THE CIA WORLD FACTBOOK 2021-2022

CENTRAL INTELLIGENCE AGENCY

Skyhorse Publishing

Copyright © 2021 by Skyhorse Publishing

Skyhorse Publishing books may be purchased in bulk at special discounts for sales promotion, corporate gifts, fund-raising, or educational purposes. Special editions can also be created to specifications. For details, contact the Special Sales Department, Skyhorse Publishing, 307 West 36th Street, 11th Floor, New York, NY 10018 or info@skyhorsepublishing.com.

Skyhorse® and Skyhorse Publishing® are registered trademarks of Skyhorse Publishing, Inc.®, a Delaware corporation.

Visit our website at www.skyhorsepublishing.com.

10 9 8 7 6 5 4 3 2 1

Library of Congress Cataloging-in-Publication Data is available on file.

Cover design by Brian Peterson

Print ISBN: 978-1-5107-6381-4
Ebook ISBN: 978-1-5107-6382-1

Printed in the United States of America

CONTENTS

INTRODUCTION

The World Factbook is prepared by the Central Intelligence Agency for the use of US Government officials, and the style, format, coverage, and content are designed to meet their specific requirements. Information is provided by Antarctic Information Program (National Science Foundation), Armed Forces Medical Intelligence Center (Department of Defense), Bureau of the Census (Department of Commerce), Bureau of Labor Statistics (Department of Labor), Central Intelligence Agency, Council of Managers of National Antarctic Programs, Defense Intelligence Agency (Department of Defense), Department of Energy, Department of State, Fish and Wildlife Service (Department of the Interior), Maritime Administration (Department of Transportation), National Geospatial-Intelligence Agency (Department of Defense), Naval Facilities Engineering Command (Department of Defense), Office of Insular Affairs (Department of the Interior), Office of Naval Intelligence (Department of Defense), US Board on Geographic Names (Department of the Interior), US Transportation Command (Department of Defense), United Nations Population Division (*World Urbanization Prospects: The 2018 Revision and World Population Prospects: The 2019 Revision* urbanization and population data used with permission),

International Telecommunication Union, International Institute for Strategic Studies, *Oil Gas Journal*, and other public and private sources.

The *Factbook* is in the public domain. Accordingly, it may be copied freely without permission of the Central Intelligence Agency (CIA). The official seal of the CIA, however, may NOT be copied without permission as required by the CIA Act of 1949 (50 U.S.C. section 403m). Misuse of the official seal of the CIA could result in civil and criminal penalties.

Citation Model

The World Factbook 2020. Washington, DC: Central Intelligence Agency, 2020.
https://www.cia.gov/the-world-factbook/

Comments and queries are welcome and may be addressed to:

Central Intelligence Agency
Attn: Office of Public Affairs
Washington, DC 20505

A BRIEF HISTORY OF BASIC INTELLIGENCE AND *THE WORLD FACTBOOK*

The Intelligence Cycle is the process by which information is acquired, converted into intelligence, and made available to policymakers. **Information** *is raw data from any source, data that may be fragmentary, contradictory, unreliable, ambiguous, deceptive, or wrong.* **Intelligence** *is information that has been collected, integrated, evaluated, analyzed, and interpreted.* **Finished intelligence** *is the final product of the Intelligence Cycle ready to be delivered to the policymaker.*

The three types of finished intelligence are: basic, current, and estimative. Basic intelligence provides the fundamental and factual reference material on a country or issue. Current intelligence reports on new developments. Estimative intelligence judges probable outcomes. The three are mutually supportive: basic intelligence is the foundation on which the other two are constructed; current intelligence continually updates the inventory of knowledge; and estimative intelligence revises overall interpretations of country and issue prospects for guidance of basic and current intelligence. *The World Factbook, The President's Daily Brief,* and the *National Intelligence Estimates* are examples of the three types of finished intelligence.

The United States has carried on foreign intelligence activities since the days of George Washington but only since World War II have they been coordinated on a government- wide basis. Three programs have highlighted the development of coordinated basic intelligence since that time: (1) *the Joint Army Navy Intelligence Studies* (JANIS), (2) *the National Intelligence Survey* (NIS), and (3) *The World Factbook.*

During World War II, intelligence consumers realized that the production of basic intelligence by different components of the US Government resulted in a great duplication of effort and conflicting information. The Japanese attack on Pearl Harbor in 1941 brought home to leaders in Congress and the executive branch the need for integrating departmental reports to national policymakers. Detailed and coordinated information was needed not only on such major powers as Germany and Japan, but also on places of little previous

interest. In the Pacific Theater, for example, the Navy and Marines had to launch amphibious operations against many islands about which information was unconfirmed or nonexistent. Intelligence authorities resolved that the United States should never again be caught unprepared.

In 1943, Gen. George B. Strong (G-2), Adm. H. C. Train (Office of Naval Intelligence—ONI), and Gen. William J. Donovan (Director of the Office of Strategic Services—OSS) decided that a joint effort should be initiated. A steering committee was appointed on 27 April 1943 that recommended the formation of a Joint Intelligence Study Publishing Board to assemble, edit, coordinate, and publish the *Joint Army Navy Intelligence Studies* (JANIS). JANIS was the first interdepartmental basic intelligence program to fulfill the needs of the US Government for an authoritative and coordinated appraisal of strategic basic intelligence. Between April 1943 and July 1947, the board published 34 JANIS studies. JANIS performed well in the war effort, and numerous letters of commendation were received, including a statement from Adm. Forrest Sherman, Chief of Staff, Pacific Ocean Areas, which said, "JANIS has become the indispensable reference work for the shore-based planners."

The need for more comprehensive basic intelligence in the postwar world was well expressed in 1946 by George S. Pettee, a noted author on national security. He wrote in *The Future of American Secret Intelligence* (Infantry Journal Press, 1946, page 46) that world leadership in peace requires even more elaborate intelligence than in war. "The conduct of peace involves all countries, all human activities—not just the enemy and his war production."

The Central Intelligence Agency was established on 26 July 1947 and officially began operating on 18 September 1947. Effective 1 October 1947, the Director of Central Intelligence assumed operational responsibility for JANIS. On 13 January 1948, the National Security Council issued Intelligence Directive (NSCID) No. 3, which authorized the *National Intelligence Survey* (NIS) program as a peacetime replacement for the wartime JANIS program. Before adequate NIS country sections could be produced, government agencies had to develop more comprehensive gazetteers and better maps. The US Board on Geographic Names (BGN) compiled the names; the Department of the Interior produced the gazetteers; and CIA produced the maps.

The Hoover Commission's Clark Committee, set up in 1954 to study the structure and administration of the CIA, reported to Congress in 1955 that: "The National Intelligence Survey is an invaluable publication which provides the essential elements of basic intelligence on all areas of the world. There will always be a continuing requirement for keeping the Survey up-to-date." The *Factbook* was created as an annual summary and update to the encyclopedic NIS studies. The first classified *Factbook* was published in August 1962, and the first unclassified version was published in June 1971. The NIS program was terminated in 1973 except for the *Factbook,* map, and gazetteer components. The 1975 *Factbook* was the first to be made available to the public with sales through the US Government Printing Office (GPO). The *Factbook* was first made available on the Internet in June 1997. The year 2019 marks the 72nd anniversary of the establishment of the Central Intelligence Agency and the 76th year of continuous basic intelligence support to the US Government by *The World Factbook* and its two predecessor programs.

The Evolution of The World Factbook

National Basic Intelligence Factbook produced semiannually until 1980. Country entries include sections on Land, Water, People, Government, Economy, Communications, and Defense Forces.

1981   Publication becomes an annual product and is renamed *The World Factbook.* A total of 165 nations are covered on 225 pages.

1983   Appendices (Conversion Factors, International Organizations) first introduced.

1984   Appendices expanded; now include: A. The United Nations, B. Selected United Nations Organizations, C. Selected International Organizations, D. Country Membership in Selected Organizations, E. Conversion Factors.

1987   A new Geography section replaces the former separate Land and Water sections. UN Organizations and Selected International Organizations appendices merged into a new International Organizations appendix. First multi-color-cover *Factbook*.

1988   More than 40 new geographic entities added to provide complete world coverage without overlap or omission. Among the new entities are Antarctica, oceans (Arctic, Atlantic, Indian, Pacific), and the World. The front-of-the-book explanatory introduction expanded and retitled to Notes, Definitions, and Abbreviations. Two new Appendices added: Weights and Measures (in place of Conversion Factors) and a Cross-Reference List of Geographic Names. *Factbook* size reaches 300 pages.

1989   Economy section completely revised and now includes an "Overview" briefly describing a country's economy. New entries added under People, Government, and Communications.

1990   The Government section revised and considerably expanded with new entries.

1991   A new International Organizations and Groups appendix added. *Factbook* size reaches 405 pages.

1992   Twenty new successor state entries replace those of the Soviet Union and Yugoslavia. New countries are respectively: Armenia, Azerbaijan, Belarus, Estonia, Georgia, Kazakhstan, Kyrgyzstan, Latvia, Lithuania, Moldova, Russia, Tajikistan, Turkmenistan, Ukraine, Uzbekistan; and Bosnia and Hercegovina, Croatia, Macedonia, Serbia and Montenegro, Slovenia. Number of nations in the *Factbook* rises to 188.

1993   Czechoslovakia's split necessitates new Czech Republic and Slovakia entries. New Eritrea entry added after it secedes from Ethiopia. Substantial enhancements made to Geography section.

1994   Two new appendices address Selected International Environmental Agreements. The gross domestic product (GDP) of most developing countries changed to a purchasing power parity (PPP) basis rather than an exchange rate basis. *Factbook* size up to 512 pages.

1995   The GDP of all countries now presented on a PPP basis. New appendix lists estimates of GDP on an exchange rate basis. Communications category split; "Railroads," "Highways," "Inland waterways," "Pipelines," "Merchant marine," and "Airports" entries now make up a new Transportation category. *The World Factbook* is first produced on CD-ROM.

1996   Maps accompanying each entry now present more detail. Flags also introduced for nearly all entities. Various new entries appear under Geography and Communications. *Factbook* abbreviations consolidated into a new Appendix A. Two new appendices present a Cross-Reference List of Country Data Codes and a Cross-Reference List of Hydrogeographic Data Codes. Geographic coordinates added to Appendix H, Cross-Reference List of Geographic Names. *Factbook* size expands by 95 pages in one year to reach 652.

1997   *The World Factbook* introduced onto the Internet. A special printed edition prepared for the CIA's 50th anniversary. A schema or Guide to Country Profiles introduced. New color maps and flags now accompany each country profile. Category headings distinguished by shaded backgrounds. Number of categories expanded to nine with the addition of an Introduction (for only a few countries) and Transnational Issues (which includes "Disputes-international" and "Illicit drugs").

1998   The Introduction category with two entries, "Current issues" and "Historical perspective," expanded to more countries. Last year for the production of CD-ROM versions of the Factbook.

1999   "Historical perspective" and "Current issues" entries in the Introduction category combined into a new "Background" statement. Several new Economy entries introduced. A new physical map of the world added to the back-of-the-book reference maps.

2000   A new "country profile" added on the Southern Ocean. The Background statements dramatically expanded to over 200 countries and possessions. A number of new Communications entries added.

2001   Background entries completed for all 267 entities in the Factbook. Several new HIV/AIDS entries introduced under the People category. Revision begun on individual country maps to include elevation extremes and a partial geographic grid. Weights and Measures appendix deleted.

2002   New entry on "Distribution of Family income—Gini index" added. Revision of individual country maps continued (process still ongoing).

2003   In the Economy category, petroleum entries added for "oil production," "consumption," "exports," "imports," and "proved reserves," as well as "natural gas proved reserves."

2004 Bi-weekly updates launched on *The World Factbook* website. Additional petroleum entries included for "natural gas production," "consumption," "exports," and "imports." In the Transportation category, under "Merchant marine," subfields added for foreign-owned vessels and those registered in other countries. Descriptions of the many forms of government mentioned in the *Factbook* incorporated into the Definitions and Notes.

2005 In the People category, a "Major infectious diseases" field added for countries deemed to pose a higher risk for travelers. In the Economy category, entries included for "Current account balance," "Investment," "Public debt," and "Reserves of foreign exchange and gold." The Transnational issues category expanded to include "Refugees and internally displaced persons." Size of the printed *Factbook* reaches 702 pages.

2006 In the Economy category, national GDP figures now presented at Official Exchange Rates (OER) in addition to GDP at purchasing power parity (PPP). Entries in the Transportation section reordered; "Highways" changed to "Roadways," and "Ports and harbors" to "Ports and terminals."

2007 In the Government category, the "Capital" entry significantly expanded with up to four subfields, including new information having to do with time. The subfields consist of the name of the capital itself, its geographic coordinates, the time difference at the capital from coordinated universal time (UTC), and, if applicable, information on daylight saving time (DST). Where appropriate, a special note is added to highlight those countries with multiple time zones. A "Trafficking in persons" entry added to the Transnational issues category. A new appendix, Weights and Measures, (re)introduced to the online version of the Factbook.

2008 In the Geography category, two fields focus on the increasingly vital resource of water: "Total renewable water resources" and "Freshwater withdrawal." In the Economy category, three fields added for: "Stock of direct foreign investment—at home," "Stock of direct foreign investment—abroad," and "Market value of publicly traded shares." Concise descriptions of all major religions included in the Definitions and Notes. Responsibility for printing of *The World Factbook* turned over to the Government Printing Office.

2009 The online *Factbook* site completely redesigned with many new features. In the People category, two new fields provide information on education in terms of opportunity and resources: "School Life Expectancy" and "Education expenditures." Additionally, the "Urbanization" entry expanded to include all countries. In the Economy category, five fields added: "Central bank discount rate," "Commercial bank prime lending rate," "Stock of narrow money," "Stock of broad money," and "Stock of domestic credit."

2010 Weekly updates inaugurated on the *The World Factbook* website. The dissolution of the Netherlands Antilles results in two new listings: Curacao and Sint Maarten. In the Communications category, a "Broadcast media" field replaces the former "Radio broadcast stations" and "TV broadcast stations" entries. In the Geography section, under "Natural hazards," a Volcanism subfield added for countries with historically active volcanoes. In the Government category, a new "National anthems" field introduced. Concise descriptions of all major Legal systems incorporated into the Definitions and Notes. In order to facilitate comparisons over time, dozens of the entries in the Economy category expanded to include two (and in some cases three) years' worth of data.

2011 The People section expanded to People and Society, incorporating ten new fields. The Economy category added "Taxes and other revenues" and "Budget surplus (+) or deficit (-)," while the

Government section introduced "International law organization participation" and "National symbols." A new African nation, South Sudan, brings the total number of countries in *The World Factbook* to 195.

2012 A new Energy category introduced with 23 energy-related fields. Several distinctive features added to *The World Factbook* website: 1) playable audio files in the Government section for the National Anthems entry, 2) online graphics in the form of a Population Pyramid feature in the People and Society category's Age Structure field, and 3) a Users Guide enabling visitors to navigate the *Factbook* more easily and efficiently. A new and distinctive Map of the World Oceans highlights an expanded array of regional and country maps. Size of the printed Factbook's 50th anniversary edition reaches 847 pages.

2013 In the People and Society section five fields introduced: "Demographic profile," "Mother's mean age at first birth," "Contraceptive prevalence rate," "Dependency ratios," and "Child labor—children ages 5-14." In the Transnational Issues category, a new *stateless persons* subfield embedded under the "Refugees and internally displaced persons" entry. In the Economy section two fields added: "GDP—composition by end use" and "Gross national saving." In the Government category the "Judicial branch" entry revised and expanded to include three new subfields: *highest court(s), judge selection and term of office, and subordinate courts.*

2014 In the Transportation category, the "Ports and terminals" field substantially expanded with subfields for *major seaport(s), river port(s), lake port(s), oil/gas terminal(s), LNG terminal(s), dry bulk cargo port(s), container port(s),* and *cruise/ferry port(s).* In the Geography section, the "Land boundaries entry" revised for all countries, including the total country border length as well as the border lengths for all *neighboring countries.*

2015 In the Government category, the first part of the "Legislative branch" field thoroughly revised, expanded, and updated for all countries under a new description heading. This subentry includes the legislative structure, the formal name(s), the number of legislative seats, the types of voting constituencies and voting systems, and the member term of office. In the Geography category, the "Land use" entry expanded to include *agricultural land, forest land, and other* uses. Area Comparison Maps introduced online for about half of the world's countries. These graphics show the size of a country in relation to a part of the United States. (More maps to follow as they become available.)

2016 In the Government section for all countries, a new "Citizenship" field added to describe policies related to the acquisition of citizenship and to the recognition of dual citizenship. Also, under the "Country name" entry, *etymologies* (historical origins) added to explain how countries acquired their names. In the Energy section, an "Electricity access" field introduced with subfields summarizing total access to electricity within an country, as well as for *urban* and *rural* populations. In the Transportation category, an expansive National air transport system field presents info on the *number of registered air carriers, number of operating aircraft, annual passenger traffic, and annual freight traffic*

2017 In the Government category, the "Constitution" entry revised and expanded with new subfields for *history* and amendments. Information on piracy moved from the Transportation category to a new "Maritime threats" field in the Military and Security category. In the Transportation section, the "Merchant marine" entry revised to not only include the total number of ships, but also the major types: bulk carrier, container ship, general cargo, oil tanker, and other. A new "Population distribution" field added to both the Geography and People and Society categories. The Government Printing Office discontinued printing The *Word Factbook*, but annual online editions may be downloaded from the *Factbook* site.

2018 One-Page Country Summaries introduced for selected countries in the *Factbook*; more to follow in the future. The Summaries highlight key information from lengthier entries and are intended for use by teachers, students, travelers, researchers, news reporters, or anyone with an interest in geography. Dozens of additional area comparison maps added; about two-thirds

of country entries now include these popular maps. In the Communications category, a "Broadband—fixed subscriptions" entry now included.

2019 The *Factbook's* World entry acquires many new Top Ten listings including those for the largest forests, largest deserts, longest mountain ranges, and climate extremes (Top Ten driest, wettest, coldest, and hottest places on earth). Also in the World entry, seven new continent area comparison maps compare their size to that of the US. In each of the five ocean entries, under the Economy section, a "Maritime fisheries" field includes info on major fishing regions, total tonnage caught, and principal fish catches.

Abbreviations This information is included in **Appendix A: Abbreviations**, which includes all abbreviations and acronyms used in the *Factbook*, with their expansions.

Acronyms An acronym is an abbreviation coined from the initial letter of each successive word in a term or phrase. In general, an acronym made up solely from the first letter of the major words in the expanded form is rendered in all capital letters (NATO from North Atlantic Treaty Organization; an exception would be ASEAN for Association of Southeast Asian Nations). In general, an acronym made up of more than the first letter of the major words in the expanded form is rendered with only an initial capital letter (Comsat from Communications Satellite Corporation; an exception would be NAM from Nonaligned Movement). Hybrid forms are sometimes used to distinguish between initially identical terms (ICC for International Chamber of Commerce and ICCt for International Criminal Court).

Administrative divisions This entry generally gives the numbers, designatory terms, and first-order administrative divisions as approved by the US Board on Geographic Names (BGN). Changes that have been reported but not yet acted on by the BGN are noted. Geographic names conform to spellings approved by the BGN with the exception of the omission of diacritical marks and special characters.

Age structure This entry provides the distribution of the population according to age. Information is included by sex and age group as follows: *0-14 years (children)*, *15-24 years (early working age)*, *25-54 years (prime working age)*, *55-64 years (mature working age)*, *65 years and over (elderly)*. The age structure of a population affects a nation's key socioeconomic issues. Countries with young populations (high percentage under age 15) need to invest more in schools, while countries with older populations (high percentage ages 65 and over) need to invest more in the health sector. The age structure can also be used to help predict potential political issues. For example, the rapid growth of a young adult population unable to find employment can lead to unrest.

Agriculture—products This entry is an ordered listing of major crops and products starting with the most important.

Airports This entry gives the total number of airports or airfields recognizable from the air. The runway(s) may be paved (concrete or asphalt surfaces) or unpaved (grass, earth, sand, or gravel surfaces) and may include closed or abandoned installations. Airports or airfields that are no longer recognizable (overgrown, no facilities, etc.) are not included. Note that not all airports have accommodations for refueling, maintenance, or air traffic control.

Airports—with paved runways This entry gives the *total* number of airports with paved runways (concrete or asphalt surfaces) by length. For airports with more than one runway, only the longest runway is included according to the following five groups— (1) *over 3,047 m* (over 10,000 ft), (2) *2,438 to 3,047 m* (8,000 to 10,000 ft), (3) *1,524 to 2,437 m* (5,000 to 8,000 ft), (4) *914 to 1,523 m* (3,000 to 5,000 ft), and (5) *under 914 m* (under 3,000 ft). Only airports with usable runways are included in this listing. Not all airports have facilities for refueling, maintenance, or air traffic control. The type aircraft capable of operating from a runway of a given length is dependent upon a number of factors including elevation of the runway, runway gradient, average maximum daily temperature at the airport, engine types, flap settings, and take-off weight of the aircraft.

Airports—with unpaved runways This entry gives the *total* number of airports with unpaved runways (grass, dirt, sand, or gravel surfaces) by length. For airports with more than one runway, only the longest runway is included according to the following five groups— (1) *over 3,047 m* (over 10,000 ft), (2) *2,438 to 3,047 m* (8,000 to 10,000 ft), (3) *1,524 to 2,437 m* (5,000 to 8,000 ft), (4) *914 to 1,523 m* (3,000 to 5,000 ft), and (5) *under 914 m* (under 3,000 ft). Only airports with usable runways are included in this listing. Not all airports have facilities for refueling, maintenance, or air traffic control. The type aircraft capable of operating from a runway of a given length is dependent upon a number of factors including elevation of the runway, runway gradient, average maximum daily temperature at the airport, engine types, flap settings, and take-off weight of the aircraft.

Appendixes This section includes *Factbook*-related material by topic.

Area This entry includes three subfields. *Total area* is the sum of all land and water areas delimited by international boundaries and/or coastlines. *Land area* is the aggregate of all surfaces delimited by international boundaries and/or coastlines, excluding inland water bodies (lakes, reservoirs, rivers). *Water area* is the sum of the surfaces of all inland water bodies, such as lakes, reservoirs, or rivers, as delimited by international boundaries and/or coastlines.

Area—comparative This entry provides an area comparison based on total area equivalents. Most entities are compared with the entire US or one of the 50 states based on area measurements (1990 revised) provided by the US Bureau of the Census. The smaller entities are compared with Washington, DC (178 sq km, 69 sq mi) or The Mall in Washington, DC (0.59 sq km, 0.23 sq mi, 146 acres).

Area—rankings This entry, which appears only in the World, Geography category, provides rankings for the earth's largest (or smallest) continents, countries, oceans, islands, mountain ranges, or other physical features.

Background This entry usually highlights major historic events and current issues and may include a statement about one or two key future trends.

Birth rate This entry gives the average annual number of births during a year per 1,000 persons in the population at midyear; also known as crude birth rate. The birth rate is usually the dominant factor in determining the rate of population growth. It depends on both the level of fertility and the age structure of the population.

Broadband—fixed subscriptions This entry gives the total number of fixed-broadband subscriptions, as well as the number of subscriptions per 100 inhabitants. Fixed broadband is a physical wired connection to the Internet (e.g., coaxial cable, optical fiber) at speeds equal to or greater than 256 kilobits/second (256 kbit/s).

Broadcast media This entry provides information on the approximate number of public and private TV and radio stations in a country, as well as basic information on the availability of satellite and cable TV services.

Budget This entry includes *revenues*, *expenditures*, and capital expenditures. These figures are calculated on an exchange rate basis, i.e., not in purchasing power parity (PPP) terms.

Budget surplus (+) or deficit (-) This entry records the difference between national government revenues and expenditures, expressed as a percent of GDP. A positive (+) number indicates that revenues exceeded expenditures (a budget surplus), while a negative (-) number indicates the reverse (a budget deficit). Normalizing the data, by dividing the budget balance by GDP, enables easy comparisons across countries and indicates whether a national government saves or borrows money. Countries with high budget deficits (relative to their GDPs) generally have more difficulty raising funds to finance expenditures, than those with lower deficits.

Capital This entry gives the *name* of the seat of government, its *geographic coordinates*, the *time difference* relative to **Coordinated Universal Time (UTC)** and the time observed in Washington, DC, and, if applicable, information on *daylight saving time* **(DST)**. Where appropriate, a special *note* has been added to highlight those countries that have multiple time zones.

Carbon dioxide emissions from consumption of energy
This entry is the total amount of carbon dioxide, measured in metric tons, released by burning fossil fuels in the process of producing and consuming energy.

Children under the age of 5 years underweight
This entry gives the percent of children under five considered to be underweight. Underweight means weight-for-age is approximately 2 kg below for standard at age one, 3 kg below standard for ages two and three, and 4 kg below standard for ages four and five. This statistic is an indicator of the nutritional status of a community. Children who suffer from growth retardation as a result of poor diets and/or recurrent infections tend to have a greater risk of suffering illness and death.

Citizenship This entry provides information related to the acquisition and exercise of citizenship; it includes four subfields:
citizenship by birth describes the acquisition of citizenship based on place of birth, known as *Jus soli*, regardless of the citizenship of parents.
citizenship by descent only describes the acquisition of citizenship based on the principle of *Jus sanguinis*, or by descent, where at least one parent is a citizen of the state and being born within the territorial limits of the state is not required.
The majority of countries adhere to this practice. In some cases, citizenship is conferred through the father or mother exclusively.
dual citizenship recognized indicates whether a state permits a citizen to simultaneously hold citizenship in another state. Many states do not permit dual citizenship and the voluntary acquisition of citizenship in another country is grounds for revocation of citizenship. Holding dual citizenship makes an individual legally obligated to more than one state and can negate the normal consular protections afforded to citizens outside their original country of citizenship.
residency requirement for naturalization lists the length of time an applicant is required to live in a state before applying for naturalization. In most countries citizenship can be acquired through the legal process of naturalization. The requirements for naturalization vary by state but generally include no criminal record, good health, economic wherewithal, and a period of authorized residency in the state. This time period can vary enormously among states and is often used to make the acquisition of citizenship difficult or impossible.

Civil aircraft registration country code prefix
This entry provides the one-or two-character alphanumeric code indicating the nationality of civil aircraft. Article 20 of the Convention on International Civil Aviation (Chicago Convention), signed in 1944, requires that all aircraft engaged in international air navigation bear appropriate nationality marks. The aircraft registration number consists of two parts: a prefix consisting of a one-or two-character alphanumeric code indicating nationality and a registration suffix of one to five characters for the specific aircraft. The prefix codes are based upon radio call-signs allocated by the International Telecommunications Union (ITU) to each country. Since 1947, the International Civil Aviation Organization (ICAO) has managed code standards and their allocation.

Climate This entry includes a brief description of typical weather regimes throughout the year; in the Word entry only, it includes four subfields that describe climate extremes:
ten driest places on earth (average annual precipitation) describes the annual average precipitation measured in both millimeters and inches for selected countries with climate extremes.
ten wettest places on earth (average annual precipitation) describes the annual average precipitation measured in both millimeters and inches for selected countries with climate extremes.
ten coldest places on earth (lowest average monthly temperature) describes temperature measured in both degrees Celsius and Fahrenheit as well as the month of the year for selected countries with climate extremes.
ten hottest places on earth (highest average monthly temperature) describes the temperature measured both in degrees Celsius and Fahrenheit as well as the month of the year for selected countries with climate extremes.

Coastline This entry gives the total length of the boundary between the land area (including islands) and the sea.

Communications This category deals with the means of exchanging information and includes the telephone, radio, television, and Internet host entries.

Communications—note This entry includes miscellaneous communications information of significance not included elsewhere.

Constitution This entry provides information on a country's constitution and includes two subfields. The history subfield includes the dates of previous constitutions and the main steps and dates in formulating and implementing the latest constitution. For countries with 1-3 previous constitutions, the years are listed; for those with 4-9 previous, the entry is listed as "several previous," and for those with 10 or more, the entry is "many previous." The amendments subfield summarizes the process of amending a country's constitution – from proposal through passage – and the dates of amendments, which are treated in the same manner as the constitution dates. Where appropriate, summaries are composed from English-language translations of non-English constitutions, which derive from official or non-official translations or machine translators.

The main steps in creating a constitution and amending it usually include the following steps: proposal, drafting, legislative and/or executive branch review and approval, public referendum, and entry into law. In many countries this process is lengthy. Terms commonly used to describe constitutional changes are "amended," "revised," or "reformed." In countries such as South Korea and Turkmenistan, sources differ as to whether changes are stated as new constitutions or are amendments/revisions to existing ones.

A few countries including Canada, Israel, and the UK have no single constitution document, but have various written and unwritten acts, statutes, common laws, and practices that, when taken together, describe a body of fundamental principles or established precedents as to how their countries are governed. Some special regions (Hong Kong, Macau) and countries (Oman, Saudi Arabia) use the term "basic law" instead of constitution.

A number of self-governing dependencies and territories such as the Cayman Islands, Bermuda, and Gibraltar (UK), Greenland and Faroe Islands (Denmark), Aruba, Curacao, and Sint Maarten (Netherlands), and Puerto Rico and the Virgin Islands (US) have their own constitutions.

Contraceptive prevalence rate This field gives the percent of women of reproductive age (15-49) who are married or in union and are using, or whose sexual partner is using, a method of contraception according to the date of the most recent available data. The contraceptive prevalence rate is an indicator of health services, development, and women's empowerment. It is also useful in understanding, past, present, and future fertility trends, especially in developing countries.

Coordinated Universal Time (UTC) UTC is the international atomic time scale that serves as the basis of timekeeping for most of the world. The hours, minutes, and seconds expressed by UTC represent the time of day at the Prime Meridian (0º longitude) located near Greenwich, England as reckoned from midnight. UTC is calculated by the Bureau International des Poids et Measures (BIPM) in Sevres, France. The BIPM averages data collected from more than 200 atomic time and frequency standards located at about 50 laboratories worldwide. UTC is the basis for all civil time, with the world divided into time zones expressed as positive or negative differences from UTC. UTC is also referred to as "Zulu time." See the Standard Time Zones of the World map included with the **Reference Maps**.

Country data codes See **Data codes**.

Country map Most versions of the *Factbook* provide a country map in color. The maps were produced from the best information available at the time of preparation. Names and/or boundaries may have changed subsequently.

Country name This entry includes all forms of the country's name approved by the US Board on Geographic Names (Italy is used as an example): *conventional long form* (Italian Republic), *conventional short form* (Italy), *local long form* (Repubblica Italiana), *local short form* (Italia), *former* (Kingdom of Italy), as well as the *abbreviation*. Also see the **Terminology** note.

Crude oil—exports This entry is the total amount of crude oil exported, in barrels per day (bbl/day).

Crude oil—imports This entry is the total amount of crude oil imported, in barrels per day (bbl/day).

Crude oil—production This entry is the total amount of crude oil produced, in barrels per day (bbl/day).

Crude oil—proved reserves This entry is the stock of proved reserves of crude oil, in barrels (bbl). Proved reserves are those quantities of petroleum which, by analysis of geological and engineering data, can be estimated with a high degree of confidence to be commercially recoverable from a given date forward, from known reservoirs and under current economic conditions.

Current account balance This entry records a country's net trade in goods and services, plus net earnings from rents, interest, profits, and dividends, and net transfer payments (such as pension funds and worker remittances) to and from the rest of the world during the period specified. These figures are calculated on an exchange rate basis, i.e., not in purchasing power parity (PPP) terms.

Current Health Expenditure Current Health Expenditure (CHE) describes the share of spending on health in each country relative to the size of its economy. It includes expenditures corresponding to the final consumption of health care goods and services and excludes investment, exports, and intermediate consumption. CHE shows the importance of the health sector in the economy and indicates the priority given to health in monetary terms. **Note:** Current Health Expenditure replaces the former Health Expenditures field and is calculated differently.

Data codes This information is presented in Appendix D: Cross-Reference List of Country Data Codes and Appendix E: Cross-Reference List of Hydrographic Data Codes.

Date of information In general, information available as of January in a given year is used in the preparation of the printed edition.

Daylight Saving Time (DST) This entry is included for those entities that have adopted a policy of adjusting the official local time forward, usually one hour, from Standard Time during summer months. Such policies are most common in mid-latitude regions.

Death rate This entry gives the average annual number of deaths during a year per 1,000 population at midyear; also known as crude death rate. The death rate, while only a rough indicator of the mortality situation in a country, accurately indicates the current mortality impact on population growth. This indicator is significantly affected by age distribution, and most countries will eventually show a rise in the overall death rate, in spite of continued decline in mortality at all ages, as declining fertility results in an aging population.

Debt—external This entry gives the total public and private debt owed to nonresidents repayable in internationally accepted currencies, goods, or services. These figures are calculated on an exchange rate basis, i.e., not in purchasing power parity (PPP) terms.

Demographic profile This entry describes a country's key demographic features and trends and how they vary among regional, ethnic, and socioeconomic sub-populations. Some of the topics addressed are population age structure, fertility, health, mortality, poverty, education, and migration.

Dependency ratios Dependency ratios are a measure of the age structure of a population. They relate the number of individuals that are likely to be economically "dependent" on the support of others. Dependency ratios contrast the ratio of youths (ages 0-14) and the elderly (ages 65+) to the number of those in the working-age group (ages 15-64). Changes in the dependency ratio provide an indication of potential social support requirements resulting from changes in population age structures. As fertility levels decline, the dependency ratio initially falls because the proportion of youths decreases while the proportion of the population of working age increases. As fertility levels continue to decline, dependency ratios eventually increase because the proportion of the population of working age starts to decline and the proportion of elderly persons continues to increase.

total dependency ratio—The total dependency ratio is the ratio of combined youth population (ages 0-14) and elderly population (ages 65+) per 100 people of working age (ages 15-64). A high total dependency ratio indicates that the working-age population and the overall economy face a greater burden to support and provide social services for youth and elderly persons, who are often economically dependent.

youth dependency ratio—The youth dependency ratio is the ratio of the youth population (ages 0-14) per 100 people of working age (ages 15-64). A high youth dependency ratio indicates that a greater investment needs to be made in schooling and other services for children.

elderly dependency ratio—The elderly dependency ratio is the ratio of the elderly population (ages 65+) per 100 people of working age (ages 15-64). Increases in the elderly dependency ratio put added pressure on governments to fund pensions and healthcare.

potential support ratio—The potential support ratio is the number of working-age people (ages 15-64) per one elderly person (ages 65+). As a population ages, the potential support ratio tends to fall, meaning there are fewer potential workers to support the elderly.

Dependency status This entry describes the formal relationship between a particular nonindependent entity and an independent state.

Dependent areas This entry contains an alphabetical listing of all nonindependent entities associated in some way with a particular independent state.

Diplomatic representation The US Government has diplomatic relations with 190 independent states, including 188 of the 193 UN members (excluded UN members are Bhutan, Cuba, Iran, North Korea, and the US itself). In addition, the US has diplomatic relations with 2 independent states that are not in the UN, the Holy See and Kosovo, as well as with the EU.

Diplomatic representation from the US This entry includes the *chief of mission, embassy* address, *mailing address, telephone* number, *FAX* number, *branch office* locations, *consulate general* locations, and *consulate* locations.

Diplomatic representation in the US This entry includes the *chief of mission, chancery address, telephone, FAX, consulate general* locations, and *consulate* locations. The use of the annotated title Appointed Ambassador refers to a new ambassador who has presented his/her credentials to the secretary of state but not the US president. Such ambassadors fulfill all diplomatic functions except meeting with or appearing at functions attended by the president until such time as they formally present their credentials at a White House ceremony.

Disputes—international This entry includes a wide variety of situations that range from traditional bilateral boundary disputes to unilateral claims of one sort or another. Information regarding disputes over international terrestrial and maritime boundaries has been reviewed by the US Department of State. References to other situations involving borders or frontiers may also be included, such as resource disputes, geopolitical questions, or irredentist issues; however, inclusion does not necessarily constitute official acceptance or recognition by the US Government.

Drinking water source This entry provides information about access to improved or unimproved drinking water sources available to segments of the population of a country. *Improved* drinking water—use of any of the following sources: piped water into dwelling, yard, or plot; public tap or standpipe; tubewell or borehole; protected dug well; protected spring; or rainwater collection. *Unimproved* drinking water—use of any of the following sources: unprotected dug well; unprotected spring; cart with small tank or drum; tanker truck; surface water, which includes rivers, dams, lakes, ponds, streams, canals or irrigation channels; or bottled water.

Economy This category includes the entries dealing with the size, development, and management of productive resources, i.e., land, labor, and capital.

Economy—overview This entry briefly describes the type of economy, including the degree of market orientation, the level of economic development, the most important natural resources, and the unique areas of specialization. It also characterizes major economic events and policy changes in the most recent 12 months and may include a statement about one or two key future macroeconomic trends.

Education expenditures This entry provides the public expenditure on education as a percent of GDP.

Electricity—consumption This entry consists of total electricity generated annually plus imports and minus exports, expressed in kilowatt-hours.

The discrepancy between the amount of electricity generated and/or imported and the amount consumed and/or exported is accounted for as loss in transmission and distribution.

Electricity—exports This entry is the total exported electricity in kilowatt-hours.

Electricity—from fossil fuels This entry measures the capacity of plants that generate electricity by burning fossil fuels (such as coal, petroleum products, and natural gas), expressed as a share of the country's total generating capacity.

Electricity—from hydroelectric plants This entry measures the capacity of plants that generate electricity by water-driven turbines, expressed as a share of the country's total generating capacity.

Electricity—from nuclear fuels This entry measures the capacity of plants that generate electricity through radioactive decay of nuclear fuel, expressed as a share of the country's total generating capacity.

Electricity—from other renewable sources This entry measures the capacity of plants that generate electricity by using renewable energy sources other than hydroelectric (including, for example, wind, waves, solar, and geothermal), expressed as a share of the country's total generating capacity.

Electricity—imports This entry is the total imported electricity in kilowatt-hours.

Electricity—installed generating capacity This entry is the total capacity of currently installed generators, expressed in kilowatts (kW), to produce electricity. A 10-kilowatt (kW) generator will produce 10 kilowatt hours (kWh) of electricity, if it runs continuously for one hour.

Electricity—production This entry is the annual electricity generated expressed in kilowatt-hours. The discrepancy between the amount of electricity generated and/or imported and the amount consumed and/or exported is accounted for as loss in transmission and distribution.

Electricity access This entry provides information on access to electricity. Electrification data – collected from industry reports, national surveys, and international sources – consists of four subfields. *Population without electricity* provides an estimate of the number of citizens that do not have access to electricity. *Electrification – total population* is the percent of a country's total population with access to electricity, *electrification – urban areas* is the percent of a country's urban population with access to electricity, while *electrification – rural areas* is the percent of a country's rural population with access to electricity. Due to differences in definitions and methodology from different sources, data quality may vary from country to country.

Elevation This entry includes the *mean elevation* and elevation extremes, *lowest point* and *highest point*.

Energy This category includes entries dealing with the production, consumption, import, and export of various forms of energy including electricity, crude oil, refined petroleum products, and natural gas.

Entities Some of the independent states, dependencies, areas of special sovereignty, and governments included in this publication are not independent, and others are not officially recognized by the US Government. "Independent state" refers to a people politically organized into a sovereign state with a definite territory. "Dependencies" and "areas of special sovereignty" refer to a broad category of political entities that are associated in some way with an independent state. "Country" names used in the table of contents or for page headings are usually the short-form names as approved by the US Board on Geographic Names and may include independent states, dependencies, and areas of special sovereignty, or other geographic entities. There are a total of 267 separate geographic entities in *The World Factbook* that may be categorized as follows:

INDEPENDENT STATES

195 Afghanistan, Albania, Algeria, Andorra, Angola, Antigua and Barbuda, Argentina, Armenia, Australia, Austria, Azerbaijan, The Bahamas, Bahrain, Bangladesh, Barbados, Belarus, Belgium, Belize, Benin, Bhutan, Bolivia, Bosnia and Herzegovina, Botswana, Brazil, Brunei, Bulgaria, Burkina Faso, Burma, Burundi, Cambodia, Cameroon, Canada, Cape Verde, Central African Republic, Chad, Chile, China, Colombia, Comoros, Democratic Republic of the Congo, Republic of the Congo, Costa Rica, Cote d'Ivoire, Croatia, Cuba, Cyprus, Czech Republic, Denmark, Djibouti, Dominica, Dominican Republic, Ecuador, Egypt, El Salvador, Equatorial Guinea, Eritrea, Estonia, Ethiopia, Fiji, Finland, France, Gabon, The Gambia, Georgia, Germany, Ghana, Greece, Grenada, Guatemala, Guinea, Guinea-Bissau, Guyana, Haiti, Holy See, Honduras, Hungary, Iceland, India, Indonesia, Iran, Iraq, Ireland, Israel, Italy, Jamaica, Japan, Jordan, Kazakhstan, Kenya, Kiribati, North Korea, South Korea, Kosovo, Kuwait, Kyrgyzstan, Laos, Latvia, Lebanon, Lesotho, Liberia, Libya, Liechtenstein, Lithuania, Luxembourg, Macedonia, Madagascar, Malawi, Malaysia, Maldives, Mali, Malta, Marshall Islands, Mauritania, Mauritius, Mexico, Federated States of Micronesia, Moldova, Monaco, Mongolia, Montenegro, Morocco, Mozambique, Namibia, Nauru, Nepal, Netherlands, NZ, Nicaragua, Niger, Nigeria, Norway, Oman, Pakistan, Palau, Panama, Papua New Guinea, Paraguay, Peru, Philippines, Poland, Portugal, Qatar, Romania, Russia, Rwanda, Saint Kitts and Nevis, Saint Lucia, Saint Vincent and the Grenadines, Samoa, San Marino, Sao Tome and Principe, Saudi Arabia, Senegal, Serbia, Seychelles, Sierra Leone, Singapore, Slovakia, Slovenia, Solomon Islands, Somalia, South Africa, South Sudan, Spain, Sri Lanka, Sudan, Suriname, Swaziland, Sweden, Switzerland, Syria, Tajikistan, Tanzania, Thailand, Timor-Leste, Togo, Tonga, Trinidad and Tobago, Tunisia, Turkey, Turkmenistan, Tuvalu, Uganda, Ukraine, UAE, UK, US, Uruguay, Uzbekistan, Vanuatu, Venezuela, Vietnam, Yemen, Zambia, Zimbabwe

OTHER

2 Taiwan, European Union

DEPENDENCIES AND AREAS OF SPECIAL SOVEREIGNTY

6 Australia—Ashmore and Cartier Islands, Christmas Island, Cocos (Keeling) Islands, Coral Sea Islands, Heard Island and McDonald Islands, Norfolk Island
2 China—Hong Kong, Macau
2 Denmark—Faroe Islands, Greenland
8 France—Clipperton Island, French Polynesia, French Southern and Antarctic Lands, New Caledonia, Saint Barthelemy, Saint Martin, Saint Pierre and Miquelon, Wallis and Futuna
3 Netherlands—Aruba, Curacao, Sint Maarten
3 New Zealand—Cook Islands, Niue, Tokelau
3 Norway—Bouvet Island, Jan Mayen, Svalbard
17 UK—Akrotiri, Anguilla, Bermuda, British Indian Ocean Territory, British Virgin Islands, Cayman Islands, Dhekelia, Falkland Islands, Gibraltar, Guernsey, Jersey, Isle of Man, Montserrat, Pitcairn Islands, Saint Helena, South Georgia and the South Sandwich Islands, Turks and Caicos Islands
14 US—American Samoa, Baker Island*, Guam, Howland Island*, Jarvis Island*, Johnston Atoll*, Kingman Reef*, Midway Islands*, Navassa Island, Northern Mariana Islands, Palmyra Atoll*, Puerto Rico, Virgin Islands, Wake Island (* consolidated in United States Pacific Island Wildlife Refuges entry)

MISCELLANEOUS

6 Antarctica, Gaza Strip, Paracel Islands, Spratly Islands, West Bank, Western Sahara

OTHER ENTITIES

5 oceans—Arctic Ocean, Atlantic Ocean, Indian Ocean, Pacific Ocean, Southern Ocean
1 World

267 total

Environment—current issues This entry lists the most pressing and important environmental problems. The following terms and abbreviations are used throughout the entry:

Acidification—the lowering of soil and water pH due to acid precipitation and deposition usually through precipitation; this process disrupts ecosystem nutrient flows and may kill freshwater fish and plants dependent on more neutral or alkaline conditions (see acid rain).

Acid rain—characterized as containing harmful levels of sulfur dioxide or nitrogen oxide; acid rain is damaging and potentially deadly to the earth's fragile ecosystems; acidity is measured using the pH scale where 7 is neutral, values greater than 7 are considered alkaline, and values below 5.6 are considered acid precipitation; note—a pH of 2.4 (the acidity of vinegar) has been measured in rainfall in New England.

Aerosol—a collection of airborne particles dispersed in a gas, smoke, or fog.

Afforestation—converting a bare or agricultural space by planting trees and plants; reforestation involves replanting trees on areas that have been cut or destroyed by fire.

Asbestos—a naturally occurring soft fibrous mineral commonly used in fireproofing materials and considered to be highly carcinogenic in particulate form.

Biodiversity—also biological diversity; the relative number of species, diverse in form and function, at the genetic, organism, community, and ecosystem level; loss of biodiversity reduces an ecosystem's ability to recover from natural or man-induced disruption.

Bio-indicators—a plant or animal species whose presence, abundance, and health reveal the general condition of its habitat.

Biomass—the total weight or volume of living matter in a given area or volume.

Carbon cycle—the term used to describe the exchange of carbon (in various forms, e.g., as carbon dioxide) between the atmosphere, ocean, terrestrial biosphere, and geological deposits.

Catchments—assemblages used to capture and retain rainwater and runoff; an important water management technique in areas with limited freshwater resources, such as Gibraltar.

DDT (dichloro-diphenyl-trichloro-ethane)—a colorless, odorless insecticide that has toxic effects on most animals; the use of DDT was banned in the US in 1972.

Defoliants—chemicals which cause plants to lose their leaves artificially; often used in agricultural practices for weed control, and may have detrimental impacts on human and ecosystem health.

Deforestation—the destruction of vast areas of forest (e.g., unsustainable forestry practices, agricultural and range land clearing, and the over exploitation of wood products for use as fuel) without planting new growth.

Desertification—the spread of desert-like conditions in arid or semi-arid areas, due to overgrazing, loss of agriculturally productive soils, or climate change.

Dredging—the practice of deepening an existing waterway; also, a technique used for collecting bottom-dwelling marine organisms (e.g., shellfish) or harvesting coral, often causing significant destruction of reef and ocean-floor ecosystems.

Drift-net fishing—done with a net, miles in extent, that is generally anchored to a boat and left to float with the tide; often results in an over harvesting and waste of large populations of non-commercial marine species (by-catch) by its effect of" sweeping the ocean clean."

Ecosystems—ecological units comprised of complex communities of organisms and their specific environments.

Effluents—waste materials, such as smoke, sewage, or industrial waste which are released into the environment, subsequently polluting it.

Endangered species—a species that is threatened with extinction either by direct hunting or habitat destruction.

Freshwater—water with very low soluble mineral content; sources include lakes, streams, rivers, glaciers, and underground aquifers.

Greenhouse gas—a gas that" traps" infrared radiation in the lower atmosphere causing surface warming; water vapor, carbon dioxide, nitrous oxide, methane, hydrofluorocarbons, and ozone are the primary greenhouse gases in the Earth's atmosphere.

Groundwater—water sources found below the surface of the earth often in naturally occurring reservoirs in permeable rock strata; the source for wells and natural springs.

Highlands Water Project—a series of dams constructed jointly by Lesotho and South Africa to redirect Lesotho's abundant water supply into a rapidly growing area in South Africa; while it is the largest infrastructure project in southern Africa, it is also the most costly and controversial; objections to the project include claims that it forces people from their homes, submerges farmlands, and squanders economic resources.

Inuit Circumpolar Conference (ICC)—represents the roughly 150,000 Inuits of Alaska, Canada, Greenland, and Russia in international environmental issues; a General Assembly convenes every three years to determine the focus of the ICC; the most current concerns are long-range transport of pollutants, sustainable development, and climate change.

Metallurgical plants—industries which specialize in the science, technology, and processing of metals; these plants produce highly concentrated and toxic wastes which can contribute to pollution of ground water and air when not properly disposed.

Noxious substances—injurious, very harmful to living beings.

Overgrazing—the grazing of animals on plant material faster than it can naturally regrow leading to the permanent loss of plant cover, a common effect of too many animals grazing limited range land.

Ozone shield—a layer of the atmosphere composed of ozone gas (O3) that resides approximately 25 miles above the Earth's surface and absorbs solar ultraviolet radiation that can be harmful to living organisms.

Poaching—the illegal killing of animals or fish, a great concern with respect to endangered or threatened species.

Pollution—the contamination of a healthy environment by man-made waste.

Potable water—water that is drinkable, safe to be consumed.

Salination—the process through which fresh (drinkable) water becomes salt (undrinkable) water; hence, desalination is the reverse process; also involves the accumulation of salts in topsoil caused by evaporation of excessive irrigation water, a process that can eventually render soil incapable of supporting crops.

Siltation—occurs when water channels and reservoirs become clotted with silt and mud, a side effect of deforestation and soil erosion.

Slash-and-burn agriculture—a rotating cultivation technique in which trees are cut down and burned in order to clear land for temporary agriculture; the land is used until its productivity declines at which point a new plot is selected and the process repeats; this practice is sustainable while population levels are low and time is permitted for regrowth of natural vegetation;

conversely, where these conditions do not exist, the practice can have disastrous consequences for the environment.

Soil degradation—damage to the land's productive capacity because of poor agricultural practices such as the excessive use of pesticides or fertilizers, soil compaction from heavy equipment, or erosion of topsoil, eventually resulting in reduced ability to produce agricultural products.

Soil erosion—the removal of soil by the action of water or wind, compounded by poor agricultural practices, deforestation, overgrazing, and desertification.

Ultraviolet (UV) radiation—a portion of the electromagnetic energy emitted by the sun and naturally filtered in the upper atmosphere by the ozone layer; UV radiation can be harmful to living organisms and has been linked to increasing rates of skin cancer in humans.

Waterborne diseases—those in which bacteria survive in, and are transmitted through, water; always a serious threat in areas with an untreated water supply.

Environment—international agreements This entry separates country participation in international environmental agreements into two levels—*party to* and *signed, but not ratified*. Agreements are listed in alphabetical order by the abbreviated form of the full name.

Environmental agreements This information is presented in Appendix C: Selected International Environmental Agreements, which includes the name, abbreviation, date opened for signature, date entered into force, objective, and parties by category.

Ethnic groups This entry provides an ordered listing of ethnic groups starting with the largest and normally includes the percent of total population.

Exchange rates This entry provides the average annual price of a country's monetary unit for the time period specified, expressed in units of local currency per US dollar, as determined by international market forces or by official fiat. The International Organization for Standardization (ISO) 4217 alphabetic currency code for the national medium of exchange is presented in parenthesis. Closing daily exchange rates are not presented in *The World Factbook*, but are used to convert stock values-e.g., the market value of publicly traded shares—to US dollars as of the specified date.

Executive branch This entry includes five subentries: *chief of state; head of government; cabinet; elections/appointments; election results. Chief of state* includes the name, title, and beginning date in office of the titular leader of the country who represents the state at official and ceremonial functions but may not be involved with the day-to-day activities of the government. *Head of government* includes the name, title of the top executive designated to manage the executive branch of the government, and the beginning date in office. *Cabinet* includes the official name of the executive branch's high-ranking body and the method of member selection. *Elections/appointments* includes the process for accession to office, date of the last election, and date of the next election. *Election results* includes each candidate's political affiliation, percent of direct popular vote or indirect legislative/parliamentary percent vote or vote count in the last election.

The executive branches in approximately 80% of the world's countries have separate chiefs of state and heads of government; for the remainder, the chief of state is also the head of government, such as in Argentina, Kenya, the Philippines, the US, and Venezuela. Chiefs of state in just over 100 countries are directly elected, most by majority popular vote; those in another 55 are indirectly elected by their national legislatures, parliaments, or electoral colleges. Another 29 countries have a monarch as the chief of state. In dependencies, territories, and collectivities of sovereign countries—except those of the US—representatives are appointed to serve as chiefs of state.

Heads of government in the majority of countries are appointed either by the president or the monarch or selected by the majority party in the legislative body. Excluding countries where the chief of state is also head of government, in only a few countries is the head of government directly elected through popular vote.

Most of the world's countries have cabinets, the majority of which are appointed by the chief of state or prime minister, many in consultation with each other or with the legislature. Cabinets in only about a dozen countries are elected solely by their legislative bodies.

Exports This entry provides the total US dollar amount of merchandise exports on an f.o.b. (free on board) basis. These figures are calculated on an exchange rate basis, i.e., not in purchasing power parity (PPP) terms.

Exports—commodities This entry provides a listing of the highest-valued exported products; it sometimes includes the percent of total dollar value.

Exports—partners This entry provides a rank ordering of trading partners starting with the most important; it sometimes includes the percent of total dollar value.

Fiscal year This entry identifies the beginning and ending months for a country's accounting period of 12 months, which often is the calendar year but which may begin in any month. All yearly references are for the calendar year (CY) unless indicated as a non-calendar fiscal year (FY).

Flag description This entry provides a written flag description produced from actual flags or the best information available at the time the entry was written. The flags of independent states are used by their dependencies unless there is an officially recognized local flag. Some disputed and other areas do not have flags.

Flag graphic Most versions of the *Factbook* include a color flag at the beginning of the country profile. The flag graphics were produced from actual flags or the best information available at the time of preparation. The flags of independent states are used by their dependencies unless there is an officially recognized local flag. Some disputed and other areas do not have flags.

GDP (official exchange rate) This entry gives the gross domestic product (GDP) or value of all final goods and services produced within a nation in a given year. A nation's GDP at official exchange rates (OER) is the home-currency-denominated annual GDP figure divided by the bilateral average US exchange rate with that country in that year. The measure is simple to compute and gives a precise measure of the value of output. Many economists prefer this measure when gauging the economic power an economy maintains vis-à-vis its neighbors, judging that an exchange rate captures the purchasing power a nation enjoys in the international marketplace. Official exchange rates, however, can be artificially fixed and/or subject to manipulation—resulting in claims of the country having an under-or over-valued currency—and are not necessarily the equivalent of a market-determined exchange rate. Moreover, even if the official exchange rate is market-determined, market exchange rates are frequently established by a relatively small set of goods and services (the ones the country trades) and may not capture the value of the larger set of goods the country produces. Furthermore, OER-converted GDP is not well suited to comparing domestic GDP over time, since appreciation/depreciation from one year to the next will make the OER GDP value rise/fall regardless of whether home-currency-denominated GDP changed.

GDP (purchasing power parity) This entry gives the gross domestic product (GDP) or value of all final goods and services produced within a nation in a given year. A nation's GDP at purchasing power parity (PPP) exchange rates is the sum value of all goods and services produced in the country valued at prices prevailing in the United States in the year noted. This is the measure most economists prefer when looking at per-capita welfare and when comparing living conditions or use of resources across countries. The measure is difficult to compute, as a US dollar value has to be assigned to all goods and services in the country regardless of whether these goods and services have a direct equivalent in the United States (for example, the value of an ox-cart or non-US military equipment); as a result, PPP estimates for some countries are based on a small and sometimes different set of goods and services. In addition, many countries do not formally participate in the World Bank's PPP project that calculates these measures, so the resulting GDP estimates for these countries may lack precision. For many developing countries, PPP-based GDP measures are multiples of the official exchange rate (OER) measure. The differences between the OER-and PPP-denominated GDP values for most of the wealthy industrialized countries are generally much smaller.

GDP—composition, by end use This entry shows who does the spending in an economy: consumers, businesses, government, and foreigners. The distribution gives the percentage contribution to total GDP of *household consumption, government consumption, investment in fixed capital, investment in inventories, exports of goods and services, and imports of goods and services,* and will total 100 percent of GDP if the data are complete.

> *household consumption* consists of expenditures by resident households, and by nonprofit institutions that serve households, on goods and services that are consumed by individuals. This includes consumption of both domestically produced and foreign goods and services.
>
> *government consumption* consists of government expenditures on goods and services. These figures exclude government transfer payments, such as interest on debt, unemployment, and social security, since such payments are not made in exchange for goods and services supplied.
>
> *investment in fixed capital* consists of total business spending on fixed assets, such as factories, machinery, equipment, dwellings, and inventories of raw materials, which provide the basis for future production. It is measured gross of the depreciation of the assets, i.e., it includes investment that merely replaces worn-out or scrapped capital. Earlier editions of *The World Factbook* referred to this concept as Investment (gross fixed) and that data now have been moved to this new field.
>
> *investment in inventories* consists of net changes to the stock of outputs that are still held by the units that produce them, awaiting further sale to an end user, such as automobiles sitting on a dealer 's lot or groceries on the store shelves. This figure may be positive or negative. If the stock of unsold output increases during the relevant time period, *investment in inventories* is positive, but, if the stock of unsold goods declines, it will be negative. *Investment in inventories* normally is an early indicator of the state of the economy. If the stock of unsold items increases unexpectedly – because people stop buying—the economy may be entering a recession; but if the stock of unsold items falls—and goods " go flying off the shelves"—businesses normally try to replace those stocks, and the economy is likely to accelerate.
>
> *exports of goods and services* consist of sales, barter, gifts, or grants of goods and services from residents to nonresidents.
>
> *imports of goods and services* consist of purchases, barter, or receipts of gifts, or grants of goods and services by residents from nonresidents. *Exports* are treated as a positive item, while

imports are treated as a negative item. In a purely accounting sense, *imports* have no direct impact on GDP, which only measures output of the domestic economy. Imports are entered as a negative item to offset the fact that the expenditure figures for consumption, investment, government, and exports also include expenditures on imports. These imports contribute directly to foreign GDP but only indirectly to domestic GDP. Because of this negative offset for imports of goods and services, the sum of the other five items, excluding imports, will always total more than 100 percent of GDP. A surplus of exports of goods and services over imports indicates an economy is investing abroad, while a deficit indicates an economy is borrowing from abroad.

GDP—composition, by sector of origin This entry shows where production takes place in an economy. The distribution gives the percentage contribution of *agriculture, industry,* and *services* to total GDP, and will total 100 percent of GDP if the data are complete. Agriculture includes farming, fishing, and forestry. Industry includes mining, manufacturing, energy production, and construction. Services cover government activities, communications, transportation, finance, and all other private economic activities that do not produce material goods.

GDP—per capita (PPP) This entry shows GDP on a purchasing power parity basis divided by population as of 1 July for the same year.

GDP—real growth rate This entry gives GDP growth on an annual basis adjusted for inflation and expressed as a percent. The growth rates are year-over-year, and not compounded.

GDP methodology In the **Economy** category, GDP dollar estimates for countries are reported both on an official exchange rate (OER) and a purchasing power parity (PPP) basis. Both measures contain information that is useful to the reader. The PPP method involves the use of standardized international dollar price weights, which are applied to the quantities of final goods and services produced in a given economy. The data derived from the PPP method probably provide the best available starting point for comparisons of economic strength and well-being between countries. In contrast, the currency exchange rate method involves a variety of international and domestic financial forces that may not capture the value of domestic output. Whereas PPP estimates for OECD countries are quite reliable, PPP estimates for developing countries are often rough approximations. In developing countries with weak currencies, the exchange rate estimate of GDP in dollars is typically one-fourth to one-half the PPP estimate. Most of the GDP estimates for developing countries are based on extrapolation of PPP numbers published by the UN International Comparison Program (UNICP) and by Professors Robert Summers and Alan Heston of the University of Pennsylvania and their colleagues. GDP derived using the OER method should be used for the purpose of calculating the share of items such as exports, imports, military expenditures, external debt, or the current account balance, because the dollar values presented in the *Factbook* for these items have been converted at official exchange rates, not at PPP. One should use the OER GDP figure to calculate the proportion of, say, Chinese defense expenditures in GDP, because that share will be the same as one calculated in local currency units. Comparison of OER GDP with PPP GDP may also indicate whether a currency is over-or under-valued. If OER GDP is smaller than PPP GDP, the official exchange rate may be undervalued, and vice versa. However, there is no strong historical evidence that market exchange rates move in the direction implied by the PPP rate, at least not in the short-or medium-term. Note: the numbers for GDP and other economic data should not be chained together from successive volumes of the *Factbook* because of changes in the US dollar measuring rod, revisions of data by statistical agencies, use of new or different sources of information, and changes in national statistical methods and practices.

Geographic coordinates This entry includes rounded latitude and longitude figures for the centroid or center point of a country expressed in degrees and minutes; it is based on the locations provided in the Geographic Names Server (GNS), maintained by the National Geospatial-Intelligence Agency on behalf of the US Board on Geographic Names.

Geographic names This information is presented in Appendix F: Cross Reference List of Geographic Names. It includes a listing of various alternate names, former names, local names, and regional names referenced to one or more related *Factbook* entries. Spellings are normally, but not always, those approved by the US Board on Geographic Names (BGN). Alternate names and additional information are included in parentheses.

Geographic overview This entry, which appears only in the World, Geography category, provides basic geographic information about the earth's oceans and continents. The entry also lists all of the countries that compose each continent.

Geography This category includes the entries dealing with the natural environment and the effects of human activity.

Geography—note This entry includes miscellaneous geographic information of significance not included elsewhere.

Gini index See entry for **Distribution of family income—Gini index**

GNP Gross national product (GNP) is the value of all final goods and services produced within a nation in a given year, plus income earned by its citizens abroad, minus income earned by foreigners from domestic production. The *Factbook,* following current practice, uses GDP rather than GNP to measure national production. However, the user must realize that in certain countries net remittances from citizens working abroad may be important to national well-being.

Government This category includes the entries dealing with the system for the adoption and administration of public policy.

Government—note This entry includes miscellaneous government information of significance not included elsewhere.

Government type This entry gives the basic form of government. Definitions of the major governmental terms are as follows. (Note that for some countries more than one definition applies.):

Absolute monarchy—a form of government where the monarch rules unhindered, i.e., without any laws, constitution, or legally organized opposition.

Anarchy—a condition of lawlessness or political disorder brought about by the absence of governmental authority.

Authoritarian—a form of government in which state authority is imposed onto many aspects of citizens' lives.

Commonwealth—a nation, state, or other political entity founded on law and united by a compact of the people for the common good.

Communist—a system of government in which the state plans and controls the economy and a single—often authoritarian—party holds power; state controls are imposed with the elimination of private ownership of property or capital while claiming to make progress toward a higher social order in which all goods are equally shared by the people (i.e., a classless society).

Confederacy (Confederation)—a union by compact or treaty between states, provinces, or territories, that creates a central government with limited powers; the constituent entities retain supreme authority over all matters except those delegated to the central government.

Constitutional—a government by or operating under an authoritative document (constitution) that sets forth the system of fundamental laws and principles that determines the nature, functions, and limits of that government.

Constitutional democracy—a form of government in which the sovereign power of the people is spelled out in a governing constitution.

Constitutional monarchy—a system of government in which a monarch is guided by a constitution whereby his/her rights, duties, and responsibilities are spelled out in written law or by custom.

Democracy—a form of government in which the supreme power is retained by the people, but which is usually exercised indirectly through a system of representation and delegated authority periodically renewed.

Democratic republic—a state in which the supreme power rests in the body of citizens entitled to vote for officers and representatives responsible to them.

Dictatorship—a form of government in which a ruler or small clique wield absolute power (not restricted by a constitution or laws).

Ecclesiastical—a government administrated by a church.

Emirate—similar to a monarchy or sultanate, but a government in which the supreme power is in the hands of an emir (the ruler of a Muslim state); the emir may be an absolute overlord or a sovereign with constitutionally limited authority.

Federal (Federation)—a form of government in which sovereign power is formally divided—usually by means of a constitution—between a central authority and a number of constituent regions (states, colonies, or provinces) so that each region retains some management of its internal affairs; differs from a confederacy in that the central government exerts influence directly upon both individuals as well as upon the regional units.

Federal republic—a state in which the powers of the central government are restricted and in which the component parts (states, colonies, or provinces) retain a degree of self-government; ultimate sovereign power rests with the voters who chose their governmental representatives.

Islamic republic—a particular form of government adopted by some Muslim states; although such a state is, in theory, a theocracy, it remains a republic, but its laws are required to be compatible with the laws of Islam.

Maoism—the theory and practice of Marxism-Leninism developed in China by Mao Zedong (Mao Tse-tung), which states that a continuous revolution is necessary if the leaders of a communist state are to keep in touch with the people.

Marxism—the political, economic, and social principles espoused by 19th century economist Karl Marx; he viewed the struggle of workers as a progression of historical forces that would proceed from a class struggle of the proletariat (workers) exploited by capitalists (business owners), to a socialist" dictatorship of the proletariat," to, finally, a classless society—Communism.

Marxism-Leninism—an expanded form of communism developed by Lenin from doctrines of Karl Marx; Lenin saw imperialism as the final stage of capitalism and shifted the focus of workers' struggle from developed to underdeveloped countries.

Monarchy—a government in which the supreme power is lodged in the hands of a monarch who reigns over a state or territory, usually for life and by hereditary right; the monarch may be either a sole absolute ruler or a sovereign—such as a king, queen, or prince—with constitutionally limited authority.

Oligarchy—a government in which control is exercised by a small group of individuals whose authority generally is based on wealth or power.

Parliamentary democracy—a political system in which the legislature (parliament) selects the government—a prime minister, premier, or chancellor along with the cabinet ministers—according to party strength as expressed in elections; by this system, the government acquires a dual responsibility: to the people as well as to the parliament.

Parliamentary government (Cabinet-Parliamentary government)—a government in which members of an executive branch (the cabinet and its leader—a prime minister, premier, or chancellor) are nominated to their positions by a legislature or parliament, and are directly responsible to It; this type of government can be dissolved at will by the parliament (legislature) by means of a no confidence vote or the leader of the cabinet may dissolve the parliament if it can no longer function.

Parliamentary monarchy—a state headed by a monarch who is not actively involved in policy formation or implementation (i.e., the exercise of sovereign powers by a monarch in a ceremonial capacity); true governmental leadership is carried out by a cabinet and its head—a prime minister, premier, or chancellor—who are drawn from a legislature (parliament).

Presidential—a system of government where the executive branch exists separately from a legislature (to which it is generally not accountable).

Republic—a representative democracy in which the people's elected deputies (representatives), not the people themselves, vote on legislation.

Socialism—a government in which the means of planning, producing, and distributing goods is controlled by a central government that theoretically seeks a more just and equitable distribution of property and labor; in actuality, most socialist governments have ended up being no more than dictatorships over workers by a ruling elite.

Sultanate—similar to a monarchy, but a government in which the supreme power is in the hands of a sultan (the head of a Muslim state); the sultan may be an absolute ruler or a sovereign with constitutionally limited authority.

Theocracy—a form of government in which a Deity is recognized as the supreme civil ruler, but the Deity's laws are interpreted by ecclesiastical authorities (bishops, mullahs, etc.); a government subject to religious authority.

Totalitarian—a government that seeks to subordinate the individual to the state by controlling not only all political and economic matters, but also the attitudes, values, and beliefs of its population.

Greenwich Mean Time (GMT) The mean solar time at the Greenwich Meridian, Greenwich, England, with the hours and days, since 1925, reckoned from midnight. GMT is now a historical term having been replaced by UTC on 1 January 1972. See **Coordinated Universal Time.**

Gross domestic product See GDP

Gross national product See GNP

Gross national saving Gross national saving is derived by deducting final consumption expenditure (household plus government) from Gross national disposable income, and consists of personal saving, plus business saving (the sum of the capital consumption allowance and retained business profits), plus government saving (the excess of tax revenues over expenditures), but excludes foreign saving (the excess of imports of goods and services over exports). The figures are presented as a percent of GDP. A negative number indicates that the economy as a whole is spending more income than it produces, thus drawing down national wealth (dissaving).

Gross world product See GWP

GWP This entry gives the gross world product (GWP) or aggregate value of all final goods and services produced worldwide in a given year.

Heliports This entry gives the total number of heliports with hard-surface runways, helipads, or landing areas that support routine sustained helicopter operations exclusively and have support facilities including one or more of the following facilities: lighting, fuel, passenger handling, or maintenance. It includes former airports used exclusively for helicopter operations but excludes heliports limited to day operations and natural clearings that could support helicopter landings and takeoffs.

HIV/AIDS—adult prevalence rate This entry gives an estimate of the percentage of adults (aged 15-49) living with HIV/AIDS. The adult prevalence rate is calculated by dividing the estimated number of adults living with HIV/AIDS at yearend by the total adult population at yearend.

HIV/AIDS—deaths This entry gives an estimate of the number of adults and children who died of AIDS during a given calendar year.

HIV/AIDS—people living with HIV/AIDS This entry gives an estimate of all people (adults and children) alive at yearend with HIV infection, whether or not they have developed symptoms of AIDS.

Hospital bed density This entry provides the number of hospital beds per 1,000 people; it serves as a general measure of inpatient service availability. Hospital beds include inpatient beds available in public, private, general, and specialized hospitals and rehabilitation centers. In most cases, beds for both acute and chronic care are included. Because the level of inpatient services required for individual countries depends on several factors—such as demographic issues and the burden of disease—there is no global target for the number of hospital beds per country. So, while 2 beds per 1,000 in one country may be sufficient, 2 beds per 1,000 in another may be woefully inadequate because of the number of people hospitalized by disease.

Household income or consumption by percentage share Data on household income or consumption come from household surveys, the results adjusted for household size. Nations use different standards and procedures in collecting and adjusting the data. Surveys based on income will normally show a more unequal distribution than surveys based on consumption. The quality of surveys is improving with time, yet caution is still necessary in making inter-country comparisons.

Hydrographic data codes See **Data codes**

Illicit drugs This entry gives information on the five categories of illicit drugs—narcotics, stimulants, depressants (sedatives), hallucinogens, and cannabis. These categories include many drugs legally produced and prescribed by doctors as well as those illegally produced and sold outside of medical channels.

    Cannabis (*Cannabis sativa*) is the common hemp plant, which provides hallucinogens with some sedative properties, and includes marijuana (pot, Acapulco gold, grass, reefer), tetrahydrocannabinol (THC, Marinol), hashish (hash), and hashish oil (hash oil).

    Coca (mostly *Erythroxylum coca*) is a bush with leaves that contain the stimulant used to make cocaine. Coca is not to be confused with cocoa, which comes from cacao seeds and is used in making chocolate, cocoa, and cocoa butter.

    Cocaine is a stimulant derived from the leaves of the coca bush.

    Depressants (sedatives) are drugs that reduce tension and anxiety and include chloral hydrate, barbiturates (Amytal, Nembutal, Seconal, phenobarbital), benzodiazepines (Librium,

Valium), methaqualone (Quaalude), glutethimide (Doriden), and others (Equanil, Placidyl, Valmid).

    Drugs are any chemical substances that effect a physical, mental, emotional, or behavioral change in an individual.

    Drug abuse is the use of any licit or illicit chemical substance that results in physical, mental, emotional, or behavioral impairment in an individual.

    Hallucinogens are drugs that affect sensation, thinking, self-awareness, and emotion. Hallucinogens include LSD (acid, microdot), mescaline and peyote (mexc, buttons, cactus), amphetamine variants (PMA, STP, DOB), phencyclidine (PCP, angel dust, hog), phencyclidine analogues (PCE, PCPy, TCP), and others (psilocybin, psilocyn).

    Hashish is the resinous exudate of the cannabis or hemp plant (*Cannabis sativa*).

    Heroin is a semisynthetic derivative of morphine.

    Mandrax is a trade name for methaqualone, a pharmaceutical depressant.

    Marijuana is the dried leaf of the cannabis or hemp plant (*Cannabis sativa*).

    Methaqualone is a pharmaceutical depressant, referred to as mandrax in Southwest Asia and Africa.

    Narcotics are drugs that relieve pain, often induce sleep, and refer to opium, opium derivatives, and synthetic substitutes. Natural narcotics include opium (paregoric, parepectolin), morphine (MS-Contin, Roxanol), codeine (Tylenol with codeine, Empirin with codeine, Robitussin AC), and thebaine. Semisynthetic narcotics include heroin (horse, smack), and hydromorphone (Dilaudid). Synthetic narcotics include meperidine or Pethidine (Demerol, Mepergan), methadone (Dolophine, Methadose), and others (Darvon, Lomotil).

    Opium is the brown, gummy exudate of the incised, unripe seedpod of the opium poppy.

    Opium poppy (*Papaver somniferum*) is the source for the natural and semisynthetic narcotics.

    Poppy straw is the entire cut and dried opium poppy-plant material, other than the seeds. Opium is extracted from poppy straw in commercial operations that produce the drug for medical use.

    Qat (kat, khat) is a stimulant from the buds or leaves of *Catha edulis* that is chewed or drunk as tea.

    Quaaludes is the North American slang term for methaqualone, a pharmaceutical depressant.

    Stimulants are drugs that relieve mild depression, increase energy and activity, and include cocaine (coke, snow, crack), amphetamines (Desoxyn, Dexedrine), ephedrine, ecstasy (clarity, essence, doctor, Adam), phenmetrazine (Preludin), methylphenidate (Ritalin), and others (Cylert, Sanorex, Tenuate).

Imports This entry provides the total US dollar amount of merchandise imports on a c. i. f. (cost, insurance, and freight) or f. o. b. (free on board) basis. These figures are calculated on an exchange rate basis, i.e., not in purchasing power parity (PPP) terms.

Imports—commodities This entry provides a listing of the highest-valued imported products; it sometimes includes the percent of total dollar value.

Imports—partners This entry provides a rank ordering of trading partners starting with the most important; it sometimes includes the percent of total dollar value.

Independence For most countries, this entry gives the date that sovereignty was achieved and from which nation, empire, or trusteeship. For the other countries, the date given may not represent "independence" in the strict sense, but rather some significant nationhood event such as the traditional founding date or the date of unification, federation, confederation, establishment, fundamental change in the form of government, or state succession. For a number of countries, the establishment of statehood was a lengthy evolutionary process occurring over decades or even centuries. In such cases, several significant dates are cited. Dependent areas include the notation "none" followed by the nature of their dependency status. Also see the **Terminology** note.

Industrial production growth rate This entry gives the annual percentage increase in industrial production (includes manufacturing, mining, and construction).

Industries This entry provides a rank ordering of industries starting with the largest by value of annual output.

Infant mortality rate This entry gives the number of deaths of infants under one year old in a given year per 1,000 live births in the same year. This rate is often used as an indicator of the level of health in a country.

Inflation rate (consumer prices) This entry furnishes the annual percent change in consumer prices compared with the previous year's consumer prices.

International disputes see Disputes—international

International law organization participation This entry includes information on a country's acceptance of jurisdiction of the International Court of Justice (ICJ) and of the International Criminal Court (ICCt); 59 countries have accepted ICJ jurisdiction with reservations and 11 have accepted ICJ jurisdiction without reservations; 122 countries have accepted ICCt jurisdiction. Appendix B: International Organizations and Groups explains the differing mandates of the ICJ and ICCt.

International organization participation This entry lists in alphabetical order by abbreviation those international organizations in which the subject country is a member or participates in some other way.

International organizations This information is presented in Appendix B: International Organizations and Groups which includes the name, abbreviation, date established, aim, and members by category.

Internet country code This entry includes the two-letter codes maintained by the International Organization for Standardization (ISO) in the ISO 3166 Alpha-2 list and used by the Internet Assigned Numbers Authority (IANA) to establish country-coded top-level domains (ccTLDs).

Internet users This entry gives the *total* number of individuals within a country who can access the Internet at home, via any device type (computer or mobile) and connection. The *percent of population* with Internet access (i.e., the penetration rate) helps gauge how widespread Internet use is within a country. Statistics vary from country to country and may include users who access the Internet at least several times a week to those who access it only once within a period of several months.

Introduction This category includes one entry, **Background.**

Investment (gross fixed) This entry records total business spending on fixed assets, such as factories, machinery, equipment, dwellings, and inventories of raw materials, which provide the basis for future production. It is measured gross of the depreciation of the assets, i.e., it includes investment that merely replaces worn-out or scrapped capital.

Irrigated land This entry gives the number of square kilometers of land area that is artificially supplied with water.

Judicial branch This entry includes three subfields. The *highest court(s)* subfield includes the name(s) of a country's highest level court(s), the number and titles of the judges, and the types of cases heard by the court, which commonly are based on civil, crimi-

nal, administrative, and constitutional law. A number of countries have separate constitutional courts. The *judge selection and term of office* subfield includes the organizations and associated officials responsible for nominating and appointing judges, and a brief description of the process. The selection process can be indicative of the independence of a country's court system from other branches of its government. Also included in this subfield are judges' tenures, which can range from a few years, to a specified retirement age, to lifelong appointments. The *subordinate courts* subfield lists the courts lower in the hierarchy of a country's court system. A few countries with federal-style governments, such as Brazil, Canada, and the US, in addition to their federal court, have separate state- or province-level court systems, though generally the two systems interact.

Labor force This entry contains the total labor force figure.

Labor force—by occupation This entry lists the percentage distribution of the labor force by sector of occupation. *Agriculture* includes farming, fishing, and forestry. *Industry* includes mining, manufacturing, energy production, and construction. *Services* cover government activities, communications, transportation, finance, and all other economic activities that do not produce material goods. The distribution will total less than 100 percent if the data are incomplete and may range from 99-101 percent due to rounding.

Land boundaries This entry contains the *total* length of all land boundaries and the individual lengths for each of the contiguous *border countries.* When available, official lengths published by national statistical agencies are used. Because surveying methods may differ, country border lengths reported by contiguous countries may differ.

Land use This entry contains the percentage shares of total land area for three different types of land use: *agricultural land, forest, and other; agricultural land* is further divided into *arable land*—land cultivated for crops like wheat, maize, and rice that are replanted after each harvest, *permanent crops*—land cultivated for crops like citrus, coffee, and rubber that are not replanted after each harvest, and includes land under flowering shrubs, fruit trees, nut trees, and vines, and *permanent pastures* and meadows – land used for at least five years or more to grow herbaceous forage, either cultivated or growing naturally; *forest* area is land spanning more than 0.5 hectare with trees higher than five meters and a canopy cover of more than 10% to include windbreaks, shelterbelts, and corridors of trees greater than 0.5 hectare and at least 20 m wide; land classified as *other* includes built-up areas, roads and other transportation features, barren land, or wasteland.

Languages This entry provides a listing of languages spoken in each country and specifies any that are official national or regional languages. When data is available, the languages spoken in each country are broken down according to the percent of the total population speaking each language as a first language. For those countries without available data, languages are listed in rank order based on prevalence, starting with the most-spoken language.

Legal system This entry provides the description of a country's legal system. A statement on judicial review of legislative acts is also included for a number of countries. The legal systems of nearly all countries are generally modeled upon elements of five main types: civil law (including French law, the Napoleonic Code, Roman law, Roman-Dutch law, and Spanish law); common law (including United State law); customary law; mixed or pluralistic law; and religious law (including Islamic law). An additional type of legal system—international law, which governs the conduct of independent nations in their relationships with one another—is also addressed below. The following list describes these legal systems, the countries or world regions where these systems are enforced, and a brief statement on the origins and major features of each.

Civil Law—The most widespread type of legal system in the world, applied in various forms in approximately 150 countries. Also referred to as European continental law, the civil law system is derived mainly from the Roman *Corpus Juris Civilus*, (Body of Civil Law), a collection of laws and legal interpretations compiled under the East Roman (Byzantine) Emperor Justinian I between A. D. 528 and 565. The major feature of civil law systems is that the laws are organized into systematic written codes. In civil law the sources recognized as authoritative are principally legislation—especially codifications in constitutions or statutes enacted by governments—and secondarily, custom. The civil law systems in some countries are based on more than one code.

Common Law—A type of legal system, often synonymous with" English common law," which is the system of England and Wales in the UK, and is also in force in approximately 80 countries formerly part of or influenced by the former British Empire. English common law reflects Biblical influences as well as remnants of law systems imposed by early conquerors including the Romans, Anglo-Saxons, and Normans. Some legal scholars attribute the formation of the English common law system to King Henry II (r. 1154-1189). Until the time of his reign, laws customary among England's various manorial and ecclesiastical (church) jurisdictions were administered locally. Henry II established the king's court and designated that laws were "common" to the entire English realm. The foundation of English common law is " legal precedent"—referred to as *stare decisis,* meaning "to stand by things decided." In the English common law system, court judges are bound in their decisions in large part by the rules and other doctrines developed—and supplemented over time—by the judges of earlier English courts.

Customary Law—A type of legal system that serves as the basis of, or has influenced, the present-day laws in approximately 40 countries—mostly in Africa, but some in the Pacific islands, Europe, and the Near East. Customary law is also referred to as "primitive law," "unwritten law," "indigenous law," and "folk law." There is no single history of customary law such as that found in Roman civil law, English common law, Islamic law, or the Napoleonic Civil Code. The earliest systems of law in human society were customary, and usually developed in small agrarian and hunter-gatherer communities. As the term implies, customary law is based upon the customs of a community. Common attributes of customary legal systems are that they are seldom written down, they embody an organized set of rules regulating social relations, and they are agreed upon by members of the community. Although such law systems include sanctions for law infractions, resolution tends to be reconciliatory rather than punitive. A number of African states practiced customary law many centuries prior to colonial influences. Following colonization, such laws were written down and incorporated to varying extents into the legal systems imposed by their colonial powers.

European Union Law—A sub-discipline of international law known as" supranational law" in which the rights of sovereign nations are limited in relation to one another. Also referred to as the Law of the European Union or Community Law, it is the unique and complex legal system that operates in tandem with the laws of the 27 member states of the European Union (EU). Similar to federal states, the EU legal system ensures compliance from the member states because of the Union's decentralized political nature. The European Court of Justice (ECJ), established in 1952 by the Treaty of Paris, has been largely responsible for the development of EU law. Fundamental principles of European Union law include:

subsidiarity—the notion that issues be handled by the smallest, lowest, or least centralized competent authority; *proportionality*—the EU may only act to the extent needed to achieve its objectives; *conferral*—the EU is a union of member states, and all its authorities are voluntarily granted by its members; *legal certainty*—requires that legal rules be clear and precise; and *precautionary principle*—a moral and political principle stating that if an action or policy might cause severe or irreversible harm to the public or to the environment, in the absence of a scientific consensus that harm would not ensue, the burden of proof falls on those who would advocate taking the action.

French Law—A type of civil law that is the legal system of France. The French system also serves as the basis for, or is mixed with, other legal systems in approximately 50 countries, notably in North Africa, the Near East, and the French territories and dependencies. French law is primarily codified or systematic written civil law. Prior to the French Revolution (1789-1799), France had no single national legal system. Laws in the northern areas of present-day France were mostly local customs based on privileges and exemptions granted by kings and feudal lords, while in the southern areas Roman law predominated. The introduction of the Napoleonic Civil Code during the reign of Napoleon I in the first decade of the 19th century brought major reforms to the French legal system, many of which remain part of France's current legal structure, though all have been extensively amended or redrafted to address a modern nation. French law distinguishes between "public law" and "private law." Public law relates to government, the French Constitution, public administration, and criminal law. Private law covers issues between private citizens or corporations. The most recent changes to the French legal system—introduced in the 1980s—were the decentralization laws, which transferred authority from centrally appointed government representatives to locally elected representatives of the people.

International Law—The law of the international community, or the body of customary rules and treaty rules accepted as legally binding by states in their relations with each other. International law differs from other legal systems in that it primarily concerns sovereign political entities. There are three separate disciplines of international law: public international law, which governs the relationship between provinces and international entities and includes treaty law, law of the sea, international criminal law, and international humanitarian law; private international law, which addresses legal jurisdiction; and supranational law—a legal framework wherein countries are bound by regional agreements in which the laws of the member countries are held inapplicable when in conflict with supranational laws. At present the European Union is the only entity under a supranational legal system. The term" international law" was coined by Jeremy Bentham in 1780 in his *Principles of Morals and Legislation*, though laws governing relations between states have been recognized from very early times (many centuries B. C.). Modern international law developed alongside the emergence and growth of the European nation-states beginning in the early 16th century. Other factors that influenced the development of international law included the revival of legal studies, the growth of international trade, and the practice of exchanging emissaries and establishing legations. The sources of International law are set out in Article 38-1 of the Statute of the International Court of Justice within the UN Charter.

Islamic Law—The most widespread type of religious law, it is the legal system enforced in over 30 countries, particularly in the Near East, but also in Central and South Asia, Africa, and Indonesia. In many countries Islamic law operates in tandem with a civil law system. Islamic law is embodied in the sharia, an Arabic word meaning" the right path." Sharia covers all aspects of public and private life and organizes them into five categories: obligatory, recommended, permitted, disliked, and forbidden. The primary sources of sharia law are the Qur'an, believed by Muslims to be the word of God revealed to the Prophet Muhammad by the angel Gabriel, and the Sunnah, the teachings of the Prophet and his works. In addition to these two primary sources, traditional Sunni Muslims recognize the consensus of Muhammad's companions and Islamic jurists on certain issues, called ijmas, and various forms of reasoning, including analogy by legal scholars, referred to as qiyas. Shia Muslims reject ijmas and qiyas as sources of sharia law.

Mixed Law—Also referred to as pluralistic law, mixed law consists of elements of some or all of the other main types of legal systems—civil, common, customary, and religious. The mixed legal systems of a number of countries came about when colonial powers overlaid their own legal systems upon colonized regions but retained elements of the colonies' existing legal systems.

Napoleonic Civil Code—A type of civil law, referred to as the Civil Code or *Code Civil des Francais*, forms part of the legal system of France, and underpins the legal systems of Bolivia, Egypt, Lebanon, Poland, and the US state of Louisiana. The Civil Code was established under Napoleon I, enacted in 1804, and officially designated the *Code Napoleon* in 1807. This legal system combined the Teutonic civil law tradition of the northern provinces of France with the Roman law tradition of the southern and eastern regions of the country. The Civil Code bears similarities in its arrangement to the Roman *Body of Civil Law* (see Civil Law above). As enacted in 1804, the Code addressed personal status, property, and the acquisition of property. Codes added over the following six years included civil procedures, commercial law, criminal law and procedures, and a penal code.

Religious Law—A legal system which stems from the sacred texts of religious traditions and in most cases professes to cover all aspects of life as a seamless part of devotional obligations to a transcendent, imminent, or deep philosophical reality. Implied as the basis of religious law is the concept of unalterability, because the word of God cannot be amended or legislated against by judges or governments. However, a detailed legal system generally requires human elaboration. The main types of religious law are sharia in Islam, halakha in Judaism, and canon law in some Christian groups. Sharia is the most widespread religious legal system (see Islamic Law), and is the sole system of law for countries including Iran, the Maldives, and Saudi Arabia. No country is fully governed by halakha, but Jewish people may decide to settle disputes through Jewish courts and be bound by their rulings. Canon law is not a divine law as such because it is not found in revelation. It is viewed instead as human law inspired by the word of God and applying the demands of that revelation to the actual situation of the church. Canon law regulates the internal ordering of the Roman Catholic Church, the Eastern Orthodox Church, and the Anglican Communion.

Roman Law—A type of civil law developed in ancient Rome and practiced from the time of the city's founding (traditionally 753 B. C.) until the fall of the Western Empire in the 5th century A. D. Roman law remained the legal system of the Byzantine (Eastern Empire) until the fall of Constantinople in 1453. Preserved fragments of the first legal text, known as the Law of the Twelve Tables, dating from the 5th century B. C., contained specific provisions designed to change the prevailing customary law. Early Roman law was drawn from custom and statutes; later, during the time of the empire, emperors asserted their authority as the ultimate source of law. The basis for Roman laws was the idea that the exact form—not the intention—of words or of actions produced legal consequences. It was only in the late 6th

century A. D. that a comprehensive Roman code of laws was published (see Civil Law above). Roman law served as the basis of law systems developed in a number of continental European countries.

Roman-Dutch Law—A type of civil law based on Roman law as applied in the Netherlands. Roman-Dutch law serves as the basis for legal systems in seven African countries, as well as Guyana, Indonesia, and Sri Lanka. This law system, which originated in the province of Holland and expanded throughout the Netherlands (to be replaced by the French Civil Code in 1809), was instituted in a number of sub-Saharan African countries during the Dutch colonial period. The Dutch jurist/philosopher Hugo Grotius was the first to attempt to reduce Roman-Dutch civil law into a system in his *Jurisprudence of Holland* (written 1619-20, commentary published 1621). The Dutch historian/lawyer Simon van Leeuwen coined the term" Roman-Dutch law" in 1652.

Spanish Law—A type of civil law, often referred to as the Spanish Civil Code, it is the present legal system of Spain and is the basis of legal systems in 12 countries mostly in Central and South America, but also in southwestern Europe, northern and western Africa, and southeastern Asia. The Spanish Civil Code reflects a complex mixture of customary, Roman, Napoleonic, local, and modern codified law. The laws of the Visigoth invaders of Spain in the 5th to 7th centuries had the earliest major influence on Spanish legal system development. The Christian Reconquest of Spain in the 11th through 15th centuries witnessed the development of customary law, which combined canon (religious) and Roman law. During several centuries of Hapsburg and Bourbon rule, systematic recompilations of the existing national legal system were attempted, but these often conflicted with local and regional customary civil laws. Legal system development for most of the 19th century concentrated on formulating a national civil law system, which was finally enacted in 1889 as the Spanish Civil Code. Several sections of the code have been revised, the most recent of which are the penal code in 1989 and the judiciary code in 2001. The Spanish Civil Code separates public and private law. Public law includes constitutional law, administrative law, criminal law, process law, financial and tax law, and international public law. Private law includes civil law, commercial law, labor law, and international private law.

United States Law—A type of common law, which is the basis of the legal system of the United States and that of its island possessions in the Caribbean and the Pacific. This legal system has several layers, more possibly than in most other countries, and is due in part to the division between federal and state law. The United States was founded not as one nation but as a union of 13 colonies, each claiming independence from the British Crown. The US Constitution, implemented in 1789, began shifting power away from the states and toward the federal government, though the states today retain substantial legal authority. US law draws its authority from four sources: *constitutional law, statutory law, administrative regulations,* and *case law.* Constitutional law is based on the US Constitution and serves as the supreme federal law. Taken together with those of the state constitutions, these documents outline the general structure of the federal and state governments and provide the rules and limits of power. US statutory law is legislation enacted by the US Congress and is codified in the United States Code. The 50 state legislatures have similar authority to enact state statutes. Administrative law is the authority delegated to federal and state executive agencies. Case law, also referred to as common law, covers areas where constitutional or statutory law is lacking. Case law is a collection of judicial decisions, customs, and general principles that began in England centuries ago, that were adopted in America at the time of the Revolution, and that continue to develop today.

Legislative branch This entry has three subfields. The *description* subfield provides the legislative structure (unicameral – single house; bicameral – an upper and a lower house); formal name(s); number of member seats; types of constituencies or voting districts (single seat, multi-seat, nationwide); electoral voting system(s); and member term of office. The elections subfield includes the dates of the last election and next election. The *election results* subfield lists *percent of vote by party/coalition* and *number of seats by party/coalition* in the last election (in bicameral legislatures, upper house results are listed first). In general, parties with less than four seats and less than 4 percent of the vote are aggregated and listed as " other," and non-party-affiliated seats are listed as " independent." Also, the entries for some countries include two sets of *percent of vote by party* and *seats by party*; the former reflects results following a formal election announcement, and the latter – following a mid-term or byelection – reflects changes in a legislature's political party composition.

Of the approximately 240 countries with legislative bodies, approximately two-thirds are unicameral, and the remainder, bicameral. The selection of legislative members is typically governed by a country's constitution and/or its electoral laws. In general, members are either directly elected by a country's eligible voters using a defined electoral system; indirectly elected or selected by its province, state, or department legislatures; or appointed by the country's executive body. Legislative members in many countries are selected both directly and indirectly, and the electoral laws of some countries reserve seats for women and various ethnic and minority groups.

Worldwide, the two predominant direct voting systems are plurality/majority and proportional representation. The most common of the several plurality/majority systems is simple majority vote, or first-past-the-post, in which the candidate receiving the most votes is elected. Countries' legislatures such as Bangladesh's Parliament, Malaysia's House of Representatives, and the United Kingdom's House of Commons use this system. Another common plurality/majority system – absolute majority or two-round – requires that candidates win at least 50 percent of the votes to be elected. If none of the candidates meets that vote threshold in the initial election, a second poll or" runoff" is held soon after for the two top vote getters, and the candidate receiving a simple vote majority is declared the winner. Examples of the two-round system are Haiti's Chamber of Deputies, Mali's National Assembly, and Uzbekistan's Legislative Chamber. Other plurality/majority voting systems, referred to as preferential voting and generally used in multi-seat constituencies, are block vote and single non-transferable vote, in which voters cast their ballots by ranking their candidate preferences from highest to lowest.

Proportional representation electoral systems – in contrast to plurality/majority systems – generally award legislative seats to political parties in approximate proportion to the number of votes each receives. For example, in a 100-member legislature, if Party A receives 50 percent of the total vote, Party B, 30 percent, and Party C, 20 percent, then Party A would be awarded 50 seats, Party B 30 seats, and Party C 20 seats. There are various forms of proportional representation and the degree of reaching proportionality varies. Some forms of proportional representation are focused solely on achieving the proportional representation of different political parties and voters cast ballots only for political parties, whereas in other forms, voters cast ballots for individual candidates within a political party.

Many countries—both unicameral and bicameral—use a mix of electoral methods, in which a portion of legislative seats are awarded using one system, such as plurality/majority, while the remaining seats are awarded by another system, such as proportional representation. Many countries with bicameral legislatures use different voting systems for the two chambers.

Life expectancy at birth This entry contains the average number of years to be lived by a group of people born in the same year, if mortality at each age remains constant in the future. Life expectancy at birth is also a measure of overall quality of life in a country and summarizes the mortality at all ages. It can also be thought of as indicating the potential return on investment in human capital and is necessary for the calculation of various actuarial measures.

Literacy This entry includes a *definition* of literacy and UNESCO's percentage estimates for populations aged 15 years and over, including *total population, males*, and *females*. There are no universal definitions and standards of literacy. Unless otherwise specified, all rates are based on the most common definition—the ability to read and write at a specified age. Detailing the standards that individual countries use to assess the ability to read and write is beyond the scope of the *Factbook*. Information on literacy, while not a perfect measure of educational results, is probably the most easily available and valid for international comparisons. Low levels of literacy, and education in general, can impede the economic development of a country in the current rapidly changing, technology-driven world.

Location This entry identifies the country's regional location, neighboring countries, and adjacent bodies of water.

Major infectious diseases This entry lists major infectious diseases likely to be encountered in countries where the risk of such diseases is assessed to be very high as compared to the United States. These infectious diseases represent risks to US government personnel traveling to the specified country for a period of less than three years. The *degree of risk* is assessed by considering the foreign nature of these infectious diseases, their severity, and the probability of being affected by the diseases present. The diseases listed do not necessarily represent the total disease burden experienced by the local population.

The risk to an individual traveler varies considerably by the specific location, visit duration, type of activities, type of accommodations, time of year, and other factors. Consultation with a travel medicine physician is needed to evaluate individual risk and recommend appropriate preventive measures such as vaccines.

Diseases are organized into the following six exposure categories shown in italics *and listed in typical descending order of risk*. Note: The sequence of exposure categories listed in individual country entries may vary according to local conditions.

food or waterborne diseases acquired through eating or drinking on the local economy:

Hepatitis A—viral disease that interferes with the functioning of the liver; spread through consumption of food or water contaminated with fecal matter, principally in areas of poor sanitation; victims exhibit fever, jaundice, and diarrhea; 15% of victims will experience prolonged symptoms over 6-9 months; vaccine available.

Hepatitis E—water-borne viral disease that interferes with the functioning of the liver; most commonly spread through fecal contamination of drinking water; victims exhibit jaundice, fatigue, abdominal pain, and dark colored urine.

Typhoid fever—bacterial disease spread through contact with food or water contaminated by fecal matter or sewage; victims exhibit sustained high fevers; left untreated, mortality rates can reach 20%.

vectorborne diseases acquired through the bite of an infected arthropod:

Malaria—caused by single-cell parasitic protozoa *Plasmodium*; transmitted to humans via the bite of the female *Anopheles* mosquito; parasites multiply in the liver attacking red blood cells resulting in cycles of fever, chills, and sweats accompanied by anemia; death due to damage to vital organs and interruption of blood supply to the brain; endemic in 100, mostly tropical, countries with 90% of cases and the majority of 0.4-0.8 million estimated annual deaths occurring in sub-Saharan Africa.

Dengue fever—mosquito-borne (*Aedes aegypti*) viral disease associated with urban environments; manifests as sudden onset of fever and severe headache; occasionally produces shock and hemorrhage leading to death in 5% of cases.

Yellow fever—mosquito-borne (in urban areas *Aedes aegypti*) viral disease; severity ranges from influenza-like symptoms to severe hepatitis and hemorrhagic fever; occurs only in tropical South America and sub-Saharan Africa, where most cases are reported; fatality rate is less than 20%.

Japanese Encephalitis—mosquito-borne (*Culex tritaeniorhynchus*) viral disease associated with rural areas in Asia; acute encephalitis can progress to paralysis, coma, and death; fatality rates 30%.

African Trypanosomiasis—caused by the parasitic protozoa *Trypanosoma*; transmitted to humans via the bite of bloodsucking Tsetse flies; infection leads to malaise and irregular fevers and, in advanced cases when the parasites invade the central nervous system, coma and death; endemic in 36 countries of sub-Saharan Africa; cattle and wild animals act as reservoir hosts for the parasites.

Cutaneous Leishmaniasis—caused by the parasitic protozoa *leishmania*; transmitted to humans via the bite of sandflies; results in skin lesions that may become chronic; endemic in 88 countries; 90% of cases occur in Iran, Afghanistan, Syria, Saudi Arabia, Brazil, and Peru; wild and domesticated animals as well as humans can act as reservoirs of infection.

Plague—bacterial disease transmitted by fleas normally associated with rats; person-to-person airborne transmission also possible; recent plague epidemics occurred in areas of Asia, Africa, and South America associated with rural areas or small towns and villages; manifests as fever, headache, and painfully swollen lymph nodes; disease progresses rapidly and without antibiotic treatment leads to pneumonic form with a death rate in excess of 50%.

Crimean-Congo hemorrhagic fever—tick-borne viral disease; infection may also result from exposure to infected animal blood or tissue; geographic distribution includes Africa, Asia, the Middle East, and Eastern Europe; sudden onset of fever, headache, and muscle aches followed by hemorrhaging in the bowels, urine, nose, and gums; mortality rate is approximately 30%.

Rift Valley fever—viral disease affecting domesticated animals and humans; transmission is by mosquito and other biting insects; infection may also occur through handling of infected meat or contact with blood; geographic distribution includes eastern and southern Africa where cattle and sheep are raised; symptoms are generally mild with fever and some liver abnormalities, but the disease may progress to hemorrhagic fever, encephalitis, or ocular disease; fatality rates are low at about 1% of cases.

Chikungunya—mosquito-borne (*Aedes aegypti*) viral disease associated with urban environments, similar to Dengue Fever; characterized by sudden onset of fever, rash, and severe joint pain usually lasting 3-7 days, some cases result in persistent arthritis.

water contact diseases acquired through swimming or wading in freshwater lakes, streams, and rivers:

Leptospirosis—bacterial disease that affects animals and humans; infection occurs through contact with water, food, or soil contaminated by animal urine; symptoms include high fever, severe headache, vomiting, jaundice, and diarrhea; untreated, the disease can result in kidney damage, liver failure, meningitis, or respiratory distress; fatality rates are low but left untreated recovery can take months.

Schistosomiasis—caused by parasitic trematode flatworm *Schistosoma*; fresh water snails act as intermediate host and release larval form of parasite that penetrates the skin of people exposed to contaminated water; worms mature and reproduce in the blood vessels, liver, kidneys, and intestines releasing eggs, which become trapped in tissues triggering an immune response; may manifest as either urinary or intestinal disease resulting in decreased work or learning capacity; mortality, while generally low, may occur in advanced cases usually due to bladder cancer; endemic in 74 developing countries with 80% of infected people living in sub-Saharan Africa; humans act as the reservoir for this parasite.

aerosolized dust or soil contact disease acquired through inhalation of aerosols contaminated with rodent urine:

Lassa fever—viral disease carried by rats of the genus *Mastomys*; endemic in portions of West Africa; infection occurs through direct contact with or consumption of food contaminated by rodent urine or fecal matter containing virus particles; fatality rate can reach 50% in epidemic outbreaks.

respiratory disease acquired through close contact with an infectious person:

Meningococcal meningitis—bacterial disease causing an inflammation of the lining of the brain and spinal cord; one of the most important bacterial pathogens is *Neisseria meningitidis* because of its potential to cause epidemics; symptoms include stiff neck, high fever, headaches, and vomiting; bacteria are transmitted from person to person by respiratory droplets and facilitated by close and prolonged contact resulting from crowded living conditions, often with a seasonal distribution; death occurs in 5-15% of cases, typically within 24-48 hours of onset of symptoms; highest burden of meningococcal disease occurs in the hyperendemic region of sub-Saharan Africa known as the "Meningitis Belt" which stretches from Senegal east to Ethiopia.

animal contact disease acquired through direct contact with local animals:

Rabies—viral disease of mammals usually transmitted through the bite of an infected animal, most commonly dogs; virus affects the central nervous system causing brain alteration and death; symptoms initially are non-specific fever and headache progressing to neurological symptoms; death occurs within days of the onset of symptoms.

Major urban areas—population This entry provides the population of the capital and up to six major cities defined as urban agglomerations with populations of at least 750,000 people. An *urban agglomeration* is defined as comprising the city or town proper and also the suburban fringe or thickly settled territory lying outside of, but adjacent to, the boundaries of the city. For smaller countries, lacking urban centers of 750,000 or more, only the population of the capital is presented.

Map references This entry includes the name of the *Factbook* reference map on which a country may be found. Note that boundary representations on these maps are not necessarily authoritative. The entry on **Geographic coordinates** may be helpful in finding some smaller countries.

Marine fisheries This entry describes the major fisheries in the world's oceans in terms of the area covered, their ranking in terms of the global catch, the main producing countries, and the principal species caught. Information provided by the Fisheries and Aquaculture Department of the UN Food and Agriculture Organization (FAO).

Maritime claims This entry includes the following claims, the definitions of which are excerpted from the United Nations Convention on the Law of the Sea (UNCLOS), which alone contains the full and definitive descriptions:

territorial sea—the sovereignty of a coastal state extends beyond its land territory and internal waters to an adjacent belt of sea, described as the territorial sea in the UNCLOS (Part II); this sovereignty extends to the air space over the territorial sea as well as its underlying seabed and subsoil; every state has the right to establish the breadth of its territorial sea up to a limit not exceeding 12 nautical miles; the normal baseline for measuring the breadth of the territorial sea is the mean low-water line along the coast as marked on large-scale charts officially recognized by the coastal state; where the coasts of two states are opposite or adjacent to each other, neither state is entitled to extend its territorial sea beyond the median line, every point of which is equidistant from the nearest points on the baseline from which the territorial seas of both states are measured; the UNCLOS describes specific rules for archipelagic states.

contiguous zone—according to the UNCLOS (Article 33), this is a zone contiguous to a coastal state's territorial sea, over which it may exercise the control necessary to: prevent infringement of its customs, fiscal, immigration, or sanitary laws and regulations within its territory or territorial sea; punish infringement of the above laws and regulations committed within its territory or territorial sea; the contiguous zone may not extend beyond 24 nautical miles from the baselines from which the breadth of the territorial sea is measured (e.g., the US has claimed a 12-nautical mile contiguous zone in addition to its 12-nautical mile territorial sea); where the coasts of two states are opposite or adjacent to each other, neither state is entitled to extend its contiguous zone beyond the median line, every point of which is equidistant from the nearest points on the baseline from which the contiguous zone of both states are measured.

exclusive economic zone (EEZ)—the UNCLOS (Part V) defines the EEZ as a zone beyond and adjacent to the territorial sea in which a coastal state has: sovereign rights for the purpose of exploring and exploiting, conserving and managing the natural resources, whether living or non-living, of the waters superjacent to the seabed and of the seabed and its subsoil, and with regard to other activities for the economic exploitation and exploration of the zone, such as the production of energy from the water, currents, and winds; jurisdiction with regard to the establishment and use of artificial islands, installations, and structures; marine scientific research; the protection and preservation of the marine environment; the outer limit of the exclusive economic zone shall not exceed 200 nautical miles from the baselines from which the breadth of the territorial sea is measured.

continental shelf—the UNCLOS (Article 76) defines the continental shelf of a coastal state as comprising the seabed and subsoil of the submarine areas that extend beyond its territorial sea throughout the natural prolongation of its land territory to the outer edge of the continental margin, or to a distance of 200 nautical miles from the baselines from which the breadth of the territorial sea is measured where the outer edge of the continental margin does not extend up to that distance; the continental margin comprises the submerged prolongation of the landmass of the coastal state, and consists of the seabed and subsoil of the shelf, the slope and the rise; wherever the continental margin extends beyond 200 nautical miles from the baseline, coastal states may extend their claim to a distance not to exceed 350

nautical miles from the baseline or 100 nautical miles from the 2,500-meter isobath, which is a line connecting points of 2,500 meters in depth; it does not include the deep ocean floor with its oceanic ridges or the subsoil thereof.

exclusive fishing zone—while this term is not used in the UNCLOS, some states (e.g., the United Kingdom) have chosen not to claim an EEZ but rather to claim jurisdiction over the living resources off their coast; in such cases, the term exclusive fishing zone is often used; the breadth of this zone is normally the same as the EEZ or 200 nautical miles.

Maritime threats This entry describes the threat of piracy, as defined in Article 101, UN Convention on the Law of the Sea (UNCLOS), or armed robbery against ships, as defined in Resolution A. 1025 (26) adopted on 2 December 2009 at the 26th Assembly Session of the International Maritime Organization. The entry includes the number of ships on the high seas or in territorial waters that were boarded or attacked by pirates, and the number of crewmen abducted or killed, as compiled by the International Maritime Bureau. Information is also supplied on the geographical range of attacks.

Maternal mortality rate The maternal mortality rate (MMR) is the annual number of female deaths per 100,000 live births from any cause related to or aggravated by pregnancy or its management (excluding accidental or incidental causes). The MMR includes deaths during pregnancy, childbirth, or within 42 days of termination of pregnancy, irrespective of the duration and site of the pregnancy, for a specified year.

Median age This entry is the age that divides a population into two numerically equal groups; that is, half the people are younger than this age and half are older. It is a single index that summarizes the age distribution of a population. Currently, the median age ranges from a low of about 15 in Niger and Uganda to 40 or more in several European countries and Japan. See the entry for "Age structure" for the importance of a young versus an older age structure and, by implication, a low versus a higher median age.

Member states This entry, which appears only in the European Union, Government category, provides a listing of all of the European Union member countries, as well as their associated overseas countries and territories.

Merchant marine This entry provides the total and the number of each type of privately or publicly owned commercial ship for each country; military ships are not included; the five ships by type include: *bulk carrier*—for cargo such as coal, grain, cement, ores, and gravel; *container ship*—for loads in truck-size containers, a transportation system called containerization; *general cargo*—also referred to as break-bulk containers—for a wide variety of packaged merchandise, such as textiles, furniture and machinery; *oil tanker*—for crude oil and petroleum products; *other*—includes chemical carriers, dredgers, liquefied natural gas (LNG) carriers, refrigerated cargo ships called reefers, tugboats, passenger vessels (cruise and ferry), and offshore supply ships

Military This category includes the entries dealing with a country's military structure, manpower, and expenditures.

Military—note This entry includes miscellaneous military information of significance not included elsewhere.

Military and security forces This entry lists the military and security forces subordinate to defense ministries or the equivalent (typically ground, naval, air, and marine forces), as well as those belonging to interior ministries or the equivalent (typically gendarmeries, border/coast guards, paramilitary police, and other internal security forces).

Military and security service personnel strengths This entry provides estimates of military and security services personnel strengths. The numbers are based on a wide-range of publicly available information. Unless otherwise noted, military estimates focus on the major services (army, navy, air force, and where applicable, gendarmeries) and do not account for activated reservists or delineate military service members assigned to joint staffs or defense ministries.

Military deployments This entry lists military forces deployed to other countries or territories abroad. *The World Factbook* defines deployed as a permanently-stationed force or a temporary deployment of greater than six months. Deployments smaller than 100 personnel or paramilitaries, police, contractors, mercenaries, or proxy forces are not included. Numbers provided are estimates only and should be considered paper strengths, not necessarily the current number of troops on the ground. In addition, some estimates, such as those by the US military, are significantly influenced by deployment policies, contingencies, or world events and may change suddenly. Where available, the organization or mission that at least some of the forces are deployed under is listed. The following terms and abbreviations are used throughout the entry:

AMISOM—Africa Union (AU) Mission in Somalia; UN-supported, AU-operated peacekeeping mission

BATUS—British Army Training Unit Suffield, Canada

BATUK—British Army Training Unit, Kenya

CSTO—Collective Security Treaty Organization

ECOMIG—ECOWUS Mission in The Gambia; Africa Union-European Union peacekeeping, stabilization, and training mission in Gambia

EUTM—European Union Training Mission

EUFOR—European Union Force Bosnia and Herzegovina (also known as Operation Althea)

EuroCorps—European multi-national corps headquartered in Strasbourg, France, consisting of troops from Belgium, France, Germany, Luxembourg, and Spain; Greece, Italy, Poland, Romania and Turkey are Associated Nations of EuroCorps

G5 Joint Force—G5 Sahel Cross-Border Joint Force comprised of troops from Burkina Faso, Chad, Mali, Mauritania, and Niger

KFOR—the Kosovo Force; a NATO-led international peacekeeping force in Kosovo

MFO—Multinational Force & Observers Sinai, headquartered in Rome

MINOSCO—United Nations Organization Stabilization Mission in the Democratic Republic of the Congo

MINUSCA—United Nations Multidimensional Integrated Stabilization Mission in the Central African Republic

MINUSMA—United Nations Multidimensional Integrated Stabilization Mission in Mali

MNJTF—Multinational Joint Task Force Against Boko Haram comprised of troops from Cameroon, Niger, and Nigeria with the mission of fighting Boko Haram in the Lake Chad Basin

NATO—North American Treaty Organization, headquartered in Brussels, Belgium

Operation Barkhane—French-led counterinsurgency and counter-terrorism mission in the Sahel alongside the G5 Joint Force; headquartered in N'Djamena, Chad and supported by Canada, Denmark, Estonia, the European Union, Germany, Spain, the United Kingdom, and the US

Operation Inherent Resolve—US-led coalition to counter the Islamic State in Iraq and Syria and provide assistance and training to Iraqi security forces

UNAFIL—United Nations Interim Force in Lebanon

UNAMID—African Union—United Nations Hybrid Operation in Darfur, Sudan

UNDOF—United Nations Disengagement Observer Force, Golan (Israel-Syria border)

UNFICYP—United Nations Peacekeeping Force in Cyprus

UNISFA—United Nations Interim Security Force for Abyei (Sudan-South Sudan border)

UNMISS—United Nations Mission in the Republic of South Sudan

UNSOM—United Nations Assistance Mission in Somalia

Military equipment inventories and acquisitions This entry provides basic information on each country's military equipment inventories, as well as how they acquire their equipment; it is intended to show broad trends in major military equipment holdings, such as tanks and other armored vehicles, air defense systems, artillery, naval ships, helicopters, and fixed-wing aircraft. Arms acquisition information is an overview of major arms suppliers over a specific period of time, including second-hand arms delivered as aid, with a focus on major weapons systems. It is based on the type and number of weapon systems ordered and delivered and the financial value of the deal. For some countries, general information on domestic defense industry capabilities is provided.

Military expenditures This entry gives estimates on spending on defense programs for the most recent year available as a percent of gross domestic product (GDP); the GDP is calculated on an exchange rate basis, i.e., not in terms of purchasing power parity (PPP). For countries with no military forces, this figure can include expenditures on public security and police.

Military service age and obligation This entry gives the required ages for voluntary or conscript military service and the length of service obligation.

Money figures All money figures are expressed in contemporaneous US dollars unless otherwise indicated.

Mother's mean age at first birth This entry provides the mean (average) age of mothers at the birth of their first child. It is a useful indicator for gauging the success of family planning programs aiming to reduce maternal mortality, increase contraceptive use—particularly among married and unmarried adolescents—delay age at first marriage, and improve the health of newborns.

National air transport system This entry includes four subfields describing the air transport system of a given country in terms of both structure and performance. The first subfield, *number of registered air carriers*, indicates the total number of air carriers registered with the country's national aviation authority and issued an air operator certificate as required by the Convention on International Civil Aviation. The second subfield, *inventory of registered aircraft operated by air carriers*, lists the total number of aircraft operated by all registered air carriers in the country. The last two subfields measure the performance of the air transport system in terms of both passengers and freight. The subfield, *annual passenger traffic on registered air carriers*, includes the total number of passengers carried by air carriers registered in the country, including both domestic and international passengers, in a given year. The last subfield, *annual freight traffic on registered air carriers*, includes the volume of freight, express, and diplomatic bags carried by registered air carriers and measured in metric tons times kilometers traveled. Freight ton-kilometers equal the sum of the products obtained by multiplying the number of tons of freight, express, and diplomatic bags carried on each flight stage by the stage distance (operation of an aircraft from takeoff to its next landing). For statistical purposes, freight includes express and diplomatic bags but not passenger baggage.

National anthem A generally patriotic musical composition—usually in the form of a song or hymn of praise—that evokes and eulogizes the history, traditions, or struggles of a nation or its people. National anthems can be officially recognized as a national song by a country's constitution or by an enacted law, or simply by tradition. Although most anthems contain lyrics, some do not.

National holiday This entry gives the primary national day of celebration—usually independence day.

National symbol(s) A national symbol is a faunal, floral, or other abstract representation—or some distinctive object—that over time has come to be closely identified with a country or entity. Not all countries have national symbols; a few countries have more than one.

Nationality This entry provides the identifying terms for citizens—*noun* and *adjective*.

Natural gas—consumption This entry is the total natural gas consumed in cubic meters (cu m). The discrepancy between the amount of natural gas produced and/or imported and the amount consumed and/or exported is due to the omission of stock changes and other complicating factors.

Natural gas—exports This entry is the total natural gas exported in cubic meters (cu m).

Natural gas—imports This entry is the total natural gas imported in cubic meters (cu m).

Natural gas—production This entry is the total natural gas produced in cubic meters (cu m). The discrepancy between the amount of natural gas produced and/or imported and the amount consumed and/or exported is due to the omission of stock changes and other complicating factors.

Natural gas—proved reserves This entry is the stock of proved reserves of natural gas in cubic meters (cu m). Proved reserves are those quantities of natural gas, which, by analysis of geological and engineering data, can be estimated with a high degree of confidence to be commercially recoverable from a given date forward, from known reservoirs and under current economic conditions.

Natural hazards This entry lists potential natural disasters. For countries where volcanic activity is common, a *volcanism* subfield highlights historically active volcanoes.

Natural resources This entry lists a country's mineral, petroleum, hydropower, and other resources of commercial importance, such as rare earth elements (REEs). In general, products appear only if they make a significant contribution to the economy, or are likely to do so in the future.

Net migration rate This entry includes the figure for the difference between the number of persons entering and leaving a country during the year per 1,000 persons (based on midyear population). An excess of persons entering the country is referred to as net immigration (e.g., 3.56 migrants/1,000 population); an excess of persons leaving the country as net emigration (e.g., -9.26 migrants/1,000 population). The net migration rate indicates the contribution of migration to the overall level of population change. The net migration rate does not distinguish between economic migrants, refugees, and other types of migrants nor does it distinguish between lawful migrants and undocumented migrants.

Obesity—adult prevalence rate This entry gives the percent of a country's population considered to be obese. Obesity is defined as an adult having a Body Mass Index (BMI) greater to or equal to 30.0. BMI is calculated by taking a person's weight in kg and dividing it by the person's squared height in meters.

People—note This entry includes miscellaneous demographic information of significance not included elsewhere.

People and Society This category includes entries dealing with national identity (including ethnicities, languages, and reli-

gions), demography (a variety of population statistics) and societal characteristics (health and education indicators).

Personal Names—Capitalization The *Factbook* capitalizes the surname or family name of individuals for the convenience of our users who are faced with a world of different cultures and naming conventions. The need for capitalization, bold type, underlining, italics, or some other indicator of the individual's surname is apparent in the following examples: MAO Zedong, Fidel CASTRO Ruz, George W. BUSH, and TUNKU SALAHUDDIN Abdul Aziz Shah ibni Al-Marhum Sultan Hisammuddin Alam Shah. By knowing the surname, a short form without all capital letters can be used with confidence as in President Castro, Chairman Mao, President Bush, or Sultan Tunku Salahuddin. The same system of capitalization is extended to the names of leaders with surnames that are not commonly used such as Queen ELIZABETH II. For Vietnamese names, the given name is capitalized because officials are referred to by their given name rather than by their surname. For example, the president of Vietnam is Tran Duc LUONG. His surname is Tran, but he is referred to by his given name—President LUONG.

Personal Names—Spelling The romanization of personal names in the *Factbook* normally follows the same transliteration system used by the US Board on Geographic Names for spelling place names. At times, however, a foreign leader expressly indicates a preference for, or the media or official documents regularly use, a romanized spelling that differs from the transliteration derived from the US Government standard. In such cases, the *Factbook* uses the alternative spelling.

Personal Names—Titles The *Factbook* capitalizes any valid title (or short form of it) immediately preceding a person's name. A title standing alone is not capitalized. Examples: President PUTIN and President OBAMA are chiefs of state. In Russia, the president is chief of state and the premier is the head of the government, while in the US, the president is both chief of state and head of government.

Petroleum See entries under **Refined petroleum products**.

Petroleum products See entries under **Refined petroleum products**.

Physicians density This entry gives the number of medical doctors (physicians), including generalist and specialist medical practitioners, per 1,000 of the population. Medical doctors are defined as doctors that study, diagnose, treat, and prevent illness, disease, injury, and other physical and mental impairments in humans through the application of modern medicine. They also plan, supervise, and evaluate care and treatment plans by other health care providers. The World Health Organization estimates that fewer than 2.3 health workers (physicians, nurses, and midwives only) per 1,000 would be insufficient to achieve coverage of primary health-care needs.

Pipelines This entry gives the lengths and types of pipelines for transporting products like natural gas, crude oil, or petroleum products.

Piracy Piracy is defined by the 1982 United Nations Convention on the Law of the Sea as any illegal act of violence, detention, or depredation directed against a ship, aircraft, persons, or property in a place outside the jurisdiction of any State. Such criminal acts committed in the territorial waters of a littoral state are generally considered to be armed robbery against ships. Information on piracy may be found, where applicable, under **Maritime threats**.

Political parties and leaders This entry includes a listing of significant political parties, coalitions, and electoral lists as of each country's last legislative election, unless otherwise noted.

Political structure This entry, which appears only in the European Union, Government category, provides a definition for the entity that is the European Union.

Population This entry gives an estimate from the US Bureau of the Census based on statistics from population censuses, vital statistics registration systems, or sample surveys pertaining to the recent past and on assumptions about future trends. The total population presents one overall measure of the potential impact of the country on the world and within its region. Note: Starting with the 1993 *Factbook*, demographic estimates for some countries (mostly African) have explicitly taken into account the effects of the growing impact of the HIV/AIDS epidemic. These countries are currently: The Bahamas, Benin, Botswana, Brazil, Burkina Faso, Burma, Burundi, Cambodia, Cameroon, Central African Republic, Democratic Republic of the Congo, Republic of the Congo, Cote d'Ivoire, Ethiopia, Gabon, Ghana, Guyana, Haiti, Honduras, Kenya, Lesotho, Malawi, Mozambique, Namibia, Nigeria, Rwanda, South Africa, Swaziland, Tanzania, Thailand, Togo, Uganda, Zambia, and Zimbabwe.

Population below poverty line National estimates of the percentage of the population falling below the poverty line are based on surveys of sub-groups, with the results weighted by the number of people in each group. Definitions of poverty vary considerably among nations. For example, rich nations generally employ more generous standards of poverty than poor nations.

Population distribution This entry provides a summary description of the population dispersion within a country. While it may suggest population density, it does not provide density figures.

Population growth rate The average annual percent change in the population, resulting from a surplus (or deficit) of births over deaths and the balance of migrants entering and leaving a country. The rate may be positive or negative. The growth rate is a factor in determining how great a burden would be imposed on a country by the changing needs of its people for infrastructure (e.g., schools, hospitals, housing, roads), resources (e.g., food, water, electricity), and jobs. Rapid population growth can be seen as threatening by neighboring countries.

Population pyramid A population pyramid illustrates the age and sex structure of a country's population and may provide insights about political and social stability, as well as economic development. The population is distributed along the horizontal axis, with males shown on the left and females on the right. The male and female populations are broken down into 5-year age groups represented as horizontal bars along the vertical axis, with the youngest age groups at the bottom and the oldest at the top. The shape of the population pyramid gradually evolves over time based on fertility, mortality, and international migration trends.

Some distinctive types of population pyramids are:

- A **youthful distribution** has a broad base and narrow peak and is characterized by a high proportion of children and low proportion of the elderly. This population distribution results from high fertility, high mortality, low life expectancy, and high population growth. It is typical of developing countries where female education and contraceptive use are low and health care and sanitation are poor.
- A **transitional distribution** is caused by declining fertility and mortality rates, increasing life expectancy, and slowing population growth. The population has a larger proportion of working-age people relative to children and the elderly and produces a barrel-shaped pyramid, where the mid-section bulges and the base and top are narrower. The large proportion of working-age

people can create a "demographic bonus" if it is educated and productively employed.
- A **mature distribution** has fairly balanced proportions of the population in the child, working-age, and elderly age groups and will gradually form an inverted triangle population pyramid as population growth continues to fall or ceases and the proportion of older people increases. Low fertility, low mortality, and high life expectancy—made possible by the availability of advanced healthcare, family planning, sanitation, and education—lead to aging populations in industrialized countries.

Ports and terminals This entry lists major ports and terminals primarily on the basis of the amount of cargo tonnage shipped through the facilities on an annual basis. In some instances, the number of containers handled or ship visits were also considered. Most ports service multiple classes of vessels including bulk carriers (dry and liquid), break bulk cargoes (goods loaded individually in bags, boxes, crates, or drums; sometimes palletized), containers, roll-on/roll-off, and passenger ships. The listing leads off with *major seaports* handling all types of cargo. Inland *river/lake ports* are listed separately along with the river or lake name. Ports configured specifically to handle bulk cargoes are designated as *oil terminals* or *dry bulk cargo ports. LNG terminals* handle liquefied natural gas (LNG) and are differentiated as either export, where the gas is chilled to a liquid state to reduce its volume for transport on specialized gas carriers, or import, where the off-loaded LNG undergoes a regasification process before entering pipelines for distribution. As break bulk cargoes are largely transported by containers today, the entry also includes a listing of major *container ports* with the corresponding throughput measured in twenty-foot equivalent units (TEUs). Some ports are significant for handling passenger traffic and are listed as *cruise/ferry ports.* In addition to commercial traffic, many seaports also provide important military infrastructure such as naval bases or dockyards.

Preliminary statement This entry, which appears only in the European Union, Introduction category, provides an explanation and justification for the inclusion of a separate European Union geographic entity.

Principality A sovereign state ruled by a monarch with the title of prince; principalities were common in the past, but today only three remain: Liechtenstein, Monaco, and the co-principality of Andorra.

Public debt This entry records the cumulative total of all government borrowings less repayments that are denominated in a country's home currency. Public debt should not be confused with external debt, which reflects the foreign currency liabilities of both the private and public sector and must be financed out of foreign exchange earnings.

Railways This entry states the *total* route length of the railway network and of its component parts by gauge, which is the measure of the distance between the inner sides of the load-bearing rails. The four typical types of gauges are: *broad, standard, narrow,* and *dual.* Other gauges are listed under *note.* Some 60% of the world's railways use the standard gauge of 1.4 m (4.7 ft). Gauges vary by country and sometimes within countries. The choice of gauge during initial construction was mainly in response to local conditions and the intent of the builder. Narrow-gauge railways were cheaper to build and could negotiate sharper curves, broad-gauge railways gave greater stability and permitted higher speeds. Standard-gauge railways were a compromise between narrow and broad gauges.

Rare earth elements Rare earth elements or REEs are 17 chemical elements that are critical in many of today's high-tech industries. They include lanthanum, cerium, praseodymium, neodymium, promethium, samarium, europium, gadolinium, terbium, dysprosium, holmium, erbium, thulium, ytterbium, lutetium, scandium, and yttrium. Typical applications for REEs include batteries in hybrid cars, fiber optic cables, flat panel displays, and permanent magnets, as well as some defense and medical products.

Reference maps This section includes world and regional maps.

Refined petroleum products—consumption This entry is the country's total consumption of refined petroleum products, in barrels per day (bbl/day). The discrepancy between the amount of refined petroleum products produced and/or imported and the amount consumed and/or exported is due to the omission of stock changes, refinery gains, and other complicating factors.

Refined petroleum products—exports This entry is the country's total exports of refined petroleum products, in barrels per day (bbl/day).

Refined petroleum products—imports This entry is the country's total imports of refined petroleum products, in barrels per day (bbl/day).

Refined petroleum products—production This entry is the country's total output of refined petroleum products, in barrels per day (bbl/day). The discrepancy between the amount of refined petroleum products produced and/or imported and the amount consumed and/or exported is due to the omission of stock changes, refinery gains, and other complicating factors.

Refugees and internally displaced persons This entry includes those persons residing in a country as *refugees, internally displaced persons (IDPs)*, or *stateless persons*. Each country's refugee entry includes only countries of origin that are the source of refugee populations of 5,000 or more. The definition of a refugee according to a UN Convention is "a person who is outside his/her country of nationality or habitual residence; has a well-founded fear of persecution because of his/her race, religion, nationality, membership in a particular social group or political opinion; and is unable or unwilling to avail himself/herself of the protection of that country, or to return there, for fear of persecution." The UN established the Office of the UN High Commissioner for Refugees (UNHCR) in 1950 to handle refugee matters worldwide. The UN Relief and Works Agency for Palestine Refugees in the Near East (UNRWA) has a different operational definition for a Palestinian refugee: "a person whose normal place of residence was Palestine during the period 1 June 1946 to 15 May 1948 and who lost both home and means of livelihood as a result of the 1948 conflict." However, UNHCR also assists some 400,000 Palestinian refugees not covered under the UNRWA definition. The term "internally displaced person" is not specifically covered in the UN Convention; it is used to describe people who have fled their homes for reasons similar to refugees, but who remain within their own national territory and are subject to the laws of that state. A **stateless person** is defined as someone who is not considered a national by any state under the operation of its law, according to UN convention.

Religions This entry is an ordered listing of religions by adherents starting with the largest group and sometimes includes the percent of total population. The core characteristics and beliefs of the world's major religions are described below.

Baha'i—Founded by Mirza Husayn-Ali (known as Baha'u'llah) in Iran in 1852, Baha'i faith emphasizes monotheism and believes in one eternal transcendent God. Its guiding focus is to encourage the unity of all peoples on the earth so that justice and peace may be achieved on earth. Baha'i revelation contends the prophets of major world religions reflect some truth or element of the divine, believes all were manifestations of God given to specific communities in specific times, and that

Baha'u'llah is an additional prophet meant to call all humankind. Bahais are an open community, located worldwide, with the greatest concentration of believers in South Asia.

Buddhism—Religion or philosophy inspired by the 5th century B.C. teachings of Siddhartha Gautama (also known as Gautama Buddha "the enlightened one"). Buddhism focuses on the goal of spiritual enlightenment centered on an understanding of Gautama Buddha's Four Noble Truths on the nature of suffering, and on the Eightfold Path of spiritual and moral practice, to break the cycle of suffering of which we are a part. Buddhism ascribes to a karmic system of rebirth. Several schools and sects of Buddhism exist, differing often on the nature of the Buddha, the extent to which enlightenment can be achieved—for one or for all, and by whom—religious orders or laity.

Basic Groupings

Theravada Buddhism: The oldest Buddhist school, Theravada is practiced mostly in Sri Lanka, Cambodia, Laos, Burma, and Thailand, with minority representation elsewhere in Asia and the West. Theravadans follow the Pali Canon of Buddha's teachings, and believe that one may escape the cycle of rebirth, worldly attachment, and suffering for oneself; this process may take one or several lifetimes.

Mahayana Buddhism, including subsets Zen and Tibetan (Lamaistic) Buddhism: Forms of Mahayana Buddhism are common in East Asia and Tibet, and parts of the West. Mahayanas have additional scriptures beyond the Pali Canon and believe the Buddha is eternal and still teaching. Unlike Theravada Buddhism, Mahayana schools maintain the Buddha-nature is present in all beings and all will ultimately achieve enlightenment. Hoa Hao: a minority tradition of Buddhism practiced in Vietnam that stresses lay participation, primarily by peasant farmers; it eschews expensive ceremonies and temples and relocates the primary practices into the home.

Christianity—Descending from Judaism, Christianity's central belief maintains Jesus of Nazareth is the promised messiah of the Hebrew Scriptures, and that his life, death, and resurrection are salvific for the world. Christianity is one of the three monotheistic Abrahamic faiths, along with Islam and Judaism, which traces its spiritual lineage to Abraham of the Hebrew Scriptures. Its sacred texts include the Hebrew Bible and the New Testament (or the Christian Gospels).

Basic Groupings

Catholicism (or Roman Catholicism): This is the oldest established western Christian church and the world's largest single religious body. It is supranational, and recognizes a hierarchical structure with the Pope, or Bishop of Rome, as its head, located at the Vatican. Catholics believe the Pope is the divinely ordered head of the Church from a direct spiritual legacy of Jesus' apostle Peter. Catholicism is comprised of 23 particular Churches, or Rites—one Western (Roman or Latin-Rite) and 22 Eastern. The Latin Rite is by far the largest, making up about 98% of Catholic membership. Eastern- Rite Churches, such as the Maronite Church and the Ukrainian Catholic Church, are in communion with Rome although they preserve their own worship traditions and their immediate hierarchy consists of clergy within their own rite. The Catholic Church has a comprehensive theological and moral doctrine specified for believers in its catechism, which makes it unique among most forms of Christianity.

Mormonism (including the Church of Jesus Christ of Latter-Day Saints): Originating in 1830 in the United States under Joseph Smith, Mormonism is not characterized as a form of Protestant Christianity because it claims additional revealed Christian scriptures after the Hebrew Bible and New Testament. The Book of Mormon maintains there was an appearance of Jesus in the New World following the Christian account of his resurrection, and that the Americas are uniquely blessed continents. Mormonism believes earlier Christian traditions, such as the Roman Catholic, Orthodox, and Protestant reform faiths, are apostasies and that Joseph Smith's revelation of the Book of Mormon is a restoration of true Christianity. Mormons have a hierarchical religious leadership structure, and actively proselytize their faith; they are located primarily in the Americas and in a number of other Western countries.

Jehovah's Witnesses structure their faith on the Christian Bible, but their rejection of the Trinity is distinct from mainstream Christianity. They believe that a Kingdom of God, the Theocracy, will emerge following Armageddon and usher in a new earthly society. Adherents are required to evangelize and to follow a strict moral code.

Orthodox Christianity: The oldest established eastern form of Christianity, the Holy Orthodox Church, has a ceremonial head in the Bishop of Constantinople (Istanbul), also known as a Patriarch, but its various regional forms (e.g., Greek Orthodox, Russian Orthodox, Serbian Orthodox, Ukrainian Orthodox) are autocephalous (independent of Constantinople's authority, and have their own Patriarchs). Orthodox churches are highly nationalist and ethnic. The Orthodox Christian faith shares many theological tenets with the Roman Catholic Church, but diverges on some key premises and does not recognize the governing authority of the Pope.

Protestant Christianity: Protestant Christianity originated in the 16th century as an attempt to reform Roman Catholicism's practices, dogma, and theology. It encompasses several forms or denominations which are extremely varied in structure, beliefs, relationship to state, clergy, and governance. Many protestant theologies emphasize the primary role of scripture in their faith, advocating individual interpretation of Christian texts without the mediation of a final religious authority such as the Roman Pope. The oldest Protestant Christianities include Lutheranism, Calvinism (Presbyterians), and Anglican Christianity (Episcopalians), which have established liturgies, governing structure, and formal clergy. Other variants on Protestant Christianity, including Pentecostal movements and independent churches, may lack one or more of these elements, and their leadership and beliefs are individualized and dynamic.

Hinduism—Originating in the Vedic civilization of India (second and first millennium B.C.), Hinduism is an extremely diverse set of beliefs and practices with no single founder or religious authority. Hinduism has many scriptures; the Vedas, the Upanishads, and the Bhagavad-Gita are among some of the most important. Hindus may worship one or many deities, usually with prayer rituals within their own home. The most common figures of devotion are the gods Vishnu, Shiva, and a mother goddess, Devi. Most Hindus believe the soul, or *atman*, is eternal, and goes through a cycle of birth, death, and rebirth (*samsara*) determined by one's positive or negative karma, or the consequences of one's actions. The goal of religious life is to learn to act so as to finally achieve liberation (*moksha*) of one's soul, escaping the rebirth cycle.

Islam—The third of the monotheistic Abrahamic faiths, Islam originated with the teachings of Muhammad in the 7th century. Muslims believe Muhammad is the final of all religious prophets (beginning with Abraham) and that the Qu'ran, which is the Islamic scripture, was revealed to him by God. Islam derives from the word submission, and obedience to God is a primary theme in this religion. In order to live an Islamic life, believers must follow the five pillars, or tenets, of Islam, which are the testimony of faith (*shahada*), daily prayer (*salah*), giving alms (*zakah*), fasting during Ramadan (*sawm*), and the pilgrimage to Mecca (*hajj*).

Basic Groupings

The two primary branches of Islam are Sunni and Shia, which split from each other over a religio-political leadership dispute about the rightful successor to Muhammad. The Shia believe Muhammad's cousin and son-in-law, Ali, was the only divinely ordained Imam (religious leader), while the Sunni maintain the first three caliphs after Muhammad were also legitimate authorities. In modern Islam, Sunnis and Shia continue to have different views of acceptable schools of Islamic jurisprudence, and who is a proper Islamic religious authority. Islam also has an active mystical branch, Sufism, with various Sunni and Shia subsets.

Sunni Islam accounts for over 75% of the world's Muslim population. It recognizes the Abu Bakr as the first caliph after Muhammad. Sunni has four schools of Islamic doctrine and law—Hanafi, Maliki, Shafi'i, and Hanbali—which uniquely interpret the *Hadith*, or recorded oral traditions of Muhammad. A Sunni Muslim may elect to follow any one of these schools, as all are considered equally valid.

Shia Islam represents 10-20% of Muslims worldwide, and its distinguishing feature is its reverence for Ali as an infallible, divinely inspired leader, and as the first Imam of the Muslim community after Muhammad. A majority of Shia are known as "Twelvers," because they believe that the 11 familial successor imams after Muhammad culminate in a 12th Imam (al-Mahdi) who is hidden in the world and will reappear at its end to redeem the righteous.

Variants

Ismaili faith: A sect of Shia Islam, its adherents are also known as "Seveners," because they believe that the rightful seventh Imam in Islamic leadership was Isma'il, the elder son of Imam Jafar al-Sadiq. Ismaili tradition awaits the return of the seventh Imam as the Mahdi, or Islamic messianic figure. Ismailis are located in various parts of the world, particularly South Asia and the Levant.

Alawi faith: Another Shia sect of Islam, the name reflects followers' devotion to the religious authority of Ali. Alawites are a closed, secretive religious group who assert they are Shia Muslims, although outside scholars speculate their beliefs may have a syncretic mix with other faiths originating in the Middle East. Alawis live mostly in Syria, Lebanon, and Turkey.

Druze faith: A highly secretive tradition and a closed community that derives from the Ismaili sect of Islam; its core beliefs are thought to emphasize a combination of Gnostic principles believing that the Fatimid caliph, al-Hakin, is the one who embodies the key aspects of goodness of the universe, which are, the intellect, the word, the soul, the preceder, and the follower. The Druze have a key presence in Syria, Lebanon, and Israel.

Jainism—Originating in India, Jain spiritual philosophy believes in an eternal human soul, the eternal universe, and a principle of "the own nature of things." It emphasizes compassion for all living things, seeks liberation of the human soul from reincarnation through enlightenment, and values personal responsibility due to the belief in the immediate consequences of one's behavior. Jain philosophy teaches non-violence and prescribes vegetarianism for monks and laity alike; its adherents are a highly influential religious minority in Indian society.

Judaism—One of the first known monotheistic religions, likely dating to between 2000-1500 B.C., Judaism is the native faith of the Jewish people, based upon the belief in a covenant of responsibility between a sole omnipotent creator God and Abraham, the patriarch of Judaism's Hebrew Bible, or *Tanakh*. Divine revelation of principles and prohibitions in the Hebrew Scriptures form the basis of Jewish law, or *halakhah*, which is a key component of the faith. While there are extensive traditions of Jewish halakhic and theological discourse, there is no final dogmatic authority in the tradition.

Local communities have their own religious leadership. Modern Judaism has three basic categories of faith: Orthodox, Conservative, and Reform/Liberal. These differ in their views and observance of Jewish law, with the Orthodox representing the most traditional practice, and Reform/Liberal communities the most accommodating of individualized interpretations of Jewish identity and faith.

Shintoism—A native animist tradition of Japan, Shinto practice is based upon the premise that every being and object has its own spirit or *kami*. Shinto practitioners worship several particular *kamis*, including the *kamis* of nature, and families often have shrines to their ancestors' *kamis*. Shintoism has no fixed tradition of prayers or prescribed dogma, but is characterized by individual ritual. Respect for the *kamis* in nature is a key Shinto value. Prior to the end of World War II, Shinto was the state religion of Japan, and bolstered the cult of the Japanese emperor.

Sikhism—Founded by the Guru Nanak (born 1469), Sikhism believes in a non-anthropomorphic, supreme, eternal, creator God; centering one's devotion to God is seen as a means of escaping the cycle of rebirth. Sikhs follow the teachings of Nanak and nine subsequent gurus. Their scripture, the Guru Granth Sahib—also known as the Adi Granth—is considered the living Guru, or final authority of Sikh faith and theology. Sikhism emphasizes equality of humankind and disavows caste, class, or gender discrimination.

Taoism—Chinese philosophy or religion based upon Lao Tzu's Tao Te Ching, which centers on belief in the Tao, or the way, as the flow of the universe and the nature of things. Taoism encourages a principle of non-force, or wu-wei, as the means to live harmoniously with the Tao. Taoists believe the esoteric world is made up of a perfect harmonious balance and nature, while in the manifest world—particularly in the body—balance is distorted. The Three Jewels of the Tao—compassion, simplicity, and humility—serve as the basis for Taoist ethics.

Zoroastrianism—Originating from the teachings of Zoroaster in about the 9th or 10th century B.C., Zoroastrianism may be the oldest continuing creedal religion. Its key beliefs center on a transcendent creator God, Ahura Mazda, and the concept of free will. The key ethical tenets of Zoroastrianism expressed in its scripture, the Avesta, are based on a dualistic worldview where one may prevent chaos if one chooses to serve God and exercises good thoughts, good words, and good deeds. Zoroastrianism is generally a closed religion and members are almost always born to Zoroastrian parents. Prior to the spread of Islam, Zoroastrianism dominated greater Iran. Today, though a minority, Zoroastrians remain primarily in Iran, India (where they are known as Parsi), and Pakistan.

Traditional beliefs

Animism: the belief that non-human entities contain souls or spirits.

Badimo: a form of ancestor worship of the Tswana people of Botswana.

Confucianism: an ideology that humans are perfectible through self-cultivation and self-creation; developed from teachings of the Chinese philosopher Confucius. Confucianism has strongly influenced the culture and beliefs of East Asian countries, including China, Japan, Korea, Singapore, Taiwan, and Vietnam.

Inuit beliefs are a form of shamanism (see below) based on animistic principles of the Inuit or Eskimo peoples.

Kirant: the belief system of the Kirat, a people who live mainly in the Himalayas of Nepal. It is primarily a form of polytheistic shamanism, but includes elements of animism and ancestor worship.

Pagan is a blanket term used to describe many unconnected belief practices throughout history, usually in reference to religions outside of the Abrahamic category (monotheistic faiths like Judaism, Christianity, and Islam).

Shamanism: beliefs and practices promoting communication with the spiritual world. Shamanistic beliefs are organized around a shaman or medicine man who—as an intermediary between the human and spirit world—is believed to be able to heal the sick (by healing their souls), communicate with the spirit world, and help souls into the afterlife through the practice of entering a trance. In shaman-based religions, the shaman is also responsible for leading sacred rites.

Spiritualism: the belief that souls and spirits communicate with the living usually through intermediaries called mediums.

Syncretic (fusion of diverse religious beliefs and practices)

Cao Dai: a nationalistic Vietnamese sect, officially established in 1926, that draws practices and precepts from Confucianism, Taoism, Buddhism, and Catholicism.

Chondogyo: or the religion of the Heavenly Way, is based on Korean shamanism, Buddhism, and Korean folk traditions, with some elements drawn from Christianity. Formulated in the 1860s, it holds that God lives in all of us and strives to convert society into a paradise on earth, populated by believers transformed into intelligent moral beings with a high social conscience.

Kimbanguist: a puritan form of the Baptist denomination founded by Simon Kimbangu in the 1920s in what is now the Democratic Republic of Congo. Adherents believe that salvation comes through Jesus' death and resurrection, like Christianity, but additionally that living a spiritually pure life following strict codes of conduct is required for salvation.

Modekngei: a hybrid of Christianity and ancient Palauan culture and oral traditions founded around 1915 on the island of Babeldaob. Adherents simultaneously worship Jesus Christ and Palauan goddesses.

Rastafarian: an afro-centrist ideology and movement based on Christianity that arose in Jamaica in the 1930s; it believes that Haile Selassie I, Emperor of Ethiopia from 1930-74, was the incarnation of the second coming of Jesus.

Santeria: practiced in Cuba, the merging of the Yoruba religion of Nigeria with Roman Catholicism and native Indian traditions. Its practitioners believe that each person has a destiny and eventually transcends to merge with the divine creator and source of all energy, Olorun.

Voodoo/Vodun: a form of spirit and ancestor worship combined with some Christian faiths, especially Catholicism. Haitian and Louisiana Voodoo, which have included more Catholic practices, are separate from West African Vodun, which has retained a focus on spirit worship.

Non-religious

Agnosticism: the belief that most things are unknowable. In regard to religion it is usually characterized as neither a belief nor non belief in a deity.

Atheism: the belief that there are no deities of any kind.

Reserves of foreign exchange and gold This entry gives the dollar value for the stock of all financial assets that are available to the central monetary authority for use in meeting a country's balance of payments needs as of the end-date of the period specified. This category includes not only foreign currency and gold, but also a country's holdings of Special Drawing Rights in the International Monetary Fund, and its reserve position in the Fund.

Roadways This entry gives the *total* length of the road network and includes the length of the *paved* and *unpaved* portions.

Sanitation facility access This entry provides information about access to improved or unimproved sanitation facilities available to segments of the population of a country. *Improved* sanitation—use of any of the following facilities: flush or pour-flush to a piped sewer system, septic tank or pit latrine; ventilated improved pit (VIP) latrine; pit latrine with slab; or a composting toilet. *Unimproved* sanitation—use of any of the following facilities: flush or pour-flush not piped to a sewer system, septic tank or pit latrine; pit latrine without a slab or open pit; bucket; hanging toilet or hanging latrine; shared facilities of any type; no facilities; or bush or field.

School life expectancy (primary to tertiary education) School life expectancy (SLE) is the total number of years of schooling (primary to tertiary) that a child can expect to receive, assuming that the probability of his or her being enrolled in school at any particular future age is equal to the current enrollment ratio at that age. Caution must be maintained when utilizing this indicator in international comparisons. For example, a year or grade completed in one country is not necessarily the same in terms of educational content or quality as a year or grade completed in another country. SLE represents the expected number of years of schooling that will be completed, including years spent repeating one or more grades.

Sex ratio This entry includes the number of males for each female in five age groups—*at birth, under 15 years, 15-64 years, 65 years and over,* and for the *total population.* Sex ratio at birth has recently emerged as an indicator of certain kinds of sex discrimination in some countries. For instance, high sex ratios at birth in some Asian countries are now attributed to sex- selective abortion and infanticide due to a strong preference for sons. This will affect future marriage patterns and fertility patterns. Eventually, it could cause unrest among young adult males who are unable to find partners.

Stateless person Statelessness is the condition whereby an individual is not considered a national by any country. Stateless people are denied basic rights, such as access to employment, housing, education, healthcare, and pensions, and they may be unable to vote, own property, open a bank account, or legally register a marriage or birth. They may also be vulnerable to arbitrary treatment and human trafficking. In at least 30 states, women cannot pass their nationality on to their children. In these countries, if a child's father is foreign, stateless, or absent, the child usually becomes stateless. Estimates of the number of stateless people are inherently imprecise because few countries have procedures to identify them; the UN approximates that there are 12 million stateless people worldwide. Stateless people are counted in a country's overall population figure if they have lived there for a year.

Suffrage This entry gives the age at enfranchisement and whether the right to vote is universal or restricted.

Taxes and other revenues This entry records total taxes and other revenues received by the national government during the time period indicated, expressed as a percent of GDP. Taxes include personal and corporate income taxes, value added taxes, excise taxes, and tariffs. Other revenues include social contributions—such as payments for social security and hospital insurance—grants, and net revenues from public enterprises. Normalizing the data, by dividing total revenues by GDP, enables easy comparisons across countries, and provides an average rate at which all income (GDP) is paid to the national level government for the supply of public goods and services.

Telecommunication systems This entry includes a brief general assessment of a country's telecommunications system with details on the domestic and international components. The following terms and abbreviations are used throughout the entry:

2G—is short for second-generation cellular network. After 2G was launched, the previous mobile wireless network systems were retroactively dubbed 1G. While radio signals on 1G networks are analog, radio signals on 2G networks are digital. Both systems use digital signaling to connect the radio towers (which listen to the devices) to the rest of the mobile system.

3G—is the third generation of wireless mobile telecommunications technology. It is the upgrade for 2.5G and 2.5G GPRS networks, for faster data transfer. This increased speed is based on a set of standards used for mobile devices and mobile telecommunications use services and networks that comply with the International Mobile Telecommunications-2000 (IMT-2000) specifications by the International Telecommunication Union. 3G finds application in wireless voice telephony, mobile Internet access, fixed wireless Internet access, video calls, and mobile TV.

4G—is the fourth generation of broadband cellular network technology, succeeding 3G. The first-release Long Term Evolution (LTE) standard was commercially deployed in Oslo, Norway, and Stockholm, Sweden in 2009, and has since been deployed throughout most parts of the world. Applications, include enhanced mobile web access, IP telephony, highdefinition mobile TV, and video conferencing.

5G—is the fifth generation technology standard for cellular networks, which cellular phone companies began deploying worldwide in 2019; it is the planned successor to the 4G networks which provide connectivity to most current cellphones. Like its predecessors, 5G networks are cellular networks, in which the service area is divided into small geographical areas called cells. All 5G wireless devices in a cell are connected to the Internet and telephone network by radio waves through a local antenna in the cell. The main advantage of the new networks is that they will have greater bandwidth, allowing higher download speeds, eventually up to 10 gigabits per second (Gbit/s). Due to the increased bandwidth, the expectation is that the new networks will not just serve cellphones like existing cellular networks, but also be used as general Internet service providers for laptops and desktop computers, competing with existing ISPs such as cable Internet. Existing 4G cellphones will not be able to use the new networks, which will require new 5G-enabled wireless devices.

ADSL—Asymmetric Digital Subscriber Line (ADSL) is a type of digital subscriber line (DSL) that allows faster data transmission via copper service phone lines to a home or business. ADSL provides an "always on" connection and higher speeds than dial-up Internet can provide. In ADSL, bandwidth and bit rate (i.e., speed) are asymmetric, meaning greater toward the customer (downstream) than the reverse (upstream).

Arabsat—Arab Satellite Communications Organization (Riyadh, Saudi Arabia).

Cellular telephone system—the telephones in this system are radio transceivers, with each instrument having its own private radio frequency and sufficient radiated power to reach the booster station in its area (cell), from which the telephone signal is fed to a telephone exchange.

Central American Microwave System—a trunk microwave radio relay system that links the countries of Central America and Mexico with each other.

Coaxial cable—a multichannel communication cable consisting of a central conducting wire, surrounded by and insulated from a cylindrical conducting shell; a large number of telephone channels can be made available within the insulated space by the use of a large number of carrier frequencies.

DSL—Digital Subscriber Line (DSL) is a family of technologies that are used to transmit digital data over telephone lines.

Eutelsat—European Telecommunications Satellite Organization (Paris).

Fiber-optic cable—a multichannel communications cable using a thread of optical glass fibers as a transmission medium in which the signal (voice, video, etc.) is in the form of a coded pulse of light.

FTTX—Fiber to the x (FTTX) is a generic term for any broadband network architecture using optical fiber to provide all or part of the local loop used for last mile telecommunications. As fiber optic cables are able to carry much more data than copper cables, especially over long distances, copper telephone networks built in the 20th century are being replaced by fiber. FTTX is a general term for several configurations of fiber deployment, broadly organized into two groups: FTTN and FTTP /H/B. Fiber to the node (FTTN), also referred to as Fiber to the neighborhood, delivers fiber to within 300m (1,000 ft) of a customer's premises. Fiber to the premises (FTTP) can be further categorized as fiber to the home (FTTH) or fiber to the building/business (FTTB). FTTN (and FTTC, fiber to the curb (to less than 300m (1,000 ft of a customer's premises)) are seen as interim steps toward full FTTP.

GPON—stands for Gigabyte Passive Optical Networks, which are networks that rely on optical cables to deliver information from a single feeding fiber from a provider—to multiple destinations—via the use of splitters. GPONs are currently the leading form of Passive Optical Networks (PON) and offer up to a 1:64 ratio on a single fiber. As opposed to a standard copper wire in most networks, GPONs are 95% more energy efficient.

GSM—a global system for mobile (cellular) communications devised by the Groupe Special Mobile of the pan-European standardization organization, Conference Europeanne des Posts et Telecommunications (CEPT) in 1982.

HF—high frequency; any radio frequency in the 3,000- to 30,000-kHz range.

HSPA—High Speed Packet Access (HSPA) is an amalgamation of two mobile protocols, High Speed Downlink Packet Access (HSDPA) and High Speed Uplink Packet Access (HSUPA), that extends and improves the performance of existing 3G mobile telecommunication networks using the WCDMA protocols. A further improved 3GPP standard, Evolved High Speed Packet Access (also known as HSPA+), was released late in 2008 with subsequent worldwide adoption beginning in 2010. The newer standard allows bit-rates to reach as high as 337 Mbit/s in the downlink and 34 Mbit/s in the uplink. However, these speeds are rarely achieved in practice.

Inmarsat—International Maritime Satellite Organization; a British satellite telecommunications company, offering global mobile services. It provides telephone and data services to users worldwide, via portable or mobile terminals that communicate with ground stations through 13 geostationary telecommunications satellites. Inmarsat's network provides communications services to a range of governments, aid agencies, media outlets, and businesses (especially in the shipping, airline, and mining industries) with a need to communicate in remote regions or where there is no reliable terrestrial network.

Intelsat—Intelsat Corporation (formerly International Telecommunications Satellite Organization, INTEL-SAT, INTEL-SAT, Intelsat) is a communications satellite services provider.

Intersputnik—International Organization of Space Communications (Moscow); first established in the former Soviet Union and the East European countries, it is now marketing its services worldwide with earth stations in North America, Africa, and East Asia.

IoT—the Internet of Things is a system of interrelated computing devices, mechanical, and digital machines provided with unique identifiers (UIDs) and the ability to transfer data over a network without requiring human-to-human or human-to-computer interaction.

Iridium—the Iridium satellite constellation provides L band voice and data information coverage to satellite phones, pagers, and integrated transceivers over the entire surface of the earth. Iridium Communications owns and operates the constellation, additionally selling equipment and access to its services.

ITU—the International Telecommunication Union (ITU) is a United Nations specialized agency that is responsible for issues that concern information and communication technologies. Founded in 1865, the ITU is the oldest global international organization. The ITU coordinates the shared global use of the radio spectrum, promotes international cooperation in assigning satellite orbits, works to improve telecommunication infrastructure in the developing world, and assists in the development and coordination of worldwide technical standards. The ITU is also active in the areas of broadband Internet, latest-generation wireless technologies, aeronautical and maritime navigation, radio astronomy, satellite-based meteorology, convergence in fixed-mobile phone, Internet access, data, voice, TV broadcasting, and next- generation networks.

IXP—an Internet exchange point (IXP) is a physical location through which Internet infrastructure companies such as Internet service providers (ISPs) and content delivery networks (CDNs) connect with each other.

Kacific 1—Kacific Broadband Satellites Group (Kacific) is a satellite operator providing high-speed broadband Internet service for the South East Asia and Pacific Islands regions. Its first Ka-band HTS satellite, Kacific1, was designed and built by Boeing and launched into geostationary orbit in December 201 9.

Landline—communication wire or cable of any sort that is installed on poles or buried in the ground.

LTE—Long -Term Evolution (LTE) is a standard for wireless broadband communication for mobile devices and data terminals Based on the GSM/EDGE and UMTS/HSPA technologies, it increases communication capacity and speed using a different radio interface together with core network improvements.

LTE Advanced—(aka LTE A) is a mobile communication standard and a major enhancement of the Long Term Evolution (LTE) standard. It was submitted as a candidate 4G in late 2009 as meeting the requirements of the IMT- Advanced standard, and was standardized by the 3rd Generation Partnership Project (3GPP) in March 2011 as 3GPP Release 10.

LTE-TDD & LTE-FDD—There are two major differences between LTE-TDD and LTE-FDD: how data is uploaded and downloaded, and what frequency spectra the networks are deployed in. While LTE-FDD uses paired frequencies to upload and download data, LTE-TDD uses a single frequency, alternating between uploading and downloading data through time. The ratio between uploads and downloads on a LTE-TDD network can be changed dynamically, depending on whether more data needs to be sent or received. LTE-TDD and LTE-FDD also operate on different frequency bands, with LTE-TDD working better at higher frequencies, and LTE-FDD working better at lower frequencies.

M-commerce—short for mobile commerce, m-commerce is the use of wireless handheld devices like cellphones and tablets to conduct commercial transactions online, including the purchase and sale of products, online banking, and paying bills.

MNO—a mobile network operator (MNO), also known as a wireless service provider, wireless carrier, cellular company, or mobile network carrier, is a provider of wireless communications services that owns or controls all the elements necessary to sell and deliver services to an end user including radio spectrum allocation, wireless network infrastructure, back haul infrastructure, billing, customer care, provisioning computer systems, and marketing and repair organizations.

MNP—mobile number portability

MVNO—a mobile virtual network operator (MVNO) does not own the wireless network infrastructure over which it provides services to its customers. A MVNO enters into a business agreement with a mobile network operator (MNO) to obtain bulk access to network services at wholesale rates, then sets retail prices independently.

Medarabtel—the Middle East Telecommunications Project of the International Telecommunications Union (ITU) providing a modern telecommunications network, primarily by microwave radio relay, linking Algeria, Djibouti, Egypt, Jordan, Libya, Morocco, Saudi Arabia, Somalia, Sudan, Syria, Tunisia, and Yemen; it was initially started in Morocco in 1970 by the Arab Telecommunications Union (ATU) and was known at that time as the Middle East Mediterranean Telecommunications Network.

Microwave radio relay—transmission of long distance telephone calls and television programs by highly directional radio microwaves that are received and sent on from one booster station to another on an optical path.

NMT—Nordic Mobile Telephone; NMT is a first generation (1G) mobile cellular phone system based on analog technology that was developed jointly by the national telecommunications authorities of the Nordic countries (Denmark, Finland, Iceland, Norway, and Sweden). NMT-450 analog networks have been replaced with digital networks using the same cellular frequencies.

Orbita—a Russian television service; also the trade name of a packet-switched digital telephone network.

PanAmSat—PanAmSat Corporation (Greenwich, CT).

Radio telephone communications—the two-way transmission and reception of sounds by broadcast radio on authorized frequencies using telephone handsets.

Satellite communication system—a communication system consisting of two or more earth stations and at least one satellite that provide long distance transmission of voice, data, and television; the system usually serves as a trunk connection between telephone exchanges; if the earth stations are in the same country, it is a domestic system.

Satellite earth station—a communications facility with a microwave radio transmitting and receiving antenna and required receiving and transmitting equipment for communicating with satellites.

Satellite link—a radio connection between a satellite and an earth station permitting communication between them, either one-way (down link from satellite to earth station—television receive-only transmission) or two-way (telephone channels).

SHF—super high frequency; any radio frequency in the 3,000- to 30,000-MHz range.

Shortwave—radio frequencies (from 1.605 to 30 MHz) that fall above the commercial broadcast band and are used for communication over long distances.

SIM card—subscriber identity/identification module card, is a small, removable integrated circuit used in a mobile phone to store data unique to the user, such as an identification number, passwords, phone numbers, and messages.

Solidaridad—geosynchronous satellites in Mexico's system of international telecommunications in the Western Hemisphere.

Spectrum—spectrum management is the allocation and regulation of the electromagnetic spectrum into radio frequency (RF) bands, a procedure normally carried out by governments in most countries. Because radio propagation does not stop at national boundaries, governments have sought to harmonise the allocation of RF bands and their standardization. A spectrum auction is a process whereby a government uses an auction system to sell the rights to transmit signals over specific bands of the electromagnetic spectrum and to assign scarce spectrum resources.

Submarine cable—a cable designed for service under water.

Telecommunication (telecom)—is the exchange of signs, signals, messages, words, images and sounds, or information of any nature by wire, radio, optical, or other electromagnetic systems (i.e., via the use of technology). Telecommunication occurs through a transmission medium, such as over physical media, for example, over electrical cable; or via electromagnetic radiation through space such as radio or light.

Teledensity—(telephone density) is the number of telephone connections for every hundred individuals living within an area. It varies widely between nations and also between urban and rural areas within a country. Telephone density correlates closely with the per capita GDP of an area, and is also used as an indicator of the purchasing power of the middle class of a country or specific region.

Telefax—facsimile service between subscriber stations via the public switched telephone network or the international Datel network.

Telegraph—a telecommunications system designed for unmodulated electric impulse transmission.

Telephony—is the field of technology involving the development, application, and deployment of telecommunication services for the purpose of electronic transmission of voice, fax, or data, between distant parties. The history of telephony is intimately linked to the invention and development of the telephone.

Telex—a communication service involving teletypewriters connected by wire through automatic exchanges.

Tropospheric scatter—a form of microwave radio transmission in which the troposphere is used to scatter and reflect a fraction of the incident radio waves back to earth. Powerful, highly directional antennas are used to transmit and receive the microwave signals. Reliable over-the-horizon communications are realized for distances up to 600 miles in a single hop; additional hops can extend the range of this system for very long distances.

Trunk network—a network of switching centers, connected by multichannel trunk lines.

UHF—ultra high frequency; any radio frequency in the 300- to 3,000-MHz range.

VHF—very high frequency; any radio frequency in the 30- to 300-MHz range.

VNO—A virtual network operator (VNO) is a management services provider and a network services reseller of other telecommunication service providers. VNOs do not possess a telecom network infrastructure; however, they provide telecom services by acquiring the required capacity from other telecom carriers. These network providers are classified as virtual because they offer network services to clients without possessing the actual network. VNOs usually lease bandwidth at agreed wholesale rates from different telecom providers and then offer solutions to their direct customers.

VOD—or video on demand is a video media distribution system that allows users to access video entertainment without a traditional video entertainment device and without the constraints of a typical static broadcasting schedule.

Voice over Internet Protocol—VoIP, also called IP telephony, refers to the delivery of voice communications and multimedia sessions over Internet Protocol (IP) networks, such as the Internet. The terms Internet telephony, broadband telephony, and broadband phone service specifically refer to the provisioning of communications services (voice, fax, text-messaging, voice-messaging) over the public Internet, rather than via the public switched telephone network (PSTN), also known as plain old telephone service (POTS).

VSAT—a VSAT (very-small-aperture terminal) is a two-way satellite ground station with a dish antenna that is smaller than

3.8 meters. The majority of VSAT antennas range from 75 cm to 1.2 m. Data rates, generally, range from 4 kbit/s up to 16 M bi t/s.

WiMAX—stands for Worldwide Interoperability for Microwave Access; it is a family of wireless broadband communication standards based on the IEEE 802.16 set of standards, which provide multiple physical layer (PHY) and Media Access Control (MAC) options.

Telephone numbers All telephone numbers in *The World Factbook* consist of the country code in brackets, the city or area code (where required) in parentheses, and the local number. The one component that is not presented is the international access code, which varies from country to country. For example, an international direct dial telephone call placed from the US to Madrid, Spain, would be as follows: 011 [34] (1) 577-xxxx, where 011 is the international access code for station-to-station calls; 01 is for calls other than station-to-station calls, [34] is the country code for Spain, (1) is the city code for Madrid, 577 is the local exchange, and xxxx is the local telephone number. An international direct dial telephone call placed from another country to the US would be as follows: international access code + [1] (202) 939-xxxx, where [1] is the country code for the US, (202) is the area code for Washington, DC, 939 is the local exchange, and xxxx is the local telephone number.

Telephones—fixed lines This entry gives the *total* number of fixed telephone lines in use, as well as the number of *subscriptions per 100 inhabitants*.

Telephones—mobile cellular This entry gives the *total* number of mobile cellular telephone subscribers, as well as the number of *subscriptions per 100 inhabitants*. Note that because of the ubiquity of mobile phone use in developed countries, the number of subscriptions per 1 00 inhabitants can exceed 1 00.

Terminology Due to the highly structured nature of the *Factbook* database, some collective generic terms have to be used. For example, the word **Country** in the **Country name** entry refers to a wide variety of dependencies, areas of special sovereignty, uninhabited islands, and other entities in addition to the traditional countries or independent states. **Military** is also used as an umbrella term for various civil defense, security, and defense activities in many entries. The **Independence** entry includes the usual colonial independence dates and former ruling states as well as other significant nationhood dates such as the traditional founding date or the date of unification, federation, confederation, establishment, or state succession that are not strictly independence dates. Dependent areas have the nature of their dependency status noted in this same entry.

Terrain This entry contains a brief description of the topography.

Terrorist group(s) This entry provides a list of US State Department designated Foreign Terrorist Organizations (FTO) that are assessed to maintain a presence in each country. This includes cases where sympathizers, supporters, or associates of designated FTOs have carried out attacks or been arrested by security forces for terrorist-type activities in the country. See Appendix T for details on each FTO.

Time difference This entry is expressed in *The World Factbook* in two ways. First, it is stated as the difference in hours between the capital of an entity and **Coordinated Universal Time** (UTC) during Standard Time. Additionally, the difference in time between the capital of an entity and that observed in Washington, D.C. is also provided. Note that the time difference assumes both locations are simultaneously observing Standard Time or Daylight Saving Time.

Time zones Ten countries (Australia, Brazil, Canada, Indonesia, Kazakhstan, Mexico, New Zealand, Russia, Spain, and the United States) and the island of Greenland observe more than one official time depending on the number of designated time zones within their boundaries. An illustration of time zones throughout the world and within countries can be seen in the Standard Time Zones of the World map included in the **Reference Maps** section of *The World Factbook*.

Total fertility rate This entry gives a figure for the average number of children that would be born per woman if all women lived to the end of their childbearing years and bore children according to a given fertility rate at each age. The total fertility rate (TFR) is a more direct measure of the level of fertility than the crude birth rate, since it refers to births per woman. This indicator shows the potential for population change in the country. A rate of two children per woman is considered the replacement rate for a population, resulting in relative stability in terms of total numbers. Rates above two children indicate populations growing in size and whose median age is declining. Higher rates may also indicate difficulties for families, in some situations, to feed and educate their children and for women to enter the labor force. Rates below two children indicate populations decreasing in size and growing older. Global fertility rates are in general decline and this trend is most pronounced in industrialized countries, especially Western Europe, where populations are projected to decline dramatically over the next 50 years.

Trafficking in persons Trafficking in persons is modern-day slavery, involving victims who are forced, defrauded, or coerced into labor or sexual exploitation. The International Labor Organization (ILO), the UN agency charged with addressing labor standards, employment, and social protection issues, estimated in 2011 that 20.9 million people worldwide were victims of forced labor, bonded labor, forced child labor, sexual servitude, and involuntary servitude. Human trafficking is a multidimensional threat, depriving people of their human rights and freedoms, risking global health, promoting social breakdown, inhibiting development by depriving countries of their human capital, and helping fuel the growth of organized crime. In 2000, the US Congress passed the Trafficking Victims Protection Act (TVPA), reauthorized in 2003 and 2005, which provides tools for the US to combat trafficking in persons, both domestically and abroad. One of the law's key components is the creation of the US Department of State's annual *Trafficking in Persons Report*, which assesses the government response (i.e., the *current situation*) in some 150 countries with a significant number of victims trafficked across their borders who are recruited, harbored, transported, provided, or obtained for forced labor or sexual exploitation. Countries in the annual report are rated in three tiers, based on government efforts to combat trafficking. The countries identified in this entry are those listed in the *2010 Trafficking in Persons Report* as 'Tier 2 Watch List' or 'Tier 3' based on the following *tier rating* definitions:

Tier 2 Watch List countries do not fully comply with the minimum standards for the elimination of trafficking but are making significant efforts to do so, and meet one of the following criteria:
they display high or significantly increasing number of victims,
they have failed to provide evidence of increasing efforts to combat trafficking in persons, or,
they have committed to take action over the next year.
Tier 3 countries neither satisfy the minimum standards for the elimination of trafficking nor demonstrate a significant effort to do so. Countries in this tier are subject to potential non-humanitarian and non-trade sanctions.

Transnational issues This category includes four entries—Disputes—international, **Refugees and internally displaced persons, Trafficking in persons,** and **Illicit drugs**—that deal with current issues going beyond national boundaries.

Transportation This category includes the entries dealing with the means for movement of people and goods.

Transportation—note This entry includes miscellaneous transportation information of significance not included elsewhere.

Unemployment rate This entry contains the percent of the labor force that is without jobs. Substantial underemployment might be noted.

Unemployment, youth ages 15-24 This entry gives the percent of the total labor force ages 15-24 unemployed during a specified year.

Union name This entry, which appears only in the European Union, Government category, provides the full name and abbreviation for the European Union.

Urbanization This entry provides two measures of the degree of urbanization of a population. The first, *urban population*, describes the percentage of the total population living in urban areas, as defined by the country. The second, *rate of urbanization*, describes the projected average rate of change of the size of the urban population over the given period of time. It is possible for a country with a 100% urban population to still display a change in the rate of urbanization (up or down). For example, a population of 100,000 that is 100% urban can change in size to 110,000 or 90,000 but remain 100% urban.

Additionally, the World entry includes a list of the *ten largest urban agglomerations*. An *urban agglomeration* is defined as comprising the city or town proper and also the suburban fringe or thickly settled territory lying outside of, but adjacent to, the boundaries of the city.

UTC (Coordinated Universal Time) See entry for Coordinated Universal Time.

Waterways This entry gives the total length of navigable rivers, canals, and other inland bodies of water.

Weights and Measures This information is presented in Appendix G: Weights and Measures and includes mathematical notations (mathematical powers and names), metric interrelationships (prefix; symbol; length, weight, or capacity; area; volume), and standard conversion factors.

Years All year references are for the calendar year (CY) unless indicated as fiscal year (FY). The calendar year is an accounting period of 12 months from 1 January to 31 December. The fiscal year is an accounting period of 12 months other than 1 January to 31 December.

GUIDE TO COUNTRY PROFILES

These are the **Categories**, **Fields**, and **subfields** of information generally recorded for each country. Links are to the Definitions and Notes about each entry.

INTRODUCTION

Background:

GEOGRAPHY

Location:
Geographic coordinates:
Map references:
Area:
total
land
water
Area—comparative:
Land boundaries:
total
border countries
regional borders
Coastline:
Maritime claims:
territorial sea
exclusive economic zone
contiguous zone
continental shelf
exclusive fishing zone
Climate:
Terrain:
Elevation:
mean elevation
lowest point
highest point
Natural resources:
Land use:
agricultural land
agricultural land: arable land
agricultural land: permanent crops
agricultural land: permanent pasture
forest
other
Irrigated land:
Population distribution:
Natural hazards:
Environment—current issues:
**Environment—international
    agreements:**
party to
signed, but not ratified
Geography—note:

PEOPLE AND SOCIETY

Population:
Nationality:
noun
adjective
Ethnic groups:
Languages:
Religions:

Demographic profile:
Age structure:
0-14 years
15-24 years
25-54 years
55-64 years
65 years and over
Dependency ratios:
total dependency ratio
youth dependency ratio
elderly dependency ratio
potential support ratio
Median age:
total
male
female
Population growth rate:
Birth rate:
Death rate:
Net migration rate:
Population distribution:
Urbanization:
urban population
rate of urbanization
Major urban areas—population:
Sex ratio:
at birth
0-14 years
15-24 years
25-54 years
55-64 years
65 years and over
total population
Mother's mean age at first birth:
Maternal mortality rate:
Infant mortality rate:
total
male
female
Life expectancy at birth:
total population
male
female
Total fertility rate:
Contraceptive prevalence rate:
Drinking water source:
improved: urban
improved: rural
improved: total
unimproved: urban
unimproved: rural
Current Health Expenditure:
Physicians density:
Hospital bed density:
Sanitation facility access:
improved: urban
improved: rural
improved: total

unimproved: urban
unimproved: rural
unimproved: total
HIV/AIDS—adult prevalence rate:
HIV/AIDS—people living with HIV/AIDS:
HIV/AIDS—deaths:
Major infectious diseases:
degree of risk
food or waterborne diseases
vectorborne diseases
water contact diseases
animal contact diseases
respiratory diseases
soil contact diseases
aerosolized dust or soil contact diseases
Obesity—adult prevalence rate:
**Children under the age of 5 years
    underweight:**
Education expenditures:
Literacy:
definition
total population
male
female
**School life expectancy (primary to
    tertiary education):**
total
male
female
Unemployment, youth ages 15-24:
total
male
female
People—note:

GOVERNMENT

Country name:
conventional long form
conventional short form
local long form
local short form
former
abbreviation
etymology
Dependency status:
Government type:
Capital:
name
geographic coordinates
time difference
daylight saving time
capital
Administrative divisions:
Dependent areas:
Independence:
National holiday:
Constitution:

history
amendments
Legal system:
International law organization participation:
Citizenship:
citizenship by birth
citizenship by descent only
dual citizenship recognized
residency requirement for naturalization
Suffrage:
Executive branch:
chief of state
head of government
cabinet
elections/appointments
election results
head of state
Legislative branch:
description
elections
election results
Judicial branch:
highest courts
judge selection and term of office
subordinate courts
Political parties and leaders:
International organization participation:
Diplomatic representation in the US:
chief of mission
chancery
telephone
FAX
consulate(s) general
consulate(s)
embassy
honorary consulate(s)
Diplomatic representation from the US:
chief of mission
telephone
embassy
mailing address
FAX
consulate(s) general
branch office(s)
consulate(s)
Flag description:
National symbol(s):
National anthem:
name
lyrics/music
Government—note:

ECONOMY

Economy—overview:
GDP (purchasing power parity):
GDP (official exchange rate):
GDP—real growth rate:
GDP—per capita (PPP):
Gross national saving:
GDP—composition, by end use:
household consumption
government consumption
investment in fixed capital
investment in inventories
exports of goods and services

imports of goods and services
GDP—composition, by sector of origin:
agriculture
industry
services
Agriculture—products:
Industries:
Industrial production growth rate:
Labor force:
Labor force—by occupation:
agriculture
industry
services
industry and services
manufacturing
construction
commerce
other services
Unemployment rate:
Population below poverty line:
Household income or consumption by percentage share:
lowest 10%
highest 10%
Budget:
revenues
expenditures
Taxes and other revenues:
Budget surplus (+) or deficit (-):
Public debt:
Fiscal year:
Inflation rate (consumer prices):
Current account balance:
Exports:
Exports—partners:
Exports—commodities:
Imports:
Imports—commodities:
Imports—partners:
Reserves of foreign exchange and gold:
Debt—external:
Exchange rates:
Marine fisheries:

ENERGY

Electricity access:
population without electricity
electrification—total population
electrification—urban areas
electrification—rural areas
Electricity—production:
Electricity—consumption:
Electricity—exports:
Electricity—imports:
Electricity—installed generating capacity:
Electricity—from fossil fuels:
Electricity—from nuclear fuels:
Electricity—from hydroelectric plants:
Electricity—from other renewable sources:
Crude oil—production:
Crude oil—exports:
Crude oil—imports:
Crude oil—proved reserves:

Refined petroleum products—production:
Refined petroleum products—consumption:
Refined petroleum products—exports:
Refined petroleum products—imports:
Natural gas—production:
Natural gas—consumption:
Natural gas—exports:
Natural gas—imports:
Natural gas—proved reserves:
Carbon dioxide emissions from consumption of energy:

COMMUNICATIONS

Telephones—fixed lines:
total subscriptions
subscriptions per 100 inhabitants
Telephones—mobile cellular:
total subscriptions
subscriptions per 100 inhabitants
Telecommunication systems:
general assessment
domestic
international
Broadcast media:
Internet country code:
Internet users:
total
percent of population
Broadband—fixed subscriptions:
total
subscriptions per 100 inhabitants
Communications—note:

TRANSPORTATION

National air transport system:
number of registered air carriers
inventory of registered aircraft operated by air carriers
annual passenger traffic on registered air carriers
annual freight traffic on registered air carriers
Civil aircraft registration country code prefix:
Airports:
total
Airports—with paved runways:
total
over 3,047 m
2,438 to 3,047 m
1,524 to 2,437 m
914 to 1,523 m
under 914 m
Airports—with unpaved runways:
total
over 3,047 m
2,438 to 3,047 m
1,524 to 2,437 m
914 to 1,523 m
under 914 m
Heliports:
Pipelines:
Railways:

total
standard gauge
narrow gauge
broad gauge
dual gauge
Roadways:
total
paved
unpaved
urban
non-urban
Waterways:
Merchant marine:
total
by type
Ports and terminals:
major seaport(s)
oil terminal(s)
container port(s) (TEUs)
lake port(s)

LNG terminal(s) (export)
LNG terminal(s) (import)
river port(s)
dry bulk cargo port(s)
bulk cargo port(s)
Transportation—note:

MILITARY AND SECURITY

Military and security forces:
Military expenditures:
Military and security service personnel strengths:
Military equipment inventories and acquisitions:
Military deployments:
Military service age and obligation:
Maritime threats:
Military—note:

TERRORISM

Terrorist group(s):
Terrorist group(s)

TRANSNATIONAL ISSUES

Disputes—international:
Refugees and internally displaced persons:
refugees (country of origin)
IDPs
stateless persons
Trafficking in persons:
current situation
tier rating
Illicit drugs:

GUIDE TO COUNTRY COMPARISONS

Country Comparison pages are presorted lists of data from selected *Factbook* data fields. Country Comparison pages are generally given in descending order - highest to lowest - such as Population and Area. The two exceptions are Unemployment Rate and Inflation Rate, which are in ascending - lowest to highest - order. Country Comparison pages are available for the following fields in seven of the ten *Factbook* categories.

GEOGRAPHY
Area:

PEOPLE AND SOCIETY
Population:
Median age:
Population growth rate:
Birth rate:
Death rate:
Net migration rate:
Maternal mortality rate:
Infant mortality rate:
Life expectancy at birth:
Total fertility rate:
HIV/AIDS—adult prevalence rate:
HIV/AIDS—people living with HIV/AIDS:
HIV/AIDS—deaths:
Obesity—adult prevalence rate:
Children under the age of 5 years underweight:
Education expenditures:
Unemployment, youth ages 15-24:

ECONOMY
GDP (purchasing power parity):
GDP—real growth rate:
GDP—per capita (PPP):
Gross national saving:

Industrial production growth rate:
Labor force:
Unemployment rate:
Taxes and other revenues:
Budget surplus (+) or deficit (-):
Public debt:
Inflation rate (consumer prices):
Current account balance:
Exports:
Imports:
Reserves of foreign exchange and gold:
Debt—external:

ENERGY
Electricity—production:
Electricity—consumption:
Electricity—exports:
Electricity—imports:
Electricity—installed generating capacity:
Electricity—from fossil fuels:
Electricity—from nuclear fuels:
Electricity—from hydroelectric plants:
Electricity—from other renewable sources:
Crude oil—production:
Crude oil—exports:
Crude oil—imports:
Crude oil—proved reserves:

Refined petroleum products—production:
Refined petroleum products—consumption:
Refined petroleum products—exports:
Refined petroleum products—imports:
Natural gas—production:
Natural gas—consumption:
Natural gas—exports:
Natural gas—imports:
Natural gas—proved reserves:
Carbon dioxide emissions from consumption of energy:

COMMUNICATIONS
Telephones—fixed lines:
Telephones—mobile cellular:
Internet users:
Broadband—fixed subscriptions:

TRANSPORTATION
Airports:
Railways:
Roadways:
Waterways:
Merchant marine:

MILITARY AND SECURITY
Military expenditures:

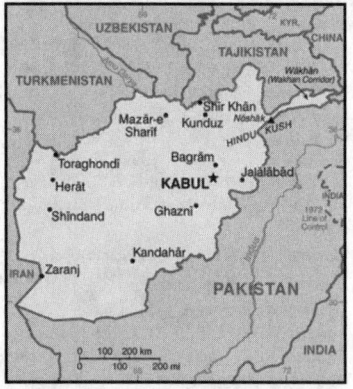

INTRODUCTION

Background: Ahmad Shah DURRANI unified the Pashtun tribes and founded Afghanistan in 1747. The country served as a buffer between the British and Russian Empires until it won independence from notional British control in 1919. A brief experiment in increased democracy ended in a 1973 coup and a 1978 communist countercoup. The Soviet Union invaded in 1979 to support the tottering Afghan communist regime, touching off a long and destructive war. The USSR withdrew in 1989 under relentless pressure by internationally supported anti-communist mujahidin rebels. A series of subsequent civil wars saw Kabul finally fall in 1996 to the Taliban, a hardline Pakistani-sponsored movement that emerged in 1994 to end the country's civil war and anarchy. Following the 11 September 2001 terrorist attacks, a US, Allied, and anti-Taliban Northern Alliance military action toppled the Taliban for sheltering Usama BIN LADIN.

A UN-sponsored Bonn Conference in 2001 established a process for political reconstruction that included the adoption of a new constitution, a presidential election in 2004, and National Assembly elections in 2005. In December 2004, Hamid KARZAI became the first democratically elected president of Afghanistan, and the National Assembly was inaugurated the following December. KARZAI was reelected in August 2009 for a second term. The 2014 presidential election was the country's first to include a runoff, which featured the top two vote-getters from the first round, Abdullah ABDULLAH and Ashraf GHANI. Throughout the summer of 2014, their campaigns disputed the results and traded accusations of fraud, leading to a US-led diplomatic intervention that included a full vote audit as well as political negotiations between the two camps. In September 2014, GHANI and ABDULLAH agreed to form the Government of National Unity, with GHANI inaugurated

as president and ABDULLAH elevated to the newly-created position of chief executive officer. The day after the inauguration, the GHANI administration signed the US-Afghan Bilateral Security Agreement and NATO Status of Forces Agreement, which provide the legal basis for the post-2014 international military presence in Afghanistan. After two postponements, the next presidential election was held in September 2019.

The Taliban remains a serious challenge for the Afghan Government in almost every province. The Taliban still considers itself the rightful government of Afghanistan, and it remains a capable and confident insurgent force fighting for the withdrawal of foreign military forces from Afghanistan, establishment of sharia law, and rewriting of the Afghan constitution. In 2019, negotiations between the US and the Taliban in Doha entered their highest level yet, building on momentum that began in late 2018. Underlying the negotiations is the unsettled state of Afghan politics, and prospects for a sustainable political settlement remain unclear.

GEOGRAPHY

Location: Southern Asia, north and west of Pakistan, east of Iran

Geographic coordinates: 33 00 N, 65 00 E

Map references: Asia

Area: *total:* 652,230 sq km
land: 652,230 sq km
water: 0 sq km
country comparison to the world: 42

Area—comparative: almost six times the size of Virginia; slightly smaller than Texas

Land boundaries: *total:* 5,987 km
border countries (6): China 91 km, Iran 921 km, Pakistan 2670 km, Tajikistan 1357 km, Turkmenistan 804 km, Uzbekistan 144 km

Coastline: 0 km (landlocked)

Maritime claims: none (landlocked)

Climate: arid to semiarid; cold winters and hot summers

Terrain: mostly rugged mountains; plains in north and southwest

Elevation: *mean elevation:* 1,884 m
lowest point: Amu Darya 258 m
highest point: Noshak 7,492 m

Natural resources: natural gas, petroleum, coal, copper, chromite, talc, barites, sulfur, lead, zinc, iron ore, salt, precious and semiprecious stones, arable land

Land use: *agricultural land:* 58.1% (2016 est.)
arable land: 11.8% (2016) / *permanent crops:* 0.3% (2016) / *permanent pasture:* 46% (2016)
forest: 2.07% (2016 est.)
other: 39% (2016)
Irrigated land: 32,080 sq km (2012)

Population distribution: populations tend to cluster in the foothills and periphery of the rugged Hindu Kush range; smaller groups are found in many of the country's interior valleys; in general, the east is more densely settled, while the south is sparsely populated

Natural hazards: damaging earthquakes occur in Hindu Kush mountains; flooding; droughts

Environment—current issues: limited natural freshwater resources; inadequate supplies of potable water; soil degradation; overgrazing; deforestation (much of the remaining forests are being cut down for fuel and building materials); desertification; air and water pollution in overcrowded urban areas

Environment—international agreements: *party to:* Biodiversity, Climate Change, Desertification, Endangered Species, Environmental Modification, Marine Dumping, Ozone Layer Protection
signed, but not ratified: Hazardous Wastes, Law of the Sea, Marine Life Conservation

Geography—note: landlocked; the Hindu Kush mountains that run northeast to southwest divide the northern provinces from the rest of the country; the highest peaks are in the northern Vakhan (Wakhan Corridor)

PEOPLE AND SOCIETY

Population: 36,643,815 (July 2020 est.)
country comparison to the world: 39

Nationality: *noun:* Afghan(s)
adjective: Afghan

Ethnic groups: Pashtun, Tajik, Hazara, Uzbek, other (includes smaller numbers of Baloch, Turkmen, Nuristani, Pamiri, Arab, Gujar, Brahui, Qizilbash, Aimaq, Pashai, and Kyrghyz) (2015)
note: current statistical data on the sensitive subject of ethnicity in Afghanistan are not available, and ethnicity data from small samples of respondents to opinion polls are not a reliable alternative; Afghanistan's 2004 constitution recognizes 14 ethnic groups: Pashtun, Tajik, Hazara, Uzbek, Baloch, Turkmen, Nuristani, Pamiri, Arab, Gujar, Brahui, Qizilbash, Aimaq, and Pashai

Languages: Afghan Persian or Dari (official) 77% (Dari functions as the lingua franca), Pashto (official) 48%, Uzbek 11%, English 6%, Turkmen 3%, Urdu 3%, Pashayi 1%, Nuristani 1%, Arabic 1%, Balochi 1% (2017 est.)
note: data represent most widely spoken languages; shares sum to more than 100% because there is much bilingualism in the country and because respondents were allowed to select more than one language
note: the Turkic languages Uzbek and Turkmen, as well as Balochi, Pashayi, Nuristani, and Pamiri are the third official languages in areas where the majority speaks them

Religions: Muslim 99.7% (Sunni 84.7—89.7%, Shia 10—15%), other 0.3% (2009 est.)

Age structure: *0-14 years:* 40.62% (male 7,562,703/female 7,321,646)
15-24 years: 21.26% (male 3,960,044/female 3,828,670)
25-54 years: 31.44% (male 5,858,675/female 5,661,887)
55-64 years: 4.01% (male 724,597/female 744,910)
65 years and over: 2.68% (male 451,852/female 528,831) (2020 est.)

Dependency ratios: *total dependency ratio:* 88.8
youth dependency ratio: 75.3
elderly dependency ratio: 4.8
potential support ratio: 21 (2020 est.)

Median age: *total:* 19.5 years
male: 19.4 years
female: 19.5 years (2020 est.)
country comparison to the world: 202

Population growth rate: 2.38% (2020 est.)
country comparison to the world: 28

Birth rate: 36.7 births/1,000 population (2020 est.)
country comparison to the world: 15

Death rate: 12.7 deaths/1,000 population (2020 est.)
country comparison to the world: 12

Net migration rate: -0.1 migrant(s)/1,000 population (2020 est.)
country comparison to the world: 99

Population distribution: populations tend to cluster in the foothills and periphery of the rugged Hindu Kush range; smaller groups are found in many of the country's interior valleys; in general, the east is more densely settled, while the south is sparsely populated

Urbanization: *urban population:* 26% of total population (2020)
rate of urbanization: 3.37% annual rate of change (2015-20 est.)

total population growth rate v. urban population growth rate, 2000-2030: Major urban areas—population: 4.222 million KABUL (capital) (2020)

Sex ratio: *at birth:* 1.05 male(s)/female
0-14 years: 1.03 male(s)/female
15-24 years: 1.03 male(s)/female
25-54 years: 1.03 male(s)/female
55-64 years: 0.97 male(s)/female
65 years and over: 0.85 male(s)/female
total population: 1.03 male(s)/female (2020 est.)

Mother's mean age at first birth: 19.9 years (2015 est.)
note: median age at first birth among women 25-29

Maternal mortality rate: 638 deaths/100,000 live births (2017 est.)
country comparison to the world: 11

Infant mortality rate: *total:* 104.3 deaths/1,000 live births
male: 111.3 deaths/1,000 live births
female: 96.9 deaths/1,000 live births (2020 est.)
country comparison to the world: 1

Life expectancy at birth: *total population:* 52.8 years
male: 51.4 years
female: 54.4 years (2020 est.)

country comparison to the world: 228

Total fertility rate: 4.82 children born/woman (2020 est.)
country comparison to the world: 16

Contraceptive prevalence rate: 18.9% (2018)
note: percent of women aged 12-49

Drinking water source:
improved:
urban: 95.9% of population
rural: 61.4% of population
total: 70.2% of population
unimproved:
urban: 3.2% of population
rural: 38.6% of population
total: 38.6% of population (2017 est.)

Current Health Expenditure: 11.8% (2017)

Physicians density: 0.28 physicians/1,000 population (2016)

Hospital bed density: 0.4 beds/1,000 population (2017)

Sanitation facility access:
improved:
urban: 83.6% of population
rural: 43% of population
total: 53.2% of population
unimproved:
urban: 16.4% of population
rural: 57% of population
total: 46.8% of population (2017 est.)

HIV/AIDS—adult prevalence rate: <.1% (2019 est.)

HIV/AIDS—people living with HIV/AIDS: 11,000 (2019 est.)
country comparison to the world: 97

HIV/AIDS—deaths: <500 (2019 est.)

Major infectious diseases: *degree of risk:* intermediate (2020)
food or waterborne diseases: bacterial diarrhea, hepatitis A, and typhoid fever
vectorborne diseases: Crimea-Congo hemorrhagic fever, malaria

Obesity—adult prevalence rate: 5.5% (2016)
country comparison to the world: 176

Children under the age of 5 years underweight: 19.1% (2018)
country comparison to the world: 25

Education expenditures: 4.1% of GDP (2017)
country comparison to the world: 95

Literacy: *definition:* age 15 and over can read and write
total population: 43%
male: 55.5%
female: 29.8% (2018)

School life expectancy (primary to tertiary education): *total:* 10 years
male: 13 years
female: 8 years (2018)

Unemployment, youth ages 15-24: *total:* 17.6%
male: 16.3%
female: 21.4% (2017)
country comparison to the world: 75

GOVERNMENT

Country name: *conventional long form:* Islamic Republic of Afghanistan
conventional short form: Afghanistan
local long form: Jamhuri-ye Islami-ye Afghanistan
local short form: Afghanistan
former: Republic of Afghanistan
etymology: the name "Afghan" originally referred to the Pashtun people (today it is understood to include all the country's ethnic groups), while the suffix "-stan" means "place of" or "country"; so Afghanistan literally means the "Land of the Afghans"

Government type: presidential Islamic republic

Capital: *name:* Kabul
geographic coordinates: 34 31 N, 69 11 E
time difference: UTC+4.5 (9.5 hours ahead of Washington, DC, during Standard Time)
daylight saving time: does not observe daylight savings time
etymology: named for the Kabul River, but the river's name is of unknown origin

Administrative divisions: 34 provinces (welayat, singular—welayat); Badakhshan, Badghis, Baghlan, Balkh, Bamyan, Daykundi, Farah, Faryab, Ghazni, Ghor, Helmand, Herat, Jowzjan, Kabul, Kandahar, Kapisa, Khost, Kunar, Kunduz, Laghman, Logar, Nangarhar, Nimroz, Nuristan, Paktika, Paktiya, Panjshir, Parwan, Samangan, Sar-e Pul, Takhar, Uruzgan, Wardak, Zabul

Independence: 19 August 1919 (from UK control over Afghan foreign affairs)

National holiday: Independence Day, 19 August (1919)

Constitution: *history:* several previous; latest drafted 14 December 2003—4 January 2004, signed 16 January 2004, ratified 26 January 2004
amendments: proposed by a commission formed by presidential decree followed by the convention of a Grand Council (Loya Jirga) decreed by the president; passage requires at least two-thirds majority vote of the Loya Jirga membership and endorsement by the president

Legal system: mixed legal system of civil, customary, and Islamic (sharia) law

International law organization participation: has not submitted an ICJ jurisdiction declaration; accepts ICCt jurisdiction

Citizenship: *citizenship by birth:* no
citizenship by descent only: at least one parent must have been born in—and continuously lived in—Afghanistan
dual citizenship recognized: no
residency requirement for naturalization: 5 years

Suffrage: 18 years of age; universal

Executive branch: *chief of state:* President of the Islamic Republic of Afghanistan Ashraf GHANI (since 29 September 2014); CEO Abdullah ABDULLAH, Dr. (since 29 September 2014); First Vice President Abdul Rashid DOSTAM (since 29 September 2014); Second Vice President

Sarwar DANESH (since 29 September 2014); First Deputy CEO Khyal Mohammad KHAN; Second Deputy CEO Mohammad MOHAQQEQ; note—the president is both chief of state and head of government

head of government: President of the Islamic Republic of Afghanistan Ashraf GHANI (since 29 September 2014); CEO Abdullah ABDULLAH, Dr. (since 29 September 2014); First Vice President Abdul Rashid DOSTAM (since 29 September 2014); Second Vice President Sarwar DANESH (since 29 September 2014); First Deputy CEO Khyal Mohammad KHAN; Second Deputy CEO Mohammad MOHAQQEQ

cabinet: Cabinet consists of 25 ministers appointed by the president, approved by the National Assembly

elections/appointments: president directly elected by absolute majority popular vote in 2 rounds if needed for a 5-year term (eligible for a second term); election last held on 28 September 2019 (next to be held in 2024)

election results: Ashraf GHANI declared winner by the Independent Election Commission on 18 February 2020; Ashraf GHANI 50.6%, Abdullah ABDULLAH, Dr. 39.5%, other 0.9%

Legislative branch: *description:* bicameral National Assembly consists of:

Meshrano Jirga or House of Elders (102 seats; 34 members indirectly elected by absolute majority vote in 2 rounds if needed by district councils to serve 3-year terms, 34 indirectly elected by absolute majority vote in 2 rounds if needed by provincial councils to serve 4-year terms, and 34 appointed by the president from nominations by civic groups, political parties, and the public, of which 17 must be women, 2 must represent the disabled, and 2 must be Kuchi nomads; presidential appointees serve 5-year terms)

Wolesi Jirga or House of People (250 seats; members directly elected in multi-seat constituencies by proportional representation vote to serve 5-year terms)

elections: Meshrano Jirga—district councils—within 5 days of installation; provincial councils—within 15 days of installation; presidential appointees—within 2 weeks after the presidential inauguration

Wolesi Jirga—last held on 20 October 2018) (next to be held in 2023)

election results: Meshrano Jirga—percent of vote by party—NA; seats by party—NA; composition—men 84, women 18, percent of women 17.6%

Wolesi Jirga—percent of vote by party NA; seats by party—NA; composition—NA

note: the constitution allows the government to convene a constitutional Loya Jirga (Grand Council) on issues of independence, national sovereignty, and territorial integrity; it consists of members of the National Assembly and chairpersons of the provincial and district councils; a Loya Jirga can amend provisions of the constitution and prosecute the president; no constitutional Loya Jirga has ever been held, and district councils

have never been elected; the president appointed 34 members of the Meshrano Jirga that the district councils should have indirectly elected

Judicial branch: *highest courts:* Supreme Court or Stera Mahkama (consists of the supreme court chief and 8 justices organized into criminal, public security, civil, and commercial divisions or dewans)

judge selection and term of office: court chief and justices appointed by the president with the approval of the Wolesi Jirga; court chief and justices serve single 10-year terms

subordinate courts: Appeals Courts; Primary Courts; Special Courts for issues including narcotics, security, property, family, and juveniles

Political parties and leaders: note—the Ministry of Justice licensed 72 political parties as of April 2019

International organization participation: ADB, CICA, CP, ECO, EITI (candidate country), FAO, G-77, IAEA, IBRD, ICAO, ICC (NGOs), ICCt, ICRM, IDA, IDB, IFAD, IFC, IFRCS, ILO, IMF, Interpol, IOC, IOM, IPU, ISO (correspondent), ITSO, ITU, ITUC (NGOs), MIGA, NAM, OIC, OPCW, OSCE (partner), SAARC, SACEP, SCO (dialogue member), UN, UNAMA, UNCTAD, UNESCO, UNHCR, UNIDO, UNWTO, UPU, WCO, WFTU (NGOs), WHO, WIPO, WMO, WTO

Diplomatic representation in the US: *chief of mission:* Ambassador Roya RAHMANI (since 24 November 2018)

chancery: 2341 Wyoming Avenue NW, Washington, DC 20008

telephone: [1] (202) 483-6410

FAX: [1] (202) 483-6488

consulate(s) general: Los Angeles, New York, Washington, DC

Diplomatic representation from the US: *chief of mission:* Ambassador (vacant); Charge d'Affaires Ross WILSON (since 18 January 2020)

telephone: [00 93] 0700 108 001

embassy: Bibi Mahru, Kabul

mailing address: U.S. Embassy Kabul, APO AE 09806

FAX: [00 93] 0700 108 564

Flag description: three equal vertical bands of black (hoist side), red, and green, with the national emblem in white centered on the red band and slightly overlapping the other 2 bands; the center of the emblem features a mosque with pulpit and flags on either side, below the mosque are Eastern Arabic numerals for the solar year 1298 (1919 in the Gregorian calendar, the year of Afghan independence from the UK); this central image is circled by a border consisting of sheaves of wheat on the left and right, in the upper-center is an Arabic inscription of the Shahada (Muslim creed) below which are rays of the rising sun over the Takbir (Arabic expression meaning "God is great"), and at bottom center is a scroll bearing the name Afghanistan; black signifies the past, red is for the blood shed for independence, and green

can represent either hope for the future, agricultural prosperity, or Islam

note: Afghanistan had more changes to its national flag in the 20th century—19 by one count—than any other country; the colors black, red, and green appeared on most of them

National symbol(s): lion; national colors: red, green, black

National anthem: *name:* "Milli Surood" (National Anthem)

lyrics/music: Abdul Bari JAHANI/Babrak WASA

note: adopted 2006; the 2004 constitution of the post-Taliban government mandated that a new national anthem should be written containing the phrase "Allahu Akbar" (God is Greatest) and mentioning the names of Afghanistan's ethnic groups

0:00 / 1:13

ECONOMY

Economy—overview: Despite improvements in life expectancy, incomes, and literacy since 2001, Afghanistan is extremely poor, landlocked, and highly dependent on foreign aid. Much of the population continues to suffer from shortages of housing, clean water, electricity, medical care, and jobs. Corruption, insecurity, weak governance, lack of infrastructure, and the Afghan Government's difficulty in extending rule of law to all parts of the country pose challenges to future economic growth. Afghanistan's living standards are among the lowest in the world. Since 2014, the economy has slowed, in large part because of the withdrawal of nearly 100,000 foreign troops that had artificially inflated the country's economic growth.

The international community remains committed to Afghanistan's development, pledging over $83 billion at ten donors' conferences between 2003 and 2016. In October 2016, the donors at the Brussels conference pledged an additional $3.8 billion in development aid annually from 2017 to 2020. Even with this help, Government of Afghanistan still faces number of challenges, including low revenue collection, anemic job creation, high levels of corruption, weak government capacity, and poor public infrastructure.

In 2017 Afghanistan's growth rate was only marginally above that of the 2014-2016 average. The drawdown of international security forces that started in 2012 has negatively affected economic growth, as a substantial portion of commerce, especially in the services sector, has catered to the ongoing international troop presence in the country. Afghan President Ashraf GHANI Ahmadzai is dedicated to instituting economic reforms to include improving revenue collection and fighting corruption. The government has implemented reforms to the budget process and in some other areas. However, many other reforms will take time to implement and Afghanistan will remain dependent on international donor support over the next several years.

GDP (purchasing power parity): $69.45 billion (2017 est.)

$67.65 billion (2016 est.)
$66.21 billion (2015 est.)
note: data are in 2017 dollars
country comparison to the world: 101

GDP (official exchange rate): $20.24 billion (2017 est.)

GDP—real growth rate: 2.7% (2017 est.)
2.2% (2016 est.)
1% (2015 est.)
country comparison to the world: 124

GDP—per capita (PPP): $2,000 (2017 est.)
$2,000 (2016 est.)
$2,000 (2015 est.)
note: data are in 2017 dollars
country comparison to the world: 209

Gross national saving: 22.7% of GDP (2017 est.)
25.8% of GDP (2016 est.)
21.4% of GDP (2015 est.)
country comparison to the world: 78

GDP—composition, by end use: household consumption: 81.6% (2016 est.)
government consumption: 12% (2016 est.)
investment in fixed capital: 17.2% (2016 est.)
investment in inventories: 30% (2016 est.)
exports of goods and services: 6.7% (2016 est.)
imports of goods and services: -47.6% (2016 est.)

GDP—composition, by sector of origin: agriculture: 23% (2016 est.)
industry: 21.1% (2016 est.)
services: 55.9% (2016 est.)
note: data exclude opium production

Agriculture—products: opium, wheat, fruits, nuts, wool, mutton, sheepskins, lambskins, poppies

Industries: small-scale production of bricks, textiles, soap, furniture, shoes, fertilizer, apparel, food products, non-alcoholic beverages, mineral water, cement; handwoven carpets; natural gas, coal, copper

Industrial production growth rate: -1.9% (2016 est.)
country comparison to the world: 181

Labor force: 8.478 million (2017 est.)
country comparison to the world: 61

Labor force—by occupation: agriculture: 44.3%
industry: 18.1%
services: 37.6% (2017 est.)

Unemployment rate: 23.9% (2017 est.)
22.6% (2016 est.)
country comparison to the world: 194

Population below poverty line: 54.5% (2017 est.)

Household income or consumption by percentage share: lowest 10%: 3.8%
highest 10%: 24% (2008)

Budget: revenues: 2.276 billion (2017 est.)
expenditures: 5.328 billion (2017 est.)

Taxes and other revenues: 11.2% (of GDP) (2017 est.)
country comparison to the world: 210

Budget surplus (+) or deficit (-): -15.1% (of GDP) (2017 est.)
country comparison to the world: 217

Public debt: 7% of GDP (2017 est.)
7.8% of GDP (2016 est.)
country comparison to the world: 202

Fiscal year: 21 December—20 December

Inflation rate (consumer prices): 5% (2017 est.)
4.4% (2016 est.)
country comparison to the world: 171

Current account balance: $1.014 billion (2017 est.)
$1.409 billion (2016 est.)
country comparison to the world: 48

Exports: $784 million (2017 est.)
$614.2 million (2016 est.)
note: not including illicit exports or reexports
country comparison to the world: 170

Exports—partners: India 56.5%, Pakistan 29.6% (2017)

Exports—commodities: opium, fruits and nuts, handwoven carpets, wool, cotton, hides and pelts, precious and semi-precious gems, and medical herbs

Imports: $7.616 billion (2017 est.)
$6.16 billion (2016 est.)
country comparison to the world: 113

Imports—commodities: machinery and other capital goods, food, textiles, petroleum products

Imports—partners: China 21%, Iran 20.5%, Pakistan 11.8%, Kazakhstan 11%, Uzbekistan 6.8%, Malaysia 5.3% (2017)

Reserves of foreign exchange and gold: $7.187 billion (31 December 2017 est.)
$6.901 billion (31 December 2015 est.)
country comparison to the world: 85

Debt—external: $284 million (FY10/11)
country comparison to the world: 185

Exchange rates: afghanis (AFA) per US dollar—
7.87 (2017 est.)
68.03 (2016 est.)
67.87 (2015)
61.14 (2014 est.)
57.25 (2013 est.)

ENERGY

Electricity access: population without electricity: 18,999,254 (2012)
electrification—total population: 84.1% (2016)
electrification—urban areas: 98% (2016)
electrification—rural areas: 79% (2016)

Electricity—production: 1.211 billion kWh (2016 est.)
country comparison to the world: 146

Electricity—consumption: 5.526 billion kWh (2016 est.)
country comparison to the world: 119

Electricity—exports: 0 kWh (2016 est.)
country comparison to the world: 96

Electricity—imports: 4.4 billion kWh (2016 est.)
country comparison to the world: 42

Electricity—installed generating capacity: 634,100 kW (2016 est.)

country comparison to the world: 138

Electricity—from fossil fuels: 45% of total installed capacity (2016 est.)
country comparison to the world: 159

Electricity—from nuclear fuels: 0% of total installed capacity (2017 est.)
country comparison to the world: 32

Electricity—from hydroelectric plants: 52% of total installed capacity (2017 est.)
country comparison to the world: 34

Electricity—from other renewable sources: 4% of total installed capacity (2017 est.)
country comparison to the world: 111

Crude oil—production: 0 bbl/day (2018 est.)
country comparison to the world: 101

Crude oil—exports: 0 bbl/day (2015 est.)
country comparison to the world: 82

Crude oil—imports: 0 bbl/day (2015 est.)
country comparison to the world: 84

Crude oil—proved reserves: 0 bbl (1 January 2018 est.)
country comparison to the world: 99

Refined petroleum products—production: 0 bbl/day (2015 est.)
country comparison to the world: 110

Refined petroleum products—consumption: 35,000 bbl/day (2016 est.)
country comparison to the world: 117

Refined petroleum products—exports: 0 bbl/day (2015 est.)
country comparison to the world: 124

Refined petroleum products—imports: 34,210 bbl/day (2015 est.)
country comparison to the world: 97

Natural gas—production: 164.2 million cu m (2017 est.)
country comparison to the world: 79

Natural gas—consumption: 164.2 million cu m (2017 est.)
country comparison to the world: 108

Natural gas—exports: 0 cu m (2017 est.)
country comparison to the world: 57

Natural gas—imports: 0 cu m (2017 est.)
country comparison to the world: 81

Natural gas—proved reserves: 49.55 billion cu m (1 January 2018 est.)
country comparison to the world: 62

Carbon dioxide emissions from consumption of energy: 9.067 million Mt (2017 est.)
country comparison to the world: 111

COMMUNICATIONS

Telephones—fixed lines: total subscriptions: 125,232
subscriptions per 100 inhabitants: less than 1 (2019 est.)
country comparison to the world: 133

Telephones—mobile cellular: total subscriptions: 21,239,280

subscriptions per 100 inhabitants: 59.36 (2019 est.)
country comparison to the world: 57

Telecommunication systems: general assessment: progress has been made on Afghanistan's first limited fixed-line telephone service and nationwide optical fiber backbone; aided by the presence of multiple providers, mobile-cellular telephone service continues to improve swiftly; the Afghan Ministry of Communications and Information claims that more than 90% of the population live in areas with access to mobile cellular service; moderate growth through 2024, assuming stable governance and improving economic environment (2020)
domestic: less than 1 per 100 for fixed-line teledensity; 59 per 100 for mobile-cellular; an increasing number of Afghans utilize mobile-cellular phone networks (2019)
international: country code—93; multiple VSAT's provide international and domestic voice and data connectivity (2019)
note: the COVID-19 outbreak is negatively impacting telecommunications production and supply chains globally; consumer spending on telecom devices and services has also slowed due to the pandemic's effect on economies worldwide; overall progress towards improvements in all facets of the telecom industry—mobile, fixed-line, broadband, submarine cable and satellite—has moderated

Broadcast media: state-owned broadcaster, Radio Television Afghanistan (RTA), operates a series of radio and television stations in Kabul and the provinces; an estimated 174 private radio stations, 83 TV stations, and about a dozen international broadcasters are available (2019)

Internet country code: .af

Internet users: total: 4,717,013

percent of population: 13.5% (July 2018 est.)
country comparison to the world: 86

Broadband—fixed subscriptions: total: 15,999
subscriptions per 100 inhabitants: less than 1 (2018 est.)
country comparison to the world: 160

TRANSPORTATION

National air transport system: number of registered air carriers: 3 (2020)
inventory of registered aircraft operated by air carriers: 13
annual passenger traffic on registered air carriers: 1,722,612 (2018)
annual freight traffic on registered air carriers: 29.56 million mt-km (2018)

Civil aircraft registration country code prefix: YA (2016)

Airports: 46 (2020)
country comparison to the world: 94

Airports—with paved runways: total: 29 (2020)
over 3,047 m: 4
2,438 to 3,047 m: 8
1,524 to 2,437 m: 12

914 to 1,523 m: 2
under 914 m: 3

Airports—with unpaved runways: total: 17 (2020)
2,438 to 3,047 m: 1
1,524 to 2,437 m: 7
914 to 1,523 m: 4
under 914 m: 5

Heliports: 1 (2020)

Pipelines: 466 km gas (2013)

Roadways: total: 34,903 km (2017)
paved: 17,903 km (2017)
unpaved: 17,000 km (2017)
country comparison to the world: 93

Waterways: 1,200 km (chiefly Amu Darya, which handles vessels up to 500 DWT) (2011)
country comparison to the world: 58

Ports and terminals: river port(s): Kheyrabad, Shir Khan

MILITARY AND SECURITY

Military and security forces: Afghan National Defense and Security Forces (ANDSF) are comprised of military, police, and other security elements: Ministry of Defense: Afghan National Army ((ANA), Afghan Air Force, Afghan Special Security Forces (includes Special Operations Forces), Afghanistan National Army Territorial Forces (ANA-TF)); Afghan Border Force (ABF); Afghan National Civil Order Force (ANCOF)

Ministry of Interior: Afghan Uniform (National) Police (AUP); Public Security Police (PSP); Afghan Border Police (ABP); Afghan Anti-Crime Police; Afghan Local Police; Afghan Public Protection Force
National Directorate of Security ((NDS), intelligence service) (2020)

Military expenditures: 1.2% of GDP (2019)
1% of GDP (2018)
0.9% of GDP (2017)
1% of GDP (2016)
1% of GDP (2015)
country comparison to the world: 99

Military and security service personnel strengths: the Afghan National Defense and Security Forces (ANDSF) have approximately 180,000 active personnel (173,000 Army; 7,000 Air); est. 150,000 Afghan National Police (2019)

Military equipment inventories and acquisitions: the Afghan Army and Air Force inventory is mostly a mix of Soviet-era and more modern US equipment; since 2010, the US is the leading supplier of arms to Afghanistan, followed by Russia (2019 est.)

Military service age and obligation: 18 is the legal minimum age for voluntary military service; no conscription (2017)

Military—note: since early 2015, the NATO-led mission in Afghanistan known as Resolute Support Mission (RSM) has focused on training, advising, and assisting Afghan government forces; as of mid-2020, RSM includes about 16,000 troops from 38 countries (June 2020)

TERRORISM

Terrorist group(s): Haqqani Taliban Network; Harakat ul-Mujahidin; Harakat ul-Jihad-i-Islami; Islamic Jihad Union; Islamic Movement of Uzbekistan; Islamic State of Iraq and ash-Sham-Khorasan Province; Islamic Revolutionary Guard Corps/Qods Force; Jaish-e-Mohammed; Jaysh al Adl (Jundallah); Lashkar i Jhangvi; Lashkar-e Tayyiba; al-Qa'ida; al-Qa'ida in the Indian Subcontinent; Tehrik-e-Taliban Pakistan (2020)
note: details about the history, aims, leadership, organization, areas of operation, tactics, targets, weapons, size, and sources of support of the group(s) appear(s) in Appendix T

TRANSNATIONAL ISSUES

Disputes—international: Afghan, Coalition, and Pakistan military meet periodically to clarify the alignment of the boundary on the ground and on maps and since 2014 have met to discuss collaboration on the Taliban insurgency and counterterrorism efforts; Afghan and Iranian commissioners have discussed boundary monument densification and resurvey; Iran protests Afghanistan's restricting flow of dammed Helmand River tributaries during drought; Pakistan has sent troops across and built fences along some remote tribal areas of its treaty-defined Durand Line border with Afghanistan which serve as bases for foreign terrorists and other illegal activities; Russia remains concerned about the smuggling of poppy derivatives from Afghanistan through Central Asian countries

Refugees and internally displaced persons: refugees (country of origin): 72,191 (Pakistan) (2019)
IDPs: 2.993 million (mostly Pashtuns and Kuchis displaced in the south and west due to natural disasters and political instability) (2019)

Illicit drugs: world's largest producer of opium; poppy cultivation increased 63 percent, to 328,304 hectares in 2017; while eradication increased slightly, it still remains well below levels achieved in 2015; the 2017 crop yielded an estimated 9,000 mt of raw opium, a 88% increase over 2016; the Taliban and other antigovernment groups participate in and profit from the opiate trade, which is a key source of revenue for the Taliban inside Afghanistan; widespread corruption and instability impede counterdrug efforts; most of the heroin consumed in Europe and Eurasia is derived from Afghan opium; Afghanistan is also struggling to respond to a burgeoning domestic opiate addiction problem; a 2015 national drug use survey found that roughly 11% of the population tested positive for one or more illicit drugs; vulnerable to drug money laundering through informal financial networks; illicit cultivation of cannabis and regional source of hashish (2018)

AKROTIRI

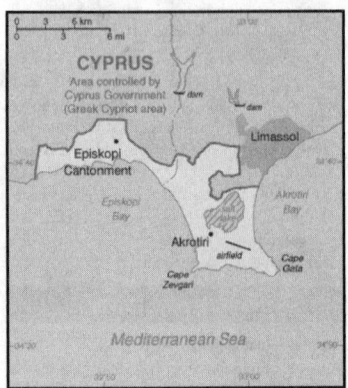

INTRODUCTION

Background: By terms of the 1960 Treaty of Establishment that created the independent Republic of Cyprus, the UK retained full sovereignty and jurisdiction over two areas of almost 254 square kilometers—Akrotiri and Dhekelia. The southernmost and smallest of these is the Akrotiri Sovereign Base Area, which is also referred to as the Western Sovereign Base Area.

GEOGRAPHY

Location: Eastern Mediterranean, peninsula on the southwest coast of Cyprus

Geographic coordinates: 34 37 N, 32 58 E

Map references: Middle East

Area: *total:* 123 sq km
note: includes a salt lake and wetlands
country comparison to the world: 224

Area—comparative: about 0.7 times the size of Washington, DC

Land boundaries: *total:* 48 km
border countries (1): Cyprus 48 km

Coastline: 56.3 km

Climate: temperate; Mediterranean with hot, dry summers and cool winters

Environment—current issues: hunting around the salt lake; note—breeding place for loggerhead and green turtles; only remaining colony of griffon vultures is on the base

Geography—note: British extraterritorial rights also extended to several small off-post sites scattered across Cyprus; of the Sovereign Base Area (SBA) land, 60% is privately owned and farmed, 20% is owned by the Ministry of Defense, and 20% is SBA Crown land

PEOPLE AND SOCIETY

Population: approximately 18,195 on the Sovereign Base Areas of Akrotiri and Dhekelia including 11,000 Cypriots and 7,195 Service and UK-based contract personnel and dependents (2020)

Languages: English, Greek

HIV/AIDS—adult prevalence rate: NA

GOVERNMENT

Country name: *conventional long form:* none
conventional short form: Akrotiri
etymology: named for the village that lies within the Western Sovereign Base Area on Cyprus

Dependency status: a special form of UK overseas territory; administered by an administrator who is also the Commander, British Forces Cyprus

Capital: *name:* Episkopi Cantonment (base administrative center for Akrotiri and Dhekelia)

geographic coordinates: 34 40 N, 32 51 E
time difference: UTC+2 (7 hours ahead of Washington, DC, during Standard Time)
daylight saving time: +1hr, begins last Sunday in March; ends last Sunday in October
etymology: "Episkopi" means "episcopal" in Greek and stems from the fact that the site previously served as the bishop's seat of an Orthodox diocese

Constitution: *history:* presented 3 August 1960, effective 16 August 1960 (The Sovereign Base Areas of Akrotiri and Dhekelia Order in Council 1960 serves as a basic legal document)
amendments: amended 1966

Legal system: laws applicable to the Cypriot population are, as far as possible, the same as the laws of the Republic of Cyprus; note—the Sovereign Base Area Administration has its own court system to deal with civil and criminal matters

Executive branch: *chief of state:* Queen ELIZABETH II (since 6 February 1952)

head of government: Administrator Major General Robert J. THOMSON (since 25 September 2019); note—administrator reports to the British Ministry of Defense; the chief officer is responsible for the day-to-day running of the civil government of the Sovereign Bases
elections/appointments: the monarchy is hereditary; administrator appointed by the monarch on the advice of the Ministry of Defense

Judicial branch: *highest courts:* Senior Judges' Court (consists of several visiting judges from England and Wales)
judge selection and term of office: see entry for United Kingdom
subordinate courts: Resident Judges' Court; Courts Martial

Diplomatic representation in the US: none (overseas territory of the UK)

Diplomatic representation from the US: none (overseas territory of the UK)

Flag description: the flag of the UK is used

National anthem: *note:* as a UK area of special sovereignty, "God Save the Queen" is official (see United Kingdom)
0:00 / 1:02

ECONOMY

Economy—overview: Economic activity is limited to providing services to the military and their families located in Akrotiri. All food and manufactured goods must be imported.

Exchange rates: *note:* uses the euro

COMMUNICATIONS

Broadcast media: British Forces Broadcast Service (BFBS) provides multi-channel satellite TV service as well as BFBS radio broadcasts to the Akrotiri Sovereign Base Area

TRANSPORTATION

Airports: 1 (2020)
country comparison to the world: 210

Airports—with paved runways: 2,438 to 3,047 m: 1 (2017)

MILITARY AND SECURITY

Military—note: defense is the responsibility of the UK; Akrotiri has a full RAF base, headquarters for British Forces Cyprus, and Episkopi Support Unit

ALBANIA

INTRODUCTION

Background: Albania declared its independence from the Ottoman Empire in 1912, but was conquered by Italy in 1939 and occupied by Germany in 1943. Communist partisans took over the country in 1944. Albania allied itself first with the USSR (until 1960), and then with China (to 1978). In the early 1990s, Albania ended 46 years of isolated communist rule and established a multiparty democracy. The transition has proven challenging as successive governments have tried

to deal with high unemployment, widespread corruption, dilapidated infrastructure, powerful organized crime networks, and combative political opponents.

Albania has made progress in its democratic development since it first held multiparty elections in 1991, but deficiencies remain. Most of Albania's post-communist elections were marred by claims of electoral fraud; however, international observers judged elections to be largely free and fair since the restoration of political stability following the collapse of pyramid schemes in 1997. Albania joined NATO in April 2009 and in June 2014 became an EU candidate. Albania in April 2017 received a European Commission recommendation to open EU accession negotiations following the passage of historic EU-mandated justice reforms in 2016. Although Albania's economy continues to grow, it has slowed, and the country is still one of the poorest in Europe. A large informal economy and a weak energy and transportation infrastructure remain obstacles.

GEOGRAPHY

Location: Southeastern Europe, bordering the Adriatic Sea and Ionian Sea, between Greece to the south and Montenegro and Kosovo to the north

Geographic coordinates: 41 00 N, 20 00 E

Map references: Europe

Area: *total:* 28,748 sq km
land: 27,398 sq km
water: 1,350 sq km
country comparison to the world: 145

Area—comparative: slightly smaller than Maryland

Land boundaries: *total:* 691 km
border countries (4): Greece 212 km, Kosovo 112 km, Macedonia 181 km, Montenegro 186 km

Coastline: 362 km

Maritime claims: *territorial sea:* 12 nm
continental shelf: 200-m depth or to the depth of exploitation

Climate: mild temperate; cool, cloudy, wet winters; hot, clear, dry summers; interior is cooler and wetter

Terrain: mostly mountains and hills; small plains along coast

Elevation: *mean elevation:* 708 m
lowest point: Adriatic Sea 0 m
highest point: Maja e Korabit (Golem Korab) 2,764 m

Natural resources: petroleum, natural gas, coal, bauxite, chromite, copper, iron ore, nickel, salt, timber, hydropower, arable land

Land use: *agricultural land:* 43.1% (2016 est.)
arable land: 22.6% (2016 est.) / *permanent crops:* 3% (2016 est.) / *permanent pasture:* 17.5% (2016 est.)
forest: 28.12% (2016 est.)
other: 28.75% (2016 est.)
Irrigated land: 3,537 sq km (2014)

Population distribution: a fairly even distribution, with somewhat higher concentrations of people in the western and central parts of the country

Natural hazards: destructive earthquakes; tsunamis occur along southwestern coast; floods; drought

Environment—current issues: deforestation; soil erosion; water pollution from industrial and domestic effluents; air pollution from industrial and power plants; loss of biodiversity due to lack of resources for sound environmental management

Environment—international agreements: *party to:* Air Pollution, Biodiversity, Climate Change, Climate Change-Kyoto Protocol, Desertification, Endangered Species, Hazardous Wastes, Law of the Sea, Ozone Layer Protection, Wetlands
signed, but not ratified: none of the selected agreements

Geography—note: strategic location along Strait of Otranto (links Adriatic Sea to Ionian Sea and Mediterranean Sea)

PEOPLE AND SOCIETY

Population: 3,074,579 (July 2020 est.)
country comparison to the world: 136

Nationality: *noun:* Albanian(s)
adjective: Albanian

Ethnic groups: Albanian 82.6%, Greek 0.9%, other 1% (including Vlach, Romani, Macedonian, Montenegrin, and Egyptian), unspecified 15.5% (2011 est.)
note: data represent population by ethnic and cultural affiliation

Languages: Albanian 98.8% (official—derived from Tosk dialect), Greek 0.5%, other 0.6% (including Macedonian, Romani, Vlach, Turkish, Italian, and Serbo-Croatian), unspecified 0.1% (2011 est.)

Religions: Muslim 56.7%, Roman Catholic 10%, Orthodox 6.8%, atheist 2.5%, Bektashi (a Sufi order) 2.1%, other 5.7%, unspecified 16.2% (2011 est.)
note: all mosques and churches were closed in 1967 and religious observances prohibited; in November 1990, Albania began allowing private religious practice

Age structure: *0-14 years:* 17.6% (male 284,636/female 256,474)
15-24 years: 15.39% (male 246,931/female 226,318)
25-54 years: 42.04% (male 622,100/female 670,307)
55-64 years: 11.94% (male 178,419/female 188,783)
65 years and over: 13.03% (male 186,335/female 214,276) (2020 est.)

Dependency ratios: *total dependency ratio:* 46.9
youth dependency ratio: 25.3
elderly dependency ratio: 21.6
potential support ratio: 4.6 (2020 est.)

Median age: *total:* 34.3 years
male: 32.9 years
female: 35.7 years (2020 est.)
country comparison to the world: 91

Population growth rate: 0.28% (2020 est.)
country comparison to the world: 173

Birth rate: 13 births/1,000 population (2020 est.)
country comparison to the world: 143

Death rate: 7.1 deaths/1,000 population (2020 est.)
country comparison to the world: 121

Net migration rate: -3.3 migrant(s)/1,000 population (2020 est.)
country comparison to the world: 179

Population distribution: a fairly even distribution, with somewhat higher concentrations of people in the western and central parts of the country

Urbanization: *urban population:* 62.1% of total population (2020)
rate of urbanization: 1.69% annual rate of change (2015-20 est.)

total population growth rate v. urban population growth rate, 2000-2030: Major urban areas—population: 494,000 TIRANA (capital) (2020)

Sex ratio: *at birth:* 1.08 male(s)/female
0-14 years: 1.11 male(s)/female
15-24 years: 1.09 male(s)/female
25-54 years: 0.93 male(s)/female
55-64 years: 0.95 male(s)/female
65 years and over: 0.87 male(s)/female
total population: 0.98 male(s)/female (2020 est.)

Mother's mean age at first birth: 24.8 years (2017/18 est.)

Maternal mortality rate: 15 deaths/100,000 live births (2017 est.)
country comparison to the world: 136

Infant mortality rate: *total:* 10.8 deaths/1,000 live births
male: 12.1 deaths/1,000 live births
female: 9.5 deaths/1,000 live births (2020 est.)
country comparison to the world: 125

Life expectancy at birth: *total population:* 79 years
male: 76.3 years
female: 81.9 years (2020 est.)
country comparison to the world: 61

Total fertility rate: 1.53 children born/woman (2020 est.)
country comparison to the world: 197

Contraceptive prevalence rate: 46% (2017/18)

Drinking water source:
improved:
urban: 96.8% of population
rural: 95.3% of population
total: 96.2% of population
unimproved:
urban: 4.7% of population
rural: 4.7% of population
total: 3.8% of population (2017 est.)

Current Health Expenditure: 6.7% (2016)

Physicians density: 1.22 physicians/1,000 population (2016)

Hospital bed density: 2.9 beds/1,000 population (2013)

Sanitation facility access:
improved:
urban: 100% of population
rural: 99.5% of population
total: 99.8% of population
unimproved:
urban: 0% of population
rural: 0.5% of population
total: 0.2% of population (2017 est.)

HIV/AIDS—adult prevalence rate: <.1 (2019 est.)

HIV/AIDS—people living with HIV/AIDS: 1,400 (2019 est.)
country comparison to the world: 139

HIV/AIDS—deaths: <100 (2019 est.)

Obesity—adult prevalence rate: 21.7% (2016)
country comparison to the world: 85

Children under the age of 5 years underweight: 1.5% (2017)
country comparison to the world: 120

Education expenditures: 4% of GDP (2016)
country comparison to the world: 101

Literacy: *definition:* age 15 and over can read and write
total population: 98.1%
male: 98.5%
female: 97.8% (2018)

School life expectancy (primary to tertiary education): *total:* 15 years
male: 14 years
female: 16 years (2019)

Unemployment, youth ages 15-24: *total:* 31.9%
male: 34.2%
female: 27.7% (2017 est.)
country comparison to the world: 27

GOVERNMENT

Country name: *conventional long form:* Republic of Albania
conventional short form: Albania
local long form: Republika e Shqiperise
local short form: Shqiperia
former: People's Socialist Republic of Albania
etymology: the English-language country name seems to be derived from the ancient Illyrian tribe of the Albani; the native name "Shqiperia" is derived from the Albanian word "Shqiponje"

("Eagle") and is popularly interpreted to mean "Land of the Eagles"

Government type: parliamentary republic

Capital: *name:* Tirana (Tirane)

geographic coordinates: 41 19 N, 19 49 E
time difference: UTC+1 (6 hours ahead of Washington, DC, during Standard Time)
daylight saving time: +1hr, begins last Sunday in March; ends last Sunday in October
etymology: the name Tirana first appears in a 1418 Venetian document; the origin of the name is unclear, but may derive from Tirkan Fortress, whose ruins survive on the slopes of Dajti mountain and which overlooks the city

Administrative divisions: 12 counties (qarqe, singular—qark); Berat, Diber, Durres, Elbasan, Fier, Gjirokaster, Korce, Kukes, Lezhe, Shkoder, Tirane, Vlore

Independence: 28 November 1912 (from the Ottoman Empire)

National holiday: Independence Day, 28 November (1912), also known as Flag Day

Constitution: *history:* several previous; latest approved by the Assembly 21 October 1998, adopted by referendum 22 November 1998, promulgated 28 November 1998
amendments: proposed by at least one-fifth of the Assembly membership; passage requires at least a two-thirds majority vote by the Assembly; referendum required only if approved by two-thirds of the Assembly; amendments approved by referendum effective upon declaration by the president of the republic; amended several times, last in 2016

Legal system: civil law system except in the northern rural areas where customary law known as the "Code of Leke" is still present

International law organization participation: has not submitted an ICJ jurisdiction declaration; accepts ICCt jurisdiction

Citizenship: *citizenship by birth:* no
citizenship by descent only: at least one parent must be a citizen of Albania
dual citizenship recognized: yes
residency requirement for naturalization: 5 years

Suffrage: 18 years of age; universal

Executive branch: *chief of state:* President of the Republic Ilir META (since 24 July 2017)

head of government: Prime Minister Edi RAMA (since 10 September 2013); Deputy Prime Minister Senida MESI (since 13 September 2017)
cabinet: Council of Ministers proposed by the prime minister, nominated by the president, and approved by the Assembly
elections/appointments: president indirectly elected by the Assembly for a 5-year term (eligible for a second term); a candidate needs three-fifths majority vote of the Assembly in 1 of 3 rounds or a simple majority in 2 additional rounds to become president; election last held in 4 rounds on 19, 20, 27, and 28 April 2017 (next election to be held in 2022); prime minister appointed by the president

on the proposal of the majority party or coalition of parties in the Assembly
election results: Ilir META elected president; Assembly vote—87—2 in fourth round

Legislative branch: *description:* unicameral Assembly or Kuvendi (140 seats; members directly elected in multi-seat constituencies by proportional representation vote to serve 4-year terms)
elections: last held on 25 June 2017 (next to be held in 2021)
election results: percent of vote by party—PS 48.3%, PD 28.9%, LSI 14.3%, PDIU 4.8%, PSD 1%, other 2.7%; seats by party—PS 74, PD 43, LSI 19, PDIU 3, PSD 1; composition—men 108, women 32, percent of women 22.9%

Judicial branch: *highest courts:* Supreme Court (consists of 19 judges, including the chief justice); Constitutional Court (consists of 9 judges, including the chairman)
judge selection and term of office: Supreme Court judges appointed by the High Judicial Council with the consent of the president to serve single 9-year terms; Supreme Court chairman is elected for a single 3-year term by the court members; appointments of Constitutional Court judges are rotated among the president, Parliament, and Supreme Court from a list of pre-qualified candidates (each institution selects 3 judges), to serve single 9-year terms; candidates are pre-qualified by a randomly selected body of experienced judges and prosecutors; Constitutional Court chairman is elected by the court members for a single, renewable 3-year term
subordinate courts: Courts of Appeal; Courts of First Instance; specialized courts: Court for Corruption and Organized Crime, Appeals Court for Corruption and Organized Crime (responsible for corruption, organized crime, and crimes of high officials)

Political parties and leaders: Democratic Party or PD [Lulzim BASHA]
Party for Justice, Integration and Unity or PDIU [Shpetim IDRIZI] (formerly part of APMI)
Social Democratic Party or PSD [Paskal MILO]
Socialist Movement for Integration or LSI [Monika KRYEMADHI]
Socialist Party or PS [Edi RAMA]

International organization participation: BSEC, CD, CE, CEI, EAPC, EBRD, EITI (compliant country), FAO, IAEA, IBRD, ICAO, ICC (national committees), ICCt, ICRM, IDA, IDB, IFAD, IFC, IFRCS, ILO, IMF, IMO, Interpol, IOC, IOM, IPU, ISO (correspondent), ITU, ITUC (NGOs), MIGA, NATO, OAS (observer), OIC, OIF, OPCW, OSCE, PCA, SELEC, UN, UNCTAD, UNESCO, UNIDO, UNWTO, UPU, WCO, WFTU (NGOs), WHO, WIPO, WMO, WTO

Diplomatic representation in the US: *chief of mission:* Ambassador Yuri KIM (since 27 January 2020)
chancery: 2100 S Street NW, Washington, DC 20008

telephone: [1] (202) 223-4942
FAX: [1] (202) 628-7342

consulate(s) general: New York

Diplomatic representation from the US: *chief of mission:* Ambassador (vacant); Charge d'Affaires Leyla MOSES-ONES (since August 2018)
telephone: [355] (4) 2247-285
embassy: Rruga e Elbasanit, 103, Tirana
mailing address: US Department of State, 9510 Tirana Place, Dulles, VA 20189-9510
FAX: [355] (4) 2232-222

Flag description: red with a black two-headed eagle in the center; the design is claimed to be that of 15th-century hero Georgi Kastrioti SKANDERBEG, who led a successful uprising against the Ottoman Turks that resulted in a short-lived independence for some Albanian regions (1443-78); an unsubstantiated explanation for the eagle symbol is the tradition that Albanians see themselves as descendants of the eagle; they refer to themselves as "Shqiptare," which translates as "sons of the eagle"

National symbol(s): black double-headed eagle; national colors: red, black

National anthem: *name:* "Hymni i Flamurit" (Hymn to the Flag)
lyrics/music: Aleksander Stavre DRENOVA/ Ciprian PORUMBESCU
note: adopted 1912
0:00 / 1:05

ECONOMY

Economy—overview: Albania, a formerly closed, centrally planned state, is a developing country with a modern open-market economy. Albania managed to weather the first waves of the global financial crisis but, the negative effects of the crisis caused a significant economic slowdown. Since 2014, Albania's economy has steadily improved and economic growth reached 3.8% in 2017. However, close trade, remittance, and banking sector ties with Greece and Italy make Albania vulnerable to spillover effects of possible debt crises and weak growth in the euro zone.

Remittances, a significant catalyst for economic growth, declined from 12-15% of GDP before the 2008 financial crisis to 5.8% of GDP in 2015, mostly from Albanians residing in Greece and Italy. The agricultural sector, which accounts for more than 40% of employment but less than one quarter of GDP, is limited primarily to small family operations and subsistence farming, because of a lack of modern equipment, unclear property rights, and the prevalence of small, inefficient plots of land. Complex tax codes and licensing requirements, a weak judicial system, endemic corruption, poor enforcement of contracts and property issues, and antiquated infrastructure contribute to Albania's poor business environment making attracting foreign investment difficult. Since 2015, Albania has launched an ambitious program to increase tax compliance and bring more businesses into the formal economy. In July 2016, Albania passed constitutional amendments reforming the

judicial system in order to strengthen the rule of law and to reduce deeply entrenched corruption.

Albania's electricity supply is uneven despite upgraded transmission capacities with neighboring countries. However, the government has recently taken steps to stem non-technical losses and has begun to upgrade the distribution grid. Better enforcement of electricity contracts has improved the financial viability of the sector, decreasing its reliance on budget support. Also, with help from international donors, the government is taking steps to improve the poor road and rail networks, a long standing barrier to sustained economic growth.

Inward foreign direct investment has increased significantly in recent years as the government has embarked on an ambitious program to improve the business climate through fiscal and legislative reforms. The government is focused on the simplification of licensing requirements and tax codes, and it entered into a new arrangement with the IMF for additional financial and technical support. Albania's three-year IMF program, an extended fund facility arrangement, was successfully concluded in February 2017. The Albanian Government has strengthened tax collection amid moderate public wage and pension increases in an effort to reduce its budget deficit. The country continues to face high public debt, exceeding its former statutory limit of 60% of GDP in 2013 and reaching 72% in 2016.

GDP (purchasing power parity): $36.01 billion (2017 est.)
$34.67 billion (2016 est.)
$33.55 billion (2015 est.)
note: data are in 2017 dollars; unreported output may be as large as 50% of official GDP
country comparison to the world: 125

GDP (official exchange rate): $13.07 billion (2017 est.)

GDP—real growth rate: 3.8% (2017 est.)
3.4% (2016 est.)
2.2% (2015 est.)
country comparison to the world: 85

GDP—per capita (PPP): $12,500 (2017 est.)
$12,100 (2016 est.)
$11,600 (2015 est.)
note: data are in 2017 dollars
country comparison to the world: 125

Gross national saving: 15.9% of GDP (2017 est.)
16.7% of GDP (2016 est.)
16.9% of GDP (2015 est.)
country comparison to the world: 130

GDP—composition, by end use: *household consumption:* 78.1% (2017 est.)
government consumption: 11.5% (2017 est.)
investment in fixed capital: 25.2% (2017 est.)
investment in inventories: 0.2% (2017 est.)
exports of goods and services: 31.5% (2017 est.)
imports of goods and services: -46.6% (2017 est.)

GDP—composition, by sector of origin:
agriculture: 21.7% (2017 est.)
industry: 24.2% (2017 est.)
services: 54.1% (2017 est.)

Agriculture—products: wheat, corn, potatoes, vegetables, fruits, olives and olive oil, grapes; meat, dairy products; sheep and goats

Industries: food; footwear, apparel and clothing; lumber, oil, cement, chemicals, mining, basic metals, hydropower

Industrial production growth rate: 6.8% (2017 est.)
country comparison to the world: 31

Labor force: 1.198 million (2017 est.)
country comparison to the world: 140

Labor force—by occupation: *agriculture:* 41.4%
industry: 18.3%
services: 40.3% (2017 est.)

Unemployment rate: 13.8% (2017 est.)
15.2% (2016 est.)
note: these official rates may not include those working at near-subsistence farming
country comparison to the world: 168

Population below poverty line: 14.3% (2012 est.)

Household income or consumption by percentage share: *lowest 10%:* 4.1%
highest 10%: 19.6% (2015 est.)

Budget: *revenues:* 3.614 billion (2017 est.)
expenditures: 3.874 billion (2017 est.)

Taxes and other revenues: 27.6% (of GDP) (2017 est.)
country comparison to the world: 99

Budget surplus (+) or deficit (-): -2% (of GDP) (2017 est.)
country comparison to the world: 103

Public debt: 71.8% of GDP (2017 est.)
73.2% of GDP (2016 est.)
country comparison to the world: 45

Fiscal year: calendar year

Inflation rate (consumer prices): 2% (2017 est.)
1.3% (2016 est.)
country comparison to the world: 102

Current account balance: -$908 million (2017 est.)
-$899 million (2016 est.)
country comparison to the world: 139

Exports: $900.7 million (2017 est.)
$789.1 million (2016 est.)
country comparison to the world: 164

Exports—partners: Italy 53.4%, Kosovo 7.7%, Spain 5.6%, Greece 4.2% (2017)

Exports—commodities: apparel and clothing, footwear; asphalt, metals and metallic ores, crude oil; cement and construction materials, vegetables, fruits, tobacco

Imports: $4.103 billion (2017 est.)
$3.67 billion (2016 est.)
country comparison to the world: 138

Imports—commodities: machinery and equipment, foodstuffs, textiles, chemicals

Imports—partners: Italy 28.5%, Turkey 8.1%, Germany 8%, Greece 8%, China 7.9%, Serbia 4% (2017)

Reserves of foreign exchange and gold: $3.59 billion (31 December 2017 est.)
$3.109 billion (31 December 2016 est.)

9

country comparison to the world: 103

Debt—external: $9.505 billion (31 December 2017 est.)
$8.421 billion (31 December 2016 est.)
country comparison to the world: 114

Exchange rates: leke (ALL) per US dollar—
121.9 (2017 est.)
124.14 (2016 est.)
124.14 (2015 est.)
125.96 (2014 est.)
105.48 (2013 est.)

ENERGY

Electricity access: electrification—total population: 100% (2016)

Electricity—production: 7.138 billion kWh (2016 est.)
country comparison to the world: 111

Electricity—consumption: 5.11 billion kWh (2016 est.)
country comparison to the world: 122

Electricity—exports: 1.869 billion kWh (2016 est.)
country comparison to the world: 46

Electricity—imports: 1.827 billion kWh (2016 est.)
country comparison to the world: 58

Electricity—installed generating capacity: 2.109 million kW (2016 est.)
country comparison to the world: 112

Electricity—from fossil fuels: 5% of total installed capacity (2016 est.)
country comparison to the world: 202

Electricity—from nuclear fuels: 0% of total installed capacity (2017 est.)
country comparison to the world: 33

Electricity—from hydroelectric plants: 95% of total installed capacity (2017 est.)
country comparison to the world: 5

Electricity—from other renewable sources: 0% of total installed capacity (2017 est.)
country comparison to the world: 172

Crude oil—production: 14,000 bbl/day (2018 est.)
country comparison to the world: 73

Crude oil—exports: 17,290 bbl/day (2015 est.)
country comparison to the world: 51

Crude oil—imports: 0 bbl/day (2015 est.)
country comparison to the world: 85

Crude oil—proved reserves: 168.3 million bbl (1 January 2018 est.)
country comparison to the world: 59

Refined petroleum products—production: 5,638 bbl/day (2015 est.)
country comparison to the world: 103

Refined petroleum products—consumption: 29,000 bbl/day (2016 est.)
country comparison to the world: 120

Refined petroleum products—exports: 3,250 bbl/day (2015 est.)
country comparison to the world: 98

Refined petroleum products—imports: 26,660 bbl/day (2015 est.)
country comparison to the world: 103

Natural gas—production: 50.97 million cu m (2017 est.)
country comparison to the world: 86

Natural gas—consumption: 50.97 million cu m (2017 est.)
country comparison to the world: 112

Natural gas—exports: 0 cu m (2017 est.)
country comparison to the world: 58

Natural gas—imports: 0 cu m (2017 est.)
country comparison to the world: 82

Natural gas—proved reserves: 821.2 million cu m (1 January 2018 est.)
country comparison to the world: 101

Carbon dioxide emissions from consumption of energy: 4.5 million Mt (2017 est.)
country comparison to the world: 136

COMMUNICATIONS

Telephones—fixed lines: total subscriptions: 258,474
subscriptions per 100 inhabitants: 8.43 (2019 est.)
country comparison to the world: 116

Telephones—mobile cellular: total subscriptions: 2,799,066
subscriptions per 100 inhabitants: 91.29 (2019 est.)
country comparison to the world: 144

Telecommunication systems: general assessment: mobile-cellular phone service has been available since 1996 and dominates over fixed-line capacity; Internet broadband services initiated in 2005 and the government continues to supports the improvement of broadband availability and access conditions; Albania has received financial aid to build its infrastructure and works towards the EU accession process, an adherence to careful scrutiny of its regulatory regime helps the telecom sector advance; Internet cafes are popular in major urban areas; 1.3 million use mobile broadband services (3G/4G) (2020)
domestic: fixed-line 8 per 100, teledensity continues to decline due to heavy use of mobile-cellular telephone services; mobile-cellular telephone use is widespread and generally effective, 91 per 100 for mobile-cellular (2019)
international: country code—355; submarine cables for the Adria 1 and Italy-Albania provide connectivity to Italy, Croatia, and Greece; a combination submarine cable and land fiber-optic system, provides additional connectivity to Bulgaria, Macedonia, and Turkey; international traffic carried by fiber-optic cable and, when necessary, by microwave radio relay from the Tirana exchange to Italy and Greece (2019)
note: the COVID-19 outbreak is negatively impacting telecommunications production and supply chains globally; consumer spending on telecom devices and services has also slowed due to the pandemic's effect on economies worldwide; overall

progress towards improvements in all facets of the telecom industry—mobile, fixed-line, broadband, submarine cable and satellite—has moderated

Broadcast media: Albania has more than 65 TV stations, including several that broadcast nationally; Albanian TV broadcasts are also available to Albanian-speaking populations in neighboring countries; many viewers have access to Italian and Greek TV broadcasts via terrestrial reception; Albania's TV stations have begun a government-mandated conversion from analog to digital broadcast; the government has pledged to provide analog-to-digital converters to low-income families affected by this decision; cable TV service is available; 2 public radio networks and roughly 78 private radio stations; several international broadcasters are available (2019)

Internet country code: .al

Internet users: total: 2,196,613
percent of population: 71.85% (July 2018 est.)
country comparison to the world: 119

Broadband—fixed subscriptions: total: 361,947
subscriptions per 100 inhabitants: 12 (2018 est.)
country comparison to the world: 94

TRANSPORTATION

National air transport system: number of registered air carriers: 2 (2020)
inventory of registered aircraft operated by air carriers: 5
annual passenger traffic on registered air carriers: 303,137 (2018)

Civil aircraft registration country code prefix: ZA (2016)

Airports: 3 (2020)
country comparison to the world: 191

Airports—with paved runways: total: 3 (2020)
2,438 to 3,047 m: 2 (2017)
1,524 to 2,437 m: 1 (2017)

Pipelines: 498 km gas (a majority of the network is in disrepair and parts of it are missing), 249 km oil (2015)

Railways: total: 677 km (447 km of major railway lines and 230 km of secondary lines) (2015)
standard gauge: 677 km 1.435-m gauge (2015)
country comparison to the world: 103

Roadways: total: 3,945 km (2018)
country comparison to the world: 156

Waterways: 41 km (on the Bojana River) (2011)
country comparison to the world: 103

Merchant marine: total: 68
by type: general cargo 49, oil tanker 1, other 18 (2019)
country comparison to the world: 104

Ports and terminals: major seaport(s): Durres, Sarande, Shengjin, Vlore

MILITARY AND SECURITY

Military and security forces: Land Forces Command, Navy Force Command (includes Coast Guard), Air Forces Command, Support

Command, Training and Doctrination Command (2019)

Military expenditures: 1.3% of GDP (2019 est.)
1.2% of GDP (2018)
1.1% of GDP (2017)
1.1% of GDP (2016)
1.2% of GDP (2015)
country comparison to the world: 91

Military and security service personnel strengths: the Albanian military has approximately 10,000 total active duty personnel (8,000 Army; 1,500 Navy; 500 Air Force; note—as many as 4,000 personnel are assigned to support forces) (2019 est.)

Military equipment inventories and acquisitions: the Albanian military was previously equipped with mostly Soviet-era weapons that were sold or destroyed; its inventory now includes a mix of mostly donated and second-hand European and US equipment; since 2010, it has received equipment from France, Germany, Italy, and the US (2019 est.)

Military deployments: 100 Afghanistan (NATO) (June 2020)

Military service age and obligation: 19 is the legal minimum age for voluntary military service; 18 is the legal minimum age in case of general/partial compulsory mobilization (2012)

TERRORISM

Terrorist group(s): Islamic Revolutionary Guard Corps/Qods Force (2020)
note: details about the history, aims, leadership, organization, areas of operation, tactics, targets, weapons, size, and sources of support of the group(s) appear(s) in Appendix T

TRANSNATIONAL ISSUES

Disputes—international: none

Refugees and internally displaced persons: *stateless persons:* 4,160 (2018)
note: 11,052 estimated refugee and migrant arrivals (January 2015-October 2020)

Illicit drugs: active transshipment point for Southwest Asian opiates, hashish, and cannabis transiting the Balkan route and—to a lesser extent—cocaine from South America destined for Western Europe; significant source country for cannabis production; ethnic Albanian narcotrafficking organizations active and expanding in Europe; vulnerable to money laundering associated with regional trafficking in narcotics, arms, contraband, and illegal aliens

ALGERIA

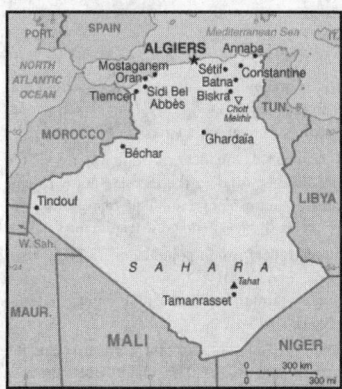

INTRODUCTION

Background: Algeria has known many empires and dynasties starting with the ancient Numidians (3rd century B.C.), Phoenicians, Carthaginians, Romans, Vandals, Byzantines, over a dozen different Arab and Berber dynasties, Spaniards, and Ottoman Turks. It was under the latter that the Barbary pirates operated from North Africa and preyed on shipping beginning in roughly 1500, peaking in the early to mid-17th century, until finally subdued by the French capture of Algiers in 1830. The French southward conquest of the entirety of Algeria proceeded throughout the 19th century and was marked by many atrocities. The country was heavily colonized by the French in the late 19th and early 20th centuries. A bloody eight-year struggle culminated in Algerian independence in 1962.

Algeria's primary political party, the National Liberation Front (FLN), was established in 1954 as part of the struggle for independence and has since largely dominated politics. The Government of Algeria in 1988 instituted a multi-party system in response to public unrest, but the surprising first round success of the Islamic Salvation Front (FIS) in the December 1991 legislative elections led the Algerian army to intervene and postpone the second round of elections to prevent what the secular elite feared would be an extremist-led government from assuming power. The army began a crackdown on the FIS that spurred FIS supporters to begin attacking government targets. Fighting escalated into an insurgency, which saw intense violence from 1992-98, resulting in over 100,000 deaths—many attributed to indiscriminate massacres of villagers by extremists. The government gained the upper hand by the late-1990s, and FIS's armed wing, the Islamic Salvation Army, disbanded in January 2000.

Abdelaziz BOUTEFLIKA, with the backing of the military, won the presidency in 1999 in an election that was boycotted by several candidates protesting alleged fraud, and won subsequent elections in 2004, 2009, and 2014. The government in 2011 introduced some political reforms in response to the Arab Spring, including lifting the 19-year-old state of emergency restrictions and increasing women's quotas for elected assemblies, while also increasing subsidies to the populace. Since 2014, Algeria's reliance on hydrocarbon revenues to fund the government and finance the large subsidies for the population has fallen under stress because of declining oil prices. Protests broke out across the country in late February 2019 against President BOUTEFLIKA's decision to seek a fifth term. BOUTEFLIKA resigned on 2 April 2019, and the speaker of the upper house of parliament, Abdelkader BENSALAH, became interim head of state on 9 April. BENSALAH remained in office beyond the 90-day constitutional limit until Algerians elected former Prime Minister Abdelmadjid TEBBOUNE as the country's new president in December 2019.

GEOGRAPHY

Location: Northern Africa, bordering the Mediterranean Sea, between Morocco and Tunisia

Geographic coordinates: 28 00 N, 3 00 E

Map references: Africa

Area: *total:* 2,381,740 sq km
land: 2,381,740 sq km
water: 0 sq km
country comparison to the world: 11

Area—comparative: slightly less than 3.5 times the size of Texas

Land boundaries: *total:* 6,734 km
border countries (7): Libya 989 km, Mali 1359 km, Mauritania 460 km, Morocco 1900 km, Niger 951 km, Tunisia 1034 km, Western Sahara 41 km

Coastline: 998 km

Maritime claims: *territorial sea:* 12 nm
exclusive fishing zone: 32-52 nm

Climate: arid to semiarid; mild, wet winters with hot, dry summers along coast; drier with cold winters and hot summers on high plateau; sirocco is a hot, dust/sand-laden wind especially common in summer

Terrain: mostly high plateau and desert; Atlas Mountains in the far north and Hoggar Mountains in the south; narrow, discontinuous coastal plain

Elevation: *mean elevation:* 800 m
lowest point: Chott Melrhir -40 m
highest point: Tahat 2,908 m

Natural resources: petroleum, natural gas, iron ore, phosphates, uranium, lead, zinc

Land use: agricultural land: 17.4% (2016 est.)
arable land: 3.1% (2016 est.) / permanent crops: 0.4% (2016 est.) / permanent pasture: 13.8% (2016 est.)

forest: 0.8% (2016 est.)
other: 81.8% (2016 est.)

Irrigated land: 13,600 sq km (2014)

Population distribution: the vast majority of the populace is found in the extreme northern part of the country along the Mediterranean Coast as shown in this population distribution map

Natural hazards: mountainous areas subject to severe earthquakes; mudslides and floods in rainy season; droughts

Environment—current issues: air pollution in major cities; soil erosion from overgrazing and other poor farming practices; desertification; dumping of raw sewage, petroleum refining wastes, and other industrial effluents is leading to the pollution of rivers and coastal waters; Mediterranean Sea, in particular, becoming polluted from oil wastes, soil erosion, and fertilizer runoff; inadequate supplies of potable water

Environment—international agreements: *party to:* Biodiversity, Climate Change, Climate Change-Kyoto Protocol, Desertification, Endangered Species,Environmental Modification, Hazardous Wastes, Law of the Sea, Ozone Layer Protection, Ship Pollution, Wetlands
signed, but not ratified: none of the selected agreements

Geography—note: largest country in Africa but 80% desert; canyons and caves in the southern Hoggar Mountains and in the barren Tassilin'Ajjer area in the southeast of the country contain numerous examples of prehistoric art—rock paintings and carvings depicting human activities and wild and domestic animals (elephants, giraffes, cattle)—that date to the African Humid Period, roughly 11,000 to 5,000 years ago, when the region was completely vegetated

PEOPLE AND SOCIETY

Population: 42,972,878 (July 2020 est.)
country comparison to the world: 35

Nationality: *noun:* Algerian(s)
adjective: Algerian

Ethnic groups: Arab-Berber 99%, European less than 1%
note: although almost all Algerians are Berber in origin (not Arab), only a minority identify themselves as primarily Berber, about 15% of the total population; these people live mostly in the mountainous region of Kabylie east of Algiers and several other communities; the Berbers are also Muslim but identify with their Berber rather than Arab cultural heritage; Berbers have long agitated, sometimes violently, for autonomy; the government is unlikely to grant autonomy but has officially recognized Berber languages and introduced them into public schools

Languages: Arabic (official), French (lingua franca), Berber or Tamazight (official); dialects include Kabyle Berber (Taqbaylit), Shawiya Berber (Tacawit), Mzab Berber, Tuareg Berber (Tamahaq)

Religions: Muslim (official; predominantly Sunni) 99%, other (includes Christian and Jewish) <1% (2012 est.)

MENA religious affiliation: Demographic profile: For the first two thirds of the 20th century, Algeria's high fertility rate caused its population to grow rapidly. However, about a decade after independence from France in 1962, the total fertility rate fell dramatically from 7 children per woman in the 1970s to about 2.4 in 2000, slowing Algeria's population growth rate by the late 1980s. The lower fertility rate was mainly the result of women's rising age at first marriage (virtually all Algerian children being born in wedlock) and to a lesser extent the wider use of contraceptives. Later marriages and a preference for smaller families are attributed to increases in women's education and participation in the labor market; higher unemployment; and a shortage of housing forcing multiple generations to live together. The average woman's age at first marriage increased from about 19 in the mid-1950s to 24 in the mid-1970s to 30.5 in the late 1990s.

Algeria's fertility rate experienced an unexpected upturn in the early 2000s, as the average woman's age at first marriage dropped slightly. The reversal in fertility could represent a temporary fluctuation in marriage age or, less likely, a decrease in the steady rate of contraceptive use.

Thousands of Algerian peasants—mainly Berber men from the Kabylia region—faced with land dispossession and economic hardship under French rule migrated temporarily to France to work in manufacturing and mining during the first half of the 20th century. This movement accelerated during World War I, when Algerians filled in for French factory workers or served as soldiers. In the years following independence, low-skilled Algerian workers and Algerians who had supported the French (known as Harkis) emigrated en masse to France. Tighter French immigration rules and Algiers' decision to cease managing labor migration to France in the 1970s limited legal emigration largely to family reunification.

Not until Algeria's civil war in the 1990s did the country again experience substantial outmigration. Many Algerians legally entered Tunisia without visas claiming to be tourists and then stayed as workers. Other Algerians headed to Europe seeking asylum, although France imposed restrictions. Sub-Saharan African migrants came to Algeria after its civil war to work in agriculture and mining. In the 2000s, a wave of educated Algerians went abroad seeking skilled jobs in a wider range of destinations, increasing their presence in North America and Spain. At the same time, legal foreign workers principally from China and Egypt came to work in Algeria's construction and oil sectors. Illegal migrants from Sub-Saharan Africa, particularly Malians, Nigeriens, and Gambians, continue to come to Algeria in search of work or to use it as a stepping stone to Libya and Europe.

Since 1975, Algeria also has been the main recipient of Sahrawi refugees from the ongoing conflict in Western Sahara. More than 1000,000

Sahrawis are estimated to be living in five refugee camps in southwestern Algeria near Tindouf.

Age structure: *0-14 years:* 29.58% (male 6,509,490/female 6,201,450)
15-24 years: 13.93% (male 3,063,972/female 2,922,368)
25-54 years: 42.91% (male 9,345,997/female 9,091,558)
55-64 years: 7.41% (male 1,599,369/female 1,585,233)
65 years and over: 6.17% (male 1,252,084/female 1,401,357) (2020 est.)

Dependency ratios: *total dependency ratio:* 60.1
youth dependency ratio: 49.3
elderly dependency ratio: 10.8
potential support ratio: 9.3 (2020 est.)

Median age: *total:* 28.9 years
male: 28.6 years
female: 29.3 years (2020 est.)
country comparison to the world: 139

Population growth rate: 1.52% (2020 est.)
country comparison to the world: 67

Birth rate: 20 births/1,000 population (2020 est.)
country comparison to the world: 76

Death rate: 4.4 deaths/ 1,000 population (2020 est.)
country comparison to the world: 208

Net migration rate: -0.9 migrant(s)/1,000 population (2020 est.)
country comparison to the world: 137

Population distribution: the vast majority of the populace is found in the extreme northern part of the country along the Mediterranean Coast as shown in this population distribution map

Urbanization: *urban population:* 73.7% of total population (2020)
rate of urbanization: 2.46% annual rate of change (2015-20 est.)

total population growth rate v. urban population growth rate, 2000-2030: Major urban areas—population: 2.768 million ALGIERS (capital), 899,000 Oran (2020)

Sex ratio: *at birth:* 1.05 male(s)/female
0-14 years: 1.05 male(s)/female
15-24 years: 1.05 male(s)/female
25-54 years: 1.03 male(s)/female
55-64 years: 1.01 male(s)/female
65 years and over: 0.89 male(s)/female
total population: 1.03 male(s)/female (2020 est.)

Maternal mortality rate: 112 deaths/100,000 live births (2017 est.)
country comparison to the world: 68

Infant mortality rate: *total:* 17.6 deaths/1,000 live births
male: 19.1 deaths/1,000 live births
female: 16 deaths/1,000 live births (2020 est.)
country comparison to the world: 85

Life expectancy at birth: *total population:* 77.5 years
male: 76.1 years
female: 79.1 years (2020 est.)
country comparison to the world: 77

Total fertility rate: 2.59 children born/woman (2020 est.)
country comparison to the world: 67

Contraceptive prevalence rate: 57.1% (2012/13)

Drinking water source:
improved:
urban: 99.2% of population
rural: 97.4% of population
total: 98.7% of population
unimproved:
urban: 0.8% of population
rural: 2.1% of population
total: 1.1% of population (2017 est.)

Current Health Expenditure: 6.4% (2017)

Physicians density: 1.79 physicians/1,000 population (2017)

Hospital bed density: 1.9 beds/1,000 population (2015)

Sanitation facility access:
improved:
urban: 96.9% of population
rural: 93.4% of population
total: 96% of population
unimproved:
urban: 3.1% of population
rural: 6.6% of population
total: 4% of population (2017 est.)

HIV/AIDS—adult prevalence rate: <.1% (2019 est.)

HIV/AIDS—people living with HIV/AIDS: 22,000 (2019 est.)
country comparison to the world: 84

HIV/AIDS—deaths: < 200 (2019 est.)

Obesity—adult prevalence rate: 27.4% (2016)
country comparison to the world: 38

Children under the age of 5 years underweight: 3% (2012)
country comparison to the world: 99

Education expenditures: NA

Literacy: definition: age 15 and over can read and write
total population: 81.4%
male: 87.4%
female: 75.3% (2018)

School life expectancy (primary to tertiary education): total: 14 years
male: 14 years
female: 15 years (2011)

Unemployment, youth ages 15-24: total: 39.3%
male: 33.1%
female: 82% (2017 est.)
country comparison to the world: 13

GOVERNMENT

Country name: *conventional long form:* People's Democratic Republic of Algeria
conventional short form: Algeria
local long form: Al Jumhuriyah al Jaza'iriyah ad Dimuqratiyah ash Sha'biyah
local short form: Al Jaza'ir
etymology: the country name derives from the capital city of Algiers

Government type: presidential republic

Capital: *name:* Algiers
geographic coordinates: 36 45 N, 3 03 E
time difference: UTC+1 (6 hours ahead of Washington, DC, during Standard Time)
etymology: name derives from the Arabic "al-Jazair" meaning "the islands" and refers to the four islands formerly off the coast but joined to the mainland since 1525

Administrative divisions: 48 provinces (wilayas, singular—wilaya); Adrar, Ain Defla, Ain Temouchent, Alger, Annaba, Batna, Bechar, Bejaia, Biskra, Blida, Bordj Bou Arreridj, Bouira, Boumerdes, Chlef, Constantine, Djelfa, El Bayadh, El Oued, El Tarf, Ghardaia, Guelma, Illizi, Jijel, Khenchela, Laghouat, Mascara, Medea, Mila, Mostaganem, M'Sila, Naama, Oran, Ouargla, Oum el Bouaghi, Relizane, Saida, Setif, Sidi Bel Abbes, Skikda, Souk Ahras, Tamanrasset, Tebessa, Tiaret, Tindouf, Tipaza, Tissemsilt, Tizi Ouzou, Tlemcen

Independence: 5 July 1962 (from France)

National holiday: Independence Day, 5 July (1962); Revolution Day, 1 November (1954)

Constitution: *history:* several previous; latest approved by referendum 23 February 1989
amendments: proposed by the president of the republic or through the president with the support of three fourths of the members of both houses of Parliament in joint session; passage requires approval by both houses, approval by referendum, and promulgation by the president; the president can forego a referendum if the Constitutional Council determines the proposed amendment does not conflict with basic constitutional principles; articles including the republican form of government, the integrity and unity of the country, and fundamental citizens' liberties and rights cannot be amended; amended 2002, 2008, 2016

Legal system: mixed legal system of French civil law and Islamic law; judicial review of legislative acts in ad hoc Constitutional Council composed of various public officials including several Supreme Court justices

International law organization participation: has not submitted an ICJ jurisdiction declaration; non-party state to the ICCt

Citizenship: *citizenship by birth:* no
citizenship by descent only: the mother must be a citizen of Algeria
dual citizenship recognized: no
residency requirement for naturalization: 7 years

Suffrage: 18 years of age; universal

Executive branch: chief of state: President Abdelmadjid TEBBOUNE (since 12 December 2019)
head of government: Abdelaziz DJERAD (since 28 December 2019)
cabinet: Cabinet of Ministers appointed by the president
elections/appointments: president directly elected by absolute majority popular vote in two rounds if needed for a 5-year term (eligible for a second term); election last held on 12 December 2019 (next to be held in 2024); prime minister nominated by the president after consultation with the majority party in Parliament
election results: Abdelmadjid TEBBOUNE (NLF) 58.1%, Abdelkader BENGRINA (Movement of National Construction) 17.4%, Ali BENFLIS (Vanguard of Freedoms) 10.6%, Azzedine MIHOUBI (RND) 7.3%, Abdelaziz BELAID (Future Front) 6.7%

Legislative branch: description: bicameral Parliament consists of: Council of the Nation (upper house with 144 seats; one-third of members appointed by the president, two-thirds indirectly elected by simple majority vote by an electoral college composed of local council members; members serve 6-year terms with one-half of the membership renewed every 3 years)
National People's Assembly (lower house with 462 seats including 8 seats for Algerians living abroad); members directly elected in multi-seat constituencies by proportional representation vote to serve 5-year terms)

elections: Council of the Nation—last held on 29 December 2018 (next to be held in December 2021) National People's Assembly—last held on 4 May 2017 (next to be held in 2022)

election results: Council of the Nation—percent of vote by party—NA; seats by party—NA; composition—men 137, women 7, percent of women 5%
National People's Assembly—percent of vote by party—NA; seats by party—FLN 164, RND 97, MSP—FC 33, TAJ 19,Ennahda—FJD 15, FFS 14, El Mostakbel 14, MPA 13, PT 11, RCD 9, ANR 8, MEN 4, other 33, independent 28; composition-men 343, women 119, percent of women 25.8%; note—total Parliament percent of women 20.8%

Judicial branch: *highest courts:* Supreme Court or Cour Suprême, (consists of 150 judges organized into 8 chambers: Civil, Commercial and Maritime, Criminal, House of Offenses and Contraventions, House of Petitions, Land, Personal Status, and Social;Constitutional Council (consists of 12 members including the court chairman and deputy chairman); note—Algeria's judicial system does not include sharia courts
judge selection and term of office: Supreme Court judges appointed by the High Council of Magistracy, an administrative body presided over by the president of the republic, and includes the republic vice-president and several members; judges appointed for life; Constitutional Council members—4 appointed by the president of the republic, 2 each by the 2 houses of Parliament, 2 by the Supreme Court, and 2 by the Council of State; Council president and members appointed for single 6-year terms with half the membership renewed every 3 years
subordinate courts: appellate or wilaya courts; first instance or daira tribunals

Political parties and leaders: Algerian National Front or FNA [Moussa TOUATI]
Algerian Popular Movement or MPA [Amara BENYOUNES]

Algerian Rally or RA [Ali ZAGHDOUD]

Algeria's Hope Rally or TAJ [Amar GHOUL]

Democratic and Social Movement or MDS [Hamid FERHI]

Dignity or El Karama [Aymene HARKATI]

Ennour El Djazairi Party (Algerian Radiance Party) or PED [Badreddine BELBAZ]

Front for Justice and Development or El Adala [Abdallah DJABALLAH]

Future Front or El Mostakbel [Abdelaziz BELAID]

Islamic Renaissance Movement or Ennahda Movement [Mohamed DOUIBI]

Justice and Development Front or FJD [Abdellah DJABALLAH]

Movement of National Construction (Harakat El-Binaa El-Watani) [Abdelkader BENGRINA]

Movement of National Understanding or MEN

Movement for National Reform or Islah [Filali GHOUINI]

Movement of Society for Peace or MSP [Abderrazak MOKRI]

National Democratic Rally (Rassemblement National Democratique) or RND [Ahmed OUYAHIA]

National Front for Social Justice or FNJS [Khaled BOUNEDJEMA]

National Liberation Front or FLN [Mohamed DJEMAI]

National Party for Solidarity and Development or PNSD [Dalila YALAQUI]

National Reform Movement or Islah [Djahid YOUNSI]

National Republican Alliance or ANR [Belkacem SAHLI]

New Dawn Party or PFJ [Tahar BENBAIBECHE]

New Generation or Jil Jadid [Soufiane DJILALI]

Oath of 1954 or Ahd 54 [Ali Fawzi REBAINE]

Party of Justice and Liberty [Mohammed SAID]

Rally for Culture and Democracy or RCD [Mohcine BELABBAS]

Socialist Forces Front or FFS [Hakim BELAHCEL]

Union for Change and Progress or UCP [Zoubida Assoul]

Union of Democratic and Social Forces or UFDS [Noureddine BAHBOUH]

Vanguard of Freedoms (Talaie El Houriat) [Ali BENFLIS]

Youth Party or PJ [Hamana BOUCHARMA]

Workers Party or PT [Louisa HANOUNE]

note: a law banning political parties based on religion was enacted in March 1997

International organization participation: ABEDA, AfDB, AFESD, AMF, AMU, AU, BIS, CAEU, CD, FAO, G-15, G-24, G-77, IAEA, IBRD, ICAO, ICC (national committees), ICRM, IDA, IDB, IFAD, IFC, IFRCS, IHO, ILO, IMF, IMO, IMSO, Interpol, IOC, IOM, IPU, ISO, ITSO, ITU,ITUC (NGOs), LAS, MIGA, MONUSCO, NAM, OAPEC, OAS (observer), OIC, OPCW, OPEC, OSCE (partner), UN, UNCTAD, UNESCO, UNHCR, UNIDO, UNITAR, UNWTO, UPU, WCO, WHO, WIPO, WMO, WTO (observer)

Diplomatic representation in the US: *chief of mission:* Ambassador Madjid BOUGUERRA (since 23 February 2015)

chancery: 2118 Kalorama Road NW, Washington, DC 20008

telephone: [1] (202) 265-2800

FAX: [1] (202) 986-5906

consulate(s) general: New York

Diplomatic representation from the US: *chief of mission:* Ambassador John P. DESROCHER (since 5 September 2017)

telephone: [213] (0) 770-08-2000

embassy: 05 Chemin Cheikh Bachir, Ibrahimi, El-Biar 16030, Alger

mailing address: B. P. 408, Alger-Gare, 16030 Algiers

FAX: [213] (0) 770-08-2064

Flag description: two equal vertical bands of green (hoist side) and white; a red, five-pointed star within a red crescent centered over the two-color boundary; the colors represent Islam (green), purity and peace (white), and liberty (red); the crescent and star are also Islamic symbols, but the crescent is more closed than those of other Muslim countries because Algerians believe the long crescent horns bring happiness

National symbol(s): five-pointed star between the extended horns of a crescent moon, fennec fox; national colors: green, white, red

National anthem: *name:* "Kassaman" (We Pledge)

lyrics/music: Mufdi ZAKARIAH/Mohamed FAWZI

note: adopted 1962; ZAKARIAH wrote "Kassaman" as a poem while imprisoned in Algiers by French colonial forces

0:00 / 1:18

ECONOMY

Economy—overview: Algeria's economy remains dominated by the state, a legacy of the country's socialist post-independence development model. In recent years the Algerian Government has halted the privatization of state-owned industries and imposed restrictions on imports and foreign involvement in its economy, pursuing an explicit import substitution policy.

Hydrocarbons have long been the backbone of the economy, accounting for roughly 30% of GDP, 60% of budget revenues, and nearly 95% of export earnings. Algeria has the 10th-largest reserves of natural gas in the world—including the 3rd-largest reserves of shale gas—and is the 6th—largest gas exporter. It ranks 16th in proven oil reserves. Hydrocarbon exports enabled Algeria to maintain macroeconomic stability, amass large foreign currency reserves, and maintain low external debt while global oil prices were high. With lower oil prices since 2014, Algeria's foreign exchange reserves have declined by more than half and its oil stabilization fund has decreased from about $20 billion at the end of 2013 to about $7 billion in 2017, which is the statutory minimum.

Declining oil prices have also reduced the government's ability to use state-driven growth to distribute rents and fund generous public subsidies, and the government has been under pressure to reduce spending. Over the past three years, the government has enacted incremental increases in some taxes, resulting in modest increases in prices for gasoline,cigarettes, alcohol, and certain imported goods, but it has refrained from reducing subsidies, particularly for education, healthcare, and housing programs.

Algiers has increased protectionist measures since 2015 to limit its import bill and encourage domestic production of non-oil and gas industries. Since 2015, the government has imposed additional restrictions on access to foreign exchange for imports, and import quotas for specific products, such as cars. In January 2018 the government imposed an indefinite suspension on the importation of roughly 850 products, subject to periodic review.

President BOUTEFLIKA announced in fall 2017 that Algeria intends to develop its non-conventional energy resources. Algeria has struggled to develop non-hydrocarbon industries because of heavy regulation and an emphasis on state-driven growth. Algeria has not increased non-hydrocarbon exports, and hydrocarbon exports have declined because of field depletion and increased domestic demand.

GDP (purchasing power parity): $630 billion (2017 est.)

$621.3 billion (2016 est.)

$602 billion (2015 est.)

note: data are in 2017 dollars

country comparison to the world: 36

GDP (official exchange rate): $167.6 billion (2017 est.)

GDP—real growth rate: 1.4% (2017 est.)

3.2% (2016 est.)

3.7% (2015 est.)

country comparison to the world: 174

GDP—per capita (PPP): $15,200 (2017 est.)

$15,200 (2016 est.)

$15,100 (2015 est.)

note: data are in 2017 dollars

country comparison to the world: 109

Gross national saving: 37.8% of GDP (2017 est.)

37.4% of GDP (2016 est.)

36.4% of GDP (2015 est.)

country comparison to the world: 13

GDP—composition, by end use: household consumption: 42.7% (2017 est.)

government consumption: 20.2% (2017 est.)

investment in fixed capital: 38.1% (2017 est.)

investment in inventories: 11.2% (2017 est.)

exports of goods and services: 23.6% (2017 est.)

imports of goods and services: -35.8% (2017 est.)

GDP—composition, by sector of origin: agriculture: 13.3% (2017 est.)

industry: 39.3% (2017 est.)

services: 47.4% (2017 est.)

Agriculture—products: wheat, barley, oats, grapes, olives, citrus, fruits; sheep, cattle

Industries: petroleum, natural gas, light industries, mining, electrical, petrochemical, food processing

Industrial production growth rate: 0.6% (2017 est.)

country comparison to the world: 164

Labor force: 11.82 million (2017 est.)
country comparison to the world: 50

Labor force—by occupation: agriculture: 10.8%
industry: 30.9%
services: 58.4% (2011 est.)

Unemployment rate: 11.7% (2017 est.)
10.5% (2016 est.)
country comparison to the world: 155

Population below poverty line: 23% (2006 est.)

Household income or consumption by percentage share: lowest 10%: 2.8%
highest 10%: 26.8% (1995)

Budget: *revenues:* 54.15 billion (2017 est.)
expenditures: 70.2 billion (2017 est.)

Taxes and other revenues: 32.3% (of GDP) (2017 est.)
country comparison to the world: 67

Budget surplus (+) or deficit (-): -9.6% (of GDP) (2017 est.)
country comparison to the world: 207

Public debt: 27.5% of GDP (2017 est.)
20.4% of GDP (2016 est.)
note: data cover central government debt as well as debt issued by subnational entities and intra-governmental debt
country comparison to the world: 170

Fiscal year: calendar year

Inflation rate (consumer prices): 5.6% (2017 est.)
6.4% (2016 est.)
country comparison to the world: 179

Current account balance: -$22.1 billion (2017 est.)
-$26.47 billion (2016 est.)
country comparison to the world: 199

Exports: $34.37 billion (2017 est.)
$29.06 billion (2016 est.)
country comparison to the world: 58

Exports—partners: Italy 17.4%, Spain 13%, France 11.9%, US 9.4%, Brazil 6.2%, Netherlands 5.5% (2017)

Exports—commodities: petroleum, natural gas, and petroleum products 97% (2009 est.)

Imports: $48.54 billion (2017 est.)
$49.43 billion (2016 est.)
country comparison to the world: 55

Imports—commodities: capital goods, foodstuffs, consumer goods

Imports—partners: China 18.2%, France 9.1%, Italy 8%, Germany 7%, Spain 6.9%, Turkey 4.4% (2017)

Reserves of foreign exchange and gold: $97.89 billion (31 December 2017 est.)
$114.7 billion (31 December 2016 est.)
country comparison to the world: 26

Debt—external: $6.26 billion (31 December 2017 est.)
$5.088 billion (31 December 2016 est.)
country comparison to the world: 128

Exchange rates: Algerian dinars (DZD) per US dollar –
108.9 (2017 est.)

109.443 (2016 est.)
109.443 (2015 est.)
100.691 (2014 est.)
80.579 (2013 est.)

ENERGY

Electricity access: *population without electricity:* 400,000 (2016)
electrification—total population: 99.4% (2016)
electrification—urban areas: 99.6% (2016)
electrification—rural areas: 99% (2016)

Electricity—production: 66.89 billion kWh (2016 est.)
country comparison to the world: 42

Electricity—consumption: 55.96 billion kWh (2016 est.)
country comparison to the world: 46

Electricity—exports: 641 million kWh (2015 est.)
country comparison to the world: 65

Electricity—imports: 257 million kWh (2016 est.)
country comparison to the world: 91

Electricity—installed generating capacity: 19.27 million kW (2016 est.)
country comparison to the world: 45

Electricity—from fossil fuels: 96% of total installed capacity (2016 est.)
country comparison to the world: 36

Electricity—from nuclear fuels: 0% of total installed capacity (2017 est.)
country comparison to the world: 34

Electricity—from hydroelectric plants: 1% of total installed capacity (2017 est.)
country comparison to the world: 144

Electricity—from other renewable sources: 2% of total installed capacity (2017 est.)
country comparison to the world: 130

Crude oil—production: 1.259 million bbl/day (2018 est.)
country comparison to the world: 18

Crude oil—exports: 756,400 bbl/day (2015 est.)
country comparison to the world: 15

Crude oil—imports: 5,340 bbl/day (2015 est.)
country comparison to the world: 75

Crude oil—proved reserves: 12.2 billion bbl (1 January 2018 est.)
country comparison to the world: 15

Refined petroleum products—production: 627,900 bbl/day (2015 est.)
country comparison to the world: 29

Refined petroleum products—consumption: 405,000 bbl/day (2016 est.)
country comparison to the world: 37

Refined petroleum products—exports: 578,800 bbl/day (2015 est.)
country comparison to the world: 15

Refined petroleum products—imports: 82,930 bbl/day (2015 est.)
country comparison to the world: 61

Natural gas—production: 93.5 billion cu m (2017 est.)
country comparison to the world: 10

Natural gas—consumption: 41.28 billion cu m (2017 est.)
country comparison to the world: 25

Natural gas—exports: 53.88 billion cu m (2017 est.)
country comparison to the world: 7

Natural gas—imports: 0 cu m (2017 est.)
country comparison to the world: 83

Natural gas—proved reserves: 4.504 trillion cu m (1 January 2018 est.)
country comparison to the world: 10

Carbon dioxide emissions from consumption of energy: 135.9 million Mt (2017 est.)
country comparison to the world: 34

COMMUNICATIONS

Telephones—fixed lines: total subscriptions: 4,558,502
subscriptions per 100 inhabitants: 10.77 (2019 est.)
country comparison to the world: 30

Telephones—mobile cellular: total subscriptions: 46,287,629
subscriptions per 100 inhabitants: 109.36 (2019 est.)
country comparison to the world: 32

Telecommunication systems: general assessment: improved international connectivity and privatization of Algeria's telecommunications sector began in 2000; three mobile-cellular licenses have been issued; LTE service growth in additional provinces and rural areas;upgrade to LTE infrastructure and migration to 5G; LTE subscriber rate up 82% in 2018; Chinese company Huawei opens smart phone assembly plant in Algeria; ending of monopolies have made broadband services more affordable; Algeria and Tunisia end roaming charges for travelers (2020)
domestic: a limited network of fixed-lines with a teledensity of less than 11 telephones per 100 persons has been offset by the rapid increase in mobile-cellular subscribership; mobile-cellular teledensity was roughly 109 telephones per 100 persons (2019)
international: country code—213; ALPAL-2 is a submarine telecommunications cable system in the Mediterranean Sea linking Algeria and the Spanish Balearic island of Majorca; ORVAL is a submarine cable to Spain; landing points for the TE North/TGN-Eurasia/SEACOM/SeaMeWe-4 fiber-optic submarine cable system that provides links to Europe, the Middle East, and Asia; MED cable connecting Algeria with France; microwave radio relay to Italy, France, Spain, Morocco, and Tunisia; Algeria part of the 4,500 Km terrestrial Trans Sahara Backbone network which connects to other fiber networks in the region; Alcomstat-1 satellite offering telemedicine network (2020)
note: the COVID-19 outbreak is negatively impacting telecommunications production and supply chains globally; consumer spending on telecom devices and services has also slowed due to the pandemic's effect on economies worldwide; overall progress towards improvements in all facets of the

15

telecom industry—mobile, fixed-line, broadband, submarine cable and satellite—has moderated

Broadcast media: state-run Radio-Television Algerienne operates the broadcast media and carries programming in Arabic, Berber dialects, and French; use of satellite dishes is widespread, providing easy access to European and Arab satellite stations; state-run radio operates several national networks and roughly 40 regional radio stations

Internet country code: .dz

Internet users: *total:* 24,819,531
percent of population: 59.58% (July 2018 est.)
country comparison to the world: 31

Broadband—fixed subscriptions: *total:* 3,067,022
subscriptions per 100 inhabitants: 7 (2018 est.)
country comparison to the world: 43

TRANSPORTATION

National air transport system: number of registered air carriers: 3 (2020)
inventory of registered aircraft operated by air carriers: 87
annual passenger traffic on registered air carriers: 6,442,442 (2018)
annual freight traffic on registered air carriers: 28.28 million mt-km (2018)

Civil aircraft registration country code prefix: 7T (2016)

Airports: 149 (2020)
country comparison to the world: 36

Airports—with paved runways: *total:* 67 (2020)
over 3,047 m: 14
2,438 to 3,047 m: 27
1,524 to 2,437 m: 18
914 to 1,523 m: 6
under 914 m: 2

Airports—with unpaved runways: *total:* 82 (2020)
2,438 to 3,047 m: 2
1,524 to 2,437 m: 16
914 to 1,523 m: 36
under 914 m: 28

Heliports: 3 (2013)

Pipelines: 2600 km condensate, 16415 km gas, 3447 km liquid petroleum gas, 7036 km oil, 144 km refined products (2013)

Railways: *total:* 3,973 km (2014)
standard gauge: 2,888 km 1.432-m gauge (283 km electrified) (2014)
narrow gauge: 1,085 km 1.055-m gauge (2014)
country comparison to the world: 50

Roadways: *total:* 104,000 km (2015)
paved: 71,656 km (2015)
unpaved: 32,344 km (2015)
country comparison to the world: 46

Merchant marine: *total:* 114
by type: bulk carrier 2, general cargo 11, oil tanker 10, other 91 (2019)
country comparison to the world: 83

Ports and terminals: major seaport(s): Algiers, Annaba, Arzew, Bejaia, Djendjene, Jijel, Mostaganem, Oran, Skikda
LNG terminal(s) (export): Arzew, Bethioua, Skikda

MILITARY AND SECURITY

Military and security forces: Algerian People's National Army (ANP): Land Forces, Naval Forces (includes coast guard), Air Forces, Territorial Air Defense Forces, Republican Guard; National Gendarmerie (subordinate to the Ministry of National Defense); Ministry of Interior: General Directorate of National Security (2020)

Military expenditures: 6% of GDP (2019)
5.5% of GDP (2018)
5.81% of GDP (2017)
6.55% of GDP (2016)
6.32% of GDP (2015)
country comparison to the world: 3

Military and security service personnel strengths: the Algerian People's National Army (ANP) has approximately 130,000 total active personnel (110,000 Army; 6,000 Navy; 14,000 Air Force); est. 40,000 Gendarmerie (2019 est.)

Military equipment inventories and acquisitions: the ANP's inventory includes mostly Russian-sourced equipment with smaller amounts from other suppliers, particularly China and Europe; since 2010, Russia is the leading supplier of armaments to Algeria, followed by China, Germany, and Italy (2020)

Military service age and obligation: 18 is the legal minimum age for voluntary military service; 19-30 years of age for compulsory service; conscript service obligation reduced from 18 to 12 months in 2014 (2019)

TERRORISM

Terrorist group(s): al-Qa'ida in the Islamic Maghreb; Islamic State of Iraq and ash-Sham

(ISIS)—Algeria; al-Mulathamun Battalion (al-Mourabitoun) (2020)

note: details about the history, aims, leadership, organization, areas of operation, tactics, targets, weapons, size, and sources of support of the group(s) appear(s) in Appendix T

TRANSNATIONAL ISSUES

Disputes—international: Algeria and many other states reject Moroccan administration of Western Sahara; the Polisario Front, exiled in Algeria, represents the "Sahrawi Arab Democratic Republic" which Algeria recognizes; the Algerian-Moroccan land border remains closed; dormant disputes include Libyan claims of about 32,000 sq km of southeastern Algeria and the National Liberation Front's (FLN) assertions of a claim to Chirac Pastures in southeastern Morocco.

Refugees and internally displaced persons: *refugees (country of origin):* more than 100,000 (Western Saharan Sahrawi, mostly living in Algerian-sponsored camps in the southwestern Algerian town of Tindouf) (2018); 7,757 (Syria) (2019)

Trafficking in persons: *current situation:* Algeria is a transit and, to a lesser extent, a destination and source country for women subjected to forced labor and sex trafficking and, to a lesser extent, men subjected to forced labor; criminal networks, sometimes extending to Sub-Saharan Africa and to Europe, are involved in human smuggling and trafficking in Algeria; Sub-Saharan adults enter Algeria voluntarily but illegally, often with the aid of smugglers, for onward travel to Europe, but some of the women are forced into prostitution, domestic service, and begging; some Sub-Saharan men, mostly from Mali, are forced into domestic servitude; some Algerian women and children are also forced into prostitution domestically

tier rating: Tier 3 – Algeria does not fully comply with the minimum standards for the elimination of trafficking and is not making significant efforts to do so: some officials denied the existence of human trafficking, hindering law enforcement efforts; the government reported its first conviction under its anti-trafficking law; one potential trafficking case was investigated in 2014, but no suspected offenders were arrested; no progress was made in identifying victims among vulnerable groups or referring them to NGO-run protection service, which left trafficking victims subject to arrest and detention; no anti-trafficking public awareness or educational campaigns were conducted (2015)

AMERICAN SAMOA

INTRODUCTION

Background: Settled as early as 1000 B.C., Samoa was not reached by European explorers until the 18th century. International rivalries in the latter half of the 19th century were settled by an 1899

treaty in which Germany and the US divided the Samoan archipelago. The US formally occupied its portion—a smaller group of eastern islands with the excellent harbor of Pago Pago—the following year.

GEOGRAPHY

Location: Oceania, group of islands in the South Pacific Ocean, about halfway between Hawaii and New Zealand

Geographic coordinates: 14 20 S, 170 00 W

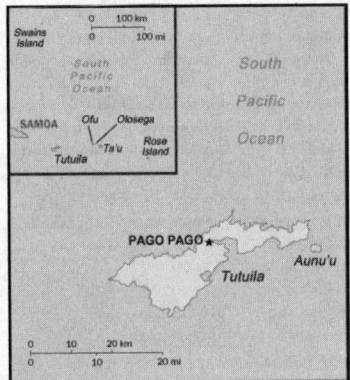

Map references: Oceania

Area: *total:* 224 sq km
land: 224 sq km
water: 0 sq km
note: includes Rose Island and Swains Island
country comparison to the world: 216

Area—comparative: slightly larger than Washington, DC

Land boundaries: 0 km

Coastline: 116 km

Maritime claims: *territorial sea:* 12 nm
exclusive economic zone: 200 nm

Climate: tropical marine, moderated by southeast trade winds; annual rainfall averages about 3 m; rainy season (November to April), dry season (May to October); little seasonal temperature variation

Terrain: five volcanic islands with rugged peaks and limited coastal plains, two coral atolls (Rose Island, Swains Island)

Elevation: *lowest point:* Pacific Ocean 0 m
highest point: Lata Mountain 964 m

Natural resources: pumice, pumicite

Land use: *agricultural land:* 21.9% (2016 est.)
arable land: 13.4% (2016 est.) / permanent crops: 8.5% (2016 est.) / permanent pasture: 0% (2016 est.)
forest: 78.1% (2016 est.)
other: 0% (2016 est.)

Irrigated land: 0 sq km (2012)

Natural hazards: cyclones common from December to March
volcanism: limited volcanic activity on the Ofu and Olosega Islands; neither has erupted since the 19th century

Environment—current issues: limited supply of drinking water; pollution; waste disposal; coastal and stream alteration; soil erosion

Geography—note: Pago Pago has one of the best natural deepwater harbors in the South Pacific Ocean, sheltered by shape from rough seas and protected by peripheral mountains from high winds; strategic location in the South Pacific Ocean

PEOPLE AND SOCIETY

Population: 49,437 (July 2020 est.)
country comparison to the world: 211

Nationality: *noun:* American Samoan(s) (US nationals)
adjective: American Samoan

Ethnic groups: Pacific Islander 92.6% (includes Samoan 88.9%, Tongan 2.9%, other .8%), Asian 3.6% (includes Filipino 2.2%, other 1.4%), mixed 2.7%, other 1.2% (2010 est.)
note: data represent population by ethnic origin or race

Languages: Samoan 88.6% (closely related to Hawaiian and other Polynesian languages), English 3.9%, Tongan 2.7%, other Pacific islander 3%, other 1.8% (2010 est.)
note: most people are bilingual

Religions: Christian 98.3%, other 1%, unaffiliated 0.7% (2010 est.)

Age structure: *0-14 years:* 27.76% (male 7,063/female 6,662)
15-24 years: 18.16% (male 4,521/female 4,458)
25-54 years: 37.49% (male 9,164/female 9,370)
55-64 years: 9.69% (male 2,341/female 2,447)
65 years and over: 6.9% (male 1,580/female 1,831) (2020 est.)

Median age: *total:* 27.2 years
male: 26.7 years
female: 27.7 years (2020 est.)
country comparison to the world: 149

Population growth rate: -1.4% (2020 est.)
country comparison to the world: 234

Birth rate: 17.8 births/1,000 population (2020 est.)
country comparison to the world: 93

Death rate: 5.9 deaths/1,000 population (2020 est.)
country comparison to the world: 168

Net migration rate: -26.1 migrant(s)/1,000 population (2020 est.)
country comparison to the world: 226

Urbanization: *urban population:* 87.2% of total population (2020)
rate of urbanization: 0.07% annual rate of change (2015-20 est.)

total population growth rate v. urban population growth rate, 2000-2030: Major urban areas—population: 49,000 PAGO PAGO (capital) (2018)

Sex ratio: *at birth:* 1.06 male(s)/female
0-14 years: 1.06 male(s)/female
15-24 years: 1.01 male(s)/female
25-54 years: 0.98 male(s)/female
55-64 years: 0.96 male(s)/female
65 years and over: 0.86 male(s)/female
total population: 1 male(s)/female (2020 est.)

Infant mortality rate: *total:* 9.9 deaths/1,000 live births
male: 11.7 deaths/1,000 live births
female: 8 deaths/1,000 live births (2020 est.)
country comparison to the world: 131

Life expectancy at birth: *total population:* 74.8 years
male: 72.3 years
female: 77.5 years (2020 est.)
country comparison to the world: 123

Total fertility rate: 2.35 children born/woman (2020 est.)
country comparison to the world: 80

Drinking water source:
improved:
total: 100% of population
unimproved:
total: 0% of population (2017 est.)

Sanitation facility access:
improved:
total: 99% of population
unimproved:
total: 1% of population (2017 est.)

HIV/AIDS—adult prevalence rate: NA

HIV/AIDS—people living with HIV/AIDS: NA

HIV/AIDS—deaths: NA

Education expenditures: NA

GOVERNMENT

Country name: *conventional long form:* American Samoa
conventional short form: American Samoa
abbreviation: AS
etymology: the meaning of Samoa is disputed; some modern explanations are that the "sa" connotes "sacred" and "moa" indicates "center," so the name can mean "Holy Center"; alternatively, some assertions state that it can mean "place of the sacred moa bird" of Polynesian mythology; the name, however, may go back to Proto-Polynesian (PPn) times (before 1000 B.C.); a plausible PPn reconstruction has the first syllable as "sa'a" meaning "tribe or people" and "moa" meaning "deep sea or ocean" to convey the meaning "people of the deep sea"

Dependency status: unincorporated unorganized territory of the US; administered by the Office of Insular Affairs, US Department of the Interior

Government type: republican form of government with separate executive, legislative, and judicial branches; unincorporated unorganized territory of the US with local self-government

Capital: *name:* Pago Pago
geographic coordinates: 14 16 S, 170 42 W
time difference: UTC-11 (6 hours behind Washington, DC, during Standard Time)
note: pronounced pahn-go pahn-go

Administrative divisions: none (territory of the US); there are no first-order administrative divisions as defined by the US Government, but there are 3 districts and 2 islands* at the second order; Eastern, Manu'a, Rose Island*, Swains Island*, Western

Independence: none (territory of the US)

National holiday: Flag Day, 17 April (1900)

Constitution: *history:* adopted 17 October 1960; revised 1 July 1967
amendments: proposed by either house of the Legislative Assembly; passage requires three-fifths majority vote by the membership of each house,

approval in a referendum, and approval by the US Secretary of the Interior; amended 1971, 1977, 1979

Legal system: mixed legal system of US common law and customary law

Citizenship: see United States
Note: in accordance with US Code Title 8, Section 1408, persons born in American Samoa are US nationals but not US citizens

Suffrage: 18 years of age; universal

Executive branch: *chief of state:* President Donald J. TRUMP (since 20 January 2017); Vice President Michael R. PENCE (since 20 January 2017)
head of government: Governor Lolo Matalasi MOLIGA (since 3 January 2013)
cabinet: Cabinet consists of 12 department directors appointed by the governor with the consent of the Legislature or Fono
elections/appointments: president and vice president indirectly elected on the same ballot by an Electoral College of 'electors' chosen from each state to serve a 4-year term (eligible for a second term); under the US Constitution, residents of unincorporated territories, such as American Samoa, do not vote in elections for US president and vice president; however, they may vote in Democratic and Republican presidential primary elections; governor and lieutenant governor directly elected on the same ballot by absolute majority popular vote in 2 rounds if needed for a 4-year term (eligible for a second term); election last held on 8 November 2016 (next to be held in November 2020)
election results: Lolo Matalasi MOLIGA reelected governor in first round; percent of vote—Lolo Matalasi MOLIGA (independent) 60.2%, Faoa Aitofele SUNIA (Democratic Party) 35.8%, Tuika TUIKA (independent) 4%

Legislative branch: *description:* bicameral Legislature or Fono consists of:
Senate (18 seats; members indirectly selected by regional governing councils to serve 4-year terms)
House of Representatives (21 seats; 20 members directly elected by simple majority vote and 1 decided by public meeting on Swains Island; members serve 2-year terms)

elections: Senate—last held on 8 November 2016 (next to be held in November 2020) House of Representatives—last held on 6 November 2018 (next to be held in November 2020)

election results: Senate—percent of vote by party—NA; seats by party—independent 18; composition—men 17, women 1, percent of women 9.5%
House of Representatives—percent of vote by party—NA; seats by party—NA; composition—men 14, women 7, percent of women 33.3%; note—total percent of women in Legislature 20.5%
note: American Samoa elects 1 member by simple majority vote to serve a 2-year term as a delegate to the US House of Representatives; the delegate can vote when serving on a committee and when the House meets as the Committee of the Whole

House, but not when legislation is submitted for a "full floor" House vote; election of delegate last held on 6 November 2018 (next to be held in November 2020)

Judicial branch: *highest courts:* High Court of American Samoa (consists of the chief justice, associate chief justice, and 6 Samoan associate judges and organized into trial, family, drug, and appellate divisions); note—American Samoa has no US federal courts
judge selection and term of office: chief justice and associate chief justice appointed by the US Secretary of the Interior to serve for life; Samoan associate judges appointed by the governor to serve for life
subordinate courts: district and village courts

Political parties and leaders: Democratic Party [Fagafaga Daniel LANGKILDE, chairman] Republican Party [William SWORD, chairman]

International organization participation: AOSIS (observer), Interpol (subbureau), IOC, PIF (observer), SPC

Diplomatic representation in the US: none (territory of the US)

Diplomatic representation from the US: none (territory of the US)

Flag description: blue, with a white triangle edged in red that is based on the fly side and extends to the hoist side; a brown and white American bald eagle flying toward the hoist side is carrying 2 traditional Samoan symbols of authority, a war club known as a "fa'alaufa'i" (upper; left talon), and a coconut-fiber fly whisk known as a "fue" (lower; right talon); the combination of symbols broadly mimics that seen on the US Great Seal and reflects the relationship between the US and American Samoa

National symbol(s): a fue (coconut fiber fly whisk; representing wisdom) crossed with a to'oto'o (staff; representing authority); national colors: red, white, blue

National anthem: *name:* "Amerika Samoa" (American Samoa)
lyrics/music: Mariota Tiumalu TUIASOSOPO/ Napoleon Andrew TUITELELEAPAGA
note: local anthem adopted 1950; as a territory of the United States, "The Star-Spangled Banner" is official (see United States)

ECONOMY

Economy—overview: American Samoa s a traditional Polynesian economy in which more than 90% of the land is communally owned. Economic activity is strongly linked to the US with which American Samoa conducts most of its commerce. Tuna fishing and processing are the backbone of the private sector with processed fish products as the primary exports. The fish processing business accounted for 15.5% of employment in 2015.
In late September 2009, an earthquake and the resulting tsunami devastated American Samoa and nearby Samoa, disrupting transportation and power generation, and resulting in about 200 deaths. The US Federal Emergency Management

Agency oversaw a relief program of nearly $25 million. Transfers from the US Government add substantially to American Samoa's economic well-being.
Attempts by the government to develop a larger and broader economy are restrained by Samoa's remote location, its limited transportation, and its devastating hurricanes. Tourism has some potential as a source of income and jobs.

GDP (purchasing power parity): $658 million (2016 est.)
$674.9 million (2015 est.)
$666.9 million (2014 est.)
note: data are in 2016 US dollars
country comparison to the world: 209

GDP (official exchange rate): $658 million (2016 est.)

GDP—real growth rate: -2.5% (2016 est.)
1.2% (2015 est.)
1% (2014 est.)
country comparison to the world: 208

GDP—per capita (PPP): $11,200 (2016 est.)
$11,300 (2015 est.)
$11,200 (2014 est.)
country comparison to the world: 134

GDP—composition, by end use: *household consumption:* 66.4% (2016 est.)
government consumption: 49.7% (2016 est.)
investment in fixed capital: 7.3% (2016 est.)
investment in inventories: 5.1% (2016 est.)
exports of goods and services: 65% (2016 est.)
imports of goods and services: -93.5% (2016 est.)

GDP—composition, by sector of origin: *agriculture:* 27.4% (2012)
industry: 12.4% (2012)
services: 60.2% (2012)

Agriculture—products: bananas, coconuts, vegetables, taro, breadfruit, yams, copra, pineapples, papayas; dairy products, livestock

Industries: tuna canneries (largely supplied by foreign fishing vessels), handicrafts

Industrial production growth rate: NA

Labor force: 17,850 (2015 est.)
country comparison to the world: 213

Labor force—by occupation: *agriculture:* NA
industry: 15.5%
services: 46.4% (2015 est.)

Unemployment rate: 29.8% (2005)
country comparison to the world: 206

Population below poverty line: NA

Household income or consumption by percentage share: *lowest 10%:* NA
highest 10%: NA

Budget: *revenues:* 249 million (2016 est.)
expenditures: 262.5 million (2016 est.)

Taxes and other revenues: 37.8% (of GDP) (2016 est.)
country comparison to the world: 53

Budget surplus (+) or deficit (-): -2.1% (of GDP) (2016 est.)
country comparison to the world: 107

Public debt: 12.2% of GDP (2016 est.)
country comparison to the world: 197

Fiscal year: 1 October—30 September

Inflation rate (consumer prices): -0.5% (2015 est.)
1.4% (2014 est.)
country comparison to the world: 5

Exports: $428 million (2016 est.)
$427 million (2015 est.)
country comparison to the world: 178

Exports—partners: Australia 25%, Ghana 19%, Indonesia 15.6%, Burma 10.4%, Portugal 5.1% (2017)

Exports—commodities: canned tuna 93%

Imports: $615 million (2016 est.)
$657 million (2015 est.)
country comparison to the world: 194

Imports—commodities: raw materials for canneries, food, petroleum products, machinery and parts

Imports—partners: Fiji 10.7%, Singapore 10.4%, NZ 10.4%, South Korea 9.3%, Samoa 8.2%, Kenya 6.4%, Australia 5.2% (2017)

Debt—external: NA

Exchange rates: the US dollar is used

ENERGY

Electricity access: *population without electricity:* 22,219 (2012)
electrification—total population: 59% (2012)
electrification—urban areas: 60% (2012)
electrification—rural areas: 45% (2012)

Electricity—production: 169 million kWh (2016 est.)
country comparison to the world: 195

Electricity—consumption: 157.2 million kWh (2016 est.)
country comparison to the world: 197

Electricity—exports: 0 kWh (2016 est.)
country comparison to the world: 97

Electricity—imports: 0 kWh (2016 est.)
country comparison to the world: 118

Electricity—installed generating capacity: 43,000 kW (2016 est.)
country comparison to the world: 195

Electricity—from fossil fuels: 98% of total installed capacity (2016 est.)
country comparison to the world: 27

Electricity—from nuclear fuels: 0% of total installed capacity (2017 est.)
country comparison to the world: 35

Electricity—from hydroelectric plants: 0% of total installed capacity (2017 est.)

country comparison to the world: 152

Electricity—from other renewable sources: 2% of total installed capacity (2017 est.)
country comparison to the world: 131

Crude oil—production: 0 bbl/day (2018 est.)
country comparison to the world: 102

Crude oil—exports: 0 bbl/day (2015 est.)
country comparison to the world: 83

Crude oil—imports: 0 bbl/day (2015 est.)
country comparison to the world: 86

Crude oil—proved reserves: 0 bbl (1 January 2018 est.)
country comparison to the world: 100

Refined petroleum products—production: 0 bbl/day (2015 est.)
country comparison to the world: 111

Refined petroleum products—consumption: 2,375 bbl/day (2016 est.)
country comparison to the world: 192

Refined petroleum products—exports: 0 bbl/day (2015 est.)
country comparison to the world: 125

Refined petroleum products—imports: 2,346 bbl/day (2015 est.)
country comparison to the world: 188

Natural gas—production: 0 cu m (2017 est.)
country comparison to the world: 97

Natural gas—consumption: 0 cu m (2017 est.)
country comparison to the world: 117

Natural gas—exports: 0 cu m (2017 est.)
country comparison to the world: 59

Natural gas—imports: 0 cu m (2017 est.)
country comparison to the world: 84

Natural gas—proved reserves: 0 cu m (1 January 2014 est.)
country comparison to the world: 103

Carbon dioxide emissions from consumption of energy: 361,100 Mt (2017 est.)
country comparison to the world: 189

COMMUNICATIONS

Telephones—fixed lines: *total subscriptions:* 8,984
subscriptions per 100 inhabitants: 17.92 (2019 est.)
country comparison to the world: 190

Telecommunication systems: *general assessment:* good telex, telegraph, facsimile, and cellular telephone services; one of the most complete and modern telecommunications systems in the South Pacific Islands; all inhabited islands have telephone connectivity
domestic: 18 per 100 fixed-line teledensity, domestic satellite system with 1 Comsat earth station (2019)
international: country code—1-684; landing points for the ASH, Southern Cross NEXT and Hawaiki providing connectivity to New Zealand, Australia, American Samoa, Hawaii, California, and SAS connecting American Samoa with Samoa; satellite earth station—1 (Intelsat-Pacific Ocean) (2019)
note: the COVID-19 outbreak is negatively impacting telecommunications production and supply chains globally; consumer spending on telecom devices and services has also slowed due to the pandemic's effect on economies worldwide; overall progress towards improvements in all facets of the telecom industry—mobile, fixed-line, broadband, submarine cable and satellite—has moderated

Broadcast media: 3 TV stations; multi-channel pay TV services are available; about a dozen radio stations, some of which are repeater stations

Internet country code: .as

Internet users: *total:* 17,000
percent of population: 31.3% (July 2016 est.)
country comparison to the world: 210

TRANSPORTATION

Airports: 3 (2020)
country comparison to the world: 192

Airports—with paved runways: *total:* 3 (2019)
over 3,047 m: 1
914 to 1,523 m: 1
under 914 m: 1

Roadways: *total:* 241 km (2016)
country comparison to the world: 205

Ports and terminals: *major seaport(s):* Pago Pago

MILITARY AND SECURITY

Military—note: defense is the responsibility of the US

TRANSNATIONAL ISSUES

Disputes—international: Tokelau included American Samoa's Swains Island (Olosega) in its 2006 draft independence constitution

ANDORRA

INTRODUCTION

Background: The landlocked Principality of Andorra is one of the smallest states in Europe, nestled high in the Pyrenees between the French and Spanish borders. For 715 years, from 1278 to 1993, Andorrans lived under a unique coprincipality, ruled by French and Spanish leaders (from 1607 onward, the French chief of state and the Bishop of Urgell). In 1993, this feudal system was modified with the introduction of a modern constitution; the co-princes remained as titular heads of state, but the government transformed into a parliamentary democracy.

Andorra has become a popular tourist destination visited by approximately 8 million people

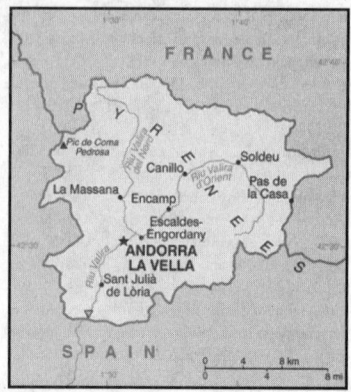

each year drawn by the winter sports, summer climate, and duty-free shopping. Andorra has also become a wealthy international commercial center because of its mature banking sector and low taxes. As part of its effort to modernize its economy, Andorra has opened to foreign investment, and engaged in other reforms, such as advancing tax initiatives aimed at supporting a broader infrastructure. Although not a member of the EU, Andorra enjoys a special relationship with the bloc that is governed by various customs and cooperation agreements and uses the euro as its national currency.

GEOGRAPHY

Location: Southwestern Europe, Pyrenees mountains, on the border between France and Spain

Geographic coordinates: 42 30 N, 1 30 E

Map references: Europe

Area: *total:* 468 sq km
land: 468 sq km
water: 0 sq km
country comparison to the world: 196

Area—comparative: 2.5 times the size of Washington, DC

Land boundaries: *total:* 118 km
border countries (2): France 55 km, Spain 63 km

Coastline: 0 km (landlocked)

Maritime claims: none (landlocked)

Climate: temperate; snowy, cold winters and warm, dry summers

Terrain: rugged mountains dissected by narrow valleys

Elevation: *mean elevation:* 1,996 m
lowest point: Riu Runer 840 m
highest point: Pic de Coma Pedrosa 2,946 m

Natural resources: hydropower, mineral water, timber, iron ore, lead

Land use: *agricultural land:* 40% (2016 est.)
arable land: 1.7% (2016 est.) / *permanent crops:* 0% (2016 est.) / *permanent pasture:* 38.3% (2016 est.)
forest: 34% (2016 est.)
other: 26% (2016 est.)

Irrigated land: 0 sq km (2012)

Population distribution: population is unevenly distributed and is concentrated in the seven urbanized valleys that make up the country's parishes (political administrative divisions)

Natural hazards: avalanches

Environment—current issues: deforestation; overgrazing of mountain meadows contributes to soil erosion; air pollution; wastewater treatment and solid waste disposal

Environment—international agreements: *party to:* Biodiversity, Desertification, Hazardous Wastes, Ozone Layer Protection
signed, but not ratified: none of the selected agreements

Geography—note: landlocked; straddles a number of important crossroads in the Pyrenees

PEOPLE AND SOCIETY

Population: 77,000 (2019 est.)
country comparison to the world: 201

Nationality: *noun:* Andorran(s)
adjective: Andorran

Ethnic groups: Andorran 48.8%, Spanish 25.1%, Portuguese 12%, French 4.4%, other 9.7% (2017 est.)
note: data represent population by nationality

Languages: Catalan (official), French, Castilian, Portuguese

Religions: Roman Catholic (predominant)

Age structure: *0-14 years:* 13.37% (male 5,901/female 5,551)
15-24 years: 10.16% (male 4,474/female 4,227)
25-54 years: 43.19% (male 18,857/female 18,131)
55-64 years: 15.91% (male 7,184/female 6,443)
65 years and over: 17.36% (male 7,544/female 7,323) (2020 est.)

Median age: *total:* 46.2 years
male: 46.3 years
female: 46.1 years (2020 est.)
country comparison to the world: 6

Population growth rate: -0.06% (2020 est.)
country comparison to the world: 200

Birth rate: 7 births/1,000 population (2020 est.)
country comparison to the world: 227

Death rate: 7.7 deaths/1,000 population (2020 est.)
country comparison to the world: 99

Net migration rate: 0 migrant(s)/1,000 population (2020 est.)
country comparison to the world: 75

Population distribution: population is unevenly distributed and is concentrated in the seven urbanized valleys that make up the country's parishes (political administrative divisions)

Urbanization: *urban population:* 87.9% of total population (2020)
rate of urbanization: -0.31% annual rate of change (2015-20 est.)

total population growth rate v. urban population growth rate, 2000-2030: Major urban

areas—population: 23,000 ANDORRA LA VELLA (capital) (2018)

Sex ratio: *at birth:* 1.07 male(s)/female
0-14 years: 1.06 male(s)/female
15-24 years: 1.06 male(s)/female
25-54 years: 1.04 male(s)/female
55-64 years: 1.12 male(s)/female
65 years and over: 1.03 male(s)/female
total population: 1.06 male(s)/female (2020 est.)

Infant mortality rate: *total:* 3.5 deaths/1,000 live births
male: 3.5 deaths/1,000 live births
female: 3.5 deaths/1,000 live births (2020 est.)
country comparison to the world: 197

Life expectancy at birth: *total population:* 83 years
male: 80.8 years
female: 85.4 years (2020 est.)
country comparison to the world: 9

Total fertility rate: 1.43 children born/woman (2020 est.)
country comparison to the world: 213

Drinking water source:
improved:
urban: 100% of population
rural: 100% of population
total: 100% of population
unimproved:
urban: 0% of population
rural: 0% of population
total: 0% of population (2017 est.)

Current Health Expenditure: 10.3% (2017)

Physicians density: 3.33 physicians/1,000 population (2015)

Hospital bed density: 2.5 beds/1,000 population (2009)

Sanitation facility access:
improved:
urban: 100% of population
rural: 100% of population
total: 100% of population
unimproved:
urban: 0% of population
rural: 0% of population
total: 0% of population (2017 est.)

HIV/AIDS—adult prevalence rate: NA

HIV/AIDS—people living with HIV/AIDS: NA

HIV/AIDS—deaths: NA

Obesity—adult prevalence rate: 25.6% (2016)
country comparison to the world: 49

Education expenditures: 3.2% of GDP (2017)
country comparison to the world: 129

Literacy: *definition:* age 15 and over can read and write
total population: 100%
male: 100%
female: 100% (2016)

GOVERNMENT

Country name: *conventional long form:* Principality of Andorra
conventional short form: Andorra

local long form: Principat d'Andorra
local short form: Andorra
etymology: the origin of the country's name is obscure; the name may derive from the Arabic "ad-darra" meaning "the forest," a reference to its location as part of the Spanish March (defensive buffer zone) against the invading Moors in the 8th century

Government type: parliamentary democracy (since March 1993) that retains its chiefs of state in the form of a co-principality; the two princes are the President of France and Bishop of Seu d'Urgell, Spain

Capital: *name:* Andorra la Vella

geographic coordinates: 42 30 N, 1 31 E
time difference: UTC+1 (6 hours ahead of Washington, DC during Standard Time)
daylight saving time: +1hr, begins last Sunday in March; ends last Sunday in October
etymology: translates as "Andorra the Old" in Catalan

Administrative divisions: 7 parishes (parroquies, singular—parroquia); Andorra la Vella, Canillo, Encamp, Escaldes-Engordany, La Massana, Ordino, Sant Julia de Loria

Independence: 1278 (formed under the joint sovereignty of the French Count of Foix and the Spanish Bishop of Urgell)

National holiday: Our Lady of Meritxell Day, 8 September (1278)

Constitution: *history:* drafted 1991, approved by referendum 14 March 1993, effective 28 April 1993
amendments: proposed by the coprinces jointly or by the General Council; passage requires at least a two-thirds majority vote by the General Council, ratification in a referendum, and sanctioning by the coprinces

Legal system: mixed legal system of civil and customary law with the influence of canon (religious) law

International law organization participation: has not submitted an ICJ jurisdiction declaration; accepts ICCt jurisdiction

Citizenship: *citizenship by birth:* no
citizenship by descent only: the mother must be an Andorran citizen or the father must have been born in Andorra and both parents maintain permanent residence in Andorra
dual citizenship recognized: no
residency requirement for naturalization: 25 years

Suffrage: 18 years of age; universal

Executive branch: *chief of state:* Co-prince Emmanuel MACRON (since 14 May 2017); represented by Patrick STROZDA (since 14 May 2017); and Co-prince Archbishop Joan-Enric VIVES i Sicilia (since 12 May 2003); represented by Josep Maria MAURI (since 20 July 2012)

head of government: Head of Government (or Cap de Govern) Xaviar Espot ZAMORA (since 16 May 2019)

cabinet: Executive Council of 12 ministers designated by the head of government
elections/appointments: head of government indirectly elected by the General Council (Andorran parliament), formally appointed by the coprinces for a 4-year term; election last held on 7 April 2019 (next to be held in April 2023); the leader of the majority party in the General Council is usually elected head of government
election results: Xaviar Espot ZAMORA (DA) elected head of government; percent of General Council vote—60.7%

Legislative branch: *description:* unicameral General Council of the Valleys or Consell General de les Valls (a minimum of 28 seats; 14 members directly elected in multi-seat constituencies (parishes) by simple majority vote and 14 directly elected in a single national constituency by proportional representation vote; members serve 4-year terms); note—voters cast two separate ballots—one for a national list and one for a parish list
elections: last held on 7 April 2019 (next to be held on April 2023)
election results: percent of vote by party—DA 35.1%, PS 30.6%, L'A 12.5%, Third Way/Lauredian Union 10.4%, other 22.4%; seats by party—DA 11, PS 7, L'A 4, Third Way/Lauredian Union 4, other 2; composition—men 14, women 14, percent of women 50%

Judicial branch: *highest courts:* Supreme Court of Justice or Tribunal Superior de la Justicia d'Andorra (consists of the court president and 8 judges organized into civil, criminal, and administrative chambers); Constitutional Court or Tribunal Constitucional (consists of 4 magistrates)
judge selection and term of office: Supreme Court president and judges appointed by the Supreme Council of Justice, a 5-member judicial policy and administrative body appointed 1 each by the coprinces, 1 by the General Council, 1 by the executive council president, and 1 by the courts; judges serve 6-year renewable terms; Constitutional magistrates—2 appointed by the coprinces and 2 by the General Council; magistrates' appointments limited to 2 consecutive 8-year terms
subordinate courts: Tribunal of Judges or Tribunal de Batlles; Tribunal of the Courts or Tribunal de Corts

Political parties and leaders: Democrats for Andorra or DA [Xaviar ESPOT ZAMORA]
Social Democratic Party or PS [Vicenc ALFY FERRER]
Liberals of Andorra or L'A [Jordi GALLARDO FERNANDEZ]
Third Way/Lauredian Union [Josep PINTAT FORNE]
Social Democracy and Progress or SDP [Victor NAUDI ZAMORA]
United for the Progress of Andorra or UPA [Alfons CLAVERA ARIZTI]
note: Andorra has several smaller parties at the parish level (one is Lauredian Union)

International organization participation: CE, FAO, ICAO, ICC (NGOs), ICCt, ICRM, IFRCS, Interpol, IOC, IPU, ITU, OIF, OPCW, OSCE, UN, UNCTAD, UNESCO, Union Latina, UNWTO, WCO, WHO, WIPO, WTO (observer)

Diplomatic representation in the US: *chief of mission:* Ambassador Elisenda VIVES BALMANA (since 2 March 2016)
chancery: 2 United Nations Plaza, 27th Floor, New York, NY 10017
telephone: [1] (212) 750-8064
FAX: [1] (212) 750-6630

Diplomatic representation from the US: the US does not have an embassy in Andorra; the US ambassador to Spain is accredited to Andorra; US interests in Andorra are represented by the US Consulate General's office in Barcelona (Spain); mailing address: Paseo Reina Elisenda de Montcada, 23, 08034 Barcelona, Spain; telephone: [34] (93) 280-2227; FAX: [34] (93) 280-6175

Flag description: three vertical bands of blue (hoist side), yellow, and red, with the national coat of arms centered in the yellow band; the latter band is slightly wider than the other 2 so that the ratio of band widths is 8:9:8; the coat of arms features a quartered shield with the emblems of (starting in the upper left and proceeding clockwise): Urgell, Foix, Bearn, and Catalonia; the motto reads VIRTUS UNITA FORTIOR (Strength United is Stronger); the flag combines the blue and red French colors with the red and yellow of Spain to show Franco-Spanish protection
note: similar to the flags of Chad and Romania, which do not have a national coat of arms in the center, and the flag of Moldova, which does bear a national emblem

National symbol(s): red cow (breed unspecified); national colors: blue, yellow, red

National anthem: *name:* "El Gran Carlemany" (The Great Charlemagne)
lyrics/music: Joan BENLLOCH i VIVO/Enric MARFANY BONS
note: adopted 1921; the anthem provides a brief history of Andorra in a first person narrative

ECONOMY

Economy—overview: Andorra has a developed economy and a free market, with per capita income above the European average and above the level of its neighbors, Spain and France. The country has developed a sophisticated infrastructure including a one-of-a- kind micro-fiber-optic network for the entire country. Tourism, retail sales, and finance comprise more than three-quarters of GDP. Duty-free shopping for some products and the country's summer and winter resorts attract millions of visitors annually. Andorra uses the euro and is effectively subject to the monetary policy of the European Central Bank. Andorra's comparative advantage as a tax haven eroded when the borders of neighboring France and Spain opened and the government eased bank secrecy laws under pressure from the EU and OECD.

Agricultural production is limited—only about 5% of the land is arable—and most food has to be imported, making the economy vulnerable to changes in fuel and food prices. The principal livestock is sheep. Manufacturing output and exports consist mainly of perfumes and cosmetic products, products of the printing industry, electrical machinery and equipment, clothing, tobacco products, and furniture. Andorra is a member of the EU Customs Union and is treated as an EU member for trade in manufactured goods (no tariffs) and as a non-EU member for agricultural products.

To provide incentives for growth and diversification in the economy, the Andorran government began sweeping economic reforms in 2006. The Parliament approved three laws to complement the first phase of economic openness: on companies (October 2007), on business accounting (December 2007), and on foreign investment (April 2008 and June 2012). From 2011 to 2015, the Parliament also approved direct taxes in the form of taxes on corporations, on individual incomes of residents and non-residents, and on capital gains, savings, and economic activities. These regulations aim to establish a transparent, modern, and internationally comparable regulatory framework, in order to attract foreign investment and businesses that offer higher value added.

GDP (purchasing power parity): $3.327 billion (2015 est.)
$3.363 billion (2014 est.)
$3.273 billion (2013 est.)
note: data are in 2012 US dollars
country comparison to the world: 186

GDP (official exchange rate): $2.712 billion (2016 est.)

GDP—real growth rate: -1.1% (2015 est.)
1.4% (2014 est.)
-0.1% (2013 est.)
country comparison to the world: 203

GDP—per capita (PPP): $49,900 (2015 est.)
$51,300 (2014 est.)
$50,300 (2013 est.)
country comparison to the world: 32

GDP—composition, by sector of origin:
agriculture: 11.9% (2015 est.)
industry: 33.6% (2015 est.)
services: 54.5% (2015 est.)

Agriculture—products: small quantities of rye, wheat, barley, oats, vegetables, tobacco; sheep, cattle

Industries: tourism (particularly skiing), banking, timber, furniture

Industrial production growth rate: NA

Labor force: 39,750 (2016)
country comparison to the world: 196

Labor force—by occupation: *agriculture:* 0.5%
industry: 4.4%
services: 95.1% (2015)

Unemployment rate: 3.7% (2016 est.)
4.1% (2015 est.)
country comparison to the world: 44

Household income or consumption by percentage share: *lowest 10%:* NA
highest 10%: NA

Budget: *revenues:* 1.872 billion (2016)
expenditures: 2.06 billion (2016)

Taxes and other revenues: 69% (of GDP) (2016)
country comparison to the world: 5

Budget surplus (+) or deficit (-): -6.9% (of GDP) (2016)
country comparison to the world: 192

Public debt: 41% of GDP (2014 est.)
41.4% of GDP (2013 est.)
country comparison to the world: 122

Fiscal year: calendar year

Inflation rate (consumer prices): -0.9% (2015 est.)
-0.1% (2014 est.)
country comparison to the world: 1

Exports: $78.71 million (2015 est.)
$79.57 million (2014 est.)
country comparison to the world: 200

Exports—commodities: tobacco products, furniture

Imports: $1.257 billion (2015 est.)
$1.264 billion (2014 est.)
country comparison to the world: 178

Imports—commodities: consumer goods, food, fuel, electricity

Debt—external: $0 (2016)
country comparison to the world: 203

Exchange rates: euros (EUR) per US dollar—
0.885 (2017 est.)
0.903 (2016 est.)
0.9214 (2015 est.)
0.885 (2014 est.)
0.7634 (2013 est.)

ENERGY

Electricity access: *electrification—total population:* 100% (2016)

Electricity—production: 99.48 million kWh (2015 est.)
country comparison to the world: 201

Electricity—consumption: 221.6 million kWh (2015 est.)
country comparison to the world: 190

Electricity—exports: 6,000 kWh (2015 est.)
country comparison to the world: 95

Electricity—imports: 471.3 million kWh (2015 est.)
country comparison to the world: 81

Electricity—installed generating capacity: 520,000 kW (2010 est.)
country comparison to the world: 147

Electricity—from fossil fuels: 61% of total installed capacity (2010 est.)
country comparison to the world: 126

Electricity—from nuclear fuels: 0% of total installed capacity (2016 est.)
country comparison to the world: 36

Electricity—from hydroelectric plants: 23% of total installed capacity (2010 est.)
country comparison to the world: 82

Electricity—from other renewable sources: 15% of total installed capacity (2010 est.)
country comparison to the world: 56

Crude oil—production: 0 bbl/day (2016)
country comparison to the world: 103

Crude oil—exports: 0 bbl/day (2016) (2016)
country comparison to the world: 84

Crude oil—imports: 0 bbl/day (2016) (2016)
country comparison to the world: 87

Crude oil—proved reserves: 0 bbl (2016) (2016)
country comparison to the world: 101

Refined petroleum products—production: 0 bbl/day (2016)
country comparison to the world: 112

Natural gas—production: 0 cu m (2016) (2016)
country comparison to the world: 98

Natural gas—consumption: 0 cu m (2016) (2016)
country comparison to the world: 118

Natural gas—exports: 0 cu m (2016) (2016)
country comparison to the world: 60

Natural gas—imports: 0 cu m (2016) (2016)
country comparison to the world: 85

Natural gas—proved reserves: 0 cu m (2016)
country comparison to the world: 104

COMMUNICATIONS

Telephones—fixed lines: total subscriptions: 44,050
subscriptions per 100 inhabitants: 51.41 (2019 est.)
country comparison to the world: 160

Telephones—mobile cellular: total subscriptions: 97,645
subscriptions per 100 inhabitants: 113.96 (2019 est.)
country comparison to the world: 193

Telecommunication systems: *general assessment:* modern automatic telephone system; broadband Internet and LTE mobile lines for both consumer and enterprise customers available (2019)
domestic: 51 per 100 fixed-line, 113 per 100 mobile-cellular (2019)
international: country code—376; landline circuits to France and Spain; modern system with microwave radio relay connections between exchanges (2019)
note: the COVID-19 outbreak is negatively impacting telecommunications production and supply chains globally; consumer spending on telecom devices and services has also slowed due to the pandemic's effect on economies worldwide; overall progress towards improvements in all facets of the telecom industry—mobile, fixed-line, broadband, submarine cable and satellite has moderated

Broadcast media: 1 public TV station and 2 public radio stations; about 10 commercial radio stations; good reception of radio and TV broadcasts from stations in France and Spain; upgraded to terrestrial digital TV broadcasting in 2007; roughly 25 international TV channels available (2019)

Internet country code: .ad

Internet users: *total:* 78,483

percent of population: 91.57% (July 2018 est.)
country comparison to the world: 182

Broadband—fixed subscriptions: *total:* 35,663
subscriptions per 100 inhabitants: 42 (2018 est.)
country comparison to the world: 139

TRANSPORTATION

Civil aircraft registration country code prefix: C3 (2016)

Roadways: *total:* 320 km (2019)
country comparison to the world: 201

MILITARY AND SECURITY

Military and security forces: no regular military forces; Police Corps of Andorra

Military—note: defense is the responsibility of France and Spain

TRANSNATIONAL ISSUES

Disputes—international: none

ANGOLA

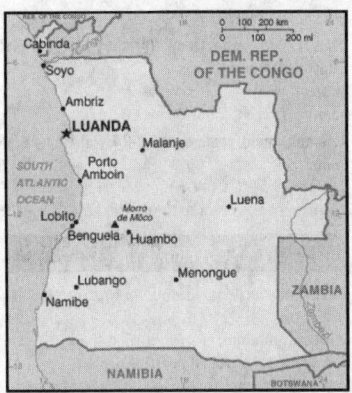

INTRODUCTION

Background: From the late 14th to the mid 19th century a Kingdom of Kongo stretched across central Africa from present-day northern Angola into the current Congo republics. It traded heavily with the Portuguese who, beginning in the 16th century, established coastal colonies and trading posts and introduced Christianity. By the 19th century, Portuguese settlement had spread to the interior; in 1914, Portugal abolished the last vestiges of the Kongo Kingdom and Angola became a Portuguese colony.

Angola scores low on human development indexes despite using its large oil reserves to rebuild since the end of a 27-year civil war in 2002. Fighting between the Popular Movement for the Liberation of Angola (MPLA), led by Jose Eduardo DOS SANTOS, and the National Union for the Total Independence of Angola (UNITA), led by Jonas SAVIMBI, followed independence from Portugal in 1975. Peace seemed imminent in 1992 when Angola held national elections, but fighting picked up again in 1993. Up to 1.5 million lives may have been lost—and 4 million people displaced—during the more than a quarter century of fighting. SAVIMBI's death in 2002 ended UNITA's insurgency and cemented the MPLA's hold on power. DOS SANTOS stepped down from the presidency in 2017, having led the

country since 1979. He pushed through a new constitution in 2010. Joao LOURENCO was elected president in August 2017 and became president of the MPLA in September 2018.

GEOGRAPHY

Location: Southern Africa, bordering the South Atlantic Ocean, between Namibia and Democratic Republic of the Congo

Geographic coordinates: 12 30 S, 18 30 E

Map references: Africa

Area: *total:* 1,246,700 sq km
land: 1,246,700 sq km
water: 0 sq km
country comparison to the world: 24

Area—comparative: about eight times the size of Georgia; slightly less than twice the size of Texas

Land boundaries: *total:* 5,369 km
border countries (4): Democratic Republic of the Congo 2646 km (of which 225 km is the boundary of discontiguous Cabinda Province), Republic of the Congo 231 km, Namibia 1427 km, Zambia 1065 km

Coastline: 1,600 km

Maritime claims: *territorial sea:* 12 nm
exclusive economic zone: 200 nm
contiguous zone: 24 nm

Climate: semiarid in south and along coast to Luanda; north has cool, dry season (May to October) and hot, rainy season (November to April)

Terrain: narrow coastal plain rises abruptly to vast interior plateau

Elevation: *mean elevation:* 1,112 m
lowest point: Atlantic Ocean 0 m
highest point: Moca 2,620 m

Natural resources: petroleum, diamonds, iron ore, phosphates, copper, feldspar, gold, bauxite, uranium

Land use: *agricultural land:* 47.5% (2016 est.)
arable land: 3.9% (2016 est.) / permanent crops: 0.3% (2016 est.) / permanent pasture: 43.3% (2016 est.)
forest: 46.3% (2016 est.)
other: 6.2% (2016 est.)

Irrigated land: 860 sq km (2014)

Population distribution: most people live in the western half of the country; urban areas account for the highest concentrations of people, particularly the capital of Luanda as shown in this population distribution map

Natural hazards: locally heavy rainfall causes periodic flooding on the plateau

Environment—current issues: overuse of pastures and subsequent soil erosion attributable to population pressures; desertification; deforestation of tropical rain forest, in response to both international demand for tropical timber and to domestic use as fuel, resulting in loss of biodiversity; soil erosion contributing to water pollution and siltation of rivers and dams; inadequate supplies of potable water

Environment—international agreements: party to: Biodiversity, Climate Change, Climate Change-Kyoto Protocol, Desertification, Hazardous Wastes, Law of the Sea, Marine Dumping, Ozone Layer Protection, Ship Pollution
signed, but not ratified: none of the selected agreements

Geography—note: the province of Cabinda is an exclave, separated from the rest of the country by the Democratic Republic of the Congo

PEOPLE AND SOCIETY

Population: 32,522,339 (July 2020 est.)
note: Angola's national statistical agency projects the country's 2017 population to be 28.4 million
country comparison to the world: 43

Nationality: *noun:* Angolan(s)
adjective: Angolan

Ethnic groups: Ovimbundu 37%, Kimbundu 25%, Bakongo 13%, mestico (mixed European and native African) 2%, European 1%, other 22%

Languages: Portuguese 71.2% (official), Umbundu 23%, Kikongo 8.2%, Kimbundu 7.8%, Chokwe 6.5%, Nhaneca 3.4%, Nganguela 3.1%, Fiote 2.4%, Kwanhama 2.3%, Muhumbi 2.1%, Luvale 1%, other 3.6% (2014 est.)
note: most widely spoken languages; shares sum to more than 100% because some respondents gave more than one answer on the census

Religions: Roman Catholic 41.1%, Protestant 38.1%, other 8.6%, none 12.3% (2014 est.)

Demographic profile: More than a decade after the end of Angola's 27-year civil war, the country still faces a variety of socioeconomic problems, including poverty, high maternal and child mortality, and illiteracy. Despite the country's rapid post-war economic growth based on oil production, about 40 percent of Angolans live below the poverty line and unemployment is widespread, especially among the large young- adult population. Only about 70% of the population is literate, and the rate drops to around 60% for women. The youthful population—about 45% are under the age of 15—is expected to continue growing rapidly with a fertility rate of more than 5 children per woman and a low rate of contraceptive use. Fewer than half of women deliver their babies with the assistance of trained health care personnel, which contributes to Angola's high maternal mortality rate.

Of the estimated 550,000 Angolans who fled their homeland during its civil war, most have returned home since 2002. In 2012, the UN assessed that conditions in Angola had been stable for several years and invoked a cessation of refugee status for Angolans. Following the cessation clause, some of those still in exile returned home voluntarily through UN repatriation programs, and others integrated into host countries.

Age structure: 0-14 *years:* 47.83% (male 7,758,636/female 7,797,869)
15-24 years: 18.64% (male 2,950,999/female 3,109,741)
25-54 years: 27.8% (male 4,301,618/female 4,740,463)
55-64 years: 3.43% (male 523,517/female 591,249)
65 years and over: 2.3% (male 312,197/female 436,050) (2020 est.)

Dependency ratios: *total dependency ratio:* 94.5
youth dependency ratio: 90.2
elderly dependency ratio: 4.3
potential support ratio: 23.5 (2020 est.)

Median age: *total:* 15.9 years
male: 15.4 years
female: 16.4 years (2020 est.)
country comparison to the world: 226

Population growth rate: 3.43% (2020 est.)
country comparison to the world: 3

Birth rate: 42.7 births/1,000 population (2020 est.)
country comparison to the world: 2

Death rate: 8.5 deaths/1,000 population (2020 est.)
country comparison to the world: 72

Net migration rate: -0.2 migrant(s)/1,000 population (2020 est.)
country comparison to the world: 105

Population distribution: most people live in the western half of the country; urban areas account for the highest concentrations of people, particularly the capital of Luanda as shown in this population distribution map

Urbanization: urban population: 66.8% of total population (2020)
rate of urbanization: 4.32% annual rate of change (2015-20 est.)

total population growth rate v. urban population growth rate, 2000-2030: Major urban areas—population: 8.330 million LUANDA (capital), 828,000 Lubango, 778,000 Cabinda (2020)

Sex ratio: *at birth:* 1.03 male(s)/female
0-14 years: 0.99 male(s)/female
15-24 years: 0.95 male(s)/female
25-54 years: 0.91 male(s)/female
55-64 years: 0.89 male(s)/female
65 years and over: 0.72 male(s)/female
total population: 0.95 male(s)/female (2020 est.)

Mother's mean age at first birth: 19.4 years (2015/16 est.)
note: median age at first birth among women 25-29

Maternal mortality rate: 241 deaths/100,000 live births (2017 est.)
country comparison to the world: 45

Infant mortality rate: *total:* 62.3 deaths/1,000 live births
male: 67.8 deaths/1,000 live births
female: 56.8 deaths/1,000 live births (2020 est.)
country comparison to the world: 11

Life expectancy at birth: *total population:* 61.3 years
male: 59.3 years
female: 63.4 years (2020 est.)
country comparison to the world: 213

Total fertility rate: 5.96 children born/woman (2020 est.)
country comparison to the world: 2

Contraceptive prevalence rate: 13.7% (2015/16)

Drinking water source:
improved:
urban: 81.7% of population
rural: 36.6% of population
total: 65.8% of population
unimproved:
urban: 18.3% of population
rural: 63.4% of population
total: 34.2% of population (2017 est.)

Current Health Expenditure: 2.8% (2017)

Physicians density: 0.21 physicians/1,000 population (2017)

Sanitation facility access:
improved:
urban: 92.2% of population
rural: 29.2% of population
total: 70.1% of population
unimproved:
urban: 7.8% of population
rural: 70.8% of population (2 est.)
total: 29.9% of population (2017 est.)

HIV/AIDS—adult prevalence rate: 1.8% (2019 est.)
country comparison to the world: 26

HIV/AIDS—people living with HIV/AIDS: 340,000 (2019 est.)
country comparison to the world: 20

HIV/AIDS—deaths: 13,000 (2019 est.)
country comparison to the world: 16

Major infectious diseases: *degree of risk:* very high (2020)

food or waterborne diseases: bacterial and protozoal diarrhea, hepatitis A, typhoid fever
vectorborne diseases: dengue fever, malaria
water contact diseases: schistosomiasis
animal contact diseases: rabies

Obesity—adult prevalence rate: 8.2% (2016)
country comparison to the world: 154

Children under the age of 5 years underweight: 19% (2016)
country comparison to the world: 27

Education expenditures: 3.4% of GDP (2010)
country comparison to the world: 124

Literacy: definition: age 15 and over can read and write
total population: 71.1%
male: 82%
female: 60.7% (2015)

School life expectancy (primary to tertiary education): *total:* 10 years
male: 12 years
female: 7 years (2011)

Unemployment, youth ages 15-24: *total:* 39.4%
male: 39%
female: 39.8% (2014 est.)
country comparison to the world: 12

GOVERNMENT

Country name: *conventional long form:* Republic of Angola
conventional short form: Angola
local long form: Republica de Angola
local short form: Angola
former: People's Republic of Angola
etymology: name derived by the Portuguese from the title "ngola" held by kings of the Ndongo (Ndongo was a kingdom in what is now northern Angola)

Government type: presidential republic

Capital: *name:* Luanda
geographic coordinates: 8 50 S, 13 13 E
time difference: UTC+ 1 (6 hours ahead of Washington, DC, during Standard Time)
daylight saving time: does not observe daylight savings time
etymology: originally named "Sao Paulo da Assuncao de Loanda" (Saint Paul of the Assumption of Loanda), which over time was shortened and corrupted to just Luanda

Administrative divisions: 18 provinces (provincias, singular—provincia); Bengo, Benguela, Bie, Cabinda, Cuando Cubango, Cuanza-Norte, Cuanza-Sul, Cunene, Huambo, Huila, Luanda, Lunda-Norte, Lunda-Sul, Malanje, Moxico, Namibe, Uige, Zaire

Independence: 11 November 1975 (from Portugal)

National holiday: Independence Day, 11 November (1975)

Constitution: history: previous 1975, 1992; latest passed by National Assembly 21 January 2010, adopted 5 February 2010
amendments: proposed by the president of the republic or supported by at least one third of the

National Assembly membership; passage requires at least two-thirds majority vote of the Assembly subject to prior Constitutional Court review if requested by the president of the republic

Legal system: civil legal system based on Portuguese civil law; no judicial review of legislation

International law organization participation: has not submitted an ICJ jurisdiction declaration; non-party state to the ICCt

Citizenship: *citizenship by birth:* no
citizenship by descent only: at least one parent must be a citizen of Angola
dual citizenship recognized: no
residency requirement for naturalization: 10 years

Suffrage: 18 years of age; universal

Executive branch: *chief of state:* President Joao Manuel Goncalves LOURENCO (since 26 September 2017); Vice President Bornito De Sousa Baltazar DIOGO (since 26 September 2017); note—the president is both chief of state and head of government
head of government: President Joao Manuel Goncalves LOURENCO (since 26 September 2017); Vice President Bornito De Sousa Baltazar DIOGO (since 26 September 2017)
cabinet: Council of Ministers appointed by the president
elections/appointments: the candidate of the winning party or coalition in the last legislative election becomes the president; president serves a 5-year term (eligible for a second consecutive or discontinuous term); last held on 23 August 2017 (next to be held in 2022)
election results: Joao Manuel Goncalves LOURENCO (MPLA) elected president by the winning party following the 23 August 2017 general election

Legislative branch: description: unicameral National Assembly or Assembleia Nacional (220 seats; members directly elected in a single national constituency and in multi-seat constituencies by closed list proportional representation vote; members serve 5-year terms)
elections: last held on 23 August 2017 (next to be held in August 2022)
election results: percent of vote by party—MPLA 61.1%, UNITA 26.7%, CASA-CE 9.5%, PRS 1.4%, FNLA 0.9%, other 0.5%; seats by party—MPLA 150, UNITA 51, CASA-CE 16, PRS 2, FNLA 1; composition—men 136, women 84, percent of women 38.2%

Judicial branch: *highest courts:* Supreme Court or Supremo Tribunal de Justica (consists of the court president, vice president, and a minimum of 16 judges); Constitutional Court or Tribunal Constitucional (consists of 11 judges)
judge selection and term of office: Supreme Court judges appointed by the president upon recommendation of the Supreme Judicial Council, an 18-member body chaired by the president; judge tenure NA; Constitutional Court judges—4 nominated by the president, 4 elected by National

Assembly, 2 elected by Supreme National Council, 1 elected by competitive submission of curricula; judges serve single 7-year terms
subordinate courts: provincial and municipal courts

Political parties and leaders: Broad Convergence for the Salvation of Angola Electoral Coalition or CASA-CE [Andre Mendes de CARVALHO]
National Front for the Liberation of Angola or FNLA; note—party has two factions; one led by Lucas NGONDA; the other by Ngola KABANGU
National Union for the Total Independence of Angola or UNITA [Isaias SAMAKUVA] (largest opposition party)
Popular Movement for the Liberation of Angola or MPLA [Joao LOURENCO]; note—Jose Eduardo DOS SANTOS stepped down 8 Sept 2018 ruling party in power since 1975
Social Renewal Party or PRS [Benedito DANIEL]

International organization participation: ACP, AfDB, AU, CEMAC, CPLP, FAO, G-77, IAEA, IBRD, ICAO, ICRM, IDA, IFAD, IFC, IFRCS, ILO, IMF, IMO, Interpol, IOC, IOM, IPU, ISO (correspondent), ITSO, ITU, ITUC (NGOs), MIGA, NAM, OAS (observer), OPEC, SADC, UN, UNCTAD, UNESCO, UNIDO, Union Latina, UNWTO, UPU, WCO, WFTU (NGOs), WHO, WIPO, WMO, WTO

Diplomatic representation in the US: *chief of mission:* Ambassador Joaquim do Espirito SANTO (since 16 September 2019)
chancery: 2100-2108 16th Street NW, Washington, DC 20009
telephone: [1] (202) 785-1156
FAX: [1] (202) 822-9049
consulate(s) general: Houston, New York

Diplomatic representation from the US: *chief of mission:* Ambassador Nina Maria FITE (since 14 February 2018)
telephone: [244] 946440977
embassy: 32 Rua Houari Boumedienne (in the Miramar area of Luanda), Luanda, C. P. 6468
mailing address: international mail: Caixa Postal 6468, Luanda; pouch: US Embassy Luanda, US Department of State, 2550 Luanda Place, Washington, DC 20521-2550
FAX: [244] (222) 64-1000

Flag description: two equal horizontal bands of red (top) and black with a centered yellow emblem consisting of a five-pointed star within half a cogwheel crossed by a machete (in the style of a hammer and sickle); red represents liberty and black the African continent; the symbols characterize workers and peasants

National symbol(s): Palanca Negra Gigante (giant black sable antelope); national colors: red, black, yellow

National anthem: *name:* "Angola Avante" (Forward Angola)
lyrics/music: Manuel Rui Alves MONTEIRO/Rui Alberto Vieira Dias MINGAO
note: adopted 1975
0:00 / 1:21

Economy—overview: Angola's economy is overwhelmingly driven by its oil sector. Oil production and its supporting activities contribute about 50% of GDP, more than 70% of government revenue, and more than 90% of the country's exports; Angola is an OPEC member and subject to its direction regarding oil production levels. Diamonds contribute an additional 5% to exports. Subsistence agriculture provides the main livelihood for most of the people, but half of the country's food is still imported.

Increased oil production supported growth averaging more than 17% per year from 2004 to 2008. A postwar reconstruction boom and resettlement of displaced persons led to high rates of growth in construction and agriculture as well. Some of the country's infrastructure is still damaged or undeveloped from the 27-year-long civil war (1975-2002). However, the government since 2005 has used billions of dollars in credit from China, Brazil, Portugal, Germany, Spain, and the EU to help rebuild Angola's public infrastructure. Land mines left from the war still mar the countryside, and as a result, the national military, international partners, and private Angolan firms all continue to remove them.

The global recession that started in 2008 stalled Angola's economic growth and many construction projects stopped because Luanda accrued billions in arrears to foreign construction companies when government revenue fell. Lower prices for oil and diamonds also resulted in GDP falling 0.7% in 2016. Angola formally abandoned its currency peg in 2009 but reinstituted it in April 2016 and maintains an overvalued exchange rate. In late 2016, Angola lost the last of its correspondent relationships with foreign banks, further exacerbating hard currency problems. Since 2013 the central bank has consistently spent down reserves to defend the kwanza, gradually allowing a 40% depreciation since late 2014. Consumer inflation declined from 325% in 2000 to less than 9% in 2014, before rising again to above 30% from 2015-2017.

Continued low oil prices, the depreciation of the kwanza, and slower than expected growth in non-oil GDP have reduced growth prospects, although several major international oil companies remain in Angola. Corruption, especially in the extractive sectors, is a major long-term challenge that poses an additional threat to the economy.

GDP (purchasing power parity): $193.6 billion (2017 est.)
$198.6 billion (2016 est.)
$203.9 billion (2015 est.)
note: data are in 2017 dollars
country comparison to the world: 65

GDP (official exchange rate): $126.5 billion (2017 est.)

GDP—real growth rate: -2.5% (2017 est.)
-2.6% (2016 est.)
0.9% (2015 est.)
country comparison to the world: 209

GDP—per capita (PPP): $6,800 (2017 est.)
$7,200 (2016 est.)
$7,600 (2015 est.)
note: data are in 2017 dollars
country comparison to the world: 160

Gross national saving: 28.6% of GDP (2017 est.)
24.5% of GDP (2016 est.)
28.5% of GDP (2015 est.)
country comparison to the world: 37

GDP—composition, by end use: *household consumption:* 80.6% (2017 est.)
government consumption: 15.6% (2017 est.)
investment in fixed capital: 10.3% (2017 est.)
investment in inventories: -1.2% (2017 est.)
exports of goods and services: 25.4% (2017 est.)
imports of goods and services: -30.7% (2017 est.)

GDP—composition, by sector of origin:
agriculture: 10.2% (2011 est.)
industry: 61.4% (2011 est.)
services: 28.4% (2011 est.)

Agriculture—products: bananas, sugarcane, coffee, sisal, corn, cotton, cassava (manioc, tapioca), tobacco, vegetables, plantains; livestock; forest products; fish

Industries: petroleum; diamonds, iron ore, phosphates, feldspar, bauxite, uranium, and gold; cement; basic metal products; fish processing; food processing, brewing, tobacco products, sugar; textiles; ship repair

Industrial production growth rate: 2.5% (2017 est.)
country comparison to the world: 115

Labor force: 12.51 million (2017 est.)
country comparison to the world: 46

Labor force—by occupation: *agriculture:* 85%
industry: 15% (2015 est.)
industry and services: 15% (2003 est.)

Unemployment rate: 6.6% (2016 est.)
country comparison to the world: 97

Population below poverty line: 36.6% (2008 est.)

Household income or consumption by percentage share: *lowest 10%:* 0.6%
highest 10%: 44.7% (2000)

Budget: *revenues:* 37.02 billion (2017 est.)
expenditures: 45.44 billion (2017 est.)

Taxes and other revenues: 29.3% (of GDP) (2017 est.)
country comparison to the world: 83

Budget surplus (+) or deficit (-): -6.7% (of GDP) (2017 est.)
country comparison to the world: 189

Public debt: 65% of GDP (2017 est.)
75.3% of GDP (2016 est.)
country comparison to the world: 59

Fiscal year: calendar year

Inflation rate (consumer prices): 29.8% (2017 est.)
30.7% (2016 est.)
country comparison to the world: 222

Current account balance: -$1.254 billion (2017 est.)
-$4.834 billion (2016 est.)

country comparison to the world: 150

Exports: $33.07 billion (2017 est.)
$31.03 billion (2016 est.)
country comparison to the world: 59

Exports—partners: China 61.2%, India 13%, US 4.2% (2017)

Exports—commodities: crude oil, diamonds, refined petroleum products, coffee, sisal, fish and fish products, timber, cotton

Imports: $19.5 billion (2017 est.)
$13.04 billion (2016 est.)
country comparison to the world: 77

Imports—commodities: machinery and electrical equipment, vehicles and spare parts; medicines, food, textiles, military goods

Imports—partners: Portugal 17.8%, China 13.5%, US 7.4%, South Africa 6.2%, Brazil 6.1%, UK 4% (2017)

Reserves of foreign exchange and gold: $17.29 billion (31 December 2017 est.)
$23.74 billion (31 December 2016 est.)
country comparison to the world: 63

Debt—external: $42.08 billion (31 December 2017 est.)
$27.14 billion (31 December 2016 est.)
country comparison to the world: 71

Exchange rates: kwanza (AOA) per US dollar - 172.6 (2017 est.)
163.656 (2016 est.)
163.656 (2015 est.)
120.061 (2014 est.)
98.303 (2013 est.)

ENERGY

Electricity access: population without electricity: 15 million (2013)
electrification—total population: 40.5% (2016)
electrification—urban areas: 70.7% (2016)
electrification—rural areas: 16% (2016)

Electricity—production: 10.2 billion kWh (2016 est.)
country comparison to the world: 102

Electricity—consumption: 9.036 billion kWh (2016 est.)
country comparison to the world: 101

Electricity—exports: 0 kWh (2016 est.)
country comparison to the world: 98

Electricity—imports: 0 kWh (2016 est.)
country comparison to the world: 119

Electricity—installed generating capacity: 2.613 million kW (2016 est.)
country comparison to the world: 103

Electricity—from fossil fuels: 34% of total installed capacity (2016 est.)
country comparison to the world: 180

Electricity—from nuclear fuels: 0% of total installed capacity (2017 est.)
country comparison to the world: 37

Electricity—from hydroelectric plants: 64% of total installed capacity (2017 est.)
country comparison to the world: 23

Electricity—from other renewable sources: 2% of total installed capacity (2017 est.)
country comparison to the world: 132

Crude oil—production: 1.593 million bbl/day (2018 est.)
country comparison to the world: 14

Crude oil—exports: 1.782 million bbl/day (2015 est.)
country comparison to the world: 7

Crude oil—imports: 0 bbl/day (2015 est.)
country comparison to the world: 88

Crude oil—proved reserves: 9.523 billion bbl (1 January 2018 est.)
country comparison to the world: 16

Refined petroleum products—production: 53,480 bbl/day (2015 est.)
country comparison to the world: 80

Refined petroleum products—consumption: 130,000 bbl/day (2016 est.)
country comparison to the world: 72

Refined petroleum products—exports: 30,340 bbl/day (2015 est.)
country comparison to the world: 62

Refined petroleum products—imports: 111,600 bbl/day (2015 est.)
country comparison to the world: 50

Natural gas—production: 3.115 billion cu m (2017 est.)
country comparison to the world: 55

Natural gas—consumption: 821.2 million cu m (2017 est.)
country comparison to the world: 95

Natural gas—exports: 3.993 billion cu m (2017 est.)
country comparison to the world: 33

Natural gas—imports: 0 cu m (2017 est.)
country comparison to the world: 86

Natural gas—proved reserves: 308.1 billion cu m (1 January 2018 est.)
country comparison to the world: 36

Carbon dioxide emissions from consumption of energy: 20.95 million Mt (2017 est.)
country comparison to the world: 85

COMMUNICATIONS

Telephones—fixed lines: *total subscriptions:* 122,566
subscriptions per 100 inhabitants: less than 1 (2019 est.)
country comparison to the world: 134

Telephones—mobile cellular: *total subscriptions:* 14,645,106
subscriptions per 100 inhabitants: 46.6 (2019 est.)
country comparison to the world: 69

Telecommunication systems: *general assessment:* progress in opening up the telecom sector to new competitors, while still retaining a 45% govt. portion of the share; slow progress in LTE network development, with only about 12% of the country covered by network infrastructure; regulator offers

4th service license to be issued for competition, cracks down on informal SIM card sales, and auctions 800MHz spectrum; M-commerce services launch pending (2020)

domestic: only about one fixed-line per 100 persons; mobile-cellular teledensity about 47 telephones per 100 persons (2019)

international: country code—244; landing points for the SAT-3/WASC, WACS, ACE and SACS fiber-optic submarine cable that provides connectivity to other countries in west Africa, Brazil, Europe and Asia; satellite earth stations—29, Angosat-2 satellite expected by 2021 (2019)

note: the COVID-19 outbreak is negatively impacting telecommunications production and supply chains globally; consumer spending on telecom devices and services has also slowed due to the pandemic's effect on economies worldwide; overall progress towards improvements in all facets of the telecom industry—mobile, fixed-line, broadband, submarine cable and satellite—has moderated

Broadcast media: state controls all broadcast media with nationwide reach; state-owned Televisao Popular de Angola (TPA) provides terrestrial TV service on 2 channels; a third TPA channel is available via cable and satellite; TV subscription services are available; state-owned Radio Nacional de Angola (RNA) broadcasts on 5 stations; about a half-dozen private radio stations broadcast locally

Internet country code: .ao

Internet users: *total:* 4,353,033
percent of population: 14.34% (July 2018 est.)
country comparison to the world: 90

Broadband—fixed subscriptions: *total:* 109,561
subscriptions per 100 inhabitants: less than 1 (2018 est.)
country comparison to the world: 120

TRANSPORTATION

National air transport system: *number of registered air carriers:* 10 (2020)

inventory of registered aircraft operated by air carriers: 55

annual passenger traffic on registered air carriers: 1,516,628 (2018)

annual freight traffic on registered air carriers: 78.16 million mt-km (2018)

Civil aircraft registration country code prefix: D2 (2016)

Airports: 102 (2020)
country comparison to the world: 54

Airports—with paved runways: *total:* 32 (2020)
over 3,047 m: 8
2,438 to 3,047 m: 8
1,524 to 2,437 m: 10
914 to 1,523 m: 6

Airports—with unpaved runways: *total:* 70 (2020)
over 3,047 m: 2
2,438 to 3,047 m: 2
1,524 to 2,437 m: 17
914 to 1,523 m: 27
under 914 m: 22

Heliports: 1 (2013)

Pipelines: 352 km gas, 85 km liquid petroleum gas, 1065 km oil, 5 km oil/gas/water (2013)

Railways: *total:* 2,852 km (2014)
narrow gauge: 2,729 km 1.067-m gauge (2014)
123 km 0.600-m gauge
country comparison to the world: 63

Roadways: *total:* 26,000 km (2018)
paved: 13,600 km (2018)
unpaved: 12,400 km (2018)
country comparison to the world: 103

Waterways: 1,300 km (2011)
country comparison to the world: 53

Merchant marine: *total:* 55
by type: general cargo 14, oil tanker 8, other 33 (2019)
country comparison to the world: 113

Ports and terminals: *major seaport(s):* Cabinda, Lobito, Luanda, Namibe
LNG terminal(s) (export): Angola Soyo

MILITARY AND SECURITY

Military and security forces: Angolan Armed Forces (Forcas Armadas Angolanas, FAA): Army, Navy (Marinha de Guerra Angola, MGA), Angolan National Air Force (Forca Aerea Nacional Angolana, FANA; under operational control of the Army); Rapid Reaction Police (paramilitary) (2019)

Military expenditures: 1.6% of GDP (2019)
1.8% of GDP (2018)
2.4% of GDP (2017)
3% of GDP (2016)
3.5% of GDP (2015)
country comparison to the world: 69

Military and security service personnel strengths: the Angolan Armed Forces (FAA) are comprised of approximately 107,000 active troops (100,000 Army; 1,000 Navy; 6,000 Air Force); est. 10,000 Rapid Reaction Police (2019)

Military equipment inventories and acquisitions: most Angolan Armed Forces weapons and equipment are of Russian, Soviet, or Warsaw Pact origin; Russia remains Angola's top supplier of military hardware, followed by Belarus and China (2019)

Military service age and obligation: 20-45 years of age for compulsory male and 18-45 years for voluntary male military service (registration at age 18 is mandatory); 20-45 years of age for voluntary female service; 2-year conscript service obligation; Angolan citizenship required; the Navy (MGA) is entirely staffed with volunteers (2019)

TRANSNATIONAL ISSUES

Disputes—international: Democratic Republic of Congo accuses Angola of shifting monuments

Refugees and internally displaced persons: refugees (country of origin): 23,395 (Democratic Republic of the Congo) (refugees and asylum seekers) (2020)

Illicit drugs: used as a transshipment point for cocaine destined for Western Europe and other African states, particularly South Africa

ANGUILLA

INTRODUCTION

Background: Colonized by English settlers from Saint Kitts in 1650, Anguilla was administered by Great Britain until the early 19th century, when the island—against the wishes of the inhabitants—was incorporated into a single British dependency along with Saint Kitts and Nevis. Several attempts at separation failed. In 1971, two years after a revolt, Anguilla was finally allowed to secede; this arrangement was formally recognized in 1980, with Anguilla becoming a separate British dependency. On 7 September 2017, the island suffered extensive damage from Hurricane Irma, particularly to communications and residential and business infrastructure.

GEOGRAPHY

Location: Caribbean, islands between the Caribbean Sea and North Atlantic Ocean, east of Puerto Rico

Geographic coordinates: 18 15 N, 63 10 W

Map references: Central America and the Caribbean

Area: *total:* 91 sq km
land: 91 sq km

water: 0 sq km
country comparison to the world: 227

Area—comparative: about one-half the size of Washington, DC

Land boundaries: 0 km

Coastline: 61 km

Maritime claims: *territorial sea:* 3 nm
exclusive fishing zone: 200 nm

Climate: tropical; moderated by northeast trade winds

Terrain: flat and low-lying island of coral and limestone

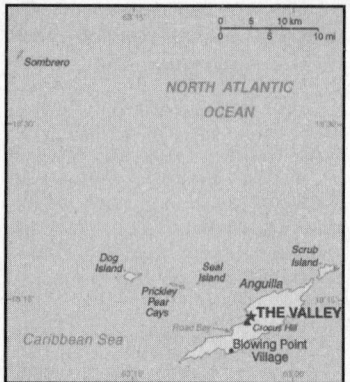

Elevation: *lowest point:* Caribbean Sea 0 m
highest point: Crocus Hill 73 m

Natural resources: salt, fish, lobster

Land use: *agricultural land:* 0% (2016 est.)
arable land: 0% (2016 est.) / permanent crops: 0%
(2016 est.) / permanent pasture: 0% (2016 est.)
forest: 61.1% (2016 est.)
other: 38.9% (2016 est.)

Irrigated land: 0 sq km (2012)

Population distribution: most of the population is
concentrated in The Valley in the center of the
island; settlmement is fairly uniform in the southwest, but rather sparce in the northeast

Natural hazards: frequent hurricanes and other
tropical storms (July to October)

Environment—current issues: supplies of potable
water sometimes cannot meet increasing demand
largely because of poor distribution system

Geography—note: the most northerly of the
Leeward Islands in the Lesser Antilles

PEOPLE AND SOCIETY

Population: 18,090 (July 2020 est.)
country comparison to the world: 220

Nationality: *noun:* Anguillan(s)
adjective: Anguillan

Ethnic groups: African/Black 85.3%, Hispanic
4.9%, mixed 3.8%, White 3.2%, East Indian/
Indian 1%, other 1.6%, unspecified 0.3% (2011
est.)

note: data represent population by ethnic origin

Languages: English (official)

Religions: Protestant 73.2% (includes Anglican
22.7%, Methodist 19.4%, Pentecostal 10.5%,
Seventh Day Adventist 8.3%, Baptist 7.1%,
Church of God 4.9%, Presbyterian 0.2%, Brethren
0.1%), Roman Catholic 6.8%, Jehovah's Witness
1.1%, other Christian 10.9%, other 3.2%, unspecified 0.3%, none 4.5% (2011 est.)

Age structure: *0-14 years:* 21.63% (male 1,991/
female 1,922)
15-24 years: 13.9% (male 1,269/female 1,246)
25-54 years: 42.27% (male 3,428/female 4,218)
55-64 years: 12.42% (male 993/female 1,254)

65 years and over: 9.78% (male 874/female 895)
(2020 est.)

Median age: *total:* 35.7 years
male: 33.7 years
female: 37.6 years (2020 est.)
country comparison to the world: 81

Population growth rate: 1.86% (2020 est.)
country comparison to the world: 54

Birth rate: 12.2 births/1,000 population (2020 est.)
country comparison to the world: 158

Death rate: 4.8 deaths/1,000 population (2020 est.)
country comparison to the world: 204

Net migration rate: 11.1 migrant(s)/1,000 population (2020 est.)
country comparison to the world: 6

Population distribution: most of the population is
concentrated in The Valley in the center of the
island; settlmement is fairly uniform in the southwest, but rather sparce in the northeast

Urbanization: *urban population:* 100% of total
population (2020)
rate of urbanization: 0.9% annual rate of change
(2015-20 est.)

**total population growth rate v. urban population
growth rate, 2000-2030:** Major urban areas—population: 1,000 THE VALLEY (capital) (2018)

Sex ratio: *at birth:* 1.03 male(s)/female
0-14 years: 1.04 male(s)/female
15-24 years: 1.02 male(s)/female
25-54 years: 0.81 male(s)/female
55-64 years: 0.79 male(s)/female
65 years and over: 0.98 male(s)/female
total population: 0.9 male(s)/female (2020 est.)

Infant mortality rate: *total:* 3.3 deaths/1,000 live
births
male: 3.6 deaths/1,000 live births
female: 2.9 deaths/1,000 live births (2020 est.)
country comparison to the world: 203

Life expectancy at birth: *total population:* 81.8
years
male: 79.2 years
female: 84.5 years (2020 est.)
country comparison to the world: 26

Total fertility rate: 1.74 children born/woman
(2020 est.)
country comparison to the world: 161

Drinking water source:
improved:
urban: 97.5% of population
total: 97.5% of population
unimproved:
urban: 2.5% of population
total: 2.5% of population (2017 est.)

Sanitation facility access:
improved:
urban: 99.1% of population
total: 99.1% of population
unimproved:
urban: 0.9% of population
total: 0.9% of population (2017 est.)

HIV/AIDS—adult prevalence rate: NA

HIV/AIDS—people living with HIV/AIDS: NA

HIV/AIDS—deaths: NA

Education expenditures: NA

GOVERNMENT

Country name: *conventional long form:* none
conventional short form: Anguilla
etymology: the name Anguilla means "eel" in
various Romance languages (Spanish, Italian,
Portuguese, French) and likely derives from the
island's lengthy shape

Dependency status: overseas territory of the UK

Government type: parliamentary democracy
(House of Assembly); self-governing overseas territory of the UK

Capital: *name:* The Valley
geographic coordinates: 18 13 N, 63 03 W
time difference: UTC-4 (1 hour ahead of
Washington, DC, during Standard Time)
etymology: name derives from the capital's location between several hills

Administrative divisions: none (overseas territory
of the UK)

Independence: none (overseas territory of the UK)

National holiday: Anguilla Day, 30 May (1967)

Constitution: *history:* several previous; latest 1
April 1982
amendments: amended 1990

Legal system: common law based on the English
model

Citizenship: see United Kingdom

Suffrage: 18 years of age; universal

Executive branch: *chief of state:* Queen
ELIZABETH II (since 6 February 1952);
represented by Governor Tim FOY (since August
2017)
head of government: Premier Dr. Ellis WEBSTER
(since 30 June 2020); note—starting in 2019, the
title of head of government was changed to premier from chief minister of Anguilla
cabinet: Executive Council appointed by the governor from among elected members of the House
of Assembly
elections/appointments: the monarchy is hereditary; governor appointed by the monarch; following legislative elections, the leader of the majority
party or majority coalition usually appointed premier by the governor

Legislative branch: *description:* unicameral House
of Assembly (11 seats; 7 members directly elected
in single-seat constituencies by simple majority
vote, 2 appointed by the governor, and 2 ex
officio members—the attorney general and deputy
governor; members serve five-year terms)
elections: last held on 29 June 2020 (next to be
held in 2025)
election results: percent of vote by party—NA;
seats by party—APM 7, AUF 4; composition—NA

Judicial branch: *highest courts:* the Eastern
Caribbean Supreme Court (ECSC) is the superior
court of the Organization of Eastern Caribbean
States; the ECSC—headquartered on St. Lucia—
consists of the Court of Appeal—headed by the

chief justice and 4 judges—and the High Court with 18 judges; the Court of Appeal is itinerant, travelling to member states on a schedule to hear appeals from the High Court and subordinate courts; High Court judges reside in the member states, though none on Anguilla

judge selection and term of office: Eastern Caribbean Supreme Court chief justice appointed by Her Majesty, Queen ELIZABETH II; other justices and judges appointed by the Judicial and Legal Services Commission; Court of Appeal justices appointed for life with mandatory retirement at age 65; High Court judges appointed for life with mandatory retirement at age 62

subordinate courts: Magistrate's Court; Juvenile Court

Political parties and leaders: Anguilla Democratic Party or ADP

Anguilla National Alliance or ANA

Anguilla Progressive Movement or APM [Dr. Ellis WEBSTER]; prior to 2019, it was known as the Anguilla United Movement or AUM

Anguilla United Front or AUF [Victor BANKS] (alliance includes ADP, ANA)

Democracy, Opportunity, Vision, and Empowerment Party or DOVE [Sutcliffe HODGE]

International organization participation: Caricom (associate), CDB, Interpol (subbureau), OECS, UNESCO (associate), UPU

Diplomatic representation in the US: none (overseas territory of the UK)

Diplomatic representation from the US: telephone: [34] [93] 280-2227

embassy: US does not have an embassy in Andorra; the US ambassador to Spain is accredited to Andorra; US interests in Andorra are represented by the US Consulate General's office in Barcelona (Spain); mailing address: Paseo Reina Elisenda de Montcada, 23, 08034 Barcelona, Spain none (overseas territory of the UK)

Flag description: blue, with the flag of the UK in the upper hoist-side quadrant and the Anguillan coat of arms centered in the outer half of the flag; the coat of arms depicts three orange dolphins in an interlocking circular design on a white background with a turquoise-blue field below; the white in the background represents peace; the blue base symbolizes the surrounding sea, as well as faith, youth, and hope; the three dolphins stand for endurance, unity, and strength

National symbol(s): dolphin

National anthem: *name:* God Bless Anguilla *lyrics/music:* Alex RICHARDSON

note: local anthem adopted 1981; as a territory of the United Kingdom, "God Save the Queen" is official (see United Kingdom)

ECONOMY

Economy—overview: Anguilla has few natural resources, is unsuited for agriculture, and the economy depends heavily on luxury tourism, offshore banking, lobster fishing, and remittances from emigrants. Increased activity in the tourism

industry has spurred the growth of the construction sector contributing to economic growth. Anguillan officials have put substantial effort into developing the offshore financial sector, which is small but growing. In the medium term, prospects for the economy will depend largely on the recovery of the tourism sector and, therefore, on revived income growth in the industrialized nations as well as on favorable weather conditions.

GDP (purchasing power parity): $175.4 million (2009 est.)

$191.7 million (2008 est.)

$108.9 million (2004 est.)

country comparison to the world: 222

GDP (official exchange rate): $175.4 million (2009 est.)

GDP—real growth rate: -8.5% (2009 est.)

country comparison to the world: 220

GDP—per capita (PPP): $12,200 (2008 est.)

country comparison to the world: 130

GDP—composition, by end use: household consumption: 74.1% (2017 est.)

government consumption: 18.3% (2017 est.)

investment in fixed capital: 26.8% (2017 est.)

investment in inventories: 0% (2017 est.)

exports of goods and services: 48.2% (2017 est.)

imports of goods and services: -67.4% (2017 est.)

GDP—composition, by sector of origin: agriculture: 3% (2017 est.)

industry: 10.5% (2017 est.)

services: 86.4% (2017 est.)

Agriculture—products: small quantities of tobacco, vegetables; cattle raising

Industries: tourism, boat building, offshore financial services

Industrial production growth rate: 4% (2017 est.)

country comparison to the world: 75

Labor force: 6,049 (2001)

country comparison to the world: 219

Labor force—by occupation: *agriculture:* 74.1%

industry: 3%

services: 18%

agriculture/fishing/forestry/mining: 4% (2000 est.)

manufacturing: 3% (2000 est.)

construction: 18% (2000 est.)

transportation and utilities: 10% (2000 est.)

commerce: 36% (2000 est.)

Unemployment rate: 8% (2002)

country comparison to the world: 115

Population below poverty line: 23% (2002 est.)

Household income or consumption by percentage share: lowest 10%: NA

highest 10%: NA

Budget: *revenues:* 81.92 million (2017 est.)

expenditures: 80.32 million (2017 est.)

Taxes and other revenues: 46.7% (of GDP) (2017 est.)

country comparison to the world: 19

Budget surplus (+) or deficit (-): 0.9% (of GDP) (2017 est.)

country comparison to the world: 34

Public debt: 20.1% of GDP (2015 est.)

20.8% of GDP (2014 est.)

country comparison to the world: 189

Fiscal year: 1 April—31 March

Inflation rate (consumer prices): 1.3% (2017 est.)

-0.6% (2016 est.)

country comparison to the world: 67

Current account balance: -$23.2 million (2017 est.)

-$25.3 million (2016 est.)

country comparison to the world: 73

Exports: $7.9 million (2017 est.)

$3.9 million (2016 est.)

country comparison to the world: 216

Exports—commodities: lobster, fish, livestock, salt, concrete blocks, rum

Imports: $186.2 million (2017 est.)

$170.1 million (2016 est.)

country comparison to the world: 211

Imports—commodities: fuels, foodstuffs, manufactures, chemicals, trucks, textiles

Reserves of foreign exchange and gold: $76.38 million (31 December 2017 est.)

$48.14 million (31 December 2015 est.)

country comparison to the world: 183

Debt—external: $41.04 million (31 December 2013)

$8.8 million (1998)

country comparison to the world: 195

Exchange rates: East Caribbean dollars (XCD) per US dollar -

2.7 (2017 est.)

2.7 (2016 est.)

2.7 (2015 est.)

2.7 (2014 est.)

2.7 (2013 est.)

COMMUNICATIONS

Telephones—fixed lines: *total subscriptions:* 7,461

subscriptions per 100 inhabitants: 42.02 (2019 est.)

country comparison to the world: 196

Telephones—mobile cellular: *total subscriptions:* 32,332

subscriptions per 100 inhabitants: 182.09 (2019 est.)

country comparison to the world: 209

Telecommunication systems: general assessment: modern internal telephone system with fiber-optic trunk lines; telecom sector provides a relatively high contribution to overall GDP; numerous competitors licensed, but small and localized; major growth sectors include the mobile telephony and data segments (2020)

domestic: fixed-line teledensity is about 42 per 100 persons; mobile-cellular teledensity is roughly 182 per 100 persons (2019)

international: country code—1-264; landing points for the SSCS, ECFS, GCN and Southern Caribbean Fiber with submarine cable links to Caribbean islands and to the US; microwave

29

radio relay to island of Saint Martin/Sint Maarten (2019)

note: the COVID-19 outbreak is negatively impacting telecommunications production and supply chains globally; consumer spending on telecom devices and services has also slowed due to the pandemic's effect on economies worldwide; overall progress towards improvements in all facets of the telecom industry—mobile, fixed-line, broadband, submarine cable and satellite—has moderated

Broadcast media: 1 private TV station; multi-channel cable TV subscription services are available; about 10 radio stations, one of which is government-owned

Internet country code: .ai

Internet users: total: 14,211
percent of population: 81.57% (July 2018 est.)

country comparison to the world: 213

TRANSPORTATION

National air transport system: *number of registered air carriers:* 2 (2020)
inventory of registered aircraft operated by air carriers: 4

Civil aircraft registration country code prefix: VP-A (2016)

Airports: 1 (2020)
country comparison to the world: 211

Airports—with paved runways: *total:* 1 (2020)
1,524 to 2,437 m: 1

Roadways: *total:* 175 km (2004)
paved: 82 km (2004)
unpaved: 93 km (2004)

country comparison to the world: 209
Merchant marine: *total:* 2
by type: other 2 (2019)
country comparison to the world: 172
Ports and terminals: *major seaport(s):* Blowing Point, Road Bay

MILITARY AND SECURITY

Military—note: defense is the responsibility of the UK

TRANSNATIONAL ISSUES

Disputes—international: none

Illicit drugs: transshipment point for South American narcotics destined for the US and Europe

ANTARCTICA

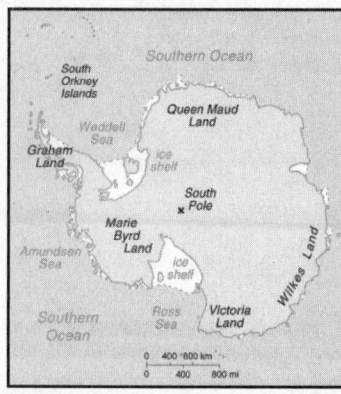

INTRODUCTION

Background: Speculation over the existence of a "southern land" was not confirmed until the early 1820s when British and American commercial operators and British and Russian national expeditions began exploring the Antarctic Peninsula region and other areas south of the Antarctic Circle. Not until 1840 was it established that Antarctica was indeed a continent and not merely a group of islands or an area of ocean. Several exploration "firsts" were achieved in the early 20th century, but generally the area saw little human activity. Following World War II, however, the continent experienced an upsurge in scientific research. A number of countries have set up a range of year-round and seasonal stations, camps, and refuges to support scientific research in Antarctica. Seven have made territorial claims, but most countries do not recognize these claims. In order to form a legal framework for the activities of nations on the continent,

an Antarctic Treaty was negotiated that neither denies nor gives recognition to existing territorial claims; signed in 1959, it entered into force in 1961. Also relevant to Antarctic governance are the Environmental Protocol to the Antarctic Treaty and the Convention for the Conservation of Antarctic Marine Living Resources.

GEOGRAPHY

Location: continent mostly south of the Antarctic Circle

Geographic coordinates: 90 00 S, 0 00 E

Map references: Antarctic Region

Area: *total:* 14.2 million sq km
land: 14.2 million sq km (285,000 sq km ice-free, 13.915 million sq km ice-covered) (est.)
note: fifth-largest continent, following Asia, Africa, North America, and South America, but larger than Australia and the subcontinent of Europe
country comparison to the world: 2

Area—comparative: slightly less than 1.5 times the size of the US

Land boundaries: 0
note: see entry on Disputes—international

Coastline: 17,968 km

Maritime claims: Australia, Chile, and Argentina claim Exclusive Economic Zone (EEZ) rights or similar over 200 nm extensions seaward from their continental claims, but like the claims themselves, these zones are not accepted by other countries; 22 of 29 Antarctic Treaty consultative parties have made no claims to Antarctic territory (although Russia and the US have reserved the right to do so); also see the Disputes—international entry

Climate: the coldest, windiest, and driest continent on Earth; severe low temperatures vary with latitude, elevation, and distance from the ocean;

East Antarctica is colder than West Antarctica because of its higher elevation; Antarctic Peninsula has the most moderate climate; higher temperatures occur in January along the coast and average slightly below freezing; summers characterized by continuous daylight, while winters bring continous darkness; persistent high pressure over the interior brings dry, subsiding air that results in very little cloud cover

Terrain: about 98% thick continental ice sheet and 2% barren rock, with average elevations between 2,000 and 4,000 m; mountain ranges up to nearly 5,000 m; ice-free coastal areas include parts of southern Victoria Land, Wilkes Land, the Antarctic Peninsula area, and parts of Ross Island on McMurdo Sound; glaciers form ice shelves along about half of the coastline, and floating ice shelves constitute 11% of the area of the continent

Elevation: *mean elevation:* 2,300 m
lowest point: Denman Glacier more than -3,500 m (-11,500 ft) below sea level
highest point: Vinson Massif 4,892 m
note: the lowest known land point in Antarctica is hidden in the Denman Galcier; at its surface is the deepest ice yet discovered and the world's lowest elevation not under seawater

Natural resources: iron ore, chromium, copper, gold, nickel, platinum and other minerals, and coal and hydrocarbons have been found in small noncommercial quantities; mineral exploitation except for scientific research is banned by the Environmental Protocol to the Antarctic Treaty; krill, icefish, toothfish, and crab have been taken by commercial fisheries, which are managed through the Commission for the Conservation of Antarctic Living Marine Resources (CCAMLR)

Land use: 0% (2015 est.)

Natural hazards: katabatic (gravity-driven) winds blow coastward from the high interior; frequent blizzards form near the foot of the plateau; cyclonic storms form over the ocean and move clockwise along the coast; volcanism on Deception Island and isolated areas of West Antarctica; other seismic activity rare and weak; large icebergs may calve from ice shelf

Environment—current issues: the discovery of a large Antarctic ozone hole in the earth's stratosphere (the ozone layer)—first announced in 1985—spurred the signing of the Montreal Protocol in 1987, an international agreement phasing out the use of ozone-depleting chemicals; the ozone layer prevents most harmful wavelengths of ultra-violet (UV) light from passing through the earth's atmosphere; ozone depletion has been shown to harm a variety of Antarctic marine plants and animals (plankton); in 2016, a gradual trend toward "healing" of the ozone hole was reported; since the 1990s, satellites have shown accelerating ice loss driven by ocean change; although considerable uncertainty remains, scientists are increasing our understanding and ability to model potential impacts of ice loss

Geography—note: the coldest, windiest, highest (on average), and driest continent; during summer, more solar radiation reaches the surface at the South Pole than is received at the Equator in an equivalent period mostly uninhabitable, 98% of the land area is covered by the Antarctic ice sheet, the largest single mass of ice on earth covering an area of 14 million sq km (5.4 million sq mi) and containing 26.5 million cu km (6.4 million cu mi) of ice (this is almost 62% of all of the world's fresh water); if all this ice were converted to liquid water, one estimate is that it would be sufficient to raise the height of the world's oceans by 58 m (190 ft)

PEOPLE AND SOCIETY

Population: no indigenous inhabitants, but there are both permanent and summer-only staffed research stations

note: 53 countries have signed the 1959 Antarctic Treaty; 30 of those operate through their National Antarctic Program a number of seasonal-only (summer) and year-round research stations on the continent and its nearby islands south of 60 degrees south latitude (the region covered by the Antarctic Treaty); the population engaging in and supporting science or managing and protecting the Antarctic region varies from approximately 4,400 in summer to 1,100 in winter; in addition, approximately 1,000 personnel, including ship's crew and scientists doing onboard research, are present in the waters of the treaty region as of 2017, peak summer (December-February) maximum capacity in scientific stations—4,877 total; Argentina 601, Australia 243, Belarus 12, Belgium 40, Brazil 66, Bulgaria 22, Chile 433, China 166, Czechia 20, Ecuador 34, Finland 17, France 90, France and Italy jointly 80, Germany 104, India 113, Italy 120, Japan 130, South Korea 130, Netherlands 10, NZ 86, Norway 70, Peru 30, Poland 40, Russia 335, South Africa 80, Spain 98, Sweden 20, Ukraine 24, UK 196, US 1,399, Uruguay 68 (2017)

winter (June-August) maximum capacity in scientific station—1,036 total; Argentina 221, Australia 52, Brazil 15, Chile 114, China 32, France 24, France and Italy jointly 13, Germany 9, India 48, Japan 40, Netherlands 10, South Korea 25, NZ 11, Norway 7, Poland 16, Russia 125, South Africa 15, Ukraine 12, UK 44, US 215, Uruguay 8 (2017)

research stations operated within the Antarctic Treaty area (south of 60 degrees south latitude) by National Antarctic Programs year-round stations—approximately 40 total; Argentina 6, Australia 3, Brazil 1, Chile 6, China 2, France 1, France and Italy jointly 1, Germany 1, India 2, Japan 1, Netherlands 1, South Korea 2, NZ 1, Norway 1, Poland 1, Russia 5, South Africa 1, Ukraine 1, UK 2, US 3, Uruguay 2 (2017)

a range of seasonal-only (summer) stations, camps, and refuges—Argentina, Australia, Belarus, Belgium, Bulgaria, Brazil, Chile, China, Czechia, Ecuador, Finland, France, Germany, India, Italy, Japan, Netherlands, South Korea, New Zealand, Norway, Peru, Poland, Russia, South Africa, Spain, Sweden, Ukraine, UK, US, and Uruguay (2017)

in addition, during the austral summer some nations have numerous occupied locations such as tent camps, summer-long temporary facilities, and mobile traverses in support of research

GOVERNMENT

Country name: *conventional long form:* none
conventional short form: Antarctica
etymology: name derived from two Greek words meaning "opposite to the Arctic" or "opposite to the north"

Government type: Antarctic Treaty Summary—the Antarctic region is governed by a system known as the Antarctic Treaty system; the system includes: 1. the Antarctic Treaty, signed on 1 December 1959 and entered into force on 23 June 1961, which establishes the legal framework for the management of Antarctica, 2. Measures, Decisions, and Resolutions adopted at Antarctic Treaty Consultative Meetings, 3. The Convention for the Conservation of Antarctic Seals (1972), 4. The Convention for the Conservation of Antarctic Marine Living Resources (1980), and 5. The Protocol on Environmental Protection to the Antarctic Treaty (1991); the Antarctic Treaty Consultative Meetings operate by consensus (not by vote) of all consultative parties at annual Treaty meetings; by January 2016, there were 53 treaty member nations: 29 consultative and 24 non-consultative; consultative (decision-making) members include the seven nations that claim portions of Antarctica as national territory (some claims overlap) and 22 non-claimant nations; the US and Russia have reserved the right to make claims; the US does not recognize the claims of others; Antarctica is administered through meetings of the consultative member nations; measures adopted at these meetings are carried out by these member nations (with respect to their own nationals and operations) in accordance with

their own national laws; the years in parentheses indicate when a consultative member-nation acceded to the Treaty and when it was accepted as a consultative member, while no date indicates the country was an original 1959 treaty signatory; claimant nations are—Argentina, Australia, Chile, France, NZ, Norway, and the UK; non-claimant consultative nations are—Belgium, Brazil (1975/1983), Bulgaria (1978/1998), China (1983/1985), Czech Republic (1962/2014), Ecuador (1987/1990), Finland (1984/1989), Germany (1979/1981), India (1983/1983), Italy (1981/1987), Japan, South Korea (1986/1989), Netherlands (1967/1990), Peru (1981/1989), Poland (1961/1977), Russia, South Africa, Spain (1982/1988), Sweden (1984/1988), Ukraine (1992/2004), Uruguay (1980/1985), and the US; non-consultative members, with year of accession in parentheses, are—Austria (1987), Belarus (2006), Canada (1988), Colombia (1989), Cuba (1984), Denmark (1965), Estonia (2001), Greece (1987), Guatemala (1991), Hungary (1984), Iceland (2015), Kazakhstan (2015), North Korea (1987), Malaysia (2011), Monaco (2008), Mongolia (2015), Pakistan (2012), Papua New Guinea (1981), Portugal (2010), Romania (1971), Slovakia (1962/1993), Switzerland (1990), Turkey (1996), and Venezuela (1999); note—Czechoslovakia acceded to the Treaty in 1962 and separated into the Czech Republic and Slovakia in 1993; Article 1—area to be used for peaceful purposes only; military activity, such as weapons testing, is prohibited, but military personnel and equipment may be used for scientific research or any other peaceful purpose; Article 2—freedom of scientific investigation and cooperation shall continue; Article 3—free exchange of information and personnel, cooperation with the UN and other international agencies; Article 4—does not recognize, dispute, or establish territorial claims and no new claims shall be asserted while the treaty is in force; Article 5—prohibits nuclear explosions or disposal of radioactive wastes; Article 6—includes under the treaty all land and ice shelves south of 60 degrees 00 minutes south and reserves high seas rights; Article 7—treaty-state observers have free access, including aerial observation, to any area and may inspect all stations, installations, and equipment; advance notice of all expeditions and of the introduction of military personnel must be given; Article 8—allows for jurisdiction over observers and scientists by their own states; Article 9—frequent consultative meetings take place among member nations; Article 10—treaty states will discourage activities by any country in Antarctica that are contrary to the treaty; Article 11—disputes to be settled peacefully by the parties concerned or, ultimately, by the ICJ; Articles 12, 13, 14—deal with upholding, interpreting, and amending the treaty among involved nations; other agreements—some 200 measures adopted at treaty consultative meetings and approved by governments; the Protocol on Environmental Protection to the Antarctic Treaty was signed 4 October 1991 and entered into force 14 January 1998; this agreement provides for the protection

THE CIA WORLD FACTBOOK

of the Antarctic environment and includes five annexes that have entered into force: 1) environmental impact assessment, 2) conservation of Antarctic fauna and flora, 3) waste disposal and waste management, 4) prevention of marine pollution, 5) area protection and management; a sixth annex addressing liability arising from environmental emergencies has yet to enter into force; the Protocol prohibits all activities relating to mineral resources except scientific research; a permanent Antarctic Treaty Secretariat was established in 2004 in Buenos Aires, Argentina

Legal system: Antarctica is administered through annual meetings—known as Antarctic Treaty Consultative Meetings—which include consultative member nations, non-consultative member nations, observer organizations, and expert organizations; decisions from these meetings are carried out by these member nations (with respect to their own nationals and operations) in accordance with their own national laws; more generally, the Antarctic Treaty area, that is to all areas between 60 and 90 degrees south latitude, is subject to a number of relevant legal instruments and procedures adopted by the states party to the Antarctic Treaty; note—US law, including certain criminal offenses by or against US nationals, such as murder, may apply extraterritoriality; some US laws directly apply to Antarctica; for example, the Antarctic Conservation Act, 16 U.S.C. section 2401 et seq., provides civil and criminal penalties for the following activities unless authorized by regulation or statute: the taking of native mammals or birds; the introduction of nonindigenous plants and animals; entry into specially protected areas; the discharge or disposal of pollutants; and the importation into the US of certain items from Antarctica; violation of the Antarctic Conservation Act carries penalties of up to $10,000 in fines and one year in prison; the National Science Foundation and Department of Justice share enforcement responsibilities; Public Law 95-541, the US Antarctic Conservation Act of 1978, as amended in 1996, requires expeditions from the US to Antarctica to notify, in advance, the Office of Oceans and Polar Affairs, Room 2665, Department of State, Washington, DC 20520, which reports such plans to other nations as required by the Antarctic Treaty; for more information, contact antarctica@state.gov

ECONOMY

Economy—overview: Scientific undertakings rather than commercial pursuits are the predominant human activity in Antarctica. Offshore fishing and tourism, both based abroad, account for Antarctica's limited economic activity.

Antarctic Fisheries, within the area covered by the Convention on Conservation of Antarctic Marine Living Resources currently target Patagonian toothfish, Antarctic toothfish, mackerel icefish and Antarctic krill. The Commission for the Conservation of Antarctic Marine Living Resources (CCAMLR) manages these fisheries using the ecosystem-based and precautionary approach. The Commission's objective is conservation of Antarctic marine living resources and it regulates the fisheries based on the level of information available, and maintaining existing ecological relationships. While Illegal, Unreported and Unregulated (IUU) fishing has declined in the Convention area since 1990, it remains a concern

A total of 51,707 tourists visited the Antarctic Treaty area in the 2017-2018 Antarctic summer, 17 percent greater than the 43,915 visitors in 2016-2017. These estimates were provided to the Antarctic Treaty by the International Association of Antarctica Tour Operators and do not include passengers on overflights. Nearly all of the tourists were passengers on commercial ships and several yachts that make trips during the summer.

ENERGY

Crude oil—production: 0 bbl/day (2018 est.)
country comparison to the world: 104

COMMUNICATIONS

Telecommunication systems: *general assessment:* local systems at some research stations (2019)
domestic: commercial cellular networks operating in a small number of locations (2019)
international: country code—none allocated; via satellite (including mobile Inmarsat and Iridium systems) to and from all research stations, ships, aircraft, and most field parties

Internet country code: .aq

Internet users: *total:* 4,400
percent of population: 100% (July 2016 est.)
country comparison to the world: 219

TRANSPORTATION

Airports: 17 (2020)
country comparison to the world: 141

Airports—with unpaved runways: *total:* 17 (2020)
over 3,047 m: 4

2,438 to 3,047 m: 2
1,524 to 2,437 m: 2
914 to 1,523 m: 5
under 914 m: 4

Heliports: 53 (2012)
note: all year-round and seasonal stations operated by National Antarctic Programs stations have some kind of helicopter landing facilities, prepared (helipads) or unprepared

Ports and terminals: most coastal stations have sparse and intermittent offshore anchorages; a few stations have basic wharf facilities

Transportation—note: US coastal stations include McMurdo (77 51 S, 166 40 E) and Palmer (64 43 S, 64 03 W); government use only; all ships at port are subject to inspection in accordance with Article 7, Antarctic Treaty; relevant legal instruments and authorization procedures adopted by the states parties to the Antarctic Treaty regulating the Antarctic Treaty area have to be complied with (see "Legal System"); The Hydrographic Commission on Antarctica (HCA), a commission of the International Hydrographic Organization (IHO), is responsible for hydrographic surveying and nautical charting matters in Antarctic Treaty area; it coordinates and facilitates provision of accurate and appropriate charts and other aids to navigation in support of safety of navigation in region; membership of HCA is open to any IHO Member State whose government has acceded to the Antarctic Treaty and which contributes resources or data to IHO Chart coverage of the area

MILITARY AND SECURITY

Military—note: the Antarctic Treaty prohibits any measures of a military nature, such as the establishment of military bases and fortifications, the carrying out of military maneuvers, or the testing of any type of weapon; it permits the use of military personnel or equipment for scientific research or for any other peaceful purposes

TRANSNATIONAL ISSUES

Disputes—international: the Antarctic Treaty freezes, and most states do not recognize, the land and maritime territorial claims made by Argentina, Australia, Chile, France, New Zealand, Norway, and the UK (some overlapping) for three-fourths of the continent; the US and Russia reserve the right to make claims

ANTIGUA AND BARBUDA

INTRODUCTION

Background: The Siboney were the first people to inhabit the islands of Antigua and Barbuda in 2400 B.C., but Arawak Indians populated the islands when COLUMBUS landed on his second voyage in 1493. Early Spanish and French settlements were succeeded by an English colony in 1667. Slavery, established to run the sugar plantations on Antigua, was abolished in 1834. The islands became an independent state within the British Commonwealth of Nations in 1981.

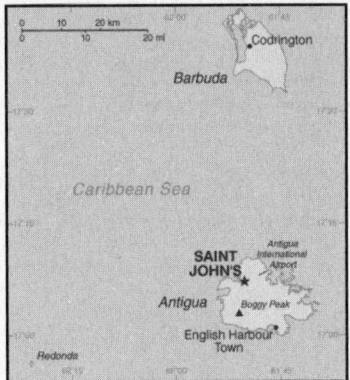

On 6 September 2017, Hurricane Irma passed over the island of Barbuda devastating the island and forcing the evacuation of the population to Antigua. Almost all the structures on Barbuda were destroyed and the vegetation stripped, but Antigua was spared the worst.

GEOGRAPHY

Location: Caribbean, islands between the Caribbean Sea and the North Atlantic Ocean, east-southeast of Puerto Rico

Geographic coordinates: 17 03 N, 61 48 W

Map references: Central America and the Caribbean

Area: *total:* 443 sq km (Antigua 280 sq km; Barbuda 161 sq km)
land: 442.6 sq km
water: 0 sq km
note: includes Redonda, 1.6 sq km
country comparison to the world: 201

Area—comparative: 2.5 times the size of Washington, DC

Land boundaries: 0 km

Coastline: 153 km

Maritime claims: *territorial sea:* 12 nm
exclusive economic zone: 200 nm
contiguous zone: 24 nm
continental shelf: 200 nm or to the edge of the continental margin

Climate: tropical maritime; little seasonal temperature variation

Terrain: mostly low-lying limestone and coral islands, with some higher volcanic areas

Elevation: *lowest point:* Caribbean Sea 0 m
highest point: Mount Obama 402 m

Natural resources: NEGL; pleasant climate fosters tourism

Land use: *agricultural land:* 20.5% (2016 est.)
arable land: 9.1% (2016 est.) / permanent crops: 2.3% (2016 est.) / permanent pasture: 9.1% (2016 est.)
forest: 22.3% (2016 est.)
other: 57.3% (2016 est.)

Irrigated land: 1.3 sq km (2012)

Population distribution: the island of Antigua is home to approximately 97% of the population; nearly the entire population of Barbuda lives in Codrington

Natural hazards: hurricanes and tropical storms (July to October); periodic droughts

Environment—current issues: water management—a major concern because of limited natural freshwater resources—is further hampered by the clearing of trees to increase crop production, causing rainfall to run off quickly

Environment—international agreements: party to: Biodiversity, Climate Change, Climate Change-Kyoto Protocol, Desertification, Endangered Species, Environmental Modification, Hazardous Wastes, Law of the Sea, Marine Dumping, Ozone Layer Protection, Ship Pollution, Wetlands, Whaling
signed, but not ratified: none of the selected agreements

Geography—note: Antigua has a deeply indented shoreline with many natural harbors and beaches; Barbuda has a large western harbor

PEOPLE AND SOCIETY

Population: 98,179 (July 2020 est.)
country comparison to the world: 197

Nationality: *noun:* Antiguan(s), Barbudan(s)
adjective: Antiguan, Barbudan

Ethnic groups: African descent 87.3%, mixed 4.7%, hispanic 2.7%, white 1.6%, other 2.7%, unspecified 0.9% (2011 est.)
note: data represent population by ethnic group

Languages: English (official), Antiguan creole

Religions: Protestant 68.3% (Anglican 17.6%, Seventh Day Adventist 12.4%, Pentecostal 12.2%, Moravian 8.3%, Methodist 5.6%, Wesleyan Holiness 4.5%, Church of God 4.1%, Baptist 3.6%), Roman Catholic 8.2%, other 12.2%, unspecified 5.5%, none 5.9% (2011 est.)

Age structure: *0-14 years:* 22.52% (male 11,243/female 10,871)
15-24 years: 16.15% (male 7,891/female 7,961)
25-54 years: 41.68% (male 18,757/female 22,167)
55-64 years: 10.74% (male 4,693/female 5,848)
65 years and over: 8.91% (male 3,736/female 5,012) (2020 est.)

Dependency ratios: *total dependency ratio:* 45.3
youth dependency ratio: 31.8
elderly dependency ratio: 13.6
potential support ratio: 7.4 (2020 est.)

Median age: *total:* 32.7 years
male: 30.7 years
female: 34.4 years (2020 est.)
country comparison to the world: 106

Population growth rate: 1.18% (2020 est.)
country comparison to the world: 92

Birth rate: 15.4 births/1,000 population (2020 est.)
country comparison to the world: 114

Death rate: 5.8 deaths/1,000 population (2020 est.)
country comparison to the world: 174

Net migration rate: 2.1 migrant(s)/1,000 population (2020 est.)
country comparison to the world: 47

Population distribution: the island of Antigua is home to approximately 97% of the population; nearly the entire population of Barbuda lives in Codrington

Urbanization: *urban population:* 24.4% of total population (2020)
rate of urbanization: 0.55% annual rate of change (2015-20 est.)

total population growth rate v. urban population growth rate, 2000-2030: Major urban areas— population: 21,000 SAINT JOHN'S (capital) (2018)

Sex ratio: *at birth:* 1.05 male(s)/female
0-14 years: 1.03 male(s)/female
15-24 years: 0.99 male(s)/female
25-54 years: 0.85 male(s)/female
55-64 years: 0.8 male(s)/female
65 years and over: 0.75 male(s)/female
total population: 0.89 male(s)/female (2020 est.)

Infant mortality rate: *total:* 11.1 deaths/1,000 live births
male: 12.7 deaths/1,000 live births
female: 9.3 deaths/1,000 live births (2020 est.)
country comparison to the world: 119

Life expectancy at birth: *total population:* 77.3 years
male: 75.1 years
female: 79.6 years (2020 est.)
country comparison to the world: 83

Total fertility rate: 1.97 children born/woman (2020 est.)
country comparison to the world: 114

Drinking water source:
improved:
total: 96.7% of population
unimproved:
total: 3.2% of population (2017 est.)

Current Health Expenditure: 4.5% (2017)

Physicians density: 2.96 physicians/1,000 population (2017)

Hospital bed density: 2.9 beds/1,000 population (2017)

Sanitation facility access:
improved:
total: 91.7% of population
unimproved:
total: 8.1% of population (2017 est.)

HIV/AIDS—adult prevalence rate: 1.1% (2018)
country comparison to the world: 41

HIV/AIDS—people living with HIV/AIDS: <1000 (2018)

HIV/AIDS—deaths: <100 (2018)

Obesity—adult prevalence rate: 18.9% (2016)
country comparison to the world: 113

Education expenditures: 2.5% of GDP (2009)
country comparison to the world: 158

Literacy: *definition:* age 15 and over has completed five or more years of schooling
total population: 99%
male: 98.4%
female: 99.4% (2015)

School life expectancy (primary to tertiary education): *total:* 15 years
male: 14 years
female: 16 years (2012)

GOVERNMENT

Country name: *conventional long form:* none
conventional short form: Antigua and Barbuda
etymology: "antiguo" is Spanish for "ancient" or "old"; the island was discovered by Christopher COLUMBUS in 1493 and, according to tradition, named by him after the church of Santa Maria la Antigua (Old Saint Mary's) in Seville; "barbuda" is Spanish for "bearded" and the adjective may refer to the alleged beards of the indigenous people or to the island's bearded fig trees

Government type: parliamentary democracy under a constitutional monarchy; a Commonwealth realm

Capital: *name:* Saint John's
geographic coordinates: 17 07 N, 61 51 W
time difference: UTC-4 (1 hour ahead of Washington, DC, during Standard Time)
etymology: named after Saint John the Apostle

Administrative divisions: 6 parishes and 2 dependencies*; Barbuda*, Redonda*, Saint George, Saint John, Saint Mary, Saint Paul, Saint Peter, Saint Philip

Independence: 1 November 1981 (from the UK)

National holiday: Independence Day, 1 November (1981)

Constitution: history: several previous; latest presented 31 July 1981, effective 31 October 1981 (The Antigua and Barbuda Constitution Order 1981)
amendments: proposed by either house of Parliament; passage of amendments to constitutional sections such as citizenship, fundamental rights and freedoms, the establishment, power, and authority of the executive and legislative branches, the Supreme Court Order, and the procedure for amending the constitution requires approval by at least two-thirds majority vote of the membership of both houses, approval by at least two-thirds majority in a referendum, and assent to by the governor general; passage of other amendments requires only two-thirds majority vote by both houses; amended 2009, 2011

Legal system: common law based on the English model

International law organization participation: has not submitted an ICJ jurisdiction declaration; accepts ICCt jurisdiction

Citizenship: *citizenship by birth:* yes
citizenship by descent only: yes
dual citizenship recognized: yes
residency requirement for naturalization: 7 years

Suffrage: 18 years of age; universal

Executive branch: *chief of state:* Queen ELIZABETH II (since 6 February 1952); represented by Governor General Rodney WILLIAMS (since 14 August 2014)
head of government: Prime Minister Gaston BROWNE (since 13 June 2014)
cabinet: Council of Ministers appointed by the governor general on the advice of the prime minister
elections/appointments: the monarchy is hereditary; governor general appointed by the monarch on the advice of the prime minister; following legislative elections, the leader of the majority party or majority coalition usually appointed prime minister by the governor general

Legislative branch: *description:* bicameral Parliament consists of: Senate (17 seats; members appointed by the governor general)
House of Representatives (18 seats; members directly elected in single-seat constituencies by simple majority vote to serve 5-year terms)
elections: Senate—last appointed on 26 March 2018 (next NA)
House of Representatives—last held on 21 March 2018 (next to be held in March 2023)
election results: Senate—composition—men 8, women 9, percent of women 52.9%
House of Representatives—percent of vote by party—ABLP 59.4%, UPP 37.2%, BPM 1.4%, other 1.9% ; seats by party—ABLP 15, UPP 1, BPM 1; composition—men 16, women 2, percent of women 11.1%; note—total Parliament percent of women 31.4%

Judicial branch: *highest courts:* the Eastern Caribbean Supreme Court (ECSC) is the superior court of the Organization of Eastern Caribbean States; the ECSC—headquartered on St. Lucia—consists of the Court of Appeal—headed by the chief justice and 4 judges—and the High Court with 18 judges; the Court of Appeal is itinerant, travelling to member states on a schedule to hear appeals from the High Court and subordinate courts; High Court judges reside in the member states, with 2 assigned to Antigua and Barbuda
judge selection and term of office: chief justice of Eastern Caribbean Supreme Court appointed by the Her Majesty, Queen ELIZABETH II; other justices and judges appointed by the Judicial and Legal Services Commission; Court of Appeal justices appointed for life with mandatory retirement at age 65; High Court judges appointed for life with mandatory retirement at age 62
subordinate courts: Industrial Court; Magistrates' Courts

Political parties and leaders: Antigua Caribbean Liberation Movement or ACLM
Antigua Labor Party or ABLP [Gaston BROWNE]
Antigua Barbuda True Labor Party or ABTLP [Sharlene SAMUEL]
Barbuda People's Movement or BPM [Trevor WALKER]
Barbuda People's Movement for Change [Arthur NIBBS]

Barbudans for a Better Barbuda [Ordrick SAMUEL]
Democratic National Alliance or DNA [Joanne MASSIAH]
Go Green for Life [Owen GEORGE]
Progressive Labor Movement or PLM
United National Democratic Party or UNDP
United Progressive Party or UPP [Harold LOVELL] (a coalition of ACLM, PLM, UNDP)

International organization participation: ACP, AOSIS, C, Caricom, CDB, CELAC, FAO, G-77, IBRD, ICAO, ICC (NGOs), ICCt, ICRM, IDA, IFAD, IFC, IFRCS, ILO, IMF, IMO, IMSO, Interpol, IOC, IOM, ISO (subscriber), ITU, ITUC (NGOs), MIGA, NAM (observer), OAS, OECS, OPANAL, OPCW, Petrocaribe, UN, UNCTAD, UNESCO, UPU, WFTU (NGOs), WHO, WIPO, WMO, WTO

Diplomatic representation in the US: *chief of mission:* Ambassador Sir Ronald SANDERS (since 17 September 2015)
chancery: 3234 Prospect Street NW, Washington, DC 20007
telephone: [1] (202) 362-5122
FAX: [1] (202) 362-5525
consulate(s) general: Miami, New York

Diplomatic representation from the US: the US does not have an embassy in Antigua and Barbuda; the US Ambassador to Barbados is accredited to Antigua and Barbuda

Flag description: red, with an inverted isosceles triangle based on the top edge of the flag; the triangle contains three horizontal bands of black (top), light blue, and white, with a yellow rising sun in the black band; the sun symbolizes the dawn of a new era, black represents the African heritage of most of the population, blue is for hope, and red is for the dynamism of the people; the "V" stands for victory; the successive yellow, blue, and white coloring is also meant to evoke the country's tourist attractions of sun, sea, and sand

National symbol(s): fallow deer; national colors: red, white, blue, black, yellow

National anthem: *name:* Fair Antigua, We Salute Thee
lyrics/music: Novelle Hamilton RICHARDS/ Walter Garnet Picart CHAMBERS
note: adopted 1967; as a Commonwealth country, in addition to the national anthem, "God Save the Queen" serves as the royal anthem (see United Kingdom)
0:00 / 1:00

ECONOMY

Economy—overview: Tourism continues to dominate Antigua and Barbuda's economy, accounting for nearly 60% of GDP and 40% of investment. The dual-island nation's agricultural production is focused on the domestic market and constrained by a limited water supply and a labor shortage stemming from the lure of higher wages in tourism and construction. Manufacturing comprises enclave-type assembly for export with major products being bedding, handicrafts, and electronic components.

Like other countries in the region, Antigua's economy was severely hit by effects of the global economic recession in 2009. The country suffered from the collapse of its largest private sector employer, a steep decline in tourism, a rise in debt, and a sharp economic contraction between 2009 and 2011. Antigua has not yet returned to its pre-crisis growth levels. Barbuda suffered significant damages after hurricanes Irma and Maria passed through the Caribbean in 2017.

Prospects for economic growth in the medium term will continue to depend on tourist arrivals from the US, Canada, and Europe and could be disrupted by potential damage from natural disasters. The new government, elected in 2014 and led by Prime Minister Gaston Browne, continues to face significant fiscal challenges. The government places some hope in a new Citizenship by Investment Program, to both reduce public debt levels and spur growth, and a resolution of a WTO dispute with the US.

GDP (purchasing power parity): $2.398 billion (2017 est.)
$2.334 billion (2016 est.)
$2.215 billion (2015 est.)
note: data are in 2017 dollars
country comparison to the world: 194

GDP (official exchange rate): $1.524 billion (2017 est.)

GDP—real growth rate: 2.8% (2017 est.)
5.3% (2016 est.)
4.1% (2015 est.)
country comparison to the world: 120

GDP—per capita (PPP): $26,400 (2017 est.)
$25,900 (2016 est.)
$24,900 (2015 est.)
note: data are in 2017 dollars
country comparison to the world: 78

Gross national saving: 17.3% of GDP (2017 est.)
24.5% of GDP (2016 est.)
30.7% of GDP (2015 est.)
country comparison to the world: 115

GDP—composition, by end use: household consumption: 53.5% (2017 est.)
government consumption: 15.2% (2017 est.)
investment in fixed capital: 23.9% (2017 est.)
investment in inventories: 0.1% (2017 est.)
exports of goods and services: 73.9% (2017 est.)
imports of goods and services: -66.5% (2017 est.)

GDP—composition, by sector of origin:
agriculture: 1.8% (2017 est.)
industry: 20.8% (2017 est.)
services: 77.3% (2017 est.)

Agriculture—products: cotton, fruits, vegetables, bananas, coconuts, cucumbers, mangoes, sugarcane; livestock

Industries: tourism, construction, light manufacturing (clothing, alcohol, household appliances)

Industrial production growth rate: 6.8% (2017 est.)
country comparison to the world: 32

Labor force: 30,000 (1991)

country comparison to the world: 204

Labor force—by occupation: agriculture: 7%
industry: 11%
services: 82% (1983 est.)

Unemployment rate: 11% (2014 est.)
country comparison to the world: 148

Population below poverty line: NA

Household income or consumption by percentage share: lowest 10%: NA
highest 10%: NA

Budget: revenues: 298.2 million (2017 est.)
expenditures: 334 million (2017 est.)

Taxes and other revenues: 19.6% (of GDP) (2017 est.)
country comparison to the world: 154

Budget surplus (+) or deficit (-): -2.4% (of GDP) (2017 est.)
country comparison to the world: 112

Public debt: 86.8% of GDP (2017 est.)
86.2% of GDP (2016 est.)
country comparison to the world: 30

Fiscal year: 1 April—31 March

Inflation rate (consumer prices): 2.5% (2017 est.)
-0.5% (2016 est.)
country comparison to the world: 122

Current account balance: -$112 million (2017 est.)
$2 million (2016 est.)
country comparison to the world: 88

Exports: $86.7 million (2017 est.)
$56.5 million (2016 est.)
country comparison to the world: 198

Exports—partners: Poland 62.2%, Cameroon 9.5%, US 5.1%, UK 4.5% (2017)

Exports—commodities: petroleum products, bedding, handicrafts, electronic components, transport equipment, food and live animals

Imports: $560 million (2017 est.)
$503.4 million (2016 est.)
country comparison to the world: 198

Imports—commodities: food and live animals, machinery and transport equipment, manufactures, chemicals, oil

Imports—partners: US 48%, Spain 4.2% (2017)

Debt—external: $441.2 million (31 December 2012)
$458 million (June 2010)
country comparison to the world: 180

Exchange rates: East Caribbean dollars (XCD) per US dollar -
2.7 (2017 est.)
2.7 (2016 est.)
2.7 (2015 est.)
2.7 (2014 est.)
2.7 (2013 est.)

ENERGY

Electricity access: population without electricity: 9,358 (2012)
electrification—total population: 97.4% (2016)
electrification—urban areas: 100% (2016)
electrification—rural areas: 96.5% (2016)

Electricity—production: 331 million kWh (2016 est.)
country comparison to the world: 179

Electricity—consumption: 307.8 million kWh (2016 est.)
country comparison to the world: 184

Electricity—exports: 0 kWh (2016 est.)
country comparison to the world: 99

Electricity—imports: 0 kWh (2016 est.)
country comparison to the world: 120

Electricity—installed generating capacity: 124,000 kW (2016 est.)
country comparison to the world: 177

Electricity—from fossil fuels: 97% of total installed capacity (2016 est.)
country comparison to the world: 31

Electricity—from nuclear fuels: 0% of total installed capacity (2017 est.)
country comparison to the world: 38

Electricity—from hydroelectric plants: 0% of total installed capacity (2017 est.)
country comparison to the world: 153

Electricity—from other renewable sources: 3% of total installed capacity (2017 est.)
country comparison to the world: 119

Crude oil—production: 0 bbl/day (2018 est.)
country comparison to the world: 105

Crude oil—exports: 0 bbl/day (2015 est.)
country comparison to the world: 85

Crude oil—imports: 0 bbl/day (2015 est.)
country comparison to the world: 89

Crude oil—proved reserves: 0 bbl (1 January 2018 est.)
country comparison to the world: 102

Refined petroleum products—production: 0 bbl/day (2015 est.)
country comparison to the world: 113

Refined petroleum products—consumption: 5,000 bbl/day (2016 est.)
country comparison to the world: 177

Refined petroleum products—exports: 91 bbl/day (2015 est.)
country comparison to the world: 119

Refined petroleum products—imports: 5,065 bbl/day (2015 est.)
country comparison to the world: 172

Natural gas—production: 0 cu m (2017 est.)
country comparison to the world: 99

Natural gas—consumption: 0 cu m (2017 est.)
country comparison to the world: 119

Natural gas—exports: 0 cu m (2017 est.)
country comparison to the world: 61

Natural gas—imports: 0 cu m (2017 est.)
country comparison to the world: 87

Natural gas—proved reserves: 0 cu m (1 January 2014 est.)
country comparison to the world: 105

Carbon dioxide emissions from consumption of energy: 740,300 Mt (2017 est.)
country comparison to the world: 175

COMMUNICATIONS

Telephones—fixed lines: *total subscriptions:* 24,403
subscriptions per 100 inhabitants: 25.15 (2019 est.)
country comparison to the world: 172

Telephones—mobile cellular: *total subscriptions:* 187,095
subscriptions per 100 inhabitants: 192.82 (2019 est.)
country comparison to the world: 184

Telecommunication systems: general assessment: good automatic telephone system with fiber-optic lines; telecom sector contributes heavily to GDP; numerous mobile network competitors licensed, but small and local; govt. to spend EC80 million in 2019 to improve state-owned telecom market competitiveness; legislative amendments extend jurisdiction of its telecom regulator in Barbuda to include mobile services (2020)
domestic: fixed-line teledensity roughly 25 per 100 persons; mobile-cellular teledensity is about 193 per 100 persons (2019)
international: country code—1-268; landing points for the ECFS and Southern Caribbean Fiber submarine cable systems with links to other islands in the eastern Caribbean; satellite earth stations—1 Intelsat (Atlantic Ocean) (2019)
note: the COVID-19 outbreak is negatively impacting telecommunications production and supply chains globally; consumer spending on telecom devices and services has also slowed due to the pandemic's effect on economies worldwide; overall progress towards improvements in all facets of the telecom industry—mobile, fixed-line, broadband, submarine cable and satellite—has moderated

Broadcast media: state-controlled Antigua and Barbuda Broadcasting Service (ABS) operates 1 TV station; multi-channel cable TV subscription services are available; ABS operates 1 radio station; roughly 15 radio stations, some broadcasting on multiple frequencies

Internet country code: .ag

Internet users: *total:* 72,870
percent of population: 76% (July 2018 est.)
country comparison to the world: 185

Broadband—fixed subscriptions: *total:* 9,261
subscriptions per 100 inhabitants: 10 (2017 est.)
country comparison to the world: 172

TRANSPORTATION

National air transport system: *number of registered air carriers:* 1 (2020)
inventory of registered aircraft operated by air carriers: 10
annual passenger traffic on registered air carriers: 580,174 (2018)
annual freight traffic on registered air carriers: 290,000 mt-km (2018)

Civil aircraft registration country code prefix: V2 (2016)

Airports: 3 (2020)
country comparison to the world: 193

Airports—with paved runways: *total:* 2 (2019)
2,438 to 3,047 m: 1
under 914 m: 1

Airports—with unpaved runways: *total:* 1 (2013)
under 914 m: 1 (2013)

Roadways: *total:* 1,170 km (2011)
paved: 386 km (2011)
unpaved: 784 km (2011)
country comparison to the world: 180

Merchant marine: *total:* 780
by type: bulk carrier 32, container ship 151, general cargo 534, oil tanker 2, other 61 (2019)
country comparison to the world: 31

Ports and terminals: *major seaport(s):* Saint John's

MILITARY AND SECURITY

Military and security forces: *Antigua and Barbuda Defense Force (ABDF):* Coast Guard and the Antigua and Barbuda Regiment (2020)

Military and security service personnel strengths: the Antigua and Barbuda Defense Force (ABDF)

has approximately 200 active personnel (2019 est.)

Military equipment inventories and acquisitions: the ABDF's equipment inventory is limited to small arms, light weapons, and soft-skin vehicles; the Coast Guard maintains ex-US patrol vessels and some smaller boats (2019 est.)

Military service age and obligation: 18 years of age for voluntary military service; no conscription; Governor-General has powers to call up men for national service and set the age at which they could be called up (2012)

TRANSNATIONAL ISSUES

Disputes—international: none

Trafficking in persons: *current situation:* Antigua and Barbuda is a destination and transit country for adults and children subjected to sex trafficking and forced labor; forced prostitution has been reported in bars, taverns, and brothels, while forced labor occurs in domestic service and the retail sector
tier rating: Tier 2 Watch List – Antigua and Barbuda does not fully comply with the minimum standards for the elimination of trafficking; however, it is making significant efforts to do so; the government made no discernible progress in convicting traffickers in 2014 but charged two individuals in separate cases; efforts to convict traffickers have been impeded by a 2014 ruling that found the 2010 anti-trafficking act was unconstitutional because jurisdiction rests with the Magistrate's Court rather than the High Court; no new prosecutions, convictions, or punishments were recorded in 2014; credible sources have raised concerns about trafficking-related complicity among some off-duty police officers, which could hinder investigations or victims willingness to report offenses; prevention efforts were sustained, but progress in protecting victims was uneven; seven victims were assisted, which was an increase over 2013 (2015)

Illicit drugs: considered a minor transshipment point for narcotics bound for the US and Europe; more significant as an offshore financial center

ARCTIC OCEAN

INTRODUCTION

Background: The Arctic Ocean is the smallest of the world's five oceans (after the Pacific Ocean, Atlantic Ocean, Indian Ocean, and the Southern Ocean). The Northwest Passage (US and Canada) and Northern Sea Route (Norway and Russia) are two important seasonal waterways. In recent years the polar ice pack has receded in the summer allowing for increased navigation and raising the possibility of future sovereignty and shipping disputes among the six countries bordering the Arctic Ocean (Canada, Denmark (Greenland), Iceland, Norway, Russia, US).

GEOGRAPHY

Location: body of water between Europe, Asia, and North America, mostly north of the Arctic Circle

Geographic coordinates: 90 00 N, 0 00 E

Map references: Arctic Region

Area: *total:* 14.056 million sq km
note: includes Baffin Bay, Barents Sea, Beaufort Sea, Chukchi Sea, East Siberian Sea, Greenland Sea, Hudson Bay, Hudson Strait, Kara Sea, Laptev Sea, Northwest Passage, and other tributary water bodies

Area—comparative: slightly less than 1.5 times the size of the US

Coastline: 45,389 km

Climate: polar climate characterized by persistent cold and relatively narrow annual temperature range; winters characterized by continuous darkness, cold and stable weather conditions, and clear skies; summers characterized by continuous daylight, damp and foggy weather, and weak cyclones with rain or snow

Terrain: central surface covered by a perennial drifting polar icepack that, on average, is about 3 m thick, although pressure ridges may be three

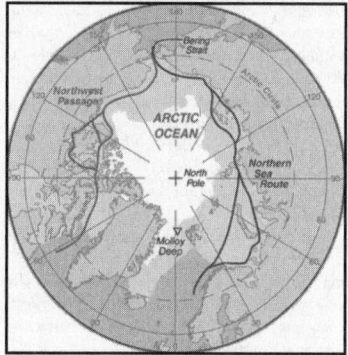

times that thickness; the icepack is surrounded by open seas during the summer, but more than doubles in size during the winter and extends to the encircling landmasses; the ocean floor is about 50% continental shelf (highest percentage of any ocean) with the remainder a central basin interrupted by three submarine ridges (Alpha Cordillera, Nansen Cordillera, and Lomonosov Ridge)

major surface currents: two major, slow-moving, wind-driven currents (drift streams) dominate: a clockwise drift pattern in the Beaufort Gyre in the western part of the Arctic Ocean and a nearly straight line Transpolar Drift Stream that moves eastward across the ocean from the New Siberian Islands (Russia) to the Fram Strait (between Greenland and Svalbard); sea ice that lies close to the center of the gyre can complete a 360 degree circle in about 2 years, while ice on the gyre periphery will complete the same circle in about 7-8 years; sea ice in the Transpolar Drift crosses the ocean in about 3 years

Elevation: *mean depth:* -1,205 m
lowest point: Molloy Deep -5,607 m
highest point: sea level

Natural resources: sand and gravel aggregates, placer deposits, polymetallic nodules, oil and gas fields, fish, marine mammals (seals and whales)

Natural hazards: ice islands occasionally break away from northern Ellesmere Island; icebergs calved from glaciers in western Greenland and extreme northeastern Canada; permafrost in islands; virtually ice locked from October to June; ships subject to superstructure icing from October to May

Environment—current issues: climate change; changes in biodiversity; use of toxic chemicals; endangered marine species include walruses and whales; fragile ecosystem slow to change and slow to recover from disruptions or damage; thinning polar icepack

Geography—note: major chokepoint is the southern Chukchi Sea (northern access to the Pacific Ocean via the Bering Strait); strategic location between North America and Russia; shortest marine link between the extremes of eastern and western Russia; floating research stations operated by the US and Russia; maximum snow cover in March or April about 20 to 50 centimeters over the frozen ocean; snow cover lasts about 10 months

PEOPLE AND SOCIETY

GOVERNMENT

Country name: *etymology:* the name Arctic comes from the Greek word "arktikos" meaning "near the bear" or "northern," and that word derives from "arktos," meaning "bear"; the name refers either to the constellation Ursa Major, the "Great Bear," which is prominent in the northern celestial sphere, or to the constellation Ursa Minor, the "Little Bear," which contains Polaris, the North (Pole) Star

ECONOMY

Economy—overview: Economic activity is limited to the exploitation of natural resources, including petroleum, natural gas, fish, and seals.

Marine fisheries: the Arctic fishery region (Region 18) is the smallest in the world with a catch of only 418 mt in 2017, although the Food and Agriculture Organization assesses that some Arctic catches are reported in adjacent regions; Russia and Canada were historically the major producers; in 2017, the five littoral states including Canada, Denmark (Greenland), Norway, Russia, and the US agreed to a 16 year ban on fishing in the Central Arctic Ocean to allow for time to study the ecological system of these waters

FAO map of world fishing regions, used with permission.:

TRANSPORTATION

Ports and terminals: *major seaport(s):* Churchill (Canada), Murmansk (Russia), Prudhoe Bay (US)

Transportation—note: sparse network of air, ocean, river, and land routes; the Northwest Passage (North America) and Northern Sea Route (Eurasia) are important seasonal waterways

TRANSNATIONAL ISSUES

Disputes—international: Canada and the US dispute how to divide the Beaufort Sea and the status of the Northwest Passage but continue to work cooperatively to survey the Arctic continental shelf; Denmark (Greenland) and Norway have made submissions to the Commission on the Limits of the Continental shelf (CLCS) and Russia is collecting additional data to augment its 2001 CLCS submission; record summer melting of sea ice in the Arctic has renewed interest in maritime shipping lanes and sea floor exploration; Norway and Russia signed a comprehensive maritime boundary agreement in 2010

ARGENTINA

INTRODUCTION

Background: In 1816, the United Provinces of the Rio Plata declared their independence from Spain. After Bolivia, Paraguay, and Uruguay went their separate ways, the area that remained became Argentina. The country's population and culture were heavily shaped by immigrants from throughout Europe, with Italy and Spain providing the largest percentage of newcomers from 1860 to 1930. Up until about the mid-20th century, much of Argentina's history was dominated by periods of internal political unrest and conflict between civilian and military factions.

After World War II, an era of Peronist populism and direct and indirect military interference

in subsequent governments was followed by a military junta that took power in 1976. Democracy returned in 1983 after a failed bid to seize the Falkland Islands (Islas Malvinas) by force, and has persisted despite numerous challenges, the most formidable of which was a severe economic crisis in 2001-02 that led to violent public protests and the successive resignations of several presidents. The years 2003-15 saw Peronist rule by Nestor and Cristina FERNANDEZ de KIRCHNER, whose policies isolated Argentina and caused economic stagnation. With the election of Mauricio MACRI in November 2015, Argentina began a period of reform and international reintegration.

GEOGRAPHY

Location: Southern South America, bordering the South Atlantic Ocean, between Chile and Uruguay

Geographic coordinates: 34 00 S, 64 00 W

Map references: South America

Area: *total:* 2,780,400 sq km
land: 2,736,690 sq km
water: 43,710 sq km
country comparison to the world: 9

Area—comparative: slightly less than three-tenths the size of the US

Land boundaries: *total:* 11,968 km

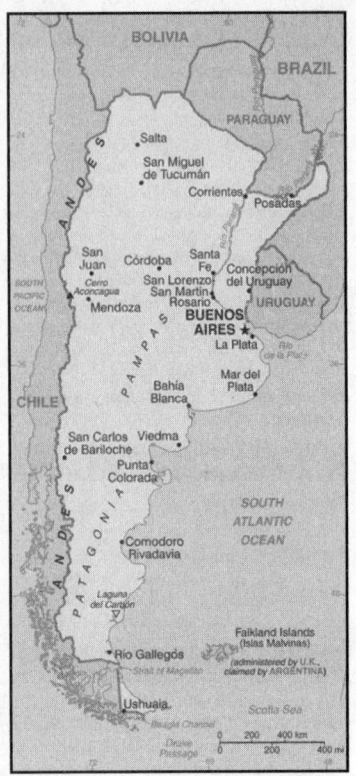

border countries (5): Bolivia 942 km, Brazil 1263 km, Chile 6691 km, Paraguay 2531 km, Uruguay 541 km

Coastline: 4,989 km

Maritime claims: *territorial sea:* 12 nm
exclusive economic zone: 200 nm
contiguous zone: 24 nm
continental shelf: 200 nm or to the edge of the continental margin

Climate: mostly temperate; arid in southeast; subantarctic in southwest

Terrain: rich plains of the Pampas in northern half, flat to rolling plateau of Patagonia in south, rugged Andes along western border

Elevation: *mean elevation:* 595 m
lowest point: Laguna del Carbon (located between Puerto San Julian and Comandante Luis Piedra Buena in the province of Santa Cruz) -105 m
highest point: Cerro Aconcagua (located in the northwestern corner of the province of Mendoza; highest point in South America) 6,962 m

Natural resources: fertile plains of the pampas, lead, zinc, tin, copper, iron ore, manganese, petroleum, uranium, arable land

Land use: *agricultural land:* 53.9% (2016 est.)
arable land: 13.9% (2016 est.) / permanent crops: 0.4% (2016 est.) / permanent pasture: 39.6% (2016 est.)
forest: 10.7% (2016 est.)

other: 35.4% (2016 est.)
Irrigated land: 23,600 sq km (2012)

Population distribution: one-third of the population lives in Buenos Aires; pockets of agglomeration occur throughout the northern and central parts of the country; Patagonia to the south remains sparsely populated

Natural hazards: San Miguel de Tucuman and Mendoza areas in the Andes subject to earthquakes; pamperos are violent windstorms that can strike the pampas and northeast; heavy flooding in some areas
volcanism: volcanic activity in the Andes Mountains along the Chilean border; Copahue (2,997 m) last erupted in 2000; other historically active volcanoes include Llullaillaco, Maipo, Planchon-Peteroa, San Jose, Tromen, Tupungatito, and Viedma

Environment—current issues: environmental problems (urban and rural) typical of an industrializing economy such as deforestation, soil degradation (erosion, salinization), desertification, air pollution, and water pollution
note: Argentina is a world leader in setting voluntary greenhouse gas targets

Environment—international agreements: *party to:* Antarctic-Environmental Protocol, Antarctic-Marine Living Resources, Antarctic Seals, Antarctic Treaty, Biodiversity, Climate Change, Climate Change-Kyoto Protocol, Desertification, Endangered Species, Environmental Modification, Hazardous Wastes, Law of the Sea, Marine Dumping, Ozone Layer Protection, Ship Pollution, Wetlands, Whaling
signed, but not ratified: Marine Life Conservation

Geography—note: second-largest country in South America (after Brazil); strategic location relative to sea lanes between the South Atlantic and the South Pacific Oceans (Strait of Magellan, Beagle Channel, Drake Passage); diverse geophysical landscapes range from tropical climates in the north to tundra in the far south; Cerro Aconcagua is the Western Hemisphere's tallest mountain, while Laguna del Carbon is the lowest point in the Western Hemisphere; shares Iguazu Falls, the world's largest waterfalls system, with Brazil

PEOPLE AND SOCIETY

Population: 45,479,118 (July 2020 est.)
country comparison to the world: 32

Nationality: *noun:* Argentine(s)
adjective: Argentine

Ethnic groups: European (mostly Spanish and Italian descent) and mestizo (mixed European and Amerindian ancestry) 97.2%, Amerindian 2.4%, African 0.4% (2010 est.)

Languages: Spanish (official), Italian, English, German, French, indigenous (Mapudungun, Quechua)

Religions: nominally Roman Catholic 92% (less than 20% practicing), Protestant 2%, Jewish 2%, other 4%

Demographic profile: Argentina's population continues to grow but at a slower rate because of its steadily declining birth rate. Argentina's fertility decline began earlier than in the rest of Latin America, occurring most rapidly between the early 20th century and the 1950s, and then becoming more gradual. Life expectancy has been improving, most notably among the young and the poor. While the population under age 15 is shrinking, the youth cohort—ages 15-24—is the largest in Argentina's history and will continue to bolster the working-age population. If this large working-age population is well-educated and gainfully employed, Argentina is likely to experience an economic boost and possibly higher per capita savings and investment. Although literacy and primary school enrollment are nearly universal, grade repetition is problematic and secondary school completion is low. Both of these issues vary widely by region and socioeconomic group.

Argentina has been primarily a country of immigration for most of its history, welcoming European immigrants (often providing needed low-skilled labor) after its independence in the 19th century and attracting especially large numbers from Spain and Italy. More than 7 million European immigrants are estimated to have arrived in Argentina between 1880 and 1930, when it adopted a more restrictive immigration policy. European immigration also began to wane in the 1930s because of the global depression. The inflow rebounded temporarily following WWII and resumed its decline in the 1950s when Argentina's military dictators tightened immigration rules and European economies rebounded. Regional migration increased, however, supplying low-skilled workers escaping economic and political instability in their home countries. As of 2015, immigrants made up almost 5% of Argentina's population, the largest share in South America. Migration from neighboring countries accounted for approximately 80% of Argentina's immigrant population in 2015.

The first waves of highly skilled Argentine emigrant workers headed mainly to the United States and Spain in the 1960s and 1970s, driven by economic decline and repressive military dictatorships. The 2008 European economic crisis drove the return migration of some Argentinean and other Latin American nationals, as well as the immigration of Europeans to South America, where Argentina was a key recipient. In 2015, Argentina received the highest number of legal migrants in Latin America and the Caribbean. The majority of its migrant inflow came from Paraguay and Bolivia.

Age structure: *0-14 years:* 24.02% (male 5,629,188/female 5,294,723)
15-24 years: 15.19% (male 3,539,021/female 3,367,321)
25-54 years: 39.6% (male 9,005,750/female 9,002,931)
55-64 years: 9.07% (male 2,000,536/female 2,122,699)
65 years and over: 12.13% (male 2,331,679/female 3,185,262) (2020 est.)

Dependency ratios: *total dependency ratio:* 56.5
youth dependency ratio: 38.1
elderly dependency ratio: 17.7
potential support ratio: 5.6 (2020 est.)

Median age: *total:* 32.4 years
male: 31.1 years
female: 33.6 years (2020 est.)
country comparison to the world: 110

Population growth rate: 0.86% (2020 est.)
country comparison to the world: 121

Birth rate: 16 births/1,000 population (2020 est.)
country comparison to the world: 110

Death rate: 7.4 deaths/1,000 population (2020 est.)
country comparison to the world: 107

Net migration rate: -0.1 migrant(s)/1,000 population (2020 est.)
country comparison to the world: 100

Population distribution: one-third of the population lives in Buenos Aires; pockets of agglomeration occur throughout the northern and central parts of the country; Patagonia to the south remains sparsely populated

Urbanization: *urban population:* 92.1% of total population (2020)
rate of urbanization: 1.07% annual rate of change (2015-20 est.)

total population growth rate v. urban population growth rate, 2000-2030: Major urban areas—population: 15.154 million BUENOS AIRES (capital), 1.573 million Cordoba, 1.532 million Rosario, 1.173 million Mendoza, 986,000 San Miguel de Tucuman, 884,000 La Plata (2020)

Sex ratio: *at birth:* 1.07 male(s)/female
0-14 years: 1.06 male(s)/female
15-24 years: 1.05 male(s)/female
25-54 years: 1 male(s)/female
55-64 years: 0.94 male(s)/female
65 years and over: 0.73 male(s)/female
total population: 0.98 male(s)/female (2020 est.)

Maternal mortality rate: 39 deaths/100,000 live births (2017 est.)
country comparison to the world: 101

Infant mortality rate: *total:* 9 deaths/1,000 live births
male: 9.9 deaths/1,000 live births
female: 8.1 deaths/1,000 live births (2020 est.)
country comparison to the world: 139

Life expectancy at birth: *total population:* 77.8 years
male: 74.7 years
female: 81.1 years (2020 est.)
country comparison to the world: 74

Total fertility rate: 2.21 children born/woman (2020 est.)
country comparison to the world: 92

Contraceptive prevalence rate: 81.3% (2013)

Drinking water source:
improved:
urban: 99% of population
rural: 100% of population
total: 99.1% of population
unimproved:

urban: 1% of population
rural: 0% of population
total: 0.9% of population (2015 est.)

Current Health Expenditure: 9.1% (2017)

Physicians density: 3.99 physicians/1,000 population (2017)

Hospital bed density: 5 beds/1,000 population (2017)

Sanitation facility access:
improved:
urban: 98.3% of population (2017 est.)
unimproved:
urban: 1.7% of population (2017 est.)

HIV/AIDS—adult prevalence rate: 0.4% (2019 est.)
country comparison to the world: 73

HIV/AIDS—people living with HIV/AIDS: 140,000 (2019 est.)
country comparison to the world: 37

HIV/AIDS—deaths: 1,400 (2019 est.)
country comparison to the world: 51

Major infectious diseases: *note:* widespread ongoing transmission of a respiratory illness caused by the novel coronavirus (COVID-19) is occurring throughout Argentina; as of 10 November 2020, Argentina has reported a total of 1,228,814 cases of COVID-19 or 27,189 cumulative cases of COVID-19 per 1 million population with 733 cumulative deaths per 1 million population

Obesity—adult prevalence rate: 28.3% (2016)
country comparison to the world: 30

Children under the age of 5 years underweight: 1.7% (2018/19)
country comparison to the world: 117

Education expenditures: 5.5% of GDP (2017)
country comparison to the world: 37

Literacy: *definition:* age 15 and over can read and write
total population: 99%
male: 98.9%
female: 99.1% (2018)

School life expectancy (primary to tertiary education): *total:* 18 years
male: 16 years
female: 19 years (2017)

Unemployment, youth ages 15-24: *total:* 23.7%
male: 20.8%
female: 27.8% (2018 est.)
country comparison to the world: 54

GOVERNMENT

Country name: *conventional long form:* Argentine Republic
conventional short form: Argentina
local long form: Republica Argentina
local short form: Argentina
etymology: originally the area was referred to as Tierra Argentina, i.e., "Land beside the Silvery River" or "silvery land," which referred to the massive estuary in the east of the country, the Rio de la Plata (River of Silver); over time the name shortened to simply Argentina or "silvery"

Government type: presidential republic

Capital: *name:* Buenos Aires

geographic coordinates: 34 36 S, 58 22 W
time difference: UTC-3 (2 hours ahead of Washington, DC, during Standard Time)
etymology: the name translates as "fair winds" in Spanish and derives from the original designation of the settlement that would become the present-day city, "Santa Maria del Buen Aire" (Saint Mary of the Fair Winds)

Administrative divisions: 23 provinces (provincias, singular—provincia) and 1 autonomous city*; Buenos Aires, Catamarca, Chaco, Chubut, Ciudad Autonoma de Buenos Aires*, Cordoba, Corrientes, Entre Rios, Formosa, Jujuy, La Pampa, La Rioja, Mendoza, Misiones, Neuquen, Rio Negro, Salta, San Juan, San Luis, Santa Cruz, Santa Fe, Santiago del Estero, Tierra del Fuego—Antartida e Islas del Atlantico Sur (Tierra del Fuego—Antarctica and the South Atlantic Islands), Tucuman
note: the US does not recognize any claims to Antarctica

Independence: 9 July 1816 (from Spain)

National holiday: Revolution Day (May Revolution Day), 25 May (1810)

Constitution: *history:* several previous; latest effective 11 May 1853
amendments: a declaration of proposed amendments requires two-thirds majority vote by both houses of the National Congress followed by approval by an ad hoc, multi-member constitutional convention; amended many times, last significant amendment in 1994

Legal system: civil law system based on West European legal systems; note—in mid-2015, Argentina adopted a new civil code, replacing the old one in force since 1871

International law organization participation: has not submitted an ICJ jurisdiction declaration; accepts ICCt jurisdiction

Citizenship: *citizenship by birth:* yes
citizenship by descent only: yes
dual citizenship recognized: yes
residency requirement for naturalization: 2 years

Suffrage: 18-70 years of age; universal and compulsory; 16-17 years of age—optional for national elections

Executive branch: *chief of state:* President Alberto Angel FERNANDEZ (since 10 December 2019); Vice President Cristina FERNANDEZ DE KIRCHNER (since 10 December 2019); note—the president is both chief of state and head of government
head of government: President Alberto Angel FERNANDEZ (since 10 December 2019); Vice President Cristina FERNANDEZ DE KIRCHNER (since 10 December 2019)
cabinet: Cabinet appointed by the president
elections/appointments: president and vice president directly elected on the same ballot by qualified majority vote (to win, a candidate must

receive at least 45% of votes or 40% of votes and a 10-point lead over the second place candidate; if neither occurs, a second round is held); the president serves a 4-year term (eligible for a second consecutive term); election last held on 27 October 2019 (next to be held in October 2023)

election results: Alberto Angel FERNANDEZ elected president; percent of vote—Alberto Angel FERNANDEZ (TODOS) 48.1%, Mauricio MACRI (PRO) 40.4%, Roberto LAVAGNA (independent) 6.2%, other 5.3%

Legislative branch: *description:* bicameral National Congress or Congreso Nacional consists of:

Senate (72 seats; members directly elected in multi-seat constituencies by simple majority vote to serve 6-year terms with one-third of the membership elected every 2 years)

Chamber of Deputies (257 seats; members directly elected in multi-seat constituencies by proportional representation vote; members serve 4-year terms with one-half of the membership renewed every 2 years)

elections: Senate—last held on 27 October 2019 (next to be held in October 2021)

Chamber of Deputies—last held on 27 October 2019 (next to be held in October 2021)

election results: Senate—percent of vote by bloc or party—NA; seats by bloc or party—TODOS 13, Cambiemos 8, FCS 2, JSRN 1;

Chamber of Deputies—percent of vote by bloc or party—NA; seats by bloc or party—TODOS 64, Cambiemos 56, CF 3, FCS 3, JSRN 1, other 3

Judicial branch: *highest courts:* Supreme Court or Corte Suprema (consists of the court president, vice president, and 5 justices)

judge selection and term of office: justices nominated by the president and approved by the Senate; justices can serve until mandatory retirement at age 75; extensions beyond 75 require renomination by the president and approval by the Senate

subordinate courts: federal level appellate, district, and territorial courts; provincial level supreme, appellate, and first instance courts

Political parties and leaders: Argentina Federal [coalition led by Pablo KOSINER]

Cambiemos [Mauricio MACRI] (coalition of CC-ARI, PRO, and UCR)

Citizen's Unity or UC [Cristina FERNANDEZ DE KIRCHNER]

Civic Coalition ARI or CC-ARI [Elisa CARRIO, Maximiliano FERRARO]

Civic Front for Santiago or FCS [Gerardo ZAMORA]

Everyone's Front (Frente de Todos) or TODOS [Alberto Angel FERNANDEZ]

Federal Consensus or CF [Roberto LAVAGNA, Juan Manuel URTUBEY]

Front for the Renewal of Concord or FRC

Front for Victory or FpV [coalition led by Cristina FERNANDEZ DE KIRCHNER and Agustin ROSSI] Generation for a National Encounter or GEN [Monica PERALTA]

Justicialist Party or PJ [Miguel Angel PICHETTO]

Radical Civic Union or UCR [Alfredo CORNEJO]

Renewal Front (Frente Renovador) or FR [Sergio MASSA]

Republican Proposal or PRO [Mauricio MACRI, Humberto SCHIAVONI]

Socialist Party or PS [Antonio BONFATTI]

Socialist Workers' Party or PTS [Jose MONTES]

Together We Are Rio Negro or JSRN [Alberto Edgardo WERETILNECK]

We Do For Cordoba (Hacemos Por Cordoba) or HC [Juan SCHIARETTI]

Workers' Party or PO [Jorge ALTAMIRA]

Worker's Socialist Movement or MST [Alejandro BODDART; Vilma RIPOLL] numerous provincial parties

International organization participation: AfDB (nonregional member), Australia Group, BCIE, BIS, CAN (associate), CD, CELAC, FAO, FATF, G-15, G-20, G-24, G- 77, IADB, IAEA, IBRD, ICAO, ICC (national committees), ICCt, ICRM, IDA, IFAD, IFC, IFRCS, IHO, ILO, IMF, IMO, IMSO, Interpol, IOC, IOM, IPU, ISO, ITSO, ITU, ITUC (NGOs), LAES, LAIA, Mercosur, MIGA, MINURSO, MINUSTAH, NAM (observer), NSG, OAS, OPANAL, OPCW, Paris Club (associate), PCA, SICA (observer), UN, UNASUR, UNCTAD, UNESCO, UNFICYP, UNHCR, UNIDO, Union Latina (observer), UNTSO, UNWTO, UPU, WCO, WFTU (NGOs), WHO, WIPO, WMO, WTO, ZC

Diplomatic representation in the US: *chief of mission:* Ambassador Jorge Martin Arturo ARGUELLO (since 6 February 2020)

chancery: 1600 New Hampshire Avenue NW, Washington, DC 20009

telephone: [1] (202) 238-6400

FAX: [1] (202) 332-3171

consulate(s) general: Atlanta, Chicago, Houston, Los Angeles, Miami, New York, Washington, DC

Diplomatic representation from the US: *chief of mission:* Ambassador Edward Charles PRADO (since 16 May 2018)

telephone: [54] (11) 5777-4533

embassy: Avenida Colombia 4300, C1425GMN Buenos Aires

mailing address: international mail: use embassy street address; APO address: US Embassy Buenos Aires, Unit 4334, APO AA 34034

FAX: [54] (11) 5777-4240

Flag description: three equal horizontal bands of sky blue (top), white, and sky blue; centered in the white band is a radiant yellow sun with a human face (delineated in brown) known as the Sun of May; the colors represent the clear skies and snow of the Andes; the sun symbol commemorates the appearance of the sun through cloudy skies on 25 May 1810 during the first mass demonstration in favor of independence; the sun features are those of Inti, the Inca god of the sun

National symbol(s): Sun of May (a sun-with-face symbol); national colors: sky blue, white

National anthem: *name:* "Himno Nacional Argentino" (Argentine National Anthem)

lyrics/music: Vicente LOPEZ y PLANES/Jose Blas PARERA

note: adopted 1813; Vicente LOPEZ was inspired to write the anthem after watching a play about the 1810 May Revolution against Spain

0:00 / 1:05

ECONOMY

Economy—overview: Argentina benefits from rich natural resources, a highly literate population, an export-oriented agricultural sector, and a diversified industrial base. Although one of the world's wealthiest countries 100 years ago, Argentina suffered during most of the 20th century from recurring economic crises, persistent fiscal and current account deficits, high inflation, mounting external debt, and capital flight.

Cristina FERNANDEZ DE KIRCHNER succeeded her husband as president in late 2007, and in 2008 the rapid economic growth of previous years slowed sharply as government policies held back exports and the world economy fell into recession. In 2010 the economy rebounded strongly, but slowed in late 2011 even as the government continued to rely on expansionary fiscal and monetary policies, which kept inflation in the double digits.

In order to deal with these problems, the government expanded state intervention in the economy: it nationalized the oil company YPF from Spain's Repsol, expanded measures to restrict imports, and further tightened currency controls in an effort to bolster foreign reserves and stem capital flight. Between 2011 and 2013, Central Bank foreign reserves dropped $21.3 billion from a high of $52.7 billion. In July 2014, Argentina and China agreed on an $11 billion currency swap; the Argentine Central Bank has received the equivalent of $3.2 billion in Chinese yuan, which it counts as international reserves.

With the election of President Mauricio MACRI in November 2015, Argentina began a historic political and economic transformation, as his administration took steps to liberalize the Argentine economy, lifting capital controls, floating the peso, removing export controls on some commodities, cutting some energy subsidies, and reforming the country's official statistics. Argentina negotiated debt payments with holdout bond creditors, continued working with the IMF to shore up its finances, and returned to international capital markets in April 2016.

In 2017, Argentina's economy emerged from recession with GDP growth of nearly 3.0%. The government passed important pension, tax, and fiscal reforms. And after years of international isolation, Argentina took on several international leadership roles, including hosting the World Economic Forum on Latin America and the World Trade Organization Ministerial Conference, and is set to assume the presidency of the G-20 in 2018.

GDP (purchasing power parity): $922.1 billion (2017 est.)

$896.5 billion (2016 est.)

$913.2 billion (2015 est.)

note: data are in 2017 dollars
country comparison to the world: 28

GDP (official exchange rate): $637.6 billion (2017 est.)

GDP—real growth rate: 2.9% (2017 est.)
-1.8% (2016 est.)
2.7% (2015 est.)
country comparison to the world: 117

GDP—per capita (PPP): $20,900 (2017 est.)
$20,600 (2016 est.)
$21,200 (2015 est.)
note: data are in 2017 dollars
country comparison to the world: 88

Gross national saving: 17.6% of GDP (2017 est.)
16.8% of GDP (2016 est.)
15.8% of GDP (2015 est.)
country comparison to the world: 114

GDP—composition, by end use: *household consumption:* 65.9% (2017 est.)
government consumption: 18.2% (2017 est.)
investment in fixed capital: 14.8% (2017 est.)
investment in inventories: 3.7% (2017 est.)
exports of goods and services: 11.2% (2017 est.)
imports of goods and services: -13.8% (2017 est.)

GDP—composition, by sector of origin:
agriculture: 10.8% (2017 est.)
industry: 28.1% (2017 est.)
services: 61.1% (2017 est.)

Agriculture—products: sunflower seeds, lemons, soybeans, grapes, corn, tobacco, peanuts, tea, wheat; livestock

Industries: food processing, motor vehicles, consumer durables, textiles, chemicals and petrochemicals, printing, metallurgy, steel

Industrial production growth rate: 2.7% (2017 est.)
note: based on private sector estimates
country comparison to the world: 111

Labor force: 18 million (2017 est.)
note: urban areas only
country comparison to the world: 33

Labor force—by occupation: *agriculture:* 5.3%
industry: 28.6%
services: 66.1% (2017 est.)

Unemployment rate: 8.4% (2017 est.)
8.5% (2016 est.)
country comparison to the world: 120

Population below poverty line: 25.7% (2017 est.)
note: data are based on private estimates

Household income or consumption by percentage share: *lowest 10%:* 1.8%
highest 10%: 31% (2017 est.)

Budget: *revenues:* 120.6 billion (2017 est.)
expenditures: 158.6 billion (2017 est.)

Taxes and other revenues: 18.9% (of GDP) (2017 est.)
country comparison to the world: 157

Budget surplus (+) or deficit (-): -6% (of GDP) (2017 est.)
country comparison to the world: 182

Public debt: 57.6% of GDP (2017 est.)
55% of GDP (2016 est.)
country comparison to the world: 77

Fiscal year: calendar year

Inflation rate (consumer prices): 25.7% (2017 est.)
26.5% (2016 est.)
note: data are derived from private estimates
country comparison to the world: 219

Current account balance: -$31.32 billion (2017 est.)
-$14.69 billion (2016 est.)
country comparison to the world: 200

Exports: $58.45 billion (2017 est.)
$57.78 billion (2016 est.)
country comparison to the world: 49

Exports—partners: Brazil 16.1%, US 7.9%, China 7.5%, Chile 4.4% (2017)

Exports—commodities: soybeans and derivatives, petroleum and gas, vehicles, corn, wheat

Imports: $63.97 billion (2017 est.)
$53.5 billion (2016 est.)
country comparison to the world: 48

Imports—commodities: machinery, motor vehicles, petroleum and natural gas, organic chemicals, plastics

Imports—partners: Brazil 26.9%, China 18.5%, US 11.3%, Germany 4.9% (2017)

Reserves of foreign exchange and gold: $55.33 billion (31 December 2017 est.)
$38.43 billion (31 December 2016 est.)
country comparison to the world: 38

Debt—external: $214.9 billion (31 December 2017 est.)
$190.2 billion (31 December 2016 est.)
country comparison to the world: 34

Exchange rates: Argentine pesos (ARS) per US dollar—
16.92 (2017 est.)
14.76 (2016 est.)
14.76 (2015 est.)
9.23 (2014 est.)
8.08 (2013 est.)

ENERGY

Electricity access: *electrification—total population:* 100% (2016)

Electricity—production: 131.9 billion kWh (2016 est.)
country comparison to the world: 30

Electricity—consumption: 121 billion kWh (2016 est.)
country comparison to the world: 30

Electricity—exports: 55 million kWh (2015 est.)
country comparison to the world: 85

Electricity—imports: 9.851 billion kWh (2016 est.)
country comparison to the world: 26

Electricity—installed generating capacity: 38.35 million kW (2016 est.)
country comparison to the world: 27

Electricity—from fossil fuels: 69% of total installed capacity (2016 est.)
country comparison to the world: 112

Electricity—from nuclear fuels: 4% of total installed capacity (2017 est.)
country comparison to the world: 23

Electricity—from hydroelectric plants: 24% of total installed capacity (2017 est.)
country comparison to the world: 79

Electricity—from other renewable sources: 3% of total installed capacity (2017 est.)
country comparison to the world: 120

Crude oil—production: 489,000 bbl/day (2018 est.)
country comparison to the world: 29

Crude oil—exports: 36,630 bbl/day (2015 est.)
country comparison to the world: 43

Crude oil—imports: 16,740 bbl/day (2015 est.)
country comparison to the world: 68

Crude oil—proved reserves: 2.162 billion bbl (1 January 2018 est.)
country comparison to the world: 32

Refined petroleum products—production: 669,800 bbl/day (2015 est.)
country comparison to the world: 26

Refined petroleum products—consumption: 806,000 bbl/day (2016 est.)
country comparison to the world: 27

Refined petroleum products—exports: 58,360 bbl/day (2015 est.)
country comparison to the world: 51

Refined petroleum products—imports: 121,400 bbl/day (2015 est.)
country comparison to the world: 49

Natural gas—production: 40.92 billion cu m (2017 est.)
country comparison to the world: 20

Natural gas—consumption: 49.04 billion cu m (2017 est.)
country comparison to the world: 17

Natural gas—exports: 76.45 million cu m (2017 est.)
country comparison to the world: 49

Natural gas—imports: 9.826 billion cu m (2017 est.)
country comparison to the world: 27

Natural gas—proved reserves: 336.6 billion cu m (1 January 2018 est.)
country comparison to the world: 35

Carbon dioxide emissions from consumption of energy: 203.7 million Mt (2017 est.)
country comparison to the world: 32

COMMUNICATIONS

Telephones—fixed lines: *total subscriptions:* 7,791,464
subscriptions per 100 inhabitants: 17.28 (2019 est.)
country comparison to the world: 21

Telephones—mobile cellular: *total subscriptions:* 59,008,618
subscriptions per 100 inhabitants: 130.87 (2019 est.)
country comparison to the world: 26

41

Telecommunication systems: *general assessment:* one of the highest broadband penetrations in Latin America, supported by operator investment and govt. programs aimed at expansion; govt. provides 20 million euros for two 5G trials, Chinese company Huawei conducts 5G trials; major networks are entirely digital and the availability of telephone service continues to improve to rural areas; Argentinians' own multiple SIM cards for work and personal use; even with numerous providers there is a lack of competition for broadband and mobile services; still Argentina is the 3rd largest in the region after Brazil and Mexico (2020)
domestic: 17 per 100 fixed-line, 131 per 100 mobile-cellular; microwave radio relay, fiber-optic cable, and a domestic satellite system with 40 earth stations serve the trunk network (2019)
international: country code—54; landing points for the UNISUR, Bicentenario, Atlantis-2, SAm-1, and SAC, Tannat, Malbec and ARBR submarine cable systems that provide links to Europe, Africa, South and Central America, and US; satellite earth stations—112 (2019)
note: the COVID-19 outbreak is negatively impacting telecommunications production and supply chains globally; consumer spending on telecom devices and services has also slowed due to the pandemic's effect on economies worldwide; overall progress towards improvements in all facets of the telecom industry—mobile, fixed-line, broadband, submarine cable and satellite—has moderated

Broadcast media: government owns a TV station and radio network; more than 2 dozen TV stations and hundreds of privately owned radio stations; high rate of cable TV subscription usage

Internet country code: .ar

Internet users: *total:* 33,203,320

percent of population: 74.29% (July 2018 est.)
country comparison to the world: 24

Broadband—fixed subscriptions: *total:* 8,473,655
subscriptions per 100 inhabitants: 19 (2018 est.)
country comparison to the world: 21

TRANSPORTATION

National air transport system: *number of registered air carriers:* 6 (2020)
inventory of registered aircraft operated by air carriers: 107
annual passenger traffic on registered air carriers: 18,081,937 (2018)
annual freight traffic on registered air carriers: 311.57 million mt-km (2018)

Civil aircraft registration country code prefix: LV (2016)

Airports: 916 (2020)
country comparison to the world: 6

Airports—with paved runways: *total:* 161 (2017)
over 3,047 m: 4 (2017)
2,438 to 3,047 m: 29 (2017)
1,524 to 2,437 m: 65 (2017)
914 to 1,523 m: 53 (2017)

under 914 m: 10 (2017)

Airports—with unpaved runways: *total:* 977 (2013)
over 3,047 m: 1 (2013)
2,438 to 3,047 m: 1 (2013)
1,524 to 2,437 m: 43 (2013)
914 to 1,523 m: 484 (2013)
under 914 m: 448 (2013)

Heliports: 2 (2013)

Pipelines: 29930 km gas, 41 km liquid petroleum gas, 6248 km oil, 3631 km refined products (2013)

Railways: *total:* 36,917 km (2014)
standard gauge: 2,745.1 km 1.435-m gauge (41.1 km electrified) (2014)
narrow gauge: 7,523.3 km 1.000-m gauge (2014)
broad gauge: 26,391 km 1.676-m gauge (149 km electrified) (2014)
258 km 0.750-m gauge
country comparison to the world: 6

Roadways: *total:* 281,290 km (2017)
paved: 117,616 km (2017)
unpaved: 163,674 km (2017)
country comparison to the world: 21

Waterways: 11,000 km (2012)
country comparison to the world: 11

Merchant marine: *total:* 192
by type: bulk carrier 1, general cargo 8, oil tanker 30, other 153 (2019)
country comparison to the world: 68

Ports and terminals: *major seaport(s):* Bahia Blanca, Buenos Aires, La Plata, Punta Colorada, Ushuaia
container port(s) (TEUs): Buenos Aires (1,851,701)
LNG terminal(s) (import): Bahia Blanca
river port(s): Arroyo Seco, Rosario, San Lorenzo-San Martin (Parana)

MILITARY AND SECURITY

Military and security forces: Armed Forces of the Argentine Republic (Fuerzas Armadas de la República Argentina): Argentine Army (Ejercito Argentino), Navy of the Argentine Republic (Armada Republica; includes naval aviation and naval infantry), Argentine Air Force (Fuerza Aerea Argentina, FAA); Ministry of Security: Gendarmerie, Prefectura Naval (coast guard) (2020)

Military expenditures: 0.7% of GDP (2019)
0.7% of GDP (2018)
0.9% of GDP (2017)
0.8% of GDP (2016)
0.9% of GDP (2015)
country comparison to the world: 131

Military and security service personnel strengths: Argentina's armed forces have approximately 75,000 (45,000 Army; 17,000 Navy; 13,000 Air Force); est. 18,000 Gendarmerie (2019 est.)

Military equipment inventories and acquisitions: the inventory of Argentina's armed forces is a mix of domestically-produced and mostly older imported weapons, largely from Europe and the US; since 2010, France and the US are the leading suppliers of equipment; Argentina has an indigenous defense industry that can produce air, land, and sea systems (2019 est.)

Military deployments: 230 Cyprus (UNFICYP) (March 2020)

Military service age and obligation: 18-24 years of age for voluntary military service (18-21 requires parental consent); no conscription; if the number of volunteers fails to meet the quota of recruits for a particular year, Congress can authorize the conscription of citizens turning 18 that year for a period not exceeding one year (2012)

Military—note: the Argentine military focuses primarily on border security and counter-narcotics operations; in 2018, the government approved a decree allowing greater latitude for the military in internal security missions, with a focus on logistics support in border areas (2019)

TERRORISM

TRANSNATIONAL ISSUES

Disputes—international: Argentina continues to assert its claims to the UK-administered Falkland Islands (Islas Malvinas), South Georgia, and the South Sandwich Islands in its constitution, forcibly occupying the Falklands in 1982, but in 1995 agreed to no longer seek settlement by force; UK continues to reject Argentine requests for sovereignty talks; territorial claim in Antarctica partially overlaps UK and Chilean claims; uncontested dispute between Brazil and Uruguay over Braziliera/Brasiliera Island in the Quarai/Cuareim River leaves the tripoint with Argentina in question; in 2010, the ICJ ruled in favor of Uruguay's operation of two paper mills on the Uruguay River, which forms the border with Argentina; the two countries formed a joint pollution monitoring regime; the joint boundary commission, established by Chile and Argentina in 2001 has yet to map and demarcate the delimited boundary in the inhospitable Andean Southern Ice Field (Campo de Hielo Sur); contraband smuggling, human trafficking, and illegal narcotic trafficking are problems in the porous areas of the border with Bolivia

Refugees and internally displaced persons: *refugees (country of origin):* 213,769 (Venezuela) (economic and political crisis; includes Venezuelans who have claimed asylum, are recognized as refugees, or have received alternative legal stay) (2020)

Illicit drugs: a transshipment country for cocaine headed for Europe, heroin headed for the US, and ephedrine and pseudoephedrine headed for Mexico; some money-laundering activity, especially in the Tri-Border Area; law enforcement corruption; a source for precursor chemicals; increasing domestic consumption of drugs in urban centers, especially cocaine base and synthetic drugs

ARMENIA

INTRODUCTION

Background: Armenia prides itself on being the first nation to formally adopt Christianity (early 4th century). Despite periods of autonomy, over the centuries Armenia came under the sway of various empires including the Roman, Byzantine, Arab, Persian, and Ottoman. During World War I in the western portion of Armenia, the Ottoman Empire instituted a policy of forced resettlement coupled with other harsh practices that resulted in at least 1 million Armenian deaths. The eastern area of Armenia was ceded by the Ottomans to Russia in 1828; this portion declared its independence in 1918, but was conquered by the Soviet Red Army in 1920.

Armenia remains involved in the protracted Nagorno-Karabakh conflict with Azerbaijan. Nagorno-Karabakh was a primarily ethnic Armenian region that Moscow recognized in 1923 as an autonomous oblast within Soviet Azerbaijan. In the late Soviet period, a separatist movement developed which sought to end Azerbaijani control over the region. Fighting over Nagorno-Karabakh began in 1988 and escalated after Armenia and Azerbaijan attained independence from the Soviet Union in 1991. By the time a ceasefire took effect in May 1994, separatists, with Armenian support, controlled Nagorno-Karabakh and seven surrounding Azerbaijani territories. The 1994 ceasefire continues to hold, although violence continues along the line of contact separating the opposing forces, as well as the Armenia-Azerbaijan international border. The final status of Nagorno-Karabakh remains the subject of international mediation by the Organization for Security and Cooperation in Europe (OSCE) Minsk Group, which works to help the sides settle the conflict peacefully. The OSCE Minsk Group is co-chaired by the US, France, and Russia.

Turkey closed the common border with Armenia in 1993 in support of Azerbaijan in its conflict with Armenia over control of Nagorno-Karabakh and surrounding areas, further hampering Armenian economic growth. In 2009, Armenia and Turkey signed Protocols normalizing relations between the two countries, but neither country ratified the Protocols, and Armenia officially withdrew from the Protocols in March 2018. In 2015, Armenia joined the Eurasian Economic Union alongside Russia, Belarus, Kazakhstan, and Kyrgyzstan. In November 2017, Armenia signed a Comprehensive and Enhanced Partnership Agreement (CEPA) with the EU. In spring 2018, Serzh SARGSIAN of the Republican Party of Armenia (RPA) stepped down and Civil Contract party leader Nikol PASHINYAN became prime minister.

GEOGRAPHY

Location: Southwestern Asia, between Turkey (to the west) and Azerbaijan; note—Armenia views itself as part of Europe; geopolitically, it can be classified as falling within Europe, the Middle East, or both

Geographic coordinates: 40 00 N, 45 00 E

Map references: Asia

Area: *total:* 29,743 sq km
land: 28,203 sq km
water: 1,540 sq km
country comparison to the world: 143

Area—comparative: slightly smaller than Maryland

Land boundaries: *total:* 1,570 km
border countries (4): Azerbaijan 996 km, Georgia 219 km, Iran 44 km, Turkey 311 km

Coastline: 0 km (landlocked)

Maritime claims: none (landlocked)

Climate: highland continental, hot summers, cold winters

Terrain: Armenian Highland with mountains; little forest land; fast flowing rivers; good soil in Aras River valley

Elevation: *mean elevation:* 1,792 m
lowest point: Debed River 400 m
highest point: Aragats Lerrnagagat' 4,090 m

Natural resources: small deposits of gold, copper, molybdenum, zinc, bauxite

Land use: *agricultural land:* 59.7% (2016 est.)
arable land: 15.8% (2016 est.) / permanent crops: 1.9% (2016 est.) / permanent pasture: 42% (2016 est.)
forest: 9.1% (2016 est.)
other: 31.2% (2016 est.)
Irrigated land: 2,740 sq km (2012)

Population distribution: most of the population is located in the northern half of the country; the capital of Yerevan is home to more than five times as many people as Gyumri, the second largest city in the country

Natural hazards: occasionally severe earthquakes; droughts

Environment—current issues: soil pollution from toxic chemicals such as DDT; deforestation; pollution of Hrazdan and Aras Rivers; the draining of Sevana Lich (Lake Sevan), a result of its use as a source for hydropower, threatens drinking water supplies; restart of Metsamor nuclear power plant in spite of its location in a seismically active zone

Environment—international agreements: *party to:* Air Pollution, Biodiversity, Climate Change, Climate Change-Kyoto Protocol, Desertification, Environmental Modification, Hazardous Wastes, Law of the Sea, Ozone Layer Protection, Wetlands *signed, but not ratified:* Air Pollution-Persistent Organic Pollutants

Geography—note: landlocked in the Lesser Caucasus Mountains; Sevana Lich (Lake Sevan) is the largest lake in this mountain range

PEOPLE AND SOCIETY

Population: 3,021,324 (July 2020 est.)
country comparison to the world: 137

Nationality: *noun:* Armenian(s)
adjective: Armenian

Ethnic groups: Armenian 98.1%, Yezidi (Kurd) 1.2%, other 0.7% (2011 est.)

Languages: Armenian (official) 97.9%, Kurdish (spoken by Yezidi minority) 1%, other 1% (2011 est.)
note: Russian is widely spoken

Religions: Armenian Apostolic 92.6%, Evangelical 1%, other 2.4%, none 1.1%, unspecified 2.9% (2011 est.)

Age structure: *0-14 years:* 18.64% (male 297,320/female 265,969)
15-24 years: 11.63% (male 184,258/female 167,197)
25-54 years: 43.04% (male 639,101/female 661,421)
55-64 years: 14.08% (male 195,754/female 229,580)
65 years and over: 12.6% (male 154,117/female 226,607) (2020 est.)

Dependency ratios: *total dependency ratio:* 48.4
youth dependency ratio: 30.9
elderly dependency ratio: 17.5
potential support ratio: 5.7 (2020 est.)

Median age: *total:* 36.6 years
male: 35.1 years
female: 38.3 years (2020 est.)
country comparison to the world: 78

Population growth rate: -0.3% (2020 est.)
country comparison to the world: 218

Birth rate: 11.9 births/1,000 population (2020 est.)
country comparison to the world: 163

Death rate: 9.5 deaths/1,000 population (2020 est.)

country comparison to the world: 44

Net migration rate: -5.5 migrant(s)/1,000 population (2020 est.)

country comparison to the world: 201

Population distribution: most of the population is located in the northern half of the country; the capital of Yerevan is home to more than five times as many people as Gyumri, the second largest city in the country

Urbanization: *urban population:* 63.3% of total population (2020)

rate of urbanization: 0.22% annual rate of change (2015-20 est.)

total population growth rate v. urban population growth rate, 2000-2030: Major urban areas—population: 1.086 million YEREVAN (capital) (2020)

Sex ratio: *at birth:* 1.1 male(s)/female
0-14 years: 1.12 male(s)/female
15-24 years: 1.1 male(s)/female
25-54 years: 0.97 male(s)/female
55-64 years: 0.85 male(s)/female
65 years and over: 0.68 male(s)/female
total population: 0.95 male(s)/female (2020 est.)

Mother's mean age at first birth: 24.8 years (2017 est.)

Maternal mortality rate: 26 deaths/100,000 live births (2017 est.)

country comparison to the world: 119

Infant mortality rate: *total:* 11.5 deaths/1,000 live births
male: 12.9 deaths/1,000 live births
female: 10 deaths/1,000 live births (2020 est.)
country comparison to the world: 112

Life expectancy at birth: *total population:* 75.6 years
male: 72.3 years
female: 79.2 years (2020 est.)
country comparison to the world: 111

Total fertility rate: 1.65 children born/woman (2020 est.)
country comparison to the world: 180

Contraceptive prevalence rate: 57.1% (2015/16)

Drinking water source:
improved:
urban: 100% of population
rural: 100% of population
total: 100% of population
unimproved:
urban: 0% of population
rural: 0% of population
total: 0% of population (2017 est.)

Current Health Expenditure: 10.4% (2017)

Physicians density: 4.4 physicians/1,000 population (2017)

Hospital bed density: 4.2 beds/1,000 population (2014)

Sanitation facility access:
improved:
urban: 100% of population
rural: 84.5% of population
total: 93.6% of population
unimproved:

urban: 0% of population
rural: 15.5% of population
total: 6.4% of population (2017 est.)

HIV/AIDS—adult prevalence rate: 0.1% (2019 est.)

country comparison to the world: 116

HIV/AIDS—people living with HIV/AIDS: 3,500 (2019 est.)

country comparison to the world: 128

HIV/AIDS—deaths: <100 (2019 est.)

Obesity—adult prevalence rate: 20.2% (2016)
country comparison to the world: 101

Children under the age of 5 years underweight: 2.6% (2016)

country comparison to the world: 105

Education expenditures: 2.7% of GDP (2017)

country comparison to the world: 150

Literacy: *definition:* age 15 and over can read and write
total population: 99.7%
male: 99.8%
female: 99.7% (2017)

School life expectancy (primary to tertiary education): *total:* 13 years
male: 13 years
female: 14 years (2019)

Unemployment, youth ages 15-24: *total:* 36.3%
male: 29.5%
female: 45.7% (2016 est.)
country comparison to the world: 17

GOVERNMENT

Country name: *conventional long form:* Republic of Armenia
conventional short form: Armenia
local long form: Hayastani Hanrapetut'yun
local short form: Hayastan
former: Armenian Soviet Socialist Republic, Armenian Republic
etymology: the etymology of the country's name remains obscure; according to tradition, the country is named after Hayk, the legendary patriarch of the Armenians and the great-great-grandson of Noah; Hayk's descendant, Aram, purportedly is the source of the name Armenia

Government type: parliamentary democracy; note—constitutional changes adopted in December 2015 transformed the government to a parliamentary system

Capital: *name:* Yerevan

geographic coordinates: 40 10 N, 44 30 E
time difference: UTC+4 (9 hours ahead of Washington, DC, during Standard Time)
etymology: name likely derives from the ancient Urartian fortress of Erebuni established on the current site of Yerevan in 782 B.C. and whose impressive ruins still survive

Administrative divisions: 11 provinces (marzer, singular—marz); Aragatsotn, Ararat, Armavir, Geghark'unik', Kotayk', Lorri, Shirak, Syunik', Tavush, Vayots' Dzor, Yerevan

Independence: 21 September 1991 (from the Soviet Union); notable earlier dates: 321 B.C. (Kingdom of Armenia established under the Orontid Dynasty), A.D. 884 (Armenian Kingdom reestablished under the Bagratid Dynasty); 1198 (Cilician Kingdom established); 28 May 1918 (Democratic Republic of Armenia declared)

National holiday: Independence Day, 21 September (1991)

Constitution: *history:* previous 1915, 1978; latest adopted 5 July 1995
amendments: proposed by the president of the republic or by the National Assembly; passage requires approval by the president, by the National Assembly, and by a referendum with at least 25% registered voter participation and more than 50% of votes; constitutional articles on the form of government and democratic procedures are not amendable; amended 2005, 2007, 2008, 2015; note—a constitutional referendum scheduled for 4 May 2020 has been postponed due to the COVID-19 pandemic

Legal system: civil law system

International law organization participation: has not submitted an ICJ jurisdiction declaration; non-party state to the ICCt

Citizenship: *citizenship by birth:* no
citizenship by descent only: at least one parent must be a citizen of Armenia
dual citizenship recognized: yes
residency requirement for naturalization: 3 years

Suffrage: 18 years of age; universal

Executive branch: *chief of state:* President Armen SARKISSIAN (since 9 April 2018)

head of government: Prime Minister Nikol PASHINYAN (since 8 May 2018); Deputy Prime Ministers Mher GRIGORYAN and Tigran AVINYAN (since 16 January 2019)
cabinet: Council of Ministers appointed by the prime minister
elections/appointments: president indirectly elected by the National Assembly in 3 rounds if needed for a single 7-year term; election last held on 2 March 2018; prime minister elected by majority vote in 2 rounds if needed by the National Assembly; election last held on 14 January 2019
election results: Armen SARKISSIAN elected president in first round; note—Armen SARKISSIAN ran unopposed and won the Assembly vote 90-10; Nikol PASHINYAN was chosen as prime minister by the parliament automatically after his party won a landslide victory in the December 2018 elections
note: After initially winning election on 8 May 2018, Nikol PASHINYAN resigned his post (but stayed on as acting prime minister) on 16 October 2018 to force a snap election (held on 9 December 2018) in which his bloc won more than 70% of the vote; PASHINYAN was reappointed prime minister on 14 January 2019

Legislative branch: *description:* unicameral National Assembly (Parliament) or Azgayin Zhoghov (minimum 101 seats, currently

132; members directly elected in single-seat constituencies by proportional representation vote; members serve 5-year terms)

elections: last held on 9 December 2018 (next elections to be held December 2023)

election results: percent of vote by party—My Step Alliance 70.4%, BHK 8.3%, Bright Armenia 6.4%, RPA 4.7%, ARF 3.9%, other 6.3%; seats by party—My Step Alliance 88, BHK 26, Bright Armenia 18; composition—men 112, women 20, percent of women 15.2%

Judicial branch: *highest courts:* Court of Cassation (consists of the Criminal Chamber with a chairman and 5 judges and the Civil and Administrative Chamber with a chairman and 10 judges – with both civil and administrative specializations); Constitutional Court (consists of 9 judges)

judge selection and term of office: Court of Cassation judges nominated by the Supreme Judicial Council, a 10-member body of selected judges and legal scholars; judges appointed by the president; judges can serve until age 65;Constitutional Court judges—4 appointed by the president, and 5 elected by the National Assembly; judges can serve until age 70

subordinate courts: criminal and civil appellate courts; administrative appellate court; first instance courts; specialized administrative and bankruptcy courts

Political parties and leaders: Armenian National Congress or ANC (bloc of independent and opposition parties) [Levon TER-PETROSSIAN]
Armenian Revolutionary Federation or ARF ("Dashnak" Party) [Hakob TER-KHACHATURYAN]
Bright Armenia [Edmon MARUKYAN]
Citizen's Decision [Suren SAHAKYAN]
Civil Contract [Nikol PASHINYAN]
Free Democrats [Khachatur KOKOBELYAN]
Heritage Party [Raffi HOVANNISIAN]
Prosperous Armenia or BHK [Gagik TSARUKYAN]
Republic [Aram SARGSYAN]
Republican Party of Armenia or RPA [Serzh SARGSIAN]
Rule of Law Party (Orinats Yerkir) or OEK [Artur BAGHDASARIAN]
Sasna Tser [Varuzhan AVETISYAN]

International organization participation: ADB, BSEC, CD, CE, CIS, CSTO, EAEC (observer), EAEU, EAPC, EBRD, FAO, GCTU, IAEA, IBRD, ICAO, ICC (NGOs), ICRM, IDA, IFAD, IFC, IFRCS, ILO, IMF, Interpol, IOC, IOM, IPU, ISO, ITSO, ITU, MIGA, NAM (observer), OAS (observer), OIF, OPCW, OSCE, PFP, UN, UNCTAD, UNESCO, UNIDO, UNIFIL, UNWTO, UPU, WCO, WFTU (NGOs), WHO, WIPO, WMO, WTO

Diplomatic representation in the US: *chief of mission:* Ambassador Varuzhan NERSESSYAN (since 11 January 2019)
chancery: 2225 R Street NW, Washington, DC 20008
telephone: [1] (202) 319-1976

FAX: [1] (202) 319-2982

consulate(s) general: Glendale (CA)

Diplomatic representation from the US: *chief of mission:* Ambassador Lynne M. TRACEY (since 5 March 2019)
telephone: [374](10) 464-700
embassy: 1 American Ave., Yerevan 0082
mailing address: American Embassy Yerevan, US Department of State, 7020 Yerevan Place, Washington, DC 20521-7020
FAX: [374](10) 464-742

Flag description: three equal horizontal bands of red (top), blue, and orange; the color red recalls the blood shed for liberty, blue the Armenian skies as well as hope, and orange the land and the courage of the workers who farm it

National symbol(s): Mount Ararat, eagle, lion; national colors: red, blue, orange

National anthem: *name:* "Mer Hayrenik" (Our Fatherland)
lyrics/music: Mikael NALBANDIAN/Barsegh KANACHYAN
note: adopted 1991; based on the anthem of the Democratic Republic of Armenia (1918-1922) but with different lyrics
0:00 / 0:29

ECONOMY

Economy—overview: Under the old Soviet central planning system, Armenia developed a modern industrial sector, supplying machine tools, textiles, and other manufactured goods to sister republics, in exchange for raw materials and energy. Armenia has since switched to small-scale agriculture and away from the large agro industrial complexes of the Soviet era. Armenia has only two open trade borders—Iran and Georgia—because its borders with Azerbaijan and Turkey have been closed since 1991 and 1993, respectively, as a result of Armenia's ongoing conflict with Azerbaijan over the separatist Nagorno-Karabakh region.

Armenia joined the World Trade Organization in January 2003. The government has made some improvements in tax and customs administration in recent years, but anti-corruption measures have been largely ineffective. Armenia will need to pursue additional economic reforms and strengthen the rule of law in order to raise its economic growth and improve economic competitiveness and employment opportunities, especially given its economic isolation from Turkey and Azerbaijan.

Armenia's geographic isolation, a narrow export base, and pervasive monopolies in important business sectors have made it particularly vulnerable to volatility in the global commodity markets and the economic challenges in Russia. Armenia is particularly dependent on Russian commercial and governmental support, as most key Armenian infrastructure is Russianowned and/or managed, especially in the energy sector. Remittances from expatriates working in Russia are equivalent to about 12-14% of GDP. Armenia joined the Russia-led Eurasian Economic Union in January 2015, but has remained interested in pursuing closer ties

with the EU as well, signing a Comprehensive and Enhanced Partnership Agreement with the EU in November 2017. Armenia's rising government debt is leading Yerevan to tighten its fiscal policies – the amount is approaching the debt to GDP ratio threshold set by national legislation.

GDP (purchasing power parity): $28.34 billion (2017 est.)
$26.37 billion (2016 est.)
$26.3 billion (2015 est.)
note: data are in 2017 dollars
country comparison to the world: 136

GDP (official exchange rate): $11.54 billion (2017 est.)

GDP—real growth rate: 7.5% (2017 est.)
0.3% (2016 est.)
3.3% (2015 est.)
country comparison to the world: 12

GDP—per capita (PPP): $9,500 (2017 est.)
$8,800 (2016 est.)
$8,800 (2015 est.)
note: data are in 2017 dollars
country comparison to the world: 142

Gross national saving: 17.8% of GDP (2017 est.)
16.6% of GDP (2016 est.)
18.4% of GDP (2015 est.)
country comparison to the world: 112

GDP—composition, by end use: *household consumption:* 76.7% (2017 est.)
government consumption: 14.2% (2017 est.)
investment in fixed capital: 17.3% (2017 est.)
investment in inventories: 4.1% (2017 est.)
exports of goods and services: 38.1% (2017 est.)
imports of goods and services: -50.4% (2017 est.)

GDP—composition, by sector of origin: *agriculture:* 16.7% (2017 est.)
industry: 28.2% (2017 est.)
services: 54.8% (2017 est.)

Agriculture—products: fruit (especially grapes and apricots), vegetables; livestock

Industries: brandy, mining, diamond processing, metal-cutting machine tools, forging and pressing machines, electric motors, knitted wear, hosiery, shoes, silk fabric, chemicals, trucks, instruments, microelectronics, jewelry, software, food processing

Industrial production growth rate: 5.4% (2017 est.)
country comparison to the world: 51

Labor force: 1.507 million (2017 est.)
country comparison to the world: 131

Labor force—by occupation: *agriculture:* 36.3%
industry: 17%
services: 46.7% (2013 est.)

Unemployment rate: 18.9% (2017 est.)
18.8% (2016 est.)
country comparison to the world: 183

Population below poverty line: 32% (2013 est.)

Household income or consumption by percentage share: *lowest 10%:* 3.5%
highest 10%: 25.7% (2014)

Budget: *revenues:* 2.644 billion (2017 est.)
expenditures: 3.192 billion (2017 est.)

Taxes and other revenues: 22.9% (of GDP) (2017 est.)
country comparison to the world: 130

Budget surplus (+) or deficit (-): -4.8% (of GDP) (2017 est.)
country comparison to the world: 167

Public debt: 53.5% of GDP (2017 est.)
51.9% of GDP (2016 est.)
country comparison to the world: 89

Fiscal year: calendar year

Inflation rate (consumer prices): 0.9% (2017 est.)
-1.4% (2016 est.)
country comparison to the world: 44

Current account balance: -$328 million (2017 est.)
-$238 million (2016 est.)
country comparison to the world: 107

Exports: $2.361 billion (2017 est.)
$1.891 billion (2016 est.)
country comparison to the world: 134

Exports—partners: Russia 24.2%, Bulgaria 12.8%, Switzerland 12%, Georgia 6.9%, Germany 5.9%, China 5.5%, Iraq 5.4%, UAE 4.6%, Netherlands 4.1% (2017)

Exports—commodities: unwrought copper, pig iron, nonferrous metals, gold, diamonds, mineral products, foodstuffs, brandy, cigarettes, energy

Imports: $3.771 billion (2017 est.)
$2.835 billion (2016 est.)
country comparison to the world: 141

Imports—commodities: natural gas, petroleum, tobacco products, foodstuffs, diamonds, pharmaceuticals, cars

Imports—partners: Russia 28%, China 11.5%, Turkey 5.5%, Germany 4.9%, Iran 4.3% (2017)

Reserves of foreign exchange and gold: $2.314 billion (31 December 2017 est.)
$2.204 billion (31 December 2016 est.)
country comparison to the world: 119

Debt—external: $10.41 billion (31 December 2017 est.)
$8.987 billion (31 December 2016 est.)
country comparison to the world: 113

Exchange rates: drams (AMD) per US dollar—
487.9 (2017 est.)
480.49 (2016 est.)
480.49 (2015 est.)
477.92 (2014 est.)
415.92 (2013 est.)

ENERGY

Electricity access: *electrification—total population:* 100% (2016)

Electricity—production: 6.951 billion kWh (2016 est.)
country comparison to the world: 112

Electricity—consumption: 5.291 billion kWh (2016 est.)
country comparison to the world: 121

Electricity—exports: 1.424 billion kWh (2015 est.)

country comparison to the world: 50

Electricity—imports: 275 million kWh (2016 est.)
country comparison to the world: 90

Electricity—installed generating capacity: 4.08 million kW (2016 est.)
country comparison to the world: 86

Electricity—from fossil fuels: 58% of total installed capacity (2016 est.)
country comparison to the world: 134

Electricity—from nuclear fuels: 9% of total installed capacity (2017 est.)
country comparison to the world: 15

Electricity—from hydroelectric plants: 32% of total installed capacity (2017 est.)
country comparison to the world: 65

Electricity—from other renewable sources: 0% of total installed capacity (2017 est.)
country comparison to the world: 173

Crude oil—production: 0 bbl/day (2018 est.)
country comparison to the world: 106

Crude oil—exports: 0 bbl/day (2015 est.)
country comparison to the world: 86

Crude oil—imports: 0 bbl/day (2015 est.)
country comparison to the world: 90

Crude oil—proved reserves: 0 bbl (1 January 2018 est.)
country comparison to the world: 103

Refined petroleum products—production: 0 bbl/day (2015 est.)
country comparison to the world: 114

Refined petroleum products—consumption: 8,000 bbl/day (2016 est.)
country comparison to the world: 162

Refined petroleum products—exports: 0 bbl/day (2015 est.)
country comparison to the world: 126

Refined petroleum products—imports: 7,145 bbl/day (2015 est.)
country comparison to the world: 158

Natural gas—production: 0 cu m (2017 est.)
country comparison to the world: 100

Natural gas—consumption: 2.35 billion cu m (2017 est.)
country comparison to the world: 80

Natural gas—exports: 0 cu m (2017 est.)
country comparison to the world: 62

Natural gas—imports: 2.35 billion cu m (2017 est.)
country comparison to the world: 48

Natural gas—proved reserves: 0 cu m (1 January 2014 est.)
country comparison to the world: 106

Carbon dioxide emissions from consumption of energy: 5.501 million Mt (2017 est.)
country comparison to the world: 131

COMMUNICATIONS

Telephones—fixed lines: *total subscriptions:* 462,725
subscriptions per 100 inhabitants: 15.27 (2019 est.)

country comparison to the world: 97

Telephones—mobile cellular: *total subscriptions:* 3,707,557
subscriptions per 100 inhabitants: 122.35 (2019 est.)
country comparison to the world: 134

Telecommunication systems: *general assessment:* telecommunications investments have made major inroads in modernizing and upgrading the outdated telecommunications network inherited from the Soviet era; now 100% privately owned and undergoing continued modernization and expansion; strong growth in mobile broadband sector and mobile services dominate over fixedline; rollout of 4G networks and falling prices due to growing competition (2020)
domestic: 15 per 100 fixed-line, 122 per 100 mobile-cellular; reliable fixed-line and mobile-cellular services are available across Yerevan and in major cities and towns; mobile-cellular coverage available in most rural areas (2019)
international: country code—374; Yerevan is connected to the Caucasus Cable System fiber-optic cable through Georgia and Iran to Europe; additional international service is available by microwave radio relay and landline connections to the other countries of the Commonwealth of Independent States, through the Moscow international switch, and by satellite to the rest of the world; satellite earth stations—3 (2019)
note: the COVID-19 outbreak is negatively impacting telecommunications production and supply chains globally; consumer spending on telecom devices and services has also slowed due to the pandemic's effect on economies worldwide; overall progress towards improvements in all facets of the telecom industry—mobile, fixed-line, broadband, submarine cable and satellite—has moderated

Broadcast media: Armenia's government-run Public Television network operates alongside 100 privately owned TV stations that provide local to near nationwide coverage; three Russian TV companies are broadcast in Armenia under interstate agreements; subscription cable TV services are available in most regions; several major international broadcasters are available, including CNN; Armenian TV completed conversion from analog to digital broadcasting in late 2016; Public Radio of Armenia is a national, state-run broadcast network that operates alongside 18 privately owned radio stations (2019)

Internet country code: .am

Internet users: *total:* 1,966,942

percent of population: 64.74% (July 2018 est.)
country comparison to the world: 122

Broadband—fixed subscriptions: *total:* 347,448
subscriptions per 100 inhabitants: 11 (2018 est.)
country comparison to the world: 97

TRANSPORTATION

National air transport system: *number of registered air carriers:* 3 (2020)
inventory of registered aircraft operated by air carriers: 5

Civil aircraft registration country code prefix: EK (2016)

Airports: 7 (2020)
country comparison to the world: 165

Airports—with paved runways: *total:* 10 (2017)
over 3,047 m: 2 (2017)
2,438 to 3,047 m: 2 (2017)
1,524 to 2,437 m: 4 (2017)
914 to 1,523 m: 2 (2017)

Airports—with unpaved runways: *total:* 1 (2013)
914 to 1,523 m: 1 (2013)

Pipelines: 3838 km gas (high and medium pressure) (2017)

Railways: *total:* 780 km (2014)

broad gauge: 780 km 1.520-m gauge (780 km electrified) (2014)
note: 726 km operational
country comparison to the world: 98

Roadways: *total:* 7,700 km (2019)
urban: 3,780 km
non-urban: 3,920 km
country comparison to the world: 140

Military and security forces: Armenian Armed Forces: Ground Forces (Armenian Army), Air Force, Air Defense; "Nagorno-Karabakh Republic": Nagorno-Karabakh Defense Army (2019)

Military expenditures: 4.9% of GDP (2019)
4.9% of GDP (2018)
3.8% of GDP (2017)
4.1% of GDP (2016)
4.2% of GDP (2015)
country comparison to the world: 7

Military and security service personnel strengths: the Armenian Armed Forces have approximately 45,000 active troops (42,000 Army; 3,000 Air Force/Air Defense) (2019 est.)

Military equipment inventories and acquisitions: the inventory of the Armenian Armed Forces (as well as the Nagorno-Karabakh Defense Army) includes mostly Russian and Soviet-era equipment; since 2010, almost all of Armenia's imported weapons have come from Russia (2019)

Military deployments: 120 Afghanistan (NATO); contributes one motorized rifle regiment (approximately 2,000 personnel) to CSTO's Rapid Reaction Force (2020)

Military service age and obligation: 18-27 years of age for voluntary or compulsory military service; 2-year conscript service obligation, which can be served as an officer upon deferment for university studies if enrolled in officer-producing program; 17 year olds are eligible to become cadets at military higher education institutes, where they are classified as military personnel (2019)

Disputes—international: the dispute over the break-away Nagorno-Karabakh region and the Armenian military occupation of surrounding lands in Azerbaijan remains the primary focus of regional instability; residents have evacuated the former Soviet-era small ethnic enclaves in Armenia and Azerbaijan; Turkish authorities have complained that blasting from quarries in Armenia might be damaging the medieval ruins of Ani, on the other side of the Arpacay valley; in 2009, Swiss mediators facilitated an accord reestablishing diplomatic ties between Armenia and Turkey, but neither side has ratified the agreement and the rapprochement effort has faltered; local border forces struggle to control the illegal transit of goods and people across the porous, undemarcated Armenian, Azerbaijani, and Georgian borders; ethnic Armenian groups in the Javakheti region of Georgia seek greater autonomy from the Georgian Government

Refugees and internally displaced persons: *refugees (country of origin):* 14,730- (Syria—ethnic Armenians) (2019)
stateless persons: 848 (2018)

Illicit drugs: illicit cultivation of small amount of cannabis for domestic consumption; minor transit point for illicit drugs—mostly opium and hashish—moving from Southwest Asia to Russia and to a lesser extent the rest of Europe

ARUBA

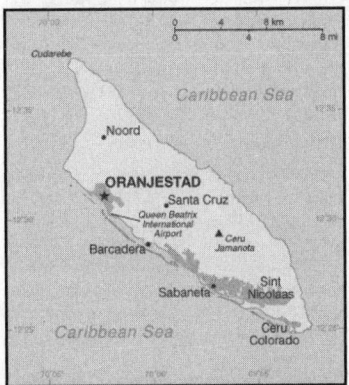

Background: Discovered and claimed for Spain in 1499, Aruba was acquired by the Dutch in 1636. The island's economy has been dominated by three main industries. A 19th century gold rush was followed by prosperity brought on by the opening in 1924 of an oil refinery. The last decades of the 20th century saw a boom in the tourism industry. Aruba seceded from the Netherlands Antilles in 1986 and became a separate, semi-autonomous member of the Kingdom of the Netherlands. Movement toward full independence was halted at Aruba's request in 1990.

Location: Caribbean, island in the Caribbean Sea, north of Venezuela

Geographic coordinates: 12 30 N, 69 58 W

Map references: Central America and the Caribbean

Area: *total:* 180 sq km
land: 180 sq km
water: 0 sq km
country comparison to the world: 218

Area—comparative: slightly larger than Washington, DC

Land boundaries: 0 km

Coastline: 68.5 km

Maritime claims: *territorial sea:* 12 nm
exclusive economic zone: 200 nm

Climate: tropical marine; little seasonal temperature variation

Terrain: flat with a few hills; scant vegetation

Elevation: *lowest point:* Caribbean Sea 0 m
highest point: Ceru Jamanota 188 m

Natural resources: NEGL; white sandy beaches foster tourism

Land use: *agricultural land:* 11.1% (2016 est.)
arable land: 11.1% (2016 est.) / permanent crops: 0% (2016 est.) / permanent pasture: 0% (2016 est.)
forest: 2.3% (2016 est.)
other: 86.6% (2016 est.)

Irrigated land: NA

Population distribution: most residents live in or around Oranjestad and San Nicolaas; most settlments tend to be located on the less mountainous western side of the island

Natural hazards: hurricanes; lies outside the Caribbean hurricane belt and is rarely threatened

Environment—current issues: difficulty in properly disposing of waste produced by large numbers of tourists; waste burning that occurs in the landfill causes air pollution and poses an environmental and health risk; ocean environmental damage due to plastic pollution

Geography—note: a flat, riverless island renowned for its white sand beaches; its tropical climate is moderated by constant trade winds from the

Atlantic Ocean; the temperature is almost constant at about 27 degrees Celsius (81 degrees Fahrenheit)

PEOPLE AND SOCIETY

Population: 119,428 (July 2020 est.)
country comparison to the world: 189

Nationality: *noun:* Aruban(s)
adjective: Aruban; Dutch

Ethnic groups: Aruban 66%, Colombian 9.1%, Dutch 4.3%, Dominican 4.1%, Venezuelan 3.2%, Curacaoan 2.2%, Haitian 1.5%, Surinamese 1.2%, Peruvian 1.1%, Chinese 1.1%, other 6.2% (2010 est.)
note: data represent population by country of birth

Languages: Papiamento (official) (a creole language that is a mixture of Portuguese, Spanish, Dutch, English, and, to a lesser extent, French, as well as elements of African languages and the language of the Arawak) 69.4%, Spanish 13.7%, English (widely spoken) 7.1%, Dutch (official) 6.1%, Chinese 1.5%, other 1.7%, unspecified 0.4% (2010 est.)

Religions: Roman Catholic 75.3%, Protestant 4.9% (includes Methodist 0.9%, Adventist 0.9%, Anglican 0.4%, other Protestant 2.7%), Jehovah's Witness 1.7%, other 12%, none 5.5%, unspecified 0.5% (2010 est.)

Age structure: 0-14 years: 17.55% (male 10,524/female 10,437)
15-24 years: 12.06% (male 7,231/female 7,175)
25-54 years: 40.54% (male 23,387/female 25,029)
55-64 years: 14.79% (male 8,285/female 9,383)
65 years and over: 15.05% (male 7,064/female 10,913) (2020 est.)

Dependency ratios: *total dependency ratio:* 47
youth dependency ratio: 25.6
elderly dependency ratio: 21.5
potential support ratio: 4.7 (2020 est.)

Median age: *total:* 39.9 years
male: 38.2 years
female: 41.5 years (2020 est.)
country comparison to the world: 53

Population growth rate: 1.19% (2020 est.)
country comparison to the world: 91

Birth rate: 12.1 births/1,000 population (2020 est.)
country comparison to the world: 160

Death rate: 8.7 deaths/1,000 population (2020 est.)
country comparison to the world: 67

Net migration rate: 8.4 migrant(s)/1,000 population (2020 est.)
country comparison to the world: 9

Population distribution: most residents live in or around Oranjestad and San Nicolaas; most settlments tend to be located on the less mountainous western side of the island

Urbanization: *urban population:* 43.7% of total population (2020)
rate of urbanization: 0.67% annual rate of change (2015-20 est.)

total population growth rate v. urban population growth rate, 2000-2030: Major urban areas—**population:** 30,000 ORANJESTAD (capital) (2018)

Sex ratio: *at birth:* 1.02 male(s)/female
0-14 years: 1.01 male(s)/female
15-24 years: 1.01 male(s)/female
25-54 years: 0.93 male(s)/female
55-64 years: 0.88 male(s)/female
65 years and over: 0.65 male(s)/female
total population: 0.9 male(s)/female (2020 est.)

Infant mortality rate: *total:* 9.8 deaths/1,000 live births
male: 12.7 deaths/1,000 live births
female: 6.8 deaths/1,000 live births (2020 est.)
country comparison to the world: 132

Life expectancy at birth: *total population:* 77.5 years
male: 74.4 years
female: 80.7 years (2020 est.)
country comparison to the world: 78

Total fertility rate: 1.83 children born/woman (2020 est.)
country comparison to the world: 143

Drinking water source:
improved:
urban: 98.1% of population
rural: 98.1% of population
total: 98.1% of population
unimproved:
urban: 1.9% of population
rural: 1.9% of population
total: 1.9% of population (2015 est.)

Sanitation facility access:
improved:
urban: 97.7% of population
rural: 97.7% of population
total: 97.7% of population
unimproved:
urban: 2.3% of population
rural: 2.3% of population
total: 2.3% of population (2015 est.)

HIV/AIDS—adult prevalence rate: NA

HIV/AIDS—people living with HIV/AIDS: NA

HIV/AIDS—deaths: NA

Education expenditures: 6.2% of GDP (2016)
country comparison to the world: 27

Literacy: *definition:* age 15 and over can read and write
total population: 97.8%
male: 97.8%
female: 97.8% (2018)

School life expectancy (primary to tertiary education): *total:* 14 years
male: 13 years
female: 14 years (2012)

GOVERNMENT

Country name: *conventional long form:* Country of Aruba
conventional short form: Aruba

local long form: Land Aruba (Dutch); Pais Aruba (Papiamento)
local short form: Aruba
etymology: the origin of the island's name is unclear; according to tradition, the name comes from the Spanish phrase "oro huba" (there was gold), but in fact no gold was ever found on the island; another possibility is the native word "oruba," which means "well-situated"

Dependency status: constituent country of the Kingdom of the Netherlands; full autonomy in internal affairs obtained in 1986 upon separation from the Netherlands Antilles; Dutch Government responsible for defense and foreign affairs

Government type: parliamentary democracy; part of the Kingdom of the Netherlands

Capital: *name:* Oranjestad
geographic coordinates: 12 31 N, 70 02 W
time difference: UTC-4 (1 hour ahead of Washington, DC, during Standard Time)
etymology: translates as "orange town" in Dutch; the city is named after William I (1533-1584), Prince of Orange, the first ruler of the Netherlands

Administrative divisions: none (part of the Kingdom of the Netherlands)
note: Aruba is one of four constituent countries of the Kingdom of the Netherlands; the other three are the Netherlands, Curacao, and Sint Maarten

Independence: *none (part of the Kingdom of the Netherlands)*

National holiday: National Anthem and Flag Day, 18 March (1976)

Constitution: *history:* previous 1947, 1955; latest drafted and approved August 1985, enacted 1 January 1986 (regulates governance of Aruba but is subordinate to the Charter for the Kingdom of the Netherlands); in 1986, Aruba became a semi-autonomous entity within the Kingdom of the Netherlands

Legal system: civil law system based on the Dutch civil code

Citizenship: see the Netherlands

Suffrage: 18 years of age; universal

Executive branch: *chief of state:* King WILLEM-ALEXANDER of the Netherlands (since 30 April 2013); represented by Governor General Alfonso BOEKHOUDT (since 1 January 2017)
head of government: Prime Minister Evelyn WEVER-CROES (since 17 November 2017)
cabinet: Council of Ministers elected by the Legislature (Staten)
elections/appointments: the monarchy is hereditary; governor general appointed by the monarch for a 6-year term; prime minister and deputy prime minister indirectly elected by the Staten for 4-year term; election last held on 27 September 2013 (next to be held by September 2017)
election results: Evelyn WEVER-CROES (MEP) elected prime minister; percent of legislative vote—NA

Legislative branch: description: unicameral Legislature or Staten (21 seats; members directly elected in a single nationwide constituency by proportional representation vote; members serve 4-year terms)

elections: last held on 22 September 2017 (next to be held in September 2021)

election results: percent of vote by party AVP 39.8%, MEP 37.6%, POR 9.4%, RED 7.1%, other 6.1%; seats by party—AVP 9, MEP 9, POR 2, RED 1; composition as of October 2018—men 14, women 7, percent of women 33.3%

Judicial branch: highest courts: Joint Court of Justice of Aruba, Curacao, Sint Maarten, and of Bonaire, Sint Eustatius and Saba or "Joint Court of Justice" (sits as a 3-judge panel); final appeals heard by the Supreme Court in The Hague, Netherlands

judge selection and term of office: Joint Court judges appointed for life by the monarch

subordinate courts: Court in First Instance

Political parties and leaders: Aruban People's Party or AVP [Michiel "Mike" EMAN]
Democratic Electoral Network or RED [L.R. CROES]
People's Electoral Movement Party or MEP [Evelyn WEVER-CROES]
Pueblo Orguyoso y Respeta or POR [O.E. ODUBER]
Real Democracy or PDR [Andin BIKKER]

International organization participation: Caricom (observer), FATF, ILO, IMF, Interpol, IOC, ITUC (NGOs), UNESCO (associate), UNWTO (associate), UPU

Diplomatic representation in the US: none (represented by the Kingdom of the Netherlands); note—Guillfred BESARIL (since 20 November 2017) is Minister Plenipotentiary of Aruba, seated with his cabinet in the Aruba House (Arubahuis) in The Hague none (represented by the Kingdom of the Netherlands) note—there is a Minister Plenipotentiary for Aruba, Rendolf "Andy" LEE, at the Embassy of the Kingdom of the Netherlands

Diplomatic representation from the US: the US does not have an embassy in Aruba; the Consul General to Curacao is accredited to Aruba

Flag description: blue, with two narrow, horizontal, yellow stripes across the lower portion and a red, four-pointed star outlined in white in the upper hoist-side corner; the star represents Aruba and its red soil and white beaches, its four points the four major languages (Papiamento, Dutch, Spanish, English) as well as the four points of a compass, to indicate that its inhabitants come from all over the world; the blue symbolizes Caribbean waters and skies; the stripes represent the island's two main "industries": the flow of tourists to the sun-drenched beaches and the flow of minerals from the earth

National symbol(s): Hooiberg (Haystack) Hill; national colors: blue, yellow, red, white

National anthem: *name:* "Aruba Deshi Tera" (Aruba Precious Country)

lyrics/music: Juan Chabaya 'Padu' LAMPE/Rufo Inocencio WEVER

note: local anthem adopted 1986; as part of the Kingdom of the Netherlands, "Het Wilhelmus" is official (see Netherlands)

0:00 / 2:06

ECONOMY

Economy—overview: Tourism, petroleum bunkering, hospitality, and financial and business services are the mainstays of the small open Aruban economy.

Tourism accounts for a majority of economic activity; as of 2017, over 2 million tourists visited Aruba annually, with the large majority (80-85%) of those from the US. The rapid growth of the tourism sector has resulted in a substantial expansion of other activities. Construction continues to boom, especially in the hospitality sector.

Aruba is heavily dependent on imports and is making efforts to expand exports to improve its trade balance. Almost all consumer and capital goods are imported, with the US, the Netherlands, and Panama being the major suppliers.

In 2016, Citgo Petroleum Corporation, an indirect wholly owned subsidiary of Petroleos de Venezuela SA, and the Government of Aruba signed an agreement to restart Valero Energy Corp.'s former 235,000-b/d refinery. Tourism and related industries have continued to grow, and the Aruban Government is working to attract more diverse industries. Aruba's banking sector continues to be a strong sector; unemployment has significantly decreased.

GDP (purchasing power parity): $4.158 billion (2017 est.)
$4.107 billion (2016 est.)
$4.112 billion (2015 est.)
country comparison to the world: 180

GDP (official exchange rate): $2.7 billion (2017 est.)

GDP—real growth rate: 1.2% (2017 est.)
-0.1% (2016 est.)
-0.4% (2015 est.)
country comparison to the world: 180

GDP—per capita (PPP): $37,500 (2017 est.)
$37,300 (2016 est.)
$37,700 (2015 est.)
country comparison to the world: 51

Gross national saving: 17% of GDP (2017 est.)
17.2% of GDP (2016 est.)
15.5% of GDP (2015 est.)
country comparison to the world: 120

GDP—composition, by end use: household consumption: 60.3% (2014 est.)
government consumption: 25.3% (2015 est.)
investment in fixed capital: 22.3% (2014 est.)
investment in inventories: 0% (2015 est.)
exports of goods and services: 70.5% (2015 est.)
imports of goods and services: -76.6% (2015 est.)

GDP—composition, by sector of origin: *agriculture:* 0.4% (2002 est.)

industry: 33.3% (2002 est.)
services: 66.3% (2002 est.)

Agriculture—products: aloes; livestock; fish

Industries: tourism, petroleum transshipment facilities, banking

Industrial production growth rate: NA

Labor force: 51,610 (2007 est.)
note: of the 51,610 workers aged 15 and over in the labor force, 32,252 were born in Aruba and 19,353 came from abroad; foreign workers are 38% of the employed population
country comparison to the world: 191

Labor force—by occupation: *agriculture:* NA
industry: NA
services: NA
note: most employment is in wholesale and retail trade, followed by hotels and restaurants

Unemployment rate: 7.7% (2016 est.)
country comparison to the world: 113

Population below poverty line: NA

Household income or consumption by percentage share: *lowest 10%:* NA
highest 10%: NA

Budget: *revenues:* 681.6 million (2017 est.)
expenditures: 755.5 million (2017 est.)

Taxes and other revenues: 25.2% (of GDP) (2017 est.)
country comparison to the world: 118

Budget surplus (+) or deficit (-): -2.7% (of GDP) (2017 est.)
country comparison to the world: 118

Public debt: 86% of GDP (2017 est.)
84.7% of GDP (2016 est.)
country comparison to the world: 31

Fiscal year: calendar year

Inflation rate (consumer prices): -0.5% (2017 est.)
-0.9% (2016 est.)
country comparison to the world: 6

Current account balance: $22 million (2017 est.)
$133 million (2016 est.)
country comparison to the world: 60

Exports: $137.1 million (2017 est.)
$283.1 million (2016 est.)
country comparison to the world: 193

Exports—partners: US 20.2%, Colombia 17.6%, Venezuela 13%, Netherlands 9.1%, Thailand 8.4%, Panama 4.8% (2017)

Exports—commodities: live animals and animal products, art and collectibles, machinery and electrical equipment, transport equipment

Imports: $1.122 billion (2017 est.)
$1.142 billion (2016 est.)
country comparison to the world: 182

Imports—commodities: machinery and electrical equipment, refined oil for bunkering and reexport, chemicals; foodstuffs

Imports—partners: US 53.7%, Netherlands 13.1% (2017)

Reserves of foreign exchange and gold: $921.8 million (31 December 2017 est.)

$828 million (31 December 2015 est.)
country comparison to the world: 134

Debt—external: $693.2 million (31 December 2014 est.)
$666.4 million (31 December 2013 est.)
country comparison to the world: 172

Exchange rates: Aruban guilders/florins per US dollar -
1.79 (2017 est.)
1.79 (2016 est.)
1.79 (2015 est.)
1.79 (2014 est.)
1.79 (2013 est.)

ENERGY

Electricity access: *population without electricity:* 11,364 (2012)
electrification—total population: 95.6% (2016)
electrification—urban areas: 100% (2016)
electrification—rural areas: 92.5% (2016)

Electricity—production: 939 million kWh (2016 est.)
country comparison to the world: 153

Electricity—consumption: 873.3 million kWh (2016 est.)
country comparison to the world: 158

Electricity—exports: 0 kWh (2016 est.)
country comparison to the world: 100

Electricity—imports: 0 kWh (2016 est.)
country comparison to the world: 121

Electricity—installed generating capacity: 296,000 kW (2016 est.)
country comparison to the world: 159

Electricity—from fossil fuels: 87% of total installed capacity (2016 est.)
country comparison to the world: 61

Electricity—from nuclear fuels: 0% of total installed capacity (2017 est.)
country comparison to the world: 39

Electricity—from hydroelectric plants: 0% of total installed capacity (2017 est.)
country comparison to the world: 154

Electricity—from other renewable sources: 13% of total installed capacity (2017 est.)
country comparison to the world: 66

Crude oil—production: 0 bbl/day (2018 est.)
country comparison to the world: 107

Crude oil—exports: 0 bbl/day (2015 est.)
country comparison to the world: 87

Crude oil—imports: 0 bbl/day (2015 est.)
country comparison to the world: 91

Crude oil—proved reserves: 0 bbl (1 January 2018 est.)
country comparison to the world: 104

Refined petroleum products—production: 0 bbl/day (2015 est.)
country comparison to the world: 115

Refined petroleum products—consumption: 8,000 bbl/day (2016 est.)
country comparison to the world: 163

Refined petroleum products—exports: 0 bbl/day (2015 est.)
country comparison to the world: 127

Refined petroleum products—imports: 7,891 bbl/day (2015 est.)
country comparison to the world: 153

Natural gas—production: 1 cu m (2017 est.)
country comparison to the world: 96

Natural gas—consumption: 1 cu m (2017 est.)
country comparison to the world: 116

Natural gas—exports: 1 cu m (2017 est.)
country comparison to the world: 56

Natural gas—imports: 1 cu m (2017 est.)
country comparison to the world: 80

Natural gas—proved reserves: 0 cu m (1 January 2014 est.)
country comparison to the world: 107

Carbon dioxide emissions from consumption of energy: 1.266 million Mt (2017 est.)
country comparison to the world: 162

COMMUNICATIONS

Telephones—fixed lines: *total subscriptions:* 39,582
subscriptions per 100 inhabitants: 33.54 (2019 est.)
country comparison to the world: 163

Telephones—mobile cellular: *total subscriptions:* 159,471
subscriptions per 100 inhabitants: 135.13 (2019 est.)
country comparison to the world: 189

Telecommunication systems: *general assessment:* modern fully automatic telecommunications system; increased competition through privatization has increased mobile-cellular teledensity; three mobile-cellular service providers are now licensed; MNO (mobile network operator) launched island-wide LTE services; MNP (mobile number portability) introduced (2018)
domestic: ongoing changes in regulations and competition improving teledensity; 34 per 100 fixed-line, 135 per 100 mobile-cellular (2019)
international: country code—297; landing points for the PAN-AM, PCCS, Deep Blue Cable, and Alonso de Ojeda submarine telecommunications cable system that extends from Trinidad and Tobago, Florida, Puerto Ricco, Jamaica, Guyana, Sint Eustatius & Saba, Suriname, Dominican Republic, BVI, USVI, Haiti, Cayman Islands, the Netherlands Antilles, through Aruba to Panama, Venezuela, Colombia, Ecuador, Peru and Chile; extensive interisland microwave radio relay links (2019)
note: the COVID-19 outbreak is negatively impacting telecommunications production and supply chains globally; consumer spending on telecom devices and services has also slowed due to the pandemic's effect on economies worldwide; overall progress towards improvements in all facets of the telecom industry—mobile, fixed-line, broadband, submarine cable and satellite—has moderated

Broadcast media: 2 commercial TV stations; cable TV subscription service provides access to foreign channels; about 19 commercial radio stations broadcast (2017)

Internet country code: .aw

Internet users: *total:* 113,277
percent of population: 97.17% (July 2018 est.)
country comparison to the world: 178

TRANSPORTATION

National air transport system: *number of registered air carriers:* 3 (2020)
inventory of registered aircraft operated by air carriers: 19
annual passenger traffic on registered air carriers: 274,280 (2018)

Civil aircraft registration country code prefix: P4 (2016)

Airports: 1 (2020)
country comparison to the world: 212

Airports—with paved runways: *total:* 1 (2019)
2,438 *to 3,047 m:* 1

Roadways: *total:* 1,000 km (2010)
country comparison to the world: 185

Ports and terminals: *major seaport(s):* Barcadera, Oranjestad
oil terminal(s): Sint Nicolaas
cruise port(s): Oranjestad

MILITARY AND SECURITY

Military and security forces: no regular military forces (2011)

Military—note: defense is the responsibility of the Netherlands; the Aruba security services focus on organized crime and terrorism

TRANSNATIONAL ISSUES

Disputes—international: none

Illicit drugs: transit point for US- and Europe-bound narcotics with some accompanying money-laundering activity; relatively high percentage of population consumes cocaine

ASHMORE AND CARTIER ISLANDS

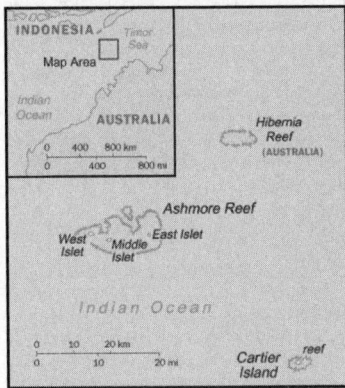

INDONESIA
Timor
Sea
Map Area
Indian
Ocean
AUSTRALIA
0   400   800 km
0   400   800 nu
Hibernia
Reef
(AUSTRALIA)
Ashmore Reef
West
Islet
Middle
Islet
East Islet
Indian Ocean
0   10   20 km
0   10   20 mi
Cartier
Island
reef

INTRODUCTION

Background: These uninhabited islands came under Australian authority in 1931; formal administration began two years later. Ashmore Reef supports a rich and diverse avian and marine habitat; in 1983, it became a National Nature Reserve. Cartier Island, a former bombing range, became a marine reserve in 2000.

GEOGRAPHY

Location: Southeastern Asia, islands in the Indian Ocean, midway between northwestern Australia and Timor island

Geographic coordinates: 12 14 S, 123 05 E

Map references: Southeast Asia

Area: *total:* 5 sq km
land: 5 sq km
water: 0 sq km
note: includes Ashmore Reef (West, Middle, and East Islets) and Cartier Island
country comparison to the world: 250

Area—comparative: about eight times the size of the National Mall in Washington, DC

Land boundaries: 0 km

Coastline: 74.1 km

Maritime claims: *territorial sea:* 12 nm
contiguous zone: 12 nm
continental shelf: 200-m depth or to the depth of exploitation
exclusive fishing zone: 200 nm

Climate: tropical

Terrain: low with sand and coral

Elevation: *lowest point:* Indian Ocean 0 m
highest point: Cartier Island 5 m

Natural resources: fish

Land use: 0% (2014 est.)

Natural hazards: surrounded by shoals and reefs that can pose maritime hazards

Environment—current issues: illegal killing of protected wildlife by traditional Indonesian fisherman, as well as fishing by non-traditional Indonesian vessels, are ongoing problems; sea level rise, changes in sea temperature, and ocean acidification are concerns; marine debris

Geography—note: Ashmore Reef National Nature Reserve established in August 1983; Cartier Island Marine Reserve established in 2000

PEOPLE AND SOCIETY

Population: no indigenous inhabitants
note: Indonesian fishermen are allowed access to the lagoon and fresh water at Ashmore Reef's West Island; access to East and Middle Islands is by permit only

GOVERNMENT

Country name: *conventional long form:* Territory of Ashmore and Cartier Islands

conventional short form: Ashmore and Cartier Islands
etymology: named after British Captain Samuel ASHMORE, who first sighted his namesake island in 1811, and after the ship Cartier, from which the second island was discovered in 1800

Dependency status: territory of Australia; administered from Canberra by the Department of Regional Australia, Local Government, Arts and Sport

Legal system: the laws of the Commonwealth of Australia and the laws of the Northern Territory of Australia, where applicable, apply

Citizenship: see Australia

Diplomatic representation in the US: none (territory of Australia)

Diplomatic representation from the US: none (territory of Australia)

Flag description: the flag of Australia is used

ECONOMY

Economy—overview: no economic activity

TRANSPORTATION

Ports and terminals: none; offshore anchorage only

MILITARY AND SECURITY

Military—note: defense is the responsibility of Australia; periodic visits by the Royal Australian Navy and Royal Australian Air Force

TRANSNATIONAL ISSUES

Disputes—international: Australia has closed parts of the Ashmore and Cartier reserve to Indonesian traditional fishing; Indonesian groups challenge Australia's claim to Ashmore Reef

ATLANTIC OCEAN

INTRODUCTION

Background: The Atlantic Ocean is the second largest of the world's five oceans (after the Pacific Ocean, but larger than the Indian Ocean, Southern Ocean, and Arctic Ocean). The Kiel Canal (Germany), Oresund (Denmark-Sweden), Bosporus (Turkey), Strait of Gibraltar (Morocco-Spain), and the Saint Lawrence Seaway (Canada-US) are important strategic access waterways. The decision by the International Hydrographic Organization in the spring of 2000 to delimit a fifth world ocean, the Southern

Ocean, removed the portion of the Atlantic Ocean south of 60 degrees south latitude.

GEOGRAPHY

Location: body of water between Africa, Europe, the Arctic Ocean, the Americas, and the Southern Ocean

Geographic coordinates: 0 00 N, 25 00 W

Map references: Political Map of the World

Area: *total:* 76.762 million sq km

note: includes Baltic Sea, Black Sea, Caribbean Sea, Davis Strait, Denmark Strait, part of the Drake Passage, Gulf of Mexico, Labrador Sea, Mediterranean Sea, North Sea, Norwegian Sea, almost all of the Scotia Sea, and other tributary water bodies

Area—comparative: about 7.5 times the size of the US

Coastline: 111, 866 km

Climate: tropical cyclones (hurricanes) develop off the coast of Africa near Cabo Verde and move westward into the Caribbean Sea; hurricanes can

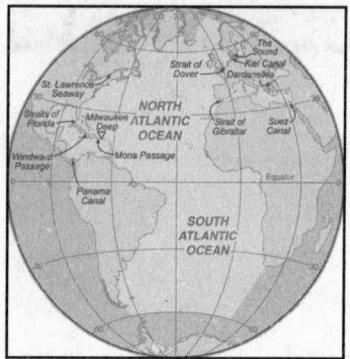

occur from May to December but are most frequent from August to November

Terrain: surface usually covered with sea ice in Labrador Sea, Denmark Strait, and coastal portions of the Baltic Sea from October to June; surface dominated by two large gyres (broad, circular systems of currents), one in the northern Atlantic and another in the southern Atlantic; the ocean floor is dominated by the Mid-Atlantic Ridge, a rugged north-south centerline for the entire Atlantic basin

major surface currents: clockwise North Atlantic Gyre consists of the northward flowing, warm Gulf Stream in the west, the eastward flowing North Atlantic Current in the north, the southward flowing cold Canary Current in the east, and the westward flowing North Equatorial Current in the south; the counterclockwise South Atlantic Gyre composed of the southward flowing warm Brazil Current in the west, the eastward flowing South Atlantic Current in the south, the northward flowing cold Benguela Current in the east, and the westward flowing South Equatorial Current in the north

Elevation: *mean depth:* -3,646 m
lowest point: Milwaukee Deep in the Puerto Rico Trench -8,605 m
highest point: sea level

Natural resources: oil and gas fields, fish, marine mammals (seals and whales), sand and gravel aggregates, placer deposits, polymetallic nodules, precious stones

Natural hazards: icebergs common in Davis Strait, Denmark Strait, and the northwestern Atlantic Ocean from February to August and have been spotted as far south as Bermuda and the Madeira Islands; ships subject to superstructure icing in extreme northern Atlantic from October to May; persistent fog can be a maritime hazard from May to September; hurricanes (May to December)

Environment—current issues: endangered marine species include the manatee, seals, sea lions, turtles, and whales; unsustainable exploitation of fisheries (over fishing, bottom trawling, drift net fishing, discards, catch of non-target species); pollution (maritime transport, discharges, offshore drilling, oil spills); municipal sludge pollution off eastern US, southern Brazil, and eastern

Argentina; oil pollution in Caribbean Sea, Gulf of Mexico, Lake Maracaibo, Mediterranean Sea, and North Sea; industrial waste and municipal sewage pollution in Baltic Sea, North Sea, and Mediterranean Sea

Geography—note: major chokepoints include the Dardanelles, Strait of Gibraltar, access to the Panama and Suez Canals; strategic straits include the Strait of Dover, Straits of Florida, Mona Passage, The Sound (Oresund), and Windward Passage; the Equator divides the Atlantic Ocean into the North Atlantic Ocean and South Atlantic Ocean

GOVERNMENT

Country name: *etymology:* name derives from the Greek description of the waters beyond the Strait of Gibraltar, Atlantis thalassa, meaning "Sea of Atlas"

ECONOMY

Economy—overview: The Atlantic Ocean provides some of the world's most heavily trafficked sea routes, between and within the Eastern and Western Hemispheres. Other economic activity includes the exploitation of natural resources, e.g., fishing, dredging of aragonite sands (The Bahamas), and production of crude oil and natural gas (Caribbean Sea, Gulf of Mexico, and North Sea).

Marine fisheries: the Atlantic Ocean fisheries are the second most important in the world accounting for 28%, or 22,434,652 mt, of the global catch in 2017; of the seven regions delineated by the Food and Agriculture Organization in the Atlantic basin, the most important include the following: Northeast Atlantic region (Region 27) is the third most important in the world producing more than 11% of the global catch or 9,309,821 mt in 2017; the region encompasses the waters north of 36° North latitude and east of 40° West longitude with the major producers including Norway (2,208,175 mt), Iceland (1,163,166 mt), Russia (1,105,548 mt), UK (717,545 mt), and Denmark (901,939 mt); the region includes the historically important fishing grounds of the North Sea, the Baltic Sea, and the Atlantic waters between Greenland, Iceland, and the British Isles; the principal catches include Atlantic cod, haddock, saithe (pollock), Blue Whiting, herring, and mackerel; not all fish caught are for human consumption, half of fish catches in the North Sea are processed as fish oil or fish meal, which are used in animal fodder

Eastern Central Atlantic region (Region 34) is the second most important Atlantic fishery, and seventh largest in the world producing more than 6% of the global catch or 5,085,264 mt in 2017; the region encompasses the waters between 36° North and 6° South latitude and east of 40° West longitude off the west coast of Africa with the major producers including Morocco (1,336,787 mt), Mauritania (779,580 mt), Nigeria (496,206 mt), Senegal (464,199 mt), Ghana (291,904 mt), Cameroon (205,190 mt), and Sierra Leone

(200,000 mt); the principal catches include pilchard, sardinellas, shad, and mackerel

Northwest Atlantic region (Region 21) is the third most important Atlantic fishery and ninth in the world producing a little more than 2% of the global catch and 1,755,861 mt in 2017; it encompasses the waters north of 35° North latitude and west of 42° West longitude including the important fishing grounds over the continental shelf of North America such as the Grand Banks, the Georges Bank, and the Flemish Cap, as well as Baffin Bay with the major producers including the US (889,668 mt), Canada (624,747 mt), and Greenland (169,830 mt); the principal catches include sea scallops, prawns, lobster, herring, and menhaden

Mediterranean and Black Sea region (Region 37) is a minor fishing region representing 1.6% or 1,348,299 mt of the world's total capture in 2017; the region encompasses all waters east of the Strait of Gibraltar with the major producers including Turkey (322,175 mt), Italy (185,067 mt), Tunisia (109,636 mt), Russia (90,883 mt), and Spain (86,342 mt); the principal catches include European anchovy, European pilchard, Gobies, and clams

FAO map of world fishing regions; used with permission.:

TRANSPORTATION

Ports and terminals: *major seaport(s):* Alexandria (Egypt), Algiers (Algeria), Antwerp (Belgium), Barcelona (Spain), Buenos Aires (Argentina), Casablanca (Morocco), Colon (Panama), Copenhagen (Denmark), Dakar (Senegal), Gdansk (Poland), Hamburg (Germany), Helsinki (Finland), Las Palmas (Canary Islands, Spain), Le Havre (France), Lisbon (Portugal), London (UK), Marseille (France), Montevideo (Uruguay), Montreal (Canada), Naples (Italy), New Orleans (US), New York (US), Oran (Algeria), Oslo (Norway), Peiraiefs or Piraeus (Greece), Rio de Janeiro (Brazil), Rotterdam (Netherlands), Saint Petersburg (Russia), Stockholm (Sweden)

Transportation—note: Kiel Canal and Saint Lawrence Seaway are two important waterways; significant domestic commercial and recreational use of Intracoastal Waterway on central and south Atlantic seaboard and Gulf of Mexico coast of US; the International Maritime Bureau reports the territorial waters of littoral states and offshore Atlantic waters as high risk for piracy and armed robbery against ships, particularly in the Gulf of Guinea off West Africa; in 2014, 41 commercial vessels were attacked in the Gulf of Guinea with 5 hijacked and 144 crew members taken hostage; hijacked vessels are often disguised and cargoes stolen; crews have been robbed and stores or cargoes stolen

MILITARY AND SECURITY

Maritime threats: West African piracy more than doubled in 2018 totaling 85 attacks, including all of the six ships highjacked during the year; 13 of the

18 vessels fired upon world-wide occurred in West African waters; Nigerian pirates are very aggresive, operating as far as 200 nm offshore and boarded 29 ships in 2018; the Maritime Administration of the US Department of Transportation has issued a Maritime Advisory (2019-010-Gulf of Guinea-Piracy/Armed Robbery/Kidnapping for Ransom) effective 19 July 2019, which states in part "Piracy, armed robbery, and kidnapping for ransom (KFR)

continue to serve as significant threats to U.S. flagged vessels transiting or operating in the Gulf of Guinea (GoG). ...According to the Office of Naval Intelligence's "Weekly Piracy Reports" 72 reported incidents of piracy and armed robbery at sea occurred in the GoG region this year as of July 9, 2019. Attacks, kidnappings for ransom (KFR), and boardings to steal valuables from the ships and crews are the most common types of incidents with

approximately 75 percent of all incidents taking place off Nigeria. During the first six months of 2019, there were 15 kidnapping and 3 hijackings in the GoG."

TRANSNATIONAL ISSUES

Disputes—international: some maritime disputes (see littoral states)

AUSTRALIA

INTRODUCTION

Background: Prehistoric settlers arrived on the continent from Southeast Asia at least 40,000 years before the first Europeans began exploration in the 17th century. No formal territorial claims were made until 1770, when Capt. James COOK took possession of the east coast in the name of Great Britain (all of Australia was claimed as British territory in 1829 with the creation of the colony of Western Australia). Six colonies were created in the late 18th and 19th centuries; they federated and became the Commonwealth of Australia in 1901. The new country took advantage of its natural resources to rapidly develop agricultural and manufacturing industries and to make a major contribution to the Allied effort in World Wars I and II.

In recent decades, Australia has become an internationally competitive, advanced market economy due in large part to economic reforms adopted in the 1980s and its location in one of the fastest growing regions of the world economy. Longterm concerns include an aging population, pressure on infrastructure, and environmental issues such as floods, droughts, and bushfires. Australia is the driest inhabited continent on earth, making it particularly vulnerable to the challenges of climate change. Australia is home to 10% of the world's biodiversity, and a great number of its flora and fauna exist nowhere else in the world.

GEOGRAPHY

Location: Oceania, continent between the Indian Ocean and the South Pacific Ocean

Geographic coordinates: 27 00 S, 133 00 E

Map references: Oceania

Area: *total:* 7,741,220 sq km
land: 7,682,300 sq km
water: 58,920 sq km
note: includes Lord Howe Island and Macquarie Island
country comparison to the world: 7

Area—comparative: slightly smaller than the US contiguous 48 states

Land boundaries: 0 km

Coastline: 25,760 km

Maritime claims: *territorial sea:* 12 nm
exclusive economic zone: 200 nm
contiguous zone: 24 nm
continental shelf: 200 nm or to the edge of the continental margin

Climate: generally arid to semiarid; temperate in south and east; tropical in north

Terrain: mostly low plateau with deserts; fertile plain in southeast

Elevation: *mean elevation:* 330 m
lowest point: Lake Eyre -15 m
highest point: Mount Kosciuszko 2,228 m

Natural resources: alumina, coal, iron ore, copper, tin, gold, silver, uranium, nickel, tungsten, rare earth elements, mineral sands, lead, zinc, diamonds, natural gas, petroleum; note—Australia is the world's largest net exporter of coal accounting for 29% of global coal exports

Land use: *agricultural land:* 52.9% (2016 est.)
arable land: 11.6% (2016 est.) / permanent crops:
0.09% (2016 est.) / permanent pasture: 88.4%
(2016 est.)
forest: 16.2% (2016 est.)
other: 30.9% (2016 est.)

Irrigated land: 25,460 sq km (2014)

Population distribution: population is primarily located on the periphery, with the highest concentration of people residing in the east and southeast; a secondary population center is located in and around Perth in the west; of the States and Territories, New South Wales has, by far, the

largest population; the interior, or "outback", has a very sparse population

Natural hazards: cyclones along the coast; severe droughts; forest fires
volcanism: volcanic activity on Heard and McDonald Islands

Environment—current issues: soil erosion from overgrazing, deforestation, industrial development, urbanization, and poor farming practices; limited natural freshwater resources; soil salinity rising due to the use of poor quality water; drought, desertification; clearing for agricultural purposes threatens the natural habitat of many unique animal and plant species; disruption of the fragile ecosystem has resulted in significant floral extinctions; the Great Barrier Reef off the northeast coast, the largest coral reef in the world, is threatened by increased shipping and its popularity as a tourist site; overfishing, pollution, and invasive species are also problems

Environment—international agreements: party to: Antarctic-Environmental Protocol, Antarctic-Marine Living Resources, Antarctic Seals, Antarctic Treaty, Biodiversity, Climate Change, Climate Change-Kyoto Protocol, Desertification, Endangered Species, Environmental Modification, Hazardous Wastes, Law of the Sea, Marine Dumping, Marine Life Conservation, Ozone Layer Protection, Ship Pollution, Tropical Timber 83, Tropical Timber 94, Wetlands, Whaling
signed, but not ratified: none of the selected agreements

Geography—note: note 1: world's smallest continent but sixth-largest country; the largest country in Oceania, the largest country entirely in the Southern Hemisphere, and the largest country without land borders; the only continent without glaciers; the invigorating sea breeze known as the "Fremantle Doctor" affects the city of Perth on the west coast and is one of the most consistent winds in the world
note 2: the Great Dividing Range that runs along eastern Australia is that continent's longest mountain range and the third-longest land-based range in the world; the term "Great Dividing Range" refers to the fact that the mountains form a watershed crest from which all of the rivers of eastern Australia flow – east, west, north, and south

PEOPLE AND SOCIETY

Population: 25,466,459 (July 2020 est.)
country comparison to the world: 55

Nationality: *noun:* Australian(s)
adjective: Australian

Ethnic groups: English 25.9%, Australian 25.4%, Irish 7.5%, Scottish 6.4%, Italian 3.3%, German 3.2%, Chinese 3.1%, Indian 1.4%, Greek 1.4%, Dutch 1.2%, other 15.8% (includes Australian aboriginal .5%), unspecified 5.4% (2011 est.)
note: data represent self-identified ancestry, over a third of respondents reported two ancestries

Languages: English 72.7%, Mandarin 2.5%, Arabic 1.4%, Cantonese 1.2%, Vietnamese 1.2%, Italian 1.2%, Greek 1%, other 14.8%, unspecified 6.5% (2016 est.)
note: data represent language spoken at home

Religions: Protestant 23.1% (Anglican 13.3%, Uniting Church 3.7%, Presbyterian and Reformed 2.3%, Baptist 1.5%, Pentecostal 1.1%, Lutheran .7%, other Protestant .5%), Roman Catholic 22.6%, other Christian 4.2%, Muslim 2.6%, Buddhist 2.4%, Orthodox 2.3% (Eastern Orthodox 2.1%, Oriental Orthodox .2%), Hindu 1.9%, other 1.3%, none 30.1%, unspecified 9.6% (2016 est.)

Age structure: 0-14 *years:* 18.72% (male 2,457,418/female 2,309,706)
15-24 years: 12.89% (male 1,710,253/female 1,572,794)
25-54 years: 41.15% (male 5,224,840/female 5,255,041)
55-64 years: 11.35% (male 1,395,844/female 1,495,806)
65 years and over: 15.88% (male 1,866,761/female 2,177,996) (2020 est.)

Dependency ratios: *total dependency ratio:* 55.1
youth dependency ratio: 29.9
elderly dependency ratio: 25.1
potential support ratio: 4 (2020 est.)

Median age: *total:* 37.5 years
male: 36.5 years
female: 38.5 years (2020 est.)
country comparison to the world: 69

Population growth rate: 1.4% (2020 est.)
country comparison to the world: 79

Birth rate: 12.4 births/1,000 population (2020 est.)
country comparison to the world: 156

Death rate: 6.9 deaths/1,000 population (2020 est.)
country comparison to the world: 128

Net migration rate: 8.1 migrant(s)/1,000 population (2020 est.)
country comparison to the world: 11

Population distribution: population is primarily located on the periphery, with the highest concentration of people residing in the east and southeast; a secondary population center is located in and around Perth in the west; of the States and Territories, New South Wales has, by far, the largest population; the interior, or "outback", has a very sparse population

Urbanization: *urban population:* 86.2% of total population (2020)
rate of urbanization: 1.43% annual rate of change (2015-20 est.)
note: data include Christmas Island, Cocos Islands, and Norfolk Island

total population growth rate v. urban population growth rate, 2000-2030: Major urban areas—population: 4.968 million Melbourne, 4.926 million Sydney, 2.406 million Brisbane, 2.042 million Perth, 1.336 million Adelaide, 457,000 CANBERRA (capital) (2020)

Sex ratio: *at birth:* 1.06 male(s)/female
0-14 years: 1.06 male(s)/female
15-24 years: 1.09 male(s)/female
25-54 years: 0.99 male(s)/female
55-64 years: 0.93 male(s)/female
65 years and over: 0.86 male(s)/female
total population: 0.99 male(s)/female (2020 est.)

Mother's mean age at first birth: 28.7 years (2014 est.)

Maternal mortality rate: 6 deaths/100,000 live births (2017 est.)
country comparison to the world: 159

Infant mortality rate: *total:* 3.1 deaths/1,000 live births
male: 3.3 deaths/1,000 live births
female: 2.9 deaths/1,000 live births (2020 est.)
country comparison to the world: 214

Life expectancy at birth: *total population:* 82.7 years
male: 80.5 years
female: 85 years (2020 est.)
country comparison to the world: 14

Total fertility rate: 1.74 children born/woman (2020 est.)
country comparison to the world: 162

Contraceptive prevalence rate: 66.9% (2015/16)
note: percent of women aged 18-45

Drinking water source:
improved:
urban: 100% of population
rural: 100% of population
total: 100% of population
unimproved:
urban: 0% of population
rural: 0% of population
total: 0% of population (2017 est.)

Current Health Expenditure: 9.2% (2017)

Physicians density: 3.68 physicians/1,000 population (2017)

Hospital bed density: 3.8 beds/1,000 population (2016)

Sanitation facility access:
improved:
total: 100% of population
unimproved:
total: 0% of population (2017 est.)

HIV/AIDS—adult prevalence rate: 0.1% (2019 est.)
country comparison to the world: 117

HIV/AIDS—people living with HIV/AIDS: 29,000 (2019 est.)
country comparison to the world: 76

HIV/AIDS—deaths: <100 (201 est.)

Obesity—adult prevalence rate: 29% (2016)
country comparison to the world: 27

Education expenditures: 5.3% of GDP (2016)
country comparison to the world: 45

School life expectancy (primary to tertiary education): *total:* 21 years
male: 20 years
female: 21 years (2018)

Unemployment, youth ages 15-24: *total:* 11.8%
male: 112.8%
female: 10.7% (2018 est.)
country comparison to the world: 110

GOVERNMENT

Country name: conventional long form: Commonwealth of Australia
conventional short form: Australia
etymology: the name Australia derives from the Latin "australis" meaning "southern"; the Australian landmass was long referred to as "Terra Australis" or the Southern Land

Government type: federal parliamentary democracy under a constitutional monarchy; a Commonwealth realm

Capital: *name:* Canberra
geographic coordinates: 35 16 S, 149 08 E
time difference: UTC+10 (15 hours ahead of Washington, DC, during Standard Time)
daylight saving time: +1hr, begins first Sunday in October; ends first Sunday in April
note: Australia has four time zones, including Lord Howe Island (UTC+10:30)
etymolgy: the name is claimed to derive from either Kambera or Camberry, which are names corrupted from the original native designation for the area "Nganbra" or "Nganbira"

Administrative divisions: 6 states and 2 territories*; Australian Capital Territory*, New South Wales, Northern Territory*, Queensland, South Australia, Tasmania, Victoria, Western Australia

Dependent areas: Ashmore and Cartier Islands, Christmas Island, Cocos (Keeling) Islands, Coral Sea Islands, Heard Island and McDonald Islands, Norfolk Island

Independence: 1 January 1901 (from the federation of UK colonies)

National holiday: Australia Day (commemorates the arrival of the First Fleet of Australian settlers), 26 January (1788); ANZAC Day (commemorates the anniversary of the landing of troops of the Australian and New Zealand Army Corps during World War I at Gallipoli, Turkey), 25 April (1915)

Constitution: *history:* approved in a series of referenda from 1898 through 1900 and became law 9 July 1900, effective 1 January 1901
amendments: proposed by Parliament; passage requires approval of a referendum bill by absolute majority vote in both houses of Parliament, approval in a referendum by a majority of voters in at least four states and in the territories, and Royal Assent; proposals that would reduce a state's representation in either house or change a state's

boundaries require that state's approval prior to Royal Assent; amended several times, last in 1977

Legal system: common law system based on the English model

International law organization participation: accepts compulsory ICJ jurisdiction with reservations; accepts ICCt jurisdiction

Citizenship: *citizenship by birth:* no
citizenship by descent only: at least one parent must be a citizen or permanent resident of Australia
dual citizenship recognized: yes
residency requirement for naturalization: 4 years

Suffrage: 18 years of age; universal and compulsory

Executive branch: chief of state: Queen ELIZABETH II (since 6 February 1952); represented by Governor General David HURLEY (since 1 July 2019)
head of government: Prime Minister Scott MORRISON (since 24 August 2018)
cabinet: Cabinet nominated by the prime minister from among members of Parliament and sworn in by the governor general
elections/appointments: the monarchy is hereditary; governor general appointed by the monarch on the recommendation of the prime minister; following legislative elections, the leader of the majority party or majority coalition is sworn in as prime minister by the governor general

Legislative branch: description: bicameral Federal Parliament consists of: Senate (76 seats; 12 members from each of the 6 states and 2 each from the 2 mainland territories; members directly elected in multi-seat constituencies by proportional representation vote; members serve 6-year terms with one-half of state membership renewed every 3 years and territory membership renewed every 3 years)

House of Representatives (151 seats; members directly elected in single-seat constituencies by majority preferential vote; members serve terms of up to 3 years)

elections: Senate—last held on 18 May 2019 (next to be held in 2022)
House of Representatives—last held on 18 May 2019 (next to be held in 2022)

election results: Senate—percent of vote by party—Liberal/National coalition 37.99%, ALP 28.79%, The Greens 10.19%, One Nation 5.4%, Centre Alliance .19%, Lambie Network .21%, other 17.23%; seats by party—Liberal/National coalition 35, ALP 26, The Greens 9, One Nation 2, Centre Alliance 2, Lambie Network 1, independents 1

House of Representatives—percent of vote by party—Liberal/National coalition 41.4%, ALP 33.3%, The Greens 10.4%, Katter's Australian Party .49%, Centre Alliance .33%, independents 3.37%, other 10.63%; seats by party—Liberal/National Coalition 77, ALP 68, The Greens 1, Katter's Australian Party 1, Centre Alliance 1, independent 3

Judicial branch: *highest courts:* High Court of Australia (consists of 7 justices, including the chief justice); note—each of the 6 states, 2 territories, and Norfolk Island has a Supreme Court; the High Court is the final appellate court beyond the state and territory supreme courts
judge selection and term of office: justices appointed by the governor-general in council for life with mandatory retirement at age 70
subordinate courts: at the federal level: Federal Court; Federal Magistrates' Courts of Australia; Family Court; at the state and territory level: Local Court—New South Wales; Magistrates' Courts – Victoria, Queensland, South Australia, Western Australia, Tasmania, Northern Territory, Australian Capital Territory; District Courts – New South Wales, Queensland, South Australia, Western Australia; County Court – Victoria; Family Court – Western Australia; Court of Petty Sessions – Norfolk Island

Political parties and leaders: Australian Greens Party [Adam BANDT]
Australian Labor Party or ALP [Anthony ALBANESE]
Country Liberal Party or CLP [Gary HIGGINS]
Liberal National Party of Queensland or LNP [Deborah FRECKLINGTON]
Liberal Party of Australia [Scott MORRISON]
The Nationals [Michael MCCORMACK]
Centre Alliance [Nick XENOPHON]
Pauline Hanson's One Nation [Pauline HANSON]

International organization participation: ADB, ANZUS, APEC, ARF, ASEAN (dialogue partner), Australia Group, BIS, C, CD, CP, EAS, EBRD, EITI (implementing country), FAO, FATF, G-20, IAEA, IBRD, ICAO, ICC (national committees), ICCt, ICRM, IDA, IEA, IFC, IFRCS, IHO, ILO, IMF, IMO, IMSO, Interpol, IOC, IOM, IPU, ISO, ITSO, ITU, ITUC (NGOs), MIGA, NEA, NSG, OECD, OPCW, OSCE (partner), Pacific Alliance (observer), Paris Club, PCA, PIF, SAARC (observer), SICA (observer), Sparteca, SPC, UN, UNCTAD, UNESCO, UNHCR, UNMISS, UNMIT, UNRWA, UNTSO, UNWTO, UPU, WCO, WFTU (NGOs), WHO, WIPO, WMO, WTO, ZC

Diplomatic representation in the US: *chief of mission:* Ambassador Arthur SINODINOS (since 6 February 2020)
chancery: 1601 Massachusetts Avenue NW, Washington, DC 20036
telephone: [1] (202) 797-3000
FAX: [1] (202) 797-3168
consulate(s) general: Atlanta, Chicago, Honolulu, Houston, Los Angeles, New York, San Francisco

Diplomatic representation from the US: *chief of mission:* Ambassador Arthur B. CULVAHOUSE (since 19 February 2019)
telephone: [61] (02) 6214-5600
embassy: Moonah Place, Yarralumla, Canberra, Australian Capital Territory 2600
mailing address: APO AP 96549
FAX: [61] (02) 6214-5970
consulate(s) general: Melbourne, Perth, Sydney

Flag description: blue with the flag of the UK in the upper hoist-side quadrant and a large seven-pointed star in the lower hoist-side quadrant known as the Commonwealth or Federation Star, representing the federation of the colonies of Australia in 1901; the star depicts one point for each of the six original states and one representing all of Australia's internal and external territories; on the fly half is a representation of the Southern Cross constellation in white with one small, five-pointed star and four larger, seven-pointed stars

National symbol(s): Commonwealth Star (seven-pointed Star of Federation), golden wattle tree (Acacia pycnantha Benth), kangaroo, emu; national colors: green, gold

National anthem: *name:* Advance Australia Fair
lyrics/music: Peter Dodds McCORMICK
note: adopted 1984; although originally written in the late 19th century, the anthem was not used for all official occasions until 1984; as a Commonwealth country, in addition to the national anthem, "God Save the Queen" serves as the royal anthem (see United Kingdom)
0:00 / 0:54

ECONOMY

Economy—overview: Australia is an open market with minimal restrictions on imports of goods and services. The process of opening up has increased productivity, stimulated growth, and made the economy more flexible and dynamic. Australia plays an active role in the WTO, APEC, the G20, and other trade forums. Australia's free trade agreement (FTA) with China entered into force in 2015, adding to existing FTAs with the Republic of Korea, Japan, Chile, Malaysia, New Zealand, Singapore, Thailand, and the US, and a regional FTA with ASEAN and New Zealand. Australia continues to negotiate bilateral agreements with Indonesia, as well as larger agreements with its Pacific neighbors and the Gulf Cooperation Council countries, and an Asia-wide Regional Comprehensive Economic Partnership that includes the 10 ASEAN countries and China, Japan, Korea, New Zealand, and India.

Australia is a significant exporter of natural resources, energy, and food. Australia's abundant and diverse natural resources attract high levels of foreign investment and include extensive reserves of coal, iron, copper, gold, natural gas, uranium, and renewable energy sources. A series of major investments, such as the US$40 billion Gorgon Liquid Natural Gas Project, will significantly expand the resources sector.

For nearly two decades up till 2017, Australia had benefited from a dramatic surge in its terms of trade. As export prices increased faster than import prices, the economy experienced continuous growth, low unemployment, contained inflation, very low public debt, and a strong and stable financial system. Australia entered 2018 facing a range of growth constraints, principally driven by the sharp fall in global prices of key export commodities. Demand for resources and energy from Asia and especially China is growing at a slower pace and sharp drops in export prices have impacted growth.

GDP (purchasing power parity): $1.248 trillion (2017 est.)
$1.221 trillion (2016 est.)
$1.19 trillion (2015 est.)
note: data are in 2017 dollars
country comparison to the world: 19

GDP (official exchange rate): $1.38 trillion (2017 est.)

GDP—real growth rate: 2.2% (2017 est.)
2.6% (2016 est.)
2.5% (2015 est.)
country comparison to the world: 144

GDP—per capita (PPP): $50,400 (2017 est.)
$50,100 (2016 est.)
$49,600 (2015 est.)
note: data are in 2017 dollars
country comparison to the world: 29

Gross national saving: 21% of GDP (2017 est.)
20.5% of GDP (2016 est.)
21.5% of GDP (2015 est.)
country comparison to the world: 88

GDP—composition, by end use: household consumption: 56.9% (2017 est.)
government consumption: 18.4% (2017 est.)
investment in fixed capital: 24.1% (2017 est.)
investment in inventories: 0.1% (2017 est.)
exports of goods and services: 21.5% (2017 est.)
imports of goods and services: -21% (2017 est.)

GDP—composition, by sector of origin: agriculture: 3.6% (2017 est.)
industry: 25.3% (2017 est.)
services: 71.2% (2017 est.)

Agriculture—products: wheat, barley, sugarcane, fruits; cattle, sheep, poultry

Industries: mining, industrial and transportation equipment, food processing, chemicals, steel

Industrial production growth rate: 1.4% (2017 est.)
country comparison to the world: 144

Labor force: 12.91 million (2017 est.)
country comparison to the world: 44

Labor force—by occupation: agriculture: 3.6%
industry: 21.1%
services: 75.3% (2009 est.)

Unemployment rate: 5.6% (2017 est.)
5.7% (2016 est.)
country comparison to the world: 81

Population below poverty line: NA

Household income or consumption by percentage share: lowest 10%: 2%
highest 10%: 25.4% (1994)

Budget: revenues: 490 billion (2017 est.)
expenditures: 496.9 billion (2017 est.)

Taxes and other revenues: 35.5% (of GDP) (2017 est.)
country comparison to the world: 61

Budget surplus (+) or deficit (-): -0.5% (of GDP) (2017 est.)
country comparison to the world: 60

Public debt: 40.8% of GDP (2017 est.)
40.6% of GDP (2016 est.)

country comparison to the world: 123

Fiscal year: 1 July—30 June

Inflation rate (consumer prices): 2% (2017 est.)
1.3% (2016 est.)
country comparison to the world: 103

Current account balance: -$36.01 billion (2017 est.)
-$41.45 billion (2016 est.)
country comparison to the world: 201

Exports: $231.6 billion (2017 est.)
$191.7 billion (2016 est.)
country comparison to the world: 22

Exports—partners: China 33.5%, Japan 14.6%, South Korea 6.6%, India 5%, Hong Kong 4% (2017)

Exports—commodities: iron ore, coal, gold, natural gas, beef, aluminum ores and conc, wheat, meat (excluding beef), wool, alumina, alcohol

Imports: $221 billion (2017 est.)
$198.7 billion (2016 est.)
country comparison to the world: 24

Imports—commodities: motor vehicles, refined petroleum, telecommunication equipment and parts; crude petroleum, medicaments, goods vehicles, gold, computers

Imports—partners: China 22.9%, US 10.8%, Japan 7.5%, Thailand 5.1%, Germany 4.9%, South Korea 4.5% (2017)

Reserves of foreign exchange and gold: $66.58 billion (31 December 2017 est.)
$55.07 billion (31 December 2016 est.)
country comparison to the world: 33

Debt—external: $1.714 trillion (31 December 2017 est.)
$1.547 trillion (31 December 2016 est.)
country comparison to the world: 11

Exchange rates: Australian dollars (AUD) per US dollar -
1.311 (2017 est.)
1.3442 (2016 est.)
1.3442 (2015 est.)
1.3291 (2014 est.)
1.1094 (2013 est.)

ENERGY

Electricity access: electrification—total population: 100% (2016)

Electricity—production: 243 billion kWh (2016 est.)
country comparison to the world: 19

Electricity—consumption: 229.4 billion kWh (2016 est.)
country comparison to the world: 19

Electricity—exports: 0 kWh (2016 est.)
country comparison to the world: 101

Electricity—imports: 0 kWh (2016 est.)
country comparison to the world: 122

Electricity—installed generating capacity: 65.56 million kW (2016 est.)
country comparison to the world: 18

Electricity—from fossil fuels: 72% of total installed capacity (2016 est.)
country comparison to the world: 101

Electricity—from nuclear fuels: 0% of total installed capacity (2017 est.)
country comparison to the world: 40

Electricity—from hydroelectric plants: 11% of total installed capacity (2017 est.)
country comparison to the world: 113

Electricity—from other renewable sources: 17% of total installed capacity (2017 est.)
country comparison to the world: 49

Crude oil—production: 284,000 bbl/day (2018 est.)
country comparison to the world: 31

Crude oil—exports: 192,500 bbl/day (2017 est.)
country comparison to the world: 31

Crude oil—imports: 341,700 bbl/day (2017 est.)
country comparison to the world: 24

Crude oil—proved reserves: 1.821 billion bbl (1 January 2018 est.)
country comparison to the world: 35

Refined petroleum products—production: 462,500 bbl/day (2017 est.)
country comparison to the world: 35

Refined petroleum products—consumption: 1.175 million bbl/day (2017 est.)
country comparison to the world: 20

Refined petroleum products—exports: 64,120 bbl/day (2017 est.)
country comparison to the world: 48

Refined petroleum products—imports: 619,600 bbl/day (2017 est.)
country comparison to the world: 12

Natural gas—production: 105.2 billion cu m (2017 est.)
country comparison to the world: 9

Natural gas—consumption: 45.25 billion cu m (2017 est.)
country comparison to the world: 19

Natural gas—exports: 67.96 billion cu m (2017 est.)
country comparison to the world: 6

Natural gas—imports: 5.776 billion cu m (2017 est.)
country comparison to the world: 33

Natural gas—proved reserves: 1.989 trillion cu m (1 January 2018 est.)
country comparison to the world: 17

Carbon dioxide emissions from consumption of energy: 439.1 million Mt (2017 est.)
country comparison to the world: 15

COMMUNICATIONS

Telephones—fixed lines: total subscriptions: 7,792,701
subscriptions per 100 inhabitants: 31.03 (2019 est.)
country comparison to the world: 20

Telephones—mobile cellular: total subscriptions: 27,780,491

subscriptions per 100 inhabitants: 110.62 (2019 est.)
country comparison to the world: 47

Telecommunication systems: general assessment: excellent domestic and international service; domestic satellite system; significant use of radiotelephone in areas of low population density; rapid growth of mobile telephones; continue to enhance 4G networks while migrating to 5G technologies; 5G connections are predicted to account for around 50—60% of total connections by 2025 (2020)
domestic: 31 per 100 fixed-line, 111 per 100 mobile-cellular; more subscribers to mobile services than there are people; 90% of all mobile device sales are now smartphones, growth in mobile traffic brisk (2019)
international: country code—61; landing points for more than 20 submarine cables including: the SeaMeWe-3 optical telecommunications submarine cable with links to Asia, the Middle East, and Europe; the INDIGO-Central, INDIGO West and ASC, North West Cable System, Australia-Papua New Guinea cable, CSCS, PPC-1, Gondwana-1, SCCN, Hawaiki, TGA, Basslink, Bass Strait-1, Bass Strait-2, JGA-S, with links to other Australian cities, New Zealand and many countries in southeast Asia, US and Europe; the H2 Cable, AJC, Telstra Endeavor, Southern Cross NEXT with links to Japan, Hong Kong, and other Pacific Ocean countries as well as the US; satellite earth stations—10 Intelsat (4 Indian Ocean and 6 Pacific Ocean), 2 Inmarsat, 2 Globalstar, 5 other (2019)
note: the COVID-19 outbreak is negatively impacting telecommunications production and supply chains globally; consumer spending on telecom devices and services has also slowed due to the pandemic's effect on economies worldwide; overall progress towards improvements in all facets of the telecom industry—mobile, fixed-line, broadband, submarine cable and satellite—has moderated

Broadcast media: the Australian Broadcasting Corporation (ABC) runs multiple national and local radio networks and TV stations, as well as Australia Network, a TV service that broadcasts throughout the Asia-Pacific region and is the main public broadcaster; Special Broadcasting Service (SBS), a second large public broadcaster, operates radio and TV networks broadcasting in multiple languages; several large national commercial TV networks, a large number of local commercial TV stations, and hundreds of commercial radio stations are accessible; cable and satellite systems are available

Internet country code: .au

Internet users: *total:* 21,419,302
percent of population: 86.55% (July 2018 est.)
country comparison to the world: 35

Broadband—fixed subscriptions: *total:* 7.64 million
subscriptions per 100 inhabitants: 31 (2018 est.)
country comparison to the world: 22

TRANSPORTATION

National air transport system: *number of registered air carriers:* 25 (2020)
inventory of registered aircraft operated by air carriers: 583
annual passenger traffic on registered air carriers: 75,667,645 (2018)
annual freight traffic on registered air carriers: 2,027,640,000 mt-km (2018)

Civil aircraft registration country code prefix: VH (2016)

Airports: 418 (2020)
country comparison to the world: 19

Airports—with paved runways: *total:* 349 (2017)
over 3,047 m: 11 (2017)
2,438 to 3,047 m: 14 (2017)
1,524 to 2,437 m: 155 (2017)
914 to 1,523 m: 155 (2017)
under 914 m: 14 (2017)

Airports—with unpaved runways: *total:* 131 (2013)
1,524 to 2,437 m: 16 (2013)
914 to 1,523 m: 101 (2013)
under 914 m: 14 (2013)

Heliports: 1 (2013)

Pipelines: 637 km condensate/gas, 30054 km gas, 240 km liquid petroleum gas, 3609 km oil, 110 km oil/gas/water, 72 km refined products (2013)

Railways: *total:* 33,343 km (2015)
standard gauge: 17,446 km 1.435-m gauge (650 km electrified) (2015)
narrow gauge: 12,318 km 1.067-m gauge (2,075.5 km electrified) (2015)
broad gauge: 3,247 km 1.600-m gauge (372 km electrified) (2015)
country comparison to the world: 8

Roadways: *total:* 873,573 km (2015)
urban: 145,928 km (2015)
non-urban: 727,645 km (2015)
country comparison to the world: 9

Waterways: 2,000 km (mainly used for recreation on Murray and Murray-Darling River systems) (2011)
country comparison to the world: 42

Merchant marine: *total:* 579
by type: bulk carrier 4general cargo 80, oil tanker 7, other 488 (2019)
country comparison to the world: 39

Ports and terminals: major seaport(s): Brisbane, Cairns, Darwin, Fremantle, Geelong, Gladstone, Hobart, Melbourne, Newcastle, Port Adelaide, Port Kembla, Sydney
container port(s) (TEUs): Melbourne (2,806,436), Sydney (2,530,122) (2017)
LNG terminal(s) (export): Australia Pacific, Barrow Island, Burrup (Pluto), Curtis Island, Darwin, Karratha, Bladin Point (Ichthys), Gladstone, Prelude (offshore FLNG), Wheatstone
dry bulk cargo port(s): Dampier (iron ore), Dalrymple Bay (coal), Hay Point (coal), Port Hedland (iron ore), Port Walcott (iron ore)

MILITARY AND SECURITY

Military and security forces: Australian Defense Force (ADF): Australian Army (includes Special Operations Command), Royal Australian Navy (includes Naval Aviation Force), Royal Australian Air Force, Joint Operations Command (JOC) (2019)

Military expenditures: 1.9% of GDP (2019)
1.9% of GDP (2018)
2% of GDP (2017)
2.1% of GDP (2016)
2% of GDP (2015)
country comparison to the world: 54

Military and security service personnel strengths: the Australian Defense Force has approximately 60,000 total active troops (30,800 Army; 14,700 Navy; 14,300 Air Force) (2019 est.)

Military equipment inventories and acquisitions: the Australian military's inventory includes a mix of domestically-produced and imported Western (mostly US-origin, particularly aircraft) weapons systems; since 2015, the US is the largest supplier of arms, followed by Spain; the Australian defense industry produces a variety of land and sea weapons platforms; the defense industry also participates in joint development and production ventures with other Western countries, including the US and Canada (2019 est.)

Military deployments: 200 Afghanistan (NATO); 750 Middle East (June 2020)

Military service age and obligation: 17 years of age for voluntary military service (with parental consent); no conscription; women allowed to serve in most combat roles (2018)

TERRORISM

Terrorist group(s): Islamic State of Iraq and ash-Sham (2020)
note: details about the history, aims, leadership, organization, areas of operation, tactics, targets, weapons, size, and sources of support of the group(s) appear(s) in Appendix-T

TRANSNATIONAL ISSUES

Disputes—international: In 2007, Australia and Timor-Leste agreed to a 50-year development zone and revenue sharing arrangement and deferred a maritime boundary; Australia asserts land and maritime claims to Antarctica; Australia's 2004 submission to the Commission on the Limits of the Continental Shelf extends its continental margins over 3.37 million square kilometers, expanding its seabed roughly 30 percent beyond its claimed EEZ; all borders between Indonesia and Australia have been agreed upon bilaterally, but a 1997 treaty that would settle the last of their maritime and EEZ boundary has yet to be ratified by Indonesia's legislature; Indonesian groups challenge Australia's claim to Ashmore Reef; Australia closed parts of the Ashmore and Cartier reserve to Indonesian traditional fishing

Refugees and internally displaced persons: *refugees (country of origin):* 13,122 (Iraq), 12,714 (Afghanistan), 12,537 (Iran), 5,578 (Pakistan) (2019)

stateless persons: 132 (2018)

Illicit drugs: Tasmania is one of the world's major suppliers of licit opiate products; government

maintains strict controls over areas of opium poppy cultivation and output of poppy straw concentrate; major consumer of cocaine and amphetamines

AUSTRIA

INTRODUCTION

Background: Once the center of power for the large Austro-Hungarian Empire, Austria was reduced to a small republic after its defeat in World War I. Following annexation by Nazi Germany in 1938 and subsequent occupation by the victorious Allies in 1945, Austria's status remained unclear for a decade. A State Treaty signed in 1955 ended the occupation, recognized Austria's independence, and forbade unification with Germany. A constitutional law that same year declared the country's "perpetual neutrality" as a condition for Soviet military withdrawal. The Soviet Union's collapse in 1991 and Austria's entry into the EU in 1995 have altered the meaning of this neutrality. A prosperous, democratic country, Austria entered the EU Economic and Monetary Union in 1999.

GEOGRAPHY

Location: Central Europe, north of Italy and Slovenia

Geographic coordinates: 47 20 N, 13 20 E

Map references: Europe

Area: *total:* 83,871 sq km
land: 82,445 sq km
water: 1,426 sq km
country comparison to the world: 115

Area—comparative: about the size of South Carolina; slightly more than two-thirds the size of Pennsylvania

Land boundaries: *total:* 2,524 km
border countries (8): Czech Republic 402 km, Germany 801 km, Hungary 321 km, Italy 404 km,

Liechtenstein 34 km, Slovakia 105 km, Slovenia 299 km, Switzerland 158 km

Coastline: 0 km (landlocked)

Maritime claims: none (landlocked)

Climate: temperate; continental, cloudy; cold winters with frequent rain and some snow in lowlands and snow in mountains; moderate summers with occasional showers

Terrain: mostly mountains (Alps) in the west and south; mostly flat or gently sloping along the eastern and northern margins

Elevation: *mean elevation:* 910 m
lowest point: Neusiedler See 115 m
highest point: Grossglockner 3,798 m

Natural resources: oil, coal, lignite, timber, iron ore, copper, zinc, antimony, magnesite, tungsten, graphite, salt, hydropower

Land use: *agricultural land:* 38.4% (2016 est.)
arable land: 16.5% (2016 est.) / permanent crops: 0.8% (2016 est.) / permanent pasture: 21.1% (2016 est.)
forest: 47.2% (2016 est.)
other: 14.4% (2016 est.)
Irrigated land: 1,170 sq km (2012)

Population distribution: the northern and eastern portions of the country are more densely populated; nearly two-thirds of the populace lives in urban areas

Natural hazards: landslides; avalanches; earthquakes

Environment—current issues: some forest degradation caused by air and soil pollution; soil pollution results from the use of agricultural chemicals; air pollution results from emissions by coal- and oil-fired power stations and industrial plants and from trucks transiting Austria between northern and southern Europe; water pollution; the Danube, as well as some of Austria's other rivers and lakes, are threatened by pollution

Environment—international agreements: *party to:* Air Pollution, Air Pollution-Nitrogen Oxides, Air Pollution-Persistent Organic Pollutants, Air Pollution-Sulfur 85, Air Pollution-Sulphur 94, Air Pollution-Volatile Organic Compounds, Antarctic Treaty, Biodiversity, Climate Change, Climate Change-Kyoto Protocol, Desertification, Endangered Species, Environmental Modification, Hazardous Wastes, Law of the Sea, Ozone Layer Protection, Ship Pollution, Tropical Timber 83, Tropical Timber 94, Wetlands, Whaling
signed, but not ratified: none of the selected agreements

Geography—note: note 1: landlocked; strategic location at the crossroads of central Europe with many easily traversable Alpine passes and valleys; major river is the Danube; population is concentrated on eastern lowlands because of steep slopes, poor soils, and low temperatures elsewhere
note 2: the world's largest and longest ice cave system at 42 km (26 mi) is the Eisriesenwelt (Ice Giants World) inside the Hochkogel mountain near Werfen, about 40 km south of Salzburg; ice caves are bedrock caves that contain year-round ice formations; they differ from glacial caves, which are transient and are formed by melting ice and flowing water within and under glaciers

PEOPLE AND SOCIETY

Population: 8,859,449 (July 2020 est.)
country comparison to the world: 97

Nationality: *noun:* Austrian(s)
adjective: Austrian

Ethnic groups: Austrian 80.8%, German 2.6%, Bosnian and Herzegovinian 1.9%, Turkish 1.8%, Serbian 1.6%, Romanian 1.3%, other 10% (2018 est.)
note: data represent population by country of birth

Languages: German (official nationwide) 88.6%, Turkish 2.3%, Serbian 2.2%, Croatian (official in Burgenland) 1.6%, other (includes Slovene, official in southern Carinthia, and Hungarian, official in Burgenland) 5.3% (2001 est.)

Religions: Catholic 57%, Eastern Orthodox 8.7%, Muslim 7.9%, Evangelical Christian 3.3%, other/none/unspecified 23.1% (2018 est.)
note: data on Muslim is a 2016 estimate; data on other/none/unspecified are from 2012-2018 estimates

Age structure: *0-14 years:* 14.01% (male 635,803/female 605,065)
15-24 years: 10.36% (male 466,921/female 451,248)
25-54 years: 41.35% (male 1,831,704/female 1,831,669)
55-64 years: 14.41% (male 635,342/female 641,389)
65 years and over: 19.87% (male 768,687/female 991,621) (2020 est.)

Dependency ratios: *total dependency ratio:* 50.6
youth dependency ratio: 21.7
elderly dependency ratio: 28.9
potential support ratio: 3.5 (2020 est.)

Median age: *total:* 44.5 years
male: 43.1 years
female: 45.8 years (2020 est.)

country comparison to the world: 14

Population growth rate: 0.35% (2020 est.)
country comparison to the world: 167

Birth rate: 9.5 births/1,000 population (2020 est.)
country comparison to the world: 195

Death rate: 9.8 deaths/1,000 population (2020 est.)
country comparison to the world: 39

Net migration rate: 3.6 migrant(s)/1,000 population (2020 est.)
country comparison to the world: 33

Population distribution: the northern and eastern portions of the country are more densely populated; nearly two-thirds of the populace lives in urban areas

Urbanization: *urban population:* 58.7% of total population (2020)
rate of urbanization: 0.59% annual rate of change (2015-20 est.)

total population growth rate v. urban population growth rate, 2000-2030: Major urban areas—population: 1.930 million VIENNA (capital) (2020)

Sex ratio: *at birth:* 1.05 male(s)/female
0-14 years: 1.05 male(s)/female
15-24 years: 1.03 male(s)/female
25-54 years: 1 male(s)/female
55-64 years: 0.99 male(s)/female
65 years and over: 0.78 male(s)/female
total population: 0.96 male(s)/female (2020 est.)

Mother's mean age at first birth: 29.3 years (2017 est.)

Maternal mortality rate: 5 deaths/100,000 live births (2017 est.)
country comparison to the world: 163

Infant mortality rate: *total:* 3.3 deaths/1,000 live births
male: 3.7 deaths/1,000 live births
female: 3 deaths/1,000 live births (2020 est.)
country comparison to the world: 204

Life expectancy at birth: *total population:* 81.9 years
male: 79.2 years
female: 84.7 years (2020 est.)
country comparison to the world: 25

Total fertility rate: 1.49 children born/woman (2020 est.)
country comparison to the world: 203

Contraceptive prevalence rate: 79% (2019)
note: percent of women aged 16-49

Drinking water source:
improved:
urban: 100% of population
rural: 100% of population
total: 100% of population
unimproved:
urban: 0% of population
rural: 0% of population
total: 0% of population (2017 est.)

Current Health Expenditure: 10.4% (2017)

Physicians density: 5.17 physicians/1,000 population (2017)

Hospital bed density: 7.4 beds/1,000 population (2017)

Sanitation facility access:
improved:
urban: 100% of population
rural: 100% of population
total: 100% of population
unimproved:
urban: 0% of population
rural: 0% of population
total: 0% of population (2017 est.)

HIV/AIDS—adult prevalence rate: 0.1% (2017 est.)
country comparison to the world: 118

HIV/AIDS—people living with HIV/AIDS: 7,400 (2017 est.)
country comparison to the world: 113

HIV/AIDS—deaths: <100 (2017 est.)

Obesity—adult prevalence rate: 20.1% (2016)
country comparison to the world: 105

Education expenditures: 5.5% of GDP (2016)
country comparison to the world: 38

School life expectancy (primary to tertiary education): *total:* 16 years
male: 16 years
female: 16 years (2018)

Unemployment, youth ages 15-24: *total:* 9.4%
male: 9.4%
female: 9.4% (2018 est.)
country comparison to the world: 129

GOVERNMENT

Country name: *conventional long form:* Republic of Austria
conventional short form: Austria
local long form: Republik Oesterreich
local short form: Oesterreich
etymology: the name Oesterreich means "eastern realm" or "eastern march" and dates to the 10th century; the designation refers to the fact that Austria was the easternmost extension of Bavaria, and, in fact, of all the Germans; the word Austria is a Latinization of the German name

Government type: federal parliamentary republic

Capital: *name:* Vienna

geographic coordinates: 48 12 N, 16 22 E
time difference: UTC+1 (6 hours ahead of Washington, DC, during Standard Time)
daylight saving time: +1hr, begins last Sunday in March; ends last Sunday in October
etymology: the origin of the name is disputed but may derive from earlier settlements of the area; a Celtic town of Vedunia, established about 500 B.C., came under Roman dominance around 15 B.C. and became known as Vindobona; archeological remains of the latter survive at many sites in the center of Vienna

Administrative divisions: 9 states (Bundeslaender, singular—Bundesland); Burgenland, Kaernten (Carinthia), Niederoesterreich (Lower Austria), Oberoesterreich (Upper Austria), Salzburg, Steiermark (Styria), Tirol (Tyrol), Vorarlberg, Wien (Vienna)

Independence: no official date of independence: 976 (Margravate of Austria established); 17 September 1156 (Duchy of Austria founded); 6 January 1453 (Archduchy of Austria acknowledged); 11 August 1804 (Austrian Empire proclaimed); 30 March 1867 (Austro-Hungarian dual monarchy established); 12 November 1918 (First Republic proclaimed); 27 April 1945 (Second Republic proclaimed)

National holiday: National Day (commemorates passage of the law on permanent neutrality), 26 October (1955)

Constitution: *history:* several previous; latest adopted 1 October 1920, revised 1929, replaced May 1934, replaced by German Weimar constitution in 1938 following German annexation, reinstated 1 May 1945
amendments: proposed through laws designated "constitutional laws" or through the constitutional process if the amendment is part of another law; approval required by at least a two-thirds majority vote by the National Assembly and the presence of one half of the members; a referendum is required only if requested by one third of the National Council or Federal Council membership; passage by referendum requires absolute majority vote; amended many times, last in 2018

Legal system: civil law system; judicial review of legislative acts by the Constitutional Court

International law organization participation: accepts compulsory ICJ jurisdiction; accepts ICCt jurisdiction

Citizenship: *citizenship by birth:* no
citizenship by descent only: at least one parent must be a citizen of Austria
dual citizenship recognized: no
residency requirement for naturalization: 10 years

Suffrage: 16 years of age; universal

Executive branch: *chief of state:* President Alexander VAN DER BELLEN (since 26 January 2017)

head of government: Sebastian KURZ elected chancellor (since 2 January 2020)
cabinet: Council of Ministers chosen by the president on the advice of the chancellor
elections/appointments: president directly elected by absolute majority popular vote in 2 rounds if needed for a 6-year term (eligible for a second term); elections last held on 24 April 2016 (first round), 22 May 2016 (second round, which was annulled), and 4 December 2016 (second round re-vote) (next election to be held in April 2022); chancellor appointed by the president but determined by the majority coalition parties in the Federal Assembly; vice chancellor appointed by the president on the advice of the chancellor
election results: Alexander VAN DER BELLEN elected in second round; percent of vote in first round—Norbert HOFER (FPOe) 35.1%, Alexander VAN DER BELLEN (independent, allied with the Greens) 21.3%, Irmgard GRISS

(independent) 18.9%, Rudolf HUNDSTORFER (SPOe) 11.3%, Andreas KHOL (OeVP) 11.1%, Richard LUGNER (independent) 2.3%; percent of vote in second round—Alexander VAN DER BELLEN 53.8%, Norbert HOFER 46.2%

Legislative branch: *description:* bicameral Federal Assembly or Bundesversammlung consists of:

Federal Council or Bundesrat (61 seats; members appointed by state parliaments with each state receiving 3 to 12 seats in proportion to its population; members serve 5- or 6-year terms)

National Council or Nationalrat (183 seats; members directly elected in single-seat constituencies by proportional representation vote; members serve 5-year terms) (e.g. 2019)

elections: Federal Council—last appointed—NA

National Council—last held on 29 September 2019 (next to be held in 2024); note—election was originally scheduled for 2022, but President VAN DER BELLEN called for an early election (e.g. 2019)

election results: Federal Council—percent of vote by party—NA; seats by party—NA; composition—men 44, women 17, percent of women 27.9%

National Council—percent of vote by party—OeVP 37.5%, SPOe 21.2%, FPOe 16.2%, The Greens 13.9%, NEOS 8.1%, other 3.1%; seats by party—OeVP 71, SPOe 40, FPOe 31, The Greens 26, NEOS 15; composition—men 115, women 68, percent of women 37.2%; note—total Federal Assembly percent of women 34.8% (e.g. 2019)

Judicial branch: *highest courts:* Supreme Court of Justice or Oberster Gerichtshof (consists of 85 judges organized into 17 senates or panels of 5 judges each); Constitutional Court or Verfassungsgerichtshof (consists of 20 judges including 6 substitutes; Administrative Court or Verwaltungsgerichtshof—2 judges plus other members depending on the importance of the case)

judge selection and term of office: Supreme Court judges nominated by executive branch departments and appointed by the president; judges serve for life; Constitutional Court judges nominated by several executive branch departments and approved by the president; judges serve for life; Administrative Court judges recommended by executive branch departments and appointed by the president; terms of judges and members determined by the president

subordinate courts: Courts of Appeal (4); Regional Courts (20); district courts (120); county courts

Political parties and leaders: Austrian People's Party or OeVP [Sebastian KURZ]

Communist Party of Austria or KPOe [Mirko MESSNER]

Freedom Party of Austria or FPOe [Heinz-Christian STRACHE]

The Greens [Werner KOGLER]

NEOS—The New Austria [Beate MEINL-REISINGER]

NOW-Pilz List (JETZT-Liste Pilz) or PILZ [Maria STERN]

Social Democratic Party of Austria or SPOe [Pamela RENDI-WAGNER]

International organization participation: ADB (nonregional member), AfDB (nonregional member), Australia Group, BIS, BSEC (observer), CD, CE, CEI, CERN, EAPC, EBRD, ECB, EIB, EMU, ESA, EU, FAO, FATF, G-9, IADB, IAEA, IBRD, ICAO, ICC (national committees), ICCt, ICRM, IDA, IEA, IFAD, IFC, IFRCS, IGAD (partners), ILO, IMF, IMO, Interpol, IOC, IOM, IPU, ISO, ITSO, ITU, ITUC (NGOs), MIGA, MINURSO, NEA, NSG, OAS (observer), OECD, OIF (observer), OPCW, OSCE, Paris Club, PCA, PFP, Schengen Convention, SELEC (observer), UN, UNCTAD, UNESCO, UNFICYP, UNHCR, UNIDO, UNIFIL, UNTSO, UNWTO, UPU, WCO, WFTU (NGOs), WHO, WIPO, WMO, WTO, ZC

Diplomatic representation in the US: *chief of mission:* Ambassador Martin WEISS (since 6 January 2020)

chancery: 3524 International Court NW, Washington, DC 20008-3035

telephone: [1] (202) 895-6700

FAX: [1] (202) 895-6750

consulate(s) general: Los Angeles, New York

consulate(s): Chicago

Diplomatic representation from the US: *chief of mission:* Ambassador Trevor TRAINA (since 24 May 2018)

telephone: [43] (1) 31339-0

embassy: Boltzmanngasse 16, A-1090, Vienna

mailing address: Boltzmanngasse 16, 1090 Vienna, Austria

FAX: [43] (1) 3100682

Flag description: three equal horizontal bands of red (top), white, and red; the flag design is certainly one of the oldest—if not the oldest—national banners in the world; according to tradition, in 1191, following a fierce battle in the Third Crusade, Duke Leopold V of Austria's white tunic became completely blood-spattered; upon removal of his wide belt or sash, a white band was revealed; the red-white-red color combination was subsequently adopted as his banner

National symbol(s): eagle, edelweiss, Alpine gentian; national colors: red, white

National anthem: *name:* "Bundeshymne" (Federal Hymn)

lyrics/music: Paula von PRERADOVIC/Wolfgang Amadeus MOZART or Johann HOLZER (disputed)

note: adopted 1947; the anthem is also known as "Land der Berge, Land am Strome" (Land of the Mountains, Land by the River); Austria adopted a new national anthem after World War II to replace the former imperial anthem composed by Franz Josef HAYDN, which had been appropriated by Germany in 1922 and was thereafter associated with the Nazi regime; a gendered version of the lyrics was adopted by the Austrian Federal Assembly in fall 2011 and became effective 1 January 2012 0:00 / 1:29

Economy—overview: Austria is a well-developed market economy with skilled labor force and high standard of living. It is closely tied to other EU economies, especially Germany's, but also the US', its third-largest trade partner. Its economy features a large service sector, a sound industrial sector, and a small, but highly developed agricultural sector.

Austrian economic growth strengthen in 2017, with a 2.9% increase in GDP. Austrian exports, accounting for around 60% of the GDP, were up 8.2% in 2017. Austria's unemployment rate fell by 0.3% to 5.5%, which is low by European standards, but still at its second highest rate since the end of World War II, driven by an increased number of refugees and EU migrants entering the labor market.

Austria's fiscal position compares favorably with other euro-zone countries. The budget deficit stood at a low 0.7% of GDP in 2017 and public debt declined again to 78.4% of GDP in 2017, after reaching a post-war high 84.6% in 2015. The Austrian government has announced it plans to balance the fiscal budget in 2019. Several external risks, such as Austrian banks' exposure to Central and Eastern Europe, the refugee crisis, and continued unrest in Russia/Ukraine, eased in 2017, but are still a factor for the Austrian economy. Exposure to the Russian banking sector and a deep energy relationship with Russia present additional risks.

Austria elected a new pro-business government in October 2017 that campaigned on promises to reduce bureaucracy, improve public sector efficiency, reduce labor market protections, and provide positive investment incentives.

GDP (purchasing power parity): $441 billion (2017 est.)

$428.1 billion (2016 est.)

$422 billion (2015 est.)

note: data are in 2017 dollars

country comparison to the world: 45

GDP (official exchange rate): $417.4 billion (2017 est.)

GDP—real growth rate: 3% (2017 est.)

1.5% (2016 est.)

1.1% (2015 est.)

country comparison to the world: 110

GDP—per capita (PPP): $50,000 (2017 est.)

$49,000 (2016 est.)

$48,900 (2015 est.)

note: data are in 2017 dollars

country comparison to the world: 31

Gross national saving: 27% of GDP (2017 est.)

26.2% of GDP (2016 est.)

25.5% of GDP (2015 est.)

country comparison to the world: 43

GDP—composition, by end use: *household consumption:* 52.1% (2017 est.)

government consumption: 19.5% (2017 est.)

investment in fixed capital: 23.5% (2017 est.)

investment in inventories: 1.6% (2017 est.)

exports of goods and services: 54.2% (2017 est.)

imports of goods and services: -50.7% (2017 est.)

GDP—composition, by sector of origin: *agriculture:* 1.3% (2017 est.)
industry: 28.4% (2017 est.)
services: 70.3% (2017 est.)

Agriculture—products: grains, potatoes, wine, fruit; dairy products, cattle, pigs, poultry; lumber and other forestry products

Industries: construction, machinery, vehicles and parts, food, metals, chemicals, lumber and paper, electronics, tourism

Industrial production growth rate: 6.5% (2017 est.)
country comparison to the world: 35

Labor force: 4.26 million (2017 est.)
country comparison to the world: 90

Labor force—by occupation: *agriculture:* 0.7%
industry: 25.2%
services: 74.1% (2017 est.)

Unemployment rate: 5.5% (2017 est.)
6% (2016 est.)
country comparison to the world: 80

Population below poverty line: 3% (2017 est.)

Household income or consumption by percentage share: *lowest 10%:* 2.8%
highest 10%: 23.5% (2012 est.)

Budget: *revenues:* 201.7 billion (2017 est.)
expenditures: 204.6 billion (2017 est.)

Taxes and other revenues: 48.3% (of GDP) (2017 est.)
country comparison to the world: 18

Budget surplus (+) or deficit (-): -0.7% (of GDP) (2017 est.)
country comparison to the world: 67

Public debt: 78.6% of GDP (2017 est.)
83.6% of GDP (2016 est.)
note: this is general government gross debt, defined in the Maastricht Treaty as consolidated general government gross debt at nominal value, outstanding at the end of the year; it covers the following categories of government liabilities (as defined in ESA95): currency and deposits (AF.2), securities other than shares excluding financial derivatives (AF.3, excluding AF.34), and loans (AF.4); the general government sector comprises the sub-sectors of central government, state government, local government and social security funds; as a percentage of GDP, the GDP used as a denominator is the gross domestic product in current year prices
country comparison to the world: 37

Fiscal year: calendar year

Inflation rate (consumer prices): 2.2% (2017 est.)
1% (2016 est.)
country comparison to the world: 111

Current account balance: $7.859 billion (2017 est.)
$8.313 billion (2016 est.)
country comparison to the world: 25

Exports: $156.7 billion (2017 est.)
$149.5 billion (2016 est.)
country comparison to the world: 32

Exports—partners: Germany 29.4%, US 6.3%, Italy 6.2%, Switzerland 5.1%, France 4.8%, Slovakia 4.8% (2017)

Exports—commodities: machinery and equipment, motor vehicles and parts, manufactured goods, chemicals, iron and steel, foodstuffs

Imports: $158.1 billion (2017 est.)
$142.3 billion (2016 est.)
country comparison to the world: 28

Imports—commodities: machinery and equipment, motor vehicles, chemicals, metal goods, oil and oil products, natural gas; foodstuffs

Imports—partners: Germany 41.8%, Italy 5.8%, Switzerland 5.5%, Czech Republic 4.4%, Netherlands 4.2% (2017)

Reserves of foreign exchange and gold: $21.57 billion (31 December 2017 est.)
$23.36 billion (31 December 2016 est.)
country comparison to the world: 57

Debt—external: $630.8 billion (31 December 2017 est.)
$679.3 billion (31 March 2015 est.)
country comparison to the world: 19

Exchange rates: euros (EUR) per US dollar—
0.885 (2017 est.)
0.903 (2016 est.)
0.9214 (2015 est.)
0.885 (2014 est.)
0.7634 (2013 est.)

ENERGY

Electricity access: *electrification—total population:* 100% (2016)

Electricity—production: 60.78 billion kWh (2016 est.)
country comparison to the world: 48

Electricity—consumption: 64.6 billion kWh (2016 est.)
country comparison to the world: 41

Electricity—exports: 19.21 billion kWh (2016 est.)
country comparison to the world: 9

Electricity—imports: 26.37 billion kWh (2016 est.)
country comparison to the world: 6

Electricity—installed generating capacity: 24.79 million kW (2016 est.)
country comparison to the world: 36

Electricity—from fossil fuels: 25% of total installed capacity (2016 est.)
country comparison to the world: 188

Electricity—from nuclear fuels: 0% of total installed capacity (2017 est.)
country comparison to the world: 41

Electricity—from hydroelectric plants: 43% of total installed capacity (2017 est.)
country comparison to the world: 45

Electricity—from other renewable sources: 31% of total installed capacity (2017 est.)
country comparison to the world: 17

Crude oil—production: 13,000 bbl/day (2018 est.)
country comparison to the world: 75

Crude oil—exports: 0 bbl/day (2017 est.)
country comparison to the world: 88

Crude oil—imports: 146,600 bbl/day (2017 est.)
country comparison to the world: 37

Crude oil—proved reserves: 41.2 million bbl (1 January 2018 est.)
country comparison to the world: 78

Refined petroleum products—production: 186,500 bbl/day (2017 est.)
country comparison to the world: 54

Refined petroleum products—consumption: 268,000 bbl/day (2017 est.)
country comparison to the world: 47

Refined petroleum products—exports: 49,960 bbl/day (2017 est.)
country comparison to the world: 55

Refined petroleum products—imports: 135,500 bbl/day (2017 est.)
country comparison to the world: 42

Natural gas—production: 1.274 billion cu m (2017 est.)
country comparison to the world: 62

Natural gas—consumption: 9.486 billion cu m (2017 est.)
country comparison to the world: 50

Natural gas—exports: 5.437 billion cu m (2017 est.)
country comparison to the world: 29

Natural gas—imports: 14.02 billion cu m (2017 est.)
country comparison to the world: 22

Natural gas—proved reserves: 6.513 billion cu m (1 January 2018 est.)
country comparison to the world: 84

Carbon dioxide emissions from consumption of energy: 63.93 million Mt (2017 est.)
country comparison to the world: 53

COMMUNICATIONS

Telephones—fixed lines: *total subscriptions:* 3,700,006
subscriptions per 100 inhabitants: 41.91 (2019 est.)
country comparison to the world: 38

Telephones—mobile cellular: *total subscriptions:* 10,574,725
subscriptions per 100 inhabitants: 119.78 (2019 est.)
country comparison to the world: 85

Telecommunication systems: *general assessment:* mobile-cellular subscribership is everywhere; cable networks are very extensive; all telephone applications and Internet services are accessible; broadband is available in all large municipalities; one company is in the process of extending fiber infrastructure to an additional 300,000 premises, affecting some 500 towns, as a consequence the number of DSL lines is likely to fall as customers move to fiber; the govt. vows to spend

61

1 billion euros to upgrade the national broadband availability; the roll-out of 5G has begun in a test phase in 2019 and is planned to be available nationwide in 2025 (2020)

domestic: developed and efficient; 41 per 100 fixed-line for households, 174 per 100 for companies; 120 per 100 mobile- cellular; broadband: 138 per 100 on smartphones; 62 per 100 fixed broadband, 54 per 100 mobile broadband (2019)

international: country code—43; earth stations available in the Astra, Intelsat, Eutelsat satellite systems (2019)

note: the COVID-19 outbreak is negatively impacting telecommunications production and supply chains globally; consumer spending on telecom devices and services has also slowed due to the pandemic's effect on economies worldwide; overall progress towards improvements in all facets of the telecom industry—mobile, fixed-line, broadband, submarine cable and satellite—has moderated

Broadcast media: worldwide cable and satellite TV are available; the public incumbent ORF competes with three other major, several regional domestic, and up to 400 international TV stations; TV coverage is in principle 100%, but only 90% use broadcast media; Internet streaming not only complements, but increasingly replaces regular TV stations (2019)

Internet country code: .at

Internet users: *total:* 7,712,665

percent of population: 87.71% (July 2018 est.)
country comparison to the world: 63

Broadband—fixed subscriptions: *total:* 2,521,100
subscriptions per 100 inhabitants: 29 (2018 est.)
country comparison to the world: 49

Communications—note: note 1: the Austrian National Library contains important collections of the Imperial Library of the Holy Roman Empire and of the Austrian Empire, as well as of the Austrian Republic; among its more than 12 million items are outstanding holdings of rare books, maps, globes, papyrus, and music; its Globe Museum is the only one in the world

note 2: on 1 October 1869, Austria-Hungary introduced the world's first postal card—postal stationery with an imprinted stamp indicating the prepayment of postage; simple and cheap (sent for a fraction of the cost of a regular letter), postal cards became an instant success, widely produced in the millions worldwide

note 3: Austria followed up with the creation of the world's first commercial picture postcards—cards bearing a picture or photo to which postage is affixed—in May 1871; sent from Vienna, the image served as a souvenir of the city; together, postal cards and post cards served as the world's e-mails of the late 19th and early 20th centuries

note 4: Austria was also an airmail pioneer; from March to October of 1918, it conducted the world's first regular (daily) airmail service—between the imperial cities of Vienna, Krakow, and Lemberg—a combined distance of some 650 km (400 mi) (earlier airmail services had been set up in a few parts of the world, but only for short stretches and none lasted beyond a few days or weeks); an expansion of the route in June of 1918 allowed private mail to be flown to Kyiv, in newly independent Ukraine, which made the route the world's first regular international airmail service (covering a distance of some 1,200 km; 750 mi)

TRANSPORTATION

National air transport system: *number of registered air carriers:* 11 (2020)
inventory of registered aircraft operated by air carriers: 130
annual passenger traffic on registered air carriers: 12,935,505 (2018)
annual freight traffic on registered air carriers: 373.51 million mt-km (2018)

Civil aircraft registration country code prefix: OE (2016)

Airports: 50 (2020)
country comparison to the world: 90

Airports—with paved runways: *total:* 24 (2017)
over 3,047 m: 1 (2017)
2,438 to 3,047 m: 5 (2017)
1,524 to 2,437 m: 1 (2017)
914 to 1,523 m: 4 (2017)
under 914 m: 13 (2017)

Airports—with unpaved runways: *total:* 28 (2013)
1,524 to 2,437 m: 1 (2013)
914 to 1,523 m: 3 (2013)
under 914 m: 24 (2013)

Heliports: 1 (2013)

Pipelines: 1888 km gas, 594 km oil, 157 km refined products (2017)

Railways: *total:* 5,800 km (2017)

standard gauge: 5,300 km 1.435-m gauge (3,826 km electrified) (2016)
country comparison to the world: 33

Roadways: *total:* 137,039 km (2018)

paved: 137,039 km (includes 2,232 km of expressways) (2018)
country comparison to the world: 39

Waterways: 358 km (2011)
country comparison to the world: 89

Ports and terminals: *river port(s):* Enns, Krems, Linz, Vienna (Danube)

MILITARY AND SECURITY

Military and security forces: Austrian Armed Forces: Land Forces Command, Air Forces Command, plus a Logistics Command and Service Support & Cyber Defence Command (2019)

Military expenditures: 0.7% of GDP (2019)
0.7% of GDP (2018)
0.8% of GDP (2017)
0.7% of GDP (2016)
0.7% of GDP (2015)
country comparison to the world: 132

Military and security service personnel strengths: the Austrian Armed Forces have approximately 23,000 total active duty personnel (13,000 Land Forces; 2,500 Air Force; 7,500 support) (2019 est.)

Military equipment inventories and acquisitions: the Austrian military's inventory includes a mix of domestically-produced and imported weapons systems from European countries and the US; since 2010, Germany and Italy are the leading suppliers of armaments to Austria; the Austrian defense industry produces a range of armored vehicles (2019 est.)

Military deployments: 150 Bosnia-Herzegovina (EUFOR); 300 Kosovo (NATO); 150 Lebanon (UNIFIL) (July 2020)

Military service age and obligation: registration requirement at age 17, the legal minimum age for voluntary military service; 18 is the legal minimum age for compulsory military service (6 months), or optionally, alternative civil/community service (9 months); males 18 to 50 years old in the militia or inactive reserve are subject to compulsory service; in a January 2012 referendum, a majority of Austrians voted in favor of retaining the system of compulsory military service (with the option of alternative/non-military service) instead of switching to a professional army system (2015)

TERRORISM

Terrorist group(s): Islamic State of Iraq and ash-Sham (ISIS) (2020)
note: details about the history, aims, leadership, organization, areas of operation, tactics, targets, weapons, size, and sources of support of the group(s) appear(s) in Appendix-T

TRANSNATIONAL ISSUES

Disputes—international: none

Refugees and internally displaced persons: *refugees (country of origin):* 51,955 (Syria), 37,276 (Afghanistan), 8,664 (Russia), 8,568 (Iraq), 7,636 (Somalia), 6,393 (Iran) (2018)
stateless persons: 1,062 (2018)

Illicit drugs: transshipment point for Southwest Asian heroin and South American cocaine destined for Western Europe; increasing consumption of European-produced synthetic drugs

AZERBAIJAN

INTRODUCTION

Background: Azerbaijan—a secular nation with a majority-Turkic and majority-Shia Muslim population—was briefly independent (from 1918 to 1920) following the collapse of the Russian Empire; it was subsequently incorporated into the Soviet Union for seven decades. Azerbaijan remains involved in the protracted Nagorno-Karabakh conflict with Armenia. Nagorno-Karabakh was a primarily ethnic Armenian region that Moscow recognized in 1923 as an autonomous oblast within Soviet Azerbaijan. In the late Soviet period, a separatist movement developed which sought to end Azerbaijani control over the region. Fighting over Nagorno-Karabakh began in 1988 and escalated after Armenia and Azerbaijan attained independence from the Soviet Union in 1991. By the time a ceasefire took effect in May 1994, separatists, with Armenian support, controlled Nagorno-Karabakh and seven surrounding Azerbaijani territories. The 1994 ceasefire continues to hold, although violence continues along the line of contact separating the opposing forces, as well as the Azerbaijan-Armenia international border. The final status of Nagorno-Karabakh remains the subject of international mediation by the Organization for Security and Cooperation in Europe (OSCE) Minsk Group, which works to help the sides settle the conflict peacefully. The OSCE Minsk Group is co-chaired by the United States, France, and Russia.

In the 25 years following its independence, Azerbaijan succeeded in significantly reducing the poverty rate and has directed revenues from its oil and gas production to develop the country's infrastructure. However, corruption remains a problem, and the government has been accused of authoritarianism. The country's leadership has remained in the Aliyev family since Heydar ALIYEV became president in 1993 and was succeeded by his son, President Ilham ALIYEV in 2003. Following two national referendums in the past several years that eliminated presidential term limits and extended presidential terms from 5 to 7 years, President ALIYEV secured a fourth term as president in April 2018 in an election that international observers noted had serious shortcomings. Reforms are underway to diversify the country's non-oil economy and additional reforms are needed to address weaknesses in government institutions, particularly in the education and health sectors, and the court system.

GEOGRAPHY

Location: Southwestern Asia, bordering the Caspian Sea, between Iran and Russia, with a small European portion north of the Caucasus range

Geographic coordinates: 40 30 N, 47 30 E

Map references: Asia

Area: *total:* 86,600 sq km
land: 82,629 sq km
water: 3,971 sq km
note: includes the exclave of Naxcivan Autonomous Republic and the Nagorno-Karabakh region; the region's autonomy was abolished by Azerbaijani Supreme Soviet on 26 November 1991
country comparison to the world: 114

Area—comparative: about three-quarters the size of Pennsylvania; slightly smaller than Maine

Land boundaries: *total:* 2,468 km
border countries (5): Armenia 996 km, Georgia 428 km, Iran 689 km, Russia 338 km, Turkey 17 km

Coastline: 0 km (landlocked); note—Azerbaijan borders the Caspian Sea (713 km)

Maritime claims: none (landlocked)

Climate: dry, semiarid steppe

Terrain: large, flat Kur-Araz Ovaligi (Kura-Araks Lowland, much of it below sea level) with Great Caucasus Mountains to the north, Qarabag Yaylasi (Karabakh Upland) to the west; Baku lies on Abseron Yasaqligi (Apsheron Peninsula) that juts into Caspian Sea

Elevation: *mean elevation:* 384 m
lowest point: Caspian Sea -28 m
highest point: Bazarduzu Dagi 4,466 m

Natural resources: petroleum, natural gas, iron ore, nonferrous metals, bauxite

Land use: *agricultural land:* 57.6% (2016 est.)
arable land: 22.8% (2016 est.) / permanent crops: 2.7% (2016 est.) / permanent pasture: 32.1% (2016 est.)
forest: 11.3% (2016 est.)
other: 31.1% (2016 est.)
Irrigated land: 14,277 sq km (2012)

Population distribution: highest population density is found in the far eastern area of the county, in and around Baku; apart from smaller urbanized areas, the rest of the country has a fairly light and evenly distributed population

Natural hazards: droughts

Environment—current issues: local scientists consider the Abseron Yasaqligi (Apsheron Peninsula) (including Baku and Sumqayit) and the Caspian Sea to be the ecologically most devastated area in the world because of severe air, soil, and water pollution; soil pollution results from oil spills, from the use of DDT pesticide, and from toxic defoliants used in the production of cotton; surface and underground water are polluted by untreated municipal and industrial wastewater and agricultural run-off

Environment—international agreements: *party to:* Air Pollution, Biodiversity, Climate Change, Climate Change-Kyoto Protocol, Desertification, Endangered Species, Hazardous Wastes, Marine Dumping, Ozone Layer Protection, Ship Pollution, Wetlands
signed, but not ratified: none of the selected agreements

Geography—note: both the main area of the country and the Naxcivan exclave are landlocked

PEOPLE AND SOCIETY

Population: 10,205,810 (July 2020 est.)
country comparison to the world: 90

Nationality: *noun:* Azerbaijani(s)
adjective: Azerbaijani

Ethnic groups: Azerbaijani 91.6%, Lezghin 2%, Russian 1.3%, Armenian 1.3%, Talysh 1.3%, other 2.4% (2009 est.)
note: the separatist Nagorno-Karabakh region is populated almost entirely by ethnic Armenians

Languages: Azerbaijani (Azeri) (official) 92.5%, Russian 1.4%, Armenian 1.4%, other 4.7% (2009 est.)
note: Russian is widely spoken

Religions: Muslim 96.9% (predominantly Shia), Christian 3%, other <0.1, unaffiliated <0.1 (2010 est.)
note: religious affiliation for the majority of Azerbaijanis is largely nominal, percentages for actual practicing adherents are probably much lower

Age structure: *0-14 years:* 22.84% (male 1,235,292/female 1,095,308)
15-24 years: 13.17% (male 714,718/female 629,494)
25-54 years: 45.29% (male 2,291,600/female 2,330,843)
55-64 years: 11.41% (male 530,046/female 634,136)
65 years and over: 7.29% (male 289,604/female 454,769) (2020 est.)

Dependency ratios: *total dependency ratio:* 43.4
youth dependency ratio: 33.7

elderly dependency ratio: 9.7
potential support ratio: 10.3 (2020 est.)

Median age: *total:* 32.6 years
male: 31.1 years
female: 34.2 years (2020 est.)
country comparison to the world: 108

Population growth rate: 0.77% (2020 est.)
country comparison to the world: 131

Birth rate: 14.5 births/1,000 population (2020 est.)
country comparison to the world: 126

Death rate: 7 deaths/1,000 population (2020 est.)
country comparison to the world: 124

Net migration rate: 0 migrant(s)/1,000 population (2020 est.)
country comparison to the world: 76

Population distribution: highest population density is found in the far eastern area of the county, in and around Baku; apart from smaller urbanized areas, the rest of the country has a fairly light and evenly distributed population

Urbanization: *urban population:* 56.4% of total population (2020)
rate of urbanization: 1.58% annual rate of change (2015-20 est.)
note: includes Nagorno-Karabakh

total population growth rate v. urban population growth rate, 2000-2030: Major urban areas—population: 2.341 million BAKU (capital) (2020)

Sex ratio: *at birth:* 1.06 male(s)/female
0-14 years: 1.13 male(s)/female
15-24 years: 1.14 male(s)/female
25-54 years: 0.98 male(s)/female
55-64 years: 0.84 male(s)/female
65 years and over: 0.64 male(s)/female
total population: 0.98 male(s)/female (2020 est.)

Mother's mean age at first birth: 23.8 years (2017 est.)

Maternal mortality rate: 26 deaths/100,000 live births (2017 est.)
country comparison to the world: 120

Infant mortality rate: *total:* 21.3 deaths/1,000 live births
male: 22.3 deaths/1,000 live births
female: 20.3 deaths/1,000 live births (2020 est.)
country comparison to the world: 72

Life expectancy at birth: *total population:* 73.6 years
male: 70.5 years
female: 76.9 years (2020 est.)
country comparison to the world: 144

Total fertility rate: 1.88 children born/woman (2020 est.)
country comparison to the world: 133

Contraceptive prevalence rate: 54.9% (2011)

Drinking water source:
improved:
urban: 100% of population
rural: 87.4% of population
total: 94.1% of population
unimproved:
urban: 0% of population
rural: 12.6% of population

total: 4.9% of population (2017 est.)

Current Health Expenditure: 6.7% (2017)

Physicians density: 3.45 physicians/1,000 population (2014)

Hospital bed density: 4.8 beds/1,000 population (2014)

Sanitation facility access:
improved:
urban: 100% of population
rural: 89.1% of population
total: 95.1% of population
unimproved:
urban: 0% of population
rural: 10.9% of population
total: 4.9% of population (2017 est.)

HIV/AIDS—adult prevalence rate: 0.1% (2019 est.)
country comparison to the world: 119

HIV/AIDS—people living with HIV/AIDS: 9,700 (2019 est.)
country comparison to the world: 104

HIV/AIDS—deaths: <500 (2019 est.)

Obesity—adult prevalence rate: 19.9% (2016)
country comparison to the world: 106

Children under the age of 5 years underweight: 4.9% (2013)
country comparison to the world: 83

Education expenditures: 2.5% of GDP (2017)
country comparison to the world: 159

Literacy: *definition:* age 15 and over can read and write
total population: 99.8%
male: 99.9%
female: 99.7% (2017)

School life expectancy (primary to tertiary education): *total:* 14 years
male: 13 years
female: 14 years (2019)

Unemployment, youth ages 15-24: *total:* 13.4%
male: 11.4%
female: 15.8% (2015 est.)
country comparison to the world: 103

GOVERNMENT

Country name: *conventional long form:* Republic of Azerbaijan
conventional short form: Azerbaijan
local long form: Azarbaycan Respublikasi
local short form: Azarbaycan
former: Azerbaijan Soviet Socialist Republic
etymology: the name translates as "Land of Fire" and refers to naturally occurring surface fires on ancient oil pools or from natural gas discharges

Government type: presidential republic

Capital: *name:* Baku (Baki, Baky)

geographic coordinates: 40 23 N, 49 52 E
time difference: UTC+4 (9 hours ahead of Washington, DC, during Standard Time)
daylight saving time: does not observe daylight savings time

etymology: the name derives from the Persian designation of the city "bad-kube" meaning "wind-pounded city" and refers to the harsh winds and severe snow storms that can hit the city
note: at approximately 28 m below sea level, Baku's elevation makes it the lowest capital city in the world

Administrative divisions: 66 rayons (ray-onlar; rayon—singular), 11 cities (saharlar; sahar—singular);

rayons: Abseron, Agcabadi, Agdam, Agdas, Agstafa, Agsu, Astara, Babak, Balakan, Barda, Beylaqan, Bilasuvar, Cabrayil, Calilabad, Culfa, Daskasan, Fuzuli, Gadabay, Goranboy, Goycay, Goygol, Haciqabul, Imisli, Ismayilli, Kalbacar, Kangarli, Kurdamir, Lacin, Lankaran, Lerik, Masalli, Neftcala, Oguz, Ordubad, Qabala, Qax, Qazax, Qobustan, Quba, Qubadli, Qusar, Saatli, Sabirabad, Sabran, Sadarak, Sahbuz, Saki, Salyan, Samaxi, Samkir, Samux, Sarur, Siyazan, Susa, Tartar, Tovuz, Ucar, Xacmaz, Xizi, Xocali, Xocavand, Yardimli, Yevlax, Zangilan, Zaqatala, Zardab

cities: Baku, Ganca, Lankaran, Mingacevir, Naftalan, Naxcivan (Nakhichevan), Saki, Sirvan, Sumqayit, Xankandi, Yevlax

Independence: 30 August 1991 (declared from the Soviet Union); 18 October 1991 (adopted by the Supreme Council of Azerbaijan)

National holiday: Republic Day (founding of the Democratic Republic of Azerbaijan), 28 May (1918)

Constitution: *history:* several previous; latest adopted 12 November 1995
amendments: proposed by the president of the republic or by at least 63 members of the National Assembly; passage requires at least 95 votes of Assembly members in two separate readings of the draft amendment six months apart and requires presidential approval after each of the two Assembly votes, followed by presidential signature; constitutional articles on the authority, sovereignty, and unity of the people cannot be amended; amended 2002, 2009, 2016

Legal system: civil law system

International law organization participation: has not submitted an ICJ jurisdiction declaration; non-party state to the ICCt

Citizenship: *citizenship by birth:* yes
citizenship by descent only: yes
dual citizenship recognized: no
residency requirement for naturalization: 5 years

Suffrage: 18 years of age; universal

Executive branch: *chief of state:* President Ilham ALIYEV (since 31 October 2003); First Vice President Mehriban ALIYEVA (since 21 February 2017)

head of government: Prime Minister Ali ASADOV (since 8 October 2019); First Deputy Prime Minister Yaqub EYYUBOV (since June 2006)

cabinet: Council of Ministers appointed by the president and confirmed by the National Assembly
elections/appointments: president directly elected by absolute majority popular vote in 2 rounds if needed for a 7-year term (eligible for unlimited terms); election last held on 11 April 2018 (next to be held in 2025); prime minister and first deputy prime minister appointed by the president and confirmed by the National Assembly; note—a constitutional amendment approved in a September 2016 referendum expanded presidential terms from 5 to 7 years; a separate constitutional amendment approved in the same referendum also introduced the post of first vice-president and additional vice-presidents, who are directly appointed by the president
election results: Ilham ALIYEV reelected president in first round; percent of vote—Ilham ALIYEV (YAP) 86%, Zahid ORUJ (independent) 3.1%, other 10.9%
note: OSCE observers noted shortcomings in the election, including a restrictive political environment, limits on fundamental freedoms, a lack of genuine competition, and ballot box stuffing

Legislative branch: *description:* unicameral National Assembly or Milli Mejlis (125 seats; members directly elected in single-seat constituencies by simple majority vote to serve 5-year terms)
elections: last held on 9 February 2020 (next to be held in 2025)
election results: percent of vote by party—NA; seats by party—YAP 70, CSP 3, AVP 1, CUP 1, Democratic Enlightenment 1, PDR 1, Great Order 1, VP 1, Whole Azerbaijan Popular Front 1, independent 41, vacant 4

Judicial branch: *highest courts:* Supreme Court (consists of the chairman, vice chairman, and 23 judges in plenum sessions and organized into civil, economic affairs, criminal, and rights violations chambers); Constitutional Court (consists of 9 judges)
judge selection and term of office: Supreme Court judges nominated by the president and appointed by the Milli Majlis; judges appointed for 10 years; Constitutional Court chairman and deputy chairman appointed by the president; other court judges nominated by the president and appointed by the Milli Majlis to serve single 15-year terms
subordinate courts: Courts of Appeal (replaced the Economic Court in 2002); district and municipal courts

Political parties and leaders: Azerbaijan Democratic Enlightenment Party
Civil Solidarity Party or CSP [Sabir RUSTAMKHANLI]
Civil Unity Party or CUP [Sabir HAJIYEV]
Great Order Party
Islamic Party of Azerbaijan [Mavsum SAMADOV]
Musavat [Arif HAJILI]
Popular Front Party [Ali KARIMLI]
Motherland Party or AVP [Fazail AGAMALI]
National Renaissance Party
Party for Democratic Reforms (PDR)
Social Democratic Party [Ayaz MUTALIBOV]

Social Prosperity Party [Khanhusein KAZIMLI]
Unity Party (VP) [Tahir KARIMLI]
Whole Azerbaijan Popular Front Party [Gudrat HASANGULIYEV]
Yeni (New) Azerbaijan Party or YAP [President Ilham ALIYEV]

International organization participation: ADB, BSEC, CD, CE, CICA, CIS, EAPC, EBRD, ECO, FAO, GCTU, GUAM, IAEA, IBRD, ICAO, ICC (NGOs), ICRM, IDA, IDB, IFAD, IFC, IFRCS, ILO, IMF, IMO, Interpol, IOC, IOM, IPU, ISO, ITSO, ITU, ITUC (NGOs), MIGA, NAM, OAS (observer), OIC, OPCW, OSCE, PFP, UN, UNCTAD, UNESCO, UNHCR, UNIDO, UNWTO, UPU, WCO, WFTU (NGOs), WHO, WIPO, WMO, WTO (observer)

Diplomatic representation in the US: *chief of mission:* Ambassador Elin SULEYMANOV (since 5 December 2011)
chancery: 2741 34th Street NW, Washington, DC 20008
telephone: [1] (202) 337-3500
FAX: [1] (202) 337-5911

consulate(s) general: Los Angeles

Diplomatic representation from the US: *chief of mission:* Ambassador Earle LITZENBERGER (since 12 March 2019)
telephone: [994] (12) 488-3300
embassy: 111 Azadliq Prospekti, Baku AZ1007
mailing address: American Embassy Baku, US Department of State, 7050 Baku Place, Washington, DC 20521-7050
FAX: [994] (12) 488-3330

Flag description: three equal horizontal bands of sky blue (top), red, and green; a vertical crescent moon and an eight-pointed star in white are centered in the red band; the blue band recalls Azerbaijan's Turkic heritage, red stands for modernization and progress, and green refers to Islam; the crescent moon and star are a Turkic insignia; the eight star points represent the eight Turkic peoples of the world

National symbol(s): flames of fire; national colors: blue, red, green

National anthem: *name:* "Azerbaijan Marsi" (March of Azerbaijan)
lyrics/music: Ahmed JAVAD/Uzeyir HAJIBEYOV
note: adopted 1992; although originally written in 1919 during a brief period of independence, "Azerbaijan Marsi" did not become the official anthem until after the dissolution of the Soviet Union
0:00 / 2:11

ECONOMY

Economy—overview: Prior to the decline in global oil prices since 2014, Azerbaijan's high economic growth was attributable to rising energy exports and to some non-export sectors. Oil exports through the Baku-Tbilisi-Ceyhan Pipeline, the Baku-Novorossiysk, and the Baku-Supsa Pipelines remain the main economic driver, but efforts to boost Azerbaijan's gas production are underway. The expected completion of the geopolitically

important Southern Gas Corridor (SGC) between Azerbaijan and Europe will open up another source of revenue from gas exports. First gas to Turkey through the SGC is expected in 2018 with project completion expected by 2020-21.

Declining oil prices caused a 3.1% contraction in GDP in 2016, and a 0.8% decline in 2017, highlighted by a sharp reduction in the construction sector. The economic decline was accompanied by higher inflation, a weakened banking sector, and two sharp currency devaluations in 2015. Azerbaijan's financial sector continued to struggle. In May 2017, Baku allowed the majority state-owned International Bank of Azerbaijan (IBA), the nation's largest bank, to default on some of its outstanding debt and file for restructuring in Azerbaijani courts; IBA also filed in US and UK bankruptcy courts to have its restructuring recognized in their respective jurisdictions.

Azerbaijan has made limited progress with market-based economic reforms. Pervasive public and private sector corruption and structural economic inefficiencies remain a drag on long-term growth, particularly in non-energy sectors. The government has, however, made efforts to combat corruption, particularly in customs and government services. Several other obstacles impede Azerbaijan's economic progress, including the need for more foreign investment in the non-energy sector and the continuing conflict with Armenia over the Nagorno-Karabakh region. While trade with Russia and the other former Soviet republics remains important, Azerbaijan has expanded trade with Turkey and Europe and is seeking new markets for non-oil/gas exports—mainly in the agricultural sector—with Gulf Cooperation Council member countries, the US, and others. It is also improving Baku airport and the Caspian Sea port of Alat for use as a regional transportation and logistics hub.

Long-term prospects depend on world oil prices, Azerbaijan's ability to develop export routes for its growing gas production, and its ability to improve the business environment and diversify the economy. In late 2016, the president approved a strategic roadmap for economic reforms that identified key non-energy segments of the economy for development, such as agriculture, logistics, information technology, and tourism. In October 2017, the long-awaited Baku-Tbilisi-Kars railway, stretching from the Azerbaijani capital to Kars in north-eastern Turkey, began limited service.

GDP (purchasing power parity): $172.2 billion (2017 est.)
$172.1 billion (2016 est.)
$177.6 billion (2015 est.)
note: data are in 2017 dollars
country comparison to the world: 73

GDP (official exchange rate): $40.67 billion (2017 est.)

GDP—real growth rate: 0.1% (2017 est.)
-3.1% (2016 est.)
0.6% (2015 est.)
country comparison to the world: 194

GDP—per capita (PPP): $17,500 (2017 est.)

$17,700 (2016 est.)
$18,500 (2015 est.)
note: data are in 2017 dollars
country comparison to the world: 101

Gross national saving: 24.6% of GDP (2017 est.)
22.7% of GDP (2016 est.)
27.3% of GDP (2015 est.)
country comparison to the world: 63

GDP—composition, by end use: *household consumption:* 57.6% (2017 est.)
government consumption: 11.5% (2017 est.)
investment in fixed capital: 23.6% (2017 est.)
investment in inventories: 0.5% (2017 est.)
exports of goods and services: 48.7% (2017 est.)
imports of goods and services: -42% (2017 est.)

GDP—composition, by sector of origin:
agriculture: 6.1% (2017 est.)
industry: 53.5% (2017 est.)
services: 40.4% (2017 est.)

Agriculture—products: fruit, vegetables, grain, rice, grapes, tea, cotton, tobacco; cattle, pigs, sheep, goats

Industries: petroleum and petroleum products, natural gas, oilfield equipment; steel, iron ore; cement; chemicals and petrochemicals; textiles

Industrial production growth rate: -3.8% (2017 est.)
country comparison to the world: 191

Labor force: 5.118 million (2017 est.)
country comparison to the world: 82

Labor force—by occupation: *agriculture:* 37%
industry: 14.3%
services: 48.9% (2014)

Unemployment rate: 5% (2017 est.)
5% (2016 est.)
country comparison to the world: 73

Population below poverty line: 4.9% (2015 est.)

Household income or consumption by percentage share: *lowest 10%:* 3.4%
highest 10%: 27.4% (2008)

Budget: *revenues:* 9.556 billion (2017 est.)
expenditures: 10.22 billion (2017 est.)

Taxes and other revenues: 23.5% (of GDP) (2017 est.)
country comparison to the world: 126

Budget surplus (+) or deficit (-): -1.6% (of GDP) (2017 est.)
country comparison to the world: 93

Public debt: 54.1% of GDP (2017 est.)
50.7% of GDP (2016 est.)
country comparison to the world: 85

Fiscal year: calendar year

Inflation rate (consumer prices): 13% (2017 est.)
12.6% (2016 est.)
country comparison to the world: 209

Current account balance: $1,685 billion (2017 est.)
-$1.363 billion (2016 est.)
country comparison to the world: 45

Exports: $15.15 billion (2017 est.)
$13.21 billion (2016 est.)
country comparison to the world: 76

Exports—partners: Italy 23.2%, Turkey 13.6%, Israel 6.1%, Russia 5.4%, Germany 5%, Czech Republic 4.6%, Georgia 4.3% (2017)

Exports—commodities: oil and gas roughly 90%, machinery, foodstuffs, cotton

Imports: $9.037 billion (2017 est.)
$9.004 billion (2016 est.)
country comparison to the world: 105

Imports—commodities: machinery and equipment, foodstuffs, metals, chemicals

Imports—partners: Russia 17.7%, Turkey 14.8%, China 9.9%, US 8.3%, Ukraine 5.3%, Germany 5.1% (2017)

Reserves of foreign exchange and gold: $6.681 billion (31 December 2017 est.)
$7.142 billion (31 December 2016 est.)
country comparison to the world: 88

Debt—external: $17.41 billion (31 December 2017 est.)
$13.83 billion (31 December 2016 est.)
country comparison to the world: 99

Exchange rates: Azerbaijani manats (AZN) per US dollar—
1.723 (2017 est.)
1.5957 (2016 est.)
1.5957 (2015 est.)
1.0246 (2014 est.)
0.7844 (2013 est.)

ENERGY

Electricity access: *electrification—total population:* 100% (2016)

Electricity—production: 23.57 billion kWh (2016 est.)
country comparison to the world: 73

Electricity—consumption: 20.24 billion kWh (2016 est.)
country comparison to the world: 71

Electricity—exports: 265 million kWh (2015 est.)
country comparison to the world: 71

Electricity—imports: 114 million kWh (2016 est.)
country comparison to the world: 97

Electricity—installed generating capacity: 7.876 million kW (2016 est.)
country comparison to the world: 71

Electricity—from fossil fuels: 84% of total installed capacity (2016 est.)
country comparison to the world: 74

Electricity—from nuclear fuels: 0% of total installed capacity (2017 est.)
country comparison to the world: 42

Electricity—from hydroelectric plants: 14% of total installed capacity (2017 est.)
country comparison to the world: 104

Electricity—from other renewable sources: 2% of total installed capacity (2017 est.)
country comparison to the world: 133

Crude oil—production: 798,000 bbl/day (2018 est.)
country comparison to the world: 23

Crude oil—exports: 718,800 bbl/day (2015 est.)

country comparison to the world: 19

Crude oil—imports: 0 bbl/day (2015 est.)
country comparison to the world: 92

Crude oil—proved reserves: 7 billion bbl (1 January 2018 est.)
country comparison to the world: 18

Refined petroleum products—production: 138,900 bbl/day (2015 est.)
country comparison to the world: 61

Refined petroleum products—consumption: 100,000 bbl/day (2016 est.)
country comparison to the world: 80

Refined petroleum products—exports: 46,480 bbl/day (2015 est.)
country comparison to the world: 57

Refined petroleum products—imports: 5,576 bbl/day (2015 est.)
country comparison to the world: 168

Natural gas—production: 16.96 billion cu m (2017 est.)
country comparison to the world: 35

Natural gas—consumption: 10.34 billion cu m (2017 est.)
country comparison to the world: 47

Natural gas—exports: 8.042 billion cu m (2017 est.)
country comparison to the world: 24

Natural gas—imports: 2.095 billion cu m (2017 est.)
country comparison to the world: 51

Natural gas—proved reserves: 991.1 billion cu m (1 January 2018 est.)
country comparison to the world: 25

Carbon dioxide emissions from consumption of energy: 35.6 million Mt (2017 est.)
country comparison to the world: 72

COMMUNICATIONS

Telephones—fixed lines: *total subscriptions:* 1,686,316
subscriptions per 100 inhabitants: 16.65 (2019 est.)
country comparison to the world: 61

Telephones—mobile cellular: *total subscriptions:* 10,835,974
subscriptions per 100 inhabitants: 106.99 (2019 est.)
country comparison to the world: 82

Telecommunication systems: *general assessment:* after the oil sector, the telecommunications sector contributes the most to the GDP; more competition will allow for falling prices and the strengthening of the 4G TD-LTE standard and the migration to 5G; Azerbaijan has moderate mobile, mobile broadband and fixed broadband penetration compared to other Asian nations (2020)
domestic: teledensity of some 17 fixed-lines per 100 persons; mobile-cellular teledensity has increased to 107 telephones per 100 persons; satellite service connects Baku to a modern switch in its exclave of Naxcivan (Nakhchivan) (2019)

international: country code—994; the TAE fiber-optic link transits Azerbaijan providing international connectivity to neighboring countries; the old Soviet system of cable and microwave is still serviceable; satellite earth stations—2 (2019) *note:* the COVID-19 outbreak is negatively impacting telecommunications production and supply chains globally; consumer spending on telecom devices and services has also slowed due to the pandemic's effect on economies worldwide; overall progress towards improvements in all facets of the telecom industry—mobile, fixed-line, broadband, submarine cable and satellite—has moderated

Broadcast media: 3 state-run and 1 public TV channels; 4 domestic commercial TV stations and about 15 regional TV stations; cable TV services are available in Baku; 1 state-run and 1 public radio network operating; a small number of private commercial radio stations broadcasting; local FM relays of Baku commercial stations are available in many localities; note—all broadcast media is pro-government, and most private broadcast media outlets are owned by entities directly linked to the government

Internet country code: .az

Internet users: *total:* 8,017,120

percent of population: 79.8% (July 2018 est.)
country comparison to the world: 60

Broadband—fixed subscriptions: *total:* 1,890,913
subscriptions per 100 inhabitants: 19 (2018 est.)
country comparison to the world: 56

TRANSPORTATION

National air transport system: *number of registered air carriers:* 42 (2020)
inventory of registered aircraft operated by air carriers: 44
annual passenger traffic on registered air carriers: 2,279,546 (2018)
annual freight traffic on registered air carriers: 44.09 million mt-km (2018)

Civil aircraft registration country code prefix: 4K (2016)

Airports: 23 (2020)
country comparison to the world: 132

Airports—with paved runways: *total:* 30 (2017)

over 3,047 m: 5 (2017)
2,438 to 3,047 m: 5 (2017)
1,524 to 2,437 m: 13 (2017)
914 to 1,523 m: 4 (2017)
under 914 m: 3 (2017)

Airports—with unpaved runways: *total:* 7 (2013)
under 914 m: 7 (2013)

Heliports: 1 (2012)

Pipelines: 89 km condensate, 3890 km gas, 2446 km oil (2013)

Railways: *total:* 2,944 km (2017)

broad gauge: 2,944.3 km 1.520-m gauge (approx. 1,767 km electrified) (2017)
country comparison to the world: 62

Roadways: *total:* 24,981 km (2013)
country comparison to the world: 105

Merchant marine: *total:* 305
by type: general cargo 40, oil tanker 48, other 217 (2019)
country comparison to the world: 53

Ports and terminals: *major seaport(s):* Baku (Baki) located on the Caspian Sea

MILITARY AND SECURITY

Military and security forces: Land Forces, Air Forces, Navy Forces; Ministry of Internal Affairs: State Border Service (includes Coast Guard), Internal Security Troops (2020)

Military expenditures: 4% of GDP (2019)
3.6% of GDP (2018)
3.8% of GDP (2017)
3.7% of GDP (2016)
5.5% of GDP (2015)
country comparison to the world: 10

Military and security service personnel strengths: the Azerbaijan military has approximately 67,000 total active personnel; 56,000 Army; 2,500 Navy; 8,500 Air Force) (2019 est.)

Military equipment inventories and acquisitions: the inventory of the Azerbaijan military includes mostly Russian and Soviet-era equipment; since 2010, Russia is the leading supplier of arms to Azerbaijan, followed by Israel and Turkey (2020)

Military deployments: 120 Afghanistan (NATO) (June 2020)

Military service age and obligation: 18-35 years of age for compulsory military service; service obligation 18 months or 12 months for university graduates; 17 years of age for voluntary service; 17 year olds are considered to be on active service at cadet military schools (2012)

TERRORISM

Terrorist group(s): Islamic State of Iraq and ash-Sham (2020)
note: details about the history, aims, leadership, organization, areas of operation, tactics, targets, weapons, size, and sources of support of the group(s) appear(s) in Appendix-T

TRANSNATIONAL ISSUES

Disputes—international: Azerbaijan, Kazakhstan, and Russia ratified the Caspian seabed delimitation treaties based on equidistance, while Iran continues to insist on a one-fifth slice of the sea; the dispute over the break-away Nagorno-Karabakh region and the Armenian military occupation of surrounding lands in Azerbaijan remains the primary focus of regional instability; residents have evacuated the former Soviet-era small ethnic enclaves in Armenia and Azerbaijan; local border forces struggle to control the illegal transit of goods and people across the porous, undemarcated Armenian, Azerbaijani, and Georgian borders; bilateral talks continue with Turkmenistan on dividing the seabed and contested oilfields in the middle of the Caspian

Refugees and internally displaced persons: *IDPs:* 351,000 (conflict with Armenia over Nagorno-Karabakh; IDPs are mainly ethnic Azerbaijanis but also include ethnic Kurds, Russians, and Turks predominantly from occupied territories around Nagorno-Karabakh; includes IDPs' descendants, returned IDPs, and people living in insecure areas and excludes people displaced by natural disasters; around half the IDPs live in the capital Baku) (2019)
stateless persons: 3,585 (2018)

Illicit drugs: limited illicit cultivation of cannabis and opium poppy, mostly for CIS consumption; small government eradication program; transit point for Southwest Asian opiates bound for Russia and to a lesser extent the rest of Europe

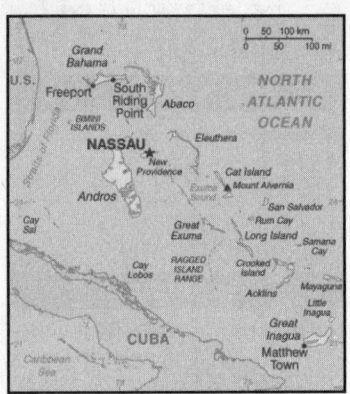

INTRODUCTION

Background: Lucayan Indians inhabited the islands when Christopher COLUMBUS first set foot in the New World on San Salvador in 1492. British settlement of the islands began in 1647; the islands became a colony in 1783. Piracy thrived in the 17th and 18th centuries because of The Bahamas close proximity to shipping lanes. Since attaining independence from the UK in 1973, The Bahamas has prospered through tourism, international banking, and investment management, which comprise up to 85% of GDP. Because of its proximity to the US—the nearest Bahamian landmass being only 80 km (50 mi) from Florida—the country is a major transshipment point for illicit trafficking, particularly to the US mainland, as well as Europe. US law enforcement agencies cooperate closely with The Bahamas, and the US Coast Guard assists Bahamian authorities in coastal defense through Operation Bahamas, Turks and Caicos, or OPBAT.

GEOGRAPHY

Location: chain of islands in the North Atlantic Ocean, southeast of Florida, northeast of Cuba

Geographic coordinates: 24 15 N, 76 00 W

Map references: Central America and the Caribbean

Area: *total:* 13,880 sq km
land: 10,010 sq km
water: 3,870 sq km
country comparison to the world: 161

Area—comparative: slightly smaller than Connecticut

Land boundaries: 0 km

Coastline: 3,542 km

Maritime claims: *territorial sea:* 12 nm
exclusive economic zone: 200 nm

Climate: tropical marine; moderated by warm waters of Gulf Stream

Terrain: long, flat coral formations with some low rounded hills

Elevation: *lowest point:* Atlantic Ocean 0 m
highest point: Mount Alvernia on Cat Island 64 m

Natural resources: salt, aragonite, timber, arable land

Land use: *agricultural land:* 1.4% (2016 est.)
arable land: 0.8% (2016 est.) / permanent crops: 0.4% (2016 est.) / permanent pasture: 0.2% (2016 est.)
forest: 51.4% (2016 est.)
other: 47.2% (2016 est.)

Irrigated land: 10 sq km (2012)

Population distribution: most of the population lives in urban areas, with two-thirds living on New Providence Island where Nassau is located

Natural hazards: hurricanes and other tropical storms cause extensive flood and wind damage

Environment—current issues: coral reef decay; solid waste disposal

Environment—international agreements: party to: Biodiversity, Climate Change, Climate Change-Kyoto Protocol, Desertification, Endangered Species, Hazardous Wastes, Law of the Sea, Ozone Layer Protection, Ship Pollution, Wetlands
signed, but not ratified: none of the selected agreements

Geography—note: strategic location adjacent to US and Cuba; extensive island chain of which 30 are inhabited

PEOPLE AND SOCIETY

Population: 337,721 (July 2020 est.)
note: estimates for this country explicitly take into account the effects of excess mortality due to AIDS; this can result in lower life expectancy, higher infant mortality, higher death rates, lower population growth rates, and changes in the distribution of population by age and sex than would otherwise be expected
country comparison to the world: 179

Nationality: *noun:* Bahamian(s)
adjective: Bahamian

Ethnic groups: black 90.6%, white 4.7%, black and white 2.1%, other 1.9%, unspecified 0.7% (2010 est.)
note: data represent population by racial group

Languages: English (official), Creole (among Haitian immigrants)

Religions: Protestant 69.9% (includes Baptist 34.9%, Anglican 13.7%, Pentecostal 8.9%, Seventh Day Adventist 4.4%, Methodist 3.6%, Church of God 1.9%, Brethren 1.6%, other Protestant .9%), Roman Catholic 12%, other Christian 13% (includes Jehovah's Witness 1.1%), other 0.6%, none 1.9%, unspecified 2.6% (2010 est.)

Age structure: *0-14 years:* 22.04% (male 37,758/female 36,668)
15-24 years: 15.39% (male 26,332/female 25,650)
25-54 years: 43.86% (male 74,485/female 73,647)
55-64 years: 10.04% (male 15,648/female 18,250)
65 years and over: 8.67% (male 11,326/female 17,957) (2020 est.)

Dependency ratios: *total dependency ratio:* 41.5
youth dependency ratio: 30.6
elderly dependency ratio: 11
potential support ratio: 9.1 (2020 est.)

Median age: *total:* 32.8 years
male: 31.7 years
female: 34 years (2020 est.)
country comparison to the world: 104

Population growth rate: 0.75% (2020 est.)
country comparison to the world: 133

Birth rate: 14.8 births/1,000 population (2020 est.)
country comparison to the world: 121

Death rate: 7.4 deaths/1,000 population (2020 est.)
country comparison to the world: 108

Net migration rate: 0 migrant(s)/1,000 population (2020 est.)
country comparison to the world: 77

Population distribution: most of the population lives in urban areas, with two-thirds living on New Providence Island where Nassau is located

Urbanization: *urban population:* 83.2% of total population (2020)
rate of urbanization: 1.13% annual rate of change (2015-20 est.)

total population growth rate v. urban population growth rate, 2000-2030: Major urban areas—population: 280,000 NASSAU (capital) (2018)

Sex ratio: *at birth:* 1.03 male(s)/female
0-14 years: 1.03 male(s)/female
15-24 years: 1.03 male(s)/female
25-54 years: 1.01 male(s)/female
55-64 years: 0.86 male(s)/female
65 years and over: 0.63 male(s)/female
total population: 0.96 male(s)/female (2020 est.)

Maternal mortality rate: 70 deaths/100,000 live births (2017 est.)
country comparison to the world: 83

Infant mortality rate: *total:* 10.6 deaths/1,000 live births
male: 10.6 deaths/1,000 live births
female: 10.6 deaths/1,000 live births (2020 est.)
country comparison to the world: 128

Life expectancy at birth: *total population:* 73.3 years
male: 70.8 years
female: 75.8 years (2020 est.)
country comparison to the world: 146

Total fertility rate: 1.92 children born/woman (2020 est.)
country comparison to the world: 126

Drinking water source:
improved:
total: 98.9% of population
unimproved:
total: 1.1% of population (2017 est.)

Current Health Expenditure: 5.8% (2017)

Physicians density: 2.01 physicians/1,000 population (2017)

Hospital bed density: 3 beds/1,000 population (2017)

Sanitation facility access:
improved:
total: 98.2% of population
unimproved:
total: 1.8% of population (2017 est.)

HIV/AIDS—adult prevalence rate: 1.8% (2018 est.)
country comparison to the world: 27

HIV/AIDS—people living with HIV/AIDS: 6,000 (2018 est.)
country comparison to the world: 118

HIV/AIDS—deaths: <200 (2018)

Obesity—adult prevalence rate: 31.6% (2016)
country comparison to the world: 21

Education expenditures: NA

Unemployment, youth ages 15-24: *total:* 25.8%
male: 20.8%
female: 31.6% (2016 est.)
country comparison to the world: 47

GOVERNMENT

Country name: *conventional long form:* Commonwealth of The Bahamas
conventional short form: The Bahamas
etymology: name derives from the Spanish "baha mar," meaning "shallow sea," which describes the shallow waters of the Bahama Banks

Government type: parliamentary democracy under a constitutional monarchy; a Commonwealth realm

Capital: *name:* Nassau
geographic coordinates: 25 05 N, 77 21 W
time difference: UTC-5 (same time as Washington, DC, during Standard Time)
daylight saving time: +1hr, begins second Sunday in March; ends first Sunday in November
etymology: named after William III (1650-1702), king of England, Scotland, and Ireland, who was a member of the House of Nassau

Administrative divisions: 31 districts; Acklins Islands, Berry Islands, Bimini, Black Point, Cat Island, Central Abaco, Central Andros, Central Eleuthera, City of Freeport, Crooked Island and Long Cay, East Grand Bahama, Exuma, Grand Cay, Harbour Island, Hope Town, Inagua, Long Island, Mangrove Cay, Mayaguana, Moore's Island, North Abaco, North Andros, North Eleuthera, Ragged Island, Rum Cay, San Salvador, South Abaco, South Andros, South Eleuthera, Spanish Wells, West Grand Bahama

Independence: 10 July 1973 (from the UK)

National holiday: Independence Day, 10 July (1973)

Constitution: history: previous 1964 (preindependence); latest adopted 20 June 1973, effective 10 July 1973
amendments: proposed as an "Act" by Parliament; passage of amendments to articles such as the organization and composition of the branches of government requires approval by at least two-thirds majority of the membership of both houses of Parliament and majority approval in a referendum; passage of amendments to constitutional articles such as fundamental rights and individual freedoms, the powers, authorities, and procedures of the branches of government, or changes to the Bahamas Independence Act 1973 requires approval by at least three-fourths majority of the membership of both houses and majority approval in a referendum; amended many times, last in 2016

Legal system: common law system based on the English model

International law organization participation: has not submitted an ICJ jurisdiction declaration; non-party state to the ICCt

Citizenship: *citizenship by birth:* no
citizenship by descent only: at least one parent must be a citizen of The Bahamas
dual citizenship recognized: no
residency requirement for naturalization: 6-9 years

Suffrage: 18 years of age; universal

Executive branch: chief of state: Queen ELIZABETH II (since 6 February 1952); represented by Governor General Cornelius A. SMITH (since 28 June 2019)
head of government: Prime Minister Hubert MINNIS (since 11 May 2017)
cabinet: Cabinet appointed by governor general on recommendation of prime minister
elections/appointments: the monarchy is hereditary; governor general appointed by the monarch on the advice of the prime minister; following legislative elections, the leader of the majority party or majority coalition usually appointed prime minister by the governor general; the prime minister recommends the deputy prime minister
note: Prime Minister Hubert MINNIS is only the fourth prime minister in Bahamian history following its independence from the UK; he is also the first prime minister in 25 years besides Perry CHRISTIE and Hubert INGRAHAM, who repeatedly traded the premiership from 1992 to 2017

Legislative branch: *description:* bicameral Parliament consists of: Senate (16 seats; members appointed by the governor general upon the advice of the prime minister and the opposition leader to serve 5-year terms)
House of Assembly (39 seats; members directly elected in single-seat constituencies by simple majority vote to serve 5- year terms)

elections: Senate—last appointments on 24 May 2017 (next appointments in 2022)

House of Assembly—last held on 10 May 2017 (next to be held by May 2022)

election results: Senate—appointed; composition—men 9, women 7, percent of women 43.8%

House of Assembly—percent of vote by party—FNM 57%, PLP 36.9%, other 6.1%; seats by party—FNM 35, PLP 4; composition—men 34, women 5, percent of women 12.8%; note—total Parliament percent of women 21.8%
note: the government may dissolve the parliament and call elections at any time

Judicial branch: *highest courts:* Court of Appeal (consists of the court president and 4 justices, organized in 3-member panels); Supreme Court (consists of the chief justice and a maximum of 11 and a minimum of 2 justices)
judge selection and term of office: Court of Appeal president and Supreme Court chief justice appointed by the governor-general on the advice of the prime minister after consultation with the leader of the opposition party; other Court of Appeal and Supreme Court justices appointed by the governor general upon recommendation of the Judicial and Legal Services Commission, a 5-member body headed by the chief justice; Court of Appeal justices appointed for life with mandatory retirement normally at age 68 but can be extended until age 70; Supreme Court justices appointed for life with mandatory retirement normally at age 65 but can be extended until age 67
subordinate courts: Industrial Tribunal; Stipendiary and Magistrates' Courts; Family Island Administrators
note: the Bahamas is a member of the 15-member Caribbean Community but is not party to the agreement establishing the Caribbean Court of Justice as its highest appellate court; the Judicial Committee of the Privy Council (in London) serves as the final court of appeal for The Bahamas

Political parties and leaders: Democratic National Alliance or DNA [Christopher MORTIMER, interim leader]
Free National Movement or FNM [Hubert MINNIS]
Progressive Liberal Party or PLP [Philip "Brave" DAVIS]

International organization participation: ACP, AOSIS, C, Caricom, CDB, CELAC, FAO, G-77, IADB, IAEA, IBRD, ICAO, ICC (NGOs), ICRM, IDA, IFAD, IFC, IFRCS, ILO, IMF, IMO, IMSO, Interpol, IOC, IOM, ISO (correspondent), ITSO, ITU, LAES, MIGA, NAM, OAS, OPANAL, OPCW, Petrocaribe, UN, UNCTAD, UNESCO, UNIDO, UNWTO, UPU, WCO, WHO, WIPO, WMO, WTO (observer)

Diplomatic representation in the US: *chief of mission:* Ambassador Sidney Stanley COLLIE (since 29 November 2017)
chancery: 2220 Massachusetts Avenue NW, Washington, DC 20008
telephone: [1] (202) 319-2660
FAX: [1] (202) 319-2668
consulate(s) general: Atlanta, Miami, New York, Washington, DC

honorary consulate(s): Aurora (CO), Chicago, Houston, Los Angeles

Diplomatic representation from the US: *chief of mission:* Ambassador (vacant); Charge d' Affaires Stephanie BOWERS (since 1 March 2018)
telephone: [1] (242) 322-1181, 328-2206 (after hours)
embassy: 42 Queen Street, Nassau, New Providence
mailing address: local or express mail address: P. O. Box N-8197, Nassau; US Department of State, 3370 Nassau Place, Washington, DC 20521-3370
FAX: [1] (242) 356-7174

Flag description: three equal horizontal bands of aquamarine (top), gold, and aquamarine, with a black equilateral triangle based on the hoist side; the band colors represent the golden beaches of the islands surrounded by the aquamarine sea; black represents the vigor and force of a united people, while the pointing triangle indicates the enterprise and determination of the Bahamian people to develop the rich resources of land and sea

National symbol(s): blue marlin, flamingo, Yellow Elder flower; national colors: aquamarine, yellow, black

National anthem: *name:* March On, Bahamaland!
lyrics/music: Timothy GIBSON
note: adopted 1973; as a Commonwealth country, in addition to the national anthem, "God Save the Queen" serves as the royal anthem (see United Kingdom)

0:00 / 1:14

ECONOMY

Economy—overview: The Bahamas has the second highest per capita GDP in the English-speaking Caribbean with an economy heavily dependent on tourism and financial services. Tourism accounts for approximately 50% of GDP and directly or indirectly employs half of the archipelago's labor force. Financial services constitute the second-most important sector of the Bahamian economy, accounting for about 15% of GDP. Manufacturing and agriculture combined contribute less than 7% of GDP and show little growth, despite government incentives aimed at those sectors. The new government led by Prime Minister Hubert MINNIS has prioritized addressing fiscal imbalances and rising debt, which stood at 75% of GDP in 2016. Large capital projects like the Baha Mar Casino and Hotel are driving growth. Public debt increased in 2017 in large part due to hurricane reconstruction and relief financing. The primary fiscal balance was a deficit of 0.4% of GDP in 2016. The Bahamas is the only country in the Western Hemisphere that is not a member of the World Trade Organization.

GDP (purchasing power parity): $12.06 billion (2017 est.)
$11.89 billion (2016 est.)
$12.09 billion (2015 est.)
note: data are in 2017 dollars

country comparison to the world: 156

GDP (official exchange rate): $12.16 billion (2017 est.)

GDP—real growth rate: 1.4% (2017 est.)
-1.7% (2016 est.)
1% (2015 est.)
country comparison to the world: 159

GDP—per capita (PPP): $32,400 (2017 est.)
$32,300 (2016 est.)
$33,200 (2015 est.)
note: data are in 2017 dollars
country comparison to the world: 62

Gross national saving: 11.4% of GDP (2017 est.)
18.2% of GDP (2016 est.)
12.3% of GDP (2015 est.)
country comparison to the world: 155

GDP—composition, by end use: *household consumption:* 68% (2017 est.)
government consumption: 13% (2017 est.)
investment in fixed capital: 26.3% (2017 est.)
investment in inventories: 0.7% (2017 est.)
exports of goods and services: 33.7% (2017 est.)
imports of goods and services: -41.8% (2017 est.)

GDP—composition, by sector of origin: *agriculture:* 2.3% (2017 est.)
industry: 7.7% (2017 est.)
services: 90% (2017 est.)

Agriculture—products: citrus, vegetables; poultry; seafood

Industries: tourism, banking, oil bunkering, maritime industries, transshipment and logistics, salt, aragonite, pharmaceuticals

Industrial production growth rate: 5.8% (2017 est.)
country comparison to the world: 44

Labor force: 196,900 (2013 est.)
country comparison to the world: 171

Labor force—by occupation: *agriculture:* 3%
industry: 11%
services: 49%
tourism: 37% (2011 est.)

Unemployment rate: 10.1% (2017 est.)
12.2% (2016 est.)
country comparison to the world: 147

Population below poverty line: 9.3% (2010 est.)

Household income or consumption by percentage share: *lowest 10%:* 1%
highest 10%: 22% (2007 est.)

Budget: *revenues:* 2.139 billion (2017 est.)
expenditures: 2.46 billion (2017 est.)

Taxes and other revenues: 17.6% (of GDP) (2017 est.)
country comparison to the world: 166

Budget surplus (+) or deficit (-): -2.6% (of GDP) (2017 est.)
country comparison to the world: 115

Public debt: 54.6% of GDP (2017 est.)
50.5% of GDP (2016 est.)
country comparison to the world: 80

Fiscal year: 1 July—30 June

Inflation rate (consumer prices): 1.4% (2017 est.)

-0.3% (2016 est.)
country comparison to the world: 75

Current account balance: -$1.909 billion (2017 est.)
-$868 million (2016 est.)
country comparison to the world: 166

Exports: $550 million (2017 est.)
$444.3 million (2016 est.)
country comparison to the world: 174

Exports—partners: US 63.9%, Namibia 19.3% (2017)

Exports—commodities: Rock lobster, aragonite, crude salt, polystyrene products

Imports: $3.18 billion (2017 est.)
$2.594 billion (2016 est.)
country comparison to the world: 146

Imports—commodities: machinery and transport equipment, manufactures, chemicals, mineral fuels; food and live animals

Imports—partners: US 83.2% (2017)

Reserves of foreign exchange and gold: $1.522 billion (31 December 2017 est.)
$1.002 billion (31 December 2016 est.)
country comparison to the world: 125

Debt—external: $17.56 billion (31 December 2013 est.)
$16.35 billion (31 December 2012 est.)
country comparison to the world: 98

Exchange rates: Bahamian dollars (BSD) per US dollar -
1 (2017 est.)
1 (2016 est.)
1 (2015 est.)
1 (2014 est.)
1 (2013 est.)

ENERGY

Electricity access: *electrification—total population:* 100% (2020)

Electricity—production: 1.778 billion kWh (2016 est.)
country comparison to the world: 140

Electricity—consumption: 1.654 billion kWh (2016 est.)
country comparison to the world: 145

Electricity—exports: 0 kWh (2016 est.)
country comparison to the world: 102

Electricity—imports: 0 kWh (2016 est.)
country comparison to the world: 123

Electricity—installed generating capacity: 577,000 kW (2016 est.)
country comparison to the world: 141

Electricity—from fossil fuels: 100% of total installed capacity (2016 est.)
country comparison to the world: 1

Electricity—from nuclear fuels: 0% of total installed capacity (2017 est.)
country comparison to the world: 43

Electricity—from hydroelectric plants: 0% of total installed capacity (2017 est.)

country comparison to the world: 155

Electricity—from other renewable sources: 0% of total installed capacity (2017 est.)
country comparison to the world: 174

Crude oil—production: 0 bbl/day (2018 est.)
country comparison to the world: 108

Crude oil—exports: 0 bbl/day (2015 est.)
country comparison to the world: 89

Crude oil—imports: 0 bbl/day (2015 est.)
country comparison to the world: 93

Crude oil—proved reserves: 0 bbl (1 January 2018 est.)
country comparison to the world: 105

Refined petroleum products—production: 0 bbl/day (2015 est.)
country comparison to the world: 116

Refined petroleum products—consumption: 20,040 bbl/day (2016 est.)
country comparison to the world: 140

Refined petroleum products—exports: 0 bbl/day (2015 est.)
country comparison to the world: 128

Refined petroleum products—imports: 19,150 bbl/day (2015 est.)
country comparison to the world: 123

Natural gas—production: 0 cu m (2017 est.)
country comparison to the world: 101

Natural gas—consumption: 48,020 cu m (2017 est.)
country comparison to the world: 115

Natural gas—exports: 0 cu m (2017 est.)
country comparison to the world: 63

Natural gas—imports: 48,020 cu m (2017 est.)
country comparison to the world: 79

Natural gas—proved reserves: 0 cu m (1 January 2009 est.)
country comparison to the world: 108

Carbon dioxide emissions from consumption of energy: 3.089 million Mt (2017 est.)
country comparison to the world: 147

COMMUNICATIONS

Telephones—fixed lines: *total subscriptions:* 78,439
subscriptions per 100 inhabitants: 23.4 (2019 est.)
country comparison to the world: 144

Telephones—mobile cellular: *total subscriptions:* 366,217
subscriptions per 100 inhabitants: 109.25 (2019 est.)
country comparison to the world: 177

Telecommunication systems: general assessment: the telecom sector across the Caribbean continues to be a growth area, contributing to the country's overall GDP; totally automatic system; highly developed; operators focus investment on mobile networks; the activation of (mobile number portability) MNP in April 2017, allowing mobile subscribers to port their numbers between competing MNO (mobile network operators) has contributed to the competition and liberalization of the market (2020)
domestic: 23 per 100 fixed-line, 109 per 100 mobile-cellular (2019)
international: country code—1-242; landing points for the ARCOS-1, BICS, Bahamas 2-US, and BDSN fiber-optic submarine cables that provide links to South and Central America, parts of the Caribbean, and the US; satellite earth stations—2; the Bahamas Domestic Submarine Network links all of the major islands; (2019)
note: the COVID-19 outbreak is negatively impacting telecommunications production and supply chains globally; consumer spending on telecom devices and services has also slowed due to the pandemic's effect on economies worldwide; overall progress towards improvements in all facets of the telecom industry—mobile, fixed-line, broadband, submarine cable and satellite—has moderated

Broadcast media: The Bahamas has 4 major TV providers that provide service to all major islands in the archipelago; 1 TV station is operated by government-owned, commercially run Broadcasting Corporation of the Bahamas (BCB) and competes freely with 4 privately owned TV stations; multi-channel cable TV subscription service is widely available; there are 32 licensed broadcast (radio) service providers, 31 are privately owned FM radio stations operating on New Providence, Grand Bahama Island, Abaco Island, and on smaller islands in the country; the BCB operates a multi-channel radio broadcasting network that has national coverage; the sector is regulated by the Utilities Regulation and Competition Authority (2019)

Internet country code: .bs

Internet users: *total:* 282,739
percent of population: 85% (July 2018 est.)
country comparison to the world: 167

Broadband—fixed subscriptions: *total:* 87,067
subscriptions per 100 inhabitants: 26 (2018 est.)
country comparison to the world: 125

TRANSPORTATION

National air transport system: *number of registered air carriers:* 5 (2020)
inventory of registered aircraft operated by air carriers: 35

annual passenger traffic on registered air carriers: 1,197,116 (2018)
annual freight traffic on registered air carriers: 160,000 mt-km (2018)

Civil aircraft registration country code prefix: C6 (2016)

Airports: 54 (2020)
country comparison to the world: 85

Airports—with paved runways: *total:* 24 (2017)
over 3,047 m: 2 (2017)
2,438 to 3,047 m: 2 (2017)
1,524 to 2,437 m: 13 (2017)
914 to 1,523 m: 7 (2017)

Airports—with unpaved runways: *total:* 37 (2013)
1,524 to 2,437 m: 4 (2013)
914 to 1,523 m: 16 (2013)
under 914 m: 17 (2013)

Heliports: 1 (2013)

Roadways: *total:* 2,700 km (2011)
paved: 1,620 km (2011)
unpaved: 1,080 km (2011)
country comparison to the world: 166

Merchant marine: *total:* 1,401
by type: bulk carrier 333, container ship 49, general cargo 78, oil tanker 268, other 673 (2019)
country comparison to the world: 20

Ports and terminals: *major seaport(s):* Freeport, Nassau, South Riding Point
cruise port(s): Nassau
container port(s) (TEUs): Freeport (1,116,272) (2011)

MILITARY AND SECURITY

Military and security forces: Royal Bahamas Defense Force: Patrol Squadron, Commando Squadron, and Air Wing (2020)

Military and security service personnel strengths: the Royal Bahamas Defense Force (RBDF) has approximately 1,300 total personnel (2019 est.)

Military equipment inventories and acquisitions: most of the RBDF's major equipment inventory is supplied by the Netherlands (2019 est.)

Military service age and obligation: 18 years of age for voluntary male and female service; no conscription (2012)

TRANSNATIONAL ISSUES

Disputes—international: disagrees with the US on the alignment of the northern axis of a potential maritime boundary

Illicit drugs: transshipment point for cocaine and marijuana bound for US and Europe; offshore financial center

BAHRAIN

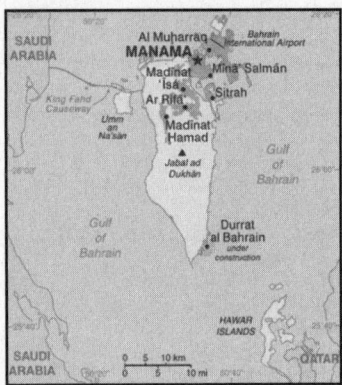

INTRODUCTION

Background: In 1783, the Sunni Al-Khalifa family took power in Bahrain. In order to secure these holdings, it entered into a series of treaties that made Bahrain a British protectorate. The archipelago attained its independence in 1971. A steady decline in oil production and reserves since 1970 prompted Bahrain to take steps to diversify its economy, in the process developing petroleum processing and refining, aluminum production, and hospitality and retail sectors. It has also endeavored to become a leading regional banking center, especially with respect to Islamic finance. Bahrain's small size, central location among Gulf countries, economic dependence on Saudi Arabia, and proximity to Iran require it to play a delicate balancing act in foreign affairs among its larger neighbors. Its foreign policy activities usually fall in line with Saudi Arabia and the UAE.

The Sunni royal family has long struggled to manage relations with its large Shia-majority population. In early 2011, amid Arab uprisings elsewhere in the region, the Bahraini Government confronted similar pro-democracy and reform protests at home with police and military action, including deploying Gulf Cooperation Council security forces to Bahrain. Failed political talks prompted opposition political societies to boycott 2014 legislative and municipal council elections. In 2018, a law preventing members of political societies dissolved by the courts from participating in elections effectively sidelined the majority of opposition figures from taking part in national elections. As a result, most members of parliament are independents. Ongoing dissatisfaction with the political status quo continues to factor into sporadic clashes between demonstrators and security forces. On 15 September 2020, Bahrain and the United Arab Emirates signed a peace accord with Israel – brokered by the US – in Washington DC. Referred to as the Abraham Accords, Bahrain

and the United Arab Emirates are the two latest Middle Eastern countries, along with Egypt and Jordan, to recognize Israel.

GEOGRAPHY

Location: Middle East, archipelago in the Persian Gulf, east of Saudi Arabia

Geographic coordinates: 26 00 N, 50 33 E

Map references: Middle East

Area: *total:* 760 sq km
land: 760 sq km
water: 0 sq km
country comparison to the world: 188

Area—comparative: 3.5 times the size of Washington, DC

Land boundaries: 0 km

Coastline: 161 km

Maritime claims: *territorial sea:* 12 nm
contiguous zone: 24 nm
continental shelf: extending to boundaries to be determined

Climate: arid; mild, pleasant winters; very hot, humid summers

Terrain: mostly low desert plain rising gently to low central escarpment

Elevation: *lowest point:* Persian Gulf 0 m
highest point: Jabal ad Dukhan 135 m

Natural resources: oil, associated and nonassociated natural gas, fish, pearls

Land use: *agricultural land:* 11.3% (2016 est.)
arable land: 2.1% (2016 est.) / permanent crops: 3.9% (2016 est.) / permanent pasture: 5.3% (2016 est.)
forest: 0.7% (2016 est.)
other: 88% (2016 est.)
Irrigated land: 40 sq km (2012)

Population distribution: smallest population of the Gulf States, but urbanization rate exceeds 90%; largest settlement concentration is found on the far northern end of the island in and around Manamah and Al Muharraq

Natural hazards: periodic droughts; dust storms

Environment—current issues: desertification resulting from the degradation of limited arable land, periods of drought, and dust storms; coastal degradation (damage to coastlines, coral reefs, and sea vegetation) resulting from oil spills and other discharges from large tankers, oil refineries, and distribution stations; lack of freshwater resources (groundwater and seawater are the only sources for all water needs); lowered water table leaves aquifers vulnerable to saline contamination; desalinization provides some 90% of the country's freshwater

Environment—international agreements: *party to:* Biodiversity, Climate Change, Climate Change-Kyoto Protocol, Desertification, Hazardous

Wastes, Law of the Sea, Ozone Layer Protection, Wetlands
signed, but not ratified: none of the selected agreements

Geography—note: close to primary Middle Eastern petroleum sources; strategic location in Persian Gulf, through which much of the Western world's petroleum must transit to reach open ocean

PEOPLE AND SOCIETY

Population: 1,505,003 (July 2020 est.)
note: immigrants make up approximately 48% of the total population, according to UN data (2017)
country comparison to the world: 154

Nationality: *noun:* Bahraini(s)
adjective: Bahraini

Ethnic groups: Bahraini 46%, Asian 45.5%, other Arab 4.7%, African 1.6%, European 1%, other 1.2% (includes Gulf Co-operative country nationals, North and South Americans, and Oceanians) (2010 est.)

Languages: Arabic (official), English, Farsi, Urdu

Religions: Muslim 73.7%, Christian 9.3%, Jewish 0.1%, other 16.9% (2017 est.)

MENA religious affiliation: *Age structure:* 0-14 years: 18.45% (male 141,039/female 136,687)
15-24 years: 15.16% (male 129,310/female 98,817)
25-54 years: 56.14% (male 550,135/female 294,778)
55-64 years: 6.89% (male 64,761/female 38,870)
65 years and over: 3.36% (male 25,799/female 24,807) (2020 est.)

Dependency ratios: *total dependency ratio:* 26.5
youth dependency ratio: 23.1
elderly dependency ratio: 3.4
potential support ratio: 29.8 (2020 est.)

Median age: *total:* 32.9 years
male: 34.4 years
female: 30.3 years (2020 est.)
country comparison to the world: 102

Population growth rate: 2.08% (2020 est.)
country comparison to the world: 43

Birth rate: 12.7 births/1,000 population (2020 est.)
country comparison to the world: 149

Death rate: 2.8 deaths/1,000 population (2020 est.)
country comparison to the world: 226

Net migration rate: 10.6 migrant(s)/1,000 population (2020 est.)
country comparison to the world: 7

Population distribution: smallest population of the Gulf States, but urbanization rate exceeds 90%; largest settlement concentration is found on the far northern end of the island in and around Manamah and Al Muharraq

Urbanization: *urban population:* 89.5% of total population (2020)

rate of urbanization: 4.38% annual rate of change (2015-20 est.)

total population growth rate v. urban population growth rate, 2000-2030: Major urban areas—population: 635,000 MANAMA (capital) (2020)

Sex ratio: *at birth:* 1.03 male(s)/female
0-14 years: 1.03 male(s)/female
15-24 years: 1.31 male(s)/female
25-54 years: 1.87 male(s)/female
55-64 years: 1.67 male(s)/female
65 years and over: 1.04 male(s)/female
total population: 1.53 male(s)/female (2020 est.)

Maternal mortality rate: 14 deaths/100,000 live births (2017 est.)
country comparison to the world: 137

Infant mortality rate: *total:* 8.3 deaths/1,000 live births
male: 9.2 deaths/1,000 live births
female: 7.4 deaths/1,000 live births (2020 est.)
country comparison to the world: 147

Life expectancy at birth: *total population:* 79.4 years
male: 77.1 years
female: 81.8 years (2020 est.)
country comparison to the world: 52

Total fertility rate: 1.69 children born/woman (2020 est.)
country comparison to the world: 176

Drinking water source:
improved:
total: 100% of population
unimproved:
total: 0% of population (2017 est.)

Current Health Expenditure: 4.7% (2017)

Physicians density: 0.93 physicians/1,000 population (2015)

Hospital bed density: 1.7 beds/1,000 population (2017)

Sanitation facility access:
improved:
total: 100% of population
unimproved:
total: 0% of population (2017 est.)

HIV/AIDS—adult prevalence rate: <.1% (2017 est.)

HIV/AIDS—people living with HIV/AIDS: <500 (2017 est.)

HIV/AIDS—deaths: <100 (2017 est.)

Obesity—adult prevalence rate: 29.8% (2016)
country comparison to the world: 25

Education expenditures: 2.3% of GDP (2017)
country comparison to the world: 163

Literacy: *definition:* age 15 and over can read and write
total population: 97.5%
male: 99.9%
female: 94.9% (2018)

School life expectancy (primary to tertiary education): *total:* 16 years
male: 16 years
female: 17 years (2019)

Unemployment, youth ages 15-24: *total:* 5.3%
male: 2.6%
female: 12.2% (2012 est.)
country comparison to the world: 164

GOVERNMENT

Country name: *conventional long form:* Kingdom of Bahrain
conventional short form: Bahrain
local long form: Mamlakat al Bahrayn
local short form: Al Bahrayn
former: Dilmun, Tylos, Awal, Mishmahig, Bahrayn, State of Bahrain
etymology: the name means "the two seas" in Arabic and refers to the water bodies surrounding the archipelago

Government type: constitutional monarchy

Capital: *name:* Manama
geographic coordinates: 26 14 N, 50 34 E
time difference: UTC+3 (8 hours ahead of Washington, DC, during Standard Time)
etymology: name derives from the Arabic "al-manama" meaning "place of rest" or "place of dreams"

Administrative divisions: 4 governorates (muhafazat, singular—muhafazah); Asimah (Capital), Janubiyah (Southern), Muharraq, Shamaliyah (Northern)
note: each governorate administered by an appointed governor

Independence: 15 August 1971 (from the UK)

National holiday: National Day, 16 December (1971); note—15 August 1971 was the date of independence from the UK, 16 December 1971 was the date of independence from British protection

Constitution: *history:* adopted 14 February 2002
amendments: proposed by the king or by at least 15 members of either chamber of the National Assembly followed by submission to an Assembly committee for review and, if approved, submitted to the government for restatement as drafts; passage requires a two-thirds majority vote by the membership of both chambers and validation by the king; constitutional articles on the state religion (Islam), state language (Arabic), and the monarchy and "inherited rule" cannot be amended; amended 2012, 2017

Legal system: mixed legal system of Islamic (sharia) law, English common law, Egyptian civil, criminal, and commercial codes; customary law

International law organization participation: has not submitted an ICJ jurisdiction declaration; non-party state to the ICCt

Citizenship: *citizenship by birth:* no
citizenship by descent only: the father must be a citizen of Bahrain
dual citizenship recognized: no
residency requirement for naturalization: 25 years; 15 years for Arab nationals

Suffrage: 20 years of age; universal

Executive branch: *chief of state:* King HAMAD bin Isa Al-Khalifa (since 6 March 1999)

head of government: Prime minister SALMAN bin Hamad Al-Khalifa (since 11 November 2020); first deputy prime minister (vacant); Deputy Prime Ministers MUHAMMAD bin Mubarak Al-Khalifa (since September 2005), Jawad bin Salim al-ARAIDH, ALI bin Khalifa bin Salman Al-Khalifa (since 11 December 2006), KHALID bin Abdallah Al-Khalifa (since November 2010); note—KHALIFA ibn Salman Al Khalifa, who served as prime minister since Bahrain's independence in 1971, died on 11 November 2020
cabinet: Cabinet appointed by the monarch
elections/appointments: the monarchy is hereditary; prime minister appointed by the monarch

Legislative branch: *description:* bicameral National Assembly consists of:
Consultative Council or Majlis al-Shura (40 seats; members appointed by the king)
Council of Representatives or Majlis al-Nuwab (40 seats; members directly elected in single-seat constituencies by absolute majority vote in 2 rounds if needed; members serve 4-year renewable terms)
elections: Consultative Council—last appointments on 12 December 2018 (next NA)
Council of Representatives—first round for 9 members held on 24 November 2018; second round for remaining 31 members held on 1 December 2018 (next to be held in 2022)
election results: Consultative Council—composition—men 31, women 9, percent of women 22.5%
Council of Representatives (for 2018 election)—percent of vote by society—NA; seats by society—Islamic Al-Asalah (Sunni Salafi) 3, Minbar al-Taqadumi (Communist) 2, National Unity Gathering (Sunni progovernment) 1, National Islamic Minbar (Sunni Muslim Brotherhood) 1, independent 33; composition—men 34, women 6, percent of women 15%; note—total National Assembly percent of women 19%

Judicial branch: *highest courts:* Court of Cassation (consists of the chairman and 3 judges); Supreme Court of Appeal (consists of the chairman and 3 judges); Constitutional Court (consists of the president and 6 members); High Sharia Court of Appeal (court sittings include the president and at least one judge)
judge selection and term of office: Court of Cassation judges appointed by royal decree and serve for a specified tenure; Constitutional Court president and members appointed by the Higher Judicial Council, a body chaired by the monarch and includes judges from the Court of Cassation, sharia law courts, and Civil High Courts of Appeal; members serve 9-year terms; High Sharia Court of Appeal member appointments by royal decree for a specified tenure
subordinate courts: Civil High Courts of Appeal; middle and lower civil courts; High Sharia Court of Appeal; Senior Sharia Court; Administrative Courts of Appeal; military courts
note: the judiciary of Bahrain is divided into civil law courts and sharia law courts; sharia courts (involving personal status and family law) are further divided into Sunni Muslim and Shia Muslim;

the Courts are supervised by the Supreme Judicial Council.

Political parties and leaders: *note:* political parties are prohibited, but political societies were legalized under a July 2005 law

International organization participation: ABEDA, AFESD, AMF, CAEU, CICA, FAO, G-77, GCC, IAEA, IBRD, ICAO, ICC (national committees), ICRM, IDA, IDB, IFC, IFRCS, IHO, ILO, IMF, IMO, IMSO, Interpol, IOC, IOM (observer), IPU, ISO, ITSO, ITU, ITUC (NGOs), LAS, MIGA, NAM, OAPEC, OIC, OPCW, PCA, UN, UNCTAD, UNESCO, UNIDO, UNWTO, UPU, WCO, WFTU (NGOs), WHO, WIPO, WMO, WTO

Diplomatic representation in the US: *chief of mission:* Ambassador Abdulla bin Rashid AL KHALIFA (since 21 July 2017)
chancery: 3502 International Drive NW, Washington, DC 20008
telephone: [1] (202) 342-1111
FAX: [1] (202) 362-2192
consulate(s) general: New York

Diplomatic representation from the US: *chief of mission:* Ambassador Justin H. SIBERELL (since November 2017)
telephone: [973] 1724-2700
embassy: Building #979, Road 3119 (next to Al-Ahli Sports Club), Block 331, Zinj District, Manama
mailing address: PSC 451, Box 660, FPO AE 09834-5100
international mail: American Embassy, Box 26431, Manama
FAX: [973] 1727-2594

Flag description: red, the traditional color for flags of Persian Gulf states, with a white serrated band (five white points) on the hoist side; the five points represent the five pillars of Islam
note: until 2002, the flag had eight white points, but this was reduced to five to avoid confusion with the Qatari flag

National symbol(s): a red field surmounted by a white serrated band with five white points; national colors: red, white

National anthem: *name:* "Bahrainona" (Our Bahrain)
lyrics/music: unknown
note: adopted 1971; although Mohamed Sudqi AYYASH wrote the original lyrics, they were changed in 2002 following the transformation of Bahrain from an emirate to a kingdom
0:00 / 0:49

ECONOMY

Economy—overview: Oil and natural gas play a dominant role in Bahrain's economy. Despite the Government's past efforts to diversify the economy, oil still comprises 85% of Bahraini budget revenues. In the last few years lower world energy prices have generated sizable budget deficits—about 10% of GDP in 2017 alone. Bahrain has few options for covering these deficits, with low foreign assets and fewer oil resources compared to its GCC neighbors. The three major US credit agencies downgraded Bahrain's sovereign debt rating to "junk" status in 2016, citing persistently low oil prices and the government's high debt levels. Nevertheless, Bahrain was able to raise about $4 billion by issuing foreign currency denominated debt in 2017.

Other major economic activities are production of aluminum—Bahrain's second biggest export after oil and gas –finance, and construction. Bahrain continues to seek new natural gas supplies as feedstock to support its expanding petrochemical and aluminum industries. In April 2018 Bahrain announced it had found a significant oil field off the country's west coast, but is still assessing how much of the oil can be extracted profitably.

In addition to addressing its current fiscal woes, Bahraini authorities face the long-term challenge of boosting Bahrain's regional competitiveness—especially regarding industry, finance, and tourism—and reconciling revenue constraints with popular pressure to maintain generous state subsidies and a large public sector. Since 2015, the government lifted subsidies on meat, diesel, kerosene, and gasoline and has begun to phase in higher prices for electricity and water. As part of its diversification plans, Bahrain implemented a Free Trade Agreement (FTA) with the US in August 2006, the first FTA between the US and a Gulf state. It plans to introduce a Value Added Tax (VAT) by the end of 2018.

GDP (purchasing power parity): $71.17 billion (2017 est.)
$68.59 billion (2016 est.)
$66.3 billion (2015 est.)
note: data are in 2017 dollars
country comparison to the world: 100

GDP (official exchange rate): $35.33 billion (2017 est.)

GDP—real growth rate: 2.49% (2019 est.)
13.89% (2018 est.)
3.85% (2017 est.)
country comparison to the world: 116

GDP—per capita (PPP): $49,000 (2017 est.)
$48,200 (2016 est.)
$48,400 (2015 est.)
note: data are in 2017 dollars
country comparison to the world: 33

Gross national saving: 19.8% of GDP (2017 est.)
21.2% of GDP (2016 est.)
22% of GDP (2015 est.)
country comparison to the world: 99

GDP—composition, by end use: *household consumption:* 45.8% (2017 est.)
government consumption: 15.5% (2017 est.)
investment in fixed capital: 26.1% (2017 est.)
investment in inventories: 0.4% (2017 est.)
exports of goods and services: 80.2% (2017 est.)
imports of goods and services: -67.9% (2017 est.)

GDP—composition, by sector of origin: *agriculture:* 0.3% (2017 est.)
industry: 39.3% (2017 est.)
services: 60.4% (2017 est.)

Agriculture—products: fruit, vegetables; poultry, dairy products; shrimp, fish

Industries: petroleum processing and refining, aluminum smelting, iron pelletization, fertilizers, Islamic and offshore banking, insurance, ship repairing, tourism

Industrial production growth rate: 0.6% (2017 est.)
country comparison to the world: 165

Labor force: 831,600 (2017 est.)
note: excludes unemployed; 44% of the population in the 15-64 age group is non-national
country comparison to the world: 144

Labor force—by occupation: *agriculture:* 1%
industry: 32%
services: 67% (2004 est.)

Unemployment rate: 3.6% (2017 est.)
3.7% (2016 est.)
note: official estimate; actual rate is higher
country comparison to the world: 49

Population below poverty line: NA

Household income or consumption by percentage share: *lowest 10%:* NA
highest 10%: NA

Budget: *revenues:* 5.854 billion (2017 est.)
expenditures: 9.407 billion (2017 est.)

Taxes and other revenues: 16.6% (of GDP) (2017 est.)
country comparison to the world: 176

Budget surplus (+) or deficit (-): -10.1% (of GDP) (2017 est.)
country comparison to the world: 211

Public debt: 88.5% of GDP (2017 est.)
81.4% of GDP (2016 est.)
country comparison to the world: 26

Fiscal year: calendar year

Inflation rate (consumer prices): 1.4% (2017 est.)
2.8% (2016 est.)
country comparison to the world: 76

Current account balance: -$1.6 billion (2017 est.)
-$1.493 billion (2016 est.)
country comparison to the world: 162

Exports: $15.38 billion (2017 est.)
$12.78 billion (2016 est.)
country comparison to the world: 75

Exports—partners: UAE 19.6%, Saudi Arabia 11.7%, US 10.8%, Oman 8.1%, China 6.5%, Qatar 5.7%, Japan 4.2% (2017)

Exports—commodities: petroleum and petroleum products, aluminum, textiles

Imports: $16.08 billion (2017 est.)
$13.59 billion (2016 est.)
country comparison to the world: 84

Imports—commodities: crude oil, machinery, chemicals

Imports—partners: China 8.8%, UAE 7.2%, US 7.1%, Australia 5.3%, Japan 4.8% (2017)

Reserves of foreign exchange and gold: $2.349 billion (31 December 2017 est.)
$3.094 billion (31 December 2016 est.)
country comparison to the world: 118

Debt—external: $52.15 billion (31 December 2017 est.)
$42.55 billion (31 December 2016 est.)
country comparison to the world: 63

Exchange rates: Bahraini dinars (BHD) per US dollar—
0.376 (2017 est.)
0.376 (2016 est.)
0.376 (2015 est.)
0.376 (2014 est.)
0.376 (2013 est.)

ENERGY

Electricity access: *electrification—total population:* 100% (2020)

Electricity—production: 26.81 billion kWh (2016 est.)
country comparison to the world: 70

Electricity—consumption: 26.11 billion kWh (2016 est.)
country comparison to the world: 67

Electricity—exports: 213 million kWh (2015 est.)
country comparison to the world: 74

Electricity—imports: 276 million kWh (2016 est.)
country comparison to the world: 89

Electricity—installed generating capacity: 3.928 million kW (2016 est.)
country comparison to the world: 90

Electricity—from fossil fuels: 100% of total installed capacity (2016 est.)
country comparison to the world: 2

Electricity—from nuclear fuels: 0% of total installed capacity (2017 est.)
country comparison to the world: 44

Electricity—from hydroelectric plants: 0% of total installed capacity (2017 est.)
country comparison to the world: 156

Electricity—from other renewable sources: 0% of total installed capacity (2017 est.)
country comparison to the world: 175

Crude oil—production: 40,000 bbl/day (2018 est.)
country comparison to the world: 58

Crude oil—exports: 0 bbl/day (2015 est.)
country comparison to the world: 90

Crude oil—imports: 226,200 bbl/day (2015 est.)
country comparison to the world: 29

Crude oil—proved reserves: 124.6 million bbl (1 January 2018 est.)
country comparison to the world: 67

Refined petroleum products—production: 274,500 bbl/day (2015 est.)
country comparison to the world: 45

Refined petroleum products—consumption: 61,000 bbl/day (2016 est.)
country comparison to the world: 94

Refined petroleum products—exports: 245,300 bbl/day (2015 est.)
country comparison to the world: 30

Refined petroleum products—imports: 14,530 bbl/day (2015 est.)
country comparison to the world: 136

Natural gas—production: 15.89 billion cu m (2017 est.)
country comparison to the world: 36

Natural gas—consumption: 15.89 billion cu m (2017 est.)
country comparison to the world: 42

Natural gas—exports: 0 cu m (2017 est.)
country comparison to the world: 64

Natural gas—imports: 0 cu m (2017 est.)
country comparison to the world: 88

Natural gas—proved reserves: 92.03 billion cu m (1 January 2018 est.)
country comparison to the world: 53

Carbon dioxide emissions from consumption of energy: 37.98 million Mt (2017 est.)
country comparison to the world: 67

COMMUNICATIONS

Telephones—fixed lines: *total subscriptions:* 246,603
subscriptions per 100 inhabitants: 16.73 (2019 est.)
country comparison to the world: 118

Telephones—mobile cellular: *total subscriptions:* 1,706,763
subscriptions per 100 inhabitants: 115.79 (2019 est.)
country comparison to the world: 156

Telecommunication systems: *general assessment:* well-developed LTE networks, 5G trials tested and deployment in the near future; mobile penetration is high compared to the region; development of its own national broadband network; competition is good and telecoms are regulated; telecom contributes 4% to the GDP (2020)
domestic: 17 per 100 fixed-line, 116 per 100 mobile-cellular; modern fiber-optic integrated services; digital network with rapidly expanding mobile-cellular telephones (2019)
international: country code—973; landing points for the FALCON, Tata TGN-Gulf, GBICS/MENA, and FOG submarine cable network that provides links to Asia, the Middle East, and Africa; tropospheric scatter to Qatar and UAE; microwave radio relay to Saudi Arabia; satellite earth station—1 (2019)
note: the COVID-19 outbreak is negatively impacting telecommunications production and supply chains globally; consumer spending on telecom devices and services has also slowed due to the pandemic's effect on economies worldwide; overall progress towards improvements in all facets of the telecom industry—mobile, fixed-line, broadband, submarine cable and satellite—has moderated

Broadcast media: state-run Bahrain Radio and Television Corporation (BRTC) operates 5 terrestrial TV networks and several radio stations; satellite TV systems provide access to international broadcasts; 1 private FM station directs broadcasts to Indian listeners; radio and TV broadcasts from countries in the region are available (2019)

Internet country code: .bh

Internet users: *total:* 1,423,039

percent of population: 98.64% (July 2018 est.)
country comparison to the world: 132

Broadband—fixed subscriptions: *total:* 184,603
subscriptions per 100 inhabitants: 13 (2018 est.)
country comparison to the world: 111

TRANSPORTATION

National air transport system: *number of registered air carriers:* 6 (2020)
inventory of registered aircraft operated by air carriers: 42
annual passenger traffic on registered air carriers: 5,877,003 (2018)
annual freight traffic on registered air carriers: 420.98 million mt-km (2018)

Civil aircraft registration country code prefix: A9C (2016)

Airports: 4 (2013)
country comparison to the world: 184

Airports—with paved runways: *total:* 4 (2017)
over 3,047 m: 3 (2017)
914 to 1,523 m: 1 (2017)

Heliports: 1 (2013)

Pipelines: 20 km gas, 54 km oil (2013)

Roadways: *total:* 4,122 km (2010)
paved: 3,392 km (2010)
unpaved: 730 km (2010)
country comparison to the world: 153

Merchant marine: *total:* 261
by type: bulk carrier 1, container ship 1, general cargo 11, oil tanker 4, other 244 (2019)
country comparison to the world: 59

Ports and terminals: *major seaport(s):* Mina' Salman, Sitrah

MILITARY AND SECURITY

Military and security forces: Bahrain Defense Force (BDF): Royal Bahraini Army (includes the Royal Guard), Royal Bahraini Navy, Royal Bahraini Air Force, Royal Bahraini Air Defense Force; Ministry of Interior security forces: National Guard, Special Security Forces Command (SSFC), Coast Guard (2019)
note: the Royal Guard is officially under the command of the Army, but exercises considerable autonomy

Military expenditures: 3.7% of GDP (2019)
4.1% of GDP (2018)
4.3% of GDP (2017)
4.7% of GDP (2016)
4.6% of GDP (2015)
country comparison to the world: 17

Military and security service personnel strengths: size assessments for the Bahrain Defense Force vary; approximately 10,000 active personnel (7,500 Army; 1,000 Navy; 1,500 Air Force); est. 2,500 National Guard
(2019 est.)

Military equipment inventories and acquisitions: the inventory of the Bahrain Defense force is comprised mostly of equipment acquired from the US along with a smaller quantity of material from

European suppliers; since 2010, Turkey and the US are the leading suppliers of arms to Bahrain (2019 est.)

Military service age and obligation: 18 years of age for voluntary military service; 15 years of age for NCOs, technicians, and cadets; no conscription (2019)

targets, weapons, size, and sources of support of the group(s) appear(s) in Appendix-T

TRANSNATIONAL ISSUES

Disputes—international: none

BANGLADESH

INTRODUCTION

Background: The huge delta region formed at the confluence of the Ganges and Brahmaputra River systems—now referred to as Bangladesh—was a loosely incorporated outpost of various empires centered on the Gangetic plain for much of the first millennium A.D. Muslim conversions and settlement in the region began in the 10th century, primarily from Arab and Persian traders and preachers. Europeans established trading posts in the area in the 16th century. Eventually the area known as Bengal, primarily Hindu in the western section and mostly Muslim in the eastern half, became part of British India. Partition in 1947 resulted in an eastern wing of Pakistan in the Muslim-majority area, which became East Pakistan. Calls for greater autonomy and animosity between the eastern and western wings of Pakistan led to a Bengali independence movement. That movement, led by the Awami League (AL) and supported by India, won the independence war for Bangladesh in 1971.

The post-independence AL government faced daunting challenges and in 1975 it was overthrown by the military, triggering a series of military coups that resulted in a military-backed government and subsequent creation of the Bangladesh Nationalist Party (BNP) in 1978. That government also ended in a coup in 1981, followed by military-backed rule until democratic elections occurred in 1991. The BNP and AL have alternated in power since 1991, with the exception of a military-backed,

emergency caretaker regime that suspended parliamentary elections planned for January 2007 in an effort to reform the political system and root out corruption. That government returned the country to fully democratic rule in December 2008 with the election of the AL and Prime Minister Sheikh HASINA. In January 2014, the incumbent AL won the national election by an overwhelming majority after the BNP boycotted the election, which extended HASINA's term as prime minister. In December 2018, HASINA secured a third consecutive term (fourth overall) with the AL coalition securing 96% of available seats, amid widespread claims of election irregularities. With the help of international development assistance, Bangladesh has reduced the poverty rate from over half of the population to less than a third, achieved Millennium Development Goals for maternal and child health, and made great progress in food security since independence. The economy has grown at an annual average of about 6% for the last two decades and the country reached World Bank lower-middle income status in 2014.

GEOGRAPHY

Location: Southern Asia, bordering the Bay of Bengal, between Burma and India

Geographic coordinates: 24 00 N, 90 00 E

Map references: Asia

Area: *total:* 148,460 sq km
land: 130,170 sq km
water: 18,290 sq km
country comparison to the world: 95

Area—comparative: slightly larger than Pennsylvania and New Jersey combined; slightly smaller than Iowa

Land boundaries: *total:* 4,413 km
border countries (2): Burma 271 km, India 4142 km

Coastline: 580 km

Maritime claims: *territorial sea:* 12 nm
exclusive economic zone: 200 nm
contiguous zone: 18 nm
continental shelf: to the outer limits of the continental margin

Climate: tropical; mild winter (October to March); hot, humid summer (March to June); humid, warm rainy monsoon (June to October)

Terrain: mostly flat alluvial plain; hilly in southeast

Elevation: *mean elevation:* 85 m
lowest point: Indian Ocean 0 m
highest point: Keokradong 1,230 m

Natural resources: natural gas, arable land, timber, coal

Land use: *agricultural land:* 70.1% (2016 est.)
arable land: 59% (2016 est.) / permanent crops: 6.5% (2016 est.) / permanent pasture: 4.6% (2016 est.)
forest: 11.1% (2016 est.)
other: 18.8% (2016 est.)
Irrigated land: 53,000 sq km (2012)

Natural hazards: droughts; cyclones; much of the country routinely inundated during the summer monsoon season

Environment—current issues: many people are landless and forced to live on and cultivate flood-prone land; waterborne diseases prevalent in surface water; water pollution, especially of fishing areas, results from the use of commercial pesticides; ground water contaminated by naturally occurring arsenic; intermittent water shortages because of falling water tables in the northern and central parts of the country; soil degradation and erosion; deforestation; destruction of wetlands; severe overpopulation with noise pollution

Environment—international agreements: *party to:* Biodiversity, Climate Change, Climate Change-Kyoto Protocol, Desertification, Endangered Species, Environmental Modification, Hazardous Wastes, Law of the Sea, Ozone Layer Protection, Ship Pollution, Wetlands
signed, but not ratified: none of the selected agreements

Geography—note: most of the country is situated on deltas of large rivers flowing from the Himalayas: the Ganges unites with the Jamuna (main channel of the Brahmaputra) and later joins the Meghna to eventually empty into the Bay of Bengal

PEOPLE AND SOCIETY

Population: 162,650,853 (July 2020 est.)
country comparison to the world: 8
Nationality: *noun:* Bangladeshi(s)
adjective: Bangladeshi

Ethnic groups: Bengali at least 98%, other indigenous ethnic groups 1.1% (2011 est.)

note: Bangladesh's government recognizes 27 indigenous ethnic groups under the 2010 Cultural Institution for Small Anthropological Groups Act; other sources estimate there are about 75 ethnic groups; critics of the 2011 census claim that it underestimates the size of Bangladesh's ethnic population

Languages: Bangla 98.8% (official, also known as Bengali), other 1.2% (2011 est.)

Religions: Muslim 89.1%, Hindu 10%, other 0.9% (includes Buddhist, Christian) (2013 est.)

Age structure: *0-14 years:* 26.48% (male 21,918,651/female 21,158,574)
15-24 years: 18.56% (male 15,186,470/female 15,001,950)
25-54 years: 40.72% (male 31,694,267/female 34,535,643)
55-64 years: 7.41% (male 5,941,825/female 6,115,856)
65 years and over: 6.82% (male 5,218,206/female 5,879,411) (2020 est.)

population pyramid: *Dependency ratios:* total dependency ratio: 47
youth dependency ratio: 39.3
elderly dependency ratio: 7.7
potential support ratio: 13 (2020 est.)

Median age: *total:* 27.9 years
male: 27.1 years
female: 28.6 years (2020 est.)
country comparison to the world: 143

Population growth rate: 0.98% (2020 est.)
country comparison to the world: 106

Birth rate: 18.1 births/1,000 population (2020 est.)
country comparison to the world: 88

Death rate: 5.5 deaths/1,000 population (2020 est.)
country comparison to the world: 182

Net migration rate: -3 migrant(s)/1,000 population (2020 est.)
country comparison to the world: 175

Urbanization: *urban population:* 38.2% of total population (2020)
rate of urbanization: 3.17% annual rate of change (2015-20 est.)

total population growth rate v. urban population growth rate, 2000-2030: Major urban areas—population: 21.006 million DHAKA (capital), 5.020 million Chittagong, 954,000 Khulna, 908,000 Rajshahi, 852,000 Sylhet (2020)

Sex ratio: *at birth:* 1.04 male(s)/female
0-14 years: 1.04 male(s)/female
15-24 years: 1.01 male(s)/female
25-54 years: 0.92 male(s)/female
55-64 years: 0.97 male(s)/female
65 years and over: 0.89 male(s)/female
total population: 0.97 male(s)/female (2020 est.)

Mother's mean age at first birth: 18.5 years (2014 est.)
note: median age at first birth among women 25-29

Maternal mortality rate: 173 deaths/100,000 live births (2017 est.)
country comparison to the world: 53

Infant mortality rate: *total:* 28.3 deaths/1,000 live births
male: 30.6 deaths/1,000 live births
female: 26 deaths/1,000 live births (2020 est.)
country comparison to the world: 60

Life expectancy at birth: *total population:* 74.2 years
male: 72 years
female: 76.5 years (2020 est.)
country comparison to the world: 133

Total fertility rate: 2.11 children born/woman (2020 est.)
country comparison to the world: 98

Contraceptive prevalence rate: 62.3% (2014)

Drinking water source:
improved:
urban: 98.9% of population
rural: 98.4% of population
total: 98.6% of population
unimproved:
urban: 1.1% of population
rural: 1.6% of population
total: 1.4% of population (2017 est.)

Current Health Expenditure: 2.3% (2017)

Physicians density: 0.54 physicians/1,000 population (2017)

Hospital bed density: 0.8 beds/1,000 population (2016)

Sanitation facility access:
improved:
urban: 82.5% of population
rural: 64.4% of population
total: 70.9% of population
unimproved:
urban: 17.5% of population
rural: 35.6% of population
total: 29.1% of population (2017 est.)

HIV/AIDS—adult prevalence rate: <.1% (2018 est.)

HIV/AIDS—people living with HIV/AIDS: 14,000 (2018 est.)
country comparison to the world: 90

HIV/AIDS—deaths: <1000 (2018 est.)

Major infectious diseases: *degree of risk:* high (2020)
food or waterborne diseases: bacterial and protozoal diarrhea, hepatitis A and E, and typhoid fever
vectorborne diseases: dengue fever and malaria are high risks in some locations
water contact diseases: leptospirosis
animal contact diseases: rabies
note: widespread ongoing transmission of a respiratory illness caused by the novel coronavirus (COVID-19) is occurring throughout Bangladesh; as of 10 November 2020, Bangladesh has reported a total of 418,764 cases of COVID-19 or 2,543 cumulative cases of COVID-19 per 1 million population with 37 cumulative deaths per 1 million population

Obesity—adult prevalence rate: 3.6% (2016)
country comparison to the world: 191

Children under the age of 5 years underweight: 21.9% (2017/18)

country comparison to the world: 17

Education expenditures: 2% of GDP (2018)
country comparison to the world: 169

Literacy: *definition:* age 15 and over can read and write
total population: 73.9%
male: 76.7%
female: 71.2% (2018)

School life expectancy (primary to tertiary education): *total:* 12 years
male: 12 years
female: 12 years (2018)

Unemployment, youth ages 15-24: *total:* 12.8%
male: 10.8%
female: 16.8% (2017 est.)
country comparison to the world: 107

GOVERNMENT

Country name: *conventional long form:* People's Republic of Bangladesh
conventional short form: Bangladesh
local long form: Gana Prajatantri Bangladesh
local short form: Bangladesh
former: East Bengal, East Pakistan
etymology: the name—a compound of the Bengali words "Bangla" (Bengal) and "desh" (country)—means "Country of Bengal"

Government type: parliamentary republic

Capital: *name:* Dhaka
geographic coordinates: 23 43 N, 90 24 E
time difference: UTC+6 (11 hours ahead of Washington, DC, during Standard Time)
etymology: the origins of the name are unclear, but some sources state that the city's site was originally called "dhakka," meaning "watchtower," and that the area served as a watch-station for Bengal rulers

Administrative divisions: 8 divisions; Barishal, Chattogram, Dhaka, Khulna, Mymensingh, Rajshahi, Rangpur, Sylhet

Independence: 16 December 1971 (from Pakistan)

National holiday: Independence Day, 26 March (1971); Victory Day, 16 December (1971); note—26 March 1971 is the date of the Awami League's declaration of an independent Bangladesh, and 16 December (Victory Day) memorializes the military victory over Pakistan and the official creation of the state of Bangladesh

Constitution: *history:* previous 1935, 1956, 1962 (preindependence); latest enacted 4 November 1972, effective 16 December 1972, suspended March 1982, restored November 1986
amendments: proposed by the House of the Nation; approval requires at least two-thirds majority vote of the House membership and assent of the president of the republic; amended many times, last in 2018

Legal system: mixed legal system of mostly English common law and Islamic law

International law organization participation: has not submitted an ICJ jurisdiction declaration; accepts ICCt jurisdiction

Citizenship: *citizenship by birth:* no
citizenship by descent only: at least one parent
must be a citizen of Bangladesh
dual citizenship recognized: yes, but limited to
select countries
residency requirement for naturalization: 5
years

Suffrage: 18 years of age; universal

Executive branch: *chief of state:* President Abdul
HAMID (since 24 April 2013); note—Abdul
HAMID served as acting president following
the death of Zillur RAHMAN in March 2013;
HAMID was subsequently indirectly elected by
the National Parliament and sworn in 24 April
2013

head of government: Prime Minister Sheikh
HASINA (since 6 January 2009)
cabinet: Cabinet selected by the prime minister,
appointed by the president
elections/appointments: president indirectly
elected by the National Parliament for a 5-year
term (eligible for a second term); election last held
on 7 February 2018 (next to be held by 2023); the
president appoints as prime minister the majority
party leader in the National Parliament
election results: President Abdul HAMID (AL)
reelected by the National Parliament unopposed
for a second term; Sheikh HASINA reappointed
prime minister as leader of the majority AL party
following parliamentary elections in 2018

Legislative branch: *description:* unicameral House
of the Nation or Jatiya Sangsad (350 seats; 300
members in single-seat territorial constituencies
directly elected by simple majority popular vote; 50
members—reserved for women only—indirectly
elected by the elected members by proportional
representation vote using single transferable vote;
all members serve 5-year terms)
elections: last held on 30 December 2018 (next to
be held in 2023)
election results: percent of vote by party—NA;
seats by party as of January 2020—AL 299, JP 27,
BNP 7, other 10, independent 4, vacant 3; composition—men 274, women 73, percent of women
21%

Judicial branch: *highest courts:* Supreme Court of
Bangladesh (organized into the Appellate Division
with 7 justices and the High Court Division with
99 justices)
judge selection and term of office: chief justice
and justices appointed by the president; justices
serve until retirement at age 67
subordinate courts: civil courts include: Assistant
Judge's Court; Joint District Judge's Court;
Additional District Judge's Court; District Judge's
Court; criminal courts include: Court of Sessions;
Court of Metropolitan Sessions; Metropolitan
Magistrate Courts; Magistrate Court; special
courts/tribunals

Political parties and leaders: Awami League or
AL [Sheikh HASINA]
Bangladesh Nationalist Front or BNF [Abdul
Kalam AZADI]

Bangladesh Nationalist Party or BNP [Khaleda
ZIA]
Bangladesh Tariqat Federation or BTF [Syed
Nozibul Bashar MAIZBHANDARI]
Jamaat-i-Islami Bangladesh or JIB (Makbul
AHMAD)
Jatiya Party or JP (Ershad faction) [Hussain
Mohammad ERSHAD]
Jatiya Party or JP (Manju faction) [Anwar Hossain
MANJU]
Liberal Democratic Party or LDP [Oli AHMED]
National Socialist Party or JSD
[KHALEQUZZAMAN]
Workers Party or WP [Rashed Khan MENON]

International organization participation: ADB,
ARF, BIMSTEC, C, CD, CICA (observer), CP,
D-8, FAO, G-77, IAEA, IBRD, ICAO, ICC
(national committees), ICRM, IDA, IDB, IFAD,
IFC, IFRCS, IHO, ILO, IMF, IMO, IMSO,
Interpol, IOC, IOM, IPU, ISO, ITSO, ITU,
ITUC (NGOs), MIGA, MINURSO, MINUSMA,
MONUSCO, NAM, OIC, OPCW, PCA,
SAARC, SACEP, UN, UNAMID, UNCTAD,
UNESCO, UNHCR, UNIDO, UNIFIL, UNMIL,
UNMISS, UNOCI, UNWTO, UPU, WCO,
WFTU (NGOs), WHO, WIPO, WMO, WTO

Diplomatic representation in the US: *chief of
mission:* Ambassador Mohammad ZIAUDDIN
(since 18 September 2014)
chancery: 3510 International Drive NW,
Washington, DC 20008
telephone: [1] (202) 244-0183
FAX: [1] (202) 244-2771
consulate(s) general: Los Angeles, New York

Diplomatic representation from the US: *chief of
mission:* Ambassador Earl Robert MILLER (since
29 November 2018)
telephone: [880] (2) 5566-2000
embassy: Madani Avenue, Baridhara, Dhaka 1212
mailing address: G. P. O. Box 323, Dhaka 1000
FAX: [880] (2) 5566-2915

Flag description: green field with a large red disk
shifted slightly to the hoist side of center; the red
disk represents the rising sun and the sacrifice to
achieve independence; the green field symbolizes
the lush vegetation of Bangladesh

National symbol(s): Bengal tiger, water lily;
national colors: green, red

National anthem: *name:* "Amar Shonar Bangla"
(My Golden Bengal)
lyrics/music: Rabindranath TAGORE
note: adopted 1971; Rabindranath TAGORE, a
Nobel laureate, also wrote India's national anthem
0:00 / 0:00

ECONOMY

Economy—overview: Bangladesh's economy has
grown roughly 6% per year since 2005 despite
prolonged periods of political instability, poor
infrastructure, endemic corruption, insufficient
power supplies, and slow implementation of economic reforms. Although more than half of GDP
is generated through the services sector, almost

half of Bangladeshis are employed in the agriculture sector, with rice as the single-most-important
product.

Garments, the backbone of Bangladesh's industrial sector, accounted for more than 80% of total
exports in FY 2016-17. The industrial sector continues to grow, despite the need for improvements
in factory safety conditions. Steady export growth
in the garment sector, combined with $13 billion
in remittances from overseas Bangladeshis, contributed to Bangladesh's rising foreign exchange
reserves in FY 2016-17. Recent improvements to
energy infrastructure, including the start of liquefied natural gas imports in 2018, represent a major
step forward in resolving a key growth bottleneck.

GDP (purchasing power parity): $690.3 billion
(2017 est.)
$642.7 billion (2016 est.)
$599.5 billion (2015 est.)
note: data are in 2017 dollars
country comparison to the world: 33

GDP (official exchange rate): $261.5 billion (2017
est.)

GDP—real growth rate: 7.4% (2017 est.)
7.2% (2016 est.)
6.8% (2015 est.)
country comparison to the world: 13

GDP—per capita (PPP): $4,200 (2017 est.)
$4,000 (2016 est.)
$3,800 (2015 est.)
note: data are in 2017 dollars
country comparison to the world: 176

Gross national saving: 30.2% of GDP (2017 est.)
30.6% of GDP (2016 est.)
30.3% of GDP (2015 est.)
country comparison to the world: 30

GDP—composition, by end use: *household
consumption:* 68.7% (2017 est.)
government consumption: 6% (2017 est.)
investment in fixed capital: 30.5% (2017 est.)
investment in inventories: 1% (2017 est.)
exports of goods and services: 15% (2017 est.)
imports of goods and services: -20.3% (2017 est.)

GDP—composition, by sector of origin:
agriculture: 14.2% (2017 est.)
industry: 29.3% (2017 est.)
services: 56.5% (2017 est.)

Agriculture—products: rice, jute, tea, wheat, sugarcane, potatoes, tobacco, pulses, oilseeds, spices,
fruit; beef, milk, poultry

Industries: jute, cotton, garments, paper, leather,
fertilizer, iron and steel, cement, petroleum products, tobacco, pharmaceuticals, ceramics, tea, salt,
sugar, edible oils, soap and detergent, fabricated
metal products, electricity, natural gas

Industrial production growth rate: 10.2% (2017
est.)
country comparison to the world: 15

Labor force: 66.64 million (2017 est.)
note: extensive migration of labor to Saudi Arabia,
Kuwait, UAE, Oman, Qatar, and Malaysia
country comparison to the world: 6

Labor force—by occupation: *agriculture:* 42.7% *industry:* 20.5% *services:* 36.9% (2016 est.)

Unemployment rate: 4.4% (2017 est.) 4.4% (2016 est.)
note: about 40% of the population is underemployed; many persons counted as employed work only a few hours a week and at low wages
country comparison to the world: 64

Population below poverty line: 24.3% (2016 est.)

Household income or consumption by percentage share: *lowest 10%:* 4% *highest 10%:* 27% (2010 est.)

Budget: *revenues:* 25.1 billion (2017 est.) *expenditures:* 33.5 billion (2017 est.)

Taxes and other revenues: 9.6% (of GDP) (2017 est.)
country comparison to the world: 214

Budget surplus (+) or deficit (-): -3.2% (of GDP) (2017 est.)
country comparison to the world: 137

Public debt: 33.1% of GDP (2017 est.) 33.3% of GDP (2016 est.)
country comparison to the world: 159

Fiscal year: 1 July—30 June

Inflation rate (consumer prices): 5.6% (2017 est.) 5.7% (2016 est.)
country comparison to the world: 180

Current account balance: -$5.322 billion (2017 est.) $1.391 billion (2016 est.)
country comparison to the world: 185

Exports: $35.3 billion (2017 est.) $34.14 billion (2016 est.)
country comparison to the world: 57

Exports—partners: Germany 12.9%, US 12.2%, UK 8.7%, Spain 5.3%, France 5.1%, Italy 4.1% (2017)

Exports—commodities: garments, knitwear, agricultural products, frozen food (fish and seafood), jute and jute goods, leather

Imports: $47.56 billion (2017 est.) $40.28 billion (2016 est.)
country comparison to the world: 56

Imports—commodities: cotton, machinery and equipment, chemicals, iron and steel, foodstuffs

Imports—partners: China 21.9%, India 15.3%, Singapore 5.7% (2017)

Reserves of foreign exchange and gold: $33.42 billion (31 December 2017 est.) $32.28 billion (31 December 2016 est.)
country comparison to the world: 49

Debt—external: $50.26 billion (31 December 2017 est.) $41.85 billion (31 December 2016 est.)
country comparison to the world: 66

Exchange rates: taka (BDT) per US dollar— 80.69 (2017 est.) 78.468 (2016 est.) 78.468 (2015 est.) 77.947 (2014 est.) 77.614 (2013 est.)

ENERGY

Electricity access: *population without electricity:* 28 million (2019)
electrification—total population: 83% (2019)
electrification—urban areas: 93% (2019)
electrification—rural areas: 77% (2019)

Electricity—production: 60.51 billion kWh (2016 est.)
country comparison to the world: 49

Electricity—consumption: 53.65 billion kWh (2016 est.)
country comparison to the world: 48

Electricity—exports: 0 kWh (2016 est.)
country comparison to the world: 103

Electricity—imports: 0 kWh (2016 est.)
country comparison to the world: 124

Electricity—installed generating capacity: 11.9 million kW (2016 est.)
country comparison to the world: 56

Electricity—from fossil fuels: 97% of total installed capacity (2016 est.)
country comparison to the world: 32

Electricity—from nuclear fuels: 0% of total installed capacity (2017 est.)
country comparison to the world: 45

Electricity—from hydroelectric plants: 2% of total installed capacity (2017 est.)
country comparison to the world: 136

Electricity—from other renewable sources: 2% of total installed capacity (2017 est.)
country comparison to the world: 134

Crude oil—production: 3,000 bbl/day (2018 est.)
country comparison to the world: 83

Crude oil—exports: 0 bbl/day (2015 est.)
country comparison to the world: 91

Crude oil—imports: 21,860 bbl/day (2015 est.)
country comparison to the world: 63

Crude oil—proved reserves: 28 million bbl (1 January 2018 est.)
country comparison to the world: 81

Refined petroleum products—production: 26,280 bbl/day (2015 est.)
country comparison to the world: 86

Refined petroleum products—consumption: 106,000 bbl/day (2016 est.)
country comparison to the world: 77

Refined petroleum products—exports: 901 bbl/ day (2015 est.)
country comparison to the world: 108

Refined petroleum products—imports: 81,570 bbl/ day (2015 est.)
country comparison to the world: 63

Natural gas—production: 29.53 billion cu m (2017 est.)
country comparison to the world: 27

Natural gas—consumption: 29.53 billion cu m (2017 est.)
country comparison to the world: 32

Natural gas—exports: 0 cu m (2017 est.)
country comparison to the world: 65

Natural gas—imports: 0 cu m (2017 est.)
country comparison to the world: 89

Natural gas—proved reserves: 185.8 billion cu m (1 January 2018 est.)
country comparison to the world: 44

Carbon dioxide emissions from consumption of energy: 79.97 million Mt (2017 est.)
country comparison to the world: 48

COMMUNICATIONS

Telephones—fixed lines: *total subscriptions:* 1,433,460
subscriptions per 100 inhabitants: less than 1 (2019 est.)
country comparison to the world: 67

Telephones—mobile cellular: *total subscriptions:* 163,559,380
subscriptions per 100 inhabitants: 101.55 (2019 est.)
country comparison to the world: 11

Telecommunication systems: *general assessment:* slow to moderate growth in mobile subscriber rate; regulator's recent budget allowance and telecoms investment in LTE infrastructure is leading the way to the migration of 5G; fixed broadband penetration in Bangladesh remains very low mainly due to the dominance of the mobile platform (2020)
domestic: fixed-line teledensity remains less than 1 per 100 persons; mobile-cellular telephone subscribership has been increasing rapidly and now exceeds 101 telephones per 100 persons; mobile subscriber growth is anticipated over the next five years to 2023; strong local competition (2019)
international: country code—880; landing points for the SeaMeWe-4 and SeaMeWe-5 fiber-optic submarine cable system that provides links to Europe, the Middle East, and Asia; satellite earth stations—6; international radiotelephone communications and landline service to neighboring countries (2019)
note: the COVID-19 outbreak is negatively impacting telecommunications production and supply chains globally; consumer spending on telecom devices and services has also slowed due to the pandemic's effect on economies worldwide; overall progress towards improvements in all facets of the telecom industry—mobile, fixed-line, broadband, submarine cable and satellite—has moderated

Broadcast media: state-owned Bangladesh Television (BTV) broadcasts throughout the country. Some channels, such as BTV World, operate via satellite. The government also owns a medium wave radio channel and some private FM radio broadcast news channels. Of the 41 Bangladesh approved TV stations, 26 are currently being used to broadcast. Of those, 23 operate under private management via cable distribution. Collectively, TV channels can reach more than 50 million people across the country.

Internet country code: .bd

Internet users: *total:* 23,917,950

percent of population: 15% (July 2018 est.)

country comparison to the world: 32

Broadband—fixed subscriptions: *total:* 10,237,003
subscriptions per 100 inhabitants: 6 (2018 est.)
country comparison to the world: 17

TRANSPORTATION

National air transport system: *number of registered air carriers:* 6 (2020)
inventory of registered aircraft operated by air carriers: 9
annual passenger traffic on registered air carriers: 5,984,155 (2018)
annual freight traffic on registered air carriers: 63.82 million mt-km (2018)

Civil aircraft registration country code prefix: S2 (2016)

Airports: 18 (2013)
country comparison to the world: 137

Airports—with paved runways: *total:* 16 (2017)
over 3,047 m: 2 (2017)
2,438 to 3,047 m: 2 (2017)
1,524 to 2,437 m: 6 (2017)
914 to 1,523 m: 1 (2017)
under 914 m: 5 (2017)

Airports—with unpaved runways: *total:* 2 (2013)
1,524 to 2,437 m: 1 (2013)
under 914 m: 1 (2013)

Heliports: 3 (2013)

Pipelines: 2950 km gas (2013)

Railways: *total:* 2,460 km (2014)
narrow gauge: 1,801 km 1.000-m gauge (2014)
broad gauge: 659 km 1.676-m gauge (2014)
country comparison to the world: 68

Roadways: *total:* 369,105 km (2018)
paved: 110,311 km (2018)
unpaved: 258,794 km (2018)
country comparison to the world: 20

Waterways: 8,370 km (includes up to 3,060 km of main cargo routes; network reduced to 5,200 km in the dry season) (2011)
country comparison to the world: 16

Merchant marine: *total:* 376

by type: bulk carrier 36, container ship 5, general cargo 97, oil tanker 136, other 102 (2019)
country comparison to the world: 47

Ports and terminals: *major seaport(s):* Chittagong
container port(s) (TEUs): Chittagong (2,566,597) (2017)
river port(s): Mongla Port (Sela River)

MILITARY AND SECURITY

Military and security forces: Bangladesh Defense Force: Bangladesh Army, Bangladesh Navy, Bangladesh Air Force; Ministry of Home Affairs: Border Guard Bangladesh (BGB), Bangladesh Coast Guard, Ansars, Village Defense Party (VDP) (2019)
note: the Ansars and VDP are paramilitary organizations for internal security

Military expenditures: 1.3% of GDP (2019)
1.3% of GDP (2018)
1.2% of GDP (2017)
1.4% of GDP (2016)
1.4% of GDP (2015)
country comparison to the world: 92

Military and security service personnel strengths: estimates of the size of the Bangladesh Defense Force vary; approximately 165,000 total active personnel (135,000 Army; 16,000 Navy; 14,000 Air Force); 38,000 Border Guards (2019 est.)

Military equipment inventories and acquisitions: the Bangladesh Defense Force inventory is comprised of mostly Chinese and Russian equipment; since 2010, China and Russia are the chief suppliers of arms to Bangladesh; Bangladesh is currently undertaking a significant defense modernization program, with a focus on naval acquisitions (2019)

Military deployments: 1,300 Central African Republic (MINUSCA); 1,650 Democratic Republic of the Congo (MONUSCO); 115 Lebanon (UNIFIL); 1,300 Mali (MINUSMA); 1,600 South Sudan (UNMISS) (2020)

Military service age and obligation: 16-21 years of age for voluntary military service; Bangladeshi nationality and 10th grade education required; officers: 1721 years of age, Bangladeshi nationality, and 12th grade education required (2018)

Maritime threats: the International Maritime Bureau reports the territorial waters of Bangladesh remain a risk for armed robbery against ships; in 2018, the number of attacks against commercial vessels increased to 12 over the 11 such incidents in 2017

TERRORISM

Terrorist group(s): Harakat ul-Jihad-i-Islami/ Bangladesh; Islamic State of Iraq and ash-Sham in Bangladesh; al-Qa'ida; al-Qa'ida in the Indian Subcontinent (2020)
note: details about the history, aims, leadership, organization, areas of operation, tactics, targets, weapons, size, and sources of support of the group(s) appear(s) in Appendix-T

TRANSNATIONAL ISSUES

Disputes—international: Bangladesh referred its maritime boundary claims with Burma and India to the International Tribunal on the Law of the Sea; Indian Prime Minister Singh's September 2011 visit to Bangladesh resulted in the signing of a Protocol to the 1974 Land Boundary Agreement between India and Bangladesh, which had called for the settlement of longstanding boundary disputes over undemarcated areas and the exchange of territorial enclaves, but which had never been implemented; Bangladesh struggles to accommodate 912,000 Rohingya, Burmese Muslim minority from Rakhine State, living as refugees in Cox's Bazar; Burmese border authorities are constructing a 200 km (124 mi) wire fence designed to deter illegal cross-border transit and tensions from the military build-up along border

Refugees and internally displaced persons: *refugees (country of origin):* 861,545 (Burma) (2020) (includes an estimated 712,152 Rohingya refugees who have fled conflict since 25 August 2017)

IDPs: 427,000 (conflict, development, human rights violations, religious persecution, natural disasters) (2019)

Illicit drugs: transit country for illegal drugs produced in neighboring countries

BARBADOS

INTRODUCTION

Background: The island was uninhabited when first settled by the British in 1627. African slaves worked the sugar plantations established on the island, which initially dominated the Caribbean sugar industry. By 1720 Barbados was no longer a dominant force within the sugar industry, having been surpassed by the Leeward Islands and Jamaica. Slavery was abolished in 1834. The Barbadian economy remained heavily dependent on sugar, rum, and molasses production through most of the 20th century. The gradual introduction of social and political reforms in the 1940s and 1950s led to complete independence from the UK in 1966. In the 1990s, tourism and manufacturing surpassed the sugar industry in economic importance. Barbados plans to remove the British monarch as its head of state by November 2021 and transition to a republic.

GEOGRAPHY

Location: Caribbean, island in the North Atlantic Ocean, northeast of Venezuela

Geographic coordinates: 13 10 N, 59 32 W

Map references: Central America and the Caribbean

Area: *total:* 430 sq km
land: 430 sq km
water: 0 sq km

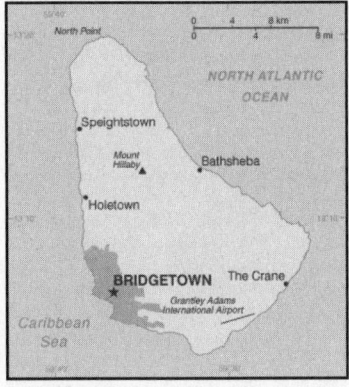

country comparison to the world: 202

Area—comparative: 2.5 times the size of Washington, DC

Land boundaries: 0 km

Coastline: 97 km

Maritime claims: *territorial sea*: 12 nm
exclusive economic zone: 200 nm

Climate: tropical; rainy season (June to October)

Terrain: relatively flat; rises gently to central highland region

Elevation: *lowest point*: Atlantic Ocean 0 m
highest point: Mount Hillaby 336 m

Natural resources: petroleum, fish, natural gas

Land use: *agricultural land*: 32.6% (2016 est.)
arable land: 25.6% (2016 est.) / *permanent crops*: 2.3% (2016 est.) / *permanent pasture*: 4.7% (2016 est.)
forest: 19.4% (2016 est.)
other: 48% (2016 est.)

Irrigated land: 50 sq km (2012)

Population distribution: most densely populated country in the eastern Caribbean; approximately one-third live in urban areas

Natural hazards: infrequent hurricanes; periodic landslides

Environment—current issues: pollution of coastal waters from waste disposal by ships; soil erosion; illegal solid waste disposal threatens contamination of aquifers

Environment—international agreements: party to: Biodiversity, Climate Change, Climate Change-Kyoto Protocol, Desertification, Endangered Species, Hazardous Wastes, Law of the Sea, Marine Dumping, Ozone Layer Protection, Ship Pollution, Wetlands
signed, but not ratified: none of the selected agreements

Geography—note: easternmost Caribbean island

PEOPLE AND SOCIETY

Population: 294,560 (July 2020 est.)
country comparison to the world: 182

Nationality: *noun*: Barbadian(s) or Bajan (colloquial)
adjective: Barbadian or Bajan (colloquial)

Ethnic groups: African descent 92.4%, mixed 3.1%, white 2.7%, East Indian 1.3%, other 0.2%, unspecified 0.3% (2010 est.)

Languages: English (official), Bajan (English-based creole language, widely spoken in informal settings)

Religions: Protestant 66.4% (includes Anglican 23.9%, other Pentecostal 19.5%, Adventist 5.9%, Methodist 4.2%, Wesleyan 3.4%, Nazarene 3.2%, Church of God 2.4%, Baptist 1.8%, Moravian 1.2%, other Protestant 0.9%), Roman Catholic 3.8%, other Christian 5.4% (includes Jehovah's Witness 2.0%, other 3.4%), Rastafarian 1%, other 1.5%, none 20.6%, unspecified 1.2% (2010 est.)

Age structure: *0-14 years*: 17.49% (male 25,762/female 25,764)
15-24 years: 12.34% (male 18,024/female 18,330)
25-54 years: 42.69% (male 62,655/female 63,093)
55-64 years: 13.91% (male 19,533/female 21,430)
65 years and over: 13.57% (male 16,398/female 23,571) (2020 est.)

Dependency ratios: *total dependency ratio*: 50.3
youth dependency ratio: 25.2
elderly dependency ratio: 25.1
potential support ratio: 4 (2020 est.)

Median age: *total*: 39.5 years
male: 38.4 years
female: 40.7 years (2020 est.)
country comparison to the world: 55

Population growth rate: 0.23% (2020 est.)
country comparison to the world: 179

Birth rate: 11.3 births/1,000 population (2020 est.)
country comparison to the world: 173

Death rate: 8.8 deaths/1,000 population (2020 est.)
country comparison to the world: 65

Net migration rate: -0.3 migrant(s)/1,000 population (2020 est.)
country comparison to the world: 112

Population distribution: most densely populated country in the eastern Caribbean; approximately one-third live in urban areas

Urbanization: *urban population*: 31.2% of total population (2020)
rate of urbanization: 0.2% annual rate of change (2015-20 est.)

total population growth rate v. urban population growth rate, 2000-2030: Major urban areas—population: 89,000 BRIDGETOWN (capital) (2018)

Sex ratio: *at birth*: 1.01 male(s)/female
0-14 years: 1 male(s)/female
15-24 years: 0.98 male(s)/female
25-54 years: 0.99 male(s)/female
55-64 years: 0.91 male(s)/female
65 years and over: 0.7 male(s)/female
total population: 0.94 male(s)/female (2020 est.)

Maternal mortality rate: 27 deaths/100,000 live births (2017 est.)
country comparison to the world: 115

Infant mortality rate: *total*: 9.6 deaths/1,000 live births
male: 10.7 deaths/1,000 live births
female: 8.4 deaths/1,000 live births (2020 est.)
country comparison to the world: 135

Life expectancy at birth: *total population*: 76 years
male: 73.6 years
female: 78.4 years (2020 est.)
country comparison to the world: 106

Total fertility rate: 1.68 children born/woman (2020 est.)
country comparison to the world: 178

Contraceptive prevalence rate: 59.2% (2012)

Drinking water source:
improved:
total: 98.5% of population
unimproved:
total: 1.5% of population (2017 est.)

Current Health Expenditure: 6.8% (2017)

Physicians density: 2.48 physicians/1,000 population (2017)

Hospital bed density: 6 beds/1,000 population (2017)

Sanitation facility access:
improved:
total: 99.2% of population
unimproved:
total: 0.8% of population (2017 est.)

HIV/AIDS—adult prevalence rate: 1.1% (2019 est.)
country comparison to the world: 42

HIV/AIDS—people living with HIV/AIDS: 2,700 (2019 est.)
country comparison to the world: 133

HIV/AIDS—deaths: <100 (2019 est.)

Obesity—adult prevalence rate: 23.1% (2016)
country comparison to the world: 67

Children under the age of 5 years underweight: 3.5% (2012)
country comparison to the world: 94

Education expenditures: 4.7% of GDP (2017)
country comparison to the world: 76

Literacy: *definition*: age 15 and over can read and write
total population: 99.6%
male: 99.6%
female: 99.6% (2014)

School life expectancy (primary to tertiary education): *total*: 15 years
male: 14 years
female: 17 years (2011)

Unemployment, youth ages 15-24: *total*: 29.6%
male: 27.9%
female: 31.5% (2016 est.)
country comparison to the world: 31

GOVERNMENT

Country name: *conventional long form*: none
conventional short form: Barbados
etymology: the name derives from the Portuguese "as barbadas," which means "the bearded ones"

and can refer either to the long, hanging roots of the island's bearded fig trees or to the alleged beards of the native Carib inhabitants

Government type: parliamentary democracy under a constitutional monarchy; a Commonwealth realm

Capital: *name:* Bridgetown
geographic coordinates: 13 06 N, 59 37 W
time difference: UTC-4 (1 hour ahead of Washington, DC, during Standard Time)
etymology: named after a bridge constructed over the swampy area (known as the Careenage) around the Constitution River that flows through the center of Bridgetown

Administrative divisions: 11 parishes and 1 city*; Bridgetown*, Christ Church, Saint Andrew, Saint George, Saint James, Saint John, Saint Joseph, Saint Lucy, Saint Michael, Saint Peter, Saint Philip, Saint Thomas

Independence: 30 November 1966 (from the UK)

National holiday: Independence Day, 30 November (1966)

Constitution: *history:* adopted 22 November 1966, effective 30 November 1966
amendments: proposed by Parliament; passage of amendments to constitutional sections such as citizenship, fundamental rights and freedoms, and the organization and authorities of the branches of government requires two-thirds majority vote by the membership of both houses of Parliament; passage of other amendments only requires a majority vote of both houses; amended several times, last in 2010

Legal system: English common law; no judicial review of legislative acts

International law organization participation: accepts compulsory ICJ jurisdiction with reservations; accepts ICCt jurisdiction

Citizenship: *citizenship by birth:* yes
citizenship by descent only: yes
dual citizenship recognized: yes
residency requirement for naturalization: 5 years

Suffrage: 18 years of age; universal

Executive branch: *chief of state:* Queen ELIZABETH II (since 6 February 1952); represented by Governor General Sandra MASON (since 8 January 2018)
head of government: Prime Minister Mia MOTTLEY (since 25 May 2018)
cabinet: Cabinet appointed by the governor general on the advice of the prime minister
elections/appointments: the monarchy is hereditary; governor general appointed by the monarch; following legislative elections, the leader of the majority party or leader of the majority coalition usually appointed prime minister by the governor general; the prime minister recommends the deputy prime minister

Legislative branch: description: bicameral Parliament consists of: Senate (21 seats; members appointed by the governor general—12 on the advice of the Prime Minister, 2 on the advice of

the opposition leader, and 7 at the discretion of the governor general)
House of Assembly (30 seats; members directly elected in single-seat constituencies by simple majority vote to serve 5- year terms)

elections: Senate—last appointments on 5 June 2018 (next appointments NA)
House of Assembly—last held on 24 May 2018 (next to be held in 2023)

election results: Senate—appointed; composition—men 16, women 5, percent of women 23.8%
House of Assembly—percent of vote by party—BLP 74.6%, DLP 22.6%, other 2.8%; seats by party—BLP 30; composition—men 24, women 6, percent of women 20%; note—total Parliament percent of women 21.6%
note: tradition dictates that the election is held within 5 years of the last election, but constitutionally it is 5 years from the first seating of Parliament plus a 90-day grace period

Judicial branch: *highest courts:* Supreme Court (consists of the High Court with 8 justices) and the Court of Appeal (consists of the High Court chief justice and president of the court and 4 justices; note—in 2005, Barbados acceded to the Caribbean Court of Justice as the final court of appeal, replacing the Judicial Committee of the Privy Council (in London)
judge selection and term of office: Supreme Court chief justice appointed by the governor general on the recommendation of the prime minister and opposition leader of Parliament; other justices appointed by the governor general on the recommendation of the Judicial and Legal Service Commission, a 5-member independent body consisting of the Supreme Court chief justice, the commission head, and governor general appointees recommended by the prime minister; justices serve until mandatory retirement at age 65
subordinate courts: Magistrates' Courts

Political parties and leaders: Bajan Free Party [Alex MITCHELL]
Barbados Integrity Movement [Neil HOLDER]
Barbados Labor Party or BLP [Mia MOTTLEY]
Democratic Labor Party or DLP [Freundel STUART]
People's Democratic Congress [Mark ADAMSON]
People's Empowerment Party or PEP [David COMISSIONG]
Solutions Barbados [Grenville PHILLIPS II]
United Progressive Party or UPP [Lynette EASTMOND]

International organization participation: ACP, AOSIS, C, Caricom, CDB, CELAC, FAO, G-77, IADB, IBRD, ICAO, ICCt, ICRM, IDA, IFAD, IFC, IFRCS, ILO, IMF, IMO, Interpol, IOC, ISO, ITSO, ITU, ITUC (NGOs), LAES, MIGA, NAM, OAS, OPANAL, OPCW, UN, UNCTAD, UNESCO, UNHCR, UNIDO, UPU, WCO, WFTU (NGOs), WHO, WIPO, WMO, WTO

Diplomatic representation in the US: *chief of mission:* Ambassador Noel Anderson LYNCH (since 11 January 2019)
chancery: 2144 Wyoming Avenue NW, Washington, DC 20008

telephone: [1] (202) 939-9200
FAX: [1] (202) 332-7467
consulate(s) general: Miami, New York

Diplomatic representation from the US: *chief of mission:* Ambassador Linda S. TAGLIALATELA (since 1 February 2016) note—also accredited to Antigua and Barbuda, Dominica, Grenada, Saint Kitts and Nevis, Saint Lucia, and Saint Vincent and the Grenadines
telephone: [1] (246) 227-4000
embassy: Wildey Business Park, Wildey, St. Michael BB 14006, W.I.
mailing address: P. O. Box 302, Bridgetown BB 11000; (Department Name) Unit 3120, DPO AA 34055
FAX: [1] (246) 431-0179

Flag description: three equal vertical bands of ultramarine blue (hoist side), gold, and ultramarine blue with the head of a black trident centered on the gold band; the band colors represent the blue of the sea and sky and the gold of the beaches; the trident head represents independence and a break with the past (the colonial coat of arms contained a complete trident)

National symbol(s): Neptune's trident, pelican, Red Bird of Paradise flower (also known as Pride of Barbados); national colors: blue, yellow, black

National anthem: *name:* The National Anthem of Barbados
lyrics/music: Irving BURGIE/C. Van Roland EDWARDS
note: adopted 1966; the anthem is also known as "In Plenty and In Time of Need"
0:00 / 1:18

ECONOMY

Economy—overview: Barbados is the wealthiest and one of the most developed countries in the Eastern Caribbean and enjoys one of the highest per capita incomes in the region. Historically, the Barbadian economy was dependent on sugarcane cultivation and related activities. However, in recent years the economy has diversified into light industry and tourism. Offshore finance and information services are important foreign exchange earners, boosted by being in the same time zone as eastern US financial centers and by a relatively highly educated workforce. Following the 2008-09 recession, external vulnerabilities such as fluctuations in international oil prices have hurt economic growth, raised Barbados' already high public debt to GDP ratio—which stood at 105% of GDP in 2016—and cut into its international reserves.

GDP (purchasing power parity): $5.218 billion (2017 est.)
$5.227 billion (2016 est.)
$5.111 billion (2015 est.)
note: data are in 2017 dollars
country comparison to the world: 178

GDP (official exchange rate): $4.99 billion (2017 est.)

GDP—real growth rate: -0.2% (2017 est.)
2.3% (2016 est.)
2.2% (2015 est.)

country comparison to the world: 196

GDP—per capita (PPP): $18,600 (2017 est.)
$18,700 (2016 est.)
$18,300 (2015 est.)
note: data are in 2017 dollars
country comparison to the world: 95

Gross national saving: 7.2% of GDP (2017 est.)
11.8% of GDP (2016 est.)
10.8% of GDP (2015 est.)
country comparison to the world: 171

GDP—composition, by end use: *household consumption:* 84.2% (2017 est.)
government consumption: 13.4% (2017 est.)
investment in fixed capital: 17.6% (2017 est.)
investment in inventories: 0.2% (2017 est.)
exports of goods and services: 31.6% (2017 est.)
imports of goods and services: -47% (2017 est.)

GDP—composition, by sector of origin:
agriculture: 1.5% (2017 est.)
industry: 9.8% (2017 est.)
services: 88.7% (2017 est.)

Agriculture—products: sugarcane, vegetables, cotton

Industries: tourism, sugar, light manufacturing, component assembly for export

Industrial production growth rate: 2.4% (2017 est.)
country comparison to the world: 119

Labor force: 144,000 (2017 est.)
country comparison to the world: 175

Labor force—by occupation: *agriculture:* 10%
industry: 15%
services: 75% (1996 est.)

Unemployment rate: 10.1% (2017 est.)
9.9% (2016 est.)
country comparison to the world: 148

Population below poverty line: NA

Household income or consumption by percentage share: *lowest 10%:* NA
highest 10%: NA

Budget: *revenues:* 1.466 billion (2013 est.) (2017 est.)
expenditures: 1.664 billion (2017 est.)

Taxes and other revenues: 29.4% (of GDP) (2017 est.)
country comparison to the world: 82

Budget surplus (+) or deficit (-): -4% (of GDP) (2017 est.)
country comparison to the world: 155

Public debt: 157.3% of GDP (2017 est.)
149.1% of GDP (2016 est.)
country comparison to the world: 3

Fiscal year: 1 April—31 March

Inflation rate (consumer prices): 4.4% (2017 est.)
1.5% (2016 est.)
country comparison to the world: 164

Current account balance: -$189 million (2017 est.)
-$206 million (2016 est.)
country comparison to the world: 97

Exports: $485.4 million (2017 est.)
$516.9 million (2016 est.)
country comparison to the world: 175

Exports—partners: US 38%, Trinidad and Tobago 10.2%, Guyana 5.5%, Jamaica 5%, China 4.8%, St. Lucia 4.6% (2017)

Exports—commodities: manufactures, sugar, molasses, rum, other foodstuffs and beverages, chemicals, electrical components

Imports: $1.52 billion (2017 est.)
$1.541 billion (2016 est.)
country comparison to the world: 174

Imports—commodities: consumer goods, machinery, foodstuffs, construction materials, chemicals, fuel, electrical components

Imports—partners: US 38.5%, Trinidad and Tobago 14.6%, China 7.1%, UK 4.7% (2017)

Reserves of foreign exchange and gold: $264.5 million (31 December 2017 est.)
$341.8 million (31 December 2016 est.)
country comparison to the world: 169

Debt—external: $4.49 billion (2010 est.)
$668 million (2003 est.)
country comparison to the world: 136

Exchange rates: Barbadian dollars (BBD) per US dollar—
2 (2017 est.)
2 (2016 est.)
2 (2015 est.)
2 (2014 est.)
2 (2013 est.)
note: the Barbadian dollar is pegged to the US dollar

ENERGY

Electricity access: *electrification—total population:* 100% (2020)

Electricity—production: 1.01 billion kWh (2016 est.)
country comparison to the world: 150

Electricity—consumption: 990 million kWh (2016 est.)
country comparison to the world: 155

Electricity—exports: 0 kWh (2017 est.)
country comparison to the world: 104

Electricity—imports: 0 kWh (2016 est.)
country comparison to the world: 125

Electricity—installed generating capacity: 269,000 kW (2016 est.)
country comparison to the world: 162

Electricity—from fossil fuels: 93% of total installed capacity (2016 est.)
country comparison to the world: 50

Electricity—from nuclear fuels: 0% of total installed capacity (2017 est.)
country comparison to the world: 46

Electricity—from hydroelectric plants: 0% of total installed capacity (2017 est.)
country comparison to the world: 157

Electricity—from other renewable sources: 7% of total installed capacity (2017 est.)
country comparison to the world: 91

Crude oil—production: 1,000 bbl/day (2018 est.)
country comparison to the world: 90

Crude oil—exports: 674 bbl/day (2015 est.)
country comparison to the world: 77

Crude oil—imports: 0 bbl/day (2015 est.)
country comparison to the world: 94

Crude oil—proved reserves: 2.534 million bbl (1 January 2018 est.)
country comparison to the world: 94

Refined petroleum products—production: 0 bbl/day (2015 est.)
country comparison to the world: 117

Refined petroleum products—consumption: 11,000 bbl/day (2016 est.)
country comparison to the world: 159

Refined petroleum products—exports: 0 bbl/day (2015 est.)
country comparison to the world: 129

Refined petroleum products—imports: 10,630 bbl/day (2015 est.)
country comparison to the world: 147

Natural gas—production: 14.16 million cu m (2017 est.)
country comparison to the world: 91

Natural gas—consumption: 19.82 million cu m (2017 est.)
country comparison to the world: 113

Natural gas—exports: 0 cu m (2017 est.)
country comparison to the world: 66

Natural gas—imports: 5.653 million cu m (2017 est.)
country comparison to the world: 78

Natural gas—proved reserves: 141.6 million cu m (1 January 2018 est.)
country comparison to the world: 102

Carbon dioxide emissions from consumption of energy: 1.76 million Mt (2017 est.)
country comparison to the world: 160

COMMUNICATIONS

Telephones—fixed lines: *total subscriptions:* 141,618
subscriptions per 100 inhabitants: 48.19 (2019 est.)
country comparison to the world: 128

Telephones—mobile cellular: *total subscriptions:* 319,177
subscriptions per 100 inhabitants: 108.61 (2019 est.)
country comparison to the world: 178

Telecommunication systems: *general assessment:* island-wide automatic telephone system; telecom sector across the Caribbean region remains one of the key growth areas and contributors to the overall GDP; numerous competitors licensed, but small and localized (2020)
domestic: fixed-line teledensity of roughly 48 per 100 persons; mobile-cellular telephone density about 109 per 100 persons (2019)
international: country code—1-246; landing points for the ECFS and Southern Caribbean Fiber submarine cable with links to 15 other islands in the eastern Caribbean extending from the British Virgin Islands to Trinidad and Puerto Ricco;

satellite earth stations—1 (Intelsat—Atlantic Ocean); tropospheric scatter to Trinidad and Saint Lucia (2019)

note: the COVID-19 outbreak is negatively impacting telecommunications production and supply chains globally; consumer spending on telecom devices and services has also slowed due to the pandemic's effect on economies worldwide; overall progress towards improvements in all facets of the telecom industry—mobile, fixed-line, broadband, submarine cable and satellite—has moderated

Broadcast media: government-owned Caribbean Broadcasting Corporation (CBC) operates the lone terrestrial TV station; CBC also operates a multi-channel cable TV subscription service; roughly a dozen radio stations, consisting of a CBC-operated network operating alongside privately owned radio stations

Internet country code: .bb

Internet users: *total:* 239,664
percent of population: 81.76% (July 2018 est.)
country comparison to the world: 171

Broadband—fixed subscriptions: *total:* 89,340
subscriptions per 100 inhabitants: 30 (2018 est.)
country comparison to the world: 124

TRANSPORTATION

Civil aircraft registration country code prefix: 8P (2016)

Airports: 1 (2020)
country comparison to the world: 213

Airports—with paved runways: *total:* 1 (2019)
over 3,047 m: 1

Pipelines: 33 km gas, 64 km oil, 6 km refined products (2013)

Roadways: *total:* 1,700 km (2015)
paved: 1,700 km (2015)
country comparison to the world: 173

Merchant marine: *total:* 132
by type: bulk carrier 25general cargo 90, other 17 (2019)
country comparison to the world: 78

Ports and terminals: *major seaport(s):* Bridgetown

MILITARY AND SECURITY

Military and security forces: Royal Barbados Defense Force: The Barbados Regiment, The Barbados Coast Guard (2020)

Military and security service personnel strengths: the Royal Barbados Defense Force (RBDF) has approximately 550 active personnel (450 Barbados Regiment; 100 Coast Guard) (2019 est.)

Military equipment inventories and acquisitions: the RBDF's major equipment inventory—maritime patrol boats—is supplied by the Netherlands (2019 est.)

Military service age and obligation: 18 years of age for voluntary military service, or earlier with parental consent; no conscription (2013)

TRANSNATIONAL ISSUES

Disputes—international: Barbados and Trinidad and Tobago abide by the April 2006 Permanent Court of Arbitration decision delimiting a maritime boundary and limiting catches of flying fish in Trinidad and Tobago's exclusive economic zone; joins other Caribbean states to counter Venezuela's claim that Aves Island sustains human habitation, a criterion under the UN Convention on the Law of the Sea, which permits Venezuela to extend its Economic Exclusion Zone/continental shelf over a large portion of the eastern Caribbean Sea

Illicit drugs: one of many Caribbean transshipment points for narcotics bound for Europe and the US; offshore financial center

BELARUS

INTRODUCTION

Background: After seven decades as a constituent republic of the USSR, Belarus attained its independence in 1991. It has retained closer political and economic ties to Russia than have any of the other former Soviet republics. Belarus and Russia signed a treaty on a two-state union on 8 December 1999 envisioning greater political and economic integration. Although Belarus agreed to a framework to carry out the accord, serious implementation has yet to take place and current negotiations on further integration have been contentious. Since his election in July 1994 as the country's first and only directly elected president, Aleksandr LUKASHENKO has steadily consolidated his power through authoritarian means and a centralized economic system. Government restrictions on political and civil freedoms, freedom of speech and the press, peaceful assembly, and religion have remained in place.

GEOGRAPHY

Location: Eastern Europe, east of Poland

Geographic coordinates: 53 00 N, 28 00 E

Map references: Europe

Area: *total:* 207,600 sq km
land: 202,900 sq km
water: 4,700 sq km
country comparison to the world: 87

Area—comparative: slightly less than twice the size of Kentucky; slightly smaller than Kansas

Land boundaries: *total:* 3,642 km
border countries (5): Latvia 161 km, Lithuania 640 km, Poland 418 km, Russia 1312 km, Ukraine 1111 km

Coastline: 0 km (landlocked)

Maritime claims: none (landlocked)

Climate: cold winters, cool and moist summers; transitional between continental and maritime

Terrain: generally flat with much marshland

Elevation: *mean elevation:* 160 m
lowest point: Nyoman River 90 m
highest point: Dzyarzhynskaya Hara 346 m

Natural resources: timber, peat deposits, small quantities of oil and natural gas, granite, dolomitic limestone, marl, chalk, sand, gravel, clay

Land use: *agricultural land:* 43.7% (2016 est.)
arable land: 27.2% (2016 est.) / permanent crops: 0.6% (2016 est.) / permanent pasture: 15.9% (2016 est.)
forest: 42.7% (2016 est.)
other: 13.6% (2016 est.)
Irrigated land: 1,140 sq km (2012)

Population distribution: a fairly even distribution throughout most of the country, with urban areas attracting larger and denser populations

Natural hazards: large tracts of marshy land

Environment—current issues: soil pollution from pesticide use; southern part of the country contaminated with fallout from 1986 nuclear reactor accident at Chornobyl' in northern Ukraine

Environment—international agreements: *party to:* Air Pollution, Air Pollution-Nitrogen Oxides, Air Pollution-Sulfur 85, Biodiversity, Climate Change, Climate Change-Kyoto Protocol, Desertification, Endangered Species, Environmental Modification, Hazardous Wastes, Law of the Sea, Marine Dumping, Ozone Layer Protection, Ship Pollution, Wetlands

signed, but not ratified: none of the selected agreements

Geography—note: landlocked; glacial scouring accounts for the flatness of Belarusian terrain and for its 11,000 lakes

PEOPLE AND SOCIETY

Population: 9,477,918 (July 2020 est.)
country comparison to the world: 94

Nationality: *noun:* Belarusian(s)
adjective: Belarusian

Ethnic groups: Belarusian 83.7%, Russian 8.3%, Polish 3.1%, Ukrainian 1.7%, other 2.4%, unspecified 0.9% (2009 est.)

Languages: Russian (official) 70.2%, Belarusian (official) 23.4%, other 3.1% (includes small Polish- and Ukrainian-speaking minorities), unspecified 3.3% (2009 est.)

Religions: Orthodox 48.3%, Catholic 7.1%, other 3.5%, non-believers 41.1% (2011 est.)

Age structure: *0-14 years:* 16.09% (male 784,231/female 740,373)
15-24 years: 9.59% (male 467,393/female 441,795)
25-54 years: 43.94% (male 2,058,648/female 2,105,910)
55-64 years: 14.45% (male 605,330/female 763,972)
65 years and over: 15.93% (male 493,055/female 1,017,211) (2020 est.)

Dependency ratios: *total dependency ratio:* 48.9
youth dependency ratio: 25.7
elderly dependency ratio: 23.2
potential support ratio: 4.3 (2020 est.)

Median age: *total:* 40.9 years
male: 38 years
female: 43.9 years (2020 est.)
country comparison to the world: 48

Population growth rate: -0.27% (2020 est.)
country comparison to the world: 215

Birth rate: 9.5 births/1,000 population (2020 est.)
country comparison to the world: 196

Death rate: 13.1 deaths/1,000 population (2020 est.)
country comparison to the world: 8

Net migration rate: 0.7 migrant(s)/1,000 population (2020 est.)
country comparison to the world: 65

Population distribution: a fairly even distribution throughout most of the country, with urban areas attracting larger and denser populations

Urbanization: *urban population:* 79.5% of total population (2020)
rate of urbanization: 0.44% annual rate of change (2015-20 est.)

total population growth rate v. urban population growth rate, 2000-2030: Major urban areas—population: 2.028 million MINSK (capital) (2020)

Sex ratio: *at birth:* 1.06 male(s)/female
0-14 years: 1.06 male(s)/female
15-24 years: 1.06 male(s)/female
25-54 years: 0.98 male(s)/female
55-64 years: 0.79 male(s)/female
65 years and over: 0.48 male(s)/female
total population: 0.87 male(s)/female (2020 est.)

Mother's mean age at first birth: 26.5 years (2017 est.)

Maternal mortality rate: 2 deaths/100,000 live births (2017 est.)
country comparison to the world: 181

Infant mortality rate: *total:* 3.5 deaths/1,000 live births
male: 3.9 deaths/1,000 live births
female: 3.1 deaths/1,000 live births (2020 est.)
country comparison to the world: 198

Life expectancy at birth: *total population:* 73.8 years
male: 68.3 years
female: 79.5 years (2020 est.)
country comparison to the world: 139

Total fertility rate: 1.5 children born/woman (2020 est.)
country comparison to the world: 201

Contraceptive prevalence rate: 72.1% (2017)
note: percent of women aged 18-49

Drinking water source:
improved:
urban: 100% of population
rural: 98.3% of population
total: 99.8% of population
unimproved:
urban: 0% of population
rural: 1.7% of population
total: 0.2% of population (2017 est.)

Current Health Expenditure: 5.9% (2017)

Physicians density: 5.19 physicians/1,000 population (2015)

Hospital bed density: 10.8 beds/1,000 population (2014)

Sanitation facility access:
improved:
urban: 99.8% of population
rural: 97.9% of population
total: 99.4% of population
unimproved:
urban: 0.2% of population
rural: 2.1% of population
total: 0.6% of population (2017 est.)

HIV/AIDS—adult prevalence rate: 0.4% (2019 est.)
country comparison to the world: 74

HIV/AIDS—people living with HIV/AIDS: 28,000 (2019 est.)
country comparison to the world: 77

HIV/AIDS—deaths: <200 (2019 est.)

Obesity—adult prevalence rate: 24.5% (2016)
country comparison to the world: 58

Education expenditures: 4.8% of GDP (2017)
country comparison to the world: 70

Literacy: *definition:* age 15 and over can read and write
total population: 99.8%
male: 99.8%
female: 99.7% (2018)

School life expectancy (primary to tertiary education): *total:* 15 years
male: 15 years
female: 16 years (2018)

Unemployment, youth ages 15-24: *total:* 10.6%
male: 12.7%
female: 8.4% (2018 est.)
country comparison to the world: 121

GOVERNMENT

Country name: *conventional long form:* Republic of Belarus
conventional short form: Belarus
local long form: Respublika Byelarus'/Respublika Belarus'
local short form: Byelarus'/Belarus'
former: Belorussian (Byelorussian) Soviet Socialist Republic
etymology: the name is a compound of the Belarusian words "bel" (white) and "Rus" (the Old East Slavic ethnic designation) to form the meaning White Rusian or White Ruthenian

Government type: presidential republic in name, although in fact a dictatorship

Capital: *name:* Minsk

geographic coordinates: 53 54 N, 27 34 E
time difference: UTC+2 (7 hours ahead of Washington, DC, during Standard Time)
etymology: the origin of the name is disputed; Minsk may originally have been located 16 km to the southwest, on the banks of Menka River; remnants of a 10th-century settlement on the banks of the Menka have been found

Administrative divisions: 6 provinces (voblastsi, singular—voblasts') and 1 municipality* (horad); Brest, Homyel' (Gomel'), Horad Minsk* (Minsk City), Hrodna (Grodno), Mahilyow (Mogilev), Minsk, Vitsyebsk (Vitebsk)
note: administrative divisions have the same names as their administrative centers; Russian spelling provided for reference when different from Belarusian

Independence: 25 August 1991 (from the Soviet Union)

National holiday: Independence Day, 3 July (1944); note—3 July 1944 was the date Minsk was liberated from German troops, 25 August 1991 was the date of independence from the Soviet Union

Constitution: *history:* several previous; latest drafted between late 1991 and early 1994, signed 15 March 1994
amendments: proposed by the president of the republic through petition to the National Assembly or by petition of at least 150,000 eligible voters; approval required by at least two-thirds majority vote in both chambers or by simple majority of votes cast in a referendum

Legal system: civil law system; note—nearly all major codes (civil, civil procedure, criminal, criminal procedure, family, and labor) were revised and came into force in 1999 and 2000

International law organization participation: has not submitted an ICJ jurisdiction declaration; non-party state to the ICCt

Citizenship: *citizenship by birth:* no
citizenship by descent only: at least one parent must be a citizen of Belarus
dual citizenship recognized: no
residency requirement for naturalization: 7 years

Suffrage: 18 years of age; universal

Executive branch: *chief of state:* President Aleksandr LUKASHENKO (since 20 July 1994)

head of government: Prime Minister Roman GOLOVCHENKO (since 4 June 2020); First Deputy Prime Minister Nikolai SNOPKOV (since 4 June 2020); Deputy Prime Ministers Vladimir KUKHAREV, Igor PETRISHENKO (since 18 August 2018), Yury NAZAROV (since 3 March 2020), Aleksander Subbotin (since 4 June 2020)
cabinet: Council of Ministers appointed by the president
elections/appointments: president directly elected by absolute majority popular vote in 2 rounds if needed for a 5-year term (no term limits); first election took place on 23 June and 10 July 1994; according to the 1994 constitution, the next election should have been held in 1999; however, Aleksandr LUKASHENKO extended his term to 2001 via a November 1996 referendum; subsequent election held on 9 September 2001; an October 2004 referendum ended presidential term limits and allowed the president to run and win in a third (19 March 2006), fourth (19 December 2010), fifth (11 October 2015), and sixth (9 August 2020); next election in 2025; prime minister and deputy prime ministers appointed by the president and approved by the National Assembly
election results: Aleksandr LUKASHENKO reelected president; percent of vote—Aleksandr LUKASHENKO (independent) 80.2%, Sviatana TSIKHANOUSKAYA (independent) 9.9%, other 9.9%; note—widespread street protests erupted following announcement of the election results amid allegations of voter fraud

Legislative branch: *description:* bicameral National Assembly or Natsionalnoye Sobraniye consists of:
Council of the Republic or Sovet Respubliki (64 seats; 56 members indirectly elected by regional and Minsk city councils and 8 members appointed by the president; members serve 4-year terms)
House of Representatives or Palata Predstaviteley (110 seats; members directly elected in single-seat constituencies by absolute majority vote in 2 rounds if needed; members serve 4-year terms)
elections: Council of the Republic—indirect election last held on 7 November 2019
House of Representatives—last held on 17 November 2019 (next to be held in 2023); OSCE observers determined that the election was neither free nor impartial and that vote counting was problematic in a number of polling stations;

proLUKASHENKO candidates won every seat; international observers determined that the previous elections, on 28 September 2008, 23 September 2012, and 11 September 2016 also fell short of democratic standards, with proLU-KASHENKO candidates winning every, or virtually every, seat
election results: Council of the Republic—percent of vote by party—NA; seats by party—NA; composition—NA
House of Representatives—percent of vote by party—NA; seats by party—KPB 11, Republican Party of Labor and Justice 6, Belarusian Patriotic Party 2, LDP 1, AP 1, independent 89; composition—men 66, women 44, percent of women 40%; note—total National Assembly percent of women—NA
note: the US does not recognize the legitimacy of the National Assembly

Judicial branch: *highest courts:* Supreme Court (consists of the chairman and deputy chairman and organized into several specialized panels, including economic and military; number of judges set by the president of the republic and the court chairman); Constitutional Court (consists of 12 judges, including a chairman and deputy chairman)
judge selection and term of office: Supreme Court judges appointed by the president with the consent of the Council of the Republic; judges initially appointed for 5 years and evaluated for life appointment; Constitutional Court judges—6 appointed by the president and 6 elected by the Council of the Republic; the presiding judge directly elected by the president and approved by the Council of the Republic; judges can serve for 11 years with an age limit of 70
subordinate courts: oblast courts; Minsk City Court; town courts; Minsk city and oblast economic courts

Political parties and leaders: pro-government parties:
Belarusian Agrarian Party or AP [Mikhail SHIMANSKY]
Belarusian Patriotic Party [Nikolai ULAKHOVICH]
Belarusian Social Sport Party [Vladimir ALEKSANDROVICH]
Communist Party of Belarus or KPB [Aleksei SOKOL]
Liberal Democratic Party or LDP [Sergey GAYDUKEVICH]
Republican Party [Vladimir BELOZOR]
Republican Party of Labor and Justice [Vasiliy ZADNEPRYANIY]
Social Democratic Party of Popular Accord [Sergei YERMAK]
opposition parties:
Belarusian Christian Democracy Party [Paval SEVIARYNETS, Volha KAVALKOVA, Vital RYMASHEWSKI] (unregistered) Belarusian Party of the Green [Anastasiya DOROFEYEVA]
Belarusian Party of the Left "Just World" [Sergey KALYAKIN]

Belarusian Popular Front or BPF [Ryhor KASTUSEU]
Belarusian Social-Democratic Assembly [Sergei CHERECHEN]
Belarusian Social Democratic Party ("Assembly") or BSDPH [Ihar BARYSAU]
Belarusian Social Democratic Party (People's Assembly) [Mikalay STATKEVICH] (unregistered)
Christian Conservative Party or BPF [Zyanon PAZNYAK]
United Civic Party or UCP [Nikolay KOZLOV]

International organization participation: BSEC (observer), CBSS (observer), CEI, CIS, CSTO, EAEC, EAEU, EAPC, EBRD, FAO, GCTU, IAEA, IBRD, ICAO, ICC (NGOs), ICRM, IDA, IFC, IFRCS, ILO, IMF, IMSO, Interpol, IOC, IOM, IPU, ISO, ITU, ITUC (NGOs), MIGA, NAM, NSG, OPCW, OSCE, PCA, PFP, SCO (dialogue member), UN, UNCTAD, UNESCO, UNIDO, UNIFIL, UNWTO, UPU, WCO, WFTU (NGOs), WHO, WIPO, WMO, WTO (observer), ZC

Diplomatic representation in the US: *chief of mission:* Ambassador (vacant; recalled by Belarus in 2008); Charge d'Affaires Dmitriy BASIK (since July 2019)
chancery: 1619 New Hampshire Avenue NW, Washington, DC 20009
telephone: [1] (202) 986-1606
FAX: [1] (202) 986-1805
consulate(s) general: New York

Diplomatic representation from the US: *chief of mission:* Ambassador (vacant; left in 2008 upon insistence of Belarusian Government); Charge d'Affaires Jenifer MOORE (since August 2018)
telephone: [375] (17) 210-1283
embassy: 46 Starovilenskaya Street, Minsk 220002
mailing address: Unit 7010 Box 100, DPO AE 09769
FAX: [375] (17) 234-7853

Flag description: red horizontal band (top) and green horizontal band one-half the width of the red band; a white vertical stripe on the hoist side bears Belarusian national ornamentation in red; the red band color recalls past struggles from oppression, the green band represents hope and the many forests of the country

National symbol(s): no clearly defined current national symbol, the mounted knight known as Pahonia (the Chaser) is the traditional Belarusian symbol; national colors: green, red, white

National anthem: *name:* "My, Bielarusy" (We Belarusians)
lyrics/music: Mikhas KLIMKOVICH and Uladzimir KARYZNA/Nester SAKALOUSKI
note: music adopted 1955, lyrics adopted 2002; after the fall of the Soviet Union, Belarus kept the music of its Soviet-era anthem but adopted new lyrics; also known as "Dziarzauny himn Respubliki Bielarus" (State Anthem of the Republic of Belarus)
0:00 / 1:55

ECONOMY

Economy—overview: As part of the former Soviet Union, Belarus had a relatively well-developed industrial base, but it is now outdated, inefficient, and dependent on subsidized Russian energy and preferential access to Russian markets. The country's agricultural base is largely dependent on government subsidies. Following the collapse of the Soviet Union, an initial burst of economic reforms included privatization of state enterprises, creation of private property rights, and the acceptance of private entrepreneurship, but by 1994 the reform effort dissipated. About 80% of industry remains in state hands, and foreign investment has virtually disappeared. Several businesses have been renationalized. State-owned entities account for 70-75% of GDP, and state banks make up 75% of the banking sector.

Economic output declined for several years following the break-up of the Soviet Union, but revived in the mid-2000s. Belarus has only small reserves of crude oil and imports crude oil and natural gas from Russia at subsidized, below market, prices. Belarus derives export revenue by refining Russian crude and selling it at market prices. Russia and Belarus have had serious disagreements over prices and quantities for Russian energy. Beginning in early 2016, Russia claimed Belarus began accumulating debt – reaching $740 million by April 2017 – for paying below the agreed price for Russian natural gas and Russia cut back its export of crude oil as a result of the debt. In April 2017, Belarus agreed to pay its gas debt and Russia restored the flow of crude.

New non-Russian foreign investment has been limited in recent years, largely because of an unfavorable financial climate. In 2011, a financial crisis lead to a nearly three-fold devaluation of the Belarusian ruble. The Belarusian economy has continued to struggle under the weight of high external debt servicing payments and a trade deficit. In mid-December 2014, the devaluation of the Russian ruble triggered a near 40% devaluation of the Belarusian ruble.

Belarus's economy stagnated between 2012 and 2016, widening productivity and income gaps between Belarus and neighboring countries. Budget revenues dropped because of falling global prices on key Belarusian export commodities. Since 2015, the Belarusian government has tightened its macro-economic policies, allowed more flexibility to its exchange rate, taken some steps towards price liberalization, and reduced subsidized government lending to state-owned enterprises. Belarus returned to modest growth in 2017, largely driven by improvement of external conditions and Belarus issued sovereign debt for the first time since 2011, which provided the country with badly-needed liquidity, and issued $600 million worth of Eurobonds in February 2018, predominantly to US and British investors.

GDP (purchasing power parity): $179.4 billion (2017 est.)
$175.1 billion (2016 est.)
$179.7 billion (2015 est.)
note: data are in 2017 dollars
country comparison to the world: 70

GDP (official exchange rate): $54.44 billion (2017 est.)

GDP—real growth rate: 1.22% (2019 est.)
3.17% (2018 est.)
2.53% (2017 est.)
country comparison to the world: 165

GDP—per capita (PPP): $18,900 (2017 est.)
$18,400 (2016 est.)
$19,000 (2015 est.)
note: data are in 2017 dollars
country comparison to the world: 94

Gross national saving: 24.5% of GDP (2017 est.)
23% of GDP (2016 est.)
25.8% of GDP (2015 est.)
country comparison to the world: 64

GDP—composition, by end use: *household consumption:* 54.8% (2017 est.)
government consumption: 14.6% (2017 est.)
investment in fixed capital: 24.9% (2017 est.)
investment in inventories: 5.7% (2017 est.)
exports of goods and services: 67% (2017 est.)
imports of goods and services: -67% (2017 est.)

GDP—composition, by sector of origin: *agriculture:* 8.1% (2017 est.)
industry: 40.8% (2017 est.)
services: 51.1% (2017 est.)

Agriculture—products: grain, potatoes, vegetables, sugar beets, flax; beef, milk

Industries: metal-cutting machine tools, tractors, trucks, earthmovers, motorcycles, synthetic fibers, fertilizer, textiles, refrigerators, washing machines and other household appliances

Industrial production growth rate: 5.6% (2017 est.)
country comparison to the world: 47

Labor force: 4.381 million (2016 est.)
country comparison to the world: 86

Labor force—by occupation: *agriculture:* 9.7%
industry: 23.4%
services: 66.8% (2015 est.)

Unemployment rate: 0.8% (2017 est.)
1% (2016 est.)
note: official registered unemployed; large number of underemployed workers
country comparison to the world: 6

Population below poverty line: 5.7% (2016 est.)

Household income or consumption by percentage share: *lowest 10%:* 3.8%
highest 10%: 21.9% (2008)

Budget: *revenues:* 22.15 billion (2017 est.)
expenditures: 20.57 billion (2017 est.)

Taxes and other revenues: 40.7% (of GDP) (2017 est.)
country comparison to the world: 35

Budget surplus (+) or deficit (-): 2.9% (of GDP) (2017 est.)
country comparison to the world: 14

Public debt: 53.4% of GDP (2017 est.)
53.5% of GDP (2016 est.)
country comparison to the world: 90

Fiscal year: calendar year

Inflation rate (consumer prices): 6% (2017 est.)
11.8% (2016 est.)
country comparison to the world: 184

Current account balance: -$931 million (2017 est.)
-$1.669 billion (2016 est.)
country comparison to the world: 143

Exports: $28.65 billion (2017 est.)
$22.98 billion (2016 est.)
country comparison to the world: 66

Exports—partners: Russia 43.9%, Ukraine 11.5%, UK 8.2% (2017)

Exports—commodities: machinery and equipment, mineral products, chemicals, metals, textiles, foodstuffs

Imports: $31.58 billion (2017 est.)
$25.61 billion (2016 est.)
country comparison to the world: 64

Imports—commodities: mineral products, machinery and equipment, chemicals, foodstuffs, metals

Imports—partners: Russia 57.2%, China 8%, Germany 5.1% (2017)

Reserves of foreign exchange and gold: $7.315 billion (31 December 2017 est.)
$4.927 billion (31 December 2016 est.)
country comparison to the world: 84

Debt—external: $39.92 billion (31 December 2017 est.)
$37.74 billion (31 December 2016 est.)
country comparison to the world: 75

Exchange rates: Belarusian rubles (BYB/BYR) per US dollar—
1.9 (2017 est.)
2 (2016 est.)
2 (2015 est.)
15,926 (2014 est.)
10,224.1 (2013 est.)

ENERGY

Electricity access: *electrification—total population:* 100% (2020)

Electricity—production: 31.58 billion kWh (2016 est.)
country comparison to the world: 63

Electricity—consumption: 31.72 billion kWh (2016 est.)
country comparison to the world: 61

Electricity—exports: 3.482 billion kWh (2015 est.)
country comparison to the world: 40

Electricity—imports: 6.319 billion kWh (2016 est.)
country comparison to the world: 32

Electricity—installed generating capacity: 10.04 million kW (2016 est.)
country comparison to the world: 59

Electricity—from fossil fuels: 96% of total installed capacity (2016 est.)
country comparison to the world: 37

Electricity—from nuclear fuels: 0% of total installed capacity (2017 est.)
country comparison to the world: 47

Electricity—from hydroelectric plants: 1% of total installed capacity (2017 est.)
country comparison to the world: 145

Electricity—from other renewable sources: 3% of total installed capacity (2017 est.)
country comparison to the world: 121

Crude oil—production: 31,000 bbl/day (2018 est.)
country comparison to the world: 62

Crude oil—exports: 31,730 bbl/day (2015 est.)
country comparison to the world: 44

Crude oil—imports: 468,400 bbl/day (2015 est.)
country comparison to the world: 21

Crude oil—proved reserves: 198 million bbl (1 January 2018 est.)
country comparison to the world: 56

Refined petroleum products—production: 477,200 bbl/day (2015 est.)
country comparison to the world: 34

Refined petroleum products—consumption: 141,000 bbl/day (2016 est.)
country comparison to the world: 68

Refined petroleum products—exports: 351,200 bbl/day (2015 est.)
country comparison to the world: 25

Refined petroleum products—imports: 14,630 bbl/day (2015 est.)
country comparison to the world: 135

Natural gas—production: 59.46 million cu m (2017 est.)
country comparison to the world: 84

Natural gas—consumption: 17.7 billion cu m (2017 est.)
country comparison to the world: 39

Natural gas—exports: 0 cu m (2017 est.)
country comparison to the world: 67

Natural gas—imports: 17.53 billion cu m (2017 est.)
country comparison to the world: 18

Natural gas—proved reserves: 2.832 billion cu m (1 January 2018 est.)
country comparison to the world: 95

Carbon dioxide emissions from consumption of energy: 56.07 million Mt (2017 est.)
country comparison to the world: 54

COMMUNICATIONS

Telephones—fixed lines: *total subscriptions:* 4,513,255
subscriptions per 100 inhabitants: 47.49 (2019 est.)
country comparison to the world: 31

Telephones—mobile cellular: *total subscriptions:* 11,682,764
subscriptions per 100 inhabitants: 122.93 (2019 est.)
country comparison to the world: 77

Telecommunication systems: *general assessment:* govt. and telecom regulator have plans to develop the telecom sector for the migration to 5G; Chinese company Huawei have started 5G trials to deliver data at 2Gb/s; fiber network reaches two million establishments; 10,000km of fiber cabling laid; August 2018 almost two million GPON connections (Gigabit Passive Optical Network, point-to-multi point access mechanism); Belarus launched its first telecoms satellite in 2016; LTE use reaches 75% of mobile subscribers (2020)
domestic: fixed-line teledensity is improving although rural areas continue to be underserved, 48 per 100 fixed-line; mobile-cellular teledensity now approaches 123 telephones per 100 persons (2019)
international: country code—375; Belarus is landlocked and therefore a member of the Trans-European Line (TEL), Trans-Asia-Europe (TAE) fiber-optic line, and has access to the Trans-Siberia Line (TSL); 3 fiber-optic segments provide connectivity to Latvia, Poland, Russia, and Ukraine; worldwide service is available to Belarus through this infrastructure; additional analog lines to Russia; Intelsat, Eutelsat, and Intersputnik earth stations; almost 31,000 base stations in service in 2019 (2020)
note: the COVID-19 outbreak is negatively impacting telecommunications production and supply chains globally; consumer spending on telecom devices and services has also slowed due to the pandemic's effect on economies worldwide; overall progress towards improvements in all facets of the telecom industry—mobile, fixed-line, broadband, submarine cable and satellite—has moderated

Broadcast media: 7 state-controlled national TV channels; Polish and Russian TV broadcasts are available in some areas; state-run Belarusian Radio operates 5 national networks and an external service; Russian and Polish radio broadcasts are available (2019)

Internet country code: .by

Internet users: *total:* 7,539,145
percent of population: 79.13% (July 2018 est.)
country comparison to the world: 67

Broadband—fixed subscriptions: *total:* 3,201,519
subscriptions per 100 inhabitants: 34 (2018 est.)
country comparison to the world: 41

TRANSPORTATION

National air transport system: *number of registered air carriers:* 2 (2020)
inventory of registered aircraft operated by air carriers: 30
annual passenger traffic on registered air carriers: 2,760,168 (2018)
annual freight traffic on registered air carriers: 1.9 million mt-km (2018)

Civil aircraft registration country code prefix: EW (2016)

Airports: 65 (2013)
country comparison to the world: 75

Airports—with paved runways: *total:* 33 (2017)
over 3,047 m: 1 (2017)
2,438 to 3,047 m: 20 (2017)
1,524 to 2,437 m: 4 (2017)
914 to 1,523 m: 1 (2017)
under 914 m: 7 (2017)

Airports—with unpaved runways: *total:* 32 (2013)
over 3,047 m: 1 (2013)
1,524 to 2,437 m: 1 (2013)
914 to 1,523 m: 2 (2013)
under 914 m: 28 (2013)

Heliports: 1 (2013)

Pipelines: 5386 km gas, 1589 km oil, 1730 km refined products (2013)

Railways: *total:* 5,528 km (2014)
standard gauge: 25 km 1.435-m gauge (2014)
broad gauge: 5,503 km 1.520-m gauge (874 km electrified) (2014)
country comparison to the world: 35

Roadways: *total:* 86,600 km (2017)
country comparison to the world: 57

Waterways: 2,500 km (major rivers are the west-flowing Western Dvina and Neman Rivers and the south-flowing Dnepr River and its tributaries, the Berezina, Sozh, and Pripyat Rivers) (2011)
country comparison to the world: 35

Merchant marine: *total:* 4
by type: other 4 (2019)
country comparison to the world: 166

Ports and terminals: *river port(s):* Mazyr (Prypyats')

MILITARY AND SECURITY

Military and security forces: Belarus Armed Forces: Army, Air and Air Defense Force, Special Operations Force; Ministry of Interior: State Border Troops, Militia, Internal Troops (2019)

Military expenditures: 1.2% of GDP (2019)
1.2% of GDP (2018)
1.2% of GDP (2017)
1.3% of GDP (2016)
1.3% of GDP (2015)
country comparison to the world: 100

Military and security service personnel strengths: the Belarus Armed Forces have approximately 45,000 active troops (29,000 Army, including Special Operations Force; 16,000 Air and Air Defense) (2020 est.)

Military equipment inventories and acquisitions: the inventory of the Belarus Armed Forces is comprised of Russian-origin equipment; Belarus's defense industry manufactures some equipment, including vehicles, guided weapons, and electronic warfare systems (2019 est.)

Military deployments: contributes about 2,000 personnel to CSTO's Rapid Reaction Force (2019 est.)

Military service age and obligation: 18-27 years of age for compulsory military or alternative service; conscript service obligation is 12-18 months, depending on academic qualifications, and 24-36 months for alternative service, depending on academic qualifications; 17 year olds are eligible to become cadets at military higher education institutes, where they are classified as military personnel (2017)

TRANSNATIONAL ISSUES

Disputes—international: boundary demarcated with Latvia and Lithuania; as a member state that forms part of the EU's external border, Poland has implemented strict Schengen border rules to restrict illegal immigration and trade along its border with Belarus

Refugees and internally displaced persons: *stateless persons:* 6,466 (2019)

Trafficking in persons: current situation: Belarus is a source, transit, and destination country for women, men, and children subjected to sex trafficking and forced labor; more victims are exploited within Belarus than abroad; Belarusians exploited abroad are primarily trafficked to

Germany, Poland, Russian, and Turkey but also other European countries, the Middle East, Japan, Kazakhstan, and Mexico; Moldovans, Russians, Ukrainians, and Vietnamese are exploited in Belarus; state-sponsored forced labor is a continuing problem; students are forced to do farm labor without pay and military conscripts are forced to perform unpaid non-military work; the government has retained a decree forbidding workers in state-owned wood processing factories from leaving their jobs without their employers' permission

tier rating: Tier 3 – Belarus does not fully comply with the minimum standards for the elimination of trafficking and was placed on Tier 3 after being on the Tier 2 Watch List for two consecutive years without making progress; government efforts to repeal state-sponsored forced labor policies and domestic trafficking were inadequate;

no trafficking offenders were convicted in 2014, and the number of investigations progressively declined from 2005-14; efforts to protect trafficking victims remain insufficient, with no identification and referral mechanism in place; care facilities were not trafficking-specific and were poorly equipped, leading most victims to seek assistance from private shelters (2015)

Illicit drugs: limited cultivation of opium poppy and cannabis, mostly for the domestic market; transshipment point for illicit drugs to and via Russia, and to the Baltics and Western Europe; a small and lightly regulated financial center; anti-moneylaundering legislation does not meet international standards and was weakened further when know-your-customer requirements were curtailed in 2008; few investigations or prosecutions of money-laundering activities

BELGIUM

INTRODUCTION

Background: Belgium became independent from the Netherlands in 1830; it was occupied by Germany during World Wars I and II. The country prospered in the past half century as a modern, technologically advanced European state and member of NATO and the EU. In recent years, political divisions between the Dutch-speaking Flemish of the north and the French-speaking Walloons of the south have led to constitutional amendments granting these regions formal recognition and autonomy. The capital city of Brussels is home to numerous international organizations including the EU and NATO.

GEOGRAPHY

Location: Western Europe, bordering the North Sea, between France and the Netherlands

Geographic coordinates: 50 50 N, 4 00 E

Map references: Europe

Area: *total:* 30,528 sq km
land: 30,278 sq km
water: 250 sq km
country comparison to the world: 141

Area—comparative: about the size of Maryland

Land boundaries: *total:* 1,297 km
border countries (4): France 556 km, Germany 133 km, Luxembourg 130 km, Netherlands 478 km

Coastline: 66.5 km

Maritime claims: *territorial sea:* 12 nm
exclusive economic zone: geographic coordinates define outer limit
contiguous zone: 24 nm
continental shelf: median line with neighbors

Climate: temperate; mild winters, cool summers; rainy, humid, cloudy

Terrain: flat coastal plains in northwest, central rolling hills, rugged mountains of Ardennes Forest in southeast

Elevation: *mean elevation:* 181 m
lowest point: North Sea 0 m
highest point: Botrange 694 m

Natural resources: construction materials, silica sand, carbonates, arable land

Land use: *agricultural land:* 44.1% (2011 est.)
arable land: 27.2% (2011 est.) / *permanent crops:* 0.8% (2011 est.) / *permanent pasture:* 16.1% (2011 est.)
forest: 22.4% (2011 est.)
other: 33.5% (2011 est.)
Irrigated land: 230 sq km (2012)

Population distribution: most of the population concentrated in the northern two-thirds of the country; the southeast is more thinly populated; considered to have one of the highest population

densities in the world; approximately 97% live in urban areas

Natural hazards: flooding is a threat along rivers and in areas of reclaimed coastal land, protected from the sea by concrete dikes

Environment—current issues: intense pressures from human activities: urbanization, dense transportation network, industry, extensive animal breeding and crop cultivation; air and water pollution also have repercussions for neighboring countries

Environment—international agreements: *party to:* Air Pollution, Air Pollution-Nitrogen Oxides, Air Pollution-Persistent Organic Pollutants, Air Pollution-Sulfur 85, Air Pollution-Sulfur 94, Air Pollution-Volatile Organic Compounds, Antarctic-Environmental Protocol, Antarctic-Marine Living Resources, Antarctic Seals, Antarctic Treaty, Biodiversity, Climate Change, Climate Change-Kyoto Protocol, Desertification, Endangered Species, Environmental Modification, Hazardous Wastes, Law of the Sea, Marine Dumping, Marine Life Conservation, Ozone Layer Protection, Ship Pollution, Tropical Timber 83, Tropical Timber 94, Wetlands, Whaling
signed, but not ratified: none of the selected agreements

Geography—note: crossroads of Western Europe; most West European capitals are within 1,000 km of Brussels, the seat of both the European Union and NATO

PEOPLE AND SOCIETY

Population: 11,720,716 (July 2020 est.)
country comparison to the world: 80

Nationality: *noun:* Belgian(s)
adjective: Belgian

Ethnic groups: Belgian 75.2%, Italian 4.1%, Moroccan 3.7%, French 2.4%, Turkish 2%, Dutch 2%, other 10.6% (2012 est.)

Languages: Dutch (official) 60%, French (official) 40%, German (official) less than 1%

Religions: Roman Catholic 50%, Protestant and other Christian 2.5%, Muslim 5%, Jewish 0.4%, Buddhist 0.3%, atheist 9.2%, none 32.6% (2009 est.)

Age structure: 0-14 years: 17.22% (male 1,033,383/female 984,624)
15-24 years: 11.2% (male 670,724/female 642,145)
25-54 years: 39.23% (male 2,319,777/female 2,278,450)
55-64 years: 13.14% (male 764,902/female 775,454)
65 years and over: 19.21% (male 988,148/female 1,263,109) (2020 est.)

Dependency ratios: *total dependency ratio:* 57
youth dependency ratio: 26.7
elderly dependency ratio: 30.2
potential support ratio: 3.3 (2020 est.)

Median age: *total:* 41.6 years
male: 40.4 years
female: 42.8 years (2020 est.)
country comparison to the world: 44

Population growth rate: 0.63% (2020 est.)
country comparison to the world: 148

Birth rate: 11.1 births/1,000 population (2020 est.)
country comparison to the world: 176

Death rate: 9.8 deaths/1,000 population (2020 est.)
country comparison to the world: 40

Net migration rate: 4.8 migrant(s)/1,000 population (2020 est.)
country comparison to the world: 26

Population distribution: most of the population concentrated in the northern two-thirds of the country; the southeast is more thinly populated; considered to have one of the highest population densities in the world; approximately 97% live in urban areas

Urbanization: *urban population:* 98.1% of total population (2020)
rate of urbanization: 0.62% annual rate of change (2015-20 est.)

total population growth rate v. urban population growth rate, 2000-2030: Major urban areas—population: 2.081 million BRUSSELS (capital), 1.042 million Antwerp (2020)

Sex ratio: *at birth:* 1.05 male(s)/female
0-14 years: 1.05 male(s)/female
15-24 years: 1.04 male(s)/female
25-54 years: 1.02 male(s)/female
55-64 years: 0.99 male(s)/female
65 years and over: 0.78 male(s)/female
total population: 0.97 male(s)/female (2020 est.)

Mother's mean age at first birth: 29 years (2018 est.)

Maternal mortality rate: 5 deaths/100,000 live births (2017 est.)
country comparison to the world: 164

Infant mortality rate: *total:* 3.3 deaths/1,000 live births
male: 3.7 deaths/1,000 live births
female: 2.9 deaths/1,000 live births (2020 est.)
country comparison to the world: 205

Life expectancy at birth: *total population:* 81.4 years
male: 78.8 years
female: 84.2 years (2020 est.)
country comparison to the world: 31

Total fertility rate: 1.77 children born/woman (2020 est.)
country comparison to the world: 152

Contraceptive prevalence rate: 66.7% (2018)

Drinking water source:
improved:
urban: 100% of population
rural: 100% of population
total: 100% of population
unimproved:
urban: 0% of population
rural: 0% of population
total: 0% of population (2017 est.)

Current Health Expenditure: 10.3% (2017)

Physicians density: 3.07 physicians/1,000 population (2017)

Hospital bed density: 5.7 beds/1,000 population (2017)

Sanitation facility access:
improved:
urban: 100% of population
rural: 100% of population
total: 100% of population
unimproved:
urban: 0% of population
rural: 0% of population
total: 0% of population (2017 est.)

HIV/AIDS—adult prevalence rate: NA

HIV/AIDS—people living with HIV/AIDS: NA

HIV/AIDS—deaths: NA

Obesity—adult prevalence rate: 22.1% (2016)
country comparison to the world: 81

Education expenditures: 6.5% of GDP (2016)
country comparison to the world: 20

School life expectancy (primary to tertiary education): *total:* 20 years
male: 19 years
female: 21 years (2018)

Unemployment, youth ages 15-24: *total:* 15.8%
male: 16.2%
female: 15.3% (2018 est.)
country comparison to the world: 86

GOVERNMENT

Country name: *conventional long form:* Kingdom of Belgium
conventional short form: Belgium
local long form: Royaume de Belgique (French)/ Koninkrijk Belgie (Dutch)/Koenigreich Belgien (German)
local short form: Belgique/Belgie/Belgien

etymology: the name derives from the Belgae, an ancient Celtic tribal confederation that inhabited an area between the English Channel and the west bank of the Rhine in the first centuries B.C.

Government type: federal parliamentary democracy under a constitutional monarchy

Capital: *name:* Brussels
geographic coordinates: 50 50 N, 4 20 E
time difference: UTC+1 (6 hours ahead of Washington, DC, during Standard Time)
daylight saving time: +1hr, begins last Sunday in March; ends last Sunday in October
etymology: may derive from the Old Dutch "bruoc/broek," meaning "marsh" and "sella/zele/ sel" signifying "home" to express the meaning "home in the marsh"

Administrative divisions: 3 regions (French: regions, singular—region; Dutch: gewesten, singular—gewest); Brussels-Capital Region, also known as Brussels Hoofdstedelijk Gewest (Dutch), Region de Bruxelles-Capitale (French long form), Bruxelles-Capitale (French short form); Flemish Region (Flanders), also known as Vlaams Gewest (Dutch long form), Vlaanderen (Dutch short form), Region Flamande (French long form), Flandre (French short form); Walloon Region (Wallonia), also known as Region Wallone (French long form), Wallonie (French short form), Waals Gewest (Dutch long form), Wallonie (Dutch short form)
note: as a result of the 1993 constitutional revision that furthered devolution into a federal state, there are now three levels of government (federal, regional, and linguistic community) with a complex division of responsibilities; the 2012 sixth state reform transferred additional competencies from the federal state to the regions and linguistic communities

Independence: 4 October 1830 (a provisional government declared independence from the Netherlands); 21 July 1831 (King LEOPOLD I ascended to the throne)

National holiday: Belgian National Day (ascension to the throne of King LEOPOLD I), 21 July (1831)

Constitution: *history:* drafted 25 November 1830, approved 7 February 1831, entered into force 26 July 1831, revised 14 July 1993 (creating a federal state)
amendments: "revisions" proposed as declarations by the federal government in accord with the king or by Parliament followed by dissolution of Parliament and new elections; adoption requires two-thirds majority vote of a two-thirds quorum in both houses of the next elected Parliament; amended many times, last in 2014

Legal system: civil law system based on the French Civil Code; note—Belgian law continues to be modified in conformance with the legislative norms mandated by the European Union; judicial review of legislative acts

International law organization participation: accepts compulsory ICJ jurisdiction with reservations; accepts ICCt jurisdiction

Citizenship: *citizenship by birth:* no

citizenship by descent only: at least one parent must be a citizen of Belgium

dual citizenship recognized: yes

residency requirement for naturalization: 5 years

Suffrage: 18 years of age; universal and compulsory

Executive branch: *chief of state:* King PHILIPPE (since 21 July 2013); Heir Apparent Princess ELISABETH (daughter of the monarch, born 25 October 2001)

head of government: Prime Minister Alexander DE CROO (since 1 October 2020); Deputy Prime Ministers Koen GEENS (27 October 2019), Didier REYNDERS (since 27 October 2019), David CLARINVAL (30 November 2019), Petra DE SUTTER (since 1 October 2020)

cabinet: Council of Ministers formally appointed by the monarch

elections/appointments: the monarchy is hereditary and constitutional; following legislative elections, the leader of the majority party or majority coalition usually appointed prime minister by the monarch and approved by Parliament

Legislative branch: *description:* bicameral Parliament consists of:

Senate or Senaat (in Dutch), Senat (in French) (60 seats; 50 members indirectly elected by the community and regional parliaments based on their election results, and 10 elected by the 50 other senators; members serve 5-year terms) Chamber of Representatives or Kamer van Volksvertegenwoordigers (in Dutch), Chambre des Representants (in French) (150 seats; members directly elected in multi-seat constituencies by proportional representation vote; members serve 5- year terms)

elections: Senate—last held 26 May 2019 (next to be held in 2024)

Chamber of Representatives—last held on 26 May 2019 (next to be held in 2024); note—elections coincided with the EU elections

election results: Senate—percent of vote by party—NA; seats by party—NA; composition men 32, women 28, percent of women 46.7%

Chamber of Representatives—percent of vote by party—N-VA 16.0%, VB 11.9%, PS 9.5%, CD&V 8.9%, PVDA+/PTB 8.62%, Open VLD 8.5%, MR 7.6%, SP.A 6.7%, Ecolo 6.1%, Groen 6.1%, CDH 3.7%, Defi 2.2%, PP 1.1%, other 20.1%; seats by party—N-VA 25, VB 18, PS 20, CD&V 12, PVDA+PTB 12, Open VLD 12, MR 14, SP.A 9, Ecolo 13, Groen 8, CDH 5, Defi 2; composition—men 86, women 64, percent of women 42.7%

note: the 1993 constitutional revision that further devolved Belgium into a federal state created three levels of government (federal, regional, and linguistic community) with a complex division of responsibilities; this reality leaves six governments, each with its own legislative assembly; changes above occurred since the sixth state reform

Judicial branch: *highest courts:* Constitutional Court or Grondwettelijk Hof (in Dutch) and Cour Constitutionelle (in French) (consists of 12 judges—6 Dutch-speaking and 6 French-speaking); Supreme Court of Justice or Hof van Cassatie (in Dutch) and Cour de Cassation (in French) (court organized into 3 chambers: civil and commercial; criminal; social, fiscal, and armed forces; each chamber includes a Dutch division and a French division, each with a chairperson and 5-6 judges)

judge selection and term of office: Constitutional Court judges appointed by the monarch from candidates submitted by Parliament; judges appointed for life with mandatory retirement at age 70; Supreme Court judges appointed by the monarch from candidates submitted by the High Council of Justice, a 44-member independent body of judicial and nonjudicial members; judges appointed for life

subordinate courts: Courts of Appeal; regional courts; specialized courts for administrative, commercial, labor, immigration, and audit issues; magistrate's courts; justices of the peace

Political parties and leaders: Flemish parties:
Christian Democratic and Flemish or CD&V [Wouter BEKE]
Flemish Liberals and Democrats or Open VLD [Gwendolyn RUTTEN]
Groen [Meyrem ALMACI] (formerly AGALEV, Flemish Greens)
New Flemish Alliance or N-VA [Bart DE WEVER]
Social Progressive Alternative or SP.A [John CROMBEZ]
Vlaams Belang (Flemish Interest) or VB [Tom VAN GRIEKEN]
Francophone parties:
Ecolo (Francophone Greens) [Jean-Marc NOLLET, Zakia KHATTABI]
Francophone Federalist Democrats or Defi [Olivier MAINGAIN]
Humanist and Democratic Center or CDH [Maxine PREVOT]
People's Party or PP [Mischael MODRIKAMEN]
Reform Movement or MR [Charles MICHEL]
Socialist Party or PS [Elio DI RUPO]
Workers' Party or PTB [Peter MERTENS] other minor parties

International organization participation: ADB (nonregional members), AfDB (nonregional members), Australia Group, Benelux, BIS, CD, CE, CERN, EAPC, EBRD, ECB, EIB, EITI (implementing country), EMU, ESA, EU, FAO, FATF, G-9, G-10, IADB, IAEA, IBRD, ICAO, ICC (national committees), ICCt, ICRM, IDA, IEA, IFAD, IFC, IFRCS, IGAD (partners), IHO, ILO, IMF, IMO, IMSO, Interpol, IOC, IOM, IPU, ISO, ITSO, ITU, ITUC (NGOs), MIGA, MONUSCO, NATO, NEA, NSG, OAS (observer), OECD, OIF, OPCW, OSCE, Pacific Alliance (observer), Paris Club, PCA, Schengen Convention, SELEC (observer), UN, UNCTAD, UNESCO, UNHCR, UNIDO, UNIFIL, UNRWA, UNTSO, UPU, WCO, WHO, WIPO, WMO, WTO, ZC

Diplomatic representation in the US: *chief of mission:* Ambassador Jean Arthur REGIBEAU (since 17 September 2020)

chancery: 3330 Garfield Street NW, Washington, DC 20008

telephone: [1] (202) 333-6900

FAX: [1] (202) 333-3079

consulate(s) general: Atlanta, Los Angeles, New York

Diplomatic representation from the US: *chief of mission:* Ambassador Ronald GIDWITZ (since 4 July 2018)

telephone: [32] (2) 811-4000

embassy: 27 Boulevard du Regent [Regentlaan], B-1000 Brussels

mailing address: PSC 82, Box 002, APO AE 09710

FAX: [32] (2) 811-4500

Flag description: three equal vertical bands of black (hoist side), yellow, and red; the vertical design was based on the flag of France; the colors are those of the arms of the duchy of Brabant (yellow lion with red claws and tongue on a black field)

National symbol(s): golden rampant lion; national colors: red, black, yellow

National anthem: *name:* "La Brabanconne" (The Song of Brabant)

lyrics/music: Louis-Alexandre DECHET[French] Victor CEULEMANS [Dutch]/Francois VAN CAMPENHOUT

note: adopted 1830; according to legend, Louis-Alexandre DECHET, an actor at the theater in which the revolution against the Netherlands began, wrote the lyrics with a group of young people in a Brussels cafe

0:00 / 0:21

ECONOMY

Economy—overview: Belgium's central geographic location and highly developed transport network have helped develop a well-diversified economy, with a broad mix of transport, services, manufacturing, and high tech. Service and high-tech industries are concentrated in the northern Flanders region while the southern region of Wallonia is home to industries like coal and steel manufacturing. Belgium is completely reliant on foreign sources of fossil fuels, and the planned closure of its seven nuclear plants by 2025 should increase its dependence on foreign energy. Its role as a regional logistical hub makes its economy vulnerable to shifts in foreign demand, particularly with EU trading partners. Roughly three-quarters of Belgium's trade is with other EU countries, and the port of Zeebrugge conducts almost half its trade with the United Kingdom alone, leaving Belgium's economy vulnerable to the outcome of negotiations on the UK's exit from the EU.

Belgium's GDP grew by 1.7% in 2017 and the budget deficit was 1.5% of GDP. Unemployment stood at 7.3%, however the unemployment rate is lower in Flanders than Wallonia, 4.4% compared to 9.4%, because of industrial differences between the regions. The economy largely recovered from the March 2016 terrorist attacks that mainly impacted the Brussels region tourist and hospitality industry. Prime Minister Charles MICHEL's center-right government has pledged to further reduce the deficit in response to EU pressure to decrease Belgium's high public debt of about 104%

of GDP, but such efforts would also dampen economic growth. In addition to restrained public spending, low wage growth and higher inflation promise to curtail a more robust recovery in private consumption.

The government has pledged to pursue a reform program to improve Belgium's competitiveness, including changes to labor market rules and welfare benefits. These changes have generally made Belgian wages more competitive regionally, but have raised tensions with trade unions, which have called for extended strikes. In 2017, Belgium approved a tax reform plan to ease corporate rates from 33% to 29% by 2018 and down to 25% by 2020. The tax plan also included benefits for innovation and SMEs, intended to spur competitiveness and private investment.

GDP (purchasing power parity): $529.2 billion (2017 est.)
$520.2 billion (2016 est.)
$513 billion (2015 est.)
note: data are in 2017 dollars
country comparison to the world: 37

GDP (official exchange rate): $493.7 billion (2017 est.)

GDP—real growth rate: 1.41% (2019 est.)
1.49% (2018 est.)
1.9% (2017 est.)
country comparison to the world: 157

GDP—per capita (PPP): $46,600 (2017 est.)
$46,000 (2016 est.)
$45,700 (2015 est.)
note: data are in 2017 dollars
country comparison to the world: 35

Gross national saving: 24.5% of GDP (2017 est.)
24% of GDP (2016 est.)
23.4% of GDP (2015 est.)
country comparison to the world: 65

GDP—composition, by end use: household consumption: 51.2% (2017 est.)
government consumption: 23.4% (2017 est.)
investment in fixed capital: 23.3% (2017 est.)
investment in inventories: 1.3% (2017 est.)
exports of goods and services: 85.1% (2017 est.)
imports of goods and services: -84.4% (2017 est.)

GDP—composition, by sector of origin: agriculture: 0.7% (2017 est.)
industry: 22.1% (2017 est.)
services: 77.2% (2017 est.)

Agriculture—products: sugar beets, fresh vegetables, fruits, grain, tobacco; beef, veal, pork, milk

Industries: engineering and metal products, motor vehicle assembly, transportation equipment, scientific instruments, processed food and beverages, chemicals, pharmaceuticals, base metals, textiles, glass, petroleum

Industrial production growth rate: 0.2% (2017 est.)
country comparison to the world: 168

Labor force: 4.122 million (2020 est.)
country comparison to the world: 88

Labor force—by occupation: agriculture: 1.3%
industry: 18.6%
services: 80.1% (2013 est.)

Unemployment rate: 5.36% (2019 est.)
5.96% (2018 est.)
country comparison to the world: 85

Population below poverty line: 15.1% (2013 est.)

Household income or consumption by percentage share: lowest 10%: 3.4%
highest 10%: 28.4% (2006)

Budget: revenues: 253.5 billion (2017 est.)
expenditures: 258.6 billion (2017 est.)

Taxes and other revenues: 51.3% (of GDP) (2017 est.)
country comparison to the world: 15

Budget surplus (+) or deficit (-): -1% (of GDP) (2017 est.)
country comparison to the world: 74

Public debt: 103.4% of GDP (2017 est.)
106% of GDP (2016 est.)
note: data cover general government debt and includes debt instruments issued (or owned) by government entities other than the treasury; the data include treasury debt held by foreign entities; the data include debt issued by subnational entities, as well as intra-governmental debt; intra-governmental debt consists of treasury borrowings from surpluses in the social funds, such as for retirement, medical care, and unemployment; debt instruments for the social funds are not sold at public auctions; general government debt is defined by the Maastricht definition and calculated by the National Bank of Belgium as consolidated gross debt; the debt is defined in European Regulation EC479/2009 concerning the implementation of the protocol on the excessive deficit procedure annexed to the Treaty on European Union (Treaty of Maastricht) of 7 February 1992; the sub-sectors of consolidated gross debt are: federal government, communities and regions, local government, and social security funds
country comparison to the world: 13

Fiscal year: calendar year

Inflation rate (consumer prices): 2.2% (2017 est.)
1.8% (2016 est.)
country comparison to the world: 112

Current account balance: $1.843 billion (2019 est.)
-$4.135 billion (2018 est.)
country comparison to the world: 41

Exports: $300.8 billion (2017 est.)
$277.7 billion (2016 est.)
country comparison to the world: 20

Exports—partners: Germany 16.6%, France 14.9%, Netherlands 12%, UK 8.4%, Italy 4.9%, US 4.8% (2017)

Exports—commodities: chemicals, machinery and equipment, finished diamonds, metals and metal products, foodstuffs

Imports: $300.4 billion (2017 est.)
$273.4 billion (2016 est.)
country comparison to the world: 17

Imports—commodities: raw materials, machinery and equipment, chemicals, raw diamonds, pharmaceuticals, foodstuffs, transportation equipment, oil products

Imports—partners: Netherlands 17.3%, Germany 13.8%, France 9.5%, US 7.1%, UK 4.9%, Ireland 4.2%, China 4.1% (2017)

Reserves of foreign exchange and gold: $26.16 billion (31 December 2017 est.)
$24.1 billion (31 December 2015 est.)
country comparison to the world: 54

Debt—external: $1.281 trillion (31 March 2016 est.)
$1.214 trillion (31 March 2015 est.)
country comparison to the world: 15

Exchange rates: euros (EUR) per US dollar—
0.885 (2017 est.)
0.903 (2016 est.)
0.9214 (2015 est.)
0.885 (2014 est.)
0.7634 (2013 est.)

ENERGY

Electricity access: electrification—total population: 100% (2020)

Electricity—production: 79.83 billion kWh (2016 est.)
country comparison to the world: 37

Electricity—consumption: 82.16 billion kWh (2016 est.)
country comparison to the world: 36

Electricity—exports: 8.465 billion kWh (2016 est.)
country comparison to the world: 25

Electricity—imports: 14.65 billion kWh (2016 est.)
country comparison to the world: 15

Electricity—installed generating capacity: 21.56 million kW (2016 est.)
country comparison to the world: 41

Electricity—from fossil fuels: 35% of total installed capacity (2016 est.)
country comparison to the world: 177

Electricity—from nuclear fuels: 28% of total installed capacity (2017 est.)
country comparison to the world: 2

Electricity—from hydroelectric plants: 1% of total installed capacity (2017 est.)
country comparison to the world: 146

Electricity—from other renewable sources: 36% of total installed capacity (2017 est.)
country comparison to the world: 8

Crude oil—production: 0 bbl/day (2018 est.)
country comparison to the world: 109

Crude oil—exports: 0 bbl/day (2017 est.)
country comparison to the world: 92

Crude oil—imports: 687,600 bbl/day (2017 est.)
country comparison to the world: 16

Crude oil—proved reserves: 0 bbl (1 January 2018 est.)
country comparison to the world: 106

Refined petroleum products—production: 731,700 bbl/day (2017 est.)
country comparison to the world: 25

Refined petroleum products—consumption: 648,600 bbl/day (2017 est.)
country comparison to the world: 31

Refined petroleum products—exports: 680,800 bbl/day (2017 est.)
country comparison to the world: 12

Refined petroleum products—imports: 601,400 bbl/day (2017 est.)
country comparison to the world: 14

Natural gas—production: 0 cu m (2017 est.)
country comparison to the world: 102

Natural gas—consumption: 17.61 billion cu m (2017 est.)
country comparison to the world: 40

Natural gas—exports: 736.2 million cu m (2017 est.)
country comparison to the world: 40

Natural gas—imports: 18.09 billion cu m (2017 est.)
country comparison to the world: 17

Natural gas—proved reserves: 0 cu m (1 January 2014 est.)
country comparison to the world: 109

Carbon dioxide emissions from consumption of energy: 134.7 million Mt (2017 est.)
country comparison to the world: 35

COMMUNICATIONS

Telephones—fixed lines: *total subscriptions:* 3,967,054
subscriptions per 100 inhabitants: 34.06 (2019 est.)
country comparison to the world: 34

Telephones—mobile cellular: *total subscriptions:* 11,616,970
subscriptions per 100 inhabitants: 99.74 (2019 est.)
country comparison to the world: 78

Telecommunication systems: *general assessment:* highly developed, technologically advanced, and completely automated domestic and international telephone and telegraph facilities; LTE availability is nearly universal in mobile sector; ongoing investments in developing applications and services for migration to 5G, operators are looking into repurposing 3G infrastructure and spectrum as they gear up for 5G; Europe-wide approach to simultaneous movement to 5G on going; 5G will be main motivation for growth and revenue in years to come; consumer are interested in quad-play/bundled services (broadband +television +telephone +wireless services) which will mean MNOs (mobile network operators) are enhancing their fixed-line offerings (2020)
domestic: 34 per 100 fixed-line, 100 per 100 mobile-cellular; nationwide mobile-cellular telephone system; extensive cable network; limited microwave radio relay network (2019)
international: country code—32; landing points for Concerto, UK-Belgium, Tangerine, and SeaMeWe-3, submarine cables that provide links to Europe, the Middle East, Australia and Asia; satellite earth stations—7 (Intelsat—3) (2019)

note: the COVID-19 outbreak is negatively impacting telecommunications production and supply chains globally; consumer spending on telecom devices and services has also slowed due to the pandemic's effect on economies worldwide; overall progress towards improvements in all facets of the telecom industry—mobile, fixed-line, broadband, submarine cable and satellite—has moderated

Broadcast media: a segmented market with the three major communities (Flemish, French, and German-speaking) each having responsibility for their own broadcast media; multiple TV channels exist for each community; additionally, in excess of 90% of households are connected to cable and can access broadcasts of TV stations from neighboring countries; each community has a public radio network coexisting with private broadcasters

Internet country code: .be

Internet users: *total:* 10,258,638

percent of population: 88.66% (July 2018 est.)
country comparison to the world: 51

Broadband—fixed subscriptions: *total:* 4,502,950
subscriptions per 100 inhabitants: 39 (2018 est.)
country comparison to the world: 31

TRANSPORTATION

National air transport system: *number of registered air carriers:* 7 (2020)
inventory of registered aircraft operated by air carriers: 117
annual passenger traffic on registered air carriers: 13,639,487 (2018)
annual freight traffic on registered air carriers: 1,285,340,000 mt-km (2018)

Civil aircraft registration country code prefix: OO (2016)

Airports: 41 (2013)
country comparison to the world: 102

Airports—with paved runways: *total:* 26 (2019)
over 3,047 m: 6
2,438 to 3,047 m: 9
1,524 to 2,437 m: 2
914 to 1,523 m: 1
under 914 m: 8

Airports—with unpaved runways: *total:* 15 (2013)
under 914 m: 15 (2013)

Heliports: 1 (2013)

Pipelines: 3139 km gas, 154 km oil, 535 km refined products (2013)

Railways: *total:* 3,592 km (2014)
standard gauge: 3,592 km 1.435-m gauge (2,960 km electrified) (2014)
country comparison to the world: 55

Roadways: *total:* 118,414 km (2015)
paved: 118,414 km (includes 1,747 km of expressways) (2015)
country comparison to the world: 41

Waterways: 2,043 km (1,528 km in regular commercial use) (2012)
country comparison to the world: 41

Merchant marine: *total:* 201

by type: bulk carrier 21 general cargo 17, oil tanker 26, other 137 (2019)
country comparison to the world: 67

Ports and terminals: *major seaport(s):* Oostende, Zeebrugge
container port(s) (TEUs): Antwerp (10,450,000) (2017)
LNG terminal(s) (import): Zeebrugge
river port(s): Antwerp, Gent (Schelde River) Brussels (Senne River) Liege (Meuse River)

MILITARY AND SECURITY

Military and security forces: Belgian Armed Forces: Land Component, Naval Component, Air Component, Medical Service (2019)

Military expenditures: 0.93% of GDP (2019)
0.93% of GDP (2018)
0.9% of GDP (2017)
0.91% of GDP (2016)
0.92% of GDP (2015)
country comparison to the world: 123

Military and security service personnel strengths: the Belgian Armed Forces have approximately 27,000 active duty personnel (10,500 Army; 1,500 Navy; 5,000 Air Force; 1,200 Medical Service; 9,000 other, including joint staff, support, and training schools) (2019 est.)

Military equipment inventories and acquisitions: the Belgian Armed Forces have a mix of weapons systems from European countries, Israel, and the US; since 2010, France, Germany, and Switzerland are the leading suppliers of armaments; Belgium has an advanced, export-focused defense industry that focuses on components and subcontracting (2019 est.)

Military deployments: 125 France (contributing member of EuroCorps); 100 Mali (EUTM/MINUSMA); est. 260 Baltic States (NATO) (2020)

Military service age and obligation: 18 years of age for male and female voluntary military service; conscription abolished in 1994 (2012)

Military—note: in 2018, the Defense Ministers of Belgium, Denmark and the Netherlands signed a Memorandum of Understanding (MOU) for the creation of a Composite Special Operations Component Command (C-SOCC); C-SOCC is scheduled to be fully operational in 2021 (2020)

TERRORISM

Terrorist group(s): Islamic Revolutionary Guard Corps/Qods Force; Islamic State of Iraq and ash-Sham (2019)
note: details about the history, aims, leadership, organization, areas of operation, tactics, targets, weapons, size, and sources of support of the group(s) appear(s) in Appendix-T

TRANSNATIONAL ISSUES

Disputes—international: none

Refugees and internally displaced persons: *refugees (country of origin):* 16,604 (Syria), 5,602 (Iraq), 5,070 (Afghanistan) (2019)
stateless persons: 10,933 (2019)

Illicit drugs: growing producer of synthetic drugs and cannabis; transit point for US-bound ecstasy; source of precursor chemicals for South American cocaine processors; transshipment point for cocaine, heroin, hashish, and marijuana entering Western Europe; despite a strengthening of legislation, the country remains vulnerable to money laundering related to narcotics, automobiles, alcohol, and tobacco; significant domestic consumption of ecstasy

BELIZE

INTRODUCTION

Background: Belize was the site of several Mayan city states until their decline at the end of the first millennium A.D. The British and Spanish disputed the region in the 17th and 18th centuries; it formally became the colony of British Honduras in 1862. Territorial disputes between the UK and Guatemala delayed the independence of Belize until 1981. Guatemala refused to recognize the new nation until 1992 and the two countries are involved in an ongoing border dispute. Both nations have voted to send the dispute for final resolution to the International Court of Justice. Tourism has become the mainstay of the economy. Current concerns include the country's heavy foreign debt burden, high crime rates, high unemployment combined with a majority youth population, growing involvement in the Mexican and South American drug trade, and one of the highest HIV/AIDS prevalence rates in Central America.

GEOGRAPHY

Location: Central America, bordering the Caribbean Sea, between Guatemala and Mexico

Geographic coordinates: 17 15 N, 88 45 W

Map references: Central America and the Caribbean

Area: *total:* 22,966 sq km
land: 22,806 sq km
water: 160 sq km
country comparison to the world: 152

Area—comparative: slightly smaller than Massachusetts

Land boundaries: *total:* 542 km
border countries (2): Guatemala 266 km, Mexico 276 km

Coastline: 386 km

Maritime claims: *territorial sea:* 12 nm in the north, 3 nm in the south; note—from the mouth of the Sarstoon River to Ranguana Cay, Belize's territorial sea is 3 nm; according to Belize's Maritime Areas Act, 1992, the purpose of this limitation is to provide a framework for negotiating a definitive agreement on territorial differences with Guatemala
exclusive economic zone: 200 nm

Climate: tropical; very hot and humid; rainy season (May to November); dry season (February to May)

Terrain: flat, swampy coastal plain; low mountains in south

Elevation: *mean elevation:* 173 m
lowest point: Caribbean Sea 0 m
highest point: Doyle's Delight 1,124 m

Natural resources: arable land potential, timber, fish, hydropower

Land use: *agricultural land:* 6.9% (2011 est.)
arable land: 3.3% (2011 est.) / permanent crops: 1.4% (2011 est.) / permanent pasture: 2.2% (2011 est.)
forest: 60.6% (2011 est.)
other: 32.5% (2011 est.)

Irrigated land: 35 sq km (2012)

Population distribution: approximately 25% to 30% of the population lives in the former capital, Belize City; over half of the overall population is rural; population density is slightly higher in the north and east

Natural hazards: frequent, devastating hurricanes (June to November) and coastal flooding (especially in south)

Environment—current issues: deforestation; water pollution, including pollution of Belize's Barrier Reef System, from sewage, industrial effluents, agricultural runoff; inability to properly dispose of solid waste

Environment—international agreements: party to: Biodiversity, Climate Change, Climate Change-Kyoto Protocol, Desertification, Endangered Species, Hazardous Wastes, Law of the Sea, Ozone Layer Protection, Ship Pollution, Wetlands, Whaling
signed, but not ratified: none of the selected agreements

Geography—note: only country in Central America without a coastline on the North Pacific Ocean

PEOPLE AND SOCIETY

Population: 399,598 (July 2020 est.)
country comparison to the world: 176

Nationality: *noun:* Belizean(s)
adjective: Belizean

Ethnic groups: mestizo 52.9%, Creole 25.9%, Maya 11.3%, Garifuna 6.1%, East Indian 3.9%, Mennonite 3.6%, white 1.2%, Asian 1%, other 1.2%, unknown 0.3% (2010 est.)
note: percentages add up to more than 100% because respondents were able to identify more than one ethnic origin

Languages: English 62.9% (official), Spanish 56.6%, Creole 44.6%, Maya 10.5%, German 3.2%, Garifuna 2.9%, other 1.8%, unknown 0.3%, none 0.2% (cannot speak) (2010 est.)
note: shares sum to more than 100% because some respondents gave more than one answer on the census

Religions: Roman Catholic 40.1%, Protestant 31.5% (includes Pentecostal 8.4%, Seventh Day Adventist 5.4%, Anglican 4.7%, Mennonite 3.7%, Baptist 3.6%, Methodist 2.9%, Nazarene 2.8%), Jehovah's Witness 1.7%, other 10.5% (includes Baha'i, Buddhist, Hindu, Mormon, Muslim, Rastafarian, Salvation Army), unspecified 0.6%, none 15.5% (2010 est.)

Demographic profile: Migration continues to transform Belize's population. About 16% of Belizeans live abroad, while immigrants constitute approximately 15% of Belize's population. Belizeans seeking job and educational opportunities have preferred to emigrate to the United States rather than former colonizer Great Britain because of the United States' closer proximity and stronger trade ties with Belize. Belizeans also emigrate to Canada, Mexico, and English-speaking Caribbean countries. The emigration of a large share of Creoles (Afro-Belizeans) and the influx of Central American immigrants, mainly Guatemalans, Salvadorans, and Hondurans, has changed Belize's ethnic composition. Mestizos have become the largest ethnic group, and Belize now has more native Spanish speakers than English or Creole speakers, despite English being the official language. In addition, Central American immigrants are establishing new communities in rural areas, which contrasts with the urbanization trend seen in neighboring countries. Recently, Chinese, European, and North American immigrants have become more frequent.

Immigration accounts for an increasing share of Belize's population growth rate, which is steadily falling due to fertility decline. Belize's declining birth rate and its increased life expectancy are creating an aging population. As the elderly population grows and nuclear families replace extended households, Belize's government will be challenged to balance a rising demand for pensions, social services, and healthcare for its senior citizens with the need to reduce poverty and social inequality and to improve sanitation.

Age structure: 0-14 *years:* 32.57% (male 66,454/female 63,700)
15-24 years: 19% (male 39,238/female 36,683)
25-54 years: 37.72% (male 73,440/female 77,300)
55-64 years: 6.18% (male 12,235/female 12,444)
65 years and over: 4.53% (male 8,781/female 9,323) (2020 est.)

Dependency ratios: *total dependency ratio:* 52
youth dependency ratio: 44.4
elderly dependency ratio: 7.6
potential support ratio: 13.1 (2020 est.)

Median age: *total:* 23.9 years
male: 23 years
female: 24.8 years (2020 est.)
country comparison to the world: 172

Population growth rate: 1.72% (2020 est.)
country comparison to the world: 61

Birth rate: 22 births/1,000 population (2020 est.)
country comparison to the world: 66

Death rate: 4.1 deaths/1,000 population (2020 est.)
country comparison to the world: 212

Net migration rate: -1 migrant(s)/1,000 population (2020 est.)
country comparison to the world: 142

Population distribution: approximately 25% to 30% of the population lives in the former capital, Belize City; over half of the overall population is rural; population density is slightly higher in the north and east

Urbanization: *urban population:* 46% of total population (2020)
rate of urbanization: 2.32% annual rate of change (2015-20 est.)
total population growth rate v. urban population growth rate, 2000-2030:
Major urban areas—population: 23,000 BELMOPAN (capital) (2018)

Sex ratio: *at birth:* 1.05 male(s)/female
0-14 years: 1.04 male(s)/female
15-24 years: 1.07 male(s)/female
25-54 years: 0.95 male(s)/female
55-64 years: 0.98 male(s)/female
65 years and over: 0.94 male(s)/female
total population: 1 male(s)/female (2020 est.)

Maternal mortality rate: 36 deaths/100,000 live births (2017 est.)
country comparison to the world: 104

Infant mortality rate: *total:* 11.2 deaths/1,000 live births

male: 12.4 deaths/1,000 live births
female: 10 deaths/1,000 live births (2020 est.)
country comparison to the world: 118

Life expectancy at birth: *total population:* 75.3 years
male: 73.7 years
female: 77 years (2020 est.)
country comparison to the world: 117

Total fertility rate: 2.7 children born/woman (2020 est.)
country comparison to the world: 64

Contraceptive prevalence rate: 51.4% (2015/16)

Drinking water source:
improved:
urban: 100% of population
rural: 98.6% of population
total: 99.2% of population
unimproved:
urban: 0% of population
rural: 1.4% of population
total: 0.8% of population (2017 est.)

Current Health Expenditure: 5.6% (2017)

Physicians density: 1.12 physicians/1,000 population (2017)

Hospital bed density: 1 beds/1,000 population (2017)

Sanitation facility access:
improved:
urban: 98.8% of population
rural: 95.3% of population
total: 96.9% of population
unimproved:
urban: 1.2% of population
rural: 4.7% of population
total: 3.1% of population (2017 est.)

HIV/AIDS—adult prevalence rate: 1.9% (2018 est.)
country comparison to the world: 24

HIV/AIDS—people living with HIV/AIDS: 4,900 (2018 est.)
country comparison to the world: 123

HIV/AIDS—deaths: <200 (2018 est.)

Obesity—adult prevalence rate: 24.1% (2016)
country comparison to the world: 60

Children under the age of 5 years underweight: 4.6% (2015)
country comparison to the world: 85

Education expenditures: 7.4% of GDP (2017)
country comparison to the world: 9

School life expectancy (primary to tertiary education): *total:* 13 years
male: 13 years
female: 13 years (2019)

Unemployment, youth ages 15-24: *total:* 15.3%
male: 9.5%
female: 24.8% (2017 est.)
country comparison to the world: 89

GOVERNMENT

Country name: *conventional long form:* none
conventional short form: Belize
former: British Honduras

etymology: may be named for the Belize River, whose name possibly derives from the Maya word "belix," meaning "muddy-watered"

Government type: parliamentary democracy (National Assembly) under a constitutional monarchy; a Commonwealth realm

Capital: *name:* Belmopan
geographic coordinates: 17 15 N, 88 46 W
time difference: UTC-6 (1 hour behind Washington, DC, during Standard Time)
etymology: the decision to move the capital of the country inland to higher and more stable land was made in the 1960s; the name chosen for the new city was formed from the union of two words: "Belize," the name of the longest river in the country, and "Mopan," one of the rivers in the area of the new capital that empties into the Belize River

Administrative divisions: 6 districts; Belize, Cayo, Corozal, Orange Walk, Stann Creek, Toledo

Independence: 21 September 1981 (from the UK)

National holiday: Battle of St. George's Caye Day (National Day), 10 September (1798); Independence Day, 21 September (1981)

Constitution: *history:* previous 1954, 1963 (preindependence); latest signed and entered into force 21 September 1981
amendments: proposed and adopted by two-thirds majority vote of the National Assembly House of Representatives except for amendments relating to rights and freedoms, changes to the Assembly, and to elections and judiciary matters, which require at least three-quarters majority vote of the House; both types of amendments require assent of the governor general; amended several times, last in 2018

Legal system: English common law

International law organization participation: has not submitted an ICJ jurisdiction declaration; accepts ICCt jurisdiction

Citizenship: *citizenship by birth:* yes
citizenship by descent only: yes
dual citizenship recognized: yes
residency requirement for naturalization: 5 years

Suffrage: 18 years of age; universal

Executive branch: *chief of state:* Queen ELIZABETH II (since 6 February 1952); represented by Governor General Sir Colville Norbert YOUNG, Sr. (since 17 November 1993)
head of government: Prime Minister Juan Antonio BRICENO (since 12 November 2020); Deputy Prime Minister Cordel HYDE (since 16 November 2020)
cabinet: Cabinet appointed by the governor general on the advice of the prime minister from among members of the National Assembly
elections/appointments: the monarchy is hereditary; governor general appointed by the monarch; following legislative elections, the leader of the majority party or majority coalition usually appointed prime minister by the governor general; prime minister recommends the deputy prime minister

Legislative branch: *description:* bicameral National Assembly consists of: Senate (14 seats, including the president); members appointed by the governor general—6 on the advice of the prime minister, 3 on the advice of the leader of the opposition, and 1 each on the advice of the Belize Council of Churches and Evangelical Association of Churches, the Belize Chamber of Commerce and Industry and the Belize Better Business Bureau, non-governmental organizations in good standing, and the National Trade Union Congress and the Civil Society Steering Committee; Senate president elected from among the Senate members or from outside the Senate; term of appointment NA

House of Representatives (31 seats; members directly elected in single-seat constituencies by simple majority vote to serve 5-year terms)

elections: Senate—last appointed 11 November 2020 (next appointments in November 2025)

House of Representatives—last held on 11 November 2020 (next to be held in November 2025)

election results: House of Representatives—percent of vote by party—PUP 59.6%, UDP 38.8%, other 1.6%; seats by party—PUP 26, UDP 5

Judicial branch: highest courts: Supreme Court of Judicature (consists of the Court of Appeal with the court president and 3 justices, and the Supreme Court with the chief justice and 10 justices); note—in 2010, Belize acceded to the Caribbean Court of Justice as the final court of appeal, replacing that of the Judicial Committee of the Privy Council in London

judge selection and term of office: Court of Appeal president and justices appointed by the governor-general upon advice of the prime minister after consultation with the National Assembly opposition leader; justices' tenures vary by terms of appointment; Supreme Court chief justice appointed by the governor-general upon the advice of the prime minister and the National Assembly opposition leader; other judges appointed by the governor-general upon the advice of the Judicial and Legal Services Section of the Public Services Commission and with the concurrence of the prime minister after consultation with the National Assembly opposition leader; judges can be appointed beyond age 65 but must retire by age 75; in 2013, the Supreme Court chief justice overturned a constitutional amendment that had restricted Court of Appeal judge appointments to as short as 1 year

subordinate courts: Magistrates' Courts; Family Court

Political parties and leaders: Belize Progressive Party or BPP [Patrick ROGERS] (formed in 2015 from a merger of the People's National Party, elements of the Vision Inspired by the People, and other smaller political groups)

People's United Party or PUP [Johnny BRICENO]
United Democratic Party or UDP [Dean Oliver BARROW]
Vision Inspired by the People or VIP [Hubert ENRIQUEZ]

International organization participation: ACP, AOSIS, C, Caricom, CD, CDB, CELAC, FAO, G-77, IADB, IAEA, IBRD, ICAO, ICC (NGOs), ICRM, IDA, IFAD, IFC, IFRCS, ILO, IMF, IMO, Interpol, IOC, IOM, ITU, LAES, MIGA, NAM, OAS, OPANAL, OPCW, PCA, Petrocaribe, SICA, UN, UNCTAD, UNESCO, UNIDO, UPU, WCO, WHO, WIPO, WMO, WTO

Diplomatic representation in the US: *chief of mission:* Ambassador Francisco Daniel GUTIEREZ (since 21 July 2017)

chancery: 2535 Massachusetts Avenue NW, Washington, DC 20008

telephone: [1] (202) 332-9636

FAX: [1] (202) 332-6888

consulate(s) general: Los Angeles

consulate(s): Miami

Diplomatic representation from the US: *chief of mission:* Ambassador (vacant); Charge d'Affaires Keith R. GILGES (since 24 July 2018)

telephone: [011] (501) 822-4011

embassy: Floral Park Road, Belmopan City, Cayo District

mailing address: P.O. Box 497, Belmopan City, Cayo District, Belize

FAX: [011] (501) 822-4012

Flag description: royal blue with a narrow red stripe along the top and the bottom edges; centered is a large white disk bearing the coat of arms; the coat of arms features a shield flanked by two workers in front of a mahogany tree with the related motto SUB UMBRA FLOREO (I Flourish in the Shade) on a scroll at the bottom, all encircled by a green garland of 50 mahogany leaves; the colors are those of the two main political parties: blue for the PUP and red for the UDP; various elements of the coat of arms—the figures, the tools, the mahogany tree, and the garland of leaves—recall the logging industry that led to British settlement of Belize

note: Belize's flag is the only national flag that depicts human beings; two British overseas territories, Montserrat and the British Virgin Islands, also depict humans

National symbol(s): Baird's tapir (a large, browsing, forest-dwelling mammal), keel-billed toucan, Black Orchid; national colors: red, blue

National anthem: *name:* Land of the Free

lyrics/music: Samuel Alfred HAYNES/Selwyn Walford YOUNG

note: adopted 1981; as a Commonwealth country, in addition to the national anthem, "God Save the Queen" serves as the royal anthem (see United Kingdom)

0:00 / 2:42

ECONOMY

Economy—overview: Tourism is the number one foreign exchange earner in this small economy, followed by exports of sugar, bananas, citrus, marine products, and crude oil.

The government's expansionary monetary and fiscal policies, initiated in September 1998, led to GDP growth averaging nearly 4% in 1999-2007, but GPD growth has averaged only 2.1% from 2007-2016, with 2.5% growth estimated for 2017. Belize's dependence on energy imports makes it susceptible to energy price shocks.

Although Belize has the third highest per capita income in Central America, the average income figure masks a huge income disparity between rich and poor, and a key government objective remains reducing poverty and inequality with the help of international donors. High unemployment, a growing trade deficit and heavy foreign debt burden continue to be major concerns. Belize faces continued pressure from rising sovereign debt, and a growing trade imbalance.

GDP (purchasing power parity): $3.218 billion (2017 est.)
$3.194 billion (2016 est.)
$3.21 billion (2015 est.)
note: data are in 2017 dollars
country comparison to the world: 187

GDP (official exchange rate): $1.854 billion (2017 est.)

GDP—real growth rate: 0.8% (2017 est.)
-0.5% (2016 est.)
3.8% (2015 est.)
country comparison to the world: 177

GDP—per capita (PPP): $8,300 (2017 est.)
$8,500 (2016 est.)
$8,800 (2015 est.)
note: data are in 2017 dollars
country comparison to the world: 149

Gross national saving: 11.3% of GDP (2017 est.)
13.3% of GDP (2016 est.)
14.2% of GDP (2015 est.)
country comparison to the world: 157

GDP—composition, by end use: household consumption: 75.1% (2017 est.)
government consumption: 15.2% (2017 est.)
investment in fixed capital: 22.5% (2017 est.)
investment in inventories: 1.2% (2017 est.)
exports of goods and services: 49.1% (2017 est.)
imports of goods and services: -63.2% (2017 est.)

GDP—composition, by sector of origin:
agriculture: 10.3% (2017 est.)
industry: 21.6% (2017 est.)
services: 68% (2017 est.)

Agriculture—products: bananas, cacao, citrus, sugar; fish, cultured shrimp; lumber

Industries: garment production, food processing, tourism, construction, oil

Industrial production growth rate: -0.6% (2017 est.)
country comparison to the world: 172

Labor force: 120,500 (2008 est.)
note: shortage of skilled labor and all types of technical personnel
country comparison to the world: 180

Labor force—by occupation: *agriculture:* 10.2%
industry: 18.1%
services: 71.7% (2007 est.)

Unemployment rate: 9% (2017 est.)
8% (2016 est.)
country comparison to the world: 137

Population below poverty line: 41% (2013 est.)

Household income or consumption by percentage share: *lowest 10%:* NA
highest 10%: NA

Budget: *revenues:* 553.5 million (2017 est.)
expenditures: 572 million (2017 est.)

Taxes and other revenues: 29.9% (of GDP) (2017 est.)
country comparison to the world: 78

Budget surplus (+) or deficit (-): -1% (of GDP) (2017 est.)
country comparison to the world: 75

Public debt: 99% of GDP (2017 est.)
95.9% of GDP (2016 est.)
country comparison to the world: 17

Fiscal year: 1 April—31 March

Inflation rate (consumer prices): 1.1% (2017 est.)
0.7% (2016 est.)
country comparison to the world: 55

Current account balance: -$143 million (2017 est.)
-$163 million (2016 est.)
country comparison to the world: 92

Exports: $457.5 million (2017 est.)
$442.7 million (2016 est.)
country comparison to the world: 177

Exports—partners: UK 33.9%, US 22%, Jamaica 6.7%, Italy 6.4%, Barbados 5.9%, Ireland 5.5%, Netherlands 4.3% (2017)

Exports—commodities: sugar, bananas, citrus, clothing, fish products, molasses, wood, crude oil

Imports: $845.9 million (2017 est.)
$916.2 million (2016 est.)
country comparison to the world: 188

Imports—commodities: machinery and transport equipment, manufactured goods; fuels, chemicals, pharmaceuticals; food, beverages, tobacco

Imports—partners: US 35.6%, China 11.2%, Mexico 11.2%, Guatemala 6.9% (2017)

Reserves of foreign exchange and gold: $312.1 million (31 December 2017 est.)
$376.7 million (31 December 2016 est.)
country comparison to the world: 167

Debt—external: $1.315 billion (31 December 2017 est.)
$1.338 billion (31 December 2016 est.)
country comparison to the world: 161

Exchange rates: Belizean dollars (BZD) per US dollar –
2 (2017 est.)
2 (2016 est.)
2 (2015 est.)
2 (2014 est.)
2 (2013 est.)

ENERGY

Electricity access: *electrification—total population:* 99.5% (2018)
electrification—urban areas: 97.1% (2016)
electrification—rural areas: 88.4% (2016)

Electricity—production: 280 million kWh (2016 est.)
country comparison to the world: 186

Electricity—consumption: 453 million kWh (2016 est.)
country comparison to the world: 172

Electricity—exports: 0 kWh (2016 est.)
country comparison to the world: 105

Electricity—imports: 243 million kWh (2016 est.)
country comparison to the world: 92

Electricity—installed generating capacity: 198,000 kW (2016 est.)
country comparison to the world: 165

Electricity—from fossil fuels: 51% of total installed capacity (2016 est.)
country comparison to the world: 147

Electricity—from nuclear fuels: 0% of total installed capacity (2017 est.)
country comparison to the world: 48

Electricity—from hydroelectric plants: 27% of total installed capacity (2017 est.)
country comparison to the world: 73

Electricity—from other renewable sources: 22% of total installed capacity (2017 est.)
country comparison to the world: 33

Crude oil—production: 2,000 bbl/day (2018 est.)
country comparison to the world: 85

Crude oil—exports: 1,220 bbl/day (2015 est.)
country comparison to the world: 73

Crude oil—imports: 0 bbl/day (2015 est.)
country comparison to the world: 95

Crude oil—proved reserves: 6.7 million bbl (1 January 2018 est.)
country comparison to the world: 93

Refined petroleum products—production: 36 bbl/day (2015 est.)
country comparison to the world: 109

Refined petroleum products—consumption: 4,000 bbl/day (2016 est.)
country comparison to the world: 182

Refined petroleum products—exports: 0 bbl/day (2015 est.)
country comparison to the world: 130

Refined petroleum products—imports: 4,161 bbl/day (2015 est.)
country comparison to the world: 176

Natural gas—production: 0 cu m (2017 est.)
country comparison to the world: 103

Natural gas—consumption: 0 cu m (2017 est.)
country comparison to the world: 120

Natural gas—exports: 0 cu m (2017 est.)
country comparison to the world: 68

Natural gas—imports: 0 cu m (2017 est.)
country comparison to the world: 90

Natural gas—proved reserves: 0 cu m (1 January 2014 est.)
country comparison to the world: 110

Carbon dioxide emissions from consumption of energy: 556,700 Mt (2017 est.)
country comparison to the world: 183

COMMUNICATIONS

Telephones—fixed lines: *total subscriptions:* 18,617
subscriptions per 100 inhabitants: 4.74 (2019 est.)
country comparison to the world: 179

Telephones—mobile cellular: *total subscriptions:* 256,479
subscriptions per 100 inhabitants: 65.3 (2019 est.)
country comparison to the world: 182

Telecommunication systems: *general assessment:* govt. telecom company, Belize Telemedia Ltd., continues to hold a monopoly in fixed-line services, mobile and broadband fixed-line teledensity; it is a small market; fixed-line teledensity and mobile penetration below regional average; lack of competition and underinvestment in telecom system, make it pricey for consumer (2020)
domestic: 5 per 100 fixed-line and mobile-cellular teledensity approaching 65 per 100 persons; mobile sector accounting for over 90% of all phone subscriptions (2019)
international: country code—501; landing points for the ARCOS and SEUL fiber-optic telecommunications submarine cable that provides links to South and Central America, parts of the Caribbean, and the US; satellite earth station—8 (Intelsat—2, unknown—6) (2019)
note: the COVID-19 outbreak is negatively impacting telecommunications production and supply chains globally; consumer spending on telecom devices and services has also slowed due to the pandemic's effect on economies worldwide; overall progress towards improvements in all facets of the telecom industry—mobile, fixed-line, broadband, submarine cable and satellite—has moderated

Broadcast media: 8 privately owned TV stations; multi-channel cable TV provides access to foreign stations; about 25 radio stations broadcasting on roughly 50 different frequencies; state-run radio was privatized in 1998 (2019)

Internet country code: .bz

Internet users: *total:* 181,660
percent of population: 47.08% (July 2018 est.)
country comparison to the world: 176

Broadband—fixed subscriptions: *total:* 24,658
subscriptions per 100 inhabitants: 6 (2018 est.)
country comparison to the world: 148

TRANSPORTATION

National air transport system: *number of registered air carriers:* 2 (2020)
inventory of registered aircraft operated by air carriers: 28
annual passenger traffic on registered air carriers: 1,297,533 (2018)
annual freight traffic on registered air carriers: 3.78 million mt-km (2018)

Civil aircraft registration country code prefix: V3 (2016)

Airports: 47 (2013)

country comparison to the world: 91

Airports—with paved runways: *total:* 6 (2017)
2,438 to 3,047 m: 1 (2017)
914 to 1,523 m: 2 (2017)
under 914 m: 3 (2017)

Airports—with unpaved runways: *total:* 41 (2013)
2,438 to 3,047 m: 1 (2013)
914 to 1,523 m: 11 (2013)
under 914 m: 29 (2013)

Roadways: *total:* 3,281 km (2017)
paved: 601 km (2017)
unpaved: 2,680 km (2017)
country comparison to the world: 160

Waterways: 825 km (navigable only by small craft) (2011)
country comparison to the world: 70

Merchant marine: *total:* 786
by type: bulk carrier 56, container ship 4, general cargo 398, oil tanker 65, other 263 (2019)
country comparison to the world: 29

Ports and terminals: *major seaport(s):* Belize City, Big Creek

MILITARY AND SECURITY

Military and security forces: *Belize Defense Force (BDF):* Army, Air Wing; Belize Coast Guard (2019)

Military expenditures: 1.2% of GDP (2019)
1.3% of GDP (2018)
1.6% of GDP (2017)

1.2% of GDP (2016)
1.1% of GDP (2015)
country comparison to the world: 101

Military and security service personnel strengths: the Belize Defense Force (BDF) has approximately 1,300 active Army personnel; 150 Belize Coast Guard (2019 est.)

Military equipment inventories and acquisitions: the BDF's inventory is limited and consists mostly of equipment from the UK and US (2019 est.)

Military service age and obligation: 18 years of age for voluntary military service; laws allow for conscription only if volunteers are insufficient; conscription has never been implemented; volunteers typically outnumber available positions by 3:1; initial service obligation 12 years (2012)

TRANSNATIONAL ISSUES

Disputes—international: Guatemala persists in its territorial claim to approximately half of Belize, but agrees to the Line of Adjacency to keep Guatemalan squatters out of Belize's forested interior; both countries agreed in April 2012 to hold simultaneous referenda, scheduled for 6 October 2013, to decide whether to refer the dispute to the ICJ for binding resolution, but this vote was suspended indefinitely; Belize and Mexico are working to solve minor border demarcation discrepancies arising from inaccuracies in the 1898 border treaty

Trafficking in persons: *current situation:* Belize is a source, destination, and transit country for men, women, and children subjected to forced labor and sex trafficking; the coerced prostitution of women and children by family members has not led to arrests; child sex tourism, involving primarily US citizens, is on the rise; sex trafficking and forced labor of Belizean and foreign women and LGBT individuals occurs in bars, nightclubs, brothels, and domestic service; workers from Central America, Mexico, and Asia may fall victim to forced labor in restaurants, shops, agriculture, and fishing
tier rating: Tier 3 – Belize does not comply fully with the minimum standards for the elimination of human trafficking and is not making significant efforts to do so; authorities did not initiate any new trafficking investigations of prosecutions, and cases from previous years remain pending; law enforcement efforts to use informal means to identify and refer victims were ineffective and draft procedures for referring victims to services are still not finalized; trafficking victims were more commonly arrested, detained, or deported based on immigration violations than provided with assistance; the government did not make progress in implementing the 2012-14 anti-trafficking national strategic plan (2015)

Illicit drugs: major transshipment point for cocaine; small-scale illicit producer of cannabis, primarily for local consumption; offshore sector money-laundering activity related to narcotics trafficking and other crimes

BENIN

INTRODUCTION

Background: Present day Benin was the site of Dahomey, a West African kingdom that rose to prominence in about 1600 and over the next two and a half centuries became a regional power, largely based on its slave trade. France began to control the coastal areas of Dahomey in the second half of the 19th century; the entire kingdom was conquered by 1894. French Dahomey achieved independence in 1960; it changed its name to the Republic of Benin in 1975.

A succession of military governments ended in 1972 with the rise to power of Mathieu KEREKOU and the establishment of a government based on Marxist-Leninist principles. A move to representative government began in 1989. Two years later, free elections ushered in former Prime Minister Nicephore SOGLO as president, marking the first successful transfer of power in Africa from a dictatorship to a democracy. KEREKOU was returned to power by elections held in 1996 and 2001, though some irregularities were alleged. KEREKOU stepped down at the end of his second term in 2006 and was succeeded by Thomas YAYI Boni, a political outsider and independent, who won a second five-year term in March 2011.

Patrice TALON, a wealthy businessman, took office in 2016 after campaigning to restore public confidence in the government.

GEOGRAPHY

Location: Western Africa, bordering the Bight of Benin, between Nigeria and Togo

Geographic coordinates: 9 30 N, 2 15 E

Map references: Africa

Area: *total:* 112,622 sq km
land: 110,622 sq km
water: 2,000 sq km
country comparison to the world: 103

Area—comparative: slightly smaller than Pennsylvania

Land boundaries: *total:* 2,123 km
border countries (4): Burkina Faso 386 km, Niger 277 km, Nigeria 809 km, Togo 651 km

Coastline: 121 km

Maritime claims: *territorial sea:* 200 nm
continental shelf: 200 nm
exclusive fishing zone: 200 nm

Climate: tropical; hot, humid in south; semiarid in north

Terrain: mostly flat to undulating plain; some hills and low mountains

Elevation: *mean elevation:* 273 m
lowest point: Atlantic Ocean 0 m
highest point: Mont Sokbaro 658 m

Natural resources: small offshore oil deposits, limestone, marble, timber

Land use: *agricultural land:* 31.3% (2011 est.)
arable land: 22.9% (2011 est.) / *permanent crops:* 3.5% (2011 est.) / *permanent pasture:* 4.9% (2011 est.)
forest: 40% (2011 est.)
other: 28.7% (2011 est.)

Irrigated land: 230 sq km (2012)

Population distribution: the population is primarily located in the south, with the highest concentration of people residing in and around the cities on the Atlantic coast; most of the north remains sparsely populated with higher concentrations of residents in the west as shown in this population distribution map

Natural hazards: hot, dry, dusty harmattan wind may affect north from December to March

Environment—current issues: inadequate supplies of potable water; water pollution; poaching

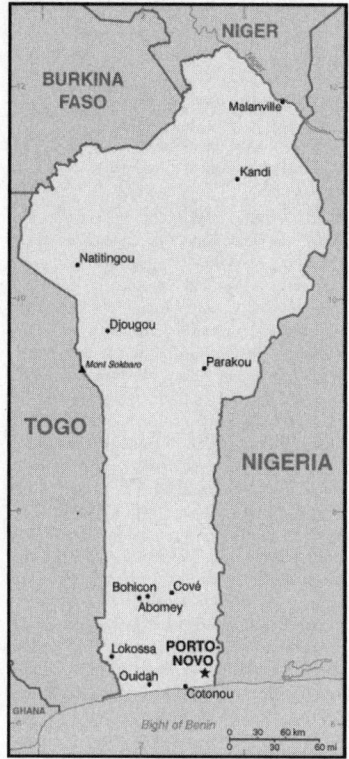

threatens wildlife populations; deforestation; desertification (the spread of the desert into agricultural lands in the north is accelerated by regular droughts)

Environment—international agreements: party to: Biodiversity, Climate Change, Climate Change-Kyoto Protocol, Desertification, Endangered Species, Environmental Modification, Hazardous Wastes, Law of the Sea, Ozone Layer Protection, Ship Pollution, Wetlands, Whaling
signed, but not ratified: none of the selected agreements

Geography—note: sandbanks create difficult access to a coast with no natural harbors, river mouths, or islands

PEOPLE AND SOCIETY

Population: 12,864,634 (July 2020 est.)
note: estimates for this country explicitly take into account the effects of excess mortality due to AIDS; this can result in lower life expectancy, higher infant mortality, higher death rates, lower population growth rates, and changes in the distribution of population by age and sex than would otherwise be expected
country comparison to the world: 74

Nationality: *noun:* Beninese (singular and plural)
adjective: Beninese

Ethnic groups: Fon and related 38.4%, Adja and related 15.1%, Yoruba and related 12%, Bariba and related 9.6%, Fulani and related 8.6%, Ottamari and related 6.1%, Yoa-Lokpa and related 4.3%, Dendi and related 2.9%, other 0.9%, foreigner 1.9% (2013 est.)

Languages: French (official), Fon and Yoruba (most common vernaculars in south), tribal languages (at least six major ones in north)

Religions: Muslim 27.7%, Roman Catholic 25.5%, Protestant 13.5% (Celestial 6.7%, Methodist 3.4%, other Protestant 3.4%), Vodoun 11.6%, other Christian 9.5%, other traditional religions 2.6%, other 2.6%, none 5.8% (2013 est.)

Demographic profile: Benin has a youthful age structure – almost 65% of the population is under the age of 25 – which is bolstered by high fertility and population growth rates. Benin's total fertility has been falling over time but remains high, declining from almost 7 children per women in 1990 to 4.8 in 2016. Benin's low contraceptive use and high unmet need for contraception contribute to the sustained high fertility rate. Although the majority of Beninese women use skilled health care personnel for antenatal care and delivery, the high rate of maternal mortality indicates the need for more access to high quality obstetric care.

Poverty, unemployment, increased living costs, and dwindling resources increasingly drive the Beninese to migrate. An estimated 4.4 million, more than 40%, of Beninese live abroad. Virtually all Beninese emigrants move to West African countries, particularly Nigeria and Cote d'Ivoire. Of the less than 1% of Beninese emigrants who settle in Europe, the vast majority live in France, Benin's former colonial ruler.

With about 40% of the population living below the poverty line, many desperate parents resort to sending their children to work in wealthy households as domestic servants (a common practice known as vidomegon), mines, quarries, or agriculture domestically or in Nigeria and other neighboring countries, often under brutal conditions. Unlike in other West African countries, where rural people move to the coast, farmers from Benin's densely populated southern and northwestern regions move to the historically sparsely populated central region to pursue agriculture. Immigrants from West African countries came to Benin in increasing numbers between 1992 and 2002 because of its political stability and porous borders.

Age structure: 0-14 *years:* 45.56% (male 2,955,396/female 2,906,079)
15-24 years: 20.36% (male 1,300,453/female 1,318,880)
25-54 years: 28.54% (male 1,735,229/female 1,935,839)
55-64 years: 3.15% (male 193,548/female 211,427)
65 years and over: 2.39% (male 140,513/female 167,270) (2020 est.)

Dependency ratios: *total dependency ratio:* 82.6
youth dependency ratio: 76.6
elderly dependency ratio: 6
potential support ratio: 16.7 (2020 est.)

Median age: *total:* 17 years

male: 16.4 years
female: 17.6 years (2020 est.)
country comparison to the world: 219

Population growth rate: 3.4% (2020 est.)
country comparison to the world: 4

Birth rate: 42.1 births/1,000 population (2020 est.)
country comparison to the world: 5

Death rate: 8.4 deaths/1,000 population (2020 est.)
country comparison to the world: 75

Net migration rate: 0.3 migrant(s)/1,000 population (2020 est.)
country comparison to the world: 69

Population distribution: the population is primarily located in the south, with the highest concentration of people residing in and around the cities on the Atlantic coast; most of the north remains sparsely populated with higher concentrations of residents in the west at shown in this population distribution map

Urbanization: *urban population:* 48.4% of total population (2020)
rate of urbanization: 3.89% annual rate of change (2015-20 est.)

total population growth rate v. urban population growth rate, 2000-2030: Major urban areas—population: 285,000 PORTO-NOVO (capital) (2018); 1.056 million Abomey-Calavi, 692,000 COTONOU (seat of government) (2020)

Sex ratio: *at birth:* 1.05 male(s)/female
0-14 years: 1.02 male(s)/female
15-24 years: 0.99 male(s)/female
25-54 years: 0.9 male(s)/female
55-64 years: 0.92 male(s)/female
65 years and over: 0.84 male(s)/female
total population: 0.97 male(s)/female (2020 est.)

Mother's mean age at first birth: 20.4 years (2017/18 est.)
note: median age at first birth among women 25-29

Maternal mortality rate: 397 deaths/ 100,000 live births (2017 est.)
country comparison to the world: 27

Infant mortality rate: *total:* 58.7 deaths/ 1,000 live births
male: 63.9 deaths/1,000 live births
female: 53.3 deaths/1,000 live births (2020 est.)
country comparison to the world: 15

Life expectancy at birth: *total population:* 61.4 years
male: 59.6 years
female: 63.3 years (2020 est.)
country comparison to the world: 212

Total fertility rate: 5.53 children born/woman (2020 est.)
country comparison to the world: 8

Contraceptive prevalence rate: 15.5% (2017/18)

Drinking water source:
improved:
urban: 81.2% of population
rural: 72.2% of population
total: 76.4% of population
unimproved:

urban: 18.8% of population
rural: 27.8% of population
total: 23.6% of population (2017 est.)

Current Health Expenditure: 3.7% (2017)

Physicians density: 0.05 physicians/ 1,000 population (2016)

Hospital bed density: 0.5 beds/ 1,000 population (2010)

Sanitation facility access:
improved:
urban: 58.7% of population
rural: 16% of population
total: 36% of population
unimproved:
urban: 41.3% of population
rural: 84% of population
total: 64% of population (2017 est.)

HIV/AIDS—adult prevalence rate: 1% (2019 est.)
country comparison to the world: 44

HIV/AIDS—people living with HIV/AIDS: 75,000 (2019 est.)
country comparison to the world: 53

HIV/AIDS—deaths: 2,300 (2019 est.)
country comparison to the world: 42

Major infectious diseases: *degree of risk:* very high (2020)
food or waterborne diseases: bacterial and protozoal diarrhea, hepatitis A, and typhoid fever
vectorborne diseases: dengue fever and malaria
animal contact diseases: rabies
respiratory diseases: meningococcal meningitis

Obesity—adult prevalence rate: 9.6% (2016)
country comparison to the world: 142

Children under the age of 5 years underweight: 16.8% (2018)
country comparison to the world: 35

Education expenditures: 4% of GDP (2016)
country comparison to the world: 102

Literacy: *definition:* age 15 and over can read and write
total population: 42.4%
male: 54%
female: 31.1% (2018)

School life expectancy (primary to tertiary education): *total:* 13 years
male: 14 years
female: 11 years (2016)

Unemployment, youth ages 15-24: *total:* 5.6%
male: 5.2%
female: 5.9% (2011 est.)
country comparison to the world: 161

GOVERNMENT

Country name: *conventional long form:* Republic of Benin
conventional short form: Benin
local long form: Republique du Benin
local short form: Benin
former: Dahomey, People's Republic of Benin
etymology: named for the Bight of Benin, the body of water on which the country lies

Government type: presidential republic

Capital: *name:* Porto-Novo (constitutional capital); Cotonou (seat of government)
geographic coordinates: 6 29 N, 2 37 E
time difference: UTC+ 1 (6 hours ahead of Washington, DC, during Standard Time)
etymology: the name Porto- Novo is Portuguese for "new port"; Cotonou means "by the river of death" in the native Fon language

Administrative divisions: 12 departments; Alibori, Atacora, Atlantique, Borgou, Collines, Couffo, Donga, Littoral, Mono, Oueme, Plateau, Zou

Independence: 1 August 1960 (from France)

National holiday: Independence Day, 1 August (1960)

Constitution: *history:* previous 1946, 1958 (preindependence); latest adopted by referendum 2 December 1990, promulgated 11 December 1990
amendments: proposed concurrently by the president of the republic (after a decision in the Council of Ministers) and the National Assembly; consideration of drafts or proposals requires at least three-fourths majority vote of the Assembly membership; passage requires approval in a referendum unless approved by at least four- fifths majority vote of the Assembly membership; constitutional articles affecting territorial sovereignty, the republican form of government, and secularity of Benin cannot be amended

Legal system: civil law system modeled largely on the French system and some customary law

International law organization participation: has not submitted an ICJ jurisdiction declaration; accepts ICCt jurisdiction

Citizenship: citizenship by birth: no
citizenship by descent only: at least one parent must be a citizen of Benin
dual citizenship recognized: yes
residency requirement for naturalization: 10 years

Suffrage: 18 years of age; universal

Executive branch: *chief of state:* President Patrice TALON (since 6 April 2016); note—the president is both chief of state and head of government
head of government: President Patrice TALON (since 6 April 2016); prime minister position abolished
cabinet: Council of Ministers appointed by the president
elections/appointments: president directly elected by absolute majority popular vote in 2 rounds if needed for a 5-year term (eligible for a second term); last held on 6 March and 20 March 2016 (next to be held in 2021)
election results: Patrice TALON elected president in second round; percent of vote in first round— Lionel ZINSOU (FCBE) 28.4%, Patrice TALON (independent) 24.8%, Sebastien AJAVON (independent) 23.%, Abdoulaye Bio TCHANE (ABT) 8.8%, Pascal KOUPAKI (NC) 5.9%, other 9.1%; percent of vote in second round—Patrice TALON 65.4%, Lionel ZINSOU 34.6%

Legislative branch: description: unicameral National Assembly or Assemblee Nationale (83 seats; members directly elected in multi-seat constituencies by proportional representation vote; members serve 4-year terms)
elections: last held on 28 April 2019 (next to be held in April 2023)
election results: percent of vote by party—Union Progressiste 56.2%, Bloc Republicain 43.8%; seats by party—Union Progressiste 47, Bloc Republicain 36; composition—men 77, women 6, percent of women 7.2%

Judicial branch: *highest courts:* Supreme Court or Cour Supreme (consists of the chief justice and 16 justices organized into an administrative division, judicial chamber, and chamber of accounts); Constitutional Court or Cour Constitutionnelle (consists of 7 members, including the court president); High Court of Justice (consists of the Constitutional Court members, 6 members appointed by the National Assembly, and the Supreme Court president); note—jurisdiction of the High Court of Justice is limited to cases of high treason by the national president or members of the government while in office
judge selection and term of office: Supreme Court president and judges appointed by the president of the republic upon the advice of the National Assembly; judges appointed for single renewable 5-year terms; Constitutional Court members— 4 appointed by the National Assembly and 3 by the president of the republic; members appointed for single renewable 5-year terms; other members of the High Court of Justice elected by the National Assembly; member tenure NA
subordinate courts: Court of Appeal or Cour d'Appel; district courts; village courts; Assize courts

Political parties and leaders: Alliance for a Triumphant Benin or ABT [Abdoulaye BIO TCHANE]
African Movement for Development and Progress or MADEP [Sefou FAGBOHOUN]
Benin Renaissance or RB [Lehady SOGLO]
Cowrie Force for an Emerging Benin or FCBE [Yayi BONI]
Democratic Renewal Party or PRD [Adrien HOUNGBEDJI]
National Alliance for Development and Democracy or AND [Valentin Aditi HOUDE]
New Consciousness Rally or NC [Pascal KOUPAKI]
Patriotic Awakening or RP [Janvier YAHOUEDEOU]
Social Democrat Party or PSD [Emmanuel GOLOU]
Sun Alliance or AS [Sacca LAFIA]
Union Makes the Nation or UN [Adrien HOUNGBEDJI] (includes PRD, MADEP)
United Democratic Forces or FDU [Mathurin NAGO]
note: approximately 20 additional minor parties

International organization participation: ACP, AfDB, AU, CD, ECOWAS, Entente, FAO, FZ, G- 77, IAEA, IBRD, ICAO, ICCt, ICRM, IDA, IDB, IFAD, IFC, IFRCS, ILO, IMF, IMO, Interpol, IOC, IOM, IPU, ISO, ITSO, ITU, ITUC (NGOs), MIGA, MINUSMA, MONUSCO,

NAM, OAS (observer), OIC, OIF, OPCW, PCA, UN, UNAMID, UNCTAD, UNESCO, UNHCR, UNIDO, UNMIL, UNMISS, UNOCI, UNWTO, UPU, WADB (regional), WAEMU, WCO, WFTU (NGOs), WHO, WIPO, WMO, WTO

Diplomatic representation in the US: *chief of mission:* Ambassador Jean Claude Felix DO REGO (since 17 July 2020)
chancery: 2124 Kalorama Road NW, Washington, DC 20008
telephone: [1] (202) 232-6656
FAX: [1] (202) 265-1996

Diplomatic representation from the US: *chief of mission:* Ambassador Patricia MAHONEY (since 18 January 2019)
telephone: [229] 21-30-06-50
embassy: Marina Avenue, 01 BP 2012, Cotonou
mailing address: 01 B.P.2012, Cotonou
FAX: [229] 21-30-03-84

Flag description: two equal horizontal bands of yellow (top) and red (bottom) with a vertical green band on the hoist side; green symbolizes hope and revival, yellow wealth, and red courage
note: uses the popular Pan- African colors of Ethiopia

National symbol(s): leopard; national colors: green, yellow, red

National anthem: *name:* "L'Aube Nouvelle" (The Dawn of a New Day)
lyrics/music: Gilbert Jean DAGNON
note: adopted 1960
0:00 / 0:00

ECONOMY

Economy—overview: The free market economy of Benin has grown consecutively for four years, though growth slowed in 2017, as its close trade links to Nigeria expose Benin to risks from volatile commodity prices. Cotton is a key export commodity, with export earnings significantly impacted by the price of cotton in the broader market. The economy began deflating in 2017, with the consumer price index falling 0.8%.

During the first two years of President TALON's administration, which began in April 2016, the government has followed an ambitious action plan to kickstart development through investments in infrastructure, education, agriculture, and governance. Electricity generation, which has constrained Benin's economic growth, has increased and blackouts have been considerably reduced. Private foreign direct investment is small, and foreign aid accounts for a large proportion of investment in infrastructure projects.

Benin has appealed for international assistance to mitigate piracy against commercial shipping in its territory, and has used equipment from donors effectively against such piracy. Pilferage has significantly dropped at the Port of Cotonou, though the port is still struggling with effective implementation of the International Ship and Port Facility Security (ISPS) Code. Projects included in Benin's $307 million Millennium Challenge Corporation (MCC) first compact (2006-11) were designed

to increase investment and private sector activity by improving key institutional and physical infrastructure. The four projects focused on access to land, access to financial services, access to justice, and access to markets (including modernization of the port). The Port of Cotonou is a major contributor to Benin's economy, with revenues projected to account for more than 40% of Benin's national budget.

Benin will need further efforts to upgrade infrastructure, stem corruption, and expand access to foreign markets to achieve its potential. In September 2015, Benin signed a second MCC Compact for $ 375 million that entered into force in June 2017 and is designed to strengthen the national utility service provider, attract private sector investment, fund infrastructure investments in electricity generation and distribution, and develop off-grid electrification for poor and unserved households. As part of the Government of Benin's action plan to spur growth, Benin passed public private partnership legislation in 2017 to attract more foreign investment, place more emphasis on tourism, facilitate the development of new food processing systems and agricultural products, encourage new information and communication technology, and establish Independent Power Producers. In April 2017, the IMF approved a three year $150.4 million Extended Credit Facility agreement to maintain debt sustainability and boost donor confidence.

GDP (purchasing power parity): $25.39 billion (2017 est.)
$24.04 billion (2016 est.)
$23.12 billion (2015 est.)
note: data are in 2017 dollars
country comparison to the world: 141

GDP (official exchange rate): $9.246 billion (2017 est.)

GDP—real growth rate: 5.6% (2017 est.)
4% (2016 est.)
2.1% (2015 est.)
country comparison to the world: 34

GDP—per capita (PPP): $2,300 (2017 est.)
$2,200 (2016 est.)
$2,200 (2015 est.)
note: data are in 2017 dollars
country comparison to the world: 201

Gross national saving: 17.3% of GDP (2017 est.)
15.2% of GDP (2016 est.)
16.6% of GDP (2015 est.)
country comparison to the world: 116

GDP—composition, by end use: *household consumption:* 70.5% (2017 est.)
government consumption: 13.1% (2017 est.)
investment in fixed capital: 27.6% (2017 est.)
investment in inventories: 0% (2017 est.)
exports of goods and services: 31.6% (2017 est.)
imports of goods and services: -43% (2017 est.)

GDP—composition, by sector of origin: *agriculture:* 26.1% (2017 est.)
industry: 22.8% (2017 est.)
services: 51.1% (2017 est.)

Agriculture—products: cotton, corn, cassava (manioc, tapioca), yams, beans, palm oil, peanuts, cashews; livestock

Industries: textiles, food processing, construction materials, cement

Industrial production growth rate: 3% (2017 est.)
country comparison to the world: 101

Labor force: 3.662 million (2007 est.)
country comparison to the world: 97

Unemployment rate: 1% (2014 est.)
country comparison to the world: 8

Population below poverty line: 36.2% (2011 est.)

Household income or consumption by percentage share: *lowest 10%:* 3.1%
highest 10%: 29% (2003)

Budget: *revenues:* 1.578 billion (2017 est.)
expenditures: 2.152 billion (2017 est.)

Taxes and other revenues: 17.1% (of GDP) (2017 est.)
country comparison to the world: 171

Budget surplus (+) or deficit (-): -6.2% (of GDP) (2017 est.)
country comparison to the world: 186

Public debt: 54.6% of GDP (2017 est.)
49.7% of GDP (2016 est.)
country comparison to the world: 81

Fiscal year: calendar year

Inflation rate (consumer prices): 0.1% (2017 est.)
-0.8% (2016 est.)
country comparison to the world: 13

Current account balance: -$1.024 billion (2017 est.)
-$808 million (2016 est.)
country comparison to the world: 147

Exports: $1.974 billion (2017 est.)
$1.588 billion (2016 est.)
country comparison to the world: 140

Exports—partners: Bangladesh 18.1%, India 10.7%, Ukraine 9%, Niger 8.1%, China 7.7%, Nigeria 7.2%, Turkey 4% (2017)

Exports—commodities: cotton, cashews, shea butter, textiles, palm products, seafood

Imports: $2.787 billion (2017 est.)
$2.443 billion (2016 est.)
country comparison to the world: 151

Imports—commodities: foodstuffs, capital goods, petroleum products

Imports—partners: Thailand 18.1%, India 15.9%, France 8.5%, China 7.5%, Togo 5.9%, Netherlands 4.3%, Belgium 4.3% (2017)

Reserves of foreign exchange and gold: $698.9 million (31 December 2017 est.)
$57.5 million (31 December 2016 est.)
country comparison to the world: 141

Debt—external: $2.804 billion (31 December 2017 est.)
$2.476 billion (31 December 2016 est.)
country comparison to the world: 144

Exchange rates: Communaute Financiere Africaine francs (XOF) per US dollar—
605.3 (2017 est.)

593.01 (2016 est.)
593.01 (2015 est.)
591.45 (2014 est.)
494.42 (2013 est.)

ENERGY

Electricity access: *population without electricity:*
8 million (2019)
electrification—total population: 33% (2019)
electrification—urban areas: 58% (2019)
electrification—rural areas: 9% (2019)

Electricity—production: 335 million kWh (2016 est.)
country comparison to the world: 177

Electricity—consumption: 1.143 billion kWh (2016 est.)
country comparison to the world: 152

Electricity—exports: 0 kWh (2016 est.)
country comparison to the world: 106

Electricity—imports: 1.088 billion kWh (2016 est.)
country comparison to the world: 68

Electricity—installed generating capacity: 321,000 kW (2016 est.)
country comparison to the world: 158

Electricity—from fossil fuels: 88% of total installed capacity (2016 est.)
country comparison to the world: 58

Electricity—from nuclear fuels: 0% of total installed capacity (2017 est.)
country comparison to the world: 49

Electricity—from hydroelectric plants: 9% of total installed capacity (2017 est.)
country comparison to the world: 116

Electricity—from other renewable sources: 2% of total installed capacity (2017 est.)
country comparison to the world: 135

Crude oil—production: 0 bbl/day (2018 est.)
country comparison to the world: 110

Crude oil—exports: 0 bbl/day (2015 est.)
country comparison to the world: 93

Crude oil—imports: 0 bbl/day (2015 est.)
country comparison to the world: 96

Crude oil—proved reserves: 8 million bbl (1 January 2018 est.)
country comparison to the world: 92

Refined petroleum products—production: 0 bbl/day (2015 est.)
country comparison to the world: 118

Refined petroleum products—consumption: 38,000 bbl/day (2016 est.)
country comparison to the world: 113

Refined petroleum products—exports: 1,514 bbl/day (2015 est.)
country comparison to the world: 107

Refined petroleum products—imports: 38,040 bbl/day (2015 est.)
country comparison to the world: 93

Natural gas—production: 0 cu m (2017 est.)
country comparison to the world: 104

Natural gas—consumption: 0 cu m (2017 est.)

country comparison to the world: 121

Natural gas—exports: 0 cu m (2017 est.)
country comparison to the world: 69

Natural gas—imports: 0 cu m (2017 est.)
country comparison to the world: 91

Natural gas—proved reserves: 1.133 billion cu m (1 January 2018 est.)
country comparison to the world: 98

Carbon dioxide emissions from consumption of energy: 5.664 million Mt (2017 est.)
country comparison to the world: 130

COMMUNICATIONS

Telephones—fixed lines: *total subscriptions:* 37,305
subscriptions per 100 inhabitants: less than 1 (2019 est.)
country comparison to the world: 164

Telephones—mobile cellular: *total subscriptions:* 10,905,559
subscriptions per 100 inhabitants: 87.7 (2019 est.)
country comparison to the world: 80

Telecommunication systems: *general assessment:* fixed-line network characterized by aging, deteriorating equipment; mobile networks account for almost all Internet connections; govt. to provide telecom services to 80% of the country, mostly via mobile infrastructure by restructuring state-owned telecom companies; (mobile number portability) MNP is available; Benin joins free roaming scheme (2019)
domestic: fixed-line teledensity only about 1 per 100 persons; spurred by the presence of multiple mobile-cellular providers, cellular telephone subscribership has increased rapidly, exceeding 88 per 100 persons (2019)
international: country code—229; landing points for the SAT-3/WASC and ACE fiber-optic submarine cable that provides connectivity to Europe, and most West African countries; satellite earth stations—7 (Intelsat- Atlantic Ocean) (2019)
note: the COVID- 19 outbreak is negatively impacting telecommunications production and supply chains globally; consumer spending on telecom devices and services has also slowed due to the pandemic's effect on economies worldwide; overall progress towards improvements in all facets of the telecom industry—mobile, fixed-line, broadband, submarine cable and satellite—has moderated

Broadcast media: state-run Office de Radiodiffusion et de Television du Benin (ORTB) operates a TV station providing a wide broadcast reach; several privately owned TV stations broadcast from Cotonou; satellite TV subscription service is available; state- owned radio, under ORTB control, includes a national station supplemented by a number of regional stations; substantial number of privately owned radio broadcast stations, transmissions of a few international broadcasters are available on FM in Cotonou (2019)

Internet country code: .bj

Internet users: *total:* 2,403,596

percent of population: 20% (July 2018 est.)
country comparison to the world: 110

Broadband—fixed subscriptions: *total:* 27,034
subscriptions per 100 inhabitants: less than 1 (2018 est.)
country comparison to the world: 145

TRANSPORTATION

National air transport system: *number of registered air carriers:* 1 (2015)
inventory of registered aircraft operated by air carriers: 1 (2015)
annual passenger traffic on registered air carriers: 112,392 (2015)
annual freight traffic on registered air carriers: 805,347 mt-km (2015)

Civil aircraft registration country code prefix: TY (2016)

Airports: 6 (2013)
country comparison to the world: 172

Airports—with paved runways: *total:* 1 (2017)
1,524 to 2,437 m: 1 (2017)

Airports—with unpaved runways: *total:* 5 (2013)
2,438 to 3,047 m: 2 (2013)
1,524 to 2,437 m: 1 (2013)
914 to 1,523 m: 2 (2013)

Pipelines: 134 km gas

Railways: *total:* 438 km (2014)
narrow gauge: 438 km 1.000-m gauge (2014)
country comparison to the world: 116

Roadways: *total:* 16,000 km (2006)
paved: 1,400 km (2006)
unpaved: 14,600 km (2006)
country comparison to the world: 121

Waterways: 150 km (seasonal navigation on River Niger along northern border) (2011)
country comparison to the world: 101

Merchant marine: *total:* 6
by type: other 6 (2019)
country comparison to the world: 162

Ports and terminals: *major seaport(s):* Cotonou
LNG terminal(s) (import): Cotonou

MILITARY AND SECURITY

Military and security forces: Benin Armed Forces (Forces Armees Beninoises, FAB): Army, Navy, Air Force; Ministry of Public Security: Republican Police (2019)

Military expenditures: 0.7% of GDP (2019)
0.86% of GDP (2018)
1.26% of GDP (2017)
1.14% of GDP (2016)
1.1% of GDP (2015)
country comparison to the world: 133

Military and security service personnel strengths: the Benin Armed Forces (FAB) are comprised of approximately 7,200 active duty troops (6,500 Army; 500 Navy; 200 Air Force); est. 5,000 Republican Police (2019)

Military equipment inventories and acquisitions: the FAB is equipped with a mix of foreign-supplied weapons; historically, France and Russia (including the former Soviet Union) have been the chief suppliers of military hardware (2019 est.)

Military deployments: 250 Mali (MINUSMA) (2020)

Military service age and obligation: 18-35 years of age for selective compulsory and voluntary military service; a higher education diploma is required; both sexes are eligible for military service; conscript tour of duty—18 months (2013)

Maritime threats: West African piracy more than doubled in 2018 to become the most dangerous area in the World; the waters off of Benin saw a dramatic increase in 2018 with five attacks reported compared with none in 2017; three ships

were boarded, two were hijacked, and 48 crew taken hostage or kidnapped

Military—note: Benin participates in the Multinational Joint Task Force (MNJTF) against Boko Haram along with Cameroon, Chad, Niger, and Nigeria; the Benin military contingent is in charge of MNJTF garrison duties (2020)

TERRORISM

Terrorist group(s): al-Qa'ida (Jama'at Nusrat al Islam wal Muslimeen); Islamic State in the Greater Sahara (2020)

note: details about the history, aims, leadership, organization, areas of operation, tactics,

targets, weapons, size, and sources of support of the group(s) appear(s) in Appendix-T

TRANSNATIONAL ISSUES

Disputes—international: talks continue between Benin and Togo on funding the Adjrala hydroelectric dam on the Mona River; Benin retains a border dispute with Burkina Faso near the town of Koualou; location of Benin-Niger-Nigeria tripoint is unresolved

Illicit drugs: transshipment point used by traffickers for cocaine destined for Western Europe; vulnerable to money laundering due to poorly enforced financial regulations

BERMUDA

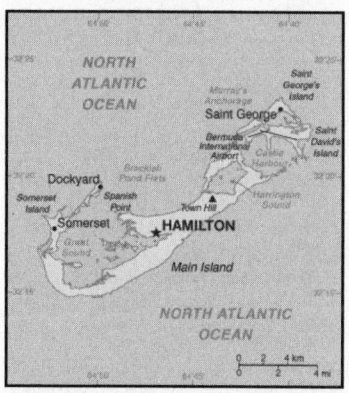

INTRODUCTION

Background: Bermuda was first settled in 1609 by shipwrecked English colonists heading for Virginia. Self-governing since 1620, Bermuda is the oldest and most populous of the British overseas territories. Vacationing to the island to escape North American winters first developed in Victorian times. Tourism continues to be important to the island's economy, although international business has overtaken it in recent years. Bermuda has also developed into a highly successful offshore financial center. A referendum on independence from the UK was soundly defeated in 1995.

GEOGRAPHY

Location: North America, group of islands in the North Atlantic Ocean, east of South Carolina (US)

Geographic coordinates: 32 20 N, 64 45 W

Map references: North America

Area: *total:* 54 sq km
land: 54 sq km

water: 0 sq km
country comparison to the world: 232

Area—comparative: about one-third the size of Washington, DC

Land boundaries: 0 km

Coastline: 103 km

Maritime claims: *territorial sea:* 12 nm
exclusive fishing zone: 200 nm

Climate: subtropical; mild, humid; gales, strong winds common in winter

Terrain: low hills separated by fertile depressions

Elevation: *lowest point:* Atlantic Ocean 0 m
highest point: Town Hill 79 m

Natural resources: limestone, pleasant climate fostering tourism

Land use: *agricultural land:* 14.8% (2011 est.)
arable land: 14.8% (2011 est.) / *permanent crops:* 0% (2011 est.) / *permanent pasture:* 0% (2011 est.)
forest: 20% (2011 est.)
other: 65.2% (2011 est.)
Irrigated land: NA

Population distribution: relatively even population distribution throughout

Natural hazards: hurricanes (June to November)

Environment—current issues: dense population and heavy vehicle traffic create serious congestion and air pollution problems; water resources scarce (most obtained as rainwater or from wells); solid waste disposal; hazardous waste disposal; sewage disposal; overfishing; oil spills

Geography—note: consists of about 138 coral islands and islets with ample rainfall, but no rivers or freshwater lakes; some land was leased by the US Government from 1941 to 1995

PEOPLE AND SOCIETY

Population: 71,750 (July 2020 est.)
country comparison to the world: 203

Nationality: *noun:* Bermudian(s)
adjective: Bermudian

Ethnic groups: African descent 53.8%, white 31%, mixed 7.5%, other 7.1%, unspecified 0.6% (2010 est.)

Languages: English (official), Portuguese

Religions: Protestant 46.2% (includes Anglican 15.8%, African Methodist Episcopal 8.6%, Seventh Day Adventist 6.7, Pentecostal 3.5%, Methodist 2.7%, Presbyterian 2.0%, Church of God 1.6%, Baptist 1.2%, Salvation Army 1.1%, Brethren 1.0%, other Protestant 2.0%), Roman Catholic 14.5%, Jehovah's Witness 1.3%, other Christian 9.1%, Muslim 1%, other 3.9%, none 17.8%, unspecified 6.2% (2010 est.)

Age structure: *0-14 years:* 16.7% (male 6,053/female 5,928)
15-24 years: 11.88% (male 4,290/female 4,235)
25-54 years: 35.31% (male 12,758/female 12,575)
55-64 years: 16.37% (male 5,560/female 6,185)
65 years and over: 19.74% (male 6,032/female 8,134) (2020 est.)

Median age: *total:* 43.6 years
male: 41.6 years
female: 45.7 years (2020 est.)
country comparison to the world: 23

Population growth rate: 0.39% (2020 est.)
country comparison to the world: 161

Birth rate: 11.2 births/1,000 population (2020 est.)
country comparison to the world: 175

Death rate: 9.1 deaths/1,000 population (2020 est.)
country comparison to the world: 57

Net migration rate: 1.6 migrant(s)/1,000 population (2020 est.)
country comparison to the world: 53

Population distribution: relatively even population distribution throughout

Urbanization: *urban population:* 100% of total population (2020)

rate of urbanization: -0.44% annual rate of change (2015-20 est.)

total population growth rate v. urban population growth rate, 2000-2030: Major urban areas—population: 10,000 HAMILTON (capital) (2018)

Sex ratio: *at birth:* 1.02 male(s)/female
0-14 years: 1.02 male(s)/female
15-24 years: 1.01 male(s)/female
25-54 years: 1.01 male(s)/female
55-64 years: 0.9 male(s)/female
65 years and over: 0.74 male(s)/female
total population: 0.94 male(s)/female (2020 est.)

Infant mortality rate: *total:* 2.5 deaths/1,000 live births
male: 2.6 deaths/1,000 live births
female: 2.4 deaths/1,000 live births (2020 est.)
country comparison to the world: 221

Life expectancy at birth: *total population:* 81.7 years
male: 78.5 years
female: 84.9 years (2020 est.)
country comparison to the world: 27

Total fertility rate: 1.91 children born/woman (2020 est.)
country comparison to the world: 128

Drinking water source:
improved:
total: 100% of population
unimproved:
total: 0% of population (2017 est.)

Sanitation facility access:
improved:
urban: 100% of population
total: 100% of population
unimproved:
urban: 0% of population
total: 0% of population (2017)

HIV/AIDS—adult prevalence rate: NA

HIV/AIDS—people living with HIV/AIDS: NA

HIV/AIDS—deaths: NA

Education expenditures: 1.5% of GDP (2017)
country comparison to the world: 171

School life expectancy (primary to tertiary education): *total:* 13 years
male: 12 years
female: 14 years (2015)

Unemployment, youth ages 15-24: *total:* 29.3%
male: 29.7%
female: 29% (2014 est.)
country comparison to the world: 35

GOVERNMENT

Country name: *conventional long form:* none
conventional short form: Bermuda
former: Somers Islands
etymology: the islands making up Bermuda are named after Juan de BERMUDEZ, an early 16th century Spanish sea captain and the first European explorer of the archipelago

Dependency status: overseas territory of the UK

Government type: parliamentary democracy; self-governing overseas territory of the UK

Capital: *name:* Hamilton

geographic coordinates: 32 17 N, 64 47 W
time difference: UTC-4 (1 hour ahead of Washington, DC, during Standard Time)
daylight saving time: +1hr, begins second Sunday in March; ends first Sunday in November
etymology: named after Henry HAMILTON (ca. 1734-1796) who served as governor of Bermuda from 1788-1794

Administrative divisions: 9 parishes and 2 municipalities*; Devonshire, Hamilton, Hamilton*, Paget, Pembroke, Saint George*, Saint George's, Sandys, Smith's, Southampton, Warwick

Independence: none (overseas territory of the UK)

National holiday: Bermuda Day, 24 May; note—formerly known as Victoria Day, Empire Day, and Commonwealth Day

Constitution: *history:* several previous (dating to 1684); latest entered into force 8 June 1968 (Bermuda Constitution Order 1968)
amendments: proposal procedure—NA; passage by an Order in Council in the UK; amended several times, last in 2012

Legal system: English common law

International law organization participation: has not submitted an ICJ jurisdiction declaration; non-party state to the ICCt

Citizenship: *citizenship by birth:* no
citizenship by descent only: at least one parent must be a citizen of the UK
dual citizenship recognized: yes
residency requirement for naturalization: 10 years

Suffrage: 18 years of age; universal

Executive branch: *chief of state:* Queen ELIZABETH II (since 6 February 1952); represented by Governor John RANKIN (since 5 December 2016)

head of government: Premier David BURT (since 19 July 2017)
cabinet: Cabinet nominated by the premier, appointed by the governor
elections/appointments: the monarchy is hereditary; governor appointed by the monarch; following legislative elections, the leader of the majority party or majority coalition usually appointed premier by the governor

Legislative branch: *description:* bicameral Parliament consists of:
Senate (11 seats; 3 members appointed by the governor, 5 by the premier, and 3 by the opposition party; members serve 5-year terms) and the House of Assembly (36 seats; members directly elected in single-seat constituencies by simple majority vote to serve up to 5-year terms)
House of Assembly (36 seats; members directly elected in single-seat constituencies by simple majority vote to serve up to 5-year terms)
elections: Senate—last appointments in August 2017 (next appointments in 2022)

House of Assembly—last held on 1 October 2020 (next to be held not later than 2025)
election results: Senate—composition—men 7, women 4, percent of women 36.4%

House of Assembly—percent of vote by party—PLP 62.1%, OBA 32.3%, other 5.4%, independent 0.2%; seats by party—PLP 30, OBA 6; composition—NA

Judicial branch: *highest courts:* Court of Appeal (consists of the court president and at least 2 justices); Supreme Court (consists of the chief justice, 4 puisne judges, and 1 associate justice); note—the Judicial Committee of the Privy Council (in London) is the court of final appeal
judge selection and term of office: Court of Appeal justice appointed by the governor; justice tenure by individual appointment; Supreme Court judges nominated by the Judicial and Legal Services Commission and appointed by the governor; judge tenure based on terms of appointment
subordinate courts: commercial court (began in 2006); magistrates' courts

Political parties and leaders: Free Democratic Movement or FDM (Marc BEAN)
One Bermuda Alliance or OBA (Craig CANNONIER)
Progressive Labor Party or PLP [Edward D. BURT]

International organization participation: Caricom (associate), ICC (NGOs), Interpol (subbureau), IOC, ITUC (NGOs), UPU, WCO

Diplomatic representation in the US: none (overseas territory of the UK)

Diplomatic representation from the US: *chief of mission:* Consul General Mary Ellen KOENIG (since 28 November 2015)
telephone: (441) 295-1342
embassy: 16 Middle Road Devonshire, DV 03
mailing address: P. O. Box HM325, Hamilton HMBX; American Consulate General Hamilton, US Department of State, 5300 Hamilton Place, Washington, DC 20520-5300
FAX: [1] (441) 295-1592, 296-9233

consulate(s) general: Crown Hill, 16 Middle Road, Devonshire DVO3

Flag description: red, with the flag of the UK in the upper hoist-side quadrant and the Bermudian coat of arms (a white shield with a red lion standing on a green grassy field holding a scrolled shield showing the sinking of the ship Sea Venture off Bermuda in 1609) centered on the outer half of the flag; it was the shipwreck of the vessel, filled with English colonists originally bound for Virginia, that led to the settling of Bermuda
note: the flag is unusual in that it is only British overseas territory that uses a red ensign, all others use blue

National symbol(s): red lion

National anthem: *name:* Hail to Bermuda
lyrics/music: Bette JOHNS
note: serves as a local anthem; as a territory of the United Kingdom, "God Save the Queen" is official (see United Kingdom)

ECONOMY

Economy—overview: International business, which consists primarily of insurance and other financial services, is the real bedrock of Bermuda's economy, consistently accounting for about 85% of the island's GDP. Tourism is the country's second largest industry, accounting for about 5% of Bermuda's GDP but a much larger share of employment. Over 80% of visitors come from the US and the sector struggled in the wake of the global recession of 2008-09. Even the financial sector has lost roughly 5,000 high-paying expatriate jobs since 2008, weighing heavily on household consumption and retail sales. Bermuda must import almost everything. Agriculture and industry are limited due to the small size of the island.

Bermuda's economy returned to negative growth in 2016, reporting a contraction of 0.1% GDP, after growing by 0.6% in 2015. Unemployment reached 7% in 2016 and 2017, public debt is growing and exceeds $2.4 billion, and the government continues to work on attracting foreign investment. Still, Bermuda enjoys one of the highest per capita incomes in the world.

GDP (purchasing power parity): $6.127 billion (2016 est.)
$6.133 billion (2015 est.)
$6.097 billion (2014 est.)
country comparison to the world: 172

GDP (official exchange rate): $6.127 billion (2016 est.)

GDP—real growth rate: -0.1% (2016 est.)
0.6% (2015 est.)
-0.3% (2014 est.)
country comparison to the world: 195

GDP—per capita (PPP): $99,400 (2016 est.)
$95,500 (2015 est.)
$87,500 (2014 est.)
country comparison to the world: 6

GDP—composition, by end use: *household consumption:* 51.3% (2017 est.)
government consumption: 15.7% (2017 est.)
investment in fixed capital: 13.7% (2017 est.)
investment in inventories: 0% (2017 est.)
exports of goods and services: 49.8% (2017 est.)
imports of goods and services: -30.4% (2017 est.)

GDP—composition, by sector of origin: *agriculture:* 0.9% (2017 est.)
industry: 5.3% (2017 est.)
services: 93.8% (2017 est.)

Agriculture—products: bananas, vegetables, citrus, flowers; dairy products, honey

Industries: international business, tourism, light manufacturing

Industrial production growth rate: 2% (2017 est.)
country comparison to the world: 129

Labor force: 33,480 (2016 est.)
country comparison to the world: 202

Labor force—by occupation: *agriculture:* 2%
industry: 13%
services: 85% (2016 est.)

Unemployment rate: 7% (2017 est.)

7% (2016 est.)
country comparison to the world: 112

Population below poverty line: 11% (2008 est.)

Household income or consumption by percentage share: *lowest 10%:* NA
highest 10%: NA

Budget: *revenues:* 999.2 million (2017 est.)
expenditures: 1.176 billion (2017 est.)

Taxes and other revenues: 16.3% (of GDP) (2017 est.)
country comparison to the world: 183

Budget surplus (+) or deficit (-): -2.9% (of GDP) (2017 est.)
country comparison to the world: 127

Public debt: 43% of GDP (FY14/15)
country comparison to the world: 117

Fiscal year: 1 April—31 March

Inflation rate (consumer prices): 1.9% (2017 est.)
1.4% (2016 est.)
country comparison to the world: 96

Current account balance: $818.6 million (2017 est.)
$763 million (2016 est.)
country comparison to the world: 52

Exports: $19 million (2017 est.)
$19 million (2016 est.)
country comparison to the world: 210

Exports—partners: Jamaica 49.1%, Luxembourg 36.1%, US 4.9% (2017)

Exports—commodities: reexports of pharmaceuticals

Imports: $1.094 billion (2017 est.)
$980 million (2016 est.)
country comparison to the world: 184

Imports—commodities: clothing, fuels, machinery and transport equipment, construction materials, chemicals, food and live animals

Imports—partners: US 72.1%, South Korea 9.7%, Canada 4.2% (2017)

Debt—external: $2.515 billion (2017 est.)
$2.435 billion (2015 est.)
country comparison to the world: 149

Exchange rates: Bermudian dollars (BMD) per US dollar—
1 (2017 est.)
1 (2016 est.)
1 (2015 est.)
1 (2014 est.)
1 (2013 est.)

ENERGY

Electricity access: *electrification—total population:* 100% (2020)

Electricity—production: 650 million kWh (2016 est.)
country comparison to the world: 159

Electricity—consumption: 604.5 million kWh (2016 est.)
country comparison to the world: 166

Electricity—exports: 0 kWh (2016 est.)
country comparison to the world: 107

Electricity—imports: 0 kWh (2016 est.)
country comparison to the world: 126

Electricity—installed generating capacity: 171,000 kW (2016 est.)
country comparison to the world: 169

Electricity—from fossil fuels: 100% of total installed capacity (2016 est.)
country comparison to the world: 3

Electricity—from nuclear fuels: 0% of total installed capacity (2017 est.)
country comparison to the world: 50

Electricity—from hydroelectric plants: 0% of total installed capacity (2017 est.)
country comparison to the world: 158

Electricity—from other renewable sources: 0% of total installed capacity (2017 est.)
note: the Tynes Bay Waste Treatment Facility turns waste to electric energy
country comparison to the world: 176

Crude oil—production: 0 bbl/day (2018 est.)
country comparison to the world: 111

Crude oil—exports: 0 bbl/day (2015 est.)
country comparison to the world: 94

Crude oil—imports: 0 bbl/day (2015 est.)
country comparison to the world: 97

Crude oil—proved reserves: 0 bbl (1 January 2018 est.)
country comparison to the world: 107

Refined petroleum products—production: 0 bbl/day (2017 est.)
country comparison to the world: 119

Refined petroleum products—consumption: 5,000 bbl/day (2016 est.)
country comparison to the world: 178

Refined petroleum products—exports: 0 bbl/day (2015 est.)
country comparison to the world: 131

Refined petroleum products—imports: 3,939 bbl/day (2015 est.)
country comparison to the world: 178

Natural gas—production: 0 cu m (2017 est.)
country comparison to the world: 105

Natural gas—consumption: 0 cu m (2017 est.)
country comparison to the world: 122

Natural gas—exports: 0 cu m (2017 est.)
country comparison to the world: 70

Natural gas—imports: 0 cu m (2017 est.)
country comparison to the world: 92

Natural gas—proved reserves: 0 cu m (1 January 2014 est.)
country comparison to the world: 111

Carbon dioxide emissions from consumption of energy: 793,700 Mt (2017 est.)
country comparison to the world: 174

COMMUNICATIONS

Telephones—fixed lines: *total subscriptions:* 24,808
subscriptions per 100 inhabitants: 34.71 (2019 est.)
country comparison to the world: 171

105

Telephones—mobile cellular: *total subscriptions:* 73,680
subscriptions per 100 inhabitants: 103.09 (2019 est.)
country comparison to the world: 199

Telecommunication systems: *general assessment:* a good, fully automatic digital telephone system with fiber-optic trunk lines; telecom sector provides a relatively high contribution to overall GDP; numerous competitors licensed, but small and localized; telecom sector a growth area across the Caribbean (2020)
domestic: the system has a high fixed-line teledensity 35 per 100, coupled with a mobile-cellular teledensity of roughly 103 per 100 persons (2019)
international: country code—1-441; landing points for the GlobeNet, Gemini Bermuda, CBUS, and the CB-1 submarine cables to the Caribbean, South America and the US; satellite earth stations—3 (2019)
note: the COVID-19 outbreak is negatively impacting telecommunications production and supply chains globally; consumer spending on telecom devices and services has also slowed due to the

pandemic's effect on economies worldwide; overall progress towards improvements in all facets of the telecom industry—mobile, fixed-line, broadband, submarine cable and satellite—has moderated

Broadcast media: 3 TV stations; cable and satellite TV subscription services are available; roughly 13 radio stations operating

Internet country code: .bm

Internet users: *total:* 70,016

percent of population: 98.37% (July 2018 est.)
country comparison to the world: 187

TRANSPORTATION

Civil aircraft registration country code prefix: VP-B (2016)

Airports: 1 (2020)
country comparison to the world: 214

Airports—with paved runways: *total:* 1 (2019)
2,438 to 3,047 m: 1

Roadways: *total:* 447 km (2010)
paved: 447 km (2010)
note: 225 km public roads; 222 km private roads

country comparison to the world: 197

Merchant marine: *total:* 148
by type: bulk carrier 2, container ship 11, oil tanker 18, other 117 (2019)
country comparison to the world: 72

Ports and terminals: *major seaport(s):* Hamilton, Ireland Island, Saint George

MILITARY AND SECURITY

Military and security forces: Royal Bermuda Regiment (2019)

Military service age and obligation: 18-45 years of age for voluntary male or female enlistment in the Bermuda Regiment; males must register at age 18 and may be subject to conscription; term of service is 38 months for volunteers or conscripts (2012)

Military—note: defense is the responsibility of the UK

TRANSNATIONAL ISSUES

Disputes—international: none

BHUTAN

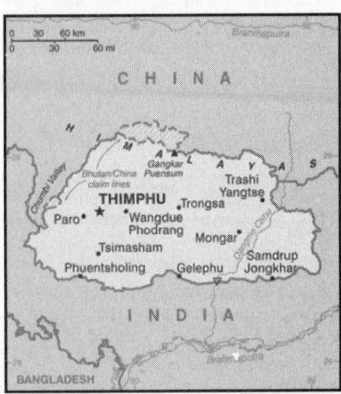

INTRODUCTION

Background: Following Britain's victory in the 1865 Duar War, Britain and Bhutan signed the Treaty of Sinchulu, under which Bhutan would receive an annual subsidy in exchange for ceding land to British India. Ugyen WANGCHUCK—who had served as the de facto ruler of an increasingly unified Bhutan and had improved relations with the British toward the end of the 19th century—was named king in 1907. Three years later, a treaty was signed whereby the British agreed not to interfere in Bhutanese internal affairs, and Bhutan allowed Britain to direct its foreign affairs. Bhutan negotiated a similar arrangement with independent India in 1949. The Indo-Bhutanese Treaty

of Friendship returned to Bhutan a small piece of the territory annexed by the British, formalized the annual subsidies the country received, and defined India's responsibilities in defense and foreign relations. Under a succession of modernizing monarchs beginning in the 1950s, Bhutan joined the UN in 1971 and slowly continued its engagement beyond its borders.
In 2005, King Jigme Singye WANGCHUCK unveiled the draft of Bhutan's first constitution—which introduced major democratic reforms—and held a national referendum for its approval. The King abdicated the throne in 2006 in favor of his son, Jigme Khesar Namgyel WANGCHUCK. In 2007, India and Bhutan renegotiated their treaty, eliminating the clause that stated that Bhutan would be "guided by" India in conducting its foreign policy, although Thimphu continues to coordinate closely with New Delhi. In 2008, Bhutan held its first parliamentary election in accordance with the constitution. Bhutan experienced a peaceful turnover of power following a parliamentary election in 2013, which resulted in the defeat of the incumbent party. In 2018, the incumbent party again lost the parliamentary election. Of the more than 100,000 ethnic Nepali—predominantly Lhotshampa—refugees who fled or were forced out of Bhutan in the 1990s, about 6,500 remain displaced in Nepal.

GEOGRAPHY

Location: Southern Asia, between China and India
Geographic coordinates: 27 30 N, 90 30 E
Map references: Asia

Area: *total:* 38,394 sq km
land: 38,394 sq km
water: 0 sq km
country comparison to the world: 137

Area—comparative: slightly larger than Maryland; about one-half the size of Indiana

Land boundaries: *total:* 1,136 km
border countries (2): China 477 km, India 659 km

Coastline: 0 km (landlocked)

Maritime claims: none (landlocked)

Climate: varies; tropical in southern plains; cool winters and hot summers in central valleys; severe winters and cool summers in Himalayas

Terrain: mostly mountainous with some fertile valleys and savanna

Elevation: *mean elevation:* 2,220 m
lowest point: Drangeme Chhu 97 m
highest point: Gangkar Puensum 7,570 m

Natural resources: timber, hydropower, gypsum, calcium carbonate

Land use: *agricultural land:* 13.6% (2011 est.)
arable land: 2.6% (2011 est.) / permanent crops: 0.3% (2011 est.) / permanent pasture: 10.7% (2011 est.)
forest: 85.5% (2011 est.)
other: 0.9% (2011 est.)
Irrigated land: 320 sq km (2012)

Natural hazards: violent storms from the Himalayas are the source of the country's Bhutanese name, which translates as Land of the Thunder Dragon; frequent landslides during the rainy season

Environment—current issues: soil erosion; limited access to potable water; wildlife conservation; industrial pollution; waste disposal

Environment—international agreements: *party to:* Biodiversity, Climate Change, Climate Change-Kyoto Protocol, Desertification, Endangered Species, Hazardous Wastes, Ozone Layer Protection

signed, but not ratified: Law of the Sea

Geography—note: landlocked; strategic location between China and India; controls several key Himalayan mountain passes

PEOPLE AND SOCIETY

Population: 782,318 (July 2020 est.)
country comparison to the world: 165

Nationality: *noun:* Bhutanese (singular and plural)
adjective: Bhutanese

Ethnic groups: Ngalop (also known as Bhote) 50%, ethnic Nepali 35% (predominantly Lhotshampas), indigenous or migrant tribes 15%

Languages: Sharchhopka 28%, Dzongkha (official) 24%, Lhotshamkha 22%, other 26% (includes foreign languages) (2005 est.)

Religions: Lamaistic Buddhist 75.3%, Indian- and Nepali-influenced Hinduism 22.1%, other 2.6% (2005 est.)

Age structure: *0-14 years:* 24.52% (male 98,113/ female 93,740)
15-24 years: 17.77% (male 70,768/female 68,211)
25-54 years: 44.72% (male 184,500/female 165,374)
55-64 years: 6.39% (male 26,714/female 23,280)
65 years and over: 6.6% (male 26,797/female 24,821) (2020 est.)

Dependency ratios: *total dependency ratio:* 45.1
youth dependency ratio: 36.1
elderly dependency ratio: 9
potential support ratio: 11.1 (2020 est.)

Median age: *total:* 29.1 years
male: 29.6 years
female: 28.6 years (2020 est.)
country comparison to the world: 136

Population growth rate: 1.02% (2020 est.)
country comparison to the world: 103

Birth rate: 16.3 births/1,000 population (2020 est.)
country comparison to the world: 106

Death rate: 6.3 deaths/1,000 population (2020 est.)
country comparison to the world: 147

Net migration rate: 0 migrant(s)/1,000 population (2020 est.)
country comparison to the world: 78

Urbanization: *urban population:* 42.3% of total population (2020)
rate of urbanization: 2.98% annual rate of change (2015-20 est.)

total population growth rate v. urban population growth rate, 2000-2030: Major urban areas—population: 203,000 THIMPHU (capital) (2018)

Sex ratio: *at birth:* 1.05 male(s)/female
0-14 years: 1.05 male(s)/female

15-24 years: 1.04 male(s)/female
25-54 years: 1.12 male(s)/female
55-64 years: 1.15 male(s)/female
65 years and over: 1.08 male(s)/female
total population: 1.08 male(s)/female (2020 est.)

Maternal mortality rate: 183 deaths/100,000 live births (2017 est.)
country comparison to the world: 51

Infant mortality rate: *total:* 27 deaths/1,000 live births
male: 27.1 deaths/1,000 live births
female: 27 deaths/1,000 live births (2020 est.)
country comparison to the world: 65

Life expectancy at birth: *total population:* 72.1 years
male: 71 years
female: 73.2 years (2020 est.)
country comparison to the world: 154

Total fertility rate: 1.82 children born/woman (2020 est.)
country comparison to the world: 147

Contraceptive prevalence rate: 65.6% (2010)

Drinking water source:
improved:
urban: 99.3% of population
rural: 100% of population
total: 99.7% of population
unimproved:
urban: 0.7% of population
rural: 0% of population
total: 0.3% of population (2017 est.)

Current Health Expenditure: 3.2% (2017)

Physicians density: 0.4 physicians/1,000 population (2017)

Hospital bed density: 1.7 beds/1,000 population (2012)

Sanitation facility access:
improved:
urban: 87.5% of population
rural: 72.1% of population
total: 78.3% of population
unimproved:
urban: 12.5% of population
rural: 27.9% of population
total: 21.7% of population (2017 est.)

HIV/AIDS—adult prevalence rate: 0.3% (2018)
country comparison to the world: 84

HIV/AIDS—people living with HIV/AIDS: 1,300 (2018)
country comparison to the world: 140

HIV/AIDS—deaths: <100 (2018)

Obesity—adult prevalence rate: 6.4% (2016)
country comparison to the world: 167

Children under the age of 5 years underweight: 12.7% (2010)
country comparison to the world: 49

Education expenditures: 6.6% of GDP (2018)
country comparison to the world: 18

Literacy: *definition:* age 15 and over can read and write
total population: 66.6%
male: 75%

female: 57.1% (2017)

School life expectancy (primary to tertiary education): *total:* 13 years
male: 13 years
female: 14 years (2018)

Unemployment, youth ages 15-24: *total:* 10.7%
male: 8.2%
female: 12.7% (2015 est.)
country comparison to the world: 119

GOVERNMENT

Country name: *conventional long form:* Kingdom of Bhutan
conventional short form: Bhutan
local long form: Druk Gyalkhap
local short form: Druk Yul
etymology: named after the Bhotia, the ethnic Tibetans who migrated from Tibet to Bhutan; "Bod" is the Tibetan name for their land; the Bhutanese name "Druk Yul" means "Land of the Thunder Dragon"

Government type: constitutional monarchy

Capital: *name:* Thimphu

geographic coordinates: 27 28 N, 89 38 E
time difference: UTC+6 (11 hours ahead of Washington, DC, during Standard Time)
etymology: the origins of the name are unclear; the traditional explanation, dating to the 14th century, is that "thim" means "dissolve" and "phu" denotes "high ground" to express the meaning of "dissolving high ground," in reference to a local deity that dissolved before a traveler's eyes, becoming a part of the rock on which the present city stands

Administrative divisions: 20 districts (dzongkhag, singular and plural); Bumthang, Chhukha, Dagana, Gasa, Haa, Lhuentse, Mongar, Paro, Pemagatshel, Punakha, Samdrup Jongkhar, Samtse, Sarpang, Thimphu, Trashigang, Trashi Yangtse, Trongsa, Tsirang, Wangdue Phodrang, Zhemgang

Independence: 17 December 1907 (became a unified kingdom under its first hereditary king); 8 August 1949 (Treaty of Friendship with India maintains Bhutanese independence)

National holiday: National Day (Ugyen WANGCHUCK became first hereditary king), 17 December (1907)

Constitution: *history:* previous governing documents were various royal decrees; first constitution drafted November 2001 to March 2005, ratified 18 July 2008
amendments: proposed as a motion by simple majority vote in a joint session of Parliament; passage requires at least a three-fourths majority vote in a joint session of the next Parliament and assent by the king

Legal system: civil law based on Buddhist religious law

International law organization participation: has not submitted an ICJ jurisdiction declaration; non-party state to the ICCt

Citizenship: *citizenship by birth:* no

107

THE CIA WORLD FACTBOOK

citizenship by descent only: the father must be a citizen of Bhutan

dual citizenship recognized: no

residency requirement for naturalization: 10 years

Suffrage: 18 years of age; universal

Executive branch: *chief of state:* King Jigme Khesar Namgyel WANGCHUCK (since 14 December 2006); note—King Jigme Singye WANGCHUCK abdicated the throne on 14 December 2006 to his son

head of government: Prime Minister Lotay TSHERING (since 7 November 2018)

cabinet: Council of Ministers or Lhengye Zhungtshog members nominated by the monarch in consultation with the prime minister and approved by the National Assembly; members serve 5-year terms

elections/appointments: the monarchy is hereditary but can be removed by a two-thirds vote of Parliament; leader of the majority party in Parliament is nominated as the prime minister, appointed by the monarch

Legislative branch: *description:* bicameral Parliament or Chi Tshog consists of:

non-partisan National Council or Gyelyong Tshogde (25 seats; 20 members directly elected in single-seat constituencies by simple majority vote and 5 members appointed by the king; members serve 5-year terms)

National Assembly or Tshogdu (47 seats; members directly elected in single-seat constituencies by proportional representation vote to serve 5-year terms)

elections: National Council election last held on 20 April 2018 (next to be held in 2023)

National Assembly—first round held on 15 September 2018 and second round held on 18 October 2018 (next to be held in 2023)

election results: National Council—seats by party—independent 20 (all candidates ran as independents); composition—men 23, women 2, percent of women 8%

National Assembly—first round—percent of vote by party—DNT 31.9%, DPT 30.9%, PDP 27.4%, BKP 9.8%; second round—percent of vote by party—NA; seats by party—DNT 30, DPT 17; composition—men 40, women 7, percent of women 14.9%; note—total Parliament percent of women 12.5%

Judicial branch: *highest courts:* Supreme Court (consists of the chief justice and 4 associate justices); note—the Supreme Court has sole jurisdiction in constitutional matters

judge selection and term of office: Supreme Court chief justice appointed by the monarch upon the advice of the National Judicial Commission, a 4-member body to include the Legislative Committee of the National Assembly, the attorney general, the Chief Justice of Bhutan and the senior Associate Justice of the Supreme Court; other judges (drangpons) appointed by the monarch from among the High Court judges selected by the National Judicial Commission; chief justice serves a 5-year term or until reaching age 65

years, whichever is earlier; the 4 other judges serve 10-year terms or until age 65, whichever is earlier

subordinate courts: High Court (first appellate court); District or Dzongkhag Courts; sub-district or Dungkhag Courts

Political parties and leaders: Bhutan Kuen-Nyam Party or BKP

Bhutan Peace and Prosperity Party (Druk Phuensum Tshogpa) or DPT [Pema GYAMTSHO] (Druk Chirwang Tshogpa or DCT merged with DPT in March 2018)

People's Democratic Party or PDP [Tshering TOBGAY]

United Party of Bhutan (Druk Nyamrup Tshogpa) or DNT [Lotay TSHERING]

International organization participation: ADB, BIMSTEC, CP, FAO, G-77, IBRD, ICAO, IDA, IFAD, IFC, IMF, Interpol, IOC, IOM (observer), IPU, ISO (correspondent), ITSO, ITU, MIGA, NAM, OPCW, SAARC, SACEP, UN, UNCTAD, UNESCO, UNIDO, UNTSO, UNWTO, UPU, WCO, WHO, WIPO, WMO, WTO (observer)

Diplomatic representation in the US: *chief of mission:* none; the Permanent Mission to the UN for Bhutan has consular jurisdiction in the US; the permanent representative to the UN is Doma TSHERING (since 13 September 2017); address: 343 East 43rd Street, New York, NY 10017; telephone [1] (212) 682-2268; FAX [1] (212) 661-0551

consulate(s) general: New York

Diplomatic representation from the US: none; frequent informal contact is maintained via the US embassy in New Delhi (India) and Bhutan's Permanent Mission to the UN

Flag description: divided diagonally from the lower hoist-side corner; the upper triangle is yellow and the lower triangle is orange; centered along the dividing line is a large black and white dragon facing away from the hoist side; the dragon, called the Druk (Thunder Dragon), is the emblem of the nation; its white color stands for purity and the jewels in its claws symbolize wealth; the background colors represent spiritual and secular powers within Bhutan: the orange is associated with Buddhism, while the yellow denotes the ruling dynasty

National symbol(s): thunder dragon known as Druk Gyalpo; national colors: orange, yellow

National anthem: *name:* "Druk tsendhen" (The Thunder Dragon Kingdom)

lyrics/music: Gyaldun Dasho Thinley DORJI/Aku TONGMI

note: adopted 1953

ECONOMY

Economy—overview: Bhutan's small economy is based largely on hydropower, agriculture, and forestry, which provide the main livelihood for more than half the population. Because rugged mountains dominate the terrain and make the building of roads and other infrastructure difficult and expensive, industrial production is primarily of

the cottage industry type. The economy is closely aligned with India's through strong trade and monetary links and is dependent on India for financial assistance and migrant laborers for development projects, especially for road construction. Bhutan signed a pact in December 2014 to expand duty-free trade with Bangladesh.

Multilateral development organizations administer most educational, social, and environment programs, and take into account the government's desire to protect the country's environment and cultural traditions. For example, the government is cautious in its expansion of the tourist sector, restricting visits to environmentally conscientious tourists. Complicated controls and uncertain policies in areas such as industrial licensing, trade, labor, and finance continue to hamper foreign investment.

Bhutan's largest export—hydropower to India—could spur sustainable growth in the coming years if Bhutan resolves chronic delays in construction. Bhutan's hydropower exports comprise 40% of total exports and 25% of the government's total revenue. Bhutan currently taps only 6.5% of its 24,000-megawatt hydropower potential and is behind schedule in

building 12 new hydropower dams with a combined capacity of 10,000 megawatts by 2020 in accordance with a deal signed in 2008 with India. The high volume of imported materials to build hydropower plants has expanded Bhutan's trade and current account deficits. Bhutan also signed a memorandum of understanding with Bangladesh and India in July 2017 to jointly construct a new hydropower plant for exporting electricity to Bangladesh.

GDP (purchasing power parity): $7.205 billion (2017 est.)

$6.71 billion (2016 est.)

$6.252 billion (2015 est.)

note: data are in 2017 dollars

country comparison to the world: 167

GDP (official exchange rate): $2.405 billion (2017 est.)

GDP—real growth rate: 7.4% (2017 est.)

7.3% (2016 est.)

6.2% (2015 est.)

country comparison to the world: 14

GDP—per capita (PPP): $9,000 (2017 est.)

$8,500 (2016 est.)

$8,000 (2015 est.)

note: data are in 2017 dollars

country comparison to the world: 145

Gross national saving: 40.4% of GDP (2017 est.)

33.3% of GDP (2016 est.)

32% of GDP (2015 est.)

country comparison to the world: 8

GDP—composition, by end use: *household consumption:* 58% (2017 est.)

government consumption: 16.8% (2017 est.)

investment in fixed capital: 47.2% (2017 est.)

investment in inventories: 0% (2017 est.)

exports of goods and services: 26% (2017 est.)

imports of goods and services: -48% (2017 est.)

GDP—composition, by sector of origin: *agriculture:* 16.2% (2017 est.)
industry: 41.8% (2017 est.)
services: 42% (2017 est.)

Agriculture—products: rice, corn, root crops, citrus; dairy products, eggs

Industries: cement, wood products, processed fruits, alcoholic beverages, calcium carbide, tourism

Industrial production growth rate: 6.3% (2017 est.)
country comparison to the world: 36

Labor force: 397,900 (2017 est.)
note: major shortage of skilled labor
country comparison to the world: 159

Labor force—by occupation: *agriculture:* 58%
industry: 20%
services: 22% (2015 est.)

Unemployment rate: 3.2% (2017 est.)
3.2% (2016 est.)
country comparison to the world: 44

Population below poverty line: 12% (2012 est.)

Household income or consumption by percentage share: *lowest 10%:* 2.8%
highest 10%: 30.6% (2012)

Budget: *revenues:* 655.3 million (2017 est.)
expenditures: 737.4 million (2017 est.)
note: the Government of India finances nearly one-quarter of Bhutan's budget expenditures

Taxes and other revenues: 27.2% (of GDP) (2017 est.)
country comparison to the world: 101

Budget surplus (+) or deficit (-): -3.4% (of GDP) (2017 est.)
country comparison to the world: 142

Public debt: 106.3% of GDP (2017 est.)
114.2% of GDP (2016 est.)
country comparison to the world: 12

Fiscal year: 1 July—30 June

Inflation rate (consumer prices): 5.8% (2017 est.)
7.6% (2016 est.)
country comparison to the world: 183

Current account balance: -$547 million (2017 est.)
-$621 million (2016 est.)
country comparison to the world: 123

Exports: $554.6 million (2017 est.)
$495.3 million (2016 est.)
country comparison to the world: 173

Exports—partners: India 95.3% (2017)

Exports—commodities: electricity (to India), ferrosilicon, cement, cardamom, calcium carbide, steel rods/bars, dolomite, gypsum

Imports: $1.025 billion (2017 est.)
$1.03 billion (2016 est.)
country comparison to the world: 185

Imports—commodities: fuel and lubricants, airplanes, machinery and parts, rice, motor vehicles

Imports—partners: India 89.5% (2017)

Reserves of foreign exchange and gold: $1.206 billion (31 December 2017 est.)
$1.127 billion (31 December 2016 est.)

country comparison to the world: 129

Debt—external: $2.671 billion (31 December 2017 est.)
$2.355 billion (31 December 2016 est.)
country comparison to the world: 146

Exchange rates: ngultrum (BTN) per US dollar—
64.97 (2017 est.)
67.2 (2016 est.)
67.2 (2015 est.)
64.15 (2014 est.)
61.03 (2013 est.)

ENERGY

Electricity access: *electrification—total population:* 100% (2020)

Electricity—production: 7.883 billion kWh (2016 est.)
country comparison to the world: 110

Electricity—consumption: 2.184 billion kWh (2016 est.)
country comparison to the world: 141

Electricity—exports: 5.763 billion kWh (2016 est.)
country comparison to the world: 32

Electricity—imports: 84 million kWh (2016 est.)
country comparison to the world: 101

Electricity—installed generating capacity: 1.632 million kW (2016 est.)
country comparison to the world: 120

Electricity—from fossil fuels: 1% of total installed capacity (2016 est.)
country comparison to the world: 211

Electricity—from nuclear fuels: 0% of total installed capacity (2017 est.)
country comparison to the world: 51

Electricity—from hydroelectric plants: 99% of total installed capacity (2017 est.)
country comparison to the world: 2

Electricity—from other renewable sources: 0% of total installed capacity (2017 est.)
country comparison to the world: 177

Crude oil—production: 0 bbl/day (2018 est.)
country comparison to the world: 112

Crude oil—exports: 0 bbl/day (2015 est.)
country comparison to the world: 95

Crude oil—imports: 0 bbl/day (2015 est.)
country comparison to the world: 98

Crude oil—proved reserves: 0 bbl (1 January 2018 est.)
country comparison to the world: 108

Refined petroleum products—production: 0 bbl/day (2017 est.)
country comparison to the world: 120

Refined petroleum products—consumption: 3,000 bbl/day (2016 est.)
country comparison to the world: 188

Refined petroleum products—exports: 0 bbl/day (2015 est.)
country comparison to the world: 132

Refined petroleum products—imports: 3,120 bbl/day (2015 est.)

country comparison to the world: 183

Natural gas—production: 0 cu m (2017 est.)
country comparison to the world: 106

Natural gas—consumption: 0 cu m (2017 est.)
country comparison to the world: 123

Natural gas—exports: 0 cu m (2017 est.)
country comparison to the world: 71

Natural gas—imports: 0 cu m (2017 est.)
country comparison to the world: 93

Natural gas—proved reserves: 0 cu m (2016 est.)
country comparison to the world: 112

Carbon dioxide emissions from consumption of energy: 604,900 Mt (2017 est.)
country comparison to the world: 181

COMMUNICATIONS

Telephones—fixed lines: *total subscriptions:* 21,916
subscriptions per 100 inhabitants: 2.83 (2019 est.)
country comparison to the world: 174

Telephones—mobile cellular: *total subscriptions:* 740,026
subscriptions per 100 inhabitants: 95.56 (2019 est.)
country comparison to the world: 165

Telecommunication systems: *general assessment:* 4G platforms now gaining traction; 4G/WiMAX networks now cover well over half of the country; fixed broadband penetration remains very low, due to the preeminence of the mobile platform; low to moderate growth is expected from this small base with a maturing mobile subscriber market (2020)
domestic: 3 to 100 fixed-line, 96 to 100 mobile cellular; domestic service inadequate, notably in rural areas (2019)
international: country code—975; international telephone and telegraph service via landline and microwave relay through India; satellite earth station—1 Intelsat
note: the COVID-19 outbreak is negatively impacting telecommunications production and supply chains globally; consumer spending on telecom devices and services has also slowed due to the pandemic's effect on economies worldwide; overall progress towards improvements in all facets of the telecom industry—mobile, fixed-line, broadband, submarine cable and satellite—has moderated

Broadcast media: state-owned TV station established in 1999; cable TV service offers dozens of Indian and other international channels; first radio station, privately launched in 1973, is now state-owned; 5 private radio stations are currently broadcasting (2012)

Internet country code: .bt

Internet users: *total:* 368,714

percent of population: 48.11% (July 2018 est.)
country comparison to the world: 160

Broadband—fixed subscriptions: *total:* 10,802
subscriptions per 100 inhabitants: 1 (2018 est.)
country comparison to the world: 169

TRANSPORTATION

National air transport system: *number of registered air carriers:* 2 (2020)
inventory of registered aircraft operated by air carriers: 6
annual passenger traffic on registered air carriers: 275,849 (2018)
annual freight traffic on registered air carriers: 690,000 mt-km (2018)

Civil aircraft registration country code prefix: A5 (2016)

Airports: 2 (2013)
country comparison to the world: 197

Airports—with paved runways: *total:* 2 (2017)
1,524 to 2,437 m: 1 (2017)
914 to 1,523 m: 1 (2017)

Airports—with unpaved runways: *total:* 1 (2012)
914 to 1,523 m: 1 (2012)

Roadways: *total:* 12,205 km (2017)
urban: 437 km (2017)
country comparison to the world: 132

MILITARY AND SECURITY

Military and security forces: Royal Bhutan Army (includes Royal Bodyguard, plus militia); Ministry of Home and Cultural Affairs: Royal Bhutan Police (2019)
note: Bhutan does not have an air force; India is responsible for military training, arms supplies, and the air defense of Bhutan

Military and security service personnel strengths: the Royal Bhutan Army has approximately 8,000 personnel (2019 est.)

Military equipment inventories and acquisitions: India has provided most of the Royal Bhutan Army's equipment, although the only recorded delivery of military equipment to Bhutan since 2010 was from France (2019 est.)

Military service age and obligation: 18 years of age for voluntary military service; no conscription; militia training is compulsory for males aged 20-25, over a 3-year period (2012)

TRANSNATIONAL ISSUES

Disputes—international: lacking any treaty describing the boundary, Bhutan and China continue negotiations to establish a common boundary alignment to resolve territorial disputes arising from substantial cartographic discrepancies, the most contentious of which lie in Bhutan's west along China's Chumbi salient

BOLIVIA

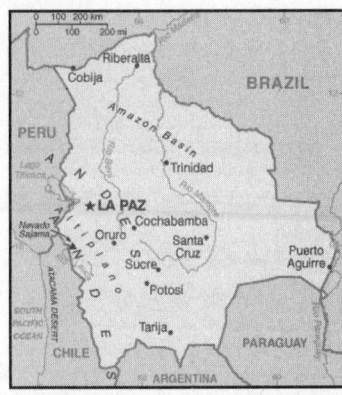

INTRODUCTION

Background: Bolivia, named after independence fighter Simon BOLIVAR, broke away from Spanish rule in 1825; much of its subsequent history has consisted of a series of coups and countercoups, with the last coup occurring in 1978. Democratic civilian rule was established in 1982, but leaders have faced difficult problems of deep-seated poverty, social unrest, and illegal drug production.

In December 2005, Bolivians elected Movement Toward Socialism leader Evo MORALES president—by the widest margin of any leader since the restoration of civilian rule in 1982—after he ran on a promise to change the country's traditional political class and empower the nation's poor, indigenous majority. In December 2009 and October 2014, President MORALES easily won reelection. His party maintained control of the legislative branch of

the government, which has allowed him to continue his process of change. In February 2016, MORALES narrowly lost a referendum to approve a constitutional amendment that would have allowed him to compete in the 2019 presidential election. However, a 2017 Supreme Court ruling stating that term limits violate human rights provided the justification for MORALES to be chosen by his party to run again in 2019. MORALES attempted to claim victory in the 20 October 2019 election, but widespread allegations of electoral fraud, rising violence, and pressure from the military ultimately forced him to flee the country. An interim government, led by President Jeanine ANEZ Chavez, prepared new elections that took place on 18 October 2020.

GEOGRAPHY

Location: Central South America, southwest of Brazil

Geographic coordinates: 17 00 S, 65 00 W

Map references: South America

Area: *total:* 1,098,581 sq km
land: 1,083,301 sq km
water: 15,280 sq km
country comparison to the world: 29

Area—comparative: slightly less than three times the size of Montana

Land boundaries: *total:* 7,252 km
border countries (5): Argentina 942 km, Brazil 3403 km, Chile 942 km, Paraguay 753 km, Peru 1212 km

Coastline: 0 km (landlocked)

Maritime claims: none (landlocked)

Climate: varies with altitude; humid and tropical to cold and semiarid

Terrain: rugged Andes Mountains with a highland plateau (Altiplano), hills, lowland plains of the Amazon Basin

Elevation: *mean elevation:* 1,192 m
lowest point: Rio Paraguay 90 m
highest point: Nevado Sajama 6,542 m

Natural resources: tin, natural gas, petroleum, zinc, tungsten, antimony, silver, iron, lead, gold, timber, hydropower

Land use: *agricultural land:* 34.3% (2011 est.)
arable land: 3.6% (2011 est.) / permanent crops: 0.2% (2011 est.) / permanent pasture: 30.5% (2011 est.)
forest: 52.5% (2011 est.)
other: 13.2% (2011 est.)
Irrigated land: 3,000 sq km (2012)

Population distribution: a high altitude plain in the west between two cordillera of the Andes, known as the Altiplano, is the focal area for most of the population; a dense settlement pattern is also found in and around the city of Santa Cruz, located on the eastern side of the Andes

Natural hazards: flooding in the northeast (March to April)
volcanism: volcanic activity in Andes Mountains on the border with Chile; historically active volcanoes in this region are Irruputuncu (5,163 m), which last erupted in 1995, and the Olca-Paruma volcanic complex (5,762 m to 5,167 m)

Environment—current issues: the clearing of land for agricultural purposes and the international demand for tropical timber are contributing to deforestation; soil erosion from overgrazing and poor cultivation methods (including slash-and-burn agriculture); desertification; loss of biodiversity; industrial pollution of water supplies used for drinking and irrigation

Environment—international agreements: *party to:* Biodiversity, Climate Change, Climate

Change-Kyoto Protocol, Desertification, Endangered Species, Hazardous Wastes, Law of the Sea, Marine Dumping, Ozone Layer Protection, Ship Pollution, Tropical Timber 83, Tropical Timber 94, Wetlands

signed, but not ratified: Environmental Modification, Marine Life Conservation

Geography—note: *note 1:* landlocked; shares control of Lago Titicaca, world's highest navigable lake (elevation 3,805 m), with Peru
note 2: the southern regions of Peru and the extreme northwestern part of Bolivia are considered to be the place of origin for the common potato

PEOPLE AND SOCIETY

Population: 11,639,909 (July 2020 est.)
country comparison to the world: 81

Nationality: *noun:* Bolivian(s)
adjective: Bolivian

Ethnic groups: mestizo (mixed white and Amerindian ancestry) 68%, indigenous 20%, white 5%, cholo/chola 2%, black 1%, other 1%, unspecified 3% ; 44% of respondents indicated feeling part of some indigenous group, predominantly Quechua or Aymara (2009 est.)
note: results among surveys vary based on the wording of the ethnicity question and the available response choices; the 2001 national census did not provide "mestizo" as a response choice, resulting in a much higher proportion of respondents identifying themselves as belonging to one of the available indigenous ethnicity choices; the use of "mestizo" and "cholo" varies among response choices in surveys, with surveys using the terms interchangeably, providing one or the other as a response choice, or providing the two as separate response choices

Languages: Spanish (official) 60.7%, Quechua (official) 21.2%, Aymara (official) 14.6%, Guarani (official) 0.6%, other native languages 0.4%, foreign languages 2.4%, none 0.1% (2001 est.)
note: Bolivia's 2009 constitution designates Spanish and all indigenous languages as official; 36 indigenous languages are specified, including a few that are extinct

Religions: Roman Catholic 76.8%, Evangelical and Pentecostal 8.1%, Protestant 7.9%, other 1.7%, none 5.5% (2012 est.)

Demographic profile: Bolivia ranks at or near the bottom among Latin American countries in several areas of health and development, including poverty, education, fertility, malnutrition, mortality, and life expectancy. On the positive side, more children are being vaccinated and more pregnant women are getting prenatal care and having skilled health practitioners attend their births.

Bolivia's income inequality is the highest in Latin America and one of the highest in the world. Public education is of poor quality, and educational opportunities are among the most unevenly distributed in Latin America, with girls and indigenous and rural children less likely to be literate or to complete primary school. The lack of access to education and family planning services helps to sustain Bolivia's high fertility rate—approximately three children per woman. Bolivia's lack of clean water and basic sanitation, especially in rural areas, contributes to health problems.

Between 7% and 16% of Bolivia's population lives abroad (estimates vary in part because of illegal migration). Emigrants primarily seek jobs and better wages in Argentina (the principal destination), the US, and Spain. In recent years, more restrictive immigration policies in Europe and the US have increased the flow of Bolivian emigrants to neighboring countries. Fewer Bolivians migrated to Brazil in 2015 and 2016 because of its recession; increasing numbers have been going to Chile, mainly to work as miners.

Age structure: *0-14 years:* 30.34% (male 1,799,925/female 1,731,565)
15-24 years: 19.21% (male 1,133,120/female 1,103,063)
25-54 years: 38.68% (male 2,212,096/female 2,289,888)
55-64 years: 6.06% (male 323,210/female 382,139)
65 years and over: 5.71% (male 291,368/female 373,535) (2020 est.)

Dependency ratios: *total dependency ratio:* 60.5
youth dependency ratio: 48.5
elderly dependency ratio: 12
potential support ratio: 8.3 (2020 est.)

Median age: *total:* 25.3 years
male: 24.5 years
female: 26 years (2020 est.)
country comparison to the world: 160

Population growth rate: 1.44% (2020 est.)
country comparison to the world: 75

Birth rate: 20.8 births/1,000 population (2020 est.)
country comparison to the world: 73

Death rate: 6.3 deaths/1,000 population (2020 est.)
country comparison to the world: 148

Net migration rate: -0.3 migrant(s)/1,000 population (2020 est.)
country comparison to the world: 113

Population distribution: a high altitude plain in the west between two cordillera of the Andes, known as the Altiplano, is the focal area for most of the population; a dense settlement pattern is also found in and around the city of Santa Cruz, located on the eastern side of the Andes

Urbanization: *urban population:* 70.1% of total population (2020)
rate of urbanization: 1.97% annual rate of change (2015-20 est.)

https://nypost.com/2020/10/15/emails-reveal-how-hunter-biden-tried-to-cash-in-big-with-chinese-firm/: Major urban areas—population: 278,000 Sucre (constitutional capital) (2018); 1.858 million LA PAZ (capital), 1.713 million Santa Cruz, 1.304 million Cochabamba (2020)

Sex ratio: *at birth:* 1.05 male(s)/female
0-14 years: 1.04 male(s)/female
15-24 years: 1.03 male(s)/female
25-54 years: 0.97 male(s)/female
55-64 years: 0.85 male(s)/female
65 years and over: 0.78 male(s)/female
total population: 0.98 male(s)/female (2020 est.)

Mother's mean age at first birth: 21.2 years (2008 est.)
note: median age at first birth among women 25-29

Maternal mortality rate: 155 deaths/100,000 live births (2017 est.)
country comparison to the world: 56

Infant mortality rate: *total:* 32.2 deaths/1,000 live births
male: 35.5 deaths/1,000 live births
female: 28.8 deaths/1,000 live births (2020 est.)
country comparison to the world: 48

Life expectancy at birth: *total population:* 70.4 years
male: 67.6 years
female: 73.4 years (2020 est.)
country comparison to the world: 165

Total fertility rate: 2.48 children born/woman (2020 est.)
country comparison to the world: 76

Contraceptive prevalence rate: 66.5% (2016)

Drinking water source:
improved:
urban: 100% of population
rural: 78.1% of population
total: 92.8% of population
unimproved:
urban: 0% of population
rural: 21.9% of population
total: 7.1% of population (2017 est.)

Current Health Expenditure: 6.4% (2017)

Physicians density: 1.59 physicians/1,000 population (2016)

Hospital bed density: 1.3 beds/1,000 population (2017)

Sanitation facility access:
improved:
urban: 94.1% of population
rural: 42.2% of population
total: 78% of population
unimproved:
urban: 5.9% of population
rural: 57.8% of population
total: 22% of population (2017 est.)

HIV/AIDS—adult prevalence rate: 0.2% (2019 est.)
country comparison to the world: 95

HIV/AIDS—people living with HIV/AIDS: 19,000 (2019 est.)
country comparison to the world: 87

HIV/AIDS—deaths: <200 (2019 est.)

Major infectious diseases: *degree of risk:* very high (2020)
food or waterborne diseases: bacterial diarrhea and hepatitis A
vectorborne diseases: dengue fever and malaria

Obesity—adult prevalence rate: 20.2% (2016)
country comparison to the world: 102

Children under the age of 5 years underweight:
3.4% (2016)
country comparison to the world: 95

Education expenditures: 7.3% of GDP (2014)
country comparison to the world: 11

Literacy: *definition:* age 15 and over can read and
write
total population: 92.5%
male: 96.5%
female: 88.6% (2015)

Unemployment, youth ages 15-24: *total:* 6.9%
male: 6.8%
female: 7.1% (2018 est.)
country comparison to the world: 151

GOVERNMENT

Country name: *conventional long form:*
Plurinational State of Bolivia
conventional short form: Bolivia
local long form: Estado Plurinacional de Bolivia
local short form: Bolivia
etymology: the country is named after Simon
BOLIVAR, a 19th-century leader in the South
American wars for independence

Government type: presidential republic

Capital: *name:* La Paz (administrative capital);
Sucre (constitutional [legislative and judicial]
capital)

geographic coordinates: 16 30 S, 68 09 W
time difference: UTC-4 (1 hour ahead of
Washington, DC, during Standard Time)
etymology: La Paz is a shortening of the original
name of the city, Nuestra Senora de La Paz (Our
Lady of Peace); Sucre is named after Antonio
Jose de Sucre (1795-1830), military hero in the
independence struggle from Spain and the second
president of Bolivia
note: at approximately 3,630 m above sea level, La
Paz's elevation makes it the highest capital city in
the world

Administrative divisions: 9 departments (depar-
tamentos, singular—departamento); Beni,
Chuquisaca, Cochabamba, La Paz, Oruro, Pando,
Potosi, Santa Cruz, Tarija

Independence: 6 August 1825 (from Spain)

National holiday: Independence Day, 6 August
(1825)

Constitution: *history:* many previous; latest drafted
6 August 2006 to 9 December 2008, approved by
referendum 25 January 2009, effective 7 February
2009; note—in late 2017, the Constitutional
Tribunal declared inapplicable provisions of
the constitution that prohibit elected officials,
including the president, from serving more than 2
consecutive terms
amendments: proposed through public petition
by at least 20% of voters or by the Plurinational
Legislative Assembly; passage requires approval by
at least two-thirds majority vote of the total membership of the Assembly and approval in a referendum; amended 2013

Legal system: civil law system with influences
from Roman, Spanish, canon (religious), French,
and indigenous law

International law organization participation: has
not submitted an ICJ jurisdiction declaration;
accepts ICCt jurisdiction

Citizenship: *citizenship by birth:* yes
citizenship by descent only: yes
dual citizenship recognized: yes
residency requirement for naturalization: 3
years

Suffrage: 18 years of age; universal and compulsory

Executive branch: *chief of state:* President Luis
Alberto ARCE Catacora (since 8 November
2020); Vice President David CHOQUEHUANCA
Cespedes (since 8 November 2020); note—
the president is both chief of state and head of
government
note: former President Juan Evo MORALES Ayma
resigned from office on 10 November 2019 over
alleged election rigging; resignations of all his
constitutionally designated successors followed,
including the Vice President, President of the
Senate, President of the Chamber of Deputies,
and First Vice President of the Senate, leaving
the Second Vice President of the Senate, Jeanine
ANEZ Chavez, the highest-ranking official still
in office; her appointment to the presidency was
endorsed by Bolivia's Constitutional Court, and
she served as interim president until the inauguration of Luis Alberto ARCE Catacora, winner of
the 18 October 2020 presidential election

head of government: President Luis Alberto
ARCE Catacora (since 8 November 2020); Vice
President David CHOQUEHUANCA Cespedes
(since 8 November 2020)
cabinet: Cabinet appointed by the president
elections/appointments: president and vice president directly elected on the same ballot one of 3
ways: candidate wins at least 50% of the vote, or at
least 40% of the vote and 10% more than the next
highest candidate; otherwise a second round is
held and the winner determined by simple majority vote; president and vice president are elected
by majority vote to serve a 5-year term; no term
limits (changed from two consecutive term limit
by Constitutional Court in late 2017); election
last held on 18 October 2020
election results: Luis Alberto ARCE Catacora
elected president; percent of vote—Luis Alberto
ARCE Catacora (MAS) 55.1%; Carlos Diego
MESA Gisbert (CC) 28.8%; Luis Fernando
CAMACHO Vaca (Creemos) 14%; other 2.1%

Legislative branch: *description:* bicameral
Plurinational Legislative Assembly or Asamblea
Legislativa Plurinacional consists of:
Chamber of Senators or Camara de Senadores
(36 seats; members directly elected in multi-seat
constituencies by proportional representation
vote; members serve 5-year terms)
Chamber of Deputies or Camara de Diputados
(130 seats; 70 members directly elected in single-seat constituencies by simple majority vote,
53 directly elected in single-seat constituencies by
proportional representation vote, and 7—apportioned to non-contiguous, rural areas in 7 of the
9 states—directly elected in single-seat constituencies by simple majority vote; members serve
5-year terms)
elections: Chamber of Senators—last held on 18
October 2020 (next to be held in 2025)
Chamber of Deputies—last held on 18 October
2020 (next to be held in 2025)
election results: Chamber of Senators—percent
of vote by party—NA; seats by party—MAS 21,
ACC 11, Creemos 4;
Chamber of Deputies—percent of vote by
party—NA; seats by party—MAS 75, ACC 39,
Creemos 16

Judicial branch: *highest courts:* Supreme Court or
Tribunal Supremo de Justicia (consists of 12 judges
or ministros organized into civil, penal, social,
and administrative chambers); Plurinational
Constitutional Tribunal (consists of 7 primary and
7 alternate magistrates); Plurinational Electoral
Organ (consists of 7 members and 6 alternates);
National Agro-Environment Court (consists of
5 primary and 5 alternate judges); Council of the
Judiciary (consists of 3 primary and 3 alternate
judges)
judge selection and term of office: Supreme
Court, Plurinational Constitutional Tribunal,
National Agro-Environmental Court, and
Council of the Judiciary candidates pre-selected
by the Plurinational Legislative Assembly and
elected by direct popular vote; judges elected for
6-year terms; Plurinational Electoral Organ judges
appointed—6 by the Legislative Assembly and 1
by the president of the republic; members serve
single 6-year terms
subordinate courts: National Electoral Court;
District Courts (in each of the 9 administrative
departments); agro- environmental lower courts

Political parties and leaders: Christian
Democratic Party or PDC [Jorge Fernando
QUIROGA Ramirez]
Community Citizen Alliance or ACC [Carlos
Diego MESA Gisbert]
Movement Toward Socialism or MAS [Juan Evo
MORALES Ayma]
National Unity or UN [Samuel DORIA MEDINA
Arana]
Social Democrat Movement or MDS [Ruben
COSTAS Aguilera]
We Believe or Creemos [Luis Fernando
CAMACHO Vaca]
note: the Democrat Unity Coalition or UD
[Samuel DORIA MEDINA Arana] was a coali-
tion comprised of several of the largest opposition
parties participating in the 2014 election, which
included the Democrats (MDS), National Unity
Front (UN), and Without Fear Movement

International organization participation: CAN,
CD, CELAC, FAO, G-77, IADB, IAEA, IBRD,
ICAO, ICC (national committees), ICCt,
ICRM, IDA, IFAD, IFC, IFRCS, ILO, IMF, IMO,
Interpol, IOC, IOM, IPU, ISO (correspondent),
ITSO, ITU, LAES, LAIA, Mercosur (associate),
MIGA, MINUSTAH, MONUSCO, NAM, OAS,

OPANAL, OPCW, PCA, UN, UN Security Council (temporary), UNAMID, UNASUR, UNCTAD, UNESCO, UNIDO, Union Latina, UNMIL, UNMISS, UNOCI, UNWTO, UPU, WCO, WFTU (NGOs), WHO, WIPO, WMO, WTO

Diplomatic representation in the US: *chief of mission:* Ambassador Walter Oscar SERRATE CUELLAR (since 2 December 2019)
chancery: 3014 Massachusetts Ave., NW, Washington, DC 20008
telephone: [1] (202) 483-4410
FAX: [1] (202) 328-3712

consulate(s) general: Houston, Los Angeles, Miami, New York, Washington, DC
note: in September 2008, the US expelled the Bolivian ambassador to the US in reciprocity for Bolivia expelling the US ambassador to Bolivia; in November 2019, the interim Bolivian Government names Oscar SERRATE Cuellar as its temporary special representative to the US

Diplomatic representation from the US: *chief of mission:* Ambassador (vacant); Charge d'Affaires Bruce WILLIAMSON (since December 2017)
telephone: [591] (2) 216-8000
embassy: Avenida Arce 2780, Casilla 425, La Paz
mailing address: 3220 La Paz Place, Dulles, VA, 20189-3220
FAX: [591] (2) 216-8111
note: in September 2008, the Bolivian Government expelled the US Ambassador to Bolivia, Philip GOLDBERG, and both countries have yet to reinstate their ambassadors

Flag description: three equal horizontal bands of red (top), yellow, and green with the coat of arms centered on the yellow band; red stands for bravery and the blood of national heroes, yellow for the nation's mineral resources, and green for the fertility of the land
note: similar to the flag of Ghana, which has a large black five-pointed star centered in the yellow band; in 2009, a presidential decree made it mandatory for a so-called wiphala—a square, multi-colored flag representing the country's indigenous peoples—to be used alongside the traditional flag

National symbol(s): llama, Andean condor, two national flowers: the cantuta and the patuju; national colors: red, yellow, green

National anthem: *name:* "Cancion Patriotica" (Patriotic Song)
lyrics/music: Jose Ignacio de SANJINES/Leopoldo Benedetto VINCENTI
note: adopted 1852
0:00 / 0:00

ECONOMY

Economy—overview: Bolivia is a resource rich country with strong growth attributed to captive markets for natural gas exports – to Brazil and Argentina. However, the country remains one of the least developed countries in Latin America because of state-oriented policies that deter investment.

Following an economic crisis during the early 1980s, reforms in the 1990s spurred private investment, stimulated economic growth, and cut poverty rates. The period 2003-05 was characterized by political instability, racial tensions, and violent protests against plans—subsequently abandoned—to export Bolivia's newly discovered natural gas reserves to large Northern Hemisphere markets. In 2005-06, the government passed hydrocarbon laws that imposed significantly higher royalties and required foreign firms then operating under risk-sharing contracts to surrender all production to the state energy company in exchange for a predetermined service fee; the laws engendered much public debate. High commodity prices between 2010 and 2014 sustained rapid growth and large trade surpluses with GDP growing 6.8% in 2013 and 5.4% in 2014. The global decline in oil prices that began in late 2014 exerted downward pressure on the price Bolivia receives for exported gas and resulted in lower GDP growth rates—4.9% in 2015 and 4.3% in 2016—and losses in government revenue as well as fiscal and trade deficits.

A lack of foreign investment in the key sectors of mining and hydrocarbons, along with conflict among social groups, pose challenges for the Bolivian economy. In 2015, President Evo MORALES expanded efforts to court international investment and boost Bolivia's energy production capacity. MORALES passed an investment law and promised not to nationalize additional industries in an effort to improve the investment climate. In early 2016, the Government of Bolivia approved the 2016-2020 National Economic and Social Development Plan aimed at maintaining growth of 5% and reducing poverty.

GDP (purchasing power parity): $83.72 billion (2017 est.)
$80.35 billion (2016 est.)
$77.07 billion (2015 est.)
note: data are in 2017 dollars
country comparison to the world: 94

GDP (official exchange rate): $37.78 billion (2017 est.)

GDP—real growth rate: 2.22% (2019 est.)
4.23% (2018 est.)
4.19% (2017 est.)
country comparison to the world: 128

GDP—per capita (PPP): $7,600 (2017 est.)
$7,400 (2016 est.)
$7,200 (2015 est.)
note: data are in 2017 dollars
country comparison to the world: 154

Gross national saving: 15.7% of GDP (2017 est.)
15.3% of GDP (2016 est.)
14.2% of GDP (2015 est.)
country comparison to the world: 132

GDP—composition, by end use: *household consumption:* 67.7% (2017 est.)
government consumption: 17% (2017 est.)
investment in fixed capital: 21.3% (2017 est.)
investment in inventories: 3.8% (2017 est.)
exports of goods and services: 21.7% (2017 est.)
imports of goods and services: -31.3% (2017 est.)

GDP—composition, by sector of origin: *agriculture:* 13.8% (2017 est.)
industry: 37.8% (2017 est.)
services: 48.2% (2017 est.)

Agriculture—products: soybeans, quinoa, Brazil nuts, sugarcane, coffee, corn, rice, potatoes, chia, coca

Industries: mining, smelting, electricity, petroleum, food and beverages, handicrafts, clothing, jewelry

Industrial production growth rate: 2.2% (2017 est.)
country comparison to the world: 123

Labor force: 5.719 million (2016 est.)
country comparison to the world: 71

Labor force—by occupation: *agriculture:* 29.4%
industry: 22%
services: 48.6% (2015 est.)

Unemployment rate: 4% (2017 est.)
4% (2016 est.)
note: data are for urban areas; widespread underemployment
country comparison to the world: 58

Population below poverty line: 38.6% (2015 est.)
note: based on percent of population living on less than the international standard of $2/day

Household income or consumption by percentage share: *lowest 10%:* 0.9%
highest 10%: 36.1% (2014 est.)

Budget: *revenues:* 15.09 billion (2017 est.)
expenditures: 18.02 billion (2017 est.)

Taxes and other revenues: 39.9% (of GDP) (2017 est.)
country comparison to the world: 39

Budget surplus (+) or deficit (-): -7.8% (of GDP) (2017 est.)
country comparison to the world: 196

Public debt: 49% of GDP (2017 est.)
44.9% of GDP (2016 est.)
note: data cover general government debt and includes debt instruments issued by government entities other than the treasury; the data include treasury debt held by foreign entities; the data include debt issued by subnational entities
country comparison to the world: 104

Fiscal year: calendar year

Inflation rate (consumer prices): 2.8% (2017 est.)
3.6% (2016 est.)
country comparison to the world: 127

Current account balance: -$2.375 billion (2017 est.)
-$1.932 billion (2016 est.)
country comparison to the world: 171

Exports: $7.746 billion (2017 est.)
$7.214 billion (2016 est.)
country comparison to the world: 97

Exports—partners: Brazil 17.9%, Argentina 16%, US 7.8%, Japan 7.3%, India 6.6%, South Korea 6.3%, Colombia 5.8%, China 5.1%, UAE 4.7% (2017)

Exports—commodities: natural gas, silver, zinc, lead, tin, gold, quinoa, soybeans and soy products

Imports: $8.601 billion (2017 est.)

$7.888 billion (2016 est.)
country comparison to the world: 107

Imports—commodities: machinery, petroleum products, vehicles, iron and steel, plastics

Imports—partners: China 21.7%, Brazil 16.8%, Argentina 12.6%, US 8.4%, Peru 6.5% (2017)

Reserves of foreign exchange and gold: $10.26 billion (31 December 2017 est.)
$10.08 billion (31 December 2016 est.)
country comparison to the world: 74

Debt—external: $12.81 billion (31 December 2017 est.)
$7.268 billion (31 December 2016 est.)
country comparison to the world: 106

Exchange rates: bolivianos (BOB) per US dollar—
6.86 (2017 est.)
6.86 (2016 est.)
6.91 (2015 est.)
6.91 (2014 est.)
6.91 (2013 est.)

ENERGY

Electricity access: *population without electricity:* 1.2 million (2013)
electrification—total population: 95.6% (2018)
electrification—urban areas: 99.3% (2016)
electrification—rural areas: 79.1% (2016)

Electricity—production: 8.951 billion kWh (2016 est.)
country comparison to the world: 107

Electricity—consumption: 7.785 billion kWh (2016 est.)
country comparison to the world: 105

Electricity—exports: 0 kWh (2017 est.)
country comparison to the world: 108

Electricity—imports: 0 kWh (2016 est.)
country comparison to the world: 127

Electricity—installed generating capacity: 2.764 million kW (2016 est.)
country comparison to the world: 101

Electricity—from fossil fuels: 76% of total installed capacity (2016 est.)
country comparison to the world: 93

Electricity—from nuclear fuels: 0% of total installed capacity (2017 est.)
country comparison to the world: 52

Electricity—from hydroelectric plants: 18% of total installed capacity (2017 est.)
country comparison to the world: 92

Electricity—from other renewable sources: 7% of total installed capacity (2017 est.)
country comparison to the world: 92

Crude oil—production: 60,000 bbl/day (2018 est.)
country comparison to the world: 50

Crude oil—exports: 1,274 bbl/day (2015 est.)
country comparison to the world: 72

Crude oil—imports: 0 bbl/day (2015 est.)
country comparison to the world: 99

Crude oil—proved reserves: 211.5 million bbl (1 January 2018 est.)
country comparison to the world: 54

Refined petroleum products—production: 65,960 bbl/day (2015 est.)
country comparison to the world: 75

Refined petroleum products—consumption: 83,000 bbl/day (2016 est.)
country comparison to the world: 86

Refined petroleum products—exports: 9,686 bbl/day (2015 est.)
country comparison to the world: 82

Refined petroleum products—imports: 20,620 bbl/day (2015 est.)
country comparison to the world: 118

Natural gas—production: 18.69 billion cu m (2017 est.)
country comparison to the world: 32

Natural gas—consumption: 3.171 billion cu m (2017 est.)
country comparison to the world: 71

Natural gas—exports: 15.46 billion cu m (2017 est.)
country comparison to the world: 15

Natural gas—imports: 0 cu m (2017 est.)
country comparison to the world: 94

Natural gas—proved reserves: 295.9 billion cu m (1 January 2018 est.)
country comparison to the world: 37

Carbon dioxide emissions from consumption of energy: 17.66 million Mt (2017 est.)
country comparison to the world: 90

COMMUNICATIONS

Telephones—fixed lines: *total subscriptions:* 719,399
subscriptions per 100 inhabitants: 6.27 (2019 est.)
country comparison to the world: 83

Telephones—mobile cellular: *total subscriptions:* 11,567,760
subscriptions per 100 inhabitants: 100.82 (2019 est.)
country comparison to the world: 79

Telecommunication systems: *general assessment:* lowest GDP in the area; much of the population live in remote valleys and telecommunications is poor; consumers pick from multiple long-distance carriers for each call; reliability, and coverage have steadily improved, but some remote areas are still underserved; operators plan to extend fiber to all 339 municipal capital cities by 2022; move from 3G to LTE available by all 3 mobile companies; 92% of all Internet is through smartphone; broadband services remain expensive by the lack of competition and that fact that Bolivia is landlocked and does not have access through submarine cables; MNP (mobile number portability) launched in October 2018; Bolivian Space Agency planning to launch a second telecom satellite after 2020 (2020)
domestic: 6 per 100 fixed-line, mobile-cellular telephone use expanding rapidly and teledensity stands at 101 per 100 persons; most telephones are concentrated in La Paz, Santa Cruz, and other capital cities (2019)

international: country code—591; Bolivia has no direct access to submarine cable networks and must therefore connect to the rest of the world either via satellite or through terrestrial links across neighboring countries; satellite earth station -1 Intelsat (Atlantic Ocean) (2019)
note: the COVID-19 outbreak is negatively impacting telecommunications production and supply chains globally; consumer spending on telecom devices and services has also slowed due to the pandemic's effect on economies worldwide; overall progress towards improvements in all facets of the telecom industry—mobile, fixed-line, broadband, submarine cable and satellite—has moderated

Broadcast media: large number of radio and TV stations broadcasting with private media outlets dominating; state-owned and private radio and TV stations generally operating freely, although both pro-government and anti-government groups have attacked media outlets in response to their reporting

Internet country code: .bo

Internet users: *total:* 4,955,569

percent of population: 43.83% (July 2018 est.)
country comparison to the world: 83

Broadband—fixed subscriptions: *total:* 504,097
subscriptions per 100 inhabitants: 4 (2018 est.)
country comparison to the world: 85

TRANSPORTATION

National air transport system: *number of registered air carriers:* 7 (2020)
inventory of registered aircraft operated by air carriers: 39
annual passenger traffic on registered air carriers: 4,122,113 (2018)
annual freight traffic on registered air carriers: 13.73 million mt-km (2018)

Civil aircraft registration country code prefix: CP (2016)

Airports: 855 (2013)
country comparison to the world: 7

Airports—with paved runways: *total:* 21 (2017)
over 3,047 m: 5 (2017)
2,438 to 3,047 m: 4 (2017)
1,524 to 2,437 m: 6 (2017)
914 to 1,523 m: 6 (2017)

Airports—with unpaved runways: *total:* 834 (2013)
over 3,047 m: 1 (2013)
2,438 to 3,047 m: 4 (2013)
1,524 to 2,437 m: 47 (2013)
914 to 1,523 m: 151 (2013)
under 914 m: 631 (2013)

Pipelines: 5457 km gas, 51 km liquid petroleum gas, 2511 km oil, 1627 km refined products (2013)

Railways: *total:* 3,960 km (2019)
narrow gauge: 3,960 km 1.000-m gauge (2014)
country comparison to the world: 51

Roadways: *total:* 90,568 km (2017)
paved: 9,792 km (2017)
unpaved: 80,776 km (2017)

country comparison to the world: 54

Waterways: 10,000 km (commercially navigable almost exclusively in the northern and eastern parts of the country) (2012)
country comparison to the world: 13

Merchant marine: total: 43
by type: general cargo 27, oil tanker 2, other 14 (2019)
country comparison to the world: 120

Ports and terminals: *river port(s):* Puerto Aguirre (Paraguay/Parana)
note: Bolivia has free port privileges in maritime ports in Argentina, Brazil, Chile, and Paraguay

Military and security forces: Bolivian Armed Forces: Bolivian Army (Ejercito Boliviano, EB), Bolivian Naval Force (Fuerza Naval Boliviana, FNB, includes Marines), Bolivian Air Force (Fuerza Aerea Boliviana, FAB); Ministry of Interior: National Police (Policía Nacional de Bolivia, PNB; includes Anti-Narcotics Special Forces (Fuerza Especial de Lucha Contra el Narcotráfico, FELCN) and other paramilitary units (2020)

Military expenditures: 1.4% of GDP (2019)
1.5% of GDP (2018)
1.5% of GDP (2017)
1.6% of GDP (2016)
1.7% of GDP (2015)
country comparison to the world: 85

Military and security service personnel strengths: size assessments for the Bolivian Armed Forces vary; approximately 39,000 total active troops

(26,000 Army; 5,500 Navy; 7,500 Air Force) (2019 est.)

Military equipment inventories and acquisitions: the Bolivian Armed Forces are equipped with a mix of mostly Brazilian, Chinese, European, and US equipment; since 2010, China and France are the leading suppliers of military hardware to Bolivia (2019 est.)

Military service age and obligation: 16-49 years of age for 12-month voluntary male and female military service; Bolivian citizenship required; minimum age for combat duty is 18; when annual number of volunteers falls short of goal, compulsory recruitment is effected, including conscription of boys as young as 14; 15-19 years of age for voluntary premilitary service, provides exemption from further military service (2017)

Disputes—international: Chile and Peru rebuff Bolivia's reactivated claim to restore the Atacama corridor, ceded to Chile in 1884, but Chile offers instead unrestricted but not sovereign maritime access through Chile for Bolivian products; contraband smuggling, human trafficking, and illegal narcotic trafficking are problems in the porous areas of its border regions with all of its neighbors (Argentina, Brazil, Chile, Paraguay, and Peru)

Trafficking in persons: current situation: Bolivia is a source country for men, women, and children subjected to forced labor and sex trafficking domestically and abroad; rural and poor Bolivians, most of whom are indigenous, and LGBT youth are particularly vulnerable; Bolivians perform forced labor domestically in mining, ranching, agriculture, and domestic service, and a significant

number are in forced labor abroad in sweatshops, agriculture, domestic service, and the informal sector; women and girls are sex trafficked within Bolivia and in neighboring countries, such as Argentina, Peru, and Chile; a limited number of women from nearby countries are sex trafficked in Bolivia

tier rating: Tier 2 Watch List – Bolivia does not comply fully with the minimum standards for the elimination of human trafficking; however, it is making significant efforts to do so; the government did not demonstrate overall increasing anti-trafficking efforts, and poor data collection made it difficult to assess the number of investigations, prosecutions, and victim identifications and referrals to care services; authorities did not adequately differentiate between human trafficking and other crimes, such as domestic violence and child abuse; law enforcement failed to implement an early detection protocol for identifying trafficking cases and lacked a formal process for identifying trafficking victims among vulnerable populations; specialized victim services were inadequately funded and virtually non-existent for adult women and male victims (2015)

Illicit drugs: world's third-largest cultivator of coca (after Colombia and Peru) with an estimated 37,500 hectares under cultivation in 2016, a 3 percent increase over 2015; third largest producer of cocaine, estimated at 275 metric tons potential pure cocaine in 2016; transit country for Peruvian and Colombian cocaine destined for Brazil, Argentina, Chile, Paraguay, and Europe; weak border controls; some money-laundering activity related to narcotics trade; major cocaine consumption

BOSNIA AND HERZEGOVINA

Background: Bosnia and Herzegovina declared sovereignty in October 1991 and independence

from the former Yugoslavia on 3 March 1992 after a referendum boycotted by ethnic Serbs. The Bosnian Serbs—supported by neighboring Serbia and Montenegro—responded with armed resistance aimed at partitioning the republic along ethnic lines and joining Serb-held areas to form a "Greater Serbia." In March 1994, Bosniaks and Croats reduced the number of warring factions from three to two by signing an agreement creating a joint Bosniak-Croat Federation of Bosnia and Herzegovina. On 21 November 1995, in Dayton, Ohio, the warring parties initialed a peace agreement that ended three years of interethnic civil strife (the final agreement was signed in Paris on 14 December 1995).

The Dayton Peace Accords retained Bosnia and Herzegovina's international boundaries and created a multiethnic and democratic government charged with conducting foreign, diplomatic, and fiscal policy. Also recognized was a second tier of government composed of

two entities roughly equal in size: the predominantly Bosniak-Bosnian Croat Federation of Bosnia and Herzegovina and the predominantly Bosnian Serb-led Republika Srpska (RS). The Federation and RS governments are responsible for overseeing most government functions. Additionally, the Dayton Accords established the Office of the High Representative to oversee the implementation of the civilian aspects of the agreement. The Peace Implementation Council at its conference in Bonn in 1997 also gave the High Representative the authority to impose legislation and remove officials, the so-called "Bonn Powers." An original NATO-led international peacekeeping force (IFOR) of 60,000 troops assembled in 1995 was succeeded over time by a smaller, NATO-led Stabilization Force (SFOR). In 2004, European Union peacekeeping troops (EUFOR) replaced SFOR. Currently, EUFOR deploys around 600 troops in theater in a security assistance and training capacity.

GEOGRAPHY

Location: Southeastern Europe, bordering the Adriatic Sea and Croatia

Geographic coordinates: 44 00 N, 18 00 E

Map references: Europe

Area: *total:* 51,197 sq km
land: 51,187 sq km
water: 10 sq km
country comparison to the world: 129

Area—comparative: slightly smaller than West Virginia

Land boundaries: *total:* 1,543 km
border countries (3): Croatia 956 km, Montenegro 242 km, Serbia 345 km

Coastline: 20 km

Maritime claims: NA

Climate: hot summers and cold winters; areas of high elevation have short, cool summers and long, severe winters; mild, rainy winters along coast

Terrain: mountains and valleys

Elevation: *mean elevation:* 500 m
lowest point: Adriatic Sea 0 m
highest point: Maglic 2,386 m

Natural resources: coal, iron ore, antimony, bauxite, copper, lead, zinc, chromite, cobalt, manganese, nickel, clay, gypsum, salt, sand, timber, hydropower

Land use: *agricultural land:* 42.2% (2011 est.)
arable land: 19.7% (2011 est.) / permanent crops: 2% (2011 est.) / permanent pasture: 20.5% (2011 est.)
forest: 42.8% (2011 est.)
other: 15% (2011 est.)
Irrigated land: 30 sq km (2012)

Population distribution: the northern and central areas of the country are the most densely populated

Natural hazards: destructive earthquakes

Environment—current issues: air pollution; deforestation and illegal logging; inadequate wastewater treatment and flood management facilities; sites for disposing of urban waste are limited; land mines left over from the 1992-95 civil strife are a hazard in some areas

Environment—international agreements: *party to:* Air Pollution, Biodiversity, Climate Change, Climate Change-Kyoto Protocol, Desertification, Hazardous Wastes, Law of the Sea, Marine Life Conservation, Ozone Layer Protection, Wetlands *signed, but not ratified:* none of the selected agreements

Geography—note: within Bosnia and Herzegovina's recognized borders, the country is divided into a joint Bosniak/Croat Federation (about 51% of the territory) and the Bosnian Serb-led Republika Srpska or RS (about 49% of the territory); the region called Herzegovina is contiguous to Croatia and Montenegro, and traditionally has been settled by an ethnic Croat majority in the west and an ethnic Serb majority in the east

PEOPLE AND SOCIETY

Population: 3,835,586 (July 2020 est.)
country comparison to the world: 131

Nationality: *noun:* Bosnian(s), Herzegovinian(s)
adjective: Bosnian, Herzegovinian

Ethnic groups: Bosniak 50.1%, Serb 30.8%, Croat 15.4%, other 2.7%, not declared/no answer 1% (2013 est.)
note: Republika Srpska authorities dispute the methodology and refuse to recognize the results; Bosniak has replaced Muslim as an ethnic term in part to avoid confusion with the religious term Muslim—an adherent of Islam

Languages: Bosnian (official) 52.9%, Serbian (official) 30.8%, Croatian (official) 14.6%, other 1.6%, no answer 0.2% (2013 est.)

Religions: Muslim 50.7%, Orthodox 30.7%, Roman Catholic 15.2%, atheist 0.8%, agnostic 0.3%, other 1.2%, undeclared/no answer 1.1% (2013 est.)

Age structure: *0-14 years:* 13.18% (male 261,430/female 244,242)
15-24 years: 10.83% (male 214,319/female 201,214)
25-54 years: 44.52% (male 859,509/female 848,071)
55-64 years: 15.24% (male 284,415/female 300,168)
65 years and over: 16.22% (male 249,624/female 372,594) (2020 est.)

Dependency ratios:
total dependency ratio: 48
youth dependency ratio: 21.5
elderly dependency ratio: 26.5
potential support ratio: 3.8 (2020 est.)

Median age: *total:* 43.3 years
male: 41.6 years
female: 44.8 years (2020 est.)
country comparison to the world: 27

Population growth rate: -0.19% (2020 est.)
country comparison to the world: 208

Birth rate: 8.6 births/1,000 population (2020 est.)
country comparison to the world: 214

Death rate: 10.2 deaths/1,000 population (2020 est.)
country comparison to the world: 33

Net migration rate: -0.4 migrant(s)/1,000 population (2020 est.)
country comparison to the world: 122

Population distribution: the northern and central areas of the country are the most densely populated

Urbanization: *urban population:* 49% of total population (2020)
rate of urbanization: 0.55% annual rate of change (2015-20 est.)

total population growth rate v urban population growth rate, 2000-2030: Major urban areas—population: 343,000 SARAJEVO (capital) (2020)

Sex ratio: *at birth:* 1.07 male(s)/female
0-14 years: 1.07 male(s)/female
15-24 years: 1.07 male(s)/female

25-54 years: 1.01 male(s)/female
55-64 years: 0.95 male(s)/female
65 years and over: 0.67 male(s)/female
total population: 0.95 male(s)/female (2020 est.)

Mother's mean age at first birth: 27.3 years (2017 est.)

Maternal mortality rate: 10 deaths/100,000 live births (2017 est.)
country comparison to the world: 143

Infant mortality rate: *total:* 5.2 deaths/1,000 live births
male: 5.3 deaths/1,000 live births
female: 5.1 deaths/1,000 live births (2020 est.)
country comparison to the world: 175

Life expectancy at birth: *total population:* 77.5 years
male: 74.5 years
female: 80.7 years (2020 est.)
country comparison to the world: 79

Total fertility rate: 1.33 children born/woman (2020 est.)
country comparison to the world: 221

Contraceptive prevalence rate: 45.8% (2011/12)

Drinking water source:
improved:
urban: 99.9% of population
rural: 100% of population
total: 99.9% of population
unimproved:
urban: 0.1% of population
rural: 0% of population
total: 0.1% of population (2017 est.)

Current Health Expenditure: 8.9% (2017)

Physicians density: 2.16 physicians/1,000 population (2015)

Hospital bed density: 3.5 beds/1,000 population (2014)

Sanitation facility access:
improved:
urban: 98.9% of population
rural: 92.1% of population
total: 95.4% of population
unimproved:
urban: 1.1% of population
rural: 7.9% of population
total: 4.5% of population (2017 est.)

HIV/AIDS—adult prevalence rate: <.1% (2018)

HIV/AIDS—people living with HIV/AIDS: <500 (2018)

HIV/AIDS—deaths: <100 (2018)

Obesity—adult prevalence rate: 17.9% (2016)
country comparison to the world: 118

Children under the age of 5 years underweight: 1.6% (2012)
country comparison to the world: 118

Education expenditures: NA

Literacy: *definition:* age 15 and over can read and write
total population: 98.5%
male: 99.5%
female: 97.5% (2015)

School life expectancy (primary to tertiary education): *total:* 14 years
male: 14 years
female: 15 years (2014)

Unemployment, youth ages 15-24: *total:* 33.8%
male: 31.3%
female: 37.9% (2019 est.)
country comparison to the world: 25

GOVERNMENT

Country name: *conventional long form:* none
conventional short form: Bosnia and Herzegovina
local long form: none
local short form: Bosna i Hercegovina
former: People's Republic of Bosnia and Herzegovina, Socialist Republic of Bosnia and Herzegovina
abbreviation: BiH
etymology: the larger northern territory is named for the Bosna River; the smaller southern section takes its name from the German word "herzog," meaning "duke," and the ending "-ovina," meaning "land," forming the combination denoting "dukedom"

Government type: parliamentary republic

Capital: *name:* Sarajevo

geographic coordinates: 43 52 N, 18 25 E
time difference: UTC+1 (6 hours ahead of Washington, DC, during Standard Time)
daylight saving time: +1hr, begins last Sunday in March; ends last Sunday in October
etymology: the name derives from the Turkish noun "saray," meaning "palace" or "mansion," and the term "ova," signifying "plain(s)," to give a meaning of "palace plains" or "the plains about the palace"

Administrative divisions: 3 first-order administrative divisions—Brcko District (Brcko Distrikt) (ethnically mixed), Federation of Bosnia and Herzegovina (Federacija Bosne i Hercegovine) (predominantly Bosniak-Croat), Republika Srpska (predominantly Serb)

Independence: 1 March 1992 (from Yugoslavia); note—referendum for independence completed on 1 March 1992; independence declared on 3 March 1992

National holiday: Independence Day, 1 March (1992) and Statehood Day, 25 November (1943)—both observed in the Federation of Bosnia and Herzegovina entity; Victory Day, 9 May (1945) and Dayton Agreement Day, 21 November (1995)—both observed in the Republika Srpska entity
note: there is no national-level holiday

Constitution: *history:* 14 December 1995 (constitution included as part of the Dayton Peace Accords); note—each of the political entities has its own constitution
amendments: decided by the Parliamentary Assembly, including a two-thirds majority vote of members present in the House of Representatives; the constitutional article on human rights and fundamental freedoms cannot be amended; amended several times, last in 2009 (2016)

Legal system: civil law system; Constitutional Court review of legislative acts

International law organization participation: has not submitted an ICJ jurisdiction declaration; accepts ICCt jurisdiction

Citizenship: *citizenship by birth:* no
citizenship by descent only: at least one parent must be a citizen of Bosnia and Herzegovina
dual citizenship recognized: yes, provided there is a bilateral agreement with the other state
residency requirement for naturalization: 8 years

Suffrage: 18 years of age, 16 if employed; universal

Executive branch: *chief of state:* Chairman of the Presidency Sefik DZAFEROVIC (chairman since 20 March 2020, presidency member since 20 November 2018—Bosniak seat); Zeljko KOMSIC (presidency member since 20 November 2018—Croat seat); Milorad DODIK (presidency member since 20 November 2018—Serb seat)

head of government: Chairman of the Council of Ministers Zoran TEGELTIJA (since 5 December 2019)
cabinet: Council of Ministers nominated by the council chairman, approved by the state-level House of Representatives
elections/appointments: 3-member presidency (1 Bosniak and 1 Croat elected from the Federation of Bosnia and Herzegovina and 1 Serb elected from the Republika Srpska) directly elected by simple majority popular vote for a 4-year term (eligible for a second term, but then ineligible for 4 years); the presidency chairpersonship rotates every 8 months with the new member of the presidency elected with the highest number of votes starting the new mandate as chair; election last held on 7 October 2018 (next to be held in October 2022); the chairman of the Council of Ministers appointed by the presidency and confirmed by the state-level House of Representatives
election results: percent of vote—Milorad DODIK (SNSD) 53.9%—Serb seat; Zeljko KOMSIC (DF) 52.6%—Croat seat; Sefik DZAFEROVIC (SDA) 36.6%—Bosniak seat
note: President of the Federation of Bosnia and Herzegovina Marinko CAVARA (since 11 February 2015); Vice Presidents Melika MAHMUTBEGOVIC (since 11 February 2015), Milan DUNOVIC (since 11 February 2015); President of the Republika Srpska Zeljka CVIJANOVIC (since 18 November 2018); Vice Presidents Ramiz SALKIC (since 24 November 2014), Josip JERKOVIC (since 24 November 2014)

Legislative branch: *description:* bicameral Parliamentary Assembly or Skupstina consists of:
House of Peoples or Dom Naroda (15 seats—5 Bosniak, 5 Croat, 5 Serb; members designated by the Federation of Bosnia and Herzegovina's House of Peoples and the Republika Srpska's National Assembly to serve 4-year terms)

House of Representatives or Predstavnicki Dom (42 seats to include 28 seats allocated to the Federation of Bosnia and Herzegovina and 14 to the Republika Srpska; members directly elected by proportional representation vote to serve 4- year terms); note—the Federation of Bosnia and Herzegovina has a bicameral legislature that consists of the House of Peoples (58 seats—17 Bosniak, 17 Croat, 17 Serb, 7 other) and the House of Representatives (98 seats; members directly elected by proportional representation vote to serve 4-year terms); Republika Srpska's unicameral legislature is the National Assembly (83 directly elected delegates serve 4-year terms)
elections: House of Peoples—last held on 18 October 2018 (next to be held in October 2022)
House of Representatives—last held on 7 October 2018 (next to be held in October 2022)
election results: House of Peoples—percent of vote by coalition/party—NA; seats by coalition/party—NA; composition—men 13, women 2, percent of women 13.3%
House of Representatives—percent of vote by coalition/party—SDA 17%, SNSD 16%, SDS/NDP/NS/SRS-VS 9.8%, SDP 9.1%, HDZ-BiH/HSS/HKDU/HSP-AS BiH/HDU BiH 9.1%, DF 5.8%, PDP 5.1%, DNS 4.2%, SBB BiH 4.2%, NS/HC 2.9%, NB 2.5%, PDA 2.3%, SP 1.9%, A-SDA 1.8%, other 17.4%; seats by coalition/party—SDA 9, SNSD 6, SDP 5, HDZ- BiH/HSS/HKDU/HSP-AS BiH/HDU BiH 5, SDS/NDP/NS/SRS-VS 3, DF 3, PDP 2, SBB BiH 2, NS/HC 2, DNS 1, NB 1 PDA 1, SP 1, A-SDA 1; composition—men 33, women 9, percent of women 21.4%; note—total Parliamentary Assembly percent of women 19.3%

Judicial branch: *highest courts:* Bosnia and Herzegovina (BiH) Constitutional Court (consists of 9 members); Court of BiH (consists of 44 national judges and 7 international judges organized into 3 divisions—Administrative, Appellate, and Criminal, which includes a War Crimes Chamber)
judge selection and term of office: BiH Constitutional Court judges—4 selected by the Federation of Bosnia and Herzegovina House of Representatives, 2 selected by the Republika Srpska's National Assembly, and 3 non-Bosnian judges selected by the president of the European Court of Human Rights; Court of BiH president and national judges appointed by the High Judicial and Prosecutorial Council; Court of BiH president appointed for renewable 6-year term; other national judges appointed to serve until age 70; international judges recommended by the president of the Court of BiH and appointed by the High Representative for Bosnia and Herzegovina; international judges appointed to serve until age 70
subordinate courts: the Federation has 10 cantonal courts plus a number of municipal courts; the Republika Srpska has a supreme court, 5 district courts, and a number of municipal courts

Political parties and leaders: Alliance for a Better Future of BiH or SBB BiH [Fahrudin RADONCIC]

Alliance of Independent Social Democrats or SNSD [Milorad DODIK]

Alternative Party for Democratic Activity or A-SDA [Nermin OGRESEVIC]

Croat Peasants' Party or HSS [Mario KARAMATIC]

Croatian Christian Democratic Union of Bosnia and Herzegovina or HKDU [Ivan MUSA]

Croatian Democratic Union of Bosnia and Herzegovina or HDU-BiH [Miro GRABOVAC-TITAN]

Croatian Democratic Union of Bosnia and Herzegovina or HDZ-BiH [Dragan COVIC]

Croatian Democratic Union 1990 or HDZ-1990 [Ilija CVITANOVIC]

Croatian Party of Rights dr. Ante Starcevic or HSP-AS Bih [Karlo STARCEVIC]

Democratic Alliance or DEMOS [Nedeljko CUBRILOVIC]

Democratic Front of DF [Zeljko KOMSIC]

Democratic Peoples' Alliance or DNS [Marko PAVIC]

Independent Bloc or NB [Senad SEPIC]

Movement for Democratic Action or PDA [Mirsad KUKIC]

Progressive Srpska or NS [Goran DORDIC]

Our Party or NS/HC [Predrag KOJOVIC]

Party for Democratic Action or SDA [Bakir IZETBEGOVIC]

Party of Democratic Progress or PDP [Branislav BORENOVIC]

People's Democratic Movement or NDP [Dragan CAVIC]

Serb Democratic Party or SDS [Vukota GOVEDARICA]

Serb Radical Party-Dr. Vojislav Seselj or SRS-VS [Vojislav SESELJ] (members joined the PDP)

Social Democratic Party or SDP [Nermin NIKSIC]

Socialist Party or SP [Petar DOKIC]

United Srpska or US [Nenad STEVANDIC]

International organization participation: BIS, CD, CE, CEI, EAPC, EBRD, FAO, G-77, IAEA, IBRD, ICAO, ICC (NGOs), ICCt, ICRM, IDA, IFAD, IFC, IFRCS, ILO, IMF, IMO, IMSO, Interpol, IOC, IOM, IPU, ISO, ITSO, ITU, ITUC (NGOs), MIGA, MINUSMA, MONUSCO, NAM (observer), OAS (observer), OIC (observer), OIF (observer), OPCW, OSCE, PFP, SELEC, UN, UNCTAD, UNESCO, UNIDO, UNWTO, UPU, WCO, WHO, WIPO, WMO, WTO (observer)

Diplomatic representation in the US: *chief of mission:* Ambassador Bojan VUJIC (since 16 September 2019)
chancery: 2109 E Street NW, Washington, DC 20037
telephone: [1] (202) 337-1500
FAX: [1] (202) 337-1502
consulate(s) general: Chicago, New York

Diplomatic representation from the US: *chief of mission:* Ambassador Eric NELSON (since 19 February 2019)
telephone: [387] (33) 704-000
embassy: 1 Robert C. Frasure Street, 71000 Sarajevo

mailing address: use embassy street address
FAX: [387] (33) 659-722

branch office(s): Banja Luka, Mostar

Flag description: a wide blue vertical band on the fly side with a yellow isosceles triangle abutting the band and the top of the flag; the remainder of the flag is blue with seven full five-pointed white stars and two half stars top and bottom along the hypotenuse of the triangle; the triangle approximates the shape of the country and its three points stand for the constituent peoples—Bosniaks, Croats, and Serbs; the stars represent Europe and are meant to be continuous (thus the half stars at top and bottom); the colors (white, blue, and yellow) are often associated with neutrality and peace, and traditionally are linked with Bosnia
note: one of several flags where a prominent component of the design reflects the shape of the country; other such flags are those of Brazil, Eritrea, and Vanuatu

National symbol(s): golden lily; national colors: blue, yellow, white

National anthem: *name:* "Drzavna himna Bosne i Hercegovine" (The National Anthem of Bosnia and Herzegovina)
lyrics/music: none officially; Dusan SESTIC and Benjamin ISOVIC/Dusan SESTIC
note: music adopted 1999; lyrics proposed in 2009 and others in 2016 were not approved; a parliamentary committee launched a new initiative for lyrics in February 2018
0:00 / 2:05

ECONOMY

Economy—overview: Bosnia and Herzegovina has a transitional economy with limited market reforms. The economy relies heavily on the export of metals, energy, textiles, and furniture as well as on remittances and foreign aid. A highly decentralized government hampers economic policy coordination and reform, while excessive bureaucracy and a segmented market discourage foreign investment. The economy is among the least competitive in the region. Foreign banks, primarily from Austria and Italy, control much of the banking sector, though the largest bank is a private domestic one. The konvertibilna marka (convertible mark)—the national currency introduced in 1998—is pegged to the euro through a currency board arrangement, which has maintained confidence in the currency and has facilitated reliable trade links with European partners. Bosnia and Herzegovina became a full member of the Central European Free Trade Agreement in September 2007. In 2016, Bosnia began a three-year IMF loan program, but it has struggled to meet the economic reform benchmarks required to receive all funding installments.

Bosnia and Herzegovina's private sector is growing slowly, but foreign investment dropped sharply after 2007 and remains low. High unemployment remains the most serious macroeconomic problem. Successful implementation of a value-added tax in 2006 provided a steady source of revenue for the government and helped rein in gray-market activity, though public perceptions of government corruption and misuse of taxpayer money has encouraged a large informal economy to persist. National-level statistics have improved over time, but a large share of economic activity remains unofficial and unrecorded.

Bosnia and Herzegovina's top economic priorities are: acceleration of integration into the EU; strengthening the fiscal system; public administration reform; World Trade Organization membership; and securing economic growth by fostering a dynamic, competitive private sector.

GDP (purchasing power parity): $44.83 billion (2017 est.)
$43.54 billion (2016 est.)
$42.19 billion (2015 est.)
note: data are in 2017 dollars
country comparison to the world: 113

GDP (official exchange rate): $18.17 billion (2017 est.)

GDP—real growth rate: 3% (2017 est.)
3.2% (2016 est.)
3.1% (2015 est.)
country comparison to the world: 111

GDP—per capita (PPP): $12,800 (2017 est.)
$12,400 (2016 est.)
$11,900 (2015 est.)
note: data are in 2017 dollars
country comparison to the world: 122

Gross national saving: 11% of GDP (2017 est.)
11.1% of GDP (2016 est.)
10.5% of GDP (2015 est.)
country comparison to the world: 158

GDP—composition, by end use: *household consumption:* 77.4% (2017 est.)
government consumption: 20% (2017 est.)
investment in fixed capital: 16.6% (2017 est.)
investment in inventories: 2.3% (2017 est.)
exports of goods and services: 38.7% (2017 est.)
imports of goods and services: -55.1% (2017 est.)

GDP—composition, by sector of origin: *agriculture:* 6.8% (2017 est.)
industry: 28.9% (2017 est.)
services: 64.3% (2017 est.)

Agriculture—products: wheat, corn, fruits, vegetables; livestock

Industries: steel, coal, iron ore, lead, zinc, manganese, bauxite, aluminum, motor vehicle assembly, textiles, tobacco products, wooden furniture, ammunition, domestic appliances, oil refining

Industrial production growth rate: 3% (2017 est.)
country comparison to the world: 102

Labor force: 1.38 million (2017 est.)
country comparison to the world: 134

Labor force—by occupation: *agriculture:* 18%
industry: 30.4%
services: 51.7% (2017 est.)

Unemployment rate: 20.5% (2017 est.)
25.4% (2016 est.)
note: official rate; actual rate is lower as many technically unemployed persons work in the gray economy

country comparison to the world: 188

Population below poverty line: 16.9% (2015 est.)

Household income or consumption by percentage share: *lowest 10%:* 2.9%
highest 10%: 25.8% (2011 est.)

Budget: *revenues:* 7.993 billion (2017 est.)
expenditures: 7.607 billion (2017 est.)

Taxes and other revenues: 44% (of GDP) (2017 est.)
country comparison to the world: 26

Budget surplus (+) or deficit (-): 2.1% (of GDP) (2017 est.)
country comparison to the world: 15

Public debt: 39.5% of GDP (2017 est.)
44.1% of GDP (2016 est.)
note: data cover general government debt and includes debt instruments issued (or owned) by government entities other than the treasury; the data include treasury debt held by foreign entities; the data include debt issued by subnational entities, as well as intra-governmental debt; intra-governmental debt consists of treasury borrowings from surpluses in the social funds, such as for retirement, medical care, and unemployment; debt instruments for the social funds are not sold at public auctions.
country comparison to the world: 130

Fiscal year: calendar year

Inflation rate (consumer prices): 1.2% (2017 est.)
-1.1% (2016 est.)
country comparison to the world: 62

Current account balance: -$873 million (2017 est.)
-$821 million (2016 est.)
country comparison to the world: 135

Exports: $5.205 billion (2017 est.)
$4.288 billion (2016 est.)
country comparison to the world: 106

Exports—partners: Germany 14.7%, Croatia 11.8%, Italy 11.1%, Serbia 10%, Slovenia 9%, Austria 8.3% (2017)

Exports—commodities: metals, clothing, wood products

Imports: $9.547 billion (2017 est.)
$8.337 billion (2016 est.)
country comparison to the world: 102

Imports—commodities: machinery and equipment, chemicals, fuels, foodstuffs

Imports—partners: Germany 11.6%, Italy 11.3%, Serbia 11.1%, Croatia 10.1%, China 6.5%, Slovenia 5%, Russia 4.7%, Turkey 4.2% (2017)

Reserves of foreign exchange and gold: $6.474 billion (31 December 2017 est.)
$5.137 billion (31 December 2016 est.)
country comparison to the world: 90

Debt—external: $10.87 billion (31 December 2017 est.)
$10.64 billion (31 December 2016 est.)
country comparison to the world: 111

Exchange rates: konvertibilna markas (BAM) per US dollar—1.729 (2017 est.)
1.7674 (2016 est.)

1.7674 (2015 est.)
1.7626 (2014 est.)
1.4718 (2013 est.)

ENERGY

Electricity access: *electrification—total population:* 100% (2016)

Electricity—production: 16.99 billion kWh (2016 est.)
country comparison to the world: 86

Electricity—consumption: 11.87 billion kWh (2016 est.)
country comparison to the world: 89

Electricity—exports: 6.007 billion kWh (2015 est.)
country comparison to the world: 31

Electricity—imports: 3.084 billion kWh (2016 est.)
country comparison to the world: 49

Electricity—installed generating capacity: 4.676 million kW (2016 est.)
country comparison to the world: 83

Electricity—from fossil fuels: 49% of total installed capacity (2016 est.)
country comparison to the world: 153

Electricity—from nuclear fuels: 0% of total installed capacity (2017 est.)
country comparison to the world: 53

Electricity—from hydroelectric plants: 51% of total installed capacity (2017 est.)
country comparison to the world: 35

Electricity—from other renewable sources: 1% of total installed capacity (2017 est.)
country comparison to the world: 148

Crude oil—production: 0 bbl/day (2018 est.)
country comparison to the world: 113

Crude oil—exports: 0 bbl/day (2015 est.)
country comparison to the world: 96

Crude oil—imports: 18,480 bbl/day (2015 est.)
country comparison to the world: 64

Crude oil—proved reserves: 0 bbl (1 January 2018 est.)
country comparison to the world: 109

Refined petroleum products—production: 0 bbl/day (2015 est.)
country comparison to the world: 121

Refined petroleum products—consumption: 32,000 bbl/day (2016 est.)
country comparison to the world: 118

Refined petroleum products—exports: 4,603 bbl/day (2015 est.)
country comparison to the world: 92

Refined petroleum products—imports: 18,280 bbl/day (2015 est.)
country comparison to the world: 129

Natural gas—production: 0 cu m (2017 est.)
country comparison to the world: 107

Natural gas—consumption: 226.5 million cu m (2017 est.)
country comparison to the world: 103

Natural gas—exports: 0 cu m (2017 est.)

country comparison to the world: 72

Natural gas—imports: 226.5 million cu m (2017 est.)
country comparison to the world: 70

Natural gas—proved reserves: 0 cu m (1 January 2014 est.)
country comparison to the world: 113

Carbon dioxide emissions from consumption of energy: 22.07 million Mt (2017 est.)
country comparison to the world: 84

COMMUNICATIONS

Telephones—fixed lines: *total subscriptions:* 920,407
subscriptions per 100 inhabitants: 23.95 (2019 est.)
country comparison to the world: 77

Telephones—mobile cellular: *total subscriptions:* 4,300,743
subscriptions per 100 inhabitants: 111.91 (2019 est.)
country comparison to the world: 126

Telecommunication systems: *general assessment:* mobile services dominate fixed-line; integration with the EU has given stability to the present economy, as an EU candidate country, the regulatory framework and telecom market has been liberalized and the regulator has given LTE license to 3 MNOs; DSL and cable are the chief platforms for fixed-line connectivity, there is a small market presence of fiber broadband; new mobile roaming fees come into effect similar to other EU countries; rural areas still suffer from insufficient connectivity (2020)
domestic: fixed-line teledensity roughly 24 per 100 persons and mobile-cellular subscribership has been increasing rapidly and stands at roughly 112 telephones per 100 persons (2019)
international: country code—387; no satellite earth stations
note: the COVID-19 outbreak is negatively impacting telecommunications production and supply chains globally; consumer spending on telecom devices and services has also slowed due to the pandemic's effect on economies worldwide; overall progress towards improvements in all facets of the telecom industry—mobile, fixed-line, broadband, submarine cable and satellite—has moderated

Broadcast media: 3 public TV broadcasters: Radio and TV of Bosnia and Herzegovina, Federation TV (operating 2 networks), and Republika Srpska Radio-TV; a local commercial network of 5 TV stations; 3 private, near-national TV stations and dozens of small independent TV broadcasting stations; 3 large public radio broadcasters and many private radio stations

Internet country code: .ba

Internet users: *total:* 2,699,544

percent of population: 70.12% (July 2018 est.)
country comparison to the world: 104

Broadband—fixed subscriptions: *total:* 693,554
subscriptions per 100 inhabitants: 18 (2018 est.)
country comparison to the world: 77

TRANSPORTATION

National air transport system: *number of registered air carriers:* 1 (2020)
inventory of registered aircraft operated by air carriers: 1
annual passenger traffic on registered air carriers: 7,070 (2015)
annual freight traffic on registered air carriers: 87 mt-km (2015)

Civil aircraft registration country code prefix: T9 (2016)

Airports: 24 (2013)
country comparison to the world: 129

Airports—with paved runways: *total:* 7 (2017)
2,438 to 3,047 m: 4 (2017)
1,524 to 2,437 m: 1 (2017)
under 914 m: 2 (2017)

Airports—with unpaved runways: *total:* 17 (2013)
1,524 to 2,437 m: 1 (2013)
914 to 1,523 m: 5 (2013)
under 914 m: 11 (2013)

Heliports: 6 (2013)

Pipelines: 147 km gas, 9 km oil (2013)

Railways: *total:* 965 km (2014)
standard gauge: 965 km 1.435-m gauge (565 km electrified) (2014)
country comparison to the world: 90

Roadways: *total:* 22,926 km (2010)
paved: 19,426 km (4,652 km of interurban roads) (2010)
unpaved: 3,500 km (2010)

country comparison to the world: 109

Waterways: (Sava River on northern border; open to shipping but use limited) (2011)
Ports and terminals: river port(s): Bosanska Gradiska, Bosanski Brod, Bosanski Samac, Brcko, Orasje (Sava River)

MILITARY AND SECURITY

Military and security forces: Armed Forces of Bosnia and Herzegovina (Oruzanih Snaga Bosne i Hercegovine, OSBiH): Operations Command (includes Army, Air, and Air Defense units), Support Command (2019)

Military expenditures: 0.9% of GDP (2019)
0.9% of GDP (2018)
0.9% of GDP (2017)
0.9% of GDP (2016)
1% of GDP (2015)
country comparison to the world: 126

Military and security service personnel strengths: the Armed Forces of Bosnia and Herzegovina have approximately 9,200 active duty personnel (2019)

Military equipment inventories and acquisitions: the inventory for the Armed Forces of Bosnia and Herzegovina includes mainly Soviet-era weapons systems with a small mix of older European and US equipment (2019 est.)

Military service age and obligation: 18 years of age for voluntary military service; mandatory retirement at age 35 or after 15 years of service for E-1 through E-4, mandatory retirement at age 50 and 30 years of service for E-5 through E-9, mandatory

retirement at age 55 and 30 years of service for all officers (2014)

TERRORISM

Terrorist group(s): Islamic Revolutionary Guard Corps/Qods Force (2019)
note: details about the history, aims, leadership, organization, areas of operation, tactics, targets, weapons, size, and sources of support of the group(s) appear(s) in Appendix-T

TRANSNATIONAL ISSUES

Disputes—international: Serbia delimited about half of the boundary with Bosnia and Herzegovina, but sections along the Drina River remain in dispute

Refugees and internally displaced persons: *refugees (country of origin):* 5,116 (Croatia) (2019)
IDPs: 99,000 (Bosnian Croats, Serbs, and Bosniaks displaced by inter-ethnic violence, human rights violations, and armed conflict during the 1992-95 war) (2019)
stateless persons: 90 (2018)
note: 68,087 estimated refugee and migrant arrivals (January 2015-October 2020)

Illicit drugs: increasingly a transit point for heroin being trafficked to Western Europe; minor transit point for marijuana; remains highly vulnerable to money-laundering activity given a primarily cash-based and unregulated economy, weak law enforcement, and instances of corruption

BOTSWANA

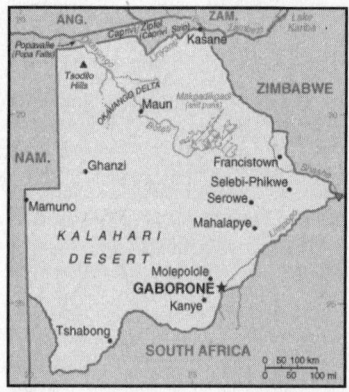

INTRODUCTION

Background: Seeking to stop the incorporation of their land into Rhodesia (Zimbabwe) or the Union of South Africa, in 1885, three tribal chiefs traveled to Great Britain and

successfully lobbied the British Government to put "Bechuanaland" under UK protection. Upon independence in 1966, the British protectorate of Bechuanaland adopted the new name of Botswana. More than five decades of uninterrupted civilian leadership, progressive social policies, and significant capital investment have created one of the most stable economies in Africa. The ruling Botswana Democratic Party has won every national election since independence; President Mokgweetsi Eric MASISI assumed the presidency in April 2018 following the retirement of former President Ian KHAMA due to constitutional term limits. MASISI won his first election as president in October 2019, and he is Botswana's fifth president since independence. Mineral extraction, principally diamond mining, dominates economic activity, though tourism is a growing sector due to the country's conservation practices and extensive nature preserves. Botswana has one of the world's highest rates of HIV/AIDS infection, but also one of Africa's most progressive and comprehensive programs for dealing with the disease.

GEOGRAPHY

Location: Southern Africa, north of South Africa

Geographic coordinates: 22 00 S, 24 00 E

Map references: Africa

Area: *total:* 581,730 sq km
land: 566,730 sq km
water: 15,000 sq km
country comparison to the world: 49

Area—comparative: slightly smaller than Texas; almost four times the size of Illinois

Land boundaries: *total:* 4,347.15 km
border countries (4): Namibia 1544 km, South Africa 1969 km, Zambia 0.15 km, Zimbabwe 834 km

Coastline: 0 km (landlocked)

Maritime claims: none (landlocked)

Climate: semiarid; warm winters and hot summers

Terrain: predominantly flat to gently rolling tableland; Kalahari Desert in southwest

Elevation: *mean elevation:* 1,013 m
lowest point: junction of the Limpopo and Shashe Rivers 513 m

highest point: Tsodilo Hills 1,489 m

Natural resources: diamonds, copper, nickel, salt, soda ash, potash, coal, iron ore, silver

Land use: *agricultural land:* 45.8% (2011 est.)
arable land: 0.6% (2011 est.) / permanent crops: 0% (2011 est.) / permanent pasture: 45.2% (2011 est.)
forest: 19.8% (2011 est.)
other: 34.4% (2011 est.)

Irrigated land: 20 sq km (2012)

Population distribution: the population is primarily concentrated in the east with a focus in and around the captial of Gaborone, and the far central-eastern city of Francistown; population density remains low in other areas in the country, especially in the Kalahari to the west as shown in this population distribution map

Natural hazards: periodic droughts; seasonal August winds blow from the west, carrying sand and dust across the country, which can obscure visibility

Environment—current issues: overgrazing; desertification; limited freshwater resources; air pollution

Environment—international agreements: party to: Biodiversity, Climate Change, Climate Change-Kyoto Protocol, Desertification, Endangered Species, Hazardous Wastes, Law of the Sea, Ozone Layer Protection, Wetlands
signed, but not ratified: none of the selected agreements

Geography—note: landlocked; population concentrated in the southern and eastern parts of the country

PEOPLE AND SOCIETY

Population: 2,317,233 (July 2020 est.)
note: estimates for this country explicitly take into account the effects of excess mortality due to AIDS; this can result in lower life expectancy, higher infant mortality, higher death rates, lower population growth rates, and changes in the distribution of population by age and sex than would otherwise be expected
country comparison to the world: 144

Nationality: *noun:* Motswana (singular), Batswana (plural)
adjective: Motswana (singular), Batswana (plural)

Ethnic groups: Tswana (or Setswana) 79%, Kalanga 11%, Basarwa 3%, other, including Kgalagadi and people of European ancestry 7%

Languages: Setswana 77.3%, Sekalanga 7.4%, Shekgalagadi 3.4%, English (official) 2.8%, Zezuru/ Shona 2%, Sesarwa 1.7%, Sembukushu 1.6%, Ndebele 1%, other 2.8% (2011 est.)

Religions: Christian 79.1%, Badimo 4.1%, other 1.4% (includes Baha'i, Hindu, Muslim, Rastafarian), none 15.2%, unspecified 0.3% (2011 est.)

Demographic profile: Botswana has experienced one of the most rapid declines in fertility in Sub-Saharan Africa. The total fertility rate has fallen from more than 5 children per woman in the mid 1980s to approximately 2.4 in 2013. The fertility reduction has been attributed to a host of factors, including higher educational attainment among women, greater participation of women in the workforce, increased contraceptive use, later first births, and a strong national family planning program. Botswana was making significant progress in several health indicators, including life expectancy and infant and child mortality rates, until being devastated by the HIV/AIDs epidemic in the 1990s.

Today Botswana has the third highest HIV/AIDS prevalence rate in the world at approximately 22%, however comprehensive and effective treatment programs have reduced HIV/AIDS-related deaths. The combination of declining fertility and increasing mortality rates because of HIV/AIDS is slowing the population aging process, with a narrowing of the youngest age groups and little expansion of the oldest age groups. Nevertheless, having the bulk of its population (about 60%) of working age will only yield economic benefits if the labor force is healthy, educated, and productively employed.

Batswana have been working as contract miners in South Africa since the 19th century. Although Botswana's economy improved shortly after independence in 1966 with the discovery of diamonds and other minerals, its lingering high poverty rate and lack of job opportunities continued to push workers to seek mining work in southern African countries. In the early 1970s, about a third of Botswana's male labor force worked in South Africa (lesser numbers went to Namibia and Zimbabwe). Not until the 1980s and 1990s, when South African mining companies had reduced their recruitment of foreign workers and Botswana's economic prospects had improved, were Batswana increasingly able to find job opportunities at home.

Most Batswana prefer life in their home country and choose cross-border migration on a temporary basis only for work, shopping, visiting family, or tourism. Since the 1970s, Botswana has pursued an open migration policy enabling it to recruit thousands of foreign workers to fill skilled labor shortages. In the late 1990s, Botswana's prosperity and political stability attracted not only skilled workers but small numbers of refugees from neighboring Angola, Namibia, and Zimbabwe.

Age structure: 0-14 *years:* 30.54% (male 357,065/female 350,550)
15-24 years: 18.31% (male 208,824/female 215,462)
25-54 years: 39.67% (male 434,258/female 484,922)
55-64 years: 5.92% (male 59,399/female 77,886)
65 years and over: 5.56% (male 53,708/female 75,159) (2020 est.)

Dependency ratios: *total dependency ratio:* 61.1
youth dependency ratio: 53.8
elderly dependency ratio: 7.3
potential support ratio: 13.8 (2020 est.)

Median age: *total:* 25.7 years
male: 24.5 years

female: 26.7 years (2020 est.)
country comparison to the world: 157

Population growth rate: 1.48% (2020 est.)
country comparison to the world: 72

Birth rate: 20.9 births/1,000 population (2020 est.)
country comparison to the world: 72

Death rate: 9.2 deaths/1,000 population (2020 est.)
country comparison to the world: 54

Net migration rate: 2.9 migrant(s)/1,000 population (2020 est.)
country comparison to the world: 38

Population distribution: the population is primarily concentrated in the east with a focus in and around the captial of Gaborone, and the far central-eastern city of Francistown; population density remains low in other areas in the country, especially in the Kalahari to the west as shown in this population distribution map

Urbanization: *urban population:* 70.9% of total population (2020)
rate of urbanization: 2.87% annual rate of change (2015-20 est.)

total population growth rate v. urban population growth rate, 2000-2030: Major urban areas—population: 269,000 GABORONE (capital) (2018)

Sex ratio: *at birth:* 1.03 male(s)/female
0-14 years: 1.02 male(s)/female
15-24 years: 0.97 male(s)/female
25-54 years: 0.9 male(s)/female
55-64 years: 0.76 male(s)/female
65 years and over: 0.71 male(s)/female
total population: 0.93 male(s)/female (2020 est.)

Maternal mortality rate: 144 deaths/100,000 live births (2017 est.)
country comparison to the world: 59

Infant mortality rate: *total:* 26.8 deaths/1,000 live births
male: 29.2 deaths/1,000 live births
female: 24.2 deaths/1,000 live births (2020 est.)
country comparison to the world: 66

Life expectancy at birth: *total population:* 64.8 years
male: 62.8 years
female: 66.9 years (2020 est.)
country comparison to the world: 197

Total fertility rate: 2.45 children born/woman (2020 est.)
country comparison to the world: 77

Contraceptive prevalence rate: 67.4% (2017)

Drinking water source:
improved:
urban: 98.2% of population
rural: 94% of population
total: 96.9% of population
unimproved:
urban: 1.8% of population
rural: 3.1% of population
total: 3.8% of population (2017 est.)

Current Health Expenditure: 6.1% (2017)

Physicians density: 0.53 physicians/1,000 population (2016)

Hospital bed density: 1.8 beds/1,000 population (2010)

Sanitation facility access:
improved:
urban: 92.9% of population
rural: 60.8% of population
total: 82.8% of population
unimproved:
urban: 7.1% of population
rural: 39.2% of population
total: 17.2% of population (2017 est.)

HIV/AIDS—adult prevalence rate: 22.2% (2019 est.)
country comparison to the world: 3

HIV/AIDS—people living with HIV/AIDS: 380,000 (2019 est.)
country comparison to the world: 19

HIV/AIDS—deaths: 5,000 (2019 est.)
country comparison to the world: 25

Major infectious diseases: *degree of risk:* high (2020)
food or waterborne diseases: bacterial diarrhea, hepatitis A, and typhoid fever
vectorborne diseases: malaria

Obesity—adult prevalence rate: 18.9% (2016)
country comparison to the world: 114

Education expenditures: 9.6% of GDP (2009)
country comparison to the world: 4

Literacy: *definition:* age 15 and over can read and write
total population: 88.5%
male: 88%
female: 88.9% (2015)

School life expectancy (primary to tertiary education): *total:* 13 years
male: 13 years
female: 13 years (2013)

Unemployment, youth ages 15-24: *total:* 36%
male: 29.6%
female: 43.5% (2010 est.)
country comparison to the world: 18

GOVERNMENT

Country name: *conventional long form:* Republic of Botswana
conventional short form: Botswana
local long form: Republic of Botswana
local short form: Botswana
former: Bechuanaland
etymology: the name Botswana means "Land of the Tswana"—referring to the country's major ethnic group

Government type: parliamentary republic

Capital: *name:* Gaborone
geographic coordinates: 24 38 S, 25 54 E
time difference: UTC+ 2 (7 hours ahead of Washington, DC, during Standard Time)
etymology: named after GABORONE (ca. 1825-1931), a revered kgosi (chief) of the Tlokwa tribe, part of the larger Tswana ethnic group

Administrative divisions: 10 districts and 6 town councils*; Central, Chobe, Francistown*,

Gaborone*, Ghanzi, Jwaneng*, Kgalagadi, Kgatleng, Kweneng, Lobatse*, North East, North West, Selebi- Phikwe*, South East, Southern, Sowa Town*

Independence: 30 September 1966 (from the UK)

National holiday: Independence Day (Botswana Day), 30 September (1966)

Constitution: *history:* previous 1960 (preindependence); latest adopted March 1965, effective 30 September 1966
amendments: proposed by the National Assembly; passage requires approval in two successive Assembly votes with at least two-thirds majority in the final vote; proposals to amend constitutional provisions on fundamental rights and freedoms, the structure and branches of government, and public services also requires approval by majority vote in a referendum and assent by the president of the republic; amended several times, last in 2006

Legal system: mixed legal system of civil law influenced by the Roman-Dutch model and also customary and common law

International law organization participation: accepts compulsory ICJ jurisdiction with reservations; accepts ICCt jurisdiction

Citizenship: *citizenship by birth:* no
citizenship by descent only: at least one parent must be a citizen of Botswana
dual citizenship recognized: no
residency requirement for naturalization: 10 years

Suffrage: 18 years of age; universal

Executive branch: chief of state: President Mokgweetse Eric MASISI (since 1 April 2018); Vice President Slumber TSOGWANE (since 4 April 2018); note—the president is both chief of state and head of government
head of government: President Mokgweetse Eric MASISI (since 1 April 2018); Vice President Slumber TSOGWANE (since 4 April 2018); note—the president is both chief of state and head of government
cabinet: Cabinet appointed by the president
elections/appointments: president indirectly elected by the National Assembly for a 5-year term (eligible for a second term); election last held on 24 October 2014 (next to be held on 31 October 2019); vice president appointed by the president
election results: President Seretse Khama Ian KHAMA (since 1 April 2008) stepped down on 1 April 2018 having completed the constitutionally mandated 10-year term limit; upon his retirement, then Vice President MASISI became president; national elections held on 23 October 2019 gave MASISI'S BPD 38 seats in the National Assembly which then selected MASISI as President

Legislative branch: description: unicameral Parliament consists of the National Assembly (63 seats; 57 members directly elected in single- seat constituencies by simple majority vote, 4 nominated by the president and indirectly elected by simple majority vote by the rest of the National Assembly, and 2 ex-officio members—the

president and attorney general; elected members serve 5-year terms); note—the House of Chiefs (Ntlo ya Dikgosi), an advisory body to the National Assembly, consists of 35 members—8 hereditary chiefs from Botswana's principal tribes, 22 indirectly elected by the chiefs, and 5 appointed by the president; the House of Chiefs consults on issues including powers of chiefs, customary courts, customary law, tribal property, and constitutional amendments
elections: last held on 23 October 2019 (next to be held in October 2024)
election results: percent of vote by party—BDP 52.7%, UDC 35.9%, BPF 4.4%, AP 5.1%, other 1.7%; seats by party—BDP 38, UDC 15, BPF 3, AP 1; composition—NA

Judicial branch: *highest courts:* Court of Appeal, High Court (each consists of a chief justice and a number of other judges as prescribed by the Parliament)
judge selection and term of office: Court of Appeal and High Court chief justices appointed by the president and other judges appointed by the president upon the advice of the Judicial Service Commission; all judges appointed to serve until age 70
subordinate courts: Industrial Court (with circuits scheduled monthly in the capital city and in 3 districts); Magistrates Courts (1 in each district); Customary Court of Appeal; Paramount Chief' s Court/Urban Customary Court; Senior Chief's Representative Court; Chief's Representative's Court; Headman's Court

Political parties and leaders: Alliance of Progressives or AP [Ndaba GAOLATHE]
Botswana Congress Party or BCP [Dumelang SALESHANDO]
Botswana Democratic Party or BDP [Mokgweetsi MASISI]
Botswana Movement for Democracy or BMD [Sidney PILANE]
Botswana National Front or BNF [Duma BOKO]
Botswana Patriotic Front or BPF [Biggie BUTALE]
Botswana Peoples Party or BPP [Motlatsi MOLAPISI]
Real Alternative Party or RAP [Gaontebale MOKGOSI]
Umbrella for Democratic Change or UDC [Duma BOKO] (various times the collation has included the BMD, BPP, BCP and BNF) (2019)

International organization participation: ACP, AfDB, AU, C, CD, FAO, G- 77, IAEA, IBRD, ICAO, ICCt, ICRM, IDA, IFAD, IFC, IFRCS, ILO, IMF, Interpol, IOC, IOM, IPU, ISO, ITSO, ITU, ITUC (NGOs), MIGA, NAM, OPCW, SACU, SADC, UN, UNCTAD, UNESCO, UNIDO, UNWTO, UPU, WCO, WFTU (NGOs), WHO, WIPO, WMO, WTO

Diplomatic representation in the US: *chief of mission:* Ambassador Onkokame Kitso MOKAILA (since 17 September 2020)
chancery: 1531-1533 New Hampshire Avenue NW, Washington, DC 20036
telephone: [1] (202) 244-4990
FAX: [1] (202) 244-4164

consulate(s) general: Atlanta

Diplomatic representation from the US: *chief of mission:* Ambassador Craig Lewis CLOUD (since 2 April 2019)
telephone: [267] 395-3982
embassy: Embassy Drive, Government Enclave (off Khama Crescent), Gaborone
mailing address: Embassy Enclave, P. O. Box 90, Gaborone
FAX: [267] 318-0232

Flag description: light blue with a horizontal white-edged black stripe in the center; the blue symbolizes water in the form of rain, while the black and white bands represent racial harmony

National symbol(s): zebra; national colors: blue, white, black

National anthem: *name:* " Fatshe leno la rona" (Our Land)
lyrics/music: Kgalemang Tumedisco MOTSETE
note: adopted 1966
0:00 / 0:56

ECONOMY

Economy—overview: Until the beginning of the global recession in 2008, Botswana maintained one of the world's highest economic growth rates since its independence in 1966. Botswana recovered from the global recession in 2010, but only grew modestly until 2017, primarily due to a downturn in the global diamond market, though water and power shortages also played a role. Through fiscal discipline and sound management, Botswana has transformed itself from one of the poorest countries in the world five decades ago into a middle-income country with a per capita GDP of approximately $18,100 in 2017.

Botswana also ranks as one of the least corrupt and best places to do business in Sub-Saharan Africa.

Because of its heavy reliance on diamond exports, Botswana's economy closely follows global price trends for that one commodity. Diamond mining fueled much of Botswana's past economic expansion and currently accounts for one-quarter of GDP, approximately 85% of export earnings, and about one- third of the government's revenues. In 2017, Diamond exports increased to the highest levels since 2013 at about 22 million carats of output, driving Botswana's economic growth to about 4.5% and increasing foreign exchange reserves to about 45% of GDP. De Beers, a major international diamond company, signed a 10-year deal with Botswana in 2012 and moved its rough stone sorting and trading division from London to Gaborone in 2013.The move was geared to support the development of Botswana's nascent downstream diamond industry.

Tourism is a secondary earner of foreign exchange and many Batswana engage in tourism-related services, subsistence farming, and cattle rearing. According to official government statistics, unemployment is around 20%, but unofficial estimates run much higher. The prevalence

of HIV/AIDS is second highest in the world and threatens the country's impressive economic gains.

GDP (purchasing power parity): $39.01 billion (2017 est.)
$38.11 billion (2016 est.)
$36.54 billion (2015 est.)
note: data are in 2017 dollars
country comparison to the world: 121

GDP (official exchange rate): $17.38 billion (2017 est.)

GDP—real growth rate: 2.4% (2017 est.)
4.3% (2016 est.)
-1.7% (2015 est.)
country comparison to the world: 119

GDP—per capita (PPP): $17,000 (2017 est.)
$16,900 (2016 est.)
$16,500 (2015 est.)
note: data are in 2017 dollars
country comparison to the world: 102

Gross national saving: 40.3% of GDP (2017 est.)
38.8% of GDP (2016 est.)
41.2% of GDP (2015 est.)
country comparison to the world: 9

GDP—composition, by end use: *household consumption:* 48.5% (2017 est.)
government consumption: 18.4% (2017 est.)
investment in fixed capital: 29% (2017 est.)
investment in inventories: -1.8% (2017 est.)
exports of goods and services: 39.8% (2017 est.)
imports of goods and services: -33.9% (2017 est.)

GDP—composition, by sector of origin: *agriculture:* 1.8% (2017 est.)
industry: 27.5% (2017 est.)
services: 70.6% (2017 est.)

Agriculture—products: livestock, sorghum, maize, millet, beans, sunflowers, groundnuts

Industries: diamonds, copper, nickel, salt, soda ash, potash, coal, iron ore, silver; beef processing; textiles

Industrial production growth rate: -4.2% (2017 est.)
country comparison to the world: 193

Labor force: 1.177 million (2017 est.)
country comparison to the world: 135

Labor force—by occupation: *agriculture:* NA
industry: NA
services: NA

Unemployment rate: 20% (2013 est.)
17.8% (2009 est.)
country comparison to the world: 188

Population below poverty line: 19.3% (2009 est.)

Household income or consumption by percentage share: *lowest 10%:* NA
highest 10%: NA

Budget: *revenues:* 5.305 billion (2017 est.)
expenditures: 5.478 billion (2017 est.)

Taxes and other revenues: 30.5% (of GDP) (2017 est.)
country comparison to the world: 75

Budget surplus (+) or deficit (-): -1% (of GDP) (2017 est.)

country comparison to the world: 76

Public debt: 14% of GDP (2017 est.)
15.6% of GDP (2016 est.)
country comparison to the world: 195

Fiscal year: 1 April–31 March

Inflation rate (consumer prices): 3.3% (2017 est.)
2.8% (2016 est.)
country comparison to the world: 136

Current account balance: $2.146 billion (2017 est.)
$2.147 billion (2016 est.)
country comparison to the world: 38

Exports: $5.934 billion (2017 est.)
$7.226 billion (2016 est.)
country comparison to the world: 102

Exports—partners: Belgium 20.3%, India 12.6%, UAE 12.4%, South Africa 11.9%, Singapore 8.7%, Israel 7%, Hong Kong 4.1%, Namibia 4.1% (2017)

Exports—commodities: diamonds, copper, nickel, soda ash, beef, textiles

Imports: $5.005 billion (2017 est.)
$5.871 billion (2016 est.)
country comparison to the world: 128

Imports—commodities: foodstuffs, machinery, electrical goods, transport equipment, textiles, fuel and petroleum products, wood and paper products, metal and metal products

Imports—partners: South Africa 66.1%, Canada 8.3%, Israel 5.3% (2017)

Reserves of foreign exchange and gold: $7.491 billion (31 December 2017 est.)
$7.189 billion (31 December 2016 est.)
country comparison to the world: 82

Debt—external: $2.187 billion (31 December 2017 est.)
$2.421 billion (31 December 2016 est.)
country comparison to the world: 150

Exchange rates: pulas (BWP) per US dollar—
10.19 (2017 est.)
10.9022 (2016 est.)
10.9022 (2015 est.)
10.1263 (2014 est.)
8.9761 (2013 est.)

ENERGY

Electricity access: *electrification—total population:* 59% (2019)
electrification—urban areas: 71% (2019)
electrification—rural areas: 29% (2019)

Electricity—production: 2.527 billion kWh (2016 est.)
country comparison to the world: 135

Electricity—consumption: 3.636 billion kWh (2016 est.)
country comparison to the world: 131

Electricity—exports: 0 kWh (2016 est.)
country comparison to the world: 109

Electricity—imports: 1.673 billion kWh (2016 est.)

country comparison to the world: 59

Electricity—installed generating capacity: 735,000 kW (2016 est.)
country comparison to the world: 135

Electricity—from fossil fuels: 100% of total installed capacity (2016 est.)
country comparison to the world: 4

Electricity—from nuclear fuels: 0% of total installed capacity (2017 est.)
country comparison to the world: 54

Electricity—from hydroelectric plants: 0% of total installed capacity (2017 est.)
country comparison to the world: 159

Electricity—from other renewable sources: 0% of total installed capacity (2017 est.)
country comparison to the world: 178

Crude oil—production: 0 bbl/ day (2018 est.)
country comparison to the world: 114

Crude oil—exports: 0 bbl/day (2015 est.)
country comparison to the world: 97

Crude oil—imports: 0 bbl/day (2015 est.)
country comparison to the world: 100

Crude oil—proved reserves: 0 bbl (1 January 2018 est.)
country comparison to the world: 110

Refined petroleum products—production: 0 bbl/day (2015 est.)
country comparison to the world: 122

Refined petroleum products—consumption: 21,000 bbl/day (2016 est.)
country comparison to the world: 135

Refined petroleum products—exports: 0 bbl/day (2015 est.)
country comparison to the world: 133

Refined petroleum products—imports: 21,090 bbl/day (2015 est.)
country comparison to the world: 116

Natural gas—production: 0 cu m (2017 est.)
country comparison to the world: 108

Natural gas—consumption: 0 cu m (2017 est.)
country comparison to the world: 124

Natural gas—exports: 0 cu m (2017 est.)
country comparison to the world: 73

Natural gas—imports: 0 cu m (2017 est.)
country comparison to the world: 95

Natural gas—proved reserves: 0 cu m (1 January 2014 est.)
country comparison to the world: 114

Carbon dioxide emissions from consumption of energy: 6.235 million Mt (2017 est.)
country comparison to the world: 127

COMMUNICATIONS

Telephones—fixed lines: *total subscriptions:* 139,735
subscriptions per 100 inhabitants: 6.12 (2019 est.)
country comparison to the world: 129

Telephones—mobile cellular: *total subscriptions:* 3,968,526
subscriptions per 100 inhabitants: 173.81 (2019 est.)
country comparison to the world: 131

Telecommunication systems: *general assessment:* the Botswana Telecommunications Corp is rolling out 4G service to over 95 sites in the country that will improve network connectivity; an effective regulatory reform has turned the Botswana's telecom market into one of the most liberalized in the region; Botswana has one of the highest mobile penetration rates in Africa; 3 MNOs have entered the underdeveloped broadband sector with the adoption of 3G, LTE and WiMAX technologies; mobile Internet remains the preferred choice; the expansion of a fully digital system with fiber-optic cables along with a system of open-wire lines links the major population centers in the east; the use of multiple SIM cards has delayed the introduction of (mobile number portability) MNP (2020)
domestic: fixed-line teledensity has declined in recent years and now stands at roughly 6 telephones per 100 persons; mobile-cellular teledensity has advanced to 174 telephones per 100 persons (2019)
international: country code—267; international calls are made via satellite, using international direct dialing; 2 international exchanges; digital microwave radio relay links to Namibia, Zambia, Zimbabwe, and South Africa; satellite earth station—1 Intelsat (Indian Ocean)
note: the COVID-19 outbreak is negatively impacting telecommunications production and supply chains globally; consumer spending on telecom devices and services has also slowed due to the pandemic's effect on economies worldwide; overall progress towards improvements in all facets of the telecom industry—mobile, fixed-line, broadband, submarine cable and satellite—has moderated

Broadcast media: 2 TV stations—1 state-owned and 1 privately owned; privately owned satellite TV subscription service is available; 2 state-owned national radio stations; 4 privately owned radio stations broadcast locally (2019)

Internet country code: .bw

Internet users: *total:* 1,057,079
percent of population: 47% (July 2018 est.)
country comparison to the world: 138

Broadband—fixed subscriptions: *total:* 40,044
subscriptions per 100 inhabitants: 2 (2018 est.)
country comparison to the world: 138

TRANSPORTATION

National air transport system: *number of registered air carriers:* 1 (2020)
inventory of registered aircraft operated by air carriers: 6
annual passenger traffic on registered air carriers: 253,417 (2018)
annual freight traffic on registered air carriers: 110,000 mt-km (2018)

Civil aircraft registration country code prefix: A2 (2016)

Airports: 74 (2013)
country comparison to the world: 70

Airports—with paved runways: *total:* 10 (2017)
over 3,047 m: 2 (2017)
2,438 to 3,047 m: 1 (2017)
1,524 to 2,437 m: 6 (2017)
914 to 1,523 m: 1 (2017)

Airports—with unpaved runways: *total:* 64 (2013)
1,524 to 2,437 m: 5 (2013)
914 to 1,523 m: 46 (2013)
under 914 m: 13 (2013)

Railways: *total:* 888 km (2014)
narrow gauge: 888 km 1.067-m gauge (2014)
country comparison to the world: 95

Roadways: *total:* 31,747 km (2017)
paved: 9,810 km (2017)
unpaved: 21,937 km (2017)
country comparison to the world: 95

MILITARY AND SECURITY

Military and security forces: Botswana Defence Force (BDF): Ground Forces Command, Air Arm Command, Defense Logistics Command (2020)

Military expenditures: 2.8% of GDP (2019)
2.8% of GDP (2018)
3% of GDP (2017)
3.4% of GDP (2016)
2.7% of GDP (2015)
country comparison to the world: 30

Military and security service personnel strengths: the Botswana Defense Force (BDF) has approximately 9,000 active personnel (8,500 Ground; 500 Air) (2019)

Military equipment inventories and acquisitions: the BDF has a mix of foreign-supplied weapons and equipment, largely from European suppliers, as well as the US; since 2010, it has received limited quantities of equipment from Canada, France, Spain, Switzerland, Ukraine, and the US (2019 est.)

Military service age and obligation: 18 is the legal minimum age for voluntary military service; no conscription (2012)

TRANSNATIONAL ISSUES

Disputes—international: none

Trafficking in persons: current situation: Botswana is a source, transit, and destination country for women and children subjected to sex trafficking and forced labor; young Batswana serving as domestic workers, sometimes sent by their parents, may be denied education and basic necessities or experience confinement and abuse indicative of forced labor; Batswana girls and women also are forced into prostitution domestically; adults and children of San ethnicity were reported to be in forced labor on farms and at cattle posts in the country's rural west *tier rating:* Tier 2 Watch List – Botswana does not fully comply with the minimum standards for the elimination of trafficking; however, it is making significant efforts to do so; an anti-trafficking act was passed at the beginning of 2014, but authorities did not investigate, prosecute, or convict any offenders or government officials complicit in trafficking or operationalize victim identification and referral procedures based on the new law; the government sponsored a radio campaign to familiarize the public with the issue of human trafficking (2015)

BOUVET ISLAND

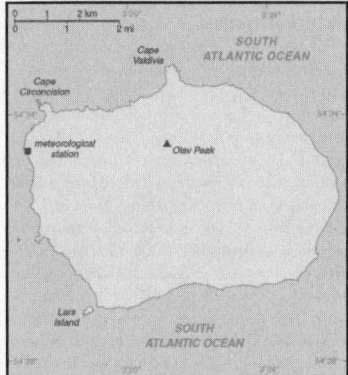

INTRODUCTION

Background: This uninhabited, volcanic, Antarctic island is almost entirely covered by glaciers making it difficult to approach; it is recognized as the most remote island on Earth. (It is furthest in distance from any other point of land, 1,639 km from Antarctica.) Bouvet Island was discovered in 1739 by a French naval officer after whom it is named. No claim was made until 1825, when the British flag was raised. A few expeditions visited the island in the late 19th century. In 1929, the UK waived its claim in favor of Norway, which had occupied the island two years previously. In 1971, Norway designated Bouvet Island and the adjacent territorial waters a nature reserve. Since 1977, Norway has run an automated meteorological station and studied foraging strategies and distribution of fur seals and penguins on the island. In February 2006, an earthquake weakened the station's foundation causing it to be blown out to sea in a winter storm. Norway erected a new research station in 2014 that can hold six people for periods of two to four months.

GEOGRAPHY

Location: island in the South Atlantic Ocean, southwest of the Cape of Good Hope (South Africa)

Geographic coordinates: 54 26 S, 3 24 E

Map references: Antarctic Region

Area: *total:* 49 sq km
land: 49 sq km
water: 0 sq km
country comparison to the world: 233

Area—comparative: about 0.3 times the size of Washington, DC

Land boundaries: 0 km

Coastline: 29.6 km

Maritime claims: *territorial sea:* 4 nm

Climate: antarctic

Terrain: volcanic; coast is mostly inaccessible

Elevation: *lowest point:* South Atlantic Ocean 0 m
highest point: Olavtoppen (Olav Peak) 780 m

Natural resources: none

Land use: *agricultural land:* 0% (2011 est.)
arable land: 0% (2011 est.) / permanent crops: 0% (2011 est.) / permanent pasture: 0% (2011 est.)
forest: 0% (2011 est.)
other: 100% (2011 est.)

Natural hazards: occasional volcanism, rock slides; harsh climate, surrounded by pack ice in winter

Environment—current issues: none; almost entirely ice covered

Geography—note: almost entirely covered by glacial ice (93%); declared a nature reserve by Norway; the distance from Bouvet Island to Norway is 12,776 km, which is almost one-third the circumference of the earth

PEOPLE AND SOCIETY

Population: uninhabited

GOVERNMENT

Country name: *conventional long form:* none
conventional short form: Bouvet Island
etymology: named after the French naval officer Jean-Baptiste Charles BOUVET who discovered the island in 1739
note: pronounced boo-vay i-land

Dependency status: territory of Norway; administered by the Polar Department of the Ministry of Justice and Oslo Police

Legal system: the laws of Norway apply where applicable

Flag description: the flag of Norway is used

ECONOMY

Economy—overview: no economic activity; declared a nature reserve

COMMUNICATIONS

Internet country code: .bv

Communications—note: has an automated meteorological station

TRANSPORTATION

Ports and terminals: none; offshore anchorage only

MILITARY AND SECURITY

Military—note: defense is the responsibility of Norway

TRANSNATIONAL ISSUES

Disputes—international: none

BRAZIL

INTRODUCTION

Background: Following more than three centuries under Portuguese rule, Brazil gained its independence in 1822, maintaining a monarchical system of government until the abolition of slavery in 1888 and the subsequent proclamation of a republic by the military in 1889. Brazilian coffee exporters politically dominated the country until populist leader Getulio VARGAS rose to power in 1930. By far the largest and most populous country in South America, Brazil underwent more than a half century of populist and military government until 1985, when the military regime peacefully ceded power to civilian rulers. Brazil continues to pursue industrial and agricultural growth and development of its interior. Having successfully weathered a period of global financial difficulty in the late

20th century, Brazil was seen as one of the world's strongest emerging markets and a contributor to global growth. The awarding of the 2014 FIFA World Cup and 2016 Summer Olympic Games, the first ever to be held in South America, was seen as symbolic of the country's rise. However, from about 2013 to 2016, Brazil was plagued by a sagging economy, high unemployment, and high inflation, only emerging from recession in 2017. Former President Dilma ROUSSEFF (2011-2016) was removed from office in 2016 by Congress for having committed impeachable acts against Brazil's budgetary laws, and her vice president, Michel TEMER, served the remainder of her second term. In October 2018, Jair BOLSONARO won the presidency with 55 percent of the vote and assumed office on 1 January 2019.

GEOGRAPHY

Location: Eastern South America, bordering the Atlantic Ocean

Geographic coordinates: 10 00 S, 55 00 W

Map references: South America

Area: total: 8,515,770 sq km
land: 8,358,140 sq km
water: 157,630 sq km
note: includes Arquipelago de Fernando de Noronha, Atol das Rocas, Ilha da Trindade, Ilhas Martin Vaz, and Penedos de Sao Pedro e Sao Paulo
country comparison to the world: 6

Area—comparative: slightly smaller than the US

Land boundaries: total: 16,145 km
border countries (10): Argentina 1263 km, Bolivia 3403 km, Colombia 1790 km, French Guiana 649 km, Guyana 1308 km, Paraguay 1371 km, Peru 2659 km, Suriname 515 km, Uruguay 1050 km, Venezuela 2137 km

Coastline: 7,491 km

Maritime claims: territorial sea: 12 nm
exclusive economic zone: 200 nm
contiguous zone: 24 nm
continental shelf: 200 nm or to edge of the continental margin

Climate: mostly tropical, but temperate in south

Terrain: mostly flat to rolling lowlands in north; some plains, hills, mountains, and narrow coastal belt

Elevation: mean elevation: 320 m
lowest point: Atlantic Ocean 0 m
highest point: Pico da Neblina 2,994 m

Natural resources: alumina, bauxite, beryllium, gold, iron ore, manganese, nickel, niobium, phosphates, platinum, tantalum, tin, rare earth elements, uranium, petroleum, hydropower, timber

Land use: agricultural land: 32.9% (2011 est.)
arable land: 8.6% (2011 est.) / permanent crops: 0.8% (2011 est.) / permanent pasture: 23.5% (2011 est.)
forest: 61.9% (2011 est.)
other: 5.2% (2011 est.)
Irrigated land: 54,000 sq km (2012)

Population distribution: the vast majority of people live along, or relatively near, the Atlantic coast in the east; the population core is in the southeast, anchored by the cities of Sao Paolo, Brasilia, and Rio de Janeiro

Natural hazards: recurring droughts in northeast; floods and occasional frost in south

Environment—current issues: deforestation in Amazon Basin destroys the habitat and endangers a multitude of plant and animal species indigenous to the area; illegal wildlife trade; illegal poaching; air and water pollution in Rio de Janeiro, Sao Paulo, and several other large cities; land degradation and water pollution caused by improper mining activities; wetland degradation; severe oil spills

Environment—international agreements: party to: Antarctic-Environmental Protocol, Antarctic-Marine Living Resources, Antarctic Seals, Antarctic Treaty, Biodiversity, Climate Change, Climate Change-Kyoto Protocol, Desertification, Endangered Species, Environmental Modification, Hazardous Wastes, Law of the Sea, Marine Dumping, Ozone Layer Protection, Ship Pollution, Tropical Timber 83, Tropical Timber 94, Wetlands, Whaling
signed, but not ratified: none of the selected agreements

Geography—note: largest country in South America and in the Southern Hemisphere; shares common boundaries with every South American country except Chile and Ecuador; most of the Pantanal, the world's largest tropical wetland, extends through the west central part of the country; shares Iguazu Falls, the world's largest waterfalls system, with Argentina

PEOPLE AND SOCIETY

Population: 211,715,973 (July 2020 est.)
country comparison to the world: 7

Nationality: noun: Brazilian(s)
adjective: Brazilian

Ethnic groups: white 47.7%, mulatto (mixed white and black) 43.1%, black 7.6%, Asian 1.1%, indigenous 0.4% (2010 est.)

Languages: Portuguese (official and most widely spoken language)

note: less common languages include Spanish (border areas and schools), German, Italian, Japanese, English, and a large number of minor Amerindian languages

Religions: Roman Catholic 64.6%, other Catholic 0.4%, Protestant 22.2% (includes Adventist 6.5%, Assembly of God 2.0%, Christian Congregation of Brazil 1.2%, Universal Kingdom of God 1.0%, other Protestant 11.5%), other Christian 0.7%, Spiritist 2.2%, other 1.4%, none 8%, unspecified 0.4% (2010 est.)

Demographic profile: Brazil's rapid fertility decline since the 1960s is the main factor behind the country's slowing population growth rate, aging population, and fast-paced demographic transition. Brasilia has not taken full advantage of its large working-age population to develop its human capital and strengthen its social and economic institutions but is funding a study abroad program to bring advanced skills back to the country. The current favorable age structure will begin to shift around 2025, with the labor force shrinking and the elderly starting to compose an increasing share of the total population. Well-funded public pensions have nearly wiped out poverty among the elderly, and Bolsa Familia and other social programs have lifted tens of millions out of poverty. More than half of Brazil's population is considered middle class, but poverty and income inequality levels remain high; the Northeast, North, and Center-West, women, and black, mixed race, and indigenous populations are disproportionately affected. Disparities in opportunities foster social exclusion and contribute to Brazil's high crime rate, particularly violent crime in cities and favelas (slums).

Brazil has traditionally been a net recipient of immigrants, with its southeast being the prime destination. After the importation of African slaves was outlawed in the mid-19th century, Brazil sought Europeans (Italians, Portuguese, Spaniards, and Germans) and later Asians (Japanese) to work in agriculture, especially coffee cultivation. Recent immigrants come mainly from Argentina, Chile, and Andean countries (many are unskilled illegal migrants) or are returning Brazilian nationals. Since Brazil's economic downturn in the 1980s, emigration to the United States, Europe, and Japan has been rising but is negligible relative to Brazil's total population. The majority of these emigrants are well-educated and middle-class. Fewer Brazilian peasants are emigrating to neighboring countries to take up agricultural work.

Age structure: 0-14 years: 21.11% (male 22,790,634/female 21,907,018)
15-24 years: 16.06% (male 17,254,363/female 16,750,581)
25-54 years: 43.83% (male 46,070,240/female 46,729,640)
55-64 years: 9.78% (male 9,802,995/female 10,911,140)
65 years and over: 9.21% (male 8,323,344/female 11,176,018) (2020 est.)

Dependency ratios: total dependency ratio: 43.5
youth dependency ratio: 29.7

elderly dependency ratio: 13.8
potential support ratio: 7.3 (2020 est.)

Median age: *total:* 33.2 years
male: 32.3 years
female: 34.1 years (2020 est.)
country comparison to the world: 101

Population growth rate: 0.67% (2020 est.)
country comparison to the world: 140

Birth rate: 13.6 births/1,000 population (2020 est.)
country comparison to the world: 137

Death rate: 6.9 deaths/1,000 population (2020 est.)
country comparison to the world: 129

Net migration rate: -0.1 migrant(s)/1,000 population (2020 est.)
country comparison to the world: 101

Population distribution: the vast majority of people live along, or relatively near, the Atlantic coast in the east; the population core is in the southeast, anchored by the cities of Sao Paolo, Brasilia, and Rio de Janeiro

Urbanization: *urban population:* 87.1% of total population (2020)
rate of urbanization: 1.05% annual rate of change (2015-20 est.)

total population growth rate v. urban population growth rate, 2000-2030: Major urban areas—population: 22.043 million Sao Paulo, 13.458 million Rio de Janeiro, 6.084 million Belo Horizonte, 4.646 million BRASILIA (capital), 4.137 million Porto Alegre, 4.127 million Recife (2020)

Sex ratio: *at birth:* 1.05 male(s)/female
0-14 years: 1.04 male(s)/female
15-24 years: 1.03 male(s)/female
25-54 years: 0.99 male(s)/female
55-64 years: 0.9 male(s)/female
65 years and over: 0.74 male(s)/female
total population: 0.97 male(s)/female (2020 est.)

Maternal mortality rate: 60 deaths/100,000 live births (2017 est.)
country comparison to the world: 88

Infant mortality rate: *total:* 15.9 deaths/1,000 live births
male: 18.8 deaths/1,000 live births
female: 12.9 deaths/1,000 live births (2020 est.)
country comparison to the world: 94

Life expectancy at birth: *total population:* 74.7 years
male: 71.2 years
female: 78.4 years (2020 est.)
country comparison to the world: 126

Total fertility rate: 1.73 children born/woman (2020 est.)
country comparison to the world: 165

Contraceptive prevalence rate: 80.2% (2013)

Drinking water source:
improved:
urban: 100% of population
rural: 91.6% of population
total: 98.2% of population
unimproved:

urban: 0% of population
rural: 8.4% of population
total: 1.6% of population (2017 est.)

Current Health Expenditure: 9.5% (2017)

Physicians density: 2.17 physicians/1,000 population (2017)

Hospital bed density: 2.1 beds/1,000 population (2017)

Sanitation facility access:
improved:
urban: 92.8% of population
rural: 60.1% of population
total: 88.3% of population
unimproved:
urban: 7.2% of population
rural: 39.9% of population
total: 11.7% of population (2017 est.)

HIV/AIDS—adult prevalence rate: 0.5% (2018 est.)
country comparison to the world: 66

HIV/AIDS—people living with HIV/AIDS: 920,000 (2019 est.)
country comparison to the world: 12

HIV/AIDS—deaths: 14,000 (2019 est.)
country comparison to the world: 12

Major infectious diseases: *degree of risk:* very high (2020)
food or waterborne diseases: bacterial diarrhea and hepatitis A
vectorborne diseases: dengue fever and malaria

water contact diseases: schistosomiasis
note: widespread ongoing transmission of a respiratory illness caused by the novel coronavirus (COVID-19) is occurring throughout Brazil; as of 10 November 2020, Brazil has reported a total of 5,631,181 cases of COVID-19 or 26,492 cumulative cases of COVID-19 per 1 million population with 762 cumulative deaths per 1 million population; the Department of Homeland Security has issued instructions requiring US passengers who have been in Brazil to travel through select airports where the US Government has implemented enhanced screening procedures

Obesity—adult prevalence rate: 22.1% (2016)
country comparison to the world: 82

Education expenditures: 6.2% of GDP (2015)
country comparison to the world: 28

Literacy: *definition:* age 15 and over can read and write
total population: 93.2%
male: 93%
female: 93.4% (2018)

School life expectancy (primary to tertiary education): *total:* 14 years
male: 14 years
female: 14 years (2011)

Unemployment, youth ages 15-24: *total:* 28.5%
male: 25.3%
female: 32.8% (2018 est.)
country comparison to the world: 39

GOVERNMENT

Country name: *conventional long form:* Federative Republic of Brazil
conventional short form: Brazil
local long form: Republica Federativa do Brasil
local short form: Brasil
etymology: the country name derives from the brazilwood tree that used to grow plentifully along the coast of Brazil and that was used to produce a deep red dye

Government type: federal presidential republic

Capital: *name:* Brasilia

geographic coordinates: 15 47 S, 47 55 W
time difference: UTC-3 (2 hours ahead of Washington, DC, during Standard Time)
daylight saving time: +1hr, begins third Sunday in October; ends third Sunday in February
note: Brazil has four time zones, including one for the Fernando de Noronha Islands
etymology: name bestowed on the new capital of Brazil upon its inauguration in 1960; previous Brazilian capitals had been Salvador from 1549 to 1763 and Rio de Janeiro from 1763 to 1960

Administrative divisions: 26 states (estados, singular—estado) and 1 federal district* (distrito federal); Acre, Alagoas, Amapa, Amazonas, Bahia, Ceara, Distrito Federal*, Espirito Santo, Goias, Maranhao, Mato Grosso, Mato Grosso do Sul, Minas Gerais, Para, Paraiba, Parana, Pernambuco, Piaui, Rio de Janeiro, Rio Grande do Norte, Rio Grande do Sul, Rondonia, Roraima, Santa Catarina, Sao Paulo, Sergipe, Tocantins

Independence: 7 September 1822 (from Portugal)

National holiday: Independence Day, 7 September (1822)

Constitution: *history:* several previous; latest ratified 5 October 1988
amendments: proposed by at least one third of either house of the National Congress, by the president of the republic, or by simple majority vote by more than half of the state legislative assemblies; passage requires at least three-fifths majority vote by both houses in each of two readings; constitutional provisions affecting the federal form of government, separation of powers, suffrage, or individual rights and guarantees cannot be amended; amended many times, last in 2017

Legal system: civil law; note—a new civil law code was enacted in 2002 replacing the 1916 code

International law organization participation: has not submitted an ICJ jurisdiction declaration; accepts ICCt jurisdiction

Citizenship: *citizenship by birth:* yes
citizenship by descent only: yes
dual citizenship recognized: yes
residency requirement for naturalization: 4 years

Suffrage: voluntary between 16 to 18 years of age, over 70, and if illiterate; compulsory between 18 to 70 years of age; note—military conscripts by law cannot vote

Executive branch: *chief of state:* President Jair BOLSONARO (since 1 January 2019); Vice President Antonio Hamilton Martins MOURAO (since 1 January 2019); note—the president is both chief of state and head of government

head of government: President Jair BOLSONARO (since 1 January 2019); Vice President Antonio Hamilton Martins MOURAO (since 1 January 2019)

cabinet: Cabinet appointed by the president

elections/appointments: president and vice president directly elected on the same ballot by absolute majority popular vote in 2 rounds if needed for a single 4-year term (eligible for a second term); election last held on 7 October 2018 with runoff on 28 October 2018 (next to be held in October 2022)

election results: Jair BOLSONARO elected president in second round; percent of vote in first round—Jair BOLSONARO (PSL) 46%, Fernando HADDAD (PT) 29.3%, Ciro GOMEZ (PDT) 12.5%, Geraldo ALCKMIN (PSDB) 4.8%, other 7.4%; percent of vote in second round—Jair BOLSONARO (PSL) 55.1%, Fernando HADDAD (PT) 44.9%

Legislative branch: *description:* bicameral National Congress or Congresso Nacional consists of:

Federal Senate or Senado Federal (81 seats; 3 members each from 26 states and 3 from the federal district directly elected in multi-seat constituencies by simple majority vote to serve 8-year terms, with one-third and two-thirds of the membership elected alternately every 4 years)

Chamber of Deputies or Camara dos Deputados (513 seats; members directly elected in multi-seat constituencies by proportional representation vote to serve 4-year terms)

elections: Federal Senate—last held on 7 October 2018 for two-thirds of the Senate (next to be held in October 2022 for one-third of the Senate)

Chamber of Deputies—last held on 7 October 2018 (next to be held in October 2022)

election results: Federal Senate—percent of vote by party—NA; seats by party—PMDB 7, PP 5, REDE 5, DEM 4, PSDB 4, PSDC 4, PSL 4, PT 4, PDT 2, PHS 2, PPS 2, PSB 2, PTB 2, Podemos 1, PR 1, PRB 1, PROS 1, PRP 1, PSC 1, SD 1; composition—men 70, women 11, percent of women 13.6%

Chamber of Deputies—percent of vote by party—NA; seats by party—PT 56, PSL 52, PP 37, PMDB 34, PSDC 34, PR 33, PSB 32, PRB 30, DEM 29, PSDB 29, PDT 28, SD 13, Podemos 11, PSOL 10, PTB 10, PCdoB 9, NOVO 8, PPS 8, PROS 8, PSC 8, Avante 7, PHS 6, Patriota 5, PRP 4, PV 4, PMN 3, PTC 2, DC 1, PPL 1, REDE 1; composition—men 462, women 51, percent of women 9.9%; total National Congress percent of women 10.4%

Judicial branch: *highest courts:* Supreme Federal Court or Supremo Tribunal Federal (consists of 11 justices)

judge selection and term of office: justices appointed by the president and approved by the Federal Senate; justices appointed to serve until mandatory retirement at age 75

subordinate courts: Tribunal of the Union, Federal Appeals Court, Superior Court of Justice, Superior Electoral Court, regional federal courts; state court system

Political parties and leaders: Avante [Luis TIBE] (formerly Labor Party of Brazil or PTdoB)

Brazilian Communist Party or PCB [Ivan Martins PINHEIRO]

Brazilian Democratic Movement Party or PMDB [Michel TEMER]

Brazilian Labor Party or PTB [Cristiane BRASIL]

Brazilian Renewal Labor Party or PRTB [Jose Levy FIDELIX da Cruz]

Brazilian Republican Party or PRB [Marcos Antonio PEREIRA]

Brazilian Social Democracy Party or PSDB [Tasso JEREISSATI]

Brazilian Socialist Party or PSB [Carlos Roberto SIQUEIRA de Barros]

Christian Democracy or DC [Jose Maria EYMAEL] (formerly Christian Social Democratic Party or PSDC)

Christian Labor Party or PTC [Daniel TOURINHO]

Communist Party of Brazil or PCdoB [Jose Renato RABELO]

Democratic Labor Party or PDT [Carlos Roberto LUPI]

The Democrats or DEM [Jose AGRIPINO] (formerly Liberal Front Party or PFL)

Free Homeland Party or PPL [Sergio RUBENS]

Green Party or PV [Jose Luiz PENNA]

Humanist Party of Solidarity or PHS [Eduardo MACHADO]

National Mobilization Party or PMN [Telma RIBEIRO dos Santos]

New Party or NOVO [Moises JARDIM]

Party of the Republic or PR [Alfredo NASCIMENTO]

Patriota [Adilson BARROSO Oliveira] (formerly National Ecologic Party or PEN)

Podemos [Renata ABREU] (formerly National Labor Party or PTN)

Popular Socialist Party or PPS [Roberto Joao Pereira FREIRE]

Progressive Party or PP [Ciro NOGUEIRA]

Progressive Republican Party or PRP [Ovasco Roma Altimari RESENDE]

Republican Social Order Party or PROS [Euripedes JUNIOR]

Social Christian Party or PSC [Vitor Jorge Abdala NOSSEIS]

Social Democratic Party or PSD [Guilherme CAMPOS]

Social Liberal Party or PSL [Luciano Caldas BIVAR]

Socialism and Freedom Party or PSOL [Luiz ARAUJO]

Solidarity or SD [Paulo PEREIRA DA SILVA]

Sustainability Network or REDE [Marina SILVA]

United Socialist Workers' Party or PSTU [Jose Maria DE ALMEIDA]

Workers' Cause Party or PCO [Rui Costa PIMENTA]

Workers' Party or PT [Gleisi HOFFMAN]

International organization participation: AfDB (nonregional member), BIS, BRICS, CAN (associate), CD, CELAC, CPLP, FAO, FATF, G-15, G-20, G-24, G-5, G-77, IADB, IAEA, IBRD, ICAO, ICC (national committees), ICCt, ICRM, IDA, IFAD, IFC, IFRCS, IHO, ILO, IMF, IMO, IMSO, Interpol, IOC, IOM, IPU, ISO, ITSO, ITU, ITUC (NGOs), LAES, LAIA, LAS (observer), Mercosur, MIGA, MINURSO, MINUSTAH, MONUSCO, NAM (observer), NSG, OAS, OECD (enhanced engagement), OPANAL, OPCW, Paris Club (associate), PCA, SICA (observer), UN, UNASUR, UNCTAD, UNESCO, UNFICYP, UNHCR, UNIDO, UNIFIL, Union Latina, UNISFA, UNITAR, UNMIL, UNMISS, UNOCI, UNRWA, UNWTO, UPU, WCO, WFTU (NGOs), WHO, WIPO, WMO, WTO

Diplomatic representation in the US: *chief of mission:* Ambassador (vacant); Charge·d'Affaires Nestor Jose FORSTER (since 11 July 2019)

chancery: 3006 Massachusetts Avenue NW, Washington, DC 20008

telephone: [1] (202) 238-2700

FAX: [1] (202) 238-2827

consulate(s) general: Atlanta, Boston, Chicago, Hartford (CT), Houston, Los Angeles, Miami, New York, San Francisco, Washington, DC

Diplomatic representation from the US: *chief of mission:* Ambassador (vacant); Charge d'Affaires William POPP (since 3 November 2018)

telephone: [55] (61) 3312-7000

embassy: Avenida das Nacoes, Quadra 801, Lote 3, Distrito Federal Cep 70403-900, Brasilia

mailing address: Unit 7500, DPO AA 34030

FAX: [55] (61) 3225-9136

consulate(s) general: Belo Horizonte, Recife, Porto Alegre, Rio de Janeiro, Sao Paulo

Flag description: green with a large yellow diamond in the center bearing a blue celestial globe with 27 white five-pointed stars; the globe has a white equatorial band with the motto ORDEM E PROGRESSO (Order and Progress); the current flag was inspired by the banner of the former Empire of Brazil (1822-1889); on the imperial flag, the green represented the House of Braganza of Pedro I, the first Emperor of Brazil, while the yellow stood for the Habsburg Family of his wife; on the modern flag the green represents the forests of the country and the yellow rhombus its mineral wealth (the diamond shape roughly mirrors that of the country); the blue circle and stars, which replaced the coat of arms of the original flag, depict the sky over Rio de Janeiro on the morning of 15 November 1889—the day the Republic of Brazil was declared; the number of stars has changed with the creation of new states and has risen from an original 21 to the current 27 (one for each state and the Federal District)

note: one of several flags where a prominent component of the design reflects the shape of the country; other such flags are those of Bosnia and Herzegovina, Eritrea, and Vanuatu

National symbol(s): Southern Cross constellation; national colors: green, yellow, blue

National anthem: *name:* "Hino Nacional Brasileiro" (Brazilian National Anthem) *lyrics/music:* Joaquim Osorio Duque ESTRADA/ Francisco Manoel DA SILVA *note:* music adopted 1890, lyrics adopted 1922; the anthem's music, composed in 1822, was used unofficially for many years before it was adopted 0:00 / 1:52

ECONOMY

Economy—overview: Brazil is the eighth-largest economy in the world, but is recovering from a recession in 2015 and 2016 that ranks as the worst in the country's history. In 2017, Brazil's GDP grew 1%, inflation fell to historic lows of 2.9%, and the Central Bank lowered benchmark interest rates from 13.75% in 2016 to 7%.

The economy has been negatively affected by multiple corruption scandals involving private companies and government officials, including the impeachment and conviction of Former President Dilma ROUSSEFF in August 2016. Sanctions against the firms involved — some of the largest in Brazil — have limited their business opportunities, producing a ripple effect on associated businesses and contractors but creating opportunities for foreign companies to step into what had been a closed market.

The succeeding TEMER administration has implemented a series of fiscal and structural reforms to restore credibility to government finances. Congress approved legislation in December 2016 to cap public spending. Government spending growth had pushed public debt to 73.7% of GDP at the end of 2017, up from over 50% in 2012. The government also boosted infrastructure projects, such as oil and natural gas auctions, in part to raise revenues. Other economic reforms, proposed in 2016, aim to reduce barriers to foreign investment, and to improve labor conditions. Policies to strengthen Brazil's workforce and industrial sector, such as local content requirements, have boosted employment, but at the expense of investment.

Brazil is a member of the Common Market of the South (Mercosur), a trade bloc that includes Argentina, Paraguay and Uruguay—Venezuela's membership in the organization was suspended In August 2017. After the Asian and Russian financial crises, Mercosur adopted a protectionist stance to guard against exposure to volatile foreign markets and it currently is negotiating Free Trade Agreements with the European Union and Canada.

GDP (purchasing power parity): $3.248 trillion (2017 est.)
$3.216 trillion (2016 est.)
$3.332 trillion (2015 est.)
note: data are in 2017 dollars
country comparison to the world: 8

GDP (official exchange rate): $2.055 trillion (2017 est.)

GDP—real growth rate: 1.13% (2019 est.)

1.2% (2018 est.)
1.62% (2017 est.)
country comparison to the world: 169

GDP—per capita (PPP): $15,600 (2017 est.)
$15,600 (2016 est.)
$16,300 (2015 est.)
note: data are in 2017 dollars
country comparison to the world: 108

Gross national saving: 15% of GDP (2017 est.)
14.1% of GDP (2016 est.)
14.1% of GDP (2015 est.)
country comparison to the world: 136

GDP—composition, by end use: *household consumption:* 63.4% (2017 est.)
government consumption: 20% (2017 est.)
investment in fixed capital: 15.6% (2017 est.)
investment in inventories: -0.1% (2017 est.)
exports of goods and services: 12.6% (2017 est.)
imports of goods and services: -11.6% (2017 est.)

GDP—composition, by sector of origin: *agriculture:* 6.6% (2017 est.)
industry: 20.7% (2017 est.)
services: 72.7% (2017 est.)

Agriculture—products: coffee, soybeans, wheat, rice, corn, sugarcane, cocoa, citrus; beef

Industries: textiles, shoes, chemicals, cement, lumber, iron ore, tin, steel, aircraft, motor vehicles and parts, other machinery and equipment

Industrial production growth rate: 0% (2017 est.)
country comparison to the world: 169

Labor force: 86.621 million (2020 est.)
country comparison to the world: 4

Labor force—by occupation: *agriculture:* 9.4%
industry: 32.1%
services: 58.5% (2017 est.)

Unemployment rate: 11.93% (2019 est.)
12.26% (2018 est.)
country comparison to the world: 163

Population below poverty line: 4.2% (2016 est.)
note: approximately 4% of the population are below the "extreme" poverty line

Household income or consumption by percentage share: *lowest 10%:* 0.8%
highest 10%: 43.4% (2016 est.)

Budget: *revenues:* 733.7 billion (2017 est.)
expenditures: 756.3 billion (2017 est.)

Taxes and other revenues: 35.7% (of GDP) (2017 est.)
country comparison to the world: 57

Budget surplus (+) or deficit (-): -1.1% (of GDP) (2017 est.)
country comparison to the world: 82

Public debt: 84% of GDP (2017 est.)
78.4% of GDP (2016 est.)
country comparison to the world: 32

Fiscal year: calendar year

Inflation rate (consumer prices): 3.4% (2017 est.)
8.7% (2016 est.)
country comparison to the world: 139

Current account balance: -$50.927 billion (2019 est.)
-$41.54 billion (2018 est.)

country comparison to the world: 203

Exports: $217.2 billion (2017 est.)
$184.5 billion (2016 est.)
country comparison to the world: 26

Exports—partners: China 21.8%, US 12.5%, Argentina 8.1%, Netherlands 4.3% (2017)

Exports—commodities: transport equipment, iron ore, soybeans, footwear, coffee, automobiles

Imports: $153.2 billion (2017 est.)
$139.4 billion (2016 est.)
country comparison to the world: 29

Imports—commodities: machinery, electrical and transport equipment, chemical products, oil, automotive parts, electronics

Imports—partners: China 18.1%, US 16.7%, Argentina 6.3%, Germany 6.1% (2017)

Reserves of foreign exchange and gold: $374 billion (31 December 2017 est.)
$367.5 billion (31 December 2016 est.)
country comparison to the world: 10

Debt—external: $547.4 billion (31 December 2017 est.)
$548.6 billion (31 December 2016 est.)
country comparison to the world: 21

Exchange rates: reals (BRL) per US dollar—
3.19 (2017 est.)
3.48 (2016 est.)
3.4901 (2015 est.)
3.3315 (2014 est.)
2.3535 (2013 est.)

ENERGY

Electricity access: *electrification—total population:* 100% (2020)

Electricity—production: 567.9 billion kWh (2016 est.)
country comparison to the world: 8

Electricity—consumption: 509.1 billion kWh (2016 est.)
country comparison to the world: 8

Electricity—exports: 219 million kWh (2015 est.)
country comparison to the world: 73

Electricity—imports: 41.31 billion kWh (2016 est.)
country comparison to the world: 3

Electricity—installed generating capacity: 150.8 million kW (2016 est.)
country comparison to the world: 7

Electricity—from fossil fuels: 17% of total installed capacity (2016 est.)
country comparison to the world: 197

Electricity—from nuclear fuels: 1% of total installed capacity (2017 est.)
country comparison to the world: 28

Electricity—from hydroelectric plants: 64% of total installed capacity (2017 est.)
country comparison to the world: 24

Electricity—from other renewable sources: 18% of total installed capacity (2017 est.)
country comparison to the world: 46

Crude oil—production: 2.587 million bbl/day (2018 est.)
country comparison to the world: 10

Crude oil—exports: 736,600 bbl/day (2015 est.)
country comparison to the world: 17

Crude oil—imports: 297,700 bbl/day (2015 est.)
country comparison to the world: 25

Crude oil—proved reserves: 12.63 billion bbl (1 January 2018 est.)
country comparison to the world: 14

Refined petroleum products—production: 2.811 million bbl/day (2015 est.)
country comparison to the world: 7

Refined petroleum products—consumption: 2.956 million bbl/day (2016 est.)
country comparison to the world: 7

Refined petroleum products—exports: 279,000 bbl/day (2015 est.)
country comparison to the world: 28

Refined petroleum products—imports: 490,400 bbl/day (2015 est.)
country comparison to the world: 17

Natural gas—production: 23.96 billion cu m (2017 est.)
country comparison to the world: 29

Natural gas—consumption: 34.35 billion cu m (2017 est.)
country comparison to the world: 28

Natural gas—exports: 134.5 million cu m (2017 est.)
country comparison to the world: 48

Natural gas—imports: 10.51 billion cu m (2017 est.)
country comparison to the world: 26

Natural gas—proved reserves: 377.4 billion cu m (1 January 2018 est.)
country comparison to the world: 34

Carbon dioxide emissions from consumption of energy: 513.8 million Mt (2017 est.)
country comparison to the world: 13

COMMUNICATIONS

Telephones—fixed lines: *total subscriptions:* 33,585,164
subscriptions per 100 inhabitants: 15.97 (2019 est.)
country comparison to the world: 6

Telephones—mobile cellular: *total subscriptions:* 207,862,093
subscriptions per 100 inhabitants: 98.84 (2019 est.)
country comparison to the world: 6

Telecommunication systems: *general assessment:* Brazil is one of the largest mobile and broadband markets in Latin America; 5G auction delayed due to interference issues; four major (mobile network operators) MNOs offering a range of voice and data services; broadband penetration only behind Chile, Argentina, and Uruguay; country is a pioneer in the region for M-commerce (electronic commerce conducted on mobile phones) (2020)

domestic: fixed-line connections have remained relatively stable in recent years and stand at about 16 per 100 persons; less-expensive mobile-cellular technology has been a major impetus broadening telephone service to the lower-income segments of the population with mobile-cellular teledensity roughly 99 per 100 persons (2019)

international: country code—55; landing points for a number of submarine cables, including Malbec, ARBR, Tamnat, SAC, SAm-1, Atlantis -2, Seabras-1, Monet, EllaLink, BRUSA, GlobeNet, AMX-1, Brazilian Festoon, Bicentenario, Unisur, Junior, Americas -II, SAE x1, SAIL, SACS and SABR that provide direct connectivity to South and Central America, the Caribbean, the US, Africa, and Europe; satellite earth stations—3 Intelsat (Atlantic Ocean), 1 Inmarsat (Atlantic Ocean region east), connected by microwave relay system to Mercosur Brazilsat B3 satellite earth station; satellites is a major communication platform, as it is almost impossible to lay fiber optic cable in the thick vegetation (2019)

note: the COVID-19 outbreak is negatively impacting telecommunications production and supply chains globally; consumer spending on telecom devices and services has also slowed due to the pandemic's effect on economies worldwide; overall progress towards improvements in all facets of the telecom industry—mobile, fixed-line, broadband, submarine cable and satellite—has moderated

Broadcast media: state-run Radiobras operates a radio and a TV network; more than 1,000 radio stations and more than 100 TV channels operating—mostly privately owned; private media ownership highly concentrated

Internet country code: .br

Internet users: *total:* 140,908,998

percent of population: 67.47% (July 2018 est.)
country comparison to the world: 4

Broadband—fixed subscriptions: *total:* 31,233,004
subscriptions per 100 inhabitants: 15 (2018 est.)
country comparison to the world: 6

TRANSPORTATION

National air transport system: *number of registered air carriers:* 9 (2020)
inventory of registered aircraft operated by air carriers: 443
annual passenger traffic on registered air carriers: 102,109,977 (2018)
annual freight traffic on registered air carriers: 1,845,650,000 mt-km (2018)

Civil aircraft registration country code prefix: PP (2016)

Airports: 4,093 (2013)
country comparison to the world: 2

Airports—with paved runways: *total:* 698 (2017)
over 3,047 m: 7 (2017)
2,438 to 3,047 m: 27 (2017)
1,524 to 2,437 m: 179 (2017)
914 to 1,523 m: 436 (2017)
under 914 m: 49 (2017)

Airports—with unpaved runways: *total:* 3,395 (2013)
1,524 to 2,437 m: 92 (2013)
914 to 1,523 m: 1,619 (2013)
under 914 m: 1,684 (2013)

Heliports: 13 (2013)

Pipelines: 5959 km refined petroleum product (1,165 km distribution, 4,794 km transport), 11696 km natural gas (2,274 km distribution, 9,422 km transport), 1985 km crude oil (distribution), 77 km ethanol/petrochemical (37 km distribution, 40 km transport) (2016)

Railways: *total:* 29,850 km (2014)
standard gauge: 194 km 1.435-m gauge (2014)
narrow gauge: 23,341.6 km 1.000-m gauge (24 km electrified) (2014)
broad gauge: 5,822.3 km 1.600-m gauge (498.3 km electrified) (2014)
dual gauge: 492 km 1.600-1.000-m gauge (2014)
country comparison to the world: 9

Roadways: *total:* 2 million km (2018)
paved: 246,000 km (2018)
unpaved: 1.754 million km (2018)
country comparison to the world: 4

Waterways: 50,000 km (most in areas remote from industry and population) (2012)
country comparison to the world: 3

Merchant marine: *total:* 864
by type: bulk carrier 12, container ship 17, general cargo 45, oil tanker 41, other 749 (2019)
country comparison to the world: 26

Ports and terminals: *major seaport(s):* Belem, Paranagua, Rio Grande, Rio de Janeiro, Santos, Sao Sebastiao, Tubarao
oil terminal(s): DTSE/Gegua oil terminal, Ilha Grande (Gebig), Guaiba Island terminal, Guamare oil terminal
container port(s) (TEUs): Santos (3,853,719) (2017)
LNG terminal(s) (import): Pecem, Rio de Janiero
river port(s): Manaus (Amazon)
dry bulk cargo port(s): Sepetiba ore terminal, Tubarao

MILITARY AND SECURITY

Military and security forces: Brazilian Armed Forces: Brazilian Army (Exercito Brasileiro, EB), Brazilian Navy (Marinha do Brasil, MB, includes Naval Aviation and Marine Corps (Corpo de Fuzileiros Navais)), Brazilian Air Force (Forca Aerea Brasileira, FAB); Public Security Forces (2020)

Military expenditures: 1.5% of GDP (2019)
1.5% of GDP (2018)
1.4% of GDP (2017)
1.4% of GDP (2016)
1.4% of GDP (2015)
country comparison to the world: 78

Military and security service personnel strengths: size assessments for the Brazilian Armed Forces vary; approximately 360,000 active personnel (215,000 Army; 75,000 Navy; 70,000 Air Force) (2019 est.)

Military equipment inventories and acquisitions: the Brazilian military's inventory consists of a mix of domestically-produced and imported weapons, largely from Europe and the US; since 2010, France, Germany, the UK, and the US are the leading suppliers of military equipment to Brazil; Brazil's defense industry is capable of designing and manufacturing equipment for all three military services and for export; it also jointly produces equipment with other countries (2019)

Military deployments: 220 Lebanon (UNIFIL) (2020)

Military service age and obligation: 18-45 years of age for compulsory military service; conscript service obligation is 10-12 months; 17-45 years of age for voluntary service; an increasing percentage of the ranks are "long-service" volunteer professionals; women were allowed to serve in the armed forces beginning in early 1980s, when the Brazilian Army became the first army in South America to accept women into career ranks; women serve in Navy and Air Force only in Women's Reserve Corps (2012)

Military—note: the military's primary role is enforcing border security, particularly in the Amazon states; it also assists with internal security operations with a focus on organized crime

Brazilian police forces are divided into Federal Police (around 15,000 personnel), Military Police (approximately 400,000 personnel), and Civil Police (approximately 125,000 personnel); the Federal Police serve under the Ministry of Justice, while the Military and Civil police are subordinate to the state governments; the National Public Security Force (Forca Nacional de Seguranca Publica or SENASP) is a national police force made up of Military Police from various states; article 144 of the Brazilian constitution states that all state Military Police are classified as reserve troops and ancillary forces of the Brazilian Army

TRANSNATIONAL ISSUES

Disputes—international: uncontested boundary dispute between Brazil and Uruguay over Braziliera/Brasiliera Island in the Quarai/Cuareim River leaves the tripoint with Argentina in question; smuggling of firearms and narcotics continues to be an issue along the Uruguay-Brazil border; Colombian-organized illegal narcotics and paramilitary activities penetrate Brazil's border region with Venezuela

Refugees and internally displaced persons: *refugees (country of origin):* 251,832 (Venezuela) (economic and political crisis; includes Venezuelans who have claimed asylum, are recognized as refugees, or received alternative legal stay) (2020)

stateless persons: 7 (2019)

Illicit drugs: second-largest consumer of cocaine in the world; illicit producer of cannabis; trace amounts of coca cultivation in the Amazon region, used for domestic consumption; government has a large-scale eradication program to control cannabis; important transshipment country for Bolivian, Colombian, and Peruvian cocaine headed for Europe; also used by traffickers as a way station for narcotics air transshipments between Peru and Colombia; upsurge in drug-related violence and weapons smuggling; important market for Colombian, Bolivian, and Peruvian cocaine; illicit narcotics proceeds are often laundered through the financial system; significant illicit financial activity in the Tri-Border Area

BRITISH INDIAN OCEAN TERRITORY

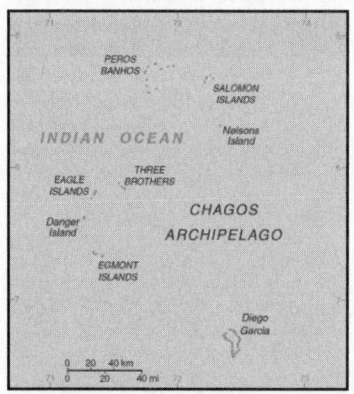

INTRODUCTION

Background: Formerly administered as part of the British Crown Colony of Mauritius, the British Indian Ocean Territory (BIOT) was established as an overseas territory of the UK in 1965. A number of the islands of the territory were later transferred to the Seychelles when it attained independence in 1976. Subsequently, BIOT has consisted only of the six main island groups comprising the Chagos Archipelago. Only Diego Garcia, the largest and most southerly of the islands, is inhabited. It contains a joint UK-US naval support facility and hosts one of four dedicated ground antennas that assist in the operation of the Global Positioning System (GPS) navigation system (the others are on Kwajalein (Marshall Islands), at Cape Canaveral, Florida (US), and on Ascension Island (Saint Helena, Ascension, and Tristan da Cunha)). The US Air Force also operates a telescope array on Diego Garcia as part of the Ground-Based Electro-Optical Deep Space Surveillance System (GEODSS) for tracking orbital debris, which can be a hazard to spacecraft and astronauts.

Between 1967 and 1973, former agricultural workers, earlier residents in the islands, were relocated primarily to Mauritius, but also to the Seychelles. Negotiations between 1971 and 1982 resulted in the establishment of a trust fund by the British Government as compensation for the displaced islanders, known as Chagossians. Beginning in 1998, the islanders pursued a series of lawsuits against the British Government seeking further compensation and the right to return to the territory. In 2006 and 2007, British court rulings invalidated the immigration policies contained in the 2004 BIOT Constitution Order that had excluded the islanders from the archipelago, but upheld the special military status of Diego Garcia. In 2008, the House of Lords, as the final court of appeal in the UK, ruled in favor of the British Government by overturning the lower court rulings and finding no right of return for the Chagossians. In March 2015, the Permanent Court of Arbitration unanimously held that the marine protected area (MPA) that the UK declared around the Chagos Archipelago in April 2010 was in violation of the UN Convention on the Law of the Sea.

In February 2019, the International Court of Justice ruled in an advisory opinion that Britain's decolonization of Mauritius was not completed lawfully because of continued Chagossian claims. A non-binding May 2019 UN General Assembly vote demanded that Britain end its "colonial administration" of the Chagos Archipelago and that it be returned to Mauritius. UK officials defend Britain's sovereignty over the islands and argue that the issue is a bilateral dispute between Mauritius and the UK that does not warrant international intervention.

GEOGRAPHY

Location: archipelago in the Indian Ocean, south of India, about halfway between Africa and Indonesia

Geographic coordinates: 6 00 S, 71 30 E;note—Diego Garcia 7 20 S, 72 25 E

Map references: Political Map of the World

Area: *total:* 60 sq km

land: 60 sq km (44 Diego Garcia)

water: 54,340 sq km

note: includes the entire Chagos Archipelago of 55 islands

country comparison to the world: 230

Area—comparative: land area is about one-third the size of Washington, DC

Land boundaries: 0 km

Coastline: 698 km

Maritime claims: *territorial sea:* 12 nm

Environment (Protection and Preservation) Zone: 200 nm

Climate: tropical marine; hot, humid, moderated by trade winds

Terrain: flat and low (most areas do not exceed two m in elevation)

Elevation: *lowest point:* Indian Ocean 0 m
highest point: ocean-side dunes on Diego Garcia 9 m

Natural resources: coconuts, fish, sugarcane

Land use: *agricultural land:* 0% (2011 est.)
arable land: 0% (2011 est.) / permanent crops: 0% (2011 est.) / permanent pasture: 0% (2011 est.)
forest: 0% (2011 est.)
other: 100% (2011 est.)

Natural hazards: none; located outside routes of Indian Ocean cyclones

Environment—current issues: wastewater discharge into the lagoon on Diego Garcia

Geography—note: *note 1:* archipelago of 55 islands; Diego Garcia, the largest and southernmost island, occupies a strategic location in the central Indian Ocean; the island is the site of a joint US-UK military facility
note 2: Diego Garcia is the only inhabited island of the BIOT and one of only two British territories where traffic drives on the right, the other being Gibraltar

PEOPLE AND SOCIETY

Population: no indigenous inhabitants
note: approximately 1,200 former agricultural workers resident in the Chagos Archipelago, often referred to as Chagossians or Ilois, were relocated to Mauritius and the Seychelles in the 1960s and 1970s; approximately 3,000 UK and US military personnel and civilian contractors living on the island of Diego Garcia (2018)

GOVERNMENT

Country name: *conventional long form:* British Indian Ocean Territory
conventional short form: none
abbreviation: BIOT
etymology: self-descriptive name specifying the territory's affiliation and location

Dependency status: overseas territory of the UK; administered by a commissioner, resident in the Foreign and Commonwealth Office in London

Legal system: the laws of the UK apply where applicable

Executive branch: *chief of state:* Queen ELIZABETH II (since 6 February 1952)
head of government: Commissioner Dr. Peter HAYES (since 17 October 2012); Administrator John MCMANUS (since April 2011); note—both reside in the UK and are represented by the officer commanding British Forces on Diego Garcia
cabinet: NA
elections/appointments: the monarchy is hereditary; commissioner and administrator appointed by the monarch

International organization participation: UPU

Diplomatic representation in the US: none (overseas territory of the UK)

Diplomatic representation from the US: none (overseas territory of the UK)

Flag description: white with six blue wavy horizontal stripes; the flag of the UK is in the upper hoist-side quadrant; the striped section bears a palm tree and yellow crown (the symbols of the territory) centered on the outer half of the flag; the wavy stripes represent the Indian Ocean; although not officially described, the six blue stripes may stand for the six main atolls of the archipelago

ECONOMY

Economy—overview: All economic activity is concentrated on the largest island of Diego Garcia, where a joint UK-US military facility is located. Construction projects and various services needed to support the military installation are performed by military and contract employees from the UK, Mauritius, the Philippines, and the US. Some of the natural resources found in this territory include coconuts, fish, and sugarcane.

Exchange rates: the US dollar is used

COMMUNICATIONS

Telecommunication systems: *general assessment:* separate facilities for military and public needs are available (2018)
domestic: all commercial telephone services are available, including connection to the Internet (2018)
international: country code (Diego Garcia)—246; landing point for the SAFE submarine cable that provides direct connectivity to Africa, Asia and near-by Indian Ocean island countries; international telephone service is carried by satellite (2019)

Broadcast media: Armed Forces Radio and Television Service (AFRTS) broadcasts over 3 separate frequencies for US and UK military personnel stationed on the islands

Internet country code: .io

Communications—note: Diego Garcia hosts one of four dedicated ground antennas that assist in the operation of the Global Positioning System (GPS) navigation system (the others are on Kwajalein (Marshall Islands), at Cape Canaveral, Florida (US), and on Ascension Island (Saint Helena, Ascension, and Tristan da Cunha))

TRANSPORTATION

Airports: 1 (2020)
country comparison to the world: 215
Airports—with paved runways: *total:* 1 (2019)
over 3,047 m: 1
Roadways: note: short section of paved road between port and airfield on Diego Garcia

Ports and terminals: *major seaport(s):* Diego Garcia

MILITARY AND SECURITY

Military and security forces: no regular military forces (2014)

Military—note: defense is the responsibility of the UK; in November 2016, the UK extended the US lease on Diego Garcia for 20 years; the lease now expires in December 2036 (2016)

TRANSNATIONAL ISSUES

Disputes—international: Mauritius and Seychelles claim the Chagos Islands; negotiations between 1971 and 1982 resulted in the establishment of a trust fund by the British Government as compensation for the displaced islanders, known as Chagossians, who were evicted between 1967-73; in 2001, the former inhabitants of the archipelago were granted UK citizenship and the right of return; in 2006 and 2007, British court rulings invalidated the immigration policies contained in the 2004 BIOT Constitution Order that had excluded the islanders from the archipelago; in 2008 a House of Lords' decision overturned lower court rulings, once again denying the right of return to Chagossians; in addition, the UK created the world's largest marine protection area around the Chagos islands prohibiting the extraction of any natural resources therein

BRITISH VIRGIN ISLANDS

INTRODUCTION

Background: First inhabited by Arawak and later by Carib Indians, the Virgin Islands were settled by the Dutch in 1648 and then annexed by the English in 1672. The islands were part of the British colony of the Leeward Islands from 1872-1960; they were granted autonomy in 1967. The economy is closely tied to the larger and more populous US Virgin Islands to the west; the US dollar is the legal currency. On 6 September 2017, Hurricane Irma devastated the island of Tortola. An estimated 80% of residential and business

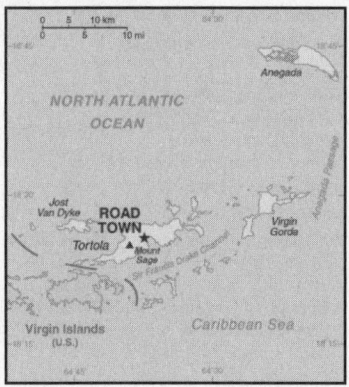

structures were destroyed or damaged, communications disrupted, and local roads rendered impassable.

GEOGRAPHY

Location: Caribbean, between the Caribbean Sea and the North Atlantic Ocean, east of Puerto Rico

Geographic coordinates: 18 30 N, 64 30 W

Map references: Central America and the Caribbean

Area: *total:* 151 sq km
land: 151 sq km
water: 0 sq km
note: comprised of 16 inhabited and more than 20 uninhabited islands; includes the islands of Tortola, Anegada, Virgin Gorda, Jost van Dyke
country comparison to the world: 220

Area—comparative: about 0.9 times the size of Washington, DC

Land boundaries: 0 km

Coastline: 80 km

Maritime claims: *territorial sea:* 3 nm
exclusive fishing zone: 200 nm

Climate: subtropical; humid; temperatures moderated by trade winds

Terrain: coral islands relatively flat; volcanic islands steep, hilly

Elevation: *lowest point:* Caribbean Sea 0 m
highest point: Mount Sage 521 m

Natural resources: NEGL; pleasant climate, beaches foster tourism

Land use: *agricultural land:* 46.7% (2011 est.)
arable land: 6.7% (2011 est.) / permanent crops: 6.7% (2011 est.) / permanent pasture: 33.3% (2011 est.)
forest: 24.3% (2011 est.)
other: 29% (2011 est.)

Irrigated land: NA

Population distribution: a fairly even distribution throughout the inhabited islands, with the largest islands of Tortola, Anegada, Virgin Gorda, and Jost Van Dyke having the largest populations

Natural hazards: hurricanes and tropical storms (July to October)

Environment—current issues: limited natural freshwater resources except for a few seasonal streams and springs on Tortola; most of the islands' water supply comes from desalination plants; sewage and mining/industry waste contribute to water pollution, threatening coral reefs

Geography—note: strong ties to nearby US Virgin Islands and Puerto Rico

PEOPLE AND SOCIETY

Population: 37,381 (July 2020 est.)
country comparison to the world: 215

Nationality: *noun:* British Virgin Islander(s)
adjective: British Virgin Islander

Ethnic groups: African/black 76.3%, Latino 5.5%, white 5.4%, mixed 5.3%, Indian 2.1%, East Indian 1.6%, other 3%, unspecified 0.8% (2010 est.)

Languages: English (official)

Religions: Protestant 70.2% (Methodist 17.6%, Church of God 10.4%, Anglican 9.5%, Seventh Day Adventist 9.0%, Pentecostal 8.2%, Baptist 7.4%, New Testament Church of God 6.9%, other Protestant 1.2%), Roman Catholic 8.9%, Jehovah's Witness 2.5%, Hindu 1.9%, other 6.2%, none 7.9%, unspecified 2.4% (2010 est.)

Age structure: *0-14 years:* 16.59% (male 3,060/female 3,142)
15-24 years: 12.53% (male 2,240/female 2,445)
25-54 years: 48.27% (male 8,424/female 9,620)
55-64 years: 12.51% (male 2,261/female 2,416)
65 years and over: 10.09% (male 1,808/female 1,965) (2020 est.)

Median age: *total:* 37.2 years
male: 37 years
female: 37.5 years (2020 est.)
country comparison to the world: 71

Population growth rate: 2.14% (2020 est.)
country comparison to the world: 39

Birth rate: 11.1 births/1,000 population (2020 est.)
country comparison to the world: 177

Death rate: 5.4 deaths/1,000 population (2020 est.)
country comparison to the world: 184

Net migration rate: 15.5 migrant(s)/1,000 population (2020 est.)
country comparison to the world: 2

Population distribution: a fairly even distribution throughout the inhabited islands, with the largest islands of Tortola, Anegada, Virgin Gorda, and Jost Van Dyke having the largest populations

Urbanization: *urban population:* 48.5% of total population (2020)
rate of urbanization: 2.42% annual rate of change (2015-20 est.)

total population growth rate v. urban population growth rate, 2000-2030: Major urban areas—population: 15,000 ROAD TOWN (capital) (2018)

Sex ratio: *at birth:* 1.05 male(s)/female
0-14 years: 0.97 male(s)/female
15-24 years: 0.92 male(s)/female
25-54 years: 0.88 male(s)/female
55-64 years: 0.94 male(s)/female
65 years and over: 0.92 male(s)/female
total population: 0.91 male(s)/female (2020 est.)

Infant mortality rate: *total:* 11 deaths/1,000 live births
male: 12.5 deaths/1,000 live births
female: 9.4 deaths/1,000 live births (2020 est.)
country comparison to the world: 122

Life expectancy at birth: *total population:* 79.2 years
male: 77.7 years
female: 80.8 years (2020 est.)
country comparison to the world: 57

Total fertility rate: 1.33 children born/woman (2020 est.)
country comparison to the world: 222

Drinking water source:
improved:
total: 100% of population
unimproved:
total: 0% of population (2017 est.)

Sanitation facility access:
improved:
urban: 97.5% of population
rural: 97.5% of population
total: 97.5% of population
unimproved:
urban: 2.5% of population
rural: 2.5% of population
total: 2.5% of population (2015 est.)

HIV/AIDS—adult prevalence rate: NA

HIV/AIDS—people living with HIV/AIDS: NA

HIV/AIDS—deaths: NA

Education expenditures: 3.2% of GDP (2017)
country comparison to the world: 130

School life expectancy (primary to tertiary education): *total:* 12 years
male: 12 years NA
female: 12 years NA (2018)

GOVERNMENT

Country name: *conventional long form:* none
conventional short form: British Virgin Islands
abbreviation: BVI
etymology: the myriad islets, cays, and rocks surrounding the major islands reminded explorer Christopher COLUMBUS in 1493 of Saint Ursula and her 11,000 virgin followers (Santa Ursula y las Once Mil Virgenes), which over time shortened to the Virgins (las Virgenes)

Dependency status: overseas territory of the UK; internal self-governing

Government type: parliamentary democracy; self-governing overseas territory of the UK

Capital: *name:* Road Town
geographic coordinates: 18 25 N, 64 37 W

time difference: UTC-4 (1 hour ahead of Washington, DC, during Standard Time)

etymology: name refers to the nautical term "roadstead" or "roads," a body of water less sheltered than a harbor but where where ships can lie reasonably safely at anchor sheltered from rip currents, spring tides, or ocean swells

Administrative divisions: none (overseas territory of the UK)

Independence: none (overseas territory of the UK)

National holiday: Territory Day, 1 July (1956)

Constitution: *history:* several previous; latest effective 15 June 2007 (The Virgin Islands Constitution Order 2007)

Legal system: English common law

Citizenship: see United Kingdom

Suffrage: 18 years of age; universal

Executive branch: *chief of state:* Queen ELIZABETH II (since 6 February 1952); represented by Governor Gus JASPERT (since 22 August 2017)

head of government: Premier Andrew FAHIE (since 26 February 2019)

cabinet: Executive Council appointed by the governor from members of the House of Assembly

elections/appointments: the monarchy is hereditary; governor appointed by the monarch; following legislative elections, the leader of the majority party or majority coalition usually appointed premier by the governor

Legislative branch: *description:* unicameral House of Assembly (15 seats; 13 members—9 in single-seat constituencies and 4 at-large seats directly elected by simple majority vote and 2 ex-officio members—the attorney general and the speaker—chosen from outside the House; members serve 4-year terms)

elections: last held on 25 February 2019 (next to be held in 2023)

election results: percent of vote by party—VIP 46.5%, NDP 28.2%, PVIM 17.4%, PU 8%; seats by party—VIP 8, NDP 3, PVIM 1, PU 1; composition—men 12, women 3, percent of women 20%

Judicial branch: *highest courts:* the Eastern Caribbean Supreme Court (ECSC) is the superior court of the Organization of Eastern Caribbean States; the ECSC—headquartered on St. Lucia—consists of the Court of Appeal—headed by the chief justice and 4 judges—and the High Court with 18 judges; the Court of Appeal is itinerant, traveling to member states on a schedule to hear appeals from the High Court and subordinate courts; High Court judges reside in the member states, with 3 in the British Virgin Islands

judge selection and term of office: Eastern Caribbean Supreme Court chief justice appointed by Her Majesty, Queen ELIZABETH II; other justices and judges appointed by the Judicial and Legal Services Commission; Court of Appeal justices appointed for life with mandatory retirement at age 65; High Court judges appointed for life with mandatory retirement at age 62

subordinate courts: Magistrates' Courts

Political parties and leaders: National Democratic Party or NDP [Myron WALWYN]
People's Empowerment Party or PEP [Alvin CHRISTOPHER]
Progressive Virgin Islands Movement or PVIM [Ronnie SKELTON]
Progressives United or PU [Julian FRASER]
Virgin Islands Party or VIP [Andrew FAHIE]

International organization participation: Caricom (associate), CDB, Interpol (subbureau), IOC, OECS, UNESCO (associate), UPU

Diplomatic representation in the US: none (overseas territory of the UK)

Diplomatic representation from the US: none (overseas territory of the UK)

Flag description: blue with the flag of the UK in the upper hoist-side quadrant and the Virgin Islander coat of arms centered in the outer half of the flag; the coat of arms depicts a woman flanked on either side by a vertical column of six oil lamps above a scroll bearing the Latin word VIGILATE (Be Watchful); the islands were named by COLUMBUS in 1493 in honor of Saint Ursula and her 11 virgin followers (some sources say 11,000) who reputedly were martyred by the Huns in the 4th or 5th century; the figure on the banner holding a lamp represents the saint; the other lamps symbolize her followers

National symbol(s): zenaida dove, white cedar flower; national colors: yellow, green, red, white, blue

National anthem: note: as a territory of the United Kingdom, "God Save the Queen" is official (see United Kingdom)
0:00 / 0:00

ECONOMY

Economy—overview: The economy, one of the most stable and prosperous in the Caribbean, is highly dependent on tourism, which generates an estimated 45% of the national income. More than 934,000 tourists, mainly from the US, visited the islands in 2008. Because of traditionally close links with the US Virgin Islands, the British Virgin Islands has used the US dollar as its currency since 1959.

Livestock raising is the most important agricultural activity; poor soils limit the islands' ability to meet domestic food requirements.

In the mid-1980s, the government began offering offshore registration to companies wishing to incorporate in the islands, and incorporation fees now generate substantial revenues. Roughly 400,000 companies were on the offshore registry by yearend 2000. The adoption of a comprehensive insurance law in late 1994, which provides a blanket of confidentiality with regulated statutory gateways for investigation of criminal offenses, made the British Virgin Islands even more attractive to international business.

GDP (purchasing power parity): $500 million (2017 est.)
$490.2 million (2016 est.)
$481.1 million (2015 est.)

country comparison to the world: 213

GDP (official exchange rate): $1.028 billion (2017 est.)

GDP—real growth rate: 2% (2017 est.)
1.9% (2016 est.)
1.8% (2015 est.)
country comparison to the world: 135

GDP—per capita (PPP): $34,200 (2017 est.)
country comparison to the world: 59

GDP—composition, by end use: household consumption: 25.1% (2017 est.)
government consumption: 7.5% (2017 est.)
investment in fixed capital: 21.7% (2017 est.)
investment in inventories: 20.4% (2017 est.)
exports of goods and services: 94.7% (2017 est.)
imports of goods and services: -69.4% (2017 est.)

GDP—composition, by sector of origin:
agriculture: 0.2% (2017 est.)
industry: 6.8% (2017 est.)
services: 93.1% (2017 est.)

Agriculture—products: fruits, vegetables; livestock, poultry; fish

Industries: tourism, light industry, construction, rum, concrete block, offshore banking center

Industrial production growth rate: 1.1% (2017 est.)
country comparison to the world: 152

Labor force: 12,770 (2004)
country comparison to the world: 215

Labor force—by occupation: *agriculture:* 0.6%
industry: 40%
services: 59.4% (2005)

Unemployment rate: 2.9% (2015 est.)
country comparison to the world: 34

Population below poverty line: NA

Household income or consumption by percentage share: *lowest 10%:* NA
highest 10%: NA

Budget: *revenues:* 400 million (2017 est.)
expenditures: 400 million (2017 est.)

Taxes and other revenues: 38.9% (of GDP) (2017 est.)
country comparison to the world: 50

Budget surplus (+) or deficit (-): 0% (of GDP) (2017 est.)
country comparison to the world: 44

Fiscal year: 1 April–31 March

Inflation rate (consumer prices): 1.1% (2017 est.)
1.1% (2016 est.)
country comparison to the world: 56

Current account balance: $362.6 million (2011 est.)
$279.8 million (2010 est.)
country comparison to the world: 55

Exports: $23 million (2017 est.)
$23 million (2015 est.)
country comparison to the world: 209

Exports—commodities: rum, fresh fish, fruits, animals; gravel, sand

Imports: $300 million NA (2017 est.)
$210 million (2016 est.)
country comparison to the world: 204

Imports—commodities: building materials, automobiles, foodstuffs, machinery

Debt—external: $36.1 million (1997)
country comparison to the world: 198

Exchange rates: the US dollar is used

ENERGY

Electricity—production: 126.3 million kWh (2016 est.)
country comparison to the world: 198

Electricity—consumption: 117.5 million kWh (2016 est.)
country comparison to the world: 200

Electricity—exports: 0 kWh (2016 est.)
country comparison to the world: 110

Electricity—imports: 0 kWh (2016 est.)
country comparison to the world: 128

Electricity—installed generating capacity: 45,200 kW (2016 est.)
country comparison to the world: 193

Electricity—from fossil fuels: 97% of total installed capacity (2016 est.)
country comparison to the world: 33

Electricity—from nuclear fuels: 0% of total installed capacity (2017 est.)
country comparison to the world: 55

Electricity—from hydroelectric plants: 0% of total installed capacity (2017 est.)
country comparison to the world: 160

Electricity—from other renewable sources: 3% of total installed capacity (2017 est.)
country comparison to the world: 122

Crude oil—production: 0 bbl/day (2018 est.)
country comparison to the world: 115

Crude oil—exports: 0 bbl/day (2015 est.)
country comparison to the world: 98

Crude oil—imports: 0 bbl/day (2015 est.)
country comparison to the world: 101

Crude oil—proved reserves: 0 bbl (1 January 2018 est.)
country comparison to the world: 111

Refined petroleum products—production: 0 bbl/day (2015 est.)
country comparison to the world: 123

Refined petroleum products—consumption: 20,000 bbl/day (2016 est.)
country comparison to the world: 141

Refined petroleum products—exports: 0 bbl/day (2015 est.)

country comparison to the world: 134

Refined petroleum products—imports: 1,227 bbl/day (2015 est.)
country comparison to the world: 200

Natural gas—production: 0 cu m (2017 est.)
country comparison to the world: 109

Natural gas—consumption: 0 cu m (2017 est.)
country comparison to the world: 125

Natural gas—exports: 0 cu m (2017 est.)
country comparison to the world: 74

Natural gas—imports: 0 cu m (2017 est.)
country comparison to the world: 96

Natural gas—proved reserves: 0 cu m (1 January 2014 est.)
country comparison to the world: 115

Carbon dioxide emissions from consumption of energy: 183,300 Mt (2017 est.)
country comparison to the world: 202

COMMUNICATIONS

Telephones—fixed lines: *total subscriptions:* 7,640
subscriptions per 100 inhabitants: 20.88 (2019 est.)
country comparison to the world: 195

Telephones—mobile cellular: *total subscriptions:* 72,589
subscriptions per 100 inhabitants: 198.38 (2019 est.)
country comparison to the world: 200

Telecommunication systems: general assessment: good overall telephone service; major expansion sectors include the mobile telephony and data segments, which continue to appeal to operator investment; several operators licensed to provide services within individual markets, most of them are small and localized; telecommunication contributes to overall GDP (2020)
domestic: fixed-line connections exceed 21 per 100 persons and mobile cellular subscribership is roughly 198 per 100 persons (2019)
international: country code—1-284; landing points for PCCS, ECFS, CBUS, Deep Blue Cable, East-West, PAN-AM, Americas-1, Southern Caribbean Fiber, Columbus- IIb, St Thomas—St Croix System, Taino-Carib, and Americas I-North via submarine cable to Caribbean, Central and South America, and US (2019)
note: the COVID-19 outbreak is negatively impacting telecommunications production and

supply chains globally; consumer spending on telecom devices and services has also slowed due to the pandemic's effect on economies worldwide; overall progress towards improvements in all facets of the telecom industry—mobile, fixed-line, broadband, submarine cable and satellite—has moderated

Broadcast media: 1 private TV station; multi-channel TV is available from cable and satellite subscription services; about a half-dozen private radio stations

Internet country code: .vg

Internet users: *total:* 27,818
percent of population: 77.7% (July 2018 est.)
country comparison to the world: 207

Broadband—fixed subscriptions: *total:* 4,715
subscriptions per 100 inhabitants: 13 (2018 est.)
country comparison to the world: 180

TRANSPORTATION

Civil aircraft registration country code prefix: VP-L (2016)

Airports: 4 (2020)
country comparison to the world: 185

Airports—with paved runways: *total:* 2 (2019)
914 to 1,523 m: 1

under 914 m: 1

Airports—with unpaved runways: *total:* 2 (2013)
914 to 1,523 m: 2 (2013)

Roadways: *total:* 200 km (2007)
paved: 200 km (2007)
country comparison to the world: 207

Merchant marine: *total:* 29
by type: general cargo 3, other 26 (2019)
country comparison to the world: 132

Ports and terminals: *major seaport(s):* Road Harbor

MILITARY AND SECURITY

Military—note: defense is the responsibility of the UK

TRANSNATIONAL ISSUES

Disputes—international: none

Illicit drugs: transshipment point for South American narcotics destined for the US and Europe; large offshore financial center makes it vulnerable to money laundering

BRUNEI

INTRODUCTION

Background: The Sultanate of Brunei's influence peaked between the 15th and 17th centuries when its control extended over coastal areas of northwest Borneo and the southern Philippines.

Brunei subsequently entered a period of decline brought on by internal strife over royal succession, colonial expansion of European powers, and piracy. In 1888, Brunei became a British protectorate; independence was achieved in 1984. The same family has ruled Brunei for over six

centuries. Brunei benefits from extensive petroleum and natural gas fields, the source of one of the highest per capita GDPs in the world. In 2017, Brunei celebrated the 50th anniversary of the Sultan Hassanal BOLKIAH's accession to the throne.

GEOGRAPHY

Location: Southeastern Asia, along the northern coast of the island of Borneo, bordering the South China Sea and Malaysia

Geographic coordinates: 4 30 N, 114 40 E

Map references: Southeast Asia

Area: *total:* 5,765 sq km
land: 5,265 sq km
water: 500 sq km
country comparison to the world: 173

Area—comparative: slightly smaller than Delaware

Land boundaries: *total:* 266 km
border countries (1): Malaysia 266 km

Coastline: 161 km

Maritime claims: *territorial sea:* 12 nm
exclusive economic zone: 200 nm or to median line

Climate: tropical; hot, humid, rainy

Terrain: flat coastal plain rises to mountains in east; hilly lowland in west

Elevation: *mean elevation:* 478 m
lowest point: South China Sea 0 m
highest point: Bukit Pagon 1,850 m

Natural resources: petroleum, natural gas, timber

Land use: *agricultural land:* 2.5% (2011 est.)
arable land: 0.8% (2011 est.) / permanent crops: 1.1% (2011 est.) / permanent pasture: 0.6% (2011 est.)
forest: 71.8% (2011 est.)
other: 25.7% (2011 est.)

Irrigated land: 10 sq km (2012)

Natural hazards: typhoons, earthquakes, and severe flooding are rare

Environment—current issues: no major environmental problems, but air pollution control is becoming a concern; seasonal trans-boundary haze from forest fires in Indonesia

Environment—international agreements: party to: Biodiversity, Climate Change, Desertification, Endangered Species, Hazardous Wastes, Law of the Sea, Ozone Layer Protection, Ship Pollution
signed, but not ratified: none of the selected agreements

Geography—note: close to vital sea lanes through South China Sea linking Indian and Pacific Oceans; two parts physically separated by Malaysia; the eastern part, the Temburong district, is an exclave and is almost an enclave within Malaysia

PEOPLE AND SOCIETY

Population: 464,478 (July 2020 est.)
country comparison to the world: 174

Nationality: *noun:* Bruneian(s)
adjective: Bruneian

Ethnic groups: Malay 65.7%, Chinese 10.3%, other 24% (2016 est.)

Languages: Malay (Bahasa Melayu) (official), English, Chinese dialects

Religions: Muslim (official) 78.8%, Christian 8.7%, Buddhist 7.8%, other (includes indigenous beliefs) 4.7% (2011 est.)

Age structure: *0-14 years:* 22.41% (male 53,653/female 50,446)
15-24 years: 16.14% (male 37,394/female 37,559)
25-54 years: 47.21% (male 103,991/female 115,291)
55-64 years: 8.34% (male 19,159/female 19,585)
65 years and over: 5.9% (male 13,333/female 14,067) (2020 est.)

population pyramid: *Dependency ratios:* total dependency ratio: 38.7
youth dependency ratio: 31
elderly dependency ratio: 7.7
potential support ratio: 12.9 (2020 est.)

Median age: *total:* 31.1 years
male: 30.5 years
female: 31.8 years (2020 est.)
country comparison to the world: 116

Population growth rate: 1.51% (2020 est.)
country comparison to the world: 70

Birth rate: 16.5 births/1,000 population (2020 est.)
country comparison to the world: 104

Death rate: 3.8 deaths/1,000 population (2020 est.)
country comparison to the world: 216

Net migration rate: 2.3 migrant(s)/1,000 population (2020 est.)
country comparison to the world: 42

Urbanization: *urban population:* 78.3% of total population (2020)
rate of urbanization: 1.66% annual rate of change (2015-20 est.)

total population growth rate v. urban population growth rate, 2000-2030: Major urban areas—population: 241,000 BANDAR SERI BEGAWAN (capital) (2011)
note: the boundaries of the capital city were expanded in 2007, greatly increasing the city area; the population of the capital increased tenfold

Sex ratio: *at birth:* 1.05 male(s)/female
0-14 years: 1.06 male(s)/female
15-24 years: 1 male(s)/female
25-54 years: 0.9 male(s)/female
55-64 years: 0.98 male(s)/female
65 years and over: 0.95 male(s)/female

total population: 0.96 male(s)/female (2020 est.)

Maternal mortality rate: 31 deaths/100,000 live births (2017 est.)
country comparison to the world: 109

Infant mortality rate: *total:* 8.8 deaths/1,000 live births
male: 10.4 deaths/1,000 live births
female: 7.1 deaths/1,000 live births (2020 est.)
country comparison to the world: 142

Life expectancy at birth: *total population:* 77.9 years
male: 75.5 years
female: 80.4 years (2020 est.)
country comparison to the world: 70

Total fertility rate: 1.75 children born/woman (2020 est.)
country comparison to the world: 159

Drinking water source:
improved:
total: 100% of population
unimproved:
total: 0% of population (2017 est.)

Current Health Expenditure: 2.4% (2017)

Physicians density: 1.61 physicians/1,000 population (2017)

Hospital bed density: 2.9 beds/1,000 population (2017)

HIV/AIDS—adult prevalence rate: NA

HIV/AIDS—people living with HIV/AIDS: NA

HIV/AIDS—deaths: NA

Obesity—adult prevalence rate: 14.1% (2016)
country comparison to the world: 129

Children under the age of 5 years underweight: 9.6% (2009)
country comparison to the world: 65

Education expenditures: 4.4% of GDP (2016)
country comparison to the world: 88

Literacy: *definition:* age 15 and over can read and write
total population: 97.2%
male: 98.1%
female: 93.4% (2018)

School life expectancy (primary to tertiary education): *total:* 14 years
male: 14 years
female: 15 years (2019)

Unemployment, youth ages 15-24: *total:* 28.9%
male: 28.4%
female: 29.5% (2017 est.)
country comparison to the world: 37

GOVERNMENT

Country name: *conventional long form:* Brunei Darussalam
conventional short form: Brunei
local long form: Negara Brunei Darussalam
local short form: Brunei
etymology: derivation of the name is unclear; according to legend, MUHAMMAD SHAH, who would become the first sultan of Brunei, upon discovering what would become Brunei exclaimed

"Baru nah," which roughly translates as "there or "that's it"

Government type: absolute monarchy or sultanate

Capital: name: Bandar Seri Begawan
geographic coordinates: 4 53 N, 114 56 E
time difference: UTC+8 (13 hours ahead of Washington, DC, during Standard Time)
etymology: named in 1970 after Sultan Omar Ali SAIFUDDIEN III (1914-1986; "The Father of Independence") who adopted the title of "Seri Begawan" (approximate meaning "honored lord") upon his abdication in 1967; "bandar" in Malay means "town" or "city"; the capital had previously been called Bandar Brunei (Brunei Town)

Administrative divisions: 4 districts (daer-ah-daerah, singular—daerah); Belait, Brunei dan Muara, Temburong, Tutong

Independence: 1 January 1984 (from the UK)

National holiday: National Day, 23 February (1984); note—1 January 1984 was the date of independence from the UK, 23 February 1984 was the date of independence from British protection; the Sultan's birthday, 15 June

Constitution: *history:* drafted 1954 to 1959, signed 29 September 1959; note—some constitutional provisions suspended since 1962 under a State of Emergency, others suspended since independence in 1984
amendments: proposed by the monarch; passage requires submission to the Privy Council for Legislative Council review and finalization takes place by proclamation; the monarch can accept or reject changes to the original proposal provided by the Legislative Council; amended several times

Legal system: mixed legal system based on English common law and Islamic law; note—in April 2019, the full sharia penal codes came into force and apply to Muslims and non-Muslims in parallel with present common law codes

International law organization participation: has not submitted an ICJ jurisdiction declaration; non-party state to the ICC

Citizenship: *citizenship by birth:* no
citizenship by descent only: the father must be a citizen of Brunei
dual citizenship recognized: no
residency requirement for naturalization: 12 years

Suffrage: 18 years of age for village elections; universal

Executive branch: *chief of state:* Sultan and Prime Minister Sir HASSANAL Bolkiah (since 5 October 1967); note—the monarch is both chief of state and head of government
head of government: Sultan and Prime Minister Sir HASSANAL Bolkiah (since 5 October 1967)
cabinet: Council of Ministers appointed and presided over by the monarch; note—4 additional advisory councils appointed by the monarch are the Religious Council, Privy Council for constitutional issues, Council of Succession, and Legislative Council

elections/appointments: none; the monarchy is hereditary

Legislative branch: *description:* unicameral Legislative Council or Majlis Mesyuarat Negara Brunei (36 seats; members appointed by the sultan including 3 ex-officio members—the speaker and first and second secretaries; members appointed for 5-year terms)
elections: appointed by the sultan
election results: NA; composition—men 33, women 3, percent of women 8.3%

Judicial branch: *highest courts:* Supreme Court (consists of the Court of Appeal and the High Court, each with a chief justice and 2 judges); Sharia Court (consists the Court of Appeals and the High Court); note—Brunei has a dual judicial system of secular and sharia (religious) courts; the Judicial Committee of Privy Council (in London) serves as the final appellate court for civil cases only
judge selection and term of office: Supreme Court judges appointed by the monarch to serve until age 65, and older if approved by the monarch; Sharia Court judges appointed by the monarch for life
subordinate courts: Intermediate Court; Magistrates' Courts; Juvenile Court; small claims courts; lower sharia courts

Political parties and leaders: National Development Party or NDP [YASSIN Affendi]
note: Brunei National Solidarity Party or PPKB [Abdul LATIF bin Chuchu] and People's Awareness Party or PAKAR [Awang Haji MAIDIN bin Haji Ahmad] were deregistered in 2007; parties are small and have limited activity

International organization participation: ADB, APEC, ARF, ASEAN, C, CP, EAS, FAO, G-77, IAEA, IBRD, ICAO, ICC (NGOs), ICRM, IDA, IFRCS, ILO, IMF, IMO, IMSO, Interpol, IOC, ISO (correspondent), ITSO, ITU, NAM, OIC, OPCW, UN, UNCTAD, UNESCO, UNIFIL, UNWTO, UPU, WCO, WHO, WIPO, WMO, WTO

Diplomatic representation in the US: *chief of mission:* Ambassador Serbini ALI (since 28 January 2016)
chancery: 3520 International Court NW, Washington, DC 20008
telephone: [1] (202) 237-1838
FAX: [1] (202) 885-0560
consulate(s): New York

Diplomatic representation from the US: *chief of mission:* Ambassador (vacant); Charge d'Affaires Scott E. WOODARD (since 20 May 2020)
telephone: [673] 238-4616
embassy: Simpang 336-52-16-9, Jalan Duta, Bandar Seri Begawan, BC4115
mailing address: Unit 4280, Box 40, FPO AP 96507; P.O. Box 2991, Bandar Seri Begawan BS8675, Negara Brunei Darussalam
FAX: [673] 238-4604

Flag description: yellow with two diagonal bands of white (top, almost double width) and black starting from the upper hoist side; the national emblem in red is superimposed at the center; yellow is the

color of royalty and symbolizes the sultanate; the white and black bands denote Brunei's chief ministers; the emblem includes five main components: a swallow-tailed flag, the royal umbrella representing the monarchy, the wings of four feathers symbolizing justice, tranquility, prosperity, and peace, the two upraised hands signifying the government's pledge to preserve and promote the welfare of the people, and the crescent moon denoting Islam, the state religion; the state motto "Always render service with God's guidance" appears in yellow Arabic script on the crescent; a ribbon below the crescent reads "Brunei, the Abode of Peace"

National symbol(s): royal parasol; national colors: yellow, white, black

National anthem: *name:* "Allah Peliharakan Sultan" (God Bless His Majesty)
lyrics/music: Pengiran Haji Mohamed YUSUF bin Pengiran Abdul Rahim/Awang Haji BESAR bin Sagap
note: adopted 1951
0:00 / 0:00

ECONOMY

Economy—overview: Brunei is an energy-rich sultanate on the northern coast of Borneo in Southeast Asia. Brunei boasts a well-educated, largely English-speaking population; excellent infrastructure; and a stable government intent on attracting foreign investment. Crude oil and natural gas production account for approximately 65% of GDP and 95% of exports, with Japan as the primary export market.

Per capita GDP is among the highest in the world, and substantial income from overseas investment supplements income from domestic hydrocarbon production. Bruneian citizens pay no personal income taxes, and the government provides free medical services and free education through the university level.

The Bruneian Government wants to diversify its economy away from hydrocarbon exports to other industries such as information and communications technology and halal manufacturing, permissible under Islamic law. Brunei's trade increased in 2016 and 2017, following its regional economic integration in the ASEAN Economic Community, and the expected ratification of the Trans-Pacific Partnership trade agreement.

GDP (purchasing power parity): $33.87 billion (2017 est.)
$33.42 billion (2016 est.)
$34.27 billion (2015 est.)
note: data are in 2017 dollars
country comparison to the world: 128

GDP (official exchange rate): $12.13 billion (2017 est.)

GDP—real growth rate: 1.3% (2017 est.)
-2.5% (2016 est.)
-0.4% (2015 est.)
country comparison to the world: 162

GDP—per capita (PPP): $78,900 (2017 est.)
$79,000 (2016 est.)
$82,200 (2015 est.)

note: data are in 2017 dollars
country comparison to the world: 9

Gross national saving: 47.5% of GDP (2017 est.)
50.1% of GDP (2016 est.)
51.9% of GDP (2015 est.)
country comparison to the world: 3

GDP—composition, by end use: *household consumption:* 25% (2017 est.)
government consumption: 24.8% (2017 est.)
investment in fixed capital: 32.6% (2017 est.)
investment in inventories: 8.5% (2017 est.)
exports of goods and services: 45.9% (2017 est.)
imports of goods and services: -36.8% (2017 est.)

GDP—composition, by sector of origin:
agriculture: 1.2% (2017 est.)
industry: 56.6% (2017 est.)
services: 42.3% (2017 est.)

Agriculture—products: rice, vegetables, fruits; chickens, water buffalo, cattle, goats, eggs

Industries: petroleum, petroleum refining, liquefied natural gas, construction, agriculture, aquaculture, transportation

Industrial production growth rate: 1.5% (2017 est.)
country comparison to the world: 142

Labor force: 203,600 (2014 est.)
country comparison to the world: 168

Labor force—by occupation: agriculture: 4.2%
industry: 62.8%
services: 33% (2008 est.)

Unemployment rate: 6.9% (2017 est.)
6.9% (2016 est.)
country comparison to the world: 108

Population below poverty line: NA

Household income or consumption by percentage share: *lowest 10%:* NA
highest 10%: NA

Budget: *revenues:* 2.245 billion (2017 est.)
expenditures: 4.345 billion (2017 est.)

Taxes and other revenues: 18.5% (of GDP) (2017 est.)
country comparison to the world: 158

Budget surplus (+) or deficit (-): -17.3% (of GDP) (2017 est.)
country comparison to the world: 218

Public debt: 2.8% of GDP (2017 est.)
3% of GDP (2016 est.)
country comparison to the world: 207

Fiscal year: 1 April—31 March

Inflation rate (consumer prices): -0.2% (2017 est.)
-0.7% (2016 est.)
country comparison to the world: 9

Current account balance: $2.021 billion (2017 est.)
$1.47 billion (2016 est.)
country comparison to the world: 40

Exports: $5.885 billion (2017 est.)
$5.023 billion (2016 est.)
country comparison to the world: 103

Exports—partners: Japan 27.8%, South Korea 12.4%, Thailand 11.5%, Malaysia 11.3%, India

9.3%, Singapore 7.7%, Switzerland 5%, China 4.7% (2017)

Exports—commodities: mineral fuels, organic chemicals

Imports: $2.998 billion (2017 est.)
$2.658 billion (2016 est.)
country comparison to the world: 147

Imports—commodities: machinery and mechanical appliance parts, mineral fuels, motor vehicles, electric machinery

Imports—partners: China 19.6%, Singapore 19%, Malaysia 18.8%, US 9.2%, Germany 5.9%, Japan 4.1%, UK 4% (2017)

Reserves of foreign exchange and gold: $3.488 billion (31 December 2017 est.)
$3.366 billion (31 December 2015 est.)
country comparison to the world: 105

Debt—external: $0 (2014)
$0 (2013)
note: public external debt only; private external debt unavailable
country comparison to the world: 204

Exchange rates: Bruneian dollars (BND) per US dollar -
1.394 (2017 est.)
1.3814 (2016 est.)
1.3814 (2015 est.)
1.3749 (2014 est.)
1.267 (2013 est.)

ENERGY

Electricity access: *electrification—total population:* 100% (2020)

Electricity—production: 4.014 billion kWh (2016 est.)
country comparison to the world: 127

Electricity—consumption: 3.771 billion kWh (2016 est.)
country comparison to the world: 129

Electricity—exports: 0 kWh (2016 est.)
country comparison to the world: 111

Electricity—imports: 0 kWh (2016 est.)
country comparison to the world: 129

Electricity—installed generating capacity: 821,000 kW (2016 est.)
country comparison to the world: 134

Electricity—from fossil fuels: 100% of total installed capacity (2016 est.)
country comparison to the world: 5

Electricity—from nuclear fuels: 0% of total installed capacity (2017 est.)
country comparison to the world: 56

Electricity—from hydroelectric plants: 0% of total installed capacity (2017 est.)
country comparison to the world: 161

Electricity—from other renewable sources: 0% of total installed capacity (2017 est.)
country comparison to the world: 179

Crude oil—production: 100,000 bbl/day (2018 est.)
country comparison to the world: 42

Crude oil—exports: 127,400 bbl/day (2015 est.)
country comparison to the world: 33

Crude oil—imports: 160 bbl/day (2015 est.)
country comparison to the world: 82

Crude oil—proved reserves: 1.1 billion bbl (1 January 2018 est.)
country comparison to the world: 39

Refined petroleum products—production: 10,310 bbl/day (2015 est.)
country comparison to the world: 100

Refined petroleum products—consumption: 18,000 bbl/day (2016 est.)
country comparison to the world: 144

Refined petroleum products—exports: 0 bbl/day (2015 est.)
country comparison to the world: 135

Refined petroleum products—imports: 6,948 bbl/day (2015 est.)
country comparison to the world: 159

Natural gas—production: 12.74 billion cu m (2017 est.)
country comparison to the world: 38

Natural gas—consumption: 3.936 billion cu m (2017 est.)
country comparison to the world: 66

Natural gas—exports: 8.268 billion cu m (2017 est.)
country comparison to the world: 23

Natural gas—imports: 0 cu m (2017 est.)
country comparison to the world: 97

Natural gas—proved reserves: 260.5 billion cu m (1 January 2018 est.)
country comparison to the world: 39

Carbon dioxide emissions from consumption of energy: 10.04 million Mt (2017 est.)
country comparison to the world: 107

COMMUNICATIONS

Telephones—fixed lines: *total subscriptions:* 91,415
subscriptions per 100 inhabitants: 19.98 (2019 est.)
country comparison to the world: 141

Telephones—mobile cellular: total subscriptions: 588,616
subscriptions per 100 inhabitants: 128.65 (2019 est.)
country comparison to the world: 171

Telecommunication systems: *general assessment:* service throughout the country is excellent; international service is good to Southeast Asia, Middle East, Western Europe, and the US; lots of investment given the high GDP per capita; launch of 5G in 2021 anticipated; while fixed-line is slowing down, mobile broadband has taken over in the advancement in the telecoms access market; broadband penetration slow to moderate growth predicted over the next five years to 2023 (2020)
domestic: every service available; 20 per 100 fixed-line, 129 per 100 mobile-cellular (2019)
international: country code—673; landing points for the SEA-ME-WE-3, SJC, AAG,

Lubuan-Brunei Submarine Cable via optical telecommunications submarine cables that provides links to Asia, the Middle East, Southeast Asia, Africa, Australia, and the US; satellite earth stations—2 Intelsat (1 Indian Ocean and 1 Pacific Ocean) (2019)

note: the COVID-19 outbreak is negatively impacting telecommunications production and supply chains globally; consumer spending on telecom devices and services has also slowed due to the pandemic's effect on economies worldwide; overall progress towards improvements in all facets of the telecom industry—mobile, fixed-line, broadband, submarine cable and satellite—has moderated

Broadcast media: state-controlled Radio Television Brunei (RTB) operates 5 channels; 3 Malaysian TV stations are available; foreign TV broadcasts are available via satellite systems; RTB operates 5 radio networks and broadcasts on multiple frequencies; British Forces Broadcast Service (BFBS) provides radio broadcasts on 2 FM stations; some radio broadcast stations from Malaysia are available via repeaters

Internet country code: .bn

Internet users: *total:* 426,234
percent of population: 94.6% (July 2018 est.)
country comparison to the world: 157

Broadband—fixed subscriptions: *total:* 49,452
subscriptions per 100 inhabitants: 11 (2018 est.)
country comparison to the world: 135

TRANSPORTATION

National air transport system: *number of registered air carriers:* 1 (2020)
inventory of registered aircraft operated by air carriers: 10
annual passenger traffic on registered air carriers: 1,234,455 (2018)
annual freight traffic on registered air carriers: 129.35 million mt-km (2018)

Civil aircraft registration country code prefix: V8 (2016)

Airports: 1 (2020)
country comparison to the world: 216

Airports—with paved runways: *total:* 1 (2019)
over 3,047 m: 1

Heliports: 3 (2013)

Pipelines: 33 km condensate, 86 km condensate/gas, 628 km gas, 492 km oil (2013)

Roadways: *total:* 2,976 km (2014)
paved: 2,559 km (2014)
unpaved: 417 km (2014)
country comparison to the world: 162

Waterways: 209 km (navigable by craft drawing less than 1.2 m; the Belait, Brunei, and Tutong Rivers are major transport links) (2012)
country comparison to the world: 96

Merchant marine: *total:* 104
by type: general cargo 18, oil tanker 2, other 84 (2019)
country comparison to the world: 89

Ports and terminals: *major seaport(s):* Muara
oil terminal(s): Lumut, Seria
LNG terminal(s) (export): Lumut

MILITARY AND SECURITY

Military and security forces: Royal Brunei Armed Forces: Royal Brunei Land Force, Royal Brunei Navy, Royal Brunei Air Force (2019)

Military expenditures: 3.3% of GDP (2019)
2.6% of GDP (2018)
2.9% of GDP (2017)
3.5% of GDP (2016)
3.3% of GDP (2015)
country comparison to the world: 21

Military and security service personnel strengths: the Royal Brunei Armed Forces is comprised of approximately 6,800 total active troops (4,500 Army; 1,200 Navy; 1,100 Air Force) (2019)

Military equipment inventories and acquisitions: the Royal Brunei Armed Forces imports nearly all of its military equipment and weapons systems; the top suppliers since 2010 are France, Germany, and the US (2019 est.)

Military service age and obligation: 17 years of age for voluntary military service; non-Malays are ineligible to serve; recruits from the army, navy, and air force all undergo 43-week initial training (2019)

Military—note: Brunei has a long-standing defense relationship with the United Kingdom and hosts a British Army garrison, which includes a Gurkha battalion and a jungle warfare school; there is also a long-term Singaporean military presence (2019)

TRANSNATIONAL ISSUES

Disputes—international: per Letters of Exchange signed in 2009, Malaysia in 2010 ceded two hydrocarbon concession blocks to Brunei in exchange for Brunei's sultan dropping claims to the Limbang corridor, which divides Brunei; nonetheless, Brunei claims a maritime boundary extending as far as a median with Vietnam, thus asserting an implicit claim to Louisa Reef

Refugees and internally displaced persons: stateless persons: 20,863 (2019); note—thousands of stateless persons, often ethnic Chinese, are permanent residents and their families have lived in Brunei for generations; obtaining citizenship is difficult and requires individuals to pass rigorous tests on Malay culture, customs, and language; stateless residents receive an International Certificate of Identity, which enables them to travel overseas; the government is considering changing the law prohibiting non-Bruneians, including stateless permanent residents, from owning land

Illicit drugs: drug trafficking and illegally importing controlled substances are serious offenses in Brunei and carry a mandatory death penalty

BULGARIA

INTRODUCTION

Background: The Bulgars, a Central Asian Turkic tribe, merged with the local Slavic inhabitants in the late 7th century to form the first Bulgarian state. In succeeding centuries, Bulgaria struggled with the Byzantine Empire to assert its place in the Balkans, but by the end of the 14th century the country was overrun by the Ottoman Turks. Northern Bulgaria attained autonomy in 1878 and all of Bulgaria became independent from the Ottoman Empire in 1908. Having fought on the losing side in both World Wars, Bulgaria fell within the Soviet sphere of influence and became a People's Republic in 1946. Communist domination ended in 1990, when Bulgaria held its first multiparty election since World War II and

began the contentious process of moving toward political democracy and a market economy while combating inflation, unemployment, corruption, and crime. The country joined NATO in 2004 and the EU in 2007.

GEOGRAPHY

Location: Southeastern Europe, bordering the Black Sea, between Romania and Turkey

Geographic coordinates: 43 00 N, 25 00 E

Map references: Europe

Area: *total:* 110,879 sq km
land: 108,489 sq km
water: 2,390 sq km
country comparison to the world: 106

Area—comparative: almost identical in size to Virginia; slightly larger than Tennessee

Land boundaries: *total:* 1,806 km
border countries (5): Greece 472 km, Macedonia 162 km, Romania 605 km, Serbia 344 km, Turkey 223 km

Coastline: 354 km

Maritime claims: *territorial sea:* 12 nm
exclusive economic zone: 200 nm
contiguous zone: 24 nm

Climate: temperate; cold, damp winters; hot, dry summers

Terrain: mostly mountains with lowlands in north and southeast

Elevation: *mean elevation:* 472 m
lowest point: Black Sea 0 m

highest point: Musala 2,925 m

Natural resources: bauxite, copper, lead, zinc, coal, timber, arable land

Land use: *agricultural land:* 46.9% (2011 est.) *arable land:* 29.9% (2011 est.) / *permanent crops:* 1.5% (2011 est.) / *permanent pasture:* 15.5% (2011 est.) *forest:* 36.7% (2011 est.) *other:* 16.4% (2011 est.) *Irrigated land:* 1,020 sq km (2012)

Population distribution: a fairly even distribution throughout most of the country, with urban areas attracting larger populations

Natural hazards: earthquakes; landslides

Environment—current issues: air pollution from industrial emissions; rivers polluted from raw sewage, heavy metals, detergents; deforestation; forest damage from air pollution and resulting acid rain; soil contamination from heavy metals from metallurgical plants and industrial wastes

Environment—international agreements: *party to:* Air Pollution, Air Pollution-Nitrogen Oxides, Air Pollution-Persistent Organic Pollutants, Air Pollution-Sulfur 85, Air Pollution-Sulfur 94, Air Pollution-Volatile Organic Compounds, Antarctic-Environmental Protocol, Antarctic-Marine Living Resources, Antarctic Treaty, Biodiversity, Climate Change, Climate Change-Kyoto Protocol, Desertification, Endangered Species, Environmental Modification, Hazardous Wastes, Law of the Sea, Marine Dumping, Ozone Layer Protection, Ship Pollution, Wetlands *signed, but not ratified:* none of the selected agreements

Geography—note: strategic location near Turkish Straits; controls key land routes from Europe to Middle East and Asia

PEOPLE AND SOCIETY

Population: 6,966,899 (July 2020 est.) *country comparison to the world:* 106

Nationality: *noun:* Bulgarian(s) *adjective:* Bulgarian

Ethnic groups: Bulgarian 76.9%, Turkish 8%, Romani 4.4%, other 0.7% (including Russian, Armenian, and Vlach), other (unknown) 10% (2011 est.)

note: Romani populations are usually underestimated in official statistics and may represent 9–11% of Bulgaria's population

Languages: Bulgarian (official) 76.8%, Turkish 8.2%, Romani 3.8%, other 0.7%, unspecified 10.5% (2011 est.)

Religions: Eastern Orthodox 59.4%, Muslim 7.8%, other (including Catholic, Protestant, Armenian Apostolic Orthodox, and Jewish) 1.7%, none 3.7%, unspecified 27.4% (2011 est.)

Age structure: *0-14 years:* 14.52% (male 520,190/female 491,506)
15-24 years: 9.4% (male 340,306/female 314,241)
25-54 years: 42.87% (male 1,538,593/female 1,448,080)
55-64 years: 13.15% (male 433,943/female 482,474)
65 years and over: 20.06% (male 562,513/female 835,053) (2020 est.)

Dependency ratios: *total dependency ratio:* 56.6 *youth dependency ratio:* 23 *elderly dependency ratio:* 33.6 *potential support ratio:* 3 (2020 est.)

Median age: *total:* 43.7 years *male:* 41.9 years *female:* 45.6 years (2020 est.) *country comparison to the world:* 20

Population growth rate: -0.65% (2020 est.) *country comparison to the world:* 228

Birth rate: 8.3 births/1,000 population (2020 est.) *country comparison to the world:* 219

Death rate: 14.6 deaths/1,000 population (2020 est.) *country comparison to the world:* 3

Net migration rate: -0.3 migrant(s)/1,000 population (2020 est.) *country comparison to the world:* 114

Population distribution: a fairly even distribution throughout most of the country, with urban areas attracting larger populations

Urbanization: *urban population:* 75.7% of total population (2020) *rate of urbanization:* -0.22% annual rate of change (2015-20 est.)

total population growth rate v. urban population growth rate, 2000-2030:

Major urban areas—population: 1.281 million SOFIA (capital) (2020)

Sex ratio: *at birth:* 1.06 male(s)/female
0-14 years: 1.06 male(s)/female
15-24 years: 1.08 male(s)/female
25-54 years: 1.06 male(s)/female
55-64 years: 0.9 male(s)/female
65 years and over: 0.67 male(s)/female
total population: 0.95 male(s)/female (2020 est.)

Mother's mean age at first birth: 27.1 years (2017 est.)

Maternal mortality rate: 10 deaths/100,000 live births (2017 est.) *country comparison to the world:* 144

Infant mortality rate: *total:* 8.1 deaths/1,000 live births *male:* 9 deaths/1,000 live births *female:* 7 deaths/1,000 live births (2020 est.) *country comparison to the world:* 149

Life expectancy at birth: *total population:* 75 years *male:* 71.8 years *female:* 78.5 years (2020 est.) *country comparison to the world:* 121

Total fertility rate: 1.49 children born/woman (2020 est.) *country comparison to the world:* 204

Drinking water source: improved: *urban:* 100% of population *rural:* 98% of population *total:* 100% of population unimproved: *urban:* 0% of population *rural:* 2% of population *total:* 0% of population (2017 est.)

Current Health Expenditure: 8.1% (2017)

Physicians density: 4.03 physicians/1,000 population (2015)

Hospital bed density: 7.5 beds/1,000 population (2017)

Sanitation facility access: improved: *urban:* 100% of population *rural:* 100% of population *total:* 100% of population unimproved: *urban:* 0% of population *rural:* 0% of population *total:* 0% of population (2017 est.)

HIV/AIDS—adult prevalence rate: <.1% (2019 est.)

HIV/AIDS—people living with HIV/AIDS: 3,300 (2019 est.) *country comparison to the world:* 131

HIV/AIDS—deaths: <100 (2019 est.)

Obesity—adult prevalence rate: 25% (2016) *country comparison to the world:* 53

Children under the age of 5 years underweight: 1.9% (2014) *country comparison to the world:* 112

Education expenditures: 4.1% of GDP (2013) *country comparison to the world:* 96

Literacy: *definition:* age 15 and over can read and write *total population:* 98.4% *male:* 98.7% *female:* 98.1% (2015)

School life expectancy (primary to tertiary education): *total:* 14 years *male:* 14 years *female:* 14 years (2018)

Unemployment, youth ages 15-24: *total:* 12.7% *male:* 13.2% *female:* 11.9% (2018 est.) *country comparison to the world:* 108

GOVERNMENT

Country name: *conventional long form:* Republic of Bulgaria
conventional short form: Bulgaria
local long form: Republika Bulgaria
local short form: Bulgaria
former: Kingdom of Bulgaria, People's Republic of Bulgaria
etymology: named after the Bulgar tribes who settled the lower Balkan region in the 7th century A.D.

Government type: parliamentary republic

Capital: *name:* Sofia
geographic coordinates: 42 41 N, 23 19 E
time difference: UTC+2 (7 hours ahead of Washington, DC, during Standard Time)
daylight saving time: +1hr, begins last Sunday in March; ends last Sunday in October
etymology: named after the Saint Sofia Church in the city, parts of which date back to the 4th century A.D.

Administrative divisions: 28 provinces (oblasti, singular—oblast); Blagoevgrad, Burgas, Dobrich, Gabrovo, Haskovo, Kardzhali, Kyustendil, Lovech, Montana, Pazardzhik, Pernik, Pleven, Plovdiv, Razgrad, Ruse, Shumen, Silistra, Sliven, Smolyan, Sofia, Sofia-Grad (Sofia City), Stara Zagora, Targovishte, Varna, Veliko Tarnovo, Vidin, Vratsa, Yambol

Independence: 3 March 1878 (as an autonomous principality within the Ottoman Empire); 22 September 1908 (complete independence from the Ottoman Empire)

National holiday: Liberation Day, 3 March (1878)

Constitution: *history:* several previous; latest drafted between late 1990 and early 1991, adopted 13 July 1991
amendments: proposed by the National Assembly or by the president of the republic; passage requires three-fourths majority vote of National Assembly members in three ballots; signed by the National Assembly chairperson; note—under special circumstances, a "Grand National Assembly" is elected with the authority to write a new constitution and amend certain articles of the constitution, including those affecting basic civil rights and national sovereignty; passage requires at least two-thirds majority vote in each of several readings; amended several times, last in 2015

Legal system: civil law

International law organization participation: accepts compulsory ICJ jurisdiction with reservations; accepts ICCt jurisdiction

Citizenship: *citizenship by birth:* no
citizenship by descent only: at least one parent must be a citizen of Bulgaria
dual citizenship recognized: yes
residency requirement for naturalization: 5 years

Suffrage: 18 years of age; universal

Executive branch: *chief of state:* President Rumen RADEV (since 22 January 2017); Vice President Iliana IOTOVA (since 22 January 2017)

head of government: Prime Minister Boyko BORISOV (since 4 May 2017); note—BORISOV served 2 previous terms as prime minister (27 July 2009-13 March 2013 and 7 November 2014-27 January 2017)
cabinet: Council of Ministers nominated by the prime minister, elected by the National Assembly
elections/appointments: president and vice president elected on the same ballot by absolute majority popular vote in 2 rounds if needed for a 5-year term (eligible for a second term); election last held on 6 and 13 November 2016 (next to be held in fall 2021); chairman of the Council of Ministers (prime minister) elected by the National Assembly; deputy prime ministers nominated by the prime minister, elected by the National Assembly
election results: Rumen RADEV elected president in second round; percent of vote—Rumen RADEV (independent, supported by Bulgarian Socialist Party) 59.4%, Tsetska TSACHEVA (GERB) 36.2%, neither 4.5%; Boyko BORISOV (GERB) elected prime minister; National Assembly vote—133 to 100

Legislative branch: *description:* unicameral National Assembly or Narodno Sabranie (240 seats; members directly elected in multi-seat constituencies by proportional representation vote to serve 4-year terms)
elections: last held on 26 March 2017 (next to be held in March 2021)
election results: percent of vote by party/coalition—GERB 32.7%, BSP 27.2%, United Patriots 9.1%, DPS 9%, Volya 4.2%, other 17.8%; seats by party/coalition—GERB 95, BSP 80, United Patriots 27, DPS 26, Volya 12; composition—men 183, women 57, percent of women 23.8%

Judicial branch: *highest courts:* Supreme Court of Cassation (consists of a chairman and approximately 72 judges organized into penal, civil, and commercial colleges); Supreme Administrative Court (organized into 2 colleges with various panels of 5 judges each); Constitutional Court (consists of 12 justices); note—Constitutional Court resides outside the judiciary
judge selection and term of office: Supreme Court of Cassation and Supreme Administrative judges elected by the Supreme Judicial Council or SJC (consists of 25 members with extensive legal experience) and appointed by the president; judges can serve until mandatory retirement at age 65; Constitutional Court justices elected by the National Assembly and appointed by the president and the SJC; justices appointed for 9-year terms with renewal of 4 justices every 3 years
subordinate courts: appeals courts; regional and district courts; administrative courts; courts martial

Political parties and leaders: Alternative for Bulgarian Revival or ABV [Rumen PETKOV]
Attack (Ataka) [Volen Nikolov SIDEROV]

Bulgarian Agrarian People's Union [Nikolay NENCHEV]
Bulgarian Socialist Party or BSP [Korneliya NINOVA]
Bulgaria of the Citizens or DBG [Dimiter DELCHEV]]
Citizens for the European Development of Bulgaria or GERB [Boyko BORISOV]
Democrats for a Strong Bulgaria or DSB [Atanas ATANASOV]
Democrats for Responsibility, Solidarity, and Tolerance or DOST [Lyutvi MESTAN]
IMRO—Bulgarian National Movement or IMRO-BNM [Krasimir KARAKACHANOV]
Movement for Rights and Freedoms or DPS [Mustafa KARADAYI]
National Front for the Salvation of Bulgaria or NFSB [Valeri SIMEONOV]
Reformist Bloc or RB (a four-party alliance including DBG and SDS)
United Patriots (alliance of IMRO-BNM, NFSB, and Attack)
Union of Democratic Forces or SDS [Bozhidar LUKARSKI]
Yes! Bulgaria [Hristo IVANOV]
Volya [Veselin MARESHKI]

International organization participation: Australia Group, BIS, BSEC, CD, CE, CEI, CERN, EAPC, EBRD, ECB, EIB, EU, FAO, G- 9, IAEA, IBRD, ICAO, ICC (national committees), ICCt, ICRM, IDA, IFC, IFRCS, IHO (pending member), ILO, IMF, IMO, IMSO, Interpol, IOC, IOM, IPU, ISO, ITU, ITUC (NGOs), MIGA, NATO, NSG, OAS (observer), OIF, OPCW, OSCE, PCA, SELEC, UN, UNCTAD, UNESCO, UNHCR, UNIDO, UNMIL, UNWTO, UPU, WCO, WFTU (NGOs), WHO, WIPO, WMO, WTO, ZC

Diplomatic representation in the US: *chief of mission:* Ambassador Tihomir Anguelov STOYTCHEV (since 27 June 2016)
chancery: 1621 22nd Street NW, Washington, DC 20008
telephone: [1] (202) 387-0174
FAX: [1] (202) 234-7973
consulate(s) general: Chicago, Los Angeles, New York

Diplomatic representation from the US: *chief of mission:* Ambassador Herro MUSTAFA (since 18 October 2019)
telephone: [359] (2) 937-5100
embassy: 16 Kozyak Street, Sofia 1408
mailing address: American Embassy Sofia, US Department of State, 5740 Sofia Place, Washington, DC 20521-5740
FAX: [359] (2) 937-5320

Flag description: three equal horizontal bands of white (top), green, and red; the pan-Slavic white-blue-red colors were modified by substituting a green band (representing freedom) for the blue
note: the national emblem, formerly on the hoist side of the white stripe, has been removed

National symbol(s): lion; national colors: white, green, red

National anthem: *name:* "Mila Rodino" (Dear Homeland)

lyrics/music: Tsvetan Tsvetkov RADOSLAVOV
note: adopted 1964; composed in 1885 by a student en route to fight in the Serbo-Bulgarian War
0:00 / 0:00

ECONOMY

Economy—overview: Bulgaria, a former communist country that entered the EU in 2007, has an open economy that historically has demonstrated strong growth, but its per-capita income remains the lowest among EU members and its reliance on energy imports and foreign demand for its exports makes its growth sensitive to external market conditions.

The government undertook significant structural economic reforms in the 1990s to move the economy from a centralized, planned economy to a more liberal, market-driven economy. These reforms included privatization of state-owned enterprises, liberalization of trade, and strengthening of the tax system—changes that initially caused some economic hardships but later helped to attract investment, spur growth, and make gradual improvements to living conditions. From 2000 through 2008, Bulgaria maintained robust, average annual real GDP growth in excess of 6%, which was followed by a deep recession in 2009 as the financial crisis caused domestic demand, exports, capital inflows and industrial production to contract, prompting the government to rein in spending. Real GDP growth remained slow—less than 2% annually—until 2015, when demand from EU countries for Bulgarian exports, plus an inflow of EU development funds, boosted growth to more than 3%. In recent years, strong domestic demand combined with low international energy prices have contributed to Bulgaria's economic growth approaching 4% and have also helped to ease inflation. Bulgaria's prudent public financial management contributed to budget surpluses both in 2016 and 2017.

Bulgaria is heavily reliant on energy imports from Russia, a potential vulnerability, and is a participant in EU-backed efforts to diversify regional natural gas supplies. In late 2016, the Bulgarian Government provided funding to Bulgaria's National Electric Company to cover the $695 million compensation owed to Russian nuclear equipment manufacturer Atomstroyexport for the cancellation of the Belene Nuclear Power Plant project, which the Bulgarian Government terminated in 2012. As of early 2018, the government was floating the possibility of resurrecting the Belene project. The natural gas market, dominated by state-owned Bulgargaz, is also almost entirely supplied by Russia. Infrastructure projects such as the Inter-Connector Greece-Bulgaria and Inter-Connector Bulgaria-Serbia, which would enable Bulgaria to have access to non-Russian gas, have either stalled or made limited progress. In 2016, the Bulgarian Government established the State eGovernment Agency. This new agency is responsible for the electronic governance, coordinating national policies with the EU, and strengthening cybersecurity.

Despite a favorable investment regime, including low, flat corporate income taxes, significant challenges remain. Corruption in public administration, a weak judiciary, low productivity, lack of transparency in public procurements, and the presence of organized crime continue to hamper the country's investment climate and economic prospects.

GDP (purchasing power parity): $153.5 billion (2017 est.)
$148.2 billion (2016 est.)
$142.6 billion (2015 est.)
note: data are in 2017 dollars
country comparison to the world: 76

GDP (official exchange rate): $56.94 billion (2017 est.)

GDP—real growth rate: 3.39% (2019 est.)
3.2% (2018 est.)
3.5% (2017 est.)
country comparison to the world: 89

GDP—per capita (PPP): $21,800 (2017 est.)
$20,900 (2016 est.)
$19,900 (2015 est.)
note: data are in 2017 dollars
country comparison to the world: 87

Gross national saving: 25.4% of GDP (2017 est.)
21.4% of GDP (2016 est.)
21.2% of GDP (2015 est.)
country comparison to the world: 58

GDP—composition, by end use:
household consumption: 61.6% (2017 est.)
government consumption: 16% (2017 est.)
investment in fixed capital: 19.2% (2017 est.)
investment in inventories: 1.7% (2017 est.)
exports of goods and services: 66.3% (2017 est.)
imports of goods and services: -64.8% (2017 est.)

GDP—composition, by sector of origin:
agriculture: 4.3% (2017 est.)
industry: 28% (2017 est.)
services: 67.4% (2017 est.)

Agriculture—products: vegetables, fruits, tobacco, wine, wheat, barley, sunflowers, sugar beets; livestock

Industries: electricity, gas, water; food, beverages, tobacco; machinery and equipment, automotive parts, base metals, chemical products, coke, refined petroleum, nuclear fuel; outsourcing centers

Industrial production growth rate: 3.6% (2017 est.)
country comparison to the world: 80

Labor force: 3.113 million (2020 est.)
note: number of employed persons
country comparison to the world: 102

Labor force—by occupation: *agriculture:* 6.8%
industry: 26.6%
services: 66.6% (2016 est.)

Unemployment rate: 5.66% (2019 est.)
6.18% (2018 est.)
country comparison to the world: 90

Population below poverty line: 23.4% (2016 est.)

Household income or consumption by percentage share: *lowest 10%:* 1.9%
highest 10%: 31.2% (2017)

Budget: *revenues:* 20.35 billion (2017 est.)
expenditures: 19.35 billion (2017 est.)

Taxes and other revenues: 35.7% (of GDP) (2017 est.)
country comparison to the world: 58

Budget surplus (+) or deficit (-): 1.8% (of GDP) (2017 est.)
country comparison to the world: 16

Public debt: 23.9% of GDP (2017 est.)
27.4% of GDP (2016 est.)
note: defined by the EU's Maastricht Treaty as consolidated general government gross debt at nominal value, outstanding at the end of the year in the following categories of government liabilities: currency and deposits, securities other than shares excluding financial derivatives, and loans; general government sector comprises the subsectors: central government, state government, local government, and social security funds
country comparison to the world: 181

Fiscal year: calendar year

Inflation rate (consumer prices): 1.2% (2017 est.)
-1.3% (2016 est.)
country comparison to the world: 63

Current account balance: $2.06 billion (2019 est.)
$611 million (2018 est.)
country comparison to the world: 39

Exports: $29.08 billion (2017 est.)
$25.37 billion (2016 est.)
country comparison to the world: 65

Exports—partners: Germany 13.5%, Italy 8.3%, Romania 8.2%, Turkey 7.7%, Greece 6.5%, Belgium 4.2%, France 4.1% (2017)

Exports—commodities: clothing, footwear, iron and steel, machinery and equipment, fuels, agriculture, tobacco, IT components

Imports: $31.43 billion (2017 est.)
$26.66 billion (2016 est.)
country comparison to the world: 66

Imports—commodities: machinery and equipment; metals and ores; chemicals and plastics; fuels, minerals, and raw materials

Imports—partners: Germany 12.3%, Russia 10.3%, Italy 7.3%, Romania 7.1%, Turkey 6.2%, Spain 5.3%, Greece 4.4% (2017)

Reserves of foreign exchange and gold: $28.38 billion (31 December 2017 est.)
$25.13 billion (31 December 2016 est.)
country comparison to the world: 51

Debt—external: $42.06 billion (31 December 2017 est.)
$35.98 billion (31 December 2016 est.)
country comparison to the world: 72

Exchange rates: leva (BGN) per US dollar—
1.63 (2017 est.)
1.86 (2016 est.)
1.768 (2015 est.)
1.7644 (2014 est.)
1.4742 (2013 est.)

ENERGY

Electricity access: *electrification—total population:* 100% (2020)

Electricity—production: 42.29 billion kWh (2016 est.)
country comparison to the world: 57

Electricity—consumption: 32.34 billion kWh (2016 est.)
country comparison to the world: 60

Electricity—exports: 9.187 billion kWh (2017 est.)
country comparison to the world: 23

Electricity—imports: 4.568 billion kWh (2016 est.)
country comparison to the world: 41

Electricity—installed generating capacity: 10.75 million kW (2016 est.)
country comparison to the world: 57

Electricity—from fossil fuels: 39% of total installed capacity (2016 est.)
country comparison to the world: 170

Electricity—from nuclear fuels: 20% of total installed capacity (2017 est.)
country comparison to the world: 8

Electricity—from hydroelectric plants: 23% of total installed capacity (2017 est.)
country comparison to the world: 83

Electricity—from other renewable sources: 19% of total installed capacity (2017 est.)
country comparison to the world: 41

Crude oil—production: 1,000 bbl/day (2018 est.)
country comparison to the world: 91

Crude oil—exports: 0 bbl/day (2015 est.)
country comparison to the world: 99

Crude oil—imports: 133,900 bbl/day (2015 est.)
country comparison to the world: 39

Crude oil—proved reserves: 15 million bbl (1 January 2018 est.)
country comparison to the world: 84
Refined petroleum products—production:

144,300 bbl/day (2015 est.)
country comparison to the world: 60

Refined petroleum products—consumption: 97,000 bbl/day (2016 est.)
country comparison to the world: 82

Refined petroleum products—exports: 92,720 bbl/day (2015 est.)
country comparison to the world: 45

Refined petroleum products—imports: 49,260 bbl/day (2015 est.)
country comparison to the world: 83

Natural gas—production: 79.28 million cu m (2017 est.)
country comparison to the world: 83

Natural gas—consumption: 3.313 billion cu m (2017 est.)
country comparison to the world: 70

Natural gas—exports: 31.15 million cu m (2017 est.)
country comparison to the world: 52

Natural gas—imports: 3.256 billion cu m (2017 est.)
country comparison to the world: 44

Natural gas—proved reserves: 5.663 billion cu m (1 January 2018 est.)
country comparison to the world: 88

Carbon dioxide emissions from consumption of energy: 46.31 million Mt (2017 est.)
country comparison to the world: 63

COMMUNICATIONS

Telephones—fixed lines: *total subscriptions:* 974,056
subscriptions per 100 inhabitants: 13.89 (2019 est.)
country comparison to the world: 75

Telephones—mobile cellular: *total subscriptions:* 8,149,389
subscriptions per 100 inhabitants: 116.21 (2019 est.)
country comparison to the world: 97

Telecommunication systems: *general assessment:* telecom sector has benefited from Bulgaria's adaptation of EU regulatory measures, more privatization and less govt. monopoly; population is moving to fiber networks for broadband; govt. investment in programs for broadband in rural areas; 5G trials by 2 operators; quality has improved with a modern digital trunk line connecting switching centers in most of the regions; remaining areas are connected by digital microwave radio relay; Bulgaria has a mature mobile market with active competition (2020)
domestic: fixed-line 14 per 100 persons, mobile-cellular teledensity, fostered by multiple service providers, is over 116 telephones per 100 persons (2019)
international: country code—359; Caucasus Cable System via submarine cable provides connectivity to Ukraine, Georgia and Russia; a combination submarine cable and land fiber-optic system provides connectivity to Italy, Albania, and Macedonia; satellite earth stations—3 (1 Intersputnik in the Atlantic Ocean region, 2 Intelsat in the Atlantic and Indian Ocean regions) (2019)
note: the COVID-19 outbreak is negatively impacting telecommunications production and supply chains globally; consumer spending on telecom devices and services has also slowed due to the pandemic's effect on economies worldwide; overall progress towards improvements in all facets of the telecom industry—mobile, fixed-line, broadband, submarine cable and satellite—has moderated

Broadcast media: 4 national terrestrial TV stations with 1 state-owned and 3 privately owned; a vast array of TV stations are available from cable and satellite TV providers; state-owned national radio broadcasts over 3 networks; large number of private radio stations broadcasting, especially in urban areas

Internet country code: .bg

Internet users: *total:* 4,571,851

percent of population: 64.78% (July 2018 est.)
country comparison to the world: 87

Broadband—fixed subscriptions: *total:* 1,903,946
subscriptions per 100 inhabitants: 27 (2018 est.)
country comparison to the world: 55

TRANSPORTATION

National air transport system:
number of registered air carriers: 8 (2020)
inventory of registered aircraft operated by air carriers: 44
annual passenger traffic on registered air carriers: 1,022,645 (2018)
annual freight traffic on registered air carriers: 1.38 million mt-km (2018)

Civil aircraft registration country code prefix: LZ (2016)

Airports: 68 (2013)
country comparison to the world: 72

Airports—with paved runways: *total:* 57 (2017)
over 3,047 m: 2 (2017)
2,438 to 3,047 m: 17 (2017)
1,524 to 2,437 m: 12 (2017)
under 914 m: 26 (2017)

Airports—with unpaved runways: *total:* 11 (2013)
914 to 1,523 m: 2 (2013)
under 914 m: 9 (2013)

Heliports: 1 (2013)

Pipelines: 2765 km gas, 346 km oil, 378 km refined products (2017)

Railways: *total:* 5,114 km (2014)
standard gauge: 4,989 km 1.435-m gauge (2,880 km electrified) (2014)
narrow gauge: 125 km 0.760-m gauge (2014)
country comparison to the world: 37

Roadways: *total:* 19,512 km (2011)
paved: 19,235 km (includes 458 km of expressways) (2011)
unpaved: 277 km (2011)
note: does not include Category IV local roads
country comparison to the world: 117

Waterways: 470 km (2009)
country comparison to the world: 83

Merchant marine: *total:* 83
by type: bulk carrier 5general cargo 16, oil tanker 8, other 54 (2019)
country comparison to the world: 98

Ports and terminals: *major seaport(s):* Burgas, Varna (Black Sea)

MILITARY AND SECURITY

Military and security forces: Bulgarian Armed Forces: Land Forces (aka Army), Naval Forces, Bulgarian Air Forces (Voennovazdushni Sili, VVS), Special Forces; Ministry of Interior: Border Guards (2020)

Military expenditures: 3.25% of GDP (2019 est.)
1.48% of GDP (2018)
1.24% of GDP (2017)
1.26% of GDP (2016)
1.26% of GDP (2015)
country comparison to the world: 22

Military and security service personnel strengths: the Bulgarian Armed Forces have approximately 35,000 active duty personnel (17,000 Army; 4,000 Navy; 7,000 Air Force; 7,000 Joint Service/Central Staff) (2019 est.)

Military equipment inventories and acquisitions: the Bulgarian Armed Forces inventory consists primarily of Soviet-era equipment, although in recent years, Bulgaria has attempted to procure more modern weapons systems from Western countries; since 2010, it has acquired limited quantities of military equipment from France, Israel, Italy, Norway, and the US (2019 est.)

Military deployments: 160 Afghanistan (NATO) (2020)

Military service age and obligation: 18-27 years of age for voluntary military service; conscription ended in January 2008; service obligation 6-9 months (2012)

TERRORISM

Terrorist group(s): Islamic Revolutionary Guard Corps/Qods Force (2019)

note: details about the history, aims, leadership, organization, areas of operation, tactics, targets, weapons, size, and sources of support of the group(s) appear(s) in Appendix-T

TRANSNATIONAL ISSUES

Disputes—international: none

Refugees and internally displaced persons: *refugees (country of origin):* 17,551 (Syria) (2019)

stateless persons: 116 (2019)

note: 58,073 estimated refugee and migrant arrivals (January 2015-October 2020); Bulgaria is predominantly a transit country and hosts approximately 992 migrants and asylum seekers as of the end of September 2018; 2,576 migrant arrivals in 2018

Trafficking in persons: current situation: Bulgaria is a source and, to a lesser extent, a transit and destination country for men, women, and children subjected to sex trafficking and forced labor; Bulgaria is one of the main sources of human trafficking in the EU; women and children are increasingly sex trafficked domestically, as well as in Europe, Russia, the Middle East, and the US; adults and children become forced laborers in agriculture, construction, and the service sector in Europe, Israel, and Zambia; Romanian girls are also subjected to sex trafficking in Bulgaria

tier rating: Tier 2 Watch List – Bulgaria does not fully comply with the minimum standards for the elimination of trafficking; however, it is making significant efforts to do so; in 2014, authorities prosecuted and convicted fewer traffickers and issued suspended sentences for the majority of those convicted; victim protection efforts declined and were minimal relative to the number of victims identified; funding for the state's two NGO-operated shelters was significantly cut, forcing them to close; specialized services for child and adult male victims were non-existent; the government took action to combat trafficking-related complicity among public officials and police officers (2015)

Illicit drugs: major European transshipment point for Southwest Asian heroin and, to a lesser degree, South American cocaine for the European market; limited producer of precursor chemicals; vulnerable to money laundering because of corruption, organized crime; some money laundering of drug-related proceeds through financial institutions

BURKINA FASO

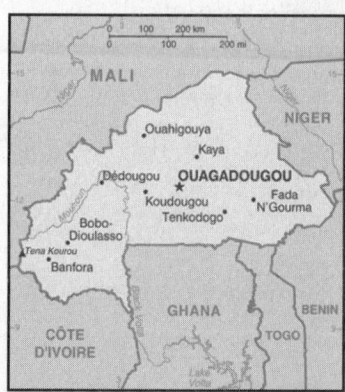

INTRODUCTION

Background: Various ethnic groups settled and established kingdoms in the area of today's Burkina Faso from medieval times onward. In the late 19th century, several European states attempted to move into the region, but it was the French who established a protectorate of Upper Volta in 1896. Independent from France in 1960, the country changed its name to Burkina Faso in 1984. Repeated military coups during the 1970s and 1980s were followed by multiparty elections in the early 1990s. Former President Blaise COMPAORE (1987-2014) resigned in late October 2014 following popular protests against his efforts to amend the constitution's two-term presidential limit. An interim administration organized presidential and legislative elections—held in November 2015—where Roch Marc Christian KABORE was elected president. The country experienced terrorist attacks in its capital in 2016, 2017, and 2018, while additional attacks in the country's northern and eastern regions resulted in more than 1,800 deaths and over 500,000 internally displaced persons in 2019. The Government of Burkina Faso has made numerous arrests of terrorist suspects, augmented the size of its special terrorism detachment *Groupement des Forces Anti-Terroristes* (GFAT) in the country's north, and joined the newly—created G5 Sahel Joint Force to fight terrorism and criminal trafficking groups with regional neighbors Chad, Mali, Mauritania, and Niger. Burkina Faso's high population growth, recurring drought, pervasive and perennial food insecurity, and limited natural resources result in poor economic prospects for the majority of its citizens. (2019)

GEOGRAPHY

Location: Western Africa, north of Ghana

Geographic coordinates: 13 00 N, 2 00 W

Map references: Africa

Area: *total:* 274,200 sq km
land: 273,800 sq km
water: 400 sq km

country comparison to the world: 76

Area—comparative: slightly larger than Colorado

Land boundaries: *total:* 3,611 km
border countries (6): Benin 386 km, Cote d'Ivoire 545 km, Ghana 602 km, Mali 1325 km, Niger 622 km, Togo 131 km

Coastline: 0 km (landlocked)

Maritime claims: none (landlocked)

Climate: three climate zones including a hot tropical savanna with a short rainy season in the southern half, a tropical hot semiarid steppe climate typical of the Sahel region in the northern half, and small area of hot desert in the very north of the country bordering the Sahara Desert

Terrain: Mostly flat to dissected, undulating plains; hills in the west and southeast. Occupies an extensive plateau with savanna that is grassy in the north and gradually gives way to sparse forests in the south. (2019)

Elevation: *mean elevation:* 297 m
lowest point: Mouhoun (Black Volta) River 200 m
highest point: Tena Kourou 749 m

Natural resources: gold, manganese, zinc, limestone, marble, phosphates, pumice, salt

Land use: *agricultural land:* 44.2% (2016 est.)
arable land: 22% (2016 est.) / permanent crops: 37% (2016 est.) / permanent pasture: 21.93% (2016 est.)
forest: 19.3% (2016 est.)
other: 36.5% (2016 est.)

Irrigated land: 550 sq km (2016)

Population distribution: Most of the population is located in the center and south. Nearly one-third of the population lives in cities. The capital and largest city is Ouagadougou (Ouaga), with a population of 1.8 million as shown in this population distribution map (2019)

Natural hazards: recurring droughts

Environment—current issues: recent droughts and desertification severely affecting agricultural activities, population distribution, and the economy; overgrazing; soil degradation; deforestation (2019)

Environment—international agreements: party to: Biodiversity, Climate Change, Climate Change-Paris Agreement, Desertification, Endangered Species, Hazardous Wastes, Law of the Sea, Marine Life Conservation, Ozone Layer Protection, Wetlands (2019)

signed, but not ratified: none of the selected agreements

Geography—note: landlocked savanna cut by the three principal rivers of the Black, Red, and White Voltas

PEOPLE AND SOCIETY

Population: 20,835,401 (July 2020 est.)

note: estimates for this country explicitly take into account the effects of excess mortality due to AIDS; this can result in lower life expectancy, higher infant mortality, higher death rates, lower population growth rates, and changes in the distribution of population by age and sex than would otherwise be expected

country comparison to the world: 61

Nationality: *noun:* Burkinabe (singular and plural)

adjective: Burkinabe

Ethnic groups: Mossi 52%, Fulani 8.4%, Gurma 7%, Bobo 4.9%, Gurunsi 4.6%, Senufo 4.5%, Bissa 3.7%, Lobi 2.4%, Dagara 2.4%, Tuareg/Bella 1.9%, Dioula 0.8%, unspecified/no answer 0.3%, other 7.2% (2010 est.)

Languages: French (official), native African languages belonging to Sudanic family spoken by 90% of the population

Religions: Muslim 61.5%, Roman Catholic 23.3%, traditional/animist 7.8%, Protestant 6.5%, other/no answer 0.2%, none 0.7% (2010 est.)

Demographic profile: Burkina Faso has a young age structure – the result of declining mortality combined with steady high fertility – and continues to experience rapid population growth, which is putting increasing pressure on the country's limited arable land. More than 65% of the population is under the age of 25, and the population is growing at 3% annually. Mortality rates, especially those of infants and children, have decreased because of improved health care, hygiene, and sanitation, but women continue to have an average of almost 6 children. Even if fertility were substantially reduced, today's large cohort entering their reproductive years would sustain high population growth for the foreseeable future. Only about a third of the population is literate

and unemployment is widespread, dampening the economic prospects of Burkina Faso's large working-age population.

Migration has traditionally been a way of life for Burkinabe, with seasonal migration being replaced by stints of up to two years abroad. Cote d'Ivoire remains the top destination, although it has experienced periods of internal conflict. Under French colonization, Burkina Faso became a main labor source for agricultural and factory work in Cote d'Ivoire. Burkinabe also migrated to Ghana, Mali, and Senegal for work between the world wars. Burkina Faso attracts migrants from Cote d'Ivoire, Ghana, and Mâli, who often share common ethnic backgrounds with the Burkinabe. Despite its food shortages and high poverty rate, Burkina Faso has become a destination for refugees in recent years and hosts about 33,500 Malians as of May 2017. (2018)

Age structure: *0-14 years:* 43.58% (male 4,606,350/female 4,473,951)

15-24 years: 20.33% (male 2,121,012/female 2,114,213)

25-54 years: 29.36% (male 2,850,621/female 3,265,926)

55-64 years: 3.57% (male 321,417/female 423,016)

65 years and over: 3.16% (male 284,838/female 374,057) (2020 est.)

population pyramid:

Dependency ratios: *total dependency ratio:* 87.9

youth dependency ratio: 83.4

elderly dependency ratio: 4.5

potential support ratio: 22.1 (2020 est.)

Median age: *total:* 17.9 years

male: 17 years

female: 18.7 years (2020 est.)

country comparison to the world: 217

Population growth rate: 2.66% (2020 est.)

country comparison to the world: 18

Birth rate: 35.1 births/1,000 population (2020 est.)

country comparison to the world: 20

Death rate: 8.2 deaths/1,000 population (2020 est.)

country comparison to the world: 84

Net migration rate: -0.6 migrant(s)/1,000 population (2020 est.)

country comparison to the world: 128

Population distribution: Most of the population is located in the center and south. Nearly one-third of the population lives in cities. The capital and largest city is Ouagadougou (Ouaga), with a population of 1.8 million as shown in this population distribution map (2019)

Urbanization: *urban population:* 30.6% of total population (2020)

rate of urbanization: 4.99% annual rate of change (2015-20 est.)

total population growth rate v. urban population growth rate, 2000-2030: Major urban areas—population: 2.780 million OUAGADOUGOU (capital), 972,000 Bobo-Dioulasso (2020)

Sex ratio: *at birth:* 1.03 male(s)/female

0-14 years: 1.03 male(s)/female

15-24 years: 1 male(s)/female

25-54 years: 0.87 male(s)/female

55-64 years: 0.76 male(s)/female

65 years and over: 0.76 male(s)/female

total population: 0.96 male(s)/female (2020 est.)

Mother's mean age at first birth: 19.4 years (2010 est.)

note: median age at first birth among women 25-29

Maternal mortality rate: 320 deaths/100,000 live births (2017 est.)

country comparison to the world: 34

Infant mortality rate: *total:* 52 deaths/1,000 live births

male: 56.4 deaths/1,000 live births

female: 47.5 deaths/1,000 live births (2020 est.)

country comparison to the world: 21

Life expectancy at birth: total population: 62.7 years

male: 60.9 years

female: 64.5 years (2020 est.)

country comparison to the world: 208

Total fertility rate: 4.51 children born/woman (2020 est.)

country comparison to the world: 23

Contraceptive prevalence rate: 32.5% (2018/19)

Drinking water source:

improved:

urban: 94.9% of population

rural: 67.9% of population

total: 75.6% of population

unimproved:

urban: 4.5% of population

rural: 32.1% of population

total: 24.4% of population (2017 est.)

Current Health Expenditure: 6.9% (2017)

Physicians density: 0.08 physicians/1,000 population (2017)

Hospital bed density: 0.4 beds/1,000 population (2010)

Sanitation facility access:

improved:

urban: 88.2% of population

rural: 30.2% of population

total: 46.9% of population

unimproved:

urban: 11.8% of population

rural: 69.8% of population

total: 53.1% of population (2017 est.)

HIV/AIDS—adult prevalence rate: 0.8% (2019 est.)

country comparison to the world: 51

HIV/AIDS—people living with HIV/AIDS: 100,000 (2019 est.)

country comparison to the world: 44

HIV/AIDS—deaths: 3,100 (2019 est.)

country comparison to the world: 33

Major infectious diseases: *degree of risk:* very high (2020)

food or waterborne diseases: bacterial and protozoal diarrhea, hepatitis A, and typhoid fever

vectorborne diseases: dengue fever and malaria

water contact diseases: schistosomiasis
animal contact diseases: rabies
respiratory diseases: meningococcal meningitis

Obesity—adult prevalence rate: 5.6% (2016)
country comparison to the world: 175

Children under the age of 5 years underweight: 17.7% (2018)
country comparison to the world: 30

Education expenditures: 4.2% of GDP (2015)
country comparison to the world: 92

Literacy: *definition:* age 15 and over can read and write
total population: 41.2%
male: 50.1%
female: 32.7% (2018)

School life expectancy (primary to tertiary education): *total:* 9 years
male: 9 years
female: 9 years (2019)

Unemployment, youth ages 15-24: *total:* 8.7%
male: 5.3%
female: 12.5% (2014)
country comparison to the world: 136

GOVERNMENT

Country name: *conventional long form:* none
conventional short form: Burkina Faso
local long form: none
local short form: Burkina Faso
former: Upper Volta, Republic of Upper Volta
etymology: name translates as "Land of the Honest (Incorruptible) Men"

Government type: presidential republic

Capital: *name:* Ouagadougou
geographic coordinates: 12 22 N, 1 31 W
time difference: UTC 0 (5 hours ahead of Washington, DC, during Standard Time)
etymology: Ouagadougou is a Francophone spelling of the native name "Wogodogo," meaning "where people get honor and respect"

Administrative divisions: 13 regions; Boucle du Mouhoun, Cascades, Centre, Centre-Est, Centre-Nord, Centre-Ouest, Centre-Sud, Est, Hauts-Bassins, Nord, Plateau-Central, Sahel, Sud-Ouest

Independence: 5 August 1960 (from France)

National holiday: Republic Day, 11 December (1958); note—commemorates the day that Upper Volta became an autonomous republic in the French Community

Constitution: *history:* several previous; latest approved by referendum 2 June 1991, adopted 11 June 1991, temporarily suspended late October to mid-November 2014; initial draft of a new constitution to usher in the new republic was completed in January 2017 and a final draft was submitted to the government in December 2017; a constitutional referendum originally scheduled for adoption in March 2019 was postponed until 2020 *amendments:* proposed by the president, by a majority of National Assembly membership, or by petition of at least 30,000 eligible voters submitted to the Assembly; passage requires at least

three-fourths majority vote in the Assembly; failure to meet that threshold requires majority voter approval in a referendum; constitutional provisions on the form of government, the multiparty system, and national sovereignty cannot be amended; amended several times, last in 2012

Legal system: civil law based on the French model and customary law; in mid-2019, the National Assembly amended the penal code

International law organization participation: has not submitted an ICJ jurisdiction declaration; accepts ICCt jurisdiction

Citizenship: *citizenship by birth:* no
citizenship by descent only: at least one parent must be a citizen of Burkina Faso
dual citizenship recognized: yes
residency requirement for naturalization: 10 years

Suffrage: 18 years of age; universal

Executive branch: *chief of state:* President Roch Marc Christian KABORE (since 29 December 2015)
head of government: Prime Minister Christophe DABIRE (since 24 January 2019)
cabinet: Council of Ministers appointed by the president on the recommendation of the prime minister
elections/appointments: president elected by absolute majority popular vote in 2 rounds if needed for a 5-year term (eligible for a second); election last held on 29 November 2015 (next to be held November 2020); prime minister appointed by the president with consent of the National Assembly
election results: Roch Marc Christian KABORE elected president in first round; percent of vote—Roch Marc Christian KABORE (MPP) 53.5%, Zephirin DIABRE (UPC) 29.6%, Tahirou BARRY (PAREN) 3.1%, Benewende Stanislas SANKARA (UNIR-MS) 2.8%, other 10.9%

Legislative branch: *description:* unicameral National Assembly (127 seats; members directly elected in multi-seat constituencies by party-list proportional representation vote to serve 5-year terms)
elections: last held on 29 November 2015 (next to be held on 22 November 2020)
election results: percent of vote by party—NA; seats by party—MPP 55, UPC 33, CDP 18, Union for Rebirth-Sankarist Party 5, ADF/ RDA 3, NTD 3, other 10; composition—men 115, women 12, percent of women 9.4%

Judicial branch: *highest courts:* Supreme Court of Appeals or Cour de Cassation (consists of NA judges); Council of State (consists of NA judges); Constitutional Council or Conseil Constitutionnel (consists of the council president and 9 members)
judge selection and term of office: Supreme Court judge appointments mostly controlled by the president of Burkina Faso; judges have no term limits; Council of State judge appointment and tenure NA; Constitutional Council judges appointed by the president of Burkina Faso upon the proposal

of the minister of justice and the president of the National Assembly; judges appointed for 9-year terms with one-third of membership renewed every 3 years
subordinate courts: Appeals Court; High Court; first instance tribunals; district courts; specialized courts relating to issues of labor, children, and juveniles; village (customary) courts

Political parties and leaders: African Democratic Rally/Alliance for Democracy and Federation or ADF/RDA [Gilbert Noel OUEDRAOGO]
African People's Movement or MAP [Victorien TOUGOUMA]
Congress for Democracy and Progress or CDP [Eddie KOMBOIGO]
Le Faso Autrement [Ablasse OUEDRAOGO]
New Alliance of the Faso or NAFA [Mahamoudou DICKO]
New Time for Democracy or NTD [Vincent DABILGOU]
Organization for Democracy and Work or ODT [Anatole BONKOUNGOU]
Party for Development and Change or PDC [Aziz SEREME]
Party for Democracy and Progress-Socialist Party or PDP-PS [Drabo TORO]
Party for Democracy and Socialism/Metba or PDS/ Metba [Philippe OUEDRAOGO]
Party for National Renaissance or PAREN [Michel BERE]
People's Movement for Progress or MPP [Simon COMPAORE]
Rally for Democracy and Socialism or RDS [Francois OUEDRAOGO]
Rally for the Development of Burkina or RDB [Celestin Saidou COMPAORE]
Rally of Ecologists of Burkina Faso or RDEB [Adama SERE]
Soleil d'Avenir [Abdoulaye SOMA]
Union for a New Burkina or UBN [Diemdioda DICKO]
Union for Progress and Change or UPC [Zephirin DIABRE]
Union for Rebirth—Sankarist Party or UNIR-MS [Benewende Stanislas SANKARA]
Union for the Republic or UPR [Toussaint Abel COULIBALY]
Youth Alliance for the Republic and Independence or AJIR [Adama KANAZOE]

International organization participation: ACP, AfDB, AU, CD, ECOWAS, EITI (compliant country), Entente, FAO, FZ, G-77, IAEA, IBRD, ICAO, ICC (NGOs), ICCt, ICRM, IDA, IDB, IFAD, IFC, IFRCS, ILO, IMF, Interpol, IOC, IOM, IPU, ISO, ITSO, ITU, ITUC (NGOs), MIGA, MINUSMA, MONUSCO, NAM, OIC, OIF, OPCW, PCA, UN, UNAMID, UNCTAD, UNESCO, UNIDO, UNISFA, UNITAR, UNWTO, UPU, WADB (regional), WAEMU, WCO, WFTU (NGOs), WHO, WIPO, WMO, WTO

Diplomatic representation in the US: *chief of mission:* Ambassador Seydou KABORE (since 18 January 2017)

chancery: 2340 Massachusetts Avenue NW, Washington, DC 20008
telephone: [1] (202) 332-5577
FAX: [1] (202) 667-1882

Diplomatic representation from the US: *chief of mission:* Ambassador Andrew YOUNG (since 1 December 2016)
telephone: [226] 25-49-53-00
embassy: Rue 15.873, Avenue Sembene Ousmane, Ouaga 2000, Secteur 15
mailing address: 01 B. P. 35, Ouagadougou 01; pouch mail—US Department of State, 2440 Ouagadougou Place, Washington, DC 20521-2440
FAX: [226] 25-49-56-28

Flag description: two equal horizontal bands of red (top) and green with a yellow five-pointed star in the center; red recalls the country's struggle for independence, green is for hope and abundance, and yellow represents the country's mineral wealth
note: uses the popular Pan-African colors of Ethiopia

National symbol(s): white stallion; national colors: red, yellow, green

National anthem: *name:* "Le Ditanye" (Anthem of Victory)
lyrics/music: Thomas SANKARA
note: adopted 1974; also known as "Une Seule Nuit" (One Single Night); written by the country's former president, an avid guitar player
0:00 / 0:00

ECONOMY

Economy—overview: Burkina Faso is a poor, landlocked country that depends on adequate rainfall. Irregular patterns of rainfall, poor soil, and the lack of adequate communications and other infrastructure contribute to the economy's vulnerability to external shocks. About 80% of the population is engaged in subsistence farming and cotton is the main cash crop. The country has few natural resources and a weak industrial base.

Cotton and gold are Burkina Faso's key exports—gold has accounted for about three-quarters of the country's total export revenues. Burkina Faso's economic growth and revenue depends largely on production levels and global prices for the two commodities. The country has seen an upswing in gold exploration, production, and exports.

In 2016, the government adopted a new development strategy, set forth in the 2016-2020 National Plan for Economic and Social Development, that aims to reduce poverty, build human capital, and to satisfy basic needs. A new three-year IMF program (2018-2020), approved in 2018, will allow the government to reduce the budget deficit and preserve critical spending on social services and priority public investments.

While the end of the political crisis has allowed Burkina Faso's economy to resume positive growth, the country's fragile security situation could put these gains at risk. Political insecurity in neighboring Mali, unreliable energy supplies, and poor transportation links pose long-term challenges.

GDP (purchasing power parity): $35.85 billion (2017 est.)
$33.69 billion (2016 est.)
$31.81 billion (2015 est.)
note: data are in 2017 dollars
country comparison to the world: 126

GDP (official exchange rate): $12.57 billion (2017 est.)

GDP—real growth rate: 6.4% (2017 est.)
5.9% (2016 est.)
3.9% (2015 est.)
country comparison to the world: 25

GDP—per capita (PPP): $1,900 (2017 est.)
$1,800 (2016 est.)
$1,800 (2015 est.)
note: data are in 2017 dollars
country comparison to the world: 211

Gross national saving: 9.3% of GDP (2017 est.)
8.5% of GDP (2016 est.)
5.3% of GDP (2015 est.)
country comparison to the world: 164

GDP—composition, by end use: *household consumption:* 56.5% (2017 est.)
government consumption: 23.9% (2017 est.)
investment in fixed capital: 24.6% (2017 est.)
investment in inventories: 1% (2017 est.)
exports of goods and services: 28.4% (2017 est.)
imports of goods and services: -34.4% (2017 est.)

GDP—composition, by sector of origin: *agriculture:* 31% (2017 est.)
industry: 23.9% (2017 est.)
services: 44.9% (2017 est.)

Agriculture—products: cotton, peanuts, shea nuts, sesame, sorghum, millet, corn, rice; livestock

Industries: cotton lint, beverages, agricultural processing, soap, cigarettes, textiles, gold

Industrial production growth rate: 10.4% (2017 est.)
country comparison to the world: 14

Labor force: 8.501 million (2016 est.)
note: a large part of the male labor force migrates annually to neighboring countries for seasonal employment
country comparison to the world: 57

Labor force—by occupation: *agriculture:* 90%
industry and services: 10% (2000 est.)

Unemployment rate: 77% (2004)
country comparison to the world: 218

Population below poverty line: 40.1% (2009 est.)

Household income or consumption by percentage share: *lowest 10%:* 2.9%
highest 10%: 32.2% (2009 est.)

Budget: *revenues:* 2.666 billion (2017 est.)
expenditures: 3.655 billion (2017 est.)

Taxes and other revenues: 21.2% (of GDP) (2017 est.)
country comparison to the world: 143

Budget surplus (+) or deficit (-): -7.9% (of GDP) (2017 est.)
country comparison to the world: 198

Public debt: 38.1% of GDP (2017 est.)
38.3% of GDP (2016 est.)

country comparison to the world: 136

Fiscal year: calendar year

Inflation rate (consumer prices): 0.4% (2017 est.)
-0.2% (2016 est.)
country comparison to the world: 22

Current account balance: -$1.019 billion (2017 est.)
-$820 million (2016 est.)
country comparison to the world: 146

Exports: $3.14 billion (2017 est.)
$2.641 billion (2016 est.)
country comparison to the world: 128

Exports—partners: Switzerland 44.9%, India 15.6%, South Africa 11.3%, Cote d'Ivoire 4.9% (2017)

Exports—commodities: gold, cotton, livestock

Imports: $3.305 billion (2017 est.)
$2.827 billion (2016 est.)
country comparison to the world: 144

Imports—commodities: capital goods, foodstuffs, petroleum

Imports—partners: China 13.2%, Cote d'Ivoire 9.5%, US 8.2%, Thailand 8.1%, France 6.5%, Ghana 4.4%, Togo 4.4%, India 4.3% (2017)

Reserves of foreign exchange and gold: $49 million (31 December 2017 est.)
$50.9 million (31 December 2016 est.)
country comparison to the world: 186

Debt—external: $3.056 billion (31 December 2017 est.)
$2.88 billion (31 December 2016 est.)
country comparison to the world: 142

Exchange rates: Communaute Financiere Africaine francs (XOF) per US dollar -
605.3 (2017 est.)
593.01 (2016 est.)
593.01 (2015 est.)
591.45 (2014 est.)
494.42 (2013 est.)

ENERGY

Electricity access: *population without electricity:* 16 million (2019)
electrification—total population: 22% (2019)
electrification—urban areas: 69% (2019)
electrification—rural areas: 2% (2019)

Electricity—production: 990 million kWh (2016 est.)
country comparison to the world: 152

Electricity—consumption: 1.551 billion kWh (2016 est.)
country comparison to the world: 148

Electricity—exports: 0 kWh (2016 est.)
country comparison to the world: 112

Electricity—imports: 630 million kWh (2016 est.)
country comparison to the world: 77

Electricity—installed generating capacity: 342,400 kW (2016 est.)
country comparison to the world: 153

Electricity—from fossil fuels: 80% of total installed capacity (2016 est.)

country comparison to the world: 82

Electricity—from nuclear fuels: 0% of total installed capacity (2017 est.)
country comparison to the world: 57

Electricity—from hydroelectric plants: 9% of total installed capacity (2017 est.)
country comparison to the world: 117

Electricity—from other renewable sources: 12% of total installed capacity (2017 est.)
country comparison to the world: 72

Crude oil—production: 0 bbl/day (2018 est.)
country comparison to the world: 116

Crude oil—exports: 0 bbl/day (2015 est.)
country comparison to the world: 100

Crude oil—imports: 0 bbl/day (2015 est.)
country comparison to the world: 102

Crude oil—proved reserves: 0 bbl (1 January 2018 est.)
country comparison to the world: 112

Refined petroleum products—production: 0 bbl/day (2015 est.)
country comparison to the world: 124

Refined petroleum products—consumption: 23,000 bbl/day (2016 est.)
country comparison to the world: 132

Refined petroleum products—exports: 0 bbl/day (2015 est.)
country comparison to the world: 136

Refined petroleum products—imports: 23,580 bbl/day (2015 est.)
country comparison to the world: 110

Natural gas—production: 0 cu m (2017 est.)
country comparison to the world: 110

Natural gas—consumption: 0 cu m (2017 est.)
country comparison to the world: 126

Natural gas—exports: 0 cu m (2017 est.)
country comparison to the world: 75

Natural gas—imports: 0 cu m (2017 est.)
country comparison to the world: 98

Natural gas—proved reserves: 0 cu m (1 January 2014 est.)
country comparison to the world: 116

Carbon dioxide emissions from consumption of energy: 3.421 million Mt (2017 est.)
country comparison to the world: 142

COMMUNICATIONS

Telephones—fixed lines: *total subscriptions:* 75,066
subscriptions per 100 inhabitants: less than 1 (2019 est.)
country comparison to the world: 146

Telephones—mobile cellular: *total subscriptions:* 20,330,657
subscriptions per 100 inhabitants: 100.21 (2019 est.)
country comparison to the world: 60

Telecommunication systems: *general assessment:* system includes microwave radio relay, open-wire, and radiotelephone communication stations; insufficient mobile spectrum, and poor condition

of fixed-line networks hinders the development of fixed-line Internet services and leaves Burkina Faso with some of the most expensive telecommunications globally; mobile telephony has experienced growth, but below the African average; govt. proposes technology-neutral licenses to boost mobile broadband connectivity and amend legislation to improve regulators and legalize the framework governing the telecom sector (2020)
domestic: fixed-line connections stand at less than 1 per 100 persons; mobile-cellular usage 100 per 100, with multiple providers there is competition and the hope for growth from a low base; Internet penetration is 11% countrywide, but higher in urban areas (2019)
international: country code—226; satellite earth station—1 Intelsat (Atlantic Ocean)
note: the COVID-19 outbreak is negatively impacting telecommunications production and supply chains globally; consumer spending on telecom devices and services has also slowed due to the pandemic's effect on economies worldwide; overall progress towards improvements in all facets of the telecom industry—mobile, fixed-line, broadband, submarine cable and satellite—has moderated

Broadcast media: since the official inauguration of Terrestrial Digital Television (TNT) in December 2017, Burkina Faso now has 14 digital TV channels among which 2 are state-owned; there are more than 140 radio stations (commercial, religious, community) available throughout the country including a national and regional state-owned network; the state-owned Radio Burkina and the private Radio Omega are among the most widespread stations and both include broadcasts in French and local languages (2019)

Internet country code: .bf

Internet users: *total:* 3,158,834
percent of population: 16% (July 2018 est.)
country comparison to the world: 96

Broadband—fixed subscriptions: *total:* 13,818
subscriptions per 100 inhabitants: less than 1 (2018 est.)
country comparison to the world: 162

TRANSPORTATION

National air transport system: *number of registered air carriers:* 1 (2020)
inventory of registered aircraft operated by air carriers: 3
annual passenger traffic on registered air carriers: 151,531 (2018)
annual freight traffic on registered air carriers: 100,000 mt-km (2018)

Civil aircraft registration country code prefix: XT (2016)

Airports: 23 (2013)
country comparison to the world: 133

Airports—with paved runways: *total:* 2 (2019)
over 3,047 m: 1
2,438 to 3,047 m: 1

Airports—with unpaved runways: *total:* 21 (2013)

1,524 to 2,437 m: 3 (2013)
914 to 1,523 m: 13 (2013)
under 914 m: 5 (2013)

Railways: *total:* 622 km (2014)
narrow gauge: 622 km 1.000-m gauge (2014)
note: another 660 km of this railway extends into Cote d'Ivoire
country comparison to the world: 108

Roadways: *total:* 15,304 km (2014)
paved: 3,642 km (2014)
unpaved: 11,662 km (2014)
country comparison to the world: 124

MILITARY AND SECURITY

Military and security forces: Armed Forces of Burkina Faso (FABF): Army of Burkina Faso (L'Armee de Terre, LAT), Air Force of Burkina Faso (Force Aerienne de Burkina Faso, FABF), National Gendarmerie, National Fire Brigade (Brigade Nationale des Sapeurs-Pompiers, BNSP) (2019)
note: the National Gendarmerie officially reports to the Ministry of Defense, but usually operates in support of the Ministry of Security and the Ministry of Justice; Gendarmerie troops are typically integrated with Army forces in anti-terrorism operations; for example, Gendarmerie, Army, and police forces were combined to form a task force known as the *Groupement des Forces Anti-Terroristes* (GFAT) to address terrorist activities along the country's northern border in 2013

Military expenditures: 2.4% of GDP (2019)
2.1% of GDP (2018)
1.4% of GDP (2017)
1.2% of GDP (2016)
1.3% of GDP (2015)
country comparison to the world: 36

Military and security service personnel strengths: the Armed Forces of Burkina Faso (FABF) have approximately 12,000 personnel (7,000 Army; 500 Air Force; 4,500 National Gendarmerie) (2019 est.)

Military equipment inventories and acquisitions: the FABF has a mix of foreign-supplied weapons; since 2010, it has received limited amounts of equipment from several countries, including donated second hand armaments; the leading suppliers are Brazil, Russia, and Turkey (2019 est.)

Military deployments: 1,100 Mali (MINUSMA) (2020)

Military service age and obligation: 18 years of age for voluntary military service; no conscription; women may serve in supporting roles (2013)

Military—note: since at least 2016, the Armed Forces of Burkina Faso have been actively engaged in combat operations with terrorist groups linked to al-Qa'ida and ISIS; military operations have occurred in the Centre-Est, Centre-Nord, Est, Nord, and Sahel administrative regions

Burkina Faso is part of a five-nation anti-jihadist task force known as the G5 Sahel Group, set up in 2014 with Chad, Mali, Mauritania, and Niger; it has committed 550 troops and 100 gendarmes to the force; the G5 force is backed by the UN, US, and France; G5 troops periodically conduct joint operations with French forces deployed to the Sahel under Operation Barkhane; in early 2020, G5 Sahel military chiefs of staff agreed to allow defense forces from each of the states to pursue terrorist fighters up to 100 km into neighboring countries (2020)

TERRORISM

Terrorist group(s): Ansarul Islam; Islamic State of Iraq and ash-Sham in the Greater Sahara; al-Mulathamun Battalion (al-Mourabitoun); Jama'at Nusrat al-Islam wal-Muslimin (2020)

note: details about the history, aims, leadership, organization, areas of operation, tactics, targets, weapons, size, and sources of support of the group(s) appear(s) in Appendix-T

TRANSNATIONAL ISSUES

Disputes—international: adding to illicit cross-border activities, Burkina Faso has issues concerning unresolved boundary alignments with its neighbors; demarcation is currently underway with Mali; the dispute with Niger was referred to the ICJ in 2010, and a dispute over several villages with Benin persists; Benin retains a border dispute with Burkina Faso around the town of Koualou

Refugees and internally displaced persons: *refugees (country of origin):* 20,951 (Mali) (2020)

IDPs: 921,471 (2020)

Trafficking in persons: current situation: Burkina Faso is a source, transit, and destination country for women and children subjected to forced labor and sex trafficking; Burkinabe children are forced to work as farm hands, gold panners and washers, street vendors, domestic servants, and beggars or in the commercial sex trade, with some transported to nearby countries; to a lesser extent, Burkinabe women are recruited for legitimate jobs in the Middle East or Europe and subsequently forced into prostitution; women from other West African countries are also lured to Burkina Faso for work and subjected to forced prostitution, forced labor in restaurants, or domestic servitude

tier rating: Tier 2 Watch List – Burkina Faso does not fully comply with the minimum standards for the elimination of trafficking; however, it is making significant efforts to do so; law enforcement efforts decreased in 2014, with a significant decline in trafficking prosecutions (none for forced begging involving Koranic school teachers – a prevalent form of trafficking) and no convictions, a 2014 law criminalizing the sale of children, child prostitution, and child pornography is undermined by a provision allowing offenders to pay a fine in lieu of serving prison time proportionate to the crime; the government sustained efforts to identify and protect a large number of child victims, relying on support from NGOs and international organizations; nationwide awareness-raising activities were sustained, but little was done to stop forced begging (2015)

BURMA

INTRODUCTION

Background: Various ethnic Burman and ethnic minority city-states or kingdoms occupied the present borders through the 19th century, and several minority ethnic groups continue to maintain independent armies and control territory within the country today, in opposition to the central government. Over a period of 62 years (1824-1886), Britain conquered Burma and incorporated all the groups within the country into its Indian Empire. Burma was administered as a province of India until 1937 when it became a separate, self-governing colony; in 1948, following major battles on its territory during World War II, Burma attained independence from the British Commonwealth. Gen. NE WIN dominated the government from 1962 to 1988, first as military ruler, then as self-appointed president, and later as political kingpin. In response to widespread civil unrest, NE WIN resigned in 1988, but within months the military crushed student-led protests and took power. Since independence, successive Burmese governments have fought on-and-off conflicts with armed ethnic groups seeking autonomy in the country's mountainous border regions.

Multiparty legislative elections in 1990 resulted in the main opposition party—the National League for Democracy (NLD)—winning a landslide victory. Instead of handing over power, the junta placed NLD leader (and 1991 Nobel Peace Prize recipient) AUNG SAN SUU KYI under house arrest from 1989 to 1995, 2000 to 2002, and from May 2003 to November 2010. In late September 2007, the ruling junta brutally suppressed protests over increased fuel prices led by prodemocracy activists and Buddhist monks, killing an unknown number of people and arresting thousands for participating in the demonstrations—popularly referred to as the Saffron Revolution. In early May 2008, Cyclone Nargis struck Burma, which left over 138,000 dead and tens of thousands injured and homeless. Despite this tragedy, the junta proceeded with its May constitutional referendum, the first vote in Burma since 1990. The 2008 constitution reserves 25% of its seats to the military. Legislative elections held in November 2010, which the NLD boycotted and many in the international community considered flawed, saw the successor ruling junta's mass organization, the Union Solidarity and Development Party garner over 75% of the contested seats.

The national legislature convened in January 2011 and selected former Prime Minister THEIN SEIN as president. Although the vast majority of national-level appointees named by THEIN SEIN were former or current military officers, the government initiated a series of political and economic reforms leading to a substantial opening of the long-isolated country. These reforms included releasing hundreds of political prisoners, signing a nationwide cease-fire with several of the country's ethnic armed groups, pursuing legal reform, and gradually reducing restrictions on freedom of the press, association, and civil society. At least due in part to these reforms, AUNG SAN SUU KYI was elected to the national legislature in April 2012 and became chair of the Committee for Rule of Law and Tranquility. Burma served as chair of the Association of Southeast Asian Nations (ASEAN) for 2014. In a flawed but largely credible national legislative election in November 2015 featuring more than 90 political parties, the NLD again won a landslide victory. Using its overwhelming majority in both houses of parliament, the NLD elected HTIN KYAW, AUNG SAN SUU KYI's confidant and long-time NLD supporter, as president. The new legislature created the position of State Counsellor, according AUNG SAN SUU KYI a formal role in the government and making her the de facto head of state. Burma's first credibly elected civilian government after more than five decades of military dictatorship was sworn into office on 30 March 2016. In March 2018, upon HTIN KYAW's resignation, parliament selected WIN MYINT, another long-time ally of AUNG SAN SUU KYI's, as president.

Attacks in October 2016 and August 2017 on security forces in northern Rakhine State by members of the Arakan Rohingya Salvation Army (ARSA), a Rohingya militant group, resulted in military crackdowns on the Rohingya population that reportedly caused thousands of deaths and human rights abuses. Following the August 2017 violence, over 740,000 Rohingya fled to neighboring Bangladesh as refugees. In November 2017, the US Department of State

determined that the August 2017 violence constituted ethnic cleansing of Rohingyas. The UN has called for Burma to allow access to a Fact Finding Mission to investigate reports of human rights violations and abuses and to work with Bangladesh to facilitate repatriation of Rohingya refugees, and in September 2018 the International Criminal Court (ICC) determined it had jurisdiction to investigate reported human rights abuses against Rohingyas. Burma has rejected charges of ethnic cleansing and genocide, and has chosen

not to work with the UN Fact Finding Mission or the ICC. In March 2018, President HTIN KYAW announced his voluntary retirement; NLD parliamentarian WIN MYINT was named by the parliament as his successor. In February 2019, the NLD announced it would establish a parliamentary committee to examine options for constitutional reform ahead of national the elections planned for 2020.

GEOGRAPHY

Location: Southeastern Asia, bordering the Andaman Sea and the Bay of Bengal, between Bangladesh and Thailand

Geographic coordinates: 22 00 N, 98 00 E

Map references: Southeast Asia

Area: *total:* 676,578 sq km
land: 653,508 sq km
water: 23,070 sq km
country comparison to the world: 41

Area—comparative: slightly smaller than Texas

Land boundaries: *total:* 6,522 km
border countries (5): Bangladesh 271 km, China 2129 km, India 1468 km, Laos 238 km, Thailand 2416 km

Coastline: 1,930 km

Maritime claims: *territorial sea:* 12 nm
exclusive economic zone: 200 nm
contiguous zone: 24 nm

continental shelf: 200 nm or to the edge of the continental margin

Climate: tropical monsoon; cloudy, rainy, hot, humid summers (southwest monsoon, June to September); less cloudy, scant rainfall, mild temperatures, lower humidity during winter (northeast monsoon, December to April)

Terrain: central lowlands ringed by steep, rugged highlands

Elevation: *mean elevation:* 702 m
lowest point: Andaman Sea/Bay of Bengal 0 m
highest point: Gamlang Razi 5,870 m

Natural resources: petroleum, timber, tin, antimony, zinc, copper, tungsten, lead, coal, marble, limestone, precious stones, natural gas, hydropower, arable land

Land use: *agricultural land:* 19.2% (2011 est.)
arable land: 16.5% (2011 est.) / *permanent crops:* 2.2% (2011 est.) / *permanent pasture:* 0.5% (2011 est.)
forest: 48.2% (2011 est.)
other: 32.6% (2011 est.)

Irrigated land: 22,950 sq km (2012)

Population distribution: population concentrated along coastal areas and in general proximity to the shores of the Irrawaddy River; the extreme north is relatively underpopulated

Natural hazards: destructive earthquakes and cyclones; flooding and landslides common during rainy season (June to September); periodic droughts

Environment—current issues: deforestation; industrial pollution of air, soil, and water; inadequate sanitation and water treatment contribute to disease; rapid depletion of the country's natural resources

Environment—international agreements: *party to:* Biodiversity, Climate Change, Climate Change-Kyoto Protocol, Desertification, Endangered Species, Hazardous Wastes, Law of the Sea, Ozone Layer Protection, Ship Pollution, Tropical Timber 83, Tropical Timber 94
signed, but not ratified: none of the selected agreements

Geography—note: strategic location near major Indian Ocean shipping lanes; the north-south flowing Irrawaddy River is the country's largest and most important commercial waterway

PEOPLE AND SOCIETY

Population: 56,590,071 (July 2020 est.)
country comparison to the world: 25

Nationality: *noun:* Burmese (singular and plural)
adjective: Burmese

Ethnic groups: Burman (Bamar) 68%, Shan 9%, Karen 7%, Rakhine 4%, Chinese 3%, Indian 2%, Mon 2%, other 5%
note: government recognizes 135 indigenous ethnic groups

Languages: Burmese (official)
note: minority ethnic groups use their own languages

Religions: Buddhist 87.9%, Christian 6.2%, Muslim 4.3%, Animist 0.8%, Hindu 0.5%, other 0.2%, none 0.1% (2014 est.)
note: religion estimate is based on the 2014 national census, including an estimate for the non-enumerated population of Rakhine State, which is assumed to mainly affiliate with the Islamic faith; as of December 2019, Muslims probably make up less than 3% of Burma's total population due to the large outmigration of the Rohingya population since 2017

Age structure: *0-14 years:* 25.97% (male 7,524,869/female 7,173,333)
15-24 years: 17% (male 4,852,122/female 4,769,412)
25-54 years: 42.76% (male 11,861,971/female 12,337,482)
55-64 years: 8.22% (male 2,179,616/female 2,472,681)
65 years and over: 6.04% (male 1,489,807/female 1,928,778) (2020 est.)

population pyramid: *Dependency ratios:* total dependency ratio: 46.5
youth dependency ratio: 37.3
elderly dependency ratio: 9.1
potential support ratio: 10.9 (2020 est.)

Median age: *total:* 29.2 years
male: 28.3 years
female: 30 years (2020 est.)
country comparison to the world: 133

Population growth rate: 0.85% (2020 est.)
country comparison to the world: 122

Birth rate: 17 births/1,000 population (2020 est.)
country comparison to the world: 99

Death rate: 7.2 deaths/1,000 population (2020 est.)
country comparison to the world: 118

Net migration rate: -1.4 migrant(s)/1,000 population (2020 est.)
country comparison to the world: 152

Population distribution: population concentrated along coastal areas and in general proximity to the shores of the Irrawaddy River; the extreme north is relatively underpopulated

Urbanization: *urban population:* 31.1% of total population (2020)
rate of urbanization: 1.74% annual rate of change (2015-20 est.)

total population growth rate v. urban population growth rate, 2000-2030: Major urban areas—population: 5.332 million RANGOON (Yangon) (capital), 1.438 million Mandalay (2020)

Sex ratio: *at birth:* 1.06 male(s)/female
0-14 years: 1.05 male(s)/female
15-24 years: 1.02 male(s)/female
25-54 years: 0.96 male(s)/female
55-64 years: 0.88 male(s)/female
65 years and over: 0.77 male(s)/female
total population: 0.97 male(s)/female (2020 est.)

Mother's mean age at first birth: 25 years (2015/16 est.)
note: median age at first birth among women 25-29

Maternal mortality rate: 250 deaths/100,000 live births (2017 est.)
country comparison to the world: 42

Infant mortality rate: *total:* 31.7 deaths/1,000 live births
male: 34.4 deaths/1,000 live births
female: 28.7 deaths/1,000 live births (2020 est.)
country comparison to the world: 50

Life expectancy at birth: *total population:* 69.3 years
male: 67.7 years
female: 71.1 years (2020 est.)
country comparison to the world: 169

Total fertility rate: 2.07 children born/woman (2020 est.)
country comparison to the world: 102

Contraceptive prevalence rate: 52.2% (2015/16)

Drinking water source:
improved:
urban: 93% of population
rural: 76.9% of population
total: 81.8% of population
unimproved:
urban: 7% of population
rural: 23.1% of population
total: 18.2% of population (2017 est.)

Current Health Expenditure: 4.7% (2017)

Physicians density: 0.86 physicians/1,000 population (2017)

Hospital bed density: 1 beds/1,000 population (2017)

Sanitation facility access:
improved:
urban: 87.6% of population
rural: 67.6% of population
total: 73.7% of population
unimproved:
urban: 12.4% of population
rural: 32.4% of population
total: 26.3% of population (2017 est.)

HIV/AIDS—adult prevalence rate: 0.6% (2019 est.)
country comparison to the world: 58

HIV/AIDS—people living with HIV/AIDS: 240,000 (2019 est.)
country comparison to the world: 24

HIV/AIDS—deaths: 7,700 (2019 est.)
country comparison to the world: 21

Major infectious diseases: *degree of risk:* very high (2020)
food or waterborne diseases: bacterial and proto-zoal diarrhea, hepatitis A, and typhoid fever
vectorborne diseases: dengue fever, malaria, and Japanese encephalitis
animal contact diseases: rabies

Obesity—adult prevalence rate: 5.8% (2016)
country comparison to the world: 172

Children under the age of 5 years underweight: 18.5% (2016)
country comparison to the world: 29

Education expenditures: 2.2% of GDP (2017)
country comparison to the world: 164

Literacy: *definition:* age 15 and over can read and write
total population: 75.6%
male: 80%
female: 71.8% (2016)

School life expectancy (primary to tertiary education): *total:* 11 years
male: 11 years
female: 11 years (2018)

Unemployment, youth ages 15-24: *total:* 2%
male: 1.8%
female: 2.2% (2018 est.)
country comparison to the world: 175

GOVERNMENT

Country name: *conventional long form:* Union of Burma
conventional short form: Burma
local long form: Pyidaungzu Thammada Myanma Naingngandaw (translated as the Republic of the Union of Myanmar)
local short form: Myanma Naingngandaw
former: Socialist Republic of the Union of Burma, Union of Myanmar
etymology: both "Burma" and "Myanmar" derive from the name of the majority Burman (Bamar) ethnic group
note: since 1989 the military authorities in Burma and the current parliamentary government have promoted the name Myanmar as a conventional name for their state; the US Government has not officially adopted the name

Government type: parliamentary republic

Capital: *name:* Rangoon (Yangon); note—Nay Pyi Taw is the administrative capital
geographic coordinates: 16 48 N, 96 09 E
time difference: UTC+6.5 (11.5 hours ahead of Washington, DC, during Standard Time)
etymology: Rangoon (Yangon) is a compound of "yan" signifying "enemies" and "koun" meaning "to run out of" and so denoting "End of Strife"; Nay Pyi Taw translates as: "Great City of the Sun" or "Abode of Kings"

Administrative divisions: 7 regions (taing-myar, singular—taing), 7 states (pyi ne-myar, singular—pyi ne), 1 union territory
regions: Ayeyarwady (Irrawaddy), Bago, Magway, Mandalay, Sagaing, Tanintharyi, Yangon (Rangoon)
states: Chin, Kachin, Kayah, Kayin, Mon, Rakhine, Shan
union territory: Nay Pyi Taw

Independence: 4 January 1948 (from the UK)

National holiday: Independence Day, 4 January (1948); Union Day, 12 February (1947)

Constitution: history: previous 1947, 1974 (suspended until 2008); latest drafted 9 April 2008, approved by referendum 29 May 2008
amendments: proposals require at least 20% approval by the Assembly of the Union membership; passage of amendments to sections of the constitution on basic principles, government structure, branches of government, state emergencies,

and amendment procedures requires 75% approval by the Assembly and approval in a referendum by absolute majority of registered voters; passage of amendments to other sections requires only 75% Assembly approval; amended 2015

Legal system: mixed legal system of English common law (as introduced in codifications designed for colonial India) and customary law

International law organization participation: has not submitted an ICJ jurisdiction declaration; non-party state to the ICCt

Citizenship: *citizenship by birth:* no
citizenship by descent only: both parents must be citizens of Burma
dual citizenship recognized: no
residency requirement for naturalization: none
note: an applicant for naturalization must be the child or spouse of a citizen

Suffrage: 18 years of age; universal

Executive branch: *chief of state:* President WIN MYINT (since 30 March 2018); Vice Presidents MYINT SWE (since 16 March 2016) and HENRY VAN THIO (since 30 March 2016); note—President HTIN KYAW (since 30 March 2016) resigned on 21 March 2018; the president is both chief of state and head of government
head of government: President WIN MYINT (since 30 March 2018); Vice Presidents MYINT SWE (since 16 March 2016) and HENRY VAN THIO (since 30 March 2016
cabinet: Cabinet appointments shared by the president and the commander-in-chief
elections/appointments: president indirectly elected by simple majority vote by the full Assembly of the Union from among 3 vice-presidential candidates nominated by the Presidential Electoral College (consists of members of the lower and upper houses and military members); the other 2 candidates become vice-presidents (president elected for a 5-year term); election last held on 28 March 2018 (next to be held in November 2020)
election results: WIN MYINT elected president; Assembly of the Union vote—WIN MYINT (NLD) 403, MYINT SWE (USDP) 211, HENRY VAN THIO (NLD) 18, 4 votes canceled (636 votes cast)
state counsellor: State Counselor AUNG SAN SUU KYI (since 6 April 2016); she concurrently serves as minister of foreign affairs and minister for the office of the president
note: a parliamentary bill creating the position of "state counsellor" was signed into law by former President HTIN KYAW on 6 April 2016; a state counsellor serves the equivalent term of the president and is similar to a prime minister in that the holder acts as a link between the parliament and the executive branch

Legislative branch: *description:* bicameral Assembly of the Union or Pyidaungsu consists of: House of Nationalities or Amyotha Hluttaw, (224 seats; 168 members directly elected in single-seat constituencies by absolute majority vote with a second round if needed and 56 appointed by the military; members serve 5-year terms) House of Representatives or Pyithu Hluttaw, (440 seats, currently 433; 330 members directly elected in single-seat constituencies by simple majority vote and 110 appointed by the military; members serve 5-year terms)

elections: House of Nationalities—last held on 8 November 2015 (next to be held on 8 November 2020)

House of Representatives—last held on 8 November 2015 (next to be held on 8 November 2020)

election results: House of Nationalities—percent of vote by party—NLD 60.3%, USDP 4.9%, ANP 4.5%, SNLD 1.3%, other 4%, military appointees 25%; seats by party—NLD 135, USDP 11, ANP 10, SNLD 3, TNP 2, ZCD 2, other 3, independent 2, military appointees 56; composition—men 201, women 23, percent of women 10.3%

House of Representatives—percent of vote by party—NLD 58%, USDP 6.8%, ANP 2.7%, SNLD 2.7%, military 25%, other 4.8%; seats by party—NLD 255, USDP 30, ANP 12, SNLD 12, PNO 3, TNP 3, LNDP 2, ZCD 2, other 3, independent 1, canceled due to insurgence 7, military appointees 110; composition—men 392, women 41, percent of women 9.5%

Judicial branch: highest courts: Supreme Court of the Union (consists of the chief justice and 7-11 judges)
judge selection and term of office: chief justice and judges nominated by the president, with approval of the Lower House, and appointed by the president; judges normally serve until mandatory retirement at age 70
subordinate courts: High Courts of the Region; High Courts of the State; Court of the Self-Administered Division; Court of the Self-Administered Zone; district and township courts; special courts (for juvenile, municipal, and traffic offenses); courts martial

Political parties and leaders: All Mon Region Democracy Party or AMRDP
Arakan National Party or ANP (formed from the 2013 merger of the Rakhine Nationalities Development Party and the Arakan League for Democracy)
National Democratic Force or NDF [KHIN MAUNG SWE]
National League for Democracy or NLD [AUNG SAN SUU KYI]
National Unity Party or NUP [THAN TIN]
Pa-O National Organization or PNO [AUNG KHAM HTI]
People's Party [KO KO GYI]
Shan Nationalities Democratic Party or SNDP [SAI AIK PAUNG]
Shan Nationalities League for Democracy or SNLD [KHUN HTUN OO]
Ta'ang National Party or TNP [AIK MONE]
Union Solidarity and Development Party or USDP [THAN HTAY]
Zomi Congress for Democracy or ZCD [PU CIN SIAN THANG]
numerous smaller parties

International organization participation: ADB, ARF, ASEAN, BIMSTEC, CP, EAS, EITI (candidate country), FAO, G-77, IAEA, IBRD, ICAO, ICRM, IDA, IFAD, IFC, IFRCS, IHO, ILO, IMF, IMO, Interpol, IOC, IOM, IPU, ISO (correspondent), ITU, ITUC (NGOs), NAM, OPCW (signatory), SAARC (observer), UN, UNCTAD, UNESCO, UNIDO, UNWTO, UPU, WCO, WHO, WIPO, WMO, WTO

Diplomatic representation in the US: *chief of mission:* Ambassador AUNG LYNN (since 16 September 2016)
chancery: 2300 S Street NW, Washington, DC 20008
telephone: [1] (202) 332-3344
FAX: [1] (202) 332-4351
consulate(s) general: Los Angeles, New York

Diplomatic representation from the US: *chief of mission:* Ambassador Scot MARCIEL (since 27 April 2016)
telephone: [95] (1) 536-509, 535-756, 538-038
embassy: 110 University Avenue, Kamayut Township, Rangoon
mailing address: Box B, APO AP 96546
FAX: [95] (1) 511-069

Flag description: design consists of three equal horizontal stripes of yellow (top), green, and red; centered on the green band is a large white five-pointed star that partially overlaps onto the adjacent colored stripes; the design revives the triband colors used by Burma from 1943-45, during the Japanese occupation

National symbol(s): chinthe (mythical lion); national colors: yellow, green, red, white

National anthem: *name:* "Kaba Ma Kyei" (Till the End of the World, Myanmar)
lyrics/music: SAYA TIN
note: adopted 1948; Burma is among a handful of non-European nations that have anthems rooted in indigenous traditions; the beginning portion of the anthem is a traditional Burmese anthem before transitioning into a Western-style orchestrated work
0:00 / 0:00

ECONOMY

Economy—overview: Since Burma began the transition to a civilian-led government in 2011, the country initiated economic reforms aimed at attracting foreign investment and reintegrating into the global economy. Burma established a managed float of the Burmese kyat in 2012, granted the Central Bank operational independence in July 2013, enacted a new anti-corruption law in September 2013, and granted licenses to 13 foreign banks in 2014-16. State Counsellor AUNG SAN SUU KYI and the ruling National League for Democracy, who took power in March 2016, have sought to improve Burma's investment climate following the US sanctions lift in October 2016 and reinstatement of Generalized System of Preferences trade benefits in November 2016. In October 2016, Burma passed a foreign investment law that consolidates investment regulations and eases rules on foreign ownership of businesses.

Burma's economic growth rate recovered from a low growth under 6% in 2011 but has been volatile between 6% and 8% between 2014 and 2018. Burma's abundant natural resources and young labor force have the potential to attract foreign investment in the energy, garment, information technology, and food and beverage sectors. The government is focusing on accelerating agricultural productivity and land reforms, modernizing and opening the financial sector, and developing transportation and electricity infrastructure. The government has also taken steps to improve transparency in the mining and oil sectors through publication of reports under the Extractive Industries Transparency Initiative (EITI) in 2016 and 2018.

Despite these improvements, living standards have not improved for the majority of the people residing in rural areas. Burma remains one of the poorest countries in Asia – approximately 26% of the country's 51 million people live in poverty. The isolationist policies and economic mismanagement of previous governments have left Burma with poor infrastructure, endemic corruption, underdeveloped human resources, and inadequate access to capital, which will require a major commitment to reverse. The Burmese Government has been slow to address impediments to economic development such as unclear land rights, a restrictive trade licensing system, an opaque revenue collection system, and an antiquated banking system.

GDP (purchasing power parity): $329.8 billion (2017 est.)
$308.7 billion (2016 est.)
$291.5 billion (2015 est.)
note: data are in 2017 dollars
country comparison to the world: 53

GDP (official exchange rate): $67.28 billion (2017 est.)

GDP—real growth rate: 6.8% (2017 est.)
5.9% (2016 est.)
7% (2015 est.)
country comparison to the world: 21

GDP—per capita (PPP): $6,300 (2017 est.)
$5,900 (2016 est.)
$5,600 (2015 est.)
note: data are in 2017 dollars
country comparison to the world: 163

Gross national saving: 17.7% of GDP (2017 est.)
17.6% of GDP (2016 est.)
18.1% of GDP (2015 est.)
country comparison to the world: 113

GDP—composition, by end use: *household consumption:* 59.2% (2017 est.)
government consumption: 13.8% (2017 est.)
investment in fixed capital: 33.5% (2017 est.)
investment in inventories: 1.5% (2017 est.)
exports of goods and services: 21.4% (2017 est.)
imports of goods and services: -28.6% (2017 est.)

GDP—composition, by sector of origin:
agriculture: 24.1% (2017 est.)
industry: 35.6% (2017 est.)
services: 40.3% (2017 est.)

Agriculture—products: rice, pulses, beans, sesame, groundnuts; sugarcane; fish and fish products; hardwood

Industries: agricultural processing; wood and wood products; copper, tin, tungsten, iron; cement, construction materials; pharmaceuticals; fertilizer; oil and natural gas; garments; jade and gems

Industrial production growth rate: 8.9% (2017 est.)
country comparison to the world: 20

Labor force: 22.3 million (2017 est.)
country comparison to the world: 24

Labor force—by occupation: agriculture: 70%
industry: 7%
services: 23% (2001 est.)

Unemployment rate: 4% (2017 est.)
4% (2016 est.)
country comparison to the world: 59

Population below poverty line: 25.6% (2016 est.)

Household income or consumption by percentage share: *lowest 10%:* 2.8%
highest 10%: 32.4% (1998)

Budget: *revenues:* 9.108 billion (2017 est.)
expenditures: 11.23 billion (2017 est.)

Taxes and other revenues: 13.5% (of GDP) (2017 est.)
country comparison to the world: 205

Budget surplus (+) or deficit (-): -3.2% (of GDP) (2017 est.)
country comparison to the world: 138

Public debt: 33.6% of GDP (2017 est.)
35.7% of GDP (2016 est.)
country comparison to the world: 156

Fiscal year: 1 April—31 March

Inflation rate (consumer prices): 4% (2017 est.)
6.8% (2016 est.)
country comparison to the world: 154

Current account balance: $240 million (2019 est.)
-$2.398 billion (2018 est.)
country comparison to the world: 56

Exports: $9.832 billion (2017 est.)
$9.085 billion (2016 est.)
note: official export figures are grossly underestimated due to the value of timber, gems, narcotics, rice, and other products smuggled to Thailand, China, and Bangladesh
country comparison to the world: 93

Exports—partners: China 36.5%, Thailand 21.8%, Japan 6.6%, Singapore 6.4%, India 5.9% (2017)

Exports—commodities: natural gas; wood products; pulses and beans; fish; rice; clothing; minerals, including jade and gems

Imports: $15.78 billion (2017 est.)
$12.81 billion (2016 est.)
note: import figures are grossly underestimated due to the value of consumer goods, diesel fuel, and other products smuggled in from Thailand, China, Malaysia, and India
country comparison to the world: 87

Imports—commodities: fabric; petroleum products; fertilizer; plastics; machinery; transport equipment; cement, construction materials; food products' edible oil

Imports—partners: China 31.4%, Singapore 15%, Thailand 11.1%, Saudi Arabia 7.5%, Malaysia 6.2%, Japan 6%, India 5.5%, Indonesia 4.5% (2017)

Reserves of foreign exchange and gold: $4.924 billion (31 December 2017 est.)
$4.63 billion (31 December 2016 est.)
country comparison to the world: 95

Debt—external: $6.594 billion (31 December 2017 est.)
$8.2 billion (31 December 2016 est.)
country comparison to the world: 125

Exchange rates: kyats (MMK) per US dollar -
1,361.9 (2017 est.)
1,234.87 (2016 est.)
1,234.87 (2015 est.)
1,162.62 (2014 est.)
984.35 (2013 est.)

ENERGY

Electricity access: *population without electricity:* 27 million (2019)
electrification—total population: 51% (2019)
electrification—urban areas: 76% (2019)
electrification—rural areas: 39% (2019)

Electricity—production: 17.32 billion kWh (2016 est.)
country comparison to the world: 83

Electricity—consumption: 14.93 billion kWh (2016 est.)
country comparison to the world: 81

Electricity—exports: 0 kWh (2016 est.)
country comparison to the world: 113

Electricity—imports: 0 kWh (2016 est.)
country comparison to the world: 130

Electricity—installed generating capacity: 5.205 million kW (2016 est.)
country comparison to the world: 79

Electricity—from fossil fuels: 39% of total installed capacity (2016 est.)
country comparison to the world: 171

Electricity—from nuclear fuels: 0% of total installed capacity (2017 est.)
country comparison to the world: 58

Electricity—from hydroelectric plants: 61% of total installed capacity (2017 est.)
country comparison to the world: 28

Electricity—from other renewable sources: 1% of total installed capacity (2017 est.)
country comparison to the world: 149

Crude oil—production: 11,000 bbl/day (2018 est.)
country comparison to the world: 77

Crude oil—exports: 1,824 bbl/day (2015 est.)
country comparison to the world: 71

Crude oil—imports: 0 bbl/day (2015 est.)
country comparison to the world: 103

Crude oil—proved reserves: 139 million bbl (1 January 2018 est.)
country comparison to the world: 63

Refined petroleum products—production: 13,330 bbl/day (2017 est.)
country comparison to the world: 97

Refined petroleum products—consumption: 123,000 bbl/day (2016 est.)
country comparison to the world: 73

Refined petroleum products—exports: 0 bbl/day (2015 est.)
country comparison to the world: 137

Refined petroleum products—imports: 102,600 bbl/day (2015 est.)
country comparison to the world: 53

Natural gas—production: 18.41 billion cu m (2017 est.)
country comparison to the world: 33

Natural gas—consumption: 4.502 billion cu m (2017 est.)
country comparison to the world: 63

Natural gas—exports: 14.07 billion cu m (2017 est.)
country comparison to the world: 16

Natural gas—imports: 0 cu m (2017 est.)
country comparison to the world: 99

Natural gas—proved reserves: 637.1 billion cu m (1 January 2018 est.)
country comparison to the world: 29

Carbon dioxide emissions from consumption of energy: 27.01 million Mt (2017 est.)
country comparison to the world: 77

COMMUNICATIONS

Telephones—fixed lines: *total subscriptions:* 544,283
subscriptions per 100 inhabitants: less than 1 (2019 est.)
country comparison to the world: 91

Telephones—mobile cellular: *total subscriptions:* 63,877,526
subscriptions per 100 inhabitants: 113.84 (2019 est.)
country comparison to the world: 25

Telecommunication systems: *general assessment:* use to claim to be one of the last underdeveloped telecom markets in Asia; the mobile market has recently experienced rapid growth, in 2014 foreign competition was allowed to compete in the market and now they have moved from 1 operator to 3; low compared to other nations in the region, but expanding nationally; moving past fixed broadband to mobile device access for Internet services; rollout of 4G to eventually 5G networks (2020)
domestic: fixed-line is 1 per 100, while mobile-cellular is 114 per 100 and shows great potential for the future (2019)
international: country code—95; landing points for the SeaMeWe-3, SeaMeWe-5, AAE-1 and Singapore-Myanmar optical telecommunications submarine cable that provides links to Asia, the Middle East, Africa, Southeast Asia, Australia and Europe; satellite earth stations—2, Intelsat (Indian Ocean) and ShinSat (2019)

note: the COVID-19 outbreak is negatively impacting telecommunications production and supply chains globally; consumer spending on telecom devices and services has also slowed due to the pandemic's effect on economies worldwide; overall progress towards improvements in all facets of the telecom industry—mobile, fixed-line, broadband, submarine cable and satellite—has moderated

Broadcast media: government controls all domestic broadcast media; 2 state-controlled TV stations with 1 of the stations controlled by the armed forces; 2 pay-TV stations are joint state-private ventures; access to satellite TV is limited; 1 state-controlled domestic radio station and 9 FM stations that are joint state-private ventures; transmissions of several international broadcasters are available in parts of Burma; the Voice of America (VOA), Radio Free Asia (RFA), BBC Burmese service, the Democratic Voice of Burma (DVB), and Radio Australia use shortwave to broadcast in Burma; VOA, RFA, and DVB produce daily TV news programs that are transmitted by satellite to audiences in Burma; in March 2017, the government granted licenses to 5 private broadcasters, allowing them digital free-to-air TV channels to be operated in partnership with government-owned Myanmar Radio and Television (MRTV) and will rely upon MRTV's transmission infrastructure (2019)

Internet country code: .mm

Internet users: *total:* 17,064,985
percent of population: 30.68% (July 2018 est.)
country comparison to the world: 39

Broadband—fixed subscriptions: *total:* 129,050
subscriptions per 100 inhabitants: less than 1 (2018 est.)
country comparison to the world: 118

TRANSPORTATION

National air transport system: *number of registered air carriers:* 8 (2020)
inventory of registered aircraft operated by air carriers: 42
annual passenger traffic on registered air carriers: 3,407,788 (2018)
annual freight traffic on registered air carriers: 4.74 million mt-km (2018)

Civil aircraft registration country code prefix: XY (2016)

Airports: 64 (2013)
country comparison to the world: 76

Airports—with paved runways: *total:* 36 (2017)
over 3,047 m: 12 (2017)
2,438 to 3,047 m: 11 (2017)
1,524 to 2,437 m: 12 (2017)
under 914 m: 1 (2017)

Airports—with unpaved runways: *total:* 28 (2013)
over 3,047 m: 1 (2013)
1,524 to 2,437 m: 4 (2013)
914 to 1,523 m: 10 (2013)
under 914 m: 13 (2013)

Heliports: 11 (2013)

Pipelines: 3739 km gas, 1321 km oil (2017)

Railways: *total:* 5,031 km (2008)
narrow gauge: 5,031 km 1.000-m gauge (2008)
country comparison to the world: 40

Roadways: *total:* 157,000 km (2013)
paved: 34,700 km (2013)
unpaved: 122,300 km (2013)
country comparison to the world: 33

Waterways: 12,800 km (2011)
country comparison to the world: 10

Merchant marine: *total:* 95
by type: bulk carrier 1general cargo 39, oil tanker 6, other 49 (2019)
country comparison to the world: 93

Ports and terminals: *major seaport(s):* Mawlamyine (Moulmein), Sittwe
river port(s): Rangoon (Yangon) (Rangoon River)

MILITARY AND SECURITY

Military and security forces: Burmese Defense Service (Tatmadaw): Army (Tatmadaw Kyi), Navy (Tatmadaw Yay), Air Force (Tatmadaw Lay), Directorate of People's Militia and Border Guard Forces (2019)

Military expenditures: 2.9% of GDP (2018)
3.2% of GDP (2017)
3.7% of GDP (2016)
4.1% of GDP (2015)
3.6% of GDP (2014)
country comparison to the world: 28

Military and security service personnel strengths: estimates of the Burmese Defense Service (Tatmadaw) vary widely; approximately 380,000 total active troops (est. 340,000 Army; 20,000 Navy; 20,000 Air Force); est. 35,000 People's Militia (2019 est.)

Military equipment inventories and acquisitions: the Burmese Defense Service's inventory is comprised mostly of older Chinese and Russian/Soviet-era equipment with a smaller mix of more modern acquisitions; since 2010, China and Russia are the leading suppliers of military hardware to Burma; other suppliers include Belarus, India, Israel, and South Korea (2019 est.)

Military service age and obligation: 18-35 years of age (men) and 18-27 years of age (women) for voluntary military service; no conscription (a 2010 law reintroducing conscription has not yet entered into force); 2-year service obligation; male (ages 18-45) and female (ages 18-35) professionals (including doctors, engineers, mechanics) serve up to 3 years; service terms may be stretched to 5 years in an officially declared emergency; Burma signed the Convention on the Rights of the Child on 15 August 1991; on 27 June 2012, the regime signed a Joint Action Plan on prevention of child recruitment; in February 2013, the military formed a new task force to address forced child conscription (2013)

Military—note: since the country's founding, the armed forces have been heavily involved in domestic politics and ran the country for five decades following a military coup in 1962; the military controls three key security ministries, one of two vice presidential appointments, and 25% of the parliamentary seats; its primary operational focus is internal security, particularly counterinsurgency operations against several insurgent groups in Kachin, Rakhine, and Shan states, such as the Arakan Army, the Kachin Independence Army, the Shan State Army, and the Tang National Liberation Army; these operations have resulted in numerous civilian casualties, human rights abuses, and internal displacement; the military is also engaged in small-scale operations against Naga insurgents along the northwestern border with India (2020)

TRANSNATIONAL ISSUES

Disputes—international: over half of Burma's population consists of diverse ethnic groups who have substantial numbers of kin in neighboring countries; Bangladesh struggles to accommodate 912,000 Rohingya, Burmese Muslim minority from Rakhine State, living as refugees in Cox's Bazar; Burmese border authorities are constructing a 200 km (124 mi) wire fence designed to deter illegal cross-border transit and tensions from the military build-up along border with Bangladesh in 2010; Bangladesh referred its maritime boundary claims with Burma and India to the International Tribunal on the Law of the Sea; Burmese forces attempting to dig in to the largely autonomous Shan State to rout local militias tied to the drug trade, prompts local residents to periodically flee into neighboring Yunnan Province in China; fencing along the India-Burma international border at Manipur's Moreh town is in progress to check illegal drug trafficking and movement of militants; over 100,000 mostly Karen refugees and asylum seekers fleeing civil strife, political upheaval, and economic stagnation in Burma were living in remote camps in Thailand near the border as of May 2017

Refugees and internally displaced persons: *IDPs:* 457,000 (government offensives against armed ethnic minority groups near its borders with China and Thailand, natural disasters, forced land evictions) (2019)
stateless persons: 600,000 (2019); note—Rohingya Muslims, living predominantly in Rakhine State, are Burma's main group of stateless people; the Burmese Government does not recognize the Rohingya as a "national race" and stripped them of their citizenship under the 1982 Citizenship Law, categorizing them as "non-nationals" or "foreign residents"; under the Rakhine State Action Plan drafted in October 2014, the Rohingya must demonstrate their family has lived in Burma for at least 60 years to qualify for a lesser naturalized citizenship and the classification of Bengali or be put in detention camps and face deportation; native-born but non-indigenous people, such as Indians, are also stateless; the Burmese Government does not grant citizenship to children born outside of the country to Burmese parents who left the country illegally or fled persecution, such as those born in Thailand; the number

of stateless persons has decreased dramatically since late 2017 because hundreds of thousands of Rohingya have fled to Bangladesh since 25 August 2017 to escape violence

note: estimate does not include stateless IDPs or stateless persons in IDP-like situations because they are included in estimates of IDPs (2017)

Trafficking in persons: *current situation:* Burma is a source country for men, women, and children trafficked for the purpose of forced labor and for women and children subjected to sex trafficking; Burmese adult and child labor migrants travel to East Asia, the Middle East, South Asia, and the US, where men are forced to work in the fishing, manufacturing, forestry, and construction industries and women and girls are forced into prostitution, domestic servitude, or forced labor in the garment sector; some Burmese economic migrants and Rohingya asylum seekers have

become forced laborers on Thai fishing boats; some military personnel and armed ethnic groups unlawfully conscript child soldiers or coerce adults and children into forced labor; domestically, adults and children from ethnic areas are vulnerable to forced labor on plantations and in mines, while children may also be subject to forced prostitution, domestic service, and begging

tier rating: Tier 2 Watch List – Burma does not fully comply with the minimum standards for the elimination of trafficking, but it is making significant efforts to do so; the government has a written plan that, if implemented, would constitute making a significant effort toward meeting the minimum standard for eliminating human trafficking; in 2014, law enforcement continued to investigate and prosecute cross-border trafficking offenses but did little to address domestic trafficking; no civilians or government officials were prosecuted or convicted for the recruitment of child soldiers,

a serious problem that is hampered by corruption and the influence of the military; victim referral and protection services remained inadequate, especially for men, and left victims vulnerable to being re-trafficked; the government coordinated antitrafficking programs as part of its five-year national action plan (2015)

Illicit drugs: world's second largest producer of illicit opium with an estimated poppy cultivation totaling 41,000 hectares in 2017, a decrease of 25% from the last survey in 2015; Shan state is the source of 91% of Burma's poppy cultivation; lack of government will to take on major narcotrafficking groups and lack of serious commitment against money laundering continues to hinder the overall antidrug effort; Burma is one of the world's largest producers of amphetamine-type stimulants, which are trafficked throughout the region, as far afield as Australia and New Zealand

BURUNDI

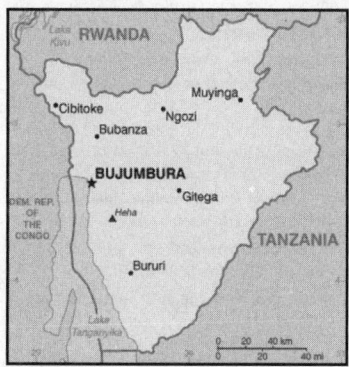

INTRODUCTION

Background: Burundi is a small country in Central-East Africa bordered by Tanzania, Rwanda, the Democratic Republic of Congo, and Lake Tanganyika. Created in the 17th century, a Burundi Kingdom was preserved under German colonial rule in the late 19th and early 20th century, and then by Belgium after World War I. Burundi gained its independence from Belgium in 1962 as the Kingdom of Burundi, but the monarchy was overthrown in 1966 and a republic established. Political violence and non-democratic transfers of power have marked much of its history; Burundi's first democratically elected president, a Hutu, was assassinated in October 1993 after only 100 days in office. The internationally brokered Arusha Agreement, signed in 2000, and subsequent ceasefire agreements with armed movements ended the 1993-2005 civil war. Burundi's second democratic elections were held in 2005. Pierre NKURUNZIZA was elected president in 2005 and 2010, and again in a controversial

election in 2015. Burundi continues to face many economic and political challenges.

GEOGRAPHY

Location: Central Africa, east of the Democratic Republic of the Congo, west of Tanzania

Geographic coordinates: 3 30 S, 30 00 E

Map references: Africa

Area: *total:* 27,830 sq km
land: 25,680 sq km
water: 2,150 sq km
country comparison to the world: 147

Area—comparative: slightly smaller than Maryland

Land boundaries: *total:* 1,140 km
border countries (3): Democratic Republic of the Congo 236 km, Rwanda 315 km, Tanzania 589 km

Coastline: 0 km (landlocked)

Maritime claims: none (landlocked)

Climate: equatorial; high plateau with considerable altitude variation (772 m to 2,670 m above sea level); average annual temperature varies with altitude from 23 to 17 degrees Celsius but is generally moderate as the average altitude is about 1,700 m; average annual rainfall is about 150 cm; two wet seasons (February to May and September to November), and two dry seasons (June to August and December to January)

Terrain: hilly and mountainous, dropping to a plateau in east, some plains

Elevation: *mean elevation:* 1,504 m
lowest point: Lake Tanganyika 772 m
highest point: Heha 2,670 m

Natural resources: nickel, uranium, rare earth oxides, peat, cobalt, copper, platinum, vanadium, arable land, hydropower, niobium, tantalum, gold, tin, tungsten, kaolin, limestone

Land use: *agricultural land:* 73.3% (2011 est.)
arable land: *38.9% (2011 est.) / permanent crops:* 15.6% (2011 est.) / permanent pasture: 18.8% (2011 est.)
forest: 6.6% (2011 est.)
other: 20.1% (2011 est.)

Irrigated land: 230 sq km (2012)

Population distribution: one of Africa's most densely populated countries; concentrations tend to be in the north and along the northern shore of Lake Tanganyika in the west; most people live on farms near areas of fertile volcanic soil as shown in this population distribution map

Natural hazards: flooding; landslides; drought

Environment—current issues: soil erosion as a result of overgrazing and the expansion of agriculture into marginal lands; deforestation (little forested land remains because of uncontrolled cutting of trees for fuel); habitat loss threatens wildlife populations

Environment—international agreements: *party to:* Biodiversity, Climate Change, Climate Change-Kyoto Protocol, Desertification, Endangered Species, Hazardous Wastes, Ozone Layer Protection, Wetlands
signed, but not ratified: Law of the Sea

Geography—note: landlocked; straddles crest of the Nile-Congo watershed; the Kagera, which drains into Lake Victoria, is the most remote headstream of the White Nile

PEOPLE AND SOCIETY

Population: 11,865,821 (July 2020 est.)
note: estimates for this country explicitly take into account the effects of excess mortality due to AIDS; this can result in lower life expectancy, higher infant mortality, higher death rates, lower population growth rates, and changes in the

distribution of population by age and sex than would otherwise be expected
country comparison to the world: 77

Nationality: *noun:* Burundian(s)
adjective: Burundian

Ethnic groups: Hutu, Tutsi, Twa (Pygmy)

Languages: Kirundi only 29.7% (official); French only .3% (official); Swahili only .2%; English only .1% (official); Kirundi and French 8.4%; Kirundi, French, and English 2.4%, other language combinations 2%, unspecified 56.9% (2008 est.)
note: data represent languages read and written by people 10 years of age or older; spoken Kirundi is nearly universal

Religions: Roman Catholic 62.1%, Protestant 23.9% (includes Adventist 2.3% and other Protestant 21.6%), Muslim 2.5%, other 3.6%, unspecified 7.9% (2008 est.)

Demographic profile: Burundi is a densely populated country with a high population growth rate, factors that combined with land scarcity and poverty place a large share of its population at risk of food insecurity. About 90% of the population relies on subsistence agriculture. Subdivision of land to sons, and redistribution to returning refugees, results in smaller, overworked, and less productive plots. Food shortages, poverty, and a lack of clean water contribute to a 60% chronic malnutrition rate among children. A lack of reproductive health services has prevented a significant reduction in Burundi's maternal mortality and fertility rates, which are both among the world's highest. With two-thirds of its population under the age of 25 and a birth rate of about 6 children per woman, Burundi's population will continue to expand rapidly for decades to come, putting additional strain on a poor country.

Historically, migration flows into and out of Burundi have consisted overwhelmingly of refugees from violent conflicts. In the last decade, more than a half million Burundian refugees returned home from neighboring countries, mainly Tanzania. Reintegrating the returnees has been problematic due to their prolonged time in exile, land scarcity, poor infrastructure, poverty, and unemployment. Repatriates and existing residents (including internally displaced persons) compete for limited land and other resources. To further complicate matters, international aid organizations reduced their assistance because they no longer classified Burundi as a post-conflict country. Conditions have deteriorated since renewed violence erupted in April 2015, causing another outpouring of refugees. In addition to refugee out—migration, Burundi has hosted thousands of refugees from neighboring countries, mostly from the Democratic Republic of the Congo and lesser numbers from Rwanda.

Age structure: *0-14 years:* 43.83% (male 2,618,868/female 2,581,597)
15-24 years: 19.76% (male 1,172,858/female 1,171,966)
25-54 years: 29.18% (male 1,713,985/female 1,748,167)
55-64 years: 4.17% (male 231,088/female 264,131)

65 years and over: 3.06% (male 155,262/female 207,899) (2020 est.)

Dependency ratios: *total dependency ratio:* 91
youth dependency ratio: 86.4
elderly dependency ratio: 4.5
potential support ratio: 22 (2020 est.)

Median age: *total:* 17.7 years
male: 17.4 years
female: 18 years (2020 est.)
country comparison to the world: 218

Population growth rate: 2.85% (2020 est.)
country comparison to the world: 11

Birth rate: 36.5 births/1,000 population (2020 est.)
country comparison to the world: 16

Death rate: 6.2 deaths/1,000 population (2020 est.)
country comparison to the world: 154

Net migration rate: -0.8 migrant(s)/1,000 population (2020 est.)
country comparison to the world: 134

Population distribution: one of Africa's most densely populated countries; concentrations tend to be in the north and along the northern shore of Lake Tanganyika in the west; most people live on farms near areas of fertile volcanic soil as shown in this population distribution map

Urbanization: *urban population:* 13.7% of total population (2020)
rate of urbanization: 5.68% annual rate of change (2015-20 est.)

total population growth rate v. urban population growth rate, 2000-2030: *Major urban areas—population:* 1,013,000 BUJUMBURA (capital) (2020)

Sex ratio: *at birth:* 1.03 male(s)/female
0-14 years: 1.01 male(s)/female
15-24 years: 1 male(s)/female
25-54 years: 0.98 male(s)/female
55-64 years: 0.87 male(s)/female
65 years and over: 0.75 male(s)/female
total population: 0.99 male(s)/female (2020 est.)

Mother's mean age at first birth: 21.3 years (2010 est.)
note: median age at first birth among women 25-29

Maternal mortality rate: 548 deaths/100,000 live births (2017 est.)
country comparison to the world: 16

Infant mortality rate: *total:* 40.1 deaths/1,000 live births
male: 44.4 deaths/1,000 live births
female: 35.7 deaths/1,000 live births (2020 est.)
country comparison to the world: 39

Life expectancy at birth: *total population:* 66.7 years
male: 64.6 years
female: 68.8 years (2020 est.)
country comparison to the world: 184

Total fertility rate: 5.28 children born/woman (2020 est.)
country comparison to the world: 12

Contraceptive prevalence rate: 28.5% (2016/17)

Drinking water source:
improved:

urban: 97.6% of population
rural: 77.8% of population
total: 80.3% of population
unimproved:
urban: -1.1% of population
rural: 22.2% of population
total: 19.7% of population (2017 est.)

Current Health Expenditure: 7.5% (2017)

Physicians density: 0.1 physicians/1,000 population (2017)

Hospital bed density: 0.8 beds/1,000 population (2014)

Sanitation facility access:
improved:
urban: 85.2% of population
rural: 53.4% of population
total: 57.4% of population
unimproved:
urban: 14.8% of population
rural: 46.6% of population
total: 42.6% of population (2017 est.)

HIV/AIDS—adult prevalence rate: 1.2% (2019 est.)
country comparison to the world: 37

HIV/AIDS—people living with HIV/AIDS: 85,000 (2019 est.)
country comparison to the world: 51

HIV/AIDS—deaths: 1,800 (2019 est.)
country comparison to the world: 47

Major infectious diseases: *degree of risk:* very high (2020)
food or waterborne diseases: bacterial and protozoal diarrhea, hepatitis A, and typhoid fever
vectorborne diseases: malaria and dengue fever
water contact diseases: schistosomiasis
animal contact diseases: rabies

Obesity—adult prevalence rate: 5.4% (2016)
country comparison to the world: 178

Children under the age of 5 years underweight: 27.2% (2018/19)
country comparison to the world: 10

Education expenditures: 4.8% of GDP (2017)
country comparison to the world: 71

Literacy: *definition:* age 15 and over can read and write
total population: 68.4%
male: 76.3%
female: 61.2% (2017)

School life expectancy (primary to tertiary education): *total:* 11 years
male: 11 years
female: 11 years (2018)

Unemployment, youth ages 15-24: *total:* 2.9%
male: 4.4%
female: 2% (2014 est.)
country comparison to the world: 173

GOVERNMENT

Country name: *conventional long form:* Republic of Burundi
conventional short form: Burundi
local long form: Republique du Burundi/Republika y'u Burundi
local short form: Burundi

former: Urundi, German East Africa, Ruanda-Urundi, Kingdom of Burundi

etymology: name derived from the pre-colonial Kingdom of Burundi (17th-19th century)

Government type: presidential republic

Capital: *name:* Gitega (political capital), Bujumbura (commercial capital); note—in January 2019, the Burundian parliament voted to make Gitega the political capital of the country while Bujumbura would remain its economic capital; all branches of the government are expected to have moved from Bujumbura to Gitega by 2021

geographic coordinates: 3 25 S, 29 55 E

time difference: UTC+ 2 (7 hours ahead of Washington, DC, during Standard Time)

etymology: the naming origins for both Gitega and Bujumbura are obscure; Bujumbura's name prior to independence in 1962 was Usumbura

Administrative divisions: 18 provinces; Bubanza, Bujumbura Mairie, Bujumbura Rural, Bururi, Cankuzo, Cibitoke, Gitega, Karuzi, Kayanza, Kirundo, Makamba, Muramvya, Muyinga, Mwaro, Ngozi, Rumonge, Rutana, Ruyigi

Independence: 1 July 1962 (from UN trusteeship under Belgian administration)

National holiday: Independence Day, 1 July (1962)

Constitution: *history:* several previous; latest ratified by referendum 28 February 2005

amendments: proposed by the president of the republic after consultation with the government or by absolute majority support of the membership in both houses of Parliament; passage requires at least two-thirds majority vote by the Senate membership and at least four-fifths majority vote by the National Assembly; the president can opt to submit amendment bills to a referendum; constitutional articles including those on national unity, the secularity of Burundi, its democratic form of government, and its sovereignty cannot be amended; amended 2018 (amendments extended the presidential term from 5 to 7 years, reintroduced the position of prime minister, and reduced the number of vice presidents from 2 to 1)

Legal system: mixed legal system of Belgian civil law and customary law

International law organization participation: has not submitted an ICJ jurisdiction declaration; withdrew from ICCt in October 2017

Citizenship: *citizenship by birth:* no

citizenship by descent only: the father must be a citizen of Burundi

dual citizenship recognized: no

residency requirement for naturalization: 10 years

Suffrage: 18 years of age; universal

Executive branch: *chief of state:* President Evariste NDAYISHIMIYE (since 18 June 2020); Vice President Prosper BAZOMBANZA (since 24 June 2020); note—the president is both chief of state and head of government

head of government: President Evariste NDAYISHIMIYE (since 18 June 2020); Vice President Prosper BAZOMBANZA (since 24 June 2020); Prime Minister Alain-Guillaume BUNYONI (since 24 June 2020)

cabinet: Council of Ministers appointed by president

elections/appointments: president directly elected by absolute majority popular vote in 2 rounds if needed for a 7-year term (eligible for a second term); election last held on 20 May 2020 (next to be held in 2025); vice presidents nominated by the president, endorsed by Parliament; note—a 2018 constitutional referendum effective for the 2020 election, increased the presidential term from 5 to 7 years with a 2-consecutive-term limit, reinstated the position of the prime minister position, and reduced the number of vice presidents from 2 to 1

election results: Evariste NDAYISHIMIYE elected president; percent of vote—Evariste NDAYISHIMIYE (CNDD-FDD) 71.5%, Agathon RWASA (CNL) 25.2%, Gaston SINDIMWO (UPRONA) 1.7%, OTHER 1.6%

Legislative branch: *description:* bicameral Parliament or Parlement consists of: Senate or Inama Nkenguzamateka (39 seats in the July 2020 election); 36 members indirectly elected by an electoral college of provincial councils using a three-round voting system, which requires a two-thirds majority vote in the first two rounds and simple majority vote for the two leading candidates in the final round; 3 seats reserved for Twas, and 30% of all votes reserved for women; members serve 5-year terms)

National Assembly or Inama Nshingamateka (123 seats in the May 2020 election; 100 members directly elected in multiseat constituencies by proportional representation vote and 23 co-opted members; 60% of seats allocated to Hutu and 40% to Tutsi; 3 seats reserved for Twas; 30% of total seats reserved for women; members serve 5-year terms)

elections: Senate—last held on 20 July 2020 (next to be held in 2025)

National Assembly—last held on 20 May 2020 (next to be held in 2025)

election results: Senate—percent of vote by party—CNDD-FDD 87.2%, Twa 7.7%, CNL 2.6%, UPRONA 2.6%; seats by party—CNDD-FDD 34, CNL 1, UPRONA 1, Twa 3; composition—men 23, women 16, percent of women 37.2%

National Assembly—percent of vote by party—CNDD-FDD 70.9%, CNL 23.4%, UPRONA 2.5%, other (co-opted Twa) 3.2%; seats by party—CNDD-FDD 86, CNL 32, UPRONA 2, Twa 3; composition—men 76, women 47, percent of women 38.2%;

note—total Parliament percent of women 38%

Judicial branch: *highest courts:* Supreme Court (consists of 9 judges and organized into judicial, administrative, and cassation chambers); Constitutional Court (consists of 7 members)

judge selection and term of office: Supreme Court judges nominated by the Judicial Service Commission, a 15-member independent body of judicial and legal profession officials), appointed by the president and confirmed by the Senate; judge tenure NA; Constitutional Court judges appointed by the president and confirmed by the Senate and serve 6-year nonrenewable terms

subordinate courts: Courts of Appeal; County Courts; Courts of Residence; Martial Court; Court Against Corruption; Commercial Court

Political parties and leaders: Front for Democracy in Burundi-Nyakuri or FRODEBU-Nyakuri [Keffa NIBIZI]

Front for Democracy in Burundi-Sahwanya or FRODEBU-Sahwanya [Pierre Claver NAHIMANA]

National Congress for Liberty or CNL [Agathon RWASA]

National Council for the Defense of Democracy—Front for the Defense of Democracy or CNDD-FDD [Evariste NDAYISHIMIYE]

National Liberation Forces or FNL [Jacques BIGITIMANA]

Union for National Progress (Union pour le Progress Nationale) or UPRONA [Abel GASHATSI]

International organization participation: ACP, AfDB, AU, CEMAC, CEPGL, CICA, COMESA, EAC, FAO, G-77, IBRD, ICAO, ICCt, ICRM, IDA, IFAD, IFC, IFRCS, ILO, IMF, Interpol, IOC, IOM, IPU, ISO (correspondent), ITU, ITUC (NGOs), MIGA, NAM, OIF, OPCW, UN, UNAMID, UNCTAD, UNESCO, UNIDO, UNISFA, UNWTO, UPU, WCO, WHO, WIPO, WMO, WTO

Diplomatic representation in the US: *chief of mission:* Ambassador S.E. Gandence SINDAYIGAYA (since 20 September 2019)

chancery: 2233 Wisconsin Avenue NW, Suite 408, Washington, DC 20007

telephone: [1] (202) 342-2574

FAX: [1] (202) 342-2578

Diplomatic representation from the US: *chief of mission:* Ambassador (vacant); Charge d'Affaires Eunice S. REDDICK (since May 2019)

telephone: [257] 22-207-000

embassy: Avenue Des Etats-Unis, Bujumbura, BP1720

mailing address: B. P. 1720, Bujumbura

FAX: [257] 22-222-926

Flag description: divided by a white diagonal cross into red panels (top and bottom) and green panels (hoist side and fly side) with a white disk superimposed at the center bearing three red six-pointed stars outlined in green arranged in a triangular design (one star above, two stars below); green symbolizes hope and optimism, white purity and peace, and red the blood shed in the struggle for independence; the three stars in the disk represent the three major ethnic groups: Hutu, Twa, Tutsi, as well as the three elements in the national motto: unity, work, progress

National symbol(s): lion; *national colors:* red, white, green

National anthem: *name:* "Burundi Bwacu" (Our Beloved Burundi)

lyrics/music: Jean-Baptiste NTAHOKAJA/Marc BARENGAYABO

note: adopted 1962

ECONOMY

Economy—overview: Burundi is a landlocked, resource-poor country with an underdeveloped manufacturing sector. Agriculture accounts for over 40% of GDP and employs more than 90% of the population. Burundi's primary exports are coffee and tea, which account for more than half of foreign exchange earnings, but these earnings are subject to fluctuations in weather and international coffee and tea prices, Burundi is heavily dependent on aid from bilateral and multilateral donors, as well as foreign exchange earnings from participation in the African Union Mission to Somalia (AMISOM). Foreign aid represented 48% of Burundi's national income in 2015, one of the highest percentages in Sub-Saharan Africa, but this figure decreased to 33.5% in 2016 due to political turmoil surrounding President NKURUNZIZA's bid for a third term. Burundi joined the East African Community (EAC) in 2009.

Burundi faces several underlying weaknesses – low governmental capacity, corruption, a high poverty rate, poor educational levels, a weak legal system, a poor transportation network, and overburdened utilities – that have prevented the implementation of planned economic reforms. The purchasing power of most Burundians has decreased as wage increases have not kept pace with inflation, which reached approximately 18% in 2017.

Real GDP growth dropped precipitously following political events in 2015 and has yet to recover to pre-conflict levels. Continued resistance by donors and the international community will restrict Burundi's economic growth as the country deals with a large current account deficit.

GDP (purchasing power parity): $8.007 billion (2017 est.)
$8.007 billion (2016 est.)
$8.091 billion (2015 est.)
note: data are in 2017 dollars
country comparison to the world: 164

GDP (official exchange rate): $3.396 billion (2017 est.)

GDP—real growth rate: 0% (2017 est.)
-1% (2016 est.)
-4% (2015 est.)
country comparison to the world: 193

GDP—per capita (PPP): $700 (2017 est.)
$800 (2016 est.)
$800 (2015 est.)
note: data are in 2017 dollars
country comparison to the world: 227

Gross national saving: -5.3% of GDP (2017 est.)
-4.1% of GDP (2016 est.)
-6.7% of GDP (2015 est.)
country comparison to the world: 183

GDP—composition, by end use: *household consumption:* 83% (2017 est.)
government consumption: 20.8% (2017 est.)
investment in fixed capital: 16% (2017 est.)
investment in inventories: 0% (2017 est.)
exports of goods and services: 5.5% (2017 est.)

imports of goods and services: -25.3% (2017 est.)

GDP—composition, by sector of origin: *agriculture:* 39.5% (2017 est.)
industry: 16.4% (2017 est.)
services: 44.2% (2017 est.)

Agriculture—products: coffee, cotton, tea, corn, beans, sorghum, sweet potatoes, bananas, cassava (manioc, tapioca); beef, milk, hides

Industries: light consumer goods (sugar, shoes, soap, beer); cement, assembly of imported components; public works construction; food processing (fruits)

Industrial production growth rate: -2% (2017 est.)
country comparison to the world: 182

Labor force: 5.012 million (2017 est.)
country comparison to the world: 78

Labor force—by occupation: *agriculture:* 93.6%
industry: 2.3%
services: 4.1% (2002 est.)

Unemployment rate: NA

Population below poverty line: 64.6% (2014 est.)

Household income or consumption by percentage share: *lowest 10%:* 4.1%
highest 10%: 28% (2006)

Budget: *revenues:* 536.7 million (2017 est.)
expenditures: 729.6 million (2017 est.)

Taxes and other revenues: 15.8% (of GDP) (2017 est.)
country comparison to the world: 186

Budget surplus (+) or deficit (-): -5.7% (of GDP) (2017 est.)
country comparison to the world: 177

Public debt: 51.7% of GDP (2017 est.)
48.4% of GDP (2016 est.)
country comparison to the world: 96

Fiscal year: calendar year

Inflation rate (consumer prices): 16.6% (2017 est.)
5.5% (2016 est.)
country comparison to the world: 214

Current account balance: -$418 million (2017 est.)
-$411 million (2016 est.)
country comparison to the world: 118

Exports: $119 million (2017 est.)
$109.7 million (2016 est.)
country comparison to the world: 195

Exports—partners: Democratic Republic of the Congo 25.5%, Switzerland 18.4%, UAE 14.9%, Belgium 6% (2017)

Exports—commodities: coffee, tea, sugar, cotton, hides

Imports: $603.8 million (2017 est.)
$527.2 million (2016 est.)
country comparison to the world: 195

Imports—commodities: capital goods, petroleum products, foodstuffs

Imports—partners: India 18.5%, China 13%, Kenya 7.9%, UAE 6.8%, Saudi Arabia 6.8%, Uganda 6%, Tanzania 5.4%, Zambia 4.6% (2017)

Reserves of foreign exchange and gold: $97.4 million (31 December 2017 est.)
$95.17 million (31 December 2016 est.)

country comparison to the world: 181

Debt—external: $610.9 million (31 December 2017 est.)
$622.4 million (31 December 2016 est.)
country comparison to the world: 173

Exchange rates: Burundi francs (BIF) per US dollar -
1,731 (2017 est.)
1,654.63 (2016 est.)
1,654.63 (2015 est.)
1,571.9 (2014 est.)
1,546.7 (2013 est.)

ENERGY

Electricity access: *population without electricity:* 10 million (2019)
electrification—total population: 11% (2019)
electrification—urban areas: 66% (2019)
electrification—rural areas: 2% (2019)

Electricity—production: 304 million kWh (2016 est.)
country comparison to the world: 183

Electricity—consumption: 382.7 million kWh (2016 est.)
country comparison to the world: 176

Electricity—exports: 0 kWh (2016 est.)
country comparison to the world: 114

Electricity—imports: 100 million kWh (2016 est.)
country comparison to the world: 100

Electricity—installed generating capacity: 68,000 kW (2016 est.)
country comparison to the world: 186

Electricity—from fossil fuels: 14% of total installed capacity (2016 est.)
country comparison to the world: 200

Electricity—from nuclear fuels: 0% of total installed capacity (2017 est.)
country comparison to the world: 59

Electricity—from hydroelectric plants: 73% of total installed capacity (2017 est.)
country comparison to the world: 14

Electricity—from other renewable sources: 14% of total installed capacity (2017 est.)
country comparison to the world: 61

Crude oil—production: 0 bbl/day (2018 est.)
country comparison to the world: 117

Crude oil—exports: 0 bbl/day (2015 est.)
country comparison to the world: 101

Crude oil—imports: 0 bbl/day (2015 est.)
country comparison to the world: 104

Crude oil—proved reserves: 0 bbl (1 January 2018 est.)
country comparison to the world: 113

Refined petroleum products—production: 0 bbl/day (2015 est.)
country comparison to the world: 125

Refined petroleum products—consumption: 1,500 bbl/day (2016 est.)
country comparison to the world: 200

Refined petroleum products—exports: 0 bbl/day (2015 est.)
country comparison to the world: 138

Refined petroleum products—imports: 1,374 bbl/day (2015 est.)
country comparison to the world: 196

Natural gas—production: 0 cu m (2017 est.)
country comparison to the world: 111

Natural gas—consumption: 0 cu m (2017 est.)
country comparison to the world: 127

Natural gas—exports: 0 cu m (2017 est.)
country comparison to the world: 76

Natural gas—imports: 0 cu m (2017 est.)
country comparison to the world: 100

Natural gas—proved reserves: 0 cu m (1 January 2014 est.)
country comparison to the world: 117

Carbon dioxide emissions from consumption of energy: 217,000 Mt (2017 est.)
country comparison to the world: 199

COMMUNICATIONS

Telephones—fixed lines: *total subscriptions:* 20,758
subscriptions per 100 inhabitants: less than 1 (2019 est.)
country comparison to the world: 175

Telephones—mobile cellular: *total subscriptions:* 6,644,833
subscriptions per 100 inhabitants: 57.62 (2019 est.)
country comparison to the world: 105

Telecommunication systems: *general assessment:* with the great population density Burundi remains one of the most alluring telecom markets in Africa for investors; the government in early 2018 began the Burundi Broadband project, which plans to deliver nationwide connectivity by 2025; mobile operators have launched 3G and LTE mobile services to capitalize on the expanding demand for Internet access; mobile penetration is at 52%, and remains low by regional standards; future plans to privatize the national telecoms (2020)
domestic: telephone density one of the lowest in the world; fixed-line connections stand at well less than 1 per 100 persons; mobile-cellular usage is 58 per 100 persons (2019)
international: country code—257; satellite earth station—1 Intelsat (Indian Ocean); the government, supported by the Word Bank, has backed a joint venture with a number of prominent telecoms to build a national fiber backbone network, offering onward connectivity to submarine cable infrastructure landings in Kenya and Tanzania (2019)
note: the COVID-19 outbreak is negatively impacting telecommunications production and supply chains globally; consumer spending on telecom devices and services has also slowed due to the pandemic's effect on economies worldwide; overall progress towards improvements in all facets of the telecom industry—mobile, fixed-line, broadband, submarine cable and satellite—has moderated

Broadcast media: state-controlled Radio Television Nationale de Burundi (RTNB) operates a TV station and a national radio network; 3 private TV stations and about 10 privately owned radio stations; transmissions of several international broadcasters are available in Bujumbura (2019)

Internet country code: .bi

Internet users: *total:* 298,684
percent of population: 2.66% (July 2018 est.)
country comparison to the world: 165

Broadband—fixed subscriptions: *total:* 3,935
subscriptions per 100 inhabitants: less than 1 (2018 est.)
country comparison to the world: 184

TRANSPORTATION

Civil aircraft registration country code prefix: 9U (2016)

Airports: 7 (2013)
country comparison to the world: 166

Airports—with paved runways: *total:* 1 (2019)
over 3,047 m: 1

Airports—with unpaved runways: *total:* 6 (2013)
914 to 1,523 m: 4 (2013)
under 914 m: 2 (2013)

Heliports: 1 (2012)

Roadways: *total:* 12,322 km (2016)
paved: 1,500 km (2016)
unpaved: 10,822 km (2016)
country comparison to the world: 130

Waterways: (mainly on Lake Tanganyika between Bujumbura, Burundi's principal port, and lake ports in Tanzania, Zambia, and the Democratic Republic of the Congo) (2011)

Ports and terminals: *lake port(s):* Bujumbura (Lake Tanganyika)

MILITARY AND SECURITY

Military and security forces: National Defense Forces (*Forces de Defense Nationale, FDN*): Army (includes maritime wing, air wing), National Police (Police Nationale du Burundi) (2019)

Military expenditures: 1.8% of GDP (2019)
1.9% of GDP (2018)
1.8% of GDP (2017)
2.2% of GDP (2016)
2.1% of GDP (2015)
country comparison to the world: 59

Military and security service personnel strengths: the National Defense Forces (FDN) have approximately 25,000 active duty Army troops (includes small air and maritime wings) (2019 est.)

Military equipment inventories and acquisitions: the FDN is armed mostly with weapons from Russia and the former Soviet Union, with some Western equipment, largely from France; since 2010, the FDN has received small amounts of mostly second-hand equipment from China, South Africa, and the US (2019)

Military deployments: 750 Central African Republic (MINUSCA); 5,400 Somalia (AMISOM) (2020)

Military service age and obligation: 18 years of age for voluntary military service; the armed forces law of 31 December 2004 did not specify a minimum age for enlistment, but the government claimed that no one younger than 18 was being recruited; mandatory retirement ages: 45 (enlisted), 50 (NCOs), 55 (officers), and 60 (officers with the rank of general) (2017)

Military—note: in addition to its foreign deployments, the FDN is focused on internal security missions, particularly against rebel groups opposed to the regime such as National Forces of Liberation (FNL), the Resistance for the Rule of Law-Tabara (aka RED Tabara), and Popular Forces of Burundi (FPB or FOREBU); the groups are based in the neighboring Democratic Republic of Congo and have carried out sporadic attacks in Burundi (2020)

TRANSNATIONAL ISSUES

Disputes—international: Burundi and Rwanda dispute two sq km (0.8 sq mi) of Sabanerwa, a farmed area in the Rukurazi Valley where the Akanyaru/Kanyaru River shifted its course southward after heavy rains in 1965; cross-border conflicts persist among Tutsi, Hutu, other ethnic groups, associated political rebels, armed gangs, and various government forces in the Great Lakes region

Refugees and internally displaced persons: *refugees (country of origin):* 77,757 (Democratic Republic of the Congo) (refugees and asylum seekers) (2020)
IDPs: 135,058 (some ethnic Tutsis remain displaced from intercommunal violence that broke out after the 1,993 coup and fighting between government forces and rebel groups; violence since April 2015) (2020)
stateless persons: 974 (2019)

Trafficking in persons: *current situation:* Burundi is a source country for children and possibly women subjected to forced labor and sex trafficking; business people recruit Burundian girls for prostitution domestically, as well as in Rwanda, Kenya, Uganda, and the Middle East, and recruit boys and girls for forced labor in Burundi and Tanzania; children and young adults are coerced into forced labor in farming, mining, informal commerce, fishing, or collecting river stones for construction; sometimes family, friends, and neighbors are complicit in exploiting children, at times luring them in with offers of educational or job opportunities
tier rating: Tier 3 – Burundi does not comply fully with the minimum standards for the elimination of human trafficking and is not making significant efforts to do so; corruption, a lack of political will, and limited resources continue to hamper efforts to combat human trafficking; in 2014, the government did not inform judicial and law enforcement officials of the enactment of an anti-trafficking law or how to implement it and approved – but did not fund – its national anti-trafficking action plan; authorities again failed to identify trafficking victims or to provide them with adequate protective services; the government has focused on transnational child trafficking but gave little attention to its domestic child trafficking problem and adult trafficking victims (2015)

INTRODUCTION

Background: The uninhabited islands were discovered and colonized by the Portuguese in the 15th century; Cabo Verde subsequently became a trading center for African slaves and later an important coaling and resupply stop for whaling and transatlantic shipping. The fusing of European and various African cultural traditions is reflected in Cabo Verde's Krioulo language, music, and pano textiles. Following independence in 1975, and a tentative interest in unification with Guinea-Bissau, a one-party system was established and maintained until multi-party elections were held in 1990. Cabo Verde continues to sustain one of Africa's most stable democratic governments and one of its most stable economies, maintaining a currency formerly pegged to the Portuguese escudo and then the euro since 1998. Repeated droughts during the second half of the 20th century caused significant hardship and prompted heavy emigration. As a result, Cabo Verde's expatriate population—concentrated in Boston and Western Europe—is greater than its domestic one. Most Cabo Verdeans have both African and Portuguese antecedents. Cabo Verde's population descends from its first permanent inhabitants in the late 15th-century—a preponderance of West African slaves, a small share of Portuguese colonists, and even fewer Italians, Spaniards, and Portuguese Jews. Among the nine inhabited islands, population distribution is variable. Islands in the east are very dry and are home to the country's growing tourism industry. The more western islands receive more precipitation and support larger populations, but agriculture and livestock grazing have damaged their soil fertility and vegetation. For centuries, the country's overall population size has fluctuated significantly, as recurring periods of famine and epidemics have caused high death tolls and emigration.

GEOGRAPHY

Location: Western Africa, group of islands in the North Atlantic Ocean, west of Senegal

Geographic coordinates: 16 00 N, 24 00 W

Map references: Africa

Area: *total:* 4,033 sq km
land: 4,033 sq km
water: 0 sq km
country comparison to the world: 176

Area—comparative: slightly larger than Rhode Island

Land boundaries: 0 km

Coastline: 965 km

Maritime claims: *territorial sea:* 12 nm
exclusive economic zone: 200 nm
contiguous zone: 24 nm
measured from claimed archipelagic baselines

Climate: temperate; warm, dry summer; precipitation meager and erratic

Terrain: steep, rugged, rocky, volcanic

Elevation: lowest point: Atlantic Ocean 0 m

highest point: Mt. Fogo (a volcano on Fogo Island) 2,829 m

Natural resources: salt, basalt rock, limestone, kaolin, fish, clay, gypsum

Land use: agricultural *land:* 18.6% (2011 est.)
arable land: 11.7% (2011 est.) / permanent crops: 0.7% (2011 est.) / permanent pasture: 6.2% (2011 est.)
forest: 21% (2011 est.)
other: 60.4% (2011 est.)
Irrigated land: 35 sq km (2012)

Population distribution: among the nine inhabited islands, population distribution is variable; islands in the east are very dry and are only sparsely settled to exploit their extensive salt deposits; the more southerly islands receive more precipitation and support larger populations, but agriculture and livestock grazing have damaged the soil fertility and vegetation; approximately half of the population lives on Sao Tiago Island, which is the location of the capital of Praia; Mindelo, on the northern island of Sao Vicente, also has a large urban population as shown in this population distribution map

Natural hazards: prolonged droughts; seasonal harmattan wind produces obscuring dust; volcanically and seismically active
volcanism: Fogo (2,829 m), which last erupted in 1995, is Cabo Verde's only active volcano

Environment—current issues: deforestation due to demand for firewood; water shortages; prolonged droughts and improper use of land (overgrazing, crop cultivation on hillsides lead to desertification and erosion); environmental damage has threatened several species of birds and reptiles; illegal beach sand extraction; overfishing

Environment—international agreements: party to: Biodiversity, Climate Change, Climate Change-Kyoto Protocol, Desertification, Endangered Species, Environmental Modification, Hazardous Wastes, Law of the Sea, Marine Dumping, Ozone Layer Protection, Ship Pollution, Wetlands

signed, but not ratified: none of the selected agreements

Geography—note: strategic location 500 km from west coast of Africa near major north-south sea routes; important communications station; important sea and air refueling site

PEOPLE AND SOCIETY

Population: 583,255 (July 2020 est.)
country comparison to the world: 173

Nationality: *noun:* Cabo Verdean(s)
adjective: Cabo Verdean

Ethnic groups: Creole (mulatto) 71%, African 28%, European 1%

Languages: Portuguese (official), Krioulo (a blend of Portuguese and West African languages)

Religions: Roman Catholic 77.3%, Protestant 4.6% (includes Church of the Nazarene 1.7%, Adventist 1.5%, Assembly of God 0.9%, Universal Kingdom of God 0.4%, and God and Love 0.1%), other Christian 3.4% (includes Christian Rationalism 1.9%, Jehovah's Witness 1%, and New Apostolic 0.5%), Muslim 1.8%, other 1.3%, none 10.8%, unspecified 0.7% (2010 est.)

Demographic profile: Cabo Verde's population descends from its first permanent inhabitants in the late 15th-century—a preponderance of West African slaves, a small share of Portuguese colonists, and even fewer Italians, Spaniards, and Portuguese Jews. Over the centuries, the country's overall population size has fluctuated significantly, as recurring periods of famine and epidemics have caused high death tolls and emigration.

Labor migration historically reduced Cabo Verde's population growth and still provides a key source of income through remittances. Expatriates probably outnumber Cabo Verde's resident population, with most families having a member abroad. Cabo Verdeans have settled in the US, Europe, Africa, and South America. The largest diaspora community in New Bedford, Massachusetts, dating to the early 1800s, is a byproduct of the transatlantic whaling industry. Cabo Verdean men fleeing poverty at home joined the crews of US whaling ships that stopped in the islands. Many settled in New Bedford and stayed in the whaling or shipping trade, worked in the textile or cranberry industries, or operated their own transatlantic packet ships that transported compatriots to the US. Increased Cabo Verdean emigration to the US coincided with the gradual and eventually complete abolition of slavery in the archipelago in 1878.

During the same period, Portuguese authorities coerced Cabo Verdeans to go to Sao Tome and Principe and other Portuguese colonies in Africa to work as indentured laborers on plantations. In the 1920s, when the US implemented immigration quotas, Cabo Verdean emigration shifted toward Portugal, West Africa (Senegal), and South America (Argentina). Growing numbers of Cabo Verdean labor migrants headed to Western Europe in the 1960s and 1970s. They filled unskilled jobs in Portugal, as many Portuguese sought out work opportunities in the more prosperous economies of northwest Europe. Cabo Verdeans eventually expanded their emigration to the Netherlands, where they worked in the shipping industry. Migration to the US resumed under relaxed migration laws. Cabo Verdean women also began migrating to southern Europe to become domestic workers, a trend that continues today and has shifted the gender balance of Cabo Verdean emigration.

Emigration has declined in more recent decades due to the adoption of more restrictive migration policies in destination countries. Reduced emigration along with a large youth population, decreased mortality rates, and increased life expectancies, has boosted population growth, putting further pressure on domestic employment and resources. In addition, Cabo Verde has attracted increasing numbers of migrants in recent decades, consisting primarily of people from West Africa, Portuguese—speaking African countries, Portugal, and China. Since the 1990s, some West African migrants have used Cabo Verde as a stepping stone for illegal migration to Europe.

Age structure: *0-14 years:* 27.95% (male 82,010/female 81,012)
15-24 years: 18.69% (male 54,521/female 54,504)
25-54 years: 40.76% (male 115,811/female 121,923)
55-64 years: 7.12% (male 18,939/female 22,597)
65 years and over: 5.48% (male 12,037/female 19,901) (2020 est.)

Dependency ratios: *total dependency ratio:* 49
youth dependency ratio: 41.8
elderly dependency ratio: 7.1
potential support ratio: 14 (2020 est.)

Median age: *total:* 26.8 years
male: 25.9 years
female: 27.6 years (2020 est.)
country comparison to the world: 151

Population growth rate: 1.28% (2020 est.)
country comparison to the world: 84

Birth rate: 19.1 births/1,000 population (2020 est.)
country comparison to the world: 79

Death rate: 5.9 deaths/1,000 population (2020 est.)
country comparison to the world: 169

Net migration rate: -0.6 migrant(s)/1,000 population (2020 est.)
country comparison to the world: 129

Population distribution: among the nine inhabited islands, population distribution is variable; islands in the east are very dry and are only sparsely settled to exploit their extensive salt deposits; the more southerly islands receive more precipitation and support larger populations, but agriculture and livestock grazing have damaged the soil fertility and vegetation; approximately half of the population lives on Sao Tiago Island, which is the location of the capital of Praia; Mindelo, on the northern island of Sao Vicente, also has a large urban population as shown in this population distribution map

Urbanization: *urban population:* 66.7% of total population (2020)
rate of urbanization: 1.97% annual rate of change (2015-20 est.)

total population growth rate v. urban population growth rate, 2000-2030: Major urban areas—population: 168,000 PRAIA (capital) (2018)

Sex ratio: *at birth:* 1.03 male(s)/female
0-14 years: 1.01 male(s)/female
15-24 years: 1 male(s)/female
25-54 years: 0.95 male(s)/female
55-64 years: 0.84 male(s)/female
65 years and over: 0.6 male(s)/female
total population: 0.95 male(s)/female (2020 est.)

Maternal mortality rate: 58 deaths/100,000 live births (2017 est.)
country comparison to the world: 91

Infant mortality rate: *total:* 19.7 deaths/1,000 live births
male: 22.7 deaths/1,000 live births
female: 16.7 deaths/1,000 live births (2020 est.)
country comparison to the world: 79

Life expectancy at birth: *total population:* 73.2 years
male: 70.8 years
female: 75.6 years (2020 est.)
country comparison to the world: 149

Total fertility rate: 2.16 children born/woman (2020 est.)
country comparison to the world: 95

Drinking water source:

improved: *urban:* 100% of population
rural: 89.1% of population
total: 96.2% of population

unimproved:

urban: *0% of population*
rural: 10.9% of population
total: 3.8% of population (2017 est.)

Current Health Expenditure: 5.2% (2017)

Physicians density: 0.78 physicians/1,000 population (2015)

Hospital bed density: 2.1 beds/1,000 population (2010)

Sanitation facility access:

improved: *urban:* 87.8% of population
rural: 64.9% of population
total: 79.8% of population

unimproved: *urban:* 12.2% of population
rural: 35.1% of population
total: 20.2% of population (2017 est.)

HIV/AIDS—adult prevalence rate: 0.6% (2019 est.)
country comparison to the world: 59

HIV/AIDS—people living with HIV/AIDS: 2,500 (2019 est.)
country comparison to the world: 135

HIV/AIDS—deaths: <100 (2019 est.)

Obesity—adult prevalence rate: 11.8% (2016)
country comparison to the world: 134

Education expenditures: 5.2% of GDP (2017)
country comparison to the world: 53

Literacy: *definition:* age 15 and over can read and write
total population: 86.8%
male: 91.7%
female: 82% (2015)

School life expectancy (primary to tertiary education): *total:* 13 years
male: 12 years
female: 13 years (2018)

Unemployment, youth ages 15-24: *total:* 27.8%
male: 24.6%
female: 31.9% (2018)
country comparison to the world: 40

GOVERNMENT

Country name: *conventional long form:* Republic of Cabo Verde
conventional short form: Cabo Verde
local long form: Republica de Cabo Verde
local short form: Cabo Verde

etymology: the name derives from Cap-Vert (Green Cape) on the Senegalese coast, the westernmost point of Africa and the nearest mainland to the islands

Government type: parliamentary republic

Capital: *name:* Praia
geographic coordinates: 14 55 N, 23 31 W
time difference: UTC-1 (4 hours ahead of Washington, DC, during Standard Time)
etymology: the earlier Portuguese name was Villa de Praia ("Village of the Beach"); it became just Praia in 1974 (prior to full independence in 1975)

Administrative divisions: 22 municipalities (concelhos, singular—concelho); Boa Vista, Brava, Maio, Mosteiros, Paul, Porto Novo, Praia, Ribeira Brava, Ribeira Grande, Ribeira Grande de Santiago, Sal, Santa Catarina, Santa Catarina do Fogo, Santa Cruz, Sao Domingos, Sao Filipe, Sao Lourenco dos Orgaos, Sao Miguel, Sao Salvador do Mundo, Sao Vicente, Tarrafal, Tarrafal de Sao Nicolau

Independence: 5 July 1975 (from Portugal)

National holiday: Independence Day, 5 July (1975)

Constitution: *history:* previous 1981; latest effective 25 September 1992

amendments: proposals require support of at least four fifths of the active National Assembly membership; amendment drafts require sponsorship of at least one third of the active Assembly membership; passage requires at least two-thirds majority vote by the Assembly membership; constitutional sections, including those on national independence, form of government, political pluralism, suffrage, and human rights and liberties, cannot be amended; revised 1995, 1999, 2010

Legal system: civil law system of Portugal

International law organization participation: has not submitted an ICJ jurisdiction declaration; accepts ICCt jurisdiction

Citizenship: *citizenship by birth:* no
citizenship by descent only: at least one parent must be a citizen of Cabo Verde

dual citizenship recognized: yes

residency requirement for naturalization: 5 years

Suffrage: 18 years of age; universal

Executive branch: *chief of state:* President Jorge Carlos FONSECA (since 9 September 2011)

head of government: Prime Minister Ulisses CORREIA E. SILVA (since 22 April 2016)

cabinet: Council of Ministers appointed by the president on the recommendation of the prime minister

elections/appointments: president directly elected by absolute majority popular vote in 2 rounds if needed for a 5-year term (eligible for a second term); election last held on 2 October 2016 (next to be held in 2021); prime minister nominated by the National Assembly and appointed by the president

election results: Jorge Carlos FONSECA reelected president; percent of vote—Jorge Carlos FONSECA (MPD) 74%, Albertino GRACA (independent) 23%, other 3%

Legislative branch: *description:* unicameral National Assembly or Assembleia Nacional (72 seats; members directly elected in multi-seat constituencies by proportional representation vote; members serve 5-year terms)

elections: last held on 20 March 2016 (next to be held in 2021)

election results: percent of vote by party MPD 54.5%, PAICV 38.2%, UCID 7%, other 0.3%; seats by party—MPD 40, PAICV 29, UCID 3; composition—men 57, women 15, percent of women 20.8%

Judicial branch: *highest courts:* Supreme Court of Justice (consists of the chief justice and at least 7 judges and organized into civil, criminal, and administrative sections)

judge selection and term of office: judge appointments—1 by the president of the republic, 1 elected by the National Assembly, and 3 by the Superior Judicial Council (SJC), a 16-member independent body chaired by the chief justice and includes the attorney general, 8 private citizens, 2 judges, 2 prosecutors, the senior legal inspector of the Attorney General's office, and a representative of the Ministry of Justice; chief justice appointed by the president of the republic from among peers of the Supreme Court of Justice and in consultation with the SJC; judges appointed for life

subordinate courts: appeals courts, first instance (municipal) courts; audit, military, and fiscal and customs courts

Political parties and leaders: rz African Party for Independence of Cabo Verde or PAICV [Janira Hopffer ALMADA] Democratic and Independent Cabo Verdean Union or UCID [Antonio MONTEIRO] Democratic Christian Party or PDC [Manuel RODRIGUES] Democratic Renovation Party or PRD [Victor FIDALGO]

Movement for Democracy or MPD [Ulisses CORREIA E SILVA] Party for Democratic Convergence or PCD [Dr. Eurico MONTEIRO] Party of Work and Solidarity or PTS [Anibal MEDINA] Social Democratic Party or PSD [Joao ALEM]

International organization participation: ACP, AfDB, AOSIS, AU, CD, CPLP, ECOWAS, FAO, G-77, IAEA, IBRD, ICAO, ICCt (signatory), ICRM, IDA, IFAD, IFC, IFRCS, ILO, IMF, IMO, Interpol, IOC, IOM, IPU, ITSO, ITU, ITUC (NGOs), MIGA, NAM, OIF, OPCW, UN, UNCTAD, UNESCO, UNIDO, Union Latina, UNWTO, UPU, WCO, WHO, WIPO, WMO, WTO

Diplomatic representation in the US: *chief of mission:* Ambassador Carlos W. VEIGA (since 18 January 2017)

chancery: 3415 Massachusetts Avenue NW, Washington, DC 20007

telephone: [1] (202) 965-6820

FAX: [1] (202) 965-1207

consulate(s) general: Boston

Diplomatic representation from the US: *chief of mission:* Ambassador John "Jeff" DAIGLE (since 28 June 2019)

telephone: [238] 260-89-00

embassy: Rua Abilio Macedo 6, Praia

mailing address: C. P. 201, Praia

FAX: [238] 261-13-55

Flag description: five unequal horizontal bands; the top—most band of blue—equal to one half the width of the flag—is followed by three bands of white, red, and white, each equal to 1/12 of the width, and a bottom stripe of blue equal to one quarter of the flag width; a circle of 10 yellow, five-pointed stars is centered on the red stripe and positioned 3/8 of the length of the flag from the hoist side; blue stands for the sea and the sky, the circle of stars represents the 10 major islands united into a nation, the stripes symbolize the road to formation of the country through peace (white) and effort (red)

National symbol(s): ten, five-pointed, yellow stars; national colors: blue, white, red, yellow

National anthem: *name:* "Cantico da Liberdade" (Song of Freedom)

lyrics/music: Amilcar Spencer LOPES/Adalberto Higino Tavares SILVA

note: adopted 1996

0:00 / 1:11

ECONOMY

Economy—overview: Cabo Verde's economy depends on development aid, foreign investment, remittances, and tourism. The economy is service-oriented with commerce, transport, tourism, and public services accounting for about three-fourths of GDP. Tourism is the mainstay of the economy and depends on conditions in the eurozone countries. Cabo Verde annually runs a high trade deficit financed by foreign aid and remittances from its large pool of emigrants; remittances

as a share of GDP are one of the highest in Sub-Saharan Africa.

Although about 40% of the population lives in rural areas, the share of food production in GDP is low. The island economy suffers from a poor natural resource base, including serious water shortages, exacerbated by cycles of long-term drought, and poor soil for growing food on several of the islands, requiring it to import most of what it consumes. The fishing potential, mostly lobster and tuna, is not fully exploited.

Economic reforms are aimed at developing the private sector and attracting foreign investment to diversify the economy and mitigate high unemployment. The government's elevated debt levels have limited its capacity to finance any shortfalls.

GDP (purchasing power parity): $3.777 billion (2017 est.)

$3.631 billion (2016 est.)

$3.468 billion (2015 est.)

note: data are in 2017 dollars

country comparison to the world: 182

GDP (official exchange rate): $1.776 billion (2017 est.)

GDP—real growth rate: 4% (2017 est.)

4.7% (2016 est.)

1% (2015 est.)

country comparison to the world: 73

GDP—per capita (PPP): $7,000 (2017 est.)

$6,800 (2016 est.)

$6,600 (2015 est.)

note: data are in 2017 dollars

country comparison to the world: 157

Gross national saving: 32.4% of GDP (2017 est.)

34.8% of GDP (2016 est.)

35.6% of GDP (2015 est.)

country comparison to the world: 25

GDP—composition, by end use: *household consumption:* 50.1% (2017 est.)

government consumption: 18.3% (2017 est.)

investment in fixed capital: 32.2% (2017 est.)

investment in inventories: 1.9% (2017 est.)

exports of goods and services: 48.6% (2017 est.)

imports of goods and services: -51.1% (2017 est.)

GDP—composition, by sector of origin: *agriculture:* 8.9% (2017 est.)

industry: 17.5% (2017 est.)

services: 73.7% (2017 est.)

Agriculture—products: bananas, corn, beans, sweet potatoes, sugarcane, coffee, peanuts; fish

Industries: food and beverages, fish processing, shoes and garments, salt mining, ship repair

Industrial production growth rate: 2.9% (2017 est.)

country comparison to the world: 107

Labor force: 196,100 (2007 est.)

country comparison to the world: 172

Unemployment rate: 9% (2017 est.)

9% (2016 est.)

country comparison to the world: 138

Population below poverty line: 30% (2000 est.)

Household income or consumption by percentage share: *lowest 10%:* 1.9%

highest 10%: 40.6% (2000)

Budget: *revenues:* 493.5 million (2017 est.)
expenditures: 546.7 million (2017 est.)

Taxes and other revenues: 27.8% (of GDP) (2017 est.)
country comparison to the world: 98

Budget surplus (+) or deficit (-): -3% (of GDP) (2017 est.)
country comparison to the world: 131

Public debt: 125.8% of GDP (2017 est.)
127.6% of GDP (2016 est.)
country comparison to the world: 8

Fiscal year: calendar year

Inflation rate (consumer prices): 0.8% (2017 est.)
-1.4% (2016 est.)
country comparison to the world: 39

Current account balance: -$109 million (2017 est.)
-$40 million (2016 est.)
country comparison to the world: 88

Exports: $189 million (2017 est.)
$148.4 million (2016 est.)
country comparison to the world: 190

Exports—partners: Spain 45.3%, Portugal 40.3%, Netherlands 8.1% (2017)

Exports—commodities: fuel (re-exports), shoes, garments, fish, hides

Imports: $836.1 million (2017 est.)
$687.3 million (2016 est.)
country comparison to the world: 189

Imports—commodities: foodstuffs, industrial products, transport equipment, fuels

Imports—partners: Portugal 43.9%, Spain 11.6%, Netherlands 6.1%, China 6.1% (2017)

Reserves of foreign exchange and gold: $617.4 million (31 December 2017 est.)
$572.7 million (31 December 2016 est.)
country comparison to the world: 145

Debt—external: $1.713 billion (31 December 2017 est.)
$1.688 billion (31 December 2016 est.)
country comparison to the world: 154

Exchange rates: Cabo Verdean escudos (CVE) per US dollar -
101.8 (2017 est.)
99.688 (2016 est.)
99.688 (2015 est.)
99.426 (2014 est.)
83.114 (2013 est.)

ENERGY

Electricity access: *electrification—total population:* 96% (2019)
electrification—urban areas: 99% (2019)
electrification—rural areas: 89% (2019)

Electricity—production: 395 million kWh (2016 est.)
country comparison to the world: 172

Electricity—consumption: 367.4 million kWh (2016 est.)
country comparison to the world: 179

Electricity—exports: 0 kWh (2016 est.)

country comparison to the world: 115

Electricity—imports: 0 kWh (2016 est.)
country comparison to the world: 131

Electricity—installed generating capacity: 162,500 kW (2016 est.)
country comparison to the world: 171

Electricity—from fossil fuels: 79% of total installed capacity (2016 est.)
country comparison to the world: 84

Electricity—from nuclear fuels: 0% of total installed capacity (2017 est.)
country comparison to the world: 60

Electricity—from hydroelectric plants: 0% of total installed capacity (2017 est.)
country comparison to the world: 162

Electricity—from other renewable sources: 21% of total installed capacity (2017 est.)
country comparison to the world: 35

Crude oil—production: 0 bbl/day (2018 est.)
country comparison to the world: 118

Crude oil—exports: 0 bbl/day (2015 est.)
country comparison to the world: 102

Crude oil—imports: 0 bbl/day (2015 est.)
country comparison to the world: 105

Crude oil—proved reserves: 0 bbl (1 January 2018 est.)
country comparison to the world: 114

Refined petroleum products—production: 0 bbl/day (2015 est.)
country comparison to the world: 126

Refined petroleum products—consumption: 5,600 bbl/day (2016 est.)
country comparison to the world: 173

Refined petroleum products—exports: 0 bbl/day (2015 est.)
country comparison to the world: 139

Refined petroleum products—imports: 5,607 bbl/day (2015 est.)
country comparison to the world: 166

Natural gas—production: 0 cu m (2017 est.)
country comparison to the world: 112

Natural gas—consumption: 0 cu m (2017 est.)
country comparison to the world: 128

Natural gas—exports: 0 cu m (2017 est.)
country comparison to the world: 77

Natural gas—imports: 0 cu m (2017 est.)
country comparison to the world: 101

Natural gas—proved reserves: 0 cu m (1 January 2016 est.)
country comparison to the world: 118

Carbon dioxide emissions from consumption of energy: 867,800 Mt (2017 est.)
country comparison to the world: 172

COMMUNICATIONS

Telephones—fixed lines: *total subscriptions:* 60,233
subscriptions per 100 inhabitants: 10.46 (2019 est.)
country comparison to the world: 154

Telephones—mobile cellular: *total subscriptions:* 623,749
subscriptions per 100 inhabitants: 108.32 (2019 est.)
country comparison to the world: 167

Telecommunication systems: *general assessment:* LTE reaches almost 40% of the population; regulator awards commercial 4G licenses and starts 5G pilot; govt. extends USD 25 million for submarine fiber-optic cable project linking Africa to Portugal and Brazil; major service provider is Cabo Verde Telecom (CVT) (2020)

domestic: 11 per 100 fixed-line and 108 per 100 mobile-cellular; fiber-optic ring, completed in 2001, links all islands providing Internet access and ISDN services; cellular service introduced in 1998; broadband services launched early in the decade (2019)

international: country code—238; landing points for the Atlantis-2, EllaLink, Cabo Verde Telecom Domestic Submarine Cable Phase 1, 2, 3 and WACS fiber-optic transatlantic telephone cable that provides links to South America, Africa, and Europe; HF radiotelephone to Senegal and Guinea-Bissau; satellite earth station—1 Intelsat (Atlantic Ocean) (2019)

note: the COVID-19 outbreak is negatively impacting telecommunications production and supply chains globally; consumer spending on telecom devices and services has also slowed due to the pandemic's effect on economies worldwide; overall progress towards improvements in all facets of the telecom industry—mobile, fixed-line, broadband, submarine cable and satellite—has moderated

Broadcast media: state-run TV and radio broadcast network plus a growing number of private broadcasters; Portuguese public TV and radio services for Africa are available; transmissions of a few international broadcasters are available (2019)

Internet country code: .cv

Internet users: *total:* 330,623

percent of population: 58.17% (July 2018 est.)
country comparison to the world: 164

Broadband—fixed subscriptions: *total:* 15,657
subscriptions per 100 inhabitants: 3 (2018 est.)
country comparison to the world: 161

TRANSPORTATION

National air transport system: *number of registered air carriers:* 2 (2020)
inventory of registered aircraft operated by air carriers: 5
annual passenger traffic on registered air carriers: 140,429 (2018)
annual freight traffic on registered air carriers: 1,728,152 mt-km (2015)

Civil aircraft registration country code prefix: D4 (2016)

Airports: 9 (2013)
country comparison to the world: 156

Airports—with paved runways: *total:* 9 (2017)

over 3,047 m: 1 (2017)
1,524 to 2,437 m: 3 (2017)
914 to 1,523 m: 3 (2017)
under 914 m: 2 (2017)

Roadways: *total:* 1,350 km (2013)
paved: 932 km (2013)
unpaved: 418 km (2013)
country comparison to the world: 176

Merchant marine: total: 44
by type: general cargo 16, oil tanker 3, other 25 (2019)
country comparison to the world: 119

Ports and terminals: *major seaport(s):* Porto Grande

MILITARY AND SECURITY

Military and security forces: Cabo Verdean Armed Forces (FACV): Army (also called the National Guard, GN), Cabo Verde Coast Guard (Guardia

Costeira de Cabo Verde, GCCV, includes naval infantry) (2013)

Military expenditures: 0.5% of GDP (2019)
0.6% of GDP (2018)
0.5% of GDP (2017)
0.6% of GDP (2016)
0.6% of GDP (2015)
country comparison to the world: 146

Military and security service personnel strengths: the Cabo Verdean Armed Forces (FACV) consist of approximately 1,100 Army (includes an air component of about 100 personnel) and 100 Coast Guard active duty troops (2019)

Military equipment inventories and acquisitions: the FACV has a limited amount of mostly dated and second-hand equipment, largely from China, European countries, and the former Soviet Union; since 2010, it has received limited quantities of equipment (naval patrol craft and air craft) from the Netherlands and Portugal (2019 est.)

Military service age and obligation: 18-35 years of age for male and female selective compulsory military service; 2-years conscript service obligation; 17 years of age for voluntary service (with parental consent) (2013)

TRANSNATIONAL ISSUES

Disputes—international: none

Refugees and internally displaced persons: *stateless persons:* 115 (2019)

Illicit drugs: used as a transshipment point for Latin American cocaine destined for Western Europe, particularly because of Lusophone links to Brazil, Portugal, and Guinea-Bissau; has taken steps to deter drug money laundering, including a 2002 anti-money laundering reform that criminalizes laundering the proceeds of narcotics trafficking and other crimes and the establishment in 2008 of a Financial Intelligence Unit

CAMBODIA

INTRODUCTION

Background: Most Cambodians consider themselves to be Khmers, descendants of the Angkor Empire that extended over much of Southeast Asia and reached its zenith between the 10th and 13th centuries. Attacks by the Thai and Cham (from present- day Vietnam) weakened the empire, ushering in a long period of decline. The king placed the country under French protection in 1863, and it became part of French Indochina in 1887. Following Japanese occupation in World War II, Cambodia gained full independence from France in 1953. In April 1975, after a seven-year struggle, communist Khmer Rouge forces captured Phnom Penh and evacuated all cities and towns. At least 1.5 million Cambodians died from execution, forced hardships, or starvation during the Khmer Rouge regime under POL POT. A December 1978 Vietnamese invasion drove

the Khmer Rouge into the countryside, began a 10-year Vietnamese occupation, and touched off 20 years of civil war.

The 1991 Paris Peace Accords mandated democratic elections and a cease-fire, which was not fully respected by the Khmer Rouge. UN-sponsored elections in 1993 helped restore some semblance of normalcy under a coalition government. Factional fighting in 1997 ended the first coalition government, but a second round of national elections in 1998 led to the formation of another coalition government and renewed political stability. The remaining elements of the Khmer Rouge surrendered in early 1999. Some of the surviving Khmer Rouge leaders were tried for crimes against humanity by a hybrid UN-Cambodian tribunal supported by international assistance. In 2018, the tribunal heard its final cases, but it remains in operation to hear appeals. Elections in July 2003 were relatively peaceful, but it took one year of negotiations between contending political parties before a coalition government was formed. In October 2004, King Norodom SIHANOUK abdicated the throne and his son, Prince Norodom SIHAMONI, was selected to succeed him. Local (Commune Council) elections were held in Cambodia in 2012, with little of the violence that preceded prior elections. National elections in July 2013 were disputed, with the opposition—the Cambodia National Rescue Party (CNRP)—boycotting the National Assembly. The political impasse was ended nearly a year later, with the CNRP agreeing to enter parliament in exchange for commitments by the ruling Cambodian People's Party (CPP) to electoral and legislative reforms. The CNRP made further gains in local commune elections in June

2017, accelerating sitting Prime Minister Hun SEN's efforts to marginalize the CNRP before national elections in 2018. Hun Sen arrested CNRP President Kem SOKHA in September 2017. The Supreme Court dissolved the CNRP in November 2017 and banned its leaders from participating in politics for at least five years. The CNRP's seats in the National Assembly were redistributed to smaller, less influential opposition parties, while all of the CNRP's 5,007 seats in the commune councils throughout the country were reallocated to the CPP. With the CNRP banned, the CPP swept the 2018 national elections, winning all 125 National Assembly seats and effectively turning the country into a one-party state.

GEOGRAPHY

Location: Southeastern Asia, bordering the Gulf of Thailand, between Thailand, Vietnam, and Laos

Geographic coordinates: 13 00 N, 105 00 E

Map references: Southeast Asia

Area: *total:* 181,035 sq km
land: 176,515 sq km
water: 4,520 sq km
country comparison to the world: 91

Area—comparative: one and a half times the size of Pennsylvania; slightly smaller than Oklahoma

Land boundaries: *total:* 2,530 km
border countries (3): Laos 555 km, Thailand 817 km, Vietnam 1158 km

Coastline: 443 km

Maritime claims: *territorial sea:* 12 nm
exclusive economic zone: 200 nm
contiguous zone: 24 nm
continental shelf: 200 nm

Climate: tropical; rainy, monsoon season (May to November); dry season (December to April); little seasonal temperature variation

Terrain: mostly low, flat plains; mountains in southwest and north

Elevation: *mean elevation:* 126 m
lowest point: Gulf of Thailand 0 m
highest point: Phnum Aoral 1,810 m

Natural resources: oil and gas, timber, gemstones, iron ore, manganese, phosphates, hydropower potential, arable land

Land use: *agricultural land:* 32.1% (2011 est.)
arable land: 22.7% (2011 est.) / permanent crops: 0.9% (2011 est.) / permanent pasture: 8.5% (2011 est.)
forest: 56.5% (2011 est.)
other: 11.4% (2011 est.)
Irrigated land: 3,540 sq km (2012)

Population distribution: population concentrated in the southeast, particularly in and around the capital of Phnom Penh; further distribution is linked closely to the Tonle Sap and Mekong Rivers

Natural hazards: monsoonal rains (June to November); flooding; occasional droughts

Environment—current issues: illegal logging activities throughout the country and strip mining for gems in the western region along the border with Thailand have resulted in habitat loss and declining biodiversity (in particular, destruction of mangrove swamps threatens natural fisheries); soil erosion; in rural areas, most of the population does not have access to potable water; declining fish stocks because of illegal fishing and overfishing; coastal ecosystems choked by sediment washed loose from deforested areas inland

Environment—international agreements: *party to:* Biodiversity, Climate Change, Climate Change-Kyoto Protocol, Desertification, Endangered Species, Hazardous Wastes, Marine Life Conservation, Ozone Layer Protection, Ship Pollution, Tropical Timber 94, Wetlands, Whaling *signed, but not ratified:* Law of the Sea

Geography—note: a land of paddies and forests dominated by the Mekong River and Tonle Sap (Southeast Asia's largest freshwater lake)

PEOPLE AND SOCIETY

Population: 16,926,984 (July 2020 est.)
country comparison to the world: 69

Nationality: *noun:* Cambodian(s)
adjective: Cambodian

Ethnic groups: Khmer 97.6%, Cham 1.2%, Chinese 0.1%, Vietnamese 0.1%, other 0.9% (2013 est.)

Languages: Khmer (official) 96.3%, other 3.7% (2008 est.)

Religions: Buddhist (official) 97.9%, Muslim 1.1%, Christian 0.5%, other 0.6% (2013 est.)

Age structure: *0-14 years:* 30.18% (male 2,582,427/female 2,525,619)
15-24 years: 17.28% (male 1,452,784/female 1,472,769)

25-54 years: 41.51% (male 3,442,051/female 3,584,592)
55-64 years: 6.44% (male 476,561/female 612,706)
65 years and over: 4.59% (male 287,021/female 490,454) (2020 est.)

Dependency ratios: *total dependency ratio:* 55.7
youth dependency ratio: 48.2
elderly dependency ratio: 7.6
potential support ratio: 13.2 (2020 est.)

Median age: *total:* 26.4 years
male: 25.6 years
female: 27.2 years (2020 est.)
country comparison to the world: 153

Population growth rate: 1.4% (2020 est.)
country comparison to the world: 80

Birth rate: 21.3 births/1,000 population (2020 est.)
country comparison to the world: 70

Death rate: 7.3 deaths/1,000 population (2020 est.)
country comparison to the world: 110

Net migration rate: -0.3 migrant(s)/1,000 population (2020 est.)
country comparison to the world: 115

Population distribution: population concentrated in the southeast, particularly in and around the capital of Phnom Penh; further distribution is linked closely to the Tonle Sap and Mekong Rivers

Urbanization: *urban population:* 24.2% of total population (2020)
rate of urbanization: 3.25% annual rate of change (2015-20 est.)
total population growth rate v. urban population growth rate, 2000-2030:

Major urban areas—population: 2.078 million PHNOM PENH (capital) (2020)

Sex ratio: *at birth:* 1.05 male(s)/female
0-14 years: 1.02 male(s)/female
15-24 years: 0.99 male(s)/female
25-54 years: 0.96 male(s)/female
55-64 years: 0.78 male(s)/female
65 years and over: 0.59 male(s)/female
total population: 0.95 male(s)/female (2020 est.)

Mother's mean age at first birth: 22.9 years (2014 est.)
note: median age at first birth among women 25-29

Maternal mortality rate: 160 deaths/100,000 live births (2017 est.)
country comparison to the world: 55

Infant mortality rate: *total:* 43.7 deaths/1,000 live births
male: 49.8 deaths/1,000 live births
female: 37.3 deaths/1,000 live births (2020 est.)
country comparison to the world: 30

Life expectancy at birth: *total population:* 65.9 years
male: 63.4 years
female: 68.6 years (2020 est.)
country comparison to the world: 189

Total fertility rate: 2.39 children born/woman (2020 est.)
country comparison to the world: 79

Contraceptive prevalence rate: 56.3% (2014)

Drinking water source:
improved:
urban: 98.4% of population
rural: 77.8% of population
total: 80.3% of population
unimproved:
urban: 1.6% of population
rural: 22.2% of population
total: 19.7% of population (2017 est.)

Current Health Expenditure: 5.9% (2017)

Physicians density: 0.19 physicians/1,000 population (2014)

Hospital bed density: 1.9 beds/1,000 population (2016)

Sanitation facility access:
improved:
urban: 100% of population
rural: 55.5% of population
total: 65.7% of population
unimproved:
urban: 0% of population
rural: 44.5% of population
total: 34.3% of population (2017 est.)

HIV/AIDS—adult prevalence rate: 0.6% (2019 est.)
country comparison to the world: 60

HIV/AIDS—people living with HIV/AIDS: 73,000 (2019 est.)
country comparison to the world: 55

HIV/AIDS—deaths: 1,300 (2019 est.)
country comparison to the world: 53

Major infectious diseases: *degree of risk:* very high (2020)
food or waterborne diseases: bacterial diarrhea, hepatitis A, and typhoid fever
vectorborne diseases: dengue fever, Japanese encephalitis, and malaria

Obesity—adult prevalence rate: 3.9% (2016)
country comparison to the world: 188

Children under the age of 5 years underweight: 24.1% (2014)
country comparison to the world: 13

Education expenditures: 1.9% of GDP (2014)
country comparison to the world: 170

Literacy: *definition:* age 15 and over can read and write
total population: 80.5%
male: 86.5%
female: 75% (2015)

School life expectancy (primary to tertiary education): *total:* 11 years
male: 11 years
female: 10 years (2008)

Unemployment, youth ages 15-24: *total:* 1.1%
male: 1%
female: 1.2% (2016 est.)
country comparison to the world: 177

GOVERNMENT

Country name: *conventional long form:* Kingdom of Cambodia

165

conventional short form: Cambodia
local long form: Preahreacheanachakr Kampuchea (phonetic transliteration)
local short form: Kampuchea
former: Khmer Republic, Democratic Kampuchea, People's Republic of Kampuchea, State of Cambodia
etymology: the English name Cambodia is an anglicization of the French Cambodge, which is the French transliteration of the native name Kampuchea

Government type: parliamentary constitutional monarchy

Capital: *name:* Phnom Penh

geographic coordinates: 11 33 N, 104 55 E
time difference: UTC+7 (12 hours ahead of Washington, DC, during Standard Time)
etymology: Phnom Penh translates as "Penh's Hill" in Khmer; the city takes its name from the present Wat Phnom (Hill Temple), the tallest religious structure in the city, whose establishment, according to legend, was inspired in the 14th century by a pious nun, Daun PENH

Administrative divisions: 24 provinces (khett, singular and plural) and 1 municipality (krong, singular and plural)
provinces: Banteay Meanchey, Battambang, Kampong Cham, Kampong Chhnang, Kampong Speu, Kampong Thom, Kampot, Kandal, Kep, Koh Kong, Kratie, Mondolkiri, Oddar Meanchey, Pailin, Preah Sihanouk, Preah Vihear, Prey Veng, Pursat, Ratanakiri, Siem Reap, Stung Treng, Svay Rieng, Takeo, Tbong Khmum
municipalities: Phnom Penh (Phnum Penh)

Independence: 9 November 1953 (from France)

National holiday: Independence Day, 9 November (1953)

Constitution: *history:* previous 1947; latest promulgated 21 September 1993
amendments: proposed by the monarch, by the prime minister, or by the president of the National Assembly if supported by one fourth of the Assembly membership; passage requires two-thirds majority of the Assembly membership; constitutional articles on the multiparty democratic form of government and the monarchy cannot be amended; amended 1999, 2008, 2014, 2018

Legal system: civil law system (influenced by the UN Transitional Authority in Cambodia) customary law, Communist legal theory, and common law

International law organization participation: accepts compulsory ICJ jurisdiction with reservations; accepts ICCt jurisdiction

Citizenship: *citizenship by birth:* no
citizenship by descent only: at least one parent must be a citizen of Cambodia
dual citizenship recognized: yes
residency requirement for naturalization: 7 years

Suffrage: 18 years of age; universal

Executive branch: *chief of state:* King Norodom SIHAMONI (since 29 October 2004)

head of government: Prime Minister HUN SEN (since 14 January 1985); Permanent Deputy Prime Minister MEN SAM AN (since 25 September 2008); Deputy Prime Ministers SAR KHENG (since 3 February 1992), TEA BANH, Gen., HOR NAMHONG, (since 16 July 2004), BIN CHHIN (since 5 September 2007), YIM CHHAI LY (since 24 September 2008), KE KIMYAN (since 12 March 2009), AUN PORNMONIROTH (since 24 September 2012), Prak SOKONN, CHEA SOPHARA (since 5 April 2016)
cabinet: Council of Ministers named by the prime minister and appointed by the monarch
elections/appointments: monarch chosen by the 9-member Royal Council of the Throne from among all eligible males of royal descent; following legislative elections, a member of the majority party or majority coalition named prime minister by the Chairman of the National Assembly and appointed by the monarch

Legislative branch: *description:* bicameral Parliament of Cambodia consists of: Senate (62 seats; 58 indirectly elected by parliamentarians and commune councils, 2 indirectly elected by the National Assembly, and 2 appointed by the monarch; members serve 6-year terms)
National Assembly (125 seats; members directly elected in multi-seat constituencies by proportional representation vote; members serve 5-year terms)
elections: Senate—last held on 25 February 2018 (next to be held in 2024); National Assembly—last held on 29 July 2018 (next to be held in 2023)
election results: Senate—percent of vote by party—CPP 96%, FUNCINPEC 2.4%, KNUP 1.6%; seats by party—CPP 58; composition—men 53, women 9, percent of women 14.5%
National Assembly—percent of vote by party—CPP 76.9%, FUNCINPEC 5.9%, LDP 4.9%, Khmer Will Party 3.4%, other 8.9%; seats by party—CPP 125; composition—men 100, women 25, percent of women 20%; note—total Parliament of Cambodia percent of women 18.2%

Judicial branch: *highest courts:* Supreme Council (organized into 5- and 9-judge panels and includes a court chief and deputy chief); Constitutional Court (consists of 9 members); note—in 1997, the Cambodian Government requested UN assistance in establishing trials to prosecute former Khmer Rouge senior leaders for crimes against humanity committed during the 1975-1979 Khmer Rouge regime; the Extraordinary Chambers of the Courts of Cambodia (also called the Khmer Rouge Tribunal) was established in 2006 and began hearings for the first case in 2009; court proceedings remain ongoing in 2019
judge selection and term of office: Supreme Court and Constitutional Council judge candidates recommended by the Supreme Council of Magistracy, a 17-member body chaired by the monarch and includes other high-level judicial officers, judges of both courts appointed by the monarch; Supreme Court judges appointed for life; Constitutional Council judges appointed for 9-year terms with one-third of the court renewed every 3 years

subordinate courts: Appellate Court; provincial and municipal courts; Military Court

Political parties and leaders: Cambodia National Rescue Party or CNRP [KHEM SOKHA] (dissolved by the Cambodian Supreme Court in November 2017; formed from a 2012 merger of the Sam Rangsi Party or SRP and the former Human Rights Party or HRP [KHEM SOKHA, also spelled KEM SOKHA])
Cambodian Nationality Party or CNP [SENG SOKHENG]
Cambodian People's Party or CPP [HUN SEN]
Khmer Economic Development Party or KEDP [HUON REACH CHAMROEUN]
Khmer National Unity Party or KNUP [NHEK BUN CHHAY]
Khmer Will Party [KONG MONIKA]
League for Democracy Party or LDP [KHEM Veasna]
National United Front for an Independent, Neutral, Peaceful, and Cooperative Cambodia or FUNCINPEC [Prince NORODOM RANARIDDH]

International organization participation: ADB, ARF, ASEAN, CICA, EAS, FAO, G-77, IAEA, IBRD, ICAO, ICRM, IDA, IFAD, IFC, IFRCS, ILO, IMF, IMO, Interpol, IOC, IOM, IPU, ISO (correspondent), ITU, MINUSMA, MIGA, NAM, OIF, OPCW, PCA, UN, UNAMID, UNCTAD, UNESCO, UNIDO, UNIFIL, UNISFA, UNMISS, UNWTO, UPU, WCO, WFTU (NGOs), WHO, WIPO, WMO, WTO

Diplomatic representation in the US: *chief of mission:* Ambassador CHUM SOUNRY (since 17 September 2018)
chancery: 4530 16th Street NW, Washington, DC 20011
telephone: [1] (202) 726-7742
FAX: [1] (202) 726-8381

Diplomatic representation from the US: *chief of mission:* Ambassador Patrick MURPHY (since 23 October 2019)
telephone: [855] (23) 728-000
embassy: #1, Street 96, Sangkat Wat Phnom, Khan Daun Penh, Phnom Penh
mailing address: Unit 8166, Box P, APO AP 96546
FAX: [855] (23) 728-600

Flag description: three horizontal bands of blue (top), red (double width), and blue with a white, three-towered temple, representing Angkor Wat, outlined in black in the center of the red band; red and blue are traditional Cambodian colors
note: only national flag to prominently incorporate an actual identifiable building into its design (a few other national flags—those of Afghanistan, San Marino, Portugal, and Spain—show small generic buildings as part of their coats of arms on the flag)

National symbol(s): Angkor Wat temple, kouprey (wild ox); national colors: red, blue

National anthem: *name:* "Nokoreach" (Royal Kingdom)

lyrics/music: CHUON NAT/F. PERRUCHOT and J. JEKYLL

note: adopted 1941, restored 1993; the anthem, based on a Cambodian folk tune, was restored after the defeat of the Communist regime

0:00 / 1:31

ECONOMY

Economy—overview: Cambodia has experienced strong economic growth over the last decade; GDP grew at an average annual rate of over 8% between 2000 and 2010 and about 7% since 2011. The tourism, garment, construction and real estate, and agriculture sectors accounted for the bulk of growth. Around 700,000 people, the majority of whom are women, are employed in the garment and footwear sector. An additional 500,000 Cambodians are employed in the tourism sector, and a further 200,000 people in construction. Tourism has continued to grow rapidly with foreign arrivals exceeding 2 million per year in 2007 and reaching 5.6 million visitors in 2017. Mining also is attracting some investor interest and the government has touted opportunities for mining bauxite, gold, iron and gems.

Still, Cambodia remains one of the poorest countries in Asia, and long-term economic development remains a daunting challenge, inhibited by corruption, limited human resources, high income inequality, and poor job prospects. According to the Asian Development Bank (ADB), the percentage of the population living in poverty decreased to 13.5% in 2016. More than 50% of the population is less than 25 years old. The population lacks education and productive skills, particularly in the impoverished countryside, which also lacks basic infrastructure.

The World Bank in 2016 formally reclassified Cambodia as a lower middle-income country as a result of continued rapid economic growth over the past several years. Cambodia's graduation from a low-income country will reduce its eligibility for foreign assistance and will challenge the government to seek new sources of financing. The Cambodian Government has been working with bilateral and multilateral donors, including the Asian Development Bank, the World Bank and IMF, to address the country's many pressing needs; more than 20% of the government budget will come from donor assistance in 2018. A major economic challenge for Cambodia over the next decade will be fashioning an economic environment in which the private sector can create enough jobs to handle Cambodia's demographic imbalance.

Textile exports, which accounted for 68% of total exports in 2017, have driven much of Cambodia's growth over the past several years. The textile sector relies on exports to the United States and European Union, and Cambodia's dependence on its comparative advantage in textile production is a key vulnerability for the economy, especially because Cambodia has continued to run a current account deficit above 9% of GDP since 2014.

GDP (purchasing power parity): $64.21 billion (2017 est.)
$60.09 billion (2016 est.)
$56.18 billion (2015 est.)
note: data are in 2017 dollars
country comparison to the world: 104

GDP (official exchange rate): $22.09 billion (2017 est.)

GDP—real growth rate: 6.9% (2017 est.)
7% (2016 est.)
7% (2015 est.)
country comparison to the world: 19

GDP—per capita (PPP): $4,000 (2017 est.)
$3,800 (2016 est.)
$3,600 (2015 est.)
note: data are in 2017 dollars
country comparison to the world: 177

Gross national saving: 13.7% of GDP (2017 est.)
14.3% of GDP (2016 est.)
13.4% of GDP (2015 est.)
country comparison to the world: 139

GDP—composition, by end use: *household consumption:* 76% (2017 est.)
government consumption: 5.4% (2017 est.)
investment in fixed capital: 21.8% (2017 est.)
investment in inventories: 1.2% (2017 est.)
exports of goods and services: 68.6% (2017 est.)
imports of goods and services: -73% (2017 est.)

GDP—composition, by sector of origin: *agriculture:* 25.3% (2017 est.)
industry: 32.8% (2017 est.)
services: 41.9% (2017 est.)

Agriculture—products: rice, rubber, corn, vegetables, cashews, cassava (manioc, tapioca), silk

Industries: tourism, garments, construction, rice milling, fishing, wood and wood products, rubber, cement, gem mining, textiles

Industrial production growth rate: 10.6% (2017 est.)
country comparison to the world: 11

Labor force: 8.913 million (2017 est.)
country comparison to the world: 52

Labor force—by occupation: *agriculture:* 48.7%
industry: 19.9%
services: 31.5% (2013 est.)

Unemployment rate: 0.3% (2017 est.)
0.2% (2016 est.)
note: high underemployment, according to official statistics
country comparison to the world: 2

Population below poverty line: 16.5% (2016 est.)

Household income or consumption by percentage share: *lowest 10%:* 2%
highest 10%: 28% (2013 est.)

Budget: *revenues:* 3.947 billion (2017 est.)
expenditures: 4.354 billion (2017 est.)

Taxes and other revenues: 17.9% (of GDP) (2017 est.)
country comparison to the world: 164

Budget surplus (+) or deficit (-): -1.8% (of GDP) (2017 est.)
country comparison to the world: 97

Public debt: 30.4% of GDP (2017 est.)
29.1% of GDP (2016 est.)
country comparison to the world: 165

Fiscal year: calendar year

Inflation rate (consumer prices): 2.9% (2017 est.)
3% (2016 est.)
country comparison to the world: 129

Current account balance: -$1.871 billion (2017 est.)
-$1.731 billion (2016 est.)
country comparison to the world: 165

Exports: $11.42 billion (2017 est.)
$10.07 billion (2016 est.)
country comparison to the world: 85

Exports—partners: US 21.5%, UK 9%, Germany 8.6%, Japan 7.6%, China 6.9%, Canada 6.7%, Spain 4.7%, Belgium 4.5% (2017)

Exports—commodities: clothing, timber, rubber, rice, fish, tobacco, footwear

Imports: $14.37 billion (2017 est.)
$12.65 billion (2016 est.)
country comparison to the world: 91

Imports—commodities: petroleum products, cigarettes, gold, construction materials, machinery, motor vehicles, pharmaceutical products

Imports—partners: China 34.1%, Singapore 12.8%, Thailand 12.4%, Vietnam 10.1% (2017)

Reserves of foreign exchange and gold: $12.2 billion (31 December 2017 est.)
$9.122 billion (31 December 2016 est.)
country comparison to the world: 69

Debt—external: $11.87 billion (31 December 2017 est.)
$10.3 billion (31 December 2016 est.)
country comparison to the world: 107

Exchange rates: riels (KHR) per US dollar—
4,055 (2017 est.)
4,058.7 (2016 est.)
4,058.7 (2015 est.)
4,067.8 (2014 est.)
4,037.5 (2013 est.)

ENERGY

Electricity access: *population without electricity:* 4 million (2019)
electrification—total population: 75% (2019)
electrification—urban areas: 100% (2019)
electrification—rural areas: 67% (2019)

Electricity—production: 5.21 billion kWh (2016 est.)
country comparison to the world: 121

Electricity—consumption: 5.857 billion kWh (2016 est.)
country comparison to the world: 117

Electricity—exports: 0 kWh (2016 est.)
country comparison to the world: 116

Electricity—imports: 1.583 billion kWh (2016 est.)
country comparison to the world: 60

Electricity—installed generating capacity: 1.697 million kW (2016 est.)

country comparison to the world: 119

Electricity—from fossil fuels: 35% of total installed capacity (2016 est.)
country comparison to the world: 178

Electricity—from nuclear fuels: 0% of total installed capacity (2017 est.)
country comparison to the world: 61

Electricity—from hydroelectric plants: 63% of total installed capacity (2017 est.)
country comparison to the world: 27

Electricity—from other renewable sources: 2% of total installed capacity (2017 est.)
country comparison to the world: 136

Crude oil—production: 0 bbl/day (2018 est.)
country comparison to the world: 119

Crude oil—exports: 0 bbl/day (2015 est.)
country comparison to the world: 103

Crude oil—imports: 0 bbl/day (2015 est.)
country comparison to the world: 106

Crude oil—proved reserves: 0 bbl (1 January 2018 est.)
country comparison to the world: 115

Refined petroleum products—production: 0 bbl/day (2015 est.)
country comparison to the world: 127

Refined petroleum products—consumption: 45,000 bbl/day (2016 est.)
country comparison to the world: 108

Refined petroleum products—exports: 0 bbl/day (2015 est.)
country comparison to the world: 140

Refined petroleum products—imports: 43,030 bbl/day (2015 est.)
country comparison to the world: 86

Natural gas—production: 0 cu m (2017 est.)
country comparison to the world: 113

Natural gas—consumption: 0 cu m (2017 est.)
country comparison to the world: 129

Natural gas—exports: 0 cu m (2017 est.)
country comparison to the world: 78

Natural gas—imports: 0 cu m (2017 est.)
country comparison to the world: 102

Natural gas—proved reserves: 0 cu m (1 January 2014 est.)
country comparison to the world: 119

Carbon dioxide emissions from consumption of energy: 10.55 million Mt (2017 est.)
country comparison to the world: 104

COMMUNICATIONS

Telephones—fixed lines: *total subscriptions:* 56,749
subscriptions per 100 inhabitants: less than 1 (2019 est.)
country comparison to the world: 156

Telephones—mobile cellular: *total subscriptions:* 21,684,767
subscriptions per 100 inhabitants: 129.92 (2019 est.)
country comparison to the world: 56

Telecommunication systems: *general assessment:* well on its way to rollout 5G services, Chinese company Huawei dealing with the infrastructure for the 5G rollout; mobile-cellular phone systems are widely used in urban areas to bypass deficiencies in the fixed-line network; mobile-phone coverage is rapidly spreading in rural areas; competition among mobile operators strong; about 50% of Cambodians own at least one smart phone; in 2018, the MPTC began a free Wi-Fi service for visitors and residents of Phnom Penh, in selected parks around the city customers can access free Wi-Fi services; fixed broadband penetration is predicted to reach over 2% by 2023; in 2021, Cambodia hopes to launch it first communications satellite into orbit (2020)
domestic: fixed-line connections stand at about 1 per 100 persons and declining; mobile-cellular usage, aided by competition among service providers, has increased to about 130 per 100 persons (2019)
international: country code—855; landing points for MCT and AAE-1 via submarine cables providing communication to Asia, the Middle East, Europe and Africa; satellite earth station—1 Intersputnik (Indian Ocean region) (2019)
note: the COVID-19 outbreak is negatively impacting telecommunications production and supply chains globally; consumer spending on telecom devices and services has also slowed due to the pandemic's effect on economies worldwide; overall progress towards improvements in all facets of the telecom industry—mobile, fixed-line, broadband, submarine cable and satellite—has moderated

Broadcast media: mixture of state-owned, joint public-private, and privately owned broadcast media; 27 TV broadcast stations with most operating on multiple channels, including 1 state-operated station broadcasting from multiple locations, 11 stations either jointly operated or privately owned with some broadcasting from several locations; multi-channel cable and satellite systems are available (2019); 84 radio broadcast stations—1 state-owned broadcaster with multiple stations and a large mixture of public and private broadcasters; one international broadcaster is available (2019) as well as one Chinese joint venture television station with the Ministry of Interior; several television and radio operators broadcast online only (often via Facebook) (2019)

Internet country code: .kh

Internet users: *total:* 6,579,808

percent of population: 40% (July 2018 est.)
country comparison to the world: 74

Broadband—fixed subscriptions: *total:* 166,200
subscriptions per 100 inhabitants: 1 (2018 est.)
country comparison to the world: 114

TRANSPORTATION

National air transport system: *number of registered air carriers:* 6 (2020)
inventory of registered aircraft operated by air carriers: 25

annual passenger traffic on registered air carriers: 1,411,059 (2018)
annual freight traffic on registered air carriers: 680,000 mt-km (2018)

Civil aircraft registration country code prefix: XU (2016)

Airports: 16 (2013)
country comparison to the world: 142

Airports—with paved runways: *total:* 6 (2019)
2,438 to 3,047 m: 3
1,524 to 2,437 m: 2
914 to 1,523 m: 1

Airports—with unpaved runways: *total:* 10 (2013)
1,524 to 2,437 m: 2 (2013)
914 to 1,523 m: 7 (2013)
under 914 m: 1 (2013)

Heliports: 1 (2013)

Railways: *total:* 642 km (2014)
narrow gauge: 642 km 1.000-m gauge (2014)
note: under restoration
country comparison to the world: 107

Roadways: *total:* 47,263 km (2013)
paved: 12,239 km (2013)
unpaved: 35,024 km (2013)
country comparison to the world: 84

Waterways: 3,700 km (mainly on Mekong River) (2012)
country comparison to the world: 28

Merchant marine: *total:* 268
by type: bulk carrier 2, general cargo 176, oil tanker 19, other 71 (2019)
country comparison to the world: 57

Ports and terminals: *major seaport(s):* Sihanoukville (Kampong Saom)
river port(s): Phnom Penh (Mekong)

MILITARY AND SECURITY

Military and security forces: *Royal Cambodian Armed Forces:* High Command Headquarters, Royal Cambodian Army, Royal Khmer Navy, Royal Cambodian Air Force; Gendarmerie Royale Khmer (military police force responsible for internal security under Ministry of Interior); the National Counter Terrorism Committee; the National Committee for Maritime Security (performs Coast Guard functions and has representation from military and civilian agencies) (2019)

Military expenditures: 2.3% of GDP (2019)
2.2% of GDP (2018)
2.1% of GDP (2017)
2% of GDP (2016)
1.8% of GDP (2015)
country comparison to the world: 38

Military and security service personnel strengths: assessments of the size of the Royal Cambodian Armed Forces vary; approximately 115,000 total active troops (110,000 Army; 3,000 Navy; 1,000 Air Force); 10,000 Gendarmerie (2019 est.)

Military equipment inventories and acquisitions: the Royal Cambodian Armed Forces are armed largely with older Chinese and Russian-origin equipment; it has received limited amounts of newer equipment since 2010 with China as the principal provider, followed by Ukraine (2019 est.)

Military deployments: 200 Central African Republic (MINUSCA); 180 Lebanon (UNIFIL); 330 Mali (MINUSMA) (2020)

Military service age and obligation: 18 is the legal minimum age for compulsory and voluntary military service (2012)

TRANSNATIONAL ISSUES

Disputes—international: Cambodia is concerned about Laos' extensive upstream dam construction; Cambodia and Thailand dispute sections of boundary; in 2011 Thailand and Cambodia resorted to arms in the dispute over the location of the boundary on the precipice surmounted by Preah Vihear Temple ruins, awarded to Cambodia by an International Court of Justice decision in 1962 and part of a UN World Heritage site; Cambodia accuses Vietnam of a wide variety of illicit cross-border activities; progress on a joint development area with Vietnam is hampered by an unresolved dispute over sovereignty of offshore islands

Refugees and internally displaced persons: *stateless persons:* 57,444 (2019)

Trafficking in persons: *current situation:* Cambodia is a source, transit, and destination country for men, women, and children subjected to forced labor and sex trafficking; Cambodian men, women, and children migrate to countries within the region and, increasingly, the Middle East for legitimate work but are subjected to sex trafficking, domestic servitude, or forced labor in fishing, agriculture, construction, and factories; Cambodian men recruited to work on Thai-owned fishing vessels are subsequently subjected to forced labor in international waters and are kept at sea for years; poor Cambodian children are vulnerable and, often with the families' complicity, are subject to forced labor, including domestic servitude and forced begging, in Thailand and Vietnam; Cambodian and ethnic Vietnamese women and girls are trafficked from rural areas to urban centers and tourist spots for sexual exploitation; Cambodian men are the main exploiters of child prostitutes, but men from other Asian countries, and the West travel to Cambodia for child sex tourism

tier rating: Tier 2 Watch List – Cambodia does not fully comply with the minimum standards for the elimination of trafficking; however, it is making significant efforts to do so; the government has a written plan that, if implemented, would constitute making significant efforts to meet the minimum standards for the elimination of trafficking; authorities made modest progress in prosecutions and convictions of traffickers in 2014 but did not provide comprehensive data; endemic corruption continued to impede law enforcement efforts, and no complicit officials were prosecuted or convicted; the government sustained efforts to identify victims and refer them to NGOs for care, but victim protection remained inadequate, particularly for assisting male victims and victims identified abroad; a new national action plan was adopted, but guidelines for victim identification and guidance on undercover investigation techniques are still pending after several years (2015)

Illicit drugs: narcotics-related corruption reportedly involving some in the government, military, and police; limited methamphetamine production; vulnerable to money laundering due to its cash-based economy and porous borders

CAMEROON

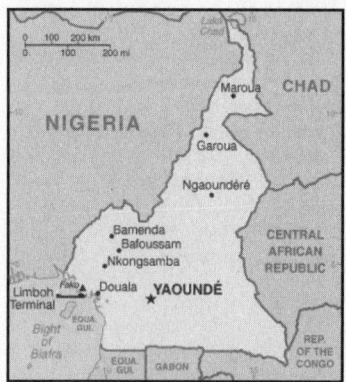

INTRODUCTION

Background: Much of the area of present-day Cameroon was ruled by powerful chiefdoms before becoming a German colony in 1884 known as Kamerun. After World War I, the territory was divided between France and the UK as League of Nations mandates. French Cameroon became independent in 1960 as the Republic of Cameroon. The following year the southern portion of neighboring British Cameroon voted to merge with the new country to form the Federal Republic of Cameroon. In 1972, a new constitution replaced the federation with a unitary state,

the United Republic of Cameroon. The country has generally enjoyed stability, which has enabled the development of agriculture, roads, and railways, as well as a petroleum industry. Despite slow movement toward democratic reform, political power remains firmly in the hands of President Paul BIYA.

GEOGRAPHY

Location: Central Africa, bordering the Bight of Biafra, between Equatorial Guinea and Nigeria

Geographic coordinates: 6 00 N, 12 00 E

Map references: Africa

Area: *total:* 475,440 sq km
land: 472,710 sq km
water: 2,730 sq km
country comparison to the world: 55

Area—comparative: slightly larger than California; about four times the size of Pennsylvania

Land boundaries: *total:* 5,018 km
border countries (6): Central African Republic 901 km, Chad 1116 km, Republic of the Congo 494 km, Equatorial Guinea 183 km, Gabon 349 km, Nigeria 1975 km
Coastline: 402 km

Maritime claims: *territorial sea:* 12 nm
contiguous zone: 24 nm

Climate: varies with terrain, from tropical along coast to semiarid and hot in north

Terrain: diverse, with coastal plain in southwest, dissected plateau in center, mountains in west, plains in north

Elevation: *mean elevation:* 667 m
lowest point: Atlantic Ocean 0 m
highest point: Fako on Mont Cameroun 4,045 m

Natural resources: petroleum, bauxite, iron ore, timber, hydropower

Land use: *agricultural land:* 20.6% (2011 est.)
arable land: 13.1% (2011 est.) / permanent crops: 3.3% (2011 est.) / permanent pasture: 4.2% (2011 est.)
forest: 41.7% (2011 est.)
other: 37.7% (2011 est.)
Irrigated land: 290 sq km (2012)

Population distribution: population concentrated in the west and north, with the interior of the country sparsely populated as shown in this <u>population distribution map</u>

Natural hazards: volcanic activity with periodic releases of poisonous gases from Lake Nyos and Lake Monoun volcanoes

volcanism: Mt. Cameroon (4,095 m), which last erupted in 2000, is the most frequently active volcano in West Africa; lakes in Oku volcanic field have released fatal levels of gas on occasion, killing some 1,700 people in 1986

Environment—current issues: waterborne diseases are prevalent; deforestation and overgrazing result in erosion, desertification, and reduced quality of pastureland; poaching; overfishing; overhunting

Environment—international agreements: *party to:* Biodiversity, Climate Change, Climate Change-Kyoto Protocol, Desertification, Endangered Species, Hazardous Wastes, Law of the Sea, Ozone Layer Protection, Tropical Timber 83, Tropical Timber 94, Wetlands, Whaling

signed, but not ratified: none of the selected agreements

Geography—note: sometimes referred to as the hinge of Africa because of its central location on the continent and its position at the west-south juncture of the Gulf of Guinea; throughout the country there are areas of thermal springs and indications of current or prior volcanic activity; Mount Cameroon, the highest mountain in Sub-Saharan west Africa, is an active volcano

PEOPLE AND SOCIETY

Population: 27,744,989 (July 2020 est.)

note: estimates for this country explicitly take into account the effects of excess mortality due to AIDS; this can result in lower life expectancy, higher infant mortality, higher death rates, lower population growth rates, and changes in the distribution of population by age and sex than would otherwise be expected
country comparison to the world: 51

Nationality: *noun:* Cameroonian(s)
adjective: Cameroonian

Ethnic groups: Bamileke-Bamu 24.3%, Beti/Bassa, Mbam 21.6%, Biu-Mandara 14.6%, Arab-Choa/Hausa/Kanuri 11%, Adamawa-Ubangi,9.8%, Grassfields 7.7%, Kako, Meka/Pygmy 3.3%, Cotier/Ngoe/Oroko 2.7%, Southwestern Bantu 0.7%, foreign/other ethnic group 4.5% (2018 est.)

Languages: 24 major African language groups, English (official), French (official)

Religions: Roman Catholic 38.3%, Protestant 25.5%, other Christian 6.9%, Muslim 24.4%, animist 2.2%, other 0.5%, none 2.2% (2018 est.)

Demographic profile: Cameroon has a large youth population, with more than 60% of the populace under the age of 25. Fertility is falling but remains at a high level, especially among poor, rural, and uneducated women, in part because of inadequate access to contraception. Life expectancy remains low at about 55 years due to the prevalence of HIV and AIDs and an elevated maternal mortality rate, which has remained flat since 1990. Cameroon, particularly the northern region, is vulnerable to food insecurity largely because of government mismanagement, corruption, high production costs, inadequate infrastructure, and natural disasters. Despite economic growth in some regions, poverty is on the rise, and is most prevalent in rural areas, which are especially affected by a shortage of jobs, declining incomes, poor school and health care infrastructure, and a lack of clean water and sanitation. Underinvestment in social safety nets and ineffective public financial management also contribute to Cameroon's high rate of poverty.

International migration has been driven by unemployment (including fewer government jobs), poverty, the search for educational opportunities, and corruption. The US and Europe are preferred destinations, but, with tighter immigration restrictions in these countries, young Cameroonians are increasingly turning to neighboring states, such as Gabon and Nigeria, South Africa, other parts of Africa, and the Near and Far East. Cameroon's limited resources make it dependent on UN support to host more than 420,000 refugees and asylum seekers as of September 2020. These refugees and asylum seekers are primarily from the Central African Republic and Nigeria.

Age structure: *0-14 years:* 42.34% (male 5,927,640/female 5,820,226)
15-24 years: 20.04% (male 2,782,376/female 2,776,873)
25-54 years: 30.64% (male 4,191,151/female 4,309,483)
55-64 years: 3.87% (male 520,771/female 552,801)
65 years and over: 3.11% (male 403,420/female 460,248) (2020 est.)

Dependency ratios: *total dependency ratio:* 81.1
*youth dependency ratio:*76.2
*elderly dependency ratio:*4.9
potential support ratio: 20.3 (2020 est.)

Median age: *total:* 18.5 years
male: 18.2 years
female: 18.8 years (2020 est.)
country comparison to the world: 210

Population growth rate: 2.78% (2020 est.)
country comparison to the world: 12

Birth rate: 36.3 births/1,000 population (2020 est.)
country comparison to the world: 17

Death rate: 8.1 deaths/1,000 population (2020 est.)
country comparison to the world: 88

Net migration rate:—0.3 migrant(s)/1,000 population (2020 est.)
country comparison to the world: 116

Population distribution: population concentrated in the west and north, with the interior of the country sparsely populated as shown in this <u>population distribution map</u>

Urbanization: *urban population:* 57.6% of total population (2020)
rate of urbanization: 3.63% annual rate of change (2015-20 est.)

total population growth rate v. urban population growth rate, 2000-2030: Major urban areas -population: 3.922 million YAOUNDE (capital), 3.663 million Douala (2020)

Sex ratio: *at birth:* 1.03 male(s)/female
0-14 years:1.02 male(s)/female
15-24 years: 1 male(s)/female
25-54 years: 0.97 male(s)/female
55-64 years: 0.94 male(s)/female
65 years and over: 0.88 male(s)/female
total population: 0.99 male(s)/female (2020 est.)

Mother's mean age at first birth: 19.7 years (2011 est.)

note: median age at first birth among women 25-29

Maternal mortality rate: 529 deaths/100,000 live births (2017 est.)
country comparison to the world: 18

Infant mortality rate: *total:* 51.5 deaths/1,000 live births
male: 56.5 deaths/1,000 live births
female: 46.3 deaths/1,000 live births (2020 est.)
country comparison to the world: 23

Life expectancy at birth: *total population:* 62.3 years
male: 60.6 years
female: 64 years (2020 est.)
country comparison to the world: 209

Total fertility rate: 4.66 children born/woman (2020 est.)
country comparison to the world: 20

Contraceptive prevalence rate: 19.3% (2018)

Drinking water source:
improved: *urban:* 94% of population
rural: 54.6% of population
total: 76.5% of population
unimproved: *urban:* 6% of population
rural: 45.3% of population
total: 23.5% of population (2017 est.)

Current Health Expenditure: 4.7% (2017)

Physicians density: 0.09 physicians/1,000 population (2011)

Hospital bed density: 1.3 beds/1,000 population (2010)

Sanitation facility access:
improved: *urban:* 83.3% of population
rural: 25.6% of population
total: 57.7% of population
unimproved: *urban:* 16.7% of population
rural: 74.4% of population
total: 42.3% of population (2017 est.)

HIV/AIDS—adult prevalence rate: 3.2% (2019 est.)
country comparison to the world: 17

HIV/AIDS—people living with HIV/AIDS: 510,000 (2019 est.)
country comparison to the world: 16

HIV/AIDS—deaths: 14,000 (2019 est.)
country comparison to the world: 13

Major infectious diseases: *degree of risk:* very high (2020)
food or waterborne diseases: bacterial and protozoal diarrhea, hepatitis A, and typhoid fever
vectorborne diseases: malaria and dengue fever
water contact diseases: schistosomiasis
animal contact diseases: rabies
respiratory diseases: meningococcal meningitis

Obesity—adult prevalence rate: 11.4% (2016)
country comparison to the world: 135

Children under the age of 5 years underweight: 11% (2018)
country comparison to the world: 60

Education expenditures: 3.1% of GDP (2017)
country comparison to the world: 133

Literacy: definition: age 15 and over can read and write
total population: 77.1%
male: 82.6%
female: 71.6% (2018)

School life expectancy (primary to tertiary education): total: 12 years
male: 13 years
female: 11 years (2016)

Unemployment, youth ages 15-24: total: 6.3%
male: 5.8%
female: 6.8% (2014 est.)
country comparison to the world: 157

GOVERNMENT

Country name: conventional long form: Republic of Cameroon
conventional short form: Cameroon
local long form: Republique du Cameroun/ Republic of Cameroon
local short form: Cameroun/Cameroon
former: Kamerun, French Cameroon, British Cameroon, Federal Republic of Cameroon, United Republic of Cameroon
etymology: in the 15th century, Portuguese explorers named the area near the mouth of the Wouri River the Rio dos Camaroes (River of Prawns) after the abundant shrimp in the water; over time the designation became Cameroon in English; this is the only instance where a country is named after a crustacean

Government type: presidential republic

Capital: name: Yaounde
geographic coordinates: 3 52 N, 11 31 E
time difference: UTC+1 (6 hours ahead of Washington, DC, during Standard Time)
etymology: founded as a German colonial settlement of Jaunde in 1888 and named after the local Yaunde (Ewondo) people

Administrative divisions: 10 regions (regions, singular—region); Adamaoua, Centre, East (Est), Far North (Extreme-Nord), Littoral, North (Nord), North-West (Nord-Ouest), West (Ouest), South (Sud), South-West (Sud-Ouest)

Independence: 1 January 1960 (from French-administered UN trusteeship)

National holiday: State Unification Day (National Day), 20 May (1972)

Constitution: history: several previous; latest effective 18 January 1996

amendments: proposed by the president of the republic or by Parliament; amendment drafts require approval of at least one third of the membership in either house of Parliament; passage requires absolute majority vote of the Parliament membership; passage of drafts requested by the president for a second reading in Parliament requires two-thirds majority vote of its membership; the president can opt to submit drafts to a referendum, in which case passage requires a simple majority; constitutional articles on Cameroon's unity and territorial integrity and its democratic principles cannot be amended; amended 2008

Legal system: mixed legal system of English common law, French civil law, and customary law

International law organization participation: accepts compulsory ICJ jurisdiction; non-party state to the ICCt

Citizenship: citizenship by birth:no
citizenship by descent only: at least one parent must be a citizen of Cameroon
dual citizenship recognized: no
residency requirement for naturalization: 5 years

Suffrage: 20 years of age; universal

Executive branch: chief of state: President Paul BIYA (since 6 November 1982)

head of government: Prime Minister Joseph Dion NGUTE (since 4 January 2019); Deputy Prime Minister Amadou ALI (since 2014)
cabinet: Cabinet proposed by the prime minister, appointed by the president
elections/appointments: president directly elected by simple majority popular vote for a 7-year term (no term limits); election last held on 7 October 2018 (next to be held in October 2025); prime minister appointed by the president
election results: Paul BIYA reelected president; percent of vote—Paul BIYA (CPDM) 71.3%, Maurice KAMTO (MRC) 14.2%, Cabral LIBII (Univers) 6.3%, other 8.2%

Legislative branch: description: bicameral Parliament or Parlement consists of:
Senate or Senat (100 seats; 70 members indirectly elected by regional councils and 30 appointed by the president; members serve 5-year terms)
National Assembly or Assemblee Nationale (180 seats; members directly elected in multi-seat constituencies by simple majority vote to serve 5-year terms)
elections: Senate—last held on 25 March 2018 (next to be held in 2023)
National Assembly—last held on 9 February 2020 (current term extended by President); note—the constitutional court has ordered a partial rerun of elections in the English speaking areas; date to be determined
election results: Senate—percent of vote by party—CDPM 81.1%, SDF 8.6%, UNDP 5.8%, UDC 1.16%, other 2.8%; seats by party -CPDM 63, SDF 7
National Assembly—percent of vote by party—NA; seats by party—CPDM 139, UNDP 7, SDF 5, PCRN 5, UDC 4, FSNC 3, MDR 2, Union of Socialist Movements 2; 13 vacant; composition—NA

Judicial branch: highest courts: Supreme Court of Cameroon (consists of 9 titular and 6 surrogate judges and organized into judicial, administrative, and audit chambers); Constitutional Council (consists of 11 members)

judge selection and term of office: Supreme Court judges appointed by the president with the advice of the Higher Judicial Council of Cameroon, a body chaired by the president and includes the minister of justice, selected magistrates, and representatives of the National Assembly; judge term NA; Constitutional Council members appointed by the president for single 9-year terms

subordinate courts: Parliamentary Court of Justice (jurisdiction limited to cases involving the

president and prime minister);appellate and first instance courts; circuit and magistrates' courts

Political parties and leaders: Alliance for Democracy and Development
Cameroon People's Democratic Movement or CPDM [Paul BIYA]
Cameroon People's Party or CPP [Edith Kah WALLA]
Cameroon Renaissance Movement or MRC [Maurice KAMTO]
Cameroonian Democratic Union or UDC [Adamou Ndam NJOYA]
Cameroonian Party for National Reconciliation or PCRN [Cabral LIBII]
Front for the National Salvation of Cameroon or FSNC [Issa Tchiroma BAKARY]
Movement for the Defense of the Republic or MDR [Dakole DAISSALA]
Movement for the Liberation and Development of Cameroon or MLDC [Marcel YONDO] National Union for Democracy and Progress or UNDP [Maigari BELLO BOUBA]
Progressive Movement or MP [Jean-Jacques EKINDI]
Social Democratic Front or SDF [John FRU NDI]
Union of Peoples of Cameroon or UPC [Provisional Management Bureau]
Union of Socialist Movements

International organization participation: ACP, AfDB, AU, BDEAC, C, CEMAC, EITI (compliant country), FAO, FZ, G-77, IAEA, IBRD, ICAO, ICRM, IDA, IDB, IFAD, IFC, IFRCS, IHO, ILO, IMF, IMO, IMSO, Interpol, IOC, IOM, IPU, ISO, ITSO, ITU, ITUC (NGOs), MIGA, MONUSCO, NAM, OIC, OIF, OPCW, PCA, UN,UNCTAD, UNESCO, UNHCR, UNIDO, UNOCI, UNWTO, UPU, WCO, WFTU (NGOs), WHO,WIPO, WMO, WTO

Diplomatic representation in the US: chief of mission: Ambassador Henri ETOUNDI ESSOMBA (since 27 June 2016)
chancery: 2349 Massachusetts Avenue, Washington, DC 20008
telephone: [1] (202) 265-8790
FAX: [1] (202) 387-3826

Diplomatic representation from the US: chief of mission: Ambassador Peter Henry BARLERIN (since 20 December 2017)
telephone: [237] 22220 1500; Consular: [237] 22220 1603
embassy: Avenue Rosa Parks, Yaoundé
mailing address: P.O.Box 817, Yaounde; pouch: American Embassy, US Department of State, Washington, DC 20521-2520
FAX: [237] 22220 1500 Ext. 4531; Consular FAX: [237] 22220 1752
branch office(s): Douala

Flag description: three equal vertical bands of green (hoist side), red, and yellow, with a yellow five-pointed star centered in the red band; the vertical tricolor recalls the flag of France; red symbolizes unity, yellow the sun, happiness, and the savannahs in the north, and green hope and the forests in the south; the star is referred to as the "star of unity"

note: uses the popular Pan-African colors of Ethiopia

National symbol(s): lion; national colors: green, red, yellow

National anthem: *name:* "O Cameroun, Berceau de nos Ancetres" (O Cameroon, Cradle of Our Forefathers)

lyrics/music: Rene Djam AFAME, Samuel Minkio BAMBA, Moise Nyatte NKO'O [French], Benard Nsokika FONLON [English]/Rene Djam AFAME

note: adopted 1957; Cameroon's anthem, also known as "Chant de Ralliement" (The Rallying Song), has been used unofficially since 1948 and officially adopted in 1957; the anthem has French and English versions whose lyrics differ
0:00 / 0.57

ECONOMY

Economy—overview: Cameroon's market-based, diversified economy features oil and gas, timber, aluminum, agriculture, mining and the service sector. Oil remains Cameroon's main export commodity, and despite falling global oil prices, still accounts for nearly 40% of exports. Cameroon's economy suffers from factors that often impact underdeveloped countries, such as stagnant per capita income, a relatively inequitable distribution of income, a top-heavy civil service, endemic corruption, continuing inefficiencies of a large parastatal system in key sectors, and a generally unfavorable climate for business enterprise.Since 1990, the government has embarked on various IMF and World Bank programs designed to spur business investment, increase efficiency in agriculture, improve trade, and recapitalize the nation's banks. The IMF continues to press for economic reforms, including increased budget transparency, privatization, and poverty reduction programs. The Government of Cameroon provides subsidies for electricity, food, and fuel that have strained the federal budget and diverted funds from education, healthcare, and infrastructure projects, as low oil prices have led to lower revenues.

Cameroon devotes significant resources to several large infrastructure projects currently under construction, including a deep seaport in Kribi and the Lom Pangar Hydropower Project. Cameroon's energy sector continues to diversify, recently opening a natural gas-powered electricity generating plant. Cameroon continues to seek foreign investment to improve its inadequate infrastructure, create jobs, and improve its economic footprint, but its unfavorable business environment remains a significant deterrent to foreign investment.

GDP (purchasing power parity): $89.54 billion (2017 est.)
$86.47 billion (2016 est.)
$82.63 billion (2015 est.)

note: data are in 2017 dollars
country comparison to the world: 88

GDP (official exchange rate): $34.99 billion (2017 est.)

GDP—real growth rate: 3.5% (2017 est.)

4.6% (2016 est.)
5.7% (2015 est.)
country comparison to the world: 84

GDP—per capita (PPP): $3,700 (2017 est.)
$3,700 (2016 est.)
$3,600 (2015 est.)

note: data are in 2017 dollars
country comparison to the world: 182

Gross national saving: 25.5% of GDP (2017 est.)
25.2% of GDP (2016 est.)
23.9% of GDP (2015 est.)
country comparison to the world: 56

GDP—composition, by end use:
household consumption: 66.3% (2017 est.)
government consumption: 11.8% (2017 est.)
investment in fixed capital: 21.6% (2017 est.)
investment in inventories: —0.3% (2017 est.)
exports of goods and services: 21.6% (2017 est.)
imports of goods and services: -20.9% (2017 est.)

GDP—composition, by sector of origin:
agriculture: 16.7% (2017 est.)
industry: 26.5% (2017 est.)
services: 56.8% (2017 est.)

Agriculture—products: coffee, cocoa, cotton, rubber, bananas, oilseed, grains, cassava (manioc, tapioca); livestock; timber

Industries: petroleum production and refining, aluminum production, food processing, light consumer goods, textiles, lumber, ship repair

Industrial production growth rate: 3.3% (2017 est.)
country comparison to the world: 94

Labor force: 9.912 million (2017 est.)
country comparison to the world: 50

Labor force—by occupation: *agriculture:* 70%
industry: 13%
services: 17% (2001 est.)

Unemployment rate: 4.3% (2014 est.)
30% (2001 est.)
country comparison to the world: 63

Population below poverty line: 30% (2001 est.)

Household income or consumption by percentage share: *lowest 10%:* 37.5%
highest 10%: 35.4% (2001)

Budget: *revenues:* 5.363 billion (2017 est.)
expenditures: 6.556 billion (2017 est.)

Taxes and other revenues: 15.3% (of GDP) (2017 est.)
country comparison to the world: 191

Budget surplus (+) or deficit (-): -3.4% (of GDP) (2017 est.)
country comparison to the world: 143

Public debt: 36.9% of GDP (2017 est.)
32.5% of GDP (2016 est.)
country comparison to the world: 143

Fiscal year: 1 July—30 June

Inflation rate (consumer prices): 0.6% (2017 est.)
0.9% (2016 est.)
country comparison to the world: 31

Current account balance: -$932 million (2017 est.)
-$1.034 billion (2016 est.)
country comparison to the world: 144

Exports: $4.732 billion (2017 est.)
$4.561 billion (2016 est.)
country comparison to the world: 109

Exports—partners: Netherlands 15.6%, France 12.6%, China 11.7%, Belgium 6.8%, Italy 6.3%, Algeria 4.8%, Malaysia 4.4% (2017)

Exports—commodities: crude oil and petroleum products, lumber, cocoa beans, aluminum, coffee, cotton

Imports: $4.812 billion (2017 est.)
$4.827 billion (2016 est.)
country comparison to the world: 132

Imports—commodities: machinery, electrical equipment, transport equipment, fuel, food

Imports—partners: China 19%, France 10.3%, Thailand 7.9%, Nigeria 4.1% (2017)

Reserves of foreign exchange and gold: $3.235 billion (31 December 2017 est.)
$2.26 billion (31 December 2016 est.)
country comparison to the world: 107

Debt—external: $9.375 billion (31 December 2017 est.)
$7.364 billion (31 December 2016 est.)
country comparison to the world: 115

Exchange rates: Cooperation Financiere en Afrique Centrale francs (XAF) per US dollar—605.3 (2017 est.)
593.01 (2016 est.)a
593.01 (2015 est.)
591.45 (2014 est.)
494.42 (2013 est.)

ENERGY

Electricity access: *population without electricity:* 8 million (2019)
electrification—total population: 70% (2019)
electrification—urban areas: 98% (2019)
electrification—rural areas: 32% (2019)

Electricity—production: 8.108 billion kWh (2016 est.)
country comparison to the world: 109

Electricity—consumption: 6.411 billion kWh (2016 est.)
country comparison to the world: 113

Electricity—exports: 0 kWh (2016 est.)
country comparison to the world: 117

Electricity—imports: 55 million kWh (2016 est.)
country comparison to the world: 105

Electricity—installed generating capacity: 1.558 million kW (2016 est.)
country comparison to the world: 122

Electricity—from fossil fuels: 52% of total installed capacity (2016 est.)
country comparison to the world: 145

Electricity—from nuclear fuels: 0% of total installed capacity (2017 est.)
country comparison to the world: 62

Electricity—from hydroelectric plants: 47% of total installed capacity (2017 est.)
country comparison to the world: 44

Electricity—from other renewable sources: 1% of total installed capacity (2017 est.)
country comparison to the world: 150

Crude oil—production: 69,000 bbl/day (2018 est.)
country comparison to the world: 47

Crude oil—exports: 96,370 bbl/day (2015 est.)
country comparison to the world: 35

Crude oil—imports: 36,480 bbl/day (2015 est.)
country comparison to the world: 59

Crude oil—proved reserves: 200 million bbl (1 January 2018 est.)
country comparison to the world: 55

Refined petroleum products—production: 39,080 bbl/day (2015 est.)
country comparison to the world: 82

Refined petroleum products—consumption: 45,000 bbl/day (2016 est.)
country comparison to the world: 109

Refined petroleum products—exports: 8,545 bbl/day (2015 est.)
country comparison to the world: 84

Refined petroleum products—imports: 14,090 bbl/day (2015 est.)
country comparison to the world: 138

Natural gas—production: 910.4 million cu m (2017 est.)
country comparison to the world: 69

Natural gas—consumption: 906.1 million cu m (2017 est.)
country comparison to the world: 93

Natural gas—exports: 0 cu m (2017 est.)
country comparison to the world: 79

Natural gas—imports: 0 cu m (2017 est.)
country comparison to the world: 103

Natural gas—proved reserves: 135.1 billion cu m (1 January 2018 est.)
country comparison to the world: 48

Carbon dioxide emissions from consumption of energy: 7.672 million Mt (2017 est.)
country comparison to the world: 119

COMMUNICATIONS

Telephones—fixed lines: *total subscriptions:* 966,035
subscriptions per 100 inhabitants: 3.58 (2019 est.)
country comparison to the world: 76

Telephones—mobile cellular: *total subscriptions:* 22,062,303
subscriptions per 100 inhabitants: 81.76 (2019 est.)
country comparison to the world: 53

Telecommunication systems: *general assessment:* 3G service and LTE service both developing given growing competition, along with a fast-developing mobile broadband sector; govt. supportive of launching programs who's aim is to improve connections nationally; about 95% of electronic transactions carried out through M-commerce services (2020)

domestic: only about 4 per 100 persons for fixed-line subscriptions; mobile-cellular usage has increased sharply, reaching a subscribership base of over 82 per 100 persons (2019)

international: country code—237; landing points for the SAT-3/WASC, SAIL, ACE, NCSCS, Ceiba-2, and WACS fiber-optic submarine cable that provides connectivity to Europe, South America, and West Africa; satellite earth stations—2 Intelsat (Atlantic Ocean) (2019)

note: the COVID-19 outbreak is negatively impacting telecommunications production and supply chains globally; consumer spending on telecom devices and services has also slowed due to the pandemic's effect on economies worldwide; overall progress towards improvements in all facets of the telecom industry—mobile, fixed—line, broadband, submarine cable and satellite—has moderated

Broadcast media: government maintains tight control over broadcast media; state-owned Cameroon Radio Television (CRTV), broadcasting on both a TV and radio network, was the only officially recognized and fully licensed broadcaster until August 2007, when the government finally issued licenses to 2 private TV broadcasters and 1 private radio broadcaster; about 70 privately owned, unlicensed radio stations operating but are subject to closure at any time; foreign news services required to partner with state- owned national station (2019)

Internet country code: .cm

Internet users: *total:* 6,089,200
percent of population: 23.2% (July 2018 est.)
country comparison to the world: 77

Broadband—fixed subscriptions: *total:* 17,987
subscriptions per 100 inhabitants: less than 1 (2018 est.)
country comparison to the world: 156

TRANSPORTATION

National air transport system: *number of registered air carriers:* 1 (2020)
inventory of registered aircraft operated by air carriers: 3
annual passenger traffic on registered air carriers: 265,136 (2018)
annual freight traffic on registered air carriers: 70,000 mt-km (2018)

Civil aircraft registration country code prefix: TJ (2016)

Airports: 33 (2013)
country comparison to the world: 111

Airports—with paved runways: *total:* 11 (2017)
over 3,047 m: 2 (2017)
2,438 to 3,047 m: 5 (2017)
1,524 to 2,437 m: 3 (2017)
914 to 1,523 m: 1 (2017)

Airports—with unpaved runways: *total:* 22 (2013)
1,524 to 2,437 m: 4 (2013)
914 to 1,523 m: 10 (2013)
under 914 m: 8 (2013)

Pipelines: 53 km gas, 5 km liquid petroleum gas, 1107 km oil, 35 km water (2013)

Railways: *total:* 987 km (2014)
narrow gauge: 987 km 1.000-m gauge (2014)
note: railway connections generally efficient but limited; rail lines connect major cities of Douala, Yaounde, Ngaoundere, and Garoua; passenger and freight service provided by CAMRAIL
country comparison to the world: 89

Roadways: *total:* 77,589 km (2016)
paved: 5,133 km (2016)
unpaved: 72,456 km (2016)
country comparison to the world: 65

Waterways: (major rivers in the south, such as the Wouri and the Sanaga, are largely non-navigable; in the north, the Benue, which connects through Nigeria to the Niger River, is navigable in the rainy season only to the port of Garoua) (2010)

Merchant marine: *total:* 29
by type: general cargo 9, other 20 (2019)
country comparison to the world: 133

Ports and terminals: *oil terminal(s):* Limboh Terminal
river port(s): Douala (Wouri)
Garoua (Benoue)

MILITARY AND SECURITY

Military and security forces: Cameroon Armed Forces (Forces Armees Camerounaises, FAC): Army (L'Armee de Terre), Navy (Marine Nationale Republique, MNR, includes naval infantry), Air Force (Armee de l'Air du Cameroun, AAC), Gendarmerie, Presidential Guard (2019)

Military expenditures: 1.1% of GDP (2019)
1.1% of GDP (2018)
1.3% of GDP (2017)
1.3% of GDP (2016)
1.3% of GDP (2015)
country comparison to the world: 110

Military and security service personnel strengths: size assessments for the Cameroon Armed Forces (FAC) vary widely; approximately 40,000 active duty troops; (25,000 Army, including the Presidential Guard; 2,000 Navy; 1,000 Air Force; 12,000 Gendarmerie) (2019 est.)

Military equipment inventories and acquisitions: the FAC inventory includes a mix of mostly older or second-hand Chinese, Russian, and Western equipment, with a limited quantity of more modern weapons; since 2010, the top suppliers to the FAC are China, Russia, Spain, and the US (2019 est.)

Military deployments: 750 Central African Republic (MINUSCA); MNJTF (approximately 2,000-2,500 troops committed; note—the national MNJTF troop contingents are deployed within their own country territories, although cross-border operations occur occasionally) (2020)

Military service age and obligation: 18-23 years of age for male and female voluntary military service; no conscription; high school graduation required;

service obligation 4 years; periodic government calls for volunteers (2012)

Military—note: the FAC is largely focused on the threat from the terror group Boko Haram along its frontiers with Nigeria and Chad (Far North region) and an insurgency from armed Anglophone separatist groups in the North-West and South-West regions (as of Feb 2020, this internal conflict has left an estimated 3,000 civilians dead and over 500,000 people displaced since fighting started in 2016); in addition, the FAC has occasionally deployed units to the border region with the Central African Republic to counter intrusions from armed militias and bandits (2020)

Terrorist group(s): Boko Haram; Islamic State of Iraq and ash-Sham—West Africa (2020)

note: details about the history, aims, leadership, organization, areas of operation, tactics, targets, weapons, size, and sources of support of the group(s) appear(s) in Appendix- T

TRANSNATIONAL ISSUES

Disputes—international: Joint Border Commission with Nigeria reviewed 2002 ICJ ruling on the entire boundary and bilaterally resolved differences, including June 2006 Greentree Agreement that immediately ceded sovereignty of the Bakassi Peninsula to Cameroon with a full phase- out of Nigerian control and patriation of residents in 2008; Cameroon and Nigeria agreed on maritime delimitation in March 2008; sovereignty dispute between Equatorial Guinea and Cameroon over an island at the mouth of the Ntem River; only Nigeria and Cameroon have heeded the Lake Chad Commission's admonition to ratify the delimitation treaty, which also includes the Chad-Niger and Niger-Nigeria boundaries

Refugees and internally displaced persons: refugees (country of origin): 310,097 (Central African Republic), 116,623 (Nigeria) (2020)

IDPs: 1,032,942 (2020) (includes far north, northwest, and southwest)

CANADA

INTRODUCTION

Background: A land of vast distances and rich natural resources, Canada became a self-governing dominion in 1867, while retaining ties to the British crown. Canada repatriated its constitution from the UK in 1982, severing a final colonial tie. Economically and technologically, the nation has developed in parallel with the US, its neighbor to the south across the world's longest international border. Canada faces the political challenges of meeting public demands for quality improvements in health care, education, social services, and economic competitiveness, as well as responding to the particular concerns of predominantly francophone Quebec. Canada also aims to develop its diverse energy resources while maintaining its commitment to the environment.

GEOGRAPHY

Location: Northern North America, bordering the North Atlantic Ocean on the east, North Pacific Ocean on the west, and the Arctic Ocean on the north, north of the conterminous US

Geographic coordinates: 60 00 N, 95 00 W

Map references: North America

Area: *total:* 9,984,670 sq km
land: 9,093,507 sq km
water: 891,163 sq km
country comparison to the world: 3

Area—comparative: slightly larger than the US

Land boundaries: *total:* 8,893 km
border countries (1): US 8893 km (includes 2477 km with Alaska)
note: Canada is the world's largest country that borders only one country

Coastline: 202,080 km
note: the Canadian Arctic Archipelago—consisting of 36,563 islands, several of them some of the world's largest—contributes to Canada easily having the longest coastline in the world

Maritime claims: *territorial sea:* 12 nm
exclusive economic zone: 200 nm
contiguous zone: 24 nm
continental shelf: 200 nm or to the edge of the continental margin

Climate: varies from temperate in south to subarctic and arctic in north

Terrain: mostly plains with mountains in west, lowlands in southeast

Elevation: *mean elevation:* 487 m
lowest point: Atlantic Ocean 0 m
highest point: Mount Logan 5,959 m

Natural resources: bauxite, iron ore, nickel, zinc, copper, gold, lead, rare earth elements, molybdenum, potash, diamonds, silver, fish, timber, wildlife, coal, petroleum, natural gas, hydropower

Land use: *agricultural land:* 6.8% (2011 est.)
arable land: 4.7% (2011 est.) / *permanent crops:* 0.5% (2011 est.) / *permanent pasture:* 1.6% (2011 est.)
forest: 34.1% (2011 est.)
other: 59.1% (2011 est.)
Irrigated land: 8,700 sq km (2012)

Population distribution: vast majority of Canadians are positioned in a discontinuous band within approximately 300 km of the southern border with the United States; the most populated province is Ontario, followed by Quebec and British Columbia

Natural hazards: continuous permafrost in north is a serious obstacle to development; cyclonic storms form east of the Rocky Mountains, a result of the mixing of air masses from the Arctic, Pacific, and North American interior, and produce most of the country's rain and snow east of the mountains
volcanism: the vast majority of volcanoes in Western Canada's Coast Mountains remain dormant

Environment—current issues: metal smelting, coal-burning utilities, and vehicle emissions impacting agricultural and forest productivity; air pollution and resulting acid rain severely affecting lakes and damaging forests; ocean waters becoming contaminated due to agricultural, industrial, mining, and forestry activities

Environment—international agreements: *party to:* Air Pollution, Air Pollution-Nitrogen Oxides, Air Pollution-Persistent Organic Pollutants, Air Pollution-Sulfur 85, Air Pollution-Sulfur 94, Antarctic-Environmental Protocol, Antarctic-Marine Living Resources, Antarctic Seals, Antarctic Treaty, Biodiversity, Climate Change, Desertification, Endangered Species, Environmental Modification, Hazardous Wastes, Law of the Sea, Marine Dumping, Ozone Layer Protection, Ship Pollution, Tropical Timber 83, Tropical Timber 94, Wetlands
signed, but not ratified: Air Pollution-Volatile Organic Compounds, Marine Life Conservation

Geography—note: *note 1:* second-largest country in world (after Russia) and largest in the Americas; strategic location between Russia and US via north polar route; approximately 90% of the population is concentrated within 160 km (100 mi) of the US border
note 2: Canada has more fresh water than any other country and almost 9% of Canadian territory is water; Canada has at least 2 million and possibly over 3 million lakes—that is more than all other countries combined

PEOPLE AND SOCIETY

Population: 37,694,085 (July 2020 est.)
country comparison to the world: 38

Nationality: *noun:* Canadian(s)
adjective: Canadian

Ethnic groups: Canadian 32.3%, English 18.3%, Scottish 13.9%, French 13.6%, Irish 13.4%, German 9.6%, Chinese 5.1%, Italian 4.6%, North American Indian 4.4%, East Indian 4%, other 51.6% (2016 est.)

note: percentages add up to more than 100% because respondents were able to identify more than one ethnic origin

Languages: English (official) 58.7%, French (official) 22%, Punjabi 1.4%, Italian 1.3%, Spanish 1.3%, German 1.3%, Cantonese 1.2%, Tagalog 1.2%, Arabic 1.1%, other 10.5% (2011 est.)

Religions: Catholic 39% (includes Roman Catholic 38.8%, other Catholic .2%), Protestant 20.3% (includes United Church 6.1%, Anglican 5%, Baptist 1.9%, Lutheran 1.5%, Pentecostal 1.5%, Presbyterian 1.4%, other Protestant 2.9%), Orthodox 1.6%, other Christian 6.3%, Muslim 3.2%, Hindu 1.5%, Sikh 1.4%, Buddhist 1.1%, Jewish 1%, other 0.6%, none 23.9% (2011 est.)

Age structure: *0-14 years:* 15.99% (male 3,094,008/female 2,931,953)
15-24 years: 11.14% (male 2,167,013/female 2,032,064)
25-54 years: 39.81% (male 7,527,554/female 7,478,737)
55-64 years: 14.08% (male 2,624,474/female 2,682,858)
65 years and over: 18.98% (male 3,274,298/female 3,881,126) (2020 est.)

Dependency ratios: *total dependency ratio:* 51.2
youth dependency ratio: 23.9
elderly dependency ratio: 27.4
potential support ratio: 3.7 (2020 est.)

Median age: *total:* 41.8 years
male: 40.6 years
female: 42.9 years (2020 est.)
country comparison to the world: 40

Population growth rate: 0.81% (2020 est.)
country comparison to the world: 127

Birth rate: 10.2 births/1,000 population (2020 est.)
country comparison to the world: 190

Death rate: 7.9 deaths/1,000 population (2020 est.)
country comparison to the world: 92

Net migration rate: 5.6 migrant(s)/1,000 population (2020 est.)
country comparison to the world: 20

Population distribution: vast majority of Canadians are positioned in a discontinuous band within approximately 300 km of the southern border with the United States; the most populated province is Ontario, followed by Quebec and British Columbia

Urbanization: *urban population:* 81.6% of total population (2020)

rate of urbanization: 0.97% annual rate of change (2015-20 est.)

total population growth rate v. urban population growth rate, 2000-2030: Major urban areas—population: 6.197 million Toronto, 4.221 million Montreal, 2.581 million Vancouver, 1.547 million Calgary, 1.461 million Edmonton, 1.393 million OTTAWA (capital) (2020)

Sex ratio: *at birth:* 1.05 male(s)/female
0-14 years: 1.06 male(s)/female
15-24 years: 1.07 male(s)/female
25-54 years: 1.01 male(s)/female
55-64 years: 0.98 male(s)/female
65 years and over: 0.84 male(s)/female
total population: 0.98 male(s)/female (2020 est.)

Mother's mean age at first birth: 29 years (2017 est.)

Maternal mortality rate: 10 deaths/100,000 live births (2017 est.)
country comparison to the world: 145

Infant mortality rate: *total:* 4.3 deaths/1,000 live births
male: 4.5 deaths/1,000 live births
female: 4.1 deaths/1,000 live births (2020 est.)
country comparison to the world: 183

Life expectancy at birth: *total population:* 83.4 years
male: 81.1 years
female: 85.9 years (2020 est.)
country comparison to the world: 6

Total fertility rate: 1.57 children born/woman (2020 est.)
country comparison to the world: 191

Drinking water source:
improved:
urban: 100% of population
rural: 98.9% of population
total: 100% of population
unimproved:
urban: 0% of population
rural: 1.1% of population
total: 0% of population (2017 est.)

Current Health Expenditure: 10.6% (2017)

Physicians density: 2.31 physicians/1,000 population (2017)

Hospital bed density: 2.5 beds/1,000 population (2017)

Sanitation facility access:
improved:
urban: 100% of population
rural: 98.7% of population
total: 100% of population
unimproved:
urban: 0% of population
rural: 1.3% of population
total: 0% of population (2017 est.)

HIV/AIDS—adult prevalence rate: NA

HIV/AIDS—people living with HIV/AIDS: NA

HIV/AIDS—deaths: NA

Obesity—adult prevalence rate: 29.4% (2016)
country comparison to the world: 26

Education expenditures: 5.3% of GDP (2011)

country comparison to the world: 46

School life expectancy (primary to tertiary education): *total:* 16 years
male: 16 years
female: 17 years (2018)

Unemployment, youth ages 15-24: *total:* 11.1%
male: 12.5%
female: 9.6% (2018 est.)
country comparison to the world: 116

GOVERNMENT

Country name: *conventional long form:* none
conventional short form: Canada
etymology: the country name likely derives from the St. Lawrence Iroquoian word "kanata" meaning village or settlement

Government type: federal parliamentary democracy (Parliament of Canada) under a constitutional monarchy; a Commonwealth realm; federal and state authorities and responsibilities regulated in constitution

Capital: *name:* Ottawa

geographic coordinates: 45 25 N, 75 42 W
time difference: UTC-5 (same time as Washington, DC, during Standard Time)
daylight saving time: +1hr, begins second Sunday in March; ends first Sunday in November
note: Canada has six time zones
etymology: the city lies on the south bank of the Ottawa River, from which it derives its name; the river name comes from the Algonquin word "adawe" meaning "to trade" and refers to the indigenous peoples who used the river as a trade highway

Administrative divisions: 10 provinces and 3 territories*; Alberta, British Columbia, Manitoba, New Brunswick, Newfoundland and Labrador, Northwest Territories*, Nova Scotia, Nunavut*, Ontario, Prince Edward Island, Quebec, Saskatchewan, Yukon*

Independence: 1 July 1867 (union of British North American colonies); 11 December 1931 (recognized by UK per Statute of Westminster)

National holiday: Canada Day, 1 July (1867)

Constitution: *history:* consists of unwritten and written acts, customs, judicial decisions, and traditions dating from 1763; the written part of the constitution consists of the Constitution Act of 29 March 1867, which created a federation of four provinces, and the Constitution Act of 17 April 1982
amendments: proposed by either house of Parliament or by the provincial legislative assemblies; there are 5 methods for passage though most require approval by both houses of Parliament, approval of at least two thirds of the provincial legislative assemblies and assent and formalization as a proclamation by the governor general in council; the most restrictive method is reserved for amendments affecting fundamental sections of the constitution, such as the office of the monarch or the governor general, and the constitutional amendment procedures, which require unanimous

approval by both houses and by all the provincial assemblies, and assent of the governor general in council; amended 11 times, last in 2011 (Fair Representation Act, 2011)

Legal system: common law system except in Quebec, where civil law based on the French civil code prevails

International law organization participation: accepts compulsory ICJ jurisdiction with reservations; accepts ICCt jurisdiction

Citizenship: *citizenship by birth:* yes
citizenship by descent only: yes
dual citizenship recognized: yes
residency requirement for naturalization: minimum of 3 of last 5 years resident in Canada

Suffrage: 18 years of age; universal

Executive branch: *chief of state:* Queen ELIZABETH II (since 6 February 1952); represented by Governor General Julie PAYETTE (since 2 October 2017)

head of government: Prime Minister Justin Pierre James TRUDEAU (Liberal Party) (since 4 November 2015)
cabinet: Federal Ministry chosen by the prime minister usually from among members of his/her own party sitting in Parliament
elections/appointments: the monarchy is hereditary; governor general appointed by the monarch on the advice of the prime minister for a 5-year term; following legislative elections, the leader of the majority party or majority coalition in the House of Commons generally designated prime minister by the governor general
note: the governor general position is largely ceremonial; Julie PAYETTE, a former space shuttle astronaut, is Canada's fourth female governor general but the first to have flown in space

Legislative branch: *description:* bicameral Parliament or Parlement consists of:
Senate or Senat (105 seats; members appointed by the governor general on the advice of the prime minister and can serve until age 75)
House of Commons or Chambre des Communes (338 seats; members directly elected in single-seat constituencies by simple majority vote with terms up to 4 years)
elections: Senate—appointed; latest appointments in December 2018
House of Commons—last held on 21 October 2019 (next to be held in October 2023)
election results: Senate—composition as of December 2018—men 51, women 54, percent of women 51.4%
House of Commons—percent of vote by party—CPC 34.4%, Liberal Party 33.1%, NDP 15.9%, Bloc Quebecois 7.7%, Greens 6.5%, other 2.4%; seats by party—Liberal Party 157, CPC 121, NDP 24, Bloc Quebecois 32, Greens 4; composition—men 240, women 98, percent of women 29%; note—total Parliament percent of women 34.3%

Judicial branch: *highest courts:* Supreme Court of Canada (consists of the chief justice and 8 judges); note—in 1949, Canada abolished all appeals

beyond its Supreme Court, which prior to that time, were heard by the Judicial Committee of the Privy Council (in London)
judge selection and term of office: chief justice and judges appointed by the prime minister in council; all judges appointed for life with mandatory retirement at age 75
subordinate courts: federal level: Federal Court of Appeal; Federal Court; Tax Court; federal administrative tribunals; Courts Martial; provincial/territorial level: provincial superior, appeals, first instance, and specialized courts; note—in 1999, the Nunavut Court—a circuit court with the power of a provincial superior court, as well as a territorial court—was established to serve isolated settlements

Political parties and leaders: Bloc Quebecois [Mario BEAULIEU]
Conservative Party of Canada or CPC [Erin O'TOOLE]
Green Party [Annamie PAUL]
Liberal Party [Justin TRUDEAU]
New Democratic Party or NDP [Jagmeet SINGH]
People's Party of Canada [Maxime BERNIER]

International organization participation: ADB (nonregional member), AfDB (nonregional member), APEC, Arctic Council, ARF, ASEAN (dialogue partner), Australia Group, BIS, C, CD, CDB, CE (observer), EAPC, EBRD, EITI (implementing country), FAO, FATF, G-7, G-8, G-10, G-20, IADB, IAEA, IBRD, ICAO, ICC (national committees), ICCt, ICRM, IDA, IEA, IFAD, IFC, IFRCS, IGAD (partners), IHO, ILO, IMF, IMO, IMSO, Interpol, IOC, IOM, IPU, ISO, ITSO, ITU, ITUC (NGOs), MIGA, MINUSTAH, MONUSCO, NAFTA, NATO, NEA, NSG, OAS, OECD, OIF, OPCW, OSCE, Pacific Alliance (observer), Paris Club, PCA, PIF (partner), UN, UNCTAD, UNESCO, UNFICYP, UNHCR, UNMISS, UNRWA, UNTSO, UPU, WCO, WFTU (NGOs), WHO, WIPO, WMO, WTO, ZC

Diplomatic representation in the US: *chief of mission:* Ambassador Kirsten HILLMAN (since 17 July 2020)
chancery: 501 Pennsylvania Avenue NW, Washington, DC 20001
telephone: [1] (202) 682-1740
FAX: [1] (202) 682-7726
consulate(s) general: Atlanta, Boston, Chicago, Dallas, Denver, Detroit, Los Angeles, Miami, Minneapolis, New York, San Francisco/Silicon Valley, Seattle
trade office(s): Houston, Palo Alto (CA), San Diego; note—there are trade offices in the Consulates General

Diplomatic representation from the US: *chief of mission:* Ambassador (vacant); Charge d'Affaires Richard M. MILLS, Jr. (since 23 August 2019)
telephone: [1] (613) 688-5335
embassy: 490 Sussex Drive, Ottawa, Ontario K1N 1G8
mailing address: P. O. Box 5000, Ogdensburg, NY 13669-0430; P.O. Box 866, Station B, Ottawa, Ontario K1P 5T1

FAX: [1] (613) 688-3082
consulate(s) general: Calgary, Halifax, Montreal, Quebec City, Toronto, Vancouver
consulate(s): Winnipeg

Flag description: two vertical bands of red (hoist and fly side, half width) with white square between them; an 11-pointed red maple leaf is centered in the white square; the maple leaf has long been a Canadian symbol

National symbol(s): maple leaf, beaver; national colors: red, white

National anthem: *name:* O Canada
lyrics/music: Adolphe-Basile ROUTHIER [French], Robert Stanley WEIR [English]/Calixa LAVALLEE
note: adopted 1980; originally written in 1880, "O Canada" served as an unofficial anthem many years before its official adoption; the anthem has French and English versions whose lyrics differ; as a Commonwealth realm, in addition to the national anthem, "God Save the Queen" serves as the royal anthem (see United Kingdom)
0:00 / 1:18

ECONOMY

Economy—overview: Canada resembles the US in its market-oriented economic system, pattern of production, and high living standards. Since World War II, the impressive growth of the manufacturing, mining, and service sectors has transformed the nation from a largely rural economy into one primarily industrial and urban. Canada has a large oil and natural gas sector with the majority of crude oil production derived from oil sands in the western provinces, especially Alberta. Canada now ranks third in the world in proved oil reserves behind Venezuela and Saudi Arabia and is the world's seventh-largest oil producer.

The 1989 Canada-US Free Trade Agreement and the 1994 North American Free Trade Agreement (which includes Mexico) dramatically increased trade and economic integration between the US and Canada. Canada and the US enjoy the world's most comprehensive bilateral trade and investment relationship, with goods and services trade totaling more than $680 billion in 2017, and two-way investment stocks of more than $800 billion. Over three-fourths of Canada's merchandise exports are destined for the US each year. Canada is the largest foreign supplier of energy to the US, including oil, natural gas, and electric power, and a top source of US uranium imports.

Given its abundant natural resources, highly skilled labor force, and modern capital stock, Canada enjoyed solid economic growth from 1993 through 2007. The global economic crisis of 2007-08 moved the Canadian economy into sharp recession by late 2008, and Ottawa posted its first fiscal deficit in 2009 after 12 years of surplus. Canada's major banks emerged from the financial crisis of 2008-09 among the strongest in the world, owing to the financial sector's tradition of conservative lending practices and strong capitalization. Canada's economy posted strong growth in 2017 at 3%, but most analysts are projecting Canada's economic growth will drop back closer to 2% in 2018.

GDP (purchasing power parity): $1.774 trillion (2017 est.)
$1.721 trillion (2016 est.)
$1.697 trillion (2015 est.)
note: data are in 2017 dollars
country comparison to the world: 17

GDP (official exchange rate): $1.653 trillion (2017 est.)

GDP—real growth rate: 1.66% (2019 est.)
2.02% (2018 est.)
3.17% (2017 est.)
country comparison to the world: 149

GDP—per capita (PPP): $48,400 (2017 est.)
$47,500 (2016 est.)
$47,400 (2015 est.)
note: data are in 2017 dollars
country comparison to the world: 34

Gross national saving: 20.8% of GDP (2017 est.)
20% of GDP (2016 est.)
20.5% of GDP (2015 est.)
country comparison to the world: 90

GDP—composition, by end use: *household consumption:* 57.8% (2017 est.)
government consumption: 20.8% (2017 est.)
investment in fixed capital: 23% (2017 est.)
investment in inventories: 0.7% (2017 est.)
exports of goods and services: 30.9% (2017 est.)
imports of goods and services: -33.2% (2017 est.)

GDP—composition, by sector of origin:
agriculture: 1.6% (2017 est.)
industry: 28.2% (2017 est.)
services: 70.2% (2017 est.)

Agriculture—products: wheat, barley, oilseed, tobacco, fruits, vegetables; dairy products; fish; forest products

Industries: transportation equipment, chemicals, processed and unprocessed minerals, food products, wood and paper products, fish products, petroleum, natural gas

Industrial production growth rate: 4.9% (2017 est.)
country comparison to the world: 60

Labor force: 18.136 million (2020 est.)
country comparison to the world: 29

Labor force—by occupation: *agriculture:* 2%
industry: 13%
services: 6%
industry and services: 76%
manufacturing: 3% (2006 est.)

Unemployment rate: 5.67% (2019 est.)
5.83% (2018 est.)
country comparison to the world: 91

Population below poverty line: 9.4% (2008 est.)
note: this figure is the Low Income Cut-Off, a calculation that results in higher figures than found in many comparable economies; Canada does not have an official poverty line

Household income or consumption by percentage share: *lowest 10%:* 2.6%
highest 10%: 24.8% (2000)

Budget: *revenues:* 649.6 billion (2017 est.)

expenditures: 665.7 billion (2017 est.)

Taxes and other revenues: 39.3% (of GDP) (2017 est.)
country comparison to the world: 48

Budget surplus (+) or deficit (-): -1% (of GDP) (2017 est.)
country comparison to the world: 77

Public debt: 89.7% of GDP (2017 est.)
91.1% of GDP (2016 est.)
note: figures are for gross general government debt, as opposed to net federal debt; gross general government debt includes both intragovernmental debt and the debt of public entities at the sub-national level
country comparison to the world: 25

Fiscal year: 1 April—31 March

Inflation rate (consumer prices): 1.6% (2017 est.)
1.4% (2016 est.)
country comparison to the world: 87

Current account balance: -$35.425 billion (2019 est.)
-$42.862 billion (2018 est.)
country comparison to the world: 201

Exports: $423.5 billion (2017 est.)
$393.5 billion (2016 est.)
country comparison to the world: 11

Exports—partners: US 76.4%, China 4.3% (2017)

Exports—commodities: motor vehicles and parts, industrial machinery, aircraft, telecommunications equipment; chemicals, plastics, fertilizers; wood pulp, timber, crude petroleum, natural gas, electricity, aluminum

Imports: $442.1 billion (2017 est.)
$413.4 billion (2016 est.)
country comparison to the world: 12

Imports—commodities: machinery and equipment, motor vehicles and parts, crude oil, chemicals, electricity, durable consumer goods

Imports—partners: US 51.5%, China 12.6%, Mexico 6.3% (2017)

Reserves of foreign exchange and gold: $86.68 billion (31 December 2017 est.)
$82.72 billion (31 December 2016 est.)
country comparison to the world: 28

Debt—external: $1.608 trillion (31 March 2016 est.)
$1.55 trillion (31 March 2015 est.)
country comparison to the world: 13

Exchange rates: Canadian dollars (CAD) per US dollar—
1.308 (2017 est.)
1.3256 (2016 est.)
1.3256 (2015 est.)
1.2788 (2014 est.)
1.0298 (2013 est.)

ENERGY

Electricity access: *electrification—total population:* 100% (2020)

Electricity—production: 649.6 billion kWh (2016 est.)

country comparison to the world: 6

Electricity—consumption: 522.2 billion kWh (2016 est.)
country comparison to the world: 7

Electricity—exports: 73.35 billion kWh (2016 est.)
country comparison to the world: 2

Electricity—imports: 2.682 billion kWh (2016 est.)
country comparison to the world: 52

Electricity—installed generating capacity: 143.5 million kW (2016 est.)
country comparison to the world: 8

Electricity—from fossil fuels: 23% of total installed capacity (2016 est.)
country comparison to the world: 191

Electricity—from nuclear fuels: 9% of total installed capacity (2017 est.)
country comparison to the world: 16

Electricity—from hydroelectric plants: 56% of total installed capacity (2017 est.)
country comparison to the world: 30

Electricity—from other renewable sources: 12% of total installed capacity (2017 est.)
country comparison to the world: 73

Crude oil—production: 4.264 million bbl/day (2018 est.)
country comparison to the world: 5

Crude oil—exports: 2.818 million bbl/day (2017 est.)
country comparison to the world: 4

Crude oil—imports: 806,700 bbl/day (2017 est.)
country comparison to the world: 14

Crude oil—proved reserves: 170.5 billion bbl (1 January 2018 est.)
country comparison to the world: 3

Refined petroleum products—production: 2.009 million bbl/day (2017 est.)
country comparison to the world: 10

Refined petroleum products—consumption: 2.445 million bbl/day (2017 est.)
country comparison to the world: 10

Refined petroleum products—exports: 1.115 million bbl/day (2017 est.)
country comparison to the world: 8

Refined petroleum products—imports: 405,700 bbl/day (2017 est.)
country comparison to the world: 21

Natural gas—production: 159.1 billion cu m (2017 est.)
country comparison to the world: 5

Natural gas—consumption: 124.4 billion cu m (2017 est.)
country comparison to the world: 6

Natural gas—exports: 83.96 billion cu m (2017 est.)
country comparison to the world: 5

Natural gas—imports: 26.36 billion cu m (2017 est.)
country comparison to the world: 13

Natural gas—proved reserves: 2.056 trillion cu m (1 January 2018 est.)
country comparison to the world: 16

Carbon dioxide emissions from consumption of energy: 640.6 million Mt (2017 est.)
country comparison to the world: 9

COMMUNICATIONS

Telephones—fixed lines: *total subscriptions:* 13,258,721
subscriptions per 100 inhabitants: 35.46 (2019 est.)
country comparison to the world: 15

Telephones—mobile cellular: *total subscriptions:* 34,597,559
subscriptions per 100 inhabitants: 92.53 (2019 est.)
country comparison to the world: 43

Telecommunication systems: *general assessment:* excellent service provided by first-rate technology; offers 99% coverage with LTE; consumer demand for mobile data services have prompted telecom companies to invest and advance LTE infrastructure, and further investment in 5G; govt. policy has aided the extension of broadband to rural and regional areas, with the result that services are almost universally accessible; govt. sets up $400 million public-private partnership to exploit benefits of 5G (2020)
domestic: 35 per 100 fixed-line; 93 per 100 mobile-cellular; comparatively low mobile penetration provides further room for growth; domestic satellite system with about 300 earth stations (2019)
international: country code—1; landing points for the Nunavut Undersea Fiber Optic Network System, Greenland Connect, Persona, GTT Atlantic, and Express, KetchCan 1 Submarine Fiber Cable system, St Pierre and Miquelon Cable submarine cables providing links to the US and Europe; satellite earth stations—7 (5 Intelsat—4 Atlantic Ocean and 1 Pacific Ocean, and 2 Intersputnik—Atlantic Ocean region) (2019)
note: the COVID-19 outbreak is negatively impacting telecommunications production and supply chains globally; consumer spending on telecom devices and services has also slowed due to the pandemic's effect on economies worldwide; overall progress towards improvements in all facets of the telecom industry—mobile, fixed-line, broadband, submarine cable and satellite—has moderated

Broadcast media: 2 public TV broadcasting networks, 1 in English and 1 in French, each with a large number of network affiliates; several private-commercial networks also with multiple network affiliates; overall, about 150 TV stations; multi-channel satellite and cable systems provide access to a wide range of stations including US stations; mix of public and commercial radio broadcasters with the Canadian Broadcasting Corporation (CBC), the public radio broadcaster, operating 4 radio networks, Radio Canada International, and radio services to indigenous populations in the north; roughly 1,119 licensed radio stations (2016)

Internet country code: .ca

Internet users: *total:* 33,743,954

percent of population: 91% (July 2018 est.)
country comparison to the world: 23

Broadband—fixed subscriptions: *total:* 14,445,606
subscriptions per 100 inhabitants: 39 (2018 est.)
country comparison to the world: 14

TRANSPORTATION

National air transport system: *number of registered air carriers:* 51 (2020)
inventory of registered aircraft operated by air carriers: 879
annual passenger traffic on registered air carriers: 89.38 million (2018)
annual freight traffic on registered air carriers: 3,434,070,000 mt-km (2018)

Civil aircraft registration country code prefix: C (2016)

Airports: 1,467 (2013)
country comparison to the world: 4

Airports—with paved runways: *total:* 523 (2017)
over 3,047 m: 21 (2017)
2,438 to 3,047 m: 19 (2017)
1,524 to 2,437 m: 147 (2017)
914 to 1,523 m: 257 (2017)
under 914 m: 79 (2017)

Airports—with unpaved runways: *total:* 944 (2013)
1,524 to 2,437 m: 75 (2013)
914 to 1,523 m: 385 (2013)
under 914 m: 484 (2013)

Heliports: 26 (2013)

Pipelines: 110000 km gas and liquid petroleum (2017)

Railways: *total:* 77,932 km (2014)
standard gauge: 77,932 km 1.435-m gauge (2014)
country comparison to the world: 4

Roadways: *total:* 1,042,300 km (2011)
paved: 415,600 km (includes 17,000 km of expressways) (2011)
unpaved: 626,700 km (2011)
country comparison to the world: 8

Waterways: 636 km (Saint Lawrence Seaway of 3,769 km, including the Saint Lawrence River of 3,058 km, shared with United States) (2011)
country comparison to the world: 77

Merchant marine: *total:* 669
by type: bulk carrier 18, container ship 1, general cargo 76, oil tanker 16, other 558 (2019)
country comparison to the world: 34

Ports and terminals: *major seaport(s):* Halifax, Saint John (New Brunswick), Vancouver
oil terminal(s): Lower Lakes terminal
container port(s) (TEUs): Montreal (1,537,669), Vancouver (3,252,225) (2017)
LNG terminal(s) (import): Saint John
river and lake port(s): Montreal, Quebec City, Sept-Isles (St. Lawrence)

dry bulk cargo port(s): Port-Cartier (iron ore and grain), Fraser River Port (Fraser) Hamilton (Lake Ontario)

MILITARY AND SECURITY

Military and security forces: Canadian Forces: Canadian Army, Royal Canadian Navy, Royal Canadian Air Force, Canadian Joint Operations Command, Canadian Special Operations Forces Command; Primary Reserve (army, air, naval reserves); Coast Guard (Department of Fisheries and Oceans) (2019)
note: the Army reserves include the Canadian Rangers, which provides a limited presence in Canada's northern, coastal, and isolated areas for sovereignty, public safety, and surveillance roles

Military expenditures: 1.31% of GDP (2019 est.)
1.31% of GDP (2018)
1.44% of GDP (2017)
1.16% of GDP (2016)
1.2% of GDP (2015)
country comparison to the world: 90

Military and security service personnel strengths: the Canadian Armed (CAF) Forces have approximately 66,000 total active personnel (23,000 Army; 8,300 Navy; 12,000 Air Force; 23,000 other uniformed personnel) (2019 est.)

Military equipment inventories and acquisitions: the CAF's inventory is a mix of domestically-produced equipment and imported weapons systems from Australia, Europe, Israel, and the US; since 2010, the leading supplier is the US; Canada's defense industry develops, maintains, and produces a range of equipment, including aircraft, combat vehicles, naval vessels, and associated components (2019 est.)

Military deployments: 540 Latvia (NATO); up to 200 Ukraine; up to 850 Middle East (multiple missions, including support to the Global Coalition to Defeat ISIS and NATO assistance mission Iraq; reduced considerably in 2020 because of COVID 19) (2020)

Military service age and obligation: 17 years of age for voluntary male and female military service (with parental consent); 16 years of age for Reserve and Military College applicants; Canadian citizenship or permanent residence status required; maximum 34 years of age; service obligation 3-9 years (2012)

TERRORISM

Terrorist group(s): Islamic State of Iraq and ash-Sham (ISIS) (2019)
note: details about the history, aims, leadership, organization, areas of operation, tactics, targets, weapons, size, and sources of support of the group(s) appear(s) in Appendix-T

TRANSNATIONAL ISSUES

Disputes—international: managed maritime boundary disputes with the US at Dixon Entrance, Beaufort Sea, Strait of Juan de Fuca, and the Gulf of Maine, including the disputed Machias Seal

Island and North Rock; Canada and the United States dispute how to divide the Beaufort Sea and the status of the Northwest Passage but continue to work cooperatively to survey the Arctic continental shelf; US works closely with Canada to intensify security measures for monitoring and controlling legal and illegal movement of people, transport, and commodities across the international border; sovereignty dispute with Denmark over Hans Island in the Kennedy Channel between Ellesmere Island and Greenland; commencing the collection of technical evidence

for submission to the Commission on the Limits of the Continental Shelf in support of claims for continental shelf beyond 200 nm from its declared baselines in the Arctic, as stipulated in Article 76, paragraph 8, of the UN Convention on the Law of the Sea

Refugees and internally displaced persons: *refugees (country of origin):* 7,356 (Colombia), 6,640 (Nigeria), 6,563 (Haiti), 6,060 (China), 5,876 (Turkey), 5,498 (Pakistan) (2018); 6,751 (Venezuela) (economic and political crisis; includes Venezuelans who have claimed asylum,

are recognized as refugees, or have received alternative legal stay) (2019)
stateless persons: 3,790 (2019)

Illicit drugs: illicit producer of cannabis for the domestic drug market and export to US; use of hydroponics technology permits growers to plant large quantities of high-quality marijuana indoors; increasing ecstasy production, some of which is destined for the US; vulnerable to narcotics money laundering because of its mature financial services sector

CAYMAN ISLANDS

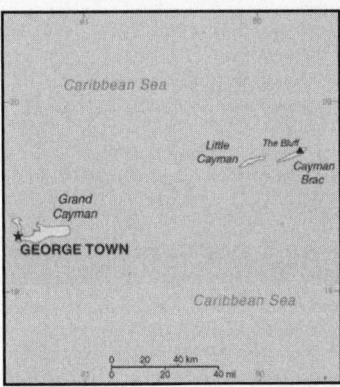

INTRODUCTION

Background: The Cayman Islands were colonized from Jamaica by the British during the 18th and 19th centuries and were administered by Jamaica after 1863. In 1959, the islands became a territory within the Federation of the West Indies. When the Federation dissolved in 1962, the Cayman Islands chose to remain a British dependency. The territory has transformed itself into a significant offshore financial center.

GEOGRAPHY

Location: Caribbean, three-island group (Grand Cayman, Cayman Brac, Little Cayman) in Caribbean Sea, 240 km south of Cuba and 268 km northwest of Jamaica

Geographic coordinates: 19 30 N, 80 30 W

Map references: Central America and the Caribbean

Area: *total:* 264 sq km
land: 264 sq km
water: 0 sq km
country comparison to the world: 211

Area—comparative: 1.5 times the size of Washington, DC

Land boundaries: 0 km

Coastline: 160 km

Maritime claims: *territorial sea:* 12 nm
exclusive fishing zone: 200 nm

Climate: tropical marine; warm, rainy summers (May to October) and cool, relatively dry winters (November to April)

Terrain: low-lying limestone base surrounded by coral reefs

Elevation: *lowest point:* Caribbean Sea 0 m

highest point: 1 km SW of The Bluff on Cayman Brac 50 m

Natural resources: fish, climate and beaches that foster tourism

Land use: *agricultural land:* 11.2% (2011 est.)
arable land: 0.8% (2011 est.) / permanent crops: 2.1% (2011 est.) / permanent pasture: 8.3% (2011 est.)
forest: 52.9% (2011 est.)
other: 35.9% (2011 est.)
Irrigated land: NA

Population distribution: majority of the population resides on Grand Cayman

Natural hazards: hurricanes (July to November)

Environment—current issues: no natural freshwater resources; drinking water supplies are met by reverse osmosis desalination plants and rainwater catchment; trash washing up on the beaches or being deposited there by residents; no recycling or waste treatment facilities; deforestation (trees being cut down to create space for commercial use)

Geography—note: important location between Cuba and Central America

PEOPLE AND SOCIETY

Population: 61,944 (July 2020 est.)
note: most of the population lives on Grand Cayman
country comparison to the world: 205

Nationality: *noun:* Caymanian(s)
adjective: Caymanian

Ethnic groups: mixed 40%, white 20%, black 20%, expatriates of various ethnic groups 20%

Languages: English (official) 90.9%, Spanish 4%, Filipino 3.3%, other 1.7%, unspecified 0.1% (2010 est.)

Religions: Protestant 67.8% (includes Church of God 22.6%, Seventh Day Adventist 9.4%, Presbyterian/United Church 8.6%, Baptist 8.3%, Pentecostal 7.1%, non-denominational 5.3%, Anglican 4.1%, Wesleyan Holiness 2.4%), Roman Catholic 14.1%, Jehovah's Witness 1.1%, other 7%, none 9.3%, unspecified 0.7% (2010 est.)

Age structure: *0-14 years:* 17.75% (male 5,535/female 5,457)
15-24 years: 11.86% (male 3,673/female 3,675)
25-54 years: 41.37% (male 12,489/female 13,140)
55-64 years: 14.78% (male 4,398/female 4,755)
65 years and over: 14.24% (male 4,053/female 4,769) (2020 est.)

Median age: *total:* 40.5 years
male: 39.7 years
female: 41.2 years (2020 est.)
country comparison to the world: 51

Population growth rate: 1.9% (2020 est.)
country comparison to the world: 51

Birth rate: 11.9 births/1,000 population (2020 est.)
country comparison to the world: 164

Death rate: 6.1 deaths/1,000 population (2020 est.)
country comparison to the world: 159

Net migration rate: 13 migrant(s)/1,000 population (2020 est.)
note: major destination for Cubans trying to migrate to the US
country comparison to the world: 4

Population distribution: majority of the population resides on Grand Cayman

Urbanization: *urban population:* 100% of total population (2020)
rate of urbanization: 1.27% annual rate of change (2015-20 est.)
total population growth rate v. urban population growth rate, 2000-2030:

Major urban areas—population: 35,000 GEORGE TOWN (capital) (2018)

Sex ratio: *at birth:* 1.02 male(s)/female

0-14 years: 1.01 male(s)/female
15-24 years: 1 male(s)/female
25-54 years: 0.95 male(s)/female
55-64 years: 0.92 male(s)/female
65 years and over: 0.85 male(s)/female
total population: 0.95 male(s)/female (2020 est.)

Infant mortality rate: *total:* 5.5 deaths/1,000 live births
male: 6.3 deaths/1,000 live births
female: 4.8 deaths/1,000 live births (2020 est.)
country comparison to the world: 172

Life expectancy at birth: *total population:* 81.6 years
male: 78.9 years
female: 84.4 years (2020 est.)
country comparison to the world: 29

Total fertility rate: 1.83 children born/woman (2020 est.)
country comparison to the world: 144

Drinking water source:
improved:
urban: 97.4% of population
total: 97.4% of population
unimproved:
urban: 2.6% of population
total: 2.6% of population (2015 est.)

Sanitation facility access:
improved:
urban: 95.6% of population
total: 95.6% of population
unimproved:
urban: 4.4% of population
total: 4.4% of population (2015 est.)

HIV/AIDS—adult prevalence rate: NA

HIV/AIDS—people living with HIV/AIDS: NA

HIV/AIDS—deaths: NA

Education expenditures: NA

Literacy: *definition:* age 15 and over has ever attended school
total population: 98.9%
male: 98.7%
female: 99% (2007)

Unemployment, youth ages 15-24: *total:* 13.8%
male: 16.4%
female: 11.4% (2015 est.)
country comparison to the world: 98

GOVERNMENT

Country name: *conventional long form:* none
conventional short form: Cayman Islands
etymology: the islands' name comes from the native Carib word "caiman," describing the marine crocodiles living there
Dependency status: overseas territory of the UK

Government type: parliamentary democracy; self-governing overseas territory of the UK

Capital: *name:* George Town (on Grand Cayman)
geographic coordinates: 19 18 N, 81 23 W
time difference: UTC-5 (same time as Washington, DC, during Standard Time)
etymology: named after English King George III (1738-1820)

Administrative divisions: 6 districts; Bodden Town, Cayman Brac and Little Cayman, East End, George Town, North Side, West Bay

Independence: none (overseas territory of the UK)

National holiday: Constitution Day, the first Monday in July (1959)

Constitution: *history:* several previous; latest approved 10 June 2009, entered into force 6 November 2009 (The Cayman Islands Constitution Order 2009)
amendments: amended several times, last in 2016

Legal system: English common law and local statutes

Citizenship: see United Kingdom

Suffrage: 18 years of age; universal

Executive branch: *chief of state:* Queen ELIZABETH II (since 6 February 1952); represented by Governor Martyn ROPER (since 29 October 2018)
head of government: Premier Alden MCLAUGHLIN (since 29 May 2013)
cabinet: Cabinet selected from the Legislative Assembly and appointed by the governor on the advice of the premier
elections/appointments: the monarchy is hereditary; governor appointed by the monarch; following legislative elections, the leader of the majority party or majority coalition appointed premier by the governor

Legislative branch: *description:* unicameral Legislative Assembly (21 seats; 19 members directly elected by majority vote and 2 ex officio members—the deputy governor and attorney general—appointed by the governor; members serve 4-year terms)
elections: last held on 24 May 2017 (next to be held in 2021)
election results: percent of vote by party—independent 44.7%, PPM 31.2%, CDP 24.1%; seats by party—independent 9, PPM 7, CDP 3; composition—men 18, women 3, percent of women 14.3%

Judicial branch: *highest courts:* Court of Appeal (consists of the court president and at least 2 judges); Grand Court (consists of the court president and at least 2 judges); note—appeals beyond the Court of Appeal are heard by the Judicial Committee of the Privy Council (in London)
judge selection and term of office: Court of Appeal and Grand Court judges appointed by the governor on the advice of the Judicial and Legal Services Commission, an 8-member independent body consisting of governor appointees, Court of Appeal president, and attorneys; Court of Appeal judges' tenure based on their individual instruments of appointment; Grand Court judges normally appointed until retirement at age 65 but can be extended until age 70
subordinate courts: Summary Court

Political parties and leaders:
People's Progressive Movement or PPM [Alden MCLAUGHLIN]
Cayman Democratic Party or CDP [McKeeva BUSH]

International organization participation: Caricom (associate), CDB, Interpol (subbureau), IOC, UNESCO (associate), UPU

Diplomatic representation in the US: none (overseas territory of the UK)

Diplomatic representation from the US: none (overseas territory of the UK); consular services provided through the US Embassy in Jamaica

Flag description: a blue field with the flag of the UK in the upper hoist-side quadrant and the Caymanian coat of arms centered on the outer half of the flag; the coat of arms includes a crest with a pineapple, representing the connection with Jamaica, and a turtle, representing Cayman's seafaring tradition, above a shield bearing a golden lion, symbolizing Great Britain, below which are three green stars (representing the three islands) surmounting white and blue wavy lines representing the sea; a scroll below the shield bears the motto HE HATH FOUNDED IT UPON THE SEAS

National symbol(s): green sea turtle

National anthem: *name:* Beloved Isle Cayman
lyrics/music: Leila E. ROSS
note: adopted 1993; served as an unofficial anthem since 1930; as a territory of the United Kingdom, in addition to the local anthem, "God Save the Queen" is official (see United Kingdom)

ECONOMY

Economy—overview: With no direct taxation, the islands are a thriving offshore financial center. More than 65,000 companies were registered in the Cayman Islands as of 2017, including more than 280 banks, 700 insurers, and 10,500 mutual funds. A stock exchange was opened in 1997. Nearly 90% of the islands' food and consumer goods must be imported. The Caymanians enjoy a standard of living comparable to that of Switzerland.

Tourism is also a mainstay, accounting for about 70% of GDP and 75% of foreign currency earnings. The tourist industry is aimed at the luxury market and caters mainly to visitors from North America. Total tourist arrivals exceeded 2.1 million in 2016, with more than three-quarters from the US.

GDP (purchasing power parity):
$2.507 billion (2014 est.)
$2.465 billion (2013 est.)
$2.435 billion (2012 est.)
country comparison to the world: 192

GDP (official exchange rate):
$2.25 billion (2008 est.)

GDP—real growth rate:
1.7% (2014 est.)
1.2% (2013 est.)
1.6% (2012 est.)
country comparison to the world: 148

GDP—per capita (PPP): $43,800 (2004 est.)
country comparison to the world: 41

GDP—composition, by end use:
household consumption: 62.3% (2017 est.)

government consumption: 14.5% (2017 est.)
investment in fixed capital: 22.1% (2017 est.)
investment in inventories: 0.1% (2017 est.)
exports of goods and services: 65.4% (2017 est.)
imports of goods and services: -64.2% (2017 est.)

GDP—composition, by sector of origin:
agriculture: 0.3% (2017 est.)
industry: 7.4% (2017 est.)
services: 92.3% (2017 est.)

Agriculture—products: vegetables, fruit; livestock; turtle farming

Industries: tourism, banking, insurance and finance, construction, construction materials, furniture

Industrial production growth rate: 2.2% (2017 est.)
country comparison to the world: 124

Labor force: 39,000 (2007 est.)
note: nearly 55% are non-nationals
country comparison to the world: 197

Labor force—by occupation:
agriculture: 1.9%
industry: 19.1%
services: 79% (2008 est.)

Unemployment rate:
4% (2008)
4.4% (2004)
country comparison to the world: 60

Population below poverty line: NA

Household income or consumption by percentage share: *lowest 10%:* NA
highest 10%: NA

Budget: *revenues:* 874.5 million (2017 est.)
expenditures: 766.6 million (2017 est.)

Taxes and other revenues: 38.9% (of GDP) (2017 est.)
country comparison to the world: 51

Budget surplus (+) or deficit (-): 4.8% (of GDP) (2017 est.)
country comparison to the world: 7

Fiscal year: 1 April—31 March

Inflation rate (consumer prices):
2% (2017 est.)
-0.6% (2016 est.)
country comparison to the world: 104

Current account balance:
-$492.6 million (2017 est.)
-$493.5 million (2016 est.)
country comparison to the world: 121

Exports:
$421.9 million (2017 est.)
$47.6 million (2016 est.)
country comparison to the world: 181

Exports—commodities: turtle products, manufactured consumer goods

Imports:
$787.3 million (2017 est.)
$810.1 million (2016 est.)
country comparison to the world: 190

Imports—commodities: foodstuffs, manufactured goods, fuels

Exchange rates: Caymanian dollars (KYD) per US dollar -
0.82 (2017 est.)
0.82 (2016 est.)
0.82 (2015 est.)
0.82 (2014 est.)
0.83 (2013 est.)

ENERGY

Electricity access: *electrification—total population:* 100% (2020)

Electricity—production: 650 million kWh (2016 est.)
country comparison to the world: 160

Electricity—consumption: 612 million kWh (2016 est.)
country comparison to the world: 165

Electricity—exports: 0 kWh (2016 est.)
country comparison to the world: 118

Electricity—imports: 0 kWh (2016 est.)
country comparison to the world: 132

Electricity—installed generating capacity: 132,000 kW (2016 est.)
country comparison to the world: 174

Electricity—from fossil fuels: 100% of total installed capacity (2016 est.)
country comparison to the world: 6

Electricity—from nuclear fuels: 0% of total installed capacity (2017 est.)
country comparison to the world: 63

Electricity—from hydroelectric plants: 0% of total installed capacity (2017 est.)
country comparison to the world: 163

Electricity—from other renewable sources: 0% of total installed capacity (2017 est.)
country comparison to the world: 180

Crude oil—production: 0 bbl/day (2018 est.)
country comparison to the world: 120

Crude oil—exports: 0 bbl/day (2015 est.)
country comparison to the world: 104

Crude oil—imports: 0 bbl/day (2015 est.)
country comparison to the world: 107

Crude oil—proved reserves: 0 bbl (1 January 2018 est.)
country comparison to the world: 116

Refined petroleum products—production: 0 bbl/day (2017 est.)
country comparison to the world: 128

Refined petroleum products—consumption: 4,400 bbl/day (2016 est.)
country comparison to the world: 181

Refined petroleum products—exports: 0 bbl/day (2015 est.)
country comparison to the world: 141

Refined petroleum products—imports: 4,285 bbl/day (2015 est.)
country comparison to the world: 175

Natural gas—production: 0 cu m (2017 est.)
country comparison to the world: 114

Natural gas—consumption: 0 cu m (2017 est.)
country comparison to the world: 130

Natural gas—exports: 0 cu m (2017 est.)
country comparison to the world: 80

Natural gas—imports: 0 cu m (2017 est.)
country comparison to the world: 104

Natural gas—proved reserves: 0 cu m (1 January 2014 est.)
country comparison to the world: 120

Carbon dioxide emissions from consumption of energy: 643,800 Mt (2017 est.)
country comparison to the world: 178

COMMUNICATIONS

Telephones—fixed lines: *total subscriptions:* 33,338
subscriptions per 100 inhabitants: 54.85 (2019 est.)
country comparison to the world: 168

Telephones—mobile cellular: *total subscriptions:* 92,691
subscriptions per 100 inhabitants: 152.5 (2019 est.)
country comparison to the world: 195

Telecommunication systems: *general assessment:* reasonably good overall telephone system with a high fixed-line teledensity; given the high dependence of tourism and activities such as fisheries and offshore financial services, the telecom sector provides a relatively high contribution to overall GDP; good competition in all sectors promotes advancement in mobile telephony and data segments (2018)
domestic: introduction of competition in the mobile-cellular market in 2004 boosted subscriptions dramatically; 55 per 100 fixed-line, 153 per 100 mobile-cellular (2019)
international: country code—1-345; landing points for the Maya-1, Deep Blue Cable, and the Cayman-Jamaica Fiber System submarine cables that provide links to the US and parts of Central and South America; satellite earth station—1 Intelsat (Atlantic Ocean) (2019)
note: the COVID-19 outbreak is negatively impacting telecommunications production and supply chains globally; consumer spending on telecom devices and services has also slowed due to the pandemic's effect on economies worldwide; overall progress towards improvements in all facets of the telecom industry—mobile, fixed-line, broadband, submarine cable and satellite—has moderated

Broadcast media: 4 TV stations; cable and satellite subscription services offer a variety of international programming; government-owned Radio Cayman operates 2 networks broadcasting on 5 stations; 10 privately owned radio stations operate alongside Radio Cayman

Internet country code: .ky

Internet users: *total:* 48,328
percent of population: 81.07% (July 2018 est.)
country comparison to the world: 198

Broadband—fixed subscriptions: *total:* 24,535
subscriptions per 100 inhabitants: 42 (2017 est.)
country comparison to the world: 150

TRANSPORTATION

National air transport system: *number of registered air carriers:* 1 (2020)
inventory of registered aircraft operated by air carriers: 6

Civil aircraft registration country code prefix: VP-C (2016)

Airports: 3 (2020)
country comparison to the world: 194

Airports—with paved runways: *total:* 3 (2017)
1,524 to 2,437 m: 2 (2017)
914 to 1,523 m: 1 (2017)

Airports—with unpaved runways: *total:* 1 (2012)
914 to 1,523 m: 1 (2012)

Roadways:
total: 785 km (2007)
paved: 785 km (2007)
country comparison to the world: 188

Merchant marine: *total:* 170
by type: bulk carrier 32 general cargo 4, oil tanker 22, other 112 (2019)
country comparison to the world: 70

Ports and terminals: *major seaport(s):* Cayman Brac, George Town

MILITARY AND SECURITY

Military and security forces: no regular military forces; Royal Cayman Islands Police Service (2019)

Military—note: defense is the responsibility of the UK

TRANSNATIONAL ISSUES

Disputes—international: none

Illicit drugs: major offshore financial center; vulnerable to drug transshipment to the US and Europe

CENTRAL AFRICAN REPUBLIC

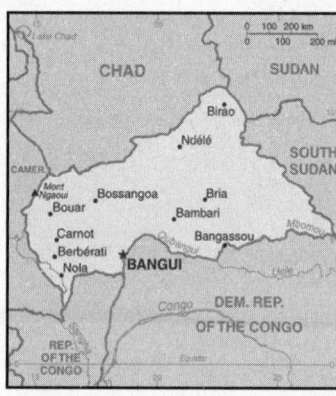

INTRODUCTION

Background: The region was the site of much slave trading activity in the centuries before becoming a French protectorate in the late 19th century, and then was heavily economically exploited in the early part of the 20th century. Upon independence in 1960, the French colony of Ubangi-Shari became the Central African Republic. After three tumultuous decades of misrule—mostly by military governments—civilian rule was established in 1993 but lasted only a decade. In March 2003, President Ange-Felix PATASSE was deposed in a military coup led by General Francois BOZIZE, who established a transitional government. Elections held in 2005 affirmed General BOZIZE as president; he was reelected in 2011 in voting widely viewed as flawed. Several rebel groups joined together in early December 2012 to launch a series of attacks that left them in control of numerous towns in the northern and central parts of the country. The rebels—unhappy with BOZIZE's government—participated in peace talks in early January 2013 which resulted in a coalition government including the rebellion's leadership. In March 2013, the coalition government dissolved, rebels seized the capital, and President BOZIZE fled the country. Rebel leader Michel DJOTODIA assumed the presidency and the following month established a National Transitional Council (CNT). In January 2014, the CNT elected Catherine SAMBA-PANZA as interim president. Elections completed in March 2016 installed independent candidate Faustin-Archange TOUADERA as president; he continues to work towards peace between the government and armed groups, and is developing a disarmament, demobilization, reintegration, and repatriation program to reintegrate the armed groups into society. Nonetheless, as of early 2020 widespread violence continued, and the government in Bangui remains unable to extend control outside the capital. Peace agreements signed in 2017 and 2019 between the government and the main armed factions have had little effect and armed groups operate openly and control large swaths—as much 80% by some estimates—of the country's territory.

GEOGRAPHY

Location: Central Africa, north of Democratic Republic of the Congo

Geographic coordinates: 7 00 N, 21 00 E

Map references: Africa

Area: *total:* 622,984 sq km
land: 622,984 sq km
water: 0 sq km
country comparison to the world: 46

Area—comparative: slightly smaller than Texas; about four times the size of Georgia

Land boundaries: *total:* 5,920 km
border countries (6): Cameroon 901 km, Chad 1556 km, Democratic Republic of the Congo 1747 km, Republic of the Congo 487 km, South Sudan 1055 km, Sudan 174 km

Coastline: 0 km (landlocked)

Maritime claims: none (landlocked)

Climate: tropical; hot, dry winters; mild to hot, wet summers

Terrain: vast, flat to rolling plateau; scattered hills in northeast and southwest

Elevation: *mean elevation:* 635 m
lowest point: Oubangui River 335 m
highest point: Mont Ngaoui 1,410 m

Natural resources: diamonds, uranium, timber, gold, oil, hydropower

Land use: agricultural *land:* 8.1% (2011 est.)
arable land: 2.9% (2011 est.) / permanent crops: 0.1% (2011 est.) / permanent pasture: 5.1% (2011 est.)
forest: 36.2% (2011 est.)
other: 55.7% (2011 est.)
Irrigated land: 10 sq km (2012)

Population distribution: majority of residents live in the western and central areas of the country, especially in and around the capital of Bangui as shown in this population distribution map

Natural hazards: hot, dry, dusty harmattan winds affect northern areas; floods are common

Environment—current issues: water pollution; tap water is not potable; poaching and mismanagement have diminished the country's reputation as one of the last great wildlife refuges; desertification; deforestation; soil erosion

Environment—international agreements: *party to:* Biodiversity, Climate Change, Climate Change-Kyoto Protocol, Desertification, Endangered Species, Hazardous Wastes, Ozone Layer Protection, Tropical Timber 94, Wetlands
signed, but not ratified: Law of the Sea

Geography—note: landlocked; almost the precise center of Africa

PEOPLE AND SOCIETY

Population: 5,990,855 (July 2020 est.)
note: estimates for this country explicitly take into account the effects of excess mortality due to AIDS; this can result in lower life expectancy, higher infant mortality, higher death rates, lower population growth rates, and changes in the

distribution of population by age and sex than would otherwise be expected

country comparison to the world: 113

Nationality: *noun:* Central African(s)
adjective: Central African

Ethnic groups: Baya 28.8%, Banda 22.9%, Mandjia 9.9%, Sara 7.9%, M'Baka-Bantu 7.9%, Arab-Fulani (Peul) 6%, Mbum 6%, Ngbanki 5.5%, Zande-Nzakara 3%, other Central African Republic ethnic groups 2%, non-Central African Republic ethnic groups. 1%

Languages: French (official), Sangho (lingua franca and national language), tribal languages

Religions: Christian 89.5%, Muslim 8.5%, folk 1%, unaffiliated 1% (2010 est.)
note: animistic beliefs and practices strongly influence the Christian majority

Demographic profile: The Central African Republic's (CAR) humanitarian crisis has worsened since a coup in March 2013. CAR's high mortality rate and low life expectancy are attributed to elevated rates of preventable and treatable diseases (including malaria and malnutrition), an inadequate health care system, precarious food security, and armed conflict. Some of the worst mortality rates are in western CAR's diamond mining region, which is impoverished because of government attempts to control the diamond trade and the fall in industrial diamond prices. To make matters worse, the government and international donors have reduced health funding in recent years. The CAR's weak educational system and low literacy rate have also suffered as a result of the country's ongoing conflict. Schools are closed, qualified teachers are scarce, infrastructure, funding, and supplies are lacking and subject to looting, and many students and teachers are displaced by violence.

Rampant poverty, human rights violations, unemployment, poor infrastructure, and a lack of security and stability have led to forced displacement internally and externally. Since the political crisis that resulted in CAR's March 2013 coup began in December 2012, approximately 600,000 people have fled to Chad, the Democratic Republic of the Congo (DRC), and other neighboring countries, while another estimated 600,000 are displaced internally as of October 2019. The UN has urged countries to refrain from repatriating CAR refugees amid the heightened lawlessness. (2019)

Age structure: *0-14 years:* 39.49% (male 1,188,682/female 1,176,958)
15-24 years: 19.89% (male 598,567/female 593,075)
25-54 years: 32.95% (male 988,077/female 986,019)
55-64 years: 4.32% (male 123,895/female 134,829)
65 years and over: 3.35% (male 78,017/female 122,736) (2020 est.)

Dependency ratios: *total dependency ratio:* 86.4
youth dependency ratio: 81.1
elderly dependency ratio: 5.2
potential support ratio: 19.2 (2020 est.)

Median age: *total:* 20 years
male: 19.7 years
female: 20.3 years (2020 est.)
country comparison to the world: 195

Population growth rate: 2.09% (2020 est.)
country comparison to the world: 41

Birth rate: 33.2 births/1,000 population (2020 est.)
country comparison to the world: 25

Death rate: 12.3 deaths/1,000 population (2020 est.)
country comparison to the world: 15

Net migration rate: 0 migrant(s)/1,000 population (2020 est.)
country comparison to the world: 79

Population distribution: majority of residents live in the western and central areas of the country, especially in and around the capital of Bangui as shown in this population distribution map

Urbanization: *urban population:* 42.2% of total population (2020)
rate of urbanization: 2.52% annual rate of change (2015-20 est.)

total population growth rate v. urban population growth rate, 2000-2030: Major urban areas—population: 889,000 BANGUI (capital) (2020)

Sex ratio: *at birth:* 1.03 male(s)/female
0-14 years: 1.01 male(s)/female
15-24 years: 1.01 male(s)/female
25-54 years: 1 male(s)/female
55-64 years: 0.92 male(s)/female
65 years and over: 0.64 male(s)/female
total population: 0.99 male(s)/female (2020 est.)

Maternal mortality rate: 829 deaths/100,000 live births (2017 est.)
country comparison to the world: 5

Infant mortality rate: *total:* 80.6 deaths/1,000 live births
male: 87.7 deaths/1,000 live births
female: 73.2 deaths/1,000 live births (2020 est.)
country comparison to the world: 3

Life expectancy at birth: *total population:* 54.2 years
male: 52.7 years
female: 55.7 years (2020 est.)
country comparison to the world: 224

Total fertility rate: 4.14 children born/woman (2020 est.)
country comparison to the world: 27

Contraceptive prevalence rate: 15.2% (2010/11)

Drinking water source:
improved:
urban: 89.6% of population
rural: 54.4% of population
total: 68.5% of population
unimproved:
urban: 10.4% of population
rural: 45.6% of population
total: 31.5% of population (2015 est.)

Current Health Expenditure: 5.8% (2017)

Physicians density: 0.07 physicians/1,000 population (2015)

Hospital bed density: 1 beds/1,000 population (2011)

Sanitation facility access:
improved:
urban: 43.6% of population
rural: 7.2% of population
total: 21.8% of population
unimproved:
urban: 56.4% of population
rural: 92.8% of population
total: 78.2% of population (2015 est.)

HIV/AIDS—adult prevalence rate: 3.6% (2019 est.)
country comparison to the world: 14

HIV/AIDS—people living with HIV/AIDS: 100,000 (2019 est.)
country comparison to the world: 45

HIV/AIDS—deaths: 3,800 (2019 est.)
country comparison to the world: 31

Major infectious diseases: *degree of risk:* very high (2020)
food or waterborne diseases: bacterial and protozoal diarrhea, hepatitis A and E, and typhoid fever
vectorborne diseases: malaria and dengue fever
water contact diseases: schistosomiasis
animal contact diseases: rabies
respiratory diseases: meningococcal meningitis

Obesity—adult prevalence rate: 7.5% (2016)
country comparison to the world: 159

Children under the age of 5 years underweight: 20.8% (2018)
country comparison to the world: 22

Education expenditures: 1.2% of GDP (2011)
country comparison to the world: 174

Literacy: *definition:* age 15 and over can read and write
total population: 37.4%
male: 49.5%
female: 25.8% (2018)

School life expectancy (primary to tertiary education): *total:* 7 years
male: 8 years
female: 6 years (2012)

GOVERNMENT

Country name: *conventional long form:* Central African Republic
conventional short form: none
local long form: Republique Centrafricaine
local short form: none
former: Ubangi-Shari, Central African Empire
abbreviation: CAR
etymology: self- descriptive name specifying the country's location on the continent; "Africa" is derived from the Roman designation of the area corresponding to present-day Tunisia "Africa terra," which meant "Land of the Afri" (the tribe resident in that area), but which eventually came to mean the entire continent

Government type: presidential republic

Capital: *name:* Bangui

geographic coordinates: 4 22 N, 18 35 E

time difference: UTC+1 (6 hours ahead of Washington, DC, during Standard Time)

etymology: established as a French settlement in 1889 and named after its location on the northern bank of the Ubangi River; the Ubangi itself was named from the native word for the "rapids" located beside the outpost, which marked the end of navigable water north from from Brazzaville

Administrative divisions: 14 prefectures (prefectures, singular—prefecture), 2 economic prefectures* (prefectures economiques, singular—prefecture economique), and 1 commune**; Bamingui-Bangoran, Bangui**, Basse-Kotto, Haute-Kotto, Haut-Mbomou, Kemo, Lobaye, Mambere-Kadei, Mbomou, Nana-Grebizi*, Nana-Mambere, Ombella-Mpoko, Ouaka, Ouham, Ouham-Pende, Sangha-Mbaere*, Vakaga

Independence: 13 August 1960 (from France)

National holiday: Republic Day, 1 December (1958)

Constitution: *history:* several previous; latest (interim constitution) approved by the Transitional Council 30 August 2015, adopted by referendum 13-14 December 2015, ratified 27 March 2016

amendments: proposals require support of the government, two thirds of the National Council of Transition, and assent by the "Mediator of the Central African" crisis; passage requires at least three-fourths majority vote by the National Council membership; non-amendable constitutional provisions include those on the secular and republican form of government, fundamental rights and freedoms, amendment procedures, or changes to the authorities of various high-level executive, parliamentary, and judicial officials

Legal system: civil law system based on the French model

International law organization participation: has not submitted an ICJ jurisdiction declaration; accepts ICCt jurisdiction

Citizenship: *citizenship by birth:* no
citizenship by descent only: least one parent must be a citizen of the Central African Republic
dual citizenship recognized: yes
residency requirement for naturalization: 35 years

Suffrage: 18 years of age; universal

Executive branch: *chief of state:* President Faustin-Archange TOUADERA (since 30 March 2016)

head of government: Prime Minister Firmin NGREBADA (since 25 February 2019)

cabinet: Council of Ministers appointed by the president

elections/appointments: under the 2015 constitution, the president is elected by universal direct suffrage for a period of 5 years (eligible for a second term); election last held 30 December 2015 with a runoff 20 February 2016 (next election scheduled to be held in December 2020)

election results: Faustin-Archange TOUADERA elected president in the second round; percent of vote in first round -Anicet-Georges DOLOGUELE (URCA) 23.7%, Faustin-Archange TOUADERA

(independent) 19.1%, Desire KOLINGBA (RDC) 12.%, Martin ZIGUELE (MLPC) 11.4%, other 33.8%; percent of vote in second round—Faustin-Archange TOUADERA 62.7%, Anicet-Georges DOLOGUELE 37.3%

note: rebel forces seized the capital in March 2013, forcing former President BOZIZE to flee the country; Interim President Michel DJOTODIA assumed the presidency, reinstated the prime minister, and established a National Transitional Council (CNT) in April 2013; the NTC elected Catherine SAMBA-PANZA interim president in January 2014 to serve until February 2015, when new elections were to be held; her term was extended because instability delayed new elections and the transition did not take place until the end of March 2016

Legislative branch: *description:* unicameral National Assembly or Assemblee Nationale (140 seats; members directly elected in single-seat constituencies by absolute majority vote with a second round if needed; members serve 5-year terms)

elections: last held 30 December 2015 (results annulled), 14 February 2016—first round and 31 March 2016—second round (next to be held on 27 December 2020)

election results: percent of vote by party—NA; seats by party—UNDP 16, URCA 11, RDC 8, MLPC 10, KNK 7, other 28, independent 60; composition—men 129, women 11, percent of women 7.9%

Judicial branch: *highest courts:* Supreme Court or Cour Supreme (consists of NA judges); Constitutional Court (consists of 9 judges, at least 3 of whom are women)

judge selection and term of office: Supreme Court judges appointed by the president; Constitutional Court judge appointments—2 by the president, 1 by the speaker of the National Assembly, 2 elected by their peers, 2 are advocates elected by their peers, and 2 are law professors elected by their peers; judges serve 7-year non-renewable terms

subordinate courts: high courts; magistrates' courts

Political parties and leaders: Action Party for Development or PAD [El Hadj Laurent NGON-BABA]
Alliance for Democracy and Progress or ADP [Clement BELIBANGA]
Central African Democratic Rally or RDC [Desire Nzanga KOLINGBA]
Movement for Democracy and Development or MDD [Louis PAPENIAH]
Movement for the Liberation of the Central African People or MLPC [Martin ZIGUELE]
National Convergence (also known as Kwa Na Kwa) or KNK [Francois BOZIZE]
National Union for Democracy and Progress or UNDP [Amine MICHEL]
New Alliance for Progress or NAP [Jean-Jacques DEMAFOUTH]
Social Democratic Party or PSD [Enoch LAKOUE]
Union for Central African Renewal or URCA [Anicet-Georges DOLOGUELE]

International organization participation: ACP, AfDB, AU, BDEAC, CEMAC, EITI (compliant country) (suspended), FAO, FZ, G-77, IAEA, IBRD, ICAO, ICCt, ICRM, IDA, IFAD, IFC, IFRCS, ILO, IMF, Interpol, IOC, IOM, ITSO, ITU, ITUC (NGOs), MIGA, NAM, OIC (observer), OIF, OPCW, UN, UNCTAD, UNESCO, UNIDO, UNWTO, UPU, WCO, WHO, WIPO, WMO, WTO

Diplomatic representation in the US: *chief of mission:* Ambassador Martial NDOUBOU (since 17 September 2018)
chancery: 2704 Ontario Road NW, Washington, DC 20009
telephone: [1] (202) 483-7800
FAX: [1] (202) 332-9893

Diplomatic representation from the US: *chief of mission:* Ambassador Lucy TAMLYN (since 6 February 2019)
telephone: [236] 21 61 0200
embassy: Avenue David Dacko, Bangui
mailing address: P.O. Box 924, Bangui
FAX: [236] 21 61 4494

Flag description: four equal horizontal bands of blue (top), white, green, and yellow with a vertical red band in center; a yellow five-pointed star to the hoist side of the blue band; banner combines the Pan-African and French flag colors; red symbolizes the blood spilled in the struggle for independence, blue represents the sky and freedom, white peace and dignity, green hope and faith, and yellow tolerance; the star represents aspiration towards a vibrant future

National symbol(s): elephant; national colors: blue, white, green, yellow, red

National anthem: *name:* "Le Renaissance" (The Renaissance)
lyrics/music: Barthelemy BOGANDA/Herbert PEPPER
note: adopted 1960; Barthelemy BOGANDA wrote the anthem's lyrics and was the first prime minister of the autonomous French territory
0:00 / 1:05

ECONOMY

Economy—overview: Subsistence agriculture, together with forestry and mining, remains the backbone of the economy of the Central African Republic (CAR), with about 60% of the population living in outlying areas. The agricultural sector generates more than half of estimated GDP, although statistics are unreliable in the conflict-prone country. Timber and diamonds account for most export earnings, followed by cotton. Important constraints to economic development include the CAR's landlocked geography, poor transportation system, largely unskilled work force, and legacy of misdirected macroeconomic policies. Factional fighting between the government and its opponents remains a drag on economic revitalization. Distribution of income is highly unequal and grants from the international community can only partially meet humanitarian needs. CAR shares a common currency with the Central African

Monetary Union. The currency is pegged to the Euro.

Since 2009, the IMF has worked closely with the government to institute reforms that have resulted in some improvement in budget transparency, but other problems remain. The government's additional spending in the run-up to the 2011 election worsened CAR's fiscal situation. In 2012, the World Bank approved $125 million in funding for transport infrastructure and regional trade, focused on the route between CAR's capital and the port of Douala in Cameroon. In July 2016, the IMF approved a three- year extended credit facility valued at $116 million; in mid-2017, the IMF completed a review of CAR's fiscal performance and broadly approved of the government's management, although issues with revenue collection, weak government capacity, and transparency remain. The World Bank in late 2016 approved a $20 million grant to restore basic fiscal management, improve transparency, and assist with economic recovery.

Participation in the Kimberley Process, a commitment to remove conflict diamonds from the global supply chain, led to a partially lifted the ban on diamond exports from CAR in 2015, but persistent insecurity is likely to constrain real GDP growth.

GDP (purchasing power parity): $3.39 billion (2017 est.)
$3.249 billion (2016 est.)
$3.108 billion (2015 est.)
note: data are in 2017 dollars
country comparison to the world: 185

GDP (official exchange rate): $1.937 billion (2017 est.)

GDP—real growth rate: 4.3% (2017 est.)
4.5% (2016 est.)
4.8% (2015 est.)
country comparison to the world: 66

GDP—per capita (PPP): $700 (2017 est.)
$700 (2016 est.)
$600 (2015 est.)
note: data are in 2017 dollars
country comparison to the world: 228

Gross national saving: 5.4% of GDP (2017 est.)
8.2% of GDP (2016 est.)
4.2% of GDP (2015 est.)
country comparison to the world: 175

GDP—composition, by end use: *household consumption:* 95.3% (2017 est.)
government consumption: 8.5% (2017 est.)
investment in fixed capital: 13.7% (2017 est.)
investment in inventories: 0% (2017 est.)
exports of goods and services: 12% (2017 est.)
imports of goods and services: -29.5% (2017 est.)

GDP—composition, by sector of origin: agriculture: 43.2% (2017 est.)
industry: 16% (2017 est.)
services: 40.8% (2017 est.)

Agriculture—products: cotton, coffee, tobacco, cassava (manioc, tapioca), yams, millet, corn, bananas; timber

Industries: gold and diamond mining, logging, brewing, sugar refining

Industrial production growth rate: 3.9% (2017 est.)
country comparison to the world: 77

Labor force: 2.242 million (2017 est.)
country comparison to the world: 119

Unemployment rate: 6.9% (2017 est.)
country comparison to the world: 109

Population below poverty line: 62% NA (2008 est.)

Household income or consumption by percentage share: *lowest 10%:* 2.1%
highest 10%: 33% (2003)

Budget: *revenues:* 282.9 million (2017 est.)
expenditures: 300.1 million (2017 est.)

Taxes and other revenues: 14.6% (of GDP) (2017 est.)
country comparison to the world: 198

Budget surplus (+) or deficit (-): -0.9% (of GDP) (2017 est.)
country comparison to the world: 71

Public debt: 52.9% of GDP (2017 est.)
56% of GDP (2016 est.)
country comparison to the world: 93

Fiscal year: calendar year

Inflation rate (consumer prices): 4.1% (2017 est.)
4.6% (2016 est.)
country comparison to the world: 159

Current account balance: -$163 million (2017 est.)
-$97 million (2016 est.)
country comparison to the world: 95

Exports: $113.7 million (2017 est.)
$101.5 million (2016 est.)
country comparison to the world: 196

Exports—partners: France 31.2%, Burundi 16.2%, China 12.5%, Cameroon 9.6%, Austria 7.8% (2017)

Exports—commodities: diamonds, timber, cotton, coffee

Imports: $393.1 million (2017 est.)
$342.2 million (2016 est.)
country comparison to the world: 200

Imports—commodities: food, textiles, petroleum products, machinery, electrical equipment, motor vehicles, chemicals, pharmaceuticals

Imports—partners: France 17.1%, US 12.3%, India 11.5%, China 8.2%, South Africa 7.4%, Japan 5.8%, Italy 5.1%, Cameroon 4.9%, Netherlands 4.6% (2017)

Reserves of foreign exchange and gold: $304.3 million (31 December 2017 est.)
$252.5 million (31 December 2016 est.)
country comparison to the world: 168

Debt—external: $779.9 million (31 December 2017 est.)
$691.5 million (31 December 2016 est.)
country comparison to the world: 170

Exchange rates: Cooperation Financiere en Afrique Centrale francs (XAF) per US dollar –
605.3 (2017 est.)
593.01 (2016 est.)
593.01 (2015 est.)
591.45 (2014 est.)
494.42 (2013 est.)

Electricity access: *population without electricity:* 5 million (2019)
electrification—total population: 3% (2019)
electrification—urban areas: 7% (2019)
electrification—rural areas: 0.4% (2019)

Electricity—production: 171.4 million kWh (2016 est.)
country comparison to the world: 194

Electricity—consumption: 159.4 million kWh (2016 est.)
country comparison to the world: 196

Electricity—exports: 0 kWh (2016 est.)
country comparison to the world: 119

Electricity—imports: 0 kWh (2016 est.)
country comparison to the world: 133

Electricity—installed generating capacity: 38,300 kW (2016 est.)
country comparison to the world: 197

Electricity—from fossil fuels: 50% of total installed capacity (2016 est.)
country comparison to the world: 151

Electricity—from nuclear fuels: 0% of total installed capacity (2017 est.)
country comparison to the world: 64

Electricity—from hydroelectric plants: 50% of total installed capacity (2017 est.)
country comparison to the world: 40

Electricity—from other renewable sources: 1% of total installed capacity (2017 est.)
country comparison to the world: 151

Crude oil—production: 0 bbl/day (2018 est.)
country comparison to the world: 121

Crude oil—exports: 0 bbl/day (2015 est.)
country comparison to the world: 105

Crude oil—imports: 0 bbl/day (2015 est.)
country comparison to the world: 108

Crude oil—proved reserves: 0 bbl (1 January 2018 est.)
country comparison to the world: 117

Refined petroleum products—production: 0 bbl/day (2017 est.)
country comparison to the world: 129

Refined petroleum products—consumption: 2,800 bbl/day (2016 est.)
country comparison to the world: 189

Refined petroleum products—exports: 0 bbl/day (2015 est.)
country comparison to the world: 142

Refined petroleum products—imports: 2,799 bbl/day (2015 est.)
country comparison to the world: 185

Natural gas—production: 0 cu m (2017 est.)
country comparison to the world: 115

Natural gas—consumption: 0 cu m (2017 est.)
country comparison to the world: 131

Natural gas—exports: 0 cu m (2017 est.)
country comparison to the world: 81

Natural gas—imports: 0 cu m (2017 est.)
country comparison to the world: 105

Natural gas—proved reserves: 0 cu m (1 January 2014 est.)
country comparison to the world: 121

Carbon dioxide emissions from consumption of energy: 413,800 Mt (2017 est.)
country comparison to the world: 187

COMMUNICATIONS

Telephones—fixed lines: *total subscriptions:* 2,934
subscriptions per 100 inhabitants: less than 1 (2019 est.)
country comparison to the world: 213

Telephones—mobile cellular: *total subscriptions:* 1,892,114
subscriptions per 100 inhabitants: 32.25 (2019 est.)
country comparison to the world: 152

Telecommunication systems: *general assessment:* network consists principally of microwave radio relay and at low-capacity; ongoing conflict has obstructed telecommunication and media development, although there are ISP (Internet service providers) and mobile phone carriers, radio is the most-popular communications medium (2018)
domestic: very limited telephone service with less than 1 fixed-line connection per 100 persons; with the presence of multiple providers mobile-cellular service has reached 33 per 100 mobile-cellular subscribers; cellular usage is increasing from a low base; most fixed-line and mobile-cellular telephone services are concentrated in Bangui (2019)
international: country code—236; satellite earth station—1 Intelsat (Atlantic Ocean)
note: the COVID-19 outbreak is negatively impacting telecommunications production and supply chains globally; consumer spending on telecom devices and services has also slowed due to the pandemic's effect on economies worldwide; overall progress towards improvements in all facets of the telecom industry—mobile, fixed-line, broadband, submarine cable and satellite—has moderated

Broadcast media: government-owned network, Radiodiffusion Television Centrafricaine, provides limited domestic TV broadcasting; state-owned radio network is supplemented by a small number of privately owned broadcast stations as well as a few community radio stations; transmissions of at least 2 international broadcasters are available (2017)

Internet country code: .cf

Internet users: *total:* 249,336

percent of population: 4.34% (July 2018 est.)
country comparison to the world: 169

Broadband—fixed subscriptions: *total:* 608
subscriptions per 100 inhabitants: less than 1 (2018 est.)
country comparison to the world: 199

TRANSPORTATION

National air transport system: *number of registered air carriers:* 2 (2020)
inventory of registered aircraft operated by air carriers: 2
annual passenger traffic on registered air carriers: 46,364 (2015)
annual freight traffic on registered air carriers: 0 mt-km (2015)

Civil aircraft registration country code prefix: TL (2016)

Airports: 39 (2013)
country comparison to the world: 106

Airports—with paved runways: *total:* 1 (2019)
2,438 to 3,047 m: 1

Airports—with unpaved runways: *total:* 37 (2013)
2,438 to 3,047 m: 1 (2013)
1,524 to 2,437 m: 11 (2013)
914 to 1,523 m: 19 (2013)
under 914 m: 6 (2013)

Roadways: *total:* 24,000 km (2018)
paved: 700 km (2018)
unpaved: 23,300 km (2018)
country comparison to the world: 106

Waterways: 2,800 km (the primary navigable river is the Ubangi, which joins the River Congo; it was the traditional route for the export of products because it connected with the Congo-Ocean railway at Brazzaville; because of the warfare on both sides of the River Congo from 1997, importers and exporters preferred routes through Cameroon) (2011)
country comparison to the world: 34
Ports and terminals: river port(s): Bangui (Oubangui) Nola (Sangha)

MILITARY AND SECURITY

Military and security forces: Central African Armed Forces (Forces Armees Centrafricaines, FACA): Ground Forces (includes Military Air Service), General Directorate of Gendarmerie Inspection (DGIG); National Police (2019)

Military expenditures: 1.5% of GDP (2019)
1.41% of GDP (2018)
1.44% of GDP (2017)
1.53% of GDP (2016)
1.69% of GDP (2015)
country comparison to the world: 79

Military and security service personnel strengths: the Central African Armed Forces (FACA) have an estimated 8,000 Army troops (including an Air Service component of about 150) and 1,500 Gendarmerie (2019 est.)

Military equipment inventories and acquisitions: the FACA is armed mostly with second-hand equipment from China, Russia, and Ukraine (2020)

Military service age and obligation: 18 years of age for military service; no conscription (2019)

Military—note: the FACA is currently assessed as unable to provide adequate internal security for the country; the military was dissolved following the 2013 rebel seizure of the government and has struggled to rebuild in the years of instability since; France, Russia, the UN, and the European Union are providing various levels of security assistance

the UN Multidimensional Integrated Stabilization Mission in the Central African Republic (MINUSCA) has operated in the country since 2014; its peacekeeping mission includes providing security, protecting civilians, facilitating humanitarian assistance, disarming and demobilizing armed groups, and supporting the country's fragile transitional government; in

November 2019, the UN Security Council extended the mandate of the MINUSCA peacekeeping mission another year; as of March 2020, MINUSCA had approximately 13,200 total personnel, including about 10,700 troops and 2,000 police

the European Union Training Mission in the Central African Republic (EUTM-RCA) has operated in the country since 2016; the EUTM-RCA contributes to the restructuring of the country's military and defense sector through advice, training, and educational programs (2020)

TRANSNATIONAL ISSUES

Disputes—international: periodic skirmishes persist over water and grazing rights among related pastoral populations along the border with southern Sudan

Refugees and internally displaced persons: *refugees (country of origin):* 5,555 (Democratic Republic of Congo) (2020)
IDPs: 684,004 (clashes between army and rebel groups since 2005; tensions between ethnic groups) (2020)

Trafficking in persons: current situation: Central African Republic (CAR) is a source, transit, and destination country for children subjected to forced labor and sex trafficking, women subjected to forced prostitution, and adults subjected to forced labor; most victims appear to be CAR citizens exploited within the country, with a smaller number transported back and forth between the CAR and nearby countries; armed groups operating in the CAR, including those aligned with the former SELEKA Government and the Lord's Resistance Army, continue to recruit and re-recruit children for military activities and labor; children are also subject to domestic servitude, commercial sexual exploitation, and forced labor in agriculture, mines, shops, and street vending; women and girls are subject to domestic servitude, sexual slavery, commercial sexual exploitation, and forced marriage

tier rating: Tier 3—the Central African Republic does not fully comply with the minimum standards for the elimination of trafficking and is not making significant efforts to do so; the government conducted a limited number of investigations and prosecutions of cases of suspected human trafficking in 2014 but did not identify, provide protection to, or refer to care providers any trafficking victims; the government did not directly provide reintegration programs for demobilized child soldiers, leaving victims vulnerable to further exploitation or retrafficking by armed groups, including those affiliated with the government; in 2014, an NGO and the government began drafting a national action plan against trafficking but no efforts were reported to establish a policy against child soldiering or to raise awareness about existing laws prohibiting the use of children in the armed forces (2015)

CHAD

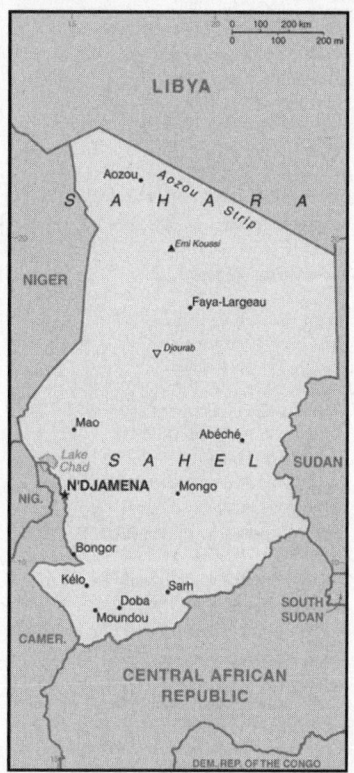

INTRODUCTION

Background: The Kanem Empire (c.700-1380) and its successor the Bornu Empire (1380s-1893) existed in Chad's southern Sahelian strip and focused on controlling the trans-Saharan trade routes that passed through the region. By 1920, France conquered the territory and incorporated it as part of French Equatorial Africa. Chad attained independence in 1960, but then endured three decades of civil warfare, as well as invasions by Libya, before peace was restored in 1990. The government eventually drafted a democratic constitution and held flawed presidential elections in 1996 and 2001. In 1998, a rebellion broke out in northern Chad, which has sporadically flared up despite several peace agreements between the government and insurgents. In June 2005, President Idriss DEBY held a referendum successfully removing constitutional term limits and won another controversial election in 2006. Sporadic rebel campaigns continued throughout 2006 and 2007. The capital experienced a significant insurrection in early 2008, but has had no significant rebel threats since then, in part due to Chad's 2010 rapprochement with Sudan, which previously used Chadian rebels as proxies. Nevertheless, a state of emergency continues to be in place in the Sila and Ouaddai regions bordering Sudan. In late 2015, the government imposed a state of emergency in the Lake Chad region following multiple attacks by the terrorist group Boko Haram throughout the year; Boko Haram also launched several bombings in N'Djamena in mid-2015. A state of emergency is also emplaced in the western Tibesti region bordering Niger where rival ethnic groups are fighting. DEBY in 2016 was reelected to his fifth term in an election that was peaceful but flawed. A new constitution promulgated in 2018 allows DEBY to run for two additional consecutive terms of six years when his current term comes to an end in 2021. As of 2020, the country continued to face multiple challenges, including widespread poverty, an economy severely weakened by the drop in international oil prices, and insurgencies led by rebel militants in the north and Boko Haram in the Lake Chad Basin. In late 2019, the government was forced to declare a state of emergency in three eastern provinces for four months to stop a cycle of interethnic violence, and the army has suffered heavy losses to Islamic terror groups in the Lake Chad area. In March 2020, Boko Haram fighters attacked a Chadian military camp in the Lake Chad region, killing nearly 100 soldiers; it was the deadliest attack in the history of the Chadian military. (2019)

GEOGRAPHY

Location: Central Africa, south of Libya

Geographic coordinates: 15 00 N, 19 00 E

Map references: Africa

Area: *total:* 1.284 million sq km
land: 1,259,200 sq km
water: 24,800 sq km
country comparison to the world: 22

Area—comparative: almost nine times the size of New York state; slightly more than three times the size of California

Land boundaries: *total:* 6,406 km
border countries (6): Cameroon 1116 km, Central African Republic 1556 km, Libya 1050 km, Niger 1196 km, Nigeria 85 km, Sudan 1403 km

Coastline: 0 km (landlocked)

Maritime claims: none (landlocked)

Climate: tropical in south, desert in north

Terrain: broad, arid plains in center, desert in north, mountains in northwest, lowlands in south

Elevation: *mean elevation:* 543 m
lowest point: Djourab 160 m
highest point: Emi Koussi 3,445 m

Natural resources: petroleum, uranium, natron, kaolin, fish (Lake Chad), gold, limestone, sand and gravel, salt

Land use: *agricultural land:* 39.6% (2011 est.)
arable land: 3.9% (2011 est.) / permanent crops: 0% (2011 est.) / permanent pasture: 35.7% (2011 est.)
forest: 9.1% (2011 est.)
other: 51.3% (2011 est.)
Irrigated land: 300 sq km (2012)

Population distribution: the population is unevenly distributed due to contrasts in climate and physical geography; the highest density is found in the southwest, particularly around Lake Chad and points south; the dry Saharan zone to the north is the least densely populated as shown in this population distribution map

Natural hazards: hot, dry, dusty harmattan winds occur in north; periodic droughts; locust plagues

Environment—current issues: inadequate supplies of potable water; improper waste disposal in rural areas and poor farming practices contribute to soil and water pollution; desertification

Environment—international agreements: *party to:* Biodiversity, Climate Change, Desertification, Endangered Species, Hazardous Wastes, Ozone Layer Protection, Wetlands
signed, but not ratified: Law of the Sea, Marine Dumping

Geography—note: *note 1:* Chad is the largest of Africa's 16 landlocked countries
note 2: not long ago—geologically speaking—what is today the Sahara was green savannah teeming with wildlife; during the African Humid Period, roughly 11,000 to 5,000 years ago, a vibrant animal community, including elephants, giraffes, hippos, and antelope lived there; the last remnant of the " Green Sahara" exists in the Lakes of Ounianga (oo-nee-ahn-ga) in northern Chad, a series of 18 interconnected freshwater, saline, and hypersaline lakes now protected as a World Heritage site
note 3: Lake Chad, the most significant water body in the Sahel, is a remnant of a former inland sea, paleolake Mega-Chad; at its greatest extent, sometime before 5000 B.C., Lake Mega-Chad was

187

the largest of four Saharan paleolakes that existed during the African Humid Period; it covered an area of about 400,000 sq km (150,000 sq mi), roughly the size of today's Caspian Sea

PEOPLE AND SOCIETY

Population: 16,877,357 (July 2020 est.)
country comparison to the world: 71

Nationality: *noun:* Chadian(s)
adjective: Chadian

Ethnic groups: Sara (Ngambaye/Sara/Madjingaye/Mbaye) 30.5%, Kanembu/Bornu/Buduma 9.8%, Arab 9.7%, Wadai/Maba/Masalit/Mimi 7%, Gorane 5.8%, Masa/Musseye/Musgum 4.9%, Bulala/Medogo/Kuka 3.7%, Marba/Lele/Mesme 3.5%, Mundang 2.7%, Bidiyo/Migaama/Kenga/Dangleat 2.5%, Dadjo/Kibet/Muro 2.4%, Tupuri/Kera 2%, Gabri/Kabalaye/Nanchere/Somrai 2%, Fulani/Fulbe/Bodore 1.8%, Karo/Zime/Peve 1.3%, Baguirmi/Barma 1.2%, Zaghawa/Bideyat/Kobe 1.1%,Tama/Assongori/Mararit 1.1%, Mesmedje/Massalat/Kadjakse 0.8%, other Chadian ethnicities 3.4%, Chadians of foreign ethnicities 0.9%, foreign nationals 0.3%, unspecified 1.7% (2014-15 est.)

Languages: French (official), Arabic (official), Sara (in south), more than 120 different languages and dialects

Religions: Muslim 52.1%, Protestant 23.9%, Roman Catholic 20%, animist 0.3%, other Christian 0.2%, none 2.8%, unspecified 0.7% (2014-15 est.)

Demographic profile: Despite the start of oil production in 2003, 40% of Chad's population lives below the poverty line. The population will continue to grow rapidly because of the country's very high fertility rate and large youth cohort—more than 65%of the populace is under the age of 25—although the mortality rate is high and life expectancy is low. Chad has the world's third highest maternal mortality rate. Among the primary risk factors are poverty, anemia, rural habitation, high fertility, poor education, and a lack of access to family planning and obstetric care. Impoverished, uneducated adolescents living in rural areas are most affected. To improve women's reproductive health and reduce fertility, Chad will need to increase women's educational attainment, job participation, and knowledge of and access to family planning. Only about a quarter of women are literate, less than 5% use contraceptives, and more than 40%undergo genital cutting.

As of October 2017, more than 320,000 refugees from Sudan and more than 75,000 from the Central African Republic strain Chad's limited resources and create tensions in host communities. Thousands of new refugees fled to Chad in 2013 to escape worsening violence in the Darfur region of Sudan. The large refugee populations are hesitant to return to their home countries because of continued instability. Chad was relatively stable in 2012 in comparison to other states in the region, but past fighting between government forces and opposition groups and inter-communal violence

have left nearly 60,000 of its citizens displaced in the eastern part of the country.

Age structure: *0-14 years:* 47.43% (male 4,050,505/female 3,954,413)
*15-24 years:*19.77% (male 1,676,495/female 1,660,417)
*25-54 years:*27.14% (male 2,208,181/female 2,371,490)
55-64 years: 3.24% (male 239,634/female 306,477)
65 years and over: 2.43% (male 176,658/female 233,087) (2020 est.)

Dependency ratios: *total dependency ratio:* 96
youth dependency ratio: 91.1
elderly dependency ratio: 4.9
potential support ratio: 20.4 (2020 est.)

Median age: *total:* 16.1 years
male: 15.6 years
female: 16.5 years (2020 est.)
country comparison to the world: 224

Population growth rate: 3.18% (2020 est.)
country comparison to the world: 7

Birth rate: 41.7 births/1,000 population (2020 est.)
country comparison to the world: 6

Death rate: 10 deaths/1,000 population (2020 est.)
country comparison to the world: 38

Net migration rate:—0.1 migrant(s)/1,000 population (2020 est.)
country comparison to the world: 102

Population distribution: the population is unevenly distributed due to contrasts in climate and physical geography; the highest density is found in the southwest, particularly around Lake Chad and points south; the dry Saharan zone to the north is the least densely populated as shown in this population distribution map

Urbanization: *urban population:* 23.5% of total population (2020)
rate of urbanization: 3.88% annual rate of change (2015-20 est.)

total population growth rate v. urban population growth rate, 2000-2030: Major urban areas—population: 1.423 million N'DJAMENA (capital) (2020)

Sex ratio: *at birth:* 1.04 male(s)/female
0-14 years: 1.02 male(s)/female
15-24 years: 1.01 male(s)/female
25-54 years: 0.93 male(s)/female
55-64 years: 0.78 male(s)/female
65 years and over: 0.76 male(s)/female
total population: 0.98 male(s)/female (2020 est.)

Mother's mean age at first birth: 17.9 years (2014/15 est.)
note: median age at first birth among women 25-29

Maternal mortality rate: 1,140 deaths/100,000 live births (2017 est.)
country comparison to the world: 2

Infant mortality rate: *total:* 68.6 deaths/1,000 live births
male: 74.5 deaths/1,000 live births
female: 62.5 deaths/1,000 live births (2020 est.)
country comparison to the world: 5

Life expectancy at birth: *total population:* 58.3 years
male: 56.5 years
female: 60.1 years (2020 est.)
country comparison to the world: 221

Total fertility rate: 5.68 children born/woman (2020 est.)
country comparison to the world: 5
Contraceptive prevalence rate: 5.7% (2014/15)

Drinking water source:
improved:
urban: 86.7% of population
rural: 46.6% of population
total: 55.7% of population
unimproved:
urban: 13.3% of population
rural: 53.4% of population
total: 44.3% of population (2017 est.)

Current Health Expenditure: 4.5% (2017)

Physicians density: 0.04 physicians/1,000 population (2017)

Sanitation facility access:
improved:
urban: 56.5% of population
rural: 3.1% of population
total: 15.3% of population
unimproved:
urban: 43.5% of population
rural: 96.9% of population
total: 84.7% of population (2017 est.)

HIV/AIDS—adult prevalence rate: 1.2% (2019 est.)
country comparison to the world: 38

HIV/AIDS—people living with HIV/AIDS: 120,000 (2019 est.)
country comparison to the world: 40

HIV/AIDS—deaths: 3,200 (2019 est.)
country comparison to the world: 32

Major infectious diseases: *degree of risk:* very high (2020)
food or waterborne diseases: bacterial and protozoal diarrhea, hepatitis A and E, and typhoid fever
vectorborne diseases: malaria and dengue fever
water contact diseases: schistosomiasis
animal contact diseases: rabies
respiratory diseases: meningococcal meningitis

Obesity—adult prevalence rate: 6.1% (2016)
country comparison to the world: 170

Children under the age of 5 years underweight: 29.4% (2015)
country comparison to the world: 7

Education expenditures: 2.9% of GDP (2013)
country comparison to the world: 140

Literacy: *definition:* age 15 and over can read and write French or Arabic
total population: 22.3%
male: 31.3%
female: 14% (2016)

School life expectancy (primary to tertiary education): *total:* 7 years
male: 9 years
female: 6 years (2015)

GOVERNMENT

Country name: *conventional long form:* Republic of Chad
conventional short form: Chad
local long form: Republique du Tchad/Jumhuriyat Tshad
local short form: Tchad/Tshad
etymology: named for Lake Chad, which lies along the country's western border; the word "tsade" means "large body of water" or "lake" in several local native languages
note: the only country whose name is composed of a single syllable with a single vowel

Government type: presidential republic

Capital: *name:* N'Djamena
geographic coordinates: 12 06 N, 15 02 E
time difference: UTC+1 (6 hours ahead of Washington, DC, during Standard Time)
etymology: name taken from the Arab name of a nearby village, Nijamina, meaning "place of rest"

Administrative divisions: 23 regions (regions, singular—region); Barh-El-Gazel, Batha, Borkou, Chari-Baguirmi, Ennedi-Est, Ennedi-Ouest, Guera, Hadjer-Lamis, Kanem, Lac, Logone Occidental, Logone Oriental, Mandoul, Mayo-Kebbi-Est, Mayo-Kebbi-Ouest, Moyen-Chari, N'Djamena, Ouaddai, Salamat, Sila, Tandjile, Tibesti, Wadi-Fira

Independence: 11 August 1960 (from France)

National holiday: Independence Day, 11 August (1960)

Constitution: *history:* several previous; latest approved 30 April 2018 by the National Assembly, entered into force 4 May 2018
amendments: proposed as a revision by the president of the republic after a Council of Ministers (cabinet) decision or by the National Assembly; approval for consideration of a revision requires at least three-fifths majority vote by the Assembly; passage requires approval by referendum or at least two-thirds majority vote by the Assembly; amended 2005, 2013

Legal system: mixed legal system of civil and customary law

International law organization participation: has not submitted an ICJ jurisdiction declaration; accepts ICCt jurisdiction

Citizenship: *citizenship by birth:* no
citizenship by descent only: both parents must be citizens of Chad
dual citizenship recognized: Chadian law does not address dual citizenship
residency requirement for naturalization: 15 years

Suffrage: 18 years of age; universal

Executive branch: *chief of state:* President Idriss DEBY Itno, Lt. Gen. (since 4 December 1990)

head of government: President Idriss DEBY Itno, Lt. Gen. (since 4 December 1990); prime minister position eliminated under the 2018 constitution
cabinet: Council of Ministers

elections/appointments: president directly elected by absolute majority popular vote in 2 rounds if needed for a 5-year term (no term limits); election last held on 10 April 2016 (next to be held in April 2021)
election results: Lt. Gen. Idriss DEBY Itno reelected president in first round; percent of vote—Lt. Gen. Idriss DEBY (MPS) 61.6%, Saleh KEBZABO (UNDR) 12.8%, Laokein Kourayo MEDAR (CTPD) 10.7%, Djimrangar DADNADJI (CAP-SUR) 5.1%, other 9.8%

Legislative branch: *description:* unicameral National Assembly (188 seats); 163 directly elected in multi-seat constituencies by proportional representation vote and 25 directly elected in single-seat constituencies by absolute majority vote with a second round if needed; members serve 4-year terms)
elections: last held on 13 February and 6 May 2011 (next originally scheduled on 13 December 2020 but postponed due to the COVID-19 pandemic)
election results: percent of vote by party—NA; seats by party—MPS 117, UNDR 10, RDP 9, RNDT/Le Reveil 8, URD 8, Viva- RNDP 5, FAR 4, CTPD 2, PDSA 2, PUR 2, UDR 2, other 19; composition—men 164, women 24, percent of women 12.8%
note: the National Assembly mandate was extended to 2020, reportedly due to a lack of funding for the scheduled 2015 election; the MPS has held a majority in the NA since 1997

Judicial branch: *highest courts:* Supreme Court (consists of the chief justice, 3 chamber presidents, and 12 judges or councilors and divided into 3 chambers); Constitutional Council (consists of 3 judges and 6 jurists)
judge selection and term of office: Supreme Court chief justice selected by the president; councilors—8 designated by the president and 7 by the speaker of the National Assembly; chief justice and councilors appointed for life; Constitutional Council judges—2 appointed by the president and 1 by the speaker of the National Assembly; jurists—3 each by the president and by the speaker of the National Assembly; judges appointed for 9-year terms
subordinate courts: High Court of Justice; Courts of Appeal; tribunals; justices of the peace

Political parties and leaders: Chadian Convention for Peace and Development or CTPD [Laoukein Kourayo MEDAR]
Federation Action for the Republic or FAR [Ngarledjy YORONGAR]
Framework of Popular Action for Solidarity and Unity of the Republic or CAP- SUR [Joseph Djimrangar DADNADJI]
National Rally for Development and Progress or Viva-RNDP [Dr. Nouradine Delwa Kassire COUMAKOYE]
National Union for Democracy and Renewal or UNDR [Saleh KEBZABO]
Party for Liberty and Development or PLD [Ahmat ALHABO]
Party for Unity and Reconciliation

Patriotic Salvation Movement or MPS [Idriss DEBY]
Rally for Democracy and Progress or RDP [Mahamat Allahou TAHER] RNDT/Le Reveil [Albert Pahimi PADACKE]
Social Democratic Party for a Change-over of Power or PDSA [Malloum YOBODA]
Union for Renewal and Democracy or URD [Felix Romadoumngar NIALBE]

International organization participation: ACP, AfDB, AU, BDEAC, CEMAC, EITI (compliant country), FAO, FZ, G-77, IAEA, IBRD, ICAO, ICCt, ICRM, IDA, IDB, IFAD, IFC, IFRCS, ILO, IMF, Interpol, IOC, IOM, IPU, ITSO, ITU, ITUC (NGOs), MIGA, MINUSMA, NAM, OIC, OIF, OPCW, UN, UNCTAD, UNESCO, UNIDO, UNOCI, UNWTO, UPU, WCO, WHO, WIPO, WMO, WTO

Diplomatic representation in the US: *chief of mission:* Ambassador Ngote Gali KOUTOU (since 22 June 2018)
chancery: 2401 Massachusetts Avenue NW, Washington, DC 20008
telephone: [1] (202) 652-1312
FAX: [1] (202) 758-0431

Diplomatic representation from the US: *chief of mission:* Ambassador (vacant); Charge d'Affaires Thomas R. GENTON (since 16 August 2019)
telephone: [235] 2251-5017
embassy: US Embassy N'Djamena, B.P. 413, N'Djamena
mailing address: B.P. 413, N'Djamena
FAX: [235] 2253-9102

Flag description: three equal vertical bands of blue (hoist side), gold, and red; the flag combines the blue and red French (former colonial) colors with the red and yellow (gold) of the Pan-African colors; blue symbolizes the sky, hope, and the south of the country, which is relatively well-watered; gold represents the sun, as well as the desert in the north of the country; red stands for progress, unity, and sacrifice
note: almost identical to the flag of Romania but with a darker shade of blue; also similar to the flags of Andorra and Moldova, both of which have a national coat of arms centered in the yellow band; design based on the flag of France

National symbol(s): goat (north), lion (south); national colors: blue, yellow, red

National anthem: *name:* "La Tchadienne" (The Chadian)
lyrics/music: Louis GIDROL and his students/ Paul VILLARD
note: adopted1960

ECONOMY

Economy—overview: Chad's landlocked location results in high transportation costs for imported goods and dependence on neighboring countries. Oil and agriculture are mainstays of Chad's economy. Oil provides about 60% of export revenues, while cotton, cattle, livestock, and gum arabic provide the bulk of Chad's non-oil export earnings. The services sector contributes less than one-third of GDP and has attracted foreign investment mostly through telecommunications and banking.

Nearly all of Chad's fuel is provided by one domestic refinery, and unanticipated shutdowns occasionally result in shortages. The country regulates the price of domestic fuel, providing an incentive for black market sales.

Although high oil prices and strong local harvests supported the economy in the past, low oil prices now stress Chad's fiscal position and have resulted in significant government cutbacks. Chad relies on foreign assistance and foreign capital for most of its public and private sector investment. Investment in Chad is difficult due to its limited infrastructure, lack of trained workers, extensive government bureaucracy, and corruption. Chad obtained a three-year extended credit facility from the IMF in 2014 and was granted debt relief under the Heavily Indebted Poor Countries Initiative in April 2015.

In 2018, economic policy will be driven by efforts that started in 2016 to reverse the recession and to repair damage to public finances and exports. The government is implementing an emergency action plan to counterbalance the drop in oil revenue and to diversify the economy. Chad's national development plan (NDP) cost just over $9 billion with a financing gap of $ 6.7 billion. The NDP emphasized the importance of private sector participation in Chad's development, as well as the need to improve the business environment, particularly in priority sectors such as mining and agriculture.

The Government of Chad reached a deal with Glencore and four other banks on the restructuring of a $1.45 billion oil-backed loan in February 2018, after a long negotiation. The new terms include an extension of the maturity to 2030 from 2022, a two-year grace period on principal repayments, and a lower interest rate of the London Inter-bank Offer Rate (Libor) plus 2%—down from Libor plus 7.5%. The original Glencore loan was to be repaid with crude oil assets, however, Chad's oil sales were hit by the downturn in the price of oil. Chad had secured a $312 million credit from the IMF in June 2017, but release of those funds hinged on restructuring the Glencore debt. Chad had already cut public spending to try to meet the terms of the IMF program, but that prompted strikes and protests in a country where nearly 40% of the population lives below the poverty line. Multinational partners, such as the African Development Bank, the EU, and the World Bank are likely to continue budget support in 2018,but Chad will remain at high debt risk, given its dependence on oil revenue and pressure to spend on subsidies and security.

GDP (purchasing power parity): $28.62 billion (2017 est.)
$29.55 billion (2016 est.)
$31.58 billion (2015 est.)
note: data are in 2017 dollars
country comparison to the world: 134

GDP (official exchange rate): $9.872 billion (2017 est.)

GDP—real growth rate: -3.1% (2017 est.)
-6.4% (2016 est.)
1.8% (2015 est.)
country comparison to the world: 211

GDP—per capita (PPP): $2,300 (2017 est.)
$2,500 (2016 est.)
$2,700 (2015 est.)
note: data are in 2017 dollars
country comparison to the world: 202

Gross national saving: 15.5% of GDP (2017 est.)
7.5% of GDP (2016 est.)
13.3% of GDP (2015 est.)
country comparison to the world: 133

GDP—composition, by end use: household consumption: 75.1% (2017 est.)
government consumption: 4.4% (2017 est.)
investment in fixed capital: 24.1% (2017 est.)
investment in inventories: 0.7% (2017 est.)
exports of goods and services: 35.1% (2017 est.)
imports of goods and services: -39.4% (2017 est.)

GDP—composition, by sector of origin: agriculture: 52.3% (2017 est.)
industry: 14.7% (2017 est.)
services: 33.1% (2017 est.)

Agriculture—products: cotton, sorghum, millet, peanuts, sesame, corn, rice, potatoes, onions, cassava (manioc, tapioca), cattle, sheep, goats, camels

Industries: oil, cotton textiles, brewing, natron (sodium carbonate), soap, cigarettes, construction materials

Industrial production growth rate: -4% (2017 est.)
country comparison to the world: 192

Labor force: 5.654 million (2017 est.)
country comparison to the world: 72

Labor force—by occupation: agriculture: 80%
industry: 20% (2006 est.)

Unemployment rate: NA

Population below poverty line: 46.7% (2011 est.)

Household income or consumption by percentage share: lowest 10%: 2.6%
highest 10%: 30.8% (2003)

Budget: revenues: 1.337 billion (2017 est.)
expenditures: 1.481 billion (2017 est.)

Taxes and other revenues: 13.5% (of GDP) (2017 est.)
country comparison to the world: 206

Budget surplus (+) or deficit (-): -1.5% (of GDP) (2017 est.)
country comparison to the world: 89

Public debt: 52.5% of GDP (2017 est.)
52.4% of GDP (2016 est.)
country comparison to the world: 94

Fiscal year: calendar year

Inflation rate (consumer prices): -0.9% (2017 est.)
-1.1% (2016 est.)
country comparison to the world: 2

Current account balance: -$558 million (2017 est.)
-$926 million (2016 est.)
country comparison to the world: 124

Exports: $2.464 billion (2017 est.)
$2.187 billion (2016 est.)
country comparison to the world: 132

Exports—partners: US 38.7%, China 16.6%, Netherlands 15.7%, UAE 12.2%, India 6.3% (2017)

Exports—commodities: oil, livestock, cotton, sesame, gum arabic, shea butter

Imports: $2.16 billion (2017 est.)
$1.997 billion (2016 est.)
country comparison to the world: 164

Imports—commodities: machinery and transportation equipment, industrial goods, foodstuffs,textiles

Imports—partners: China 19.9%, Cameroon 17.2%, France 17%, US 5.4%, India 4.9%, Senegal 4.5% (2017)

Reserves of foreign exchange and gold: $22.9 million (31 December 2017 est.)
$20.92 million (31 December 2016 est.)
country comparison to the world: 190

Debt—external: $1.724 billion (31 December 2017 est.)
$1.281 billion (31 December 2016 est.)
country comparison to the world: 153

Exchange rates: Cooperation Financiere en Afrique Centrale francs (XAF) per US dollar - 605.3 (2017 est.)
593.01 (2016 est.)
593.01 (2015 est.)
591.45 (2014 est.)
494.42 (2013 est.)

ENERGY

Electricity access: population without electricity: 15 million (2019)
electrification—total population: 9% (2019)
electrification—urban areas: 32% (2019)
electrification—rural areas: 1% (2019)

Electricity—production: 224.3 million kWh (2016 est.)
country comparison to the world:190

Electricity—consumption: 208.6 million kWh (2016 est.)
country comparison to the world: 192

Electricity—exports: 0 kWh (2016 est.)
country comparison to the world: 120

Electricity—imports: 0 kWh (2016 est.)
country comparison to the world: 134

Electricity—installed generating capacity: 48,200 kW (2016 est.)
country comparison to the world: 192

Electricity—from fossil fuels: 98% of total installed capacity (2016 est.)
country comparison to the world: 28

Electricity—from nuclear fuels: 0% of total installed capacity (2017 est.)
country comparison to the world: 65

Electricity—from hydroelectric plants: 0% of total installed capacity (2017 est.)
country comparison to the world: 164

Electricity—from other renewable sources: 3% of total installed capacity (2017 est.)

*country comparison to the world:*123

Crude oil—production: 132,000 bbl/day (2018 est.)
country comparison to the world: 40

Crude oil—exports: 70,440 bbl/day (2015 est.)
country comparison to the world: 37

Crude oil—imports: 0 bbl/day (2015 est.)
country comparison to the world: 109

Crude oil—proved reserves: 1.5 billion bbl (1 January 2018 est.)
country comparison to the world: 38

Refined petroleum products—production: 0 bbl/day (2015 est.)
country comparison to the world: 130

Refined petroleum products—consumption: 2,300 bbl/day (2016 est.)
country comparison to the world: 193

Refined petroleum products—exports: 0 bbl/day (2015 est.)
country comparison to the world: 143

Refined petroleum products—imports: 2,285 bbl/day (2015 est.)
*country comparison to the world:*189

Natural gas—production: 0 cu m (2017 est.)
*country comparison to the world:*116

Natural gas—consumption: 0 cu m (2017 est.)
country comparison to the world: 132

Natural gas—exports: 0 cu m (2017 est.)
*country comparison to the world:*82

Natural gas—imports: 0 cu m (2017 est.)
country comparison to the world: Natural gas—proved reserves:106

0 cu m (1 January 2014 est.)
*country comparison to the world:*122

Carbon dioxide emissions from consumption of energy: 342,200 Mt (2017 est.)
country comparison to the world: 190

COMMUNICATIONS

Telephones—fixed lines: *total subscriptions:* 6,540

subscriptions per 100 inhabitants: less than 1 (2019 est.)
country comparison to the world: 201

Telephones—mobile cellular: *total subscriptions:* 7,857,758

subscriptions per 100 inhabitants: 48.06 (2019 est.)
country comparison to the world: 99

Telecommunication systems: *general assessment:* inadequate system of radio telephone communication stations with high maintenance costs and low telephone density; Chad remains one of the least developed on the African continent, telecom infrastructure is particularly low, with penetration rates in all sectors—fixed, mobile and Internet—well below African averages; low usage also due to 18% excise duty tax on telecom services and a negative impact on operator revenue (2020) *domestic:* fixed-line connections less than 1 per 100 persons, with mobile-cellular subscribership base of about 48 per 100 persons (2019)

international: country code—235; satellite earth station—1 Intelsat (Atlantic Ocean)
note: the COVID-19 outbreak is negatively impacting telecommunications production and supply chains globally; consumer spending on telecom devices and services has also slowed due to the pandemic's effect on economies worldwide; overall progress towards improvements in all facets of the telecom industry—mobile, fixed-line, broadband, submarine cable and domestic networks—has moderated

Broadcast media: 1 state-owned TV station; 2 privately-owned TV stations; state-owned radio network, Radiodiffusion Nationale Tchadienne (RNT), operates national and regional stations; over 10 private radio stations; some stations rebroadcast programs from international broadcasters (2017)

Internet country code: . td

Internet users: *total:* 1,029,153

percent of population: 6.5% (July 2018 est.)
country comparison to the world: 140

Broadband—fixed subscriptions: *total:* 334
subscriptions per 100 inhabitants: less than 1 (2018 est.)
country comparison to the world: 202

TRANSPORTATION

National air transport system: *number of registered air carriers:* 2 (2020)
inventory of registered aircraft operated by air carriers: 3

Airports: 59 (2013)
country comparison to the world: 80

Airports—with paved runways: *total:* 9 (2017)
over 3,047 m: 2 (2017)
2,438 to 3,047 m: 4 (2017)
1,524 to 2,437 m: 2 (2017)
under 914 m: 1 (2017)

Airports—with unpaved runways: *total:* 50 (2013)
over 3,047 m: 1 (2013)
2,438 to 3,047 m: 2 (2013)
1,524 to 2,437 m: 14 (2013)
914 to 1,523 m: 22 (2013)
under 914 m: 11 (2013)

Pipelines: 582 km oil (2013)

Roadways: *total:* 40,000 km (2018)
note: consists of 25,000 km of national and regional roads and 15,000 km of local roads; 206 km of urban roads are paved
country comparison to the world: 89

Waterways: (Chari and Legone Rivers are navigable only in wet season) (2012)

MILITARY AND SECURITY

Military and security forces: Chadian National Army (Armee Nationale du Tchad, ANT): Ground Forces (l'Armee de Terre, AdT), Chadian Air Force (l'Armee de l'Air Tchadienne, AAT), General Direction of the Security Services of State Institutions (Direction Generale des Services de Securite des Institutions de l'Etat, GDSSIE);

National Gendarmerie; National Nomadic Guard of Chad (GNNT) (2019)
note(s): the GDSSIE, formerly known as the Republican Guard, is the presidential guard force and considered an elite military unit; it is comprised of men from President DEBY's own Zaghawa ethnic group; the Chadian Army also includes the US-trained and equipped Special Anti-Terrorist Group (SATG)

Military expenditures: 2.2% of GDP (2019)
2.3% of GDP (2018)
2.2% of GDP (2017)
1.8% of GDP (2016)
2% of GDP (2015)
country comparison to the world: 42

Military and security service personnel strengths: the Chadian National Army (ANT) has approximately 34,000 active personnel (29,000 Ground Forces; 300 Air Force; 4,500 General Direction of the Security Services of State Institutions); 5,000 National Gendarmerie; 3,500 National Nomadic Guard of Chad (2019 est.)

Military equipment inventories and acquisitions: the ANT is mostly armed with older or second-hand equipment from Belgium, France, Russia, and the former Soviet Union; since 2010, the leading suppliers are China, Italy, and Ukraine; the US has also donated equipment (2019)

Military deployments: 1,450 Mali (MINUSMA) (2020)

Military service age and obligation: 20 is the legal minimum age for compulsory military service, with a 3-year service obligation; 18 is the legal minimum age for voluntary service; no minimum age restriction for volunteers with consent from a parent or guardian; women are subject to 1 year of compulsory military or civic service at age 21; while provisions for military service have not been repealed, they have never been fully implemented (2015)

Military—note: the ANT is chiefly focused on counterinsurgency/counter-terrorist operations against Boko Haram (BH) and the Islamic State in West Africa (ISWA) in the Lake Chad Basin area (primarily the Lac Province) and countering the terrorist threat in the Sahel; in 2020, it conducted a large military operation against BH in the Lake Chad region; also in 2020, Chad sent troops to the tri-border area with Burkina Faso, Mali, and Niger to combat ISWA militants

Chad is part of a five-nation anti-jihadist task force known as the G5 Sahel Group, set up in 2014 with Burkina Faso, Mali, Mauritania, and Niger; Chad has committed 550 troops and 100 gendarmes to the force; in early 2020, G5 Sahel military chiefs of staff agreed to allow defense forces from each of the states to pursue terrorist fighters up to 100 km into neighboring countries; the G5 force is backed by the UN, US, and France; G5 troops periodically conduct joint operations with French forces deployed to the Sahel under Operation Barkhane; Chad hosts the headquarters of Operation Barkhane in N'Djamena

Chad has committed approximately 1,000-1,500 troops to the Multinational Joint Task

Force (MNJTF) against Boko Haram; national MNJTF troop contingents are deployed within their own territories, although cross-border operations are conducted periodically; in 2019, Chad sent more than 1,000 troops to Nigeria's Borno State to fight BH as part of the MNJTF mission (2020)

TERRORISM

Terrorist group(s): Boko Haram; Islamic State of Iraq and ash- Sham—West Africa (2020)

note: details about the history, aims, leadership, organization, areas of operation, tactics, targets, weapons, size, and sources of support of the group(s) appear(s) in Appendix-T

TRANSNATIONAL ISSUES

Disputes—international: since 2003, ad hoc armed militia groups and the Sudanese military have driven hundreds of thousands of Darfur residents into Chad; Chad wishes to be a helpful mediator in resolving the Darfur conflict, and in 2010

established a joint border monitoring force with Sudan, which has helped to reduce cross-border banditry and violence; only Nigeria and Cameroon have heeded the Lake Chad Commission's admonition to ratify the delimitation treaty, which also includes the Chad-Niger and Niger-Nigeria boundaries

Refugees and internally displaced persons: refugees (country of origin): 361,945 (Sudan), 95,051 (Central African Republic), 15,843 (Nigeria) (2020)
IDPs: 236,426 (majority are in the east) (2020)

CHILE

INTRODUCTION

Background: Prior to the arrival of the Spanish in the 16th century, the Inca ruled northern Chile for nearly a century while an indigenous people, the Mapuche, inhabited central and southern Chile. Although Chile declared its independence in 1810, it did not achieve decisive victory over the Spanish until 1818. In the War of the Pacific (1879-83), Chile defeated Peru and Bolivia to win its present northern regions. In the 1880s, the

Chilean central government gained control over the central and southern regions inhabited by the Mapuche. After a series of elected governments, the three-year-old Marxist government of Salvador ALLENDE was overthrown in 1973 by a military coup led by General Augusto PINOCHET, who ruled until a democratically-elected president was inaugurated in 1990. Economic reforms, maintained consistently since the 1980s, contributed to steady growth, reduced poverty rates by over half, and helped secure the country's commitment to democratic and representative government. Chile has increasingly assumed regional and international leadership roles befitting its status as a stable, democratic nation.

GEOGRAPHY

Location: Southern South America, bordering the South Pacific Ocean, between Argentina and Peru

Geographic coordinates: 30 00 S, 71 00 W

Map references: South America

Area: *total:* 756,102 sq km
land: 743,812 sq km
water: 12,290 sq km
note: includes Easter Island (Isla de Pascua) and Isla Sala y Gomez
country comparison to the world: 39

Area—comparative: slightly smaller than twice the size of Montana

Land boundaries: *total:* 7,801 km
border countries (3): Argentina 6691 km, Bolivia 942 km, Peru 168 km

Coastline: 6,435 km

Maritime claims: *territorial sea:* 12 nm
exclusive economic zone: 200 nm
contiguous zone: 24 nm
continental shelf: 200/350 nm

Climate: temperate; desert in north; Mediterranean in central region; cool and damp in south

Terrain: low coastal mountains, fertile central valley, rugged Andes in east

Elevation: *mean elevation:* 1,871 m
lowest point: Pacific Ocean 0 m
highest point: Nevado Ojos del Salado 6,880 m

Natural resources: copper, timber, iron ore, nitrates, precious metals, molybdenum, hydropower

Land use: *agricultural land:* 21.1% (2011 est.)
arable land: 1.7% (2011 est.) / permanent crops: 0.6% (2011 est.) / permanent pasture: 18.8% (2011 est.)
forest: 21.9% (2011 est.)
other: 57% (2011 est.)
Irrigated land: 11,100 sq km (2012)

Population distribution: 90% of the population is located in the middle third of the country around the capital of Santiago; the far north (anchored by the Atacama Desert) and the extreme south are relatively underpopulated

Natural hazards: severe earthquakes; active volcanism; tsunamis
volcanism: significant volcanic activity due to more than three-dozen active volcanoes along the Andes Mountains; Lascar (5,592 m), which last erupted in 2007, is the most active volcano in the northern Chilean Andes; Llaima (3,125 m) in central Chile, which last erupted in 2009, is another of the country's most active; Chaiten's 2008 eruption forced major evacuations; other notable historically active volcanoes include Cerro Hudson, Calbuco, Copahue, Guallatiri, Llullaillaco, Nevados de Chillan, Puyehue, San Pedro, and Villarrica; see note 2 under "Geography—note"

Environment—current issues: air pollution from industrial and vehicle emissions; water pollution from raw sewage; noise pollution; improper garbage disposal; soil degradation; widespread deforestation and mining threaten the environment; wildlife conservation

Environment—international agreements: *party to:* Antarctic-Environmental Protocol, Antarctic-Marine Living Resources, Antarctic Seals, Antarctic Treaty, Biodiversity, Climate Change, Climate Change-Kyoto Protocol, Desertification, Endangered Species, Environmental Modification, Hazardous Wastes, Law of the Sea, Marine Dumping, Ozone Layer Protection, Ship Pollution, Wetlands, Whaling
signed, but not ratified: none of the selected agreements

Geography—note: *note 1:* the longest north-south trending country in the world, extending across 39 degrees of latitude; strategic location relative to sea lanes between the Atlantic and Pacific Oceans (Strait of Magellan, Beagle Channel, Drake Passage)

note 2: Chile is one of the countries along the Ring of Fire, a belt of active volcanoes and earthquake epicenters bordering the Pacific Ocean; up to 90% of the world's earthquakes and some 75% of the world's volcanoes occur within the Ring of Fire

note 3: the Atacama Desert—the driest desert in the world—spreads across the northern part of the country; Ojos del Salado (6,893 m) in the Atacama Desert is the highest active volcano in the world, Chile's tallest mountain, and the second highest in the Western Hemisphere and the Southern Hemisphere—its small crater lake (at 6,390 m) is the world's highest lake

PEOPLE AND SOCIETY

Population: 18,186,770 (July 2020 est.)
country comparison to the world: 65

Nationality: *noun:* Chilean(s)
adjective: Chilean

Ethnic groups: white and non-indigenous 88.9%, Mapuche 9.1%, Aymara 0.7%, other indigenous groups 1% (includes Rapa Nui, Likan Antai, Quechua, Colla, Diaguita, Kawesqar, Yagan or Yamana), unspecified 0.3% (2012 est.)

Languages: Spanish 99.5% (official), English 10.2%, indigenous 1% (includes Mapudungun, Aymara, Quechua, Rapa Nui), other 2.3%, unspecified 0.2% (2012 est.)

note: shares sum to more than 100% because some respondents gave more than one answer on the census

Religions: Roman Catholic 66.7%, Evangelical or Protestant 16.4%, Jehovah's Witness 1%, other 3.4%, none 11.5%, unspecified 1.1% (2012 est.)

Demographic profile: Chile is in the advanced stages of demographic transition and is becoming an aging society—with fertility below replacement level, low mortality rates, and life expectancy on par with developed countries. Nevertheless, with its dependency ratio nearing its low point, Chile could benefit from its favorable age structure. It will need to keep its large working-age population productively employed, while preparing to provide for the needs of its growing proportion of elderly people, especially as women—the traditional caregivers—increasingly enter the workforce. Over the last two decades, Chile has made great strides in reducing its poverty rate, which is now lower than most Latin American countries. However, its severe income inequality ranks as the worst among members of the Organization for Economic Cooperation and Development. Unequal access to quality education perpetuates this uneven income distribution.

Chile has historically been a country of emigration but has slowly become more attractive to immigrants since transitioning to democracy in 1990 and improving its economic stability (other regional destinations have concurrently experienced deteriorating economic and political conditions). Most of Chile's small but growing foreign-born population consists of transplants from other Latin American countries, especially Peru.

Age structure: *0-14 years:* 19.79% (male 1,836,240/female 1,763,124)
15-24 years: 13.84% (male 1,283,710/female 1,233,238)
25-54 years: 42.58% (male 3,882,405/female 3,860,700)
55-64 years: 11.98% (male 1,034,049/female 1,145,022)
65 years and over: 11.81% (male 902,392/female 1,245,890) (2020 est.)

Dependency ratios: *total dependency ratio:* 45.9
youth dependency ratio: 28.1
elderly dependency ratio: 17.9
potential support ratio: 5.6 (2020 est.)

Median age: *total:* 35.5 years
male: 34.3 years
female: 36.7 years (2020 est.)
country comparison to the world: 83

Population growth rate: 0.71% (2020 est.)
country comparison to the world: 138

Birth rate: 13.1 births/1,000 population (2020 est.)
country comparison to the world: 142

Death rate: 6.5 deaths/1,000 population (2020 est.)
country comparison to the world: 142

Net migration rate: 0.3 migrant(s)/1,000 population (2020 est.)
country comparison to the world: 70

Population distribution: 90% of the population is located in the middle third of the country around the capital of Santiago; the far north (anchored by the Atacama Desert) and the extreme south are relatively underpopulated

Urbanization: *urban population:* 87.7% of total population (2020)
rate of urbanization: 0.87% annual rate of change (2015-20 est.)

total population growth rate v. urban population growth rate, 2000-2030: Major urban areas—population: 6.767 million SANTIAGO (capital), 984,000 Valparaiso, 881,000 Concepcion (2020)

Sex ratio: *at birth:* 1.04 male(s)/female
0-14 years: 1.04 male(s)/female
15-24 years: 1.04 male(s)/female
25-54 years: 1.01 male(s)/female
55-64 years: 0.9 male(s)/female
65 years and over: 0.72 male(s)/female
total population: 0.97 male(s)/female (2020 est.)

Maternal mortality rate: 13 deaths/100,000 live births (2017 est.)
country comparison to the world: 138

Infant mortality rate: *total:* 6.2 deaths/1,000 live births
male: 6.6 deaths/1,000 live births
female: 5.7 deaths/1,000 live births (2020 est.)
country comparison to the world: 164

Life expectancy at birth: *total population:* 79.4 years
male: 76.3 years
female: 82.5 years (2020 est.)
country comparison to the world: 53

Total fertility rate: 1.77 children born/woman (2020 est.)
country comparison to the world: 153

Contraceptive prevalence rate: 76.3% (2015/16)

Drinking water source:
improved:
urban: 100% of population
rural: 100% of population
total: 100% of population
unimproved:
urban: 0% of population
rural: 0% of population
total: 0% of population (2017 est.)

Current Health Expenditure: 9% (2017)

Physicians density: 2.44 physicians/1,000 population (2017)

Hospital bed density: 2.1 beds/1,000 population (2017)

Sanitation facility access:
improved:
urban: 100% of population
rural: 100% of population
total: 100% of population
unimproved:
urban: 0% of population
rural: 0% of population
total: 0% of population (2017 est.)

HIV/AIDS—adult prevalence rate: 0.5% (2019 est.)
country comparison to the world: 67

HIV/AIDS—people living with HIV/AIDS: 74,000 (201 est.)
country comparison to the world: 54

HIV/AIDS—deaths: <1000 (2018)

Obesity—adult prevalence rate: 28% (2016)
country comparison to the world: 32

Children under the age of 5 years underweight: 0.5% (2014)
country comparison to the world: 129

Education expenditures: 5.4% of GDP (2017)
country comparison to the world: 41

Literacy: *definition:* age 15 and over can read and write
total population: 96.4%
male: 96.3%
female: 96.3% (2017)

School life expectancy (primary to tertiary education): *total:* 17 years
male: 16 years
female: 17 years (2018)

Unemployment, youth ages 15-24: *total:* 18.1%
male: 16.7%
female: 20.2% (2018 est.)
country comparison to the world: 73

GOVERNMENT

Country name: *conventional long form:* Republic of Chile

conventional short form: Chile
local long form: Republica de Chile
local short form: Chile
etymology: derivation of the name is unclear, but it may come from the Mapuche word "chilli" meaning "limit of the earth" or from the Quechua "chiri" meaning "cold"

Government type: presidential republic

Capital: *name:* Santiago; note—Valparaiso is the seat of the national legislature

geographic coordinates: 33 27 S, 70 40 W
time difference: UTC-3 (2 hours ahead of Washington, DC, during Standard Time)
daylight saving time: +1hr, begins second Sunday in August; ends second Sunday in May; note—Punta Arenas observes DST throughout the year
note: Chile has three time zones: the continental portion at UTC-3; the southern Magallanes region, which does not use daylight savings time and remains at UTC-3 for the summer months; and Easter Island at UTC-5
etymology: Santiago is named after the biblical figure Saint James (ca. A.D. 3-44), patron saint of Spain, but especially revered in Galicia; "Santiago" derives from the local Galician evolution of the Vulgar Latin "Sanctu Iacobu"; Valparaiso derives from the Spanish "Valle Paraiso" meaning "Paradise Valley"

Administrative divisions: 16 regions (regiones, singular—region); Aysen, Antofagasta, Araucania, Arica y Parinacota, Atacama, Biobio, Coquimbo, Libertador General Bernardo O'Higgins, Los Lagos, Los Rios, Magallanes y de la Antartica Chilena (Magallanes and Chilean Antarctica), Maule, Nuble, Region Metropolitana (Santiago), Tarapaca, Valparaiso
note: the US does not recognize any claims to Antarctica

Independence: 18 September 1810 (from Spain)

National holiday: Independence Day, 18 September (1810)

Constitution: *history:* many previous; latest adopted 11 September 1980, effective 11 March 1981; a referendum held in late October 2020 approved a referendum on forming a convention to draft a new constittion
amendments: proposed by members of either house of the National Congress or by the president of the republic; passage requires at least three-fifths majority vote of the membership in both houses and approval by the president; passage of amendments to constitutional articles, such as the republican form of government, basic rights and freedoms, the Constitutional Tribunal, electoral justice, the Council of National Security, or the constitutional amendment process, requires at least two-third majority vote by both houses of Congress and approval by the president; the president can opt to hold a referendum when Congress and the president disagree on an amendment; amended many times, last in 2020; note—a referendum on a new constitution scheduled for 26 April 2020 has been postponed due to the COVID-19 pandemic

Legal system: civil law system influenced by several West European civil legal systems; judicial review of legislative acts by the Constitutional Tribunal

International law organization participation: has not submitted an ICJ jurisdiction declaration; accepts ICCt jurisdiction

Citizenship: *citizenship by birth:* yes
citizenship by descent only: yes
dual citizenship recognized: yes
residency requirement for naturalization: 5 years

Suffrage: 18 years of age; universal

Executive branch: *chief of state:* President Sebastian PINERA Echenique (since 11 March 2018); note—the president is both chief of state and head of government

head of government: President Sebastian PINERA Echenique (since 11 March 2018)
cabinet: Cabinet appointed by the president
elections/appointments: president directly elected by absolute majority popular vote in 2 rounds if needed for a single 4-year term; election last held on 19 November 2017 with a runoff held 17 December 2017 (next to be held in November 2021)
election results: Sebastian PINERA Echenique elected president in second round; percent of vote in first round Sebastian PINERA Echenique (independent) 36.6%; Alejandro GUILLIER (independent) 22.7%; Beatriz SANCHEZ (independent) 20.3%; Jose Antonio KAST (independent) 7.9%; Carolina GOIC (PDC) 5.9%; Marco ENRIQUEZ-OMINAMI (PRO) 5.7%; other 0.9%; percent of vote in second round—Sebastian PINERA Echenique 54.6%, Alejandro GUILLIER 45.4%

Legislative branch: *description:* bicameral National Congress or Congreso Nacional consists of:
Senate or Senado (43 seats following the 2017 election; to increase to 50 in 2021); members directly elected in multi-seat constituencies by open party-list proportional representation vote to serve 8-year terms with one-half of the membership renewed every 4 years)
Chamber of Deputies or Camara de Diputados (155 seats; members directly elected in multi-seat constituencies by oen party-list proportional representation vote to serve 4-year terms)
elections: Senate—last held on 19 November 2017 (next to be held in 2021)
Chamber of Deputies—last held on 19 November 2017 (next to be held in 2021)
election results: Senate—percent of vote by party—NA; seats by party—New Majority Coalition (formerly known as Concertacion) 19 (PDC 6, PS 6, PPD 6, MAS 1), Let's Go Chile Coalition (formerly known as the Coalition for Change and the Alianza coalition) 15 (RN 6, UDI 8, Amplitude Party 1), independent 4; composition—men 33, women 10, percent of women 23.3% Chamber of Deputies—percent of vote by party—NA; seats by party—New Majority 68 (PDC 21, PS 16, PPD 14, PC 6, PRSD 6, Citizen Left 1, independent 4), Coalition for Change 47

(UDI 29, RN 14, independent 3, EP 1), Liberal Party 1, independent 4; composition -men 120, women 35, percent of women 22.6%; note—total National Congress percent of women 22.7%

Judicial branch: *highest courts:* Supreme Court or Corte Suprema (consists of a court president and 20 members or ministros); Constitutional Court (consists of 10 members); Elections Qualifying Court (consists of 5 members)
judge selection and term of office: Supreme Court president and judges (ministers) appointed by the president of the republic and ratified by the Senate from lists of candidates provided by the court itself; judges appointed for life with mandatory retirement at age 70; Constitutional Court members appointed—3 by the Supreme Court, 3 by the president of the republic, 2 by the Chamber of Deputies, and 2 by the Senate; members serve 9-year terms with partial membership replacement every 3 years (the court reviews constitutionality of legislation); Elections Qualifying Court members appointed by lottery—1 by the former president or vice president of the Senate and 1 by the former president or vice president of the Chamber of Deputies, 2 by the Supreme Court, and 1 by the Appellate Court of Valparaiso; members appointed for 4-year terms
subordinate courts: Courts of Appeal; oral criminal tribunals; military tribunals; local police courts; specialized tribunals and courts in matters such as family, labor, customs, taxes, and electoral affairs

Political parties and leaders: Amplitude (Amplitud) [Lily PEREZ]
Broad Front Coalition (Frente Amplio) or FA (includes RD, PL, PH, PEV, Igualdad, and Poder) [Beatriz SANCHEZ]
Broad Social Movement of Leftist Citizens (includes former MAS and Izquierda Ciudadana) [Fernando ZAMORANO]
Christian Democratic Party or PDC [Fuad CHAHIN]
Citizen Power (Poder) [Karina OLIVA]
Communist Party of Chile or PC [Guillermo TEILLIER del Valle]
Democratic Revolution or RD [Rodrigo ECHECOPAR]
Equality Party (Igualdad) [Guillermo GONZALEZ]
Green Ecological Party or PEV [Felix GONZALEZ]
Humanist Party or PH [Octavio GONZALEZ]
Independent Democratic Union or UDI [Jacqueline VAN RYSSELBERGHE Herrera]
Independent Regionalist Democratic Party or PRI [Hugo ORTIZ de Filippi]
Let's Go Chile Coalition (Chile Vamos) [Sebastian PINERA]
(includes EVOPOLI, PRI, RN, UDI) Liberal Party (Partido Liberal de Chile) or PL [Luis Felipe RAMOS] National Renewal or RN [Mario DESBORDES]
New Majority Coalition (Nueva Mayoria) [Michelle BACHELET] (includes PDC, PC, PPD, PRSD, PS); note—dissolved in March 2018
Party for Democracy or PPD [Heraldo MUNOZ]
Political Evolution or EVOPOLI [Hernan LARRAIN MATTE]

Progressive Party or PRO [Camilo LAGOS]
Radical Social Democratic Party or PRSD [Carlos
MALDONADO Curti], Socialist Party or PS
[Alvaro ELIZALDE Soto] (formerly known as
Concertacion)

International organization participation: APEC,
BIS, CAN (associate), CD, CELAC, FAO,
G-15, G-77, IADB, IAEA, IBRD, ICAO, ICC
(national committees), ICCt, ICRM, IDA,
IFAD, IFC, IFRCS, IHO, ILO, IMF, IMO, IMSO,
Interpol, IOC, IOM, IPU, ISO, ITSO, ITU,
ITUC (NGOs), LAES, LAIA, Mercosur (associ-
ate), MIGA, MINUSTAH, NAM, OAS, OECD
(enhanced engagement), OPANAL, OPCW,
Pacific Alliance, PCA, SICA (observer), UN,
UNASUR, UNCTAD, UNESCO, UNFICYP,
UNHCR, UNIDO, Union Latina, UNMOGIP,
UNTSO, UNWTO, UPU, WCO, WFTU
(NGOs), WHO, WIPO, WMO, WTO

Diplomatic representation in the US: *chief of
mission:* Ambassador Oscar Alfonso Sebastian
SILVA Navarro (since 17 September 2018)
chancery: 1732 Massachusetts Avenue NW,
Washington, DC 20036
telephone: [1] (202) 785-1746
FAX: [1] (202) 887-5579
consulate(s) general: Chicago, Houston, Los
Angeles, Miami, New York, San Francisco

Diplomatic representation from the US: *chief of
mission:* Ambassador (vacant); Charge d'Affaires
Richard H. GLENN (since August 2020)
telephone: [56] (2) 2330-3000
embassy: Avenida Andres Bello 2800, Las Condes,
Santiago
mailing address: APO AA 34033
FAX: [56] (2) 2330-3710, 2330-3160

Flag description: two equal horizontal bands of
white (top) and red; a blue square the same height
as the white band at the hoist-side end of the
white band; the square bears a white five-pointed
star in the center representing a guide to progress
and honor; blue symbolizes the sky, white is for the
snow-covered Andes, and red represents the blood
spilled to achieve independence
note: design influenced by the US flag

National symbol(s): huemul (mountain deer),
Andean condor; national colors: red, white, blue

National anthem: *name:* "Himno Nacional de
Chile" (National Anthem of Chile)
lyrics/music: Eusebio LILLO Robles and Bernardo
DE VERA y Pintado/Ramon CARNICER y Battle
note: music adopted 1828, original lyrics adopted
1818, adapted lyrics adopted 1847; under Augusto
PINOCHET's military rule, a verse glorifying
the army was added; however, as a protest, some
citizens refused to sing this verse; it was removed
when democracy was restored in 1990
0:00 / 2:02

ECONOMY

Economy—overview: Chile has a market-oriented
economy characterized by a high level of foreign
trade and a reputation for strong financial institutions and sound policy that have given it the

strongest sovereign bond rating in South America.
Exports of goods and services account for approximately one-third of GDP, with commodities
making up some 60% of total exports. Copper is
Chile's top export and provides 20% of government revenue.

From 2003 through 2013, real growth averaged
almost 5% per year, despite a slight contraction in
2009 that resulted from the global financial crisis.
Growth slowed to an estimated 1.4% in 2017. A
continued drop in copper prices prompted Chile
to experience its third consecutive year of slow
growth.

Chile deepened its longstanding commitment
to trade liberalization with the signing of a free
trade agreement with the US, effective 1 January
2004. Chile has 26 trade agreements covering
60 countries including agreements with the EU,
Mercosur, China, India, South Korea, and Mexico.
In May 2010, Chile signed the OECD Convention,
becoming the first South American country to
join the OECD. In October 2015, Chile signed
the Trans-Pacific Partnership trade agreement,
which was finalized as the Comprehensive and
Progressive Trans-Pacific Partnership (CPTPP)
and signed at a ceremony in Chile in March 2018.

The Chilean Government has generally followed a countercyclical fiscal policy, under which
it accumulates surpluses in sovereign wealth funds
during periods of high copper prices and economic
growth, and generally allows deficit spending only
during periods of low copper prices and growth.
As of 31 October 2016, those sovereign wealth
funds—kept mostly outside the country and separate from Central Bank reserves—amounted to
more than $23.5 billion. Chile used these funds
to finance fiscal stimulus packages during the 2009
economic downturn.

In 2014, then-President Michelle BACHELET
introduced tax reforms aimed at delivering her
campaign promise to fight inequality and to provide access to education and health care. The
reforms are expected to generate additional tax
revenues equal to 3% of Chile's GDP, mostly by
increasing corporate tax rates to OECD averages.

GDP (purchasing power parity): $452.1 billion
(2017 est.)
$445.5 billion (2016 est.)
$439.9 billion (2015 est.)
note: data are in 2017 dollars
country comparison to the world: 44

GDP (official exchange rate): $277 billion (2017
est.)

GDP—real growth rate: 1.03% (2019 est.)
4% (2018 est.)
1.41% (2017 est.)
country comparison to the world: 171

GDP—per capita (PPP): $24,600 (2017 est.)
$24,500 (2016 est.)
$24,400 (2015 est.)
note: data are in 2017 dollars
country comparison to the world: 82

Gross national saving: 20.5% of GDP (2017 est.)
20.9% of GDP (2016 est.)
21.4% of GDP (2015 est.)

country comparison to the world: 94

GDP—composition, by end use: *household
consumption:* 62.3% (2017 est.)
government consumption: 14% (2017 est.)
investment in fixed capital: 21.5% (2017 est.)
investment in inventories: 0.5% (2017 est.)
exports of goods and services: 28.7% (2017 est.)
imports of goods and services: -27% (2017 est.)

GDP—composition, by sector of origin:
agriculture: 4.2% (2017 est.)
industry: 32.8% (2017 est.)
services: 63% (2017 est.)

Agriculture—products: grapes, apples, pears,
onions, wheat, corn, oats, peaches, garlic, asparagus, beans; beef, poultry, wool; fish; timber

Industries: copper, lithium, other minerals, foodstuffs, fish processing, iron and steel, wood and
wood products, transport equipment, cement,
textiles

Industrial production growth rate: -0.4% (2017
est.)
country comparison to the world: 170

Labor force: 7.249 million (2020 est.)
country comparison to the world: 63

Labor force—by occupation: *agriculture:* 9.2%
industry: 23.7%
services: 67.1% (2013)

Unemployment rate: 7.22% (2019 est.)
7.33% (2018 est.)
country comparison to the world: 115

Population below poverty line: 14.4% (2013)

**Household income or consumption by percentage
share:** *lowest 10%:* 1.7%
highest 10%: 41.5% (2013 est.)

Budget: *revenues:* 57.75 billion (2017 est.)
expenditures: 65.38 billion (2017 est.)

Taxes and other revenues: 20.8% (of GDP) (2017
est.)
country comparison to the world: 145

Budget surplus (+) or deficit (-): -2.8% (of GDP)
(2017 est.)
country comparison to the world: 124

Public debt: 23.6% of GDP (2017 est.)
21% of GDP (2016 est.)
country comparison to the world: 182

Fiscal year: calendar year

Inflation rate (consumer prices): 2.2% (2017 est.)
3.8% (2016 est.)
country comparison to the world: 113

Current account balance: -$10.933 billion (2019
est.)
-$10.601 billion (2018 est.)
country comparison to the world: 193

Exports: $69.23 billion (2017 est.)
$60.6 billion (2016 est.)
country comparison to the world: 42

Exports—partners: China 27.5%, US 14.5%,
Japan 9.3%, South Korea 6.2%, Brazil 5% (2017)

Exports—commodities: copper, fruit, fish products, paper and pulp, chemicals, wine

Imports: $61.31 billion (2017 est.)

$55.29 billion (2016 est.)
country comparison to the world: 49

Imports—commodities: petroleum and petroleum products, chemicals, electrical and telecommunications equipment, industrial machinery, vehicles, natural gas

Imports—partners: China 23.9%, US 18.1%, Brazil 8.6%, Argentina 4.5%, Germany 4% (2017)

Reserves of foreign exchange and gold: $38.98 billion (31 December 2017 est.)
$40.49 billion (31 December 2016 est.)
country comparison to the world: 44

Debt—external: $183.4 billion (31 December 2017 est.)
$158.1 billion (31 December 2016 est.)
country comparison to the world: 37

Exchange rates: Chilean pesos (CLP) per US dollar—
653.9 (2017 est.)
676.94 (2016 est.)
676.94 (2015 est.)
658.93 (2014 est.)
570.37 (2013 est.)

ENERGY

Electricity access: *electrification—total population:* 100% (2020)

Electricity—production: 76.09 billion kWh (2016 est.)
country comparison to the world: 39

Electricity—consumption: 73.22 billion kWh (2016 est.)
country comparison to the world: 38

Electricity—exports: 0 kWh (2016 est.)
country comparison to the world: 121

Electricity—imports: 0 kWh (2016 est.)
country comparison to the world: 135

Electricity—installed generating capacity: 24.53 million kW (2016 est.)
country comparison to the world: 37

Electricity—from fossil fuels: 59% of total installed capacity (2016 est.)
country comparison to the world: 133

Electricity—from nuclear fuels: 0% of total installed capacity (2017 est.)
country comparison to the world: 66

Electricity—from hydroelectric plants: 26% of total installed capacity (2017 est.)
country comparison to the world: 75

Electricity—from other renewable sources: 15% of total installed capacity (2017 est.)
country comparison to the world: 57

Crude oil—production: 3,000 bbl/day (2018 est.)
country comparison to the world: 84

Crude oil—exports: 0 bbl/day (2017 est.)
country comparison to the world: 106

Crude oil—imports: 169,600 bbl/day (2017 est.)
country comparison to the world: 33

Crude oil—proved reserves: 150 million bbl (1 January 2018 est.)
country comparison to the world: 60

Refined petroleum products—production: 216,200 bbl/day (2017 est.)
country comparison to the world: 49

Refined petroleum products—consumption: 354,500 bbl/day (2017 est.)
country comparison to the world: 39

Refined petroleum products—exports: 7,359 bbl/day (2017 est.)
country comparison to the world: 87

Refined petroleum products—imports: 166,400 bbl/day (2017 est.)
country comparison to the world: 38

Natural gas—production: 1.218 billion cu m (2017 est.)
country comparison to the world: 65

Natural gas—consumption: 5.125 billion cu m (2017 est.)
country comparison to the world: 58

Natural gas—exports: 277.5 million cu m (2017 est.)
country comparison to the world: 43

Natural gas—imports: 4.446 billion cu m (2017 est.)
country comparison to the world: 38

Natural gas—proved reserves: 97.97 billion cu m (1 January 2018 est.)
country comparison to the world: 52

Carbon dioxide emissions from consumption of energy: 88.23 million Mt (2017 est.)
country comparison to the world: 47

COMMUNICATIONS

Telephones—fixed lines: *total subscriptions:* 2,620,195
subscriptions per 100 inhabitants: 14.51 (2019 est.)
country comparison to the world: 48

Telephones—mobile cellular: *total subscriptions:* 23,870,679
subscriptions per 100 inhabitants: 132.19 (2019 est.)
country comparison to the world: 52

Telecommunication systems: *general assessment:* most advanced telecommunications infrastructure in South America; although Chile has one of the highest mobile penetration rates in the region, the number of subscribers has fallen due to subscribers ending multiple SIM card use; the country ranks second highest in South and Central America in terms of available broadband speeds; effective competition in the broadband and mobile sectors; LTE infrastructure is extensive but national plan for 5G services awaits spectrum auctions; during the COVID-19 pandemic Chile provided free access to educational content for about 3 million school pupils (2020)
domestic: number of fixed-line connections have stagnated to 15 per 100 in recent years as mobile-cellular usage continues to increase, reaching 132 telephones per 100 persons; domestic satellite system with 3 earth stations (2019)
international: country code—56; landing points for the Pan-Am, Prat, SAm-1, American

Movil-Telxius West Coast Cable, FOS Quellon-Chacabuco, Fibra Optical Austral, SAC and Curie submarine cables providing links to the US, Caribbean and to Central and South America; satellite earth stations—2 Intelsat (Atlantic Ocean) (2019)
note: the COVID-19 outbreak is negatively impacting telecommunications production and supply chains globally; consumer spending on telecom devices and services has also slowed due to the pandemic's effect on economies worldwide; overall progress towards improvements in all facets of the telecom industry—mobile, fixed-line, broadband, submarine cable and satellite—has moderated

Broadcast media: national and local terrestrial TV channels, coupled with extensive cable TV networks; the state-owned Television Nacional de Chile (TVN) network is self-financed through commercial advertising revenues and is not under direct government control; large number of privately owned TV stations; about 250 radio stations

Internet country code: .cl

Internet users: *total:* 14,757,868

percent of population: 82.33% (July 2018 est.)
country comparison to the world: 45

Broadband—fixed subscriptions: *total:* 3,250,678
subscriptions per 100 inhabitants: 18 (2018 est.)
country comparison to the world: 39

TRANSPORTATION

National air transport system: *number of registered air carriers:* 9 (2020)
inventory of registered aircraft operated by air carriers: 173
annual passenger traffic on registered air carriers: 19,517,185 (2018)
annual freight traffic on registered air carriers: 1,226,440,000 mt-km (2018)

Civil aircraft registration country code prefix: CC (2016)

Airports: 481 (2013)
country comparison to the world: 14

Airports—with paved runways: *total:* 90 (2017)
over 3,047 m: 5 (2017)
2,438 to 3,047 m: 7 (2017)
1,524 to 2,437 m: 23 (2017)
914 to 1,523 m: 31 (2017)
under 914 m: 24 (2017)

Airports—with unpaved runways: *total:* 391 (2013)
2,438 to 3,047 m: 5 (2013)
1,524 to 2,437 m: 11 (2013)
914 to 1,523 m: 56 (2013)
under 914 m: 319 (2013)

Heliports: 1 (2013)

Pipelines: 3160 km gas, 781 km liquid petroleum gas, 985 km oil, 722 km refined products (2013)

Railways: *total:* 7,282 km (2014)
narrow gauge: 3,853.5 km 1.000-m gauge (2014)
broad gauge: 3,428 km 1.676-m gauge (1,691 km electrified) (2014)
country comparison to the world: 30

Roadways: *total:* 77,801 km (2016)
country comparison to the world: 63

Merchant marine: *total:* 221
by type: bulk carrier 7, container ship 5, general cargo 51, oil tanker 14, other 144 (2019)
country comparison to the world: 63

Ports and terminals: *major seaport(s):* Coronel, Huasco, Lirquen, Puerto Ventanas, San Antonio, San Vicente, Valparaiso
container port(s) (TEUs): San Antonio (1,296,890), Valparaiso (1,073,734) (2017)
LNG terminal(s) (import): Mejillones, Quintero

MILITARY AND SECURITY

Military and security forces: Armed Forces of Chile (Fuerzas Armadas de Chile): Chilean Army, Chilean Navy (Armada de Chile, includes Naval Aviation, Marine Corps, and Maritime Territory and Merchant Marine Directorate (Directemar)), Chilean Air Force (Fuerza Aerea de Chile, FACh); Carabineros de Chile (National Police Force) (2020)
note: Carabineros de Chile are responsible to both the Ministry of Defense and the Ministry of Interior

Military expenditures: 1.8% of GDP (2019)
1.9% of GDP (2018)

1.9% of GDP (2017)
1.9% of GDP (2016)
1.9% of GDP (2015)
country comparison to the world: 60

Military and security service personnel strengths: the Armed Forces of Chile have approximately 80,000 active personnel (45,000 Army; 22,000 Navy; 13,000 Air Force); approximately 45,000 Carabineros (2019 est.)

Military equipment inventories and acquisitions: the Chilean military inventory is comprised of a mix of mostly European and US equipment and a limited number of domestically-produced systems; since 2010, France, Germany, the Netherlands, and the US are the leading suppliers; Chile's defense industry produces some military vehicles and naval craft (2019 est.)

Military service age and obligation: 18-45 years of age for voluntary male and female military service, although the right to compulsory recruitment of males 18-45 is retained; service obligation is 12 months for Army and 22 months for Navy and Air Force (2015)

TRANSNATIONAL ISSUES

Disputes—international: Chile and Peru rebuff Bolivia's reactivated claim to restore the Atacama

corridor, ceded to Chile in 1884, but Chile has offered instead unrestricted but not sovereign maritime access through Chile to Bolivian natural gas; Chile rejects Peru's unilateral legislation to change its latitudinal maritime boundary with Chile to an equidistance line with a southwestern axis favoring Peru; in October 2007, Peru took its maritime complaint with Chile to the ICJ; territorial claim in Antarctica (Chilean Antarctic Territory) partially overlaps Argentine and British claims; the joint boundary commission, established by Chile and Argentina in 2001, has yet to map and demarcate the delimited boundary in the inhospitable Andean Southern Ice Field (Campo de Hielo Sur)

Refugees and internally displaced persons: *refugees (country of origin):* 475,688 (Venezuela) (economic and political crisis; includes Venezuelans who have claimed asylum or have received alternative legal stay) (2020)

Illicit drugs: transshipment country for cocaine destined for Europe and the region; some money laundering activity, especially through the Iquique Free Trade Zone; imported precursors passed on to Bolivia; domestic cocaine consumption is rising, making Chile a significant consumer of cocaine

CHINA

1990s, China has increased its global outreach and participation in international organizations.

GEOGRAPHY

Location: Eastern Asia, bordering the East China Sea, Korea Bay, Yellow Sea, and South China Sea, between North Korea and Vietnam

Geographic coordinates: 35 00 N, 105 00 E

Map references: Asia

Area: *total:* 9,596,960 sq km
land: 9,326,410 sq km
water: 270,550 sq km
country comparison to the world: 5

Area—comparative: slightly smaller than the US

Land boundaries: *total:* 22,457 km
border countries (15): Afghanistan 91 km, Bhutan 477 km, Burma 2129 km, India 2659 km, Kazakhstan 1765 km, North Korea 1352 km, Kyrgyzstan 1063 km, Laos 475 km, Mongolia 4630 km, Nepal 1389 km, Pakistan 438 km, Russia (northeast) 4133 km, Russia (northwest) 46 km, Tajikistan 477 km, Vietnam 1297 km

Coastline: 14,500 km

Maritime claims: *territorial sea:* 12 nm
exclusive economic zone: 200 nm
contiguous zone: 24 nm
continental shelf: 200 nm or to the edge of the continental margin

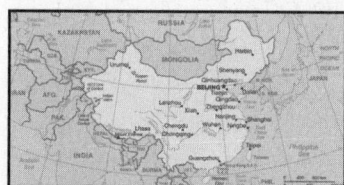

INTRODUCTION

Background: China's historical civilization dates from at least 1200 B.C.; from the 3rd century B.C. and for the next two millennia, China alternated between periods of unity and disunity under a succession of imperial dynasties. In the 19th and early 20th centuries, the country was beset by civil unrest, major famines, military defeats, and foreign occupation. After World War II, the Chinese Communist Party under MAO Zedong established an autocratic socialist system that, while ensuring China's sovereignty, imposed strict controls over everyday life and cost the lives of tens of millions of people. After 1978, MAO's successor DENG Xiaoping and other leaders focused on market-oriented economic development and by 2000 output had quadrupled. For much of the population, living standards have improved dramatically but political controls remain tight. Since the early

Climate: extremely diverse; tropical in south to subarctic in north

Terrain: mostly mountains, high plateaus, deserts in west; plains, deltas, and hills in east

Elevation: *mean elevation:* 1,840 m
lowest point: Turpan Pendi -154 m
highest point: Mount Everest (highest peak in Asia and highest point on earth above sea level) 8,848 m

Natural resources: coal, iron ore, helium, petroleum, natural gas, arsenic, bismuth, cobalt, cadmium, ferrosilicon, gallium, germanium, hafnium, indium, lithium, mercury, tantalum, tellurium, tin, titanium, tungsten, antimony, manganese, magnesium, molybdenum, selenium, strontium, vanadium, magnetite, aluminum, lead, zinc, rare earth elements, uranium, hydropower potential (world's largest), arable land

Land use: *agricultural land:* 54.7% (2011 est.)
arable land: 11.3% (2011 est.) / permanent crops: 1.6% (2011 est.) / permanent pasture: 41.8% (2011 est.)
forest: 22.3% (2011 est.)
other: 23% (2011 est.)
Irrigated land: 690,070 sq km (2012)

Population distribution: overwhelming majority of the population is found in the eastern half of the country; the west, with its vast mountainous and desert areas, remains sparsely populated; though

197

ranked first in the world in total population, overall density is less than that of many other countries in Asia and Europe; high population density is found along the Yangtze and Yellow River valleys, the Xi Jiang River delta, the Sichuan Basin (around Chengdu), in and around Beijing, and the industrial area around Shenyang

Natural hazards: frequent typhoons (about five per year along southern and eastern coasts); damaging floods; tsunamis; earthquakes; droughts; land subsidence

volcanism: China contains some historically active volcanoes including Changbaishan (also known as Baitoushan, Baegdu, or P'aektu-san), Hainan Dao, and Kunlun although most have been relatively inactive in recent centuries

Environment—current issues: air pollution (greenhouse gases, sulfur dioxide particulates) from reliance on coal produces acid rain; China is the world's largest single emitter of carbon dioxide from the burning of fossil fuels; water shortages, particularly in the north; water pollution from untreated wastes; coastal destruction due to land reclamation, industrial development, and aquaculture; deforestation and habitat destruction; poor land management leads to soil erosion, landslides, floods, droughts, dust storms, and desertification; trade in endangered species

Environment—international agreements: *party to:* Antarctic-Environmental Protocol, Antarctic Treaty, Biodiversity, Climate Change, Climate Change-Kyoto Protocol, Desertification, Endangered Species, Environmental Modification, Hazardous Wastes, Law of the Sea, Marine Dumping, Ozone Layer Protection, Ship Pollution, Tropical Timber 83, Tropical Timber 94, Wetlands, Whaling
signed, but not ratified: none of the selected agreements

Geography—note: *note 1:* world's fourth largest country (after Russia, Canada, and US) and largest country situated entirely in Asia; Mount Everest on the border with Nepal is the world's tallest peak above sea level
note 2: the largest cave chamber in the world is the Miao Room, in the Gebihe cave system at China's Ziyun Getu He Chuandong National Park, which encloses some 10.78 million cu m (380.7 million cu ft) of volume
note 3: China appears to have been the center of domestication for two of the world's leading cereal crops: millet in the north along the Yellow River and rice in the south along the lower or middle Yangtze River

PEOPLE AND SOCIETY

Population: 1,394,015,977 (July 2020 est.)
country comparison to the world: 1

Nationality: *noun:* Chinese (singular and plural)
adjective: Chinese

Ethnic groups: Han Chinese 91.6%, Zhuang 1.3%, other (includes Hui, Manchu, Uighur, Miao, Yi, Tujia, Tibetan, Mongol, Dong, Buyei, Yao, Bai, Korean, Hani, Li, Kazakh, Dai, and other nationalities) 7.1% (2010 est.)

note: the Chinese Government officially recognizes 56 ethnic groups

Languages: Standard Chinese or Mandarin (official; Putonghua, based on the Beijing dialect), Yue (Cantonese), Wu (Shanghainese), Minbei (Fuzhou), Minnan (Hokkien-Taiwanese), Xiang, Gan, Hakka dialects, minority languages (see Ethnic groups entry)
note: Zhuang is official in Guangxi Zhuang, Yue is official in Guangdong, Mongolian is official in Nei Mongol, Uighur is official in Xinjiang Uygur, Kyrgyz is official in Xinjiang Uygur, and Tibetan is official in Xizang (Tibet)

Religions: Buddhist 18.2%, Christian 5.1%, Muslim 1.8%, folk religion 21.9%, Hindu < 0.1%, Jewish < 0.1%, other 0.7% (includes Daoist (Taoist)), unaffiliated 52.2% (2010 est.)
note: officially atheist

Age structure: *0-14 years:* 17.29% (male 129,296,339/female 111,782,427)
15-24 years: 11.48% (male 86,129,841/female 73,876,148)
25-54 years: 46.81% (male 333,789,731/female 318,711,557)
55-64 years: 12.08% (male 84,827,645/female 83,557,507)
65 years and over: 12.34% (male 81,586,490/female 90,458,292) (2020 est.)

Dependency ratios: *total dependency ratio:* 42.2
youth dependency ratio: 25.2
elderly dependency ratio: 17
potential support ratio: 5.9 (2020 est.)
data do not include Hong Kong, Macau, and Taiwan

Median age: *total:* 38.4 years
male: 37.5 years
female: 39.4 years (2020 est.)
country comparison to the world: 62

Population growth rate: 0.32% (2020 est.)
country comparison to the world: 170

Birth rate: 11.6 births/1,000 population (2020 est.)
country comparison to the world: 168

Death rate: 8.2 deaths/1,000 population (2020 est.)
country comparison to the world: 85

Net migration rate: -0.4 migrant(s)/1,000 population (2020 est.)
country comparison to the world: 123

Population distribution: overwhelming majority of the population is found in the eastern half of the country; the west, with its vast mountainous and desert areas, remains sparsely populated; though ranked first in the world in total population, overall density is less than that of many other countries in Asia and Europe; high population density is found along the Yangtze and Yellow River valleys, the Xi Jiang River delta, the Sichuan Basin (around Chengdu), in and around Beijing, and the industrial area around Shenyang

Urbanization: *urban population:* 61.4% of total population (2020)
rate of urbanization: 2.42% annual rate of change (2015-20 est.)

note: data do not include Hong Kong and Macau total population growth rate v. urban population growth rate, 2000-2030:

Major urban areas—population: 27.058 million Shanghai, 20.463 million BEIJING (capital), 15.872 million Chongqing, 13.589 million Tianjin, 13.302 million Guangzhou, 12.357 million Shenzhen (2020)

Sex ratio: *at birth:* 1.11 male(s)/female
0-14 years: 1.16 male(s)/female
15-24 years: 1.17 male(s)/female
25-54 years: 1.05 male(s)/female
55-64 years: 1.02 male(s)/female
65 years and over: 0.9 male(s)/female
total population: 1.06 male(s)/female (2020 est.)

Maternal mortality rate: 29 deaths/100,000 live births (2017 est.)
country comparison to the world: 111

Infant mortality rate: *total:* 11.4 deaths/1,000 live births
male: 11.9 deaths/1,000 live births
female: 10.9 deaths/1,000 live births (2020 est.)
country comparison to the world: 115

Life expectancy at birth: *total population:* 76.1 years
male: 74 years
female: 78.4 years (2020 est.)
country comparison to the world: 104

Total fertility rate: 1.6 children born/woman (2020 est.)
country comparison to the world: 184
Contraceptive prevalence rate: 84.5% (2017)

Drinking water source:
improved:
urban: 97.7% of population
rural: 87.8% of population
total: 92.8% of population
unimproved:
urban: 2.3% of population
rural: 12.2% of population
total: 7.2% of population (2017 est.)

Current Health Expenditure: 5.2% (2017)

Physicians density: 1.98 physicians/1,000 population (2017)

Hospital bed density: 4.3 beds/1,000 population (2017)

Sanitation facility access:
improved:
urban: 97.1% of population
rural: 82% of population
total: 90.7% of population
unimproved:
urban: 2.4% of population
rural: 18% of population
total: 9.3% of population (2017 est.)

HIV/AIDS—adult prevalence rate: NA

HIV/AIDS—people living with HIV/AIDS: NA

HIV/AIDS—deaths: NA

Major infectious diseases: *degree of risk:* high (2020)
food or waterborne diseases: bacterial diarrhea, hepatitis A, and typhoid fever

vectorborne diseases: Crimean-Congo hemorrhagic fever, Japanese encephalitis

soil contact diseases: hantaviral hemorrhagic fever with renal syndrome (HFRS)

note: a new coronavirus is causing an outbreak of respiratory illness (COVID-19) in China; illness with this virus has ranged from mild to severe with fatalities reported; the US Department of State has issued a do not travel advisory for China due to COVID-19; the Centers for Disease Control and Prevention has also recommended against travel to China and published additional guidance at https://wwwnc.cdc.gov/travel/notices/warning/novel-coronavirus-china; the US Department of Homeland Security has issued instructions requiring US passengers who have been in China to travel through select airports where the US Government has implemented enhanced screening procedures; as of 10 November 2020, China has reported 92,195 confirmed cases of COVID-19 with 4,748 deaths to the World Health Organization

Obesity—adult prevalence rate: 6.2% (2016)
country comparison to the world: 169

Children under the age of 5 years underweight: 2.4% (2013)
country comparison to the world: 108

Education expenditures: NA

Literacy: *definition:* age 15 and over can read and write
total population: 96.8%
male: 98.5%
female: 95.2% (2018)

School life expectancy (primary to tertiary education): *total:* 14 years
male: 14 years
female: 14 years (2015)

People—note: in October 2015, the Chinese Government announced that it would change its rules to allow all couples to have two children, loosening a 1979 mandate that restricted many couples to one child; the new policy was implemented on 1 January 2016 to address China's rapidly aging population and future economic needs

GOVERNMENT

Country name: *conventional long form:* People's Republic of China
conventional short form: China
local long form: Zhonghua Renmin Gongheguo
local short form: Zhongguo
abbreviation: PRC
etymology: English name derives from the Qin (Chin) rulers of the 3rd century B.C., who comprised the first imperial dynasty of ancient China; the Chinese name Zhongguo translates as "Central Nation" or "Middle Kingdom"

Government type: communist party-led state

Capital: *name:* Beijing

geographic coordinates: 39 55 N, 116 23 E
time difference: UTC+8 (13 hours ahead of Washington, DC, during Standard Time)

note: China is the largest country (in terms of area) with just one time zone; before 1949 it was divided into five
etymology: the Chinese meaning is "Northern Capital"

Administrative divisions: 23 provinces (sheng, singular and plural), 5 autonomous regions (zizhiqu, singular and plural), and 4 municipalities (shi, singular and plural)
provinces: Anhui, Fujian, Gansu, Guangdong, Guizhou, Hainan, Hebei, Heilongjiang, Henan, Hubei, Hunan, Jiangsu, Jiangxi, Jilin, Liaoning, Qinghai, Shaanxi, Shandong, Shanxi, Sichuan, Yunnan, Zhejiang; (see note on Taiwan)
autonomous regions: Guangxi, Nei Mongol (Inner Mongolia), Ningxia, Xinjiang Uyghur, Xizang (Tibet)
municipalities: Beijing, Chongqing, Shanghai, Tianjin
note: China considers Taiwan its 23rd province; see separate entries for the special administrative regions of Hong Kong and Macau

Independence: 1 October 1949 (People's Republic of China established); notable earlier dates: 221 B.C. (unification under the Qin Dynasty); 1 January 1912 (Qing Dynasty replaced by the Republic of China)

National holiday: National Day (anniversary of the founding of the People's Republic of China), 1 October (1949)

Constitution: *history:* several previous; latest promulgated 4 December 1982
amendments: proposed by the Standing Committee of the National People's Congress or supported by more than one fifth of the National People's Congress membership; passage requires more than two-thirds majority vote of the Congress membership; amended several times, last in 2018

Legal system: civil law influenced by Soviet and continental European civil law systems; legislature retains power to interpret statutes; note—on 28 May 2020, the National People's Congress adopted the PRC Civil Code, which codifies personal relations and property matters

International law organization participation: has not submitted an ICJ jurisdiction declaration; non-party state to the ICCt

Citizenship: *citizenship by birth:* no
citizenship by descent only: least one parent must be a citizen of China
dual citizenship recognized: no
residency requirement for naturalization: while naturalization is theoretically possible, in practical terms it is extremely difficult; residency is required but not specified

Suffrage: 18 years of age; universal

Executive branch: *chief of state:* President XI Jinping (since 14 March 2013); Vice President WANG Qishan (since 17 March 2018)

head of government: Premier LI Keqiang (since 16 March 2013); Executive Vice Premiers HAN Zheng (since 19 March 2018), SUN Chunlan

(since 19 March 2018), LIU He (since 19 March 2018), HU Chunhua (since 19 March 2018)
cabinet: State Council appointed by National People's Congress
elections/appointments: president and vice president indirectly elected by National People's Congress for a 5-year term (unlimited terms); election last held on 17 March 2018 (next to be held in March 2023); premier nominated by president, confirmed by National People's Congress
election results: XI Jinping reelected president; National People's Congress vote—2,970 (unanimously); WANG Qishan elected vice president with 2,969 votes

Legislative branch: *description:* unicameral National People's Congress or Quanguo Renmin Daibiao Dahui (maximum of 3,000 seats; members indirectly elected by municipal, regional, and provincial people's congresses, and the People's Liberation Army; members serve 5-year terms); note—in practice, only members of the Chinese Communist Party (CCP), its 8 allied independent parties, and CCP-approved independent candidates are elected
elections: last held in December 2017-February 2018 (next to be held in late 2022 to early 2023)
election results: percent of vote—NA; seats by party—NA; composition—men 2,238, women 742, percent of women 24.9%

Judicial branch: *highest courts:* Supreme People's Court (consists of over 340 judges, including the chief justice and 13 grand justices organized into a civil committee and tribunals for civil, economic, administrative, complaint and appeal, and communication and transportation cases)
judge selection and term of office: chief justice appointed by the People's National Congress (NPC); limited to 2 consecutive 5-year-terms; other justices and judges nominated by the chief justice and appointed by the Standing Committee of the NPC; term of other justices and judges determined by the NPC
subordinate courts: Higher People's Courts; Intermediate People's Courts; District and County People's Courts; Autonomous Region People's Courts; International Commercial Courts; Special People's Courts for military, maritime, transportation, and forestry issues
note: in late 2014, China unveiled a multi-year judicial reform program; progress continued in 2018

Political parties and leaders: Chinese Communist Party or CCP [XI Jinping]
note: China has 8 nominally independent small parties controlled by the CCP

International organization participation: ADB, AfDB (nonregional member), APEC, Arctic Council (observer), ARF, ASEAN (dialogue partner), BIS, BRICS, CDB, CICA, EAS, FAO, FATF, G-20, G-24 (observer), G-5, G-77, IADB, IAEA, IBRD, ICAO, ICC (national committees), ICRM, IDA, IFAD, IFC, IFRCS, IHO, ILO, IMF, IMO, IMSO, Interpol, IOC, IOM (observer), IPU, ISO, ITSO, ITU, LAIA (observer), MIGA, MINURSO, MINUSMA, MONUSCO, NAM (observer), NSG, OAS (observer), OPCW, Pacific Alliance (observer), PCA, PIF (partner), SAARC (observer), SCO, SICA (observer), UN, UNAMID, UNCTAD, UNESCO, UNFICYP, UNHCR, UNIDO, UNIFIL, UNMIL, UNMISS, UNOCI, UN Security Council (permanent), UNTSO, UNWTO, UPU, WCO, WHO, WIPO, WMO, WTO, ZC

Diplomatic representation in the US: *chief of mission:* Ambassador CUI Tiankai (since 3 April 2013)
chancery: 3505 International Place NW, Washington, DC 20008
telephone: [1] (202) 495-2266
FAX: [1] (202) 495-2138
consulate(s) general: Chicago, Los Angeles, New York, San Francisco; note—the US ordered closure of the Houston consulate in late July 2020

Diplomatic representation from the US: *chief of mission:* Ambassador Terry BRANSTAD (since 12 July 2017)
telephone: [86] (10) 8531-3000
embassy: 55 An Jia Lou Lu, 100600 Beijing
mailing address: PO AP 96521
FAX: [86] (10) 8531-3300
consulate(s) general: Guangzhou, Shanghai, Shenyang, Wuhan; note—the Chinese Government ordered closure of the US consulate in Chengdu in late July 2020

Flag description: red with a large yellow five-pointed star and four smaller yellow five-pointed stars (arranged in a vertical arc toward the middle of the flag) in the upper hoist-side corner; the color red represents revolution, while the stars symbolize the four social classes—the working class, the peasantry, the urban petty bourgeoisie, and the national bourgeoisie (capitalists)—united under the Communist Party of China

National symbol(s): dragon, giant panda; national colors: red, yellow

National anthem: *name:* "Yiyongjun Jinxingqu" (The March of the Volunteers)
lyrics/music: TIAN Han/NIE Er
note: adopted 1949; the anthem, though banned during the Cultural Revolution, is more commonly known as "Zhongguo Guoge" (Chinese National Song); it was originally the theme song to the 1935 Chinese movie, "Sons and Daughters in a Time of Storm"
0:00 / 0:43

ECONOMY

Economy—overview: Since the late 1970s, China has moved from a closed, centrally planned system to a more market-oriented one that plays a major global role. China has implemented reforms in a gradualist fashion, resulting in efficiency gains that have contributed to a more than tenfold increase in GDP since 1978. Reforms began with the phase-out of collectivized agriculture, and expanded to include the gradual liberalization of prices, fiscal decentralization, increased autonomy for state enterprises, growth of the private sector, development of stock markets and a modern banking system, and opening to foreign trade and investment. China continues to pursue an industrial policy, state support of key sectors, and a restrictive investment regime. From 2013 to 2017, China had one of the fastest growing economies in the world, averaging slightly more than 7% real growth per year. Measured on a purchasing power parity (PPP) basis that adjusts for price differences, China in 2017 stood as the largest economy in the world, surpassing the US in 2014 for the first time in modern history. China became the world's largest exporter in 2010, and the largest trading nation in 2013. Still, China's per capita income is below the world average.

In July 2005 moved to an exchange rate system that references a basket of currencies. From mid-2005 to late 2008, the renminbi (RMB) appreciated more than 20% against the US dollar, but the exchange rate remained virtually pegged to the dollar from the onset of the global financial crisis until June 2010, when Beijing announced it would resume a gradual appreciation. From 2013 until early 2015, the renminbi held steady against the dollar, but it depreciated 13% from mid-2015 until end-2016 amid strong capital outflows; in 2017 the RMB resumed appreciating against the dollar – roughly 7% from end-of-2016 to end-of-2017. In 2015, the People's Bank of China announced it would continue to carefully push for full convertibility of the renminbi, after the currency was accepted as part of the IMF's special drawing rights basket. However, since late 2015 the Chinese Government has strengthened capital controls and oversight of overseas investments to better manage the exchange rate and maintain financial stability.

The Chinese Government faces numerous economic challenges including: (a) reducing its high domestic savings rate and correspondingly low domestic household consumption; (b) managing its high corporate debt burden to maintain financial stability; (c) controlling off-balance sheet local government debt used to finance infrastructure stimulus; (d) facilitating higher-wage job opportunities for the aspiring middle class, including rural migrants and college graduates, while maintaining competitiveness; (e) dampening speculative investment in the real estate sector without sharply slowing the economy; (f) reducing industrial overcapacity; and (g) raising productivity growth rates through the more efficient allocation of capital and state-support for innovation. Economic development has progressed further in coastal provinces than in the interior, and by 2016 more than 169.3 million migrant workers and their dependents had relocated to urban areas to find work. One consequence of China's population control policy known as the "one-child policy"—which was relaxed in 2016 to permit all families to have two children—is that China is now one of the most rapidly aging countries in the world. Deterioration in the environment—notably air pollution, soil erosion, and the steady fall of the water table, especially in the North—is another long-term problem. China continues to lose arable land because of erosion and urbanization. The Chinese Government is seeking to add energy production capacity from sources other than coal and oil, focusing on natural gas, nuclear, and clean energy development. In 2016, China ratified the Paris Agreement, a multilateral agreement to combat climate change, and committed to peak its carbon dioxide emissions between 2025 and 2030.

The government's 13th Five-Year Plan, unveiled in March 2016, emphasizes the need to increase innovation and boost domestic consumption to make the economy less dependent on government investment, exports, and heavy industry. However, China has made more progress on subsidizing innovation than rebalancing the economy. Beijing has committed to giving the market a more decisive role in allocating resources, but the Chinese Government's policies continue to favor state-owned enterprises and emphasize stability. Chinese leaders in 2010 pledged to double China's GDP by 2020, and the 13th Five Year Plan includes annual economic growth targets of at least 6.5% through 2020 to achieve that goal. In recent years, China has renewed its support for state-owned enterprises in sectors considered important to "economic security," explicitly looking to foster globally competitive industries. Chinese leaders also have undermined some market-oriented reforms by reaffirming the "dominant" role of the state in the economy, a stance that threatens to discourage private initiative and make the economy less efficient over time. The slight acceleration in economic growth in 2017—the first such uptick since 2010—gives Beijing more latitude to pursue its economic reforms, focusing on financial sector deleveraging and its Supply-Side Structural Reform agenda, first announced in late 2015.

GDP (purchasing power parity): $25.36 trillion (2018)
$23.21 trillion (2017 est.)
$21.72 trillion (2016 est.)
note: data are in 2017 dollars
country comparison to the world: 1

GDP (official exchange rate): $12.01 trillion (2017 est.)
note: because China's exchange rate is determined by fiat rather than by market forces, the official exchange rate measure of GDP is not an accurate measure of China's output; GDP at the official exchange rate substantially understates the actual level of China's output vis-a-vis the rest of the world; in China's situation, GDP at purchasing power parity provides the best measure for comparing output across countries

GDP—real growth rate: 6.14% (2019 est.)
6.75% (2018 est.)
6.92% (2017 est.)
country comparison to the world: 27

GDP—per capita (PPP): $18,200 (2018)
$16,700 (2017 est.)
$15,700 (2016 est.)
note: data are in 2017 dollars
country comparison to the world: 96

Gross national saving: 45.8% of GDP (2017 est.)
45.9% of GDP (2016 est.)
47.5% of GDP (2015 est.)

country comparison to the world: 6

GDP—composition, by end use: *household consumption:* 39.1% (2017 est.)
government consumption: 14.5% (2017 est.)
investment in fixed capital: 42.7% (2017 est.)
investment in inventories: 1.7% (2017 est.)
exports of goods and services: 20.4% (2017 est.)
imports of goods and services: -18.4% (2017 est.)

GDP—composition, by sector of origin: *agriculture:* 7.9% (2017 est.)
industry: 40.5% (2017 est.)
services: 51.6% (2017 est.)

Agriculture—products: world leader in gross value of agricultural output; rice, wheat, potatoes, corn, tobacco, peanuts, tea, apples, cotton, pork, mutton, eggs; fish, shrimp

Industries: world leader in gross value of industrial output; mining and ore processing, iron, steel, aluminum, and other metals, coal; machine building; armaments; textiles and apparel; petroleum; cement; chemicals; fertilizer; consumer products (including footwear, toys, and electronics); food processing; transportation equipment, including automobiles, railcars and locomotives, ships, aircraft; telecommunications equipment, commercial space launch vehicles, satellites

Industrial production growth rate: 6.1% (2017 est.)
country comparison to the world: 40

Labor force: 774.71 million (2019 est.)
note: by the end of 2012, China's working age population (15-64 years) was 1.004 billion
country comparison to the world: 1

Labor force—by occupation: *agriculture:* 27.7%
industry: 28.8%
services: 43.5% (2016 est.)

Unemployment rate: 3.64% (2019 est.)
3.84% (2018 est.)
note: data are for registered urban unemployment, which excludes private enterprises and migrants
country comparison to the world: 51

Population below poverty line: 3.3% (2016 est.)
note: in 2011, China set a new poverty line at RMB 2300 (approximately US $400)

Household income or consumption by percentage share: *lowest 10%:* 2.1%
highest 10%: 31.4% (2012)
note: data are for urban households only

Budget: *revenues:* 2.553 trillion (2017 est.)
expenditures: 3.008 trillion (2017 est.)

Taxes and other revenues: 21.3% (of GDP) (2017 est.)
country comparison to the world: 141

Budget surplus (+) or deficit (-): -3.8% (of GDP) (2017 est.)
country comparison to the world: 152

Public debt: 47% of GDP (2017 est.)
44.2% of GDP (2016 est.)
note: official data; data cover both central and local government debt, including debt officially recognized by China's National Audit Office report in 2011; data exclude policy bank bonds, Ministry of Railway debt, and China Asset Management Company debt

country comparison to the world: 111

Fiscal year: calendar year

Inflation rate (consumer prices): 1.6% (2017 est.)
2% (2016 est.)
country comparison to the world: 88

Current account balance: $141.335 billion (2019 est.)
$25.499 billion (2018 est.)
country comparison to the world: 3

Exports: $2.49 trillion (2018)
$2.216 trillion (2017 est.)
$1.99 trillion (2016 est.)
country comparison to the world: 1

Exports—partners: US 19.2%, Hong Kong 12.2%, Japan 5.9%, South Korea 4.4% (2018)

Exports—commodities: electrical and other machinery, including computers and telecommunications equipment, apparel, furniture, textiles

Imports: $2.14 trillion (2018)
$1.74 trillion (2017 est.)
$1.501 trillion (2016 est.)
country comparison to the world: 2

Imports—commodities: electrical and other machinery, including integrated circuits and other computer components, oil and mineral fuels; optical and medical equipment, metal ores, motor vehicles; soybeans

Imports—partners: South Korea 9.7%, Japan 8.6%, US 7.3%, Germany 5%, Australia 4.9% (2018)

Reserves of foreign exchange and gold: $3.236 trillion (31 December 2017 est.)
$3.098 trillion (31 December 2016 est.)
country comparison to the world: 1

Debt—external: $1.598 trillion (31 December 2017 est.)
$1.429 trillion (31 December 2016 est.)
country comparison to the world: 14

Exchange rates: Renminbi yuan (RMB) per US dollar—
7.76 (2017 est.)
6.6446 (2016 est.)
6.2275 (2015 est.)
6.1434 (2014 est.)
6.1958 (2013 est.)

ENERGY

Electricity access: *electrification—total population:* 100% (2020)

Electricity—production: 5.883 trillion kWh (2016 est.)
country comparison to the world: 1

Electricity—consumption: 5.564 trillion kWh (2016 est.)
country comparison to the world: 1

Electricity—exports: 18.91 billion kWh (2016 est.)
country comparison to the world: 10

Electricity—imports: 6.185 billion kWh (2016 est.)
country comparison to the world: 33

Electricity—installed generating capacity: 1.653 billion kW (2016 est.)
country comparison to the world: 1

Electricity—from fossil fuels: 62% of total installed capacity (2016 est.)
country comparison to the world: 124

Electricity—from nuclear fuels: 2% of total installed capacity (2016 est.)
country comparison to the world: 25

Electricity—from hydroelectric plants: 18% of total installed capacity (2017 est.)
country comparison to the world: 93

Electricity—from other renewable sources: 18% of total installed capacity (2017 est.)
country comparison to the world: 47

Crude oil—production: 3.773 million bbl/day (2018 est.)
country comparison to the world: 7

Crude oil—exports: 57,310 bbl/day (2015 est.)
country comparison to the world: 40

Crude oil—imports: 6.71 million bbl/day (2015 est.)
country comparison to the world: 2

Crude oil—proved reserves: 25.63 billion bbl (1 January 2018 est.)
country comparison to the world: 12

Refined petroleum products—production: 11.51 million bbl/day (2015 est.)
country comparison to the world: 2

Refined petroleum products—consumption: 12.47 million bbl/day (2016 est.)
country comparison to the world: 2

Refined petroleum products—exports: 848,400 bbl/day (2015 est.)
country comparison to the world: 9

Refined petroleum products—imports: 1.16 million bbl/day (2015 est.)
country comparison to the world: 4

Natural gas—production: 145.9 billion cu m (2017 est.)
country comparison to the world: 6

Natural gas—consumption: 238.6 billion cu m (2017 est.)
country comparison to the world: 3

Natural gas—exports: 3.37 billion cu m (2017 est.)
country comparison to the world: 35

Natural gas—imports: 97.63 billion cu m (2017 est.)
country comparison to the world: 3

Natural gas—proved reserves: 5.44 trillion cu m (1 January 2018 est.)
country comparison to the world: 9

Carbon dioxide emissions from consumption of energy: 11.67 billion Mt (2017 est.)
country comparison to the world: 1

COMMUNICATIONS

Telephones—fixed lines: *total subscriptions:* 185,097,221

subscriptions per 100 inhabitants: 13.32 (2019 est.)

country comparison to the world: 1

Telephones—mobile cellular: *total subscriptions:* 1,672,545,161

subscriptions per 100 inhabitants: 120.36 (2019 est.)

country comparison to the world: 1

Telecommunication systems: *general assessment:* the largest Internet market in the world, with the majority, 98.6% of users accessing the Internet through mobile devices; moderate growth is predicted over the next five years in the fixed broadband segment; one of the biggest drivers of commercial growth is its increasing urbanization rate as rural residents move to cities; China will be the world's largest 5G market; the Chinese mobile market to reach penetration of 134% by 2024; maintains the largest M2M market in the world (2020)

domestic: 13 per 100 fixed line and 120 per 100 mobile-cellular; a domestic satellite system with several earth stations has been in place since 2018 (2019)

international: country code—86; landing points for the RJCN, EAC-C2C, TPE, APCN-2, APG, NCP, TEA, SeaMeWe-3, SJC2, Taiwan Strait Express-1, AAE-1, APCN-2, AAG, FEA, FLAG and TSE submarine cables providing connectivity to Asia, the Middle East, Europe, and the US; satellite earth stations—7 (5 Intelsat—4 Pacific Ocean and 1 Indian Ocean; 1 Intersputnik—Indian Ocean region; and 1 Inmarsat—Pacific and Indian Ocean regions) (2019)

note: the COVID-19 outbreak is negatively impacting telecommunications production and supply chains globally; consumer spending on telecom devices and services has also slowed due to the pandemic's effect on economies worldwide; overall progress towards improvements in all facets of the telecom industry—mobile, fixed-line, broadband, submarine cable and satellite—has moderated

Broadcast media: all broadcast media are owned by, or affiliated with, the Communist Party of China or a government agency; no privately owned TV or radio stations; state-run Chinese Central TV, provincial, and municipal stations offer more than 2,000 channels; the Central Propaganda Department sends directives to all domestic media outlets to guide its reporting with the government maintaining authority to approve all programming; foreign-made TV programs must be approved prior to broadcast; increasingly, Chinese turn to online and satellite television to access Chinese and international films and television shows (2019)

Internet country code: .cn

Internet users: *total:* 751,886,119

percent of population: 54.3% (July 2018 est.)

country comparison to the world: 1

Broadband—fixed subscriptions: *total:* 407.382 million

subscriptions per 100 inhabitants: 29 (2018 est.)

country comparison to the world: 1

TRANSPORTATION

National air transport system: *number of registered air carriers:* 56 (2020)

inventory of registered aircraft operated by air carriers: 2,890

annual passenger traffic on registered air carriers: 436,183,969 (2018)

annual freight traffic on registered air carriers: 611,439,830 mt-km (2018)

Civil aircraft registration country code prefix: B (2016)

Airports: 507 (2013)

country comparison to the world: 13

Airports—with paved runways: *total:* 510 (2019)

over 3,047 m: 87

2,438 to 3,047 m: 187

1,524 to 2,437 m: 109

914 to 1,523 m: 43

under 914 m: 84

Airports—with unpaved runways: *total:* 23 (2019)

over 3,047 m: 2

2,438 to 3,047 m: 0

1,524 to 2,437 m: 1

914 to 1,523 m: 7

under 914 m: 13

Heliports: 39 (2019)

Pipelines: 76000 km gas, 30400 km crude oil, 27700 km refined petroleum products, 797000 km water (2018)

Railways: *total:* 131,000 km 1.435-m gauge (80,000 km electrified); 102,000 traditional, 29,000 high-speed (2018)

country comparison to the world: 2

Roadways: *total:* 4,960,600 km (2017)

paved: 4,338,600 km (includes 136,500 km of expressways) (2017)

unpaved: 622,000 km (2017)

country comparison to the world: 2

Waterways: 110,000 km (navigable waterways) (2011)

country comparison to the world: 1

Merchant marine: *total:* 5,594

by type: bulk carrier 1,231, container ship 262, general cargo 846, oil tanker 777, other 2,478 (2019)

country comparison to the world: 3

Ports and terminals: *major seaport(s):* Dalian, Ningbo, Qingdao, Qinhuangdao, Shanghai, Shenzhen, Tianjin container port(s) (TEUs): Dalian (9,707,000), Guangzhou (18,858,000), Ningbo (24,607,000), Qingdao (18,262,000), Shanghai (40,233,000), Shenzhen (25,208,000), Tianjin (15,040,000) (2017)

LNG terminal(s) (import): Fujian, Guangdong, Jiangsu, Shandong, Shanghai, Tangshan, Zhejiang *river port(s):* Guangzhou (Pearl)

Transportation—note: seven of the world's ten largest container ports are in China

MILITARY AND SECURITY

Military and security forces: *People's Liberation Army (PLA):* Ground Forces, Navy (PLAN, includes marines and naval aviation), Air Force (PLAAF, includes airborne forces), Rocket Force (strategic missile force), and Strategic Support Force (information warfare, cyber, space forces); People's Armed Police (PAP, includes Coast Guard, Border Defense Force, Internal Security Forces); PLA Reserve Force (2020)

Military expenditures: 1.9% of GDP (2019)

1.9% of GDP (2018)

1.9% of GDP (2017)

1.9% of GDP (2016)

1.9% of GDP (2015)

country comparison to the world: 55

Military and security service personnel strengths: assessments of the size of the People's Liberation Army (PLA) vary; approximately 2 million total active duty troops (approximately 1.0 million Ground; 250,000 Navy/Marines; 350,000 Air Force; 120,000 Rocket Forces; 150,000 Strategic Support Forces); 650,000 People's Armed Police (2019)

Military equipment inventories and acquisitions: the PLA is outfitted primarily with a mix of older and modern domestically-produced systems heavily influenced by technology derived from other countries; Russia is the top supplier of foreign military equipment since 2010, followed by France and Ukraine (2019)

Military deployments: 425 Mali (MINUSMA); 220 Democratic Republic of the Congo (MONUSCO); 360 Sudan (UNAMID); 410 Lebanon (UNIFIL); 1,050 South Sudan (UNMISS); est. 250 Djibouti (2020)

Military service age and obligation: 18-22 years of age for selective compulsory military service, with a 2-year service obligation; no minimum age for voluntary service (all officers are volunteers); 18-19 years of age for women high school graduates who meet requirements for specific military jobs (2018)

TRANSNATIONAL ISSUES

Disputes—international: China and India continue their security and foreign policy dialogue started in 2005 related to a number of boundary disputes across the 2,000 mile shared border; India does not recognize Pakistan's 1964 ceding to China of the Aksai Chin, a territory designated as part of the princely state of Kashmir by the British Survey of India in 1865; China claims most of the Indian state Arunachal Pradesh to the base of the Himalayas; the US recognizes the state of Arunachal Pradesh as Indian territory; Bhutan and China continue negotiations to establish a common boundary alignment to resolve territorial disputes arising from substantial cartographic discrepancies, the most contentious of which lie in Bhutan's west along China's Chumbi salient; Chinese maps show an international boundary symbol (the so-called "nine-dash line") off the coasts of the littoral states of the South China Sea, where China has interrupted Vietnamese hydrocarbon exploration; China asserts sovereignty over Scarborough Reef along with the Philippines and Taiwan, and over the Spratly Islands together with

Malaysia, the Philippines, Taiwan, Vietnam, and Brunei; the 2002 Declaration on the Conduct of Parties in the South China Sea eased tensions in the Spratlys, and in 2017 China and ASEAN began confidential negotiations for an updated Code of Conduct for the South China Sea designed not to settle territorial disputes but establish rules and norms in the region; this still is not the legally binding code of conduct sought by some parties; Vietnam and China continue to expand construction of facilities in the Spratlys and in early 2018 China began deploying advanced military systems to disputed Spratly outposts; China occupies some of the Paracel Islands also claimed by Vietnam and Taiwan; the Japanese-administered Senkaku Islands are also claimed by China and Taiwan; certain islands in the Yalu and Tumen Rivers are in dispute with North Korea; North Korea and China seek to stem illegal migration to China by North Koreans, fleeing privation and oppression; China and Russia have demarcated the once disputed islands at the Amur and Ussuri confluence and in the Argun River in accordance with their 2004 Agreement; China and Tajikistan have begun demarcating the revised boundary agreed to in the delimitation of 2002; the decade-long demarcation of the China-Vietnam land boundary was completed in 2009; citing environmental, cultural, and social concerns, China has reconsidered construction of 13 dams on the Salween River, but energy-starved Burma, with backing

from Thailand, continues to consider building five hydro-electric dams downstream despite regional and international protests

Refugees and internally displaced persons: *refugees (country of origin):* 303,095 (Vietnam), undetermined (North Korea) (2019)
IDPs: undetermined (2014)

Trafficking in persons: *current situation:* China is a source, transit, and destination country for men, women, and children subjected to sex trafficking and forced labor; Chinese adults and children are forced into prostitution and various forms of forced labor, including begging and working in brick kilns, coal mines, and factories; women and children are recruited from rural areas and taken to urban centers for sexual exploitation, often lured by criminal syndicates or gangs with fraudulent job offers; state-sponsored forced labor, where detainees work for up to four years often with no remuneration, continues to be a serious concern; Chinese men, women, and children also may be subjected to conditions of sex trafficking and forced labor worldwide, particularly in overseas Chinese communities; women and children are trafficked to China from neighboring countries, as well as Africa and the Americas, for forced labor and prostitution

tier rating: Tier 2 Watch List—China does not fully comply with the minimum standards for the elimination of trafficking; however, it is making

significant efforts to do so; official data for 2014 states that 194 alleged traffickers were arrested and at least 35 were convicted, but the government's conflation of human trafficking with other crimes makes it difficult to assess law enforcement efforts to investigate and to prosecute trafficking offenses according to international law; despite reports of complicity, no government officials were investigated, prosecuted, or convicted for their roles in trafficking offenses; authorities did not adequately protect victims and did not provide the data needed to ascertain the number of victims identified or assisted or the services provided; the National People's Congress ratified a decision to abolish "reform through labor" in 2013, but some continued to operate as state-sponsored drug detention or "custody and education" centers that force inmates to perform manual labor; some North Korean refugees continued to be forcibly repatriated as illegal economic migrants, despite reports that some were trafficking victims (2015)

Illicit drugs: major transshipment point for heroin produced in the Golden Triangle region of Southeast Asia; growing domestic consumption of synthetic drugs, and heroin from Southeast and Southwest Asia; source country for methamphetamine and heroin chemical precursors, despite new regulations on its large chemical industry; more people believed to be convicted and executed for drug offences than anywhere else in the world, according to NGOs

CHRISTMAS ISLAND

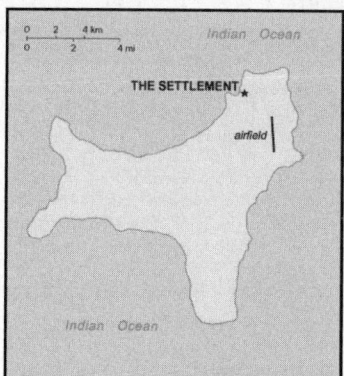

under the jurisdiction of the new British Colony of Singapore. The island existed as a separate Crown colony from 1 January 1958 to 1 October 1958 when its transfer to Australian jurisdiction was finalized. That date is still celebrated on the first Monday in October as Territory Day. Almost two-thirds of the island has been declared a national park.

GEOGRAPHY

Location: Southeastern Asia, island in the Indian Ocean, south of Indonesia

Geographic coordinates: 10 30 S, 105 40 E

Map references: Southeast Asia

Area: *total:* 135 sq km
land: 135 sq km
water: 0 sq km
country comparison to the world: 222

Area—comparative: about three-quarters the size of Washington, DC

Land boundaries: *0 km*

Coastline: 138.9 km

Maritime claims: *territorial sea:* 12 nm
contiguous zone: 12 nm
exclusive fishing zone: 200 nm

Climate: tropical with a wet season (December to April) and dry season; heat and humidity moderated by trade winds

Terrain: steep cliffs along coast rise abruptly to central plateau

Elevation: *lowest point:* Indian Ocean 0 m
highest point: Murray Hill 361 m

Natural resources: phosphate, beaches

Land use: *agricultural land:* 0% (2011 est.)
arable land: 0% (2011 est.) / permanent crops: 0% (2011 est.) / permanent pasture: 0% (2011 est.)
other: 100% (2011 est.)
Irrigated land: NA

Population distribution: majority of the population lives on the northern tip of the island

Natural hazards: the narrow fringing reef surrounding the island can be a maritime hazard

Environment—current issues: loss of rainforest; impact of phosphate mining

Geography—note: located along major sea lanes of the Indian Ocean

PEOPLE AND SOCIETY

Population: 2,205 (2016 est.)
country comparison to the world: 231

INTRODUCTION

Background: Although Europeans had sighted the island at least as early as 1615, it was only named in 1643 for the day of its rediscovery. The island was annexed and settlement began by the UK in 1888 with the discovery of the island's phosphate deposits. Following the Second World War, Christmas Island came

Nationality: *noun:* Christmas Islander(s)
adjective: Christmas Island

Ethnic groups: Chinese 70%, European 20%, Malay 10% (2001)
note: no indigenous population

Languages: English (official) 27.6%, Mandarin 17.2%, Malay 17.1%, Cantonese 3.9%, Min Nan 1.6%, Tagalog 1%, other 4.5%, unspecified 27.1% (2016 est.)
note: data represent language spoken at home

Religions: Muslim 19.4%, Buddhist 18.3%, Roman Catholic 8.8%, Protestant 6.5% (includes Anglican 3.6%, Uniting Church 1.2%, other 1.7%), other Christian 3.3%, other 0.6%, none 15.3%, unspecified 27.7% (2016 est.)

Age structure: 0-14 *years:* 12.79% (male 147/female 135) (2017 est.)
15-24 years: 12.2% (male 202/female 67) (2017 est.)
25-54 years: 57.91% (male 955/female 322) (2017 est.)
55-64 years: 11.66% (male 172/female 85) (2017 est.)
65 years and over: 5.44% (male 84/female 36) (2017 est.)

Population growth rate: 1.11% (2014 est.)
country comparison to the world: 95

Population distribution: majority of the population lives on the northern tip of the island

Sex ratio: NA

Infant mortality rate: *total:* NA (2018)
male: NA
female: NA

Life expectancy at birth: *total population:* NA (2017 est.)
male: NA
female: NA

Total fertility rate: NA

HIV/AIDS—adult prevalence rate: NA

HIV/AIDS—people living with HIV/AIDS: NA

HIV/AIDS—deaths: NA

GOVERNMENT

Country name: *conventional long form:* Territory of Christmas Island
conventional short form: Christmas Island
etymology: named by English Captain William MYNORS for the day of its rediscovery, Christmas Day (25 December 1643); the island had been sighted by Europeans as early as 1615

Dependency status: non-self governing territory of Australia; administered from Canberra by the Department of Regional Australia, Local Government, Arts and Sport

Government type: non-self-governing overseas territory of Australia

Capital: *name:* The Settlement (Flying Fish Cove)

geographic coordinates: 10 25 S, 105 43 E
time difference: UTC+7 (12 hours ahead of Washington, DC, during Standard Time)

etymology: self-descriptive name for the main locus of population

Administrative divisions: none (territory of Australia)

Independence: none (territory of Australia)

National holiday: Australia Day (commemorates the arrival of the First Fleet of Australian settlers), 26 January (1788)

Constitution: *history:* 1 October 1958 (Christmas Island Act 1958)
amendments: amended many times, last in 2016

Legal system: legal system is under the authority of the governor general of Australia and Australian law

Citizenship: see Australia

Suffrage: 18 years of age

Executive branch: *chief of state:* Queen ELIZABETH II (since 6 February 1952); represented by Governor General of the Commonwealth of Australia General Sir Peter COSGROVE (since 28 March 2014)

head of government: Administrator Natasha GRIGGS (since 5 October 2018)
elections/appointments: the monarchy is hereditary; governor general appointed by the monarch on the recommendation of the Australian prime minister; administrator appointed by the governor general of Australia for a 2year term and represents the monarch and Australia

Legislative branch: *description:* unicameral Christmas Island Shire Council (9 seats; members directly elected by simple majority vote to serve 4-year terms with a portion of the membership renewed every 2 years)
elections: held every 2 years with half the members standing for election; last held on 21 October 2017 (next to be held in October 2019)
election results: percent of vote—NA; seats by party—independent 9; composition as of 17 October 2015—men 7, women 2, percent of women 22.2%

Judicial branch: under the terms of the Territorial Law Reform Act 1992, Western Australia provides court services as needed for the island, including the Supreme Court and subordinate courts (District Court, Magistrate Court, Family Court, Children's Court, and Coroners' Court)

Political parties and leaders: none

International organization participation: none

Diplomatic representation in the US: none (territory of Australia)

Diplomatic representation from the US: none (territory of Australia)

Flag description: territorial flag; divided diagonally from upper hoist to lower fly; the upper triangle is green with a yellow image of the Golden Bosun Bird superimposed; the lower triangle is blue with the Southern Cross constellation, representing Australia, superimposed; a centered yellow disk displays a green map of the island
note: the flag of Australia is used for official purposes

National symbol(s): golden bosun bird

National anthem: note: as a territory of Australia, "Advance Australia Fair" remains official as the national anthem, while "God Save the Queen" serves as the royal anthem (see Australia)
0:00/ 0:54

ECONOMY

Economy—overview: The main economic activities on Christmas Island are the mining of low grade phosphate, limited tourism, the provision of government services and, since 2005, the construction and operation of the Immigration Detention Center. The government sector includes administration, health, education, policing, customs, quarantine, and defense.

GDP (purchasing power parity): NA

Agriculture—products: NA

Industries: tourism, phosphate extraction (near depletion)

Labor force: NA

Budget: *revenues:* NA
expenditures: NA

Fiscal year: 1 July—30 June

Exports: NA

Exports—commodities: phosphate

Imports: NA

Imports—commodities: consumer goods

Exchange rates: Australian dollars (AUD) per US dollar—
1.311 (2017 est.)
1.3442 (2016 est.)
1.3442 (2015)
1.3291 (2014 est.)
1.1094 (2013 est.)

COMMUNICATIONS

Telecommunication systems: *general assessment:* service provided by the Australian network
domestic: local area code—08; GSM mobile-cellular telephone service is provided by Telstra as part of the Australian network
international: international code—61 8; ASC submarine cable to Singapore and Australia; satellite earth station—1 (Intelsat provides telephone and telex service) (2019)

Broadcast media: 1 community radio station; satellite broadcasts of several Australian radio and TV stations (2017)

Internet country code: .cx

Internet users: *total:* 790

percent of population: 35.8% (July 2016 est.)
country comparison to the world: 227

TRANSPORTATION

Airports: 1 (2020)
country comparison to the world: 217

Airports—with paved runways: *total:* 1 (2019)
1,524 to 2,437 m: 1

Railways: *total:* 18 km (2017)

standard gauge: 18 km 1.435-m (not in operation) (2017)

note: the 18-km Christmas Island Phosphate Company Railway between Flying Fish Cove and South Point was decommissioned in 1987; some tracks and scrap remain in place
country comparison to the world: 134

Roadways: *total:* 140 km (2011)

paved: 30 km (2011)
unpaved: 110 km (2011)
country comparison to the world: 210

Ports and terminals: *major seaport(s):* Flying Fish Cove

MILITARY AND SECURITY

Military—note: defense is the responsibility of Australia

TRANSNATIONAL ISSUES

Disputes—international: none

CLIPPERTON ISLAND

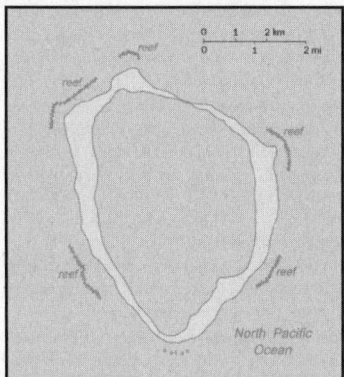

INTRODUCTION

Background: This isolated atoll was named for John CLIPPERTON, an English pirate who was rumored to have made it his hideout early in the 18th century. Annexed by France in 1855 and claimed by the US, it was seized by Mexico in 1897. Arbitration eventually awarded the island to France in 1931, which took possession in 1935.

GEOGRAPHY

Location: Middle America, atoll in the North Pacific Ocean, 1,120 km southwest of Mexico

Geographic coordinates: 10 17 N, 109 13 W

Map references: Political Map of the World

Area: *total:* 6 sq km

land: 6 sq km
water: 0 sq km
country comparison to the world: 248

Area—comparative: about 12 times the size of The Mall in Washington, DC

Land boundaries: 0 km

Coastline: 11.1 km

Maritime claims: *territorial sea:* 12 nm
exclusive economic zone: 200 nm

Climate: tropical; humid, average temperature 20-32 degrees Celsius, wet season (May to October)

Terrain: coral atoll

Elevation: *lowest point:* Pacific Ocean 0 m
highest point: Rocher Clipperton 29 m

Natural resources: fish

Land use: *agricultural land:* 0% (2011 est.)
arable land: 0% (2011 est.) / permanent crops: 0% (2011 est.) / permanent pasture: 0% (2011 est.)
forest: 0% (2011 est.)
other: 100% (2011 est.)

Natural hazards: subject to tropical storms and hurricanes from May to October

Environment—current issues: no natural resources, guano deposits depleted; the ring-shaped atoll encloses a stagnant fresh-water lagoon

Geography—note: the atoll reef is approximately 12 km (7.5 mi) in circumference; an attempt to colonize the atoll in the early 20th century ended in disaster and was abandoned in 1917

PEOPLE AND SOCIETY

Population: uninhabited

GOVERNMENT

Country name: *conventional long form:* none
conventional short form: Clipperton Island
local long form: none
local short form: Ile Clipperton
former: sometimes referred to as Ile de la Passion or Atoll Clipperton
etymology: named after an 18th-century English pirate who supposedly used the island as a base

Dependency status: possession of France; administered directly by the Minister of Overseas France

Legal system: the laws of France apply

Flag description: the flag of France is used

ECONOMY

Economy—overview: Although 115 species of fish have been identified in the territorial waters of Clipperton Island, tuna fishing is the only economically viable species.

TRANSPORTATION

Ports and terminals: none; offshore anchorage only

MILITARY AND SECURITY

Military—note: defense is the responsibility of France

TRANSNATIONAL ISSUES

Disputes—international: none

COCOS (KEELING) ISLANDS

INTRODUCTION

Background: There are 27 coral islands in the group. Captain William KEELING discovered the islands in 1609, but they remained uninhabited until the 19th century. From the 1820s to 1978, members of the CLUNIES-ROSS family controlled the islands and the copra produced from local coconuts. Annexed by the UK in 1857, the Cocos Islands were transferred to the Australian Government in 1955. Apart from North Keeling Island, which lies 30 kilometers north of the main group, the islands form a horseshoe-shaped atoll surrounding a lagoon. North Keeling Island was declared a national park in 1995 and is administered by Parks Australia. The population on the two inhabited islands generally is split between the ethnic Europeans on West Island and the ethnic Malays on Home Island.

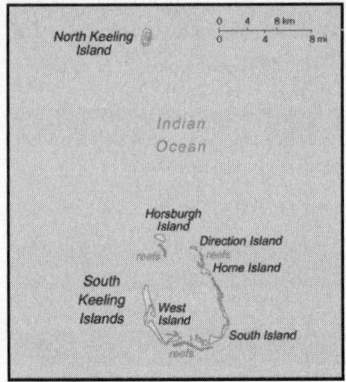

GEOGRAPHY

Location: Southeastern Asia, group of islands in the Indian Ocean, southwest of Indonesia, about halfway between Australia and Sri Lanka

Geographic coordinates: 12 30 S, 96 50 E

Map references: Southeast Asia

Area: *total:* 14 sq km
land: 14 sq km
water: 0 sq km
note: includes the two main islands of West Island and Home Island
country comparison to the world: 241

Area—comparative: about 24 times the size of The Mall in Washington, DC

Land boundaries: 0 km

Coastline: 26 km

Maritime claims: *territorial sea:* 12 nm
exclusive fishing zone: 200 nm

Climate: tropical with high humidity, moderated by the southeast trade winds for about nine months of the year

Terrain: flat, low-lying coral atolls

Elevation: *lowest point:* Indian Ocean 0 m
highest point: South Point on South Island 9 m

Natural resources: fish

Land use: *agricultural land:* 0% (2011 est.)
arable land: 0% (2011 est.) / *permanent crops:* 0% (2011 est.) / *permanent pasture:* 0% (2011 est.)
forest: 0% (2011 est.)
other: 100% (2011 est.)
Irrigated land: NA

Population distribution: only Home Island and West Island are populated

Natural hazards: cyclone season is October to April

Environment—current issues: freshwater resources are limited to rainwater accumulations in natural underground reservoirs; illegal fishing a concern

Geography—note: islands are thickly covered with coconut palms and other vegetation; site of a World War I naval battle in November 1914 between the Australian light cruiser HMAS Sydney and the German raider SMS Emden;

after being heavily damaged in the engagement, the Emden was beached by her captain on North Keeling Island

PEOPLE AND SOCIETY

Population: 596 (July 2014 est.)
country comparison to the world: 237

Nationality: *noun:* Cocos Islander(s)
adjective: Cocos Islander

Ethnic groups: Europeans, Cocos Malays

Languages: English 22.3%, Malay (Cocos dialect) 68.8%, unspecified 8.9% (2016 est.)
note: data represent language spoken at home

Religions: Muslim (predominantly Sunni) 75%, Anglican 3.5%, Roman Catholic 2.2%, none 12.9%, unspecified 6.3% (2016 est.)

Population growth rate: 0% (2014 est.)
country comparison to the world: 194

Population distribution: only Home Island and West Island are populated

Infant mortality rate: *total:* NA (2018)
male: NA
female: NA

Life expectancy at birth: *total population:* NA (2017 est.)
male: NA
female: NA

Total fertility rate: NA

HIV/AIDS—adult prevalence rate: NA

HIV/AIDS—people living with HIV/AIDS: NA

HIV/AIDS—deaths: NA

GOVERNMENT

Country name: *conventional long form:* Territory of Cocos (Keeling) Islands
conventional short form: Cocos (Keeling) Islands
etymology: the name refers to the abundant coconut trees on the islands and to English Captain William KEELING, the first European to sight the islands in 1609

Dependency status: non-self governing territory of Australia; administered from Canberra by the Department of Regional Australia, Local Government, Arts and Sport

Government type: non-self-governing overseas territory of Australia

Capital: *name:* West Island
geographic coordinates: 12 10 S, 96 50 E
time difference: UTC+6.5 (11.5 hours ahead of Washington, DC, during Standard Time)

Administrative divisions: none (territory of Australia)

Independence: none (territory of Australia)

National holiday: Australia Day (commemorates the arrival of the First Fleet of Australian settlers), 26 January (1788)

Constitution: *history:* 23 November 1955 (Cocos (Keeling) Islands Act 1955)
amendments: amended many times, last in 2016

Legal system: common law based on the Australian model

Citizenship: see Australia

Suffrage: 18 years of age

Executive branch: *chief of state:* Queen ELIZABETH II (since 6 February 1952); represented by Governor General of the Commonwealth of Australia General Sir Peter COSGROVE (since 28 March 2014)

head of government: Administrator Natasha GRIGGS (since 5 October 2018)
cabinet: NA
elections/appointments: the monarchy is hereditary; governor general appointed by the monarch on the recommendation of the Australian prime minister; administrator appointed by the governor general for a 2-year term and represents the monarch and Australia

Legislative branch: *description:* unicameral Cocos (Keeling) Islands Shire Council (7 seats; members directly elected by simple majority vote to serve 4-year terms with half the membership renewed every 2 years)
elections: last held in October 2017 (next to be held on 31 October 2019)
election results: percent of vote by party—NA; seats by party—NA; composition—men 5, women 2, percent of women 28.6%

Judicial branch: under the terms of the Territorial Law Reform Act 1992, Western Australia provides court services as needed for the island including the Supreme Court and subordinate courts (District Court, Magistrate Court, Family Court, Children's Court, and Coroners' Court)

Political parties and leaders: none

International organization participation: none

Diplomatic representation in the US: none (territory of Australia)

Diplomatic representation from the US: none (territory of Australia)

Flag description: the flag of Australia is used

National anthem: *note:* as a territory of Australia, "Advance Australia Fair" remains official as the national anthem, while "God Save the Queen" serves as the royal anthem (see Australia)
0:00/ 0:54

ECONOMY

Economy—overview: Coconuts, grown throughout the islands, are the sole cash crop. Small local gardens and fishing contribute to the food supply, but additional food and most other necessities must be imported from Australia. There is a small tourist industry.

GDP (purchasing power parity): NA

GDP—real growth rate: 1% (2003)
country comparison to the world: 172

Agriculture—products: vegetables, bananas, pawpaws, coconuts

Industries: copra products, tourism

Labor force: NA

Labor force—by occupation: *note:* the Cocos Islands Cooperative Society Ltd. employs construction workers, stevedores, and lighterage workers; tourism is the other main source of employment

Unemployment rate: 0.1% (2011)
60% (2000 est.)
country comparison to the world: 1

Budget: *revenues:* NA
expenditures: NA

Fiscal year: 1 July—30 June

Exports: NA

Exports—commodities: copra

Imports: NA

Imports—commodities: foodstuffs

Exchange rates: Australian dollars (AUD) per US dollar—
1.311 (2017 est.)
1.3442 (2016 est.)
1.3442 (2015)
1.3291 (2014)
1.1094 (2013)

Telecommunication systems: *general assessment:* telephone service is part of the Australian network; an operational local mobile-cellular network available; wireless Internet connectivity available
domestic: local area code—08
international: international code—61 8; telephone, telex, and facsimile communications with Australia and elsewhere via satellite; satellite earth station—1 (Intelsat)

Broadcast media: 1 local radio station staffed by community volunteers; satellite broadcasts of several Australian radio and TV stations available (2017)

Internet country code: .cc

TRANSPORTATION

Airports: 1 (2020)
country comparison to the world: 218

Airports—with paved runways: *total:* 1 (2019)
2,438 to 3,047 m: 1

Roadways: *total:* 22 km (2007)
paved: 10 km (2007)
unpaved: 12 km (2007)
country comparison to the world: 221

Ports and terminals: *major seaport(s):* Port Refuge

MILITARY AND SECURITY

Military—note: defense is the responsibility of Australia; the territory has a five-person police force

TRANSNATIONAL ISSUES

Disputes—international: none

COLOMBIA

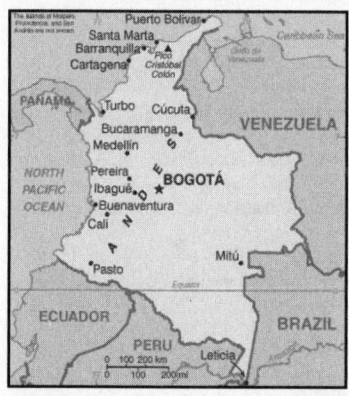

INTRODUCTION

Background: Colombia was one of the three countries that emerged after the dissolution of Gran Colombia in 1830 (the others are Ecuador and Venezuela). A decades-long conflict between government forces, paramilitaries, and antigovernment insurgent groups heavily funded by the drug trade, principally the Revolutionary Armed Forces of Colombia (FARC), escalated during the 1990s. More than 31,000 former United Self Defense Forces of Colombia (AUC) paramilitaries demobilized by the end of 2006, and the AUC as a formal organization ceased to operate. In the wake of the paramilitary demobilization, illegal armed groups arose, whose members include some former paramilitaries. After four years of formal peace negotiations, the Colombian Government signed a final peace accord with the FARC in November 2016, which was subsequently ratified by the Colombian Congress. The accord calls for members of the FARC to demobilize, disarm, and reincorporate into society and politics. The accord also committed the Colombian Government to create three new institutions to form a 'comprehensive system for truth, justice, reparation, and non-repetition,' to include a truth commission, a special unit to coordinate the search for those who disappeared during the conflict, and a 'Special Jurisdiction for Peace' to administer justice for conflict-related crimes. The Colombian Government has stepped up efforts to expand its presence into every one of its administrative departments. Despite decades of internal conflict and drug- related security challenges, Colombia maintains relatively strong democratic institutions characterized by peaceful, transparent elections and the protection of civil liberties.

GEOGRAPHY

Location: Northern South America, bordering the Caribbean Sea, between Panama and Venezuela, and bordering the North Pacific Ocean, between Ecuador and Panama

Geographic coordinates: 4 00 N, 72 00 W

Map references: South America

Area: *total:* 1,138,910 sq km
land: 1,038,700 sq km
water: 100,210 sq km
note: includes Isla de Malpelo, Roncador Cay, and Serrana Bank
country comparison to the world: 27

Area—comparative: slightly less than twice the size of Texas

Land boundaries: *total:* 6,672 km

border countries (5): Brazil 1790 km, Ecuador 708 km, Panama 339 km, Peru 1494 km, Venezuela 2341 km

Coastline: 3,208 km (Caribbean Sea 1,760 km, North Pacific Ocean 1,448 km)

Maritime claims: *territorial sea:* 12 nm
exclusive economic zone: 200 nm
continental shelf: 200-m depth or to the depth of exploitation

Climate: tropical along coast and eastern plains; cooler in highlands

Terrain: flat coastal lowlands, central highlands, high Andes Mountains, eastern lowland plains (Llanos)

Elevation: *mean elevation:* 593 m
lowest point: Pacific Ocean 0 m
highest point: Pico Cristobal Colon 5,730 m

Natural resources: petroleum, natural gas, coal, iron ore, nickel, gold, copper, emeralds, hydropower

Land use: *agricultural land:* 37.5% (2011 est.)
arable land: 1.4% (2011 est.) / permanent crops: 1.6% (2011 est.) / permanent pasture: 34.5% (2011 est.)
forest: 54.4% (2011 est.)
other: 8.1% (2011 est.)
Irrigated land: 10,900 sq km (2012)

Population distribution: the majority of people live in the north and west where agricultural opportunities and natural resources are found; the vast grasslands of the llanos to the south and east, which make up approximately 60% of the country, are sparsely populated

Natural hazards: highlands subject to volcanic eruptions; occasional earthquakes; periodic droughts *volcanism:* Galeras (4,276 m) is one of Colombia's most active volcanoes, having erupted in 2009 and 2010 causing major evacuations; it has been deemed a Decade Volcano by the International Association of Volcanology and Chemistry of the Earth's Interior, worthy of study due to its explosive history and close proximity to human populations; Nevado del Ruiz (5,321 m), 129 km (80 mi) west of Bogota, erupted in 1985 producing lahars (mudflows) that killed 23,000 people; the volcano last erupted in 1991; additionally, after 500 years of dormancy, Nevado del Huila reawakened in 2007 and has experienced frequent eruptions since then; other historically active volcanoes include Cumbal, Dona Juana, Nevado del Tolima, and Purace

Environment—current issues: deforestation resulting from timber exploitation in the jungles of the Amazon and the region of Chocó; illicit drug crops grown by peasants in the national parks; soil erosion; soil and water quality damage from overuse of pesticides; air pollution, especially in Bogota, from vehicle emissions

Environment—international agreements: *party to:* Antarctic Treaty, Biodiversity, Climate Change, Climate Change-Kyoto Protocol, Desertification, Endangered Species, Hazardous Wastes, Marine Life Conservation, Ozone Layer Protection, Ship Pollution, Tropical Timber 83, Tropical Timber 94, Wetlands
signed, but not ratified: Law of the Sea

Geography—note: only South American country with coastlines on both the North Pacific Ocean and Caribbean Sea

PEOPLE AND SOCIETY

Population: 49,084,841 (July 2020 est.)
country comparison to the world: 30

Nationality: *noun:* Colombian(s)
adjective: Colombian

Ethnic groups: mestizo and white 87.6%, Afro-Colombian (includes mulatto, Raizal, and Palenquero) 6.8%, Amerindian 4.3%, unspecified 1.4% (2018 est.)

Languages: Spanish (official)

Religions: Roman Catholic 79%, Protestant 14% (includes Pentecostal 6%, mainline Protestant 2%, other 6%), other 2%, unspecified 5% (2014 est.)

Demographic profile: Colombia is in the midst of a demographic transition resulting from steady declines in its fertility, mortality, and population growth rates. The birth rate has fallen from more than 6 children per woman in the 1960s to just above replacement level today as a result of increased literacy, family planning services, and urbanization. However, income inequality is among the worst in the world, and more than a third of the population lives below the poverty line.

Colombia experiences significant legal and illegal economic emigration and refugee outflows. Large-scale labor emigration dates to the 1960s;

the United States and, until recently, Venezuela have been the main host countries. Emigration to Spain picked up in the 1990s because of its economic growth, but this flow has since diminished because of Spain's ailing economy and high unemployment. Colombia has been the largest source of Latin American refugees in Latin America, nearly 400,000 of whom live primarily in Venezuela and Ecuador. Venezuela's political and economic crisis since 2015, however, has created a reverse flow, consisting largely of Colombians returning home.

Forced displacement continues to be prevalent because of violence among guerrillas, paramilitary groups, and Colombian security forces. Afro-Colombian and indigenous populations are disproportionately affected. Even with the Colombian Government's December 2016 peace agreement with the Revolutionary Armed Forces of Colombia (FARC), the risk of displacement remains as other rebel groups fill the void left by the FARC. Between 1985 and September 2017, nearly 7.6 million persons have been internally displaced, the highest total in the world. These estimates may undercount actual numbers because many internally displaced persons are not registered. Historically, Colombia also has one of the world's highest levels of forced disappearances. About 30,000 cases have been recorded over the last four decades—although the number is likely to be much higher—including human rights activists, trade unionists, Afro-Colombians, indigenous people, and farmers in rural conflict zones.

Because of political violence and economic problems, Colombia received limited numbers of immigrants during the 19th and 20th centuries, mostly from the Middle East, Europe, and Japan. More recently, growth in the oil, mining, and South America—The World Factbook—Central Intelligence Agency manufacturing sectors has attracted increased labor migration; the primary source countries are Venezuela, the US, Mexico, and Argentina. Colombia has also become a transit area for illegal migrants from Africa, Asia, and the Caribbean—especially Haiti and Cuba—who are en route to the US or Canada.

Age structure: *0-14 years:* 23.27% (male 5,853,351/female 5,567,196)
15-24 years: 16.38% (male 4,098,421/female 3,939,870)
25-54 years: 42.04% (male 10,270,516/female 10,365,423)
55-64 years: 9.93% (male 2,307,705/female 2,566,173)
65 years and over: 8.39% (male 1,725,461/female 2,390,725) (2020 est.)

Dependency ratios: *total dependency ratio:* 45.4
youth dependency ratio: 32.3
elderly dependency ratio: 13.2
potential support ratio: 7.6 (2020 est.)

Median age: *total:* 31.2 years
male: 30.2 years
female: 32.2 years (2020 est.)
country comparison to the world: 115

Population growth rate: 0.93% (2020 est.)
country comparison to the world: 114

Birth rate: 15.4 births/1,000 population (2020 est.)
country comparison to the world: 115

Death rate: 5.6 deaths/1,000 population (2020 est.)
country comparison to the world: 180

Net migration rate: -0.6 migrant(s)/1,000 population (2020 est.)
country comparison to the world: 130

Population distribution: the majority of people live in the north and west where agricultural opportunities and natural resources are found; the vast grasslands of the llanos to the south and east, which make up approximately 60% of the country, are sparsely populated

Urbanization: *urban population:* 81.4% of total population (2020)
rate of urbanization: 1.22% annual rate of change (2015-20 est.)

total population growth rate v. urban population growth rate, 2000-2030: Major urban areas—population: 10.978 million BOGOTA (capital), 4.000 million Medellin, 2.782 million Cali, 2.273 million Barranquilla, 1.331 million Bucaramanga, 1.063 million Cartagena (2020)

Sex ratio: *at birth:* 1.06 male(s)/female
0-14 years: 1.05 male(s)/female
15-24 years: 1.04 male(s)/female
25-54 years: 0.99 male(s)/female
55-64 years: 0.9 male(s)/female
65 years and over: 0.72 male(s)/female
total population: 0.98 male(s)/female (2020 est.)

Mother's mean age at first birth: 21.7 years (2015 est.)
note: median age at first birth among women 25-29

Maternal mortality rate: 83 deaths/100,000 live births (2017 est.)
country comparison to the world: 78

Infant mortality rate: *total:* 12.3 deaths/1,000 live births
male: 14.9 deaths/1,000 live births
female: 9.5 deaths/1,000 live births (2020 est.)
country comparison to the world: 106

Life expectancy at birth: *total population:* 76.6 years
male: 73.5 years
female: 80 years (2020 est.)
country comparison to the world: 92

Total fertility rate: 1.94 children born/woman (2020 est.)
country comparison to the world: 122

Contraceptive prevalence rate: 81% (2015/16)

Drinking water source:
improved:
urban: 100% of population
rural: 86.4% of population
total: 97.3% of population
unimproved:
urban: 0% of population
rural: 13.6% of population
total: 2.7% of population (2017 est.)

Current Health Expenditure: 7.2% (2017)

Physicians density: 2.11 physicians/1,000 population (2017)

Hospital bed density: 1.7 beds/1,000 population (2017)

Sanitation facility access:
improved:
urban: 98.3% of population
rural: 80.1% of population
total: 94.7% of population
unimproved:
urban: 1.7% of population
rural: 19.9% of population
total: 5.3% of population (2017 est.)

HIV/AIDS—adult prevalence rate: 0.5% (2019 est.)
country comparison to the world: 68

HIV/AIDS—people living with HIV/AIDS: 200,000 (2019 est.)
country comparison to the world: 29

HIV/AIDS—deaths: 4,100 (2019 est.)
country comparison to the world: 29

Major infectious diseases: *degree of risk:* high (2020)
food or waterborne diseases: bacterial diarrhea
vectorborne diseases: dengue fever, malaria, and yellow fever
note: widespread ongoing transmission of a respiratory illness caused by the novel coronavirus (COVID-19) is occurring throughout Colombia; as of 10 November 2020, Colombia has reported a total of 1,127,733 cases of COVID-19 or 22,163 cumulative cases of COVID-19 per 1 million population with 637 cumulative deaths per 1 million population

Obesity—adult prevalence rate: 22.3% (2016)
country comparison to the world: 78

Children under the age of 5 years underweight: 3.7% (2015/16)
country comparison to the world: 92

Education expenditures: 4.5% of GDP (2017)
country comparison to the world: 86

Literacy: *definition:* age 15 and over can read and write
total population: 95.1%
male: 94.9%
female: 95.3% (2018)

School life expectancy (primary to tertiary education): *total:* 14 years
male: 14 years
female: 15 years (2018)

Unemployment, youth ages 15-24: *total:* 18.5%
male: 14.4%
female: 24% (2018 est.)
country comparison to the world: 71

GOVERNMENT

Country name: *conventional long form:* Republic of Colombia
conventional short form: Colombia
local long form: Republica de Colombia
local short form: Colombia

etymology: the country is named after explorer Christopher COLUMBUS

Government type: presidential republic

Capital: *name:* Bogota

geographic coordinates: 4 36 N, 74 05 W
time difference: UTC-5 (same time as Washington, DC, during Standard Time)
etymology: originally referred to as "Bacata," meaning "enclosure outside of the farm fields," by the indigenous Muisca

Administrative divisions: 32 departments (departamentos, singular—departamento) and 1 capital district* (distrito capital); Amazonas, Antioquia, Arauca, Atlantico, Bogota*, Bolivar, Boyaca, Caldas, Caqueta, Casanare, Cauca, Cesar, Choco, Cordoba, Cundinamarca, Guainia, Guaviare, Huila, La Guajira, Magdalena, Meta, Narino, Norte de Santander, Putumayo, Quindio, Risaralda, Archipielago de San Andres, Providencia y Santa Catalina (colloquially San Andres y Providencia), Santander, Sucre, Tolima, Valle del Cauca, Vaupes, Vichada

Independence: 20 July 1810 (from Spain)

National holiday: Independence Day, 20 July (1810)

Constitution: *history:* several previous; latest promulgated 4 July 1991
amendments: proposed by the government, by Congress, by a constituent assembly, or by public petition; passage requires a majority vote by Congress in each of two consecutive sessions; passage of amendments to constitutional articles on citizen rights, guarantees, and duties also require approval in a referendum by over one half of voters and participation of over one fourth of citizens registered to vote; amended many times, last in 2020

Legal system: civil law system influenced by the Spanish and French civil codes

International law organization participation: has not submitted an ICJ jurisdiction declaration; accepts ICCt jurisdiction

Citizenship: *citizenship by birth:* no
citizenship by descent only: least one parent must be a citizen or permanent resident of Colombia
dual citizenship recognized: yes
residency requirement for naturalization: 5 years

Suffrage: 18 years of age; universal

Executive branch: *chief of state:* President Ivan DUQUE Marquez (since 7 August 2018); Vice President Marta Lucia RAMIREZ Blanco (since 7 August 2018); the president is both chief of state and head of government

head of government: President Ivan DUQUE Marquez (since 7 August 2018); Vice President Marta Lucia RAMIREZ Blanco (since 7 August 2018)
cabinet: Cabinet appointed by the president
elections/appointments: president directly elected by absolute majority vote in 2 rounds if needed for a single 4-year term; election last held on 27 May 2018 with a runoff held on 17 June

2018 (next to be held in 2022); note—political reform in 2015 eliminated presidential reelection
election results: Ivan DUQUE Marquez elected president in second round; percent of vote—Ivan DUQUE Marquez (CD) 54%, Gustavo PETRO (Humane Colombia) 41.8%, other/blank/invalid 4.2%

Legislative branch: *description:* bicameral Congress or Congreso consists of:
Senate or Senado (108 seats; 100 members elected in a single nationwide constituency by party-list proportional representation vote, 2 members elected in a special nationwide constituency for indigenous communities, 5 members of the People's Alternative Revolutionary Force (FARC) political party for the 2018 and 2022 elections only as per the 2016 peace accord, and 1 seat reserved for the runner-up presidential candidate in the recent election; all members serve 4year terms)
Chamber of Representatives or Camara de Representantes (172 seats; 165 members elected in multi-seat constituencies by party-list proportional representation vote, 5 members of the FARC for the 2018 and 2022 elections only as per the 2016 peace accord, and 1 seat reserved for the runner-up vice presidential candidate in the recent election; all members serve 4-year terms)
elections: Senate—last held on 11 March 2018 (next to be held in March 2022)
Chamber of Representatives—last held on 11 March 2018 (next to be held in March 2022)
election results: Senate—percent of vote by party—NA; seats by party—CD 19, CR 16, PC 15, PL 14, U Party 14, Green Alliance 10, PDA 5, other 9; composition—men 77, women 31, percent of women 28.7%
Chamber of Representatives—percent of vote by party—NA; seats by party—PL 35, CD 32, CR 30, U Party 25, PC 21, Green Alliance 9, other 13; composition—men 147, women 25, percent of women 14.5%; total Congress percent of women 20%

Judicial branch: *highest courts:* Supreme Court of Justice or Corte Suprema de Justicia (consists of the Civil-Agrarian and Labor Chambers each with 7 judges, and the Penal Chamber with 9 judges); Constitutional Court (consists of 9 magistrates); Council of State (consists of 27 judges); Superior Judiciary Council (consists of 13 magistrates)
judge selection and term of office: Supreme Court judges appointed by the Supreme Court members from candidates submitted by the Superior Judiciary Council; judges elected for individual 8-year terms; Constitutional Court magistrates— nominated by the president, by the Supreme Court, and elected by the Senate; judges elected for individual 8-year terms; Council of State members appointed by the State Council plenary from lists nominated by the Superior Judiciary Council
subordinate courts: Superior Tribunals (appellate courts for each of the judicial districts); regional courts; civil municipal courts; Superior Military Tribunal; first instance administrative courts

Political parties and leaders: Alternative Democratic Pole or PDA [Jorge Enrique ROBLEDO]

Citizens Option (Opcion Ciudadana) or OC [Angel ALIRIO Moreno] (formerly known as the National Integration Party or PIN)

Conservative Party or PC [Hernan ANDRADE]

Democratic Center Party or CD [Alvaro URIBE Velez]

Green Alliance [Claudia LOPEZ Hernandez]

Humane Colombia [Gustavo PETRO]

Liberal Party or PL [Cesar GAVIRIA]

People's Alternative Revolutionary Force or FARC [Rodrigo LONDONO Echeverry]

Radical Change or CR [Rodrigo LARA Restrepo]

Social National Unity Party or U Party [Roy BARRERAS]

note: Colombia has numerous smaller political movements

International organization participation: BCIE, BIS, CAN, Caricom (observer), CD, CDB, CELAC, EITI (candidate country), FAO, G-3, G-24, G-77, IADB, IAEA, IBRD, ICAO, ICC (national committees), ICCt, ICRM, IDA, IFAD, IFC, IFRCS, IHO, ILO, IMF, IMO, IMSO, Interpol, IOC, IOM, IPU, ISO, ITSO, ITU, ITUC (NGOs), LAES, LAIA, Mercosur (associate), MIGA, NAM, OAS, OPANAL, OPCW, Pacific Alliance, PCA, UN, UNASUR, UNCTAD, UNESCO, UNHCR, UNIDO, Union Latina, UNWTO, UPU, WCO, WFTU (NGOs), WHO, WIPO, WMO, WTO

Diplomatic representation in the US: *chief of mission:* Ambassador Francisco SANTOS Calderon (since 17 September 2018)

chancery: 1724 Massachusetts Avenue NW, Washington, DC 20036

telephone: [1] (202) 387-8338

FAX: [1] (202) 232-8643

consulate(s) general: Atlanta, Houston, Los Angeles, Miami, New York, Newark (NJ), Orlando, San Juan (Puerto Rico)

consulate(s): Boston, Chicago, San Francisco

Diplomatic representation from the US: *chief of mission:* Ambassador Philip S. GOLDBERG (since 19 September 2019)

telephone: [57] (1) 275-2000

embassy: Carrera 45, No. 24B-27, Bogota

mailing address: Carrera 45 No. 24B-27, Bogota, D.C.

FAX: [57] (1) 275-4600

Flag description: three horizontal bands of yellow (top, double-width), blue, and red; the flag retains the three main colors of the banner of Gran Colombia, the short-lived South American republic that broke up in 1830; various interpretations of the colors exist and include: yellow for the gold in Colombia's land, blue for the seas on its shores, and red for the blood spilled in attaining freedom; alternatively, the colors have been described as representing more elemental concepts such as sovereignty and justice (yellow), loyalty and vigilance (blue), and valor and generosity (red); or simply the principles of liberty, equality, and fraternity

note: similar to the flag of Ecuador, which is longer and bears the Ecuadorian coat of arms superimposed in the center

National symbol(s): Andean condor; national colors: yellow, blue, red

National anthem: *name:* "Himno Nacional de la Republica de Colombia" (National Anthem of the Republic of Colombia)

lyrics/music: Rafael NUNEZ/Oreste SINDICI

note: adopted 1920; the anthem was created from an inspirational poem written by President Rafael NUNEZ

0:00 / 2:40

ECONOMY

Economy—overview: Colombia heavily depends on energy and mining exports, making it vulnerable to fluctuations in commodity prices. Colombia is Latin America's fourth largest oil producer and the world's fourth largest coal producer, third largest coffee exporter, and second largest cut flowers exporter. Colombia's economic development is hampered by inadequate infrastructure, poverty, narcotrafficking, and an uncertain security situation, in addition to dependence on primary commodities (goods that have little value-added from processing or labor inputs).

Colombia's economy slowed in 2017 because of falling world market prices for oil and lower domestic oil production due to insurgent attacks on pipeline infrastructure. Although real GDP growth averaged 4.7% during the past decade, it fell to an estimated 1.8% in 2017. Declining oil prices also have contributed to reduced government revenues. In 2016, oil revenue dropped below 4% of the federal budget and likely remained below 4% in 2017. A Western credit rating agency in December 2017 downgraded Colombia's sovereign credit rating to BBB-, because of weaker-than-expected growth and increasing external debt. Colombia has struggled to address local referendums against foreign investment, which have slowed its expansion, especially in the oil and mining sectors. Colombia's FDI declined by 3% to \$10.2 billion between January and September 2017.

Colombia has signed or is negotiating Free Trade Agreements (FTA) with more than a dozen countries; the US-Colombia FTA went into effect in May 2012. Colombia is a founding member of the Pacific Alliance—a regional trade block formed in 2012 by Chile, Colombia, Mexico, and Peru to promote regional trade and economic integration. The Colombian government took steps in 2017 to address several bilateral trade irritants with the US, including those on truck scrappage, distilled spirits, pharmaceuticals, ethanol imports, and labor rights. Colombia hopes to accede to the Organization for Economic Cooperation and Development

GDP (purchasing power parity): \$711.6 billion (2017 est.)

\$699.1 billion (2016 est.)

\$685.6 billion (2015 est.)

note: data are in 2017 dollars

country comparison to the world: 31

GDP (official exchange rate): \$314.5 billion (2017 est.)

GDP—real growth rate: 3.26% (2019 est.)

2.51% (2018 est.)

1.36% (2017 est.)

country comparison to the world: 92

GDP—per capita (PPP): \$14,400 (2017 est.)

\$14,300 (2016 est.)

\$14,200 (2015 est.)

note: data are in 2017 dollars

country comparison to the world: 116

Gross national saving: 18.9% of GDP (2017 est.)

19% of GDP (2016 est.)

17.4% of GDP (2015 est.)

country comparison to the world: 104

GDP—composition, by end use: *household consumption:* 68.2% (2017 est.)

government consumption: 14.8% (2017 est.)

investment in fixed capital: 22.2% (2017 est.)

investment in inventories: 0.2% (2017 est.)

exports of goods and services: 14.6% (2017 est.)

imports of goods and services: -19.7% (2017 est.)

GDP—composition, by sector of origin: *agriculture:* 7.2% (2017 est.)

industry: 30.8% (2017 est.)

services: 62.1% (2017 est.)

Agriculture—products: coffee, cut flowers, bananas, rice, tobacco, corn, sugarcane, cocoa beans, oilseed, vegetables; shrimp; forest products

Industries: textiles, food processing, oil, clothing and footwear, beverages, chemicals, cement; gold, coal, emeralds

Industrial production growth rate: -2.2% (2017 est.)

country comparison to the world: 185

Labor force: 19.309 million (2020 est.)

country comparison to the world: 27

Labor force—by occupation: *agriculture:* 17%

industry: 21%

South America—The World Factbook—Central Intelligence Agency

services: 62% (2011 est.)

Unemployment rate: 10.5% (2019 est.)

9.68% (2018 est.)

country comparison to the world: 152

Population below poverty line: 28% (2017 est.)

Household income or consumption by percentage share: *lowest 10%:* 1.2%

highest 10%: 39.6% (2015 est.)

Budget: *revenues:* 83.35 billion (2017 est.)

expenditures: 91.73 billion (2017 est.)

Taxes and other revenues: 26.5% (of GDP) (2017 est.)

country comparison to the world: 109

Budget surplus (+) or deficit (-): -2.7% (of GDP) (2017 est.)

country comparison to the world: 119

Public debt: 49.4% of GDP (2017 est.)

49.8% of GDP (2016 est.)

note: data cover general government debt, and includes debt instruments issued (or owned) by government entities other than the treasury; the

data include treasury debt held by foreign entities; the data include debt issued by subnational entities
country comparison to the world: 102

Fiscal year: calendar year

Inflation rate (consumer prices): 4.3% (2017 est.) 7.5% (2016 est.)
country comparison to the world: 163

Current account balance: -$13.748 billion (2019 est.)
-$13.118 billion (2018 est.)
country comparison to the world: 196

Exports: $39.48 billion (2017 est.)
$31.39 billion (2016 est.)
country comparison to the world: 55

Exports—partners: US 28.5%, Panama 8.6%, China 5.1% (2017)

Exports—commodities: petroleum, coal, emeralds, coffee, nickel, cut flowers, bananas, apparel

Imports: $44.24 billion (2017 est.)
$43.24 billion (2016 est.)
country comparison to the world: 57

Imports—commodities: industrial equipment, transportation equipment, consumer goods, chemicals, paper products, fuels, electricity

Imports—partners: US 26.3%, China 19.3%, Mexico 7.5%, Brazil 5%, Germany 4.1% (2017)

Reserves of foreign exchange and gold: $47.13 billion (31 December 2017 est.)

$46.18 billion (31 December 2016 est.)
country comparison to the world: 42

Debt—external: $124.6 billion (31 December 2017 est.)
$115 billion (31 December 2016 est.)
country comparison to the world: 46

Exchange rates: Colombian pesos (COP) per US dollar—
2,957 (2017 est.)
3,055.3 (2016 est.)
3,055.3 (2015 est.)
2,001 (2014 est.)
2,001.1 (2013 est.)

ENERGY

Electricity access: *population without electricity:* 1 million (2017)
electrification—total population: 99% (2016)
electrification—urban areas: 100% (2016)
electrification—rural areas: 95.7% (2016)

Electricity—production: 74.92 billion kWh (2016 est.)
country comparison to the world: 41

Electricity—consumption: 68.25 billion kWh (2016 est.)
country comparison to the world: 40

Electricity—exports: 460 million kWh (2015 est.)
country comparison to the world: 69

Electricity—imports: 378 million kWh (2016 est.)
country comparison to the world: 82

Electricity—installed generating capacity: 16.89 million kW (2016 est.)
country comparison to the world: 49

Electricity—from fossil fuels: 29% of total installed capacity (2016 est.)
country comparison to the world: 184

Electricity—from nuclear fuels: 0% of total installed capacity (2017 est.)
country comparison to the world: 67

Electricity—from hydroelectric plants: 69% of total installed capacity (2017 est.)
country comparison to the world: 17

Electricity—from other renewable sources: 2% of total installed capacity (2017 est.)
country comparison to the world: 137

Crude oil—production: 863,000 bbl/day (2018 est.)
country comparison to the world: 22

Crude oil—exports: 726,700 bbl/day (2015 est.)
country comparison to the world: 18

Crude oil—imports: 0 bbl/day (2015 est.)
country comparison to the world: 110

Crude oil—proved reserves: 1.665 billion bbl (1 January 2018 est.)
country comparison to the world: 36

Refined petroleum products—production: 303,600 bbl/day (2015 est.)
country comparison to the world: 41

Refined petroleum products—consumption: 333,000 bbl/day (2016 est.)
country comparison to the world: 40

Refined petroleum products—exports: 56,900 bbl/day (2015 est.)
country comparison to the world: 52

Refined petroleum products—imports: 57,170 bbl/day (2015 est.)
country comparison to the world: 74

Natural gas—production: 10.02 billion cu m (2017 est.)
country comparison to the world: 41

Natural gas—consumption: 10.08 billion cu m (2017 est.)
country comparison to the world: 48

Natural gas—exports: 0 cu m (2017 est.)
country comparison to the world: 83

Natural gas—imports: 48.14 million cu m (2017 est.)
country comparison to the world: 76

Natural gas—proved reserves: 113.9 billion cu m (1 January 2018 est.)
country comparison to the world: 49

Carbon dioxide emissions from consumption of energy: 95.59 million Mt (2017 est.)
country comparison to the world: 45

COMMUNICATIONS

Telephones—fixed lines: *total subscriptions:* 6,774,363
subscriptions per 100 inhabitants: 13.93 (2019 est.)
country comparison to the world: 23

Telephones—mobile cellular: *total subscriptions:* 64,033,049

subscriptions per 100 inhabitants: 131.67 (2019 est.)
country comparison to the world: 24

Telecommunication systems: *general assessment:* fastest growing sector is mobile broadband with LTE infrastructure and investment in 5G; strong demand in rural areas for mobile broadband, potential is high while penetration is low; fiber-optic network linking 50 cities; the cable sector commands about half of the market by subscribers, with DSL having a declining share while fiberbased broadband is developing strongly; competition among the MVNO (mobile virtual network operator) sector has promoted 2.9 million subscribers as of mid-2018; most infrastructure is primarily in high-density urban areas; growing popularity of bundled services (2020)
domestic: fixed-line connections stand at about 14 per 100 persons; mobile cellular telephone subscribership is about 132 per 100 persons; competition among cellular service providers is resulting in falling local and international calling rates and contributing to the steep decline in the market share of fixed-line services; domestic satellite system with 41 earth stations (2019)
international: country code—57; landing points for the SAC, Maya-1, SAIT, ACROS, AMX-1, CFX-1, PCCS, Deep Blue Cable, Globe Net, PAN-AM, SAm-1 submarine cable systems providing links to the US, parts of the Caribbean, and Central and South America; satellite earth stations—10 (6 Intelsat, 1 Inmarsat, 3 fully digitalized international switching centers) (2019)
note: the COVID-19 outbreak is negatively impacting telecommunications production and supply chains globally; consumer spending on telecom devices and services has also slowed due to the pandemic's effect on economies worldwide; overall progress towards improvements in all facets of the telecom industry—mobile, fixed-line, broadband, submarine cable and satellite—has moderated

Broadcast media: combination of state-owned and privately owned broadcast media provide service; more than 500 radio stations and many national, regional, and local TV stations (2019)

Internet country code: .co

Internet users: total: 29,990,017

percent of population: 62.26% (July 2018 est.)
country comparison to the world: 27

Broadband—fixed subscriptions: *total:* 6,678,543
subscriptions per 100 inhabitants: 14 (2018 est.)
country comparison to the world: 25

TRANSPORTATION

National air transport system: *number of registered air carriers:* 12 (2020)
inventory of registered aircraft operated by air carriers: 157
annual passenger traffic on registered air carriers: 33,704,037 (2018)
annual freight traffic on registered air carriers: 1,349,450,000 mt-km (2018)

Civil aircraft registration country code prefix: HJ, HK (2016)

Airports: 836 (2013)
country comparison to the world: 8

Airports—with paved runways: *total:* 121 (2017)
over 3,047 m: 2 (2017)
2,438 to 3,047 m: 9 (2017)
1,524 to 2,437 m: 39 (2017)
914 to 1,523 m: 53 (2017)
under 914 m: 18 (2017)

Airports—with unpaved runways: *total:* 715 (2013)
over 3,047 m: 1 (2013)
1,524 to 2,437 m: 25 (2013)
914 to 1,523 m: 201 (2013)
under 914 m: 488 (2013)

Heliports: 3 (2013)

Pipelines: 4991 km gas, 6796 km oil, 3429 km refined products (2013)

Railways: *total:* 2,141 km (2015)
standard gauge: 150 km 1.435-m gauge (2015)
narrow gauge: 1,991 km 0.914-m gauge (2015)
country comparison to the world: 72

Roadways: *total:* 206,500 km (2016)
country comparison to the world: 26

Waterways: 24,725 km (18,300 km navigable; the most important waterway, the River Magdalena, of which 1,488 km is navigable, is dredged regularly to ensure safe passage of cargo vessels and container barges) (2012)
country comparison to the world: 6

Merchant marine: *total:* 115
by type: general cargo 21, oil tanker 9, other 85 (2019)
country comparison to the world: 82

Ports and terminals: *major seaport(s):* Atlantic Ocean (Caribbean)—Cartagena, Santa Marta, Turbo
oil terminal(s): Covenas offshore terminal
container port(s) (TEUs): Cartagena (2,663,415) (2017)
river port(s): Barranquilla (Rio Magdalena)
dry bulk cargo port(s): Puerto Bolivar (coal)

Pacific Ocean—Buenaventura

MILITARY AND SECURITY

Military and security forces: Military Forces of Colombia (Fuerzas Militares de Colombia): National Army (Ejercito Nacional), Republic of Colombia Navy (Armada Republica de Colombia, ARC; includes Coast Guard), Colombian Air Force (Fuerza Aerea de Colombia, FAC);
Colombian National Police (civilian force that is part of the Ministry of Defense) (2020)

Military expenditures: 3.2% of GDP (2019)
3.1% of GDP (2018 est.)
3.2% of GDP (2017)
3.1% of GDP (2016)
3.1% of GDP (2015)
country comparison to the world: 23

Military and security service personnel strengths: size estimates for the Military Forces of Colombia (FMC) vary; approximately 295,000 total active troops (235,000 Army; 45,000 Navy, including about 22,000 marines; 14,000 Air Force) (2019)

Military equipment inventories and acquisitions: the Colombian military inventory includes a wide mix of equipment from a variety of suppliers, including Brazil, Canada, Europe, Israel, South Korea, and the US; Germany, Israel, and the US are the leading suppliers of military hardware since 2010; Colombia's defense industry is active in producing air, land, and naval platforms (2019 est.)

Military deployments: 275 Egypt (MFO) (Dec. 2019)

Military service age and obligation: 18-24 years of age for compulsory and voluntary military service; service obligation is 18 months (2012)

Military—note: the Colombian Armed Forces are primarily focused on internal security, particularly counter-narcotics, counter-terrorism, and counterinsurgency operations against drug traffickers, militants from the Revolutionary Armed Forces of Colombia (FARC) and National Liberation Army (ELN) terrorist/guerrilla organizations, and other illegal armed groups; the Colombian Government signed a peace agreement with the FARC in 2016, but some former members (known as dissidents) have returned to fighting; the Colombian military resumed operations against FARC dissidents and their successor paramilitary groups in late 2019; in 2017, the Colombian Government initiated formal peace talks with the ELN, but in January 2019, the government ended the peace talks shortly after the ELN exploded a car bomb at the National Police Academy in Bogotá; the military is also focused on the security challenges posed by its neighbor, Venezuela (2020)

TERRORISM

Terrorist group(s): National Liberation Army; Revolutionary Armed Forces of Colombia (2020)
note: details about the history, aims, leadership, organization, areas of operation, tactics, targets, weapons, size, and sources of support of the group(s) appear(s) in Appendix-T

TRANSNATIONAL ISSUES

Disputes—international: in December 2007, ICJ allocated San Andres, Providencia, and Santa Catalina islands to Colombia under 1928 Treaty but did not rule on 82 degrees W meridian as maritime boundary with Nicaragua; managed dispute with Venezuela over maritime boundary and Venezuelan-administered Los Monjes Islands near the Gulf of Venezuela; Colombian-organized illegal narcotics, guerrilla, and paramilitary activities penetrate all neighboring borders and have caused Colombian citizens to flee mostly into neighboring countries; Colombia, Honduras, Nicaragua, Jamaica, and the US assert various claims to Bajo Nuevo and Serranilla Bank

Refugees and internally displaced persons: *refugees (country of origin):* 768,714 (Venezuela) (economic and political crisis; includes Venezuelans who have claimed asylum, are recognized as refugees, or received alternative legal stay)(2020)
IDPs: 7,967,965 (conflict between government and illegal armed groups and drug traffickers since 1985; about 300,000 new IDPs each year since 2000) (2020)
stateless persons: 11 (2019)

Illicit drugs: illicit producer of coca, opium poppy, and cannabis; world's leading coca cultivator with 188,000 hectares in coca cultivation in 2016, a 18% increase over 2015, producing a potential of 710 mt of pure cocaine; the world's largest producer of coca derivatives; supplies cocaine to nearly all of the US market and the great majority of other international drug markets; in 2016, the Colombian government reported manual eradication of 17,642 hectares; Colombia suspended aerial eradication in October 2015 making 2016 the first full year without aerial eradication; a significant portion of narcotics proceeds are either laundered or invested in Colombia through the black market peso exchange; Colombia probably remains the second largest supplier of heroin to the US market; opium poppy cultivation was estimated to be 1,100 hectares in 2015, sufficient to potentially produce three metric tons of pure heroin

COMOROS

INTRODUCTION

Background: The archipelago of the Comoros in the Indian Ocean, composed of the islands of Mayotte, Anjouan, Moheli, and Grande Comore declared independence from France
on 6 July 1975. Residents of Mayotte voted to remain in France, and France now has classified it as a department of France. Since independence, Comoros has endured political instability through realized and attempted coups. In 1997, the islands of Anjouan and Moheli declared independence
from Comoros. In 1999, military chief Col. AZALI Assoumani seized power of the entire government in a bloodless coup; he initiated the 2000 Fomboni Accords, a power-sharing agreement in which the federal presidency rotates among the three islands, and each island maintains its local government.

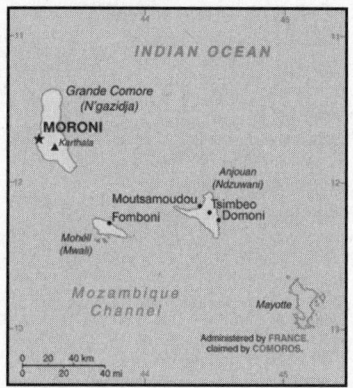

AZALI won the 2002 federal presidential election as president of the Union of the Comoros from Grande Comore Island, which held the first four-year term. AZALI stepped down in 2006 and President Ahmed Abdallah Mohamed SAMBI was elected to office as president from Anjouan. In 2007, Mohamed BACAR effected Anjouan's de-facto secession from the Union of the Comoros, refusing to step down when Comoros' other islands held legitimate elections in July. The African Union (AU) initially attempted to resolve the political crisis by applying sanctions and a naval blockade to Anjouan, but in March 2008 the AU and Comoran soldiers seized the island. The island's inhabitants generally welcomed the move. In 2009, the Comorian population approved a constitutional referendum extending the term of the president from four years to five years. In May 2011, Ikililou DHOININE won the presidency in peaceful elections widely deemed to be free and fair. In closely contested elections in 2016, former President AZALI Assoumani won a second term, when the rotating presidency returned to Grande Comore. A new July 2018 constitution removed the presidential term limits and the requirement for the presidency to rotate between the three main islands. In August 2018, President AZALI formed a new government and subsequently ran and was elected president in March 2019.

GEOGRAPHY

Location: Southern Africa, group of islands at the northern mouth of the Mozambique Channel, about two-thirds of the way between northern Madagascar and northern Mozambique

Geographic coordinates: 12 10 S, 44 15 E

Map references: Africa

Area: *total:* 2,235 sq km
land: 2,235 sq km
water: 0 sq km
country comparison to the world: 180

Area—comparative: slightly more than 12 times the size of Washington, DC

Land boundaries: 0 km

Coastline: 340 km

Maritime claims: *territorial sea:* 12 nm

exclusive economic zone: 200 nm

Climate: tropical marine; rainy season (November to May)

Terrain: volcanic islands, interiors vary from steep mountains to low hills

Elevation: *Lowest point:* Indian Ocean 0 m highest point: Karthala 2,360 m
Natural resources: fish

Land use: *agricultural land:* 84.4% (2011 est.)
arable land: 46.7% (2011 est.) / *permanent crops:* 29.6% (2011 est.) / *permanent pasture:* 8.1% (2011 est.)
forest: 1.4% (2011 est.)
other: 14.2% (2011 est.)
Irrigated land: 1.3 sq km (2012)

Population distribution: the capital city of Maroni, located on the western side of the island of Grande Comore, is the country's largest city; however, of the three islands that comprise Comoros, it is Anjouan that is the most densely populated as shown in this population distribution map

Natural hazards: cyclones possible during rainy season (December to April); volcanic activity on Grand Comore

volcanism: Karthala (2,361 m) on Grand Comore Island last erupted in 2007; a 2005 eruption forced thousands of people to be evacuated and produced a large ash cloud

Environment—current issues: deforestation; soil degradation and erosion results from forest loss and from crop cultivation on slopes without proper terracing; marine biodiversity affected as soil erosion leads to the silting of coral reefs

Environment—international agreements: *party to:* Biodiversity, Climate Change, Climate Change-Kyoto Protocol, Desertification, Endangered Species, Hazardous Wastes, Law of the Sea, Ozone Layer Protection, Ship Pollution, Wetlands
signed, but not ratified: none of the selected agreements

Geography—note: important location at northern end of Mozambique Channel

PEOPLE AND SOCIETY

Population: 846,281 (July 2020 est.)
country comparison to the world: 163

Nationality: *noun:* Comoran(s)
adjective: Comoran

Ethnic groups: Antalote, Cafre, Makoa, Oimatsaha, Sakalava

Languages: Arabic (official), French (official), Shikomoro (official; a blend of Swahili and Arabic) (Comorian)

Religions: Sunni Muslim 98%, other (including Shia Muslim, Roman Catholic, Jehovah's Witness, Protestant) 2%
note: Sunni Islam is the state religion

Demographic profile: Comoros' population is a melange of Arabs, Persians, Indonesians, Africans, and Indians, and the much smaller number of Europeans that settled on the islands between the 8th and 19th centuries, when they served as

a regional trade hub. The Arab and Persian influence is most evident in the islands' overwhelmingly Muslim majority—about 98% of Comorans are Sunni Muslims. The country is densely populated, averaging nearly 350 people per square mile, although this varies widely among the islands, with Anjouan being the most densely populated.

Given the large share of land dedicated to agriculture and Comoros' growing population, habitable land is becoming increasingly crowded. The combination of increasing population pressure on limited land and resources, widespread poverty, and poor job prospects motivates thousands of Comorans each year to attempt to illegally migrate using small fishing boats to the neighboring island of Mayotte, which is a French territory. The majority of legal Comoran migration to France came after Comoros' independence from France in 1975, with the flow peaking in the mid-1980s.

At least 150,000 to 200,000 people of Comoran citizenship or descent live abroad, mainly in France, where they have gone seeking a better quality of life, job opportunities, higher education (Comoros has no universities), advanced health care, and to finance elaborate traditional wedding ceremonies (aada). Remittances from the diaspora are an economic mainstay, in 2013 representing approximately 25% of Comoros' GDP and significantly more than the value of its exports of goods and services (only 15% of GDP). Grand Comore, Comoros' most populous island, is both the primary source of emigrants and the main recipient of remittances. Most remittances are spent on private consumption, but this often goes toward luxury goods and the aada and does not contribute to economic development or poverty reduction. Although the majority of the diaspora is now French-born with more distant ties to Comoros, it is unclear whether they will sustain the current level of remittances.

Age structure: *0-14 years:* 36.68% (male 154,853/female 155,602)
15-24 years: 20.75% (male 85,208/female 90,422)
25-54 years: 33.99% (male 136,484/female 151,178)
55-64 years: 4.49% (male 17,237/female 20,781)
65 years and over: 4.08% (male 15,437/female 19,079) (2020 est.)

Dependency ratios: *total dependency ratio:* 75.5
youth dependency ratio: 67.4
elderly dependency ratio: 5.4
potential support ratio: 18.6 (2020 est.)

Median age: *total:* 20.9 years
male: 20.2 years
female: 21.5 years (2020 est.)
country comparison to the world: 189

Population growth rate: 1.47% (2020 est.)
country comparison to the world: 73

Birth rate: 23.6 births/1,000 population (2020 est.)
country comparison to the world: 52

Death rate: 6.9 deaths/1,000 population (2020 est.)
country comparison to the world: 130

Net migration rate: -2.3 migrant(s)/1,000 population (2020 est.)

country comparison to the world: 169

Population distribution: the capital city of Maroni, located on the western side of the island of Grande Comore, is the country' s largest city; however, of the three islands that comprise Comoros, it is Anjouan that is the most densely populated as shown in this population distribution map

Urbanization: *urban population:* 29.4% of total population (2020)
rate of urbanization: 2.87% annual rate of change (2015-20 est.)

total population growth rate v. urban population growth rate, 2000-2030: Major urban areas -population: 62,000 MORONI (capital) (2018)

Sex ratio: *at birth:* 1.03 male(s)/female
0-14 years: 1male(s)/female
15-24 years: 0.94 male(s)/female
25-54 years: 0.9 male(s)/female
55-64 years: 0.83 male(s)/female
65 years and over: 0.81 male(s)/female
total population: 0.94 male(s)/female (2020 est.)

Mother's mean age at first birth: 24.6 years (2012 est.)

note: median age at first birth among women 25-29

Maternal mortality rate: 273 deaths/100,000 live births (2017 est.)
country comparison to the world: 40

Infant mortality rate: *total:* 55 deaths/1,000 live births
male: 64.8 deaths/1,000 live births
female: 45.1 deaths/1,000 live births (2020 est.)
country comparison to the world: 17

Life expectancy at birth: *total population:* 65.7 years
male: 63.3 years
female: 68.1 years (2020 est.)
country comparison to the world: 191

Total fertility rate: 2.95 children born/ woman (2020 est.)
country comparison to the world: 53

Contraceptive prevalence rate: 19.4% (2012)

Drinking water source:

improved: *urban:* 97.4% of population
rural: 88.5% of population
total: 91% of population

unimproved: urban:2.6%of population
rural:11.5%of population
total: 8.9% of population (2017 est.)

Current Health Expenditure: 7.4% (2017)

Physicians density: 0.27 physicians/1,000 population (2016)

Hospital bed density: 2.2 beds/1,000 population (2010)

Sanitation facility access:

improved: *urban:* 62.4% of population
rural: 43.6% of population
total: 49% of population

unimproved: *urban:* 37.6% of population
rural: 56.4% of population
total: 51% of population (2017 est.)

HIV/AIDS—adult prevalence rate: <.1% (2019 est.)

HIV/AIDS—people living with HIV/AIDS: <200 (2019 est.)

HIV/AIDS—deaths: <100 (2019 est.)

Obesity—adult prevalence rate: 7.8% (2016)
country comparison to the world: 157

Children under the age of 5 years underweight: 16.9% (2012)
country comparison to the world: 34

Education expenditures: 2.5% of GDP (2015)
country comparison to the world: 160

Literacy: *definition:* age 15 and over can read and write
total population: 58.8%
male: 64.6%
female: 53% (2018)

School life expectancy (primary to tertiary education): *total:* 11 years
male: 11 years
female: 11 years (2014)

Unemployment, youth ages 15-24: *total:* 19.5%
male: 20%
female: 18.8% (2018)
country comparison to the world: 69

GOVERNMENT

Country name: *conventional long form:* Union of the Comoros
conventional short form: Comoros
local long form: Udzima wa Komori (Comorian), Union des Comores (French), Jumhuriyat al Qamar al Muttahidah (Arabic)
local short form: Komori (Comorian), Comores (French), Juzur al Qamar (Arabic)
etymology: name derives from the Arabic designation "Juzur al Qamar" meaning "Islands of the Moon"

Government type: federal presidential republic

Capital: *name:* Moroni
geographic coordinates: 11 42 S, 43 14 E
time difference: UTC+3 (8 hours ahead of Washington, DC, during Standard Time)
etymology: Moroni derives from "mroni," which means "at the river" in Shingazidja, the Comorian language spoken on Grande Comore (N'gazidja)

Administrative divisions: 3 islands; Anjouan (Ndzuwani), Grande Comore (N'gazidja), Moheli (Mwali)

Independence: 6 July 1975 (from France)

National holiday: Independence Day, 6 July (1975)

Constitution: *history:* previous 1996, 2001; newest adopted 30 July 2018
amendments: proposed by the president of the union or supported by at least one third of the Assembly of the Union membership; adoption requires approval by at three- quarters majority of the total Assembly membership or approval in a referendum
note: a referendum held on 30 July 2018—boycotted by the opposition—overwhelmingly approved a new constitution that allows for 2 consecutive

5-year presidential terms and revises the rotating presidency within the islands

Legal system: mixed legal system of Islamic religious law, the French civil code of 1975, and customary law

International law organization participation: has not submitted an ICJ jurisdiction declaration; accepts ICCt jurisdiction

Citizenship: *citizenship by birth:* no
citizenship by descent only: at least one parent must be a citizen of the Comoros
dual citizenship recognized: no
residency requirement for naturalization: 10 years

Suffrage: 18 years of age; universal

Executive branch: *chief of state:* President AZALI Assoumani (since 26 May 2016); note—the president is both chief of state and head of government; note—AZALI takes oath of office 2 June 2019 after 24 March 2019 reelection (2019)

head of government: President AZALI Assoumani (since 26 May 2016)
cabinet: Council of Ministers appointed by the president
elections/appointments: president directly elected by simple majority popular vote in 2 rounds for a 5-year term (eligible for a second term); election last held on 24 March 2019 (next to be held in 2024)
election results: AZALI Assoumani (CRC) elected president in first round; with a 59% of the vote;—AZALI Assoumani (CRC) 60.8%, Ahamada MAHAMOUDOU (PJ) 14.6%, and Mouigni Baraka Said SOILIHI (Independent) 5.6%

Legislative branch: *description:* unicameral Assembly of the Union (33 seats; 24 members directly elected by absolute majority vote in 2 rounds if needed and 9 members indirectly elected by the 3 island assemblies; members serve 5-year terms) (2017)
elections: last held on 19 January 2020 with a runoff on 23 February 2020 (next to be held in 2025) (2020)
election results: seats by party -1st round—Boycotting parties 16, Independent 3, CRC 2, RDC 2, RADHI 1, Orange party 0; note—9 additional seats filled by the 3 island assemblies; 2nd round—CRC 20, Orange Party 2, Independents 2;composition as of 23 January 2020 men 20, women 4, percent of women 16.7% (2019)

Judicial branch: *highest courts:* Supreme Court or Cour Supreme (consists of 7 judges)
judge selection and term of office: Supreme Court judges—selection and term of office NA
subordinate courts: Court of Appeals (in Moroni); Tribunal de premiere instance; island village (community) courts; religious courts

Political parties and leaders: Convention for the Renewal of the Comoros or CRC [AZALI Assoumani]
Democratic Rally of the Comoros or RDC [Mouigni BARAKA]
Independent Party [N/A]

Juwa Party or PJ [[Ahmed Abdallah SAMBI, Mahamoudou AHAMADA]

Orange Party [Mohamed DAOUDOU]

Party for the Comorian Agreement (Partie Pour l'Entente Commorienne) or PEC [Fahmi Said IBRAHIM]

Rally for an Alternative of Harmonious and Integrated Development or RADHI [Houmed MSAIDIE, Abdou SOEFO]

Rally with a Development Initiative for Enlightened Youth or RIDJA [Said LARIFOU]

Union for the Development of the Comoros or UPDC [Mohamed HALIFA] (2018)

International organization participation: ACP, AfDB, AMF, AOSIS, AU, CAEU (candidates), COMESA, FAO, FZ, G-77, IBRD, ICAO, ICCt, ICRM, IDA, IDB, IFAD, IFC, IFRCS,ILO, IMF, IMO, IMSO, InOC, Interpol, IOC, IOM, ITSO, ITU, ITUC (NGOs), LAS, MIGA, NAM, OIC, OIF,OPCW, UN, UNCTAD, UNESCO, UNIDO, UPU, WCO, WHO, WIPO, WMO, WTO (observer)

Diplomatic representation in the US: *chief of mission:* Ambassador Eric ANDRIAMIHAJA Robson, since March 2018

chancery: Mission to the US, 866 United Nations Plaza, Suite 418, New York, NY 10017

telephone: [1] (212) 750—1637

FAX: [1] (212) 750—1657

Diplomatic representation from the US: the US does not have an embassy in Comoros; the US Ambassador to Madagascar is accredited to Comoros

Flag description: four equal horizontal bands of yellow (top), white, red, and blue, with a green isosceles triangle based on the hoist; centered within the triangle is a vertical white crescent moon with the convex side facing the hoist and four white, five-pointed stars placed vertically in a line between the points of the crescent; the horizontal bands and the four stars represent the four main islands of the archipelago—Mwali, N'gazidja, Ndzuwani, and Mahore (Mayotte—department of France, but claimed by Comoros)

note: the crescent, stars, and color green are traditional symbols of Islam

National symbol(s): four five-pointed stars and crescent moon; national colors: green, white

National anthem: *name:* "Udzima wa ya Masiwa" (The Union of the Great Islands)

lyrics/music: Said Hachim SIDI ABDEREMANE/ Said Hachim SIDI ABDEREMANE and Kamildine ABDALLAH

note: adopted 1978

0:00 / 1:36

ECONOMY

Economy—overview: One of the world's poorest and smallest economies, the Comoros is made up of three islands that are hampered by inadequate transportation links, a young and rapidly increasing population, and few natural resources. The low educational level of the labor force contributes to a subsistence level of economic activity and a heavy dependence on foreign grants and technical assistance. Agriculture, including fishing, hunting, and forestry, accounts for about 50% of GDP, employs a majority of the labor force, and provides most of the exports. Export income is heavily reliant on the three main crops of vanilla, cloves, and ylang ylang (perfume essence); and the Comoros' export earnings are easily disrupted by disasters such as fires and extreme weather. Despite agriculture's importance to the economy, the country imports roughly 70% of its food; rice, the main staple, and other dried vegetables account for more than 25% of imports. Remittances from about 300,000 Comorans contribute about 25% of the country's GDP. France, Comoros's colonial power, remains a key trading partner and bilateral donor.

Comoros faces an education system in need of upgrades, limited opportunities for private commercial and industrial enterprises, poor health services, limited exports, and a high population growth rate. Recurring political instability, sometimes initiated from outside the country, and an ongoing electricity crisis have inhibited growth. The government, elected in mid-2016, has moved to improve revenue mobilization, reduce expenditures, and improve electricity access, although the public sector wage bill remains one of the highest in Sub-Saharan Africa. In mid-2017, Comoros joined the Southern African Development Community with 15 other regional member states.

GDP (purchasing power parity): $1.319 billion (2017 est.)

$1.284 billion (2016 est.)

$1.257 billion (2015 est.)

note: data are in 2017 dollars

country comparison to the world: 201

GDP (official exchange rate): $652 million (2017 est.)

GDP—real growth rate: 2.7% (2017 est.)

2.2% (2016 est.)

1% (2015 est.)

country comparison to the world: 107

GDP—per capita (PPP): $1,600 (2017 est.)

$1,600 (2016 est.)

$1,600 (2015 est.)

note: data are in 2017 dollars

country comparison to the world: 216

Gross national saving: 17.3% of GDP (2017 est.)

13.6% of GDP (2016 est.)

18% of GDP (2015 est.)

country comparison to the world: 117

GDP—composition, by end use: *household consumption:* 92.6% (2017 est.)

government consumption: 20.4% (2017 est.)

investment in fixed capital: 20% (2017 est.)

investment in inventories: -3.1% (2017 est.)

exports of goods and services: 17.2% (2017 est.)

imports of goods and services:—47.1% (2017 est.)

GDP—composition, by sector of origin: *agriculture:* 47.7% (2017 est.)

industry: 11.8% (2017 est.)

services: 40.5% (2017 est.)

Agriculture—products: vanilla, cloves, ylang-ylang (perfume essence), coconuts, bananas, cassava (manioc)

Industries: fishing, tourism, perfume distillation

Industrial production growth rate: 1% (2017 est.)

country comparison to the world: 154

Labor force: 278,500 (2016 est.)

country comparison to the world: 165

Labor force—by occupation: *agriculture:* 80%

industry: 20% (1996 est.)

industry and services: 20% (1996 est.)

Unemployment rate: 6.5% (2014 est.)

country comparison to the world: 101

Population below poverty line: 44.8% (2004 est.)

Household income or consumption by percentage share: *lowest 10%:* 0.9%

highest 10%: 55.2% (2004)

Budget: *revenues:* 165.2 million (2017 est.)

expenditures: 207.3 million (2017 est.)

Taxes and other revenues: 25.3% (of GDP) (2017 est.)

country comparison to the world: 117

Budget surplus (+) or deficit (-):

-6.5% (of GDP) (2017 est.)

country comparison to the world: 188

Public debt:

32.4% of GDP (2017 est.)

27.7% of GDP (2016 est.)

country comparison to the world: 160

Fiscal year: calendar year

Inflation rate (consumer prices): 1% (2017 est.)

1.8% (2016 est.)

country comparison to the world: 50

Current account balance: -$27 million (2017 est.)

-$45 million (2016 est.)

country comparison to the world: 74

Exports: $18.9 million (2017 est.)

$17.9 million (2016 est.)

country comparison to the world: 212

Exports—partners: France 36.5%, India 12.2%, Germany 8.2%, Pakistan 6.3%, Switzerland 5.8%, South Korea 4.7%, Russia 4.3% (2017)

Exports—commodities: vanilla, ylang-ylang (perfume essence), cloves

Imports: $207.8 million (2017 est.)

$189.9 million (2016 est.)

country comparison to the world: 209

Imports—commodities: rice and other foodstuffs, consumer goods, petroleum products, cement and construction materials, transport equipment

Imports—partners: UAE 32.8%, France 17.3%, China 13.2%, Madagascar 6.1%, Pakistan 4.5%, India 4.3% (2017)

Reserves of foreign exchange and gold: $208 million (31 December 2017 est.)

$159.5 million (31 December 2016 est.)

country comparison to the world: 173

Debt—external: $199.8 million (31 December 2017 est.)

$132 million (31 December 2016 est.)

country comparison to the world: 189

Exchange rates: Comoran francs (KMF) per US dollar –
458.2 (2017 est.)
444.76 (2016 est.)
444.76 (2015 est.)
443.6 (2014 est.)
370.81 (2013 est.)

ENERGY

Electricity access: *electrification—total population:* 70% (2019)
electrification—urban areas: 89% (2019)
electrification—rural areas: 62% (2019)

Electricity—production: 42 million kWh (2016 est.)
country comparison to the world: 207

Electricity—consumption: 39.06 million kWh (2016 est.)
country comparison to the world: 207

Electricity—exports: 0 kWh (2016 est.)
country comparison to the world: 122

Electricity—imports: 0 kWh (2016 est.)
country comparison to the world: 136

Electricity—installed generating capacity: 27,000 kW (2016 est.)
country comparison to the world: 203

Electricity—from fossil fuels: 96% of total installed capacity (2016 est.)
country comparison to the world: 38

Electricity—from nuclear fuels: 0% of total installed capacity (2017 est.)
country comparison to the world: 68

Electricity—from hydroelectric plants: 4% of total installed capacity (2017 est.)
country comparison to the world: 131

Electricity—from other renewable sources: 0% of total installed capacity (2017 est.)
country comparison to the world: 181

Crude oil—production: 0 bbl/day (2018 est.)
country comparison to the world: 122

Crude oil -exports: 0 bbl/day (2015 est.)
country comparison to the world: 107

Crude oil -imports: 0 bbl/day (2015 est.)
country comparison to the world: 111

Crude oil—proved reserves: 0 bbl (1 January 2018 est.)
country comparison to the world: 118

Refined petroleum products—production: 0 bbl/day (2015 est.)
country comparison to the world: 131

Refined petroleum products—consumption: 1,300 bbl/day (2016 est.)
country comparison to the world: 202

Refined petroleum products—exports: 0 bbl/day (2015 est.)
country comparison to the world: 144

Refined petroleum products—imports: 1,241 bbl/day (2015 est.)
country comparison to the world: 198

Natural gas—production: 0 cu m (2017 est.)
country comparison to the world: 117

Natural gas—consumption: 0 cu m (2017 est.)
country comparison to the world: 133

Natural gas—exports: 0 cu m (2017 est.)
country comparison to the world: 84

Natural gas—imports: 0 cu m (2017 est.)
country comparison to the world: 107

Natural gas—proved reserves: 0 cu m (1 January 2014 est.)
country comparison to the world: 123

Carbon dioxide emissions from consumption of energy: 193,600 Mt (2017 est.)
country comparison to the world: 201

COMMUNICATIONS

Telephones—fixed lines: *total subscriptions:* 9,840
subscriptions per 100 inhabitants: 1.18 (2019 est.)
country comparison to the world: 189

Telephones—mobile cellular: *total subscriptions:* 563,722
subscriptions per 100 inhabitants: 67.6 (2019 est.)
country comparison to the world: 172

Telecommunication systems: *general assessment:* Qatar launched a special program for the construction of a wireless network to inter connect the 3 islands of the archipelago; telephone service limited to the islands' few towns (2020)
domestic: fixed-line connections only about 1 per 100 persons; mobile-cellular usage over 68 per 100 persons; two companies provide domestic and international mobile service and wireless data (2019)
international: country code-269; landing point for the EASSy, Comoros Domestic Cable System, Avassa, and FLY- LION 3 fiber-optic submarine cable system connecting East Africa with Europe; HF radiotelephone communications to Madagascar and Reunion (2019)
note: the COVID- 19 outbreak is negatively impacting telecommunications production and supply chains globally; consumer spending on telecom devices and services has also slowed due to the pandemic's effect on economies worldwide; overall progress towards improvements in all facets of the telecom industry—mobile, fixed-line, broadband, submarine cable and satellite—has moderated

Broadcast media: national state-owned TV station and a TV station run by Anjouan regional government; national state-owned radio; regional governments on the islands of Grande Comore and Anjouan each operate a radio station; a few independent and small community radio stations operate on the islands of Grande Comore and Moheli, and these two islands have access to Mayotte Radio and French TV

Internet country code: .km

Internet users: *total:* 69,635

percent of population: 8.48% (July 2018 est.)
country comparison to the world: 188

Broadband—fixed subscriptions: *total:* 1,531

subscriptions per 100 inhabitants: less than 1 (2018 est.)
country comparison to the world: 191

TRANSPORTATION

National air transport system: *number of registered air carriers:* 2 (2020)
inventory of registered aircraft operated by air carriers: 9

Civil aircraft registration country code prefix: D6 (2016)

Airports: 4 (2013)
country comparison to the world: 186

Airports—with paved runways: *total:* 4 (2017)
2,438 to 3,047 m: 1 (2017)
914 to 1,523 m: 3 (2017)

Roadways: *total:* 880 km (2002)
paved: 673 km (2002)
unpaved: 207 km (2002)
country comparison to the world: 187

Merchant marine: *total:* 230
by type: bulk carrier 7, container ship 5, general cargo 109, oil tanker 27, other 82 (2019)
country comparison to the world: 62

Ports and terminals: *major seaport(s):* Moroni, Moutsamoudou

MILITARY AND SECURITY

Military and security forces: National Army for Development (l'Armee Nationale de Developpement, AND): Comoran Security Force (also called Comoran Defense Force (Force Comorienne de Defense, FCD), includes Gendarmerie), Comoran Coast Guard, Comoran Federal Police (2017)

Military service age and obligation: 18 years of age for 2-year voluntary male and female military service; no conscription (2015)

TRANSNATIONAL ISSUES

Disputes—international: claims French- administered Mayotte and challenges France's and Madagascar's claims to Banc du Geyser, a drying reef in the Mozambique Channel; in May 2008, African Union forces assisted the Comoros military recapture Anjouan Island from rebels who seized it in 2001

Trafficking in persons: *current situation:* Comoros is a source country for children subjected to forced labor and, reportedly, sex trafficking domestically, and women and children are subjected to forced labor in Mayotte; it is possibly a transit and destination country for Malagasy women and girls and a transit country for East African women and girls exploited in domestic service in the Middle East; Comoran children are forced to labor in domestic service, roadside and street vending, baking, fishing, and agriculture; some Comoran students at Koranic schools are exploited for forced agricultural or domestic labor, sometimes being subjected to physical and sexual abuse; Comoros may be particularly vulnerable to transnational trafficking because of inadequate border controls, government corruption, and the presence of international criminal networks

tier rating: Tier 3—Comoros does not fully comply with the minimum standards for the elimination of trafficking and was placed on Tier 3 after being on the Tier 2 Watch List for two consecutive

years without making progress; Parliament passed revisions to the penal code in 2014, including anti- trafficking provisions and enforcement guidelines, but these amendments have not yet been passed approved by the President and put into effect; a new child labor law was passed in 2015 prohibiting child trafficking, but existing laws do not criminalize the forced prostitution of

adults; authorities did not investigate, prosecute, or convict alleged trafficking offenders, including complicit officials; the government lacked victim identification and care referral procedures, did not assist any victims during 2014, and provided minimal support to NGOs offering victims psychosocial services (2015)

CONGO, DEMOCRATIC REPUBLIC OF THE

INTRODUCTION

Background: The Kingdom of Kongo ruled the area around the mouth of the Congo River from the 14th to 19th centuries. To the center and east, the Kingdoms of Luba and Lunda ruled from the 16th and 17th centuries to the 19th century. in the 1870s, European exploration of the Congo Basin, sponsored by King Leopold II of Belgium, eventually allowed the ruler to acquire rights to the Congo territory and to make it his private property under the name of the Congo Free State. During the Free State, the king's colonial military forced the local population to produce rubber. From 1885 to 1908, millions of Congolese people died as a result of disease and exploitation. International condemnation finally forced Leopold to cede the land to Belgium, creating the Belgian Congo.

The Republic of the Congo gained its independence from Belgium in 1960, but its early years were marred by political and social instability. Col. Joseph MOBUTU seized power and declared himself president in a November 1965 coup. He subsequently changed his name—to MOBUTU Sese Seko—as well as that of the country—to Zaire. MOBUTU retained his position for 32 years through several sham elections, as well as through brutal force. Ethnic strife and civil war, touched off by a massive inflow of refugees in 1994 from conflict in Rwanda and Burundi, led in May 1997 to the toppling of the MOBUTU regime by a rebellion backed by Rwanda and Uganda and fronted

by Laurent KABILA. KABILA renamed the country the Democratic Republic of the Congo (DRC), but in August 1998 his regime was itself challenged by a second insurrection again backed by Rwanda and Uganda. Troops from Angola, Chad, Namibia, Sudan, and Zimbabwe intervened to support KABILA's regime. In January 2001, KABILA was assassinated and his son, Joseph KABILA, was named head of state. In October 2002, the new president was successful in negotiating the withdrawal of Rwandan forces occupying the eastern DRC; two months later, the Pretoria Accord was signed by all remaining warring parties to end the fighting and establish a government of national unity. Presidential, National Assembly, and provincial legislatures took place in 2006, with Joseph KABILA elected to office.

National elections were held in November 2011 and disputed results allowed Joseph KABILA to be reelected to the presidency. While the DRC constitution barred President KABILA from running for a third term, the DRC Government delayed national elections originally slated for November 2016, to 30 December 2018. This failure to hold elections as scheduled fueled significant civil and political unrest, with sporadic street protests by KABILA's opponents and exacerbation of tensions in the tumultuous eastern DRC regions. Presidential, legislative, and provincial elections were held in late December 2018 and early 2019 across most of the country. The DRC Government canceled presidential elections in the cities of Beni and Butembo (citing concerns over an ongoing Ebola outbreak in the region) as well as Yumbi (which had recently experienced heavy violence).

Opposition candidate Felix TSHISEKEDI was announced the election winner on 10 January 2019 and inaugurated two weeks later. This was the first transfer of power to an opposition candidate without significant violence or a coup since the DRC's independence.

The DRC, particularly in the East, continues to experience violence perpetrated by more than 100 armed groups active in the region, including the Allied Democratic Forces (ADF), the Democratic Forces for the Liberation of Rwanda (FDLR), and assorted Mai Mai militias. The UN Organization Stabilization Mission in the DRC (MONUSCO) has operated in the region since 1999 and is the

largest and most expensive UN peacekeeping mission in the world.

GEOGRAPHY

Location: Central Africa, northeast of Angola

Geographic coordinates: 0 00 N, 25 00 E

Map references: Africa

Area: *total:* 2,344,858 sq km
land: 2,267,048 sq km
water: 77,810 sq km
country comparison to the world: 12

Area—comparative: slightly less than one-fourth the size of the US

Land boundaries: *total:* 10,481 km
border countries (9): Angola 2646 km (of which 225 km is the boundary of Angola's discontiguous Cabinda Province), Burundi 236 km, Central African Republic 1747 km, Republic of the Congo 1229 km, Rwanda 221 km, South Sudan 714 km, Tanzania 479 km, Uganda 877 km, Zambia 2332 km

Coastline: 37 km

Maritime claims: *territorial sea:* 12 nm
exclusive economic zone: since 2011 the DRC has a Common Interest Zone agreement with Angola for the mutual development of off- shore resources

Climate: tropical; hot and humid in equatorial river basin; cooler and drier in southern highlands; cooler and wetter in eastern highlands; north of Equator—wet season (April to October), dry season (December to February); south of Equator—wet season (November to March), dry season (April to October)

Terrain: vast central basin is a low-lying plateau; mountains in east

Elevation: *mean elevation:* 726 m
lowest point: Atlantic Ocean 0 m
highest point: Pic Marguerite on Mont Ngaliema (Mount Stanley) 5,110 m

Natural resources: cobalt, copper, niobium, tantalum, petroleum, industrial and gem diamonds, gold, silver, zinc, manganese, tin, uranium, coal, hydropower, timber

Land use: *agricultural land:* 11.4% (2011 est.)
arable land: 3.1% (2011 est.) / permanent crops: 0.3% (2011 est.) / permanent pasture: 8% (2011 est.)

forest: 67.9% (2011 est.)
other: 20.7% (2011 est.)
Irrigated land: 110 sq km (2012)

Population distribution: urban clusters are spread throughout the country, particularly in the northeast along the boarder with Uganda, Rwanda, and Burundi; the largest city is the capital, Kinshasha, located in the west along the Congo River; the south is least densely populated as shown in this population distribution map

Natural hazards: periodic droughts in south; Congo River floods (seasonal); active volcanoes in the east along the Great Rift Valley
volcanism: Nyiragongo (3,470 m), which erupted in 2002 and is experiencing ongoing activity, poses a major threat to the city of Goma, home to a quarter million people; the volcano produces unusually fast-moving lava, known to travel up to 100 km /hr; Nyiragongo has been deemed a Decade Volcano by the International Association of Volcanology and Chemistry of the Earth's Interior, worthy of study due to its explosive history and close proximity to human populations; its neighbor, Nyamuragira, which erupted in 2010, is Africa's most active volcano; Visoke is the only other historically active volcano

Environment—current issues: poaching threatens wildlife populations; water pollution; deforestation (forests endangered by fires set to clean the land for agricultural purposes; forests also used as a source of fuel); soil erosion; mining (diamonds, gold, coltan—a mineral used in creating capacitors for electronic devices) causing environmental damage

Environment—international agreements: *party to:* Biodiversity, Climate Change, Climate Change-Kyoto Protocol, Desertification, Endangered Species, Hazardous Wastes, Law of the Sea, Marine Dumping, Ozone Layer Protection, Tropical Timber 83, Tropical Timber 94, Wetlands *signed, but not ratified:* Environmental Modification

Geography—note: *note 1:* second largest country in Africa (after Algeria) and largest country in Sub-Saharan Africa; straddles the equator; dense tropical rain forest in central river basin and eastern highlands; the narrow strip of land that controls the lower Congo River is the DRC's only outlet to the South Atlantic Ocean
note 2: because of its speed, cataracts, rapids, and turbulence the Congo River, most of which flows through the DRC, has never been accurately measured along much of its length; nonetheless, it is conceded to be the deepest river in the world; estimates of its greatest depth vary between 220 and 250 meters

PEOPLE AND SOCIETY

Population: 101,780,263 (July 2020 est.)
note: estimates for this country explicitly take into account the effects of excess mortality due to AIDS; this can result in lower life expectancy, higher infant mortality, higher death rates, lower population growth rates, and changes in the

distribution of population by age and sex than would otherwise be expected
country comparison to the world: 15

Nationality: *noun:* Congolese (singular and plural)
adjective: Congolese or Congo

Ethnic groups: more than 200 African ethnic groups of which the majority are Bantu; the four largest tribes—Mongo, Luba, Kongo (all Bantu), and the Mangbetu-Azande (Hamitic)—make up about 45% of the population

Languages: French (official), Lingala (a lingua franca trade language), Kingwana (a dialect of Kiswahili or Swahili), Kikongo, Tshiluba

Religions: Roman Catholic 29.9%, Protestant 26.7%, Kimbanguist 2.8%, other Christian 36.5%, Muslim 1.3%, other (includes syncretic sects and indigenous beliefs) 1.2%, none 1.3%, unspecified .2% (2014 est.)

Demographic profile: Despite a wealth of fertile soil, hydroelectric power potential, and mineral resources, the Democratic Republic of the Congo (DRC) struggles with many socioeconomic problems, including high infant and maternal mortality rates, malnutrition, poor vaccination coverage, lack of access to improved water sources and sanitation, and frequent and early fertility. Ongoing conflict, mismanagement of resources, and a lack of investment have resulted in food insecurity; almost 30 percent of children under the age of 5 are malnourished. The overall coverage of basic public services—education, health, sanitation, and potable water—is very limited and piecemeal, with substantial regional and rural/urban disparities. Fertility remains high at almost 5 children per woman and is likely to remain high because of the low use of contraception and the cultural preference for larger families.
   The DRC is a source and host country for refugees. Between 2012 and 2014, more than 119,000 Congolese refugees returned from the Republic of Congo to the relative stability of northwest DRC, but more than 540,000 Congolese refugees remained abroad as of year- end 2015. In addition, an estimated 3.9 million Congolese were internally displaced as of October 2017, the vast majority fleeing violence between rebel group and Congolese armed forces. Thousands of refugees have come to the DRC from neighboring countries, including Rwanda, the Central African Republic, and Burundi.

Age structure: *0-14 years:* 46.38% (male 23,757,297/female 23,449,057)
15-24 years: 19.42% (male 9,908,686/female 9,856,841)
25-54 years: 28.38% (male 14,459,453/female 14,422,912)
55-64 years: 3.36% (male 1,647,267/female 1,769,429)
65 years and over: 2.47% (male 1,085,539/female 1,423,782) (2020 est.)

Dependency ratios: *total dependency ratio:* 95.4
youth dependency ratio: 89.5
elderly dependency ratio: 5.9
potential support ratio: 17 (2020 est.)

Median age: *total:* 16.7 years
male: 16.5 years
female: 16.8 years (2020 est.)
country comparison to the world: 223

Population growth rate: 3.18% (2020 est.)
country comparison to the world: 8

Birth rate: 41 births/1,000 population (2020 est.)
country comparison to the world: 7

Death rate: 8.4 deaths/1,000 population (2020 est.)
country comparison to the world: 76

Net migration rate: -0.9 migrant(s)/1,000 population (2020 est.)
country comparison to the world: 138

Population distribution: urban clusters are spread throughout the country, particularly in the northeast along the boarder with Uganda, Rwanda, and Burundi; the largest city is the capital, Kinshasha, located in the west along the Congo River; the south is least densely populated as shown in this population distribution map

Urbanization: *urban population:* 45.6% of total population (2020)
rate of urbanization: 4.53% annual rate of change (2015-20 est.)

total population growth rate v. urban population growth rate, 2000-2030: Major urban areas—population: 14.342 million KINSHASA (capital), 2.525 million Mbuji-Mayi, 2.478 million Lubumbashi, 1.458 million Kananga, 1.261 million Kisangani, 1.078 million Bukavu (2020)

Sex ratio: *at birth:* 1.03 male(s)/female
0-14 years: 1.01 male(s)/female
15-24 years: 1.01 male(s)/female
25-54 years: 1 male(s)/female
55-64 years: 0.93 male(s)/female
65 years and over: 0.76 male(s)/female
total population: 1 male(s)/female (2020 est.)

Mother's mean age at first birth: 19.9 years (2013/14 est.)
note: median age at first birth among women 25-29

Maternal mortality rate: 473 deaths/100,000 live births (2017 est.)
country comparison to the world: 23

Infant mortality rate: *total:* 64.5 deaths/1,000 live births
male: 70.3 deaths/1,000 live births
female: 58.4 deaths/1,000 live births (2020 est.)
country comparison to the world: 8

Life expectancy at birth: *total population:* 61 years
male: 59.3 years
female: 62.8 years (2020 est.)
country comparison to the world: 216

Total fertility rate: 5.77 children born/woman (2020 est.)
country comparison to the world: 3
Contraceptive prevalence rate: 20.4% (2013/14)

Drinking water source:
improved:
urban: 84.3% of population
rural: 32.4% of population
total: 55.2% of population

unimproved:
urban: 15.7% of population
rural: 67.6% of population
total: 44.8% of population (2017 est.)

Current Health Expenditure: 4% (2017)

Physicians density: 0.07 physicians/1,000 population (201)

Sanitation facility access:
improved:
urban: 54.7% of population
rural: 29.8% of population
total: 40.7% of population
unimproved:
urban: 44.5% of population
rural: 70.2% of population
total: 59.3% of population (2017 est.)

HIV/AIDS—adult prevalence rate: 0.8% (2019 est.)
country comparison to the world: 52

HIV/AIDS—people living with HIV/AIDS: 520,000 (2019 est.)
country comparison to the world: 15

HIV/AIDS—deaths: 15,000 (2019 est.)
country comparison to the world: 11

Major infectious diseases: degree of risk: very high (2020)
food or waterborne diseases: bacterial and protozoal diarrhea, hepatitis A, and typhoid fever
vectorborne diseases: malaria, dengue fever, and trypanosomiasis-gambiense (African sleeping sickness)
water contact diseases: schistosomiasis
animal contact diseases: rabies
note: on 18 October 2019, the Centers for Disease Control and Prevention issued a Travel Health Notice for an Ebola outbreak in the South Kivu (Kivu Sud), North Kivu (Kivu Nord), and Ituri provinces in the northeastern part of the Democratic Republic of the Congo; travelers to this area could be infected with Ebola if they come into contact with an infected person's blood or other body fluids; travelers should seek medical care immediately if they develop fever, muscle pain, sore throat, diarrhea, weakness, vomiting, stomach pain, or unexplained bleeding or bruising during or after travel

Obesity—adult prevalence rate: 6.7% (2016)
country comparison to the world: 164

Children under the age of 5 years underweight: 23.4% (2013)
country comparison to the world: 14

Education expenditures: 1.5% of GDP (2017)
country comparison to the world: 172

Literacy: definition: age 15 and over can read and write French, Lingala, Kingwana, or Tshiluba
total population: 77%
male: 88.5%
female: 66.5% (2016)

School life expectancy (primary to tertiary education): total: 11 years
male: 10 years
female: 9 years (2013)

Unemployment, youth ages 15-24: total: 8.7%
male: 11.3%

female: 6.8% (2012 est.)
country comparison to the world: 137

GOVERNMENT

Country name: conventional long form: Democratic Republic of the Congo
conventional short form: DRC
local long form: Republique Democratique du Congo
local short form: RDC
former: Congo Free State, Belgian Congo, Congo/Leopoldville, Congo/Kinshasa, Zaire
abbreviation: DRC (or DROC)
etymology: named for the Congo River, most of which lies within the DRC; the river name derives from Kongo, a Bantu kingdom that occupied its mouth at the time of Portuguese discovery in the late 15th century and whose name stems from its people the Bakongo, meaning "hunters"

Government type: semi-presidential republic

Capital: name: Kinshasa
geographic coordinates: 4 19 S, 15 18 E
time difference: UTC+ 1 (6 hours ahead of Washington, DC, during Standard Time)
note: the DRC has two time zones
etymology: founded as a trading post in 1881 and named Leopoldville in honor of King Leopold II of the Belgians, who controlled the Congo Free State, the vast central African territory that became the Democratic Republic of the Congo in 1960; in 1966, Leopoldville was renamed Kinshasa, after a village of that name that once stood near the site

Administrative divisions: 26 provinces (provinces, singular—province); Bas-Uele (Lower Uele), Equateur, Haut-Katanga (Upper Katanga), Haut-Lomami (Upper Lomami), Haut-Uele (Upper Uele), Ituri, Kasai, Kasai-Central, Kasai-Oriental (East Kasai), Kinshasa, Kongo Central, Kwango, Kwilu, Lomami, Lualaba, Mai-Ndombe, Maniema, Mongala, Nord-Kivu (North Kivu), Nord-Ubangi (North Ubangi), Sankuru, Sud-Kivu (South Kivu), Sud-Ubangi (South Ubangi), Tanganyika, Tshopo, Tshuapa

Independence: 30 June 1960 (from Belgium)

National holiday: Independence Day, 30 June (1960)

Constitution: history: several previous; latest adopted 13 May 2005, approved by referendum 18-19 December 2005, promulgated 18 February 2006
amendments: proposed by the president of the republic, by the government, by either house of Parliament, or by public petition; agreement on the substance of a proposed bill requires absolute majority vote in both houses; passage requires a referendum only if both houses in joint meeting fail to achieve three-fifths majority vote; constitutional articles, including the form of government, universal suffrage, judicial independence, political pluralism, and personal freedoms, cannot be amended; amended 2011

Legal system: civil law system primarily based on Belgian law, but also customary and tribal law

International law organization participation: accepts compulsory ICJ jurisdiction with reservations; accepts ICCt jurisdiction

Citizenship: citizenship by birth: no
citizenship by descent only: at least one parent must be a citizen of the Democratic Republic of the Congo
dual citizenship recognized: no
residency requirement for naturalization: 5 years

Suffrage: 18 years of age; universal and compulsory

Executive branch: chief of state: President Felix TSHISEKEDI (since 24 January 2019)
head of government: Prime Minister Sylvestre ILUNGA Ilunkamba (since 20 May 2019); Deputy Prime Ministers Jose MAKILA, Leonard She OKITUNDU, Henri MOVA Sankanyi (since February 2018)
cabinet: Ministers of State appointed by the president
elections/appointments: president directly elected by simple majority vote for a 5-year term (eligible for a second term); election last held on 30 December 2018 (next to be held in December 2023); prime minister appointed by the president
election results: Felix TSHISEKEDI elected president; percent of vote—Felix TSHISEKEDI (UDPS) 38.6%, Martin FAYULU (Lamuka coalition) 34.8%, Emmanuel Ramazani SHADARY (PPRD) 23.9%, other 2.7%; note—election marred by serious voting irregularities

Legislative branch: description: bicameral Parliament or Parlement consists of:
Senate (108 seats; members indirectly elected by provincial assemblies by proportional representation vote; members serve 5-year terms)
National Assembly (500 seats; 439 members directly elected in multi-seat constituencies by proportional representation vote and 61 directly elected in single-seat constituencies by simple majority vote; members serve 5-year terms)
elections: Senate—last held on 19 January 2007 (follow-on election has been delayed) National Assembly—last held on 30 December 2018
election results: Senate—percent of vote by party—NA; seats by party—PPRD 22, MLC 14, FR 7, RCD 7, PDC 6, CDC 3, MSR 3, PALU 2, other 18, independent 26; composition—men 103, women 5, percent of women 4.6%
National Assembly—percent of vote by party—NA; seats by party—PPRD 62, UDPS 41, PPPD 29, MSR 27, MLC 22, PALU 19, UNC 17, ARC 16, AFDC 15, ECT 11, RRC 11, other 214 (includes numerous political parties that won 10 or fewer seats and 2 constituencies where voting was halted), independent 16; composition—men 456, women 44, percent of women 8.8%; total Parliament percent of women 8.1%; note—the November 2011 election was marred by violence including the destruction of ballots in 2 constituencies resulting in the closure of polling sites; election results were delayed 3 months, strongly contested, and continue to be unresolved

Judicial branch: highest courts: Court of Cassation or Cour de Cassation (consists of 26 justices and organized into legislative and judiciary sections); Constitutional Court (consists of 9 judges)

judge selection and term of office: Court of Cassation judges nominated by the Judicial Service Council, an independent body of public prosecutors and selected judges of the lower courts; judge tenure NA; Constitutional Court judges—3 nominated by the president, 3 by the Judicial Service Council, and 3 by the legislature; judges appointed by the president to serve 9-year non-renewable terms with one-third of the membership renewed every 3 years

subordinate courts: State Security Court; Court of Appeals (organized into administrative and judiciary sections); Tribunal de Grande; magistrates' courts; customary courts

Political parties and leaders: Christian Democrat Party or PDC [Jose ENDUNDO]
Congolese Rally for Democracy or RCD [Azarias RUBERWA]
Convention of Christian Democrats or CDC
Engagement for Citizenship and Development or ECiDe [Martin FAYULU]
Forces of Renewal or FR [Mbusa NYAMWISI]
Lamuka coalition [Martin FAYULU] (includes ECiDe, MLC, Together for Change, CNB, and, Nouvel Elan)
Movement for the Liberation of the Congo or MLC [Jean- Pierre BEMBA]
Nouvel Elan [Adolphe MUZITO]
Our Congo or CNB ("Congo Na Biso") [Freddy MATUNGULU]
People's Party for Reconstruction and Democracy or PPRD [Henri MOVA Sakanyi]
Social Movement for Renewal or MSR [Pierre LUMBI]
Together for Change (Ensemble") [Moise KATUMBI]
Unified Lumumbist Party or PALU [Antoine GIZENGA]
Union for the Congolese Nation or UNC [Vital KAMERHE]
Union for Democracy and Social Progress or UDPS [Felix TSHISEKEDI]

International organization participation: ACP, AfDB, AU, CEMAC, CEPGL, COMESA, EITI (compliant country), FAO, G- 24, G- 77, IAEA, IBRD, ICAO, ICC (NGOs), ICCt, ICRM, IDA, IFAD, IFC, IFRCS, IHO, ILO, IMF, IMO, Interpol, IOC, IOM, IPU, ISO, ITSO, ITU, ITUC (NGOs), MIGA, NAM, OIF, OPCW, PCA, SADC, UN, UNCTAD, UNESCO, UNHCR, UNIDO, UNWTO, UPU, WCO, WFTU (NGOs), WHO, WIPO, WMO, WTO

Diplomatic representation in the US: *chief of mission:* Ambassador Francois Nkuna BALUMUENE (since 23 September 2015)
chancery: 1100 Connecticut Avenue NW, Suite 725, Washington DC 20036
telephone: [1] (202) 234-7690 through 7691
FAX: [1] (202) 234-2609
representative office: New York New York

Diplomatic representation from the US: *chief of mission:* Ambassador Michael A. HAMMER (since 22 December 2018)

telephone: [243] 081 556-0151
embassy: 310 Avenue des Aviateurs, Kinshasa, Gombe
mailing address: Unit 2220, DPO AE 09828
FAX: [243] 81 556-0175

Flag description: sky blue field divided diagonally from the lower hoist corner to upper fly corner by a red stripe bordered by two narrow yellow stripes; a yellow, five- pointed star appears in the upper hoist corner; blue represents peace and hope, red the blood of the country's martyrs, and yellow the country's wealth and prosperity; the star symbolizes unity and the brilliant future for the country

National symbol(s): *leopard; national colors: sky blue, red, yellow*

National anthem: *name:* "Debout Congolaise" (Arise Congolese)
lyrics/music: Joseph LUTUMBA/Simon- Pierre BOKA di Mpasi Londi
note: adopted 1960; replaced when the country was known as Zaire; but readopted in 1997

ECONOMY

Economy—overview: The economy of the Democratic Republic of the Congo—a nation endowed with vast natural resource wealth—continues to perform poorly. Systemic corruption since independence in 1960, combined with countrywide instability and intermittent conflict that began in the early-90s, has reduced national output and government revenue, and increased external debt. ith the installation of a transitional government in 2003 after peace accords, economic conditions slowly began to improve as the government reopened relations with international financial institutions and international donors, and President KABILA began implementing reforms. Progress on implementing substantive economic reforms remains slow because of political instability, bureaucratic inefficiency, corruption, and patronage, which also dampen international investment prospects.

Renewed activity in the mining sector, the source of most export income, boosted Kinshasa's fiscal position and GDP growth until 2015, but low commodity prices have led to slower growth, volatile inflation, currency depreciation, and a growing fiscal deficit. An uncertain legal framework, corruption, and a lack of transparency in government policy are long-term problems for the large mining sector and for the economy as a whole. Much economic activity still occurs in the informal sector and is not reflected in GDP data.

Poverty remains widespread in DRC, and the country failed to meet any Millennium Development Goals by 2015. DRC also concluded its program with the IMF in 2015. The price of copper—the DRC's primary export—plummeted in 2015 and remained at record lows during 2016-17, reducing government revenues, expenditures, and foreign exchange reserves, while inflation reached nearly 50% in mid-2017—its highest level since the early 2000s.

GDP (purchasing power parity): $68.6 billion (2017 est.)
$66.33 billion (2016 est.)
$64.78 billion (2015 est.)
note: data are in 2017 dollars
country comparison to the world: 103

GDP (official exchange rate): $41.44 billion (2017 est.)

GDP—real growth rate: 3.4% (2017 est.)
2.4% (2016 est.)
6.9% (2015 est.)
country comparison to the world: 87

GDP—per capita (PPP): $800 (2017 est.)
$800 (2016 est.)
$800 (2015 est.)
note: data are in 2017 dollars
country comparison to the world: 226

Gross national saving: 11.5% of GDP (2017 est.)
8.7% of GDP (2016 est.)
16.5% of GDP (2015 est.)
country comparison to the world: 154

GDP—composition, by end use: *household consumption:* 78.5% (2017 est.)
government consumption: 12.7% (2017 est.)
investment in fixed capital: 15.9% (2017 est.)
investment in inventories: 0% (2017 est.)
exports of goods and services: 25.7% (2017 est.)
imports of goods and services: -32.8% (2017 est.)

GDP—composition, by sector of origin: *agriculture:* 19.7% (2017 est.)
industry: 43.6% (2017 est.)
services: 36.7% (2017 est.)

Agriculture—products: coffee, sugar, palm oil, rubber, tea, cotton, cocoa, quinine, cassava (manioc, tapioca), bananas, plantains, peanuts, root crops, corn, fruits; wood products

Industries: mining (copper, cobalt, gold, diamonds, coltan, zinc, tin, tungsten), mineral processing, consumer products (textiles, plastics, footwear, cigarettes), metal products, processed foods and beverages, timber, cement, commercial ship repair

Industrial production growth rate: 1.6% (2017 est.)
country comparison to the world: 140

Labor force: 20.692 million (2012 est.)
country comparison to the world: 25

Labor force—by occupation: *agriculture:* NA
industry: NA
services: NA

Unemployment rate: NA

Population below poverty line: 63% (2014 est.)

Household income or consumption by percentage share: *lowest 10%:* 2.3%
highest 10%: 34.7% (2006)

Budget: *revenues:* 4.634 billion (2017 est.)
expenditures: 5.009 billion (2017 est.)

Taxes and other revenues: 11.2% (of GDP) (2017 est.)
country comparison to the world: 211

Budget surplus (+) or deficit (-): -0.9% (of GDP) (2017 est.)
country comparison to the world: 72

Public debt: 18.1% of GDP (2017 est.)
19.3% of GDP (2016 est.)
country comparison to the world: 192

Fiscal year: calendar year

Inflation rate (consumer prices): 41.5% (2017 est.)
18.2% (2016 est.)
country comparison to the world: 224

Current account balance: -$200 million (2017 est.)
-$1.215 billion (2016 est.)
country comparison to the world: 99

Exports: $10.98 billion (2017 est.)
$8.228 billion (2016 est.)
country comparison to the world: 89

Exports—partners: China 41.4%, Zambia 22.7%, South Korea 7.2%, Finland 6.2% (2017)

Exports—commodities: diamonds, copper, gold, cobalt, wood products, crude oil, coffee

Imports: $10.82 billion (2017 est.)
$10.21 billion (2016 est.)
country comparison to the world: 100

Imports—commodities: foodstuffs, mining and other machinery, transport equipment, fuels

Imports—partners: China 19.9%, South Africa 18%, Zambia 10.4%, Belgium 9.1%, India 4.3%, Tanzania 4.2% (2017)

Reserves of foreign exchange and gold: $457.5 million (31 December 2017 est.)
$708.2 million (31 December 2016 est.)
country comparison to the world: 155

Debt—external: $4.963 billion (31 December 2017 est.)
$5.35 billion (31 December 2016 est.)
country comparison to the world: 134

Exchange rates: Congolese francs (CDF) per US dollar—
1,546.8 (2017 est.)
1,010.3 (2016 est.)
1,010.3 (2015 est.)
925.99 (2014 est.)
925.23 (2013 est.)

ENERGY

Electricity access: *population without electricity:* 79 million (2019)
electrification—total population: 9% (2019)
electrification—urban areas: 19% (2019)
electrification—rural areas: 0.4% (2019)

Electricity—production: 9.046 billion kWh (2016 est.)
country comparison to the world: 106

Electricity—consumption: 7.43 billion kWh (2016 est.)
country comparison to the world: 106

Electricity—exports: 422 million kWh (2015 est.)
country comparison to the world: 70

Electricity—imports: 20 million kWh (2016 est.)
country comparison to the world: 113

Electricity—installed generating capacity: 2.587 million kW (2016 est.)

country comparison to the world: 105

Electricity—from fossil fuels: 2% of total installed capacity (2016 est.)
country comparison to the world: 210

Electricity—from nuclear fuels: 0% of total installed capacity (2017 est.)
country comparison to the world: 69

Electricity—from hydroelectric plants: 98% of total installed capacity (2017 est.)
country comparison to the world: 4

Electricity—from other renewable sources: 0% of total installed capacity (2017 est.)
country comparison to the world: 182

Crude oil—production: 17,000 bbl/day (2018 est.)
country comparison to the world: 68

Crude oil—exports: 20,000 bbl/day (2015 est.)
country comparison to the world: 49

Crude oil—imports: 0 bbl/day (2015 est.)
country comparison to the world: 112

Crude oil—proved reserves: 180 million bbl (1 January 2018 est.)
country comparison to the world: 58

Refined petroleum products—production: 0 bbl/day (2017 est.)
country comparison to the world: 132

Refined petroleum products—consumption: 21,000 bbl/day (2016 est.)
country comparison to the world: 136

Refined petroleum products—exports: 0 bbl/day (2015 est.)
country comparison to the world: 145

Refined petroleum products—imports: 21,140 bbl/day (2015 est.)
country comparison to the world: 115

Natural gas—production: 0 cu m (2017 est.)
country comparison to the world: 118

Natural gas—consumption: 0 cu m (2017 est.)
country comparison to the world: 134

Natural gas—exports: 0 cu m (2017 est.)
country comparison to the world: 85

Natural gas—imports: 0 cu m (2017 est.)
country comparison to the world: 108

Natural gas—proved reserves: 991.1 million cu m (1 January 2018 est.)
country comparison to the world: 99

Carbon dioxide emissions from consumption of energy: 3.146 million Mt (2017 est.)
country comparison to the world: 145

COMMUNICATIONS

Telephones—fixed lines: *total subscriptions:* 0 NA
subscriptions per 100 inhabitants: less than 1 (2018 est.)
country comparison to the world: 222

Telephones—mobile cellular: *total subscriptions:* 42,166,976
subscriptions per 100 inhabitants: 42.77 (2019 est.)
country comparison to the world: 35

Telecommunication systems: *general assessment:* poorly developed national and international infrastructure; bandwidth is limited; Internet pricing is expensive; domestic satellite system with 14 earth stations; wars and social upheaval have not promoted advancement; a revised Telecommunications Act adopted in May 2018; govt. only loosely regulates the telecom sector, much of the investment is from donor countries (specifically China) (2020)
domestic: fixed-line connections less than 1 per 100 persons; given the backdrop of a wholly inadequate fixed-line infrastructure, the use of mobile-cellular services is over 43 per 100 persons (2019)
international: country code—243; ACE and WACS submarine cables to West and South Africa and Europe; satellite earth station—1 Intelsat (Atlantic Ocean) (2019)
note: the COVID-19 outbreak is negatively impacting telecommunications production and supply chains globally; consumer spending on telecom devices and services has also slowed due to the pandemic's effect on economies worldwide; overall progress towards improvements in all facets of the telecom industry—mobile, fixed-line, broadband, submarine cable and satellite—has moderated

Broadcast media: state-owned TV broadcast station with near national coverage; more than a dozen privately owned TV stations—2 with near national coverage; 2 state-owned radio stations are supplemented by more than 100 private radio stations; transmissions of at least 2 international broadcasters are available

Internet country code: .cd

Internet users: *total:* 8,231,357
percent of population: 8.62% (July 2018 est.)
country comparison to the world: 59

Broadband—fixed subscriptions: *total:* 4,620
subscriptions per 100 inhabitants: less than 1 (2018 est.)
country comparison to the world: 181

TRANSPORTATION

National air transport system: *number of registered air carriers:* 8 (2020)
inventory of registered aircraft operated by air carriers: 13
annual passenger traffic on registered air carriers: 932,043 (2018)
annual freight traffic on registered air carriers: 890,000 mt-km (2018)

Civil aircraft registration country code prefix: 9Q (2016)

Airports: 198 (2013)
country comparison to the world: 27

Airports—with paved runways: *total:* 26 (2017)
over 3,047 m: 3 (2017)
2,438 to 3,047 m: 3 (2017)
1,524 to 2,437 m: 17 (2017)
914 to 1,523 m: 2 (2017)
under 914 m: 1 (2017)

Airports—with unpaved runways: *total:* 172 (2013)
1,524 to 2,437 m: 20 (2013)
914 to 1,523 m: 87 (2013)
under 914 m: 65 (2013)

Heliports: 1 (2013)

Pipelines: 62 km gas, 77 km oil, 756 km refined products (2013)

Railways: *total:* 4,007 km (2014)
narrow gauge: 3,882 km 1.067-m gauge (858 km electrified) (2014)
125 1.000-m gauge
country comparison to the world: 48

Roadways: *total:* 152,373 km (2015)
paved: 3,047 km (2015)
unpaved: 149,326 km (2015)
urban: 7,400 km (2015)
non-urban: 144,973 km
country comparison to the world: 34

Waterways: 15,000 km (including the Congo River, its tributaries, and unconnected lakes) (2011)
country comparison to the world: 8

Merchant marine: *total:* 21
by type: general cargo 4, oil tanker 2, other 15 (2019)
country comparison to the world: 142

Ports and terminals: *major seaport(s):* Banana
river or lake port(s): Boma, Bumba, Kinshasa, Kisangani, Matadi, Mbandaka (Congo) Kindu (Lualaba) Bukavu, Goma (Lake Kivu) Kalemie (Lake Tanganyika)

MILITARY AND SECURITY

Military and security forces: *Armed Forces of the Democratic Republic of the Congo (Forces d'Armees de la Republique Democratique du Congo, FARDC):* Land Forces, National Navy (La Marine Nationale), Congolese Air Force (Force Aerienne Congolaise, FAC); Republican Guard (responsible for presidential security) (2019)

Military expenditures: 0.7% of GDP (2019)
0.7% of GDP (2018)
0.7% of GDP (2017)
1.3% of GDP (2016)
1.4% of GDP (2015)
country comparison to the world: 134

Military and security service personnel strengths: size estimates for the Armed Forces of the Democratic Republic of Congo (FARDC) vary widely because of inconsistent and unreliable data, as well as the ongoing integration of various non- state armed groups/militias; approximately 100,000 active troops (80,000 Army; 7,000 Navy; 2,000 Air Force; 10,000 Republican Guard) (2019 est.)

Military equipment inventories and acquisitions: the FARDC is equipped mostly with a mix of second-hand Russian and Soviet-era weapons acquired from Ukraine and other former Warsaw

Pact nations, as well as some equipment provided by Brazil and France; most equipment was acquired between 1970 and 2000; since 2010, Ukraine is the largest supplier of arms to the FARDC (2019 est.)

Military service age and obligation: 18-45 years of age for voluntary and compulsory military service (2012)

Military—note: the modern FARDC was created out of the armed factions of the two Congo wars of 1996-1997 and 1998-2003; as part of the peace accords that ended the last war, the largest rebel groups were incorporated into the FARDC; many armed groups (at least 70 and by some recent estimates more than 100), however, continue to fight; as of September 2020, the FARDC is actively engaged in combat operations against numerous armed groups inside the country, particularly in the eastern provinces of Ituri, North Kivu, and South Kivu, although violence also continues in Maniema, Kasai, Kasai Central, and Tanganyika provinces; the military is widely assessed as being unable to provide adequate security throughout the country due to insufficient training, poor morale and leadership, ill-discipline and corruption, low equipment readiness, a fractious ethnic makeup, and the sheer size of the country and diversity of armed rebel groups

MONUSCO, the United Nations peacekeeping and stabilization force in the Democratic Republic of Congo, has operated in the central and eastern parts of the country since 1999; as of March 2020, MONUSCO comprised around 18,500 personnel, including nearly 14,000 military troops; in December 2019, the UN extended MONUSCO's s mandate until 20 December 2020; MONUSCO includes a Force Intervention Brigade (FIB; 3 infantry battalions), the first ever UN peacekeeping force specifically tasked to carry out targeted offensive operations to neutralize and disarm groups considered a threat to state authority and civilian security (2020)

TERRORISM

Terrorist group(s): Islamic State of Iraq and ash-Sham—Central Arica (2020)
note: details about the history, aims, leadership, organization, areas of operation, tactics, targets, weapons, size, and sources of support of the group(s) appear(s) in Appendix-T

TRANSNATIONAL ISSUES

Disputes—international: heads of the Great Lakes states and UN pledged in 2004 to abate tribal, rebel, and militia fighting in the region, including northeast Congo, where the UN Organization Mission in the Democratic Republic of the Congo (MONUC), organized in 1999, maintains over 16,500 uniformed peacekeepers; members of Uganda's Lord's Resistance Army forces continue to seek refuge in Congo's Garamba National Park as peace talks with the Uganda Government evolve; the location of the boundary in the broad

Congo River with the Republic of the Congo is indefinite except in the Pool Malebo/Stanley Pool area; Uganda and DRC dispute Rukwanzi Island in Lake Albert and other areas on the Semliki River with hydrocarbon potential; boundary commission continues discussions over Congolese- administered triangle of land on the right bank of the Lunkinda River claimed by Zambia near the DRC village of Pweto; DRC accuses Angola of shifting monuments

Refugees and internally displaced persons: *refugees (country of origin):* 172,234 (Central African Republic), 214,777 (Rwanda) (refugees and asylum seekers), 89,401 (South Sudan) (refugees and asylum seekers), 48,824 (Burundi) (2020)
IDPs: 5.512 million (fighting between government forces and rebels since mid-1990s; conflict in Kasai region since 2016) (2019)

Trafficking in persons: *current situation:* The Democratic Republic of the Congo is a source, destination, and possibly a transit country for men, women, and children subjected to forced labor and sex trafficking; the majority of this trafficking is internal, and much of it is perpetrated by armed groups and rogue government forces outside official control in the country's unstable eastern provinces; Congolese adults are subjected to forced labor, including debt bondage, in unlicensed mines, and women may be forced into prostitution; Congolese women and girls are subjected to forced marriages where they are vulnerable to domestic servitude or sex trafficking, while children are forced to work in agriculture, mining, mineral smuggling, vending, portering, and begging; Congolese women and children migrate to countries in Africa, the Middle East, and Europe where some are subjected to forced prostitution, domestic servitude, and forced labor in agriculture and diamond mining; indigenous and foreign armed groups, including the Lord's Resistance Army, abduct and forcibly recruit Congolese adults and children to serve as laborers, porters, domestics, combatants, and sex slaves; some elements of the Congolese national army (FARDC) also forced adults to carry supplies, equipment, and looted goods, but no cases of the FARDC recruiting child soldiers were reported in 2014—a significant change

tier rating: Tier 2 Watch List—The Democratic Republic of the Congo does not fully comply with the minimum standards for the elimination of trafficking; however, it is making significant efforts to do so; the government took significant steps to hold military and police officials complicit in human trafficking accountable with convictions for sex slavery and arrests of armed group commanders for the recruitment and use of child soldiers; the government appears to have ceased the recruitment of child soldiers through the implementation of a UN-backed action plan; little effort was made to address labor and sex trafficking crimes committed by persons other than officials, or to identify the victims, or to provide or refer the victims to care services; awareness of various forms of trafficking is limited among law enforcement personnel and training and resources are inadequate to conduct investigations (2015)

Illicit drugs: traffickers exploit lax shipping controls to transit pseudoephedrine through the capital; while rampant corruption and inadequate supervision leave the banking system vulnerable to money laundering, the lack of a well-developed financial system limits the country's utility as a money-laundering center

CONGO, REPUBLIC OF THE

INTRODUCTION

Background: Upon independence in 1960, the former French region of Middle Congo became the Republic of the Congo. A quarter century of experimentation with Marxism was abandoned in 1990 and a democratically elected government took office in 1992. A two-year civil war that ended in 1999 restored former Marxist President Denis SASSOU- Nguesso, who had ruled from 1979 to 1992, and sparked a short period of ethnic and political unrest that was resolved by a peace agreement in late 1999. A new constitution adopted three years later provided for a multiparty system and a seven-year presidential term, and elections arranged shortly thereafter installed SASSOU-Nguesso. Following a year of renewed fighting, President SASSOU-Nguesso and southern-based rebel groups agreed to a final peace accord in March 2003. SASSOU-Nguesso was reelected in 2009 and, after passing a referendum allowing him to run for a third term, was reelected again in 2016. The Republic of Congo is one of Africa's largest petroleum producers, but with declining production it will need new offshore oil finds to sustain its oil earnings over the long term.

GEOGRAPHY

Location: Central Africa, bordering the South Atlantic Ocean, between Angola and Gabon

Geographic coordinates: 1 00 S, 15 00 E

Map references: Africa

Area: *total:* 342,000 sq km
land: 341,500 sq km
water: 500 sq km

country comparison to the world: 65

Area—comparative: slightly smaller than Montana; about twice the size of Florida

Land boundaries: *total:* 5,008 km
border countries (5): Angola 231 km, Cameroon 494 km, Central African Republic 487 km, Democratic Republic of the Congo 1229 km, Gabon 2567 km

Coastline: 169 km

Maritime claims: *territorial sea:* 12 nm
exclusive economic zone: 200 nm
contiguous zone: 24 nm

Climate: tropical; rainy season (March to June); dry season (June to October); persistent high temperatures and humidity; particularly enervating climate astride the Equator

Terrain: coastal plain, southern basin, central plateau, northern basin

Elevation: *mean elevation:* 430 m
lowest point: Atlantic Ocean 0 m
highest point: Mount Berongou 903 m

Natural resources: petroleum, timber, potash, lead, zinc, uranium, copper, phosphates, gold, magnesium, natural gas, hydropower

Land use: *agricultural land:* 31.1% (2011 est.)
arable land: 1.6% (2011 est.) / *permanent crops:* 0.2% (2011 est.) / *permanent pasture:* 29.3% (2011 est.)
forest: 65.6% (2011 est.)
other: 3.3% (2011 est.)
Irrigated land: 20 sq km (2012)

Population distribution: the population is primarily located in the south, in and around the capital of Brazzaville as shown in this population distribution map

Natural hazards: seasonal flooding

Environment—current issues: air pollution from vehicle emissions; water pollution from raw sewage; tap water is not potable; deforestation; wildlife protection

Environment—international agreements: *party to:* Biodiversity, Climate Change, Climate Change-Kyoto Protocol, Desertification, Endangered Species, Hazardous Wastes, Law of the Sea, Ozone Layer Protection, Ship Pollution, Tropical Timber 83, Tropical Timber 94, Wetlands
signed, but not ratified: none of the selected agreements

Geography—note: about 70% of the population lives in Brazzaville, Pointe- Noire, or along the railroad between them

PEOPLE AND SOCIETY

Population: 5,293,070 (July 2020 est.)
note: estimates for this country explicitly take into account the effects of excess mortality due to AIDS; this can result in lower life expectancy, higher infant mortality, higher death rates, lower population growth rates, and changes in the distribution of population by age and sex than would otherwise be expected
country comparison to the world: 121

Nationality: *noun:* Congolese (singular and plural)
adjective: Congolese or Congo

Ethnic groups: Kongo 40.5%, Teke 16.9%, Mbochi 13.1%, foreigner 8.2%, Sangha 5.6%, Mbere/Mbeti/Kele 4.4%, Punu 4.3%, Pygmy 1.6%, Oubanguiens 1.6%, Duma 1.5%, Makaa 1.3%, other and unspecified 1% (2014-15 est.)

Languages: French (official), French Lingala and Monokutuba (lingua franca trade languages), many local languages and dialects (of which Kikongo is the most widespread)

Religions: Roman Catholic 33.1%, Awakening Churches/Christian Revival 22.3%, Protestant 19.9%, Salutiste 2.2%, Muslim 1.6%, Kimbanguiste 1.5%, other 8.1%, none 11.3% (2010 est.)

Age structure: *0-14 years:* 41.57% (male 1,110,484/female 1,089,732)
15-24 years: 17.14% (male 454,981/female 452,204)
25-54 years: 33.5% (male 886,743/female 886,312)
55-64 years: 4.59% (male 125,207/female 117,810)
65 years and over: 3.2% (male 75,921/female 93,676) (2020 est.)

Dependency ratios: *total dependency ratio:* 78.7
youth dependency ratio: 73.7
elderly dependency ratio: 4.9
potential support ratio: 20.3 (2020 est.)

Median age: *total:* 19.5 years
male: 19.3 years
female: 19.7 years (2020 est.)
country comparison to the world: 203

Population growth rate: 2.26% (2020 est.)
country comparison to the world: 33

Birth rate: 32.6 births/1,000 population (2020 est.)
country comparison to the world: 26

Death rate: 8.7 deaths/1,000 population (2020 est.) c
country comparison to the world: 68

Net migration rate: -0.9 migrant(s)/1,000 population (2020 est.)

223

country comparison to the world: 139

Population distribution: the population is primarily located in the south, in and around the capital of Brazzaville as shown in this population distribution map

Urbanization: *urban population:* 67.8% of total population (2020)

rate of urbanization: 3.28% annual rate of change (2015-20 est.)

total population growth rate v. urban population growth rate, 2000-2030: Major urban areas—population: 2.388 million BRAZZAVILLE (capital), 1.214 million Pointe-Noire (2020)

Sex ratio: at birth: 1.03 male(s)/female

0-14 years: 1.02 male(s)/female

15-24 years: 1.01 male(s)/female

25-54 years: 1 male(s)/female

55-64 years: 1.06 male(s)/female

65 years and over: 0.81 male(s)/female

total population: 1.01 male(s)/female (2020 est.)

Mother's mean age at first birth: 19.8 years (2011/12 est.)

note: median age at first birth among women 25-29

Maternal mortality rate: 378 deaths/100,000 live births (2017 est.)

country comparison to the world: 29

Infant mortality rate: *total:* 50.7 deaths/1,000 live births

male: 55.3 deaths/1,000 live births

female: 45.9 deaths/1,000 live births (2020 est.)

country comparison to the world: 24

Life expectancy at birth: *total population:* 61.3 years

male: 59.9 years

female: 62.7 years (2020 est.)

country comparison to the world: 214

Total fertility rate: 4.45 children born/woman (2020 est.)

country comparison to the world: 24

Contraceptive prevalence rate: 30.1% (2014/15)

Drinking water source:

improved:

urban: 97.5% of population

rural: 56.4% of population

total: 83.7% of population

unimproved:

urban: 2.5% of population

rural: 43.6% of population

total: 16.3% of population (2017 est.)

Current Health Expenditure: 2.9% (2017)

Physicians density: 0.16 physicians/1,000 population (2011)

Sanitation facility access:

improved:

urban: 73.4% of population

rural: 15.1% of population

total: 53.9% of population

unimproved:

urban: 26.6% of population

rural: 84.9% of population

total: 46.1% of population (2017 est.)

HIV/AIDS—adult prevalence rate: 3.1% (2019 est.)

country comparison to the world: 18

HIV/AIDS—people living with HIV/AIDS: 100,000 (2019 est.)

country comparison to the world: 46

HIV/AIDS—deaths: 4,500 (2019 est.)

country comparison to the world: 28

Major infectious diseases: *degree of risk:* very high (2020)

food or waterborne diseases: bacterial and protozoal diarrhea, hepatitis A, and typhoid fever

vectorborne diseases: malaria and dengue fever

water contact diseases: schistosomiasis

animal contact diseases: rabies

Obesity—adult prevalence rate: 9.6% (2016)

country comparison to the world: 143

Children under the age of 5 years underweight: 12.3% (2015)

country comparison to the world: 52

Education expenditures: 4.6% of GDP (2015)

country comparison to the world: 82

Literacy: *definition:* age 15 and over can read and write

total population: 80.3%

male: 86.1%

female: 74.6% (2018)

School life expectancy (primary to tertiary education): *total:* 11 years

male: 11 years

female: 11 years (2012)

GOVERNMENT

Country name: *conventional long form:* Republic of the Congo

conventional short form: Congo (Brazzaville)

local long form: Republique du Congo

local short form: Congo

former: French Congo, Middle Congo, People's Republic of the Congo, Congo/Brazzaville

etymology: named for the Congo River, which makes up much of the country's eastern border; the river name derives from Kongo, a Bantu kingdom that occupied its mouth at the time of Portuguese discovery in the late 15th century and whose name stems from its people the Bakongo, meaning "hunters"

Government type: presidential republic

Capital: *name:* Brazzaville

geographic coordinates: 4 15 S, 15 17 E

time difference: UTC+ 1 (6 hours ahead of Washington, DC, during Standard Time)

etymology: named after the Italian-born French explorer and humanitarian, Pierre Savorgnan de BRAZZA (1852-1905), who promoted French colonial interests in central Africa and worked against slavery and the abuse of African laborers

Administrative divisions: 12 departments (departments, singular—department), Bouenza, Brazzaville, Cuvette, Cuvette-Ouest, Kouilou, Lekoumou, Likouala, Niari, Plateaux, Pointe-Noire, Pool, Sangha

Independence: 15 August 1960 (from France)

National holiday: Independence Day, 15 August (1960)

Constitution: *history:* several previous; latest approved by referendum 25 October 2015

amendments: proposed by the president of the republic or by Parliament; passage of presidential proposals requires Supreme Court review followed by approval in a referendum; such proposals may also be submitted directly to Parliament, in which case passage requires at least three-quarters majority vote of both houses in joint session; proposals by Parliament require three-fourths majority vote of both houses in joint session; constitutional articles including those affecting the country's territory, republican form of government, and secularity of the state are not amendable

Legal system: mixed legal system of French civil law and customary law

International law organization participation: has not submitted an ICJ jurisdiction declaration; accepts ICCt jurisdiction

Citizenship: *citizenship by birth:* no

citizenship by descent only: at least one parent must be a citizen of the Republic of the Congo

dual citizenship recognized: no

residency requirement for naturalization: 10 years

Suffrage: 18 years of age; universal

Executive branch: *chief of state:* President Denis SASSOU-Nguesso (since 25 October 1997)

head of government: Prime Minister Clement MOUAMBA (since 24 April 2016); note—a constitutional referendum held in 2015 approved the change of the head of government from the president to the prime minister (2019)

cabinet: Council of Ministers appointed by the president

elections/appointments: president directly elected by absolute majority popular vote in 2 rounds if needed for a 5-year term (eligible for 2 additional terms); election last held on 20 March 2016 (next to be held in 2021)

election results: Denis SASSOU-Nguesso reelected president in the first round; percent of vote—Denis SASSOU- Nguesso (PCT) 60.4%, Guy Price Parfait KOLELAS (MCDDI) 15.1%, Jean- Marie MOKOKO (independent) 13.9%, Pascal Tsaty MABIALA (UPADS) 4.4%, other 6.2%

Legislative branch: *description:* bicameral Parliament or Parlement consists of:

Senate (72 seats; members indirectly elected by regional councils by simple majority vote to serve 6-year terms with one- half of membership renewed every 3 years)

National Assembly (151 seats; members directly elected in single-seat constituencies by absolute majority popular vote in 2 rounds if needed; members serve 5-year terms)

elections: Senate—last held on 31 August 2017 for expiry of half the seats (next to be held in 2020) National Assembly—last held on 16 and 30 July 2017 (next to be held in July 2022) *election results:* Senate—percent of vote by party—NA; seats by party—PCT 46, independent 12, MAR 2, RDPS 2, UPADS 2, DRD 1, FP 1, MCDDI 1, PRL 1, Pulp 1, PUR 1, RC 1; composition—men 58, women 14, percent of women 19.4%

National Assembly—percent of vote by party—NA; seats by party—PCT 96, UPADS 8, MCDDI 4, other 23 (less than 4 seats) independent 20; composition—men 134, women 17, percent of women 11.3%; note—total Parliament percent of women 13.9%

Judicial branch: *highest courts:* Supreme Court or Cour Supreme (consists of NA judges); Constitutional Court (consists of 9 members); note—a High Court of Justice, outside the judicial authority, tries cases involving treason by the president of the republic

judge selection and term of office: Supreme Court judges elected by Parliament and serve until age 65; Constitutional Court members appointed by the president of the republic—3 directly by the president and 6 nominated by Parliament; members appointed for renewable 9-year terms with one-third of the membership renewed every 3 years

subordinate courts: Court of Audit and Budgetary Discipline; courts of appeal; regional and district courts; employment tribunals; juvenile courts

Political parties and leaders: Action Movement for Renewal or MAR [Roland BOUITI-VIAUDO] Citizen's Rally or RC [Claude Alphonse NSILOU] Congolese Labour Party or PCT [Denis SASSOU-NGUESSO] Congolese Movement for Democracy and Integral Development or MCDDI [Guy Price Parfait KOLELAS] Movement for Unity, Solidarity, and Work or MUST [Claudine MUNARI] Pan-African Union for Social Development or UPADS [Pascal Tsaty MABIALA] Party for the Unity of the Republic or PUR Patriotic Union for Democracy and Progress or UPDP [Auguste-Celestin GONGARD NKOUA] Prospects and Realities Club or CPR Rally for Democracy and Social Progress or RDPS [Bernard BATCHI] Rally of the Presidential Majority or RMP Republican and Liberal Party or PRL [Bonaventure MIZIDY] Union for the Republic or UR Union of Democratic Forces or UDF Union for Democracy and Republic or UDR many smaller parties

International organization participation: ACP, AfDB, AU, BDEAC, CEMAC, EITI (compliant country), FAO, FZ, G-77, IAEA, IBRD, ICAO, ICCt, ICRM, IDA, IFAD, IFC, IFRCS, ILO, IMF, IMO, Interpol, IOC, IOM, IPU, ISO (correspondent), ITSO, ITU, ITUC (NGOs), MIGA, NAM, OIF, OPCW, UN, UNCTAD, UNESCO, UNHCR, UNIDO, UNITAR, UNWTO, UPU, WCO, WFTU (NGOs), WHO, WIPO, WMO, WTO

Diplomatic representation in the US: *chief of mission:* Ambassador Serge MOMBOULI (since 31 July 2001)
chancery: 1720 16th Street NW, Washington, DC 20009
telephone: [1] (202) 726-5500
FAX: [1] (202) 726-1860

Diplomatic representation from the US: *chief of mission:* Ambassador Todd P. HASKELL (since July 2017)
telephone: [242] 06 612-2000
embassy: 70-83 Section D, Maya-Maya Boulevard, Brazzaville
mailing address: B. P. 1015, Brazzaville

Flag description: divided diagonally from the lower hoist side by a yellow band; the upper triangle (hoist side) is green and the lower triangle is red; green symbolizes agriculture and forests, yellow the friendship and nobility of the people, red is unexplained but has been associated with the struggle for independence
note: uses the popular Pan-African colors of Ethiopia

National symbol(s): lion, elephant; national colors: green, yellow, red

National anthem: *name:* "La Congolaise" (The Congolese)
lyrics/music: Jacques TONDRA and Georges KIBANGHI/Jean ROYER and Joseph SPADILIERE
note: originally adopted 1959, restored 1991
0:00 / 1:28

ECONOMY

Economy—overview: The Republic of the Congo's economy is a mixture of subsistence farming, an industrial sector based largely on oil and support services, and government spending. Oil has supplanted forestry as the mainstay of the economy, providing a major share of government revenues and exports. Natural gas is increasingly being converted to electricity rather than being flared, greatly improving energy prospects. New mining projects, particularly iron ore, which entered production in late 2013, may add as much as $1 billion to annual government revenue. The Republic of the Congo is a member of the Central African Economic and Monetary Community (CEMAC) and shares a common currency—the Central African Franc—with five other member states in the region.

The current administration faces difficult economic challenges of stimulating recovery and reducing poverty. The drop in oil prices that began in 2014 has constrained government spending; lower oil prices forced the government to cut more than $1 billion in planned spending. The fiscal deficit amounted to 11% of GDP in 2017. The government's inability to pay civil servant salaries has resulted in multiple rounds of strikes by many groups, including doctors, nurses, and teachers. In the wake of a multi-year recession, the country reached out to the IMF in 2017 for a new program; the IMF noted that the country's continued

dependence on oil, unsustainable debt, and significant governance weakness are key impediments to the country's economy. In 2018, the country's external debt level will approach 120% of GDP. The IMF urged the government to renegotiate debts levels to sustainable levels before it agreed to a new macroeconomic adjustment package.

GDP (purchasing power parity): $29.39 billion (2017 est.)
$30.33 billion (2016 est.)
$31.22 billion (2015 est.)
note: data are in 2017 dollars
country comparison to the world: 133

GDP (official exchange rate): $8.718 billion (2017 est.)

GDP—real growth rate: -3.1% (2017 est.)
-2.8% (2016 est.)
2.6% (2015 est.)
country comparison to the world: 212

GDP—per capita (PPP): $6,800 (2017 est.)
$7,200 (2016 est.)
$7,500 (2015 est.)
note: data are in 2017 dollars
country comparison to the world: 161

Gross national saving: 19.5% of GDP (2017 est.)
-12.8% of GDP (2016 est.)
6.6% of GDP (2015 est.)
country comparison to the world: 101

GDP—composition, by end use: *household consumption:* 47.6% (2017 est.)
government consumption: 9.6% (2017 est.)
investment in fixed capital: 42.5% (2017 est.)
investment in inventories: 0.1% (2017 est.)
exports of goods and services: 62.9% (2017 est.)
imports of goods and services: -62.7% (2017 est.)

GDP—composition, by sector of origin: *agriculture:* 9.3% (2017 est.)
industry: 51% (2017 est.)
services: 39.7% (2017 est.)

Agriculture—products: cassava (manioc, tapioca), sugar, rice, corn, peanuts, vegetables, coffee, cocoa; forest products

Industries: petroleum extraction, cement, lumber, brewing, sugar, palm oil, soap, flour, cigarettes

Industrial production growth rate: -3% (2017 est.)
country comparison to the world: 187

Labor force: 2.055 million (2016 est.)
country comparison to the world: 121

Labor force—by occupation: *agriculture:* 35.4%
industry: 20.6%
services: 44% (2005 est.)

Unemployment rate: 36% (2014 est.)
country comparison to the world: 211

Population below poverty line: 46.5% (2011 est.)

Household income or consumption by percentage share: *lowest 10%:* 2.1%
highest 10%: 37.1% (2005)

Budget: *revenues:* 1.965 billion (2017 est.)
expenditures: 2.578 billion (2017 est.)

Taxes and other revenues: 22.5% (of GDP) (2017 est.)
country comparison to the world: 132

Budget surplus (+) or deficit (-): -7% (of GDP) (2017 est.)
country comparison to the world: 194

Public debt: 130.8% of GDP (2017 est.)
128.7% of GDP (2016 est.)
country comparison to the world: 7

Fiscal year: calendar year

Inflation rate (consumer prices): 0.5% (2017 est.)
3.2% (2016 est.)
country comparison to the world: 26

Current account balance: -$1.128 billion (2017 est.)
-$5.735 billion (2016 est.)
country comparison to the world: 148

Exports: $4.193 billion (2017 est.)
$4.116 billion (2016 est.)
country comparison to the world: 114

Exports—partners: China 53.8%, Angola 6.2%, Gabon 5.7%, Italy 5.4%, Spain 5.4%, Australia 4.8% (2017)

Exports—commodities: petroleum, lumber, plywood, sugar, cocoa, coffee, diamonds

Imports: $2.501 billion (2017 est.)
$5.639 billion (2016 est.)
country comparison to the world: 157

Imports—commodities: capital equipment, construction materials, foodstuffs

Imports—partners: France 15%, China 14%, Belgium 12.2%, Norway 8.1% (2017)

Reserves of foreign exchange and gold: $505.7 million (31 December 2017 est.)
$727.1 million (31 December 2016 est.)
country comparison to the world: 151

Debt—external: $4.605 billion (31 December 2017 est.)
$4.721 billion (31 December 2016 est.)
country comparison to the world: 135

Exchange rates: Cooperation Financiere en Afrique Centrale francs (XAF) per US dollar - 579.8 (2017 est.)
593.01 (2016 est.)
593.01 (2015 est.)
591.45 (2014 est.)
494.42 (2013 est.)

ENERGY

Electricity access: *population without electricity:* 2 million (2019)
electrification—total population: 72% (2019)
electrification—urban areas: 89% (2019)
electrification—rural areas: 36% (2019)

Electricity—production: 1.696 billion kWh (2016 est.)
country comparison to the world: 143

Electricity—consumption: 912 million kWh (2016 est.)
country comparison to the world: 157

Electricity—exports: 22 million kWh (2015 est.)
country comparison to the world: 90

Electricity—imports: 18 million kWh (2016 est.)
country comparison to the world: 115

Electricity—installed generating capacity: 591,500 kW (2016 est.)
country comparison to the world: 139

Electricity—from fossil fuels: 64% of total installed capacity (2016 est.)
country comparison to the world: 121

Electricity—from nuclear fuels: 0% of total installed capacity (2017 est.)
country comparison to the world: 70

Electricity—from hydroelectric plants: 36% of total installed capacity (2017 est.)
country comparison to the world: 59

Electricity—from other renewable sources: 0% of total installed capacity (2017 est.)
country comparison to the world: 183

Crude oil—production: 340,000 bbl/day (2018 est.)
country comparison to the world: 30

Crude oil—exports: 254,100 bbl/day (2015 est.)
country comparison to the world: 28

Crude oil—imports: 0 bbl/day (2015 est.)
country comparison to the world: 113

Crude oil—proved reserves: 1.6 billion bbl (1 January 2018 est.)
country comparison to the world: 37

Refined petroleum products—production: 15,760 bbl/day (2015 est.)
country comparison to the world: 93

Refined petroleum products—consumption: 17,000 bbl/day (2016 est.)
country comparison to the world: 149

Refined petroleum products—exports: 5,766 bbl/day (2015 est.)
country comparison to the world: 89

Refined petroleum products—imports: 7,162 bbl/day (2015 est.)
country comparison to the world: 156

Natural gas—production: 1.387 billion cu m (2017 est.)
country comparison to the world: 61

Natural gas—consumption: 1.387 billion cu m (2017 est.)
country comparison to the world: 85

Natural gas—exports: 0 cu m (2017 est.)
country comparison to the world: 86

Natural gas—imports: 0 cu m (2017 est.)
country comparison to the world: 109

Natural gas—proved reserves: 90.61 billion cu m (1 January 2018 est.)
country comparison to the world: 54

Carbon dioxide emissions from consumption of energy: 5.239 million Mt (2017 est.)
country comparison to the world: 134

COMMUNICATIONS

Telephones—fixed lines: *total subscriptions:* 17,076

subscriptions per 100 inhabitants: less than 1 (2019 est.)
country comparison to the world: 181

Telephones—mobile cellular: *total subscriptions:* 4,933,529
subscriptions per 100 inhabitants: 95.34 (2019 est.)
country comparison to the world: 120

Telecommunication systems: *general assessment:* primary network consists of microwave radio relay and coaxial cable with services barely adequate for government use; key exchanges are in Brazzaville, Pointe-Noire, and Loubomo; intercity lines frequently out of order; youth are seeking the Internet more than their parents and often gaining access in cyber cafes, only the most affluent have Internet access in their homes (2020)
domestic: fixed-line infrastructure inadequate, providing less than 1 fixed-line connection per 100 persons; in the absence of an adequate fixed-line infrastructure, mobile-cellular subscribership has surged to 95 per 100 persons (2019)
international: country code—242; WACS submarine cables to Europe and Western and South Africa; satellite earth station—1 Intelsat (Atlantic Ocean) (2019)
note: the COVID-19 outbreak is negatively impacting telecommunications production and supply chains globally; consumer spending on telecom devices and services has also slowed due to the pandemic's effect on economies worldwide; overall progress towards improvements in all facets of the telecom industry—mobile, fixed-line, broadband, submarine cable and satellite—has moderated

Broadcast media: 1 state-owned TV and 3 state-owned radio stations; several privately owned TV and radio stations; satellite TV service is available; rebroadcasts of several international broadcasters are available

Internet country code: .cg

Internet users: *total:* 437,865
percent of population: 8.65% (July 2018 est.)
country comparison to the world: 156

TRANSPORTATION

National air transport system: *number of registered air carriers:* 3 (2020)
inventory of registered aircraft operated by air carriers: 12
annual passenger traffic on registered air carriers: 333,899 (2018)
annual freight traffic on registered air carriers: 4.6 million mt-km (2018)

Civil aircraft registration country code prefix: TN (2016)

Airports: 27 (2013)
country comparison to the world: 123

Airports—with paved runways: *total:* 8 (2017)
over 3,047 m: 2 (2017)
2,438 to 3,047 m: 1 (2017)
1,524 to 2,437 m: 5 (2017)

Airports—with unpaved runways: *total:* 19 (2013)
1,524 to 2,437 m: 8 (2013)
914 to 1,523 m: 9 (2013)
under 914 m: 2 (2013)

Pipelines: 232 km gas, 4 km liquid petroleum gas, 982 km oil (2013)

Railways: *total:* 510 km (2014)
narrow gauge: 510 km 1.067-m gauge (2014)
country comparison to the world: 112

Roadways: *total:* 23,324 km (2017)
paved: 3,111 km (2017)
unpaved: 20,213 km (2017)
note: road network in Congo is composed of 23,324 km of which 17,000 km are classified as national, departmental, and routes of local interest: 6,324 km are non-classified routes
country comparison to the world: 108

Waterways: 1,120 km (commercially navigable on Congo and Oubanqui Rivers above Brazzaville; there are many ferries across the river to Kinshasa; the Congo south of Brazzaville- Kinshasa to the coast is not navigable because of rapids, necessitating a rail connection to Pointe Noire; other rivers are used for local traffic only) (2011)
country comparison to the world: 61

Merchant marine: *total:* 11
by type: general cargo 1, oil tanker 1, other 9 (2019)
country comparison to the world: 152

Ports and terminals: *major seaport(s):*
Pointe-Noire
oil terminal(s): Djeno
river port(s): Brazzaville (Congo)
Impfondo (Oubangi) Ouesso (Sangha) Oyo (Alima)

Military and security forces: *Congolese Armed Forces (Forces Armees Congolaises, FAC):* Army (Armee de Terre), Navy, Congolese Air Force (Armee de l'Air Congolaise); Gendarmerie; Presidential Guard (2019)

Military expenditures: 2.7% of GDP (2019)
2.5% of GDP (2018)
4.3% of GDP (2017)
6.4% of GDP (2016)
country comparison to the world: 32

Military and security service personnel strengths: the Congolese Armed Forces (FAC) have an estimated 12,000 active duty troops (8,000 Army; 800 Navy; 1-1,200 Air Force; 2,000 Gendarmerie) (2019 est.)

Military equipment inventories and acquisitions: the FAC is armed with mostly ageing Russian/former Soviet Union weapons, with some French and South African equipment; the leading suppliers of arms to the FAC since 2010 are Russia and South Africa (2019 est.)

Military service age and obligation: 18 years of age for voluntary military service; women may serve in the Armed Forces (2013)

Disputes—international: the location of the boundary in the broad Congo River with the Democratic Republic of the Congo is undefined except in the Pool Malebo/Stanley Pool area

Refugees and internally displaced persons: *refugees (country of origin):* 20,700 (Central African Republic), 19,780 (Democratic Republic

of the Congo) (refugees and asylum seekers) (2020)
IDPs: 304,430 (multiple civil wars since 1992) (2020)

Trafficking in persons: *current situation:* the Republic of the Congo is a source and destination country for children, men, and women, subjected to forced labor and sex trafficking; most trafficking victims are from Benin, the Democratic Republic of the Congo (DRC) and, to a lesser extent, other neighboring countries and are subjected to domestic servitude and market vending by West African and Congolese nationals; adults and children, the majority from the DRC, are also sex trafficked in Congo, mainly Brazzaville; internal trafficking victims, often from rural areas, are exploited as domestic servants or forced to work in quarries, bakeries, fishing, and agriculture

tier rating: Tier 2 Watch List—the Republic of the Congo does not fully comply with the minimum standards for the elimination of trafficking; however, it is making significant efforts to do so; the country drafted an action plan based on anti- trafficking legislation, which remains pending in the Supreme Court; the government made minimal anti-trafficking law enforcement efforts in 2014, failing to prosecute or convict suspected traffickers from cases dating back to 2010; serious allegations of official complicity continue to be reported; the government lacks a systematic means of identifying victims and relies on NGOs and international organizations to identify victims and NGOs and foster families to provide care to victims; the quality of care varied widely because the foster care system was allegedly undermined by inadequate security and official complicity (2015)

COOK ISLANDS

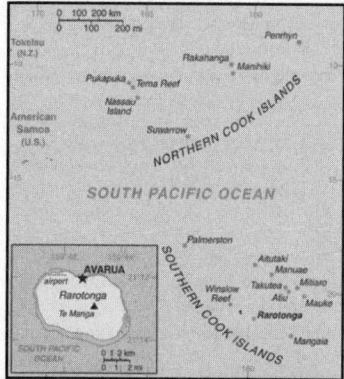

Background: The Cook Islands, named after Captain James Cook who landed in 1773, became

a British protectorate in 1888 and was later annexed by proclamation in 1900. The Cook Islands was first included within the boundaries of New Zealand in 1901, and in 1965, residents chose self-government in free association with New Zealand. The Cook Islands' economy relies on tourism, fisheries, and foreign aid. More recently a growing offshore financial sector exposed the country to vulnerabilities which the government has addressed with legislation and regulations for the oversight of all banks and financial institutions, and with enforcement measures. The Cook Islands continues to face challenges with the emigration of skilled workers, government deficits, inadequate infrastructure, and natural resource depletion. The Cook Islands is expected to graduate to the high-income threshold set by the World Bank, which will limit the country's access to Official Development Assistance under OECD guidelines.

Location: Oceania, group of islands in the South Pacific Ocean, about halfway between Hawaii and New Zealand

Geographic coordinates: 21 14 S, 159 46 W

Map references: Oceania

Area: *total:* 236 sq km
land: 236 sq km
water: 0 sq km
country comparison to the world: 215

Area—comparative: 1.3 times the size of Washington, DC

Land boundaries: *0 km*

Coastline: 120 km

Maritime claims: *territorial sea:* 12 nm
exclusive economic zone: 200 nm
continental shelf: 200 nm or to the edge of the continental margin

Climate: tropical oceanic; moderated by trade winds; a dry season from April to November and a more humid season from December to March

Terrain: low coral atolls in north; volcanic, hilly islands in south

Elevation: *lowest point:* Pacific Ocean 0 m
highest point: Te Manga 652 m

Natural resources: coconuts (copra)

Land use: *agricultural land:* 8.4% (2011 est.)
arable land: 4.2% (2011 est.) / permanent crops: 4.2% (2011 est.) / permanent pasture: 0% (2011 est.)
forest: 64.6% (2011 est.)
other: 27% (2011 est.)
Irrigated land: NA

Population distribution: most of the population is found on the island of Rarotonga

Natural hazards: tropical cyclones (November to March)

Environment—current issues: limited land presents solid and liquid waste disposal problems; soil destruction and deforestation; environmental degradation due to indiscriminant use of pesticides; improper disposal of pollutants; overfishing and destructive fishing practices; over dredging of lagoons and coral rubble beds; unregulated building

Environment—international agreements: *party to:* Biodiversity, Climate Change, Climate Change-Kyoto Protocol, Desertification, Hazardous Wastes, Law of the Sea, Ozone Layer Protection

Geography—note: the northern Cook Islands are seven low-lying, sparsely populated, coral atolls; the southern Cook Islands, where most of the population lives, consist of eight elevated, fertile, volcanic isles, including the largest, Rarotonga, at 67 sq km

PEOPLE AND SOCIETY

Population: 8,574 (July 2020 est.)
note: the Cook Islands' Ministry of Finance & Economic Management estimated the resident population to have been
11,700 in September 2016
country comparison to the world: 224

Nationality: *noun:* Cook Islander(s)
adjective: Cook Islander

Ethnic groups: Cook Island Maori (Polynesian) 81.3%, part Cook Island Maori 6.7%, other 11.9% (2011 est.)

Languages: English (official) 86.4%, Cook Islands Maori (Rarotongan) (official) 76.2%, other 8.3% (2011 est.)
note: shares sum to more than 100% because some respondents gave more than one answer on the census

Religions: Protestant 62.8% (Cook Islands Christian Church 49.1%, Seventh Day Adventist 7.9%, Assemblies of God 3.7%, Apostolic Church 2.1%), Roman Catholic 17%, Mormon 4.4%,

other 8%, none 5.6%, no response 2.2% (2011 est.)

Age structure: *0-14 years:* 19.93% (male 901/female 808)
15-24 years: 14.89% (male 684/female 593)
25-54 years: 37.66% (male 1,595/female 1,634)
55-64 years: 14.15% (male 674/female 539)
65 years and over: 13.37% (male 555/female 591) (2020 est.)

Median age: *total:* 38.3 years
male: 37.8 years
female: 38.7 years (2020 est.)
country comparison to the world: 64

Population growth rate: -2.59% (2020 est.)
country comparison to the world: 236

Birth rate: 13.3 births/1,000 population (2020 est.)
country comparison to the world: 140

Death rate: 9 deaths/1,000 population (2020 est.)
country comparison to the world: 61

Net migration rate: -29.9 migrant(s)/1,000 population (2020 est.)
country comparison to the world: 227

Population distribution: most of the population is found on the island of Rarotonga
Urbanization: urban population: 75.5% of total population (2020)
rate of urbanization: 0.37% annual rate of change (2015-20 est.)

total population growth rate v. urban population growth rate, 2000-2030: Sex ratio: at birth: 1.05 male(s)/female
0-14 years: 1.12 male(s)/female
15-24 years: 1.15 male(s)/female
25-54 years: 0.98 male(s)/female
55-64 years: 1.25 male(s)/female
65 years and over: 0.94 male(s)/female

total population: 1.06 male(s)/female (2020 est.)

Infant mortality rate: *total:* 11.9 deaths/1,000 live births
male: 14.4 deaths/1,000 live births
female: 9.2 deaths/1,000 live births (2020 est.)
country comparison to the world: 108

Life expectancy at birth: *total population:* 76.6 years
male: 73.8 years
female: 79.6 years (2020 est.)
country comparison to the world: 93

Total fertility rate: 2.12 children born/woman (2020 est.)
country comparison to the world: 97

Drinking water source:
improved:
total: 100% of population
unimproved:
total: 0% of population (2017 est.)

Current Health Expenditure: 3.3% (2017)

Physicians density: 1.41 physicians/1,000 population (2014)

Sanitation facility access:
improved:
total: 97.6% of population

unimproved:
total: 2.4% of population (2017 est.)

HIV/AIDS—adult prevalence rate: NA

HIV/AIDS—people living with HIV/AIDS: NA

HIV/AIDS—deaths: NA

Major infectious diseases: *degree of risk:* high (2020)
food or waterborne diseases: bacterial diarrhea
vectorborne diseases: malaria

Obesity—adult prevalence rate: 55.9% (2016)
country comparison to the world: 2
Education expenditures: 4.7% of GDP (2016)
country comparison to the world: 77

School life expectancy (primary to tertiary education): *total:* 1,516 years
male: 15 years
female: 14 years (2012)

GOVERNMENT

Country name: *conventional long form:* none
conventional short form: Cook Islands
former: Hervey Islands
etymology: named after Captain James COOK, the British explorer who visited the islands in 1773 and 1777

Dependency status: self-governing in free association with New Zealand; Cook Islands is fully responsible for internal affairs; New Zealand retains responsibility for external affairs and defense in consultation with the Cook Islands

Government type: parliamentary democracy

Capital: *name:* Avarua

geographic coordinates: 21 12 S, 159 46 W
time difference: UTC-10 (5 hours behind Washington, DC, during Standard Time)
etymology: translates as "two harbors" in Maori

Administrative divisions: none

Independence: none (became self-governing in free association with New Zealand on 4 August 1965 with the right at any time to move to full independence by unilateral action)

National holiday: Constitution Day, the first Monday in August (1965)

Constitution: *history:* 4 August 1965 (Cook Islands Constitution Act 1964)
amendments: proposed by Parliament; passage requires at least two-thirds majority vote by the Parliament membership in each of several readings and assent of the chief of state's representative; passage of amendments relating to the chief of state also requires two-thirds majority approval in a referendum; amended many times, last in 2004

Legal system: common law similar to New Zealand common law

International law organization participation: has not submitted an ICJ jurisdiction declaration (New Zealand normally retains responsibility for external affairs); accepts ICCt jurisdiction

Suffrage: 18 years of age; universal

Executive branch: *chief of state:* Queen ELIZABETH II (since 6 February 1952); represented by Sir Tom J. MARSTERS (since 9 August 2013); New Zealand Acting High Commissioner Ms Rachel BENNETT (since 9 December 2019)

head of government: Prime Minister Mark BROWN (since 1 October 2020)

cabinet: Cabinet chosen by the prime minister

elections/appointments: the monarchy is hereditary; UK representative appointed by the monarch; New Zealand high commissioner appointed by the New Zealand Government; following legislative elections, the leader of the majority party or majority coalition usually becomes prime minister

Legislative branch: *description:* unicameral Parliament, formerly the Legislative Assembly (24 seats; members directly elected in single-seat constituencies by simple majority vote to serve 4-year terms); note—the House of Ariki, a 24-member parliamentary body of traditional leaders appointed by the Queen's representative serves as a consultative body to the Parliament

elections: last held on 14 June 2018 (next to be held by 2022)

election results: percent of vote by party—NA; seats by party—Demo 11, CIP 10, One Cook Islands Movement 1, independent 2; composition—men 15, women 9, percent of women 37.5%

Judicial branch: *highest courts:* Court of Appeal (consists of the chief justice and 3 judges of the High Court); High Court (consists of the chief justice and at least 4 judges and organized into civil, criminal, and land divisions); note—appeals beyond the Cook Islands Court of Appeal are heard by the Judicial Committee of the Privy Council (in London)

judge selection and term of office: High Court chief justice appointed by the Queen's Representative on the advice of the Executive Council tendered by the prime minister; other judges appointed by the Queen's Representative, on the advice of the Executive Council tendered by the chief justice, High Court chief justice, and the minister of justice; chief justice and judges appointed for 3-year renewable terms

subordinate courts: justices of the peace

Political parties and leaders: Cook Islands Party or CIP [Henry PUNA]
Democratic Party or Demo [Tina BROWNE]
One Cook Islands Movement [Teina BISHOP]

International organization participation: ACP, ADB, AOSIS, FAO, ICAO, ICCt, ICRM, IFAD, IFRCS, IMO, IMSO, IOC, ITUC (NGOs), OPCW, PIF, Sparteca, SPC, UNESCO, UPU, WHO, WMO

Diplomatic representation in the US: none (self-governing in free association with New Zealand)

Diplomatic representation from the US: none (self-governing in free association with New Zealand)

Flag description: blue with the flag of the UK in the upper hoist-side quadrant and a large circle of 15 white five-pointed stars (one for every island) centered in the outer half of the flag

National symbol(s): a circle of 15, five-pointed, white stars on a blue field, Tiare maori (Gardenia taitensis) flower; national colors: green, white

National anthem: *name:* "Te Atua Mou E" (To God Almighty)

lyrics/music: Tepaeru Te RITO/Thomas DAVIS

note: adopted 1982; as prime minister, Sir Thomas DAVIS composed the anthem; his wife, a tribal chief, wrote the lyrics

0:00/ 1:05

ECONOMY

Economy—overview: Like many other South Pacific island nations, the Cook Islands' economic development is hindered by the isolation of the country from foreign markets, the limited size of domestic markets, lack of natural resources, periodic devastation from natural disasters, and inadequate infrastructure. Agriculture, employing more than one-quarter of the working population, provides the economic base with major exports of copra and citrus fruit. Black pearls are the Cook Islands' leading export. Manufacturing activities are limited to fruit processing, clothing, and handicrafts. Trade deficits are offset by remittances from emigrants and by foreign aid overwhelmingly from New Zealand. In the 1980s and 1990s, the country became overextended, maintaining a bloated public service and accumulating a large foreign debt. Subsequent reforms, including the sale of state assets, the strengthening of economic management, the encouragement of tourism, and a debt restructuring agreement, have rekindled investment and growth. The government is targeting fisheries and seabed mining as sectors for future economic growth.

GDP (purchasing power parity): $299.9 million (2016 est.)
$183.2 million (2005 est.)
country comparison to the world: 216

GDP (official exchange rate): $299.9 million (2016 est.)

GDP—real growth rate: 0.1% (2005 est.)
country comparison to the world: 190

GDP—per capita (PPP): $16,700 (2016 est.)
$9,100 (2005 est.)
country comparison to the world: 106

GDP—composition, by sector of origin:
agriculture: 5.1% (2010 est.)
industry: 12.7% (2010 est.)
services: 82.1% (2010 est.)

Agriculture—products: copra, citrus, pineapples, tomatoes, beans, pawpaws, bananas, yams, taro, coffee; pigs, poultry

Industries: fishing, fruit processing, tourism, clothing, handicrafts

Industrial production growth rate: 1% (2002)
country comparison to the world: 155

Labor force: 6,820 (2001)
country comparison to the world: 218

Labor force—by occupation: agriculture: 29% industry: 15%
services: 56% (1995)

Unemployment rate: 13.1% (2005)
country comparison to the world: 168

Population below poverty line: NA

Household income or consumption by percentage share: *lowest 10%:* NA
highest 10%: NA

Budget: revenues: 86.9 million (2010)
expenditures: 77.9 million (2010)

Taxes and other revenues: 29% (of GDP) (2010 est.)
country comparison to the world: 86

Budget surplus (+) or deficit (-): 3% (of GDP) (2010 est.)
country comparison to the world: 13

Fiscal year: 1 April–31 March

Inflation rate (consumer prices): 2.2% (2011 est.)
country comparison to the world: 114

Current account balance: $26.67 million (2005)
country comparison to the world: 58

Exports: $3.125 million (2011 est.)
$5.163 million (2010 est.)
country comparison to the world: 219

Exports—commodities: fish; copra, papayas, fresh and canned citrus fruit, coffee; pearls and pearl shells; clothing

Imports: $109.3 million (2011 est.)
$90.62 million (2010 est.)
country comparison to the world: 214

Imports—commodities: foodstuffs, textiles, fuels, timber, capital goods

Debt—external: $141 million (1996 est.)
country comparison to the world: 191

Exchange rates: NZ dollars (NZD) per US dollar—
1.416 (2017 est.)
1.4341 (2016 est.)
1.4341 (2015 est.)
1.441 (2014 est.)
1.4279 (2013 est.)

ENERGY

Electricity—production: 34 million kWh (2016 est.)
country comparison to the world: 209

Electricity—consumption: 31.62 million kWh (2016 est.)
country comparison to the world: 209

Electricity—exports: 0 kWh (2016 est.)
country comparison to the world: 123

Electricity—imports: 0 kWh (2016 est.)
country comparison to the world: 137

Electricity—installed generating capacity: 14,000 kW (2016 est.)
country comparison to the world: 207

Electricity—from fossil fuels: 79% of total installed capacity (2016 est.)
country comparison to the world: 85

Electricity—from nuclear fuels: 0% of total installed capacity (2017 est.)
country comparison to the world: 71

Electricity—from hydroelectric plants: 0% of total installed capacity (2017 est.)
country comparison to the world: 165

Electricity—from other renewable sources: 21% of total installed capacity (2017 est.)
country comparison to the world: 36

Crude oil—production: 0 bbl/day (2018 est.)
country comparison to the world: 123

Crude oil—exports: 0 bbl/day (2015 est.)
country comparison to the world: 108

Crude oil—imports: 0 bbl/day (2015 est.)
country comparison to the world: 114

Crude oil—proved reserves: 0 bbl (1 January 2018 est.)
country comparison to the world: 119

Refined petroleum products—production: 0 bbl/day (2015 est.)
country comparison to the world: 133

Refined petroleum products—consumption: 600 bbl/day (2016 est.)
country comparison to the world: 209

Refined petroleum products—exports: 0 bbl/day (2015 est.)
country comparison to the world: 146

Refined petroleum products—imports: 611 bbl/day (2015 est.)
country comparison to the world: 205

Natural gas—production: 0 cu m (2017 est.)
country comparison to the world: 119

Natural gas—consumption: 0 cu m (2017 est.)
country comparison to the world: 135

Natural gas—exports: 0 cu m (2017 est.)
country comparison to the world: 87

Natural gas—imports: 0 cu m (2017 est.)
country comparison to the world: 110

Natural gas—proved reserves: 0 cu m (1 January 2014 est.)

country comparison to the world: 124

Carbon dioxide emissions from consumption of energy: 88,810 Mt (2017 est.)
country comparison to the world: 207

COMMUNICATIONS

Telephones—fixed lines: *total subscriptions:* 3,305
subscriptions per 100 inhabitants: 37.56 (2019 est.)
country comparison to the world: 210

Telephones—mobile cellular: *total subscriptions:* 7,308
subscriptions per 100 inhabitants: 83.05 (2019 est.)
country comparison to the world: 216

Telecommunication systems: *general assessment:* demand for mobile broadband is increasing due to mobile services being the primary and most widespread source for Internet access across the region; Telecom Cook Islands offers international direct dialing, Internet, email, and fax; individual islands are connected by a combination of satellite earth stations, microwave systems, and VHF and HF radiotelephone (2020)
domestic: service is provided by small exchanges connected to subscribers by open-wire, cable, and fiber-optic cable; 38 per 100 fixed-line, 83 per 100 mobile-cellular (2019)
international: country code—682; the Manatua submarine cable to surrounding islands of Niue, Samoa, French Polynesia and other Cook Islands, the topography of the South Pacific region has made Internet connectivity a serious issue for many of the remote islands; submarine fiber-optic networks are expensive to build and maintain; satellite earth station—1 Intelsat (Pacific Ocean) (2019)
note: the COVID-19 outbreak is negatively impacting telecommunications production and supply chains globally; consumer spending on telecom devices and services has also slowed due to the pandemic's effect on economies worldwide; overall progress towards improvements in all facets of the telecom industry—mobile, fixed-line, broadband, submarine cable and satellite—has moderated

Broadcast media: 1 privately owned TV station broadcasts from Rarotonga providing a mix of local news and overseas-sourced programs (2019)

Internet country code: .ck

Internet users: *total:* 4,881

percent of population: 54% (July 2018 est.)
country comparison to the world: 217

TRANSPORTATION

National air transport system: *number of registered air carriers:* 1 (2020)
inventory of registered aircraft operated by air carriers: 6

Civil aircraft registration country code prefix: E5 (2016)

Airports: 11 (2013)
country comparison to the world: 153

Airports—with paved runways: *total:* 1 (2019)
1,524 to 2,437 m: 1

Airports—with unpaved runways: *total:* 10 (2013)
1,524 to 2,437 m: 2 (2013)
914 to 1,523 m: 7 (2013)
under 914 m: 1 (2013)

Roadways: *total:* 295 km (2018)
paved: 207 km (2018)
unpaved: 88 km (2018)
country comparison to the world: 202

Merchant marine: *total:* 205
by type: bulk carrier 21, container ship 3, general cargo 85, oil tanker 33, other 63 (2019)
country comparison to the world: 65

Ports and terminals: *major seaport(s):* Avatiu

MILITARY AND SECURITY

Military and security forces: no regular military forces; Cook Islands Police Service. (2018)

Military—note: defense is the responsibility of New Zealand in consultation with the Cook Islands and at its request

TRANSNATIONAL ISSUES

Disputes—international: none

CORAL SEA ISLANDS

INTRODUCTION

Background: Scattered over more than three-quarters of a million square kilometers of ocean, the Coral Sea Islands were declared a territory of Australia in 1969. They are uninhabited except for a small meteorological staff on the Willis Islets. Automated weather stations, beacons, and a lighthouse occupy many other islands and reefs. The Coral Sea Islands Act 1969 was amended in 1997 to extend the boundaries of the Coral Sea Islands Territory around Elizabeth and Middleton Reefs.

GEOGRAPHY

Location: Oceania, islands in the Coral Sea, northeast of Australia

Geographic coordinates: 18 00 S, 152 00 E

Map references: Oceania

Area: *total:* 3 sq km less than
land: 3 sq km less than
water: 0 sq km
note: includes numerous small islands and reefs scattered over a sea area of about 780,000 sq km

(300,000 sq mi) with the Willis Islets the most important
country comparison to the world: 253

Area—comparative: about four times the size of the National Mall in Washington, DC

Land boundaries: *0 km*

Coastline: 3,095 km

Maritime claims: *territorial sea:* 3 nm
exclusive fishing zone: 200 nm

Climate: tropical

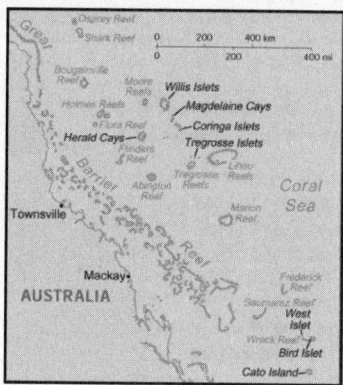

Terrain: sand and coral reefs and islands (cays)

Elevation: *lowest point:* Pacific Ocean 0 m
highest point: unnamed location on Cato Island 9 m

Natural resources: fish

Land use: *agricultural land:* 0% (2011 est.)
arable land: 0% (2011 est.) / permanent crops: 0% (2011 est.) / permanent pasture: 0% (2011 est.)
forest: 0% (2011 est.)
other: 100% (2011 est.)

Natural hazards: occasional tropical cyclones

Environment—current issues: no permanent freshwater resources; damaging activities include coral mining, destructive fishing practices (over-fishing, blast fishing)

Geography—note: important nesting area for birds and turtles

PEOPLE AND SOCIETY

Population: no indigenous inhabitants (2017 est.)
note: there is a staff of four at the meteorological station on Willis Island

GOVERNMENT

Country name: *conventional long form:* Coral Sea Islands Territory
conventional short form: Coral Sea Islands
etymology: self-descriptive name to reflect the islands' position in the Coral Sea off the north-eastern coast of Australia

Dependency status: territory of Australia; admin-istered from Canberra by the Department of Regional Australia, Local Government, Arts and Sport

Legal system: the common law legal system of Australia applies where applicable

Citizenship: see Australia

Diplomatic representation in the US: none (terri-tory of Australia)

Diplomatic representation from the US: none (ter-ritory of Australia)

Flag description: the flag of Australia is used

ECONOMY

Economy—overview: no economic activity

COMMUNICATIONS

Communications—note: automatic weather sta-tions on many of the isles and reefs relay data to the mainland

TRANSPORTATION

Ports and terminals: none; offshore anchorage only

MILITARY AND SECURITY

Military—note: defense is the responsibility of Australia

TRANSNATIONAL ISSUES

Disputes—international: none

COSTA RICA

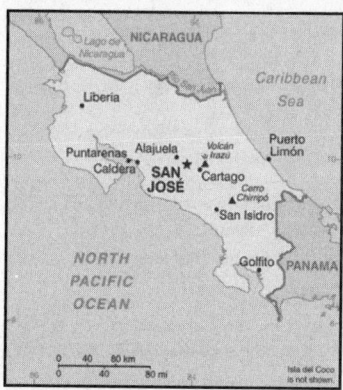

INTRODUCTION

Background: Although explored by the Spanish early in the 16th century, initial attempts at col-onizing Costa Rica proved unsuccessful due to a combination of factors, including disease from mosquito-infested swamps, brutal heat, resistance by natives, and pirate raids. It was not until 1563 that a permanent settlement of Cartago was established in the cooler, fertile central high-lands. The area remained a colony for some two and a half centuries. In 1821, Costa Rica became one of several Central American provinces that jointly declared their independence from Spain. Two years later it joined the United Provinces of Central America, but this federation disintegrated in 1838, at which time Costa Rica proclaimed its sovereignty and independence. Since the late 19th century, only two brief periods of violence have marred the country's democratic development. On 1 December 1948, Costa Rica dissolved its armed forces. Although it still maintains a large agricultural sector, Costa Rica has expanded its economy to include strong technology and tourism industries. The standard of living is relatively high. Land ownership is widespread.

GEOGRAPHY

Location: Central America, bordering both the Caribbean Sea and the North Pacific Ocean, between Nicaragua and Panama

Geographic coordinates: 10 00 N, 84 00 W

Map references: Central America and the Caribbean

Area: *total:* 51,100 sq km
land: 51,060 sq km
water: 40 sq km
note: includes Isla del Coco
country comparison to the world: 130

Area—comparative: slightly smaller than West Virginia

Land boundaries: *total:* 661 km
border countries (2): Nicaragua 313 km, Panama 348 km

Coastline: 1,290 km

Maritime claims: *territorial sea:* 12 nm
exclusive economic zone: 200 nm
continental shelf: 200 nm

Climate: tropical and subtropical; dry season (December to April); rainy season (May to November); cooler in highlands

Terrain: coastal plains separated by rugged moun-tains including over 100 volcanic cones, of which several are major active volcanoes

Elevation: *mean elevation:* 746 m
lowest point: Pacific Ocean 0 m
highest point: Cerro Chirripo 3,819 m

Natural resources: hydropower

Land use: *agricultural land:* 37.1% (2011 est.)
arable land: 4.9% (2011 est.) / permanent crops: 6.7% (2011 est.) / permanent pasture: 25.5% (2011 est.)
forest: 51.5% (2011 est.)
other: 11.4% (2011 est.)
Irrigated land: 1,015 sq km (2012)

Population distribution: roughly half of the nation's population resides in urban areas; the capital of San Jose is the largest city and home to approximately one-fifth of the population

Natural hazards: occasional earthquakes, hurricanes along Atlantic coast; frequent flooding of lowlands at onset of rainy season and landslides; active volcanoes

volcanism: Arenal (1,670 m), which erupted in 2010, is the most active volcano in Costa Rica; a 1968 eruption destroyed the town of Tabacon; Irazu (3,432 m), situated just east of San Jose, has the potential to spew ash over the capital city as it did between 1963 and 1965; other historically active volcanoes include Miravalles, Poas, Rincon de la Vieja, and Turrialba

Environment—current issues: deforestation and land use change, largely a result of the clearing of land for cattle ranching and agriculture; soil erosion; coastal marine pollution; fisheries protection; solid waste management; air pollution

Environment—international agreements: party to: Biodiversity, Climate Change, Climate Change-Kyoto Protocol, Desertification, Endangered Species, Environmental Modification, Hazardous Wastes, Law of the Sea, Marine Dumping, Ozone Layer Protection, Wetlands, Whaling
signed, but not ratified: Marine Life Conservation

Geography—note: four volcanoes, two of them active, rise near the capital of San Jose in the center of the country; one of the volcanoes, Irazu, erupted destructively in 1963-65

PEOPLE AND SOCIETY

Population: 5,097,988 (July 2020 est.)
country comparison to the world: 123

Nationality: *noun:* Costa Rican(s)
adjective: Costa Rican

Ethnic groups: white or mestizo 83.6%, mulatto 6.7%, indigenous 2.4%, black of African descent 1.1%, other 1.1%, none 2.9%, unspecified 2.2% (2011 est.)

Languages: Spanish (official), English

Religions: Roman Catholic 71.8%, Evangelical and Pentecostal 12.3%, other Protestant 2.6%, Jehovah's Witness 0.5%, other 2.4%, none 10.4% (2016 est.)

Demographic profile: Costa Rica's political stability, high standard of living, and well-developed social benefits system set it apart from its Central American neighbors. Through the government's sustained social spending—almost 20% of GDP annually—Costa Rica has made tremendous progress toward achieving its goal of providing universal access to education, healthcare, clean water, sanitation, and electricity. Since the 1970s, expansion of these services has led to a rapid decline in infant mortality, an increase in life expectancy at birth, and a sharp decrease in the birth rate. The average number of children born per women has fallen from about 7 in the 1960s to 3.5 in the early 1980s to below replacement level today. Costa Rica's poverty rate is lower than in most Latin American countries, but it has stalled at around 20% for almost two decades.

Costa Rica is a popular regional immigration destination because of its job opportunities and social programs. Almost 9% of the population is foreign-born, with Nicaraguans comprising nearly three-quarters of the foreign population. Many Nicaraguans who perform unskilled seasonal labor enter Costa Rica illegally or overstay their visas, which continues to be a source of tension. Less than 3% of Costa Rica's population lives abroad. The overwhelming majority of expatriates have settled in the United States after completing a university degree or in order to work in a highly skilled field.

Age structure: *0-14 years:* 22.08% (male 575,731/female 549,802)
15-24 years: 15.19% (male 395,202/female 379,277)
25-54 years: 43.98% (male 1,130,387/female 1,111,791)
55-64 years: 9.99% (male 247,267/female 261,847)
65 years and over: 8.76% (male 205,463/female 241,221) (2020 est.)

Dependency ratios: *total dependency ratio:* 45.1
youth dependency ratio: 30.2
elderly dependency ratio: 14.9
potential support ratio: 6.7 (2020 est.)

Median age: *total:* 32.6 years
male: 32.1 years
female: 33.1 years (2020 est.)
country comparison to the world: 109

Population growth rate: 1.08% (2020 est.)
country comparison to the world: 99

Birth rate: 14.8 births/1,000 population (2020 est.)
country comparison to the world: 122

Death rate: 4.9 deaths/1,000 population (2020 est.)
country comparison to the world: 198

Net migration rate: 0.8 migrant(s)/1,000 population (2020 est.)
country comparison to the world: 63

Population distribution: roughly half of the nation's population resides in urban areas; the capital of San Jose is the largest city and home to approximately one-fifth of the population

Urbanization: *urban population:* 80.8% of total population (2020)
rate of urbanization: 1.5% annual rate of change (2015-20 est.)
total population growth rate v. urban population growth rate, 2000-2030:

Major urban areas—population: 1.400 million SAN JOSE (capital) (2020)

Sex ratio: *at birth:* 1.05 male(s)/female
0-14 years: 1.05 male(s)/female
15-24 years: 1.04 male(s)/female
25-54 years: 1.02 male(s)/female
55-64 years: 0.94 male(s)/female
65 years and over: 0.85 male(s)/female
total population: 1 male(s)/female (2020 est.)

Maternal mortality rate: 27 deaths/100,000 live births (2017 est.)
country comparison to the world: 116

Infant mortality rate: *total:* 7.5 deaths/1,000 live births
male: 8.2 deaths/1,000 live births
female: 6.7 deaths/1,000 live births (2020 est.)
country comparison to the world: 153

Life expectancy at birth: *total population:* 79.2 years
male: 76.5 years
female: 82 years (2020 est.)
country comparison to the world: 58

Total fertility rate: 1.87 children born/woman (2020 est.)
country comparison to the world: 136

Contraceptive prevalence rate: 70.9% (2018)

Drinking water source:
improved:
urban: 100% of population
rural: 100% of population
total: 100% of population
unimproved:
urban: 0% of population
rural: 0% of population
total: 0% of population (2017 est.)

Current Health Expenditure: 7.3% (2017)

Physicians density: 2.95 physicians/1,000 population (2017)

Hospital bed density: 1.1 beds/1,000 population (2017)

Sanitation facility access:
improved:
urban: 98.4% of population
rural: 95.8% of population
total: 97.8% of population
unimproved:
urban: 1.6% of population
rural: 4.2% of population
total: 2.2% of population (2017 est.)

HIV/AIDS—adult prevalence rate: 0.4% (2019 est.)
country comparison to the world: 75

HIV/AIDS—people living with HIV/AIDS: 14,000 (2019 est.)
country comparison to the world: 91

HIV/AIDS—deaths: <200 (2019 est.)

Major infectious diseases: *degree of risk:* intermediate (2020)
food or waterborne diseases: bacterial diarrhea
vectorborne diseases: dengue fever

Obesity—adult prevalence rate: 25.7% (2016)
country comparison to the world: 48

Education expenditures: 7.4% of GDP (2017)
country comparison to the world: 10

Literacy: *definition:* age 15 and over can read and write
total population: 97.9%
male: 97.8%
female: 97.9% (2018)

School life expectancy (primary to tertiary education): *total:* 16 years
male: 17 years
female: 17 years (2019)

Unemployment, youth ages 15-24: *total:* 20.6%

male: 17.6%
female: 25.9% (2017 est.)
country comparison to the world: 63

GOVERNMENT

Country name: *conventional long form:* Republic of Costa Rica
conventional short form: Costa Rica
local long form: Republica de Costa Rica
local short form: Costa Rica
etymology: the name means "rich coast" in Spanish and was first applied in the early colonial period of the 16th century

Government type: presidential republic

Capital: *name:* San Jose
geographic coordinates: 9 56 N, 84 05 W
time difference: UTC-6 (1 hour behind Washington, DC, during Standard Time)
etymology: named in honor of Saint Joseph

Administrative divisions: 7 provinces (provincias, singular—provincia); Alajuela, Cartago, Guanacaste, Heredia, Limon, Puntarenas, San Jose

Independence: 15 September 1821 (from Spain)

National holiday: Independence Day, 15 September (1821)

Constitution: *history:* many previous; latest effective 8 November 1949
amendments: proposals require the signatures of at least 10 Legislative Assembly members or petition of at least 5% of qualified voters; consideration of proposals requires two-thirds majority approval in each of three readings by the Assembly, followed by preparation of the proposal as a legislative bill and its approval by simple majority of the Assembly; passage requires at least two-thirds majority vote of the Assembly membership; a referendum is required only if approved by at least two thirds of the Assembly; amended many times, last in 2015

Legal system: civil law system based on Spanish civil code; judicial review of legislative acts in the Supreme Court

International law organization participation: accepts compulsory ICJ jurisdiction; accepts ICCt jurisdiction

Citizenship: *citizenship by birth:* yes
citizenship by descent only: yes
dual citizenship recognized: yes
residency requirement for naturalization: 7 years

Suffrage: 18 years of age; universal and compulsory

Executive branch: *chief of state:* President Carlos ALVARADO Quesada (since 8 May 2018); First Vice President Epsy CAMPBELL Barr (since 8 May 2018); Second Vice President Marvin RODRIGUEZ Cordero (since 8 May 2018); note—the president is both chief of state and head of government
head of government: President Carlos ALVARADO Quesada (since 8 May 2018); First Vice President Epsy CAMPBELL Barr (since

8 May 2018); Second Vice President Marvin RODRIGUEZ Cordero (since 8 May 2018)
cabinet: Cabinet selected by the president
elections/appointments: president and vice presidents directly elected on the same ballot by modified majority popular vote (40% threshold) for a 4-year term (eligible for non-consecutive terms); election last held on 4 February 2018 with a runoff on 1 April 2018 (next to be held in February 2022)
election results: Carlos ALVARADO Quesada elected president in second round; percent of vote in first round—Fabricio ALVARADO Munoz (PRN) 25%; Carlos ALVARADO Quesada (PAC) 21.6%; Antonio ALVAREZ (PLN) 18.6%; Rodolfo PIZA (PUSC) 16%; Juan Diego CASTRO (PIN) 9.5%; Rodolfo HERNANDEZ (PRS) 4.9%, other 4.4%; percent of vote in second round—Carlos ALVARADO Quesada (PAC) 60.7%; Fabricio ALVARADO Munoz (PRN) 39.3%

Legislative branch: *description:* unicameral Legislative Assembly or Asamblea Legislativa (57 seats; members directly elected in multiseat constituencies—corresponding to the country's 7 provinces—by closed list proportional representation vote; members serve 4-year terms)
elections: last held on 4 February 2018 (next to be held in February 2022)
election results: percent of vote by party—PLN 19.5%, PRN 18.2%, PAC 16.3%, PUSC 14.6%, PIN 7.7%, PRS 4.2%, PFA 4%, ADC 2.5%, ML 2.3%, PASE 2.3%, PNG 2.2%, other 6.2%; seats by party—PLN 17, PRN 14, PAC 10, PUSC 9, PIN 4, PRS 2, PFA 1; composition—men 31, women 26, percent of women 45.6%

Judicial branch: *highest courts:* Supreme Court of Justice (consists of 22 judges organized into 3 cassation chambers each with 5 judges and the Constitutional Chamber with 7 judges)
judge selection and term of office: Supreme Court of Justice judges elected by the National Assembly for 8-year terms with renewal decided by the National Assembly
subordinate courts: appellate courts; trial courts; first instance and justice of the peace courts; Superior Electoral Tribunal

Political parties and leaders:
Accessibility Without Exclusion or PASE [Oscar Andres LOPEZ Arias]
Broad Front (Frente Amplio) or PFA [Ana Patricia MORA Castellanos]
Christian Democratic Alliance or ADC [Mario REDONDO Poveda]
Citizen Action Party or PAC [Marta Eugenia SOLANO Arias]
Costa Rican Renewal Party or PRC [Justo OROZCO Alvarez]
Libertarian Movement Party or ML [Victor Danilo CUBERO Corrales]
National Integration Party or PIN [Walter MUNOZ Cespedes]
National Liberation Party or PLN [Jorge Julio PATTONI Saenz]
National Restoration Party or PRN [Carlos Luis AVENDANO Calvo]
New Generation or PNG [Sergio MENA]

Patriotic Alliance [Jorge ARAYA Westover]
Social Christian Republican Party or PRS [Dragos DOLANESCU Valenciano]
Social Christian Unity Party or PUSC [Pedro MUNOZ Fonseca]

International organization participation: BCIE, CACM, CD, CELAC, FAO, G-77, IADB, IAEA, IBRD, ICAO, ICC (national committees), ICCt, ICRM, IDA, IFAD, IFC, IFRCS, ILO, IMF, IMO, IMSO, Interpol, IOC, IOM, IPU, ISO, ITSO, ITU, ITUC (NGOs), LAES, LAIA (observer), MIGA, NAM (observer), OAS, OIF (observer), OPANAL, OPCW, Pacific Alliance (observer), PCA, SICA, UN, UNCTAD, UNESCO, UNHCR, UNIDO, Union Latina, UNWTO, UPU, WCO, WFTU (NGOs), WHO, WIPO, WMO, WTO

Diplomatic representation in the US: chief of mission: Ambassador Fernando LLORCA Castro (since 17 September 2018)
chancery: 2114 S Street NW, Washington, DC 20008
telephone: [1] (202) 499-2980
FAX: [1] (202) 265-4795
consulate(s) general: Atlanta, Chicago, Houston, Los Angeles, Miami, New York, Washington DC
consulate(s): Saint Paul (MN), San Juan (Puerto Rico), Tucson (AZ)

Diplomatic representation from the US: chief of mission: Ambassador Sharon DAY (since 5 October 2017)
telephone: [506] 2519-2000
embassy: Calle 98 Via 104, Pavas, San Jose
mailing address: APO AA 34020
FAX: [506] 2519-2305

Flag description: five horizontal bands of blue (top), white, red (double width), white, and blue, with the coat of arms in a white elliptical disk placed toward the hoist side of the red band; Costa Rica retained the earlier blue-white-blue flag of Central America until 1848 when, in response to revolutionary activity in Europe, it was decided to incorporate the French colors into the national flag and a central red stripe was added; today the blue color is said to stand for the sky, opportunity, and perseverance, white denotes peace, happiness, and wisdom, while red represents the blood shed for freedom, as well as the generosity and vibrancy of the people
note: somewhat resembles the flag of North Korea; similar to the flag of Thailand but with the blue and red colors reversed

National symbol(s): *yiguirro (clay-colored robin); national colors:* blue, white, red

National anthem: *name:* "Himno Nacional de Costa Rica" (National Anthem of Costa Rica)
lyrics/music: Jose Maria ZELEDON Brenes/ Manuel Maria GUTIERREZ
note: adopted 1949; the anthem's music was originally written for an 1853 welcome ceremony for diplomatic missions from the US and UK; the lyrics were added in 1903
0:00 / 0:00

ECONOMY

Economy—overview: Since 2010, Costa Rica has enjoyed strong and stable economic growth—3.8% in 2017. Exports of bananas, coffee, sugar, and beef are the backbone of its commodity exports. Various industrial and processed agricultural products have broadened exports in recent years, as have high value-added goods, including medical devices. Costa Rica's impressive biodiversity also makes it a key destination for ecotourism.

Foreign investors remain attracted by the country's political stability and relatively high education levels, as well as the incentives offered in the free-trade zones; Costa Rica has attracted one of the highest levels of foreign direct investment per capita in Latin America. The US-Central American-Dominican Republic Free Trade Agreement (CAFTA-DR), which became effective for Costa Rica in 2009, helped increase foreign direct investment in key sectors of the economy, including insurance and telecommunication. However, poor infrastructure, high energy costs, a complex bureaucracy, weak investor protection, and uncertainty of contract enforcement impede greater investment.

Costa Rica's economy also faces challenges due to a rising fiscal deficit, rising public debt, and relatively low levels of domestic revenue. Poverty has remained around 20-25% for nearly 20 years, and the government's strong social safety net has eroded due to increased constraints on its expenditures. Costa Rica's credit rating was downgraded from stable to negative in 2015 and again in 2017, upping pressure on lending rates—which could hurt small business, on the budget deficit—which could hurt infrastructure development, and on the rate of return on investment—which could soften foreign direct investment (FDI). Unlike the rest of Central America, Costa Rica is not highly dependent on remittances—which represented just 1 % of GDP in 2016, but instead relies on FDI—which accounted for 5.1% of GDP.

GDP (purchasing power parity):
$83.94 billion (2017 est.)
$81.27 billion (2016 est.)
$77.96 billion (2015 est.)
note: data are in 2017 dollars
country comparison to the world: 93

GDP (official exchange rate):
$58.27 billion (2017 est.)

GDP—real growth rate:
3.3% (2017 est.)
4.2% (2016 est.)
3.6% (2015 est.)
country comparison to the world: 90

GDP—per capita (PPP):
$16,900 (2017 est.)
$16,600 (2016 est.)
$16,100 (2015 est.)
note: data are in 2017 dollars
country comparison to the world: 105

Gross national saving:
15.1% of GDP (2017 est.)
16.1% of GDP (2016 est.)

15% of GDP (2015 est.)
country comparison to the world: 135

GDP—composition, by end use:
household consumption: 64.2% (2017 est.)
government consumption: 17.3% (2017 est.)
investment in fixed capital: 17.1% (2017 est.)
investment in inventories: 1% (2017 est.)
exports of goods and services: 33.3% (2017 est.)
imports of goods and services: -32.9% (2017 est.)

GDP—composition, by sector of origin:
agriculture: 5.5% (2017 est.)
industry: 20.6% (2017 est.)
services: 73.9% (2017 est.)

Agriculture—products: bananas, pineapples, coffee, melons, ornamental plants, sugar, corn, rice, beans, potatoes; beef, poultry, dairy; timber

Industries: medical equipment, food processing, textiles and clothing, construction materials, fertilizer, plastic products

Industrial production growth rate: 1.3% (2017 est.)
country comparison to the world: 147

Labor force: 1.843 million (2020 est.)
note: official estimate; excludes Nicaraguans living in Costa Rica
country comparison to the world: 123

Labor force—by occupation:
agriculture: 14%
industry: 22%
services: 64% (2006 est.)

Unemployment rate:
8.1% (2017 est.)
9.5% (2016 est.)
country comparison to the world: 126

Population below poverty line: 21.7% (2014 est.)

Household income or consumption by percentage share: lowest 10%: 1.5%
highest 10%: 36.9% (2014 est.)

Budget: revenues: 8.357 billion (2017 est.)
expenditures: 11.92 billion (2017 est.)

Taxes and other revenues: 14.3% (of GDP) (2017 est.)
country comparison to the world: 200

Budget surplus (+) or deficit (-): -6.1% (of GDP) (2017 est.)
country comparison to the world: 185

Public debt:
48.9% of GDP (2017 est.)
44.9% of GDP (2016 est.)
country comparison to the world: 105

Fiscal year: calendar year

Inflation rate (consumer prices):
1.6% (2017 est.)
0% (2016 est.)
country comparison to the world: 89

Current account balance:
-$1.692 billion (2017 est.)
-$1.326 billion (2016 est.)
country comparison to the world: 163

Exports:
$10.81 billion (2017 est.)
$10.15 billion (2016 est.)
country comparison to the world: 90

Exports—partners: US 40.9%, Belgium 6.3%, Panama 5.6%, Netherlands 5.6%, Nicaragua 5.1%, Guatemala 5% (2017)

Exports—commodities: bananas, pineapples, coffee, melons, ornamental plants, sugar; beef; seafood; electronic components, medical equipment

Imports:
$15.15 billion (2017 est.)
$14.53 billion (2016 est.)
country comparison to the world: 89

Imports—commodities: raw materials, consumer goods, capital equipment, petroleum, construction materials

Imports—partners: US 38.1%, China 13.1%, Mexico 7.3% (2017)

Reserves of foreign exchange and gold:
$7.15 billion (31 December 2017 est.)
$7.574 billion (31 December 2016 est.)
country comparison to the world: 86

Debt—external:
$26.83 billion (31 December 2017 est.)
$24.3 billion (31 December 2016 est.)
country comparison to the world: 86

Exchange rates: Costa Rican colones (CRC) per US dollar—
573.5 (2017 est.)
544.74 (2016 est.)
544.74 (2015 est.)
534.57 (2014 est.)
538.32 (2013 est.)

ENERGY

Electricity access: electrification—total population: 100% (2020)

Electricity—production: 10.79 billion kWh (2016 est.)
country comparison to the world: 100

Electricity—consumption: 9.812 billion kWh (2016 est.)
country comparison to the world: 98

Electricity—exports: 643 million kWh (2015 est.)
country comparison to the world: 64

Electricity—imports: 807 million kWh (2016 est.)
country comparison to the world: 72

Electricity—installed generating capacity: 3.584 million kW (2016 est.)
country comparison to the world: 94

Electricity—from fossil fuels: 18% of total installed capacity (2016 est.)
country comparison to the world: 196

Electricity—from nuclear fuels: 0% of total installed capacity (2017 est.)
country comparison to the world: 72

Electricity—from hydroelectric plants: 64% of total installed capacity (2017 est.)
country comparison to the world: 25

Electricity—from other renewable sources: 18% of total installed capacity (2017 est.)
country comparison to the world: 48

Crude oil—production: 0 bbl/day (2018 est.)
country comparison to the world: 124

Crude oil—exports: 0 bbl/day (2015 est.)
country comparison to the world: 109

Crude oil—imports: 0 bbl/day (2015 est.)
country comparison to the world: 115

Crude oil—proved reserves: 0 bbl (1 January 2018 est.)
country comparison to the world: 120

Refined petroleum products—production: 0 bbl/day (2015 est.)
country comparison to the world: 134

Refined petroleum products—consumption: 53,000 bbl/day (2016 est.)
country comparison to the world: 100

Refined petroleum products—exports: 0 bbl/day (2015 est.)
country comparison to the world: 147

Refined petroleum products—imports: 51,320 bbl/day (2015 est.)
country comparison to the world: 80

Natural gas—production: 0 cu m (2017 est.)
country comparison to the world: 120

Natural gas—consumption: 0 cu m (2017 est.)
country comparison to the world: 136

Natural gas—exports: 0 cu m (2017 est.)
country comparison to the world: 88

Natural gas—imports: 0 cu m (2017 est.)
country comparison to the world: 111

Natural gas—proved reserves: 0 cu m (1 January 2014 est.)
country comparison to the world: 125

Carbon dioxide emissions from consumption of energy: 7.653 million Mt (2017 est.)
country comparison to the world: 120

COMMUNICATIONS

Telephones—fixed lines: *total subscriptions:* 630,386
subscriptions per 100 inhabitants: 12.5 (2019 est.)
country comparison to the world: 89

Telephones—mobile cellular: *total subscriptions:* 8,163,744
subscriptions per 100 inhabitants: 161.88 (2019 est.)
country comparison to the world: 96

Telecommunication systems: *general assessment:* good domestic telephone service in terms of breadth of coverage; in recent years growth has been achieved from liberalization of the telecom sector and has seen substantial expansion in all sectors; Costa Rica's broadband market is the most advanced in Central America, with the highest broadband penetration for this sub-region; broadband penetration does lag behind many South American countries; with the implementation of number portability there is greater opportunity for increased competition in the future (2018)
domestic: 13 per 100 fixed-line, 162 per 100 mobile-cellular; point-to-point and point-to-multi-point microwave, fiberoptic, and coaxial

cable link rural areas; Internet service is available (2019)
international: country code—506; landing points for the ARCOS-1, MAYA-1, and the PAC submarine cables that provide links to South and Central America, parts of the Caribbean, and the US; connected to Central American Microwave System; satellite earth stations—2 Intelsat (Atlantic Ocean) (2019)
note: the COVID-19 outbreak is negatively impacting telecommunications production and supply chains globally; consumer spending on telecom devices and services has also slowed due to the pandemic's effect on economies worldwide; overall progress towards improvements in all facets of the telecom industry—mobile, fixed-line, broadband, submarine cable and satellite—has moderated

Broadcast media: multiple privately owned TV stations and 1 publicly owned TV station; cable network services are widely available; more than 100 privately owned radio stations and a public radio network (2017)

Internet country code: .cr

Internet users: *total:* 3,694,974
percent of population: 74.09% (July 2018 est.)
country comparison to the world: 94

Broadband—fixed subscriptions: *total:* 834,784
subscriptions per 100 inhabitants: 17 (2018 est.)
country comparison to the world: 73

TRANSPORTATION

National air transport system: *number of registered air carriers:* 1 (2020)
inventory of registered aircraft operated by air carriers: 39
annual passenger traffic on registered air carriers: 1,948,546 (2018)
annual freight traffic on registered air carriers: 11.13 million mt-km (2018)

Civil aircraft registration country code prefix: TI (2016)

Airports: 161 (2013)
country comparison to the world: 34

Airports—with paved runways: *total:* 47 (2017)
2,438 to 3,047 m: 2 (2017)
1,524 to 2,437 m: 2 (2017)
914 to 1,523 m: 27 (2017)
under 914 m: 16 (2017)

Airports—with unpaved runways: *total:* 114 (2013)
914 to 1,523 m: 18 (2013)
under 914 m: 96 (2013)

Pipelines: 662 km refined products (2013)

Railways: *total:* 278 km (2014)
narrow gauge: 278 km 1.067-m gauge (2014)
note: the entire rail network fell into disrepair and out of use at the end of the 20th century; since 2005, certain sections of rail have been rehabilitated
country comparison to the world: 123

Roadways: *total:* 5,035 km (2017)
country comparison to the world: 147

Waterways: 730 km (seasonally navigable by small craft) (2011)
country comparison to the world: 74

Merchant marine: *total:* 11

by type: other 11 (2019)
country comparison to the world: 153

Ports and terminals: *major seaport(s):* Atlantic Ocean (Caribbean)—Puerto Limon

Pacific Ocean—Caldera

MILITARY AND SECURITY

Military and security forces: no regular military forces; Ministry of Public Security commands the Public Forces of Costa Rica, which includes the Public Force (National Police), Anti-Drug Police, and National Coast Guard Service (2020)
note: Costa Rica's armed forces were constitutionally abolished in 1949

Military and security service personnel strengths: the Public Forces of Costa Rica have approximately 12,000 personnel (2019 est.)

Military equipment inventories and acquisitions: the Public Forces' inventory includes mostly second-hand US equipment; since 2000, the only reported major equipment deliveries were from the US (light helicopters in 2012 and 2014 and second-hand coast guard cutters in 2018) (2019 est.)

TRANSNATIONAL ISSUES

Disputes—international: Costa Rica and Nicaragua regularly file border dispute cases over the delimitations of the San Juan River and the northern tip of Calero Island to the International Court of Justice (ICJ); in 2009, the ICJ ruled that Costa Rican vessels carrying out police activities could not use the river, but official Costa Rican vessels providing essential services to riverside inhabitants and Costa Rican tourists could travel freely on the river; in 2011, the ICJ provisionally ruled that both countries must remove personnel from the disputed area; in 2013, the ICJ rejected Nicaragua's 2012 suit to halt Costa Rica's construction of a highway paralleling the river on the grounds of irreparable environmental damage; in 2013, the ICJ, regarding the disputed territory, ordered that Nicaragua should refrain from dredging or canal construction and refill and repair damage caused by trenches connecting the river to the Caribbean and upheld its 2010 ruling that Nicaragua must remove all personnel; in early 2014, Costa Rica brought Nicaragua to the ICJ over offshore oil concessions in the disputed region

Refugees and internally displaced persons: *refugees (country of origin):* 13,517 (Venezuela) (economic and political crisis; includes Venezuelans who have claimed asylum, are recognized as refugees, or received alternative legal stay) (2020)
stateless persons: 231 (2019)

Trafficking in persons: *current situation:* Costa Rica is a source, transit, and destination country for men, women, and children subjected to sex trafficking and forced labor; Costa Rican

235

women and children, as well as those from Nicaragua, the Dominican Republic, and other Latin American countries, are sex trafficked in Costa Rica; child sex tourism is a particular problem with offenders coming from the US and Europe; men and children from Central America, including indigenous Panamanians, and Asia are exploited in agriculture, construction, fishing, and commerce; Nicaraguans transit Costa Rica to reach Panama, where some are subjected to forced labor or sex trafficking

tier rating: Tier 2 Watch List—Costa Rica does not fully comply with the minimum standards

for the elimination of trafficking; however, it is making significant efforts to do so; anti-trafficking law enforcement efforts declined in 2014, with fewer prosecutions and no convictions and no actions taken against complicit government personnel; some officials conflated trafficking with smuggling, and authorities reported the diversion of funds to combat smuggling hindered antitrafficking efforts; the government identified more victims than the previous year but did not make progress in ensuring that victims received adequate protective services; specialized services were limited and mostly provided by NGOs

without government support, even from a dedicated fund for anti-trafficking efforts; victims services were virtually nonexistent outside of the capital (2015)

Illicit drugs: transshipment country for cocaine and heroin from South America; illicit production of cannabis in remote areas; domestic cocaine consumption, particularly crack cocaine, is rising; significant consumption of amphetamines; seizures of smuggled cash in Costa Rica and at the main border crossing to enter Costa Rica from Nicaragua have risen in recent years

COTE D'IVOIRE

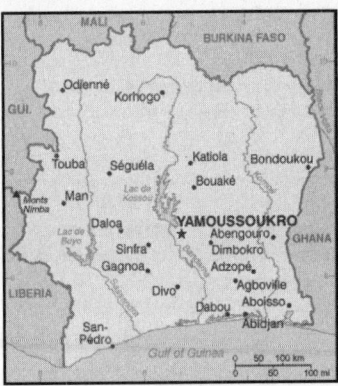

INTRODUCTION

Background: Close ties to France following independence in 1960, the development of cocoa production for export, and foreign investment all made Cote d'Ivoire one of the most prosperous of the West African states but did not protect it from political turmoil. In December 1999, a military coup—the first ever in Cote d'Ivoire's history—overthrew the government. Junta leader Robert GUEI attempted to rig the elections held in late 2000 and declared himself the winner. Popular protest forced him to step aside and an election brought Laurent GBAGBO into power. Ivoirian dissidents and disaffected members of the military launched a failed coup attempt in September 2002 that developed into a rebellion and then a civil war. In 2003, a cease-fire resulted in the country being divided with the rebels holding the north, the government the south, and peacekeeping forces a buffer zone between the two. In March 2007, President GBAGBO and former New Forces rebel leader Guillaume SORO signed an agreement in which SORO joined GBAGBO's government as prime minister and the two agreed to reunite the country by dismantling the buffer zone, integrating rebel forces into the national

armed forces, and holding elections. Difficulties in preparing electoral registers delayed balloting until 2010. In November 2010, Alassane Dramane OUATTARA won the presidential election over GBAGBO, but GBAGBO refused to hand over power, resulting in a five- month resumption of violent conflict. In April 2011, after widespread fighting, GBAGBO was formally forced from office by armed OUATTARA supporters with the help of UN and French forces. OUATTARA won a second term in 2015 and is focused on rebuilding the country's economy and infrastructure while reforming the security forces. The UN peacekeeping mission departed in June 2017. GBAGBO was in The Hague on trial for crimes against humanity, but was acquitted in January 2019. Côte d'Ivoire is scheduled to hold presidential elections in November 2020.

GEOGRAPHY

Location: Western Africa, bordering the North Atlantic Ocean, between Ghana and Liberia

Geographic coordinates: 8 00 N, 5 00 W

Map references: Africa

Area: *total:* 322,463 sq km
land: 318,003 sq km
water: 4,460 sq km
country comparison to the world: 70

Area—comparative: slightly larger than New Mexico

Land boundaries: *total:* 3,458 km
border countries (5): Burkina Faso 545 km, Ghana 720 km, Guinea 816 km, Liberia 778 km, Mali 599 km

Coastline: 515 km

Maritime claims: *territorial sea:* 12 nm
exclusive economic zone: 200 nm
continental shelf: 200 nm

Climate: tropical along coast, semiarid in far north; three seasons—warm and dry (November to March), hot and dry (March to May), hot and wet (June to October)

Terrain: mostly flat to undulating plains; mountains in northwest

Elevation: *mean elevation:* 250 m
lowest point: Gulf of Guinea 0 m
highest point: Monts Nimba 1,752 m

Natural resources: petroleum, natural gas, diamonds, manganese, iron ore, cobalt, bauxite, copper, gold, nickel, tantalum, silica sand, clay, cocoa beans, coffee, palm oil, hydropower

Land use: *agricultural land:* 64.8% (2011 est.)
arable land: 9.1% (2011 est.) / permanent crops: 14.2% (2011 est.) / permanent pasture: 41.5% (2011 est.)
forest: 32.7% (2011 est.)
other: 2.5% (2011 est.)
Irrigated land: 730 sq km (2012)

Population distribution: the population is primarily located in the forested south, with the highest concentration of people residing in and around the cities on the Atlantic coast; most of the northern savanna remains sparsely populated with higher concentrations located along transportation corridors as shown in this population distribution map

Natural hazards: coast has heavy surf and no natural harbors; during the rainy season torrential flooding is possible

Environment—current issues: deforestation (most of the country's forests—once the largest in West Africa—have been heavily logged); water pollution from sewage, and from industrial, mining, and agricultural effluents

Environment—international agreements: *party to:* Biodiversity, Climate Change, Climate Change-Kyoto Protocol, Desertification, Endangered Species, Hazardous Wastes, Law of the Sea, Marine Dumping, Ozone Layer Protection, Ship Pollution, Tropical Timber 83, Tropical Timber 94, Wetlands, Whaling
signed, but not ratified: none of the selected agreements

Geography—note: most of the inhabitants live along the sandy coastal region; apart from the capital area, the forested interior is sparsely populated

PEOPLE AND SOCIETY

Population: 27,481,086 (July 2020 est.)
note: estimates for this country explicitly take into account the effects of excess mortality due to AIDS; this can result in lower life expectancy, higher infant mortality, higher death rates, lower population growth rates, and changes in the distribution of population by age and sex than would otherwise be expected
country comparison to the world: 52

Nationality: *noun:* Ivoirian(s)
adjective: Ivoirian

Ethnic groups: Akan 28.9%, Voltaique or Gur 16.1%, Northern Mande 14.5%, Kru 8.5%, Southern Mande 6.9%, unspecified 0.9%, non-Ivoirian 24.2% (2014 est.)

Languages: French (official), 60 native dialects of which Dioula is the most widely spoken

Religions: Muslim 42.9%, Catholic 17.2%, Evangelical 11.8%, Methodist 1.7%, other Christian 3.2%, animist 3.6%, other religion 0.5%, none 19.1% (2014 est.)
note: the majority of foreign migrant workers are Muslim (72.7%) and Christian (17.7%)

Demographic profile: Cote d'Ivoire's population is likely to continue growing for the foreseeable future because almost 60% of the populace is younger than 25, the total fertility rate is holding steady at about 3.5 children per woman, and contraceptive use is under 20%. The country will need to improve education, health care, and gender equality in order to turn its large and growing youth cohort into human capital. Even prior to 2010 unrest that shuttered schools for months, access to education was poor, especially for women. As of 2015, only 53% of men and 33% of women were literate. The lack of educational attainment contributes to Cote d'Ivoire's high rates of unskilled labor, adolescent pregnancy, and HIV/AIDS prevalence.

Following its independence in 1960, Cote d'Ivoire's stability and the blossoming of its labor-intensive cocoa and coffee industries in the southwest made it an attractive destination for migrants from other parts of the country and its neighbors, particularly Burkina Faso. The HOUPHOUET-BOIGNY administration continued the French colonial policy of encouraging labor immigration by offering liberal land ownership laws. Foreigners from West Africa, Europe (mainly France), and Lebanon composed about 25% of the population by 1998.

Ongoing economic decline since the 1980s and the power struggle after HOUPHOUET-BOIGNY's death in 1993 ushered in the politics of "Ivoirite," institutionalizing an Ivoirian identity that further marginalized northern Ivoirians and scapegoated immigrants. The hostile Muslim north-Christian south divide snowballed into a 2002 civil war, pushing tens of thousands of foreign migrants, Liberian refugees, and Ivoirians to flee to war-torn Liberia or other regional countries and more than a million people to be internally displaced. Subsequently, violence following the contested 2010 presidential election prompted some 250,000 people to seek refuge in Liberia and other neighboring countries and again internally displaced as many as a million people. By July 2012, the majority had returned home, but ongoing inter-communal tension and armed conflict continue to force people from their homes.

Age structure: *0-14 years:* 38.53% (male 5,311,971/female 5,276,219)
15-24 years: 20.21% (male 2,774,374/female 2,779,012)
25-54 years: 34.88% (male 4,866,957/female 4,719,286)
55-64 years: 3.53% (male 494,000/female 476,060)
65 years and over: 2.85% (male 349,822/female 433,385) (2020 est.)

Dependency ratios: *total dependency ratio:* 79.8
youth dependency ratio: 74.6
elderly dependency ratio: 5.2
potential support ratio: 19.3 (2020 est.)

Median age: *total:* 20.3 years
male: 20.3 years
female: 20.3 years (2020 est.)
country comparison to the world: 191

Population growth rate: 2.26% (2020 est.)
country comparison to the world: 34

Birth rate: 29.1 births/1,000 population (2020 est.)
country comparison to the world: 35

Death rate: 7.9 deaths/1,000 population (2020 est.)
country comparison to the world: 93

Net migration rate: 1.2 migrant(s)/1,000 population (2020 est.)
country comparison to the world: 57

Population distribution: the population is primarily located in the forested south, with the highest concentration of people residing in and around the cities on the Atlantic coast; most of the northern savanna remains sparsely populated with higher concentrations located along transportation corridors as shown in this population distribution map

Urbanization: *urban population:* 51.7% of total population (2020)
rate of urbanization: 3.38% annual rate of change (2015-20 est.)

total population growth rate v. urban population growth rate, 2000-2030: *Major urban areas—population:* 231,000 YAMOUSSOUKRO (capital) (2018), 5.203 million ABIDJAN (seat of government) (2020)

Sex ratio: *at birth:* 1.03 male(s)/female
0-14 years: 1.01 male(s)/female
15-24 years: 1 male(s)/female
25-54 years: 1.03 male(s)/female
55-64 years: 1.04 male(s)/female
65 years and over: 0.81 male(s)/female
total population: 1.01 male(s)/female (2020 est.)

Mother's mean age at first birth: 19.8 years (2011/12 est.)
note: median age at first birth among women 25-29

Maternal mortality rate: 617 deaths/100,000 live births (2017 est.)
country comparison to the world: 12

Infant mortality rate: *total:* 59.1 deaths/1,000 live births
male: 66.7 deaths/1,000 live births
female: 51.2 deaths/1,000 live births (2020 est.)
country comparison to the world: 14

Life expectancy at birth: *total population:* 61.3 years
male: 59.2 years
female: 63.6 years (2020 est.)
country comparison to the world: 215

Total fertility rate: 3.67 children born/woman (2020 est.)
country comparison to the world: 36

Contraceptive prevalence rate: 23.3% (2018)

Drinking water source:
improved:
urban: 90.4% of population
rural: 67.8% of population
total: 79.2% of population
unimproved:
urban: 9.6% of population
rural: 32.2% of population
total: 20.8% of population (2017 est.)

Current Health Expenditure: 4.5% (2017)

Physicians density: 0.23 physicians/1,000 population (2014)

Sanitation facility access:
improved:
urban: 75.9% of population
rural: 32.7% of population
total: 54.5% of population
unimproved:
urban: 24.1% of population
rural: 67.3% of population
total: 45.5% of population (2017 est.)

HIV/AIDS—adult prevalence rate: 2.7% (2019 est.)
country comparison to the world: 20

HIV/AIDS—people living with HIV/AIDS: 430,000 (2019 est.)
country comparison to the world: 18

HIV/AIDS—deaths: 13,000 (2019 est.)
country comparison to the world: 17

Major infectious diseases: *degree of risk:* very high (2020)
food or waterborne diseases: bacterial diarrhea, hepatitis A, and typhoid fever
vectorborne diseases: malaria, dengue fever, and yellow fever
water contact diseases: schistosomiasis
animal contact diseases: rabies
respiratory diseases: meningococcal meningitis

Obesity—adult prevalence rate: 10.3% (2016)
country comparison to the world: 138

Children under the age of 5 years underweight: 12.8% (2016)
country comparison to the world: 48

Education expenditures: 5.1% of GDP (2017)
country comparison to the world: 58

Literacy: *definition:* age 15 and over can read and write
total population: 47.2%
male: 53.7%
female: 40.5% (2018)

School life expectancy (primary to tertiary education): total: 11 years
male: 10 years
female: 9 years (2017)

Unemployment, youth ages 15-24: total: 5.5%
male: 4.7%
female: 6.5% (2017 est.)
country comparison to the world: 163

GOVERNMENT

Country name: *conventional long form:* Republic of Cote d'Ivoire
conventional short form: Cote d'Ivoire
local long form: Republique de Cote d'Ivoire
local short form: Cote d'Ivoire
former: Ivory Coast
etymology: name reflects the intense ivory trade that took place in the region from the 15th to 17th centuries
note: pronounced coat-div-whar

Government type: presidential republic

Capital: *name:* Yamoussoukro (legislative capital), Abidjan (administrative capital); note—although Yamoussoukro has been the official capital since 1983, Abidjan remains the administrative capital as well as the officially designated economic capital; the US, like other countries, maintains its Embassy in Abidjan
geographic coordinates: 6 49 N, 5 16 W
time difference: UTC 0 (5 hours ahead of Washington, DC, during Standard Time)
etymology: Yamoussoukro is named after Queen YAMOUSSOU, who ruled in the village of N'Gokro in 1929 at the time of French colonization; the village was renamed Yamoussoukro, the suffix "-kro" meaning "town" in the native Baoule language; Abidjan's name supposedly comes from a misunderstanding; tradition states that an old man carrying branches met a European explorer who asked for the name of the nearest village; the man, not understanding and terrified by this unexpected encounter, fled shouting "min-chan m'bidjan," which in the Ebrie language means: "I return from cutting leaves"; the explorer, thinking that his question had been answered, recorded the name of the locale as Abidjan; a different version has the first colonists asking native women the name of the place and getting a similar response

Administrative divisions: 12 districts and 2 autonomous districts*; Abidjan*, Bas-Sassandra, Comoe, Denguele, Goh-Djiboua, Lacs, Lagunes, Montagnes, Sassandra-Marahoue, Savanes, Vallee du Bandama, Woroba, Yamoussoukro*, Zanzan

Independence: 7 August 1960 (from France)

National holiday: Independence Day, 7 August (1960)

Constitution: *history:* previous 1960, 2000; latest draft completed 24 September 2016, approved by the National Assembly 11 October 2016, approved by referendum 30 October 2016, promulgated 8 November 2016
amendments: proposed by the president of the republic or by Parliament; consideration of drafts or proposals requires an absolute majority

vote by the parliamentary membership; passage of amendments affecting presidential elections, presidential term of office and vacancies, and amendment procedures requires approval by absolute majority in a referendum; passage of other proposals by the president requires at least four-fifths majority vote by Parliament; constitutional articles on the sovereignty of the state and its republican and secular form of government cannot be amended

Legal system: civil law system based on the French civil code; judicial review of legislation held in the Constitutional Chamber of the Supreme Court

International law organization participation: accepts compulsory ICJ jurisdiction with reservations; accepts ICCt jurisdiction

Citizenship: *citizenship by birth:* no
citizenship by descent only: at least one parent must be a citizen of Cote d'Ivoire
dual citizenship recognized: no
residency requirement for naturalization: 5 years

Suffrage: 18 years of age; universal

Executive branch: *chief of state:* President Alassane Dramane OUATTARA (since 4 December 2010); Vice President (vacant); note—Vice President Daniel Kablan DUNCAN resigned 8 July 2020; note—the 2016 constitution calls for the establishment of the position of vice-president

head of government: Prime Minister Hamed BAKAYOKO (since 30 July 2020); note—Prime Minister Amadou Gon COULIBALY died on 8 July 2020 after a Council of Ministers meeting
cabinet: Council of Ministers appointed by the president
elections/appointments: president directly elected by absolute majority popular vote in 2 rounds if needed for a single renewable 5-year term; election last held on 31 October 2020 (next to be held in October 2025); vice president elected on same ballot as president; prime minister appointed by the president; note—because President OUATTARA promulgated the new constitution during his second term, he has claimed that the clock is reset on term limits, allowing him to run for up to two additional terms
election results: Alassane OUATTARA reelected president; percent of vote—Alassane OUATTARA (RDR) 94.3%, Kouadio Konan BERTIN (PDCI-RDA) 2.0%, other 3.7%

Legislative branch: *description:* bicameral Parliament consists of:
Senate or Senat (99 seats; 66 members indirectly elected by the National Assembly and members of municipal, autonomous districts, and regional councils, and 33 members appointed by the president; members serve 5-year terms)
National Assembly (255 seats; members directly elected in single-and multi-seat constituencies by simple majority vote to serve 5-year terms)
elections: Senate—first ever held on 25 March 2018 (next to be held in 2023)
National Assembly—last held on 18 December 2016 (next to be held in 2021)

election results: Senate—percent by party NA; seats by party—RHDP 50, independent 16; composition—men 80, women 19, percent of women 19.2%
National Assembly—percent of vote by party—RHDP 50.3%, FPI 5.8%, UDPCI 1%, other 1.4%, independent 38.5%; seats by party—RHDP, 167, UDPCI 6, FPI 3, UPCI 3, independent 76; composition—men 228, women 27, percent of women 10.6%; note—total Parliament percent of women 13%

Judicial branch: *highest courts:* Supreme Court or Cour Supreme (organized into Judicial, Audit, Constitutional, and Administrative Chambers; consists of the court president, 3 vice presidents for the Judicial, Audit, and Administrative chambers, and 9 associate justices or magistrates)
judge selection and term of office: judges nominated by the Superior Council of the Magistrature, a 7-member body consisting of the national president (chairman), 3 "bench" judges, and 3 public prosecutors; judges appointed for life
subordinate courts: Courts of Appeal (organized into civil, criminal, and social chambers); first instance courts; peace courts

Political parties and leaders: Democratic Party of Cote d'Ivoire or PDCI [Henri Konan BEDIE]
Ivorian Popular Front or FPI [former pres. Laurent GBAGBO]
Liberty and Democracy for the Republic or LIDER [Mamadou KOULIBALY]
Movement of the Future Forces or MFA [Innocent Augustin ANAKY KOBENA]
Rally of Houphouetists for Democracy and Peace or RHDP [Alassane OUATTARA] (alliance includes MFA, PDCI, RDR, UDPCI, UPCI)
Rally of the Republicans or RDR [Henriette DIABATE]
Union for Cote d'Ivoire or UPCI [Gnamien KONAN]
Union for Democracy and Peace in Cote d'Ivoire or UDPCI [Albert Toikeusse MABRI]

International organization participation: ACP, AfDB, AU, ECOWAS, EITI (compliant country), Entente, FAO, FZ, G-24, G-77, IAEA, IBRD, ICAO, ICC, ICCt, ICRM, IDA, IDB, IFAD, IFC, IFRCS, ILO, IMF, IMO, Interpol, IOC, IOM, IPU, ISO, ITSO, ITU, ITUC (NGOs), MIGA, MINUSMA, MONUSCO, NAM, OIC, OIF, OPCW, UN, UNCTAD, UNESCO, UNHCR, UNIDO, Union Latina, UN Security Council (temporary), UNWTO, UPU, WADB (regional), WAEMU, WCO, WFTU (NGOs), WHO, WIPO, WMO, WTO

Diplomatic representation in the US: *chief of mission:* Ambassador Mamadou HAIDARA (since 28 March 2018)
chancery: 2424 Massachusetts Avenue NW, Washington, DC 20008
telephone: [1] (202) 797-0300
FAX: [1] (202) 462-9444

Diplomatic representation from the US: *chief of mission:* Ambassador Richard K. BELL (since 3 September 2019)
telephone: [225] 22 49 40 00

embassy: Cocody Riviéra Golf, 01 BP 1712 Abidjan 01, Abidjan
mailing address: B. P. 1712, Abidjan 01
FAX: [225] 22 49 43 23

Flag description: three equal vertical bands of orange (hoist side), white, and green; orange symbolizes the land (savannah) of the north and fertility, white stands for peace and unity, green represents the forests of the south and the hope for a bright future
note: similar to the flag of Ireland, which is longer and has the colors reversed—green (hoist side), white, and orange; also similar to the flag of Italy, which is green (hoist side), white, and red; design was based on the flag of France

National symbol(s): *elephant; national colors:* orange, white, green

National anthem: *name:* "L'Abidjanaise" (Song of Abidjan)
lyrics/music: Mathieu EKRA, Joachim BONY, and Pierre Marie COTY/Pierre Marie COTY and Pierre Michel PANGO
note: adopted 1960; although the nation's capital city moved from Abidjan to Yamoussoukro in 1983, the anthem still owes its name to the former capital
0:00 / 1:09

ECONOMY

Economy—overview: For the last 5 years Cote d'Ivoire's growth rate has been among the highest in the world. Cote d'Ivoire is heavily dependent on agriculture and related activities, which engage roughly two-thirds of the population. Cote d'Ivoire is the world's largest producer and exporter of cocoa beans and a significant producer and exporter of coffee and palm oil. Consequently, the economy is highly sensitive to fluctuations in international prices for these products and to climatic conditions. Cocoa, oil, and coffee are the country's top export revenue earners, but the country has targeted agricultural processing of cocoa, cashews, mangoes, and other commodities as a high priority. Mining gold and exporting electricity are growing industries outside agriculture.

Following the end of more than a decade of civil conflict in 2011, Cote d'Ivoire has experienced a boom in foreign investment and economic growth. In June 2012, the IMF and the World Bank announced $4.4 billion in debt relief for Cote d'Ivoire under the Highly Indebted Poor Countries Initiative.

GDP (purchasing power parity): $97.16 billion (2017 est.)
$90.12 billion (2016 est.)
$83.19 billion (2015 est.)
note: data are in 2017 dollars
country comparison to the world: 86

GDP (official exchange rate): $40.47 billion (2017 est.)

GDP—real growth rate: 7.8% (2017 est.)
8.3% (2016 est.)
8.8% (2015 est.)
country comparison to the world: 10

GDP—per capita (PPP): $3,900 (2017 est.)
$3,700 (2016 est.)
$3,500 (2015 est.)
note: data are in 2017 dollars
country comparison to the world: 179

Gross national saving: 15.9% of GDP (2017 est.)
19.2% of GDP (2016 est.)
19.5% of GDP (2015 est.)
country comparison to the world: 131

GDP—composition, by end use: *household consumption:* 61.7% (2017 est.)
government consumption: 14.9% (2017 est.)
investment in fixed capital: 22.4% (2017 est.)
investment in inventories: 0.3% (2017 est.)
exports of goods and services: 30.8% (2017 est.)
imports of goods and services:—30.1% (2017 est.)

GDP—composition, by sector of origin: *agriculture:* 20.1% (2017 est.)
industry: 26.6% (2017 est.)
services: 53.3% (2017 est.)

Agriculture—products: coffee, cocoa beans, bananas, palm kernels, corn, rice, cassava (manioc, tapioca), sweet potatoes, sugar, cotton, rubber; timber

Industries: foodstuffs, beverages; wood products, oil refining, gold mining, truck and bus assembly, textiles, fertilizer, building materials, electricity

Industrial production growth rate: 4.2% (2017 est.)
country comparison to the world: 71

Labor force: 8.747 million (2017 est.)
country comparison to the world: 55

Labor force—by occupation: *agriculture:* 68% (2007 est.)

Unemployment rate: 9.4% (2013 est.)
country comparison to the world: 141

Population below poverty line: 46.3% (2015 est.)

Household income or consumption by percentage share: *lowest 10%:* 2.2%
highest 10%: 31.8% (2008)

Budget: *revenues:* 7.749 billion (2017 est.)
expenditures: 9.464 billion (2017 est.)

Taxes and other revenues:
19.1% (of GDP) (2017 est.)
country comparison to the world: 156

Budget surplus (+) or deficit (-): -4.2% (of GDP) (2017 est.)
country comparison to the world: 159

Public debt:
47% of GDP (2017 est.)
47% of GDP (2016 est.)
country comparison to the world: 112

Fiscal year: calendar year

Inflation rate (consumer prices): 0.8% (2017 est.)
0.7% (2016 est.)
country comparison to the world: 40

Current account balance:
-$1.86 billion (2017 est.)
-$414 million (2016 est.)
country comparison to the world: 164

Exports: $11.74 billion (2017 est.)
$11.77 billion (2016 est.)

country comparison to the world: 82

Exports—partners: Netherlands 11.8%, US 7.9%, France 6.4%, Belgium 6.4%, Germany 5.8%, Burkina Faso 4.5%, India 4.4%, Mali 4.2% (2017)

Exports—commodities: cocoa, coffee, timber, petroleum, cotton, bananas, pineapples, palm oil, fish

Imports: $9.447 billion (2017 est.)
$7.81 billion (2016 est.)
country comparison to the world: 104

Imports—commodities: fuel, capital equipment, foodstuffs

Imports—partners: Nigeria 15%, France 13.4%, China 11.3%, US 4.3% (2017)

Reserves of foreign exchange and gold: $6.257 billion (31 December 2017 est.)
$4.935 billion (31 December 2016 est.)
country comparison to the world: 91

Debt—external: $13.07 billion (31 December 2017 est.)
$11.02 billion (31 December 2016 est.)
country comparison to the world: 105

Exchange rates: Communaute Financiere Africaine francs (XOF) per US dollar -
594.3 (2017 est.)
593.01 (2016 est.)
593.01 (2015 est.)
591.45 (2014 est.)
494.42 (2013 est.)

ENERGY

Electricity access: *population without electricity:* 6 million (2019)
electrification—total population: 76% (2019)
electrification—urban areas: 99% (2019)
electrification—rural areas: 51% (2019)

Electricity—production: 9.73 billion kWh (2016 est.)
country comparison to the world: 104

Electricity—consumption: 6.245 billion kWh (2016 est.)
country comparison to the world: 114

Electricity—exports: 872 million kWh (2015 est.)
country comparison to the world: 61

Electricity—imports: 19 million kWh (2016 est.)
country comparison to the world: 114

Electricity—installed generating capacity: 1.914 million kW (2016 est.)
country comparison to the world: 114

Electricity—from fossil fuels: 60% of total installed capacity (2016 est.)
country comparison to the world: 130

Electricity—from nuclear fuels: 0% of total installed capacity (2017 est.)
country comparison to the world: 73

Electricity—from hydroelectric plants: 40% of total installed capacity (2017 est.)
country comparison to the world: 51

Electricity—from other renewable sources: 0% of total installed capacity (2017 est.)
country comparison to the world: 184

Crude oil—production: 52,000 bbl/day (2018 est.)
country comparison to the world: 52

Crude oil—exports: 26,700 bbl/day (2015 est.)
country comparison to the world: 47

Crude oil—imports: 62,350 bbl/day (2015 est.)
country comparison to the world: 52

Crude oil—proved reserves: 100 million bbl (1 January 2018 est.)
country comparison to the world: 69

Refined petroleum products—production: 69,360 bbl/day (2017 est.)
country comparison to the world: 72

Refined petroleum products—consumption: 51,000 bbl/day (2016 est.)
country comparison to the world: 104

Refined petroleum products—exports: 31,450 bbl/day (2015 est.)
country comparison to the world: 61

Refined petroleum products—imports: 7,405 bbl/day (2015 est.)
country comparison to the world: 154

Natural gas—production: 2.322 billion cu m (2017 est.)
country comparison to the world: 59

Natural gas—consumption: 2.322 billion cu m (2017 est.)
country comparison to the world: 82

Natural gas—exports: 0 cu m (2017 est.)
country comparison to the world: 89

Natural gas—imports: 0 cu m (2017 est.)
country comparison to the world: 112

Natural gas—proved reserves: 28.32 billion cu m (1 January 2018 est.)
country comparison to the world: 68

Carbon dioxide emissions from consumption of energy: 11.54 million Mt (2017 est.)
country comparison to the world: 101

COMMUNICATIONS

Telephones—fixed lines: *total subscriptions:* 284,799
subscriptions per 100 inhabitants: 1.06 (2019 est.)
country comparison to the world: 110

Telephones—mobile cellular: *total subscriptions:* 39,049,743
subscriptions per 100 inhabitants: 145.34 (2019 est.)
country comparison to the world: 39

Telecommunication systems: *general assessment:* strongest sector in the overall market is the mobile sector; fixed internet and broadband sectors have remained underdeveloped; country 90% digitalized; Côte d'Ivoire continues to benefit from strong economic growth; progress has been made in building out the national backbone network and connecting in 2019 to the MainOne submarine cable; this development puts the country in a better position to develop its broadband market and work on its digital economy; government further tightens SIM card registration rules (2020)

domestic: less than 1 per 100 fixed-line, with multiple mobile- cellular service providers competing in the market, usage has increased to about 145 per 100 persons (2019)

international: country code—225; landing point for the SAT- 3/WASC, ACE, MainOne, and WACS fiber-optic submarine cable that provides connectivity to Europe and South and West Africa; satellite earth stations—2 Intelsat (1 Atlantic Ocean and 1 Indian Ocean) (2019)

note: the COVID-19 outbreak is negatively impacting telecommunications production and supply chains globally; consumer spending on telecom devices and services has also slowed due to the pandemic's effect on economies worldwide; overall progress towards improvements in all facets of the telecom industry—mobile, fixed-line, broadband, submarine cable and satellite—has moderated

Broadcast media: state-controlled Radiodiffusion Television Ivoirienne (RTI) is made up of 2 radios stations (Radio Cote d'Ivoire and Frequence2) and 2 television stations (RTI1 and RTI2), with nationwide coverage, broadcasts mainly in French; after 2011 post- electoral crisis, President OUATTARA's administration reopened RTI Bouake', the broadcaster's office in Cote d'Ivoire's 2nd largest city, where facilities were destroyed during the 2002 rebellion; Cote d'Ivoire is also home to 178 proximity radios stations, 16 religious radios stations, 5 commercial radios stations, and 5 international radios stations, according to the Haute Autorite' de la Communication Audiovisuelle (HACA); govt now runs radio UNOCIFM, a radio station previously owned by the UN Operation in Cote d'Ivoire; in Dec 2016, the govt announced 4 companies had been granted licenses to operate—Live TV, Optimum Media Cote d'Ivoire, the Audiovisual Company of Cote d'Ivoire (Sedaci), and Sorano-CI, out of the 4 companies only one has started operating (2019)

Internet country code: .ci

Internet users: *total:* 12,295,204
percent of population: 46.82% (July 2018 est.)
country comparison to the world: 48

Broadband—fixed subscriptions: total: 175,918
subscriptions per 100 inhabitants: 1 (2018 est.)
country comparison to the world: 112

TRANSPORTATION

National air transport system: *number of registered air carriers:* 1 (2020)
inventory of registered aircraft operated by air carriers: 10
annual passenger traffic on registered air carriers: 779,482 (2018)
annual freight traffic on registered air carriers: 5.8 million mt-km (2018)

Civil aircraft registration country code prefix: TU (2016)

Airports: 27 (2013)
country comparison to the world: 124

Airports—with paved runways: *total:* 7 (2017)

over 3,047 m: 1 (2017)
2,438 to 3,047 m: 2 (2017)
1,524 to 2,437 m: 4 (2017)

Airports—with unpaved runways: *total:* 20 (2013)
1,524 to 2,437 m: 6 (2013)
914 to 1,523 m: 11 (2013)
under 914 m: 3 (2013)

Heliports: 1 (2013)

Pipelines: 101 km condensate, 256 km gas, 118 km oil, 5 km oil/gas/water, 7 km water (2013)

Railways: *total:* 660 km (2008)
narrow gauge: 660 km 1.000-m gauge (2008)
note: an additional 622 km of this railroad extends into Burkina Faso
country comparison to the world: 104

Roadways: *total:* 81,996 km (2007)
paved: 6,502 km (2007)
unpaved: 75,494 km (2007)
note: includes intercity and urban roads; another 20,000 km of dirt roads are in poor condition and 150,000 km of dirt roads are impassable
country comparison to the world: 62

Waterways: 980 km (navigable rivers, canals, and numerous coastal lagoons) (2011)
country comparison to the world: 66

Merchant marine: *total:* 15
by type: oil tanker 2, other 13 (2019)
country comparison to the world: 147

Ports and terminals: *major seaport(s):* Abidjan, San-Pedro
oil terminal(s): Espoir Offshore Terminal

MILITARY AND SECURITY

Military and security forces: Armed Forces of Cote d'Ivoire (Forces Armees de Cote d'Ivoire, FACI; aka Republican Forces of Ivory Coast, FRCI): Army (Armee de Terre), Navy (Marine Nationale), Cote Air Force (Force Aerienne Cote), Special Forces (Forces Speciale)

other security services include the National Gendarmerie (under the Ministry of Defense), the National Police (under the Ministry of Security and Civil Protection), and the Coordination Center for Operational Decisions (a mix of police, gendarmerie, and FACI personnel for assisting police in providing security in some large cities) (2019)

Military expenditures: 1.1% of GDP (2019)
1.4% of GDP (2018)
1.3% of GDP (2017)
1.7% of GDP (2016)
1.7% of GDP (2015)
country comparison to the world: 111

Military and security service personnel strengths: the Armed Forces of Cote d'Ivoire have approximately 25,000 active troops (23,000 Army; 1,000 Navy; 1,000 Air Force) (2019 est.)

Military equipment inventories and acquisitions: the FACI is mostly equipped with second-hand weapons and equipment of Russian origin; the leading suppliers since 2000 are Belarus, Bulgaria, and Romania (2019 est.)

Military deployments: 800 Mali (MINUSMA) (2020)

Military service age and obligation: 18-25 years of age for compulsory and voluntary male and female military service; conscription is not enforced; voluntary recruitment of former rebels into the new national army is restricted to ages 22-29 (2012)

Military—note: the military has mutinied several times since the late 1990s, most recently in 2017, and has had a large role in the country's political turmoil; currently, the FACI is focused on internal security and the growing threat posed by Islamic militants associated with the al-Qa'ida in the Islamic Maghreb (AQIM) terrorist group operating across the border in southern Burkina Faso; AQIM militants conducted significant attacks in the country in 2016 and 2020; Côte d'Ivoire since 2016 has stepped up border security and built a joint terrorism training center with France near Abidjan in 2018

the UN maintained a 9,000-strong peacekeeping force in Cote d'Ivoire (UNOCI) from 2004 until 2017 (2020)

TERRORISM

Terrorist group(s): al-Qa'ida in the Islamic Maghreb (2020)
note: details about the history, aims, leadership, organization, areas of operation, tactics, targets, weapons, size, and sources of support of the group(s) appear(s) in Appendix- T

TRANSNATIONAL ISSUES

Disputes—international: disputed maritime border between Cote d'Ivoire and Ghana

Refugees and internally displaced persons: *IDPs:* 303,000 (post- election conflict in 2010-11, as well as civil war from 2002-04; land disputes; most pronounced in western and southwestern regions) (2019)
stateless persons: 955,399 (2019); note—many Ivoirians lack documentation proving their nationality, which prevent them from accessing education and healthcare; birth on Ivorian soil does not automatically result in citizenship;

disputes over citizenship and the associated rights of the large population descended from migrants from neighboring countries is an ongoing source of tension and contributed to the country's 2002 civil war; some observers believe the government's mass naturalizations of thousands of people over the last couple of years is intended to boost its electoral support base; the government in October 2013 acceded to international conventions on statelessness and in August 2013 reformed its nationality law, key steps to clarify the nationality of thousands of residents; since the adoption of the Abidjan Declaration to eradicate statelessness in West Africa in February 2015, 6,400 people have received nationality papers

Illicit drugs: illicit producer of cannabis, mostly for local consumption; utility as a narcotic transshipment point to Europe reduced by ongoing political instability; while rampant corruption and inadequate supervision leave the banking system vulnerable to money laundering, the lack of a developed financial system limits the country's utility as a major money-laundering center

CROATIA

INTRODUCTION

Background: The lands that today comprise Croatia were part of the Austro-Hungarian Empire until the close of World War I. In 1918, the Croats, Serbs, and Slovenes formed a kingdom known after 1929 as Yugoslavia. Following World War II, Yugoslavia became a federal independent communist state consisting of six socialist republics under the strong hand of Marshal Josip Broz, aka TITO. Although Croatia declared its independence from Yugoslavia in 1991, it took four years of sporadic, but often bitter, fighting before occupying Yugoslav forces, dominated by Serb officers, were mostly cleared from Croatian lands, along with a majority of Croatia's ethnic Serb population. Under UN supervision, the last Serb-held enclave

in eastern Slavonia was returned to Croatia in 1998. The country joined NATO in April 2009 and the EU in July 2013.

GEOGRAPHY

Location: Southeastern Europe, bordering the Adriatic Sea, between Bosnia and Herzegovina and Slovenia

Geographic coordinates: 45 10 N, 15 30 E

Map references: Europe

Area: *total:* 56,594 sq km
land: 55,974 sq km
water: 620 sq km
country comparison to the world: 128

Area—comparative: slightly smaller than West Virginia

Land boundaries: *total:* 2,237 km
border countries (5): Bosnia and Herzegovina 956 km, Hungary 348 km, Montenegro 19 km, Serbia 314 km, Slovenia 600 km

Coastline: 5,835 km (mainland 1,777 km, islands 4,058 km)

Maritime claims: *territorial sea:* 12 nm
continental shelf: 200-m depth or to the depth of exploitation

Climate: Mediterranean and continental; continental climate predominant with hot summers and cold winters; mild winters, dry summers along coast

Terrain: geographically diverse; flat plains along Hungarian border, low mountains and highlands near Adriatic coastline and islands

Elevation: *mean elevation:* 331 m

lowest point: Adriatic Sea 0 m
highest point: Dinara 1,831 m

Natural resources: oil, some coal, bauxite, low-grade iron ore, calcium, gypsum, natural asphalt, silica, mica, clays, salt, hydropower

Land use: *agricultural land:* 23.7% (2011 est.)
arable land: 16% (2011 est.) / permanent crops: 1.5% (2011 est.) / permanent pasture: 6.2% (2011 est.)
forest: 34.4% (2011 est.)
other: 41.9% (2011 est.)
Irrigated land: 240 sq km (2012)

Population distribution: more of the population lives in the northern half of the country, with approximately a quarter of the populace residing in and around the capital of Zagreb; many of the islands are sparsely populated

Natural hazards: destructive earthquakes

Environment—current issues: air pollution improving but still a concern in urban settings and in emissions arriving from neighboring countries; surface water pollution in the Danube River Basin

Environment—international agreements: *party to:* Air Pollution, Air Pollution-Nitrogen Oxides, Air Pollution-Persistent Organic Pollutants, Air Pollution-Sulfur 94, Air Pollution-Volatile Organic Compounds, Biodiversity, Climate Change, Climate Change-Kyoto Protocol, Desertification, Endangered Species, Hazardous Wastes, Law of the Sea, Marine Dumping, Ozone Layer Protection, Ship Pollution, Wetlands, Whaling
signed, but not ratified: none of the selected agreements

Geography—note: controls most land routes from Western Europe to Aegean Sea and Turkish Straits; most Adriatic Sea islands lie off the coast of Croatia—some 1,200 islands, islets, ridges, and rocks

PEOPLE AND SOCIETY

Population: 4,227,746 (July 2020 est.)
country comparison to the world: 127

Nationality: *noun:* Croat(s), Croatian(s)
adjective: Croatian
note: the French designation of "Croate" to Croatian mercenaries in the 17th century eventually became "Cravate" and later came to be applied to the soldiers' scarves—the cravat; Croatia celebrates Cravat Day every 18 October

Ethnic groups: Croat 90.4%, Serb 4.4%, other 4.4% (including Bosniak, Hungarian, Slovene, Czech, and Romani), unspecified 0.8% (2011 est.)

Languages: Croatian (official) 95.6%, Serbian 1.2%, other 3% (including Hungarian, Czech, Slovak, and Albanian), unspecified 0.2% (2011 est.)

Religions: Roman Catholic 86.3%, Orthodox 4.4%, Muslim 1.5%, other 1.5%, unspecified 2.5%, not religious or atheist 3.8% (2011 est.)

Age structure: *0-14 years:* 14.16% (male 308,668/female 289,996)
15-24 years: 10.76% (male 233,602/female 221,495)
25-54 years: 39.77% (male 841,930/female 839,601)
55-64 years: 14.24% (male 290,982/female 310,969)
65 years and over: 21.06% (male 364,076/female 526,427) (2020 est.)

Dependency ratios: *total dependency ratio:* 55.7
youth dependency ratio: 22.6
elderly dependency ratio: 33.1
potential support ratio: 3 (2020 est.)

Median age: *total:* 43.9 years
male: 42 years
female: 45.9 years (2020 est.)
country comparison to the world: 18

Population growth rate: -0.5% (2020 est.)
country comparison to the world: 225

Birth rate: 8.7 births/1,000 population (2020 est.)
country comparison to the world: 211

Death rate: 12.8 deaths/1,000 population (2020 est.)
country comparison to the world: 11

Net migration rate: -1 migrant(s)/1,000 population (2020 est.)
country comparison to the world: 143

Population distribution: more of the population lives in the northern half of the country, with approximately a quarter of the populace residing in and around the capital of Zagreb; many of the islands are sparsely populated

Urbanization: *urban population:* 57.6% of total population (2020)

rate of urbanization: -0.08% annual rate of change (2015-20 est.)

total population growth rate v. urban population growth rate, 2000-2030: Major urban areas—population: 685,000 ZAGREB (capital) (2020)

Sex ratio: *at birth:* 1.06 male(s)/female
0-14 years: 1.06 male(s)/female
15-24 years: 1.05 male(s)/female
25-54 years: 1 male(s)/female
55-64 years: 0.94 male(s)/female
65 years and over: 0.69 male(s)/female
total population: 0.93 male(s)/female (2020 est.)

Mother's mean age at first birth: 28.9 years (2017 est.)

Maternal mortality rate: 8 deaths/100,000 live births (2017 est.)
country comparison to the world: 150

Infant mortality rate: *total:* 8.6 deaths/1,000 live births
male: 8.4 deaths/1,000 live births
female: 8.9 deaths/1,000 live births (2020 est.)
country comparison to the world: 145

Life expectancy at birth: *total population:* 76.7 years
male: 73.6 years
female: 80.1 years (2020 est.)
country comparison to the world: 88

Total fertility rate: 1.42 children born/woman (2020 est.)
country comparison to the world: 215

Drinking water source:
improved:
urban: 100% of population
rural: 100% of population
total: 100% of population
unimproved:
urban: 0% of population
rural: 0% of population
total: 0% of population (2017 est.)

Current Health Expenditure: 6.8% (2017)

Physicians density: 3 physicians/1,000 population (2016)

Hospital bed density: 5.5 beds/1,000 population (2017)

Sanitation facility access:
improved:
urban: 99.5% of population
rural: 98.4% of population
total: 99% of population
unimproved:
urban: 0.5% of population
rural: 1.6% of population
total: 1% of population (2017 est.)

HIV/AIDS—adult prevalence rate: <.1% (2019 est.)

HIV/AIDS—people living with HIV/AIDS: 1,600 (2019 est.)
country comparison to the world: 137

HIV/AIDS—deaths: <100 (2019 est.)

Major infectious diseases:
degree of risk: intermediate (2020)
vectorborne diseases: tickborne encephalitis

Obesity—adult prevalence rate: 24.4% (2016)
country comparison to the world: 59

Education expenditures: 4.6% of GDP (2013)
country comparison to the world: 83

Literacy: *definition:* age 15 and over can read and write
total population: 99.3%
male: 99.7%
female: 98.9% (2015)

School life expectancy (primary to tertiary education): *total:* 15 years
male: 15 years
female: 16 years (2018)

Unemployment, youth ages 15-24: *total:* 23.7%
male: 19.6%
female: 29.4% (2018 est.)
country comparison to the world: 55

GOVERNMENT

Country name: *conventional long form:* Republic of Croatia
conventional short form: Croatia
local long form: Republika Hrvatska
local short form: Hrvatska
former: People's Republic of Croatia, Socialist Republic of Croatia
etymology: name derives from the Croats, a Slavic tribe who migrated to the Balkans in the 7th century A.D.

Government type: parliamentary republic

Capital: *name:* Zagreb
geographic coordinates: 45 48 N, 16 00 E
time difference: UTC+1 (6 hours ahead of Washington, DC, during Standard Time)
daylight saving time: +1hr, begins last Sunday in March; ends last Sunday in October
etymology: the name seems to be related to "digging"; archeologists suggest that the original settlement was established beyond a water-filled hole or "graba" and that the name derives from this; "za" in Slavic means "beyond"; the overall meaning may be "beyond the trench (fault, channel, ditch)"

Administrative divisions: 20 counties (zupanije, zupanija—singular) and 1 city* (grad—singular) with special county status; Bjelovarsko-Bilogorska (Bjelovar-Bilogora), Brodsko-Posavska (Brod-Posavina), Dubrovacko-Neretvanska (Dubrovnik-Neretva), Istarska (Istria), Karlovacka (Karlovac), Koprivnicko-Krizevacka (Koprivnica-Krizevci), Krapinsko-Zagorska (Krapina-Zagorje), Licko-Senjska (Lika-Senj), Medimurska (Medimurje), Osjecko-Baranjska (Osijek-Baranja), Pozesko-Slavonska (Pozega- Slavonia), Primorsko-Goranska (Primorje-Gorski Kotar), Sibensko-Kninska (Sibenik-Knin), Sisacko-Moslavacka (Sisak- Moslavina), Splitsko-Dalmatinska (Split-Dalmatia), Varazdinska (Varazdin), Viroviticko-Podravska (Virovitica-Podravina), Vukovarsko-Srijemska (Vukovar-Syrmia), Zadarska (Zadar), Zagreb*, Zagrebacka (Zagreb county)

Independence: 25 June 1991 (from Yugoslavia); notable earlier dates: ca. 925 (Kingdom of Croatia established); 1 December 1918 (Kingdom of Serbs, Croats, and Slovenes (Yugoslavia) established)

National holiday: Independence Day, 8 October (1991) and Statehood Day, 25 June (1991); note—25 June 1991 was the day the Croatian parliament voted for independence; following a three-month moratorium to allow the European Community to solve the Yugoslav crisis peacefully, parliament adopted a decision on 8 October 1991 to sever constitutional relations with Yugoslavia

Constitution: *history:* several previous; latest adopted 22 December 1990

amendments: proposed by at least one fifth of the Assembly membership, by the president of the republic, by the Government of Croatia, or through petition by at least 10% of the total electorate; proceedings to amend require majority vote by the Assembly; passage requires two-thirds majority vote by the Assembly; passage by petition requires a majority vote in a referendum and promulgation by the Assembly; amended several times, last in 2014

Legal system: civil law system influenced by legal heritage of Austria-Hungary; note—Croatian law was fully harmonized with the European Community acquis as of the June 2010 completion of EU accession negotiations

International law organization participation: has not submitted an ICJ jurisdiction declaration; accepts ICCt jurisdiction

Citizenship: *citizenship by birth:* no

citizenship by descent only: at least one parent must be a citizen of Croatia

dual citizenship recognized: yes

residency requirement for naturalization: 5 years

Suffrage: 18 years of age; universal

Executive branch: *chief of state:* President Zoran MILANOVIC (since 18 February 2020)

head of government: Prime Minister Andrej PLENKOVIC (since 19 October 2016); Deputy Prime Ministers Damir KRSTICEVIC (since 19 October 2016), Predrag STROMAR (since 9 June 2017), Marija Pejcinovic BURIC (since 19 June 2017), and Tomislav TOLUSIC (since 25 May 2018)

cabinet: Council of Ministers named by the prime minister and approved by the Assembly

elections/appointments: president directly elected by absolute majority popular vote in 2 rounds if needed for a 5-year term (eligible for a second term); election last held on 22 December 2019 with a runoff on 5 January 2020 (next to be held in 2024); the leader of the majority party or majority coalition usually appointed prime minister by the president and approved by the Assembly

election results: Zoran MILANOVIC elected president in second round; percent of vote—Zoran MILANOVIC (SDP) 52.7%, Kolinda GRABAR-KITAROVIC (HDZ) 47.3%

Legislative branch: *description:* unicameral Assembly or Hrvatski Sabor (151 seats; 140 members in 10 multi-seat constituencies and 3

members in a single constituency for Croatian diaspora directly elected by proportional representation vote using the D'Hondt method with a 5% threshold; an additional 8 members elected from a nationwide constituency by simple majority by voters belonging to minorities recognized by Croatia; the Serb minority elects 3 Assembly members, the Hungarian and Italian minorities elect 1 each, the Czech and Slovak minorities elect 1 jointly, and all other minorities elect 2; all members serve 4-year terms

elections: early election held on 5 July 2020 (next to be held by 2024)

election results: percent of vote by coalition/party—HDZ-led coalition 37.3%, Restart coalition 24.9%, DPMS-led coalition 10.9%, MOST 7.4%, Green-Left coalition 7%, P-F-SSIP 4%, HNS-LD 1.3%, People's Party—Reformists 1%, other 6.2%; number of seats by coalition/party—HDZ-led coalition 66, Restart coalition 41, DPMS-led coalition 16, MOST 8, Green-Left coalition 7, P-F-SSIP 3, HNS-LD 1, People's Party—Reformists—1, national minorities 8; composition—men 116, women 35, percent of women 23.2%

note: seats by party as of June 2019—HDZ 55, SDP 29, MOST-NL 10, HNS 4, HSS 4, GLAS 4, IDS 3, SDSS 3, BM365- SRS 3, Human Shield 2, HDS 2, NHR 2, other 8, independent 21

Judicial branch: *highest courts:* Supreme Court (consists of the court president and vice president, 25 civil department justices, and 16 criminal department justices)

judge selection and term of office: president of Supreme Court nominated by the president of Croatia and elected by the Sabor for a 4-year term; other Supreme Court justices appointed by the National Judicial Council; all judges serve until age 70

subordinate courts: Administrative Court; county, municipal, and specialized courts; note—there is an 11-member Constitutional Court with jurisdiction limited to constitutional issues but is outside of the judicial system

Political parties and leaders: Bloc for Croatia or BZH [Zlatko HASANBEGOVIC]

Bridge of Independent Lists or Most [Bozo PETROV]

Civic Liberal Alliance or GLAS [Ankar Mrak TARITAS]

Croatian Christian Democratic Party or HDS [Goran DODIG]

Croatian Conservative Party or HKS [Marijan PAVLICEK]

Croatian Democratic Congress of Slavonia and Baranja or HDSSB [Branimir GLAVAS]

Croatian Democratic Union or HDZ [Andrej PLENKOVIC]

Croatian Democratic Union-led coalition (includes HSLS, HDS, HDSSB)

Croatian Peasant Party or HSS [Kreso BELJAK]

Croatian Pensioner Party or HSU [Silvano HRELJA]

Croatian People's Party—Liberal Democrats or HNS-LD [Ivan VRDOLJAK]

Croatian Social Liberal Party or HSLS [Dario HREBAK]

Croatian Sovereignists coalition (includes HK, HRAST)

FOKUS [Davor NADI]

Green-Left coalition (includes MOZEMO!, RF, NL)

Homeland Movement or DPMS [Miloslav SKORO]

Homeland Movement-led coalition (includes DPMS, Croatian Sovereignists coalition, BZH)

Istrian Democratic Assembly or IDS [Boris MILETIC]

Movement for Successful Croatia or HRAST [Ladislav ILCIC]

New Left or NL [Dragan MARKOVINA]

Pametno [Marijana PULJAK]

Pametno, FOKUS, SSIP coalition

Party with a First and Last Name or SSIP [Ivan KOVACIC]

People's Party—Reformists [Radimir CACIC]

Restart Coalition (includes HSLS, HDS, HDSSB)

Social Democratic Party of Croatia or SDP [Zlatko KOMADINA, acting leader]

We Can! or MOZEMO! [collective leadership]

Workers' Front or RF [collective leadership]

International organization participation: Australia Group, BIS, BSEC (observer), CD, CE, CEI, EAPC, EBRD, ECB, EMU, EU, FAO, G-11, IADB, IAEA, IBRD, ICAO, ICC (national committees), ICCt, ICRM, IDA, IFAD, IFC, IFRCS, IHO, ILO, IMF, IMO, IMSO, Interpol, IOC, IOM, IPU, ISO, ITSO, ITU, ITUC (NGOs), MIGA, MINURSO, NAM (observer), NATO, NSG, OAS (observer), OIF (observer), OPCW, OSCE, PCA, SELEC, UN, UNCTAD, UNESCO, UNFICYP, UNHCR, UNIDO, UNIFIL, UNMIL, UNMOGIP, UNWTO, UPU, WCO, WHO, WIPO, WMO, WTO, ZC

Diplomatic representation in the US: *chief of mission:* Ambassador Pjer SIMUNOVIC (since 8 September 2017)

chancery: 2343 Massachusetts Avenue NW, Washington, DC 20008

telephone: [1] (202) 588-5899

FAX: [1] (202) 588-8936

consulate(s) general: Chicago, Los Angeles, New York

Diplomatic representation from the US: *chief of mission:* Ambassador W. Robert KOHORST (since 12 January 2018)

telephone: [385] (1) 661-2200

embassy: 2 Thomas Jefferson Street, 10010 Zagreb

mailing address: use embassy street address

FAX: [385] (1) 661-2373

Flag description: three equal horizontal bands of red (top), white, and blue—the Pan-Slav colors—superimposed by the Croatian coat of arms; the coat of arms consists of one main shield (a checkerboard of 13 red and 12 silver (white) fields) surmounted by five smaller shields that form a crown over the main shield; the five small shields represent five historic regions (from left to right): Croatia, Dubrovnik, Dalmatia, Istria, and Slavonia *note:* the Pan-Slav colors were inspired by the 19th-century flag of Russia

National symbol(s): red-white checkerboard; national colors: red, white, blue

National anthem: *name:* "Lijepa nasa domovino" (Our Beautiful Homeland)

lyrics/music: Antun MIHANOVIC/Josip RUNJANIN

note: adopted in 1972 while still part of Yugoslavia; "Lijepa nasa domovino," whose lyrics were written in 1835, served as an unofficial anthem beginning in 1891

0:00 / 0:59

ECONOMY

Economy—overview: Though still one of the wealthiest of the former Yugoslav republics, Croatia's economy suffered badly during the 1991-95 war. The country's output during that time collapsed, and Croatia missed the early waves of investment in Central and Eastern Europe that followed the fall of the Berlin Wall. Between 2000 and 2007, however, Croatia's economic fortunes began to improve with moderate but steady GDP growth between 4% and 6%, led by a rebound in tourism and credit-driven consumer spending. Inflation over the same period remained tame and the currency, the kuna, stable.

Croatia experienced an abrupt slowdown in the economy in 2008; economic growth was stagnant or negative in each year between 2009 and 2014, but has picked up since the third quarter of 2014, ending 2017 with an average of 2.8% growth. Challenges remain including uneven regional development, a difficult investment climate, an inefficient judiciary, and loss of educated young professionals seeking higher salaries elsewhere in the EU. In 2016, Croatia revised its tax code to stimulate growth from domestic consumption and foreign investment. Income tax reduction began in 2017, and in 2018 various business costs were removed from income tax calculations. At the start of 2018, the government announced its economic reform plan, slated for implementation in 2019.

Tourism is one of the main pillars of the Croatian economy, comprising 19.6% of Croatia's GDP. Croatia is working to become a regional energy hub, and is undertaking plans to open a floating liquefied natural gas (LNG) regasification terminal by the end of 2019 or early in 2020 to import LNG for re-distribution in southeast Europe.

Croatia joined the EU on July 1, 2013, following a decade-long accession process. Croatia has developed a plan for Eurozone accession, and the government projects Croatia will adopt the Euro by 2024. In 2017, the Croatian government decreased public debt to 78% of GDP, from an all-time high of 84% in 2014, and realized a 0.8% budget surplus—the first surplus since independence in 1991. The government has also sought to accelerate privatization of non-strategic assets with mixed success. Croatia's economic recovery is still somewhat fragile; Croatia's largest private company narrowly avoided collapse in 2017, thanks to a capital infusion from an American investor. Restructuring is ongoing, and projected to finish by mid-July 2018.

GDP (purchasing power parity): $102.1 billion (2017 est.)
$99.37 billion (2016 est.)
$95.97 billion (2015 est.)
note: data are in 2017 dollars
country comparison to the world: 85

GDP (official exchange rate): $54.76 billion (2017 est.)

GDP—real growth rate: 2.94% (2019 est.)
2.7% (2018 est.)
3.14% (2017 est.)
country comparison to the world: 101

GDP—per capita (PPP): $24,700 (2017 est.)
$23,800 (2016 est.)
$22,800 (2015 est.)
note: data are in 2017 dollars
country comparison to the world: 81

Gross national saving: 24.7% of GDP (2017 est.)
23.4% of GDP (2016 est.)
24.5% of GDP (2015 est.)
country comparison to the world: 62

GDP—composition, by end use: *household consumption:* 57.3% (2017 est.)
government consumption: 19.5% (2017 est.)
investment in fixed capital: 20% (2017 est.)
investment in inventories: 0% (2017 est.)
exports of goods and services: 51.1% (2017 est.)
imports of goods and services: -48.8% (2017 est.)

GDP—composition, by sector of origin: *agriculture:* 3.7% (2017 est.)
industry: 26.2% (2017 est.)
services: 70.1% (2017 est.)

Agriculture—products: arable crops (wheat, corn, barley, sugar beet, sunflower, rapeseed, alfalfa, clover); vegetables (potatoes, cabbage, onion, tomato, pepper); fruits (apples, plum, mandarins, olives), grapes for wine; livestock (cattle, cows, pigs); dairy products

Industries: chemicals and plastics, machine tools, fabricated metal, electronics, pig iron and rolled steel products, aluminum, paper, wood products, construction materials, textiles, shipbuilding, petroleum and petroleum refining, food and beverages, tourism

Industrial production growth rate: 1.2% (2017 est.)
country comparison to the world: 148

Labor force: 1.656 million (2020 est.)
country comparison to the world: 126

Labor force—by occupation: *agriculture:* 1.9%
industry: 27.3%
services: 70.8% (2017 est.)

Unemployment rate: 8.07% (2019 est.)
9.86% (2018 est.)
country comparison to the world: 125

Population below poverty line: 19.5% (2015 est.)

Household income or consumption by percentage share: *lowest 10%:* 2.7%
highest 10%: 23% (2015 est.)

Budget: *revenues:* 25.24 billion (2017 est.)
expenditures: 24.83 billion (2017 est.)

Taxes and other revenues: 46.1% (of GDP) (2017 est.)
country comparison to the world: 21

Budget surplus (+) or deficit (-): 0.8% (of GDP) (2017 est.)
country comparison to the world: 35

Public debt: 77.8% of GDP (2017 est.)
82.3% of GDP (2016 est.)
country comparison to the world: 38

Fiscal year: calendar year

Inflation rate (consumer prices): 1.1% (2017 est.)
-1.1% (2016 est.)
country comparison to the world: 57

Current account balance: $1.597 billion (2019 est.)
$1 billion (2018 est.)
country comparison to the world: 45

Exports: $13.15 billion (2017 est.)
$13.88 billion (2016 est.)
country comparison to the world: 80

Exports—partners: Italy 13.4%, Germany 12.2%, Slovenia 10.6%, Bosnia and Herzegovina 9.8%, Austria 6.2%, Serbia 4.8% (2017)

Exports—commodities: transport equipment, machinery, textiles, chemicals, foodstuffs, fuels

Imports: $22.34 billion (2017 est.)
$19.76 billion (2016 est.)
country comparison to the world: 71

Imports—commodities: machinery, transport and electrical equipment; chemicals, fuels and lubricants; foodstuffs

Imports—partners: Germany 15.7%, Italy 12.9%, Slovenia 10.7%, Hungary 7.5%, Austria 7.5% (2017)

Reserves of foreign exchange and gold: $18.82 billion (31 December 2017 est.)
$14.24 billion (31 December 2016 est.)
country comparison to the world: 60

Debt—external: $48.1 billion (31 December 2017 est.)
$46.96 billion (31 December 2016 est.)
country comparison to the world: 67

Exchange rates: kuna (HRK) per US dollar—
6.62 (2017 est.)
6.8 (2016 est.)
6.806 (2015 est.)
6.8583 (2014 est.)
5.7482 (2013 est.)

ENERGY

Electricity access: *electrification—total population:* 100% (2020)

Electricity—production: 12.2 billion kWh (2016 est.)
country comparison to the world: 95

Electricity—consumption: 15.93 billion kWh (2016 est.)

country comparison to the world: 76

Electricity—exports: 3.2 billion kWh (2016 est.)
country comparison to the world: 42

Electricity—imports: 8.702 billion kWh (2016 est.)
country comparison to the world: 28

Electricity—installed generating capacity: 4.921 million kW (2016 est.)
country comparison to the world: 80

Electricity—from fossil fuels: 45% of total installed capacity (2016 est.)
country comparison to the world: 160

Electricity—from nuclear fuels: 0% of total installed capacity (2017 est.)
country comparison to the world: 74

Electricity—from hydroelectric plants: 40% of total installed capacity (2017 est.)
country comparison to the world: 52

Electricity—from other renewable sources: 16% of total installed capacity (2017 est.)
country comparison to the world: 50

Crude oil—production: 14,000 bbl/day (2018 est.)
country comparison to the world: 74

Crude oil—exports: 0 bbl/day (2015 est.)
country comparison to the world: 110

Crude oil—imports: 55,400 bbl/day (2015 est.)
country comparison to the world: 55

Crude oil—proved reserves: 71 million bbl (1 January 2018 est.)
country comparison to the world: 74

Refined petroleum products—production: 74,620 bbl/day (2015 est.)
country comparison to the world: 70

Refined petroleum products—consumption: 73,000 bbl/day (2016 est.)
country comparison to the world: 91

Refined petroleum products—exports: 40,530 bbl/day (2015 est.)
country comparison to the world: 58

Refined petroleum products—imports: 35,530 bbl/day (2015 est.)
country comparison to the world: 94

Natural gas—production: 1.048 billion cu m (2017 est.)
country comparison to the world: 67

Natural gas—consumption: 2.577 billion cu m (2017 est.)
country comparison to the world: 77

Natural gas—exports: 172.7 million cu m (2017 est.)
country comparison to the world: 46

Natural gas—imports: 1.841 billion cu m (2017 est.)
country comparison to the world: 54

Natural gas—proved reserves: 24.92 billion cu m (1 January 2018 est.)
country comparison to the world: 71

Carbon dioxide emissions from consumption of energy: 17.96 million Mt (2017 est.)
country comparison to the world: 89

COMMUNICATIONS

Telephones—fixed lines: *total subscriptions:* 1,371,999
subscriptions per 100 inhabitants: 32.29 (2019 est.)
country comparison to the world: 68

Telephones—mobile cellular: *total subscriptions:* 4,531,122
subscriptions per 100 inhabitants: 106.64 (2019 est.)
country comparison to the world: 124

Telecommunication systems: *general assessment:* the mobile market has one of the highest penetration rates in the Balkans region; covering much of what were once inaccessible areas; local lines are digital; telecom market in Croatia has been shaped by Croatia becoming part of the European Union in 2013, a process which opened up the market and the creation of a regulatory environment leading to competition in mobile and broadband; investment among operators has led to a relatively high broadband penetration in the region; trials for 5G technologies underway (2020)

domestic: fixed-line teledensity has dropped somewhat to about 32 per 100 persons; mobile-cellular telephone subscriptions 107 per 100 (2019)

international: country code—385; the ADRIA-1 submarine cable provides connectivity to Albania and Greece; digital international service is provided through the main switch in Zagreb; Croatia participates in the Trans-Asia-Europe fiberoptic project, which consists of 2 fiber-optic trunk connections with Slovenia and a fiber-optic trunk line from Rijeka to Split and Dubrovnik (2019)

note: the COVID-19 outbreak is negatively impacting telecommunications production and supply chains globally; consumer spending on telecom devices and services has also slowed due to the pandemic's effect on economies worldwide; overall progress towards improvements in all facets of the telecom industry—mobile, fixed-line, broadband, submarine cable and satellite—has moderated

Broadcast media: the national state-owned public broadcaster, Croatian Radiotelevision, operates 4 terrestrial TV networks, a satellite channel that rebroadcasts programs for Croatians living abroad, and 6 regional TV centers; 2 private broadcasters operate national terrestrial networks; 29 privately owned regional TV stations; multi-channel cable and satellite TV subscription services are available; state-owned public broadcaster operates 4 national radio networks and 23 regional radio stations; 2 privately owned national radio networks and 117 local radio stations (2019)

Internet country code: .hr

Internet users: *total:* 3,104,212

percent of population: 72.69% (July 2018 est.)
country comparison to the world: 98

Broadband—fixed subscriptions: *total:* 1,127,591
subscriptions per 100 inhabitants: 26 (2018 est.)
country comparison to the world: 67

TRANSPORTATION

National air transport system: *number of registered air carriers:* 2 (2020)
inventory of registered aircraft operated by air carriers: 18
annual passenger traffic on registered air carriers: 2,093,577 (2018)
annual freight traffic on registered air carriers: 530,000 mt-km (2018)

Civil aircraft registration country code prefix: 9A (2016)

Airports: 69 (2013)
country comparison to the world: 71

Airports—with paved runways: *total:* 24 (2017)
over 3,047 m: 2 (2017)
2,438 to 3,047 m: 6 (2017)
1,524 to 2,437 m: 3 (2017)
914 to 1,523 m: 3 (2017)
under 914 m: 10 (2017)

Airports—with unpaved runways: *total:* 45 (2013)
1,524 to 2,437 m: 1 (2013)
914 to 1,523 m: 6 (2013)
under 914 m: 38 (2013)

Heliports: 1 (2013)

Pipelines: 2410 km gas, 610 km oil (2011)

Railways: *total:* 2,722 km (2014)
standard gauge: 2,722 km 1.435-m gauge (980 km electrified) (2014)
country comparison to the world: 64

Roadways: *total:* 26,958 km (includes 1,416 km of expressways) (2015)
country comparison to the world: 101

Waterways: 785 km (2009)
country comparison to the world: 73

Merchant marine: *total:* 336
by type: bulk carrier 15, general cargo 32, oil tanker 20, other 269 (2019)
country comparison to the world: 52

Ports and terminals: *major seaport(s):* Ploce, Rijeka, Sibenik, Split

oil terminal(s): Omisalj
river port(s): Vukovar (Danube)

MILITARY AND SECURITY

Military and security forces: Armed Forces of the Republic of Croatia (Oruzane Snage Republike Hrvatske, OSRH) consists of five major commands directly subordinate to a General Staff: Ground Forces (Hrvatska Kopnena Vojska, HKoV), Naval Forces (Hrvatska Ratna Mornarica, HRM, includes Coast Guard), Air Force and Air Defense Command (Hrvatsko Ratno Zrakoplovstvo I Protuzracna Obrana), Joint Education and Training Command, Logistics Command; Military Police Force supports each of the three Croatian military forces (2019)

Military expenditures: 1.68% of GDP (2019 est.)
1.59% of GDP (2018)
1.67% of GDP (2017)

1.62% of GDP (2016)
1.78% of GDP (2015)
country comparison to the world: 67

Military and security service personnel strengths: the Armed Forces of the Republic of Croatia have approximately 15,000 active duty personnel (10,000 Army; 1,500 Navy; 1,500 Air force; 2,000 other) (2019 est.)

Military equipment inventories and acquisitions: the inventory of the Croatian Armed Forces consists mostly of Soviet-era equipment, although in recent years, it has attempted to acquire more modern weapon systems from Western suppliers; since 2010, the leading suppliers of military equipment to Croatia are Finland, Germany, and the US (2019 est.)

Military service age and obligation: 18-27 years of age for voluntary military service; conscription abolished in 2008 (2017)

TRANSNATIONAL ISSUES

Disputes—international: dispute remains with Bosnia and Herzegovina over several small sections of the boundary related to maritime access that hinders ratification of the 1999 border agreement; since the breakup of Yugoslavia in the early 1990s, Croatia and Slovenia have each claimed sovereignty over Piranski Bay and four villages, and Slovenia has objected to Croatia's claim of an exclusive economic zone in the Adriatic Sea; in 2009, however Croatia and Slovenia signed a binding international arbitration agreement to define their disputed land and maritime borders, which led to Slovenia lifting its objections to Croatia joining the EU; Slovenia continues to impose a hard border Schengen regime with Croatia, which joined the EU in 2013 but has not yet fulfilled Schengen requirements

Refugees and internally displaced persons: *stateless persons:* 2,886 (2019)
note: 713,772 estimated refugee and migrant arrivals (January 2015-October 2020); flows slowed considerably in 2017; Croatia is predominantly a transit country and hosts about 340 asylum seekers as of the end of June 2018

Illicit drugs: primarily a transit country along the Balkan route for maritime shipments of South American cocaine bound for Western Europe and other illicit drugs and chemical precursors to and from Western Europe; no significant domestic production of illicit drugs

CUBA

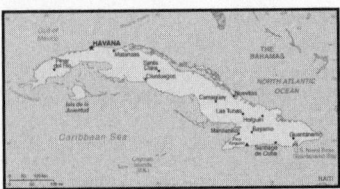

INTRODUCTION

Background: The native Amerindian population of Cuba began to decline after the European discovery of the island by Christopher COLUMBUS in 1492 and following its development as a Spanish colony during the next several centuries. Large numbers of African slaves were imported to work the coffee and sugar plantations, and Havana became the launching point for the annual treasure fleets bound for Spain from Mexico and Peru. Spanish rule eventually provoked an independence movement and occasional rebellions were harshly suppressed. US intervention during the Spanish-American War in 1898 assisted the Cubans in overthrowing Spanish rule. The Treaty of Paris established Cuban independence from Spain in 1898 and, following three-and-a-half years of subsequent US military rule, Cuba became an independent republic in 1902 after which the island experienced a string of governments mostly dominated by the military and corrupt politicians. Fidel CASTRO led a rebel army to victory in 1959; his authoritarian rule held the subsequent regime together for nearly five decades. He stepped down as president in February 2008 in favor of his younger brother Raul CASTRO. Cuba's communist revolution, with Soviet support, was exported throughout Latin America and Africa during the 1960s, 1970s, and 1980s. Miguel DIAZ-CANEL Bermudez, hand-picked by Raul CASTRO to succeed him, was approved as president by the National Assembly and took office on 19 April 2018.

The country faced a severe economic downturn in 1990 following the withdrawal of former Soviet subsidies worth $4-6 billion annually. Cuba traditionally and consistently portrays the US embargo, in place since 1961, as the source of its difficulties. As a result of efforts begun in December 2014 to re-establish diplomatic relations with the Cuban Government, which were severed in January 1961, the US and Cuba reopened embassies in their respective countries in July 2015. The embargo remains in place, and the relationship between the US and Cuba remains tense.

Illicit migration of Cuban nationals to the US via maritime and overland routes has been a longstanding challenge. On 12 January 2017, the US and Cuba signed a Joint Statement ending the so-called "wet-foot, dry-foot" policy – by which Cuban nationals who reached US soil were permitted to stay. Illicit Cuban migration by sea has since dropped significantly, but land border crossings continue. In FY 2018, the US Coast Guard interdicted 312 Cuban nationals at sea. Also in FY 2018, 7,249 Cuban migrants presented themselves at various land border ports of entry throughout the US.

GEOGRAPHY

Location: Caribbean, island between the Caribbean Sea and the North Atlantic Ocean, 150 km south of Key West, Florida

Geographic coordinates: 21 30 N, 80 00 W

Map references: Central America and the Caribbean

Area: *total:* 110,860 sq km
land: 109,820 sq km
water: 1,040 sq km

country comparison to the world: 107

Area—comparative: slightly smaller than Pennsylvania

Land boundaries: *total:* 28.5 km
border countries (1): US Naval Base at Guantanamo Bay 28.5 km
note: Guantanamo Naval Base is leased by the US and remains part of Cuba

Coastline: 3,735 km

Maritime claims: *territorial sea:* 12 nm
exclusive economic zone: 200 nm
contiguous zone: 24 nm

Climate: tropical; moderated by trade winds; dry season (November to April); rainy season (May to October)

Terrain: mostly flat to rolling plains, with rugged hills and mountains in the southeast

Elevation: *mean elevation:* 108 m
lowest point: Caribbean Sea 0 m
highest point: Pico Turquino 1,974 m

Natural resources: cobalt, nickel, iron ore, chromium, copper, salt, timber, silica, petroleum, arable land

Land use: *agricultural land:* 60.3% (2011 est.)
arable land: 33.8% (2011 est.) / permanent crops: 3.6% (2011 est.) / permanent pasture: 22.9% (2011 est.)
forest: 27.3% (2011 est.)
other: 12.4% (2011 est.)
Irrigated land: 8,700 sq km (2012)

Population distribution: large population clusters found throughout the country, the more significant ones being in the larger towns and cities, particularly the capital of Havana

Natural hazards: the east coast is subject to hurricanes from August to November (in general, the country averages about one hurricane every other year); droughts are common

Environment—current issues: soil degradation and desertification (brought on by poor farming techniques and natural disasters) are the main environmental problems; biodiversity loss; deforestation; air and water pollution

Environment—international agreements: party to: Antarctic Treaty, Biodiversity, Climate Change, Climate Change-Kyoto Protocol, Desertification, Endangered Species, Environmental Modification, Hazardous Wastes, Law of the Sea, Marine Dumping, Ozone Layer Protection, Ship Pollution, Wetlands

signed, but not ratified: Marine Life Conservation

Geography—note: largest country in Caribbean and westernmost island of the Greater Antilles

PEOPLE AND SOCIETY

Population: 11,059,062 (July 2020 est.)
country comparison to the world: 83

Nationality: *noun:* Cuban(s)
adjective: Cuban

Ethnic groups: white 64.1%, mulatto or mixed 26.6%, black 9.3% (2012 est.)
note: data represent racial self-identification from Cuba's 2012 national census

Languages: Spanish (official)

Religions: Christian 59.2%, folk 17.4%, other .4%, none 23% (2010 est.)
note: folk religion includes religions of African origin, spiritualism, and others intermingled with Catholicism or Protestantism; data is estimative because no authoritative source on religious affiliation exists in Cuba

Age structure: *0-14 years:* 16.34% (male 929,927/female 877,035)
15-24 years: 11.81% (male 678,253/female 627,384)
25-54 years: 41.95% (male 2,335,680/female 2,303,793)
55-64 years: 14.11% (male 760,165/female 799,734)
65 years and over: 15.8% (male 794,743/female 952,348) (2020 est.)

Dependency ratios: *total dependency ratio:* 46.7
youth dependency ratio: 23.3
elderly dependency ratio: 23.3
potential support ratio: 4.3 (2020 est.)

Median age: *total:* 42.1 years
male: 40.2 years
female: 43.8 years (2020 est.)
country comparison to the world: 37

Population growth rate: -0.25% (2020 est.)
country comparison to the world: 212

Birth rate: 10.4 births/1,000 population (2020 est.)
country comparison to the world: 188

Death rate: 9.1 deaths/1,000 population (2020 est.)
country comparison to the world: 58

Net migration rate: -3.7 migrant(s)/1,000 population (2020 est.)
country comparison to the world: 185

Population distribution: large population clusters found throughout the country, the more significant ones being in the larger towns and cities, particularly the capital of Havana

Urbanization: *urban population:* 77.2% of total population (2020)
rate of urbanization: 0.14% annual rate of change (2015-20 est.)
total population growth rate v. urban population growth rate, 2000-2030:

Major urban areas—population: 2.140 million HAVANA (capital) (2020)

Sex ratio: *at birth:* 1.06 male(s)/female
0-14 years: 1.06 male(s)/female
15-24 years: 1.08 male(s)/female
25-54 years: 1.01 male(s)/female
55-64 years: 0.95 male(s)/female
65 years and over: 0.83 male(s)/female
total population: 0.99 male(s)/female (2020 est.)

Maternal mortality rate: 36 deaths/100,000 live births (2017 est.)
country comparison to the world: 105

Infant mortality rate: *total:* 4.3 deaths/1,000 live births
male: 4.8 deaths/1,000 live births
female: 3.8 deaths/1,000 live births (2020 est.)
country comparison to the world: 184

Life expectancy at birth: *total population:* 79.2 years
male: 76.8 years
female: 81.7 years (2020 est.)
country comparison to the world: 59

Total fertility rate: 1.71 children born/woman (2020 est.)
country comparison to the world: 169

Contraceptive prevalence rate: 73.7% (2014)

Drinking water source:
improved:
urban: 98.2% of population
rural: 94.5% of population
total: 97.4% of population
unimproved:
urban: 1.8% of population
rural: 5.5% of population
total: 2.6% of population (2017 est.)

Current Health Expenditure: 11.7% (2017)

Physicians density: 8.3 physicians/1,000 population (2017)

Hospital bed density: 5.3 beds/1,000 population (2017)

Sanitation facility access:
improved:
urban: 96.1% of population
rural: 94.8% of population
total: 95.8% of population
unimproved:
urban: 3.9% of population
rural: 5.2% of population
total: 4.2% of population (2017 est.)

HIV/AIDS—adult prevalence rate: 0.3% (2019 est.)
country comparison to the world: 85

HIV/AIDS—people living with HIV/AIDS: 32,000 (2019 est.)
country comparison to the world: 72

HIV/AIDS—deaths: <500 (2019 est.)

Major infectious diseases: *degree of risk:* intermediate (2020)
food or waterborne diseases: bacterial diarrhea and hepatitis A
vectorborne diseases: dengue fever

Obesity—adult prevalence rate: 24.6% (2016)
country comparison to the world: 56

Education expenditures: 12.8% of GDP (2010)
country comparison to the world: 1

Literacy: *definition:* age 15 and over can read and write
total population: 99.8%
male: 99.9%
female: 99.8% (2015)

School life expectancy (primary to tertiary education): *total:* 14 years
male: 14 years
female: 15 years (2018)

Unemployment, youth ages 15-24: *total:* 6.1%
male: 6.4%
female: 5.6% (2010 est.)
country comparison to the world: 159

People—note: illicit emigration is a continuing problem; Cubans attempt to depart the island and enter the US using homemade rafts, alien smugglers, direct flights, or falsified visas; Cubans also use non-maritime routes to enter the US including direct flights to Miami and overland via the southwest border; the number of Cubans migrating to the US surged after the announcement of normalization of US-Cuban relations in late December 2014 but has decreased since the end of the so-called "wet-foot, dry-foot" policy on 12 January 2017

GOVERNMENT

Country name: *conventional long form:* Republic of Cuba
conventional short form: Cuba
local long form: Republica de Cuba
local short form: Cuba
etymology: name derives from the Taino Indian designation for the island "coabana" meaning "great place"

Government type: communist state

Capital: *name:* Havana
geographic coordinates: 23 07 N, 82 21 W
time difference: UTC-5 (same time as Washington, DC, during Standard Time)

daylight saving time: +1hr, begins second Sunday in March; ends first Sunday in November; note—Cuba has been known to alter the schedule of DST on short notice in an attempt to conserve electricity for lighting
etymology: the sites of Spanish colonial cities often retained their original Taino names; Habana, the Spanish name for the city, may be based on the name of a local Taino chief, HABAGUANEX

247

Administrative divisions: 15 provinces (provincias, singular—provincia) and 1 special municipality* (municipio especial); Artemisa, Camaguey, Ciego de Avila, Cienfuegos, Granma, Guantanamo, Holguin, Isla de la Juventud*, La Habana, Las Tunas, Matanzas, Mayabeque, Pinar del Rio, Sancti Spiritus, Santiago de Cuba, Villa Clara

Independence: 20 May 1902 (from Spain 10 December 1898; administered by the US from 1898 to 1902); not acknowledged by the Cuban Government as a day of independence

National holiday: Triumph of the Revolution (Liberation Day), 1 January (1959)

Constitution: *history:* several previous; latest drafted 14 July 2018, approved by the National Assembly 22 December 2018, approved by referendum 24 February 2019
amendments: proposed by the National Assembly of People's Power; passage requires approval of at least two-thirds majority of the National Assembly membership; amendments to constitutional articles on the authorities of the National Assembly, Council of State, or any rights and duties in the constitution also require approval in a referendum; constitutional articles on the Cuban political, social, and economic system cannot be amended

Legal system: civil law system based on Spanish civil code

International law organization participation: has not submitted an ICJ jurisdiction declaration; non-party state to the ICCt

Citizenship: *citizenship by birth:* yes
citizenship by descent only: yes
dual citizenship recognized: no
residency requirement for naturalization: unknown

Suffrage: 16 years of age; universal

Executive branch: *chief of state:* President Miguel DIAZ-CANEL Bermudez (since 10 October 2019); Vice President Salvador Antonio VALDES Mesa (since 10 October 2019); note—the president is both chief of state and head of government
head of government: Prime Minister Manuel MARRERO Cruz (since 21 December 2019); Deputy Prime Ministers Ramiro VALDES Menendez, Roberto MORALES Ojeda, Ines Maria CHAPMAN Waugh, Jorge Luis TAPIA Fonseca, Alejandro GIL Fernandez, Ricardo CABRISAS Ruiz (since 21 December 2019)
cabinet: Council of Ministers proposed by the president and appointed by the National Assembly; it is subordinate to the 21-member Council of State, which is elected by the Assembly to act on its behalf when it is not in session
elections/appointments: president and vice president indirectly elected by the National Assembly for a 5-year term (may be reelected for another 5-year term); election last held on 10 October 2019 (next to be held in 2024)
election results: Miguel DIAZ-CANEL Bermudez (PCC) elected president; percent of National Assembly vote—98.8%; Salvador Antonio

VALDES Mesa (PCC) elected vice president; percent of National Assembly vote—98.1%
note—on 19 April 2018, DIAZ-CANEL succeeded Raul CASTRO as president of the Council of State; on 10 October 2019 he was elected to the newly created position of President of the Republic, which replaced the position of President of the Council of State

Legislative branch: *description:* unicameral National Assembly of People's Power or Asamblea Nacional del Poder Popular (605 seats; members directly elected by absolute majority vote; members serve 5-year terms); note 1—the National Candidature Commission submits a slate of approved candidates; to be elected, candidates must receive more than 50% of valid votes otherwise the seat remains vacant or the Council of State can declare another election; note 2—in july 2019, the National Assembly passed a law which reduces the number of members from 605 to 474, effective with the 2023 general election
elections: last held on 11 March 2018 (next to be held in early 2023)
election results: Cuba's Communist Party is the only legal party, and officially sanctioned candidates run unopposed; composition—men 283, women 322, percent of women 53.2%

Judicial branch: *highest courts:* People's Supreme Court (consists of court president, vice president, 41 professional justices, and NA lay judges); organization includes the State Council, criminal, civil, administrative, labor, crimes against the state, and military courts)
judge selection and term of office: professional judges elected by the National Assembly are not subject to a specific term; lay judges nominated by workplace collectives and neighborhood associations and elected by municipal or provincial assemblies; lay judges appointed for 5-year terms and serve up to 30 days per year
subordinate courts: People's Provincial Courts; People's Regional Courts; People's Courts

Political parties and leaders: Cuban Communist Party or PCC [Raul CASTRO Ruz]

International organization participation: ACP, ALBA, AOSIS, CELAC, FAO, G-77, IAEA, ICAO, ICC (national committees), ICRM, IFAD, IFRCS, IHO, ILO, IMO, IMSO, Interpol, IOC, IOM (observer), IPU, ISO, ITSO, ITU, LAES, LAIA, NAM, OAS (excluded from formal participation since 1962), OPANAL, OPCW, PCA, Petrocaribe, PIF (partner), UN, UNCTAD, UNESCO, UNIDO, Union Latina, UNWTO, UPU, WCO, WFTU (NGOs), WHO, WIPO, WMO, WTO

Diplomatic representation in the US: chief of mission: Ambassador Jose Ramon CABANAS Rodriguez (since 17 September 2015)
chancery: 2630 16th Street NW, Washington, DC 20009
telephone: [1] (202) 797-8518

Diplomatic representation from the US: chief of mission: Ambassador (vacant); Charge d'Affaires Mara TEKACH (since 20 June 2018)

telephone: [53] (7) 839-4100
embassy: Calzada between L & M Streets, Vedado, Havana
mailing address: use embassy street address
FAX: NA

Flag description: five equal horizontal bands of blue (top, center, and bottom) alternating with white; a red equilateral triangle based on the hoist side bears a white, five-pointed star in the center; the blue bands refer to the three old divisions of the island: central, occidental, and oriental; the white bands describe the purity of the independence ideal; the triangle symbolizes liberty, equality, and fraternity, while the red color stands for the blood shed in the independence struggle; the white star, called La Estrella Solitaria (the Lone Star) lights the way to freedom and was taken from the flag of Texas
note: design similar to the Puerto Rican flag, with the colors of the bands and triangle reversed

National symbol(s): *royal palm; national colors:* red, white, blue

National anthem: *name:* "La Bayamesa" (The Bayamo Song)
lyrics/music: Pedro FIGUEREDO
note: adopted 1940; Pedro FIGUEREDO first performed "La Bayamesa" in 1868 during the Ten Years War against the Spanish; a leading figure in the uprising, FIGUEREDO was captured in 1870 and executed by a firing squad; just prior to the fusillade he is reputed to have shouted, "Morir por la Patria es vivir" (To die for the country is to live), a line from the anthem
0:00 / 1:10

ECONOMY

Economy—overview: The government continues to balance the need for loosening its socialist economic system against a desire for firm political control. In April 2011, the government held the first Cuban Communist Party Congress in almost 13 years, during which leaders approved a plan for wide-ranging economic changes. Since then, the government has slowly and incrementally implemented limited economic reforms, including allowing Cubans to buy electronic appliances and cell phones, stay in hotels, and buy and sell used cars. The government has cut state sector jobs as part of the reform process, and it has opened up some retail services to "self-employment," leading to the rise of so-called "cuentapropistas" or entrepreneurs. More than 500,000 Cuban workers are currently registered as self-employed.

The Cuban regime has updated its economic model to include permitting the private ownership and sale of real estate and new vehicles, allowing private farmers to sell agricultural goods directly to hotels, allowing the creation of non-agricultural cooperatives, adopting a new foreign investment law, and launching a "Special Development Zone" around the Mariel port.

Since 2016, Cuba has attributed slowed economic growth in part to problems with petroleum product deliveries from Venezuela. Since late

2000, Venezuela provided petroleum products to Cuba on preferential terms, supplying at times nearly 100,000 barrels per day. Cuba paid for the oil, in part, with the services of Cuban personnel in Venezuela, including some 30,000 medical professionals.

GDP (purchasing power parity):
$137 billion (2017 est.)
$134.8 billion (2016 est.)
$134.2 billion (2015 est.)
note: data are in 2016 US dollars
country comparison to the world: 79

GDP (official exchange rate):
$93.79 billion (2017 est.)
note: data are in Cuban Pesos at 1 CUP = 1 US$; official exchange rate

GDP—real growth rate:
1.6% (2017 est.)
0.5% (2016 est.)
4.4% (2015 est.)
country comparison to the world: 151

GDP—per capita (PPP):
$12,300 (2016 est.)
$12,200 (2015 est.)
$12,100 (2014 est.)
note: data are in 2016 US dollars
country comparison to the world: 128

Gross national saving:
11.4% of GDP (2017 est.)
12.3% of GDP (2016 est.)
12.1% of GDP (2015 est.)
country comparison to the world: 156

GDP—composition, by end use:
household consumption: 57% (2017 est.)
government consumption: 31.6% (2017 est.)
investment in fixed capital: 9.6% (2017 est.)
investment in inventories: 0% (2017 est.)
exports of goods and services: 14.6% (2017 est.)
imports of goods and services: -12.7% (2017 est.)

GDP—composition, by sector of origin:
agriculture: 4% (2017 est.)
industry: 22.7% (2017 est.)
services: 73.4% (2017 est.)

Agriculture—products: sugar, tobacco, citrus, coffee, rice, potatoes, beans; livestock

Industries: petroleum, nickel, cobalt, pharmaceuticals, tobacco, construction, steel, cement, agricultural machinery, sugar

Industrial production growth rate: -1.2% (2017 est.)
country comparison to the world: 179

Labor force: 4.691 million (2017 est.)
note: state sector 72.3%, non-state sector 27.7%
country comparison to the world: 83

Labor force—by occupation:
agriculture: 18%
industry: 10%
services: 72% (2016 est.)

Unemployment rate:
2.6% (2017 est.)
2.4% (2016 est.)

note: data are official rates; unofficial estimates are about double
country comparison to the world: 29

Population below poverty line: NA

Household income or consumption by percentage share: lowest 10%: NA
highest 10%: NA

Budget: revenues: 54.52 billion (2017 est.)
expenditures: 64.64 billion (2017 est.)

Taxes and other revenues: 58.1% (of GDP) (2017 est.)
country comparison to the world: 8

Budget surplus (+) or deficit (-): -10.8% (of GDP) (2017 est.)
country comparison to the world: 214

Public debt:
47.7% of GDP (2017 est.)
42.7% of GDP (2016 est.)
country comparison to the world: 110

Fiscal year: calendar year

Inflation rate (consumer prices):
5.5% (2017 est.)
4.5% (2016 est.)
country comparison to the world: 178

Current account balance:
$985.4 million (2017 est.)
$2.008 billion (2016 est.)
country comparison to the world: 50

Exports:
$2.63 billion (2017 est.)
$2.546 billion (2016 est.)
country comparison to the world: 131

Exports—partners: Venezuela 17.8%, Spain 12.2%, Russia 7.9%, Lebanon 6.1%, Indonesia 4.5%, Germany 4.3% (2017)

Exports—commodities: petroleum, nickel, medical products, sugar, tobacco, fish, citrus, coffee

Imports:
$11.06 billion (2017 est.)
$10.28 billion (2016 est.)
country comparison to the world: 98

Imports—commodities: petroleum, food, machinery and equipment, chemicals

Imports—partners: China 22%, Spain 14%, Russia 5%, Brazil 5%, Mexico 4.9%, Italy 4.8%, US 4.5% (2017)

Reserves of foreign exchange and gold:
$11.35 billion (31 December 2017 est.)
$12.3 billion (31 December 2016 est.)
country comparison to the world: 72

Debt—external:
$30.06 billion (31 December 2017 est.)
$29.89 billion (31 December 2016 est.)
country comparison to the world: 80

Exchange rates: Cuban pesos (CUP) per US dollar—
1 (2017 est.)
1 (2016 est.)
1 (2015 est.)
1 (2014 est.)
22.7 (2013 est.)

ENERGY

Electricity access: electrification—total population: 100% (2020)

Electricity—production: 19.28 billion kWh (2016 est.)
country comparison to the world: 76

Electricity—consumption: 16.16 billion kWh (2016 est.)
country comparison to the world: 75

Electricity—exports: 0 kWh (2016 est.)
country comparison to the world: 124

Electricity—imports: 0 kWh (2016 est.)
country comparison to the world: 138

Electricity—installed generating capacity: 6.998 million kW (2016 est.)
country comparison to the world: 74

Electricity—from fossil fuels: 91% of total installed capacity (2016 est.)
country comparison to the world: 53

Electricity—from nuclear fuels: 0% of total installed capacity (2017 est.)
country comparison to the world: 75

Electricity—from hydroelectric plants: 1% of total installed capacity (2017 est.)
country comparison to the world: 147

Electricity—from other renewable sources: 8% of total installed capacity (2017 est.)
country comparison to the world: 85

Crude oil—production: 50,000 bbl/day (2018 est.)
country comparison to the world: 53

Crude oil—exports: 0 bbl/day (2015 est.)
country comparison to the world: 111

Crude oil—imports: 112,400 bbl/day (2015 est.)
country comparison to the world: 41

Crude oil—proved reserves: 124 million bbl (1 January 2018 est.)
country comparison to the world: 68

Refined petroleum products—production: 104,100 bbl/day (2015 est.)
country comparison to the world: 67

Refined petroleum products—consumption: 175,000 bbl/day (2016 est.)
country comparison to the world: 60

Refined petroleum products—exports: 24,190 bbl/day (2015 est.)
country comparison to the world: 69

Refined petroleum products—imports: 52,750 bbl/day (2015 est.)
country comparison to the world: 78

Natural gas—production: 1.189 billion cu m (2017 est.)
country comparison to the world: 66

Natural gas—consumption: 1.189 billion cu m (2017 est.)
country comparison to the world: 90

Natural gas—exports: 0 cu m (2017 est.)
country comparison to the world: 90

Natural gas—imports: 0 cu m (2017 est.)
country comparison to the world: 113

Natural gas—proved reserves: 70.79 billion cu m (1 January 2018 est.)
country comparison to the world: 57

Carbon dioxide emissions from consumption of energy: 26.94 million Mt (2017 est.)
country comparison to the world: 78

COMMUNICATIONS

Telephones—fixed lines: *total subscriptions:* 1,475,679
subscriptions per 100 inhabitants: 13.31 (2019 est.)
country comparison to the world: 64

Telephones—mobile cellular: *total subscriptions:* 5,911,586
subscriptions per 100 inhabitants: 53.32 (2019 est.)
country comparison to the world: 112

Telecommunication systems: *general assessment:* lowest mobile phone and Internet penetration rates in the region, fixed-line teledensity is also low; fixed-line and mobile services run by the state-run ETESCA; mobile-cellular telephone service is expensive and must be paid in convertible pesos; the Cuban Government has opened several hundred Wi-Fi hotspots around the island, which are expensive, and launched a new residential Internet pilot in Havana and other provinces; as of 2018, 3G mobile service is available, if limited (2020)
domestic: fixed-line density remains low at about 13 per 100 inhabitants; mobile-cellular service is expanding to about 53 per 100 persons (2019)
international: country code—53; the ALBA-1, GTMO-1, and GTMO-PR fiber-optic submarine cables link Cuba, Jamaica, and Venezuela; satellite earth station—1 Intersputnik (Atlantic Ocean region) (2019)
note: the COVID-19 outbreak is negatively impacting telecommunications production and supply chains globally; consumer spending on telecom devices and services has also slowed due to the pandemic's effect on economies worldwide; overall progress towards improvements in all facets of the telecom industry—mobile, fixed-line, broadband, submarine cable and satellite—has moderated

Broadcast media: Government owns and controls all broadcast media: five national TV channels (Cubavision, Tele Rebelde, Multivision, Educational Channel 1 and 2,) 2 international channels (Cubavision Internacional and Caribe,) 16 regional TV stations, 6 national radio networks and multiple regional stations; the Cuban government beams over the Radio-TV Marti signal; although private ownership of electronic media is prohibited, several online independent news sites exist; those that are not openly critical of the government are often tolerated; the others are blocked by the government; there are no independent TV channels, but several outlets have created strong audiovisual content (El Toque, for example); a community of young Youtubers is also growing, mostly with channels about sports, technology and fashion; Christian denominations

are creating original video content to distribute via social media (2019)

Internet country code: .cu

Internet users: *total:* 6,353,020
percent of population: 57.15% (July 2018 est.)
note: private citizens are prohibited from buying computers or accessing the Internet without special authorization; foreigners may access the Internet in large hotels but are subject to firewalls; some Cubans buy illegal passwords on the black market or take advantage of public outlets to access limited email and the government-controlled "intranet"
country comparison to the world: 76

Broadband—fixed subscriptions: *total:* 98,838
subscriptions per 100 inhabitants: 1 less than 1 (2018 est.)
country comparison to the world: 122

TRANSPORTATION

National air transport system: *number of registered air carriers:* 4 (2020)
inventory of registered aircraft operated by air carriers: 18
annual passenger traffic on registered air carriers: 560,754 (2018)
annual freight traffic on registered air carriers: 17.76 million mt-km (2018)

Civil aircraft registration country code prefix: CU (2016)

Airports: 133 (2017)
country comparison to the world: 41

Airports—with paved runways: *total:* 64 (2017)
over 3,047 m: 7 (2017)
2,438 to 3,047 m: 10 (2017)
1,524 to 2,437 m: 16 (2017)
914 to 1,523 m: 4 (2017)
under 914 m: 27 (2017)

Airports—with unpaved runways: *total:* 69 (2013)
914 to 1,523 m: 11 (2013)
under 914 m: 58 (2013)

Pipelines: 41 km gas, 230 km oil (2013)

Railways: *total:* 8,367 km (2017)
standard gauge: 8,195 km 1.435-m gauge (124 km electrified) (2017)
narrow gauge: 172 km 1.000-m gauge (2017)
note: 82 km of standard gauge track is not for public use
country comparison to the world: 26

Roadways:
total: 60,000 km (2015)
paved: 20,000 km (2001)
unpaved: 40,000 km (2001)
country comparison to the world: 76

Waterways: 240 km (almost all navigable inland waterways are near the mouths of rivers) (2011)
country comparison to the world: 94

Merchant marine: *total:* 52
by type: general cargo 12, oil tanker 3, other 37 (2019)
country comparison to the world: 116

Ports and terminals: *major seaport(s):* Antilla, Cienfuegos, Guantanamo, Havana, Matanzas, Mariel, Nuevitas Bay, Santiago de Cuba

MILITARY AND SECURITY

Military and security forces: Revolutionary Armed Forces (Fuerzas Armadas Revolucionarias, FAR): Revolutionary Army (Ejercito Revolucionario, ER), Revolutionary Navy (Marina de Guerra Revolucionaria, MGR, includes Marine Corps), Revolutionary Air and Air Defense Forces (Defensas Anti-Aereas y Fuerza Aerea Revolucionaria, DAAFAR); Paramilitary forces: Youth Labor Army (Ejercito Juvenil del Trabajo, EJT), Territorial Militia Troops (Milicia de Tropas de Territoriales, MTT), Civil Defense Force; Ministry of Interior: Border Guards, State Security (2020)

Military expenditures:
2.9% of GDP (2018)
2.9% of GDP (2017)
3.1% of GDP (2016)
3.1% of GDP (2015)
3.5% of GDP (2014)
country comparison to the world: 29

Military and security service personnel strengths: the Revolutionary Armed Forces (FAR) of Cuba have approximately 50,000 active personnel (39,000 Army; 3,000 Navy; 8,0 Air Force) (2019 est.)

Military equipment inventories and acquisitions: the Cuban military inventory is comprised of Russian and Soviet-era equipment; the last recorded arms delivery to Cuba was by Russia in 2004 (2019 est.)

Military service age and obligation: 17-28 years of age for compulsory military service; 2-year service obligation for males, optional for females (2017)

Military—note: the FAR remains well trained and professional in nature, but the collapse of the Soviet Union deprived the Cuban military of its major economic and logistic support and had a significant impact on the state of equipment; the lack of replacement parts for its existing equipment has increasingly affected operational capabilities (2019)

TRANSNATIONAL ISSUES

Disputes—international: US Naval Base at Guantanamo Bay is leased to US and only mutual agreement or US abandonment of the facility can terminate the lease

Trafficking in persons: *current situation:* Cuba is a source country for adults and children subjected to sex trafficking and forced labor; child sex trafficking and child sex tourism occur in Cuba, while some Cubans are forced into prostitution in South America and the Caribbean; allegations have been made that some Cubans have been forced or coerced to work at Cuban medical missions abroad; assessing the scope of trafficking within Cuba is difficult because of the lack of information

tier rating: Tier 2 Watch List—Cuba does not fully comply with the minimum standards for the elimination of trafficking; however, it is making significant efforts to do so; Cuba's penal code does not criminalize all forms of human trafficking, but the government reported that it is in the process of amending its criminal code to comply with the 2000 UN TIP Protocol, to which it acceded in 2013; the government in 2014 prosecuted and convicted 13 sex traffickers and provided services to the victims in those cases but does not have shelters specifically for trafficking victims; the government did not recognize forced labor as a problem and took no action to address it; state media produced newspaper articles and TV and radio programs to raise public awareness about sex trafficking (2015)

Illicit drugs: territorial waters and air space serve as transshipment zone for US- and European-bound drugs; established the death penalty for certain drug-related crimes in 1999

CURACAO

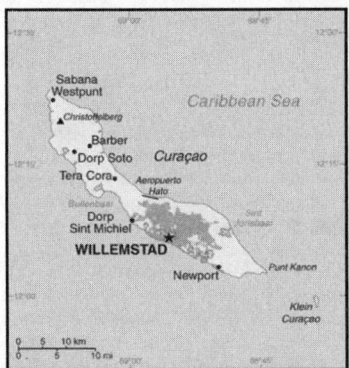

INTRODUCTION

Background: The original Arawak Indian settlers who arrived on the island from South America in about 1000, were largely enslaved by the Spanish early in the 16th century and forcibly relocated to other colonies where labor was needed. Curacao was seized by the Dutch from the Spanish in 1634. Once the center of the Caribbean slave trade, Curacao was hard hit economically by the abolition of slavery in 1863. Its prosperity (and that of neighboring Aruba) was restored in the early 20th century with the construction of the Isla Refineria to service the newly discovered Venezuelan oil fields. In 1954, Curacao and several other Dutch Caribbean possessions were reorganized as the Netherlands Antilles, part of the Kingdom of the Netherlands. In referenda in 2005 and 2009, the citizens of Curacao voted to become a self-governing country within the Kingdom of the Netherlands. The change in status became effective in October 2010 with the dissolution of the Netherlands Antilles.

GEOGRAPHY

Location: Caribbean, an island in the Caribbean Sea, 55 km off the coast of Venezuela

Geographic coordinates: 12 10 N, 69 00 W

Map references: Central America and the Caribbean

Area: *total:* 444 sq km
land: 444 sq km
water: 0 sq km

country comparison to the world: 200

Area—comparative: more than twice the size of Washington, DC

Land boundaries: 0

Coastline: 364 km

Maritime claims: *territorial sea:* 12 nm
exclusive economic zone: 200 nm

Climate: tropical marine climate, ameliorated by northeast trade winds, results in mild temperatures; semiarid with average rainfall of 60 cm/year

Terrain: generally low, hilly terrain

Elevation: *lowest point:* Caribbean Sea 0 m
highest point: Mt. Christoffel 372 m

Natural resources: calcium phosphates, aloes, sorghum, peanuts, vegetables, tropical fruit

Land use: *agricultural land:* 10% (2011 est.)
arable land: 10% / permanent crops: 0% / permanent pasture:* 0% (2011 est.)
forest: 0% (2011 est.)
other: 90% (2011 est.)
Irrigated land: NA

Population distribution: largest concentration on the island is Willemstad; smaller settlements near the coast can be found throughout the island, particularly in the northwest

Natural hazards: Curacao is south of the Caribbean hurricane belt and is rarely threatened

Environment—current issues: problems in waste management that threaten environmental sustainability on the island include pollution of marine areas from domestic sewage, inadequate sewage treatment facilities, industrial effluents and agricultural runoff, the mismanagement of toxic substances, and ineffective regulations; the refinery in Sint Anna Bay, at the eastern edge of Willemstad's large natural harbor, processes heavy crude oil from Venezuela; it has caused significant environmental damage to the surrounding area because of neglect and a lack of strict environmental controls; the release of noxious fumes and potentially hazardous particles causes schools downwind to regularly close

Geography—note: Curacao is a part of the Windward Islands (southern) group in the Lesser Antilles

PEOPLE AND SOCIETY

Population: 151,345 (July 2020 est.)

country comparison to the world: 188

Nationality: *noun:* Curacaoan
adjective: Curacaoan; Dutch

Ethnic groups: Curacaoan 75.4%, Dutch 6%, Dominican 3.6%, Colombian 3%, Bonairean, Sint Eustatian, Saban 1.5%, Haitian 1.2%, Surinamese 1.2%, Venezuelan 1.1%, Aruban 1.1%, other 5%, unspecified 0.9% (2011 est.)

Languages: Papiamento (official) (a creole language that is a mixture of Portuguese, Spanish, Dutch, English, and, to a lesser extent, French, as well as elements of African languages and the language of the Arawak) 79.9%, Dutch (official) 8.8%, Spanish 5.6%, English (official) 3.1%, other 2.9%, unspecified .3% (2001 census)
note: data represent most spoken language in household

Religions: Roman Catholic 72.8%, Pentecostal 6.6%, Protestant 3.2%, Adventist 3%, Jehovah's Witness 2%, Evangelical 1.9%, other 3.8%, none 6%, unspecified 0.6% (2011 est.)

Age structure: 0-14 years: 19.68% (male 15,227/female 14,553)
15-24 years: 13.38% (male 10,438/female 9,806)
25-54 years: 36.55% (male 27,733/female 27,589)
55-64 years: 13.88% (male 9,130/female 11,873)
65 years and over: 16.52% (male 10,127/female 14,869) (2020 est.)

Dependency ratios: *total dependency ratio:* 55.9
youth dependency ratio: 28.3
elderly dependency ratio: 27.5
potential support ratio: 3.6 (2020 est.)

Median age: *total:* 36.7 years
male: 34.4 years
female: 39.5 years (2020 est.)
country comparison to the world: 77

Population growth rate: 0.35% (2020 est.)
country comparison to the world: 168

Birth rate: 13.4 births/1,000 population (2020 est.)
country comparison to the world: 139

Death rate: 8.7 deaths/1,000 population (2020 est.)
country comparison to the world: 69

Net migration rate: -1.3 migrant(s)/1,000 population (2020 est.)
country comparison to the world: 148

Population distribution: largest concentration on the island is Willemstad; smaller settlements near the coast can be found throughout the island, particularly in the northwest

Urbanization: *urban population:* 89.1% of total population (2020) (2018)
rate of urbanization: 0.62% annual rate of change (2015-20 est.) (2015-20 est.)
total population growth rate v. urban population growth rate, 2000-2030:

Major urban areas—population: 144000 WILLEMSTAD (capital) (2018)

Sex ratio: *at birth:* 1.05 male(s)/female
0-14 years: 1.05 male(s)/female
15-24 years: 1.06 male(s)/female
25-54 years: 1.01 male(s)/female
55-64 years: 0.77 male(s)/female
65 years and over: 0.68 male(s)/female
total population: 0.92 male(s)/female (2020 est.)

Infant mortality rate: *total:* 7 deaths/1,000 live births
male: 7.5 deaths/1,000 live births
female: 6.5 deaths/1,000 live births (2020 est.)
country comparison to the world: 160

Life expectancy at birth: *total population:* 79 years
male: 76.6 years
female: 81.4 years (2020 est.)
country comparison to the world: 62

Total fertility rate: 2 children born/woman (2020 est.)
country comparison to the world: 113

Drinking water source:
improved:
total: 100% of population
unimproved:
total: 0% of population (2017 est.)

Sanitation facility access:
improved:
total: 100% of population
unimproved:
total: 0% of population (2017)

HIV/AIDS—adult prevalence rate: NA

HIV/AIDS—people living with HIV/AIDS: NA

HIV/AIDS—deaths: NA

Education expenditures: 4.9% of GDP (2013)
country comparison to the world: 65

School life expectancy (primary to tertiary education): *total:* 17 years
male: 18 years
female: 18 years (2013)

Unemployment, youth ages 15-24: *total:* 29.3%
male: 25.4% NA
female: 34.5% NA (2018 est.)
country comparison to the world: 36

GOVERNMENT

Country name: *conventional long form:* Country of Curacao
conventional short form: Curacao
local long form: Land Curacao (Dutch); Pais Korsou (Papiamento)
local short form: Curacao (Dutch); Korsou (Papiamento)

former: Netherlands Antilles; Curacao and Dependencies
etymology: the most plausible name derivation is that the island was designated Isla de la Curacion (Spanish meaning "Island of the Cure" or "Island of Healing") or Ilha da Curacao (Portuguese meaning the same) to reflect the locale's function as a recovery stop for sick crewmen

Dependency status: constituent country within the Kingdom of the Netherlands; full autonomy in internal affairs granted in 2010; Dutch Government responsible for defense and foreign affairs

Government type: parliamentary democracy

Capital: *name:* Willemstad
geographic coordinates: 12 06 N, 68 55 W
time difference: UTC-4 (1 hour ahead of Washington, DC, during Standard Time)
etymology—named after Prince William II of Orange (1626-1650), who served as stadtholder (Dutch head of state) from 1647 to 1650, shortly after the the Dutch captured Curacao from the Spanish in 1634

Administrative divisions: none (part of the Kingdom of the Netherlands)
note: Curacao is one of four constituent countries of the Kingdom of the Netherlands; the other three are the Netherlands, Aruba, and Sint Maarten

Independence: none (part of the Kingdom of the Netherlands)

National holiday: King's Day (birthday of King WILLEM-ALEXANDER), 27 April (1967); note—King's or Queen's Day are observed on the ruling monarch's birthday; celebrated on 26 April if 27 April is a Sunday

Constitution: *history:* previous 1947, 1955; latest adopted 5 September 2010, entered into force 10 October 2010 (regulates governance of Curacao but is subordinate to the Charter for the Kingdom of the Netherlands); note—in October 2010, with the dissolution of the Netherlands Antilles, Curacao became a semi-autonomous entity within the Kingdom of the Netherlands

Legal system: based on Dutch civil law

Citizenship: *see the Netherlands*

Suffrage: 18 years of age; universal

Executive branch: *chief of state:* King WILLEM-ALEXANDER of the Netherlands (since 30 April 2013); represented by Governor Lucille A. GEORGE-WOUT (since 4 November 2013)
head of government: Prime Minister Ivar ASJES (since 7 June 2013)
cabinet: Cabinet sworn-in by the governor
elections/appointments: the monarch is hereditary; governor appointed by the monarch; following legislative elections, the leader of the majority party usually elected prime minister by the Parliament of Curacao; next election scheduled for 2016

Legislative branch: *description:* unicameral Parliament of Curacao (21 seats; members directly

elected by proportional representation vote to serve 4-year terms)
elections: last held on 28 April 2017 (next to be held in 2021); early elections were held after Prime Minister Hensley KOEIMAN resigned on 12 February 2017, when the coalition government lost its majority
election results: percent of vote by party—PAR 23.3%, MAN 20.4%, MFK 19.9%, KdnT 9.4%, PIN 5.3%, PS 5.1%, MP 4.9%, other 11.7%; seats by party—PAR 6, MAN 5, MFK 5, KdnT 2, PIN 1, PS 1, MP 1; composition—men 15, women 6, percent of women 28.6%

Judicial branch: *highest courts:* Joint Court of Justice of Aruba, Curacao, Sint Maarten, and of Bonaire, Sint Eustatius and Saba or "Joint Court of Justice" (sits as a 3-judge panel); final appeals heard by the Supreme Court, in The Hague, Netherlands
judge selection and term of office: Joint Court judges appointed by the monarch for life
subordinate courts: first instance courts, appeals court; specialized courts

Political parties and leaders:
Korsou di Nos Tur or KdnT [Amparo dos SANTOS]
Mayors for Liberec Region (Starostove pro Liberecky Kraj) or SLK [Martin PUTA]
Movementu Futuro Korsou or MFK [Gerrit SCHOTTE]
Movementu Progresivo or MP [Marylin MOSES]
Movishon Antia Nobo or MAN [Hensley KOEIMAN]
Partido Antia Restruktura or PAR [Eugene RHUGGENAATH]
Partido Inovashon Nashonal or PIN [Suzanne CAMELIA-ROMER]
Partido pa Adelanto I Inovashon Soshal or PAIS [Alex ROSARIA]
Partido Nashonal di Pueblo or PNP [Humphrey DAVELAAR]
Pueblo Soberano or PS
Un Korsou Hustu [Omayra LEEFLANG]

International organization participation: Caricom (observer), FATF, ILO, ITU, UNESCO (associate), UPU

Diplomatic representation in the US: none (represented by the Kingdom of the Netherlands)

Diplomatic representation from the US: chief of mission: Consul General Allen GREENBERG (since June 2019); note—also accredited to Aruba and Sint Maarten
telephone: [599] (9) 4613066
mailing address: P. O. Box 158, J.B. Gorsiraweg #1
FAX: [599] (9) 4616489

Flag description: on a blue field a horizontal yellow band somewhat below the center divides the flag into proportions of 5:1:2; two five-pointed white stars—the smaller above and to the left of the larger—appear in the canton; the blue of the upper and lower sections symbolizes the sky and sea respectively; yellow represents the sun; the stars symbolize Curacao and its uninhabited smaller sister island of Klein Curacao; the five star points signify the five continents from which Curacao's people derive

National symbol(s): *laraha (citrus tree); national colors:* blue, yellow, white

National anthem: *name:* Himmo di Korsou (Anthem of Curacao)
lyrics/music: Guillermo ROSARIO, Mae HENRIQUEZ, Enrique MULLER, Betty DORAN/Frater Candidus NOWENS, Errol "El Toro" COLINA
note: adapted 1978; the lyrics, originally written in 1899, were rewritten in 1978 to make them less colonial in nature

ECONOMY

Economy—overview: Most of Curacao's GDP results from services. Tourism, petroleum refining and bunkering, offshore finance, and transportation and communications are the mainstays of this small island economy, which is closely tied to the outside world. Curacao has limited natural resources, poor soil, and inadequate water supplies, and budgetary problems complicate reform of the health and education systems. Although GDP grew only slightly during the past decade, Curacao enjoys a high per capita income and a well-developed infrastructure compared to other countries in the region.

Curacao has an excellent natural harbor that can accommodate large oil tankers, and the port of Willemstad hosts a free trade zone and a dry dock. Venezuelan state-owned oil company PdVSA, under a contract in effect until 2019, leases the single refinery on the island from the government, directly employing some 1,000 people. Most of the oil for the refinery is imported from Venezuela and most of the refined products are exported to the US and Asia. Almost all consumer and capital goods are imported, with the US, the Netherlands, and Venezuela being the major suppliers.

The government is attempting to diversify its industry and trade. Curacao is an Overseas Countries and Territories (OCT) of the European Union. Nationals of Curacao are citizens of the European Union, even though it is not a member. Based on its OCT status, products that originate in Curacao have preferential access to the EU and are exempt from import duties. Curacao is a beneficiary of the Caribbean Basin Initiative and, as a result, products originating in Curacao can be imported tax free into the US if at least 35% has been added to the value of these products in Curacao. The island has state-of-the-art information and communication technology connectivity with the rest of the world, including a Tier IV datacenter. With several direct satellite and submarine optic fiber cables, Curacao has one of the best Internet speeds and reliability in the Western Hemisphere.

GDP (purchasing power parity):
$3.128 billion (2012 est.)
$3.02 billion (2011 est.)
$2.96 billion (2010 est.)
note: data are in 2012 US dollars
country comparison to the world: 189

GDP (official exchange rate):
$5.6 billion (2012 est.)

GDP—real growth rate:
3.6% (2012 est.)
2% (2011 est.)
0.1% (2010 est.)
country comparison to the world: 82

GDP—per capita (PPP):
$15,000 (2004 est.)
country comparison to the world: 112

GDP—composition, by end use:
household consumption: 66.9% (2016 est.)
government consumption: 33.6% (2016 est.)
investment in fixed capital: 19.4% (2016 est.)
investment in inventories: 0% (2016 est.)
exports of goods and services: 17.5% (2016 est.)
imports of goods and services: -37.5% (2016 est.)

GDP—composition, by sector of origin:
agriculture: 0.7% (2012 est.)
industry: 15.5% (2012 est.)
services: 83.8% (2012 est.)

Agriculture—products: aloe, sorghum, peanuts, vegetables, tropical fruit

Industries: tourism, petroleum refining, petroleum transshipment, light manufacturing, financial and business services

Industrial production growth rate: NA

Labor force: 73,010 (2013)
country comparison to the world: 185

Labor force—by occupation:
agriculture: 1.2%
industry: 16.9%
services: 81.8% (2008 est.)

Unemployment rate:
13% (2013 est.)
9.8% (2011 est.)
country comparison to the world: 167

Taxes and other revenues: 16.6% (of GDP) (2012 est.)
country comparison to the world: 177

Budget surplus (+) or deficit (-): -0.4% (of GDP) (2012 est.)
country comparison to the world: 56

Public debt:
33.2% of GDP (2012 est.)
40.6% of GDP (2011 est.)
country comparison to the world: 158

Inflation rate (consumer prices):
2.6% (2013 est.)
2.8% (2012 est.)
country comparison to the world: 124

Current account balance:
-$400 million (2011 est.)
-$600 million (2010 est.)
country comparison to the world: 116

Exports:
$839.7 million (2017 est.)
$1.44 billion (2010 est.)
country comparison to the world: 166

Exports—commodities: petroleum products

Imports:
$540.3 billion (2018 est.)

$453.8 billion (2017 est.)
country comparison to the world: 8

Imports—commodities: crude petroleum, food, manufactures

Reserves of foreign exchange and gold:
$0 (31 December 2017 est.)
country comparison to the world: 191

Exchange rates: Netherlands Antillean guilders (ANG) per US dollar—
1.79 (2017 est.)
1.79 (2016 est.)
1.79 (2015 est.)
1.79 (2014 est.)
1.79 (2013 est.)

ENERGY

Electricity access: *electrification—total population:* 100% (2020)

Electricity—production: 1.785 billion kWh (2012 est.)
country comparison to the world: 139

Electricity—consumption: 968 million kWh (2008 est.)
country comparison to the world: 156

Electricity—exports: 0 kWh (2009 est.)
country comparison to the world: 125

Electricity—imports: 0 kWh (2009 est.)
country comparison to the world: 139

Crude oil—production: 0 bbl/day (2017 est.)
country comparison to the world: 125

Crude oil—exports: 0 bbl/day (2015 est.)
country comparison to the world: 112

Crude oil—imports: 191,300 bbl/day (2015 est.)
country comparison to the world: 31

Crude oil—proved reserves: 0 bbl (1 January 2011 est.)
country comparison to the world: 121

Refined petroleum products—production: 189,800 bbl/day (2015 est.)
country comparison to the world: 53

Refined petroleum products—consumption: 70,000 bbl/day (2016 est.)
country comparison to the world: 93

Refined petroleum products—exports: 167,500 bbl/day (2015 est.)
country comparison to the world: 33

Refined petroleum products—imports: 45,800 bbl/day (2015 est.)
country comparison to the world: 85

Natural gas—production: 0 cu m (2009 est.)
country comparison to the world: 121

Natural gas—consumption: 0 cu m (2009 est.)
country comparison to the world: 137

Natural gas—exports: 0 cu m (2009 est.)
country comparison to the world: 91

Natural gas—imports: 0 cu m (2009 est.)
country comparison to the world: 114

Natural gas—proved reserves: 0 cu m (1 January 2011 est.)
country comparison to the world: 126

COMMUNICATIONS

Telephones—fixed lines: *total subscriptions:* 57,443
subscriptions per 100 inhabitants: 38.09 (2019 est.)
country comparison to the world: 155

Telephones—mobile cellular: *total subscriptions:* 174,260
subscriptions per 100 inhabitants: 115.55 (2019 est.)
country comparison to the world: 186

Telecommunication systems: *general assessment:* fully automatic modern telecommunications system; telecom sector across the Caribbean region continues to be one of the growth areas; given the lack of economic diversity in the region, with a high dependence on tourism and activities such as fisheries and offshore financial services the telecom sector contributes greatly to the GDP (2020)
domestic: 39 per 100 users for fixed-line and 116 per 100 users for cellular-mobile, majority of the islanders have Internet; market revenue has been affected in recent quarters as a result of competition and regulatory measures on termination rates and roaming tariffs (2020)

international: country code—+599, PCCS submarine cable system to US, Caribbean and Central and South America (2019)
note: the COVID-19 outbreak is negatively impacting telecommunications production and supply chains globally; consumer spending on telecom devices and services has also slowed due to the pandemic's effect on economies worldwide; overall progress towards improvements in all facets of the telecom industry—mobile, fixed-line, broadband, submarine cable and satellite—has moderated

Broadcast media: government-run TeleCuracao operates a TV station and a radio station; 2 other privately owned TV stations and several privately owned radio stations (2019)

Internet country code: .cw

Internet users: *total:* 102,359
percent of population: 68.13% (July 2018 est.)
country comparison to the world: 179

Broadband—fixed subscriptions: *total:* 51,836
subscriptions per 100 inhabitants: 35 (2018 est.)
country comparison to the world: 134

TRANSPORTATION

National air transport system: *number of registered air carriers:* 2 (2020)

inventory of registered aircraft operated by air carriers: 11

Civil aircraft registration country code prefix: PJ (2016)

Airports: 1 (2020)
country comparison to the world: 219

Airports—with paved runways: *total:* 1 (2019)
over 3,047 m: 1

Roadways: *total:* 550 km
country comparison to the world: 193

Merchant marine: *total:* 74
by type: general cargo 9, oil tanker 1, other 64 (2019)
country comparison to the world: 100

Ports and terminals: *major seaport(s):* Willemstad
oil terminal(s): Bullen Baai (Curacao Terminal)
bulk cargo port(s): Fuik Bay (phosphate rock)

MILITARY AND SECURITY

Military and security forces: no regular military forces; the Dutch Government controls foreign and defense policy; the Dutch Caribbean Coast Guard (DCCG) provides maritime security (2019)

Military service age and obligation: no conscription (2010)

Military—note: defense is the responsibility of the Kingdom of the Netherlands (2019)

CYPRUS

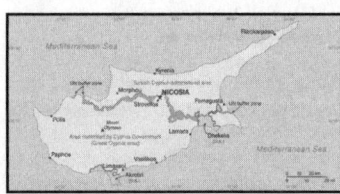

INTRODUCTION

Background: A former British colony, Cyprus became independent in 1960 following years of resistance to British rule. Tensions between the Greek Cypriot majority and Turkish Cypriot minority communities came to a head in December 1963, when violence broke out in the capital of Nicosia. Despite the deployment of UN peacekeepers in 1964, sporadic intercommunal violence continued, forcing most Turkish Cypriots into enclaves throughout the island. In 1974, a Greek Government-sponsored attempt to overthrow the elected president of Cyprus was met by military intervention from Turkey, which soon controlled more than a third of the island. In 1983, the Turkish Cypriot administered area declared itself the "Turkish Republic of Northern Cyprus" ("TRNC"), but it is recognized only by Turkey. An UN-mediated agreement, the Annan Plan, failed

to win approval by both communities in 2004. In February 2014, after a hiatus of nearly two years, the leaders of the two communities resumed formal discussions under UN auspices aimed at reuniting the divided island. The most recent round of negotiations to reunify the island were suspended in July 2017 after failure to achieve a breakthrough. The entire island entered the EU on 1 May 2004, although the EU acquis—the body of common rights and obligations—applies only to the areas under the internationally recognized government, and is suspended in the "TRNC." However, individual Turkish Cypriots able to document their eligibility for Republic of Cyprus citizenship legally enjoy the same rights accorded to other citizens of EU states.

GEOGRAPHY

Location: Middle East, island in the Mediterranean Sea, south of Turkey; note—Cyprus views itself as part of Europe; geopolitically, it can be classified as falling within Europe, the Middle East, or both

Geographic coordinates: 35 00 N, 33 00 E

Map references: Middle East

Area: *total:* 9,251 sq km (of which 3,355 sq km are in north Cyprus)
land: 9,241 sq km
water: 10 sq km

country comparison to the world: 170

Area—comparative: about 0.6 times the size of Connecticut

Land boundaries: *total:* 156 km

border sovereign base areas: Akrotiri 48 km, Dhekelia 108 km

Coastline: 648 km

Maritime claims: *territorial sea:* 12 nm
contiguous zone: 24 nm
continental shelf: 200-m depth or to the depth of exploitation

Climate: temperate; Mediterranean with hot, dry summers and cool winters

Terrain: central plain with mountains to north and south; scattered but significant plains along southern coast

Elevation: *mean elevation:* 91 m
lowest point: Mediterranean Sea 0 m
highest point: Mount Olympus 1,951 m

Natural resources: copper, pyrites, asbestos, gypsum, timber, salt, marble, clay earth pigment

Land use: *agricultural land:* 13.4% (2011 est.)
arable land: 9.8% (2011 est.) / permanent crops: 3.2% (2011 est.) / permanent pasture: 0.4% (2011 est.)
forest: 18.8% (2011 est.)
other: 67.8% (2011 est.)

Irrigated land: 460 sq km (2012)

Population distribution: population concentrated in central Nicosia and in the major cities of the south: Paphos, Limassol, and Larnaca

Natural hazards: moderate earthquake activity; droughts

Environment—current issues: water resource problems (no natural reservoir catchments, seasonal disparity in rainfall, sea water intrusion to island's largest aquifer, increased salination in the north); water pollution from sewage, industrial wastes, and pesticides; coastal degradation; erosion; loss of wildlife habitats from urbanization

Environment—international agreements: *party to:* Air Pollution, Air Pollution-Nitrogen Oxides, Air Pollution-Persistent Organic Pollutants, Air Pollution-Sulfur 94, Biodiversity, Climate Change, Climate Change-Kyoto Protocol, Desertification, Endangered Species, Environmental Modification, Hazardous Wastes, Law of the Sea, Marine Dumping, Ozone Layer Protection, Ship Pollution, Wetlands
signed, but not ratified: none of the selected agreements

Geography—note: the third largest island in the Mediterranean Sea (after Sicily and Sardinia); several small Cypriot enclaves exist within the Dhekelia Sovereign Base Area

PEOPLE AND SOCIETY

Population: 1,266,676 (July 2020 est.)
country comparison to the world: 157

Nationality: *noun:* Cypriot(s)
adjective: Cypriot

Ethnic groups: Greek 98.8%, other 1% (includes Maronite, Armenian, Turkish-Cypriot), unspecified 0.2% (2011 est.)
note: data represent only the Greek-Cypriot citizens in the Republic of Cyprus

Languages: Greek (official) 80.9%, Turkish (official) 0.2%, English 4.1%, Romanian 2.9%, Russian 2.5%, Bulgarian 2.2%, Arabic 1.2%, Filipino 1.1%, other 4.3%, unspecified 0.6% (2011 est.)
note: data represent only the Republic of Cyprus

Religions: Orthodox Christian 89.1%, Roman Catholic 2.9%, Protestant/Anglican 2%, Muslim 1.8%, Buddhist 1%, other (includes Maronite, Armenian Church, Hindu) 1.4%, unknown 1.1%, none/atheist 0.6% (2011 est.)
note: data represent only the government-controlled area of Cyprus

Age structure: *0-14 years:* 15.69% (male 102,095/female 96,676)
15-24 years: 12.29% (male 84,832/female 70,879)
25-54 years: 47.12% (male 316,710/female 280,167)
55-64 years: 11.92% (male 72,476/female 78,511)
65 years and over: 12.97% (male 71,053/female 93,277) (2020 est.)

Dependency ratios: *total dependency ratio:* 44.9
youth dependency ratio: 24
elderly dependency ratio: 20.9

potential support ratio: 4.8 (2020 est.)
note: data represent the whole country

Median age: *total:* 37.9 years
male: 36.7 years
female: 39.4 years (2020 est.)
country comparison to the world: 65

Population growth rate: 1.15% (2020 est.)
country comparison to the world: 94

Birth rate: 10.9 births/1,000 population (2020 est.)
country comparison to the world: 180

Death rate: 7 deaths/1,000 population (2020 est.)
country comparison to the world: 125

Net migration rate: 7.6 migrant(s)/1,000 population (2020 est.)
country comparison to the world: 13

Population distribution: population concentrated in central Nicosia and in the major cities of the south: Paphos, Limassol, and Larnaca

Urbanization: *urban population:* 66.8% of total population (2020)
rate of urbanization: 0.75% annual rate of change (2015-20 est.)

total population growth rate v. urban population growth rate, 2000-2030: Major urban areas—population: 269,000 NICOSIA (capital) (2018)

Sex ratio: *at birth:* 1.05 male(s)/female
0-14 years: 1.06 male(s)/female
15-24 years: 1.2 male(s)/female
25-54 years: 1.13 male(s)/female
55-64 years: 0.92 male(s)/female
65 years and over: 0.76 male(s)/female
total population: 1.05 male(s)/female (2020 est.)

Mother's mean age at first birth: 29.2 years (2017 est.)
note: data represent only government-controlled areas

Maternal mortality rate: 6 deaths/100,000 live births (2017 est.)
country comparison to the world: 160

Infant mortality rate: *total:* 7.4 deaths/1,000 live births
male: 8.6 deaths/1,000 live births
female: 6.1 deaths/1,000 live births (2020 est.)
country comparison to the world: 155

Life expectancy at birth: *total population:* 79.3 years
male: 76.4 years
female: 82.2 years (2020 est.)
country comparison to the world: 55

Total fertility rate: 1.48 children born/woman (2020 est.)
country comparison to the world: 206

Drinking water source:
improved:
urban: 100% of population
rural: 100% of population
total: 100% of population
unimproved:
urban: 0% of population
rural: 0% of population
total: 0% of population (2017 est.)

Current Health Expenditure: 6.7% (2017)

Physicians density: 1.95 physicians/1,000 population (2016)

Hospital bed density: 3.4 beds/1,000 population (2017)

Sanitation facility access:
improved:
urban: 100% of population
rural: 98.4% of population
total: 99% of population
unimproved:
urban: 0% of population
rural: 1.6% of population
total: 1% of population (2017 est.)

HIV/AIDS—adult prevalence rate: 0.1% (2017 est.)
country comparison to the world: 120

HIV/AIDS—people living with HIV/AIDS: <1000 (2017 est.)

HIV/AIDS—deaths: <100 (2017 est.)

Obesity—adult prevalence rate: 21.8% (2016)
country comparison to the world: 84

Education expenditures: 6.4% of GDP (2015)
country comparison to the world: 22

Literacy: *definition:* age 15 and over can read and write
total population: 99.1%
male: 99.5%
female: 98.7% (2015)

School life expectancy (primary to tertiary education): *total:* 15 years
male: 15 years
female: 15 years (2018)

Unemployment, youth ages 15-24: *total:* 20.2%
male: 25%
female: 16.2% (2018 est.)
country comparison to the world: 67

People—note: demographic data for Cyprus represent the population of the government-controlled area and the area administered by Turkish Cypriots, unless otherwise indicated

GOVERNMENT

Country name: *conventional long form:* Republic of Cyprus
conventional short form: Cyprus
local long form: Kypriaki Dimokratia/Kibris Cumhuriyeti
local short form: Kypros/Kibris
etymology: the derivation of the name "Cyprus" is unknown, but the extensive mining of copper metal on the island in antiquity gave rise to the Latin word "cuprum" for copper
note: the Turkish Cypriot community, which administers the northern part of the island, refers to itself as the "Turkish Republic of Northern Cyprus" or "TRNC" ("Kuzey Kibris Turk Cumhuriyeti" or "KKTC")

Government type: Republic of Cyprus—presidential republic; "Turkish Republic of Northern Cyprus" (self-declared)—parliamentary republic with enhanced presidency *note:* a separation of the two main ethnic communities inhabiting the island began following the outbreak of communal strife in 1963; this separation was further solidified when a Greek military-junta-supported coup attempt prompted the Turkish military intervention in July 1974 that gave the Turkish Cypriots de facto control in the north; Greek Cypriots control the only internationally recognized government on the island; on 15 November 1983, then Turkish Cypriot "President" Rauf DENKTAS declared independence and the formation of the "TRNC," which is recognized only by Turkey

Capital: *name:* Nicosia (Lefkosia/Lefkosa)

geographic coordinates: 35 10 N, 33 22 E

time difference: UTC+2 (7 hours ahead of Washington, DC, during Standard Time)

daylight saving time: +1hr, begins last Sunday in March; ends last Sunday in October

etymology: a mispronunciation of the city's Greek name Lefkosia and its Turkish name Lefkosa, both of which mean "White City"; the Greek name may derive from the Greek phrase "leuke ousia" ("white estate")

Administrative divisions: 6 districts; Ammochostos (Famagusta); (all but a small part located in the Turkish Cypriot community), Keryneia (Kyrenia; the only district located entirely in the Turkish Cypriot community), Larnaka (Larnaca; with a small part located in the Turkish Cypriot community), Lefkosia (Nicosia; a small part administered by Turkish Cypriots), Lemesos (Limassol), Pafos (Paphos); note—the 5 "districts" of the "TRNC" are Gazimagusa (Famagusta), Girne (Kyrenia), Guzelyurt (Morphou), Iskele (Trikomo), Lefkosa (Nicosia)

Independence: 16 August 1960 (from the UK); note—Turkish Cypriots proclaimed self-rule on 13 February 1975 and independence in 1983, but these proclamations are recognized only by Turkey

National holiday: Independence Day, 1 October (1960); note—Turkish Cypriots celebrate 15 November (1983) as "Republic Day"

Constitution: *history:* ratified 16 August 1960; note—in 1963, the constitution was partly suspended as Turkish Cypriots withdrew from the government; Turkish-held territory in 1983 was declared the "Turkish Republic of Northern Cyprus" ("TRNC"); in 1985, the "TRNC" approved its own constitution

amendments: constitution of the Republic of Cyprus—proposed by the House of Representatives; passage requires at least two-thirds majority vote of the total membership of the "Greek Community" and the "Turkish Community"; however, all seats of Turkish Cypriot members have remained vacant since 1964; amended 10 times, last in 2016 constitution of the "Turkish Republic of Northern Cyprus"—proposed by at least 10 members of the "Assembly of the Republic"; passage requires at least two-thirds majority vote of the total Assembly membership and approval by referendum; amended 2014

Legal system: mixed legal system of English common law and civil law with European law supremacy

International law organization participation: accepts compulsory ICJ jurisdiction with reservations; accepts ICC jurisdiction

Citizenship: *citizenship by birth:* no

citizenship by descent only: at least one parent must be a citizen of Cyprus

dual citizenship recognized: yes

residency requirement for naturalization: 7 years

Suffrage: 18 years of age; universal

Executive branch: *chief of state:* President Nikos ANASTASIADIS (since 28 February 2013); the president is both chief of state and head of government; note—vice presidency reserved for a Turkish Cypriot, but vacant since 1974 because Turkish Cypriots do not participate in the Republic of Cyprus Government

head of government: President Nikos ANASTASIADIS (since 28 February 2013)

cabinet: Council of Ministers appointed by the president; note—under the 1960 constitution, 3 of the ministerial posts reserved for Turkish Cypriots, appointed by the vice president; positions currently filled by Greek Cypriots

elections/appointments: president directly elected by absolute majority popular vote in 2 rounds if needed for a 5-year term; election last held on 28 January 2018 with a runoff on 4 February 2018 (next to be held in February 2023)

election results: Nikos ANASTASIADIS reelected president in second round; percent of vote in first round—Nikos ANASTASIADIS (DISY) 35.5%, Stavros MALAS (AKEL) 30.2%, Nicolas PAPADOPOULOS (DIKO) 25.7%, other 8.6%; percent of vote in second round—Nikos ANASTASIADIS 56%, Savros MALAS 44%

note: the first round of the TRNC presidential election, originally scheduled for 26 April 2020, was postponed to 11 October 202 due to the COVID-19 pandemic; results—Ersin TATAR (UBP) 32.4%, Mustafa AKINCI (independent) 29.8%, Tufan ERHURMAN (RTP) 21.7%, Kudret OZERSAY (independent) 5.7%, Erhan ARIKLI (YDP) 5.4%, Serdar DENKTAS (independent) 4.2%, other 0.8%; the second round to be held on 18 October

Legislative branch: *description:* area under government control: unicameral House of Representatives or Vouli

Antiprosopon (80 seats; 56 assigned to Greek Cypriots, 24 to Turkish Cypriots, but only those assigned to Greek Cypriots are filled; members directly elected by both proportional representation and preferential vote; members serve 5-year terms); area administered by Turkish Cypriots: unicameral "Assembly of the Republic" or Cumhuriyet Meclisi (50 seats; members directly elected to 5- year terms by proportional representation system using a hybrid d'Hondt method with voter preferences for individual candidates

elections: area under government control: last held on 22 May 2016 (next to be held in May

2021); area administered by Turkish Cypriots: last held on 7 January 2018 (next to be held in 2023, unless early election called)

election results: area under government control: House of Representatives—percent of vote by party—DISY 30.7%, AKEL 25.7%, DIKO 14.5%, KS-EDEK 6.2%, SP 6% Solidarity Movement 5.2%, other 11.7%; seats by party—DISY 18, AKEL 16, DIKO 9, KS-EDEK 3, Citizen's Alliance 3 (2 left the party in 2017 and 2018 due to disagreements over the party's policy regarding the presidential election campaign; one joined DIKO and the other became an independent MP), Solidarity Movement 3, other 4; area administered by Turkish Cypriots: "Assembly of the Republic"—percent of vote by party—UBP 35.6%, CTP 20.9%, HP 17.1%, TDP 8.6%, DP 7.8%, YDP 7%, 3%; seats by party—UBP 21, CTP 12, HP 9, DP 3, TDP 3, YDP 2

Judicial branch: *highest courts:* Supreme Court of Cyprus (consists of 13 judges, including the court president); note—the highest court in the "TRNC" is the "Supreme Court" (consists of 8 "judges," including the "court president")

judge selection and term of office: Republic of Cyprus Supreme Court judges appointed by the president of the republic upon the recommendation of the Supreme Court judges; judges can serve until age 68; "TRNC Supreme Court" judges appointed by the "Supreme Council of Judicature," a 12-member body of judges, the attorney general, appointees by the president of the "TRNC," and by the "Legislative Assembly," and members elected by the bar association; judge tenure NA

subordinate courts: Republic of Cyprus district courts; Assize Courts; Administrative Court; specialized courts for issues relating to family, industrial disputes, the military, and rent control; "TRNC Assize Courts"; "district and family courts"

Political parties and leaders: area under government control:

Citizens' Alliance or SP [Giorgos LILLIKAS]

Democratic Party or DIKO [Nicolas PAPADOPOULOS]

Democratic Rally or DISY [Averof NEOPHYTOU]

Movement of Ecologists and Environmentalists or KOP (Green party) [Giorgos PERDIKIS]

I, the Citizen or EOP [Georgios KOUNTOURIS]

Movement of Social Democrats EDEK [Marinos SIZOPOULOS]

National Popular Front or ELAM [Christos CHRISTOU]

Progressive Party of the Working People or AKEL (Communist party) [Andros KYPRIANOU]

Solidarity Movement [Eleni THEOCHAROUS]

United Democrats or EDI [Praxoula ANTONIADOU]

Democratic Front or DIPA [Marios GAROYIAN]

Animal Party Cyprus or APC [Kyriacos KYRIACOU]

area administered by Turkish Cypriots:

Communal Democracy Party or TDP [Cemal OZYIGIT]

Communal Liberation Party-New Forces or TKP-YG [Mehmet CAKICI]

Cyprus Socialist Party or KSP [Mehmet BIRINCI]

Democratic Party or DP [Serdar DENKTAS]
National Democratic Party or NDP [Buray BUSKUVUTCU]
National Unity Party or UBP [Ersin TATAR]
New Cyprus Party or YKP [Murat KANATLI]
People's Party or HP [Kudret OZERSAY]
Rebirth Party or YDP [Erhan ARIKLI]
Republican Turkish Party or CTP [Tufan ERHURMAN]
United Cyprus Party or BKP [Izzet IZCAN]

International organization participation: Australia Group, C, CD, CE, EBRD, ECB, EIB, EMU, EU, FAO, IAEA, IBRD, ICAO, ICC (national committees), ICCt, ICRM, IDA, IFAD, IFC, IFRCS, IHO, ILO, IMF, IMO, IMSO, Interpol, IOC, IOM, IPU, ISO, ITSO, ITU, ITUC (NGOs), MIGA, NAM, NSG, OAS (observer), OIF, OPCW, OSCE, PCA, UN, UNCTAD, UNESCO, UNHCR, UNIDO, UNIFIL, UNWTO, UPU, WCO, WFTU (NGOs), WHO, WIPO, WMO, WTO

Diplomatic representation in the US: *chief of mission:* Ambassador Marios LYSIOTIS (since 17 September 2018)
chancery: 2211 R Street NW, Washington, DC 20008
telephone: [1] (202) 462-5772, 462-0873
FAX: [1] (202) 483-6710
consulate(s) general: New York
note: representative of the Turkish Cypriot community in the US is Mustafa LAKADAMYALI; office at 1667 K Street NW, Washington, DC; telephone [1] (202) 887-6198

Diplomatic representation from the US: *chief of mission:* Ambassador Judith Gail GARBER (since 18 March 2019)
telephone: [357] (22) 393939
embassy: corner of Metochiou and Ploutarchou Streets, 2407 Engomi, Nicosia
mailing address: P. O. Box 24536, 1385 Nicosia
FAX: [357] (22) 393344

Flag description: centered on a white field is a copper-colored silhouette of the island (the island has long been famous for its copper deposits) above two olive-green-colored, crossed olive branches; the branches symbolize the hope for peace and reconciliation between the Greek and Turkish communities
note: one of only two national flags that uses a map as a design element; the flag of Kosovo is the other
note: the "Turkish Republic of Northern Cyprus" flag retains the white field of the Cyprus national flag but displays narrow horizontal red stripes positioned a small distance from the top and bottom edges between which are centered a red crescent and a red five-pointed star; the banner is modeled after the Turkish national flag but with the colors reversed

National symbol(s): Cypriot mouflon (wild sheep), white dove; national colors: blue, white

National anthem: *name:* "Ymnos eis tin Eleftherian" (Hymn to Liberty)
lyrics/music: Dionysios SOLOMOS/Nikolaos MANTZAROS

note: adopted 1960; Cyprus adopted the Greek national anthem as its own; the Turkish Cypriot community in Cyprus uses the anthem of Turkey

ECONOMY

Economy—overview: The area of the Republic of Cyprus under government control has a market economy dominated by a services sector that accounts for more than four-fifths of GDP. Tourism, finance, shipping, and real estate have traditionally been the most important services. Cyprus has been a member of the EU since May 2004 and adopted the euro as its national currency in January 2008.

During the first five years of EU membership, the Cyprus economy grew at an average rate of about 4%, with unemployment between 2004 and 2008 averaging about 4%. However, the economy tipped into recession in 2009 as the ongoing global financial crisis and resulting low demand hit the tourism and construction sectors. An overextended banking sector with excessive exposure to Greek debt added to the contraction. Cyprus' biggest two banks were among the largest holders of Greek bonds in Europe and had a substantial presence in Greece through bank branches and subsidiaries. Following numerous downgrades of its credit rating, Cyprus lost access to international capital markets in May 2011. In July 2012, Cyprus became the fifth euro-zone government to request an economic bailout program from the European Commission, European Central Bank and the International Monetary Fund—known collectively as the "Troika."

Shortly after the election of President Nikos ANASTASIADES in February 2013, Cyprus reached an agreement with the Troika on a $13 billion bailout that triggered a two-week bank closure and the imposition of capital controls that remained partially in place until April 2015. Cyprus' two largest banks merged and the combined entity was recapitalized through conversion of some large bank deposits to shares and imposition of losses on bank bondholders. As with other EU countries, the Troika conditioned the bailout on passing financial and structural reforms and privatizing state-owned enterprises. Despite downsizing and restructuring, the Cypriot financial sector remains burdened by the largest stock of non-performing loans in the euro zone, equal to nearly half of all loans. Since the bailout, Cyprus has received positive appraisals by the Troika and outperformed fiscal targets but has struggled to overcome political opposition to bailout-mandated legislation, particularly regarding privatizations. The rate of non-performing loans (NPLs) is still very high at around 49%, and growth would accelerate if Cypriot banks could increase the pace of resolution of the NPLs.

In October 2013, a US-Israeli consortium completed preliminary appraisals of hydrocarbon deposits in Cyprus' exclusive economic zone (EEZ), which estimated gross mean reserves of about 130 billion cubic meters. Though exploration continues in Cyprus' EEZ, no additional

commercially exploitable reserves have been identified. Developing offshore hydrocarbon resources remains a critical component of the government's economic recovery efforts, but development has been delayed as a result of regional developments and disagreements about exploitation methods.

GDP (purchasing power parity): $31.78 billion (2017 est.)
$30.59 billion (2016 est.)
$29.58 billion (2015 est.)
note: data are in 2017 dollars
country comparison to the world: 129

GDP (official exchange rate): $21.7 billion (2017 est.)

GDP—real growth rate: 3.08% (2019 est.)
5.25% (2018 est.)
5.16% (2017 est.)
country comparison to the world: 97

GDP—per capita (PPP): $37,200 (2017 est.)
$36,100 (2016 est.)
$34,900 (2015 est.)
note: data are in 2017 dollars
country comparison to the world: 53

Gross national saving: 13.7% of GDP (2017 est.)
11.9% of GDP (2016 est.)
12.8% of GDP (2015 est.)
country comparison to the world: 140

GDP—composition, by end use:
household consumption: 68.7% (2017 est.)
government consumption: 14.9% (2017 est.)
investment in fixed capital: 21.1% (2017 est.)
investment in inventories: -0.7% (2017 est.)
exports of goods and services: 63.8% (2017 est.)
imports of goods and services: -67.8% (2017 est.)

GDP—composition, by sector of origin:
agriculture: 2% (2017 est.)
industry: 12.5% (2017 est.)
services: 85.5% (2017 est.)

Agriculture—products: citrus, vegetables, barley, grapes, olives, vegetables; poultry, pork, lamb; dairy, cheese

Industries: tourism, food and beverage processing, cement and gypsum, ship repair and refurbishment, textiles, light chemicals, metal products, wood, paper, stone and clay products

Industrial production growth rate: 13.4% (2017 est.)
country comparison to the world: 5

Labor force: 416,000 (2019 est.)
country comparison to the world: 158

Labor force—by occupation: *agriculture:* 3.8%
industry: 15.2%
services: 81% (2014 est.)

Unemployment rate: 7.07% (2019 est.)
8.37% (2018 est.)
country comparison to the world: 114

Population below poverty line: NA

Household income or consumption by percentage share: *lowest 10%:* 3.3%
highest 10%: 28.8% (2014)

Budget: *revenues:* 8.663 billion (2017 est.)
expenditures: 8.275 billion (2017 est.)

Taxes and other revenues: 39.9% (of GDP) (2017 est.)

country comparison to the world: 40

Budget surplus (+) or deficit (-): 1.8% (of GDP) (2017 est.)

country comparison to the world: 17

Public debt: 97.5% of GDP (2017 est.)
106.6% of GDP (2016 est.)

note: data cover general government debt and include debt instruments issued (or owned) by government entities other than the treasury; the data include treasury debt held by foreign entities; the data exclude debt issued by subnational entities, as well as intragovernmental debt; intragovernmental debt consists of treasury borrowings from surpluses in the social funds, such as for retirement, medical care, and unemployment
country comparison to the world: 19

Fiscal year: calendar year

Inflation rate (consumer prices): 0.7% (2017 est.)
-1.2% (2016 est.)
country comparison to the world: 35

Current account balance: -$1.578 billion (2019 est.)
-$958 million (2018 est.)
country comparison to the world: 161

Exports: $2.805 billion (2017 est.)
$2.7 billion (2016 est.)
country comparison to the world: 130

Exports—partners: Libya 9.4%, Greece 7.7%, Norway 6.7%, UK 5.3%, Germany 4.1% (2017)

Exports—commodities: citrus, potatoes, pharmaceuticals, cement, clothing

Imports: $7.935 billion (2017 est.)
$7.153 billion (2016 est.)
country comparison to the world: 110

Imports—commodities: consumer goods, petroleum and lubricants, machinery, transport equipment

Imports—partners: Greece 19%, Italy 7.5%, China 7.4%, South Korea 7.3%, Germany 7%, Netherlands 5.1%, UK 5%, Israel 4.1% (2017)

Reserves of foreign exchange and gold: $888.2 million (31 December 2017 est.)
$817.7 million (31 December 2016 est.)
country comparison to the world: 136

Debt—external: $95.28 billion (31 December 2013 est.)
$103.5 billion (31 December 2012 est.)
country comparison to the world: 50

Exchange rates: euros (EUR) per US dollar—
0.885 (2017 est.)
0.903 (2016 est.)
0.9214 (2015 est.)
0.885 (2014 est.)
0.7634 (2013 est.)

Economy of the area administered by Turkish Cypriots: Economy—overview: Even though the whole of the island is part of the EU, implementation of the EU "acquis communautaire" has been suspended in the area administered by Turkish Cypriots, known locally as the "Turkish Republic

of Northern Cyprus" ("TRNC"), until political conditions permit the reunification of the island. The market-based economy of the "TRNC" is roughly one-fifth the size of its southern neighbor and is likewise dominated by the service sector with a large portion of the population employed by the government. In 2012—the latest year for which data are available—the services sector, which includes the public sector, trade, tourism, and education, contributed 58.7% to economic output. In the same year, light manufacturing and agriculture contributed 2.7% and 6.2%, respectively. Manufacturing is limited mainly to food and beverages, furniture and fixtures, construction materials, metal and non-metal products, textiles and clothing. The "TRNC" maintains few economic ties with the Republic of Cyprus outside of trade in construction materials. Since its creation, the "TRNC" has heavily relied on financial assistance from Turkey, which supports the "TRNC" defense, telecommunications, water and postal services. The Turkish Lira is the preferred currency, though foreign currencies are widely accepted in business transactions. The "TRNC" remains vulnerable to the Turkish market and monetary policy because of its use of the Turkish Lira. The "TRNC" weathered the European financial crisis relatively unscathed—compared to the Republic of Cyprus—because of the lack of financial sector development, the health of the Turkish economy, and its separation from the rest of the island. The "TRNC" economy experienced growth estimated at 2.8% in 2013 and 2.3% in 2014 and is projected to grow 3.8% in 2015.;

GDP (purchasing power parity): $1.829 billion (2007 est.);

GDP—real growth rate: 2.3% (2014 est.);
2.8% (2013 est.);

GDP—per capita: $11,700 (2007 est.);

GDP—composition by sector: *agriculture:* 6.2%,; industry: 35.1%,; services: 58.7% (2012 est.);

Labor force: 95,030 (2007 est.);

Labor force—by occupation: *agriculture:* 14.5%,; industry: 29%,; services: 56.5% (2004);

Unemployment rate: 9.4% (2005 est.);

Population below poverty line: %NA;

Inflation rate: 11.4% (2006);

Budget: *revenues:* $2.5 billion,; expenditures: $2.5 billion (2006);

Agriculture—products: citrus fruit, dairy, potatoes, grapes, olives, poultry, lamb;

Industries: foodstuffs, textiles, clothing, ship repair, clay, gypsum, copper, furniture;

Industrial production growth rate: -0.3% (2007 est.);

Electricity production: 998.9 million kWh (2005);

Electricity consumption: 797.9 million kWh (2005);

Exports: $68.1 million, f.o.b. (2007 est.);

Export—commodities: citrus, dairy, potatoes, textiles;

Export—partners: Turkey 40%; direct trade between the area administered by Turkish Cypriots and the area under government control remains limited;

Imports: $1.2 billion, f.o.b. (2007 est.);

Import—commodities: vehicles, fuel, cigarettes, food, minerals, chemicals, machinery;

Import—partners: Turkey 60%; direct trade between the area administered by Turkish Cypriots and the area under government control remains limited;

Reserves of foreign exchange and gold: NA;

Debt—external: NA;

Currency (code): Turkish new lira (YTL);

Exchange rates: Turkish new lira per US dollar:; 1.9 (2013); 1.8 (2012); 1.668 (2011); 1.5026 (2010); 1.55 (2009);

ENERGY

Electricity access: *electrification—total population:* 100% (2020)

Electricity—production: 4.618 billion kWh (2016 est.)
country comparison to the world: 123

Electricity—consumption: 4.355 billion kWh (2016 est.)
country comparison to the world: 126

Electricity—exports: 0 kWh (2016 est.)
country comparison to the world: 126

Electricity—imports: 0 kWh (2016 est.)
country comparison to the world: 140

Electricity—installed generating capacity: 1.77 million kW (2016 est.)
country comparison to the world: 117

Electricity—from fossil fuels: 85% of total installed capacity (2016 est.)
country comparison to the world: 70

Electricity—from nuclear fuels: 0% of total installed capacity (2017 est.)
country comparison to the world: 76

Electricity—from hydroelectric plants: 0% of total installed capacity (2017 est.)
country comparison to the world: 166

Electricity—from other renewable sources: 15% of total installed capacity (2017 est.)
country comparison to the world: 58

Crude oil—production: 0 bbl/day (2018 est.)
country comparison to the world: 126

Crude oil—exports: 0 bbl/day (2015 est.)
country comparison to the world: 113

Crude oil—imports: 0 bbl/day (2015 est.)
country comparison to the world: 116

Crude oil—proved reserves: 0 bbl (1 January 2018 est.)
country comparison to the world: 122

Refined petroleum products—production: 0 bbl/day (2015 est.)
country comparison to the world: 135

Refined petroleum products—consumption: 49,000 bbl/day (2016 est.)
country comparison to the world: 106

Refined petroleum products—exports: 500 bbl/day (2015 est.)
country comparison to the world: 110

Refined petroleum products—imports: 49,240 bbl/day (2015 est.)
country comparison to the world: 84

Natural gas—production: 0 cu m (2017 est.)
country comparison to the world: 122

Natural gas—consumption: 0 cu m (2017 est.)
country comparison to the world: 138

Natural gas—exports: 0 cu m (2017 est.)
country comparison to the world: 92

Natural gas—imports: 0 cu m (2017 est.)
country comparison to the world: 115

Natural gas—proved reserves: 141.6 billion cu m (1 January 2014 est.)
country comparison to the world: 47

Carbon dioxide emissions from consumption of energy: 7.72 million Mt (2017 est.)
country comparison to the world: 118

COMMUNICATIONS

Telephones—fixed lines: *total subscriptions:* 469,305
subscriptions per 100 inhabitants: 37.48 (2019 est.)
country comparison to the world: 96

Telephones—mobile cellular: *total subscriptions:* 1,801,213
subscriptions per 100 inhabitants: 143.85 (2019 est.)
country comparison to the world: 155

Telecommunication systems: *general assessment:* broadband market steadily developing with one of the highest penetrations rates in the region; despite the growth of Cyprus's telecom sector, the market overall continues to be dominated by the incumbent, Cyprus Telecommunications Authority (CyTA), which is still fully-owned by the state, but it is losing ground to its competition annually; improved regulatory circumstances, especially in relation to network interconnection and access, has given competing operators the certainty to invest in network infrastructure, and to launch competing services; fiber infrastructure in the early days and DSL remains the dominate access platform (2020)
domestic: fixed-line is 37 per 100, and 144 per 100 for mobile-cellular; open-wire, fiber-optic cable, and microwave radio relay (2019)
international: country code—357 (area administered by Turkish Cypriots uses the country code of Turkey—90); a number of submarine cables, including the SEA-ME-WE-3, CADMOS, MedNautilus Submarine System, POSEIDON, TE North/TGN-Eurasia/SEACOM/Alexandros/Medes, UGARIT, Aphrodite2, Hawk, Lev Submarine System, and Tamares combine to provide connectivity to Europe, the Middle East, Africa, Asia, Australia, and Southeast Asia; Turcyos-1 and Turcyos-2 submarine cable in Turkish North Cyprus link to Turkey; tropospheric scatter; satellite earth stations—8 (3 Intelsat—1 Atlantic Ocean and 2 Indian Ocean, 2 Eutelsat, 2 Intersputnik, and 1 Arabsat) (2019)
note: the COVID-19 outbreak is negatively impacting telecommunications production and supply chains globally; consumer spending on telecom devices and services has also slowed due to the pandemic's effect on economies worldwide; overall progress towards improvements in all facets of the telecom industry—mobile, fixed-line, broadband, submarine cable and satellite—has moderated

Broadcast media: mixture of state and privately run TV and radio services; the public broadcaster operates 2 TV channels and 4 radio stations; 6 private TV broadcasters, satellite and cable TV services including telecasts from Greece and Turkey, and a number of private radio stations are available; in areas administered by Turkish Cypriots, there are 2 public TV stations, 4 public radio stations, and 7 privately owned TV and 21 radio broadcast stations plus 6 radio and 4 TV channels of local universities, plus 1 radio station of military, security forces and 1 radio station of civil defense cooperation, as well as relay stations from Turkey (2019)

Internet country code: .cy

Internet users: *total:* 1,044,473
percent of population: 84.43% (July 2018 est.)
country comparison to the world: 139

Broadband—fixed subscriptions: *total:* 313,462
subscriptions per 100 inhabitants: 25 (2018 est.)
country comparison to the world: 101

TRANSPORTATION

National air transport system: *number of registered air carriers:* 2 (2020)
inventory of registered aircraft operated by air carriers: 6
annual passenger traffic on registered air carriers: 401,408 (2018)
annual freight traffic on registered air carriers: 20,000 mt-km (2018)

Civil aircraft registration country code prefix: 5B (2016)

Airports: 15 (2013)
country comparison to the world: 145

Airports—with paved runways: *total:* 13 (2017)
2,438 to 3,047 m: 7 (2017)
1,524 to 2,437 m: 2 (2017)
914 to 1,523 m: 3 (2017)
under 914 m: 1 (2017)

Airports—with unpaved runways: *total:* 2 (2013)
under 914 m: 2 (2013)

Heliports: 9 (2013)

Pipelines: 0 km oil

Roadways: *total:* 19,901 km (2016)
government control: 12,901 km (includes 272 km of expressways) (2016)
paved: 8,631 km (2016)
unpaved: 4,270 km (2016)
Turkish Cypriot control: 7,000 km (2011)

country comparison to the world: 114

Merchant marine: *total:* 1,039
by type: bulk carrier 304, container ship 190, general cargo 183, oil tanker 42, other 320 (2019)
country comparison to the world: 25

Ports and terminals: *major seaport(s):* area under government control: Larnaca, Limassol, Vasilikos area administered by Turkish Cypriots: Famagusta, Kyrenia

MILITARY AND SECURITY

Military and security forces: *Republic of Cyprus:* Cypriot National Guard (Ethniki Froura, EF, includes Army Land Forces, Naval Command, Air Command) (2020)

Military expenditures: 1.6% of GDP (2019)
1.8% of GDP (2018)
1.6% of GDP (2017)
1.4% of GDP (2016)
1.7% of GDP (2015)
country comparison to the world: 70

Military and security service personnel strengths: the Cypriot National Guard has approximately 13-15,000 total active duty personnel (2019 est.)

Military equipment inventories and acquisitions: the inventory of the Cypriot National Guard is a mix of Soviet-era and some more modern weapons systems; since 2010, it has received equipment from France, Israel, Italy, Oman, and Russia (2019 est.)

Military service age and obligation: Cypriot National Guard (CNG): 18-50 years of age for compulsory military service for all Greek Cypriot males; 17 years of age for voluntary service; 12-month service obligation (2019)

Military—note: the United Nations Peacekeeping Force in Cyprus (UNICYP) was set up in 1964 to prevent further fighting between the Greek Cypriot and Turkish Cypriot communities on the island and bring about a return to normal conditions; as of March 2020, the UNICYP mission consisted of about 830 personnel (2020)

TRANSNATIONAL ISSUES

Disputes—international: hostilities in 1974 divided the island into two de facto autonomous entities, the internationally recognized Cypriot Government and a Turkish-Cypriot community (north Cyprus); the 1,000-strong UN Peacekeeping Force in Cyprus (UNFICYP) has served in Cyprus since 1964 and maintains the buffer zone between north and south; on 1 May 2004, Cyprus entered the EU still divided, with the EU's body of legislation and standards (acquis communitaire) suspended in the north; Turkey protests Cypriot Government creating hydrocarbon blocks and maritime boundary with Lebanon in March 2007

Refugees and internally displaced persons: *refugees (country of origin):* 7,372 (Syria) (2019)

259

IDPs: 228,000 (both Turkish and Greek Cypriots; many displaced since 1974) (2019)
note: 10,690 estimated refugee and migrant arrivals (January 2015-November 2019)

Illicit drugs: minor transit point for heroin and hashish via air routes and container traffic to Europe, especially from Lebanon and Turkey; some cocaine transits as well; despite a strengthening of anti-money-laundering legislation, remains vulnerable to money laundering; reporting of suspicious transactions in offshore sector remains weak

CZECHIA

INTRODUCTION

Background: At the close of World War I, the Czechs and Slovaks of the former Austro-Hungarian Empire merged to form Czechoslovakia. During the interwar years, having rejected a federal system, the new country's predominantly Czech leaders were frequently preoccupied with meeting the increasingly strident demands of other ethnic minorities within the republic, most notably the Slovaks, the Sudeten Germans, and the Ruthenians (Ukrainians). On the eve of World War II, Nazi Germany occupied the territory that today comprises Czechia, and Slovakia became an independent state allied with Germany. After the war, a reunited but truncated Czechoslovakia (less Ruthenia) fell within the Soviet sphere of influence. In 1968, an invasion by Warsaw Pact troops ended the efforts of the country's leaders to liberalize communist rule and create "socialism with a human face," ushering in a period of repression known as "normalization." The peaceful "Velvet Revolution" swept the Communist Party from power at the end of 1989 and inaugurated a return to democratic rule and a market economy. On 1 January 1993, the country underwent a nonviolent "velvet divorce" into its two national components, the Czech Republic and Slovakia. The Czech Republic joined NATO in 1999 and the European Union in 2004. The country added the short-form name Czechia in 2016, while continuing to use the full form name, Czech Republic.

GEOGRAPHY

Location: Central Europe, between Germany, Poland, Slovakia, and Austria

Geographic coordinates: 49 45 N, 15 30 E

Map references: Europe

Area: *total:* 78,867 sq km
land: 77,247 sq km
water: 1,620 sq km
country comparison to the world: 117

Area—comparative: about two-thirds the size of Pennsylvania; slightly smaller than South Carolina

Land boundaries: *total:* 2,143 km
border countries (4): Austria 402 km, Germany 704 km, Poland 796 km, Slovakia 241 km

Coastline: 0 km (landlocked)

Maritime claims: none (landlocked)

Climate: temperate; cool summers; cold, cloudy, humid winters

Terrain: Bohemia in the west consists of rolling plains, hills, and plateaus surrounded by low mountains; Moravia in the east consists of very hilly country

Elevation: *mean elevation:* 433 m
lowest point: Labe (Elbe) River 115 m
highest point: Snezka 1,602 m

Natural resources: hard coal, soft coal, kaolin, clay, graphite, timber, arable land

Land use: *agricultural land:* 54.8% (2011 est.)
arable land: 41% (2011 est.) / *permanent crops:* 1% (2011 est.) / *permanent pasture:* 12.8% (2011 est.)
forest: 34.4% (2011 est.)
other: 10.8% (2011 est.)
Irrigated land: 320 sq km (2012)

Population distribution: a fairly even distribution throughout most of the country, but the northern and eastern regions tend to have larger urban concentrations

Natural hazards: flooding

Environment—current issues: air and water pollution in areas of northwest Bohemia and in northern Moravia around Ostrava present health risks; acid rain damaging forests; land pollution caused by industry, mining, and agriculture

Environment—international agreements: *party to:* Air Pollution, Air Pollution-Nitrogen Oxides, Air Pollution-Persistent Organic Pollutants, Air Pollution-Sulfur 85, Air Pollution-Sulfur 94, Air Pollution-Volatile Organic Compounds, Antarctic-Environmental Protocol, Antarctic Treaty, Biodiversity, Climate Change, Climate Change-Kyoto Protocol, Desertification, Endangered Species, Environmental Modification, Hazardous Wastes, Law of the Sea, Ozone Layer Protection, Ship Pollution, Wetlands, Whaling

signed, but not ratified: none of the selected agreements

Geography—note: *note 1:* landlocked; strategically located astride some of oldest and most significant land routes in Europe; Moravian Gate is a traditional military corridor between the North European Plain and the Danube in central Europe
note 2: the Hranice Abyss in Czechia is the world's deepest surveyed underwater cave at 404 m (1,325 ft); its survey is not complete and it could end up being some 800-1,200 m deep

PEOPLE AND SOCIETY

Population: 10,702,498 (July 2020 est.)
country comparison to the world: 85

Nationality: *noun:* Czech(s)
adjective: Czech

Ethnic groups: Czech 64.3%, Moravian 5%, Slovak 1.4%, other 1.8%, unspecified 27.5% (2011 est.)

Languages: Czech (official) 95.4%, Slovak 1.6%, other 3% (2011 census)

Religions: Roman Catholic 10.4%, Protestant (includes Czech Brethren and Hussite) 1.1%, other and unspecified 54%, none 34.5% (2011 est.)

Age structure: *0-14 years:* 15.17% (male 834,447/female 789,328)
15-24 years: 9.2% (male 508,329/female 475,846)
25-54 years: 43.29% (male 2,382,899/female 2,249,774)
55-64 years: 12.12% (male 636,357/female 660,748)
65 years and over: 20.23% (male 907,255/female 1,257,51 5) (2020 est.)

Dependency ratios: *total dependency ratio:* 56
youth dependency ratio: 24.6
elderly dependency ratio: 31.4
potential support ratio: 3.2 (2020 est.)

Median age: *total:* 43.3 years
male: 42 years
female: 44.7 years (2020 est.)
country comparison to the world: 28

Population growth rate: 0.06% (2020 est.)
country comparison to the world: 188

Birth rate: 8.9 births/1,000 population (2020 est.)
country comparison to the world: 205

Death rate: 10.7 deaths/1,000 population (2020 est.)
country comparison to the world: 26

Net migration rate: 2.3 migrant(s)/1,000 population (2020 est.)
country comparison to the world: 43

Population distribution: a fairly even distribution throughout most of the country, but the northern and eastern regions tend to have larger urban concentrations

Urbanization: *urban population:* 74.1% of total population (2020)
rate of urbanization: 0.21% annual rate of change (2015-20 est.)

total population growth rate v. urban population growth rate, 2000-2030:

Major urban areas—population: 1.306 million PRAGUE (capital) (2020)

Sex ratio: *at birth:* 1.05 male(s)/female
0-14 years: 1.06 male(s)/female
15-24 years: 1.07 male(s)/female
25-54 years: 1.06 male(s)/female
55-64 years: 0.96 male(s)/female
65 years and over: 0.72 male(s)/female
total population: 0.97 male(s)/female (2020 est.)

Mother's mean age at first birth: 28.4 years (2018 est.)

Maternal mortality rate: 3 deaths/100,000 live births (2017 est.)
country comparison to the world: 176

Infant mortality rate: *total:* 2.6 deaths/1,000 live births
male: 2.8 deaths/1,000 live births
female: 2.5 deaths/1,000 live births (2020 est.)
country comparison to the world: 218

Life expectancy at birth: *total population:* 79.3 years
male: 76.3 years
female: 82.4 years (2020 est.)
country comparison to the world: 56

Total fertility rate: 1.48 children born/woman (2020 est.)
country comparison to the world: 207

Drinking water source:
improved:
urban: 100% of population
rural: 100% of population
total: 100% of population
unimproved:
urban: 0% of population
rural: 0% of population
total: 0% of population (2017 est.)

Current Health Expenditure: 7.2% (2017)

Physicians density: 4.07 physicians/1,000 population (2017)

Hospital bed density: 6.6 beds/1,000 population (2017)

Sanitation facility access:
improved:
urban: 100% of population
rural: 100% of population
total: 100% of population
unimproved:
urban: 0% of population
rural: 0% of population
total: 0% of population (2017 est.)

HIV/AIDS—adult prevalence rate: <.1% (2018 est.)

HIV/AIDS—people living with HIV/AIDS: 4,400 (2018 est.)
country comparison to the world: 124

HIV/AIDS—deaths: <100 (2018 est.)

Obesity—adult prevalence rate: 26% (2016)
country comparison to the world: 46

Education expenditures: 5.6% of GDP (2016)
country comparison to the world: 36

Literacy: *definition:* NA
total population: 99%
male: 99%
female: 99% (2011)

School life expectancy (primary to tertiary education): *total:* 16 years
male: 16 years
female: 17 years (2018)

Unemployment, youth ages 15-24: *total:* 6.7%
male: 6.4%
female: 7.2% (2018 est.)
country comparison to the world: 155

GOVERNMENT

Country name: *conventional long form:* Czech Republic
conventional short form: Czechia
local long form: Ceska republika
local short form: Cesko
etymology: name derives from the Czechs, a West Slavic tribe who rose to prominence in the late 9th century A.D.; the country officially adopted the English short-form name of Czechia on 1 July 2016

Government type: parliamentary republic

Capital: *name:* Prague
geographic coordinates: 50 05 N, 14 28 E
time difference: UTC + 1 (6 hours ahead of Washington, DC, during Standard Time)
daylight saving time: +1hr, begins last Sunday in March; ends last Sunday in October
etymology: the name may derive from an old Slavic root "praga" or "prah", meaning "ford", and refer to the city's origin at a crossing point of the Vltava (Moldau) River

Administrative divisions: 13 regions (kraje, singular—kraj) and 1 capital city* (hlavni mesto); Jihocesky (South Bohemia), Jihomoravsky (South Moravia), Karlovarsky (Karlovy Vary), Kralovehradecky (Hradec Kralove), Liberecky (Liberec), Moravskoslezsky (Moravia- Silesia), Olomoucky (Olomouc), Pardubicky (Pardubice), Plzensky (Pilsen), Praha (Prague)*, Stredocesky (Central Bohemia), Ustecky (Usti), Vysocina (Highlands), Zlinsky (Zlin)

Independence: 1 January 1993 (Czechoslovakia split into the Czech Republic and Slovakia); note—although 1 January is the day the Czech Republic came into being, the Czechs commemorate 28 October 1918, the day the former Czechoslovakia declared its independence from the Austro-Hungarian Empire, as their independence day

National holiday: Czechoslovak Founding Day, 28 October (1918)

Constitution: *history:* previous 1960; latest ratified 16 December 1992, effective 1 January 1993
amendments: passage requires at least three-fifths concurrence of members present in both houses of Parliament; amended several times, last in 2013

Legal system: new civil code enacted in 2014, replacing civil code of 1964—based on former Austro-Hungarian civil codes and socialist theory—and reintroducing former Czech legal terminology

International law organization participation: has not submitted an ICJ jurisdiction declaration; accepts ICCt jurisdiction

Citizenship: *citizenship by birth:* no
citizenship by descent only: at least one parent must be a citizen of Czechia
dual citizenship recognized: no
residency requirement for naturalization: 5 years

Suffrage: 18 years of age; universal

Executive branch: *chief of state:* President Milos ZEMAN (since 8 March 2013)
head of government: Prime Minister Andrej BABIS (since 13 December 2017); First Deputy Prime Minister Jan HAMACEK (since 27 June 2018), Deputy Prime Minister Alena SCHILLEROVA (since 30 April 2019)
cabinet: Cabinet appointed by the president on the recommendation of the prime minister
elections/appointments: president directly elected by absolute majority popular vote in 2 rounds if needed for a 5-year term (limited to 2 consecutive terms); elections last held on 12-13 January 2018 with a runoff on 26-27 January 2018 (next to be held in January 2023); prime minister appointed by the president for a 4-year term
election results: Milos ZEMAN reelected president in the second round; percent of vote—Milos ZEMAN (SPO) 51.4%, Jiri DRAHOS (independent) 48.6%

Legislative branch: *description:* bicameral Parliament or Parlament consists of:
Senate or Senat (81 seats; members directly elected in single-seat constituencies by absolute majority vote in 2 rounds if needed; members serve 6-year terms with one-third of the membership renewed every 2 years)
Chamber of Deputies or Poslanecka Snemovna (200 seats; members directly elected in 14 multiseat constituencies by proportional representation vote with a 5% threshold required to fill a seat; members serve 4-year terms)
elections: Senate—last held in 2 rounds on 2-3 and 9-10 October 2020 (next to be held in October 2022)
Chamber of Deputies—last held on 20-21 October 2017 (next to be held by October 2021)
election results: Senate—percent of vote by party—NA; seats by party—STAN 19, ODS 18, KDU-CSL 12, ANO 5, TOP 09 5, CSSD 3, SEN 21 3, Pirates 2, SZ 1 , minor parties with one seat each 9, independents 4
Chamber of Deputies—percent of vote by party—ANO 29.6%, ODS 11.3%, Pirates 10.8%,

SPD 10.6%, KSCM 7.8%, CSSD 7.3%, KDU-CSL 5.8%, TOP 09 5.3%, STAN 5.2%, other 6.3%; seats by party—ANO 78, ODS 25, Pirates 22, SPD 22, CSSD 15, KSCM 15, KDU-CSL 10, TOP 09 7, STAN 6; composition—men 155, women 45, percent of women 24%; note—total Parliament percent of women 20.6%

Judicial branch: *highest courts:* Supreme Court (organized into Civil Law and Commercial Division, and Criminal Division each with a court chief justice, vice justice, and several judges); Constitutional Court (consists of 15 justices); Supreme Administrative Court (consists of 36 judges, including the court president and vice president, and organized into 6-, 7-, and 9-member chambers)
judge selection and term of office: Supreme Court judges proposed by the Chamber of Deputies and appointed by the president; judges appointed for life; Constitutional Court judges appointed by the president and confirmed by the Senate; judges appointed for 10-year, renewable terms; Supreme Administrative Court judges selected by the president of the Court; unlimited terms
subordinate courts: High Court; regional and district courts

Political parties and leaders: Christian Democratic Union-Czechoslovak People's Party or KDU-CSL [Pavel BELOBRADEK]
Civic Democratic Party or ODS [Petr FIALA]
Communist Party of Bohemia and Moravia or KSCM [Vojtech FILIP]
Czech Social Democratic Party or CSSD [Jan HAMACEK]
Freedom and Direct Democracy or SPD [Tomio OKAMURA]
Green Party or SZ [Petr STEPANEK]
Mayors and Independents or STAN [Petr GAZDIK]
Movement of Dissatisfied Citizens or ANO [Andrej BABIS]
Party of Civic Rights or SPO [Lubomir NECAS]
Pirate Party or Pirates [Ivan BARTOS]
Tradition Responsibility Prosperity 09 or TOP 09 [Jiri POSPISIL]

International organization participation: Australia Group, BIS, BSEC (observer), CD, CE, CEI, CERN, EAPC, EBRD, ECB, EIB, ESA, EU, FAO, IAEA, IBRD, ICAO, ICC (national committees), ICCt, ICRM, IDA, IEA, IFC, IFRCS, ILO, IMF, IMO, IMSO, Interpol, IOC, IOM, IPU, ISO, ITSO, ITU, ITUC (NGOs), MIGA, MONUSCO, NATO, NEA, NSG, OAS (observer), OECD, OIF (observer), OPCW, OSCE, PCA, Schengen Convention, SELEC, UN, UNCTAD, UNESCO, UNHCR, UNIDO, UNWTO, UPU, WCO, WFTU (NGOs), WHO, WIPO, WMO, WTO, ZC

Diplomatic representation in the US: *chief of mission:* Ambassador Hynek KMONICEK (since 24 April 2017)
chancery: 3900 Spring of Freedom Street NW, Washington, DC 20008
telephone: [1] (202) 274-9100
FAX: [1] (202) 966-8540

consulate(s) general: Chicago, Los Angeles, New York

Diplomatic representation from the US: *chief of mission:* Ambassador Stephen B. KING (since 6 December 2017)
telephone: [420] 257 022 000
embassy: Trziste 15, 118 01 Prague 1—Mala Strana
mailing address: use embassy street address
FAX: [420] 257 022 809

Flag description: two equal horizontal bands of white (top) and red with a blue isosceles triangle based on the hoist side
note: combines the white and red colors of Bohemia with blue from the arms of Moravia; is identical to the flag of the former Czechoslovakia

National symbol(s): *silver (or white), double-tailed, rampant lion; national colors:* white, red, blue

National anthem: *name:* "Kde domov muj?" (Where is My Home?)
lyrics/music: Josef Kajetan TYL/Frantisek Jan SKROUP
note: adopted 1993; the anthem was originally written as incidental music to the play "Fidlovacka" (1834), it soon became very popular as an unofficial anthem of the Czech nation; its first verse served as the official Czechoslovak anthem beginning in 1918, while the second verse (Slovak) was dropped after the split of Czechoslovakia in 1993 0:00/ 0:00

ECONOMY

Economy—overview: Czechia is a prosperous market economy that boasts one of the highest GDP growth rates and lowest unemployment levels in the EU, but its dependence on exports makes economic growth vulnerable to contractions in external demand. Czechia's exports comprise some 80% of GDP and largely consist of automobiles, the country's single largest industry. Czechia acceded to the EU in 2004 but has yet to join the eurozone. While the flexible koruna helps Czechia weather external shocks, it was one of the world's strongest performing currencies in 2017, appreciating approximately 16% relative to the US dollar after the central bank (Czech National Bank—CNB) ended its cap on the currency's value in early April 2017, which it had maintained since November 2013. The CNB hiked rates in August and November 2017—the first rate changes in nine years—to address rising inflationary pressures brought by strong economic growth and a tight labor market.

Since coming to power in 2014, the new government has undertaken some reforms to try to reduce corruption, attract investment, and improve social welfare programs, which could help increase the government's revenues and improve living conditions for Czechs. The government introduced in December 2016 an online tax reporting system intended to reduce tax evasion and increase revenues. The government also plans to remove labor market rigidities to improve the

business climate, bring procurement procedures in line with EU best practices, and boost wages. The country's low unemployment rate has led to steady increases in salaries, and the government is facing pressure from businesses to allow greater migration of qualified workers, at least from Ukraine and neighboring Central European countries.

Long-term challenges include dealing with a rapidly aging population, a shortage of skilled workers, a lagging education system, funding an unsustainable pension and health care system, and diversifying away from manufacturing and toward a more high-tech, services-based, knowledge economy.

GDP (purchasing power parity): $375.9 billion (2017 est.)
$360.5 billion (2016 est.)
$351.9 billion (2015 est.)
note: data are in 2017 dollars
country comparison to the world: 49

GDP (official exchange rate): $215.8 billion (2017 est.)

GDP—real growth rate: 2.27% (2019 est.)
3.18% (2018 est.)
5.35% (2017 est.)
country comparison to the world: 124

GDP—per capita (PPP): $35,500 (2017 est.)
$34,200 (2016 est.)
$33,400 (2015 est.)
note: data are in 2017 dollars
country comparison to the world: 57

Gross national saving: 26.9% of GDP (2017 est.)
27.5% of GDP (2016 est.)
28.2% of GDP (2015 est.)
country comparison to the world: 45

GDP—composition, by end use: *household consumption:* 47.4% (2017 est.)
government consumption: 19.2% (2017 est.)
investment in fixed capital: 24.7% (2017 est.)
investment in inventories: 1.1% (2017 est.)
exports of goods and services: 79.9% (2017 est.)
imports of goods and services: -72.3% (2017 est.)

GDP—composition, by sector of origin:
agriculture: 2.3% (2017 est.)
industry: 36.9% (2017 est.)
services: 60.8% (2017 est.)

Agriculture—products: wheat, potatoes, sugar beets, hops, fruit; pigs, poultry

Industries: motor vehicles, metallurgy, machinery and equipment, glass, armaments

Industrial production growth rate: 7.5% (2017 est.)
country comparison to the world: 27

Labor force: 5.222 million (2020 est.)
country comparison to the world: 75

Labor force—by occupation: *agriculture:* 2.8%
industry: 38%
services: 59.2% (2015)

Unemployment rate: 2.8% (2019 est.)
3.18% (2018 est.)
country comparison to the world: 32

Population below poverty line: 9.7% (2015 est.)

Household income or consumption by percentage share: *lowest 10%:* 4.1%

highest 10%: 21.7% (2015 est.)

Budget: *revenues:* 87.37 billion (2017 est.)
expenditures: 83.92 billion (2017 est.)

Taxes and other revenues: 40.5% (of GDP) (2017 est.)
country comparison to the world: 37

Budget surplus (+) or deficit (-): 1.6% (of GDP) (2017 est.)
country comparison to the world: 19

Public debt: 34.7% of GDP (2017 est.)
36.8% of GDP (2016 est.)
country comparison to the world: 153

Fiscal year: calendar year

Inflation rate (consumer prices): 2.4% (2017 est.)
0.7% (2016 est.)
country comparison to the world: 118

Current account balance: -$678 million (2019 est.)
$1.259 billion (2018 est.)
country comparison to the world: 130

Exports: $144.8 billion (2017 est.)
$131.1 billion (2016 est.)
country comparison to the world: 33

Exports—partners: Germany 32.8%, Slovakia 7.8%, Poland 6.1%, France 5.1%, UK 4.9%, Austria 4.4%, Italy 4.1% (2017)

Exports—commodities: machinery and transport equipment, raw materials, fuel, chemicals

Imports: $134.7 billion (2017 est.)
$120.5 billion (2016 est.)
country comparison to the world: 32

Imports—commodities: machinery and transport equipment, raw materials and fuels, chemicals

Imports—partners: Germany 29.8%, Poland 9.1%, China 7.4%, Slovakia 5.8%, Netherlands 5.3%, Italy 4% (2017)

Reserves of foreign exchange and gold: $148 billion (31 December 2017 est.)
$85.73 billion (31 December 2016 est.)
country comparison to the world: 18

Debt—external: $205.2 billion (31 December 2017 est.)
$138 billion (31 December 2016 est.)
country comparison to the world: 35

Exchange rates: koruny (CZK) per US dollar—
23.34 (2017 est.)
24.44 (2016 est.)
24.44 (2015 est.)
24.599 (2014 est.)
20.758 (2013 est.)

ENERGY

Electricity access: *electrification—total population:* 100% (2020)

Electricity—production: 77.39 billion kWh (2016 est.)
country comparison to the world: 38

Electricity—consumption: 62.34 billion kWh (2016 est.)
country comparison to the world: 42

Electricity—exports: 24.79 billion kWh (2016 est.)
country comparison to the world: 7

Electricity—imports: 13.82 billion kWh (2016 est.)
country comparison to the world: 18

Electricity—installed generating capacity: 21.63 million kW (2016 est.)
country comparison to the world: 40

Electricity—from fossil fuels:
60% of total installed capacity (2016 est.)
country comparison to the world: 131

Electricity—from nuclear fuels:
19% of total installed capacity (2017 est.)
country comparison to the world: 10

Electricity—from hydroelectric plants: 5% of total installed capacity (2017 est.)
country comparison to the world: 130

Electricity—from other renewable sources: 16% of total installed capacity (2017 est.)
country comparison to the world: 51

Crude oil—production: 2,000 bbl/day (2018 est.)
country comparison to the world: 86

Crude oil—exports: 446 bbl/day (2017 est.)
country comparison to the world: 78

Crude oil—imports: 155,900 bbl/day (2017 est.)
country comparison to the world: 36

Crude oil—proved reserves: 15 million bbl (1 January 2018 est.)
country comparison to the world: 85

Refined petroleum products—production: 177,500 bbl/day (2017 est.)
country comparison to the world: 56

Refined petroleum products—consumption: 213,700 bbl/day (2017 est.)
country comparison to the world: 56

Refined petroleum products—exports: 52,200 bbl/day (2017 est.)
country comparison to the world: 54

Refined petroleum products—imports: 83,860 bbl/day (2017 est.)
country comparison to the world: 60

Natural gas—production: 229.4 million cu m (2017 est.)
country comparison to the world: 78

Natural gas—consumption:
8.721 billion cu m (2017 est.)
country comparison to the world: 51

Natural gas—exports: 0 cu m (2017 est.)
country comparison to the world: 93

Natural gas—imports: 8.891 billion cu m (2017 est.)
country comparison to the world: 28

Natural gas—proved reserves: 3.964 billion cu m (1 January 2018 est.)
country comparison to the world: 93

Carbon dioxide emissions from consumption of energy: 115.8 million Mt (2017 est.)
country comparison to the world: 39

COMMUNICATIONS

Telephones—fixed lines: *total subscriptions:* 1,473,846
subscriptions per 100 inhabitants: 13.78 (2019 est.)
country comparison to the world: 65

Telephones—mobile cellular: *total subscriptions:* 13,213,279
subscriptions per 100 inhabitants: 123.54 (2019 est.)
country comparison to the world: 71

Telecommunication systems: *general assessment:* good telephone and Internet service; the Czech Republic has a sophisticated telecom market; mobile sector showing steady growth, but perhaps without enough competition, regulator makes progress for 5G services; the govt. trying to stimulate competition, improve end-users pricing and step up quality; strong growth in cable and fiber sectors; fixed wireless broadband remains strong, with penetration among the highest in the EU (2020)
domestic: 14 per 100 fixed-line and mobile telephone usage increased to 124 per 100 mobile-cellular, the number of cellular telephone subscriptions now greatly exceeds the population (2019)
international: country code—420; satellite earth stations—6 (2 Intersputnik—Atlantic and Indian Ocean regions, 1 Intelsat, 1 Eutelsat, 1 Inmarsat, 1 Globalstar) (2019)
note: the COVID-19 outbreak is negatively impacting telecommunications production and supply chains globally; consumer spending on telecom devices and services has also slowed due to the pandemic's effect on economies worldwide; overall progress towards improvements in all facets of the telecom industry—mobile, fixed-line, broadband, submarine cable and satellite—has moderated

Broadcast media: 22 TV stations operate nationally, with 17 of them in private hands; publicly operated Czech Television has 5 national channels; throughout the country, there are some 350 TV channels in operation, many through cable, satellite, and IPTV subscription services; 63 radio broadcasters are registered, operating over 80 radio stations, including 7 multiregional radio stations or networks; publicly operated broadcaster Czech Radio operates 4 national, 14 regional, and 4 Internet stations; both Czech Radio and Czech Television are partially financed through a license fee (2019)

Internet country code: .cz

Internet users: *total:* 8,622,750
percent of population: 80.69% (July 2018 est.)
country comparison to the world: 58

Broadband—fixed subscriptions: *total:* 3,222,835
subscriptions per 100 inhabitants: 30 (2018 est.)
country comparison to the world: 40

TRANSPORTATION

National air transport system: *number of registered air carriers:* 4 (2020)

inventory of registered aircraft operated by air carriers: 48

annual passenger traffic on registered air carriers: 5,727,200 (2018)

annual freight traffic on registered air carriers: 25.23 million mt-km (2018)

Civil aircraft registration country code prefix: OK (2016)

Airports: 128 (2013)
country comparison to the world: 45

Airports—with paved runways: *total:* 41 (2017)
over 3,047 m: 2 (2017)
2,438 to 3,047 m: 9 (2017)
1,524 to 2,437 m: 12 (2017)
914 to 1,523 m: 2 (2017)
under 914 m: 16 (2017)

Airports—with unpaved runways: *total:* 87 (2013)
1,524 to 2,437 m: 1 (2013)
914 to 1,523 m: 25 (2013)
under 914 m: 61 (2013)

Heliports: 1 (2013)

Pipelines: 7,160 km gas, 675 km oil, 94 km refined products (2016)

Railways: *total:* 9,408 km (2017)
standard gauge: 9,385 km 1.435-m gauge (3,218 km electrified) (2017)

narrow gauge: 23 km 0.760-m gauge (2017)
country comparison to the world: 24

Roadways: *total:* 55,744 km (includes urban and category I, II, III roads) (2019)
paved: 55,744 km (includes 1,252 km of expressways) (2019)
country comparison to the world: 82

Waterways: 664 km (principally on Elbe, Vltava, Oder, and other navigable rivers, lakes, and canals) (2010)
country comparison to the world: 76

Ports and terminals: *river port(s):* Prague (Vltava) Decin, Usti nad Labem (Elbe)

MILITARY AND SECURITY

Military and security forces: *Ministry of Defense and Armed Forces:* Land Forces; Air Forces; Cyber Forces; Special Forces Directorate (2020)

Military expenditures: 1.19% of GDP (2019 est.)
1.13% of GDP (2018)
1.04% of GDP (2017)
0.96% of GDP (2016)
1.03% of GDP (2015)
country comparison to the world: 109

Military and security service personnel strengths: the Czech military has approximately 25,000 active personnel (20,000 Army; 5,000 Air Force) (2019)

Military equipment inventories and acquisitions: the Czech military has a mix of Soviet-era and more modern equipment, mostly of European origin; since 2010, the leading suppliers of military equipment to Czechia are Austria and Spain (2019)

Military service age and obligation: 18-28 years of age for male and female voluntary military service; no conscription (2012)

TRANSNATIONAL ISSUES

Disputes—international: none

Refugees and internally displaced persons: *stateless persons:* 1,394 (2019)

Illicit drugs: transshipment point for Southwest Asian heroin and minor transit point for Latin American cocaine to Western Europe; producer of synthetic drugs for local and regional markets; susceptible to money laundering related to drug trafficking, organized crime; significant consumer of ecstasy

INTRODUCTION

Background: Once the seat of Viking raiders and later a major north European power, Denmark has evolved into a modern, prosperous nation that is participating in the general political and economic integration of Europe. It joined NATO in 1949 and the EEC (now the EU) in 1973. However, the country has opted out of certain elements of the EU's Maastricht Treaty, including the European Economic and Monetary Union, European defense cooperation, and issues concerning certain justice and home affairs.

GEOGRAPHY

Location: Northern Europe, bordering the Baltic Sea and the North Sea, on a peninsula north of Germany (Jutland); also includes several major islands (Sjaelland, Fyn, and Bornholm)

Geographic coordinates: 56 00 N, 10 00 E

Map references: Europe

Area: *total:* 43,094 sq km

land: 42,434 sq km

water: 660 sq km

note: includes the island of Bornholm in the Baltic Sea and the rest of metropolitan Denmark (the Jutland Peninsula, and the major islands of Sjaelland and Fyn), but excludes the Faroe Islands and Greenland

country comparison to the world: 134

Area—comparative: slightly less than twice the size of Massachusetts; about two-thirds the size of West Virginia

Land boundaries: *total:* 140 km
border countries (1): Germany 140 km

Coastline: 7,314 km

Maritime claims:
territorial sea: 12 nm
exclusive economic zone: 200 nm
contiguous zone: 24 nm

continental shelf: 200-m depth or to the depth of exploitation

Climate: temperate; humid and overcast; mild, windy winters and cool summers

Terrain: low and flat to gently rolling plains

Elevation: *mean elevation:* 34 m
lowest point: Lammefjord -7 m
highest point: Mollehoj/Ejer Bavnehoj 171 m

Natural resources: petroleum, natural gas, fish, arable land, salt, limestone, chalk, stone, gravel and sand

Land use: *agricultural land:* 63.4% (2011 est.)
arable land: 58.9% (2011 est.) / permanent crops: 0.1% (2011 est.) / permanent pasture: 4.4% (2011 est.)
forest: 12.9% (2011 est.)
other: 23.7% (2011 est.)
note: highest percentage of arable land for any country in the world

Irrigated land: 4,350 sq km (2012)

Population distribution: with excellent access to the North Sea, Skagerrak, Kattegat, and the Baltic Sea, population centers tend to be along coastal areas, particularly in Copenhagen and the eastern side of the country's mainland

Natural hazards: flooding is a threat in some areas of the country (e.g., parts of Jutland, along the southern coast of the island of Lolland) that are protected from the sea by a system of dikes

Environment—current issues: air pollution, principally from vehicle and power plant emissions; nitrogen and phosphorus pollution of the North Sea; drinking and surface water becoming polluted from animal wastes and pesticides; much of country's household and industrial waste is recycled

Environment—international agreements: *party to:* Air Pollution, Air Pollution-Nitrogen Oxides, Air Pollution-Persistent Organic Pollutants, Air Pollution-Sulfur 85, Air Pollution-Sulfur 94, Air Pollution-Volatile Organic Compounds, Antarctic Treaty, Biodiversity, Climate Change, Climate Change-Kyoto Protocol, Desertification, Endangered Species, Environmental Modification, Hazardous Wastes, Law of the Sea, Marine Dumping, Marine Life Conservation, Ozone Layer Protection, Ship Pollution, Tropical Timber 83, Tropical Timber 94, Wetlands, Whaling
signed, but not ratified: none of the selected agreements

Geography—note: composed of the Jutland Peninsula and a group of more than 400 islands (Danish Archipelago); controls Danish Straits (Skagerrak and Kattegat) linking Baltic and North Seas; about one-quarter of the population lives in greater Copenhagen

PEOPLE AND SOCIETY

Population: 5,869,410 (July 2020 est.)
country comparison to the world: 115

Nationality: *noun:* Dane(s)
adjective: Danish

Ethnic groups: Danish (includes Greenlandic (who are predominantly Inuit) and Faroese) 86.3%, Turkish 1.1%, other 12.6% (largest groups are Polish, Syrian, German, Iraqi, and Romanian) (2018 est.)
note: data represent population by ancestry

Languages: Danish, Faroese, Greenlandic (an Inuit dialect), German (small minority)
note: English is the predominant second language

Religions: Evangelical Lutheran (official) 74.7%, Muslim 5.5%, other/none/unspecified (denominations of less than 1% each in descending order of size include Roman Catholic, Jehovah's Witness, Serbian Orthodox Christian, Jewish, Baptist, Buddhist, Mormon, Pentecostal, and nondenominational Christian) 19.8% (2019 est.)

Age structure: *0-14 years:* 16.42% (male 494,806/female 469,005)
15-24 years: 12.33% (male 370,557/female 352,977)
25-54 years: 38.71% (male 1,149,991/female 1,122,016)
55-64 years: 12.63% (male 370,338/female 371,149)
65 years and over: 1 9.91% (male 538,096/female 630,475) (2020 est.)

Dependency ratios: *total dependency ratio:* 57.3
youth dependency ratio: 25.6
elderly dependency ratio: 31.7
potential support ratio: 3.2 (2020 est.)

Median age: *total:* 42 years
male: 40.9 years
female: 43.1 years (2020 est.)
country comparison to the world: 38

Population growth rate: 0.48% (2020 est.)
country comparison to the world: 157

Birth rate: 11.1 births/1,000 population (2020 est.)
country comparison to the world: 178

Death rate: 9.5 deaths/1,000 population (2020 est.)
country comparison to the world: 45

Net migration rate: 2.8 migrant(s)/1,000 population (2020 est.)
country comparison to the world: 39

Population distribution: with excellent access to the North Sea, Skagerrak, Kattegat, and the Baltic Sea, population centers tend to be along coastal areas, particularly in Copenhagen and the eastern side of the country's mainland

Urbanization: *urban population:* 88.1% of total population (2020)
rate of urbanization: 0.51% annual rate of change (2015-20 est.)
total population growth rate v. urban population growth rate, 2000-2030:

Major urban areas—Population: 1.346 million COPENHAGEN (capital) (2020)

Sex ratio: *at birth:* 1.07 male(s)/female
0-14 years: 1.06 male(s)/female
15-24 years: 1.05 male(s)/female
25-54 years: 1.02 male(s)/female
55-64 years: 1 male(s)/female
65 years and over: 0.85 male(s)/female
total population: 0.99 male(s)/female (2020 est.)

Mother's mean age at first birth:
29.2 years (2017 est.)

Maternal mortality rate: 4 deaths/100,000 live births (2017 est.)
country comparison to the world: 172

Infant mortality rate: *total:* 3.2 deaths/1,000 live births
male: 3.6 deaths/1,000 live births
female: 2.7 deaths/1,000 live births (2020 est.)
country comparison to the world: 210
Life expectancy at birth: total population: 81.2 years
male: 79.3 years
female: 83.3 years (2020 est.)
country comparison to the world: 35

Total fertility rate: 1.78 children born/woman (2020 est.)
country comparison to the world: 151

Drinking water source:
improved:
urban: 100% of population
rural: 100% of population
total: 100% of population
unimproved:
urban: 0% of population
rural: 0% of population
total: 0% of population (2017 est.)

Current Health Expenditure: 10.1% (2017)

Physicians density: 4.01 physicians/1,000 population (2016)

Hospital bed density: 2.6 beds/1,000 population (2017)

Sanitation facility access:
improved:
urban: 100% of population
rural: 100% of population
total: 100% of population
unimproved:
urban: 0% of population
rural: 0% of population
total: 0% of population (2017 est.)

HIV/AIDS—adult prevalence rate: 0.1% (2018 est.)
country comparison to the world: 121

HIV/AIDS—people living with HIV/AIDS: 6,200 (2018 est.)
country comparison to the world: 117

HIV/AIDS—deaths: <100 (2018 est.)

Obesity—adult prevalence rate:
19.7% (2016)
country comparison to the world: 109

Education expenditures: 7.6% of GDP (2014)
country comparison to the world: 7

School life expectancy (primary to tertiary education): *total:* 18 years

male: 19 years
female: 19 years (2018)

Unemployment, youth ages 15-24: *total:* 9.4%
male: 10.5%
female: 8.2% (2018 est.)
country comparison to the world: 130

GOVERNMENT

Country name: *conventional long form:* Kingdom of Denmark
conventional short form: Denmark
local long form: Kongeriget Danmark
local short form: Danmark
etymology: the name derives from the words "Dane(s)" and "mark"; the latter referring to a march (borderland) or forest

Government type: parliamentary constitutional monarchy

Capital: *name:* Copenhagen
geographic coordinates: 55 40 N, 12 35 E
time difference: UTC + 1 (6 hours ahead of Washington, DC, during Standard Time)
daylight saving time: +1hr, begins last Sunday in March; ends last Sunday in October; note—applies to continental Denmark only, not to its North Atlantic components
etymology: name derives from the city's Danish appellation Kobenhavn, meaning "Merchant's Harbor"

Administrative divisions: metropolitan Denmark—5 regions (regioner, singular—region); Hovedstaden (Capital), Midtjylland (Central Jutland), Nordjylland (North Jutland), Sjaelland (Zealand), Syddanmark (Southern Denmark)

Independence: ca. 965 (unified and Christianized under HARALD I Gormsson); 5 June 1849 (became a parliamentary constitutional monarchy)

National holiday: Constitution Day, 5 June (1849); note—closest equivalent to a national holiday

Constitution: *history:* several previous; latest adopted 5 June 1953
amendments: proposed by the Folketing with consent of the government; passage requires approval by the next Folketing following a general election, approval by simple majority vote of at least 40% of voters in a referendum, and assent of the chief of state; changed several times, last in 2009 (Danish Act of Succession)

Legal system: civil law; judicial review of legislative acts

International law organization participation: accepts compulsory ICJ jurisdiction with reservations; accepts ICCt jurisdiction

Citizenship: *citizenship by birth:* no
citizenship by descent only: at least one parent must be a citizen of Denmark
dual citizenship recognized: yes
residency requirement for naturalization: 7 years

Suffrage: 18 years of age; universal

Executive branch: *chief of state:* Queen MARGRETHE II (since 14 January 1972); Heir

Apparent Crown Prince FREDERIK (elder son of the monarch, born on 26 May 1968)
head of government: Prime Minister Mette FREDERIKSEN (since 27 June 2019)
cabinet: Council of State appointed by the monarch
elections/appointments: the monarchy is hereditary; following legislative elections, the leader of the majority party or majority coalition usually appointed prime minister by the monarch

Legislative branch: *description:* unicameral People's Assembly or Folketing (179 seats, including 2 each representing Greenland and the Faroe Islands; members directly elected in multi-seat constituencies by proportional representation vote; members serve 4-year terms unless the Folketing is dissolved earlier)
elections: last held on 5 June 2019 (next to be held on June 2023)
election results: percent of vote by party—SDP 25.9%, V 23.4%, DF 8.7%, SLP 8.6%, SF 7.7%, EL 6.9%, C 6.6%, A 3.0%, NB 2.4%, LA 2.3%; seats by party—SDP 48, V 43, DF 16, SLP 16, SF 14, EL 13, C 12, A 5, NB 4, LA 4; composition—men 109, women 70 (includes 2 from Greenland), percent of women 39.1%

Judicial branch: *highest courts:* Supreme Court (consists of the court president and 18 judges)

judge selection and term of office: judges appointed by the monarch upon the recommendation of the Minister of Justice, with the advice of the Judicial Appointments Council, a 6-member independent body of judges and lawyers; judges appointed for life with retirement at age 70
subordinate courts: Special Court of Indictment and Revision; 2 High Courts; Maritime and Commercial Court; county courts

Political parties and leaders:

The Alternative A or AP (vacant)
Conservative People's Party or DKF or C [Soren PAPE POULSEN]
Danish People's Party or DF or O [Kristian THULESEN DAHL]
Liberal Alliance or LA [Alex VANOPSLAGH]
Liberal Party (Venstre) or V [Jakob ELLEMAN-JENSEN]
New Right Party or D or NB [Pemille VERMUND]
Red-Green Alliance (Unity List) or EL [collective leadership, Pemille SKIPPER, spokesperson]
Social Democrats or A or SDP [Mette FREDERIKSEN]
Social Liberal Party or B or SLP [Sofie CARSTEN]
Socialist People's Party or SF [Pia OLSEN DYHR]

International organization participation: ADB (nonregional member), AfDB (nonregional member), Arctic Council, Australia Group, BIS, CBSS, CD, CE, CERN, EAPC, EBRD, ECB, EIB, EITI (implementing country), ESA, EU, FAO, FATF, G-9, IADB, IAEA, IBRD, ICAO, ICC (national committees), ICCt, ICRM, IDA, IEA, IFAD, IFC, IFRCS, IGAD (partners), IHO, ILO, IMF, IMO, IMSO, Interpol, IOC, IOM, IPU, ISO, ITSO, ITU, ITUC (NGOs), MIGA, MINUSMA, NATO, NC, NEA, NIB, NSG, OAS (observer), OECD, OPCW, OSCE, Paris Club,

PCA, Schengen Convention, UN, UNCTAD, UNESCO, UNHCR, UNIDO, UNMIL, UNMISS, UNRWA, UNTSO, UPU, WCO, WHO, WIPO, WMO, WTO, ZC

Diplomatic representation in the US: *chief of mission:* Ambassador Lone Dencker WISBORG (since 17 September 2015)
chancery: 3200 Whitehaven Street NW, Washington, DC 20008
telephone: [1] (202) 234-4300
FAX: [1] (202) 328-1470
consulate(s) general: Chicago, Houston, New York, Palo Alto (CA)

Diplomatic representation from the US: *chief of mission:* Ambassador Carla SANDS (since 15 December 2017)
telephone: [45] 33 41 71 00
embassy: Dag Hammarskjolds Alle 24, 2100 Copenhagen 0
mailing address: Unit 5280 ODC, DPO AE 09716
FAX: [45] 35 43 02 23

Flag description: red with a white cross that extends to the edges of the flag; the vertical part of the cross is shifted to the hoist side; the banner is referred to as the Dannebrog (Danish flag) and is one of the oldest national flags in the world; traditions as to the origin of the flag design vary, but the best known is a legend that the banner fell from the sky during an early-13th century battle; caught up by the Danish king before it ever touched the earth, this heavenly talisman inspired the royal army to victory; in actuality, the flag may derive from a crusade banner or ensign
note: the shifted cross design element was subsequently adopted by the other Nordic countries of Finland, Iceland, Norway, and Sweden

National symbol(s): *lion, mute swan; national colors:* red, white

National anthem: *name:* "Der er et yndigt land" (There is a Lovely Country); "Kong Christian" (King Christian)
lyrics/music: Adam Gottlob OEHLENSCHLAGER/Hans Ernst KROYER; Johannes EWALD/unknown
note: Denmark has two national anthems with equal status; "Der er et yndigt land," adopted 1844, is a national anthem, while "Kong Christian," adopted 1780, serves as both a national and royal anthem; "Kong Christian" is also known as "Kong Christian stod ved hojen mast" (King Christian Stood by the Lofty Mast) and "Kongesangen" (The King's Anthem); within Denmark, the royal anthem is played only when royalty is present and is usually followed by the national anthem; when royalty is not present, only the national anthem is performed; outside Denmark, the royal anthem is played, unless the national anthem is requested 0:00/ 1:19

ECONOMY

Economy—overview: This thoroughly modern market economy features advanced industry with world-leading firms in pharmaceuticals, maritime shipping, and renewable energy, and a high-tech agricultural sector. Danes enjoy a high standard of living, and the Danish economy is characterized by extensive government welfare measures and an equitable distribution of income. An aging population will be a long-term issue.

Denmark's small open economy is highly dependent on foreign trade, and the government strongly supports trade liberalization. Denmark is a net exporter of food, oil, and gas and enjoys a comfortable balance of payments surplus, but depends on imports of raw materials for the manufacturing sector.

Denmark is a member of the EU but not the eurozone. Despite previously meeting the criteria to join the European Economic and Monetary Union, Denmark has negotiated an opt-out with the EU and is not required to adopt the euro.

Denmark is experiencing a modest economic expansion. The economy grew by 2.0% in 2016 and 2.1% in 2017. The expansion is expected to decline slightly in 2018. Unemployment stood at 5.5% in 2017, based on the national labor survey. The labor market was tight in 2017, with corporations experiencing some difficulty finding appropriately-skilled workers to fill billets. The Danish Government offers extensive programs to train unemployed persons to work in sectors that need qualified workers.

Denmark maintained a healthy budget surplus for many years up to 2008, but the global financial crisis swung the budget balance into deficit. Since 2014 the balance has shifted between surplus and deficit. In 2017 there was a surplus of 1.0%. The government projects a lower deficit in 2018 and 2019 of 0.7%, and public debt (EMU debt) as a share of GDP is expected to decline to 35.6% in 2018 and 34.8% in 2019. The Danish Government plans to address increasing municipal, public housing and integration spending in 2018.

GDP (purchasing power parity): $287.8 billion (2017 est.)
$281.4 billion (2016 est.)
$276 billion (2015 est.)
note: data are in 2017 dollars
country comparison to the world: 60

GDP (official exchange rate): $325.6 billion (2017 est.)

GDP—real growth rate:
2.85% (2019 est.)
2.18% (2018 est.)
2.83% (2017 est.)
country comparison to the world: 102

GDP—per capita (PPP): $50,100 (2017 est.)
$49,300 (2016 est.)
$48,800 (2015 est.)
note: data are in 2017 dollars
country comparison to the world: 30

Gross national saving:
28.8% of GDP (2017 est.)
28.3% of GDP (2016 est.)
28.7% of GDP (2015 est.)
country comparison to the world: 35

GDP—composition, by end use:
household consumption: 48% (2017 est.)
government consumption: 25.2% (2017 est.)
investment in fixed capital: 20% (2017 est.)
investment in inventories: -0.2% (2017 est.)
exports of goods and services: 54.5% (2017 est.)
imports of goods and services: -47.5% (2017 est.)

GDP—composition, by sector of origin:
agriculture: 1.3% (2017 est.)
industry: 22.9% (2017 est.)
services: 75.8% (2017 est.)

Agriculture—products: barley, wheat, potatoes, sugar beets; pork, dairy products; fish

Industries: wind turbines, pharmaceuticals, medical equipment, shipbuilding and refurbishment, iron, steel, nonferrous metals, chemicals, food processing, machinery and transportation equipment, textiles and clothing, electronics, construction, furniture and other wood products

Industrial production growth rate: 2.5% (2017 est.)
country comparison to the world: 116

Labor force:
2.736 million (2020 est.)
country comparison to the world: 107

Labor force—by occupation: *agriculture:* 2.4%
industry: 18.3%
services: 79.3% (2016 est.)

Unemployment rate:
3.05% (2019 est.)
3.07% (2018 est.)
country comparison to the world: 39

Population below poverty line: 13.4% (2011 est.)
note: excludes students

Household income or consumption by percentage share: *lowest 10%:* 9%
highest 10%: 23.4% (2016 est.)

Budget: *revenues:* 172.5 billion (2017 est.)
expenditures: 168.9 billion (2017 est.)

Taxes and other revenues: 53% (of GDP) (2017 est.)
country comparison to the world: 12

Budget surplus (+) or deficit (-): 1.1% (of GDP) (2017 est.)
country comparison to the world: 30

Public debt:
35.3% of GDP (2017 est.)
37.9% of GDP (2016 est.)
note: data cover general government debt and include debt instruments issued (or owned) by government entities other than the treasury; the data include treasury debt held by foreign entities; the data include debt issued by subnational entities, as well as intra-governmental debt; intragovernmental debt consists of treasury borrowings from surpluses in the social funds, such as for retirement, medical care, and unemployment; debt instruments for the social funds are not sold at public auctions
country comparison to the world: 151

Fiscal year: calendar year

Inflation rate (consumer prices):
1.1% (2017 est.)
0.3% (2016 est.)
country comparison to the world: 58

Current account balance:
$30.935 billion (2019 est.)
$24.821 billion (2018 est.)
country comparison to the world: 12

Exports:
$113.6 billion (2017 est.)
$103.6 billion (2016 est.)
country comparison to the world: 34

Exports—partners: Germany 15.5%, Sweden 11.6%, UK 8.2%, US 7.5%, Norway 6%, China 4.4%, Netherlands 4.4% (2017)

Exports—commodities: wind turbines, pharmaceuticals, machinery and instruments, meat and meat products, dairy products, fish, furniture and design

Imports:
$94.93 billion (2017 est.)
$86.81 billion (2016 est.)
country comparison to the world: 37

Imports—commodities: machinery and equipment, raw materials and semimanufactures for industry, chemicals, grain and foodstuffs, consumer goods

Imports—partners: Germany 21.3%, Sweden 11.9%, Netherlands 7.8%, China 7.1%, Norway 6.3%, Poland 4% (2017)

Reserves of foreign exchange and gold:
$75.25 billion (31 December 2017 est.)
$64.25 billion (31 December 2016 est.)
country comparison to the world: 30

Debt—external:
$484.8 billion (31 March 2016 est.)
$519.8 billion (31 March 2015 est.)
country comparison to the world: 25

Exchange rates: Danish kroner (DKK) per US dollar –
6.586 (2017 est.)
6.7309 (2016 est.)
6.7309 (2015 est.)
6.7236 (2014 est.)
5.6125 (2013 est.)

ENERGY

Electricity access: electrification—total population: 100% (2020)

Electricity—production: 29.84 billion kWh (2016 est.)
country comparison to the world: 65

Electricity—consumption: 33.02 billion kWh (2016 est.)
country comparison to the world: 59

Electricity—exports: 9.919 billion kWh (2016 est.)
country comparison to the world: 20

Electricity—imports: 14.98 billion kWh (2016 est.)
country comparison to the world: 14

Electricity—installed generating capacity: 14.34 million kW (2016 est.)
country comparison to the world: 52

Electricity—from fossil fuels: 46% of total installed capacity (2016 est.)

country comparison to the world: 158

Electricity—from nuclear fuels: 0% of total installed capacity (2017 est.)
country comparison to the world: 77

Electricity—from hydroelectric plants: 0% of total installed capacity (2017 est.)
country comparison to the world: 167

Electricity—from other renewable sources: 54% of total installed capacity (2017 est.)
country comparison to the world: 3

Crude oil—production: 115,000 bbl/day (2018 est.)
country comparison to the world: 41

Crude oil—exports:
82,980 bbl/day (2017 est.)
country comparison to the world: 36

Crude oil—imports:
98,240 bbl/day (2017 est.)
country comparison to the world: 44

Crude oil—proved reserves: 439 million bbl (1 January 2018 est.)
country comparison to the world: 46

Refined petroleum products—production: 183,900 bbl/day (2017 est.)
country comparison to the world: 55

Refined petroleum products—consumption: 158,500 bbl/day (2017 est.)
country comparison to the world: 64

Refined petroleum products—exports: 133,700 bbl/day (2017 est.)
country comparison to the world: 38

Refined petroleum products—imports: 109,700 bbl/day (2017 est.)
country comparison to the world: 51

Natural gas—production: 4.842 billion cu m (2017 est.)
country comparison to the world: 52

Natural gas—consumption: 3.115 billion cu m (2017 est.)
country comparison to the world: 73

Natural gas—exports: 2.237 billion cu m (2017 est.)
country comparison to the world: 37

Natural gas—imports: 509.7 million cu m (2017 est.)
country comparison to the world: 65

Natural gas—proved reserves: 12.86 billion cu m (1 January 2018 est.)
country comparison to the world: 77

Carbon dioxide emissions from consumption of energy: 37.45 million Mt (2017 est.)
country comparison to the world: 70

COMMUNICATIONS

Telephones—fixed lines: *total subscriptions:* 1,017,009
subscriptions per 100 inhabitants: 17.41 (2019 est.)
country comparison to the world: 74

Telephones—mobile cellular: *total subscriptions:* 7,331,110

subscriptions per 100 inhabitants: 125.5 (2019 est.)
country comparison to the world: 102

Telecommunication systems: *general assessment:* excellent telephone and Internet services; Denmark's competitive telecom market has led to the country having the second highest broadband penetration rate in Europe; the fixed-line sector continues to see a decline in revenue while customers move to VoIP (Voice over Internet Protocol) and mobile alternatives; comprehensive LTE coverage and a fast-developing 5G segment; the government is able to offer broadband coverage in rural areas (2020)

domestic: fixed-line 17 per 100, 126 per 100 for mobile-cellular (2019)

international: country code—45; landing points for the NSC, COBRAcable, CANTAT-3, DANICE, Havfrue/AEC-2, TAT-14m Denmark-Norway-5 & 6, Skagenfiber West & East, GC1, GC2, GC3, GC-KPN, Kattegat 1 & 2 & 3, Energinet Lyngsa- Laeso, Energinet Laeso-Varberg, Fehmarn Balt, Baltica, German-Denmark 2 & 3, Ronne-Rodvig, Denmark-Sweden 15 & 16 & 17 & 18, IP-Only Denmark-Sweden, Scandinavian South, Scandinavian Ring North, Danica North, 34 series of fiberoptic submarine cables link Denmark with Canada, Faroe Islands, Germany, Iceland, Netherlands, Norway, Poland,

Russia, Sweden, US and UK; satellite earth stations—18 (6 Intelsat, 10 Eutelsat, 1 Orion, 1 Inmarsat (Blaavand-Atlantic- East)); note—the Nordic countries (Denmark, Finland, Iceland, Norway, and Sweden) share the Danish earth station and the Eik, Norway, station for worldwide Inmarsat access (2019)

note: the COVID-19 outbreak is negatively impacting telecommunications production and supply chains globally; consumer spending on telecom devices and services has also slowed due to the pandemic's effect on economies worldwide; overall progress towards improvements in all facets of the telecom industry—mobile, fixed-line, broadband, submarine cable and satellite—has moderated

Broadcast media: strong public-sector TV presence with state-owned Danmarks Radio (DR) operating 6 channels and publicly owned TV2 operating roughly a half-dozen channels; broadcasts of privately owned stations are available via satellite and cable feed;

DR operates 4 nationwide FM radio stations, 10 digital audio broadcasting stations, and 14 web-based radio stations; 140 commercial and 187 community (non-commercial) radio stations (2019)

Internet country code: .dk

Internet users: *total:* 5,672,398
percent of population: 97.64% (July 2018 est.)
country comparison to the world: 79

Broadband—fixed subscriptions: *total:* 2,534,348
subscriptions per 100 inhabitants: 44 (2018 est.)
country comparison to the world: 48

TRANSPORTATION

National air transport system: *number of registered air carriers:* 10 (2020)
inventory of registered aircraft operated by air carriers: 76
annual passenger traffic on registered air carriers: 582,011 (2015)
annual freight traffic on registered air carriers: 0 mt-km (2015)

Civil aircraft registration country code prefix: OY (2016)

Airports: 80 (2013)
country comparison to the world: 68

Airports—with paved runways: *total:* 28 (2017)
over 3,047 m: 2 (2017)
2,438 to 3,047 m: 7 (2017)
1,524 to 2,437 m: 5 (2017)
914 to 1,523 m: 12 (2017)
under 914 m: 2 (2017)

Airports—with unpaved runways: *total:* 52 (2013)
914 to 1,523 m: 5 (2013)
under 914 m: 47 (2013)

Pipelines: 1536 km gas, 330 km oil (2015)

Railways: *total:* 3,476 km (2017)
standard gauge: 3,476 km 1.435-m gauge (1,756 km electrified) (2017)
country comparison to the world: 57

Roadways: *total:* 74,558 km (2017)
paved: 74,558 km (includes 1,205 km of expressways) (2017)
country comparison to the world: 67

Waterways: 400 km (2010)
country comparison to the world: 87

Merchant marine: *total:* 682
by type: bulk carrier 7, container ship 149, general cargo 59, oil tanker 77, other 390 (2019)
country comparison to the world: 32

Ports and terminals: *major seaport(s):* Baltic Sea—Aarhus, Copenhagen, Fredericia, Kalundborg
cruise port(s): Copenhagen
river port(s): Aalborg (Langerak)
dry bulk cargo port(s): Ensted (coal)
North Sea—Esbjerg,

MILITARY AND SECURITY

Military and security forces: Royal Danish Army, Royal Danish Navy, Royal Danish Air Force, Danish Home Guard (Reserves) (2020)
note: the Danish military also maintains a Joint Arctic Command

Military expenditures:
1.32% of GDP (2019)
1.3% of GDP (2018)
1.15% of GDP (2017)
1.15% of GDP (2016)
1.11% of GDP (2015)
country comparison to the world: 89

Military and security service personnel strengths: the Danish military has approximately 16,000 active duty personnel (8,500 Army; 2,500 Navy; 3,000 Air Force; 2,000 joint service, other) (2019 est.)

Military equipment inventories and acquisitions: the Danish military inventory is comprised of a mix of modern European, US, and domestically-produced equipment; the US is the largest supplier of military equipment to Denmark since 2010, followed by Germany and the Netherlands; the Danish defense industry is mainly active in the production of naval vessels, defense electronics, and subcomponents of larger weapons systems, such as the US F-35 fighter aircraft (2019 est.)

Military deployments: 110 Afghanistan (NATO); 130 Middle East/Iraq (NATO/Operation Inherent Resolve) (2020)

Military service age and obligation: 18 years of age for compulsory and voluntary military service; conscripts serve an initial training period that varies from 4 to 12 months depending on specialization; former conscripts are assigned to mobilization units; women eligible to volunteer for military service; in addition to full time employment, the Danish Military offers reserve contracts in all three branches (2016)

Military—note: in 2018, the Defense Ministers of Belgium, Denmark and the Netherlands signed a Memorandum of Understanding (MOU) for the creation of a Composite Special Operations Component Command (C-SOCC); C-SOCC is scheduled to be fully operational in 2021 (2020)

TERRORISM

Terrorist group(s): Islamic Revolutionary Guard Corps/Qods Force (2019)
note: details about the history, aims, leadership, organization, areas of operation, tactics, targets, weapons, size, and sources of support of the group(s) appear(s) in Appendix-T

TRANSNATIONAL ISSUES

Disputes—international: Iceland, the UK, and Ireland dispute Denmark's claim that the Faroe Islands' continental shelf extends beyond 200 nm; sovereignty dispute with Canada over Hans Island in the Kennedy Channel between Ellesmere Island and Greenland; Denmark (Greenland) and Norway have made submissions to the Commission on the Limits of the Continental Shelf (CLCS) and Russia is collecting additional data to augment its 2001 CLCS submission

Refugees and internally displaced persons:
refugees (country of origin): 20,046 (Syria), 5,320 (Eritrea) (2019)
stateless persons: 8,672 (2019)

DHEKELIA

INTRODUCTION

Background: By terms of the 1960 Treaty of Establishment that created the independent Republic of Cyprus, the UK retained full sovereignty and jurisdiction over two areas of almost 254 square kilometers—Akrotiri and Dhekelia. The larger of these is the Dhekelia Sovereign Base Area, which is also referred to as the Eastern Sovereign Base Area.

GEOGRAPHY

Location: Eastern Mediterranean, on the southeast coast of Cyprus near Famagusta

Geographic coordinates: 34 59 N, 33 45 E

Map references: Middle East

Area: *total:* 131 sq km

note: area surrounds three Cypriot enclaves
country comparison to the world: 223

Area—comparative: about three-quarters the size of Washington, DC

Land boundaries: *total:* 108 km
border countries (1): Cyprus 108 km

Coastline: 27.5 km

Climate: temperate; Mediterranean with hot, dry summers and cool winters

Environment—current issues: netting and trapping of small migrant songbirds in the spring and autumn

Geography—note: British extraterritorial rights also extended to several small off-post sites scattered across Cyprus; several small Cypriot enclaves exist within the Sovereign Base Area (SBA); of the SBA land, 60% is privately owned and farmed, 20% is owned by the Ministry of Defense, and 20% is SBA Crown land

PEOPLE AND SOCIETY

Population: approximately 15,500 on the Sovereign Base Areas of Akrotiri and Dhekelia including 9,700 Cypriots and 5,800 Service and UK-based contract personnel and dependents

Languages: English, Greek

HIV/AIDS—adult prevalence rate: NA

GOVERNMENT

Country name: *conventional long form:* none
conventional short form: Dhekelia

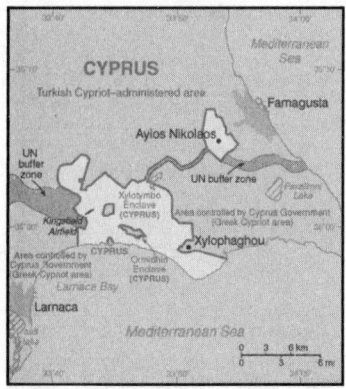

Dependency status: a special form of UK overseas territory; administered by an administrator who is also the Commander, British Forces Cyprus

Capital: *name:* Episkopi Cantonment (base administrative center for Akrotiri and Dhekelia); located in Akrotiri
geographic coordinates: 34 40 N, 32 51 E
time difference: UTC+2 (7 hours ahead of Washington, DC, during Standard Time)
daylight saving time: +1hr, begins last Sunday in March; ends last Sunday in October
etymology: "Episkopi" means "episcopal" in Greek and stems from the fact that the site previously served as the bishop's seat of an Orthodox diocese

Constitution: *history:* presented 3 August 1960, effective 16 August 1960 (The Sovereign Base Areas of Akrotiri and Dhekelia Order in Council 1960, serves as a basic legal document); amended 1966

Legal system: laws applicable to the Cypriot population are, as far as possible, the same as the laws of the Republic of Cyprus; note—the Sovereign Base Area Administration has its own court system to deal with civil and criminal matters

Executive branch: *chief of state:* Queen ELIZABETH II (since 6 February 1952)
head of government: Administrator Major General Robert J. THOMSON (since 25 September 2019); note—administrator reports to the British Ministry of Defense; the chief officer is responsible for the day-to-day running of the civil government of the Sovereign Bases
elections/appointments: the monarchy is hereditary; administrator appointed by the monarch on the advice of the Ministry of Defense

Judicial branch: *highest courts:* Senior Judges' Court (consists of several visiting judges from England and Wales)
judge selection and term of office: see entry for United Kingdom
subordinate courts: Resident Judges' Court; military courts

Diplomatic representation in the US: none (overseas territory of the UK)

Diplomatic representation from the US: none (overseas territory of the UK)

Flag description: the flag of the UK is used

National anthem: *note:* as a United Kingdom area of special sovereignty, "God Save the Queen" is official (see United Kingdom)
0:00/ 1:02

ECONOMY

Economy—overview: Economic activity is limited to providing services to the military and their families located in Dhekelia. All food and manufactured goods must be imported.

Industries: none

Exchange rates:
note: uses the euro

COMMUNICATIONS

Broadcast media: British Forces Broadcast Service (BFBS) provides multi-channel satellite TV service as well as BFBS radio broadcasts to the Dhekelia Sovereign Base

TRANSPORTATION

MILITARY AND SECURITY

Military—note: defense is the responsibility of the UK; includes Dhekelia Garrison and Ayios Nikolaos Station connected by a roadway

DJIBOUTI

INTRODUCTION

Background: The region of present-day Djibouti was the site of the medieval Ifat and Adal Sultanates. In the late 19th century, treaties signed by the ruling Somali and Afar sultans with the French allowed the latter to establish the colony of French Somaliland. The designation continued in use until 1967, when the name was changed to the French Territory of the Afars and the Issas. Upon independence in 1977, the country was named after its capital city of Djibouti. Hassan Gouled APTIDON installed an authoritarian one- party state and proceeded to serve as president until 1999. Unrest among the Afar minority during the 1990s led to a civil war that ended in 2001 with a peace accord between Afar rebels and the Somali Issa-dominated government. In 1999, Djibouti's first multiparty presidential election resulted in the election of Ismail Omar GUELLEH as president; he was reelected to a second term in 2005 and extended his tenure in office via a constitutional amendment, which allowed him to serve a third term in 2011 and begin a fourth term in 2016. Djibouti occupies a strategic geographic location at the intersection of the Red Sea and the Gulf of Aden. Its ports handle 95% of Ethiopia's trade. Djibouti's ports also service transshipments between Europe, the Middle East, and Asia. The government holds longstanding ties to France, which maintains a military presence in the country, as does the US, Japan, Italy, Germany, Spain, and China.

GEOGRAPHY

Location: Eastern Africa, bordering the Gulf of Aden and the Red Sea, between Eritrea and Somalia

Geographic coordinates: 11 30 N, 43 00 E

Map references: Africa

Area: *total:* 23,200 sq km
land: 23,180 sq km
water: 20 sq km
country comparison to the world: 151

Area—comparative: slightly smaller than New Jersey

Land boundaries: *total:* 528 km
border countries (3): Eritrea 125 km, Ethiopia 342 km, Somalia 61 km

Coastline: 314 km

Maritime claims: *territorial sea:* 12 nm
exclusive economic zone: 200 nm

contiguous zone: 24 nm

Climate: desert; torrid, dry

Terrain: coastal plain and plateau separated by central mountains

Elevation: *mean elevation:* 430 m
lowest point: Lac Assal—155 m
highest point: Moussa Ali 2,021 m

Natural resources: potential geothermal power, gold, clay, granite, limestone, marble, salt, diatomite, gypsum, pumice, petroleum

Land use: *agricultural land:* 73.4% (2011 est.)
arable land: 0.1% (2011 est.) /permanent crops: 0% (2011 est.) / permanent pasture: 73.3% (2011 est.)
forest: 0.2% (2011 est.)
other: 26.4% (2011 est.)
Irrigated land: 10 sq km (2012)

Population distribution: most densely populated areas are in the east; the largest city is Djibouti, with a population over 600,000; no other city in the country has a total population over 50,000 as shown in this population distribution map

Natural hazards: earthquakes; droughts; occasional cyclonic disturbances from the Indian Ocean bring heavy rains and flash floods

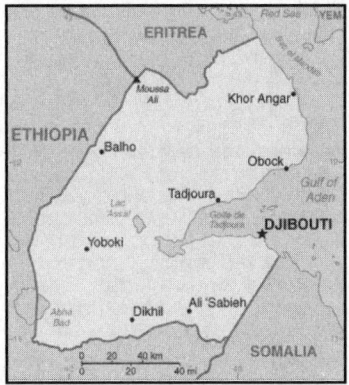

volcanism: experiences limited volcanic activity; Ardoukoba (298 m) last erupted in 1978; Manda-Inakir, located along the Ethiopian border, is also historically active

Environment—current issues: inadequate supplies of potable water; water pollution; limited arable land; deforestation (forests threatened by agriculture and the use of wood for fuel); desertification; endangered species

Environment—international agreements: *party to:* Biodiversity, Climate Change, Climate Change-Kyoto Protocol, Desertification, Endangered Species, Hazardous Wastes, Law of the Sea, Ozone Layer Protection, Ship Pollution, Wetlands
signed, but not ratified: none of the selected agreements

Geography—note: strategic location near world's busiest shipping lanes and close to Arabian oilfields; terminus of rail traffic into Ethiopia; mostly wasteland; Lac Assal (Lake Assal) is the lowest point in Africa and the saltiest lake in the world

PEOPLE AND SOCIETY

Population: 921,804 (July 2020 est.)
country comparison to the world: 162

Nationality: *noun:* Djiboutian(s)
adjective: Djiboutian

Ethnic groups: Somali 60%, Afar 35%, other 5% (mostly Yemeni Arab, also French, Ethiopian, and Italian)

Languages: French (official), Arabic (official), Somali, Afar

Religions: Sunni Muslim 94% (nearly all Djiboutians), Christian 6% (mainly foreign-born residents)

Demographic profile: Djibouti is a poor, predominantly urban country, characterized by high rates of illiteracy, unemployment, and childhood malnutrition. More than 75% of the population lives in cities and towns (predominantly in the capital, Djibouti). The rural population subsists primarily on nomadic herding. Prone to droughts and floods, the country has few natural resources and must import more than 80% of its food from neighboring countries or Europe. Health care, particularly

outside the capital, is limited by poor infrastructure, shortages of equipment and supplies, and a lack of qualified personnel. More than a third of health care recipients are migrants because the services are still better than those available in their neighboring home countries. The nearly universal practice of female genital cutting reflects Djibouti's lack of gender equality and is a major contributor to obstetrical complications and its high rates of maternal and infant mortality. A 1995 law prohibiting the practice has never been enforced.

Because of its political stability and its strategic location at the confluence of East Africa and the Gulf States along the Gulf of Aden and the Red Sea, Djibouti is a key transit point for migrants and asylum seekers heading for the Gulf States and beyond. Each year some hundred thousand people, mainly Ethiopians and some Somalis, journey through Djibouti, usually to the port of Obock, to attempt a dangerous sea crossing to Yemen. However, with the escalation of the ongoing Yemen conflict, Yemenis began fleeing to Djibouti in March 2015, with almost 20,000 arriving by August 2017. Most Yemenis remain unregistered and head for Djibouti City rather than seeking asylum at one of Djibouti's three spartan refugee camps. Djibouti has been hosting refugees and asylum seekers, predominantly Somalis and lesser numbers of Ethiopians and Eritreans, at camps for 20 years, despite lacking potable water, food shortages, and unemployment.

Age structure: *0-14 years:* 29.97% (male 138,701/female 137,588)
15-24 years: 20.32% (male 88,399/female 98,955)
25-54 years: 40.73% (male 156,016/female 219,406)
55-64 years: 5.01% (male 19,868/female 26,307)
65 years and over: 3.97% (male 16,245/female 20,319) (2020 est.)

Dependency ratios: *total dependency ratio:* 50.6
youth dependency ratio: 43.6
elderly dependency ratio: 7.1
potential support ratio: 14.1 (2020 est.)

Median age: *total:* 24.9 years
male: 23 years
female: 26.4 years (2020 est.)
country comparison to the world: 163

Population growth rate: 2.07% (2020 est.)
country comparison to the world: 44

Birth rate: 22.7 births/1,000 population (2020 est.)
country comparison to the world: 62

Death rate: 7.3 deaths/1,000 population (2020 est.)
country comparison to the world: 111

Net migration rate: 5.1 migrant(s)/1,000 population (2020 est.)
country comparison to the world: 23

Population distribution: most densely populated areas are in the east; the largest city is Djibouti, with a population over 600,000; no other city in the country has a total population over 50,000 as shown in this population distribution map

Urbanization: *urban population:* 78.1% of total population (2020)

rate of urbanization: 1.67% annual rate of change (2015-20 est.)

total population growth rate v. urban population growth rate, 2000-2030: Major urban areas—population: 576,000 DJIBOUTI (capital) (2020)

Sex ratio: *at birth:* 1.03 male(s)/female
0-14 years: 1.01 male(s)/female
15-24 years: 0.89 male(s)/female
25-54 years: 0.71 male(s)/female
55-64 years: 0.76 male(s)/female
65 years and over: 0.8 male(s)/female
total population: 0.83 male(s)/female (2020 est.)

Maternal mortality rate: 248 deaths/100,000 live births (2017 est.)
country comparison to the world: 43

Infant mortality rate: *total:* 41.6 deaths/1,000 live births
male: 47.9 deaths/1,000 live births
female: 35.2 deaths/1,000 live births (2020 est.)
country comparison to the world: 37

Life expectancy at birth: *total population:* 64.7 years
male: 62.1 years
female: 67.4 years (2020 est.)
country comparison to the world: 199

Total fertility rate: 2.19 children born/woman (2020 est.)
country comparison to the world: 93

Contraceptive prevalence rate: 19% (2012)

Drinking water source:

improved: *urban:* 99.3% of population
rural: 59.1% of population
total: 90.3% of population

unimproved: *urban:* 0.7% of population
rural: 40.9% of population
total: 9.7% of population (2017 est.)

Current Health Expenditure: 3.3% (2017)

Physicians density: 0.22 physicians/1,000 population (2014)

Hospital bed density: 1.4 beds/1,000 population (2017)

Sanitation facility access:

improved: *urban:* 84% of population
rural: 21.5% of population
total: 70.1% of population

unimproved: *urban:* 16% of population
rural: 78.5% of population
total: 29.9% of population (2017 est.)

HIV/AIDS—adult prevalence rate: 0.9% (2019 est.)
country comparison to the world: 46

HIV/AIDS—people living with HIV/AIDS: 6,800 (2019 est.)
country comparison to the world: 115

HIV/AIDS—deaths: <500 (2019 est.)

Major infectious diseases: degree of risk: high (2020)

food or waterborne diseases: bacterial and protozoal diarrhea, hepatitis A, and typhoid fever

vectorborne diseases: dengue fever

Obesity—adult prevalence rate: 13.5% (2016)
country comparison to the world: 131

Children under the age of 5 years underweight: 29.9% (2012)

country comparison to the world: 6

Education expenditures: 4.5% of GDP (2010)

country comparison to the world: 87

School life expectancy (primary to tertiary education): *total:* 7 years

male: 7 years

female: 67 years (2011)

GOVERNMENT

Country name: *conventional long form:* Republic of Djibouti

conventional short form: Djibouti

local long form: Republique de Djibouti/ Jumhuriyat Jibuti

local short form: Djibouti/Jibuti

former: French Somaliland, French Territory of the Afars and Issas

etymology: the country name derives from the capital city of Djibouti

Government type: presidential republic

Capital: *name:* Djibouti

geographic coordinates: 11 35 N, 43 09 E

time difference: UTC+ 3 (8 hours ahead of Washington, DC, during Standard Time)

etymology: the origin of the name is disputed; multiple descriptions, possibilities, and theories have been proposed

Administrative divisions: 6 districts (cercles, singular—cercle); Ali Sabieh, Arta, Dikhil, Djibouti, Obock, Tadjourah

Independence: 27 June 1977 (from France)

National holiday: Independence Day, 27 June (1977)

Constitution: *history:* approved by referendum 4 September 1992

amendments: proposed by the president of the republic or by the National Assembly; Assembly consideration of proposals requires assent at least one third of the membership; passage requires a simple majority vote by the Assembly and approval by simple majority vote in a referendum; the president can opt to bypass a referendum if adopted by at least two- thirds majority vote of the Assembly; constitutional articles on the sovereignty of Djibouti, its republican form of government, and its pluralist form of democracy cannot by amended; amended 2006, 2008, 2010

Legal system: mixed legal system based primarily on the French civil code (as it existed in 1997), Islamic religious law (in matters of family law and successions), and customary law

International law organization participation: accepts compulsory ICJ jurisdiction with reservations; accepts ICCt jurisdiction

Citizenship: *citizenship by birth:* no

citizenship by descent only: the mother must be a citizen of Djibouti

dual citizenship recognized: no

residency requirement for naturalization: 10 years

Suffrage: 18 years of age; universal

Executive branch: *chief of state:* President Ismail Omar GUELLEH (since 8 May 1999)

head of government: Prime Minister Abdoulkader Kamil MOHAMED (since 1 April 2013)

cabinet: Council of Ministers appointed by the prime minister

elections/appointments: president directly elected by absolute majority popular vote in 2 rounds if needed for a 5-year term; election last held on 8 April 2016 (next to be held by 2021); prime minister appointed by the president

election results: Ismail Omar GUELLEH reelected president for a fourth term; percent of vote— Ismail Omar GUELLEH (RPP) 87%, Omar Elmi KHAIREH (CDU) 7.3%, other 5.6%

Legislative branch: *description:* unicameral National Assembly or Assemblee Nationale, formerly the Chamber of Deputies (65 seats; members directly elected in multi-seat constituencies by party- list proportional representation vote; members serve 5-year terms)

elections: last held on 23 February 2018 (next to be held in February 2023)

election results: percent of vote by party—NA; seats by party—UMP 57, UDJ- PDD 7, CDU 1; composition—men 47, women 18, percent of women 26.7%

Judicial branch: *highest courts:* Supreme Court or Cour Supreme (consists of NA magistrates); Constitutional Council (consists of 6 magistrates)

judge selection and term of office: Supreme Court magistrates appointed by the president with the advice of the Superior Council of the Magistracy CSM, a 10-member body consisting of 4 judges, 3 members (non parliamentarians and judges) appointed by the president, and 3 appointed by the National Assembly president or speaker; magistrates appointed for life with retirement at age 65; Constitutional Council magistrate appointments—2 by the president of the republic, 2 by the president of the National Assembly, and 2 by the CSM; magistrates appointed for 8-year, non-renewable terms

subordinate courts: High Court of Appeal; 5 Courts of First Instance; customary courts; State Court (replaced sharia courts in 2003)

Political parties and leaders: Center for United Democrats or CDU [Ahmed Mohamed YOUSSOUF, chairman]

Democratic Renewal Party or PRD [Abdillahi HAMARITEH]

Djibouti Development Party or PDD [Mohamed Daoud CHEHEM]

Front for Restoration of Unity and Democracy (Front pour la Restauration de l'Unite Democratique) or FRUD [Ali Mohamed DAOUD]

Movement for Democratic Renewal and Development [Daher Ahmed FARAH]

Movement for Development and Liberty or MoDel [Ismail Ahmed WABERI]

National Democratic Party or PND [Aden Robleh AWALEH]

People's Rally for Progress or RPP [Ismail Omar GUELLEH] (governing party)

Peoples Social Democratic Party or PPSD [Hasna Moumin BAHDON]

Republican Alliance for Democracy or ARD [Aden Mohamed ABDOU, interim president]

Union for a Presidential Majority or UMP (coalition includes RPP, FRUD, PND, PPSD)

Union for Democracy and Justice or UDJ [Ilya Ismail GUEDI Hared]

International organization participation: ACP, AfDB, AFESD, AMF, AU, CAEU (candidates), COMESA, FAO, G- 77, IBRD, ICAO, ICCt, ICRM, IDA, IDB, IFAD, IFC, IFRCS, IGAD, ILO, IMF, IMO, Interpol, IOC, IOM, IPU, ITU, ITUC (NGOs), LAS, MIGA, MINURSO, NAM, OIC, OIF, OPCW, UN, UNCTAD, UNESCO, UNHCR, UNIDO, UNWTO, UPU, WCO, WFTU (NGOs), WHO, WIPO, WMO, WTO

Diplomatic representation in the US: *chief of mission:* Ambassador Mohamed Said DOUALEH (28 December 2016)

chancery: 1156 15th Street NW, Suite 515, Washington, DC 20005

telephone: [1] (202) 331-0270

FAX: [1] (202) 331-0302

Diplomatic representation from the US: *chief of mission:* Ambassador Larry Edward ANDRE, Jr. (since 20 November 2017)

telephone: [253] 21 45 30 00

embassy: Lot 350-B, Haramouss B. P. 185

mailing address: B. P. 185, Djibouti

FAX: [253] 21 45 31 29

Flag description: two equal horizontal bands of light blue (top) and light green with a white isosceles triangle based on the hoist side bearing a red five- pointed star in the center; blue stands for sea and sky and the Issa Somali people; green symbolizes earth and the Afar people; white represents peace; the red star recalls the struggle for independence and stands for unity

National symbol(s): red star; national colors: light blue, green, white, red

National anthem: *name:* "Jabuuti" (Djibouti)

lyrics/music: Aden ELMI/Abdi ROBLEH

note: adopted 1977

0:00 / 0:47

ECONOMY

Economy—overview: Djibouti's economy is based on service activities connected with the country's strategic location as a deepwater port on the Red Sea. Three- fourths of Djibouti's inhabitants live in the capital city; the remainder are mostly nomadic herders. Scant rainfall and less than 4% arable land limits crop production to small quantities of fruits and vegetables, and most food must be imported.

Djibouti provides services as both a transit port for the region and an international transshipment and refueling center. Imports, exports, and reexports represent 70% of port activity at Djibouti's container terminal. Reexports consist primarily of coffee from landlocked neighbor Ethiopia. Djibouti has few natural resources and little industry. The nation is, therefore, heavily dependent on foreign assistance to support its balance of payments and to finance development projects. An official unemployment rate of nearly 40%—with youth unemployment near 80%—continues to be a major problem. Inflation was a modest 3% in 2014-2017, due to low international food prices and a decline in electricity tariffs.

Djibouti's reliance on diesel-generated electricity and imported food and water leave average consumers vulnerable to global price shocks, though in mid-2015 Djibouti passed new legislation to liberalize the energy sector. The government has emphasized infrastructure development for transportation and energy and Djibouti—with the help of foreign partners, particularly China—has begun to increase and modernize its port capacity. In 2017, Djibouti opened two of the largest projects in its history, the Doraleh Port and Djibouti- Addis Ababa Railway, funded by China as part of the "Belt and Road Initiative," which will increase the country's ability to capitalize on its strategic location.

GDP (purchasing power parity): $3.64 billion (2017 est.)
$3.411 billion (2016 est.)
$3.203 billion (2015 est.)
note: data are in 2017 dollars
country comparison to the world: 183

GDP (official exchange rate): $2.029 billion (2017 est.)

GDP—real growth rate: 6.7% (2017 est.)
6.5% (2016 est.)
6.5% (2015 est.)
country comparison to the world: 23

GDP—per capita (PPP): $3,600 (2017 est.)
$3,400 (2016 est.)
$3,300 (2015 est.)
note: data are in 2017 dollars
country comparison to the world: 185

Gross national saving: 22.3% of GDP (2017 est.)
38.1% of GDP (2016 est.)
19% of GDP (2015 est.)
country comparison to the world: 81

GDP—composition, by end use: household consumption: 56.5% (2017 est.)
government consumption: 29.2% (2017 est.)
investment in fixed capital: 41.8% (2017 est.)
investment in inventories: 0.3% (2017 est.)
exports of goods and services: 38.6% (2017 est.)
imports of goods and services: -66.4% (2017 est.)

GDP—composition, by sector of origin:
agriculture: 2.4% (2017 est.)
industry: 17.3% (2017 est.)
services: 80.2% (2017 est.)

Agriculture—products: fruits, vegetables; goats, sheep, camels, animal hides

Industries: construction, agricultural processing, shipping

Industrial production growth rate: 2.7% (2017 est.)
country comparison to the world: 112

Labor force: 294,600 (2012)
country comparison to the world: 163

Labor force—by occupation: agriculture: NA
industry: NA
services: NA

Unemployment rate: 40% (2017 est.)
60% (2014 est.)
country comparison to the world: 213

Population below poverty line: 23% (2015 est.)
note: percent of population below $1.25 per day at purchasing power parity

Household income or consumption by percentage share: lowest 10%: 2.4%
highest 10%: 30.9% (2002)

Budget: revenues: 717 million (2017 est.)
expenditures: 899.2 million (2017 est.)

Taxes and other revenues: 35.3% (of GDP) (2017 est.)
country comparison to the world: 62

Budget surplus (+) or deficit (-): -9% (of GDP) (2017 est.)
country comparison to the world: 205

Public debt: 31.8% of GDP (2017 est.)
33.7% of GDP (2016 est.)
country comparison to the world: 161

Fiscal year: calendar year

Inflation rate (consumer prices): 0.7% (2017 est.)
2.7% (2016 est.)
country comparison to the world: 36

Current account balance: -$280 million (2017 est.)
-$178 million (2016 est.)
country comparison to the world: 105

Exports: $139.9 million (2017 est.)
(2016)
country comparison to the world: 192

Exports—partners: Ethiopia 38.8%, Somalia 17.1%, Qatar 9.1%, Brazil 8.9%, Yemen 4.9%, US 4.6% (2017)

Exports—commodities: reexports, hides and skins, scrap metal

Imports: $726.4 million (2017 est.)
$705.2 million (2016 est.)
country comparison to the world: 192

Imports—commodities: foods, beverages, transport equipment, chemicals, petroleum products, clothing

Imports—partners: UAE 25%, France 15.2%, Saudi Arabia 11%, China 9.6%, Ethiopia 6.8%, Yemen 4.6% (2017)

Reserves of foreign exchange and gold: $547.7 million (31 December 2017 est.)
$398.5 million (31 December 2016 est.)
country comparison to the world: 148

Debt—external: $1.954 billion (31 December 2017 est.)
$1.519 billion (31 December 2016 est.)
country comparison to the world: 152

Exchange rates: Djiboutian francs (DJF) per US dollar -
177.7 (2017 est.)
177.72 (2016 est.)
177.72 (2015 est.)
177.72 (2014 est.)
177.72 (2013 est.)

ENERGY

Electricity access: population without electricity: 400,000 (2019)
electrification—total population: 42% (2019)
electrification—urban areas: 54% (2019)

electrification—rural areas: 1% (2019)

Electricity—production: 405.5 million kWh (2016 est.)
country comparison to the world: 170

Electricity—consumption: 377.1 million kWh (2016 est.)
country comparison to the world: 177

Electricity—exports: 0 kWh (2016 est.)
country comparison to the world: 127

Electricity—imports: 0 kWh (2016 est.)
country comparison to the world: 141

Electricity—installed generating capacity: 130,300 kW (2016 est.)
country comparison to the world: 175

Electricity—from fossil fuels: 100% of total installed capacity (2016 est.)
country comparison to the world: 7

Electricity—from nuclear fuels: 0% of total installed capacity (2017 est.)
country comparison to the world: 78

Electricity—from hydroelectric plants: 0% of total installed capacity (2017 est.)
country comparison to the world: 168

Electricity—from other renewable sources: 0% of total installed capacity (2017 est.)
country comparison to the world: 185

Crude oil—production: 0 bbl/day (2018 est.)
country comparison to the world: 127

Crude oil—exports: 0 bbl/day (2015 est.)
country comparison to the world: 114

Crude oil—imports: 0 bbl/day (2015 est.)
country comparison to the world: 117

Crude oil—proved reserves: 0 bbl (1 January 2018 est.)
country comparison to the world: 123

Refined petroleum products—production: 0 bbl/day (2015 est.)
country comparison to the world: 136

Refined petroleum products—consumption: 6,360 bbl/day (2016 est.)
country comparison to the world: 170

Refined petroleum products—exports: 403 bbl/day (2015 est.)
country comparison to the world: 112

Refined petroleum products—imports: 6,692 bbl/day (2015 est.)
country comparison to the world: 161

Natural gas—production: 0 cu m (2017 est.)
country comparison to the world: 123

Natural gas—consumption: 0 cu m (2017 est.)
country comparison to the world: 139

Natural gas—exports: 0 cu m (2017 est.)
country comparison to the world: 94

Natural gas—imports: 0 cu m (2017 est.)
country comparison to the world: 116

Natural gas—proved reserves: 0 cu m (1 January 2014 est.)
country comparison to the world: 127

Carbon dioxide emissions from consumption of energy: 950,200 Mt (2017 est.)
country comparison to the world: 171

COMMUNICATIONS

Telephones—fixed lines: *total subscriptions:* 34,671
subscriptions per 100 inhabitants: 3.84 (2019 est.)
country comparison to the world: 167

Telephones—mobile cellular: *total subscriptions:* 371,992
subscriptions per 100 inhabitants: 41.2 (2019 est.)
country comparison to the world: 175

Telecommunication systems: *general assessment:* telephone facilities in the city of Djibouti are adequate, as are the microwave radio relay connections to outlying areas of the country; Djibouti is one of the few remaining countries in which the national telco, Djibouti Telecom (DT), has a monopoly on all telecom services, including fixed lines, mobile, Internet and broadband; the lack of competition has meant that the market has not lived up to its potential; broadband's growth held back by the expense and mobile and Internet markets need foreign investment (2020)
domestic: 4 per 100 fixed-line and 41 per 100 mobile-cellular; Djibouti Telecom (DT) is the sole provider of telecommunications services and utilizes mostly a microwave radio relay network; fiber-optic cable is installed in the capital; rural areas connected via wireless local loop radio systems; mobile cellular coverage is primarily limited to the area in and around Djibouti city (2019)
international: country code—253; landing points for the SEA-ME-WE-3 & 5, EASSy, Aden-Djibouti, Africa-1, DARE-1, EIG, MENA, Bridge International, PEACE Cable, and SEACOM fiber-optic submarine cable systems providing links to Asia, the Middle East, Europe, Southeast Asia, Australia and Africa; satellite earth stations—2 (1 Intelsat—Indian Ocean and 1 Arabsat) (2019)
note: the COVID-19 outbreak is negatively impacting telecommunications production and supply chains globally; consumer spending on telecom devices and services has also slowed due to the pandemic's effect on economies worldwide; overall progress towards improvements in all facets of the telecom industry—mobile, fixed-line, broadband, submarine cable and satellite—has moderated

Broadcast media: state-owned Radiodiffusion-Television de Djibouti operates the sole terrestrial TV station, as well as the only 2 domestic radio networks; no private TV or radio stations; transmissions of several international broadcasters are available (2019)

Internet country code: .dj

Internet users: *total:* 492,221
percent of population: 55.68% (July 2018 est.)
country comparison to the world: 153

Broadband—fixed subscriptions: *total:* 25,508
subscriptions per 100 inhabitants: 3 (2018 est.)
country comparison to the world: 147

TRANSPORTATION

National air transport system: *number of registered air carriers:* 2 (2020)
inventory of registered aircraft operated by air carriers: 4

Civil aircraft registration country code prefix: J2 (2016)

Airports: 13 (2013)
country comparison to the world: 151

Airports—with paved runways: *total:* 3 (2017)
over 3,047 m: 1 (2017)
2,438 to 3,047 m: 1 (2017)
1,524 to 2,437 m: 1 (2017)

Airports—with unpaved runways: *total:* 10 (2013)
1,524 to 2,437 m: 1 (2013)
914 to 1,523 m: 7 (2013)
under 914 m: 2 (2013)

Railways: *total:* 97 km (Djibouti segment of the 756 km Addis Ababa-Djibouti railway) (2017)
standard gauge: 97 km 1.435-m gauge (2017)
country comparison to the world: 127

Roadways: total: 2,893 km (2013)
country comparison to the world: 163

Merchant marine: *total:* 20
by type: general cargo 1, other 19 (2019)
country comparison to the world: 143

Ports and terminals: *major seaport(s):* Djibouti

MILITARY AND SECURITY

Military and security forces: Djibouti Armed Forces (FAD): Djibouti National Army (includes Navy, Djiboutian Air Force, National Gendarmerie); Djibouti Coast Guard (2019)

Military and security service personnel strengths: the Djibouti Armed Forces (FAD) have approximately 10,500 active troops (8,000 Army; 250 Naval; 250 Air; 2,000 Gendarmerie); 150 Coast Guard (2019 est.)

Military equipment inventories and acquisitions: the FAD is armed mostly with older French and Soviet-era weapons systems; since 2010, it has received limited amounts of newer equipment, with China and the US as the largest suppliers (2019 est.)

Military deployments: 960 Somalia (AMISOM) (2020)

Military service age and obligation: 18 years of age for voluntary military service; 16-25 years of age for voluntary military training; no conscription (2012)

Maritime threats: the International Maritime Bureau reports offshore waters in the Red Sea and Gulf of Aden remain a high risk for piracy; the presence of several naval task forces in the Gulf of Aden and additional anti-piracy measures on the part of ship operators, including the use of on-board armed security teams, contributed to the drop in incidents; there was one incident in the Gulf of Aden and none in the Red Sea in 2018; Operation Ocean Shield, the NATO/EUNAVFOR naval task force established in 2009 to combat Somali piracy, concluded its operations in December 2016 as a result of the drop in reported incidents over the last few years; the EU naval mission, Operation ATALANTA, continues its operations in the Gulf of Aden and Indian Ocean through 2020; naval units from Japan, India, and China also operate in conjunction with EU forces; China has established a logistical base in Djibouti to support its deployed naval units in the Horn of Africa

TERRORISM

Terrorist group(s): al-Shabaab (2019)
note: details about the history, aims, leadership, organization, areas of operation, tactics, targets, weapons, size, and sources of support of the group(s) appear(s) in Appendix-T

TRANSNATIONAL ISSUES

Disputes—international: Djibouti maintains economic ties and border accords with "Somaliland" leadership while maintaining some political ties to various factions in Somalia; Kuwait is chief investor in the 2008 restoration and upgrade of the Ethiopian-Djibouti rail link; in 2008, Eritrean troops moved across the border on Ras Doumera peninsula and occupied Doumera Island with undefined sovereignty in the Red Sea

Refugees and internally displaced persons: *refugees (country of origin):* 12,139 (Somalia) (2020)

Trafficking in persons: *current situation:* Djibouti is a transit, source, and destination country for men, women, and children subjected to forced labor and sex trafficking; economic migrants from East Africa en route to Yemen and other Middle East locations are vulnerable to exploitation in Djibouti; some women and girls may be forced into domestic servitude or prostitution after reaching Djibouti City, the Ethiopia-Djibouti trucking corridor, or Obock—the main crossing point into Yemen; Djiboutian and foreign children may be forced to beg, to work as domestic servants, or to commit theft and other petty crimes

tier rating: Tier 2 Watch List—Djibouti does not fully comply with the minimum standards for the elimination of trafficking; however, it is making significant efforts to do so; in 2014, Djibouti was granted a waiver from an otherwise required downgrade to Tier 3 because its government has a written plan that, if implemented would constitute making significant efforts to bring itself into compliance with the minimum standards for the elimination of trafficking; one forced labor trafficker was convicted in 2014 but received a suspended sentence inadequate to deter trafficking; authorities did not investigate or prosecute any other forced labor crimes, any sex trafficking offenses, or any officials complicit in human trafficking, and remained limited in their ability to recognize or protect trafficking victims; official round-ups, detentions, and deportations of non-Djiboutian residents, including children without screening for trafficking victims remained routine; the government did not provide care to victims but supported local NGOs operating centers that assisted victims (2015)

DOMINICA

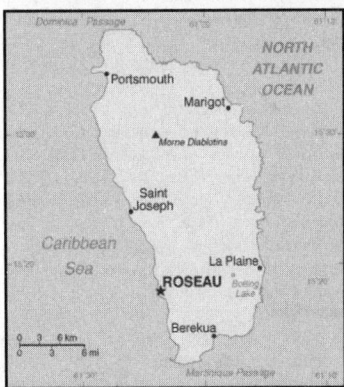

Maritime claims: *territorial sea:* 12 nm
exclusive economic zone: 200 nm
contiguous zone: 24 nm

Climate: tropical; moderated by northeast trade winds; heavy rainfall

Terrain: rugged mountains of volcanic origin

Elevation: *lowest point:* Caribbean Sea 0 m
highest point: Morne Diablotins 1,447 m

Natural resources: timber, hydropower, arable land

Land use: *agricultural land:* 34.7% (2011 est.)
arable land: 8% (2011 est.) / *permanent crops:*
24% (2011 est.) / *permanent pasture:* 2.7% (2011 est.)
forest: 59.2% (2011 est.)
other: 6.1% (2011 est.)

Irrigated land: NA

Population distribution: population is mosly clustered along the coast, with roughly a third living in the parish of St. George, in or around the capital of Roseau; the volcanic interior is sparsely populated

Natural hazards: flash floods are a constant threat; destructive hurricanes can be expected during the late summer months
volcanism: Dominica was the last island to be formed in the Caribbean some 26 million years ago, it lies in the middle of the volcanic island arc of the Lesser Antilles that extends from the island of Saba in the north to Grenada in the south; of the 16 volcanoes that make up this arc, five are located on Dominica, more than any other island in the Caribbean: Morne aux Diables (861 m), Morne Diablotins (1,430 m), Morne Trois Pitons (1,387 m), Watt Mountain (1,224 m), which last erupted in 1997, and Morne Plat Pays (940 m); the two best known volcanic features on Dominica, the Valley of Desolation and the Boiling Lake thermal areas, lie on the flanks of Watt Mountain and both are popular tourist destinations

Environment—current issues: water shortages a continuing concern; pollution from agrochemicals and from untreated sewage; forests endangered by the expansion of farming; soil erosion; pollution of the coastal zone by agricultural and industrial chemicals, and untreated sewage

Environment—international agreements: party to: Biodiversity, Climate Change, Climate Change-Kyoto Protocol, Desertification, Endangered Species, Environmental Modification, Hazardous Wastes, Law of the Sea, Ozone Layer Protection, Ship Pollution, Whaling
signed, but not ratified: none of the selected agreements

Geography—note: known as "The Nature Island of the Caribbean" due to its spectacular, lush, and varied flora and fauna, which are protected by an extensive natural park system; the most mountainous of the Lesser Antilles, its volcanic peaks are cones of lava craters and include Boiling Lake, the second-largest, thermally active lake in the world

INTRODUCTION

Background: Dominica was the last of the Caribbean islands to be colonized by Europeans due chiefly to the fierce resistance of the native Caribs. France ceded possession to Great Britain in 1763, which colonized the island in 1805. Slavery ended in 1833 and in 1835 the first three men of African descent were elected to the legislative assembly of Dominica. In 1871, Dominica became part first of the British Leeward Islands and then the British Windward Islands until 1958. In 1967 Dominica became an associated state of the UK, and formally took responsibility for its internal affairs. In 1980, two years after independence, Dominica's fortunes improved when a corrupt and tyrannical administration was replaced by that of Mary Eugenia CHARLES, the first female prime minister in the Caribbean, who remained in office for 15 years. On 18 September 2017, Hurricane Maria passed over the island causing extensive damage to structures, roads, communications, and the power supply, and largely destroying critical agricultural areas.

GEOGRAPHY

Location: Caribbean, island between the Caribbean Sea and the North Atlantic Ocean, about halfway between Puerto Rico and Trinidad and Tobago

Geographic coordinates: 15 25 N, 61 20 W

Map references: Central America and the Caribbean

Area: *total:* 751 sq km
land: 751 sq km
water: NEGL
country comparison to the world: 189

Area—comparative: slightly more than four times the size of Washington, DC

Land boundaries: 0 km

Coastline: 148 km

PEOPLE AND SOCIETY

Population: 74,243 (July 2020 est.)
country comparison to the world: 202

Nationality: *noun:* Dominican(s)
adjective: Dominican

Ethnic groups: African descent 86.6%, mixed 9.1%, indigenous 2.9%, other 1.3%, unspecified 0.2% (2001 est.)

Languages: English (official), French patois

Religions: Roman Catholic 61.4%, Protestant 28.6% (includes Evangelical 6.7%, Seventh Day Adventist 6.1%, Pentecostal 5.6%, Baptist 4.1%, Methodist 3.7%, Church of God 1.2%, other 1.2%), Rastafarian 1.3%, Jehovah's Witness 1.2%, other 0.3%, none 6.1%, unspecified 1.1% (2001 est.)

Age structure: *0-14 years:* 21.41% (male 8,135/ female 7,760)
15-24 years: 13.15% (male 5,017/female 4,746)
25-54 years: 42.79% (male 16,133/female 15,637)
55-64 years: 10.53% (male 4,089/female 3,731)
65 years and over: 12.12% (male 4,128/female 4,867) (2020 est.)

Median age: *total:* 34.9 years
male: 34.4 years
female: 35.5 years (2020 est.)
country comparison to the world: 87

Population growth rate: 0.13% (2020 est.)
country comparison to the world: 184

Birth rate: 14.5 births/1,000 population (2020 est.)
country comparison to the world: 127

Death rate: 8 deaths/1,000 population (2020 est.)
country comparison to the world: 91

Net migration rate: -5.3 migrant(s)/1,000 population (2020 est.)
country comparison to the world: 199

Population distribution: population is mosly clustered along the coast, with roughly a third living in the parish of St. George, in or around the capital of Roseau; the volcanic interior is sparsely populated

Urbanization: *urban population:* 71.1% of total population (2020)
rate of urbanization: 0.94% annual rate of change (2015-20 est.)
total population growth rate v. urban population growth rate, 2000-2030:

Major urban areas—population: 15,000 ROSEAU (capital) (2018)

Sex ratio: *at birth:* 1.05 male(s)/female
0-14 years: 1.05 male(s)/female
15-24 years: 1.06 male(s)/female
25-54 years: 1.03 male(s)/female
55-64 years: 1.1 male(s)/female
65 years and over: 0.85 male(s)/female
total population: 1.02 male(s)/female (2020 est.)

Infant mortality rate: *total:* 9.7 deaths/1,000 live births

male: 12.7 deaths/1,000 live births
female: 6.5 deaths/1,000 live births (2020 est.)
country comparison to the world: 134

Life expectancy at birth: *total population:* 77.7 years
male: 74.7 years
female: 80.9 years (2020 est.)
country comparison to the world: 76

Total fertility rate: 2.02 children born/woman (2020 est.)
country comparison to the world: 110

Drinking water source:
improved:
urban: 95.7% of population
unimproved:
urban: 4.3% of population

Current Health Expenditure: 5.9% (2017)

Physicians density: 1.12 physicians/1,000 population (2017)

Hospital bed density: 3.8 beds/1,000 population (2010)

HIV/AIDS—adult prevalence rate: 0.6% (2018)
country comparison to the world: 61

HIV/AIDS—people living with HIV/AIDS: <500 (2018)

HIV/AIDS—deaths: <100 (2018)

Obesity—adult prevalence rate: 27.9% (2016)
country comparison to the world: 33

Education expenditures: 3.4% of GDP (2015)
country comparison to the world: 125

People—note: 3,000-3,500 Kalinago (Carib) still living on Dominica are the only pre-Columbian population remaining in the Caribbean; only 70-100 may be "pure" Kalinago because of years of integration into the broader population

GOVERNMENT

Country name: *conventional long form:* Commonwealth of Dominica
conventional short form: Dominica
etymology: the island was named by explorer Christopher COLUMBUS for the day of the week on which he spotted it, Sunday ("Domingo" in Latin), 3 November 1493

Government type: parliamentary republic

Capital: *name:* Roseau
geographic coordinates: 15 18 N, 61 24 W
time difference: UTC-4 (1 hour ahead of Washington, DC, during Standard Time)
etymology: the name is French for "reed"; the first settlement was named after the river reeds that grew in the area

Administrative divisions: 10 parishes; Saint Andrew, Saint David, Saint George, Saint John, Saint Joseph, Saint Luke, Saint Mark, Saint Patrick, Saint Paul, Saint Peter

Independence: 3 November 1978 (from the UK)

National holiday: Independence Day, 3 November (1978)

Constitution: *history:* previous 1967 (preindependence); latest presented 25 July 1978, entered into force 3 November 1978
amendments: proposed by the House of Assembly; passage of amendments to constitutional sections such as fundamental rights and freedoms, the government structure, and constitutional amendment procedures requires approval by three fourths of the Assembly membership in the final reading of the amendment bill, approval by simple majority in a referendum, and assent of the president; amended several times, last in 2015

Legal system: common law based on the English model

International law organization participation: accepts compulsory ICJ jurisdiction; accepts ICCt jurisdiction

Citizenship: *citizenship by birth:* yes
citizenship by descent only: yes
dual citizenship recognized: yes
residency requirement for naturalization: 5 years

Suffrage: 18 years of age; universal

Executive branch: *chief of state:* President Charles A. SAVARIN (since 2 October 2013)
head of government: Prime Minister Roosevelt SKERRIT (since 8 January 2004)
cabinet: Cabinet appointed by the president on the advice of the prime minister
elections/appointments: president nominated by the prime minister and leader of the opposition party and elected by the House of Assembly for a 5-year term (eligible for a second term); election last held on 1 October 2018 (next to be held in October 2023); prime minister appointed by the president
election results: Charles A. SAVARIN (DLP) reelected president unopposed

Legislative branch: *description:* unicameral House of Assembly (32 seats; 21 representatives directly elected in single-seat constituencies by simple majority vote, 9 senators appointed by the president—5 on the advice of the prime minister, and 4 on the advice of the leader of the opposition party, plus 2 ex-officio members—the house speaker and the attorney general; members serve 5-year terms)
elections: last held on 6 December 2019 (next to be held in 2024); note—tradition dictates that the election is held within 5 years of the last election, but technically it is 5 years from the first seating of parliament plus a 90-day grace period
election results: percent of vote by party—DLP 59.0%, UWP 41.0%; seats by party—DLP 18, UWP 3

Judicial branch: *highest courts:* the Eastern Caribbean Supreme Court (ECSC) is the superior court of the Organization of Eastern Caribbean States; the ECSC—headquartered on St. Lucia—consists of the Court of Appeal—headed by the chief justice and 4 judges—and the High Court with 18 judges; the Court of Appeal is itinerant, traveling to member states on a schedule to hear appeals from the High Court and subordinate courts; High Court judges reside in the member states, with 2 in Dominica; note—in 2015, Dominica acceded to the Caribbean Court of Justice as final court of appeal, replacing that of the Judicial Committee of the Privy Council, in London
judge selection and term of office: chief justice of Eastern Caribbean Supreme Court appointed by the Her Majesty, Queen ELIZABETH II; other justices and judges appointed by the Judicial and Legal Services Commission, an independent body of judicial officials; Court of Appeal justices appointed for life with mandatory retirement at age 65; High Court judges appointed for life with mandatory retirement at age 62
subordinate courts: Court of Summary Jurisdiction; magistrates' courts

Political parties and leaders:
Dominica Freedom Party or DFP [Judith PESTAINA]
Dominica Labor Party or DLP [Roosevelt SKERRIT]
Dominica United Workers Party or UWP [Lennox LINTON]

International organization participation: ACP, AOSIS, C, Caricom, CD, CDB, CELAC, Commonwealth of Nations, ECCU, FAO, G-77, IAEA, IBRD, ICCt, ICRM, IDA, IFAD, IFC, IFRCS, ILO, IMF, IMO, Interpol, IOC, ISO (correspondent), ITU, ITUC (NGOs), MIGA, NAM, OAS, OECS, OIF, OPANAL, OPCW, Petrocaribe, UN, UNCTAD, UNESCO, UNIDO, UPU, WFTU, WHO, WIPO, WMO, WTO

Diplomatic representation in the US: *chief of mission:* Ambassador Vince HENDERSON (since 18 January 2017)
chancery: 3216 New Mexico Avenue NW, Washington, DC 20016
telephone: [1] (202) 364-6781
FAX: [1] (202) 364-6791
consulate(s) general: New York

Diplomatic representation from the US: the US does not have an embassy in Dominica; the US Ambassador to Barbados is accredited to Dominica

Flag description: green with a centered cross of three equal bands—the vertical part is yellow (hoist side), black, and white and the horizontal part is yellow (top), black, and white; superimposed in the center of the cross is a red disk bearing a Sisserou parrot, unique to Dominica, encircled by 10 green, five-pointed stars edged in yellow; the 10 stars represent the 10 administrative divisions (parishes); green symbolizes the island's lush vegetation; the triple-colored cross represents the Christian Trinity; the yellow color denotes sunshine, the main agricultural products (citrus and bananas), and the native Carib Indians; black is for the rich soil and the African heritage of most citizens; white signifies rivers, waterfalls, and the purity of aspirations; the red disc stands for social justice

National symbol(s): *Sisserou parrot, Carib Wood flower; national colors:* green, yellow, black, white, red

National anthem: *name:* Isle of Beauty
lyrics/music: Wilfred Oscar Morgan POND/ Lemuel McPherson CHRISTIAN
note: adopted 1967
0:00 / 0:48

ECONOMY

Economy—overview: The Dominican economy was dependent on agriculture—primarily bananas—in years past, but increasingly has been driven by tourism, as the government seeks to promote Dominica as an "ecotourism" destination. However, Hurricane Maria, which passed through the island in September 2017, destroyed much of the country's agricultural sector and caused damage to all of the country's transportation and physical infrastructure. Before Hurricane Maria, the government had attempted to foster an offshore financial industry and planned to sign agreements with the private sector to develop geothermal energy resources. At a time when government finances are fragile, the government's focus has been to get the country back in shape to service cruise ships. The economy contracted in 2015 and recovered to positive growth in 2016 due to a recovery of agriculture and tourism. Dominica suffers from high debt levels, which increased from 67% of GDP in 2010 to 77% in 2016. Dominica is one of five countries in the East Caribbean that have citizenship by investment programs whereby foreigners can obtain passports for a fee and revenue from this contribute to government budgets.

GDP (purchasing power parity):
$783 million (2017 est.)
$821.5 million (2016 est.)
$800.4 million (2015 est.)
note: data are in 2017 dollars
country comparison to the world: 206

GDP (official exchange rate):
$557 million (2017 est.)

GDP—real growth rate:
-4.7% (2017 est.)
2.6% (2016 est.)
-3.7% (2015 est.)
country comparison to the world: 217

GDP—per capita (PPP):
$11,000 (2017 est.)
$11,600 (2016 est.)
$11,300 (2015 est.)
note: data are in 2017 dollars
country comparison to the world: 136

Gross national saving:
10.8% of GDP (2017 est.)
20% of GDP (2016 est.)
14.3% of GDP (2015 est.)
country comparison to the world: 160

GDP—composition, by end use:
household consumption: 60.6% (2017 est.)
government consumption: 26.2% (2017 est.)
investment in fixed capital: 21.5% (2017 est.)

investment in inventories: 0% (2017 est.)
exports of goods and services: 54.4% (2017 est.)
imports of goods and services: -62.7% (2017 est.)

GDP—composition, by sector of origin:
agriculture: 22.3% (2017 est.)
industry: 12.6% (2017 est.)
services: 65.1% (2017 est.)

Agriculture—products: bananas, citrus, mangos, root crops, coconuts, cocoa
note: forest and fishery potential not exploited

Industries: soap, coconut oil, tourism, copra, furniture, cement blocks, shoes

Industrial production growth rate: -13% (2017 est.)
country comparison to the world: 199

Labor force: 25,000 (2000 est.)
country comparison to the world: 208

Labor force—by occupation:
agriculture: 40%
industry: 32%
services: 28% (2002 est.)

Unemployment rate: 23% (2000 est.)
country comparison to the world: 192

Population below poverty line: 29% (2009 est.)

Household income or consumption by percentage share: *lowest 10%:* NA
highest 10%: NA

Budget: *revenues:* 227.8 million (2017 est.)
expenditures: 260.4 million (2017 est.)

Taxes and other revenues: 40.9% (of GDP) (2017 est.)
country comparison to the world: 34

Budget surplus (+) or deficit (-): -5.9% (of GDP) (2017 est.)
country comparison to the world: 181

Public debt:
82.7% of GDP (2017 est.)
71.7% of GDP (2016 est.)
country comparison to the world: 33

Fiscal year: 1 July—30 June

Inflation rate (consumer prices):
0.6% (2017 est.)
0% (2016 est.)
country comparison to the world: 32

Current account balance:
-$70 million (2017 est.)
$5 million (2016 est.)
country comparison to the world: 83

Exports:
$28 million (2017 est.)
$43.7 million (2016 est.)
country comparison to the world: 206

Exports—partners: Saudi Arabia 42.6%, Trinidad and Tobago 9.3%, Jamaica 8.1%, St. Kitts and Nevis 7.1%, Guyana 6.7% (2017)

Exports—commodities: bananas, soap, bay oil, vegetables, grapefruit, oranges

Imports:
$206.6 million (2017 est.)
$188.4 million (2016 est.)

country comparison to the world: 210

Imports—commodities: manufactured goods, machinery and equipment, food, chemicals

Imports—partners: US 61.3%, Trinidad and Tobago 9.8% (2017)

Reserves of foreign exchange and gold:
$212.3 million (31 December 2017 est.)
$221.9 million (31 December 2016 est.)
country comparison to the world: 172

Debt—external:
$280.4 million (31 December 2017 est.)
$314.2 million (31 December 2015 est.)
country comparison to the world: 186

Exchange rates: East Caribbean dollars (XCD) per US dollar -
2.7 (2017 est.)
2.7 (2016 est.)
2.7 (2015 est.)
2.7 (2014 est.)
2.7 (2013 est.)

ENERGY

Electricity access: *electrification—total population:* 100% (2020)

Electricity—production: 111.4 million kWh (2016 est.)
country comparison to the world: 199

Electricity—consumption: 103.6 million kWh (2016 est.)
country comparison to the world: 201

Electricity—exports: 0 kWh (2016 est.)
country comparison to the world: 128

Electricity—imports: 0 kWh (2016 est.)
country comparison to the world: 142

Electricity—installed generating capacity: 27,800 kW (2016 est.)
country comparison to the world: 201

Electricity—from fossil fuels: 72% of total installed capacity (2016 est.)
country comparison to the world: 102

Electricity—from nuclear fuels: 0% of total installed capacity (2017 est.)
country comparison to the world: 79

Electricity—from hydroelectric plants: 25% of total installed capacity (2017 est.)
country comparison to the world: 76

Electricity—from other renewable sources: 3% of total installed capacity (2017 est.)
country comparison to the world: 124

Crude oil—production: 0 bbl/day (2018 est.)
country comparison to the world: 128

Crude oil—exports: 0 bbl/day (2015 est.)
country comparison to the world: 115

Crude oil—imports: 0 bbl/day (2015 est.)
country comparison to the world: 118

Crude oil—proved reserves: 0 bbl (1 January 2018 est.)
country comparison to the world: 124

Refined petroleum products—production: 0 bbl/day (2015 est.)

277

country comparison to the world: 137

Refined petroleum products—consumption: 1,300 bbl/day (2016 est.)
country comparison to the world: 203

Refined petroleum products—exports: 0 bbl/day (2015 est.)
country comparison to the world: 148

Refined petroleum products—imports: 1,237 bbl/day (2015 est.)
country comparison to the world: 199

Natural gas—production: 0 cu m (2017 est.)
country comparison to the world: 124

Natural gas—consumption: 0 cu m (2017 est.)
country comparison to the world: 140

Natural gas—exports: 0 cu m (2017 est.)
country comparison to the world: 95

Natural gas—imports: 0 cu m (2017 est.)
country comparison to the world: 117

Natural gas—proved reserves: 0 cu m (1 January 2014 est.)
country comparison to the world: 128

Carbon dioxide emissions from consumption of energy: 199,600 Mt (2017 est.)
country comparison to the world: 200

COMMUNICATIONS

Telephones—fixed lines: *total subscriptions:* 2,751
subscriptions per 100 inhabitants: 3.71 (2019 est.)
country comparison to the world: 214

Telephones—mobile cellular: *total subscriptions:* 78,437
subscriptions per 100 inhabitants: 105.79 (2019 est.)
country comparison to the world: 198

Telecommunication systems: *general assessment:* fully automatic network; there are multiple competing operators licensed to provide services, most of them are small and localized; the telecom

sector across the Caribbean region remains one of the key growth areas; (2020)
domestic: fixed-line connections continue to decline slowly with only two active operators providing about 4 fixed-line connections per 100 persons; subscribership among the three mobile-cellular providers is about 106 per 100 persons (2019)
international: country code—1-767; landing points for the ECFS and the Southern Caribbean Fiber submarine cables providing connectivity to other islands in the eastern Caribbean extending from the British Virgin Islands to Trinidad and to the US; microwave radio relay and SHF radiotelephone links to Martinique and Guadeloupe; VHF and UHF radiotelephone links to Saint Lucia (2019)
note: the COVID-19 outbreak is negatively impacting telecommunications production and supply chains globally; consumer spending on telecom devices and services has also slowed due to the pandemic's effect on economies worldwide; overall progress towards improvements in all facets of the telecom industry—mobile, fixed-line, broadband, submarine cable and satellite—has moderated

Broadcast media: no terrestrial TV service available; subscription cable TV provider offers some locally produced programming plus channels from the US, Latin America, and the Caribbean; state-operated radio broadcasts on 6 stations; privately owned radio broadcasts on about 15 stations (2019)

Internet country code: .dm

Internet users: *total:* 51,538
percent of population: 69.62% (July 2018 est.)
country comparison to the world: 196

Broadband—fixed subscriptions: *total:* 11,514
subscriptions per 100 inhabitants: 16 (2018 est.)
country comparison to the world: 167

TRANSPORTATION

Civil aircraft registration country code prefix: J7 (2016)

Airports: 2 (2020)
country comparison to the world: 198

Airports—with paved runways: *total:* 2 (2019)
1,524 to 2,437 m: 1
914 to 1,523 m: 1

Roadways: *total:* 1,512 km (2018)
paved: 762 km (2018)
unpaved: 750 km (2018)
country comparison to the world: 174

Merchant marine: *total:* 108
by type: general cargo 29, oil tanker 29, other 50 (2019)
country comparison to the world: 85

Ports and terminals: *major seaport(s):* Portsmouth, Roseau

MILITARY AND SECURITY

Military and security forces: no regular military forces; Commonwealth of Dominica Police Force (includes Coast Guard) (2019)

Military—note: Dominica participates in the Regional Security System (RSS) an international agreement for the defense and security of the eastern Caribbean region (2019)

TRANSNATIONAL ISSUES

Disputes—international: Dominica is the only Caribbean state to challenge Venezuela's sovereignty claim over Aves Island and joins the other island nations in challenging whether the feature sustains human habitation, a criterion under the UN Convention on the Law of the Sea, which permits Venezuela to extend its EEZ and continental shelf claims over a large portion of the eastern Caribbean Sea

Illicit drugs: transshipment point for narcotics bound for the US and Europe; minor cannabis producer

DOMINICAN REPUBLIC

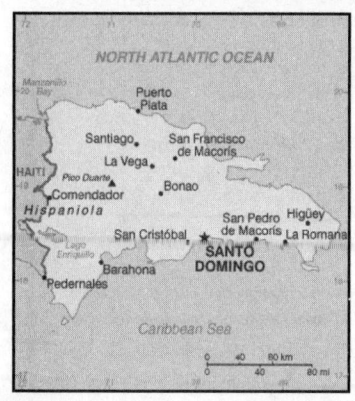

INTRODUCTION

Background: The Taino—indigenous inhabitants of Hispaniola prior to the arrival of the Europeans—divided the island into five chiefdoms and territories. Christopher COLUMBUS explored and claimed the island on his first voyage in 1492; it became a springboard for Spanish conquest of the Caribbean and the American mainland. In 1697, Spain recognized French dominion over the western third of the island, which in 1804 became Haiti. The remainder of the island, by then known as Santo Domingo, sought to gain its own independence in 1821 but was conquered and ruled by the Haitians for 22 years; it finally attained independence as the Dominican

Republic in 1844. In 1861, the Dominicans voluntarily returned to the Spanish Empire, but two years later they launched a war that restored independence in 1865. A legacy of unsettled, mostly non-representative rule followed, capped by the dictatorship of Rafael Leonidas TRUJILLO from 1930 to 1961. Juan BOSCH was elected president in 1962 but was deposed in a military coup in 1963. In 1965, the US led an intervention in the midst of a civil war sparked by an uprising to restore BOSCH. In 1966, Joaquin BALAGUER defeated BOSCH in the presidential election. BALAGUER maintained a tight grip on power for most of the next 30 years when international reaction to flawed elections forced him to curtail his term in 1996. Since then, regular competitive elections have been held in which opposition candidates have won the presidency. Former President Leonel FERNANDEZ Reyna (first term 1996-2000) won election to a new term in 2004 following a constitutional amendment allowing presidents to serve more than one term, and was later reelected to a second consecutive term. In 2012, Danilo MEDINA Sanchez became president; he was reelected in 2016.

GEOGRAPHY

Location: Caribbean, eastern two-thirds of the island of Hispaniola, between the Caribbean Sea and the North Atlantic Ocean, east of Haiti

Geographic coordinates: 19 00 N, 70 40 W

Map references: Central America and the Caribbean

Area: *total:* 48,670 sq km
land: 48,320 sq km
water: 350 sq km
country comparison to the world: 132

Area—comparative: slightly more than twice the size of New Jersey

Land boundaries: *total:* 376 km
border countries (1): Haiti 376 km

Coastline: 1,288 km

Maritime claims: *territorial sea:* 12 nm
exclusive economic zone: 200 nm
contiguous zone: 24 nm
continental shelf: 200 nm or to the edge of the continental margin measured from claimed archipelagic straight baselines

Climate: tropical maritime; little seasonal temperature variation; seasonal variation in rainfall

Terrain: rugged highlands and mountains interspersed with fertile valleys

Elevation: *mean elevation:* 424 m
lowest point: Lago Enriquillo -46 m
highest point: Pico Duarte 3,098 m

Natural resources: nickel, bauxite, gold, silver, arable land

Land use: *agricultural land:* 51.5% (2011 est.)
arable land: 16.6% (2011 est.) / permanent crops: 10.1% (2011 est.) / permanent pasture: 24.8% (2011 est.)
forest: 40.8% (2011 est.)
other: 7.7% (2011 est.)

Irrigated land: 3,070 sq km (2012)

Population distribution: coastal development is significant, especially in the southern coastal plains and the Cibao Valley, where population density is highest; smaller population clusters exist in the interior mountains (Cordillera Central)

Natural hazards: lies in the middle of the hurricane belt and subject to severe storms from June to October; occasional flooding; periodic droughts

Environment—current issues: water shortages; soil eroding into the sea damages coral reefs; deforestation

Environment—international agreements: party to: Biodiversity, Climate Change, Climate Change-Kyoto Protocol, Desertification, Endangered Species, Hazardous Wastes, Marine Dumping, Marine Life Conservation, Ozone Layer Protection, Ship Pollution, Wetlands
signed, but not ratified: Law of the Sea

Geography—note: shares island of Hispaniola with Haiti (eastern two-thirds makes up the Dominican Republic, western one-third is Haiti); the second largest country in the Antilles (after Cuba); geographically diverse with the Caribbean's tallest mountain, Pico Duarte, and lowest elevation and largest lake, Lago Enriquillo

PEOPLE AND SOCIETY

Population: 10,499,707 (July 2020 est.)
country comparison to the world: 88

Nationality: *noun:* Dominican(s)
adjective: Dominican

Ethnic groups: mixed 70.4% (mestizo/indio 58%, mulatto 12.4%), black 15.8%, white 13.5%, other 0.3% (2014 est.)
note: respondents self-identified their race; the term "indio" in the Dominican Republic is not associated with people of indigenous ancestry but people of mixed ancestry or skin color between light and dark

Languages: Spanish (official)

Religions: Roman Catholic 47.8%, Protestant 21.3%, other 2.2%, none 28%, don't know/no response .7% (2017 est.)

Age structure: *0-14 years:* 26.85% (male 1,433,166/female 1,385,987)
15-24 years: 18.15% (male 968,391/female 937,227)
25-54 years: 40.54% (male 2,168,122/female 2,088,926)
55-64 years: 8.17% (male 429,042/female 428,508)
65 years and over: 6.29% (male 310,262/female 350,076) (2020 est.)

Dependency ratios: *total dependency ratio:* 53.8
youth dependency ratio: 42.2
elderly dependency ratio: 11.6
potential support ratio: 8.6 (2020 est.)

Median age: *total:* 27.9 years
male: 27.8 years
female: 28.1 years (2020 est.)
country comparison to the world: 144

Population growth rate: 0.95% (2020 est.)
country comparison to the world: 112

Birth rate: 18.5 births/1,000 population (2020 est.)
country comparison to the world: 83

Death rate: 6.3 deaths/1,000 population (2020 est.)
country comparison to the world: 149

Net migration rate: -2.7 migrant(s)/1,000 population (2020 est.)
country comparison to the world: 172

Population distribution: coastal development is significant, especially in the southern coastal plains and the Cibao Valley, where population density is highest; smaller population clusters exist in the interior mountains (Cordillera Central)

Urbanization: *urban population:* 82.5% of total population (2020)
rate of urbanization: 2.06% annual rate of change (2015-20 est.)
total population growth rate v. urban population growth rate, 2000-2030:

Major urban areas—population: 3.318 million SANTO DOMINGO (capital) (2020)

Sex ratio: *at birth:* 1.04 male(s)/female
0-14 years: 1.03 male(s)/female
15-24 years: 1.03 male(s)/female
25-54 years: 1.04 male(s)/female
55-64 years: 1 male(s)/female
65 years and over: 0.89 male(s)/female
total population: 1.02 male(s)/female (2020 est.)

Mother's mean age at first birth: 21.3 years (2013 est.)
note: median age at first birth among women 25-29

Maternal mortality rate: 95 deaths/100,000 live births (2017 est.)
country comparison to the world: 71

Infant mortality rate: *total:* 20.9 deaths/1,000 live births
male: 23.1 deaths/1,000 live births
female: 18.7 deaths/1,000 live births (2020 est.)
country comparison to the world: 73

Life expectancy at birth: *total population:* 72 years
male: 70.3 years
female: 73.8 years (2020 est.)
country comparison to the world: 155

Total fertility rate: 2.24 children born/woman (2020 est.)
country comparison to the world: 89

Contraceptive prevalence rate: 69.5% (2014)

Drinking water source:
improved:
urban: 98.3% of population
rural: 92% of population
total: 96.7% of population
unimproved:
urban: 1.7% of population
rural: 8% of population
total: 3.3% of population (2017 est.)

Current Health Expenditure: 6.1% (2017)

Physicians density: 1.56 physicians/1,000 population (2017)

Hospital bed density: 1.6 beds/1,000 population (2017)

Sanitation facility access:
improved:
urban: 96.3% of population
rural: 89.5% of population
total: 95% of population
unimproved:
urban: 13.8% of population
rural: 3.7% of population
total: 5% of population (2017 est.)

HIV/AIDS—adult prevalence rate: 0.9% (2019 est.)

country comparison to the world: 47

HIV/AIDS—people living with HIV/AIDS: 72,000 (2019 est.)
country comparison to the world: 56

HIV/AIDS—deaths: 1,900 (2019 est.)
country comparison to the world: 45

Major infectious diseases: *degree of risk:* high (2020)
food or waterborne diseases: bacterial diarrhea, hepatitis A, and typhoid fever
vectorborne diseases: dengue fever

Obesity—adult prevalence rate: 27.6% (2016)
country comparison to the world: 37

Children under the age of 5 years underweight: 4% (2013)
country comparison to the world: 89

Education expenditures: NA

Literacy: *definition:* age 15 and over can read and write
total population: 93.8%
male: 93.8%
female: 93.8% (2016)

School life expectancy (primary to tertiary education): *total:* 14 years
male: 14 years
female: 15 years (2017)

Unemployment, youth ages 15-24: *total:* 13.5%
male: 9.9%
female: 19.7% (2017 est.)
country comparison to the world: 102

GOVERNMENT

Country name: *conventional long form:* Dominican Republic
conventional short form: The Dominican
local long form: Republica Dominicana
local short form: La Dominicana
etymology: the country name derives from the capital city of Santo Domingo (Saint Dominic)

Government type: presidential republic

Capital: *name:* Santo Domingo
geographic coordinates: 18 28 N, 69 54 W
time difference: UTC-4 (1 hour ahead of Washington, DC, during Standard Time)
etymology: named after Saint Dominic de Guzman (1170-1221), founder of the Dominican Order

Administrative divisions: 10 regions (regiones, singular—region); Cibao Nordeste, Cibao Noroeste, Cibao Norte, Cibao Sur, El Valle, Enriquillo, Higuamo, Ozama, Valdesia, Yuma

Independence: 27 February 1844 (from Haiti)

National holiday: Independence Day, 27 February (1844)

Constitution: *history:* many previous (38 total); latest proclaimed 13 June 2015
amendments: proposed by a special session of the National Congress called the National Revisory Assembly; passage requires at least two-thirds majority approval by at least one half of those present in both houses of the Assembly; passage

of amendments to constitutional articles, such as fundamental rights and guarantees, territorial composition, nationality, or the procedures for constitutional reform, also requires approval in a referendum; amended many times, last in 2017

Legal system: civil law system based on the French civil code; Criminal Procedures Code modified in 2004 to include important elements of an accusatory system

International law organization participation: accepts compulsory ICJ jurisdiction; accepts ICCt jurisdiction

Citizenship: *citizenship by birth:* no
citizenship by descent only: at least one parent must be a citizen of the Dominican Republic
dual citizenship recognized: yes
residency requirement for naturalization: 2 years

Suffrage: 18 years of age; universal and compulsory; married persons regardless of age can vote; note—members of the armed forces and national police by law cannot vote

Executive branch: *chief of state:* President Danilo MEDINA Sanchez (since 16 August 2012); Vice President Margarita CEDENO DE FERNANDEZ (since 16 August 2012); note—the president is both chief of state and head of government
head of government: President Danilo MEDINA Sanchez (since 16 August 2012); Vice President Margarita CEDENO DE FERNANDEZ (since 16 August 2012)
cabinet: Cabinet nominated by the president
elections/appointments: president and vice president directly elected on the same ballot by absolute vote in 2 rounds if needed for a 4-year term (eligible for a maximum of two consecutive terms); election last held on 15 May 2016 (rescheduled from 17 May to 5 July 2020 due to COVID-19 pandemic)
election results: Danilo MEDINA Sanchez reelected president in first round; percent of vote— Danilo MEDINA Sanchez (PLD) 61.7%, Luis Rodolfo ABINADER Corona (PRM) 35%, other 3.3%; Margarita CEDENO DE FERNANDEZ (PLD) reelected vice president

Legislative branch: *description:* bicameral National Congress or Congreso Nacional consists of:
Senate or Senado (32 seats; note—electoral system changes by the Central Election Commission are being challenged by the ruling party and opposition)
House of Representatives or Camara de Diputados (190 seats; members directly elected in multi-seat constituencies by proportional representation vote; members serve 4-year terms)
elections: Senate—last held on 15 May 2016 (rescheduled from 17 May to 5 July 2020 due to COVID-19 pandemic)
House of Representatives—last held on 15 May 2016 (rescheduled from 17 May to 5 July 2020 due to COVID-19 pandemic)
election results: Senate—percent of vote by party—NA; seats by party—PLD 26, PRM 2, BIS

1, PLRD 1, PRD 1, PRSC 1; composition as of 2018—men 29, women 3, percent of women 9.4%
House of Representatives—percent of vote by party—NA; seats by party—PLD 106, PRM 42, PRSC 18, PRD 16, PLRD 3, other 5; composition as of 2018—men 139, women 51, percent of women 26.8%; note—total National Congress percent of women 24.3%

Judicial branch: *highest courts:* Supreme Court of Justice or Suprema Corte de Justicia (consists of a minimum of 16 magistrates); Constitutional Court or Tribunal Constitucional (consists of 13 judges); note—the Constitutional Court was established in 2010 by constitutional amendment
judge selection and term of office: Supreme Court and Constitutional Court judges appointed by the National Council of the Judiciary comprised of the president, the leaders of both chambers of congress, the president of the Supreme Court, and a non-governing party congressional representative; Supreme Court judges appointed for 7-year terms; Constitutional Court judges appointed for 9-year terms
subordinate courts: courts of appeal; courts of first instance; justices of the peace; special courts for juvenile, labor, and land cases; Contentious Administrative Court for cases filed against the government

Political parties and leaders:
Dominican Liberation Party or PLD [Leonel FERNANDEZ Reyna]
Dominican Revolutionary Party or PRD [Miguel VARGAS Maldonado]
Institutional Social Democratic Bloc or BIS
Liberal Reformist Party or PRL (formerly the Liberal Party of the Dominican Republic or PLRD)
Modern Revolutionary Party or PRM [Jose Ignacio PALIZA]
National Progressive Front or FNP [Vinicio CASTILLO, Pelegrin CASTILLO]
Social Christian Reformist Party or PRSC [Federico ANTUN]

International organization participation: ACP, AOSIS, BCIE, Caricom (observer), CD, CELAC, FAO, G-77, IADB, IAEA, IBRD, ICAO, ICC (national committees), ICCt, ICRM, IDA, IFAD, IFC, IFRCS, IHO, ILO, IMF, IMO, Interpol, IOC, IOM, IPU, ISO (correspondent), ITSO, ITU, ITUC (NGOs), LAES, LAIA, MIGA, MINUSMA, NAM, OAS, OIF (observer), OPANAL, OPCW, Pacific Alliance (observer), PCA, Petrocaribe, SICA (associated member), UN, UNCTAD, UNESCO, UNIDO, Union Latina, UNWTO, UPU, WCO, WFTU (NGOs), WHO, WIPO, WMO, WTO

Diplomatic representation in the US: chief of mission: Ambassador Jose Tomas PEREZ Vazquez (since 23 February 2015)
chancery: 1715 22nd Street NW, Washington, DC 20008
telephone: [1] (202) 332-6280, 660-2263
FAX: [1] (202) 265-8057
consulate(s) general: Boston, Chicago, Los Angeles, Mayaguez (Puerto Rico), Miami, New Orleans, New York, San Juan (Puerto Rico)

consulate(s): San Francisco

Diplomatic representation from the US: chief of mission: Ambassador Robin BERNSTEIN (since 6 September 2018)

telephone: [1] (809) 567-7775

embassy: Av. Republica de Colombia # 57, Santo Domingo

mailing address: Unit 5500, APO AA 34041-5500

FAX: [1] (809) 686-7437

Flag description: a centered white cross that extends to the edges divides the flag into four rectangles—the top ones are ultramarine blue (hoist side) and vermilion red, and the bottom ones are vermilion red (hoist side) and ultramarine blue; a small coat of arms featuring a shield supported by a laurel branch (left) and a palm branch (right) is at the center of the cross; above the shield a blue ribbon displays the motto, DIOS, PATRIA, LIBERTAD (God, Fatherland, Liberty), and below the shield, REPUBLICA DOMINICANA appears on a red ribbon; in the shield a bible is opened to a verse that reads "Y la verdad nos hara libre" (And the truth shall set you free); blue stands for liberty, white for salvation, and red for the blood of heroes

National symbol(s): *palmchat (bird); national colors:* red, white, blue

National anthem: *name:* "Himno Nacional" (National Anthem)

lyrics/music: Emilio PRUD'HOMME/Jose REYES

note: adopted 1934; also known as "Quisqueyanos valientes" (Valient Sons of Quisqueye); the anthem never refers to the people as Dominican but rather calls them "Quisqueyanos," a reference to the indigenous name of the island

0:00/ 1:36

ECONOMY

Economy—overview: The Dominican Republic was for most of its history primarily an exporter of sugar, coffee, and tobacco, but over the last three decades the economy has become more diversified as the service sector has overtaken agriculture as the economy's largest employer, due to growth in construction, tourism, and free trade zones. The mining sector has also played a greater role in the export market since late 2012 with the commencement of the extraction phase of the Pueblo Viejo Gold and Silver mine, one of the largest gold mines in the world.

For the last 20 years, the Dominican Republic has been one of the fastest growing economies in Latin America. The economy rebounded from the global recession in 2010-16, and the fiscal situation is improving. A tax reform package passed in November 2012, a reduction in government spending, and lower energy costs helped to narrow the central government budget deficit from 6.6% of GDP in 2012 to 2.6% in 2016, and public debt is declining. Marked income inequality, high unemployment, and underemployment remain important long-term challenges; the poorest half of the population receives less than one-fifth of GDP, while the richest 10% enjoys nearly 40% of GDP.

The economy is highly dependent upon the US, the destination for approximately half of exports and the source of 40% of imports. Remittances from the US amount to about 7% of GDP, equivalent to about a third of exports and two-thirds of tourism receipts. The Central America-Dominican Republic Free Trade Agreement came into force in March 2007, boosting investment and manufacturing exports.

GDP (purchasing power parity):
$173 billion (2017 est.)
$165.4 billion (2016 est.)
$155.2 billion (2015 est.)
note: data are in 2017 dollars
country comparison to the world: 72

GDP (official exchange rate):
$76.09 billion (2017 est.)

GDP—real growth rate:
4.6% (2017 est.)
6.6% (2016 est.)
7% (2015 est.)
country comparison to the world: 57

GDP—per capita (PPP):
$17,000 (2017 est.)
$16,400 (2016 est.)
$15,500 (2015 est.)
note: data are in 2017 dollars
country comparison to the world: 103

Gross national saving:
21.6% of GDP (2017 est.)
20.8% of GDP (2016 est.)
20.7% of GDP (2015 est.)
country comparison to the world: 84

GDP—composition, by end use:
household consumption: 69.3% (2017 est.)
government consumption: 12.2% (2017 est.)
investment in fixed capital: 21.9% (2017 est.)
investment in inventories: -0.1% (2017 est.)
exports of goods and services: 24.8% (2017 est.)
imports of goods and services: -28.1% (2017 est.)

GDP—composition, by sector of origin:
agriculture: 5.6% (2017 est.)
industry: 33% (2017 est.)
services: 61.4% (2017 est.)

Agriculture—products: cocoa, tobacco, sugarcane, coffee, cotton, rice, beans, potatoes, corn, bananas; cattle, pigs, dairy products, beef, eggs

Industries: tourism, sugar processing, gold mining, textiles, cement, tobacco, electrical components, medical devices

Industrial production growth rate: 3.1% (2017 est.)
country comparison to the world: 99

Labor force: 4.732 million (2017 est.)
country comparison to the world: 81

Labor force—by occupation:
agriculture: 14.4%
industry: 20.8% (2014)
services: 64.7% (2014 est.)

Unemployment rate:
5.1% (2017 est.)
5.5% (2016 est.)
country comparison to the world: 80

Population below poverty line: 30.5% (2016 est.)

Household income or consumption by percentage share: *lowest 10%:* 1.9%
highest 10%: 37.4% (2013 est.)

Budget: *revenues:* 11.33 billion (2017 est.)
expenditures: 13.62 billion (2017 est.)

Taxes and other revenues: 14.9% (of GDP) (2017 est.)
country comparison to the world: 194

Budget surplus (+) or deficit (-): -3% (of GDP) (2017 est.)
country comparison to the world: 132

Public debt:
37.2% of GDP (2017 est.)
34.6% of GDP (2016 est.)
country comparison to the world: 140

Fiscal year: calendar year

Inflation rate (consumer prices):
3.3% (2017 est.)
1.6% (2016 est.)
country comparison to the world: 137

Current account balance:
-$165 million (2017 est.)
-$815 million (2016 est.)
country comparison to the world: 96

Exports:
$10.12 billion (2017 est.)
$9.86 billion (2016 est.)
country comparison to the world: 91

Exports—partners: US 50.3%, Haiti 9.1%, Canada 8.2%, India 5.6% (2017)

Exports—commodities: gold, silver, cocoa, sugar, coffee, tobacco, meats, consumer goods

Imports:
$17.7 billion (2017 est.)
$17.4 billion (2016 est.)
country comparison to the world: 82

Imports—commodities: petroleum, foodstuffs, cotton and fabrics, chemicals and pharmaceuticals

Imports—partners: US 41.4%, China 13.9%, Mexico 4.5%, Brazil 4.3% (2017)

Reserves of foreign exchange and gold:
$6.873 billion (31 December 2017 est.)
$6.134 billion (31 December 2016 est.)
country comparison to the world: 87

Debt—external:
$29.16 billion (31 December 2017 est.)
$27.7 billion (31 December 2016 est.)
country comparison to the world: 83

Exchange rates: Dominican pesos (DOP) per US dollar –
47.042 (2017 est.)
46.078 (2016 est.)
46.078 (2015 est.)
45.052 (2014 est.)
43.556 (2013 est.)

ENERGY

Electricity access: *electrification—total population:* 100% (2020)

Electricity—production: 18.03 billion kWh (2016 est.)

country comparison to the world: 81

Electricity—consumption: 15.64 billion kWh (2016 est.)
country comparison to the world: 78

Electricity—exports: 0 kWh (2016 est.)
country comparison to the world: 129

Electricity—imports: 0 kWh (2016 est.)
country comparison to the world: 143

Electricity—installed generating capacity: 3.839 million kW (2016 est.)
country comparison to the world: 91

Electricity—from fossil fuels: 77% of total installed capacity (2016 est.)
country comparison to the world: 92

Electricity—from nuclear fuels: 0% of total installed capacity (2017 est.)
country comparison to the world: 80

Electricity—from hydroelectric plants: 16% of total installed capacity (2017 est.)
country comparison to the world: 99

Electricity—from other renewable sources: 7% of total installed capacity (2017 est.)
country comparison to the world: 93

Crude oil—production: 0 bbl/day (2018 est.)
country comparison to the world: 129

Crude oil—exports: 0 bbl/day (2015 est.)
country comparison to the world: 116

Crude oil—imports: 16,980 bbl/day (2015 est.)
country comparison to the world: 67

Crude oil—proved reserves: 0 bbl (1 January 2018 est.)
country comparison to the world: 125

Refined petroleum products—production: 16,060 bbl/day (2015 est.)
country comparison to the world: 92

Refined petroleum products—consumption: 134,000 bbl/day (2016 est.)
country comparison to the world: 70

Refined petroleum products—exports: 0 bbl/day (2015 est.)
country comparison to the world: 149

Refined petroleum products—imports: 108,500 bbl/day (2015 est.)
country comparison to the world: 52

Natural gas—production: 0 cu m (2017 est.)
country comparison to the world: 125

Natural gas—consumption: 1.161 billion cu m (2017 est.)
country comparison to the world: 92

Natural gas—exports: 0 cu m (2017 est.)
country comparison to the world: 96

Natural gas—imports: 1.161 billion cu m (2017 est.)
country comparison to the world: 60

Natural gas—proved reserves: 0 cu m (1 January 2014 est.)
country comparison to the world: 129

Carbon dioxide emissions from consumption of energy: 23.79 million Mt (2017 est.)
country comparison to the world: 81

COMMUNICATIONS

Telephones—fixed lines: *total subscriptions:* 1,172,083
subscriptions per 100 inhabitants: 11.27 (2019 est.)
country comparison to the world: 71

Telephones—mobile cellular: *total subscriptions:* 8,665,302
subscriptions per 100 inhabitants: 83.32 (2019 est.)
country comparison to the world: 94

Telecommunication systems: *general assessment:* there are multiple operators licensed to provide services, most of them are small and localized; the telecom sector across the Caribbean region remains one of the key growth areas; fixed-line teledensity well-below Latin America averages; development of LTE and HSPA (high speed packet access) services, mobile broadband has taken off; income inequalities seen in telephone accesses (2020)
domestic: fixed-line teledensity is about 11 per 100 persons; multiple providers of mobile-cellular service with a subscribership of 83 per 100 persons (2019)
international: country code—1-809; 1-829; 1-849; landing point for the ARCOS-1, Antillas 1, AMX-1, SAm-1, East-West, Deep Blue Cable and the Fibralink submarine cables that provide links to South and Central America, parts of the Caribbean, and US; satellite earth station—1 Intelsat (Atlantic Ocean) (2019)
note: the COVID-19 outbreak is negatively impacting telecommunications production and supply chains globally; consumer spending on telecom devices and services has also slowed due to the pandemic's effect on economies worldwide; overall progress towards improvements in all facets of the telecom industry—mobile, fixed-line, broadband, submarine cable and satellite—has moderated

Broadcast media: combination of state-owned and privately owned broadcast media; 1 state-owned TV network and a number of private TV networks; networks operate repeaters to extend signals throughout country; combination of state-owned and privately owned radio stations with more than 300 radio stations operating (2019)

Internet country code: .do

Internet users: *total:* 7,705,529
percent of population: 74.82% (July 2018 est.)
country comparison to the world: 64

Broadband—fixed subscriptions: *total:* 794,788
subscriptions per 100 inhabitants: 8 (2018 est.)
country comparison to the world: 74

TRANSPORTATION

National air transport system: *number of registered air carriers:* 1 (2020)

inventory of registered aircraft operated by air carriers: 6

Civil aircraft registration country code prefix: HI (2016)

Airports: 36 (2013)
country comparison to the world: 108

Airports—with paved runways: *total:* 16 (2017)
over 3,047 m: 3 (2017)
2,438 to 3,047 m: 4 (2017)
1,524 to 2,437 m: 4 (2017)
914 to 1,523 m: 4 (2017)
under 914 m: 1 (2017)

Airports—with unpaved runways: *total:* 20 (2013)
1,524 to 2,437 m: 1 (2013)
914 to 1,523 m: 1 (2013)
under 914 m: 18 (2013)

Heliports: 1 (2013)

Pipelines: 27 km gas, 103 km oil (2013)

Railways: *total:* 496 km (2014)
standard gauge: 354 km 1.435-m gauge (2014)
narrow gauge: 142 km 0.762-m gauge (2014)
country comparison to the world: 114

Roadways:
total: 19,705 km (2002)
paved: 9,872 km (2002)
unpaved: 9,833 km (2002)
country comparison to the world: 115

Merchant marine: *total:* 37
by type: bulk carrier 1, general cargo 2, oil tanker 1, other 33 (2019)
country comparison to the world: 125

Ports and terminals: *major seaport(s):* Puerto Haina, Puerto Plata, Santo Domingo
oil terminal(s): Punta Nizao oil terminal

LNG terminal(s) (import): Andres LNG terminal (Boca Chica)

MILITARY AND SECURITY

Military and security forces: Armed Forces of the Dominican Republic: Army (Ejercito Nacional, EN), Navy (Marina de Guerra, MdG, includes naval infantry), Dominican Air Force (Fuerza Aerea Dominicana, FAD) (2020)
note: in addition to the military, the Ministry of Armed Forces directs the Airport Security Authority and Civil Aviation, Port Security Authority, and Border Security Corps

Military expenditures:
0.7% of GDP (2019)
0.7% of GDP (2018)
0.7% of GDP (2017)
0.7% of GDP (2016)
0.7% of GDP (2015)
country comparison to the world: 135

Military and security service personnel strengths: the Armed Forces of the Dominican Republic have approximately 62,000 active personnel (33,000 Army; 12,000 Navy; 17,000 Air Force) (2019 est.)

Military equipment inventories and acquisitions: the military's inventory consists mostly of older US equipment with limited quantities of Brazilian, European, and Israeli material; since 2010, Brazil and Israel are the leading suppliers of armaments to the Dominican Republic (2019 est.)

Military service age and obligation: 17-21 years of age for voluntary military service; recruits must have completed primary school and be Dominican Republic citizens; women may volunteer (2012)

TRANSNATIONAL ISSUES

Disputes—international: Haitian migrants cross the porous border into the Dominican Republic to find work; illegal migrants from the Dominican Republic cross the Mona Passage each year to Puerto Rico to find better work

Refugees and internally displaced persons: *refugees (country of origin):* 8,119 (Venezuela) (economic and political crisis; includes Venezuelans who have claimed asylum or have received alternative legal stay) (2019)
stateless persons: 133,770 (2016); note—a September 2013 Constitutional Court ruling revoked the citizenship of those born after 1929 to immigrants without proper documentation, even though the constitution at the time automatically granted citizenship to children born in the Dominican Republic and the 2010 constitution provides that constitutional provisions cannot be applied retroactively; the decision overwhelmingly affected people of Haitian descent whose relatives had come to the Dominican Republic since the 1890s as a cheap source of labor for sugar plantations; a May 2014 law passed by the Dominican Congress regularizes the status of those with birth certificates but will require those without them to prove they were born in the Dominican Republic and to apply for naturalization; the government has issued documents to thousands of individuals who may claim citizenship under this law, but no official estimate has been released
note: revised estimate includes only individuals born to parents who were both born abroad; it does not include individuals born in the country to one Dominican-born and one foreign-born parent or subsequent generations of individuals of foreign descent; the estimate, as such, does not include all stateless persons (2015)

Illicit drugs: transshipment point for South American drugs destined for the US and Europe; has become a transshipment point for ecstasy from the Netherlands and Belgium destined for US and Canada; substantial money laundering activity in particular by Colombian narcotics traffickers; significant amphetamine consumption

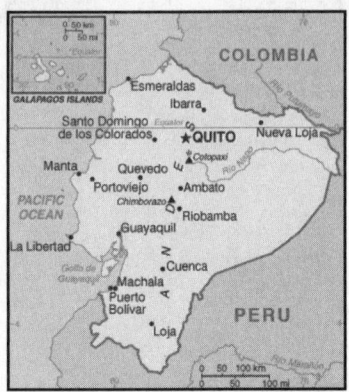

INTRODUCTION

Background: What is now Ecuador formed part of the northern Inca Empire until the Spanish conquest in 1533. Quito became a seat of Spanish colonial government in 1563 and part of the Viceroyalty of New Granada in 1717. The territories of the Viceroyalty—New Granada (Colombia), Venezuela, and Quito—gained their independence between 1819 and 1822 and formed a federation known as Gran Colombia. When Quito withdrew in 1830, the traditional name was changed in favor of the "Republic of the Equator." Between 1904 and 1942, Ecuador lost territories in a series of conflicts with its neighbors. A border war with Peru that flared in 1995 was resolved in 1999. Although Ecuador marked 30 years of civilian governance in 2004, the period was marred by political instability. Protests in Quito contributed to the mid-term ouster of three of Ecuador's last four democratically elected presidents. In late 2008, voters approved a new constitution, Ecuador's 20th since gaining independence. General elections were held in April 2017, and voters elected President Lenin MORENO.

GEOGRAPHY

Location: Western South America, bordering the Pacific Ocean at the Equator, between Colombia and Peru

Geographic coordinates: 2 00 S, 77 30 W

Map references: South America

Area: *total:* 283,561 sq km
land: 276,841 sq km
water: 6,720 sq km
note: includes Galapagos Islands
country comparison to the world: 75

Area—comparative: slightly smaller than Nevada

Land boundaries: *total:* 2,237 km
border countries (2): Colombia 708 km, Peru 1529 km

Coastline: 2,237 km

Maritime claims: *territorial sea:* 200 nm
exclusive economic zone: 200 nm
continental shelf: 200 nm
note: Ecuador has declared its right to extend its continental shelf to 350nm measured from the baselines of the Galapagos Archipelago

Climate: tropical along coast, becoming cooler inland at higher elevations; tropical in Amazonian jungle lowlands

Terrain: coastal plain (costa), inter-Andean central highlands (sierra), and flat to rolling eastern jungle (oriente)

Elevation: *mean elevation:* 1,117 m
lowest point: Pacific Ocean 0 m
highest point: Chimborazo 6,267
note: because the earth is not a perfect sphere and has an equatorial bulge, the highest point on the planet farthest from its center is Mount Chimborazo not Mount Everest, which is merely the highest peak above sea level

Natural resources: petroleum, fish, timber, hydropower

Land use: *agricultural land:* 29.7% (2011 est.)

arable land: 4.7% (2011 est.) / permanent crops: 5.6% (2011 est.) / permanent pasture: 19.4% (2011 est.)
forest: 38.9% (2011 est.)
other: 31.4% (2011 est.)

Irrigated land: 15,000 sq km (2012)

Population distribution: nearly half of the population is concentrated in the interior in the Andean intermontane basins and valleys, with large concentrations also found along the western coastal strip; the rainforests of the east remain sparsely populated

Natural hazards: frequent earthquakes; landslides; volcanic activity; floods; periodic droughts
volcanism: volcanic activity concentrated along the Andes Mountains; Sangay (5,230 m), which erupted in 2010, is mainland Ecuador's most active volcano; other historically active volcanoes in the Andes include Antisana, Cayambe, Chacana, Cotopaxi, Guagua Pichincha, Reventador, Sumaco, and Tungurahua; Fernandina (1,476 m), a shield volcano that last erupted in 2009, is the most active of the many Galapagos volcanoes; other historically active Galapagos volcanoes include Wolf, Sierra Negra, Cerro Azul, Pinta, Marchena, and Santiago

Environment—current issues: deforestation; soil erosion; desertification; water pollution; pollution from oil production wastes in ecologically sensitive areas of the Amazon Basin and Galapagos Islands

Environment—international agreements:
party to: Antarctic-Environmental Protocol, Antarctic Treaty, Biodiversity, Climate Change, Climate Change-Kyoto Protocol, Desertification, Endangered Species, Hazardous Wastes, Ozone Layer Protection, Ship Pollution, Tropical Timber 83, Tropical Timber 94, Wetlands

signed, but not ratified: none of the selected agreements

Geography—note: note 1: Cotopaxi in Andes is highest active volcano in world
note 2: according to the latest archeological research, the cacao tree, whose seeds are used to make chocolate and which was long thought to have originated in Mesoamerica, was first domesticated in the upper Amazon region of northwest South America—present-day Ecuador—about 3,300 B.C. (2020)

PEOPLE AND SOCIETY

Population: 16,904,867 (July 2020 est.)
country comparison to the world: 70

Nationality: *noun:* Ecuadorian(s)
adjective: Ecuadorian

Ethnic groups: mestizo (mixed Amerindian and white) 71.9%, Montubio 7.4%, Amerindian 7%, white 6.1%, Afroecuadorian 4.3%, mulatto 1.9%, black 1%, other 0.4% (2010 est.)

Languages: Spanish (Castilian) 93% (official), Quechua 4.1%, other indigenous 0.7%, foreign 2.2% (2010 est.)
note: (Quechua and Shuar are official languages of intercultural relations; other indigenous languages are in official use by indigenous peoples in the areas they inhabit)

Religions: Roman Catholic 74%, Evangelical 10.4%, Jehovah's Witness 1.2%, other 6.4% (includes Mormon, Buddhist, Jewish, Spiritualist, Muslim, Hindu, indigenous, African American, Pentecostal), atheist 7.9%, agnostic 0.1% (2012 est.)
note: data represent persons at least 16 years of age from five Ecuadoran cities

Demographic profile: Ecuador's high poverty and income inequality most affect indigenous, mixed race, and rural populations. The government has increased its social spending to ameliorate these problems, but critics question the efficiency and implementation of its national development plan. Nevertheless, the conditional cash transfer program, which requires participants' children to attend school and have medical check-ups, has helped improve educational attainment and healthcare among poor children. Ecuador is stalled at above replacement level fertility and the population most likely will keep growing rather than stabilize.

An estimated 2 to 3 million Ecuadorians live abroad, but increased unemployment in key receiving countries—Spain, the United States, and Italy—is slowing emigration and increasing the likelihood of returnees to Ecuador. The first large-scale emigration of Ecuadorians occurred between 1980 and 2000, when an economic crisis drove Ecuadorians from southern provinces to New York City, where they had trade contacts. A second, nationwide wave of emigration in the late 1990s was caused by another economic downturn, political instability, and a currency crisis. Spain was the logical destination because of its shared language and the wide availability of low-skilled, informal jobs at a time when increased border surveillance made illegal migration to the US difficult. Ecuador has a small but growing immigrant population and is Latin America's top recipient of refugees; 98% are neighboring Colombians fleeing violence in their country.

Age structure: *0-14 years:* 25.82% (male 2,226,240/female 2,138,219)

15-24 years: 17.8% (male 1,531,545/female 1,478,222)

25-54 years: 40.31% (male 3,333,650/female 3,480,262)

55-64 years: 7.92% (male 647,718/female 691,759)

65 years and over: 8.15% (male 648,761/female 728,491) (2020 est.)

Dependency ratios: *total dependency ratio:* 53.8
youth dependency ratio: 42.1
elderly dependency ratio: 11.7
potential support ratio: 8.6 (2020 est.)

Median age: *total:* 28.8 years
male: 28 years
female: 29.6 years (2020 est.)
country comparison to the world: 140

Population growth rate: 1.2% (2020 est.)
country comparison to the world: 89

Birth rate: 17 births/1,000 population (2020 est.)
country comparison to the world: 100

Death rate: 5.2 deaths/1,000 population (2020 est.)
country comparison to the world: 194

Net migration rate: 0 migrant(s)/1,000 population (2020 est.)
country comparison to the world: 80

Population distribution: nearly half of the population is concentrated in the interior in the Andean intermontane basins and valleys, with large concentrations also found along the western coastal strip; the rainforests of the east remain sparsely populated

Urbanization: *urban population:* 64.2% of total population (2020)
rate of urbanization: 1.66% annual rate of change (2015-20 est.)
total population growth rate v. urban population growth rate, 2000-2030:

Major urban areas—Population: 2.994 million Guayaquil, 1.874 million QUITO (capital) (2020)

Sex ratio: *at birth:* 1.05 male(s)/female
0-14 years: 1.04 male(s)/female
15-24 years: 1.04 male(s)/female
25-54 years: 0.96 male(s)/female
55-64 years: 0.94 male(s)/female
65 years and over: 0.89 male(s)/female
total population: 0.99 male(s)/female (2020 est.)

Maternal mortality rate: 59 deaths/100,000 live births (2017 est.)
country comparison to the world: 90

Infant mortality rate: *total:* 15 deaths/1,000 live births
male: 17.8 deaths/1,000 live births

female: 12 deaths/1,000 live births (2020 est.)
country comparison to the world: 97
Life expectancy at birth: total population: 77.5 years
male: 74.5 years
female: 80.6 years (2020 est.)
country comparison to the world: 80

Total fertility rate: 2.09 children born/woman (2020 est.)
country comparison to the world: 99

Contraceptive prevalence rate: 80.1% (2007/12)

Drinking water source:
improved:
urban: 100% of population
rural: 83.5% of population
total: 94% of population
unimproved:
urban: 0% of population
rural: 16.2% of population
total: 6% of population (2017 est.)

Current Health Expenditure: 8.3% (2017)

Physicians density: 2.04 physicians/1,000 population (2016)

Hospital bed density: 1.4 beds/1,000 population (2016)

Sanitation facility access:
improved:
urban: 100% of population (2015 est.)
rural: 91.9% of population
total: 97.1% of population
unimproved:
urban: 0% of population
rural: 8.1% of population
total: 2.1% of population (2017 est.)

HIV/AIDS—adult prevalence rate: 0.4% (2019 est.)
country comparison to the world: 76

HIV/AIDS—people living with HIV/AIDS: 47,000 (2019 est.)
country comparison to the world: 62

HIV/AIDS—deaths: <1000 (2019 est.)

Major infectious diseases: *degree of risk:* high (2020)
food or waterborne diseases: bacterial diarrhea, hepatitis A, and typhoid fever
vectorborne diseases: dengue fever and malaria

Obesity—adult prevalence rate:
19.9% (2016)
country comparison to the world: 107

Children under the age of 5 years underweight: 5.1% (2014)
country comparison to the world: 81

Education expenditures: 5% of GDP (2015)
country comparison to the world: 62

Literacy: definition: age 15 and over can read and write
total population: 92.8%
male: 93.8%
female: 92.1% (2017)

School life expectancy (primary to tertiary education): *total:* 15 years
male: 15 years

female: 16 years (2015)

Unemployment, youth ages 15-24: *total:* 7.9%
male: 6.4%
female: 10.6% (2018 est.)
country comparison to the world: 143

GOVERNMENT

Country name: *conventional long form:* Republic of Ecuador
conventional short form: Ecuador
local long form: Republica del Ecuador
local short form: Ecuador
etymology: the country's position on the globe, straddling the Equator, accounts for its Spanish name

Government type: presidential republic

Capital: *name:* Quito
geographic coordinates: 0 13 S, 78 30 W
time difference: UTC-5 (same time as Washington, DC, during Standard Time)
note: Ecuador has two time zones, including the Galapagos Islands (UTC-6)
etymology: named after the Quitus, a Pre-Columbian indigenous people credited with founding the city

Administrative divisions: 24 provinces (provincias, singular—provincia); Azuay, Bolivar, Canar, Carchi, Chimborazo, Cotopaxi, El Oro, Esmeraldas, Galapagos, Guayas, Imbabura, Loja, Los Rios, Manabi, Morona-Santiago, Napo, Orellana, Pastaza, Pichincha, Santa Elena, Santo Domingo de los Tsachilas, Sucumbios, Tungurahua, Zamora-Chinchipe

Independence: 24 May 1822 (from Spain)

National holiday: Independence Day (independence of Quito), 10 August (1809)

Constitution: *history:* many previous; latest approved 20 October 2008
amendments: proposed by the president of the republic through a referendum, by public petition of at least 1% of registered voters, or by agreement of at least one-third membership of the National Assembly; passage requires two separate readings a year apart and approval by at least two-thirds majority vote of the Assembly, and approval by absolute majority in a referendum; amendments such as changes to the structure of the state, constraints on personal rights and guarantees, or constitutional amendment procedures are not allowed; amended 2011, 2015, 2018; note—a 2015 constitutional amendment lifting presidential term limits was overturned by a February 2018 referendum

Legal system: civil law based on the Chilean civil code with modifications; traditional law in indigenous communities

International law organization participation: has not submitted an ICJ jurisdiction declaration; accepts ICCt jurisdiction

Citizenship: citizenship by birth: yes
citizenship by descent only: yes
dual citizenship recognized: no
residency requirement for naturalization: 3 years

Suffrage: 18-65 years of age; universal and compulsory; 16-18, over 65, and other eligible voters, voluntary

Executive branch: *chief of state:* President Lenin MORENO Garces (since 24 May 2017); Vice President María Alejandra MUNOZ (since 17 July 2020); the president is both chief of state and head of government

head of government: President Lenin MORENO Garces (since 24 May 2017); Vice President (vacant)

cabinet: Cabinet appointed by the president

elections/appointments: president and vice president directly elected on the same ballot by absolute majority popular vote in 2 rounds if needed for a 4-year term (eligible for a second term); election last held on 19 February 2017 with a runoff on 2 April 2017 (next to be held in 2021)

election results: Lenin MORENO Garces elected president in second round; percent of vote—Lenin MORENO Garces (Alianza PAIS Movement) 51.1%, Guillermo LASSO (CREO) 48.9%

Legislative branch: *description:* unicameral National Assembly or Asamblea Nacional (137 seats; 116 members directly elected in single-seat constituencies by simple majority vote, 15 members directly elected in a single nationwide constituency by proportional representation vote, and 6 directly elected in multi-seat constituencies for Ecuadorians living abroad by simple majority vote; members serve 4-year terms)

elections: last held on 19 February 2017 (next to be held on 7 February 2021)

election results: percent of vote by party—PAIS 39.1%, CREO-SUMA 20.1%, PSC 15.9%, ID 3.8%, MUPP 2.7%, other 10.7; seats by party—PAIS 74, CREO-SUMA 34, PSC 15, ID 4, MUPP 4, PSP 2, Fuerza Ecuador 1, independent 3; composition—men 85, women 52, percent of women 38%; note—defections by members of National Assembly are commonplace, resulting in frequent changes in the numbers of seats held by the various parties

Judicial branch: *highest courts:* National Court of Justice or Corte Nacional de Justicia (consists of 21 judges, including the chief justice and organized into 5 specialized chambers); Constitutional Court or Corte Constitucional (consists of 9 judges)

judge selection and term of office: justices of National Court of Justice elected by the Judiciary Council, a 9-member independent body of law professionals; judges elected for 9-year, non-renewable terms, with one-third of the membership renewed every 3 years; Constitutional Court judges appointed by the executive, legislative, and Citizen Participation branches of government; judges appointed for 9-year non-renewable terms with one-third of the membership renewed every 3 years

subordinate courts: Fiscal Tribunal, Election Dispute Settlement Courts, provincial courts (one for each province); cantonal courts

Political parties and leaders:

Alianza PAIS movement [Lenin Voltaire MORENO Garces]

Avanza Party or AVANZA [Ramiro GONZALEZ]

Citizen Revolution Movement or MRC [Rafael CORREA]

Creating Opportunities Movement or CREO [Guillermo LASSO]

Democratic Left or ID

Forward Ecuador Movement [Alvaro NOBOA]

Fuerza Ecuador [Abdala BUCARAM] (successor to Roldosist Party)

Pachakutik Plurinational Unity Movement or MUPP [Marlon Rene SANTI Gualinga]

Patriotic Society Party or PSP [Gilmar GUTIERREZ Borbua]

Popular Democracy Movement or MPD [Luis VILLACIS]

Social Christian Party or PSC [Pascual DEL CIOPPO]

Socialist Party [Patricio ZABRANO]

Society United for More Action or SUMA [Mauricio RODAS]

International organization participation: CAN, CD, CELAC, FAO, G-11, G-77, IADB, IAEA, IBRD, ICAO, ICC (national committees), ICCt, ICRM, IDA, IFAD, IFC, IFRCS, IHO, ILO, IMF, IMO, Interpol, IOC, IOM, IPU, ISO, ITSO, ITU, ITUC (NGOs), LAES, LAIA, Mercosur (associate), MIGA, MINUSTAH, NAM, OAS, OPANAL, OPCW, OPEC, Pacific Alliance (observer), PCA, SICA (observer), UN, UNAMID, UNASUR, UNCTAD, UNESCO, UNHCR, UNIDO, Union Latina, UNISFA, UNMIL, UNMISS, UNOCI, UNWTO, UPU, WCO, WFTU (NGOs), WHO, WIPO, WMO, WTO

Diplomatic representation in the US: *chief of mission:* Ambassador Francisco Benjamin Esteban CARRION Mena (since 24 January 2018)

chancery: 2535 15th Street NW, Washington, DC 20009

telephone: [1] (202) 234-7200

FAX: [1] (202) 667-3482

consulate(s) general: Atlanta, Chicago, Houston, Los Angeles, Miami, Minneapolis, New Haven (CT), New Orleans, New York, Newark (NJ), Phoenix, San Francisco

Diplomatic representation from the US: *chief of mission:* Ambassador Michael J. FITZPATRICK (since 18 June 2019)

telephone: [593] (2) 398-5000

embassy: Avenida Avigiras E12-170 y Avenida Eloy Alfaro, Quito

mailing address: Avenida Guayacanes N52-205 y Avenida Avigiras

FAX: [593] (2) 398-5100

consulate(s) general: Guayaquil

Flag description: three horizontal bands of yellow (top, double width), blue, and red with the coat of arms superimposed at the center of the flag; the flag retains the three main colors of the banner of Gran Colombia, the South American republic that broke up in 1830; the yellow color represents sunshine, grain, and mineral wealth, blue the sky, sea, and rivers, and red the blood of patriots spilled in the struggle for freedom and justice

note: similar to the flag of Colombia, which is shorter and does not bear a coat of arms

National symbol(s): *Andean condor; national colors:* yellow, blue, red

National anthem: *name:* "Salve, Oh Patria!" (We Salute You, Our Homeland)

lyrics/music: Juan Leon MERA/Antonio NEUMANE

note: adopted 1948; Juan Leon MERA wrote the lyrics in 1865; only the chorus and second verse are sung

0:00 / 3:18

ECONOMY

Economy—overview: Ecuador is substantially dependent on its petroleum resources, which accounted for about a third of the country's export earnings in 2017. Remittances from overseas Ecuadorian are also important.

In 1999/2000, Ecuador's economy suffered from a banking crisis that lead to some reforms, including adoption of the US dollar as legal tender. Dollarization stabilized the economy, and positive growth returned in most of the years that followed. China has become Ecuador's largest foreign lender since 2008 and now accounts for 77.7% of the Ecuador's bilateral debt. Various economic policies under the CORREA administration, such as an announcement in 2017 that Ecuador would terminate 13 bilateral investment treaties—including one with the US, generated economic uncertainty and discouraged private investment.

Faced with a 2013 trade deficit of $1.1 billion, Ecuador imposed tariff surcharges from 5% to 45% on an estimated 32% of imports. Ecuador's economy fell into recession in 2015 and remained in recession in 2016. Declining oil prices and exports forced the CORREA administration to cut government outlays. Foreign investment in Ecuador is low as a result of the unstable regulatory environment and weak rule of law.

n April of 2017, Lenin MORENO was elected President of Ecuador by popular vote. His immediate challenge was to reengage the private sector to improve cash flow in the country. Ecuador's economy returned to positive, but sluggish, growth. In early 2018, the MORENO administration held a public referendum on seven economic and political issues in a move counter to CORREA-administration policies, reduce corruption, strengthen democracy, and revive employment and the economy. The referendum resulted in repeal of taxes associated with recovery from the earthquake of 2016, reduced restrictions on metal mining in the Yasuni Intangible Zone—a protected area, and several political reforms.

GDP (purchasing power parity): $193 billion (2017 est.)

$188.6 billion (2016 est.)

$190.9 billion (2015 est.)

note: data are in 2017 dollars

country comparison to the world: 66

GDP (official exchange rate): $104.3 billion (2017 est.)

GDP—real growth rate:
0.06% (2019 est.)
1.29% (2018 est.)
2.37% (2017 est.)
country comparison to the world: 192

GDP—per capita (PPP): $11,500 (2017 est.)
$11,400 (2016 est.)
$11,700 (2015 est.)
note: data are in 2017 dollars
country comparison to the world: 132

Gross national saving:
25.9% of GDP (2017 est.)
26.4% of GDP (2016 est.)
24.7% of GDP (2015 est.)
country comparison to the world: 51

GDP—composition, by end use:
household consumption: 60.7% (2017 est.)
government consumption: 14.4% (2017 est.)
investment in fixed capital: 24.3% (2017 est.)
investment in inventories: 1% (2017 est.)
exports of goods and services: 20.8% (2017 est.)
imports of goods and services: -21.3% (2017 est.)

GDP—composition, by sector of origin:
agriculture: 6.7% (2017 est.)
industry: 32.9% (2017 est.)
services: 60.4% (2017 est.)

Agriculture—products: bananas, coffee, cocoa, rice, potatoes, cassava (manioc, tapioca), plantains, sugarcane; cattle, sheep, pigs, beef, pork, dairy products; fish, shrimp; balsa wood

Industries: petroleum, food processing, textiles, wood products, chemicals

Industrial production growth rate: -0.6% (2017 est.)
note: excludes oil refining
country comparison to the world: 173

Labor force:
8.086 million (2017 est.)
country comparison to the world: 59

Labor force—by occupation: *agriculture:* 26.1%
industry: 18.4%
services: 55.5% (2017 est.)

Unemployment rate:
5.71% (2019 est.)
5.26% (2018 est.)
country comparison to the world: 93

Population below poverty line: 21.5% (December 2017 est.)

Household income or consumption by percentage share: *lowest 10%:* 1.4%
highest 10%: 35.4% (2012 est.)
note: data are for urban households only

Budget: *revenues:* 33.43 billion (2017 est.)
expenditures: 38.08 billion (2017 est.)

Taxes and other revenues: 32% (of GDP) (2017 est.)
country comparison to the world: 69

Budget surplus (+) or deficit (-): -4.5% (of GDP) (2017 est.)
country comparison to the world: 164

Public debt:
45.4% of GDP (2017 est.)
43.2% of GDP (2016 est.)
country comparison to the world: 114

Fiscal year: calendar year

Inflation rate (consumer prices):
0.4% (2017 est.)
1.7% (2016 est.)
country comparison to the world: 23

Current account balance:
-$53 million (2019 est.)
-$1.328 billion (2018 est.)
country comparison to the world: 79

Exports:
$19.62 billion (2017 est.)
$16.8 billion (2016 est.)
country comparison to the world: 70

Exports—partners: US 31.5%, Vietnam 7.6%, Peru 6.7%, Chile 6.5%, Panama 4.9%, Russia 4.4%, China 4% (2017)

Exports—commodities: petroleum, bananas, cut flowers, shrimp, cacao, coffee, wood, fish

Imports:
$19.31 billion (2017 est.)
$15.86 billion (2016 est.)
country comparison to the world: 78

Imports—commodities: industrial materials, fuels and lubricants, nondurable consumer goods

Imports—partners: US 22.8%, China 15.4%, Colombia 8.7%, Panama 6.4%, Brazil 4.4%, Peru 4.2% (2017)

Reserves of foreign exchange and gold:
$2.395 billion (31 December 2017 est.)
$4.259 billion (31 December 2016 est.)
country comparison to the world: 116

Debt—external:
$39.29 billion (31 December 2017 est.)
$38.14 billion (31 December 2016 est.)
country comparison to the world: 77

Exchange rates: the US dollar became Ecuador's currency in 2001

ENERGY

Electricity access: *population without electricity:* 500,000 (2016)
electrification—total population: 99.9% (2016)
electrification—urban areas: 100% (2016)
electrification—rural areas: 99.8% (2016)

Electricity—production: 26.5 billion kWh (2016 est.)
country comparison to the world: 71

Electricity—consumption: 22.68 billion kWh (2016 est.)
country comparison to the world: 70

Electricity—exports: 211 million kWh (2015 est.)
country comparison to the world: 75

Electricity—imports: 82 million kWh (2016 est.)
country comparison to the world: 102

Electricity—installed generating capacity: 8.192 million kW (2016 est.)
country comparison to the world: 69

Electricity—from fossil fuels: 43% of total installed capacity (2016 est.)

country comparison to the world: 163

Electricity—from nuclear fuels: 0% of total installed capacity (2017 est.)
country comparison to the world: 81

Electricity—from hydroelectric plants: 54% of total installed capacity (2017 est.)
country comparison to the world: 32

Electricity—from other renewable sources: 2% of total installed capacity (2017 est.)
country comparison to the world: 138

Crude oil—production: 517,000 bbl/day (2018 est.)
country comparison to the world: 28

Crude oil—exports:
383,500 bbl/day (2017 est.)
country comparison to the world: 22

Crude oil—imports:
0 bbl/day (2015 est.)
country comparison to the world: 119

Crude oil—proved reserves: 8.273 billion bbl (1 January 2018 est.)
country comparison to the world: 17

Refined petroleum products—production: 137,400 bbl/day (2015 est.)
country comparison to the world: 62

Refined petroleum products—consumption: 265,000 bbl/day (2016 est.)
country comparison to the world: 48

Refined petroleum products—exports: 25,870 bbl/day (2015 est.)
country comparison to the world: 66

Refined petroleum products—imports: 153,900 bbl/day (2015 est.)
country comparison to the world: 40

Natural gas—production: 477.8 million cu m (2017 est.)
country comparison to the world: 73

Natural gas—consumption: 453.1 million cu m (2017 est.)
country comparison to the world: 100

Natural gas—exports: 0 cu m (2017 est.)
country comparison to the world: 97

Natural gas—imports: 0 cu m (2017 est.)
country comparison to the world: 118

Natural gas—proved reserves: 10.9 billion cu m (1 January 2018 est.)
country comparison to the world: 78

Carbon dioxide emissions from consumption of energy: 37.54 million Mt (2017 est.)
country comparison to the world: 69

COMMUNICATIONS

Telephones—fixed lines: *total subscriptions:* 2,111,291
subscriptions per 100 inhabitants: 12.64 (2019 est.)
country comparison to the world: 52

Telephones—mobile cellular: *total subscriptions:* 15,241,719
subscriptions per 100 inhabitants: 91.25 (2019 est.)

country comparison to the world: 67

Telecommunication systems: *general assessment:* much of the country's fixed-line structure is influenced by topographical challenges associated with the Andes Mountains; Ecuador has a small telecom market with a dominant mobile sector; the state-owned incumbent CNT dominates the fixed-line market, and therefore the DSL broadband market as well; mobile broadband market growing and expanding LTE services (2020)

domestic: fixed-line services with digital networks provided by multiple telecommunications operators; fixed-line teledensity stands at about 13 per 100 persons and mobile-cellular use has surged and subscribership has reached 91 per 100 persons (2019)

international: country code—593; landing points for the PAN-AM, PCCS, America Movil-Telxius West Coast Cable and SAm-1 submarine cables that provide links to South and Central America, and extending onward to the Caribbean and the US; satellite earth station—1 Intelsat (Atlantic Ocean) (2019)

note: the COVID-19 outbreak is negatively impacting telecommunications production and supply chains globally; consumer spending on telecom devices and services has also slowed due to the pandemic's effect on economies worldwide; overall progress towards improvements in all facets of the telecom industry—mobile, fixed-line, broadband, submarine cable and satellite—has moderated

Broadcast media: about 60 media outlets are recognized as national; the Ecuadorian Government controls 12 national outlets and multiple radio stations; there are multiple TV networks and many local channels, as well as more than 300 radio stations; many TV and radio stations are privately owned; broadcast media is required by law to give the government free airtime to broadcast programs produced by the state; the Ecuadorian Government is the biggest advertiser and grants advertising contracts to outlets that provide favorable coverage; an antimonopoly law and communication law limit ownership and investment in the media by non-media businesses (2019)

Internet country code: .ec

Internet users: *total:* 9,448,692

percent of population: 57.27% (July 2018 est.)
country comparison to the world: 55

Broadband—fixed subscriptions: *total:* 1,953,607
subscriptions per 100 inhabitants: 12 (2018 est.)
country comparison to the world: 54

TRANSPORTATION

National air transport system: *number of registered air carriers:* 7 (2020)
inventory of registered aircraft operated by air carriers: 35
annual passenger traffic on registered air carriers: 5,365,261 (2018)
annual freight traffic on registered air carriers: 64.2 million mt-km (2018)

Civil aircraft registration country code prefix: HC (2016)

Airports: 432 (2013)
country comparison to the world: 18

Airports—with paved runways: *total:* 104 (2017)
over 3,047 m: 4 (2017)
2,438 to 3,047 m: 5 (2017)
1,524 to 2,437 m: 18 (2017)
914 to 1,523 m: 26 (2017)
under 914 m: 51 (2017)

Airports—with unpaved runways: *total:* 328 (2013)
914 to 1,523 m: 37 (2013)
under 914 m: 291 (2013)

Heliports: 2 (2013)

Pipelines: 485 km extra heavy crude, 123 km gas, 2131 km oil, 1526 km refined products (2017)

Railways: *total:* 965 km (2017)
narrow gauge: 965 km 1.067-m gauge (2017)
note: passenger service limited to certain sections of track, mostly for tourist trains
country comparison to the world: 91

Roadways: *total:* 43,216 km (2015)
paved: 8,161 km (2015)
unpaved: 35,055 km (2015)
country comparison to the world: 87

Waterways: 1,500 km (most inaccessible) (2012)
country comparison to the world: 52

Merchant marine: *total:* 137
by type: bulk carrier 1, general cargo 7, oil tanker 28, other 101 (2019)
country comparison to the world: 76

Ports and terminals: *major seaport(s):* Esmeraldas, Manta, Puerto Bolivar
container port(s) (TEUs): Guayaquil (1,871,591) (2017)
river port(s): Guayaquil (Guayas)

MILITARY AND SECURITY

Military and security forces: *Ecuadorian Armed Forces:* Ecuadorian Land Force (Fuerza Terrestre Ecuatoriana, FTE), Ecuadorian Navy (Fuerza Naval del Ecuador, FNE, includes naval infantry, naval aviation, coast guard), Ecuadorian Air Force (Fuerza Aerea Ecuatoriana, FAE) (2020)

Military expenditures:
2.3% of GDP (2019)
2.4% of GDP (2018)
2.4% of GDP (2017)
2.5% of GDP (2016)
2.6% of GDP (2015)
country comparison to the world: 39

Military and security service personnel strengths: the Ecuadorian Armed Forces have approximately 40,000 active personnel (25,000 Army; 9,000 Navy; 6,000 Air Force) (2019 est.)

Military equipment inventories and acquisitions: the military's equipment inventory is mostly older and derived from a wide variety of sources; since 2010, the leading suppliers of military hardware are Brazil, the Netherlands, South Africa, and Spain (2019 est.)

Military service age and obligation: 18 years of age for selective conscript military service; conscription has been suspended; 18 years of age for voluntary military service; Air Force 18-22 years of age, Ecuadorian birth requirement; 1-year service obligation (2013)

Maritime threats: the International Maritime Bureau continues to report the territorial and offshore waters as at risk for piracy and armed robbery against ships; vessels, including commercial shipping and pleasure craft, have been attacked and hijacked both at anchor and while underway; crews have been robbed and stores or cargoes stolen; after several years with no incidents, there has been an increase over the last two years with four attacks reported in 2018

TRANSNATIONAL ISSUES

Disputes—international: organized illegal narcotics operations in Colombia penetrate across Ecuador's shared border

Refugees and internally displaced persons: *refugees (country of origin):* 102,928 (Colombia) (2019); 207,324 (Venezuela) (economic and political crisis; includes Venezuelans who have claimed asylum, are recognized as refugees, or have received alternative legal stay) (2020)

Illicit drugs: significant transit country for cocaine originating in Colombia and Peru, with much of the US-bound cocaine passing through Ecuadorian Pacific waters; importer of precursor chemicals used in production of illicit narcotics; attractive location for cash-placement by drug traffickers laundering money because of dollarization and weak anti-moneylaundering regime; increased activity on the northern frontier by trafficking groups and Colombian insurgents

EGYPT

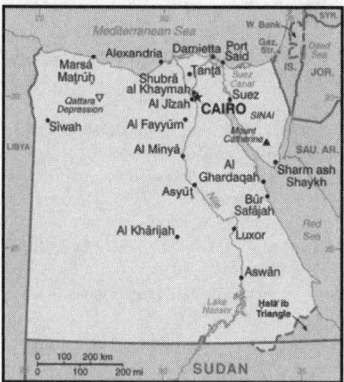

INTRODUCTION

Background: The regularity and richness of the annual Nile River flood, coupled with semi-isolation provided by deserts to the east and west, allowed for the development of one of the world's great civilizations. A unified kingdom arose circa 3200 B.C., and a series of dynasties ruled in Egypt for the next three millennia. The last native dynasty fell to the Persians in 341 B.C., who in turn were replaced by the Greeks, Romans, and Byzantines. It was the Arabs who introduced Islam and the Arabic language in the 7th century and who ruled for the next six centuries. A local military caste, the Mamluks took control about 1250 and continued to govern after the conquest of Egypt by the Ottoman Turks in 1517. Completion of the Suez Canal in 1869 elevated Egypt as an important world transportation hub. Ostensibly to protect its investments, Britain seized control of Egypt's government in 1882, but nominal allegiance to the Ottoman Empire continued until 1914. Partially independent from the UK in 1922, Egypt acquired full sovereignty from Britain in 1952. The completion of the Aswan High Dam in 1971 and the resultant Lake Nasser have reaffirmed the time- honored place of the Nile River in the agriculture and ecology of Egypt. A rapidly growing population (the largest in the Arab world), limited arable land, and dependence on the Nile all continue to overtax resources and stress society. The government has struggled to meet the demands of Egypt's fast-growing population as it implements far- reaching economic reforms, including the reduction of select subsidies, large-scale infrastructure projects, energy cooperation, and foreign direct investment appeals.

Inspired by the 2010 Tunisian revolution, Egyptian opposition groups led demonstrations and labor strikes countrywide, culminating in President Hosni MUBARAK's ouster in 2011. Egypt's military assumed national leadership until

a new legislature was in place in early 2012; later that same year, Muhammad MURSI won the presidential election. Following protests throughout the spring of 2013 against MURSI's government and the Muslim Brotherhood, the Egyptian Armed Forces intervened and removed MURSI from power in July 2013 and replaced him with interim president Adly MANSOUR. Simultaneously, the government began enacting laws to limit freedoms of assembly and expression. In January 2014, voters approved a new constitution by referendum and in May 2014 elected former defense minister Abdelfattah ELSISI president. Egypt elected a new legislature in December 2015, its first Hose of Representatives since 2012. ELSISI was reelected to a second four-year term in March 2018. In April 2019, Egypt approved via national referendum a set of constitutional amendments extending ELSISI's term in office through 2024 and possibly through 2030 if re-elected for a third term. The amendments would also allow future presidents up to two consecutive six- year terms in office, re-establish an upper legislative house, allow for one or more vice presidents, establish a 25% quota for female legislators, reaffirm the military's role as guardian of Egypt, and expand presidential authority to appoint the heads of judicial councils.

GEOGRAPHY

Location: Northern Africa, bordering the Mediterranean Sea, between Libya and the Gaza Strip, and the Red Sea north of Sudan, and includes the Asian Sinai Peninsula

Geographic coordinates: 27 00 N, 30 00 E

Map references: Africa

Area: *total:* 1,001,450 sq km
land: 995,450 sq km
water: 6,000 sq km
country comparison to the world: 31

Area—comparative: more than eight times the size of Ohio; slightly more than three times the size of New Mexico

Land boundaries: *total:* 2,612 km
border countries (4): Gaza Strip 13 km, Israel 208 km, Libya 1115 km, Sudan 1276 km

Coastline: 2,450 km

Maritime claims: *territorial sea:* 12 nm
exclusive economic zone: 200 nm or the equidistant median line with Cyprus
contiguous zone: 24 nm
continental shelf: 200 nm

Climate: desert; hot, dry summers with moderate winters

Terrain: vast desert plateau interrupted by Nile valley and delta

Elevation: *mean elevation:* 321 m
lowest point: Qattara Depression—133 m
highest point: Mount Catherine 2,629 m

Natural resources: petroleum, natural gas, iron ore, phosphates, manganese, limestone, gypsum, talc, asbestos, lead, rare earth elements, zinc

Land use: *agricultural land:* 3.6% (2011 est.)
arable land: 2.8% (2011 est.) / permanent crops: 0.8% (2011 est.) / permanent pasture: 0% (2011 est.)
forest: 0.1% (2011 est.)
other: 96.3% (2011 est.)
Irrigated land: 36,500 sq km (2012)

Population distribution: approximately 95% of the population lives within 20 km of the Nile River and its delta; vast areas of the country remain sparsely populated or uninhabited as shown in this population distribution map

Natural hazards: periodic droughts; frequent earthquakes; flash floods; landslides; hot, driving windstorms called khamsin occur in spring; dust storms; sandstorms

Environment—current issues: agricultural land being lost to urbanization and windblown sands; increasing soil salination below Aswan High Dam; desertification; oil pollution threatening coral reefs, beaches, and marine habitats; other water pollution from agricultural pesticides, raw sewage, and industrial effluents; limited natural freshwater resources away from the Nile, which is the only perennial water source; rapid growth in population overstraining the Nile and natural resources

Environment—international agreements: *party to:* Biodiversity, Climate Change, Climate Change-Kyoto Protocol, Desertification, Endangered Species, Environmental Modification, Hazardous Wastes, Law of the Sea, Marine Dumping, Ozone Layer Protection, Ship Pollution, Tropical Timber 83, Tropical Timber 94, Wetlands
signed, but not ratified: none of the selected agreements

Geography—note: controls Sinai Peninsula, the only land bridge between Africa and remainder of Eastern Hemisphere; controls Suez Canal, a sea link between Indian Ocean and Mediterranean Sea; size, and juxtaposition to Israel, establish its major role in Middle Eastern geopolitics; dependence on upstream neighbors; dominance of Nile basin issues; prone to influxes of refugees from Sudan and the Palestinian territories

PEOPLE AND SOCIETY

Population: 104,124,440 (July 2020 est.)
country comparison to the world: 14

Nationality: *noun:* Egyptian(s)
adjective: Egyptian

Ethnic groups: Egyptian 99.7%, other 0.3% (2006 est.)
note: data represent respondents by nationality

Languages: Arabic (official), Arabic, English, and French widely understood by educated classes

Religions: Muslim (predominantly Sunni) 90%, Christian (majority Coptic Orthodox, other Christians include Armenian Apostolic, Catholic, Maronite, Orthodox, and Anglican) 10% (2015 est.)

MENA religious affiliation:

Demographic profile: Egypt is the most populous country in the Arab world and the third most populous country in Africa, behind Nigeria and Ethiopia. Most of the country is desert, so about 95% of the population is concentrated in a narrow strip of fertile land along the Nile River, which represents only about 5% of Egypt's land area. Egypt's rapid population growth—46% between 1994 and 2014—stresses limited natural resources, jobs, housing, sanitation, education, and health care.

Although the country's total fertility rate (TFR) fell from roughly 5.5 children per woman in 1980 to just over 3 in the late 1990s, largely as a result of state-sponsored family planning programs, the population growth rate dropped more modestly because of decreased mortality rates and longer life expectancies. During the last decade, Egypt's TFR decline stalled for several years and then reversed, reaching 3.6 in 2011, and has plateaued the last few years. Contraceptive use has held steady at about 60%, while preferences for larger families and early marriage may have strengthened in the wake of the recent 2011 revolution. The large cohort of women of or nearing childbearing age will sustain high population growth for the foreseeable future (an effect called population momentum). Nevertheless, post-MUBARAK governments have not made curbing population growth a priority. To increase contraceptive use and to prevent further overpopulation will require greater government commitment and substantial social change, including encouraging smaller families and better educating and empowering women. Currently, literacy, educational attainment, and labor force participation rates are much lower for women than men. In addition, the prevalence of violence against women, the lack of female political representation, and the perpetuation of the nearly universal practice of female genital cutting continue to keep women from playing a more significant role in Egypt's public sphere.

Population pressure, poverty, high unemployment, and the fragmentation of inherited land holdings have historically motivated Egyptians, primarily young men, to migrate internally from rural and smaller urban areas in the Nile Delta region and the poorer rural south to Cairo, Alexandria, and other urban centers in the north, while a much smaller number migrated to the Red Sea and Sinai areas. Waves of forced internal migration also resulted from the 1967 Arab- Israeli War and the floods caused by the completion of the Aswan High Dam in 1970. Limited numbers of students and professionals emigrated temporarily prior to the early 1970s, when economic problems and high unemployment pushed the Egyptian Government to lift restrictions on labor migration. At the same time, high oil revenues enabled Saudi Arabia, Iraq, and other Gulf states, as well as Libya and Jordan, to fund development projects, creating a demand for unskilled labor (mainly in construction), which attracted tens of thousands of young Egyptian men.

Between 1970 and 1974 alone, Egyptian migrants in the Gulf countries increased from approximately 70,000 to 370,000. Egyptian officials encouraged legal labor migration both to alleviate unemployment and to generate remittance income (remittances continue to be one of Egypt's largest sources of foreign currency and GDP). During the mid-1980s, however, depressed oil prices resulting from the Iran-Iraq War, decreased demand for low-skilled labor, competition from less costly South Asian workers, and efforts to replace foreign workers with locals significantly reduced Egyptian migration to the Gulf States. The number of Egyptian migrants dropped from a peak of almost 3.3 million in 1983 to about 2.2 million at the start of the 1990s, but numbers gradually recovered.

In the 2000s, Egypt began facilitating more labor migration through bilateral agreements, notably with Arab countries and Italy, but illegal migration to Europe through overstayed visas or maritime human smuggling via Libya also rose. The Egyptian Government estimated there were 6.5 million Egyptian migrants in 2009, with roughly 75% being temporary migrants in other Arab countries (Libya, Saudi Arabia, Jordan, Kuwait, and the United Arab Emirates) and 25% being predominantly permanent migrants in the West (US, UK, Italy, France, and Canada).

During the 2000s, Egypt became an increasingly important transit and destination country for economic migrants and asylum seekers, including Palestinians, East Africans, and South Asians and, more recently, Iraqis and Syrians. Egypt draws many refugees because of its resettlement programs with the West; Cairo has one of the largest urban refugee populations in the world. Many East African migrants are interned or live in temporary encampments along the Egypt-Israel border, and some have been shot and killed by Egyptian border guards.

Age structure: *0-14 years:* 33.62% (male 18,112,550/female 16,889,155)
15-24 years: 18.01% (male 9,684,437/female 9,071,163)
25-54 years: 37.85% (male 20,032,310/female 19,376,847)
55-64 years: 6.08% (male 3,160,438/female 3,172,544)
65 years and over: 4.44% (male 2,213,539/female 2,411,457) (2020 est.)

population pyramid: *Dependency ratios:* total dependency ratio: 64.6
youth dependency ratio: 55.8
elderly dependency ratio: 8.8
potential support ratio: 11.4 (2020 est.)

Median age: *total:* 24.1 years
male: 23.8 years
female: 24.5 years (2020 est.)
country comparison to the world: 166

Population growth rate: 2.28% (2020 est.)

country comparison to the world: 31

Birth rate: 27.2 births/1,000 population (2020 est.)
country comparison to the world: 42

Death rate: 4.4 deaths/1,000 population (2020 est.)
country comparison to the world: 209

Net migration rate: -0.3 migrant(s)/1,000 population (2020 est.)
country comparison to the world: 117

Population distribution: approximately 95% of the population lives within 20 km of the Nile River and its delta; vast areas of the country remain sparsely populated or uninhabited as shown in this population distribution map

Urbanization: *urban population:* 42.8% of total population (2020)
rate of urbanization: 1.86% annual rate of change (2015-20 est.)

total population growth rate v. urban population growth rate, 2000-2030: Major urban areas—population: 20.901 million CAIRO (capital), 5.281 million Alexandria (2020)

Sex ratio: *at birth:* 1.06 male(s)/female
0-14 years: 1.07 male(s)/female
15-24 years: 1.07 male(s)/female
25-54 years: 1.03 male(s)/female
55-64 years: 1 male(s)/female
65 years and over: 0.92 male(s)/female
total population: 1.05 male(s)/female (2020 est.)

Mother's mean age at first birth: 22.7 years (2014 est.)
note: median age at first birth among women 25-29

Maternal mortality rate: 37 deaths/100,000 live births (2017 est.)
country comparison to the world: 102

Infant mortality rate: *total:* 17.1 deaths/1,000 live births
male: 18.2 deaths/1,000 live births
female: 15.8 deaths/1,000 live births (2020 est.)
country comparison to the world: 87

Life expectancy at birth: *total population:* 73.7 years
male: 72.3 years
female: 75.3 years (2020 est.)
country comparison to the world: 140

Total fertility rate: 3.29 children born/woman (2020 est.)
country comparison to the world: 45

Contraceptive prevalence rate: 58.5% (2014)

Drinking water source:

improved: *urban:* 100% of population
rural: 98.8% of population
total: 100% of population

unimproved: *urban:* 0% of population
rural: 1.2% of population
total: 0.6% of population (2017 est.)

Current Health Expenditure: 5.3% (2017)

Physicians density: 0.8 physicians/1,000 population (2017)

Hospital bed density: 1.4 beds/1,000 population (2017)

Sanitation facility access:

improved: *urban:* 99.8% of population
rural: 97.6% of population
total: 98.5% of population
unimproved: urban: 0.2% of population
rural: 2.4% of population
total: 1.5% of population (2017 est.)

HIV/AIDS—adult prevalence rate: <.1% (2019 est.)

HIV/AIDS—people living with HIV/AIDS: 26,000 (2019 est.)
country comparison to the world: 80

HIV/AIDS—deaths: <500 (2019 est.)

Major infectious diseases: *degree of risk:* intermediate (2020)
food or waterborne diseases: bacterial diarrhea, hepatitis A, and typhoid fever
water contact diseases: schistosomiasis
note: clusters of cases of a respiratory illness caused by the novel coronavirus (COVID-19) are occurring throughout Egypt; as of 10 November 2020, Egypt has reported a total of 108,962 cases of COVID-19 or 1,065 cumulative cases of COVID-19 per 1 million population with 62 cumulative deaths per 1 million population

Obesity—adult prevalence rate: 32% (2016)
country comparison to the world: 18

Children under the age of 5 years underweight: 7% (2014)
country comparison to the world: 73

Education expenditures: NA

Literacy: *definition:* age 15 and over can read and write
total population: 71.2%
male: 76.5%
female: 65.5% (2017)

School life expectancy (primary to tertiary education): *total:* 13 years
male: 13 years
female: 13 years (2017)

Unemployment, youth ages 15-24: *total:* 29.6%
male: 25.7%
female: 38.3% (2017 est.)
country comparison to the world: 32

GOVERNMENT

Country name: *conventional long form:* Arab Republic of Egypt
conventional short form: Egypt
local long form: Jumhuriyat Misr al-Arabiyah
local short form: Misr
former: United Arab Republic (with Syria)
etymology: the English name "Egypt" derives from the ancient Greek name for the country "Aigyptos"; the Arabic name "Misr" can be traced to the ancient Akkadian "misru" meaning border or frontier

Government type: presidential republic

Capital: *name:* Cairo
geographic coordinates: 30 03 N, 31 15 E
time difference: UTC+ 2 (7 hours ahead of Washington, DC, during Standard Time)

etymology: from the Arabic "al-Qahira," meaning "the victorious"

Administrative divisions: 27 governorates (muhafazat, singular—muhafazat); Ad Daqahliyah, Al Bahr al Ahmar (Red Sea), Al Buhayrah, Al Fayyum, Al Gharbiyah, Al Iskandariyah (Alexandria), Al Isma'iliyah (Ismailia), Al Jizah (Giza), Al Minufiyah, Al Minya, Al Qahirah (Cairo), Al Qalyubiyah, Al Uqsur (Luxor), Al Wadi al Jadid (New Valley), As Suways (Suez), Ash Sharqiyah, Aswan, Asyut, Bani Suwayf, Bur Sa'id (Port Said), Dumyat (Damietta), Janub Sina' (South Sinai), Kafr ash Shaykh, Matruh, Qina, Shamal Sina' (North Sinai), Suhaj

Independence: 28 February 1922 (from UK protectorate status; the military-led revolution that began on 23 July 1952 led to a republic being declared on 18 June 1953 and all British troops withdrawn on 18 June 1956); note—it was ca. 3200 B.C. that the Two Lands of Upper (southern) and Lower (northern) Egypt were first united politically

National holiday: Revolution Day, 23 July (1952)

Constitution: *history:* several previous; latest approved by a constitutional committee in December 2013, approved by referendum held on 14-15 January 2014, ratified by interim president on 19 January 2014
amendments: proposed by the president of the republic or by one fifth of the House of Representatives members; a decision to accept the proposal requires majority vote by House members; passage of amendment requires a two-thirds majority vote by House members and passage by majority vote in a referendum; articles of reelection of the president and principles of freedom are not amendable unless the amendment "brings more guarantees" amended 2019

Legal system: mixed legal system based on Napoleonic civil and penal law, Islamic religious law, and vestiges of colonial-era laws; judicial review of the constitutionality of laws by the Supreme Constitutional Court

International law organization participation: accepts compulsory ICJ jurisdiction with reservations; non- party state to the ICCt

Citizenship: *citizenship by birth:* no
citizenship by descent only: if the father was born in Egypt
dual citizenship recognized: only with prior permission from the government
residency requirement for naturalization: 10 years

Suffrage: 18 years of age; universal and compulsory

Executive branch: *chief of state:* President Abdelfattah ELSISI (since 8 June 2014)

head of government: Prime Minister Mostafa MADBOULY (since 7 June 2018)
cabinet: Cabinet ministers nominated by the executive branch and approved by the House of Representatives
elections/appointments: president elected by absolute majority popular vote in 2 rounds if needed for a 6-year term (eligible for 3

consecutive terms); election last held on 26-28 March 2018 (next to be held in 2024); prime minister appointed by the president, approved by the House of Representatives; note—following a constitutional amendment approved by referendum in April 2019, the presidential term was extended from 4 to 6 years and eligibility extended to 3 consecutive terms
election results: Abdelfattah ELSISI reelected president in first round; percent of valid votes cast—Abdelfattah ELSISI (independent) 97.1%, Moussa Mostafa MOUSSA (El Ghad Party) 2.9%; note—more than 7% of ballots cast were deemed invalid

Legislative branch: *description:* bicameral Parliament consists of:
Senate (Majlis Al-Shiyoukh) (300 seats; 100 members elected in single seat constituencies, 100 elected by closed party-list system, and 100 appointed by the president; note—the upper house, previously the Shura Council, was eliminated in the 2014 constitution, reestablished as the Senate, following passage in a 2019 constitutional referendum and approved by the House of Representatives in June 2020
House of Representatives (Majlis Al-Nowaab) (596 seats; 448 members directly elected by individual candidacy system, 120 members—with quotas for women, youth, Christians and workers—elected in party-list constituencies by simple majority popular vote, and 28 members appointed by the president; members of both houses serve 5-year terms
elections: Senate—first round held on 11-12 August 2020 (9-10 August for diaspora); second round to be held on 8-9 September (6-7 September for diaspora) (next to be held in 2025)
House of Representatives—last held from 17 October to 2 December 2015 (next to be held 24-25 October and 7-8 November 2020)
election results: Senate first round results—percent of vote by party—NA; seats by party—Nation's Future Party 100, independent 100; composition—NA
House of Representatives (2015)—percent of vote by party—NA; seats by party—Free Egyptians Party 65, Future of the Nation 53, New Wafd Party 36, Homeland's Protector Party 18, Republican People's Party 13, Congress Party 12, Al-Nour Party 11, Conservative Party 6, Democratic Peace Party 5, Egyptian National Movement 4, Egyptian Social Democratic Party 4, Modern Egypt Party 4, Freedom Party 3, My Homeland Egypt Party 3, Reform and Development Party 3, National Progressive Unionist Party 2, Arab Democratic Nasserist Party 1, El Serh El Masry el Hor 1, Revolutionary Guards Party 1, independent 351; composition—men 507, women 89, percent of women 14.9%

Judicial branch: *highest courts:* Supreme Constitutional Court (SCC) (consists of the court president and 10 justices); the SCC serves as the final court of arbitration on the constitutionality of laws and conflicts between lower courts regarding jurisdiction and rulings; Court of Cassation (CC) (consists of the court president and 550 judges

291

organized in circuits with cases heard by panels of 5 judges); the CC is the highest appeals body for civil and criminal cases, also known as "ordinary justices"; Supreme Administrative Court (SAC) (consists of the court president and NA judges and organized in circuits with cases heard by panels of 5 judges); the SAC is the highest court of the State Council

judge selection and term of office: under the 2014 constitution, all judges and justices selected and appointed by the Supreme Judiciary Council and approved as a formality by the president of the Republic; judges appointed for life; under the 2019 amendments, the president has the power to appoint heads of judiciary authorities and courts, the prosecutor general, and the head of the Supreme Constitutional Court

subordinate courts: Courts of Appeal; Courts of First Instance; courts of limited jurisdiction; Family Court (established in 2004)

Political parties and leaders: Al-Nour [Yunis MAKHYUN]

Arab Democratic Nasserist Party [Dr. Mohamed ABDUL ELLA]

Congress Party [Omar Al-Mokhtar SEMIDA]

Conservative Party [Akmal KOURTAM]

Democratic Peace Party [Ahmed FADALY]

Egyptian National Movement Party [Gen. Raouf EL SAYED]

Egyptian Social Democratic Party [Farid ZAHRAN]

El Ghad Party [Moussa Mostafa MOUSSA]

El Serh El Masry el Hor [Tarek Ahmed Abbas NADIM] Freedom Party [Salah HASSABALAH]

Free Egyptians Party [Essam KHALIL]

Homeland's Protector Party [Lt. Gen. (retired) Galal AL-HARIDI]

Modern Egypt Party [Nabil DEIBIS]

Nation's Future Party (Mostaqbal Watan) [Mohamed Ashraf RASHAD]

My Homeland Egypt Party [Gen. Seif El Islam ABDEL BARY]

National Progressive Unionist (Tagammu) Party [Sayed Abdel AAL]

Reform and Development Party [Mohamad Anwar al- SADAT]

Republican People's Party [Hazim AMR]

Revolutionary Guards Party [Magdy EL-SHARIF] Wafd Party note—party chairman Bahaa ABU SHOKA resigned in late September 2020

International organization participation: ABEDA, AfDB, AFESD, AMF, AU, BSEC (observer), CAEU, CD, CICA, COMESA, D-8, EBRD, FAO, G-15, G-24, G-77, IAEA, IBRD, ICAO, ICC (national committees), ICRM, IDA, IDB, IFAD, IFC, IFRCS, IHO, ILO, IMF, IMO, IMSO, Interpol, IOC, IOM, IPU, ISO, ITSO, ITU, LAS, MIGA, MINURSO, MINUSMA, MONUSCO, NAM, OAPEC, OAS (observer), OIC, OIF, OSCE (partner), PCA, UN, UNAMID, UNCTAD, UNESCO, UNHCR, UNIDO, UNMISS, UNOCI, UNRWA, UNWTO, UPU, WCO, WFTU (NGOs), WHO, WIPO, WMO, WTO

Diplomatic representation in the US: *chief of mission:* Ambassador Motaz Mounir ZAHRAN (since 17 September 2020)

chancery: 3521 International Court NW, Washington, DC 20008

telephone: [1] (202) 895-5400

FAX: [1] (202) 244-5131

consulate(s) general: Chicago, Houston, Los Angeles, New York

Diplomatic representation from the US: *chief of mission:* Ambassador Jonathan R. COHEN (since 17 November 2019)

telephone: [20-2] 2797-3300

embassy: 5 Tawfik Diab St., Garden City, Cairo

mailing address: Unit 64900, Box 15, APO AE 09839-4900; 5 Tawfik Diab Street, Garden City, Cairo

FAX: [20-2] 2797-3200

Flag description: three equal horizontal bands of red (top), white, and black; the national emblem (a gold Eagle of Saladin facing the hoist side with a shield superimposed on its chest above a scroll bearing the name of the country in Arabic) centered in the white band; the band colors derive from the Arab Liberation flag and represent oppression (black), overcome through bloody struggle (red), to be replaced by a bright future (white)

note: similar to the flag of Syria, which has two green stars in the white band, Iraq, which has an Arabic inscription centered in the white band, and Yemen, which has a plain white band

National symbol(s): golden eagle, white lotus; national colors: red, white, black

National anthem: *name:* "Bilady, Bilady, Bilady" (My Homeland, My Homeland, My Homeland)

lyrics/music: Younis-al QADI/Sayed DARWISH

note: adopted 1979; the current anthem, less militaristic than the previous one, was created after the signing of the 1979 peace treaty with Israel; Sayed DARWISH, commonly considered the father of modern Egyptian music, composed the anthem 0:00 / 0:00

ECONOMY

Economy—overview: Occupying the northeast corner of the African continent, Egypt is bisected by the highly fertile Nile valley where most economic activity takes place. Egypt's economy was highly centralized during the rule of former President Gamal Abdel NASSER but opened up considerably under former Presidents Anwar EL-SADAT and Mohamed Hosni MUBARAK. Agriculture, hydrocarbons, manufacturing, tourism, and other service sectors drove the country's relatively diverse economic activity.

Despite Egypt's mixed record for attracting foreign investment over the past two decades, poor living conditions and limited job opportunities have contributed to public discontent. These socioeconomic pressures were a major factor leading to the January 2011 revolution that ousted MUBARAK. The uncertain political, security, and policy environment since 2011 has restricted economic growth and failed to alleviate persistent unemployment, especially among the young.

In late 2016, persistent dollar shortages and waning aid from its Gulf allies led Cairo to turn to the IMF for a 3-year, $12 billion loan program. To secure the deal, Cairo floated its currency, introduced new taxes, and cut energy subsidies—all of which pushed inflation above 30% for most of 2017, a high that had not been seen in a generation. Since the currency float, foreign investment in Egypt's high interest treasury bills has risen exponentially, boosting both dollar availability and central bank reserves. Cairo will be challenged to obtain foreign and local investment in manufacturing and other sectors without a sustained effort to implement a range of business reforms.

GDP (purchasing power parity): $1.204 trillion (2017 est.)

$1.155 trillion (2016 est.)

$1.107 trillion (2015 est.)

note: data are in 2017 dollars

country comparison to the world: 21

GDP (official exchange rate): $236.5 billion (2017 est.)

GDP—real growth rate: 4.2% (2017 est.)

4.3% (2016 est.)

4.4% (2015 est.)

country comparison to the world: 68

GDP—per capita (PPP): $12,700 (2017 est.)

$12,800 (2016 est.)

$12,400 (2015 est.)

note: data are in 2017 dollars

country comparison to the world: 124

Gross national saving: 9% of GDP (2017 est.)

9.1% of GDP (2016 est.)

10.6% of GDP (2015 est.)

country comparison to the world: 166

GDP—composition, by end use: *household consumption:* 86.8% (2017 est.)

government consumption: 10.1% (2017 est.)

investment in fixed capital: 14.8% (2017 est.)

investment in inventories: 0.5% (2017 est.)

exports of goods and services: 16.3% (2017 est.)

imports of goods and services:—28.5% (2017 est.)

GDP—composition, by sector of origin: *agriculture:* 11.7% (2017 est.)

industry: 34.3% (2017 est.)

services: 54% (2017 est.)

Agriculture—products: cotton, rice, corn, wheat, beans, fruits, vegetables; cattle, water buffalo, sheep, goats

Industries: textiles, food processing, tourism, chemicals, pharmaceuticals, hydrocarbons, construction, cement, metals, light manufactures

Industrial production growth rate: 3.5% (2017 est.)

country comparison to the world: 84

Labor force: 24.113 million (2020 est.)

country comparison to the world: 22

Labor force—by occupation: *agriculture:* 25.8%

industry: 25.1%

services: 49.1% (2015 est.)

Unemployment rate: 7.86% (2019 est.)

12.7% (2016 est.)
country comparison to the world: 121

Population below poverty line: 27.8% (2016 est.)

Household income or consumption by percentage share: *lowest 10%:* 4%
highest 10%: 26.6% (2008)

Budget: *revenues:* 42.32 billion (2017 est.)
expenditures: 62.61 billion (2017 est.)

Taxes and other revenues: 17.9% (of GDP) (2017 est.)
country comparison to the world: 165

Budget surplus (+) or deficit (-): -8.6% (of GDP) (2017 est.)
country comparison to the world: 202

Public debt: 103% of GDP (2017 est.)
96.8% of GDP (2016 est.)
note: data cover central government debt and include debt instruments issued (or owned) by government entities other than the treasury; the data include treasury debt held by foreign entities; the data include debt issued by subnational entities, as well as intragovernmental debt; intragovernmental debt consists of treasury borrowings from surpluses in the social funds, such as for retirement, medical care, and unemployment; debt instruments for the social funds are sold at public auctions
country comparison to the world: 14

Fiscal year: 1 July—30 June

Inflation rate (consumer prices): 23.5% (2017 est.)
10.2% (2016 est.)
country comparison to the world: 217

Current account balance: -$8.915 billion (2019 est.)
-$7.682 billion (2018 est.)
country comparison to the world: 190

Exports: $23.3 billion (2017 est.)
$20.02 billion (2016 est.)
country comparison to the world: 67

Exports—partners: UAE 10.9%, Italy 10%, US 7.4%, UK 5.7%, Turkey 4.4%, Germany 4.3%, India 4.3% (2017)

Exports—commodities: crude oil and petroleum products, fruits and vegetables, cotton, textiles, metal products, chemicals, processed food

Imports: $59.78 billion (2017 est.)
$57.84 billion (2016 est.)
country comparison to the world: 50

Imports—commodities: machinery and equipment, foodstuffs, chemicals, wood products, fuels

Imports—partners: China 7.9%, UAE 5.2%, Germany 4.8%, Saudi Arabia 4.6%, US 4.4%, Russia 4.3% (2017)

Reserves of foreign exchange and gold: $35.89 billion (31 December 2017 est.)
$23.2 billion (31 December 2016 est.)
country comparison to the world: 47

Debt—external: $77.47 billion (31 December 2017 est.)
$62.38 billion (31 December 2016 est.)
country comparison to the world: 56

Exchange rates: Egyptian pounds (EGP) per US dollar—
18.05 (2017 est.)
8.8 (2016 est.)
10.07 (2015 est.)
7.7133 (2014 est.)
7.08 (2013 est.)

ENERGY

Electricity access: *electrification—total population:* 100% (2020)

Electricity—production: 183.5 billion kWh (2016 est.)
country comparison to the world: 22

Electricity—consumption: 159.7 billion kWh (2016 est.)
country comparison to the world: 23

Electricity—exports: 1.158 billion kWh (2015 est.)
country comparison to the world: 57

Electricity—imports: 54 million kWh (2016 est.)
country comparison to the world: 106

Electricity—installed generating capacity: 45.12 million kW (2016 est.)
country comparison to the world: 23

Electricity—from fossil fuels: 91% of total installed capacity (2016 est.)
country comparison to the world: 54

Electricity—from nuclear fuels: 0% of total installed capacity (2017 est.)
country comparison to the world: 82

Electricity—from hydroelectric plants: 6% of total installed capacity (2017 est.)
country comparison to the world: 129

Electricity—from other renewable sources: 2% of total installed capacity (2017 est.)
country comparison to the world: 139

Crude oil—production: 639,000 bbl/day (2018 est.)
country comparison to the world: 27

Crude oil—exports: 246,500 bbl/day (2017 est.)
country comparison to the world: 29

Crude oil—imports: 64,760 bbl/day (2015 est.)
country comparison to the world: 51

Crude oil—proved reserves: 4.4 billion bbl (1 January 2018 est.)
country comparison to the world: 24

Refined petroleum products—production: 547,500 bbl/day (2015 est.)
country comparison to the world: 31

Refined petroleum products—consumption: 878,000 bbl/day (2016 est.)
country comparison to the world: 25

Refined petroleum products—exports: 47,360 bbl/day (2015 est.)
country comparison to the world: 56

Refined petroleum products—imports: 280,200 bbl/day (2015 est.)
country comparison to the world: 26

Natural gas—production: 50.86 billion cu m (2017 est.)

country comparison to the world: 16

Natural gas—consumption: 57.71 billion cu m (2017 est.)
country comparison to the world: 13

Natural gas—exports: 212.4 million cu m (2017 est.)
country comparison to the world: 45

Natural gas—imports: 7.079 billion cu m (2017 est.)
country comparison to the world: 29

Natural gas—proved reserves: 2.186 trillion cu m (1 January 2018 est.)
country comparison to the world: 15

Carbon dioxide emissions from consumption of energy: 232.7 million Mt (2017 est.)
country comparison to the world: 30

COMMUNICATIONS

Telephones—fixed lines: *total subscriptions:* 8,885,103
subscriptions per 100 inhabitants: 8.73 (2019 est.)
country comparison to the world: 19

Telephones—mobile cellular: *total subscriptions:* 96,657,295
subscriptions per 100 inhabitants: 94.97 (2019 est.)
country comparison to the world: 17

Telecommunication systems: *general assessment:* one of the biggest fixed-line systems in Africa and the Arab region; one of the largest mobile telecom markets in North Africa; penetration rate of about 94%; LTE launch in late 2017, which greatly helped the capabilities of mobile broadband services, and the beginning of developing the 5G network; recent govt. efforts to fund next generation networks, develop technology parks and extend broadband availability (2020)
domestic: fixed-line 9 per 100, mobile-cellular 95 per 100 (2019)
international: country code—20; landing points for Aletar, Africa-1, FEA, Hawk, IMEWE, and SEA-ME-WE-3 & 4 submarine cable networks linking to Asia, Africa, the Middle East, and Australia; satellite earth stations—4 (2 Intelsat—Atlantic Ocean and Indian Ocean, 1 Arabsat, and 1 Inmarsat); tropospheric scatter to Sudan; microwave radio relay to Israel; a participant in Medarabtel (2019)
note: the COVID-19 outbreak is negatively impacting telecommunications production and supply chains globally; consumer spending on telecom devices and services has also slowed due to the pandemic's effect on economies worldwide; overall progress towards improvements in all facets of the telecom industry—mobile, fixed-line, broadband, submarine cable and satellite—has moderated

Broadcast media: mix of state-run and private broadcast media; state-run TV operates 2 national and 6 regional terrestrial networks, as well as a few satellite channels; dozens of private satellite channels and a large number of Arabic satellite channels are available for free; some limited satellite services are also available via subscription; state-run radio operates about 30 stations belonging to 8 networks; privately-owned radio includes 8 major stations, 4 of which belong to 1 network (2019)

Internet country code: .eg

Internet users: *total:* 46,644,728

percent of population: 46.92% (July 2018 est.) *country comparison to the world:* 18

Broadband—fixed subscriptions: *total:* 6,579,762 *subscriptions per 100 inhabitants:* 7 (2018 est.) *country comparison to the world:* 26

Communications—note: one of the largest and most famous libraries in the ancient world was the Great Library of Alexandria in Egypt (founded about 295 B.C., it may have survived in some form into the 5th century A. D.); seeking to resurrect the great center of learning and communication, the Egyptian Government in 2002 inaugurated the Bibliotheca Alexandrina, an Egyptian National Library on the site of the original Great Library, which commemorates the original archive and also serves as a center of cultural and scientific excellence

TRANSPORTATION

National air transport system: *number of registered air carriers:* 14 (2020) *inventory of registered aircraft operated by air carriers:* 101 *annual passenger traffic on registered air carriers:* 12,340,32 (2018) *annual freight traffic on registered air carriers:* 437.63 million mt-km (2018)

Civil aircraft registration country code prefix: SU (2016)

Airports: 83 (2013) *country comparison to the world:* 65

Airports—with paved runways: *total:* 72 (2017) *over 3,047 m:* 15 (2017) *2,438 to 3,047 m:* 36 (2017) *1,524 to 2,437 m:* 15 (2017) *under 914 m:* 6 (2017)

Airports—with unpaved runways: *total:* 11 (2013) *2,438 to 3,047 m:* 1 (2013) *1,524 to 2,437 m:* 3 (2013) *914 to 1,523 m:* 4 (2013) *under 914 m:* 3 (2013)

Heliports: 7 (2013)

Pipelines: 486 km condensate, 74 km condensate/gas, 7986 km gas, 957 km liquid petroleum gas, 5225 km oil, 37 km oil/gas/water, 895 km refined products, 65 km water (2013)

Railways: *total:* 5,085 km (2014) *standard gauge:* 5,085 km 1.435-m gauge (62 km electrified) (2014) *country comparison to the world:* 39

Roadways: total: 65,050 km (2017) *paved:* 48,000 km (2017) *unpaved:* 17,050 km (2017) *country comparison to the world:* 74

Waterways: 3,500 km (includes the Nile River, Lake Nasser, Alexandria-Cairo Waterway, and numerous smaller canals in Nile Delta; the Suez Canal (193.5 km including approaches) is navigable by oceangoing vessels drawing up to 17.68 m) (2011) *country comparison to the world:* 29

Merchant marine: *total:* 393 *by type:* bulk carrier 13, container ship 7, general cargo 28, oil tanker 36, other 309 (2019) *country comparison to the world:* 45

Ports and terminals: major seaport(s): Mediterranean Sea—Alexandria, Damietta, El Dekheila, Port Said

oil terminal(s): Ain Sukhna terminal, Sidi Kerir terminal *container port(s) (TEUs):* Alexandria (1,613,000), Port Said (East) (2,968,308) (2017) *LNG terminal(s) (export):* Damietta, Idku (Abu Qir Bay) Gulf of Suez—Suez

MILITARY AND SECURITY

Military and security forces: Egyptian Armed Forces (EAF): Army (includes surface-to-surface missile forces, special forces, Republican Guard), Navy (includes coastal defense, Coast Guard), Air Force, Air Defense Command; Ministry of Interior: Central Security Forces, National Police (2019) *note:* some tribal militias in the Sinai Peninsula cooperate with the Egyptian military against insurgent/terrorist groups such as the Islamic State

Military expenditures: 1.2% of GDP (2019) 1.2% of GDP (2018) 1.4% of GDP (2017) 1.7% of GDP (2016) 1.7% of GDP (2015) *country comparison to the world:* 102

Military and security service personnel strengths: estimates of the size of the Egyptian Armed Forces (EAF) vary; approximately 450,000 total active personnel (325,000 Army; 18,500 Navy; 30,000 Air Force; 75,000 Air Defense Command) (2019 est.)

Military equipment inventories and acquisitions: the EAF's inventory is comprised of a mix of domestically produced, Soviet-era, and more modern, particularly US, weapons systems; in recent years, the EAF has embarked on an extensive equipment modernization program with major purchases from a variety of suppliers; since 2010, the leading suppliers of military hardware to Egypt are France, Germany, Russia, and the US; Egypt has an established defense industry that produces a range of products from small arms to armored vehicles and naval vessels; it also has licensed and co-production agreements with several countries, including France (naval frigates) and the US (tanks) (2019)

Military deployments: 1,000 Central African Republic (MINUSCA); 1,050 Mali (MINUSMA); 150 Sudan (UNAMID) (2020)

Military service age and obligation: 18-30 years of age for male conscript military service; service obligation—18-36 months, followed by a 9-year reserve obligation; voluntary enlistment possible from age 15 (2017)

Military—note: since 2011, the Egyptian Armed Forces, police, and other security forces have been actively engaged in counterinsurgency and counter- terrorism operations in the North Sinai governorate against several militant groups, particularly the Islamic State of Iraq and ash- Sham—Sinai Province; as of early 2020, Egypt reportedly had over 40,000 troops plus thousands of police and other security personnel deployed to the Sinai for internal security duties where more than 1,000 have been killed the military has a large stake in the civilian economy, including running businesses, producing consumer and industrial goods, importing commodities, and building and managing infrastructure projects, such as bridges, roads, hospitals, and housing developments

the Multinational Force & Observers (MFO) has operated in the Sinai since 1982 as a peacekeeping and monitoring force to supervise the implementation of the security provisions of the 1979 Egyptian-Israeli Treaty of Peace; the MFO is an independent international organization, created by agreement between the Arab Republic of Egypt and the State of Israel; it is composed of about 1,150 troops from 13 countries (2020)

TERRORISM

Terrorist group(s): Army of Islam; Islamic State of Iraq and ash- Sham—Sinai Province; Mujahidin Shura Council in the Environs of Jerusalem; al-Qa'ida (2019) *note:* details about the history, aims, leadership, organization, areas of operation, tactics, targets, weapons, size, and sources of support of the group(s) appear(s) in Appendix- T

TRANSNATIONAL ISSUES

Disputes—international: Sudan claims but Egypt de facto administers security and economic development of Halaib region north of the 22nd parallel boundary; Egypt no longer shows its administration of the Bir Tawil trapezoid in Sudan on its maps; Gazan breaches in the security wall with Egypt in January 2008 highlight difficulties in monitoring the Sinai border; Saudi Arabia claims Egyptian-administered islands of Tiran and Sanafir

Refugees and internally displaced persons: *refugees (country of origin):* 70,010 (West Bank and Gaza Strip) (2019); 130,085 (Syria) (refugees and asylum seekers), 49,290 (Sudan) (refugees and asylum seekers), 19,814 (South Sudan) (refugees and asylum seekers), 19,200 (Eritrea) (refugees and asylum seekers), 16,181 (Ethiopia) (refugees and asylum seekers), 9,259 (Yemen) (refugees and asylum seekers), 6,824 (Iraq) (refugees and asylum seekers), 6,755 (Somalia) (refugees and asylum seekers) (2020) *IDPs:* 97,000 (2019) *stateless persons:* 5 (2019)

Trafficking in persons: *current situation:* Egypt is a source, transit, and destination country for men, women, and children subjected to sex trafficking and forced labor; Egyptian children, including the large population of street children are vulnerable to forced labor in domestic service, begging and agriculture or may be victims of sex trafficking or child sex tourism, which occurs in Cairo, Alexandria, and Luxor; some Egyptian women and girls are sold into "temporary" or "summer" marriages with Gulf men, through the complicity of their parents or marriage brokers, and are exploited for prostitution or forced labor; Egyptian men are subject to forced labor in neighboring countries, while adults from South and Southeast Asia and East Africa—and increasingly Syrian refugees—are forced to work in domestic service, construction, cleaning, and begging in Egypt; women and girls, including migrants and refugees,

from Asia, Sub- Saharan Africa, and the Middle East are sex trafficked in Egypt; the Egyptian military cracked down on criminal group's smuggling, abducting, trafficking, and extorting African migrants in the Sinai Peninsula, but the practice has reemerged along Egypt's western border with Libya

tier rating: Tier 2 Watch List—Egypt does not fully comply with the minimum standards for the elimination of trafficking; however, it is making significant efforts to do so; the government gathered data nationwide on trafficking cases to better allocate and prioritize anti- trafficking efforts, but overall it did not demonstrate increased progress; prosecutions increased in 2014, but no offenders were

convicted for the second consecutive year; fewer trafficking victims were identified in 2014, which represents a significant and ongoing decrease from the previous two reporting periods; the government relied on NGOs and international organizations to identify and refer victims to protective services, and focused on Egyptian victims and refused to provide some services to foreign victims, at times including shelter (2015)

Illicit drugs: transit point for cannabis, heroin, and opium moving to Europe, Israel, and North Africa; transit stop for Nigerian drug couriers; concern as money laundering site due to lax enforcement of financial regulations

EL SALVADOR

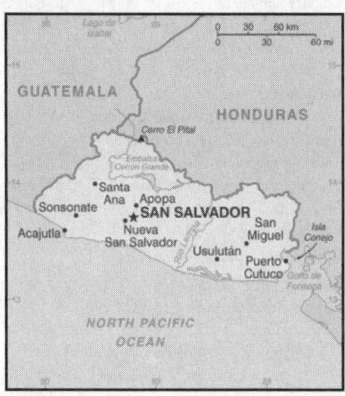

INTRODUCTION

Background: El Salvador achieved independence from Spain in 1821 and from the Central American Federation in 1839. war, which cost about 75,000 lives, was brought to a close in 1992 when the government and leftist rebels that provided for military and political reforms. El Salvador is beset by one of the world's highest homicide pervasive criminal gangs.

GEOGRAPHY

Location: Central America, bordering the North Pacific Ocean, between Guatemala and Honduras

Geographic coordinates: 13 50 N, 88 55 W

Map references: Central America and the Caribbean

Area: *total:* 21,041 sq km
land: 20,721 sq km
water: 320 sq km
country comparison to the world: 154

Area—comparative: about the same size as New Jersey

Land boundaries: *total:* 590 km
border countries (2): Guatemala 199 km, Honduras 391 km

Coastline: 307 km

Maritime claims: *territorial sea:* 12 nm
exclusive economic zone: 200 nm
contiguous zone: 24 nm

Climate: tropical; rainy season (May to October); dry season (November to April); tropical on coast; temperate in uplands

Terrain: mostly mountains with narrow coastal belt and central plateau

Elevation: *mean elevation:* 442 m
lowest point: Pacific Ocean 0 m
highest point: Cerro El Pital 2,730 m

Natural resources: hydropower, geothermal power, petroleum, arable land

Land use: *agricultural land:* 74.7% (2011 est.)
arable land: 33.1% (2011 est.) / permanent crops: 10.9% (2011 est.) / permanent pasture: 30.7% (2011 est.)
forest: 13.6% (2011 est.)
other: 11.7% (2011 est.)

Irrigated land: 452 sq km (2012)

Population distribution: athough it is the smallest country in land area in Central America, El Salvador has a population that is 18 times larger than Belize; at least 20% of the population lives abroad; high population density country-wide, with particular concentration around the capital of San Salvador

Natural hazards: known as the Land of Volcanoes; frequent and sometimes destructive earthquakes and volcanic activity; extremely susceptible to hurricanes
volcanism: significant volcanic activity; San Salvador (1,893 m), which last erupted in 1917, has the potential to cause major harm to the country's capital, which lies just below the volcano's slopes; San Miguel (2,130 m), which last erupted in 2002, is one of the most active volcanoes in

the country; other historically active volcanoes include Conchaguita, Ilopango, Izalco, and Santa Ana

Environment—current issues: deforestation; soil erosion; water pollution; contamination of soils from disposal of toxic wastes

Environment—international agreements: party to: Biodiversity, Climate Change, Climate Change-Kyoto Protocol, Desertification, Endangered Species, Hazardous Wastes, Ozone Layer Protection, Wetlands
signed, but not ratified: Law of the Sea

Geography—note: smallest Central American country and only one without a coastline on the Caribbean Sea

PEOPLE AND SOCIETY

Population: 6,481,102 (July 2020 est.)
country comparison to the world: 109

Nationality: *noun:* Salvadoran(s)
adjective: Salvadoran

Ethnic groups: mestizo 86.3%, white 12.7%, Amerindian 0.2% (includes Lenca, Kakawira, Nahua-Pipil), black 0.1%, other 0.6% (2007 est.)

Languages: Spanish (official), Nawat (among some Amerindians)

Religions: Roman Catholic 50%, Protestant 36%, other 2%, none 12% (2014 est.)

Demographic profile: El Salvador is the smallest and most densely populated country in Central America. It is well into its demographic transition, experiencing slower population growth, a decline in its number of youths, and the gradual aging of its population. The increased use of family planning has substantially lowered El Salvador's fertility rate, from approximately 6 children per woman in the 1970s to replacement level today. A 2008 national family planning survey showed that female sterilization remained the most common contraception method in El Salvador—its sterilization rate is among the highest in Latin America and the Caribbean—but that the use of injectable contraceptives is growing. Fertility differences between rich and poor and urban and rural women are narrowing.

Salvadorans fled during the 1979 to 1992 civil war mainly to the United States but also to Canada and to neighboring Mexico, Guatemala, Honduras, Nicaragua, and Costa Rica. Emigration to the United States increased again in the 1990s and 2000s as a result of deteriorating economic conditions, natural disasters (Hurricane Mitch in 1998 and earthquakes in 2001), and family reunification. At least 20% of El Salvador's population lives abroad. The remittances they send home account for close to 20% of GDP, are the second largest source of external income after exports, and have helped reduce poverty.

Age structure: *0-14 years:* 25.83% (male 857,003/female 817,336)
15-24 years: 18.82% (male 619,368/female 600,501)
25-54 years: 40.51% (male 1,221,545/female 1,404,163)
55-64 years: 7.23% (male 198,029/female 270,461)
65 years and over: 7.6% (male 214,717/female 277,979) (2020 est.)

Dependency ratios: *total dependency ratio:* 54.4
youth dependency ratio: 41.1
elderly dependency ratio: 13.4
potential support ratio: 7.5 (2020 est.)

Median age: *total:* 27.7 years
male: 26.2 years
female: 29.3 years (2020 est.)
country comparison to the world: 145

Population growth rate: 0.83% (2020 est.)
country comparison to the world: 126

Birth rate: 18.6 births/1,000 population (2020 est.)
country comparison to the world: 82

Death rate: 5.9 deaths/1,000 population (2020 est.)
country comparison to the world: 170

Net migration rate: -4.8 migrant(s)/1,000 population (2020 est.)
country comparison to the world: 195

Population distribution: athough it is the smallest country in land area in Central America, El Salvador has a population that is 18 times larger than Belize; at least 20% of the population lives abroad; high population density country-wide, with particular concentration around the capital of San Salvador

Urbanization: *urban population:* 73.4% of total population (2020)
rate of urbanization: 1.57% annual rate of change (2015-20 est.)
total population growth rate v. urban population growth rate, 2000-2030:

Major urban areas—population: 1.106 million SAN SALVADOR (capital) (2020)

Sex ratio: *at birth:* 1.05 male(s)/female
0-14 years: 1.05 male(s)/female
15-24 years: 1.03 male(s)/female
25-54 years: 0.87 male(s)/female

55-64 years: 0.73 male(s)/female
65 years and over: 0.77 male(s)/female
total population: 0.92 male(s)/female (2020 est.)

Mother's mean age at first birth: 20.8 years (2008 est.)
note: median age at first birth among women 25-29

Maternal mortality rate: 46 deaths/100,000 live births (2017 est.)
country comparison to the world: 95

Infant mortality rate: *total:* 11.8 deaths/1,000 live births
male: 13.4 deaths/1,000 live births
female: 10.2 deaths/1,000 live births (2020 est.)
country comparison to the world: 109

Life expectancy at birth: *total population:* 74.8 years
male: 71.3 years
female: 78.6 years (2020 est.)
country comparison to the world: 124

Total fertility rate: 2.09 children born/woman (2020 est.)
country comparison to the world: 100

Contraceptive prevalence rate: 71.9% (2014)

Drinking water source:
improved:
urban: 100% of population
rural: 92.2% of population
total: 97.4% of population
unimproved:
urban: 0% of population
rural: 7.8% of population
total: 2.6% of population (2015 est.)

Current Health Expenditure: 7.2% (2017)

Physicians density: 1.57 physicians/1,000 population (2016)

Hospital bed density: 1.2 beds/1,000 population (2017)

Sanitation facility access:
improved:
urban: 99.8% of population
rural: 94.7% of population
total: 98.3% of population
unimproved:
urban: 0.2% of population
rural: 5.3% of population
total: 1.7% of population (2017 est.)

HIV/AIDS—adult prevalence rate: 0.6% (2019 est.)
country comparison to the world: 62

HIV/AIDS—people living with HIV/AIDS: 27,000 (2019 est.)
country comparison to the world: 79

HIV/AIDS—deaths: <1000 (2019 est.)

Major infectious diseases: *degree of risk:* high (2020)
food or waterborne diseases: bacterial and protozoal diarrhea
vectorborne diseases: dengue fever

Obesity—adult prevalence rate: 24.6% (2016)
country comparison to the world: 57

Children under the age of 5 years underweight: 5% (2014)
country comparison to the world: 82

Education expenditures: 3.8% of GDP (2017)
country comparison to the world: 110

Literacy: *definition:* age 15 and over can read and write
total population: 88.5%
male: 90.6%
female: 86.7% (2017)

School life expectancy (primary to tertiary education): *total:* 12 years
male: 12 years
female: 12 years (2018)

Unemployment, youth ages 15-24: *total:* 9.6%
male: 8.4%
female: 11.7% (2018)
country comparison to the world: 128

GOVERNMENT

Country name: *conventional long form:* Republic of El Salvador
conventional short form: El Salvador
local long form: Republica de El Salvador
local short form: El Salvador
etymology: name is an abbreviation of the original Spanish conquistador designation for the area "Provincia de Nuestro Senor Jesus Cristo, el Salvador del Mundo" (Province of Our Lord Jesus Christ, the Saviour of the World), which became simply "El Salvador" (The Savior)

Government type: presidential republic

Capital: *name:* San Salvador
geographic coordinates: 13 42 N, 89 12 W
time difference: UTC-6 (1 hour behind Washington, DC, during Standard Time)
etymology: Spanish for "Holy Savior" (referring to Jesus Christ)

Administrative divisions: 14 departments (departamentos, singular—departamento); Ahuachapan, Cabanas, Chalatenango, Cuscatlan, La Libertad, La Paz, La Union, Morazan, San Miguel, San Salvador, San Vicente, Santa Ana, Sonsonate, Usulutan

Independence: 15 September 1821 (from Spain)

National holiday: Independence Day, 15 September (1821)

Constitution: *history:* many previous; latest drafted 16 December 1983, enacted 23 December 1983
amendments: proposals require agreement by absolute majority of the Legislative Assembly membership; passage requires at least two-thirds majority vote of the Assembly; constitutional articles on basic principles, and citizen rights and freedoms cannot be amended; amended many times, last in 2018

Legal system: civil law system with minor common law influence; judicial review of legislative acts in the Supreme Court

International law organization participation: has not submitted an ICJ jurisdiction declaration; non-party state to the ICCt

Citizenship: *citizenship by birth:* yes
citizenship by descent only: yes
dual citizenship recognized: yes
residency requirement for naturalization: 5 years

Suffrage: 18 years of age; universal

Executive branch: *chief of state:* President Nayib Armando BUKELE Ortez (since 1 June 2019); Vice President Felix Augusto Antonio ULLOA Garay (since 1 June 2019); note—the president is both chief of state and head of government
head of government: President Nayib Armando BUKELE Ortez (since 1 June 2019); Vice President Felix Augusto Antonio ULLOA Garay (since 1 June 2019)
cabinet: Council of Ministers selected by the president
elections/appointments: president and vice president directly elected on the same ballot by absolute majority popular vote in 2 rounds if needed for a single 5-year term; election last held on 3 February 2019 (next to be held on February 2024)
election results: Nayib Armando BUKELE Ortez elected president—Nayib Armando BUKELE Ortez (GANA) 53.1%, Carlos CALLEJA Hakker (ARENA) 31.72%, Hugo MARTINEZ (FMLN) 14.41%, other 0.77%

Legislative branch: *description:* unicameral Legislative Assembly or Asamblea Legislativa (84 seats; members directly elected in multi-seat constituencies and a single nationwide constituency by proportional representation vote to serve 3-year terms)
elections: last held on 4 March 2018 (next to be held in March 2021)
election results: percent of vote by party—ARENA 42.3%, FMLN 24.4%, GANA 11.5%, PCN 10.8%, PDC 3.2%, CD 0.9%, Independent 0.7%, other 6.2%; seats by party—ARENA 37, FMLN 23, GANA 11, PCN 8, PDC 3, CD 1, independent 1; composition -men 58, women 26, percent of women 31%

Judicial branch: *highest courts:* Supreme Court or Corte Suprema de Justicia (consists of 16 judges and 16 substitutes judges organized into Constitutional, Civil, Penal, and Administrative Conflict Chambers)
judge selection and term of office: judges elected by the Legislative Assembly on the recommendation of both the National Council of the Judicature, an independent body elected by the Legislative Assembly, and the Bar Association; judges elected for 9-year terms, with renewal of one-third of membership every 3 years; consecutive reelection is allowed
subordinate courts: Appellate Courts; Courts of First Instance; Courts of Peace

Political parties and leaders:
Christian Democratic Party or PDC [Rodolfo Antonio PARKER Soto]
Democratic Change (Cambio Democratico) or CD [Douglas AVILES] (formerly United Democratic Center or CDU)
Farabundo Marti National Liberation Front or FMLN [Medardo GONZALEZ]
Great Alliance for National Unity or GANA [Jose Andres ROVIRA Caneles]
National Coalition Party or PCN [Manuel RODRIGUEZ]
Nationalist Republican Alliance or ARENA [Mauricio INTERIANO]
Nuevas Ideas [Federico Gerardo ANLIKER]

International organization participation: BCIE, CACM, CD, CELAC, FAO, G-11, G-77, IADB, IAEA, IBRD, ICAO, ICC (national committees), ICRM, IDA, IFAD, IFC, IFRCS, ILO, IMF, IMO, Interpol, IOC, IOM, IPU, ISO (correspondent), ITSO, ITU, ITUC (NGOs), LAES, LAIA (observer), MIGA, MINURSO, MINUSTAH, NAM (observer), OAS, OPANAL, OPCW, Pacific Alliance (observer), PCA, Petrocaribe, SICA, UN, UNCTAD, UNESCO, UNIDO, UNIFIL, Union Latina, UNISFA, UNMISS, UNOCI, UNWTO, UPU, WCO, WFTU (NGOs), WHO, WIPO, WMO, WTO

Diplomatic representation in the US: chief of mission: Ambassador (vacant); Charge d'Affaires Werner Matias ROMERO Guerra (since 9 June 2019)
chancery: 1400 16th Street NW, Suite 100, Washington, DC 20036
telephone: [1] (202) 595-7500
FAX: [1] (202) 232-1928
consulate(s) general: Atlanta, Boston, Brentwood (NY), Chicago, Dallas, Doral (FL), Doraville (GA), Houston, Las Vegas (NV), Los Angeles, McAllen (TX), New York, Nogales (AZ), San Francisco, Silver Spring (MD), Tucson (AZ), Washington, DC, Woodbridge (VA)
consulate(s): Elizabeth (NJ), Newark (NJ), Seattle, Woodbridge (VA)

Diplomatic representation from the US: chief of mission: Ambassador Ronald D. JOHNSON (since 6 December 2019)
telephone: [503] 2501-2999
embassy: Final Boulevard Santa Elena, Antiguo Cuscatlan, La Libertad, San Salvador
mailing address: Unit 3450, APO AA 34023; 3450 San Salvador Place, Washington, DC 20521 -3450
FAX: [503] 2501-2150

Flag description: three equal horizontal bands of cobalt blue (top), white, and cobalt blue with the national coat of arms centered in the white band; the coat of arms features a round emblem encircled by the words REPUBLICA DE EL SALVADOR EN LA AMERICA CENTRAL; the banner is based on the former blue-white-blue flag of the Federal Republic of Central America; the blue bands symbolize the Pacific Ocean and the Caribbean Sea, while the white band represents the land between the two bodies of water, as well as peace and prosperity
note: similar to the flag of Nicaragua, which has a different coat of arms centered in the white band; also similar to the flag of Honduras, which has five blue stars arranged in an X pattern centered in the white band

National symbol(s): *turquoise-browed motmot (bird); national colors:* blue, white

National anthem: *name:* "Himno Nacional de El Salvador" (National Anthem of El Salvador)
lyrics/music: Juan Jose CANAS/Juan ABERLE
note: officially adopted 1953, in use since 1879; at 4:20 minutes, the anthem of El Salvador is one of the world's longest
0:00 / 4:20

ECONOMY

Economy—overview: The smallest country in Central America geographically, El Salvador has the fourth largest economy in the region. With the global recession, real GDP contracted in 2009 and economic growth has since remained low, averaging less than 2% from 2010 to 2014, but recovered somewhat in 2015-17 with an average annual growth rate of 2.4%. Remittances accounted for approximately 18% of GDP in 2017 and were received by about a third of all households.

In 2006, El Salvador was the first country to ratify the Dominican Republic-Central American Free Trade Agreement, which has bolstered the export of processed foods, sugar, and ethanol, and supported investment in the apparel sector amid increased Asian competition. In September 2015, El Salvador kicked off a five-year $277 million second compact with the Millennium Challenge Corporation—a US Government agency aimed at stimulating economic growth and reducing poverty—to improve El Salvador's competitiveness and productivity in international markets.

The Salvadoran Government maintained fiscal discipline during reconstruction and rebuilding following earthquakes in 2001 and hurricanes in 1998 and 2005, but El Salvador's public debt, estimated at 59.3% of GDP in 2017, has been growing over the last several years.

GDP (purchasing power parity):
$51.17 billion (2017 est.)
$50.01 billion (2016 est.)
$48.75 billion (2015 est.)
note: data are in 2017 dollars
country comparison to the world: 109

GDP (official exchange rate):
$24.81 billion (2017 est.)

GDP—real growth rate:
2.3% (2017 est.)
2.6% (2016 est.)
2.4% (2015 est.)
country comparison to the world: 121

GDP—per capita (PPP):
$8,000 (2017 est.)
$7,900 (2016 est.)
$7,700 (2015 est.)
note: data are in 2017 dollars
country comparison to the world: 152

Gross national saving:
14.9% of GDP (2017 est.)
13% of GDP (2016 est.)
12.4% of GDP (2015 est.)
country comparison to the world: 137

GDP—composition, by end use:
household consumption: 84.5% (2017 est.)
government consumption: 15.8% (2017 est.)

297

investment in fixed capital: 16.9% (2017 est.)
investment in inventories: 0% (2017 est.)
exports of goods and services: 27.6% (2017 est.)
imports of goods and services: -44.9% (2017 est.)

GDP—composition, by sector of origin:
agriculture: 12% (2017 est.)
industry: 27.7% (2017 est.)
services: 60.3% (2017 est.)

Agriculture—products: coffee, sugar, corn, rice, beans, oilseed, cotton, sorghum; beef, dairy products

Industries: food processing, beverages, petroleum, chemicals, fertilizer, textiles, furniture, light metals

Industrial production growth rate: 3.6% (2017 est.)
country comparison to the world: 81

Labor force: 2.908 million (2019 est.)
country comparison to the world: 105

Labor force—by occupation:
agriculture: 21%
industry: 20%
services: 58% (2011 est.)

Unemployment rate:
7% (2017 est.)
6.9% (2016 est.)
note: data are official rates; but underemployment is high
country comparison to the world: 113

Population below poverty line: 32.7% (2016 est.)

Household income or consumption by percentage share: *lowest 10%:* 2.2%
highest 10%: 32.3% (2014 est.)

Budget: *revenues:* 5.886 billion (2017 est.)
expenditures: 6.517 billion (2017 est.)

Taxes and other revenues: 23.7% (of GDP) (2017 est.)
country comparison to the world: 123

Budget surplus (+) or deficit (-): -2.5% (of GDP) (2017 est.)
country comparison to the world: 114

Public debt:
67.9% of GDP (2017 est.)
66.4% of GDP (2016 est.)
note: El Salvador's total public debt includes non-financial public sector debt, financial public sector debt, and central bank debt
country comparison to the world: 54

Fiscal year: calendar year

Inflation rate (consumer prices):
1% (2017 est.)
0.6% (2016 est.)
country comparison to the world: 51

Current account balance:
-$501 million (2017 est.)
-$500 million (2016 est.)
country comparison to the world: 122

Exports:
$4.662 billion (2017 est.)
$5.42 billion (2016 est.)
country comparison to the world: 111

Exports—partners: US 45.7%, Honduras 13.9%, Guatemala 13.5%, Nicaragua 6.7%, Costa Rica 4.6% (2017)

Exports—commodities: offshore assembly exports, coffee, sugar, textiles and apparel, ethanol, chemicals, electricity, iron and steel manufactures

Imports:
$9.499 billion (2017 est.)
$8.954 billion (2016 est.)
country comparison to the world: 103

Imports—commodities: raw materials, consumer goods, capital goods, fuels, foodstuffs, petroleum, electricity

Imports—partners: US 36.7%, Guatemala 10.5%, China 8.7%, Mexico 7.4%, Honduras 6.7% (2017)

Reserves of foreign exchange and gold:
$3.567 billion (31 December 2017 est.)
$3.238 billion (31 December 2016 est.)
country comparison to the world: 104

Debt—external:
$15.51 billion (31 December 2017 est.)
$16.32 billion (31 December 2016 est.)
country comparison to the world: 102

Exchange rates: *note:* the US dollar is used as a medium of exchange and circulates freely in the economy 1 (2017 est.)

ENERGY

Electricity access: *population without electricity:* 400,000 (2016)
electrification—total population: 98.6% (2016)
electrification—urban areas: 98.6% (2016)
electrification—rural areas: 98.8% (2016)

Electricity—production: 5.83 billion kWh (2016 est.)
country comparison to the world: 116

Electricity—consumption: 5.928 billion kWh (2016 est.)
country comparison to the world: 116

Electricity—exports: 89.6 million kWh (2017 est.)
country comparison to the world: 82

Electricity—imports: 1.066 billion kWh (2016 est.)
country comparison to the world: 70

Electricity—installed generating capacity: 1.983 million kW (2016 est.)
country comparison to the world: 113

Electricity—from fossil fuels: 49% of total installed capacity (2016 est.)
country comparison to the world: 154

Electricity—from nuclear fuels: 0% of total installed capacity (2017 est.)
country comparison to the world: 83

Electricity—from hydroelectric plants: 23% of total installed capacity (2017 est.)
country comparison to the world: 84

Electricity—from other renewable sources: 29% of total installed capacity (2017 est.)
country comparison to the world: 18

Crude oil—production: 0 bbl/day (2018 est.)
country comparison to the world: 130

Crude oil—exports: 0 bbl/day (2015 est.)
country comparison to the world: 117

Crude oil—imports: 0 bbl/day (2015 est.)
country comparison to the world: 120

Crude oil—proved reserves: 0 bbl (1 January 2018 est.)
country comparison to the world: 126

Refined petroleum products—production: 0 bbl/day (2015 est.)
country comparison to the world: 138

Refined petroleum products—consumption: 52,000 bbl/day (2016 est.)
country comparison to the world: 103

Refined petroleum products—exports: 347 bbl/day (2015 est.)
country comparison to the world: 115

Refined petroleum products—imports: 49,280 bbl/day (2015 est.)
country comparison to the world: 82

Natural gas—production: 0 cu m (2017 est.)
country comparison to the world: 126

Natural gas—consumption: 0 cu m (2017 est.)
country comparison to the world: 141

Natural gas—exports: 0 cu m (2017 est.)
country comparison to the world: 98

Natural gas—imports: 0 cu m (2017 est.)
country comparison to the world: 119

Natural gas—proved reserves: 0 cu m (1 January 2017 est.)
country comparison to the world: 130

Carbon dioxide emissions from consumption of energy: 7.331 million Mt (2017 est.)
country comparison to the world: 124

COMMUNICATIONS

Telephones—fixed lines: *total subscriptions:* 882,498
subscriptions per 100 inhabitants: 13.73 (2019 est.)
country comparison to the world: 79

Telephones—mobile cellular: *total subscriptions:* 9,442,667
subscriptions per 100 inhabitants: 146.91 (2019 est.)
country comparison to the world: 91

Telecommunication systems: *general assessment:* multiple mobile-cellular operators began rolling out (Long Term Evolution) LTE data services in late-2016; Internet usage grew almost 400% between 2007 and 2015; 6% of phones are fixed-line, while 94% are mobilecellular; as of March 2019, the regulator launched a public dialog that allowed mobile network operators to improve the reach and quality of service; telecom legislation encourages competition and foreign investment; only 1 DSL market leader retaining a monopoly; govt. increases tax on telecom services to 18% (2020)
domestic: growth in fixed-line services 14 per 100, has slowed in the face of mobile-cellular competition at 147 per 100 (2019)
international: country code—503; satellite earth station—1 Intelsat (Atlantic Ocean); connected to Central American Microwave System (2019)

note: the COVID-19 outbreak is negatively impacting telecommunications production and supply chains globally; consumer spending on telecom devices and services has also slowed due to the pandemic's effect on economies worldwide; overall progress towards improvements in all facets of the telecom industry—mobile, fixed-line, broadband, submarine cable and satellite—has moderated

Broadcast media: multiple privately owned national terrestrial TV networks, supplemented by cable TV networks that carry international channels; hundreds of commercial radio broadcast stations and 1 government-owned radio broadcast station; transition to digital transmission to begin in 2018 along with adaptation of the Japanese-Brazilian Digital Standard (ISDB-T)

Internet country code: .sv

Internet users: *total:* 2,153,776
percent of population: 33.82% (July 2018 est.)
country comparison to the world: 120

Broadband—fixed subscriptions: *total:* 492,265
subscriptions per 100 inhabitants: 8 (2018 est.)
country comparison to the world: 86

TRANSPORTATION

National air transport system: *number of registered air carriers:* 1 (2020)
inventory of registered aircraft operated by air carriers: 13
annual passenger traffic on registered air carriers: 2,545,105 (2018)
annual freight traffic on registered air carriers: 10.73 million mt-km (2018)

Civil aircraft registration country code prefix: YS (2016)

Airports: 68 (2013)
country comparison to the world: 73

Airports—with paved runways: *total:* 5 (2017)
over 3,047 m: 1 (2017)

1,524 to 2,437 m: 1 (2017)
914 to 1,523 m: 2 (2017)
under 914 m: 1 (2017)

Airports—with unpaved runways: *total:* 63 (2013)
1,524 to 2,437 m: 1 (2013)
914 to 1,523 m: 11 (2013)
under 914 m: 51 (2013)

Heliports: 2 (2013)

Railways: *total:* 13 km (2014)
narrow gauge: 12.5 km 0.914-m gauge (2014)
country comparison to the world: 135

Roadways: *total:* 9,012 km (2017)
paved: 5,341 km (2017)
unpaved: 3,671 km (2017)
country comparison to the world: 138

Waterways: (Rio Lempa River is partially navigable by small craft) (2011)

Merchant marine: *total:* 2
by type: other 2 (2019)
country comparison to the world: 173

Ports and terminals: *major seaport(s):* Puerto Cutuco
oil terminal(s): Acajutla offshore terminal

MILITARY AND SECURITY

Military and security forces: Armed Force of El Salvador (Fuerza Armada de El Salvador, FAES): Army of El Salvador (Ejercito de El Salvador, ES), Navy of El Salvador (Fuerza Naval de El Salvador, FNES), Salvadoran Air Force (Fuerza Aerea Salvadorena, FAS) (2020)
note: supporting the National Police (Ministry of Interior) in countering gang violence and drug trafficking is a primary mission for the FAES

Military expenditures:
1.2% of GDP (2019)
1% of GDP (2018)
0.9% of GDP (2017)

0.9% of GDP (2016)
0.95% of GDP (2015)
country comparison to the world: 103

Military and security service personnel strengths: the Armed Force of El Salvador (FAES) has approximately 22,000 active troops (18,000 Army; 2,000 Navy; 2,000 Air Force) (2019 est.)

Military equipment inventories and acquisitions: the FAES is dependent on a mix of imported Cold War-era platforms, largely from the US; since 2000, the FAES has received limited amounts of equipment from Chile, Israel, and the US (2019 est.)

Military deployments: 200 Mali (MINUSMA) (2020)

Military service age and obligation: 18 years of age for selective compulsory military service; 16-22 years of age for voluntary male or female service; service obligation is 12 months, with 11 months for officers and NCOs (2012)

TRANSNATIONAL ISSUES

Disputes—international: International Court of Justice (ICJ) ruled on the delimitation of "bolsones" (disputed areas) along the El Salvador-Honduras boundary, in 1992, with final agreement by the parties in 2006 after an Organization of American States survey and a further ICJ ruling in 2003; the 1992 ICJ ruling advised a tripartite resolution to a maritime boundary in the Gulf of Fonseca advocating Honduran access to the Pacific; El Salvador continues to claim tiny Conejo Island, not identified in the ICJ decision, off Honduras in the Gulf of Fonseca

Refugees and internally displaced persons: *IDPs:* 71,500 (2018)

Illicit drugs: transshipment point for cocaine; small amounts of marijuana produced for local consumption; significant use of cocaine

EQUATORIAL GUINEA

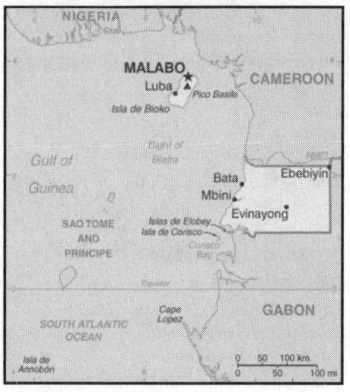

INTRODUCTION

Background: Equatorial Guinea gained independence in 1968 after 190 years of Spanish rule; it is one of the smallest countries in Africa consisting of a mainland territory and five inhabited islands. The capital of Malabo is located on the island of Bioko, approximately 25 km from the Cameroonian coastline in the Gulf of Guinea. Between 1968 and 1979, autocratic President Francisco MACIAS NGUEMA virtually destroyed all of the country's political, economic, and social institutions before being deposed by his nephew Teodoro OBIANG NGUEMA MBASOGO in a coup. President OBIANG has ruled since October 1979. He has been elected several times since 1996, and was most recently reelected in 2016. Although nominally a

constitutional democracy since 1991, presidential and legislative elections since 1996 have generally been labeled as flawed. The president exerts almost total control over the political system and has placed legal and bureaucratic barriers that hinder political opposition. Equatorial Guinea experienced rapid economic growth in the early years of the 21st century due to the discovery of large offshore oil reserves in 1996. Production peaked in late 2004 and has slowly declined since, although aggressive searches for new oil fields continue. Despite the country's economic windfall from oil production, resulting in massive increases in government revenue in past years, the drop in global oil prices as of 2014 has placed significant strain on the state budget and pushed the country into recession. Oil revenues have mainly been used for the development of infrastructure and there have been limited improvements in the population's living standards. Equatorial Guinea continues to seek to diversify its economy and to increase foreign investment. The country hosts major regional and international conferences and continues to seek a greater role in international affairs, and leadership in the sub-region.

GEOGRAPHY

Location: Central Africa, bordering the Bight of Biafra, between Cameroon and Gabon

Geographic coordinates: 2 00 N, 10 00 E

Map references: Africa

Area: *total:* 28,051 sq km
land: 28,051 sq km
water: 0 sq km
country comparison to the world: 146

Area—comparative: slightly smaller than Maryland

Land boundaries: *total:* 528 km
border countries (2): Cameroon 183 km, Gabon 345 km

Coastline: 296 km

Maritime claims: *territorial sea:* 12 nm
exclusive economic zone: 200 nm

Climate: tropical; always hot, humid

Terrain: coastal plains rise to interior hills; islands are volcanic

Elevation: *mean elevation:* 577 m
lowest point: Atlantic Ocean 0 m
highest point: Pico Basile 3,008 m

Natural resources: petroleum, natural gas, timber, gold, bauxite, diamonds, tantalum, sand and gravel, clay

Land use: *agricultural land:* 10.1% (2011 est.)
arable land: 4.3% (2011 est.) / permanent crops: 2.1% (2011 est.) / permanent pasture: 3.7% (2011 est.)
forest: 57.5% (2011 est.)
other: 32.4% (2011 est.)
Irrigated land: NA

Population distribution: only two large cities over 30,000 people (Bata on the mainland, and the capital Malabo on the island of Bioko); small communities are scattered throughout the mainland and the five inhabited islands as shown in this population distribution map

Natural hazards: violent windstorms; flash floods
volcanism: Santa Isabel (3,007 m), which last erupted in 1923, is the country's only historically active volcano; Santa Isabel, along with two dormant volcanoes, form Bioko Island in the Gulf of Guinea

Environment—current issues: deforestation (forests are threatened by agricultural expansion, fires, and grazing); desertification; water pollution (tap water is non- potable); wildlife preservation

Environment—international agreements: *party to:* Biodiversity, Climate Change, Climate Change-Kyoto Protocol, Desertification, Endangered Species, Hazardous Wastes, Law of the Sea, Marine Dumping, Ozone Layer Protection, Ship Pollution, Wetlands
signed, but not ratified: none of the selected agreements

Geography—note: insular and continental regions widely separated; despite its name, no part of the Equator passes through Equatorial Guinea; the mainland part of the country is located just north of the Equator

PEOPLE AND SOCIETY

Population: 836,178 (July 2020 est.)
country comparison to the world: 164

Nationality: *noun:* Equatorial Guinean(s) or Equatoguinean(s)
adjective: Equatorial Guinean or Equatoguinean

Ethnic groups: Fang 85.7%, Bubi 6.5%, Mdowe 3.6%, Annobon 1.6%, Bujeba 1.1%, other 1.4% (1994 census)

Languages: Spanish (official) 67.6%, other (includes Fang, Bubi, Portuguese (official), French (official)) 32.4% (1994 census)

Religions: nominally Christian and predominantly Roman Catholic, Muslim, Baha'i, animist, indigenous

Demographic profile: Equatorial Guinea is one of the smallest and least populated countries in continental Africa and is the only independent African country where Spanish is an official language. Despite a boom in oil production in the 1990s, authoritarianism, corruption, and resource mismanagement have concentrated the benefits among a small elite. These practices have perpetuated income inequality and unbalanced development, such as low public spending on education and health care.

Unemployment remains problematic because the oil-dominated economy employs a small labor force dependent on skilled foreign workers. The agricultural sector, Equatorial Guinea's main employer, continues to deteriorate because of a lack of investment and the migration of rural workers to urban areas. About three-quarters of the population lives below the poverty line.

Equatorial Guinea's large and growing youth population—about 60% are under the age of 25—is particularly affected because job creation in the non- oil sectors is limited, and young people often do not have the skills needed in the labor market. Equatorial Guinean children frequently enter school late, have poor attendance, and have high dropout rates. Thousands of Equatorial Guineans fled across the border to Gabon in the 1970s to escape the dictatorship of MACIAS NGUEMA; smaller numbers have followed in the decades since. Continued inequitable economic growth and high youth unemployment increases the likelihood of ethnic and regional violence.

Age structure: *0-14 years:* 38.73% (male 164,417/female 159,400)
15-24 years: 19.94% (male 84,820/female 81,880)
25-54 years: 32.72% (male 137,632/female 135,973)
55-64 years: 4.69% (male 17,252/female 22,006)
65 years and over: 3.92% (male 13,464/female 19,334) (2020 est.)

Dependency ratios: *total dependency ratio:* 64.4
youth dependency ratio: 60.5
elderly dependency ratio: 3.9
potential support ratio: 25.5 (2020 est.)

Median age: *total:* 20.3 years
male: 19.9 years
female: 20.7 years (2020 est.)
country comparison to the world: 192

Population growth rate: 2.35% (2020 est.)
country comparison to the world: 29

Birth rate: 30.7 births/1,000 population (2020 est.)
country comparison to the world: 31

Death rate: 7.3 deaths/1,000 population (2020 est.)
country comparison to the world: 112

Net migration rate: 0 migrant(s)/1,000 population (2020 est.)
country comparison to the world: 81

Population distribution: only two large cities over 30,000 people (Bata on the mainland, and the capital Malabo on the island of Bioko); small communities are scattered throughout the mainland and the five inhabited islands as shown in this population distribution map

Urbanization: *urban population:* 73.1% of total population (2020)
rate of urbanization: 4.28% annual rate of change (2015-20 est.)

total population growth rate v. urban population growth rate, 2000-2030: Major urban areas—population: 297,000 MALABO (capital) (2018)

Sex ratio: *at birth:* 1.03 male(s)/female
0-14 years: 1.03 male(s)/female
15-24 years: 1.04 male(s)/female
25-54 years: 1.01 male(s)/female
55-64 years: 0.78 male(s)/female
65 years and over: 0.7 male(s)/female
total population: 1 male(s)/female (2020 est.)

Maternal mortality rate: 301 deaths/100,000 live births (2017 est.)
country comparison to the world: 37

Infant mortality rate: *total:* 59.7 deaths/1,000 live births
male: 60.8 deaths/1,000 live births
female: 58.6 deaths/1,000 live births (2020 est.)
country comparison to the world: 13

Life expectancy at birth: *total population:* 65.7 years
male: 64.4 years
female: 66.9 years (2020 est.)
country comparison to the world: 192

Total fertility rate: 4.11 children born/woman (2020 est.)
country comparison to the world: 29

Contraceptive prevalence rate: 12.6% (2011)

Drinking water source:
improved: *urban:* 81.7% of population
rural: 32.1% of population
total: 67.6% of population

unimproved: *urban:* 18.3% of population
rural: 67.9% of population

total: 32.4% of population (2017 est.)

Current Health Expenditure: 3.1% (2017)

Physicians density: 0.4 physicians/1,000 population (2017)

Hospital bed density: 2.1 beds/1,000 population (2010)

Sanitation facility access:

improved: *urban:* 81.2% of population
rural: 63.4% of population
total: 76.2% of population

unimproved: *urban:* 18.8% of population
rural: 36.6% of population
total: 23.8% of population (2017 est.)

HIV/AIDS—adult prevalence rate: 7% (2019 est.)
country comparison to the world: 10

HIV/AIDS—people living with HIV/AIDS: 65,000 (2019 est.)
country comparison to the world: 57

HIV/AIDS—deaths: 1,800 (2019 est.)
country comparison to the world: 48

Major infectious diseases: *degree of risk:* very high (2020)
food or waterborne diseases: bacterial and protozoal diarrhea, hepatitis A, and typhoid fever
vectorborne diseases: malaria and dengue fever
animal contact diseases: rabies

Obesity—adult prevalence rate: 8% (2016)
country comparison to the world: 156

Children under the age of 5 years underweight: 5.6% (2011)
country comparison to the world: 80

Education expenditures: NA

Literacy: *definition:* age 15 and over can read and write
total population: 95.3%
male: 97.4%
female: 93% (2015)

GOVERNMENT

Country name: *conventional long form:* Republic of Equatorial Guinea
conventional short form: Equatorial Guinea
local long form: Republica de Guinea Ecuatorial/Republique de Guinee Equatoriale
local short form: Guinea Ecuatorial/Guinee Equatoriale
former: Spanish Guinea
etymology: the country is named for the Guinea region of West Africa that lies along the Gulf of Guinea and stretches north to the Sahel; the "equatorial" refers to the fact that the country lies just north of the Equator

Government type: presidential republic

Capital: *name:* Malabo; note—a new capital of Cuidad de la Paz (formerly referred to as Oyala) is being built on the mainland near Djibloho; Malabo is on the island of Bioko
geographic coordinates: 3 45 N, 8 47 E
time difference: UTC+ 1 (6 hours ahead of Washington, DC, during Standard Time)

etymology: named after Malabo Lopelo Melaka (1837–1937), the last king of the Bubi, the ethnic group indigenous to the island of Bioko; the name of the new capital, Cuidad de la Paz, translates to "City of Peace" in Spanish

Administrative divisions: 7 provinces (provincias, singular—provincia); Annobon, Bioko Norte, Bioko Sur, Centro Sur, Kie-Ntem, Litoral, Wele-Nzas

Independence: 12 October 1968 (from Spain)

National holiday: Independence Day, 12 October (1968)

Constitution: *history:* previous 1968, 1973, 1982; approved by referendum 17 November 1991
amendments: proposed by the president of the republic or supported by three fourths of the membership in either house of the National Assembly; passage requires three- fourths majority vote by both houses of the Assembly and approval in a referendum if requested by the president; amended several times, last in 2012

Legal system: mixed system of civil and customary law

International law organization participation: accepts compulsory ICJ jurisdiction; accepts ICCt jurisdiction

Citizenship: *citizenship by birth:* no
citizenship by descent only: at least one parent must be a citizen of Equatorial Guinea
dual citizenship recognized: no
residency requirement for naturalization: 10 years

Suffrage: 18 years of age; universal

Executive branch: *chief of state:* President Brig. Gen. (Ret.) Teodoro OBIANG Nguema Mbasogo (since 3 August 1979 when he seized power in a military coup); Vice President Teodoro Nguema OBIANG Mangue(since 2012)
head of government: Prime Minister Francisco Pascual Eyegue OBAMA Asue (since 23 June 2016); First Deputy Prime Minister Clemente Engonga NGUEMA Onguene (since 23 June 2016); Second Deputy Prime Minister Angel MESIE Mibuy (since 5 February 2018); Third Deputy Prime Minister Alfonso Nsue MOKUY (since 23 June 2016)
cabinet: Council of Ministers appointed by the president and overseen by the prime minister
elections/appointments: president directly elected by simple majority popular vote for a 7-year term (eligible for a second term); election last held on 24 April 2016 (next to be held in 2023); prime minister and deputy prime ministers appointed by the president
election results: Teodoro OBIANG Nguema Mbasogo reelected president; percent of vote— Teodoro OBIANG Nguema Mbasogo (PDGE) 93.5%, other 6.5%

Legislative branch: *description:* bicameral National Assembly or Asemblea Nacional consists of:

Senate or Senado (70 seats; 55 members directly elected in multi-seat constituencies by

closed party- list proportional representation vote and 15 appointed by the president)

Chamber of Deputies or Camara de los Diputados (100 seats; members directly elected in multi- seat constituencies by closed paryt-list proportional representation vote to serve 5-year terms)
elections: Senate—last held on 12 November 2017 (next to be held in 2022/2023)

Chamber of Deputies—last held on 12 November 2017 (next to be held in 2022/2023)
election results: Senate—percent of vote by party—NA; seats by party—PDGE and aligned coalition 70; composition—men 60, women 10, percent of women 14.3%

Chamber of Deputies—percent of vote by party—NA; seats by party—PDGE 99, CI 1; composition—men 78, women 22, percent of women 22%; note—total National Assembly percent of women 18.8%

Judicial branch: *highest courts:* Supreme Court of Justice (consists of the chief justice—who is also chief of state—and 9 judges organized into civil, criminal, commercial, labor, administrative, and customary sections); Constitutional Court (consists of the court president and 4 members)
judge selection and term of office: Supreme Court judges appointed by the president for 5-year terms; Constitutional Court members appointed by the president, 2 of whom are nominated by the Chamber of Deputies; note—judges subject to dismissal by the president at any time
subordinate courts: Court of Guarantees; military courts; Courts of Appeal; first instance tribunals; district and county tribunals

Political parties and leaders: Citizens for Innovation or CI [Gabriel Nse Obiang OBONO]
Convergence Party for Social Democracy or CPDS [Andres ESONO ONDO]
Democratic Party for Equatorial Guinea or PDGE [Teodoro Obiang NGUEMA MBASOGO]
Electoral Coalition or EC
Juntos Podemos (coalition includes CPDS, FDR, UDC)
National Congress of Equatorial Guinea [Agustin MASOKO ABEGUE]
National Democratic Party [Benedicto OBIANG MANGUE]
National Union for Democracy [Thomas MBA MONABANG]
Popular Action of Equatorial Guinea or APGE [Carmelo MBA BACALE]
Popular Union or UP [Daniel MARTINEZ AYECABA]
Union for the Center right or UDC [Avelino MOCACHE MEHENGA]
not officially registered parties:
Democratic Republican Force or FDR [Guillermo NGUEMA ELA]
Party for Progress of Equatorial Guinea or PPGE [Severo MOTO]

International organization participation: ACP, AfDB, AU, BDEAC, CEMAC, CPLP (associate), FAO, FZ, G- 77, IBRD, ICAO, ICRM, IDA, IFAD, IFC, IFRCS, ILO, IMF, IMO, Interpol, IOC, IPU,

ITSO, ITU, MIGA, NAM, OAS (observer), OIF, OPCW, UN, UNCTAD, UNESCO, UNIDO, UN Security Council (temporary), UNWTO, UPU, WHO, WIPO, WTO (observer)

Diplomatic representation in the US: *chief of mission:* Ambassador Miguel Ntutumu EVUNA ANDEME (since 23 February 2015)
chancery: 2020 16th Street NW, Washington, DC 20009
telephone: [1] (202) 518-5700
FAX: [1] (202) 518-5252
consulate(s) general: Houston

Diplomatic representation from the US: *chief of mission:* Ambassador Susan N. STEVENSON (since 7 May 2019)
telephone: [240] 333 09 57 41 or 1-301-985-8750
embassy: Malabo II Highway (between the Headquarters of Sonagas and the offices of the United Nations), Malabo
mailing address: US Embassy Malabo, 2320 Malabo Place, Washington, DC 20521-2520

Flag description: three equal horizontal bands of green (top), white, and red, with a blue isosceles triangle based on the hoist side and the coat of arms centered in the white band; the coat of arms has six yellow six-pointed stars (representing the mainland and five offshore islands) above a gray shield bearing a silk-cotton tree and below which is a scroll with the motto UNIDAD, PAZ, JUSTICIA (Unity, Peace, Justice); green symbolizes the jungle and natural resources, blue represents the sea that connects the mainland to the islands, white stands for peace, and red recalls the fight for independence

National symbol(s): *silk cotton tree; national colors:* green, white, red, blue

National anthem: *name:* "Caminemos pisando la senda" (Let Us Tread the Path)
lyrics/music: Atanasio Ndongo MIYONO/ Atanasio Ndongo MIYONO or Ramiro Sanchez LOPEZ (disputed)
note: adopted 1968

ECONOMY

Economy—overview: Exploitation of oil and gas deposits, beginning in the 1990s, has driven economic growth in Equatorial Guinea; a recent rebasing of GDP resulted in an upward revision of the size of the economy by approximately 30%. Forestry and farming are minor components of GDP. Although preindependence Equatorial Guinea counted on cocoa production for hard currency earnings, the neglect of the rural economy since independence has diminished the potential for agriculture-led growth. Subsistence farming is the dominant form of livelihood. Declining revenue from hydrocarbon production, high levels of infrastructure expenditures, lack of economic diversification, and corruption have pushed the economy into decline in recent years and limited improvements in the general population's living conditions. Equatorial Guinea's real GDP growth has been weak in recent years, averaging -0.5%

per year from 2010 to 2014, because of a declining hydrocarbon sector. Inflation remained very low in 2016, down from an average of 4% in 2014.

As a middle income country, Equatorial Guinea is now ineligible for most low-income World Bank and the IMF funding. The government has been widely criticized for its lack of transparency and misuse of oil revenues and has attempted to address this issue by working toward compliance with the Extractive Industries Transparency Initiative. US foreign assistance to Equatorial Guinea is limited in part because of US restrictions pursuant to the Trafficking Victims Protection Act.

Equatorial Guinea hosted two economic diversification symposia in 2014 that focused on attracting investment in five sectors: agriculture and animal ranching, fishing, mining and petrochemicals, tourism, and financial services. Undeveloped mineral resources include gold, zinc, diamonds, columbite-tantalite, and other base metals. In 2017 Equatorial Guinea signed a preliminary agreement with Ghana to sell liquefied natural gas (LNG); as oil production wanes, the government believes LNG could provide a boost to revenues, but it will require large investments and long lead times to develop.

GDP (purchasing power parity): $31.52 billion (2017 est.)
$32.57 billion (2016 est.)
$35.62 billion (2015 est.)
note: data are in 2017 dollars
country comparison to the world: 130

GDP (official exchange rate): $12.49 billion (2017 est.)

GDP—real growth rate: -3.2% (2017 est.)
-8.6% (2016 est.)
-9.1% (2015 est.)
country comparison to the world: 213

GDP—per capita (PPP): $37,400 (2017 est.)
$39,700 (2016 est.)
$44,600 (2015 est.)
note: data are in 2017 dollars
country comparison to the world: 52

Gross national saving: 6.1% of GDP (2017 est.)
3.6% of GDP (2016 est.)
8.5% of GDP (2015 est.)
country comparison to the world: 173

GDP—composition, by end use: *household consumption:* 50% (2017 est.)
government consumption: 21.8% (2017 est.)
investment in fixed capital: 10.2% (2017 est.)
investment in inventories: 0.1% (2017 est.)
exports of goods and services: 56.9% (2017 est.)
imports of goods and services: -39% (2017 est.)

GDP—composition, by sector of origin: *agriculture:* 2.5% (2017 est.)
industry: 54.6% (2017 est.)
services: 42.9% (2017 est.)

Agriculture—products: coffee, cocoa, rice, yams, cassava (manioc, tapioca), bananas, palm oil nuts; livestock; timber

Industries: petroleum, natural gas, sawmilling

Industrial production growth rate: -6.9% (2017 est.)
country comparison to the world: 197

Labor force: 195,200 (2007 est.)
country comparison to the world: 173

Unemployment rate: 8.6% (2014 est.)
22.3% (2009 est.)
country comparison to the world: 131

Population below poverty line: 44% (2011 est.)

Household income or consumption by percentage share: *lowest 10%:* NA
highest 10%: NA

Budget: *revenues:* 2.114 billion (2017 est.)
expenditures: 2.523 billion (2017 est.)

Taxes and other revenues: 16.9% (of GDP) (2017 est.)
country comparison to the world: 173

Budget surplus (+) or deficit (-): -3.3% (of GDP) (2017 est.)
country comparison to the world: 141

Public debt: 37.4% of GDP (2017 est.)
43.3% of GDP (2016 est.)
country comparison to the world: 139

Fiscal year: calendar year

Inflation rate (consumer prices): 0.7% (2017 est.)
1.4% (2016 est.)
country comparison to the world: 37

Current account balance: -$738 million (2017 est.)
-$1.457 billion (2016 est.)
country comparison to the world: 136

Exports: $6.118 billion (2017 est.)
$5.042 billion (2016 est.)
country comparison to the world: 101

Exports—partners: China 28%, India 11.8%, South Korea 10.3%, Portugal 8.7%, US 6.9%, Spain 4.9% (2017)

Exports—commodities: petroleum products, timber

Imports: $2.577 billion (2017 est.)
$2.915 billion (2016 est.)
country comparison to the world: 155

Imports—commodities: petroleum sector equipment, other equipment, construction materials, vehicles

Imports—partners: Spain 20.5%, China 19.4%, US 13%, Cote dIvoire 6.2%, Netherlands 4.7% (2017)

Reserves of foreign exchange and gold: $45.5 million (31 December 2017 est.)
$62.31 million (31 December 2016 est.)
country comparison to the world: 188

Debt—external: $1.211 billion (31 December 2017 est.)
$1.074 billion (31 December 2016 est.)
country comparison to the world: 162

Exchange rates: Cooperation Financiere en Afrique Centrale francs (XAF) per US dollar - 605.3 (2017 est.)
593.01 (2016 est.)
593.01 (2015 est.)

591.45 (2014 est.)
494.42 (2013 est.)

ENERGY

Electricity access: *electrification—total population:* 67% (2019)
electrification—urban areas: 75% (2019)
electrification—rural areas: 45% (2019)

Electricity—production: 500 million kWh (2016 est.)
country comparison to the world: 166

Electricity—consumption: 465 million kWh (2016 est.)
country comparison to the world: 171

Electricity—exports: 0 kWh (2016 est.)
country comparison to the world: 130

Electricity—imports: 0 kWh (2016 est.)
country comparison to the world: 144

Electricity—installed generating capacity: 331,000 kW (2016 est.)
country comparison to the world: 156

Electricity—from fossil fuels: 61% of total installed capacity (2016 est.)
country comparison to the world: 127

Electricity—from nuclear fuels: 0% of total installed capacity (2017 est.)
country comparison to the world: 84

Electricity—from hydroelectric plants: 38% of total installed capacity (2017 est.)
country comparison to the world: 54

Electricity—from other renewable sources: 2% of total installed capacity (2017 est.)
country comparison to the world: 140

Crude oil—production: 172,000 bbl/day (2018 est.)
country comparison to the world: 38

Crude oil—exports: 308,700 bbl/day (2017 est.)
country comparison to the world: 26

Crude oil—imports: 0 bbl/day (2015 est.)
country comparison to the world: 121

Crude oil—proved reserves: 1.1 billion bbl (1 January 2018 est.)
country comparison to the world: 40

Refined petroleum products—production: 0 bbl/day (2015 est.)
country comparison to the world: 139

Refined petroleum products—consumption: 5,200 bbl/day (2016 est.)
country comparison to the world: 176

Refined petroleum products—exports: 0 bbl/day (2015 est.)
country comparison to the world: 150

Refined petroleum products—imports: 5,094 bbl/day (2015 est.)
country comparison to the world: 171

Natural gas—production: 6.069 billion cu m (2017 est.)
country comparison to the world: 46

Natural gas—consumption: 1.189 billion cu m (2017 est.)
country comparison to the world: 91

Natural gas—exports: 4.878 billion cu m (2017 est.)
country comparison to the world: 30

Natural gas—imports: 0 cu m (2017 est.)
country comparison to the world: 120

Natural gas—proved reserves: 36.81 billion cu m (1 January 2018 est.)
country comparison to the world: 66

Carbon dioxide emissions from consumption of energy: 3.062 million Mt (2017 est.)
country comparison to the world: 148

COMMUNICATIONS

Telephones—fixed lines: *total subscriptions:* 6,779
subscriptions per 100 inhabitants: less than 1 (2019 est.)
country comparison to the world: 199

Telephones—mobile cellular: *total subscriptions:* 368,920
subscriptions per 100 inhabitants: 45.17 (2019 est.)
country comparison to the world: 176

Telecommunication systems: *general assessment:* digital fixed-line network in most major urban areas and decent mobile cellular coverage; 3G technology has allowed for estimated 9.5% of growth during 2016—2021; mobile data will be the fastest- growing segment 2016-2021 (2018)
domestic: fixed-line density is about 1 per 100 persons and mobile-cellular subscribership is 45 per 100 (2019)
international: country code—240; landing points for the ACE, Ceiba-1, and Ceiba-2 submarine cables providing communication from Bata and Malabo, Equatorial Guinea to numerous Western African and European countries; satellite earth station—1 Intelsat (Indian Ocean) (2019)
note: the COVID-19 outbreak is negatively impacting telecommunications production and supply chains globally; consumer spending on telecom devices and services has also slowed due to the pandemic's effect on economies worldwide; overall progress towards improvements in all facets of the telecom industry—mobile, fixed-line, broadband, submarine cable and satellite—has moderated

Broadcast media: the state maintains control of broadcast media with domestic broadcast media limited to 1 state-owned TV station, 1 private TV station owned by the president's eldest son (who is the Vice President), 1 state-owned radio station, and 1 private radio station owned by the president's eldest son; satellite TV service is available; transmissions of multiple international broadcasters are generally accessible (2019)

Internet country code: .gq

Internet users: *total:* 209,253
percent of population: 26.24% (July 2018 est.)
country comparison to the world: 175

Broadband—fixed subscriptions: *total:* 1,620
subscriptions per 100 inhabitants: less than 1 (2018 est.)
country comparison to the world: 188

TRANSPORTATION

National air transport system: *number of registered air carriers:* 6 (2020)
inventory of registered aircraft operated by air carriers: 15
annual passenger traffic on registered air carriers: 466,435 (2018)
annual freight traffic on registered air carriers: 350,000 mt-km (2018)

Civil aircraft registration country code prefix: 3C (2016)

Airports: 7 (2013)
country comparison to the world: 167

Airports—with paved runways: *total:* 6 (2019)
over 3,047 m: 1
2,438 to 3,047 m: 2
1,524 to 2,437 m: 1
under 914 m: 2

Airports—with unpaved runways: *total:* 1 (2013)
2,438 to 3,047 m: 1 (2013)

Pipelines: 42 km condensate, 5 km condensate/gas, 79 km gas, 71 km oil (2013)

Roadways: *total:* 2,880 km (2017)
country comparison to the world: 164

Merchant marine: *total:* 38
by type: bulk carrier 1, general cargo 7, oil tanker 6, other 24 (2019)
country comparison to the world: 124

Ports and terminals: *major seaport(s):* Bata, Luba, Malabo
LNG terminal(s) (export): Bioko Island

MILITARY AND SECURITY

Military and security forces: *Equatorial Guinea Armed Forces (FAGE):* Equatorial Guinea National Guard (Guardia Nacional de Guinea Ecuatorial, GNGE (Army), Navy, Air Force; Guardia Civil (paramilitary force for internal security) (2019)

Military expenditures: 1.1% of GDP (2018)
1.1% of GDP (2017)
1.2% of GDP (2016)
1% of GDP (2015)
country comparison to the world: 112

Military and security service personnel strengths: the Equatorial Guinea Armed Forces (FAGE) have approximately 1,400 active duty troops (1,100 Army; 200 Navy; 100 Air Force) (2019)

Military equipment inventories and acquisitions: the FAGE is armed with mostly second-hand Russian and Soviet-era weapons; Ukraine is the leading provider of equipment since 2010 followed by Israel (2019 est.)

Military service age and obligation: 18 years of age for selective compulsory military service, although conscription is rare in practice; 2-year service obligation; women hold only administrative positions in the Navy (2013)

Disputes—international: in 2002, ICJ ruled on an equidistance settlement of Cameroon- Equatorial Guinea-Nigeria maritime boundary in the Gulf of Guinea, but a dispute between Equatorial Guinea and Cameroon over an island at the mouth of the Ntem River and imprecisely defined maritime coordinates in the ICJ decision delayed final delimitation; UN' urged Equatorial Guinea and Gabon to resolve the sovereignty dispute over Gabon- occupied Mbane and lesser islands and to create a maritime boundary in the hydrocarbon-rich Corisco Bay

Trafficking in persons: *current situation:* Equatorial Guinea is a source country for children subjected to sex trafficking and destination country for men, women, and children subjected to forced labor; Equatorial Guinean girls may be encouraged by their parents to engage in the sex trade in urban centers to receive groceries, gifts, housing, and money; children are also trafficked from nearby countries for work as domestic servants, market laborers, ambulant vendors, and launderers; women are trafficked to Equatorial Guinea from Cameroon, Benin, other neighboring countries, and China for forced labor or prostitution

tier rating: Tier 3—Equatorial Guinea does not fully comply with the minimum standards on the elimination of trafficking and is not making significant efforts to do so; in 2014, the government made no efforts to investigate or prosecute any suspected trafficking offenders or to identify or protect victims, despite its 2004 law prohibiting all forms of trafficking and mandating the provision of services to victims; undocumented migrants continued to be deported without being screened to assess whether any were trafficking victims; authorities did not undertake any trafficking awareness campaigns, implement any programs to address forced child labor, or make any other efforts to prevent trafficking (2015)

ERITREA

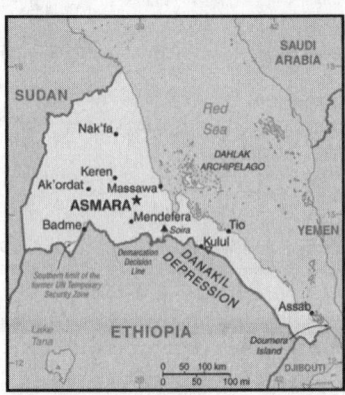

INTRODUCTION

Background: After independence from Italian colonial control in 1941 and 10 years of British administrative control, the UN established Eritrea as an autonomous region within the Ethiopian federation in 1952. Ethiopia's full annexation of Eritrea as a province 10 years later sparked a violent 30-year struggle for independence that ended in 1991 with Eritrean rebels defeating government forces. Eritreans overwhelmingly approved independence in a 1993 referendum. ISAIAS Afwerki has been Eritrea's only president since independence; his rule, particularly since 2001, has been highly autocratic and repressive. His government has created a highly militarized society by pursuing an unpopular program of mandatory conscription into national service—divided between military and civilian service—of indefinite length. A two-and-a-half-year border war with Ethiopia that erupted in 1998 ended under UN auspices in December 2000. A subsequent 2007 Eritrea-Ethiopia Boundary Commission (EEBC) demarcation was rejected by Ethiopia. More than a decade of a tense "no peace, no war" stalemate ended in 2018 after the newly elected Ethiopian prime

minister accepted the EEBC's 2007 ruling, and the two countries signed declarations of peace and friendship. Following the July 2018 peace agreement with Ethiopia, Eritrean leaders engaged in intensive diplomacy around the Horn of Africa, bolstering regional peace, security, and cooperation, as well as brokering rapprochements between governments and opposition groups. In November 2018, the UN Security Council lifted an arms embargo that had been imposed on Eritrea since 2009, after the UN Somalia-Eritrea Monitoring Group reported they had not found evidence of Eritrean support in recent years for Al- Shabaab. The country's rapprochement with Ethiopia has led to a steady resumption of economic ties, with increased air transport, trade, tourism, and port activities, but the economy remains agriculture- dependent, and Eritrea is still one of Africa's poorest nations. Despite the country's improved relations with its neighbors, ISAIAS has not let up on repression and conscription and militarization continue.

GEOGRAPHY

Location: Eastern Africa, bordering the Red Sea, between Djibouti and Sudan

Geographic coordinates: 15 00 N, 39 00 E

Map references: Africa

Area: *total:* 117,600 sq km
land: 101,000 sq km
water: 16,600 sq km
country comparison to the world: 102

Area—comparative: slightly smaller than Pennsylvania

Land boundaries: *total:* 1,840 km
border countries (3): Djibouti 125 km, Ethiopia 1033 km, Sudan 682 km

Coastline: 2,234 km (mainland on Red Sea 1,151 km, islands in Red Sea 1,083 km)

Maritime claims: *territorial sea:* 12 nm

Climate: hot, dry desert strip along Red Sea coast; cooler and wetter in the central highlands (up

to 61 cm of rainfall annually, heaviest June to September); semiarid in western hills and lowlands

Terrain: dominated by extension of Ethiopian north-south trending highlands, descending on the east to a coastal desert plain, on the northwest to hilly terrain and on the southwest to flat-to-rolling plains

Elevation: *mean elevation:* 853 m
lowest point: near Kulul within the Danakil Depression—75 m
highest point: Soira 3,018 m

Natural resources: gold, potash, zinc, copper, salt, possibly oil and natural gas, fish

Land use: *agricultural land:* 75.1% (2011 est.)
arable land: 6.8% (2011 est.) / permanent crops: 0% (2011 est.) / permanent pasture: 68.3% (2011 est.)
forest: 15.1% (2011 est.)
other: 9.8% (2011 est.)
Irrigated land: 210 sq km (2012)

Population distribution: density is highest in the center of the country in and around the cities of Asmara (capital) and Keren; smaller settlements exist in the north and south as shown in this population distribution map

Natural hazards: frequent droughts, rare earthquakes and volcanoes; locust swarms
volcanism: Dubbi (1,625 m), which last erupted in 1861, was the country's only historically active volcano until Nabro (2,218 m) came to life on 12 June 2011

Environment—current issues: deforestation; desertification; soil erosion; overgrazing

Environment—international agreements: *party to:* Biodiversity, Climate Change, Climate Change-Kyoto Protocol, Desertification, Endangered Species, Hazardous Wastes, Ozone Layer Protection
signed, but not ratified: none of the selected agreements

Geography—note: strategic geopolitical position along world's busiest shipping lanes; Eritrea retained the entire coastline of Ethiopia along the Red Sea upon de jure independence from Ethiopia on 24 May 1993

PEOPLE AND SOCIETY

Population: 6,081,196 (July 2020 est.)
country comparison to the world: 112

Nationality: *noun:* Eritrean(s)
adjective: Eritrean

Ethnic groups: Tigrinya 55%, Tigre 30%, Saho 4%, Kunama 2%, Rashaida 2%, Bilen 2%, other (Afar, Beni Amir, Nera) 5% (2010 est.)

note: data represent Eritrea's nine recognized ethnic groups

Languages: Tigrinya (official), Arabic (official), English (official), Tigre, Kunama, Afar, other Cushitic languages

Religions: Sunni Muslim, Coptic Christian, Roman Catholic, Protestant

Demographic profile: Eritrea is a persistently poor country that has made progress in some socioeconomic categories but not in others. Education and human capital formation are national priorities for facilitating economic development and eradicating poverty. To this end, Eritrea has made great strides in improving adult literacy—doubling the literacy rate over the last 20 years—in large part because of its successful adult education programs. The overall literacy rate was estimated to be almost 74% in 2015; more work needs to be done to raise female literacy and school attendance among nomadic and rural communities. Subsistence farming fails to meet the needs of Eritrea's growing population because of repeated droughts, dwindling arable land, overgrazing, soil erosion, and a shortage of farmers due to conscription and displacement. The government's emphasis on spending on defense over agriculture and its lack of foreign exchange to import food also contribute to food insecurity.

Eritrea has been a leading refugee source country since at least the 1960s, when its 30-year war for independence from Ethiopia began. Since gaining independence in 1993, Eritreans have continued migrating to Sudan, Ethiopia, Yemen, Egypt, or Israel because of a lack of basic human rights or political freedom, educational and job opportunities, or to seek asylum because of militarization. Eritrea's large diaspora has been a source of vital remittances, funding its war for independence and providing 30% of the country's GDP annually since it became independent.

In the last few years, Eritreans have increasingly been trafficked and held hostage by Bedouins in the Sinai Desert, where they are victims of organ harvesting, rape, extortion, and torture. Some Eritrean trafficking victims are kidnapped after being smuggled to Sudan or Ethiopia, while others are kidnapped from within or around refugee camps or crossing Eritrea's borders. Eritreans composed approximately 90% of the conservatively estimated 25,000-30,000 victims of Sinai trafficking from 2009-2013, according to a 2013 consultancy firm report.

Age structure: *0-14 years:* 38.23% (male 1,169,456/female 1,155,460)
15-24 years: 20.56% (male 622,172/female 627,858)
25-54 years: 33.42% (male 997,693/female 1,034,550)
55-64 years: 3.8% (male 105,092/female 125,735)
65 years and over: 4% (male 99,231/female 143,949) (2020 est.)

Dependency ratios: *total dependency ratio:* 83.9
youth dependency ratio: 75.6
elderly dependency ratio: 8.3
potential support ratio: 12.1 (2020 est.)

Median age: *total:* 20.3 years
male: 19.7 years
female: 20.8 years (2020 est.)
country comparison to the world: 193

Population growth rate: 0.93% (2020 est.)
country comparison to the world: 115

Birth rate: 27.9 births/1,000 population (2020 est.)
country comparison to the world: 39

Death rate: 6.9 deaths/1,000 population (2020 est.)
country comparison to the world: 131

Net migration rate: -11.6 migrant(s)/1,000 population (2020 est.)
country comparison to the world: 220

Population distribution: density is highest in the center of the country in and around the cities of Asmara (capital) and Keren; smaller settlements exist in the north and south as shown in this population distribution map

Urbanization: *urban population:* 41.3% of total population (2020)
rate of urbanization: 3.86% annual rate of change (2015-20 est.)

total population growth rate v. urban population growth rate, 2000-2030: *Major urban areas—population:* 963,000 ASMARA (capital) (2020)

Sex ratio: *at birth:* 1.03 male(s)/female
0-14 years: 1.01 male(s)/female
15-24 years: 0.99 male(s)/female
25-54 years: 0.96 male(s)/female
55-64 years: 0.84 male(s)/female
65 years and over: 0.69 male(s)/female
total population: 0.97 male(s)/female (2020 est.)

Mother's mean age at first birth: 21.3 years (2010 est.)
note: median age at first birth among women 25-29

Maternal mortality rate: 480 deaths/100,000 live births (2017 est.)
country comparison to the world: 21

Infant mortality rate: *total:* 43.3 deaths/1,000 live births
male: 50.3 deaths/1,000 live births
female: 36.1 deaths/1,000 live births (2020 est.)
country comparison to the world: 31

Life expectancy at birth: *total population:* 66.2 years
male: 63.6 years
female: 68.8 years (2020 est.)
country comparison to the world: 188

Total fertility rate: 3.73 children born/woman (2020 est.)
country comparison to the world: 35

Contraceptive prevalence rate: 8.4% (2010)

Drinking water source:

improved: *urban:* 73.2% of population
rural: 53.3% of population
total: 57.8% of population

unimproved: *urban:* 26.8% of population
rural: 46.7% of population
total: 42.2% of population (2015 est.)

Current Health Expenditure: 2.9% (2017)

Physicians density: 0.06 physicians/1,000 population (2016)

Hospital bed density: 0.7 beds/1,000 population (2011)

Sanitation facility access:

improved: *urban:* 44.5% of population
rural: 7.3% of population
total: 15.7% of population

unimproved: *urban:* 55.5% of population
rural: 92.7% of population
total: 84.3% of population (2015 est.)

HIV/AIDS—adult prevalence rate: 0.7% (2019 est.)
country comparison to the world: 55

HIV/AIDS—people living with HIV/AIDS: 14,000 (2019 est.)
country comparison to the world: 92

HIV/AIDS—deaths: <500 (2019 est.)

Major infectious diseases: *degree of risk:* high (2020)
food or waterborne diseases: bacterial diarrhea, hepatitis A, and typhoid fever
vectorborne diseases: malaria and dengue fever

Obesity—adult prevalence rate: 5% (2016)
country comparison to the world: 183

Children under the age of 5 years underweight: 39.4% (2010)
country comparison to the world: 2

Education expenditures: NA

Literacy: *definition:* age 15 and over can read and write
total population: 76.6%
male: 84.4%
female: 68.9% (2018)

School life expectancy (primary to tertiary education): *total:* 59 years
male: 8 years
female: 7 years (2015)

GOVERNMENT

Country name: *conventional long form:* State of Eritrea
conventional short form: Eritrea
local long form: Hagere Ertra
local short form: Ertra
former: Eritrea Autonomous Region in Ethiopia
etymology: the country name derives from the ancient Greek appellation "Erythra Thalassa" meaning Red Sea, which is the major water body bordering the country

305

Government type: presidential republic

Capital: *name:* Asmara (Asmera)
geographic coordinates: 15 20 N, 38 56 E
time difference: UTC+ 3 (8 hours ahead of Washington, DC, during Standard Time)
etymology: the name means "they [women] made them unite," which according to Tigrinya oral tradition refers to the women of the four clans in the Asmara area who persuaded their menfolk to unite and defeat their common enemy; the name has also been translated as "live in peace"

Administrative divisions: 6 regions (zobatat, singular—zoba); Anseba, Debub (South), Debubawi K'eyih Bahri (Southern Red Sea), Gash Barka, Ma'akel (Central), Semenawi K'eyih Bahri (Northern Red Sea)

Independence: 24 May 1993 (from Ethiopia)

National holiday: Independence Day, 24 May (1991)

Constitution: *history:* ratified by the Constituent Assembly 23 May 1997 (not fully implemented)
amendments: proposed by the president of Eritrea or by assent of at least one half of the National Assembly membership; passage requires at least an initial three- quarters majority vote by the Assembly and, after one year, final passage by at least four-fifths majority vote by the Assembly

Legal system: mixed legal system of civil, customary, and Islamic religious law

International law organization participation: has not submitted an ICJ jurisdiction declaration; non-party state to the ICCt

Citizenship: *citizenship by birth:* no
citizenship by descent only: at least one parent must be a citizen of Eritrea
dual citizenship recognized: no
residency requirement for naturalization: 20 years

Suffrage: 18 years of age; universal

Executive branch: *chief of state:* President ISAIAS Afwerki (since 8 June 1993); note—the president is both chief of state and head of government and is head of the State Council and National Assembly
head of government: President ISAIAS Afwerki (since 8 June 1993)
cabinet: State Council appointed by the president
elections/appointments: president indirectly elected by the National Assembly for a 5-year term (eligible for a second term); the only election was held on 8 June 1993, following independence from Ethiopia (next election postponed indefinitely)
election results: ISAIAS Afwerki elected president by the transitional National Assembly; percent of National Assembly vote—ISAIAS Afwerki (PFDJ) 95%, other 5%

Legislative branch: *description:* unicameral National Assembly (Hagerawi Baito) (150 seats; 75 members indirectly elected by the ruling party and 75 directly elected by simple majority vote; members serve 5-year terms)
elections: in May 1997, following the adoption of the new constitution, 75 members of the PFDJ

Central Committee (the old Central Committee of the EPLF), 60 members of the 527-member Constituent Assembly, which had been established in 1997 to discuss and ratify the new constitution, and 15 representatives of Eritreans living abroad were formed into a Transitional National Assembly to serve as the country's legislative body until countrywide elections to form a National Assembly were held; although only 75 of 150 members of the Transitional National Assembly were elected, the constitution stipulates that once past the transition stage, all members of the National Assembly will be elected by secret ballot of all eligible voters; National Assembly elections scheduled for December 2001 were postponed indefinitely due to the war with Ethiopia, and as of May 2019, there was no sitting legislative body
election results: NA

Judicial branch: *highest courts:* High Court (consists of 20 judges and organized into civil, commercial, criminal, labor, administrative, and customary sections)
judge selection and term of office: High Court judges appointed by the president
subordinate courts: regional/zonal courts; community courts; special courts; sharia courts (for issues dealing with Muslim marriage, inheritance, and family); military courts

Political parties and leaders: People's Front for Democracy and Justice or PFDJ [ISAIAS Afwerki] (the only party recognized by the government)

International organization participation: ACP, AfDB, AU, COMESA, FAO, G- 77, IAEA, IBRD, ICAO, ICC (NGOs), IDA, IFAD, IFC, IFRCS (observer), ILO, IMF, IMO, Interpol, IOC, ISO (correspondent), ITU, ITUC (NGOs), LAS (observer), MIGA, NAM, OPCW, PCA, UN, UNCTAD, UNESCO, UNIDO, UNWTO, UPU, WCO, WFTU (NGOs), WHO, WIPO, WMO

Diplomatic representation in the US: *chief of mission:* Ambassador (vacant); Charge d'Affaires BERHANE Gebrehiwet Solomon (since 15 March 2011)
chancery: 1708 New Hampshire Avenue NW, Washington, DC 20009
telephone: [1] (202) 319-1991
FAX: [1] (202) 319-1304

Diplomatic representation from the US: *chief of mission:* Ambassador (vacant); Charge d'Affaires Natalie E. BROWN (since September 2016)
telephone: [291] (1) 120004
embassy: 179 Ala Street, Asmara
mailing address: P. O. Box 211, Asmara
FAX: [291] (1) 127584

Flag description: red isosceles triangle (based on the hoist side) dividing the flag into two right triangles; the upper triangle is green, the lower one is blue; a gold wreath encircling a gold olive branch is centered on the hoist side of the red triangle; green stands for the country's agriculture economy, red signifies the blood shed in the fight for freedom, and blue symbolizes the bounty of the sea; the wreath-olive branch symbol is similar to that on the first flag of Eritrea from 1952; the shape of

the red triangle broadly mimics the shape of the country
note: one of several flags where a prominent component of the design reflects the shape of the country; other such flags are those of Bosnia and Herzegovina, Brazil, and Vanuatu

National symbol(s): *camel; national colors:* green, red, blue

National anthem: *name:* "Ertra, Ertra, Ertra" (Eritrea, Eritrea, Eritrea)
lyrics/music: SOLOMON Tsehaye Beraki/Isaac Abraham MEHAREZGI and ARON Tekle Tesfatsion
note: adopted 1993; upon independence from Ethiopia
0:00 / 1:59

ECONOMY

Economy—overview: Since formal independence from Ethiopia in 1993, Eritrea has faced many economic problems, including lack of financial resources and chronic drought. Eritrea has a command economy under the control of the sole political party, the People's Front for Democracy and Justice. Like the economies of many African nations, a large share of the population—nearly 80% in Eritrea—is engaged in subsistence agriculture, but the sector only produces a small share of the country's total output. Mining accounts for the lion's share of output.

The government has strictly controlled the use of foreign currency by limiting access and availability; new regulations in 2013 aimed at relaxing currency controls have had little economic effect. Few large private enterprises exist in Eritrea and most operate in conjunction with government partners, including a number of large international mining ventures, which began production in 2013. In late 2015, the Government of Eritrea introduced a new currency, retaining the name nakfa, and restricted the amount of hard currency individuals could withdraw from banks per month. The changeover has resulted in exchange fluctuations and the scarcity of hard currency available in the market.

While reliable statistics on Eritrea are difficult to obtain, erratic rainfall and the large percentage of the labor force tied up in military service continue to interfere with agricultural production and economic development. Eritrea's harvests generally cannot meet the food needs of the country without supplemental grain purchases. Copper, potash, and gold production are likely to continue to drive limited economic growth and government revenue over the next few years, but military spending will continue to compete with development and investment plans.

GDP (purchasing power parity): $9.402 billion (2017 est.)
$8.953 billion (2016 est.)
$8.791 billion (2015 est.)
note: data are in 2017 dollars
country comparison to the world: 161

GDP (official exchange rate): $5.813 billion (2017 est.)

GDP—real growth rate: 5% (2017 est.)
1.9% (2016 est.)
2.6% (2015 est.)
country comparison to the world: 46

GDP—per capita (PPP): $1,600 (2017 est.)
$1,500 (2016 est.)
$1,500 (2015 est.)
note: data are in 2017 dollars
country comparison to the world: 217

Gross national saving: 5.5% of GDP (2017 est.)
6% of GDP (2016 est.)
6.8% of GDP (2015 est.)
country comparison to the world: 174

GDP—composition, by end use: *household consumption:* 80.9% (2017 est.)
government consumption: 24.3% (2017 est.)
investment in fixed capital: 6.4% (2017 est.)
investment in inventories: 0.1% (2017 est.)
exports of goods and services: 10.9% (2017 est.)
*imports of goods and services:—22.5% (2017 est.)

GDP—composition, by sector of origin:
agriculture: 11.7% (2017 est.)
industry: 29.6% (2017 est.)
services: 58.7% (2017 est.)

Agriculture—products: sorghum, lentils, vegetables, corn, cotton, tobacco, sisal; livestock, goats; fish

Industries: food processing, beverages, clothing and textiles, light manufacturing, salt, cement

Industrial production growth rate: 5.4% (2017 est.)
country comparison to the world: 52

Labor force: 2.71 million (2017 est.)
country comparison to the world: 108

Labor force—by occupation: *agriculture:* 80%
industry: 20% (2004 est.)

Unemployment rate: 5.8% (2017 est.)
10% (2016 est.)
country comparison to the world: 94

Population below poverty line: 50% (2004 est.)

Household income or consumption by percentage share: *lowest 10%:* NA
highest 10%: NA
Budget: NA
revenues: 2.029 billion (2017 est.)
expenditures: 2.601 billion (2017 est.)

Taxes and other revenues: 34.9% (of GDP) (2017 est.)
country comparison to the world: 64

Budget surplus (+) or deficit (-): -9.8% (of GDP) (2017 est.)
country comparison to the world: 209

Public debt: 131.2% of GDP (2017 est.)
132.8% of GDP (2016 est.)
country comparison to the world: 6

Fiscal year: calendar year

Inflation rate (consumer prices): 9% (2017 est.)
9% (2016 est.)
country comparison to the world: 201

Current account balance: -$137 million (2017 est.)

-$105 million (2016 est.)
country comparison to the world: 91

Exports: $624.3 million (2017 est.)
$485.4 million (2016 est.)
country comparison to the world: 171

Exports—partners: China 62%, South Korea 28.3% (2017)

Exports—commodities: gold and other minerals, livestock, sorghum, textiles, food, small industry manufactures

Imports: $1.127 billion (2017 est.)
$1.048 billion (2016 est.)
country comparison to the world: 181

Imports—commodities: machinery, petroleum products, food, manufactured goods

Imports—partners: UAE 14.5%, China 13.2%, Saudi Arabia 13.2%, Italy 12.9%, Turkey 5.6%, South Africa 4.6% (2017)

Reserves of foreign exchange and gold: $236.7 million (31 December 2017 est.)
$218.4 million (31 December 2016 est.)
country comparison to the world: 171

Debt—external: $792.7 million (31 December 2017 est.)
$875.6 million (31 December 2016 est.)
country comparison to the world: 169

Exchange rates: nakfa (ERN) per US dollar -
15.38 (2017 est.)
15.375 (2016 est.)
15.375 (2015 est.)
15.375 (2014 est.)
15.375 (2013 est.)

ENERGY

Electricity access: *population without electricity:* 3 million (2019)
electrification—total population: 47% (2019)
electrification—urban areas: 95% (2019)
electrification—rural areas: 13% (2019)

Electricity—production: 415.9 million kWh (2016 est.)
country comparison to the world: 168

Electricity—consumption: 353.9 million kWh (2016 est.)
country comparison to the world: 180

Electricity—exports: 0 kWh (2016 est.)
country comparison to the world: 131

Electricity—imports: 0 kWh (2016 est.)
country comparison to the world: 145

Electricity—installed generating capacity: 160,700 kW (2016 est.)
country comparison to the world: 172

Electricity—from fossil fuels: 99% of total installed capacity (2016 est.)
country comparison to the world: 23

Electricity—from nuclear fuels: 0% of total installed capacity (2017 est.)
country comparison to the world: 85

Electricity—from hydroelectric plants: 0% of total installed capacity (2017 est.)
country comparison to the world: 169

Electricity—from other renewable sources: 1% of total installed capacity (2017 est.)
country comparison to the world: 152

Crude oil—production: 0 bbl/day (2018 est.)
country comparison to the world: 131

Crude oil—exports: 0 bbl/day (2015 est.)
country comparison to the world: 118

Crude oil—imports: 0 bbl/day (2015 est.)
country comparison to the world: 122

Crude oil—proved reserves: 0 bbl (1 January 2018 est.)
country comparison to the world: 127

Refined petroleum products—production: 0 bbl/day (2015 est.)
country comparison to the world: 140

Refined petroleum products—consumption: 4,000 bbl/day (2016 est.)
country comparison to the world: 183

Refined petroleum products—exports: 0 bbl/day (2015 est.)
country comparison to the world: 151

Refined petroleum products—imports: 3,897 bbl/day (2015 est.)
country comparison to the world: 179

Natural gas—production: 0 cu m (2017 est.)
country comparison to the world: 127

Natural gas—consumption: 0 cu m (2017 est.)
country comparison to the world: 142

Natural gas—exports: 0 cu m (2017 est.)
country comparison to the world: 99

Natural gas—imports: 0 cu m (2017 est.)
country comparison to the world: 121

Natural gas—proved reserves: 0 cu m (1 January 2014 est.)
country comparison to the world: 131

Carbon dioxide emissions from consumption of energy: 597,100 Mt (2017 est.)
country comparison to the world: 182

COMMUNICATIONS

Telephones—fixed lines: *total subscriptions:* 116,882
subscriptions per 100 inhabitants: 1.94 (2019 est.)
country comparison to the world: 137

Telephones—mobile cellular: *total subscriptions:* 1,226,660
subscriptions per 100 inhabitants: 20.36 (2019 est.)
country comparison to the world: 159

Telecommunication systems: *general assessment:* woefully inadequate service provided by state-owned telecom monopoly; most fixed-line telephones are in Asmara; cell phone use is limited by government control of SIM card issuance; no data service; only about 4% of households having computers with 2% Internet; untapped market ripe for competition; direct phone service between Eritrea and Ethiopia was restored in September 2018; government telco working on roll-out of 3G network; in 2019 11% mobile penetration (2020) *domestic:* fixed-line subscribership is less than 2 per 100 person and mobile-cellular 20 per 100 (2019)

international: country code—291 (2019)

note: the COVID-19 outbreak is negatively impacting telecommunications production and supply chains globally; consumer spending on telecom devices and services has also slowed due to the pandemic's effect on economies worldwide; overall progress towards improvements in all facets of the telecom industry—mobile, fixed-line, broadband, submarine cable and satellite—has moderated

Broadcast media: government controls broadcast media with private ownership prohibited; 1 state-owned TV station; state- owned radio operates 2 networks; purchases of satellite dishes and subscriptions to international broadcast media are permitted (2019)

Internet country code: .er

Internet users: *total:* 78,215
percent of population: 1.31% (July 2018 est.)
country comparison to the world: 183

Broadband—fixed subscriptions: *total:* 600
subscriptions per 100 inhabitants: less than 1 (2017 est.)
country comparison to the world: 201

TRANSPORTATION

National air transport system: *number of registered air carriers:* 1 (2020)
inventory of registered aircraft operated by air carriers: 1
annual passenger traffic on registered air carriers: 102,729 (2018)

Civil aircraft registration country code prefix: E3 (2016)

Airports: 13 (2020)
country comparison to the world: 152

Airports—with paved runways: *total:* 4 (2019)
over 3,047 m: 2
2,438 to 3,047 m: 2

Airports—with unpaved runways: *total:* 9 (2013)

over 3,047 m: 1 (2013)
2,438 to 3,047 m: 1 (2013)
1,524 to 2,437 m: 5 (2013)
914 to 1,523 m: 2 (2013)

Heliports: 1 (2013)

Railways: *total:* 306 km (2018)
narrow gauge: 306 km 0.950-m gauge (2018)
country comparison to the world: 121

Roadways: *total:* 16,000 km (2018)
paved: 1,600 km (2000)
unpaved: 14,400 km (2000)
country comparison to the world: 122

Merchant marine: *total:* 9
by type: general cargo 4, oil tanker 1, other 4 (2019)
country comparison to the world: 156

Ports and terminals: *major seaport(s):* Assab, Massawa

MILITARY AND SECURITY

Military and security forces: *Eritrean Defense Forces:* Eritrean Ground Forces, Eritrean Navy, Eritrean Air Force (includes Air Defense Force) (2019)

Military and security service personnel strengths: the Eritrean Defense Forces are comprised of an estimated 200,000 personnel, including about 2,000 in the naval and air forces; note—includes significant numbers of conscripts; it is unclear how many of the EDF's 200,000 are on active duty; many conscripts are reportedly not under arms (2019)

Military equipment inventories and acquisitions: the Eritrean Defense Forces inventory is comprised primarily of Soviet- era systems; Eritrea was under a UN arms embargo from 2009 to 2018; prior to 2009, Belarus, Bulgaria, and Russia were the leading arms suppliers (2019 est.)

Military service age and obligation: 18-40 years of age for male and female voluntary and compulsory military service; 18-month conscript service obligation (2019)

TRANSNATIONAL ISSUES

Disputes—international: Eritrea and Ethiopia agreed to abide by 2002 Ethiopia- Eritrea Boundary Commission's (EEBC) delimitation decision, but neither party responded to the revised line detailed in the November 2006 EEBC Demarcation Statement; Sudan accuses Eritrea of supporting eastern Sudanese rebel groups; in 2008, Eritrean troops moved across the border on Ras Doumera peninsula and occupied Doumera Island with undefined sovereignty in the Red Sea

Trafficking in persons: *current situation:* Eritrea is a source country for men, women, and children trafficked for the purposes of forced labor domestically and, to a lesser extent, sex and labor trafficking abroad; the country's national service program is often abused, with conscripts detained indefinitely and subjected to forced labor; Eritrean migrants, often fleeing national service, face strict exit control procedures and limited access to passports and visas, making them vulnerable to trafficking; Eritrean secondary school children are required to take part in public works projects during their summer breaks and must attend military and educational camp in their final year to obtain a high school graduation certificate and to gain access to higher education and some jobs; some Eritreans living in or near refugee camps, particularly in Sudan, are kidnapped by criminal groups and held for ransom in the Sinai Peninsula and Libya, where they are subjected to forced labor and abuse

tier rating: Tier 3—Eritrea does not fully comply with the minimum standards for the elimination of trafficking and is not making significant efforts to do so; the government failed to investigate or prosecute any trafficking offenses or to identify or protect any victims; while the government continued to warn citizens of the dangers of human trafficking through awareness-raising events and poster campaigns, authorities lacked an understanding of the crime, conflating trafficking with transnational migration; Eritrea is not a party to the 2000 UN TIP Protocol (2015)

ESTONIA

INTRODUCTION

Background: After centuries of Danish, Swedish, German, and Russian rule, Estonia attained independence in 1918. Forcibly incorporated into the USSR in 1940—an action never recognized by the US and many other countries—it regained its freedom in 1991 with the collapse of the Soviet Union. Since the last Russian troops left in 1994, Estonia has been free to promote economic and political ties with the West. It joined both NATO and the EU in the spring of 2004, formally joined the OECD in late 2010, and adopted the euro as its official currency on 1 January 2011.

GEOGRAPHY

Location: Eastern Europe, bordering the Baltic Sea and Gulf of Finland, between Latvia and Russia

Geographic coordinates: 59 00 N, 26 00 E

Map references: Europe

Area: *total:* 45,228 sq km
land: 42,388 sq km
water: 2,840 sq km
note: includes 1,520 islands in the Baltic Sea
country comparison to the world: 133

Area—comparative: about twice the size of New Jersey

Land boundaries: *total:* 657 km
border countries (2): Latvia 333 km, Russia 324 km

Coastline: 3,794 km

Maritime claims: *territorial sea:* 12 nm
exclusive economic zone: limits as agreed to by Estonia, Finland, Latvia, Sweden, and Russia

Climate: maritime; wet, moderate winters, cool summers

Terrain: marshy, lowlands; flat in the north, hilly in the south

Elevation: *mean elevation:* 61 m
lowest point: Baltic Sea 0 m

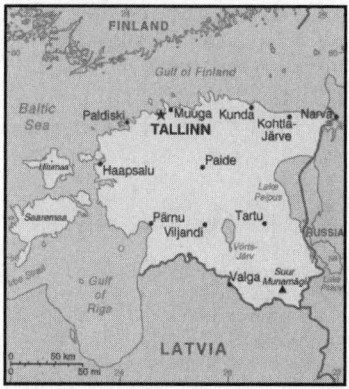

highest point: Suur Munamagi 318 m

Natural resources: oil shale, peat, rare earth elements, phosphorite, clay, limestone, sand, dolomite, arable land, sea mud

Land use: *agricultural land:* 22.2% (2011 est.)
arable land: 14.9% (2011 est.) / *permanent crops:* 0.1% (2011 est.) / *permanent pasture:* 7.2% (2011 est.)
forest: 52.1% (2011 est.)
other: 25.7% (2011 est.)

Irrigated land: 40 sq km (2012)

Population distribution: a fairly even distribution throughout most of the country, with urban areas attracting larger and denser populations

Natural hazards: sometimes flooding occurs in the spring

Environment—current issues: air polluted with sulfur dioxide from oil-shale burning power plants in northeast; however, the amounts of pollutants emitted to the air have fallen dramatically and the pollution load of wastewater at purification plants has decreased substantially due to improved technology and environmental monitoring; Estonia has more than 1,400 natural and manmade lakes, the smaller of which in agricultural areas need to be monitored; coastal seawater is polluted in certain locations

Environment—international agreements: *party to:* Air Pollution, Air Pollution-Nitrogen Oxides, Air Pollution-Persistent Organic Pollutants, Air Pollution-Sulfur 85, Air Pollution-Volatile Organic Compounds, Antarctic Treaty, Biodiversity, Climate Change, Climate Change-Kyoto Protocol, Endangered Species, Hazardous Wastes, Law of the Sea, Ozone Layer Protection, Ship Pollution, Wetlands
signed, but not ratified: none of the selected agreements

Geography—note: the mainland terrain is flat, boggy, and partly wooded; offshore lie more than 1,500 islands

PEOPLE AND SOCIETY

Population: 1,228,624 (July 2020 est.)
country comparison to the world: 158

Nationality: *noun:* Estonian(s)
adjective: Estonian

Ethnic groups: Estonian 68.7%, Russian 24.8%, Ukrainian 1.7%, Belarusian 1%, Finn 0.6%, other 1.6%, unspecified 1.6% (2011 est.)

Languages: Estonian (official) 68.5%, Russian 29.6%, Ukrainian 0.6%, other 1.2%, unspecified 0.1% (2011 est.)

Religions: Orthodox 16.2%, Lutheran 9.9%, other Christian (including Methodist, Seventh-Day Adventist, Roman Catholic, Pentecostal) 2.2%, other 0.9%, none 54.1%, unspecified 16.7% (2011 est.)

Age structure: *0-14 years:* 16.22% (male 102,191/female 97,116)
15-24 years: 8.86% (male 56,484/female 52,378)
25-54 years: 40.34% (male 252,273/female 243,382)
55-64 years: 13.58% (male 76,251/female 90,576)
65 years and over: 21% (male 89,211/female 168,762) (2020 est.)

Dependency ratios: *total dependency ratio:* 58.4
youth dependency ratio: 26.1
elderly dependency ratio: 32.3
potential support ratio: 3.1 (2020 est.)

Median age: *total:* 43.7 years
male: 40.4 years
female: 47 years (2020 est.)
country comparison to the world: 21

Population growth rate: -0.65% (2020 est.)
country comparison to the world: 229

Birth rate: 9.3 births/1,000 population (2020 est.)
country comparison to the world: 200

Death rate: 12.9 deaths/1,000 population (2020 est.)
country comparison to the world: 9

Net migration rate: -3.1 migrant(s)/1,000 population (2020 est.)
country comparison to the world: 177

Population distribution: a fairly even distribution throughout most of the country, with urban areas attracting larger and denser populations

Urbanization: *urban population:* 69.2% of total population (2020)
rate of urbanization: 0.01% annual rate of change (2015-20 est.)
total population growth rate v. urban population growth rate, 2000-2030:

Major urban areas—Population: 445,000 TALLINN (capital) (2020)

Sex ratio: *at birth:* 1.05 male(s)/female
0-14 years: 1.05 male(s)/female
15-24 years: 1.08 male(s)/female
25-54 years: 1.04 male(s)/female
55-64 years: 0.84 male(s)/female
65 years and over: 0.53 male(s)/female
total population: 0.88 male(s)/female (2020 est.)

Mother's mean age at first birth: 27.4 years (2017 est.)

Maternal mortality rate: 9 deaths/100,000 live births (2017 est.)
country comparison to the world: 147

Infant mortality rate: *total:* 3.7 deaths/1,000 live births
male: 3.6 deaths/1,000 live births
female: 3.8 deaths/1,000 live births (2020 est.)
country comparison to the world: 193

Life expectancy at birth: *total population:* 77.4 years
male: 72.7 years
female: 82.3 years (2020 est.)
country comparison to the world: 82

Total fertility rate: 1.61 children born/woman (2020 est.)
country comparison to the world: 183

Drinking water source:
improved:
urban: 100% of population
rural: 100% of population
total: 100% of population
unimproved:
urban: 0% of population
rural: 0% of population
total: 0% of population (2017 est.)

Current Health Expenditure: 6.4% (2017)

Physicians density: 3.46 physicians/1,000 population (2017)

Hospital bed density: 4.7 beds/1,000 population (2017)

Sanitation facility access:
improved:
urban: 100% of population
rural: 100% of population
total: 100% of population
unimproved:
urban: 0% of population
rural: 0% of population
total: 0% of population (2017 est.)

HIV/AIDS—adult prevalence rate: 0.9% (2018 est.)
country comparison to the world: 48

HIV/AIDS—people living with HIV/AIDS: 7,400 (2018 est.)
country comparison to the world: 114

HIV/AIDS—deaths: <100 (2018 est.)

Major infectious diseases: *degree of risk:* intermediate (2020)
vectorborne diseases: tickborne encephalitis

Obesity—adult prevalence rate:
21.2% (2016)
country comparison to the world: 92

Education expenditures: 5.2% of GDP (2016)
country comparison to the world: 54

Literacy: *definition:* age 15 and over can read and write
total population: 99.8%
male: 99.8%
female: 99.8% (2015)

School life expectancy (primary to tertiary education): *total:* 15 years
male: 16 years
female: 17 years (2018)

Unemployment, youth ages 15-24: *total:* 11.8%
male: 12.3%
female: 11.4% (2018 est.)

country comparison to the world: 111

Country name: *conventional long form:* Republic of Estonia
conventional short form: Estonia
local long form: Eesti Vabariik
local short form: Eesti
former: Estonian Soviet Socialist Republic
etymology: the country name may derive from the Aesti, an ancient people who lived along the eastern Baltic Sea in the first centuries A.D.

Government type: parliamentary republic

Capital: *name:* Tallinn
geographic coordinates: 59 26 N, 24 43 E
time difference: UTC+2 (7 hours ahead of Washington, DC, during Standard Time)
daylight saving time: +1hr, begins last Sunday in March; ends last Sunday in October
etymology: the Estonian name is generally believed to be derived from "Taani-linn" (originally meaning "Danish castle", now "Danish town") after a stronghold built in the area by the Danes; it could also have come from "tali-linn" ("winter castle" or "winter town") or "talu-linn" ("home castle" or "home town")

Administrative divisions: 15 urban municipalities (linnad, singular—linn), 64 rural municipalities (vallad, singular vald)
urban municipalities: Haapsalu, Keila, Kohtla-Jarve, Loksa, Maardu, Narva, Narva-Joesuu, Paide, Parnu, Rakvere, Sillamae, Tallinn, Tartu, Viljandi, Voru
rural municipalities: Alutaguse, Anija, Antsla, Elva, Haademeeste, Haljala, Harku, Hiiumaa, Jarva, Joelahtme, Jogeva, Johvi, Kadrina, Kambja, Kanepi, Kastre, Kehtna, Kihnu, Kiili, Kohila, Kose, Kuusalu, Laane-Harju, Laane-Nigula, Laaneranna, Luganuse, Luunja, Marjamaa, Muhu, Mulgi, Mustvee, Noo, Otepaa, Peipsiaare, Pohja-Parnumaa, PohjaSakala, Poltsamaa, Polva, Raasiku, Rae, Rakvere, Rapina, Rapla, Rouge, Ruhnu, Saarde, Saaremaa, Saku, Saue, Setomaa, Tapa, Tartu, Toila, Tori, Torva, Turi, Vaike-Maarja, Valga, Viimsi, Viljandi, Vinni, Viru-Nigula, Vormsi, Voru

Independence: 24 February 1918 (from Soviet Russia); 20 August 1991 (declared from the Soviet Union); 6 September 1991 (recognized by the Soviet Union)

National holiday: Independence Day, 24 February (1918); note—24 February 1918 was the date Estonia declared its independence from Soviet Russia and established its statehood; 20 August 1991 was the date it declared its independence from the Soviet Union restoring its statehood

Constitution: *history:* several previous; latest adopted 28 June 1992
amendments: proposed by at least one-fifth of Parliament members or by the president of the republic; passage requires three readings of the proposed amendment and a simple majority vote in two successive memberships of Parliament; passage of amendments to the "General Provisions"

and "Amendment of the Constitution" chapters requires at least three- fifths majority vote by Parliament to conduct a referendum and majority vote in a referendum; amended several times, last in 2015

Legal system: civil law system

International law organization participation: accepts compulsory ICJ jurisdiction with reservations; accepts ICCt jurisdiction

Citizenship: *citizenship by birth:* no
citizenship by descent only: at least one parent must be a citizen of Estonia
dual citizenship recognized: no
residency requirement for naturalization: 5 years

Suffrage: 18 years of age; universal; age 16 for local elections

Executive branch: *chief of state:* President Kersti KALJULAID (since 10 October 2016)
head of government: Juri RATAS (since 23 November 2016)
cabinet: Cabinet appointed by the prime minister, approved by Parliament
elections/appointments: president indirectly elected by Parliament for a 5-year term (eligible for a second term); if a candidate does not secure two-thirds of the votes after 3 rounds of balloting, then an electoral college consisting of Parliament members and local council members elects the president, choosing between the 2 candidates with the highest number of votes; election last held on 29-30 August 2016, but three rounds were inconclusive; two electoral college votes on 24 September 2016 were also indecisive, so the election passed back to Parliament; on 3 October the Parliament elected Kersti KALJULAID as president; prime minister nominated by the president and approved by Parliament
election results: Kersti KALJULAID elected president; Parliament vote—Kersti KALJULAID (independent) 81 of 98 votes; note—KALJULAID is Estonia's first female president

Legislative branch: *description:* unicameral Parliament or Riigikogu (101 seats; members directly elected in multi-seat constituencies by proportional representation vote to serve 4-year terms)
elections: last held on 3 March 2019 (next to be held in March 2023)
election results: percent of vote by party—RE 28.9%, K 23.1%, EKRE 17.8%, Pro Patria 11.4%, SDE 9.8%, other 9%; seats by party—RE 34, K 26, EKRE 19, Pro Patria 12, SDE 10; composition—men 72, women 29, percent of women 28.7%

Judicial branch: *highest courts:* Supreme Court (consists of 19 justices, including the chief justice, and organized into civil, criminal, administrative, and constitutional review chambers)
judge selection and term of office: the chief justice is proposed by the president of the republic and appointed by the Riigikogu; other justices proposed by the chief justice and appointed by the Riigikogu; justices appointed for life

subordinate courts: circuit (appellate) courts; administrative, county, city, and specialized courts

Political parties and leaders:
Center Party of Estonia (Keskerakond) or K [Juri RATAS]
Estonia 200 [Kristina KALLAS]
Estonian Conservative People's Party (Konservatiivne Rahvaerakond) or EKRE [Mart HELME]
Estonian Reform Party (Reformierakond) or RE [Kaja KALLAS]
Free Party or EV [Andres HERKEL]
Pro Patria (Isamaa) [Helir-Valdor SEEDER]
Social Democratic Party or SDE [Jevgeni OSSINOVSKI]

International organization participation: Australia Group, BA, BIS, CBSS, CD, CE, EAPC, EBRD, ECB, EIB, EMU, ESA (cooperating state), EU, FAO, IAEA, IBRD, ICAO, ICC (national committees), ICCt, ICRM, IDA, IEA, IFAD, IFC, IFRCS, IHO, ILO, IMF, IMO, Interpol, IOC, IOM, IPU, ISO, ITSO, ITU, ITUC (NGOs), MIGA, MINUSMA, NATO, NIB, NSG, OAS (observer), OECD, OIF (observer), OPCW,OSCE, PCA, Schengen Convention, UN, UNCTAD, UNESCO, UNHCR, UNTSO, UPU, WCO, WHO, WIPO, WMO, WTO

Diplomatic representation in the US: *chief of mission:* Ambassador Jonatan VSEVIOV (since 17 September 2018)
chancery: 2131 Massachusetts Avenue NW, Washington, DC 20008
telephone: [1] (202) 588-0101
FAX: [1] (202) 588-0108
consulate(s) general: New York

Diplomatic representation from the US: *chief of mission:* Ambassador (vacant); Charge d'Affaires Brian RORAFF (since July 2019)
telephone: [372] 668-8100
embassy: Kentmanni 20, 15099 Tallinn
mailing address: use embassy street address
FAX: [372] 668-8265

Flag description: three equal horizontal bands of blue (top), black, and white; various interpretations are linked to the flag colors; blue represents faith, loyalty, and devotion, while also reminiscent of the sky, sea, and lakes of the country; black symbolizes the soil of the country and the dark past and suffering endured by the Estonian people; white refers to the striving towards enlightenment and virtue, and is the color of birch bark and snow, as well as summer nights illuminated by the midnight sun

National symbol(s): *barn swallow, cornflower;* *national colors:* blue, black, white

National anthem: *name:* "Mu isamaa, mu onn ja room" (My Native Land, My Pride and Joy)
lyrics/music: Johann Voldemar JANNSEN/ Fredrik PACIUS
note: adopted 1920, though banned between 1940 and 1990 under Soviet occupation; the anthem, used in Estonia since 1869, shares the same melody as Finland's but has different lyrics
0:00/ 0:00

ECONOMY

Economy—overview: Estonia, a member of the EU since 2004 and the euro zone since 2011, has a modern market-based economy and one of the higher per capita income levels in Central Europe and the Baltic region, but its economy is highly dependent on trade, leaving it vulnerable to external shocks. Estonia's successive governments have pursued a free market, pro-business economic agenda, and sound fiscal policies that have resulted in balanced budgets and the lowest debt-to-GDP ratio in the EU.

The economy benefits from strong electronics and telecommunications sectors and strong trade ties with Finland, Sweden, Germany, and Russia. The economy's 4.9% GDP growth in 2017 was the fastest in the past six years, leaving the Estonian economy in its best position since the financial crisis 10 years ago. For the first time in many years, labor productivity increased faster than labor costs in 2017. Inflation also rose in 2017 to 3.5% alongside increased global prices for food and energy, which make up a large share of Estonia's consumption.

Estonia is challenged by a shortage of labor, both skilled and unskilled, although the government has amended its immigration law to allow easier hiring of highly qualified foreign workers, and wage growth that outpaces productivity gains. The government is also pursuing efforts to boost productivity growth with a focus on innovations that emphasize technology start-ups and e-commerce.

GDP (purchasing power parity): $41.65 billion (2017 est.)
$39.72 billion (2016 est.)
$38.92 billion (2015 est.)
note: data are in 2017 dollars
country comparison to the world: 116

GDP (official exchange rate): $25.97 billion (2017 est.)

GDP—real growth rate:
5% (2019 est.)
4.36% (2018 est.)
5.51% (2017 est.)
country comparison to the world: 48

GDP—per capita (PPP): $31,700 (2017 est.)
$30,200 (2016 est.)
$29,600 (2015 est.)
note: data are in 2017 dollars
country comparison to the world: 64

Gross national saving:
27% of GDP (2017 est.)
24.6% of GDP (2016 est.)
25.8% of GDP (2015 est.)
country comparison to the world: 44

GDP—composition, by end use:
household consumption: 50.3% (2017 est.)
government consumption: 20.4% (2017 est.)
investment in fixed capital: 24% (2017 est.)
investment in inventories: 2.2% (2017 est.)
exports of goods and services: 77.2% (2017 est.)
imports of goods and services: -74% (2017 est.)

GDP—composition, by sector of origin:
agriculture: 2.8% (2017 est.)
industry: 29.2% (2017 est.)
services: 68.1% (2017 est.)

Agriculture—products: grain, potatoes, vegetables; livestock and dairy products; fish

Industries: food, engineering, electronics, wood and wood products, textiles; information technology, telecommunications

Industrial production growth rate: 9.5% (2017 est.)
country comparison to the world: 17

Labor force: 648,000 (2020 est.)
country comparison to the world: 151

Labor force—by occupation: agriculture: 2.7%
industry: 20.5%
services: 76.8% (2017 est.)

Unemployment rate:
4.94% (2019 est.)
4.73% (2018 est.)
country comparison to the world: 73

Population below poverty line: 21.1% (2016 est.)

Household income or consumption by percentage share: lowest 10%: 2.3%
highest 10%: 25.6% (2015)

Budget: revenues: 10.37 billion (2017 est.)
expenditures: 10.44 billion (2017 est.)

Taxes and other revenues: 39.9% (of GDP) (2017 est.)
country comparison to the world: 41

Budget surplus (+) or deficit (-): -0.3% (of GDP) (2017 est.)
country comparison to the world: 52

Public debt:
9% of GDP (2017 est.)
9.4% of GDP (2016 est.)
note: data cover general government debt and include debt instruments issued (or owned) by government entities, including sub-sectors of central government, state government, local government, and social security funds
country comparison to the world: 199

Fiscal year: calendar year

Inflation rate (consumer prices):
3.7% (2017 est.)
0.8% (2016 est.)
country comparison to the world: 145

Current account balance:
$616 million (2019 est.)
$280 million (2018 est.)
country comparison to the world: 53

Exports:
$13.44 billion (2017 est.)
$12.36 billion (2016 est.)
country comparison to the world: 79

Exports—partners: Finland 16.2%, Sweden 13.5%, Latvia 9.2%, Russia 7.3%, Germany 6.9%, Lithuania 5.9% (2017)

Exports—commodities: machinery and electrical equipment 30%, food products and beverages 9%, mineral fuels 6%, wood and wood products 14%, articles of base metals 7%, furniture and bedding 11%, vehicles and parts 3%, chemicals 4% (2016 est.)

Imports:
$14.42 billion (2017 est.)
$13.23 billion (2016 est.)
country comparison to the world: 90

Imports—commodities: machinery and electrical equipment 28%, mineral fuels 11%, food and food products 10%, vehicles 9%, chemical products 8%, metals 8% (2015 est.)

Imports—partners: Finland 14%, Germany 10.7%, Lithuania 8.9%, Sweden 8.5%, Latvia 8.2%, Poland 7.2%, Russia 6.7%, Netherlands 5.9%, China 4.7% (2017)

Reserves of foreign exchange and gold:
$345 million (31 December 2017 est.)
$352.2 million (31 December 2016 est.)
country comparison to the world: 164

Debt—external:
$19.05 billion (31 December 2016 est.)
$18.3 billion (31 December 2015 est.)
country comparison to the world: 93

Exchange rates: euros (EUR) per US dollar—0.92 (2017 est.)
0.9 (2016 est.)
0.9214 (2015 est.)
0.885 (2014 est.)
0.7634 (2013 est.)

ENERGY

Electricity access: electrification—total population: 100% (2020)

Electricity—production: 11.55 billion kWh (2016 est.)
Country comparison to the world: 97

Electricity—consumption: 8.795 billion kWh (2016 est.)
country comparison to the world: 102

Electricity—exports: 5.613 billion kWh (2016 est.)
country comparison to the world: 33

Electricity—imports: 3.577 billion kWh (2016 est.)
country comparison to the world: 46

Electricity—installed generating capacity: 2.578 million kW (2016 est.)
country comparison to the world: 106

Electricity—from fossil fuels: 72% of total installed capacity (2016 est.)
country comparison to the world: 103

Electricity—from nuclear fuels: 0% of total installed capacity (2017 est.)
country comparison to the world: 86

Electricity—from hydroelectric plants: 0% of total installed capacity (2017 est.)
country comparison to the world: 170

Electricity—from other renewable sources: 28% of total installed capacity (2017 est.)
country comparison to the world: 22

Crude oil—production: 0 bbl/day (2018 est.)
country comparison to the world: 132

Crude oil—exports: 0 bbl/day (2015 est.)

country comparison to the world: 119

Crude oil—imports: 0 bbl/day (2017 est.)
country comparison to the world: 123

Crude oil—proved reserves: 0 bbl (1 January 2018 est.)
country comparison to the world: 128

Refined petroleum products—production: 0 bbl/day (2017 est.)
country comparison to the world: 141

Refined petroleum products—consumption: 28,300 bbl/day (2017 est.)
country comparison to the world: 121

Refined petroleum products—exports: 27,150 bbl/day (2017 est.)
country comparison to the world: 64

Refined petroleum products—imports: 35,520 bbl/day (2017 est.)
country comparison to the world: 95

Natural gas—production: 0 cu m (2017 est.)
country comparison to the world: 128

Natural gas—consumption: 481.4 million cu m (2017 est.)
country comparison to the world: 98

Natural gas—exports: 0 cu m (2017 est.)
country comparison to the world: 100

Natural gas—imports: 481.4 million cu m (2017 est.)
country comparison to the world: 67

Natural gas—proved reserves: 0 cu m (2016 est.)
country comparison to the world: 132

Carbon dioxide emissions from consumption of energy: 5.306 million Mt (2017 est.)
country comparison to the world: 133

COMMUNICATIONS

Telephones—fixed lines: total subscriptions: 302,606
subscriptions per 100 inhabitants: 24.47 (2019 est.)
country comparison to the world: 108

Telephones—mobile cellular: total subscriptions: 1,820,088
subscriptions per 100 inhabitants: 147.18 (2019 est.)
country comparison to the world: 154

Telecommunication systems: general assessment: a range of regulatory measures, competition and foreign investment in the form of joint business ventures has greatly improved telephone service with a wide range of high-quality voice, data, and Internet services; one of the most advanced mobile markets in Europe; one of the highest broadband penetration in Europe; govt. commits 20 million euro to rural broadband program; regulator auctions spectrum in the 2.6GHz band for LTE and 5G services (2020)

domestic: 25 per 100 for fixed-line and 147 per 100 for mobile-cellular; substantial fiber-optic cable systems carry telephone, TV, and radio traffic in the digital mode; Internet services are widely available; schools and libraries are connected to the Internet, a large percentage of the population

files income tax returns online, and online voting—in local and parliamentary elections—has climbed steadily since first being introduced in 2005; a large percent of Estonian households have broadband access (2019)

international: country code—372; landing points for the EE-S-1, EESF-3, Baltic Sea Submarine Cable, FEC and EESF-2 fiber-optic submarine cables to other Estonia points, Finland, and Sweden; 2 international switches are located in Tallinn (2019)

note: the COVID-19 outbreak is negatively impacting telecommunications production and supply chains globally; consumer spending on telecom devices and services has also slowed due to the pandemic's effect on economies worldwide; overall progress towards improvements in all facets of the telecom industry—mobile, fixed-line, broadband, submarine cable and satellite—has moderated

Broadcast media: the publicly owned broadcaster, Eesti Rahvusringhaaling (ERR), operates 3 TV channels and 5 radio networks; growing number of private commercial radio stations broadcasting nationally, regionally, and locally; fully transitioned to digital television in 2010; national private TV channels expanding service; a range of channels are aimed at Russian-speaking viewers; in 2016, there were 42 on-demand services available in Estonia, including 19 pay TVOD and SVOD services; roughly 85% of households accessed digital television services

Internet country code: .ee

Internet users: total: 1,111,896

percent of population: 89.36% (July 2018 est.)
country comparison to the world: 136

Broadband—fixed subscriptions: total: 441,167
subscriptions per 100 inhabitants: 35 (2018 est.)
country comparison to the world: 87

TRANSPORTATION

National air transport system: number of registered air carriers: 3 (2020)
inventory of registered aircraft operated by air carriers: 14
annual passenger traffic on registered air carriers: 31,981 (2018)

Civil aircraft registration country code prefix: ES (2016)

Airports: 18 (2013)
country comparison to the world: 138

Airports—with paved runways: total: 13 (2017)
over 3,047 m: 2 (2017)
2,438 to 3,047 m: 8 (2017)
1,524 to 2,437 m: 2 (2017)
914 to 1,523 m: 1 (2017)

Airports—with unpaved runways: total: 5 (2013)
1,524 to 2,437 m: 1 (2013)
914 to 1,523 m: 1 (2013)
under 914 m: 3 (2013)

Heliports: 1 (2012)

Pipelines: 2360 km gas (2016)

Railways: total: 2,146 km (2016)

broad gauge: 2,146 km 1.520-m and 1.524-m gauge (132 km electrified) (2016)
note: includes 1,510 km public and 636 km non-public railway
country comparison to the world: 71

Roadways: total: 58,412 km (includes urban roads) (2011)
paved: 10,427 km (includes 115 km of expressways) (2011)
unpaved: 47,985 km (2011)
country comparison to the world: 79

Waterways: 335 km (320 km are navigable year-round) (2011)
country comparison to the world: 90

Merchant marine: total: 69
by type: general cargo 1, oil tanker 6, other 62 (2019)
country comparison to the world: 103

Ports and terminals: major seaport(s): Kuivastu, Kunda, Muuga, Parnu Reid, Sillamae, Tallinn

MILITARY AND SECURITY

Military and security forces: Estonian Defense Forces: Land Forces, Navy, Air Force, Estonian Defence League (Reserves); Ministry of Interior: Border Guards (2019)

Military expenditures:
2.14% of GDP (2019 est.)
2% of GDP (2018)
2.03% of GDP (2017)
2.07% of GDP (2016)
2.02% of GDP (2015)
country comparison to the world: 44

Military and security service personnel strengths: the Estonian Defense Forces have approximately 6,000 active duty personnel (5,000 Army; 400 Navy; 500 Air Force); est. 15,000 Estonian Defense League (2020 est.)

Military equipment inventories and acquisitions: the Estonian Defense Forces have a limited inventory of Soviet-era and more modern Western weapons systems; France and the Netherlands are the leading suppliers of armaments to Estonia since 2010 (2019 est.)

Military deployments: approximately 100 Mali (Operation Barkhane/MINUSMA/EUTM) (2020)

Military service age and obligation: 18-27 for compulsory military or governmental service, conscript service requirement 8-11 months depending on education; NCOs, reserve officers, and specialists serve 11 months (2016)

TRANSNATIONAL ISSUES

Disputes—international: Russia and Estonia in May 2005 signed a technical border agreement, but Russia in June 2005 recalled its signature after the Estonian parliament added to its domestic ratification act a historical preamble referencing the Soviet occupation and Estonia's pre-war borders under the 1920 Treaty of Tartu; Russia contends that the preamble allows Estonia to make territorial claims on Russia in the future, while Estonian officials deny that the preamble has any legal impact on the treaty text; Russia demands better treatment of the Russian-speaking population in Estonia; as a member state that forms part of the EU's external border, Estonia implements strict Schengen border rules with Russia

Refugees and internally displaced persons:
stateless persons: 75,599 (2019); note—following independence in 1991, automatic citizenship was

restricted to those who were Estonian citizens prior to the 1940 Soviet occupation and their descendants; thousands of ethnic Russians remained stateless when forced to choose between passing Estonian language and citizenship tests or applying for Russian citizenship; one reason for demurring on Estonian citizenship was to retain the right of visa-free travel to Russia; stateless residents can vote in local elections but not general elections; stateless parents who have been lawful residents of Estonia for at least five years can

apply for citizenship for their children before they turn 15 years old

Illicit drugs: growing producer of synthetic drugs; increasingly important transshipment zone for cannabis, cocaine, opiates, and synthetic drugs since joining the European Union and the Schengen Accord; potential money laundering related to organized crime and drug trafficking is a concern, as is possible use of the gambling sector to launder funds; major use of opiates and ecstasy

ESWATINI

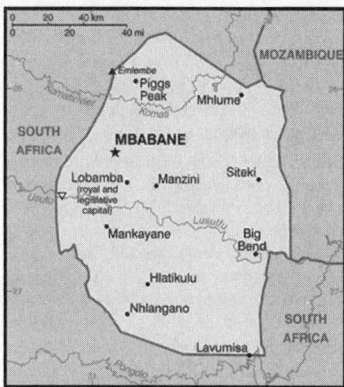

INTRODUCTION

Background: Autonomy for Eswatini was guaranteed by the British in the late 19th century; independence was granted in 1968. A new constitution came into effect in 2006, which included provisions for a more independent parliament and judiciary, but the legal status of political parties remains unclear. King MSWATI III renamed the country from Swaziland to Eswatini in April 2018. Despite its classification as a lower- middle income country, Eswatini suffers from severe poverty and high unemployment. Eswatini has the world's highest HIV/AIDS prevalence rate, although recent years have shown marked declines in new infections.

GEOGRAPHY

Location: Southern Africa, between Mozambique and South Africa

Geographic coordinates: 26 30 S, 31 30 E

Map references: Africa

Area: *total:* 17,364 sq km
land: 17,204 sq km
water: 160 sq km
country comparison to the world: 159

Area—comparative: slightly smaller than New Jersey

Land boundaries: *total:* 546 km
border countries (2): Mozambique 108 km, South Africa 438 km

Coastline: 0 km (landlocked)

Maritime claims: none (landlocked)

Climate: varies from tropical to near temperate

Terrain: mostly mountains and hills; some moderately sloping plains

Elevation: *mean elevation:* 305 m
lowest point: Great Usutu River 21 m
highest point: Emlembe 1,862 m

Natural resources: asbestos, coal, clay, cassiterite, hydropower, forests, small gold and diamond deposits, quarry stone, and talc

Land use: *agricultural land:* 68.3% (2011 est.)
arable land: 9.8% (2011 est.) / permanent crops: 0.8% (2011 est.) / permanent pasture: 57. 7% (2011 est.)
forest: 31.7% (2011 est.)
other: 0% (2011 est.)
Irrigated land: 500 sq km (2012)

Population distribution: because of its mountainous terrain, the population distribution is uneven throughout the country, concentrating primarily in valleys and plains as shown in this population distribution map

Natural hazards: drought

Environment—current issues: limited supplies of potable water; wildlife populations being depleted because of excessive hunting; population growth, deforestation, and overgrazing lead to soil erosion and soil degradation

Environment—international agreements: *party to:* Biodiversity, Climate Change, Climate Change-Kyoto Protocol, Desertification, Endangered Species, Hazardous Wastes, Ozone Layer Protection
signed, but not ratified: Law of the Sea

Geography—note: landlocked; almost completely surrounded by South Africa

PEOPLE AND SOCIETY

Population: 1,104,479 (July 2020 est.)
note: estimates for this country explicitly take into account the effects of excess mortality due

to AIDS; this can result in lower life expectancy, higher infant mortality, higher death rates, lower population growth rates, and changes in the distribution of population by age and sex than would otherwise be expected
country comparison to the world: 160

Nationality: *noun:* liSwati (singular), emaSwati (plural); note—former term, Swazi(s), still used among English speakers
adjective: Swati; note—former term, Swazi, still used among English speakers

Ethnic groups: predominantly Swazi; smaller populations of other African ethnic groups, including the Zulu, as well as people of European ancestry

Languages: English (official, used for government business), siSwati (official)

Religions: Christian 90% (Zionist—a blend of Christianity and indigenous ancestral worship—40%, Roman Catholic 20%, other 30%—includes Anglican, Methodist, Mormon, Jehovah'fs Witness), Muslim 2%, other 8% (includes Baha'i, Buddhist, Hindu, indigenous, Jewish) (2015 est.)

Demographic profile: Eswatini, a small, predominantly rural, landlocked country surrounded by South Africa and Mozambique, suffers from severe poverty and the world's highest HIV/AIDS prevalence rate. A weak and deteriorating economy, high unemployment, rapid population growth, and an uneven distribution of resources all combine to worsen already persistent poverty and food insecurity, especially in rural areas. Erratic weather (frequent droughts and intermittent heavy rains and flooding), overuse of small plots, the overgrazing of cattle, and outdated agricultural practices reduce crop yields and further degrade the environment, exacerbating Eswatini's poverty and subsistence problems. Eswatini's extremely high HIV/AIDS prevalence rate—more than 28% of adults have the disease—compounds these issues. Agricultural production has declined due to HIV/AIDS, as the illness causes households to lose manpower and to sell livestock and other assets to pay for medicine and funerals.

Swazis, mainly men from the country's rural south, have been migrating to South Africa to work in coal, and later gold, mines since the late

19th century. Although the number of miners abroad has never been high in absolute terms because of Eswatini's small population, the outflow has had important social and economic repercussions. The peak of mining employment in South Africa occurred during the 1980s. Cross-border movement has accelerated since the 1990s, as increasing unemployment has pushed more Swazis to look for work in South Africa (creating a "brain drain" in the health and educational sectors); southern Swazi men have continued to pursue mining, although the industry has downsized. Women now make up an increasing share of migrants and dominate cross-border trading in handicrafts, using the proceeds to purchase goods back in Eswatini. Much of today's migration, however, is not work- related but focuses on visits to family and friends, tourism, and shopping.

Age structure: 0-14 years: 33.63% (male 185,640/female 185,808)
15-24 years: 18.71% (male 98,029/female 108,654)
25-54 years: 39.46% (male 202,536/female 233,275)
55-64 years: 4.36% (male 20,529/female 27,672)
65 years and over: 3.83% (male 15,833/female 26,503) (2020 est.)

Dependency ratios: *total dependency ratio:* 70.8
youth dependency ratio: 64
elderly dependency ratio: 6.9
potential support ratio: 14.6 (2020 est.)

Median age: *total:* 23.7 years
male: 22.5 years
female: 24.7 years (2020 est.)
country comparison to the world: 174

Population growth rate: 0.77% (2020 est.)
country comparison to the world: 132

Birth rate: 24.5 births/1,000 population (2020 est.)
country comparison to the world: 50

Death rate: 10.1 deaths/1,000 population (2020 est.)
country comparison to the world: 36

Net migration rate: -6.8 migrant(s)/1,000 population (2020 est.)
country comparison to the world: 209

Population distribution: because of its mountainous terrain, the population distribution is uneven throughout the country, concentrating primarily in valleys and plains as shown in this population distribution map

Urbanization: *urban population:* 24.2% of total population (2020)
rate of urbanization: 2.46% annual rate of change (2015-20 est.)

total population growth rate v. urban population growth rate, 2000-2030: *Major urban areas—population:* 68,000 MBABANE (capital) (2018)

Sex ratio: *at birth:* 1.03 male(s)/female
0-14 years: 1 male(s)/female
15-24 years: 0.9 male(s)/female
25-54 years: 0.87 male(s)/female
55-64 years: 0.74 male(s)/female
65 years and over: 0.6 male(s)/female
total population: 0.9 male(s)/female (2020 est.)

Maternal mortality rate: 437 deaths/100,000 live births (2017 est.)
country comparison to the world: 25

Infant mortality rate: *total:* 42.8 deaths/1,000 live births
male: 47.3 deaths/1,000 live births
female: 38.2 deaths/1,000 live births (2020 est.)
country comparison to the world: 32

Life expectancy at birth: *total population:* 58.6 years
male: 56.5 years
female: 60.7 years (2020 est.)
country comparison to the world: 220

Total fertility rate: 2.52 children born/woman (2020 est.)
country comparison to the world: 71

Contraceptive prevalence rate: 66.1% (2014)

Drinking water source:

improved: *urban:* 96.8% of population
rural: 72.3% of population
total: 78.3% of population

unimproved: *urban:* 3.2% of population
rural: 27.7% of population
total: 21.7% of population (2017 est.)

Current Health Expenditure: 6.9% (2017)

Physicians density: 0.33 physicians/1,000 population (2016)

Hospital bed density: 2.1 beds/1,000 population (2011)

Sanitation facility access:

improved: *urban:* 93.5% of population
rural: 82.4% of population
total: 85% of population

unimproved: *urban:* 6.5% of population
rural: 17.6% of population
total: 15% of population (2017 est.)

HIV/AIDS—adult prevalence rate: 27.1% (2019 est.)
country comparison to the world: 1

HIV/AIDS—people living with HIV/AIDS: 200,000 (2019 est.)
country comparison to the world: 30

HIV/AIDS—deaths: 2,300 (2019 est.)
country comparison to the world: 43

Major infectious diseases: *degree of risk:* intermediate (2020)
food or waterborne diseases: bacterial diarrhea, hepatitis A, and typhoid fever
vectorborne diseases: malaria
water contact diseases: schistosomiasis

Obesity—adult prevalence rate: 16.5% (2016)
country comparison to the world: 124

Children under the age of 5 years underweight: 5.8% (2014)
country comparison to the world: 77

Education expenditures: 7.1% of GDP (2014)
country comparison to the world: 13

Literacy: *definition:* age 15 and over can read and write
total population: 88.4%
male: 88.3%

female: 88.5% (2015)

School life expectancy (primary to tertiary education): *total:* 13 years
male: 13 years
female: 12 years (2013)

Unemployment, youth ages 15-24: *total:* 47.1%
male: 44.2%
female: 50.1% (2016)
country comparison to the world: 5

GOVERNMENT

Country name: *conventional long form:* Kingdom of Eswatini
conventional short form: Eswatini
local long form: Umbuso weSwatini
local short form: eSwatini
former: Swaziland
etymology: the country name derives from 19th century King MSWATI II, under whose rule Swati territory was expanded and unified
note: pronounced ay-swatini or eh-swatini

Government type: absolute monarchy

Capital: *name:* Mbabane (administrative capital); Lobamba (royal and legislative capital)
geographic coordinates: 26 19 S, 31 08 E
time difference: UTC+ 2 (7 hours ahead of Washington, DC, during Standard Time)
etymology: named after a Swati chief, Mbabane Kunene, who lived in the area at the onset of British settlement

Administrative divisions: 4 regions; Hhohho, Lubombo, Manzini, Shiselweni

Independence: 6 September 1968 (from the UK)

National holiday: Independence Day (Somhlolo Day), 6 September (1968)

Constitution: *history:* previous 1968, 1978; latest signed by the king 26 July 2005, effective 8 February 2006
amendments: proposed at a joint sitting of both houses of Parliament; passage requires majority vote by both houses and/or majority vote in a referendum, and assent of the king; passage of amendments affecting " specially entrenched" constitutional provisions requires at least three-fourths majority vote by both houses, passage by simple majority vote in a referendum, and assent of the king; passage of " entrenched" provisions requires at least two- thirds majority vote of both houses, passage in a referendum, and assent of the king

Legal system: mixed legal system of civil, common, and customary law

International law organization participation: accepts compulsory ICJ jurisdiction with reservations; non-party state to the ICCt

Citizenship: *citizenship by birth:* no
citizenship by descent only: both parents must be citizens of Eswatini
dual citizenship recognized: no
residency requirement for naturalization: 5 years

Suffrage: 18 years of age

Executive branch: *chief of state:* King MSWATI III (since 25 April 1986)

head of government: Prime Minister Ambrose Mandvulo DLAMINI (since 27 October 2018); Deputy Prime Minister Themba MASUKU (since 6 November 2018)
cabinet: Cabinet recommended by the prime minister, confirmed by the monarch; at least one- half of the cabinet membership must be appointed from among elected members of the House of Assembly
elections/appointments: the monarchy is hereditary; prime minister appointed by the monarch from among members of the House of Assembly

Legislative branch: *description:* bicameral Parliament (Libandla) consists of:

Senate (30 seats; 20 members appointed by the monarch and 10 indirectly elected by simple majority vote by the House of Assembly; members serve 5-year terms)

House of Assembly (73 seats; 59 members directly elected in single-seat constituencies or tinkhundla by absolute majority vote in 2 rounds if needed, 10 members appointed by the monarch, 4 women elected by the members if representation of elected women is less than 30%; members serve 5-year terms)
elections: Senate—last held on 23 October 2018 (next to be held—31 October 2023)

House of Assembly—last held on 21 September 2018 (next to be held in 2023)
election results: Senate—percent of seats by party—NA; seats by party—NA; composition—men 20, women 10, percent of women 33.3%

House of Assembly—percent of vote by party—NA; seats by party—independent 59; composition—men 60, women 5, percent of women 7.7%; note—total Parliament percent of women 15.8%

Judicial branch: *highest courts:* Supreme Court (consists of the chief justice and at least 4 justices) and the High Court (consists of the chief justice—ex officio—and 4 justices); note—the Supreme Court has jurisdiction in all constitutional matters
judge selection and term of office: justices of the Supreme Court and High Court appointed by the monarch on the advice of the Judicial Service Commission (JSC), a judicial advisory body consisting of the Supreme Court Chief Justice, 4 members appointed by the monarch, and the chairman of the Civil Service Commission; justices of both courts eligible for retirement at age 65 with mandatory retirement at age 75
subordinate courts: magistrates' courts; National Swazi Courts for administering customary/traditional laws (jurisdiction restricted to customary law for Swazi citizens)

Political parties and leaders: political parties exist, but conditions for their operations, particularly in elections, are undefined, legally unclear, or culturally restricted; the following are considered political associations:
African United Democratic Party or AUDP [Sibusiso DLAMINI]
Ngwane National Liberatory Congress or NNLC [Dr. Alvit DLAMINI]
People's United Democratic Movement or PUDEMO [Mario MASUKU]

Swazi Democratic Party or SWADEPA [Jan SITHOLE]

International organization participation: ACP, AfDB, AU, C, COMESA, FAO, G- 77, IAEA, IBRD, ICAO, ICRM, IDA, IFAD, IFC, IFRCS, ILO, IMF, IMO, Interpol, IOC, IOM, ISO (correspondent), ITSO, ITU, ITUC (NGOs), MIGA, NAM, OPCW, PCA, SACU, SADC, UN, UNCTAD, UNESCO, UNIDO, UNWTO, UPU, WCO, WHO, WIPO, WMO, WTO

Diplomatic representation in the US: *chief of mission:* Ambassador Njabuliso Busisiwe Sikhulile GWEBU (since 24 April 2017)
chancery: 1712 New Hampshire Avenue, NW, Washington, DC 20009
telephone: [1] (202) 234-5002
FAX: [1] (202) 234-8254

Diplomatic representation from the US: *chief of mission:* Ambassador Lisa J. PETERSON (since February 2016)
telephone: (268) 404-6441; EMER: +(268) 7602-8414
embassy: 7th Floor, Central Bank Building, Mahlokohla Street, Mbabane
mailing address: PO Box 199, Mbabane, Eswatini
FAX: [268] 2416-3344

Flag description: three horizontal bands of blue (top), red (triple width), and blue; the red band is edged in yellow; centered in the red band is a large black and white shield covering two spears and a staff decorated with feather tassels, all placed horizontally; blue stands for peace and stability, red represents past struggles, and yellow the mineral resources of the country; the shield, spears, and staff symbolize protection from the country's enemies, while the black and white of the shield are meant to portray black and white people living in peaceful coexistence

National symbol(s): *lion, elephant; national colors:* blue, yellow, red

National anthem: *name:* "Nkulunkulu Mnikati wetibusiso temaSwati" (Oh God, Bestower of the Blessings of the Swazi)
lyrics/music: Andrease Enoke Fanyana SIMELANE/David Kenneth RYCROFT
note: adopted 1968; uses elements of both ethnic Swazi and Western music styles
0:00 / 1:07

ECONOMY

Economy—overview: A small, landlocked kingdom, Eswatini is bordered in the north, west and south by the Republic of South Africa and by Mozambique in the east. Eswatini depends on South Africa for a majority of its exports and imports. Eswatini's currency is pegged to the South African rand, effectively relinquishing Eswatini's monetary policy to South Africa. The government is dependent on customs duties from the Southern African Customs Union (SACU) for almost half of its revenue. Eswatini is a lower middle income country. As of 2017, more than one- quarter of the adult population was infected with HIV/AIDS; Eswatini has the world's highest HIV prevalence

rate, a financial strain and source of economic instability.

The manufacturing sector diversified in the 1980s and 1990 s, but manufacturing has grown little in the last decade. Sugar and soft drink concentrate are the largest foreign exchange earners, although a drought in 2015-16 decreased sugar production and exports. Overgrazing, soil depletion, drought, and floods are persistent problems. Mining has declined in importance in recent years. Coal, gold, diamond, and quarry stone mines are small scale, and the only iron ore mine closed in 2014. With an estimated 28% unemployment rate, Eswatini's need to increase the number and size of small and medium enterprises and to attract foreign direct investment is acute.

Eswatini's national development strategy, which expires in 2022, prioritizes increases in infrastructure, agriculture production, and economic diversification, while aiming to reduce poverty and government spending. Eswatini's revenue from SACU receipts are likely to continue to decline as South Africa pushes for a new distribution scheme, making it harder for the government to maintain fiscal balance without introducing new sources of revenue.

GDP (purchasing power parity): $11.6 billion (2017 est.)
$11.41 billion (2016 est.)
$11.26 billion (2015 est.)
note: data are in 2017 dollars
country comparison to the world: 157

GDP (official exchange rate): $4.417 billion (2017 est.)

GDP—real growth rate: 1.6% (2017 est.)
1.4% (2016 est.)
0.4% (2015 est.)
country comparison to the world: 152

GDP—per capita (PPP): $10,100 (2017 est.)
$10,100 (2016 est.)
$10,100 (2015 est.)
note: data are in 2017 dollars
country comparison to the world: 139

Gross national saving: 25.4% of GDP (2017 est.)
29.7% of GDP (2016 est.)
23.3% of GDP (2015 est.)
country comparison to the world: 59

GDP—composition, by end use: *household consumption:* 64% (2017 est.)
government consumption: 21.3% (2017 est.)
investment in fixed capital: 13.4% (2017 est.)
investment in inventories:—0.1% (2017 est.)
exports of goods and services: 47.9% (2017 est.)
imports of goods and services:—46.3% (2017 est.)

GDP—composition, by sector of origin: *agriculture:* 6.5% (2017 est.)
industry: 45% (2017 est.)
services: 48.6% (2017 est.)

Agriculture—products: sugarcane, corn, cotton, citrus, pineapples, cattle, goats

Industries: soft drink concentrates, coal, forestry, sugar processing, textiles, and apparel

Industrial production growth rate: 5.6% (2017 est.)

315

country comparison to the world: 48

Labor force: 427,900 (2016 est.)
country comparison to the world: 157

Labor force—by occupation: *agriculture:* 10.7%
industry: 30.4%
services: 58.9% (2014 est.)

Unemployment rate: 28% (2014 est.)
28% (2013 est.)
country comparison to the world: 201

Population below poverty line: 63% (2010 est.)

**Household income or consumption by percentage
share:** *lowest 10%:* 1.7%
highest 10%: 40.1% (2010 est.)

Budget: *revenues:* 1.263 billion (2017 est.)
expenditures: 1.639 billion (2017 est.)

Taxes and other revenues: 28.6% (of GDP) (2017
est.)
country comparison to the world: 94

Budget surplus (+) or deficit (-): -8. 5% (of GDP)
(2017 est.)
country comparison to the world: 201

Public debt: 28.4% of GDP (2017 est.)
25.5% of GDP (2016 est.)
country comparison to the world: 168

Fiscal year: 1 April–31 March

Inflation rate (consumer prices): 6.2% (2017 est.)
7.8% (2016 est.)
country comparison to the world: 188

Current account balance: $604 million (2017 est.)
$642 million (2016 est.)
country comparison to the world: 54

Exports: $1.83 billion (2017 est.)
$1.577 billion (2016 est.)
country comparison to the world: 145

Exports—partners: South Africa 94% (2017)

Exports—commodities: soft drink concentrates,
sugar, timber, cotton yarn, refrigerators, citrus, and
canned fruit

Imports: $1.451 billion (2017 est.)
$1.266 billion (2016 est.)
country comparison to the world: 175

Imports—commodities: motor vehicles, machinery, transport equipment, foodstuffs, petroleum
products, chemicals

Imports—partners: South Africa 81.6%, China
5.2% (2017)

Reserves of foreign exchange and gold: $563.1
million (31 December 2017 est.)
$564.4 million (31 December 2016 est.)
country comparison to the world: 147

Debt—external: $526.3 million (31 December
2017 est.)
$468.9 million (31 December 2016 est.)
country comparison to the world: 177

Exchange rates: emalangeni per US dollar—
14.44 (2017 est.)
14.6924 (2016 est.)
14.6924 (2015 est.)
12.7581 (2014 est.)
10.8469 (2013 est.)

ENERGY

Electricity access: *electrification—total
population:* 90% (2019)
electrification—urban areas: 98% (2019)
electrification—rural areas: 87% (2019)

Electricity—production: 381 million kWh (2016
est.)
country comparison to the world: 173

Electricity—consumption: 1.431 billion kWh
(2016 est.)
country comparison to the world: 149

Electricity—exports: 0 kWh (2016)
country comparison to the world: 132

Electricity—imports: 1.077 billion kWh (2016
est.)
country comparison to the world: 69

Electricity—installed generating capacity:
295,900 kW (2016 est.)
country comparison to the world: 160

Electricity—from fossil fuels: 39% of total
installed capacity (2016 est.)
country comparison to the world: 172

Electricity—from nuclear fuels: 0% of total
installed capacity (2017 est.)
country comparison to the world: 87

Electricity—from hydroelectric plants:
20% of total installed capacity (2017 est.)
country comparison to the world: 87

Electricity—from other renewable sources: 41%
of total installed capacity (2017 est.)
country comparison to the world: 6

Crude oil—production: 0 bbl/day (2018 est.)
country comparison to the world: 133

Crude oil—exports: 0 bbl/day (2015 est.)
country comparison to the world: 120

Crude oil—imports: 0 bbl/day (2015 est.)
country comparison to the world: 124

Crude oil—proved reserves: 0 bbl (1 January
2018)
country comparison to the world: 129

Refined petroleum products—production:
0 bbl/day (2015 est.)
country comparison to the world: 142

Refined petroleum products—consumption:
5,300 bbl/day (2016 est.)
country comparison to the world: 175

Refined petroleum products—exports:
0 bbl/day (2015 est.)
country comparison to the world: 152

Refined petroleum products—imports: 5,279 bbl/
day (2015 est.)
country comparison to the world: 169

Natural gas—production: 0 cu m (2017 est.)
country comparison to the world: 129

Natural gas—consumption: 0 cu m (2017 est.)
country comparison to the world: 143

Natural gas—exports: 0 cu m (2017 est.)
country comparison to the world: 101

Natural gas—imports: 0 cu m (2017 est.)

country comparison to the world: 122

Natural gas—proved reserves: 0 cu m (1 January
2014 est.)
country comparison to the world: 133

**Carbon dioxide emissions from consumption of
energy:** 1.14 million Mt (2017 est.)
country comparison to the world: 166

COMMUNICATIONS

Telephones—fixed lines: *total subscriptions:*
40,003
subscriptions per 100 inhabitants: 3.65 (2019
est.)
country comparison to the world: 162

Telephones—mobile cellular: *total subscriptions:*
1,025,061
subscriptions per 100 inhabitants: 93.53 (2019
est.)
country comparison to the world: 162

Telecommunication systems: *general assessment:*
earlier government monopoly in telecommunications hindered its growth; new regulatory authority established in 2013 has aided expansion in
the telecom sector; 2G, 3G, 4G and LTE services
(2019)
domestic: Eswatini has 2 mobile- cellular providers; communication infrastructure has a geographic
coverage of about 90% and a rising subscriber base;
fixed-line stands at 4 per 100 and mobile-cellular
teledensity roughly 94 telephones per 100 persons;
telephone system consists of carrier-equipped,
open-wire lines and low-capacity, microwave radio
relay (2019)
international: country code—268; satellite earth
station—1 Intelsat (Atlantic Ocean)
note: the COVID-19 outbreak is negatively
impacting telecommunications production and
supply chains globally; consumer spending on telecom devices and services has also slowed due to the
pandemic's effect on economies worldwide; overall
progress towards improvements in all facets of the
telecom industry—mobile, fixed-line, broadband,
submarine cable and satellite—has moderated

Broadcast media: 1 state-owned TV station; satellite dishes are able to access South African providers; state- owned radio network with 3 channels; 1
private radio station (2019)

Internet country code: .sz

Internet users: *total:* 510,984
percent of population: 47% (July 2018 est.)
country comparison to the world: 151

Broadband—fixed subscriptions: *total:* 7,000
subscriptions per 100 inhabitants: less than 1
(2017 est.)
country comparison to the world: 176

TRANSPORTATION

Civil aircraft registration country code prefix: 3
(2016)

Airports: 14 (2013)
country comparison to the world: 148

Airports—with paved runways: *total:* 2 (2019)

over 3,047 m: 1
2,438 to 3,047 m: 1

Airports—with unpaved runways: *total:* 12 (2013)
914 to 1,523 m: 5 *(2013)*
under 914 m: 7 *(2013)*

Railways: *total:* 301 km (2014)
narrow gauge: 301 km 1.067-m gauge (2014)
country comparison to the world: 122

Roadways: *total:* 3,769 km (2019)
country comparison to the world: 157

MILITARY AND SECURITY

Military and security forces: *Umbutfo Eswatini Defense Force (UEDF):* Ground Force (includes Air Wing (no operational aircraft)) (2019)

Military expenditures: 1.8% of GDP (2019)
1.9% of GDP (2018)
1.9% of GDP (2017)
2% of GDP (2016)
1.8% of GDP (2015)
country comparison to the world: 61

Military and security service personnel strengths: the Umbutfo Eswatini Defense Force has approximately 3,100 active personnel (3,000 Army; 100 Air Force) (2020 est.)

Military equipment inventories and acquisitions: the inventory of the UEDF consists mostly of equipment from South Africa; the only publicly recorded military acquisitions since 2010 were two secondhand helicopters from Taiwan in 2019 (2020)

Military service age and obligation: 18-30 years of age for male and female voluntary military service; no conscription; compulsory HIV testing required, only HIV-negative applicants accepted (2013)

TRANSNATIONAL ISSUES

Disputes—international: in 2006, Swati king advocated resorting to ICJ to claim parts of Mpumalanga and KwaZulu- Natal from South Africa

ETHIOPIA

INTRODUCTION

Background: Unique among African countries, the ancient Ethiopian monarchy maintained its freedom from colonial rule with the exception of a short-lived Italian occupation from 1936-41. In 1974, a military junta, the Derg, deposed Emperor Haile SELASSIE (who had ruled since 1930) and established a socialist state. Torn by bloody coups, uprisings, wide-scale drought, and massive refugee problems, the regime was finally toppled in 1991 by a coalition of rebel forces, the Ethiopian People's Revolutionary Democratic Front (EPRDF). A constitution was adopted in 1994, and Ethiopia's first multiparty elections were held in 1995.

A border war with Eritrea in the late 1990s ended with a peace treaty in December 2000. In November 2007, the Eritrea-Ethiopia Border Commission (EEBC) issued specific coordinates as virtually demarcating the border and pronounced its work finished. Alleging that the EEBC acted beyond its mandate in issuing the coordinates, Ethiopia did not accept them and maintained troops in previously contested

areas pronounced by the EEBC as belonging to Eritrea. This intransigence resulted in years of heightened tension between the two countries. In August 2012, longtime leader Prime Minister MELES Zenawi died in office and was replaced by his Deputy Prime Minister HAILEMARIAM Desalegn, marking the first peaceful transition of power in decades. Following a wave of popular dissent and anti- government protest that began in 2015, HAILEMARIAM resigned in February 2018 and ABIY Ahmed Ali took office in April 2018 as Ethiopia's first ethnic Oromo prime minister. In June 2018, ABIY announced Ethiopia would accept the border ruling of 2000, prompting rapprochement between Ethiopia and Eritrea that was marked with a peace agreement in July 2018 and a reopening of the border in September 2018. In November 2019, Ethiopia's nearly 30-year ethnic-based ruling coalition—the EPRDF—merged into a single unity party called the Prosperity Party, however, one of the four constituent parties refused to join.

GEOGRAPHY

Location: Eastern Africa, west of Somalia

Geographic coordinates: 8 00 N, 38 00 E

Map references: Africa

Area: *total:* 1,104,300 sq km
land: 1,096,570 sq km
water: 7,730 sq km
note: area numbers are approximate since a large portion of the Ethiopia-Somalia border is undefined
country comparison to the world: 28

Area—comparative: slightly less than twice the size of Texas

Land boundaries: *total:* 5,925 km
border countries (6): Djibouti 342 km, Eritrea 1033 km, Kenya 867 km, Somalia 1640 km, South Sudan 1299 km, Sudan 744 km

Coastline: 0 km (landlocked)

Maritime claims: none (landlocked)

Climate: tropical monsoon with wide topographic-induced variation

Terrain: high plateau with central mountain range divided by Great Rift Valley

Elevation: *mean elevation:* 1,330 m
lowest point: Danakil Depression—125 m
highest point: Ras Dejen 4,550 m

Natural resources: small reserves of gold, platinum, copper, potash, natural gas, hydropower

Land use: *agricultural land:* 36.3% (2011 est.)
arable land: 15.2% (2011 est.) / *permanent crops:* 1.1% (2011 est.) / *permanent pasture:* 20% (2011 est.)
forest: 12.2% (2011 est.)
other: 51.5% (2011 est.)
Irrigated land: 2,900 sq km (2012)

Population distribution: highest density is found in the highlands of the north and middle areas of the country, particularly around the centrally located capital city of Addis Ababa; the far east and southeast are sparsely populated as shown in this population distribution map

Natural hazards: geologically active Great Rift Valley susceptible to earthquakes, volcanic eruptions; frequent droughts
volcanism: volcanic activity in the Great Rift Valley; Erta Ale (613 m), which has caused frequent lava flows in recent years, is the country's most active volcano; Dabbahu became active in 2005, forcing evacuations; other historically active volcanoes include Alayta, Dalaffilla, Dallol, Dama Ali, Fentale, Kone, Manda Hararo, and Manda-Inakir

Environment—current issues: deforestation; overgrazing; soil erosion; desertification; loss of biodiversity; water shortages in some areas from water-intensive farming and poor management; industrial pollution and pesticides contribute to air, water, and soil pollution

317

Environment—international agreements: *party to:* Biodiversity, Climate Change, Climate Change-Kyoto Protocol, Desertification, Endangered Species, Hazardous Wastes, Ozone Layer Protection

signed, but not ratified: Environmental Modification, Law of the Sea

Geography—note: *note 1:* landlocked—entire coastline along the Red Sea was lost with the de jure independence of Eritrea on 24 May 1993; Ethiopia is, therefore, the most populous landlocked country in the world; the Blue Nile, the chief headstream of the Nile by water volume, rises in T'ana Hayk (Lake Tana) in northwest Ethiopia

note 2: three major crops are believed to have originated in Ethiopia: coffee, grain sorghum, and castor bean

PEOPLE AND SOCIETY

Population: 108,113,150 (July 2020 est.)

note: estimates for this country explicitly take into account the effects of excess mortality due to AIDS; this can result in lower life expectancy, higher infant mortality, higher death rates, lower population growth rates, and changes in the distribution of population by age and sex than would otherwise be expected

country comparison to the world: 13

Nationality: *noun:* Ethiopian(s)

adjective: Ethiopian

Ethnic groups: Oromo 34.9%, Amhara (Amara) 27.9%, Tigray (Tigrinya) 7.3%, Sidama 4.1%, Welaita 3%, Gurage 2.8%, Somali (Somalie) 2.7%, Hadiya 2.2%, Afar (Affar) .6%, other 12.6% (2016 est.)

Languages: Oromo (official working language in the State of Oromiya) 33.8%, Amharic (official national language) 29.3%, Somali (official working language of the State of Sumale) 6.2%, Tigrigna (Tigrinya) (official working language of the State of Tigray) 5.9%, Sidamo 4%, Wolaytta 2.2%, Gurage 2%, Afar (official working language of the State of Afar) 1.7%, Hadiyya 1.7%, Gamo 1.5%, Gedeo 1.3%, Opuuo 1.2%, Kafa 1.1%, other 8.1%, English (major foreign language taught in schools), Arabic (2007 est.)

Religions: Ethiopian Orthodox 43.8%, Muslim 31.3%, Protestant 22.8%, Catholic 0.7%, traditional .6%, other 0.8% (2016 est.)

Demographic profile: Ethiopia is a predominantly agricultural country—more than 80% of the population lives in rural areas—that is in the early stages of demographic transition. Infant, child, and maternal mortality have fallen sharply over the past decade, but the total fertility rate has declined more slowly and the population continues to grow. The rising age of marriage and the increasing proportion of women remaining single have contributed to fertility reduction. While the use of modern contraceptive methods among married women has increased significantly from 6 percent in 2000 to 27 percent in 2012, the overall rate is still quite low.

Ethiopia's rapid population growth is putting increasing pressure on land resources, expanding environmental degradation, and raising vulnerability to food shortages. With more than 40 percent of the population below the age of 15 and a fertility rate of over 5 children per woman (and even higher in rural areas), Ethiopia will have to make further progress in meeting its family planning needs if it is to achieve the age structure necessary for reaping a demographic dividend in the coming decades.

Poverty, drought, political repression, and forced government resettlement have driven Ethiopia's internal and external migration since the 1960s. Before the 1974 revolution, only small numbers of the Ethiopian elite went abroad to study and then returned home, but under the brutal Derg regime thousands fled the country, primarily as refugees. Between 1982 and 1991 there was a new wave of migration to the West for family reunification. Since the defeat of the Derg in 1991, Ethiopians have migrated to escape violence among some of the country's myriad ethnic groups or to pursue economic opportunities. Internal and international trafficking of women and children for domestic work and prostitution is a growing problem.

Age structure: *0-14 years:* 39.81% (male 21,657,152/female 21,381,628)

15-24 years: 19.47% (male 10,506,144/female 10,542,128)

25-54 years: 32.92% (male 17,720,540/female 17,867,298)

55-64 years: 4.42% (male 2,350,606/female 2,433,319)

65 years and over: 3.38% (male 1,676,478/female 1,977,857) (2020 est.)

Dependency ratios: *total dependency ratio:* 76.8

youth dependency ratio: 70.6

elderly dependency ratio: 6.3

potential support ratio: 16 (2020 est.)

Median age: *total:* 19.8 years

male: 19.6 years

female: 20.1 years (2020 est.)

country comparison to the world: 198

Population growth rate: 2.56% (2020 est.)

country comparison to the world: 20

Birth rate: 31.6 births/1,000 population (2020 est.)

country comparison to the world: 30

Death rate: 5.9 deaths/1,000 population (2020 est.)

country comparison to the world: 171

Net migration rate: -0.2 migrant(s)/1,000 population (2020 est.)

country comparison to the world: 106

Population distribution: highest density is found in the highlands of the north and middle areas of the country, particularly around the centrally located capital city of Addis Ababa; the far east and southeast are sparsely populated as shown in this population distribution map

Urbanization: *urban population:* 21.7% of total population (2020)

rate of urbanization: 4.63% annual rate of change (2015-20 est.)

total population growth rate v. urban population growth rate, 2000-2030: Major urban areas—population: 4.794 million ADDIS ABABA (capital) (2020)

Sex ratio: *at birth:* 1.03 male(s)/female

0-14 years: 1.01 male(s)/female

15-24 years: 1 male(s)/female

25-54 years: 0.99 male(s)/female

55-64 years: 0.97 male(s)/female

65 years and over: 0.85 male(s)/female

total population: 1 male(s)/female (2020 est.)

Mother's mean age at first birth: 20 years (2016 est.)

note: median age at first birth among women 25-29

Maternal mortality rate: 401 deaths/100,000 live births (2017 est.)

country comparison to the world: 26

Infant mortality rate: *total:* 35.8 deaths/1,000 live births

male: 40.8 deaths/1,000 live births

female: 30.5 deaths/1,000 live births (2020 est.)

country comparison to the world: 44

Life expectancy at birth: *total population:* 67.5 years

male: 65.5 years

female: 69.7 years (2020 est.)

country comparison to the world: 180

Total fertility rate: 4.14 children born/woman (2020 est.)

country comparison to the world: 28

Contraceptive prevalence rate: 40.1% (2018)

Drinking water source:

improved: *urban:* 97% of population

rural: 61.7% of population

total: 68.9% of population

unimproved: *urban:* 3% of population

rural: 38.3% of population

total: 31.1% of population (2017 est.)

Current Health Expenditure: 3.5% (2017)

Physicians density: 0.1 physicians/1,000 population (2017)

Hospital bed density: 0.3 beds/1,000 population (2016)

Sanitation facility access:

improved: *urban:* 49.7% of population

rural: 5.7% of population

total: 14.7% of population

unimproved: *urban:* 50.3% of population

rural: 94.3% of population

total: 85.3% of population (2017 est.)

HIV/AIDS—adult prevalence rate: 1.1% (2019 est.)

country comparison to the world: 43

HIV/AIDS—people living with HIV/AIDS: 670,000 (2019 est.)

country comparison to the world: 13

HIV/AIDS—deaths: 12,000 (2019 est.)

country comparison to the world: 19

Major infectious diseases: *degree of risk:* very high (2020)

food or waterborne diseases: bacterial and proto-zoal diarrhea, hepatitis A, and typhoid fever
vectorborne diseases: malaria and dengue fever
water contact diseases: schistosomiasis
animal contact diseases: rabies
respiratory diseases: meningococcal meningitis

Obesity—adult prevalence rate: 4.5% (2016)
country comparison to the world: 185

Children under the age of 5 years underweight: 21.1% (2019)
country comparison to the world: 20

Education expenditures: 4.7% of GDP (2015)
country comparison to the world: 78

Literacy: *definition:* age 15 and over can read and write
total population: 51.8%
male: 57.2%
female: 44.4% (2017)

School life expectancy (primary to tertiary education): *total:* 9 years
male: 8 years
female: 8 years (2012)

Unemployment, youth ages 15-24: *total:* 25.2%
male: 17.1%
female: 30.9% (2016 est.)
country comparison to the world: 49

GOVERNMENT

Country name: *conventional long form:* Federal Democratic Republic of Ethiopia
conventional short form: Ethiopia
local long form: Ityop'iya Federalawi Demokrasiyawi Ripeblik
local short form: Ityop'iya
former: Abyssinia, Italian East Africa
abbreviation: FDRE
etymology: the country name derives from the Greek word "Aethiopia," which in classical times referred to lands south of Egypt in the Upper Nile region

Government type: federal parliamentary republic

Capital: *name:* Addis Ababa

geographic coordinates: 9 02 N, 38 42 E
time difference: UTC+3 (8 hours ahead of Washington, DC, during Standard Time)
etymology: the name in Amharic means "new flower" and was bestowed on the city in 1889, three years after its founding

Administrative divisions: 9 ethnically based regional states (kililoch, singular—kilil) and 2 self-governing administrations* (astedaderoch, singular—astedader), Adis Abeba* (Addis Ababa), Afar, Amara (Amhara), Binshangul Gumuz, Dire Dawa*, Gambela Hizboch (Gambela Peoples), Hareri Hizb (Harari People), Oromiya (Oromia), Sumale (Somali), Tigray, Ye Debub Biheroch Bihereseboch na Hizboch (Southern Nations, Nationalities and Peoples)

Independence: oldest independent country in Africa and one of the oldest in the world—at least 2,000 years (may be traced to the Aksumite Kingdom, which coalesced in the first century B.C.)

National holiday: Derg Downfall Day (defeat of MENGISTU regime), 28 May (1991)

Constitution: *history:* several previous; latest drafted June 1994, adopted 8 December 1994, entered into force 21 August 1995
amendments: proposals submitted for discussion require two-thirds majority approval in either house of Parliament or majority approval of one-third of the State Councils; passage of amendments other than constitutional articles on fundamental rights and freedoms and the initiation and amendment of the constitution requires two-thirds majority vote in a joint session of Parliament and majority vote by two thirds of the State Councils; passage of amendments affecting rights and freedoms and amendment procedures requires two-thirds majority vote in each house of Parliament and majority vote by all the State Councils

Legal system: civil law system

International law organization participation: has not submitted an ICJ jurisdiction declaration; non-party state to the ICCt

Citizenship: *citizenship by birth:* no
citizenship by descent only: at least one parent must be a citizen of Ethiopia
dual citizenship recognized: no
residency requirement for naturalization: 4 years

Suffrage: 18 years of age; universal

Executive branch: *chief of state:* President SAHLE-WORK Zewde (since 25 October 2018)

head of government: Prime Minister ABIY Ahmed (since 2 April 2018); Deputy Prime Minister DEMEKE Mekonnen Hassen (since 29 November 2012); note—Prime Minister HAILEMARIAM Desalegn (since 21 September 2012) resigned on 15 February 2018 and continued as caretaker until the new prime minister was sworn into office on 2 April 2018
cabinet: Council of Ministers selected by the prime minister and approved by the House of People's Representatives
elections/appointments: president indirectly elected by both chambers of Parliament for a 6-year term (eligible for a second term); snap election held on 25 October 2018 due to resignation of President MULATA Teshome (next election postponed by Prime Minister ABIY due to the COVID-19 pandemic); prime minister designated by the majority party following legislative elections
election results: SAHLE-WORK Zewde elected president; Parliament vote—659 (unanimous)
note: SAHLE-WORK Zewde is the first female elected head of state in Ethiopia; she is currently the only female president in Africa. Former President Dr. Mulatu TESHOME resigned on 25 October 2018, one year ahead of finishing his six-year term.

Legislative branch: *description:* bicameral Parliament consists of:
House of Federation or Yefedereshein Mikir Bete (153 seats; members indirectly elected by state assemblies to serve 5-year terms)

House of People's Representatives or Yehizb Tewokayoch Mekir Bete (547 seats; members directly elected in single-seat constituencies by simple majority vote; 22 seats reserved for minorities; all members serve 5-year terms)
elections: House of Federation—last held 24 May 2015 (next originally scheduled on 29 August 2020 but postponed a year due to the COVID-19 pandemic)
House of People's Representatives—last held on 24 May 2015 (next originally scheduled on 29 August 2020 but postponed to 2021 due to the COVID-19 pandemic)
election results: House of Federation—percent of vote by coalition/party—NA; seats by coalition/party—NA; composition-men 104, women 49, percent of women 32%
House of Representatives—percent of vote by coalition/party—NA; seats by coalition/ party—EPRDF 501, SPDP 24, BGPDUP 9, ANDP 8, GPUDM 3, APDO 1, HNL 1; composition—men 335, women 212, percent of women 38.8%; note—total Parliament percent of women 37.3%
note: House of Federation is responsible for interpreting the constitution and federal-regional issues and the House of People's Representatives is responsible for passing legislation

Judicial branch: *highest courts:* Federal Supreme Court (consists of 11 judges); note—the House of Federation has jurisdiction for all constitutional issues
judge selection and term of office: president and vice president of Federal Supreme Court recommended by the prime minister and appointed by the House of People's Representatives; other Supreme Court judges nominated by the Federal Judicial Administrative Council (a 10-member body chaired by the president of the Federal Supreme Court) and appointed by the House of People's Representatives; judges serve until retirement at age 60
subordinate courts: federal high courts and federal courts of first instance; state court systems (mirror structure of federal system); sharia courts and customary and traditional courts

Political parties and leaders: Afar National Democratic Party or ANDP [Taha AHMED]
Argoba People Democratic Organization or APDO
Benishangul Gumuz People's Democratic Unity Party or BGPDUP
Ethiopian Federal Democratic Unity Forum or MEDREK or FORUM [Beyene PETROS] (includes ESD-SCUP, OFC, SLM, and UTDS)
Ethiopia Citizens for Social Justice or ECSJ Party (formed in May 2019 from 7 other parties, including Patriotic Genbot 7, Ethiopian Democratic Party (EDP), All Ethiopian Democratic Party (AEDP), Semayawi Party, New Generation Party, Gambella Regional Movement (GRM), Unity for Democracy and Justice (UDJ) Party [Berhanu Negu])
Prosperity Party or PP [ABIY Ahmed] (created in November 2019 from member parties of the former Ethiopian People's Revolutionary Democratic Front or EPRDF, which included the Amhara

319

National Democratic Movement (ANDM), Oromo People's Democratic Organization (OPDO), Southern Ethiopian People's Democratic Movement (SEPDM), Tigray People's Liberation Front (TPLF), plus other ERPRF allies

Ethiopian Social Democracy-Southern Coalition Unity Party or ESD-SCUP

Gambella Peoples Unity Democratic Movement or GPUDM

Harari National League or HNL [Murad ABDULHADI]

Oromo Fderalist Congress or OFC

Sidama Liberaton Movement or SLM

Somali People's Democratic Party or SPDP

Union of Tigraians for Democracy & Sovergnty or UTDS

Tigray Independence Party [Girmay BERHE] (2020)

International organization participation: ACP, AfDB, AU, COMESA, EITI (candidate country), FAO, G-24, G-77, IAEA, IBRD, ICAO, ICRM, IDA, IFAD, IFC, IFRCS, IGAD, ILO, IMF, IMO, Interpol, IOC, IOM, IPU, ISO, ITSO, ITU, ITUC (NGOs), MIGA, NAM, OPCW, PCA, UN, UNAMID, UNCTAD, UNESCO, UNHCR, UNIDO, UNISFA, UNMIL, UN Security Council (temporary), UNOCI, UNWTO, UPU, WCO, WFTU (NGOs), WHO, WIPO, WMO, WTO (observer)

Diplomatic representation in the US: *chief of mission:* Ambassador Ato FITSUM Arega (since 9 April 2019)

chancery: 3506 International Drive NW, Washington, DC 20008

telephone: [1] (202) 364-1200

FAX: [1] (202) 587-0195

consulate(s) general: Los Angeles, Seattle

consulate(s): Houston, New York

Diplomatic representation from the US: *chief of mission:* Ambassador Michael RAYNOR (since 3 October 2017)

telephone: [251] 11 130-6000

embassy: Entoto Street, P.O. Box 1014, Addis Ababa

mailing address: P.O. Box 1014, Addis Ababa

FAX: [251] 11 124-2401

Flag description: three equal horizontal bands of green (top), yellow, and red, with a yellow pentagram and single yellow rays emanating from the angles between the points on a light blue disk centered on the three bands; green represents hope and the fertility of the land, yellow symbolizes justice and harmony, while red stands for sacrifice and heroism in the defense of the land; the blue of the disk symbolizes peace and the pentagram represents the unity and equality of the nationalities and peoples of Ethiopia

note: Ethiopia is the oldest independent country in Africa, and the three main colors of her flag (adopted ca. 1895) were so often appropriated by other African countries upon independence that they became known as the Pan-African colors; the emblem in the center of the current flag was added in 1996

National symbol(s): Abyssinian lion (traditional), yellow pentagram with five rays of light on a blue field (promoted by current government); national colors: green, yellow, red

National anthem: *name:* "Whedefit Gesgeshi Woud Enat Ethiopia" (March Forward, Dear Mother Ethiopia)

lyrics/music: DEREJE Melaku Mengesha/ SOLOMON Lulu

note: adopted 1992

0:00 / 1:23

ECONOMY

Economy—overview: Ethiopia—the second most populous country in Africa—is a one-party state with a planned economy. For more than a decade before 2016, GDP grew at a rate between 8% and 11% annually—one of the fastest growing states among the 188 IMF member countries. This growth was driven by government investment in infrastructure, as well as sustained progress in the agricultural and service sectors. More than 70% of Ethiopia's population is still employed in the agricultural sector, but services have surpassed agriculture as the principal source of GDP.

Ethiopia has the lowest level of income-inequality in Africa and one of the lowest in the world, with a Gini coefficient comparable to that of the Scandinavian countries. Yet despite progress toward eliminating extreme poverty, Ethiopia remains one of the poorest countries in the world, due both to rapid population growth and a low starting base. Changes in rainfall associated with world-wide weather patterns resulted in the worst drought in 30 years in 2015-16, creating food insecurity for millions of Ethiopians.

The state is heavily engaged in the economy. Ongoing infrastructure projects include power production and distribution, roads, rails, airports and industrial parks. Key sectors are state-owned, including telecommunications, banking and insurance, and power distribution. Under Ethiopia's constitution, the state owns all land and provides long-term leases to tenants. Title rights in urban areas, particularly Addis Ababa, are poorly regulated, and subject to corruption.

Ethiopia's foreign exchange earnings are led by the services sector—primarily the state-run Ethiopian Airlines—followed by exports of several commodities. While coffee remains the largest foreign exchange earner, Ethiopia is diversifying exports, and commodities such as gold, sesame, khat, livestock and horticulture products are becoming increasingly important. Manufacturing represented less than 8% of total exports in 2016, but manufacturing exports should increase in future years due to a growing international presence.

The banking, insurance, telecommunications, and micro-credit industries are restricted to domestic investors, but Ethiopia has attracted roughly $8.5 billion in foreign direct investment (FDI), mostly from China, Turkey, India and the EU; US FDI is $567 million. Investment has been primarily in infrastructure, construction, agriculture/horticulture, agricultural processing, textiles, leather and leather products.

To support industrialization in sectors where Ethiopia has a comparative advantage, such as textiles and garments, leather goods, and processed agricultural products, Ethiopia plans to increase installed power generation capacity by 8,320 MW, up from a capacity of 2,000 MW, by building three more major dams and expanding to other sources of renewable energy. In 2017, the government devalued the birr by 15% to increase exports and alleviate a chronic foreign currency shortage in the country.

GDP (purchasing power parity): $200.6 billion (2017 est.)

$181 billion (2016 est.)

$167.6 billion (2015 est.)

note: data are in 2017 dollars

country comparison to the world: 64

GDP (official exchange rate): $80.87 billion (2017 est.)

GDP—real growth rate: 10.9% (2017 est.)

8% (2016 est.)

10.4% (2015 est.)

country comparison to the world: 5

GDP—per capita (PPP): $2,200 (2017 est.)

$2,000 (2016 est.)

$1,900 (2015 est.)

note: data are in 2017 dollars

country comparison to the world: 204

Gross national saving: 32.1% of GDP (2017 est.)

32.7% of GDP (2016 est.)

32.4% of GDP (2015 est.)

country comparison to the world: 26

GDP—composition, by end use:

household consumption: 69.6% (2017 est.)

government consumption: 10% (2017 est.)

investment in fixed capital: 43.5% (2017 est.)

investment in inventories: -0.1% (2017 est.)

exports of goods and services: 8.1% (2017 est.)

imports of goods and services: -31.2% (2017 est.)

GDP—composition, by sector of origin:

agriculture: 34.8% (2017 est.)

industry: 21.6% (2017 est.)

services: 43.6% (2017 est.)

Agriculture—products: cereals, coffee, oilseed, cotton, sugarcane, vegetables, khat, cut flowers; hides, cattle, sheep, goats; fish

Industries: food processing, beverages, textiles, leather, garments, chemicals, metals processing, cement

Industrial production growth rate: 10.5% (2017 est.)

country comparison to the world: 13

Labor force: 52.82 million (2017 est.)

country comparison to the world: 11

Labor force—by occupation: agriculture: 72.7%

industry: 7.4%

services: 19.9% (2013 est.)

Unemployment rate: 17.5% (2012 est.)

18% (2011 est.)

country comparison to the world: 183

Population below poverty line: 29.6% (2014 est.)

Household income or consumption by percentage share: *lowest 10%:* 4.1%
highest 10%: 25.6% (2005)

Budget: *revenues:* 11.24 billion (2017 est.)
expenditures: 13.79 billion (2017 est.)

Taxes and other revenues: 13.9% (of GDP) (2017 est.)
country comparison to the world: 203

Budget surplus (+) or deficit (-): -3.2% (of GDP) (2017 est.)
country comparison to the world: 139

Public debt: 54.2% of GDP (2017 est.)
53.2% of GDP (2016 est.)
country comparison to the world: 83

Fiscal year: 8 July–7 July

Inflation rate (consumer prices): 9.9% (2017 est.)
7.3% (2016 est.)
country comparison to the world: 203

Current account balance: -$6.551 billion (2017 est.)
-$6.574 billion (2016 est.)
country comparison to the world: 186

Exports: $3.23 billion (2017 est.)
$2.814 billion (2016 est.)
country comparison to the world: 126

Exports—partners: Sudan 23.3%, Switzerland 10.2%, China 8.1%, Somalia 6.6%, Netherlands 6.2%, US 4.7%, Germany 4.7%, Saudi Arabia 4.6%, UK 4.6% (2017)

Exports—commodities: coffee (27%, by value), oilseeds (17%), edible vegetables including khat (17%), gold (13%), flowers (7%), live animals (7%), raw leather products (3%), meat products (3%)

Imports: $15.59 billion (2017 est.)
$14.69 billion (2016 est.)
country comparison to the world: 88

Imports—commodities: machinery and aircraft (14%, by value), metal and metal products, (14%), electrical materials, (13%), petroleum products (12%), motor vehicles, (10%), chemicals and fertilizers (4%)

Imports—partners: China 24.1%, Saudi Arabia 10.1%, India 6.4%, Kuwait 5.3%, France 5.2% (2017)

Reserves of foreign exchange and gold: $3.013 billion (31 December 2017 est.)
$3.022 billion (31 December 2016 est.)
country comparison to the world: 110

Debt—external: $26.05 billion (31 December 2017 est.)
$24.82 billion (31 December 2016 est.)
country comparison to the world: 87

Exchange rates: birr (ETB) per US dollar—
25 (2017 est.)
21.732 (2016 est.)
21.732 (2015 est.)
21.55 (2014 est.)
19.8 (2013 est.)

ENERGY

Electricity access: *population without electricity:* 60 million (2019)
electrification—total population: 47% (2019)
electrification—urban areas: 96% (2019)
electrification—rural areas: 34% (2019)

Electricity—production: 11.15 billion kWh (2016 est.)
country comparison to the world: 99

Electricity—consumption: 9.062 billion kWh (2016 est.)
country comparison to the world: 100

Electricity—exports: 166 million kWh (2015 est.)
country comparison to the world: 78

Electricity—imports: 0 kWh (2016 est.)
country comparison to the world: 146

Electricity—installed generating capacity: 2.784 million kW (2016 est.)
country comparison to the world: 99

Electricity—from fossil fuels: 3% of total installed capacity (2016 est.)
country comparison to the world: 207

Electricity—from nuclear fuels: 0% of total installed capacity (2017 est.)
country comparison to the world: 88

Electricity—from hydroelectric plants: 86% of total installed capacity (2017 est.)
country comparison to the world: 11

Electricity—from other renewable sources: 11% of total installed capacity (2017 est.)
country comparison to the world: 76

Crude oil—production: 0 bbl/day (2018 est.)
country comparison to the world: 134

Crude oil—exports: 0 bbl/day (2015 est.)
country comparison to the world: 121

Crude oil—imports: 0 bbl/day (2015 est.)
country comparison to the world: 125

Crude oil—proved reserves: 428,000 bbl (1 January 2018 est.)
country comparison to the world: 98

Refined petroleum products—production: 0 bbl/day (2017 est.)
country comparison to the world: 143

Refined petroleum products—consumption: 74,000 bbl/day (2016 est.)
country comparison to the world: 89

Refined petroleum products—exports: 0 bbl/day (2015 est.)
country comparison to the world: 153

Refined petroleum products—imports: 69,970 bbl/day (2015 est.)
country comparison to the world: 67

Natural gas—production: 0 cu m (2017 est.)
country comparison to the world: 130

Natural gas—consumption: 0 cu m (2017 est.)
country comparison to the world: 144

Natural gas—exports: 0 cu m (2017 est.)
country comparison to the world: 102

Natural gas—imports: 0 cu m (2017 est.)
country comparison to the world: 123

Natural gas—proved reserves: 24.92 billion cu m (1 January 2018 est.)
country comparison to the world: 72

Carbon dioxide emissions from consumption of energy: 12.18 million Mt (2017 est.)
country comparison to the world: 99

COMMUNICATIONS

Telephones—fixed lines: *total subscriptions:* 1,095,946
subscriptions per 100 inhabitants: 1.04 (2019 est.)
country comparison to the world: 73

Telephones—mobile cellular: *total subscriptions:* 38,147,361
subscriptions per 100 inhabitants: 36.2 (2019 est.)
country comparison to the world: 41

Telecommunication systems: *general assessment:* Ethio Telecom maintained a monopoly over telecommunication services until recently and is now part-private; new expansion of LTE services; in 2019 govt. approved legislations which opened the market to competition and provides much needed foreign investment; one of the tech companies is Chinese company Huawei; govt. reduces tariffs by up to 50% in 2018, the result is an increase in data and voice traffic; govt. launches mobile app as part of e-govt initiative to build tech city (2020)
domestic: fixed-line subscriptions at 1 per 100 while mobile-cellular stands at 36 per 100; the number of mobile telephones is increasing steadily (2019)
international: country code—251; open- wire to Sudan and Djibouti; microwave radio relay to Kenya and Djibouti; 2 domestic satellites provide the national trunk service; satellite earth stations—3 Intelsat (1 Atlantic Ocean and 2 Pacific Ocean) (2016)
note: the COVID-19 outbreak is negatively impacting telecommunications production and supply chains globally; consumer spending on telecom devices and services has also slowed due to the pandemic's effect on economies worldwide; overall progress towards improvements in all facets of the telecom industry—mobile, fixed-line, broadband, submarine cable and satellite—has moderated

Broadcast media: 6 public TV stations broadcasting nationally and 10 public radio broadcasters; 7 private radio stations and 19 community radio stations (2017)

Internet country code: .et

Internet users: *total:* 19,118,470

percent of population: 18.62% (July 2018 est.)
country comparison to the world: 37

Broadband—fixed subscriptions: *total:* 580,120
subscriptions per 100 inhabitants: 1 (2017 est.)
country comparison to the world: 81

TRANSPORTATION

National air transport system: *number of registered air carriers:* 1 (2020)
inventory of registered aircraft operated by air carriers: 75
annual passenger traffic on registered air carriers: 11,501,244 (2018)
annual freight traffic on registered air carriers: 2,089,280,000 mt-km (2018)

Civil aircraft registration country code prefix: ET (2016)

Airports: 57 (2013)
country comparison to the world: 81

Airports—with paved runways: *total:* 17 (2017)
over 3,047 m: 3 (2017)
2,438 to 3,047 m: 8 (2017)
1,524 to 2,437 m: 4 (2017)
under 914 m: 2 (2017)

Airports—with unpaved runways: *total:* 40 (2013)
2,438 to 3,047 m: 3 (2013)
1,524 to 2,437 m: 9 (2013)
914 to 1,523 m: 20 (2013)
under 914 m: 8 (2013)

Railways: *total:* 659 km (Ethiopian segment of the 756 km Addis Ababa- Djibouti railroad) (2017)
standard gauge: 659 km 1.435-m gauge (2017)
note: electric railway with redundant power supplies; under joint control of Djibouti and Ethiopia and managed by a Chinese contractor
country comparison to the world: 105

Roadways: *total:* 120,171 km (2018)
country comparison to the world: 40

Merchant marine: *total:* 11
by type: general cargo 9, oil tanker 2 (2019)
country comparison to the world: 154

Ports and terminals: Ethiopia is landlocked and uses the ports of Djibouti in Djibouti and Berbera in Somalia

MILITARY AND SECURITY

Military and security forces: Ethiopian National Defense Force (ENDF): Ground Forces, Ethiopian Air Force (Ye Ityopya Ayer Hayl, ETAF) (2020)

note: in January 2020 the Ethiopian Government announced it had re-established a navy, which was disbanded in 1996; in March 2019 Ethiopia signed a defense cooperation agreement with France which stipulated that France would support the establishment of an Ethiopian navy in 2018, Ethiopia established a Republican Guard for protecting senior officials; the Republican Guard is a military unit accountable to the Prime Minister

Military expenditures: 0.7% of GDP (2019)
0.7% of GDP (2018)
0.7% of GDP (2017)
0.7% of GDP (2016)
0.7% of GDP (2015)
country comparison to the world: 136

Military and security service personnel strengths: estimates for the size of the Ethiopian National Defense Force (ENDF) vary; approximately 150,000 active duty troops, including about 3,000 Air Force personnel (no personnel numbers available for the newly-reestablished Navy) (2020)

Military equipment inventories and acquisitions: the ENDF's inventory is comprised mostly of Soviet-era equipment; since 2010, Russia and Ukraine are the leading suppliers of largely second-hand weapons and equipment to the ENDF, followed by China and Hungary; Ethiopia has a modest industrial defense base centered on small arms and licensed production of light-armored vehicles (2019 est.)

Military deployments: 15-20,000 Somalia (includes about 4,400 under AMISOM); 800 Sudan (UNAMID); 3,600 Sudan (UNISFA); 2,100 South Sudan (UNMISS) (2020)

Military service age and obligation: 18 years of age for voluntary military service; no compulsory military service, but the military can conduct callups when necessary and compliance is compulsory (2013)

Military—note: each of the nine states has a regional, a special police force, or both that report to regional civilian authorities; local militias operate across the country in loose and varying coordination with these regional police, the Ethiopian Federal Police (EFP), and the military; the EFP reports to the Ministry of Peace, which was created in October of 2018 (2019)

TERRORISM

TRANSNATIONAL ISSUES

Disputes—international: Eritrea and Ethiopia agreed to abide by the 2002 Eritrea- Ethiopia Boundary Commission's (EEBC) delimitation decision, but neither party responded to the revised line detailed in the November 2006 EEBC Demarcation Statement; the undemarcated former British administrative line has little meaning as a political separation to rival clans within Ethiopia's Ogaden and southern Somalia's Oromo region; Ethiopian forces invaded southern Somalia and routed Islamist courts from Mogadishu in January 2007; "Somaliland" secessionists provide port facilities in Berbera and trade ties to landlocked Ethiopia; civil unrest in eastern Sudan has hampered efforts to demarcate the porous boundary with Ethiopia; Ethiopia's construction of a large dam (the Grand Ethiopian Renaissance Dam) on the Blue Nile since 2011 has become a focal point of relations with Egypt and Sudan; as of 2020, four years of three-way talks between the three capitals over operating the dam and filling its reservoir had made little progress; Ethiopia plans to start filling the dam in July 2020

Refugees and internally displaced persons: *refugees (country of origin):* 362,787 (South Sudan), 201,465 (Somalia), 178,559 (Eritrea), 43,729 (Sudan) (2020)
IDPs: 1,735,481 (includes conflict-and climate-induced IDPs, excluding unverified estimates from the Amhara region; border war with Eritrea from 1998-2000; ethnic clashes; and ongoing fighting between the Ethiopian military and separatist rebel groups in the Somali and Oromia regions; natural disasters; intercommunal violence; most IDPs live in Sumale state) (2019)

Illicit drugs: transit hub for heroin originating in Southwest and Southeast Asia and destined for Europe, as well as cocaine destined for markets in southern Africa; cultivates qat (khat) for local use and regional export, principally to Djibouti and Somalia (legal in all three countries); the lack of a well-developed financial system limits the country's utility as a money laundering center

EUROPEAN UNION

INTRODUCTION

Preliminary statement: The evolution of what is today the European Union (EU) from a regional economic agreement among six neighboring states in 1951 to today's hybrid intergovernmental and supranational organization of 27 countries across the European continent stands as an unprecedented phenomenon in the annals of history. Dynastic unions for territorial consolidation were long the norm in Europe; on a few occasions even country-level unions were arranged—the Polish-Lithuanian Commonwealth and the Austro-Hungarian Empire were examples. But for such a large number of nation-states to cede some of their sovereignty to an overarching entity is unique.

Although the EU is not a federation in the strict sense, it is far more than a free-trade association such as ASEAN or Mercosur, and it has certain attributes associated with independent nations: its own flag, currency (for some members), and law-making abilities, as well as diplomatic representation and a common foreign and security policy in its dealings with external partners.

Thus, inclusion of basic intelligence on the EU has been deemed appropriate as a separate entity in The World Factbook. However, because of the EU's special status, this description is placed after the regular country entries.

Background: Following the two devastating World Wars in the first half of the 20th century, a number of far-sighted European leaders in the late 1940s sought a response to the overwhelming desire for peace and reconciliation on the continent. In 1950, the French Foreign Minister Robert SCHUMAN proposed pooling the production of coal and steel in Western Europe and setting up an organization for that purpose that would bring France and the Federal Republic of Germany together and would be open to other countries as well. The following year, the European Coal and Steel Community (ECSC) was set up when six members—Belgium, France, West Germany, Italy,

Luxembourg, and the Netherlands—signed the Treaty of Paris.

The ECSC was so successful that within a few years the decision was made to integrate other elements of the countries' economies. In 1957, envisioning an "ever closer union," the Treaties of Rome created the European Economic Community (EEC) and the European Atomic Energy Community (Euratom), and the six member states undertook to eliminate trade barriers among themselves by forming a common market. In 1967, the institutions of all three communities were formally merged into the European Community (EC), creating a single Commission, a single Council of Ministers, and the body known today as the European Parliament. Members of the European Parliament were initially selected by national parliaments, but in 1979 the first direct elections were undertaken and have been held every five years since.

In 1973, the first enlargement of the EC took place with the addition of Denmark, Ireland, and the UK. The 1980s saw further membership expansion with Greece joining in 1981 and Spain and Portugal in 1986. The 1992 Treaty of Maastricht laid the basis for further forms of cooperation in foreign and defense policy, in judicial and internal affairs, and in the creation of an economic and monetary union—including a common currency. This further integration created the European Union (EU), at the time standing alongside the EC. In 1995, Austria, Finland, and Sweden joined the EU/EC, raising the membership total to 15.

A new currency, the euro, was launched in world money markets on 1 January 1999; it became the unit of exchange for all EU member states except Denmark, Sweden, and the UK. In 2002, citizens of those 12 countries began using euro banknotes and coins. Ten new countries joined the EU in 2004—Cyprus, the Czech Republic, Estonia, Hungary, Latvia, Lithuania, Malta, Poland, Slovakia, and Slovenia. Bulgaria and Romania joined in 2007 and Croatia in 2013, but the UK withdrew in 2020. Current membership stands at 27. (Seven of the new countries—Cyprus, Estonia, Latvia, Lithuania, Malta, Slovakia, and Slovenia—have now adopted the euro, bringing total euro-zone membership to 19.)

In an effort to ensure that the EU could function efficiently with an expanded membership, the Treaty of Nice (concluded in 2000; entered into force in 2003) set forth rules to streamline the size and procedures of EU institutions. An effort to establish a "Constitution for Europe," growing out of a Convention held in 2002-2003, foundered when it was rejected in referenda in France and the Netherlands in 2005. A subsequent effort in 2007 incorporated many of the features of the rejected draft Constitutional Treaty while also making a number of substantive and symbolic changes. The new treaty, referred to as the Treaty of Lisbon, sought to amend existing treaties rather

than replace them. The treaty was approved at the EU intergovernmental conference of the then 27 member states held in Lisbon in December 2007, after which the process of national ratifications began. In October 2009, an Irish referendum approved the Lisbon Treaty (overturning a previous rejection) and cleared the way for an ultimate unanimous endorsement. Poland and the Czech Republic ratified soon after. The Lisbon Treaty came into force on 1 December 2009 and the EU officially replaced and succeeded the EC. The Treaty's provisions are part of the basic consolidated versions of the Treaty on European Union (TEU) and the Treaty on the Functioning of the European Union (TFEU) now governing what remains a very specific integration project.

UK citizens on 23 June 2016 narrowly voted to leave the EU; the formal exit took place on 31 January 2020. The EU and UK have negotiated and ratified a Withdrawal Agreement that includes a status quo transition period through December 2020, which can be extended if both sides agree.

GEOGRAPHY

Location: Europe between the North Atlantic Ocean in the west and Russia, Belarus, and Ukraine to the east

Map references: Europe

Area: *total:* 4,236,351 sq km

rank by area (sq km):
1. France (includes five overseas regions) 643,801
2. Spain 505,370
3. Sweden 450,295
4. Germany 357,022
5. Finland 338,145
6. Poland 312,685
7. Italy 301,340
8. Romania 238,391
9. Greece 131,957
10. Bulgaria 110,879
11. Hungary 93,028
12. Portugal 92,090 1 3
13. Austria 83,871
14. Czechia 78,867
15. Ireland 70,273
16. Lithuania 65,300
17. Latvia 64,589
18. Croatia 56,594
19. Slovakia 49,035
20. Estonia 45,228
21. Denmark 43,094
22. Netherlands 41,543
23. Belgium 30,528
24. Slovenia 20,273
25. Cyprus 9,251
26. Luxembourg 2,586
27. Malta 316

Area—comparative: less than one-half the size of the US

Land boundaries: total: 13,770 km

border countries (19): Albania 212 km, Andorra 118 km, Belarus 1176 km, Bosnia and Herzegovina 956 km, Holy See 3 km, Liechtenstein 34 km, Macedonia 396 km, Moldova 683 km, Monaco 6 km, Montenegro 19 km, Norway 2375 km, Russia 2435 km, San Marino 37 km, Serbia 1353 km, Switzerland 1729 km, Turkey 415 km, United Kingdom 499 km, Ukraine 1324 km; note—the Brexit Withdrawal Agreement (2020) commits the United Kingdom (UK) to maintain an open border in Ireland, so the border between Northern Ireland (UK) and the Republic of Ireland is only de jure and is not a hard border; the de facto border is the Irish Sea between the islands of Ireland and Great Britain

note: data for European continent only

Coastline: 53,563.9 km

Climate: cold temperate; potentially subarctic in the north to temperate; mild wet winters; hot dry summers in the south

Terrain: fairly flat along Baltic and Atlantic coasts; mountainous in the central and southern areas

Elevation: *lowest point:* Zuidplaspolder, Netherlands -7 m

highest point: Mont Blanc, France 4,810 m

Natural resources: iron ore, natural gas, petroleum, coal, copper, lead, zinc, bauxite, uranium, potash, salt, hydropower, arable land, timber, fish

Irrigated land: 154,539.82 sq km (2011 est.)

Population distribution: population distribution varies considerably from country to country, but tends to follow a pattern of coastal and river settlement, with urban agglomerations forming large hubs facilitating large scale housing, industry, and commerce; the area in and around the Netherlands, Belgium, and Luxembourg (known collectively as Benelux), is the most densely populated area in the EU

Natural hazards: flooding along coasts; avalanches in mountainous area; earthquakes in the south; volcanic eruptions in Italy; periodic droughts in Spain; ice floes in the Baltic

Environment—current issues: various forms of air, soil, and water pollution; see individual country entries

Environment—international agreements: *party to:* Air Pollution, Air Pollution-Nitrogen Oxides, Air Pollution-Persistent Organic Pollutants, Air Pollution-Sulphur 94, Antarctic-Marine Living Resources, Biodiversity, Climate Change, Climate Change-Kyoto Protocol, Desertification, Hazardous Wastes, Law of the Sea, Ozone Layer Protection, Tropical Timber 83, Tropical Timber 94 *signed, but not ratified:* Air Pollution-Volatile Organic Compounds

PEOPLE AND SOCIETY

Population: 453,007,803

rank by population: Germany—80,159,662; France—67,848,156; Italy—62,402,659; Spain—50,015,792; Poland—38,282,325;

Romania—21,302,893; Netherlands—17,280,397; Belgium—11,720,716; Czechia—10,702,498; Greece—10,607,051; Portugal—10,302,674; Sweden—10,202,491; Hungary—9,771,827; Austria—8,859,449; Bulgaria—6,966,899; Denmark—5,869,410; Finland—5,571,665; Slovakia—5,440,602; Ireland—5,176,569; Croatia—4,227,746; Lithuania—2,731,464; Slovenia—2,102,678; Latvia—1,881,232; Cyprus—1,266,676; Estonia—1,228,624; Luxembourg - 628,381; Malta—457,267 (July 2020 est.)

Languages: Bulgarian, Croatian, Czech, Danish, Dutch, English, Estonian, Finnish, French, German, Greek, Hungarian, Irish, Italian, Latvian, Lithuanian, Maltese, Polish, Portuguese, Romanian, Slovak, Slovene, Spanish, Swedish *note:* only the 24 official languages are listed; German, the major language of Germany, Austria, and Switzerland, is the most widely spoken mother tongue—about 16% of the EU population; English is the most widely spoken foreign language-about 29% of the EU population is conversant with it (2020)

Religions: Roman Catholic 48%, Protestant 12%, Orthodox 8%, other Christian 4%, Muslim 2%, other 1% (includes Jewish, Sikh, Buddhist, Hindu), atheist 7%, non-believer/agnostic 16%, unspecified 2% (2012 est.)

Age structure: *0-14 years:* 15.05% (male 34,978,216/female 33,217,600)

15-24 years: 10.39% (male 24,089,260/female 22,990,579)

25-54 years: 40.54% (male 92,503,000/female 91,144,596)

55-64 years: 13.52% (male 29,805,200/female 31,424,172)

65 years and over: 20.5% (male 39,834,507/female 53,020,673) (2020 est.)

Median age: *total:* 44 years

male: 42.6 years

female: 45.5 years (2020 est.)

Population growth rate: 0.10% (2020 est.)

Birth rate: 9.5 births/1,000 population (2020 est.)

Death rate: 10.7 deaths/1,000 population (2020 est.)

Net migration rate: 2.2 migrant(s)/1,000 population (2020 est.)

Population distribution: population distribution varies considerably from country to country, but tends to follow a pattern of coastal and river settlement, with urban agglomerations forming large hubs facilitating large scale housing, industry, and commerce; the area in and around the Netherlands, Belgium, and Luxembourg (known collectively as Benelux), is the most densely populated area in the EU

Sex ratio: *at birth:* 1.06 male(s)/female

0-14 years: 1.05 male(s)/female

15-24 years: 1.05 male(s)/female

25-54 years: 1.01 male(s)/female

55-64 years: 0.95 male(s)/female

65 years and over: 0.75 male(s)/female

total population: 0.95 male(s)/female (2020 est.)

Infant mortality rate: *total:* 3.7 deaths/1,000 live births

male: 4 deaths/1,000 live births

female: 3.3 deaths/1,000 live births (2020 est.)

Life expectancy at birth:

total population: 80.9 years

male: 78 years

female: 83.9 years (2020 est.)

Total fertility rate: 1.62 children born/woman (2020 est.)

Current Health Expenditure: 9.9% (2016)

HIV/AIDS—adult prevalence rate: note—see individual entries of member states

HIV/AIDS—people living with HIV/AIDS: note—see individual entries of member states

HIV/AIDS—deaths: note—see individual entries of member states

Major infectious diseases:

note: widespread ongoing transmission of a respiratory illness caused by the novel coronavirus (COVID-19) is occurring regionally; the US Department of Homeland Security has issued instructions requiring US passengers who have been in the European Union's Schengen Area (comprised of the following 26 European states: Austria, Belgium, Czech Republic, Denmark, Estonia, Finland, France, Germany, Greece, Hungary, Iceland, Italy, Latvia, Liechtenstein, Lithuania, Luxembourg, Malta, Netherlands, Norway, Poland, Portugal, Slovakia, Slovenia, Spain, Sweden, and Switzerland) to travel through select airports where the US Government has implemented enhanced screening procedures

Education expenditures: 4.6% of GDP (2017)

Unemployment, youth ages 15-24: *total:* 17.1%

male: 17.3%

female: 16.9% (2018 est.)

GOVERNMENT

Union name: *conventional long form:* European Union

abbreviation: EU

Political structure: a hybrid and unique intergovernmental and supranational organization

Capital: *name:* Brussels (Belgium), Strasbourg (France), Luxembourg, Frankfurt (Germany); note—the European Council, a gathering of the EU heads of state and/or government, and the Council of the European Union, a ministerial-level body of ten formations, meet in Brussels, Belgium, except for Council meetings held in Luxembourg in April, June, and October; the European Parliament meets in Brussels and Strasbourg, France, and has administrative offices in Luxembourg; the Court of Justice of the European Union is located in Luxembourg; and the European Central Bank is located in Frankfurt, Germany

geographic coordinates: (Brussels) 50 50 N, 4 20 E

time difference: UTC + 1 (6 hours ahead of Washington, DC, during Standard Time)

daylight saving time: +1hr, begins last Sunday in March; ends last Sunday in October

note: the 27 European Union countries spread across three time zones; a proposal has been put forward to do away with daylight savings time in all EU countries

Member states:

27 countries: Austria, Belgium, Bulgaria, Croatia, Cyprus, Czechia, Denmark, Estonia, Finland, France, Germany, Greece, Hungary, Ireland, Italy, Latvia, Lithuania, Luxembourg, Malta, Netherlands, Poland, Portugal, Romania, Slovakia, Slovenia, Spain, Sweden; note—candidate countries: Albania, Montenegro, North Macedonia, Serbia, Turkey there are 13 overseas countries and territories (OCTs) (1 with Denmark [Greenland], 6 with France [French Polynesia; French Southern and Antarctic Lands; New Caledonia; Saint Barthelemy; Saint Pierre and Miquelon; Wallis and Futuna], and 6 with the Netherlands [Aruba, Bonaire, Curacao, Saba, Sint Eustatius, Sint Maarten]), all are part of the Overseas Countries and Territories Association (OCTA)

note: there are non-European OCTs having special relations with Denmark, France, and the Netherlands (list is annexed to the Treaty on the Functioning of the European Union), that are associated with the EU to promote their economic and social development; member states apply to their trade with OCTs the same treatment as they accord each other pursuant to the treaties; OCT nationals are in principle EU citizens, but these countries are neither part of the EU, nor subject to the EU

Independence: 7 February 1992 (Maastricht Treaty signed establishing the European Union); 1 November 1993 (Maastricht Treaty entered into force)

note: the Treaties of Rome, signed on 25 March 1957 and subsequently entered into force on 1 January 1958, created the European Economic Community and the European Atomic Energy Community; a series of subsequent treaties have been adopted to increase efficiency and transparency, to prepare for new member states, and to introduce new areas of cooperation—such as a single currency; the Treaty of Lisbon, signed on 13 December 2007 and entered into force on 1 December 2009 is the most recent of these treaties and is intended to make the EU more democratic, more efficient, and better able to address global problems with one voice

National holiday: Europe Day (also known as Schuman Day), 9 May (1950); note—the day in 1950 that Robert SCHUMAN proposed the creation of what became the European Coal and Steel Community, the progenitor of today's European Union, with the aim of achieving a united Europe

Constitution: *history:* none; note—the EU legal order relies primarily on two consolidated texts encompassing all provisions as amended from a series of past treaties: the Treaty on European Union (TEU), as modified by the 2009 Lisbon

Treaty states in Article 1 that "the HIGH CONTRACTING PARTIES establish among themselves a EUROPEAN UNION ... on which the Member States confer competences to attain objectives they have in common"; Article 1 of the TEU states further that the EU is "founded on the present Treaty and on the Treaty on the Functioning of the European Union (hereinafter referred to as 'the Treaties')," both possessing the same legal value; Article 6 of the TEU provides that a separately adopted Charter of Fundamental Rights of the European Union "shall have the same legal value as the Treaties"

amendments: European Union treaties can be amended in several ways: 1) Ordinary Revision Procedure (for key amendments to the treaties); initiated by an EU country's government, by the European Parliament, or by the European Commission; following adoption of the proposal by the European Council, a convention is formed of national government

representatives to review the proposal and subsequently a conference of government representatives also reviews the proposal; passage requires ratification by all EU countries; 2) Simplified Revision Procedure (for amendment of EU internal policies and actions); passage of a proposal requires unanimous European Council vote following European Council consultation with the European Commission, the European Parliament, and the European Central Bank (if the amendment concerns monetary matters) and requires ratification by all EU countries; 3) Passerelle Clause (allows the alteration of a legislative procedure without a formal amendment of the treaties); 4) Flexibility Clause (permits the EU to decide in subject areas where EU competences have not been explicitly granted in the Treaties but are necessary to the attainment of the objectives set out in the Treaty); note—the Treaty of Lisbon (signed in December 2007 and effective in December 2009) amended the two treaties that formed the EU—the Maastricht Treaty (1993) and the Treaty of Rome (1958), known in updated form as the Treaty on the Functioning of the European Union

Legal system: unique supranational law system in which, according to an interpretive declaration of member-state governments appended to the Treaty of Lisbon, "the Treaties and the law adopted by the Union on the basis of the Treaties have primacy over the law of Member States" under conditions laid down in the case law of the Court of Justice; key principles of EU law include fundamental rights as guaranteed by the Charter of Fundamental Rights and as resulting from constitutional traditions common to the EU's 27-member states; EU law is divided into 'primary' and 'secondary' legislation; primary legislation is derived from the consolidated versions of the Treaty on European Union and the Treaty on the Functioning of the European Union and are the basis for all EU action; secondary legislation—which includes directives, regulations, and decisions—is derived from the principles and objectives set out in the treaties

Suffrage: 18 years of age (16 years in Austria); universal; voting for the European Parliament is permitted in each member state

Executive branch: under the EU treaties there are three distinct institutions , each of which conducts functions that may be regarded as executive in nature:

European Council—brings together heads of state and government, along with the president of the European Commission, and meets at least four times a year; its aim is to provide the impetus for the development of the Union and to issue general policy guidelines; the Treaty of Lisbon established the position of "permanent" (full-time) president of the European Council; leaders of the EU member states appoint the president for a 2 1/2 year term, renewable once; the president's responsibilities include chairing the EU summits and providing policy and organizational continuity; the current president is Donald TUSK (Poland), since 1 December 2014, succeeding Herman VAN ROMPUY (Belgian; 2009-14) Council of the European Union—consists of ministers of each EU member state and meets regularly in 10 different configurations depending on the subject matter; it conducts policymaking and coordinating functions as well as legislative functions; ministers of EU member states chair meetings of the Council of the EU based on a 6-month rotating presidency except for the meetings of EU Foreign Ministers in the Foreign Affairs Council that are chaired by the High Representative for Foreign Affairs and Security Policy

European Commission—headed by a College of Commissioners comprised of 28 members (one from each member country) including the president; each commissioner is responsible for one or more policy areas; the Commission's main responsibilities include the sole right to initiate EU legislation (except for foreign and security/defense policy), promoting the general interest of the EU, acting as "guardian of the Treaties" by monitoring the application of EU law, implementing/executing the EU budget, managing programs, negotiating on the EU's behalf in core policy areas such as trade, and ensuring the Union's external representation in some policy areas; its current president is Jean-Claude JUNCKER (Luxembourg) elected on 15 July 2014 (took office on 1 November 2014); the president of the European Commission is nominated by the European Council and formally "elected" by the European Parliament; the Commission president allocates specific responsibilities among the members of the College (appointed by common accord of the member state governments in consultation with the president-elect); the European Parliament confirms the entire Commission for a 5-year term; President JUNCKER reorganized the structure of the College around clusters or project teams coordinated by 7 vice presidents in line with the current Commission's main political priorities and appointed Frans TIMMERMANS (Netherlands) to act as his first vice president; the confirmation process for the next Commission expected be held in the fall of 2019

note: for external representation and foreign policy making, leaders of the EU member states appointed Federica MOGHERINI (Italy) as the High Representative of the European Union for Foreign Affairs and Security Policy; MOGHERINI took office on 1 November 2014, succeeding Catherine ASHTON (UK) (2009-14); the High Representative's concurrent appointment as Vice President of the European Commission was meant to bring more coherence to the EU's foreign policy (horizontally, between policies managed by the Commission that are particularly relevant for EU external relations, such as trade, humanitarian aid and crisis management, neighborhood policy and enlargement; and vertically, between national capitals and the EU); the High Representative helps develop and implement the EU's Common Foreign and Security Policy and Common Security and Defense Policy components, chairs the Foreign Affairs Council, represents and acts for the Union in many international contexts, and oversees the European External Action Service, the diplomatic corps of the EU, established on 1 December 2010

Legislative branch: *description:* two legislative bodies consisting of the Council of the European Union (27 seats; ministers representing the 27 member states) and the European Parliament (705 seats; seats allocated among member states roughly in proportion to population size; members elected by proportional representation to serve 5-year terms); note—the European Parliament President, Antonio TAJANI (Italian center-right), was elected in January 2017 by a majority of fellow members of the European Parliament (MEPs) and represents the Parliament within the EU and internationally; the Council of the EU and the MEPs share responsibilities for adopting the bulk of EU legislation, normally acting in co-decision on Commission proposals (but not in the area of Common Foreign and Security Policy, which is governed by consensus of the EU member state governments)

elections: last held on 23-26 May 2019 (next to be held May 2024)

election results: percent of vote—NA; seats by party (as of 31 January 2020)—EPP 187, S&D 148, ALDE/EDP 97, ID 76, Greens/EFA 67, ECR 59, GUE-NGL 40, non-inscripts 31; composition—NA

Judicial branch: *highest courts:* Court of Justice of the European Union, which includes the Court of Justice (informally known as the European Court of Justice or ECJ) and the General Court (consists of 27 judges, one drawn from each member state; the ECJ includes 11 Advocates General while the General Court can include additional judges; both the ECJ and the General Court may sit in a "Grand Chamber" of 15 judges in special cases but usually in chambers of 3 to 5 judges

judge selection and term of office: judges appointed by the common consent of the member states to serve 6-year renewable terms

note: the ECJ is the supreme judicial authority of the EU; it ensures that EU law is interpreted and applied uniformly throughout the EU, resolves disputed issues among the EU institutions and with member states, and reviews issues and opinions regarding questions of EU law referred by member state courts

Political parties and leaders:
Alliance of Liberals and Democrats for Europe or ALDE [Guy VERHOFSTADT]
European United Left-Nordic Green Left or GUE/NGL [Gabriele ZIMMER]
Europe of Freedom and Direct Democracy or EFDD [Nigel FARAGE]
Europe of Nations and Freedom or ENF or ENL [Nicolas BAY and Marcel DE GRAAFF]
European Conservatives and Reformists or ECR [Syed KAMALL and Ryszard LEGUTKO]
European Greens/European Free Alliance or Greens/EFA [Ska KELLER, Philippe LAMBERTS]
European People's Party or EPP [Manfred WEBER]
Identity and Democracy Party [Marco ZANNI]
Progressive Alliance of Socialists and Democrats or S&D [Udo BULLMANN]

International organization participation: ARF, ASEAN (dialogue member), Australian Group, BIS, BSEC (observer), CBSS, CERN, EBRD, FAO, FATF, G-8, G-10, G-20, IDA, IEA, IGAD (partners), LAIA (observer), NSG (observer), OAS (observer), OECD, PIF (partner), SAARC (observer), SICA (observer), UN (observer), UNRWA (observer), WCO, WTO, ZC (observer)

Diplomatic representation in the US: *chief of mission:* Ambassador David O'SULLIVAN (since 18 November 2014)
chancery: 2175 K Street, NW, Suite 800, Washington, DC 20037
telephone: [1] (202) 862-9500
FAX: [1] (202) 429-1766

Diplomatic representation from the US: *chief of mission:* Ambassador Gordon SONDLAND (since 9 July 2018)
telephone: [32] (2) 811-4100
embassy: 13 Zinnerstraat/Rue Zinner, B-1000 Brussels
mailing address: use embassy street address
FAX: [32] (2) 811-5154

Flag description: a blue field with 12 five-pointed gold stars arranged in a circle in the center; blue represents the sky of the Western world, the stars are the peoples of Europe in a circle, a symbol of unity; the number of stars is fixed

National symbol(s): a circle of 12, five-pointed, golden yellow stars on a blue field; union colors: blue, yellow

National anthem: name: Ode to Joy
lyrics/music: no lyrics/Ludwig VAN BEETHOVEN, arranged by Herbert VON KARAJAN
note: official EU anthem since 1985; the anthem is meant to represent all of Europe rather than just the organization, conveying ideas of peace, freedom, and unity
0:00/ 1:00

ECONOMY

Economy—overview: The 27 member states that make up the EU have adopted an internal single market with free movement of goods, services, capital, and labor. The EU, which is also a customs union, aims to bolster Europe's trade position and its political and economic weight in international affairs.

Despite great differences in per capita income among member states (from $28,000 to $109,000) and in national attitudes toward issues like inflation, debt, and foreign trade, the EU has achieved a high degree of coordination of monetary and fiscal policies. A common currency — the euro — circulates among 19 of the member states that make up the European Economic and Monetary Union (EMU). Eleven member states introduced the euro as their common currency on 1 January 1999 (Greece did so two years later). Since 2004, 13 states acceded to the EU. Of the 13, Slovenia (2007), Cyprus and Malta (2008), Slovakia (2009), Estonia (2011), Latvia (2014), and Lithuania (2015) have adopted the euro; seven other member states—excluding Denmark, which has a formal opt-out—are required by EU treaties to adopt the common currency upon meeting fiscal and monetary convergence criteria.

The EU economy posted moderate GDP growth for 2014 through 2017, capping five years of sustained growth since the 2008-09 global economic crisis and the ensuing sovereign debt crisis in the euro zone in 2011. However, the bloc's recovery was uneven. Some EU member states (Czechia, Ireland, Malta, Romania, Sweden, and Spain) recorded strong growth, others (Italy) experienced modest expansion, and Greece finally ended its EU rescue program in August 2018. Overall, the EU's recovery was buoyed by lower commodities prices and accommodative monetary policy, which lowered interest rates and stimulated demand. The euro zone, which makes up about 70% of the total EU economy, performed well, achieving a growth rate not seen in a decade. In October 2017 the European Central Bank (ECB) announced it would extend its bond-buying program through September 2018, and possibly beyond that date, to keep the euro zone recovery on track. The ECB's efforts to spur more lending and investment through its asset-buying program, negative interest rates, and long-term loan refinancing programs have not yet raised inflation in line with the ECB's statutory target of just under 2%.

Despite its performance, high unemployment in some member states, high levels of public and private debt, muted productivity, an incomplete single market in services, and an aging population remain sources of potential drag on the EU's future growth. Moreover, the EU economy remains vulnerable to a slowdown of global trade and bouts of political and financial turmoil. In June 2016, the UK voted to withdraw from the EU, the first member country ever to attempt to secede.

Continued uncertainty about the implications of the UK's exit from the EU (concluded January

2020) could hurt consumer and investor confidence and dampen EU growth, particularly if trade and cross-border investment significantly declines. Political disagreements between EU member states on reforms to fiscal and economic policy also may impair the EU's ability to bolster its crisis-prevention and resolution mechanisms. International investors' fears of a broad dissolution of the single currency area have largely dissipated, but these concerns could resurface if elected leaders implement policies that contravene euro-zone budget or banking rules. State interventions in ailing banks, including rescue of banks in Italy and resolution of banks in Spain, have eased financial vulnerabilities in the European banking sector even though some banks are struggling with low profitability and a large stock of bad loans, fragilities that could precipitate localized crises. Externally, the EU has continued to pursue comprehensive free trade agreements to expand EU external market share, particularly with Asian countries; EU and Japanese leaders reached a political-level agreement on a free trade agreement in July 2017, and agreement with Mexico in April 2018 on updates to an existing free trade agreement.

GDP (purchasing power parity): $20.85 trillion (2017 est.)
$20.38 trillion (2016 est.)
$19.98 trillion (2015 est.)
note: data are in 2017 dollars

GDP (official exchange rate): $17.11 trillion (2017 est.)

GDP—real growth rate:
2.3% (2017 est.)
2% (2016 est.)
2.3% (2015 est.)

GDP—per capita (PPP): $40,900 (2017 est.)
$39,400 (2016 est.)
$38,200 (2015 est.)
note: data are in 2017 dollars

Gross national saving:
22.7% of GDP (2017 est.)
22.2% of GDP (2016 est.)
22% of GDP (2015 est.)

GDP—composition, by end use:
household consumption: 54.4% (2016 est.)
government consumption: 20.4% (2016 est.)
investment in fixed capital: 19.8% (2016 est.)
investment in inventories: 0.4% (2016 est.)
exports of goods and services: 43.9% (2016 est.)
imports of goods and services: -40.5% (2016 est.)

GDP—composition, by sector of origin:
agriculture: 1.6% (2017 est.)
industry: 25.1% (2017 est.)
services: 70.9% (2017 est.)

Agriculture—products: wheat, barley, oilseeds, sugar beets, wine, grapes; dairy products, cattle, sheep, pigs, poultry; fish

Industries: among the world's largest and most technologically advanced regions, the EU industrial base includes: ferrous and nonferrous metal production and processing, metal products, petroleum, coal, cement, chemicals, pharmaceuticals, aerospace, rail transportation equipment, passenger and commercial vehicles, construction equipment, industrial equipment, shipbuilding, electrical power equipment, machine tools and automated manufacturing systems, electronics and telecommunications equipment, fishing, food and beverages, furniture, paper, textiles

Industrial production growth rate: 3.5% (2017 est.)

Labor force: 238.9 million (2016 est.)

Labor force—by occupation: *agriculture:* 5%
industry: 21.9%
services: 73.1% (2014 est.)

Unemployment rate:
8.6% (2016 est.)
9.4% (2015 est.)

Population below poverty line: 9.8% (2013 est.)
note: see individual country entries of member states

Household income or consumption by percentage share: *lowest 10%:* 2.8%
highest 10%: 23.8% (2016 est.)

Taxes and other revenues: 45.2% (of GDP) (2014)

Budget surplus (+) or deficit (-): -3% (of GDP) (2014)

Public debt:
86.8% of GDP (2014)
85.5% of GDP (2013)

Fiscal year: NA

Inflation rate (consumer prices):
1.5% (2017 est.)
1.1% (2016 est.)

Current account balance:
$404.9 billion (2017 est.)
$359.7 billion (2016 est.)

Exports:
$1.929 trillion (2016 est.)
$1.985 trillion (2015 est.)
note: external exports, excluding intra-EU trade

Exports—partners: United States 20.7%, China 9.6%, Switzerland 8.1%, Turkey 4.4%, Russia 4.1% (2016 est.)

Exports—commodities: machinery, motor vehicles, pharmaceuticals and other chemicals, fuels, aircraft, plastics, iron and steel, wood pulp and paper products, alcoholic beverages, furniture

Imports:
$1.895 trillion (2016 est.)
$1.92 trillion (2015 est.)
note: external imports, excluding intra-EU trade

Imports—commodities: fuels and crude oil, machinery, vehicles, pharmaceuticals and other chemicals, precious gemstones, textiles, aircraft, plastics, metals, ships

Imports—partners: China 20.1%, United States 14.5%, Switzerland 7.1%, Russia 6.3% (2016 est.)

Reserves of foreign exchange and gold:
$740.9 billion (31 December 2014 est.)
$746.9 billion (31 December 2013)
note: data are for the European Central Bank

Debt—external:
$29.27 trillion (31 December 2016 est.)
$28.68 trillion (31 December 2015 est.)

Exchange rates: euros per US dollar—0.885 (2017 est.)
0.903 (2016 est.)
0.9214 (2015 est.)
0.885 (2014 est.)
0.7634 (2013 est.)

ENERGY

Electricity—production: 3.043 trillion kWh (2015 est.)

Electricity—consumption: 2.845 trillion kWh (2015 est.)

Electricity—exports: 390 billion kWh (2015 est.)

Electricity—imports: 397 billion kWh (2015 est.)

Electricity—installed generating capacity: 975 million kW (2015 est.)

Electricity—from fossil fuels: 44% of total installed capacity (2015 est.)

Electricity—from nuclear fuels: 12% of total installed capacity (2015 est.)

Electricity—from hydroelectric plants: 11% of total installed capacity (2015 est.)

Electricity—from other renewable sources: 44% of total installed capacity (2015 est.)

Crude oil—production: 1.488 million bbl/day (2016 est.)

Crude oil—proved reserves: 5.1 billion bbl (2016 est.)

Refined petroleum products—production: 11.66 million bbl/day (2016 est.)

Refined petroleum products—consumption: 12.89 million bbl/day (2015 est.)

Refined petroleum products—exports: 2.196 million bbl/day (2017 est.)

Refined petroleum products—imports: 8.613 million bbl/day (2017 est.)

Natural gas—production: 118.2 billion cu m (2016 est.)

Natural gas—consumption: 428.8 billion cu m (2016 est.)

Natural gas—exports: 93.75 billion cu m (2010 est.)

Natural gas—imports: 420.6 billion cu m (2010 est.)

Natural gas—proved reserves: 1.3 trillion cu m (1 January 2017 est.)

Carbon dioxide emissions from consumption of energy: 3.475 billion Mt (2015 est.)

COMMUNICATIONS

Telephones—fixed lines: *total subscriptions:* 210,621,546
subscriptions per 100 inhabitants: 4 (2017 est.)

Telephones—mobile cellular: *total subscriptions:* 625,000,799
subscriptions per 100 inhabitants: 121 (2017 est.)

Telecommunication systems: *note—see individual country entries of member states*

Internet country code: eu; note—see country entries of member states for individual country codes

Internet users: *total:* 398.1 million (2018 est.) *percent of population:* 85%

Broadband—fixed subscriptions: *total:* 174,634,171 *subscriptions per 100 inhabitants:* 3 (2017)

TRANSPORTATION

National air transport system: *annual passenger traffic on registered air carriers:* 636,860,155 (2018) *annual freight traffic on registered air carriers:* 31,730,660,000 (2018)

Airports—with paved runways: *total:* 1,882 (2017) over 3,047 m: 120 (2017) 2,438 to 3,047 m: 341 (2017) 1,524 to 2,437 m: 507 (2017) 914 to 1,523 m: 425 (2017) under 914 m: 489 (2017)

Airports—with unpaved runways: *total:* 1,244 (2013) over 3,047 m: 1 (2013) 2.438 to 3,047 m: 1 (2013) 1.524 to 2,437 m: 15 (2013) 914 to 1,523 m: 245 (2013) under 914 m: 982 (2013)

Heliports: 90 (2013)

Railways: *total:* 230,548 km (2013)

Roadways: *total:* 10,582,653 km (2013)

Waterways: 53,384 km (2013)

Ports and terminals: *major port(s):* Antwerp (Belgium), Barcelona (Spain), Braila (Romania), Bremen (Germany), Burgas (Bulgaria) Constanta (Romania), Copenhagen (Denmark), Galati (Romania), Gdansk (Poland), Hamburg (Germany), Helsinki (Finland), Las Palmas (Canary Islands, Spain), Le Havre (France), Lisbon (Portugal), Marseille (France), Naples (Italy), Peiraiefs or Piraeus (Greece), Riga (Latvia), Rotterdam (Netherlands), Split (Croatia), Stockholm (Sweden), Talinn (Estonia), Tulcea (Romania), Varna (Bulgaria)

MILITARY AND SECURITY

Military expenditures: 1.5% of GDP (2018) 1.49% of GDP (2017) 1.48% of GDP (2016) 1.48% of GDP (2015) 1.5% of GDP (2014)

Military deployments: 180 Central African Republic (EUTM); 600 Bosnia-Herzegovina (EUTM); 700 Mali (EUTM); 200 Somalia (EUTM) (2020)

Military—note: the current five-nation Eurocorps, formally established in 1992 and activated the following year, began in 1987 as a French-German Brigade; Belgium (1993), Spain (1994), and Luxembourg (1996) joined over the next few years; five additional countries participate in Eurocorps as associated nations: Greece, Poland, and Turkey (since 2002), Italy and Romania (joined in 2009 and 2016 respectively); Eurocorps consists of approximately 1,000 troops at its headquarters in Strasbourg, France and the 5,000-man Franco-German Brigade; Eurocorps has deployed troops and police on NATO peacekeeping missions to Bosnia-Herzegovina (1998-2000), Kosovo (2000), and Afghanistan (2004-05 and 2012); Eurocorps has been involved in EU operations to Mali (2015) and the Central African Republic (201 6-1 7) (2019)

TERRORISM

Terrorist group(s): Islamic Revolutionary Guard Corps/Qods Force; Islamic State of Iraq and ash-Sham (ISIS); al-Qa'ida (2019)

note: details about the history, aims, leadership, organization, areas of operation, tactics, targets, weapons, size, and sources of support of the group(s) appear(s) in Appendix-T

TRANSNATIONAL ISSUES

Disputes—international: as a political union, the EU has no border disputes with neighboring countries, but Estonia has no land boundary agreements with Russia, Slovenia disputes its land and maritime boundaries with Croatia, and Spain has territorial and maritime disputes with Morocco and with the UK over Gibraltar; the EU has set up a Schengen area—consisting of 22 EU member states that have signed the convention implementing the Schengen agreements or "acquis" (1985 and 1990) on the free movement of persons and the harmonization of border controls in Europe; these agreements became incorporated into EU law with the implementation of the 1997 Treaty of Amsterdam on 1 May 1999; in addition, non-EU states Iceland and Norway (as part of the Nordic Union) have been included in the Schengen area since 1996 (full members in 2001), Switzerland since 2008, and Liechtenstein since 2011 bringing the total current membership to 26; the UK (since 2000) and Ireland (since 2002) take part in only some aspects of the Schengen area, especially with respect to police and criminal matters; nine of the 13 new member states that joined the EU since 2004 joined Schengen on 21 December 2007; of the four remaining EU states, Romania, Bulgaria, and Croatia are obligated to eventually join, while Cyprus' entry is held up by the ongoing Cyprus dispute

FALKLAND ISLANDS (ISLAS MALVINAS)

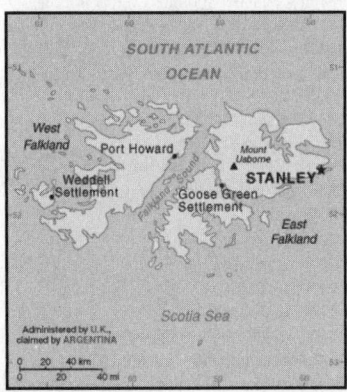

INTRODUCTION

Background: Although first sighted by an English navigator in 1592, the first landing (English) did not occur until almost a century later in 1690, and the first settlement (French) was not established until 1764. The colony was turned over to Spain two years later and the islands have since been the subject of a territorial dispute, first between Britain and Spain, then between Britain and Argentina. The UK asserted its claim to the islands by establishing a naval garrison there in 1833. Argentina invaded the islands on 2 April 1982. The British responded with an expeditionary force that landed seven weeks later and after fierce fighting forced an Argentine surrender on 14 June 1982. With hostilities ended and Argentine forces withdrawn, UK administration resumed. In response to renewed calls from Argentina for Britain to relinquish control of the islands, a referendum was held in March 2013, which resulted in 99.8% of the population voting to remain a part of the UK.

GEOGRAPHY

Location: Southern South America, islands in the South Atlantic Ocean, about 500 km east of southern Argentina

Geographic coordinates: 51 45 S, 59 00 W

Map references: South America

Area: *total:* 12,173 sq km
land: 12,173 sq km
water: 0 sq km
note: includes the two main islands of East and West Falkland and about 200 small islands
country comparison to the world: 164

Area—comparative: slightly smaller than Connecticut

Land boundaries: 0 km

Coastline: 1,288 km

Maritime claims: *territorial sea:* 12 nm
continental shelf: 200 nm
exclusive fishing zone: 200 nm

Climate: cold marine; strong westerly winds, cloudy, humid; rain occurs on more than half of days in year; average annual rainfall is 60 cm in Stanley; occasional snow all year, except in January and February, but typically does not accumulate

Terrain: rocky, hilly, mountainous with some boggy, undulating plains

Elevation: *lowest point:* Atlantic Ocean 0 m
highest point: Mount Usborne 705 m

Natural resources: fish, squid, wildlife, calcified seaweed, sphagnum moss

Land use: *agricultural land:* 92.4% (2011 est.)
arable land: 0% (2011 est.) / *permanent crops:* 0% (2011 est.) / *permanent pasture:* 92.4% (2011 est.)
forest: 0% (2011 est.)
other: 7.6% (2011 est.)

Irrigated land: NA

Population distribution: a very small population, with most residents living in and around Stanley

Natural hazards: strong winds persist throughout the year

Environment—current issues: overfishing by unlicensed vessels is a problem; reindeer—introduced to the islands in 2001 from South Georgia—are part of a farming effort to produce specialty meat and diversify the islands' economy; this is the only commercial reindeer herd in the world unaffected by the 1986 Chornobyl disaster; grazing threatens important habitats including tussac grass and its ecosystem with penguins and sea lions; soil erosion from fires

Geography—note: deeply indented coast provides good natural harbors; short growing season

PEOPLE AND SOCIETY

Population: 3,198 (2016 est.)
note: data include all persons usually resident in the islands at the time of the 2016 census
country comparison to the world: 229

Nationality: *noun:* Falkland Islander(s)
adjective: Falkland Island

Ethnic groups: Falkland Islander 48.3%, British 23.1%, St. Helenian 7.5%, Chilean 4.6%, mixed 6%, other 8.5%, unspecified 2% (2016 est.)

Languages: English 89%, Spanish 7.7%, other 3.3% (2006 est.)

Religions: Christian 57.1%, other 1.6%, none 35.4%, unspecified 6% (2016 est.)

Population growth rate: 0.01% (2014 est.)
country comparison to the world: 190

Birth rate: 10.9 births/1,000 population (2012 est.)
country comparison to the world: 181

Death rate: 4.9 deaths/1,000 population (2012 est.)
country comparison to the world: 199

Net migration rate: NA

Population distribution: a very small population, with most residents living in and around Stanley

Urbanization: *urban population:* 78.5% of total population (2020)
rate of urbanization: 0.76% annual rate of change (2015-20 est.)
total population growth rate v. urban population growth rate, 2000-2030:

Major urban areas—Population: 2,000 STANLEY (capital) (2018)

Sex ratio: 1.12 male(s)/female (2016 est.)
note: sex ratio is somewhat skewed by the high proportion of males at the Royal Air Force station, Mount Pleasant Airport (MPA); excluding MPA, the sex ratio of the total population would be 1.04

Infant mortality rate: *total:* NA (2018)
male: NA
female: NA
Life expectancy at birth: total population: 77.9 (2017 est.)
male: 75.6
female: 79.6
Total fertility rate: NA

Drinking water source:
improved:
urban: 100% of population
rural: 78.2% of population
total: 92% of population
unimproved:
urban: 0% of population
rural: 21.8% of population
total: 5% of population (2017 est.)

Sanitation facility access:
improved:
urban: 95% of population
rural: 78% of population
total: 100% of population
unimproved:
urban: 5% of population
rural: 22% of population
total: 0% of population (2017)

HIV/AIDS—adult prevalence rate: NA

HIV/AIDS—people living with HIV/AIDS: NA

HIV/AIDS—deaths: NA

GOVERNMENT

Country name: *conventional long form:* none
conventional short form: Falkland Islands (Islas Malvinas)
etymology: the archipelago takes its name from the Falkland Sound, the strait separating the two main islands; the channel itself was named after the Viscount of Falkland, who sponsored

an expedition to the islands in 1690; the Spanish name for the archipelago derives from the French "Iles Malouines," the name applied to the islands by French explorer Louis-Antoine de BOUGAINVILLE in 1764

Dependency status: overseas territory of the UK; also claimed by Argentina

Government type: parliamentary democracy (Legislative Assembly); self-governing overseas territory of the UK

Capital: *name:* Stanley
geographic coordinates: 51 42 S, 57 51 W
time difference: UTC-4 (1 hour ahead of Washington, DC, during Standard Time)
etymology: named after Edward SMITH-STANLEY (1799-1869), the 14th Earl of Derby, a British statesman and three-time prime minister of the UK who never visited the islands

Administrative divisions: none (overseas territory of the UK; also claimed by Argentina)

Independence: none (overseas territory of the UK; also claimed by Argentina)

National holiday: Liberation Day, 14 June (1982)

Constitution: *history:* previous 1985; latest entered into force 1 January 2009 (The Falkland Islands Constitution Order 2008)
amendments: NA

Legal system: English common law and local statutes

Citizenship: see United Kingdom

Suffrage: 18 years of age; universal

Executive branch: *chief of state:* Queen ELIZABETH II (since 6 February 1952); represented by Governor Nigel PHILLIPS (since 12 September 2017)
head of government: Chief Executive Barry ROWLAND (since 3 October 2016)
cabinet: Executive Council elected by the Legislative Council
elections/appointments: the monarchy is hereditary; governor appointed by the monarch; chief executive appointed by the governor

Legislative branch: *description:* unicameral Legislative Assembly, formerly the Legislative Council (10 seats; 8 members directly elected by majority vote and 2 appointed ex-officio members—the chief executive, appointed by the governor, and the financial secretary; members serve 4-year terms)
elections: last held on 9 November 2017 (next to be held in November 2021)
election results: percent of vote—NA; seats—independent 8; composition -men 8, women 2, percent of women 20%

Judicial branch: *highest courts:* Court of Appeal (consists of the court president, the chief justice as an ex officio, non-resident member, and 2 justices of appeal); Supreme Court (consists of the chief justice); note—appeals beyond the Court of

Appeal are referred to the Judicial Committee of the Privy Council (in London)
judge selection and term of office: all justices appointed by the governor; tenure specified in each justice's instrument of appointment
subordinate courts: Magistrate's Court (senior magistrate presides over civil and criminal divisions); Court of Summary Jurisdiction

Political parties and leaders: none; all independents

International organization participation: UPU

Diplomatic representation in the US: none (overseas territory of the UK)

Diplomatic representation from the US: none (overseas territory of the UK; also claimed by Argentina)

Flag description: blue with the flag of the UK in the upper hoist-side quadrant and the Falkland Island coat of arms centered on the outer half of the flag; the coat of arms contains a white ram (sheep raising was once the major economic activity) above the sailing ship Desire (whose crew discovered the islands) with a scroll at the bottom bearing the motto DESIRE THE RIGHT

National symbol(s): ram

National anthem: *name:* Song of the Falklands"
lyrics/music: Christopher LANHAM
note: adopted 1930s; the song is the local unofficial anthem; as a territory of the United Kingdom, "God Save the Queen" is official (see United Kingdom)

ECONOMY

Economy—overview: The economy was formerly based on agriculture, mainly sheep farming, but fishing and tourism currently comprise the bulk of economic activity. In 1987, the government began selling fishing licenses to foreign trawlers operating within the Falkland Islands' exclusive fishing zone. These license fees net more than $40 million per year, which help support the island's health, education, and welfare system. The waters around the Falkland Islands are known for their squid, which account for around 75% of the annual 200,000-ton catch.

Dairy farming supports domestic consumption; crops furnish winter fodder. Foreign exchange earnings come from shipments of high-grade wool to the UK and from the sale of postage stamps and coins.

Tourism, especially ecotourism, is increasing rapidly, with about 69,000 visitors in 2009 and adds approximately $5.5 million to the Falkland's annual GDP. The British military presence also provides a sizable economic boost. The islands are now self-financing except for defense.

In 1993, the British Geological Survey announced a 200-mile oil exploration zone around the islands, and early seismic surveys suggest substantial reserves capable of producing 500,000

barrels per day. Political tensions between the UK and Argentina remain high following the start of oil drilling activities in the waters. In May 2010 the first commercial oil discovery was made, signaling the potential for the development of a long term hydrocarbon industry in the Falkland Islands.

GDP (purchasing power parity): $206.4 million (2015 est.)
$164.5 million (2014 est.)
$167.5 million (2013 est.)
country comparison to the world: 220

GDP (official exchange rate): $206.4 million (2015 est.)

GDP—real growth rate:
25.5% (2015 est.)
-1.8% (2014 est.)
-20.4% (2013 est.)
country comparison to the world: 3

GDP—per capita (PPP): $70,800 (2015 est.)
$63,000 (2014 est.)
country comparison to the world: 12

GDP—composition, by sector of origin:
agriculture: 41% (2015 est.)
industry: 20.6% NA (2015 est.)
services: 38.4% NA (2015 est.)

Agriculture—products: fodder and vegetable crops; venison, sheep, dairy products; fish, squid

Industries: fish and wool processing; tourism

Industrial production growth rate: NA

Labor force:
1,850 (2016 est.)
country comparison to the world: 227

Labor force—by occupation: *agriculture:* 41%
industry: 24.5%
services: 34.5% (2015 est.)

Unemployment rate: 1% (2016 est.)
country comparison to the world: 9

Population below poverty line: NA

Household income or consumption by percentage share: *lowest 10%:* NA
highest 10%: NA

Budget: *revenues:* 67.1 million (FY09/10)
expenditures: 75.3 million (FY09/10)

Taxes and other revenues: 32.5% (of GDP) (FY09/10)
country comparison to the world: 66

Budget surplus (+) or deficit (-): -4% (of GDP) (FY09/10)
country comparison to the world: 156

Public debt: 0% of GDP (2015 est.)
country comparison to the world: 209

Fiscal year: 1 April—31 March

Inflation rate (consumer prices): 1.4% (2014 est.)
country comparison to the world: 77

Exports:
$257.3 million (2015 est.)
$125 million (2004 est.)
country comparison to the world: 186

Exports—partners: Spain 74.4%, Namibia 10.4%, US 5% (2017)

Exports—commodities: wool, hides, meat, venison, fish, squid

Imports: $90 million (2004 est.)
country comparison to the world: 218

Imports—commodities: fuel, food and drink, building materials, clothing

Imports—partners: UK 47.8%, Spain 28.4%, Greece 10.2%, Netherlands 5.7%, Cote d'Ivoire 4.3% (2017)

Debt—external:
$0 (2017 est.)
$0 (2016 est.)
country comparison to the world: 205

Exchange rates: Falkland pounds (FKP) per US dollar -
0.7836 (2017 est.)
0.6542 (2016 est.)
0.6542 (2015)
0.6542 (2014 est.)
0.6391 (2013 est.)

ENERGY

Electricity—production: 19 million kWh (2016 est.)
country comparison to the world: 213

Electricity—consumption: 17.67 million kWh (2016 est.)
country comparison to the world: 213

Electricity—exports: 0 kWh (2016 est.)
country comparison to the world: 133

Electricity—imports: 0 kWh (2016 est.)
country comparison to the world: 147

Electricity—installed generating capacity: 12,100 kW (2016 est.)
country comparison to the world: 208

Electricity—from fossil fuels: 74% of total installed capacity (2016 est.)
country comparison to the world: 96

Electricity—from nuclear fuels: 0% of total installed capacity (2017 est.)
country comparison to the world: 89

Electricity—from hydroelectric plants: 0% of total installed capacity (2017 est.)
country comparison to the world: 171

Electricity—from other renewable sources: 26% of total installed capacity (2017 est.)
country comparison to the world: 26

Crude oil—production: 0 bbl/day (2018 est.)
country comparison to the world: 135

Crude oil—exports:

0 bbl/day (2015 est.)
country comparison to the world: 122

Crude oil—imports:
0 bbl/day (2015 est.)
country comparison to the world: 126

Crude oil—proved reserves: 0 bbl (1 January 2018 est.)
country comparison to the world: 130

Refined petroleum products—production: 0 bbl/day (2017 est.)
country comparison to the world: 144

Refined petroleum products—consumption: 290 bbl/day (2016 est.)
country comparison to the world: 213

Refined petroleum products—exports: 0 bbl/day (2015 est.)
country comparison to the world: 154

Refined petroleum products—imports: 286 bbl/day (2015 est.)
country comparison to the world: 209

Natural gas—production: ``0 cu m (2017 est.)
country comparison to the world: 131

Natural gas—consumption: 0 cu m (2017 est.)
country comparison to the world: 145

Natural gas—exports: 0 cu m (2017 est.)
country comparison to the world: 103

Natural gas—imports: 0 cu m (2017 est.)
country comparison to the world: 124

Natural gas—proved reserves: 0 cu m (1 January 2014 est.)
country comparison to the world: 134

Carbon dioxide emissions from consumption of energy: 44,070 Mt (2017 est.)
country comparison to the world: 211

COMMUNICATIONS

Telephones—fixed lines: *total subscriptions:* 2,255
subscriptions per 100 inhabitants: 77 (July 2016 est.)
country comparison to the world: 216

Telephones—mobile cellular: *total subscriptions:* 4,674
subscriptions per 100 inhabitants: 146 (July 2016 est.)
country comparison to the world: 219

Telecommunication systems: *general assessment:* government-operated radiotelephone and private VHF/CB radiotelephone networks provide effective service to almost all points on both islands
domestic: fixed-line subscriptions 77 per 100, 146 per 100 for mobile-cellular (2019)

international: country code—500; satellite earth station—1 Intelsat (Atlantic Ocean) with links through London to other countries (2015)

Broadcast media: TV service provided by a multi-channel service provider; radio services provided by the public broadcaster, Falkland Islands Radio Service, broadcasting on both AM and FM frequencies, and by the British Forces Broadcasting Service (BFBS) (2007)

Internet country code: .fk

Internet users: *total:* 3,000
percent of population: 98.3% (July 2016 est.)
country comparison to the world: 221

Broadband—fixed subscriptions: *total:* 1,610
subscriptions per 100 inhabitants: 50 (2017 est.)
country comparison to the world: 189

TRANSPORTATION

National air transport system: *number of registered air carriers:* 1 (2020)
inventory of registered aircraft operated by air carriers: 5

Civil aircraft registration country code prefix: VP-F (2016)

Airports: 7 (2020)
country comparison to the world: 168

Airports—with paved runways: *total:* 2 (2019)
2,438 to 3,047 m: 1
914 to 1,523 m: 1

Airports—with unpaved runways: *total:* 5 (2013)
under 914 m: 5 (2013)

Roadways: *total:* 440 km (2008)
paved: 50 km (2008)
unpaved: 390 km (2008)
country comparison to the world: 198

Merchant marine: *total:* 3
by type: general cargo 1, other 2 (2019)
country comparison to the world: 170

Ports and terminals: *major seaport(s):* Stanley

MILITARY AND SECURITY

Military and security forces: no regular military forces

Military—note: defense is the responsibility of the UK

TRANSNATIONAL ISSUES

Disputes—international: Argentina, which claims the islands in its constitution and briefly occupied them by force in 1982, agreed in 1995 to no longer seek settlement by force; UK continues to reject Argentine requests for sovereignty talks

FAROE ISLANDS

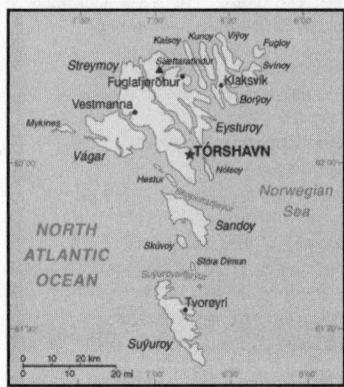

INTRODUCTION

Background: The population of the Faroe Islands is largely descended from Viking settlers who arrived in the 9th century. The islands have been connected politically to Denmark since the 14th century. A high degree of self-government was granted the Faroese in 1948, who have autonomy over most internal affairs while Denmark is responsible for justice, defense, and foreign affairs. The Faroe Islands are not part of the European Union.

GEOGRAPHY

Location: Northern Europe, island group between the Norwegian Sea and the North Atlantic Ocean, about halfway between Iceland and Norway

Geographic coordinates: 62 00 N, 7 00 W

Map references: Europe

Area: *total:* 1,393 sq km
land: 1,393 sq km
water: 0 sq km (some lakes and streams)
country comparison to the world: 183

Area—comparative: eight times the size of Washington, DC

Land boundaries: 0 km

Coastline: 1,117 km

Maritime claims:
territorial sea: 12 nm
continental shelf: 200 nm or agreed boundaries or median line
exclusive fishing zone: 200 nm or agreed boundaries or median line

Climate: mild winters, cool summers; usually overcast; foggy, windy

Terrain: rugged, rocky, some low peaks; cliffs along most of coast

Elevation: *lowest point:* Atlantic Ocean 0 m
highest point: Slaettaratindur 882 m

Natural resources: fish, whales, hydropower, possible oil and gas

Land use: *agricultural land:* 2.1% (2011 est.)
arable land: 2.1% (2011 est.) / permanent crops: 0% (2011 est.) / permanent pasture: 0% (2011 est.)
forest: 0.1% (2011 est.)
other: 97.8% (2011 est.)

Population distribution: the island of Streymoy is by far the most populous with over 40% of the population; it has approximately twice as many inhabitants as Eysturoy, the second most populous island; seven of the inhabited islands have fewer than 100 people

Natural hazards: strong winds and heavy rains can occur throughout the year

Environment—current issues: coastal erosion, landslides and rockfalls, flash flooding, wind storms; oil spills

Environment—international agreements: *party to:* Marine Dumping—associate member to the London Convention and Ship Pollution

Geography—note: archipelago of 17 inhabited islands and one uninhabited island, and a few uninhabited islets; strategically located along important sea lanes in northeastern Atlantic; precipitous terrain limits habitation to small coastal lowlands

PEOPLE AND SOCIETY

Population: 51,628 (July 2020 est.)
country comparison to the world: 209

Nationality: *noun:* Faroese (singular and plural)
adjective: Faroese

Ethnic groups: Faroese 87.6% (Scandinavian and Anglo-Saxon descent), Danish 7.8%, other Nordic 1.4%, other 3.2% (includes Filipino, Thai, British) (2018 est.)
note: data represent respondents by country of birth

Languages: Faroese 93.8% (derived from Old Norse), Danish 3.2%, other 3% (2011 est.)
note: data represent population by primary language

Religions: Christian 89.3% (predominantly Evangelical Lutheran), other 1%, none 3.8%, unspecified 6% (2011 est.)

Age structure: *0-14 years:* 19.69% (male 5,247/female 4,920)
15-24 years: 13.89% (male 3,708/female 3,465)
25-54 years: 37.01% (male 10,277/female 8,828)
55-64 years: 12% (male 3,199/female 2,996)
65 years and over: 17.41 % (male 4,352/female 4,636) (2020 est.)

Median age: *total:* 37.2 years
male: 36.9 years
female: 37.7 years (2020 est.)
country comparison to the world: 72

Population growth rate: 0.6% (2020 est.)
country comparison to the world: 150

Birth rate: 14.9 births/1,000 population (2020 est.)
country comparison to the world: 119

Death rate: 8.8 deaths/1,000 population (2020 est.)
country comparison to the world: 66

Net migration rate: 0 migrant(s)/1,000 population (2020 est.)
country comparison to the world: 82

Population distribution: the island of Streymoy is by far the most populous with over 40% of the population; it has approximately twice as many inhabitants as Eysturoy, the second most populous island; seven of the inhabited islands have fewer than 100 people

Urbanization: *urban population:* 42.4% of total population (2020)
rate of urbanization: 0.74% annual rate of change (2015-20 est.)
total population growth rate v. urban population growth rate, 2000-2030:

Major urban areas—Population: 21,000 TORSHAVN (capital) (2018)

Sex ratio: *at birth:* 1.07 male(s)/female
0-14 years: 1.07 male(s)/female
15-24 years: 1.07 male(s)/female
25-54 years: 1.16 male(s)/female
55-64 years: 1.07 male(s)/female
65 years and over: 0.94 male(s)/female
total population: 1.08 male(s)/female (2020 est.)

Infant mortality rate: *total:* 5.1 deaths/1,000 live births
male: 5.4 deaths/1,000 live births
female: 4.8 deaths/1,000 live births (2020 est.)
country comparison to the world: 176
Life expectancy at birth: total population: 80.8 years
male: 78.3 years
female: 83.6 years (2020 est.)
country comparison to the world: 42

Total fertility rate: 2.31 children born/woman (2020 est.)
country comparison to the world: 82

Drinking water source:
improved:
total: 99% of population
unimproved:
total: 0% of population (2017 est.)

Physicians density: 2.62 physicians/1,000 population (2016)

Hospital bed density: 4.1 beds/1,000 population (2015)

Sanitation facility access:
improved:
total: 100% of population
unimproved:
total: 1% of population (2017)

HIV/AIDS—adult prevalence rate: NA

HIV/AIDS—people living with HIV/AIDS: NA

HIV/AIDS—deaths: NA

GOVERNMENT

Country name: conventional long form: none
conventional short form: Faroe Islands
local long form: none
local short form: Foroyar
etymology: the archipelago's name may derive from the Old Norse word "faer," meaning sheep

Dependency status: part of the Kingdom of Denmark; self-governing overseas administrative division of Denmark since 1948

Government type: parliamentary democracy (Faroese Parliament); part of the Kingdom of Denmark

Capital: *name:* Torshavn
geographic coordinates: 62 00 N, 6 46 W
time difference: UTC 0 (5 hours ahead of Washington, DC, during Standard Time)
daylight saving time: +1hr, begins last Sunday in March; ends last Sunday in October
etymology: the meaning in Danish is Thor's harbor

Administrative divisions: part of the Kingdom of Denmark; self-governing overseas administrative division of Denmark; there are 29 first-order municipalities (kommunur, singular—kommuna) Eidhis, Eystur, Famjins, Fuglafjardhar, Fugloyar, Hovs, Husavikar, Hvalbiar, Hvannasunds, Klaksvikar, Kunoyar, Kvivik, Nes, Porkeris, Runavikar, Sands, Sjovar, Skalavikar, Skopunar, Skuvoyar, Sorvags, Sumbiar, Sunda, Torshavnar, Tvoroyrar, Vaga, Vags, Vestmanna, Vidhareidhis

Independence: none (part of the Kingdom of Denmark; self-governing overseas administrative division of Denmark)

National holiday: Olaifest (Olavsoka) (commemorates the death in battle of King OLAF II of Norway, later St. OLAF), 29 July (1030)

Constitution: *history:* 5 June 1953 (Danish Constitution), 23 March 1948 (Home Rule Act), and 24 June 2005 (Takeover Act) serve as the Faroe Islands' constitutional position in the Unity of the Realm
amendments: see entry for Denmark

Legal system: the laws of Denmark apply where applicable

Citizenship: see Denmark

Suffrage: 18 years of age; universal

Executive branch: *chief of state:* Queen MARGRETHE II of Denmark (since 14 January 1972), represented by High Commissioner Lene Moyell JOHANSEN, chief administrative officer (since 15 May 2017)
head of government: Prime Minister Bardhur A STEIG NIELSEN (since 16 September 2019)
cabinet: Landsstyri appointed by the prime minister
elections/appointments: the monarchy is hereditary; high commissioner appointed by the monarch; following legislative elections, the leader of the majority party or majority coalition usually elected prime minister by the Faroese Parliament;

election last held on 31 August 2019 (next to be held in 2023)
election results: Bardhur A STEIGNIELSEN elected prime minister; Parliament vote—NA

Legislative branch: *description:* unicameral Faroese Parliament or Logting (33 seats); members directly elected in a single nationwide constituency by proportional representation vote; members serve 4-year terms) the Faroe Islands elect 2 members to the Danish Parliament to serve 4-year terms
elections: Faroese Parliament—last held on 31 August 2019 (next to be held in 2023) Faroese seats in the Danish Parliament last held on 5 June 2019 (next to be held no later than June 2023)
election results: Faroese Parliament percent of vote by party—People's Party 24.5%, JF 22.1%, Union Party 20.3%, Republic 18.1%, Center Party 5.4%, Progressive Party 4.6%, New Self-Government Party 3.4%, other 1.4%, seats by party—People's Party 8, JF 7, Union Party 7, Republic 6, Center Party 2, Progressive Party 2, New Self-Government Party 1, composition—men 25, women 8; percent of women 24.2% Faroese seats in Danish Parliament—percent of vote by party—NA; seats by party—Social Democratic Party 1, Republican Party 1; composition—2 men

Judicial branch: *highest courts:* Faroese Court or Raett (Rett—Danish) decides both civil and criminal cases; the Court is part of the Danish legal system
subordinate courts: Court of the First Instance or Tribunal de Premiere Instance; Court of Administrative Law or Tribunal Administratif; Mixed Commercial Court; Land Court

Political parties and leaders:
Center Party (Midflokkurin) [Jenis av RANA]
Self-Government Party (Sjalvstyri or Sjalvstyrisflokkurin) [Jogvan SKORHEIM]
People's Party (Folkaflokkurin) [Jorgen NICLASEN]
Progressive Party (Framsokn) [Poul MICHELSEN
Republic (Tjodveldi) [Hogni HOYDAL] (formerly the Republican Party)
Social Democratic Party (Javnadarflokkurin) or JF [Aksel V. JOHANNESEN]
Union Party (Sambandsflokkurin) [Bardhur A STEIG NIELSEN]

International organization participation: Arctic Council, IMO (associate), NC, NIB, UNESCO (associate), UPU

Diplomatic representation in the US: none (self-governing overseas administrative division of Denmark)

Diplomatic representation from the US: none (self-governing overseas administrative division of Denmark)

Flag description: white with a red cross outlined in blue extending to the edges of the flag; the vertical part of the cross is shifted toward the hoist side in the style of the Dannebrog (Danish flag); referred to as Merkid, meaning "the banner" or "the mark,"

the flag resembles those of neighboring Iceland and Norway, and uses the same three colors—but in a different sequence; white represents the clear Faroese sky, as well as the foam of the waves; red and blue are traditional Faroese colors
note: the blue on the flag is a lighter blue (azure) than that found on the flags of Iceland or Norway

National symbol(s): *ram; national colors:* red, white, blue

National anthem: *name:* "Mitt alfagra land" (My Fairest Land)
lyrics/music: Simun av SKAROI/Peter ALBERG
note: adopted 1948; the anthem is also known as "Tu alfagra land mitt" (Thou Fairest Land of Mine); as a self-governing overseas administrative division of Denmark, the Faroe Islands are permitted their own national anthem

ECONOMY

Economy—overview: The Faroese economy has experienced a period of significant growth since 2011, due to higher fish prices and increased salmon farming and catches in the pelagic fisheries. Fishing has been the main source of income for the Faroe Islands since the late 19th century, but dependence on fishing makes the economy vulnerable to price fluctuations. Nominal GDP, measured in current prices, grew 5.6% in 2015 and 6.8% in 2016. GDP growth was forecast at 6.2% in 2017, slowing to 0.5% in 2018, due to lower fisheries quotas, higher oil prices and fewer farmed salmon combined with lower salmon prices. The fisheries sector accounts for about 97% of exports, and half of GDP. Unemployment is low, estimated at 2.1% in early 2018. Aided by an annual subsidy from Denmark, which amounts to about 11% of Faroese GDP, Faroese have a standard of living equal to that of Denmark. The Faroe Islands have bilateral free trade agreements with the EU, Iceland, Norway, Switzerland, and Turkey.

For the first time in 8 years, the Faroe Islands managed to generate a public budget surplus in 2016, a trend which continued in 2017. The local government intends to use this to reduce public debt, which reached 38% of GDP in 2015. A fiscal sustainability analysis of the Faroese economy shows that a long-term tightening of fiscal policy of 5% of GDP is required for fiscal sustainability.

Increasing public infrastructure investments are likely to lead to continued growth in the short term, and the Faroese economy is becoming somewhat more diversified. Growing industries include financial services, petroleum-related businesses, shipping, maritime manufacturing services, civil aviation, IT, telecommunications, and tourism.

GDP (purchasing power parity): $2.001 billion (2014 est.)
$1.89 billion (2013 est.)
$1.608 billion (2012 est.)
country comparison to the world: 197

GDP (official exchange rate): $2.765 billion (2014 est.)

GDP—real growth rate:
5.9% (2017 est.)

7.5% (2016 est.)
2.4% (2015 est.)
country comparison to the world: 31

GDP—per capita (PPP): $40,000 (2014 est.)
country comparison to the world: 45

Gross national saving:
25.7% of GDP (2012 est.)
25.2% of GDP (2011 est.)
25.9% of GDP (2010 est.)
country comparison to the world: 53

GDP—composition, by end use: household consumption: 52% (2013)
government consumption: 29.6% (2013)
investment in fixed capital: 18.4% (2013)

GDP—composition, by sector of origin:
agriculture: 18% (2013 est.)
industry: 39% (2013 est.)
services: 43% (2013 est.)

Agriculture—products: milk, potatoes, vegetables, sheep, salmon, herring, mackerel and other fish

Industries: fishing, fish processing, tourism, small ship repair and refurbishment, handicrafts

Industrial production growth rate: 3.4% (2009 est.)
country comparison to the world: 91

Labor force: 27,540 (2017 est.)
country comparison to the world: 206

Labor force—by occupation: *agriculture:* 15%
industry: 15%
services: 70% (December 2016 est.)

Unemployment rate:
2.2% (2017 est.)
3.4% (2016 est.)
country comparison to the world: 21

Population below poverty line: 10% (2015 est.)

Household income or consumption by percentage share: *lowest 10%:* NA
highest 10%: NA

Budget: *revenues:* 835.6 million (2014 est.)
expenditures: 883.8 million (2014)
note: Denmark supplies the Faroe Islands with almost one-third of its public funds

Taxes and other revenues: 30.2% (of GDP) (2014 est.)
country comparison to the world: 76

Budget surplus (+) or deficit (-): -1.7% (of GDP) (2014 est.)
country comparison to the world: 95

Public debt:
35% of GDP (2014 est.)
country comparison to the world: 152

Fiscal year: calendar year

Inflation rate (consumer prices):
-0.3% (2016)
-1.7% (2015)
country comparison to the world: 8

Exports:
$1.184 billion (2016 est.)
$1.019 billion (2015 est.)
country comparison to the world: 153

Exports—partners: Russia 26.4%, UK 14.1%, Germany 8.4%, China 7.9%, Spain 6.8%, Denmark 6.2%, US 4.7%, Poland 4.4%, Norway 4.1% (2017)

Exports—commodities: fish and fish products (97%) (2017 est.)

Imports:
$978.4 million (2016 est.)
$906.1 million (2015 est.)
country comparison to the world: 186

Imports—commodities: goods for household consumption, machinery and transport equipment, fuels, raw materials and semi-manufactures, cars

Imports—partners: Denmark 33%, China 10.7%, Germany 7.6%, Poland 6.8%, Norway 6.7%, Ireland 5%, Chile 4.3% (2017)

Debt—external:
$387.6 million (2012)
$274.5 million (2010)
country comparison to the world: 181

Exchange rates: Danish kroner (DKK) per US dollar –
6.586 (2017 est.)
6.7269 (2016 est.)
6.7269 (2015 est.)
6.7236 (2014 est.)
5.6125 (2013 est.)

ENERGY

Electricity access: electrification—total population: 100% (2020)

Electricity—production: 307 million kWh (2016 est.)
country comparison to the world: 180

Electricity—consumption: 285.5 million kWh (2016 est.)
country comparison to the world: 185

Electricity—exports: 0 kWh (2016 est.)
country comparison to the world: 134

Electricity—imports: 0 kWh (2016 est.)
country comparison to the world: 148

Electricity—installed generating capacity: 128,300 kW (2016 est.)
country comparison to the world: 176

Electricity—from fossil fuels: 54% of total installed capacity (2016 est.)
country comparison to the world: 142

Electricity—from nuclear fuels: 0% of total installed capacity (2017 est.)
country comparison to the world: 90

Electricity—from hydroelectric plants: 31% of total installed capacity (2017 est.)
country comparison to the world: 66

Electricity—from other renewable sources: 16% of total installed capacity (2017 est.)
country comparison to the world: 52

Crude oil—production: 0 bbl/day (2018 est.)
country comparison to the world: 136

Crude oil—exports: 0 bbl/day (2015 est.)
country comparison to the world: 123

Crude oil—imports: 0 bbl/day (2015 est.)
country comparison to the world: 127

Crude oil—proved reserves: 0 bbl (1 January 2018 est.)
country comparison to the world: 131

Refined petroleum products—production: 0 bbl/day (2015 est.)
country comparison to the world: 145

Refined petroleum products—consumption: 4,600 bbl/day (2016 est.)
country comparison to the world: 180

Refined petroleum products—exports: 0 bbl/day (2015 est.)
country comparison to the world: 155

Refined petroleum products—imports: 4,555 bbl/day (2015 est.)
country comparison to the world: 174

Natural gas—production: 0 cu m (2017 est.)
country comparison to the world: 132

Natural gas—consumption: 0 cu m (2017 est.)
country comparison to the world: 146

Natural gas—exports: 0 cu m (2017 est.)
country comparison to the world: 104

Natural gas—imports: 0 cu m (2017 est.)
country comparison to the world: 125

Natural gas—proved reserves: 0 cu m (1 January 2014 est.)
country comparison to the world: 135

Carbon dioxide emissions from consumption of energy: 739,300 Mt (2017 est.)
country comparison to the world: 176

COMMUNICATIONS

Telephones—fixed lines: *total subscriptions:* 19,137
subscriptions per 100 inhabitants: 37.29 (2019 est.)
country comparison to the world: 178

Telephones—mobile cellular: *total subscriptions:* 59,771
subscriptions per 100 inhabitants: 116.47 (2019 est.)
country comparison to the world: 204

Telecommunication systems: *general assessment:* good international and domestic communications; telecommunications network of high standards with excellent coverage throughout most parts of the country and at competitive prices (2020)
domestic: 37 per 100 for fixed-line and 116 per 100 for mobile-cellular; both NMT (analog) and GSM (digital) mobile telephone systems are installed (2019)
international: country code—298; landing points for the SHEFA-2, FARICE-1, and CANTAT-3 fiber-optic submarine cables from the Faeroe Islands, to Denmark, Germany, UK and Iceland; satellite earth stations—1 Orion; (2019)
note: the COVID-19 outbreak is negatively impacting telecommunications production and supply chains globally; consumer spending on telecom devices and services has also slowed due to the pandemic's effect on economies worldwide; overall progress towards improvements in all facets of the

telecom industry—mobile, fixed-line, broadband, submarine cable and satellite—has moderated

Broadcast media: 1 publicly owned TV station; the Faroese telecommunications company distributes local and international channels through its digital terrestrial network; publicly owned radio station supplemented by 3 privately owned stations broadcasting over multiple frequencies

Internet country code: .fo

Internet users: total: 49,783
percent of population: 97.58% (July 2018 est.)
country comparison to the world: 197

Broadband—fixed subscriptions: total: 18,181
subscriptions per 100 inhabitants: 36 (2018 est.)
country comparison to the world: 155

TRANSPORTATION

National air transport system: *number of registered air carriers:* 1 (registered in Denmark) (2020)

inventory of registered aircraft operated by air carriers: 3 (registered in Denmark)

Civil aircraft registration country code prefix: OY-H (2016)

Airports: 1 (2020)
country comparison to the world: 220

Airports—with paved runways: *total:* 1 (2019)
1,524 to 2,437 m: 1

Roadways: total: 960 km (2017)
*paved:*500 km (2017)
unpaved: 460 km (2017)
note: those islands not connected by roads (bridges or tunnels) are connected by seven different ferry links operated by the nationally owned company SSL; 28 km of tunnels
country comparison to the world: 186

Merchant marine: total: 107
by type: bulk carrier 6, general cargo 46, oil tanker 1, other 54 (2019)
country comparison to the world: 86

Ports and terminals: *major seaport(s):* Fuglafjordur, Torshavn, Vagur

MILITARY AND SECURITY

Military and security forces: no regular military forces or conscription; the Government of Denmark has responsibility for defense; as such, the Danish military's Joint Arctic Command in Nuuk, Greenland is responsible for territorial defense of the Faroe Islands; the Joint Arctic Command has a contact element in the capital of Torshavn (2019)

Military—note: defense is the responsibility of Denmark

TRANSNATIONAL ISSUES

Disputes—international: because anticipated offshore hydrocarbon resources have not been realized, earlier Faroese proposals for full independence have been deferred; Iceland, the UK, and Ireland dispute Denmark's claim to UNCLOS that the Faroe Islands' continental shelf extends beyond 200 nm

FIJI

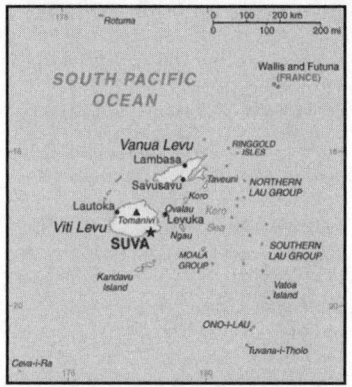

INTRODUCTION

Background: Fiji became independent in 1970 after nearly a century as a British colony. Democratic rule was interrupted by two military coups in 1987 caused by concern over a government perceived as dominated by the Indian community (descendants of contract laborers brought to the islands by the British in the 19th century). The coups and a 1990 constitution that cemented native Melanesian control of Fiji led to heavy Indian emigration; the population loss resulted in economic difficulties, but ensured that Melanesians became the majority. A new constitution enacted in 1997 was more equitable. Free and peaceful elections in 1999 resulted in a government led by an Indo-Fijian, but a civilian-led coup in 2000 ushered in a prolonged

period of political turmoil. Parliamentary elections held in 2001 provided Fiji with a democratically elected government led by Prime Minister Laisenia QARASE. Reelected in May 2006, QARASE was ousted in a December 2006 military coup led by Commodore Voreqe BAINIMARAMA, who initially appointed himself acting president but in January 2007 became interim prime minister. Following years of political turmoil, long-delayed legislative elections were held in September 2014 that were deemed "credible" by international observers and that resulted in BAINIMARAMA being reelected. He was reelected in November 2018 in elections deemed free and fair.

GEOGRAPHY

Location: Oceania, island group in the South Pacific Ocean, about two-thirds of the way from Hawaii to New Zealand

Geographic coordinates: 18 00 S, 175 00 E

Map references: Oceania

Area: total: 18,274 sq km
land: 18,274 sq km
water: 0 sq km
country comparison to the world: 157

Area—comparative: slightly smaller than New Jersey

Land boundaries: 0 km

Coastline: 1,129 km

Maritime claims: *territorial sea:* 12 nm
exclusive economic zone: 200 nm
contiguous zone: 24 nm

continental shelf: 200-m depth or to the depth of exploitation
measured from claimed archipelagic straight baselines

Climate: tropical marine; only slight seasonal temperature variation

Terrain: mostly mountains of volcanic origin

Elevation: *lowest point:* Pacific Ocean 0 m
highest point: Tomanivi 1,324 m

Natural resources: timber, fish, gold, copper, offshore oil potential, hydropower

Land use: *agricultural land:* 23.3% (2011 est.)
arable land: 9% (2011 est.) / permanent crops: 4.7% (2011 est.) / permanent pasture: 9.6% (2011 est.)
forest: 55.7% (2011 est.)
other: 21% (2011 est.)

Irrigated land: 40 sq km (2012)

Population distribution: approximately 70% of the population lives on the island of Viti Levu; roughly half of the population lives in urban areas

Natural hazards: cyclonic storms can occur from November to January

Environment—current issues: the widespread practice of waste incineration is a major contributor to air pollution in the country, as are vehicle emissions in urban areas; deforestation and soil erosion are significant problems; a contributory factor to erosion is clearing of land by bush burning, a widespread practie that threatens biodiversity

Environment—international agreements: *party to:* Biodiversity, Climate Change, Climate

335

Change-Kyoto Protocol, Desertification, Endangered Species, Law of the Sea, Marine Life Conservation, Ozone Layer Protection, Tropical Timber 83, Tropical Timber 94, Wetlands
signed, but not ratified: none of the selected agreements

Geography—note: includes 332 islands; approximately 110 are inhabited

PEOPLE AND SOCIETY

Population: 935,974 (July 2020 est.)
country comparison to the world: 161

Nationality: *noun:* Fijian(s)

adjective: Fijian

Ethnic groups: iTaukei 56.8% (predominantly Melanesian with a Polynesian admixture), Indo-Fijian 37.5%, Rotuman 1.2%, other 4.5% (European, part European, other Pacific Islanders, Chinese) (2007 est.)
note: a 2010 law replaces 'Fijian' with 'iTaukei' when referring to the original and native settlers of Fiji

Languages: English (official), Fijian (official), Hindustani

Religions: Protestant 45% (Methodist 34.6%, Assembly of God 5.7%, Seventh Day Adventist 3.9%, and Anglican 0.8%), Hindu 27.9%, other Christian 10.4%, Roman Catholic 9.1%, Muslim 6.3%, Sikh 0.3%, other 0.3%, none 0.8% (2007 est.)

Age structure: *0-14 years:* 26.86% (male 128,499/female 122,873)
15-24 years: 15.51% (male 73,993/female 71,139)
25-54 years: 41.05% (male 196,932/female 187,270)
55-64 years: 9.25% (male 43,813/female 42,763)
65 years and over: 7.34% (male 31,556/female 37,136) (2020 est.)

Dependency ratios: *total dependency ratio:* 53.4
youth dependency ratio: 44.5
elderly dependency ratio: 8.9
potential support ratio: 11.2 (2020 est.)

Median age: *total:* 29.9 years
male: 29.7 years
female: 30.1 years (2020 est.)
country comparison to the world: 125

Population growth rate: 0.5% (2020 est.)
country comparison to the world: 155

Birth rate: 17.4 births/1,000 population (2020 est.)
country comparison to the world: 96

Death rate: 6.3 deaths/1,000 population (2020 est.)
country comparison to the world: 150

Net migration rate: -6.2 migrant(s)/1,000 population (2020 est.)
country comparison to the world: 207

Population distribution: approximately 70% of the population lives on the island of Viti Levu; roughly half of the population lives in urban areas

Urbanization: *urban population:* 57.2% of total population (2020)

rate of urbanization: 1.62% annual rate of change (2015-20 est.)
total population growth rate v. urban population growth rate, 2000-2030:

Major urban areas—Population: 178,000 SUVA (capital) (2018)

Sex ratio: *at birth:* 1.05 male(s)/female
0-14 years: 1.05 male(s)/female
15-24 years: 1.04 male(s)/female
25-54 years: 1.05 male(s)/female
55-64 years: 1.02 male(s)/female
65 years and over: 0.85 male(s)/female
total population: 1.03 male(s)/female (2020 est.)

Maternal mortality rate: 34 deaths/100,000 live births (2017 est.)
country comparison to the world: 107

Infant mortality rate: *total:* 8.8 deaths/1,000 live births
male: 9.7 deaths/1,000 live births
female: 7.9 deaths/1,000 live births (2020 est.)
country comparison to the world: 143

Life expectancy at birth: total population: 73.7 years
male: 71 years
female: 76.6 years (2020 est.)
country comparison to the world: 141

Total fertility rate: 2.31 children born/woman (2020 est.)
country comparison to the world: 83

Drinking water source:
improved:
urban: 97.8% of population
rural: 88.7% of population
total: 93.8% of population
unimproved:
urban: 2.2% of population
rural: 11.3% of population
total: 6.2% of population (2017 est.)

Current Health Expenditure: 3.5% (2017)

Physicians density: 0.86 physicians/1,000 population (2015)

Hospital bed density: 2 beds/1,000 population (2016)

Sanitation facility access:
improved:
urban: 94% of population
rural: 89% of population
total: 98% of population
unimproved:
urban: 6% of population
rural: 11% of population
total: 2% of population (2017 est.)

HIV/AIDS—adult prevalence rate: 0.2% (2019 est.)
country comparison to the world: 96

HIV/AIDS—people living with HIV/AIDS: 1,000 (2019 est.)
country comparison to the world: 145

HIV/AIDS—deaths: <100 (2019 est.)

Major infectious diseases: *degree of risk:* high (2020)
food or waterborne diseases: bacterial diarrhea
vectorborne diseases: malaria

Obesity—adult prevalence rate: 30.2% (2016)
country comparison to the world: 24

Education expenditures: 3.9% of GDP (2013)
country comparison to the world: 106

Literacy:
total population: 99.1%
male: 99.1%
female: 99.1% (2018)

Unemployment, youth ages 15-24: *total:* 15.4%
male: 11.9%
female: 22.4% (2016 est.)
country comparison to the world: 87

GOVERNMENT

Country name: *conventional long form:* Republic of Fiji
conventional short form: Fiji
local long form: Republic of Fiji/Matanitu ko Viti
local short form: Fiji/Viti
etymology: the Fijians called their home Viti, but the neighboring Tongans called it Fisi, and in the Anglicized spelling of the Tongan pronunciation—promulgated by explorer Captain James COOK—the designation became Fiji

Government type: parliamentary republic

Capital: *name:* Suva (on Viti Levu)
geographic coordinates: 18 08 S, 178 25 E
time difference: UTC+12 (17 hours ahead of Washington, DC, during Standard Time)
daylight saving time: +1hr, begins first Sunday in November; ends second Sunday in January

Administrative divisions: 14 provinces and 1 dependency*; Ba, Bua, Cakaudrove, Kadavu, Lau, Lomaiviti, Macuata, Nadroga and Navosa, Naitasiri, Namosi, Ra, Rewa, Rotuma*, Serua, Tailevu

Independence: 10 October 1970 (from the UK)

National holiday: Fiji (Independence) Day, 10 October (1970)

Constitution: *history:* several previous; latest signed into law 6 September 2013

Legal system: common law system based on the English model

International law organization participation: has not submitted an ICJ jurisdiction declaration; accepts ICCt jurisdiction

Citizenship: *citizenship by birth:* no
citizenship by descent only: at least one parent must be a citizen of Fiji
dual citizenship recognized: yes
residency requirement for naturalization: at least 5 years residency out of the 10 years preceding application

Suffrage: 18 years of age; universal

Executive branch: *chief of state:* President Jioji Konousi KONROTE (since 12 November 2015)
head of government: Prime Minister Voreqe "Frank" BAINIMARAMA (since 22 September 2014)

cabinet: Cabinet appointed by the prime minister from among members of Parliament and is responsible to Parliament

elections/appointments: president elected by Parliament for a 3-year term (eligible for a second term); election last held on 31 August 2018 (next to be held in 2021); prime minister endorsed by the president

election results: Jioji Konousi KONROTE reelected president (unopposed)

Legislative branch: *description:* unicameral Parliament (51 seats; members directly elected in a nationwide, multi-seat constituency by openlist proportional representation vote to serve 4-year terms)

elections: last held on 14 November 2018 (next to be held in 2022)

election results: percent of vote by party—FijiFirst 50%, SODELPA 39.6%, NFP 7.4%; seats by party—FijiFirst 27, SODELPA 21, NFP 3; composition—men 41, women 10, percent of women 19.6%

Judicial branch: *highest courts:* Supreme Court (consists of the chief justice, all justices of the Court of Appeal, and judges appointed specifically as Supreme Court judges); Court of Appeal (consists of the court president, all puisne judges of the High Court, and judges specifically appointed to the Court of Appeal); High Court (chaired by the chief justice and includes a minimum of 10 puisne judges; High Court organized into civil, criminal, family, employment, and tax divisions)

judge selection and term of office: chief justice appointed by the president of Fiji on the advice of the prime minister following consultation with the parliamentary leader of the opposition; judges of the Supreme Court, the president of the Court of Appeal, the justices of the Court of Appeal, and puisne judges of the High Court appointed by the president of Fiji upon the nomination of the Judicial Service Commission after consulting with the cabinet minister and the committee of the House of Representatives responsible for the administration of justice; the chief justice, Supreme Court judges and justices of Appeal generally required to retire at age 70, but this requirement may be waived for one or more sessions of the court; puisne judges appointed for not less than 4 years nor more than 7 years, with mandatory retirement at age 65

subordinate courts: Magistrates' Court (organized into civil, criminal, juvenile, and small claims divisions)

Political parties and leaders:
FijiFirst [Veroqe "Frank" BAINIMARAMA]
Fiji Labor Party or FLP [Mahendra CHAUDHRY]
Fiji United Freedon Party or FUFP [Jagath KARUNARATNE]
National Federation Party or NFP [Biman PRASAD] (primarily Indian)
Peoples Democratic Party or PDP [Lynda TABUYA]
Social Democratic Liberal Party or SODELPA
Unity Fiji [Adi QORO]

International organization participation: ACP, ADB, AOSIS, C, CP, FAO, G-77, IAEA, IBRD, ICAO, ICCt, ICRM, IDA, IFAD, IFC, IFRCS, IHO, ILO, IMF, IMO, Interpol, IOC, IOM, ISO, ITSO, ITU, ITUC (NGOs), MIGA, OPCW, PCA, PIF, Sparteca (suspended), SPC, UN, UNCTAD, UNDOF, UNESCO, UNIDO, UNMISS, UNWTO, UPU, WCO, WFTU (NGOs), WHO, WIPO, WMO, WTO

Diplomatic representation in the US: *chief of mission:* Ambassador (vacant); Charge d'Affaires Akuila VUIRA

chancery: 2000 M Street NW, Suite 710, Washington, DC 20036

telephone: [1] (202) 466-8320

FAX: [1] (202) 466-8325

Diplomatic representation from the US: *chief of mission:* Ambassador Joseph James CELLA (since 23 December 2019); note—also accredited to Kiribati, Nauru, Tonga, and Tuvalu

telephone: [679] 331-4466

embassy: 158 Princes Rd, Tamavua

mailing address: P. O. Box 218, Suva

FAX: [679] 330-8685

Flag description: light blue with the flag of the UK in the upper hoist-side quadrant and the Fijian shield centered on the outer half of the flag; the blue symbolizes the Pacific Ocean and the Union Jack reflects the links with Great Britain; the shield—taken from Fiji's coat of arms—depicts a yellow lion, holding a coconut pod between its paws, above a white field quartered by the cross of Saint George; the four quarters depict stalks of sugarcane, a palm tree, a banana bunch, and a white dove of peace

National symbol(s): *Fijian canoe; national color:* light blue

National anthem: *name:* God Bless Fiji

lyrics/music: Michael Francis Alexander PRESCOTT/C. Austin MILES (adapted by Michael Francis Alexander PRESCOTT)

note: adopted 1970; known in Fijian as "Meda Dau Doka" (Let Us Show Pride); adapted from the hymn, "Dwelling in Beulah Land," the anthem's English lyrics are generally sung, although they differ in meaning from the official Fijian lyrics 0:00/ 0:00

ECONOMY

Economy—overview: Fiji, endowed with forest, mineral, and fish resources, is one of the most developed and connected of the Pacific island economies. Earnings from the tourism industry, with an estimated 842,884 tourists visiting in 2017, and remittances from Fijian's working abroad are the country's largest foreign exchange earners.

Bottled water exports to the US is Fiji's largest domestic export. Fiji's sugar sector remains a significant industry and a major export, but crops and one of the sugar mills suffered damage during Cyclone Winston in 2016. Fiji's trade imbalance continues to widen with increased imports and sluggish performance of domestic exports.

The return to parliamentary democracy and successful elections in September 2014 improved investor confidence, but increasing bureaucratic regulation, new taxes, and lack of consultation with relevant stakeholders brought four consecutive years of decline for Fiji on the World Bank Ease of Doing Business index. Private sector investment in 2017 approached 20% of GDP, compared to 13% in 2013.

GDP (purchasing power parity): $8.629 billion (2017 est.)
$8.376 billion (2016 est.)
$8.321 billion (2015 est.)
note: data are in 2017 dollars
country comparison to the world: 163

GDP (official exchange rate): $4.891 billion (2017 est.)

GDP—real growth rate:
3% (2017 est.)
0.7% (2016 est.)
3.8% (2015 est.)
country comparison to the world: 99

GDP—per capita (PPP): $9,800 (2017 est.)
$9,600 (2016 est.)
$9,600 (2015 est.)
note: data are in 2017 dollars
country comparison to the world: 140

Gross national saving:
12.7% of GDP (2017 est.)
13.4% of GDP (2016 est.)
16.1% of GDP (2015 est.)
country comparison to the world: 146

GDP—composition, by end use:
household consumption: 81.3% (2017 est.)
government consumption: 24.4% (2017 est.)
investment in fixed capital: 16.9% (2017 est.)
investment in inventories: 0% (2017 est.)
exports of goods and services: 29% (2017 est.)
imports of goods and services: -51.6% (2017 est.)

GDP—composition, by sector of origin:
agriculture: 13.5% (2017 est.)
industry: 17.4% (2017 est.)
services: 69.1% (2017 est.)

Agriculture—products: sugarcane, copra, ginger, tropical fruits, vegetables; beef, pork, chicken, fish

Industries: tourism, sugar processing, clothing, copra, gold, silver, lumber

Industrial production growth rate: 2.8% (2017 est.)
country comparison to the world: 108

Labor force: 353,100 (2017 est.)
country comparison to the world: 161

Labor force—by occupation: *agriculture:* 44.2%
industry: 14.3%
services: 41.6% (2011)

Unemployment rate:
4.5% (2017 est.)
5.5% (2016 est.)
country comparison to the world: 66

Population below poverty line: 31% (2009 est.)

Household income or consumption by percentage share: *lowest 10%:* 2.6%
highest 10%: 34.9% (2009 est.)

Budget: *revenues:* 1.454 billion (2017 est.) *expenditures:* 1.648 billion (2017 est.)

Taxes and other revenues: 29.7% (of GDP) (2017 est.)
country comparison to the world: 80

Budget surplus (+) or deficit (-): -4% (of GDP) (2017 est.)
country comparison to the world: 157

Public debt:
48.9% of GDP (2017 est.)
47.5% of GDP (2016 est.)
country comparison to the world: 106

Fiscal year: calendar year

Inflation rate (consumer prices):
3.4% (2017 est.)
3.9% (2016 est.)
country comparison to the world: 140

Current account balance:
-$277 million (2017 est.)
-$131 million (2016 est.)
country comparison to the world: 104

Exports:
$908.2 million (2017 est.)
$709 million (2016 est.)
country comparison to the world: 163

Exports—partners: US 20.8%, Australia 14.9%, NZ 7.7%, Tonga 5%, Vanuatu 4.6%, China 4.5%, Spain 4.3%, UK 4.3%, Kiribati 4.1% (2017)

Exports—commodities: fuel, including oil, fish, beverages, gems, sugar, garments, gold, timber, fish, molasses, coconut oil, mineral water

Imports:
$1.911 billion (2017 est.)
$1.761 billion (2016 est.)
country comparison to the world: 169

Imports—commodities: manufactured goods, machinery and transport equipment, petroleum products, food and beverages, chemicals, tobacco

Imports—partners: Australia 19.2%, NZ 17.2%, Singapore 17%, China 13.8% (2017)

Reserves of foreign exchange and gold:
$1.116 billion (31 December 2017 est.)
$908.6 million (31 December 2016 est.)
country comparison to the world: 130

Debt—external:
$1.022 billion (31 December 2017 est.)
$696.4 million (31 December 2016 est.)
country comparison to the world: 165

Exchange rates: Fijian dollars (FJD) per US dollar -
2.075 (2017 est.)
2.0947 (2016 est.)
2.0947 (2015 est.)
2.0976 (2014 est.)
1.8874 (2013 est.)

ENERGY

Electricity access: electrification—total population: 98.6% (2016)
electrification—urban areas: 99.2% (2016)
electrification—rural areas: 98% (2016)

Electricity—production: 914 million kWh (2016 est.)
country comparison to the world: 155

Electricity—consumption: 850 million kWh (2016 est.)
country comparison to the world: 159

Electricity—exports: 0 kWh (2016 est.)
country comparison to the world: 135

Electricity—imports: 0 kWh (2016 est.)
country comparison to the world: 149

Electricity—installed generating capacity: 338,000 kW (2016 est.)
country comparison to the world: 154

Electricity—from fossil fuels: 34% of total installed capacity (2016 est.)
country comparison to the world: 181

Electricity—from nuclear fuels: 0% of total installed capacity (2017 est.)
country comparison to the world: 91

Electricity—from hydroelectric plants: 38% of total installed capacity (2017 est.)
country comparison to the world: 55

Electricity—from other renewable sources: 27% of total installed capacity (2017 est.)
country comparison to the world: 24

Crude oil—production: 0 bbl/day (2018 est.)
country comparison to the world: 137

Crude oil—exports: 0 bbl/day (2015 est.)
country comparison to the world: 124

Crude oil—imports: 0 bbl/day (2015 est.)
country comparison to the world: 128

Crude oil—proved reserves: 0 bbl (1 January 2018 est.)
country comparison to the world: 132

Refined petroleum products—production: 0 bbl/day (2015 est.)
country comparison to the world: 146

Refined petroleum products—consumption: 16,000 bbl/day (2016 est.)
country comparison to the world: 151

Refined petroleum products—exports: 0 bbl/day (2015 est.)
country comparison to the world: 156

Refined petroleum products—imports: 17,460 bbl/day (2015 est.)
country comparison to the world: 131

Natural gas—production: 0 cu m (2017 est.)
country comparison to the world: 133

Natural gas—consumption: 0 cu m (2017 est.)
country comparison to the world: 147

Natural gas—exports: 0 cu m (2017 est.)
country comparison to the world: 105

Natural gas—imports: 0 cu m (2017 est.)
country comparison to the world: 126

Natural gas—proved reserves: 0 cu m (1 January 2014 est.)
country comparison to the world: 136

Carbon dioxide emissions from consumption of energy: 2.369 million Mt (2017 est.)
country comparison to the world: 155

COMMUNICATIONS

Telephones—fixed lines: *total subscriptions:* 80,650
subscriptions per 100 inhabitants: 8.66 (2019 est.)
country comparison to the world: 143

Telephones—mobile cellular: *total subscriptions:* 1,097,345
subscriptions per 100 inhabitants: 117.83 (2019 est.)
country comparison to the world: 161

Telecommunication systems: *general assessment:* local, interisland, and international telecommunications; subject to occasional devastating cyclones; Fiji is a leader in the Pacific region in terms of development of its ICT (Information & Communications Technology) sector and investment in telecoms infrastructure; mobile services the primary source of Internet access across the region; most advanced economy in the Pacific island region as well as hosting the highest mobile Internet penetration; initial progress towards 5G readiness (2020)
domestic: fixed-line 9 per 100 persons and mobile-cellular teledensity roughly 118 per 100 persons (2019)
international: country code—679; landing points for the ICN1, SCCN, Southern Cross NEXT, Tonga Cable and Tui-Samoa submarine cable links to US, NZ, Australia and Pacific islands of Fiji, Vanuatu, Kiribati, Samoa, Tokelau, Tonga, Fallis & Futuna, and American Samoa; satellite earth stations—2 Inmarsat (Pacific Ocean) (2019)
note: the COVID-19 outbreak is negatively impacting telecommunications production and supply chains globally; consumer spending on telecom devices and services has also slowed due to the pandemic's effect on economies worldwide; overall progress towards improvements in all facets of the telecom industry—mobile, fixed-line, broadband, submarine cable and satellite—has moderated

Broadcast media: Fiji TV, a publicly traded company, operates a free-to-air channel; Digicel Fiji operates the Sky Fiji and Sky Pacific multichannel pay-TV services; state-owned commercial company, Fiji Broadcasting Corporation, Ltd, operates 6 radio stations—2 public broadcasters and 4 commercial broadcasters with multiple repeaters; 5 radio stations with repeaters operated by Communications Fiji, Ltd; transmissions of multiple international broadcasters are available

Internet country code: .fj

Internet users: *total:* 462,860

percent of population: 49.97% (July 2018 est.)
country comparison to the world: 154

Broadband—fixed subscriptions: *total:* 13,033
subscriptions per 100 inhabitants: 1 (2018 est.)
country comparison to the world: 165

TRANSPORTATION

National air transport system: *number of registered air carriers:* 2 (2020)

inventory of registered aircraft operated by air carriers: 16
annual passenger traffic on registered air carriers: 1,670,216 (2018)
annual freight traffic on registered air carriers: 106.83 million mt-km (2018)

Civil aircraft registration country code prefix: DQ (2016)

Airports: 28 (2013)
country comparison to the world: 120

Airports—with paved runways: *total:* 4 (2017)
over 3,047 m: 1 (2017)
1,524 to 2,437 m: 1 (2017)
914 to 1,523 m: 2 (2017)

Airports—with unpaved runways: *total:* 24 (2013)
914 to 1,523 m: 5 (2013)
under 914 m: 19 (2013)

Railways: *total:* 597 km (2008)
narrow gauge: 597 km 0.600-m gauge (2008)
note: belongs to the government-owned Fiji Sugar Corporation; used to haul sugarcane during the harvest season, which runs from May to December
country comparison to the world: 109

Roadways: *total:* 3,440 km (2011)
paved: 1,686 km (2011)
unpaved: 1,754 km (2011)
country comparison to the world: 159

Waterways: 203 km (122 km are navigable by motorized craft and 200-metric-ton barges) (2012)
country comparison to the world: 97

Merchant marine: *total:* 64
by type: general cargo 19, oil tanker 4, other 41 (2019)
country comparison to the world: 106

Ports and terminals: *major seaport(s):* Lautoka, Levuka, Suva

MILITARY AND SECURITY

Military and security forces: *Republic of Fiji Military Forces (RFMF):* Land Force Command, Maritime Command (2019)

Military expenditures:
1.6% of GDP (2019)
1.6% of GDP (2018)
1.5% of GDP (2017)

1.2% of GDP (2016)
1% of GDP (2015)
country comparison to the world: 71

Military and security service personnel strengths: the Republic of Fiji Military Forces (RFMF) have about 3,500 personnel (3,200 Land Force; 300 Maritime Command) (2019)

Military equipment inventories and acquisitions: the RFMF's small inventory is a mix of equipment from Australia, New Zealand, Russia, Singapore, South Korea, the UK, and the US; since 2010, the only recorded arms deliveries were from Australia; China has donated some non-lethal material since 2018 (2019 est.)

Military deployments: 170 Egypt (MFO); 170 Iraq (UNAMI); 130 Golan Heights (UNDOF) (2020)

Military service age and obligation: 18 years of age for voluntary military service; mandatory retirement at age 55 (2013)

TRANSNATIONAL ISSUES

Disputes—international: maritime boundary dispute with Tonga

FINLAND

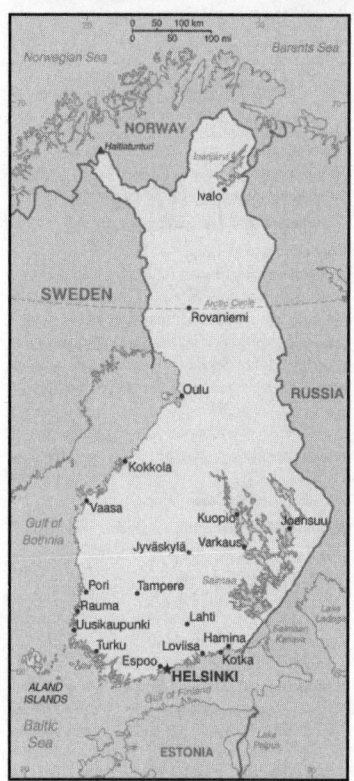

INTRODUCTION

Background: Finland was a province and then a grand duchy under Sweden from the 12th to the 19th centuries, and an autonomous grand duchy of Russia after 1809. It gained complete independence in 1917. During World War II, Finland successfully defended its independence through cooperation with Germany and resisted subsequent invasions by the Soviet Union—albeit with some loss of territory. In the subsequent half century, Finland transformed from a farm/forest economy to a diversified modern industrial economy; per capita income is among the highest in Western Europe. A member of the EU since 1995, Finland was the only Nordic state to join the euro single currency at its initiation in January 1999. In the 21st century, the key features of Finland's modern welfare state are high quality education, promotion of equality, and a national social welfare system—currently challenged by an aging population and the fluctuations of an export-driven economy.

GEOGRAPHY

Location: Northern Europe, bordering the Baltic Sea, Gulf of Bothnia, and Gulf of Finland, between Sweden and Russia

Geographic coordinates: 64 00 N, 26 00 E

Map references: Europe

Area: *total:* 338,145 sq km
land: 303,815 sq km
water: 34,330 sq km
country comparison to the world: 66

Area—comparative: slightly more than two times the size of Georgia; slightly smaller than Montana

Land boundaries: *total:* 2,563 km
border countries (3): Norway 709 km, Sweden 545 km, Russia 1309 km

Coastline: 1,250 km

Maritime claims: *territorial sea:* 12 nm (in the Gulf of Finland—3 nm)
contiguous zone: 24 nm
continental shelf: 200 m depth or to the depth of exploitation
exclusive fishing zone: 12 nm; extends to continental shelf boundary with Sweden, Estonia, and Russia

Climate: cold temperate; potentially subarctic but comparatively mild because of moderating influence of the North Atlantic Current, Baltic Sea, and more than 60,000 lakes

Terrain: mostly low, flat to rolling plains interspersed with lakes and low hills

Elevation: *mean elevation:* 164 m
lowest point: Baltic Sea 0 m
highest point: Halti (alternatively Haltia, Haltitunturi, Haltiatunturi) 1,328 m

Natural resources: timber, iron ore, copper, lead, zinc, chromite, nickel, gold, silver, limestone

Land use: *agricultural land:* 7.5% (2011 est.)
arable land: 7.4% (2011 est.) / permanent crops: 0% (2011 est.) / permanent pasture: 0.1% (2011 est.)
forest: 72.9% (2011 est.)
other: 19.6% (2011 est.)

Irrigated land: 690 sq km (2012)

Population distribution: the vast majority of people are found in the south; the northern interior areas remain sparsely populated

Natural hazards: severe winters in the north

Environment—current issues: limited air pollution in urban centers; some water pollution from industrial wastes, agricultural chemicals; habitat loss threatens wildlife populations

Environment—international agreements: *party to:* Air Pollution, Air Pollution-Nitrogen Oxides, Air Pollution-Persistent Organic Pollutants, Air Pollution-Sulfur 85, Air Pollution-Sulfur 94, Air Pollution-Volatile Organic Compounds, Antarctic-Environmental Protocol, Antarctic-Marine Living Resources, Antarctic Treaty, Biodiversity, Climate Change, Climate Change-Kyoto Protocol, Desertification, Endangered Species, Environmental Modification, Hazardous Wastes, Law of the Sea, Marine Dumping, Marine Life Conservation, Ozone Layer Protection, Ship Pollution, Tropical Timber 83, Tropical Timber 94, Wetlands, Whaling
signed, but not ratified: none of the selected agreements

Geography—note: long boundary with Russia; Helsinki is northernmost national capital on European continent; population concentrated on small southwestern coastal plain

PEOPLE AND SOCIETY

Population: 5,571,665 (July 2020 est.)
country comparison to the world: 116

Nationality: *noun:* Finn(s)
adjective: Finnish

Ethnic groups: Finn, Swede, Russian, Estonian, Romani, Sami

Languages: Finnish (official) 87.6%, Swedish (official) 5.2%, Russian 1.4%, other 5.8% (2018 est.)

Religions: Lutheran 69.8%, Greek Orthodox 1.1%, other 1.7%, unspecified 27.4% (2018 est.)

Age structure: *0-14 years:* 16.41% (male 467,220/female 447,005)
15-24 years: 10.95% (male 312,179/female 297,717)
25-54 years: 37.37% (male 1,064,326/female 1,017,545)
55-64 years: 13.02% (male 357,687/female 367,610)
65 years and over: 22.26% (male 543,331/female 697,045) (2020 est.)

Dependency ratios: *total dependency ratio:* 62.4
youth dependency ratio: 25.8
elderly dependency ratio: 36.6
potential support ratio: 2.7 (2020 est.)

Median age: *total:* 42.8 years
male: 41.3 years
female: 44.4 years (2020 est.)
country comparison to the world: 31

Population growth rate: 0.3% (2020 est.)
country comparison to the world: 171

Birth rate: 10.6 births/1,000 population (2020 est.)

country comparison to the world: 186

Death rate: 10.3 deaths/1,000 population (2020 est.)
country comparison to the world: 31

Net migration rate: 2.6 migrant(s)/1,000 population (2020 est.)
country comparison to the world: 40

Population distribution: the vast majority of people are found in the south; the northern interior areas remain sparsely populated

Urbanization: *urban population:* 85.5% of total population (2020)
rate of urbanization: 0.42% annual rate of change (2015-20 est.)
total population growth rate v. urban population growth rate, 2000-2030

Major urban areas—Population: 1.305 million HELSINKI (capital) (2020)

Sex ratio: *at birth:* 1.05 male(s)/female
0-14 years: 1.05 male(s)/female
15-24 years: 1.05 male(s)/female
25-54 years: 1.05 male(s)/female
55-64 years: 0.97 male(s)/female
65 years and over: 0.78 male(s)/female
total population: 0.97 male(s)/female (2020 est.)

Mother's mean age at first birth: 29.2 years (2017 est.)

Maternal mortality rate: 3 deaths/100,000 live births (2017 est.)
country comparison to the world: 177

Infant mortality rate: *total:* 2.5 deaths/1,000 live births
male: 2.7 deaths/1,000 live births
female: 2.4 deaths/1,000 live births (2020 est.)
country comparison to the world: 222
Life expectancy at birth: total population: 81.3 years
male: 78.4 years
female: 84.4 years (2020 est.)
country comparison to the world: 33

Total fertility rate: 1.74 children born/woman (2020 est.)
country comparison to the world: 163

Contraceptive prevalence rate: 85.5% (2015)
note: percent of women aged 18-49

Drinking water source:
improved:
urban: 100% of population
rural: 100% of population
total: 100% of population
unimproved:
urban: 0% of population
rural: 0% of population
total: 0% of population (2017 est.)

Current Health Expenditure: 9.2% (2017)

Physicians density: 3.81 physicians/1,000 population (2016)

Hospital bed density: 3.3 beds/1,000 population (2017)

Sanitation facility access:
improved:
urban: 100% of population

rural: 100% of population
total: 100% of population
unimproved:
urban: 0% of population
rural: 0% of population
total: 0% of population (2017 est.)

HIV/AIDS—adult prevalence rate: 0.1% (2018)
country comparison to the world: 122

HIV/AIDS—people living with HIV/AIDS: 4,000 (2018)
country comparison to the world: 125

HIV/AIDS—deaths: <100 (2018)

Obesity—adult prevalence rate: 22.2% (2016)
country comparison to the world: 80

Education expenditures: 6.9% of GDP (2016)
country comparison to the world: 14

School life expectancy (primary to tertiary education): *total:* 19 years
male: 20 years
female: 20 years (2018)

Unemployment, youth ages 15-24: *total:* 17%
male: 17.3%
female: 16.8% (2018 est.)
country comparison to the world: 77

GOVERNMENT

Country name: *conventional long form:* Republic of Finland
conventional short form: Finland
local long form: Suomen tasavalta/Republiken Finland
local short form: Suomi/Finland
etymology: name may derive from the ancient Fenni peoples who are first described as living in northeastern Europe in the first centuries A.D.

Government type: parliamentary republic

Capital: *name:* Helsinki
geographic coordinates: 60 10 N, 24 56 E
time difference: UTC+2 (7 hours ahead of Washington, DC, during Standard Time)
daylight saving time: +1hr, begins last Sunday in March; ends last Sunday in October
etymology: the name may derive from the Swedish "helsing," an archaic name for "neck" ("hals"), and which may refer to a narrowing of the Vantaa River that flows into the Gulf of Finland at Helsinki; "fors" refers to "rapids," so "helsing fors" meaning becomes "the narrows' rapids"

Administrative divisions: 19 regions (maakunnat, singular—maakunta (Finnish); landskapen, singular—landskapet (Swedish)); Aland (Swedish), Ahvenanmaa (Finnish); Etela-Karjala (Finnish), Sodra Karelen (Swedish) [South Karelia]; Etela-Pohjanmaa (Finnish), Sodra Osterbotten (Swedish) [South Ostrobothnia]; Etela-Savo (Finnish), Sodra Savolax (Swedish) [South Savo]; Kanta-Hame (Finnish), Egentliga Tavastland (Swedish); Kainuu (Finnish), Kajanaland (Swedish); Keski-Pohjanmaa (Finnish), Mellersta Osterbotten (Swedish) [Central Ostrobothnia]; Keski-Suomi (Finnish), Mellersta Finland

(Swedish) [Central Finland]; Kymenlaakso (Finnish), Kymmenedalen (Swedish); Lappi (Finnish), Lappland (Swedish); Paijat-Hame (Finnish), Paijanne-Tavastland (Swedish); Pirkanmaa (Finnish), Birkaland (Swedish) [Tampere]; Pohjanmaa (Finnish), Osterbotten (Swedish) [Ostrobothnia]; Pohjois-Karjala (Finnish), Norra Karelen (Swedish) [North Karelia]; Pohjois-Pohjanmaa (Finnish), Norra Osterbotten (Swedish) [North Ostrobothnia]; Pohjois-Savo (Finnish), Norra Savolax (Swedish) [North Savo]; Satakunta (Finnish and Swedish); Uusimaa (Finnish), Nyland (Swedish) [Newland]; Varsinais-Suomi (Finnish), Egentliga Finland (Swedish) [Southwest Finland]

Independence: 6 December 1917 (from Russia)

National holiday: Independence Day, 6 December (1917)

Constitution: *history:* previous 1906, 1919; latest drafted 17 June 1997, approved by Parliament 11 June 1999, entered into force 1 March 2000
amendments: proposed by Parliament; passage normally requires simple majority vote in two readings in the first parliamentary session and at least two-thirds majority vote in a single reading by the newly elected Parliament; proposals declared "urgent" by five-sixths of Parliament members can be passed by at least two-thirds majority vote in the first parliamentary session only; amended several times, last in 2012

Legal system: civil law system based on the Swedish model

International law organization participation: accepts compulsory ICJ jurisdiction with reservations; accepts ICCt jurisdiction

Citizenship: citizenship by birth: no
citizenship by descent only: at least one parent must be a citizen of Finland
dual citizenship recognized: yes
residency requirement for naturalization: 6 years

Suffrage: 18 years of age; universal

Executive branch: *chief of state:* President Sauli NIINISTO (since 1 March 2012)
head of government: Prime Minister Sanna MARIN (since 10 December 2019)
cabinet: Council of State or Valtioneuvosto appointed by the president, responsible to Parliament
elections/appointments: president directly elected by absolute majority popular vote in 2 rounds if needed for a 6-year term (eligible for a second term); election last held on 28 January 2018 (next to be held in January 2024); prime minister appointed by Parliament
election results: Sauli NIINISTO reelected president; percent of vote Sauli NIINISTO (independent) 62.7%, Pekka HAAVISTO (Vihr) 12.4%, Laura HUHTASAARI (PS) 6.9%, Paavo VAYRYNEN (independent) 6.2%, Matti VANHANEN (Kesk) 4.1%, other 7.7%

Legislative branch: *description:* unicameral Parliament or Eduskunta (200 seats; 199 members directly elected in single- and multi-seat

constituencies by proportional representation vote and 1 member in the province of Aland directly elected by simple majority vote; members serve 4-year terms) (e.g. 201 9)
elections: last held on 14 April 2019 (next to be held on April 2023) (e.g. 2019)
election results: percent of vote by party/coalition—SDP 17.7%, Finn Party 17.5%, Kok 17.0%. Centre Party 13.8%, Green League 11.5%, Left Alliance 8.2%; seats by party/coalition -SDP 40, Finn Party 39, Kok 38, Centre Party 31, Green League 20, Left Alliance 16; composition men 107, women 93, percent of women 46.5% (e.g. 2019)

Judicial branch: *highest courts:* Supreme Court or Korkein Oikeus (consists of the court president and 18 judges); Supreme Administrative Court (consists of 21 judges, including the court president and organized into 3 chambers); note— Finland has a dual judicial system—courts with civil and criminal jurisdiction and administrative courts with jurisdiction for litigation between individuals and administrative organs of the state and communities
judge selection and term of office: Supreme Court and Supreme Administrative Court judges appointed by the president of the republic; judges serve until mandatory retirement at age 68
subordinate courts: 6 Courts of Appeal; 8 regional administrative courts; 27 district courts; special courts for issues relating to markets, labor, insurance, impeachment, land, tenancy, and water rights

Political parties and leaders:
Aland Coalition (a coalition of several political parties on the Aland Islands)
Center Party or Kesk [Katri KULMUNI]
Christian Democrats or KD [Sari ESSAYAH]
Finns Party or PS [Jussi HALLA-AHO]
Green League or Vihr [Pekka HAAVISTO]
Left Alliance or Vas [Li ANDERSSON]
National Coalition Party or Kok [Petteri ORPO]
Social Democratic Party or SDP [Antti RINNE]
Swedish People's Party or SFP [Anna-Maja HENRIKSSON]

International organization participation: ADB (nonregional member), AfDB (nonregional member), Arctic Council, Australia Group, BIS, CBSS, CD, CE, CERN, EAPC, EBRD, ECB, EIB, EITI (implementing country), EMU, ESA, EU, FAO, FATF, G-9, IADB, IAEA, IBRD, ICAO, ICC (national committees), ICCt, ICRM, IDA, IEA, IFAD, IFC, IFRCS, IHO, ILO, IMF, IMO, IMSO, Interpol, IOC, IOM, IPU, ISO, ITSO, ITU, ITUC (NGOs), MIGA, MINUSMA, NC, NEA, NIB, NSG, OAS (observer), OECD, OPCW, OSCE, Pacific Alliance (observer), Paris Club, PCA, PFP, Schengen Convention, UN, UNCTAD, UNESCO, UNHCR, UNIDO, UNIFIL, UNMIL, UNMOGIP, UNRWA, UNTSO, UPU, WCO, WFTU (NGOs), WHO, WIPO, WMO, WTO, ZC

Diplomatic representation in the US: *chief of mission:* Ambassador Mikko Tapani HAUTALA (since 17 September 2020)

chancery: 3301 Massachusetts Avenue NW, Washington, DC 20008
telephone: [1] (202) 298-5800
FAX: [1] (202) 298-6030
consulate(s) general: Los Angeles, New York

Diplomatic representation from the US: *chief of mission:* Ambassador Robert "Bob" Frank PENCE (since 24 May 2018)
telephone: [358] (9) 6162-50
embassy: Itainen Puistotie 14B, 00140 Helsinki
mailing address: APO AE 09723
FAX: [358] (9) 6162-5135

Flag description: white with a blue cross extending to the edges of the flag; the vertical part of the cross is shifted to the hoist side in the style of the Dannebrog (Danish flag); the blue represents the thousands of lakes scattered across the country, while the white is for the snow that covers the land in winter

National symbol(s): *lion; national colors:* blue, white

National anthem: *name:* "Maamme" (Our Land)
lyrics/music: Johan Ludvig RUNEBERG/Fredrik PACIUS
note: in use since 1848; although never officially adopted by law, the anthem has been popular since it was first sung by a student group in 1848; Estonia's anthem uses the same melody as that of Finland
0:00/ 0:46

ECONOMY

Economy—overview: Finland has a highly industrialized, largely free-market economy with per capita GDP almost as high as that of Austria and the Netherlands and slightly above that of Germany and Belgium. Trade is important, with exports accounting for over one-third of GDP in recent years. The government is open to, and actively takes steps to attract, foreign direct investment.

Finland is historically competitive in manufacturing, particularly in the wood, metals, engineering, telecommunications, and electronics industries. Finland excels in export of technology as well as promotion of startups in the information and communications technology, gaming, cleantech, and biotechnology sectors. Except for timber and several minerals, Finland depends on imports of raw materials, energy, and some components for manufactured goods. Because of the cold climate, agricultural development is limited to maintaining self-sufficiency in basic products. Forestry, an important export industry, provides a secondary occupation for the rural population.

Finland had been one of the best performing economies within the EU before 2009 and its banks and financial markets avoided the worst of global financial crisis. However, the world slowdown hit exports and domestic demand hard in that year, causing Finland's economy to contract from 2012 to 2014. The recession affected general government finances and the debt ratio. The economy returned to growth in 2016, posting a

1.9% GDP increase before growing an estimated 3.3% in 2017, supported by a strong increase in investment, private consumption, and net exports. Finnish economists expect GDP to grow a rate of 2-3% in the next few years.

Finland's main challenges will be reducing high labor costs and boosting demand for its exports. In June 2016, the government enacted a Competitiveness Pact aimed at reducing labor costs, increasing hours worked, and introducing more flexibility into the wage bargaining system. As a result, wage growth was nearly flat in 2017. The Government was also seeking to reform the health care system and social services. In the long term, Finland must address a rapidly aging population and decreasing productivity in traditional industries that threaten competitiveness, fiscal sustainability, and economic growth.

GDP (purchasing power parity): $244.9 billion (2017 est.)
$238.2 billion (2016 est.)
$232.4 billion (2015 est.)
note: data are in 2017 dollars
country comparison to the world: 62

GDP (official exchange rate): $252.8 billion (2017 est.)

GDP—real growth rate:
1.15% (2019 est.)
1.52% (2018 est.)
3.27% (2017 est.)
country comparison to the world: 168

GDP—per capita (PPP): $44,500 (2017 est.)
$43,400 (2016 est.)
$42,500 (2015 est.)
note: data are in 2017 dollars
country comparison to the world: 38

Gross national saving:
23.3% of GDP (2017 est.)
21.7% of GDP (2016 est.)
20% of GDP (2015 est.)
country comparison to the world: 73

GDP—composition, by end use:
household consumption: 54.4% (2017 est.)
government consumption: 22.9% (2017 est.)
investment in fixed capital: 22.1% (2017 est.)
investment in inventories: 0.4% (2017 est.)
exports of goods and services: 38.5% (2017 est.)
imports of goods and services: -38.2% (2017 est.)

GDP—composition, by sector of origin:
agriculture: 2.7% (2017 est.)
industry: 28.2% (2017 est.)
services: 69.1% (2017 est.)

Agriculture—products: barley, wheat, sugar beets, potatoes; dairy cattle; fish

Industries: metals and metal products, electronics, machinery and scientific instruments, shipbuilding, pulp and paper, foodstuffs, chemicals, textiles, clothing

Industrial production growth rate: 6.2% (2017 est.)
country comparison to the world: 39

Labor force: 2.52 million (2020 est.)
country comparison to the world: 113

Labor force—by occupation: agriculture: 4%

industry: 20.7%
services: 75.3% (2017 est.)

Unemployment rate:
6.63% (2019 est.)
7.38% (2018 est.)
country comparison to the world: 105

Household income or consumption by percentage share: lowest 10%: 6.7%
highest 10%: 45.2% (2013)

Budget: revenues: 134.2 billion (2017 est.)
expenditures: 135.6 billion (2017 est.)
note: Central Government Budget data; these numbers represent a significant reduction from previous official reporting

Taxes and other revenues: 53.1% (of GDP) (2017 est.)
country comparison to the world: 11

Budget surplus (+) or deficit (-): -0.6% (of GDP) (2017 est.)
country comparison to the world: 64

Public debt:
61.3% of GDP (2017 est.)
62.9% of GDP (2016 est.)
note: data cover general government debt and include debt instruments issued (or owned) by government entities other than the treasury; the data include treasury debt held by foreign entities; the data include debt issued by subnational entities, as well as intragovernmental debt; intragovernmental debt consists of treasury borrowings from surpluses in the social funds, such as for retirement, medical care, and unemployment; debt instruments for the social funds are not sold at public auctions
country comparison to the world: 72

Fiscal year: calendar year

Inflation rate (consumer prices):
0.8% (2017 est.)
0.4% (2016 est.)
country comparison to the world: 41

Current account balance:
-$603 million (2019 est.)
-$4.908 billion (2018 est.)
country comparison to the world: 127

Exports:
$67.73 billion (2017 est.)
$51.9 billion (2016 est.)
country comparison to the world: 43

Exports—partners: Germany 14.2%, Sweden 10.1%, US 7%, Netherlands 6.8%, China 5.7%, Russia 5.7%, UK 4.5% (2017)

Exports—commodities: electrical and optical equipment, machinery, transport equipment, paper and pulp, chemicals, basic metals; timber

Imports:
$65.26 billion (2017 est.)
$58.18 billion (2016 est.)
country comparison to the world: 47

Imports—commodities: foodstuffs, petroleum and petroleum products, chemicals, transport equipment, iron and steel, machinery, computers, electronic industry products, textile yarn and fabrics, grains

Imports—partners: Germany 17.7%, Sweden 15.8%, Russia 13.1%, Netherlands 8.7% (2017)

Reserves of foreign exchange and gold:
$10.51 billion (31 December 2017 est.)
$11.2 billion (31 December 2016 est.)
country comparison to the world: 73

Debt—external:
$150.6 billion (31 December 2016 est.)
$147.8 billion (31 December 2015 est.)
country comparison to the world: 42

Exchange rates: euros (EUR) per US dollar –
0.885 (2017 est.)
0.903 (2016 est.)
0.9214 (2015 est.)
0.885 (2014 est.)
0.7634 (2013 est.)

ENERGY

Electricity access: electrification—total population: 100% (2020)

Electricity—production: 66.54 billion kWh (2016 est.)
country comparison to the world: 43

Electricity—consumption: 82.79 billion kWh (2016 est.)
country comparison to the world: 35

Electricity—exports: 3.159 billion kWh (2016 est.)
country comparison to the world: 43

Electricity—imports: 22.11 billion kWh (2016 est.)
country comparison to the world: 8

Electricity—installed generating capacity: 16.27 million kW (2016 est.)
country comparison to the world: 50

Electricity—from fossil fuels: 41% of total installed capacity (2016 est.)
country comparison to the world: 165

Electricity—from nuclear fuels: 17% of total installed capacity (2017 est.)
country comparison to the world: 12

Electricity—from hydroelectric plants: 20% of total installed capacity (2017 est.)
country comparison to the world: 88

Electricity—from other renewable sources: 23% of total installed capacity (2017 est.)
country comparison to the world: 30

Crude oil—production: 0 bbl/day (2018 est.)
country comparison to the world: 138

Crude oil—exports: 0 bbl/day (2015 est.)
country comparison to the world: 125

Crude oil—imports: 236,700 bbl/day (2017 est.)
country comparison to the world: 27

Crude oil—proved reserves: 0 bbl (1 January 2018 est.)
country comparison to the world: 133

Refined petroleum products—production: 310,600 bbl/day (2017 est.)
country comparison to the world: 40

Refined petroleum products—consumption: 217,100 bbl/day (2017 est.)

country comparison to the world: 55

Refined petroleum products—exports: 166.200 bbl/day (2017 est.)
country comparison to the world: 34

Refined petroleum products—imports: 122.200 bbl/day (2017 est.)
country comparison to the world: 48

Natural gas—production: 0 cu m (2017 est.)
country comparison to the world: 134

Natural gas—consumption: 2.35 billion cu m (2017 est.)
country comparison to the world: 81

Natural gas—exports: 4 million cu m (2017 est.)
country comparison to the world: 54

Natural gas—imports: 2.322 billion cu m (2017 est.)
country comparison to the world: 49

Natural gas—proved reserves: NA cu m (1 January 2016 est.)

Carbon dioxide emissions from consumption of energy: 46.01 million Mt (2017 est.)
country comparison to the world: 64

COMMUNICATIONS

Telephones—fixed lines: *total subscriptions:* 269,980
subscriptions per 100 inhabitants: 4.86 (2019 est.)
country comparison to the world: 113

Telephones—mobile cellular: *total subscriptions:* 7,179,481
subscriptions per 100 inhabitants: 129.24 (2019 est.)
country comparison to the world: 103

Telecommunication systems: *general assessment:* excellent service; one of the most progressive in Europe; one of the highest broadband and mobile penetrations rates in the region; for 2025 and 2030 FttP (fiber to the home) and DOCSIS3.1 (new generation of cable services for high speed connections) technologies; subscribers are migrating from 3G to LTE and 5G networks; astute regulatory measures have encouraged market competition and company investment (2020)
domestic: fixed-line 5 per 100 subscription and 129 per 100 mobile-cellular (2019)
international: country code—358; landing points for Botnia, BCS North-1 & 2, SFL, SFS-4, C-Lion1, Eastern Lights,
Baltic Sea Submarine Cable, FEC, and EESF-2 & 3 submarine cables that provide links to many Finland points, Estonia, Sweden, Germany, and Russia; satellite earth stations—access to Intelsat transmission service via a Swedish satellite earth station, 1 Inmarsat (Atlantic and Indian Ocean regions); note—Finland shares the Inmarsat earth station with the other Nordic countries (Denmark, Iceland, Norway, and Sweden) (2019)
note: the COVID-19 outbreak is negatively impacting telecommunications production and supply chains globally; consumer spending on telecom devices and services has also slowed due to the pandemic's effect on economies worldwide; overall progress towards improvements in all facets of the telecom industry—mobile, fixed-line, broadband, submarine cable and satellite—has moderated

Broadcast media: a mix of 3 publicly operated TV stations and numerous privately owned TV stations; several free and special-interest payTV channels; cable and satellite multi-channel subscription services are available; all TV signals are broadcast digitally; Internet television, such as Netflix and others, is available; public broadcasting maintains a network of 13 national and 25 regional radio stations; a large number of private radio broadcasters and access to Internet radio

Internet country code: .fi
note—Aland Islands assigned .ax

Internet users: total: 4,922,163
percent of population: 88.89% (July 2018 est.)
country comparison to the world: 84

Broadband—fixed subscriptions: *total:* 1.737 million
subscriptions per 100 inhabitants: 31 (2018 est.)
country comparison to the world: 58

TRANSPORTATION

National air transport system: *number of registered air carriers:* 3 (2020)
inventory of registered aircraft operated by air carriers: 77
annual passenger traffic on registered air carriers: 13,364,839 (2018)
annual freight traffic on registered air carriers: 957.64 million mt-km (2018)

Civil aircraft registration country code prefix: OH (2016)

Airports: 148 (2013)
country comparison to the world: 37

Airports—with paved runways: *total:* 74 (2017)
over 3,047 m: 3 (2017)
2,438 to 3,047 m: 26 (2017)
1,524 to 2,437 m: 10 (2017)
914 to 1,523 m: 21 (2017)
under 914 m: 14 (2017)

Airports—with unpaved runways: *total:* 74 (2013)
914 to 1,523 m: 3 (2013)
under 914 m: 71 (2013)

Pipelines: 1288 km gas transmission pipes, 1976 km distribution pipes (2016)

Railways: *total:* 5,926 km (2016)
broad gauge: 5,926 km 1.524-m gauge (3,270 km electrified) (2016)
country comparison to the world: 32

Roadways: *total:* 454,000 km (2012)
highways: 78,000 km (50,000 paved, including 700 km of expressways; 28,000 unpaved) (2012)
private and forest roads: 350,000 km (2012)
urban: 26,000 km (2012)
country comparison to the world: 16

Waterways: 8,000 km (includes Saimaa Canal system of 3,577 km; southern part leased from Russia; water transport used frequently in the summer and widely replaced with sledges on the ice in winter; there are 187,888 lakes in Finland that cover 31,500 km); Finland also maintains 8,200 km of coastal fairways (2013)
country comparison to the world: 17

Merchant marine: *total:* 269
by type: bulk carrier 8, container ship 1, general cargo 79, oil tanker 4, other 177 (2019)
country comparison to the world: 56

Ports and terminals: *major seaport(s):* Helsinki, Kotka, Naantali, Porvoo, Raahe, Rauma

MILITARY AND SECURITY

Military and security forces: *Finnish Defense Forces (FDF):* Army (Maavoimat), Navy (Merivoimat), Air Force (Ilmavoimat); Ministry of the Interior: Border Guard (Rajavartiolaitos) (2019)
note: the Border Guard becomes part of the FDF in wartime

Military expenditures:
1.5% of GDP (2019)
1.4% of GDP (2018)
1.4% of GDP (2017)
1.4% of GDP (2016)
1.5% of GDP (2015)
country comparison to the world: 80

Military and security service personnel strengths: estimates for the size of the Finnish Defense Forces (FDF) vary; approximately 23,000 total active duty personnel (16,000 Army; 4,000 Navy; 3,000 Air Force) (2019 est.)

Military equipment inventories and acquisitions: the inventory of the Finnish Defense Forces consists of a wide mix of mostly modern Western and domestically-produced weapons systems, as well as a limited quantity of Soviet-era equipment, particularly artillery and armored personnel carriers; since 2010, France, Italy, the Netherlands, Norway, and the US are the leading foreign suppliers of armaments to Finland; the Finish defense industry produces a variety of military equipment, including wheeled armored vehicles and naval vessels (2019 est.)

Military deployments: 200 Lebanon (UNIFIL) (2020)

Military service age and obligation: at age 18, all Finnish men are obligated to serve 6-12 months of service within a branch of the military or the Border Guard, and women may volunteer for service; after completing their initial conscript obligation, individuals enter the reserves and remain eligible for mobilization until the age of 60 (2019)

TRANSNATIONAL ISSUES

Disputes—international: various groups in Finland advocate restoration of Karelia and other areas ceded to the former Soviet Union, but the Finnish Government asserts no territorial demands

Refugees and internally displaced persons: *refugees (country of origin):* 8,862 (Iraq) (2019)
stateless persons: 2,801 (2019)

343

FRANCE

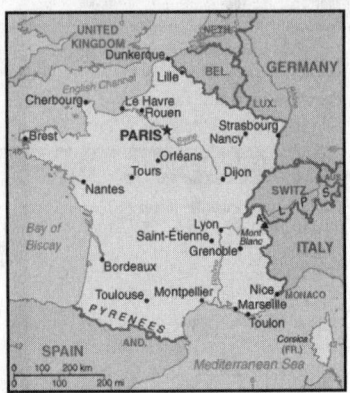

INTRODUCTION

Background: France today is one of the most modern countries in the world and is a leader among European nations. It plays an influential global role as a permanent member of the United Nations Security Council, NATO, the G-7, the G-20, the EU, and other multilateral organizations. France rejoined NATO's integrated military command structure in 2009, reversing DE GAULLE's 1966 decision to withdraw French forces from NATO. Since 1958, it has constructed a hybrid presidential-parliamentary governing system resistant to the instabilities experienced in earlier, more purely parliamentary administrations. In recent decades, its reconciliation and cooperation with Germany have proved central to the economic integration of Europe, including the introduction of a common currency, the euro, in January 1999. In the early 21st century, five French overseas entities—French Guiana, Guadeloupe, Martinique, Mayotte, and Reunion—became French regions and were made part of France proper.

GEOGRAPHY

Location: metropolitan France: Western Europe, bordering the Bay of Biscay and English Channel, between Belgium and Spain, southeast of the UK; bordering the Mediterranean Sea, between Italy and Spain;

French Guiana: Northern South America, bordering the North Atlantic Ocean, between Brazil and Suriname;

Guadeloupe: Caribbean, islands between the Caribbean Sea and the North Atlantic Ocean, southeast of Puerto Rico;

Martinique: Caribbean, island between the Caribbean Sea and North Atlantic Ocean, north of Trinidad and Tobago;

Mayotte: Southern Indian Ocean, island in the Mozambique Channel, about halfway between northern Madagascar and northern Mozambique;

Reunion: Southern Africa, island in the Indian Ocean, east of Madagascar

Geographic coordinates: metropolitan France: 46 00 N, 2 00 E;

French Guiana: 4 00 N, 53 00 W;

Guadeloupe: 16 15 N, 61 35 W;

Martinique: 14 40 N, 61 00 W;

Mayotte: 12 50 S, 45 10 E;

Reunion: 21 06 S, 55 36 E

Map references: *metropolitan France:* Europe;

French Guiana: South America;

Guadeloupe: Central America and the Caribbean;

Martinique: Central America and the Caribbean;

Mayotte: Africa;

Reunion: World

Area: *total:* 643,801 sq km ; 551,500 sq km (metropolitan France)
land: 640,427 sq km ; 549,970 sq km (metropolitan France)
water: 3,374 sq km ; 1,530 sq km (metropolitan France)
note: the first numbers include the overseas regions of French Guiana, Guadeloupe, Martinique, Mayotte, and Reunion
country comparison to the world: 44

Area—comparative: slightly more than four times the size of Georgia; slightly less than the size of Texas

Land boundaries: *border countries (8):* Andorra 55 km, Belgium 556 km, Germany 418 km, Italy 476 km, Luxembourg 69 km, Monaco 6 km, Spain 646 km, Switzerland 525 km

metropolitan France—total: 2751

French Guiana—total: 1205

Coastline: 4,853 km

metropolitan France: 3,427 km

Maritime claims:
territorial sea: 12 nm
exclusive economic zone: 200 nm (does not apply to the Mediterranean Sea)
contiguous zone: 24 nm
continental shelf: 200-m depth or to the depth of exploitation

Climate: metropolitan France: generally cool winters and mild summers, but mild winters and hot summers along the Mediterranean; occasional strong, cold, dry, north-to-northwesterly wind known as the mistral;

French Guiana: tropical; hot, humid; little seasonal temperature variation;

Guadeloupe and Martinique: subtropical tempered by trade winds; moderately high humidity; rainy

season (June to October); vulnerable to devastating cyclones (hurricanes) every eight years on average;

Mayotte: tropical; marine; hot, humid, rainy season during northeastern monsoon (November to May); dry season is cooler (May to November);

Reunion: tropical, but temperature moderates with elevation; cool and dry (May to November), hot and rainy (November to April)

Terrain: metropolitan France: mostly flat plains or gently rolling hills in north and west; remainder is mountainous, especially Pyrenees in south, Alps in east;

French Guiana: low-lying coastal plains rising to hills and small mountains;

Guadeloupe: Basse-Terre is volcanic in origin with interior mountains; Grande-Terre is low limestone formation; most of the seven other islands are volcanic in origin;

Martinique: mountainous with indented coastline; dormant volcano;

Mayotte: generally undulating, with deep ravines and ancient volcanic peaks;

Reunion: mostly rugged and mountainous; fertile lowlands along coast

Elevation: *mean elevation:* 375 m
lowest point: Rhone River delta -2 m
highest point: Mont Blanc 4,810
note: to assess the possible effects of climate change on the ice and snow cap of Mont Blanc, its surface and peak have been extensively measured in recent years; these new peak measurements have exceeded the traditional height of 4,807 m and have varied between 4,808 m and 4,811 m; the actual rock summit is 4,792 m and is 40 m away from the ice- covered summit

Natural resources: metropolitan France: coal, iron ore, bauxite, zinc, uranium, antimony, arsenic, potash, feldspar, fluorspar, gypsum, timber, arable land, fish, French Guiana, gold deposits, petroleum, kaolin, niobium, tantalum, clay

Land use: *agricultural land:* 52.7% (2011 est.)
arable land: 33.4% (2011 est.) / permanent crops: 1.8% (2011 est.) / permanent pasture: 17.5% (2011 est.)
forest: 29.2% (2011 est.)
other: 18.1% (2011 est.)

Irrigated land: 26,420 sq km 26,950 sq km (2012)
metropolitan France: 26,000 sq km (2012)

Population distribution: much of the population is concentrated in the north and southeast; although there are many urban agglomerations throughout the country, Paris is by far the largest city, with Lyon ranked a distant second

Natural hazards: metropolitan France: flooding; avalanches; midwinter windstorms; drought; forest fires in south near the Mediterranean;

overseas departments: hurricanes (cyclones); flooding;

volcanism: Montagne Pelee (1,394 m) on the island of Martinique in the Caribbean is the most active volcano of the Lesser Antilles arc, it last erupted in 1932; a catastrophic eruption in May 1902 destroyed the city of St. Pierre, killing an estimated 30,000 people;; La Soufriere (1,467 m) on the island of Guadeloupe in the Caribbean last erupted from July 1976 to March 1977;; these volcanoes are part of the volcanic island arc of the Lesser Antilles that extends from Saba in the north to Grenada in the south

Environment—current issues: some forest damage from acid rain; air pollution from industrial and vehicle emissions; water pollution from urban wastes, agricultural runoff

Environment—international agreements: party to: Air Pollution, Air Pollution-Nitrogen Oxides, Air Pollution-Persistent Organic Pollutants, Air Pollution-Sulfur 85, Air Pollution-Sulfur 94, Air Pollution-Volatile Organic Compounds, Antarctic-Environmental Protocol, Antarctic-Marine Living Resources, Antarctic Seals, Antarctic Treaty, Biodiversity, Climate Change, Climate Change-Kyoto Protocol, Desertification, Endangered Species, Hazardous Wastes, Law of the Sea, Marine Dumping, Marine Life Conservation, Ozone Layer Protection, Ship Pollution, Tropical Timber 83, Tropical Timber 94, Wetlands, Whaling

signed, but not ratified: none of the selected agreements

Geography—note: largest West European nation; most major French rivers—the Meuse, Seine, Loire, Charente, Dordogne, and Garonne—flow northward or westward into the Atlantic Ocean, only the Rhone flows southward into the Mediterranean Sea

PEOPLE AND SOCIETY

Population: 67,848,156 (July 2020 est.)
note: the above figure is for metropolitan France and five overseas regions; the metropolitan France population is 62,814,233
country comparison to the world: 21

Nationality: noun: Frenchman(men), Frenchwoman(women)

adjective: French

Ethnic groups: Celtic and Latin with Teutonic, Slavic, North African, Indochinese, Basque minorities
note: overseas departments: black, white, mulatto, East Indian, Chinese, Amerindian

Languages: French (official) 100%, declining regional dialects and languages (Provencal, Breton, Alsatian, Corsican, Catalan, Basque, Flemish, Occitan, Picard)
note: overseas departments: French, Creole patois, Mahorian (a Swahili dialect)

Religions: Christian (overwhelmingly Roman Catholic) 63-66%, Muslim 7-9%, Buddhist 0.5-0.75%, Jewish 0.5-0.75%, other 0.5-1.0%, none 23-28% (2015 est.)
note: France maintains a tradition of secularism and has not officially collected data on religious affiliation since the 1872 national census, which complicates assessments of France's religious composition; an 1872 law prohibiting state authorities from collecting data on individuals' ethnicity or religious beliefs was reaffirmed by a 1978 law emphasizing the prohibition of the collection or exploitation of personal data revealing an individual's race, ethnicity, or political, philosophical, or religious opinions; a 1905 law codified France's separation of church and state

Age structure: 0-14 years: 18.36% (male 6,368,767/female 6,085,318)
15-24 years: 11.88% (male 4,122,981/female 3,938,938)
25-54 years: 36.83% (male 12,619,649/female 12,366,120)
55-64 years: 12.47% (male 4,085,564/female 4,376,272)
65 years and over: 20.46% (male 6,029,303/female 7,855,244) (2020 est.)

Dependency ratios: total dependency ratio: 62.4
youth dependency ratio: 28.7
elderly dependency ratio: 33.7
potential support ratio: 3 (2020 est.)

Median age: total: 41.7 years
male: 40 years
female: 43.4 years (2020 est.)
country comparison to the world: 43

Population growth rate: 0.35% (2020 est.)
country comparison to the world: 169

Birth rate: 11.9 births/1,000 population (2020 est.)
country comparison to the world: 165

Death rate: 9.6 deaths/1,000 population (2020 est.)
country comparison to the world: 42

Net migration rate: 1.1 migrant(s)/1,000 population (2020 est.)
country comparison to the world: 59

Population distribution: much of the population is concentrated in the north and southeast; although there are many urban agglomerations throughout the country, Paris is by far the largest city, with Lyon ranked a distant second

Urbanization: urban population: 81% of total population (2020)
rate of urbanization: 0.72% annual rate of change (2015-20 est.)
total population growth rate v. urban population growth rate, 2000-2030:

Major urban areas—Population: 11.017 million PARIS (capital), 1.719 million Lyon, 1.608 million Marseille-Aix-en-Provence, 1.063 million Lille, 1.024 million Toulouse, 969,000 Bordeaux (2020)

Sex ratio: at birth: 1.05 male(s)/female
0-14 years: 1.05 male(s)/female
15-24 years: 1.05 male(s)/female
25-54 years: 1.02 male(s)/female
55-64 years: 0.93 male(s)/female
65 years and over: 0.77 male(s)/female
total population: 0.96 male(s)/female (2020 est.)

Mother's mean age at first birth: 28.7 years (2018 est.)

Maternal mortality rate: 8 deaths/100,000 live births (2017 est.)
country comparison to the world: 151

Infant mortality rate: total: 3.2 deaths/1,000 live births
male: 3.5 deaths/1,000 live births
female: 2.9 deaths/1,000 live births (2020 est.)
country comparison to the world: 211
Life expectancy at birth: total population: 82.2 years
male: 79.1 years
female: 85.4 years (2020 est.)
country comparison to the world: 19

Total fertility rate: 2.06 children born/woman (2020 est.)
country comparison to the world: 104

Contraceptive prevalence rate: 78.4% (2010/11)

Drinking water source:
improved:
urban: 100% of population
rural: 100% of population
total: 100% of population
unimproved:
urban: 0% of population
rural: 0% of population
total: 0% of population (2017 est.)

Current Health Expenditure: 11.3% (2017)

Physicians density: 3.26 physicians/1,000 population (2017)

Hospital bed density: 6 beds/1,000 population (2017)

Sanitation facility access:
improved:
urban: 100% of population
rural: 100% of population
total: 100% of population
unimproved:
urban: 0% of population
rural: 0% of population
total: 0% of population (2017 est.)

HIV/AIDS—adult prevalence rate: 0.3% (2019 est.)
country comparison to the world: 86

HIV/AIDS—people living with HIV/AIDS: 190,000 (2019 est.)
country comparison to the world: 31

HIV/AIDS—deaths: <500 (2019 est.)

Major infectious diseases:
note: widespread ongoing transmission of a respiratory illness caused by the novel coronavirus (COVID-19) is occurring throughout France; as of 10 November 2020, France has reported a total of 1,714,361 cases of COVID-19 or 26,264 cumulative cases of COVID-19 per 1 million population with 610 cumulative deaths per 1 million population

Obesity—adult prevalence rate: 21.6% (2016)
country comparison to the world: 87

Education expenditures: 5.4% of GDP (2016)
country comparison to the world: 42

School life expectancy (primary to tertiary education): *total:* 16 years
male: 16 years
female: 16 years (201 8)

Unemployment, youth ages 15-24: *total:* 20.8%
male: 21.4%
female: 20% (2018 est.)
country comparison to the world: 61

GOVERNMENT

Country name: *conventional long form:* French Republic
conventional short form: France
local long form: Republique francaise
local short form: France
etymology: name derives from the Latin "Francia" meaning "Land of the Franks"; the Franks were a group of Germanic tribes located along the middle and lower Rhine River in the 3rd century A.D. who merged with Gallic-Roman populations in succeeding centuries and to whom they passed on their name

Government type: semi-presidential republic

Capital: *name:* Paris
geographic coordinates: 48 52 N, 2 20 E
time difference: UTC + 1 (6 hours ahead of Washington, DC, during Standard Time)
daylight saving time: +1hr, begins last Sunday in March; ends last Sunday in October
note: applies to metropolitan France only; for its overseas regions the time difference is UTC-4 for Guadeloupe and Martinique, UTC-3 for French Guiana, UTC+3 for Mayotte, and UTC+4 for Reunion
etymology: name derives from the Parisii, a Celtic tribe that inhabited the area from the 3rd century B.C., but who were conquered by the Romans in the 1st century B.C.; the Celtic settlement became the Roman town of Lutetia Parisiorum (Lutetia of the Parisii); over subsequent centuries it became Parisium and then just Paris

Administrative divisions: 18 regions (regions, singular—region); Auvergne-Rhone-Alpes, Bourgogne-Franche-Comte (Burgundy-Free County), Bretagne (Brittany), Centre-Val de Loire (Center-Loire Valley), Corse (Corsica), Grand Est (Grand East), Guadeloupe, Guyane (French Guiana), Hauts-de-France (Upper France), Ile-de-France, Martinique, Mayotte, Normandie (Normandy), Nouvelle-Aquitaine (New Aquitaine), Occitanie (Occitania), Pays de la Loire (Lands of the Loire), Provence-Alpes-Cote d'Azur, Reunion
note: France is divided into 13 metropolitan regions (including the "collectivity" of Corse or Corsica) and 5 overseas regions (French Guiana, Guadeloupe, Martinique, Mayotte, and Reunion) and is subdivided into 96 metropolitan departments and 5 overseas departments (which are the same as the overseas regions)

Dependent areas: Clipperton Island, French Polynesia, French Southern and Antarctic Lands, New Caledonia, Saint Barthelemy, Saint Martin, Saint Pierre and Miquelon, Wallis and Futuna
note: the US Government does not recognize claims to Antarctica; New Caledonia has been considered a "sui generis" collectivity of France since 1998, a unique status falling between that of an independent country and a French overseas department

Independence: no official date of independence: 486 (Frankish tribes unified under Merovingian kingship); 10 August 843 (Western Francia established from the division of the Carolingian Empire); 14 July 1789 (French monarchy overthrown); 22 September 1792 (First French Republic founded); 4 October 1958 (Fifth French Republic established)

National holiday: Fete de la Federation, 14 July (1790); note—although often incorrectly referred to as Bastille Day, the celebration actually commemorates the holiday held on the first anniversary of the storming of the Bastille (on 14 July 1789) and the establishment of a constitutional monarchy; other names for the holiday are Fete Nationale (National Holiday) and quatorze juillet (14th of July)

Constitution: *history:* many previous; latest effective 4 October 1958
amendments: proposed by the president of the republic (upon recommendation of the prime minister and Parliament) or by Parliament; proposals submitted by Parliament members require passage by both houses followed by approval in a referendum; passage of proposals submitted by the government can bypass a referendum if submitted by the president to Parliament and passed by at least three-fifths majority vote by Parliament's National Assembly; amended many times, last in 2008; note—in May 2018, the prime minister submitted a bill to the National Assembly to amend several provisions of the constitution

Legal system: civil law; review of administrative but not legislative acts

International law organization participation: has not submitted an ICJ jurisdiction declaration; accepts ICCt jurisdiction

Citizenship: citizenship by birth: no
citizenship by descent only: at least one parent must be a citizen of France
dual citizenship recognized: yes
residency requirement for naturalization: 5 years

Suffrage: 18 years of age; universal

Executive branch: *chief of state:* President Emmanuel MACRON (since 14 May 2017)
head of government: Prime Minister Jean CASTEX (since 3 July 2020)
cabinet: Council of Ministers appointed by the president at the suggestion of the prime minister
elections/appointments: president directly elected by absolute majority popular vote in 2 rounds if needed for a 5-year term (eligible for a second term); election last held on 23 April with a runoff on 7 May 2017 (next to be held in April 2022); prime minister appointed by the president

election results: Emmanuel MACRON elected president in second round; percent of vote in first round—Emmanuel MACRON (EM) 24.%, Marine LE PEN (FN) 21.3%, Francois FILLON (LR) 20.%, Jean-Luc MELENCHON (FI) 19.6%, Benoit HAMON (PS) 6.4%, other 8.7%; percent of vote in second round—MACRON 66.1%, LE PEN 33.9%

Legislative branch: *description:* bicameral Parliament or Parlement consists of:
Senate or Senat (348 seats—328 for metropolitan France and overseas departments and regions of Guadeloupe, Martinique, French Guiana, Reunion, and Mayotte, 2 for New Caledonia, 2 for French Polynesia, 1 for Saint-Pierre and Miquelon, 1 for Saint-Barthelemy, 1 for Saint-Martin, 1 for Wallis and Futuna, and 12 for French nationals abroad; members indirectly elected by departmental electoral colleges using absolute majority vote in 2 rounds if needed for departments with 1-3 members and proportional representation vote in departments with 4 or more members; members serve 6-year terms with one-half of the membership renewed every 3 years)
National Assembly or Assemblee Nationale (577 seats—556 for metropolitan France, 10 for overseas departments, and 11 for citizens abroad; members directly elected by absolute majority vote in 2 rounds if needed to serve 5-year terms)
elections: Senate—last held on 24 September 2017 (next to be held on 24 September 2020)
National Assembly—last held on 11 and 18 June 2017 (next to be held in June 2022)
election results: Senate—percent of vote by party—NA; seats by political caucus (party or group of parties)—LR 144, PS 73, UC 51.
   LREM 23, RDSE 22, CRCE 16, RTLI 13, other 6; composition—men 246, women 102, percent of women 29.3%
   National Assembly—percent of vote by party first round—LREM 28.2%, LR 15.8%. FN 13.2%, FI 11%, PS 7.4%, other 24.4%; percent of vote by party second round—LREM 43.1%, LR 22.2%, FN 8.8%, MoDEM 6.1%, PS 5.7%. FI 4.9%, other 9.2%; seats by political caucus (party or group of parties)—LREM 306, LR 104, MoDEM 46, UDI/ Agir 29, PS 29, UDI 18,
   FI 17, Liberties and Territories 16, PCF 16, other 14; composition—men 349, women 228, percent of women 39.5%; note—total Parliament percent of women 35.7%

Judicial branch: *highest courts:* Court of Cassation or Cour de Cassation (consists of the court president, 6 divisional presiding judges, 120 trial judges, and 70 deputy judges organized into 6 divisions—3 civil, 1 commercial, 1 labor, and 1 criminal); Constitutional Council (consists of 9 members)
judge selection and term of office: Court of Cassation judges appointed by the president of the republic from nominations from the High Council of the Judiciary, presided over by the Court of Cassation and 15 appointed members; judges appointed for life; Constitutional Council members—3 appointed by the president of the republic and 3 each by the National Assembly and Senate

presidents; members serve 9-year, non-renewable terms with one-third of the membership renewed every 3 years

subordinate courts: appellate courts or Cour d'Appel; regional courts or Tribunal de Grande Instance; first instance courts or Tribunal d'instance; administrative courts

note: in April 2018, the French Government announced its intention to reform the country's judicial system

Political parties and leaders:
Presidential majority Parties [Edouard PHILIPPE]
Democratic Movement or MoDem [Francois BAYROU]
La Republique en Marche! or LREM [Richard FERRAND]
Movement of Progressives or MDP Robert HUE]
Parliamentary right Parties [Francois BAROIN]
Hunting, Fishing, Nature and Tradition or CPNT [Eddie PUYJAION]
The Republicans or LR [Annie GENEVARD]
Union of Democrats and Independents or UDI [Jean-Christophe CAMBADELIS]
Parliamentary left Parties [Bernard CAZENEUVE]
Sociatlist Party or PS [Jean-Christophe CAMBADEMAND]
Radical Party of the Left or PRG [Sylvia PINEL]
Citizen and Republican Movement or MRC [Jean-Luc LAURENT]
Martinican Progressive Party or PPM [Aiem CESAIRE]
Debout la France or DLF [Nicolas DUPONT-AIGNAN]
Ecology Democracy Solidarity or EDS [Paula FORTEZA, Matthieu ORPHELIN (splinter party formed in May 2020 by defectors of LREM)]
Europe Ecologists—the Greens or EELV [David CORMAND]
French Communist Party or PCF [Pierre LAURENT]
La France Insoumise or FI [Jean-Luc MELENCHONLIS]
National Front or FN [Marine LE PEN]

International organization participation: ADB (nonregional member), AfDB (nonregional member), Arctic Council (observer), Australia Group, BDEAC, BIS, BSEC (observer), CBSS (observer), CE, CERN, EAPC, EBRD, ECB, EIB, EITI (implementing country), EMU, ESA, EU, FAO, FATF, FZ, G-5, G-7, G-8, G-10, G-20, IADB, IAEA, IBRD, ICAO, ICC (national committees), ICCt, ICRM, IDA, IEA, IFAD, IFC, IFRCS, IGAD (partners), IHO, ILO, IMF, IMO, IMSO, InOC, Interpol, IOC, IOM, IPU, ISO, ITSO, ITU, ITUC (NGOs), MIGA, MINURSO, MINUSMA, MINUSTAH, MONUSCO, NATO, NEA, NSG, OAS (observer), OECD, OIF, OPCW, OSCE, Pacific Alliance (observer), Paris Club, PCA, PIF (partner), Schengen Convention, SELEC (observer), SPC, UN, UNCTAD, UNESCO, UNHCR, UNIDO, UNIFIL, Union Latina, UNMIL, UNOCI, UNRWA, UN Security Council (permanent), UNTSO, UNWTO, UPU, WCO, WFTU (NGOs), WHO, WIPO, WMO, WTO, ZC

Diplomatic representation in the US: *chief of mission:* Ambassador Philippe ETIENNE (since 8 July 2019)
chancery: 4101 Reservoir Road NW, Washington, DC 20007
telephone: [1] (202) 944-6000
FAX: [1] (202) 944-6166
consulate(s) general: Atlanta, Boston, Chicago, Houston, Los Angeles, Miami, New Orleans, New York, San Francisco, Washington, DC

Diplomatic representation from the US: *chief of mission:* Ambassador Jamie D. McCOURT (since 18 December 2017); note—also accredited to Monaco
telephone: [33] (1) 43-12-22-22
embassy: 2 Avenue Gabriel, 75008 Paris
mailing address: PSC 116, APO AE 09777
FAX: [33] (1) 42 66 97 83
consulate(s) general: Marseille, Strasbourg
consulate(s): Bordeaux, Lyon, Rennes

Flag description: three equal vertical bands of blue (hoist side), white, and red; known as the "Le drapeau tricolore" (French Tricolor), the origin of the flag dates to 1790 and the French Revolution when the "ancient French color" of white was combined with the blue and red colors of the Parisian militia; the official flag for all French dependent areas
note: the design and/or colors are similar to a number of other flags, including those of Belgium, Chad, Cote d'Ivoire, Ireland, Italy, Luxembourg, and Netherlands

National symbol(s): Gallic rooster, fleur-de-lis, Marianne (female personification); national colors: blue, white, red

National anthem: *name:* "La Marseillaise" (The Song of Marseille)
lyrics/music: Claude-Joseph ROUGET de Lisle
note: adopted 1795, restored 1870; originally known as "Chant de Guerre pour l'Armee du Rhin" (War Song for the Army of the Rhine), the National Guard of Marseille made the song famous by singing it while marching into Paris in 1792 during the French Revolutionary Wars
0:00/ 1:19

ECONOMY

Economy—overview: The French economy is diversified across all sectors. The government has partially or fully privatized many large companies, including Air France, France Telecom, Renault, and Thales. However, the government maintains a strong presence in some sectors, particularly power, public transport, and defense industries. France is the most visited country in the world with 89 million foreign tourists in 2017. France's leaders remain committed to a capitalism in which they maintain social equity by means of laws, tax policies, and social spending that mitigate economic inequality.

France's real GDP grew by 1.9% in 2017, up from 1.2% the year before. The unemployment rate (including overseas territories) increased from 7.8% in 2008 to 10.2% in 2015, before falling to 9.0% in 2017. Youth unemployment in metropolitan France decreased from 24.6% in the fourth quarter of 2014 to 20.6% in the fourth quarter of 2017.

France's public finances have historically been strained by high spending and low growth. In 2017, the budget deficit improved to 2.7% of GDP, bringing it in compliance with the EU-mandated 3% deficit target. Meanwhile, France's public debt rose from 89.5% of GDP in 2012 to 97% in 2017.

Since entering office in May 2017, President Emmanuel MACRON launched a series of economic reforms to improve competitiveness and boost economic growth. President MACRON campaigned on reforming France's labor code and in late 2017 implemented a range of reforms to increase flexibility in the labor market by making it easier for firms to hire and fire and simplifying negotiations between employers and employees. In addition to labor reforms, President MACRON's 2018 budget cuts public spending, taxes, and social security contributions to spur private investment and increase purchasing power. The government plans to gradually reduce corporate tax rate for businesses from 33.3% to 25% by 2022.

GDP (purchasing power parity): $2.856 trillion (2017 est.)
$2.791 trillion (2016 est.)
$2.761 trillion (2015 est.)
note: data are in 2017 dollars
country comparison to the world: 10

GDP (official exchange rate):
$2.588 trillion (2017 est.)

GDP—real growth rate:
1.49% (2019 est.)
1.81% (2018 est.)
2.42% (2017 est.)
country comparison to the world: 155

GDP—per capita (PPP): $44,100 (2017 est.)
$43,200 (2016 est.)
$42,900 (2015 est.)
note: data are in 2017 dollars
country comparison to the world: 40

Gross national saving:
22.9% of GDP (2017 est.)
21.9% of GDP (2016 est.)
22.3% of GDP (2015 est.)
country comparison to the world: 77

GDP—composition, by end use:
household consumption: 54.1% (2017 est.)
government consumption: 23.6% (2017 est.)
investment in fixed capital: 22.5% (2017 est.)
investment in inventories: 0.9% (2017 est.)
exports of goods and services: 30.9% (2017 est.)
imports of goods and services: -32% (2017 est.)

GDP—composition, by sector of origin:
agriculture: 1.7% (2017 est.)
industry: 19.5% (2017 est.)
services: 78.8% (2017 est.)

Agriculture—products: wheat, cereals, sugar beets, potatoes, wine grapes; beef, dairy products; fish

Industries: machinery, chemicals, automobiles, metallurgy, aircraft, electronics; textiles, food processing; tourism

Industrial production growth rate: 2% (2017 est.)
country comparison to the world: 130

Labor force:
27.742 million (2020 est.)
country comparison to the world: 18

Labor force—by occupation: *agriculture:* 2.8%
(2016 est.)
industry: 20% (2016 est.)
services: 77.2% (2016 est.)

Unemployment rate:
8.12% (2019 est.)
8.69% (2018 est.)
note: includes overseas territories
country comparison to the world: 128

Population below poverty line: 14.2% (2015 est.)

Household income or consumption by percentage share: *lowest 10%:* 3.6%
highest 10%: 25.4% (2013)

Budget: *revenues:* 1.392 trillion (2017 est.)
expenditures: 1.459 trillion (2017 est.)

Taxes and other revenues: 53.8% (of GDP) (2017 est.)
country comparison to the world: 10

Budget surplus (+) or deficit (-): -2.6% (of GDP) (2017 est.)
country comparison to the world: 116

Public debt:
96.8% of GDP (2017 est.)
96.6% of GDP (2016 est.)
note: data cover general government debt and include debt instruments issued (or owned) by government entities other than the treasury; the data include treasury debt held by foreign entities; the data include debt issued by subnational entities, as well as intragovernmental debt; intragovernmental debt consists of treasury borrowings from surpluses in the social funds, such as for retirement, medical care, and unemployment; debt instruments for the social funds are not sold at public auctions
country comparison to the world: 20

Fiscal year: calendar year

Inflation rate (consumer prices):
1.2% (2017 est.)
0.3% (2016 est.)
country comparison to the world: 64

Current account balance:
-$18.102 billion (2019 est.)
-$16.02 billion (2018 est.)
country comparison to the world: 197

Exports:
$549.9 billion (2017 est.)
$507 billion (2016 est.)
country comparison to the world: 7

Exports—partners: Germany 14.8%, Spain 7.7%, Italy 7.5%, US 7.2%, Belgium 7%, UK 6.7% (2017)

Exports—commodities: machinery and transportation equipment, aircraft, plastics, chemicals, pharmaceutical products, iron and steel, beverages

Imports:
$601.7 billion (2017 est.)

$536.7 billion (2016 est.)
country comparison to the world: 7

Imports—commodities: machinery and equipment, vehicles, crude oil, aircraft, plastics, chemicals

Imports—partners: Germany 18.5%, Belgium 10.2%, Netherlands 8.3%, Italy 7.9%, Spain 7.1%, UK 5.3%, US 5.2%, China 5.1% (2017)

Reserves of foreign exchange and gold:
$156.4 billion (31 December 2017 est.)
$138.2 billion (31 December 2015 est.)
country comparison to the world: 15

Debt—external:
$5.36 trillion (31 March 2016 est.)
$5.25 trillion (31 March 2015 est.)
country comparison to the world: 3

Exchange rates: euros (EUR) per US dollar –
0.885 (2017 est.)
0.903 (2016 est.)
0.9214 (2015 est.)
0.885 (2014 est.)
0.7634 (2013 est.)

ENERGY

Electricity access: electrification—total population: 100% (2020)

Electricity—production: 529.1 billion kWh (2016 est.)
country comparison to the world: 9

Electricity—consumption: 450.8 billion kWh (2016 est.)
country comparison to the world: 10

Electricity—exports: 61.41 billion kWh (2016 est.)
country comparison to the world: 3

Electricity—imports: 19.9 billion kWh (2016 est.)
country comparison to the world: 10

Electricity—installed generating capacity: 130.8 million kW (2016 est.)
country comparison to the world: 9

Electricity—from fossil fuels: 17% of total installed capacity (2016 est.)
country comparison to the world: 198

Electricity—from nuclear fuels: 50% of total installed capacity (2017 est.)
country comparison to the world: 1

Electricity—from hydroelectric plants: 15% of total installed capacity (2017 est.)
country comparison to the world: 102

Electricity—from other renewable sources: 19% of total installed capacity (2017 est.)
country comparison to the world: 42

Crude oil—production: 16,000 bbl/day (2018 est.)
country comparison to the world: 71

Crude oil—exports: 0 bbl/day (2015 est.)
country comparison to the world: 126

Crude oil—imports: 1.147 million bbl/day (2017 est.)
country comparison to the world: 9

Crude oil—proved reserves: 65.97 million bbl (1 January 2018 est.)

country comparison to the world: 75

Refined petroleum products—production: 1.311 million bbl/day (2017 est.)
country comparison to the world: 15

Refined petroleum products—consumption: 1.705 million bbl/day (2017 est.)
country comparison to the world: 13

Refined petroleum products—exports: 440,600 bbl/day (2017 est.)
country comparison to the world: 19

Refined petroleum products—imports: 886,800 bbl/day (2017 est.)
country comparison to the world: 8

Natural gas—production: 16.99 million cu m (2017 est.)
country comparison to the world: 90

Natural gas—consumption: 41.88 billion cu m (2017 est.)
country comparison to the world: 24

Natural gas—exports: 6.031 billion cu m (2017 est.)
country comparison to the world: 26

Natural gas—imports: 48.59 billion cu m (2017 est.)
country comparison to the world: 10

Natural gas—proved reserves: 8.41 billion cu m (1 January 2018 est.)
country comparison to the world: 81

Carbon dioxide emissions from consumption of energy: 341.2 million Mt (2017 est.)
country comparison to the world: 22

COMMUNICATIONS

Telephones—fixed lines: *total subscriptions.*
39,234,941
subscriptions per 100 inhabitants: 58.03 (2019 est.)
country comparison to the world: 4

Telephones—mobile cellular: *total subscriptions:*
74,791,818
subscriptions per 100 inhabitants: 110.62 (2019 est.)
country comparison to the world: 22

Telecommunication systems: *general assessment:* one of the largest mobile phone markets in Europe, worth 13 billion annually; LTE has universal coverage with extensive 5G launching any day, one of the largest broadband subscriber bases in Europe; regional govt. and telecom companies have invested in higher bandwidth w/ fiber infrastructure improvements, an investment more than 20 billion euros (2020)
domestic: 58 per 100 persons for fixed-line and 111 per 100 for mobile-cellular subscriptions (2019)
international: country code—33; landing points for Circe South, TAT-14, INGRID, FLAG Atlantic-1, Apollo, HUGO, IFC-1, ACE, SeaMeWe-3 & 4, Dunant, Africa-1, AAE-1, Atlas Offshore, Hawk, IMEWE, Med Cable, PEACE Cable, and TE North/TGN-Eurasia/SEACOM/Alexandros/Medex submarine cables providing links throughout Europe, Asia, Australia, the

Middle East, Southeast Asia, Africa and US; satellite earth stations—more than 3 (2 Intelsat (with total of 5 antennas—2 for Indian Ocean and 3 for Atlantic Ocean), NA Eutelsat, 1 Inmarsat—Atlantic Ocean region); HF radiotelephone communications with more than 20 countries (2019)

overseas departments: country codes: French Guiana—594; landing points for Ella Link, Kanawa, Americas II to South America, Europe, Caribbean and US; Guadeloupe—590; landing points for GCN, Southern Caribbean Fiber, and ECFS around the Caribbean and US; Martinique—596; landing points for Americas II, ECFS, and Southern Caribbean Fiber to South America, US and around the Caribbean; Mayotte—262; landing points for FLY-LION3 and LION2 to East Africa and East African Islands in Indian Ocean; Reunion—262; landing points for SAFE, METISS, and LION submarine cables to Asia, South and East Africa, Southeast Asia and nearby Indian Ocean Island countries of Mauritius, and Madagascar (2019)

note: the COVID-19 outbreak is negatively impacting telecommunications production and supply chains globally; consumer spending on telecom devices and services has also slowed due to the pandemic's effect on economies worldwide; overall progress towards improvements in all facets of the telecom industry—mobile, fixed-line, broadband, submarine cable and satellite—has moderated

Broadcast media: a mix of both publicly operated and privately owned TV stations; state-owned France television stations operate 4 networks, one of which is a network of regional stations, and has part-interest in several thematic cable/satellite channels and international channels; a large number of privately owned regional and local TV stations; multi-channel satellite and cable services provide a large number of channels; public broadcaster Radio France operates 7 national networks, a series of regional networks, and operates services for overseas territories and foreign audiences; Radio France Internationale, under the Ministry of Foreign Affairs, is a leading international broadcaster; a large number of commercial FM stations, with many of them consolidating into commercial networks

Internet country code: metropolitan France—.fr; French Guiana—.gf; Guadeloupe—.gp; Martinique—.mq; Mayotte—.yt; Reunion—.re

Internet users: *total:* 55,265,718
percent of population: 82.04% (July 2018 est.)
country comparison to the world: 16

Broadband—fixed subscriptions: total: 29.1 million
subscriptions per 100 inhabitants: 43 (2018 est.)
country comparison to the world: 7

TRANSPORTATION

National air transport system: *number of registered air carriers:* 19 (2020)
inventory of registered aircraft operated by air carriers: 553

annual passenger traffic on registered air carriers: 70,188,028 (2018)
annual freight traffic on registered air carriers: 4,443,790,000 mt-km (2018)

Civil aircraft registration country code prefix: F (2016)

Airports: 464 (2013)
country comparison to the world: 15

Airports—with paved runways: *total:* 294 (2017)
over 3,047 m: 14 (2017)
2,438 to 3,047 m: 25 (2017)
1,524 to 2,437 m: 97 (2017)
914 to 1,523 m: 83 (2017)
under 914 m: 75 (2017)

Airports—with unpaved runways: *total:* 170 (2013)
1,524 to 2,437 m: 1 (2013)
914 to 1,523 m: 64 (2013)
under 914 m: 105 (2013)

Heliports: 1 (2013)

Pipelines: 15322 km gas, 2939 km oil, 5084 km refined products (2013)

Railways: *total:* 29,640 km (2014)
standard gauge: 29,473 km 1.435-m gauge (15,561 km electrified) (2014)
narrow gauge: 167 km 1.000-m gauge (63 km electrified) (2014)
country comparison to the world: 10

Roadways: *total:* 1,053,215 km (2011)
urban: 654,201 km (2011)
non-urban: 399,014 km (2011)
country comparison to the world: 7

Waterways: metropolitan France: 8,501 km (1,621 km navigable by craft up to 3,000 metric tons) (2010)

Merchant marine: *total:* 552
by type: bulk carrier 25, general cargo 53, oil tanker 29, other 445 (2019)
note: includes Monaco
country comparison to the world: 40

Ports and terminals: *major seaport(s):* Brest, Calais, Dunkerque, Le Havre, Marseille, Nantes, *container port(s) (TEUs):* Le Havre (2,870,000) (2017)

LNG terminal(s) (import): Fos Cavaou, Fos Tonkin, Montoir de Bretagne
river port(s): Paris, Rouen (Seine)
cruise/ferry port(s): Calais, Cherbourg, Le Havre Strasbourg (Rhine) Bordeaux (Garronne)

Transportation—note: begun in 1988 and completed in 1994, the Channel Tunnel (nicknamed the Chunnel) is a 50.5-km (31.4-mi) rail tunnel beneath the English Channel at the Strait of Dover that runs from Folkestone, Kent, England to Coquelles, Pas-de- Calais in northern France; it is the only fixed link between the island of Great Britain and mainland Europe

MILITARY AND SECURITY

Military and security forces: Army (Armee de Terre; includes Foreign Legion), Navy (Marine Nationale), Air Force (Armee de l'Air (AdlA);

includes Air Defense), National Guard (Reserves), National Gendarmerie (paramilitary police force that is a branch of the Armed Forces but under the jurisdiction of the Ministry of the Interior; also has additional duties to the Ministry of Defense and the Ministry of Justice) (2019)

Military expenditures:
1.84% of GDP (2019 est.)
1.82% of GDP (2018)
1.78% of GDP (2017)
1.79% of GDP (2016)
1.78% of GDP (2015)
country comparison to the world: 58

Military and security service personnel strengths: the French military has approximately 205,000 active duty troops (114,500 Army; 35,000 Navy; 40,500 Air Force; 15,000 other, such as joint staffs, medical service, etc.); approximately 100,000 National Gendarmerie (2019)

Military equipment inventories and acquisitions: the French military's inventory consists almost entirely of domestically-produced weapons systems, including some jointly- produced with other European countries; there is a limited mix of armaments from other Western countries, particularly the US; since 2010, the US is the leading foreign supplier of military hardware to France; France has a defense industry capable of manufacturing the full spectrum of air, land, and naval military weapons systems (2019 est.)

Military deployments: 5,100 Burkina Faso/Chad/Mali/Niger (Operation Barkhane); 900 Cote D'Ivoire; 1,450 Djibouti; 300 Baltics (NATO); 2,000 French Guyana; 900 French Polynesia; 1,000 French West Indies; 350 Gabon; est. 500 Middle East (Iraq/Jordan/Syria); 780 Lebanon (UNIFIL); 1,400-1,500 New Caledonia; 1,700 Reunion Island; 350 Senegal; 650 United Arab Emirates; note—France has been a contributing member of the EuroCorps since 1992 (2020)

Military service age and obligation: 18-25 years of age for male and female voluntary military service; no conscription; 1-year service obligation; women serve in noncombat posts (2013)

TERRORISM

Terrorist group(s): Islamic Revolutionary Guard Corps/Qods Force; Islamic State of Iraq and ash-Sham; al-Qa'ida 2019)
note: details about the history, aims, leadership, organization, areas of operation, tactics, targets, weapons, size, and sources of support of the group(s) appear(s) in Appendix-T

TRANSNATIONAL ISSUES

Disputes—international: Madagascar claims the French territories of Bassas da India, Europa Island, Glorioso Islands, and Juan de Nova Island; Comoros claims Mayotte; Mauritius claims Tromelin Island; territorial dispute between Suriname and the French overseas department of French Guiana; France asserts a territorial claim in Antarctica (Adelie Land); France and Vanuatu claim Matthew and Hunter Islands, east of New Caledonia

Refugees and internally displaced persons: *refugees (country of origin):* 24,293 (Afghanistan), 23,821 (Sri Lanka), 18,473 (Sudan), 18,244 (Syria), 17,512 (Democratic Republic of the Congo), 16,412 (Russia), 14,141 (Serbia and Kosovo), 11,863 (Turkey), 11,038 (Guinea), 11,021 (Cambodia), 8,829 (Iraq), 7,735 (Vietnam), 6,918 (China), 6,464 (Laos), 6,372 (Eritrea), 6,156 (Bangladesh), 5,675 (Mauritania), 5,652 (Cote d'Ivoire), 5,169 (Mali) (2019) *stateless persons:* 1,521 (2019)

Illicit drugs: metropolitan France: transshipment point for South American cocaine, Southwest Asian heroin, and European synthetics;

French Guiana: small amount of marijuana grown for local consumption; minor transshipment point to Europe;

Martinique: transshipment point for cocaine and marijuana bound for the US and Europe

FRENCH POLYNESIA

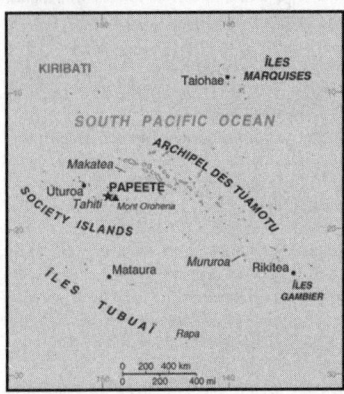

INTRODUCTION

Background: The French annexed various Polynesian island groups during the 19th century. In 1966, the French Government began testing nuclear weapons on the uninhabited Mururoa Atoll; following mounting opposition, the tests were moved underground in 1975. In September 1995, France stirred up widespread protests by resuming nuclear testing after a three-year moratorium. The tests were halted in January 1996. In recent years, French Polynesia's autonomy has been considerably expanded.

GEOGRAPHY

Location: Oceania, five archipelagoes (Archipel des Tuamotu, Iles Gambier, Iles Marquises, Iles Tubuai, Society Islands) in the South Pacific Ocean about halfway between South America and Australia

Geographic coordinates: 15 00 S, 140 00 W

Map references: Oceania

Area: *total:* 4,167 sq km (118 islands and atolls; 67 are inhabited) *land:* 3,827 sq km *water:* 340 sq km *country comparison to the world:* 175

Area—comparative: slightly less than one-third the size of Connecticut

Land boundaries: 0 km

Coastline: 2,525 km

Maritime claims: *territorial sea:* 12 nm *exclusive economic zone:* 200 nm

Climate: tropical, but moderate

Terrain: mixture of rugged high islands and low islands with reefs

Elevation: *lowest point:* Pacific Ocean 0 m *highest point:* Mont Orohena 2,241 m

Natural resources: timber, fish, cobalt, hydropower

Land use: *agricultural land:* 12.5% (2011 est.) *arable land:* 0.7% (2011 est.) / permanent crops: 6.3% (2011 est.) / permanent pasture: 5.5% (2011 est.) *forest:* 43.7% (2011 est.) *other:* 43.8% (2011 est.)

Irrigated land: 10 sq km (2012)

Population distribution: the majority of the population lives in the Society Islands, one of five archipelagos that includes the most populous island—Tahiti—with approximately 70% of the nation's population

Natural hazards: occasional cyclonic storms in January

Environment—current issues: sea level rise; extreme weather events (cyclones, storms, and tsunamis producing floods, landslides, erosion, and reef damage); droughts; fresh water scarcity

Geography—note: includes five archipelagoes: four volcanic (Iles Gambier, Iles Marquises, Iles Tubuai, Society Islands) and one coral (Archipel des Tuamotu); the Tuamotu Archipelago forms the largest group of atolls in the world—78 in total, 48 inhabited; Makatea in the Tuamotu Archipelago is one of the three great phosphate rock islands in the Pacific Ocean—the others are Banaba (Ocean Island) in Kiribati and Nauru

PEOPLE AND SOCIETY

Population: 295,121 (July 2020 est.) *country comparison to the world:* 181

Nationality: *noun:* French Polynesian(s) *adjective:* French Polynesian

Ethnic groups: Polynesian 78%, Chinese 12%, local French 6%, metropolitan French 4%

Languages: French (official) 70%, Polynesian (official) 28.2%, other 1.8% (2012 est.)

Religions: Protestant 54%, Roman Catholic 30%, other 10%, no religion 6%

Age structure: *0-14 years:* 21.69% (male 32,920/female 31,100) *15-24 years:* 14.72% (male 22,640/female 20,793) *25-54 years:* 44.24% (male 66,921/female 63,636) *55-64 years:* 10.31% (male 15,610/female 14,823) *65 years and over:* 9.04% (male 12,854/female 13,824) (2020 est.)

Dependency ratios: *total dependency ratio:* 45.5 *youth dependency ratio:* 32.3 *elderly dependency ratio:* 13.2 *potential support ratio:* 7.6 (2020 est.)

Median age: *total:* 33.3 years *male:* 33 years *female:* 33.5 years (2020 est.) *country comparison to the world:* 98

Population growth rate: 0.79% (2020 est.) *country comparison to the world:* 128

Birth rate: 14 births/1,000 population (2020 est.) *country comparison to the world:* 135

Death rate: 5.5 deaths/1,000 population (2020 est.) *country comparison to the world:* 183

Net migration rate: -0.7 migrant(s)/1,000 population (2020 est.) *country comparison to the world:* 131

Population distribution: the majority of the population lives in the Society Islands, one of five archipelagos that includes the most populous island—Tahiti—with approximately 70% of the nation's population

Urbanization: *urban population:* 62% of total population (2020) *rate of urbanization:* 1.01% annual rate of change (2015-20 est.) total population growth rate v. urban population growth rate, 2000-2030:

Major urban areas—Population: 136,000 PAPEETE (capital) (2018)

Sex ratio: *at birth:* 1.05 male(s)/female *0-14 years:* 1.06 male(s)/female *15-24 years:* 1.09 male(s)/female *25-54 years:* 1.05 male(s)/female *55-64 years:* 1.05 male(s)/female *65 years and over:* 0.93 male(s)/female *total population:* 1.05 male(s)/female (2020 est.)

Infant mortality rate: *total:* 4.5 deaths/1,000 live births
male: 5 deaths/1,000 live births
female: 4 deaths/1,000 live births (2020 est.)
country comparison to the world: 182
Life expectancy at birth: total population: 77.9 years
male: 75.6 years
female: 80.4 years (2020 est.)
country comparison to the world: 71

Total fertility rate: 1.83 children born/woman (2020 est.)
country comparison to the world: 145

Drinking water source:
improved:
total: 100% of population
unimproved:
total: 0% of population (2017 est.)

Physicians density: 2.13 physicians/1,000 population (2009)

Sanitation facility access:
improved:
total: 96.9% of population
unimproved:
total: 3.1% of population (2017 est.)

HIV/AIDS—adult prevalence rate: NA

HIV/AIDS—people living with HIV/AIDS: NA

HIV/AIDS—deaths: NA

Major infectious diseases: *degree of risk:* high (2020)
food or waterborne diseases: bacterial diarrhea
vectorborne diseases: malaria

Unemployment, youth ages 15-24: *total:* 56.7%
male: 54.5%
female: 59.7% (2012 est.)
country comparison to the world: 1

GOVERNMENT

Country name: *conventional long form:* Overseas Lands of French Polynesia
conventional short form: French Polynesia
local long form: Pays d'outre-mer de la Polynesie Francaise
local short form: Polynesie Francaise
former: Establishments in Oceania, French Establishments in Oceania
etymology: the term "Polynesia" is an 18th-century construct composed of two Greek words, "poly" (many) and "nesoi" (islands), and refers to the more than 1,000 islands scattered over the central and southern Pacific Ocean

Dependency status: overseas country of France; note—overseas territory of France from 1946-2003; overseas collectivity of France since 2003, though it is often referred to as an overseas country due to its degree of autonomy

Government type: parliamentary democracy (Assembly of French Polynesia); an overseas collectivity of France

Capital: *name:* Papeete (located on Tahiti)
geographic coordinates: 17 32 S, 149 34 W

time difference: UTC-10 (5 hours behind Washington, DC, during Standard Time)
etymology: the name means "water basket" and refers to the fact that the islanders originally used calabashes enclosed in baskets to fetch water at a spring in the area

Administrative divisions: 5 administrative subdivisions (subdivisions administratives, singular—subdivision administrative): Iles Australes (Austral Islands), Iles du Vent (Windward Islands), Iles Marquises (Marquesas Islands), Iles Sous-le-Vent (Leeward Islands), Iles Tuamotu-Gambier; note—the Leeward Islands and the Windward Islands together make up the Society Islands (Iles de la Societe)

Independence: none (overseas lands of France)

National holiday: Fete de la Federation, 14 July (1790); note—the local holiday is Internal Autonomy Day, 29 June (1880)

Constitution: *history:* 4 October 1958 (French Constitution)
amendments: French constitution amendment procedures apply

Legal system: the laws of France, where applicable, apply

Citizenship: see France

Suffrage: 18 years of age; universal

Executive branch: *chief of state:* President Emmanuel MACRON (since 14 May 2017), represented by High Commissioner of the Republic Dominique SORAIN (since 10 July 2019)
head of government: President of French Polynesia Edouard FRITCH (since 12 September 2014)
cabinet: Council of Ministers approved by the Assembly from a list of its members submitted by the president
elections/appointments: French president directly elected by absolute majority popular vote in 2 rounds if needed for a 5-year term (eligible for a second term); high commissioner appointed by the French president on the advice of the French Ministry of Interior; French Polynesia president indirectly elected by Assembly of French Polynesia for a 5-year term (no term limits)

Legislative branch: *description:* unicameral Assembly of French Polynesia or Assemblée de la Polynésie française (57 seats; elections held in 2 rounds; in the second round, 38 members directly elected in multi-seat constituencies by a closed-list proportional representation vote; the party receiving the most votes gets an additional 19 seats; members serve 5-year terms)
French Polynesia indirectly elects 2 senators to the French Senate via an electoral college by absolute majority vote for 6year terms with one-half the membership renewed every 3 years and directly elects 3 deputies to the French National Assembly by absolute majority vote in 2 rounds if needed for 5-year terms
elections: Assembly of French Polynesia—last held on 22 April 2018 and 6 May 2018 (next to be held in 2023)

French Senate—last held in September 2017 (next to be held in September 2020)
French National Assembly—last held in 2 rounds on 3 and 17 June 2017 (next to be held in 2022)
election results: Assembly of French Polynesia—percent of vote by party—Tapura Huiraatira 45.1%, Popular Rally 29.3%, Tavini Huiraatira 25.6%; seats by party—Tapura Huiraatira 38, Popular Rally 11, Tavini Huiraatira 8; composition—men 27, women 30, percent of women 52.6%
French Senate—percent of vote by party—NA; seats by party—Popular Rally 1, People's Servant Party 1; composition—men 246, women 102, percent of women 29.3%
French National Assembly—percent of vote by party—NA; seats by party—Tapura Huiraactura 2, Tavini Huiraatura 1; composition—men 353, women 224, percent of women 38.8%; note—total Parliament percent of women 20%

Judicial branch: *highest courts:* Court of Appeal or Cour d'Appel (composition NA); note—appeals beyond the French Polynesia Court of Appeal are heard by the Court of Cassation (in Paris)
judge selection and term of office: judges assigned from France normally for 3 years
subordinate courts: Court of the First Instance or Tribunal de Premiere Instance; Court of Administrative Law or Tribunal Administratif

Political parties and leaders:
A Tia Porinetia [Teva ROHFRITSCH]
Alliance for a New Democracy or ADN (includes The New Star [Philip SCHYLE], This Country is Yours [Nicole BOUTEAU])
New Fatherland Party (Ai'a Api) [Emile VERNAUDON]
Our Home alliance
People's Servant Party (Tavini Huiraatira) [Oscar TEMARU]
Popular Rally (Tahoeraa Huiraatira) [Gaston FLOSSE]
Tapura Huiraatira [Edouard FRITCH]
Tavini Huiraatira [James CHANCELOR]
Union for Democracy alliance or UPD [Oscar TEMARU]

International organization participation: ITUC (NGOs), PIF (associate member), SPC, UPU, WMO

Diplomatic representation in the US: none (overseas lands of France)

Diplomatic representation from the US: none (overseas lands of France)

Flag description: two red horizontal bands encase a wide white band in a 1:2:1 ratio; centered on the white band is a disk with a blue and white wave pattern depicting the sea on the lower half and a gold and white ray pattern depicting the sun on the upper half; a Polynesian canoe rides on the wave pattern; the canoe has a crew of five represented by five stars that symbolize the five island groups; red and white are traditional Polynesian colors
note: identical to the red-white-red flag of Tahiti, the largest and most populous of the islands in French Polynesia, but which has no emblem in the

351

white band; the flag of France is used for official occasions

National symbol(s): outrigger canoe, Tahitian gardenia (Gardenia taitensis) flower; national colors: red, white

National anthem: *name:* "Ia Ora 'O Tahiti Nui" (Long Live Tahiti Nui)

lyrics/music: Maeva BOUGES, Irmine TEHEI, Angele TEROROTUA, Johanna NOUVEAU, Patrick AMARU, Louis MAMATUI, and Jean-Pierre CELESTIN (the compositional group created both the lyrics and music)

note: adopted 1993; serves as a local anthem; as a territory of France, "La Marseillaise" is official (see France)

Government—note: under certain acts of France, French Polynesia has acquired autonomy in all areas except those relating to police, monetary policy, tertiary education, immigration, and defense and foreign affairs; the duties of its president are fashioned after those of the French prime minister

ECONOMY

Economy—overview: Since 1962, when France stationed military personnel in the region, French Polynesia has changed from a subsistence agricultural economy to one in which a high proportion of the work force is either employed by the military or supports the tourist industry. With the halt of French nuclear testing in 1996, the military contribution to the economy fell sharply.

After growing at an average yearly rate of 4.2% from 1997-2007, the economic and financial crisis in 2008 marked French Polynesia's entry into recession. However, since 2014, French Polynesia has shown signs of recovery. Business turnover reached 1.8% year-on-year in September 2016, tourism increased 1.8% in 2015, and GDP grew 2.0% in 2015.

French Polynesia's tourism-dominated service sector accounted for 85% of total value added for the economy in 2012. Tourism employs 17% of the workforce. Pearl farming is the second biggest industry, accounting for 54% of exports in 2015; however, the output has decreased to 12.5 tons – the lowest level since 2008. A small manufacturing sector predominantly processes commodities from French Polynesia's primary sector—8% of total economy in 2012—including agriculture and fishing.

France has agreed to finance infrastructure, marine businesses, and cultural and ecological sites at roughly $80 million per year between 2015 and 2020. Japan, the US, and China are French Polynesia's three largest trade partners.

GDP (purchasing power parity): $5.49 billion (2015 est.)
$5.383 billion (2014 est.)
$6.963 billion (2010 est.)
country comparison to the world: 177

GDP (official exchange rate): $4.795 billion (2015 est.)

GDP—real growth rate:
2% (2015 est.)

-2.7% (2014 est.)
-2.5% (2010 est.)
country comparison to the world: 136

GDP—per capita (PPP): $17,000 (2015 est.)
$20,100 (2014 est.)
$22,700 (2010)
country comparison to the world: 104

GDP—composition, by end use:
household consumption: 66.9% (2014 est.)
government consumption: 33.6% (2014 est.)
investment in fixed capital: 19.4% (2014 est.)
investment in inventories: 0.1% (2014 est.)
exports of goods and services: 17.5% (2014 est.)
imports of goods and services: -37.5% (2014 est.)

GDP—composition, by sector of origin:
agriculture: 2.5% (2009)
industry: 13% (2009)
services: 84.5% (2009)

Agriculture—products: coconuts, vanilla, vegetables, fruits, coffee; poultry, beef, dairy products; fish

Industries: tourism, pearls, agricultural processing, handicrafts, phosphates

Industrial production growth rate: NA

Labor force: 126,300 (2016 est.)
country comparison to the world: 179

Labor force—by occupation: *agriculture:* 13%
industry: 19%
services: 68% (2013 est.)

Unemployment rate:
21.8% (2012)
11.7% (2010)
country comparison to the world: 191

Population below poverty line: 19.7% (2009 est.)

Household income or consumption by percentage share: *lowest 10%:* NA
highest 10%: NA

Budget: *revenues:* 1.891 billion (2012)
expenditures: 1.833 billion (2011)

Taxes and other revenues: 39.4% (of GDP) (2012)
country comparison to the world: 46

Budget surplus (+) or deficit (-): 1.2% (of GDP) (2012)
country comparison to the world: 28

Fiscal year: calendar year

Inflation rate (consumer prices):
0% (2015 est.)
0.3% (2014 est.)
country comparison to the world: 10

Current account balance:
$207.7 million (2014 est.)
$158.8 million (2013 est.)
country comparison to the world: 57

Exports:
$1.245 billion (2014 est.)
$1.168 billion (2013 est.)
country comparison to the world: 152

Exports—partners: Japan 23.1%, Hong Kong 21.5%, Kyrgyzstan 15.9%, US 15.9%, France 12.4% (2017)

Exports—commodities: cultured pearls, coconut products, mother-of-pearl, vanilla, shark meat

Imports:
$2.235 billion (2014 est.)
$2.271 billion (2013 est.)
country comparison to the world: 162

Imports—commodities: fuels, foodstuffs, machinery and equipment

Imports—partners: France 27.9%, South Korea 12.1%, US 10.1%, China 7.3%, NZ 6.7%, Singapore 4.2% (2017)

Debt—external: NA

Exchange rates: Comptoirs Francais du Pacifique francs (XPF) per US dollar -
110.2 (2017 est.)
107.84 (2016 est.)
107.84 (2015 est.)
89.85 (2014 est.)
90.56 (2013 est.)

ENERGY

Electricity access: *electrification—total population:* 100% (2020)

Electricity—production: 677.3 million kWh (2016 est.)
country comparison to the world: 158

Electricity—consumption: 629.9 million kWh (2016 est.)
country comparison to the world: 164

Electricity—exports: 0 kWh (2016 est.)
country comparison to the world: 136

Electricity—imports: 0 kWh (2016 est.)
country comparison to the world: 150

Electricity—installed generating capacity: 253,000 kW (2016 est.)
country comparison to the world: 163

Electricity—from fossil fuels: 70% of total installed capacity (2016 est.)
country comparison to the world: 108

Electricity—from nuclear fuels: 0% of total installed capacity (2017 est.)
country comparison to the world: 92

Electricity—from hydroelectric plants: 19% of total installed capacity (2017 est.)
country comparison to the world: 89

Electricity—from other renewable sources: 11% of total installed capacity (2017 est.)
country comparison to the world: 77

Crude oil—production: 0 bbl/day (2018 est.)
country comparison to the world: 139

Crude oil—exports: 0 bbl/day (2015 est.)
country comparison to the world: 127

Crude oil—imports: 0 bbl/day (2015 est.)
country comparison to the world: 129

Crude oil—proved reserves: 0 bbl (1 January 2018 est.)
country comparison to the world: 134

Refined petroleum products—production: 0 bbl/day (2015 est.)
country comparison to the world: 147

Refined petroleum products—consumption: 6,600 bbl/day (2016 est.)
country comparison to the world: 168

Refined petroleum products—exports: 0 bbl/day (2015 est.)
country comparison to the world: 157

Refined petroleum products—imports: 6,785 bbl/day (2015 est.)
country comparison to the world: 160

Natural gas—production: 0 cu m (2017 est.)
country comparison to the world: 135

Natural gas—consumption: 0 cu m (2017 est.)
country comparison to the world: 148

Natural gas—exports: 0 cu m (2017 est.)
country comparison to the world: 106

Natural gas—imports: 0 cu m (2017 est.)
country comparison to the world: 127

Natural gas—proved reserves: 0 cu m (1 January 2014 est.)
country comparison to the world: 137

Carbon dioxide emissions from consumption of energy: 1.03 million Mt (2017 est.)
country comparison to the world: 168

COMMUNICATIONS

Telephones—fixed lines: *total subscriptions:* 63,769
subscriptions per 100 inhabitants: 21.78 (2019 est.)
country comparison to the world: 152

Telephones—mobile cellular: *total subscriptions:* 305,233
subscriptions per 100 inhabitants: 104.25 (2019 est.)
country comparison to the world: 179

Telecommunication systems: *general assessment:* one of the most advanced telecom infrastructures for the Pacific island region; 85% mobile broadband coverage; 40% of its mobile connections using 3G and the rest using emerging 4G LTE technology; 100% mobile penetration; uses Uplink systems of the Galileo satellite network; and with the launch of the Kacific-1 satellite in 2019, it will allow speedy access to the Internet for Pacific islands (2020)
domestic: fixed-line subscriptions 22 per 100 persons and mobile-cellular density is roughly 104 per 100 persons (2019)
international: country code—689; landing points for the NATITUA, Manatua, and Honotua submarine cables to other

French Polynesian Islands, Cook Islands, Niue, Samoa and US; satellite earth station—1 Intelsat (Pacific Ocean) (2019)
note: the COVID-19 outbreak is negatively impacting telecommunications production and supply chains globally; consumer spending on telecom devices and services has also slowed due to the pandemic's effect on economies worldwide; overall progress towards improvements in all facets of the telecom industry—mobile, fixed-line, broadband, submarine cable and satellite—has moderated

Broadcast media: French public overseas broadcaster Reseau Outre-Mer provides 2 TV channels and 1 radio station; 1 government-owned TV station; a small number of privately owned radio stations (2019)

Internet country code: .pf

Internet users: *total:* 211,101
percent of population: 72.7% (July 2018 est.)
country comparison to the world: 174

Broadband—fixed subscriptions: *total:* 59,790
subscriptions per 100 inhabitants: 21 (2018 est.)
country comparison to the world: 132

TRANSPORTATION

National air transport system: *number of registered air carriers:* 2 (registered in France) (2020)

inventory of registered aircraft operated by air carriers: 19 (registered in France)

Civil aircraft registration country code prefix: F-OH (2016)

Airports: 54 (2013)
country comparison to the world: 86

Airports—with paved runways: *total:* 45 (2017)
over 3,047 m: 2 (2017)
1,524 to 2,437 m: 5 (2017)
914 to 1,523 m: 33 (2017)
under 914 m: 5 (2017)

Airports—with unpaved runways: *total:* 9 (2013)
914 to 1,523 m: 4 (2013)
under 914 m: 5 (2013)

Heliports: 1 (2013)

Roadways: *total:* 2,590 km (1999)
paved: 1,735 km (1999)
unpaved: 855 km (1999)
country comparison to the world: 167

Merchant marine: *total:* 17
by type: general cargo 10, other 7 (2019)
country comparison to the world: 145

Ports and terminals: *major seaport(s):* Papeete

MILITARY AND SECURITY

Military and security forces: no regular military forces (2019)

Military—note: defense is the responsibility of France and France maintains forces in French Polynesia (2019)

TRANSNATIONAL ISSUES

Disputes—international: none

FRENCH SOUTHERN AND ANTARCTIC LANDS

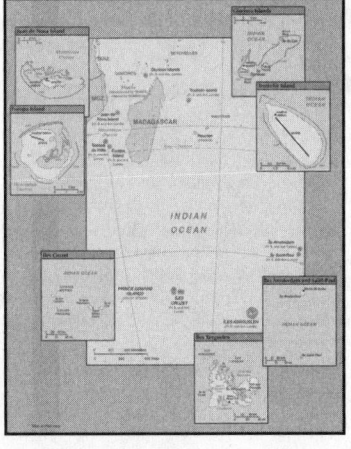

INTRODUCTION

Background: In February 2007, the Iles Eparses became an integral part of the French Southern and Antarctic Lands (TAAF). The Southern Lands are now divided into five administrative districts, two of which are archipelagos, Iles Crozet and Iles Kerguelen; the third is a district composed of two volcanic islands, Ile Saint-Paul and Ile Amsterdam; the fourth, Iles Eparses, consists of five scattered tropical islands around Madagascar. They contain no permanent inhabitants and are visited only by researchers studying the native fauna, scientists at the various scientific stations, fishermen, and military personnel. The fifth district is the Antarctic portion, which consists of "Adelie Land," a thin slice of the Antarctic continent discovered and claimed by the French in 1840.

Ile Amsterdam: Discovered but not named in 1522 by the Spanish, the island subsequently received the appellation of Nieuw Amsterdam from a Dutchman; it was claimed by France in 1843. A short-lived attempt at cattle farming began in 1871. A French meteorological station established on the island in 1949 is still in use.;
Ile Saint Paul: Claimed by France since 1893, the island was a fishing industry center from 1843 to 1914. In 1928, a spiny lobster cannery was established, but when the company went bankrupt in 1931, seven workers were abandoned. Only two survived until 1934 when rescue finally arrived.;
Iles Crozet: A large archipelago formed from the Crozet Plateau, Iles Crozet is divided into two main groups:

L'Occidental (the West), which includes Ile aux Cochons, Ilots des Apotres, Ile des Pingouins, and the reefs Brisants de l'Heroine; and L'Oriental (the East), which includes Ile d'Est and Ile de la Possession (the largest island of the Crozets). Discovered and claimed by France in 1772, the islands were used for seal hunting and as a base for whaling. Originally administered as a dependency of Madagascar, they became part of the TAAF in 1955.;

Iles Kerguelen: This island group, discovered in 1772, consists of one large island (Ile Kerguelen) and about 300 smaller islands. A permanent group of 50 to 100 scientists resides at the main base at Port-aux-Francais.;

Adelie Land: The only non-insular district of the TAAF is the Antarctic claim known as "Adelie Land." The US Government does not recognize it as a French dependency.;

Bassas da India: A French possession since 1897, this atoll is a volcanic rock surrounded by reefs and is awash at high tide.;

Europa Island: This heavily wooded island has been a French possession since 1897; it is the site of a small military garrison that staffs a weather station.;

Glorioso Islands: A French possession since 1892, the Glorioso Islands are composed of two lushly vegetated coral islands (Ile Glorieuse and Ile du Lys) and three rock islets. A military garrison operates a weather and radio station on Ile Glorieuse.;

Juan de Nova Island: Named after a famous 15th-century Spanish navigator and explorer, the island has been a French possession since 1897. It has been exploited for its guano and phosphate. Presently a small military garrison oversees a meteorological station.;

Tromelin Island: First explored by the French in 1776, the island came under the jurisdiction of Reunion in 1814. At present, it serves as a sea turtle sanctuary and is the site of an important meteorological station.

GEOGRAPHY

Location: southeast and east of Africa, islands in the southern Indian Ocean, some near Madagascar and others about equidistant between Africa, Antarctica, and Australia; note—French Southern and Antarctic Lands include Ile Amsterdam, Ile Saint- Paul, Iles Crozet, Iles Kerguelen, Bassas da India, Europa Island, Glorioso Islands, Juan de Nova Island, and Tromelin Island in the southern Indian Ocean, along with the French-claimed sector of Antarctica, "Adelie Land"; the US does not recognize the French claim to "Adelie Land"

Geographic coordinates: Ile Amsterdam (Ile Amsterdam et Ile Saint-Paul): 37 50 S, 77 32 E;
Ile Saint-Paul (Ile Amsterdam et Ile Saint-Paul): 38 72 S, 77 53 E;
Iles Crozet: 46 25 S, 51 00 E;
Iles Kerguelen: 49 15 S, 69 35 E;
Bassas da India (Iles Eparses): 21 30 S, 39 50 E;
Europa Island (Iles Eparses): 22 20 S, 40 22 E;
Glorioso Islands (Iles Eparses): 11 30 S, 47 20 E;
Juan de Nova Island (Iles Eparses): 17 03 S, 42 45 E;
Tromelin Island (Iles Eparses): 15 52 S, 54 25 E

Map references: Antarctic RegionAfrica

Area: Ile Amsterdam (Ile Amsterdam et Ile Saint-Paul): total—55 sq km; land—55 sq km; water—0 sq km
Ile Saint-Paul (Ile Amsterdam et Ile Saint-Paul): total—7 sq km; land—7 sq km; water—0 sq km
Iles Crozet: total—352 sq km; land—352 sq km; water—0 sq km
Iles Kerguelen: total—7,215 sq km; land—7,215 sq km; water—0 sq km
Bassas da India (Iles Eparses): total—80 sq km; land—0.2 sq km; water—79.8 sq km (lagoon)
Europa Island (Iles Eparses): total—28 sq km; land—28 sq km; water—0 sq km
Glorioso Islands (Iles Eparses): total—5 sq km; land—5 sq km; water—0 sq km
Juan de Nova Island (Iles Eparses): total—4.4 sq km; land—4.4 sq km; water—0 sq km
Tromelin Island (Iles Eparses): total—1 sq km; land—1 sq km; water—0 sq km
note: excludes "Adelie Land" claim of about 500,000 sq km in Antarctica that is not recognized by the US

Area—comparative: Ile Amsterdam (Ile Amsterdam et Ile Saint-Paul): less than one-half the size of Washington, DC;
Ile Saint-Paul (Ile Amsterdam et Ile Saint-Paul): more than 10 times the size of the National Mall in Washington, DC;
Iles Crozet: about twice the size of Washington, DC;
Iles Kerguelen: slightly larger than Delaware;
Bassas da India (Iles Eparses): land area about one-third the size of the National Mall in Washington, DC;
Europa Island (Iles Eparses): about one-sixth size of Washington, DC;
Glorioso Islands (Iles Eparses): about eight times the size of the National Mall in Washington, DC;
Juan de Nova Island (Iles Eparses): about seven times the size of the National Mall in Washington, DC;
Tromelin Island (Iles Eparses): about 1.7 times the size of the National Mall in Washington, DC

Land boundaries: *0 km*

Coastline: Ile Amsterdam (Ile Amsterdam et Ile Saint-Paul): 28 km
Ile Saint-Paul (Ile Amsterdam et Ile Saint-Paul):
Iles Kerguelen: 2,800 km
Bassas da India (Iles Eparses): 35.2 km
Europa Island (Iles Eparses): 22.2 km
Glorioso Islands (Iles Eparses): 35.2 km
Juan de Nova Island (Iles Eparses): 24.1 km
Tromelin Island (Iles Eparses): 3.7 km

Maritime claims: *territorial sea:* 12 nm
exclusive economic zone: 200 nm from Iles Kerguelen and Iles Eparses (does not include the rest of French Southern and Antarctic Lands); Juan de Nova Island and Tromelin Island claim a continental shelf of 200-m depth or to the depth of exploitation

Climate: Ile Amsterdam et Ile Saint-Paul: oceanic with persistent westerly winds and high humidity;
Iles Crozet: windy, cold, wet, and cloudy;

Iles Kerguelen: oceanic, cold, overcast, windy;
Iles Eparses: tropical

Terrain: Ile Amsterdam (Ile Amsterdam et Ile Saint-Paul): a volcanic island with steep coastal cliffs; the center floor of the volcano is a large plateau;
Ile Saint-Paul (Ile Amsterdam et Ile Saint-Paul): triangular in shape, the island is the top of a volcano, rocky with steep cliffs on the eastern side; has active thermal springs;
Iles Crozet: a large archipelago formed from the Crozet Plateau is divided into two groups of islands;
Iles Kerguelen: the interior of the large island of Ile Kerguelen is composed of high mountains, hills, valleys, and plains with peninsulas stretching off its coasts;
Bassas da India (Iles Eparses): atoll, awash at high tide; shallow (15 m) lagoon;
Europa Island, Glorioso Islands, Juan de Nova Island: low, flat, and sandy;
Tromelin Island (Iles Eparses): low, flat, sandy; likely volcanic seamount

Elevation: *lowest point:* Indian Ocean 0 m
highest point: Mont de la Dives on Ile Amsterdam (Ile Amsterdam et Ile Saint-Paul) 867 m
highest points throughout the French Southern and Antarctic Lands: unnamed location on Ile Saint-Paul (Ile Amsterdam et Ile Saint-Paul) 272 m; Pic Marion-Dufresne in Iles Crozet 1090 m; Mont Ross in Iles Kerguelen 1850 m; unnamed location on Bassas de India (Iles Eparses) 2.4 m;24 unnamed location on Europa Island (Iles Eparses) 24 m; unnamed location on Glorioso Islands (Iles Eparses) 12 m; unnamed location on Juan de Nova Island (Iles Eparses) 10 m; unnamed location on Tromelin Island (Iles Eparses) 7 m

Natural resources: fish, crayfish, note, Glorioso Islands and Tromelin Island (Iles Eparses) have guano, phosphates, and coconuts
note—in the 1950's and 1960's, several species of trout were introduced to Iles Kerguelen of which two, Brown trout and Brook trout, survived to establish wild populations; reindeer were also introduced to Iles Kerguelen in 1956 as a source of fresh meat for whaling crews, the herd today, one of two in the Southern Hemisphere, is estimated to number around 4,000

Natural hazards: Ile Amsterdam and Ile Saint-Paul are inactive volcanoes; Iles Eparses subject to periodic cyclones; Bassas da India is a maritime hazard since it is under water for a period of three hours prior to and following the high tide and surrounded by reefs
volcanism: Reunion Island—Piton de la Fournaise (2,632 m), which has erupted many times in recent years including 2010, 2015, and 2017, is one of the world's most active volcanoes; although rare, eruptions outside the volcano's caldera could threaten nearby cities

Environment—current issues: introduction of foreign species on Iles Crozet has caused severe damage to the original ecosystem; overfishing of

Patagonian toothfish around Iles Crozet and Iles Kerguelen

Geography—note: islands' component is widely scattered across remote locations in the southern Indian Ocean

Bassas da India (Iles Eparses): atoll is a circular reef atop a long-extinct, submerged volcano;

Europa Island and Juan de Nova Island (Iles Eparses): wildlife sanctuary for seabirds and sea turtles;

Glorioso Island (Iles Eparses): islands and rocks are surrounded by an extensive reef system;

Tromelin Island (Iles Eparses): climatologically important location for forecasting cyclones in the western Indian Ocean; wildlife sanctuary (seabirds, tortoises)

PEOPLE AND SOCIETY

Population: no indigenous inhabitants

Ile Amsterdam (Ile Amsterdam et Ile Saint-Paul): uninhabited but has a meteorological station

Ile Saint-Paul (Ile Amsterdam et Ile Saint-Paul): uninhabited but is frequently visited by fishermen and has a scientific research cabin for short stays

Iles Crozet: uninhabited except for 18 to 30 people staffing the Alfred Faure research station on Ile del la Possession

Iles Kerguelen: 50 to 100 scientists are located at the main base at Port-aux-Francais on Ile Kerguelen

Bassas da India (Iles Eparses): uninhabitable

Europa Island, Glorioso Islands, Juan de Nova Island (Iles Eparses): a small French military garrison and a few meteorologists on each possession; visited by scientists

Tromelin Island (Iles Eparses): uninhabited, except for visits by scientists

GOVERNMENT

Country name: *conventional long form:* Territory of the French Southern and Antarctic Lands
conventional short form: French Southern and Antarctic Lands
local long form: Territoire des Terres Australes et Antarctiques Francaises
local short form: Terres Australes et Antarctiques Francaises
abbreviation: TAAF
etymology: self-descriptive name specifying the territories' affiliation and location in the Southern Hemisphere

Dependency status: overseas territory of France since 1955

Administrative divisions: none (overseas territory of France); there are no first-order administrative divisions as defined by the US Government, but there are 5 administrative districts named Iles Crozet, Iles Eparses, Iles Kerguelen, Ile Saint-Paul et Ile Amsterdam; the fifth district is the "Adelie Land" claim in Antarctica that is not recognized by the US

Legal system: the laws of France, where applicable, apply

Citizenship: see France

Executive branch: President Emmanuel MACRON (since 14 May 2017), represented by Prefect Cecile POZZO DI BORGO (since 13 October 2014)

International organization participation: UPU

Diplomatic representation in the US: none (overseas territory of France)

Diplomatic representation from the US: none (overseas territory of France)

Flag description: the flag of France is used

National anthem: note: as a territory of France, "La Marseillaise" is official (see France)

0:00/ 1:19

ECONOMY

Economy—overview: Economic activity is limited to servicing meteorological and geophysical research stations, military bases, and French and other fishing fleets. The fish catches landed on Iles Kerguelen by foreign ships are exported to France and Reunion.

COMMUNICATIONS

Internet country code: .tf

Communications—note: has one or more meteorological stations on each possession

TRANSPORTATION

Airports: 4 (2020)
country comparison to the world: 187

Ports and terminals: none; offshore anchorage only

MILITARY AND SECURITY

Military—note: defense is the responsibility of France

TRANSNATIONAL ISSUES

Disputes—international: French claim to "Adelie Land" in Antarctica is not recognized by the US; *Bassas da India, Europa Island, Glorioso Islands, Juan de Nova Island (Iles Eparses):* ; claimed by Madagascar; the vegetated drying cays of Banc du Geyser, which were claimed by Madagascar in 1976, also fall within the EEZ claims of the Comoros and France (Glorioso Islands); ;
Tromelin Island (Iles Eparses): ; claimed by Mauritius

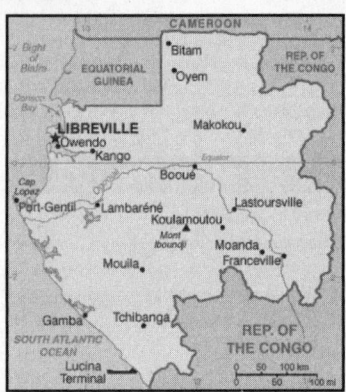

INTRODUCTION

Background: Following, independence from France in 1960, El Hadj Omar BONGO Ondimba—one of the longest-ruling heads of state in the world—dominated the country's political scene for four decades (1967-2009). President BONGO introduced a nominal multiparty system and a new constitution in the early 1990s. However, allegations of electoral fraud during local elections in December 2002 and the presidential election in 2005 exposed the weaknesses of formal political structures in Gabon. Following President BONGO's death in 2009, a new election brought his son, Ali BONGO Ondimba, to power. Despite constrained political conditions, Gabon's small population, abundant natural resources, and considerable foreign support have helped make it one of the more stable African countries.

President Ali BONGO Ondimba's controversial August 2016 reelection sparked unprecedented opposition protests that resulted in the burning of the parliament building. The election was contested by the opposition after fraudulent results were f lagged by international election observers. Gabon's Constitutional Court reviewed the election results but ruled in favor of President BONGO, upholding his win and extending his mandate to 2023.

GEOGRAPHY

Location: Central Africa, bordering the Atlantic Ocean at the Equator, between Republic of the Congo and Equatorial Guinea

Geographic coordinates: 1 00 S, 11 45 E

Map references: Africa

Area: *total:* 267,667 sq km
land: 257,667 sq km
water: 10,000 sq km
country comparison to the world: 78

Area—comparative: slightly smaller than Colorado

Land boundaries: *total:* 3,261 km

border countries (3): Cameroon 349 km, Republic of the Congo 2567 km, Equatorial Guinea 345 km

Coastline: 885 km

Maritime claims: *territorial sea:* 12 nm
exclusive economic zone: 200 nm
contiguous zone: 24 nm

Climate: tropical; always hot, humid

Terrain: narrow coastal plain; hilly interior; savanna in east and south

Elevation: *mean elevation:* 377 m
lowest point: Atlantic Ocean 0 m
highest point: Mont Iboundji 1,575 m

Natural resources: petroleum, natural gas, diamond, niobium, manganese, uranium, gold, timber, iron ore, hydropower

Land use: *agricultural land:* 19% (2011 est.)
arable land: 1.2% (2011 est.) / permanent crops: 0.6% (2011 est.) / permanent pasture: 17.2% (2011 est.)
forest: 81% (2011 est.)
other: 0% (2011 est.)
Irrigated land: 40 sq km (2012)

Population distribution: the relatively small population is spread in pockets throughout the country; the largest urban center is the capital of Libreville, located along the Atlantic coast in the northwest as shown in this population distribution map

Natural hazards: none

Environment—current issues: deforestation (the forests that cover three-quarters of the country are threatened by excessive logging); burgeoning population exacerbating disposal of solid waste; oil industry contributing to water pollution; wildlife poaching

Environment—international agreements: *party to:* Biodiversity, Climate Change, Climate Change-Kyoto Protocol, Desertification, Endangered Species, Hazardous Wastes, Law of the Sea, Marine Dumping, Ozone Layer Protection, Ship Pollution, Tropical Timber 83, Tropical Timber 94, Wetlands, Whaling
signed, but not ratified: none of the selected agreements

Geography—note: a small population and oil and mineral reserves have helped Gabon become one of Africa's wealthier countries; in general, these circumstances have allowed the country to maintain and conserve its pristine rain forest and rich biodiversity

PEOPLE AND SOCIETY

Population: 2,230,908 (July 2020 est.)
note: estimates for this country explicitly take into account the effects of excess mortality due to AIDS; this can result in lower life expectancy, higher infant mortality, higher death rates, lower population growth rates, and changes in the distribution of population by age and sex than would otherwise be expected

country comparison to the world: 145

Nationality: *noun:* Gabonese (singular and plural)
adjective: Gabonese

Ethnic groups: Gabonese-born 80.1% (includes Fang 23.2%, Shira-Punu/Vili 18.9%, Nzabi-Duma 11.3%, Mbede-Teke 6.9%, Myene 5%, Kota-Kele 4.9%, Okande-Tsogo 2.1%, Pygmy .3%, other 7.5%), Cameroonian 4.6%, Malian 2.4%, Beninese 2.1%, acquired Gabonese nationality 1.6%, Togolese 1.6%, Senegalese 1.1%, Congolese (Brazzaville) 1%, other 5.5% (includes Congolese (Kinshasa), Equatorial Guinean, Nigerian) (2012)

Languages: French (official), Fang, Myene, Nzebi, Bapounou/Eschira, Bandjabi

Religions: Roman Catholic 42.3%, Protestant 12.3%, other Christian 27.4%, Muslim 9.8%, animist 0.6%, other 0.5%, none/no answer 7.1% (2012 est.)

Demographic profile: Gabon's oil revenues have given it one of the highest per capita income levels in Sub-Saharan Africa, but the wealth is not evenly distributed and poverty is widespread. Unemployment is especially prevalent among the large youth population; more than 60% of the population is under the age of 25. With a fertility rate still averaging more than 4 children per woman, the youth population will continue to grow and further strain the mismatch between Gabon's supply of jobs and the skills of its labor force.

Gabon has been a magnet to migrants from neighboring countries since the 1960s because of the discovery of oil, as well as the country's political stability and timber, mineral, and natural gas resources. Nonetheless, income inequality and high unemployment have created slums in Libreville full of migrant workers from Senegal, Nigeria, Cameroon, Benin, Togo, and elsewhere in West Africa. In 2011, Gabon declared an end to refugee status for 9,500 remaining Congolese nationals to whom it had granted asylum during the Republic of the Congo's civil war between 1997 and 2003. About 5,400 of these refugees received permits to reside in Gabon.

Age structure: *0-14 years:* 36.45% (male 413,883/female 399,374)
15-24 years: 21.9% (male 254,749/female 233,770)
25-54 years: 32.48% (male 386,903/female 337,776)
55-64 years: 5.19% (male 58,861/female 56,843)
65 years and over: 3.98% (male 44,368/female 44,381) (2020 est.)

Dependency ratios: *total dependency ratio:* 68.9
youth dependency ratio: 62.9
elderly dependency ratio: 6
potential support ratio: 16.8 (2020 est.)

Median age: *total:* 21 years
male: 21.4 years
female: 20.6 years (2020 est.)
country comparison to the world: 187

Population growth rate: 2.5% (2020 est.)

country comparison to the world: 25

Birth rate: 26.3 births/1,000 population (2020 est.)
country comparison to the world: 45

Death rate: 5.9 deaths/1,000 population (2020 est.)
country comparison to the world: 172

Net migration rate: 3.9 migrant(s)/1,000 population (2020 est.)
country comparison to the world: 30

Population distribution: the relatively small population is spread in pockets throughout the country; the largest urban center is the capital of Libreville, located along the Atlantic coast in the northwest as shown in this population distribution map

Urbanization: *urban population*: 90.1% of total population (2020)
rate of urbanization: 2.61% annual rate of change (2015-20 est.)

total population growth rate v. urban population growth rate, 2000-2030: Major urban areas—population: 834,000 LIBREVILLE (capital) (2020)

Sex ratio: *at birth*: 1.03 male(s)/female
0-14 years: 1.04 male(s)/female
15-24 years: 1.09 male(s)/female
25-54 years: 1.15 male(s)/female
55-64 years: 1.04 male(s)/female
65 years and over: 1 male(s)/female
total population: 1.08 male(s)/female (2020 est.)

Mother's mean age at first birth: 20.3 years (2012 est.)
note: median age at first birth among women 25-29

Maternal mortality rate: 252 deaths/100,000 live births (2017 est.)
country comparison to the world: 41

Infant mortality rate: *total*: 30.4 deaths/1,000 live births
male: 33.6 deaths/1,000 live births
female: 27 deaths/1,000 live births (2020 est.)
country comparison to the world: 54

Life expectancy at birth: *total population*: 69 years
male: 67.3 years
female: 70.8 years (2020 est.)
country comparison to the world: 172

Total fertility rate: 3.41 children born/woman (2020 est.)
country comparison to the world: 43

Contraceptive prevalence rate: 31.1% (2012)

Drinking water source:

improved: *urban*: 97% of population
rural: 68% of population
total: 93.8% of population

unimproved: *urban*: 0.3% of population
rural: 32% of population
total: 6.2% of population (2017 est.)

Current Health Expenditure: 2.8% (2017)

Physicians density: 0.68 physicians/1,000 population (2017)

Hospital bed density: 6.3 beds/1,000 population (2010)

Sanitation facility access:

improved: *urban*: 77.7% of population
rural: 51.9% of population
total: 74.8% of population

unimproved: *urban*: 22.3% of population
rural: 48.1% of population
total: 25.2% of population (2017 est.)

HIV/AIDS—adult prevalence rate: 3.6% (2019 est.)
country comparison to the world: 15

HIV/AIDS—people living with HIV/AIDS: 51,000 (2019 est.)
country comparison to the world: 60

HIV/AIDS—deaths: 1,100 (2019 est.)
country comparison to the world: 56

Major infectious diseases: *degree of risk*: very high (2020)
food or waterborne diseases: bacterial diarrhea, hepatitis A, and typhoid fever
vectorborne diseases: malaria and dengue fever
water contact diseases: schistosomiasis
animal contact diseases: rabies

Obesity—adult prevalence rate: 15% (2016)
country comparison to the world: 127

Children under the age of 5 years underweight: 6.4% (2012)
country comparison to the world: 75

Education expenditures: 2.7% of GDP (2014)
country comparison to the world: 151

Literacy: *definition*: age 15 and over can read and write
total population: 84.7%
male: 85.9%
female: 83.4% (2018)

Unemployment, youth ages 15-24: *total*: 35.7%
male: 30.5%
female: 41.9% (2010 est.)
country comparison to the world: 20

GOVERNMENT

Country name: *conventional long form*: Gabonese Republic
conventional short form: Gabon
local long form: Republique Gabonaise
local short form: Gabon
etymology: name originates from the Portuguese word "gabao" meaning "cloak," which is roughly the shape that the early explorers gave to the estuary of the Komo River by the capital of Libreville

Government type: presidential republic

Capital: *name*: Libreville

geographic coordinates: 0 23 N, 9 27 E
time difference: UTC+ 1 (6 hours ahead of Washington, DC, during Standard Time)
etymology: original site settled by freed slaves and the name means "free town" in French; named in imitation of Freetown, the capital of Sierra Leone

Administrative divisions: 9 provinces; Estuaire, Haut-Ogooue, Moyen-Ogooue, Ngounie, Nyanga, Ogooue-Ivindo, Ogooue-Lolo, Ogooue-Maritime, Woleu-Ntem

Independence: 17 August 1960 (from France)

National holiday: Independence Day, 17 August (1960)

Constitution: *history*: previous 1961; latest drafted May 1990, adopted 15 March 1991, promulgated 26 March 1991
amendments: proposed by the president of the republic, by the Council of Ministers, or by one third of either house of Parliament; passage requires Constitutional Court evaluation, at least two-thirds majority vote of two thirds of the Parliament membership convened in joint session, and approval in a referendum; constitutional articles on Gabon's democratic form of government cannot be amended; amended several times, last in 2011

Legal system: mixed legal system of French civil law and customary law

International law organization participation: has not submitted an ICJ jurisdiction declaration; accepts ICCt jurisdiction

Citizenship: *citizenship by birth*: no
citizenship by descent only: at least one parent must be a citizen of Gabon
dual citizenship recognized: no
residency requirement for naturalization: 10 years

Suffrage: 18 years of age; universal

Executive branch: *chief of state*: President Ali BONGO Ondimba (since 16 October 2009)

head of government: Prime Minister Rose Christiane Ossouka RAPONDA (since 16 July 2020)
cabinet: Council of Ministers appointed by the prime minister in consultation with the president
elections/appointments: president directly elected by simple majority popular vote for a 7-year term (no term limits); election last held on 27 August 2016 (next to be held in August 2023); prime minister appointed by the president
election results: Ali BONGO Ondimba reelected president; percent of vote—Ali BONGO Ondimba (PDG) 49.8%, Jean PING (UFC) 48.2%, other 2.0%

Legislative branch: *description*: bicameral Parliament or Parlement consists of:
Senate or Senat (102 seats; members indirectly elected by municipal councils and departmental assemblies by absolute majority vote in 2 rounds if needed; members serve 6-year terms)
National Assembly or Assemblee Nationale (143 seats; members elected in single-seat constituencies by absolute majority vote in 2 rounds if needed; members serve 5-year terms)
elections: Senate—last held on 13 December 2014 (next to be held on 31 December 2020)
National Assembly—held in 2 rounds on 6 and 27 October 2018 (next to be held in 2023)
election results: Senate—percent of vote by party—NA; seats by party—PDG 81, CLR 7, PSD 2, ADERE- UPG 1, UPG 1, PGCI 1, independent 7; composition—men 84, women 18, percent of women 17.6%
National Assembly—percent of vote by party—NA; seats by party—PDG 98, The Democrats or LD 11, RV 8, Social Democrats of Gabon 5, RH&M 4, other 9, independent 8;

357

composition—men 123, women 20, percent of women 14%; note—total Parliament percent of women 15.5%

Judicial branch: highest courts: Supreme Court (consists of 4 permanent specialized supreme courts—Supreme Court or Cour de Cassation, Administrative Supreme Court or Conseil d'Etat, Accounting Supreme Court or Cour des Comptes, Constitutional Court or Cour Constitutionnelle, and the non-permanent Court of State Security, initiated only for cases of high treason by the president and criminal activity by executive branch officials)

judge selection and term of office: appointment and tenure of Supreme, Administrative, Accounting, and State Security courts NA; Constitutional Court judges appointed—3 by the national president, 3 by the president of the Senate, and 3 by the president of the National Assembly; judges serve single renewable 7-year terms

subordinate courts: Courts of Appeal; county courts; military courts

Political parties and leaders: Circle of Liberal Reformers or CLR [Gen. Jean- Boniface ASSELE] Democratic and Republican Alliance or ADERE [DIDJOB Divungui di Ndinge] Gabonese Democratic Party or PDG [Ali BONGO Ondimba] Independent Center Party of Gabon or PGCI [Luccheri GAHILA] Legacy and Modernity Party or RH&M Rally for Gabon or RPG Restoration of Republican Values or RV Social Democratic Party or PSD [Pierre Claver MAGANGA-MOUSSAVOU] Social Democrats of Gabon The Democrats or LD Union for the New Republic or UPRN [Louis Gaston MAYILA] Union of Gabonese People or UPG [Richard MOULOMBA] Union of Forces for Change or UFC [Jean PING]

International organization participation: ACP, AfDB, AU, BDEAC, CEMAC, FAO, FZ, G- 24, G- 77, IAEA, IBRD, ICAO, ICCt, ICRM, IDA, IDB, IFAD, IFC, IFRCS, ILO, IMF, IMO, IMSO, Interpol, IOC, IOM, IPU, ISO, ITSO, ITU, ITUC (NGOs), MIGA, NAM, OIC, OIF, OPCW, UN, UNCTAD, UNESCO, UNIDO, UNWTO, UPU, WCO, WHO, WIPO, WMO, WTO

Diplomatic representation in the US: *chief of mission:* Ambassador Michael MOUSSA-ADAMO (since September 9, 2011) *chancery:* 2034 20th Street NW, Suite 200, Washington, DC 20009 *telephone:* [1] (202) 797-1000 *FAX:* [1] (301) 332-0668

Diplomatic representation from the US: *chief of mission:* Ambassador (vacant); Charge d'Affaires Robert E. WHITEHEAD (since March 2019); note—also accredited to Sao Tome and Principe *telephone:* [241] 01-45-71-00 *embassy:* Sabliere, B.P. 4000, Libreville

mailing address: Centre Ville, B.P. 4000, Libreville; pouch: 2270 Libreville Place, Washington, DC 20521-2270 *FAX:* [241] 01-74-55-07

Flag description: three equal horizontal bands of green (top), yellow, and blue; green represents the country's forests and natural resources, gold represents the equator (which transects Gabon) as well as the sun, blue represents the sea

National symbol(s): black panther; national colors: green, yellow, blue

National anthem: *name:* "La Concorde" (The Concorde) *lyrics/music:* Georges Aleka DAMAS *note:* adopted 1960 0:00 / 1:44

ECONOMY

Economy—overview: Gabon enjoys a per capita income four times that of most Sub-Saharan African nations, but because of high income inequality, a large proportion of the population remains poor. Gabon relied on timber and manganese exports until oil was discovered offshore in the early 1970s. From 2010 to 2016, oil accounted for approximately 80% of Gabon's exports, 45% of its GDP, and 60% of its state budget revenues.

Gabon faces fluctuating international prices for its oil, timber, and manganese exports. A rebound of oil prices from 2001 to 2013 helped growth, but declining production, as some fields passed their peak production, has hampered Gabon from fully realizing potential gains. GDP grew nearly 6% per year over the 2010-14 period, but slowed significantly from 2014 to just 1% in 2017 as oil prices declined. Low oil prices also weakened government revenue and negatively affected the trade and current account balances. In the wake of lower revenue, Gabon signed a 3-year agreement with the IMF in June 2017.

Despite an abundance of natural wealth, poor fiscal management and over-reliance on oil has stifled the economy. Power cuts and water shortages are frequent. Gabon is reliant on imports and the government heavily subsidizes commodities, including food, but will be hard pressed to tamp down public frustration with unemployment and corruption.

GDP (purchasing power parity): $36.66 billion (2017 est.) $36.5 billion (2016 est.) $35.75 billion (2015 est.) *note:* data are in 2017 dollars *country comparison to the world:* 123

GDP (official exchange rate): $14.93 billion (2017 est.)

GDP—real growth rate: 0.5% (2017 est.) 2.1% (2016 est.) 3.9% (2015 est.) *country comparison to the world:* 185

GDP—per capita (PPP): $18,100 (2017 est.) $18,400 (2016 est.) $18,500 (2015 est.) *note:* data are in 2017 dollars

country comparison to the world: 98

Gross national saving: 25.6% of GDP (2017 est.) 24.3% of GDP (2016 est.) 29.2% of GDP (2015 est.) *country comparison to the world:* 55

GDP—composition, by end use: *household consumption:* 37.6% (2017 est.) *government consumption:* 14.1% (2017 est.) *investment in fixed capital:* 29% (2017 est.) *investment in inventories:* -0.6% (2016 est.) *exports of goods and services:* 46.7% (2017 est.) *imports of goods and services:* -26.8% (2017 est.)

GDP—composition, by sector of origin: *agriculture:* 5% (2017 est.) *industry:* 44.7% (2017 est.) *services:* 50.4% (2017 est.)

Agriculture—products: cocoa, coffee, sugar, palm oil, rubber; cattle; okoume (a tropical softwood); fish

Industries: petroleum extraction and refining; manganese, gold; chemicals, ship repair, food and beverages, textiles, lumbering and plywood, cement

Industrial production growth rate: 1.8% (2017 est.) *country comparison to the world:* 135

Labor force: 557,800 (2017 est.) *country comparison to the world:* 153

Labor force—by occupation: *agriculture:* 64% *industry:* 12% *services:* 24% (2005 est.)

Unemployment rate: 28% (2015 est.) 20.4% (2014 est.) *country comparison to the world:* 202

Population below poverty line: 34.3% (2015 est.)

Household income or consumption by percentage share: *lowest 10%:* 2.5% *highest 10%:* 32.7% (2005)

Budget: *revenues:* 2.634 billion (2017 est.) *expenditures:* 2.914 billion (2017 est.)

Taxes and other revenues: 17.6% (of GDP) (2017 est.) *country comparison to the world:* 167

Budget surplus (+) or deficit (-): -1.9% (of GDP) (2017 est.) *country comparison to the world:* 101

Public debt: 62.7% of GDP (2017 est.) 64.2% of GDP (2016 est.) *country comparison to the world:* 68

Fiscal year: calendar year

Inflation rate (consumer prices): 2.7% (2017 est.) 2.1% (2016 est.) *country comparison to the world:* 125

Current account balance: -$725 million (2017 est.) -$1.389 billion (2016 est.) *country comparison to the world:* 135

Exports: $5.564 billion (2017 est.) $4.364 billion (2016 est.) *country comparison to the world:* 105

Exports—partners: China 36.4%, US 10%, Ireland 8.5%, Netherlands 6.3%, South Korea 5.1%, Australia 5%, Italy 4.6% (2017)

Exports—commodities: crude oil, timber, manganese, uranium

Imports: $2.829 billion (2017 est.)
$2.652 billion (2016 est.)
country comparison to the world: 150

Imports—commodities: machinery and equipment, foodstuffs, chemicals, construction materials

Imports—partners: France 23.6%, Belgium 19.6%, China 15.2% (2017)

Reserves of foreign exchange and gold: $981.6 million (31 December 2017 est.)
$804.1 million (31 December 2016 est.)
country comparison to the world: 133

Debt—external: $6.49 billion (31 December 2017 est.)
$5.321 billion (31 December 2016 est.)
country comparison to the world: 127

Exchange rates: Cooperation Financiere en Afrique Centrale francs (XAF) per US dollar—
605.3 (2017 est.)
593.01 (2016 est.)
593.01 (2015 est.)
591.45 (2014 est.)
494.42 (2013 est.)

ENERGY

Electricity access: *electrification—total population:* 92% (2019)
electrification—urban areas: 99% (2019)
electrification—rural areas: 39% (2019)

Electricity—production: 2.244 billion kWh (2016 est.)
country comparison to the world: 137

Electricity—consumption: 2.071 billion kWh (2016 est.)
country comparison to the world: 143

Electricity—exports: 0 kWh (2016 est.)
country comparison to the world: 137

Electricity—imports: 344 million kWh (2016 est.)
country comparison to the world: 85

Electricity—installed generating capacity: 671,000 kW (2016 est.)
country comparison to the world: 137

Electricity—from fossil fuels: 51% of total installed capacity (2016 est.)
country comparison to the world: 148

Electricity—from nuclear fuels: 0% of total installed capacity (2017 est.)
country comparison to the world: 93

Electricity—from hydroelectric plants: 49% of total installed capacity (2017 est.)
country comparison to the world: 41

Electricity—from other renewable sources: 0% of total installed capacity (2017 est.)
country comparison to the world: 186

Crude oil—production: 196,000 bbl/day (2018 est.)
country comparison to the world: 35

Crude oil—exports: 214,200 bbl/day (2017 est.)
country comparison to the world: 30

Crude oil—imports: 0 bbl/day (2015 est.)

country comparison to the world: 130

Crude oil—proved reserves: 2 billion bbl (1 January 2018 est.)
country comparison to the world: 34

Refined petroleum products—production: 16,580 bbl/day (2017 est.)
country comparison to the world: 91

Refined petroleum products—consumption: 24, 000 bbl/day (2016 est.)
country comparison to the world: 129

Refined petroleum products—exports: 4,662 bbl/day (2015 est.)
country comparison to the world: 91

Refined petroleum products—imports: 10,680 bbl/day (2015 est.)
country comparison to the world: 146

Natural gas—production: 401 million cu m (2017 est.)
country comparison to the world: 74

Natural gas—consumption: 401 million cu m (2017 est.)
country comparison to the world: 101

Natural gas—exports: 0 cu m (2017 est.)
country comparison to the world: 107

Natural gas—imports: 0 cu m (2017 est.)
country comparison to the world: 128

Natural gas—proved reserves: 28.32 billion cu m (1 January 2018 est.)
country comparison to the world: 69

Carbon dioxide emissions from consumption of energy: 4.293 million Mt (2017 est.)
country comparison to the world: 137

COMMUNICATIONS

Telephones—fixed lines: *total subscriptions:* 22,412
subscriptions per 100 inhabitants: 1.03 (2019 est.)
country comparison to the world: 173

Telephones—mobile cellular: *total subscriptions:* 3, 008, 814
subscriptions per 100 inhabitants: 138.28 (2019 est.)
country comparison to the world: 140

Telecommunication systems: *general assessment:* fixed-line and Internet sectors have remained underdeveloped due to the lack of competition and high prices; sufficient international bandwidth due to submarine cable systems, but monopolized by Gabon Telecom; 3 G and mobile LTE services and mobile broadband available; govt. commits to XAF 150 billion in backbone infrastructure work through 2020 ; efforts towards new legal and regulatory improvements (2020)
domestic: fixed-line is 1 per 100 subscriptions; a growing mobile cellular network with multiple providers is making telephone service more widely available with mobile cellular teledensity at 138 per 100 persons (2019)
international: country code—241; landing points for the SAT-3/WASC, ACE and Libreville-Port Gentil Cable fiber- optic submarine cable that

provides connectivity to Europe and West Africa; satellite earth stations—3 Intelsat (Atlantic Ocean) (2019)
note: the COVID-19 outbreak is negatively impacting telecommunications production and supply chains globally; consumer spending on telecom devices and services has also slowed due to the pandemic's effect on economies worldwide; overall progress towards improvements in all facets of the telecom industry—mobile, fixed-line, broadband, submarine cable and satellite—has moderated

Broadcast media: state owns and operates 2 TV stations and 2 radio broadcast stations; a few private radio and TV stations; transmissions of at least 2 international broadcasters are accessible; satellite service subscriptions are available

Internet country code: .ga

Internet users: *total:* 1, 313, 802

percent of population: 62% (July 2018 est.)
country comparison to the world: 133

Broadband—fixed subscriptions: *total:* 29,099
subscriptions per 100 inhabitants: 1 (2018 est.)
country comparison to the world: 143

TRANSPORTATION

National air transport system: *number of registered air carriers:* 3 (2020)
inventory of registered aircraft operated by air carriers: 8

Civil aircraft registration country code prefix: TR (2016)

Airports: 44 (2013)
country comparison to the world: 97

Airports—with paved runways: *total:* 14 (2019)
over 3,047 m: 1
2,438 to 3,047 m: 2
1,524 to 2,437 m: 9
914 to 1,523 m: 1
under 914 m: 1

Airports—with unpaved runways: *total:* 30 (2013)
1,524 to 2,437 m: 7 (2013)
914 to 1,523 m: 9 (2013)
under 914 m: 14 (2013)

Pipelines: 807 km gas, 1639 km oil, 3 km water (2013)

Railways: *total:* 649 km (2014)
standard gauge: 649 km 1.435-Fm gauge (2014)
country comparison to the world: 106

Roadways: *total:* 14,300 km (2001)
paved: 900 km (2001)
unpaved: 13,400 km (2001)
country comparison to the world: 127

Waterways: 1,600 km (310 km on Ogooue River) (2010)
country comparison to the world: 48

Merchant marine: *total:* 28
by type: general cargo 9, oil tanker 1, other 18 (2019)
country comparison to the world: 135

Ports and terminals: *major seaport(s):* Libreville, Owendo, Port-Gentil
oil terminal(s): Gamba, Lucina

MILITARY AND SECURITY

Military and security forces: Gabonese Defense Forces (Forces de Defense Gabonaise): Land Force (Force Terrestre), Gabonese Navy (Marine Gabonaise), Gabonese Air Forces (Forces Aerienne Gabonaises, FAG), Gabonese National Gendarmerie (2019)

Military expenditures: 1.6% of GDP (2019)
1.5% of GDP (2018)
1.8% of GDP (2017)
1.4% of GDP (2016)
1.2% of GDP (2015)
country comparison to the world: 72

Military and security service personnel strengths: the Gabonese Defense Forces (FDG) are comprised of approximately 6,500 active duty troops (3,000 Land Forces; 500 Navy; 1,000 Air Force; 2,000 Gendarmerie) (2019)

Military equipment inventories and acquisitions: the FDG's inventory is comprised mostly of Brazilian, French, and South African equipment; since 2010, the leading suppliers are France and South Africa (2019 est.)

Military deployments: 450 Central African Republic (MINUSCA) (2020)

Military service age and obligation: 20 years of age for voluntary military service; no conscription (2013)

TRANSNATIONAL ISSUES

Disputes—international: UN urges Equatorial Guinea and Gabon to resolve the sovereignty dispute over Gabon-occupied Mbane Island and lesser islands and to establish a maritime boundary in hydrocarbon-rich Corisco Bay

Trafficking in persons: *current situation:* Gabon is primarily a destination and transit country for adults and children from West and Central African countries subjected to forced labor and sex trafficking; boys are forced to work as street vendors, mechanics, or in the fishing sector, while girls are subjected to domestic servitude or forced to work in markets or roadside restaurants; West African women are forced into domestic servitude or prostitution; men are reportedly forced to work on cattle farms; some foreign adults end up in forced labor in Gabon after initially seeking the help of human smugglers to help them migrate

clandestinely; traffickers operate in loose, ethnic-based criminal networks, with female traffickers recruiting and facilitating the transport of victims from source countries; in some cases, families turn child victims over to traffickers, who promise paid jobs in Gabon

tier rating: Tier 2 Watch List—Gabon does not fully comply with the minimum standards for the elimination of trafficking; however, it is making significant efforts to do so; Gabon's existing laws do not prohibit all forms of trafficking, and the government failed to pass a legal amendment drafted in 2013 to criminalize the trafficking of adults; anti-trafficking law enforcement decreased in 2014, dropping from 50 investigations to 16, and the only defendant to face prosecution fled the country; government efforts to identify and refer victims to protective services declined from 50 child victims in 2013 to just 3 in 2014, none of whom was referred to a care facility; the government provided support to four centers offering services to orphans and vulnerable children—14 child victims identified by an NGO received government assistance; no adult victims have been identified since 2009 (2015)

GAMBIA, THE

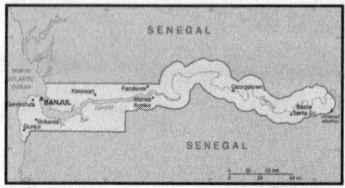

INTRODUCTION

Background: The Gambia gained its independence from the UK in 1965.Geographically surrounded by Senegal, it formed a short-lived Confederation of Senegambia between 1982 and 1989.In 1991, the two nations signed a friendship and cooperation treaty, although tensions flared up intermittently during the regime of Yahya JAMMEH. JAMMEH led a military coup in 1994 that overthrew the president and banned political activity. A new constitution and presidential election in 1996, followed by parliamentary balloting in 1997, completed a nominal return to civilian rule. JAMMEH was elected president in all subsequent elections including most recently in late 2011. After 22 years of increasingly authoritarian rule, President JAMMEH was defeated in free and fair elections in December 2016. Due to The Gambia's poor human rights record under JAMMEH, international development partners had distanced themselves, and substantially reduced aid to the country. These channels have now reopened

under the administration of President Adama BARROW, who took office in January 2017. The US and The Gambia currently enjoy improved relations. US assistance to the country has supported military education and training programs, as well as various capacity building and democracy strengthening activities.

GEOGRAPHY

Location: Western Africa, bordering the North Atlantic Ocean and Senegal

Geographic coordinates: 13 28 N, 16 34 W

Map references: Africa

Area: *total:* 11,300 sq km
land: 10,120 sq km
water: 1,180 sq km
country comparison to the world: 166

Area—comparative: slightly less than twice the size of Delaware

Land boundaries: *total:* 749 km
border countries (1): Senegal 749 km

Coastline: 80 km

Maritime claims: *territorial sea:* 12 nm
contiguous zone: 18 nm
continental shelf: extent not specified
exclusive fishing zone: 200 nm

Climate: tropical; hot, rainy season (June to November); cooler, dry season (November to May)

Terrain: flood plain of the Gambia River flanked by some low hills

Elevation: *mean elevation:* 34 m
lowest point: Atlantic Ocean 0 m
highest point: unnamed elevation 53 m

Natural resources: fish, clay, silica sand, titanium (rutile and ilmenite), tin, zircon

Land use: *agricultural land:* 56.1% (2011 est.)
arable land: 41% (2011 est.) / permanent crops: 0.5% (2011 est.) / permanent pasture: 14.6% (2011 est.)
forest: 43.9% (2011 est.)
other: 0% (2011 est.)
Irrigated land: 50 sq km (2012)

Population distribution: settlements are found scattered along the Gambia River; the largest communities, including the capital of Banjul, and the country's largest city, Serekunda, are found at the mouth of the Gambia River along the Atlantic coast as shown in this population distribution map

Natural hazards: droughts

Environment—current issues: deforestation due to slash-and-burn agriculture; desertification; water pollution; water-borne diseases

Environment—international agreements: *party to:* Biodiversity, Climate Change, Climate Change-Kyoto Protocol, Desertification, Endangered Species, Hazardous Wastes, Law of the Sea, Ozone Layer Protection, Ship Pollution, Wetlands, Whaling signed, *but not ratified*: none of the selected agreements

Geography—note: almost an enclave of Senegal; smallest country on the African mainland

PEOPLE AND SOCIETY

Population: 2,173,999 (July 2020 est.)
country comparison to the world: 146

Nationality: *noun:* Gambian(s)
adjective: Gambian

Ethnic groups: Mandinka/Jahanka 34%, Fulani/Tukulur/Lorobo 22.4%, Wolof 12.6%, Jola/Karoninka 10.7%, Serahuleh 6.6%, Serer 3.2%, Manjago 2.1%, Bambara 1%, Creole/Aku Marabout 0.7%, other 0.9%, non-Gambian 5.2%, no answer 0.6% (2013 est.)

Languages: English (official), Mandinka, Wolof, Fula, other indigenous vernaculars

Religions: Muslim 95.7%, Christian 4.2%, none 0.1%, no response 0.1% (2013 est.)

Demographic profile: The Gambia's youthful age structure—almost 60% of the population is under the age of 25—is likely to persist because the country's total fertility rate remains strong at nearly 4 children per woman. The overall literacy rate is around 55%, and is significantly lower for women than for men. At least 70% of the populace are farmers who are reliant on rain-fed agriculture and cannot afford improved seeds and fertilizers. Crop failures caused by droughts between 2011 and 2013 have increased poverty, food shortages, and malnutrition.

The Gambia is a source country for migrants and a transit and destination country for migrants and refugees. Since the 1980s, economic deterioration, drought, and high unemployment, especially among youths, have driven both domestic migration (largely urban) and migration abroad (legal and illegal). Emigrants are largely skilled workers, including doctors and nurses, and provide a significant amount of remittances. The top receiving countries for Gambian emigrants are Spain, the US, Nigeria, Senegal, and the UK. While the Gambia and Spain do not share historic, cultural, or trade ties, rural Gambians have migrated to Spain in large numbers because of its proximity and the availability of jobs in its underground economy (this flow slowed following the onset of Spain's late 2007 economic crisis).

The Gambia's role as a host country to refugees is a result of wars in several of its neighboring West African countries. Since 2006, refugees from the Casamance conflict in Senegal have replaced their pattern of flight and return with permanent settlement in The Gambia, often moving in with relatives along the Senegal-Gambia border. The strain of providing for about 7,400 Casamance refugees has increased poverty among Gambian villagers.

Age structure: *0-14 years:* 35.96% (male 392,714/female 389,027)
15-24 years: 20.09% (male 216,307/female 220,514)
25-54 years: 35.85% (male 382,138/female 397,324)

55-64 years: 4.4% (male 45,614/female 50,143)
65 years and over: 3.69% (male 36,773/female 43,445) (2020 est.)

Dependency ratios: *total dependency ratio:* 86.9
youth dependency ratio: 82.1
elderly dependency ratio: 4.7
potential support ratio: 21.1 (2020 est.)

Median age: *total:* 21.8 years
male: 21.5 years
female: 22.2 years (2020 est.)
country comparison to the world: 182

Population growth rate: 1.87% (2020 est.)
country comparison to the world: 52

Birth rate: 27 births/1,000 population (2020 est.)
country comparison to the world: 44

Death rate: 6.7 deaths/1,000 population (2020 est.)
country comparison to the world: 136

Net migration rate: -1.6 migrant(s)/1,000 population (2020 est.)
country comparison to the world: 155

Population distribution: settlements are found scattered along the Gambia River; the largest communities, including the capital of Banjul, and the country's largest city, Serekunda, are found at the mouth of the Gambia River along the Atlantic coast as shown in this population distribution map

Urbanization: *urban population:* 62.6% of total population (2020)
rate of urbanization: 4.07% annual rate of change (2015-20 est.)

total population growth rate v. urban population growth rate, 2000-2030: Major urban areas—population: 451,000 BANJUL (capital) (2020)
note: includes the local government areas of Banjul and Kanifing

Sex ratio: *at birth:* 1.03 male(s)/female
0-14 years: 1.01 male(s)/female
15-24 years: 0.98 male(s)/female
25-54 years: 0.96 male(s)/female
55-64 years: 0.91 male(s)/female
65 years and over: 0.85 male(s)/female
total population: 0.98 male(s)/female (2020 est.)

Mother's mean age at first birth: 20.9 years (2013 est.)
note: median age at first birth among women 25-29

Maternal mortality rate: 597 deaths/100,000 live births (2017 est.)
country comparison to the world: 13

Infant mortality rate: *total:* 54.9 deaths/1,000 live births
male: 60.1 deaths/1,000 live births
female: 49.6 deaths/1,000 live births (2020 est.)
country comparison to the world: 18

Life expectancy at birth: *total population:* 65.8 years
male: 63.5 years
female: 68.3 years (2020 est.)
country comparison to the world: 190

Total fertility rate: 3.21 children born/woman (2020 est.)
country comparison to the world: 46

Contraceptive prevalence rate: 16.8% (2018)
note: percent of women aged 15-50

Drinking water source:
improved: *urban:* 91.4% of population
rural: 80.4% of population
total: 87.1% of population

unimproved: *urban:* 8.6% of population
rural: 19.6% of population
total: 12.9% of population (2017 est.)

Current Health Expenditure: 3.3% (2017)

Physicians density: 0.1 physicians/1,000 population (2015)

Hospital bed density: 1.1 beds/1,000 population (2011)

Sanitation facility access:
improved: *urban:* 80.4% of population
rural: 44.5% of population
total: 66.3% of population

unimproved: *urban:* 19.6% of population
rural: 55.5% of population
total: 33.7% of population (2017 est.)

HIV/AIDS—adult prevalence rate: 2% (2019 est.)
country comparison to the world: 23

HIV/AIDS—people living with HIV/AIDS: 28,000 (2019 est.)
country comparison to the world: 78

HIV/AIDS—deaths: 1,100 (2019 est.)
country comparison to the world: 57

Major infectious diseases: *degree of risk:* very high (2020)
food or waterborne diseases: bacterial and protozoal diarrhea, hepatitis A, and typhoid fever
vectorborne diseases: malaria and dengue fever
water contact diseases: schistosomiasis
animal contact diseases: rabies
respiratory diseases: meningococcal meningitis

Obesity—adult prevalence rate: 10.3% (2016)
country comparison to the world: 139

Children under the age of 5 years underweight: 10.3% (2018)
country comparison to the world: 63

Education expenditures: 2.1% of GDP (2016)
country comparison to the world: 167

Literacy: *definition:* age 15 and over can read and write
total population: 50.8%
male: 61.8%
female: 41.6% (2015)

School life expectancy (primary to tertiary education): *total:* 9 years
male: 9 years
female: 9 years (2010)

Unemployment, youth ages 15-24: *total:* 13.1%
male: 9.1%
female: 17.2% (2012 est.)
country comparison to the world: 106

GOVERNMENT

Country name: *conventional long form:* Republic of The Gambia
conventional short form: The Gambia

etymology: named for the Gambia River that flows through the heart of the country

Government type: presidential republic

Capital: *name:* Banjul

geographic coordinates: 13 27 N, 16 34 W

time difference: UTC 0 (5 hours ahead of Washington, DC, during Standard Time)

etymology: Banjul is located on Saint Mary's Island at the mouth of the Gambia River; the Mandinka used to gather fibrous plants on the island for the manufacture of ropes; "bang julo" is Mandinka for "rope fiber"; mispronunciation over time caused the term became the word Banjul

Administrative divisions: 5 regions, 1 city*, and 1 municipality**; Banjul*, Central River, Kanifing**, Lower River, North Bank, Upper River, West Coast

Independence: 18 February 1965 (from the UK)

National holiday: Independence Day, 18 February (1965)

Constitution: *history:* previous 1965 (Independence Act), 1970; latest adopted 8 April 1996, approved by referendum 8 August 1996, effective 16 January 1997; note—referendum on new constitution planned over the next 2 years

amendments: proposed by the National Assembly; passage requires at least three-fourths majority vote by the Assembly membership in each of several readings and approval by the president of the republic; a referendum is required for amendments affecting national sovereignty, fundamental rights and freedoms, government structures and authorities, taxation, and public funding; passage by referendum requires participation of at least 50% of eligible voters and approval by at least 75% of votes cast; amended 2001, 2004, 2010

Legal system: mixed legal system of English common law, Islamic law, and customary law

International law organization participation: accepts compulsory ICJ jurisdiction with reservations; accepts ICCt jurisdiction

Citizenship: *citizenship by birth:* yes
citizenship by descent only: yes
dual citizenship recognized: no
residency requirement for naturalization: 5 years

Suffrage: 18 years of age; universal

Executive branch: *chief of state:* President Adama BARROW (since 19 January 2017); Vice President Isatou TOURAY (since 15 March 2019); note—the president is both chief of state and head of government

head of government: President Adama BARROW (since 19 January 2017); Vice President Isatou TOURAY (since 15 March 2019)

cabinet: Cabinet appointed by the president

elections/appointments: president directly elected by simple majority popular vote for a 5-year term (no term limits); election last held on 1 December 2016 (next to be held in 2021); vice president appointed by the president

election results: Adama BARROW elected president; percent of vote—Adama BARROW (Coalition 2016) 43.3%, Yahya JAMMEH (APRC) 39.6%, Mamma KANDEH (GDC) 17.1%

Legislative branch: *description:* unicameral National Assembly (58 seats; 53 members directly elected in single-seat constituencies by simple majority vote and 5 appointed by the president; members serve 5-year terms)

elections: last held on 6 April 2017 (next to be held in 2022)

election results: percent of vote by party—UDP 37.5%, GDC 17.4%, APRC 16%, PDOIS 9%, NRP 6.3%, PPP 2.5%, other 1.7%, independent 9.6%; seats by party—UDP 31, APRC 5, GDC 5, NRP 5, PDOIS 4, PPP 2, independent 1; composition—men 52, women 6, percent of women 10.3%

Judicial branch: *highest courts:* Supreme Court of The Gambia (consists of the chief justice and 6 justices; court sessions held with 5 justices)

judge selection and term of office: justices appointed by the president after consultation with the Judicial Service Commission, a 6-member independent body of high-level judicial officials, a presidential appointee, and a National Assembly appointee; justices appointed for life or until mandatory retirement at age 75

subordinate courts: Court of Appeal; High Court; Special Criminal Court; Khadis or Muslim courts; district tribunals; magistrates courts; cadi courts

Political parties and leaders: Alliance for Patriotic Reorientation and Construction or APRC [Fabakary JATTA]
Coalition 2016 [collective leadership] (electoral coalition includes UDP, PDOIS, NRP, GMC, GDC, PPP, and GPDP)
Gambia Democratic Congress or GDC [Mama KANDEH]
Gambia Moral Congress or GMC [Mai FATTY]
Gambia Party for Democracy and Progress or GPDP [Sarja JARJOU]
National Convention Party or NCP [Yaya SANYANG and Majanko SAMUSA (both claiming leadership)]
National Democratic Action Movement or NDAM [Lamin Yaa JUARA]
National People's Party or NPP [Adama BARROW]
National Reconciliation Party or NRP [Hamat BAH]
People's Democratic Organization for Independence and Socialism or PDOIS [Sidia JATTA]
People's Progressive Party or PPP [Yaya CEESAY]
United Democratic Party or UDP [Ousainou DARBOE]

International organization participation: ACP, AfDB, AU, ECOWAS, FAO, G-77, IBRD, ICAO, ICCt, ICRM, IDA, IDB, IFAD, IFC, IFRCS, ILO, IMF, IMO, Interpol, IOC, IOM, IPU, ISO (correspondent), ITSO, ITU, ITUC (NGOs), MIGA, MINUSMA, NAM, OIC, OPCW, UN, UNAMID, UNCTAD, UNESCO, UNIDO,

UNMIL, UNOCI, UNWTO, UPU, WCO, WFTU (NGOs), WHO, WIPO, WMO, WTO

Diplomatic representation in the US: *chief of mission:* Ambassador Dawda D.FADERA (since 24 January 2018)

chancery: 5630 16th Street NW, Washington, DC 20011

telephone: [1] (202) 785-1399

FAX: [1] (202) 342-0240

Diplomatic representation from the US: *chief of mission:* Ambassador Richard "Carl" PASCHALL (since 9 April 2019)

telephone: [220] 439-2856

embassy: Kairaba Avenue, Fajara, P.M.B.19, Banjul

mailing address: P.M.B.19, Banjul

FAX: [220] 439-2475

Flag description: three equal horizontal bands of red (top), blue with white edges, and green; red stands for the sun and the savannah, blue represents the Gambia River, and green symbolizes forests and agriculture; the white stripes denote unity and peace

National symbol(s): lion; national colors: red, blue, green, white

National anthem: *name:* For The Gambia, Our Homeland

lyrics/music: Virginia Julie HOWE/adapted by Jeremy Frederick HOWE

note: adopted 1965; the music is an adaptation of the traditional Mandinka song "Foday Kaba Dumbuya"

ECONOMY

Economy—overview: The government has invested in the agriculture sector because three-quarters of the population depends on the sector for its livelihood and agriculture provides for about one-third of GDP, making The Gambia largely reliant on sufficient rainfall. The agricultural sector has untapped potential—less than half of arable land is cultivated and agricultural productivity is low. Small-scale manufacturing activity features the processing of cashews, groundnuts, fish, and hides. The Gambia's reexport trade accounts for almost 80% of goods exports and China has been its largest trade partner for both exports and imports for several years.

The Gambia has sparse natural resource deposits. It relies heavily on remittances from workers overseas and tourist receipts. Remittance inflows to The Gambia amount to about one-fifth of the country's GDP. The Gambia's location on the ocean and proximity to Europe has made it one of the most frequented tourist destinations in West Africa, boosted by private sector investments in eco-tourism and facilities. Tourism normally brings in about 20% of GDP, but it suffered in 2014 from tourists' fears of Ebola virus in neighboring West African countries. Unemployment and underemployment remain high.

Economic progress depends on sustained bilateral and multilateral aid, on responsible government economic management, and on continued technical assistance from multilateral and bilateral donors. International donors and lenders were concerned about the quality of fiscal management under the administration of former President Yahya JAMMEH, who reportedly stole hundreds of millions of dollars of the country's funds during his 22 years in power, but anticipate significant improvements under the new administration of President Adama BARROW, who assumed power in early 2017.As of April 2017, the IMF, the World Bank, the European Union, and the African Development Bank were all negotiating with the new government of The Gambia to provide financial support in the coming months to ease the country's financial crisis.

The country faces a limited availability of foreign exchange, weak agricultural output, a border closure with Senegal, a slowdown in tourism, high inflation, a large fiscal deficit, and a high domestic debt burden that has crowded out private sector investment and driven interest rates to new highs. The government has committed to taking steps to reduce the deficit, including through expenditure caps, debt consolidation, and reform of state-owned enterprises.

GDP (purchasing power parity): $5.556 billion (2017 est.)
$5.314 billion (2016 est.)
$5.292 billion (2015 est.)
note: data are in 2017 dollars
country comparison to the world: 176

GDP (official exchange rate): $1.482 billion (2017 est.)

GDP—real growth rate: 4.6% (2017 est.)
0.4% (2016 est.)
5.9% (2015 est.)
country comparison to the world: 58

GDP—per capita (PPP): $2,600 (2017 est.)
$2,600 (2016 est.)
$2,700 (2015 est.)
note: data are in 2017 dollars
country comparison to the world: 197

Gross national saving: 6.8% of GDP (2017 est.)
7.1% of GDP (2016 est.)
3.7% of GDP (2015 est.)
country comparison to the world: 172

GDP—composition, by end use:
household consumption: 90.7% (2017 est.)
government consumption: 12% (2017 est.)
investment in fixed capital: 19.2% (2017 est.)
investment in inventories: -2.7% (2017 est.)
exports of goods and services: 20.8% (2017 est.)
imports of goods and services: -40% (2017 est.)

GDP—composition, by sector of origin:
agriculture: 20.4% (2017 est.)
industry: 14.2% (2017 est.)
services: 65.4% (2017 est.)

Agriculture—products: rice, millet, sorghum, peanuts, corn, sesame, cassava (manioc, tapioca), palm kernels; cattle, sheep, goats

Industries: peanuts, fish, hides, tourism, beverages, agricultural machinery assembly, woodworking, metalworking, clothing

Industrial production growth rate: -0.8% (2017 est.)
country comparison to the world: 175

Labor force: 777,100 (2007 est.)
country comparison to the world: 147

Labor force—by occupation: *agriculture:* 75%
industry: 19%
services: 6% (1996 est.)

Unemployment rate: NA

Population below poverty line: 48.4% (2010 est.)

Household income or consumption by percentage share: *lowest 10%:* 2%
highest 10%: 36.9% (2003)

Budget: *revenues:* 300.4 million (2017 est.)
expenditures: 339 million (2017 est.)

Taxes and other revenues: 20.3% (of GDP) (2017 est.)
country comparison to the world: 148

Budget surplus (+) or deficit (-): -2.6% (of GDP) (2017 est.)
country comparison to the world: 117

Public debt: 88% of GDP (2017 est.)
82.3% of GDP (2016 est.)
country comparison to the world: 28

Fiscal year: calendar year

Inflation rate (consumer prices): 8% (2017 est.)
7.2% (2016 est.)
country comparison to the world: 196

Current account balance: -$194 million (2017 est.)
-$85 million (2016 est.)
country comparison to the world: 98

Exports: $72.9 million (2017 est.)
$106.6 million (2016 est.)
country comparison to the world: 201

Exports—partners: Guinea-Bissau 51.9%, Vietnam 14.6%, Senegal 8.8%, Mali 7.2% (2017)

Exports—commodities: peanut products, fish, cotton lint, palm kernels

Imports: $376.9 million (2017 est.)
$310.5 million (2016 est.)
country comparison to the world: 201

Imports—commodities: foodstuffs, manufactures, fuel, machinery and transport equipment

Imports—partners: Cote dIvoire 11.5%, Brazil 10.6%, Spain 10.2%, China 7.8%, Russia 6.4%, Netherlands 5.3%, India 5% (2017)

Reserves of foreign exchange and gold: $170 million (31 December 2017 est.)
$87.64 million (31 December 2016 est.)
country comparison to the world: 179

Debt—external: $586.8 million (31 December 2017 est.)
$571.2 million (31 December 2016 est.)
country comparison to the world: 174

Exchange rates: dalasis (GMD) per US dollar—
49.74 (2017 est.)
43.8846 (2016 est.)
43.8846 (2015 est.)
41.89 (2014 est.)
41.733 (2013 est.)

Electricity access: *population without electricity:* 1 million (2019)
electrification—total population: 49% (2019)
electrification—urban areas: 69% (2019)
electrification—rural areas: 16% (2019)

Electricity—production: 304.1 million kWh (2016 est.)
country comparison to the world: 182

Electricity—consumption: 282.8 million kWh (2016 est.)
country comparison to the world: 186

Electricity—exports: 0 kWh (2016 est.)
country comparison to the world: 138

Electricity—imports: 0 kWh (2016 est.)
country comparison to the world: 151

Electricity—installed generating capacity: 117,000 kW (2016 est.)
country comparison to the world: 178

Electricity—from fossil fuels: 97% of total installed capacity (2016 est.)
country comparison to the world: 34

Electricity—from nuclear fuels: 0% of total installed capacity (2017 est.)
country comparison to the world: 94

Electricity—from hydroelectric plants: 0% of total installed capacity (2017 est.)
country comparison to the world: 172

Electricity—from other renewable sources: 3% of total installed capacity (2017 est.)
country comparison to the world: 125

Crude oil—production: 0 bbl/day (2018 est.)
country comparison to the world: 140

Crude oil—exports: 0 bbl/day (2015 est.)
country comparison to the world: 128

Crude oil—imports: 0 bbl/day (2015 est.)
country comparison to the world: 131

Crude oil—proved reserves: 0 bbl (1 January 2018 est.)
country comparison to the world: 135

Refined petroleum products—production: 0 bbl/day (2017 est.)
country comparison to the world: 148

Refined petroleum products—consumption: 3,800 bbl/day (2016 est.)
country comparison to the world: 185

Refined petroleum products—exports: 42 bbl/day (2015 est.)
country comparison to the world: 122

Refined petroleum products—imports: 3,738 bbl/day (2015 est.)
country comparison to the world: 181

Natural gas—production: 0 cu m (2017 est.)
country comparison to the world: 136

Natural gas—consumption: 0 cu m (2017 est.)
country comparison to the world: 149

Natural gas—exports: 0 cu m (2017 est.)
country comparison to the world: 108

Natural gas—imports:

0 cu m (2017 est.)
country comparison to the world: 129

Natural gas—proved reserves:
0 cu m (1 January 2014 est.)
country comparison to the world: 138

Carbon dioxide emissions from consumption of energy: 607,300 Mt (2017 est.)
country comparison to the world: 180

COMMUNICATIONS

Telephones—fixed lines: *total subscriptions:* 41,179
subscriptions per 100 inhabitants: 1.93 (2019 est.)
country comparison to the world: 161

Telephones—mobile cellular: *total subscriptions:* 2,977,068
subscriptions per 100 inhabitants: 139.53 (2019 est.)
country comparison to the world: 141

Telecommunication systems: *general assessment:* state-owned Gambia Telecommunications partially privatized but still retaining a monopoly with fixed-line service; multiple mobile networks offering effective competition; three licensed ISPs which serve local area without much competition; mobile penetrations above the African average; lack of availability of fixed-line services in many rural areas of the country; govt. started a National Broadband Network program aimed at closing the digital divide but not funded by Parliament in 2018; the Chinese company Huawei helping in the telecommunications sector (2020)
domestic: fixed-line stands at 2 per 100 subscriptions with one dominant company and mobile-cellular teledensity, aided by multiple mobile-cellular providers, is over 140 per 100 persons (2019)
international: country code—220; landing point for the ACE submarine cable to West Africa and Europe; microwave radio relay links to Senegal and Guinea-Bissau; satellite earth station—1 Intelsat (Atlantic Ocean) (2019)
note: the COVID-19 outbreak is negatively impacting telecommunications production and supply chains globally; consumer spending on telecom devices and services has also slowed due to the pandemic's effect on economies worldwide; overall progress towards improvements in all facets of the telecom industry—mobile, fixed-line, broadband, submarine cable and satellite—has moderated

Broadcast media: 1 state-run TV-channel; one privately-owned TV-station; 1 Online TV-station; three state-owned radio station and 31 privately owned radio stations; eight community radio stations; transmissions of multiple international broadcasters are available, some via shortwave radio; cable and satellite TV subscription services are obtainable in some parts of the country (2019)

Internet country code: .gm

Internet users: *total:* 406,918
percent of population: 19.84% (July 2017 est.)
country comparison to the world: 158

Broadband—fixed subscriptions: *total:* 4,433
subscriptions per 100 inhabitants: less than 1 (2018 est.)
country comparison to the world: 182

TRANSPORTATION

National air transport system: *number of registered air carriers:* 2 (2020)
inventory of registered aircraft operated by air carriers: 6
annual passenger traffic on registered air carriers: 53,735 (2018)

Civil aircraft registration country code prefix: C5 (2016)

Airports: 1 (2020)
country comparison to the world: 221

Airports—with paved runways: *total:* 1 (2019)
over 3,047 m: 1

Roadways: *total:* 2,977 km (2011)
paved: 518 km (2011)
unpaved: 2,459 km (2011)
country comparison to the world: 161

Waterways: 390 km (on River Gambia; small oceangoing vessels can reach 190 km) (2010)
country comparison to the world: 88

Merchant marine: *total:* 8
by type: other 8 (2019)
country comparison to the world: 158

Ports and terminals: *major seaport(s):* Banjul

MILITARY AND SECURITY

Military and security forces: *Gambia Armed Forces:* the Gambian National Army (GNA, includes an air wing); Gambia Navy; Republican National Guard (2020)

Military expenditures: 0.8% of GDP (2019)
0.7% of GDP (2018)
1% of GDP (2015)
1.2% of GDP (2014)
0.8% of GDP (2013)
country comparison to the world: 129

Military and security service personnel strengths: estimates for the size of the Gambian National Army (GNA) vary; approximately 3,000 active troops (2019 est.)

Military equipment inventories and acquisitions: the GNA has a limited equipment inventory; the only reported weapons deliveries to the GNA since 2000 are second-hand patrol boats from Taiwan (2009) and one aircraft from Georgia (2004) (2019 est.)

Military deployments: 130 Sudan (UNAMID) (2020)

Military service age and obligation: 18 years of age for male and female voluntary military service; no conscription; service obligation 6 months (2012)

TRANSNATIONAL ISSUES

Disputes—international: attempts to stem refugees, cross-border raids, arms smuggling, and other illegal activities by separatists from southern Senegal's Casamance region, as well as from conflicts in other west African states

Trafficking in persons: *current situation:* The Gambia is a source and destination country for women and children subjected to forced labor and sex trafficking; Gambian women, girls, and, to a lesser extent, boys are exploited for prostitution and domestic servitude; women, girls, and boys from West African countries are trafficked to The Gambia for commercial sexual exploitation, particularly by European sex tourists; boys in some Koranic schools are forced into street vending or begging; some Gambian children have been identified as victims of forced labor in neighboring West African countries

tier rating: Tier 3—The Gambia does not fully comply with the minimum standards for the elimination of trafficking and is not making significant efforts to do so; the government demonstrated minimal anti-trafficking law enforcement efforts, investigating one trafficking case but not prosecuting or convicting any offenders in 2014; authorities did not investigate, prosecute, or convict any government employees complicit in trafficking, although corruption was a serious problem; the government identified and repatriated 19 Gambian girls subjected to domestic servitude in Lebanon but did not identify or provide protective services to any trafficking victims in The Gambia; a government program continued to provide resources and financial support to 12 Koranic schools on the condition that their students were not forced to beg (2015)

GAZA STRIP

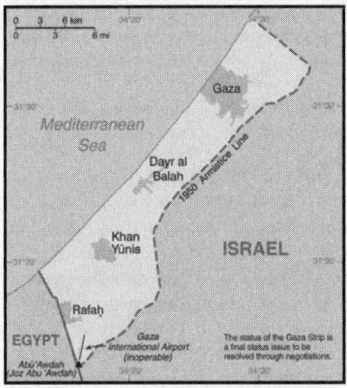

INTRODUCTION

Background: Inhabited since at least the 15th century B.C., the Gaza Strip has been dominated by many different peoples and empires throughout its history; it was incorporated into the Ottoman Empire in the early 16th century. The Gaza Strip fell to British forces during World War I, becoming a part of the British Mandate of Palestine. Following the 1948 Arab-Israeli War, Egypt administered the newly formed Gaza Strip; Israel captured it in the Six-Day War in 1967. Under a series of agreements known as the Oslo accords signed between 1993 and 1999, Israel transferred to the newly-created Palestinian Authority (PA) security and civilian responsibility for many Palestinian-populated areas of the Gaza Strip as well as the West Bank. In 2000, a violent intifada or uprising began, and in 2001 negotiations to determine the permanent status of the West bank and Gaza Strip stalled. Subsequent attempts to re-start negotiations have not resulted in progress toward determining final status of the Israeli-Palestinian conflict.

Israel by late 2005 unilaterally withdrew all of its settlers and soldiers and dismantled its military facilities in the Gaza Strip, but it continues to control the Gaza Strip's land and maritime borders and airspace. In early 2006, the Islamic Resistance Movement (HAMAS) won a majority in the Palestinian Legislative Council election. Attempts to form a unity government between Fatah, the dominant Palestinian political faction in the West Bank, and HAMAS failed, leading to violent clashes between their respective supporters and HAMAS's violent seizure of all military and governmental institutions in the Gaza Strip in June 2007. Since HAMAS's takeover, Israel and Egypt have enforced tight restrictions on movement and access of goods and individuals into and out of the territory. Fatah and HAMAS have since reached a series of agreements aimed at restoring political unity between the Gaza Strip and the West Bank but have struggled to enact them; a reconciliation agreement signed in October 2017 remains unimplemented.

In July 2014, HAMAS and other Gaza-based militant groups engaged in a 51-day conflict with Israel culminating in late August with an open-ended truce. Since 2014, Palestinian militants and the Israel Defense Forces have exchanged projectiles and air strikes respectively, sometimes lasting multiple days and resulting in multiple deaths on both sides. Egypt, Qatar, and the UN Special Coordinator for the Middle East Peace Process have negotiated multiple ceasefires to avert a broader conflict. Since March 2018, HAMAS has coordinated weekly demonstrations along the Gaza security fence, many of which have turned violent, resulting in one Israeli soldier death and several Israeli soldier injuries as well as more than 200 Palestinian deaths and thousands of injuries.

GEOGRAPHY

Location: Middle East, bordering the Mediterranean Sea, between Egypt and Israel

Geographic coordinates: 31 25 N, 34 20 E

Map references: Middle East

Area: *total:* 360 sq km
land: 360 sq km
water: 0 sq km
country comparison to the world: 207

Area—comparative: slightly more than twice the size of Washington, DC

Land boundaries: *total:* 72 km
border countries (2): Egypt 13 km, Israel 59 km

Coastline: 40 km

Maritime claims: see entry for Israel
note: effective 3 January 2009, the Gaza maritime area is closed to all maritime traffic and is under blockade imposed by Israeli Navy until further notice

Climate: temperate, mild winters, dry and warm to hot summers

Terrain: flat to rolling, sand- and dune-covered coastal plain

Elevation: *lowest point:* Mediterranean Sea 0 m
highest point: Abu 'Awdah (Joz Abu 'Awdah) 105 m

Natural resources: arable land, natural gas

Irrigated land: 240 sq km; note—includes the West Bank (2012)

Population distribution: population concentrated in major cities, particularly Gaza City in the north

Natural hazards: droughts

Environment—current issues: soil degradation; desertification; water pollution from chemicals and pesticides; salination of fresh water; improper sewage treatment; water-borne disease; depletion and contamination of underground water resources

Geography—note: strategic strip of land along Mideast-North African trade routes has experienced an incredibly turbulent history; the town of Gaza itself has been besieged countless times in its history; there are no Israeli settlements in the Gaza Strip; the Gaza Strip settlements were evacuated in 2005

PEOPLE AND SOCIETY

Population: 1,918,221 (July 2020 est.)
country comparison to the world: 152

Nationality: *noun:* NA
adjective: NA

Ethnic groups: Palestinian Arab

Languages: Arabic, Hebrew (spoken by many Palestinians), English (widely understood)

Religions: Muslim 98.0—99.0% (predominantly Sunni), Christian <1.0%, other, unaffiliated, unspecified <1.0% (2012 est.)
note: dismantlement of Israeli settlements was completed in September 2005; Gaza has had no Jewish population since then

MENA religious affiliation:

Age structure: *0-14 years:* 42.53% (male 418,751/female 397,013)
15-24 years: 21.67% (male 210,240/female 205,385)
25-54 years: 29.47% (male 275,976/female 289,277)
55-64 years: 3.66% (male 36,409/female 33,731)
65 years and over: 2.68% (male 27,248/female 24,191) (2020 est.)

Dependency ratios: *total dependency ratio:* 71.2
youth dependency ratio: 65.7
elderly dependency ratio: 5.5
potential support ratio: 18.2 (2020 est.)
note: data represent Gaza Strip and the West Bank

Median age: *total:* 18 years
male: 17.7 years
female: 18.4 years (2020 est.)
country comparison to the world: 214

Population growth rate: 2.13% (2020 est.)
country comparison to the world: 40

Birth rate: 28.6 births/1,000 population (2020 est.)
country comparison to the world: 37

Death rate: 3 deaths/1,000 population (2020 est.)
country comparison to the world: 225

Net migration rate: -4.7 migrant(s)/1,000 population (2020 est.)
country comparison to the world: 194

Population distribution: population concentrated in major cities, particularly Gaza City in the north

Urbanization: *urban population:* 76.7% of total population (2020)
rate of urbanization: 3% annual rate of change (2015-20 est.)
note: data represent Gaza Strip and the West Bank

365

total population growth rate v. urban population growth rate, 2000-2030:

Sex ratio: *at birth:* 1.06 male(s)/female
0-14 years: 1.05 male(s)/female
15-24 years: 1.02 male(s)/female
25-54 years: 0.95 male(s)/female
55-64 years: 1.08 male(s)/female
65 years and over: 1.13 male(s)/female
total population: 1.02 male(s)/female (2020 est.)

Maternal mortality rate: 27 deaths/100,000 live births (2017 est.)
note: data represent Gaza Strip and the West Bank
country comparison to the world: 117

Infant mortality rate: *total:* 14.9 deaths/1,000 live births
male: 16 deaths/1,000 live births
female: 13.9 deaths/1,000 live births (2020 est.)
country comparison to the world: 98
Life expectancy at birth: total population: 74.9 years
male: 73.1 years
female: 76.7 years (2020 est.)
country comparison to the world: 122

Total fertility rate: 3.64 children born/woman (2020 est.)
country comparison to the world: 39

Contraceptive prevalence rate: 57.2% (2014)
note: includes Gaza Strip and West Bank

Drinking water source:
improved:
urban: 97.1% of population
rural: 97.1% of population
total: 96.8% of population
unimproved:
urban: 2.9% of population
rural: 2.9% of population
total: 3.2% of population (2017 est.)
note: includes Gaza Strip and the West Bank

Physicians density: 2.77 physicians/1,000 population (2018)

Hospital bed density: 1.3 beds/1,000 population (2018)

Sanitation facility access:
improved:
urban: 100% of population
rural: 99.3% of population
total: 99.8% of population
unimproved:
urban: 0% of population
rural: 0.7% of population
total: 0.2% of population (2017 est.)
note: note includes Gaza Strip and the West Bank

HIV/AIDS—adult prevalence rate: NA

HIV/AIDS—people living with HIV/AIDS: NA

HIV/AIDS—deaths: NA

Children under the age of 5 years underweight: 1.4% (2014)
note: estimate is for Gaza Strip and the West Bank
country comparison to the world: 122

Education expenditures: 5.3% of GDP (2017)
note: includes Gaza Strip and the West Bank
country comparison to the world: 47

Literacy: *definition:* age 15 and over can read and write
total population: 97.2%
male: 98.7%
female: 95.7% (2018)
note: estimates are for Gaza and the West Bank

School life expectancy (primary to tertiary education): *total:* 13 years
male: 13 years
female: 14 years (2013)
note: data represent Gaza Strip and the West Bank

Unemployment, youth ages 15-24: *total:* 42.2%
male: 37%
female: 69.4% (2018 est.) note: includes the West Bank
country comparison to the world: 8

GOVERNMENT

Country name: conventional long form: none
conventional short form: Gaza Strip
local long form: none
local short form: Qita' Ghazzah
etymology: named for the largest city in the region, Gaza, whose settlement can be traced back to at least the 15th century B.C. (as "Ghazzat")

ECONOMY

Economy—overview: Movement and access restrictions, violent attacks, and the slow pace of post-conflict reconstruction continue to degrade economic conditions in the Gaza Strip, the smaller of the two areas comprising the Palestinian territories. Israeli controls became more restrictive after HAMAS seized control of the territory in June 2007. Under Hamas control, Gaza has suffered from rising unemployment, elevated poverty rates, and a sharp contraction of the private sector, which had relied primarily on export markets.

Since April 2017, the Palestinian Authority has reduced payments for electricity supplied to Gaza and cut salaries for its employees there, exacerbating poor economic conditions. Since 2014, Egypt's crackdown on the Gaza Strip's extensive tunnel-based smuggling network has exacerbated fuel, construction material, and consumer goods shortages in the territory. Donor support for reconstruction following the 51-day conflict in 2014 between Israel and HAMAS and other Gaza-based militant groups has fallen short of post-conflict needs.

GDP (purchasing power parity): see entry for the West Bank

GDP (official exchange rate): $2.938 billion (2014 est.)
note: excludes the West Bank

GDP—real growth rate:
-15.2% (2014 est.)
5.6% (2013 est.)
7% (2012 est.)
note: excludes the West Bank
country comparison to the world: 223

GDP—per capita (PPP): see entry for the the West Bank

GDP—composition, by end use:
household consumption: 88.6% (2017 est.)
government consumption: 26.3% (2017 est.)
investment in fixed capital: 22.4% (2017 est.)
investment in inventories: 0% (2017 est.)
exports of goods and services: 18.6% (2017 est.)
imports of goods and services: -55.6% (2017 est.)
note: data exclude the West Bank

GDP—composition, by sector of origin:
agriculture: 3% (2017 est.)
industry: 21.1% (2017 est.)
services: 75% (2017 est.)
note: data exclude the West Bank

Agriculture—products: olives, fruit, vegetables, flowers; beef, dairy products

Industries: textiles, food processing, furniture

Industrial production growth rate: 2.2% (2017 est.)
note: see entry for the West Bank
country comparison to the world: 125

Labor force:
1.24 million (2017 est.)
note: excludes the West Bank
country comparison to the world: 138

Labor force—by occupation: *agriculture:* 5.2%
industry: 10%
services: 84.8% (2015 est.)
note: data exclude the West Bank

Unemployment rate:
27.9% (2017 est.)
27% (2016 est.)
note: data exclude the West Bank
country comparison to the world: 201

Population below poverty line: 30% (2011 est.)
note: data exclude the West Bank

Budget: *see entry for the West Bank Fiscal year:* calendar year

Inflation rate (consumer prices):
0.2% (2017 est.)
-0.2% (2016 est.)
note: excludes the West Bank
country comparison to the world: 16

Current account balance:
-$1.444 billion (2017 est.)
-$1.348 billion (2016 est.)
note: excludes the West Bank
country comparison to the world: 152

Exports:
$1.955 billion (2017 est.)
$1.827 billion (2016 est.)
country comparison to the world: 141

Exports—commodities: strawberries, carnations, vegetables, fish (small and irregular shipments, as permitted to transit the Israeli-controlled Kerem Shalom crossing)

Imports:
$8.59 billion (2018 est.)
$7.852 billion (2017 est.)
see entry for the West Bank
country comparison to the world: 108

Imports—commodities: food, consumer goods, fuel

Reserves of foreign exchange and gold:

$446.3 million (31 December 2017 est.)
$583 million (31 December 2015 est.)
country comparison to the world: 156

Debt—external: see entry for the West Bank

Exchange rates: see entry for the West Bank

ENERGY

Electricity access: *population without electricity:* 80,930 (2012)
electrification—total population: 98% (2012)
electrification—urban areas: 99% (2012)
electrification—rural areas: 93% (2012)
note: data for Gaza Strip and West Bank combined

Electricity—production: 51,000 kWh (2011 est.)
country comparison to the world: 218

Electricity—consumption: 202,000 kWh (2009 est.)
country comparison to the world: 216

Electricity—exports: 0 kWh (2011 est.)
country comparison to the world: 139

Electricity—imports: 193,000 kWh (2011 est.)
country comparison to the world: 117

Crude oil—proved reserves: 0 bbl (1 January 2010 est.)
country comparison to the world: 136

COMMUNICATIONS

Telephones—fixed lines: *total subscriptions:* 472,293 (includes the West Bank); (July 2016 est.)
subscriptions per 100 inhabitants: 9 (includes the West Bank); (July 2016 est.)
country comparison to the world: 92

Telephones—mobile cellular: *total subscriptions:* 4,135,363 (includes the West Bank)
subscriptions per 100 inhabitants: 76 (includes the West Bank) (2017 est.)
country comparison to the world: 128

Telecommunication systems: *general assessment:* Israel has final say in allocating frequencies in the Gaza Strip and does not permit anything beyond a 2G network (2018)
domestic: Israeli company BEZEK and the Palestinian company PALTEL are responsible for fixed-line services; the Palestinian JAWWAL

company provides cellular services; a slow 2G network allows calls and limited data transmission; fixed-line 9 per 100 and mobile-cellular 76 per 100 (includes West Bank)
international: country code 970 or 972 (2018)
note: the COVID-19 outbreak is negatively impacting telecommunications production and supply chains globally; consumer spending on telecom devices and services has also slowed due to the pandemic's effect on economies worldwide; overall progress towards improvements in all facets of the telecom industry—mobile, fixed-line, broadband, submarine cable and satellite—has moderated

Broadcast media: 1 TV station and about 10 radio stations; satellite TV accessible

Internet country code: .ps*note—same as the West Bank*

Internet users: *total:* 2.673 million (includes the West Bank)
percent of population: 57.4% (July 2016 est.)
country comparison to the world: 105

Broadband—fixed subscriptions: *total:* 320,500
subscriptions per 100 inhabitants: 14 (2016 est.)
note: includes West Bank
country comparison to the world: 100

TRANSPORTATION

Airports: 1 (2013)
country comparison to the world: 222

Airports—with paved runways: *total:* 1 (2019)
under 914 m: 1
note—non-operational

Heliports: 1 (2013)

Roadways: *note:* see entry for the West Bank

Ports and terminals: *major seaport(s):* Gaza

MILITARY AND SECURITY

Military and security forces: HAMAS does not have a conventional military in the Gaza Strip but maintains security forces in addition to its military wing, the 'Izz al-Din al-Qassam Brigades; the military wing reports to the HAMAS Political Bureau leadership; there are several other militant groups operating in Gaza, most notably the

Al-Quds Brigades of Palestine Islamic Jihad, which are usually but not always beholden to HAMAS's authority (2019)

Military and security service personnel strengths: the military wing of HAMAS, the Izz al-Din al-Qassam Brigades, has an estimated 15-25,000 fighters (2019 est.)

Military equipment inventories and acquisitions: the military wing of HAMAS is armed with light weapons, including an inventory of improvised rocket, anti-tank missile, and mortar capabilities; HAMAS acquires its weapons through smuggling or local construction; Iran provides military support to HAMAS (2019 est.)

TERRORISM

Terrorist group(s): Army of Islam; Abdallah Azzam Brigades; al-Aqsa Martyrs Brigade; HAMAS; Islamic Revolutionary Guard Corps/ Qods Force; Islamic State of Iraq and ash-Sham (ISIS)-Sinai Province; Mujahidin Shura Council in the Environs of Jerusalem; Palestine Islamic Jihad; Palestine Liberation Front; PFLP-General Command; Popular Front for the Liberation of Palestine (2019)
note: details about the history, aims, leadership, organization, areas of operation, tactics, targets, weapons, size, and sources of support of the group(s) appear(s) in Appendix-T

TRANSNATIONAL ISSUES

Disputes—international: the status of the Gaza Strip is a final status issue to be resolved through negotiations; Israel removed settlers and military personnel from Gaza Strip in September 2005

Refugees and internally displaced persons: *refugees (country of origin):* 1,460,315 (Palestinian refugees) (2020)

IDPs: 243,000 (includes persons displaced within the Gaza Strip due to the intensification of the Israeli-Palestinian conflict since June 2014 and other Palestinian IDPs in the Gaza Strip and West Bank who fled as long ago as 1967, although confirmed cumulative data do not go back beyond 2006) (2019)

GEORGIA

INTRODUCTION

Background: The region of present day Georgia contained the ancient kingdoms of Colchis and Kartli-Iberia. The area came under Roman influence in the first centuries A.D., and Christianity became the state religion in the 330s. Domination by Persians, Arabs, and Turks was followed by a Georgian golden age (11th-13th centuries) that was cut short by the Mongol invasion of 1236. Subsequently, the Ottoman and Persian empires

competed for influence in the region. Georgia was absorbed into the Russian Empire in the 19th century. Independent for three years (1918-1921) following the Russian revolution, it was forcibly incorporated into the USSR in 1921 and regained its independence when the Soviet Union dissolved in 1991.

Mounting public discontent over rampant corruption and ineffective government services, followed by an attempt by the incumbent Georgian Government to manipulate parliamentary elections in November 2003, touched off widespread protests that led to the resignation of Eduard SHEVARDNADZE, president since 1995. In the aftermath of that popular movement, which became known as the "Rose Revolution," new elections in early 2004 swept Mikheil SAAKASHVILI into power along with his United National Movement (UNM) party. Progress on market reforms and democratization has been made in the years since independence, but this progress has been complicated by Russian assistance and support to the separatist regions of Abkhazia and South Ossetia. Periodic flare-ups in tension and violence culminated in a five-day conflict in August 2008 between Russia and Georgia, including the invasion of large portions of undisputed Georgian territory. Russian troops pledged to pull back from most occupied Georgian territory, but in late August 2008 Russia unilaterally recognized the independence of Abkhazia and South Ossetia, and Russian military forces remain in those regions.

Billionaire Bidzina IVANISHVILI's unexpected entry into politics in October 2011 brought the divided opposition together under his Georgian Dream coalition, which won a majority of seats in the October 2012 parliamentary elections and removed UNM from power. Conceding defeat, SAAKASHVILI named IVANISHVILI as prime minister and allowed Georgian Dream to create a new government. Giorgi MARGVELASHVILI was inaugurated as president on 17 November 2013, ending a tense year of power-sharing between SAAKASHVILI and IVANISHVILI. At the time, these changes in leadership represented unique examples of a former Soviet state that emerged to conduct democratic and peaceful government transitions of power. IVANISHVILI voluntarily resigned from office after the presidential succession, and Georgia's legislature on 20 November 2013 confirmed Irakli GARIBASHVILI as his replacement. GARIBASHVILI was replaced by Giorgi KVIRIKASHVILI in December 2015. KVIRIKASHVILI remained prime minister following Georgian Dream's success in the October 2016 parliamentary elections, where the party won a constitutional majority. IVANISHVILI reemerged as Georgian Dream party chairman in April 2018. KVIRIKASHVILI resigned in June 2018 and was replaced by Mamuka BAKHTADZE. In September 2019, BAKHTADZE resigned and Giorgi GAKHARIA was named the country's new head of government, Georgia's fifth prime minister in seven years. Popular and government support for integration with the West is high in Georgia. Joining the EU and NATO are among the country's top foreign policy goals.

GEOGRAPHY

Location: Southwestern Asia, bordering the Black Sea, between Turkey and Russia, with a sliver of land north of the Caucasus extending into Europe; note—Georgia views itself as part of Europe; geopolitically, it can be classified as falling within Europe, the Middle East, or both

Geographic coordinates: 42 00 N, 43 30 E

Map references: Asia

Area: *total:* 69,700 sq km
land: 69,700 sq km
water: 0 sq km
note: approximately 12,560 sq km, or about 18% of Georgia's area, is Russian occupied; the seized area includes all of Abkhazia and the breakaway region of South Ossetia, which consists of the northern part of Shida Kartli, eastern slivers of the Imereti region and Racha-Lechkhumi and Kvemo Svaneti, and part of western Mtskheta-Mtianeti
country comparison to the world: 122

Area—comparative: slightly smaller than South Carolina; slightly larger than West Virginia

Land boundaries: *total:* 1,814 km
border countries (4): Armenia 219 km, Azerbaijan 428 km, Russia 894 km, Turkey 273 km

Coastline: 310 km

Maritime claims: *territorial sea:* 12 nm
exclusive economic zone: 200 nm

Climate: warm and pleasant; Mediterranean-like on Black Sea coast

Terrain: largely mountainous with Great Caucasus Mountains in the north and Lesser Caucasus Mountains in the south; Kolkhet'is Dablobi (Kolkhida Lowland) opens to the Black Sea in the west; Mtkvari River Basin in the east; fertile soils in river valley flood plains and foothills of Kolkhida Lowland

Elevation: *mean elevation:* 1,432 m
lowest point: Black Sea 0 m
highest point: Mt'a Shkhara 5,193 m

Natural resources: timber, hydropower, manganese deposits, iron ore, copper, minor coal and oil deposits; coastal climate and soils allow for important tea and citrus growth

Land use: *agricultural land:* 35.5% (2011 est.)
arable land: 5.8% (2011 est.) / permanent crops: 1.8% (2011 est.) / permanent pasture: 27.9% (2011 est.)
forest: 39.4% (2011 est.)
other: 25.1% (2011 est.)

Irrigated land: 4,330 sq km (2012)

Population distribution: settlement concentrated in the central valley, particularly in the capital city of Tbilisi in the east; smaller urban agglomerations dot the Black Sea coast, with Bat'umi being the largest

Natural hazards: earthquakes

Environment—current issues: air pollution, particularly in Rust'avi; heavy water pollution of Mtkvari River and the Black Sea; inadequate supplies of potable water; soil pollution from toxic chemicals; land and forest degradation; biodiversity loss; waste management

Environment—international agreements: *party to:* Air Pollution, Biodiversity, Climate Change, Climate Change-Kyoto Protocol, Desertification, Endangered Species, Hazardous Wastes, Law of the Sea, Ozone Layer Protection, Ship Pollution, Wetlands

signed, but not ratified: none of the selected agreements

Geography—note: note 1: strategically located east of the Black Sea; Georgia controls much of the Caucasus Mountains and the routes through them
note 2: the world's four deepest caves are all in Georgia, including two that are the only known caves on earth deeper than 2,000 m: Krubera Cave at -2,197 m (-7,208 ft; reached in 2012) and Veryovkina Cave at -2,212 (-7,257 ft; reached in 2018)

PEOPLE AND SOCIETY

Population: 3.997 million (2019 est. est.)
country comparison to the world: 129

Nationality: *noun:* Georgian(s)
adjective: Georgian

Ethnic groups: Georgian 86.8%, Azeri 6.3%, Armenian 4.5%, other 2.3% (includes Russian, Ossetian, Yazidi, Ukrainian, Kist, Greek) (2014 est.)

Languages: Georgian (official) 87.6%, Azeri 6.2%, Armenian 3.9%, Russian 1.2%, other 1% (2014 est.)
note: Abkhaz is the official language in Abkhazia

Religions: Orthodox (official) 83.4%, Muslim 10.7%, Armenian Apostolic 2.9%, other 1.2% (includes Catholic, Jehovah's Witness, Yazidi, Protestant, Jewish), none 0.5%, unspecified/no answer 1.2% (2014 est.)

Age structure: *0-14 years:* 18.42% (male 472,731/ female 435,174)
15-24 years: 10.9% (male 286,518/female 250,882)
25-54 years: 40.59% (male 984,942/female 1,016,353)
55-64 years: 13.24% (male 288,650/female 364,117)
65 years and over: 16.85% (male 326,219/female 504,444) (2020 est.)

Dependency ratios: *total dependency ratio:* 55
youth dependency ratio: 31.3
elderly dependency ratio: 23.6
potential support ratio: 4.2 (2020 est.)

Median age: *total:* 38.6 years
male: 35.9 years
female: 41.4 years (2020 est.)
country comparison to the world: 60

Population growth rate: 0.05% (2020 est.)
country comparison to the world: 189

Birth rate: 11.6 births/1,000 population (2020 est.)
country comparison to the world: 169

Death rate: 11 deaths/1,000 population (2020 est.)
country comparison to the world: 21

Net migration rate: 0.1 migrant(s)/1,000 population (2020 est.)
country comparison to the world: 74

Population distribution: settlement concentrated in the central valley, particularly in the capital city of Tbilisi in the east; smaller urban agglomerations dot the Black Sea coast, with Bat'umi being the largest

Urbanization: *urban population:* 59.5% of total population (2020)

rate of urbanization: 0.42% annual rate of change (2015-20 est.)

note: data include Abkhazia and South Ossetia total population growth rate v. urban population growth rate, 2000-2030:

Major urban areas—Population: 1.078 million TBILISI (capital) (2020)

Sex ratio: *at birth:* 1.05 male(s)/female

0-14 years: 1.09 male(s)/female

15-24 years: 1.14 male(s)/female

25-54 years: 0.97 male(s)/female

55-64 years: 0.79 male(s)/female

65 years and over: 0.65 male(s)/female

total population: 0.92 male(s)/female (2020 est.)

Mother's mean age at first birth: 25.4 years (2017 est.)

note: data do not cover Abkhazia and South Ossetia

Maternal mortality rate: 25 deaths/100,000 live births (2017 est.)

country comparison to the world: 121

Infant mortality rate: *total:* 13.8 deaths/1,000 live births

male: 15.8 deaths/1,000 live births

female: 11.7 deaths/1,000 live births (2020 est.)

country comparison to the world: 101

Life expectancy at birth: total population: 77 years

male: 72.9 years

female: 81.3 years (2020 est.)

country comparison to the world: 85

Total fertility rate: 1.75 children born/woman (2020 est.)

country comparison to the world: 160

Contraceptive prevalence rate: 40.6% (2018)

Drinking water source:

improved:

urban: 100% of population

rural: 96.2% of population

total: 98.4% of population

unimproved:

urban: 0% of population

rural: 3.8% of population

total: 1.6% of population (2017 est.)

Current Health Expenditure: 7.6% (2017)

Physicians density: 6.13 physicians/1,000 population (2017)

Hospital bed density: 2.9 beds/1,000 population (2014)

Sanitation facility access:

improved:

urban: 97% of population

rural: 82.7% of population

total: 91.1% of population

unimproved:

urban: 3% of population

rural: 17.3% of population

total: 8.9% of population (2017 est.)

HIV/AIDS—adult prevalence rate: 0.3% (2019 est.)

country comparison to the world: 87

HIV/AIDS—people living with HIV/AIDS: 9,100 (2019 est.)

country comparison to the world: 107

HIV/AIDS—deaths: <100 (2019 est.)

Obesity—adult prevalence rate: 21.7% (2016)

country comparison to the world: 86

Children under the age of 5 years underweight: 1.1% (2009)

country comparison to the world: 126

Education expenditures: 3.8% of GDP (2017)

country comparison to the world: 111

Literacy: *definition:* age 15 and over can read and write

total population: 99.4%

male: 99.4%

female: 99.3% (2017)

School life expectancy (primary to tertiary education): *total:* 15 years

male: 16 years

female: 16 years (2019)

Unemployment, youth ages 15-24: *total:* 29.9%

male: 26.7%

female: 35.3% (2018 est.)

country comparison to the world: 29

GOVERNMENT

Country name: *conventional long form:* none

conventional short form: Georgia

local long form: none

local short form: Sak'art'velo

former: Georgian Soviet Socialist Republic

etymology: the Western name may derive from the Persian designation "gurgan" meaning "Land of the Wolves"; the native name "Sak'art'velo" means "Land of the Kartvelians" and refers to the core central Georgian region of Kartli

Government type: semi-presidential republic

Capital: *name:* Tbilisi

geographic coordinates: 41 41 N, 44 50 E

time difference: UTC+4 (9 hours ahead of Washington, DC, during Standard Time)

etymology: the name in Georgian means "warm place," referring to the numerous sulfuric hot springs in the area

Administrative divisions: 9 regions (mkharebi, singular—mkhare), 1 city (kalaki), and 2 autonomous republics (avtomnoy respubliki, singular—avtom respublika)

regions: Guria, Imereti, Kakheti, Kvemo Kartli, Mtskheta Mtianeti, Racha-Lechkhumi and Kvemo Svaneti, Samegrelo and Zemo Svaneti, Samtskhe-Javakheti, Shida Kartli; note—the breakaway region of South Ossetia consists of the northern part of Shida Kartli, eastern slivers of the Imereti region and Racha-Lechkhumi and Kvemo Svaneti, and part of western Mtskheta—Mtianeti

city: Tbilisi

autonomous republics: Abkhazia or Ap'khazet'is Avtonomiuri Respublika (Sokhumi), Ajaria or Acharis Avtonomiuri Respublika (Bat'umi)

note 1: the administrative centers of the two autonomous republics are shown in parentheses

note 2: the United States recognizes the breakaway regions of Abkhazia and South Ossetia to be part of Georgia

Independence: 9 April 1991 (from the Soviet Union); notable earlier date: A.D. 1008 (Georgia unified under King BAGRAT III)

National holiday: Independence Day, 26 May (1918); note—26 May 1918 was the date of independence from Soviet Russia, 9 April 1991 was the date of independence from the Soviet Union

Constitution: *history:* previous 1921, 1978 (based on 1977 Soviet Union constitution); latest approved 24 August 1995, effective 17 October 1995

amendments: proposed as a draft law supported by more than one half of the Parliament membership or by petition of at least 200,000 voters; passage requires support by at least three fourths of the Parliament membership in two successive sessions three months apart and the signature and promulgation by the president of Georgia; amended several times, last in 2020 (legislative electoral system revised)

Legal system: civil law system

International law organization participation: accepts compulsory ICJ jurisdiction; accepts ICCt jurisdiction

Citizenship: citizenship by birth: no

citizenship by descent only: at least one parent must be a citizen of Georgia

dual citizenship recognized: no

residency requirement for naturalization: 10 years

Suffrage: 18 years of age; universal

Executive branch: *chief of state:* President Salome ZOURABICHVILI (since 16 December 2018)

head of government: Prime Minister Giorgi GAKHARIA (since 8 September 2019)

cabinet: Cabinet of Ministers

elections/appointments: president directly elected by absolute majority popular vote in 2 rounds if needed for a 5-year term (eligible for a second term); election last held on 28 November 2018 (next to be held in 2024); prime minister nominated by Parliament, appointed by the president note—2017 constitutional amendments made the 2018 election the last where the president was directly elected; future presidents will be elected by a 300-member College of Electors; in light of these changes, ZOURABICHVILI was allowed a six-year term

election results: Salome ZOURABICHVILI elected president in runoff; percent of vote—Salome ZOURABICHVILI (independent, backed by Georgian Dream) 59.5%, Grigol VASHADZE (UNM) 40.5%; Giorgi GAKHARIA approved as prime minister by Parliamentary vote 98-0

Legislative branch: *description:* unicameral Parliament or Sakartvelos Parlamenti (150 seats; 120 members directly elected in a single nationwide constituency by closed, party-list proportional representation vote and 30 directly elected in single-seat constituencies by at least 50% majority vote, with a runoff if needed; no party earning less than 40% of total votes may claim a majority; members serve 4-year terms)

elections: last held on 31 October and 21 November 2020 (next to be held in October 2024)

election results: percent of vote by party—Georgian Dream 48.2%, UNM 27.2%, European Georgia 3.8%, Lelo 3.2%, Strategy 3.2%, Alliance of Patriots 3.1%, Girchi 2.9%, Citizens 1.3%, Labor 1%; seats by party—Georgian Dream 90, UNM 36, European Georgia 5, Lelo 4, Strategy 4, Alliance of Patriots 4, Girchi 4, Citizens 2, Labor 1

Judicial branch: *highest courts:* Supreme Court (consists of 28 judges organized into several specialized judicial chambers; number of judges determined by the president of Georgia); Constitutional Court (consists of 9 judges); note—the Abkhazian and Ajarian Autonomous republics each have a supreme court and a hierarchy of lower courts

judge selection and term of office: Supreme Court judges nominated by the High Council of Justice (a 14-member body consisting of the Supreme Court chairperson, common court judges, and appointees of the president of Georgia) and appointed by Parliament; judges appointed for life; Constitutional Court judges appointed 3 each by the president, by Parliament, and by the Supreme Court judges; judges appointed for 10-year terms

subordinate courts: Courts of Appeal; regional (town) and district courts

Political parties and leaders:

Alliance of Patriots [Irma INASHVILI]
Democratic Movement-United Georgia [Nino BURJANADZE]
Citizens Party
Development Movement [Davit USPASHVILI]
European Georgia-Movement for Liberty [Davit BAKRADZE]
For Justice Party [Eka BESELIA]
Free Democrats or FD [Shalva SHAVGULIDZE]
Georgian Dream-Democratic Georgia [Bidzina IVANISHVILI]
Girchi (Pinecone) [Zurab JAPARIDZE]
Industry Will Save Georgia (Industrialists) or IWSG [Giorgi TOPADZE]
Labor Party [Shalva NATELASHVILI]
Lelo for Georgia [Mamuka KHAZARADZE]
New Georgia [Giorgi VASHADZE]
Republican Party [Khatuna SAMNIDZE]
Strategy Aghmashenebeli [Giorgi VASHADZE]
United National Movement or UNM [Grigol VASHADZE]

International organization participation: ADB, BSEC, CD, CE, CPLP (associate), EAPC, EBRD, FAO, G-11, GCTU, GUAM, IAEA, IBRD, ICAO, ICC (national committees), ICCt, ICRM, IDA, IFAD, IFC, IFRCS, ILO, IMF, IMO, Interpol, IOC, IOM, IPU, ISO (correspondent), ITSO, ITU, ITUC (NGOs), MIGA, OAS (observer), OIF (observer), OPCW, OSCE, PFP, SELEC (observer), UN, UNCTAD, UNESCO, UNIDO, UNWTO, UPU, WCO, WHO, WIPO, WMO, WTO

Diplomatic representation in the US: *chief of mission:* Ambassador David BAKRADZE (since 18 January 2017)

chancery: 1824 R Street NW, Washington, DC 20009

telephone: [1] (202) 387-2390

FAX: [1] (202) 387-0864

consulate(s) general: New York

Diplomatic representation from the US: *chief of mission:* Ambassador Kelly C. DEGNAN (since 31 January 2020)

telephone: [995] (32) 227-70-00

embassy: 11 George Balanchine Street, Tbilisi, 0131

mailing address: 7060 T'bilisi Place, Washington, DC 20521-7060

FAX: [995] (32) 253-23-10

Flag description: white rectangle with a central red cross extending to all four sides of the flag; each of the four quadrants displays a small red bolnur-katskhuri cross; sometimes referred to as the Five-Cross Flag; although adopted as the official Georgian flag in 2004, the five-cross design appears to date back to the 14th century

National symbol(s): Saint George, lion; national colors: red, white

National anthem: *name:* "Tavisupleba" (Liberty)

lyrics/music: Davit MAGRADSE/Zakaria PALIASHVILI (adapted by Joseb KETSCHAKMADSE)

note: adopted 2004; after the Rose Revolution, a new anthem with music based on the operas "Abesalom da Eteri" and "Daisi" was adopted

0:00/ 0:00

ECONOMY

Economy—overview: Georgia's main economic activities include cultivation of agricultural products such as grapes, citrus fruits, and hazelnuts; mining of manganese, copper, and gold; and producing alcoholic and nonalcoholic beverages, metals, machinery, and chemicals in small-scale industries. The country imports nearly all of its needed supplies of natural gas and oil products. It has sizeable hydropower capacity that now provides most of its electricity needs.

Georgia has overcome the chronic energy shortages and gas supply interruptions of the past by renovating hydropower plants and by increasingly relying on natural gas imports from Azerbaijan instead of from Russia. Construction of the Baku-Tbilisi-Ceyhan oil pipeline, the South Caucasus gas pipeline, and the Baku-Tbilisi-Kars railroad are part of a strategy to capitalize on Georgia's strategic location between Europe and Asia and develop its role as a transit hub for gas, oil, and other goods.

Georgia's economy sustained GDP growth of more than 10% in 2006-07, based on strong inflows of foreign investment, remittances, and robust government spending. However, GDP growth slowed following the August 2008 conflict with Russia, and sank to negative 4% in 2009 as foreign direct investment and workers' remittances declined in

the wake of the global financial crisis. The economy rebounded in the period 2010-17, but FDI inflows, the engine of Georgian economic growth prior to the 2008 conflict, have not recovered fully. Unemployment remains persistently high.

The country is pinning its hopes for faster growth on a continued effort to build up infrastructure, enhance support for entrepreneurship, simplify regulations, and improve professional education, in order to attract foreign investment and boost employment, with a focus on transportation projects, tourism, hydropower, and agriculture. Georgia had historically suffered from a chronic failure to collect tax revenues; however, since 2004 the government has simplified the tax code, increased tax enforcement, and cracked down on petty corruption, leading to higher revenues. The government has received high marks from the World Bank for improvements in business transparency. Since 2012, the Georgian Dream-led government has continued the previous administration's low-regulation, low-tax, free market policies, while modestly increasing social spending and amending the labor code to comply with International Labor Standards. In mid-2014, Georgia concluded an association agreement with the EU, paving the way to free trade and visa-free travel. In 2017, Georgia signed Free Trade Agreement (FTA) with China as part of Tbilisi's efforts to diversify its economic ties. Georgia is seeking to develop its Black Sea ports to further facilitate East-West trade.

GDP (purchasing power parity): $39.85 billion (2017 est.)

$37.96 billion (2016 est.)

$36.91 billion (2015 est.)

note: data are in 2017 dollars

country comparison to the world: 119

GDP (official exchange rate): $15.16 billion (2017 est.)

GDP—real growth rate:

5% (2017 est.)

2.8% (2016 est.)

2.9% (2015 est.)

country comparison to the world: 47

GDP—per capita (PPP): $10,700 (2017 est.)

$10,300 (2016 est.)

$9,900 (2015 est.)

note: data are in 2017 dollars

country comparison to the world: 138

Gross national saving:

23% of GDP (2017 est.)

19.9% of GDP (2016 est.)

19.5% of GDP (2015 est.)

country comparison to the world: 75

GDP—composition, by end use:

household consumption: 62.8% (2017 est.)

government consumption: 17.1% (2017 est.)

investment in fixed capital: 29.5% (2017 est.)

investment in inventories: 2.4% (2017 est.)

exports of goods and services: 50.4% (2017 est.)

imports of goods and services: -62.2% (2017 est.)

GDP—composition, by sector of origin:

agriculture: 8.2% (2017 est.)

industry: 23.7% (2017 est.)
services: 67.9% (2017 est.)

Agriculture—products: citrus, grapes, tea, hazelnuts, vegetables; livestock

Industries: steel, machine tools, electrical appliances, mining (manganese, copper, gold), chemicals, wood products, wine

Industrial production growth rate: 6.7% (2017 est.)
country comparison to the world: 34

Labor force: 686,000 (2019 est.)
country comparison to the world: 150

Labor force—by occupation: *agriculture:* 55.6%
industry: 8.9%
services: 35.5% (2006 est.)

Unemployment rate:
NA% (2017 est.)
11.8% (2016 est.)
country comparison to the world: 160

Population below poverty line: 9.2% (2010 est.)

Household income or consumption by percentage share: *lowest 10%:* 2%
highest 10%: 31.3% (2008)

Budget: *revenues:* 4.352 billion (2017 est.)
expenditures: 4.925 billion (2017 est.)

Taxes and other revenues: 28.7% (of GDP) (2017 est.)
country comparison to the world: 91

Budget surplus (+) or deficit (-): -3.8% (of GDP) (2017 est.)
country comparison to the world: 153

Public debt:
44.9% of GDP (2017 est.)
44.4% of GDP (2016 est.)
note: data cover general government debt and include debt instruments issued (or owned) by government entities other than the treasury; the data include treasury debt held by foreign entities; the data include debt issued by subnational entities; Georgia does not maintain intragovernmental debt or social funds
country comparison to the world: 116

Fiscal year: calendar year

Inflation rate (consumer prices):
6% (2017 est.)
2.1% (2016 est.)
country comparison to the world: 185

Current account balance:
-$1.348 billion (2017 est.)
-$1.84 billion (2016 est.)
country comparison to the world: 156

Exports:
$3.566 billion (2017 est.)
$2.831 billion (2016 est.)
country comparison to the world: 121

Exports—partners: Russia 14.5%, Azerbaijan 10%, Turkey 7.9%, Armenia 7.7%, China 7.6%, Bulgaria 6.6%, Ukraine 4.6%, US 4.5% (2017)

Exports—commodities: vehicles, ferro-alloys, fertilizers, nuts, scrap metal, gold, copper ores

Imports:
$7.415 billion (2017 est.)
$6.747 billion (2016 est.)

country comparison to the world: 114

Imports—commodities: fuels, vehicles, machinery and parts, grain and other foods, pharmaceuticals

Imports—partners: Turkey 17.2%, Russia 9.9%, China 9.2%, Azerbaijan 7.6%, Ukraine 5.6%, Germany 5.4% (2017)

Reserves of foreign exchange and gold:
$3.039 billion (31 December 2017 est.)
$2.756 billion (31 December 2016 est.)
country comparison to the world: 108

Debt—external:
$16.99 billion (31 December 2017 est.)
$14.08 billion (31 December 2016 est.)
country comparison to the world: 100

Exchange rates: laris (GEL) per US dollar—
2.535 (2017 est.)
2.3668 (2016 est.)
2.3668 (2015 est.)
2.2694 (2014 est.)
1.7657 (2013 est.)

ENERGY

Electricity access: electrification—total population: 100% (2020)

Electricity—production: 13.24 billion kWh (2016 est.)
country comparison to the world: 91

Electricity—consumption: 12.37 billion kWh (2016 est.)
country comparison to the world: 87

Electricity—exports: 560 million kWh (2016 est.)
country comparison to the world: 66

Electricity—imports: 1.329 billion kWh (2016 est.)
country comparison to the world: 63

Electricity—installed generating capacity: 4.641 million kW (2016 est.)
country comparison to the world: 84

Electricity—from fossil fuels: 35% of total installed capacity (2016 est.)
country comparison to the world: 179

Electricity—from nuclear fuels: 0% of total installed capacity (2017 est.)
country comparison to the world: 95

Electricity—from hydroelectric plants: 65% of total installed capacity (2017 est.)
country comparison to the world: 22

Electricity—from other renewable sources: 0% of total installed capacity (2017 est.)
country comparison to the world: 187

Crude oil—production: 400 bbl/day (2018 est.)
country comparison to the world: 93

Crude oil—exports: 3,006 bbl/day (2017 est.)
country comparison to the world: 68

Crude oil—imports: 2,660 bbl/day (2015 est.)
country comparison to the world: 78

Crude oil—proved reserves: 35 million bbl (1 January 2018 est.)
country comparison to the world: 80

Refined petroleum products—production: 247 bbl/day (2017 est.)

country comparison to the world: 106

Refined petroleum products—consumption: 27,000 bbl/day (2016 est.)
country comparison to the world: 122

Refined petroleum products—exports: 2,052 bbl/day (2015 est.)
country comparison to the world: 104

Refined petroleum products—imports: 28,490 bbl/day (2015 est.)
country comparison to the world: 101

Natural gas—production: 7.363 million cu m (2017 est.)
country comparison to the world: 95

Natural gas—consumption: 2.294 billion cu m (2017 est.)
country comparison to the world: 83

Natural gas—exports: 0 cu m (2017 est.)
country comparison to the world: 109

Natural gas—imports: 2.294 billion cu m (2017 est.)
country comparison to the world: 50

Natural gas—proved reserves: 8.495 billion cu m (1 January 2018 est.)
country comparison to the world: 80

Carbon dioxide emissions from consumption of energy: 9.912 million Mt (2017 est.)
country comparison to the world: 109

COMMUNICATIONS

Telephones—fixed lines: *total subscriptions:* 638,092
subscriptions per 100 inhabitants: 12.95 (2019 est.)
country comparison to the world: 88

Telephones—mobile cellular: *total subscriptions:* 6,638,125
subscriptions per 100 inhabitants: 134.72 (2019 est.)
country comparison to the world: 106

Telecommunication systems: *general assessment:* telecommunications fastest growing area of Georgia's economy; LTE services now cover the vast majority of the population; fixed-line telecommunications network has limited coverage outside Tbilisi; multiple mobilecellular providers provide services to an increasing subscribership throughout the country; broadband subscribers steadily increasing; with the recent investment in infrastructure customers are moving from copper to fiber networks (2020)
domestic: fixed-line 13 per 100, cellular telephone networks cover the entire country; mobile-cellular teledensity roughly 135 per 100 persons; intercity facilities include a fiber-optic line between T'bilisi and K'ut'aisi (2019)
international: country code—995; landing points for the Georgia-Russia, Diamond Link Global, and Caucasus Cable System fiber-optic submarine cable that provides connectivity to Russia, Romania and Bulgaria; international service is available by microwave, landline, and satellite through the Moscow switch; international electronic mail and telex service are available (2019)

371

note: the COVID-19 outbreak is negatively impacting telecommunications production and supply chains globally; consumer spending on telecom devices and services has also slowed due to the pandemic's effect on economies worldwide; overall progress towards improvements in all facets of the telecom industry—mobile, fixed-line, broadband, submarine cable and satellite—has moderated

Broadcast media: The Tbilisi-based Georgian Public Broadcaster (GPB) includes Channel 1, Channel 2 as well as the Batumi-based Adjara TV, and the State Budget funds all three; there are also a number of independent commercial television broadcasters, such as Imedi, Rustavi 2, Pirveli TV, Maestro, Kavkasia, Georgian Dream Studios (GDS), Obiektivi, Mtavari Arkhi, and a small Russian language operator TOK TV; Tabula and Post TV are web-based television outlets; all of these broadcasters and web-based television outlets, except GDS, carry the news; the Georgian Orthodox Church also operates a satellite-based television station called Unanimity; there are 26 regional television broadcasters across Georgia that are members of the Georgian Association of Regional Broadcasters and/or the Alliance of Georgian Broadcasters; the broadcaster organizations seek to strengthen the regional media's capacities and distribution of regional products; a nationwide digital switchover occurred in 2015; there are several dozen private radio stations; GPB operates 2 radio stations (2019)

Internet country code: .ge

Internet users: total: 3,151,218
percent of population: 63.97% (July 2018 est.)
country comparison to the world: 97

Broadband—fixed subscriptions: total: 840,603
subscriptions per 100 inhabitants: 17 (2018 est.)
country comparison to the world: 72

TRANSPORTATION

National air transport system: *number of registered air carriers:* 4 (2020)
inventory of registered aircraft operated by air carriers: 12

annual passenger traffic on registered air carriers: 516,034 (2018)
annual freight traffic on registered air carriers: 750,000 mt-km (2018)
Civil aircraft registration country code prefix: 4L (2016)

Airports: 22 (2013)
country comparison to the world: 134

Airports—with paved runways: *total:* 18 (2017)
over 3,047 m: 1 (2017)
2,438 to 3,047 m: 7 (2017)
1,524 to 2,437 m: 3 (2017)
914 to 1,523 m: 5 (2017)
under 914 m: 2 (2017)

Airports—with unpaved runways: *total:* 4 (2013)
1,524 to 2,437 m: 1 (2013)
914 to 1,523 m: 2 (2013)
under 914 m: 1 (2013)

Heliports: 2 (2013)

Pipelines: 1596 km gas, 1175 km oil (2013)

Railways: *total:* 1,363 km (2014)
narrow gauge: 37 km 0.912-m gauge (37 km electrified) (2014)
broad gauge: 1,326 km 1.520-m gauge (1,251 km electrified) (2014)
country comparison to the world: 84

Roadways: *total:* 20,295 km (2018)
country comparison to the world: 113

Merchant marine: *total:* 82
by type: bulk carrier 3 general cargo 22, oil tanker 2, other 55 (2019)
country comparison to the world: 99

Ports and terminals: *major seaport(s):* Black Sea, Bat'umi, P'ot'i

MILITARY AND SECURITY

Military and security forces: *Georgian Defense Forces:* Land Forces (includes Aviation and Air Defense Forces); Special Operations Forces; National Guard; Ministry of the Interior: Border Police, Coast Guard (includes Georgian naval forces, which were merged with the Coast Guard in 2009) (2020)

Military expenditures:
2% of GDP (2019)
2% of GDP (2018)
2.1% of GDP (2017)
2.2% of GDP (2016)
2.1% of GDP (2015)
country comparison to the world: 50

Military and security service personnel strengths: estimates for the size of the Georgian Defense Forces vary; approximately 25,000 active troops, including the National Guard (2019 est.)

Military equipment inventories and acquisitions: the Georgian Defense Forces are equipped mostly with older Russian and Soviet-era weapons; since 2010, it has received limited quantities of equipment from Bulgaria, France, and the US (2019)

Military deployments: 860 Afghanistan (NATO) (2020)

Military service age and obligation: conscription reinstated in 2017; 18 to 27 years of age for compulsory and voluntary active duty military service; conscript service obligation is 12 months (2019)

Military—note: Georgia does not have any military stationed in the separatist territories of Abkhazia and South Ossetia, but large numbers of Russian servicemen have been stationed in these regions since the 2008 Russia-Georgia War (2019)

TRANSNATIONAL ISSUES

Disputes—international: Russia's military support and subsequent recognition of Abkhazia and South Ossetia independence in 2008 continue to sour relations with Georgia

Refugees and internally displaced persons: *IDPs:* 301,000 (displaced in the 1990s as a result of armed conflict in the breakaway republics of Abkhazia and South Ossetia; displaced in 2008 by fighting between Georgia and Russia over South Ossetia) (2019)
stateless persons: 559 (2019)

Illicit drugs: limited cultivation of cannabis and opium poppy, mostly for domestic consumption; used as transshipment point for opiates via Central Asia to Western Europe and Russia

GERMANY

INTRODUCTION

Background: As Europe's largest economy and second most populous nation (after Russia), Germany is a key member of the continent's economic, political, and defense organizations. European power struggles immersed Germany in two devastating world wars in the first half of the 20th century and left the country occupied by the victorious Allied powers of the US, UK, France, and the Soviet Union in 1945. With the advent of the Cold War, two German states were formed in 1949: the western Federal Republic of Germany

(FRG) and the eastern German Democratic Republic (GDR). The democratic FRG embedded itself in key western economic and security organizations, the EC (now the EU) and NATO, while the communist GDR was on the front line of the Soviet-led Warsaw Pact. The decline of the USSR and the end of the Cold War allowed for German reunification in 1990. Since then, Germany has expended considerable funds to bring eastern productivity and wages up to western standards. In January 1999, Germany and 10 other EU countries introduced a common European exchange currency, the euro.

GEOGRAPHY

Location: Central Europe, bordering the Baltic Sea and the North Sea, between the Netherlands and Poland, south of Denmark

Geographic coordinates: 51 00 N, 9 00 E

Map references: Europe

Area: *total:* 357,022 sq km
land: 348,672 sq km
water: 8,350 sq km
country comparison to the world: 64

Area—comparative: three times the size of Pennsylvania; slightly smaller than Montana

Land boundaries: *total:* 3,714 km
border countries (9): Austria 801 km, Belgium 133 km, Czech Republic 704 km, Denmark 140 km, France 418 km, Luxembourg 128 km, Netherlands 575 km, Poland 467 km, Switzerland 348 km

Coastline: 2,389 km

Maritime claims: *territorial sea:* 12 nm
exclusive economic zone: 200 nm
continental shelf: 200-m depth or to the depth of exploitation

Climate: temperate and marine; cool, cloudy, wet winters and summers; occasional warm mountain (foehn) wind

Terrain: lowlands in north, uplands in center, Bavarian Alps in south

Elevation: *mean elevation:* 263 m
lowest point: Neuendorf bei Wilster -3.5 m
highest point: Zugspitze 2,963 m

Natural resources: coal, lignite, natural gas, iron ore, copper, nickel, uranium, potash, salt, construction materials, timber, arable land

Land use: *agricultural land:* 48% (2011 est.)
arable land: 34.1% (2011 est.) / *permanent crops:* 0.6% (2011 est.) / *permanent pasture:* 13.3% (2011 est.)
forest: 31.8% (2011 est.)
other: 20.2% (2011 est.)

Irrigated land: 6,500 sq km (2012)

Population distribution: most populous country in Europe; a fairly even distribution throughout most of the country, with urban areas attracting larger and denser populations, particularly in the far western part of the industrial state of North Rhine-Westphalia

Natural hazards: flooding

Environment—current issues: emissions from coal-burning utilities and industries contribute to air pollution; acid rain, resulting from sulfur dioxide emissions, is damaging forests; pollution in the Baltic Sea from raw sewage and industrial effluents from rivers in eastern Germany; hazardous waste disposal; government established a mechanism for ending the use of nuclear power by 2022;

government working to meet EU commitment to identify nature preservation areas in line with the EU's Flora, Fauna, and Habitat directive

Environment—international agreements: *party to:* Air Pollution, Air Pollution-Nitrogen Oxides, Air Pollution-Persistent Organic Pollutants, Air Pollution-Sulfur 85, Air Pollution-Sulfur 94, Air Pollution-Volatile Organic Compounds, Antarctic-Environmental Protocol, Antarctic-Marine Living Resources, Antarctic Seals, Antarctic Treaty, Biodiversity, Climate Change, Climate Change-Kyoto Protocol, Desertification, Endangered Species, Environmental Modification, Hazardous Wastes, Law of the Sea, Marine Dumping, Ozone Layer Protection, Ship Pollution, Tropical Timber 83, Tropical Timber 94, Wetlands, Whaling
signed, but not ratified: none of the selected agreements

Geography—note: strategic location on North European Plain and along the entrance to the Baltic Sea; most major rivers in Germany—the Rhine, Weser, Oder, Elbe—flow northward; the Danube, which originates in the Black Forest, flows eastward

PEOPLE AND SOCIETY

Population: 80,159,662 (July 2020 est.)
country comparison to the world: 19

Nationality: *noun:* German(s)
adjective: German

Ethnic groups: German 87.2%, Turkish 1.8%, Polish 1%, Syrian 1%, other 9% (2017 est.)
note: data represent population by nationality
Languages: German (official)
note: Danish, Frisian, Sorbian, and Romani are official minority languages; Low German, Danish, North Frisian, Sater Frisian, Lower Sorbian, Upper Sorbian, and Romani are recognized as regional languages under the European Charter for Regional or Minority Languages

Religions: Roman Catholic 27.7%, Protestant 25.5%, Muslim 5.1%, Orthodox 1.9%, other Christian 1.1%, other .9%, none 37.8% (2018 est.)

Age structure: *0-14 years:* 12.89% (male 5,302,850/female 5,025,863)
15-24 years: 9.81% (male 4,012,412/female 3,854,471)
25-54 years: 38.58% (male 15,553,328/female 15,370,417)
55-64 years: 15.74% (male 6,297,886/female 6,316,024)
65 years and over: 22.99% (male 8,148,873/female 10,277,538) (2020 est.)

Dependency ratios: *total dependency ratio:* 55.4
youth dependency ratio: 21.7
elderly dependency ratio: 33.7
potential support ratio: 3 (2020 est.)

Median age: *total:* 47.8 years
male: 46.5 years
female: 49.1 years (2020 est.)
country comparison to the world: 4

Population growth rate: -0.19% (2020 est.)
country comparison to the world: 209

Birth rate: 8.6 births/1,000 population (2020 est.)
country comparison to the world: 215

Death rate: 12.1 deaths/1,000 population (2020 est.)
country comparison to the world: 16

Net migration rate: 1.5 migrant(s)/1,000 population (2020 est.)
country comparison to the world: 54

Population distribution: most populous country in Europe; a fairly even distribution throughout most of the country, with urban areas attracting larger and denser populations, particularly in the far western part of the industrial state of North Rhine-Westphalia

Urbanization: *urban population:* 77.5% of total population (2020)
rate of urbanization: 0.27% annual rate of change (2015-20 est.)
total population growth rate v. urban population growth rate, 2000-2030:

Major urban areas—Population: 3.562 million BERLIN (capital), 1.790 million Hamburg, 1.538 million Munich, 1.119 million Cologne (2020)

Sex ratio: *at birth:* 1.05 male(s)/female
0-14 years: 1.06 male(s)/female
15-24 years: 1.04 male(s)/female
25-54 years: 1.01 male(s)/female
55-64 years: 1 male(s)/female
65 years and over: 0.79 male(s)/female
total population: 0.96 male(s)/female (2020 est.)

Mother's mean age at first birth: 29.6 years (2017 est.)

Maternal mortality rate: 7 deaths/100,000 live births (2017 est.)
country comparison to the world: 154

Infant mortality rate: *total:* 3.3 deaths/1,000 live births
male: 3.6 deaths/1,000 live births
female: 3 deaths/1,000 live births (2020 est.)
country comparison to the world: 206
Life expectancy at birth: total population: 81.1 years
male: 78.7 years
female: 83.6 years (2020 est.)
country comparison to the world: 37

Total fertility rate: 1.47 children born/woman (2020 est.)
country comparison to the world: 208

Contraceptive prevalence rate: 80.3% (2011)
note: percent of women aged 18-49

Drinking water source:
improved:
urban: 100% of population
rural: 100% of population
total: 100% of population
unimproved:
urban: 0% of population
rural: 0% of population
total: 0% of population (2017 est.)

Current Health Expenditure: 11.2% (2017)

Physicians density: 4.25 physicians/1,000 population (2017)

Hospital bed density: 8 beds/1,000 population (2017)

Sanitation facility access:
improved:
urban: 100% of population
rural: 100% of population
total: 100% of population
unimproved:
urban: 0% of population
rural: 0% of population
total: 0% of population (2017 est.)

HIV/AIDS—adult prevalence rate: 0.1% (2018 est.)
country comparison to the world: 123

HIV/AIDS—people living with HIV/AIDS: 87,000 (2018 est.)
country comparison to the world: 49

HIV/AIDS—deaths: <500 (2018 est.)

Obesity—adult prevalence rate: 22.3% (2016)
country comparison to the world: 79

Children under the age of 5 years underweight: 0.5% (2014/17)
country comparison to the world: 130

Education expenditures: 4.8% of GDP (2016)
country comparison to the world: 72

School life expectancy (primary to tertiary education): *total:* 17 years
male: 17 years
female: 17 years (2018)

Unemployment, youth ages 15-24: *total:* 6.2%
male: 7.1%
female: 5.1% (2018 est.)
country comparison to the world: 158

GOVERNMENT

Country name: *conventional long form:* Federal Republic of Germany
conventional short form: Germany
local long form: Bundesrepublik Deutschland
local short form: Deutschland
former: German Reich
etymology: the Gauls (Celts) of Western Europe may have referred to the newly arriving Germanic tribes who settled in neighboring areas east of the Rhine during the first centuries B.C. as "Germani," a term the Romans adopted as "Germania"; the native designation "Deutsch" comes from the Old High German "diutisc" meaning "of the people"

Government type: federal parliamentary republic

Capital: *name:* Berlin
geographic coordinates: 52 31 N, 13 24 E
time difference: UTC+1 (6 hours ahead of Washington, DC, during Standard Time)
daylight saving time: +1hr, begins last Sunday in March; ends last Sunday in October
etymology: the origin of the name is unclear but may be related to the old West Slavic (Polabian) word "berl" or "birl," meaning "swamp"

Administrative divisions: 16 states (Laender, singular—Land); Baden-Wuerttemberg, Bayern (Bavaria), Berlin, Brandenburg, Bremen, Hamburg, Hessen (Hesse), Mecklenburg-Vorpommern (Mecklenburg-Western Pomerania), Niedersachsen (Lower Saxony), Nordrhein-Westfalen (North Rhine-Westphalia), Rheinland-Pfalz (Rhineland-Palatinate), Saarland, Sachsen (Saxony), Sachsen-Anhalt (Saxony-Anhalt), Schleswig-Holstein, Thueringen (Thuringia); note—Bayern, Sachsen, and Thueringen refer to themselves as free states (Freistaaten, singular—Freistaat), while Bremen calls itself a Free Hanseatic City (Freie Hansestadt) and Hamburg considers itself a Free and Hanseatic City (Freie und Hansestadt)

Independence: 18 January 1871 (establishment of the German Empire); divided into four zones of occupation (UK, US, USSR, and France) in 1945 following World War II; Federal Republic of Germany (FRG or West Germany) proclaimed on 23 May 1949 and included the former UK, US, and French zones; German Democratic Republic (GDR or East Germany) proclaimed on 7 October 1949 and included the former USSR zone; West Germany and East Germany unified on 3 October 1990; all four powers formally relinquished rights on 15 March 1991; notable earlier dates: 10 August 843 (Eastern Francia established from the division of the Carolingian Empire); 2 February 962 (crowning of OTTO I, recognized as the first Holy Roman Emperor)

National holiday: German Unity Day, 3 October (1990)

Constitution: *history:* previous 1919 (Weimar Constitution); latest drafted 10-23 August 1948, approved 12 May 1949, promulgated 23 May 1949, entered into force 24 May 1949
amendments: proposed by Parliament; passage and enactment into law require two-thirds majority vote by both the Bundesrat (upper house) and the Bundestag (lower house) of Parliament; articles including those on basic human rights and freedoms cannot be amended; amended many times, last in 2017

Legal system: civil law system

International law organization participation: accepts compulsory ICJ jurisdiction with reservations; accepts ICCt jurisdiction

Citizenship: citizenship by birth: no
citizenship by descent only: at least one parent must be a German citizen or a resident alien who has lived in Germany at least 8 years
dual citizenship recognized: yes, but requires prior permission from government
residency requirement for naturalization: 8 years

Suffrage: 18 years of age; universal; age 16 for some state and municipal elections

Executive branch: *chief of state:* President Frank-Walter STEINMEIER (since 19 March 2017)
head of government: Chancellor Angela MERKEL (since 22 November 2005)
cabinet: Cabinet or Bundesminister (Federal Ministers) recommended by the chancellor, appointed by the president

elections/appointments: president indirectly elected by a Federal Convention consisting of all members of the Federal Parliament (Bundestag) and an equivalent number of delegates indirectly elected by the state parliaments; president serves a 5-year term (eligible for a second term); election last held on 12 February 2017 (next to be held in February 2022); following the most recent Federal Parliament election, the party or coalition with the most representatives usually elects the chancellor (Angela Merkel since 2005) and appointed by the president to serve a renewable 4-year term; Federal Parliament vote for chancellor last held on 14 March 2018 (next to be held after the Bundestag elections in 2021)
election results: Frank-Walter STEINMEIER elected president; Federal Convention vote count—Frank-Walter STEINMEIER (SPD) 931, Christopher BUTTERWEGGE (The Left) 128, Albrecht GLASER (Alternative for Germany AfD) 42, Alexander HOLD (BVB/FW) 25, Engelbert SONNEBORN (Pirates) 10; Angela MERKEL (CDU) reelected chancellor; Federal Parliament vote—364 to 315

Legislative branch: *description:* bicameral Parliament or Parlament consists of:
Federal Council or Bundesrat (69 seats; members appointed by each of the 16 state governments)
Federal Diet or Bundestag (709 seats—total seats can vary each electoral term; approximately one-half of members directly elected in multi-seat constituencies by proportional representation vote and approximately one-half directly elected in single-seat constituencies by simple majority vote; members serve 4-year terms)
elections: Bundesrat—none; composition is determined by the composition of the state-level governments; the composition of the Bundesrat has the potential to change any time one of the 16 states holds an election
Bundestag—last held on 24 September 2017 (next to be held in 2021 at the latest); most postwar German governments have been coalitions
election results: Bundesrat—composition—men 50, women 19, percent of women 27.5%
Bundestag—percent of vote by party—CDU/CSU 33%, SPD 20.5%, AfD 12.6%, FDP 10.7%, The Left 9.2%, Alliance '90/Greens 8.9%, other 5%; seats by party—CDU/CSU 246, SPD 152, AfD 91, FDP 80, The Left 69, Alliance '90/Greens 67; composition—men 490, women 219, percent of women 30.5%; note—total Parliament percent of women 30.5%

Judicial branch: *highest courts:* Federal Court of Justice (court consists of 127 judges, including the court president, vice presidents, presiding judges, other judges and organized into 25 Senates subdivided into 12 civil panels, 5 criminal panels, and 8 special panels); Federal Constitutional Court or Bundesverfassungsgericht (consists of 2 Senates each subdivided into 3 chambers, each with a chairman and 8 members)
judge selection and term of office: Federal Court of Justice judges selected by the Judges Election Committee, which consists of the Secretaries of Justice from each of the 16 federated states and

16 members appointed by the Federal Parliament; judges appointed by the president; judges serve until mandatory retirement at age 65; Federal Constitutional Court judges—one-half elected by the House of Representatives and one-half by the Senate; judges appointed for 12-year terms with mandatory retirement at age 68

subordinate courts: Federal Administrative Court; Federal Finance Court; Federal Labor Court; Federal Social Court; each of the 16 federated states or Land has its own constitutional court and a hierarchy of ordinary (civil, criminal, family) and specialized (administrative, finance, labor, social) courts

Political parties and leaders:
Alliance '90/Greens [Annalena BAERBOCK and Robert HABECK]
Alternative for Germany or AfD [Alexander GAULAND and Joerg MEUTHEN]
Christian Democratic Union or CDU [Annegret KRAMP-KARRENBAUER]
Christian Social Union or CSU [Markus SOEDER]
Free Democratic Party or FDP [Christian LINDNER]
The Left or Die Linke [Katja KIPPING and Bernd RIEXINGER]
Social Democratic Party or SPD [Andrea NAHLES]

International organization participation: ADB (nonregional member), AfDB (nonregional member), Arctic Council (observer), Australia Group, BIS, BSEC (observer), CBSS, CD, CDB, CE, CERN, EAPC, EBRD, ECB, EIB, EITI (implementing country), EMU, ESA, EU, FAO, FATF, G-5, G-7, G-8, G-10, G-20, IADB, IAEA, IBRD, ICAO, ICC (national committees), ICCt, ICRM, IDA, IEA, IFAD, IFC, IFRCS, IGAD (partners), IHO, ILO, IMF, IMO, IMSO, Interpol, IOC, IOM, IPU, ISO, ITSO, ITU, ITUC (NGOs), MIGA, MINURSO, MINUSMA, NATO, NEA, NSG, OAS (observer), OECD, OPCW, OSCE, Pacific Alliance (observer), Paris Club, PCA, Schengen Convention, SELEC (observer), SICA (observer), UN, UNAMID, UNCTAD, UNESCO, UNHCR, UNIDO, UNIFIL, UNMISS, UNRWA, UNWTO, UPU, WCO, WHO, WIPO, WMO, WTO, ZC

Diplomatic representation in the US: *chief of mission:* Ambassador Emily Margarethe HABER (since 22 June 2018)
chancery: 4645 Reservoir Road NW, Washington, DC 20007
telephone: [1] (202) 298-4000
FAX: [1] (202) 298-4249
consulate(s) general: Atlanta, Boston, Chicago, Houston, Los Angeles, Miami, New York, San Francisco

Diplomatic representation from the US: *chief of mission:* Ambassador Richard GRENELL (since 8 May 2018)
telephone: [49] (30) 8305-0
embassy: Clayallee 170, 14191 Berlin
mailing address: Clayallee 170, 14191 Berlin
FAX: [49] (30) 8305-1215

consulate(s) general: Dusseldorf, Frankfurt am Main, Hamburg, Leipzig, Munich

Flag description: three equal horizontal bands of black (top), red, and gold; these colors have played an important role in German history and can be traced back to the medieval banner of the Holy Roman Emperor—a black eagle with red claws and beak on a gold field

National symbol(s): *eagle; national colors:* black, red, yellow

National anthem: *name:* "Das Lied der Deutschen" (Song of the Germans)
lyrics/music: August Heinrich HOFFMANN VON FALLERSLEBEN/Franz Joseph HAYDN
note: adopted 1922; the anthem, also known as "Deutschlandlied" (Song of Germany), was originally adopted for its connection to the March 1848 liberal revolution; following appropriation by the Nazis of the first verse, specifically the phrase, "Deutschland, Deutschland ueber alles" (Germany, Germany above all) to promote nationalism, it was banned after 1945; in 1952, its third verse was adopted by West Germany as its national anthem; in 1990, it became the national anthem for the reunited Germany
0:00 / 1:13

ECONOMY

Economy—overview: The German economy—the fifth largest economy in the world in PPP terms and Europe's largest—is a leading exporter of machinery, vehicles, chemicals, and household equipment. Germany benefits from a highly skilled labor force, but, like its Western European neighbors, faces significant demographic challenges to sustained long-term growth. Low fertility rates and a large increase in net immigration are increasing pressure on the country's social welfare system and necessitate structural reforms.

Reforms launched by the government of Chancellor Gerhard SCHROEDER (1998-2005), deemed necessary to address chronically high unemployment and low average growth, contributed to strong economic growth and falling unemployment. These advances, as well as a government subsidized, reduced working hour scheme, help explain the relatively modest increase in unemployment during the 2008-09 recession— the deepest since World War II. The German Government introduced a minimum wage in 2015 that increased to $9.79 (8.84 euros) in January 2017.

Stimulus and stabilization efforts initiated in 2008 and 2009 and tax cuts introduced in Chancellor Angela MERKEL's second term increased Germany's total budget deficit—including federal, state, and municipal—to 4.1% in 2010, but slower spending and higher tax revenues reduced the deficit to 0.8% in 2011 and in 2017 Germany reached a budget surplus of 0.7%. A constitutional amendment approved in 2009 limits the federal government to structural deficits of no more than 0.35% of GDP per annum as of 2016, though the target was already reached in 2012.

Following the March 2011 Fukushima nuclear disaster, Chancellor Angela MERKEL announced in May 2011 that eight of the country's 17 nuclear reactors would be shut down immediately and the remaining plants would close by 2022. Germany plans to replace nuclear power largely with renewable energy, which accounted for 29.5% of gross electricity consumption in 2016, up from 9% in 2000. Before the shutdown of the eight reactors, Germany relied on nuclear power for 23% of its electricity generating capacity and 46% of its baseload electricity production.

The German economy suffers from low levels of investment, and a government plan to invest 15 billion euros during 2016-18, largely in infrastructure, is intended to spur needed private investment. Domestic consumption, investment, and exports are likely to drive German GDP growth in 2018, and the country's budget and trade surpluses are likely to remain high.

GDP (purchasing power parity): $4.199 trillion (2017 est.)
$4.099 trillion (2016 est.)
$4.012 trillion (2015 est.)
note: data are in 2017 dollars
country comparison to the world: 5

GDP (official exchange rate): $3.701 trillion (2017 est.)

GDP—real growth rate:
0.59% (2019 est.)
1.3% (2018 est.)
2.91% (2017 est.)
country comparison to the world: 184

GDP—per capita (PPP): $50,800 (2017 est.)
$49,800 (2016 est.)
$49,100 (2015 est.)
note: data are in 2017 dollars
country comparison to the world: 27

Gross national saving:
28% of GDP (2017 est.)
28.2% of GDP (2016 est.)
28.1% of GDP (2015 est.)
country comparison to the world: 40

GDP—composition, by end use:
household consumption: 53.1% (2017 est.)
government consumption: 19.5% (2017 est.)
investment in fixed capital: 20.4% (2017 est.)
investment in inventories: -0.5% (2017 est.)
exports of goods and services: 47.3% (2017 est.)
imports of goods and services: -39.7% (2017 est.)

GDP—composition, by sector of origin:
agriculture: 0.7% (2017 est.)
industry: 30.7% (2017 est.)
services: 68.6% (2017 est.)

Agriculture—products: potatoes, wheat, barley, sugar beets, fruit, cabbages; milk products; cattle, pigs, poultry

Industries: among the world's largest and most technologically advanced producers of iron, steel, coal, cement, chemicals, machinery, vehicles, machine tools, electronics, automobiles, food and beverages, shipbuilding, textiles

Industrial production growth rate: 3.3% (2017 est.)
country comparison to the world: 95

Labor force: 44.585 million (2020 est.)
country comparison to the world: 13

Labor force—by occupation: *agriculture:* 1.4%
industry: 24.2%
services: 74.3% (2016)

Unemployment rate:
4.98% (2019 est.)
5.19% (2018 est.)
country comparison to the world: 75

Population below poverty line: 16.7% (2015 est.)

Household income or consumption by percentage share: *lowest 10%:* 3.6%
highest 10%: 24% (2000)

Budget: *revenues:* 1.665 trillion (2017 est.)
expenditures: 1.619 trillion (2017 est.)

Taxes and other revenues: 45% (of GDP) (2017 est.)
country comparison to the world: 22

Budget surplus (+) or deficit (-): 1.3% (of GDP) (2017 est.)
country comparison to the world: 25

Public debt:
63.9% of GDP (2017 est.)
67.9% of GDP (2016 est.)
note: general government gross debt is defined in the Maastricht Treaty as consolidated general government gross debt at nominal value, outstanding at the end of the year in the following categories of government liabilities (as defined in ESA95): currency and deposits (AF.2), securities other than shares excluding financial derivatives (AF.3, excluding AF.34), and loans (AF.4); the general government sector comprises the sub-sectors of central government, state government, local government and social security funds; the series are presented as a percentage of GDP and in millions of euros; GDP used as a denominator is the gross domestic product at current market prices; data expressed in national currency are converted into euro using end-of-year exchange rates provided by the European Central Bank
country comparison to the world: 61

Fiscal year: calendar year

Inflation rate (consumer prices):
1.7% (2017 est.)
0.4% (2016 est.)
country comparison to the world: 92

Current account balance:
$280.238 billion (2019 est.)
$297.434 billion (2018 est.)
country comparison to the world: 1

Exports:
$1.434 trillion (2017 est.)
$1.322 trillion (2016 est.)
country comparison to the world: 3

Exports—partners: US 8.8%, France 8.2%, China 6.8%, Netherlands 6.7%, UK 6.6%, Italy 5.1%, Austria 4.9%, Poland 4.7%, Switzerland 4.2% (2017)

Exports—commodities: motor vehicles, machinery, chemicals, computer and electronic products, electrical equipment, pharmaceuticals, metals,
transport equipment, foodstuffs, textiles, rubber and plastic products

Imports:
$1.135 trillion (2017 est.)
$1.022 trillion (2016 est.)
country comparison to the world: 3

Imports—commodities: machinery, data processing equipment, vehicles, chemicals, oil and gas, metals, electric equipment, pharmaceuticals, foodstuffs, agricultural products

Imports—partners: Netherlands 13.8%, China 7%, France 6.6%, Belgium 5.9%, Italy 5.4%, Poland 5.4%, Czechia 4.8%, US 4.5%, Austria 4.3%, Switzerland 4.2% (2017)

Reserves of foreign exchange and gold:
$200.1 billion (31 December 2017 est.)
$173.7 billion (31 December 2015 est.)
country comparison to the world: 13

Debt—external:
$5.326 trillion (31 March 2016 est.)
$5.21 trillion (31 March 2015 est.)
country comparison to the world: 4

Exchange rates: euros (EUR) per US dollar -
0.885 (2017 est.)
0.903 (2016 est.)
0.9214 (2015 est.)
0.885 (2014 est.)
0.7634 (2013 est.)

ENERGY

Electricity access: electrification—total population: 100% (2020)

Electricity—production: 612.8 billion kWh (2016 est.)
country comparison to the world: 7

Electricity—consumption: 536.5 billion kWh (2016 est.)
country comparison to the world: 6

Electricity—exports: 78.86 billion kWh (2016 est.)
country comparison to the world: 1

Electricity—imports: 28.34 billion kWh (2016 est.)
country comparison to the world: 5

Electricity—installed generating capacity: 208.5 million kW (2016 est.)
country comparison to the world: 6

Electricity—from fossil fuels: 41% of total installed capacity (2016 est.)
country comparison to the world: 166

Electricity—from nuclear fuels: 5% of total installed capacity (2017 est.)
country comparison to the world: 21

Electricity—from hydroelectric plants: 2% of total installed capacity (2017 est.)
country comparison to the world: 137

Electricity—from other renewable sources: 52% of total installed capacity (2017 est.)
country comparison to the world: 4

Crude oil—production: 41,000 bbl/day (2018 est.)
country comparison to the world: 56

Crude oil—exports: 6,569 bbl/day (2017 est.)
country comparison to the world: 63

Crude oil—imports: 1.836 million bbl/day (2017 est.)
country comparison to the world: 6

Crude oil—proved reserves: 129.6 million bbl (1 January 2018 est.)
country comparison to the world: 65

Refined petroleum products—production: 2.158 million bbl/day (2017 est.)
country comparison to the world: 9

Refined petroleum products—consumption: 2.46 million bbl/day (2017 est.)
country comparison to the world: 9

Refined petroleum products—exports: 494,000 bbl/day (2017 est.)
country comparison to the world: 17

Refined petroleum products—imports: 883,800 bbl/day (2017 est.)
country comparison to the world: 9

Natural gas—production: 7.9 billion cu m (2017 est.)
country comparison to the world: 45

Natural gas—consumption: 93.36 billion cu m (2017 est.)
country comparison to the world: 8

Natural gas—exports: 34.61 billion cu m (2017 est.)
country comparison to the world: 11

Natural gas—imports: 119.5 billion cu m (2017 est.)
country comparison to the world: 1

Natural gas—proved reserves: 39.5 billion cu m (1 January 2018 est.)
country comparison to the world: 64

Carbon dioxide emissions from consumption of energy: 847.6 million Mt (2017 est.)
country comparison to the world: 6

COMMUNICATIONS

Telephones—fixed lines: *total subscriptions:* 38,847,530
subscriptions per 100 inhabitants: 48.37 (2019 est.)
country comparison to the world: 5

Telephones—mobile cellular: *total subscriptions:* 103,090,116
subscriptions per 100 inhabitants: 128.36 (2019 est.)
country comparison to the world: 16

Telecommunication systems: *general assessment:* one of the world's most technologically advanced telecommunications systems; as a result of intensive capital expenditures since reunification, the formerly backward system of the eastern part of the country, dating back to World War II, has been modernized and integrated with that of the western part; universal 3G infrastructure available and LTE networks; mobile market the largest in Europe 107.5 million as of 2019; available reach of 5G services in 5 cities; 98% LTE coverage; penetration in broadband and mobile sectors average for region; Hamburg develops smart city concept (2020)

domestic: extensive system of automatic telephone exchanges connected by modern networks of fiber-optic cable, coaxial cable, microwave radio relay, and a domestic satellite system; cellular telephone service is widely available, expanding rapidly, and includes roaming service to many foreign countries; 48 per 100 for fixed-line and 128 per 100 for mobile-cellular (2019)

international: country code—49; landing points for SeaMeWe-3, TAT-14, AC-1, CONTACT-3, Fehmarn Balt, C-Lion1, GC1, GlobalConnect-KPN, and Germany-Denmark 2 & 3—submarine cables to Europe, Africa, the Middle East, Asia, Southeast Asia and Australia; as well as earth stations in the Inmarsat, Intelsat, Eutelsat, and Intersputnik satellite systems (2019)

note: the COVID-19 outbreak is negatively impacting telecommunications production and supply chains globally; consumer spending on telecom devices and services has also slowed due to the pandemic's effect on economies worldwide; overall progress towards improvements in all facets of the telecom industry—mobile, fixed-line, broadband, submarine cable and satellite—has moderated

Broadcast media: a mixture of publicly operated and privately owned TV and radio stations; 70 national and regional public broadcasters compete with nearly 400 privately owned national and regional TV stations; more than 90% of households have cable or satellite TV; hundreds of radio stations including multiple national radio networks, regional radio networks, and a large number of local radio stations

Internet country code: .de

Internet users: *total:* 72,202,773
percent of population: 89.74% (July 2018 est.)
country comparison to the world: 10

Broadband—fixed subscriptions: *total:* 34,174,900
subscriptions per 100 inhabitants: 42 (2018 est.)
country comparison to the world: 4

TRANSPORTATION

National air transport system: *number of registered air carriers:* 20 (2020)
inventory of registered aircraft operated by air carriers: 1,113
annual passenger traffic on registered air carriers: 109,796,202 (2018)
annual freight traffic on registered air carriers: 7,969,860,000 mt-km (2018)

Civil aircraft registration country code prefix: D (2016)

Airports: 539 (2013)
country comparison to the world: 12

Airports—with paved runways: *total:* 318 (2017)
over 3,047 m: 14 (2017)

2,438 to 3,047 m: 49 (2017)
1,524 to 2,437 m: 60 (2017)
914 to 1,523 m: 70 (2017)
under 914 m: 125 (2017)

Airports—with unpaved runways: *total:* 221 (2013)
1,524 to 2,437 m: 1 (2013)
914 to 1,523 m: 35 (2013)
under 914 m: 185 (2013)

Heliports: 23 (2013)

Pipelines: 37 km condensate, 26985 km gas, 2400 km oil, 4479 km refined products, 8 km water (2013)

Railways: *total:* 33,590 km (2017)
standard gauge: 33,331 km 1.435-m gauge (19,973 km electrified) (2015)
narrow gauge: 220 km 1.000-m gauge (79 km electrified)
15 km 0.900-m gauge, 24 km 0.750-m gauge (2015)
country comparison to the world: 7

Roadways: *total:* 625,000 km (2017)
paved: 625,000 km (includes 12,996 km of expressways) (2017)
note: includes local roads
country comparison to the world: 12

Waterways: 7,467 km (Rhine River carries most goods; Main-Danube Canal links North Sea and Black Sea) (2012)
country comparison to the world: 18

Merchant marine: *total:* 609
by type: bulk carrier 1, container ship 87, general cargo 83, oil tanker 36, other 402 (2019)
country comparison to the world: 38

Ports and terminals: *major seaport(s):* Baltic Sea—Rostock
oil terminal(s): Brunsbuttel Canal terminals
container port(s) (TEUs): Bremen/Bremerhaven (5,510,000), Hamburg (8,860,000) (2017)

LNG terminal(s) (import): Hamburg
river port(s): Bremen (Weser)
North Sea—Wilhelmshaven Bremerhaven (Geeste) Duisburg, Karlsruhe, Neuss-Dusseldorf (Rhine) Brunsbuttel, Hamburg (Elbe) Lubeck (Wakenitz)

MILITARY AND SECURITY

Military and security forces: *Federal Armed Forces (Bundeswehr):* Army (Heer), Navy (Deutsche Marine, includes naval air arm), Air Force (Luftwaffe, includes air defense), Joint Support Service (Streitkraeftebasis, SKB), Central Medical Service (Zentraler Sanitaetsdienst, ZSanDstBw), Cyber and Information Space Command (Kommando Cyber- und Informationsraum, Kdo CIR) (2020)

Military expenditures:
1.38% of GDP (2019 est.)

1.24% of GDP (2018)
1.23% of GDP (2017)
1.19% of GDP (2016)
1.18% of GDP (2015)
country comparison to the world: 87

Military and security service personnel strengths: the German Federal Armed Forces have approximately 180,000 active duty personnel (62,000 Army; 16,000 Navy; 28,000 Air Force; 27,000 Joint Support Service; 20,000 Medical Service, 13,000 Cyber and Information Space Command; 14,000 other) (2019 est.)

Military equipment inventories and acquisitions: the German Federal Armed Forces inventory is mostly comprised of weapons systems produced domestically or jointly with other European countries; since 2010, the US is the leading foreign supplier of armaments to Germany, followed by the Netherlands and Switzerland; Germany's defense industry is capable of manufacturing the full spectrum of air, land, and naval military weapons systems (2019 est.)

Military deployments: 1,300 Afghanistan (NATO); approximately 100-200 Middle East (NATO/Counter-ISIS campaign); 110 Lebanon (UNIFIL); 500 Lithuania (NATO); 400 Mali (MINUSMA); 350 Mali (EUTM); note—Germany is a contributing member of the EuroCorps (2020)

Military service age and obligation: 17-23 years of age for male and female voluntary military service; conscription ended 1 July 2011; service obligation 8-23 months or 12 years; women have been eligible for voluntary service in all military branches and positions since 2001 (2013)

TERRORISM

Terrorist group(s): Islamic Revolutionary Guard Corps/Qods Force; Islamic State of Iraq and ash-Sham (2019)
note: details about the history, aims, leadership, organization, areas of operation, tactics, targets, weapons, size, and sources of support of the group(s) appear(s) in Appendix-T

TRANSNATIONAL ISSUES

Disputes—international: none

Refugees and internally displaced persons: *refugees (country of origin):* 572,818 (Syria), 141,650 (Iraq), 140,366 (Afghanistan), 58,569 (Eritrea), 43,244 (Iran), 28,470 (Turkey), 26,015 (Somalia), 8,722 (Russia), 8,639 (Serbia and Kosovo), 8,125 (Pakistan), 7,828 (Nigeria) (2019)
stateless persons: 14,947 (2019)

Illicit drugs: source of precursor chemicals for South American cocaine processors; transshipment point for and consumer of Southwest Asian heroin, Latin American cocaine, and European-produced synthetic drugs; major financial center

GHANA

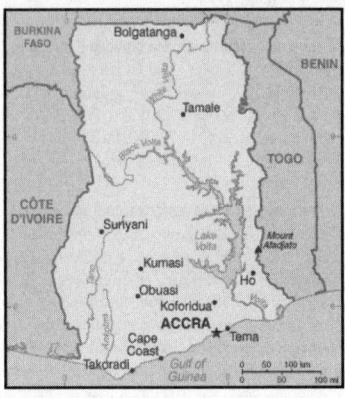

INTRODUCTION

Background: Formed from the merger of the British colony of the Gold Coast and the Togoland trust territory, Ghana in 1957 became the first Sub-Saharan country in colonial Africa to gain its independence. Ghana endured a series of coups before Lt. Jerry RAWLINGS took power in 1981 and banned political parties. After approving a new constitution and restoring multiparty politics in 1992, RAWLINGS won presidential elections in 1992 and 1996 but was constitutionally prevented from running for a third term in 2000. John KUFUOR of the opposition New Patriotic Party (NPP) succeeded him and was reelected in 2004. John Atta MILLS of the National Democratic Congress won the 2008 presidential election and took over as head of state. MILLS died in July 2012 and was constitutionally succeeded by his vice president, John Dramani MAHAMA, who subsequently won the December 2012 presidential election. In 2016, Nana Addo Dankwa AKUFO-ADDO of the NPP defeated MAHAMA, marking the third time that Ghana's presidency has changed parties since the return to democracy.

GEOGRAPHY

Location: Western Africa, bordering the Gulf of Guinea, between Cote d'Ivoire and Togo

Geographic coordinates: 8 00 N, 2 00 W

Map references: Africa

Area: *total:* 238,533 sq km
land: 227,533 sq km
water: 11,000 sq km
country comparison to the world: 83

Area—comparative: slightly smaller than Oregon

Land boundaries: *total:* 2,420 km
border countries (3): Burkina Faso 602 km, Cote d'Ivoire 720 km, Togo 1098 km

Coastline: 539 km

Maritime claims: *territorial sea:* 12 nm
exclusive economic zone: 200 nm
contiguous zone: 24 nm
continental shelf: 200 nm

Climate: tropical; warm and comparatively dry along southeast coast; hot and humid in southwest; hot and dry in north

Terrain: mostly low plains with dissected plateau in south-central area

Elevation: *mean elevation:* 190 m
lowest point: Atlantic Ocean 0 m
highest point: Mount Afadjato 885 m

Natural resources: gold, timber, industrial diamonds, bauxite, manganese, fish, rubber, hydropower, petroleum, silver, salt, limestone

Land use: *agricultural land:* 69.1% (2011 est.)
arable land: 20.7% (2011 est.) / *permanent crops:* 11.9% (2011 est.) / *permanent pasture:* 36.5% (2011 est.)
forest: 21.2% (2011 est.)
other: 9.7% (2011 est.)
Irrigated land: 340 sq km (2012)

Population distribution: population is concentrated in the southern half of the country, with the highest concentrations being on or near the Atlantic coast as shown in this population distribution map

Natural hazards: dry, dusty, northeastern harmattan winds from January to March; droughts

Environment—current issues: recurrent drought in north severely affects agricultural activities; deforestation; overgrazing; soil erosion; poaching and habitat destruction threaten wildlife populations; water pollution; inadequate supplies of potable water

Environment—international agreements: *party to:* Biodiversity, Climate Change, Climate Change-Kyoto Protocol, Desertification, Endangered Species, Environmental Modification, Hazardous Wastes, Law of the Sea, Ozone Layer Protection, Ship Pollution, Tropical Timber 83, Tropical Timber 94, Wetlands
signed, but not ratified: Marine Life Conservation

Geography—note: Lake Volta is the world's largest artificial lake (manmade reservoir) by surface area (8,482 sq km; 3,275 sq mi); the lake was created following the completion of the Akosombo Dam in 1965, which holds back the White Volta and Black Volta Rivers

PEOPLE AND SOCIETY

Population: 29,340,248 (July 2020 est.)
note: estimates for this country explicitly take into account the effects of excess mortality due to AIDS; this can result in lower life expectancy, higher infant mortality, higher death rates, lower population growth rates, and changes in the distribution of population by age and sex than would otherwise be expected
country comparison to the world: 49

Nationality: *noun:* Ghanaian(s)
adjective: Ghanaian

Ethnic groups: Akan 47.5%, Mole-Dagbon 16.6%, Ewe 13.9%, Ga-Dangme 7.4%, Gurma 5.7%, Guan 3.7%, Grusi 2.5%, Mande 1.1%, other 1.4% (2010 est.)

Languages: Asante 16%, Ewe 14%, Fante 11.6%, Boron (Brong) 4.9%, Dagomba 4.4%, Dangme 4.2%, Dagarte (Dagaba) 3.9%, Kokomba 3.5%, Akyem 3.2%, Ga 3.1%, other 31.2% (2010 est.)
note: English is the official language

Religions: Christian 71.2% (Pentecostal/Charismatic 28.3%, Protestant 18.4%, Catholic 13.1%, other 11.4%), Muslim 17.6%, traditional 5.2%, other 0.8%, none 5.2% (2010 est.)

Demographic profile: Ghana has a young age structure, with approximately 57% of the population under the age of 25. Its total fertility rate fell significantly during the 1980s and 1990s but has stalled at around four children per woman for the last few years. Fertility remains higher in the northern region than the Greater Accra region. On average, desired fertility has remained stable for several years; urban dwellers want fewer children than rural residents. Increased life expectancy, due to better health care, nutrition, and hygiene, and reduced fertility have increased Ghana's share of elderly persons; Ghana's proportion of persons aged 60+ is among the highest in Sub-Saharan Africa. Poverty has declined in Ghana, but it remains pervasive in the northern region, which is susceptible to droughts and floods and has less access to transportation infrastructure, markets, fertile farming land, and industrial centers. The northern region also has lower school enrollment, higher illiteracy, and fewer opportunities for women.

Ghana was a country of immigration in the early years after its 1957 independence, attracting labor migrants largely from Nigeria and other neighboring countries to mine minerals and harvest cocoa—immigrants composed about 12% of Ghana's population in 1960. In the late 1960s, worsening economic and social conditions discouraged immigration, and hundreds of thousands of immigrants, mostly Nigerians, were expelled.

During the 1970s, severe drought and an economic downturn transformed Ghana into a country of emigration; neighboring Cote d'Ivoire was the initial destination. Later, hundreds of thousands of Ghanaians migrated to Nigeria to work in its booming oil industry, but most were deported in 1983 and 1985 as oil prices plummeted. Many Ghanaians then turned to more distant destinations, including other parts of Africa, Europe, and North America, but the majority continued to migrate within West Africa. Since the 1990s, increased emigration of skilled Ghanaians, especially to the US and the UK, drained the country of its health care and education professionals. Internally, poverty and other developmental disparities continue to drive Ghanaians from the north to the south, particularly to its urban centers.

Age structure: *0-14 years:* 37.44% (male 5,524,932/female 5,460,943)
15-24 years: 18.64% (male 2,717,481/female 2,752,601)
25-54 years: 34.27% (male 4,875,985/female 5,177,959)
55-64 years: 5.21% (male 743,757/female 784,517)
65 years and over: 4.44% (male 598,387/female 703,686) (2020 est.)

Dependency ratios: *total dependency ratio:* 67.4
youth dependency ratio: 62.2
elderly dependency ratio: 5.3
potential support ratio: 17.1 (2020 est.)

Median age: *total:* 21.4 years
male: 21 years
female: 21.9 years (2020 est.)
country comparison to the world: 185

Population growth rate: 2.15% (2020 est.)
country comparison to the world: 38

Birth rate: 29.6 births/1,000 population (2020 est.)
country comparison to the world: 34

Death rate: 6.6 deaths/1,000 population (2020 est.)
country comparison to the world: 138

Net migration rate: -1.6 migrant(s)/1,000 population (2020 est.)
country comparison to the world: 156

Population distribution: population is concentrated in the southern half of the country, with the highest concentrations being on or near the Atlantic coast as shown in this population distribution map

Urbanization: *urban population:* 57.3% of total population (2020)
rate of urbanization: 3.34% annual rate of change (2015-20 est.)

total population growth rate v. urban population growth rate, 2000-2030: Major urban areas—population: 3.348 million Kumasi, 2.514 million ACCRA (capital), 946,000 Sekondi Takoradi (2020)

Sex ratio: *at birth:* 1.03 male(s)/female
0-14 years: 1.01 male(s)/female
15-24 years: 0.99 male(s)/female
25-54 years: 0.94 male(s)/female
55-64 years: 0.95 male(s)/female
65 years and over: 0.85 male(s)/female
total population: 0.97 male(s)/female (2020 est.)

Mother's mean age at first birth: 22.3 years (2017 est.)
note: median age at first birth among women 25-29

Maternal mortality rate: 308 deaths/100,000 live births (2017 est.)
country comparison to the world: 36

Infant mortality rate: *total:* 32.1 deaths/1,000 live births

male: 35.9 deaths/1,000 live births
female: 28.2 deaths/1,000 live births (2020 est.)
country comparison to the world: 49

Life expectancy at birth: *total population:* 68.2 years
male: 65.6 years
female: 70.8 years (2020 est.)
country comparison to the world: 176

Total fertility rate: 3.9 children born/woman (2020 est.)
country comparison to the world: 32

Contraceptive prevalence rate: 30.8% (2017)

Drinking water source:

improved: *urban:* 97.4% of population
rural: 80.6% of population
total: 89.9% of population

unimproved: *urban:* 2.6% of population
rural: 19.4% of population
total: 10.1% of population (2017 est.)

Current Health Expenditure: 3.3% (2017)

Physicians density: 0.14 physicians/1,000 population (2017)

Hospital bed density: 0.9 beds/1,000 population (2011)

Sanitation facility access:

improved: *urban:* 84.2% of population
rural: 49.5% of population
total: 68.7% of population

unimproved: *urban:* 15.8% of population
rural: 50.5% of population
total: 31.3% of population (2017 est.)

HIV/AIDS—adult prevalence rate: 1.7% (2019 est.)
country comparison to the world: 28

HIV/AIDS—people living with HIV/AIDS: 340,000 (2019 est.)
country comparison to the world: 21

HIV/AIDS—deaths: 14,000 (2019 est.)
country comparison to the world: 14

Major infectious diseases: *degree of risk:* very high (2020)
food or waterborne diseases: bacterial and protozoal diarrhea, hepatitis A, and typhoid fever
vectorborne diseases: malaria, dengue fever, and yellow fever
water contact diseases: schistosomiasis
animal contact diseases: rabies
respiratory diseases: meningococcal meningitis

Obesity—adult prevalence rate: 10.9% (2016)
country comparison to the world: 136

Children under the age of 5 years underweight: 12.6% (2017/18)
country comparison to the world: 50

Education expenditures: 3.6% of GDP (2017)
country comparison to the world: 119

Literacy: *definition:* age 15 and over can read and write
total population: 76.6%
male: 82%
female: 71.4% (2015)

School life expectancy (primary to tertiary education): *total:* 12 years

male: 12 years
female: 12 years (2019)

Unemployment, youth ages 15-24: *total:* 9.1%
male: 9.4%
female: 8.7% (2017 est.)
country comparison to the world: 132

Country name: *conventional long form:* Republic of Ghana
conventional short form: Ghana
former: Gold Coast
etymology: named for the medieval West African kingdom of the same name but whose location was actually further north than the modern country

Government type: presidential republic

Capital: *name:* Accra
geographic coordinates: 5 33 N, 0 13 W
time difference: UTC 0 (5 hours ahead of Washington, DC, during Standard Time)
etymology: the name derives from the Akan word "nkran" meaning "ants," and refers to the numerous anthills in the area around the capital

Administrative divisions: 16 regions; Ahafo, Ashanti, Bono, Bono East, Central, Eastern, Greater Accra, North East, Northern, Oti, Savannah, Upper East, Upper West, Volta, Western, Western North

Independence: 6 March 1957 (from the UK)

National holiday: Independence Day, 6 March (1957)

Constitution: *history:* several previous; latest drafted 31 March 1992, approved and promulgated 28 April 1992, entered into force 7 January 1993
amendments: proposed by Parliament; consideration requires prior referral to the Council of State, a body of prominent citizens who advise the president of the republic; passage of amendments to "entrenched" constitutional articles (including those on national sovereignty, fundamental rights and freedoms, the structure and authorities of the branches of government, and amendment procedures) requires approval in a referendum by at least 40% participation of eligible voters and at least 75% of votes cast, followed by at least two-thirds majority vote in Parliament, and assent of the president; amendments to non-entrenched articles do not require referenda; amended 1996

Legal system: mixed system of English common law and customary law

International law organization participation: has not submitted an ICJ jurisdiction declaration; accepts ICCt jurisdiction

Citizenship: *citizenship by birth:* no
citizenship by descent only: at least one parent or grandparent must be a citizen of Ghana
dual citizenship recognized: yes
residency requirement for naturalization: 5 years

Suffrage: 18 years of age; universal

Executive branch: *chief of state:* President Nana Addo Dankwa AKUFO-ADDO (since 7 January

2017); Vice President Mahamudu BAWUMIA (since 7 January 2017); the president is both chief of state and head of government

head of government: President Nana Addo Dankwa AKUFO-ADDO (since 7 January 2017); Vice President Mahamudu BAWUMIA (since 7 January 2017)

cabinet: Council of Ministers; nominated by the president, approved by Parliament

elections/appointments: president and vice president directly elected on the same ballot by absolute majority popular vote in 2 rounds if needed for a 4-year term (eligible for a second term); election last held on 7 December 2016 (next to be held in December 2020)

election results: Nana Addo Dankwa AKUFO-ADDO elected president in the first round; percent of vote—Nana Addo Dankwa AKUFO-ADDO (NPP) 53.7%, John Dramani MAHAMA (NDC) 44.5%, other 1.8%

Legislative branch: *description:* unicameral Parliament (275 seats; members directly elected in single-seat constituencies by simple majority vote to serve 4-year terms)

elections: last held on 7 December 2016 (next to be held on 7 December 2020)

election results: percent of vote by party—NPP 54%, NDC 44%, other 2%; seats by party—NPP 171, NDC 104; composition—men 240, women 35, percent of women 12.7%

Judicial branch: *highest courts:* Supreme Court (consists of the chief justice and 13 justices)

judge selection and term of office: chief justice appointed by the president in consultation with the Council of State (a small advisory body of prominent citizens) and with the approval of Parliament; other justices appointed by the president upon the advice of the Judicial Council (an 18-member independent body of judicial, military and police officials, and presidential nominees) and on the advice of the Council of State; justices can retire at age 60, with compulsory retirement at age 70

subordinate courts: Court of Appeal; High Court; Circuit Court; District Court; regional tribunals

Political parties and leaders:

note: Ghana has more than 20 registered parties; included are 5 of the more popular parties as of May 2017

International organization participation: ACP, AfDB, AU, C, ECOWAS, EITI (compliant country), FAO, G-24, G-77, IAEA, IBRD, ICAO, ICC (national committees), ICCt, ICRM, IDA, IFAD, IFC, IFRCS, ILO, IMF, IMO, IMSO, Interpol, IOC, IOM, IPU, ISO, ITSO, ITU, ITUC (NGOs), MIGA, MINURSO, MINUSMA, MONUSCO, NAM, OAS (observer), OIF, OPCW, UN, UNAMID, UNCTAD, UNESCO, UNHCR, UNIDO, UNIFIL, UNISFA, UNMIL, UNMISS, UNOCI, UNWTO, UPU, WCO, WFTU (NGOs), WHO, WIPO, WMO, WTO

Diplomatic representation in the US: *chief of mission:* Ambassador Barfour ADJEI-BARWUAH (since 21 July 2017)

chancery: 3512 International Drive NW, Washington, DC 20008

telephone: [1] (202) 686-4520

FAX: [1] (202) 686-4527

consulate(s) general: New York

Diplomatic representation from the US: *chief of mission:* Ambassador Stephanie S. SULLIVAN (since 30 November 2018)

telephone: [233] 030-274-1000

embassy: 24 Fourth Circular Rd., Cantonments, Accra

mailing address: P.O. Box 194, Accra

FAX: [233] 030-274-1389

Flag description: three equal horizontal bands of red (top), yellow, and green, with a large black five-pointed star centered in the yellow band; red symbolizes the blood shed for independence, yellow represents the country's mineral wealth, while green stands for its forests and natural wealth; the black star is said to be the lodestar of African freedom

note: uses the popular Pan-African colors of Ethiopia; similar to the flag of Bolivia, which has a coat of arms centered in the yellow band

National symbol(s): black star, golden eagle; national colors: red, yellow, green, black

National anthem: *name:* God Bless Our Homeland Ghana

lyrics/music: unknown/Philip GBEHO

note: music adopted 1957, lyrics adopted 1966; the lyrics were changed twice, in 1960 when a republic was declared and after a 1966 coup

0:00 / 0:41

ECONOMY

Economy—overview: Ghana has a market-based economy with relatively few policy barriers to trade and investment in comparison with other countries in the region, and Ghana is endowed with natural resources. Ghana's economy was strengthened by a quarter century of relatively sound management, a competitive business environment, and sustained reductions in poverty levels, but in recent years has suffered the consequences of loose fiscal policy, high budget and current account deficits, and a depreciating currency.

Agriculture accounts for about 20% of GDP and employs more than half of the workforce, mainly small landholders. Gold, oil, and cocoa exports, and individual remittances, are major sources of foreign exchange. Expansion of Ghana's nascent oil industry has boosted economic growth, but the fall in oil prices since 2015 reduced by half Ghana's oil revenue. Production at Jubilee, Ghana's first commercial offshore oilfield, began in mid-December 2010. Production from two more fields, TEN and Sankofa, started in 2016 and 2017 respectively. The country's first gas processing plant at Atuabo is also producing natural gas from the Jubilee field, providing power to several of Ghana's thermal power plants.

As of 2018, key economic concerns facing the government include the lack of affordable electricity, lack of a solid domestic revenue base, and the high debt burden. The AKUFO-ADDO administration has made some progress by committing to fiscal consolidation, but much work is still to be done. Ghana signed a $920 million extended credit facility with the IMF in April 2015 to help it address its growing economic crisis. The IMF fiscal targets require Ghana to reduce the deficit by cutting subsidies, decreasing the bloated public sector wage bill, strengthening revenue administration, boosting tax revenues, and improving the health of Ghana's banking sector. Priorities for the new administration include rescheduling some of Ghana's $31 billion debt, stimulating economic growth, reducing inflation, and stabilizing the currency. Prospects for new oil and gas production and follow through on tighter fiscal management are likely to help Ghana's economy in 2018.

GDP (purchasing power parity): $134 billion (2017 est.)

$123.6 billion (2016 est.)

$119.2 billion (2015 est.)

note: data are in 2017 dollars

country comparison to the world: 80

GDP (official exchange rate): $47.02 billion (2017 est.)

GDP—real growth rate: 8.4% (2017 est.)

3.7% (2016 est.)

3.8% (2015 est.)

country comparison to the world: 7

GDP—per capita (PPP): $4,700 (2017 est.)

$4,500 (2016 est.)

$4,400 (2015 est.)

note: data are in 2017 dollars

country comparison to the world: 172

Gross national saving: 9% of GDP (2017 est.)

7.8% of GDP (2016 est.)

9% of GDP (2015 est.)

country comparison to the world: 167

GDP—composition, by end use:

household consumption: 80.1% (2017 est.)

government consumption: 8.6% (2017 est.)

investment in fixed capital: 13.7% (2017 est.)

investment in inventories: 1.1% (2017 est.)

exports of goods and services: 43% (2017 est.)

imports of goods and services: -46.5% (2017 est.)

GDP—composition, by sector of origin:

agriculture: 18.3% (2017 est.)

industry: 24.5% (2017 est.)

services: 57.2% (2017 est.)

Agriculture—products: cocoa, rice, cassava (manioc, tapioca), peanuts, corn, shea nuts, bananas; timber

Industries: mining, lumbering, light manufacturing, aluminum smelting, food processing, cement, small commercial ship building, petroleum

Industrial production growth rate: 16.7% (2017 est.)

country comparison to the world: 2

Labor force: 12.49 million (2017 est.)

country comparison to the world: 45

Labor force—by occupation: *agriculture:* 44.7%

industry: 14.4%

services: 40.9% (2013 est.)

Unemployment rate: 11.9% (2015 est.)
5.2% (2013 est.)
country comparison to the world: 162

Population below poverty line: 24.2% (2013 est.)

Household income or consumption by percentage share: *lowest 10%:* 2%
highest 10%: 32.8% (2006)

Budget: *revenues:* 9.544 billion (2017 est.)
expenditures: 12.36 billion (2017 est.)

Taxes and other revenues: 20.3% (of GDP) (2017 est.)
country comparison to the world: 149

Budget surplus (+) or deficit (-): -6% (of GDP) (2017 est.)
country comparison to the world: 183

Public debt: 71.8% of GDP (2017 est.)
73.4% of GDP (2016 est.)
country comparison to the world: 46

Fiscal year: calendar year

Inflation rate (consumer prices): 12.4% (2017 est.)
17.5% (2016 est.)
country comparison to the world: 206

Current account balance: -$2.131 billion (2017 est.)
-$2.86 billion (2016 est.)
country comparison to the world: 169

Exports: $13.84 billion (2017 est.)
$11.14 billion (2016 est.)
country comparison to the world: 77

Exports—partners: India 23.8%, UAE 13.4%, China 10.8%, Switzerland 10.1%, Vietnam 5.2%, Burkina Faso 4% (2017)

Exports—commodities: oil, gold, cocoa, timber, tuna, bauxite, aluminum, manganese ore, diamonds, horticultural products

Imports: $12.65 billion (2017 est.)
$12.91 billion (2016 est.)
country comparison to the world: 92

Imports—commodities: capital equipment, refined petroleum, foodstuffs

Imports—partners: China 16.8%, US 8%, UK 6.2%, Belgium 5.9%, India 4.1% (2017)

Reserves of foreign exchange and gold: $7.555 billion (31 December 2017 est.)
$6.162 billion (31 December 2016 est.)
country comparison to the world: 81

Debt—external: $22.14 billion (31 December 2017 est.)
$16.5 billion (31 December 2016 est.)
country comparison to the world: 90

Exchange rates: cedis (GHC) per US dollar—
4.385 (2017 est.)
3.909 (2016 est.)
3.909 (2015 est.)
3.712 (2014 est.)
2.895 (2013 est.)

ENERGY

Electricity access: *population without electricity:* 5 million (2019)

electrification—total population: 85% (2019)
electrification—urban areas: 93% (2019)
electrification—rural areas: 75% (2019)

Electricity—production: 12.52 billion kWh (2016 est.)
country comparison to the world: 94

Electricity—consumption: 9.363 billion kWh (2016 est.)
country comparison to the world: 99

Electricity—exports: 187 million kWh (2016 est.)
country comparison to the world: 76

Electricity—imports: 511 million kWh (2016 est.)
country comparison to the world: 79

Electricity—installed generating capacity: 3.801 million kW (2016 est.)
country comparison to the world: 92

Electricity—from fossil fuels: 58% of total installed capacity (2016 est.)
country comparison to the world: 135

Electricity—from nuclear fuels: 0% of total installed capacity (2017 est.)
country comparison to the world: 96

Electricity—from hydroelectric plants: 42% of total installed capacity (2017 est.)
country comparison to the world: 48

Electricity—from other renewable sources: 1% of total installed capacity (2017 est.)
country comparison to the world: 153

Crude oil—production: 173,000 bbl/day (2018 est.)
country comparison to the world: 37

Crude oil—exports: 104,000 bbl/day (2017 est.)
country comparison to the world: 34

Crude oil—imports: 6,220 bbl/day (2015 est.)
country comparison to the world: 74

Crude oil—proved reserves: 660 million bbl (1 January 2018 est.)
country comparison to the world: 41

Refined petroleum products—production: 2,073 bbl/day (2015 est.)
country comparison to the world: 104

Refined petroleum products—consumption: 90,000 bbl/day (2016 est.)
country comparison to the world: 83

Refined petroleum products—exports: 2,654 bbl/day (2015 est.)
country comparison to the world: 100

Refined petroleum products—imports: 85,110 bbl/day (2015 est.)
country comparison to the world: 59

Natural gas—production: 914.4 million cu m (2017 est.)
country comparison to the world: 68

Natural gas—consumption: 1.232 billion cu m (2017 est.)
country comparison to the world: 87

Natural gas—exports: 0 cu m (2017 est.)
country comparison to the world: 110

Natural gas—imports: 317.4 million cu m (2017 est.)
country comparison to the world: 68

Natural gas—proved reserves: 22.65 billion cu m (1 January 2018 est.)
country comparison to the world: 73

Carbon dioxide emissions from consumption of energy: 13.67 million Mt (2017 est.)
country comparison to the world: 96

COMMUNICATIONS

Telephones—fixed lines: *total subscriptions:* 272,801
subscriptions per 100 inhabitants: less than 1 (2019 est.)
country comparison to the world: 112

Telephones—mobile cellular: *total subscriptions:* 38,571,189
subscriptions per 100 inhabitants: 134.32 (2019 est.)
country comparison to the world: 40

Telecommunication systems: *general assessment:* highly competitive Internet market; govt. helped fund programs for telecom services nationally; mobile accounts for how people access the Internet; LTE service launched in 2019; the government invested in fiber infrastructure and set up 600 additional towers to provide basic mobile services; m-money inter-operability launched; international submarine cables and new terrestrial cables have improved Internet capacity and reduced price for end-users; one of the most active mobile markets in Africa (2020)
domestic: fixed-line 1 per 100 subscriptions; competition among multiple mobile-cellular providers has spurred growth with a subscribership of more than 134 per 100 persons and rising (2019)
international: country code—233; landing points for the SAT-3/WASC, MainOne, ACE, WACS and GLO-1 fiber-optic submarine cables that provide connectivity to South and West Africa, and Europe; satellite earth stations—4 Intelsat (Atlantic Ocean); microwave radio relay link to Panaftel system connects Ghana to its neighbors; Ghana-1 satellite launched in 2020 (2019)
note: the COVID-19 outbreak is negatively impacting telecommunications production and supply chains globally; consumer spending on telecom devices and services has also slowed due to the pandemic's effect on economies worldwide; overall progress towards improvements in all facets of the telecom industry—mobile, fixed-line, broadband, submarine cable and satellite—has moderated

Broadcast media: state-owned TV station, 2 state-owned radio networks; several privately owned TV stations and a large number of privately owned radio stations; transmissions of multiple international broadcasters are accessible; several cable and satellite TV subscription services are obtainable

Internet country code: .gh

Internet users: *total:* 10,959,964
percent of population: 39% (July 2018 est.)
country comparison to the world: 49

Broadband—fixed subscriptions: *total:* 62,320
subscriptions per 100 inhabitants: less than 1 (2018 est.)
country comparison to the world: 130

TRANSPORTATION

National air transport system: *number of registered air carriers:* 3 (2020)
inventory of registered aircraft operated by air carriers: 21
annual passenger traffic on registered air carriers: 467,438 (2018)

Civil aircraft registration country code prefix: 9G (2016)

Airports: 10 (2013)
country comparison to the world: 154

Airports—with paved runways: *total:* 7 (2017)
over 3,047 m: 1 (2017)
2,438 to 3,047 m: 1 (2017)
1,524 to 2,437 m: 3 (2017)
914 to 1,523 m: 2 (2017)

Airports—with unpaved runways: *total:* 3 (2013)
914 to 1,523 m: 3 (2013)

Pipelines: 394 km gas, 20 km oil, 361 km refined products (2013)

Railways: *total:* 947 km (2014)
narrow gauge: 947 km 1.067-m gauge (2014)
country comparison to the world: 92

Roadways: *total:* 109,515 km (2009)
paved: 13,787 km (2009)
unpaved: 95,728 km (2009)
country comparison to the world: 45

Waterways: 1,293 km (168 km for launches and lighters on Volta, Ankobra, and Tano Rivers; 1,125 km of arterial and feeder waterways on Lake Volta) (2011)
country comparison to the world: 56

Merchant marine: *total:* 48
by type: general cargo 6, oil tanker 3, other 39 (2019)
country comparison to the world: 118

Ports and terminals: *major seaport(s):* Takoradi, Tema

MILITARY AND SECURITY

Military and security forces: *Ghana Armed Forces:* Army, Navy, Air Force (2019)

Military expenditures: 0.4% of GDP (2019)
0.41% of GDP (2018)
0.4% of GDP (2017)
0.38% of GDP (2016)
0.52% of GDP (2015)
country comparison to the world: 151

Military and security service personnel strengths: the Ghana Armed Forces consists of approximately 14,000 active personnel (10,000 Army; 2,000 Navy; 2,000 Air Force) (2019)

Military equipment inventories and acquisitions: the inventory of the Ghana Armed Forces is a mix of Russian, Chinese, and Western equipment; the top suppliers of armaments since 2010 are China, Germany, Spain, and Russia (2019 est.)

Military deployments: 140 Mali (MINUSMA); 180 Democratic Republic of the Congo (MONUSCO); 875 Lebanon (UNIFIL); 850 South Sudan (UNMISS) (2020)
note: Ghana has pledged to maintain about 1,000 military personnel in readiness for UN peacekeeping missions

Military service age and obligation: 18-26 years of age for voluntary military service, with basic education certificate; no conscription (2019)

Maritime threats: West African piracy more than doubled in 2018 to become the most dangerous area in the World; the waters off of Ghana saw a dramatic increase with 10 attacks reported in 2018 compared with only one in 2017; eight ships were boarded, one hijacked, and 47 crew taken hostage or kidnapped

TRANSNATIONAL ISSUES

Disputes—international: disputed maritime border between Ghana and Cote d'Ivoire

Refugees and internally displaced persons: *refugees (country of origin):* 6,406 (Cote d'Ivoire) (flight from 2010 post-election fighting) (2020)

Trafficking in persons: *current situation:* Ghana is a source, transit, and destination country for men, women, and children subjected to forced labor and sex trafficking; the trafficking of Ghanians, particularly children, internally is more common than the trafficking of foreign nationals; Ghanian children are subjected to forced labor in fishing, domestic service, street hawking, begging, portering, mining, quarrying, herding, and agriculture, with girls, and to a lesser extent boys, forced into prostitution; Ghanian women, sometimes lured with legitimate job offers, and girls are sex trafficked in West Africa, the Middle East, and Europe; Ghanian men fraudulently recruited for work in the Middle East are subjected to forced labor or prostitution, and a few Ghanian adults have been identified as victims of false labor in the US; women and girls from Vietnam, China, and neighboring West African countries are sex trafficked in Ghana; the country is also a transit point for sex trafficking from West Africa to Europe

tier rating: Tier 2 Watch List—Ghana does not fully comply with the minimum standards for the elimination of trafficking; however, it is making significant efforts to do so; Ghana continued to investigate and prosecute trafficking offenses but was unable to ramp up its anti-trafficking efforts in 2014 because the government failed to provide law enforcement or protection agencies with operating budgets; victim protection efforts decreased in 2014, with significantly fewer victims identified; most child victims were referred to NGO-run facilities, but care for adults was lacking because the government did not provide any support to the country's Human Trafficking Fund for victim services or its two shelters; anti-trafficking prevention measures increased modestly, including reconvening of the Human Trafficking Management Board, public awareness campaigns on child labor and trafficking, and anti-trafficking TV and radio programs (2015)

Illicit drugs: illicit producer of cannabis for the international drug trade; major transit hub for Southwest and Southeast Asian heroin and, to a lesser extent, South American cocaine destined for Europe and the US; widespread crime and money-laundering problem, but the lack of a well-developed financial infrastructure limits the country's utility as a money- laundering center; significant domestic cocaine and cannabis use

GIBRALTAR

INTRODUCTION

Background: Strategically important, Gibraltar was reluctantly ceded to Great Britain by Spain in the 1713 Treaty of Utrecht; the British garrison was formally declared a colony in 1830. In a referendum held in 1967, Gibraltarians voted overwhelmingly to remain a British dependency. The subsequent granting of autonomy in 1969 by the UK led Spain to close the border and sever all communication links. Between 1997 and 2002, the UK and Spain held a series of talks on establishing temporary joint sovereignty over Gibraltar. In response to these talks, the Gibraltar Government called a referendum in late 2002 in which the majority of citizens voted overwhelmingly against any sharing of sovereignty with Spain. Since late 2004, Spain, the UK, and Gibraltar have held tripartite talks with the aim of cooperatively resolving problems that affect the local population, and work continues on cooperation agreements in areas such as taxation and financial services; communications and maritime security; policy, legal and customs services; environmental protection; and education and visa services. A new noncolonial constitution came into force in 2007, and the European Court of First Instance recognized Gibraltar's right to regulate its own tax regime in December 2008. The UK retains responsibility for defense, foreign relations, internal security, and financial stability.

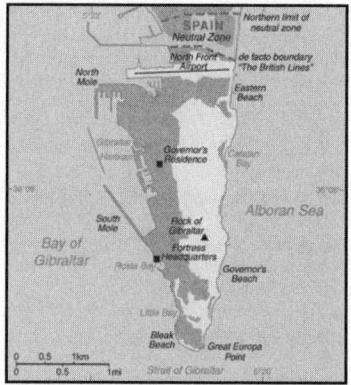

Spain and the UK continue to spar over the territory. Throughout 2009, a dispute over Gibraltar's claim to territorial waters extending out three miles gave rise to periodic non-violent maritime confrontations between Spanish and UK naval patrols and in 2013, the British reported a record number of entries by Spanish vessels into waters claimed by Gibraltar following a dispute over Gibraltar's creation of an artificial reef in those waters. Spain renewed its demands for an eventual return of Gibraltar to Spanish control after the UK's June 2016 vote to leave the EU, but London has dismissed any connection between the vote and its continued sovereignty over Gibraltar. The EU has said that Gibraltar will be ouside the territorial scope of any future UK-EU trade deal and that separate agreements between the EU and UK regarding Gibraltar would require Spain's prior approval.

GEOGRAPHY

Location: Southwestern Europe, bordering the Strait of Gibraltar, which links the Mediterranean Sea and the North Atlantic Ocean, on the southern coast of Spain

Geographic coordinates: 36 08 N, 5 21 W

Map references: Europe

Area: *total:* 7 sq km
land: 6.5 sq km
water: 0 sq km
country comparison to the world: 245

Area—comparative: more than 10 times the size of The National Mall in Washington, D.C.

Land boundaries: *total:* 1.2 km
border countries (I): Spain 1.2 km

Coastline: 12 km

Maritime claims: *territorial sea:* 3 nm

Climate: Mediterranean with mild winters and warm summers

Terrain: a narrow coastal lowland borders the Rock of Gibraltar

Elevation: *lowest point:* Mediterranean Sea 0 m
highest point: Rock of Gibraltar 426 m

Natural resources: none

Land use: *agricultural land:* 0% (2011 est.)
arable land: 0% (2011 est.) / permanent crops: 0% (2011 est.) / permanent pasture: 0% (2011 est.)
forest: 0% (2011 est.)
other: 100% (2011 est.)

Irrigated land: NA

Natural hazards: occasional droughts; no streams or large bodies of water on the peninsula (all potable water comes from desalination)

Environment—current issues: limited natural freshwater resources: more than 90% of drinking water supplied by desalination, the remainder from stored rainwater; a separate supply of saltwater used for sanitary services

Geography—note: *note 1:* strategic location on Strait of Gibraltar that links the North Atlantic Ocean and Mediterranean Sea
note 2: one of only two British territories where traffic drives on the right, the other being the island of Diego Garcia in the British Indian Ocean Territory

PEOPLE AND SOCIETY

Population: 29,581 (July 2020 est.)
country comparison to the world: 218

Nationality: *noun:* Gibraltarian(s)
adjective: Gibraltar

Ethnic groups: Gibraltarian 79%, other British 13.2%, Spanish 2.1%, Moroccan 1.6%, other EU 2.4%, other 1.6% (2012 est.)
note: data represent population by nationality

Languages: English (used in schools and for official purposes), Spanish, Italian, Portuguese

Religions: Roman Catholic 72.1%, Church of England 7.7%, other Christian 3.8%, Muslim 3.6%, Jewish 2.4%, Hindu 2%, other 1.1%, none 7.1%, unspecified 0.1% (2012 est.)

Age structure: *0-14 years:* 20.24% (male 3,080/ female 2,907)
15-24 years: 13.07% (male 2,000/female 1,866)
25-54 years: 41.28% (male 6,289/female 5,922)
55-64 years: 8.71% (male 1,082/female 1,495)
65 years and over: 16.7% (male 2,378/female 2,562) (2020 est.)

Median age: *total:* 35.5 years
male: 34.4 years
female: 36.6 years (2020 est.)
country comparison to the world: 84

Population growth rate: 0.2% (2020 est.)
country comparison to the world: 180

Birth rate: 13.8 births/1,000 population (2020 est.)
country comparison to the world: 136

Death rate: 8.6 deaths/1,000 population (2020 est.)
country comparison to the world: 71

Net migration rate: -3.3 migrant(s)/1,000 population (2020 est.)
country comparison to the world: 180

Urbanization: *urban population:* 100% of total population (2020)

rate of urbanization: 0.45% annual rate of change (2015-20 est.)
total population growth rate v. urban population growth rate, 2000-2030:

Major urban areas—Population: 35,000 GIBRALTAR (capital) (2018)

Sex ratio: *at birth:* 1.07 male(s)/female
0-14 years: 1.06 male(s)/female
15-24 years: 1.07 male(s)/female
25-54 years: 1.06 male(s)/female
55-64 years: 0.72 male(s)/female
65 years and over: 0.93 male(s)/female
total population: 1.01 male(s)/female (2020 est.)

Infant mortality rate: *total:* 5.6 deaths/1,000 live births
male: 6.2 deaths/1,000 live births
female: 5 deaths/1,000 live births (2020 est.)
country comparison to the world: 168

Life expectancy at birth: *total population:* 80 years
male: 77.1 years
female: 83 years (2020 est.)
country comparison to the world: 49

Total fertility rate: 1.9 children born/woman (2020 est.)
country comparison to the world: 129

Drinking water source:
improved:
urban: 100% of population
total: 100% of population
unimproved:
urban: 0% of population
total: 0% of population (2017 est.)

Sanitation facility access:
improved:
urban: 100% of population
total: 100% of population
unimproved:
urban: 0% of population
total: 0% of population (2017)

HIV/AIDS—adult prevalence rate: NA

HIV/AIDS—people living with HIV/AIDS: NA

HIV/AIDS—deaths: NA

Education expenditures: NA

GOVERNMENT

Country name: *conventional long form:* none
conventional short form: Gibraltar
etymology: from the Spanish derivation of the Arabic "Jabal Tariq," which means "Mountain of Tariq" and which refers to the Rock of Gibraltar

Dependency status: overseas territory of the UK

Government type: parliamentary democracy (Parliament); self-governing overseas territory of the UK

Capital: *name:* Gibraltar
geographic coordinates: 36 08 N, 5 21 W
time difference: UTC+1 (6 hours ahead of Washington, DC, during Standard Time)
daylight saving time: +1hr, begins last Sunday in March; ends last Sunday in October

etymology: from the Spanish derivation of the Arabic "Jabal Tariq," which means "Mountain of Tariq" and which refers to the Rock of Gibraltar

Administrative divisions: none (overseas territory of the UK)

Independence: none (overseas territory of the UK)

National holiday: National Day, 10 September (1967); note—day of the national referendum to decide whether to remain with the UK or join Spain

Constitution: *history*: previous 1969; latest passed by referendum 30 November 2006, entered into effect 14 December 2006, entered into force 2 January 2007

amendments: proposed by Parliament and require prior consent of the British monarch (through the Secretary of State); passage requires at least three-fourths majority vote in Parliament followed by simple majority vote in a referendum; note – only sections 1 through 15 in Chapter 1 (Protection of Fundamental Rights and Freedoms) can be amended by Parliament

Legal system: the laws of the UK, where applicable, apply

Citizenship: see United Kingdom

Suffrage: 18 years of age; universal; and British citizens with six months residence or more

Executive branch: *chief of state*: Queen ELIZABETH II (since 6 February 1952); represented by Governor Lt. Gen. Edward DAVIS (since 19 January 2016)

head of government: Chief Minister Fabian PICARDO (since 9 December 2011)

cabinet: Council of Ministers appointed from among the 17 elected members of Parliament by the governor in consultation with the chief minister

elections/appointments: the monarchy is hereditary; governor appointed by the monarch; following legislative elections, the leader of the majority party or majority coalition usually appointed chief minister by the governor

Legislative branch: *description*: unicameral Parliament (18 seats; 17 members directly elected in a single nationwide constituency by majority vote and 1 appointed by Parliament as speaker; members serve 4-year terms) (e.g. 2019)

elections: last held on 17 October 2019 (next to be held in 2023) (e.g. 2019)

election results: percent of vote by party—GSLP-Liberal Alliance 52.5% (GSLP 37.0%, LPG 15.5%), GSD 25.6%; seats by party—GSLP-Liberal Alliance 10 (GSLP 7, LPG 3), GSD 6; composition of elected members—men 15, women 2, percent of women 11.8% (e.g. 2019)

Judicial branch: *highest courts*: Court of Appeal (consists of at least 3 judges, including the court president); Supreme Court of Gibraltar (consists of the chief justice and 3 judges); note—appeals beyond the Court of Appeal are heard by the Judicial Committee of the Privy Council (in London)

judge selection and term of office: Court of Appeal and Supreme Court judges appointed by the governor upon the advice of the Judicial Service Commission, a 7-member body of judges and appointees of the governor; tenure of the Court of Appeal president based on terms of appointment; Supreme Court chief justice and judges normally appointed until retirement at age 67 but tenure can be extended 3 years

subordinate courts: Court of First Instance; Magistrates' Court; specialized tribunals for issues relating to social security, taxes, and employment

Political parties and leaders:
Gibraltar Liberal Party or Liberal Party of Gibraltar or LPG [Joseph GARCIA]
Gibraltar Social Democrats or GSD [Keith AZOPARDI]
Gibraltar Socialist Labor Party or GSLP [Fabian PICARDO]
GSLP-Liberal Alliance (includes GSLP and LPG)
Together Gibraltar or TG [Marlene HASSAN-NAHON]

International organization participation: ICC (NGOs), Interpol (subbureau), UPU

Diplomatic representation in the US: *none (overseas territory of the UK)*

Diplomatic representation from the US: none (overseas territory of the UK)

Flag description: two horizontal bands of white (top, double width) and red with a three-towered red castle in the center of the white band; hanging from the castle gate is a gold key centered in the red band; the design is that of Gibraltar's coat of arms granted on 10 July 1502 by King Ferdinand and Queen Isabella of Spain; the castle symbolizes Gibraltar as a fortress, while the key represents Gibraltar's strategic importance—the key to the Mediterranean

National symbol(s): *Barbary macaque; national colors*: red, white, yellow

National anthem: *name*: Gibraltar Anthem
lyrics/music: Peter EMBERLEY
note: adopted 1994; serves as a local anthem; as a territory of the United Kingdom, "God Save the Queen" is official (see United Kingdom)

ECONOMY

Economy—overview: Self-sufficient Gibraltar benefits from an extensive shipping trade, offshore banking, and its position as an international conference center. Tax rates are low to attract foreign investment. The British military presence has been sharply reduced and now contributes about 7% to the local economy, compared with 60% in 1984. In recent years, Gibraltar has seen major structural change from a public to a private sector economy, but changes in government spending still have a major impact on the level of employment.

The financial sector, tourism (over 11 million visitors in 2012), gaming revenues, shipping services fees, and duties on consumer goods also generate revenue. The financial sector, tourism, and the shipping sector contribute 30%, 30%, and 25%, respectively, of GDP. Telecommunications, e-commerce, and e-gaming account for the remaining 15%.

GDP (purchasing power parity): $2.044 billion (2014 est.)
$1.85 billion (2013 est.)
$2 billion (2012 est.)
note: data are in 2014 dollars
country comparison to the world: 196

GDP (official exchange rate): $2.044 billion (2014 est.)

GDP—per capita (PPP): $61,700 (2014 est.)
$43,000 (2008 est.)
$41,200 (2007 est.)
country comparison to the world: 18

GDP—composition, by sector of origin:
agriculture: 0% (2016 est.)
industry: 0% (2008 est.)
services: 100% (2016 est.)

Agriculture—products: none

Industries: tourism, banking and finance, ship repairing, tobacco

Industrial production growth rate: NA

Labor force: 24,420 (2014 est.)
country comparison to the world: 209

Labor force—by occupation: *agriculture*: NEGL
industry: 1.8%
services: 98.2% (2014 est.)

Unemployment rate: 1% (2016 est.)
country comparison to the world: 10

Population below poverty line: NA

Household income or consumption by percentage share: *lowest 10%*: NA
highest 10%: NA

Budget: *revenues*: 475.8 million (2008 est.)
expenditures: 452.3 million (2008 est.)

Taxes and other revenues: 23.3% (of GDP) (2008 est.)
country comparison to the world: 127

Budget surplus (+) or deficit (-): 1.1% (of GDP) (2008 est.)
country comparison to the world: 31

Public debt:
7.5% of GDP (2008 est.)
8.4% of GDP (2006 est.)
country comparison to the world: 200

Fiscal year: 1 July—30 June

Inflation rate (consumer prices):
2.5% (2013 est.)
2.2% (2012 est.)
country comparison to the world: 123

Exports:
$202.3 million (2014 est.)

$271 million (2004 est.)
country comparison to the world: 189

Exports—partners: Spain 27.1%, Germany 20.4%, Netherlands 10.8%, Poland 8.6%, France 6.6%, Italy 5.7%, Cote dIvoire 4.5% (2017)

Exports—commodities: (principally reexports) petroleum 51%, manufactured goods (2010 est.)

Imports:
$2.967 billion (2004 est.)
country comparison to the world: 148

Imports—commodities: fuels, manufactured goods, foodstuffs

Imports—partners: Spain 15.6%, Italy 13.4%, US 13.3%, Netherlands 10.9%, Greece 8.5%, Russia 6.6%, UK 5.8%, Belgium 4.4% (2017)

Debt—external: NA

Exchange rates: Gibraltar pounds (GIP) per US dollar -
0.885 (2017 est.)
0.903 (2016 est.)
0.9214 (2015 est.)
0.885 (2014 est.)
0.7634 (2013 est.)

ENERGY

Electricity access: electrification—total population: 100% (2020)

Electricity—production: 238.8 million kWh (2016 est.)
country comparison to the world: 187

Electricity—consumption: 230.8 million kWh (2016 est.)
country comparison to the world: 189

Electricity—exports: 0 kWh (2016 est.)
country comparison to the world: 140

Electricity—imports: 0 kWh (2016 est.)
country comparison to the world: 152

Electricity—installed generating capacity: 43,000 kW (2016 est.)
country comparison to the world: 196

Electricity—from fossil fuels: 100% of total installed capacity (2016 est.)
country comparison to the world: 8

Electricity—from nuclear fuels: 0% of total installed capacity (2017 est.)
country comparison to the world: 97

Electricity—from hydroelectric plants: 0% of total installed capacity (2017 est.)
country comparison to the world: 173

Electricity—from other renewable sources: 0% of total installed capacity (2017 est.)
country comparison to the world: 188

Crude oil—production: 0 bbl/day (2018 est.)
country comparison to the world: 141

Crude oil—exports: 0 bbl/day (2015 est.)
country comparison to the world: 129

Crude oil—imports: 0 bbl/day (2015 est.)
country comparison to the world: 132

Crude oil—proved reserves: 0 bbl (1 January 2018 est.)
country comparison to the world: 137

Refined petroleum products—production: 0 bbl/day (2017 est.)
country comparison to the world: 149

Refined petroleum products—consumption: 78,000 bbl/day (2016 est.)
country comparison to the world: 88

Refined petroleum products—exports: 0 bbl/day (2015 est.)
country comparison to the world: 158

Refined petroleum products—imports: 74,200 bbl/day (2015 est.)
country comparison to the world: 66

Natural gas—production: 0 cu m (2017 est.)
country comparison to the world: 137

Natural gas—consumption: 0 cu m (2017 est.)
country comparison to the world: 150

Natural gas—exports: 0 cu m (2017 est.)
country comparison to the world: 111

Natural gas—imports: 0 cu m (2017 est.)
country comparison to the world: 130

Natural gas—proved reserves: 0 cu m (1 January 2014 est.)
country comparison to the world: 139

Carbon dioxide emissions from consumption of energy: 13.34 million Mt (2017 est.)
country comparison to the world: 98

COMMUNICATIONS

Telephones—fixed lines: *total subscriptions:* 14,865
subscriptions per 100 inhabitants: 50.35 (2019 est.)
country comparison to the world: 183

Telephones—mobile cellular: *total subscriptions:* 35,510
subscriptions per 100 inhabitants: 120.28 (2019 est.)
country comparison to the world: 208

Telecommunication systems: *general assessment:* adequate, automatic domestic system and adequate international facilities (2018)
domestic: automatic exchange facilities; 50 per 100 fixed-line and 120 per 100 mobile-cellular (2019)

international: country code—350; landing point for the EIG to Europe, Asia, Africa and the Middle East via submarine cables; radiotelephone; microwave radio relay; satellite earth station—1 Intelsat (Atlantic Ocean) (2019)

Broadcast media: Gibraltar Broadcasting Corporation (GBC) provides TV and radio broadcasting services via 1 TV station and 4 radio stations; British Forces Broadcasting Service (BFBS) operates 1 radio station; broadcasts from Spanish radio and TV stations are accessible

Internet country code: .gi

Internet users: *total:* 27,823
percent of population: 94.44% (July 2018 est.)
country comparison to the world: 206

Broadband—fixed subscriptions: *total:* 19,232
subscriptions per 100 inhabitants: 65 (2018 est.)
country comparison to the world: 154

TRANSPORTATION

Civil aircraft registration country code prefix: VP-G (2016)

Airports: 1 (2013)
country comparison to the world: 223

Airports—with paved runways: *total:* 1 (2017)
1,524 to 2,437 m: 1 (2017)

Roadways: *total:* 29 km (2007)
paved: 29 km (2007)
country comparison to the world: 220

Merchant marine: *total:* 232
by type: bulk carrier 10, container ship 24, general cargo 71, oil tanker 24, other 103 (2019)
country comparison to the world: 61

Ports and terminals: *major seaport(s):* Gibraltar

MILITARY AND SECURITY

Military and security forces: Royal Gibraltar Regiment (2019)

Military—note: defense is the responsibility of the UK; the Royal Gibraltar Regiment replaced the last British regular infantry forces in 1991 (2019)

TRANSNATIONAL ISSUES

Disputes—international: in 2002, Gibraltar residents voted overwhelmingly by referendum to reject any "shared sovereignty" arrangement; the Government of Gibraltar insists on equal participation in talks between the UK and Spain; Spain disapproves of UK plans to grant Gibraltar even greater autonomy

GREECE

INTRODUCTION

Background: Greece achieved independence from the Ottoman Empire in 1830. During the second half of the 19th century and the first half of the 20th century, it gradually added neighboring islands and territories, most with Greek-speaking populations. In World War II, Greece was first invaded by Italy (1940) and subsequently occupied by Germany (1941-44); fighting endured in a protracted civil war between supporters of the king and other anti-communist and communist rebels. Following the latter's defeat in 1949, Greece joined NATO in 1952. In 1967, a group of military officers seized power, establishing a military dictatorship that suspended many political liberties and forced the king to flee the country. In 1974 following the collapse of the dictatorship, democratic elections and a referendum created a parliamentary republic and abolished the monarchy. In 1981, Greece joined the EC (now the EU); it became the 12th member of the European Economic and Monetary Union (EMU) in 2001. Greece has suffered a severe economic crisis since late 2009, due to nearly a decade of chronic overspending and structural rigidities. Beginning in 2010, Greece entered three bailout agreements—with the European Commission, the European Central Bank (ECB), the IMF, and the third in 2015 with the European Stability Mechanism (ESM)—worth in total about $300 billion. The Greek Government formally exited the third bailout in August 2018.

GEOGRAPHY

Location: Southern Europe, bordering the Aegean Sea, Ionian Sea, and the Mediterranean Sea, between Albania and Turkey

Geographic coordinates: 39 00 N, 22 00 E

Map references: Europe

Area: *total:* 131,957 sq km
land: 130,647 sq km

water: 1,310 sq km
country comparison to the world: 98

Area—comparative: slightly smaller than Alabama

Land boundaries: *total:* 1,110 km
border countries (4): Albania 212 km, Bulgaria 472 km, Macedonia 234 km, Turkey 192 km

Coastline: 13,676 km

Maritime claims: *territorial sea:* 12 nm
continental shelf: 200-m depth or to the depth of exploitation

Climate: temperate; mild, wet winters; hot, dry summers

Terrain: mountainous with ranges extending into the sea as peninsulas or chains of islands

Elevation: *mean elevation:* 498 m
lowest point: Mediterranean Sea 0 m
highest point: Mount Olympus 2,917
note: Mount Olympus actually has 52 peaks but its highest point, Mytikas (meaning "nose"), rises to 2,917 meters; in Greek mythology, Olympus' Mytikas peak was the home of the Greek gods

Natural resources: lignite, petroleum, iron ore, bauxite, lead, zinc, nickel, magnesite, marble, salt, hydropower potential

Land use: *agricultural land:* 63.4% (2011 est.)
arable land: 19.7% (2011 est.) / permanent crops: 8.9% (2011 est.) / permanent pasture: 34.8% (2011 est.)
forest: 30.5% (2011 est.)
other: 6.1% (2011 est.)

Irrigated land: 15,550 sq km (2012)

Population distribution: one-third of the population lives in and around metropolitan Athens; the remainder of the country has moderate population density mixed with sizeable urban clusters

Natural hazards: severe earthquakes

volcanism: Santorini (367 m) has been deemed a Decade Volcano by the International Association of Volcanology and Chemistry of the Earth's Interior, worthy of study due to its explosive history and close proximity to human populations; although there have been very few eruptions in recent centuries, Methana and Nisyros in the Aegean are classified as historically active

Environment—current issues: air pollution; air emissions from transport and electricity power stations; water pollution; degradation of coastal zones; loss of biodiversity in terrestrial and marine ecosystems; increasing municipal and industrial waste

Environment—international agreements: *party to:* Air Pollution, Air Pollution-Nitrogen Oxides, Air Pollution-Sulfur 94, Antarctic-Environmental Protocol, Antarctic-Marine Living Resources, Antarctic Treaty, Biodiversity, Climate Change, Climate Change-Kyoto Protocol, Desertification, Endangered Species, Environmental Modification, Hazardous Wastes, Law of the Sea, Marine Dumping, Ozone Layer Protection, Ship Pollution, Tropical Timber 83, Tropical Timber 94, Wetlands
signed, but not ratified: Air Pollution-Persistent Organic Pollutants, Air Pollution-Volatile Organic Compounds

Geography—note: strategic location dominating the Aegean Sea and southern approach to Turkish Straits; a peninsular country, possessing an archipelago of about 2,000 islands

PEOPLE AND SOCIETY

Population: 10,607,051 (July 2020 est.)
country comparison to the world: 86

Nationality: *noun:* Greek(s)
adjective: Greek

Ethnic groups: Greek 91.6%, Albanian 4.4%, other 4% (2011)
note: data represent citizenship; Greece does not collect data on ethnicity

Languages: Greek (official) 99%, other (includes English and French) 1%

Religions: Greek Orthodox (official) 81-90%, Muslim 2%, other 3%, none 4-15%, unspecified 1% (2015 est.)

Age structure: 0-14 years: 14.53% (male 794,918/female 745,909)
15-24 years: 10.34% (male 577,134/female 519,819)
25-54 years: 39.6% (male 2,080,443/female 2,119,995)
55-64 years: 13.1% (male 656,404/female 732,936)
65 years and over: 22.43% (male 1,057,317/female 1,322,176) (2020 est.)

Dependency ratios: *total dependency ratio:* 56.1
youth dependency ratio: 21.3
elderly dependency ratio: 34.8
potential support ratio: 2.9 (2020 est.)

Median age: *total:* 45.3 years
male: 43.7 years
female: 46.8 years (2020 est.)
country comparison to the world: 9

Population growth rate: -0.31% (2020 est.)
country comparison to the world: 220

Birth rate: 7.8 births/1,000 population (2020 est.)
country comparison to the world: 225

Death rate: 12 deaths/1,000 population (2020 est.)
country comparison to the world: 17

Net migration rate: 0.9 migrant(s)/1,000 population (2020 est.)
country comparison to the world: 61

Population distribution: one-third of the population lives in and around metropolitan Athens; the remainder of the country has moderate population density mixed with sizeable urban clusters

Urbanization: *urban population:* 79.7% of total population (2020)
rate of urbanization: 0.22% annual rate of change (2015-20 est.)

total population growth rate v. urban population growth rate, 2000-2030

Major urban areas—Population: 3.153 million ATHENS (capital), 812,000 Thessaloniki (2020)

Sex ratio: *at birth:* 1.07 male(s)/female
0-14 years: 1.07 male(s)/female
15-24 years: 1.11 male(s)/female
25-54 years: 0.98 male(s)/female
55-64 years: 0.9 male(s)/female
65 years and over: 0.8 male(s)/female
total population: 0.95 male(s)/female (2020 est.)

Mother's mean age at first birth: 29.9 years (2017 est.)

Maternal mortality rate: 3 deaths/100,000 live births (2017 est.)
country comparison to the world: 178

Infant mortality rate: *total:* 3.7 deaths/1,000 live births
male: 4 deaths/1,000 live births
female: 3.3 deaths/1,000 live births (2020 est.)
country comparison to the world: 194
Life expectancy at birth: total population: 81.1 years
male: 78.5 years
female: 83.8 years (2020 est.)
country comparison to the world: 38

Total fertility rate: 1.38 children born/woman (2020 est.)
country comparison to the world: 217

Drinking water source:
improved:
urban: 100% of population
rural: 100% of population
total: 100% of population
unimproved:
urban: 0% of population
rural: 0% of population
total: 0% of population (2017 est.)

Current Health Expenditure: 8% (2017)

Physicians density: 5.48 physicians/1,000 population (2017)

Hospital bed density: 4.2 beds/1,000 population (2017)

Sanitation facility access:
improved:
urban: 100% of population
rural: 100% of population
total: 100% of population
unimproved:
urban: 0% of population
rural: 0% of population
total: 0% of population (2017 est.)

HIV/AIDS—adult prevalence rate: 0.2% (2017 est.)
country comparison to the world: 97

HIV/AIDS—people living with HIV/AIDS: 14,000 (2017 est.)
country comparison to the world: 93

HIV/AIDS—deaths: <100 (2017 est.)

Obesity—adult prevalence rate:
24.9% (2016)
country comparison to the world: 54

Education expenditures: NA

Literacy: *definition:* age 15 and over can read and write
total population: 97.7%
male: 98.5%
female: 96.9% (2015)

School life expectancy (primary to tertiary education): *total:* 20 years
male: 20 years
female: 20 years (2018)

Unemployment, youth ages 15-24: *total:* 39.9%
male: 36.4%
female: 43.9% (2018 est.)
country comparison to the world: 11

GOVERNMENT

Country name: *conventional long form:* Hellenic Republic
conventional short form: Greece
local long form: Elliniki Dimokratia
local short form: Ellas or Ellada
former: Hellenic State, Kingdom of Greece
etymology: the English name derives from the Roman (Latin) designation "Graecia," meaning "Land of the Greeks"; the Greeks call their country "Hellas" or "Ellada"

Government type: parliamentary republic

Capital: *name:* Athens
geographic coordinates: 37 59 N, 23 44 E
time difference: UTC+2 (7 hours ahead of Washington, DC, during Standard Time)
daylight saving time: +1hr, begins last Sunday in March; ends last Sunday in October
etymology: Athens is the oldest European capital city; according to tradition, the city is named after Athena, the Greek goddess of wisdom; in actuality, the appellation probably derives from a lost name in a pre-Hellenic language

Administrative divisions: 13 regions (perifereies, singular—perifereia) and 1 autonomous monastic state* (aftonomi monastiki politeia); Agion Oros* (Mount Athos), Anatoliki Makedonia kai Thraki (East Macedonia and Thrace), Attiki (Attica), Dytiki Ellada (West Greece), Dytiki Makedonia (West Macedonia), Ionia Nisia (Ionian Islands), Ipeiros (Epirus), Kentriki Makedonia (Central Macedonia), Kriti (Crete), Notio Aigaio (South Aegean), Peloponnisos (Peloponnese), Sterea Ellada (Central Greece), Thessalia (Thessaly), Voreio Aigaio (North Aegean)

Independence: 3 February 1830 (from the Ottoman Empire); note—25 March 1821, outbreak of the national revolt against the Ottomans; 3 February 1830, signing of the London Protocol recognizing Greek independence by Great Britain, France, and Russia

National holiday: Independence Day, 25 March (1821)

Constitution: *history:* many previous; latest entered into force 11 June 1975
amendments: proposed by at least 50 members of Parliament and agreed by three-fifths majority vote in two separate ballots at least 30 days apart;

passage requires absolute majority vote by the next elected Parliament; entry into force finalized through a "special parliamentary resolution"; articles on human rights and freedoms and the form of government cannot be amended; amended 1986, 2001, 2008

Legal system: civil legal system based on Roman law

International law organization participation: accepts compulsory ICJ jurisdiction with reservations; accepts ICCt jurisdiction

Citizenship: citizenship by birth: no
citizenship by descent only: at least one parent must be a citizen of Greece
dual citizenship recognized: yes
residency requirement for naturalization: 10 years

Suffrage: 17 years of age; universal and compulsory

Executive branch: *chief of state:* President Ekaterini SAKELLAROPOULOU (since 13 March 2020)
head of government: Prime Minister Kyriakos MITSOTAKIS (since 8 July 2019)
cabinet: Cabinet appointed by the president on the recommendation of the prime minister
elections/appointments: president elected by Hellenic Parliament for a 5-year term (eligible for a second term); election last held on 22 January 2020 (next to be held by February 2025); president appoints as prime minister the leader of the majority party or coalition in the Hellenic Parliament
election results: Katerina SAKELLAROPOULOU (independent) elected president by Parliament—261 of 300 votes; note—SAKELLAROPOULOU is Greece's first woman president

Legislative branch: *description:* unicameral Hellenic Parliament or Vouli ton Ellinon (300 seats; 280 members in multi-seat constituencies and 12 members in a single nationwide constituency directly elected by open party-list proportional representation vote; 8 members in single-seat constituencies elected by simple majority vote; members serve up to 4 years); note—only parties surpassing a 3% threshold are entitled to parliamentary seats; parties need 10 seats to become formal parliamentary groups but can retain that status if the party participated in the last election and received the minimum 3% threshold
elections: last held on 7 July 2019 (next to be held by July 2023)
election results: percent of vote by party—ND 39.9%, SYRIZA 31.5%, KINAL 8.1%, KKE 5.3%, Greek Solution 3.7%, MeRA25 3.4%, other 8.1%; seats by party—ND 158, SYRIZA 86, KINAL 22, KKE 15, Greek Solution 10, MeRA25 9; composition—men 244, women 56, percent of women 18.7%

Judicial branch: *highest courts:* Supreme Civil and Criminal Court or Areios Pagos (consists of 56 judges, including the court presidents); Council of State (supreme administrative court) (consists of the president, 7 vice presidents, 42

privy councilors, 48 associate councilors and 50 reporting judges, organized into six 5- and 7-member chambers; Court of Audit (government audit and enforcement) consists of the president, 5 vice presidents, 20 councilors, and 90 associate and reporting judges

judge selection and term of office: Supreme Court judges appointed by presidential decree on the advice of the Supreme Judicial Council (SJC), which includes the president of the Supreme Court, other judges, and the prosecutor of the Supreme Court; judges appointed for life following a 2-year probationary period; Council of State president appointed by the Greek Cabinet to serve a 4-year term; other judge appointments and tenure NA; Court of Audit president appointed by decree of the president of the republic on the advice of the SJC; court president serves a 4-year term or until age 67; tenure of vice presidents, councilors, and judges NA

subordinate courts: Courts of Appeal and Courts of First Instance (district courts)

Political parties and leaders:

Anticapitalist Left Cooperation for the Overthrow or ANTARSYA [collective leadership]

Coalition of the Radical Left or SYRIZA [Alexios (Alexis) TSIPRAS]

Communist Party of Greece or KKE [Dimitrios KOUTSOUMBAS]

Democratic Left or DIMAR [Athanasios (Thanasis) THEOCHAROPOULOS]

European Realistic Disobedience Front or MeRA25 [Yanis VAROUFAKIS]

Greek Solution [Kyriakos VELOPOULOS]

Independent Greeks or ANEL [Panagiotis (Panos) KAMMENOS]

Movement for Change or KINAL [Foteini (Fofi) GENIMMATA]

New Democracy or ND [Kyriakos MITSOTAKIS]

People's Association-Golden Dawn [Nikolaos MICHALOLIAKOS]

Popular Unity or LAE [Panagiotis LAFAZANIS]

The River (To Potami) [Stavros THEODORAKIS]

Union of Centrists or EK [Vasileios (Vasilis) LEVENTIS]

International organization participation: Australia Group, BIS, BSEC, CD, CE, CERN, EAPC, EBRD, ECB, EIB, EMU, ESA, EU, FAO, FATF, IAEA, IBRD, ICAO, ICC (national committees), ICCt, ICRM, IDA, IEA, IFAD, IFC, IFRCS, IGAD (partners), IHO, ILO, IMF, IMO, IMSO, Interpol, IOC, IOM, IPU, ISO, ITSO, ITU, ITUC (NGOs), MIGA, NATO, NEA, NSG, OAS (observer), OECD, OIF, OPCW, OSCE, PCA, Schengen Convention, SELEC, UN, UNCTAD, UNESCO, UNHCR, UNIDO, UNIFIL, UNWTO, UPU, WCO, WFTU (NGOs), WHO, WIPO, WMO, WTO, ZC

Diplomatic representation in the US: *chief of mission:* Ambassador Theocharis LALAKOS (since 27 June 2016)

chancery: 2217 Massachusetts Avenue NW, Washington, DC 20008

telephone: [1] (202) 939-1300

FAX: [1] (202) 939-1324

consulate(s) general: Boston, Chicago, Los Angeles, New York, Tampa (FL), San Francisco

consulate(s): Atlanta, Houston

Diplomatic representation from the US: *chief of mission:* Ambassador Geoffrey R. PYATT (since 24 October 2016)

telephone: [30] (210) 721-2951

embassy: 91 Vasillisis Sophias Avenue, 10160 Athens

mailing address: PSC 108, APO AE 09842-0108

FAX: [30] (210) 645-6282

consulate(s) general: Thessaloniki

Flag description: nine equal horizontal stripes of blue alternating with white; a blue square bearing a white cross appears in the upper hoist-side corner; the cross symbolizes Greek Orthodoxy, the established religion of the country; there is no agreed upon meaning for the nine stripes or for the colors

note: Greek legislation states that the flag colors are cyan and white, but cyan can mean "blue" in Greek, so the exact shade of blue has never been set and has varied from a light to a dark blue over time; in general, the hue of blue normally encountered is a form of azure

National symbol(s): *Greek cross (white cross on blue field, arms equal length); national colors:* blue, white

National anthem: *name:* "Ymnos eis tin Eleftherian" (Hymn to Liberty)

lyrics/music: Dionysios SOLOMOS/Nikolaos MANTZAROS

note: adopted 1864; the anthem is based on a 158-stanza poem by the same name, which was inspired by the Greek Revolution of 1821 against the Ottomans (only the first two stanzas are used); Cyprus also uses "Hymn to Liberty" as its anthem 0:00 / 0:45

ECONOMY

Economy—overview: Greece has a capitalist economy with a public sector accounting for about 40% of GDP and with per capita GDP about two-thirds that of the leading euro-zone economies. Tourism provides 18% of GDP. Immigrants make up nearly one-fifth of the work force, mainly in agricultural and unskilled jobs. Greece is a major beneficiary of EU aid, equal to about 3.3% of annual GDP.

The Greek economy averaged growth of about 4% per year between 2003 and 2007, but the economy went into recession in 2009 as a result of the world financial crisis, tightening credit conditions, and Athens' failure to address a growing budget deficit. By 2013, the economy had contracted 26%, compared with the pre-crisis level of 2007. Greece met the EU's Growth and Stability Pact budget deficit criterion of no more than 3% of GDP in 2007-08, but violated it in 2009, when the deficit reached 15% of GDP. Deteriorating public finances, inaccurate and misreported statistics, and consistent underperformance on reforms prompted major credit rating agencies to downgrade Greece's international debt rating in late 2009 and led the country into a financial crisis. Under intense pressure from the EU and international market

participants, the government accepted a bailout program that called on Athens to cut government spending, decrease tax evasion, overhaul the civil-service, health-care, and pension systems, and reform the labor and product markets. Austerity measures reduced the deficit to 1.3% in 2017. Successive Greek governments, however, failed to push through many of the most unpopular reforms in the face of widespread political opposition, including from the country's powerful labor unions and the general public.

In April 2010, a leading credit agency assigned Greek debt its lowest possible credit rating, and in May 2010, the IMF and euro-zone governments provided Greece emergency short- and medium-term loans worth $147 billion so that the country could make debt repayments to creditors. Greece, however, struggled to meet the targets set by the EU and the IMF, especially after Eurostat—the EU's statistical office—revised upward Greece's deficit and debt numbers for 2009 and 2010. European leaders and the IMF agreed in October 2011 to provide Athens a second bailout package of $169 billion. The second deal called for holders of Greek government bonds to write down a significant portion of their holdings to try to alleviate Greece's government debt burden. However, Greek banks, saddled with a significant portion of sovereign debt, were adversely affected by the write down and $60 billion of the second bailout package was set aside to ensure the banking system was adequately capitalized.

In 2014, the Greek economy began to turn the corner on the recession. Greece achieved three significant milestones: balancing the budget—not including debt repayments; issuing government debt in financial markets for the first time since 2010; and generating 0.7% GDP growth — the first economic expansion since 2007.

Despite the nascent recovery, widespread discontent with austerity measures helped propel the far-left Coalition of the Radical Left (SYRIZA) party into government in national legislative elections in January 2015. Between January and July 2015, frustrations grew between the SYRIZA-led government and Greece's EU and IMF creditors over the implementation of bailout measures and disbursement of funds. The Greek government began running up significant arrears to suppliers, while Greek banks relied on emergency lending, and Greece's future in the euro zone was called into question. To stave off a collapse of the banking system, Greece imposed capital controls in June 2015, then became the first developed nation to miss a loan payment to the IMF, rattling international financial markets. Unable to reach an agreement with creditors, Prime Minister Alexios TSIPRAS held a nationwide referendum on 5 July on whether to accept the terms of Greece's bailout, campaigning for the ultimately successful "no" vote. The TSIPRAS government subsequently agreed, however, to a new $96 billion bailout in order to avert Greece's exit from the monetary bloc. On 20 August 2015, Greece signed its third bailout, allowing it to cover significant debt payments to its EU and IMF creditors and to ensure the banking sector retained access to emergency liquidity. The TSIPRAS government — which retook office on 20 September 2015 after calling new elections in late August — successfully secured disbursal of two delayed tranches of bailout funds. Despite the economic turmoil, Greek GDP did not contract as sharply as feared, boosted in part by a strong tourist season.

In 2017, Greece saw improvements in GDP and unemployment. Unfinished economic reforms, a massive non-performing loan problem, and ongoing uncertainty regarding the political direction of the country hold the economy back. Some estimates put Greece's black market at 20- to 25% of GDP, as more people have stopped reporting their income to avoid paying taxes that, in some cases, have risen to 70% of an individual's gross income.

GDP (purchasing power parity): $299.3 billion (2017 est.)
$295.3 billion (2016 est.)
$296 billion (2015 est.)
note: data are in 2017 dollars
country comparison to the world: 56

GDP (official exchange rate): $200.7 billion (2017 est.)

GDP—real growth rate:
1.87% (2019 est.)
1.91% (2018 est.)
1.44% (2017 est.)
country comparison to the world: 145

GDP—per capita (PPP): $27,800 (2017 est.)
$27,400 (2016 est.)
$27,300 (2015 est.)
note: data are in 2017 dollars
country comparison to the world: 75

Gross national saving:
10.9% of GDP (2017 est.)
9.5% of GDP (2016 est.)
9.6% of GDP (2015 est.)
country comparison to the world: 159

GDP—composition, by end use:
household consumption: 69.6% (2017 est.)
government consumption: 20.1% (2017 est.)
investment in fixed capital: 12.5% (2017 est.)
investment in inventories: -1% (2017 est.)
exports of goods and services: 33.4% (2017 est.)
imports of goods and services: -34.7% (2017 est.)

GDP—composition, by sector of origin:
agriculture: 4.1% (2017 est.)
industry: 16.9% (2017 est.)
services: 79.1% (2017 est.)

Agriculture—products: wheat, corn, barley, sugar beets, olives, tomatoes, wine, tobacco, potatoes; beef, dairy products

Industries: tourism, food and tobacco processing, textiles, chemicals, metal products; mining, petroleum

Industrial production growth rate: 3.5% (2017 est.)
country comparison to the world: 85

Labor force: 4 million (2020 est.)
country comparison to the world: 90

Labor force—by occupation: *agriculture:* 12.6%

industry: 15%
services: 72.4% (30 October 2015 est.)

Unemployment rate:
17.3% (2019 est.)
19.34% (2018 est.)
country comparison to the world: 182

Population below poverty line: 36% (2014 est.)

Household income or consumption by percentage share: *lowest 10%:* 1.7%
highest 10%: 26.7% (2015 est.)

Budget: *revenues:* 97.99 billion (2017 est.)
expenditures: 96.35 billion (2017 est.)

Taxes and other revenues: 48.8% (of GDP) (2017 est.)
country comparison to the world: 17

Budget surplus (+) or deficit (-): 0.8% (of GDP) (2017 est.)
country comparison to the world: 36

Public debt:
181.8% of GDP (2017 est.)
183.5% of GDP (2016 est.)
country comparison to the world: 2

Fiscal year: calendar year

Inflation rate (consumer prices):
1.1% (2017 est.)
0% (2016 est.)
country comparison to the world: 59

Current account balance:
-$3.114 billion (2019 est.)
-$6.245 billion (2018 est.)
country comparison to the world: 175

Exports:
$31.54 billion (2017 est.)
$27.1 billion (2016 est.)
country comparison to the world: 63

Exports—partners: Italy 10.6%, Germany 7.1%, Turkey 6.8%, Cyprus 6.5%, Bulgaria 4.9%, Lebanon 4.3% (2017)

Exports—commodities: food and beverages, manufactured goods, petroleum products, chemicals, textiles

Imports:
$52.27 billion (2017 est.)
$45.45 billion (2016 est.)
country comparison to the world: 52

Imports—commodities: machinery, transport equipment, fuels, chemicals

Imports—partners: Germany 10.4%, Italy 8.2%, Russia 6.8%, Iraq 6.3%, South Korea 6.1%, China 5.4%, Netherlands 5.3%, France 4.3% (2017)

Reserves of foreign exchange and gold:
$7.807 billion (31 December 2017 est.)
$6.026 billion (31 December 2015 est.)
country comparison to the world: 80

Debt—external:
$506.6 billion (31 March 2016 est.)
$468.2 billion (31 March 2015 est.)
country comparison to the world: 23

Exchange rates: euros (EUR) per US dollar -
0.885 (2017 est.)
0.903 (2016 est.)
0.9214 (2015 est.)

0.885 (2014 est.)
0.7634 (2013 est.)

ENERGY

Electricity access: electrification—total population: 100% (2020)

Electricity—production: 52.05 billion kWh (2016 est.)
country comparison to the world: 53

Electricity—consumption: 56.89 billion kWh (2016 est.)
country comparison to the world: 45

Electricity—exports: 1.037 billion kWh (2016 est.)
country comparison to the world: 58

Electricity—imports: 9.833 billion kWh (2016 est.)
country comparison to the world: 27

Electricity—installed generating capacity: 19.17 million kW (2016 est.)
country comparison to the world: 46

Electricity—from fossil fuels: 57% of total installed capacity (2016 est.)
country comparison to the world: 137

Electricity—from nuclear fuels: 0% of total installed capacity (2017 est.)
country comparison to the world: 98

Electricity—from hydroelectric plants: 14% of total installed capacity (2017 est.)
country comparison to the world: 105

Electricity—from other renewable sources: 29% of total installed capacity (2017 est.)
country comparison to the world: 19

Crude oil—production: 4,100 bbl/day (2018 est.)
country comparison to the world: 80

Crude oil—exports: 3,229 bbl/day (2017 est.)
country comparison to the world: 67

Crude oil—imports: 484,300 bbl/day (2017 est.)
country comparison to the world: 20

Crude oil—proved reserves: 10 million bbl (1 January 2018 est.)
country comparison to the world: 90

Refined petroleum products—production: 655,400 bbl/day (2017 est.)
country comparison to the world: 28

Refined petroleum products—consumption: 304,100 bbl/day (2017 est.)
country comparison to the world: 43

Refined petroleum products—exports: 371,900 bbl/day (2017 est.)
country comparison to the world: 22

Refined petroleum products—imports: 192,200 bbl/day (2017 est.)
country comparison to the world: 35

Natural gas—production: 8 million cu m (2017 est.)
country comparison to the world: 93

Natural gas—consumption: 4.927 billion cu m (2017 est.)
country comparison to the world: 61

Natural gas—exports: 0 cu m (2017 est.)

country comparison to the world: 112

Natural gas—imports: 4.984 billion cu m (2017 est.)
country comparison to the world: 36

Natural gas—proved reserves: 991.1 million cu m (1 January 2018 est.)
country comparison to the world: 100

Carbon dioxide emissions from consumption of energy: 69.37 million Mt (2017 est.)
country comparison to the world: 51

COMMUNICATIONS

Telephones—fixed lines: *total subscriptions:* 5,080,386
subscriptions per 100 inhabitants: 47.75 (2019 est.)
country comparison to the world: 29

Telephones—mobile cellular: *total subscriptions:* 12,070,571
subscriptions per 100 inhabitants: 113.45 (2019 est.)
country comparison to the world: 75

Telecommunication systems: *general assessment:* good mobile telephone and international services; 3 mobile network operators; broadband penetration developing steadily despite rough economic conditions; plans to repurpose 3G network for LTE and 5G by 2022 (2020)
domestic: microwave radio relay trunk system; extensive open-wire connections; submarine cable to offshore islands; 48 per 100 for fixed-line and 114 per 100 for mobile-cellular (2019)
international: country code—30; landing points for the SEA-ME-WE-3, Adria-1, Italy-Greece 1, OTEGLOBE, MedNautilus Submarine System, Aphrodite 2, AAE-1 and Silphium optical telecommunications submarine cable that provides links to Europe, the Middle East, Africa, Southeast Asia, Asia and Australia; tropospheric scatter; satellite earth stations—4 (2 Intelsat—1 Atlantic Ocean and 1 Indian Ocean, 1 Eutelsat, and 1 Inmarsat—Indian Ocean region) (2019)
note: the COVID-19 outbreak is negatively impacting telecommunications production and supply chains globally; consumer spending on telecom devices and services has also slowed due to the pandemic's effect on economies worldwide; overall progress towards improvements in all facets of the telecom industry—mobile, fixed-line, broadband, submarine cable and satellite—has moderated

Broadcast media: broadcast media dominated by the private sector; roughly 150 private TV channels, about 10 of which broadcast nationwide; 1 government-owned terrestrial TV channel with national coverage; 3 privately owned satellite channels; multi-channel satellite and cable TV services available; upwards of 1,500 radio stations, all of them privately owned; government-owned broadcaster has 2 national radio stations

Internet country code: .gr

Internet users: *total:* 7,783,381
percent of population: 72.95% (July 2018 est.)
country comparison to the world: 61

Broadband—fixed subscriptions: *total:* 3,961,864
subscriptions per 100 inhabitants: 37 (2018 est.)
country comparison to the world: 35

TRANSPORTATION

National air transport system: *number of registered air carriers:* 11 (2020)
inventory of registered aircraft operated by air carriers: 97
annual passenger traffic on registered air carriers: 15,125,933 (2018)
annual freight traffic on registered air carriers: 21.91 million mt-km (2018)

Civil aircraft registration country code prefix: SX (2016)

Airports: 77 (2013)
country comparison to the world: 69

Airports—with paved runways: *total:* 68 (2017)
over 3,047 m: 6 (2017)
2,438 to 3,047 m: 15 (2017)
1,524 to 2,437 m: 19 (2017)
914 to 1,523 m: 18 (2017)
under 914 m: 10 (2017)

Airports—with unpaved runways: *total:* 9 (2013)
914 to 1,523 m: 2 (2013)
under 914 m: 7 (2013)

Heliports: 9 (2013)

Pipelines: 1329 km gas, 94 km oil (2013)

Railways: *total:* 2,548 km (2014)
standard gauge: 1,565 km 1.435-m gauge (764 km electrified) (2014)
narrow gauge: 961 km 1.000-m gauge (2014)
22 0.750-m gauge
country comparison to the world: 67

Roadways: *total:* 117,000 km (2018)
country comparison to the world: 42

Waterways: 6 km (the 6-km-long Corinth Canal crosses the Isthmus of Corinth; it shortens a sea voyage by 325 km) (2012)
country comparison to the world: 106

Merchant marine: *total:* 1,308
by type: bulk carrier 180, container ship 6, general cargo 95, oil tanker 375, other 652 (2019)
country comparison to the world: 22

Ports and terminals: *major seaport(s):* Aspropyrgos, Pachi, Piraeus, Thessaloniki
oil terminal(s): Agioi Theodoroi
container port(s) (TEUs): Piraeus (4,145,079) (2017)

LNG terminal(s) (import): Revithoussa

MILITARY AND SECURITY

Military and security forces: *Hellenic Armed Forces:* Hellenic Army (Ellinikos Stratos, ES; includes National Guard reserves), Hellenic Navy (Elliniko Polemiko Navtiko, EPN), Hellenic Air Force (Elliniki Polemiki Aeroporia, EPA; includes air defense) (2019)

Military expenditures:
2.28% of GDP (2019 est.)
2.48% of GDP (2018)
2.34% of GDP (2017)

2.38% of GDP (2016)
2.3% of GDP (2015)
country comparison to the world: 41

Military and security service personnel strengths: the Hellenic Armed Forces have approximately 141,000 active duty personnel (90,000 Army; 16,000 Navy; 25,000 Air Force; 10,000 joint service, support, staff); approximately 35,000 National Guard (2019 est.)

Military equipment inventories and acquisitions: the inventory of the Hellenic Armed Forces consists mostly of a mix of imported weapons from Europe and the US, as well as a limited number of domestically produced systems, particularly naval vessels; Germany is the leading supplier of weapons systems to Greece since 2010, followed by France and the US; Greece's defense industry is capable of producing naval vessels and associated subsystems (2019 est.)

Military deployments: est. 1,000 Cyprus; 110 Kosovo (NATO); 140 Lebanon (UNIFIL) (2020)

Military service age and obligation: 19-45 years of age for compulsory military service; during wartime the law allows for recruitment beginning January of the year of inductee's 18th birthday, thus including 17 year olds; 18 years of age for volunteers; conscript service obligation is 1 year for the Army and 9 months for the Air Force and Navy; women are eligible for voluntary military service (2014)

TERRORISM

Terrorist group(s): Revolutionary Struggle (2019)
note: details about the history, aims, leadership, organization, areas of operation, tactics, targets, weapons, size, and sources of support of the group(s) appear(s) in Appendix-T

TRANSNATIONAL ISSUES

Disputes—international: Greece and Turkey continue discussions to resolve their complex maritime, air, territorial, and boundary disputes in the Aegean Sea; the mass migration of unemployed Albanians still remains a problem for developed countries, chiefly Greece and Italy

Refugees and internally displaced persons: *refugees (country of origin):* 26,696 (Syria), 17,685 (Afghanistan), 9,614 (Afghanistan) (2019)
stateless persons: 4,734 (2019)
note: 1,203,437 estimated refugee and migrant arrivals (January 2015-November 2020); as of the end of December 2019, an estimated 112,300 migrants and refugees were stranded in Greece since 2015-16; 50,215 migrant arrivals in 2018

Illicit drugs: a gateway to Europe for traffickers smuggling cannabis and heroin from the Middle East and Southwest Asia to the West and precursor chemicals to the East; some South American cocaine transits or is consumed in Greece; money laundering related to drug trafficking and organized crime

GREENLAND

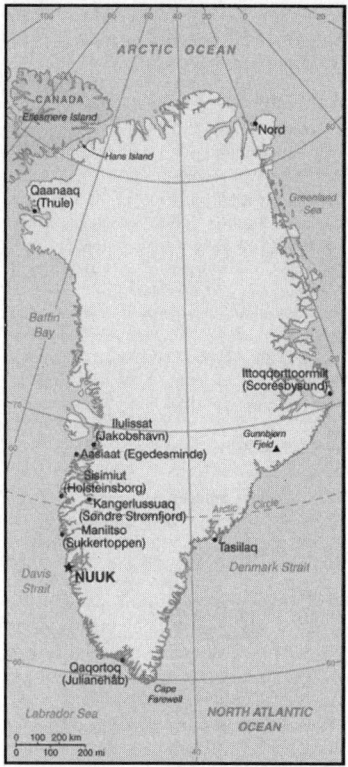

INTRODUCTION

Background: Greenland, the world's largest island, is about 80% ice-capped. Vikings reached the island in the 10th century from Iceland; Danish colonization began in the 18th century, and Greenland became an integral part of the Danish Realm in 1953. It joined the European Community (now the EU) with Denmark in 1973 but withdrew in 1985 over a dispute centered on stringent fishing quotas. Greenland remains a member of the Overseas Countries and Territories Association of the EU. Greenland was granted self-government in 1979 by the Danish parliament; the law went into effect the following year. Greenland voted in favor of increased self-rule in November 2008 and acquired greater responsibility for internal affairs when the Act on Greenland Self-Government was signed into law in June 2009. Denmark, however, continues to exercise control over several policy areas on behalf of Greenland, including foreign affairs, security, and financial policy in consultation with Greenland's Self-Rule Government.

GEOGRAPHY

Location: Northern North America, island between the Arctic Ocean and the North Atlantic Ocean, northeast of Canada

Geographic coordinates: 72 00 N, 40 00 W

Map references: Arctic Region

Area: *total:* 2,166,086 sq km
land: 2,166,086 sq km (approximately 1,710,000 sq km ice-covered)
country comparison to the world: 13

Area—comparative: slightly more than three times the size of Texas

Land boundaries: 0 km

Coastline: 44,087 km

Maritime claims: *territorial sea:* 3 nm
continental shelf: 200 nm or agreed boundaries or median line
exclusive fishing zone: 200 nm or agreed boundaries or median line

Climate: arctic to subarctic; cool summers, cold winters

Terrain: flat to gradually sloping icecap covers all but a narrow, mountainous, barren, rocky coast

Elevation: *mean elevation:* 1,792 m
lowest point: Atlantic Ocean 0 m
highest point: Gunnbjorn Fjeld 3,694 m

Natural resources: coal, iron ore, lead, zinc, molybdenum, diamonds, gold, platinum, niobium, tantalite, uranium, fish, seals, whales, hydropower, possible oil and gas

Land use: *agricultural land:* 0.6% (2011 est.)
arable land: 0% (2011 est.) / permanent crops: 0% (2011 est.) / permanent pasture: 0.6% (2011 est.)
forest: 0% (2011 est.)
other: 99.4% (2011 est.)

Irrigated land: NA

Population distribution: settlement concentrated on the southwest shoreline, with limited settlements scattered along the remaining coast; interior is uninhabited

Natural hazards: continuous permafrost over northern two-thirds of the island

Environment—current issues: especially vulnerable to climate change and disruption of the Arctic environment; preservation of the Inuit traditional way of life, including whaling and seal hunting

Geography—note: dominates North Atlantic Ocean between North America and Europe; sparse population confined to small settlements along coast; close to one-quarter of the population lives in the capital, Nuuk; world's second largest ice sheet after that of Antarctica covering an area of 1.71 million sq km (660,000 sq mi) or about 79% of the island, and containing 2.85 million cu km (684 thousand cu mi) of ice (this is almost 7% of all of the world's fresh water); if all this ice were converted to liquid water, one estimate is that it would be sufficient to raise the height of the world's oceans by 7.2 m (24 ft)

PEOPLE AND SOCIETY

Population: 57,616 (July 2020 est.)
country comparison to the world: 206

Nationality: *noun:* Greenlander(s)
adjective: Greenlandic

Ethnic groups: Greenlandic 89.7%, Danish 7.8%, other Nordic 1.1%, and other 1.4% (2018 est.)
note: data represent population by country of birth

Languages: Greenlandic (West Greenlandic or Kalaallisut is the official language), Danish, English

Religions: Evangelical Lutheran, traditional Inuit spiritual beliefs

Age structure: *0-14 years:* 20.82% (male 6,079/female 5,916)
15-24 years: 14.45% (male 4,186/female 4,137)
25-54 years: 39.72% (male 11,962/female 10,921)
55-64 years: 14.66% (male 4,561/female 3,886)
65 years and over: 10.36% (male 3,170/female 2,798) (2020 est.)

Median age: *total:* 34.3 years
male: 35.1 years
female: 33.4 years (2020 est.)
country comparison to the world: 92

Population growth rate: -0.08% (2020 est.)
country comparison to the world: 202

Birth rate: 14.1 births/1,000 population (2020 est.)
country comparison to the world: 133

Death rate: 9 deaths/1,000 population (2020 est.)
country comparison to the world: 62

Net migration rate: -6 migrant(s)/1,000 population (2020 est.)
country comparison to the world: 204

Population distribution: settlement concentrated on the southwest shoreline, with limited settlements scattered along the remaining coast; interior is uninhabited

Urbanization: *urban population:* 87.3% of total population (2020)
rate of urbanization: 0.42% annual rate of change (2015-20 est.)
total population growth rate v. urban population growth rate, 2000-2030:

Major urban areas—Population: 18,000 NUUK (capital) (2018)

Sex ratio: *at birth:* 1.05 male(s)/female
0-14 years: 1.03 male(s)/female
15-24 years: 1.01 male(s)/female
25-54 years: 1.1 male(s)/female

55-64 years: 1.17 male(s)/female
65 years and over: 1.13 male(s)/female
total population: 1.08 male(s)/female (2020 est.)

Infant mortality rate: *total:* 8.3 deaths/1,000 live births
male: 9.5 deaths/1,000 live births
female: 7.1 deaths/1,000 live births (2020 est.)
country comparison to the world: 148
Life expectancy at birth: total population: 73.4 years
male: 70.7 years
female: 76.3 years (2020 est.)
country comparison to the world: 145

Total fertility rate: 1.94 children born/woman (2020 est.)
country comparison to the world: 123

Drinking water source:
improved:
urban: 100% of population
rural: 100% of population
total: 100% of population
unimproved:
urban: 0% of population
rural: 0% of population
total: 0% of population (2017 est.)

Physicians density: 1.87 physicians/1,000 population (2016)

Hospital bed density: 14 beds/1,000 population (2016)

Sanitation facility access:
improved:
urban: 100% of population
rural: 100% of population
total: 100% of population
unimproved:
urban: 0% of population
rural: 0% of population
total: 0% of population (2017 est.)

HIV/AIDS—adult prevalence rate: NA

HIV/AIDS—people living with HIV/AIDS: NA

HIV/AIDS—deaths: NA

Education expenditures: NA

Literacy: *definition:* age 15 and over can read and write
total population: 100%
male: 100%
female: 100% (2015)

GOVERNMENT

Country name: *conventional long form:* none
conventional short form: Greenland
local long form: none
local short form: Kalaallit Nunaat
note: named by Norwegian adventurer Erik THORVALDSSON (Erik the Red) in A.D. 985 in order to entice settlers to the island

Dependency status: part of the Kingdom of Denmark; self-governing overseas administrative division of Denmark since 1979

Government type: parliamentary democracy (Parliament of Greenland or Inatsisartut)

Capital: *name:* Nuuk (Godthaab)

geographic coordinates: 64 11 N, 51 45 W
time difference: UTC-3 (2 hours ahead of Washington, DC, during Standard Time)
daylight saving time: +1hr, begins last Sunday in March; ends last Sunday in October
note: Greenland has four time zones
etymology: "nuuk" is the Inuit word for "cape" and refers to the city's position at the end of the Nuup Kangerlua fjord

Administrative divisions: 5 municipalities (kommuner, singular kommune); Avannaata, Kujalleq, Qeqertalik, Qeqqata, Sermersooq
note: Northeast Greenland National Park (Kalaallit Nunaanni Nuna Eqqissisimatitaq) and the Thule Air Base in Pituffik (in northwest Greenland) are two unincorporated areas; the national park's 972,000 sq km—about 46% of the island—makes it the largest national park in the world and also the most northerly

Independence: none (extensive self-rule as part of the Kingdom of Denmark; foreign affairs is the responsibility of Denmark, but Greenland actively participates in international agreements relating to Greenland)

National holiday: National Day, June 21; note—marks the summer solstice and the longest day of the year in the Northern Hemisphere

Constitution: *history:* previous 1953 (Greenland established as a constituency in the Danish constitution), 1979 (Greenland Home Rule Act); latest 21 June 2009 (Greenland Self-Government Act)

Legal system: the laws of Denmark apply where applicable and Greenlandic law applies to other areas

Citizenship: see Denmark

Suffrage: 18 years of age; universal

Executive branch: *chief of state:* Queen MARGRETHE II of Denmark (since 14 January 1972), represented by High Commissioner Mikaela ENGELL (since April 2011)
head of government: Premier Kim KIELSEN (since 30 September 2014)
cabinet: Self-rule Government (Naalakkersuisut) elected by the Parliament (Inatsisartut) on the basis of the strength of parties
elections/appointments: the monarchy is hereditary; high commissioner appointed by the monarch; premier indirectly elected by Parliament for a 4-year term
election results: Kim KIELSEN elected premier; Parliament vote—Kim KIELSEN (S) 27.2%, Sara OLSVIG (IA) 25.5%, Randi Vestergaard EVALDSEN (D) 19.5%, other 27.8%

Legislative branch: *description:* unicameral Parliament or Inatsisartut (31 seats; members directly elected in multi-seat constituencies by proportional representation vote to serve 4-year terms)
Greenland elects 2 members to the Danish Parliament to serve 4-year terms
elections: Greenland Parliament—last held on 24 April 2018 (next to be held by 2022)

Greenland members to Danish Parliament—last held on 5 June 2019(next to be held by 4 June 2023)
election results: Greenland Parliament percent of vote by party—S 27.2%, IA 25.5%, D 19.5%, PN 13.4%, A 5.9%, SA 4.1%, NQ 3.4% other 1%; seats by party—S 9, IA 8, D 6, PN 4, A 2, SA 1, NQ 1; composition—men 19, women 12, percent of women 38.7%
Greenland members in Danish Parliament—percent of vote by party—NA; seats by party—IA 1, S 1; composition—2 women

Judicial branch: *highest courts:* High Court of Greenland (consists of the presiding professional judge and 2 lay assessors); note—appeals beyond the High Court of Greenland can be heard by the Supreme Court (in Copenhagen)
judge selection and term of office: judges appointed by the monarch upon the recommendation of the Judicial Appointments Council, a 6-member independent body of judges and lawyers; judges appointed for life with retirement at age 70
subordinate courts: Court of Greenland; 18 district or magistrates' courts

Political parties and leaders:
Cooperation Party (Suleqatigiissitsisut or Samarbejdspartiet) or SA [Michael ROSING]
Democrats Party (Demokraatit) or D [Niels THOMSEN]
Forward Party (Siumut) or S [Kim KIELSEN]
Inuit Community (Inuit Ataqatigiit) or IA [Sara OLSVIG]
Our Country's Future (Nunatta Qitornai) or NQ [Vittus QUJAUKITSOQ]
Signpost Party (Partii Naleraq) or PN [Hans ENOKSEN] Fellowship Party (Atassut) or A [Siverth Karl HEILMANN]

International organization participation: Arctic Council, ICC, NC, NIB, UPU

Diplomatic representation in the US: none (self-governing overseas administrative division of Denmark); note—Greenland has an office in the Danish Embassy in the US; it also has offices in the Danish consulates in Chicago and New York

Diplomatic representation from the US: none (self-governing overseas administrative division of Denmark)

Flag description: two equal horizontal bands of white (top) and red with a large disk slightly to the hoist side of center—the top half of the disk is red, the bottom half is white; the design represents the sun reflecting off a field of ice; the colors are the same as those of the Danish flag and symbolize Greenland's links to the Kingdom of Denmark

National symbol(s): polar bear; *national colors:* red, white

National anthem: *name:* "Nunarput utoqqarsuanngoravit" ("Our Country, Who's Become So Old" also translated as "You Our Ancient Land")
lyrics/music: Henrik LUND/Jonathan PETERSEN

note: adopted 1916; the government also recognizes "Nuna asiilasooq" as a secondary anthem

ECONOMY

Economy—overview: Greenland's economy depends on exports of shrimp and fish, and on a substantial subsidy from the Danish Government. Fish account for over 90% of its exports, subjecting the economy to price fluctuations. The subsidy from the Danish Government is budgeted to be about $535 million in 2017, more than 50% of government revenues, and 25% of GDP.

The economy is expanding after a period of decline. The economy contracted between 2012 and 2014, grew by 1.7% in 2015 and by 7.7%in 2016. The expansion has been driven by larger quotas for shrimp, the predominant Greenlandic export, and also by increased activity in the construction sector, especially in Nuuk, the capital. Private consumption and tourism also are contributing to GDP growth more than in previous years. Tourism in Greenland grew annually around 20% in 2015 and 2016, largely a result of increasing numbers of cruise lines now operating in Greenland's western and southern waters during the peak summer tourism season.

The public sector, including publicly owned enterprises and the municipalities, plays a dominant role in Greenland's economy. During the last decade the Greenland Self Rule Government pursued conservative fiscal and monetary policies, but public pressure has increased for better schools, health care, and retirement systems. The budget was in deficit in 2014 and 2016, but public debt remains low at about 5% of GDP. The government plans a balanced budget for the 2017– 20 period.

Significant challenges face the island, including low levels of qualified labor, geographic dispersion, lack of industry diversification, the long-term sustainability of the public budget, and a declining population due to emigration.

Hydrocarbon exploration has ceased with declining oil prices. The island has potential for natural resource exploitation with rare-earth, uranium, and iron ore mineral projects proposed, but a lack of infrastructure hinders development.

GDP (purchasing power parity): $2.413 billion (2015 est.)
$2.24 billion (2014 est.)
$2.203 billion (2013 est.)
note: data are in 2015 US dollars
country comparison to the world: 193

GDP (official exchange rate): $2.221 billion (2015 est.)

GDP—real growth rate:
7.7% (2016 est.)
1.7% (2015 est.)
-0.8% (2014 est.)
country comparison to the world: 11

GDP—per capita (PPP): $41,800 (2015 est.)
$38,800 (2014 est.)
$38,500 (2013 est.)
country comparison to the world: 44

GDP—composition, by end use:

household consumption: 68.1% (2015 est.)
government consumption: 28% (2015 est.)
investment in fixed capital: 14.3% (2015 est.)
investment in inventories: -13.9% (2015 est.)
exports of goods and services: 18.2% (2015 est.)
imports of goods and services: -28.6% (2015 est.)

GDP—composition, by sector of origin:
agriculture: 15.9% (2015 est.)
industry: 10.1% (2015 est.)
services: 73.9% (2015)

Agriculture—products: sheep, cattle, reindeer, fish, shellfish

Industries: fish processing (mainly shrimp and Greenland halibut); anorthosite and ruby mining; handicrafts, hides and skins, small shipyards

Industrial production growth rate: NA

Labor force: 26,840 (2015 est.)
country comparison to the world: 207

Labor force—by occupation: *agriculture:* 15.9%
industry: 10.1%
services: 73.9% (2015 est.)

Unemployment rate:
9.1% (2015 est.)
10.3% (2014 est.)
country comparison to the world: 139

Population below poverty line: 16.2% (2015 est.)

Household income or consumption by percentage share: *lowest 10%:* NA
highest 10%: NA

Budget: *revenues:* 1.719 billion (2016 est.)
expenditures: 1.594 billion (2016 est.)

Taxes and other revenues: 77.4% (of GDP) (2016 est.)
country comparison to the world: 3

Budget surplus (+) or deficit (-): 5.6% (of GDP) (2016 est.)
country comparison to the world: 5

Public debt:
13% of GDP (2015 est.)
country comparison to the world: 196

Fiscal year: calendar year

Inflation rate (consumer prices):
0.3% (January 2017 est.)
1.2% (January 2016 est.)
country comparison to the world: 19

Exports:
$407.1 million (2015 est.)
$599.7 million (2014 est.)
country comparison to the world: 182

Exports—partners: Denmark 82.5%, Iceland 4.4% (2017)

Exports—commodities: fish and fish products 91% (2015 est.)

Imports:
$783.5 million (2015 est.)
$866.1 million (2014 est.)
country comparison to the world: 191

Imports—commodities: machinery and transport equipment, manufactured goods, food, petroleum products

Imports—partners: Denmark 69.7%, Sweden 10.6% (2017)

Debt—external:
$36.4 million (2010)
$58 million (2009)
country comparison to the world: 197

Exchange rates: Danish kroner (DKK) per US dollar -
6.586 (2017 est.)
6.7309 (2016 est.)
6.7309 (2015 est.)
6.7326 (2014 est.)
5.6125 (2013 est.)

ENERGY

Electricity access: electrification—total population: 100% (2020)

Electricity—production: 538 million kWh (2016 est.)
country comparison to the world: 163

Electricity—consumption: 468 million kWh (2016 est.)
country comparison to the world: 170

Electricity—exports: 0 kWh (2016 est.)
country comparison to the world: 141

Electricity—imports: 0 kWh (2016 est.)
country comparison to the world: 153

Electricity—installed generating capacity: 187,000 kW (2016 est.)
country comparison to the world: 167

Electricity—from fossil fuels: 51% of total installed capacity (2016 est.)
country comparison to the world: 149

Electricity—from nuclear fuels: 0% of total installed capacity (2017 est.)
country comparison to the world: 99

Electricity—from hydroelectric plants: 49% of total installed capacity (2017 est.)
country comparison to the world: 42

Electricity—from other renewable sources: 0% of total installed capacity (2017 est.)
country comparison to the world: 189

Crude oil—production: 0 bbl/day (2018 est.)
country comparison to the world: 142

Crude oil—exports: 0 bbl/day (2015 est.)
country comparison to the world: 130

Crude oil—imports: 0 bbl/day (2015 est.)
country comparison to the world: 133

Crude oil—proved reserves: 0 bbl (1 January 2018 est.)
country comparison to the world: 138

Refined petroleum products—production: 0 bbl/day (2015 est.)
country comparison to the world: 150

Refined petroleum products—consumption: 4,000 bbl/day (2016 est.)
country comparison to the world: 184

Refined petroleum products—exports: 0 bbl/day (2015 est.)
country comparison to the world: 159

Refined petroleum products—imports: 3,973 bbl/day (2015 est.)
country comparison to the world: 177

Natural gas—production: 0 cu m (2017 est.)
country comparison to the world: 138

Natural gas—consumption: 0 cu m (2017 est.)
country comparison to the world: 151

Natural gas—exports: 0 cu m (2017 est.)
country comparison to the world: 113

Natural gas—imports: 0 cu m (2017 est.)
country comparison to the world: 131

Natural gas—proved reserves: 0 cu m (1 January 2014 est.)
country comparison to the world: 140

Carbon dioxide emissions from consumption of energy: 613,800 Mt (2017 est.)
country comparison to the world: 179

COMMUNICATIONS

Telephones—fixed lines: *total subscriptions:* 7,259
subscriptions per 100 inhabitants: 12.59 (2019 est.)
country comparison to the world: 197

Telephones—mobile cellular: *total subscriptions:* 66,009
subscriptions per 100 inhabitants: 114.48 (2019 est.)
country comparison to the world: 202

Telecommunication systems: *general assessment:* adequate domestic and international service provided by satellite, cables, and microwave radio relay; the fundamental telecommunications infrastructure consists of a digital radio link from Nanortalik in south Greenland to Uummannaq in north Greenland; satellites cover north and east Greenland for domestic and foreign telecommunications; a marine cable connects south and west Greenland to the rest of the world, extending from Nuuk and Qaqortoq to Canada and Iceland (2018)
domestic: 13 per 100 for fixed-line subscriptions and 115 per 100 for mobile-cellular (2019)
international: country code—299; landing points for Greenland Connect, Greenland Connect North, Nunavut Undersea Fiber System submarine cables to Greenland, Iceland, and Canada; satellite earth stations—15 (12 Intelsat, 1 Eutelsat, 2 Americom GE-2 (all Atlantic Ocean)) (2019)
note: the COVID-19 outbreak is negatively impacting telecommunications production and supply chains globally; consumer spending on telecom devices and services has also slowed due to the pandemic's effect on economies worldwide; overall progress towards improvements in all facets of the telecom industry—mobile, fixed-line, broadband, submarine cable and satellite—has moderated

Broadcast media: the Greenland Broadcasting Company provides public radio and TV services throughout the island with a broadcast station and a series of repeaters; a few private local TV and radio stations; Danish public radio rebroadcasts are available (2019)

Internet country code: .gl

Internet users: *total:* 40,084
percent of population: 69.48% (July 2018 est.)
country comparison to the world: 201

Broadband—fixed subscriptions: *total:* 13,192
subscriptions per 100 inhabitants: 23 (2018 est.)
country comparison to the world: 164

TRANSPORTATION

National air transport system: *number of registered air carriers:* 1 (registered in Denmark) (2020)
inventory of registered aircraft operated by air carriers: 8 (registered in Denmark)

Civil aircraft registration country code prefix: OY-H (2016)

Airports: 15 (2013)
country comparison to the world: 146

Airports—with paved runways: *total:* 10 (2019)
2,438 to 3,047 m: 2
1,524 to 2,437 m: 1
914 to 1,523 m: 1
under 914 m: 6

Airports—with unpaved runways: *total:* 5 (2013)
1,524 to 2,437 m: 1 (2013)
914 to 1,523 m: 2 (2013)
under 914 m: 2 (2013)

Roadways: *note:* although there are short roads in towns, there are no roads between towns; inter-urban transport is either by sea or by air

Merchant marine: *total:* 8
by type: other 8 (2019)
country comparison to the world: 159

Ports and terminals: *major seaport(s):* Sisimiut

MILITARY AND SECURITY

Military and security forces: no regular military forces or conscription. (2019)

Military—note: The Danish military's Joint Arctic Command in Nuuk is responsible for territorial defense of Greenland (2019)

TRANSNATIONAL ISSUES

Disputes—international: managed dispute between Canada and Denmark over Hans Island in the Kennedy Channel between Canada's Ellesmere Island and Greenland; Denmark (Greenland) and Norway have made submissions to the Commission on the Limits of the Continental Shelf (CLCS) and Russia is collecting additional data to augment its 2001 CLCS submission

GRENADA

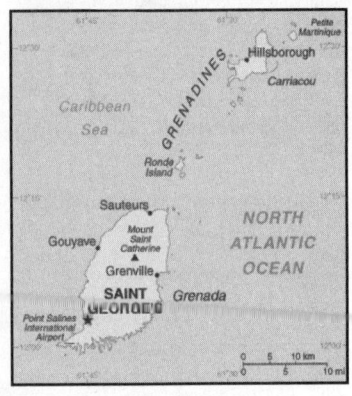

INTRODUCTION

Background: Carib Indians inhabited Grenada when Christopher COLUMBUS discovered the island in 1498, but it remained uncolonized for more than a century. The French settled Grenada in the 17th century, established sugar estates, and imported large numbers of African slaves. Britain took the island in 1762 and vigorously expanded sugar production. In the 19th century, cacao eventually surpassed sugar as the main export crop; in the 20th century, nutmeg became the leading export. In 1967, Britain gave Grenada autonomy over its internal affairs. Full independence was attained in 1974 making Grenada one of the smallest independent countries in the Western Hemisphere. In 1979, a leftist New Jewel Movement seized power under Maurice BISHOP ushering in the Grenada Revolution. On 19 October 1983, factions within the revolutionary government overthrew and killed BISHOP and members of his party. Six days later the island was invaded by US forces and those of six other Caribbean nations, which quickly captured the ringleaders and their hundreds of Cuban advisers. The rule of law was restored and democratic elections were reinstituted the following year and have continued since then.

GEOGRAPHY

Location: Caribbean, island between the Caribbean Sea and Atlantic Ocean, north of Trinidad and Tobago

Geographic coordinates: 12 07 N, 61 40 W

Map references: Central America and the Caribbean

Area: *total:* 344 sq km
land: 344 sq km
water: 0 sq km
country comparison to the world: 208

Area—comparative: twice the size of Washington, DC

Land boundaries: 0 km

Coastline: 121 km

Maritime claims: *territorial sea:* 12 nm

exclusive economic zone: 200 nm

Climate: tropical; tempered by northeast trade winds

Terrain: volcanic in origin with central mountains

Elevation: *lowest point:* Caribbean Sea 0 m
highest point: Mount Saint Catherine 840 m

Natural resources: timber, tropical fruit

Land use: *agricultural land:* 32.3% (2011 est.)
arable land: 8.8% (2011 est.) / permanent crops: 20.6% (2011 est.) / permanent pasture: 2.9% (2011 est.)
forest: 50% (2011 est.)
other: 17.7% (2011 est.)

Irrigated land: 20 sq km (2012)

Population distribution: approximately one-third of the population is found in the capital of St. George's; the island's population is concentrated along the coast

Natural hazards: lies on edge of hurricane belt; hurricane season lasts from June to November
volcanism: Mount Saint Catherine (840 m) lies on the island of Grenada; Kick 'em Jenny, an active submarine volcano (seamount) on the Caribbean Sea floor, lies about 8 km north of the island of Grenada; these two volcanoes are at the southern end of the volcanic island arc of the Lesser Antilles that extends up to the Netherlands dependency of Saba in the north

Environment—current issues: deforestation causing habitat destruction and species loss; coastal erosion and contamination; pollution and sedimentation; inadequate solid waste management

Environment—international agreements: party to: Biodiversity, Climate Change, Climate Change-Kyoto Protocol, Desertification, Endangered Species, Law of the Sea, Ozone Layer Protection, Whaling
signed, but not ratified: none of the selected agreements

Geography—note: the administration of the islands of the Grenadines group is divided between Saint Vincent and the Grenadines and Grenada

PEOPLE AND SOCIETY

Population: 113,094 (July 2020 est.)
country comparison to the world: 190

Nationality: *noun:* Grenadian(s)
adjective: Grenadian

Ethnic groups: African descent 82.4%, mixed 13.3%, East Indian 2.2%, other 1.3%, unspecified 0.9% (2011 est.)

Languages: English (official), French patois

Religions: Protestant 49.2% (includes Pentecostal 17.2%, Seventh Day Adventist 13.2%, Anglican 8.5%, Baptist 3.2%, Church of God 2.4%, Evangelical 1.9%, Methodist 1.6%, other 1.2%), Roman Catholic 36%, Jehovah's Witness 1.2%, Rastafarian 1.2%, other 5.5%, none 5.7%, unspecified 1.3% (2011 est.)

Age structure: *0-14 years:* 23.23% (male 13,709/female 12,564)
15-24 years: 14.14% (male 8,034/female 7,959)
25-54 years: 40.05% (male 23,104/female 22,187)
55-64 years: 11.69% (male 6,734/female 6,490)
65 years and over: 10.89% (male 5,774/female 6,539) (2020 est.)

Dependency ratios: *total dependency ratio:* 50.5
youth dependency ratio: 35.8
elderly dependency ratio: 14.7
potential support ratio: 6.8 (2020 est.)

Median age: *total:* 33.3 years
male: 33.1 years
female: 33.4 years (2020 est.)
country comparison to the world: 99

Population growth rate: 0.38% (2020 est.)
country comparison to the world: 164

Birth rate: 14.6 births/1,000 population (2020 est.)
country comparison to the world: 125

Death rate: 8.3 deaths/1,000 population (2020 est.)
country comparison to the world: 78

Net migration rate: -2.6 migrant(s)/1,000 population (2020 est.)
country comparison to the world: 171

Population distribution: approximately one-third of the population is found in the capital of St. George's; the island's population is concentrated along the coast

Urbanization: *urban population:* 36.5% of total population (2020)
rate of urbanization: 0.76% annual rate of change (2015-20 est.)
total population growth rate v. urban population growth rate, 2000-2030:

Major urban areas—population: 39,000 SAINT GEORGE'S (capital) (2018)

Sex ratio: *at birth:* 1.1 male(s)/female
0-14 years: 1.09 male(s)/female
15-24 years: 1.01 male(s)/female
25-54 years: 1.04 male(s)/female
55-64 years: 1.04 male(s)/female
65 years and over: 0.88 male(s)/female
total population: 1.03 male(s)/female (2020 est.)

Maternal mortality rate: 25 deaths/100,000 live births (2017 est.)
country comparison to the world: 122

Infant mortality rate: *total:* 8.9 deaths/1,000 live births
male: 8.6 deaths/1,000 live births
female: 9.4 deaths/1,000 live births (2020 est.)
country comparison to the world: 141

Life expectancy at birth: *total population:* 75.2 years
male: 72.6 years
female: 78.1 years (2020 est.)
country comparison to the world: 119

Total fertility rate: 1.96 children born/woman (2020 est.)
country comparison to the world: 116

Drinking water source:
improved:
total: 96.8% of population
unimproved:
total: 3.2% of population (2017 est.)

Current Health Expenditure: 4.8% (2017)

Physicians density: 1.41 physicians/1,000 population (2017)

Hospital bed density: 3.6 beds/1,000 population (2017)

Sanitation facility access:
improved:
total: 93.7% of population
unimproved:
total: 6.3% of population (2017 est.)

HIV/AIDS—adult prevalence rate: 0.5% (2018)
country comparison to the world: 69

HIV/AIDS—people living with HIV/AIDS: <500 (2018)

HIV/AIDS—deaths: <100 (2018)

Obesity—adult prevalence rate: 21.3% (2016)
country comparison to the world: 90

Education expenditures: 3.2% of GDP (2017)
country comparison to the world: 131

Literacy: *definition:* age 15 and over can read and write
total population: 98.6%
male: 98.6%
female: 98.6% (2014 est.)

School life expectancy (primary to tertiary education): *total:* 19 years
male: 18 years
female: 19 years (2018)

GOVERNMENT

Country name: *conventional long form:* none
conventional short form: Grenada
etymology: derivation of the name remains obscure; some sources attribute the designation to Spanish influence (most likely named for the Spanish city of Granada), with subsequent French and English interpretations resulting in the present-day Grenada; in Spanish "granada" means "pomegranate"

Government type: parliamentary democracy under a constitutional monarchy; a Commonwealth realm

Capital: *name:* Saint George's
geographic coordinates: 12 03 N, 61 45 W
time difference: UTC-4 (1 hour ahead of Washington, DC, during Standard Time)
etymology: the 1763 Treaty of Paris transferred possession of Grenada from France to Great Britain; the new administration renamed Ville de

Fort Royal (Fort Royal Town) to Saint George's Town, after the patron saint of England; eventually the name became simply Saint George's

Administrative divisions: 6 parishes and 1 dependency*; Carriacou and Petite Martinique*, Saint Andrew, Saint David, Saint George, Saint John, Saint Mark, Saint Patrick

Independence: 7 February 1974 (from the UK)

National holiday: Independence Day, 7 February (1974)

Constitution: *history:* previous 1967; latest presented 19 December 1973, effective 7 February 1974, suspended 1979 following a revolution but restored in 1983

amendments: proposed by either house of Parliament; passage requires two-thirds majority vote by the membership in both houses and assent of the governor general; passage of amendments to constitutional sections, such as personal rights and freedoms, the structure, authorities, and procedures of the branches of government, the delimitation of electoral constituencies, or the procedure for amending the constitution, also requires two-thirds majority approval in a referendum; amended 1991, 1992

Legal system: common law based on English model

International law organization participation: has not submitted an ICJ jurisdiction declaration; accepts ICCt jurisdiction

Citizenship: *citizenship by birth:* yes
citizenship by descent only: yes
dual citizenship recognized: yes
residency requirement for naturalization: 7 years for persons from a non-Caribbean state and 4 years for a person from a Caribbean state

Suffrage: 18 years of age; universal

Executive branch: *chief of state:* Queen ELIZABETH II (since 6 February 1952); represented by Governor General Cecile LA GRENADE (since 7 May 2013)
head of government: Prime Minister Keith MITCHELL (since 20 February 2013)
cabinet: Cabinet appointed by the governor general on the advice of the prime minister
elections/appointments: the monarchy is hereditary; governor general appointed by the monarch; following legislative elections, the leader of the majority party or majority coalition usually appointed prime minister by the governor general

Legislative branch: *description:* bicameral Parliament consists of:
Senate (13 seats; members appointed by the governor general—10 on the advice of the prime minister and 3 on the advice of the leader of the opposition party; members serve 5-year terms)
House of Representatives (15 seats; members directly elected in single-seat constituencies by simple majority vote to serve 5-year terms)
elections: Senate—last appointments on 27 April 2018 (next no later than2023)
House of Representatives—last held on 13 March 2018 (next no later than 2023)

election results: Senate—percent by party—NA; seats by party—NA; composition—men 11, women 2 percent of women 15.4%
House of Representatives—percent of vote by party—NNP 58.9%, NDC 40.5%; other 0.6% seats by party—NNP 15; composition—men 8, women 7, percent of women 46.7%; note—total Parliament percent of women 32.1%

Judicial branch: *highest courts:* regionally, the Eastern Caribbean Supreme Court (ECSC) is the superior court of the Organization of Eastern Caribbean States; the ECSC—headquartered on St. Lucia—consists of the Court of Appeal—headed by the chief justice and 4 judges—and the High Court with 18 judges; the Court of Appeal is itinerant, traveling to member states on a schedule to hear appeals from the High Court and subordinate courts; High Court judges reside in the member states, with 2 in Grenada; appeals beyond the ECSC in civil and criminal matters are heard by the Judicial Committee of the Privy Council (in London)
judge selection and term of office: chief justice of Eastern Caribbean Supreme Court appointed by Her Majesty, Queen ELIZABETH II; other justices and judges appointed by the Judicial and Legal Services Commission, and independent body of judicial officials; Court of Appeal justices appointed for life with mandatory retirement at age 65; High Court judges appointed for life with mandatory retirement at age 62
subordinate courts: magistrates' courts; Court of Magisterial Appeals

Political parties and leaders:
National Democratic Congress or NDC [Nazim BURKE]
New National Party or NNP [Keith MITCHELL]

International organization participation: ACP, AOSIS, C, Caricom, CDB, CELAC, FAO, G-77, IBRD, ICAO, ICCt (signatory), ICRM, IDA, IFAD, IFC, IFRCS, ILO, IMF, IMO, Interpol, IOC, ITU, ITUC, LAES, MIGA, NAM, OAS, OECS, OPANAL, OPCW, Petrocaribe, UN, UNCTAD, UNESCO, UN IDO, UPU, WHO, WIPO, WTO

Diplomatic representation in the US: chief of mission: Ambassador Yolande Yvonne SMITH (since 8 April 2019)
chancery: 1701 New Hampshire Avenue NW, Washington, DC 20009
telephone: [1] (202) 265-2561
FAX: [1] (202) 265-2468
consulate(s) general: Miami

Diplomatic representation from the US: chief of mission: the US does not have an official embassy in Grenada; the US Ambassador to Barbados is accredited to Grenada
telephone: [1] (473) 444-1173 through 1176
embassy: Lance-aux-Epines Stretch, Saint George's
mailing address: P. O. Box 54, Saint George's
FAX: [1] (473) 444-4820

Flag description: a rectangle divided diagonally into yellow triangles (top and bottom) and green triangles (hoist side and outer side), with a red

border around the flag; there are seven yellow, five-pointed stars with three centered in the top red border, three centered in the bottom red border, and one on a red disk superimposed at the center of the flag; there is also a symbolic nutmeg pod on the hoist-side triangle (Grenada is a leading nutmeg producer); the seven stars stand for the seven administrative divisions, with the central star denoting the capital, St. George's; yellow represents the sun and the warmth of the people, green stands for vegetation and agriculture, and red symbolizes harmony, unity, and courage

National symbol(s): *Grenada dove, bougainvillea flower; national colors:* red, yellow, green

National anthem: *name:* Hail Grenada
lyrics/music: Irva Merle BAPTISTE/Louis Arnold MASANTO
note: adopted 1974

ECONOMY

Economy—overview: Grenada relies on tourism and revenue generated by St. George's University—a private university offering degrees in medicine, veterinary medicine, public health, the health sciences, nursing, arts and sciences, and business—as its main source of foreign exchange. In the past two years the country expanded its sources of revenue, including from selling passports under its citizenship by investment program. These projects produced a resurgence in the construction and manufacturing sectors of the economy.
In 2017, Grenada experienced its fifth consecutive year of growth and the government successfully marked the completion of its five-year structural adjustment program that included among other things austerity measures, increased tax revenue and debt restructuring. Public debt-to-GDP was reduced from 100% of GDP in 2013 to 71.8% in 2017.

GDP (purchasing power parity):
$1.634 billion (2017 est.)
$1.555 billion (2016 est.)
$1.5 billion (2015 est.)
note: data are in 2017 dollars
country comparison to the world: 198

GDP (official exchange rate):
$1.119 billion (2017 est.)

GDP—real growth rate:
5.1% (2017 est.)
3.7% (2016 est.)
6.4% (2015 est.)
country comparison to the world: 46

GDP—per capita (PPP):
$15,100 (2017 est.)
$14,500 (2016 est.)
$14,000 (2015 est.)
note: data are in 2017 dollars
country comparison to the world: 110

Gross national saving:
11.7% of GDP (2017 est.)
17% of GDP (2016 est.)
13.9% of GDP (2015 est.)
country comparison to the world: 153

GDP—composition, by end use:
household consumption: 63% (2017 est.)
government consumption: 12% (2017 est.)
investment in fixed capital: 20% (2017 est.)
investment in inventories: -0.1% (2017 est.)
exports of goods and services: 60% (2017 est.)
imports of goods and services: -55% (2017 est.)

GDP—composition, by sector of origin:
agriculture: 6.8% (2017 est.)
industry: 15.5% (2017 est.)
services: 77.7% (2017 est.)

Agriculture—products: bananas, cocoa, nutmeg, mace, soursop, citrus, avocados, root crops, corn, vegetables, fish

Industries: food and beverages, textiles, light assembly operations, tourism, construction, education, call-center operations

Industrial production growth rate: 10% (2017 est.)
country comparison to the world: 16

Labor force: 55,270 (2017 est.)
country comparison to the world: 189

Labor force—by occupation:
agriculture: 11%
industry: 20%
services: 69% (2008 est.)

Unemployment rate:
24% (2017 est.)
28.2% (2016 est.)
country comparison to the world: 195

Population below poverty line: 38% (2008 est.)

Household income or consumption by percentage share: *lowest 10%:* NA
highest 10%: NA

Budget: *revenues:* 288.4 million (2017 est.)
expenditures: 252.3 million (2017 est.)

Taxes and other revenues: 25.8% (of GDP) (2017 est.)
country comparison to the world: 116

Budget surplus (+) or deficit (-): 3.2% (of GDP) (2017 est.)
country comparison to the world: 12

Public debt:
70.4% of GDP (2017 est.)
82% of GDP (2016 est.)
country comparison to the world: 50

Fiscal year: calendar year

Inflation rate (consumer prices):
0.9% (2017 est.)
1.7% (2016 est.)
country comparison to the world: 45

Current account balance:
-$77 million (2017 est.)
-$34 million (2016 est.)
country comparison to the world: 83

Exports:
$39.9 million (2017 est.)
$44.2 million (2016 est.)
country comparison to the world: 205

Exports—partners: US 25.3%, Japan 10.1%, Guyana 8.7%, Dominica 6.6%, St. Lucia 6.4%, Netherlands 4.7%, Barbados 4.1%, St. Kitts and Nevis 4% (2017)

Exports—commodities: nutmeg, bananas, cocoa, fruit and vegetables, clothing, mace, chocolate, fish

Imports:
$316 million (2017 est.)
$314.7 million (2016 est.)
country comparison to the world: 203

Imports—commodities: food, manufactured goods, machinery, chemicals, fuel

Imports—partners: US 31.7%, Trinidad and Tobago 24.9%, China 6.7% (2017)

Reserves of foreign exchange and gold:
$199.1 million (31 December 2017 est.)
$198 million (31 December 2015 est.)
country comparison to the world: 175

Debt—external:
$793.5 million (2017 est.)
$682.3 million (2016 est.)
country comparison to the world: 168

Exchange rates: East Caribbean dollars (XCD) per US dollar -
2.7 (2017 est.)
2.7 (2016 est.)
2.7 (2015 est.)
2.7 (2014 est.)
2.7 (2013 est.)

ENERGY

Electricity access: *electrification—total population:* 92.3% (2016)
electrification—urban areas: 92.3% (2016)
electrification—rural areas: 92.3% (2016)

Electricity—production: 202.1 million kWh (2016 est.)
country comparison to the world: 192

Electricity—consumption: 185.1 million kWh (2016 est.)
country comparison to the world: 194

Electricity—exports: 0 kWh (2016 est.)
country comparison to the world: 142

Electricity—imports: 0 kWh (2016 est.)
country comparison to the world: 154

Electricity—installed generating capacity: 51,100 kW (2016 est.)
country comparison to the world: 191

Electricity—from fossil fuels: 96% of total installed capacity (2016 est.)
country comparison to the world: 39

Electricity—from nuclear fuels: 0% of total installed capacity (2017 est.)
country comparison to the world: 100

Electricity—from hydroelectric plants: 0% of total installed capacity (2017 est.)
country comparison to the world: 174

Electricity—from other renewable sources: 4% of total installed capacity (2017 est.)
country comparison to the world: 112

Crude oil—production: 0 bbl/day (2018 est.)
country comparison to the world: 143

Crude oil—exports: 0 bbl/day (2015 est.)
country comparison to the world: 131

Crude oil—imports: 0 bbl/day (2015 est.)
country comparison to the world: 134

Crude oil—proved reserves: 0 bbl (1 January 2018 est.)
country comparison to the world: 139

Refined petroleum products—production: 0 bbl/day (2015 est.)
country comparison to the world: 151

Refined petroleum products—consumption: 2,000 bbl/day (2016 est.)
country comparison to the world: 194

Refined petroleum products—exports: 0 bbl/day (2015 est.)
country comparison to the world: 160

Refined petroleum products—imports: 1,886 bbl/day (2015 est.)
country comparison to the world: 191

Natural gas—production: 0 cu m (2017 est.)
country comparison to the world: 139

Natural gas—consumption: 0 cu m (2017 est.)
country comparison to the world: 152

Natural gas—exports: 0 cu m (2017 est.)
country comparison to the world: 114

Natural gas—imports: 0 cu m (2017 est.)
country comparison to the world: 132

Natural gas—proved reserves: 0 cu m (1 January 2014 est.)
country comparison to the world: 141

Carbon dioxide emissions from consumption of energy: 283,600 Mt (2017 est.)
country comparison to the world: 193

COMMUNICATIONS

Telephones—fixed lines: *total subscriptions:* 33,011
subscriptions per 100 inhabitants: 29.3 (2019 est.)
country comparison to the world: 170

Telephones—mobile cellular: *total subscriptions:* 115,008
subscriptions per 100 inhabitants: 102.08 (2019 est.)
country comparison to the world: 192

Telecommunication systems: *general assessment:* adequate, island-wide telephone system; lack of local competition, but telecoms are a high contributors to overall GDP; growth sectors include the mobile telephony and data segments (2020)
domestic: interisland VHF and UHF radiotelephone links; 29 per 100 for fixed-line and 102 per 100 for mobile-cellular (2019)
international: country code—1-473; landing points for the ECFS, Southern Caribbean Fiber and CARCIP submarine cables with links to 13 Caribbean islands extending from the British Virgin Islands to Trinidad & Tobago including Puerto Rico and Barbados; SHF radiotelephone links to Trinidad and Tobago and Saint Vincent; VHF and UHF radio links to Trinidad (2019)
note: the COVID-19 outbreak is negatively impacting telecommunications production and

397

supply chains globally; consumer spending on telecom devices and services has also slowed due to the pandemic's effect on economies worldwide; overall progress towards improvements in all facets of the telecom industry—mobile, fixed-line, broadband, submarine cable and satellite—has moderated

Broadcast media: multiple publicly and privately owned television and radio stations; Grenada Information Service (GIS) is governmentowned and provides television and radio services; the Grenada Broadcasting Network, jointly owned by the government and the Caribbean Communications Network of Trinidad and Tobago, operates a TV station and 2 radio stations; Meaningful Television (MTV) broadcasts island-wide and is part of a locally-owned media house, Moving Target Company, that also includes an FM radio station and a weekly newspaper; multi-channel cable TV subscription service is provided by Columbus Communications Grenada (FLOW GRENADA) and is available island wide;

approximately 25 private radio stations also broadcast throughout the country (2019)

internet country code: .gd

Internet users: *total:* 66,281
percent of population: 59.07% (July 2018 est.)
country comparison to the world: 191

Broadband—fixed subscriptions: *total:* 22,235
subscriptions per 100 inhabitants: 20 (2017 est.)
country comparison to the world: 151

TRANSPORTATION

Civil aircraft registration country code prefix: J3 (2016)

Airports: 3 (2013)
country comparison to the world: 195

Airports—with paved runways: *total:* 3 (2017)
2,438 to 3,047 m: 1 (2017)
1,524 to 2,437 m: 1 (2017)
under 914 m: 1 (2017)

Roadways: *total:* 1,127 km (2017)

*paved:*902 km (2017)
unpaved: 225 km (2017)
country comparison to the world: 182

Merchant marine: *total:* 6
by type: general cargo 3, other 3 (2019)
country comparison to the world: 163

Ports and terminals: *major seaport(s):* Saint George's

MILITARY AND SECURITY

Military and security forces: no regular military forces; Royal Grenada Police Force (includes Coast Guard) (2019)

TRANSNATIONAL ISSUES

Disputes—international: none

Illicit drugs: small-scale cannabis cultivation; lesser transshipment point for marijuana and cocaine to US

GUAM

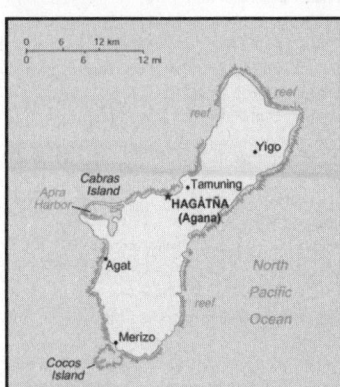

INTRODUCTION

Background: Spain ceded Guam to the US in 1898. Captured by the Japanese in 1941, it was retaken by the US three years later. The military installations on the island are some of the most strategically important US bases in the Pacific; they also constitute the island's most important source of income and economic stability.

GEOGRAPHY

Location: Oceania, island in the North Pacific Ocean, about three-quarters of the way from Hawaii to the Philippines

Geographic coordinates: 13 28 N, 144 47 E

Map references: Oceania

Area: *total:* 544 sq km

land: 544 sq km
water: 0 sq km
country comparison to the world: 195

Area—comparative: three times the size of Washington, DC

Land boundaries: 0 km

Coastline: 125.5 km

Maritime claims: *territorial sea:* 12 nm
exclusive economic zone: 200 nm

Climate: tropical marine; generally warm and humid, moderated by northeast trade winds; dry season (January to June), rainy season (July to December); little seasonal temperature variation

Terrain: volcanic origin, surrounded by coral reefs; relatively flat coralline limestone plateau (source of most fresh water), with steep coastal cliffs and narrow coastal plains in north, low hills in center, mountains in south

Elevation: *lowest point:* Pacific Ocean 0 m
highest point: Mount Lamlam 406 m

Natural resources: aquatic wildlife (supporting tourism); fishing (largely undeveloped)

Land use: *agricultural land:* 33.4% (2011 est.)
arable land: 1.9% (2011 est.) / permanent crops: 16.7% (2011 est.) / permanent pasture: 14.8% (2011 est.)
forest: 47.9% (2011 est.)
other: 18.7% (2011 est.)

Irrigated land: 2 sq km (2012)

Population distribution: no large cities exist on the island, though large villages (municipalities) attract much of the population; the largest of these is Dededo

Natural hazards: frequent squalls during rainy season; relatively rare but potentially destructive typhoons (June to December)

Environment—current issues: fresh water scarcity; reef damage; inadequate sewage treatment; extermination of native bird populations by the rapid proliferation of the brown tree snake, an exotic, invasive species

Geography—note: largest and southernmost island in the Mariana Islands archipelago and the largest island in Micronesia; strategic location in western North Pacific Ocean

PEOPLE AND SOCIETY

Population: 168,485 (July 2020 est.)
country comparison to the world: 186

Nationality: *noun:* Guamanian(s) (US citizens)
adjective: Guamanian

Ethnic groups: Chamorro 37.3%, Filipino 26.3%, white 7.1%, Chuukese 7%, Korean 2.2%, other Pacific Islander 2%, other Asian 2%, Chinese 1.6%, Palauan 1.6%, Japanese 1.5%, Pohnpeian 1.4%, mixed 9.4%, other 0.6% (2010 est.)

Languages: English 43.6%, Filipino 21.2%, Chamorro 17.8%, other Pacific island languages 10%, Asian languages 6.3%, other 1.1% (2010 est.)

Religions: Roman Catholic 85%, other 15% (1999 est.)

Age structure: *0-14 years:* 27.22% (male 23,748/ female 22,122)
15-24 years: 16.08% (male 14,522/female 12,572)
25-54 years: 36.65% (male 31,880/female 29,871)
55-64 years: 10.5% (male 9,079/female 8,610)

65 years and over: 9.54% (male 7,504/female 8,577) (2020 est.)

Dependency ratios: *total dependency ratio:* 52.4
youth dependency ratio: 36.4
elderly dependency ratio: 16.1
potential support ratio: 6.2 (2020 est.)

Median age: *total:* 29.4 years
male: 28.7 years
female: 30.2 years (2020 est.)
country comparison to the world: 130

Population growth rate: 0.2% (2020 est.)
country comparison to the world: 181

Birth rate: 18.9 births/1,000 population (2020 est.)
country comparison to the world: 80

Death rate: 6 deaths/1,000 population (2020 est.)
country comparison to the world: 163

Net migration rate: -11 migrant(s)/1,000 population (2020 est.)
country comparison to the world: 217

Population distribution: no large cities exist on the island, though large villages (municipalities) attract much of the population; the largest of these is Dededo

Urbanization: *urban population:* 94.9% of total population (2020)
rate of urbanization: 0.92% annual rate of change (2015-20 est.)
total population growth rate v. urban population growth rate, 2000-2030:

Major urban areas—Population: 147,000 HAGATNA (capital) (2018)

Sex ratio: *at birth:* 1.07 male(s)/female
0-14 years: 1.07 male(s)/female
15-24 years: 1.16 male(s)/female
25-54 years: 1.07 male(s)/female
55-64 years: 1.05 male(s)/female
65 years and over: 0.87 male(s)/female
total population: 1.06 male(s)/female (2020 est.)

Infant mortality rate: *total:* 10.8 deaths/1,000 live births
male: 10.7 deaths/1,000 live births
female: 10.8 deaths/1,000 live births (2020 est.)
country comparison to the world: 126
Life expectancy at birth: total population: 77 years
male: 74.6 years
female: 79.6 years (2020 est.)
country comparison to the world: 86

Total fertility rate: 2.84 children born/woman (2020 est.)
country comparison to the world: 59

Drinking water source:
improved:
total: 100% of population
unimproved:
total: 0% of population (2017 est.)

Sanitation facility access:
improved:
urban: 89.8% of population (2015 est.)
rural: 89.8% of population (2015 est.)
total: 89.8% of population (2015 est.)
unimproved:
urban: 10.2% of population (2015 est.)

rural: 10.2% of population (2015 est.)
total: 10.2% of population (2015 est.)

HIV/AIDS—adult prevalence rate: NA

HIV/AIDS—people living with HIV/AIDS: NA

HIV/AIDS—deaths: NA

Education expenditures: NA

Unemployment, youth ages 15-24: *total:* 29.4%
male: 29.7%
female: 28.9% (2011 est.)
country comparison to the world: 33

GOVERNMENT

Country name: *conventional long form:* none
conventional short form: Guam
local long form: none
local short form: Guahan
abbreviation: GU
etymology: the native Chamorro name for the island "Guahan" (meaning "we have" or "ours") was changed to Guam in the 1898 Treaty of Paris, whereby Spain relinquished Guam, Cuba, Puerto Rico, and the Philippines to the US

Dependency status: unincorporated organized territory of the US with policy relations between Guam and the federal government under the jurisdiction of the Office of Insular Affairs, US Department of the Interior

Government type: republican form of government with separate executive, legislative, and judicial branches; unincorporated organized territory of the US with local self-government

Capital: *name:* Hagatna (Agana)
geographic coordinates: 13 28 N, 144 44 E
time difference: UTC+10 (15 hours ahead of Washington, DC, during Standard Time)
etymology: the name is derived from the Chamoru word "haga," meaning "blood", and may refer to the bloodlines of the various families that established the original settlement

Administrative divisions: none (territory of the US)

Independence: none (territory of the US)

National holiday: Discovery Day (or Magellan Day), first Monday in March (1521)

Constitution: *history:* effective 1 July 1950 (Guam Act of 1950 serves as a constitution)
amendments: amended many times, last in 2015

Legal system: common law modeled on US system; US federal laws apply

Citizenship: see United States

Suffrage: 18 years of age; universal; note—Guamanians are US citizens but do not vote in US presidential elections

Executive branch: *chief of state:* President Donald J. TRUMP (since 20 January 2017); Vice President Michael R. PENCE (since 20 January 2017)
head of government: Governor Lourdes LEON GUERRERO (since 7 January 2019); Lieutenant Governor Josh TENORIO (since 7 January 2019)
cabinet: Cabinet appointed by the governor with the consent of the Legislature

elections/appointments: president and vice president indirectly elected on the same ballot by an Electoral College of 'electors' chosen from each state to serve a 4-year term (eligible for a second term); under the US Constitution, residents of unincorporated territories, such as Guam, do not vote in elections for US president and vice president; however, they may vote in Democratic and Republican presidential primary elections; governor and lieutenant governor elected on the same ballot by absolute majority vote in 2 rounds if needed for a 4-year term (eligible for 2 consecutive terms); election last held on 6 November 2018 (next to be held in November 2022)
election results: Lourdes LEON GUERRERO elected governor; percent of vote—Lourdes LEON GUERRERO (Democratic Party) 50.7%, Ray TENORIO (Republican Party) 26.4%; Josh TENORIO (Democratic Party) elected lieutenant governor

Legislative branch: *description:* unicameral Legislature of Guam or Liheslaturan Guahan (15 seats; members elected in a single countrywide constituency by simple majority vote to serve 2-year terms)
elections: last held on 6 November 2018 (next to be held on 3 November 2020)
election results: percent of vote by party—NA; seats by party—Democratic Party 10, Republican Party 5; composition—men 5, women 10, percent of women 66.7%
note: Guam directly elects 1 member by simple majority vote to serve a 2-year term as a delegate to the US House of Representatives; the delegate can vote when serving on a committee and when the House meets as the Committee of the Whole House, but not when legislation is submitted for a "full floor" House vote; election of delegate last held on 6 November 2018 (next to be held on 3 November 2020); election results—seat by party—Democratic Party 1; composition 1 man

Judicial branch: *highest courts:* Supreme Court of Guam (consists of 3 justices); note—appeals beyond the Supreme Court of Guam are referred to the US Supreme Court
judge selection and term of office: justices appointed by the governor and confirmed by the Guam legislature; justices appointed for life subject to retention election every 10 years
subordinate courts: Superior Court of Guam—includes several divisions; US Federal District Court for the District of Guam (a US territorial court; appeals beyond this court are heard before the US Court of Appeals for the Ninth Circuit)

Political parties and leaders:
Democratic Party [Joaquin "Kin" PEREZ]
Republican Party [Jerry CRISOSTOMO]

International organization participation: AOSIS (observer), IOC, PIF (observer), SPC, UPU

Diplomatic representation in the US: none (territory of the US)

Diplomatic representation from the US: none (territory of the US)

Flag description: territorial flag is dark blue with a narrow red border on all four sides; centered is a red-bordered, pointed, vertical ellipse containing a beach scene, a proa or outrigger canoe with sail, and a palm tree with the word GUAM superimposed in bold red letters; the proa is sailing in Agana Bay with the promontory of Punta Dos Amantes, near the capital, in the background; the shape of the central emblem is that of a Chamorro sling stone, used as a weapon for defense or hunting; blue represents the sea and red the blood shed in the struggle against oppression
note: the US flag is the national flag

National symbol(s): *coconut tree; national colors:* deep blue, red

National anthem: *name:* "Fanohge Chamoru" (Stand Ye Guamanians)
lyrics/music: Ramon Manalisay SABLAN [English], Lagrimas UNTALAN [Chamoru]/ Ramon Manalisay SABLAN
note: adopted 1919; the local anthem is also known as "Guam Hymn"; as a territory of the United States, "The Star- Spangled Banner," which generally follows the playing of "Stand Ye Guamanians," is official (see United States)

ECONOMY

Economy—overview: US national defense spending is the main driver of Guam's economy, followed closely by tourism and other services. Guam serves as a forward US base for the Western Pacific and is home to thousands of American military personnel. Total federal spending (defense and non-defense) amounted to $1.988 billion in 2016, or 34.2 of Guam's GDP. Of that total, federal grants and cover-over payments amounted to $3444.1 million in 2016, or 35.8% of Guam's total revenues for the fiscal year. In 2016, Guam's economy grew 0.3%. Despite slow growth, Guam's economy has been stable over the last decade. National defense spending cushions the island's economy against fluctuations in tourism. Service exports, mainly spending by foreign tourists in Guam, amounted to over $1 billion for the first time in 2016, or 17.8% of GDP.

GDP (purchasing power parity): $5.793 billion (2016 est.)
$5.697 billion (2015 est.)
$5.531 billion (2014 est.)
country comparison to the world: 174

GDP (official exchange rate): $5.793 billion (2016 est.)

GDP—real growth rate:
0.4% (2016 est.)
0.5% (2015 est.)
1.6% (2014 est.)
country comparison to the world: 186

GDP—per capita (PPP): $35,600 (2016 est.)
$35,200 (2015 est.)
$34,400 (2014 est.)
country comparison to the world: 56

GDP—composition, by end use:
household consumption: 56.2% (2016 est.)
government consumption: 55% (2016 est.)

investment in fixed capital: 20.6% (2016 est.)
investment in inventories: NA (2016 est.)
exports of goods and services: 19.4% (2016 est.)
imports of goods and services: -51.2% (2016 est.)

GDP—composition, by sector of origin:
agriculture: NA
industry: NA
services: 58.4% NA (2015 est.)

Agriculture—products: fruits, copra, vegetables; eggs, pork, poultry, beef

Industries: national defense, tourism, construction, transshipment services, concrete products, printing and publishing, food processing, textiles

Industrial production growth rate: NA

Labor force: 73,210 (2016 est.)
note: includes only the civilian labor force
country comparison to the world: 184

Labor force—by occupation: *agriculture:* 0.3%
industry: 21.6%
services: 78.1% (2013 est.)

Unemployment rate:
4.5% (2017 est.)
3.9% (2016 est.)
country comparison to the world: 67

Population below poverty line: 23% (2001 est.)

Household income or consumption by percentage share: *lowest 10%:* NA
highest 10%: NA

Budget: *revenues:* 1.24 billion (2016 est.)
expenditures: 1.299 billion (2016 est.)

Taxes and other revenues: 21.4% (of GDP) (2016 est.)
country comparison to the world: 139

Budget surplus (+) or deficit (-): -1% (of GDP) (2016 est.)
country comparison to the world: 78

Public debt:
22.1% of GDP (2016 est.)
32.1% of GDP (2013)
country comparison to the world: 184

Fiscal year: 1 October—30 September

Inflation rate (consumer prices):
1% (2017 est.)
0% (2016 est.)
country comparison to the world: 52

Exports:
$1.124 billion (2016 est.)
$1.046 billion (2015 est.)
country comparison to the world: 156

Exports—partners: Palau 13.6% (2017)

Exports—commodities: transshipments of refined petroleum products, construction materials, fish, foodstuffs and beverages

Imports:
$2.964 billion (2016 est.)
$3.054 billion (2015 est.)
country comparison to the world: 149

Imports—commodities: petroleum and petroleum products, food, manufactured goods

Imports—partners: Singapore 41.7%, Japan 30.6%, Hong Kong 10.6% (2017)

Debt—external: NA

Exchange rates: the US dollar is used

ENERGY

Electricity access: electrification—total population: 100% (2020)

Electricity—production: 1.722 billion kWh (2016 est.)
country comparison to the world: 141

Electricity—consumption: 1.601 billion kWh (2016 est.)
country comparison to the world: 146

Electricity—exports: 0 kWh (2016 est.)
country comparison to the world: 143

Electricity—imports: 0 kWh (2016 est.)
country comparison to the world: 155

Electricity—install ed generating capacity: 560,000 kW (2016 est.)
country comparison to the world: 143

Electricity—from fossil fuels: 94% of total installed capacity (2016 est.)
country comparison to the world: 46

Electricity—from nuclear fuels: 0% of total installed capacity (2017 est.)
country comparison to the world: 101

Electricity—from hydroelectric plants: 0% of total installed capacity (2017 est.)
country comparison to the world: 175

Electricity—from other renewable sources: 6% of total installed capacity (2017 est.)
country comparison to the world: 98

Crude oil—production: 0 bbl/day (2018 est.)
country comparison to the world: 144

Crude oil—exports: 0 bbl/day (2015 est.)
country comparison to the world: 132

Crude oil—imports: 0 bbl/day (2015 est.)
country comparison to the world: 135

Crude oil—proved reserves: 0 bbl (1 January 2018 est.)
country comparison to the world: 140

Refined petroleum products—production: 0 bbl/day (2015 est.)
country comparison to the world: 152

Refined petroleum products—consumption: 14,000 bbl/day (2016 est.)
country comparison to the world: 153

Refined petroleum products—exports: 0 bbl/day (2015 est.)
country comparison to the world: 161

Refined petroleum products—imports: 13,500 bbl/day (2015 est.)
country comparison to the world: 141

Natural gas—production: 0 cu m (2017 est.)
country comparison to the world: 140

Natural gas—consumption: 0 cu m (2017 est.)
country comparison to the world: 153

Natural gas—exports: 0 cu m (2017 est.)
country comparison to the world: 115

Natural gas—imports: 0 cu m (2017 est.)
country comparison to the world: 133

Natural gas—proved reserves: 0 cu m (1 January 2014 est.)
country comparison to the world: 142

Carbon dioxide emissions from consumption of energy: 2.214 million Mt (2017 est.)
country comparison to the world: 157

COMMUNICATIONS

Telephones—fixed lines: *total subscriptions:* 70,639
subscriptions per 100 inhabitants: 42.01 (2019 est.)
country comparison to the world: 149

Telephones—mobile cellular: *total subscriptions:* 181,000
subscriptions per 100 inhabitants: 113 (July 2016 est.)
country comparison to the world: 185

Telecommunication systems: *general assessment:* integrated with US facilities for direct dialing, including free use of 800 numbers (2020)
domestic: three major companies provide both fixed-line and mobile services, as well as access to the Internet; fixed-line 42 per 100 and 113 per 100 for mobile-cellular (2019)

international: country code—1-671; major landing points for Atisa, HANTRU1, HK-G, JGA-N, JGA-S, PIPE-1, SEA-US, SxS, Tata TGN-Pacific, AJC, GOKI, AAG, AJC and Mariana-Guam Cable submarine cables between Asia, Australia, and the US (Guam is a transpacific communications hub for major carriers linking the US and Asia); satellite earth stations—2 Intelsat (Pacific Ocean) (2019)

note: the COVID-19 outbreak is negatively impacting telecommunications production and supply chains globally; consumer spending on telecom devices and services has also slowed due to the pandemic's effect on economies worldwide; overall progress towards improvements in all facets of the telecom industry—mobile, fixed-line, broadband, submarine cable and satellite—has moderated

Broadcast media: about a dozen TV channels, including digital channels; multi-channel cable TV services are available; roughly 20 radio stations

Internet country code: .gu

Internet users: *total:* 135,073
percent of population: 80.51% (July 2018 est.)
country comparison to the world: 177

TRANSPORTATION

Airports: 5 (2013)
country comparison to the world: 180

Airports—with paved runways: *total:* 4 (2017)
over 3,047 m: 2 (2017)
2,438 to 3,047 m: 1 (2017)
914 to 1,523 m: 1 (2017)

Airports—with unpaved runways: *total:* 1 (2013)
under 914 m: 1 (2013)

Roadways: *total:* 1,045 km (2008)
country comparison to the world: 184

Merchant marine: *total:* 3
by type: other 3 (2019)
country comparison to the world: 171

Ports and terminals: *major seaport(s):* Apra Harbor

MILITARY AND SECURITY

Military—note: defense is the responsibility of the US

TRANSNATIO NAL ISSUES

Disputes—international: none

GUATEMALA

INTRODUCTION

Background: The Maya civilization flourished in Guatemala and surrounding regions during the first millennium A.D. After almost three centuries as a Spanish colony, Guatemala won its independence in 1821. During the second half of the 20th century, it experienced a variety of military and civilian governments, as well as a 36-year guerrilla war. In 1996, the government signed a peace agreement formally ending the internal conflict.

GEOGRAPHY

Location: Central America, bordering the North Pacific Ocean, between El Salvador and Mexico, and bordering the Gulf of Honduras (Caribbean Sea) between Honduras and Belize

Geographic coordinates: 15 30 N, 90 15 W

Map references: Central America and the Caribbean

Area: *total:* 108,889 sq km
land: 107,159 sq km
water: 1,730 sq km
country comparison to the world: 108

Area—comparative: slightly smaller than Pennsylvania

Land boundaries: *total:* 1,667 km
border countries (4): Belize 266 km, El Salvador 199 km, Honduras 244 km, Mexico 958 km

Coastline: 400 km

Maritime claims: *territorial sea:* 12 nm
exclusive economic zone: 200 nm
continental shelf: 200-m depth or to the depth of exploitation

Climate: tropical; hot, humid in lowlands; cooler in highlands

Terrain: two east-west trending mountain chains divide the country into three regions: the mountainous highlands, the Pacific coast south of mountains, and the vast northern Peten lowlands

Elevation: *mean elevation:* 759 m
lowest point: Pacific Ocean 0 m
highest point: Volcan Tajumulco (highest point in Central America) 4,220 m

Natural resources: petroleum, nickel, rare woods, fish, chicle, hydropower

Land use: *agricultural land:* 41.2% (2011 est.)
arable land: 14.2% (2011 est.) / permanent crops: 8.8% (2011 est.) / permanent pasture: 18.2% (2011 est.)
forest: 33.6% (2011 est.)
other: 25.2% (2011 est.)

Irrigated land: 3,375 sq km (2012)

Population distribution: the vast majority of the populace resides in the southern half of the country, particularly in the mountainous regions; more than half of the population lives in rural areas

Natural hazards: numerous volcanoes in mountains, with occasional violent earthquakes; Caribbean coast extremely susceptible to hurricanes and other tropical storms
volcanism: significant volcanic activity in the Sierra Madre range; Santa Maria (3,772 m) has been deemed a Decade Volcano by the International Association of Volcanology and Chemistry of the Earth's Interior, worthy of study due to its explosive history and close proximity to human populations; Pacaya (2,552 m), which erupted in May 2010 causing an ashfall on Guatemala City and prompting evacuations, is

one of the country's most active volcanoes with frequent eruptions since 1965; other historically active volcanoes include Acatenango, Almolonga, Atitlan, Fuego, and Tacana; see note 2 under "Geography—note"

Environment—current issues: deforestation in the Peten rainforest; soil erosion; water pollution

Environment—international agreements: party to: Antarctic Treaty, Biodiversity, Climate Change, Climate Change-Kyoto Protocol, Desertification, Endangered Species, Environmental Modification, Hazardous Wastes, Law of the Sea, Marine Dumping, Ozone Layer Protection, Ship Pollution, Wetlands, Whaling
signed, but not ratified: none of the selected agreements

Geography—note: *note 1:* despite having both eastern and western coastlines (Caribbean Sea and Pacific Ocean respectively), there are no natural harbors on the west coast
note 2: Guatemala is one of the countries along the Ring of Fire, a belt of active volcanoes and earthquake epicenters bordering the Pacific Ocean; up to 90% of the world's earthquakes and some 75% of the world's volcanoes occur within the Ring of Fire

PEOPLE AND SOCIETY

Population: 17,153,288 (July 2020 est.)
country comparison to the world: 68

Nationality: *noun:* Guatemalan(s)
adjective: Guatemalan

Ethnic groups: mestizo (mixed Amerindian-Spanish—in local Spanish called Ladino) 56%, Maya 41.7%, Xinca (indigenous, non-Maya) 1.8%, African descent .2%, Garifuna (mixed West and Central African, Island Carib, and Arawak) .1%, foreign .2% (2018 est.)

Languages: Spanish (official) 69.9%, Maya languages 29.7% (Q'eqchi' 8.3%, K'iche 7.8%, Mam 4.4%, Kaqchikel 3%, Q'anjob'al 1.2%, Poqomchi' 1%, other 4%), other 0.4% (includes Xinca and Garifuna) (2018 est.)
note: the 2003 Law of National Languages officially recognized 23 indigenous languages, including 21 Maya languages, Xinca, and Garifuna

Religions: Roman Catholic, Protestant, indigenous Maya

Demographic profile: Guatemala is a predominantly poor country that struggles in several areas of health and development, including infant, child, and maternal mortality, malnutrition, literacy, and contraceptive awareness and use. The country's large indigenous population is disproportionately affected. Guatemala is the most populous country in Central America and has the highest fertility rate in Latin America. It also has the highest population growth rate in Latin America, which is likely to continue because of its large reproductive-age population and high birth rate. Almost half of Guatemala's population is under age 19, making it the youngest population in Latin America. Guatemala's total fertility rate has slowly declined during the last few decades due in part

to limited government-funded health programs. However, the birth rate is still more close to three children per woman and is markedly higher among its rural and indigenous populations.

Guatemalans have a history of emigrating legally and illegally to Mexico, the United States, and Canada because of a lack of economic opportunity, political instability, and natural disasters. Emigration, primarily to the United States, escalated during the 1960 to 1996 civil war and accelerated after a peace agreement was signed. Thousands of Guatemalans who fled to Mexico returned after the war, but labor migration to southern Mexico continues.

Age structure: *0-14 years:* 33.68% (male 2,944,145/female 2,833,432)
15-24 years: 19.76% (male 1,705,730/female 1,683,546)
25-54 years: 36.45% (male 3,065,933/female 3,186,816)
55-64 years: 5.41% (male 431,417/female 496,743)
65 years and over: 4.7% (male 363,460/female 442,066) (2020 est.)

Dependency ratios: *total dependency ratio:* 62.3
youth dependency ratio: 54.1
elderly dependency ratio: 8.2
potential support ratio: 12.2 (2020 est.)

Median age: *total:* 23.2 years
male: 22.6 years
female: 23.8 years (2020 est.)
country comparison to the world: 178

Population growth rate: 1.68% (2020 est.)
country comparison to the world: 62

Birth rate: 23.3 births/1,000 population (2020 est.)
country comparison to the world: 55

Death rate: 4.9 deaths/1,000 population (2020 est.)
country comparison to the world: 200

Net migration rate: -1.7 migrant(s)/1,000 population (2020 est.)
country comparison to the world: 158

Population distribution: the vast majority of the populace resides in the southern half of the country, particularly in the mountainous regions; more than half of the population lives in rural areas

Urbanization: *urban population:* 51.8% of total population (2020)
rate of urbanization: 2.68% annual rate of change (2015-20 est.)
total population growth rate v. urban population growth rate, 2000-2030:

Major urban areas—population: 2.935 million GUATEMALA CITY (capital) (2020)

Sex ratio: *at birth:* 1.05 male(s)/female
0-14 years: 1.04 male(s)/female
15-24 years: 1.01 male(s)/female
25-54 years: 0.96 male(s)/female
55-64 years: 0.87 male(s)/female
65 years and over: 0.82 male(s)/female
total population: 0.99 male(s)/female (2020 est.)

Mother's mean age at first birth: 21.2 years (2014/15 est.)
note: median age at first birth among women 25-29

Maternal mortality rate: 95 deaths/100,000 live births (2017 est.)
country comparison to the world: 72

Infant mortality rate: *total:* 21.8 deaths/1,000 live births
male: 24 deaths/1,000 live births
female: 19.6 deaths/1,000 live births (2020 est.)
country comparison to the world: 71

Life expectancy at birth: *total population:* 72.4 years
male: 70.3 years
female: 74.5 years (2020 est.)
country comparison to the world: 153

Total fertility rate: 2.72 children born/woman (2020 est.)
country comparison to the world: 63

Contraceptive prevalence rate: 60.6% (2014/15)

Drinking water source:
improved:
urban: 97.9% of population
rural: 92.2% of population
total: 95.2% of population
unimproved:
urban: 2.1% of population
rural: 7.8% of population
total: 4.8% of population (2017 est.)

Current Health Expenditure: 5.8% (2017)

Physicians density: 0.36 physicians/1,000 population (2018)

Hospital bed density: 0.4 beds/1,000 population (2017)

Sanitation facility access:
improved:
urban: 91.4% of population
rural: 61.7% of population
total: 76.7% of population
unimproved:
urban: 8.6% of population
rural: 38.3% of population
total: 23.3% of population (2017 est.)

HIV/AIDS—adult prevalence rate: 0.3% (2019 est.)
country comparison to the world: 88

HIV/AIDS—people living with HIV/AIDS: 36,000 (2019 est.)
country comparison to the world: 69

HIV/AIDS—deaths: 1,200 (2019 est.)
country comparison to the world: 54

Major infectious diseases: *degree of risk:* high (2020)
food or waterborne diseases: bacterial diarrhea, hepatitis A, and typhoid fever
vectorborne diseases: dengue fever and malaria

Obesity—adult prevalence rate: 21.2% (2016)
country comparison to the world: 93

Children under the age of 5 years underweight: 12.4% (2015)
country comparison to the world: 51

Education expenditures: 2.8% of GDP (2017)
country comparison to the world: 145

Literacy: *definition:* age 15 and over can read and write

total population: 81.5%
male: 87.4%
female: 76.3% (2015)

School life expectancy (primary to tertiary education): *total:* 11 years
male: 11 years
female: 11 years (2015)

Unemployment, youth ages 15-24: *total:* 5%
male: 3.7%
female: 8% (2017 est.)
country comparison to the world: 166

GOVERNMENT

Country name: *conventional long form:* Republic of Guatemala
conventional short form: Guatemala
local long form: Republica de Guatemala
local short form: Guatemala
etymology: the Spanish conquistadors used many native Americans as allies in their conquest of Guatemala; the site of their first capital (established in 1524), a former Maya settlement, was called "Quauhtemallan" by their Nahuatl-speaking Mexican allies, a name that means "land of trees" or "forested land", but which the Spanish pronounced "Guatemala"; the Spanish applied that name to a re founded capital city three years later and eventually it became the name of the country

Government type: presidential republic

Capital: *name:* Guatemala City
geographic coordinates: 14 37 N, 90 31 W
time difference: UTC-6 (1 hour behind Washington, DC, during Standard Time)
etymology: the Spanish conquistadors used many native Americans as allies in their conquest of Guatemala; the site of their first capital (established in 1524), a former Maya settlement, was called "Quauhtemallan" by their Nahuatl-speaking Mexican allies, a name that means "land of trees" or "forested land", but which the Spanish pronounced "Guatemala"; the Spanish applied that name to a re founded capital city three years later and eventually it became the name of the country

Administrative divisions: 22 departments (departamentos, singular—departamento); Alta Verapaz, Baja Verapaz, Chimaltenango, Chiquimula, El Progreso, Escuintla, Guatemala, Huehuetenango, Izabal, Jalapa, Jutiapa, Peten, Quetzaltenango, Quiche, Retalhuleu, Sacatepequez, San Marcos, Santa Rosa, Solola, Suchitepequez, Totonicapan, Zacapa

Independence: 15 September 1821 (from Spain)

National holiday: Independence Day, 15 September (1821)

Constitution: *history:* several previous; latest adopted 31 May 1985, effective 14 January 1986; suspended and reinstated in 1994
amendments: proposed by the president of the republic, by agreement of 10 or more deputies of Congress, by the Constitutional Court, or by public petition of at least 5,000 citizens; passage

requires at least two-thirds majority vote by the Congress membership and approval by public referendum, referred to as "popular consultation"; constitutional articles such as national sovereignty, the republican form of government, limitations on those seeking the presidency, or presidential tenure cannot be amended; amended 1994

Legal system: civil law system; judicial review of legislative acts

International law organization participation: has not submitted an ICJ jurisdiction declaration; accepts ICCt jurisdiction

Citizenship: *citizenship by birth:* yes
citizenship by descent only: yes
dual citizenship recognized: yes
residency requirement for naturalization: 5 years with no absences of six consecutive months or longer or absences totaling more than a year

Suffrage: 18 years of age; universal; note—active duty members of the armed forces and police by law cannot vote and are restricted to their barracks on election day

Executive branch: *chief of state:* President Alejandro GIAMMATTEI (since 14 January 2020); Vice President Cesar Guillermo CASTILLO Reyes (since 14 January 2020); note—the president is both chief of state and head of government
head of government: President Alejandro GIAMMATTEI (since 14 January 2020); Vice President Cesar Guillermo CASTILLO Reyes (since 14 January 2020)
cabinet: Council of Ministers appointed by the president
elections/appointments: president and vice president directly elected on the same ballot by absolute majority popular vote in 2 rounds if needed for a 4-year term (not eligible for consecutive terms); election last held on 16 June 2019 with a runoff on 11 August 2019 (next to be held in June 2023)
election results: Alejandro GIAMMATTEI elected president; percent of vote in first round—Sandra TORRES (UNE) 25.54%, Alejandro GIAMMATTEI (VAMOS) 13.95%, Edmond MULET (PHG) 11.21%, Thelma CABRERA (MLP) 10.37%, Roberto ARZU (PAN-PODEMOS) 6.08%; percent of vote in second round—Alejandro GIAMMATTEI (VAMOS) 58%, Sandra TORRES (UNE) 42%

Legislative branch: *description:* unicameral Congress of the Republic or Congreso de la Republica (158 seats; 127 members directly elected in multi-seat constituencies in the country's 22 departments by simple majority vote and 31 directly elected in a single nationwide constituency by closed-list, proportional representation vote; members serve 4-year terms); note—two additional seats will be added to the new congress when it is seated in January 2020
elections: last held on 16 June 2019 (next to be held on June 2023)
election results: percent of vote by party—NA; seats by party—UNE 53, VAMOS 16, UCN 12, VALOR 9, BIEN 8, FCNNACION 8, SEMILLA 7, TODOS 7, VIVA 7, CREO 6, PHG 6,

VICTORIA 4, Winaq 4, PC 3, PU 3, URNG 3, PAN 2, MLP 1, PODEMOS 1
note: current seats by party as of 1 June 2019—FCN 37, UNE 32, MR 20, TODOS 17, AC 12, EG 7, UCN 6, CREO 5, LIDER 5, VIVA 4, Convergence 3, PAN 3, PP 2, FUERZA 1 , PU 1 , URNG 1 , Winaq 1 , independent 1 ; composition—men 136, women 22, percent of women 13.9%

Judicial branch: *highest courts:* Supreme Court of Justice or Corte Suprema de Justicia (consists of 13 magistrates, including the court president and organized into 3 chambers); note—the court president also supervises trial judges countrywide; Constitutional Court or Corte de Constitucionalidad (consists of 5 titular magistrates and 5 substitute magistrates)
judge selection and term of office: Supreme Court magistrates elected by the Congress of the Republic from candidates proposed by the Postulation Committee, an independent body of deans of the country's university law schools, representatives of the country's law associations, and representatives of the Courts of Appeal; magistrates elected for concurrent, renewable 5-year terms; Constitutional Court judges—1 elected by the Congress of the Republic, 1 by the Supreme Court, 1 by the president of the republic, 1 by the (public) University of San Carlos, and 1 by the Assembly of the College of Attorneys and Notaries; judges elected for renewable, consecutive 5-year terms; the presidency of the court rotates among the magistrates for a single 1 -year term
subordinate courts: numerous first instance and appellate courts

Political parties and leaders:
Bienestar Nacional or BIEN [Alfonso PORTILLO and Evelyn MORATAYA]
Citizen Alliance or AC
Citizen Prosperity or PC [Dami Anita Elizabeth KRISTENSON Sales]
Commitment, Renewal, and Order or CREO [Roberto GONZALEZ Diaz-Duran]
Convergence [Sandra MORAN]
Encounter for Guatemala or EG [Nineth MONTENEGRO Cottom]
Everyone Together for Guatemala or TODOS [Felipe ALEJOS]
Force or FUERZA [Mauricio RADFORD]
Guatemalan National Revolutionary Unity or URNG-MAIZ or URNG [Gregorio CHAY Laynez]
Humanist Party of Guatemala or PHG [Edmond MULET]
Movement for the Liberation of Peoples or MLP [Thelma CABRERA]
Movimiento Semilla or SEMILLA [Thelma ALDANA]
National Advancement Party or PAN [Harald JOHANNESSEN]
National Convergence Front or FCN-NACION or FCN [Jimmy MORALES]
National Unity for Hope or UNE [Sandra TORRES]
Nationalist Change Union or UCN [Mario ESTRADA]

Patriotic Party or PP

PODEMOS [Jose Raul VIRGIL Arias]

Political Movement Winaq or Winaq [Sonia GUTIERREZ Raguay]

Reform Movement or MR

Renewed Democratic Liberty or LIDER (dissolved mid-February 2016)

TODOS [Felipe ALEJOS]

Unionista Party or PU [Alvaro ARZU Escobar]

Value or VALOR [Zury RIOS]

Vamos por una Guatemala Diferente or VAMOS [Alejandro GIAMMATTEI]

Victory or VICTORIA [Amilcar RIVERA]

Vision with Values or VIVA [Armando Damian CASTILLO Alvarado]

note: parties represented in the last election, but have since dissolved—FCN (2017), LIDER (2016), and PP (2017)

International organization participation: BCIE, CACM, CD, CELAC, EITI (compliant country), FAO, G-24, G-77, IADB, IAEA, IBRD, ICAO, ICC (national committees), ICCt (signatory), ICRM, IDA, IFAD, IFC, IFRCS, IHO, ILO, IMF, IMO, Interpol, IOC, IOM, IPU, ISO (correspondent), ITSO, ITU, ITUC (NGOs), LAES, LAIA (observer), MIGA, MINUSTAH, MONUSCO, NAM, OAS, OPANAL, OPCW, Pacific Alliance (observer), PCA, Petrocaribe, SICA, UN, UNCTAD, UNESCO, UNIDO, UNIFIL, Union Latina, UNISFA, UNITAR, UNMISS, UNOCI, UNWTO, UPU, WCO, WFTU (NGOs), WHO, WIPO, WMO, WTO

Diplomatic representation in the US: *chief of mission:* Ambassador Alfonso Jose QUINONEZ LEMUS (since 17 July 2020)

chancery: 2220 R Street NW, Washington, DC 20008

telephone: [1] (202) 745-4952

FAX: [1] (202) 745-1908

consulate(s) general: Atlanta, Chicago, Del Rio (TX), Denver, Houston, Los Angeles, McAllen (TX), Miami, New York, Oklahoma City, Philadelphia, Phoenix, Providence (RI), Raleigh (NC), San Bernardino (CA), San Francisco, Seattle

consulate(s): Lake Worth (FL), Tucson (AZ)

Diplomatic representation from the US: *chief of mission:* Ambassador Luis E. ARREAGA (since 4 October 2017)

telephone: [502] 2326-4000

embassy: 7-01 Avenida Reforma, Zone 10, Guatemala City

mailing address: DPO AA 34024

FAX: [502] 2326-4654

Flag description: three equal vertical bands of light blue (hoist side), white, and light blue, with the coat of arms centered in the white band; the coat of arms includes a green and red quetzal (the national bird) representing liberty and a scroll bearing the inscription LIBERTAD 15 DE SEPTIEMBRE DE 1821 (the original date of independence from Spain) all superimposed on a pair of crossed rifles signifying Guatemala's willingness to defend itself and a pair of crossed swords representing honor and framed by a laurel wreath

symbolizing victory; the blue bands represent the Pacific Ocean and Caribbean Sea; the white band denotes peace and purity

note: one of only two national flags featuring a firearm, the other is Mozambique

National symbol(s): *quetzal (bird); national colors:* blue, white

National anthem: *name:* "Himno Nacional de Guatemala" (National Anthem of Guatemala)

lyrics/music: Jose Joaquin PALMA/Rafael Alvarez OVALLE

note: adopted 1897, modified lyrics adopted 1934; Cuban poet Jose Joaquin PALMA anonymously submitted lyrics to a public contest calling for a national anthem; his authorship was not discovered until 1911

0:00/ 1:37

ECONOMY

Economy—overview: Guatemala is the most populous country in Central America with a GDP per capita roughly half the average for Latin America and the Caribbean. The agricultural sector accounts for 13.5% of GDP and 31% of the labor force; key agricultural exports include sugar, coffee, bananas, and vegetables. Guatemala is the top remittance recipient in Central America as a result of Guatemala's large expatriate community in the US. These inflows are a primary source of foreign income, equivalent to two-thirds of the country's exports and about a tenth of its GDP.

The 1996 peace accords, which ended 36 years of civil war, removed a major obstacle to foreign investment, and Guatemala has since pursued important reforms and macroeconomic stabilization. The Dominican Republic-Central America Free Trade Agreement (CAFTA-DR) entered into force in July 2006, spurring increased investment and diversification of exports, with the largest increases in ethanol and non-traditional agricultural exports. While CAFTA-DR has helped improve the investment climate, concerns over security, the lack of skilled workers, and poor infrastructure continue to hamper foreign direct investment.

The distribution of income remains highly unequal with the richest 20% of the population accounting for more than 51% of Guatemala's overall consumption. More than half of the population is below the national poverty line, and 23% of the population lives in extreme poverty. Poverty among indigenous groups, which make up more than 40% of the population, averages 79%, with 40% of the indigenous population living in extreme poverty. Nearly one-half of Guatemala's children under age five are chronically malnourished, one of the highest malnutrition rates in the world.

GDP (purchasing power parity):

$138.1 billion (2017 est.)

$134.4 billion (2016 est.)

$130.4 billion (2015 est.)

note: data are in 2017 dollars

country comparison to the world: 77

GDP (official exchange rate):

$75.62 billion (2017 est.)

GDP—real growth rate:

2.8% (2017 est.)

3.1% (2016 est.)

4.1% (2015 est.)

country comparison to the world: 104

GDP—per capita (PPP):

$8,200 (2017 est.)

$8,100 (2016 est.)

$8,000 (2015 est.)

note: data are in 2017 dollars

country comparison to the world: 150

Gross national saving:

13.6% of GDP (2017 est.)

14.4% of GDP (2016 est.)

13.5% of GDP (2015 est.)

country comparison to the world: 141

GDP—composition, by end use:

household consumption: 86.3% (2017 est.)

government consumption: 9.7% (2017 est.)

investment in fixed capital: 12.3% (2017 est.)

investment in inventories: -0.2% (2017 est.)

exports of goods and services: 18.8% (2017 est.)

imports of goods and services: -26.9% (2017 est.)

GDP—composition, by sector of origin:

agriculture: 13.3% (2017 est.)

industry: 23.4% (2017 est.)

services: 63.2% (2017 est.)

Agriculture—products: sugarcane, corn, bananas, coffee, beans, cardamom; cattle, sheep, pigs, chickens

Industries: sugar, textiles and clothing, furniture, chemicals, petroleum, metals, rubber, tourism

Industrial production growth rate: 1.8% (2017 est.)

country comparison to the world: 136

Labor force: 6.664 million (2017 est.)

country comparison to the world: 67

Labor force—by occupation:

agriculture: 31.4%

industry: 12.8%

services: 55.8% (2017 est.)

Unemployment rate:

2.3% (2017 est.)

2.4% (2016 est.)

country comparison to the world: 23

Population below poverty line: 59.3% (2014 est.)

Household income or consumption by percentage share: *lowest 10%:* 1.6%

highest 10%: 38.4% (2014)

Budget: *revenues:* 8.164 billion (2017 est.)

expenditures: 9.156 billion (2017 est.)

Taxes and other revenues: 10.8% (of GDP) (2017 est.)

country comparison to the world: 212

Budget surplus (+) or deficit (-): -1.3% (of GDP) (2017 est.)

country comparison to the world: 85

Public debt:

24.7% of GDP (2017 est.)

24.5% of GDP (2016 est.)

country comparison to the world: 175

Fiscal year: calendar year

Inflation rate (consumer prices):
4.4% (2017 est.)
4.4% (2016 est.)
country comparison to the world: 165

Current account balance:
$1.134 billion (2017 est.)
$1.023 billion (2016 est.)
country comparison to the world: 48

Exports:
$11.12 billion (2017 est.)
$10.58 billion (2016 est.)
country comparison to the world: 88

Exports—partners: US 33.8%, El Salvador 11.1%, Honduras 8.8%, Nicaragua 5.1%, Mexico 4.7% (2017)

Exports—commodities: sugar, coffee, petroleum, apparel, bananas, fruits and vegetables, cardamom, manufacturing products, precious stones and metals, electricity

Imports:
$17.11 billion (2017 est.)
$15.77 billion (2016 est.)
country comparison to the world: 83

Imports—commodities: fuels, machinery and transport equipment, construction materials, grain, fertilizers, electricity, mineral products, chemical products, plastic materials and products

Imports—partners: US 39.8%, China 10.7%, Mexico 10.7%, El Salvador 5.3% (2017)

Reserves of foreign exchange and gold:
$11.77 billion (31 December 2017 est.)
$9.156 billion (31 December 2016 est.)
country comparison to the world: 71

Debt—external:
$22.92 billion (31 December 2017 est.)
$21.45 billion (31 December 2016 est.)
country comparison to the world: 89

Exchange rates: quetzales (GTQ) per US dollar—7.323 (2017 est.)
7.5999 (2016 est.)
7.5999 (2015 est.)
7.6548 (2014 est.)
7.7322 (2013 est.)

ENERGY

Electricity access: *population without electricity:* 1 million (2017)
electrification—total population: 91.8% (2016)
electrification—urban areas: 96.8% (2016)
electrification—rural areas: 86.4% (2016)

Electricity—production: 12.12 billion kWh (2016 est.)
country comparison to the world: 96

Electricity—consumption: 10.1 billion kWh (2016 est.)
country comparison to the world: 96

Electricity—exports: 1.858 billion kWh (2017 est.)
country comparison to the world: 47

Electricity—imports: 747 million kWh (2016 est.)
country comparison to the world: 75

Electricity—installed generating capacity: 4.605 million kW (2016 est.)
country comparison to the world: 85

Electricity—from fossil fuels: 41% of total installed capacity (2016 est.)
country comparison to the world: 167

Electricity—from nuclear fuels: 0% of total installed capacity (2017 est.)
country comparison to the world: 102

Electricity—from hydroelectric plants: 31% of total installed capacity (2017 est.)
country comparison to the world: 67

Electricity—from other renewable sources: 28% of total installed capacity (2017 est.)
country comparison to the world: 23

Crude oil—production: 9,600 bbl/day (2018 est.)
country comparison to the world: 78

Crude oil—exports: 9,383 bbl/day (2017 est.)
country comparison to the world: 59

Crude oil—imports: 0 bbl/day (2015 est.)
country comparison to the world: 136

Crude oil—proved reserves: 83.07 million bbl (1 January 2018 est.)
country comparison to the world: 71

Refined petroleum products—production: 1,162 bbl/day (2015 est.)
country comparison to the world: 105

Refined petroleum products—consumption: 89,000 bbl/day (2016 est.)
country comparison to the world: 84

Refined petroleum products—exports: 10,810 bbl/day (2015 est.)
country comparison to the world: 80

Refined petroleum products—imports: 97,900 bbl/day (2015 est.)
country comparison to the world: 55

Natural gas—production: 0 cu m (2017 est.)
country comparison to the world: 141

Natural gas—consumption: 0 cu m (2017 est.)
country comparison to the world: 154

Natural gas—exports: 0 cu m (2017 est.)
country comparison to the world: 116

Natural gas—imports: 0 cu m (2017 est.)
country comparison to the world: 134

Natural gas—proved reserves: 2.96 billion cu m (1 January 2006 est.)
country comparison to the world: 94

Carbon dioxide emissions from consumption of energy: 17.15 million Mt (2017 est.)
country comparison to the world: 91

COMMUNICATIONS

Telephones—fixed lines: *total subscriptions:* 1,894,179
subscriptions per 100 inhabitants: 11.23 (2019 est.)
country comparison to the world: 58

Telephones—mobile cellular: *total subscriptions:* 20,026,347
subscriptions per 100 inhabitants: 118.73 (2019 est.)

country comparison to the world: 61

Telecommunication systems: *general assessment:* network centered in the city of Guatemala; one of the lowest teledensities in the region especially in the country, rural areas have no fixed-line access so mobile services adopted as necessary; state-owned telecommunications company privatized in the late 1990s opened the way for competition; steady improvement of fixedline which has also spurred growth in mobile-cellular and broadband; open regulatory framework coupled with competition and greater disposable household revenue spurs growth (2020)
domestic: fixed-line teledensity roughly 11 per 100 persons; fixed-line investments are concentrating on improving rural connectivity; mobile-cellular teledensity about 119 per 100 persons (2019)
international: country code—502; landing points for the ARCOS, AMX-1, American Movil-Texius West Coast Cable and the SAm-1 fiber-optic submarine cable system that, together, provide connectivity to South and Central America, parts of the Caribbean, and the US; connected to Central American Microwave System; satellite earth station—1 Intelsat (Atlantic Ocean) (2019)
note: the COVID-19 outbreak is negatively impacting telecommunications production and supply chains globally; consumer spending on telecom devices and services has also slowed due to the pandemic's effect on economies worldwide; overall progress towards improvements in all facets of the telecom industry—mobile, fixed-line, broadband, submarine cable and satellite—has moderated

Broadcast media: 4 privately owned national terrestrial TV channels dominate TV broadcasting; multi-channel satellite and cable services are available; 1 government-owned radio station and hundreds of privately owned radio stations (2019)

Internet country code: gt

Internet users: total: 10,777,827
percent of population: 65% (July 2018 est.)
country comparison to the world: 50

Broadband—fixed subscriptions: total: 506,000
subscriptions per 100 inhabitants: 3 (2017 est.)
country comparison to the world: 84

TRANSPORTATION

National air transport system: *number of registered air carriers:* 3 (2020)
inventory of registered aircraft operated by air carriers: 5
annual passenger traffic on registered air carriers: 145,795 (2018)
annual freight traffic on registered air carriers: 110,000 mt-km (2018)

Civil aircraft registration country code prefix: TG (2016)

Airports: 291 (2013)
country comparison to the world: 23

Airports—with paved runways: total: 16 (2017)
2,438 to 3,047 m: 2 (2017)
1,524 to 2,437 m: 4 (2017)
914 to 1,523 m: 6 (2017)

under 914 m: 4 (2017)

Airports—with unpaved runways: *total:* 275 (2013)
2,438 to 3,047 m: 1 (2013)
1,524 to 2,437 m: 2 (2013)
914 to 1,523 m: 77 (2013)
under 914 m: 195 (2013)

Heliports: 1 (2013)

Pipelines: 480 km oil (2013)

Railways: *total:* 800 km (2018)
narrow gauge: 800 km 0.914-m gauge (2018)
note: despite the existence of a railway network, all rail service was suspended in 2007 and no passenger or freight train currently runs in the country (2018)
country comparison to the world: 97

Roadways:
total: 17,621 km (2016)
paved: 7,489 km (2016)
unpaved: 10,132 km (includes 4,960 km of rural roads) (2016)
country comparison to the world: 119

Waterways: 990 km (260 km navigable year round; additional 730 km navigable during high-water season) (2012)
country comparison to the world: 65

Merchant marine: *total:* 9
by type: oil tanker 1, other 8 (2019)
country comparison to the world: 157

Ports and terminals: *major seaport(s):* Puerto Quetzal, Santo Tomas de Castilla

MILITARY AND SECURITY

Military and security forces: Army of Guatemala (Ejercito de Guatemala): Land Forces (Fuerzas de Tierra), Naval Forces (Fuerza de Mar), and Air Force (Fuerza de Aire); Ministry of Interior: National Civil Police (Policia Nacional Civil; includes paramilitary units) (2020)

Military expenditures:
0.4% of GDP (2019)
0.4% of GDP (2018)
0.4% of GDP (2017)
0.4% of GDP (2016)
0.4% of GDP (2015)
country comparison to the world: 152

Military and security service personnel strengths: assessments of the size of the Army of Guatemala vary; approximately 21,500 active personnel (19,000 Land Forces; 1,500 Naval Forces; 1,000 Air Forces); approximately 30,000 National Civil Police (2019 est.)

Military equipment inventories and acquisitions: the Guatemalan military inventory is small and mostly comprised of older US equipment; since 2010, Guatemala has received limited amounts of equipment from Canada, Colombia, Spain, Taiwan, and the US (2019 est.)

Military deployments: 150 Democratic Republic of the Congo (MONUSCO) (2020)

Military service age and obligation: all male citizens between the ages of 18 and 50 are eligible for military service; in practice, most of the force is volunteer, however, a selective draft system is employed, resulting in a small portion of 17-21 year-olds conscripted; conscript service obligation varies from 1 to 2 years; women can serve as officers (2013)

TRANSNATIONAL ISSUES

Disputes—international: annual ministerial meetings under the Organization of American States-initiated Agreement on the Framework for Negotiations and Confidence Building Measures continue to address Guatemalan land and maritime claims in Belize and the Caribbean Sea; Guatemala persists in its territorial claim to half of Belize, but agrees to Line of Adjacency to keep Guatemalan squatters out of Belize's forested interior; both countries agreed in April 2012 to hold simultaneous referenda, scheduled for 6 October 2013, to decide whether to refer the dispute to the ICJ for binding resolution, but this vote was suspended indefinitely; Mexico must deal with thousands of impoverished Guatemalans and other Central Americans who cross the porous border looking for work in Mexico and the US

Refugees and internally displaced persons: *IDPs:* 242,000 (more than three decades of internal conflict that ended in 1996 displaced mainly the indigenous Maya population and rural peasants; ongoing drug cartel and gang violence) (2019)

Illicit drugs: major transit country for cocaine and heroin; it is estimated that 1,000 mt of cocaine are smuggled through the country each year, primarily destined for the US market; in 2016, the Guatamalan government estimated that an average of 4,500 hectares of opium poppy were being cultivated; marijuana cultivation for mostly domestic consumption; proximity to Mexico makes Guatemala a major staging area for drugs (particularly for cocaine); money laundering is a serious problem; corruption is a major problem

GUERNSEY

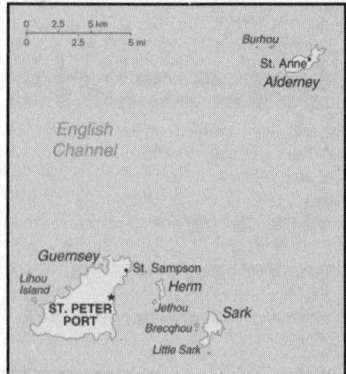

INTRODUCTION

Background: Guernsey and the other Channel Islands represent the last remnants of the medieval Duchy of Normandy, which held sway in both France and England. The islands were the only British soil occupied by German troops in World War II. The Bailiwick of Guernsey is a self-governing British Crown dependency that is not part of the United Kingdom. However, the UK Government is constitutionally responsible for its defense and international representation. The Bailiwick of Guernsey consists of the main island of Guernsey and a number of smaller islands including Alderney, Sark, Herm, Jethou, Brecqhou, and Lihou.

GEOGRAPHY

Location: Western Europe, islands in the English Channel, northwest of France

Geographic coordinates: 49 28 N, 2 35 W

Map references: Europe

Area: *total:* 78 sq km
land: 78 sq km
water: 0 sq km
note: includes Alderney, Guernsey, Herm, Sark, and some other smaller islands
country comparison to the world: 228

Area—comparative: about one-half the size of Washington, DC

Land boundaries: 0 km

Coastline: 50 km

Maritime claims: *territorial sea:* 3 nm
exclusive fishing zone: 12 nm

Climate: temperate with mild winters and cool summers; about 50% of days are overcast

Terrain: mostly flat with low hills in southwest

Elevation: *lowest point:* English Channel 0 m
highest point: Le Moulin on Sark 114 m

Natural resources: cropland

Irrigated land: NA

Natural hazards: very large tidal variation and fast currents can make local waters dangerous

Environment—current issues: coastal erosion, coastal flooding; declining biodiversity due to land abandonment and succession to scrub or woodland

Geography—note: large, deepwater harbor at Saint Peter Port

PEOPLE AND SOCIETY

Population: 67,052 (July 2020 est.)
country comparison to the world: 204

Nationality: *noun:* Channel Islander(s)
adjective: Channel Islander

Ethnic groups: Guernsey 52%, UK and Ireland 23.7%, Portugal 2.1%, Latvia 1.5%, other 6.7%, unspecified 14.1%
note: data represent population by country of birth; the native population is of British and Norman-French descent

Languages: English, French, Norman-French dialect spoken in country districts

Religions: Protestant (Anglican, Presbyterian, Baptist, Congregational, Methodist), Roman Catholic

Age structure: *0-14 years:* 14.5% (male 5,008/ female 4,712)
15-24 years: 10.58% (male 3,616/female 3,476)
25-54 years: 40.73% (male 13,821/female 13,492)
55-64 years: 13.96% (male 4,635/female 4,728)
65 years and over: 20.23% (male 6,229/female 7,335) (2020 est.)

Dependency ratios: *total dependency ratio:* 49
youth dependency ratio: 22.3
elderly dependency ratio: 26.7
potential support ratio: 3.7 (2020 est.)
note: data represent the Guernsey and Jersey

Median age: *total:* 44.3 years
male: 43 years
female: 45.6 years (2020 est.)
country comparison to the world: 17

Population growth rate: 0.26% (2020 est.)
country comparison to the world: 176

Birth rate: 9.8 births/1,000 population (2020 est.)
country comparison to the world: 193

Death rate: 9.2 deaths/1,000 population (2020 est.)
country comparison to the world: 55

Net migration rate: 1.9 migrant(s)/1,000 population (2020 est.)
country comparison to the world: 49

Urbanization: *urban population:* 31% of total population (2020)
rate of urbanization: 0.46% annual rate of change (2015-20 est.)
note: data represent Guernsey and Jersey
total population growth rate v. urban population growth rate, 2000-2030

Major urban areas—Population: 16,000 SAINT PETER PORT (capital) (2018)

Sex ratio: *at birth:* 1.05 male(s)/female
0-14 years: 1.06 male(s)/female
15-24 years: 1.04 male(s)/female
25-54 years: 1.02 male(s)/female
55-64 years: 0.98 male(s)/female

65 years and over: 0.85 male(s)/female
total population: 0.99 male(s)/female (2020 est.)

Infant mortality rate: *total:* 3.3 deaths/1,000 live births
male: 3.6 deaths/1,000 live births
female: 3 deaths/1,000 live births (2020 est.)
country comparison to the world: 207

Life expectancy at birth: total population: 82.8 years
male: 80.1 years
female: 85.7 years (2020 est.)
country comparison to the world: 11

Total fertility rate: 1.57 children born/woman (2020 est.)
country comparison to the world: 192

Drinking water source:
improved:
total: 94.2% of population
unimproved:
total: 5.9% of population (2017 est.)
note: includes data for Jersey

Sanitation facility access:
improved:
total: 98% of population
unimproved:
total: 1.2% of population (2017 est.)
note: data represent Guernsey and Jersey

HIV/AIDS—adult prevalence rate: NA

HIV/AIDS—people living with HIV/AIDS: NA

HIV/AIDS—deaths: NA

Education expenditures: NA

GOVERNMENT

Country name: *conventional long form:* Bailiwick of Guernsey
conventional short form: Guernsey
former: Norman Isles
etymology: the name is of Old Norse origin, but the meaning of the root "Guern(s)" is uncertain; the "-ey" ending means "island"

Dependency status:
British crown dependency

Government type: parliamentary democracy (States of Deliberation)

Capital: *name:* Saint Peter Port
geographic coordinates: 49 27 N, 2 32 W
time difference: UTC 0 (5 hours ahead of Washington, DC, during Standard Time)
daylight saving time: +1hr, begins last Sunday in March; ends last Sunday in October
etymology: Saint Peter Port is the name of the town and its surrounding parish; the "port" distinguishes this parish from that of Saint Peter on the other side of the island

Administrative divisions: none (British Crown dependency); there are no first-order administrative divisions as defined by the US Government, but there are 10 parishes: Castel, Forest, Saint Andrew, Saint Martin, Saint Peter Port, Saint Pierre du Bois, Saint Sampson, Saint Saviour, Torteval, Vale

note: two additional parishes for Guernsey are sometimes listed—Saint Anne on the island of Alderney and Saint Peter on the island of Sark—but they are generally not included in the enumeration of parishes

Independence: none (British Crown dependency)

National holiday: Liberation Day, 9 May (1945)

Constitution: *history:* unwritten; includes royal charters, statutes, and common law and practice
amendments: new laws or changes to existing laws are initiated by the States of Deliberation; passage requires majority vote

Legal system: customary legal system based on Norman customary law; includes elements of the French civil code and English common law

Citizenship: see United Kingdom

Suffrage: 16 years of age; universal

Executive branch: *chief of state:* Queen ELIZABETH II (since 6 February 1952); represented by Lieutenant-Governor Vice Admiral Ian CORDER (since 14 March 2016)
head of government: Chief Minister Gavin ST PIER (since 6 May 2016); Bailiff Sir Richard COLLAS (since 23 March 2012); note—the chief minister is the president of the Policy and Resources Committee and is the de facto head of government; the Policy and Resources Committee, elected by the States of Deliberation, functions as the executive; the 5 members all have equal voting rights
cabinet: none
elections/appointments: the monarchy is hereditary; lieutenant governor and bailiff appointed by the monarch; chief minister, who is the president of the Policy and Resources Committee indirectly elected by the States of Deliberation for a 4-year term; last held on 6 May 2016 (next to be held in June 2020)
election results: Gavin ST PIER (independent) elected president of the Policy and Resources Committee and chief minister

Legislative branch: *description:* unicameral States of Deliberation (40 seats; 38 People's Deputies and 2 representatives of the States of Alderney; members directly elected by majority vote to serve 4-year terms); note—non-voting members include the bailiff (presiding officer), attorney-general, and solicitor-general
elections: last held on 27 April 2016 (next to be held in June 2020)
election results: percent of vote—NA; seats—independent 38; composition—men 27, women 13, percent of women 32.5%

Judicial branch: *highest courts:* Guernsey Court of Appeal (consists of the Bailiff of Guernsey, who is the ex-officio president of the Guernsey Court of Appeal, and at least 12 judges); Royal Court (organized into 3 divisions—Full Court sits with 1 judge and 7 to 12 jurats acting as judges of fact, Ordinary Court sits with 1 judge and normally 3 jurats, and Matrimonial Causes Division sits with 1 judge and 4 jurats); note—appeals beyond Guernsey courts are heard by the Judicial Committee of the Privy Council (in London) *judge selection and term of office:* Royal Court Bailiff, Deputy Bailiff, and Court of Appeal justices appointed by the British Crown and hold office at Her Majesty's pleasure; jurats elected by the States of Election, a body chaired by the Bailiff and a number of jurats

subordinate courts: Court of Alderney; Court of the Seneschal of Sark; Magistrates' Court (includes Juvenile Court); Contracts Court; Ecclesiastical Court; Court of Chief Pleas

Political parties and leaders: none; all independents

International organization participation: UPU

Diplomatic representation in the US: none (British crown dependency)

Diplomatic representation from the US: none (British crown dependency)

Flag description: white with the red cross of Saint George (patron saint of England) extending to the edges of the flag and a yellow equalarmed cross of William the Conqueror superimposed on the Saint George cross; the red cross represents the old ties with England and the fact that Guernsey is a British Crown dependency; the gold cross is a replica of the one used by Duke William of Normandy at the Battle of Hastings in 1066

National symbol(s): *Guernsey cow, donkey; national colors:* red, white, yellow

National anthem: *name:* "Sarnia Cherie" (Guernsey Dear)
lyrics/music: George DEIGHTON/Domencio SANTANGELO
note: adopted 1911; serves as a local anthem; as a British crown dependency "God Save the Queen" is official (see United Kingdom)

ECONOMY

Economy—overview: Financial services accounted for about 21% of employment and about 32% of total income in 2016 in this tiny, prosperous Channel Island economy. Construction, manufacturing, and horticulture, mainly tomatoes and cut flowers, have been declining. Financial services, professional services, tourism, retail, and the public sector have been growing. Light tax and death duties make Guernsey a popular offshore financial center.

GDP (purchasing power parity): $3.465 billion (2015 est.)
$3.451 billion (2014 est.)
$3.42 billion (2013 est.)
note: data are in 2015 dollars
country comparison to the world: 184

GDP (official exchange rate): $2.742 billion (2005 est.)

GDP—real growth rate:
0.4% (2015 est.)
1.2% (2014 est.)

4.2% (2012 est.)
country comparison to the world: 187

GDP—per capita (PPP): $52,500 (2014 est.)
country comparison to the world: 24

GDP—composition, by sector of origin:
agriculture: 3% (2000)
industry: 10% (2000)
services: 87% (2000)

Agriculture—products: tomatoes, greenhouse flowers, sweet peppers, eggplant, fruit; Guernsey cattle

Industries: tourism, banking

Industrial production growth rate: NA

Labor force: 31,470 (March 2006)
country comparison to the world: 203

Unemployment rate: 1.2% (2016 est.)
country comparison to the world: 14

Population below poverty line: NA

Household income or consumption by percentage share: *lowest 10%:* NA
highest 10%: NA

Budget: *revenues:* 563.6 million (2005)
expenditures: 530.9 million (2005 est.)

Taxes and other revenues: 20.6% (of GDP) (2005)
country comparison to the world: 146

Budget surplus (+) or deficit (-): 1.2% (of GDP) (2005)
country comparison to the world: 29

Fiscal year: calendar year

Inflation rate (consumer prices): 3.4% (June 2006 est.)
country comparison to the world: 141

Exports: NA

Exports—commodities: tomatoes, flowers and ferns, sweet peppers, eggplant, other vegetables

Imports: NA

Imports—commodities: coal, gasoline, oil, machinery, and equipment

Debt—external: NA

Exchange rates: Guernsey pound per US dollar
0.7836 (2017 est.)
0.738 (2016 est.)
0.738 (2015)
0.6542 (2014)
0.607 (2013)

ENERGY

Electricity access: electrification—total population: 100% (2020)

COMMUNICATIONS

Telephones—fixed lines: *total subscriptions:* 36,547
subscriptions per 100 inhabitants: 60 (July 2016 est.)

country comparison to the world: 165

Telephones—mobile cellular: *total subscriptions:* 71,249
subscriptions per 100 inhabitants: 113 (July 2016 est.)
country comparison to the world: 201

Telecommunication systems: *general assessment:* high performance global connections with quality service; connections to major cities around the world to rival and attract future investment and future needs of islanders and businesses (2018)
domestic: fixed-line 60 per 100 and mobile-cellular 113 per 100 persons (2019)
international: country code—44; landing points for Guernsey-Jersey, HUGO, INGRID, Channel Islands -9 Liberty and UK-Channel Islands-7 submarine cable to UK and France (2019)
note: the COVID-19 outbreak is negatively impacting telecommunications production and supply chains globally; consumer spending on telecom devices and services has also slowed due to the pandemic's effect on economies worldwide; overall progress towards improvements in all facets of the telecom industry—mobile, fixed-line, broadband, submarine cable and satellite—has moderated

Broadcast media: multiple UK terrestrial TV broadcasts are received via a transmitter in Jersey with relays in Jersey, Guernsey, and Alderney; satellite packages are available; BBC Radio Guernsey and 1 other radio station operating

Internet country code: .gg

Internet users: *total:* 55,050
percent of population: 83.3% (July 2016 est.)
country comparison to the world: 195

TRANSPORTATION

National air transport system: number of registered air carriers: 1 (registered in UK) (2020)
inventory of registered aircraft operated by air carriers: 9 (registered in UK)

Airports: 2 (2013)
country comparison to the world: 199

Airports—with paved runways: *total:* 2 (2019)
1,524 to 2,437 m: 1
under 914 m: 1

Roadways: *total:* 260 km (2017)
country comparison to the world: 204

Ports and terminals: *major seaport(s):* Braye Bay, Saint Peter Port

MILITARY AND SECURITY

Military—note: defense is the responsibility of the UK

TRANSNATIONAL ISSUES

Disputes—international: none

GUINEA

INTRODUCTION

Background: Guinea is at a turning point after decades of authoritarian rule since gaining its independence from France in 1958. Sekou TOURE ruled the country as president from independence to his death in 1984. Lansana CONTE came to power in 1984 when the military seized the government after TOURE's death. Gen. CONTE organized and won presidential elections in 1993, 1998, and 2003, though results were questionable due to a lack in transparency and neutrality in the electoral process. Upon CONTE's death in December 2008, Capt. Moussa Dadis CAMARA led a military coup, seizing power and suspending the constitution. His unwillingness to yield to domestic and international pressure to step down led to heightened political tensions that peaked in September 2009 when presidential guards opened fire on an opposition rally killing more than 150 people. In early December 2009, CAMARA was wounded in an assassination attempt and exiled to Burkina Faso. A transitional government led by Gen. Sekouba KONATE paved the way for Guinea's transition to a fledgling democracy. The country held its first free and competitive democratic presidential and legislative elections in 2010 and 2013 respectively, and in October 2015 held a second consecutive presidential election. Alpha CONDE was reelected to a second five-year term as president in 2015, and the National Assembly was seated in January 2014.CONDE's first cabinet is the first all-civilian government in Guinea. The country held a successful political dialogue in August and September 2016 that brought together the government and opposition to address long-standing tensions. Local elections were held in February 2018, and disputed results in some of the races resulted in ongoing protests against CONDE's government.

GEOGRAPHY

Location: Western Africa, bordering the North Atlantic Ocean, between Guinea- Bissau and Sierra Leone

Geographic coordinates: 11 00 N, 10 00 W

Map references: Africa

Area: *total:* 245,857 sq km
land: 245,717 sq km
water: 140 sq km
country comparison to the world: 80

Area—comparative: slightly smaller than Oregon; slightly larger than twice the size of Pennsylvania

Land boundaries: *total:* 4,046 km

border countries (6): Cote d'Ivoire 816 km, Guinea-Bissau 421 km, Liberia 590 km, Mali 1062 km, Senegal 363 km, Sierra Leone 794 km

Coastline: 320 km

Maritime claims: *territorial sea:* 12 nm

exclusive economic zone: 200 nm

Climate: generally hot and humid; monsoon-al-type rainy season (June to November) with southwesterly winds; dry season (December to May) with northeasterly harmattan winds

Terrain: generally flat coastal plain, hilly to mountainous interior

Elevation: *mean elevation:* 472 m
lowest point: Atlantic Ocean 0 m
highest point: Mont Nimba 1,752 m

Natural resources: bauxite, iron ore, diamonds, gold, uranium, hydropower, fish, salt

Land use: *agricultural land:* 58.1% (2011 est.)
arable land: 11.8% (2011 est.) / permanent crops: 2.8% (2011 est.) / permanent pasture: 43.5% (2011 est.)
forest: 26.5% (2011 est.)
other: 15.4% (2011 est.)

Irrigated land: 950 sq km (2012)

Population distribution: areas of highest density are in the west and south; interior is sparsely populated as shown in this population distribution map

Natural hazards: hot, dry, dusty harmattan haze may reduce visibility during dry season

Environment—current issues: deforestation; inadequate potable water; desertification; soil contamination and erosion; overfishing, overpopulation in forest region; poor mining practices lead to environmental damage; water pollution; improper waste disposal

Environment—international agreements: *party to:* Biodiversity, Climate Change, Climate Change-Kyoto Protocol, Desertification, Endangered Species, Hazardous Wastes, Law of the Sea, Ozone Layer Protection, Ship Pollution, Wetlands, Whaling
signed, but not ratified: none of the selected agreements

Geography—note: the Niger and its important tributary the Milo River have their sources in the Guinean highlands

PEOPLE AND SOCIETY

Population: 12,527,440 (July 2020 est.)
country comparison to the world: 76

Nationality: *noun:* Guinean(s)
adjective: Guinean

Ethnic groups: Fulani (Peuhl) 33.4%, Malinke 29.4%, Susu 21.2%, Guerze 7.8%, Kissi 6.2%, Toma 1.6%, other/foreign.4% (2018 est.)

Languages: French (official), Pular, Maninka, Susu, other native languages
note: about 40 languages are spoken; each ethnic group has its own language

Religions: Muslim 89.1%, Christian 6.8%, animist 1.6%, other.1%, none 2.4% (2014 est.)

Demographic profile: Guinea's strong population growth is a result of declining mortality rates and sustained elevated fertility. The population growth rate was somewhat tempered in the 2000s because of a period of net outmigration. Although life expectancy and mortality rates have improved over the last two decades, the nearly universal practice of female genital cutting continues to contribute to high infant and maternal mortality rates. Guinea's total fertility remains high at about 5 children per woman because of the ongoing preference for larger families, low contraceptive usage and availability, a lack of educational attainment and empowerment among women, and poverty. A lack of literacy and vocational training programs limit job prospects for youths, but even those with university degrees often have no option but to work in the informal sector. About 60% of the country's large youth population is unemployed.

Tensions and refugees have spilled over Guinea's borders with Sierra Leone, Liberia, and Cote d'Ivoire. During the 1990s Guinea harbored as many as half a million refugees from Sierra Leone and Liberia, more refugees than any other African country for much of that decade. About half sought refuge in the volatile "Parrot's Beak" region of southwest Guinea, a wedge of land jutting into Sierra Leone near the Liberian border. Many were relocated within Guinea in the early 2000s because the area suffered repeated cross-border attacks from various government and rebel forces, as well as anti-refugee violence.

Age structure: *0-14 years:* 41.2% (male 2,601,221/female 2,559,918)
15-24 years: 19.32% (male 1,215,654/female 1,204,366)
25-54 years: 30.85% (male 1,933,141/female 1,930,977)
55-64 years: 4.73% (male 287,448/female 305,420)

65 years and over: 3.91% (male 218,803/female 270,492) (2020 est.)

Dependency ratios: *total dependency ratio:* 85.2
youth dependency ratio: 79.7
elderly dependency ratio: 5.5
potential support ratio: 18.3 (2020 est.)

Median age: *total:* 19.1 years
male: 18.9 years
female: 19.4 years (2020 est.)
country comparison to the world: 206

Population growth rate: 2.76% (2020 est.)
country comparison to the world: 13

Birth rate: 36.1 births/1,000 population (2020 est.)
country comparison to the world: 18

Death rate: 8.4 deaths/1,000 population (2020 est.)
country comparison to the world: 77

Net migration rate: 0 migrant(s)/1,000 population (2020 est.)
country comparison to the world: 83

Population distribution: areas of highest density are in the west and south; interior is sparsely populated as shown in this population distribution map

Urbanization: *urban population:* 36.5% of total population (2020)
rate of urbanization: 3.54% annual rate of change (2015-20 est.)
total population growth rate v. urban population growth rate, 2000-2030:

Major urban areas—population: 1.938 million CONAKRY (capital) (2020)

Sex ratio: *at birth:* 1.03 male(s)/female
0-14 years: 1.02 male(s)/female
15-24 years: 1.01 male(s)/female
25-54 years: 1 male(s)/female
55-64 years: 0.94 male(s)/female
65 years and over: 0.81 male(s)/female
total population: 1 male(s)/female (2020 est.)

Mother's mean age at first birth: 19.5 years (2018 est.)
note: median age at first birth among women 25-29

Maternal mortality rate: 576 deaths/100,000 live births (2017 est.)
country comparison to the world: 14

Infant mortality rate: *total:* 52.4 deaths/1,000 live births
male: 57.3 deaths/1,000 live births
female: 47.3 deaths/1,000 live births (2020 est.)
country comparison to the world: 19

Life expectancy at birth: *total population:* 63.2 years
male: 61.3 years
female: 65 years (2020 est.)
country comparison to the world: 204

Total fertility rate: 4.92 children born/woman (2020 est.)
country comparison to the world: 14

Contraceptive prevalence rate: 10.9% (2018)

Drinking water source:
improved:
urban: 97.9% of population

rural: 69.8% of population
total: 79.9% of population
unimproved:
urban: 2.1% of population
rural: 27.6% of population
total: 20.1% of population (2017 est.)

Current Health Expenditure: 4.1% (2017)

Physicians density: 0.08 physicians/1,000 population (2016)

Hospital bed density: 0.3 beds/1,000 population (2011)

Sanitation facility access:
improved:
urban: 85.6% of population
rural: 34.8% of population
total: 53% of population
unimproved:
urban: 14.4% of population
rural: 65.2% of population
total: 47% of population (2017 est.)

HIV/AIDS—adult prevalence rate: 1.4% (2019 est.)
country comparison to the world: 32

HIV/AIDS—people living with HIV/AIDS: 110,000 (2019 est.)
country comparison to the world: 42

HIV/AIDS—deaths: 3,100 (2019 est.)
country comparison to the world: 34

Major infectious diseases: *degree of risk:* very high (2020)
food or waterborne diseases: bacterial and protozoal diarrhea, hepatitis A, and typhoid fever
vectorborne diseases: malaria, dengue fever, and yellow fever
water contact diseases: schistosomiasis
animal contact diseases: rabies
aerosolized dust or soil contact diseases: Lassa fever (2016)

Obesity—adult prevalence rate: 7.7% (2016)
country comparison to the world: 158

Children under the age of 5 years underweight: 16.3% (2018)
country comparison to the world: 36

Education expenditures: 2.2% of GDP (2017)
country comparison to the world: 165

Literacy: *definition:* age 15 and over can read and write
total population: 30.4%
male: 38.1%
female: 22.8% (2015)

School life expectancy (primary to tertiary education): *total:* 9 years
male: 10 years
female: 8 years (2014)

Unemployment, youth ages 15-24: *total:* 1%
male: 1.5%
female: 0.6% (2012 est.)
country comparison to the world: 178

COUNTRY NAME: GOVERNMENT

conventional long form: Republic of Guinea
conventional short form: Guinea

local long form: Republique de Guinee
local short form: Guinee
former: French Guinea
etymology: the country is named after the Guinea region of West Africa that lies along the Gulf of Guinea and stretches north to the Sahel

Government type: presidential republic

Capital: *name:* Conakry

geographic coordinates: 9 30 N, 13 42 W
time difference: UTC 0 (5 hours ahead of Washington, DC, during Standard Time)

etymology: according to tradition, the name derives from the fusion of the name "Cona," a Baga wine and cheese producer who lived on Tombo Island (the original site of the present-day capital), and the word "nakiri," which in Susu means "the other bank" or "the other side"; supposedly, Baga's palm grove produced the best wine on the island and people traveling to sample his vintage, would say: " I am going to Cona, on the other bank (Cona- nakiri)," which over time became Conakry

Administrative divisions: 7 regions administrative and 1 gouvenorat*; Boke, Conakry*, Faranah, Kankan, Kindia, Labe, Mamou, N'Zerekore

Independence: 2 October 1958 (from France)

National holiday: Independence Day, 2 October (1958)

Constitution: *history:* previous 1958, 1990; latest promulgated 19 April 2010, approved 7 May 2010; note—in late December 2019, President CONDE announced a new draft constitution
amendments: proposed by the National Assembly or by the president of the republic; consideration of proposals requires approval by simple majority vote by the Assembly; passage requires approval in referendum; the president can opt to submit amendments directly to the Assembly, in which case approval requires at least two- thirds majority vote; amended 2020

Legal system: civil law system based on the French model

International law organization participation: accepts compulsory ICJ jurisdiction with reservations; accepts ICCt jurisdiction

Citizenship: *citizenship by birth:* no
citizenship by descent only: at least one parent must be a citizen of Guinea
dual citizenship recognized: no
residency requirement for naturalization: na

Suffrage: 18 years of age; universal

Executive branch: *chief of state:* President Alpha CONDE (since 21 December 2010)

head of government: Prime Minister Ibrahima FOFANA (since 22 May 2018)
cabinet: Council of Ministers appointed by the president
elections/appointments: president directly elected by absolute majority popular vote in 2 rounds if needed for a 5-year term (eligible for a second term); election last held on 18 October 2020 (next to be held in October 2025); prime minister appointed by the president

election results: Alpha CONDE reelected president in the first round; percent of vote—Alpha CONDE (RPG) 59.5%, Cellou Dalein DIALLO (UFDG) 33.5%, other 7%

Legislative branch: *description:* unicameral People's National Assembly or Assemblee Nationale Populaire (114 seats; 76 members directly elected in a single nationwide constituency by proportional representation vote and 38 directly elected in single-seat constituencies by simple majority vote; members serve 5-year terms) *elections:* last held on 28 September 2013 (next to be held 1 March 2020)
election results: percent of vote by party—NA; seats by party—RPG 53, UFDG 37, UFR 10, PEDN 2, UPG 2, other 10; composition—men 89, women 25, percent of women 21.9%

Judicial branch: *highest courts:* Supreme Court or Cour Supreme (organized into Administrative Chamber and Civil, Penal, and Social Chamber; court consists of the first president, 2 chamber presidents, 10 councilors, the solicitor general, and NA deputies); Constitutional Court (consists of 9 members)
judge selection and term of office: Supreme Court first president appointed by the national president after consultation with the National Assembly; other members appointed by presidential decree; members serve until age 65; Constitutional Court member appointments—2 by the National Assembly and the president of the republic, 3 experienced judges designated by their peers, 1 experienced lawyer, 1 university professor with expertise in public law designated by peers, and 2 experienced representatives of the Independent National Institution of Human Rights; members serve single 9-year terms
subordinate courts: Court of Appeal or Cour d'Appel; High Court of Justice or Cour d'Assises; Court of Account (Court of Auditors); Courts of First Instance (Tribunal de Premiere Instance); labor court; military tribunal; justices of the peace; specialized courts

Political parties and leaders:
Bloc Liberal or BL [Faya MILLIMONO]
National Party for Hope and Development or PEDN [Lansana KOUYATE]
Rally for the Guinean People or RPG [Alpha CONDE]
Union for the Progress of Guinea or UPG
Union of Democratic Forces of Guinea or UFDG [Cellou Dalein DIALLO]
Union of Republican Forces or UFR [Sidya TOURE]
Ruling party Rally of the Guinean People (*Rassemblement du Peuple Guinéen, RPG*) Opposition parties African Democratic Party of Guinea (*Parti démocratique africain de Guinée*) Party of Unity and Progress (Parti de l'Unité et du Progrès, PUP) Union for Progress and Renewal (*Union pour le Progrès et le Renouveau, UPR*) Union for Progress of Guinea (*Union pour le Progrès de la Guinée, UPG*) Democratic Party of Guinea-African Democratic Rally (*Parti Démocratique de*

Guinée-Rassemblement Démocratique Africain, PDG- RDA) National Alliance for Progress (*Alliance Nationale pour le Progrès, ANP*) Party of the Union for Development (*Parti de l'Union pour le Développement, PUD*) Union of Democratic Forces of Guinea (*Union des Forces Démocratiques de Guinée, UFDG*), led by Cellou Dalein Diallo Union of Republican Forces (*Union des Forces Républicaines, UFR*) the Party of Democrats for Hope ("PADES") Led by Dr Ousmane Kaba

International organization participation: ACP, AfDB, AU, ECOWAS, EITI (compliant country), FAO, G- 77, IBRD, ICAO, ICCt, ICRM, IDA, IDB, IFAD, IFC, IFRCS, ILO, IMF, IMO, Interpol, IOC, IOM, IPU, ISO (correspondent), ITSO, ITU, ITUC (NGOs), MIGA, MINURSO, MINUSMA, MONUSCO, NAM, OIC, OIF, OPCW, UN, UNCTAD, UNESCO, UNHCR, UNIDO, UNISFA, UNMISS, UNOCI, UNWTO, UPU, WCO, WFTU (NGOs), WHO, WIPO, WMO, WTO

Diplomatic representation in the US: *chief of mission:* Ambassador Kerfalla YANSANE (since 24 January 2018)
chancery: 2112 Leroy Place NW, Washington, DC 20008
telephone: [1] (202) 986-4300
FAX: [1] (202) 986-3800

Diplomatic representation from the US: *chief of mission:* Ambassador Simon HENSHAW (since 4 March 2019)
telephone: [224] 655-10-40-00
embassy: Transversale # 2, Center Administratif de Koloma, Commune de Ratoma, Conakry
mailing address: P. O. Box 603, Transversale No. 2, Centre Administratif de Koloma, Commune de Ratoma, Conakry
FAX: [224] 655-10-42-97

Flag description: three equal vertical bands of red (hoist side), yellow, and green; red represents the people's sacrifice for liberation and work; yellow stands for the sun, for the riches of the earth, and for justice; green symbolizes the country's vegetation and unity
note: uses the popular Pan- African colors of Ethiopia; the colors from left to right are the reverse of those on the flags of neighboring Mali and Senegal

National symbol(s): *elephant; national colors:* red, yellow, green

National anthem: *name:* "Liberte" (Liberty)
lyrics/music: unknown/Fodeba KEITA
note: adopted 1958

ECONOMY

Economy—overview: Guinea is a poor country of approximately 12.9 million people in 2016 that possesses the world's largest reserves of bauxite and largest untapped high-grade iron ore reserves, as well as gold and diamonds. In addition, Guinea has fertile soil, ample rainfall, and is the source of several West African rivers, including the Senegal,

Niger, and Gambia. Guinea's hydro potential is enormous and the country could be a major exporter of electricity. The country also has tremendous agriculture potential. Gold, bauxite, and diamonds are Guinea's main exports. International investors have shown interest in Guinea's unexplored mineral reserves, which have the potential to propel Guinea's future growth.

Following the death of long-term President Lansana CONTE in 2008 and the coup that followed, international donors, including the G-8, the IMF, and the World Bank, significantly curtailed their development programs in Guinea. However, the IMF approved a 3-year Extended Credit Facility arrangement in 2012, following the December 2010 presidential elections. In September 2012, Guinea achieved Heavily Indebted Poor Countries completion point status. Future access to international assistance and investment will depend on the government's ability to be transparent, combat corruption, reform its banking system, improve its business environment, and build infrastructure. In April 2013, the government amended its mining code to reduce taxes and royalties. In 2014, Guinea complied with requirements of the Extractive Industries Transparency Initiative by publishing its mining contracts. Guinea completed its program with the IMF in October 2016 even though some targeted reforms have been delayed. Currently Guinea is negotiating a new IMF program which will be based on Guinea's new five-year economic plan, focusing on the development of higher value-added products, including from the agro- business sector and development of the rural economy.

Political instability, a reintroduction of the Ebola virus epidemic, low international commodity prices, and an enduring legacy of corruption, inefficiency, and lack of government transparency are factors that could impact Guinea's future growth. Economic recovery will be a long process while the government adjusts to lower inflows of international donor aid following the surge of Ebola- related emergency support. Ebola stalled promising economic growth in the 2014-15 period and impeded several projects, such as offshore oil exploration and the Simandou iron ore project. The economy, however, grew by 6.6% in 2016 and 6.7% in 2017, mainly due to growth from bauxite mining and thermal energy generation as well as the resiliency of the agricultural sector. The 240-megawatt Kaleta Dam, inaugurated in September 2015, has expanded access to electricity for residents of Conakry. An combined with fears of Ebola virus, continue to undermine Guinea's economic viability.

Guinea's iron ore industry took a hit in 2016 when investors in the Simandou iron ore project announced plans to divest from the project. In 2017, agriculture output and public investment boosted economic growth, while the mining sector continued to play a prominent role in economic performance.

Successive governments have failed to address the country's crumbling infrastructure. Guinea suffers from chronic electricity shortages; poor roads, rail lines and bridges; and a lack of access to clean water—all of which continue to plague economic development. The present government, led by President Alpha CONDE, is working to create an environment to attract foreign investment and hopes to have greater participation from western countries and firms in Guinea's economic development.

GDP (purchasing power parity):
$27.97 billion (2017 est.)
$25.84 billion (2016 est.)
$23.39 billion (2015 est.)
note: data are in 2017 dollars
country comparison to the world: 138

GDP (official exchange rate):
$10.25 billion (2017 est.)

GDP—real growth rate:
8.2% (2017 est.)
10.5% (2016 est.)
3.8% (2015 est.)
country comparison to the world: 8

GDP—per capita (PPP):
$2,200 (2017 est.)
$2,000 (2016 est.)
$1,900 (2015 est.)
note: data are in 2017 dollars
country comparison to the world: 205

Gross national saving:
5.1% of GDP (2017 est.)
-6.3% of GDP (2016 est.)
-5.3% of GDP (2015 est.)
country comparison to the world: 176

GDP—composition, by end use:
household consumption: 80.8% (2017 est.)
government consumption: 6.6% (2017 est.)
investment in fixed capital: 9.1% (2017 est.)
investment in inventories: 18.5% (2017 est.)
exports of goods and services: 21.9% (2017 est.)
imports of goods and services: -36.9% (2017 est.)

GDP—composition, by sector of origin:
agriculture: 19.8% (2017 est.)
industry: 32.1% (2017 est.)
services: 48.1% (2017 est.)

Agriculture—products: rice, coffee, pineapples, mangoes, palm kernels, cocoa, cassava (manioc, tapioca), bananas, potatoes, sweet potatoes; cattle, sheep, goats; timber

Industries: bauxite, gold, diamonds, iron ore; light manufacturing, agricultural processing

Industrial production growth rate: 11% (2017 est.)
country comparison to the world: 9

Labor force:
5.558 million (2017 est.)
country comparison to the world: 73

Labor force—by occupation: *agriculture:* 76%
industry: 24% (2006 est.)

Unemployment rate: 2.7% (2017 est.)
2.8% (2016 est.)
country comparison to the world: 30

Population below poverty line: 47% (2006 est.)

Household income or consumption by percentage share: *lowest 10%:* 2.7%

highest 10%: 30.3% (2007)

Budget: *revenues:* 1.7 billion (2017 est.)
expenditures: 1.748 billion (2017 est.)

Taxes and other revenues: 16.6% (of GDP) (2017 est.)
country comparison to the world: 178

Budget surplus (+) or deficit (-): -0.5% (of GDP) (2017 est.)
country comparison to the world: 61

Public debt: 37.9% of GDP (2017 est.)
41.8% of GDP (2016 est.)
country comparison to the world: 137

Fiscal year: calendar year

Inflation rate (consumer prices): 8.9% (2017 est.)
8.2% (2016 est.)
country comparison to the world: 200

Current account balance: -$705 million (2017 est.)
-$2.705 billion (2016 est.)
country comparison to the world: 132

Exports: $3.514 billion (2017 est.)
$1.954 billion (2016 est.)
country comparison to the world: 123

Exports—partners: China 35.8%, Ghana 20.1%, UAE 11.6%, India 4.3% (2017)

Exports—commodities: bauxite, gold, diamonds, coffee, fish, agricultural products

Imports: $4.799 billion (2017 est.)
$4.43 billion (2016 est.)
country comparison to the world: 133

Imports—commodities: petroleum products, metals, machinery, transport equipment, textiles, grain and other foodstuffs

Imports—partners: Netherlands 17.2%, China 13.2%, India 11.8%, Belgium 10%, France 6.9%, UAE 4.5% (2017)

Reserves of foreign exchange and gold: $331.8 million (31 December 2017 est.)
$383.4 million (31 December 2016 est.)
country comparison to the world: 165

Debt—external: $1.458 billion (31 December 2017 est.)
$1.462 billion (31 December 2016 est.)
country comparison to the world: 159

Exchange rates: Guinean francs (GNF) per US dollar –
9,230 (2017 est.)
9,085 (2016 est.)
9,085 (2015 est.)
7,485.5 (2014 est.)
7,014.1 (2013 est.)

ENERGY

Electricity access: *population without electricity:* 7 million (2019)
electrification—total population: 46% (2019)
electrification—urban areas: 84% (2019)
electrification—rural areas: 24% (2019)

Electricity—production: 598 million kWh (2016 est.)
country comparison to the world: 162

Electricity—consumption: 556.1 million kWh (2016 est.)
country comparison to the world: 168

Electricity—exports: 0 kWh (2016 est.)
country comparison to the world: 144

Electricity—imports: 0 kWh (2016 est.)
country comparison to the world: 156

Electricity—installed generating capacity: 550,000 kW (2016 est.)
country comparison to the world: 145

Electricity—from fossil fuels: 33% of total installed capacity (2016 est.)
country comparison to the world: 182

Electricity—from nuclear fuels: 0% of total installed capacity (2017 est.)
country comparison to the world: 103

Electricity—from hydroelectric plants: 67% of total installed capacity (2017 est.)
country comparison to the world: 20

Electricity—from other renewable sources: 0% of total installed capacity (2017 est.)
country comparison to the world: 190

Crude oil—production: 0 bbl/day (2018 est.)
country comparison to the world: 145

Crude oil—exports: 0 bbl/day (2015 est.)
country comparison to the world: 133

Crude oil—imports: 0 bbl/day (2015 est.)
country comparison to the world: 137

Crude oil—proved reserves: 0 bbl (1 January 2018 est.)
country comparison to the world: 141

Refined petroleum products—production: 0 bbl/day (2017 est.)
country comparison to the world: 153

Refined petroleum products—consumption: 19,000 bbl/day (2016 est.)
country comparison to the world: 143

Refined petroleum products—exports: 0 bbl/day (2015 est.)
country comparison to the world: 162

Refined petroleum products—imports: 18,460 bbl/day (2015 est.)
country comparison to the world: 128

Natural gas—production: 0 cu m (2017 est.)
country comparison to the world: 142

Natural gas—consumption: 0 cu m (2017 est.)
country comparison to the world: 155

Natural gas—exports: 0 cu m (2017 est.)
country comparison to the world: 117

Natural gas—imports: 0 cu m (2017 est.)
country comparison to the world: 135

Natural gas—proved reserves: 0 cu m (1 January 2014 est.)
country comparison to the world: 143

Carbon dioxide emissions from consumption of energy: 2.794 million Mt (2017 est.)
country comparison to the world: 149

COMMUNICATIONS

Telephones—fixed lines: *total subscriptions:* 0

subscriptions per 100 inhabitants: less than 1 (2018 est.)
country comparison to the world: 223

Telephones—mobile cellular: *total subscriptions:* 12,283,911

subscriptions per 100 inhabitants: 100.8 (2019 est.)

country comparison to the world: 74

Telecommunication systems: *general assessment:* huge improvement over the last ten years; in May 2019, 4G Wi-Fi was launched in the capital; the regional administrative centers all have 3G access; the 2018 set up of an IXP (Internet Exchange Point) reduced the cost of Internet bandwidth and improved infrastructure; a National Backbone Network is nearing completion to connect administrative centers (2020)

domestic: there is national coverage and Conakry is reasonably well- served; coverage elsewhere remains inadequate but is improving; fixed-line teledensity is less than 1 per 100 persons; mobile-cellular subscribership is expanding rapidly and now 101 per 100 persons (2019)

international: country code—224; ACE submarine cable connecting Guinea with 20 landing points in Western and South Africa and Europe; satellite earth station—1 Intelsat (Atlantic Ocean (2019)

note: the COVID- 19 outbreak is negatively impacting telecommunications production and supply chains globally; consumer spending on telecom devices and services has also slowed due to the pandemic's effect on economies worldwide; overall progress towards improvements in all facets of the telecom industry—mobile, fixed-line, broadband, submarine cable and satellite—has moderated

Broadcast media: government maintains marginal control over broadcast media; single state- run TV station; state- run radio broadcast station also operates several stations in rural areas; a dozen private television stations; a steadily increasing number of privately owned radio stations, nearly all in Conakry, and about a dozen community radio stations; foreign TV programming available via satellite and cable subscription services (2019)

Internet country code: .gn

Internet users: *total:* 2,133,974

percent of population: 18% (July 2018 est.)
country comparison to the world: 121

Broadband—fixed subscriptions:
total: 1,213
subscriptions per 100 inhabitants: less than 1 (2018 est.)
country comparison to the world: 194

TRANSPORTATION

Civil aircraft registration country code prefix: 3X (2016)

Airports: 16 (2013)
country comparison to the world: 143

Airports—with paved runways:
total: 4 (2019)
over 3,047 m: 1
1,524 to 2,437 m: 3

Airports—with unpaved runways:
total: 12 (2013)
1,524 to 2,437 m: 7 (2013)
914 to 1,523 m: 3 (2013)
under 914 m: 2 (2013)

Railways:
total: 1,086 km (2017)
standard gauge: 279 km 1.435-m gauge (2017)
narrow gauge: 807 km 1.000-m gauge (2017)
country comparison to the world: 88

Roadways:
total: 44,301 km (2018)
paved: 3,346 km (2018)
unpaved: 40,955 km (2018)
country comparison to the world: 85

Waterways: 1,300 km (navigable by shallow-draft native craft in the northern part of the Niger River system) (2011)
country comparison to the world: 54

Merchant marine: *total:* 2
by type: other 2 (2019)
country comparison to the world: 174

Ports and terminals: *major seaport(s):* Conakry, Kamsar

MILITARY AND SECURITY

Military and security forces: *National Armed Forces:* Army, Guinean Navy (Armee de Mer or Marine Guineenne, includes Marines), Guinean Air Force (Force Aerienne de Guinee), Presidential Security Battalion (Battailon Autonome de la Sécurité Presidentielle, BASP), Gendarmerie, People's Militia (Reserves) (2019)

Military expenditures:
2% of GDP (2019)
2.3% of GDP (2018)
2.5% of GDP (2017)
2.5% of GDP (2016)
3.3% of GDP (2015)
country comparison to the world: 51

Military and security service personnel strengths: Guinean National Armed Forces are comprised of approximately 13,000 active personnel (est. 9,000 Army; 400 Navy; 800 Air Force; 1,400 Gendarmerie; 1,600 Republican Guard) (2019)

Military equipment inventories and acquisitions: the inventory of the Guinean military consists largely of ageing and outdated (mostly Soviet-era) equipment; since 2010, it has received a limited amount of equipment from France, Russia, and South Africa (2019 est.)

Military deployments: 1,500 Mali (MINUSMA) (2020)

Military service age and obligation: no compulsory military service (2017)

TRANSNATIONAL ISSUES

Disputes—international: Sierra Leone considers Guinea's definition of the flood plain limits to define the left bank boundary of the Makona and Moa Rivers excessive and protests Guinea's continued occupation of these lands, including the hamlet of Yenga, occupied since 1998

Trafficking in persons: *current situation:* Guinea is a source, transit, and, to a lesser extent, a destination country for men, women, and children subjected to forced labor and sex trafficking; the majority of trafficking victims are Guinean children, and trafficking is more prevalent among Guineans than foreign national migrants; Guinean girls are subjected to domestic servitude and commercial sexual exploitation, while boys are forced to beg or to work as street vendors, shoe shiners, or miners; Guinea is a source country and transit point for West African children forced to work as miners in the region; Guinean women and girls are subjected to domestic servitude and sex trafficking in West Africa, the Middle East, the US, and increasingly Europe, while Thai, Chinese, and Vietnamese women are forced into prostitution and some West Africans are forced into domestic servitude in Guinea

tier rating: Tier 2 Watch List – Guinea does not fully comply with the minimum standards for the elimination of trafficking; however, it is making significant efforts to do so; in 2014, Guinea was granted a waiver from an otherwise required downgrade to Tier 3 because its government has a written plan that, if implemented would constitute making significant efforts to bring itself into compliance with the minimum standards for the elimination of trafficking; no new investigations were conducted in 2014, and the one ongoing case led to the prosecution of four offenders for forced child labor, three of whom were convicted but given inadequate sentences for the crime; the government did not identify or provide protective services to victims and did not support NGOs that assisted victims but continued to refer child victims to NGOs on an ad hoc basis; Guinean law does not prohibit all forms of trafficking, excluding, for example, debt bondage; the 2014 Ebolavirus outbreak negatively affected Guinea's ability to address human trafficking (2015)

GUINEA-BISSAU

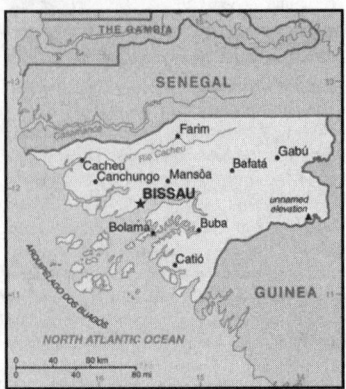

INTRODUCTION

Background: Since independence from Portugal in 1974, Guinea-Bissau has experienced considerable political and military upheaval. In 1980, a military coup established authoritarian General Joao Bernardo 'Nino' VIEIRA as president. Despite eventually setting a path to a market economy and multiparty system, VIEIRA's regime was characterized by the suppression of political opposition and the purging of political rivals. Several coup attempts through the 1980s and early 1990s failed to unseat him. In 1994 VIEIRA was elected president in the country's first free, multiparty election. A military mutiny and resulting civil war in 1998 eventually led to VIEIRA's ouster in May 1999. In February 2000, a transitional government turned over power to opposition leader Kumba YALA after he was elected president in transparent polling. In September 2003, after only three years in office, YALA was overthrown in a bloodless military coup, and businessman Henrique ROSA was sworn in as interim president. In 2005, former President VIEIRA was reelected, pledging to pursue economic development and national reconciliation; he was assassinated in March 2009. Malam Bacai SANHA was elected in an emergency election held in June 2009, but he passed away in January 2012 from a long-term illness. A military coup in April 2012 prevented Guinea-Bissau's second-round presidential election—to determine SANHA's successor—from taking place. Following mediation by the Economic Community of Western African States, a civilian transitional government assumed power in 2012 and remained until Jose Mario VAZ won a free and fair election in 2014. Beginning in 2015, a political dispute between factions in the ruling PAIGC party brought government gridlock. It was not until April 2018 that a consensus prime minister could be appointed, the national legislature reopened (having been closed for two years), and a new government formed under Prime Minister

Aristides GOMES. In March 2019, the government held legislative elections, voting in the PAIGC as the ruling party; however, President VAZ continues to perpetuate a political stalemate by refusing to name PAICG President Domingos SIMOES PEREIRA Prime Minister.

GEOGRAPHY

Location: Western Africa, bordering the North Atlantic Ocean, between Guinea and Senegal

Geographic coordinates: 12 00 N, 15 00 W

Map references:

Area: *total:* 36,125 sq km
land: 28,120 sq km
water: 8,005 sq km
country comparison to the world: 138

Area—comparative: slightly less than three times the size of Connecticut

Land boundaries: *total:* 762 km
border countries (2): Guinea 421 km, Senegal 341 km

Coastline: 350 km

Maritime claims: *territorial sea:* 12 nm
exclusive economic zone: 200 nm

Climate: tropical; generally hot and humid; monsoonal-type rainy season (June to November) with southwesterly winds; dry season (December to May) with northeasterly harmattan winds

Terrain: mostly low-lying coastal plain with a deeply indented estuarine coastline rising to savanna in east; numerous off-shore islands including the Arquipelago Dos Bijagos consisting of 18 main islands and many small islets

Elevation: *mean elevation:* 70 m
lowest point: Atlantic Ocean 0 m
highest point: unnamed elevation in the eastern part of the country 300 m

Natural resources: fish, timber, phosphates, bauxite, clay, granite, limestone, unexploited deposits of petroleum

Land use: *agricultural land:* 44.8% (2011 est.)
arable land: 8.2% (2011 est.) / permanent crops: 6.9% (2011 est.) / permanent pasture: 29.7% (2011 est.)
forest: 55.2% (2011 est.)
other: 0% (2011 est.)

Irrigated land: 250 sq km (2012)

Population distribution: approximately one-fifth of the population lives in the capital city of Bissau along the Atlantic coast; the remainder is distributed among the eight other, mainly rural, regions as shown in this population distribution map

Natural hazards: hot, dry, dusty harmattan haze may reduce visibility during dry season; brush fires

Environment—current issues: deforestation (rampant felling of trees for timber and agricultural purposes); soil erosion; overgrazing; overfishing

Environment—international agreements: *party to:* Biodiversity, Climate Change, Climate Change-Kyoto Protocol, Desertification, Endangered Species, Hazardous Wastes, Law of the Sea, Ozone Layer Protection, Wetlands

signed, but not ratified: none of the selected agreements

Geography—note: this small country is swampy along its western coast and low-lying inland

PEOPLE AND SOCIETY

Population: 1,927,104 (July 2020 est.)
country comparison to the world: 151

Nationality: *noun:* Bissau-Guinean(s)
adjective: Bissau-Guinean

Ethnic groups: Fulani 28.5%, Balanta 22.5%, Mandinga 14.7%, Papel 9.1%, Manjaco 8.3%, Beafada 3.5%, Mancanha 3.1%, Bijago 2.1%, Felupe 1.7%, Mansoanca 1.4%, Balanta Mane 1%, other 1.8%, none 2.2% (2008 est.)

Languages: Crioulo (lingua franca), Portuguese (official; largely used as a second or third language), Pular (a Fula language), Mandingo

Religions: Muslim 45.1%, Christian 22.1%, animist 14.9%, none 2%, unspecified 15.9% (2008 est.)

Demographic profile: Guinea-Bissau's young and growing population is sustained by high fertility; approximately 60% of the population is under the age of 25. Its large reproductive-age population and total fertility rate of more than 4 children per woman offsets the country's high infant and maternal mortality rates. The latter is among the world's highest because of the prevalence of early childbearing, a lack of birth spacing, the high percentage of births outside of health care facilities, and a shortage of medicines and supplies.

Guinea-Bissau's history of political instability, a civil war, and several coups (the latest in 2012) have resulted in a fragile state with a weak economy, high unemployment, rampant corruption, widespread poverty, and thriving drug and child trafficking. With the country lacking educational infrastructure, school funding and materials, and qualified teachers, and with the cultural emphasis placed on religious education, parents frequently send boys to study in residential Koranic schools (daaras) in Senegal and The Gambia. They often are extremely deprived and are forced into street begging or agricultural work by marabouts (Muslim religious teachers), who enrich themselves at the expense of the children. Boys who leave their marabouts often end up on the streets of Dakar or other large Senegalese towns and are vulnerable to even worse abuse. Some young men lacking in education and job prospects become involved in the flourishing international drug trade. Local drug use and associated violent crime are growing.

Age structure: *0-14 years:* 43.17% (male 417,810/female 414,105)
15-24 years: 20.38% (male 192,451/female 200,370)
25-54 years: 30.24% (male 275,416/female 307,387)
55-64 years: 3.12% (male 29,549/female 30,661)
65 years and over: 3.08% (male 25,291/female 34,064) (2020 est.)

Dependency ratios: *total dependency ratio:* 81.2
youth dependency ratio: 76
elderly dependency ratio: 5.2
potential support ratio: 19.1 (202 est.)

Median age: *total:* 18 years
male: 17.4 years
female: 18.6 years (2020 est.)
country comparison to the world: 215

Population growth rate: 2.51% (2020 est.)
country comparison to the world: 24

Birth rate: 36.9 births/1,000 population (2020 est.)
country comparison to the world: 14

Death rate: 7.9 deaths/1,000 population (2020 est.)
country comparison to the world: 94

Net migration rate:—3.8 migrant(s)/1,000 population (2020 est.)
country comparison to the world: 186

Population distribution: approximately one-fifth of the population lives in the capital city of Bissau along the Atlantic coast; the remainder is distributed among the eight other, mainly rural, regions as shown in this population distribution map

Urbanization: *urban population:* 44.2% of total population (2020)
rate of urbanization: 3.41% annual rate of change (2015-20 est.)
total population growth rate v. urban population growth rate, 2000-2030:

Major urban areas—population: 600,000 BISSAU (capital) (2020)

Sex ratio: *at birth:* 1.03 male(s)/female
0-14 years: 1.01 male(s)/female
15-24 years: 0.96 male(s)/female
25-54 years: 0.9 male(s)/female
55-64 years: 0.96 male(s)/female
65 years and over: 0.74 male(s)/female
total population: 0.95 male(s)/female (2020 est.)

Maternal mortality rate: 667 deaths/100,000 live births (2017 est.)
country comparison to the world: 8

Infant mortality rate: *total:* 51.9 deaths/1,000 live births
male: 57.9 deaths/1,000 live births
female: 45.7 deaths/1,000 live births (2020 est.)
country comparison to the world: 22

Life expectancy at birth: *total population:* 62.8 years
male: 60.6 years
female: 65.1 years (2020 est.)
country comparison to the world: 207

Total fertility rate: 4.75 children born/woman (2020 est.)
country comparison to the world: 17

Contraceptive prevalence rate: 16% (2014)

Drinking water source:
improved:
urban: 91.2% of population
rural: 60.3% of population
total: 73.5% of population
unimproved:
urban: 8.5% of population
rural: 39.7% of population
total: 26.5% of population (2017 est.)

Current Health Expenditure: 7.2% (2017)

Physicians density: 0.13 physicians/1,000 population (2016)

Hospital bed density: 1 beds/1,000 population (2009)

Sanitation facility access:
improved:
urban: 66.5% of population
rural: 13.4% of population
total: 36.2% of population
unimproved:
urban: 33.5% of population
rural: 86.6% of population
total: 63.8% of population (2017 est.)

HIV/AIDS—adult prevalence rate: 3.4% (2019 est.)
country comparison to the world: 16

HIV/AIDS—people living with HIV/AIDS: 40,000 (2019 est.)
country comparison to the world: 67

HIV/AIDS—deaths: 1,500 (2019 est.)
country comparison to the world: 50

Major infectious diseases: *degree of risk:* very high (2020)
food or waterborne diseases: bacterial and protozoal diarrhea, hepatitis A, and typhoid fever
vectorborne diseases: malaria, dengue fever, and yellow fever
water contact diseases: schistosomiasis
animal contact diseases: rabies

Obesity—adult prevalence rate: 9.5% (2016)
country comparison to the world: 144

Children under the age of 5 years underweight: 17% (2014)
country comparison to the world: 33

Education expenditures: 2.1% of GDP (2013)
country comparison to the world: 168

Literacy: *definition:* age 15 and over can read and write
total population: 59.9%
male: 71.8%
female: 48.3% (2015)

GOVERNMENT

Country name: *conventional long form:* Republic of Guinea-Bissau
conventional short form: Guinea-Bissau
local long form: Republica da Guine-Bissau
local short form: Guine-Bissau
former: Portuguese Guinea
etymology: the country is named after the Guinea region of West Africa that lies along the Gulf of Guinea and stretches north to the Sahel; "Bissau," the name of the capital city, distinguishes the country from neighboring Guinea

Government type: semi-presidential republic

Capital: *name:* Bissau
geographic coordinates: 11 51 N, 15 35 W
time difference: UTC 0 (5 hours ahead of Washington, DC, during Standard Time)
etymology: the meaning of Bissau is uncertain, it might be an alternative name for the Papel people who live in the area of the city of Bissau

Administrative divisions: 9 regions (regioes, singular—regiao); Bafata, Biombo, Bissau, Bolama/Bijagos, Cacheu, Gabu, Oio, Quinara, Tombali

Independence: 24 September 1973 (declared); 10 September 1974 (from Portugal)

National holiday: Independence Day, 24 September (1973)

Constitution: *history:* promulgated 16 May 1984; note—constitution suspended following military coup in April 2012 and restored in 2014
amendments: proposed by the National People's Assembly if supported by at least one third of its members, by the Council of State (a presidential consultant body), or by the government; passage requires approval by at least two-thirds majority vote of the Assembly; constitutional articles on the republican and secular form of government and national sovereignty cannot be amended; amended 1991, 1993, 1996

Legal system: mixed legal system of civil law, which incorporated Portuguese law at independence and influenced by Economic Community of West African States (ECOWAS), West African Economic and Monetary Union (UEMOA), African Francophone Public Law, and customary law

International law organization participation: accepts compulsory ICJ jurisdiction; non- party state to the ICCt

Citizenship: *citizenship by birth:* yes
citizenship by descent only: yes
dual citizenship recognized: no
residency requirement for naturalization: 5 years

Suffrage: 18 years of age; universal

Executive branch: *chief of state:* President Umaro Cissoko EMBALO (since 27 February 2020); note—President EMBALO was declared winner of the 29 December 2019 runoff presidential election by the electoral commission; however, on 28 February 2020, Cipriano CASSAMA was appointed as interim president by the parliament until the Supreme Court rules on the legitimacy of the elections due to alleged irregularities in voting; CASSAMA resigned the following day stating he had received death threats

head of government: Prime Minister Nuno NABIAM (since 27 February 2020)
cabinet: Cabinet nominated by the prime minister, appointed by the president
elections/appointments: president directly elected by absolute majority popular vote in 2

rounds if needed for a 5-year term; election last held on 24 November 2019 with a runoff on 29 December 2019 (next to be held in 2024); prime minister appointed by the president after consultation with party leaders in the National People's Assembly; note—the president cannot apply for a third consecutive term, nor during the 5 years following the end of the second term

election results: Umaro Sissoco EMBALO elected president in second round; percent of vote in first round—Domingos Simoes PEREIRA (PAIGC) 40.1%, Umaro Sissoco EMBALO (Madem G15) 27.7%, Nuno Gomez NABIAM (APU-PDGB) 13.2%, Jose Mario VAZ (independent) 12.4%, other 6.6%; percent of vote in second round— Umaro Sissoco EMBALO 53.6%, Domingos Simoes PEREIRA 46.5%

Legislative branch: *description:* unicameral National People's Assembly or Assembleia Nacional Popular (102 seats; 100 members directly elected in 27 multi- seat constituencies by closed party-list proportional representation vote and 2 elected in single- seat constituencies for citizens living abroad (1 for Africa, 1 for Europe); all members serve 4-year terms)

elections: last held on 10 March 2019 (next to be held in March 2023)

election results: percent of vote by party—PAIGC 35.2%, Madem G-15 21.1%, PRS 21.1%, other 22.6%; seats by party—PAIGC 47, Madem G-15 27, PRS 21, other 7; composition—men 88, women 14, percent of women 13.7%

Judicial branch: *highest courts:* Supreme Court or Supremo Tribunal de Justica (consists of 9 judges and organized into Civil, Criminal, and Social and Administrative Disputes Chambers); note—the Supreme Court has both appellate and constitutional jurisdiction

judge selection and term of office: judges nominated by the Higher Council of the Magistrate, a major government organ responsible for judge appointments, dismissals, and judiciary discipline; judges appointed by the president for life

subordinate courts: Appeals Court; regional (first instance) courts; military court

Political parties and leaders:

African Party for the Independence of Guinea-Bissau and Cabo Verde or PAIGC [Domingos SIMOES PEREIRA]

Democratic Convergence Party or PCD [Vicente FERNANDES]

Movement for Democratic Alternation Group of 15 or MADEM-G15 [Braima CAMARA]

National People's Assembly – Democratic Party of Guinea Bissau or APU- PDGB [Nuno Gomes NABIAM]

New Democracy Party or PND [Mamadu Iaia DJALO]

Party for Social Renewal or PRS [Alberto NAMBEIA]

Republican Party for Independence and Development or PRID [Aristides GOMES] Union for Change or UM [Agnelo REGALA]

International organization participation: ACP, AfDB, AOSIS, AU, CPLP, ECOWAS, FAO, FZ, G- 77, IBRD, ICAO, ICRM, IDA, IDB, IFAD, IFC, IFRCS, ILO, IMF, IMO, Interpol, IOC, IOM, IPU, ITSO, ITU, ITUC (NGOs), MIGA, MINUSMA, NAM, OIC, OIF, OPCW, UN, UNCTAD, UNESCO, UNIDO, UNWTO, UPU, WADB (regional), WAEMU, WCO, WFTU (NGOs), WHO, WIPO, WMO, WTO

Diplomatic representation in the US: none; note— Guinea-Bissau does not have official representation in Washington, DC

Diplomatic representation from the US: the US Embassy suspended operations on 14 June 1998; the US Ambassador to Senegal is accredited to Guinea-Bissau

Flag description: two equal horizontal bands of yellow (top) and green with a vertical red band on the hoist side; there is a black five-pointed star centered in the red band; yellow symbolizes the sun; green denotes hope; red represents blood shed during the struggle for independence; the black star stands for African unity

note: uses the popular Pan-African colors of Ethiopia; the flag design was heavily influenced by the Ghanaian flag

National symbol(s): *black star; national colors:* red, yellow, green, black

National anthem: *name:* " Esta e a Nossa Patria Bem Amada" (This Is Our Beloved Country)

lyrics/music: Amilcar Lopes CABRAL/XIAO He

note: adopted 1974; a delegation from then Portuguese Guinea visited China in 1963 and heard music by XIAO He; Amilcar Lopes CABRAL, the leader of Guinea-Bissau's independence movement, asked the composer to create a piece that would inspire his people to struggle for independence

ECONOMY

Economy—overview: Guinea-Bissau is highly dependent on subsistence agriculture, cashew nut exports, and foreign assistance. Two out of three Bissau- Guineans remain below the absolute poverty line. The legal economy is based on cashews and fishing. Illegal logging and trafficking in narcotics also play significant roles. The combination of limited economic prospects, weak institutions, and favorable geography have made this West African country a way station for drugs bound for Europe.

Guinea-Bissau has substantial potential for development of mineral resources, including phosphates, bauxite, and mineral sands. Offshore oil and gas exploration has begun. The country's climate and soil make it feasible to grow a wide range of cash crops, fruit, vegetables, and tubers; however, cashews generate more than 80% of export receipts and are the main source of income for many rural communities.

The government was deposed in August 2015, and since then, a political stalemate has resulted in weak governance and reduced donor support.

The country is participating in a three- year, IMF extended credit facility program that was suspended because of a planned bank bailout. The program was renewed in 2017, but the major donors of direct budget support (the EU, World Bank, and African Development Bank) have halted their programs indefinitely. Diversification of the economy remains a key policy goal, but Guinea-Bissau's poor infrastructure and business' climate will constrain this effort.

GDP (purchasing power parity):
$3.171 billion (2017 est.)
$2.994 billion (2016 est.)
$2.817 billion (2015 est.)
note: data are in 2017 dollars
country comparison to the world: 188

GDP (official exchange rate):
$1.35 billion (2017 est.)
GDP—real growth rate:
5.9% (2017 est.)
6.3% (2016 est.)
6.1% (2015 est.)
country comparison to the world: 32

GDP—per capita (PPP):
$1,900 (2017 est.)
$1,800 (2016 est.)
$1,700 (2015 est.)
note: data are in 2017 dollars
country comparison to the world: 212

Gross national saving:
8.6% of GDP (2017 est.)
10.1% of GDP (2016 est.)
10.5% of GDP (2015 est.)
country comparison to the world: 168

GDP—composition, by end use:
household consumption: 83.9% (2017 est.)
government consumption: 12% (2017 est.)
investment in fixed capital: 4.1% (2017 est.)
investment in inventories: 0.2% (2017 est.)
exports of goods and services: 26.4% (2017 est.)
imports of goods and services: -26.5% (2017 est.)

GDP—composition, by sector of origin:
agriculture: 50% (2017 est.)
industry: 13.1% (2017 est.)
services: 36.9% (2017 est.)

Agriculture—products: rice, corn, beans, cassava (manioc, tapioca), cashew nuts, peanuts, palm kernels, cotton; timber; fish

Industries: agricultural products processing, beer, soft drinks

Industrial production growth rate: 2.5% (2017 est.)
country comparison to the world: 117

Labor force: 731,300 (2013 est.)
country comparison to the world: 148

Labor force—by occupation: *agriculture:* 82%

industry and services: 18% (2000 est.)

Unemployment rate: NA

Population below poverty line: 67% (2015 est.)

Household income or consumption by percentage share: *lowest 10%:* 2.9%
highest 10%: 28% (2002)

Budget: *revenues:* 246.2 million (2017 est.)
expenditures: 263.5 million (2017 est.)

Taxes and other revenues: 18.2% (of GDP) (2017 est.)

country comparison to the world: 162

Budget surplus (+) or deficit (-): -1.3% (of GDP) (2017 est.)
country comparison to the world: 86

Public debt:
53.9% of GDP (2017 est.)
57.9% of GDP (2016 est.)
country comparison to the world:

Fiscal year: calendar year

Inflation rate (consumer prices): 1.1% (2017 est.)
1.5% (2016 est.)
country comparison to the world: 60

Current account balance: -$27 million (2017 est.)
$16 million (2016 est.)
country comparison to the world: 75
Exports: $328.1 million (2017 est.)
$278.6 million (2016 est.)
country comparison to the world:

Exports—partners: India 67.1%, Vietnam 21.1% (2017)

Exports—commodities: fish, shrimp; cashews, peanuts, palm kernels, raw and sawn lumber

Imports: $283.5 million (2017 est.)
$136.5 million (2016 est.)
country comparison to the world: 206

Imports—commodities: foodstuffs, machinery and transport equipment, petroleum products

Imports—partners: Portugal 47.8%, Senegal 12.1%, China 10.4%, Netherlands 8.1%, Pakistan 5.4% (2017)

Reserves of foreign exchange and gold:
$356.4 million (31 December 2017 est.)
$349.4 million (31 December 2016 est.)
country comparison to the world: 163

Debt—external:
$1.095 billion (31 December 2010 est.)
$941.5 million (31 December 2000 est.)
country comparison to the world: 163

Exchange rates: Communaute Financiere Africaine francs (XOF) per US dollar - 605.3 (2017 est.)
593.01 (2016 est.)
593.01 (2015 est.)
591.45 (2014 est.)
494.42 (2013 est.)

ENERGY

Electricity access: *population without electricity:* 1 million (2019)
electrification—total population: 28% (2019)
electrification—urban areas: 56% (2019)
electrification—rural areas: 7% (2019)

Electricity—production: 39 million kWh (2016 est.)
country comparison to the world: 208

Electricity—consumption: 36.27 million kWh (2016 est.)
country comparison to the world: 208

Electricity—exports: 0 kWh (2016 est.)
country comparison to the world: 145

Electricity—imports: 0 kWh (2016 est.)

country comparison to the world: 157

Electricity—installed generating capacity: 28,300 kW (2016 est.)
country comparison to the world: 200

Electricity—from fossil fuels: 99% of total installed capacity (2016 est.)
country comparison to the world: 24

Electricity—from nuclear fuels: 0% of total installed capacity (2017 est.)
country comparison to the world: 104

Electricity—from hydroelectric plants: 0% of total installed capacity (2017 est.)
country comparison to the world: 176

Electricity—from other renewable sources: 1% of total installed capacity (2017 est.)
country comparison to the world: 154

Crude oil—production: 0 bbl/day (2018 est.)
country comparison to the world: 146

Crude oil—exports: 0 bbl/day (2015 est.)
country comparison to the world: 134

Crude oil—imports: 0 bbl/day (2015 est.)
country comparison to the world: 138

Crude oil—proved reserves: 0 bbl (1 January 2018 est.)
country comparison to the world: 142

Refined petroleum products—production: 0 bbl/day (2015 est.)
country comparison to the world: 154

Refined petroleum products—consumption: 2,700 bbl/day (2016 est.)
country comparison to the world: 190

Refined petroleum products—exports: 0 bbl/day (2015 est.)
country comparison to the world: 163

Refined petroleum products—imports: 2,625 bbl/day (2015 est.)
country comparison to the world: Natural gas—production:
0 cu m (2017 est.)
country comparison to the world: 143

Natural gas—consumption: 0 cu m (2017 est.)
country comparison to the world: 156

Natural gas—exports: 0 cu m (2017 est.)
country comparison to the world: 118

Natural gas—imports: 0 cu m (2017 est.)
country comparison to the world: 136

Natural gas—proved reserves: cu m (1 January 2014 est.)
country comparison to the world: 144

Carbon dioxide emissions from consumption of energy: 397,900 Mt (2017 est.)
country comparison to the world: 188

COMMUNICATIONS

Telephones—fixed lines: *total subscriptions:* 0
subscriptions per 100 inhabitants: less than 1 (2018 est.)
country comparison to the world: 224

Telephones—mobile cellular:
total subscriptions: 1,555,961

subscriptions per 100 inhabitants: 82.79 (2019 est.)
country comparison to the world: 157

Telecommunication systems: *general assessment:* small system including a combination of microwave radio relay, open-wire lines, radiotelephone, and mobile cellular communications; 2 mobile network operators; one of the poorest countries in the world and this is reflected in the countries telecommunications development; radio is the most important source of information for the public (2020)
domestic: fixed-line teledensity less than 1 per 100 persons; mobile cellular teledensity is roughly 83 per 100 persons (2019)
international: country code—245; ACE submarine cable connecting Guinea-Bissau with 20 landing points in Western and South Africa and Europe (2019)
note: the COVID-19 outbreak is negatively impacting telecommunications production and supply chains globally; consumer spending on telecom devices and services has also slowed due to the pandemic's effect on economies worldwide; overall progress towards improvements in all facets of the telecom industry—mobile, fixed-line, broadband, submarine cable and satellite—has moderated

Broadcast media: 1 state-owned TV station, Televisao da Guine-Bissau (TGB) and a second station, Radio e Televisao de Portugal (RTP) Africa, is operated by Portuguese public broadcaster (RTP); 1 state-owned radio station, several private radio stations, and some community radio stations; multiple international broadcasters are available (2019)

Internet country code: .gw

Internet users: *total:* 72,047

percent of population: 3.93% (July 2018 est.)
country comparison to the world: 186

Broadband—fixed subscriptions: *total:* 1,204
subscriptions per 100 inhabitants: less than 1 (2018 est.)
country comparison to the world: 195

TRANSPORTATION

Civil aircraft registration country code prefix: J5 (2016)

Airports: 8 (2013)
country comparison to the world: 159

Airports—with paved runways:
total: 2 (2019)
over 3,047 m: 1
1,524 to 2,437 m: 1

Airports—with unpaved runways:
total: 6 (2013)
1,524 to 2,437 m: 1 (2013)
914 to 1,523 m: 2 (2013)
under 914 m: 3 (2013)

Roadways: *total:* 4,400 km (2018)
paved: 453 km (2018)
unpaved: 3,947 km (2018)
country comparison to the world: 150

Waterways: (rivers are partially navigable; many inlets and creeks provide shallow- water access to much of interior) (2012)

Merchant marine: *total:* 8
by type: general cargo 5, other 3 (2019)
country comparison to the world: 160

Ports and terminals: *major seaport(s):* Bissau, Buba, Cacheu, Farim

MILITARY AND SECURITY

Military and security forces: People's Revolutionary Armed Force (FARP): Army, Navy, National Air Force (Forca Aerea Nacional); Guard Nacional (Ministry of Internal Administration) (2020)

Military expenditures:
1.4% of GDP (2017)
1.3% of GDP (2016)
1.6% of GDP (2015)
2% of GDP (2014)
2.1% of GDP (2013)
country comparison to the world: 86

Military and security service personnel strengths: the People's Revolutionary Armed Force (FARP) has approximately 4,400 active troops (4,000 Army; 300 Navy; 100 Air Force) (2019)

Military equipment inventories and acquisitions: the inventory of the FARP consists of Soviet-era equipment; the only reported deliveries of military equipment to Guinea Bissau since 2015 were patrol boats from Spain in 2017 and non-lethal equipment from China in 2015 (2019 est.)

Military service age and obligation: 18-25 years of age for selective compulsory military service (Air Force service is voluntary); 16 years of age or younger, with parental consent, for voluntary service (2013)

TRANSNATIONAL ISSUES

Disputes—international: a longstanding low-grade conflict continues in parts of Casamance, in Senegal across the border; some rebels use Guinea-Bissau as a safe haven

Refugees and internally displaced persons: *refugees (country of origin):* 7,696 (Senegal) (2020)

Trafficking in persons: *current situation:* Guinea-Bissau is a source country for children subjected to forced labor and sex trafficking; the extent to which adults are trafficked for forced labor or forced prostitution is unclear; boys are forced into street vending in Guinea-Bissau and manual labor, agriculture, and mining in Senegal, while girls may be forced into street vending, domestic service, and, to a lesser extent, prostitution in Guinea and Senegal; some Bissau- Guinean boys at Koranic schools are forced into begging by religious teachers

tier rating: Tier 3—Guinea-Bissau does not fully comply with the minimum standards for the elimination of trafficking and is not making significant efforts to do so; despite enacting an anti- trafficking law and adopting a national action plan in 2011, the country failed to demonstrate any notable anti- trafficking efforts for the third consecutive year; existing laws prohibiting all forms of trafficking were not used to prosecute any trafficking offenders in 2014, and only one case of potential child labor trafficking was under investigation; authorities continued to rely entirely on NGOs and international organizations to provide victims with protective services; no trafficking prevention activities were conducted (2015)

Illicit drugs: increasingly important transit country for South American cocaine en route to Europe; enabling environment for trafficker operations due to pervasive corruption; archipelago- like geography near the capital facilitates drug smuggling

GUYANA

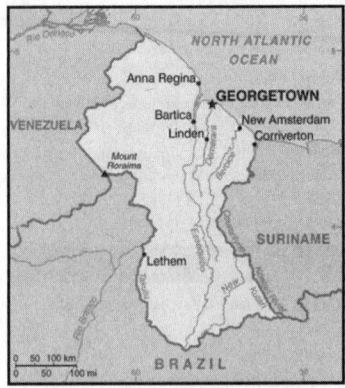

INTRODUCTION

Background: Originally a Dutch colony in the 17th century, by 1815 Guyana had become a British possession. The abolition of slaveryled to settlement of urban areas by former slaves and the importation of indentured servants from India to work the sugarplantations. The resulting ethnocultural divide has persisted and has led to turbulent politics. Guyana achieved independence from the UK in 1966, and since then it has been ruled mostly by socialist-oriented governments. In 1992, Cheddi JAGAN was elected president in what is considered the country's first free and fair election since independence. After his death five years later, his wife, Janet JAGAN, became president but resigned in 1999 due to poor health. Her successor, Bharrat JAGDEO, was elected in 2001 and again in 2006. Early elections held in May 2015 resulted in the first change in governing party and the replacement of President Donald RAMOTAR by current President David GRANGER. After a December 2018 no-confidence vote against the GRANGER government, national elections will be held before the scheduled spring 2020 date.

GEOGRAPHY

Location: Northern South America, bordering the North Atlantic Ocean, between Suriname and Venezuela

Geographic coordinates: 5 00 N, 59 00 W

Map references: South America

Area: *total:* 214,969 sq km
land: 196,849 sq km
water: 18,120 sq km
country comparison to the world: 86

Area—comparative: slightly smaller than Idaho; almost twice the size of Tennessee

Land boundaries: *total:* 2,933 km
border countries (3): Brazil 1308 km, Suriname 836 km, Venezuela 789 km

Coastline: 459 km

Maritime claims: *territorial sea:* 12 nm
exclusive economic zone: 200 nm

continental shelf: 200 nm or to the outer edge of the continental margin

Climate: tropical; hot, humid, moderated by northeast trade winds; two rainy seasons (May to August, November to January)

Terrain: mostly rolling highlands; low coastal plain; savanna in south

Elevation: *mean elevation:* 207 m
lowest point: Atlantic Ocean 0 m
highest point: Laberintos del Norte on Mount Roraima 2,775 m

Natural resources: bauxite, gold, diamonds, hardwood timber, shrimp, fish

Land use: *agricultural land:* 8.4% (2011 est.)
arable land: 2.1% (2011 est.) / permanent crops: 0.1% (2011 est.) / permanent pasture: 6.2% (2011 est.)
forest: 77.4% (2011 est.)
other: 14.2% (2011 est.)

Irrigated land: 1,430 sq km (2012)

Population distribution: population is heavily concentrated in the northeast in and around Georgetown, with noteable concentrations along the Berbice River to the east; the remainder of the country is sparsely populated

Natural hazards: flash flood threat during rainy seasons

Environment—current issues: water pollution from sewage and agricultural and industrial chemicals; deforestation

Environment—international agreements: *party to:* Biodiversity, Climate Change, Climate Change-Kyoto Protocol, Desertification, Endangered Species, Hazardous Wastes, Law of the Sea, Ozone Layer Protection, Ship Pollution, Tropical Timber 83, Tropical Timber 94

signed, but not ratified: none of the selected agreements

Geography—note: the third-smallest country in South America after Suriname and Uruguay; substantial portions of its western and eastern territories are claimed by Venezuela and Suriname respectively; contains some of the largest unspoiled rainforests on the continent

PEOPLE AND SOCIETY

Population: 750,204 (July 2020 est.)
note: estimates for this country explicitly take into account the effects of excess mortality due to AIDS; this can result in lower life expectancy, higher infant mortality, higher death rates, lower population growth rates, and changes in the distribution of population by age and sex than would otherwise be expected
country comparison to the world: 166

Nationality: *noun:* Guyanese (singular and plural)
adjective: Guyanese

Ethnic groups: East Indian 39.8%, African descent 29.3%, mixed 19.9%, Amerindian 10.5%, other 0.5% (includes Portuguese, Chinese, white) (2012 est.)

Languages: English (official), Guyanese Creole, Amerindian languages (including Caribbean and Arawak languages), Indian languages (including Caribbean Hindustani, a dialect of Hindi), Chinese (2014 est.)

Religions: Protestant 34.8% (Pentecostal 22.8%, Seventh Day Adventist 5.4%, Anglican 5.2%, Methodist 1.4%), Hindu 24.8%, Roman Catholic 7.1%, Muslim 6.8%, Jehovah's Witness 1.3%, Rastafarian 0.5%, other Christian 20.8%, other 0.9%, none 3.1% (2012 est.)

Demographic profile: Guyana is the only English-speaking country in South America and shares cultural and historical bonds with the Anglophone Caribbean. Guyana's two largest ethnic groups are the Afro-Guyanese (descendants of African slaves) and the Indo-Guyanese (descendants of Indian indentured laborers), which together comprise about three quarters of Guyana's population. Tensions periodically have boiled over between the two groups, which back ethnically based political parties and vote along ethnic lines. Poverty reduction has stagnated since the late 1990s. About one-third of the Guyanese population lives below the poverty line; indigenous people are disproportionately affected. Although Guyana's literacy rate is reported to be among the highest in the Western Hemisphere, the level of functional literacy is considerably lower, which has been attributed to poor education quality, teacher training, and infrastructure.

Guyana's emigration rate is among the highest in the world—more than 55% of its citizens reside abroad—and it is one of the largest recipients of remittances relative to GDP among Latin American and Caribbean counties. Although remittances are a vital source of income for most citizens, the pervasive emigration of skilled workers deprives Guyana of professionals in healthcare and other key sectors. More than 80% of Guyanese nationals with tertiary level educations have emigrated. Brain drain and the concentration of limited medical resources in Georgetown hamper Guyana's ability to meet the health needs of its predominantly rural population. Guyana has one of the highest HIV prevalence rates in the region and continues to rely on international support for its HIV treatment and prevention programs.

Age structure: *0-14 years:* 23.91% (male 91,317/female 88,025)
15-24 years: 21.23% (male 81,294/female 77,987)
25-54 years: 39.48% (male 154,825/female 141,385)
55-64 years: 8.37% (male 29,385/female 33,386)
65 years and over: 7.01% (male 21,325/female 31,275) (2020 est.)

Dependency ratios: *total dependency ratio:* 53.2
youth dependency ratio: 42.5
elderly dependency ratio: 10.7
potential support ratio: 9.3 (2020 est.)

Median age: *total:* 27.5 years
male: 27.2 years
female: 27.9 years (2020 est.)
country comparison to the world: 146

Population growth rate: 0.72% (2020 est.)
country comparison to the world: 135

Birth rate: 15.5 births/1,000 population (2020 est.)
country comparison to the world: 113

Death rate: 7.5 deaths/1,000 population (2020 est.)
country comparison to the world: 103

Net migration rate: 0 migrant(s)/1,000 population (2020 est.)
country comparison to the world: 84

Population distribution: population is heavily concentrated in the northeast in and around Georgetown, with noteable concentrations along the Berbice River to the east; the remainder of the country is sparsely populated

Urbanization: *urban population:* 26.8% of total population (2020)
rate of urbanization: 0.83% annual rate of change (2015-20 est.)
total population growth rate v. urban population growth rate, 2000-2030:

Major urban areas—Population: 110,000 GEORGETOWN (capital) (2018)

Sex ratio: *at birth:* 1.05 male(s)/female
0-14 years: 1.04 male(s)/female
15-24 years: 1.04 male(s)/female
25-54 years: 1.1 male(s)/female
55-64 years: 0.88 male(s)/female
65 years and over: 0.68 male(s)/female
total population: 1.02 male(s)/female (2020 est.)

Mother's mean age at first birth: 20.8 years (2009 est.)

note: median age at first birth among women 25-29

Maternal mortality rate: 667 deaths/100,000 live births (2017 est.)
country comparison to the world: 9

Infant mortality rate: *total:* 27.6 deaths/1,000 live births
male: 31.3 deaths/1,000 live births
female: 23.8 deaths/1,000 live births (2020 est.)
country comparison to the world: 64
Life expectancy at birth: total population: 69.5 years
male: 66.5 years
female: 72.6 years (2020 est.)
country comparison to the world: 168

Total fertility rate: 1.89 children born/woman (2020 est.)
country comparison to the world: 131

Contraceptive prevalence rate: 33.9% (2014)

Drinking water source:
improved:
urban: 100% of population
rural: 95.6% of population
total: 96.7% of population
unimproved:
urban: 0% of population
rural: 38.7% of population
total: 26.5% of population (2017 est.)

Current Health Expenditure: 4.9% (2017)

Physicians density: 0.8 physicians/1,000 population (2018)

Hospital bed density: 1.7 beds/1,000 population (2016)

Sanitation facility access:
improved:
urban: 97.8% of population
rural: 95.4% of population
total: 96% of population
unimproved:
urban: 2.2% of population
rural: 4.6% of population
total: 4% of population (2017 est.)

HIV/AIDS—adult prevalence rate: 1.4% (2019 est.)
country comparison to the world: 33

HIV/AIDS—people living with HIV/AIDS: 8,700 (2019 est.)
country comparison to the world: 109

HIV/AIDS—deaths: <200 (2019 est.)

Major infectious diseases: *degree of risk:* very high (2020)
food or waterborne diseases: bacterial and protozoal diarrhea, hepatitis A, and typhoid fever
vectorborne diseases: dengue fever and malaria

Obesity—adult prevalence rate: 20.2% (2016)
country comparison to the world: 103

Children under the age of 5 years underweight: 8.2% (2014)
country comparison to the world: 70

Education expenditures: 6.3% of GDP (2017)
country comparison to the world: 25

Literacy: *definition:* age 15 and over has ever attended school

total population: 88.5%
male: 87.2%
female: 89.8% (2015)

School life expectancy (primary to tertiary education): *total:* 11 years
male: 11 years
female: 12 years (2012)

Unemployment, youth ages 15-24: *total:* 21.5%
male: 17.3%
female: 27.7% (2017 est.)
country comparison to the world: 58

GOVERNMENT

Country name: *conventional long form:* Cooperative Republic of Guyana
conventional short form: Guyana
former: British Guiana
etymology: the name is derived from Guiana, the original name for the region that included British Guiana, Dutch Guiana, and French Guiana; ultimately the word is derived from an indigenous Amerindian language and means "Land of Many Waters" (referring to the area's multitude of rivers and streams)

Government type: parliamentary republic

Capital: *name:* Georgetown
geographic coordinates: 6 48 N, 58 09 W
time difference: UTC-4 (1 hour ahead of Washington, DC, during Standard Time)
etymology: when the British took possession of the town from the Dutch in 1812, they renamed it Georgetown in honor of King George III (1738-1820)

Administrative divisions: 10 regions; Barima-Waini, Cuyuni-Mazaruni, Demerara-Mahaica, East Berbice-Corentyne, Essequibo Islands-West Demerara, Mahaica-Berbice, Pomeroon-Supenaam, Potaro-Siparuni, Upper Demerara-Berbice, Upper Takutu-Upper Essequibo

Independence: 26 May 1966 (from the UK)

National holiday: Republic Day, 23 February (1970)

Constitution: *history:* several previous; latest promulgated 6 October 1980
amendments: proposed by the National Assembly; passage of amendments affecting constitutional articles, such as national sovereignty, government structure and powers, and constitutional amendment procedures, requires approval by the Assembly membership, approval in a referendum, and assent of the president; other amendments only require Assembly approval; amended many times, last in 2016

Legal system: common law system, based on the English model, with some Roman-Dutch civil law influence

International law organization participation: has not submitted an ICJ jurisdiction declaration; accepts ICCt jurisdiction

Citizenship: citizenship by birth: yes
citizenship by descent only: yes
dual citizenship recognized: no
residency requirement for naturalization: na

Suffrage: 18 years of age; universal

Executive branch: *chief of state:* President Mohammed Irfaan ALI (since 2 August 2020); First Vice President Mark PHILLIPS (since 20 May 2015); Vice Presidents Bharrat JAGDEO (since 20 May 2015), Sydney ALLICOCK (since 2 August 2020), Khemraj RAMJATTAN (since 2 August 2020); note—the president is both chief of state and head of government
head of government: President Mohammed Irfaan ALI (since 2 August 2020); First Vice President Mark PHILLIPS (since 20 May 2015); Vice Presidents Bharrat JAGDEO (since 20 May 2015), Sydney ALLICOCK (since 2 August 2020), Khemraj RAMJATTAN (since 2 August 2020)
cabinet: Cabinet of Ministers appointed by the president, responsible to the National Assembly
elections/appointments: the predesignated candidate of the winning party in the last National Assembly election becomes president for a 5-year term (no term limits); election last held on 2 March 2020 (next to be held in 2025); prime minister appointed by the president
election results: Mohammed Irfaan ALI (PPP/C) designated president by the majority party in the National Assembly

Legislative branch: *description:* unicameral National Assembly (65 seats; 40 members directly elected in a single nationwide constituency and 25 directly elected in multi-seat constituencies—all by closed list proportional representation vote; members serve 5year terms)
elections: last held on 2 March 2020 (next to be held in 2025)
election results: percent of vote by party—PPP/C 50.69%, APNU-AFC 47.34%, LJP 0.58%, ANUG 0.5%, TNM 0.05%, other 0.84%; seats by party—PPP/C 33, APNU-AFC 31, LJP-ANUG-TNM 1; composition—men 43, women 22, percent of women 33.8%; note—the initial results were declared invalid and a partial recount was conducted from 6 May to 8 June 2020, in which PPP/C was declared the winner

Judicial branch: *highest courts:* Supreme Court of Judicature (consists of the Court of Appeal with a chief justice and 3 justices, and the High Court with a chief justice and 10 justices organized into 3- or 5-judge panels); note—in 2009, Guyana acceded to the Caribbean Court of Justice as the final court of appeal in civil and criminal cases, replacing that of the Judicial Committee of the Privy Council (in London)
judge selection and term of office: Court of Appeal and High Court chief justices appointed by the president; other judges of both courts appointed by the Judicial Service Commission, a body appointed by the president; judges appointed for life with retirement at age 65
subordinate courts: Land Court; magistrates' courts

Political parties and leaders:
A New and United Guyana or ANUG [Ralph RAMKARRAN]
A Partnership for National Unity or APNU [David A. GRANGER]

Alliance for Change or AFC [Raphael TROTMAN]
Justice for All Party [C.N. SHARMA]
Liberty and Justice Party or LJP [Lenox SHUMAN]
National Independent Party or NIP [Saphier Husain SUBEDAR]
People's Progressive Party/Civic or PPP/C [Bharrat JAGDEO]
The New Movement or TNM [joint leadership of several medical doctors]
The United Force or TUF [Manzoor NADIR]
United Republican Party or URP [Vishnu BANDHU]

International organization participation: ACP, AOSIS, C, Caricom, CD, CDB, CELAC, FAO, G-77, IADB, IBRD, ICAO, ICCt, ICRM, IDA, IFAD, IFC, IFRCS, ILO, IMF, IMO, Interpol, IOC, IOM, ISO (correspondent), ITU, LAES, MIGA, NAM, OAS, OIC, OPANAL, OPCW, PCA, Petrocaribe, UN, UNASUR, UNCTAD, UNESCO, UNIDO, UPU, WCO, WFTU (NGOs), WHO, WIPO, WMO, WTO

Diplomatic representation in the US: *chief of mission:* Ambassador Riyad David INSANALLY (since 16 Sept 2016)
chancery: 2490 Tracy Place NW, Washington, DC 20008
telephone: [1] (202) 265-6900
FAX: [1] (202) 232-1297
consulate(s) general: New York

Diplomatic representation from the US: *chief of mission:* Ambassador Sarah-Ann LYNCH (since 13 March 2019)
telephone: [592] 225-4900 through 4909
embassy: US Embassy, 100 Young and Duke Streets, Kingston, Georgetown
mailing address: P. O. Box 10507, Georgetown; US Embassy, 3170 Georgetown Place, Washington DC 20521-3170
FAX: [592] 225-8497

Flag description: green with a red isosceles triangle (based on the hoist side) superimposed on a long, yellow arrowhead; there is a narrow, black border between the red and yellow, and a narrow, white border between the yellow and the green; green represents forest and foliage; yellow stands for mineral resources and a bright future; white symbolizes Guyana's rivers; red signifies zeal and the sacrifice of the people; black indicates perseverance; also referred to by its nickname The Golden Arrowhead

National symbol(s): *Canje pheasant (hoatzin), jaguar, Victoria Regia water lily; national colors:* red, yellow, green, black, white

National anthem: *name:* Dear Land of Guyana, of Rivers and Plains
lyrics/music: Archibald Leonard LUKERL/Robert Cyril Gladstone POTTER
note: adopted 1966
0:00 /0:53

ECONOMY

Economy—overview: The Guyanese economy exhibited moderate economic growth in recent years and is based largely on agriculture and extractive industries. The economy is heavily dependent upon the export of six commodities—sugar, gold, bauxite, shrimp, timber, and rice—which represent nearly 60% of the country's GDP and are highly susceptible to adverse weather conditions and fluctuations in commodity prices. Guyana closed or consolidated several sugar estates in 2017, reducing production of sugar to a forecasted 147,000 tons in 2018, less than half of 2017 production. Much of Guyana's growth in recent years has come from a surge in gold production. With a record-breaking 700,000 ounces of gold produced in 2016, Gold production in Guyana has offset the economic effects of declining sugar production. In January 2018, estimated 3.2 billion barrels of oil were found offshore and Guyana is scheduled to become a petroleum producer by March 2020.

Guyana's entrance into the Caricom Single Market and Economy in January 2006 broadened the country's export market, primarily in the raw materials sector. Guyana has experienced positive growth almost every year over the past decade. Inflation has been kept under control. Recent years have seen the government's stock of debt reduced significantly—with external debt now less than half of what it was in the early 1990s. Despite these improvements, the government is still juggling a sizable external debt against the urgent need for expanded public investment. In March 2007, the Inter-American Development Bank, Guyana's principal donor, canceled Guyana's nearly $470 million debt, equivalent to 21% of GDP, which along with other Highly Indebted Poor Country debt forgiveness, brought the debt-to-GDP ratio down from 183% in 2006 to 52% in 2017. Guyana had become heavily indebted as a result of the inward-looking, state-led development model pursued in the 1970s and 1980s. Chronic problems include a shortage of skilled labor and a deficient infrastructure.

GDP (purchasing power parity): $6.301 billion (2017 est.)
$6.169 billion (2016 est.)
$5.969 billion (2015 est.)
note: data are in 2017 dollars
country comparison to the world: 171

GDP (official exchange rate): $3.561 billion (2017 est.)

GDP—real growth rate:
2.1% (2017 est.)
3.4% (2016 est.)
3.1% (2015 est.)
country comparison to the world: 146

GDP—per capita (PPP): $8,100 (2017 est.)
$8,000 (2016 est.)
$7,800 (2015 est.)
note: data are in 2017 dollars
country comparison to the world: 151

Gross national saving:
10.5% of GDP (2017 est.)
15% of GDP (2016 est.)
8.8% of GDP (2015 est.)
country comparison to the world: 161

GDP—composition, by end use:

household consumption: 71.1% (2017 est.)
government consumption: 18.2% (2017 est.)
investment in fixed capital: 25.4% (2017 est.)
investment in inventories: 0% (2017 est.)
exports of goods and services: 47.8% (2017 est.)
imports of goods and services: -63% (2017 est.)

GDP—composition, by sector of origin:
agriculture: 15.4% (2017 est.)
industry: 15.3% (2017 est.)
services: 69.3% (2017 est.)

Agriculture—products: sugarcane, rice, edible oils; beef, pork, poultry; shrimp, fish

Industries: bauxite, sugar, rice milling, timber, textiles, gold mining

Industrial production growth rate: -5% (2017 est.)
country comparison to the world: 196

Labor force: 313,800 (2013 est.)
country comparison to the world: 162

Labor force—by occupation: *agriculture:* NA
industry: NA
services: NA

Unemployment rate:
11.1% (2013)
11.3% (2012)
country comparison to the world: 151

Population below poverty line: 35% (2006 est.)

Household income or consumption by percentage share: *lowest 10%:* 1.3%
highest 10%: 33.8% (1999)

Budget: *revenues:* 1.002 billion (2017 est.)
expenditures: 1.164 billion (2017 est.)

Taxes and other revenues: 28.1% (of GDP) (2017 est.)
country comparison to the world: 96

Budget surplus (+) or deficit (-): -4.5% (of GDP) (2017 est.)
country comparison to the world: 165

Public debt:
52.2% of GDP (2017 est.)
50.7% of GDP (2016 est.)
country comparison to the world: 95

Fiscal year: calendar year

Inflation rate (consumer prices):
2% (2017 est.)
0.8% (2016 est.)
country comparison to the world: 105

Current account balance:
-$237 million (2017 est.)
$13 million (2016 est.)
country comparison to the world: 100

Exports:
$1.439 billion (2017 est.)
$1.38 billion (2016 est.)
country comparison to the world: 149

Exports—partners: Canada 24.9%, US 16.5%, Panama 9.6%, UK 7.7%, Jamaica 5.1%, Trinidad and Tobago 5% (2017)

Exports—commodities: sugar, gold, bauxite, alumina, rice, shrimp, molasses, rum, timber

Imports:
$1.626 billion (2017 est.)

$1.341 billion (2016 est.)
country comparison to the world: 173

Imports—commodities: manufactures, machinery, petroleum, food

Imports—partners: Trinidad and Tobago 27.5%, US 26.5%, China 8.9%, Suriname 6.1% (2017)

Reserves of foreign exchange and gold:
$565.4 million (31 December 2017 est.)
$581 million (31 December 2016 est.)
country comparison to the world: 146

Debt—external:
$1.69 billion (31 December 2017 est.)
$1.542 billion (31 December 2016 est.)
country comparison to the world: 156

Exchange rates: Guyanese dollars (GYD) per US dollar -
207 (2017 est.)
206.5 (2016 est.)
206.5 (2015 est.)
206.5 (2014 est.)
206.45 (2013 est.)

ENERGY

Electricity access: electrification—total population: 84.2% (2016)
electrification—urban areas: 90.2% (2016)
electrification—rural areas: 81.9% (2016)

Electricity—production: 1.01 billion kWh (2016 est.)
country comparison to the world: 151

Electricity—consumption: 790.1 million kWh (2016 est.)
country comparison to the world: 161

Electricity—exports: 0 kWh (2016 est.)
country comparison to the world: 146

Electricity—imports: 0 kWh (2016 est.)
country comparison to the world: 158

Electricity—installed generating capacity: 428,000 kW (2016 est.)
country comparison to the world: 151

Electricity—from fossil fuels: 89% of total installed capacity (2016 est.)
country comparison to the world: 57

Electricity—from nuclear fuels: 0% of total installed capacity (2017 est.)
country comparison to the world: 105

Electricity—from hydroelectric plants: 0% of total installed capacity (2017 est.)
country comparison to the world: 177

Electricity—from other renewable sources: 11% of total installed capacity (2017 est.)
country comparison to the world: 78

Crude oil—production: 0 bbl/day (2018 est.)
country comparison to the world: 147

Crude oil—exports: 0 bbl/day (2015 est.)
country comparison to the world: 135

Crude oil—imports: 0 bbl/day (2015 est.)
country comparison to the world: 139

Crude oil—proved reserves: 0 bbl (1 January 2018 est.)
country comparison to the world: 143

Refined petroleum products—production: 0 bbl/day (2015 est.)
country comparison to the world: 155

Refined petroleum products—consumption: 14,000 bbl/day (2016 est.)
country comparison to the world: 154

Refined petroleum products—exports: 0 bbl/day (2015 est.)
country comparison to the world: 164

Refined petroleum products—imports: 13,720 bbl/day (2015 est.)
country comparison to the world: 140

Natural gas—production: 0 cu m (2017 est.)
country comparison to the world: 144

Natural gas—consumption: 0 cu m (2017 est.)
country comparison to the world: 157

Natural gas—exports: 0 cu m (2017 est.)
country comparison to the world: 119

Natural gas—imports: 0 cu m (2017 est.)
country comparison to the world: 137

Natural gas—proved reserves: 0 cu m (1 January 2014 est.)
country comparison to the world: 145

Carbon dioxide emissions from consumption of energy: 2.131 million Mt (2017 est.)
country comparison to the world: 158

COMMUNICATIONS

Telephones—fixed lines: *total subscriptions:* 130,497
subscriptions per 100 inhabitants: 17.52 (2019 est.)
country comparison to the world: 131

Telephones—mobile cellular: *total subscriptions:* 617,998
subscriptions per 100 inhabitants: 82.97 (2019 est.)
country comparison to the world: 169

Telecommunication systems: *general assessment:* reliable international long distance service; 100% digital network; national transmission supported by fiber optic cable and rural network by microwaves; more than 150,000 lines; many areas still lack fixed-line telephone services; 2019 budget allocates funds for ICT (Information and Communications Technology) development; broadband subscribers remains small and end-users incur expense to use (2020)
domestic: fixed-line teledensity is about 18 per 100 persons; mobile-cellular teledensity about 83 per 100 persons (2019)
international: country code—592; landing point for the SG-SCS submarine cable to Suriname, and the Caribbean; satellite earth station—1 Intelsat (Atlantic Ocean) (2019)
note: the COVID-19 outbreak is negatively impacting telecommunications production and supply chains globally; consumer spending on telecom devices and services has also slowed due to the pandemic's effect on economies worldwide; overall progress towards improvements in all facets of the telecom industry—mobile, fixed-line,

broadband, submarine cable and satellite—has moderated

Broadcast media: government-dominated broadcast media; the National Communications Network (NCN) TV is state-owned; a few private TV stations relay satellite services; the state owns and operates 2 radio stations broadcasting on multiple frequencies capable of reaching the entire country; government limits on licensing of new private radio stations has constrained competition in broadcast media

Internet country code: .gy

Internet users: *total:* 276,498
percent of population: 37.33% (July 2018 est.)
country comparison to the world: 168

Broadband—fixed subscriptions: *total:* 64,889
subscriptions per 100 inhabitants: 9 (2017 est.)
country comparison to the world: 129

TRANSPORTATION

Civil aircraft registration country code prefix: 8R (2016)

Airports: 117 (2013)
country comparison to the world: 48

Airports—with paved runways: *total:* 11 (2017)
1,524 to 2,437 m: 2 (2017)
914 to 1,523 m: 1 (2017)
under 914 m: 8 (2017)

Airports—with unpaved runways: *total:* 106 (2013)
1,524 to 2,437 m: 1 (2013)
914 to 1,523 m: 16 (2013)
under 914 m: 89 (2013)

Roadways: *total:* 3,995 km (2019)
paved: 799 km (2019)
unpaved: 3,196 km (2019)
country comparison to the world: 155

Waterways: 330 km (the Berbice, Demerara, and Essequibo Rivers are navigable by oceangoing vessels for 150 km, 100 km, and 80 km respectively) (2012)
country comparison to the world: 91

Merchant marine: *total:* 56
by type: general cargo 28, oil tanker 6, other 22 (2019)
country comparison to the world: 112

Ports and terminals: *major seaport(s):* Georgetown

MILITARY AND SECURITY

Military and security forces: *Guyana Defense Force:* Army, Air Corps, Coast Guard (2019)

Military expenditures:
1.7% of GDP (2019)
1.6% of GDP (2018)
1.6% of GDP (2017)
1.5% of GDP (2016)
1.5% of GDP (2015)
country comparison to the world: 65

Military and security service personnel strengths: the Guyana Defense Force has approximately 3,000 active personnel (2019 est.)

Military equipment inventories and acquisitions: the Guyana Defense Force's limited inventory is mostly comprised of second-hand platforms from a variety of foreign suppliers, including Brazil, China, the former Soviet Union, the UK, and the US; since 2000, Guyana has received limited amounts of military equipment from Brazil, China, Costa Rica, and the UK (2019 est.)

Military service age and obligation: 18 years of age or older for voluntary military service; no conscription (2014)

TRANSNATIONAL ISSUES

Disputes—international: all of the area west of the Essequibo River is claimed by Venezuela preventing any discussion of a maritime boundary; Guyana has expressed its intention to join Barbados in asserting claims before UN Convention on the Law of the Sea (UNCLOS) that Trinidad and Tobago's maritime boundary with Venezuela extends into their waters; Suriname claims a triangle of land between the New and Kutari/Koetari Rivers in a historic dispute over the headwaters of the Courantyne

Trafficking in persons: *current situation:* Guyana is a source and destination country for men, women, and children subjected to sex trafficking and forced labor – children are particularly vulnerable; women and girls from Guyana, Venezuela, Suriname, Brazil, and the Dominican Republic are forced into prostitution in Guyana's interior mining communities and urban areas; forced labor is reported in mining, agriculture, forestry, domestic service, and shops; Guyanese nationals are also trafficked to Suriname, Jamaica, and other Caribbean countries for sexual exploitation and forced labor

tier rating: Tier 2 Watch List – Guyana does not fully comply with the minimum standards for the elimination of trafficking; however, it is making significant efforts to do so; in 2014, Guyana was granted a waiver from an otherwise required downgrade to Tier 3 because its government has a written plan that, if implemented would constitute making significant efforts to bring itself into compliance with the minimum standards for the elimination of trafficking; the government released its anti-trafficking action plan in June 2014 but made uneven efforts to implement it; law enforcement was weak, investigating seven trafficking cases, prosecuting four alleged traffickers, and convicting one trafficker – a police officer – who was released on bail pending appeal; in 2014, as in previous years, Guyanese courts dismissed the majority of ongoing trafficking prosecutions; the government referred some victims to care services, which were provided by NGOs with little or no government support (2015)

Illicit drugs: transshipment point for narcotics from South America—primarily Venezuela—to Europe and the US; producer of cannabis; rising money laundering related to drug trafficking and human smuggling

HAITI

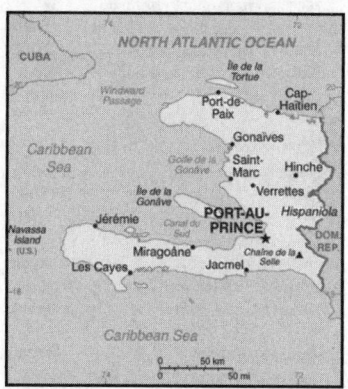

INTRODUCTION

Background: The native Taino—who inhabited the island of Hispaniola when Christopher COLUMBUS first landed on it in 1492—were virtually wiped out by Spanish settlers within 25 years. In the early 17th century, the French established a presence on Hispaniola. In 1697, Spain ceded to the French the western third of the island, which later became Haiti. The French colony, based on forestry and sugar-related industries, became one of the wealthiest in the Caribbean but relied heavily on the forced labor of enslaved Africans and environmentally degrading practices. In the late 18th century, Toussaint L'OUVERTURE led a revolution of Haiti's nearly half a million slaves that ended France's rule on the island. After a prolonged struggle, and under the leadership of Jean-Jacques DESSALINES, Haiti became the first country in the world led by former slaves after declaring its independence in 1804, but it was forced to pay an indemnity to France for more than a century and was shunned by other countries for nearly 40 years. After the US occupied Haiti from 1915-1934, Francois "Papa Doc" DUVALIER and then his son Jean-Claude "Baby Doc" DUVALIER led repressive and corrupt regimes that ruled Haiti from 1957-1971 and 1971-1986, respectively. A massive magnitude 7.0 earthquake struck Haiti in January 2010 with an epicenter about 25 km (15 mi) west of the capital, Port-au-Prince. Estimates are that over 300,000 people were killed and some 1.5 million left homeless. The earthquake was assessed as the worst in this region over the last 200 years. On 4 October 2016, Hurricane Matthew made landfall in Haiti, resulting in over 500 deaths and causing extensive damage to crops, houses, livestock, and infrastructure. Currently the poorest country in the Western Hemisphere, Haiti continues to experience bouts of political instability.

GEOGRAPHY

Location: Caribbean, western one-third of the island of Hispaniola, between the Caribbean Sea and the North Atlantic Ocean, west of the Dominican Republic

Geographic coordinates: 19 00 N, 72 25 W

Map references: Central America and the Caribbean

Area: *total:* 27,750 sq km
land: 27,560 sq km
water: 190 sq km
country comparison to the world: 148

Area—comparative: slightly smaller than Maryland

Land boundaries: *total:* 376 km
border countries (1): Dominican Republic 376 km

Coastline: 1,771 km

Maritime claims: *territorial sea:* 12 nm
exclusive economic zone: 200 nm
contiguous zone: 24 nm
continental shelf: to depth of exploitation

Climate: tropical; semiarid where mountains in east cut off trade winds

Terrain: mostly rough and mountainous

Elevation: *mean elevation:* 470 m
lowest point: Caribbean Sea 0 m
highest point: Chaine de la Selle 2,680 m

Natural resources: bauxite, copper, calcium carbonate, gold, marble, hydropower, arable land

Land use: *agricultural land:* 66.4% (2011 est.)
arable land: 38.5% (2011 est.) / *permanent crops:* 10.2% (2011 est.) / *permanent pasture:* 17.7% (2011 est.)
forest: 3.6% (2011 est.)
other: 30% (2011 est.)

Irrigated land: 970 sq km (2012)

Population distribution: fairly even distribution; largest concentrations located near coastal areas

Natural hazards: lies in the middle of the hurricane belt and subject to severe storms from June to October; occasional flooding and earthquakes; periodic droughts

Environment—current issues: extensive deforestation (much of the remaining forested land is being cleared for agriculture and used as fuel); soil erosion; overpopulation leads to inadequate supplies of potable water and and a lack of sanitation; natural disasters

Environment—international agreements: *party to:* Biodiversity, Climate Change, Climate Change-Kyoto Protocol, Desertification, Law of the Sea, Marine Dumping, Marine Life Conservation, Ozone Layer Protection
signed, but not ratified: Hazardous Wastes

Geography—note: shares island of Hispaniola with Dominican Republic (western one-third is Haiti, eastern two-thirds is the Dominican Republic); it is the most mountainous nation in the Caribbean

PEOPLE AND SOCIETY

Population: 11,067,777 (July 2020 est.)
note: estimates for this country explicitly take into account the effects of excess mortality due to AIDS; this can result in lower life expectancy, higher infant mortality, higher death rates, lower population growth rates, and changes in the distribution of population by age and sex than would otherwise be expected
country comparison to the world: 82

Nationality: *noun:* Haitian(s)
adjective: Haitian

Ethnic groups: black 95%, mixed and white 5%

Languages: French (official), Creole (official)

Religions: Roman Catholic 54.7%, Protestant 28.5% (Baptist 15.4%, Pentecostal 7.9%, Adventist 3%, Methodist 1.5%, other 0.7%), Vodou 2.1%, other 4.6%, none 10.2% (2003 est.)
note: many Haitians practice elements of Vodou in addition to another religion, most often Roman Catholicism; Vodou was recognized as an official religion in 2003

Age structure: *0-14 years:* 31.21% (male 1,719,961/female 1,734,566)
15-24 years: 20.71% (male 1,145,113/female 1,146,741)
25-54 years: 38.45% (male 2,110,294/female 2,145,209)
55-64 years: 5.3% (male 280,630/female 305,584)
65 years and over: 4.33% (male 210,451/female 269,228) (2020 est.)

Dependency ratios: *total dependency ratio:* 60.4
youth dependency ratio: 52.1
elderly dependency ratio: 8.3
potential support ratio: 13.3 (2020 est.)

Median age: *total:* 24.1 years
male: 23.8 years
female: 24.3 years (2020 est.)
country comparison to the world: 167

Population growth rate: 1.26% (2020 est.)
country comparison to the world: 87

Birth rate: 21.7 births/1,000 population (2020 est.)
country comparison to the world: 69

Death rate: 7.4 deaths/1,000 population (2020 est.)
country comparison to the world: 109

Net migration rate: -1.9 migrant(s)/1,000 population (2020 est.)
country comparison to the world: 165

Population distribution: fairly even distribution; largest concentrations located near coastal areas

Urbanization: urban population: 57.1% of total population (2020)
rate of urbanization: 2.9% annual rate of change (2015-20 est.)

total population growth rate v. urban population growth rate, 2000-2030:

Major urban areas—population: 2.774 million PORT-AU-PRINCE (capital) (2020)

Sex ratio: *at birth:* 1.01 male(s)/female
0-14 years: 0.99 male(s)/female
15-24 years: 1 male(s)/female
25-54 years: 0.98 male(s)/female
55-64 years: 0.92 male(s)/female
65 years and over: 0.78 male(s)/female
total population: 0.98 male(s)/female (2020 est.)

Mother's mean age at first birth: 22.8 years (2016/7 est.)
note: median age at first birth among women 25-29

Maternal mortality rate: 480 deaths/100,000 live births (2017 est.)
country comparison to the world: 22

Infant mortality rate: *total:* 42.6 deaths/1,000 live births
male: 48.5 deaths/1,000 live births
female: 36.6 deaths/1,000 live births (2020 est.)
country comparison to the world: 33

Life expectancy at birth: *total population:* 65.3 years
male: 62.6 years
female: 68 years (2020 est.)
country comparison to the world: 194

Total fertility rate: 2.52 children born/woman (2020 est.)
country comparison to the world: 72

Contraceptive prevalence rate: 34.3% (2016/17)

Drinking water source:
improved:
urban: 91.5% of population
rural: 55.4% of population
total: 75% of population
unimproved:
urban: 8.5% of population
rural: 44.6% of population
total: 25% of population (2017 est.)

Current Health Expenditure: 8% (2017)

Physicians density: 0.23 physicians/1,000 population (2018)

Hospital bed density: 0.7 beds/1,000 population (2013)

Sanitation facility access:
improved:
urban: 80.6% of population
rural: 40% of population
total: 62.1% of population
unimproved:
urban: 19.4% of population
rural: 60% of population
total: 37.9% of population (2017 est.)

HIV/AIDS—adult prevalence rate: 1.9% (2019 est.)
country comparison to the world: 25

HIV/AIDS—people living with HIV/AIDS: 160,000 (2019 est.)
country comparison to the world: 35

HIV/AIDS—deaths: 2,700 (2019 est.)
country comparison to the world: 38

Major infectious diseases: *degree of risk:* very high (2020)
food or waterborne diseases: bacterial and protozoal diarrhea, hepatitis A and E, and typhoid fever
vectorborne diseases: dengue fever and malaria

Obesity—adult prevalence rate: 22.7% (2016)
country comparison to the world: 72

Children under the age of 5 years underweight: 9.5% (2017)
country comparison to the world: 67

Education expenditures: 2.4% of GDP (2016)
country comparison to the world: 162

Literacy: *definition:* age 15 and over can read and write
total population: 61.7%
male: 65.3%
female: 58.3% (2016)

GOVERNMENT

Country name: *conventional long form:* Republic of Haiti
conventional short form: Haiti
local long form: Republique d'Haiti/Repiblik d Ayiti
local short form: Haiti/Ayiti
etymology: the native Taino name means "Land of High Mountains" and was originally applied to the entire island of Hispaniola

Government type: semi-presidential republic

Capital: *name:* Port-au-Prince
geographic coordinates: 18 32 N, 72 20 W
time difference: UTC-5 (same time as Washington, DC, during Standard Time)
daylight saving time: +1hr, begins second Sunday in March; ends first Sunday in November
etymology: according to tradition, in 1706, a Captain de Saint-Andre named the bay and its surrounding area after his ship Le Prince; the name of the town that grew there means, "the Port of The Prince"

Administrative divisions: 10 departments (departements, singular—departement); Artibonite, Centre, Grand'Anse, Nippes, Nord, Nord-Est, Nord- Ouest, Ouest, Sud, Sud-Est

Independence: 1 January 1804 (from France)

National holiday: Independence Day, 1 January (1804)

Constitution: *history:* many previous; latest adopted 10 March 1987
amendments: proposed by the executive branch or by either the Senate or the Chamber of Deputies; consideration of proposed amendments requires support by at least two-thirds majority of both houses; passage requires at least twothirds majority of the membership present and at least two-thirds majority of the votes cast; approved amendments enter into force after installation of the next president of the republic; constitutional articles on the democratic and republican form of government cannot be amended; amended 2011, 2012

Legal system: civil law system strongly influenced by Napoleonic Code

International law organization participation: accepts compulsory ICJ jurisdiction; non-party state to the ICCt

Citizenship: *citizenship by birth:* no
citizenship by descent only: at least one parent must be a native-born citizen of Haiti
dual citizenship recognized: no
residency requirement for naturalization: 5 years

Suffrage: 18 years of age; universal

Executive branch: *chief of state:* President Jovenel MOISE (since 7 February 2017)
head of government: Prime Minister Joseph JOUTHE (since since 4 March 2020)
cabinet: Cabinet chosen by the prime minister in consultation with the president; parliament must ratify the Cabinet and Prime Minister's governing policy
elections/appointments: president directly elected by absolute majority popular vote in 2 rounds if needed for a 5-year term (eligible for a single non-consecutive term); last election originally scheduled for 9 October 2016 but postponed until 20 November 2016 due to Hurricane Matthew
election results: Jovenel MOISE elected president in first round; percent of vote—Jovenel MOISE (PHTK) 55.6%, Jude CELESTIN (LAPEH) 19.6%, Jean-Charles MOISE (PPD) 11%, Maryse NARCISSE (FL) 9%; other 4.8%

Legislative branch: *description:* bicameral legislature or le Corps l'egislatif ou le Parlement consists of: le S'enat or Senate (30 seats, 29 filled as of June 2019; members directly elected in multi-seat constituencies by absolute majority vote in 2 rounds if needed; members serve 6-year terms with one-third of the membership renewed every 2 years) la Chambre de deput'es or Chamber of Deputies (119 seats; 116 filled as of June 2019; members directly elected in single-seat constituencies by absolute majority vote in 2 rounds if needed; members serve 4-year terms); note—when the 2 chambers meet collectively it is known as L'Assembl'ee nationale or the National Assembly and is convened for specific purposes spelled out in the constitution
elections: Senate—last held on 20 November 2016 with runoff on 29 January 2017 (next scheduled for 27 October 2019)
Chamber of Deputies—last held on 9 August 2015 with runoff on 25 October 2015 and 20 November 2016 (next scheduled for 27 October 2019)
election results: Senate—percent of vote by party—NA; seats by party—NA; composition—men 27, women 1, percent of women 3.6%
Chamber of Deputies—percent of vote by party—NA; seats by party—NA; composition—men 115, women 3, percent of women 2.5%; note—total legislature percent of women 2.7%

Judicial branch: *highest courts:* Supreme Court or Cour de cassation (consists of a chief judge and other judges); note—Haiti is a member of the Caribbean Court of Justice
judge selection and term of office: judges appointed by the president from candidate lists submitted by the Senate of the National Assembly;

note—Article 174 of Haiti's constitution states that judges of the Supreme Court are appointed for 10 years, whereas Article 177 states that judges of the Supreme Court are appointed for life

subordinate courts: Courts of Appeal; Courts of First Instance; magistrate's courts; land, labor, and children's courts

note: the Superior Council of the Judiciary or Conseil Superieur du Pouvoir Judiciaire is a 9-member body charged with the administration and oversight of the judicial branch of government

Political parties and leaders:

Alternative League for Haitian Progress and Empowerment or LAPEH [Jude CELESTIN]

Christian Movement for a New Haiti or MCNH [Luc MESADIEU]

Christian National Movement for the Reconstruction of Haiti or UNCRH [Chavannes JEUNE]

Convention for Democratic Unity or KID [Evans PAUL]

Cooperative Action to Rebuild Haiti or KONBA [Jean William JEANTY]

December 16 Platform or Platfom 16 Desanm [Dr. Gerard BLOT]

Democratic Alliance Party or ALYANS [Evans PAUL] (coalition includes KID and PPRH)

Democratic Centers' National Council or CONACED [Osner FEVRY]

Dessalinian Patriotic and Popular Movement or MOPOD [Jean Andre VICTOR]

Effort and Solidarity to Create an Alternative for the People or ESKAMP [Joseph JASME]

Fanmi Lavalas or FL [Jean-Bertrand ARISTIDE]

For Us All or PONT [Jean-Marie CHERESTAL]

Fusion of Haitian Social Democrats or FHSD [Edmonde Supplice BEAUZILE]

Grouping of Citizens for Hope or RESPE [Charles-Henri BAKER]

Haitians for Haiti [Yvon NEPTUNE]

Haitian Tet Kale Party or PHTK [Ann Valerie Timothee MILFORT]

Haiti in Action or AAA [Youri LATORTUE]

Independent Movement for National Reconstruction or MIRN [Luc FLEURINORD]

Konbit Pou refe Ayiti or KONBIT

Lavni Organization or LAVNI [Yves CRISTALIN]

Liberal Party of Haiti or PLH [Jean Andre VICTOR]

Love Haiti or Renmen Ayiti [Jean-Henry CEANT, Camille LEBLANC]

Mobilization for National Development or MDN [Hubert de RONCERAY]

New Christian Movement for a New Haiti or MOCHRENA [Luc MESADIEU]

Organization for the Advancement of Haiti and Haitians or OLAHH

Party for the Integral Advancement of the Haitian People or PAIPH

Patriotic Unity or IP [Marie Denise CLAUDE]

Peasant's Response or Repons Peyizan [Michel MARTELLY]

Platform Alternative for Progress and Democracy or ALTENATIV [Victor BENOIT and Evans PAUL]

Platform of Haitian Patriots or PLAPH [Dejean BELISAIRE, Himmler REBU]

Platform Pitit Desaline or PPD [Jean-Charles MOISE]

Pont

Popular Party for the Renewal of Haiti or PPRH [Claude ROMAIN]

PPG18

Rally of Progressive National Democrats or RDNP [Mirlande MANIGAT]

Renmen Ayiti or RA [Jean-Henry CEANT]

Reseau National Bouclier or Bouclier Respect or RESPE

Strength in Unity or Ansanm Nou Fo [Leslie VOLTAIRE]

Struggling People's Organization or OPL [Jacques-Edouard ALEXIS]

Truth (Verite)

Union [Chavannes JEUNE]

Unity or Inite [Levaillant LOUIS-JEUNE]

Vigilance or Veye Yo [Lavarice GAUDIN]

International organization participation: ACP, AOSIS, Caricom, CD, CDB, CELAC, FAO, G-77, IADB, IAEA, IBRD, ICAO, ICC (NGOs), ICRM, IDA, IFAD, IFC, IFRCS, ILO, IMF, IMO, Interpol, IOC, IOM, IPU, ITSO, ITU, ITUC (NGOs), LAES, MIGA, NAM, OAS, OIF, OPANAL, OPCW, PCA, Petrocaribe, UN, UNCTAD, UNESCO, UNIDO, Union Latina, UNWTO, UPU, WCO, WFTU (NGOs), WHO, WIPO, WMO, WTO

Diplomatic representation in the US: *chief of mission:* Charge d'Affaires Herve DENIS (since 7 March 2019)

chancery: 2311 Massachusetts Avenue NW, Washington, DC 20008

telephone: [1] (202) 332-4090

FAX: [1] (202) 745-7215

consulate(s) general: Atlanta, Boston, Chicago, Miami, Orlando (FL), New York, San Juan (Puerto Rico)

Diplomatic representation from the US: *chief of mission:* Ambassador Michele SISON (since 21 February 2018)

telephone: [509] 229-8000

embassy: Tabarre 41, Route de Tabarre, Port-au-Prince

mailing address: (in Haiti) P.O. Box 1634, Port-au-Prince, Haiti; (from abroad) 3400 Port-au-Prince, State Department, Washington, DC 20521 -3400

FAX: [509] 229-8028

Flag description: two equal horizontal bands of blue (top) and red with a centered white rectangle bearing the coat of arms, which contains a palm tree flanked by flags and two cannons above a scroll bearing the motto L'UNION FAIT LA FORCE (Union Makes Strength); the colors are taken from the French Tricolor and represent the union of blacks and mulattoes

National symbol(s): *Hispaniolan trogon (bird), hibiscus flower; national colors:* blue, red

National anthem: *name:* "La Dessalinienne" (The Dessalines Song)

lyrics/music: Justin LHERISSON/Nicolas GEFFRARD

note: adopted 1904; named for Jean-Jacques DESSALINES, a leader in the Haitian Revolution and first ruler of an independent Haiti

0:00/ 0:36

ECONOMY

Economy—overview: Haiti is a free market economy with low labor costs and tariff-free access to the US for many of its exports. Two-fifths of all Haitians depend on the agricultural sector, mainly small-scale subsistence farming, which remains vulnerable to damage from frequent natural disasters. Poverty, corruption, vulnerability to natural disasters, and low levels of education for much of the population represent some of the most serious impediments to Haiti's economic growth. Remittances are the primary source of foreign exchange, equivalent to more than a quarter of GDP, and nearly double the combined value of Haitian exports and foreign direct investment.

Currently the poorest country in the Western Hemisphere, with close to 60% of the population living under the national poverty line, Haiti's GDP growth rose to 5.5% in 2011 as the Haitian economy began recovering from the devastating January 2010 earthquake that destroyed much of its capital city, Port-au-Prince, and neighboring areas. However, growth slowed to below 2% in 2015 and 2016 as political uncertainty, drought conditions, decreasing foreign aid, and the depreciation of the national currency took a toll on investment and economic growth. Hurricane Matthew, the fiercest Caribbean storm in nearly a decade, made landfall in Haiti on 4 October 2016, with 140 mile-per-hour winds, creating a new humanitarian emergency. An estimated 2.1 million people were affected by the category 4 storm, which caused extensive damage to crops, houses, livestock, and infrastructure across Haiti's southern peninsula.

US economic engagement under the Caribbean Basin Trade Partnership Act (CBTPA) and the 2008 Haitian Hemispheric Opportunity through Partnership Encouragement Act (HOPE II) have contributed to an increase in apparel exports and investment by providing duty-free access to the US. The Haiti Economic Lift Program (HELP) Act of 2010 extended the CBTPA and HOPE II until 2020, while the Trade Preferences Extension Act of 2015 extended trade benefits provided to Haiti in the HOPE and HELP Acts through September 2025. Apparel sector exports in 2016 reached approximately $850 million and account for over 90% of Haitian exports and more than 10% of the GDP.

Investment in Haiti is hampered by the difficulty of doing business and weak infrastructure, including access to electricity. Haiti's outstanding external debt was cancelled by donor countries following the 2010 earthquake, but has since risen to $2.6 billion as of December 2017, the majority of which is owed to Venezuela under the PetroCaribe program. Although the government has increased its revenue collection, it continues to rely on formal international economic assistance for fiscal sustainability, with over 20% of its annual budget coming from foreign aid or direct budget support.

GDP (purchasing power parity):
$19.97 billion (2017 est.)
$19.74 billion (2016 est.)
$19.46 billion (2015 est.)
note: data are in 2017 dollars
country comparison to the world: 150

GDP (official exchange rate):
$8.608 billion (2017 est.)

GDP—real growth rate:
1.2% (2017 est.)
1.5% (2016 est.)
1.2% (2015 est.)
country comparison to the world: 181

GDP—per capita (PPP):
$1,800 (2017 est.)
$1,800 (2016 est.)
$1,800 (2015 est.)
note: data are in 2017 dollars
country comparison to the world: 213

Gross national saving:
24.9% of GDP (2017 est.)
29.5% of GDP (2016 est.)
29.3% of GDP (2015 est.)
country comparison to the world: 61

GDP—composition, by end use:
household consumption: 99.1% (2017 est.)
government consumption: 10% (2016 est.)
investment in fixed capital: 32.6% (2016 est.)
investment in inventories: -1.4% (2017 est.)
exports of goods and services: 20% (2017 est.)
imports of goods and services: -60.3% (2017 est.)
note: figure for household consumption also includes government consumption

GDP—composition, by sector of origin:
agriculture: 22.1% (2017 est.)
industry: 20.3% (2017 est.)
services: 57.6% (2017 est.)

Agriculture—products: coffee, mangoes, cocoa, sugarcane, rice, corn, sorghum; wood, vetiver
Industries: textiles, sugar refining, flour milling, cement, light assembly using imported parts

Industrial production growth rate: 0.9% (2017 est.)
country comparison to the world: 161

Labor force: 4.594 million (2014 est.)
note: shortage of skilled labor; unskilled labor abundant
country comparison to the world: 88

Labor force—by occupation:
agriculture: 38.1%
industry: 11.5%
services: 50.4% (2010)

Unemployment rate: 40.6% (2010 est.)
note: widespread unemployment and underemployment; more than two-thirds of the labor force do not have formal jobs
country comparison to the world: 215

Population below poverty line: 58.5% (2012 est.)
Household income or consumption by percentage share: *lowest 10%:* 0.7%
highest 10%: 47.7% (2001)

Budget: *revenues:* 1.567 billion (2017 est.)

expenditures: 1.65 billion (2017 est.)
Taxes and other revenues: 18.2% (of GDP) (2017 est.)
country comparison to the world: 163

Budget surplus (+) or deficit (-): -1% (of GDP) (2017 est.)
country comparison to the world: 79

Public debt:
31.1% of GDP (2017 est.)
33.9% of GDP (2016 est.)
country comparison to the world: 164

Fiscal year: 1 October—30 September

Inflation rate (consumer prices):
14.7% (2017 est.)
13.4% (2016 est.)
country comparison to the world: 211

Current account balance:
-$348 million (2017 est.)
-$83 million (2016 est.)
country comparison to the world: 108

Exports:
$980.2 million (2017 est.)
$995 million (2016 est.)
country comparison to the world: 160

Exports—partners: US 80.6%, Dominican Republic 4.9% (2017)

Exports—commodities: apparel, manufactures, oils, cocoa, mangoes, coffee

Imports:
$3.618 billion (2017 est.)
$3.183 billion (2016 est.)
country comparison to the world: 143

Imports—commodities: food, manufactured goods, machinery and transport equipment, fuels, raw materials

Imports—partners: US 20.7%, China 18.8%, Netherlands Antilles 15.7%, Indonesia 8.5% (2017)

Reserves of foreign exchange and gold:
$2.361 billion (31 December 2017 est.)
$2.11 billion (31 December 2016 est.)
country comparison to the world: 117

Debt—external:
$2.762 billion (31 December 2017 est.)
$2.17 billion (31 December 2016 est.)
country comparison to the world: 145

Exchange rates: gourdes (HTG) per US dollar -
65.21 (2017 est.)
63.34 (2016 est.)
63.34 (2015 est.)
50.71 (2014 est.)
45.22 (2013 est.)

ENERGY

Electricity access: *population without electricity:* 8 million (2017)
electrification—total population: 38.7% (2016)
electrification—urban areas: 65.4% (2016)
electrification—rural areas: 0.5% (2016)

Electricity—production: 1.023 billion kWh (2016 est.)
country comparison to the world: 149

Electricity—consumption: 406.2 million kWh (2016 est.)
country comparison to the world: 173

Electricity—exports: 0 kWh (2016 est.)
country comparison to the world: 147

Electricity—imports: 0 kWh (2016 est.)
country comparison to the world: 159

Electricity—installed generating capacity: 332,000 kW (2016 est.)
country comparison to the world: 155

Electricity—from fossil fuels: 82% of total installed capacity (2016 est.)
country comparison to the world: 78

Electricity—from nuclear fuels: 0% of total installed capacity (2017 est.)
country comparison to the world: 106

Electricity—from hydroelectric plants: 18% of total installed capacity (2017 est.)
country comparison to the world: 94

Electricity—from other renewable sources: 0% of total installed capacity (2017 est.)
country comparison to the world: 191

Crude oil—production: 0 bbl/day (2018 est.)
country comparison to the world: 148

Crude oil—exports: 0 bbl/day (2015 est.)
country comparison to the world: 136

Crude oil—imports: 0 bbl/day (2015 est.)
country comparison to the world: 140

Crude oil—proved reserves: 0 bbl (1 January 2018 est.)
country comparison to the world: 144

Refined petroleum products—production: 0 bbl/day (2015 est.)
country comparison to the world: 156

Refined petroleum products—consumption: 21,000 bbl/day (2016 est.)
country comparison to the world: 137

Refined petroleum products—exports: 0 bbl/day (2015 est.)
country comparison to the world: 165

Refined petroleum products—imports: 20,030 bbl/day (2015 est.)
country comparison to the world: 122

Natural gas—production: 0 cu m (2017 est.)
country comparison to the world: 145

Natural gas—consumption: 0 cu m (2017 est.)
country comparison to the world: 158

Natural gas—exports: 0 cu m (2017 est.)
country comparison to the world: 120

Natural gas—imports: 0 cu m (2017 est.)
country comparison to the world: 138

Natural gas—proved reserves: 0 cu m (1 January 2014 est.)
country comparison to the world: 146

Carbon dioxide emissions from consumption of energy: 3.595 million Mt (2017 est.)
country comparison to the world: 141

COMMUNICATIONS

Telephones—fixed lines: *total subscriptions:* 5,464

subscriptions per 100 inhabitants: less than 1 (2019 est.)
country comparison to the world: 204

Telephones—mobile cellular: *total subscriptions:* 6,287,411
subscriptions per 100 inhabitants: 57.53 (2019 est.)
country comparison to the world: 110

Telecommunication systems: *general assessment:* telecommunications infrastructure is among the least-developed in Latin America and the Caribbean; domestic cell service is functional; Hurricane Matthew in 2016 caused $35 million worth of damage to telecoms infrastructure; some expansion of LTE services (2020)
domestic: fixed-line is less than 1 per 100; mobile-cellular telephone services have expanded greatly in the last decade due to low-cost GSM (Global Systems for Mobile) phones and pay-as-you-go plans; mobile-cellular teledensity is 58 per 100 persons (2019)
international: country code—509; landing points for the BDSNi and Fibralink submarine cables to 14 points in the Bahamas and Dominican Republic; satellite earth station—1 Intelsat (Atlantic Ocean) (2019)
note: the COVID-19 outbreak is negatively impacting telecommunications production and supply chains globally; consumer spending on telecom devices and services has also slowed due to the pandemic's effect on economies worldwide; overall progress towards improvements in all facets of the telecom industry—mobile, fixed-line, broadband, submarine cable and satellite—has moderated

Broadcast media: 98 television stations throughout the country, including 1 government-owned; cable TV subscription service available; 850 radio stations (of them, only 346 are licensed), including 1 government-owned; more than 100 community radio stations; over 64 FM stations in Port-au-Prince alone; VOA Creole Service broadcasts daily on 30 affiliate stations (2016)

Internet country code: .ht

Internet users: *total:* 3,503,006
percent of population: 32.47% (July 2018 est.)
country comparison to the world: 95

Broadband—fixed subscriptions: *total:* 31,100
subscriptions per 100 inhabitants: less than 1 (2018 est.)
country comparison to the world: 142

TRANSPORTATION

National air transport system: *number of registered air carriers:* 1 (2020)
inventory of registered aircraft operated by air carriers: 1

Civil aircraft registration country code prefix: HH (2016)

Airports: 14 (2013)
country comparison to the world: 149

Airports—with paved runways: *total:* 4 (2019)
2,438 to 3,047 m: 2
914 to 1,523 m: 2

Airports—with unpaved runways: *total:* 10 (2013)
914 to 1,523 m: 2 (2013)
under 914 m: 8 (2013)

Roadways: *total:* 4,266 km (2009)
paved: 768 km (2009)
unpaved: 3,498 km (2009)
country comparison to the world: 152

Merchant marine: total: 4
by type: general cargo 3 other 1 (2019)
country comparison to the world: 167

Ports and terminals: *major seaport(s):* Cap-Haitien, Gonaives, Jacmel, Port-au-Prince

MILITARY AND SECURITY

Military and security forces: the Haitian Armed Forces (FAdH), disbanded in 1995, began to be reconstituted in 2017 to assist with natural disaster relief, border security, and combating transnational crime; it established an Army command in 2018; the small Coast Guard is not part of the military, but rather the Haitian National Police (2020)

Military and security service personnel strengths: the country's army is planned to eventually number around 5,000 personnel (2019 est.)

Military equipment inventories and acquisitions: N/A

TRANSNATIONAL ISSUES

Disputes—international: since 2004, peacekeepers from the UN Stabilization Mission in Haiti have assisted in maintaining civil order in Haiti; the mission currently includes 6,685 military, 2,607 police, and 443 civilian personnel; despite efforts to control illegal migration, Haitians cross into the Dominican Republic and sail to neighboring countries; Haiti claims US-administered Navassa Island

Refugees and internally displaced persons: *IDPs:* 34,508 (includes only IDPs from the 2010 earthquake living in camps or camp-like situations; information is lacking about IDPs living outside of camps or who have left camps) (2019)
stateless persons: 2,992 (2018); note—individuals without a nationality who were born in the Dominican Republic prior to January 2010

Trafficking in persons: *current situation:* Haiti is a source, transit, and destination country for men, women, and children subjected to forced labor and sex trafficking; most of Haiti's trafficking cases involve children in domestic servitude vulnerable to physical and sexual abuse; dismissed and runaway child domestic servants often end up in prostitution, begging, or street crime; other exploited populations included low-income Haitians, child laborers, and women and children living in IDP camps dating to the 2010 earthquake; Haitian adults are vulnerable to fraudulent labor recruitment abroad and, along with children, may be subjected to forced labor in the Dominican Republic, elsewhere in the Caribbean, South America, and the US; Dominicans are exploited in sex trafficking and forced labor in Haiti

tier rating: Tier 2 Watch List — Haiti does not fully comply with the minimum standards for the elimination of trafficking; however, it is making significant efforts to do so; in 2014, Haiti was granted a waiver from an otherwise required downgrade to Tier 3 because its government has a written plan that, if implemented would constitute making significant efforts to bring itself into compliance with the minimum standards for the elimination of trafficking; in 2014, Haiti developed a national anti-trafficking action plan and enacted a law prohibiting all forms of human trafficking, although judicial corruption hampered its implementation; progress was made in investigating and prosecuting suspected traffickers, but no convictions were made; the government sustained limited efforts to identify and refer victims to protective services, which were provided mostly by NGOs without government support; ampaigns to raise awareness about child labor and child trafficking continued (2015)

Illicit drugs: Caribbean transshipment point for cocaine en route to the US and Europe; substantial bulk cash smuggling activity; Colombian narcotics traffickers favor Haiti for illicit financial transactions; pervasive corruption; significant consumer of cannabis

HEARD ISLAND AND MCDONALD ISLANDS

INTRODUCTION

Background: The UK transferred these uninhabited, barren, sub-Antarctic islands to Australia in 1947. Populated by large numbers of seal and bird species, the islands have been designated a nature preserve.

GEOGRAPHY

Location: islands in the Indian Ocean, about two-thirds of the way from Madagascar to Antarctica

Geographic coordinates: 53 06 S, 72 31 E

Map references: Antarctic Region

Area: *total:* 412 sq km
land: 412 sq km
water: 0 sq km
country comparison to the world: 203

Area—comparative: slightly more than two times the size of Washington, DC

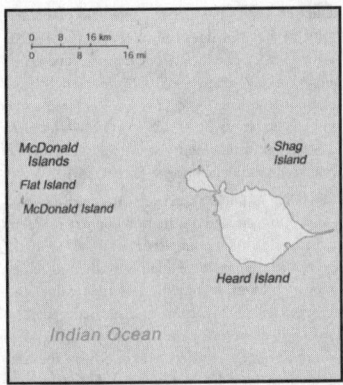

Land boundaries: *0 km*

Coastline: 101.9 km

Maritime claims: *territorial sea:* 12 nm
exclusive fishing zone: 200 nm

Climate: antarctic

Terrain: Heard Island—80% ice-covered, bleak and mountainous, dominated by a large massif (Big Ben) and an active volcano (Mawson Peak); McDonald Islands—small and rocky

Elevation: *lowest point:* Indian Ocean 0 m
highest point: Mawson Peak on Big Ben volcano 2,745 m

Natural resources: fish

Land use: *agricultural land:* 0% (2011 est.)

arable land: 0% (2011 est.) / *permanent crops:* 0% (2011 est.) / *permanent pasture:* 0% (2011 est.)
forest: 0% (2011 est.)
other: 100% (2011 est.)

Natural hazards: Mawson Peak, an active volcano, is on Heard Island

Environment—current issues: none; uninhabited and mostly ice covered

Geography—note: Mawson Peak on Heard Island is the highest Australian mountain (at 2,745 meters, it is taller than Mt. Kosciuszko in Australia proper), and one of only two active volcanoes located in Australian territory, the other being McDonald Island; in 1992, McDonald Island broke its dormancy and began erupting; it has erupted several times since, most recently in 2005

PEOPLE AND SOCIETY

Population: uninhabited

GOVERNMENT

Country name: *conventional long form:* Territory of Heard Island and McDonald Islands
conventional short form: Heard Island and McDonald Islands
abbreviation: HIMI
etymology: named after American Captain John HEARD, who sighted the island on 25 November 1853, and American Captain William McDONALD, who discovered the islands on 4 January 1854

Dependency status: territory of Australia; administered from Canberra by the Department of

Sustainability, Environment, Water, Population and Communities (Australian Antarctic Division)

Legal system: the laws of Australia apply where applicable

Diplomatic representation in the US: none (territory of Australia)

Diplomatic representation from the US: none (territory of Australia)

Flag description: the flag of Australia is used

ECONOMY

Economy—overview: The islands have no indigenous economic activity, but the Australian Government allows limited fishing in the surrounding waters. Visits to Heard Island typically focus on terrestrial and marine research and infrequent private expeditions.

COMMUNICATIONS

Internet country code: .hm

TRANSPORTATION

Ports and terminals: none; offshore anchorage only

MILITARY AND SECURITY

Military—note: defense is the responsibility of Australia; Australia conducts fisheries patrols

TRANSNATIONAL ISSUES

Disputes—international: none

HOLY SEE (Vatican City)

INTRODUCTION

Background: Popes in their secular role ruled portions of the Italian peninsula for more than a thousand years until the mid-19th century, when many

of the Papal States were seized by the newly united Kingdom of Italy. In 1870, the pope's holdings were further circumscribed when Rome itself was annexed. Disputes between a series of "prisoner" popes and Italy were resolved in 1929 by three Lateran Treaties, which established the independent state of Vatican City and granted Roman Catholicism special status in Italy. In 1984, a concordat between the Holy See and Italy modified certain of the earlier treaty provisions, including the primacy of Roman Catholicism as the Italian state religion. Present concerns of the Holy See include religious freedom, threats against minority Christian communities in Africa and the Middle East, the plight of refugees and migrants, sexual misconduct by clergy, international development, interreligious dialogue and reconciliation, and the application of church doctrine in an era of rapid change and globalization. About 1.3 billion people worldwide profess Catholicism—the world's largest Christian faith.

GEOGRAPHY

Location: Southern Europe, an enclave of Rome (Italy)

Geographic coordinates: 41 54 N, 12 27 E

Map references: Europe

Area: *total:* 0 sq km
land: 0.44 sq km
water: 0 sq km
country comparison to the world: 258

Area—comparative: about 0.7 times the size of the National Mall in Washington, DC

Land boundaries: *total:* 3.4 km
border countries (1): Italy 3.4 km

Coastline: 0 km (landlocked)

Maritime claims: none (landlocked)

Climate: temperate; mild, rainy winters (September to May) with hot, dry summers (May to September)

Terrain: urban; low hill

Elevation: *lowest point:* Saint Peter's Square 19 m

highest point: Vatican Gardens (Vatican Hill) 78 m

Natural resources: none

Land use: *agricultural land:* 0% (2011 est.)

arable land: 0% (2011 est.) / permanent crops: 0% (2011 est.) / permanent pasture: 0% (2011 est.)

forest: 0% (2011 est.)

other: 100% (2011 est.)

Natural hazards: occasional earthquakes

Environment—current issues: some air pollution from the surrounding city of Rome

Environment—international agreements: *party to:* Ozone Layer Protection

signed, but not ratified: Air Pollution, Environmental Modification

Geography—note: landlocked; an enclave in Rome, Italy; world's smallest state; beyond the territorial boundary of Vatican City, the Lateran Treaty of 1929 grants the Holy See extraterritorial authority over 23 sites in Rome and five outside of Rome, including the Pontifical Palace at Castel Gandolfo (the Pope's summer residence)

PEOPLE AND SOCIETY

Population: 1,000 (2019 est.)

country comparison to the world: 236

Nationality: *noun:* none

adjective: none

Ethnic groups: Italian, Swiss, Argentinian, and other nationalities from around the world (2017)

Languages: Italian, Latin, French, various other languages

Religions: Roman Catholic

Population growth rate: 0% (2014 est.)

country comparison to the world: 193

Urbanization: *urban population:* 100% of total population (2020)

rate of urbanization: -0.05% annual rate of change (2015-20 est.)

total population growth rate v. urban population growth rate, 2000-2030: Major urban areas—population: 1,000 VATICAN CITY (capital) (2018)

Drinking water source:

improved: *total:* 100% of population

unimproved: *total:* 0% of population (2017 est.)

HIV/AIDS—adult prevalence rate: NA

HIV/AIDS—people living with HIV/AIDS: NA

HIV/AIDS—deaths: NA

Education expenditures: NA

GOVERNMENT

Country name: *conventional long form:* The Holy See (Vatican City State)

conventional short form: Holy See (Vatican City)

local long form: La Santa Sede (Stato della Citta del Vaticano)

local short form: Santa Sede (Citta del Vaticano)

etymology: "holy" comes from the Greek word "hera" meaning "sacred"; "see" comes from the Latin word "sedes" meaning "seat," and refers to the episcopal chair; the term "Vatican" derives from the hill Mons Vaticanus on which the Vatican is located and which comes from the Latin "vaticinari" (to prophesy), referring to the fortune tellers and soothsayers who frequented the area in Roman times

Government type: ecclesiastical elective monarchy; self-described as an "absolute monarchy"

Capital: *name:* Vatican City

geographic coordinates: 41 54 N, 12 27 E

time difference: UTC+1 (6 hours ahead of Washington, DC, during Standard Time)

daylight saving time: +1hr, begins last Sunday in March; ends last Sunday in October

Administrative divisions: none

Independence: 11 February 1929; note—the three treaties signed with Italy on 11 February 1929 acknowledged, among other things, the full sovereignty of the Holy See and established its territorial extent; however, the origin of the Papal States, which over centuries varied considerably in extent, may be traced back to A.D. 754

National holiday: Election Day of Pope FRANCIS, 13 March (2013)

Constitution: *history:* previous 1929, 1963; latest adopted 26 November 2000, effective 22 February 2001 (Fundamental Law of Vatican City State); note—in October 2013, Pope Francis instituted a 9-member Council of Cardinal Advisors to reform the administrative apparatus of the Holy See (Roman Curia) to include writing a new constitution

amendments: note—although the Fundamental Law of Vatican City State makes no mention of amendments, Article Four (drafting laws), states that this legislative responsibility resides with the Pontifical Commission for Vatican City State; draft legislation is submitted through the Secretariat of State and considered by the pope

Legal system: religious legal system based on canon (religious) law

International law organization participation: has not submitted an ICJ jurisdiction declaration; non-party state to the ICCt

Citizenship: *citizenship by birth:* no

citizenship by descent only: no

dual citizenship recognized: no

residency requirement for naturalization: not applicable

note: in the Holy See, citizenship is acquired by law, ex iure, or by adminstrative decision; in the first instance, citizenship is a function of holding office within the Holy See as in the case of cardinals resident in Vatican City or diplomats of the Holy See; in the second instance, citizenship may be requested in a limited set of circumstances for those who reside within Vatican City under papal authorization, as a function of their office or service, or as the spouses and children of current citizens; citizenship is lost once an individual no longer permanently resides in Vatican City, normally reverting to the citizenship previously held

Suffrage: election of the pope is limited to cardinals less than 80 years old

Executive branch: *chief of state:* Pope FRANCIS (since 13 March 2013)

head of government: Secretary of State Cardinal Pietro PAROLIN (since 15 October 2013); note—Head of Government of Vatican City is President Cardinal Giuseppe BERTELLO (since 1 October 2011)

cabinet: Pontifical Commission for the State of Vatican City appointed by the pope

elections/appointments: pope elected by the College of Cardinals, usually for life or until voluntary resignation; election last held on 13 March 2013 (next to be held after the death or resignation of the current pope); Secretary of State appointed by the pope

election results: Jorge Mario BERGOGLIO, former Archbishop of Buenos Aires, elected Pope FRANCIS

Legislative branch: *description:* unicameral Pontifical Commission for Vatican City State or Pontificia Commissione per lo Stato della Citta del Vaticano (7 seats; members appointed by the pope to serve 5-year terms)

elections: last held on 11 July 2018

election results: composition—men 7, women 0

Judicial branch: *highest courts:* Supreme Court or Supreme Tribunal of the Apostolic Signatura (consists of the cardinal prefect, who serves as ex-officio president of the court, and 2 other cardinals of the Prefect Signatura); note—judicial duties were established by the Motu Proprio, papal directive, of Pope PIUS XII on 1 May 1946; most Vatican City criminal matters are handled by the Republic of Italy courts

judge selection and term of office: cardinal prefect appointed by the pope; the other 2 cardinals of the court appointed by the cardinal prefect on a yearly basis

subordinate courts: Appellate Court of Vatican City; Tribunal of Vatican City

Political parties and leaders: none

International organization participation: CE (observer), IAEA, Interpol, IOM, ITSO, ITU, ITUC (NGOs), OAS (observer), OPCW, OSCE, Schengen Convention (de facto member), SICA (observer), UN (observer), UNCTAD, UNHCR, Union Latina (observer), UNWTO (observer), UPU, WIPO, WTO (observer)

Diplomatic representation in the US: *chief of mission:* Apostolic Nuncio Archbishop Christophe PIERRE (since 27 June 2016)

chancery: 3339 Massachusetts Avenue NW, Washington, DC 20008

telephone: [1] (202) 333-7121

FAX: [1] (202) 337-4036

Diplomatic representation from the US: *chief of mission:* Ambassador Callista GINGRICH (since 22 December 2017)

telephone: [39] (06) 4674-1

429

embassy: American Embassy to the Holy See, Via Sallustiana, 49, 00187 Rome

mailing address: Unit 5660, Box 66, DPO AE 09624-0066

FAX: [39] (06) 4674-3412

Flag description: two vertical bands of yellow (hoist side) and white with the arms of the Holy See, consisting of the crossed keys of Saint Peter surmounted by the three-tiered papal tiara, centered in the white band; the yellow color represents the pope's spiritual power, the white his worldly power

National symbol(s): *crossed keys beneath a papal tiara; national colors*: yellow, white

National anthem: *name*: "Inno e Marcia Pontificale" (Hymn and Pontifical March); often called The Pontifical Hymn

lyrics/music: Raffaello LAVAGNA/Charles-Francois GOUNOD

note: adopted 1950

ECONOMY

Economy—overview: The Holy See is supported financially by a variety of sources, including investments, real estate income, and donations from Catholic individuals, dioceses, and institutions; these help fund the Roman Curia (Vatican bureaucracy), diplomatic missions, and media outlets. Moreover, an annual collection taken up in dioceses and from direct donations go to a nonbudgetary fund, known as Peter's Pence, which is used directly by the pope for charity, disaster relief, and aid to churches in developing nations.

The separate Vatican City State budget includes the Vatican museums and post office and is supported financially by the sale of stamps, coins, medals, and tourist mementos as well as fees for admission to museums and publication sales. Revenues increased between 2010 and 2011 because of expanded operating hours and a growing number of visitors. However, the Holy See did not escape the financial difficulties experienced by other European countries; in 2012, it started a spending review to determine where to cut costs to reverse its 2011 budget deficit of $20 million. The Holy See generated a modest surplus in 2012 before recording a $32 million deficit in 2013, driven primarily by the decreasing value of gold. The incomes and living standards of lay workers

are comparable to those of counterparts who work in the city of Rome so most public expenditures go to wages and other personnel costs;. In February 2014, Pope FRANCIS created the Secretariat of the Economy to oversee financial and administrative operations of the Holy See, part of a broader campaign to reform the Holy See's finances.

GDP (purchasing power parity): NA

Industries: printing; production of coins, medals, postage stamps; mosaics, staff uniforms; worldwide banking and financial activities

Labor force: 4,822 (2016)

country comparison to the world: 221

Labor force—by occupation: *note*: essentially services with a small amount of industry; nearly all dignitaries, priests, nuns, guards, and the approximately 3,000 lay workers live outside the Vatican

Population below poverty line: NA

Budget: *revenues*: 315 million (2013)

expenditures: 348 million (2013)

Taxes and other revenues: NA

Budget surplus (+) or deficit (-): NA

Fiscal year: calendar year

Exchange rates: euros (EUR) per US dollar - 0.885 (2017 est.)
0.903 (2016 est.)
0.9214 (2015 est.)
0.885 (2014 est.)
0.7634 (2013 est.)

ENERGY

Electricity access: *electrification—total population*: 100% (2020)

COMMUNICATIONS

Telecommunication systems: *general assessment*: automatic digital exchange (2018)

domestic: connected via fiber-optic cable to Telecom Italia network (2018)

international: country code—39; uses Italian system

note: the COVID-19 outbreak is negatively impacting telecommunications production and supply chains globally; consumer spending on telecom devices and services has also slowed due to the

pandemic's effect on economies worldwide; overall progress towards improvements in all facets of the telecom industry—mobile, fixed-line, broadband, submarine cable and satellite—has moderated

Broadcast media: the Vatican Television Center (CTV) transmits live broadcasts of the Pope's Sunday and Wednesday audiences, as well as the Pope's public celebrations; CTV also produces documentaries; Vatican Radio is the Holy See's official broadcasting service broadcasting via shortwave, AM and FM frequencies, and via satellite and Internet connections

Internet country code: .va

Communications—note: the Vatican Apostolic Library is one of the world's oldest libraries, formally established in 1475, but actually much older; it holds a significant collection of historic texts including 1.1 million printed books and 75,000 codices (manuscript books with handwritten contents); it serves as a research library for history, law, philosophy, science, and theology; the library's collections have been described as "the world's greatest treasure house of the writings at the core of Western tradition"

TRANSPORTATION

MILITARY AND SECURITY

Military and security forces: Pontifical Swiss Guard Corps (Corpo della Guardia Svizzera Pontificia); the Gendarmerie Corps of Vatican City is a police force that helps augment the Pontifical Swiss Guard during the Pope's appearances, as well as providing general security, traffic direction, and investigative duties for the Vatican City State (2019)

Military service age and obligation: *Pontifical Swiss Guard Corps (Corpo della Guardia Svizzera Pontificia)*: 19-30 years of age for voluntary military service; no conscription; must be Roman Catholic, a single male, and a Swiss citizen, with a secondary education (2019)

Military—note: defense is the responsibility of Italy

TRANSNATIONAL ISSUES

Disputes—international: none

HONDURAS

INTRODUCTION

Background: Once part of Spain's vast empire in the New World, Honduras became an independent nation in 1821. After two and a half decades of mostly military rule, a freely elected civilian government came to power in 1982. During the 1980s, Honduras proved a haven for anti-Sandinista contras fighting the Marxist Nicaraguan Government and an ally to Salvadoran Government forces

fighting leftist guerrillas. The country was devastated by Hurricane Mitch in 1998, which killed about 5,600 people and caused approximately $2 billion in damage. Since then, the economy has slowly rebounded.

GEOGRAPHY

Location: Central America, bordering the Caribbean Sea, between Guatemala and Nicaragua

and bordering the Gulf of Fonseca (North Pacific Ocean), between El Salvador and Nicaragua

Geographic coordinates: 15 00 N, 86 30 W

Map references: Central America and the Caribbean

Area: *total*: 112,090 sq km

land: 111,890 sq km

water: 200 sq km

country comparison to the world: 104

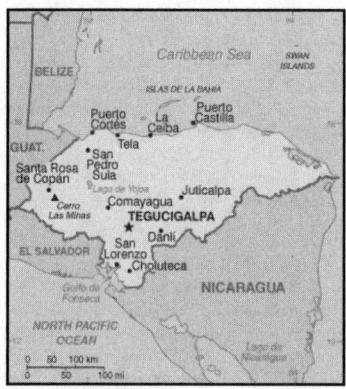

Area—comparative: slightly larger than Tennessee

Land boundaries: *total:* 1,575 km
border countries (3): Guatemala 244 km, El Salvador 391 km, Nicaragua 940 km

Coastline: 823 km (Caribbean Sea 669 km, Gulf of Fonseca 163 km)

Maritime claims: *territorial sea:* 12 nm
exclusive economic zone: 200 nm
contiguous zone: 24 nm
continental shelf: natural extension of territory or to 200 nm

Climate: subtropical in lowlands, temperate in mountains

Terrain: mostly mountains in interior, narrow coastal plains

Elevation: *mean elevation:* 684 m
lowest point: Caribbean Sea 0 m
highest point: Cerro Las Minas 2,870 m

Natural resources: timber, gold, silver, copper, lead, zinc, iron ore, antimony, coal, fish, hydropower

Land use: *agricultural land:* 28.8% (2011 est.)
arable land: 9.1% (2011 est.) / permanent crops: 4% (2011 est.) / permanent pasture: 15.7% (2011 est.)
forest: 45.3% (2011 est.)
other: 25.9% (2011 est.)

Irrigated land: 900 sq km (2012)

Population distribution: most residents live in the mountainous western half of the country; unlike other Central American nations, Honduras is the only one with an urban population that is distributed between two large centers—the capital of Tegucigalpa and the city of San Pedro Sula; the Rio Ulua valley in the north is the only densely populated lowland area

Natural hazards: frequent, but generally mild, earthquakes; extremely susceptible to damaging hurricanes and floods along the Caribbean coast

Environment—current issues: urban population expanding; deforestation results from logging and the clearing of land for agricultural purposes; further land degradation and soil erosion hastened by uncontrolled development and improper land use practices such as farming of marginal lands; mining

activities polluting Lago de Yojoa (the country's largest source of fresh water), as well as several rivers and streams, with heavy metals

Environment—international agreements: party to: Biodiversity, Climate Change, Climate Change-Kyoto Protocol, Desertification, Endangered Species, Hazardous Wastes, Law of the Sea, Marine Dumping, Ozone Layer Protection, Ship Pollution, Tropical Timber 83, Tropical Timber 94, Wetlands *signed, but not ratified:* none of the selected agreements

Geography—note: has only a short Pacific coast but a long Caribbean shoreline, including the virtually uninhabited eastern Mosquito Coast

PEOPLE AND SOCIETY

Population: 9,235,340 (July 2020 est.)
note: estimates for this country explicitly take into account the effects of excess mortality due to AIDS; this can result in lower life expectancy, higher infant mortality, higher death rates, lower population growth rates, and changes in the distribution of population by age and sex than would otherwise be expected
country comparison to the world: 95

Nationality: *noun:* Honduran(s)
adjective: Honduran

Ethnic groups: mestizo (mixed Amerindian and European) 90%, Amerindian 7%, black 2%, white 1%

Languages: Spanish (official), Amerindian dialects

Religions: Roman Catholic 46%, Protestant 41%, atheist 1%, other 2%, none 9% (2014 est.)

Demographic profile: Honduras is one of the poorest countries in Latin America and has one of the world's highest murder rates. More than half of the population lives in poverty and per capita income is one of the lowest in the region. Poverty rates are higher among rural and indigenous people and in the south, west, and along the eastern border than in the north and central areas where most of Honduras' industries and infrastructure are concentrated. The increased productivity needed to break Honduras' persistent high poverty rate depends, in part, on further improvements in educational attainment. Although primary-school enrollment is near 100%, educational quality is poor, the drop-out rate and grade repetition remain high, and teacher and school accountability is low.

Honduras' population growth rate has slowed since the 1990s, but it remains high at nearly 2% annually because the birth rate averages approximately three children per woman and more among rural, indigenous, and poor women.

Consequently, Honduras' young adult population—ages 15 to 29—is projected to continue growing rapidly for the next three decades and then stabilize or slowly shrink. Population growth and limited job prospects outside of agriculture will continue to drive emigration. Remittances represent about a fifth of GDP.

Age structure: *0-14 years:* 30.2% (male 1,411,537/female 1,377,319)
15-24 years: 21.03% (male 969,302/female 972,843)
25-54 years: 37.79% (male 1,657,260/female 1,832,780)
55-64 years: 5.58% (male 233,735/female 281,525)
65 years and over: 5.4% (male 221,779/female 277,260) (2020 est.)

Dependency ratios: *total dependency ratio:* 55.2
youth dependency ratio: 47.5
elderly dependency ratio: 7.7
potential support ratio: 13 (2020 est.)

Median age: *total:* 24.4 years
male: 23.5 years
female: 25.2 years (2020 est.)
country comparison to the world: 165

Population growth rate: 1.27% (2020 est.)
country comparison to the world: 85

Birth rate: 18.5 births/1,000 population (2020 est.)
country comparison to the world: 84

Death rate: 4.7 deaths/1,000 population (2020 est.)
country comparison to the world: 205

Net migration rate: -1.4 migrant(s)/1,000 population (2020 est.)
country comparison to the world: 153

Population distribution: most residents live in the mountainous western half of the country; unlike other Central American nations, Honduras is the only one with an urban population that is distributed between two large centers—the capital of Tegucigalpa and the city of San Pedro Sula; the Rio Ulua valley in the north is the only densely populated lowland area

Urbanization: *urban population:* 58.4% of total population (2020)
rate of urbanization: 2.75% annual rate of change (2015-20 est.)
total population growth rate v. urban population growth rate, 2000-2030

Major urban areas—population: 1.444 million TEGUCIGALPA (capital), 903,000 San Pedro Sula (2020)

Sex ratio: *at birth:* 1.03 male(s)/female
0-14 years: 1.02 male(s)/female
15-24 years: 1 male(s)/female
25-54 years: 0.9 male(s)/female
55-64 years: 0.83 male(s)/female
65 years and over: 0.8 male(s)/female
total population: 0.95 male(s)/female (2020 est.)

Mother's mean age at first birth: 20.4 years (2011/12 est.)
note: median age a first birth among women 25-29

Maternal mortality rate: 65 deaths/100,000 live births (2017 est.)
country comparison to the world: 86

Infant mortality rate: *total:* 14.6 deaths/1,000 live births
male: 16.6 deaths/1,000 live births
female: 12.4 deaths/1,000 live births (2020 est.)
country comparison to the world: 100

Life expectancy at birth: *total population:* 74.6 years
male: 71.1 years
female: 78.3 years (2020 est.)
country comparison to the world: 129

Total fertility rate: 2.09 children born/woman (2020 est.)
country comparison to the world: 101

Contraceptive prevalence rate: 73.2% (2011/12)

Drinking water source:
improved:
urban: 100% of population
rural: 88.9% of population
total: 94.8% of population
unimproved:
urban: 0% of population
rural: 11.1% of population
total: 5.2% of population (2017 est.)

Current Health Expenditure: 7.9% (2017)

Physicians density: 0.31 physicians/1,000 population (2017)

Hospital bed density: 0.6 beds/1,000 population (2017)

Sanitation facility access:
improved:
urban: 95.4% of population
rural: 83.5% of population
total: 90.2% of population
unimproved:
urban: 4.6% of population
rural: 16.5% of population (2015 est.)
total: 9.8% of population (2017 est.)

HIV/AIDS—adult prevalence rate: 0.4% (2019 est.)
country comparison to the world: 77

HIV/AIDS—people living with HIV/AIDS: 25,000 (2019 est.)
country comparison to the world: 82

HIV/AIDS—deaths: <1000 (2019 est.)

Major infectious diseases: *degree of risk:* high (2020)
food or waterborne diseases: bacterial diarrhea, hepatitis A, and typhoid fever
vectorborne diseases: dengue fever and malaria

Obesity—adult prevalence rate: 21.4% (2016)
country comparison to the world: 89

Children under the age of 5 years underweight: 7.1% (2012)
country comparison to the world: 72

Education expenditures: 6% of GDP (2017)
country comparison to the world: 32

Literacy: *definition:* age 15 and over can read and write
total population: 87.2%
male: 87.1%
female: 87.3% (2016)

School life expectancy (primary to tertiary education): *total:* 10 years
male: 10 years
female: 11 years (2017)

Unemployment, youth ages 15-24: *total:* 10.7%
male: 7.7%

female: 16.3% (2018 est.)
country comparison to the world: 120

GOVERNMENT

Country name: *conventional long form:* Republic of Honduras
conventional short form: Honduras
local long form: Republica de Honduras
local short form: Honduras
etymology: the name means "depths" in Spanish and refers to the deep anchorage in the northern Bay of Trujillo

Government type: presidential republic

Capital: *name:* Tegucigalpa; note—article eight of the Honduran constitution states that the twin cities of Tegucigalpa and Comayaguela, jointly, constitute the capital of the Republic of Honduras; however, virtually all governmental institutions are on the Tegucigalpa side, which in practical terms makes Tegucigalpa the capital
geographic coordinates: 14 06 N, 87 13 W
time difference: UTC-6 (1 hour behind Washington, DC during Standard Time)
etymology: while most sources agree that Tegucigalpa is of Nahuatl derivation, there is no consensus on its original meaning

Administrative divisions: 18 departments (departamentos, singular—departamento); Atlantida, Choluteca, Colon, Comayagua, Copan, Cortes, El Paraiso, Francisco Morazan, Gracias a Dios, Intibuca, Islas de la Bahia, La Paz, Lempira, Ocotepeque, Olancho, Santa Barbara, Valle, Yoro

Independence: 15 September 1821 (from Spain)

National holiday: Independence Day, 15 September (1821)

Constitution: *history:* several previous; latest approved 11 January 1982, effective 20 January 1982
amendments: proposed by the National Congress with at least two-thirds majority vote of the membership; passage requires at least two-thirds majority vote of Congress in its next annual session; constitutional articles, such as the form of government, national sovereignty, the presidential term, and the procedure for amending the constitution, cannot be amended; amended many times, last in 2015; note—the 2015 amendment struck down several constitutional articles on presidential term limits

Legal system: civil law system

International law organization participation: accepts compulsory ICJ jurisdiction with reservations; accepts ICCt jurisdiction

Citizenship: *citizenship by birth:* yes
citizenship by descent only: yes
dual citizenship recognized: yes
residency requirement for naturalization: 1 to 3 years

Suffrage: 18 years of age; universal and compulsory

Executive branch: *chief of state:* President Juan Orlando HERNANDEZ Alvarado (since 27 January 2014); Vice Presidents Ricardo ALVAREZ, Maria RIVERA, and Olga ALVARADO (since 26

January 2018); note—the president is both chief of state and head of government
head of government: President Juan Orlando HERNANDEZ Alvarado (since 27 January 2014); Vice Presidents Ricardo ALVAREZ, Maria RIVERA, and Olga ALVARADO (since 26 January 2018)
cabinet: Cabinet appointed by president
elections/appointments: president directly elected by simple majority popular vote for a 4-year term; election last held on 26 November 2017 (next to be held in November 2021); note—in 2015, the Constitutional Chamber of the Honduran Supreme Court struck down the constitutional provisions on presidential term limits
election results: Juan Orlando HERNANDEZ Alvarado reelected president; percent of vote Juan Orlando HERNANDEZ Alvarado (PNH) 43%, Salvador NASRALLA (Alianza de Oposicion conta la Dictadura) 41.4%, Luis Orlando ZELAYA Medrano (PL) 14.7%, other .9%

Legislative branch: *description:* unicameral National Congress or Congreso Nacional (128 seats; members directly elected in multi-seat constituencies by closed, party-list proportional representation vote; members serve 4-year terms)
elections: last held on 27 November 2017 (next to be held on 28 November 2021)
election results: percent of vote by party—PNH 47.7%, LIBRE 23.4%, PL 20.3%, AP 3.1%, PINU 3.1%, DC 0.8%, PAC 0.8%, UD 0.8%; seats by party—PNH 61, LIBRE 30, PL 26, AP 4, PINU 4, DC 1, PAC 1, UD 1; composition—men 101, women 27, percent of women 21.1%

Judicial branch: *highest courts:* Supreme Court of Justice or Corte Suprema de Justicia (15 principal judges, including the court president, and 7 alternates; court organized into civil, criminal, constitutional, and labor chambers); note—the court has both judicial and constitutional jurisdiction
judge selection and term of office: court president elected by his peers; judges elected by the National Congress from candidates proposed by the Nominating Board, a diverse 7-member group of judicial officials and other government and non-government officials nominated by each of their organizations; judges elected by Congress for renewable, 7-year terms
subordinate courts: courts of appeal; courts of first instance; justices of the peace

Political parties and leaders:
Alliance against the Dictatorship or Alianza de Oposicion conta la Dictadura [Salvador NASRALLA] (electoral coalition) Anti-Corruption Party or PAC [Marlene ALVARENGA]
Christian Democratic Party or DC [Lucas AGUILERA]
Democratic Unification Party or UD [Alfonso DIAZ]
Freedom and Refoundation Party or LIBRE [Jose Manuel ZELAYA Rosales]
Honduran Patriotic Alliance or AP [Romeo VASQUEZ Velasquez]

Liberal Party or PL [Luis Orlando ZELAYA Medrano]
National Party of Honduras or PNH [Reinaldo SANCHEZ Rivera]
Innovation and Unity Party or PINU [Guillermo VALLE]

International organization participation: BCIE, CACM, CD, CELAC, EITI (candidate country), FAO, G-11, G-77, IADB, IAEA, IBRD, ICAO, ICCt, ICRM, IDA, IFAD, IFC, IFRCS, ILO, IMF, IMO, Interpol, IOC (suspended), IOM, IPU, ISO (subscriber), ITSO, ITU, ITUC (NGOs), LAES, LAIA (observer), MIGA, MINURSO, MINUSTAH, NAM, OAS, OPANAL, OPCW, Pacific Alliance (observer), PCA, Petrocaribe, SICA, UN, UNCTAD, UNESCO, UNIDO, Union Latina, UNWTO, UPU, WCO (suspended), WFTU (NGOs), WHO, WIPO, WMO, WTO

Diplomatic representation in the US: chief of mission: Ambassador Luis Fernando SUAZO BARAHONA (since 17 September 2020)
chancery: Suite 700, 1250 Connecticut Avenue NW, Washington, DC 20036
telephone: [1] (202) 966-7702
FAX: [1] (202) 966-9751
consulate(s) general: Atlanta, Chicago, Houston, Los Angeles, Miami, New Orleans, New York, San Francisco
consulate(s): Dallas, McAllen (TX)

Diplomatic representation from the US: chief of mission: Ambassador (vacant); Charge d'Affaires Colleen A. HOEY (since August 2019)
telephone: [504] 2236-9320, 2238-5114
embassy: Avenida La Paz, Tegucigalpa M.D.C.
mailing address: American Embassy, APO AA 34022, Tegucigalpa
FAX: [504] 2236-9037

Flag description: three equal horizontal bands of cerulean blue (top), white, and cerulean blue, with five cerulean, five-pointed stars arranged in an X pattern centered in the white band; the stars represent the members of the former Federal Republic of Central America: Costa Rica, El Salvador, Guatemala, Honduras, and Nicaragua; the blue bands symbolize the Pacific Ocean and the Caribbean Sea; the white band represents the land between the two bodies of water and the peace and prosperity of its people
note: similar to the flag of El Salvador, which features a round emblem encircled by the words REPUBLICA DE EL SALVADOR EN LA AMERICA CENTRAL centered in the white band; also similar to the flag of Nicaragua, which features a triangle encircled by the words REPUBLICA DE NICARAGUA on top and AMERICA CENTRAL on the bottom, centered in the white band

National symbol(s): *scarlet macaw, white-tailed deer; national colors:* blue, white

National anthem: *name:* "Himno Nacional de Honduras" (National Anthem of Honduras)
lyrics/music: Augusto Constancio COELLO/ Carlos HARTLING

note: adopted 1915; the anthem's seven verses chronicle Honduran history; on official occasions, only the chorus and last verse are sung
0:00 / 2:46

ECONOMY

Economy—overview: Honduras, the second poorest country in Central America, suffers from extraordinarily unequal distribution of income, as well as high underemployment. While historically dependent on the export of bananas and coffee, Honduras has diversified its export base to include apparel and automobile wire harnessing.

Honduras's economy depends heavily on US trade and remittances. The US-Central America-Dominican Republic Free Trade Agreement came into force in 2006 and has helped foster foreign direct investment, but physical and political insecurity, as well as crime and perceptions of corruption, may deter potential investors; about 15% of foreign direct investment is from US firms.

The economy registered modest economic growth of 3.1%-4.0% from 2010 to 2017, insufficient to improve living standards for the nearly 65% of the population in poverty. In 2017, Honduras faced rising public debt, but its economy has performed better than expected due to low oil prices and improved investor confidence. Honduras signed a three-year standby arrangement with the IMF in December 2014, aimed at easing Honduras's poor fiscal position.

GDP (purchasing power parity):
$46.3 billion (2017 est.)
$44.18 billion (2016 est.)
$42.58 billion (2015 est.)
note: data are in 2017 dollars
country comparison to the world: 112

GDP (official exchange rate):
$22.98 billion (2017 est.)

GDP—real growth rate:
4.8% (2017 est.)
3.8% (2016 est.)
3.8% (2015 est.)
country comparison to the world: 56

GDP—per capita (PPP):
$5,600 (2017 est.)
$5,400 (2016 est.)
$5,300 (2015 est.)
note: data are in 2017 dollars
country comparison to the world: 170

Gross national saving:
22.1% of GDP (2017 est.)
20.6% of GDP (2016 est.)
20.5% of GDP (2015 est.)
country comparison to the world: 83

GDP—composition, by end use:
household consumption: 77.7% (2017 est.)
government consumption: 13.8% (2017 est.)
investment in fixed capital: 23.1% (2017 est.)
investment in inventories: 0.7% (2017 est.)
exports of goods and services: 43.6% (2017 est.)
imports of goods and services: -58.9% (2017 est.)

GDP—composition, by sector of origin:
agriculture: 14.2% (2017 est.)

industry: 28.8% (2017 est.)
services: 57% (2017 est.)

Agriculture—products: bananas, coffee, citrus, corn, African palm; beef; timber; shrimp, tilapia, lobster, sugar, oriental vegetables
Industries: sugar processing, coffee, woven and knit apparel, wood products, cigars

Industrial production growth rate: 4.5% (2017 est.)
country comparison to the world: 65

Labor force: 3.735 million (2017 est.)
country comparison to the world: 96

Labor force—by occupation:
agriculture: 39.2%
industry: 20.9%
services: 39.8% (2005 est.)

Unemployment rate:
5.6% (2017 est.)
6.3% (2016 est.)
note: about one-third of the people are underemployed
country comparison to the world: 82

Population below poverty line: 29.6% (2014)

Household income or consumption by percentage share: *lowest 10%:* 1.2%
highest 10%: 38.4% (2014)

Budget: *revenues:* 4.658 billion (2017 est.)
expenditures: 5.283 billion (2017 est.)

Taxes and other revenues: 20.3% (of GDP) (2017 est.)
country comparison to the world: 150

Budget surplus (+) or deficit (-): -2.7% (of GDP) (2017 est.)
country comparison to the world: 120

Public debt:
39.5% of GDP (2017 est.)
38.5% of GDP (2016 est.)
country comparison to the world: 131

Fiscal year: calendar year

Inflation rate (consumer prices):
3.9% (2017 est.)
2.7% (2016 est.)
country comparison to the world: 152

Current account balance:
-$380 million (2017 est.)
-$587 million (2016 est.)
country comparison to the world: 110

Exports:
$8.675 billion (2017 est.)
$7.841 billion (2016 est.)
country comparison to the world: 94

Exports—partners: US 34.5%, Germany 8.9%, Belgium 7.7%, El Salvador 7.3%, Netherlands 7.2%, Guatemala 5.2%, Nicaragua 4.8% (2017)

Exports—commodities: coffee, apparel, coffee, shrimp, automobile wire harnesses, cigars, bananas, gold, palm oil, fruit, lobster, lumber

Imports:
$11.32 billion (2017 est.)
$10.56 billion (2016 est.)
country comparison to the world: 96

Imports—commodities: communications equipment, machinery and transport, industrial raw materials, chemical products, fuels, foodstuffs

Imports—partners: US 40.3%, Guatemala 10.5%, China 8.5%, Mexico 6.2%, El Salvador 5.7%, Panama 4.4%, Costa Rica 4.2% (2017)

Reserves of foreign exchange and gold:
$4.708 billion (31 December 2017 est.)
$3.814 billion (31 December 2016 est.)
country comparison to the world: 96

Debt—external:
$8.625 billion (31 December 2017 est.)
$7.852 billion (31 December 2016 est.)
country comparison to the world: 118

Exchange rates: lempiras (HNL) per US dollar –
23.74 (2017 est.)
22.995 (2016 est.)
22.995 (2015 est.)
22.098 (2014 est.)
21.137 (2013 est.)

ENERGY

Electricity access: *population without electricity:*
2 million (2017)
electrification—total population: 87.6% (2016)
electrification—urban areas: 100% (2016)
electrification—rural areas: 72.2% (2016)

Electricity—production: 8.501 billion kWh (2016 est.)
country comparison to the world: 108

Electricity—consumption: 7.22 billion kWh (2016 est.)
country comparison to the world: 107

Electricity—exports: 536 million kWh (2015 est.)
country comparison to the world: 67

Electricity—imports: 195 million kWh (2016 est.)
country comparison to the world: 94

Electricity—installed generating capacity: 2.546 million kW (2016 est.)
country comparison to the world: 108

Electricity—from fossil fuels: 40% of total installed capacity (2016 est.)
country comparison to the world: 169

Electricity—from nuclear fuels: 0% of total installed capacity (2017 est.)
country comparison to the world: 107

Electricity—from hydroelectric plants: 25% of total installed capacity (2017 est.)
country comparison to the world: 77

Electricity—from other renewable sources: 34% of total installed capacity (2017 est.)
country comparison to the world: 11

Crude oil—production: 0 bbl/day (2018 est.)
country comparison to the world: 149

Crude oil—exports: 0 bbl/day (2015 est.)
country comparison to the world: 137

Crude oil—imports: 0 bbl/day (2015 est.)
country comparison to the world: 141

Crude oil—proved reserves: 0 bbl (1 January 2018 est.)
country comparison to the world: 145

Refined petroleum products—production: 0 bbl/day (2017 est.)
country comparison to the world: 157

Refined petroleum products—consumption: 59,000 bbl/day (2016 est.)
country comparison to the world: 97

Refined petroleum products—exports: 12,870 bbl/day (2015 est.)
country comparison to the world: 77

Refined petroleum products—imports: 56,120 bbl/day (2015 est.)
country comparison to the world: 75

Natural gas—production: 0 cu m (2017 est.)
country comparison to the world: 146

Natural gas—consumption: 0 cu m (2017 est.)
country comparison to the world: 159

Natural gas—exports: 0 cu m (2017 est.)
country comparison to the world: 121

Natural gas—imports: 0 cu m (2017 est.)
country comparison to the world: 139

Natural gas—proved reserves: 0 cu m (1 January 2014 est.)
country comparison to the world: 147

Carbon dioxide emissions from consumption of energy: 9.436 million Mt (2017 est.)
country comparison to the world: 110

COMMUNICATIONS

Telephones—fixed lines: *total subscriptions:* 458,696
subscriptions per 100 inhabitants: 5.03 (2019 est.)
country comparison to the world: 98

Telephones—mobile cellular: *total subscriptions:* 6,633,309
subscriptions per 100 inhabitants: 72.74 (2019 est.)
country comparison to the world: 107

Telecommunication systems: *general assessment:* fixed-line connections are increasing but still limited; competition among multiple providers of mobile-cellular services and international investment has contributed to a sharp increase in subscribership; demand for broadband increasing and some investment needed in network upgrades; mobile penetration below regional average; free access to the Internet in public schools (2020) *domestic:* private sub-operators allowed to provide fixed lines in order to expand telephone coverage contributing to a small increase in fixed-line teledensity 5 per 100; mobile-cellular subscribership is roughly 73 per 100 persons (2019) *international:* country code—504; landing points for both the ARCOS and the MAYA-1 fiber-optic submarine cable systems that together provide connectivity to South and Central America, parts of the Caribbean, and the US; satellite earth stations 2 Intelsat (Atlantic Ocean); connected to Central American Microwave System (2019) *note:* the COVID-19 outbreak is negatively impacting telecommunications production and supply chains globally; consumer spending on telecom devices and services has also slowed due to the pandemic's effect on economies worldwide; overall progress towards improvements in all facets of the telecom industry—mobile, fixed-line, broadband, submarine cable and satellite—has moderated

Broadcast media: multiple privately owned terrestrial TV networks, supplemented by multiple cable TV networks; Radio Honduras is the lone government-owned radio network; roughly 300 privately owned radio stations

Internet country code: .hn

Internet users: *total:* 2,853,505
percent of population: 31.7% (July 2018 est.)
country comparison to the world: 101

Broadband—fixed subscriptions: *total:* 354,861
subscriptions per 100 inhabitants: 4 (2018 est.)
country comparison to the world: 96

TRANSPORTATION

National air transport system: *number of registered air carriers:* 4 (2020)
inventory of registered aircraft operated by air carriers: 26
annual passenger traffic on registered air carriers: 251,149 (2018)
annual freight traffic on registered air carriers: 450,000 mt-km (2018)

Civil aircraft registration country code prefix: HR (2016)

Airports: 103 (2013)
country comparison to the world: 53

Airports—with paved runways: *total:* 13 (2017)
2,438 to 3,047 m: 3 (2017)
1,524 to 2,437 m: 3 (2017)
914 to 1,523 m: 4 (2017)
under 914 m: 3 (2017)

Airports—with unpaved runways: *total:* 90 (2013)
1,524 to 2,437 m: 1 (2013)
914 to 1,523 m: 16 (2013)
under 914 m: 73 (2013)

Railways: *total:* 699 km (2014)
narrow gauge: 164 km 1.067-m gauge (2014)
115 km 1.057-m gauge
420 km 0.91 4-m gauge
country comparison to the world: 101

Roadways:
total: 14,742 km (2012)
paved: 3,367 km (2012)
unpaved: 11,375 km (1,543 km summer only) (2012)
note: an additional 8,951 km of non-official roads used by the coffee industry
country comparison to the world: 126

Waterways: 465 km (most navigable only by small craft) (2012)
country comparison to the world: 84

Merchant marine: *total:* 527
by type: general cargo 247, oil tanker 83, other 197 (2019)
country comparison to the world: 41

Ports and terminals: major seaport(s): La Ceiba, Puerto Cortes, San Lorenzo, Tela

MILITARY AND SECURITY

Military and security forces: Honduran Armed Forces (Fuerzas Armadas de Honduras, FFAA): Army, Honduran Naval Force (FNH; includes marines), Honduran Air Force (Fuerza Aerea Hondurena, FAH), Honduran Public Order Military Police (PMOP); Ministry of Public Security and Defense: Public Security Forces (includes paramilitary units) (2020)

Military expenditures:
1.6% of GDP (2019)
1.6% of GDP (2018)
1.7% of GDP (2017)
1.7% of GDP (2016)
1.7% of GDP (2015)
country comparison to the world: 73

Military and security service personnel strengths: the Honduran Armed Forces (FFAA) have approximately 15,500 active personnel (7,500 Army; 1,500 Navy; 2,500 Air Force; 4,000 Public Order Military Police) (2019)

Military equipment inventories and acquisitions: the FFAA's inventory is comprised of mostly older imported equipment from Israel, the UK, and the US; since 2010, Honduras has received limited amounts of military equipment from Colombia, Israel, Netherlands, Taiwan, and the US (2019 est.)

Military service age and obligation: 18 years of age for voluntary 2- to 3-year military service; no conscription (2012)

TRANSNATIONAL ISSUES

Disputes—international: International Court of Justice (ICJ) ruled on the delimitation of "bolsones" (disputed areas) along the El Salvador-Honduras border in 1992 with final settlement by the parties in 2006 after an Organization of American States survey and a further ICJ ruling in 2003; the 1992 ICJ ruling advised a tripartite resolution to a maritime boundary in the Gulf of Fonseca with consideration of Honduran access to the Pacific; El Salvador continues to claim tiny Conejo Island, not mentioned in the ICJ ruling, off Honduras in the Gulf of Fonseca; Honduras claims the Belizean-administered Sapodilla Cays off the coast of Belize in its constitution, but agreed to a joint ecological park around the cays should Guatemala consent to a maritime corridor in the Caribbean under the OAS-sponsored 2002 Belize-Guatemala Differendum

Refugees and internally displaced persons: *IDPs:* 247,000 (violence, extortion, threats, forced recruitment by urban gangs between 2004 and 2018) (2019)

Illicit drugs: transshipment point for drugs and narcotics; illicit producer of cannabis, cultivated on small plots and used principally for local consumption; corruption is a major problem; some money-laundering activity

HONG KONG

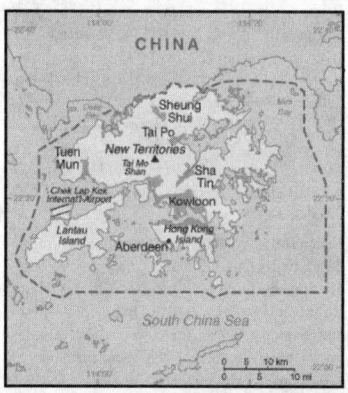

CHINA

Sheung Shui
Tai Po
Tuen Mun
New Territories
Tai Mo Shan
Sha Tin
Chek Lap Kok Internat'l Airport
Kowloon
Lantau Island
Hong Kong Island
Aberdeen
South China Sea

INTRODUCTION

Background: Occupied by the UK in 1841, Hong Kong was formally ceded by China the following year; various adjacent lands were added later in the 19th century. Pursuant to an agreement signed by China and the UK on 19 December 1984, Hong Kong became the Hong Kong Special Administrative Region of the People's Republic of China on 1 July 1997. In this agreement, China promised that, under its "one country, two systems" formula, China's socialist economic system would not be imposed on Hong Kong and that Hong Kong would enjoy a "high degree of autonomy" in all matters except foreign and defense affairs for the subsequent 50 years.

GEOGRAPHY

Location: Eastern Asia, bordering the South China Sea and China

Geographic coordinates: 22 15 N, 114 10 E

Map references: Southeast Asia

Area: *total:* 1,108 sq km
land: 1,073 sq km
water: 35 sq km
country comparison to the world: 184

Area—comparative: six times the size of Washington, DC

Land boundaries: *total:* 33 km
regional borders (1): China 33 km

Coastline: 733 km

Maritime claims: *territorial sea:* 12 nm

Climate: subtropical monsoon; cool and humid in winter, hot and rainy from spring through summer, warm and sunny in fall

Terrain: hilly to mountainous with steep slopes; lowlands in north

Elevation: *lowest point:* South China Sea 0 m
highest point: Tai Mo Shan 958 m

Natural resources: outstanding deepwater harbor, feldspar

Land use: *agricultural land:* 5% (2011 est.)
arable land: 3.2% (2011 est.) / permanent crops: 0.9% (2011 est.) / permanent pasture: 0.9% (2011 est.)
forest: 0% (2011 est.)
other: 95% (2011 est.)
Irrigated land: 10 sq km (2012)

Population distribution: population fairly evenly distributed

Natural hazards: occasional typhoons

Environment—current issues: air and water pollution from rapid urbanization; urban waste pollution; industrial pollution

Environment—international agreements: *party to:* Marine Dumping (associate member), Ship Pollution (associate member)

Geography—note: consists of a mainland area (the New Territories) and more than 200 islands

PEOPLE AND SOCIETY

Population: 7,249,907 (July 2020 est.)
country comparison to the world: 103

Nationality: *noun:* Chinese/Hong Konger
adjective: Chinese/Hong Kong

Ethnic groups: Chinese 92%, Filipino 2.5%, Indonesian 2.1%, other 3.4% (2016 est.)

Languages: Cantonese (official) 88.9%, English (official) 4.3%, Mandarin (official) 1.9%, other Chinese dialects 3.1%, other 1.9% (2016 est.)

Religions: Buddhist or Taoist 27.9%, Protestant 6.7%, Roman Catholic 5.3%, Muslim 4.2%, Hindu 1.4%, Sikh 0.2%, other or none 54.3% (2016 est.)
note: many people practice Confucianism, regardless of their religion or not having a religious affiliation

Age structure: *0-14 years:* 12.81% (male 490,477/female 437,971)
15-24 years: 8.81% (male 334,836/female 303,897)
25-54 years: 42.66% (male 1,328,529/female 1,763,970)
55-64 years: 17.24% (male 582,047/female 668,051)
65 years and over: 18.48% (male 625,453/female 714,676) (2020 est.)

Dependency ratios: *total dependency ratio:* 44.7
youth dependency ratio: 18.3
elderly dependency ratio: 26.3
potential support ratio: 3.8 (2020 est.)

435

Median age: *total:* 45.6 years
male: 44.2 years
female: 46.5 years (2020 est.)
country comparison to the world: 7

Population growth rate: 0.24% (2020 est.)
country comparison to the world: 178

Birth rate: 8.4 births/1,000 population (2020 est.)
country comparison to the world: 217

Death rate: 7.9 deaths/1,000 population (2020 est.)
country comparison to the world: 95

Net migration rate: 1.7 migrant(s)/1,000 population (2020 est.)
country comparison to the world: 51

Population distribution: population fairly evenly distributed

Urbanization: *urban population:* 100% of total population (2020)
rate of urbanization: 0.82% annual rate of change (2015-20 est.)
total population growth rate v. urban population growth rate, 2000-2030:

Major urban areas—population: 7.548 million Hong Kong (2020)

Sex ratio: *at birth:* 1.06 male(s)/female
0-14 years: 1.12 male(s)/female
15-24 years: 1.1 male(s)/female
25-54 years: 0.75 male(s)/female
55-64 years: 0.87 male(s)/female
65 years and over: 0.88 male(s)/female
total population: 0.86 male(s)/female (2020 est.)

Mother's mean age at first birth: 29.8 years (2008 est.)

Infant mortality rate: *total:* 2.7 deaths/1,000 live births
male: 2.9 deaths/1,000 live births
female: 2.5 deaths/1,000 live births (2020 est.)
country comparison to the world: 217

Life expectancy at birth: *total population:* 83.2 years
male: 80.5 years
female: 86.1 years (2020 est.)
country comparison to the world: 8

Total fertility rate: 1.21 children born/woman (2020 est.)
country comparison to the world: 225

Contraceptive prevalence rate: 66.7% (2017)

Drinking water source:
improved:
urban: 100% of population
total: 100% of population
unimproved:
urban: -1% of population
total: 0% of population (2017 est.)

Physicians density: 1.96 physicians/1,000 population (2018)

Hospital bed density: 5.4 beds/1,000 population (2018)

Sanitation facility access:
improved:
urban: 96.4% of population
total: 96.4% of population

unimproved:
urban: 3.6% of population
total: 3.6% of population (2017)

HIV/AIDS—adult prevalence rate: NA

HIV/AIDS—people living with HIV/AIDS: NA

HIV/AIDS—deaths: NA

Education expenditures: 3.3% of GDP (2018)
country comparison to the world: 128

School life expectancy (primary to tertiary education): *total:* 17 years
male: 17 years
female: 18 years (2019)

Unemployment, youth ages 15-24: *total:* 8.7%
male: 9.3%
female: 8.2% (2017 est.)
country comparison to the world: 138

GOVERNMENT

Country name: *conventional long form:* Hong Kong Special Administrative Region
conventional short form: Hong Kong
local long form: Heung Kong Takpit Hangching Ku (Eitel/Dyer-Ball)
local short form: Heung Kong (Eitel/Dyer-Ball)
abbreviation: HK
etymology: probably an imprecise phonetic rendering of the Cantonese name meaning "fragrant harbor"

Dependency status: special administrative region of the People's Republic of China

Government type: presidential limited democracy; a special administrative region of the People's Republic of China

Administrative divisions: none (special administrative region of the People's Republic of China)

Independence: none (special administrative region of China)

National holiday: National Day (Anniversary of the Founding of the People's Republic of China), 1 October (1949); note—1 July (1997) is celebrated as Hong Kong Special Administrative Region Establishment Day

Constitution: *history:* several previous (governance documents while under British authority); latest drafted April 1988 to February 1989, approved March 1990, effective 1 July 1997 (Basic Law of the Hong Kong Special Administrative Region of the People's Republic of China serves as the constitution); note—since 1990, China's National People's Congress has interpreted specific articles of the Basic Law
amendments: proposed by the Standing Committee of the National People's Congress (NPC), the People's Republic of China State Council, and the Special Administrative Region of Hong Kong; submittal of proposals to the NPC requires two-thirds majority vote by the Legislative Council of Hong Kong, approval by two thirds of Hong Kong's deputies to the NPC, and approval by the Hong Kong chief executive; final passage requires approval by the NPC

Legal system: mixed legal system of common law based on the English model and Chinese customary law (in matters of family and land tenure)

Citizenship: see China

Suffrage: 18 years of age in direct elections for half of the Legislative Council seats and all of the seats in 18 district councils; universal for permanent residents living in the territory of Hong Kong for the past 7 years; note—in indirect elections, suffrage is limited to about 220,000 members of functional constituencies for the other half of the legislature and a 1,200member election committee for the chief executive drawn from broad sectoral groupings, central government bodies, municipal organizations, and elected Hong Kong officials

Executive branch: *chief of state:* President of China XI Jinping (since 14 March 2013)

head of government: Chief Executive Carrie LAM (since 1 July 2017)
cabinet: Executive Council or ExCo appointed by the chief executive
elections/appointments: president indirectly elected by National People's Congress for a 5-year term (eligible for a second term); election last held on 17 March 2018 (next to be held in March 2023); chief executive indirectly elected by the Election Committee and appointed by the PRC Government for a 5-year term (eligible for a second term); election last held on 26 March 2017 (next to be held in 2022)
election results: Carrie LAM elected chief executive; Election Committee vote—Carrie LAM 777, John TSANG 365, WOO Kwok-hing 21, invalid 23
note: the Legislative Council voted in June 2010 to expand the Election Committee to 1,200 members

Legislative branch: *description:* unicameral Legislative Council or LegCo (70 seats; 35 members directly elected in multi-seat constituencies by party-list proportional representation vote; 30 members indirectly elected by the approximately 220,000 members of various functional constituencies based on a variety of methods; 5 at large "super-seat" members directly elected by all of Hong Kong's eligible voters who do not participate in a functional constituency; members serve 4-year terms)
elections: last held on 4 September 2016; (scheduled for September 2020, but delayed until 2021); note—byelection held on 11 March and 25 November 2018 to fill 5 seats left vacant after 5 legislators were removed from office
election results: percent of vote by block—pro-democracy 36%; pro-Beijing 40.2%, localist 19%, other 4.8%; seats by block/party—pro-Beijing 40 (DAB 12, BPA 7, FTU 5, Liberal Party 4, NPP 3, other 9); pro-democracy 23 (Democratic Party 7, Civic Party 6, PP-LSD 2, Professional Commons 2, Labor 1, NWSC 1, PTU 1, other democrats 3), localists 6 (ALLinHK 2, CP-PPI-HKRO 1, Demosisto 1, Democracy Groundwork 1, other localist 1), non-aligned independent 1; composition—men 59, women 11, percent of women 15.7%; note—2 localists were barred from taking

office in November 2016 and 4 pro-democracy legislators were removed in July 2017; two pan-democratic, two DAB, and one pro-establishment candidates won the byelections in 2018 to fill the seats vacated by the 5 legislators removed from office; one prodemocracy seat remains unfilled pending a court appeal; percent of vote by block as of March 2019—pro-Beijing 62% prodemocracy 38%; seats by block/party as of March 2019—pro-Beijing 43 (DAB 13, BPA 7, FTU 5, Liberal Party 4, NPP 3, other 11); pro-democracy 26 (Democratic Party 7, Civic Party 5, Professional Commons 2, Civic Passion 1, Labor 1 PTU 1, Council Front 6, independent 3); composition as of March 2019—men 58, women 11; percent of women 15.7%

Judicial branch: *highest courts:* Court of Final Appeal (consists of the chief justice, 3 permanent judges, and 20 non-permanent judges); note—a sitting bench consists of the chief justice, 3 permanent judges, and 1 non-permanent judge *judge selection and term of office:* all judges appointed by the Hong Kong Chief Executive upon the recommendation of the Judicial Officers Recommendation Commission, an independent body consisting of the Secretary for Justice, other judges, and judicial and legal professionals; permanent judges serve until normal retirement at age 65, but term can be extended; non-permanent judges appointed for renewable 3-year terms without age limit

subordinate courts: High Court (consists of the Court of Appeal and Court of First Instance); District Courts (includes Family and Land Courts); magistrates' courts; specialized tribunals

Political parties and leaders: *parties:* ALLinHK (alliance of 6 localist groups)
Business and Professional Alliance or BPA [LO Wai-kwok]
Civic Party [Alvin YEUNG]
Civic Passion or CP [CHENG Chung-tai] (part of Civic Passion-Proletariat Political Institute-Hong Kong Resurgence Order alliance or CP-PPI-HKRO that dissolved after the 2016 election)
Democracy Groundwork [LAU Siu-lai]
Democratic Alliance for the Betterment and Progress of Hong Kong or DAB [Starry LEE Wai-king]
Democratic Party [WU Chi-wai]
Demosisto [Ivan LAM] (announced cessation of all operations, 30 June 2020)
Federation of Trade Unions or FTU [Stanley NG Chau-pei]
Labor Party [Steven KWOK Wing-kin]
League of Social Democrats or LSD [Avery NG Man-yuen]
Liberal Party [Felix CHUNG Kwok-pan]
Neighborhood and Workers Service Center or NWSC [LEUNG Yui-chung]
New People's Party or NPP [Regina IP Lau Su-yee]
People Power or PP [Raymond CHAN]
Youngspiration [Sixtus "Baggio" LEUNG Chung-hang]
other:
Professional Commons [Charles Peter MOK] (think tank)
Professional Teachers Union or PTU

note: political blocks include: pro-democracy—Civic Party, Democratic Party, Labor Party, LSD, NWSC, PP, Professional Commons, PTU; pro-Beijing—DAB, FTU, Liberal Party, NPP, BPA; localist—ALLinHK, CP, Democracy Groundwork, Demosisto; there is no political party ordinance, so there are no registered political parties; politically active groups register as societies or companies

International organization participation: ADB, APEC, BIS, FATF, ICC (national committees), IHO, IMF, IMO (associate), Interpol (subbureau), IOC, ISO (correspondent), ITUC (NGOs), UNWTO (associate), UPU, WCO, WMO, WTO

Diplomatic representation in the US: *chief of mission:* none (Special Administrative Region of China); Hong Kong Economic and Trade Office (HKETO) carries out normal liaison activities and communication with the US Government and other US entities; Eddie MAK, JP (since 3 July 2018) is the Hong Kong Commissioner to the US Government of the Hong Kong Special Administrative Region; address: 1520 18th Street NW, Washington, DC 20036; telephone: [1] 202 331-8947; FAX: [1] 202 331-8958

HKETO offices: New York, San Francisco

Diplomatic representation from the US: *chief of mission:* Consul General Hanscom SMITH (since July 2019); note—also accredited to Macau
telephone: [852] 2523-9011
embassy: U. S. Consulate General Hong Kong and Macau
26 Garden Road
Central Hong Kong
mailing address: Unit 8000, Box 1, DPO AP 96521-0006
FAX: [852] 2845-1598
consulate(s) general: 26 Garden Road, Hong Kong

Flag description: red with a stylized, white, five-petal Bauhinia flower in the center; each petal contains a small, red, five-pointed star in its middle; the red color is the same as that on the Chinese flag and represents the motherland; the fragrant Bauhinia—developed in Hong Kong the late 19th century—has come to symbolize the region; the five stars echo those on the flag of China

National symbol(s): *orchid tree flower; national colors:* red, white

National anthem: *note:* as a Special Administrative Region of China, "Yiyongjun Jinxingqu" is the official anthem (see China)
0:00 / 0:43

ECONOMY

Economy—overview: Hong Kong has a free market economy, highly dependent on international trade and finance—the value of goods and services trade, including the sizable share of reexports, is about four times GDP. Hong Kong has no tariffs on imported goods, and it levies excise duties on only four commodities, whether imported or produced locally: hard alcohol, tobacco, oil, and methyl alcohol. There are no quotas or dumping laws.

Hong Kong continues to link its currency closely to the US dollar, maintaining an arrangement established in 1983.

Excess liquidity, low interest rates and a tight housing supply have caused Hong Kong property prices to rise rapidly. The lower and middle-income segments of the population increasingly find housing unaffordable.

Hong Kong's open economy has left it exposed to the global economic situation. Its continued reliance on foreign trade and investment makes it vulnerable to renewed global financial market volatility or a slowdown in the global economy.

Mainland China has long been Hong Kong's largest trading partner, accounting for about half of Hong Kong's total trade by value. Hong Kong's natural resources are limited, and food and raw materials must be imported. As a result of China's easing of travel restrictions, the number of mainland tourists to the territory surged from 4.5 million in 2001 to 47.3 million in 2014, outnumbering visitors from all other countries combined. After peaking in 2014, overall tourist arrivals dropped 2.5% in 2015 and 4.5% in 2016. The tourism sector rebounded in 2017, with visitor arrivals rising 3.2% to 58.47 million. Travelers from Mainland China totaled 44.45 million, accounting for 76% of the total.

The Hong Kong Government is promoting the Special Administrative Region (SAR) as the preferred business hub for renminbi (RMB) internationalization. Hong Kong residents are allowed to establish RMB-denominated savings accounts, RMB-denominated corporate and Chinese government bonds have been issued in Hong Kong, RMB trade settlement is allowed, and investment schemes such as the Renminbi Qualified Foreign Institutional Investor (RQFII) Program was first launched in Hong Kong. Offshore RMB activities experienced a setback, however, after the People's Bank of China changed the way it set the central parity rate in August 2015. RMB deposits in Hong Kong fell from 1.0 trillion RMB at the end of 2014 to 559 billion RMB at the end of 2017, while RMB trade settlement handled by banks in Hong Kong also shrank from 6.8 trillion RMB in 2015 to 3.9 trillion RMB in 2017.

Hong Kong has also established itself as the premier stock market for Chinese firms seeking to list abroad. In 2015, mainland Chinese companies constituted about 50% of the firms listed on the Hong Kong Stock Exchange and accounted for about 66% of the exchange's market capitalization.

During the past decade, as Hong Kong's manufacturing industry moved to the mainland, its service industry has grown rapidly. In 2014, Hong Kong and China signed a new agreement on achieving basic liberalization of trade in services in Guangdong Province under the Closer Economic Partnership Agreement (CEPA), adopted in 2003 to forge closer ties between Hong Kong and the mainland. The new measures, which took effect in March 2015, cover a negative list and a most-favored treatment provision. On the basis of the Guangdong Agreement, the Agreement on Trade in Services signed in November 2015 further enhanced liberalization, including extending the

437

implementation of the majority of Guangdong pilot liberalization measures to the whole Mainland, reducing the restrictive measures in the negative list, and adding measures in the positive lists for cross-border services as well as cultural and telecommunications services. In June 2017, the Investment Agreement and the Agreement on Economic and Technical Cooperation (Ecotech Agreement) were signed under the framework of CEPA.

Hong Kong's economic integration with the mainland continues to be most evident in the banking and finance sector. Initiatives like the Hong Kong-Shanghai Stock Connect, the Hong Kong- Shenzhen Stock Connect the Mutual Recognition of Funds, and the Bond Connect scheme are all important steps towards opening up the Mainland's capital markets and have reinforced Hong Kong's role as China's leading offshore RMB market. Additional connect schemes such as ETF Connect (for exchange-traded fund products) are also under exploration by Hong Kong authorities. In 2017, Chief Executive Carrie LAM announced plans to increase government spending on research and development, education, and technological innovation with the aim of spurring continued economic growth through greater sector diversification.

GDP (purchasing power parity):
$480.5 billion (2018)
$455.9 billion (2017 est.)
$439.2 billion (2016 est.)
note: data are in 2017 dollars
country comparison to the world: 42

GDP (official exchange rate):
$341.4 billion (2017 est.)

GDP—real growth rate: -
1.25% (2019 est.)
2.86% (2018 est.)
3.8% (2017 est.)
country comparison to the world: 202

GDP—per capita (PPP): $64,500 (2018)
$61,500 (2017 est.)
$59,500 (2016 est.)
note: data are in 2017 dollars
country comparison to the world: 16

Gross national saving:
26.6% of GDP (2017 est.)
25.5% of GDP (2016 est.)
24.9% of GDP (2015 est.)
country comparison to the world: 47

GDP—composition, by end use:
household consumption: 67% (2017 est.)
government consumption: 9.9% (2017 est.)
investment in fixed capital: 21.8% (2017 est.)
investment in inventories: 0.4% (2017 est.)
exports of goods and services: 188% (2017 est.)
imports of goods and services: -187.1% (2017 est.)

GDP—composition, by sector of origin:
agriculture: 0.1% (2017 est.)
industry: 7.6% (2017 est.)
services: 92.3% (2017 est.)

Agriculture—products: fresh vegetables and fruit; poultry, pork; fish
Industries: trading and logistics, financial services, professional services, tourism, cultural and creative, clothing and textiles, shipping, electronics, toys, clocks and watches

Industrial production growth rate: 1.7% (2017 est.)
country comparison to the world: 139

Labor force: 3.627 million (2020 est.)
country comparison to the world: 98

Labor force—by occupation: *agriculture:* 3.8% (2013 est.)
industry: 2% (2016 est.)
services: 54.5% (2016 est.)
industry and services: 12.5% (2013 est.)
agriculture/fishing/forestry/mining: 10.1% (2013)
manufacturing: 17.1% (2013 est.)
note: above data exclude public sector

Unemployment rate: 2.93% (2019 est.)
2.83% (2018 est.)
country comparison to the world: 36

Population below poverty line: 19.9% (2016 est.)

Household income or consumption by percentage share: *lowest 10%:* 1.8% NA
highest 10%: 38.1% NA (2016)

Budget: *revenues:* 79.34 billion (2017 est.)
expenditures: 61.64 billion (2017 est.)

Taxes and other revenues: 23.2% (of GDP) (2017 est.)
country comparison to the world: 128

Budget surplus (+) or deficit (-): 5.2% (of GDP) (2017 est.)
country comparison to the world: 6

Public debt: 0.1% of GDP (2017 est.)
0.1% of GDP (2016 est.)
country comparison to the world: 208

Fiscal year: 1 April—31 March

Inflation rate (consumer prices): 1.5% (2017 est.)
2.4% (2016 est.)
country comparison to the world: 81

Current account balance: $22.469 billion (2019 est.)
$13.516 billion (2018 est.)
country comparison to the world: 15

Exports: $530.6 billion (2018 est.)
$537.8 billion (2017 est.)
$460 billion (2016 est.)
country comparison to the world: 8

Exports—partners: China 55%, US 8.6% (2018 est.)

Exports—commodities: electrical machinery and appliances, textiles, apparel, watches and clocks, toys, "jewelry, goldsmiths' and silversmiths' wares, and other articles of precious or semi-precious materials"; Hong Kong plays an important role as entrepot to the Chinese mainland; in 2017, 58% of Hong Kong's re-exports originated in mainland China, and 54% were destined for the Chinese mainland

Imports: $602.4 billion (2018 est.)
$561.8 billion (2017 est.)

$518.2 billion (2016 est.)
country comparison to the world: 6

Imports—commodities: raw materials and semi-manufactures, consumer goods, capital goods, foodstuffs, fuel (most is reexported)

Imports—partners: China 46.3%, Singapore 6.4%, South Korea 5.9%, Japan 5.5%, US 4.9% (2018 est.)

Reserves of foreign exchange and gold: $431.4 billion (31 December 2017 est.)
$386.2 billion (31 December 2016 est.)
country comparison to the world: 7

Debt—external: $633.6 billion (31 December 2017 est.)
$1.349 trillion (31 December 2016 est.)
country comparison to the world: 18

Exchange rates: Hong Kong dollars (HKD) per US dollar -
7.82 (2017 est.)
7.76 (2016 est.)
7.762 (2015 est.)
7.752 (2014 est.)
7.754 (2013 est.)

ENERGY

Electricity access: *electrification—total population:* 100% (2020)

Electricity—production: 35.97 billion kWh (2016 est.)
country comparison to the world: 60

Electricity—consumption: 41.84 billion kWh (2016 est.)
country comparison to the world: 54

Electricity—exports: 1.205 billion kWh (2016 est.)
country comparison to the world: 55

Electricity—imports: 11.62 billion kWh (2016 est.)
country comparison to the world: 21

Electricity—installed generating capacity: 12.63 million kW (2016 est.)
country comparison to the world: 55

Electricity—from fossil fuels: 100% of total installed capacity (2016 est.)
country comparison to the world: 9

Electricity—from nuclear fuels: 0% of total installed capacity (2017 est.)
country comparison to the world: 108

Electricity—from hydroelectric plants: 0% of total installed capacity (2017 est.)
country comparison to the world: 178

Electricity—from other renewable sources: 0% of total installed capacity (2017 est.)
country comparison to the world: 192

Crude oil—production: 0 bbl/day (2018 est.)
country comparison to the world: 150

Crude oil—exports: 0 bbl/day (2015 est.)
country comparison to the world: 138

Crude oil—imports: 0 bbl/day (2015 est.)
country comparison to the world: 142

Crude oil—proved reserves: 0 bbl (1 January 2018 est.)

country comparison to the world: 146

Refined petroleum products—production: 0 bbl/day (2015 est.)
country comparison to the world: 158

Refined petroleum products—consumption: 403,100 bbl/day (2016 est.)
country comparison to the world: 38

Refined petroleum products—exports: 13,570 bbl/day (2015 est.)
country comparison to the world: 76

Refined petroleum products—imports: 402,100 bbl/day (2015 est.)
country comparison to the world: 22

Natural gas—production: 0 cu m (2017 est.)
country comparison to the world: 147

Natural gas—consumption: 3.37 billion cu m (2017 est.)
country comparison to the world: 69

Natural gas—exports: 0 cu m (2017 est.)
country comparison to the world: 122

Natural gas—imports: 3.37 billion cu m (2017 est.)
country comparison to the world: 43

Natural gas—proved reserves: 0 cu m (1 January 2016 est.)
country comparison to the world: 148

Carbon dioxide emissions from consumption of energy: 102.5 million Mt (2017 est.)
country comparison to the world: 43

COMMUNICATIONS

Telephones—fixed lines: *total subscriptions:* 3,942,605
subscriptions per 100 inhabitants: 54.51 (2019 est.)
country comparison to the world: 35

Telephones—mobile cellular: *total subscriptions:* 20,868,827
subscriptions per 100 inhabitants: 288.53 (2019 est.)
country comparison to the world: 59

Telecommunication systems: *general assessment:* excellent domestic and international services; some of the highest peak average broadband speeds in the world; HK aims to be among the earliest adopters of 5G mobile technology; almost all households have access to high-speed broadband connectivity; HK broadband penetration rate is among the highest in the world; in the next

five years the government has organized the development of 'smart cities' in six areas—"smart mobility", "smart living", "smart environment", "smart people", "smart government", and "smart economy" by 2024 (2020)

domestic: microwave radio relay links and extensive fiber-optic network; fixed-line is 55 per 100 and mobile-cellular is 289 per 100 (2019)

international: country code—852; landing points for the APG, ASE, EAC-C2C, HK-G, Bay-to-Bay Express Cable System, H2 Cable, HKA, SJC, SJC2, PLCN, SeaMeWe-3, TGN-IA, APCN-2, AAG, FLAG and FEA submarine cables that provide connections to Asia, US, Australia, the Middle East, and Europe; satellite earth stations—3 Intelsat (1 Pacific Ocean and 2 Indian Ocean); coaxial cable to Guangzhou, China (2019)

note: the COVID-19 outbreak is negatively impacting telecommunications production and supply chains globally; consumer spending on telecom devices and services has also slowed due to the pandemic's effect on economies worldwide; overall progress towards improvements in all facets of the telecom industry—mobile, fixed-line, broadband, submarine cable and satellite—has moderated

Broadcast media: 4 commercial terrestrial TV networks each with multiple stations; multi-channel satellite and cable TV systems available; 3 licensed broadcasters of terrestrial radio, one of which is government funded, operate about 12 radio stations; note—4 digital radio broadcasters operated in Hong Kong from 2010 to 2017, but all digital radio services were terminated in September 2017 due to weak market demand (2019)

Internet country code: .hk

Internet users: *total:* 6,450,167

percent of population: 89.42% (July 2018 est.)
country comparison to the world: 75

Broadband—fixed subscriptions: *total:* 2,714,679
subscriptions per 100 inhabitants: 38 (2018 est.)
country comparison to the world: 45

TRANSPORTATION

National air transport system: *number of registered air carriers:* 12 (registered in China) (2020)
inventory of registered aircraft operated by air carriers: 275 (registered in China)
annual passenger traffic on registered air carriers: 47,101,822 (2018)

annual freight traffic on registered air carriers: 12,676,720,000 mt-km (2018)

Civil aircraft registration country code prefix: B-H (2016)

Airports: 2 (2013)
country comparison to the world: 200

Airports—with paved runways: *total:* 2 (2019)
over 3,047 m: 1
1,524 to 2,437 m: 1

Heliports: 9 (2013)

Roadways: *total:* 2,107 km (2017)
paved: 2,107 km (2017)
country comparison to the world: 170

Merchant marine: *total:* 2,701
by type: bulk carrier 1,164, container ship 540, general cargo 201, oil tanker 377, other 419 (2019)
country comparison to the world: 10

Ports and terminals: *major seaport(s):* Hong Kong
container port(s) (TEUs): Hong Kong (20,770,000) (2017)

MILITARY AND SECURITY

Military and security forces: no regular indigenous military forces; Hong Kong Police Force; Hong Kong garrison of China's People's Liberation Army (PLA) includes elements of the PLA Army, PLA Navy, and PLA Air Force; these forces are under the direct leadership of the Central Military Commission in Beijing and under administrative control of the adjacent Southern Theater Command (2019)

Military—note: defense is the responsibility of China

TRANSNATIONAL ISSUES

Disputes—international: Hong Kong plans to reduce its 2,800-hectare Frontier Closed Area (FCA) to 400 hectares by 2015; the FCA established in 1951 as a buffer zone between Hong Kong and mainland China to prevent illegal migration from and the smuggling of goods

Illicit drugs: despite strenuous law enforcement efforts, faces difficult challenges in controlling transit of heroin and methamphetamine to regional and world markets; modern banking system provides conduit for money laundering; rising indigenous use of synthetic drugs, especially among young people

HUNGARY

INTRODUCTION

Background: Hungary became a Christian kingdom in A.D. 1000 and for many centuries served as a bulwark against Ottoman Turkish expansion in Europe. The kingdom eventually became part

of the polyglot Austro-Hungarian Empire, which collapsed during World War I. The country fell under communist rule following World War II. In 1956, a revolt and an announced withdrawal from the Warsaw Pact were met with a massive military intervention by Moscow. Under the leadership of

Janos KADAR in 1968, Hungary began liberalizing its economy, introducing so-called "Goulash Communism." Hungary held its first multiparty elections in 1990 and initiated a free market economy. It joined NATO in 1999 and the EU five years later.

GEOGRAPHY

Location: Central Europe, northwest of Romania

Geographic coordinates: 47 00 N, 20 00 E

Map references: Europe

Area: *total:* 93,028 sq km
land: 89,608 sq km
water: 3,420 sq km
country comparison to the world: 111

Area—comparative: slightly smaller than Virginia; about the same size as Indiana

Area comparison map: *[INSERT IMAGE: Hungary-Area comparison map]*

Land boundaries: *total:* 2,106 km
border countries (7): Austria 321 km, Croatia 348 km, Romania 424 km, Serbia 164 km, Slovakia 627 km, Slovenia 94 km, Ukraine 128 km

Coastline: 0 km (landlocked)

Maritime claims: none (landlocked)

Climate: temperate; cold, cloudy, humid winters; warm summers

Terrain: mostly flat to rolling plains; hills and low mountains on the Slovakian border

Elevation: *mean elevation:* 143 m
lowest point: Tisza River 78 m
highest point: Kekes 1,014 m

Natural resources: bauxite, coal, natural gas, fertile soils, arable land

Land use: *agricultural land:* 58.9% (2011 est.)
arable land: 48.5% (2011 est.) / permanent crops: 2% (2011 est.) / permanent pasture: 8.4% (2011 est.)
forest: 22.5% (2011 est.)
other: 18.6% (2011 est.)

Irrigated land: 1,721 sq km (2012)

Population distribution: a fairly even distribution throughout most of the country, with urban areas attracting larger and denser populations

Environment—current issues: air and water pollution are some of Hungary's most serious environmental problems; water quality in the Hungarian part of the Danube has improved but is still plagued by pollutants from industry and large-scale agriculture; soil pollution

Environment—international agreements: *party to:* Air Pollution, Air Pollution-Nitrogen Oxides, Air Pollution-Persistent Organic Pollutants, Air Pollution-Sulfur 85, Air Pollution-Sulfur 94, Air Pollution-Volatile Organic Compounds, Antarctic Treaty, Biodiversity, Climate Change, Climate Change-Kyoto Protocol, Desertification, Endangered Species, Environmental Modification, Hazardous Wastes, Law of the Sea, Marine Dumping, Ozone Layer Protection, Ship Pollution, Wetlands, Whaling

signed, but not ratified: none of the selected agreements

Geography—note: landlocked; strategic location astride main land routes between Western Europe and Balkan Peninsula as well as between Ukraine and Mediterranean basin; the north-south flowing Duna (Danube) and Tisza Rivers divide the country into three large regions

PEOPLE AND SOCIETY

Population: 9,771,827 (July 2020 est.)
country comparison to the world: 93

Nationality: *noun:* Hungarian(s)
adjective: Hungarian

Ethnic groups: Hungarian 85.6%, Romani 3.2%, German 1.9%, other 2.6%, unspecified 14.1% (2011 est.)
note: percentages add up to more than 100% because respondents were able to identify more than one ethnic group;
Romani populations are usually underestimated in official statistics and may represent 5–10% of Hungary's population

Languages: Hungarian (official) 99.6%, English 16%, German 11.2%, Russian 1.6%, Romanian 1.3%, French 1.2%, other 4.2% (2011 est.)
note: shares sum to more than 100% because some respondents gave more than one answer on the census; Hungarian is the mother tongue of 98.9% of Hungarian speakers

Religions: Roman Catholic 37.2%, Calvinist 11.6%, Lutheran 2.2%, Greek Catholic 1.8%, other 1.9%, none 18.2%, no response 27.2% (2011 est.)

Age structure: *0-14 years:* 14.54% (male 731,542/female 689,739)
15-24 years: 10.43% (male 526,933/female 492,388)
25-54 years: 42.17% (male 2,075,763/female 2,044,664)
55-64 years: 12.17% (male 552,876/female 636,107)
65 years and over: 20.69% (male 773,157/female 1,248,658) (2020 est.)

Dependency ratios: *total dependency ratio:* 46.9
youth dependency ratio: 22
elderly dependency ratio: 30.8
potential support ratio: 3.2 (2020 est.)

Median age: *total:* 43.6 years *male:* 41.5 years
female: 45.5 years (2020 est.)
country comparison to the world: 24

Population growth rate: -0.28% (2020 est.)
country comparison to the world: 217

Birth rate: 8.8 births/1,000 population (2020 est.)
country comparison to the world: 208

Death rate: 12.9 deaths/1,000 population (2020 est.)
country comparison to the world: 10

Net migration rate: 1.3 migrant(s)/1,000 population (2020 est.)
country comparison to the world: 56

Population distribution: a fairly even distribution throughout most of the country, with urban areas attracting larger and denser populations

Urbanization: *urban population:* 71.9% of total population (2020)
rate of urbanization: 0.07% annual rate of change (2015-20 est.)
total population growth rate v. urban population growth rate, 2000-2030: Major urban areas—population: 1.768 million BUDAPEST (capital) (2020)

Sex ratio: *at birth:* 1.06 male(s)/female
0-14 years: 1.06 male(s)/female
15-24 years: 1.07 male(s)/female
25-54 years: 1.02 male(s)/female
55-64 years: 0.87 male(s)/female
65 years and over: 0.62 male(s)/female
total population: 0.91 male(s)/female (2020 est.)

Mother's mean age at first birth: 28.6 years (2017 est.)

Maternal mortality rate: 12 deaths/100,000 live births (2017 est.)
country comparison to the world: 139

Infant mortality rate: *total:* 4.7 deaths/1,000 live births
male: 5 deaths/1,000 live births
female: 4.4 deaths/1,000 live births (2020 est.)
country comparison to the world: 180

Life expectancy at birth: *total population:* 76.7 years
male: 73 years
female: 80.6 years (2020 est.)
country comparison to the world: 89

Total fertility rate: 1.47 children born/woman (2020 est.)
country comparison to the world: 209

Drinking water source:
improved: urban: 100% of population
rural: 100% of population
total: 100% of population
unimproved: urban: 0% of population
rural: 0% of population
total: 0% of population (2017 est.)

Current Health Expenditure: 6.9% (2017)

Physicians density: 3.34 physicians/1,000 population (2017)

Hospital bed density: 7 beds/1,000 population (2017)

Sanitation facility access:
improved: urban: 100% of population
rural: 100% of population
total: 100% of population
unimproved: urban: 0% of population
rural: 0% of population
total: 0% of population (2017 est.)

HIV/AIDS—adult prevalence rate: <.1% (2018 est.)

HIV/AIDS—people living with HIV/AIDS: 3,700 (2018 est.)
country comparison to the world: 126

HIV/AIDS—deaths: <100 (2018 est.)

Major infectious diseases: *degree of risk:* intermediate (2016)

vectorborne diseases: tickborne encephalitis (2016)

Obesity—adult prevalence rate: 26.4% (2016)
country comparison to the world: 41

Education expenditures: 4.7% of GDP (2016)
country comparison to the world: 79

Literacy: *definition:* age 15 and over can read and write
total population: 99.1%
male: 99.1%
female: 99% (2015)

School life expectancy (primary to tertiary education): *total:* 15 years
male: 15 years
female: 15 years (2018)

Unemployment, youth ages 15-24: *total:* 10.2%
male: 9.8%
female: 10.7% (2018 est.)
country comparison to the world: 123

GOVERNMENT

Country name: *conventional long form:* none
conventional short form: Hungary
local long form: none
local short form: Magyarorszag
former: Kingdom of Hungary, Hungarian People's Republic, Hungarian Soviet Republic, Hungarian Republic
etymology: the Byzantine Greeks refered to the tribes that arrived on the steppes of Eastern Europe in the 9th century as the "Oungroi," a name that was later Latinized to "Ungri" and which became "Hungari"; the name originally meant an " [alliance of] ten tribes"; the Hungarian name "Magyarorszag" means "Country of the Magyars"; the term may derive from the most prominent of the Hungarian tribes, the Megyer

Government type: parliamentary republic

Capital: *name:* Budapest
geographic coordinates: 47 30 N, 19 05 E
time difference: UTC+1 (6 hours ahead of Washington, DC, during Standard Time)
daylight saving time: +1hr, begins last Sunday in March; ends last Sunday in October
etymology: the Hungarian capital city was formed in 1873 from the merger of three cities on opposite banks of the Danube: Buda and Obuda (Old Buda) on the western shore and Pest on the eastern; the origins of the original names are obscure, but according to the second century A.D. geographer, Ptolemy, the settlement that would become Pest was called "Pession" in ancient times; "Buda" may derive from either a Slavic or Turkic personal name

Administrative divisions: 19 counties (megyek, singular—megye), 23 cities with county rights (megyei jogu varosok, singular—megyei jogu varos), and 1 capital city (fovaros)
counties: Bacs-Kiskun, Baranya, Bekes, Borsod-Abauj-Zemplen, Csongrad, Fejer, Gyor-Moson-Sopron, Hajdu-Bihar, Heves,

Jasz-Nagykun-Szolnok, Komarom-Esztergom, Nograd, Pest, Somogy, Szabolcs-Szatmar-Bereg, Tolna, Vas, Veszprem, Zala
cities with county rights: Bekescsaba, Debrecen, Dunaujvaros, Eger, Erd, Gyor, Hodmezovasarhely, Kaposvar, Kecskemet, Miskolc, Nagykanizsa, Nyiregyhaza, Pecs, Salgotarjan, Sopron, Szeged, Szekesfehervar, Szekszard, Szolnok, Szombathely, Tatabanya, Veszprem, Zalaegerszeg
capital city: Budapest

Independence: 16 November 1918 (republic proclaimed); notable earlier dates: 25 December 1000 (crowning of King STEPHEN I, traditional founding date); 30 March 1867 (Austro-Hungarian dual monarchy established)

National holiday: Saint Stephen's Day, 20 August (1083); note—commemorates his canonization and the transfer of his remains to Buda (now Budapest) in 1083

Constitution: *history:* previous 1949 (heavily amended in 1989 following the collapse of communism); latest approved 18 April 2011, signed 25 April 2011, effective 1 January 2012
amendments: proposed by the president of the republic, by the government, by parliamentary committee, or by Parliament members; passage requires two-thirds majority vote of Parliament members and approval by the president; amended several times, last in 2018

Legal system: civil legal system influenced by the German model

International law organization participation: accepts compulsory ICJ jurisdiction with reservations; accepts ICC jurisdiction

Citizenship: *citizenship by birth:* no
citizenship by descent only: at least one parent must be a citizen of Hungary
dual citizenship recognized: yes
residency requirement for naturalization: 8 years

Suffrage: 18 years of age, 16 if married and marriage is registered in Hungary; universal

Executive branch: *chief of state:* President Janos ADER (since 10 May 2012)
head of government: Prime Minister Viktor ORBAN (since 29 May 2010)
cabinet: Cabinet of Ministers proposed by the prime minister and appointed by the president
elections/appointments: president indirectly elected by the National Assembly with two-thirds majority vote in first round or simple majority vote in second round for a 5-year term (eligible for a second term); election last held on 13 March 2017 (next to be held spring 2022); prime minister elected by the National Assembly on the recommendation of the president; election last held on 10 May 2018 (next to be held by spring 2022)
election results: Janos ADER (Fidesz) reelected president; National Assembly vote—131 to 39; Viktor ORBAN (Fidesz) reelected prime minister; National Assembly vote—134 to 28

Legislative branch: *description:* unicameral National Assembly or Orszaggyules (199 seats; 106 members directly elected in single-member

constituencies by simple majority vote and 93 members directly elected in a single nationwide constituency by party list proportional representation vote; members serve 4-year terms)
elections: last held on 8 April 2018 (next to be held in April 2022)
election results: percent of vote by party list—Fidesz-KDNP 49.3%, Jobbik 19.1%, MSZP-PM 11.9%, LMP 7.1%, DK 5.4%, Momentum Movement 3.1%, Together 0.7%, LdU 0.5%, other 2.9%; seats by party—Fidesz 117, Jobbik 26, KDNP 16, MSZP 15, DK 9, LMP 8, PM 5, Together 1, LdU 1, independent 1; composition—men 174, women 25, percent of women 12.6%

Judicial branch: *highest courts:* Curia or Supreme Judicial Court (consists of the president, vice president, department heads, and approximately 91 judges and is organized into civil, criminal, and administrative-labor departments; Constitutional Court (consists of 15 judges, including the court president and vice president)
judge selection and term of office: Curia president elected by the National Assembly on the recommendation of the president of the republic; other Curia judges appointed by the president upon the recommendation of the National Judicial Council, a separate 15-member administrative body; judge tenure based on interim evaluations until normal retirement at age 62; Constitutional Court judges, including the president of the court, elected by the National Assembly; court vice president elected by the court itself; members serve 12-year terms with mandatory retirement at age 62
subordinate courts: 5 regional courts of appeal; 19 regional or county courts (including Budapest Metropolitan Court); 20 administrative-labor courts; 111 district or local courts

Political parties and leaders: Christian Democratic People's Party or KDNP [Zsolt SEMJEN]
Democratic Coalition or DK [Ferenc GYURCSANY]
Dialogue for Hungary (Parbeszed) or PM [Gergely KARACSONY, Timea SZABO]
Fidesz-Hungarian Civic Alliance or Fidesz [Viktor ORBAN]
Hungarian Socialist Party or MSZP [Bertalan TOTH]
Momentum Movement (Momentum Mozgalom) [Andras FEKETE-GYOR]
Movement for a Better Hungary or Jobbik [Tamas SNEIDER]
National Self-Government of Germans in Hungary or LdU [Olivia SCHUBERT]
Politics Can Be Different or LMP [Marta DEMETER, Laszlo LORANT-KERESZTES]
Together (Egyutt)

International organization participation: Australia Group, BIS, CD, CE, CEI, CERN, EAPC, EBRD, ECB, EIB, ESA (cooperating state), EU, FAO, G-9, IAEA, IBRD, ICAO, ICC (national committees), ICCt, ICRM, IDA, IEA, IFAD, IFC, IFRCS, ILO, IMF, IMO, IMSO, Interpol, IOC, IOM, IPU, ISO, ITSO, ITU, ITUC (NGOs), MIGA, MINURSO, NATO, NEA, NSG, OAS (observer), OECD, OIF (observer), OPCW, OSCE, PCA, Schengen Convention, SELEC, UN, UNCTAD, UNESCO, UNFICYP, UNHCR, UNIDO, UNIFIL, UNWTO, UPU, WCO, WFTU (NGOs), WHO, WIPO, WMO, WTO, ZC

Diplomatic representation in the US: chief of mission: Charge d'Affaires Dora ZOMBORI (since 14 April 2020)
chancery: 3910 Shoemaker Street NW, Washington, DC 20008
telephone: [1] (202) 362-6730
FAX: [1] (202) 966-8135
consulate(s) general: Chicago, Los Angeles, New York

Diplomatic representation from the US: chief of mission: Ambassador David B. CORNSTEIN (since 25 June 2018)
telephone: [36] (1) 475-4400
embassy: Szabadsag ter 12, H-1054 Budapest

mailing address: pouch: American Embassy Budapest, 5270 Budapest Place, US Department of State, Washington, DC 20521-5270
FAX: [36] (1) 475-4248

Flag description: three equal horizontal bands of red (top), white, and green; the flag dates to the national movement of the 18th and 19th centuries, and fuses the medieval colors of the Hungarian coat of arms with the revolutionary tricolor form of the French flag; folklore attributes virtues to the colors: red for strength, white for faithfulness, and green for hope; alternatively, the red is seen as being for the blood spilled in defense of the land, white for freedom, and green for the pasturelands that make up so much of the country

National symbol(s): Holy Crown of Hungary (Crown of Saint Stephen); national colors: red, white, green

National anthem: name: "Himnusz" (Hymn)
lyrics/music: Ferenc KOLCSEY/Ferenc ERKEL
note: adopted 1844

ECONOMY

Economy—overview: Hungary has transitioned from a centrally planned to a market-driven economy with a per capita income approximately two thirds of the EU-28 average; however, in recent years the government has become more involved in managing the economy. Budapest has implemented unorthodox economic policies to boost household consumption and has relied on EU-funded development projects to generate growth.

Following the fall of communism in 1990, Hungary experienced a drop-off in exports and financial assistance from the former Soviet Union. Hungary embarked on a series of economic reforms, including privatization of state-owned enterprises and reduction of social spending programs, to shift from a centrally planned to a market-driven economy, and to reorient its economy towards trade with the West. These efforts helped to spur growth, attract investment, and

reduce Hungary's debt burden and fiscal deficits. Despite these reforms, living conditions for the average Hungarian initially deteriorated as inflation increased and unemployment reached double digits. Conditions slowly improved over the 1990s as the reforms came to fruition and export growth accelerated. Economic policies instituted during that decade helped position Hungary to join the European Union in 2004. Hungary has not yet joined the euro-zone. Hungary suffered a historic economic contraction as a result of the global economic slowdown in 2008-09 as export demand and domestic consumption dropped, prompting it to take an IMF-EU financial assistance package.

Since 2010, the government has backpedaled on many economic reforms and taken a more populist approach towards economic management. The government has favored national industries and government-linked businesses through legislation, regulation, and public procurements. In 2011 and 2014, Hungary nationalized private pension funds, which squeezed financial service providers out of the system, but also helped Hungary curb its public debt and lower its budget deficit to below 3% of GDP, as subsequent pension contributions have been channeled into the state-managed pension fund. Hungary's public debt (at 74.5% of GDP) is still high compared to EU peers in Central Europe. Real GDP growth has been robust in the past few years due to increased EU funding, higher EU demand for Hungarian exports, and a rebound in domestic household consumption. To further boost household consumption ahead of the 2018 election, the government embarked on a six-year phased increase to minimum wages and public sector salaries, decreased taxes on foodstuffs and services, cut the personal income tax from 16% to 15%, and implemented a uniform 9% business tax for small and medium-sized enterprises and large companies. Real GDP growth slowed in 2016 due to a cyclical decrease in EU funding, but increased to 3.8% in 2017 as the government pre-financed EU funded projects ahead of the 2018 election. Systemic economic challenges include pervasive corruption, labor shortages driven by demographic declines and migration, widespread poverty in rural areas, vulnerabilities to changes in demand for exports, and a heavy reliance on Russian energy imports.

GDP (purchasing power parity): $289.6 billion (2017 est.)
$278.5 billion (2016 est.)
$272.5 billion (2015 est.)
note: data are in 2017 dollars
country comparison to the world: 59

GDP (official exchange rate): $139.2 billion (2017 est.)

GDP—real growth rate: 4.58% (2019 est.)
5.44% (2018 est.)
4.45% (2017 est.)
country comparison to the world: 60

GDP—per capita (PPP): $29,600 (2017 est.)
$28,300 (2016 est.)
$27,600 (2015 est.)
note: data are in 2017 dollars

country comparison to the world: 68

Gross national saving: 25.7% of GDP (2017 est.)
25.8% of GDP (2016 est.)
25.3% of GDP (2015 est.)
country comparison to the world: 54

GDP—composition, by end use: household consumption: 49.6% (2017 est.)
government consumption: 20% (2017 est.)
investment in fixed capital: 21.6% (2017 est.)
investment in inventories: 1% (2017 est.)
exports of goods and services: 90.2% (2017 est.)
imports of goods and services: -82.4% (2017 est.)

GDP—composition, by sector of origin: agriculture: 3.9% (2017 est.)
industry: 31.3% (2017 est.)
services: 64.8% (2017 est.)

Agriculture—products: wheat, corn, sunflower seed, potatoes, sugar beets; pigs, cattle, poultry, dairy products
Industries: mining, metallurgy, construction materials, processed foods, textiles, chemicals (especially pharmaceuticals), motor vehicles

Industrial production growth rate: 7.4% (2017 est.)
country comparison to the world: 29

Labor force: 4.414 million (2020 est.)
country comparison to the world: 85

Labor force—by occupation: agriculture: 4.9%
industry: 30.3%
services: 64.5% (2015 est.)

Unemployment rate: 3.45% (2019 est.)
3.71% (2018 est.)
country comparison to the world: 47

Population below poverty line: 14.9% (2015 est.)

Household income or consumption by percentage share: lowest 10%: 3.3%
highest 10%: 22.4% (2015)

Budget: revenues: 61.98 billion (2017 est.)
expenditures: 64.7 billion (2017 est.)

Taxes and other revenues: 44.5% (of GDP) (2017 est.)
country comparison to the world: 23

Budget surplus (+) or deficit (-): -2% (of GDP) (2017 est.)
note: Hungary has been under the EU Excessive Deficit Procedure since it joined the EU in 2004; in March 2012, the EU elevated its Excessive Deficit Procedure against Hungary and proposed freezing 30% of the country's Cohesion Funds because 2011 deficit reductions were not achieved in a sustainable manner; in June 2012, the EU lifted the freeze, recognizing that steps had been taken to reduce the deficit; the Hungarian deficit increased above 3% both in 2013 and in 2014 due to sluggish growth and the government's fiscal tightening
country comparison to the world: 104

Public debt: 73.6% of GDP (2017 est.)
76% of GDP (2016 est.)
note: general government gross debt is defined in the Maastricht Treaty as consolidated general government gross debt at nominal value, outstanding at the end of the year in the following categories of government liabilities: currency and deposits,

securities other than shares excluding financial derivatives, and national, state, and local government and social security funds.
country comparison to the world: 43

Fiscal year: calendar year

Inflation rate (consumer prices): 2.4% (2017 est.) 0.4% (2016 est.)
country comparison to the world: 119

Current account balance: -$392 million (2019 est.)
$510 million (2018 est.)
country comparison to the world: 115

Exports: $98.74 billion (2017 est.)
$91.6 billion (2016 est.)
country comparison to the world: 38

Exports—partners: Germany 27.7%, Romania 5.4%, Italy 5.1%, Austria 5%, Slovakia 4.8%, France 4.4%, Czech Republic 4.4%, Poland 4.3% (2017)

Exports—commodities: machinery and equipment (55.8%), other manufactures (32.7%), food products (6.8%), raw materials (2.4%), fuels and electricity (2.3%) (2017 est.)

Imports: $96.3 billion (2017 est.)
$83.5 billion (2016 est.)
country comparison to the world: 35

Imports—commodities: machinery and equipment 45.4%, other manufactures 34.3%, fuels and electricity 12.6%, food products 5.3%, raw materials 2.5% (2012)

Imports—partners: Germany 26.2%, Austria 6.3%, China 5.9%, Poland 5.5%, Slovakia 5.3%, Netherlands 5%, Czech Republic 4.8%, Italy 4.7%, France 4% (2017)

Reserves of foreign exchange and gold: $28 billion (31 December 2017 est.)
$25.82 billion (31 December 2016 est.)
country comparison to the world: 52

Debt—external: $138.1 billion (31 December 2017 est.)
$131.3 billion (31 December 2016 est.)
country comparison to the world: 43

Exchange rates: forints (HUF) per US dollar -
279.5 (2017 est.)
281.52 (2016 est.)
281.52 (2015 est.)
279.33 (2014 est.)
232.6 (2013 est.)

ENERGY

Electricity access: *electrification—total population:* 100% (2020)

Electricity—production: 30.22 billion kWh (2016 est.)
country comparison to the world: 64

Electricity—consumption: 39.37 billion kWh (2016 est.)
country comparison to the world: 56

Electricity—exports: 5.24 billion kWh (2016 est.)
country comparison to the world: 34

Electricity—imports: 17.95 billion kWh (2016 est.)

country comparison to the world: 13

Electricity—installed generating capacity: 8.639 million kW (2016 est.)
country comparison to the world: 67

Electricity—from fossil fuels: 64% of total installed capacity (2016 est.)
country comparison to the world: 122

Electricity—from nuclear fuels: 22% of total installed capacity (2017 est.)
country comparison to the world: 5

Electricity—from hydroelectric plants: 1% of total installed capacity (2017 est.)
country comparison to the world: 148

Electricity—from other renewable sources: 13% of total installed capacity (2017 est.)
country comparison to the world: 67

Crude oil—production: 16,000 bbl/day (2018 est.)
country comparison to the world: 72

Crude oil—exports: 2,713 bbl/day (2017 est.)
country comparison to the world: 69

Crude oil—imports: 121,000 bbl/day (2017 est.)
country comparison to the world: 40

Crude oil—proved reserves: 24 million bbl (1 January 2018 est.)
country comparison to the world: 82

Refined petroleum products—production: 152,400 bbl/day (2017 est.)
country comparison to the world: 59

Refined petroleum products—consumption: 167,700 bbl/day (2017 est.)
country comparison to the world: 62

Refined petroleum products—exports: 58,720 bbl/day (2017 est.)
country comparison to the world: 50

Refined petroleum products—imports: 82,110 bbl/day (2017 est.)
country comparison to the world: 62

Natural gas—production: 1.812 billion cu m (2017 est.)
country comparison to the world: 60

Natural gas—consumption: 10.39 billion cu m (2017 est.)
country comparison to the world: 46

Natural gas—exports: 3.52 billion cu m (2017 est.)
country comparison to the world: 34

Natural gas—imports: 13.37 billion cu m (2017 est.)
country comparison to the world: 24

Natural gas—proved reserves: 6.598 billion cu m (1 January 2018 est.)
country comparison to the world: 83

Carbon dioxide emissions from consumption of energy: 51.28 million Mt (2017 est.)
country comparison to the world: 59

COMMUNICATIONS

Telephones—fixed lines: *total subscriptions:* 3,084,836
subscriptions per 100 inhabitants: 31.48 (2019 est.)

country comparison to the world: 43

Telephones—mobile cellular: *total subscriptions:* 10,394,172
subscriptions per 100 inhabitants: 106.07 (2019 est.)
country comparison to the world: 86

Telecommunication systems: *general assessment:* telephone system is digital and highly automated; broadband penetration is the highest in Eastern Europe; replacement of all copper infrastructure with fiber nationally; govt. expands e-payment systems; regulator makes preparations for 5G service (2020)
domestic: competition among mobile-cellular service providers has led to a sharp increase in the use of mobile-cellular phones, and a decrease in the number of fixed-line connections, 31 per 100 persons, while mobile-cellular is 106 per 100 (2019)
international: country code—36; Hungary has fiber-optic cable connections with all neighboring countries; the international switch is in Budapest; satellite earth stations—2 Intelsat (Atlantic Ocean and Indian Ocean regions), 1 Inmarsat, 1 (very small aperture terminal) VSAT system of ground terminals
note: the COVID-19 outbreak is negatively impacting telecommunications production and supply chains globally; consumer spending on telecom devices and services has also slowed due to the pandemic's effect on economies worldwide; overall progress towards improvements in all facets of the telecom industry—mobile, fixed-line, broadband, submarine cable and satellite—has moderated

Broadcast media: mixed system of state-supported public service broadcast media and private broadcasters; the 5 publicly owned TV channels and the 2 main privately owned TV stations are the major national broadcasters; a large number of special interest channels; highly developed market for satellite and cable TV services with about two-thirds of viewers utilizing their services; 4 state-supported public-service radio networks; a large number of local stations including commercial, public service, nonprofit, and community radio stations; digital transition completed at the end of 2013; government-linked businesses have greatly consolidated ownership in broadcast and print media

Internet country code: .hu

Internet users: *total:* 7,474,413
percent of population: 76.07% (July 2018 est.)
country comparison to the world: 68

Broadband—fixed subscriptions: *total:* 3,079,549
subscriptions per 100 inhabitants: 31 (2018 est.)
country comparison to the world: 42

TRANSPORTATION

National air transport system: *number of registered air carriers:* 5 (2020)
inventory of registered aircraft operated by air carriers: 145
annual passenger traffic on registered air carriers: 31,226,848 (2018)

Civil aircraft registration country code prefix: HA (2016)

Airports: 41 (2013)
country comparison to the world: 103

Airports—with paved runways: *total:* 20 (2017)
over 3,047 m: 2 (2017)
2,438 to 3,047 m: 6 (2017)
1,524 to 2,437 m: 6 (2017)
914 to 1,523 m: 5 (2017)
under 914 m: 1 (2017)

Airports—with unpaved runways: *total:* 21 (2013)
1,524 to 2,437 m: 2 (2013)
914 to 1,523 m: 8 (2013)
under 914 m: 11 (2013)

Heliports: 3 (2013)

Pipelines: 5874 km gas (high-pressure transmission system), 83732 km gas (low-pressure distribution network), 850 km oil, 1200 km refined products (2016)

Railways: *total:* 8,049 km (2014)
standard gauge: 7,794 km 1.435-m gauge (2,889 km electrified) (2014)
narrow gauge: 219 km 0.760-m gauge (2014)
broad gauge: 36 km 1.524-m gauge (2014)
country comparison to the world: 28

Roadways: *total:* 203,601 km (2014)
paved: 77,087 km (includes 1,582 km of expressways) (2014)
unpaved: 126,514 km (2014)
country comparison to the world: 27

Waterways: 1,622 km (most on Danube River) (2011)
country comparison to the world: 47

Ports and terminals: *river port(s):* Baja, Csepel (Budapest), Dunaujvaros, Gyor-Gonyu, Mohacs (Danube)

MILITARY AND SECURITY

Military and security forces: *Hungarian Defense Forces:* Ground Forces and Hungarian Air Force (2019)
note: the Hungarian Defense Forces are organized into a joint force structure with ground, air, and logistic components

Military expenditures: 1.21% of GDP (2019 est.)
1.15% of GDP (2018)
1.05% of GDP (2017)
1.02% of GDP (2016)
0.92% of GDP (2015)
country comparison to the world: 98

Military and security service personnel strengths: the Hungarian Defense Forces have approximately 29,000 active duty troops (18,000 Army; 5,000 Air Force; 6,000 other) (2019 est.)

Military equipment inventories and acquisitions: the inventory of the Hungarian Defense Forces consists largely of Soviet-era weapons, with a smaller mix of more modern European and US equipment; since 2010, Hungary has received limited quantities of equipment from Czechia, Finland, France, Germany, Russia, Sweden, and the US (2019 est.)

Military deployments: 160 Bosnia-Herzegovina (EUFOR stabilization force); 150 Iraq (counter-ISIS coalition); 400 Kosovo (NATO) (2020)

Military service age and obligation: 18-25 years of age for voluntary military service; no conscription; 6-month service obligation (2012)

TRANSNATIONAL ISSUES

Disputes—international: bilateral government, legal, technical and economic working group negotiations continue in 2006 with Slovakia over Hungary's failure to complete its portion of the Gabcikovo-Nagymaros hydroelectric dam project along the Danube; as a member state that forms part of the EU's external border, Hungary has implemented the strict Schengen border rules

Refugees and internally displaced persons: *refugees (country of origin):* 5,950 applicants for forms of legal stay other than asylum (Ukraine) (2015)
stateless persons: 76 (2019)
note: 432,744 estimated refugee and migrant arrivals (January 2015-December 2018); Hungary is predominantly a transit country and hosts 137 migrants and asylum seekers as of the end of June 2018; 1,626 migrant arrivals in 2017

Illicit drugs: transshipment point for Southwest Asian heroin and cannabis and for South American cocaine destined for Western Europe; limited producer of precursor chemicals, particularly for amphetamine and methamphetamine; efforts to counter money laundering, related to organized crime and drug trafficking are improving but remain vulnerable; significant consumer of ecstasy

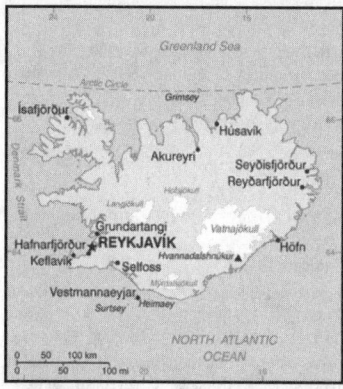

INTRODUCTION

Background: Settled by Norwegian and Celtic (Scottish and Irish) immigrants during the late 9th and 10th centuries A.D., Iceland boasts the world's oldest functioning legislative assembly, the Althingi, established in 930. Independent for over 300 years, Iceland was subsequently ruled by Norway and Denmark. Fallout from the Askja volcano of 1875 devastated the Icelandic economy and caused widespread famine. Over the next quarter century, 20% of the island's population emigrated, mostly to Canada and the US. Denmark granted limited home rule in 1874 and complete independence in 1944. The second half of the 20th century saw substantial economic growth driven primarily by the fishing industry. The economy diversified greatly after the country joined the European Economic Area in 1994, but Iceland was especially hard hit by the global financial crisis in the years following 2008. The economy is now on an upward trajectory, fueled primarily by a tourism and construction boom. Literacy, longevity, and social cohesion are first rate by world standards.

GEOGRAPHY

Location: Northern Europe, island between the Greenland Sea and the North Atlantic Ocean, northwest of the United Kingdom

Geographic coordinates: 65 00 N, 18 00 W

Map references: Arctic Region

Area: *total:* 103,000 sq km
land: 100,250 sq km
water: 2,750 sq km
country comparison to the world: 109

Area—comparative: slightly smaller than Pennsylvania; about the same size as Kentucky

Land boundaries: 0 km

Coastline: 4,970 km

Maritime claims: *territorial sea:* 12 nm
exclusive economic zone: 200 nm

continental shelf: 200 nm or to the edge of the continental margin

Climate: temperate; moderated by North Atlantic Current; mild, windy winters; damp, cool summers

Terrain: mostly plateau interspersed with mountain peaks, icefields; coast deeply indented by bays and fiords

Elevation: *mean elevation:* 557 m
lowest point: Atlantic Ocean 0 m
highest point: Hvannadalshnukur (at Vatnajokull Glacier) 2,110 m

Natural resources: fish, hydropower, geothermal power, diatomite

Land use: *agricultural land:* 18.7% (2011 est.)
arable land: 1.2% (2011 est.) / permanent crops: 0% (2011 est.) / permanent pasture: 17.5% (2011 est.)
forest: 0.3% (2011 est.)
other: 81% (2011 est.)

Irrigated land: NA

Population distribution: Iceland is almost entirely urban with half of the population located in and around the capital of Reykjavik; smaller clusters are primarily found along the coast in the north and west

Natural hazards: earthquakes and volcanic activity
volcanism: Iceland, situated on top of a hotspot, experiences severe volcanic activity; Eyjafjallajokull (1,666 m) erupted in 2010, sending ash high into the atmosphere and seriously disrupting European air traffic; scientists continue to monitor nearby Katla (1,512 m), which has a high probability of eruption in the very near future, potentially disrupting air traffic; Grimsvoetn and Hekla are Iceland's most active volcanoes; other historically active volcanoes include Askja, Bardarbunga, Brennisteinsfjoll, Esjufjoll, Hengill, Krafla, Krisuvik, Kverkfjoll, Oraefajokull, Reykjanes, Torfajokull, and Vestmannaeyjar

Environment—current issues: water pollution from fertilizer runoff

Environment—international agreements: *party to:* Air Pollution, Air Pollution-Persistent Organic Pollutants, Biodiversity, Climate Change, Climate Change-Kyoto Protocol, Desertification, Endangered Species, Hazardous Wastes, Kyoto Protocol, Law of the Sea, Marine Dumping, Ozone Layer Protection, Ship Pollution, Transboundary Air Pollution, Wetlands, Whaling
signed, but not ratified: Environmental Modification, Marine Life Conservation

Geography—note: strategic location between Greenland and Europe; westernmost European country; Reykjavik is the northernmost national capital in the world; more land covered by glaciers than in all of continental Europe

PEOPLE AND SOCIETY

Population: 350,734 (July 2020 est.)
country comparison to the world: 178

Nationality: *noun:* Icelander(s)
adjective: Icelandic

Ethnic groups: homogeneous mixture of descendants of Norse and Celts 81%, population with foreign background 19% (2018 est.)
note: population with foreign background includes immigrants and persons having at least one parent who was born abroad

Languages: Icelandic, English, Nordic languages, German

Religions: Evangelical Lutheran Church of Iceland (official) 67.2%, Roman Catholic 3.9%, Reykjavik Free Church 2.8%, Hafnarfjordur Free Church 2%, Asatru Association 1.2%, The Independent Congregation .9%, other religions 4% (includes Zuist and Pentecostal), none 6.7%, other or unspecified 11.3% (2018 est.)

Age structure: *0-14 years:* 20.31% (male 36,394/female 34,837)
15-24 years: 12.85% (male 22,748/female 22,317)
25-54 years: 39.44% (male 70,227/female 68,095)
55-64 years: 11.94% (male 20,762/female 21,111)
65 years and over: 15.47% (male 25,546/female 28,697) (2020 est.)

Dependency ratios: *total dependency ratio:* 54
youth dependency ratio: 29.9
elderly dependency ratio: 24.1
potential support ratio: 4.2 (2020 est.)

Median age: *total:* 37.1 years
male: 36.6 years
female: 37.7 years (2020 est.)
country comparison to the world: 74

Population growth rate: 1.02% (2020 est.)
country comparison to the world: 104

Birth rate: 13.3 births/1,000 population (2020 est.)
country comparison to the world: 141

Death rate: 6.6 deaths/1,000 population (2020 est.)
country comparison to the world: 139

Net migration rate: 3.3 migrant(s)/1,000 population (2020 est.)
country comparison to the world: 34

Population distribution: Iceland is almost entirely urban with half of the population located in and around the capital of Reykjavik; smaller clusters are primarily found along the coast in the north and west

Urbanization: *urban population:* 93.9% of total population (2020)
rate of urbanization: 0.81% annual rate of change (2015-20 est.)
total population growth rate v. urban population growth rate, 2000-2030: Major urban areas—population: 216,000 REYKJAVIK (capital) (2018)

Sex ratio: *at birth:* 1.05 male(s)/female
0-14 years: 1.04 male(s)/female
15-24 years: 1.02 male(s)/female
25-54 years: 1.03 male(s)/female
55-64 years: 0.98 male(s)/female

65 years and over: 0.89 male(s)/female
total population: 1 male(s)/female (2020 est.)

Mother's mean age at first birth: 27.8 years (2017 est.)

Maternal mortality rate: 4 deaths/100,000 live births (2017 est.)
country comparison to the world: 173

Infant mortality rate: *total:* 2.1 deaths/1,000 live births
male: 2.3 deaths/1,000 live births
female: 2 deaths/1,000 live births (2020 est.)
country comparison to the world: 225

Life expectancy at birth: *total population:* 83.3 years
male: 81 years
female: 85.6 years (2020 est.)
country comparison to the world: 7

Total fertility rate: 1.97 children born/woman (2020 est.)
country comparison to the world: 115

Drinking water source:

improved: *urban:* 100% of population
rural: 100% of population
total: 100% of population

unimproved: *urban:* 0% of population
rural: 0% of population
total: 0% of population (2017 est.)

Current Health Expenditure: 8.3% (2017)

Physicians density: 3.98 physicians/1,000 population (2017)

Hospital bed density: 3.1 beds/1,000 population (2017)

Sanitation facility access:

improved: *urban:* 100% of population
rural: 100% of population
total: 100% of population

unimproved: *urban:* 0% of population
rural: 0% of population
total: 0% of population (2017 est.)

HIV/AIDS—adult prevalence rate: 0.1% (2018)
country comparison to the world: 124

HIV/AIDS—people living with HIV/AIDS: <500 (2018)

HIV/AIDS—deaths: <100 (2018)

Obesity—adult prevalence rate: 21.9% (2016)
country comparison to the world: 83

Education expenditures: 7.5% of GDP (2016)
country comparison to the world: 8

School life expectancy (primary to tertiary education): *total:* 19 years
male: 18 years
female: 20 years (2018)

Unemployment, youth ages 15-24: *total:* 6.1%
male: 6.5%
female: 5.6% (2018 est.)
country comparison to the world: 160

GOVERNMENT

Country name: *conventional long form:* Republic of Iceland

conventional short form: Iceland
local long form: Lydveldid Island
local short form: Island
etymology: Floki VILGERDARSON, an early explorer of the island (9th century), applied the name "Land of Ice" after spotting a fjord full of drift ice to the north and spending a bitter winter on the island; he eventually settled on the island, however, after he saw how it greened up in the summer and that it was, in fact, habitable

Government type: unitary parliamentary republic

Capital: *name:* Reykjavik
geographic coordinates: 64 09 N, 21 57 W
time difference: UTC 0 (5 hours ahead of Washington, DC, during Standard Time)
etymology: the name means "smoky bay" in Icelandic and refers to the steamy, smoke-like vapors discharged by hot springs in the area

Administrative divisions: 72 municipalities (sveitarfelog, singular—sveitarfelagidh); Akrahreppur, Akraneskaupstadhur, Akureyrarkaupstadhur, Arneshreppur, Asahreppur, Blaskogabyggdh, Blonduosbaer, Bolungarvikurkaupstadhur, Borgarbyggdh, Borgarfjardharhreppur, Dalabyggdh, Dalvikurbyggdh, Djupavogshreppur, Eyjafjardharsveit, Eyja-og Miklaholtshreppur, Fjallabyggdh, Fjardhabyggdh, Fljotsdalsheradh, Fljotsdalshreppur, Floahreppur, Gardhabaer, Grimsnes-og Grafningshreppur, Grindavikurbaer, Grundarfjardharbaer, Grytubakkahreppur, Hafnarfjardharkaupstadhur, Helgafellssveit, Horgarsveit, Hrunamannahreppur, Hunathing Vestra, Hunavatnshreppur, Hvalfjardharsveit, Hveragerdhisbaer, Isafjardharbaer, Kaldrananeshreppur, Kjosarhreppur, Kopavogsbaer, Langanesbyggdh, Mosfellsbaer, Myrdalshreppur, Nordhurthing, Rangarthing Eystra, Rangarthing Ytra, Reykholahreppur, Reykjanesbaer, Reykjavikurborg, Seltjarnarnesbaer, Seydhisfjardharkaupstadhur, Skaftarhreppur, Skagabyggdh, Skeidha-og Gnupverjahreppur, Skorradalshreppur, Skutustadhahreppur, Snaefellsbaer, Strandabyggdh, Stykkisholmsbaer, Sudhavikurhreppur, Sudhurnesjabaer, Svalbardhshreppur, Svalbardhsstrandarhreppur, Sveitarfelagidh Arborg, Sveitarfelagidh Hornafjordhur, Sveitarfelagidh Olfus, Sveitarfelagidh Skagafjordhur, Sveitarfelagidh Skagastrond, Sveitarfelagidh Vogar, Talknafjardharhreppur, Thingeyjarsveit, Tjorneshreppur, Vestmannaeyjabaer, Vesturbyggdh, Vopnafjardharhreppur

Independence: 1 December 1918 (became a sovereign state under the Danish Crown); 17 June 1944 (from Denmark; birthday of Jon SIGURDSSON, leader of Iceland's 19th Century independence movement)

National holiday: Independence Day, 17 June (1944)

Constitution: *history:* several previous; latest ratified 16 June 1944, effective 17 June 1944 (at independence)
amendments: proposed by the Althingi; passage requires approval by the Althingi and by the next elected Althingi, and confirmation by the president of the republic; proposed amendments to Article 62 of the constitution – that the

Evangelical Lutheran Church shall be the state church of Iceland – also require passage by referendum; amended many times, last in 2013

Legal system: civil law system influenced by the Danish model

International law organization participation: has not submitted an ICJ jurisdiction declaration; accepts ICCt jurisdiction

Citizenship: *citizenship by birth:* no
citizenship by descent only: at least one parent must be a citizen of Iceland
dual citizenship recognized: yes
residency requirement for naturalization: 3 to 7 years

Suffrage: 18 years of age; universal

Executive branch: *chief of state:* President Gudni Thorlacius JOHANNESSON (since 1 August 2016)
head of government: Prime Minister Katrin JAKOBSDOTTIR (since 30 November 2017)
cabinet: Cabinet appointed by the president upon the recommendation of the prime minister
elections/appointments: president directly elected by simple majority popular vote for a 4-year term (no term limits); election last held on 27 June 2020 (next to be held in 2024); following legislative elections, the leader of the majority party or majority coalition becomes prime minister
election results: Gudni Thorlacius JOHANNESSON reelected president; percent of vote—Gudni Thorlacius JOHANNESSON (independent) 92.2%, Gudmundur Franklin JONSSON (independent) 7.8%

Legislative branch: *description:* unicameral Althingi or Parliament (63 seats; members directly elected in multi-seat constituencies by proportional representation vote to serve 4-year terms)
elections: last held on 28 October 2017 (next to be held in 2021)
election results: percent of vote by party—IP 25.2%, LGM 16.9%, SDA 12.1%, CP 10.9%, PP 10.7%, Pirate Party 9.2%, People's Party 6.9%, Reform Party 6.7%. other 1.5%; seats by party—IP 16, LGM 11, SDA 7, CP 7, PP 8, Pirate Party 6, Reform Party 4, People's Party 4

Judicial branch: *highest courts:* Supreme Court or Haestirettur (consists of 9 judges)
judge selection and term of office: judges proposed by Ministry of Interior selection committee and appointed by the president; judges appointed for an indefinite period
subordinate courts: Appellate Court or Landsrettur; 8 district courts; Labor Court

Political parties and leaders: Centrist Party (Midflokkurinn) or CP [Sigmundur David GUNNLAUGSSON]
Independence Party (Sjalfstaedisflokkurinn) or IP [Bjarni BENEDIKTSSON]
Left-Green Movement (Vinstrihreyfingin-graent frambod) or LGM [Katrin JAKOBSDOTTIR]
People's Party (Flokkur Folksins) [Inga SAELAND]
Pirate Party (Piratar) [rotating leadership]

Progressive Party (Framsoknarflokkurinn) or PP [Sigurdur Ingi JOHANNSSON]
Reform Party (Vidreisn) [Thorgerdur Katrin GUNNARSDOTTIR]
Social Democratic Alliance (Samfylkingin) or SDA [Logi Mar EINARSSON]

International organization participation: Arctic Council, Australia Group, BIS, CBSS, CD, CE, EAPC, EBRD, EFTA, FAO, FATF, IAEA, IBRD, ICAO, ICC (national committees), ICCt, ICRM, IDA, IFAD, IFC, IFRCS, IHO, ILO, IMF, IMO, IMSO, Interpol, IOC, IOM, IPU, ISO, ITSO, ITU, ITUC (NGOs), MIGA, NATO, NC, NEA, NIB, NSG, OAS (observer), OECD, OPCW, OSCE, PCA, Schengen Convention, UN, UNCTAD, UNESCO, UPU, WCO, WHO, WIPO, WMO, WTO

Diplomatic representation in the US: *chief of mission:* Ambassador Geir Hilmar HAARDE (since 23 February 2015)
chancery: House of Sweden, 2900 K Street NW, #509, Washington, DC 20007
telephone: [1] (202) 265-6653
FAX: [1] (202) 265-6656
consulate(s) general: New York

Diplomatic representation from the US: *chief of mission:* Ambassador Jeffrey Ross GUNTER (since 2 July 2019)
telephone: [354] 595-2200
embassy: Laufasvegur 21, 101 Reykjavik
mailing address: US Department of State, 5640 Reykjavik Place, Washington, D.C. 20521-5640
FAX: [354] 562-9118

Flag description: *blue with a red cross outlined in white extending to the edges of the flag; the vertical part of the cross is shifted to the hoist side in the style of the Dannebrog (Danish flag); the colors represent three of the elements that make up the island:* red is for the island's volcanic fires, white recalls the snow and ice fields of the island, and blue is for the surrounding ocean

National symbol(s): *gyrfalcon; national colors:* blue, white, red

National anthem: *name:* "Lofsongur" (Song of Praise)
lyrics/music: Matthias JOCHUMSSON/ Sveinbjorn SVEINBJORNSSON
note: adopted 1944; also known as "O, Gud vors lands" (O, God of Our Land), the anthem was originally written and performed in 1874

ECONOMY

Economy—overview: Iceland's economy combines a capitalist structure and free-market principles with an extensive welfare system. Except for a brief period during the 2008 crisis, Iceland has in recent years achieved high growth, low unemployment, and a remarkably even distribution of income. Iceland's economy has been diversifying into manufacturing and service industries in the last decade, particularly within the fields of tourism, software production, and biotechnology. Abundant geothermal and hydropower sources have attracted substantial foreign investment in the aluminum sector, boosted economic growth, and sparked some interest from high-tech firms looking to establish data centers using cheap green energy.

Tourism, aluminum smelting, and fishing are the pillars of the economy. For decades the Icelandic economy depended heavily on fisheries, but tourism has now surpassed fishing and aluminum as Iceland's main export industry. Tourism accounted for 8.6% of Iceland's GDP in 2016, and 39% of total exports of merchandise and services. From 2010 to 2017, the number of tourists visiting Iceland increased by nearly 400%. Since 2010, tourism has become a main driver of Icelandic economic growth, with the number of tourists reaching 4.5 times the Icelandic population in 2016. Iceland remains sensitive to fluctuations in world prices for its main exports, and to fluctuations in the exchange rate of the Icelandic Krona.

Following the privatization of the banking sector in the early 2000s, domestic banks expanded aggressively in foreign markets, and consumers and businesses borrowed heavily in foreign currencies. Worsening global financial conditions throughout 2008 resulted in a sharp depreciation of the krona vis-a-vis other major currencies. The foreign exposure of Icelandic banks, whose loans and other assets totaled nearly nine times the country's GDP, became unsustainable. Iceland's three largest banks collapsed in late 2008. GDP fell 6.8% in 2009, and unemployment peaked at 9.4% in February 2009. Three new banks were established to take over the domestic assets of the collapsed banks. Two of them have majority ownership by the state, which intends to re-privatize them.

Since the collapse of Iceland's financial sector, government economic priorities have included stabilizing the krona, implementing capital controls, reducing Iceland's high budget deficit, containing inflation, addressing high household debt, restructuring the financial sector, and diversifying the economy. Capital controls were lifted in March 2017, but some financial protections, such as reserve requirements for specified investments connected to new inflows of foreign currency, remain in place.

GDP (purchasing power parity): $18.18 billion (2017 est.)
$17.48 billion (2016 est.)
$16.29 billion (2015 est.)
note: data are in 2017 dollars
country comparison to the world: 153

GDP (official exchange rate): $24.48 billion (2017 est.)

GDP—real growth rate: 1.94% (2019 est.)
3.88% (2018 est.)
4.57% (2017 est.)
country comparison to the world: 142

GDP—per capita (PPP): $52,200 (2017 est.)
$51,700 (2016 est.)
$48,900 (2015 est.)
note: data are in 2017 dollars
country comparison to the world: 25

Gross national saving: 25.8% of GDP (2017 est.)
29.1% of GDP (2016 est.)
24.5% of GDP (2015 est.)

country comparison to the world: 52

GDP—composition, by end use: *household consumption:* 50.4% (2017 est.)
government consumption: 23.3% (2017 est.)
investment in fixed capital: 22.1% (2017 est.)
investment in inventories: 0% (2017 est.)
exports of goods and services: 47% (2017 est.)
imports of goods and services: -42.8% (2017 est.)

GDP—composition, by sector of origin: *agriculture:* 5.8% (2017 est.)
industry: 19.7% (2017 est.)
services: 74.6% (2017 est.)

Agriculture—products: potatoes, carrots, green vegetables, tomatoes, cucumbers; mutton, chicken, pork, beef, dairy products; fish
Industries: tourism, fish processing; aluminum smelting;; geothermal power, hydropower; medical/pharmaceutical products

Industrial production growth rate: 2.4% (2017 est.)
country comparison to the world: 120

Labor force: 200,000 (2020 est.)
country comparison to the world: 170

Labor force—by occupation: *agriculture:* 4.8%
industry: 22.2%
services: 73% (2008)

Unemployment rate: 3.62% (2019 est.)
2.73% (2018 est.)
country comparison to the world: 50

Population below poverty line: NA
note: 332,100 families (2011 est.)

Household income or consumption by percentage share: *lowest 10%:* NA
highest 10%: NA

Budget: *revenues:* 10.39 billion (2017 est.)
expenditures: 10.02 billion (2017 est.)

Taxes and other revenues: 42.4% (of GDP) (2017 est.)
country comparison to the world: 31

Budget surplus (+) or deficit (-): 1.5% (of GDP) (2017 est.)
country comparison to the world: 22

Public debt: 40% of GDP (2017 est.)
51.7% of GDP (2016 est.)
country comparison to the world: 126

Fiscal year: calendar year

Inflation rate (consumer prices): 1.8% (2017 est.)
1.7% (2016 est.)
country comparison to the world: 93

Current account balance: $1.496 billion (2019 est.)
$814 million (2018 est.)
country comparison to the world: 47

Exports: $4.957 billion (2017 est.)
$4.483 billion (2016 est.)
country comparison to the world: 108

Exports—partners: Netherlands 25.5%, Spain 13.6%, UK 9.4%, Germany 7.6%, US 7%, France 6.3%, Norway 4.9% (2017)

Exports—commodities: fish and fish products (42%), aluminum (38%), agricultural products, medicinal and medical products, ferro-silicon (2015)

Imports: $6.525 billion (2017 est.)
$5.315 billion (2016 est.)
country comparison to the world: 118

Imports—commodities: machinery and equipment, petroleum products, foodstuffs, textiles

Imports—partners: Germany 10.7%, Norway 9.2%, China 7%, Netherlands 6.7%, US 6.4%, Denmark 6.2%, UK 5.7%, Sweden 4.1% (2017)

Reserves of foreign exchange and gold: $6.567 billion (31 December 2017 est.)
$7.226 billion (31 December 2016 est.)
country comparison to the world: 89 Debt—external: $21.7 billion (31 December 2017 est.)
$25.02 billion (31 December 2016 est.)
country comparison to the world: 91

Exchange rates: Icelandic kronur (ISK) per US dollar -
111.7 (2017 est.)
120.81 (2016 est.)
120.81 (2015 est.)
131.92 (2014 est.)
116.77 (2013 est.)

ENERGY

Electricity access: *electrification—total population:* 100% (2020)

Electricity—production: 18.17 billion kWh (2016 est.)
country comparison to the world: 80

Electricity—consumption: 17.68 billion kWh (2016 est.)
country comparison to the world: 73

Electricity—exports: 0 kWh (2016 est.)
country comparison to the world: 140

Electricity—imports: 0 kWh (2016 est.)
country comparison to the world: 160

Electricity—installed generating capacity: 2.772 million kW (2016 est.)
country comparison to the world: 100

Electricity—from fossil fuels: 4% of total installed capacity (2016 est.)
country comparison to the world: 206

Electricity—from nuclear fuels: 0% of total installed capacity (2017 est.)
country comparison to the world: 109

Electricity—from hydroelectric plants: 71% of total installed capacity (2017 est.)
country comparison to the world: 16

Electricity—from other renewable sources: 25% of total installed capacity (2017 est.)
country comparison to the world: 29

Crude oil—production: 0 bbl/day (2018 est.)
country comparison to the world: 151

Crude oil—exports: 0 bbl/day (2017 est.)
country comparison to the world: 139

Crude oil—imports: 0 bbl/day (2017 est.)
country comparison to the world: 143

Crude oil—proved reserves: 0 bbl (1 January 2018 est.)
country comparison to the world: 147

Refined petroleum products—production: 0 bbl/day (2017 est.)
country comparison to the world: 159

Refined petroleum products—consumption: 20,850 bbl/day (2017 est.)
country comparison to the world: 139

Refined petroleum products—exports: 2,530 bbl/day (2017 est.)
country comparison to the world: 101

Refined petroleum products—imports: 20,220 bbl/day (2017 est.)
country comparison to the world: 120

Natural gas—production: 0 cu m (2017 est.)
country comparison to the world: 148

Natural gas—consumption: 0 cu m (2017 est.)
country comparison to the world: 160

Natural gas—exports: 0 cu m (2017 est.)
country comparison to the world: 123

Natural gas—imports: 0 cu m (2017 est.)
country comparison to the world: 140

Natural gas—proved reserves: 0 cu m (1 January 2014 est.)
country comparison to the world: 149

Carbon dioxide emissions from consumption of energy: 3.228 million Mt (2017 est.)
country comparison to the world: 144

COMMUNICATIONS

Telephones—fixed lines: *total subscriptions:* 128,597
subscriptions per 100 inhabitants: 37.04 (2019 est.)
country comparison to the world: 132

Telephones—mobile cellular: *total subscriptions:* 423,390
subscriptions per 100 inhabitants: 121.95 (2019 est.)
country comparison to the world: 174

Telecommunication systems: *general assessment:* telecommunications infrastructure is modern and fully digitized, with satellite-earth stations, fiberoptic cables, and an extensive broadband network; LTE licenses providing 99% population coverage; small but most progressive telecom market in Europe; good competition among mobile and broadband markets (2020)
domestic: liberalization of the telecommunications sector beginning in the late 1990s has led to increased competition especially in the mobile services segment of the market; 37 per 100 for fixed line and 122 per 100 for mobile-cellular subscriptions (2019)
international: country code—354; landing points for the CANTAT-3, FARICE-1, Greenland Connect and DANICE submarine cable system that provides connectivity to Canada, the Faroe Islands, Greenland, UK, Denmark, and Germany; satellite earth stations—2 Intelsat (Atlantic Ocean), 1 Inmarsat (Atlantic and Indian Ocean regions); note—Iceland shares the Inmarsat earth station with the other Nordic countries (Denmark, Finland, Norway, and Sweden) (2019)

note: the COVID-19 outbreak is negatively impacting telecommunications production and supply chains globally; consumer spending on telecom devices and services has also slowed due to the pandemic's effect on economies worldwide; overall progress towards improvements in all facets of the telecom industry—mobile, fixed-line, broadband, submarine cable and satellite—has moderated

Broadcast media: state-owned public TV broadcaster (RUV) operates 21 TV channels nationally (RUV and RUV 2, though RUV 2 is used less frequently); RUV broadcasts nationally, every household in Iceland is required to have RUV as it doubles as the emergency broadcast network; RUV also operates stringer offices in the north (Akureyri) and the east (Egilsstadir) but operations are all run out of RUV headquarters in Reykjavik; there are 3 privately owned TV stations; Stod 2 (Channel 2) is owned by Syn, following 365 Media and Vodafone merger, and is headquartered in Reykjavik; Syn also operates 4 sports channels under Stod 2; N4 is the only television station headquartered outside of Reykjavik, in Akureyri, with local programming for the north, south, and east of Iceland; Hringbraut is the newest station and is headquartered in Reykjavik; all of these television stations have nationwide penetration as 100% of households have multi-channel access though digital and/or fiber-optic connections

RUV operates 3 radio stations (RAS 1, RAS2, and Rondo) as well as 4 regional stations (but they mostly act as range extenders for RUV radio broadcasts nationwide); there is 1 privately owned radio conglomerate, Syn (4 stations), that broadcasts nationwide, and 3 other radio stations that broadcast to the most densely populated regions of the country. In addition there are upwards of 20 radio stations that operate regionally (2019)

Internet country code: .is

Internet users: *total:* 340,117
percent of population: 99.01% (July 2018 est.)
country comparison to the world: 163

Broadband—fixed subscriptions: *total:* 136,556
subscriptions per 100 inhabitants: 40 (2018 est.)
country comparison to the world: 116

TRANSPORTATION

National air transport system: *number of registered air carriers:* 6 (2020)
inventory of registered aircraft operated by air carriers: 63
annual passenger traffic on registered air carriers: 7,819,740 (2018)
annual freight traffic on registered air carriers: 163.65 million mt-km (2018)

Civil aircraft registration country code prefix: TF (2016)

Airports: 96 (2013)
country comparison to the world: 59

Airports—with paved runways: *total:* 7 (2017)
over 3,047 m: 1 (2017)
1,524 to 2,437 m: 3 (2017)

914 to 1,523 m: 3 (2017)

Airports—with unpaved runways: *total:* 89 (2013)
1,524 to 2,437 m: 3 (2013)
914 to 1,523 m: 26 (2013)
under 914 m: 60 (2013)

Roadways: *total:* 12,898 km (2012)
paved/oiled gravel: 5,647 km (excludes urban roads) (2012)
unpaved: 7,251 km (2012)
country comparison to the world: 129

Merchant marine: *total:* 37
by type: general cargo 5, oil tanker 2, other 30 (2019)
country comparison to the world: 126

Ports and terminals: *major seaport(s):* Grundartangi, Hafnarfjordur, Reykjavik

MILITARY AND SECURITY

Military and security forces: no regular military forces; Icelandic Coast Guard; Icelandic National Police (2019)

Military expenditures: 0.3% of GDP (2018)
0.3% of GDP (2017)
0.3% of GDP (2016)
0.3% of GDP (2015)
0.5% of GDP (2014)
country comparison to the world: 155

Military and security service personnel strengths: the Icelandic Coast Guard has approximately 250 personnel (2019 est.)

Military equipment inventories and acquisitions: the Icelandic Coast Guard's inventory consists of equipment from European suppliers (2019 est.)

Military—note: Iceland is the only NATO member that has no standing military force; defense of Iceland remains a NATO commitment and NATO maintains an air policing presence in Icelandic airspace; Iceland participates in international peacekeeping missions with the civilian-manned Icelandic Crisis Response Unit (ICRU) (2019)

TRANSNATIONAL ISSUES

Disputes—international: Iceland, the UK, and Ireland dispute Denmark's claim that the Faroe Islands' continental shelf extends beyond 200 nm; the European Free Trade Association Surveillance Authority filed a suit against Iceland, claiming the country violated the Agreement on the European Economic Area in failing to pay minimum compensation to Icesave depositors

Refugees and internally displaced persons: *stateless persons:* 48 (2019)

INDIA

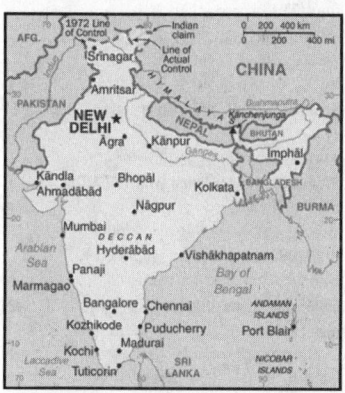

INTRODUCTION

Background: The Indus Valley civilization, one of the world's oldest, flourished during the 3rd and 2nd millennia B.C. and extended into northwestern India. Aryan tribes from the northwest infiltrated the Indian subcontinent about 1500 B.C.; their merger with the earlier Dravidian inhabitants created the classical Indian culture. The Maurya Empire of the 4th and 3rd centuries B.C.—which reached its zenith under ASHOKA—united much of South Asia. The Golden Age ushered in by the Gupta dynasty (4th to 6th centuries A.D.) saw a flowering of Indian science, art, and culture. Islam spread across the subcontinent over a period of 700 years. In the 10th and 11th centuries, Turks and Afghans invaded India and established the Delhi Sultanate. In the early 16th century, the Emperor BABUR established the Mughal Dynasty, which ruled India for more than three centuries;

European explorers began establishing footholds in India during the 16th century.

By the 19th century, Great Britain had become the dominant political power on the subcontinent and India was seen as the "Jewel in the Crown" of the British Empire. The British Indian Army played a vital role in both World Wars. Years of nonviolent resistance to British rule, led by Mohandas GANDHI and Jawaharlal NEHRU, eventually resulted in Indian independence in 1947. Large-scale communal violence took place before and after the subcontinent partition into two separate states—India and Pakistan. The neighboring countries have fought three wars since independence, the last of which was in 1971 and resulted in East Pakistan becoming the separate nation of Bangladesh. India's nuclear weapons tests in 1998 emboldened Pakistan to conduct its own tests that same year. In November 2008, terrorists originating from Pakistan conducted a series of coordinated attacks in Mumbai, India's financial capital. India's economic growth following the launch of economic reforms in 1991, a massive youthful population, and a strategic geographic location have contributed to India's emergence as a regional and global power. However, India still faces pressing problems such as environmental degradation, extensive poverty, and widespread corruption, and its restrictive business climate is dampening economic growth expectations.

GEOGRAPHY

Location: Southern Asia, bordering the Arabian Sea and the Bay of Bengal, between Burma and Pakistan

Geographic coordinates: 20 00 N, 77 00 E

Map references: Asia

Area: *total:* 3,287,263 sq km
land: 2,973,193 sq km
water: 314,070 sq km
country comparison to the world: 8

Area—comparative: slightly more than one-third the size of the US

Land boundaries: *total:* 13,888 km
border countries (6): Bangladesh 4142 km, Bhutan 659 km, Burma 1468 km, China 2659 km, Nepal 1770 km, Pakistan 3190 km

Coastline: 7,000 km

Maritime claims: *territorial sea:* 12 nm
exclusive economic zone: 200 nm
contiguous zone: 24 nm
continental shelf: 200 nm or to the edge of the continental margin

Climate: varies from tropical monsoon in south to temperate in north

Terrain: upland plain (Deccan Plateau) in south, flat to rolling plain along the Ganges, deserts in west, Himalayas in north

Elevation: *mean elevation:* 160 m
lowest point: Indian Ocean 0 m
highest point: Kanchenjunga 8,586 m

Natural resources: coal (fourth-largest reserves in the world), antimony, iron ore, lead, manganese, mica, bauxite, rare earth elements, titanium ore, chromite, natural gas, diamonds, petroleum, limestone, arable land

Land use: *agricultural land:* 60.5% (2011 est.)
arable land: 52.8% (2011 est.) / permanent crops: 4.2% (2011 est.) / permanent pasture: 3.5% (2011 est.)
forest: 23.1% (2011 est.)
other: 16.4% (2011 est.)

Irrigated land: 667,000 sq km (2012)

Population distribution: with the notable exception of the deserts in the northwest, including the Thar Desert, and the mountain fringe in the north, a very high population density exists throughout most of the country; the core of the population is in the north along the banks of the Ganges, with other river valleys and southern coastal areas also having large population concentrations

Natural hazards: droughts; flash floods, as well as widespread and destructive flooding from monsoonal rains; severe thunderstorms; earthquakes
volcanism: Barren Island (354 m) in the Andaman Sea has been active in recent years

Environment—current issues: deforestation; soil erosion; overgrazing; desertification; air pollution from industrial effluents and vehicle emissions; water pollution from raw sewage and runoff of agricultural pesticides; tap water is not potable throughout the country; huge and growing population is overstressing natural resources; preservation and quality of forests; biodiversity loss

Environment—international agreements: *party to:* Antarctic-Environmental Protocol, Antarctic-Marine Living Resources, Antarctic Treaty, Biodiversity, Climate Change, Climate Change-Kyoto Protocol, Desertification, Endangered Species, Environmental Modification, Hazardous Wastes, Law of the Sea, Ozone Layer Protection, Ship Pollution, Tropical Timber 83, Tropical Timber 94, Wetlands, Whaling
signed, but not ratified: none of the selected agreements

Geography—note: dominates South Asian subcontinent; near important Indian Ocean trade routes; Kanchenjunga, third tallest mountain in the world, lies on the border with Nepal

PEOPLE AND SOCIETY

Population: 1,326,093,247 (July 2020 est.)
country comparison to the world: 2

Nationality: *noun:* Indian(s)
adjective: Indian

Ethnic groups: Indo-Aryan 72%, Dravidian 25%, Mongoloid and other 3% (2000)

Languages: Hindi 43.6%, Bengali 8%, Marathi 6.9%, Telugu 6.7%, Tamil 5.7%, Gujarati 4.6%, Urdu 4.2%, Kannada 3.6%, Odia 3.1%, Malayalam 2.9%, Punjabi 2.7%, Assamese 1.3%, Maithili 1.1%, other 5.6% (2011 est.)
note: English enjoys the status of subsidiary official language but is the most important language for national, political, and commercial communication; there are 22 other officially recognized languages: Assamese, Bengali, Bodo, Dogri, Gujarati, Hindi, Kannada, Kashmiri, Konkani, Maithili, Malayalam, Manipuri, Nepali, Odia, Punjabi, Sanskrit, Santali, Sindhi, Tamil, Telugu, Urdu; Hindustani is a popular variant of Hindi/Urdu spoken widely throughout northern India but is not an official language

Religions: Hindu 79.8%, Muslim 14.2%, Christian 2.3%, Sikh 1.7%, other and unspecified 2% (2011 est.)

Age structure: *0-14 years:* 26.31% (male 185,017,089/female 163,844,572)

15-24 years: 17.51% (male 123,423,531/female 108,739,780)

25-54 years: 41.56% (male 285,275,667/female 265,842,319)

55-64 years: 7.91% (male 52,444,817/female 52,447,038)

65 years and over: 6.72% (male 42,054,459/female 47,003,975) (2020 est.)

Dependency ratios: *total dependency ratio:* 48.7
youth dependency ratio: 38.9
elderly dependency ratio: 9.8
potential support ratio: 10.2 (2020 est.)

Median age: *total:* 28.7 years
male: 28 years
female: 29.5 years (2020 est.)
country comparison to the world: 141

Population growth rate: 1.1% (2020 est.)
country comparison to the world: 96

Birth rate: 18.2 births/1,000 population (2020 est.)
country comparison to the world: 87

Death rate: 7.3 deaths/1,000 population (2020 est.)
country comparison to the world: 113

Net migration rate: 0 migrant(s)/1,000 population (2020 est.)
country comparison to the world: 85

Population distribution: with the notable exception of the deserts in the northwest, including the Thar Desert, and the mountain fringe in the north, a very high population density exists throughout most of the country; the core of the population is in the north along the banks of the Ganges, with other river valleys and southern coastal areas also having large population concentrations

Urbanization: *urban population:* 34.9% of total population (2020)
rate of urbanization: 2.37% annual rate of change (2015-20 est.)
total population growth rate v. urban population growth rate, 2000-2030: Major urban areas—population: 30.291 million NEW DELHI (capital), 20.411 million Mumbai, 14.850 million Kolkata, 1.237 million Bangalore, 10.971 million Chennai, 10.004 million Hyderabad (2020)

Sex ratio: *at birth:* 1.11 male(s)/female

0-14 years: 1.13 male(s)/female

15-24 years: 1.14 male(s)/female

25-54 years: 1.07 male(s)/female

55-64 years: 1 male(s)/female

65 years and over: 0.89 male(s)/female
total population: 1.08 male(s)/female (2020 est.)

Maternal mortality rate: 145 deaths/100,000 live births (2017 est.)
country comparison to the world: 57

Infant mortality rate: *total:* 35.4 deaths/1,000 live births
male: 34.4 deaths/1,000 live births
female: 36.5 deaths/1,000 live births (2020 est.)
country comparison to the world: 45

Life expectancy at birth: *total population:* 69.7 years
male: 68.4 years
female: 71.2 years (2020 est.)
country comparison to the world: 167

Total fertility rate: 2.35 children born/woman (2020 est.)
country comparison to the world: 81

Contraceptive prevalence rate: 53.5% (2015/16)

Drinking water source:

improved: *urban:* 96% of population
rural: 91% of population
total: 92.7% of population

unimproved: *urban:* 4% of population
rural: 9% of population
total: 7.2% of population (2017 est.)

Current Health Expenditure: 3.5% (2017)

Physicians density: 0.78 physicians/1,000 population (2017)

Hospital bed density: 0.5 beds/1,000 population (2017)

Sanitation facility access:

improved: *urban:* 93.7% of population
rural: 61.1% of population
total: 72% of population

unimproved: *urban:* 6.3% of population
rural: 38.9% of population
total: 28% of population (2017 est.)

HIV/AIDS—adult prevalence rate: 0.2% (2017 est.)
country comparison to the world: 98

HIV/AIDS—people living with HIV/AIDS: 2.1 million (2017 est.)
country comparison to the world: 3

HIV/AIDS—deaths: 69,000 (2017 est.)
country comparison to the world: 2

Major infectious diseases: *degree of risk:* very high (2020)
food or waterborne diseases: bacterial diarrhea, hepatitis A and E, and typhoid fever
vectorborne diseases: dengue fever, Crimean-Congo hemorrhagic fever, Japanese encephalitis, and malaria
water contact diseases: leptospirosis
animal contact diseases: rabies
note: clusters of cases of a respiratory illness caused by the novel coronavirus (COVID-19) are being reported across 27 States and Union Territories in India; as of 10 November 2020, India has reported a total of 8,507,754 cases of COVID-19 or 6,165 cumulative cases of COVID-19 per 1 million population with 91 cumulative deaths per 1 million population; on 16 March 2020, the government proposed extensive social distancing measures, including closure of all schools, museums, and cultural and social centers; prohibited gatherings of more than 50 people; and called on the public to avoid all nonessential travel; international commercial passenger flights remain suspended

Obesity—adult prevalence rate: 3.9% (2016)
country comparison to the world: 189

Children under the age of 5 years underweight: 33.4% (2016/18)
country comparison to the world: 4

Education expenditures: 3.8% of GDP (2013)
country comparison to the world: 112

Literacy: *definition:* age 15 and over can read and write
total population: 74.4%
male: 82.4%
female: 65.8% (2018)

School life expectancy (primary to tertiary education): *total:* 12 years
male: 11 years
female: 12 years (2019)

Unemployment, youth ages 15-24: *total:* 22.5%
male: 22.2%
female: 24.2% (2018 est.)
country comparison to the world: 56

GOVERNMENT

Country name: *conventional long form:* Republic of India
conventional short form: India
local long form: Republic of India/Bharatiya Ganarajya
local short form: India/Bharat
etymology: the English name derives from the Indus River; the Indian name "Bharat" may derive from the "Bharatas" tribe mentioned in the Vedas of the second millennium B.C.; the name is also associated with Emperor Bharata, the legendary conqueror of all of India

Government type: federal parliamentary republic

Capital: *name:* New Delhi
geographic coordinates: 28 36 N, 77 12 E
time difference: UTC+5.5 (10.5 hours ahead of Washington, DC, during Standard Time)
etymology: the city's name is associated with various myths and legends; the original name for the city may have been Dhilli or Dhillika; alternatively, the name could be a corruption of the Hindustani words "dehleez" or "dehali"—both terms meaning "threshold" or "gateway"—and indicative of the city as a gateway to the Gangetic Plain; after the British decided to move the capital of their Indian Empire from Calcutta to Delhi in 1911, they created a new governmental district south of the latter designated as New Delhi; the new capital was not formally inaugurated until 1931

Administrative divisions: 28 states and 8 union territories*; Andaman and Nicobar Islands*, Andhra Pradesh, Arunachal Pradesh, Assam, Bihar, Chandigarh*, Chhattisgarh, Dadra and Nagar Haveli and Daman and Diu*, Delhi*, Goa, Gujarat, Haryana, Himachal Pradesh, Jammu and Kashmir*, Jharkhand, Karnataka, Kerala, Ladakh*, Lakshadweep*, Madhya Pradesh, Maharashtra, Manipur, Meghalaya, Mizoram, Nagaland, Odisha, Puducherry*, Punjab, Rajasthan, Sikkim, Tamil Nadu, Telangana, Tripura, Uttar Pradesh, Uttarakhand, West Bengal

note: although its status is that of a union territory, the official name of Delhi is National Capital Territory of Delhi

Independence: 15 August 1947 (from the UK)

National holiday: Republic Day, 26 January (1950)

Constitution: *history:* previous 1935 (preindependence); latest draft completed 4 November 1949, adopted 26 November 1949, effective 26 January 1950
amendments: proposed by either the Council of States or the House of the People; passage requires majority participation of the total membership in each house and at least two-thirds majority of voting members of each house, followed by assent of the president of India; proposed amendments to the constitutional amendment procedures also must be ratified by at least one half of the India state legislatures before presidential assent; amended many times, last in 2019

Legal system: common law system based on the English model; separate personal law codes apply to Muslims, Christians, and Hindus; judicial review of legislative acts; note—in late 2019 the Government of India began discussions to overhaul its penal code, which dates to the British colonial period

International law organization participation: accepts compulsory ICJ jurisdiction with reservations; non-party state to the ICCt

Citizenship: *citizenship by birth:* no
citizenship by descent only: at least one parent must be a citizen of India
dual citizenship recognized: no
residency requirement for naturalization: 5 years

Suffrage: 18 years of age; universal

Executive branch: *chief of state:* President Ram Nath KOVIND (since 25 July 2017); Vice President M. Venkaiah NAIDU (since 11 August 2017)
head of government: Prime Minister Narendra MODI (since 26 May 2014)
cabinet: Union Council of Ministers recommended by the prime minister, appointed by the president
elections/appointments: president indirectly elected by an electoral college consisting of elected members of both houses of Parliament for a 5-year term (no term limits); election last held on 17 July 2017 (next to be held in July 2022); vice president indirectly elected by an electoral college consisting of elected members of both houses of Parliament for a 5-year term (no term limits); election last held on 5 August 2017 (next to be held in August 2022); following legislative elections, the prime minister is elected by Lok Sabha members of the majority party
election results: Ram Nath KOVIND elected president; percent of electoral college vote—Ram Nath KOVIND (BJP) 65.7% Meira KUMAR (INC) 34.3%; M. Venkaiah NAIDU elected vice president; electoral college vote—M. Venkaiah NAIDU (BJP) 516, Gopalkrishna GANDHI (independent) 244

Legislative branch: *description:* bicameral Parliament or Sansad consists of:
Council of States or Rajya Sabha (245 seats; 233 members indirectly elected by state and territorial assemblies by proportional representation vote and 12 members appointed by the president; members serve 6-year terms)
House of the People or Lok Sabha (545 seats; 543 members directly elected in single-seat constituencies by simple majority vote and 2 appointed by the president; members serve 5-year terms)
elections: Council of States—last held by state and territorial assemblies at various dates in 2019 (next originally scheduled for March, June, and November 2020 but were postponed due to the COVID-19 pandemic)
House of the People—last held April-May 2019 in 7 phases (next to be held in 2024)
election results: Council of States—percent of vote by party—NA; seats by party—BJP 83, INC 46, AITC 13, DMK 11, SP, other 77, independent 6; composition—men 220, women 25, percent of women 10.2%
House of the People—percent of vote by party—BJP 55.8%, INC 9.6%, AITC 4.4%, YSRC 4.4%, DMK 4.2%, SS 3.3%, JDU 2.9%, BJD 2.2%, BSP 1.8%, TRS 1.7%, LJP 1.1%, NCP 0.9%, SP 0.9%, other 6.4%, independent 0.7%; seats by party—BJP 303, INC 52, DMK 24, AITC 22, YSRC 22, SS 18, JDU 16, BJD 12, BSP 10, TRS 9, LJP 6, NCP 5, SP 5, other 35, independent 4, vacant 2; composition—men 465, women 78, percent of women 14.3%; note—total Parliament percent of women 11.3%

Judicial branch: *highest courts:* Supreme Court (consists of 28 judges, including the chief justice)
judge selection and term of office: justices appointed by the president to serve until age 65
subordinate courts: High Courts; District Courts; Labour Court
note: in mid-2011, India's Cabinet approved the "National Mission for Justice Delivery and Legal Reform" to eliminate judicial corruption and reduce the backlog of cases

Political parties and leaders: Aam Aadmi Party or AAP [Arvind KEJRIWAL]
All India Anna Dravida Munnetra Kazhagam or AIADMK [Edappadi PALANISWAMY, Occhaathevar PANNEERSELVAM]
All India Trinamool Congress or AITC [Mamata BANERJEE]
Bahujan Samaj Party or BSP [MAYAWATI]
Bharatiya Janata Party or BJP [Amit SHAH]
Biju Janata Dal or BJD [Naveen PATNAIK]
Communist Party of India-Marxist or CPI(M) [Sitaram YECHURY]
Indian National Congress or INC
Lok Janshakti Party (LJP) [Ram Vilas PASWAN]
Nationalist Congress Party or NCP [Sharad PAWAR]
Rashtriya Janata Dal or RJD [Lalu Prasad YADAV]
Samajwadi Party or SP [Akhilesh YADAV]
Shiromani Akali Dal or SAD [Sukhbir Singh BADAL]
Shiv Sena or SS [Uddhav THACKERAY]

Telegana Rashtra Samithi or TRS [K. Chandrashekar RAO]

Telugu Desam Party or TDP [Chandrababu NAIDU]

YSR Congress or YSRC [Jagan Mohan REDDY]

note: India has dozens of national and regional political parties

International organization participation: ADB, AfDB (nonregional member), Arctic Council (observer), ARF, ASEAN (dialogue partner), BIMSTEC, BIS, BRICS, C, CD, CERN (observer), CICA, CP, EAS, FAO, FATF, G-15, G-20, G-24, G-5, G-77, IAEA, IBRD, ICAO, ICC (national committees), ICRM, IDA, IFAD, IFC, IFRCS, IHO, ILO, IMF, IMO, IMSO, Interpol, IOC, IOM, IPU, ISO, ITSO, ITU, ITUC (NGOs), LAS (observer), MIGA, MINURSO, MONUSCO, NAM, OAS (observer), OECD, OPCW, Pacific Alliance (observer), PCA, PIF (partner), SAARC, SACEP, SCO (observer), UN, UNCTAD, UNDOF, UNESCO, UNHCR, UNIDO, UNIFIL, UNISFA, UNITAR, UNMISS, UNOCI, UNWTO, UPU, WCO, WFTU (NGOs), WHO, WIPO, WMO, WTO

Diplomatic representation in the US: *chief of mission:* Ambassador Taranjit Singh SANDHU (since 6 February 2020)

chancery: 2107 Massachusetts Avenue NW, Washington, DC 20008; Consular Wing located at 2536 Massachusetts Avenue NW, Washington, DC 20008

telephone: [1] (202) 939-7000

FAX: [1] (202) 265-4351

consulate(s) general: Atlanta, Chicago, Houston, New York, San Francisco

Diplomatic representation from the US: *chief of mission:* Ambassador Kenneth I. JUSTER (since 23 November 2017)

telephone: [91] (11) 2419-8000

embassy: Shantipath, Chanakyapuri, New Delhi 110021

mailing address: use embassy street address

FAX: [91] (11) 2419-0017

consulate(s) general: Chennai (Madras), Hyderabad, Kolkata (Calcutta), Mumbai (Bombay)

Flag description: three equal horizontal bands of saffron (subdued orange) (top), white, and green, with a blue chakra (24-spoked wheel) centered in the white band; saffron represents courage, sacrifice, and the spirit of renunciation; white signifies purity and truth; green stands for faith and fertility; the blue chakra symbolizes the wheel of life in movement and death in stagnation

note: similar to the flag of Niger, which has a small orange disk centered in the white band

National symbol(s): *the Lion Capital of Ashoka, which depicts four Asiatic lions standing back to back mounted on a circular abacus, is the official emblem; Bengal tiger; lotus flower; national colors:* saffron, white, green

National anthem: *name:* "Jana-Gana-Mana" (Thou Art the Ruler of the Minds of All People) *lyrics/music:* Rabindranath TAGORE

note: adopted 1950; Rabindranath TAGORE, a Nobel laureate, also wrote Bangladesh's national anthem

ECONOMY

Economy—overview: India's diverse economy encompasses traditional village farming, modern agriculture, handicrafts, a wide range of modern industries, and a multitude of services. Slightly less than half of the workforce is in agriculture, but services are the major source of economic growth, accounting for nearly two-thirds of India's output but employing less than one-third of its labor force. India has capitalized on its large educated English-speaking population to become a major exporter of information technology services, business outsourcing services, and software workers. Nevertheless, per capita income remains below the world average. India is developing into an open-market economy, yet traces of its past autarkic policies remain. Economic liberalization measures, including industrial deregulation, privatization of state-owned enterprises, and reduced controls on foreign trade and investment, began in the early 1990s and served to accelerate the country's growth, which averaged nearly 7% per year from 1997 to 2017.

India's economic growth slowed in 2011 because of a decline in investment caused by high interest rates, rising inflation, and investor pessimism about the government's commitment to further economic reforms and about slow world growth. Investors' perceptions of India improved in early 2014, due to a reduction of the current account deficit and expectations of post-election economic reform, resulting in a surge of inbound capital flows and stabilization of the rupee. Growth rebounded in 2014 through 2016. Despite a high growth rate compared to the rest of the world, India's government-owned banks faced mounting bad debt, resulting in low credit growth. Rising macroeconomic imbalances in India and improving economic conditions in Western countries led investors to shift capital away from India, prompting a sharp depreciation of the rupee through 2016.

The economy slowed again in 2017, due to shocks of "demonetizaton" in 2016 and introduction of GST in 2017. Since the election, the government has passed an important goods and services tax bill and raised foreign direct investment caps in some sectors, but most economic reforms have focused on administrative and governance changes, largely because the ruling party remains a minority in India's upper house of Parliament, which must approve most bills.

India has a young population and corresponding low dependency ratio, healthy savings and investment rates, and is increasing integration into the global economy. However, long-term challenges remain significant, including: India's discrimination against women and girls, an inefficient power generation and distribution system, ineffective enforcement of intellectual property rights, decades-long civil litigation dockets, inadequate transport and agricultural infrastructure, limited non-agricultural employment opportunities, high spending and poorly targeted subsidies, inadequate availability of quality basic and higher education, and accommodating rural-to-urban migration.

GDP (purchasing power parity): $9.474 trillion (2017 est.)

$8.88 trillion (2016 est.)

$8.291 trillion (2015 est.)

note: data are in 2017 dollars

country comparison to the world: 3

GDP (official exchange rate): $2.602 trillion (2017 est.)

GDP—real growth rate: 4.86% (2019 est.)

6.78% (2018 est.)

6.55% (2017 est.)

country comparison to the world: 52

GDP—per capita (PPP): $7,200 (2017 est.)

$6,800 (2016 est.)

$6,500 (2015 est.)

note: data are in 2017 dollars

country comparison to the world: 156

Gross national saving: 28.8% of GDP (2017 est.)

29.7% of GDP (2016 est.)

30.7% of GDP (2015 est.)

country comparison to the world: 36

GDP—composition, by end use: *household consumption:* 59.1% (2017 est.)

government consumption: 11.5% (2017 est.)

investment in fixed capital: 28.5% (2017 est.)

investment in inventories: 3.9% (2017 est.)

exports of goods and services: 19.1% (2017 est.)

imports of goods and services: -22% (2017 est.)

GDP—composition, by sector of origin: *agriculture:* 15.4% (2016 est.)

industry: 23% (2016 est.)

services: 61.5% (2016 est.)

Agriculture—products: rice, wheat, oilseed, cotton, jute, tea, sugarcane, lentils, onions, potatoes; dairy products, sheep, goats, poultry; fish

Industries: textiles, chemicals, food processing, steel, transportation equipment, cement, mining, petroleum, machinery, software, pharmaceuticals

Industrial production growth rate: 5.5% (2017 est.)

country comparison to the world: 49

Labor force: 521.9 million (2017 est.)

country comparison to the world: 2

Labor force—by occupation: *agriculture:* 47%

industry: 22%

services: 31% (FY 2014 est.)

Unemployment rate: 8.5% (2017 est.)

8.5% (2016 est.)

country comparison to the world: 130

Population below poverty line: 21.9% (2011 est.)

Household income or consumption by percentage share: *lowest 10%:* 3.6%

highest 10%: 29.8% (2011)

Budget: *revenues:* 238.2 billion (2017 est.)

expenditures: 329 billion (2017 est.)

Taxes and other revenues: 9.2% (of GDP) (2017 est.)

country comparison to the world: 215

Budget surplus (+) or deficit (-): -3.5% (of GDP) (2017 est.)
country comparison to the world: 146

Public debt: 71.2% of GDP (2017 est.)
69.5% of GDP (2016 est.)
note: data cover central government debt, and exclude debt instruments issued (or owned) by government entities other than the treasury; the data include treasury debt held by foreign entities; the data exclude debt issued by subnational entities, as well as intragovernmental debt; intragovernmental debt consists of treasury borrowings from surpluses in the social funds, such as for retirement, medical care, and unemployment; debt instruments for the social funds are not sold at public auctions
country comparison to the world: 47

Fiscal year: 1 April—31 March

Inflation rate (consumer prices): 3.6% (2017 est.)
4.5% (2016 est.)
country comparison to the world: 143

Current account balance: -$29.748 billion (2019 est.)
-$65.939 billion (2018 est.)
country comparison to the world: 199

Exports: $304.1 billion (2017 est.)
$268.6 billion (2016 est.)
country comparison to the world: 19

Exports—partners: US 15.6%, UAE 10.2%, Hong Kong 4.9%, China 4.3% (2017)

Exports—commodities: petroleum products, precious stones, vehicles, machinery, iron and steel, chemicals, pharmaceutical products, cereals, apparel

Imports: $452.2 billion (2017 est.)
$376.1 billion (2016 est.)
country comparison to the world: 11

Imports—commodities: crude oil, precious stones, machinery, chemicals, fertilizer, plastics, iron and steel

Imports—partners: China 16.3%, US 5.5%, UAE 5.2%, Saudi Arabia 4.8%, Switzerland 4.7% (2017)

Reserves of foreign exchange and gold: $409.8 billion (31 December 2017 est.)
$359.7 billion (31 December 2016 est.)
country comparison to the world: 8

Debt—external: $501.6 billion (31 December 2017 est.)
$456.4 billion (31 December 2016 est.)
country comparison to the world: 24

Exchange rates: Indian rupees (INR) per US dollar -
65.17 (2017 est.)
67.195 (2016 est.)
67.195 (2015 est.)
64.152 (2014 est.)
61.03 (2013 est.)

ENERGY

Electricity access: *population without electricity:* 6 million (2019)

electrification—total population: 99% (2019)
electrification—urban areas: 99% (2019)
electrification—rural areas: 99% (2019)

Electricity—production: 1.386 trillion kWh (2016 est.)
country comparison to the world: 3

Electricity—consumption: 1.137 trillion kWh (2016 est.)
country comparison to the world: 3

Electricity—exports: 5.15 billion kWh (2015 est.)
country comparison to the world: 36

Electricity—imports: 5.617 billion kWh (2016 est.)
country comparison to the world: 35

Electricity—installed generating capacity: 367.8 million kW (2016 est.)
country comparison to the world: 3

Electricity—from fossil fuels: 71% of total installed capacity (2016 est.)
country comparison to the world: 104

Electricity—from nuclear fuels: 2% of total installed capacity (2017 est.)
country comparison to the world: 26

Electricity—from hydroelectric plants: 2% of total installed capacity (2017 est.)
country comparison to the world: 111

Electricity—from other renewable sources: 16% of total installed capacity (2017 est.)
country comparison to the world: 53

Crude oil—production: 709,000 bbl/day (2018 est.)
country comparison to the world: 25

Crude oil—exports: 0 bbl/day (2015 est.)
country comparison to the world: 140

Crude oil—imports: 4.057 million bbl/day (2015 est.)
country comparison to the world: 3

Crude oil—proved reserves: 4.495 billion bbl (1 January 2018 est.)
country comparison to the world: 23

Refined petroleum products—production: 4.897 million bbl/day (2015 est.)
country comparison to the world: 4

Refined petroleum products—consumption: 4.521 million bbl/day (2016 est.)
country comparison to the world: 3

Refined petroleum products—exports: 1.305 million bbl/day (2015 est.)
country comparison to the world: 7

Refined petroleum products—imports: 653,300 bbl/day (2015 est.)
country comparison to the world: 11

Natural gas—production: 31.54 billion cu m (2017 est.)
country comparison to the world: 25

Natural gas—consumption: 55.43 billion cu m (2017 est.)
country comparison to the world: 14

Natural gas—exports: 76.45 million cu m (2017 est.)
country comparison to the world: 50

Natural gas—imports: 23.96 billion cu m (2017 est.)
country comparison to the world: 14

Natural gas—proved reserves: 1.29 trillion cu m (1 January 2018 est.)
country comparison to the world: 22

Carbon dioxide emissions from consumption of energy: 2.383 billion Mt (2017 est.)
country comparison to the world: 3

COMMUNICATIONS

Telephones—fixed lines: *total subscriptions:* 20,198,012
subscriptions per 100 inhabitants: 1.54 (2019 est.)
country comparison to the world: 13

Telephones—mobile cellular: *total subscriptions:* 1,105,250,941
subscriptions per 100 inhabitants: 84.27 (2019 est.)
country comparison to the world: 2

Telecommunication systems: *general assessment:* supported by deregulation and liberalization of telecommunication laws and policies, India has emerged as one of the fastest-growing telecom markets in the world; implementation of 4G/LTE services shift to data services across the country; highly competitive mobile market with price wars and value-added-services of mobile data; potential to become one of the largest five data center markets globally; steps taken towards 5G services; fixed broadband penetration is expected to grow at a moderate rate over the next five years to 2023 (2020)
domestic: fixed-line subscriptions stands at 2 per 100 and mobile-cellular at 84 per 100; mobile cellular service introduced in 1994 and organized nationwide into four metropolitan areas and 19 telecom circles, each with multiple private service providers and one or more state-owned service providers; in recent years significant trunk capacity added in the form of fiber-optic cable and one of the world's largest domestic satellite systems, the Indian National Satellite system (INSAT), with 6 satellites supporting 33,000 (very small aperture terminals) VSAT (2019)
international: country code—91; a number of major international submarine cable systems, including SEA-ME-WE-3 & 4, AAE-1, BBG, EIG, FALCON, FEA, GBICS, MENA, IMEWE, SEACOM/ Tata TGN-Eurasia, SAFE, WARF, Bharat Lanka Cable System, IOX, Chennai-Andaman & Nicobar Island Cable, SAEx2, Tata TGN-Tata Indicom and i2icn that provide connectivity to Europe, Africa, Asia, the Middle East, South East Asia, numerous Indian Ocean islands including Australia ; satellite earth stations—8 Intelsat (Indian Ocean) and 1 Inmarsat (Indian Ocean region) (2019)
note: the COVID-19 outbreak is negatively impacting telecommunications production and supply chains globally; consumer spending on telecom devices and services has also slowed due to the pandemic's effect on economies worldwide; overall

progress towards improvements in all facets of the telecom industry—mobile, fixed-line, broadband, submarine cable and satellite—has moderated

Broadcast media: Doordarshan, India's public TV network, has a monopoly on terrestrial broadcasting and operates about 20 national, regional, and local services; a large and increasing number of privately owned TV stations are distributed by cable and satellite service providers; in 2015, more than 230 million homes had access to cable and satellite TV offering more than 700 TV channels; government controls AM radio with All India Radio operating domestic and external networks; news broadcasts via radio are limited to the All India Radio Network; since 2000, privately owned FM stations have been permitted and their numbers have increased rapidly

Internet country code: .in

Internet users: total: 446,759,327
percent of population: 34.45% (July 2018 est.)
country comparison to the world: 2

Broadband—fixed subscriptions: total: 18.17 million
subscriptions per 100 inhabitants: 1 (2018 est.)
country comparison to the world: 11

TRANSPORTATION

National air transport system: number of registered air carriers: 14 (2020)
inventory of registered aircraft operated by air carriers: 485
annual passenger traffic on registered air carriers: 164,035,637 (2018)
annual freight traffic on registered air carriers: 2,703,960,000 mt-km (2018)

Civil aircraft registration country code prefix: VT (2016)

Airports: 346 (2013)
country comparison to the world: 21

Airports—with paved runways: total: 253 (2017)
over 3,047 m: 22 (2017)
2,438 to 3,047 m: 59 (2017)
1,524 to 2,437 m: 76 (2017)
914 to 1,523 m: 82 (2017)
under 914 m: 14 (2017)

Airports—with unpaved runways: total: 93 (2013)
over 3,047 m: 1 (2013)
2,438 to 3,047 m: 3 (2013)
1,524 to 2,437 m: 6 (2013)
914 to 1,523 m: 38 (2013)
under 914 m: 45 (2013)

Heliports: 45 (2013)

Pipelines: 9 km condensate/gas, 13581 km gas, 2054 km liquid petroleum gas, 8943 km oil, 20 km oil/gas/water, 11069 km refined products (2013)

Railways: total: 68,525 km (2014)
narrow gauge: 9,499 km 1.000-m gauge (2014)
broad gauge: 58,404 km 1.676-m gauge (23,654 electrified) (2014)
622 0.762-m gauge
country comparison to the world: 5

Roadways: total: 4,699,024 km (2015)

note: includes 96,214 km of national highways and expressways, 147,800 km of state highways, and 4,455,010 km of other roads
country comparison to the world: 3

Waterways: 14,500 km (5,200 km on major rivers and 485 km on canals suitable for mechanized vessels) (2012)
country comparison to the world: 9

Merchant marine: total: 1,731
by type: bulk carrier 67, container ship 25, general cargo 579, oil tanker 128, other 932 (2019)
country comparison to the world: 16

Ports and terminals: major seaport(s): Chennai, Jawaharal Nehru Port, Kandla, Kolkata (Calcutta), Mumbai (Bombay), Sikka, Vishakhapatnam
container port(s) (TEUs): Chennai (1,549,457), Jawaharal Nehru Port (4,833,397), Mundra (4,240,260) (2017)

LNG terminal(s) (import): Dabhol, Dahej, Hazira

MILITARY AND SECURITY

Military and security forces: Indian Armed Forces: Army, Navy (includes marines), Air Force, Coast Guard; Defense Security Corps (paramilitary forces); Ministry of Home Affairs paramilitary forces: Central Armed Police Force (includes Assam Rifles, Border Security Force, Central Industrial Security Force, Central Reserve Police Force, Indo-Tibetan Border Police, National Security Guards, Sashastra Seema Bal) (2019)

Military expenditures: 2.4% of GDP (2019)
2.4% of GDP (2018)
2.5% of GDP (2017)
2.5% of GDP (2016)
2.4% of GDP (2015)
country comparison to the world: 37

Military and security service personnel strengths: assessments of the size of the Indian Armed Forces vary; approximately 1.45 million active personnel (est. 1.25 million Army; 66,000 Navy; 140,000 Air Force; 11,000 Coast Guard); est. 1.5 million paramilitary forces (Ministry of Defense and Ministry of Home Affairs) (2019)

Military equipment inventories and acquisitions: the inventory of the Indian Armed Forces consists mostly of Russian-origin equipment, along with a smaller mix of Western and domestically-produced arms; since 2010, Russia is the leading supplier of arms to India, followed by France, Israel, the UK, and the US; India's defense industry is capable of producing a range of air, land, missile, and naval weapons systems (2019)

Military deployments: 1,900 Democratic Republic of the Congo (MONUSCO); 190 Golan Heights (UNDOF); 780 Lebanon (UNIFIL); 2,350 South Sudan (UNMISS) (2020)

Military service age and obligation: 16-18 years of age for voluntary military service (Army 17 1/2, Air Force 17, Navy 16 1/2); no conscription; women may join as officers, currently serve in combat roles as pilots, and under consideration for Army combat roles (2019)

TERRORISM

Terrorist group(s): Harakat ul-Mujahidin; Harakat ul-Jihad-i-Islami; Hizbul Mujahideen; Indian Mujahedeen; Islamic State of Iraq and ashSham – India; Jaish-e-Mohammed; Lashkar-e Tayyiba; al-Qa'ida; al-Qa'ida in the Indian Subcontinent (2019)
note: details about the history, aims, leadership, organization, areas of operation, tactics, targets, weapons, size, and sources of support of the group(s) appear(s) in Appendix-T

TRANSNATIONAL ISSUES

Disputes—international: since China and India launched a security and foreign policy dialogue in 2005, consolidated discussions related to the dispute over most of their rugged, militarized boundary, regional nuclear proliferation, Indian claims that China transferred
missiles to Pakistan, and other matters continue; Kashmir remains the site of the world's largest and most militarized
territorial dispute with portions under the de facto administration of China (Aksai Chin), India (Jammu and Kashmir), and Pakistan (Azad Kashmir and Northern Areas); India and Pakistan resumed bilateral dialogue in February 2011 after a two- year hiatus, have maintained the 2003 cease-fire in Kashmir, and continue to have disputes over water sharing of the Indus River and its tributaries; UN Military Observer Group in India and Pakistan has maintained a small group of peacekeepers since 1949; India does not recognize Pakistan's ceding historic Kashmir lands to China in 1964; to defuse tensions and prepare for discussions on a maritime boundary, India and Pakistan seek technical resolution of the disputed boundary in Sir Creek estuary at the mouth of the Rann of Kutch in the Arabian Sea; Pakistani maps continue to show its Junagadh claim in Indian Gujarat State; Prime Minister Singh's September 2011 visit to Bangladesh resulted in the signing of a Protocol to the 1974 Land Boundary Agreement between India and Bangladesh, which had called for the settlement of longstanding boundary disputes over undemarcated areas and the exchange of territorial enclaves, but which had never been implemented; Bangladesh referred its maritime boundary claims with Burma and India to the International Tribunal on the Law of the Sea; Joint Border Committee with Nepal continues to examine contested boundary sections, including the 400 sq km dispute over the source of the Kalapani River; India maintains a strict border regime to keep out Maoist insurgents and control illegal cross-border activities from Nepal

Refugees and internally displaced persons: refugees (country of origin): 108,008 (Tibet/China), 59,428 (Sri Lanka), 18,813 (Burma), 7,470 (Afghanistan) (2019)

IDPs: 470,000 (armed conflict and intercommunal violence) (2019)
stateless persons: 17,730 (2019)

Illicit drugs: world's largest producer of licit opium for the pharmaceutical trade, but an undetermined quantity of opium is diverted to illicit international drug markets; transit point for illicit narcotics produced in neighboring countries and throughout Southwest Asia; illicit producer of methaqualone; vulnerable to narcotics money laundering through the hawala system; licit ketamine and precursor production

INDIAN OCEAN

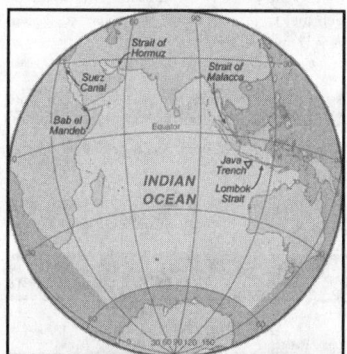

INTRODUCTION

Background: The Indian Ocean is the third largest of the world's five oceans (after the Pacific Ocean and Atlantic Ocean, but larger than the Southern Ocean and Arctic Ocean). Four critically important access waterways are the Suez Canal (Egypt), Bab el Mandeb (Djibouti-Yemen), Strait of Hormuz (Iran-Oman), and Strait of Malacca (Indonesia-Malaysia).The decision by the International Hydrographic Organization in the spring of 2000 to delimit a fifth ocean, the Southern Ocean, removed the portion of the Indian Ocean south of 60 degrees south latitude.

GEOGRAPHY

Location: body of water between Africa, the Southern Ocean, Asia, and Australia

Geographic coordinates: 20 00 S, 80 00 E

Map references: Political Map of the World

Area: *total:* 68.556 million sq km
note: includes Andaman Sea, Arabian Sea, Bay of Bengal, Flores Sea, Great Australian Bight, Gulf of Aden, Gulf of Oman, Java Sea, Mozambique Channel, Persian Gulf, Red Sea, Savu Sea, Strait of Malacca, Timor Sea, and other tributary water bodies

Area—comparative: almost 7 times the size of the US

Coastline: 66,526 km

Climate: northeast monsoon (December to April), southwest monsoon (June to October); tropical cyclones occur during May/June and October/ November in the northern Indian Ocean and January/February in the southern Indian Ocean

Terrain: surface dominated by a major gyre (broad, circular system of currents) in the southern Indian Ocean and a unique reversal of surface currents in the northern Indian Ocean; ocean floor is dominated by the Mid-Indian Ocean Ridge and subdivided by the Southeast Indian Ocean Ridge, Southwest Indian Ocean Ridge, and Ninetyeast Ridge
major surface currents: the counterclockwise Indian Ocean Gyre comprised of the southward flowing warm Agulhas and East Madagascar Currents in the west, the eastward flowing South Indian Current in the south, the northward flowing cold West Australian Current in the east, and the westward flowing South Equatorial Current in the north; a distinctive annual reversal of surface currents occurs in the northern Indian Ocean; low atmospheric pressure over southwest Asia from hot, rising, summer air results in the southwest monsoon and southwest-to-northeast winds and clockwise currents, while high pressure over northern Asia from cold, falling, winter air results in the northeast monsoon and northeast-to-southwest winds and counterclockwise currents

Elevation: *mean depth:* -3,741 m
lowest point: Java Trench -7,258 m
highest point: sea level

Natural resources: oil and gas fields, fish, shrimp, sand and gravel aggregates, placer deposits, polymetallic nodules

Natural hazards: occasional icebergs pose navigational hazard in southern reaches

Environment—current issues: marine pollution caused by ocean dumping, waste disposal, and oil spills; deep sea mining; oil pollution in Arabian Sea, Persian Gulf, and Red Sea; coral reefs threatened due to climate change, direct human pressures, and inadequate governance, awareness, and political will; loss of biodiversity; endangered marine species include the dugong, seals, turtles, and whales

Geography—note: major chokepoints include Bab el Mandeb, Strait of Hormuz, Strait of Malacca, southern access to the Suez Canal, and the Lombok Strait

GOVERNMENT

Country name: etymology: named for the country of India, which makes up much of its northern border

ECONOMY

Economy—overview: The Indian Ocean provides major sea routes connecting the Middle East, Africa, and East Asia with Europe and the Americas. It carries a particularly heavy traffic of petroleum and petroleum products from the oilfields of the Persian Gulf and Indonesia. Its fish are of great and growing importance to the bordering countries for domestic consumption and export. Fishing fleets from Russia, Japan, South Korea, and Taiwan also exploit the Indian Ocean, mainly for shrimp and tuna. Large reserves of hydrocarbons are being tapped in the offshore areas of Saudi Arabia, Iran, India, and western Australia. An estimated 40% of the world's offshore oil production comes from the Indian Ocean. Beach sands rich in heavy minerals and offshore placer deposits are actively exploited by bordering countries, particularly India, South Africa, Indonesia, Sri Lanka, and Thailand.

Marine fisheries: the Indian Ocean fisheries are the third most important in the world accounting for 15%, or 12,311,688 mt of the global catch in 2017; tuna, small pelagic fish, and shrimp are important species in these regions; the Food and Agriculture Organization delineated two fishing regions in the Indian Ocean:
Eastern Indian Ocean region (Region 57) is the most important region and the fifth largest producing region in the world with more than 8%, or 6,966,875 mt, of the global catch in 2017; the region encompasses the waters north of 55º South latitude and east of 80º East longitude including the Bay of Bengal and Andaman Sea with the major producers including Indonesia (1,940,190 mt), India (1,431,700 mt), Burma (1,263,080 mt), Bangladesh (637,476 mt), and Sri Lanka (422,842 mt); the principal catches include shad, Skipjack tuna, mackerel, shrimp, and sardinellas
Western Indian Ocean region (Region 51) is the world's sixth largest producing region with more than 6% or 5,344,813 mt of the global catch in 2017; this region encompasses the waters north of 40º South latitude and west of 80º East longitude including the western Indian Ocean, Arabian Sea, Persian Gulf, and Red Sea as well as the waters along the east coast of Africa and Madagascar, the south coast of the Arabian Peninsula, and the west coast of India with major producers including India (2,402,878 mt), Pakistan (382,768 mt), Oman (347,539 mt), and Mozambique (232,299 mt); the principal catches include Skipjack and Yellowfin tuna, mackerel, sardines, shrimp, and cephalopods

FAO map of world fishing regions; used with permission.: Transportation

Ports and terminals: *major seaport(s):* Chennai (Madras, India); Colombo (Sri Lanka); Durban (South Africa); Jakarta (Indonesia); Kolkata (Calcutta, India); Melbourne (Australia); Mumbai (Bombay, India); Richards Bay (South Africa)

MILITARY AND SECURITY

Maritime threats: the International Maritime Bureau continues to report the territorial waters of littoral states and offshore waters as high risk for piracy and armed robbery against ships, particularly in the Gulf of Aden, along the east coast of Africa, the Bay of Bengal, and the Strait of Malacca; the presence of several naval task forces in the Gulf of Aden and additional anti-piracy measures on the part of ship operators, including the use of on-board armed security teams, have reduced incidents of piracy; in response, Somali-based pirates, using hijacked fishing trawlers as "mother ships" to extend their range, shifted operations as far south as the Mozambique Channel, eastward to the vicinity of the Maldives, and northeastward to the Strait of Hormuz; 2018 saw a slight decrease in attacks over 2017, with one incident in the Gulf of Aden, none in the Red Sea, and two off the coast of Somalia; Operation Ocean Shield, the NATO naval task force established in 2009 to combat Somali piracy, concluded its operations in December 2016 as a result of the drop in reported incidents over the last few years; the EU naval mission, Operation ATALANTA, continues its operations in the Gulf of Aden and Indian Ocean through 2020; naval units from Japan, India, and China also operate in conjunction with EU forces; China has established a logistical base in Djibouti to support its deployed naval units in the Horn of Africa

the Maritime Administration of the US Department of Transportation has issued a Maritime Advisory (2019-012-Persian Gulf, Strait of Hormuz, Gulf of Oman, Arabian Sea, Red Sea-Threats to US and International Shipping from Iran) effective 7 August 2019, which states in part that "heightened military activities and increased political tensions in this region continue to present risk to commercial shipping...there is a continued possibility that Iran and/or its regional proxies could take actions against US and partner interests in the region;" at present, Iran has seized two foreign-flagged tankers in the Persian Gulf; the US and UK navies have established Operation Sentinel to provide escorts for commercial shipping transiting the Persian Gulf, Strait of Hormuz, and Gulf of Oman

TRANSNATIONAL ISSUES

Disputes—international: some maritime disputes (see littoral states)

INDONESIA

INTRODUCTION

Background: The archipelago gradually adopted Islam between the 13th and 16th centuries. The Dutch began to colonize Indonesia in the early 17th century; Japan occupied the islands from 1942 to 1945. Indonesia declared its independence shortly before Japan's surrender, but it required four years of sometimes brutal fighting, intermittent negotiations, and UN mediation before the Netherlands agreed to transfer sovereignty in 1949. A period of sometimes unruly parliamentary democracy ended in 1957 when President SOEKARNO declared martial law and instituted "Guided Democracy." After an abortive coup in 1965 by alleged communist sympathizers, SOEKARNO was gradually eased from power. From 1967 until 1998, President SUHARTO ruled Indonesia with his "New Order" government. After street protests toppled SUHARTO in 1998, free and fair legislative elections took place in 1999. Indonesia is now the world's third most populous democracy, the world's largest archipelagic state, and the world's largest Muslim-majority nation. Current issues include: alleviating poverty, improving education, preventing terrorism, consolidating democracy after four decades of authoritarianism, implementing economic and financial reforms, stemming corruption, reforming the criminal justice system, addressing climate change, and controlling infectious diseases, particularly those of global and regional importance. In 2005, Indonesia reached a historic peace agreement with armed separatists in Aceh, which led to democratic elections in Aceh in December 2006. Indonesia continues to face low intensity armed resistance in Papua by the separatist Free Papua Movement.

GEOGRAPHY

Location: Southeastern Asia, archipelago between the Indian Ocean and the Pacific Ocean

Geographic coordinates: 5 00 S, 120 00 E

Map references: Southeast Asia

Area: *total:* 1,904,569 sq km
land: 1,811,569 sq km
water: 93,000 sq km
country comparison to the world: 16

Area—comparative: slightly less than three times the size of Texas

Land boundaries: *total:* 2,958 km
border countries (3): Malaysia 1881 km, Papua New Guinea 824 km, Timor-Leste 253 km

Coastline: 54,716 km

Maritime claims: *territorial sea:* 12 nm

exclusive economic zone: 200 nm
measured from claimed archipelagic straight baselines

Climate: tropical; hot, humid; more moderate in highlands

Terrain: mostly coastal lowlands; larger islands have interior mountains

Elevation: *mean elevation:* 367 m
lowest point: Indian Ocean 0 m
highest point: Puncak Jaya 4,884 m

Natural resources: petroleum, tin, natural gas, nickel, timber, bauxite, copper, fertile soils, coal, gold, silver

Land use: *agricultural land:* 31.2% (2011 est.)
arable land: 13% (2011 est.) / permanent crops: 12.1% (2011 est.) / permanent pasture: 6.1% (2011 est.)
forest: 51.7% (2011 est.)
other: 17.1% (2011 est.)
Irrigated land: 67,220 sq km (2012)

Population distribution: major concentration on the island of Java, which is considered one of the most densely populated places on earth; of the outer islands (those surrounding Java and Bali), Sumatra contains some of the most significant clusters, particularly in the south near the Selat Sunda, and along the northeastern coast near Medan; the cities of Makasar (Sulawesi), Banjarmasin (Kalimantan) are also heavily populated

Natural hazards: occasional floods; severe droughts; tsunamis; earthquakes; volcanoes; forest fires

volcanism: Indonesia contains the most volcanoes of any country in the world—some 76 are historically active; significant volcanic activity occurs on Java, Sumatra, the Sunda Islands, Halmahera Island, Sulawesi Island, Sangihe Island, and in the Banda Sea; Merapi (2,968 m), Indonesia's most active volcano and in eruption since 2010, has been deemed a Decade Volcano by the International Association of Volcanology and Chemistry of the Earth's Interior, worthy of study due to its explosive history and close proximity to human populations; other notable historically active volcanoes include Agung, Awu, Karangetang, Krakatau (Krakatoa), Makian, Raung, and Tambora; see note 2 under "Geography—note"

Environment—current issues: large-scale deforestation (much of it illegal) and related wildfires cause heavy smog; over-exploitation of marine resources; environmental problems associated with rapid urbanization and economic development, including air pollution, traffic congestion, garbage management, and reliable water and waste water services; water pollution from industrial wastes, sewage

Environment—international agreements: *party to:* Biodiversity, Climate Change, Climate Change-Kyoto Protocol, Desertification, Endangered Species, Hazardous Wastes, Law of the Sea, Ozone Layer Protection, Ship Pollution, Tropical Timber 83, Tropical Timber 94, Wetlands
signed, but not ratified: Marine Life Conservation

Geography—note: *note 1:* according to Indonesia's National Coordinating Agency for Survey and Mapping, the total number of islands in the archipelago is 13,466, of which 922 are permanently inhabited (Indonesia is the world's largest country comprised solely of islands); the country straddles the equator and occupies a strategic location astride or along major sea lanes from the Indian Ocean to the Pacific Ocean
note 2: Indonesia is one of the countries along the Ring of Fire, a belt of active volcanoes and earthquake epicenters bordering the Pacific Ocean; up to 90% of the world's earthquakes and some 75% of the world's volcanoes occur within the Ring of Fire
note 3: despite having the fourth largest population in the world, Indonesia is the most heavily forested region on earth after the Amazon

PEOPLE AND SOCIETY

Population: 267,026,366 (July 2020 est.)
country comparison to the world: 4

Nationality: *noun:* Indonesian(s)
adjective: Indonesian

Ethnic groups: Javanese 40.1%, Sundanese 15.5%, Malay 3.7%, Batak 3.6%, Madurese 3%, Betawi 2.9%, Minangkabau 2.7%, Buginese 2.7%, Bantenese 2%, Banjarese 1.7%, Balinese 1.7%, Acehnese 1.4%, Dayak 1.4%, Sasak 1.3%, Chinese 1.2%, other 15% (2010 est.)

Languages: Bahasa Indonesia (official, modified form of Malay), English, Dutch, local dialects (of which the most widely spoken is Javanese)
note: more than 700 languages are used in Indonesia

Religions: Muslim 87.2%, Protestant 7%, Roman Catholic 2.9%, Hindu 1.7%, other 0.9% (includes Buddhist and Confucian), unspecified 0.4% (2010 est.)

Age structure: *0-14 years:* 23.87% (male 32,473,246/female 31,264,034)
15-24 years: 16.76% (male 22,786,920/female 21,960,130)
25-54 years: 42.56% (male 58,249,570/female 55,409,579)
55-64 years: 8.99% (male 11,033,838/female 12,968,005)
65 years and over: 7.82% (male 9,099,773/female 11,781,271) (2020 est.)

Dependency ratios: *total dependency ratio:* 47.5
youth dependency ratio: 38.3
elderly dependency ratio: 9.2

potential support ratio: 10.8 (2020 est.)

Median age: *total:* 31.1 years
male: 30.5 years
female: 31.8 years (2020 est.)
country comparison to the world: 117

Population growth rate: 0.79% (2020 est.)
country comparison to the world: 129

Birth rate: 15.4 births/1,000 population (2020 est.)
country comparison to the world: 116

Death rate: 6.6 deaths/1,000 population (2020 est.)
country comparison to the world: 140

Net migration rate: -1.1 migrant(s)/1,000 population (2020 est.)
country comparison to the world: 145

Population distribution: major concentration on the island of Java, which is considered one of the most densely populated places on earth; of the outer islands (those surrounding Java and Bali), Sumatra contains some of the most significant clusters, particularly in the south near the Selat Sunda, and along the northeastern coast near Medan; the cities of Makasar (Sulawesi), Banjarmasin (Kalimantan) are also heavily populated

Urbanization: *urban population:* 56.6% of total population (2020)
rate of urbanization: 2.27% annual rate of change (2015-20 est.)
total population growth rate v. urban population growth rate, 2000-2030:

Major urban areas—population: 10.770 million JAKARTA (capital), 3.394 million Bekasi, 2.944 million Surabaya, 2.580 million Bandung, 2.339 million Tangerang, 2.338 million Medan (2020)

Sex ratio: *at birth:* 1.05 male(s)/female
0-14 years: 1.04 male(s)/female
15-24 years: 1.04 male(s)/female
25-54 years: 1.05 male(s)/female
55-64 years: 0.85 male(s)/female
65 years and over: 0.77 male(s)/female
total population: 1 male(s)/female (2020 est.)

Mother's mean age at first birth: 22.8 years (2012 est.)
note: median age at first birth among women 25-29

Maternal mortality rate: 177 deaths/100,000 live births (2017 est.)
country comparison to the world: 52

Infant mortality rate: *total:* 20.4 deaths/1,000 live births
male: 24 deaths/1,000 live births
female: 16.7 deaths/1,000 live births (2020 est.)
country comparison to the world: 74

Life expectancy at birth: *total population:* 73.7 years
male: 71.1 years
female: 76.5 years (2020 est.)
country comparison to the world: 142

Total fertility rate: 2.04 children born/woman (2020 est.)
country comparison to the world: 106

Contraceptive prevalence rate: 55.5% (2018)

Drinking water source:
improved:
urban: 96.6% of population
rural: 83.7% of population
total: 90.8% of population
unimproved:
urban: 3.4% of population
rural: 16.3% of population
total: 9.2% of population (2017 est.)

Current Health Expenditure: 3% (2017)

Physicians density: 0.38 physicians/1,000 population (2017)

Hospital bed density: 1 beds/1,000 population (2017)

Sanitation facility access:
improved:
urban: 92.5% of population
rural: 76.8% of population
total: 85.4% of population
unimproved:
urban: 7.5% of population
rural: 23.2% of population
total: 14.6% of population (2017 est.)

HIV/AIDS—adult prevalence rate: 0.4% (2018 est.)
country comparison to the world: 78

HIV/AIDS—people living with HIV/AIDS: 640,000 (2018 est.)
country comparison to the world: 14

HIV/AIDS—deaths: 38,000 (2018 est.)
country comparison to the world: 5

Major infectious diseases: *degree of risk:* very high (2020)

food or waterborne diseases: bacterial diarrhea, hepatitis A, and typhoid fever

vectorborne diseases: dengue fever and malaria

Obesity—adult prevalence rate: 6.9% (2016)
country comparison to the world: 162

Children under the age of 5 years underweight: 17.7% (2018)
country comparison to the world: 31

Education expenditures: 3.6% of GDP (2015)
country comparison to the world: 120

Literacy: *definition:* age 15 and over can read and write
total population: 95.7%
male: 97.3%
female: 94% (2018)

School life expectancy (primary to tertiary education): *total:* 14 years
male: 14 years
female: 14 years (2018)

Unemployment, youth ages 15-24: *total:* 16.5%
male: 16.5%
female: 16.5% (2018 est.)
country comparison to the world: 82

GOVERNMENT

Country name: *conventional long form:* Republic of Indonesia

conventional short form: Indonesia
local long form: Republik Indonesia
local short form: Indonesia
former: Netherlands East Indies, Dutch East Indies
etymology: the name is an 18th-century construct of two Greek words, "Indos" (India) and "nesoi" (islands), meaning "Indian islands"

Government type: presidential republic

Capital: *name:* Jakarta

geographic coordinates: 6 10 S, 106 49 E
time difference: UTC+7 (12 hours ahead of Washington, DC, during Standard Time)
note: Indonesia has three time zones
etymology: "Jakarta" derives from the Sanscrit "Jayakarta" meaning "victorious city" and refers to a successful defeat and expulsion of the Portuguese in 1527; previously the port had been named "Sunda Kelapa"

Administrative divisions: 31 provinces (provinsi-provinsi, singular—provinsi), 1 autonomous province*, 1 special region** (daerah-daerah istimewa, singular—daerah istimewa), and 1 national capital district*** (daerah khusus ibukota); Aceh*, Bali, Banten, Bengkulu, Gorontalo, Jakarta***, Jambi, Jawa Barat (West Java), Jawa Tengah (Central Java), Jawa Timur (East Java), Kalimantan Barat (West Kalimantan), Kalimantan Selatan (South Kalimantan), Kalimantan Tengah (Central Kalimantan), Kalimantan Timur (East Kalimantan), Kalimantan Utara (North Kalimantan), Kepulauan Bangka Belitung (Bangka Belitung Islands), Kepulauan Riau (Riau Islands), Lampung, Maluku, Maluku Utara (North Maluku), Nusa Tenggara Barat (West Nusa Tenggara), Nusa Tenggara Timur (East Nusa Tenggara), Papua, Papua Barat (West Papua), Riau, Sulawesi Barat (West Sulawesi), Sulawesi Selatan (South Sulawesi), Sulawesi Tengah (Central Sulawesi), Sulawesi Tenggara (Southeast Sulawesi), Sulawesi Utara (North Sulawesi), Sumatera Barat (West Sumatra), Sumatera Selatan (South Sumatra), Sumatera Utara (North Sumatra), Yogyakarta**
note: following the implementation of decentralization beginning on 1 January 2001, regencies and municipalities have become the key administrative units responsible for providing most government services

Independence: 17 August 1945 (declared independence from the Netherlands)

National holiday: Independence Day, 17 August (1945)

Constitution: *history:* drafted July to August 1945, effective 18 August 1945, abrogated by 1949 and 1950 constitutions; 1945 constitution restored 5 July 1959
amendments: proposed by the People's Consultative Assembly, with at least two thirds of its members present; passage requires simple majority vote by the Assembly membership; constitutional articles on the unitary form of the state cannot be amended; amended several times, last in 2002

Legal system: civil law system based on the Roman-Dutch model and influenced by customary law
International law organization participation: has not submitted an ICJ jurisdiction declaration; non-party state to the ICCt

Citizenship: *citizenship by birth:* no
citizenship by descent only: at least one parent must be a citizen of Indonesia
dual citizenship recognized: no
residency requirement for naturalization: 5 continuous years

Suffrage: 17 years of age; universal and married persons regardless of age

Executive branch: *chief of state:* President Joko WIDODO (since 20 October 2014, reelected 17 April 2019, inauguration 19 October 2019); Vice President Ma'ruf AMIN (since 20 October 2019); note—the president is both chief of state and head of government (2019)

head of government: President Joko WIDODO (since 20 October 2014); Vice President Ma'ruf AMIN (since 20 October 2019) (2019)
cabinet: Cabinet appointed by the president
elections/appointments: president and vice president directly elected by absolute majority popular vote for a 5-year term (eligible for a second term); election last held on 17 April 2019 (next election 2024)
election results: Joko WIDODO elected president; percent of vote—Joko WIDODO (PDI-P) 55.5%, PRABOWO Subianto Djojohadikusumo (GERINDRA) 44.5%

Legislative branch: *description:* bicameral People's Consultative Assembly or Majelis Permusyawaratan Rakyat consists of:
Regional Representative Council or Dewan Perwakilan Daerah (136 seats; non-partisan members directly elected in multiseat constituencies—4 each from the country's 34 electoral districts—by proportional representation vote to serve 5-year terms); note—the Regional Representative Council has no legislative authority
House of Representatives or Dewan Perwakilan Rakyat (575 seats; members directly elected in multi-seat constituencies by single non-transferable vote to serve 5-year terms) (2019)
elections: Regional Representative Council—last held 17 April 2019 (next to be held 2024)
House of Representatives—last held on 17 April 2019 (next to be held 2024) (2019)
election results: Regional Representative Council—all seats elected on a non-partisan basis; composition—NA
 House of Representatives—percent of vote by party—PDI-P 19.3%, Gerindra 12.6%, Golkar 12.3%, PKB 9.7%, Nasdem 9.1%, PKS 8.2%, PD 7.8%, PAN 6.8%, PPP 4.5%, other 9.6%; seats by party—PDI-P 128, Golkar 85, Gerindra 78, Nasdem 59, PKB 58, PD 54, PKS 50, PAN 44, PPP 19; composition—men 475, women 100, percent of women 17.9%; total People's Consultative Assembly percent of women NA (2019)

Judicial branch: *highest courts:* Supreme Court or Mahkamah Agung (51 judges divided into 8 chambers); Constitutional Court or Mahkamah Konstitusi (consists of 9 judges)
judge selection and term of office: Supreme Court judges nominated by Judicial Commission, appointed by president with concurrence of parliament; judges serve until retirement at age 65; Constitutional Court judges—3 nominated by president, 3 by Supreme Court, and 3 by parliament; judges appointed by the president; judges serve until mandatory retirement at age 70
subordinate courts: High Courts of Appeal, district courts, religious courts

Political parties and leaders: Democrat Party or PD [Susilo Bambang YUDHOYONO]
Functional Groups Party or GOLKAR [Airlangga HARTARTO]
Great Indonesia Movement Party or GERINDRA [PRABOWO Subianto Djojohadikusumo]
Indonesia Democratic Party-Struggle or PDI-P [MEGAWATI Sukarnoputri]
National Awakening Party or PKB [Muhaiman ISKANDAR]
National Democratic Party or NasDem [Surya PALOH]
National Mandate Party or PAN [Zulkifli HASAN]
Party of the Functional Groups or Golkar [Airlangga HARTARTO]
People's Conscience Party or HANURA [Oesman Sapta ODANG]
Prosperous Justice Party or PKS [Muhammad Sohibul IMAN]
United Development Party or PPP [Muhammad ROMAHURMUZIY] (2019)

International organization participation: ADB, APEC, ARF, ASEAN, BIS, CD, CICA (observer), CP, D-8, EAS, EITI (compliant country), FAO, G-11, G-15, G-20, G- 77, IAEA, IBRD, ICAO, ICC (national committees), ICRM, IDA, IDB, IFAD, IFC, IFRCS, IHO, ILO, IMF, IMO, IMSO, Interpol, IOC, IOM (observer), IORA, IPU, ISO, ITSO, ITU, ITUC (NGOs), MIGA, MINURSO, MINUSTAH, MONUSCO, MSG (associate member), NAM, OECD (enhanced engagement), OIC, OPCW, PIF (partner), UN, UNAMID, UNCTAD, UNESCO, UNIDO, UNIFIL, UNISFA, UNMIL, UNWTO, UPU, WCO, WFTU (NGOs), WHO, WIPO, WMO, WTO

Diplomatic representation in the US: *chief of mission:* Ambassador Muhammad LUTFI (since 17 September 2020)
chancery: 2020 Massachusetts Avenue NW, Washington, DC 20036
telephone: [1] (202) 775-5200
FAX: [1] (202) 775-5365
consulate(s) general: Chicago, Houston, Los Angeles, New York, San Francisco

Diplomatic representation from the US: *chief of mission:*
Charge d'Affaires Heather VARIAVA (14 February 2020)
telephone: [62] (21) 3435-9000 (2020)
embassy: Jalan Medan Merdeka Selatan 3-5, Jakarta 10110

mailing address: Unit 8129, Box 1, FPO AP 96520
FAX: [62] (21) 2395-1697 (2018)
consulate(s) general: Surabaya
consulate(s): Medan

Flag description: two equal horizontal bands of red (top) and white; the colors derive from the banner of the Majapahit Empire of the 13th- 15th centuries; red symbolizes courage, white represents purity
note: similar to the flag of Monaco, which is shorter; also similar to the flag of Poland, which is white (top) and red

National symbol(s): *garuda (mythical bird);* *national colors:* red, white

National anthem: *name:* "Indonesia Raya" (Great Indonesia)
lyrics/music: Wage Rudolf SOEPRATMAN
note: adopted 1945
0:00 /1:48

ECONOMY

Economy—overview: Indonesia, the largest economy in Southeast Asia, has seen a slowdown in growth since 2012, mostly due to the end of the commodities export boom. During the global financial crisis, Indonesia outperformed its regional neighbors and joined China and India as the only G20 members posting growth. Indonesia's annual budget deficit is capped at 3% of GDP, and the Government of Indonesia lowered its debt-to-GDP ratio from a peak of 100% shortly after the Asian financial crisis in 1999 to 34% today. In May 2017 Standard & Poor's became the last major ratings agency to upgrade Indonesia's sovereign credit rating to investment grade.

Poverty and unemployment, inadequate infrastructure, corruption, a complex regulatory environment, and unequal resource distribution among its regions are still part of Indonesia's economic landscape. President Joko WIDODO—elected in July 2014 – seeks to develop Indonesia's maritime resources and pursue other infrastructure development, including significantly increasing its electrical power generation capacity. Fuel subsidies were significantly reduced in early 2015, a move which has helped the government redirect its spending to development priorities. Indonesia, with the nine other ASEAN members, will continue to move towards participation in the ASEAN Economic Community, though full implementation of economic integration has not yet materialized.

GDP (purchasing power parity):
$3.25 trillion (2017 est.)
$3.093 trillion (2016 est.)
$2.945 trillion (2015 est.)
note: data are in 2017 dollars
country comparison to the world: 7

GDP (official exchange rate):
$1.015 trillion (2017 est.)

GDP—real growth rate: 5.03% (2019 est.)
5.17% (2018 est.)
5.07% (2017 est.)
country comparison to the world: 45

GDP—per capita (PPP): $12,400 (2017 est.)
$12,000 (2016 est.)
$11,500 (2015 est.)
note: data are in 2017 dollars
country comparison to the world: 127

Gross national saving:
31.7% of GDP (2017 est.)
32% of GDP (2016 est.)
32% of GDP (2015 est.)
country comparison to the world: 27

GDP—composition, by end use:
household consumption: 57.3% (2017 est.)
government consumption: 9.1% (2017 est.)
investment in fixed capital: 32.1% (2017 est.)
investment in inventories: 0.3% (2017 est.)
exports of goods and services: 20.4% (2017 est.)
imports of goods and services: -19.2% (2017 est.)

GDP—composition, by sector of origin:
agriculture: 13.7% (2017 est.)
industry: 41% (2017 est.)
services: 45.4% (2017 est.)

Agriculture—products: rubber and similar products, palm oil, poultry, beef, forest products, shrimp, cocoa, coffee, medicinal herbs, essential oil, fish and its similar products, and spices
Industries: petroleum and natural gas, textiles, automotive, electrical appliances, apparel, footwear, mining, cement, medical instruments and appliances, handicrafts, chemical fertilizers, plywood, rubber, processed food, jewelry, and tourism

Industrial production growth rate: 4.1% (2017 est.)
country comparison to the world: 73

Labor force: 129.366 million (2019 est.)
country comparison to the world: 3

Labor force—by occupation: *agriculture:* 32%
industry: 21%
services: 47% (2016 est.)

Unemployment rate: 5.31% (2018 est.)
5.4% (2017 est.)
country comparison to the world: 84

Population below poverty line: 10.9% (2016 est.)

Household income or consumption by percentage share: *lowest 10%:* 3.4%
highest 10%: 28.2% (2010)

Budget: *revenues:* 131.7 billion (2017 est.)
expenditures: 159.6 billion (2017 est.)

Taxes and other revenues: 13% (of GDP) (2017 est.)
country comparison to the world: 208

Budget surplus (+) or deficit (-): -2.7% (of GDP) (2017 est.)
country comparison to the world: 121

Public debt: 28.8% of GDP (2017 est.)
28.3% of GDP (2016 est.)
country comparison to the world: 166

Fiscal year: calendar year

Inflation rate (consumer prices): 3.8% (2017 est.)
3.5% (2016 est.)
country comparison to the world: 150

Current account balance: -$30.359 billion (2019 est.)
-$30.633 billion (2018 est.)

country comparison to the world: 200

Exports: $168.9 billion (2017 est.)
$144.4 billion (2016 est.)
country comparison to the world: 29

Exports—partners: China 13.6%, US 10.6%, Japan 10.5%, India 8.4%, Singapore 7.6%, Malaysia 5.1%, South Korea 4.8% (2017)

Exports—commodities: mineral fuels, animal or vegetable fats (includes palm oil), electrical machinery, rubber, machinery and mechanical appliance parts

Imports: $150.1 billion (2017 est.)
$129.2 billion (2016 est.)
country comparison to the world: 31

Imports—commodities: mineral fuels, boilers, machinery, and mechanical parts, electric machinery, iron and steel, foodstuffs

Imports—partners: China 23.2%, Singapore 10.9%, Japan 10%, Thailand 6%, Malaysia 5.6%, South Korea 5.3%, US 5.2% (2017)
Reserves of foreign exchange and gold:
$130.2 billion (31 December 2017 est.)
country comparison to the world: 19

Debt—external: $344.4 billion (31 December 2017 est.)
country comparison to the world: 30

Exchange rates: Indonesian rupiah (IDR) per US dollar -
13,385 (2017 est.)
13,308.3 (2016 est.)
13,308.3 (2015 est.)
13,389.4 (2014 est.)
11,865.2 (2013 est.)

ENERGY

Electricity access: *population without electricity:* 2 million (2019)
electrification—total population: 99% (2019)
electrification—urban areas: 100% (2019)
electrification—rural areas: 99% (2019)

Electricity—production: 235.4 billion kWh (2016 est.)
country comparison to the world: 20

Electricity—consumption: 213.4 billion kWh (2016 est.)
country comparison to the world: 20

Electricity—exports: 0 kWh (2017 est.)
country comparison to the world: 149

Electricity—imports: 693 million kWh (2016 est.)
country comparison to the world: 76

Electricity—installed generating capacity: 61.43 million kW (2016 est.)
country comparison to the world: 19

Electricity—from fossil fuels: 85% of total installed capacity (2016 est.)
country comparison to the world: 71

Electricity—from nuclear fuels: 0% of total installed capacity (2017 est.)
country comparison to the world: 110

Electricity—from hydroelectric plants: 9% of total installed capacity (2017 est.)

country comparison to the world: 118

Electricity—from other renewable sources: 6% of total installed capacity (2017 est.)
country comparison to the world: 99

Crude oil—production: 772,000 bbl/day (2018 est.)
country comparison to the world: 24

Crude oil—exports: 302,300 bbl/day (2015 est.)
country comparison to the world: 27

Crude oil—imports: 498,500 bbl/day (2015 est.)
country comparison to the world: 18

Crude oil—proved reserves: 3.31 billion bbl (1 January 2018 est.)
country comparison to the world: 28

Refined petroleum products—production: 950,000 bbl/day (2015 est.)
country comparison to the world: 18

Refined petroleum products—consumption: 1.601 million bbl/day (2016 est.)
country comparison to the world: 14

Refined petroleum products—exports: 79,930 bbl/day (2015 est.)
country comparison to the world: 47

Refined petroleum products—imports: 591,500 bbl/day (2015 est.)
country comparison to the world: 15

Natural gas—production: 72.09 billion cu m (2017 est.)
country comparison to the world: 12

Natural gas—consumption: 42.32 billion cu m (2017 est.)
country comparison to the world: 23

Natural gas—exports: 29.78 billion cu m (2017 est.)
country comparison to the world: 12

Natural gas—imports: 0 cu m (2017 est.)
country comparison to the world: 141

Natural gas—proved reserves: 2.866 trillion cu m (1 January 2018 est.)
country comparison to the world: 12

Carbon dioxide emissions from consumption of energy: 540.7 million Mt (2017 est.)
country comparison to the world: 12

COMMUNICATIONS

Telephones—fixed lines: *total subscriptions:* 9,272,754
subscriptions per 100 inhabitants: 3.5 (2019 est.)
country comparison to the world: 18

Telephones—mobile cellular: *total subscriptions:* 337,766,682
subscriptions per 100 inhabitants: 127.49 (2019 est.)
country comparison to the world: 4

Telecommunication systems: *general assessment:* international service good; Indonesia has very low fixed line and fixed broadband penetration, high mobile penetration and moderate mobile broadband penetration; 4G mobile services are relatively advanced, 7 operators compete for revenue in the Indonesian market; Chinese company Huawei working on the development of 5G technology in the country; mobile broadband market still in early stages of development; data center market has experienced significant growth; Kacific-1 satellite launched in 2019 to significantly improve telecommunications (2020)

domestic: fixed-line 4 per 100 and mobile-cellular 127 per 100 persons; coverage provided by existing network has been expanded by use of over 200,000 telephone kiosks many located in remote areas; mobile-cellular subscribership growing rapidly (2019)

international: country code—62; landing points for the SEA-ME-WE-3 & 5, DAMAI, JASUKA, BDM, Dumai-Melaka Cable System, IGG, JIBA, Link 1, 3, 4, & 5, PGASCOM, B3J2, Tanjung Pandam-Sungai Kakap Cable System, JAKABARE, JAYABAYA, INDIGO-West, Matrix Cable System, ASC, SJJK, Jaka2LaDeMa, S-U-B Cable System, JBCS, MKCS, BALOK, Palapa Ring East, West and Middle, SMPCS Packet-1 and 2, LTCS, TSCS, SEA-US and Kamal Domestic Submarine Cable System, 35 submarine cable networks that provide links throughout Asia, the Middle East, Australia, Southeast Asia, Africa and Europe; satellite earth stations—2 Intelsat (1 Indian Ocean and 1 Pacific Ocean) (2019)
note: the COVID-19 outbreak is negatively impacting telecommunications production and supply chains globally; consumer spending on telecom devices and services has also slowed due to the pandemic's effect on economies worldwide; overall progress towards improvements in all facets of the telecom industry—mobile, fixed-line, broadband, submarine cable and satellite—has moderated

Broadcast media: mixture of about a dozen national TV networks—1 public broadcaster, the remainder private broadcasters—each with multiple transmitters; more than 100 local TV stations; widespread use of satellite and cable TV systems; public radio broadcaster operates 6 national networks, as well as regional and local stations; overall, more than 700 radio stations with more than 650 privately operated (2019)

Internet country code: .id

Internet users: *total:* 104,563,108
percent of population: 39.79% (July 2018 est.)
country comparison to the world: 7

Broadband—fixed subscriptions: *total:* 8,874,116
subscriptions per 100 inhabitants: 3 (2018 est.)
country comparison to the world: 20

TRANSPORTATION

National air transport system: *number of registered air carriers:* 25 (2020)
inventory of registered aircraft operated by air carriers: 611
annual passenger traffic on registered air carriers: 115,154,100 (2018)
annual freight traffic on registered air carriers: 1,131,910,000 mt-km (2018)

Civil aircraft registration country code prefix: PK (2016)

Airports: 673 (2013)
country comparison to the world: 10

Airports—with paved runways:
total: 186 (2017)
over 3,047 m: 5 (2017)
2,438 to 3,047 m: 21 (2017)
1,524 to 2,437 m: 51 (2017)
914 to 1,523 m: 72 (2017)
under 914 m: 37 (2017)

Airports—with unpaved runways:
total: 487 (2013)
1,524 to 2,437 m: 4 (2013)
914 to 1,523 m: 23 (2013)
under 914 m: 460 (2013)

Heliports: 76 (2013)

Pipelines: 1064 km condensate, 150 km condensate/gas, 11702 km gas, 119 km liquid petroleum gas, 7767 km oil, 77 km oil/gas/water, 728 km refined products, 53 km unknown, 44 km water (2013)

Railways: *total:* 8,159 km (2014)
narrow gauge: 8,159 km 1.067-m gauge (565 km electrified) (2014)
note: 4,816 km operational
country comparison to the world: 27

Roadways: *total:* 496,607 km (2011)
paved: 283,102 km (2011)
unpaved: 213,505 km (2011)
country comparison to the world: 14

Waterways: 21,579 km (2011)
country comparison to the world: 7

Merchant marine: *total:* 9,879
by type: bulk carrier 109, container ship 217, general cargo 2,198, oil tanker 622, other 6,733 (2019)
country comparison to the world: 1

Ports and terminals: *major seaport(s):* Banjarmasin, Belawan, Kotabaru, Krueg Geukueh, Palembang, Panjang, Sungai Pakning, Tanjung Perak, Tanjung Priok
container port(s) (TEUs): Tanjung Perak (3,553,370), Tanjung Priok (6,090,000) (2017)
LNG terminal(s) (export): Bontang, Tangguh
LNG terminal(s) (import): Arun, Lampung, West Java

MILITARY AND SECURITY

Military and security forces: Indonesian National Armed Forces (Tentara Nasional Indonesia, TNI): Army (TNI-Angkatan Darat (TNI-AD)), Navy (TNI- Angkatan Laut (TNI-AL), includes marines (Korps Marinir, KorMar), naval air arm), Air Force (TNI-Angkatan Udara (TNI- AU)), National Air Defense Command (Komando Pertahanan Udara Nasional (Kohanudnas)), Armed Forces Special Operations Command (Koopssus), Strategic Reserve Command (Kostrad)
Indonesian Sea and Coast Guard (Kesatuan Penjagaan Laut dan Pantai, KPLP) is under the Ministry of Transportation (2019)
note: the Indonesian National Police includes a paramilitary Mobile Brigade Corps (BRIMOB)

Military expenditures: 0.7% of GDP (2019)
0.7% of GDP (2018)
0.9% of GDP (2017)
0.8% of GDP (2016)
0.9% of GDP (2015)
country comparison to the world: 137

Military and security service personnel strengths:
the Indonesian National Armed Forces have an
estimated 395,000 active duty troops (300,000
Army; 65,000 Navy; 30,000 Air Force); the Police
Mobile Brigade Corps (BRIMOB) has an estimated 14,000 personnel (2019)

Military equipment inventories and acquisitions:
the Indonesian military inventory is comprised of
equipment from a wide variety of sources; since
2010, the top suppliers are China, Germany, the
Netherlands, Russia, South Korea, the UK, and
the US (2019)

Military deployments: 200 Central African
Republic (MINUSCA); 1,025 Democratic
Republic of the Congo (MONUSCO); 1,250
Lebanon (UNIFIL) (2020)

Military service age and obligation: 18-45 years
of age for voluntary military service, with selective
conscription authorized; 2-year service obligation, with reserve obligation to age 45 (officers);
Indonesian citizens only (2013)

Maritime threats: The International Maritime
Bureau continues to report the territorial and off-
shore waters in the Strait of Malacca and South
China Sea as high risk for piracy and armed

robbery against ships; attacks declined for the
third year in a row from 43 incidents in 2016 to
36 in 2018 due to aggressive maritime patrolling
by regional authorities; in 2018, 29 commercial
vessels were boarded and three crew members were
taken hostage; hijacked vessels are often disguised
and cargo diverted to ports in East Asia (2018)

TERRORISM

Terrorist group(s): Islamic State of Iraq and ash-
Sham (aka Jemaah Anshorut Daulah); Jemaah
Islamiyah (2019)

note: details about the history, aims, leadership,
organization, areas of operation, tactics, targets, weapons, size, and sources of support of the
group(s) appear(s) in Appendix-T

TRANSNATIONAL ISSUES

Disputes—international: Indonesia has a stated
foreign policy objective of establishing stable fixed
land and maritime boundaries with all of its neighbors; three stretches of land borders with Timor-
Leste have yet to be delimited, two of which are
in the Oecussi exclave area, and no maritime or
Exclusive Economic Zone (EEZ) boundaries have
been established between the countries; all borders
between Indonesia and Australia have been agreed
upon bilaterally, but a 1997 treaty that would settle the last of their maritime and EEZ boundary
has yet to be ratified by Indonesia's legislature;
Indonesian groups challenge Australia's claim

to Ashmore Reef; Australia has closed parts of
the Ashmore and Cartier Reserve to Indonesian
traditional fishing and placed restrictions on
certain catches; land and maritime negotiations
with Malaysia are ongoing, and disputed areas
include the controversial Tanjung Datu and
Camar Wulan border area in Borneo and the maritime boundary in the Ambalat oil block in the
Celebes Sea; Indonesia and Singapore continue to
work on finalizing their 1973 maritime boundary
agreement by defining unresolved areas north of
Indonesia's Batam Island; Indonesian secessionists,
squatters, and illegal migrants create repatriation
problems for Papua New Guinea; maritime delimitation talks continue with Palau; EEZ negotiations
with Vietnam are ongoing, and the two countries
in Fall 2011 agreed to work together to reduce illegal fishing along their maritime boundary

Refugees and internally displaced persons:
refugees (country of origin): 6,098 (Afghanistan)
(2018)
IDPs: 40,000 (inter-communal, inter-faith, and
separatist violence between 1998 and 2004 in
Aceh and Papua; religious attacks and land
conflicts in 2007 and 2013; most IDPs in Aceh,
Maluku, East Nusa Tenggara) (2019)

stateless persons: 582 (2019)

Illicit drugs: illicit producer of cannabis largely for
domestic use; producer of methamphetamine and
ecstasy; President WIDODO's war on drugs has led
to an increase in death sentences and executions,
particularly of foreign drug traffickers

IRAN

INTRODUCTION

Background: Known as Persia until 1935, Iran
became an Islamic republic in 1979 after the
ruling monarchy was overthrown and Shah
Mohammad Reza PAHLAVI was forced into exile.
Conservative clerical forces led by Ayatollah
Ruhollah KHOMEINI established a theocratic

system of government with ultimate political
authority vested in a learned religious scholar
referred to commonly as the Supreme Leader who,
according to the constitution, is accountable only
to the Assembly of Experts (AOE)—a popularly
elected 88-member body of clerics. US-Iranian
relations became strained when a group of Iranian
students seized the US Embassy in Tehran in
November 1979 and held embassy personnel
hostages until mid- January 1981. The US cut
off diplomatic relations with Iran in April 1980.
During the period 1980-88, Iran fought a bloody,
indecisive war with Iraq that eventually expanded
into the Persian Gulf and led to clashes between
US Navy and Iranian military forces. Iran has
been designated a state sponsor of terrorism and
was subject to US, UN, and EU economic sanctions and export controls because of its continued
involvement in terrorism and concerns over possible military dimensions of its nuclear program until
Joint Comprehensive Plan of Action (JCPOA)
Implementation Day in 2016. The US began gradually re-imposing sanctions on Iran after the US
withdrawal from JCPOA in May 2018.

Following the election of reformer Hojjat
ol-Eslam Mohammad KHATAMI as president in
1997 and a reformist Majles (legislature) in 2000,

a campaign to foster political reform in response
to popular dissatisfaction was initiated. The movement floundered as conservative politicians, supported by the Supreme Leader, unelected institutions of authority like the Council of Guardians,
and the security services reversed and blocked
reform measures while increasing security repression. Starting with nationwide municipal elections
in 2003 and continuing through Majles elections
in 2004, conservatives reestablished control over
Iran's elected government institutions, which
culminated with the August 2005 inauguration
of hardliner Mahmud AHMADI-NEJAD as president. His controversial reelection in June 2009
sparked nationwide protests over allegations of
electoral fraud, but the protests were quickly suppressed. Deteriorating economic conditions due
primarily to government mismanagement and
international sanctions prompted at least two
major economically based protests in July and
October 2012, but Iran's internal security situation
remained stable. President AHMADI-NEJAD's
independent streak angered regime establishment
figures, including the Supreme Leader, leading to
conservative opposition to his agenda for the last
year of his presidency, and an alienation of his
political supporters. In June 2013 Iranians elected

461

a centrist cleric Dr. Hasan Fereidun ROHANI to the presidency. He is a longtime senior member in the regime, but has made promises of reforming society and Iran's foreign policy. The UN Security Council has passed a number of resolutions calling for Iran to suspend its uranium enrichment and reprocessing activities and comply with its IAEA obligations and responsibilities, and in July 2015 Iran and the five permanent members, plus Germany (P5+1) signed the JCPOA under which Iran agreed to restrictions on its nuclear program in exchange for sanctions relief. Iran held elections in 2016 for the AOE and Majles, resulting in a conservative-controlled AOE and a Majles that many Iranians perceive as more supportive of the ROHANI administration than the previous, conservative-dominated body. ROHANI was reelected president in May 2017. Economic concerns once again led to nationwide protests in December 2017 and January 2018 but they were contained by Iran's security services. Additional widespread economic protests broke out in November 2019 in response to the raised price of subsidized gasoline.

GEOGRAPHY

Location: Middle East, bordering the Gulf of Oman, the Persian Gulf, and the Caspian Sea, between Iraq and Pakistan

Geographic coordinates: 32 00 N, 53 00 E

Map references: Middle East

Area: *total:* 1,648,195 sq km
land: 1,531,595 sq km
water: 116,600 sq km
country comparison to the world: 19

Area—comparative: almost 2.5 times the size of Texas; slightly smaller than Alaska

Land boundaries:
total: 5,894 km
border countries (7): Afghanistan 921 km, Armenia 44 km, Azerbaijan 689 km, Iraq 1599 km, Pakistan 959 km, Turkey 534 km, Turkmenistan 1148 km

Coastline: *2,440 km—note:* Iran also borders the Caspian Sea (740 km)

Maritime claims:
territorial sea: 12 nm
exclusive economic zone: bilateral agreements or median lines in the Persian Gulf
contiguous zone: 24 nm
continental shelf: natural prolongation

Climate: mostly arid or semiarid, subtropical along Caspian coast

Terrain: rugged, mountainous rim; high, central basin with deserts, mountains; small, discontinuous plains along both coasts

Elevation: *mean elevation:* 1,305 m
lowest point: Caspian Sea -28 m
highest point: Kuh-e Damavand 5,625 m

Natural resources: petroleum, natural gas, coal, chromium, copper, iron ore, lead, manganese, zinc, sulfur

Land use: *agricultural land:* 30.1% (2011 est.)
arable land: 10.8% (2011 est.) / permanent crops: 1.2% (2011 est.) / permanent pasture: 18.1% (2011 est.)
forest: 6.8% (2011 est.)
other: 63.1% (2011 est.)

Irrigated land: *Area comparison map:* 95,530 sq km (2012)

Population distribution: population is concentrated in the north, northwest, and west, reflecting the position of the Zagros and Elburz Mountains; the vast dry areas in the center and eastern parts of the country, around the deserts of the Dasht-e Kavir and Dasht-e Lut, have a much lower population density

Natural hazards: periodic droughts, floods; dust storms, sandstorms; earthquakes

Environment—current issues: air pollution, especially in urban areas, from vehicle emissions, refinery operations, and industrial effluents; deforestation; overgrazing; desertification; oil pollution in the Persian Gulf; wetland losses from drought; soil degradation (salination); inadequate supplies of potable water; water pollution from raw sewage and industrial waste; urbanization

Environment—international agreements: *party to:* Biodiversity, Climate Change, Climate Change-Kyoto Protocol, Desertification, Endangered Species, Hazardous Wastes, Marine Dumping, Ozone Layer Protection, Ship Pollution, Wetlands *signed, but not ratified:* Environmental Modification, Law of the Sea, Marine Life Conservation

Geography—note: strategic location on the Persian Gulf and Strait of Hormuz, which are vital maritime pathways for crude oil transport

PEOPLE AND SOCIETY

Population: 84,923,314 (July 2020 est.)
country comparison to the world: 17

Nationality: *noun:* Iranian(s)
adjective: Iranian

Ethnic groups: Persian, Azeri, Kurd, Lur, Baloch, Arab, Turkmen and Turkic tribes

Languages: Persian Farsi (official), Azeri and other Turkic dialects, Kurdish, Gilaki and Mazandarani, Luri, Balochi, Arabic

Religions: Muslim (official) 99.4% (Shia 90-95%, Sunni 5-10%), other (includes Zoroastrian, Jewish, and Christian) 0.3%, unspecified 0.4% (2011 est.)

MENA religious affiliation: *Age structure:*
0-14 years: 24.11% (male 10,472,844/female 10,000,028)
15-24 years: 13.36% (male 5,806,034/female 5,537,561)
25-54 years: 48.94% (male 21,235,038/female 20,327,384)
55-64 years: 7.72% (male 3,220,074/female 3,337,420)
65 years and over: 5.87% (male 2,316,677/female 2,670,254) (2020 est.)

Dependency ratios: *total dependency ratio:* 45.6

youth dependency ratio: 36
elderly dependency ratio: 9.6
potential support ratio: 14.2 (2020 est.)

Median age: *total:* 31.7 years
male: 31.5 years
female: 32 years (2020 est.)
country comparison to the world: 113

Population growth rate: 1.1% (2020 est.)
country comparison to the world: 97

Birth rate: 16.3 births/1,000 population (2020 est.)
country comparison to the world: 107

Death rate: 5.3 deaths/1,000 population (2020 est.)
country comparison to the world: 189

Net migration rate: -0.3 migrant(s)/1,000 population (2020 est.)
country comparison to the world: 118

Population distribution: population is concentrated in the north, northwest, and west, reflecting the position of the Zagros and Elburz Mountains; the vast dry areas in the center and eastern parts of the country, around the deserts of the Dasht-e Kavir and Dasht-e Lut, have a much lower population density

Urbanization: *urban population:* 75.9% of total population (2020)
rate of urbanization: 1.71% annual rate of change (2015-20 est.)
total population growth rate v. urban population growth rate, 2000-2030: Major urban areas—population: 9.135 million TEHRAN (capital), 3.152 million Mashhad, 2.086 million Esfahan, 1.628 million Shiraz, 1.581 million Karaj, 1.596 million Tabriz (2020)

Sex ratio: *at birth:* 1.05 male(s)/female
0-14 years: 1.05 male(s)/female
15-24 years: 1.05 male(s)/female
25-54 years: 1.04 male(s)/female
55-64 years: 0.96 male(s)/female
65 years and over: 0.87 male(s)/female
total population: 1.03 male(s)/female (2020 est.)

Maternal mortality rate: 16 deaths/100,000 live births (2017 est.)
country comparison to the world: 135

Infant mortality rate: *total:* 14.9 deaths/1,000 live births
male: 15.8 deaths/1,000 live births
female: 13.8 deaths/1,000 live births (2020 est.)
country comparison to the world: 99

Life expectancy at birth: *total population:* 74.5 years
male: 73.1 years
female: 76 years (2020 est.)
country comparison to the world: 131

Total fertility rate: 1.94 children born/woman (2020 est.)
country comparison to the world: 124

Contraceptive prevalence rate: 77.4% (2010/11)

Drinking water source:

improved: *urban:* 98.6% of population
rural: 93.1% of population
total: 97.2% of population

unimproved: *urban:* 1.4% of population
rural: 6.9% of population
total: 2.8% of population (2017 est.)

Current Health Expenditure: 8.7% (2017)

Physicians density: 1.13 physicians/1,000 population (2017)

Hospital bed density: 1.6 beds/1,000 population (2017)

Sanitation facility access:

improved: *urban:* 98.9% of population
rural: 95.7% of population
total: 98.1% of population

unimproved: *urban:* 1.1% of population (2015 est.)
rural: 4.3% of population
total: 1.9% of population (2017 est.)

HIV/AIDS—adult prevalence rate: <.1% (2019 est.)

HIV/AIDS—people living with HIV/AIDS: 59,000 (2019 est.)
country comparison to the world: 58

HIV/AIDS—deaths: 2,500 (2019 est.)
country comparison to the world: 41

Major infectious diseases: *degree of risk:* intermediate (2020)
food or waterborne diseases: bacterial diarrhea
vectorborne diseases: Crimean-Congo hemorrhagic fever
note: a new coronavirus is causing sustained community spread of respiratory illness (COVID-19) in Iran; sustained community spread means that people have been infected with the virus, but how or where they became infected is not known, and the spread is ongoing; illness with this virus has ranged from mild to severe with fatalities reported; as of 10 November 2020, Iran has reported a total of 673,250 cases of COVID-19 or 8,016 cumulative cases of COVID-19 per 1 million population with 450 cumulative deaths per 1 million population

Obesity—adult prevalence rate: 25.8% (2016)
country comparison to the world: 47

Children under the age of 5 years underweight: 4.1% (2011)
country comparison to the world: 88

Education expenditures: 4% of GDP (2018)
country comparison to the world: 103

Literacy: *definition:* age 15 and over can read and write
total population: 85.5%
male: 90.4%
female: 80.8% (2016)

School life expectancy (primary to tertiary education):
total: 15 years
male: 15 years
female: 15 years (2017)

Unemployment, youth ages 15-24: *total:* 27.6%
male: 24.3%
female: 39.9% (2018 est.)
country comparison to the world: 41

GOVERNMENT

Country name: *conventional long form:* Islamic Republic of Iran
conventional short form: Iran
local long form: Jomhuri-ye Eslami-ye Iran
local short form: Iran
former: Persia
etymology: name derives from the Avestan term "aryanam" meaning "Land of the Noble [Ones]"

Government type: theocratic republic

Capital: *name:* Tehran
geographic coordinates: 35 42 N, 51 25 E
time difference: UTC+3.5 (8.5 hours ahead of Washington, DC, during Standard Time)
daylight saving time: +1hr, begins fourth Wednesday in March; ends fourth Friday in September
etymology: various explanations of the city's name have been proffered, but the most plausible states that it derives from the Persian words "tah" meaning "end or bottom" and "ran" meaning "[mountain] slope" to signify "bottom of the mountain slope"; Tehran lies at the bottom slope of the Elburz Mountains

Administrative divisions: 31 provinces (ostanha, singular—ostan); Alborz, Ardabil, Azarbayjan-e Gharbi (West Azerbaijan), Azarbayjan-e Sharqi (East Azerbaijan), Bushehr, Chahar Mahal va Bakhtiari, Esfahan, Fars, Gilan, Golestan, Hamadan, Hormozgan, Ilam, Kerman, Kermanshah, Khorasan-e Jonubi (South Khorasan), Khorasan-e Razavi (Razavi Khorasan), Khorasan-e Shomali (North Khorasan), Khuzestan, Kohgiluyeh va Bowyer Ahmad, Kordestan, Lorestan, Markazi, Mazandaran, Qazvin, Qom, Semnan, Sistan va Baluchestan, Tehran, Yazd, Zanjan

Independence: *1 April 1979 (Islamic Republic of Iran proclaimed); notable earlier dates:* ca. 550 B.C. (Achaemenid (Persian) Empire established); A.D. 1501 (Iran reunified under the Safavid Dynasty); 1794 (beginning of Qajar Dynasty); 12 December 1925 (modern Iran established under the PAHLAVI Dynasty)

National holiday: Republic Day, 1 April (1979)

Constitution: *history:* previous 1906; latest adopted 24 October 1979, effective 3 December 1979
amendments: proposed by the supreme leader – after consultation with the Exigency Council – and submitted as an edict to the "Council for Revision of the Constitution," a body consisting of various executive, legislative, judicial, and academic leaders and members; passage requires absolute majority vote in a referendum and approval of the supreme leader; articles including Iran's political system, its religious basis, and its form of government cannot be amended; amended 1989

Legal system: religious legal system based on secular and Islamic law

International law organization participation: has not submitted an ICJ jurisdiction declaration; non-party state to the ICCt

Citizenship: *citizenship by birth:* no

citizenship by descent only: the father must be a citizen of Iran
dual citizenship recognized: no
residency requirement for naturalization: 5 years

Suffrage: 18 years of age; universal

Executive branch: *chief of state:* Supreme Leader Ali Hoseini-KHAMENEI (since 4 June 1989)
head of government: President Hasan Fereidun ROHANI (since 3 August 2013); First Vice President Eshagh JAHANGIRI (since 5 August 2013)
cabinet: Council of Ministers selected by the president with legislative approval; the supreme leader has some control over appointments to several ministries
elections/appointments: supreme leader appointed for life by Assembly of Experts; president directly elected by absolute majority popular vote in 2 rounds if needed for a 4-year term (eligible for a second term and an additional nonconsecutive term); election last held on 19 May 2017 (next to be held in 2021)
election results: Hasan Fereidun ROHANI reelected president; percent of vote— Hasan Fereidun ROHANI (Moderation and Development Party) 58.8%, Ebrahim RAI'SI (Combat Clergy Association) 39.4% , Mostafa MIR-SALIM Islamic Coalition Party) 1.2%, Mostafa HASHEMITABA (Executives of Construction Party) 0.5%
note: 3 oversight bodies are also considered part of the executive branch of government

Legislative branch: *description:* unicameral Islamic Consultative Assembly or Majles-e Shura-ye Eslami or Majles (290 seats; 285 members directly elected in single- and multi-seat constituencies by 2-round vote, and 1 seat each for Zoroastrians, Jews, Assyrian and Chaldean Christians, Armenians in the north of the country and Armenians in the south; members serve 4-year terms); note—all candidates to the Majles must be approved by the Council of Guardians, a 12-member group of which 6 are appointed by the supreme leader and 6 are jurists nominated by the judiciary and elected by the Majles
elections: first round held on 21 February 2020 and second round for 11 remaining seats held on 11 September 2020 (next full Majles election to be held in 2024)
election results: percent of vote by coalition (first round)—NA; seats by coalition (first round)— conservatives 219, reformists 20, independents 35, religious minorities 5; remaining 11 seats to be decided in April 2020

Judicial branch: *highest courts:* Supreme Court (consists of the chief justice and organized into 42 two-bench branches, each with a justice and a judge)
judge selection and term of office: Supreme Court president appointed by the head of the High Judicial Council (HJC), a 5-member body to include the Supreme Court chief justice, the prosecutor general, and 3 clergy, in consultation with judges of the Supreme Court; president appointed

for a single, renewable 5-year term; other judges appointed by the HJC; judge tenure NA

subordinate courts: Penal Courts I and II; Islamic Revolutionary Courts; Courts of Peace; Special Clerical Court (functions outside the judicial system and handles cases involving clerics); military courts

Political parties and leaders: Combatant Clergy Association

Council for Coordinating the Reforms Front

Executives of Construction Party

Followers of the Guardianship of the Jurisprudent [Ali LARIJANI]

Front of Islamic Revolutionary Stability [Morteza AGHA-TEHRANI, general secretary]

ISLAMIC COALITION PARTY

Islamic Iran Participation Front [associated with former President Mohammed KHATAMI]

Militant Clerics Society Moderation and Development Party National Trust Party National Unity Party

Pervasive Coalition of Reformists [Ali SUFI, chairman] (includes Council for Coordinating the Reforms Front, National Trust Party, Union of Islamic Iran People Party, Moderation and Development Party)

Principlists Grand Coalition [Ali Reza ZAKANI] (includes Combatant Clergy Association and Islamic Coalition Party,

Society of Devotees and Pathseekers of the Islamic Revolution, Front of Islamic Revolution Stability)

Progress, Welfare, and Justice Front

Progress and Justice Population of Islamic Iran or PJP [Hosein GHORBANZADEH, general secretary]

Resistance Front of Islamic Iran [Yadollah HABIBI, general secretary]

STEADFASTNESS FRONT

Union of Islamic Iran People's Party
Wayfarers of the Islamic Revolution

International organization participation: CICA, CP, D-8, ECO, FAO, G-15, G-24, G-77, IAEA, IBRD, ICAO, ICC (national committees), ICRM, IDA, IDB, IFAD, IFC, IFRCS, IHO, ILO, IMF, IMO, IMSO, Interpol, IOC, IOM, IPU, ISO, ITSO, ITU, MIGA, NAM, OIC, OPCW, OPEC, PCA, SAARC (observer), SCO (observer), UN, UNAMID, UNCTAD, UNESCO, UNHCR, UNIDO, UNITAR, UNWTO, UPU, WCO, WFTU (NGOs), WHO, WIPO, WMO, WTO (observer)

Diplomatic representation in the US: none; *Iran has an Interests Section in the Pakistani Embassy; address:* Iranian Interests Section, Pakistani Embassy, 2209 Wisconsin Avenue NW, Washington, DC 20007; telephone: [1] (202) 965-4990; FAX [1] (202) 965-1073

Diplomatic representation from the US: none; the US Interests Section is located in the Embassy of Switzerland; Embassy of Switzerland, US

Foreign Interests Section No. 39, Shahid Mousavi (Golestan 5th), Pasdaran Ave., Tehran, Iran

Flag description: three equal horizontal bands of green (top), white, and red; the national emblem (a stylized representation of the word Allah in the shape of a tulip, a symbol of martyrdom) in red is centered in the white band; ALLAH AKBAR (God is Great) in white Arabic script is repeated 11 times along the bottom edge of the green band and 11 times along the top edge of the red band; green is the color of Islam and also represents growth, white symbolizes honesty and peace, red stands for bravery and martyrdom

National symbol(s): *lion; national colors:* green, white, red

National anthem: *name:* "Soroud-e Melli-ye Jomhouri-ye Eslami-ye Iran" (National Anthem of the Islamic Republic of Iran)

lyrics/music: multiple authors/Hassan RIAHI

note 1: adopted 1990; Iran has had six national anthems; the first, entitled Salam-e Shah (Royal Salute) was in use from 1873-1909; next came Salamati-ye Dowlat-e Elliye-ye Iran (Salute of the Sublime State of Persia, 1909-1933); it was followed by Sorud-e melli (The Imperial Anthem of Iran; 1933-1979), which chronicled the exploits of the Pahlavi Dynasty; Ey Iran (Oh Iran) functioned unofficially as the national anthem for a brief period between the ouster of the Shah in 1979 and the early days of the Islamic Republic in 1980; Payandeh Bada Iran (Long Live Iran) was used between 1980 and 1990 during the time of Ayatollah KHOMEINI

note 2: a recording of the current Iranian national anthem is unavailable since the US Navy Band does not record anthems for countries from which the US does not anticipate official visits; the US does not have diplomatic relations with Iran

ECONOMY

Economy—overview: Iran's economy is marked by statist policies, inefficiencies, and reliance on oil and gas exports, but Iran also possesses significant agricultural, industrial, and service sectors. The Iranian government directly owns and operates hundreds of state-owned enterprises and indirectly controls many companies affiliated with the country's security forces. Distortions—including corruption, price controls, subsidies, and a banking system holding billions of dollars of non-performing loans—weigh down the economy, undermining the potential for private-sector-led growth.

Private sector activity includes small-scale workshops, farming, some manufacturing, and services, in addition to mediumscale construction, cement production, mining, and metalworking. Significant informal market activity flourishes and corruption is widespread.

The lifting of most nuclear-related sanctions under the Joint Comprehensive Plan of Action (JCPOA) in January 2016 sparked a restoration of Iran's oil production and revenue that drove rapid GDP growth, but economic growth declined in 2017 as oil production plateaued. The economy continues to suffer from low levels of investment

and declines in productivity since before the JCPOA, and from high levels of unemployment, especially among women and college-educated Iranian youth.

In May 2017, the re-election of President Hasan RUHANI generated widespread public expectations that the economic benefits of the JCPOA would expand and reach all levels of society. RUHANI will need to implement structural reforms that strengthen the banking sector and improve Iran's business climate to attract foreign investment and encourage the growth of the private sector. Sanctions that are not related to Iran's nuclear program remain in effect, and these—plus fears over the possible re-imposition of nuclear-related sanctions—will continue to deter foreign investors from engaging with Iran.

GDP (purchasing power parity): $1.64 trillion (2017 est.)

$1.581 trillion (2016 est.)

$1.405 trillion (2015 est.)

note: data are in 2017 dollars

country comparison to the world: 18

GDP (official exchange rate): $430.7 billion (2017 est.)

GDP—real growth rate: 3.7% (2017 est.)

12.5% (2016 est.)

-1.6% (2015 est.)

country comparison to the world: 78

GDP—per capita (PPP): $20,100 (2017 est.)

$19,600 (2016 est.)

$17,700 (2015 est.)

note: data are in 2017 dollars

country comparison to the world: 89

Gross national saving: 37.9% of GDP (2017 est.)

37.6% of GDP (2016 est.)

35.2% of GDP (2015 est.)

country comparison to the world: 12

GDP—composition, by end use: *household consumption:* 49.7% (2017 est.)

government consumption: 14% (2017 est.)

investment in fixed capital: 20.6% (2017 est.)

investment in inventories: 14.5% (2017 est.)

exports of goods and services: 26% (2017 est.)

imports of goods and services: -24.9% (2017 est.)

GDP—composition, by sector of origin:

agriculture: 9.6% (2016 est.)

industry: 35.3% (2016 est.)

services: 55% (2017 est.)

Agriculture—products: wheat, rice, other grains, sugar beets, sugarcane, fruits, nuts, cotton; dairy products, wool; caviar

Industries: petroleum, petrochemicals, gas, fertilizer, caustic soda, textiles, cement and other construction materials, food processing (particularly sugar refining and vegetable oil production), ferrous and nonferrous metal fabrication, armaments

Industrial production growth rate: 3% (2017 est.)

country comparison to the world: 103

Labor force: 30.5 million (2017 est.)

note: shortage of skilled labor

country comparison to the world: 17

Labor force—by occupation: *agriculture:* 16.3%

industry: 35.1%
services: 48.6% (2013 est.)

Unemployment rate: 11.8% (2017 est.)
12.4% (2016 est.)
note: data are Iranian Government numbers
country comparison to the world: 161

Population below poverty line: 18.7% (2007 est.)

Household income or consumption by percentage share:
lowest 10%: 2.6%
highest 10%: 29.6% (2005)

Budget: revenues: 74.4 billion (2017 est.)
expenditures: 84.45 billion (2017 est.)

Taxes and other revenues: 17.3% (of GDP) (2017 est.)
country comparison to the world: 170

Budget surplus (+) or deficit (-): -2.3% (of GDP) (2017 est.)
country comparison to the world: 110

Public debt: 39.5% of GDP (2017 est.)
47.5% of GDP (2016 est.)
note: includes publicly guaranteed debt
country comparison to the world: 132

Fiscal year: 21 March—20 March

Inflation rate (consumer prices): 9.6% (2017 est.)
9.1% (2016 est.)
note: official Iranian estimate
country comparison to the world: 202

Current account balance: $9.491 billion (2017 est.)
$16.28 billion (2016 est.)
country comparison to the world: 25

Exports: $101.4 billion (2017 est.)
$83.98 billion (2016 est.)
country comparison to the world: 37

Exports—partners: China 27.5%, India 15.1%, South Korea 11.4%, Turkey 11.1%, Italy 5.7%, Japan 5.3% (2017)

Exports—commodities: petroleum 60%, chemical and petrochemical products, fruits and nuts, carpets, cement, ore Imports: $76.39 billion (2017 est.)
$63.14 billion (2016 est.)
country comparison to the world: 44

Imports—commodities: industrial supplies, capital goods, foodstuffs and other consumer goods, technical services

Imports—partners: UAE 29.8%, China 12.7%, Turkey 4.4%, South Korea 4%, Germany 4% (2017)

Reserves of foreign exchange and gold: $120.6 billion (31 December 2017 est.)
$133.7 billion (31 December 2016 est.)
country comparison to the world: 21

Debt—external: $7.995 billion (31 December 2017 est.)
$8.196 billion (31 December 2016 est.)
country comparison to the world: 122

Exchange rates: Iranian rials (IRR) per US dollar—
32,769.7 (2017 est.)
30,914.9 (2016 est.)

30,914.9 (2015 est.)
29,011.5 (2014 est.)
25,912 (2013 est.)

ENERGY

Electricity access: electrification—total population: 100% (2020)

Electricity—production: 272.3 billion kWh (2016 est.)
country comparison to the world: 15

Electricity—consumption: 236.3 billion kWh (2016 est.)
country comparison to the world: 17

Electricity—exports: 6.822 billion kWh (2015 est.)
country comparison to the world: 28

Electricity—imports: 4.221 billion kWh (2016 est.)
country comparison to the world: 44

Electricity—installed generating capacity: 77.6 million kW (2016 est.)
country comparison to the world: 16

Electricity—from fossil fuels: 84% of total installed capacity (2016 est.)
country comparison to the world: 75

Electricity—from nuclear fuels: 1% of total installed capacity (2017 est.)
country comparison to the world: 29

Electricity—from hydroelectric plants: 15% of total installed capacity (2017 est.)
country comparison to the world: 103

Electricity—from other renewable sources: 0% of total installed capacity (2017 est.)
country comparison to the world: 193

Crude oil—production: 4.251 million bbl/day (2018 est.)
country comparison to the world: 6

Crude oil—exports: 750,200 bbl/day (2015 est.)
country comparison to the world: 16

Crude oil—imports: 0 bbl/day (2015 est.)
country comparison to the world: 144

Crude oil—proved reserves: 157.2 billion bbl (1 January 2018 est.)
country comparison to the world: 4

Refined petroleum products—production: 1.764 million bbl/day (2015 est.)
country comparison to the world: 11

Refined petroleum products—consumption: 1.804 million bbl/day (2016 est.)
country comparison to the world: 12

Refined petroleum products—exports: 397,200 bbl/day (2015 est.)
country comparison to the world: 21

Refined petroleum products—imports: 64,160 bbl/day (2015 est.)
country comparison to the world: 72

Natural gas—production: 214.5 billion cu m (2017 est.)
country comparison to the world: 3

Natural gas—consumption: 206.9 billion cu m (2017 est.)

country comparison to the world: 4

Natural gas—exports: 11.64 billion cu m (2017 est.)
country comparison to the world: 18

Natural gas—imports: 3.993 billion cu m (2017 est.)
country comparison to the world: 40

Natural gas—proved reserves: 33.72 trillion cu m (1 January 2018 est.)
country comparison to the world: 2

Carbon dioxide emissions from consumption of energy: 638.3 million Mt (2017 est.)
country comparison to the world: 10

COMMUNICATIONS

Telephones—fixed lines:
total subscriptions: 29,330,454
subscriptions per 100 inhabitants: 34.92 (2019 est.)
country comparison to the world: 9

Telephones—mobile cellular: total subscriptions: 119,598,034
subscriptions per 100 inhabitants: 142.39 (2019 est.)
country comparison to the world: 15

Telecommunication systems: general assessment: opportunities for telecoms growth, but disadvantaged by the lack of significant investment; one of the largest populations in the Middle East with a huge demand for services; mobile penetration is high with over 90% accessing 4G LTE coverage; Iranian-net, is currently expanding a fiber network to reach 8 million customers; govt. is proactively preparing regulations for 5G development (2020)
domestic: 35 per 100 for fixed-line and 142 per 100 for mobile-cellular subscriptions; investment by Iran's state-owned telecom company has greatly improved and expanded both the fixed-line and mobile cellular networks; a huge percentage of the cell phones in the market have been smuggled into the country (2019)
international: country code—98; landing points for Kuwait-Iran, GBICS & MENA, FALCON, OMRAN/3PEG Cable System, POI and UAE-Iran submarine fiber-optic cable to the Middle East, Africa and India; (TAE) fiber-optic line runs from Azerbaijan through the northern portion of Iran to Turkmenistan with expansion to Georgia and Azerbaijan; HF radio and microwave radio relay to Turkey, Azerbaijan, Pakistan, Afghanistan, Turkmenistan, Syria, Kuwait, Tajikistan, and Uzbekistan; satellite earth stations—13 (9 Intelsat and 4 Inmarsat) (2019)
note: the COVID-19 outbreak is negatively impacting telecommunications production and supply chains globally; consumer spending on telecom devices and services has also slowed due to the pandemic's effect on economies worldwide; overall progress towards improvements in all facets of the telecom industry—mobile, fixed-line, broadband, submarine cable and satellite—has moderated

Broadcast media: state-run broadcast media with no private, independent broadcasters; Islamic

Republic of Iran Broadcasting (IRIB), the state-run TV broadcaster, operates 19 nationwide channels including a news channel, about 34 provincial channels, and several international channels; about 20 foreign Persian-language TV stations broadcasting on satellite TV are capable of being seen in Iran; satellite dishes are illegal and, while their use is subjectively tolerated, authorities confiscate satellite dishes from time to time; IRIB operates 16 nationwide radio networks, a number of provincial stations, and an external service; most major international broadcasters transmit to Iran (2019)

Internet country code: .ir

Internet users:
total: 58,117,322
percent of population: 70% (July 2018 est.)
country comparison to the world: 14

Broadband—fixed subscriptions:
total: 9,806,123
subscriptions per 100 inhabitants: 12 (2018 est.)
country comparison to the world: 18

TRANSPORTATION

National air transport system:
number of registered air carriers: 22 (2020)
inventory of registered aircraft operated by air carriers: 237
annual passenger traffic on registered air carriers: 25,604,871 (2018)
annual freight traffic on registered air carriers: 290.74 million mt-km (2018)

Civil aircraft registration country code prefix: EP (2016)

Airports: 319 (2013)
country comparison to the world: 22

Airports—with paved runways:
total: 140 (2019)
over 3,047 m: 42
2,438 to 3,047 m: 29
1,524 to 2,437 m: 26
914 to 1,523 m: 36
under 914 m: 7

Airports—with unpaved runways:
total: 179 (2013)
over 3,047 m: 1 (2013)
2,438 to 3,047 m: 2 (2013)
1,524 to 2,437 m: 9 (2013)
914 to 1,523 m: 135 (2013)
under 914 m: 32 (2013)

Heliports: 26 (2013)

Pipelines: 7 km condensate, 973 km condensate/gas, 20794 km gas, 570 km liquid petroleum gas, 8625 km oil, 7937 km refined products (2013)

Railways: total: 8,484 km (2014)
standard gauge: 8,389.5 km 1.435-m gauge (189.5 km electrified) (2014)
broad gauge: 94 km 1.676-m gauge (2014)
country comparison to the world: 25

Roadways: total: 223,485 km (2018)
paved: 195,485 km (2018)
unpaved: 28,000 km (2018)
country comparison to the world: 23

Waterways: 850 km (on Karun River; some navigation on Lake Urmia) (2012)

COUNTRY COMPARISON TO THE

world: 69 Merchant marine: total: 785
by type: bulk carrier 31, container ship 26, general cargo 361, oil tanker 17, other 350 (2019)
country comparison to the world: 30

Ports and terminals: major seaport(s): Bandar-e Asaluyeh, Bandar Abbas, Bandar Emam
container port(s) (TEUs): Bandar Abbas (2,607,000) (2017)

MILITARY AND SECURITY

Military and security forces: Islamic Republic of Iran Regular Forces (Artesh): Ground Forces, Navy (includes marines), Air Force, Air Defense Forces; Islamic Revolutionary Guard Corps (Sepah, IRGC): Ground Forces, Navy (includes marines), Aerospace Force (controls strategic missile force), Qods Force (special operations), Cyber Command, Basij Paramilitary Forces (Popular Mobilization Army); Law Enforcement Forces (border and security troops, assigned to the armed forces in wartime) (2019)

Military expenditures: 3.8% of GDP (2019 est.)
6.1% of GDP (2018)
5.3% of GDP (2017 est.)
4.1% of GDP (2016 est.)
4.3% of GDP (2015 est.)
(Estimates)
country comparison to the world: 16

Military and security service personnel strengths: assessments of the size of the armed forces of Iran vary; approximately 600,000 total active personnel including 410,000 Islamic Republic of Iran Regular Forces (350,000 Ground Forces; 18,000 Navy; 45,000 Air Force/Air Defense Forces) and 190,000 Islamic Revolutionary Guard Corps (150,000 Ground Forces; 20,000 Navy; 15,000 Aerospace Force; 5,000 Qods Force); est. 90,000 active Basij Paramilitary Forces; est. 50,000 Law Enforcement Forces
(2019 est.)
Military equipment inventories and acquisitions: the Iranian military's inventory includes a mix of domestically-produced and mostly older foreign equipment largely of Chinese, Russian, Soviet, and US origin (US equipment acquired prior to the Islamic Revolution in 1979); weapons imports from Western countries are restricted by international sanctions; since 2010, Iran has received equipment from
Belarus, China, and Russia; Iran has a defense industry with the capacity to develop, produce, support, and sustain air, land, missile, and naval weapons programs (2019 est.)

Military deployments: est. 1,000 Syria (2020)
note: Iran has recruited, trained, and funded thousands of Syrian and foreign fighters to support the ASAD regime during the Syrian civil war

Military service age and obligation: 18 years of age for compulsory military service; 16 years of age for volunteers; 17 years of age for Law Enforcement Forces; 15 years of age for Basij Forces (Popular Mobilization Army); conscript military service obligation is 18-24 months; women exempt from military service (2019)

Maritime threats: the Maritime Administration of the US Department of Transportation has issued a Maritime Advisory (2019-012-Persian Gulf, Strait of Hormuz, Gulf of Oman, Arabian Sea, Red Sea-Threats to US and International Shipping from Iran) effective 7 August 2019, which states in part that "heightened military activities and increased political tensions in this region continue to present risk to commercial shipping...there is a continued possibility that Iran and/or its regional proxies could take actions against US and partner interests in the region;" at present, Iran has seized two foreign-flagged tankers in the Persian Gulf; the US and UK navies have established Operation Sentinel to provide escorts for commercial shipping transiting the Persian Gulf, Strait of Hormuz, and Gulf of Oman

TERRORISM

Terrorist group(s): Islamic Revolutionary Guard Corps/Qods Force; Jaysh al Adl (Jundallah); Kurdistan Workers' Party; al-Qa'ida (2019)
note: details about the history, aims, leadership, organization, areas of operation, tactics, targets, weapons, size, and sources of support of the group(s) appear(s) in Appendix-T

TRANSNATIONAL ISSUES

Disputes—international: Iran protests Afghanistan's limiting flow of dammed Helmand River tributaries during drought; Iraq's lack of a maritime boundary with Iran prompts jurisdiction disputes beyond the mouth of the Shatt al Arab in the Persian Gulf; Iran and UAE dispute Tunb Islands and Abu Musa Island, which are occupied by Iran; Azerbaijan, Kazakhstan, and Russia ratified Caspian seabed delimitation treaties based on equidistance, while Iran continues to insist on a one-fifth slice of the sea; Afghan and Iranian commissioners have discussed boundary monument densification and resurvey

Refugees and internally displaced persons: refugees (country of origin): 2.5-3.0 (1 million registered, 1.5-2.0 million undocumented) (Afghanistan) (2015); 28,268 (Iraq) (2019)

Trafficking in persons: current situation: Iran is a source, transit, and destination country for men, women, and children subjected to sex trafficking and forced labor; organized groups sex traffic Iranian women and children in Iran and to the UAE and Europe; the transport of girls from and through Iran en route to the Gulf for sexual exploitation or forced marriages is on the rise; Iranian children are also forced to work as beggars, street vendors, and in domestic workshops; Afghan boys forced to work in construction or agriculture are vulnerable to sexual abuse by their employers; Pakistani and Afghan migrants being smuggled to Europe often are subjected to forced labor, including debt bondage

tier rating: Tier 3 – Iran does not comply with the minimum standards for the elimination of trafficking, and is not making significant efforts to do so; the government does not share information on its anti-trafficking efforts, but publically available information from NGOs, the media, and international organizations indicates that Iran is not taking adequate measures to address its trafficking problems, particularly protecting victims; Iranian law does not prohibit all forms of human trafficking; female victims find it extremely difficult to get justice because Iranian courts accord women's testimony half the weight of men's, and female victims of sexual abuse, including trafficking, are likely to be prosecuted for adultery; the government did not identify or provide protection services to any victims and continued to punish victims for unlawful acts committed as a direct result of being trafficked; the government made some effort to cooperate with neighboring governments and an international organization to combat human trafficking and other crimes (2015)

Illicit drugs: despite substantial interdiction efforts and considerable control measures along the border with Afghanistan, Iran remains one of the primary transshipment routes for Southwest Asian heroin to Europe; suffers one of the highest opiate addiction rates in the world, and has an increasing problem with synthetic drugs; regularly enforces the death penalty for drug offences; lacks anti-money laundering laws; has reached out to neighboring countries to share counter-drug intelligence

IRAQ

INTRODUCTION

Background: Formerly part of the Ottoman Empire, Iraq was occupied by the United Kingdom during World War I and was declared a League of Nations mandate under UK administration in 1920. Iraq attained its independence as a kingdom in 1932. It was proclaimed a "republic" in 1958 after a coup overthrew the monarchy, but in actuality, a series of strongmen ruled the country until 2003. The last was SADDAM Husayn from 1979 to 2003. Territorial disputes with Iran led to an inconclusive and costly eight-year war (1980-88). In August 1990, Iraq seized Kuwait but was expelled by US-led UN coalition forces during the Gulf War of January-February 1991. After Iraq's expulsion, the UN Security Council (UNSC) required Iraq to scrap all weapons of mass destruction and long-range missiles and to allow UN verification inspections. Continued Iraqi noncompliance with UNSC resolutions led to the Second Gulf War in March 2003 and the ouster of the SADDAM Husayn regime by US-led forces.

In October 2005, Iraqis approved a constitution in a national referendum and, pursuant to this document, elected a 275- member Council of Representatives (COR) in December 2005. The COR approved most cabinet ministers in May 2006, marking the transition to Iraq's first

constitutional government in nearly a half century. Iraq held elections for provincial councils in all governorates in January 2009 and April 2013 and postponed the next provincial elections, originally planned for April 2017, until 2019. Iraq has held three national legislative elections since 2005, most recently in May 2018 when 329 legislators were elected to the COR. Adil ABD AL-MAHDI assumed the premiership in October 2018 as a consensus and independent candidate—the first prime minister who is not an active member of a major political bloc. However, widespread protests that began in October 2019 demanding more employment opportunities and an end to corruption prompted ABD AL-MAHDI to announce his resignation on 20 November 2019.

Between 2014 and 2017, Iraq was engaged in a military campaign against the Islamic State of Iraq and ash-Sham (ISIS) to recapture territory lost in the western and northern portion of the country. Iraqi and allied forces recaptured Mosul, the country's second-largest city, in 2017 and drove ISIS out of its other urban strongholds. In December 2017, then-Prime Minister Haydar al-ABADI publicly declared victory against ISIS while continuing operations against the group's residual presence in rural areas. Also in late 2017, ABADI responded to an independence referendum held by the Kurdistan Regional Government by ordering Iraqi forces to take control of disputed territories across central and northern Iraq that were previously occupied and governed by Kurdish forces.

GEOGRAPHY

Location: Middle East, bordering the Persian Gulf, between Iran and Kuwait

Geographic coordinates: 33 00 N, 44 00 E

Map references: Middle East

Area: *total:* 438,317 sq km
land: 437,367 sq km
water: 950 sq km
country comparison to the world: 60

Area—comparative: slightly more than three times the size of New York state

Land boundaries: *total:* 3,809 km

border countries (6): Iran 1599 km, Jordan 179 km, Kuwait 254 km, Saudi Arabia 811 km, Syria 599 km, Turkey 367 km

Coastline: 58 km

Maritime claims: *territorial sea:* 12 nm
continental shelf: not specified

Climate: mostly desert; mild to cool winters with dry, hot, cloudless summers; northern mountainous regions along Iranian and Turkish borders experience cold winters with occasionally heavy snows that melt in early spring, sometimes causing extensive flooding in central and southern Iraq

Terrain: mostly broad plains; reedy marshes along Iranian border in south with large flooded areas; mountains along borders with Iran and Turkey

Elevation: *mean elevation:* 312 m
lowest point: Persian Gulf 0 m
highest point: Cheekha Dar (Kurdish for "Black Tent") 3,611 m

Natural resources: petroleum, natural gas, phosphates, sulfur

Land use: *agricultural land:* 18.1% (2011 est.)
arable land: 8.4% (2011 est.) / permanent crops: 0.5% (2011 est.) / permanent pasture: 9.2% (2011 est.)
forest: 1.9% (2011 est.)
other: 80% (2011 est.)

Irrigated land: 35,250 sq km (2012)

Population distribution: population is concentrated in the north, center, and eastern parts of the country, with many of the larger urban agglomerations found along extensive parts of the Tigris and Euphrates Rivers; much of the western and southern areas are either lightly populated or uninhabited

Natural hazards: dust storms; sandstorms; floods

Environment—current issues: government water control projects drained most of the inhabited marsh areas east of An Nasiriyah by drying up or diverting the feeder streams and rivers; a once sizable population of Marsh Arabs, who inhabited these areas for thousands of years, has been displaced; furthermore, the destruction of the natural habitat poses serious threats to the area's wildlife populations; inadequate supplies of potable water;

467

soil degradation (salination) and erosion; desertification; military and industrial infrastructure has released heavy metals and other hazardous substances into the air, soil, and groundwater; major sources of environmental damage are effluents from oil refineries, factory and sewage discharges into rivers, fertilizer and chemical contamination of the soil, and industrial air pollution in urban areas

Environment—international agreements: *party to:* Biodiversity, Hazardous Wastes, Law of the Sea, Ozone Layer Protection
signed, but not ratified: Environmental Modification

Geography—note: strategic location on Shatt al Arab waterway and at the head of the Persian Gulf

PEOPLE AND SOCIETY

Population: 38,872,655 (July 2020 est.)
country comparison to the world: 36

Nationality: *noun:* Iraqi(s)
adjective: Iraqi

Ethnic groups: Arab 75-80%, Kurdish 15-20%, other 5% (includes Turkmen, Yezidi, Shabak, Kaka'i, Bedouin, Romani, Assyrian, Circassian, Sabaean-Mandaean, Persian)
note: data is a 1987 government estimate; no more recent reliable numbers are available

Languages: Arabic (official), Kurdish (official), Turkmen (a Turkish dialect), Syriac (Neo-Aramaic), and Armenian are official in areas where native speakers of these languages constitute a majority of the population

Religions: Muslim (official) 95-98% (Shia 64-69%, Sunni 29-34%), Christian 1% (includes Catholic, Orthodox, Protestant, Assyrian Church of the East), other 1-4% (2015 est.)
note: while there has been voluntary relocation of many Christian families to northern Iraq, the overall Christian population has decreased at least 50% and perhaps as high as 90% since the fall of the SADDAM Husayn regime in 2003, according to US Embassy estimates, with many fleeing to Syria, Jordan, and Lebanon

MENA religious affiliation:

Age structure: *0-14 years:* 37.02% (male 7,349,868/female 7,041,405)
15-24 years: 19.83% (male 3,918,433/female 3,788,157)
25-54 years: 35.59% (male 6,919,569/female 6,914,856)
55-64 years: 4.23% (male 805,397/female 839,137)
65 years and over: 3.33% (male 576,593/female 719,240) (2020 est.)

Dependency ratios:
total dependency ratio: 69.9
youth dependency ratio: 64.1
elderly dependency ratio: 5.9
potential support ratio: 17.1 (2020 est.)

Median age: *total:* 21.2 years
male: 20.8 years
female: 21.6 years (2020 est.)
country comparison to the world: 186

Population growth rate: 2.16% (2020 est.)
country comparison to the world: 37

Birth rate: 25.7 births/1,000 population (2020 est.)
country comparison to the world: 47

Death rate: 3.9 deaths/1,000 population (2020 est.)
country comparison to the world: 215

Net migration rate: -0.5 migrant(s)/1,000 population (2020 est.)
country comparison to the world: 127

Population distribution: population is concentrated in the north, center, and eastern parts of the country, with many of the larger urban agglomerations found along extensive parts of the Tigris and Euphrates Rivers; much of the western and southern areas are either lightly populated or uninhabited

Urbanization: *urban population:* 70.9% of total population (2020)
rate of urbanization: 3.06% annual rate of change (2015-20 est.)
total population growth rate v. urban population growth rate, 2000-2030: Major urban areas—population: 7.144 million BAGHDAD (capital), 1.630 million Mosul, 1.352 million Basra, 1.013 million Kirkuk, 874,000 Najaf, 846,000 Erbil (2020)

Sex ratio: *at birth:* 1.05 male(s)/female
0-14 years: 1.04 male(s)/female
15-24 years: 1.03 male(s)/female
25-54 years: 1 male(s)/female
55-64 years: 0.96 male(s)/female
65 years and over: 0.8 male(s)/female
total population: 1.01 male(s)/female (2020 est.)

Maternal mortality rate: 79 deaths/100,000 live births (2017 est.)
country comparison to the world: 80

Infant mortality rate: *total:* 19.5 deaths/1,000 live births
male: 20.9 deaths/1,000 live births
female: 18.2 deaths/1,000 live births (2020 est.)
country comparison to the world: 80

Life expectancy at birth: *total population:* 72.6 years
male: 70.7 years
female: 74.6 years (2020 est.)
country comparison to the world: 152

Total fertility rate: 3.39 children born/woman (2020 est.)
country comparison to the world: 44

Contraceptive prevalence rate: 52.8% (2018)

Drinking water source:
improved: *urban:* 98.8% of population
rural: 95% of population
total: 97.9% of population

unimproved: *urban:* 1.2% of population
rural: 5% of population
total: 2.1% of population (2017 est.)

Current Health Expenditure: 4.2% (2017)

Physicians density: 0.84 physicians/1,000 population (2017)

Hospital bed density: 1.3 beds/1,000 population (2017)

Sanitation facility access:
improved: *urban:* 96.7% of population
rural: 89.7% of population
total: 95.2% of population

unimproved: *urban:* 3.3% of population
rural: 10.3% of population
total: 4.8% of population (2017 est.)

HIV/AIDS—adult prevalence rate: NA

HIV/AIDS—people living with HIV/AIDS: NA

HIV/AIDS—deaths: NA

Major infectious diseases:
degree of risk: intermediate (2020)
food or waterborne diseases: bacterial diarrhea, hepatitis A, and typhoid fever
note: widespread ongoing transmission of a respiratory illness caused by the novel coronavirus (COVID-19) is occurring throughout Iraq; as of 10 November 2020, Iraq has reported a total of 496,019 cases of COVID-19 or 12,332 cumulative cases of COVID-19 per 1 million population with 281 cumulative deaths per 1 million population

Obesity—adult prevalence rate: 30.4% (2016)
country comparison to the world: 23

Children under the age of 5 years underweight: 3.9% (2018)
country comparison to the world: 91

Education expenditures: NA

Literacy: *definition:* age 15 and over can read and write
total population: 50.1%
male: 56.2%
female: 44% (2018)

Unemployment, youth ages 15-24:
total: 25.6%
male: 22%
female: 63.3% (2017)
country comparison to the world: 48

GOVERNMENT

Country name: *conventional long form:* Republic of Iraq
conventional short form: Iraq
local long form: Jumhuriyat al-Iraq/Komar-i Eraq
local short form: Al Iraq/Eraq
former: Mesopotamia, Mandatory Iraq, Hashemite Kingdom of Iraq
etymology: the name probably derives from "Uruk" (Biblical "Erech"), the ancient Sumerian and Babylonian city on the Euphrates River

Government type: federal parliamentary republic

Capital: *name:* Baghdad
geographic coordinates: 33 20 N, 44 24 E
time difference: UTC+3 (8 hours ahead of Washington, DC, during Standard Time)
although the origin of the name is disputed, it likely has compound Persian roots with "bagh" and "dad" meaning "god" and "given" respectively to create the meaning of "bestowed by God"

Administrative divisions: 18 governorates (muhafazat, singular—muhafazah (Arabic); parezgakan, singular—parezga (Kurdish)) and 1 region*; Al Anbar; Al Basrah; Al Muthanna; Al Qadisiyah (Ad Diwaniyah); An Najaf; Arbil (Erbil) (Arabic), Hewler (Kurdish); As Sulaymaniyah (Arabic), Slemani (Kurdish); Babil; Baghdad; Dahuk (Arabic), Dihok (Kurdish); Dhi Qar; Diyala; Karbala'; Kirkuk; Kurdistan Regional Government*; Maysan; Ninawa; Salah ad Din; Wasit

Independence: 3 October 1932 (from League of Nations mandate under British administration); note—on 28 June 2004 the Coalition Provisional Authority transferred sovereignty to the Iraqi Interim Government

National holiday: Independence Day, 3 October (1932); Republic Day, 14 July (1958)

Constitution: *history:* several previous; latest adopted by referendum 15 October 2005

amendments: proposed by the president of the republic and the Council of Minsters collectively, or by one fifth of the Council of Representatives members; passage requires at least two-thirds majority vote by the Council of Representatives, approval by referendum, and ratification by the president; passage of amendments to articles on citizen rights and liberties requires two-thirds majority vote of Council of Representatives members after two successive electoral terms, approval in a referendum, and ratification by the president

Legal system: mixed legal system of civil and Islamic law

International law organization participation: has not submitted an ICJ jurisdiction declaration; non-party state to the ICCt

Citizenship: *citizenship by birth:* no

citizenship by descent only: at least one parent must be a citizen of Iraq

dual citizenship recognized: yes

residency requirement for naturalization: 10 years

Suffrage: 18 years of age; universal

Executive branch: *chief of state:* President Barham SALIH (since 2 October 2018); vice presidents (vacant)

head of government: Prime Minister Mustafa al-KADHIMI (since 7 May 2020)

cabinet: Council of Ministers proposed by the prime minister, approved by Council of Representatives

elections/appointments: president indirectly elected by Council of Representatives (COR) to serve a 4-year term (eligible for a second term); COR election last held on 12 May 2018 (next NA)

election results: COR vote in first round—Barham SALIH (PUK) 165, Fuad HUSAYN (KDP) 90; Barham SALIH elected president in second round—Barham SALIH 219, Fuad HUSAYN 22; note—the COR vote on 1 October 2018 failed due to a lack of quorum, and a new session was held on 2 October

Legislative branch: *description:* unicameral Council of Representatives or Majlis an-Nuwwab al-Iraqiyy (329 seats; 320 members directly elected in 83 multi-seat constituencies by simple majority vote and 9 seats at the national level reserved for minorities—5 for Christians, 1 each for Sabaean-Mandaeans, Yazidis, Shabaks, Fayli Kurds; 25% of seats allocated to women; members serve 4-year terms); note—in early November 2020, the president ratified a new electoral law—approved by the Council of Representatives in late October—that eliminates the proportional representation electoral system

elections: last held on 12 May 2018 (next originally scheduled for May 2022, but rescheduled earlier to 6 June 2021)

election results: percent of vote by party/coalition—NA; seats by party/coalition—Sa'irun Alliance 54, Al Fatah Alliance 48, Al Nasr Alliance 42, KDP 25, State of Law Coalition 25, Wataniyah 21, National Wisdom Trend 19, PUK 18, Iraqi Decision Alliance 14, Anbar Our Identity 6, Goran Movement 5, New Generation 4, other 48; composition—men 245, women 84, percent of women 25.5%

Judicial branch: *highest courts:* Federal Supreme Court or FSC (consists of 9 judges); note—court jurisdiction limited to constitutional issues and disputes between regions or governorates and the central government; Court of Cassation (consists of a court president, 5 vice presidents, and at least 24 judges)

judge selection and term of office: Federal Supreme Court and Court of Cassation judges selected by the president of the republic from nominees selected by the Higher Judicial Council (HJC), a 25-member committee of judicial officials that manages the judiciary and prosecutors; FSC members appointed for life; Court of Cassation judges appointed by the HJC and confirmed by the Council of Representatives to serve until retirement nominally at age 63

subordinate courts: Courts of Appeal (governorate level); civil courts, including first instance, personal status, labor, and customs; criminal courts including felony, misdemeanor, investigative, major crimes, juvenile, and traffic courts

Political parties and leaders: Al Fatah Alliance [Hadi al-AMIRI]

Al Nasr Alliance [Haydar al-ABADI]

Al Sadiqun Bloc [Adnan al-DULAYMI]

Al Sa'irun Alliance [Muqtda al-SADR]

Badr Organization [Hadi al-AMIRI]

Da Āwa Party [Nuri al-MALIKI]

Fadilah Party [Muhammad al-YAQUBI]

Goran Movement [Omar SAYYID ALI]

Iraqi Communist Party [Hamid Majid MUSA]

Iraq Decision Alliance [Khamis al-KHANJAR, Usama al-NUJAYFI]

Islamic Supreme Council of Iraq or ISCI [Humam HAMMUDI]

Kurdistan Democratic Party or KDP [Masoud BARZANI]

National Wisdom Trend [Ammar al-HAKIM]

New Generation Movement [SHASWAR Abd al-Wahid Qadir]

Our Identity [Muhammad al-HALBUSI]

Patriotic Union of Kurdistan or PUK [KOSRAT Rasul Ali, acting]

State of Law Coalition [Nuri al MALIKI Wataniyah coalition [Ayad ALLAWI]

numerous smaller religious, local, tribal, and minority parties

International organization participation: ABEDA, AFESD, AMF, CAEU, CICA, EITI (compliant country), FAO, G-77, IAEA, IBRD, ICAO, ICRM, IDA, IDB, IFAD, IFC, IFRCS, ILO, IMF, IMO, IMSO, Interpol, IOC, IPU, ISO, ITSO, ITU, LAS, MIGA, NAM, OAPEC, OIC, OPCW, OPEC, PCA, UN, UNCTAD, UNESCO, UNIDO, UNWTO, UPU, WCO, WFTU (NGOs), WHO, WIPO, WMO, WTO (observer)

Diplomatic representation in the US: chief of mission: Ambassador Farid YASIN (since 18 January 2017)

chancery: 3421 Massachusetts Avenue, NW, Washington, DC 20007

telephone: [1] (202) 742-1600

FAX: [1] (202) 333-1129

consulate(s) general: Detroit, Los Angeles

Diplomatic representation from the US: *chief of mission:* Ambassador Matthew TUELLER (since 9 June 2019)

telephone: 0760-030-3000

embassy: Al-Kindi Street, International Zone, Baghdad; note—consulate in Al Basrah closed as of 28 September 2018

mailing address: APO AE 09316

FAX: NA

Flag description: three equal horizontal bands of red (top), white, and black; the Takbir (Arabic expression meaning "God is great") in green Arabic script is centered in the white band; the band colors derive from the Arab Liberation flag and represent oppression (black), overcome through bloody struggle (red), to be replaced by a bright future (white); the Council of Representatives approved this flag in 2008 as a compromise replacement for the Ba'thist SADDAM-era flag

note: similar to the flag of Syria, which has two stars but no script; Yemen, which has a plain white band; and that of Egypt, which has a golden Eagle of Saladin centered in the white band

National symbol(s): *golden eagle; national colors:* red, white, black

National anthem: *name:* "Mawtini" (My Homeland)

lyrics/music: Ibrahim TOUQAN/Mohammad FLAYFEL

note: adopted 2004; following the ouster of SADDAM Husayn, Iraq adopted "Mawtini," a popular folk song throughout the Arab world; also serves as an unofficial anthem of the Palestinian people

ECONOMY

Economy—overview: Iraq's GDP growth slowed to 1.1% in 2017, a marked decline compared to the previous two years as domestic consumption and investment fell because of civil violence and a sluggish oil market. The Iraqi Government received its third tranche of funding from its 2016 Stand-By Arrangement (SBA) with the IMF in

469

August 2017, which is intended to stabilize its finances by encouraging improved fiscal management, needed economic reform, and expenditure reduction. Additionally, in late 2017 Iraq received more than $1.4 billion in financing from international lenders, part of which was generated by issuing a $1 billion bond for reconstruction and rehabilitation in areas liberated from ISIL. Investment and key sector diversification are crucial components to Iraq's long-term economic development and require a strengthened business climate with enhanced legal and regulatory oversight to bolster private-sector engagement. The overall standard of living depends on global oil prices, the central government passage of major policy reforms, a stable security environment post-ISIS, and the resolution of civil discord with the Kurdish Regional Government (KRG).

Iraq's largely state-run economy is dominated by the oil sector, which provides roughly 85% of government revenue and 80% of foreign exchange earnings, and is a major determinant of the economy's fortunes. Iraq's contracts with major oil companies have the potential to further expand oil exports and revenues, but Iraq will need to make significant upgrades to its oil processing, pipeline, and export infrastructure to enable these deals to reach their economic potential.

In 2017, Iraqi oil exports from northern fields were disrupted following a KRG referendum that resulted in the Iraqi Government reasserting federal control over disputed oil fields and energy infrastructure in Kirkuk. The Iraqi government and the KRG dispute the role of federal and regional authorities in the development and export of natural resources. In 2007, the KRG passed an oil law to develop IKR oil and gas reserves independent of the federal government. The KRG has signed about 50 contracts with foreign energy companies to develop its reserves, some of which lie in territories taken by Baghdad in October 2017. The KRG is able to unilaterally export oil from the fields it retains control of through its own pipeline to Turkey, which Baghdad claims is illegal. In the absence of a national hydrocarbons law, the two sides have entered into five provisional oil- and revenue-sharing deals since 2009, all of which collapsed.

Iraq is making slow progress enacting laws and developing the institutions needed to implement economic policy, and political reforms are still needed to assuage investors' concerns regarding the uncertain business climate. The Government of Iraq is eager to attract additional foreign direct investment, but it faces a number of obstacles, including a tenuous political system and concerns about security and societal stability. Rampant corruption, outdated infrastructure, insufficient essential services, skilled labor shortages, and antiquated commercial laws stifle investment and continue to constrain growth of private, nonoil sectors. Under the Iraqi constitution, some competencies relevant to the overall investment climate are either shared by the federal government and the regions or are devolved entirely to local governments. Investment in the IKR operates within the framework of the Kurdistan Region Investment Law (Law 4 of 2006) and the Kurdistan Board of Investment, which is designed to provide incentives to help economic development in areas under the authority of the KRG.

Inflation has remained under control since 2006. However, Iraqi leaders remain hard-pressed to translate macroeconomic gains into an improved standard of living for the Iraqi populace. Unemployment remains a problem throughout the country despite a bloated public sector. Overregulation has made it difficult for Iraqi citizens and foreign investors to start new businesses. Corruption and lack of economic reforms—such as restructuring banks and developing the private sector – have inhibited the growth of the private sector.

GDP (purchasing power parity): $649.3 billion (2017 est.)
$662.9 billion (2016 est.)
$586.3 billion (2015 est.)
note: data are in 2017 dollars
country comparison to the world: 34

GDP (official exchange rate): $192.4 billion (2017 est.)

GDP—real growth rate: -2.1% (2017 est.)
13.1% (2016 est.)
2.5% (2015 est.)
country comparison to the world: 206

GDP—per capita (PPP): $16,700 (2017 est.)
$17,500 (2016 est.)
$15,900 (2015 est.)
note: data are in 2017 dollars
country comparison to the world: 107

Gross national saving: 19% of GDP (2017 est.)
13.1% of GDP (2016 est.)
18.4% of GDP (2015 est.)
country comparison to the world: 103

GDP—composition, by end use:
household consumption: 50.4% (2013 est.)
government consumption: 22.9% (2016 est.)
investment in fixed capital: 20.6% (2016 est.)
investment in inventories: 0% (2016 est.)
exports of goods and services: 32.5% (2016 est.)
imports of goods and services: -40.9% (2016 est.)

GDP—composition, by sector of origin:
agriculture: 3.3% (2017 est.)
industry: 51% (2017 est.)
services: 45.8% (2017 est.)

Agriculture—products: wheat, barley, rice, vegetables, dates, cotton; cattle, sheep, poultry
Industries: petroleum, chemicals, textiles, leather, construction materials, food processing, fertilizer, metal fabrication/processing

Industrial production growth rate: 0.7% (2017 est.)
country comparison to the world: 163

Labor force: 8.9 million (2010 est.)
country comparison to the world: 54

Labor force—by occupation: *agriculture:* 21.6%
industry: 18.7%
services: 59.8% (2008 est.)

Unemployment rate: 16% (2012 est.)
15% (2010 est.)

country comparison to the world: 178

Population below poverty line: 23% (2014 est.)

Household income or consumption by percentage share:
lowest 10%: 3.6%
highest 10%: 25.7% (2007 est.)

Budget: *revenues:* 68.71 billion (2017 est.)
expenditures: 76.82 billion (2017 est.)

Taxes and other revenues: 35.7% (of GDP) (2017 est.)
country comparison to the world: 59

Budget surplus (+) or deficit (-): -4.2% (of GDP) (2017 est.)
country comparison to the world: 160

Public debt: 59.7% of GDP (2017 est.)
66% of GDP (2016 est.)
country comparison to the world: 74

Fiscal year: calendar year

Inflation rate (consumer prices): 0.1% (2017 est.)
0.5% (2016 est.)
country comparison to the world: 14

Current account balance: $4.344 billion (2017 est.)
-$13.38 billion (2016 est.)
country comparison to the world: 31

Exports: $61.4 billion (2017 est.)
$41.72 billion (2016 est.)
country comparison to the world: 46

Exports—partners: India 21.2%, China 20.2%, US 15.8%, South Korea 9.4%, Greece 5.3%, Netherlands 4.8%, Italy 4.7% (2017)

Exports—commodities: crude oil 99%, crude materials excluding fuels, food, live animals

Imports: $39.47 billion (2017 est.)
$19.57 billion (2016 est.)
country comparison to the world: 60

Imports—commodities:
food, medicine, manufactures

Imports—partners: Turkey 27.8%, China 25.7%, South Korea 4.7%, Russia 4.3% (2017)

Reserves of foreign exchange and gold: $48.88 billion (31 December 2017 est.)
$45.36 billion (31 December 2016 est.)
country comparison to the world: 41

Debt—external: $73.02 billion (31 December 2017 est.)
$64.16 billion (31 December 2016 est.)
country comparison to the world: 59

Exchange rates: Iraqi dinars (IQD) per US dollar—
1,184 (2017 est.)
1,182 (2016 est.)
1,182 (2015 est.)
1,167.63 (2014 est.)
1,213.72 (2013 est.)

ENERGY

Electricity access: *electrification—total population:* 100% (2020)

Electricity—production: 75.45 billion kWh (2016 est.)

country comparison to the world: 40

Electricity—consumption: 38.46 billion kWh (2016 est.)
country comparison to the world: 57

Electricity—exports: 0 kWh (2016 est.)
country comparison to the world: 150

Electricity—imports: 11.97 billion kWh (2016 est.)
country comparison to the world: 20

Electricity—installed generating capacity: 27.09 million kW (2016 est.)
country comparison to the world: 34

Electricity—from fossil fuels: 91% of total installed capacity (2016 est.)
country comparison to the world: 55

Electricity—from nuclear fuels: 0% of total installed capacity (2017 est.)
country comparison to the world: 111

Electricity—from hydroelectric plants: 9% of total installed capacity (2017 est.)
country comparison to the world: 119

Electricity—from other renewable sources: 0% of total installed capacity (2017 est.)
country comparison to the world: 194

Crude oil—production: 4.613 million bbl/day (2018 est.)
country comparison to the world: 4

Crude oil—exports: 3.092 million bbl/day (2015 est.)
country comparison to the world: 3

Crude oil—imports: 0 bbl/day (2015 est.)
country comparison to the world: 145

Crude oil—proved reserves: 148.8 billion bbl (1 January 2018 est.)
country comparison to the world: 5

Refined petroleum products—production: 398,000 bbl/day (2015 est.)
country comparison to the world: 37

Refined petroleum products—consumption: 826,000 bbl/day (2016 est.)
country comparison to the world: 26

Refined petroleum products—exports: 8,284 bbl/day (2015 est.)
country comparison to the world: 86

Refined petroleum products—imports: 255,100 bbl/day (2015 est.)
country comparison to the world: 28

Natural gas—production: 1.274 billion cu m (2017 est.)
country comparison to the world: 63

Natural gas—consumption: 2.633 billion cu m (2017 est.)
country comparison to the world: 76

Natural gas—exports: 0 cu m (2017 est.)
country comparison to the world: 124

Natural gas—imports: 1.359 billion cu m (2017 est.)
country comparison to the world: 56

Natural gas—proved reserves: 3.82 trillion cu m (1 January 2018 est.)
country comparison to the world: 11

Carbon dioxide emissions from consumption of energy: 117.9 million Mt (2017 est.)
country comparison to the world: 37

COMMUNICATIONS

Telephones—fixed lines:
total subscriptions: 2,678,046
subscriptions per 100 inhabitants: 7.04 (2019 est.)
country comparison to the world: 46

Telephones—mobile cellular:
total subscriptions: 36,092,758
subscriptions per 100 inhabitants: 94.88 (2019 est.)
country comparison to the world: 42

Telecommunication systems: *general assessment:* the 2003 liberation of Iraq severely disrupted telecommunications throughout Iraq; widespread government efforts to rebuild domestic and international communications have slowed due to political unrest; 2018 showed signs of stability and installations of new fiber-optic cables and growth in mobile broadband subscribers; the most popular plans are pre-paid; 3 major operators in mobile sector preparing 4G and even 5G technologies; operators focused on fixing and replacing networks damaged during civil war (2020)
domestic: the mobile cellular market continues to expand; 3G services offered by three major mobile operators; 4G offered by one operator in Iraqi; conflict has destroyed infrastructure in areas; 7 per 100 for fixed-line and 95 per 100 for mobile-cellular subscriptions (2019)
international: country code—964; landing points for FALCON, and GBICS/MENA submarine cables providing connections to the Middle East, Africa and India; satellite earth stations—4 (2 Intelsat—1 Atlantic Ocean and 1 Indian Ocean, 1 Intersputnik—Atlantic Ocean region, and 1 Arabsat (inoperative)); local microwave radio relay connects border regions to Jordan, Kuwait, Syria, and Turkey (2019)
note: the COVID-19 outbreak is negatively impacting telecommunications production and supply chains globally; consumer spending on telecom devices and services has also slowed due to the pandemic's effect on economies worldwide; overall progress towards improvements in all facets of the telecom industry—mobile, fixed-line, broadband, submarine cable and satellite—has moderated

Broadcast media: the number of private radio and TV stations has increased rapidly since 2003; government-owned TV and radio stations are operated by the publicly funded Iraqi Media Network; private broadcast media are mostly linked to political, ethnic, or religious groups; satellite TV is available to an estimated 70% of viewers and many of the broadcasters are based abroad; transmissions of multiple international radio broadcasters are accessible (2019)

Internet country code: .iq

Internet users: *total:* 18,364,390
percent of population: 49.36% (July 2018 est.)
country comparison to the world: 38

Broadband—fixed subscriptions:
total: 4,492,328
subscriptions per 100 inhabitants: 12 (2018 est.)
country comparison to the world: 32

TRANSPORTATION

National air transport system:
number of registered air carriers: 4 (2020)
inventory of registered aircraft operated by air carriers: 34
annual passenger traffic on registered air carriers: 2,075,065 (2018)
annual freight traffic on registered air carriers: 16.2 million mt-km (2018)

Civil aircraft registration country code prefix: YI (2016)

Airports: 102 (2013)
country comparison to the world: 55

Airports—with paved runways:
total: 72 (2017)
over 3,047 m: 20 (2017)
2,438 to 3,047 m: 34 (2017)
1,524 to 2,437 m: 4 (2017)
914 to 1,523 m: 7 (2017)
under 914 m: 7 (2017)

Airports—with unpaved runways:
total: 30 (2013)
over 3,047 m: 3 (2013)
2,438 to 3,047 m: 5 (2013)
1,524 to 2,437 m: 3 (2013)
914 to 1,523 m: 13 (2013)
under 914 m: 6 (2013)

Heliports: 16 (2013)

Pipelines: 2455 km gas, 913 km liquid petroleum gas, 5432 km oil, 1637 km refined products (2013)

Railways: *total:* 2,272 km (2014)
standard gauge: 2,272 km 1.435-m gauge (2014)
country comparison to the world: 69

Roadways: *total:* 59,623 km (2012)
paved: 59,623 km (includes Kurdistan region) (2012)
country comparison to the world: 77

Waterways: 5,279 km (the Euphrates River (2,815 km), Tigris River (1,899 km), and Third River (565 km) are the principal waterways) (2012)
country comparison to the world: 22

Merchant marine: *total:* 73
by type: general cargo 1, oil tanker 6, other 66 (2019)
country comparison to the world: 101

Ports and terminals: *river port(s):* Al Basrah (Shatt al Arab); Khawr az Zubayr, Umm Qasr (Khawr az Zubayr waterway)

MILITARY AND SECURITY

Military and security forces: *Ministry of Defense:* Iraqi Army, Army Aviation Command, Iraqi Navy, Iraqi Air Force, Iraqi Air Defense Command, Special Forces Command; National-Level Security Forces: Iraqi Counterterrorism Service (CTS; a Special Forces Division aka the

"Golden Division"), Prime Minister's Special Forces Division, Presidential Brigades; Ministry of Interior: Federal Police Forces Command, Border Guard Forces Command, Federal Intelligence and Investigations Agency, Emergency Response Division, Facilities Protection Directorate, and Energy Police Directorate; Popular Mobilization Commission and Affiliated Forces (PMF); Ministry of Pershmerga (Kurdistan Regional Government) (2020)

note: the PMF is a collection of approximately 50 paramilitary militias of different sizes and with varying political interests

Military expenditures: 3.5% of GDP (2019)
2.9% of GDP (2018)
3.9% of GDP (2017)
3.5% of GDP (2016)
5.4% of GDP (2015)
country comparison to the world: 18

Military and security service personnel strengths: assessments of the size of the Iraqi military, security services, and associated militia forces vary widely; the military and the security services are rebuilding after suffering considerable losses in personnel and equipment fighting the ISIS terrorist group (see note) and are also attempting to incorporate local militia groups; approximately 190,000 active personnel (180,000 Army; 3,000 Navy; 5,000 Air Force); National-Level Security Forces: est.

10,000 Iraqi Counterterrorism Service; est. 10,000 Presidential Brigades; est. 6,000 Prime Minister's Special Forces Division; other: est. 100-150,000 Popular Mobilization Forces; est. 150,000-200,000 Peshmerga Forces (2019)

note: Iraqi Army strength reportedly fell from about 200,000 personnel in 2009 to around 50,000 in 2016

Military equipment inventories and acquisitions: the Iraqi military inventory is comprised of Russian and Soviet-era equipment combined with newer European- and US-sourced platforms; since 2010, Russia and the US are the leading suppliers of military hardware to Iraq (2019 est.)

Military service age and obligation: 18-40 years of age for voluntary military service; no conscription (2019)

TERRORISM

Terrorist group(s): Ansar al-Islam; Asa'ib Ahl al-Haq; Islamic Revolutionary Guard Corps/Qods Force; Islamic State of Iraq and ash-Sham; Jaysh Rijal al-Tariq al-Naqshabandi; Kata'ib Hizballah; Kurdistan Workers' Party (2019)

note: details about the history, aims, leadership, organization, areas of operation, tactics, targets, weapons, size, and sources of support of the group(s) appear(s) in Appendix-T

TRANSNATIONAL ISSUES

Disputes—international: Iraq's lack of a maritime boundary with Iran prompts jurisdiction disputes beyond the mouth of the Shatt al Arab in the Persian Gulf; Turkey has expressed concern over the autonomous status of Kurds in Iraq

Refugees and internally displaced persons: *refugees (country of origin):* 15,167 (Turkey), 7,858 (West Bank and Gaza Strip), 5,041 (Iran) (2018); 241,738 (Syria) (2020)

IDPs: 1,389,540 (displacement in central and northern Iraq since January 2014) (2020)

stateless persons: 47,253 (2019); note—in the 1970s and 1980s under SADDAM Husayn's regime, thousands of Iraq's Faili Kurds, followers of Shia Islam, were stripped of their Iraqi citizenship, had their property seized by the government, and many were deported; some Faili Kurds had their citizenship reinstated under the 2,006 Iraqi Nationality Law, but others lack the documentation to prove their Iraqi origins; some Palestinian refugees persecuted by the SADDAM regime remain stateless

note: estimate revised to reflect the reduction of statelessness in line with Law 26 of 2006, which allows stateless persons to apply for nationality in certain circumstances; more accurate studies of statelessness in Iraq are pending (2015)

IRELAND

INTRODUCTION

Background: Celtic tribes arrived on the island between 600 and 150 B.C. Invasions by Norsemen that began in the late 8th century were finally ended when King Brian BORU defeated the Danes in 1014. Norman invasions began in the 12th century and set off more than seven centuries of Anglo-Irish struggle marked by fierce rebellions and harsh repressions. The Irish famine of the mid-19th century was responsible for a drop in

the island's population by more than one quarter through starvation, disease, and emigration. For more than a century afterward, the population of the island continued to fall only to begin growing again in the 1960s. Over the last 50 years, Ireland's high birthrate has made it demographically one of the youngest populations in the EU.

The modern Irish state traces its origins to the failed 1916 Easter Monday Uprising that touched off several years of guerrilla warfare resulting in independence from the UK in 1921 for 26 southern counties; six northern (Ulster) counties remained part of the UK. Deep sectarian divides between the Catholic and Protestant populations and systemic discrimination in Northern Ireland erupted into years of violence known as the "Troubles" that began in the 1960s. The Government of Ireland was part of a process along with the UK and US Governments that helped broker the Good Friday Agreement in Northern Ireland in 1998. This initiated a new phase of cooperation between the Irish and British Governments. Ireland was neutral in World War II and continues its policy of military neutrality. Ireland joined the European Community in 1973 and the euro-zone currency union in 1999. The economic boom years of the Celtic Tiger (1995-2007) saw rapid economic growth, which came to an abrupt end in 2008 with the meltdown of the Irish banking system. Today

the economy is recovering, fueled by large and growing foreign direct investment, especially from US multinationals.

GEOGRAPHY

Location: Western Europe, occupying five-sixths of the island of Ireland in the North Atlantic Ocean, west of Great Britain

Geographic coordinates: 53 00 N, 8 00 W

Map references: Europe

Area: *total:* 70,273 sq km
land: 68,883 sq km
water: 1,390 sq km
country comparison to the world: 121

Area—comparative: slightly larger than West Virginia

Land boundaries: *total:* 490 km
border countries (1): UK 490 km

Coastline: 1,448 km

Maritime claims: *territorial sea:* 12 nm
exclusive fishing zone: 200 nm

Climate: temperate maritime; modified by North Atlantic Current; mild winters, cool summers; consistently humid; overcast about half the time

Terrain: mostly flat to rolling interior plain surrounded by rugged hills and low mountains; sea cliffs on west coast

Elevation: *mean elevation:* 118 m
lowest point: Atlantic Ocean 0 m
highest point: Carrauntoohil 1,041 m

Natural resources: natural gas, peat, copper, lead, zinc, silver, barite, gypsum, limestone, dolomite

Land use: *agricultural land:* 66.1% (2011 est.)
arable land: 15.4% (2011 est.) / permanent crops: 0% (2011 est.) / permanent pasture: 50.7% (2011 est.)
forest: 10.9% (2011 est.)
other: 23% (2011 est.)

Irrigated land: 0 sq km (2012)

Population distribution: population distribution is weighted to the eastern side of the island, with the largest concentration being in and around Dublin; populations in the west are small due to mountainous land, poorer soil, lack of good transport routes, and fewer job opportunities

Natural hazards: rare extreme weather events

Environment—current issues: water pollution, especially of lakes, from agricultural runoff; acid rain kills plants, destroys soil fertility, and contributes to deforestation

Environment—international agreements: *party to:* Air Pollution, Air Pollution-Nitrogen Oxides, Air Pollution-Sulfur 94, Biodiversity, Climate Change, Climate Change-Kyoto Protocol, Desertification, Endangered Species, Environmental Modification, Hazardous Wastes, Law of the Sea, Marine Dumping, Ozone Layer Protection, Ship Pollution, Tropical Timber 83, Tropical Timber 94, Wetlands, Whaling
signed, but not ratified: Air Pollution-Persistent Organic Pollutants, Marine Life Conservation
Geography—note: strategic location on major air and sea routes between North America and northern Europe; over 40% of the population resides within 100 km of Dublin

PEOPLE AND SOCIETY

Population: 5,176,569 (July 2020 est.)
country comparison to the world: 122

Nationality: *noun:* Irishman(men), Irishwoman(women), Irish (collective plural)
adjective: Irish

Ethnic groups: Irish 82.2%, Irish travelers 0.7%, other white 9.5%, Asian 2.1%, black 1.4%, other 1.5%, unspecified 2.6% (2016 est.)

Languages: English (official, the language generally used), Irish (Gaelic or Gaeilge) (official, spoken by approximately 39.8% of the population as of 2016; mainly spoken in areas along Ireland's western coast known as gaeltachtai, which are officially recognized regions where Irish is the predominant language)

Religions: Roman Catholic 78.3%, Church of Ireland 2.7%, other Christian 1.6%, Orthodox 1.3%, Muslim 1.3%, other 2.4%, none 9.8%, unspecified 2.6% (2016 est.)

Age structure: *0-14 years:* 21.15% (male 560,338/female 534,570)

15-24 years: 12.08% (male 316,239/female 308,872)
25-54 years: 42.19% (male 1,098,058/female 1,085,794)
55-64 years: 10.77% (male 278,836/female 278,498)
65 years and over: 13.82% (male 331,772/female 383,592) (2020 est.)

Dependency ratios: *total dependency ratio:* 54.8
youth dependency ratio: 32.3
elderly dependency ratio: 22.6
potential support ratio: 4.4 (2020 est.)

Median age: *total:* 37.8 years
male: 37.4 years
female: 38.2 years (2020 est.)
country comparison to the world: 66
Population growth rate: 1.04% (2020 est.)
country comparison to the world: 101

Birth rate: 13 births/1,000 population (2020 est.)
country comparison to the world: 144

Death rate: 6.8 deaths/1,000 population (2020 est.)
country comparison to the world: 134

Net migration rate: 3.9 migrant(s)/1,000 population (2020 est.)
country comparison to the world: 31

Population distribution: population distribution is weighted to the eastern side of the island, with the largest concentration being in and around Dublin; populations in the west are small due to mountainous land, poorer soil, lack of good transport routes, and fewer job opportunities

Urbanization: *urban population:* 63.7% of total population (2020)
rate of urbanization: 1.14% annual rate of change (2015-20 est.)
total population growth rate v. urban population growth rate, 2000-2030: Major urban areas—population: 1.228 million DUBLIN (capital) (2020)

Sex ratio: *at birth:* 1.06 male(s)/female
0-14 years: 1.05 male(s)/female
15-24 years: 1.02 male(s)/female
25-54 years: 1.01 male(s)/female
55-64 years: 1 male(s)/female
65 years and over: 0.86 male(s)/female
total population: 1 male(s)/female (2020 est.)

Mother's mean age at first birth: 30.5 years (2018 est.)

Maternal mortality rate: 5 deaths/100,000 live births (2017 est.)
country comparison to the world: 165

Infant mortality rate: *total:* 3.6 deaths/1,000 live births
male: 3.9 deaths/1,000 live births
female: 3.2 deaths/1,000 live births (2020 est.)
country comparison to the world: 195

Life expectancy at birth: *total population:* 81.2 years
male: 78.9 years
female: 83.7 years (2020 est.)
country comparison to the world: 36

Total fertility rate: 1.94 children born/woman (2020 est.)
country comparison to the world: 125

Contraceptive prevalence rate: 73.3% (2010)
note: percent of women aged 18-45

Drinking water source:
improved: *urban:* 97% of population
rural: 98.1% of population
total: 97.4% of population
unimproved: *urban:* 3% of population
rural: 1.9% of population
total: 2.6% of population (2017 est.)

Current Health Expenditure: 7.2% (2017)

Physicians density: 3.29 physicians/1,000 population (2017)

Hospital bed density: 3 beds/1,000 population (2017)

Sanitation facility access:
improved: *urban:* 97.7% of population
rural: 99% of population
total: 98.2% of population
unimproved: *urban:* 2.3% of population
rural: 1% of population
total: 1.8% of population (2017 est.)

HIV/AIDS—adult prevalence rate: 0.2% (2019 est.)
country comparison to the world: 99

HIV/AIDS—people living with HIV/AIDS: 7,500 (2019 est.)
country comparison to the world: 112

HIV/AIDS—deaths: <100 (2019 est.)

Obesity—adult prevalence rate: 25.3% (2016)
country comparison to the world: 51

Education expenditures: 3.7% of GDP (2016)
country comparison to the world: 117

School life expectancy (primary to tertiary education): *total:* 20 years
male: 19 years
female: 20 years (2018)

Unemployment, youth ages 15-24: *total:* 13.8%
male: 14.8%
female: 12.6% (2018 est.)
country comparison to the world: 99

GOVERNMENT

Country name: *conventional long form:* none
conventional short form: Ireland
local long form: none
local short form: Eire
etymology: the modern Irish name "Eire" evolved from the Gaelic "Eriu," the name of the matron goddess of Ireland (goddess of the land); the names "Ireland" in English and "Eire" in Irish are direct translations of each other

Government type: parliamentary republic

Capital: *name:* Dublin
geographic coordinates: 53 19 N, 6 14 W
time difference: UTC 0 (5 hours ahead of Washington, DC, during Standard Time)
daylight saving time: +1hr, begins last Sunday in March; ends last Sunday in October

etymology: derived from Irish "dubh" and "lind" meaning respectively "black, dark" and "pool" and which referred to the dark tidal pool where the River Poddle entered the River Liffey; today the area is the site of the castle gardens behind Dublin Castle

Administrative divisions: 28 counties and 3 cities*; Carlow, Cavan, Clare, Cork, Cork*, Donegal, Dublin*, Dun Laoghaire-Rathdown, Fingal, Galway, Galway*, Kerry, Kildare, Kilkenny, Laois, Leitrim, Limerick, Longford, Louth, Mayo, Meath, Monaghan, Offaly, Roscommon, Sligo, South Dublin, Tipperary, Waterford, Westmeath, Wexford, Wicklow

Independence: 6 December 1921 (from the UK by the Anglo-Irish Treaty, which ended British rule); 6 December 1922 (Irish Free State established); 18 April 1949 (Republic of Ireland Act enabled)

National holiday: Saint Patrick's Day, 17 March; note—marks the traditional death date of Saint Patrick, patron saint of Ireland, during the latter half of the fifth century A.D. (most commonly cited years are c. 461 and c. 493); although Saint Patrick's feast day was celebrated in Ireland as early as the ninth century, it only became an official public holiday in Ireland in 1903

Constitution: *history:* previous 1922; latest drafted 14 June 1937, adopted by plebiscite 1 July 1937, effective 29 December 1937

amendments: proposed as bills by Parliament; passage requires majority vote by both the Senate and House of Representatives, majority vote in a referendum, and presidential signature; amended many times, last in 2019

Legal system: common law system based on the English model but substantially modified by customary law; judicial review of legislative acts by Supreme Court

International law organization participation: accepts compulsory ICJ jurisdiction with reservations; accepts ICCt jurisdiction

Citizenship: *citizenship by birth:* no, unless a parent of a child born in Ireland has been legally resident in Ireland for at least three of the four years prior to the birth of the child

citizenship by descent only: yes

dual citizenship recognized: yes

residency requirement for naturalization: 4 of the previous 8 years

Suffrage: 18 years of age; universal

Executive branch: *chief of state:* President Michael D. HIGGINS (since 11 November 2011)

head of government: Taoiseach (Prime Minister) Micheál MARTIN (since 27 June 2020); note—MARTIN will serve through December 2022 and will then be succeeded by Leo VARADKAR

cabinet: Cabinet nominated by the prime minister, appointed by the president, approved by the Dáil Eireann (lower house of Parliament)

elections/appointments: president directly elected by majority popular vote for a 7-year term (eligible for a second term); election last held on 26 October 2018 (next to be held no later than November 2025); taoiseach (prime minister)

nominated by the House of Representatives (Dail Eireann), appointed by the president

election results: Michael D. HIGGINS reelected president; percent of vote—Michael D. HIGGINS (independent) 55.8%, Peter CASEY (independent) 23.3%, Sean GALLAGHER (independent) 6.4%, Liadh NI RIADA (Sinn Fein) 6.4%, Joan FREEMAN (independent) 6%, Gavin DUFFY (independent) 2.2%

Legislative branch: *description:* bicameral Parliament or Oireachtas consists of: Senate or Seanad Eireann (60 seats; 43 members indirectly elected from 5 vocational panels of nominees by an electoral college consisting of members from the House of Representatives, outgoing Senate members, and city and county council members, 11 appointed by the prime minister, and 6 elected by 2 university constituencies—3 each from the University of Dublin (Trinity College) and the National University of Ireland)

House of Representatives or Dail Eireann (158 seats; members directly elected in multi-seat constituencies by proportional representation vote; all Parliament members serve 5-year terms)

elections: Senate—last held in April and May 2016 (next to be held no later than 2021)

House of Representatives—last held on 8 February 2020 (next to be held no later than 2025)

election results: Senate—percent of vote by party—NA; seats by party—Fine Gael 19, Fianna Fail 14, Sinn Fein 7, Labor Party 5, Green Party 1, independent 14; composition—men 42, women 18, percent of women 30%

House of Representatives—percent of vote by party—Sinn Fein 23%, Fianna Fail 23%, Fine Gael 22%, Green Party 8%, Labor Party 4%, Social Democrats 4%, AAA-PBD 3%, Aontu 0.6%, Independents for Change 0.6%, Ceann Comhairle 0.6%, Independents 12%; seats by party—Sinn Fein 37, Fianna Fail 37, Fine Gael 35, Green Party 12, Labor Party 6, Social Democrats 6, AAA-PBD 5, Aontu 1, Independents for Change 1, Ceann Comhairle 1, Independents 19; composition—men 123, women 35, percent of women 22.2%; note—total Parliament percent of women 24.3%

Judicial branch: *highest courts:* Supreme Court of Ireland (consists of the chief justice, 9 judges, 2 ex-officio members—the presidents of the High Court and Court of Appeal—and organized in 3-, 5-, or 7-judge panels, depending on the importance or complexity of an issue of law)

judge selection and term of office: judges nominated by the prime minister and Cabinet and appointed by the president; chief justice serves in the position for 7 years; judges can serve until age 70

subordinate courts: High Court, Court of Appeal; circuit and district courts; criminal courts

Political parties and leaders: Solidarity-People Before Profit or AAAS-PBP [collective leadership]
Fianna Fail [Micheal MARTIN]
Fine Gael [Leo VARADKAR]
Green Party [Eamon RYAN]
Labor (Labour) Party (vacant)

Renua Ireland (vacant)
Sinn Fein [Mary Lou MCDONALD]
Social Democrats [Catherine MURPHY, Roisin SHORTALL]
Socialist Party [collective leadership]
The Workers' Party [Michael DONNELLY]

International organization participation: ADB (nonregional member), Australia Group, BIS, CD, CE, EAPC, EBRD, ECB, EIB, EMU, ESA, EU, FAO, FATF, IAEA, IBRD, ICAO, ICC (national committees), ICCt, ICRM, IDA, IEA, IFAD, IFC, IFRCS, IGAD (partners), IHO, ILO, IMF, IMO, Interpol, IOC, IOM, IPU, ISO, ITSO, ITU, ITUC (NGOs), MIGA, MINURSO, MONUSCO, NEA, NSG, OAS (observer), OECD, OPCW, OSCE, Paris Club, PCA, PFP, UN, UNCTAD, UNDOF, UNESCO, UNHCR, UNIDO, UNIFIL, UNOCI, UNRWA, UNTSO, UPU, WCO, WHO, WIPO, WMO, WTO, ZC

Diplomatic representation in the US: *chief of mission:* Ambassador Daniel Gerard MULHALL (since 8 September 2017)

chancery: 2234 Massachusetts Avenue NW, Washington, DC 20008

telephone: [1] (202) 462-3939

FAX: [1] (202) 232-5993

consulate(s) general: Atlanta, Austin (TX), Boston, Chicago, New York, San Francisco

Diplomatic representation from the US: *chief of mission:* Ambassador Edward F. CRAWFORD (since 1 July 2019)

telephone: [353] (1) 668-8777

embassy: 42 Elgin Road, Ballsbridge, Dublin 4

mailing address: use embassy street address

FAX: [353] (1) 688-9946

Flag description: three equal vertical bands of green (hoist side), white, and orange; officially the flag colors have no meaning, but a common interpretation is that the green represents the Irish nationalist (Gaelic) tradition of Ireland; orange represents the Orange tradition (minority supporters of William of Orange); white symbolizes peace (or a lasting truce) between the green and the orange

note: similar to the flag of Cote d'Ivoire, which is shorter and has the colors reversed—orange (hoist side), white, and green; also similar to the flag of Italy, which is shorter and has colors of green (hoist side), white, and red

National symbol(s): harp, shamrock (trefoil); *national colors:* blue, green

National anthem: *name:* "Amhran na bhFiann" (The Soldier's Song)

lyrics/music: Peadar KEARNEY [English], Liam O RINN [Irish]/Patrick HEENEY and Peadar KEARNEY

note: adopted 1926; instead of "Amhran na bhFiann," the song "Ireland's Call" is often used at athletic events where citizens of Ireland and Northern Ireland compete as a unified team

ECONOMY

Economy—overview: Ireland is a small, modern, trade-dependent economy. It was among the initial group of 12 EU nations that began circulating the euro on 1 January 2002. GDP growth averaged 6% in 1995-2007, but economic activity dropped sharply during the world financial crisis and the subsequent collapse of its domestic property market and construction industry during 2008-11. Faced with sharply reduced revenues and a burgeoning budget deficit from efforts to stabilize its fragile banking sector, the Irish Government introduced the first in a series of draconian budgets in 2009. These measures were not sufficient to stabilize Ireland's public finances. In 2010, the budget deficit reached 32.4% of GDP—the world's largest deficit, as a percentage of GDP. In late 2010, the former COWEN government agreed to a $92 billion loan package from the EU and IMF to help Dublin recapitalize Ireland's banking sector and avoid defaulting on its sovereign debt. In March 2011, the KENNY government intensified austerity measures to meet the deficit targets under Ireland's EU-IMF bailout program.

In late 2013, Ireland formally exited its EU-IMF bailout program, benefiting from its strict adherence to deficit-reduction targets and success in refinancing a large amount of banking-related debt. In 2014, the economy rapidly picked up. In late 2014, the government introduced a fiscally neutral budget, marking the end of the austerity program. Continued growth of tax receipts has allowed the government to lower some taxes and increase public spending while keeping to its deficitreduction targets. In 2015, GDP growth exceeded 26%. The magnitude of the increase reflected one-off statistical revisions, multinational corporate restructurings in intellectual property, and the aircraft leasing sector, rather than real gains in the domestic economy, which was still growing. Growth moderated to around 4.1% in 2017, but the recovering economy assisted lowering the deficit to 0.6% of GDP.

In the wake of the collapse of the construction sector and the downturn in consumer spending and business investment during the 2008-11 economic crisis, the export sector, dominated by foreign multinationals, has become an even more important component of Ireland's economy. Ireland's low corporation tax of 12.5% and a talented pool of high-tech laborers have been some of the key factors in encouraging business investment. Loose tax residency requirements made Ireland a common destination for international firms seeking to pay less tax or, in the case of U.S. multinationals, defer taxation owed to the United States. In 2014, amid growing international pressure, the Irish government announced it would phase in more stringent tax laws, effectively closing a commonly used loophole. The Irish economy continued to grow in 2017 and is forecast to do so through 2019, supported by a strong export sector, robust job growth, and low inflation, to the point that the Government must now address concerns about overheating and potential loss of competitiveness. The greatest risks to the economy are the UK's scheduled departure from the European Union ("Brexit") in March 2019, possible changes

to international taxation policies that could affect Ireland's revenues, and global trade pressures.

GDP (purchasing power parity): $353.3 billion (2017 est.)
$329.5 billion (2016 est.)
$314.1 billion (2015 est.)
note: data are in 2017 dollars
country comparison to the world: 51

GDP (official exchange rate): $331.5 billion (2017 est.)

GDP—real growth rate: 5.86% (2019 est.)
9.42% (2018 est.)
9.49% (2017 est.)
country comparison to the world: 33

GDP—per capita (PPP): $73,200 (2017 est.)
$69,100 (2016 est.)
$66,600 (2015 est.)
note: data are in 2017 dollars
country comparison to the world: 10

Gross national saving: 33.1% of GDP (2017 est.)
33.7% of GDP (2016 est.)
29% of GDP (2015 est.)
country comparison to the world: 23

GDP—composition, by end use: *household consumption:* 34% (2017 est.)
government consumption: 10.1% (2017 est.)
investment in fixed capital: 23.4% (2017 est.)
investment in inventories: 1.2% (2017 est.)
exports of goods and services: 119.9% (2017 est.)
imports of goods and services: -89.7% (2017 est.)

GDP—composition, by sector of origin: *agriculture:* 1.2% (2017 est.)
industry: 38.6% (2017 est.)
services: 60.2% (2017 est.)

Agriculture—products: barley, potatoes, wheat; beef, dairy products
Industries: pharmaceuticals, chemicals, computer hardware and software, food products, beverages and brewing; medical devices

Industrial production growth rate: 7.8% (2017 est.)
country comparison to the world: 25

Labor force: 2.289 million (2020 est.)
country comparison to the world: 117

Labor force—by occupation: *agriculture:* 5%
industry: 11%
services: 84% (2015 est.)

Unemployment rate: 4.98% (2019 est.)
5.78% (2018 est.)
country comparison to the world: 74

Population below poverty line: 8.2% (2013 est.)

Household income or consumption by percentage share: *lowest 10%:* 2.9%
highest 10%: 27.2% (2000)

Budget: *revenues:* 86.04 billion (2017 est.)
expenditures: 87.19 billion (2017 est.)

Taxes and other revenues: 26% (of GDP) (2017 est.)
country comparison to the world: 115

Budget surplus (+) or deficit (-): -0.3% (of GDP) (2017 est.)
country comparison to the world: 53

Public debt: 68.6% of GDP (2017 est.)

73.6% of GDP (2016 est.)
note: data cover general government debt and include debt instruments issued (or owned) by government entities other than the treasury; the data include treasury debt held by foreign entities; the data include debt issued by subnational entities, as well as intragovernmental debt; intragovernmental debt consists of treasury borrowings from surpluses in the social funds, such as for retirement, medical care, and unemployment; debt instruments for the social funds are not sold at public auctions
country comparison to the world: 53

Fiscal year: calendar year

Inflation rate (consumer prices): 0.3% (2017 est.)
-0.2% (2016 est.)
country comparison to the world: 20

Current account balance: -$44.954 billion (2019 est.)
$24.154 billion (2018 est.)
country comparison to the world: 202

Exports: $219.7 billion (2017 est.)
$206 billion (2016 est.)
country comparison to the world: 25

Exports—partners: US 27.1%, UK 13.4%, Belgium 11%, Germany 8.1%, Switzerland 5.1%, Netherlands 4.9%, France 4.3% (2017)

Exports—commodities: machinery and equipment, computers, chemicals, medical devices, pharmaceuticals; foodstuffs, animal products

Imports: $98.13 billion (2017 est.)
$92.09 billion (2016 est.)
country comparison to the world: 34

Imports—commodities: data processing equipment, other machinery and equipment, chemicals, petroleum and petroleum products, textiles, clothing

Imports—partners: UK 29%, US 18.9%, France 12.1%, Germany 9.6%, Netherlands 4.1% (2017)

Reserves of foreign exchange and gold: $4.412 billion (31 December 2017 est.)
$2.203 billion (31 December 2015 est.)
country comparison to the world: 99

Debt—external: $2.47 trillion (31 March 2016 est.)
$2.35 trillion (31 March 2015 est.)
country comparison to the world: 8

Exchange rates: euros (EUR) per US dollar -
0.885 (2017 est.)
0.903 (2016 est.)
0.9214 (2015 est.)
0.885 (2014 est.)
0.7634 (2013 est.)

ENERGY

Electricity access: *electrification—total population:* 100% (2020)

Electricity—production: 28.53 billion kWh (2016 est.)
country comparison to the world: 69

Electricity—consumption: 25.68 billion kWh (2016 est.)

country comparison to the world: 68

Electricity—exports: 1.583 billion kWh (2016 est.)

country comparison to the world: 48

Electricity—imports: 871 million kWh (2016 est.)

country comparison to the world: 71

Electricity—installed generating capacity: 9.945 million kW (2016 est.)

country comparison to the world: 61

Electricity—from fossil fuels: 65% of total installed capacity (2016 est.)

country comparison to the world: 117

Electricity—from nuclear fuels: 0% of total installed capacity (2017 est.)

country comparison to the world: 112

Electricity—from hydroelectric plants: 2% of total installed capacity (2017 est.)

country comparison to the world: 138

Electricity—from other renewable sources: 33% of total installed capacity (2017 est.)

country comparison to the world: 12

Crude oil—production: 0 bbl/day (2018 est.)

country comparison to the world: 152

Crude oil—exports: 5,900 bbl/day (2017 est.)

country comparison to the world: 64

Crude oil—imports: 66,210 bbl/day (2017 est.)

country comparison to the world: 50

Crude oil—proved reserves: 0 bbl (1 January 2018 est.)

country comparison to the world: 148

Refined petroleum products—production: 64,970 bbl/day (2017 est.)

country comparison to the world: 76

Refined petroleum products—consumption: 153,700 bbl/day (2017 est.)

country comparison to the world: 66

Refined petroleum products—exports: 37,040 bbl/day (2017 est.)

country comparison to the world: 59

Refined petroleum products—imports: 126,600 bbl/day (2017 est.)

country comparison to the world: 47

Natural gas—production: 3.511 billion cu m (2017 est.)

country comparison to the world: 54

Natural gas—consumption: 5.238 billion cu m (2017 est.)

country comparison to the world: 55

Natural gas—exports: 0 cu m (2017 est.)

country comparison to the world: 125

Natural gas—imports: 1.642 billion cu m (2017 est.)

country comparison to the world: 55

Natural gas—proved reserves: 9.911 billion cu m (1 January 2018 est.)

country comparison to the world: 79

Carbon dioxide emissions from consumption of energy: 36.91 million Mt (2017 est.)

country comparison to the world: 71

COMMUNICATIONS

Telephones—fixed lines: *total subscriptions:* 1,854,605

subscriptions per 100 inhabitants: 36.2 (2019 est.)

country comparison to the world: 59

Telephones—mobile cellular: *total subscriptions:* 5,398,848

subscriptions per 100 inhabitants: 105.38 (2019 est.)

country comparison to the world: 117

Telecommunication systems: *general assessment:* a previous depressed economic climate has changed to one with Ireland having one of the highest GDP growth rates in Europe, which translates to mean spending among telecom consumers; introduction of flat-rate plans; upgraded LTE technologies in rural areas; govt. intends to spend millions on the National Broadband Plan (NBP) initiative to change the broadband landscape; plans to auction spectrum suitable for 5G services; broadband market seen steady development; 20 towns see commercial 5G services (2020)

domestic: increasing levels of broadband access particularly in urban areas; fixed-line 36 per 100 and mobile-cellular 105 per 100 subscriptions; digital system using cable and microwave radio relay (2019)

international: country code—353; landing point for the AEConnect -1, Celtic-Norse, Havfrue/AEC-2, GTT Express, Celtic, ESAT-1, IFC-1, Solas, Pan European Crossing, ESAT-2, CeltixConnect -1 & 2, GTT Atlantic, Sirius South, Emerald Bridge Fibres and Geo Eirgrid submarine cable with links to the US, Canada, Norway, Isle of Man and UK; satellite earth stations—81 (2019)

note: the COVID-19 outbreak is negatively impacting telecommunications production and supply chains globally; consumer spending on telecom devices and services has also slowed due to the pandemic's effect on economies worldwide; overall progress towards improvements in all facets of the telecom industry—mobile, fixed-line, broadband, submarine cable and satellite—has moderated

Broadcast media: publicly owned broadcaster Radio Telefis Eireann (RTE) operates 4 TV stations; commercial TV stations are available; about 75% of households utilize multi-channel satellite and TV services that provide access to a wide range of stations; RTE operates 4 national radio stations and has launched digital audio broadcasts on several stations; a number of commercial broadcast stations operate at the national, regional, and local levels (2019)

Internet country code: .ie

Internet users: *total:* 4,283,516

percent of population: 84.52% (July 2018 est.)

country comparison to the world: 92

Broadband—fixed subscriptions: *total:* 1,430,160

subscriptions per 100 inhabitants: 28 (2018 est.)

country comparison to the world: 65

TRANSPORTATION

National air transport system: *number of registered air carriers:* 9 (2020)

inventory of registered aircraft operated by air carriers: 450

annual passenger traffic on registered air carriers: 167,598,633 (2018)

annual freight traffic on registered air carriers: 168.71 million mt-km (2018)

Civil aircraft registration country code prefix: EI (2016)

Airports: 40 (2013)

country comparison to the world: 105

Airports—with paved runways: *total:* 16 (2019)

over 3,047 m: 1

2,438 to 3,047 m: 1

1,524 to 2,437 m: 4

914 to 1,523 m: 5

under 914 m: 5

Airports—with unpaved runways: *total:* 24 (2013)

2,438 to 3,047 m: 1 (2013)

914 to 1,523 m: 2 (2013)

under 914 m: 21 (2013)

Pipelines: 2,427 km gas (2017)

Railways: *total:* 4,301 km (2018)

narrow gauge: 1,930 km 0.914-m gauge (operated by the Irish Peat Board to transport peat to power stations and briquetting plants) (2018)

broad gauge: 2,371 km 1.600-m gauge (53 km electrified) (2018)

country comparison to the world: 44

Roadways: *total:* 99,830 km (2018)

paved: 99,830 km (includes 2,717 km of expressways) (2018)

country comparison to the world: 48

Waterways: 956 km (pleasure craft only) (2010)

country comparison to the world: 67

Merchant marine: *total:* 93

by type: bulk carrier 9 general cargo 37, oil tanker 1, other 46 (2019)

country comparison to the world: 95

Ports and terminals: *major seaport(s):* Dublin, Shannon Foynes

cruise port(s): Cork, Dublin

container port(s) (TEUs): Dublin (529,563) (2016)

river port(s): Cork (Lee), Waterford (Suir)

MILITARY AND SECURITY

Military and security forces: *Irish Defence Forces (Oglaigh na h-Eireannn):* Army (includes Army Reserve), Naval Service (includes Naval Service Reserves), Air Corps (2019)

Military expenditures: 0.3% of GDP (2019)

0.3% of GDP (2018)

0.3% of GDP (2017)

0.3% of GDP (2016)

0.3% of GDP (2015)

country comparison to the world: 156

Military and security service personnel strengths: the Irish Defence Forces have approximately 8,700 active duty personnel (7,000 Army; 1,000 Navy; 700 Air Force) (2019 est.)

Military equipment inventories and acquisitions: the Irish Defense Forces have a small inventory of imported weapons systems from a variety of European countries, as well as South Africa and the US; the UK is the leading supplier of military hardware to Ireland since 2010 (2019 est.)

Military deployments: 130 Golan Heights (UNDOF); 340 Lebanon (UNIFIL) (2020)

Military service age and obligation: 18-25 years of age for male and female voluntary military service recruits to the Defence Forces (18-27 years of age for the Naval Service); 18-26 for cadetship (officer) applicants; 12-year service (5 active,

7 reserves); Irish citizen, European Economic Area citizenship, or refugee status (2019)

TERRORISM

Terrorist group(s): Continuity Irish Republican Army; New Irish Republican Army (2019)
note: details about the history, aims, leadership, organization, areas of operation, tactics, targets, weapons, size, and sources of support of the group(s) appear(s) in Appendix-T

TRANSNATIONAL ISSUES

Disputes—international: Ireland, Iceland, and the UK dispute Denmark's claim that the Faroe Islands' continental shelf extends beyond 200 nm

Refugees and internally displaced persons: *stateless persons:* 99 (2019)

Illicit drugs: transshipment point for and consumer of hashish from North Africa to the UK and Netherlands and of European-produced synthetic drugs; increasing consumption of South American cocaine; minor transshipment point for heroin and cocaine destined for Western Europe; despite recent legislation, narcotics-related money laundering—using bureaux de change, trusts, and shell companies involving the offshore financial community—remains a concern

ISLE OF MAN

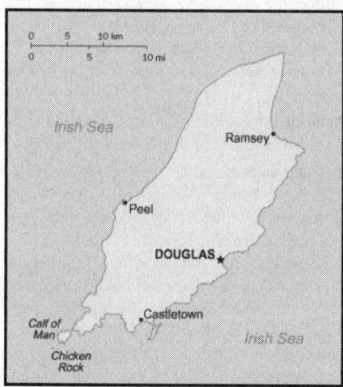

INTRODUCTION

Background: Part of the Norwegian Kingdom of the Hebrides until the 13th century when it was ceded to Scotland, the isle came under the British crown in 1765. Current concerns include reviving the almost extinct Manx Gaelic language. The Isle of Man is a British Crown dependency, which makes it a self-governing possession of the British Crown that is not part of the UK. The UK Government, however, remains constitutionally responsible for its defense and international representation.

GEOGRAPHY

Location: Western Europe, island in the Irish Sea, between Great Britain and Ireland

Geographic coordinates: 54 15 N, 4 30 W

Map references: Europe

Area: *total:* 572 sq km
land: 572 sq km
water: 0 sq km
country comparison to the world: 194

Area—comparative: slightly more than three times the size of Washington, DC

Land boundaries: 0 km

Coastline: 160 km

Maritime claims: *territorial sea:* 12 nm
exclusive fishing zone: 12 nm

Climate: temperate; cool summers and mild winters; overcast about a third of the time

Terrain: hills in north and south bisected by central valley

Elevation: *lowest point:* Irish Sea 0 m
highest point: Snaefell 621 m

Natural resources: none

Land use: *agricultural land:* 74.7% (2011 est.)
arable land: 43.8% (2011 est.) / *permanent crops:* 0% (2011 est.) / *permanent pasture:* 30.9% (2011 est.)
forest: 6.1% (2011 est.)
other: 19.2% (2011 est.)

Irrigated land: 0 sq km (2012)

Population distribution: most people concentrated in cities and large towns of which Douglas, in the southeast, is the largest

Natural hazards: occasional high winds and rough seas

Environment—current issues: air pollution, marine pollution; waste disposal (both household and industrial)

Geography—note: one small islet, the Calf of Man, lies to the southwest and is a bird sanctuary

PEOPLE AND SOCIETY

90,499 (July 2020 est.)
country comparison to the world: 199

Nationality: *noun:* Manxman(men), Manxwoman(women)
adjective: Manx

Ethnic groups: white 96.5%, Asian/Asian British 1.9%, other 1.5% (2011 est.)

Languages: English, Manx Gaelic (about 2% of the population has some knowledge)

Religions: Protestant (Anglican, Methodist, Baptist, Presbyterian, Society of Friends), Roman Catholic

Age structure: *0-14 years:* 16.28% (male 7,688/female 7,046)
15-24 years: 11.02% (male 5,328/female 4,642)
25-54 years: 37.8% (male 17,080/female 17,131)
55-64 years: 13.82% (male 6,284/female 6,219)
65 years and over: 21.08% (male 9,023/female 10,058) (2020 est.)

Median age: *total:* 44.6 years
male: 43.6 years
female: 45.6 years (2020 est.)
country comparison to the world: 12

Population growth rate: 0.59% (2020 est.)
country comparison to the world: 151

Birth rate: 10.8 births/1,000 population (2020 est.)
country comparison to the world: 182

Death rate: 10.4 deaths/1,000 population (2020 est.)
country comparison to the world: 29

Net migration rate: 5.2 migrant(s)/1,000 population (2020 est.)
country comparison to the world: 21

Population distribution: most people concentrated in cities and large towns of which Douglas, in the southeast, is the largest

Urbanization: *urban population:* 52.9% of total population (2020)
rate of urbanization: 0.89% annual rate of change (2015-20 est.)
total population growth rate v. urban population growth rate, 2000-2030: Major urban areas—*population:* 27,000 DOUGLAS (capital) (2018)

Sex ratio: *at birth:* 1.08 male(s)/female
0-14 years: 1.09 male(s)/female
15-24 years: 1.15 male(s)/female

25-54 years: 1 male(s)/female
55-64 years: 1.01 male(s)/female
65 years and over: 0.9 male(s)/female
total population: 1.01 male(s)/female (2020 est.)

Infant mortality rate: *total:* 3.9 deaths/1,000 live births
male: 3.9 deaths/1,000 live births
female: 3.9 deaths/1,000 live births (2020 est.)
country comparison to the world: 191

Life expectancy at birth: *total population:* 81.6 years
male: 79.8 years
female: 83.6 years (2020 est.)
country comparison to the world: 30

Total fertility rate: 1.9 children born/woman (2020 est.)
country comparison to the world: 130

Drinking water source:

improved: *total:* 100% of population

unimproved: *total:* 0% of population (2017 est.)

HIV/AIDS—adult prevalence rate: NA

HIV/AIDS—people living with HIV/AIDS: NA

HIV/AIDS—deaths: NA

Education expenditures: NA

Unemployment, youth ages 15-24: *total:* 10.1%
male: 11.8%
female: 8.2% (2011 est.)
country comparison to the world: 126

GOVERNMENT

Country name: *conventional long form:* none
conventional short form: Isle of Man
abbreviation: I.O.M.
etymology: the name "man" may be derived from the Celtic word for "mountain"

Dependency status: British crown dependency

Government type: parliamentary democracy (Tynwald)

Capital: *name:* Douglas
geographic coordinates: 54 09 N, 4 29 W
time difference: UTC 0 (5 hours ahead of Washington, DC, during Standard Time)
daylight saving time: +1hr, begins last Sunday in March; ends last Sunday in October
etymology: name derives from the Dhoo and Glass Rivers, which flow through the valley in which the town is located and which in Manx mean the «dark» and the «light» rivers respectively

Administrative divisions: none; there are no first-order administrative divisions as defined by the US Government, but there are 24 local authorities each with its own elections

Independence: none (British Crown dependency)

National holiday: Tynwald Day, 5 July (1417); date Tynwald Day was first recorded

Constitution: *history:* development of the Isle of Man constitution dates to at least the 14th century
amendments: proposed as a bill in the House of Keys, by the «Government,» by a «Member of the House,» or through petition to the House or Legislative Council; passage normally requires three separate readings and approval of at least 13 House members; following both House and Council agreement, assent is required by the lieutenant governor on behalf of the Crown; the constitution has been expanded and amended many times, last in 2019

Legal system: the laws of the UK apply where applicable and include Manx statutes

Citizenship: see United Kingdom

Suffrage: 16 years of age; universal

Executive branch: *chief of state:* Lord of Mann Queen ELIZABETH II (since 6 February 1952); represented by Lieutenant Governor Sir Richard GOZNEY (since 27 May 2016)
head of government: Chief Minister Howard QUAYLE (since 4 October 2016)
cabinet: Council of Ministers appointed by the lieutenant governor
elections/appointments: the monarchy is hereditary; lieutenant governor appointed by the monarch; chief minister indirectly elected by the Tynwald for a 5-year term (eligible for second term); election last held on 4 October 2016 (next to be held in 2021)
election results: Howard QUAYLE (independent) elected chief minister; Tynwald vote—21 of 33

Legislative branch: *description:* bicameral Tynwald or the High Court of Tynwald consists of: Legislative Council (11 seats; includes the President of Tynwald, 2 ex-officio members—the Lord Bishop of Sodor and Man and the attorney general (non-voting)—and 8 members indirectly elected by the House of Keys with renewal of 4 members every 2 years; elected members serve 4-year terms)
House of Keys (24 seats; 2 members directly elected by simple majority vote from 12 constituencies to serve 5-year terms)
elections: Legislative Council—last held 28 February 2018 (next to be held 12 March 2020)
House of Keys—last held on 22 September 2016 (next to be held on 23 September 2021)
election results: Legislative Council—composition—men 6, women 5, percent of women 45.5%
House of Keys—percent of vote by party—Liberal Vannin 6.4%, independent 92.3%, other 1.3%; seats by party—Liberal Vannin 3, independent 21; composition—men 19, women 5, percent of women 20.8%; note—total Tynwald percent of women 28.6%
note: as of January 2019, seats by party—Liberal Vannin 2, independent 22

Judicial branch: *highest courts:* Isle of Man High Court of Justice (consists of 3 permanent judges or "deemsters" and 1 judge of appeal; organized into the Staff of Government Division or Court of Appeal and the Civil Division); the Court of General Gaol Delivery is not formally part of the High Court but is administered as though part of the High Court and deals with serious criminal cases; note—appeals beyond the Court of Appeal are referred to the Judicial Committee of the Privy Council (in London)
judge selection and term of office: deemsters appointed by the Lord Chancellor of England on the nomination of the lieutenant governor; deemsters can serve until age 70
subordinate courts: High Court; Court of Summary Gaol Delivery; Summary Courts; Magistrate's Court; specialized courts

Political parties and leaders: Liberal Vannin Party [Kate BEECROFT]

MANX LABOR PARTY

Mec Vannin [Mark KERMODE] (sometimes referred to as the Manx Nationalist Party)
note: most members sit as independents

International organization participation: UPU

Diplomatic representation in the US: none (British crown dependency)

Diplomatic representation from the US: none (British crown dependency)

Flag description: red with the Three Legs of Man emblem (triskelion), in the center; the three legs are joined at the thigh and bent at the knee; in order to have the toes pointing clockwise on both sides of the flag, a two-sided emblem is used; the flag is based on the coat of arms of the last recognized Norse King of Mann, Magnus III (r. 1252-65); the triskelion has its roots in an early Celtic sun symbol

National symbol(s): *triskelion (a motif of three legs); national colors:* red, white

National anthem: *name:* "Arrane Ashoonagh dy Vannin" (O Land of Our Birth)
lyrics/music: William Henry GILL [English], John J. KNEEN [Manx]/traditional
note: adopted 2003, in use since 1907; serves as a local anthem; as a British Crown dependency, "God Save the Queen" is official (see United Kingdom) and is played when the sovereign, members of the royal family, or the lieutenant governor are present

ECONOMY

Economy—overview: Financial services, manufacturing, and tourism are key sectors of the economy. The government offers low taxes and other incentives to high-technology companies and financial institutions to locate on the island; this has paid off in expanding employment opportunities in high-income industries. As a result, agriculture and fishing, once the mainstays of the economy, have declined in their contributions to GDP. The Isle of Man also attracts online gambling sites and the film industry. Online gambling sites provided about 10% of the islands income in 2014. The Isle of Man currently enjoys free access to EU markets and trade is mostly with the UK. The Isle of Man's trade relationship with the EU derives from the United Kingdom's EU membership and will need to be renegotiated in light of the United Kingdom's decision to withdraw from the bloc. A transition period is expected to allow the free movement of goods and agricultural products to the EU until the end of 2020 or until a new settlement is negotiated.

GDP (purchasing power parity): $6.792 billion (2015 est.)
$7.428 billion (2014 est.)
$6.298 billion (2013 est.)
note: data are in 2014 US dollars
country comparison to the world: 169

GDP (official exchange rate): $6.792 billion (2015 est.)

GDP—real growth rate: -8.6% (2015 est.)
17.9% (2014 est.)
2.1% (2010 est.)
country comparison to the world: 221

GDP—per capita (PPP): $84,600 (2014 est.)
$86,200 (2013 est.)
$73,700 (2012 est.)
country comparison to the world: 8

GDP—composition, by sector of origin:
agriculture: 1% (FY12/13 est.)
industry: 13% (FY12/13 est.)
services: 86% (FY12/13 est.)

Agriculture—products: cereals, vegetables; cattle, sheep, pigs, poultry
Industries: financial services, light manufacturing, tourism

Labor force: 41,790 (2006)
country comparison to the world: 195

Labor force—by occupation: *manufacturing:* 5% (2006 est.)
construction: 8% (2006 est.)
tourism: 1% (2006 est.)
transport and communications: 9% (2006 est.)
agriculture, forestry, and fishing: 2% (2006 est.)
gas, electricity, and water: 1% (2006 est.)
wholesale and retail distribution: 11% (2006 est.)
professional and scientific services: 20% (2006 est.)
public administration: 7% (2006 est.)
banking and finance: 23% (2006 est.)
entertainment and catering: 5% (2006 est.)
miscellaneous services: 8% (2006 est.)

Unemployment rate: 1.1% (2017 est.)
2% (April 2011 est.)

country comparison to the world: 11

Population below poverty line: NA

Household income or consumption by percentage share: *lowest 10%:* NA
highest 10%: NA

Budget: *revenues:* 965 million (FY05/06 est.)
expenditures: 943 million (FY05/06 est.)

Taxes and other revenues: 14.2% (of GDP) (FY05/06 est.)
country comparison to the world: 201

Budget surplus (+) or deficit (-): 0.3% (of GDP) (FY05/06 est.)
country comparison to the world: 40

Fiscal year: 1 April—31 March

Inflation rate (consumer prices): 4.1% (2017 est.)
1% (2016 est.)
country comparison to the world: 160

Exports: NA

Exports—commodities: tweeds, herring, processed shellfish, beef, lamb

Imports: NA

Imports—commodities: timber, fertilizers, fish

Debt—external: NA

Exchange rates: Manx pounds (IMP) per US dollar -
0.7836 (2017 est.)
0.738 (2016 est.)
0.738 (2015)
0.6542 (2014)
0.6472 (2013 est.)

ENERGY

Electricity access: *electrification—total population:* 100% (2020)

COMMUNICATIONS

Telecommunication systems: *domestic:* landline, telefax, mobile cellular telephone system
international: country code—44; fiber-optic cable, microwave radio relay, satellite earth station, submarine cable

note: the COVID-19 outbreak is negatively impacting telecommunications production and supply chains globally; consumer spending on telecom devices and services has also slowed due to the pandemic's effect on economies worldwide; overall progress towards improvements in all facets of the telecom industry—mobile, fixed-line, broadband, submarine cable and satellite—has moderated

Broadcast media: national public radio broadcasts over 3 FM stations and 1 AM station; 2 commercial broadcasters operating with 1 having multiple FM stations; receives radio and TV services via relays from British TV and radio broadcasters

Internet country code: .im

TRANSPORTATION

Civil aircraft registration country code prefix: M (2016)

Airports: 1 (2013)
country comparison to the world: 224

Airports—with paved runways: *total:* 1 (2019)
1,524 to 2,437 m: 1

Railways: *total:* 63 km (2008)
narrow gauge: 6 km 1.076-m gauge (6 km electrified) (2008)
57 0.914-m gauge (29 km electrified)
note: primarily summer tourist attractions
country comparison to the world: 130

Roadways: *total:* 500 km (2008)
country comparison to the world: 196

Ports and terminals: *major seaport(s):* Douglas, Ramsey

MILITARY AND SECURITY

Military—note: defense is the responsibility of the UK

TRANSNATIONAL ISSUES

Disputes—international: none

ISRAEL

INTRODUCTION

Background: The State of Israel was declared in 1948, after Britain withdrew from its mandate of Palestine. The UN proposed partitioning the area into Arab and Jewish states, and Arab armies that rejected the UN plan were defeated. Israel was admitted as a member of the UN in 1949 and saw rapid population growth, primarily due to migration from Europe and the Middle East, over the following years. Israel fought wars against its Arab neighbors in 1967 and 1973, followed by peace treaties with Egypt in 1979 and Jordan in 1994. Israel took control of the West Bank and Gaza

Strip in the 1967 war, and subsequently administered those territories through military authorities. Israel and Palestinian officials signed a number of interim agreements in the 1990s that created an interim period of Palestinian self-rule in the West Bank and Gaza. Israel withdrew from Gaza in 2005. While the most recent formal efforts to negotiate final status issues occurred in 2013-2014, the US continues its efforts to advance peace. Immigration to Israel continues, with 28,600 new immigrants, mostly Jewish, in 2016. The Israeli economy has undergone a dramatic transformation in the last 25 years, led by cutting-edge, high-tech sectors. Offshore gas discoveries in the

Mediterranean, most notably in the Tamar and Leviathan gas fields, place Israel at the center of a potential regional natural gas market. However, longer-term structural issues such as low labor force participation among minority populations, low workforce productivity, high costs for housing and consumer staples, and a lack of competition, remain a concern for many Israelis and an important consideration for Israeli politicians. Prime Minister Benjamin NETANYAHU has led the Israeli Government since 2009; he formed a center-right coalition following the 2015 elections. Three Knesset elections held in April and September 2019 and March 2020 all failed

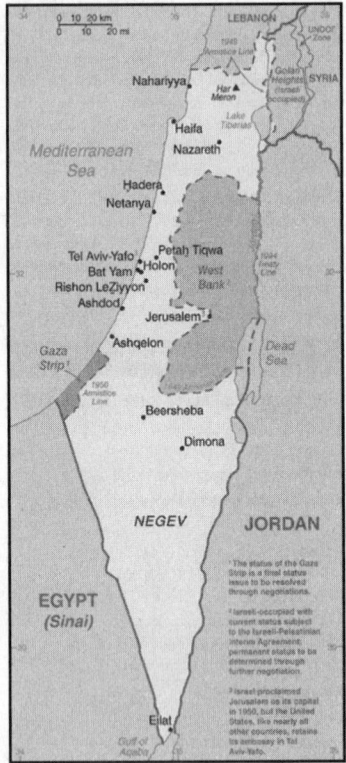

Dead Sea), Lebanon 107 km, Syria 79 km, West Bank 278 km

Coastline: 273 km

Maritime claims: *territorial sea:* 12 nm *continental shelf:* to depth of exploitation

Climate: temperate; hot and dry in southern and eastern desert areas

Terrain: Negev desert in the south; low coastal plain; central mountains; Jordan Rift Valley

Elevation: *mean elevation:* 508 m note—does not include elevation data from the Golan Heights *lowest point:* Dead Sea -431 m *highest point:* Mitspe Shlagim 2,224 m; note—this is the highest named point, the actual highest point is an unnamed dome slightly to the west of Mitspe Shlagim at 2,236 m; both points are on the northeastern border of Israel, along the southern end of the Anti-Lebanon mountain range

Natural resources: timber, potash, copper ore, natural gas, phosphate rock, magnesium bromide, clays, sand

Land use: *agricultural land:* 23.8% (2011 est.) *arable land:* 13.7% (2011 est.) / permanent crops: 3.8% (2011 est.) / permanent pasture: 6.3% (2011 est.) *forest:* 7.1% (2011 est.) *other:* 69.1% (2011 est.)

Irrigated land: 2,250 sq km (2012)

Population distribution: population concentrated in and around Tel-Aviv, as well as around the Sea of Galilee; the south remains sparsely populated with the exception of the shore of the Gulf of Aqaba

Natural hazards: sandstorms may occur during spring and summer; droughts; periodic earthquakes

Environment—current issues: limited arable land and restricted natural freshwater resources; desertification; air pollution from industrial and vehicle emissions; groundwater pollution from industrial and domestic waste, chemical fertilizers, and pesticides

Environment—international agreements: *party to:* Biodiversity, Climate Change, Climate Change-Kyoto Protocol, Desertification, Endangered Species, Hazardous Wastes, Ozone Layer Protection, Ship Pollution, Wetlands, Whaling *signed, but not ratified:* Marine Life Conservation

Geography—note: *note 1:* Lake Tiberias (Sea of Galilee) is an important freshwater source; the Dead Sea is the second saltiest body of water in the world (after Lake Assal in Djibouti) *note 2:* the Malham Cave in Mount Sodom is the world's longest salt cave at 10 km (6 mi); its survey is not complete and its length will undoubtedly increase; Mount Sodom is actually a hill some 220 m (722 ft) high that is 80% salt (multiple salt layers covered by a veneer of rock) *note 3:* in March 2019, there were 380 Israeli settlements, to include 213 settlements and 132 outposts in the West Bank, and 35 settlements in East Jerusalem; there are no Israeli settlements in the Gaza Strip, as all were evacuated in 2005 (2019)

to form a new government. The political stalemate was finally resolved in April 2020 when NETANYAHU and Blue and White party leader Benny GANTZ signed an agreement to form a coalition government. Under the terms of the agreement, NETANYAHU would remain as prime minister until October 2021 when GANTZ would succeed him. On 15 September 2020, Israel signed a peace agreement, the Abraham Accords – brokered by the US – with Bahrain and the United Arab Emirates in Washington DC. Israel signed similar peace agreements with Egypt (1979) and Jordan (1994).

GEOGRAPHY

Location: Middle East, bordering the Mediterranean Sea, between Egypt and Lebanon

Geographic coordinates: 31 30 N, 34 45 E

Map references: Middle East

Area: *total:* 21,937 sq km *land:* 21,497 sq km *water:* 440 sq km *country comparison to the world:* 153

Area—comparative: slightly larger than New Jersey

Land boundaries: *total:* 1,065 km *border countries (6):* Egypt 206 km, Gaza Strip 59 km, Jordan 336 km (20 km are within the

PEOPLE AND SOCIETY

Population: 8,675,475 (includes populations of the Golan Heights or Golan Sub-District and also East Jerusalem, which was annexed by Israel after 1967) (July 2020 est.) *note:* approximately 22,900 Israeli settlers live in the Golan Heights (2018); approximately 215,900 Israeli settlers live in East Jerusalem (2017) *country comparison to the world:* 98

Nationality: *noun:* Israeli(s) *adjective:* Israeli

Ethnic groups: Jewish 74.4% (of which Israel-born 76.9%, Europe/America/Oceania-born 15.9%, Africa-born 4.6%, Asia-born 2.6%), Arab 20.9%, other 4.7% (2018 est.)

Languages: Hebrew (official), Arabic (special status under Israeli law), English (most commonly used foreign language)

Religions: Jewish 74.3%, Muslim 17.8%, Christian 1.9%, Druze 1.6%, other 4.4% (2018 est.)

MENA religious affiliation: *Age structure:* 0-14 years: 26.76% (male 1,187,819/female 1,133,365) 15-24 years: 15.67% (male 694,142/female 665,721) 25-54 years: 37.2% (male 1,648,262/female 1,579,399) 55-64 years: 8.4% (male 363,262/female 365,709) 65 years and over: 11.96% (male 467,980/female 569,816) (2020 est.)

Dependency ratios: *total dependency ratio:* 67.3 *youth dependency ratio:* 46.6 *elderly dependency ratio:* 20.8 *potential support ratio:* 4.8 (2020 est.)

Median age: *total:* 30.4 years *male:* 29.8 years *female:* 31 years (2020 est.) *country comparison to the world:* 121

Population growth rate: 1.46% (2020 est.) *country comparison to the world:* 74

Birth rate: 17.6 births/1,000 population (2020 est.) *country comparison to the world:* 94

Death rate: 5.3 deaths/1,000 population (2020 est.) *country comparison to the world:* 190

Net migration rate: 2.1 migrant(s)/1,000 population (2020 est.) *country comparison to the world:* 48

Population distribution: population concentrated in and around Tel-Aviv, as well as around the Sea of Galilee; the south remains sparsely populated with the exception of the shore of the Gulf of Aqaba

Urbanization: *urban population:* 92.6% of total population (2020) *rate of urbanization:* 1.64% annual rate of change (2015-20 est.) *total population growth rate v. urban population growth rate, 2000-2030:* Major urban areas—population: 4.181 million Tel Aviv-Yafo, 1.147 million Haifa, 932,000 JERUSALEM (capital) (2020)

Sex ratio: *at birth:* 1.05 male(s)/female
0-14 years: 1.05 male(s)/female
15-24 years: 1.04 male(s)/female
25-54 years: 1.04 male(s)/female
55-64 years: 0.99 male(s)/female
65 years and over: 0.82 male(s)/female
total population: 1.01 male(s)/female (2020 est.)

Mother's mean age at first birth: 27.6 years (2017 est.)

Maternal mortality rate: 3 deaths/100,000 live births (2017 est.)
country comparison to the world: 179

Infant mortality rate: *total:* 3.3 deaths/1,000 live births
male: 3.3 deaths/1,000 live births
female: 3.3 deaths/1,000 live births (2020 est.)
country comparison to the world: 208

Life expectancy at birth: *total population:* 83 years
male: 81.1 years
female: 85 years (2020 est.)
country comparison to the world: 10

Total fertility rate: 2.59 children born/woman (2020 est.)
country comparison to the world: 68

Drinking water source:

improved: *urban:* 100% of population
rural: 100% of population
total: 100% of population

unimproved: *urban:* 0% of population
rural: 0% of population
total: 0% of population (2017 est.)

Current Health Expenditure: 7.4% (2017)

Physicians density: 3.48 physicians/1,000 population (2017)

Hospital bed density: 3 beds/1,000 population (2017)

Sanitation facility access:

improved: *urban:* 100% of population
rural: 100% of population
total: 100% of population

unimproved: *urban:* 0% of population
rural: 0% of population
total: 0% of population (2017 est.)

HIV/AIDS—adult prevalence rate: 0.2% (2018)
country comparison to the world: 100

HIV/AIDS—people living with HIV/AIDS: 9,000 (2018)
country comparison to the world: 108

HIV/AIDS—deaths: <100 (2018)

Obesity—adult prevalence rate: 26.1% (2016)
country comparison to the world: 44

Education expenditures: 5.8% of GDP (2016)
country comparison to the world: 33

Literacy: *definition:* age 15 and over can read and write
total population: 97.8%
male: 98.7%
female: 96.8% (2011)

School life expectancy (primary to tertiary education):

total: 16 years
male: 16 years
female: 17 years (2018)

Unemployment, youth ages 15-24:
total: 7.2%
male: 6.9%
female: 7.4% (2018 est.)
country comparison to the world: 149

GOVERNMENT

Country name: *conventional long form:* State of Israel
conventional short form: Israel
local long form: Medinat Yisra'el
local short form: Yisra'el
etymology: named after the ancient Kingdom of Israel; according to Biblical tradition, the Jewish patriarch Jacob received the name "Israel" ("He who struggles with God") after he wrestled an entire night with an angel of the Lord; Jacob's 12 sons became the ancestors of the Israelites, also known as the Twelve Tribes of Israel, who formed the Kingdom of Israel

Government type: parliamentary democracy

Capital: *name:* Jerusalem; note—the US recognized Jerusalem as Israel's capital in December 2017 without taking a position on the specific boundaries of Israeli sovereignty
geographic coordinates: 31 46 N, 35 14 E
time difference: UTC+2 (7 hours ahead of Washington, DC, during Standard Time)
daylight saving time: +1hr, Friday before the last Sunday in March; ends the last Sunday in October
etymology: Jerusalem's settlement may date back to 2800 B.C.; it is named Urushalim in Egyptian texts of the 14th century B.C.; "uru-shalim" likely means "foundation of [by] the god Shalim", and derives from Hebrew/Semitic "yry", "to found or lay a cornerstone", and Shalim, the Canaanite god of dusk and the nether world; Shalim was associated with sunset and peace and the name is based on the same S-L-M root from which Semitic words for "peace" are derived (Salam or Shalom in modern Arabic and Hebrew); this confluence has thus led to naming interpretations such as "The City of Peace" or "The Abode of Peace"

Administrative divisions: 6 districts (mehozot, singular—mehoz); Central, Haifa, Jerusalem, Northern, Southern, Tel Aviv

Independence: 14 May 1948 (following League of Nations mandate under British administration)

National holiday: Independence Day, 14 May (1948); note— Israel declared independence on 14 May 1948, but the Jewish calendar is lunar and the holiday may occur in April or May

Constitution: *history:* no formal constitution; some functions of a constitution are filled by the Declaration of Establishment (1948), the Basic Laws, and the Law of Return (as amended)
amendments: proposed by Government of Israel ministers or by the Knesset; passage requires a majority vote of Knesset members and subject to Supreme Court judicial review; 11 of the 13 Basic

Laws have been amended at least once, latest in 2020

Legal system: mixed legal system of English common law, British Mandate regulations, and Jewish, Christian, and Muslim religious laws

International law organization participation: has not submitted an ICJ jurisdiction declaration; withdrew acceptance of ICCt jurisdiction in 2002

Citizenship: *citizenship by birth:* no
citizenship by descent only: at least one parent must be a citizen of Israel
dual citizenship recognized: yes, but naturalized citizens are not allowed to maintain dual citizenship
residency requirement for naturalization: 3 out of the 5 years preceding the application for naturalization
note: Israeli law (Law of Return, 5 July 1950) provides for the granting of citizenship to any Jew—defined as a person being born to a Jewish mother or having converted to Judaism while renouncing any other religion—who immigrates to and expresses a desire to settle in Israel on the basis of the Right of aliyah; the 1970 amendment of this act extended the right to family members including the spouse of a Jew, any child or grandchild, and the spouses of children and grandchildren

Suffrage: 18 years of age; universal; 17 years of age for municipal elections

Executive branch: *chief of state:* President Reuben RIVLIN (since 27 July 2014)
head of government: Prime Minister Binyamin NETANYAHU (since 31 March 2009)
cabinet: Cabinet selected by prime minister and approved by the Knesset
elections/appointments: president indirectly elected by the Knesset for a single 7-year term; election last held on 10 June 2014 (next to be held in 2021); following legislative elections, the president, in consultation with party leaders, tasks a Knesset member (usually the member of the largest party) with forming a government
election results: Reuven RIVLIN elected president in second round; Knesset vote—Reuven RIVLIN (Likud) 63, Meir SHEETRIT (The Movement) 53, other/invalid 4; note—on 20 May 2020 – after three national elections, each ending in failed bids by Prime Minister NETANYAHU and Blue and White party leader Benny GANTZ to form a coalition government, both signed an agreement on the formation of a national emergency government in which NETANYAHU continues as prime minister for 18 months when GANTZ will replace him

Legislative branch: *description:* unicameral Knesset (120 seats; members directly elected in a single nationwide constituency by closed-list proportional representation vote, with a 3.25% threshold to gain representation; members serve 4-year terms)
elections: last held on 2 March 2020 (next to be held in 2024)
election results: percent by party (preliminary)—Likud 29.2%, Blue and White 26.4%, Joint List 13.1%, Shas 7.7%, United Torah Judaism 6.2%,

Yisrael Beiteinu 5.9%, Labor-Gesher-Meretz 5.7%, Yamina 5%, other 0.8%; seats by party (preliminary)—Likud 36, Blue and White 33, Joint List 15, Shas 9, United Torah Judaism 7, Yisrael Beiteinu 7, Labor- Gesher Meretz 7, Yamina 6; composition—NA

Judicial branch: *highest courts:* Supreme Court (consists of the president, deputy president, 13 justices, and 2 registrars) and normally sits in panels of 3 justices; in special cases, the panel is expanded with an uneven number of justices

judge selection and term of office: judges selected by the 9-member Judicial Selection Committee, consisting of the Minister of Justice (chair), the president of the Supreme Court, two other Supreme Court justices, 1 other Cabinet minister, 2 Knesset members, and 2 representatives of the Israel Bar Association; judges can serve up to mandatory retirement at age 70

subordinate courts: district and magistrate courts; national and regional labor courts; family and juvenile courts; special and religious courts

Political parties and leaders: Democratic Union [Nitzan HOROWITZ] (alliance includes Democratic Israel, Meretz, Green Movement)
Joint List [Ayman ODEH] (alliance includes Hadash, Ta'al, United Arab List, Balad)
Kahol Lavan [Benny GANTZ] (alliance includes Israeli Resilience, Yesh Atid, Telem)
Labor-Gesher [Amir PERETZ]
Likud [Binyamin NETANYAHU]
Otzma Yehudit [Itamar BEN-GVIR]
SHAS [Arye DERI]
United Torah Judaism, or UTJ [Yaakov LITZMAN] (alliance includes Agudat Israel and Degel HaTorah)
Yamina [Ayelet SHAKED]
Yisrael Beiteinu [Avigdor LIEBERMAN]
Zehut [Moshe FEIGLIN]

International organization participation: BIS, BSEC (observer), CE (observer), CERN, CICA, EBRD, FAO, IADB, IAEA, IBRD, ICAO, ICC (national committees), ICRM, IDA, IFAD, IFC, IFRCS, ILO, IMF, IMO, IMSO, Interpol, IOC, IOM, IPU, ISO, ITSO, ITU, ITUC (NGOs), MIGA, OAS (observer), OECD, OPCW (signatory), OSCE (partner), Pacific Alliance (observer), Paris Club, PCA, SELEC (observer), UN, UNCTAD, UNESCO, UNHCR, UNIDO, UNWTO, UPU, WCO, WHO, WIPO, WMO, WTO

Diplomatic representation in the US: *chief of mission:* Ambassador Ron DERMER (since 3 December 2013)
chancery: 3514 International Drive NW, Washington, DC 20008
telephone: [1] (202) 364-5500
FAX: [1] (202) 364-5607
consulate(s) general: Atlanta, Boston, Chicago, Houston, Los Angeles, Miami, New York, Philadelphia, San Francisco

Diplomatic representation from the US: *chief of mission:* Ambassador David M. FRIEDMAN (since 23 May 2017)
telephone: [972] (2) 630-4000

embassy: David Flusser St.14, Jerusalem, 9378322
FAX: NA
note: on 14 May 2018, the US Embassy relocated to Jerusalem from Tel Aviv; on 4 March 2019, Consulate General Jerusalem merged into US Embassy Jerusalem to form a single diplomatic mission

Flag description: white with a blue hexagram (six-pointed linear star) known as the Magen David (Star of David or Shield of David) centered between two equal horizontal blue bands near the top and bottom edges of the flag; the basic design resembles a traditional Jewish prayer shawl (tallit), which is white with blue stripes; the hexagram as a Jewish symbol dates back to medieval times
note: the Israeli flag proclamation states that the flag colors are sky blue and white, but the exact shade of blue has never been set and can vary from a light to a dark blue

National symbol(s): *Star of David (Magen David), menorah (seven-branched lampstand); national colors:* blue, white

National anthem: *name:* "Hatikvah" (The Hope)
lyrics/music: Naftali Herz IMBER/traditional, arranged by Samuel COHEN
note: adopted 2004, unofficial since 1948; used as the anthem of the Zionist movement since 1897; the 1888 arrangement by Samuel COHEN is thought to be based on the Romanian folk song "Carul cu boi" (The Ox Driven Cart)

ECONOMY

Economy—overview: Israel has a technologically advanced free market economy. Cut diamonds, high-technology equipment, and pharmaceuticals are among its leading exports. Its major imports include crude oil, grains, raw materials, and military equipment. Israel usually posts sizable trade deficits, which are offset by tourism and other service exports, as well as significant foreign investment inflows.

Between 2004 and 2013, growth averaged nearly 5% per year, led by exports. The global financial crisis of 2008-09 spurred a brief recession in Israel, but the country entered the crisis with solid fundamentals, following years of prudent fiscal policy and a resilient banking sector. Israel's economy also weathered the 2011 Arab Spring because strong trade ties outside the Middle East insulated the economy from spillover effects.

Slowing domestic and international demand and decreased investment resulting from Israel's uncertain security situation reduced GDP growth to an average of roughly 2.8% per year during the period 2014-17. Natural gas fields discovered off Israel's coast since 2009 have brightened Israel's energy security outlook. The Tamar and Leviathan fields were some of the world's largest offshore natural gas finds in the last decade. Political and regulatory issues have delayed the development of the massive Leviathan field, but production from Tamar provided a 0.8% boost to Israel's GDP in 2013 and a 0.3% boost in 2014. One of the most carbon intense OECD countries, Israel generates

about 57% of its power from coal and only 2.6% from renewable sources.

Income inequality and high housing and commodity prices continue to be a concern for many Israelis. Israel's income inequality and poverty rates are among the highest of OECD countries, and there is a broad perception among the public that a small number of "tycoons" have a cartel-like grip over the major parts of the economy. Government officials have called for reforms to boost the housing supply and to increase competition in the banking sector to address these public grievances. Despite calls for reforms, the restricted housing supply continues to impact younger Israelis seeking to purchase homes. Tariffs and non-tariff barriers, coupled with guaranteed prices and customs tariffs for farmers kept food prices high in 2016. Private consumption is expected to drive growth through 2018, with consumers benefitting from low inflation and a strong currency.

In the long term, Israel faces structural issues including low labor participation rates for its fastest growing social segments—the ultraorthodox and Arab-Israeli communities. Also, Israel's progressive, globally competitive, knowledge-based technology sector employs only about 8% of the workforce, with the rest mostly employed in manufacturing and services—sectors which face downward wage pressures from global competition. Expenditures on educational institutions remain low compared to most other OECD countries with similar GDP per capita.

GDP (purchasing power parity): $317.1 billion (2017 est.)
$307 billion (2016 est.)
$295.3 billion (2015 est.)
note: data are in 2017 dollars
country comparison to the world: 54

GDP (official exchange rate): $350.7 billion (2017 est.)

GDP—real growth rate: 3.28% (2019 est.)
3.69% (2018 est.)
3.63% (2017 est.)
country comparison to the world: 91

GDP—per capita (PPP): $36,400 (2017 est.)
$35,900 (2016 est.)
$35,200 (2015 est.)
note: data are in 2017 dollars
country comparison to the world: 55

Gross national saving: 23.6% of GDP (2017 est.)
24.2% of GDP (2016 est.)
25% of GDP (2015 est.)
country comparison to the world: 72

GDP—composition, by end use:
household consumption: 55.1% (2017 est.)
government consumption: 22.8% (2017 est.)
investment in fixed capital: 20.1% (2017 est.)
investment in inventories: 0.7% (2017 est.)
exports of goods and services: 28.9% (2017 est.)
imports of goods and services: -27.5% (2017 est.)

GDP—composition, by sector of origin:
agriculture: 2.4% (2017 est.)
industry: 26.5% (2017 est.)
services: 69.5% (2017 est.)

Agriculture—products: citrus, vegetables, cotton; beef, poultry, dairy products

Industries: high-technology products (including aviation, communications, computer-aided design and manufactures, medical electronics, fiber optics), wood and paper products, potash and phosphates, food, beverages, and tobacco, caustic soda, cement, pharmaceuticals, construction, metal products, chemical products, plastics, cut diamonds, textiles, footwear

Industrial production growth rate: 3.5% (2017 est.)
country comparison to the world: 86

Labor force: 3.893 million (2020 est.)
country comparison to the world: 91

Labor force—by occupation: *agriculture:* 1.1% *industry:* 17.3%
services: 81.6% (2015 est.)

Unemployment rate: 3.81% (2019 est.)
4% (2018 est.)
country comparison to the world: 56

Population below poverty line: 22% (2014 est.) (2014 est.)
note: Israel's poverty line is $7.30 per person per day

Household income or consumption by percentage share:
lowest 10%: 1.7%
highest 10%: 31.3% (2010)

Budget: *revenues:* 93.11 billion (2017 est.)
expenditures: 100.2 billion (2017 est.)

Taxes and other revenues: 26.5% (of GDP) (2017 est.)
country comparison to the world: 110

Budget surplus (+) or deficit (-): -2% (of GDP) (2017 est.)
country comparison to the world: 105

Public debt: 60.9% of GDP (2017 est.)
62.3% of GDP (2016 est.)
country comparison to the world: 73

Fiscal year: calendar year

Inflation rate (consumer prices): 0.2% (2017 est.)
-0.5% (2016 est.)
country comparison to the world: 17

Current account balance: $13.411 billion (2019 est.)
$7.888 billion (2018 est.)
country comparison to the world: 20

Exports: $58.67 billion (2017 est.)
$56.17 billion (2016 est.)
country comparison to the world: 48

Exports—partners: US 28.8%, UK 8.2%, Hong Kong 7%, China 5.4%, Belgium 4.5% (2017)

Exports—commodities: machinery and equipment, software, cut diamonds, agricultural products, chemicals, textiles and apparel

Imports: $68.61 billion (2017 est.)
$63.9 billion (2016 est.)
country comparison to the world: 46

Imports—commodities: raw materials, military equipment, investment goods, rough diamonds, fuels, grain, consumer goods

Imports—partners: US 11.7%, China 9.5%, Switzerland 8%, Germany 6.8%, UK 6.2%, Belgium 5.9%, Netherlands 4.2%, Turkey 4.2%, Italy 4% (2017)

Reserves of foreign exchange and gold: $113 billion (31 December 2017 est.)
$95.45 billion (31 December 2016 est.)
country comparison to the world: 23

Debt—external: $88.66 billion (31 December 2017 est.)
$87.96 billion (31 December 2016 est.)
country comparison to the world: 54

Exchange rates: new Israeli shekels (ILS) per US dollar—
3.606 (2017 est.)
3.8406 (2016 est.)
3.8406 (2015 est.)
3.8869 (2014 est.)
3.5779 (2013 est.)

ENERGY

Electricity access: *electrification—total population:* 100% (2020)

Electricity—production: 63.09 billion kWh (2016 est.)
country comparison to the world: 46

Electricity—consumption: 55 billion kWh (2016 est.)
country comparison to the world: 47

Electricity—exports: 5.2 billion kWh (2016 est.)
country comparison to the world: 35

Electricity—imports: 0 kWh (2016 est.)
country comparison to the world: 161

Electricity—installed generating capacity: 17.59 million kW (2016 est.)
country comparison to the world: 48

Electricity—from fossil fuels: 95% of total installed capacity (2016 est.)
country comparison to the world: 44

Electricity—from nuclear fuels: 0% of total installed capacity (2017 est.)
country comparison to the world: 113

Electricity—from hydroelectric plants: 0% of total installed capacity (2017 est.)
country comparison to the world: 179

Electricity—from other renewable sources: 5% of total installed capacity (2017 est.)
country comparison to the world: 107

Crude oil—production: 390 bbl/day (2018 est.)
country comparison to the world: 94

Crude oil—exports: 0 bbl/day (2017 est.)
country comparison to the world: 141

Crude oil—imports: 231,600 bbl/day (2017 est.)
country comparison to the world: 28

Crude oil—proved reserves: 12.73 million bbl (1 January 2018 est.)
country comparison to the world: 87

Refined petroleum products—production: 294,300 bbl/day (2017 est.)
country comparison to the world: 42

Refined petroleum products—consumption: 242,200 bbl/day (2017 est.)
country comparison to the world: 52

Refined petroleum products—exports: 111,700 bbl/day (2017 est.)
country comparison to the world: 39

Refined petroleum products—imports: 98,860 bbl/day (2017 est.)
country comparison to the world: 54

Natural gas—production: 9.826 billion cu m (2017 est.)
country comparison to the world: 42

Natural gas—consumption: 9.995 billion cu m (2017 est.)
country comparison to the world: 49

Natural gas—exports: 0 cu m (2017 est.)
country comparison to the world: 126

Natural gas—imports: 509.7 million cu m (2017 est.)
country comparison to the world: 66

Natural gas—proved reserves: 176 billion cu m (1 January 2018 est.)
country comparison to the world: 45

Carbon dioxide emissions from consumption of energy: 73.82 million Mt (2017 est.)
country comparison to the world: 49

COMMUNICATIONS

Telephones—fixed lines:
total subscriptions: 3,050,693
subscriptions per 100 inhabitants: 35.68 (2019 est.)
country comparison to the world: 44

Telephones—mobile cellular:
total subscriptions: 10,839,024
subscriptions per 100 inhabitants: 126.77 (2019 est.)
country comparison to the world: 81

Telecommunication systems: *general assessment:* one of the most highly developed system in the Middle East; mobile broadband 100% population penetration; consumers enjoy inexpensive 3G and 4G cellular service; fixed broadband available to 99% of all households; 6 mobile operators in fierce competition; in 2019 govt. began process of 5G licensing (2020)

domestic: good system of coaxial cable and microwave radio relay; all systems are digital; competition among both fixed-line and mobile cellular providers results in good coverage countrywide; fixed-line 36 per 100 and 127 per 100 for mobile-cellular subscriptions (2019)

international: country code—972; landing points for the MedNautilus Submarine System, Tameres North, Jonah and Lev Submarine System, submarine cables that provide links to Europe, Cyprus, and parts of the Middle East; satellite earth stations—3 Intelsat (2 Atlantic Ocean and 1 Indian Ocean) (2019)

note: the COVID-19 outbreak is negatively impacting telecommunications production and supply chains globally; consumer spending on telecom devices and services has also slowed due to the pandemic's effect on economies worldwide; overall progress towards improvements in all facets of the telecom industry—mobile, fixed-line, broadband, submarine cable and satellite—has moderated

Broadcast media: the Israel Broadcasting Corporation (est 2015) broadcasts on 3 channels, two in Hebrew and the other in Arabic; multi-channel satellite and cable TV packages provide access to foreign channels; the Israeli Broadcasting Corporation broadcasts on 8 radio networks with multiple repeaters and Israel Defense Forces Radio broadcasts over multiple stations; about 15 privately owned radio stations; overall more than 100 stations and repeater stations (2019)

Internet country code: .il

Internet users: *total:* 6,873,037
percent of population: 81.58% (July 2018 est.)
country comparison to the world: 73

Broadband—fixed subscriptions:
total: 2.41 million
subscriptions per 100 inhabitants: 29 (2018 est.)
country comparison to the world: 51

TRANSPORTATION

National air transport system:
number of registered air carriers: 6 (2020)
inventory of registered aircraft operated by air carriers: 64
annual passenger traffic on registered air carriers: 7,404,373 (2018)
annual freight traffic on registered air carriers: 994.54 million mt-km (2018)

Civil aircraft registration country code prefix: 4X (2016)

Airports: 42 (2020)
country comparison to the world: 100

Airports—with paved runways: *total:* 33 (2019)
over 3,047 m: 3
2,438 to 3,047 m: 5
1,524 to 2,437 m: 5
914 to 1,523 m: 12
under 914 m: 8

Airports—with unpaved runways:
total: 9 (2020)
914 to 1,523 m: 3
under 914 m: 6

Heliports: 3 (2013)

Pipelines: 763 km gas, 442 km oil, 261 km refined products (2013)

Railways: *total:* 1,384 km (2014)

standard gauge: 1,384 km 1.435-m gauge (2014)
country comparison to the world: 83

Roadways: *total:* 19,555 km (2017)
paved: 19,555 km (includes 449 km of expressways) (2017)
country comparison to the world: 116

Merchant marine: *total:* 40
by type: bulk carrier 5, general cargo 3, oil tanker 3, other 29 (2019)
country comparison to the world: 121

Ports and terminals: *major seaport(s):* Ashdod, Elat (Eilat), Hadera, Haifa
container port(s) (TEUs): Ashdod (1,443,000) (2016)

MILITARY AND SECURITY

Military and security forces: Israel Defense Forces (IDF): Ground Forces, Israel Naval Force (IN, includes commandos), Israel Air Force (IAF, includes air defense); Ministry of Public Security: Border Police (2019)
note: the Border Police is a unit within the Israel Police with its own organizational and command structure; it works both independently as well as in cooperation with or in support of the Israel Police and Israel Defense Force

Military expenditures: 5% of GDP (2019)
5% of GDP (2018)
5.5% of GDP (2017)
5.5% of GDP (2016)
5.5% of GDP (2015)
country comparison to the world: 6

Military and security service personnel strengths: the Israel Defense Forces (IDF) have approximately 173,000 active personnel (130,000 Ground Forces; 9,500 Naval; 34,000 Air Force) (2019)

Military equipment inventories and acquisitions: the majority of the IDF's inventory is comprised of weapons that are domestically-produced or imported from Europe and the US; since 2010, Germany and the US are the leading suppliers of weapons to Israel; Israel has a broad defense industrial base that can develop, produce, support, and sustain a wide variety of weapons systems for both domestic use and export, particularly armored vehicles, unmanned aerial systems, air defense, and guided missiles (2019 est.)

Military service age and obligation: 18 years of age for compulsory (Jews, Druze) military service; 17 years of age for voluntary (Christians, Muslims,

Circassians) military service; both sexes are obligated to military service; conscript service obligation—32 months for enlisted men and about 24 months for enlisted women (varies based on military occupation), 48 months for officers; pilots commit to 9-year service; reserve obligation to age 41-51 (men), age 24 (women) (2015)

Military—note: the United Nations Disengagement Observer Force (UNDOF) has operated in the Golan between Israel and Syria since 1974 to monitor the ceasefire following the 1973 Arab-Israeli War and supervise the areas of separation between the two countries; as of March 2020, UNDOF consisted of about 1,000 personnel (2020)

TERRORISM

Terrorist group(s): Kahane Chai; Popular Front for the Liberation of Palestine; Palestinian Islamic Jihad (2019)
note: details about the history, aims, leadership, organization, areas of operation, tactics, targets, weapons, size, and sources of support of the group(s) appear(s) in Appendix-T

TRANSNATIONAL ISSUES

Disputes—international: West Bank and Gaza Strip are Israeli-occupied with current status subject to the Israeli-Palestinian Interim Agreement—permanent status to be determined through further negotiation; Israel continues construction of a "seam line" separation barrier along parts of the Green Line and within the West Bank; Israel withdrew its settlers and military from the Gaza Strip and from four settlements in the West Bank in August 2005; Golan Heights is Israeli-controlled (Lebanon claims the Shab'a Farms area of Golan Heights); since 1948, about 350 peacekeepers from the UN Truce Supervision Organization headquartered in Jerusalem monitor ceasefires, supervise armistice agreements, prevent isolated incidents from escalating, and assist other UN personnel in the region

Refugees and internally displaced persons: *refugees (country of origin):* 12,181 (Eritrea), 5,061 (Ukraine) (2019)
stateless persons: 42 (2019)

Illicit drugs: increasingly concerned about ecstasy, cocaine, and heroin abuse; drugs arrive in country from Lebanon and, increasingly, from Jordan; money-laundering center

ITALY

INTRODUCTION

Background: Italy became a nation-state in 1861 when the regional states of the peninsula, along with Sardinia and Sicily, were united under King Victor EMMANUEL II. An era of parliamentary

government came to a close in the early 1920s when Benito MUSSOLINI established a Fascist dictatorship. His alliance with Nazi Germany led to Italy's defeat in World War II. A democratic republic replaced the monarchy in 1946 and economic revival followed. Italy is a charter

member of NATO and the European Economic Community (EEC) and its subsequent successors the EC and the EU. It has been at the forefront of European economic and political unification, joining the Economic and Monetary Union in 1999. Persistent problems include sluggish economic

growth, high youth and female unemployment, organized crime, corruption, and economic disparities between southern Italy and the more prosperous north.

GEOGRAPHY

Location: Southern Europe, a peninsula extending into the central Mediterranean Sea, northeast of Tunisia

Geographic coordinates: 42 50 N, 12 50 E

Map references: Europe

Area: *total:* 301,340 sq km
land: 294,140 sq km
water: 7,200 sq km
note: includes Sardinia and Sicily
country comparison to the world: 73

Area—comparative: almost twice the size of Georgia; slightly larger than Arizona

Land boundaries:
total: 1,836.4 km
border countries (6): Austria 404 km, France 476 km, Holy See (Vatican City) 3.4 km, San Marino 37 km, Slovenia 218 km, Switzerland 698 km

Coastline: 7,600 km

Maritime claims:
territorial sea: 12 nm
continental shelf: 200-m depth or to the depth of exploitation

Climate: predominantly Mediterranean; alpine in far north; hot, dry in south

Terrain: mostly rugged and mountainous; some plains, coastal lowlands

Elevation: *mean elevation:* 538 m
lowest point: Mediterranean Sea 0 m
highest point: Mont Blanc (Monte Bianco) de Courmayeur (a secondary peak of Mont Blanc) 4,748 m

Natural resources: coal, antimony, mercury, zinc, potash, marble, barite, asbestos, pumice, fluorspar, feldspar, pyrite (sulfur), natural gas and crude oil reserves, fish, arable land

Land use: *agricultural land:* 47.1% (2011 est.)

arable land: 22.8% (2011 est.) / permanent crops: 8.6% (2011 est.) / permanent pasture: 15.7% (2011 est.)
forest: 31.4% (2011 est.)
other: 21.5% (2011 est.)

Irrigated land: 39,500 sq km (2012)

Population distribution: despite a distinctive pattern with an industrial north and an agrarian south, a fairly even population distribution exists throughout most of the country, with coastal areas, the Po River Valley, and urban centers (particularly Milan, Rome, and Naples), attracting larger and denser populations

Natural hazards: regional risks include landslides, mudflows, avalanches, earthquakes, volcanic eruptions, flooding; land subsidence in Venice
volcanism: significant volcanic activity; Etna (3,330 m), which is in eruption as of 2010, is Europe's most active volcano; flank eruptions pose a threat to nearby Sicilian villages; Etna, along with the famous Vesuvius, which remains a threat to the millions of nearby residents in the Bay of Naples area, have both been deemed Decade Volcanoes by the International Association of Volcanology and Chemistry of the Earth's Interior, worthy of study due to their explosive history and close proximity to human populations; Stromboli, on its namesake island, has also been continuously active with moderate volcanic activity; other historically active volcanoes include Campi Flegrei, Ischia, Larderello, Pantelleria, Vulcano, and Vulsini

Environment—current issues: air pollution from industrial emissions such as sulfur dioxide; coastal and inland rivers polluted from industrial and agricultural effluents; acid rain damaging lakes; inadequate industrial waste treatment and disposal facilities

Environment—international agreements: *party to:* Air Pollution, Air Pollution-Nitrogen Oxides, Air Pollution-Persistent Organic Pollutants, Air Pollution-Sulfur 85, Air Pollution-Sulfur 94, Air Pollution-Volatile Organic Compounds, Antarctic-Environmental Protocol, Antarctic-Marine Living Resources, Antarctic Seals, Antarctic Treaty, Biodiversity, Climate Change, Climate Change-Kyoto Protocol, Desertification, Endangered Species, Environmental Modification, Hazardous Wastes, Law of the Sea, Marine Dumping, Ozone Layer Protection, Ship Pollution, Tropical Timber 83, Tropical Timber 94, Wetlands, Whaling
signed, but not ratified: none of the selected agreements

Geography—note: strategic location dominating central Mediterranean as well as southern sea and air approaches to Western Europe

PEOPLE AND SOCIETY

Population: 62,402,659 (July 2020 est.)
country comparison to the world: 23

Nationality: *noun:* Italian(s)
adjective: Italian

Ethnic groups: Italian (includes small clusters of German-, French-, and Slovene-Italians in the north and Albanian-Italians and Greek- Italians in the south)

Languages: Italian (official), German (parts of Trentino-Alto Adige region are predominantly German speaking), French (small French-speaking minority in Valle d'Aosta region), Slovene (Slovene-speaking minority in the Trieste-Gorizia area)

Religions: Christian 83.3% (overwhelmingly Roman Catholic with very small groups of Jehovah's Witnesses and Protestants), Muslim 3.7%, unaffiliated 12.4%, other 0.6% (2010 est.)

Age structure: *0-14 years:* 13.45% (male 4,292,431/female 4,097,732)
15-24 years: 9.61% (male 3,005,402/female 2,989,764)
25-54 years: 40.86% (male 12,577,764/female 12,921,614)
55-64 years: 14% (male 4,243,735/female 4,493,581)
65 years and over: 22.08% (male 5,949,560/female 7,831,076) (2020 est.)

Dependency ratios:
total dependency ratio: 57
youth dependency ratio: 20.4
elderly dependency ratio: 36.6
potential support ratio: 2.7 (2020 est.)

Median age:
total: 46.5 years
male: 45.4 years
female: 47.5 years (2020 est.)
country comparison to the world: 5

Population growth rate: 0.11% (2020 est.)
country comparison to the world: 186

Birth rate: 8.4 births/1,000 population (2020 est.)
country comparison to the world: 218

Death rate: 10.7 deaths/1,000 population (2020 est.)
country comparison to the world: 27

Net migration rate: 3.2 migrant(s)/1,000 population (2020 est.)
country comparison to the world: 36

Population distribution: despite a distinctive pattern with an industrial north and an agrarian south, a fairly even population distribution exists throughout most of the country, with coastal areas, the Po River Valley, and urban centers (particularly Milan, Rome, and Naples), attracting larger and denser populations

Urbanization: *urban population:* 71% of total population (2020)
rate of urbanization: 0.29% annual rate of change (2015-20 est.)
total population growth rate v. urban population growth rate, 2000-2030:

Major urban areas—population: 4.257 million ROME (capital), 3.140 million Milan, 2.187 million Naples, 1.792 million Turin, 892,000 Bergamo, 851,000 Palermo (2020)

Sex ratio: *at birth:* 1.06 male(s)/female
0-14 years: 1.05 male(s)/female

15-24 years: 1.01 male(s)/female
25-54 years: 0.97 male(s)/female
55-64 years: 0.94 male(s)/female
65 years and over: 0.76 male(s)/female
total population: 0.93 male(s)/female (2020 est.)

Mother's mean age at first birth: 31.1 years (2017 est.)

Maternal mortality rate: 2 deaths/100,000 live births (2017 est.)
country comparison to the world: 182

Infant mortality rate: *total:* 3.2 deaths/1,000 live births
male: 3.4 deaths/1,000 live births
female: 3 deaths/1,000 live births (2020 est.)
country comparison to the world: 212

Life expectancy at birth: *total population:* 82.5 years
male: 79.8 years
female: 85.3 years (2020 est.)
country comparison to the world: 17

Total fertility rate: 1.47 children born/woman (2020 est.)
country comparison to the world: 210

Contraceptive prevalence rate: 65.1% (2013)
note: percent of women aged 18-49

Drinking water source:

improved: *urban:* 100% of population
rural: 100% of population
total: 100% of population

unimproved: *urban:* 0% of population
rural: 0% of population
total: 0% of population (2017 est.)

Current Health Expenditure: 8.8% (2017)

Physicians density: 3.98 physicians/1,000 population (2017)

Hospital bed density: 3.2 beds/1,000 population (2017)

Sanitation facility access:

improved: *urban:* 98.8% of population
rural: 98.6% of population
total: 98.8% of population

unimproved: *urban:* 1.2% of population
rural: 1.4% of population
total: 1.2% of population (2017 est.)

HIV/AIDS—adult prevalence rate: 0.3% (2019 est.)
country comparison to the world: 89

HIV/AIDS—people living with HIV/AIDS: 130,000 (2019 est.)
country comparison to the world: 39

HIV/AIDS—deaths: <1000 (2019 est.)

Major infectious diseases: Covid-19 (see note) (2020)

note: a new coronavirus is causing sustained community spread of respiratory illness (COVID-19) in Italy; sustained community spread means that people have been infected with the virus, but how or where they became infected is not known, and the spread is ongoing; illness with this virus has ranged from mild to severe with fatalities reported; as of 10 November 2020, Italy has reported a total of 902,490 cases of COVID-19 or 14,927 cumulative cases of COVID-19 per 1 million population with 679 cumulative deaths per 1 million population; the US Department of State has issued a Travel Advisory to reconsider travel to Italy due to the recent outbreak of COVID-19; the Centers for Disease Control and Prevention has also recommended postponing nonessential travel to Italy at this time and published additional guidance at https://wwwnc.cdc.gov/travel/notices/alert/coronavirus-italy; the US Department of Homeland Security has issued instructions requiring US passengers who have been in Italy to travel through select airports where the US Government has implemented enhanced screening procedures

Obesity—adult prevalence rate: 19.9% (2016)
country comparison to the world: 108

Education expenditures: 3.8% of GDP (2016)
country comparison to the world: 113

Literacy: *definition:* age 15 and over can read and write
total population: 99.2%
male: 99.4%
female: 99% (2018)

School life expectancy (primary to tertiary education):
total: 16 years
male: 16 years
female: 17 years (2018)

Unemployment, youth ages 15-24:
total: 32.2%
male: 30.4%
female: 34.8% (2018 est.)
country comparison to the world: 26

GOVERNMENT

Country name: *conventional long form:* Italian Republic
conventional short form: Italy
local long form: Repubblica Italiana
local short form: Italia
former: Kingdom of Italy
etymology: derivation is unclear, but the Latin "Italia" may come from the Oscan "Viteliu" meaning "[Land] of Young Cattle" (the bull was a symbol of southern Italic tribes)

Government type: parliamentary republic

Capital: *name:* Rome
geographic coordinates: 41 54 N, 12 29 E
time difference: UTC+1 (6 hours ahead of Washington, DC, during Standard Time)
daylight saving time: +1hr, begins last Sunday in March; ends last Sunday in October
etymology: by tradition, named after Romulus, one of the legendary founders of the city and its first king

Administrative divisions: 15 regions (regioni, singular—regione) and 5 autonomous regions (regioni autonome, singular—regione autonoma)
regions: Abruzzo, Basilicata, Calabria, Campania, Emilia-Romagna, Lazio (Latium), Liguria, Lombardia, Marche, Molise, Piemonte (Piedmont), Puglia (Apulia), Toscana (Tuscany), Umbria, Veneto;
autonomous regions: Friuli Venezia Giulia, Sardegna (Sardinia), Sicilia (Sicily), Trentino-Alto Adige (Trentino-South Tyrol) or Trentino-Suedtirol (German), Valle d'Aosta (Aosta Valley) or Vallee d'Aoste (French)

Independence: 17 March 1861 (Kingdom of Italy proclaimed; Italy was not finally unified until 1871)

National holiday: Republic Day, 2 June (1946)

Constitution: *history:* previous 1848 (originally for the Kingdom of Sardinia and adopted by the Kingdom of Italy in 1861); latest enacted 22 December 1947, adopted 27 December 1947, entered into force 1 January 1948
amendments: proposed by both houses of Parliament; passage requires two successive debates and approval by absolute majority of each house on the second vote; a referendum is only required when requested by one fifth of the members of either house, by voter petition, or by five Regional Councils (elected legislative assemblies of the 15 first-level administrative regions and 5 autonomous regions of Italy); referendum not required if an amendment has been approved by a two-thirds majority in each house in the second vote; amended many times, last in 2012; note—a referendum held on 4 December 2016 on constitutional amendments was defeated

Legal system: civil law system; judicial review of legislation under certain conditions in Constitutional Court

International law organization participation: accepts compulsory ICJ jurisdiction with reservations; accepts ICCt jurisdiction

Citizenship: *citizenship by birth:* no
citizenship by descent only: at least one parent must be a citizen of Italy
dual citizenship recognized: yes
residency requirement for naturalization: 4 years for EU nationals, 5 years for refugees and specified exceptions, 10 years for all others

Suffrage: 18 years of age; universal except in senatorial elections, where minimum age is 25

Executive branch: *chief of state:* President Sergio MATTARELLA (since 3 February 2015)
head of government: Prime Minister Giuseppe CONTE (since 1 June 2018); the prime minister's official title is President of the Council of Ministers; note—CONTE resigned on 20 August 2019 but returned as prime minister after PD and M5S agreed to form a new coalition government on 28 August 2019
cabinet: Council of Ministers proposed by the prime minister, known officially as the President of the Council of Ministers and locally as the Premier; nominated by the president; the current deputy prime ministers, known officially as vice-presidents of the Council of Ministers, are Matteo Salvini (L) and Luigi Di Maio (M5S) (since 1 June 2018)
elections/appointments: president indirectly elected by an electoral college consisting of both

houses of Parliament and 58 regional representatives for a 7-year term (no term limits); election last held on 31 January 2015 (next to be held in 2022); prime minister appointed by the president, confirmed by parliament

election results: Sergio MATTARELLA (independent) elected president; electoral college vote count in fourth round—665 out of 1,009 (505-vote threshold)

Legislative branch: *description:* bicameral Parliament or Parlamento consists of: Senate or Senato della Repubblica (321 seats; 116 members directly elected in single-seat constituencies by simple majority vote, 193 members in multi-seat constituencies and 6 members in multi-seat constituencies abroad directly elected by party-list proportional representation vote to serve 5-year terms and 6 ex-officio members appointed by the president of the Republic to serve for life)

Chamber of Deputies or Camera dei Deputati (630 seats; 629 members directly elected in single- and multi-seat constituencies by proportional representation vote and 1 member from Valle d'Aosta elected by simple majority vote; members serve 5-year terms); note—a 29 March 2020 referendum on the proposed reduction of Parliament membership has been postponed indefinitely due to the COVID-19 pandemic

elections: Senate—last held on 4 March 2018 (next to be held in March 2023)

Chamber of Deputies—last held on 4 March 2018 (next to be held in March 2023)

election results: Senate—percent of vote by party—center-right coalition 37.5% (L 17.6%, FI 14.4%, FdI 4.3%, UdC 1.2%), M5S 32.2%, center-left coalition (PD 19.1%, +E 2.3%, I 0.5%, CP 0.5%, SVP-PATT 0.4%), LeU 3.3%; seats by party—center-right coalition 77(L 37, FI 33, FdI 7), M5S 68, center-left coalition 44(PD 43, SVP-PATT 1), LeU 4; composition—men 208, women 113, percent of women 35.2%

Chamber of Deputies—percent of vote by party—center-right coalition 37% (L 17.4%, FI 14%, FdI 4.4%, UdC 1.3%), M5S 33%, center-left coalition 22.9% (PD 18.8%, E+ 2.6%, I 0.6%, CP 0.5%, SVP-PATT 0.4%); seats by party—center-right coalition 151 (L73, FI 59, FdI 19), M5S 133, center-left coalition 88 (PD 86, SVP 2), LeU 14; composition—men 405, women 225, percent of women 35.7%; note—total Parliament percent of women 35.5%

Note: in October 2019, Italy's Parliament voted to reduce the number of Senate seats from 315 to 200 and the number of Chamber of Deputies seats from 630 to 400; the law is subject to a referendum to be held between 15 April and 15 June 2020; changes will be effective for the 2023 election if the law is adopted

Judicial branch: *highest courts:* Supreme Court of Cassation or Corte Suprema di Cassazione (consists of the first president (chief justice), deputy president, 54 justices presiding over 6 civil and 7 criminal divisions, and 288 judges; an additional 30 judges of lower courts serve as supporting judges; cases normally heard by 5-judge panels; more complex cases heard by 9-judge panels); Constitutional Court or Corte Costituzionale (consists of the court president and 14 judges)

judge selection and term of office: Supreme Court judges appointed by the High Council of the Judiciary, headed by the president of the republic; judges may serve for life; Constitutional Court judges—5 appointed by the president, 5 elected by Parliament, 5 elected by select higher courts; judges serve up to 9 years

subordinate courts: various lower civil and criminal courts (primary and secondary tribunals and courts of appeal)

Political parties and leaders: *Governing Coalition:* Northern League (Lega Nord) or Lega [Matteo SALVINI]

Five Star Movement or M5S [Vito CRIMI, acting leader]

Left-center-right opposition: Democratic Party or PD [Nicola ZINGARETTI]

Forza Italia or FI [Silvio BERLUSCONI]

Brothers of Italy [Giorgia MELONI]

Free and Equal (Liberi e Uguali) or LeU [Pietro GRASSO]

More Europe or +EU [Emma BONINO]

Popular Civic List or CP [Beatrice LORENZIN]

Other parties and parliamentary groups: Possible [Beatrice BRIGNONE]

Us with Italy [Raffaele FITTO]

South Tyrolean People's Party or SVP [Philipp ACHAMMER]

Trentino Tyrolean Autonomist Party (Partito Autonomista Trentino Tirolese) or PATT [Franco PANIZZA, secretary]

Article One or Art.1-MDP [Roberto SPERANZA]

International organization participation: ADB (nonregional member), AfDB (nonregional member), Arctic Council (observer), Australia Group, BIS, BSEC (observer), CBSS (observer), CD, CDB, CE, CEI, CERN, EAPC, EBRD, ECB, EIB, EITI (implementing country), EMU, ESA, EU, FAO, FATF, G-7, G-8, G-10, G-20, IADB, IAEA, IBRD, ICAO, ICC (national committees), ICCt, ICRM, IDA, IEA, IFAD, IFC, IFRCS, IGAD (partners), IHO, ILO, IMF, IMO, IMSO, Interpol, IOC, IOM, IPU, ISO, ITSO, ITU, ITUC (NGOs), LAIA (observer), MIGA, MINURSO, MINUSMA, NATO, NEA, NSG, OAS (observer), OECD, OPCW, OSCE, Pacific Alliance (observer), Paris Club, PCA, PIF (partner), Schengen Convention, SELEC (observer), SICA (observer), UN, UNCTAD, UNESCO, UNHCR, UNIDO, UNIFIL, Union Latina, UNMOGIP, UNRWA, UNTSO, UNWTO, UPU, WCO, WHO, WIPO, WMO, WTO, ZC

Diplomatic representation in the US: *chief of mission:* Ambassador Armando VARRICCHIO (since 2 March 2016)

chancery: 3000 Whitehaven Street NW, Washington, DC 20008

telephone: [1] (202) 612-4400

FAX: [1] (202) 518-2151

consulate(s) general: Boston, Chicago, Detroit, Houston, Miami, New York, Los Angeles, Philadelphia, San Francisco

consulate(s): Charlotte (NC), Cleveland (OH), Detroit (MI), Hattiesburg (MS), Honolulu (HI), New Orleans, Newark (NJ), Norfolk (VA), Pittsburgh (PA), Portland (OR), Seattle

Diplomatic representation from the US: *chief of mission:* Ambassador Lewis EISENBERG (since 4 October 2017); note—also accredited to San Marino

telephone: [39] 06-4674-1

embassy: Via Vittorio Veneto 121, 00187-Rome

mailing address: PSC 59, Box 100, APO AE 09624

FAX: [39] 06-488-2672

consulate(s) general: Florence, Milan, Naples

Flag description: three equal vertical bands of green (hoist side), white, and red; design inspired by the French flag brought to Italy by Napoleon in 1797; colors are those of Milan (red and white) combined with the green uniform color of the Milanese civic guard

note: similar to the flag of Mexico, which is longer, uses darker shades of green and red, and has its coat of arms centered on the white band; Ireland, which is longer and is green (hoist side), white, and orange; also similar to the flag of the Cote d'Ivoire, which has the colors reversed—orange (hoist side), white, and green

National symbol(s): *white, five-pointed star (Stella d'Italia); national colors:* red, white, green

National anthem: *name:* "Il Canto degli Italiani" (The Song of the Italians)

lyrics/music: Goffredo MAMELI/Michele NOVARO

note: adopted 1946; the anthem, originally written in 1847, is also known as "L'Inno di Mameli" (Mameli's Hymn), and "Fratelli D'Italia" (Brothers of Italy)

ECONOMY

Economy—overview: Italy's economy comprises a developed industrial north, dominated by private companies, and a less-developed, highly subsidized, agricultural south, with a legacy of unemployment and underdevelopment. The Italian economy is driven in large part by the manufacture of high-quality consumer goods produced by small and medium-sized enterprises, many of them family-owned. Italy also has a sizable underground economy, which by some estimates accounts for as much as 17% of GDP. These activities are most common within the agriculture, construction, and service sectors.

Italy is the third-largest economy in the euro zone, but its exceptionally high public debt and structural impediments to growth have rendered it vulnerable to scrutiny by financial markets. Public debt has increased steadily since 2007, reaching 131% of GDP in 2017. Investor concerns about Italy and the broader euro-zone crisis eased in 2013, bringing down Italy's borrowing costs on sovereign government debt from euro-era records. The government still faces pressure from investors and European partners to sustain its efforts to address Italy's longstanding structural economic problems, including labor market inefficiencies, a sluggish judicial system, and a weak banking sector. Italy's economy returned to modest growth in late 2014 for the first time since 2011. In 2015-16, Italy's economy grew at about 1% each year, and in 2017 growth accelerated to 1.5% of GDP. In 2017, overall unemployment was 11.4%, but youth unemployment remained high at 37.1%. GDP growth is projected to slow slightly in 2018.

GDP (purchasing power parity): $2.317 trillion (2017 est.)
$2.282 trillion (2016 est.)
$2.263 trillion (2015 est.)
note: data are in 2017 dollars
country comparison to the world: 12

GDP (official exchange rate): $1.939 trillion (2017 est.)

GDP—real growth rate: 0.34% (2019 est.)
0.83% (2018 est.)
1.73% (2017 est.)
country comparison to the world: 188

GDP—per capita (PPP): $38,200 (2017 est.)
$37,600 (2016 est.)
$37,200 (2015 est.)
note: data are in 2017 dollars
country comparison to the world: 50

Gross national saving: 20.3% of GDP (2017 est.)
19.7% of GDP (2016 est.)
18.8% of GDP (2015 est.)
country comparison to the world: 95

GDP—composition, by end use:
household consumption: 61% (2017 est.)
government consumption: 18.6% (2017 est.)
investment in fixed capital: 17.5% (2017 est.)
investment in inventories: -0.2% (2017 est.)
exports of goods and services: 31.4% (2017 est.)
imports of goods and services: -28.3% (2017 est.)

GDP—composition, by sector of origin:
agriculture: 2.1% (2017 est.)
industry: 23.9% (2017 est.)
services: 73.9% (2017 est.)

Agriculture—products: fruits, vegetables, grapes, potatoes, sugar beets, soybeans, grain, olives; beef, dairy products; fish
Industries: tourism, machinery, iron and steel, chemicals, food processing, textiles, motor vehicles, clothing, footwear, ceramics

Industrial production growth rate: 2.1% (2017 est.)
country comparison to the world: 128

Labor force: 22.92 million (2020 est.)
country comparison to the world: 23

Labor force—by occupation: *agriculture:* 3.9%
industry: 28.3%
services: 67.8% (2011)

Unemployment rate: 9.88% (2019 est.)
10.63% (2018 est.)
country comparison to the world: 145

Population below poverty line: 29.9% (2012 est.)

Household income or consumption by percentage share:
lowest 10%: 2.3%
highest 10%: 26.8% (2000)

Budget: *revenues:* 903.3 billion (2017 est.)
expenditures: 948.1 billion (2017 est.)

Taxes and other revenues: 46.6% (of GDP) (2017 est.)
country comparison to the world: 20

Budget surplus (+) or deficit (-): -2.3% (of GDP) (2017 est.)
country comparison to the world: 111

Public debt: 131.8% of GDP (2017 est.)
132% of GDP (2016 est.)
note: Italy reports its data on public debt according to guidelines set out in the Maastricht Treaty; general government gross debt is defined in the Maastricht Treaty as consolidated general government gross debt at nominal value, outstanding at the end of the year, in the following categories of government liabilities (as defined in ESA95): currency and deposits (AF.2), securities other than shares excluding financial derivatives (AF.3, excluding AF.34), and loans (AF.4); the general government sector comprises central, state, and local government and social security funds
country comparison to the world: 5

Fiscal year: calendar year

Inflation rate (consumer prices): 1.3% (2017 est.)
-0.1% (2016 est.)
country comparison to the world: 68

Current account balance: $59.517 billion (2019 est.)
$51.735 billion (2018 est.)
country comparison to the world: 10

Exports: $496.3 billion (2017 est.)
$454.1 billion (2016 est.)
country comparison to the world: 9

Exports—partners: Germany 12.5%, France 10.3%, US 9%, Spain 5.2%, UK 5.2%, Switzerland 4.6% (2017)

Exports—commodities: engineering products, textiles and clothing, production machinery, motor vehicles, transport equipment, chemicals; foodstuffs, beverages, and tobacco; minerals, non-ferrous metals

Imports: $432.9 billion (2017 est.)
$389.8 billion (2016 est.)
country comparison to the world: 13

Imports—commodities: engineering products, chemicals, transport equipment, energy products, minerals and nonferrous metals, textiles and clothing; food, beverages, tobacco

Imports—partners: Germany 16.3%, France 8.8%, China 7.1%, Netherlands 5.6%, Spain 5.3%, Belgium 4.5% (2017)

Reserves of foreign exchange and gold: $151.2 billion (31 December 2017 est.)
$130.6 billion (31 December 2015 est.)
country comparison to the world: 16

Debt—external: $2.444 trillion (31 March 2016 est.)

$2.3 trillion (31 March 2015 est.)
country comparison to the world: 9

Exchange rates: euros (EUR) per US dollar—
0.885 (2017 est.)
0.903 (2016 est.)
0.9214 (2015 est.)
0.885 (2014 est.)
0.7634 (2013 est.)

ENERGY

Electricity access: *electrification—total population:* 100% (2020)

Electricity—production: 275.3 billion kWh (2016 est.)
country comparison to the world: 14

Electricity—consumption: 293.5 billion kWh (2016 est.)
country comparison to the world: 13

Electricity—exports: 6.155 billion kWh (2016 est.)
country comparison to the world: 30

Electricity—imports: 43.18 billion kWh (2016 est.)
country comparison to the world: 2

Electricity—installed generating capacity: 114.2 million kW (2016 est.)
country comparison to the world: 10

Electricity—from fossil fuels: 54% of total installed capacity (2016 est.)
country comparison to the world: 143

Electricity—from nuclear fuels: 0% of total installed capacity (2017 est.)
country comparison to the world: 114

Electricity—from hydroelectric plants: 14% of total installed capacity (2017 est.)
country comparison to the world: 106

Electricity—from other renewable sources: 32% of total installed capacity (2017 est.)
country comparison to the world: 14

Crude oil—production: 90,000 bbl/day (2018 est.)
country comparison to the world: 44

Crude oil—exports: 13,790 bbl/day (2017 est.)
country comparison to the world: 57

Crude oil—imports: 1.341 million bbl/day (2017 est.)
country comparison to the world: 7

Crude oil—proved reserves: 487.8 million bbl (1 January 2018 est.)
country comparison to the world: 45

Refined petroleum products—production: 1.607 million bbl/day (2017 est.)
country comparison to the world: 12

Refined petroleum products—consumption: 1.236 million bbl/day (2017 est.)
country comparison to the world: 19

Refined petroleum products—exports: 615,900 bbl/day (2017 est.)
country comparison to the world: 13

Refined petroleum products—imports: 422,500 bbl/day (2017 est.)

country comparison to the world: 19

Natural gas—production: 5.55 billion cu m (2017 est.)
country comparison to the world: 50

Natural gas—consumption: 75.15 billion cu m (2017 est.)
country comparison to the world: 11

Natural gas—exports: 271.8 million cu m (2017 est.)
country comparison to the world: 44

Natural gas—imports: 69.66 billion cu m (2017 est.)
country comparison to the world: 5

Natural gas—proved reserves: 38.11 billion cu m (1 January 2018 est.)
country comparison to the world: 65

Carbon dioxide emissions from consumption of energy: 351 million Mt (2017 est.)
country comparison to the world: 20

COMMUNICATIONS

Telephones—fixed lines:
total subscriptions: 20,196,475
subscriptions per 100 inhabitants: 32.4 (2019 est.)
country comparison to the world: 14

Telephones—mobile cellular:
total subscriptions: 82,955,151
subscriptions per 100 inhabitants: 133.08 (2019 est.)
country comparison to the world: 19

Telecommunication systems: *general assessment:* well-developed, fully automated telephone, and data services; highest mobile penetration rates in Europe, benefitted from progressive govt. programs aimed at developing fiber in broadband sector; leading edge of development with 5G in 6 cities; regulator consults on extending 3.5 Gz licensing; fiber network reaches 60% of population (2020)
domestic: high-capacity cable and microwave radio relay trunks; 32 per 100 for fixed-line and 133 per 100 for mobilecellular subscriptions (2019)
international: country code—39; landing points for Italy-Monaco, Italy-Libya, Italy-Malta, Italy-Greece-1, Italy-Croatia, BlueMed, Janna, FEA, SeaMeWe-3 & 4 & 5, Trapani-Kelibia, Columbus-III, Didon, GO-1, HANNIBAL System, MENA, Bridge International, Malta-Italy Interconnector, Melita1, IMEWE, VMSCS, AAE-1, and OTEGLOBE, submarine cables that provide links to Asia, the Middle East, Europe, North Africa, Southeast Asia, Australia and US; satellite earth stations—3 Intelsat (with a total of 5 antennas—3 for Atlantic Ocean and 2 for Indian Ocean) (2019)
note: the COVID-19 outbreak is negatively impacting telecommunications production and supply chains globally; consumer spending on telecom devices and services has also slowed due to the pandemic's effect on economies worldwide; overall progress towards improvements in all facets of the

telecom industry—mobile, fixed-line, broadband, submarine cable and satellite—has moderated

Broadcast media: two Italian media giants dominate—the publicly owned Radiotelevisione Italiana (RAI) with 3 national terrestrial stations and privately owned Mediaset with 3 national terrestrial stations; a large number of private stations and Sky Italia—a satellite TV network; RAI operates 3 AM/FM nationwide radio stations; about 1,300 commercial radio stations

Internet country code: .it

Internet users:
total: 46,305,301
percent of population: 74.39% (July 2018 est.)
country comparison to the world: 19

Broadband—fixed subscriptions:
total: 17,060,505
subscriptions per 100 inhabitants: 27 (2018 est.)
country comparison to the world: 12

TRANSPORTATION

National air transport system:
number of registered air carriers: 9 (2020)
inventory of registered aircraft operated by air carriers: 180
annual passenger traffic on registered air carriers: 27,630,435 (2018)
annual freight traffic on registered air carriers: 1.418 billion mt-km (2018)

Civil aircraft registration country code prefix: I (2016)

Airports: 129 (2013)
country comparison to the world: 44

Airports—with paved runways:
total: 98 (2017)
over 3,047 m: 9 (2017)
2,438 to 3,047 m: 31 (2017)
1,524 to 2,437 m: 18 (2017)
914 to 1,523 m: 29 (2017)
under 914 m: 11 (2017)

Airports—with unpaved runways:
total: 31 (2013)
1,524 to 2,437 m: 1 (2013)
914 to 1,523 m: 10 (2013)
under 914 m: 20 (2013)

Heliports: 5 (2013)

Pipelines: 20223 km gas, 1393 km oil, 1574 km refined products (2013)

Railways: *total:* 20,182 km (2014)
standard gauge: 18,770.1 km 1.435-m gauge (12,893.6 km electrified) (2014)
narrow gauge: 122.3 km 1.000-m gauge (122.3 km electrified) (2014)
1289.3 0.950-m gauge (151.3 km electrified)
country comparison to the world: 15

Roadways: *total:* 487,700 km (2007)
paved: 487,700 km (includes 6,700 km of expressways) (2007)
country comparison to the world: 15

Waterways: 2,400 km (used for commercial traffic; of limited overall value compared to road and rail) (2012)
country comparison to the world: 36

Merchant marine:
total: 1,353
by type: bulk carrier 48, container ship 9, general cargo 116, oil tanker 120, other 1,060 (2019)
country comparison to the world: 21

Ports and terminals: *major seaport(s):* Augusta, Cagliari, Genoa, Livorno, Taranto, Trieste, Venice
oil terminal(s): Melilli (Santa Panagia) oil terminal, Sarroch oil terminal
container port(s) (TEUs): Genoa (2,622,200), Gioia Tauro (2,448,600) (2017)

LNG terminal(s) (import): La Spezia, Panigaglia, Porto Levante

MILITARY AND SECURITY

Military and security forces: *Italian Armed Forces:* Army (Esercito Italiano, EI), Navy (Marina Militare Italiana, MMI; includes aviation, marines), Italian Air Force (Aeronautica Militare Italiana, AMI), Carabinieri Corps (Arma dei Carabinieri, CC) (2019)
note(s): the Financial Guard (Guardia di Finanza) under the Ministry of Economy and Finance is a force with military status and nationwide remit for financial crime investigations, including narcotics trafficking, smuggling, and illegal immigration

Military expenditures: 1.22% of GDP (2019 est.)
1.21% of GDP (2018)
1.21% of GDP (2017)
1.18% of GDP (2016)
1.07% of GDP (2015)
country comparison to the world: 97

Military and security service personnel strengths: the Italian Armed Forces have approximately 271,000 active personnel, including the Carabinieri (96,000 Army; 28,000 Navy; 40,000 Air Force; 107,000 Carabinieri) (2019 est.)

Military equipment inventories and acquisitions: the Italian Armed Forces' inventory includes a mix of domestically-produced, jointly-produced, and imported European and US weapons systems; the US is the leading supplier of weapons to Italy since 2010, followed by Germany; the Italian defense industry is capable of producing equipment across all the military domains with particular strengths in naval vessels and aircraft; it also participates in joint development and production of advanced weapons systems with other European countries and the US (2019)

Military deployments: 900 Afghanistan (NATO); 120 Djibouti; 1,100 Middle East/Iraq/Kuwait (NATO, counter-ISIS campaign, European Assistance Mission Iraq); 620 Kosovo (NATO); 200 Latvia (NATO); 1,050 Lebanon (UNIFIL); 400 Libya; 290 Niger; 150 Somalia (EUTM); 100 United Arab Emirates (April 2020)

Military service age and obligation: 18-25 years of age for voluntary military service; women may serve in any military branch; Italian citizenship required; 1- year service obligation (2013)

TRANSNATIONAL ISSUES

Disputes—international: Italy's long coastline and developed economy entices tens of thousands of illegal immigrants from southeastern Europe and northern Africa

Refugees and internally displaced persons: *refugees (country of origin):* 25,241 (Nigeria), 20,063 (Pakistan), 17,849 (Afghanistan), 15,842 (Mali), 14,029 (Somalia), 12,968 (Gambia), 8,974 (Bangladesh), 7,659 (Cote d'Ivoire), 7,644 (Senegal), 7,118 (Eritrea), 6,995 (Iraq), 6,353 (Ukraine), 5,953 (Ghana) (2019); note—estimate for Ukraine represents asylum applicants since the beginning of the Ukraine crisis in 2014 to July 2018

stateless persons: 15,822 (2019)

note: 520,474 estimated refugee and migrant arrivals by sea (January 2015-November 2020); hosts an estimated 96,862 migrants and asylum seekers as of the end of October 2019; 23,370 arrivals in 2018

Illicit drugs: important gateway for and consumer of Latin American cocaine and Southwest Asian heroin entering the European market; money laundering by organized crime and from smuggling

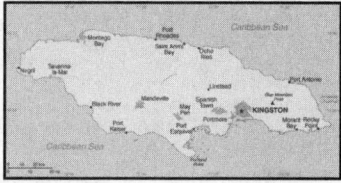

INTRODUCTION

Background: The island—discovered by Christopher COLUMBUS in 1494—was settled by the Spanish early in the 16th century. The native Taino, who had inhabited Jamaica for centuries, were gradually exterminated and replaced by African slaves. England seized the island in 1655 and established a plantation economy based on sugar, cocoa, and coffee. The abolition of slavery in 1834 freed a quarter million slaves, many of whom became small farmers. Jamaica gradually increased its independence from Britain. In 1958 it joined other British Caribbean colonies in forming the Federation of the West Indies. Jamaica withdrew from the Federation in 1961 and gained full independence in 1962. Deteriorating economic conditions during the 1970s led to recurrent violence as rival gangs affiliated with the major political parties evolved into powerful organized crime networks involved in international drug smuggling and money laundering. Violent crime, drug trafficking, and poverty pose significant challenges to the government today. Nonetheless, many rural and resort areas remain relatively safe and contribute substantially to the economy.

GEOGRAPHY

Location: Caribbean, island in the Caribbean Sea, south of Cuba

Geographic coordinates: 18 15 N, 77 30 W

Map references: Central America and the Caribbean

Area: total: 10,991 sq km
land: 10,831 sq km
water: 160 sq km
country comparison to the world: 167

Area—comparative: about half the size of New Jersey; slightly smaller than Connecticut

Land boundaries: 0 km

Coastline: 1,022 km

Maritime claims: territorial sea: 12 nm
exclusive economic zone: 200 nm
contiguous zone: 24 nm
continental shelf: 200 nm or to edge of the continental margin
measured from claimed archipelagic straight baselines

Climate: tropical; hot, humid; temperate interior

Terrain: mostly mountains, with narrow, discontinuous coastal plain

Elevation: mean elevation: 18 m
lowest point: Caribbean Sea 0 m
highest point: Blue Mountain Peak 2,256 m

Natural resources: bauxite, alumina, gypsum, limestone

Land use: agricultural land: 41.4% (2011 est.)
arable land: 11.1% (2011 est.) / permanent crops: 9.2% (2011 est.) / permanent pasture: 21.1% (2011 est.)
forest: 31.1% (2011 est.)
other: 27.5% (2011 est.)

Irrigated land: 250 sq km (2012)

Population distribution: population density is high throughout, but increases in and around Kingston, Montego Bay, and Port Esquivel

Natural hazards: hurricanes (especially July to November)

Environment—current issues: heavy rates of deforestation; coastal waters polluted by industrial waste, sewage, and oil spills; damage to coral reefs; air pollution in Kingston from vehicle emissions; land erosion

Environment—international agreements: *party to:* Biodiversity, Climate Change, Climate Change-Kyoto Protocol, Desertification, Endangered Species, Hazardous Wastes, Law of the Sea, Marine Dumping, Marine Life Conservation, Ozone Layer Protection, Ship Pollution, Wetlands
signed, but not ratified: none of the selected agreements

Geography—note: third largest island in the Caribbean (after Cuba and Hispaniola); strategic location between Cayman Trench and Jamaica Channel, the main sea lanes for the Panama Canal

PEOPLE AND SOCIETY

Population: 2,808,570 (July 2020 est.)
country comparison to the world: 140

Nationality: noun: Jamaican(s)
adjective: Jamaican

Ethnic groups: black 92.1%, mixed 6.1%, East Indian 0.8%, other 0.4%, unspecified 0.7% (2011 est.)

Languages: English, English patois

Religions: Protestant 64.8% (includes Seventh Day Adventist 12.0%, Pentecostal 11.0%, Other Church of God 9.2%, New Testament Church of God 7.2%, Baptist 6.7%, Church of God in Jamaica 4.8%, Church of God of Prophecy 4.5%, Anglican 2.8%, United Church 2.1%, Methodist 1.6%, Revived 1.4%, Brethren 0.9%, and Moravian 0.7%), Roman Catholic 2.2%, Jehovah's Witness 1.9%, Rastafarian 1.1%, other 6.5%, none 21.3%, unspecified 2.3% (2011 est.)

Age structure: 0-14 years: 25.2% (male 360,199/female 347,436)

15-24 years: 17.95% (male 255,102/female 248,927)
25-54 years: 38.06% (male 518,583/female 550,410)
55-64 years: 9.63% (male 133,890/female 136,442)
65 years and over: 9.17% (male 121,969/female 135,612) (2020 est.)

Dependency ratios: total dependency ratio: 48
youth dependency ratio: 34.6
elderly dependency ratio: 13.4
potential support ratio: 7.4 (2020 est.)

Median age: total: 29.4 years
male: 28.6 years
female: 30.1 years (2020 est.)
country comparison to the world: 131

Population growth rate: -0.07% (2020 est.)
country comparison to the world: 201

Birth rate: 16.1 births/1,000 population (2020 est.)
country comparison to the world: 108

Death rate: 7.5 deaths/1,000 population (2020 est.)
country comparison to the world: 104

Net migration rate: -9.4 migrant(s)/1,000 population (2020 est.)
country comparison to the world: 216

Population distribution: population density is high throughout, but increases in and around Kingston, Montego Bay, and Port Esquivel

Urbanization: urban population: 56.3% of total population (2020)
rate of urbanization: 0.82% annual rate of change (2015-20 est.)
total population growth rate v. urban population growth rate, 2000-2030: Major urban areas—population: 591,000 KINGSTON (capital) (2020)

Sex ratio: at birth: 1.05 male(s)/female
0-14 years: 1.04 male(s)/female
15-24 years: 1.02 male(s)/female
25-54 years: 0.94 male(s)/female
55-64 years: 0.98 male(s)/female
65 years and over: 0.9 male(s)/female
total population: 0.98 male(s)/female (2020 est.)
Mother's mean age at first birth: 21.2 years (2008 est.)
note: median age at first birth among women 25-29

Maternal mortality rate: 80 deaths/100,000 live births (2017 est.)
country comparison to the world: 79

Infant mortality rate: total: 11.6 deaths/1,000 live births
male: 13 deaths/1,000 live births
female: 10.1 deaths/1,000 live births (2020 est.)
country comparison to the world: 111

Life expectancy at birth: *total population:* 75.2 years
male: 73.4 years
female: 77.1 years (2020 est.)

country comparison to the world: 120

Total fertility rate: 2.07 children born/woman (2020 est.)

country comparison to the world: 103

Drinking water source:

improved: *urban:* 98.5% of population
rural: 93% of population
total: 96% of population

unimproved: *urban:* 1.5% of population
rural: 7% of population
total: 4% of population (2017 est.)

Current Health Expenditure: 6% (2017)

Physicians density: 1.31 physicians/1,000 population (2017)

Hospital bed density: 1.7 beds/1,000 population (2017)

Sanitation facility access: *improved:* urban: 98.5% of population
rural: 99.5% of population
total: 99% of population
unimproved: urban: 1.5% of population
rural: 0.5% of population
total: 1% of population (2017 est.)

HIV/AIDS—adult prevalence rate: 1.4% (2019 est.)

country comparison to the world: 34

HIV/AIDS—people living with HIV/AIDS: 32,000 (2019 est.)

country comparison to the world: 73

HIV/AIDS—deaths: 1,000 (2019 est.)

country comparison to the world: 59

Obesity—adult prevalence rate: 24.7% (2016)
country comparison to the world: 55

Children under the age of 5 years underweight: 2.2% (2014)

country comparison to the world: 109

Education expenditures: 5.4% of GDP (2018)

country comparison to the world: 43

Literacy: definition: age 15 and over has ever attended school
total population: 88.7%
male: 84%
female: 93.1% (2015)

School life expectancy (primary to tertiary education): *total:* 12 years
male: 11 years
female: 13 years (2015)

Unemployment, youth ages 15-24: *total:* 24.2%
male: 20%
female: 29.3% (2018 est.)
country comparison to the world: 51

GOVERNMENT

Country name: conventional long form: none
conventional short form: Jamaica
etymology: from the native Taino word "haymaca" meaning "Land of Wood and Water" or possibly "Land of Springs"

Government type: parliamentary democracy (Parliament) under a constitutional monarchy; a Commonwealth realm

Capital: name: Kingston

geographic coordinates: 18 00 N, 76 48 W
time difference: UTC-5 (same time as Washington, DC, during Standard Time)
etymology: the name is a blending of the words "king's" and "town"; the English king at the time of the city's founding in 1692 was William III (r. 1689-1702)

Administrative divisions: 14 parishes; Clarendon, Hanover, Kingston, Manchester, Portland, Saint Andrew, Saint Ann, Saint Catherine, Saint Elizabeth, Saint James, Saint Mary, Saint Thomas, Trelawny, Westmoreland
note: for local government purposes, Kingston and Saint Andrew were amalgamated in 1923 into the present single corporate body known as the Kingston and Saint Andrew Corporation

Independence: 6 August 1962 (from the UK)

National holiday: Independence Day, 6 August (1962)

Constitution: history: several previous (preindependence); latest drafted 1961-62, submitted to British Parliament 24 July 1962, entered into force 6 August 1962 (at independence)
amendments: proposed by Parliament; passage of amendments to "non-entrenched" constitutional sections, such as lowering the voting age, requires majority vote by the Parliament membership; passage of amendments to "entrenched" sections, such as fundamental rights and freedoms, requires two-thirds majority vote of Parliament; passage of amendments to "specially entrenched" sections such as the dissolution of Parliament or the executive authority of the monarch requires two-thirds approval by Parliament and approval in a referendum; amended many times, last in 2017

Legal system: common law system based on the English model

International law organization participation: has not submitted an ICJ jurisdiction declaration; non-party state to the ICCt

Citizenship: citizenship by birth: yes
citizenship by descent only: yes
dual citizenship recognized: yes
residency requirement for naturalization: 4 out of the previous 5 years

Suffrage: 18 years of age; universal

Executive branch: chief of state: Queen ELIZABETH II (since 6 February 1952); represented by Governor General Sir Patrick L. ALLEN (since 26 February 2009)
head of government: Prime Minister Andrew HOLNESS (since 3 March 2016)
cabinet: Cabinet appointed by the governor general on the advice of the prime minister
elections/appointments: the monarchy is hereditary; governor general appointed by the monarch on the recommendation of the prime minister; following legislative elections, the leader of the majority party or majority coalition in the House of Representatives is appointed prime minister by the governor general

Legislative branch: description: bicameral Parliament consists of: Senate (21 seats; members appointed by the governor general on the recommendation of the prime minister and the opposition leader—13 seats allocated to the ruling party and 8 to the opposition party; members serve 5-year terms or until Parliament is dissolved)

House of Representatives (63 seats; members directly elected in single-seat constituencies by simple majority vote to serve 5-year terms or until Parliament is dissolved)

elections: Senate—last full slate of appointments on 10 March 2016 (next full slate early on 3 September 2020, following dissolution in mid-August)

House of Representatives—last held on 3 September 2020 (next to be held in 2025)
election results: Senate—percent by party—NA; seats by party—NA; composition—men 16, women 5, percent of women 23.8%

House of Representatives—percent of vote by party—JLP 57%, PNP 42.8%, independent 0.2%; seats by party—JLP 48, PNP 15; composition—men 45, women 18; percent of women 28.6%; note—total Parliament percent of women 27.4%

Judicial branch: highest courts: Court of Appeal (consists of president of the court and a minimum of 4 judges); Supreme Court (40 judges organized in specialized divisions); note—appeals beyond Jamaica's highest courts are referred to the Judicial Committee of the Privy Council (in London) rather than to the Caribbean Court of Justice (the appellate court for member states of the Caribbean Community)
judge selection and term of office: chief justice of the Supreme Court and president of the Court of Appeal appointed by the governor-general on the advice of the prime minister; other judges of both courts appointed by the governor-general on the advice of the Judicial Service Commission; judges of both courts serve till age 70
subordinate courts: resident magistrate courts, district courts, and petty sessions courts

Political parties and leaders: Jamaica Labor Party or JLP [Andrew Michael HOLNESS]
People's National Party or PNP [Dr. Peter David PHILLIPS]
National Democratic Movement or NDM [Peter TOWNSEND]

International organization participation: ACP, AOSIS, C, Caricom, CDB, CELAC, FAO, G-15, G-77, IADB, IAEA, IBRD, ICAO, ICC (NGOs), ICRM, IDA, IFAD, IFC, IFRCS, IHO, ILO, IMF, IMO, Interpol, IOC, IOM, ISO, ITSO, ITU, LAES, MIGA, NAM, OAS, OPANAL, OPCW, Petrocaribe, UN, UNCTAD, UNESCO, UNIDO, UNITAR, UNWTO, UPU, WCO, WFTU (NGOs), WHO, WIPO, WMO, WTO

Diplomatic representation in the US: *chief of mission:* Ambassador Audrey Patrice MARKS (since 18 January 2017)
chancery: 1520 New Hampshire Avenue NW, Washington, DC 20036
telephone: [1] (202) 452-0660
FAX: [1] (202) 452-0036
consulate(s) general: Miami, New York
consulate(s): Atlanta, Boston, Chicago, Concord (MA), Houston, Los Angeles, Philadelphia, Richmond (VA), San Francisco, Seattle

Diplomatic representation from the US: *chief of mission:* Ambassador Donald R. TAPIA (since 11 September 2019)
telephone: [1] (876) 702-6000 (2018)
embassy: 142 Old Hope Road, Kingston 6
mailing address: P.O. Box 541, Kingston 5
FAX: [1] (876) 702-6001 (2018)

Flag description: diagonal yellow cross divides the flag into four triangles—green (top and bottom) and black (hoist side and fly side); green represents hope, vegetation, and agriculture, black reflects hardships overcome and to be faced, and yellow recalls golden sunshine and the island's natural resources

National symbol(s): *green-and-black streamertail (bird), Guaiacum officinale (Guaiacwood); national colors:* green, yellow, black

National anthem: name: Jamaica, Land We Love
lyrics/music: Hugh Braham SHERLOCK/Robert Charles LIGHTBOURNE
note: adopted 1 962

ECONOMY

Economy—overview: The Jamaican economy is heavily dependent on services, which accounts for more than 70% of GDP. The country derives most of its foreign exchange from tourism, remittances, and bauxite/alumina. Earnings from remittances and tourism each account for 14% and 20% of GDP, while bauxite/alumina exports have declined to less than 5% of GDP.

Jamaica's economy has grown on average less than 1% a year for the last three decades and many impediments remain to growth: a bloated public sector which crowds out spending on important projects; high crime and corruption; red-tape; and a high debt-to-GDP ratio. Jamaica, however, has made steady progress in reducing its debt-to-GDP ratio from a high of almost 150% in 2012 to less than 110% in 2017, in close collaboration with the International Monetary Fund (IMF). The current IMF Stand-By Agreement requires Jamaica to produce an annual primary surplus of 7%, in an attempt to reduce its debt burden below 60% by 2025.

Economic growth reached 1.6% in 2016, but declined to 0.9% in 2017 after intense rainfall, demonstrating the vulnerability of the economy to weather-related events. The HOLNESS administration therefore faces the difficult prospect of maintaining fiscal discipline to reduce the debt load while simultaneously implementing growth inducing policies and attacking a serious crime problem. High unemployment exacerbates the crime problem, including gang violence fueled by advanced fee fraud (lottery scamming) and the drug trade.

GDP (purchasing power parity): $26.06 billion (2017 est.)
$25.89 billion (2016 est.)
$25.51 billion (2015 est.)
note: data are in 2017 dollars
country comparison to the world: 140

GDP (official exchange rate): $14.77 billion (2017 est.)

GDP—real growth rate: 0.7% (2017 est.)
1.5% (2016 est.)
0.9% (2015 est.)
country comparison to the world: 181

GDP—per capita (PPP): $9,200 (2017 est.)
$9,200 (2016 est.)
$9,100 (2015 est.)
note: data are in 2017 dollars
country comparison to the world: 143

Gross national saving: 18.3% of GDP (2017 est.)
20.6% of GDP (2016 est.)
18% of GDP (2015 est.)
country comparison to the world: 109

GDP—composition, by end use: *household consumption:* 81.9% (2017 est.)
government consumption: 13.7% (2017 est.)
investment in fixed capital: 21.3% (2017 est.)
investment in inventories: 0.1% (2017 est.)
exports of goods and services: 30.1% (2017 est.)
imports of goods and services: -47.1% (2017 est.)

GDP—composition, by sector of origin: *agriculture:* 7% (2017 est.
industry: 21.1% (2017 est.)
services: 71.9% (2017 est.)

Agriculture—products: sugar cane, bananas, coffee, citrus, yams, ackees, vegetables; poultry, goats, milk; shellfish
Industries: agriculture, mining, manufacture, construction, financial and insurance services, tourism, telecommunications

Industrial production growth rate: 0.9% (2017 est.)
country comparison to the world: 162

Labor force: 1.113 million (2020 est.)
country comparison to the world: 138

Labor force—by occupation: *agriculture:* 16.1%
industry: 16%
services: 67.9% (2017)

Unemployment rate: 7.72% (2019 est.)
9.13% (2018 est.)
country comparison to the world: 120

Population below poverty line: 17.1% (2016 est.)

Household income or consumption by percentage share: *lowest 10%:* 2.6%
highest 10%: 29.3% (2015)

Budget: revenues: 4.382 billion (2017 est.)
expenditures: 4.314 billion (2017 est.)

Taxes and other revenues: 29.7% (of GDP) (2017 est.)
country comparison to the world: 81

Budget surplus (+) or deficit (-): 0.5% (of GDP) (2017 est.)
country comparison to the world: 37

Public debt: 101% of GDP (2017 est.)
113.6% of GDP (2016 est.)
country comparison to the world: 16

Fiscal year: 1 April–31 March
Inflation rate (consumer prices): 4.4% (2017 est.)
2.3% (2016 est.)
country comparison to the world: 166

Current account balance: -$298 million (2019 est.)
-$288 million (2018 est.)
country comparison to the world: 107

Exports: $1.296 billion (2017 est.)
$1.195 billion (2016 est.)
country comparison to the world: 151
Exports—partners: US 39.1%, Netherlands 12.3%, Canada 8.4% (2017)
Exports—commodities: alumina, bauxite, chemicals, coffee, mineral fuels, waste and scrap metals, sugar, yams

Imports: $5.151 billion (2017 est.)
$4.169 billion (2016 est.)
country comparison to the world: 126

Imports—commodities: food and other consumer goods, industrial supplies, fuel, parts and accessories of capital goods, machinery and transport equipment, construction materials

Imports—partners: US 40.6%, Colombia 6.8%, Japan 5.8%, China 5.8%, Trinidad and Tobago 4.7% (2017)

Reserves of foreign exchange and gold: $3.781 billion (31 December 2017 est.)
$2.719 billion (31 December 2016 est.)
country comparison to the world: 100

Debt—external: $14.94 billion (31 December 2017 est.)
$10.24 billion (31 December 2016 est.)
country comparison to the world: 103

Exchange rates: Jamaican dollars (JMD) per US dollar –
128.36 (2017 est.)
125.14 (2016 est.)
125.126 (2015 est.)
116.898 (2014 est.)
110.935 (2013 est.)

ENERGY

Electricity access: electrification—total population: 98.2% (2016)
electrification—urban areas: 100% (2016)
electrification—rural areas: 96% (2016)

Electricity—production: 4.007 billion kWh (2016 est.)
country comparison to the world: 128

Electricity—consumption: 2.847 billion kWh (2016 est.)
country comparison to the world: 137

Electricity—exports: 0 kWh (2016 est.)
country comparison to the world: 151

Electricity—imports: 0 kWh (2016 est.)
country comparison to the world: 162

Electricity—installed generating capacity: 1.078 million kW (2016 est.)
country comparison to the world: 126

Electricity—from fossil fuels: 83% of total installed capacity (2016 est.)
country comparison to the world: 76

Electricity—from nuclear fuels: 0% of total installed capacity (2017 est.)
country comparison to the world: 115

Electricity—from hydroelectric plants: 3% of total installed capacity (2017 est.)
country comparison to the world: 134

Electricity—from other renewable sources: 15% of total installed capacity (2017 est.)
country comparison to the world: 59

Crude oil—production: 0 bbl/day (2018 est.)
country comparison to the world: 153

Crude oil—exports: 0 bbl/day (2015 est.)
country comparison to the world: 142

Crude oil—imports: 24,360 bbl/day (2015 est.)
country comparison to the world: 61

Crude oil—proved reserves: 0 bbl (1 January 2018 est.)
country comparison to the world: 149

Refined petroleum products—production: 24,250 bbl/day (2017 est.)
country comparison to the world: 87

Refined petroleum products—consumption: 55,000 bbl/day (2016 est.)
country comparison to the world: 99

Refined petroleum products—exports: 823 bbl/day (2015 est.)
country comparison to the world: 109

Refined petroleum products—imports: 30,580 bbl/day (2015 est.)
country comparison to the world: 100

Natural gas—production: 0 cu m (2017 est.)
country comparison to the world: 149

Natural gas—consumption: 198.2 million cu m (2017 est.)
country comparison to the world: 104

Natural gas—exports: 0 cu m (2017 est.)
country comparison to the world: 127

Natural gas—imports: 198.2 million cu m (2017 est.)
country comparison to the world: 71

Natural gas—proved reserves: 0 cu m (1 January 2014 est.)
country comparison to the world: 150

Carbon dioxide emissions from consumption of energy: 8.9 million Mt (2017 est.)
country comparison to the world: 112

COMMUNICATIONS

Telephones—fixed lines: *total subscriptions:* 379,420
subscriptions per 100 inhabitants: 13.5 (2019 est.)
country comparison to the world: 103

Telephones—mobile cellular: *total subscriptions:* 2,882,469
subscriptions per 100 inhabitants: 102.56 (2019 est.)
country comparison to the world: 143

Telecommunication systems: general assessment: good domestic and international service; mobile sector dominates, accounting for 82% of the Internet connections; extensive LTE networks providing coverage to 90% of the island population (2020)

domestic: while the number of fixed-lines, 14 per 100, subscriptions has declined, cellular-mobile has grown 103 per 100 subscriptions (2019)
international: country code—1-876 and 1-658; landing points for the ALBA-1, CFX-1, Fibralink, East-West, and Cayman- Jamaican Fiber System submarine cables providing connections to South America, parts of the Caribbean, Central America and the US; satellite earth stations—2 Intelsat (Atlantic Ocean) (2019)
note: the COVID-19 outbreak is negatively impacting telecommunications production and supply chains globally; consumer spending on telecom devices and services has also slowed due to the pandemic's effect on economies worldwide; overall progress towards improvements in all facets of the telecom industry—mobile, fixed-line, broadband, submarine cable and satellite—has moderated

Broadcast media: 3 free-to-air TV stations, subscription cable services, and roughly 30 radio stations (2019)

Internet country code: .jm

Internet users: total: 1,548,618
percent of population: 55.07% (July 2018 est.)
country comparison to the world: 130

Broadband—fixed subscriptions: *total:* 284,756
subscriptions per 100 inhabitants: 10 (2018 est.)
country comparison to the world: 103

TRANSPORTATION

National air transport system: *number of registered air carriers:* 0 (2020)

Civil aircraft registration country code prefix: 6Y (2016)

Airports: 28 (2013)
country comparison to the world: 121

Airports—with paved runways: *total:* 11 (2017)
2,438 to 3,047 m: 2 (2017)
914 to 1,523 m: 4 (2017)
under 914 m: 5 (2017)

Airports—with unpaved runways: *total:* 17 (2013)
914 to 1,523 m: 1 (2013)
under 914 m: 16 (2013)

Roadways: *total:* 22,121 km (includes 44 km of expressways) (2011)
paved: 16,148 km (2011)
unpaved: 5,973 km (2011)
country comparison to the world: 110

Merchant marine: total: 39
by type: bulk carrier 1, container ship 8, general cargo 6, other 24 (2019)
country comparison to the world: 122

Ports and terminals: *major seaport(s):* Discovery Bay (Port Rhoades), Kingston, Montego Bay, Port Antonio, Port Esquivel, Port Kaiser, Rocky Point
container port(s) (TEUs): Kingston (1,681,706) (2017)

MILITARY AND SECURITY

Military and security forces: *Jamaica Defense Force (JDF):* Jamaica Regiment (Ground Forces), Maritime-Air-Cyber Command (2020)

Military expenditures: 1.6% of GDP (2019)
1.3% of GDP (2018)
1% of GDP (2017)
1% of GDP (2016)
1% of GDP (2015)
country comparison to the world: 74

Military and security service personnel strengths: assessments of the size of the the Jamaica Defense Forces vary; approximately 3,500 active personnel (3,000 Ground Forces; 300 Coast Guard; 200 Air Wing) (2019)

Military equipment inventories and acquisitions: the Jamaica Defense Force's inventory is limited and features mostly older equipment imported from a variety of foreign suppliers, including the UK and US; since 2010, Jamaica has received limited quantities of military equipment from Australia, Austria, the Netherlands, and the US (2019 est.)

Military service age and obligation: 17 1/2 is the legal minimum age for voluntary military service; no conscription (2012)

TRANSNATIONAL ISSUES

Disputes—international: none

Trafficking in persons: *current situation:* Jamaica is a source and destination country for children and adults subjected to sex trafficking and forced labor; sex trafficking of children and adults occurs on the street, in night clubs, bars, massage parlors, and private homes; child sex tourism is a problem in resort areas; Jamaicans have been subjected to sexual exploitation or forced labor in the Caribbean, Canada, the US, and the UK, while foreigners have endured conditions of forced labor in Jamaica or aboard foreign-flagged fishing vessels operating in Jamaican waters; a high number of Jamaican children are reported missing
tier rating: Tier 2 Watch List — Jamaica does not fully comply with the minimum standards for the elimination of trafficking; however, it is making significant efforts to do so; in 2014, the government made significant efforts to raise public awareness of human trafficking, and named a national trafficking-in-persons rapporteur — the first in the region; authorities initiated more new trafficking investigations than in 2013 and concluded a trafficking case in the Supreme Court, but chronic delays impeded prosecutions and no offenders were convicted for the sixth consecutive year; more adult trafficking victims were identified than in previous years, but only one child victim was identified, which was exceptionally low relative to the number of vulnerable children (2015)

Illicit drugs: transshipment point for cocaine from South America to North America and Europe; illicit cultivation and consumption of cannabis; government has an active manual cannabis eradication program; corruption is a major concern; substantial money-laundering activity; Colombian narcotics traffickers favor Jamaica for illicit financial transactions

JAN MAYEN

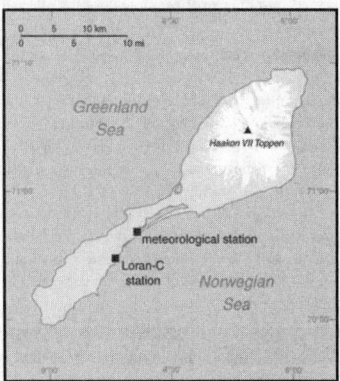

INTRODUCTION

Background: This desolate, arctic, mountainous island was named after a Dutch whaling captain who indisputably discovered it in 1614 (earlier claims are inconclusive). Visited only occasionally by seal hunters and trappers over the following centuries, the island came under Norwegian sovereignty in 1929. The long dormant Beerenberg volcano, the northernmost active volcano on earth, resumed activity in 1970 and the most recent eruption occurred in 1985.

GEOGRAPHY

Location: Northern Europe, island between the Greenland Sea and the Norwegian Sea, northeast of Iceland

Geographic coordinates: 71 00 N, 8 00 W

Map references:

Arctic Region Area: *total:* 377 sq km
land: 377 sq km
water: 0 sq km
country comparison to the world: 206

Area—comparative: slightly more than twice the size of Washington, DC

Land boundaries: 0 km

Coastline: 124.1 km

Maritime claims: *territorial sea:* 12 nm

exclusive economic zone: 200 nm
contiguous zone: 24 nm
continental shelf: 200-m depth or to the depth of exploitation

Climate: arctic maritime with frequent storms and persistent fog

Terrain: volcanic island, partly covered by glaciers

Elevation: *lowest point:* Norwegian Sea 0 m
highest point: Haakon VII Toppen on Beerenberg 2,277
note: Beerenberg volcano has numerous peaks; the highest point on the volcano rim is named Haakon VII Toppen, after Norway's first king following the reestablishment of Norwegian independence in 1905

Natural resources: none

Land use: *agricultural land:* 0% (2011 est.)
arable land: 0% (2011 est.) / permanent crops: 0% (2011 est.) / permanent pasture: 0% (2011 est.)
forest: 0% (2011 est.)
other: 100% (2011 est.)

Irrigated land: 0 sq km (2012)

Natural hazards: dominated by the volcano Beerenberg
volcanism: Beerenberg (2,227 m) is Norway's only active volcano; volcanic activity resumed in 1970; the most recent eruption occurred in 1985

Environment—current issues: pollutants transported from southerly latitudes by winds, ocean currents, and rivers accumulate in the food chains of native animals; climate change

Geography—note: *barren volcanic spoon-shaped island with some moss and grass flora; island consists of two parts:* a larger northeast Nord-Jan (the spoon "bowl") and the smaller Sor-Jan (the "handle"), linked by a 2.5 km-wide isthmus (the "stem") with two large lakes, Sorlaguna (South Lagoon) and Nordlaguna (North Lagoon)

PEOPLE AND SOCIETY

Population: no indigenous inhabitants
note: military personnel operate the the weather and coastal services radio station

GOVERNMENT

Country name: *conventional long form:* none

conventional short form: Jan Mayen
etymology: named after Dutch Captain Jan Jacobszoon MAY, one of the first explorers to reach the island in 1614

Dependency status: territory of Norway; since August 1994, administered from Oslo through the county governor (fylkesmann) of Nordland; however, authority has been delegated to a station commander of the Norwegian Defense Communication Service; in 2010, Norway designated the majority of Jan Mayen as a nature reserve

Legal system: the laws of Norway apply where applicable

Flag description: the flag of Norway is used

ECONOMY

Economy—overview: Jan Mayen is a volcanic island with no exploitable natural resources, although surrounding waters contain substantial fish stocks and potential untapped petroleum resources. Economic activity is limited to providing services for employees of Norway's radio and meteorological stations on the island.

COMMUNICATIONS

Broadcast media: a coastal radio station has been remotely operated since 1994

TRANSPORTATION

Airports: 1 (2013)
country comparison to the world: 225

Airports—with unpaved runways:
total: 1 (2013)
1,524 to 2,437 m: 1 (2013)

Ports and terminals: none; offshore anchorage only

MILITARY AND SECURITY

Military—note: defense is the responsibility of Norway

TRANSNATIONAL ISSUES

Disputes—international: none

JAPAN

INTRODUCTION

Background: In 1603, after decades of civil warfare, the Tokugawa shogunate (a military-led, dynastic government) ushered in a long period of relative political stability and isolation from foreign influence. For more than two centuries this policy enabled Japan to enjoy a flowering of its indigenous culture. Japan opened its ports after signing the Treaty of Kanagawa with the US in 1854 and began to intensively modernize and industrialize. During the late 19th and early 20th centuries, Japan became a regional power that was able to defeat the forces of both China and Russia. It occupied Korea, Formosa (Taiwan), and southern Sakhalin Island. In 1931-32 Japan

occupied Manchuria, and in 1937 it launched a full-scale invasion of China. Japan attacked US forces in 1941—triggering America's entry into World War II—and soon occupied much of East and Southeast Asia. After its defeat in World War II, Japan recovered to become an economic power and an ally of the US. While the emperor retains his throne as a symbol of national unity, elected politicians hold actual decisionmaking power. Following three decades of unprecedented growth, Japan's economy experienced a major slowdown starting in the 1990s, but the country remains an economic power. In March 2011, Japan's strongest-ever earthquake, and an accompanying tsunami, devastated the northeast part of Honshu island, killed thousands, and damaged several nuclear power plants. Prime Minister Shinzo ABE was reelected to office in December 2012, and has since embarked on ambitious economic and security reforms to improve Japan's economy and bolster the country's international standing. In November 2019, ABE became Japan's longest-serving post-war prime minister.

GEOGRAPHY

Location: Eastern Asia, island chain between the North Pacific Ocean and the Sea of Japan, east of the Korean Peninsula

Geographic coordinates: 36 00 N, 138 00 E

Map references: Asia

Area: *total:* 377,915 sq km
land: 364,485 sq km
water: 13,430 sq km
note: includes Bonin Islands (Ogasawara-gunto), Daito-shoto, Minami-jima, Okino-tori-shima, Ryukyu Islands (Nansei- shoto), and Volcano Islands (Kazan-retto)
country comparison to the world: 63

Area—comparative: slightly smaller than California

Land boundaries: 0 km

Coastline: 29,751 km

Maritime claims: *territorial sea:* 12 nm; between 3 nm and 12 nm in the international straits—La Perouse or Soya, Tsugaru, Osumi, and Eastern and Western Channels of the Korea or Tsushima Strait
exclusive economic zone: 200 nm
contiguous zone: 24 nm

Climate: varies from tropical in south to cool temperate in north

Terrain: mostly rugged and mountainous

Elevation: *mean elevation:* 438 m
lowest point: Hachiro-gata -4 m
highest point: Mount Fuji 3,776 m

Natural resources: negligible mineral resources, fish, note, with virtually no natural energy resources, Japan is the world's largest importer of coal and liquefied natural gas, as well as the second largest importer of oil

Land use: *agricultural land:* 12.5% (2011 est.)
arable land: 11.7% (2011 est.) / permanent crops: 0.8% (2011 est.) / permanent pasture: 0% (2011 est.)
forest: 68.5% (2011 est.)
other: 19% (2011 est.)
Irrigated land: 24,690 sq km (2012)

Population distribution: all primary and secondary regions of high population density lie on the coast; one-third of the population resides in and around Tokyo on the central plain (Kanto Plain)

Natural hazards: many dormant and some active volcanoes; about 1,500 seismic occurrences (mostly tremors but occasional severe earthquakes) every year; tsunamis; typhoons
volcanism: both Unzen (1,500 m) and Sakurajima (1,117 m), which lies near the densely populated city of Kagoshima, have been deemed Decade Volcanoes by the International Association of Volcanology and Chemistry of the Earth's Interior, worthy of study due to their explosive history and close proximity to human populations; other notable historically active volcanoes include Asama, Honshu Island's most active volcano, Aso, Bandai, Fuji, Iwo-Jima, Kikai, Kirishima, Komaga-take, Oshima, Suwanosejima, Tokachi, Yake-dake, and Usu; see note 2 under "Geography—note"

Environment—current issues: air pollution from power plant emissions results in acid rain; acidification of lakes and reservoirs degrading water quality and threatening aquatic life; Japan is one of the largest consumers of fish and tropical timber, contributing to the depletion of these resources in Asia and elsewhere; following the 2011 Fukushima nuclear disaster, Japan originally planned to phase out nuclear power, but it has now implemented a new policy of seeking to restart nuclear power plants that meet strict new safety standards; waste management is an ongoing issue; Japanese municipal facilities used to burn high volumes of trash, but air pollution issues forced the government to adopt an aggressive recycling policy

Environment—international agreements: *party to:* Antarctic-Environmental Protocol, Antarctic-Marine Living Resources, Antarctic Seals, Antarctic Treaty, Biodiversity, Climate Change, Climate Change-Kyoto Protocol, Desertification, Endangered Species, Environmental Modification, Hazardous Wastes, Law of the Sea, Marine Dumping, Ozone Layer Protection, Ship Pollution, Tropical Timber 83, Tropical Timber 94, Wetlands, Whaling
signed, but not ratified: none of the selected agreements

Geography—note: *note 1:* strategic location in northeast Asia; composed of four main islands—from north: Hokkaido, Honshu (the largest and most populous), Shikoku, and Kyushu (the "Home Islands")—and 6,848 smaller islands and islets
note 2: Japan annually records the most earthquakes in the world; it is one of the countries along the Ring of Fire, a belt of active volcanoes and earthquake epicenters bordering the Pacific Ocean; up to 90% of the world's earthquakes and some 75% of the world's volcanoes occur within the Ring of Fire

PEOPLE AND SOCIETY

Population: 125,507,472 (July 2020 est.)
country comparison to the world: 11

Nationality: *noun:* Japanese (singular and plural)
adjective: Japanese

Ethnic groups: Japanese 98.1%, Chinese 0.5%, Korean 0.4%, other 1% (includes Filipino, Vietnamese, and Brazilian) (2016 est.)
note: data represent population by nationality; up to 230,000 Brazilians of Japanese origin migrated to Japan in the 1990s to work in industries; some have returned to Brazil

Languages: Japanese

Religions: Shintoism 70.4%, Buddhism 69.8%, Christianity 1.5%, other 6.9% (2015 est.)
note: total adherents exceeds 100% because many people practice both Shintoism and Buddhism

Age structure: *0-14 years:* 12.49% (male 8,047,183/female 7,623,767)
15-24 years: 9.47% (male 6,254,352/female 5,635,377)
25-54 years: 36.8% (male 22,867,385/female 23,317,140)
55-64 years: 12.06% (male 7,564,067/female 7,570,732)
65 years and over: 29.18% (male 16,034,973/female 20,592,496) (2020 est.)

Dependency ratios: *total dependency ratio:* 69
youth dependency ratio: 21
elderly dependency ratio: 48
potential support ratio: 2.1 (2020 est.)

Median age: *total:* 48.6 years
male: 47.2 years
female: 50 years (2020 est.)
country comparison to the world: 2

Population growth rate: -0.27% (2020 est.)
country comparison to the world: 216

Birth rate: 7.3 births/1,000 population (2020 est.)
country comparison to the world: 226

Death rate: 10.2 deaths/1,000 population (2020 est.)
country comparison to the world: 34

Net migration rate: 0 migrant(s)/1,000 population (2020 est.)

country comparison to the world: 86

Population distribution: all primary and secondary regions of high population density lie on the coast; one-third of the population resides in and around Tokyo on the central plain (Kanto Plain)

Urbanization: *urban population:* 91.8% of total population (2020)

rate of urbanization: -0.14% annual rate of change (2015-20 est.)

total population growth rate v. urban population growth rate, 2000-2030:

Major urban areas—population: 37.393 million TOKYO (capital), 19.165 million Osaka, 9.552 million Nagoya, 5.29 million Kitakyushu-Fukuoka, 2.922 million Shizuoka-Hamamatsu, 2.670 million Sapporo (2020)

Sex ratio: *at birth:* 1.06 male(s)/female
0-14 years: 1.06 male(s)/female
15-24 years: 1.11 male(s)/female
25-54 years: 0.98 male(s)/female
55-64 years: 1 male(s)/female
65 years and over: 0.78 male(s)/female
total population: 0.94 male(s)/female (2020 est.)

Mother's mean age at first birth: 30.7 years (2015 est.)

Maternal mortality rate: 5 deaths/100,000 live births (2017 est.)
country comparison to the world: 166

Infant mortality rate: *total:* 1.9 deaths/1,000 live births
male: 2.1 deaths/1,000 live births
female: 1.7 deaths/1,000 live births (2020 est.)
country comparison to the world: 226

Life expectancy at birth:
total population: 86 years
male: 82.7 years
female: 89.5 years (2020 est.)
country comparison to the world: 2

Total fertility rate: 1.43 children born/woman (2020 est.)
country comparison to the world: 214

Contraceptive prevalence rate: 39.8% (2015)
note: percent of women aged 20-49

Drinking water source:
improved:
total: 100% of population
unimproved:
total: 0% of population (2017 est.)

Current Health Expenditure: 10.9% (2017)

Physicians density: 2.41 physicians/1,000 population (2016)

Hospital bed density: 13.1 beds/1,000 population (2017)

Sanitation facility access: *improved:*
total: 100% of population
unimproved:
total: 0% of population (2017 est. est.)

HIV/AIDS—adult prevalence rate: <.1% (2018 est.)

HIV/AIDS—people living with HIV/AIDS: 30,000 (2018 est.)

country comparison to the world: 74

HIV/AIDS—deaths: <200 (2017 est.)

Major infectious diseases: Covid-19 (see note) (2020)

note: clusters of cases of respiratory illness caused by a new coronavirus (COVID-19) in Japan; illness with this virus has ranged from mild to severe with fatalities reported; as of 10 November 2020, Japan has reported 107,086 confirmed cases of COVID19 with 1,812 deaths; the US Department of State has issued a Travel Advisory recommending increased caution in Japan due to the recent outbreak of COVID-19; the Centers for Disease Control and Prevention has recommended postponing nonessential international travel at this time and published additional guidance at https://wwwnc.cdc.gov/travel/notices/alert/; on 25 May 2020, Japan ended its state of emergency

Obesity—adult prevalence rate: 4.3% (2016)
country comparison to the world: 186

Children under the age of 5 years underweight: 3.4% (2010)
country comparison to the world: 96

Education expenditures: 3.5% of GDP (2016)
country comparison to the world: 122

School life expectancy (primary to tertiary education): *total:* 15 years
male: 15 years
female: 15 years (2016)

Unemployment, youth ages 15-24:
total: 3.6%
male: 4.1%
female: 3.1% (2018 est.)
country comparison to the world: 171

GOVERNMENT

Country name: *conventional long form:* none
conventional short form: Japan
local long form: Nihon-koku/Nippon-koku
local short form: Nihon/Nippon
etymology: the English word for Japan comes via the Chinese name for the country "Cipangu"; both Nihon and Nippon mean "where the sun originates" and are frequently translated as "Land of the Rising Sun"

Government type: parliamentary constitutional monarchy

Capital: *name:* Tokyo

geographic coordinates: 35 41 N, 139 45 E

time difference: UTC+9 (14 hours ahead of Washington, DC, during Standard Time)
etymology: originally known as Edo, meaning "estuary" in Japanese, the name was changed to Tokyo, meaning "eastern capital," in 1868

Administrative divisions: 47 prefectures; Aichi, Akita, Aomori, Chiba, Ehime, Fukui, Fukuoka, Fukushima, Gifu, Gunma, Hiroshima, Hokkaido, Hyogo, Ibaraki, Ishikawa, Iwate, Kagawa, Kagoshima, Kanagawa, Kochi, Kumamoto, Kyoto, Mie, Miyagi, Miyazaki, Nagano, Nagasaki, Nara, Niigata, Oita, Okayama, Okinawa, Osaka, Saga, Saitama, Shiga, Shimane, Shizuoka, Tochigi,

Tokushima, Tokyo, Tottori, Toyama, Wakayama, Yamagata, Yamaguchi, Yamanashi

Independence: 3 May 1947 (current constitution adopted as amendment to Meiji Constitution); notable earlier dates: 11 February 660 B.C. (mythological date of the founding of the nation by Emperor JIMMU); 29 November 1890 (Meiji Constitution provides for constitutional monarchy)

National holiday: Birthday of Emperor NARUHITO, 23 February (1960); note—celebrates the birthday of the current emperor

Constitution: *history:* previous 1890; latest approved 6 October 1946, adopted 3 November 1946, effective 3 May 1947
amendments: proposed by the Diet; passage requires approval by at least two-thirds majority of both houses of the Diet and approval by majority in a referendum; note—the constitution has not been amended since its enactment in 1947

Legal system: civil law system based on German model; system also reflects Anglo-American influence and Japanese traditions; judicial review of legislative acts in the Supreme Court

International law organization participation: accepts compulsory ICJ jurisdiction with reservations; accepts ICCt jurisdiction

Citizenship: *citizenship by birth:* no
citizenship by descent only: at least one parent must be a citizen of Japan
dual citizenship recognized: no
residency requirement for naturalization: 5 years

Suffrage: 18 years of age; universal

Executive branch: *chief of state:* Emperor NARUHITO (since 1 May 2019); note—succeeds his father who abdicated on 30 April 2019

head of government: Prime Minister Yoshihide SUGA (since 16 September 2020); Deputy Prime Minister Taro ASO (since 26 December 2012)
cabinet: Cabinet appointed by the prime minister
elections/appointments: the monarchy is hereditary; the leader of the majority party or majority coalition in the House of Representatives usually becomes prime minister

Legislative branch: *description:* bicameral Diet or Kokkai consists of:
House of Councillors or Sangi-in (242 seats; 146 members directly elected in multi-seat districts by simple majority vote and 96 directly elected in a single national constituency by proportional representation vote; members serve 6-year terms with half the membership renewed every 3 years)

House of Representatives or Shugi-in (465 seats; 289 members directly elected in single-seat districts by simple majority vote and 176 directly elected in multi-seat districts by party-list proportional representation vote; members serve 4-year terms)

elections: House of Councillors—last held on 10 July 2016 (next to be held in July 2019)

House of Representatives—last held on 22 October 2017 (next to be held by 21 October 2021) *election results:* House of Councillors—percent of vote by party—NA; seats by party—LDP 55, DP 32, Komeito 14, JCP 6, Osaka Ishin no Kai (Initiatives from Osaka) 7, PLPTYF 1, SDP 1, independent 5

House of Representatives—percent of vote by party—NA; seats by party—LDP 284, CDP 55, Party of Hope 50, Komeito 29, JCP 12, JIP 11, SDP 2, independent 22

note: the Diet in June 2017 redrew Japan's electoral district boundaries and reduced from 475 to 465 seats in the House of Representatives; the amended electoral law, which cuts 6 seats in single-seat districts and 4 in multi-seat districts, was reportedly intended to reduce voting disparities between densely and sparsely populated voting districts

Judicial branch: *highest courts:* Supreme Court or Saiko saibansho (consists of the chief justice and 14 associate justices); note—the Supreme Court has jurisdiction in constitutional issues

judge selection and term of office: Supreme Court chief justice designated by the Cabinet and appointed by the monarch; associate justices appointed by the Cabinet and confirmed by the monarch; all justices are reviewed in a popular referendum at the first general election of the House of Representatives following each judge's appointment and every 10 years afterward

subordinate courts: 8 High Courts (Koto-saibansho), each with a Family Court (Katei-saiban-sho); 50 District Courts (Chiho saibansho), with 203 additional branches; 438 Summary Courts (Kani saibansho)

Political parties and leaders: Constitutional Democratic Party of Japan or CDP [Yukio EDANO]

Democratic Party of Japan or DPJ [Kohei OTSUKA]

Group of Reformists [Sakihito OZAWA]

Initiatives from Osaka (Osaka Ishin no kai) [Ichiro MATSUI]

Japan Communist Party or JCP [Kazuo SHII]

Japan Innovation Party or JIP [Ichiro MATSUI]

Party of Hope or Kibo no To [Yuichiro TAMAKI]

Komeito [Natsuo YAMAGUCHI]

Liberal Democratic Party or LDP [Yoshihide SUGA]

Liberal Party [Ichiro OZAWA] (formerly People's Life Party & Taro Yamamoto and Friends or PLPTYF)New Renaissance Party [Hiroyuki ARAI]

Party for Japanese Kokoro or PJK [Masashi NAKANO]Social Democratic Party or SDP [Tadatomo YOSHIDA]The Assembly to Energize Japan and the Independents [Kota MATSUDA]

International organization participation: ADB, AfDB (nonregional member), APEC, Arctic Council (observer), ARF, ASEAN (dialogue partner), Australia Group, BIS, CD, CE (observer), CERN (observer), CICA (observer), CP, CPLP (associate), EAS, EBRD, EITI (implementing country), FAO, FATF, G-5, G-7, G-8, G-10, G-20, IADB, IAEA, IBRD, ICAO, ICC (national committees), ICCt, ICRM, IDA, IEA, IFAD, IFC, IFRCS, IGAD (partners), IHO, ILO, IMF, IMO,

IMSO, Interpol, IOC, IOM, IPU, ISO, ITSO, ITU, ITUC (NGOs), LAIA (observer), MIGA, NEA, NSG, OAS (observer), OECD, OPCW, OSCE (partner), Pacific Alliance (observer), Paris Club, PCA, PIF (partner), SAARC (observer), SELEC (observer), SICA (observer), UN, UNCTAD, UNESCO, UNHCR, UNIDO, UNMISS, UNRWA, UNWTO, UPU, WCO, WFTU (NGOs), WHO, WIPO, WMO, WTO, ZC

Diplomatic representation in the US: *chief of mission:* Ambassador Shinsuke SUGIYAMA (since 28 March 2018) (2018)

chancery: 2520 Massachusetts Avenue NW, Washington, DC 20008

telephone: [1] (202) 238-6700

FAX: [1] (202) 328-2187

consulate(s) general: Anchorage (AK), Atlanta, Boston, Chicago, Dallas, Denver (CO), Detroit (MI), Honolulu, Houston, Las Vegas (NV), Los Angeles, Miami, Nashville (TN), New Orleans, New York, Oklahoma City (OK), Orlando (FL), Philadelphia, Phoenix (AZ), Portland (OR), San Francisco, Seattle, Saipan (Northern Mariana Islands), Tamuning (Guam)

Diplomatic representation from the US: *chief of mission:* Ambassador (vacant); Charge d'Affaires Joseph M. YOUNG (since 22 July 2019)

telephone: [81] (03) 3224-5000

embassy: 1-10-5 Akasaka, Minato-ku, Tokyo 107-8420

mailing address: Unit 9800, Box 300, APO AP 96303-0300

FAX: [81] (03) 3505-1862

consulate(s) general: Naha (Okinawa), Osaka-Kobe, Sapporo

consulate(s): Fukuoka, Nagoya

Flag description: white with a large red disk (representing the sun without rays) in the center

National symbol(s): *red sun disc, chrysanthemum; national colors:* red, white

National anthem: *name:* "Kimigayo" (The Emperor"s Reign)

lyrics/music: unknown/Hiromori HAYASHI

note: adopted 1999; unofficial national anthem since 1883; oldest anthem lyrics in the world, dating to the 10th century or earlier; there is some opposition to the anthem because of its association with militarism and worship of the emperor

0:00 / 0:58

ECONOMY

Economy—overview: Over the past 70 years, government-industry cooperation, a strong work ethic, mastery of high technology, and a comparatively small defense allocation (slightly less than 1% of GDP) have helped Japan develop an advanced economy. Two notable characteristics of the post-World War II economy were the close interlocking structures of manufacturers, suppliers, and distributors, known as keiretsu, and the guarantee of lifetime employment for a substantial portion of the urban labor force. Both features have significantly eroded under the dual pressures

of global competition and domestic demographic change.

Measured on a purchasing power parity basis that adjusts for price differences, Japan in 2017 stood as the fourth-largest economy in the world after first-place China, which surpassed Japan in 2001, and third-place India, which edged out Japan in 2012. For three postwar decades, overall real economic growth was impressive—averaging 10% in the 1960s, 5% in the 1970s, and 4% in the 1980s. Growth slowed markedly in the 1990s, averaging just 1.7%, largely because of the aftereffects of inefficient investment and the collapse of an asset price bubble in the late 1980s, which resulted in several years of economic stagnation as firms sought to reduce excess debt, capital, and labor. Modest economic growth continued after 2000, but the economy has fallen into recession four times since 2008.

Japan enjoyed an uptick in growth since 2013, supported by Prime Minister Shinzo ABE's "Three Arrows" economic revitalization agenda—dubbed "Abenomics"—of monetary easing, "flexible" fiscal policy, and structural reform. Led by the Bank of Japan's aggressive monetary easing, Japan is making modest progress in ending deflation, but demographic decline – a low birthrate and an aging, shrinking population – poses a major long-term challenge for the economy. The government currently faces the quandary of balancing its efforts to stimulate growth and institute economic reforms with the need to address its sizable public debt, which stands at 235% of GDP. To help raise government revenue, Japan adopted legislation in 2012 to gradually raise the consumption tax rate. However, the first such increase, in April 2014, led to a sharp contraction, so Prime Minister ABE has twice postponed the next increase, which is now scheduled for October 2019. Structural reforms to unlock productivity are seen as central to strengthening the economy in the long-run.

Scarce in critical natural resources, Japan has long been dependent on imported energy and raw materials. After the complete shutdown of Japan's nuclear reactors following the earthquake and tsunami disaster in 2011, Japan's industrial sector has become even more dependent than before on imported fossil fuels. However, ABE's government is seeking to restart nuclear power plants that meet strict new safety standards and is emphasizing nuclear energy's importance as a base-load electricity source. In August 2015, Japan successfully restarted one nuclear reactor at the Sendai Nuclear Power Plant in Kagoshima prefecture, and several other reactors around the country have since resumed operations; however, opposition from local governments has delayed several more restarts that remain pending. Reforms of the electricity and gas sectors, including full liberalization of Japan's energy market in April 2016 and gas market in April 2017, constitute an important part of Prime Minister Abe's economic program.

Under the Abe Administration, Japan's government sought to open the country's economy to greater foreign competition and create new export opportunities for Japanese businesses, including by joining 11 trading partners in the Trans-Pacific Partnership (TPP). Japan became the first country to ratify the TPP in December 2016, but the United States signaled its withdrawal from the agreement in January 2017. In November 2017 the remaining 11 countries agreed on the core elements of a modified agreement, which they renamed the Comprehensive and Progressive Agreement for Trans-Pacific Partnership (CPTPP). Japan also reached agreement with the European Union on an Economic Partnership Agreement in July 2017, and is likely seek to ratify both agreements in the Diet this year.

GDP (purchasing power parity):
$5.443 trillion (2017 est.)
$5.35 trillion (2016 est.)
$5.299 trillion (2015 est.)
note: data are in 2017 dollars
country comparison to the world: 4

GDP (official exchange rate):
$4.873 trillion (2017 est.)

GDP—real growth rate: 0.7% (2019 est.)
0.29% (2018 est.)
2.19% (2017 est.)
country comparison to the world: 183

GDP—per capita (PPP): $42,900 (2017 est.)
$42,100 (2016 est.)
$41,700 (2015 est.)
note: data are in 2017 dollars
country comparison to the world: 42

Gross national saving:
28% of GDP (2017 est.)
27.5% of GDP (2016 est.)
27.1% of GDP (2015 est.)
country comparison to the world: 41

GDP—composition, by end use:
household consumption: 55.5% (2017 est.)
government consumption: 19.6% (2017 est.)
investment in fixed capital: 24% (2017 est.)
investment in inventories: 0% (2017 est.)
exports of goods and services: 17.7% (2017 est.)
imports of goods and services: -16.8% (2017 est.)

GDP—composition, by sector of origin:
agriculture: 1.1% (2017 est.)
industry: 30.1% (2017 est.)
services: 68.7% (2017 est.)
Agriculture—products: vegetables, rice, fish, poultry, fruit, dairy products, pork, beef, flowers, potatoes/taros/yams, sugarcane, tea, legumes, wheat and barley
Industries: among world's largest and most technologically advanced producers of motor vehicles, electronic equipment, machine tools, steel and nonferrous metals, ships, chemicals, textiles, processed foods

Industrial production growth rate: 1.4% (2017 est.)
country comparison to the world: 145

Labor force: 66.54 million (2020 est.)
country comparison to the world: 7

Labor force—by occupation: *agriculture:* 2.9%
industry: 26.2%
services: 70.9% (February 2015 est.)

Unemployment rate: 2.36% (2019 est.)

2.44% (2018 est.)
country comparison to the world: 25

Population below poverty line: 16.1% (2013 est.)

Household income or consumption by percentage share: *lowest 10%:* 2.7%
highest 10%: 24.8% (2008)

Budget: *revenues:* 1.714 trillion (2017 est.)
expenditures: 1.885 trillion (2017 est.)

Taxes and other revenues: 35.2% (of GDP) (2017 est.)
country comparison to the world: 63

Budget surplus (+) or deficit (-): -3.5% (of GDP) (2017 est.)
country comparison to the world: 147

Public debt:
237.6% of GDP (2017 est.)
235.6% of GDP (2016 est.)
country comparison to the world: 1

Fiscal year: 1 April—31 March

Inflation rate (consumer prices): 0.5% (2017 est.)
-0.1% (2016 est.)
country comparison to the world: 27

Current account balance: $185.644 billion (2019 est.)
$177.08 billion (2018 est.)
country comparison to the world: 2

Exports: $688.9 billion (2017 est.)
$634.9 billion (2016 est.)
country comparison to the world: 4

Exports—partners: US 19.4%, China 19%, South Korea 7.6%, Hong Kong 5.1%, Thailand 4.2% (2017)

Exports—commodities: 14.9 motor vehicles5.4 iron and steel products5 semiconductors4.8 auto parts3.5 power generating machinery3.3 plastic materials (2014 est.)

Imports: $644.7 billion (2017 est.)
$584.7 billion (2016 est.)
country comparison to the world: 4

Imports—commodities: 16.1 petroleum9.1 liquid natural gas3.8 clothing3.3 semiconductors2.4 coal1.4 audio and visual apparatus (2014 est.)

Imports—partners: China 24.5%, US 11%, Australia 5.8%, South Korea 4.2%, Saudi Arabia 4.1% (2017)

Reserves of foreign exchange and gold: $1.264 trillion (31 December 2017 est.)
$1.233 trillion (31 December 2015 est.)
country comparison to the world: 2

Debt—external: $3.24 trillion (31 March 2016 est.)
$2.83 trillion (31 March 2015 est.)
country comparison to the world: 7

Exchange rates: yen (JPY) per US dollar -
111.1 (2017 est.)
108.76 (2016 est.)
108.76 (2015 est.)
121.02 (2014 est.)
97.44 (2013 est.)

ENERGY

Electricity access: *electrification—total population:* 100% (2020)

Electricity—production: 989.3 billion kWh (2016 est.)
country comparison to the world: 5

Electricity—consumption: 943.7 billion kWh (2016 est.)
country comparison to the world: 4

Electricity—exports:
0 kWh (2016 est.)
country comparison to the world: 152

Electricity—imports: 0 kWh (2016 est.)
country comparison to the world: 163

Electricity—installed generating capacity: 295.9 million kW (2016 est.)
country comparison to the world: 4

Electricity—from fossil fuels: 71% of total installed capacity (2016 est.)
country comparison to the world: 105

Electricity—from nuclear fuels: 1% of total installed capacity (2017 est.)
country comparison to the world: 30

Electricity—from hydroelectric plants: 8% of total installed capacity (2017 est.)
country comparison to the world: 121

Electricity—from other renewable sources: 20% of total installed capacity (2017 est.)
country comparison to the world: 38

Crude oil—production: 3,200 bbl/day (2018 est.)
country comparison to the world: 82

Crude oil—exports: 0 bbl/day (2017 est.)
country comparison to the world: 143

Crude oil—imports: 3.208 million bbl/day (2017 est.)
country comparison to the world: 4

Crude oil—proved reserves: 44.12 million bbl (1 January 2018 est.)
country comparison to the world: 77

Refined petroleum products—production: 3.467 million bbl/day (2017 est.)
country comparison to the world: 5

Refined petroleum products—consumption: 3.894 million bbl/day (2017 est.)
country comparison to the world: 4

Refined petroleum products—exports: 370,900 bbl/day (2017 est.)
country comparison to the world: 24

Refined petroleum products—imports: 1.1 million bbl/day (2017 est.)
country comparison to the world: 5

Natural gas—production: 3.058 billion cu m (2017 est.)
country comparison to the world: 57

Natural gas—consumption: 127.2 billion cu m (2017 est.)
country comparison to the world: 5

Natural gas—exports: 169.9 million cu m (2017 est.)
country comparison to the world: 47

Natural gas—imports: 116.6 billion cu m (2017 est.)
country comparison to the world: 2

Natural gas—proved reserves: 20.9 billion cu m (1 January 2018 est.)
country comparison to the world: 74

Carbon dioxide emissions from consumption of energy: 1.268 billion Mt (2017 est.)
country comparison to the world: 5

COMMUNICATIONS

Telephones—fixed lines: *total subscriptions:* 62,775,494
subscriptions per 100 inhabitants: 49.88 (2019 est.)
country comparison to the world: 3

Telephones—mobile cellular:
total subscriptions: 175,187,425
subscriptions per 100 inhabitants: 139.2 (2019 est.)
country comparison to the world: 8

Telecommunication systems: *general assessment:* excellent domestic and international service; exceedingly high mobile, mobile broadband and fixed broadband penetration; strong govt. policies for over a decade see over 90% of households with FttX; one of Japan's largest e-commerce companies planning to build its own nationwide stand-alone 5G mobile network; govt. to implement a telecom tax to pay for rural 5G network; FttH will continue to increase its share of total fixed broadband subscriptions as DSL is phased out; mature telecom system will show slow growth in the next few years to 2024 (2020)

domestic: high level of modern technology and excellent service of every kind; 50 per 100 for fixed-line and 140 per 100 for mobile-cellular subscriptions (2019)

international: country code—81; numerous submarine cables with landing points for HSCS, JIH, RJCN, APCN-2, JUS, EAC-C2C, PC-1, Tata TGN-Pacific, FLAG North Asia Loop/REACH North Asia Loop, APCN-2, FASTER, SJC, SJC2, Unity/EAC-Pacific, JGA-N, APG, ASE, AJC, JUPITER, MOC, Okinawa Cellular Cable, KJCN, GOKI, KJCN, and SeaMeWE- 3, submarine cables provide links throughout Asia, Australia, the Middle East, Europe, Southeast Asia, Africa and US; satellite earth stations—7 Intelsat (Pacific and Indian Oceans), 1 Intersputnik (Indian Ocean region), 2 Inmarsat (Pacific and Indian Ocean regions), and 8 SkyPerfect JSAT (2019)
note: the COVID-19 outbreak is negatively impacting telecommunications production and supply chains globally; consumer spending on telecom devices and services has also slowed due to the pandemic's effect on economies worldwide; overall progress towards improvements in all facets of the telecom industry mobile, fixed-line, broadband, submarine cable and satellite—has moderated

Broadcast media: a mixture of public and commercial broadcast TV and radio stations; 6 national terrestrial TV networks including 1 public broadcaster; the large number of radio and TV stations

available provide a wide range of choices; satellite and cable services provide access to international channels (2019)

Internet country code: .jp

Internet users:
total: 106,725,643
percent of population: 84.59% (July 2018 est.)
country comparison to the world: 6

Broadband—fixed subscriptions: *total:* 41,496,293
subscriptions per 100 inhabitants: 33 (2018 est.)
country comparison to the world: 3

TRANSPORTATION

National air transport system:
number of registered air carriers: 22 (2020)
inventory of registered aircraft operated by air carriers: 673
annual passenger traffic on registered air carriers: 126,387,527 (2018)
annual freight traffic on registered air carriers: 9,420,660,000 mt-km (2018)

Civil aircraft registration country code prefix: JA (2016)

Airports: 175 (2013)
country comparison to the world: 32

Airports—with paved runways: *total:* 142 (2017)
over 3,047 m: 6 (2017)
2,438 to 3,047 m: 45 (2017)
1,524 to 2,437 m: 38 (2017)
914 to 1,523 m: 28 (2017)
under 914 m: 25 (2017)

Airports—with unpaved runways:
total: 33 (2013)
914 to 1,523 m: 5 (2013)
under 914 m: 28 (2013)

Heliports: 16 (2013)

Pipelines: 4456 km gas, 174 km oil, 104 km oil/gas/water (2013)

Railways: *total:* 27,311 km (2015)
standard gauge: 4,800 km 1.435-m gauge (4,800 km electrified) (2015)
narrow gauge: 124 km 1.372-m gauge (124 km electrified) (2015)
dual gauge: 132 km 1.435-1.067-m gauge (132 km electrified) (2015)
22,207 km 1.067-m gauge (15,430 km electrified)
48 km 0.762-m gauge (48 km electrified)
country comparison to the world: 11

Roadways: *total:* 1,218,772 km (2015)
paved: 992,835 km (includes 8,428 km of expressways) (2015)
unpaved: 225,937 km (2015)
country comparison to the world: 6

Waterways: 1,770 km (seagoing vessels use inland seas) (2010)
country comparison to the world: 44

Merchant marine: *total:* 5,017
by type: bulk carrier 158, container ship 37, general cargo 1,767, oil tanker 661, other 2,394 (2019)
country comparison to the world: 4

Ports and terminals: *major seaport(s):* Chiba, Kawasaki, Kobe, Mizushima, Moji, Nagoya, Osaka, Tokyo, Tomakomai, Yokohama

container port(s) (TEUs): Kobe (2,924,179), Nagoya (2,784,109), Osaka (2,326,852), Tokyo (4,500,156), Yokohama (2,926,698) (2017)

LNG terminal(s) (import): Chita, Fukwoke, Futtsu, Hachinone, Hakodate, Hatsukaichi, Higashi Ohgishima, Higashi Niigata, Himeiji, Joetsu, Kagoshima, Kawagoe, Kita Kyushu, Mizushima, Nagasaki, Naoetsu, Negishi, Ohgishima, Oita, Sakai, Sakaide, Senboku, Shimizu, Shin Minato, Sodegaura, Tobata, Yanai, Yokkaichi Okinawa—Nakagusuku

MILITARY AND SECURITY

Military and security forces: *Japan Self-Defense Force (JSDF):* Ground Self-Defense Force (Rikujou Jieitai, GSDF; includes aviation), Maritime SelfDefense Force (Kaijou Jieitai, MSDF; includes naval aviation), Air Self-Defense Force (Koukuu Jieitai, ASDF); Japan Coast Guard (Ministry of Land, Transport, Infrastructure and Tourism) (2019)

Military expenditures: 0.93% of GDP (2019)
0.93% of GDP (2018)
0.93% of GDP (2017)
0.94% of GDP (2016)
0.94% of GDP (2015)
country comparison to the world: 124

Military and security service personnel strengths: the Japanese Self Defense Force (JSDF) is comprised of approximately 240,000 active personnel (145,000 Ground; 45,000 Maritime; 45,000 Air; 4,000 Joint Forces); 14,000 Coast Guard; 56,000 reserves (2019 est.)

Military equipment inventories and acquisitions: the JSDF is equipped with a mix of imported and domestically-produced equipment; Japan is capable of producing a wide range of air, ground, and naval weapons systems; the majority of its weapons imports are from the US and some domestically-produced weapons are US-origin and manufactured under license (2019)

Military deployments: approximately 170 Djibouti (2020)

Military service age and obligation: 18 years of age for voluntary military service; no conscription; mandatory retirement at age 53 for senior enlisted personnel and at 62 years for senior service officers (2012)

TERRORISM

Terrorist group(s): Aum Shinrikyo (AUM/Aleph) (2019)
note: details about the history, aims, leadership, organization, areas of operation, tactics, targets, weapons, size, and sources of support of the group(s) appear(s) in Appendix-T

Disputes—international: the sovereignty dispute over the islands of Etorofu, Kunashiri, and Shikotan, and the Habomai group, known in Japan as the "Northern Territories" and in Russia as the "Southern Kuril Islands," occupied by the Soviet Union in 1945, now administered by Russia and claimed by Japan, remains the primary sticking point to signing a peace treaty formally ending World War II hostilities; Japan and South Korea claim Liancourt Rocks (Take-shima/Tok-do) occupied by South Korea since 1954; the Japanese-administered Senkaku Islands are also claimed by China and Taiwan

Refugees and internally displaced persons: *stateless persons:* 687 (2019)

JERSEY

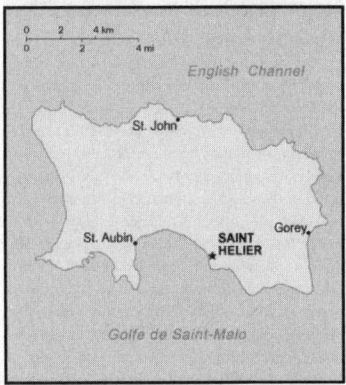

INTRODUCTION

Background: Jersey and the other Channel Islands represent the last remnants of the medieval Duchy of Normandy that held sway in both France and England. These islands were the only British soil occupied by German troops in World War II. The Bailiwick of Jersey is a British Crown dependency, which means that it is not part of the UK but is rather a self-governing possession of the British Crown. However, the UK Government is constitutionally responsible for its defense and international representation.

GEOGRAPHY

Location: Western Europe, island in the English Channel, northwest of France

Geographic coordinates: 49 15 N, 2 10 W

Map references: Europe

Area: *total:* 116 sq km
land: 116 sq km
water: 0 sq km
country comparison to the world: 225

Area—comparative: about two-thirds the size of Washington, DC

Land boundaries: 0 km

Coastline: 70 km

Maritime claims:
territorial sea: 12 nm
exclusive fishing zone: 12 nm

Climate: temperate; mild winters and cool summers

Terrain: gently rolling plain with low, rugged hills along north coast

Elevation: *lowest point:* English Channel 0 m
highest point: Les Platons 136 m

Natural resources:
arable land

Land use: *agricultural land:* 66% (2011 est.)
arable land: 66% (2011 est.) / permanent crops: 0% (2011 est.) / permanent pasture: 0% (2011 est.)
forest: 0% (2011 est.)
other: 34% (2011 est.)

Irrigated land: NA

Population distribution: fairly even distribution; no notable trends

Natural hazards: very large tidal variation can be hazardous to navigation

Environment—current issues: habitat and species depletion due to human encroachment; water pollution; improper solid waste disposal

Geography—note: largest and southernmost of Channel Islands; about 30% of population concentrated in Saint Helier

PEOPLE AND SOCIETY

Population: 101,073 (July 2020 est.)
country comparison to the world: 196

Nationality: *noun:* Channel Islander(s)
adjective: Channel Islander

Ethnic groups: Jersey 46.4%, British 32.7%, Portuguese/Madeiran 8.2%, Polish 3.3%, Irish, French, and other white 7.1%, other 2.4% (2011 est.)

Languages: English (official) 94.5%, Portuguese 4.6%, other .9% (includes French (official) and Jerriais) (2001 est.)
note: data represent main spoken language; the traditional language of Jersey is Jerriais or Jersey French (a Norman language), which was spoken by fewer than 3,000 people as of 2001; two-thirds of Jerriais speakers are aged 60 and over

Religions: Protestant (Anglican, Baptist, Congregational New Church, Methodist, Presbyterian), Roman Catholic

Age structure: *0-14 years:* 16.63% (male 8,689/female 8,124)
15-24 years: 12.98% (male 6,764/female 6,354)
25-54 years: 40.12% (male 20,499/female 20,054)
55-64 years: 13.22% (male 6,515/female 6,844)
65 years and over: 17.05% (male 7,324/female 9,906) (2020 est.)

Dependency ratios:
total dependency ratio: 49
youth dependency ratio: 22.3
elderly dependency ratio: 26.7
potential support ratio: 3.7 (2020 est.)
note: data represent the Guernsey and Jersey

Median age: *total:* 37.5 years
male: 36 years
female: 39.5 years (2020 est.)
country comparison to the world: 70

Population growth rate: 0.72% (2020 est.)
country comparison to the world: 136

Birth rate: 12.7 births/1,000 population (2020 est.)
country comparison to the world: 150

Death rate: 7.9 deaths/1,000 population (2020 est.)
country comparison to the world: 96

Net migration rate: 2.3 migrant(s)/1,000 population (2020 est.)
country comparison to the world: 44

Population distribution: fairly even distribution; no notable trends

Urbanization: *urban population:* 31% of total population (2020)
rate of urbanization: 0.46% annual rate of change (2015-20 est.)
note: data represent Guernsey and Jersey
total population growth rate v. urban population growth rate, 2000-2030: Major urban areas—population: 34,000 SAINT HELIER (capital) (2018)

Sex ratio: *at birth:* 1.06 male(s)/female
0-14 years: 1.07 male(s)/female
15-24 years: 1.06 male(s)/female
25-54 years: 1.02 male(s)/female
55-64 years: 0.95 male(s)/female
65 years and over: 0.74 male(s)/female
total population: 0.97 male(s)/female (2020 est.)

Infant mortality rate: *total:* 3.6 deaths/1,000 live births
male: 3.9 deaths/1,000 live births
female: 3.4 deaths/1,000 live births (2020 est.)
country comparison to the world: 196

Life expectancy at birth: *total population:* 82.2 years
male: 79.7 years
female: 84.9 years (2020 est.)
country comparison to the world: 20

Total fertility rate: 1.67 children born/woman (2020 est.)
country comparison to the world: 179

Drinking water source:

improved: *total:* 94.2% of population

unimproved: *total:* 5.9% of population (2017 est.)
note: includes data for Guernsey

Sanitation facility access:

improved: *total:* 98.5% of population

unimproved: *total:* 1.5% of population (2017)

HIV/AIDS—adult prevalence rate: NA

HIV/AIDS—people living with HIV/AIDS: NA

HIV/AIDS—deaths: NA

Education expenditures: NA

GOVERNMENT

Country name: *conventional long form:* Bailiwick of Jersey
conventional short form: Jersey
former: Norman Isles
etymology: the name is of Old Norse origin, but the meaning of the root "Jer(s)" is uncertain; the "-ey" ending means "island"

Dependency status: British crown dependency

Government type: parliamentary democracy (Assembly of the States of Jersey)

Capital: *name:* Saint Helier
geographic coordinates: 49 11 N, 2 06 W
time difference: UTC 0 (5 hours ahead of Washington, DC, during Standard Time)
daylight saving time: +1hr, begins last Sunday in March; ends last Sunday in October
etymology: named after Saint Helier, the patron saint of Jersey, who was reputedly martyred on the island in A.D. 555

Administrative divisions: none (British crown dependency); there are no first-order administrative divisions as defined by the US Government, but there are 12 parishes; Grouville, Saint Brelade, Saint Clement, Saint Helier, Saint John, Saint Lawrence, Saint Martin, Saint Mary, Saint Ouen, Saint Peter, Saint Saviour, and Trinity

Independence: none (British Crown dependency)

National holiday: Liberation Day, 9 May (1945)

Constitution: *history:* unwritten; partly statutes, partly common law and practice
amendments: proposed by a government minister to the Assembly of the States of Jersey, by an Assembly member, or by an elected parish head; passage requires several Assembly readings, a majority vote by the Assembly, review by the UK Ministry of Justice, and approval of the British monarch (Royal Assent)

Legal system: the laws of the UK apply where applicable; includes local statutes

Citizenship: see United Kingdom

Suffrage: 16 years of age; universal

Executive branch: *chief of state:* Queen ELIZABETH II (since 6 February 1952); represented by Lieutenant Governor Sir Stephen DALTON (since 13 March 2017)

head of government: Chief Minister John LE FONDRE (since 8 June 2018); Bailiff William BAILHACHE (since 29 January 2015)
cabinet: Council of Ministers appointed individually by the states
elections/appointments: the monarchy is hereditary; Council of Ministers, including the chief minister, indirectly elected by the Assembly of States; lieutenant governor and bailiff appointed by the monarch

Legislative branch: *description:* unicameral Assembly of the States of Jersey (49 elected members; 8 senators to serve 4-year terms, and 29 deputies and 12 connetables, or heads of parishes, to serve 4-year terms; 5 non-voting members appointed by the monarch include the bailiff, lieutenant governor, dean of Jersey, attorney general, and the solicitor general)
elections: last held on 16 May 2018 (next to be held on 16 May 2022)
election results: percent of vote—NA; seats—independents 49; composition—men 36, women 13, percent of 26.5%

Judicial branch: *highest courts:* Jersey Court of Appeal (consists of the bailiff, deputy bailiff, and 12 judges); Royal Court (consists of the bailiff, deputy bailiff, 6 commissioners and lay people referred to as jurats, and is organized into Heritage, Family, Probate, and Samedi Divisions); appeals beyond the Court of Appeal are heard by the Judicial Committee of the Privy Council (in London)
judge selection and term of office: Jersey Court of Appeal bailiffs and judges appointed by the Crown upon the advice of the Secretary of State for Justice; bailiffs and judges appointed for "extent of good behavior;" Royal Court bailiffs appointed by the Crown upon the advice of the Secretary of State for Justice; commissioners appointed by the bailiff; jurats appointed by the Electoral College; bailiffs and commissioners appointed for "extent of good behavior;" jurats appointed until retirement at age 72
subordinate courts: Magistrate's Court; Youth Court; Petty Debts Court; Parish Hall Enquires (a process of preliminary investigation into youth and minor adult offenses to determine need for presentation before a court)

Political parties and leaders: *one registered party:* Reform Jersey [Sam MEZEC]
note: most senators and deputies sit as independents

International organization participation: UPU

Diplomatic representation in the US: none (British Crown dependency)
none (British Crown dependency)
Diplomatic representation from the US:
none (British Crown dependency)

Flag description: white with a diagonal red cross extending to the corners of the flag; in the upper quadrant, surmounted by a yellow crown, a red shield with three lions in yellow; according to tradition, the ships of Jersey—in an attempt to differentiate themselves from English ships flying the horizontal cross of St. George—rotated the cross to the "X" (saltire) configuration; because this

arrangement still resembled the Irish cross of St. Patrick, the yellow Plantagenet crown and Jersey coat of arms were added

National symbol(s): *Jersey cow; national colors:* red, white

National anthem: *name:* "Isle de Siez Nous" (Island Home)
lyrics/music: Gerard LE FEUVRE
note: adopted 2008; serves as a local anthem; as a British Crown dependency, "God Save the Queen" is official (see United Kingdom)

ECONOMY

Economy—overview: Jersey's economy is based on international financial services, agriculture, and tourism. In 2016, the financial services sector accounted for about 41% of the island's output. Agriculture represented about 1% of Jersey's economy in 2016. Potatoes are an important export crop, shipped mostly to the UK. The Jersey breed of dairy cattle originated on the island and is known worldwide. The dairy industry remains important to the island with approximately $8.8 million gallons of milk produced in 2015. Tourism accounts for a significant portion of Jersey's economy, with more than 700,000 total visitors in 2015. Living standards come close to those of the UK. All raw material and energy requirements are imported as well as a large share of Jersey's food needs. Light taxes and death duties make the island a popular offshore financial center. Jersey maintains its relationship with the EU through the UK. Therefore, in light of the UK's decision to leave the EU, Jersey will also need to renegotiate its ties to the EU.

GDP (purchasing power parity): $5.569 billion (2016 est.)
$5.514 billion (2015 est.)
$4.98 billion (2014)
note: data are in 2015 US dollars
country comparison to the world: 175

GDP (official exchange rate): $5.004 billion (2015 est.)

GDP—real growth rate: 1% (2016 est.)
10.7% (2015 est.)
country comparison to the world: 173

GDP—per capita (PPP): $56,600 (2016 est.)
$49,500 (2015 est.)
country comparison to the world: 21

GDP—composition, by sector of origin:
agriculture: 2% (2010)
industry: 2% (2010)
services: 96% (2010)

Agriculture—products: potatoes, cauliflower, tomatoes; beef, dairy products
Industries: tourism, banking and finance, dairy, electronics

Industrial production growth rate: NA

Labor force: 39,930 (2017 est.)
country comparison to the world: 187

Labor force—by occupation: *agriculture:* 3%
industry: 12%
services: 85% (2014 est.)

Unemployment rate: 4% (2015 est.)

4.6% (2014 est.)
country comparison to the world: 61

Population below poverty line: NA

Household income or consumption by percentage share:
lowest 10%: NA
highest 10%: NA

Budget: *revenues:* 829 million (2005)
expenditures: 851 million (2005)

Taxes and other revenues: 16.6% (of GDP) (2005)
country comparison to the world: 179

Budget surplus (+) or deficit (-): -0.4% (of GDP) (2005)
country comparison to the world: 57

Fiscal year: 1 April—31 March

Inflation rate (consumer prices): 3.7% (2006)
country comparison to the world: 146

Exports: NA

Exports—commodities: light industrial and electrical goods, dairy cattle, foodstuffs, textiles, flowers

Imports: NA

Imports—commodities: machinery and transport equipment, manufactured goods, foodstuffs, mineral fuels, chemicals

Debt—external: NA

Exchange rates: Jersey pounds (JEP) per US dollar
0.7836 (2017 est.)
0.738 (2016 est.)
0.738 (2015)
0.6542 (2012)
0.6391 (2011 est.)

ENERGY

Electricity access: *electrification—total population:* 100% (2020)

Electricity—production: NA (2017)

Electricity—consumption: 630.1 million kWh (2004 est.)
country comparison to the world: 163

Carbon dioxide emissions from consumption of energy: 450,000 Mt (2012 est.)
country comparison to the world: 185

COMMUNICATIONS

Telephones—fixed lines:
total subscriptions: 55,938
subscriptions per 100 inhabitants: 58 (July 2016 est.)
country comparison to the world: 157

Telephones—mobile cellular:
total subscriptions: 122,668
subscriptions per 100 inhabitants: 119 (July 2016 est.)
country comparison to the world: 191

Telecommunication systems: *general assessment:* good system with broadband access (2018)
domestic: fixed-line and mobile-cellular services widely available; fixed-line 58 per 100 and mobile-cellular 119 per 100 subscriptions (2018)
international: country code—44; landing points for the INGRID, UK-Channel Islands-8, and Guernsey-Jersey-4, submarine cable connectivity to Guernsey, the UK, and France (2019)
note: the COVID-19 outbreak is negatively impacting telecommunications production and supply chains globally; consumer spending on telecom devices and services has also slowed due to the pandemic's effect on economies worldwide; overall

progress towards improvements in all facets of the telecom industry—mobile, fixed-line, broadband, submarine cable and satellite—has moderated

Broadcast media: multiple UK terrestrial TV broadcasts are received via a transmitter in Jersey; satellite packages available; BBC Radio Jersey and 1 other radio station operating

Internet country code: .je

Internet users:
total: 58,000
percent of population: 59.6% (July 2016 est.)
country comparison to the world: 193

TRANSPORTATION

National air transport system: *number of registered air carriers:* 1 (registered in UK) (2020)
inventory of registered aircraft operated by air carriers: 4 (registered in UK)

Airports: 1 (2013)
country comparison to the world: 226

Airports—with paved runways:
total: 1 (2019)
1,524 to 2,437 m: 1

Roadways: *total:* 576 km (2010)
country comparison to the world: 192

Ports and terminals: *major seaport(s):* Gorey, Saint Aubin, Saint Helier

MILITARY AND SECURITY

Military—note: defense is the responsibility of the UK

TRANSNATIONAL ISSUES

Disputes—international: none

JORDAN

INTRODUCTION

Background: Following World War I and the dissolution of the Ottoman Empire, the League of Nations awarded Britain the mandate to govern much of the Middle East. Britain demarcated a semi-autonomous region of Transjordan from Palestine in the early 1920s. The area gained its independence in 1946 and thereafter became The Hashemite Kingdom of Jordan. The country's long-time ruler, King HUSSEIN (1953-99), successfully navigated competing pressures from the major powers (US, USSR, and UK), various Arab states, Israel, and a large internal Palestinian population. Jordan lost the West Bank to Israel in the 1967 Six-Day War. King HUSSEIN in 1988 permanently relinquished Jordanian claims to the West Bank; in 1994 he signed a peace treaty with Israel. King ABDALLAH II, King HUSSEIN's eldest son, assumed the throne following his

father's death in 1999. He has implemented modest political reforms, including the passage of a new electoral law in early 2016 and an effort to devolve some authority to governorate- and municipal-level councils following subnational elections in 2017. In 2016, the Islamic Action Front, which is the political arm of the Jordanian Muslim Brotherhood, returned to the National Assembly with 15 seats after boycotting the previous two elections in 2010 and 2013.

GEOGRAPHY

Location: Middle East, northwest of Saudi Arabia, between Israel (to the west) and Iraq

Geographic coordinates: 31 00 N, 36 00 E

Map references: Middle East

Area: *total:* 89,342 sq km
land: 88,802 sq km

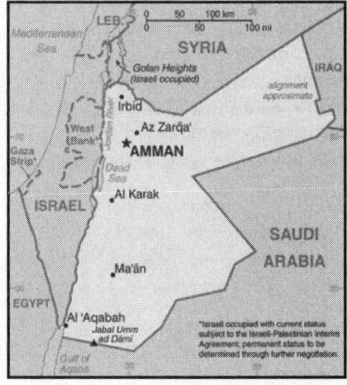

water: 540 sq km
country comparison to the world: 113

Area—comparative: about three-quarters the size of Pennsylvania; slightly smaller than Indiana

Land boundaries: *total:* 1,744 km
border countries (5): Iraq 179 km, Israel 307 km, Saudi Arabia 731 km, Syria 379 km, West Bank 148 km

Coastline: 26 km

Maritime claims: *territorial sea:* 3 nm

Climate: mostly arid desert; rainy season in west (November to April)

Terrain: mostly arid desert plateau; a great north-south geological rift along the west of the country is the dominant topographical feature and includes the Jordan River Valley, the Dead Sea, and the Jordanian Highlands

Elevation: *mean elevation:* 812 m
lowest point: Dead Sea -431 m
highest point: Jabal Umm ad Dami 1,854 m

Natural resources: phosphates, potash, shale oil

Land use: *agricultural land:* 11.4% (2011 est.)
arable land: 2% (2011 est.) / permanent crops: 1% (2011 est.) / permanent pasture: 8.4% (2011 est.)
forest: 1.1% (2011 est.)
other: 87.5% (2011 est.)

Irrigated land: 964 sq km (2012)

Population distribution: population heavily concentrated in the west, and particularly the northwest, in and around the capital of Amman; a sizeable, but smaller population is located in the southwest along the shore of the Gulf of Aqaba

Natural hazards: droughts; periodic earthquakes; flash floods

Environment—current issues: limited natural freshwater resources; declining water table; salinity; deforestation; overgrazing; soil erosion; desertification; biodiversity and ecosystem damage/loss

Environment—international agreements: *party to:* Biodiversity, Climate Change, Climate Change-Kyoto Protocol, Desertification, Endangered Species, Hazardous Wastes, Law of the Sea, Marine Dumping, Ozone Layer Protection, Wetlands
signed, but not ratified: none of the selected agreements

Geography—note: strategic location at the head of the Gulf of Aqaba and as the Arab country that shares the longest border with Israel and the occupied West Bank; the Dead Sea, the lowest point in Asia and the second saltiest body of water in the world (after Lac Assal in Djibouti), lies on Jordan's western border with Israel and the West Bank; Jordan is almost landlocked but does have a 26 km southwestern coastline with a single port, Al 'Aqabah (Aqaba)

PEOPLE AND SOCIETY

Population: 10,820,644 (July 2020 est.)
note: increased estimate reflects revised assumptions about the net migration rate due to the increased flow of Syrian refugees

country comparison to the world: 84

Nationality: *noun:* Jordanian(s)
adjective: Jordanian

Ethnic groups: Jordanian 69.3%, Syrian 13.3%, Palestinian 6.7%, Egyptian 6.7%, Iraqi 1.4%, other 2.6% (includes Armenian, Circassian) (2015 est.)
note: data represent population by self-identified nationality

Languages: Arabic (official), English (widely understood among upper and middle classes)

Religions: Muslim 97.2% (official; predominantly Sunni), Christian 2.2% (majority Greek Orthodox, but some Greek and Roman Catholics, Syrian Orthodox, Coptic Orthodox, Armenian Orthodox, and Protestant denominations), Buddhist 0.4%, Hindu 0.1%, Jewish <0.1, folk <0.1, unaffiliated <0.1, other <0.1 (2010 est.)

MENA religious affiliation: *Age structure:* 0-14 years: 33.05% (male 1,837,696/female 1,738,935)
15-24 years: 19.77% (male 1,126,567/female 1,012,812)
25-54 years: 38.39% (male 2,250,328/female 1,903,996)
55-64 years: 5.11% (male 290,633/female 262,827)
65 years and over: 3.67% (male 194,464/female 202,386) (2020 est.)

Dependency ratios: *total dependency ratio:* 58.2
youth dependency ratio: 52
elderly dependency ratio: 6.3
potential support ratio: 16 (2020 est.)

Median age: *total:* 23.5 years
male: 23.9 years
female: 22.9 years (2020 est.)
country comparison to the world: 175

Population growth rate: 1.4% (2020 est.)
country comparison to the world: 81

Birth rate: 23 births/1,000 population (2020 est.)
country comparison to the world: 58

Death rate: 3.4 deaths/1,000 population (2020 est.)
country comparison to the world: 220

Net migration rate: -11.3 migrant(s)/1,000 population (2020 est.)
country comparison to the world: 218

Population distribution: population heavily concentrated in the west, and particularly the northwest, in and around the capital of Amman; a sizeable, but smaller population is located in the southwest along the shore of the Gulf of Aqaba

Urbanization: *urban population:* 91.4% of total population (2020)
rate of urbanization: 2.43% annual rate of change (2015-20 est.)
total population growth rate v. urban population growth rate, 2000-2030: Major urban areas—population: 2.148 million AMMAN (capital) (2020)

Sex ratio: *at birth:* 1.06 male(s)/female
0-14 years: 1.06 male(s)/female
15-24 years: 1.11 male(s)/female
25-54 years: 1.18 male(s)/female
55-64 years: 1.11 male(s)/female
65 years and over: 0.96 male(s)/female

total population: 1.11 male(s)/female (2020 est.)

Mother's mean age at first birth: 24.8 years (2017/18 est.)
note: median age at first birth among women 30-34

Maternal mortality rate: 46 deaths/100,000 live births (2017 est.)
country comparison to the world: 96

Infant mortality rate: *total:* 12.8 deaths/1,000 live births
male: 13.6 deaths/1,000 live births
female: 12 deaths/1,000 live births (2020 est.)
country comparison to the world: 103

Life expectancy at birth: *total population:* 75.5 years
male: 74 years
female: 77.1 years (2020 est.)
country comparison to the world: 114

Total fertility rate: 3.04 children born/woman (2020 est.)
country comparison to the world: 51

Contraceptive prevalence rate: 51.8% (2017/18)

Drinking water source:

improved: *urban:* 100% of population
rural: 97.7% of population
total: 98.9% of population

unimproved: *urban:* 0% of population
rural: 2.2% of population
total: 1.1% of population (2017 est.)

Current Health Expenditure: 8.1% (2017)

Physicians density: 2.32 physicians/1,000 population (2017)

Hospital bed density: 1.5 beds/1,000 population (2017)

Sanitation facility access:

improved: *urban:* 98.6% of population
rural: 96.6% of population
total: 98.5% of population

unimproved: *urban:* 1.4% of population
rural: 3.7% of population
total: 1.5% of population (2017 est.)

HIV/AIDS—adult prevalence rate: <.1% (2018 est.)

HIV/AIDS—people living with HIV/AIDS: <500 (2018 est.)

HIV/AIDS—deaths: <100 (2018 est.)

Obesity—adult prevalence rate: 35.5% (2016)
country comparison to the world: 13

Children under the age of 5 years underweight: 3% (2012)
country comparison to the world: 100

Education expenditures: 3.6% of GDP (2017)
country comparison to the world: 121

Literacy: *definition:* age 15 and over can read and write
total population: 98.2%
male: 98.6%
female: 97.8% (2018)

School life expectancy (primary to tertiary education): *total:* 11 years
male: 11 years

female: 11 years (2012)

Unemployment, youth ages 15-24: *total:* 35.6%
male: 31.5%
female: 57% (2016 est.)
country comparison to the world: 21

GOVERNMENT

Country name: *conventional long form:* Hashemite Kingdom of Jordan
conventional short form: Jordan
local long form: Al Mamlakah al Urduniyah al Hashimiyah
local short form: Al Urdun
former: Transjordan
etymology: named for the Jordan River, which makes up part of Jordan's northwest border

Government type: parliamentary constitutional monarchy

Capital: *name:* Amman
geographic coordinates: 31 57 N, 35 56 E
time difference: UTC+2 (7 hours ahead of Washington, DC, during Standard Time)
daylight saving time: +1hr, begins last Friday in March; ends last Friday in October
etymology: in the 13th century B.C., the Ammonites named their main city "Rabbath Ammon"; "rabbath" designated "capital," so the name meant "The Capital of [the] Ammon[ites]"; over time, the "Rabbath" came to be dropped and the city became known simply as "Ammon" and then "Amman"

Administrative divisions: 12 governorates (muhafazat, singular—muhafazah); 'Ajlun, Al 'Aqabah, Al Balqa', Al Karak, Al Mafraq, Al 'Asimah (Amman), At Tafilah, Az Zarqa', Irbid, Jarash, Ma'an, Madaba

Independence: 25 May 1946 (from League of Nations mandate under British administration)

National holiday: Independence Day, 25 May (1946)

Constitution: *history:* previous 1928 (preindependence); latest initially adopted 28 November 1947, revised and ratified 1 January 1952
amendments: constitutional amendments require at least a two-thirds majority vote of both the Senate and the House and ratification by the king; no amendment of the constitution affecting the rights of the king and the succession to the throne is permitted during the regency period; amended several times, last in 2016

Legal system: mixed system developed from codes instituted by the Ottoman Empire (based on French law), British common law, and Islamic law

International law organization participation: has not submitted an ICJ jurisdiction declaration; accepts ICC jurisdiction

Citizenship: *citizenship by birth:* no
citizenship by descent only: the father must be a citizen of Jordan
dual citizenship recognized: yes
residency requirement for naturalization: 15 years

Suffrage: 18 years of age; universal

Executive branch: *chief of state:* King ABDALLAH II (since 7 February 1999); Crown Prince HUSSEIN (born 28 June 1994), eldest son of King ABDALLAH II
head of government: Prime Minister Bisher AL-KHASAWNEH (since 7 October 2020)
cabinet: Cabinet appointed by the prime minister in consultation with the monarch
elections/appointments: the monarchy is hereditary; prime minister appointed by the monarch

Legislative branch: *description:* bicameral National Assembly or Majlis al-'Umma consists of: Senate or the House of Notables or Majlis al-Ayan (65 seats; members appointed by the monarch to serve 4-year terms) Chamber of Deputies or House of Representatives or Majlis al-Nuwaab (130 seats; 115 members directly elected in 23 multi-seat constituencies by open-list proportional representation vote and 15 seats for women; 12 of the 115 seats reserved for Christian, Chechen, and Circassian candidates; members serve 4-year terms)
elections: Chamber of Deputies—last held on 10 November 2020 (next to be held in November 2024)
election results: Chamber of Deputies—percent of vote by party—NA; seats by party—NA

Judicial branch: *highest courts:* Court of Cassation or Supreme Court (consists of 15 members, including the chief justice); Constitutional Court (consists of 9 members)
judge selection and term of office: Supreme Court chief justice appointed by the king; other judges nominated by the Judicial Council, an 11-member judicial policymaking body consisting of high-level judicial officials and judges, and approved by the king; judge tenure generally not limited; Constitutional Court members appointed by the king for 6-year non-renewable terms with one-third of the membership renewed every 2 years
subordinate courts: Courts of Appeal; Great Felonies Court; religious courts; military courts; juvenile courts; Land Settlement Courts; Income Tax Court; Higher Administrative Court; Customs Court; special courts including the State Security Court

Political parties and leaders: Ahrar al-Urdun (Free People of Jordan) Party [Samir al-ZU'BI]
Al-Awn al-Watani (National Aid) Party [Faysal al-AWAR]
Al-Balad al-Amin Party [Khalil al-SAYED]
Al-Itijah al-Watani (National Trend Party) [Ahmad al-KAYED]
Al-Mustaqbal (Future) Party [Salah al-QUDAH]
Al-Nida' Party [Abd-al-Majid ABU-KHALID]
Al-Rayah Party (Flag Party) [Bilal DHEISAT]
Al-Shahama Party [Mashhour ZREIQAT]
Al-Shura Party [Firas al-ABBADI]
Arab Socialist Ba'th Party [Zyad AL-HOMSI]
Conservatives Party [Hasan RASHID]
Democratic Popular Unity Party [Sa'eed DHIYAB]
Democratic Sha'b Party (HASHD) [Abla ABU-OLBEH]
Freedom and Equality Party [Hamad Abu ZEID]

Islamic Action Front [Murad AL-ADAYLAH]
Islamic Centrist Party [Madallah AL-TARAWNEH]
Jordanian Al-Ansar Party [Awni al-RJOUB]
Jordanian Al-Hayah Party [Abd-al-Fattah al-KILANI]
Jordanian Communist Party [Faraj ITMIZYEH]
Jordanian Democratic Socialist Party [Jamil al-NIMRI]
Jordanian Democratic Tabiy'ah (Nature) Party [Ali ASFOUR]
Jordanian Equality Party [Zuhair al-SHURAFA]
Jordanian Fursan (Cavaliers Party) [Ali al-DHWEIB]
Jordanian Justice and Development Party [Ali al-SHURAFA]
Jordanian National Action Party [Abd-al-Hadi al-MAHARMAH]
Jordanian National Constitutional Party [Ahmad al-SHUNNAQ]
Jordanian National Democratic Grouping Party [Shakir al-ABBADI]
Jordanian National Party [Muna ABU-BAKR]
Jordanian National Union Party [Zeid ABU-ZEID]
Jordanian Progressive Ba'th Party [Fu'ad DABBOUR]
Jordanian Promise Party [Mahmoud al-KHALILI]
Jordanian Reform Party [Eid DHAYYAT]
Jordanian Social Justice Party [Abd-al-Fattah al-NSOUR]
Jordanian Wafa' (Loyalty) Party [Mazin al-QADI]
Justice and Reform Party [Sa'eed Nathir ARABIYAT]
Modernity and Change Party [Nayef al-HAMAYDEH]
National Congress Party [Irhayil GHARAYBEH] (formerly the Zamzam party)
National Renaissance Front Party [Isma'il KHATATBEH]
National Unity Party [Muhammad al-ZBOUN]
Pan Arab Movement Party [Dayfallah FARRAJ]
Partnership and Salvation Party [Muhammad al-HAMMOURI]
Reform and Renewal Party [Mazin RYAL]
Risalah Party [Hazim QASHOU']
Stronger Jordan Party [Rula al-HROUB]
Unified Jordanian Front Party [Farouq AL-ABBADI]

International organization participation: ABEDA, AFESD, AMF, CAEU, CD, CICA, EBRD, FAO, G-11, G-77, IAEA, IBRD, ICAO, ICC (national committees), ICCt, ICRM, IDA, IDB, IFAD, IFC, IFRCS, ILO, IMF, IMO, IMSO, Interpol, IOC, IOM, IPU, ISO, ITSO, ITU, ITUC (NGOs), LAS, MIGA, MINUSTAH, MINUSMA, MONUSCO, NAM, OIC, OPCW, OSCE (partner), PCA, UN, UNAMID, UNCTAD, UNESCO, UNHCR, UNIDO, UNMIL, UNMISS, UNOCI, UNRWA, UNWTO, UPU, WCO, WFTU (NGOs), WHO, WIPO, WMO, WTO

Diplomatic representation in the US: *chief of mission:* Ambassador Dina Khalil Tawiq KAWAR (since 27 June 2016)
chancery: 3504 International Drive NW, Washington, DC 20008
telephone: [1] (202) 966-2664

FAX: [1] (202) 966-3110

Diplomatic representation from the US: *chief of mission:* Ambassador (vacant); Charge d'Affaires Mike HANKEY (since 4 August 2019)
telephone: [962] (6) 590-6000
embassy: Abdoun, Al-Umawyeen St., Amman
mailing address: P. O. Box 354, Amman 11118 Jordan; Unit 70200, Box 5, DPO AE 09892-0200
FAX: [962] (6) 592-0163

Flag description: three equal horizontal bands of black (top), representing the Abbassid Caliphate, white, representing the Ummayyad Caliphate, and green, representing the Fatimid Caliphate; a red isosceles triangle on the hoist side, representing the Great Arab Revolt of 1916, and bearing a small white seven-pointed star symbolizing the seven verses of the opening Sura (Al-Fatiha) of the Holy Koran; the seven points on the star represent faith in One God, humanity, national spirit, humility, social justice, virtue, and aspirations; design is based on the Arab Revolt flag of World War I

National symbol(s): *eagle; national colors:* black, white, green, red

National anthem: *name:* "As-salam al-malaki al-urdoni" (Long Live the King of Jordan)
lyrics/music: Abdul-Mone'm al-RIFAI'/Abdul-Qader al-TANEER
note: adopted 1946; the shortened version of the anthem is used most commonly, while the full version is reserved for special occasions

ECONOMY

Economy—overview: Jordan's economy is among the smallest in the Middle East, with insufficient supplies of water, oil, and other natural resources, underlying the government's heavy reliance on foreign assistance. Other economic challenges for the government include chronic high rates of unemployment and underemployment, budget and current account deficits, and government debt.

King ABDALLAH, during the first decade of the 2000s, implemented significant economic reforms, such as expanding foreign trade and privatizing state-owned companies that attracted foreign investment and contributed to average annual economic growth of 8% for 2004 through 2008. The global economic slowdown and regional turmoil contributed to slower growth from 2010 to 2017—with growth averaging about 2.5% per year—and hurt export-oriented sectors, construction/real estate, and tourism. Since the onset of the civil war in Syria and resulting refugee crisis, one of Jordan's most pressing socioeconomic challenges has been managing the influx of approximately 660,000 UN-registered refugees, more than 80% of whom live in Jordan's urban areas. Jordan's own official census estimated the refugee number at 1.3 million Syrians as of early 2016.

Jordan is nearly completely dependent on imported energy—mostly natural gas—and energy consistently makes up 25-30% of Jordan's imports. To diversify its energy mix, Jordan has secured several contracts for liquefied and pipeline natural gas, developed several major renewables projects, and is currently exploring nuclear power generation and exploitation of abundant oil shale reserves. In August 2016, Jordan and the IMF agreed to a $723 million Extended Fund Facility that aims to build on the three-year, $2.1 billion IMF program that ended in August 2015 with the goal of helping Jordan correct budgetary and balance of payments imbalances.

GDP (purchasing power parity): $89 billion (2017 est.)
$87.28 billion (2016 est.)
$85.56 billion (2015 est.)
note: data are in 2017 dollars
country comparison to the world: 90

GDP (official exchange rate): $40.13 billion (2017 est.)

GDP—real growth rate: 2% (2019 est.)
1.94% (2018 est.)
2.12% (2017 est.)
country comparison to the world: 137

GDP—per capita (PPP): $9,200 (2017 est.)
$9,200 (2016 est.)
$9,300 (2015 est.)
note: data are in 2017 dollars
country comparison to the world: 144

Gross national saving: 9.1% of GDP (2017 est.)
9.3% of GDP (2016 est.)
10.2% of GDP (2015 est.)
country comparison to the world: 165

GDP—composition, by end use: *household consumption:* 80.5% (2017 est.)
government consumption: 19.8% (2017 est.)
investment in fixed capital: 22.8% (2017 est.)
investment in inventories: 0.7% (2017 est.)
exports of goods and services: 34.2% (2017 est.)
imports of goods and services: -58% (2017 est.)

GDP—composition, by sector of origin: *agriculture:* 4.5% (2017 est.)
industry: 28.8% (2017 est.)
services: 66.6% (2017 est.)

Agriculture—products: citrus, tomatoes, cucumbers, olives, strawberries, stone fruits; sheep, poultry, dairy
Industries: tourism, information technology, clothing, fertilizer, potash, phosphate mining, pharmaceuticals, petroleum refining, cement, inorganic chemicals, light manufacturing

Industrial production growth rate: 1.4% (2017 est.)
country comparison to the world: 146

Labor force: 731,000 (2020 est.)
country comparison to the world: 149

Labor force—by occupation: *agriculture:* 2%
industry: 20%
services: 78% (2013 est.)

Unemployment rate: 19.1% (2019 est.)
18.61% (2018 est.)
note: official rate; unofficial rate is approximately 30%
country comparison to the world: 186

Population below poverty line: 14.2% (2002 est.)

Household income or consumption by percentage share: *lowest 10%:* 3.4%

highest 10%: 28.7% (2010 est.)

Budget: *revenues:* 9.462 billion (2017 est.)
expenditures: 11.51 billion (2017 est.)

Taxes and other revenues: 23.6% (of GDP) (2017 est.)
country comparison to the world: 124

Budget surplus (+) or deficit (-): -5.1% (of GDP) (2017 est.)
country comparison to the world: 170

Public debt: 95.9% of GDP (2017 est.)
95.1% of GDP (2016 est.)
note: data cover central government debt and include debt instruments issued (or owned) by government entities other than the treasury; the data include treasury debt held by foreign entities; the data exclude debt issued by subnational entities, as well as intragovernmental debt; intragovernmental debt consists of treasury borrowings from surpluses in the social funds, such as for retirement, medical care, and unemployment; debt instruments for the social funds are not sold at public auctions
country comparison to the world: 22

Fiscal year: calendar year

Inflation rate (consumer prices): 3.3% (2017 est.)
-0.8% (2016 est.)
country comparison to the world: 138

Current account balance: -$1.222 billion (2019 est.)
-$2.964 billion (2018 est.)
country comparison to the world: 152

Exports: $7.511 billion (2017 est.)
$7.509 billion (2016 est.)
country comparison to the world: 98

Exports—partners: US 24.9%, Saudi Arabia 12.8%, India 8.2%, Iraq 8.7%, Kuwait 5.4%, UAE 4.6% (2017)

Exports—commodities: textiles, fertilizers, potash, phosphates, vegetables, pharmaceuticals

Imports: $18.21 billion (2017 est.)
$17.14 billion (2016 est.)
country comparison to the world: 81

Imports—commodities: crude oil, refined petroleum products, machinery, transport equipment, iron, cereals

Imports—partners: China 13.6%, Saudi Arabia 13.6%, US 9.9%, UAE 4.9%, Germany 4.4% (2017)

Reserves of foreign exchange and gold: $15.56 billion (31 December 2017 est.)
$15.54 billion (31 December 2016 est.)
country comparison to the world: 67

Debt—external: $29.34 billion (31 December 2017 est.)
$26.38 billion (31 December 2016 est.)
country comparison to the world: 82

Exchange rates: Jordanian dinars (JOD) per US dollar -
0.71 (2017 est.)
0.71 (2016 est.)
0.71 (2015 est.)
0.71 (2014 est.)

0.71 (2013 est.)

ENERGY

Electricity access: *electrification—total population:* 100% (2020)

Electricity—production: 18.6 billion kWh (2016 est.)
country comparison to the world: 77

Electricity—consumption: 16.82 billion kWh (2016 est.)
country comparison to the world: 74

Electricity—exports: 50 million kWh (2015 est.)
country comparison to the world: 88

Electricity—imports: 334 million kWh (2016 est.)
country comparison to the world: 86

Electricity—installed generating capacity: 4.764 million kW (2016 est.)
country comparison to the world: 82

Electricity—from fossil fuels: 87% of total installed capacity (2016 est.)
country comparison to the world: 62

Electricity—from nuclear fuels: 0% of total installed capacity (2017 est.)
country comparison to the world: 116

Electricity—from hydroelectric plants: 0% of total installed capacity (2017 est.)
country comparison to the world: 180

Electricity—from other renewable sources: 12% of total installed capacity (2017 est.)
country comparison to the world: 74

Crude oil—production: 22 bbl/day (2018 est.)
country comparison to the world: 99

Crude oil—exports: 0 bbl/day (2015 est.)
country comparison to the world: 144

Crude oil—imports: 67,980 bbl/day (2015 est.)
country comparison to the world: 49

Crude oil—proved reserves: 1 million bbl (1 January 2018 est.)
country comparison to the world: 96

Refined petroleum products—production: 67,240 bbl/day (2015 est.)
country comparison to the world: 73

Refined petroleum products—consumption: 139,000 bbl/day (2016 est.)
country comparison to the world: 69

Refined petroleum products—exports: 0 bbl/day (2015 est.)
country comparison to the world: 166

Refined petroleum products—imports: 68,460 bbl/day (2015 est.)
country comparison to the world: 68

Natural gas—production: 121.8 million cu m (2017 est.)
country comparison to the world: 80

Natural gas—consumption: 5.238 billion cu m (2017 est.)
country comparison to the world: 56

Natural gas—exports: 1.359 billion cu m (2017 est.)
country comparison to the world: 38

Natural gas—imports: 6.456 billion cu m (2017 est.)
country comparison to the world: 31

Natural gas—proved reserves: 6.031 billion cu m (1 January 2018 est.)
country comparison to the world: 87

Carbon dioxide emissions from consumption of energy: 27.39 million Mt (2017 est.)
country comparison to the world: 76

COMMUNICATIONS

Telephones—fixed lines: *total subscriptions:* 375,576
subscriptions per 100 inhabitants: 3.52 (2019 est.)
country comparison to the world: 104

Telephones—mobile cellular: *total subscriptions:* 8,215,735
subscriptions per 100 inhabitants: 77 (2019 est.)
country comparison to the world: 95

Telecommunication systems: *general assessment:* microwave radio relay transmission and coaxial and fiber-optic cable are employed on trunk lines; growing mobile-cellular usage in both urban and rural areas is reducing use of fixed-line services; recent influx of refugees putting burden on country's economy, infrastructure and society; mobile broadband is area of growth with 4G services; govt. recently launched Ministry of Digital Economy & Entrepreneurship; preparing for next wave of development with 5G and IoT/MsM services (2020)
domestic: 1995 a telecommunications law opened all non-fixed-line services to private competition; in 2005, the monopoly over fixed-line services terminated and the entire telecommunications sector was opened to competition; currently fixed-line 4 per 100 persons and multiple mobile-cellular providers with subscribership up to 77 per 100 persons (2019)
international: country code—962; landing point for the FEA and Taba-Aqaba submarine cable networks providing connectivity to Europe, the Middle East, Southeast Asia and Asia; satellite earth stations—33 (3 Intelsat, 1 Arabsat, and 29 land and maritime Inmarsat terminals (2019)
note: the COVID-19 outbreak is negatively impacting telecommunications production and supply chains globally; consumer spending on telecom devices and services has also slowed due to the pandemic's effect on economies worldwide; overall progress towards improvements in all facets of the telecom industry—mobile, fixed-line, broadband, submarine cable and satellite—has moderated

Broadcast media: radio and TV dominated by the government-owned Jordan Radio and Television Corporation (JRTV) that operates a main network, a sports network, a film network, and a satellite channel; first independent TV broadcaster aired in 2007; international satellite TV and Israeli and Syrian TV broadcasts are available; roughly 30 radio stations with JRTV operating the main government-owned station; transmissions of multiple international radio broadcasters are available

Internet country code: .jo

Internet users: *total:* 6,985,174
percent of population: 66.79% (July 2018 est.)
country comparison to the world: 71

Broadband—fixed subscriptions: *total:* 399,596
subscriptions per 100 inhabitants: 4 (2018 est.)
country comparison to the world: 90

TRANSPORTATION

National air transport system: *number of registered air carriers:* 4 (2020)
inventory of registered aircraft operated by air carriers: 54
annual passenger traffic on registered air carriers: 3,383,805 (2018)
annual freight traffic on registered air carriers: 175.84 million mt-km (2018)

Civil aircraft registration country code prefix: JY (2016)

Airports: 18 (2013)
country comparison to the world: 139

Airports—with paved runways: *total:* 16 (2017)
over 3,047 m: 8 (2017)
2,438 to 3,047 m: 5 (2017)
1,524 to 2,437 m: 2 (2017)
914 to 1,523 m: 1 (2017)

Airports—with unpaved runways: *total:* 2 (2013)
under 914 m: 2 (2013)

Heliports: 1 (2012)

Pipelines: 473 km gas, 49 km oil (2013)

Railways: *total:* 509 km (2014)
narrow gauge: 509 km 1.050-m gauge (2014)
country comparison to the world: 113

Roadways: *total:* 7,203 km (2011)
paved: 7,203 km (2011)
country comparison to the world: 141

Merchant marine: *total:* 32
by type: general cargo 7, oil tanker 1, other 24 (2019)
country comparison to the world: 129

Ports and terminals: *major seaport(s):* Al 'Aqabah

MILITARY AND SECURITY

Military and security forces: *Jordanian Armed Forces (JAF):* Royal Jordanian Army (includes Special Operations Forces, Border Guards, Royal Guard), Royal Jordanian Navy, Royal Jordanian Air Force; Ministry of Interior: General Directorate of Gendarmerie Forces, Public Security Directorate (2020)

Military expenditures: 4.7% of GDP (2019)
4.7% of GDP (2018)
4.8% of GDP (2017)
4.6% of GDP (2016)
4.3% of GDP (2015)
country comparison to the world: 8

Military and security service personnel strengths: the Jordanian Armed Forces (JAF) have approximately 101,000 active personnel (87,000 Army; 500 Navy; 14,000 Air Force); est. 15,000 Gendarmerie Forces (2019)

507

Military equipment inventories and acquisitions: the JAF inventory is comprised of a wide mix of imported weapons, mostly second-hand equipment from Europe and the US; some of the equipment is received from third-party suppliers such as the United Arab Emirates; since 2010, the Netherlands and the US are the leading suppliers of military hardware to Jordan (2019)

Military service age and obligation: 17 years of age for voluntary male military service; initial service term 2 years, with option to reenlist for 18 years; conscription at age 18 suspended in 1999; women are not conscripted, but can volunteer to serve in noncombat military positions in the Royal Jordanian Arab Army Women's Corps and RJAF (2013)

TRANSNATIONAL ISSUES

Disputes—international: 2004 Agreement settles border dispute with Syria pending demarcation

Refugees and internally displaced persons: *refugees (country of origin):* 2,272,411 (Palestinian refugees), 661,997 (Syria), 66,835 (Iraq), 14,640 (Yemen), 6,098 Sudan (2020)

KAZAKHSTAN

INTRODUCTION

Background: Ethnic Kazakhs, a mix of Turkic and Mongol nomadic tribes with additional Persian cultural influences, migrated to the region in the 15th century. The area was conquered by Russia in the 18th and 19th centuries, and Kazakhstan became a Soviet Republic in 1925. Repression and starvation associated with forced agricultural collectivization led to a massive number of deaths in the 1930s. During the 1950s and 1960s, the agricultural "Virgin Lands" program led to an influx of settlers (mostly ethnic Russians, but also other nationalities) and at the time of Kazakhstan's independence in 1991, ethnic Kazakhs were a minority. Non-Muslim ethnic minorities departed Kazakhstan in large numbers from the mid-1990s through the mid-2000s and a national program has repatriated about a million ethnic Kazakhs (from Uzbekistan, Tajikistan, Mongolia, and the Xinjiang region of China) back to Kazakhstan. As a result of this shift, the ethnic Kazakh share of the population now exceeds two-thirds.

Kazakhstan's economy is the largest in the Central Asian states, mainly due to the country's vast natural resources. Current issues include: diversifying the economy, obtaining membership in global and regional international economic institutions, enhancing Kazakhstan's economic competitiveness, and strengthening relations with neighboring states and foreign powers.

GEOGRAPHY

Location: Central Asia, northwest of China; a small portion west of the Ural (Zhayyq) River in easternmost Europe

Geographic coordinates: 48 00 N, 68 00 E

Map references: Asia

Area: *total:* 2,724,900 sq km
land: 2,699,700 sq km
water: 25,200 sq km
country comparison to the world: 10

Area—comparative: slightly less than four times the size of Texas

Land boundaries: *total:* 13,364 km
border countries (5): China 1765 km, Kyrgyzstan 1212 km, Russia 7644 km, Turkmenistan 413 km, Uzbekistan 2330 km

Coastline: 0 km (landlocked); note—Kazakhstan borders the Aral Sea, now split into two bodies of water (1,070 km), and the Caspian Sea (1,894 km)

Maritime claims: none (landlocked)

Climate: continental, cold winters and hot summers, arid and semiarid

Terrain: vast flat steppe extending from the Volga in the west to the Altai Mountains in the east and from the plains of western Siberia in the north to oases and deserts of Central Asia in the south

Elevation: *mean elevation:* 387 m
lowest point: Vpadina Kaundy -132 m
highest point: Khan Tangiri Shyngy (Pik Khan-Tengri) 6,995 m

Natural resources: major deposits of petroleum, natural gas, coal, iron ore, manganese, chrome ore, nickel, cobalt, copper, molybdenum, lead, zinc, bauxite, gold, uranium

Land use: *agricultural land:* 77.4% (2011 est.)
arable land: 8.9% (2011 est.) / permanent crops: 0% (2011 est.) / permanent pasture: 68.5% (2011 est.)
forest: 1.2% (2011 est.)
other: 21.4% (2011 est.)

Irrigated land: 20,660 sq km (2012)

Population distribution: most of the country displays a low population density, particularly the interior; population clusters appear in urban agglomerations in the far northern and southern portions of the country

Natural hazards: earthquakes in the south; mudslides around Almaty

Environment—current issues: radioactive or toxic chemical sites associated with former defense industries and test ranges scattered throughout the country pose health risks for humans and animals; industrial pollution is severe in some cities; because the two main rivers that flowed into the Aral Sea have been diverted for irrigation, it is drying up and leaving behind a harmful layer of chemical pesticides and natural salts; these substances are then picked up by the wind and blown into noxious dust storms; pollution in the Caspian Sea; desertification; soil pollution from overuse of agricultural chemicals and salination from poor infrastructure and wasteful irrigation practices

Environment—international agreements: *party to:* Air Pollution, Biodiversity, Climate Change, Desertification, Endangered Species, Environmental Modification, Hazardous Wastes, Ozone Layer Protection, Ship Pollution, Wetlands
signed, but not ratified: Climate Change-Kyoto Protocol

Geography—note: world's largest landlocked country and one of only two landlocked countries in the world that extends into two continents (the other is Azerbaijan); Russia leases approximately 6,000 sq km of territory enclosing the Baykonur

Cosmodrome; in January 2004, Kazakhstan and Russia extended the lease to 2050

PEOPLE AND SOCIETY

Population: 19,091,949 (July 2020 est.)
country comparison to the world: 64

Nationality: *noun:* Kazakhstani(s)
adjective: Kazakhstani

Ethnic groups: Kazakh (Qazaq) 68%, Russian 19.3%, Uzbek 3.2%, Ukrainian 1.5%, Uighur 1.5%, Tatar 1.1%, German 1%, other 4.4% (2019 est.)

Languages: Kazakh (official, Qazaq) 83.1% (understand spoken language) and trilingual (Kazakh, Russian, English) 22.3% (2017 est.); Russian (official, used in everyday business, designated the "language of interethnic communication") 94.4% (understand spoken language) (2009 est.)

Religions: Muslim 70.2%, Christian 26.2% (mainly Russian Orthodox), other 0.2%, atheist 2.8%, unspecified 0.5% (2009 est.)

Age structure: *0-14 years:* 26.13% (male 2,438,148/female 2,550,535)
15-24 years: 12.97% (male 1,262,766/female 1,212,645)
25-54 years: 42.23% (male 3,960,188/female 4,102,845)
55-64 years: 10.25% (male 856,180/female 1,099,923)
65 years and over: 8.43% (male 567,269/female 1,041,450) (2020 est.)

Dependency ratios: *total dependency ratio:* 58.8
youth dependency ratio: 46.3
elderly dependency ratio: 12.6
potential support ratio: 8 (2020 est.)

Median age: *total:* 31.6 years
male: 30.3 years
female: 32.8 years (2020 est.)
country comparison to the world: 114

Population growth rate: 0.89% (2020 est.)
country comparison to the world: 117

Birth rate: 16.4 births/1,000 population (2020 est.)
country comparison to the world: 105

Death rate: 8.2 deaths/1,000 population (2020 est.)
country comparison to the world: 86

Net migration rate: 0.4 migrant(s)/1,000 population (2020 est.)
country comparison to the world: 67

Population distribution: most of the country displays a low population density, particularly the interior; population clusters appear in urban agglomerations in the far northern and southern portions of the country

Urbanization: *urban population:* 57.7% of total population (2020)

rate of urbanization: 1.29% annual rate of change (2015-20 est.)

total population growth rate v. urban population growth rate, 2000-2030: Major urban areas - population: 1.896 million Almaty, 1.896 million NUR-SULTAN (capital), 1.058 million Shimkent (2020)

Sex ratio: *at birth:* 0.94 male(s)/female
0-14 years: 0.96 male(s)/female
15-24 years: 1.04 male(s)/female
25-54 years: 0.97 male(s)/female
55-64 years: 0.78 male(s)/female
65 years and over: 0.54 male(s)/female
total population: 0.91 male(s)/female (2020 est.)

Mother's mean age at first birth: 28.5 years (2017 est.)

Maternal mortality rate: 10 deaths/100,000 live births (2017 est.)
country comparison to the world: 146

Infant mortality rate: *total:* 17.9 deaths/1,000 live births
male: 20.4 deaths/1,000 live births
female: 15.5 deaths/1,000 live births (2020 est.)
country comparison to the world: 83

Life expectancy at birth: *total population:* 72 years
male: 66.8 years
female: 76.8 years (2020 est.)
country comparison to the world: 156

Total fertility rate: 2.16 children born/woman (2020 est.)
country comparison to the world: 96

Contraceptive prevalence rate: 53% (2018)
note: percent of women aged 18-49

Drinking water source:

improved: urban: 100% of population
rural: 93.8% of population
total: 97.4% of population

unimproved: urban: 0% of population
rural: 6.2% of population
total: 2.6% of population (2017 est.)

Current Health Expenditure: 3.1% (2017)

Physicians density: 3.98 physicians/1,000 population (2014)

Hospital bed density: 6.1 beds/1,000 population (2014)

Sanitation facility access:

improved: urban: 99.9% of population
rural: 100% of population
total: 99.9% of population

unimproved: urban: 0.1% of population
rural: 0% of population
total: 0.1% of population (2017 est.)

HIV/AIDS—adult prevalence rate: 0.2% (2019 est.)
country comparison to the world: 101

HIV/AIDS—people living with HIV/AIDS: 33,000 (2019 est.)
country comparison to the world: 70

HIV/AIDS—deaths: <500 (2019 est.)

Obesity—adult prevalence rate: 21% (2016)

country comparison to the world: 94

Children under the age of 5 years underweight: 2% (2015)
country comparison to the world: 111

Education expenditures: 2.8% of GDP (2017)
country comparison to the world: 146

Literacy: *definition:* age 15 and over can read and write
total population: 99.8%
male: 99.8%
female: 99.8% (2015)

School life expectancy (primary to tertiary education): *total:* 16 years
male: 15 years
female: 16 years (2019)

Unemployment, youth ages 15-24: *total:* 3.8%
male: 3.6%
female: 4% (2016 est.)
country comparison to the world: 169

GOVERNMENT

Country name: *conventional long form:* Republic of Kazakhstan
conventional short form: Kazakhstan
local long form: Qazaqstan Respublikasy
local short form: Qazaqstan
former: Kazakh Soviet Socialist Republic
etymology: the name "Kazakh" derives from the Turkic word "kaz" meaning "to wander," recalling the Kazakh's nomadic lifestyle; the Persian suffix "-stan" means "place of" or "country," so the word Kazakhstan literally means "Land of the Wanderers"

Government type: presidential republic

Capital: *name:* Nur-Sultan
geographic coordinates: 51 10 N, 71 25 E
time difference: UTC+6 (11 hours ahead of Washington, DC, during Standard Time)
note: Kazakhstan has two time zones
etymology: on 20 March 2019, Kazakhstan changed the name of its capital city from Astana to Nur-Sultan in honor of its long-serving, recently retired president, Nursultan NAZARBAYEV; this was not the first time the city had its name changed; founded in 1830 as Akmoly, it became Akmolinsk in 1832, Tselinograd in 1961, Akmola (Aqmola) in 1992, and Astana in 1998

Administrative divisions: 14 provinces (oblyslar, singular—oblys) and 4 cities* (qalalar, singular—qala); Almaty (Taldyqorghan), Almaty*, Aqmola (Kokshetau), Aqtobe, Astana*, Atyrau, Batys Qazaqstan [West Kazakhstan] (Oral), Bayqongyr*, Mangghystau (Aqtau), Pavlodar, Qaraghandy, Qostanay, Qyzylorda, Shyghys Qazaqstan [East Kazakhstan] (Oskemen), Shymkent*, Soltustik Qazaqstan [North Kazakhstan] (Petropavl), Turkistan, Zhambyl (Taraz)
note: administrative divisions have the same names as their administrative centers (exceptions have the administrative center name following in parentheses); in 1995, the Governments of Kazakhstan and Russia entered into an agreement whereby Russia would lease for a period of 20 years

an area of 6,000 sq km enclosing the Baikonur space launch facilities and the city of Bayqongyr (Baikonur, formerly Leninsk); in 2004, a new agreement extended the lease to 2050

Independence: 16 December 1991 (from the Soviet Union)

National holiday: Independence Day, 16 December (1991)

Constitution: *history:* previous 1937, 1978 (preindependence), 1993; latest approved by referendum 30 August 1995, effective 5 September 1995
amendments: introduced by a referendum initiated by the president of the republic, on the recommendation of Parliament, or by the government; the president has the option of submitting draft amendments to Parliament or directly to a referendum; passage of amendments by Parliament requires four-fifths majority vote of both houses and the signature of the president; passage by referendum requires absolute majority vote by more than one half of the voters in at least two thirds of the oblasts, major cities, and the capital, followed by the signature of the president; amended several times, last in 2019

Legal system: civil law system influenced by Roman-Germanic law and by the theory and practice of the Russian Federation

International law organization participation: has not submitted an ICJ jurisdiction declaration; non-party state to the ICCt

Citizenship: *citizenship by birth:* no
citizenship by descent only: at least one parent must be a citizen of Kazakhstan
dual citizenship recognized: no
residency requirement for naturalization: 5 years

Suffrage: 18 years of age; universal

Executive branch: *chief of state:* President Kasym-Zhomart TOKAYEV (since 20 March 2019); note - Nursultan NAZARBAYEV, who was president since 24 April 1990 (and in power since 22 June 1989 under the Soviet period), resigned on 20 March 2019; NAZARBAYEV retained the title and powers of "First President"; TOKAYEV completed NAZARBAYEV's term, which was shortened due to the early election of 9 June 2019, and then continued as president following his election victory
head of government: Prime Minister Askar MAMIN (since 25 February 2019); First Deputy Prime Minister Alikhan SMAILOV (since 25 February 2019); Deputy Prime Ministers Berdibek SAPARBAYEV and Roman SKLYAR (since 18 September 2019)
cabinet: the president appoints ministers after consultations with the Chair of the Security Council (NAZARBAYEV) who has veto power over all appointments except for the ministers of defense, internal affairs, and foreign affairs; however, the president is required to discuss these three offices with the National Security Committee, which NAZARBAYEV chairs under a lifetime appointment

elections/appointments: president directly elected by simple majority popular vote for a 5-year term (eligible for a second consecutive term); election last held on 9 June 2019 (next to be held in 2024); prime minister and deputy prime ministers appointed by the president, approved by the Mazhilis

election results: Kasym-Zhomart TOKAYEV elected president; percent of vote—Kassym-Jomart TOKAYEV (Nur Otan) 71%, Amirzhan KOSANOV (Ult Tagdyry) 16.2%, Daniya YESPAYEVA (Ak Zhol) 5.1%, other 7.7%

Legislative branch: *description:* bicameral Parliament consists of: Senate (49 seats; 34 members indirectly elected by majority 2-round vote by the oblast-level assemblies and 15 members appointed by decree of the president; members serve 6-year terms, with one-half of the membership renewed every 3 years)

Mazhilis (107 seats; 98 members directly elected in a single national constituency by proportional representation vote to serve 5-year terms and 9 indirectly elected by the Assembly of People of Kazakhstan, a 350-member, presidentially appointed advisory body designed to represent the country's ethnic minorities)

elections: Senate—last held on 12 August 2020 (next to be held in 2026)

Mazhilis—last held on 20 March 2016 (next to be held by 2021)

election results: Senate—percent of vote by party - NA; seats by party—NA; composition—men 42, women 5, percent of women 10.6% Mazhilis—percent of vote by party—Nur Otan 82.2%, Ak Zhol 7.2%, Communist People's Party 7.1%, other 3.5%; seats by party—Nur Otan 84, Ak Zhol 7, Communist People's Party 7; composition—men 78, women 29, percent of women 27.1%; note—total Parliament percent of women 22.1%

Judicial branch: *highest courts:* Supreme Court of the Republic (consists of 44 members); Constitutional Council (consists of the chairman and 6 members)

judge selection and term of office: Supreme Court judges proposed by the president of the republic on recommendation of the Supreme Judicial Council and confirmed by the Senate; judges normally serve until age 65 but can be extended to age 70; Constitutional Council—the president of the republic, the Senate chairperson, and the Mazhilis chairperson each appoints 2 members for a 6-year term; chairman of the Constitutional Council appointed by the president for a 6-year term

subordinate courts: regional and local courts

Political parties and leaders: Ak Zhol (Bright Path) Party or Democratic Party of Kazakhstan Ak Zhol [Azat PERUASHEV]
Birlik (Unity) Party [Serik SULTANGALI]
Communist People's Party of Kazakhstan [informal leader Aikyn KONUROV]
National Social Democratic Party or NSDP [Zharmakhan TUYAKBAY]
Nur Otan (Radiant Fatherland) Democratic People's Party [Nursultan NAZARBAYEV]

People's Democratic (Patriotic) Party "Auyl" [Ali BEKTAYEV]
Ult Tagdyry (Conscience of the Nation)

International organization participation: ADB, CICA, CIS, CSTO, EAEU, EAPC, EBRD, ECO, EITI (compliant country), FAO, GCTU, IAEA, IBRD, ICAO, ICC (NGOs), ICRM, IDA, IDB, IFAD, IFC, IFRCS, ILO, IMF, IMO, Interpol, IOC, IOM, IPU, ISO, ITSO, ITU, MIGA, MINURSO, NAM (observer), NSG, OAS (observer), OIC, OPCW, OSCE, PFP, SCO, UN, UNCTAD, UNESCO, UNIDO, UN Security Council (temporary), UNWTO, UPU, WCO, WFTU (NGOs), WHO, WIPO, WMO, WTO (observer), ZC

Diplomatic representation in the US: *chief of mission:* Ambassador Yerzhan KAZYKHANOV (since 24 April 2017)

chancery: 1401 16th Street NW, Washington, DC 20036

telephone: [1] (202) 232-5488

FAX: [1] (202) 232-5845

consulate(s) general: New York

Diplomatic representation from the US: *chief of mission:* Ambassador William MOSER (since 27 March 2019)

telephone: [7] (7172) 70-21-00

embassy: Rakhymzhan Koshkarbayev Ave. No 3, Astana 010010

mailing address: use embassy street address

FAX: [7] (7172) 54-09-14

consulate(s) general: Almaty

Flag description: a gold sun with 32 rays above a soaring golden steppe eagle, both centered on a sky blue background; the hoist side displays a national ornamental pattern "koshkar-muiz" (the horns of the ram) in gold; the blue color is of religious significance to the Turkic peoples of the country, and so symbolizes cultural and ethnic unity; it also represents the endless sky as well as water; the sun, a source of life and energy, exemplifies wealth and plenitude; the sun's rays are shaped like grain, which is the basis of abundance and prosperity; the eagle has appeared on the flags of Kazakh tribes for centuries and represents freedom, power, and the flight to the future

National symbol(s): *golden eagle; national colors:* blue, yellow

National anthem: *name:* "Menin Qazaqstanim" (My Kazakhstan)

lyrics/music: Zhumeken NAZHIMEDENOV and Nursultan NAZARBAYEV/Shamshi KALDAYAKOV

note: adopted 2006; President Nursultan NAZARBAYEV played a role in revising the lyrics

ECONOMY

Economy—overview: Kazakhstan's vast hydrocarbon and mineral reserves form the backbone of its economy. Geographically the largest of the former Soviet republics, excluding Russia, Kazakhstan, g possesses substantial fossil fuel reserves and other minerals and metals, such as uranium, copper, and zinc. It also has a large agricultural sector featuring livestock and grain. The government realizes that

its economy suffers from an overreliance on oil and extractive industries and has made initial attempts to diversify its economy by targeting sectors like transport, pharmaceuticals, telecommunications, petrochemicals and food processing for greater development and investment. It also adopted a Subsoil Code in December 2017 with the aim of increasing exploration and investment in the hydrocarbon, and particularly mining, sectors.

Kazakhstan's oil production and potential is expanding rapidly. A $36.8 billion expansion of Kazakhstan's premiere Tengiz oil field by Chevron-led Tengizchevroil should be complete in 2022. Meanwhile, the super-giant Kashagan field finally launched production in October 2016 after years of delay and an estimated $55 billion in development costs. Kazakhstan's total oil production in 2017 climbed 10.5%.

Kazakhstan is landlocked and depends on Russia to export its oil to Europe. It also exports oil directly to China. In 2010, Kazakhstan joined Russia and Belarus to establish a Customs Union in an effort to boost foreign investment and improve trade. The Customs Union evolved into a Single Economic Space in 2012 and the Eurasian Economic Union (EAEU) in January 2015. Supported by rising commodity prices, Kazakhstan's exports to EAEU countries increased 30.2% in 2017. Imports from EAEU countries grew by 24.1%.

The economic downturn of its EAEU partner, Russia, and the decline in global commodity prices from 2014 to 2016 contributed to an economic slowdown in Kazakhstan. In 2014, Kazakhstan devalued its currency, the tenge, and announced a stimulus package to cope with its economic challenges. In the face of further decline in the ruble, oil prices, and the regional economy, Kazakhstan announced in 2015 it would replace its currency band with a floating exchange rate, leading to a sharp fall in the value of the tenge. Since reaching a low of 391 to the dollar in January 2016, the tenge has modestly appreciated, helped by somewhat higher oil prices. While growth slowed to about 1% in both 2015 and 2016, a moderate recovery in oil prices, relatively stable inflation and foreign exchange rates, and the start of production at Kashagan helped push 2017 GDP growth to 4%.

Despite some positive institutional and legislative changes in the last several years, investors remain concerned about corruption, bureaucracy, and arbitrary law enforcement, especially at the regional and municipal levels. An additional concern is the condition of the country's banking sector, which suffers from poor asset quality and a lack of transparency. Investors also question the potentially negative effects on the economy of a contested presidential succession as Kazakhstan's first president, Nursultan NAZARBAYEV, turned 77 in 2017.

GDP (purchasing power parity): $478.6 billion (2017 est.)
$460.3 billion (2016 est.)
$455.3 billion (2015 est.)
note: data are in 2017 dollars

country comparison to the world: 43

GDP (official exchange rate): $159.4 billion (2017 est.)

GDP—real growth rate: 6.13% (2019 est.)
4.41% (2018 est.)
4.38% (2017 est.)
country comparison to the world: 28

GDP—per capita (PPP): $26,300 (2017 est.)
$25,700 (2016 est.)
$25,800 (2015 est.)
note: data are in 2017 dollars
country comparison to the world: 79

Gross national saving: 23.7% of GDP (2017 est.)
21.4% of GDP (2016 est.)
25.1% of GDP (2015 est.)
country comparison to the world: 71

GDP—composition, by end use: *household consumption:* 53.2% (2017 est.)
government consumption: 11.1% (2017 est.)
investment in fixed capital: 22.5% (2017 est.)
investment in inventories: 4.8% (2017 est.)
exports of goods and services: 35.4% (2017 est.)
imports of goods and services: -27.1% (2017 est.)

GDP—composition, by sector of origin: *agriculture:* 4.7% (2017 est.)
industry: 34.1% (2017 est.)
services: 61.2% (2017 est.)

Agriculture—products: grain (mostly spring wheat and barley), potatoes, vegetables, melons; livestock
Industries: oil, coal, iron ore, manganese, chromite, lead, zinc, copper, titanium, bauxite, gold, silver, phosphates, sulfur, uranium, iron and steel; tractors and other agricultural machinery, electric motors, construction materials

Industrial production growth rate: 5.8% (2017 est.)
country comparison to the world: 45

Labor force: 8.685 million (2020 est.)
country comparison to the world: 56

Labor force—by occupation: *agriculture:* 18.1%
industry: 20.4%
services: 61.6% (2017 est.)

Unemployment rate: 4.8% (2019 est.)
4.85% (2018 est.)
country comparison to the world: 70

Population below poverty line: 2.6% (2016 est.)

Household income or consumption by percentage share: *lowest 10%:* 4.2%
highest 10%: 23.3% (2016)

Budget: *revenues:* 35.48 billion (2017 est.)
expenditures: 38.3 billion (2017 est.)

Taxes and other revenues: 22.3% (of GDP) (2017 est.)
country comparison to the world: 134

Budget surplus (+) or deficit (-): -1.8% (of GDP) (2017 est.)
country comparison to the world: 98

Public debt: 20.8% of GDP (2017 est.)
19.7% of GDP (2016 est.)
country comparison to the world: 187

Fiscal year: calendar year

Inflation rate (consumer prices): 7.4% (2017 est.)

14.6% (2016 est.)
country comparison to the world: 194

Current account balance: -$7.206 billion (2019 est.)
-$138 million (2018 est.)
country comparison to the world: 189

Exports: $49.29 billion (2017 est.)
$37.26 billion (2016 est.)
country comparison to the world: 51

Exports—partners: Italy 17.9%, China 11.9%, Netherlands 9.8%, Russia 9.3%, Switzerland 6.4%, France 5.9% (2017)

Exports—commodities: oil and oil products, natural gas, ferrous metals, chemicals, machinery, grain, wool, meat, coal

Imports: $31.85 billion (2017 est.)
$28.07 billion (2016 est.)
country comparison to the world: 63

Imports—commodities: machinery and equipment, metal products, foodstuffs

Imports—partners: Russia 38.9%, China 16.1%, Germany 5.1%, US 4.3% (2017)

Reserves of foreign exchange and gold: $30.75 billion (31 December 2017 est.)
$29.53 billion (31 December 2016 est.)
country comparison to the world: 50

Debt—external: $167.5 billion (31 December 2017 est.)
$163.6 billion (31 December 2016 est.)
country comparison to the world: 40

Exchange rates: tenge (KZT) per US dollar - 326.3 (2017 est.)
342.13 (2016 est.)
342.13 (2015 est.)
221.73 (2014 est.)
179.19 (2013 est.)

ENERGY

Electricity access: *electrification—total population:* 100% (2020)

Electricity—production: 100.8 billion kWh (2016 est.)
country comparison to the world: 35

Electricity—consumption: 94.23 billion kWh (2016 est.)
country comparison to the world: 33

Electricity—exports: 5.1 billion kWh (2017 est.)
country comparison to the world: 37

Electricity—imports: 1.318 billion kWh (2016 est.)
country comparison to the world: 64

Electricity—installed generating capacity: 20.15 million kW (2016 est.)
country comparison to the world: 44

Electricity—from fossil fuels: 86% of total installed capacity (2016 est.)
country comparison to the world: 66

Electricity—from nuclear fuels: 0% of total installed capacity (2017 est.)
country comparison to the world: 117

Electricity—from hydroelectric plants: 14% of total installed capacity (2017 est.)
country comparison to the world: 107

Electricity—from other renewable sources: 1% of total installed capacity (2017 est.)
country comparison to the world: 155

Crude oil—production: 1.856 million bbl/day (2018 est.)
country comparison to the world: 12

Crude oil—exports: 1.409 million bbl/day (2015 est.)
country comparison to the world: 9

Crude oil—imports: 1,480 bbl/day (2015 est.)
country comparison to the world: 79

Crude oil—proved reserves: 30 billion bbl (1 January 2018 est.)
country comparison to the world: 11

Refined petroleum products—production: 290,700 bbl/day (2015 est.)
country comparison to the world: 44

Refined petroleum products—consumption: 274,000 bbl/day (2016 est.)
country comparison to the world: 46

Refined petroleum products—exports: 105,900 bbl/day (2015 est.)
country comparison to the world: 41

Refined petroleum products—imports: 39,120 bbl/day (2015 est.)
country comparison to the world: 90

Natural gas—production: 22.41 billion cu m (2017 est.)
country comparison to the world: 30

Natural gas—consumption: 15.37 billion cu m (2017 est.)
country comparison to the world: 43

Natural gas—exports: 12.8 billion cu m (2017 est.)
country comparison to the world: 17

Natural gas—imports: 5.748 billion cu m (2017 est.)
country comparison to the world: 34

Natural gas—proved reserves: 2.407 trillion cu m (1 January 2018 est.)
country comparison to the world: 14

Carbon dioxide emissions from consumption of energy: 304.6 million Mt (2017 est.)
country comparison to the world: 23

COMMUNICATIONS

Telephones—fixed lines: *total subscriptions:* 3,275,584
subscriptions per 100 inhabitants: 17.31 (2019 est.)
country comparison to the world: 39

Telephones—mobile cellular: *total subscriptions:* 26,223,595
subscriptions per 100 inhabitants: 138.58 (2019 est.)
country comparison to the world: 48

Telecommunication systems: *general assessment:* one of the most progressive telecoms sectors in

Central Asia; vast 4G network; low fixed-line and fixed-broadband penetration, moderate mobile broadband penetration and high mobile penetration; mobile market highly competitive and slow growth due to saturation (2020)

domestic: intercity by landline and microwave radio relay; number of fixed-line connections is 17 per 100 persons; mobile-cellular usage increased rapidly and the subscriber base approaches 139 per 100 persons (2019)

international: country code—7; international traffic with other former Soviet republics and China carried by landline and microwave radio relay and with other countries by satellite and by the TAE fiber-optic cable; satellite earth stations—2 Intelsat

note: the COVID-19 outbreak is negatively impacting telecommunications production and supply chains globally; consumer spending on telecom devices and services has also slowed due to the pandemic's effect on economies worldwide; overall progress towards improvements in all facets of the telecom industry—mobile, fixed-line, broadband, submarine cable and satellite—has moderated

Broadcast media: the state owns nearly all radio and TV transmission facilities and operates national TV and radio networks; there are 96 TV channels, many of which are owned by the government, and 4 state-run radio stations; some former state-owned media outlets have been privatized; households with satellite dishes have access to foreign media; a small number of commercial radio stations operate along with state-run radio stations; recent legislation requires all media outlets to register with the government and all TV providers to broadcast in digital format by 2018; broadcasts reach some 99% of the population as well as neighboring countries

Internet country code: .kz

Internet users: *total:* 14,789,448
percent of population: 78.9% (July 2018 est.)
country comparison to the world: 44

Broadband—fixed subscriptions: *total:* 2,462,900
subscriptions per 100 inhabitants: 13 (2018 est.)
country comparison to the world: 50

TRANSPORTATION

National air transport system: *number of registered air carriers:* 12 (2020)

inventory of registered aircraft operated by air carriers: 84
annual passenger traffic on registered air carriers: 7,143,797 (2018)
annual freight traffic on registered air carriers: 50.22 million mt-km (2018)

Civil aircraft registration country code prefix: UP (2016)

Airports: 96 (2013)
country comparison to the world: 60

Airports—with paved runways: *total:* 63 (2017)
over 3,047 m: 10 (2017)
2,438 to 3,047 m: 25 (2017)
1,524 to 2,437 m: 15 (2017)
914 to 1,523 m: 5 (2017)
under 914 m: 8 (2017)

Airports—with unpaved runways: *total:* 33 (2013)
over 3,047 m: 5 (2013)
2,438 to 3,047 m: 7 (2013)
1,524 to 2,437 m: 3 (2013)
914 to 1,523 m: 5 (2013)
under 914 m: 13 (2013)

Heliports: 3 (2013)

Pipelines: 658 km condensate, 15,256 km gas (2017), 8,013 km oil (2017), 1,095 km refined products, 1,975 km water (2016) (2017)

Railways: *total:* 16,614 km (2017)
broad gauge: 16,614 km 1.520-m gauge (4,200 km electrified) (2017)
country comparison to the world: 18

Roadways: *total:* 95,409 km (2017)
paved: 81,814 km (2017)
unpaved: 13,595 km (2017)
country comparison to the world: 51

Waterways: 4,000 km (on the Ertis (Irtysh) River (80%) and Syr Darya (Syrdariya) River) (2010)
country comparison to the world: 25

Merchant marine: *total:* 124
by type: general cargo 2, oil tanker 5, other 117 (2019)
country comparison to the world: 79

Ports and terminals: *major seaport(s):* Caspian Sea—Aqtau (Shevchenko), Atyrau (Gur'yev)
river port(s): Oskemen (Ust-Kamenogorsk), Pavlodar, Semey (Semipalatinsk) (Irtysh River)

MILITARY AND SECURITY

Military and security forces: *Armed Forces of the Republic of Kazakhstan:* Land Forces, Navy, Air and Air Defense Force; Ministry of Internal Affairs: National Guard, Border Service (includes Coast Guard), State Security Service (2019)

Military expenditures: 1.1% of GDP (2019)
0.9% of GDP (2018)
0.9% of GDP (2017)
0.9% of GDP (2016)
1.1% of GDP (2015)
country comparison to the world: 113

Military and security service personnel strengths: estimates of the size of the Armed Forces of Kazakhstan vary; approximately 45,000 active duty personnel (25,000 Army; 3,000 Navy; 14,000 Air and Air Defense; 3,000 other) (2019 est.)

Military equipment inventories and acquisitions: the Kazakh military's inventory is comprised of mostly older Russian and Soviet-era equipment; since 2010, Russia remains by far the leading supplier of weapons systems, but Kazakhstan has also received weapons systems from China, Germany, Israel, South Africa, Turkey, Ukraine, and the US (2019 est.)

Military deployments: 120 Lebanon (UNIFIL); as of mid-2019, Kazakhstan contributed a brigade to CSTO's Rapid Reaction Force (2020)

Military service age and obligation: All men 18-27 are required to serve in the military for at least one year. (2019)

TRANSNATIONAL ISSUES

Disputes—international: in January 2019, the Kyrgyz Republic ratified the demarcation agreement of the Kazakh-Kyrgyz border; the demarcation of the Kazakh-Uzbek borders is ongoing; the ongoing demarcation with Russia began in 2007; demarcation with China completed in 2002

Refugees and internally displaced persons: *stateless persons:* 8,386 (2019)

Illicit drugs: significant illicit cultivation of cannabis for CIS markets, as well as limited cultivation of opium poppy and ephedra (for the drug ephedrine); limited government eradication of illicit crops; transit point for Southwest Asian narcotics bound for Russia and the rest of Europe; significant consumer of opiates

KENYA

INTRODUCTION

Background: Founding president and liberation struggle icon Jomo KENYATTA led Kenya from independence in 1963 until his death in 1978, when Vice President Daniel Arap MOI took power in a constitutional succession. The country was a de facto one-party state from 1969 until

1982, after which time the ruling Kenya African National Union (KANU) changed the constitution to make itself the sole legal party in Kenya. MOI acceded to internal and external pressure for political liberalization in late 1991. The ethnically fractured opposition failed to dislodge KANU from power in elections in 1992 and 1997, which

were marred by violence and fraud. President MOI stepped down in December 2002 following fair and peaceful elections. Mwai KIBAKI, running as the candidate of the multiethnic, united opposition group, the National Rainbow Coalition (NARC), defeated KANU candidate Uhuru KENYATTA, the son of founding president Jomo KENYATTA,

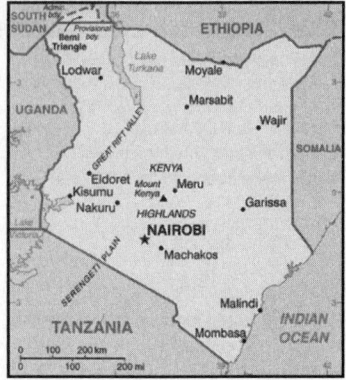

and assumed the presidency following a campaign centered on an anticorruption platform.

KIBAKI's reelection in December 2007 brought charges of vote rigging from Orange Democratic Movement (ODM) candidate Raila ODINGA and unleashed two months of violence in which approximately 1,100 people died. African Union-sponsored mediation led by former UN Secretary General Kofi ANNAN in late February 2008 resulted in a power-sharing accord bringing ODINGA into the government in the restored position of prime minister. The power sharing accord included a broad reform agenda, the centerpiece of which was constitutional reform. In August 2010, Kenyans overwhelmingly adopted a new constitution in a national referendum. The new constitution introduced additional checks and balances to executive power and devolved power and resources to 47 newly created counties. It also eliminated the position of prime minister. Uhuru KENYATTA won the first presidential election under the new constitution in March 2013, and was sworn into office the following month; he began a second term in November 2017 following a contentious, repeat election.

GEOGRAPHY

Location: Eastern Africa, bordering the Indian Ocean, between Somalia and Tanzania

Geographic coordinates: 1 00 N, 38 00 E

Map references: Africa

Area: *total:* 580,367 sq km
land: 569,140 sq km
water: 11,227 sq km
country comparison to the world: 50

Area—comparative: five times the size of Ohio; slightly more than twice the size of Nevada

Land boundaries: *total:* 3,457 km
border countries (5): Ethiopia 867 km, Somalia 684 km, South Sudan 317 km, Tanzania 775 km, Uganda 814 km

Coastline: 536 km

Maritime claims: *territorial sea:* 12 nm
exclusive economic zone: 200 nm
continental shelf: 200-m depth or to the depth of exploitation

Climate: varies from tropical along coast to arid in interior

Terrain: low plains rise to central highlands bisected by Great Rift Valley; fertile plateau in west

Elevation: *mean elevation:* 762 m
lowest point: Indian Ocean 0 m
highest point: Mount Kenya 5,199 m

Natural resources: limestone, soda ash, salt, gemstones, fluorspar, zinc, diatomite, gypsum, wildlife, hydropower

Land use: *agricultural land:* 48.1% (2011 est.)
arable land: 9.8% (2011 est.) / *permanent crops:* 0.9% (2011 est.) / *permanent pasture:* 37.4% (2011 est.)
forest: 6.1% (2011 est.)
other: 45.8% (2011 est.)
Irrigated land: 1,030 sq km (2012)

Population distribution: population heavily concentrated in the west along the shore of Lake Victoria; other areas of high density include the capital of Nairobi, and in the southeast along the Indian Ocean coast as shown in this population distribution map

Natural hazards: recurring drought; flooding during rainy seasons

volcanism: limited volcanic activity; the Barrier (1,032 m) last erupted in 1921; South Island is the only other historically active volcano

Environment—current issues: water pollution from urban and industrial wastes; water shortage and degraded water quality from increased use of pesticides and fertilizers; flooding; water hyacinth infestation in Lake Victoria; deforestation; soil erosion; desertification; poaching

Environment—international agreements: *party to:* Biodiversity, Climate Change, Climate Change-Kyoto Protocol, Desertification, Endangered Species, Hazardous Wastes, Law of the Sea, Marine Dumping, Marine Life Conservation, Ozone Layer Protection, Ship Pollution, Wetlands, Whaling
signed, but not ratified: none of the selected agreements

Geography—note: the Kenyan Highlands comprise one of the most successful agricultural production regions in Africa; glaciers are found on Mount Kenya, Africa's second highest peak; unique physiography supports abundant and varied wildlife of scientific and economic value; Lake Victoria, the world's largest tropical lake and the second largest fresh water lake, is shared among three countries: Kenya, Tanzania, and Uganda

PEOPLE AND SOCIETY

Population: 53,527,936 (July 2020 est.)
note: estimates for this country explicitly take into account the effects of excess mortality due to AIDS; this can result in lower life expectancy, higher infant mortality, higher death rates, lower population growth rates, and changes in the distribution of population by age and sex than would otherwise be expected
country comparison to the world: 27

Nationality: *noun:* Kenyan(s)
adjective: Kenyan

Ethnic groups: Kikuyu 17.1%, Luhya 14.3%, Kalenjin 13.4%, Luo 10.7%, Kamba 9.8%, Somali 5.8%, Kisii 5.7%, Mijikenda 5.2%, Meru 4.2%, Maasai 2.5%, Turkana 2.1%, non-Kenyan 1%, other 8.2% (2019 est.)

Languages: English (official), Kiswahili (official), numerous indigenous languages

Religions: Christian 85.5% (Protestant 33.4%, Catholic 20.6%, Evangelical 20.4%, African Instituted Churches 7%, other Christian 4.1%), Muslim 10.9%, other 1.8%, none 1.6%, don't know/no answer 0.2% (2019 est.)

Demographic profile: Kenya has experienced dramatic population growth since the mid-20th century as a result of its high birth rate and its declining mortality rate. More than 40% of Kenyans are under the age of 15 because of sustained high fertility, early marriage and childbearing, and an unmet need for family planning. Kenya's persistent rapid population growth strains the labor market, social services, arable land, and natural resources. Although Kenya in 1967 was the first Sub-Saharan country to launch a nationwide family planning program, progress in reducing the birth rate has largely stalled since the late 1990s, when the government decreased its support for family planning to focus on the HIV epidemic. Government commitment and international technical support spurred Kenyan contraceptive use, decreasing the fertility rate (children per woman) from about 8 in the late 1970s to less than 5 children twenty years later, but it has plateaued at just over 3 children today.

Kenya is a source of emigrants and a host country for refugees. In the 1960s and 1970s, Kenyans pursued higher education in the UK because of colonial ties, but as British immigration rules tightened, the US, the then Soviet Union, and Canada became attractive study destinations. Kenya's stagnant economy and political problems during the 1980s and 1990s led to an outpouring of Kenyan students and professionals seeking permanent opportunities in the West and southern Africa. Nevertheless, Kenya's relative stability since its independence in 1963 has attracted hundreds of thousands of refugees escaping violent conflicts in neighboring countries; Kenya shelters more than 300,000 Somali refugees as of April 2017.

Age structure: *0-14 years:* 38.71% (male 10,412,321/female 10,310,908)
15-24 years: 20.45% (male 5,486,641/female 5,460,372)
25-54 years: 33.75% (male 9,046,946/female 9,021,207)
55-64 years: 4.01% (male 1,053,202/female 1,093,305)
65 years and over: 3.07% (male 750,988/female 892,046) (2020 est.)

Dependency ratios: *total dependency ratio:* 69.8
youth dependency ratio: 65.5
elderly dependency ratio: 4.3
potential support ratio: 23.5 (2020 est.)

Median age: *total:* 20 years
male: 19.9 years
female: 20.1 years (2020 est.)
country comparison to the world: 196

Population growth rate: 2.2% (2020 est.)
country comparison to the world: 36

Birth rate: 27.2 births/1,000 population (2020 est.)
country comparison to the world: 43

Death rate: 5.2 deaths/1,000 population (2020 est.)
country comparison to the world: 195

Net migration rate: -0.2 migrant(s)/1,000 population (2020 est.)
country comparison to the world: 107

Population distribution: population heavily concentrated in the west along the shore of Lake Victoria; other areas of high density include the capital of Nairobi, and in the southeast along the Indian Ocean coast as shown in this population distribution map

Urbanization: *urban population:* 28% of total population (2020)
rate of urbanization: 4.23% annual rate of change (2015-20 est.)
total population growth rate v. urban population growth rate, 2000-2030:

Major urban areas—population: 4.735 million NAIROBI (capital), 1.296 million Mombassa (2020)

Sex ratio: *at birth:* 1.02 male(s)/female
0-14 years: 1.01 male(s)/female
15-24 years: 1 male(s)/female
25-54 years: 1 male(s)/female
55-64 years: 0.96 male(s)/female
65 years and over: 0.84 male(s)/female
total population: 1 male(s)/female (2020 est.)

Mother's mean age at first birth: 20.3 years (2014 est.)
note: median age at first birth among women 25-29

Maternal mortality rate: 342 deaths/100,000 live births (2017 est.)
country comparison to the world: 32

Infant mortality rate: *total:* 29.8 deaths/1,000 live births
male: 33 deaths/1,000 live births
female: 26.6 deaths/1,000 live births (2020 est.)
country comparison to the world: 57

Life expectancy at birth: *total population:* 69 years
male: 67.3 years
female: 70.6 years (2020 est.)
country comparison to the world: 173

Total fertility rate: 3.43 children born/woman (2020 est.)
country comparison to the world: 42

Contraceptive prevalence rate: 60.5% (2017)

Drinking water source:
improved:
urban: 89% of population
rural: 60.4% of population
total: 68% of population
unimproved:

urban: 11% of population
rural: 39.6% of population
total: 32% of population (2017 est.)

Current Health Expenditure: 4.8% (2017)

Physicians density: 0.2 physicians/1,000 population (2014)

Hospital bed density: 1.4 beds/1,000 population (2010)

Sanitation facility access:
improved:
urban: 78.8% of population
rural: 41.2% of population
total: 51.2% of population
unimproved:
urban: 21.2% of population
rural: 58.8% of population
total: 48.8% of population (2017 est.)

HIV/AIDS—adult prevalence rate: 4.8% (2019 est.)
country comparison to the world: 13

HIV/AIDS—people living with HIV/AIDS: 1.5 million (2019 est.)
country comparison to the world: 6

HIV/AIDS—deaths: 21,000 (2019 est.)
country comparison to the world: 7

Major infectious diseases: *degree of risk:* very high (2020)
food or waterborne diseases: bacterial and protozoal diarrhea, hepatitis A, and typhoid fever
vectorborne diseases: malaria, dengue fever, and Rift Valley fever
water contact diseases: schistosomiasis
animal contact diseases: rabies

Obesity—adult prevalence rate: 7.1% (2016)
country comparison to the world: 161

Children under the age of 5 years underweight: 11.2% (2014)
country comparison to the world: 58

Education expenditures: 5.2% of GDP (2017)
country comparison to the world: 55

Literacy: *definition:* age 15 and over can read and write
total population: 81.5%
male: 85%
female: 78.2% (2018)

School life expectancy (primary to tertiary education): *total:* 11 years
male: 11 years
female: 11 years (2009)

Unemployment, youth ages 15-24: *total:* 7.4%
male: 7.3%
female: 7.4% (2016)
country comparison to the world: 146

GOVERNMENT

Country name: *conventional long form:* Republic of Kenya
conventional short form: Kenya
local long form: Republic of Kenya/Jamhuri ya Kenya
local short form: Kenya
former: British East Africa

etymology: named for Mount Kenya; the meaning of the name is unclear but may derive from the Kikuyu, Embu, and Kamba words "kirinyaga," "kirenyaa," and "kiinyaa"—all of which mean "God's resting place"

Government type: presidential republic

Capital: *name:* Nairobi
geographic coordinates: 1 17 S, 36 49 E
time difference: UTC+ 3 (8 hours ahead of Washington, DC, during Standard Time)
etymology: the name derives from the Maasai expression meaning "cool waters" and refers to a cold water stream that flowed through the area in the late 19th century

Administrative divisions: 47 counties; Baringo, Bomet, Bungoma, Busia, Elgeyo/Marakwet, Embu, Garissa, Homa Bay, Isiolo, Kajiado, Kakamega, Kericho, Kiambu, Kilifi, Kirinyaga, Kisii, Kisumu, Kitui, Kwale, Laikipia, Lamu, Machakos, Makueni, Mandera, Marsabit, Meru, Migori, Mombasa, Murang'a, Nairobi City, Nakuru, Nandi, Narok, Nyamira, Nyandarua, Nyeri, Samburu, Siaya, Taita/Taveta, Tana River, Tharaka-Nithi, Trans Nzoia, Turkana, Uasin Gishu, Vihiga, Wajir, West Pokot

Independence: 12 December 1963 (from the UK)

National holiday: Jamhuri Day (Independence Day), 12 December (1963); note—Madaraka Day, 1 June (1963) marks the day Kenya attained internal self-rule

Constitution: *history:* previous 1963, 1969; latest drafted 6 May 2010, passed by referendum 4 August 2010, promulgated 27 August 2010
amendments: proposed by either house of Parliament or by petition of at least one million eligible voters; passage of amendments by Parliament requires approval by at least two-thirds majority vote of both houses in each of two readings, approval in a referendum by majority of votes cast by at least 20% of eligible voters in at least one half of Kenya's counties, and approval by the president; passage of amendments introduced by petition requires approval by a majority of county assemblies, approval by majority vote of both houses, and approval by the president

Legal system: mixed legal system of English common law, Islamic law, and customary law; judicial review in the new Supreme Court established by the new constitution

International law organization participation: accepts compulsory ICJ jurisdiction with reservations; accepts ICCt jurisdiction

Citizenship: *citizenship by birth:* no
citizenship by descent only: at least one parent must be a citizen of Kenya
dual citizenship recognized: yes
residency requirement for naturalization: 4 out of the previous 7 years

Suffrage: 18 years of age; universal

Executive branch: *chief of state:* President Uhuru KENYATTA (since 9 April 2013); Deputy President William RUTO (since 9 April 2013);

note—the president is both chief of state and head of government

head of government: President Uhuru KENYATTA (since 9 April 2013); Deputy President William RUTO (since 9 April 2013); note—position of the prime minister was abolished after the March 2013 elections

cabinet: Cabinet appointed by the president, subject to confirmation by the National Assembly

elections/appointments: president and deputy president directly elected on the same ballot by qualified majority popular vote for a 5-year term (eligible for a second term); in addition to receiving an absolute majority popular vote, the presidential candidate must also win at least 25% of the votes cast in at least 24 of the 47 counties to avoid a runoff; election last held on 26 October 2017 (next to be held in 2022)

election results: Uhuru KENYATTA reelected president; percent of vote—Uhuru KENYATTA (Jubilee Party) 98.3%, Raila ODINGA (ODM) 1%, other 0.7%; note—Kenya held a previous presidential election on 8 August 2017, but Kenya's Supreme Court on 1 September 2017 nullified the results, citing irregularities; the political opposition boycotted the October vote

Legislative branch: *description:* bicameral Parliament consists of:
Senate (67 seats; 47 members directly elected in single-seat constituencies by simple majority vote and 20 directly elected by proportional representation vote—16 women, 2 representing youth, and 2 representing the disabled; members serve 5-year terms)
National Assembly (349 seats; 290 members directly elected in single- seat constituencies by simple majority vote, 47 women in single-seat constituencies elected by simple majority vote, and 12 members nominated by the National Assembly—6 representing youth and 6 representing the disabled; members serve 5-year terms)

elections: Senate—last held on 8 August 2017 (next to be held in August 2022) National Assembly—last held on 8 August 2017 (next to be held in August 2022)

election results: Senate—percent of vote by party/coalition—NA; seats by party/coalition—Jubilee Party 24; National Super Alliance 28, other 14, independent 1; composition—men 46, women 41, percent of women is 31.3%
National Assembly—percent of vote by party/coalition—NA; seats by party/coalition—Jubilee Party 165, National Super Alliance 119, other 51, independent 13; composition—men 273, women 76, percent of women 21.8%; note—total Parliament percent of women is 23%

Judicial branch: *highest courts:* Supreme Court (consists of chief and deputy chief justices and 5 judges)

judge selection and term of office: chief and deputy chief justices nominated by Judicial Service Commission (JSC) and appointed by the president with approval of the National Assembly; other judges nominated by the JSC and appointed by president; chief justice serves a nonrenewable

10-year term or until age 70, whichever comes first; other judges serve until age 70

subordinate courts: High Court; Court of Appeal; military courts; magistrates' courts; religious courts

Political parties and leaders:
Alliance Party of Kenya or APK [Kiraitu MURUNGI]
Amani National Congress or ANC [Musalia MUDAVADI] Federal Party of Kenya or FPK [Cyrus JIRONGA]
Forum for the Restoration of Democracy- Kenya or FORD-K [Moses WETANGULA]
Forum for the Restoration of Democracy- People or FORD- P [Henry OBWOCHA]
Jubilee Party [Uhuru KENYATTA]
Kenya African National Union or KANU [Gideon MOI]
National Rainbow Coalition or NARC [Charity NGILU]
Orange Democratic Movement Party of Kenya or ODM [Raila ODINGA]
Wiper Democratic Movement-K or WDM-K (formerly Orange Democratic Movement- Kenya or ODM-K) [Kalonzo MUSYOKA]

International organization participation: ACP, AfDB, AU, C, CD, COMESA, EAC, EADB, FAO, G-15, G- 77, IAEA, IBRD, ICAO, ICCt, ICRM, IDA, IFAD, IFC, IFRCS, IGAD, ILO, IMF, IMO, IMSO, Interpol, IOC, IOM, IPU, ISO, ITSO, ITU, ITUC (NGOs), MIGA, MINUSMA, MONUSCO, NAM, OPCW, PCA, UN, UNAMID, UNCTAD, UNESCO, UNHCR, UNIDO, UNIFIL, UNMIL, UNMISS, UNWTO, UPU, WCO, WHO, WMO, WTO

Diplomatic representation in the US: *chief of mission:* Ambassador Lazarus Ombai AMAYO (since 17 July 2020)

chancery: 2249 R Street NW, Washington, DC 20008

telephone: [1] (202) 387-6101
FAX: [1] (202) 462-3829
consulate(s) general: Los Angeles
consulate(s): New York

Diplomatic representation from the US: *chief of mission:* Ambassador Kyle MCCARTER (since 12 March 2019)

telephone: [254] (20) 363-6000
embassy: United Nations Avenue, Nairobi; P. O. Box 606 Village Market, Nairobi 00621
mailing address: American Embassy Nairobi, U. S. Department of State, Washington, DC 20521-8900
FAX: [254] (20) 363-6157

Flag description: three equal horizontal bands of black (top), red, and green; the red band is edged in white; a large Maasai warrior's shield covering crossed spears is superimposed at the center; black symbolizes the majority population, red the blood shed in the struggle for freedom, green stands for natural wealth, and white for peace; the shield and crossed spears symbolize the defense of freedom

National symbol(s): lion; national colors: black, red, green, white

National anthem: name: "Ee Mungu Nguvu Yetu" (Oh God of All Creation)

lyrics/music: Graham HYSLOP, Thomas KALUME, Peter KIBUKOSYA, Washington OMONDI, and George W. SENOGA- ZAKE/ traditional, adapted by Graham HYSLOP, Thomas KALUME, Peter KIBUKOSYA, Washington OMONDI, and George W. SENOGA-ZAKE

note: adopted 1963; based on a traditional Kenyan folk song

ECONOMY

Economy—overview: Kenya is the economic, financial, and transport hub of East Africa. Kenya's real GDP growth has averaged over 5% for the last decade. Since 2014, Kenya has been ranked as a lower middle income country because its per capita GDP crossed a World Bank threshold. While Kenya has a growing entrepreneurial middle class and steady growth, its economic development has been impaired by weak governance and corruption. Although reliable numbers are hard to find, unemployment and under-employment are extremely high, and could be near 40% of the population. In 2013, the country adopted a devolved system of government with the creation of 47 counties, and is in the process of devolving state revenues and responsibilities to the counties.

Agriculture remains the backbone of the Kenyan economy, contributing one-third of GDP. About 75% of Kenya's population of roughly 48.5 million work at least part-time in the agricultural sector, including livestock and pastoral activities. Over 75% of agricultural output is from small-scale, rain-fed farming or livestock production. Tourism also holds a significant place in Kenya's economy. In spite of political turmoil throughout the second half of 2017, tourism was up 20%, showcasing the strength of this sector. Kenya has long been a target of terrorist activity and has struggled with instability along its northeastern borders. Some high visibility terrorist attacks during 2013-2015 (e. g., at Nairobi's Westgate Mall and Garissa University) affected the tourism industry severely, but the sector rebounded strongly in 2016-2017 and appears poised to continue growing.

Inadequate infrastructure continues to hamper Kenya's efforts to improve its annual growth so that it can meaningfully address poverty and unemployment. The KENYATTA administration has been successful in courting external investment for infrastructure development. International financial institutions and donors remain important to Kenya's growth and development, but Kenya has also successfully raised capital in the global bond market issuing its first sovereign bond offering in mid-2014, with a second occurring in February 2018. The first phase of a Chinese-financed and constructed standard gauge railway connecting Mombasa and Nairobi opened in May 2017.

In 2016 the government was forced to take over three small and undercapitalized banks when underlying weaknesses were exposed. The government also enacted legislation that limits interest rates banks can charge on loans and set a rate that banks must pay their depositors. This measure led to a sharp shrinkage of credit in the economy. A prolonged election cycle in 2017 hurt the economy, drained government resources, and slowed GDP growth. Drought-like conditions in parts of the country pushed 2017 inflation above 8%, but the rate had fallen to 4.5% in February 2018.

The economy, however, is well placed to resume its decade-long 5%-6% growth rate. While fiscal deficits continue to pose risks in the medium term, other economic indicators, including foreign exchange reserves, interest rates, current account deficits, remittances and FDI are positive. The credit and drought- related impediments were temporary. Now In his second term, President KENYATTA has pledged to make economic growth and development a centerpiece of his second administration, focusing on his "Big Four" initiatives of universal healthcare, food security, affordable housing, and expansion of manufacturing.

GDP (purchasing power parity):
$163.7 billion (2017 est.)
$156 billion (2016 est.)
$147.4 billion (2015 est.)
note: data are in 2017 dollars
country comparison to the world: 74

GDP (official exchange rate):
$79.22 billion (2017 est.)

GDP—real growth rate:
5.39% (2019 est.)
6.32% (2018 est.)
4.79% (2017 est.)
country comparison to the world: 39

GDP—per capita (PPP):
$3,500 (2017 est.)
$3,400 (2016 est.)
$3,300 (2015 est.)
note: data are in 2017 dollars
country comparison to the world: 187

Gross national saving:
10.4% of GDP (2017 est.)
11% of GDP (2016 est.)
11.4% of GDP (2015 est.)
country comparison to the world: 162

GDP—composition, by end use:
household consumption: 79.5% (2017 est.)
government consumption: 14.3% (2017 est.)
investment in fixed capital: 18.9% (2017 est.)
investment in inventories: —1% (2017 est.)
exports of goods and services: 13.9% (2017 est.)
imports of goods and services: -25.5% (2017 est.)

GDP—composition, by sector of origin:
agriculture: 34.5% (2017 est.)
industry: 17.8% (2017 est.)
services: 47.5% (2017 est.)

Agriculture—products: tea, coffee, corn, wheat, sugarcane, fruit, vegetables; dairy products, beef, fish, pork, poultry, eggs
Industries: small-scale consumer goods (plastic, furniture, batteries, textiles, clothing, soap, cigarettes, flour), agricultural products, horticulture, oil refining; aluminum, steel, lead; cement, commercial ship repair, tourism, information technology

Industrial production growth rate: 3.6% (2017 est.)
country comparison to the world: 82

Labor force: 19.6 million (2017 est.)
country comparison to the world: 26

Labor force—by occupation: *agriculture:* 61.1%
industry: 6.7%
services: 32.2% (2005 est.)

Unemployment rate:
40% (2013 est.)
40% (2001 est.)
country comparison to the world: 214

Population below poverty line: 36.1% (2016 est.)

Household income or consumption by percentage share: *lowest 10%:* 1.8%
highest 10%: 37.8% (2005)

Budget: *revenues:* 13.95 billion (2017 est.)
expenditures: 19.24 billion (2017 est.)

Taxes and other revenues: 17.6% (of GDP) (2017 est.)
country comparison to the world: 168

Budget surplus (+) or deficit (-): -6.7% (of GDP) (2017 est.)
country comparison to the world: 190

Public debt:
54.2% of GDP (2017 est.)
53.2% of GDP (2016 est.)
country comparison to the world: 84

Fiscal year: 1 July—30 June

Inflation rate (consumer prices):
8% (2017 est.)
6.3% (2016 est.)
country comparison to the world: 197

Current account balance:
-$57.594 billion (2019 est.)
-$56.194 billion (2018 est.)
country comparison to the world: 204

Exports:
$5.792 billion (2017 est.)
$5.695 billion (2016 est.)
country comparison to the world: 104

Exports—partners: Uganda 10.8%, Pakistan 10.6%, US 8.1%, Netherlands 7.3%, UK 6.4%, Tanzania 4.8%, UAE 4.4% (2017)

Exports—commodities: tea, horticultural products, coffee, petroleum products, fish, cement, apparel

Imports:
$15.99 billion (2017 est.)
$13.41 billion (2016 est.)
country comparison to the world: 85

Imports—commodities: machinery and transportation equipment, oil, petroleum products, motor vehicles, iron and steel, resins and plastics

Imports—partners: China 22.5%, India 9.9%, UAE 8.7%, Saudi Arabia 5.1%, Japan 4.5% (2017)

Reserves of foreign exchange and gold:
$7.354 billion (31 December 2017 est.)
$7.256 billion (31 December 2016 est.)
country comparison to the world: 83

Debt—external:
$27.59 billion (31 December 2017 est.)
$37.7 billion (31 December 2016 est.)
country comparison to the world: 85

Exchange rates: Kenyan shillings (KES) per US dollar –
102.1 (2017 est.)
101.5 (2016 est.)
101.504 (2015 est.)
98.179 (2014 est.)
87.921 (2013 est.)

ENERGY

Electricity access: *population without electricity:* 8 million (2019)
electrification—total population: 85% (2019)
electrification—urban areas: 99% (2019)
electrification—rural areas: 79% (2019)

Electricity—production: 9.634 billion kWh (2016 est.)
country comparison to the world: 105

Electricity—consumption: 7.863 billion kWh (2016 est.)
country comparison to the world: 104

Electricity—exports: 39.1 million kWh (2016 est.)
country comparison to the world: 89

Electricity—imports: 184 million kWh (2016 est.)
country comparison to the world: 95

Electricity—installed generating capacity: 2.401 million kW (2016 est.)
country comparison to the world: 109

Electricity—from fossil fuels: 33% of total installed capacity (2016 est.)
country comparison to the world: 183

Electricity—from nuclear fuels: 0% of total installed capacity (2017 est.)
country comparison to the world: 118

Electricity—from hydroelectric plants: 34% of total installed capacity (2017 est.)
country comparison to the world: 62

Electricity—from other renewable sources: 33% of total installed capacity (2017 est.)
country comparison to the world: 13

Crude oil—production: 0 bbl/day (2018 est.)
country comparison to the world: 154

Crude oil—exports: 0 bbl/day (2015 est.)
country comparison to the world: 145

Crude oil—imports: 12,550 bbl/day (2015 est.)
country comparison to the world: 71

Crude oil—proved reserves: 0 bbl (1 January 2018 est.)
country comparison to the world: 150

Refined petroleum products—production: 13,960 bbl/day (2015 est.)
country comparison to the world: 96

Refined petroleum products—consumption: 109,000 bbl/day (2016 est.)
country comparison to the world: 76

Refined petroleum products—exports: 173 bbl/day (2015 est.)

country comparison to the world: 118

Refined petroleum products—imports: 90,620 bbl/day (2015 est.)
country comparison to the world: 57

Natural gas—production: 0 cu m (2017 est.)
country comparison to the world: 150

Natural gas—consumption: 0 cu m (2017 est.)
country comparison to the world: 161

Natural gas—exports: 0 cu m (2017 est.)
country comparison to the world: 128

Natural gas—imports: 0 cu m (2017 est.)
country comparison to the world: 142

Natural gas—proved reserves: 0 cu m (1 January 2014 est.)
country comparison to the world: 151

Carbon dioxide emissions from consumption of energy: 17.98 million Mt (2017 est.)
country comparison to the world: 88

COMMUNICATIONS

Telephones—fixed lines: *total subscriptions:* 68,072
subscriptions per 100 inhabitants: less than 1 (2019 est.)
country comparison to the world: 151

Telephones—mobile cellular: *total subscriptions:* 54,336,841
subscriptions per 100 inhabitants: 103.77 (2019 est.)
country comparison to the world: 29

Telecommunication systems: *general assessment:* the mobile- cellular system is generally good with a mobile subscriber base of 47 million, especially in urban areas; fixed-line telephone system is small and inefficient; trunks are primarily microwave radio relay; to encourage advancement of the LTE services the govt. has fostered an open-access approach and pushed for a national broadband strategy; more licensing being awarded has led to competition which is good for growth; govt. commits KE 300 million to its free Wi-Fi project (2020)
domestic: fixed-line subscriptions stand at less than 1 per 100 persons; multiple providers in the mobile- cellular segment of the market fostering a boom in mobile-cellular telephone usage with teledensity reaching 104 per 100 persons (2019)
international: country code—254; landing point for the EASSy, TEAMS, LION2, DARE1, PEACE Cable, and SEACOM fiber-optic submarine cable systems covering East, North and South Africa, Europe, the Middle East, and Asia; satellite earth stations—4 Intelsat; launched first micro satellites in 2018 (2019)
note: the COVID- 19 outbreak is negatively impacting telecommunications production and supply chains globally; consumer spending on telecom devices and services has also slowed due to the pandemic's effect on economies worldwide; overall progress towards improvements in all facets of the telecom industry—mobile, fixed-line, broadband, submarine cable and satellite—has moderated

Broadcast media: about a half-dozen large-scale privately owned media companies with TV and radio stations, as well as a state- owned TV broadcaster, provide service nationwide; satellite and cable TV subscription services available; state-owned radio broadcaster operates 2 national radio channels and provides regional and local radio services in multiple languages; many private radio stations broadcast on a national level along with over 100 private and non- profit regional stations broadcasting in local languages; TV transmissions of all major international broadcasters available, mostly via paid subscriptions; direct radio frequency modulation transmissions available for several foreign government-owned broadcasters (2019)

Internet country code: .ke

Internet users: *total:* 9,129,243

percent of population: 17.83% (July 2018 est.)
country comparison to the world: 57

Broadband—fixed subscriptions:
total: 371,498
subscriptions per 100 inhabitants: 1 less than 1 (2018 est.)
country comparison to the world: 92

TRANSPORTATION

National air transport system:
number of registered air carriers: 25 (2020)
inventory of registered aircraft operated by air carriers: 188
annual passenger traffic on registered air carriers: 5,935,831 (2018)
annual freight traffic on registered air carriers: 294.97 million mt-km (2018)

Civil aircraft registration country code prefix: 5Y (2016)

Airports: 197 (2013)
country comparison to the world: 28

Airports—with paved runways: *total:* 16 (2017)
over 3,047 m: 5 (2017)
2,438 to 3,047 m: 2 (2017)
1,524 to 2,437 m: 2 (2017)
914 to 1,523 m: 6 (2017)
under 914 m: 1 (2017)

Airports—with unpaved runways: *total:* 181 (2013)
1,524 to 2,437 m: 14 (2013)
914 to 1,523 m: 107 (2013)
under 914 m: 60 (2013)

Pipelines: 4 km oil, 1,432 km refined products (2018)

Railways: *total:* 3,819 km (2018)
standard gauge: 485 km 1.435-m gauge (2018)
narrow gauge: 3,334 km 1.000-m gauge (2018)
country comparison to the world: 52

Roadways: *total:* 177,800 km (2018)
paved: 14,420 km (8,500 km highways, 1,872 urban roads, and 4,048 rural roads) (2017)
unpaved: 147,032 km (2017)
country comparison to the world: 31

Waterways: none specifically; the only significant inland waterway is the part of Lake Victoria within the boundaries of Kenya; Kisumu is the main port and has ferry connections to Uganda and Tanzania (2011)

Merchant marine: *total:* 24
by type: oil tanker 2, other 22 (2019)
country comparison to the world: 140

Ports and terminals: *major seaport(s):* Kisumu, Mombasa
LNG terminal(s) (import): Mombasa

MILITARY AND SECURITY

Military and security forces: *Kenya Defence Forces:* Kenya Army, Kenya Navy, Kenya Air Force (2019)
note: the National Police Service includes a para-military General Service Unit

Military expenditures:
1.3% of GDP (2019)
1.2% of GDP (2018)
1.3% of GDP (2017)
1.3% of GDP (2016)
1.3% of GDP (2015)
country comparison to the world: 93

Military and security service personnel strengths: the Kenyan Defense Forces (KDF) are comprised of approximately 24,000 personnel (20,000 Army; 1,500 Navy; 2,500 Air Force); 5,000 Police General Services Unit (2019 est.)

Military equipment inventories and acquisitions: the KDF's inventory traditionally carried mostly older or second- hand Western weapons systems, particularly from France, the UK, and the US; however, since the 2000s it has sought to modernize and diversify its imports; top suppliers since 2010 include China, Italy, Jordan, Serbia, South Africa, Spain, Ukraine, and the US (2019 est.)

Military deployments: 3,600 Somalia (AMISOM) (2020)

Military service age and obligation: 18-26 years of age for male and female voluntary service (under 18 with parental consent), with a 9-year obligation (7 years for Kenyan Navy) and subsequent 3-year reenlistments; applicants must be Kenyan citizens and provide a national identity card (obtained at age 18) and a school-leaving certificate, and undergo a series of mental and physical examinations; women serve under the same terms and conditions as men; mandatory retirement at age 55 but personnel leaving before this age remain in a reserve status until they reach age 55 unless they were removed for disciplinary reasons; there is no active military reserve, although the Ministry of Defence has stated its desire to create one as recently as 2017 (2019)

Maritime threats: The International Maritime Bureau reports that shipping in territorial and off-shore waters in the Indian Ocean remain at risk for piracy and armed robbery against ships, especially as Somali-based pirates extend their activities south; numerous commercial vessels have been attacked and hijacked both at anchor and while underway; crews have been robbed and stores or cargoes stolen.

Military—note: the Kenya Coast Guard Service (established 2018) is under the Ministry of Interior, but led by a military officer and comprised of personnel from the military, as well as the National Police Service, intelligence services, and other government agencies (2019)

TERRORISM

Terrorist group(s): al-Shabaab; Islamic Revolutionary Guard Corps/Qods Force (2019) *note:* details about the history, aims, leadership, organization, areas of operation, tactics, targets, weapons, size, and sources of support of the group(s) appear(s) in Appendix- T

TRANSNATIONAL ISSUES :: KENYA

Disputes—international: Kenya served as an important mediator in brokering Sudan's north-south separation in February 2005; as of March 2019, Kenya provides shelter to nearly 475,000 refugees and asylum seekers, including Ugandans who flee across the border periodically to seek protection from Lord's Resistance Army rebels; Kenya works hard to prevent the clan and militia fighting in Somalia from spreading across the border, which has long been open to nomadic pastoralists; the boundary that separates Kenya's and Sudan's sovereignty is unclear in the "Ilemi Triangle," which Kenya has administered since colonial times; in 2018, Kenya signed an MoU with Uganda and South Sudan to help demarcate their borders

Refugees and internally displaced persons: *refugees (country of origin):* 266,074 (Somalia) (refugees and asylum seekers), 122,256 (South Sudan) (refugees and asylum seekers), 44,836 (Democratic Republic of the Congo) (refugees and asylum seekers), 28,836 (Ethiopia) (refugees and asylum seekers), 16,010 (Burundi) (refugees and asylum seekers), 10,007 (Sudan) (refugees and asylum seekers) (2020)

IDPs: 162,000 (election- related violence, inter-communal violence, resource conflicts, al-Shabaab attacks in 2017 and 2018) (2019) *stateless persons:* 18,500 (2019); note—the stateless population consists of Nubians, Kenyan Somalis, and coastal Arabs; the Nubians are descendants of Sudanese soldiers recruited by the British to fight for them in East Africa more than a century ago; Nubians did not receive Kenyan citizenship when the country became independent in 1963; only recently have Nubians become a formally recognized tribe and had less trouble obtaining national IDs; Galjeel and other Somalis who have lived in Kenya for decades are included with more recent Somali refugees and denied ID cards

Illicit drugs: widespread harvesting of small plots of marijuana; transit country for South Asian heroin destined for Europe and North America; Indian methaqualone also transits on way to South Africa; significant potential for money-laundering activity given the country's status as a regional financial center; massive corruption, and relatively high levels of narcotics-associated activities

KIRIBATI

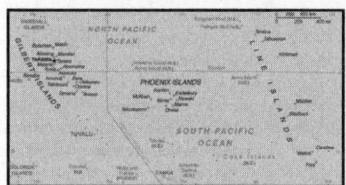

INTRODUCTION

Background: The Gilbert Islands became a British protectorate in 1892 and a colony in 1915; they were captured by the Japanese in the Pacific War in 1941. The islands of Makin and Tarawa were the sites of major US amphibious victories over entrenched Japanese garrisons in 1943. The Gilbert Islands were granted self-rule by the UK in 1971 and complete independence in 1979 under the new name of Kiribati. The US relinquished all claims to the sparsely inhabited Phoenix and Line Island groups in a 1979 treaty of friendship with Kiribati. Kiribati joined the UN in 1999 and has been an active participant in international efforts to combat climate change.

GEOGRAPHY

Location: Oceania, group of 32 coral atolls and one raised coral island in the Pacific Ocean, straddling the Equator; the capital Tarawa is about halfway between Hawaii and Australia

Geographic coordinates: 1 25 N, 173 00 E

Map references: Oceania

Area: *total:* 811 sq km

land: 811 sq km
water: 0 sq km
note: includes three island groups—Gilbert Islands, Line Islands, and Phoenix Islands—dispersed over about 3.5 million sq km (1.35 million sq mi)
country comparison to the world: 187

Area—comparative: four times the size of Washington, DC

Land boundaries: 0 km

Coastline: 1,143 km

Maritime claims: *territorial sea:* 12 nm *exclusive economic zone:* 200 nm

Climate: tropical; marine, hot and humid, moderated by trade winds

Terrain: mostly low-lying coral atolls surrounded by extensive reefs

Elevation: *mean elevation:* 2 m *lowest point:* Pacific Ocean 0 m *highest point:* unnamed elevation on Banaba 81 m m

Natural resources: phosphate (production discontinued in 1979), coconuts (copra), fish

Land use: *agricultural land:* 42% (2011 est.) *arable land:* 2.5% (2011 est.) / permanent crops: 39.5% (2011 est.) / permanent pasture: 0% (2011 est.)
forest: 15% (2011 est.)
other: 43% (2011 est.)

Irrigated land: 0 sq km (2012)

Population distribution: consists of three achipelagos spread out over an area roughly the size of India; the eastern Line Islands and central Phoenix

Islands are sparsely populated, but the western Gilbert Islands are some of the most densely settled places on earth, with the main island of South Tarawa boasting a population density similar to Tokyo or Hong Kong

Natural hazards: typhoons can occur any time, but usually November to March; occasional tornadoes; low level of some of the islands make them sensitive to changes in sea level

Environment—current issues: heavy pollution in lagoon of south Tarawa atoll due to overcrowding mixed with traditional practices such as lagoon latrines and open-pit dumping; ground water at risk; potential for water shortages, disease; coastal erosion

Environment—international agreements: *party to:* Biodiversity, Climate Change, Climate Change-Kyoto Protocol, Desertification, Hazardous Wastes, Law of the Sea, Marine Dumping, Ozone Layer Protection, Whaling *signed, but not ratified:* none of the selected agreements

Geography—note: 21 of the 33 islands are inhabited; Banaba (Ocean Island) in Kiribati is one of the three great phosphate rock islands in the Pacific Ocean—the others are Makatea in French Polynesia, and Nauru; Kiribati is the only country in the world to fall into all four hemispheres (northern, southern, eastern, and western)

PEOPLE AND SOCIETY

Population: 111, 796 (July 2020 est.) *country comparison to the world:* 191

Nationality: *noun:* I-Kiribati (singular and plural)

adjective: I-Kiribati

Ethnic groups: I-Kiribati 96.2%, I-Kiribati/mixed 1.8%, Tuvaluan 0.2%, other 1.8% (2015 est.)

Languages: I-Kiribati, English (official)

Religions: Roman Catholic 57.3%, Kiribati Uniting Church 31.3%, Mormon 5.3%, Baha'i 2.1%, Seventh Day Adventist 1.9%, other 2.1% (2015 est.)

Age structure: *0-14 years:* 28.47% (male 16,223/female 15,604)
15-24 years: 20.24% (male 11,171/female 11,459)
25-54 years: 40.05% (male 21,530/female 23,249)
55-64 years: 6.65% (male 3,350/female 4,084)
65 years and over: 4.59% (male 2,004/female 3,122) (2020 est.)

Dependency ratios: *total dependency ratio:* 67
youth dependency ratio: 60
elderly dependency ratio: 7
potential support ratio: 14.2 (2020 est.)

Median age: *total:* 25.7 years
male: 24.8 years
female: 26.6 years (2020 est.)
country comparison to the world: 158

Population growth rate: 1.09% (2020 est.)
country comparison to the world: 98

Birth rate: 20.5 births/1,000 population (2020 est.)
country comparison to the world: 75

Death rate: 6.9 deaths/1,000 population (2020 est.)
country comparison to the world: 132

Net migration rate: -2.8 migrant(s)/1,000 population (2020 est.)
country comparison to the world: 173

Population distribution: consists of three archipelagos spread out over an area roughly the size of India; the eastern Line Islands and central Phoenix Islands are sparsely populated, but the western Gilbert Islands are some of the most densely settled places on earth, with the main island of South Tarawa boasting a population density similar to Tokyo or Hong Kong

Urbanization: *urban population:* 55.6% of total population (2020)
rate of urbanization: 3.19% annual rate of change (2015-20 est.)
total population growth rate v. urban population growth rate, 2000-2030: Major urban areas—population: 64,000 TARAWA (capital) (2018)

Sex ratio: *at birth:* 1.05 male(s)/female
0-14 years: 1.04 male(s)/female
15-24 years: 0.97 male(s)/female
25-54 years: 0.93 male(s)/female
55-64 years: 0.82 male(s)/female
65 years and over: 0.64 male(s)/female
total population: 0.94 male(s)/female (2020 est.)

Mother's mean age at first birth: 23.1 years (2009 est.)
note: median age at first birth among women 25-29

Maternal mortality rate: 92 deaths/100,000 live births (2017 est.)

country comparison to the world: 73

Infant mortality rate: *total:* 29.2 deaths/1,000 live births
male: 30.3 deaths/1,000 live births
female: 27.9 deaths/1,000 live births (2020 est.)
country comparison to the world: 58

Life expectancy at birth: *total population:* 67.5 years
male: 65 years
female: 70.2 years (2020 est.)
country comparison to the world: 181

Total fertility rate: 2.25 children born/woman (2020 est.)
country comparison to the world: 88

Drinking water source:

improved: *total:* 71.6% of population

unimproved: *total:* 28.4% of population (2017 est.)

Current Health Expenditure: 10.8% (2017)

Physicians density: 0.2 physicians/1,000 population (2013)

Hospital bed density: 1.9 beds/1,000 population (2016)

Sanitation facility access:

improved: *total:* 61.1% of population

unimproved: *total:* 38.9% of population (2017 est.)

HIV/AIDS—adult prevalence rate: NA

HIV/AIDS—people living with HIV/AIDS: NA

HIV/AIDS—deaths: NA

Major infectious diseases: *degree of risk:* high (2020)
food or waterborne diseases: bacterial diarrhea
vectorborne diseases: malaria

Obesity—adult prevalence rate: 46% (2016)
country comparison to the world: 9

Children under the age of 5 years underweight: 14.9% (2009)
country comparison to the world: 40

Education expenditures: NA

School life expectancy (primary to tertiary education): *total:* 12 years
male: 11 years
female: 12 years (2008)

Unemployment, youth ages 15-24: *total:* 17.1%
male: 22.2%
female: 7.4% (2015 est.)
country comparison to the world: 76

Country name: Government
conventional long form: Republic of Kiribati
conventional short form: Kiribati
local long form: Republic of Kiribati
local short form: Kiribati
former: Gilbert Islands
etymology: the name is the local pronunciation of "Gilberts," the former designation of the islands; originally named after explorer Thomas GILBERT, who mapped many of the islands in 1788
note: pronounced keer-ree-bahss

Government type: presidential republic

Capital: *name:* Tarawa
geographic coordinates: 1 21 N, 173 02 E
time difference: UTC+12 (17 hours ahead of Washington, DC, during Standard Time)
note: Kiribati has three time zones: the Gilbert Islands group at UTC+12, the Phoenix Islands at UTC+13, and the Line Islands at UTC+14
etymology: in Kiribati creation mythology, "tarawa" was what the spider Nareau named the land to distinguish it from "karawa" (the sky) and "marawa" (the ocean)

Administrative divisions: *3 geographical units:* Gilbert Islands, Line Islands, Phoenix Islands; note—there are no first-order administrative divisions, but there are 6 districts (Banaba, Central Gilberts, Line Islands, Northern Gilberts, Southern Gilberts, Tarawa) and 21 island councils—one for each of the inhabited islands (Abaiang, Abemama, Aranuka, Arorae, Banaba, Beru, Butaritari, Kanton, Kiritimati, Kuria, Maiana, Makin, Marakei, Nikunau, Nonouti, Onotoa, Tabiteuea, Tabuaeran, Tamana, Tarawa, Teraina)

Independence: 12 July 1979 (from the UK)

National holiday: Independence Day, 12 July (1979)

Constitution: *history:* The Gilbert and Ellice Islands Order in Council 1915, The Gilbert Islands Order in Council 1975 (preindependence); latest promulgated 12 July 1979 (at independence) *amendments:* proposed by the House of Assembly; passage requires two-thirds majority vote by the Assembly membership; passage of amendments affecting the constitutional section on amendment procedures and parts of the constitutional chapter on citizenship requires deferral of the proposal to the next Assembly meeting where approval is required by at least two-thirds majority vote of the Assembly membership and support of the nominated or elected Banaban member of the Assembly; amendments affecting the protection of fundamental rights and freedoms also requires approval by at least two-thirds majority in a referendum; amended 1995, 2013

Legal system: English common law supplemented by customary law

International law organization participation: has not submitted an ICJ jurisdiction declaration; non-party state to the ICCt

Citizenship: *citizenship by birth:* no
citizenship by descent only: at least one parent must be a native-born citizen of Kiribati
dual citizenship recognized: no
residency requirement for naturalization: 7 years

Suffrage: 18 years of age; universal

Executive branch: *chief of state:* President Taneti MAAMAU (since 11 March 2016); Vice President Kourabi NENEM (since 17 March 2016); note—the president is both chief of state and head of government
head of government: President Taneti MAAMAU (since 11 March 2016); Vice President Kourabi NENEM (since 17 March 2016)

cabinet: Cabinet appointed by the president from among House of Assembly members

elections/appointments: president directly elected by simple majority popular vote following nomination of candidates from among House of Assembly members; term is 4 years (eligible for 2 additional terms); election last held on 22 June 2020 (next to be held in 2024); vice president appointed by the president

election results: Taneti MAAMAU reelected president; percent of vote—Taneti MAAMAU (TKB) 59.3%, Banuera BERINA (BKM) 40.7%.

Legislative branch: *description:* unicameral House of Assembly or Maneaba Ni Maungatabu (46 seats; 44 members directly elected in single- and multi-seat constituencies by absolute majority vote in two-rounds if needed; 1 member appointed by the Rabi Council of Leaders—representing Banaba Island, and 1 ex officio member—the attorney general; members serve 4-year terms)

elections: legislative elections originally scheduled to be held in two rounds on 7 and 15 April 2020 but rescheduled for 14 and 21 April (next to be held in 2024)

election results: percent of vote by party (second round)—NA; seats by party (second round)—NA

Judicial branch: *highest courts:* High Court (consists of a chief justice and other judges as prescribed by the president); note—the High Court has jurisdiction on constitutional issues

judge selection and term of office: chief justice appointed by the president on the advice of the cabinet in consultation with the Public Service Commission (PSC); other judges appointed by the president on the advice of the chief justice along with the PSC

subordinate courts: Court of Appeal; magistrates' courts

Political parties and leaders: Boutokaan Kiribati Moa Party (BKM) [Tessie LAMBOURNE]

Boutokaan Te Koaua Party or BTK or Pillars of Truth [Anote TONG]

Kamaeuraoan Te I-Kiribati Party or KTK [Tetaua TAITAI]

Maurin Kiribati Pati or MKP [Rimeta BENIAMINA]

Tobwaan Kiribati Party or TKP [Taneti MAAMAU]

note: there is no tradition of formally organized political parties in Kiribati; they more closely resemble factions or interest groups because they have no party headquarters, formal platforms, or party structures

International organization participation: ABEDA, ACP, ADB, AOSIS, C, FAO, IBRD, ICAO, ICRM, IDA, IFAD, IFC, IFRCS, ILO, IMF, IMO, IOC, ITU, ITUC (NGOs), OPCW, PIF, Sparteca, SPC, UN, UNCTAD, UNESCO, UPU, WHO, WIPO, WMO

Diplomatic representation in the US: *chief of mission:* Ambassador Teburoro TITO (since 24 January 2018)

chancery: 800 Second Avenue, Suite 400A, New York, NY 10017

telephone: [1](212)867-3310

FAX: [1](212)867-3320

note—the Kiribati Permanent Mission to the UN serves as the embassy

Diplomatic representation from the US: the US does not have an embassy in Kiribati; the US Ambassador to Fiji is accredited to Kiribati

Flag description: the upper half is red with a yellow frigatebird flying over a yellow rising sun, and the lower half is blue with three horizontal wavy white stripes to represent the Pacific ocean; the white stripes represent the three island groups—the Gilbert, Line, and Phoenix Islands; the 17 rays of the sun represent the 16 Gilbert Islands and Banaba (formerly Ocean Island); the frigatebird symbolizes authority and freedom

National symbol(s): *frigatebird; national colors:* red, white, blue, yellow

National anthem: *name:* "Teirake kaini Kiribati" (Stand Up, Kiribati)

lyrics/music: Urium Tamuera IOTEBA

note: adopted 1979

ECONOMY

Economy—overview: A remote country of 33 scattered coral atolls, Kiribati has few natural resources and is one of the least developed Pacific Island countries. Commercially viable phosphate deposits were exhausted by the time of independence from the United Kingdom in 1979. Earnings from fishing licenses and seafarer remittances are important sources of income. Although the number of seafarers employed declined due to changes in global shipping demands, remittances are expected to improve with more overseas temporary and seasonal work opportunities for Kiribati nationals.

Economic development is constrained by a shortage of skilled workers, weak infrastructure, and remoteness from international markets. The public sector dominates economic activity, with ongoing capital projects in infrastructure including road rehabilitation, water and sanitation projects, and renovations to the international airport, spurring some growth. Public debt increased from 23% of GDP at the end of 2015 to 25.8% in 2016.

Kiribati is dependent on foreign aid, which was estimated to have contributed over 32.7% in 2016 to the government's finances. The country's sovereign fund, the Revenue Equalization Reserve Fund (RERF), which is held offshore, had an estimated balance of $855.5 million in late July 2016. The RERF seeks to avoid exchange rate risk by holding investments in more than 20 currencies, including the Australian dollar, US dollar, the Japanese yen, and the Euro. Drawdowns from the RERF helped finance the government's annual budget.

GDP (purchasing power parity): $227 million (2017 est.)

$220.2 million (2016 est.)

$217.7 million (2015 est.)

note: data are in 2017 dollars

country comparison to the world: 219

GDP (official exchange rate): $197 million (2017 est.)

GDP—real growth rate: 3.1% (2017 est.)

1.1% (2016 est.)

10.3% (2015 est.)

country comparison to the world: 96

GDP—per capita (PPP): $2,000 (2017 est.)

$2,000 (2016 est.)

$2,000 (2015 est.)

note: data are in 2017 dollars

country comparison to the world: 210

GDP—composition, by sector of origin:

agriculture: 23% (2016 est.)

industry: 7% (2016 est.)

services: 70% (2016 est.)

Agriculture—products: copra, breadfruit, fish

Industries: fishing, handicrafts

Industrial production growth rate: 1.1% (2012 est.)

country comparison to the world: 153

Labor force: 39,000 (2010 est.)

note: economically active, not including subsistence farmers

country comparison to the world: 198

Labor force—by occupation: *agriculture:* 15%

industry: 10%

services: 75% (2010)

Unemployment rate: 30.6% (2010 est.)

6.1% (2005)

country comparison to the world: 208

Population below poverty line: NA

Household income or consumption by percentage share: *lowest 10%:* NA

highest 10%: NA

Budget: *revenues:* 151.2 million (2017 est.)

expenditures: 277.5 million (2017 est.)

Taxes and other revenues: 76.8% (of GDP) (2017 est.)

country comparison to the world: 4

Budget surplus (+) or deficit (-): -64.1% (of GDP) (2017 est.)

country comparison to the world: 221

Public debt: 26.3% of GDP (2017 est.)

22.9% of GDP (2016 est.)

country comparison to the world: 172

Fiscal year: NA

Inflation rate (consumer prices): 0.4% (2017 est.)

1.9% (2016 est.)

country comparison to the world: 24

Current account balance: $18 million (2017 est.)

$35 million (2016 est.)

country comparison to the world: 61

Exports: $84.75 million (2013 est.)

$62.31 million (2012 est.)

country comparison to the world: 199

Exports—partners: Philippines 50.8%, Malaysia 17.2%, US 11.4%, Bangladesh 5.8%, Fiji 5.4% (2017)

Exports—commodities: fish, coconut products

Imports: $107.1 million (2016 est.)

$182.2 million (2013 est.)

country comparison to the world: 215

Imports—commodities: food, machinery and equipment, miscellaneous manufactured goods, fuel

Imports—partners: Australia 29.3%, Fiji 17.3%, NZ 10.7%, China 5.8%, US 5.8%, Singapore 5.1%, Japan 4.6%, Thailand 4.1% (2017)

Reserves of foreign exchange and gold: $0 (31 December 2017 est.)
$8.37 million (31 December 2010 est.)
country comparison to the world: 192

Debt—external: $40.9 million (2016 est.)
$32.3 million (2015 est.)
country comparison to the world: 196

Exchange rates: Australian dollars (AUD) per US dollar -
1.31 (2017 est.)
1.34 (2016 est.)
1.34 (2015 est.)
1.33 (2014 est.)
1.11 (2013 est.)
note: the Australian dollar circulates as legal tender

ENERGY

Electricity access: *electrification—total population:* 100% (2020)

Electricity—production: 29 million kWh (2016 est.)
country comparison to the world: 210

Electricity—consumption: 26.97 million kWh (2016 est.)
country comparison to the world: 210

Electricity—exports: 0 kWh (2016 est.)
country comparison to the world: 153

Electricity—imports: 0 kWh (2016 est.)
country comparison to the world: 164

Electricity—installed generating capacity: 11,000 kW (2016 est.)
country comparison to the world: 209

Electricity—from fossil fuels: 73% of total installed capacity (2016 est.)
country comparison to the world: 99

Electricity—from nuclear fuels: 0% of total installed capacity (2017 est.)
country comparison to the world: 119

Electricity—from hydroelectric plants: 0% of total installed capacity (2017 est.)
country comparison to the world: 181

Electricity—from other renewable sources: 27% of total installed capacity (2017 est.)
country comparison to the world: 25

Crude oil—production: 0 bbl/day (2018 est.)
country comparison to the world: 155

Crude oil—exports: 0 bbl/day (2015 est.)
country comparison to the world: 146

Crude oil—imports: 0 bbl/day (2015 est.)
country comparison to the world: 146

Crude oil—proved reserves: 0 bbl (1 January 2018 est.)

country comparison to the world: 151

Refined petroleum products—production: 0 bbl/day (2015 est.)
country comparison to the world: 160

Refined petroleum products—consumption: 400 bbl/day (2016 est.)
country comparison to the world: 211

Refined petroleum products—exports: 0 bbl/day (2015 est.)
country comparison to the world: 167

Refined petroleum products—imports: 420 bbl/day (2015 est.)
country comparison to the world: 207

Natural gas—production: 0 cu m (2017 est.)
country comparison to the world: 151

Natural gas—consumption: 0 cu m (2017 est.)
country comparison to the world: 162

Natural gas—exports: 0 cu m (2017 est.)
country comparison to the world: 129

Natural gas—imports: 0 cu m (2017 est.)
country comparison to the world: 143

Natural gas—proved reserves: 0 cu m (1 January 2014 est.)
country comparison to the world: 152

Carbon dioxide emissions from consumption of energy: 58,850 Mt (2017 est.)
country comparison to the world: 209

COMMUNICATIONS

Telephones—fixed lines: *total subscriptions:* 22 *subscriptions per 100 inhabitants:* less than 1 (2019 est.)
country comparison to the world: 221

Telephones—mobile cellular: *total subscriptions:* 51,401
subscriptions per 100 inhabitants: 46.48 (2019 est.)
country comparison to the world: 205

Telecommunication systems: *general assessment:* generally good national and international service; wireline service available on Tarawa and Kiritimati (Christmas Island); connections to outer islands by HF/VHF radiotelephone; recently formed (mobile network operator) MNO is implementing the first phase of improvements with 3G and 4G upgrades on some islands; islands are connected to each other and the rest of the world via satellite; launch of Kacific-1 in December 2019 will improve telecommunication for Kiribati (2020)
domestic: fixed-line 1 per 100 and mobile-cellular 46 per 100 subscriptions (2019)
international: country code—686; landing point for the Southern Cross NEXT submarine cable system from Australia, 7 Pacific Ocean island countries to the US; satellite earth station—1 Intelsat (Pacific Ocean) (2019)
note: the COVID-19 outbreak is negatively impacting telecommunications production and supply chains globally; consumer spending on telecom devices and services has also slowed due to the

pandemic's effect on economies worldwide; overall progress towards improvements in all facets of the telecom industry—mobile, fixed-line, broadband, submarine cable and satellite—has moderated

Broadcast media: multi-channel TV packages provide access to Australian and US stations; 1 government-operated radio station broadcasts on AM, FM, and shortwave (2017)

Internet country code: .ki

Internet users: *total:* 15,946
percent of population: 14.58% (July 2018 est.)
country comparison to the world: 212

Broadband—fixed subscriptions: *total:* 884
subscriptions per 100 inhabitants: 1 less than 1 (2018 est.)
country comparison to the world: 198

TRANSPORTATION

National air transport system: *number of registered air carriers:* 2 (2020)
inventory of registered aircraft operated by air carriers: 8
annual passenger traffic on registered air carriers: 66,567 (2018)

Civil aircraft registration country code prefix: T3 (2016)

Airports: 19 (2013)
country comparison to the world: 136

Airports—with paved runways: *total:* 4 (2017)
1,524 to 2,437 m: 4 (2017)

Airports—with unpaved runways: *total:* 15 (2013)
914 to 1,523 m: 10 (2013)
under 914 m: 5 (2013)

Roadways: *total:* 670 km (2017)
country comparison to the world: 190

Waterways: 5 km (small network of canals in Line Islands) (2012)
country comparison to the world: 107

Merchant marine: *total:* 89
by type: bulk carrier 1 general cargo 34, oil tanker 11, other 43 (2019)
country comparison to the world: 96

Ports and terminals: *major seaport(s):* Betio (Tarawa Atoll), Canton Island, English Harbor

MILITARY AND SECURITY

Military and security forces: no regular military forces (establishment prevented by the constitution); Police Force (2011)

Military—note: Kiribati does not have military forces; defense assistance is provided by Australia and NZ

TRANSNATIONAL ISSUES

Disputes—international: none

KOREA, NORTH

INTRODUCTION

Background: An independent kingdom for much of its long history, Korea was occupied by Japan beginning in 1905 following the Russo-Japanese War. Five years later, Japan formally annexed the entire peninsula. Following World War II, Korea was split with the northern half coming under Soviet-sponsored communist control. After failing in the Korean War (1950-53) to conquer the US-backed Republic of Korea (ROK) in the southern portion by force, North Korea (DPRK), under its founder President KIM Il Sung, adopted a policy of ostensible diplomatic and economic "self-reliance" as a check against outside influence. The DPRK demonized the US as the ultimate threat to its social system through state-funded propaganda, and molded political, economic, and military policies around the core ideological objective of eventual unification of Korea under Pyongyang's control. KIM Il Sung's son, KIM Jong Il, was officially designated as his father's successor in 1980, assuming a growing political and managerial role until the elder KIM's death in 1994. Under KIM Jong Il's rein, the DPRK continued developing nuclear weapons and ballistic missiles. KIM Jong Un was publicly unveiled as his father's successor in 2010. Following KIM Jong Il's death in 2011, KIM Jong Un quickly assumed power and has since occupied the regime's highest political and military posts.

After decades of economic mismanagement and resource misallocation, the DPRK since the mid-1990s has faced chronic food shortages. In recent years, the North's domestic agricultural production has increased, but still falls far short of producing sufficient food to provide for its entire population. The DPRK began to ease restrictions to allow semi-private markets, starting in 2002, but has made few other efforts to meet its goal of improving the overall standard of living. North Korea's history of regional military provocations; proliferation of military-related items; long-range missile development; WMD programs including tests of nuclear devices in 2006, 2009, 2013, 2016, and 2017; and massive conventional armed forces are of major concern to the international community and have limited the DPRK's international engagement, particularly economically. In 2013, the DPRK declared a policy of simultaneous development of its nuclear weapons program and economy. In late 2017, KIM Jong Un declared the North's nuclear weapons development complete. In 2018, KIM announced a pivot towards diplomacy, including a re-prioritization of economic development, a pause in missile testing beginning in late 2017, and a refrain from anti-US rhetoric starting in June 2018. Since 2018, KIM has participated in four meetings with Chinese President XI Jinping, three with ROK President MOON Jae-in, and three with US President TRUMP. Since July 2019, North Korea has restarted its short-range missile tests and issued statements condemning the US.

GEOGRAPHY

Location: Eastern Asia, northern half of the Korean Peninsula bordering the Korea Bay and the Sea of Japan, between China and South Korea

Geographic coordinates: 40 00 N, 127 00 E

Map references: Asia

Area: *total:* 120,538 sq km
land: 120,408 sq km
water: 130 sq km
country comparison to the world: 100

Area—comparative: slightly larger than Virginia; slightly smaller than Mississippi

Land boundaries: *total:* 1,607 km
border countries (3): China 1352 km, South Korea 237 km, Russia 18 km

Coastline: 2,495 km

Maritime claims: *territorial sea:* 12 nm
exclusive economic zone: 200 nm
note: military boundary line 50 nm in the Sea of Japan and the exclusive economic zone limit in the Yellow Sea where all foreign vessels and aircraft without permission are banned

Climate: temperate, with rainfall concentrated in summer; long, bitter winters

Terrain: mostly hills and mountains separated by deep, narrow valleys; wide coastal plains in west, discontinuous in east

Elevation: *mean elevation:* 600 m
lowest point: Sea of Japan 0 m
highest point: Paektu-san 2,744 m

Natural resources: coal, iron ore, limestone, magnesite, graphite, copper, zinc, lead, precious metals, hydropower

Land use: *agricultural land:* 21.8% (2011 est.)
arable land: 19.5% (2011 est.) / permanent crops: 1.9% (2011 est.) / permanent pasture: 0.4% (2011 est.)
forest: 46% (2011 est.)

other: 32.2% (2011 est.)
Irrigated land: 14,600 sq km (2012)

Population distribution: population concentrated in the plains and lowlands; least populated regions are the mountainous provinces adjacent to the Chinese border; largest concentrations are in the western provinces, particularly the municipal district of Pyongyang, and around Hungnam and Wonsan in the east

Natural hazards: late spring droughts often followed by severe flooding; occasional typhoons during the early fall

volcanism: Changbaishan (2,744 m) (also known as Baitoushan, Baegdu or P'aektu-san), on the Chinese border, is considered historically active

Environment—current issues: water pollution; inadequate supplies of potable water; waterborne disease; deforestation; soil erosion and degradation

Environment—international agreements: *party to:* Antarctic Treaty, Biodiversity, Climate Change, Climate Change-Kyoto Protocol, Desertification, Environmental Modification, Hazardous Wastes, Ozone Layer Protection, Ship Pollution
signed, but not ratified: Law of the Sea

Geography—note: strategic location bordering China, South Korea, and Russia; mountainous interior is isolated and sparsely populated

PEOPLE AND SOCIETY

Population: 25,643,466 (July 2020 est.)
country comparison to the world: 54

Nationality: *noun:* Korean(s)
adjective: Korean

Ethnic groups: racially homogeneous; there is a small Chinese community and a few ethnic Japanese

Languages: Korean

Religions: traditionally Buddhist and Confucianist, some Christian and syncretic Chondogyo (Religion of the Heavenly Way)
note: autonomous religious activities now almost nonexistent; government-sponsored religious groups exist to provide illusion of religious freedom

Age structure: *0-14 years:* 20.47% (male 2,677,578/female 2,571,118)
15-24 years: 14.68% (male 1,894,091/female 1,869,799)
25-54 years: 44% (male 5,659,446/female 5,624,034)
55-64 years: 11.2% (male 1,369,199/female 1,503,086)
65 years and over: 9.65% (male 859,151/female 1,615,964) (2020 est.)

Dependency ratios: *total dependency ratio:* 41.2
youth dependency ratio: 28
elderly dependency ratio: 13.2
potential support ratio: 7.6 (2020 est.)

Median age: *total:* 34.6 years
male: 33.2 years

female: 36.2 years (2020 est.)
country comparison to the world: 89

Population growth rate: 0.51% (2020 est.)
country comparison to the world: 154

Birth rate: 14.5 births/1,000 population (2020 est.)
country comparison to the world: 128

Death rate: 9.4 deaths/1,000 population (2020 est.)
country comparison to the world: 48

Net migration rate: 0 migrant(s)/1,000 population (2020 est.)
country comparison to the world: 87

Population distribution: population concentrated in the plains and lowlands; least populated regions are the mountainous provinces adjacent to the Chinese border; largest concentrations are in the western provinces, particularly the municipal district of Pyongyang, and around Hungnam and Wonsan in the east

Urbanization: *urban population:* 62.4% of total population (2020)
rate of urbanization: 0.82% annual rate of change (2015-20 est.)
total population growth rate v. urban population growth rate, 2000-2030:

Major urban areas—population: 3.084 million PYONGYANG (capital) (2020)

Sex ratio: *at birth:* 1.06 male(s)/female
0-14 years: 1.04 male(s)/female
15-24 years: 1.01 male(s)/female
25-54 years: 1.01 male(s)/female
55-64 years: 0.91 male(s)/female
65 years and over: 0.53 male(s)/female
total population: 0.95 male(s)/female (2020 est.)

Maternal mortality rate: 89 deaths/100,000 live births (2017 est.)
country comparison to the world: 74

Infant mortality rate: *total:* 20 deaths/1,000 live births
male: 22.3 deaths/1,000 live births
female: 17.6 deaths/1,000 live births (2020 est.)
country comparison to the world: 76

Life expectancy at birth: *total population:* 71.6 years
male: 67.7 years
female: 75.6 years (2020 est.)
country comparison to the world: 161

Total fertility rate: 1.92 children born/woman (2020 est.)
country comparison to the world: 127

Contraceptive prevalence rate: 70.2% (2017)

Drinking water source:
improved:
urban: 97.2% of population
rural: 90.2% of population
total: 94.5% of population
unimproved:
urban: 2.8% of population
rural: 9.8% of population
total: 5.5% of population (2017 est.)

Physicians density: 3.68 physicians/1,000 population (2017)

Hospital bed density: 13.2 beds/1,000 population (2012)

Sanitation facility access:
improved:
urban: 91.9% of population
rural: 72.3% of population
total: 84.5% of population
unimproved:
urban: 8.1% of population
rural: 27.7% of population
total: 15.5% of population (2017 est.)

HIV/AIDS—adult prevalence rate: NA

HIV/AIDS—people living with HIV/AIDS: NA

HIV/AIDS—deaths: NA

Obesity—adult prevalence rate: 6.8% (2016)
country comparison to the world: 163

Children under the age of 5 years underweight: 9.3% (2017)
country comparison to the world: 68

Education expenditures: NA

Literacy: *definition:* age 15 and over can read and write
total population: 100%
male: 100%
female: 100% (2015)

School life expectancy (primary to tertiary education): *total:* 11 years
male: 11 years
female: 11 years (2015)

GOVERNMENT

Country name: *conventional long form:* Democratic People's Republic of Korea
conventional short form: North Korea
local long form: Choson-minjujuui-inmin-konghwaguk
local short form: Choson
abbreviation: DPRK
etymology: derived from the Chinese name for Goryeo, which was the Korean dynasty that united the peninsula in the 10th century A.D.; the North Korean name "Choson" means "[Land of the] Morning Calm"

Government type: dictatorship, single-party state; official state ideology of "Juche" or "national self-reliance"

Capital: *name:* Pyongyang

geographic coordinates: 39 01 N, 125 45 E

time difference: UTC+9 (14 hours ahead of Washington, DC, during Standard Time)
note: on 5 May 2018, North Korea reverted to UTC+9, the same time zone as South Korea
etymology: the name translates as "flat land" in Korean

Administrative divisions: 9 provinces (do, singular and plural) and 3 cities (si, singular and plural)

provinces: Chagang, Hambuk (North Hamgyong), Hamnam (South Hamgyong), Hwangbuk (North Hwanghae), Hwangnam (South Hwanghae), Kangwon, P'yongbuk (North Pyongan), P'yongnam (South Pyongan), Ryanggang

major cities: Nampo, P'yongyang, Rason
note: Nampo is sometimes designated as a metropolitan city, P'yongyang as a directly controlled city, and Rason as a city

Independence: 15 August 1945 (from Japan)

National holiday: Founding of the Democratic People's Republic of Korea (DPRK), 9 September (1948)

Constitution: *history:* previous 1948, 1972; latest adopted 1998 (during KIM Jong Il era)
amendments: proposed by the Supreme People's Assembly (SPA); passage requires more than two-thirds majority vote of the total SPA membership; revised 2009, 2012, 2013, 2016, 2019

Legal system: civil law system based on the Prussian model; system influenced by Japanese traditions and Communist legal theory

International law organization participation: has not submitted an ICJ jurisdiction declaration; non-party state to the ICC

Citizenship: *citizenship by birth:* no
citizenship by descent only: at least one parent must be a citizen of North Korea
dual citizenship recognized: no
residency requirement for naturalization: unknown

Suffrage: 17 years of age; universal and compulsory

Executive branch: *chief of state:* Supreme People's Assembly President CHOE Ryong Hae (since 11 April 2019); note—functions as the technical head of state and performs related duties, such as receiving ambassadors' credentials

head of government: State Affairs Commission Chairman KIM Jong Un (since 17 December 2011); note—functions as the commander-in-chief and chief executive
cabinet: Cabinet or Naegak members appointed by the Supreme People's Assembly except the Minister of People's Armed Forces
elections/appointments: chief of state and premier indirectly elected by the Supreme People's Assembly; election last held on 10 March 2019 (next election March 2024)
election results: KIM Jong In reelected unopposed
note: the Korean Workers' Party continues to list deceased leaders KIM Il Sung and KIM Jong Il as Eternal President and Eternal General Secretary respectively

Legislative branch: *description:* unicameral Supreme People's Assembly or Ch'oego Inmin Hoeui (687 seats; members directly elected by majority vote in 2 rounds if needed to serve 5-year terms); note—the Korean Workers' Party selects all candidates
elections: last held on 10 March 2019 (next to be held March 2024)
election results: percent of vote by party—NA; seats by party—KWP 607, KSDP 50, Chondoist Chongu Party 22, General Association of Korean Residents in Japan (Chongryon) 5, religious associations 3; ruling party approves a list of candidates who are elected without opposition;

composition—men 575, women 112, percent of women 16.3%

note: KWP, KSDP, Chondoist Chongu Party, and Chongryon are under the KWP's control; a token number of seats reserved for minor parties

Judicial branch: *highest courts:* Supreme Court or Central Court (consists of one judge and 2 "People's Assessors" or, for some cases, 3 judges)

judge selection and term of office: judges elected by the Supreme People's Assembly for 5-year terms

subordinate courts: lower provincial courts as determined by the Supreme People's Assembly

Political parties and leaders: *major parties:* Korean Workers' Party or KWP [KIM Jong Un] General Association of Korean Residents in Japan (Chongryon)

minor parties: Chondoist Chongu Party (under KWP control) Social Democratic Party or KSDP [KIM Yong Dae] (under KWP control)

International organization participation: ARF, FAO, G-77, ICAO, ICRM, IFAD, IFRCS, IHO, IMO, IMSO, IOC, IPU, ISO, ITSO, ITU, NAM, UN, UNCTAD, UNESCO, UNIDO, UNWTO, UPU, WFTU (NGOs), WHO, WIPO, WMO

Diplomatic representation in the US: none; North Korea has a Permanent Mission to the UN in New York

Diplomatic representation from the US: none; the Swedish Embassy in Pyongyang represents the US as consular protecting power

Flag description: three horizontal bands of blue (top), red (triple width), and blue; the red band is edged in white; on the hoist side of the red band is a white disk with a red five-pointed star; the broad red band symbolizes revolutionary traditions; the narrow white bands stand for purity, strength, and dignity; the blue bands signify sovereignty, peace, and friendship; the red star represents socialism

National symbol(s): *red star, chollima (winged horse); national colors:* red, white, blue

National anthem: *name:* "Aegukka" (Patriotic Song) *lyrics/music:* PAK Se Yong/KIM Won Gyun *note:* adopted 1947; both North Korea's and South Korea's anthems share the same name and have a vaguely similar melody but have different lyrics; the North Korean anthem is also known as "Ach'imun pinnara" (Let Morning Shine)

ECONOMY

Economy—overview: North Korea, one of the world's most centrally directed and least open economies, faces chronic economic problems. Industrial capital stock is nearly beyond repair as a result of years of underinvestment, shortages of spare parts, and poor maintenance. Large-scale military spending and development of its ballistic missile and nuclear programs severely draws off resources needed for investment and civilian consumption. Industrial and power outputs have stagnated for years at a fraction of pre-1990 levels. Frequent weather-related crop failures aggravated

chronic food shortages caused by on-going systemic problems, including a lack of arable land, collective farming practices, poor soil quality, insufficient fertilization, and persistent shortages of tractors and fuel.

The mid 1990s through mid-2000s were marked by severe famine and widespread starvation. Significant food aid was provided by the international community through 2009. Since that time, food assistance has declined significantly. In the last few years, domestic corn and rice production has improved, although domestic production does not fully satisfy demand. A large portion of the population continues to suffer from prolonged malnutrition and poor living conditions. Since 2002, the government has allowed semi-private markets to begin selling a wider range of goods, allowing North Koreans to partially make up for diminished public distribution system rations. It also implemented changes in the management process of communal farms in an effort to boost agricultural output.

In December 2009, North Korea carried out a redenomination of its currency, capping the amount of North Korean won that could be exchanged for the new notes, and limiting the exchange to a one-week window. A concurrent crackdown on markets and foreign currency use yielded severe shortages and inflation, forcing Pyongyang to ease the restrictions by February 2010. In response to the sinking of the South Korean warship Cheonan and the shelling of Yeonpyeong Island in 2010, South Korea's government cut off most aid, trade, and bilateral cooperation activities. In February 2016, South Korea ceased its remaining bilateral economic activity by closing the Kaesong Industrial Complex in response to North Korea's fourth nuclear test a month earlier. This nuclear test and another in September 2016 resulted in two United Nations Security Council Resolutions that targeted North Korea's foreign currency earnings, particularly coal and other mineral exports. Throughout 2017, North Korea's continued nuclear and missile tests led to a tightening of UN sanctions, resulting in full sectoral bans on DPRK exports and drastically limited key imports. Over the last decade, China has been North Korea's primary trading partner.

The North Korean Government continues to stress its goal of improving the overall standard of living, but has taken few steps to make that goal a reality for its populace. In 2016, the regime used two mass mobilizations — one totaling 70 days and another 200 days — to spur the population to increase production and complete construction projects quickly. The regime released a five-year economic development strategy in May 2016 that outlined plans for promoting growth across sectors. Firm political control remains the government's overriding concern, which likely will inhibit formal changes to North Korea's current economic system.

GDP (purchasing power parity): $40 billion (2015 est.) $40 billion (2014 est.) $40 billion (2013 est.)

note: data are in 2015 US dollars

North Korea does not publish reliable National Income Accounts data; the data shown are derived from purchasing power parity (PPP) GDP estimates that were made by Angus MADDISON in a study conducted for the OECD; his figure for 1999 was extrapolated to 2015 using estimated real growth rates for North Korea's GDP and an inflation factor based on the US GDP deflator; the results were rounded to the nearest $10 billion. *country comparison to the world:* 118

GDP (official exchange rate): $28 billion (2013 est.)

GDP—real growth rate: -1.1% (2015 est.) 1% (2014 est.) 1.1% (2013 est.) *country comparison to the world:* 201

GDP—per capita (PPP): $1,700 (2015 est.) $1,800 (2014 est.) $1,800 (2013 est.) *note:* data are in 2015 US dollars *country comparison to the world:* 214

Gross national saving: NA

GDP—composition, by end use: *household consumption:* NA (2014 est.) *government consumption:* NA (2014 est.) *investment in fixed capital:* NA (2014 est.) *investment in inventories:* NA (2014 est.) *exports of goods and services:* 5.9% (2016 est.) *imports of goods and services:* -11.1% (2016 est.)

GDP—composition, by sector of origin: *agriculture:* 22.5% (2017 est.) *industry:* 47.6% (2017 est.) *services:* 29.9% (2017 est.)

Agriculture—products: rice, corn, potatoes, wheat, soybeans, pulses, beef, pork, eggs, fruit, nuts *Industries:* military products; machine building, electric power, chemicals; mining (coal, iron ore, limestone, magnesite, graphite, copper, zinc, lead, and precious metals); metallurgy; textiles; food processing; tourism

Industrial production growth rate: 1% (2017 est.) *country comparison to the world:* 156

Labor force: 14 million (2014 est.) *note:* estimates vary widely *country comparison to the world:* 38

Labor force—by occupation: *agriculture:* 37% *industry:* 63% (2008 est.)

Unemployment rate: 25.6% (2013 est.) 25.5% (2012 est.) *country comparison to the world:* 197

Population below poverty line: NA

Household income or consumption by percentage share: *lowest 10%:* NA *highest 10%:* NA

Budget: *revenues:* 3.2 billion (2007 est.) *expenditures:* 3.3 billion (2007 est.)

Taxes and other revenues: 11.4% (of GDP) (2007 est.) *note:* excludes earnings from state-operated enterprises *country comparison to the world:* 209

Budget surplus (+) or deficit (-): -0.4% (of GDP) (2007 est.)
country comparison to the world: 58

Fiscal year: calendar year

Inflation rate (consumer prices): NA

Exports: $222 million (2018)
$4.582 billion (2017 est.)
$2.908 billion (2015 est.)
country comparison to the world: 188

Exports—partners: China 86.3% (2017)

Exports—commodities: minerals, metallurgical products, manufactures (including armaments), textiles, agricultural and fishery products

Imports: $2.32 billion (2018 est.)
$3.86 billion (2016 est.)
country comparison to the world: 160

Imports—commodities: petroleum, coking coal, machinery and equipment, textiles, grain

Imports—partners: China 91.9% (2017)

Debt—external: $5 billion (2013 est.)
country comparison to the world: 132

Exchange rates: North Korean won (KPW) per US dollar (average market rate)
135 (2017 est.)
130 (2016 est.)
130 (2015 est.)
98.5 (2013 est.)
155.5 (2012 est.)

ENERGY

Electricity access: *population without electricity:* 19 million (2019)
electrification—total population: 26% (2019)
electrification—urban areas: 36% (2019)
electrification—rural areas: 11% (2019)

Electricity—production: 16.57 billion kWh (2016 est.)
country comparison to the world: 87

Electricity—consumption: 13.89 billion kWh (2016 est.)
country comparison to the world: 83

Electricity—exports: 0 kWh (2016 est.)
country comparison to the world: 154

Electricity—imports: 0 kWh (2016 est.)
country comparison to the world: 165

Electricity—installed generating capacity: 10.01 million kW (2016 est.)
country comparison to the world: 60

Electricity—from fossil fuels: 45% of total installed capacity (2016 est.)
country comparison to the world: 161

Electricity—from nuclear fuels: 0% of total installed capacity (2017 est.)
country comparison to the world: 120

Electricity—from hydroelectric plants: 55% of total installed capacity (2017 est.)
country comparison to the world: 31

Electricity—from other renewable sources: 0% of total installed capacity (2017 est.)
country comparison to the world: 195

Crude oil—production: 0 bbl/day (2018 est.)
country comparison to the world: 156

Crude oil—exports: 0 bbl/day (2015 est.)
country comparison to the world: 147

Crude oil—imports: 10,640 bbl/day (2015 est.)
country comparison to the world: 72

Crude oil—proved reserves: 0 bbl (1 January 2018 est.)
country comparison to the world: 152

Refined petroleum products—production: 11,270 bbl/day (2015 est.)
country comparison to the world: 99

Refined petroleum products—consumption: 18,000 bbl/day (2016 est.)
country comparison to the world: 145

Refined petroleum products—exports: 0 bbl/day (2015 est.)
country comparison to the world: 168

Refined petroleum products—imports: 8,260 bbl/day (2015 est.)
country comparison to the world: 151

Natural gas—production: 0 cu m (2017 est.)
country comparison to the world: 152

Natural gas—consumption: 0 cu m (2017 est.)
country comparison to the world: 163

Natural gas—exports: 0 cu m (2017 est.)
country comparison to the world: 130

Natural gas—imports: 0 cu m (2017 est.)
country comparison to the world: 144

Natural gas—proved reserves: 0 cu m (1 January 2014 est.)
country comparison to the world: 153

Carbon dioxide emissions from consumption of energy: 27.83 million Mt (2017 est.)
country comparison to the world: 74

COMMUNICATIONS

Telephones—fixed lines: *total subscriptions:* 1,183,806
subscriptions per 100 inhabitants: 4.64 (2019 est.)
country comparison to the world: 70

Telephones—mobile cellular: *total subscriptions:* 3,821,857
subscriptions per 100 inhabitants: 14.98 (2019 est.)
country comparison to the world: 132

Telecommunication systems: *general assessment:* nationwide fiber-optic network; mobile-cellular service expanded beyond Pyongyang; infrastructure underdeveloped yet growing mobile penetration by means of foreign investment; Chinese services being increasingly favored and FaceBook and Instagram actions dropped and now absent; low broadband penetration; mobile penetration in North Korea believed to stay well below other Asian nations due to government restrictions; 3G network deployed among universal population (2020)

domestic: fiber-optic links installed down to the county level; telephone directories unavailable; mobile service launched in late 2008 for the Pyongyang area and considerable progress in expanding to other parts of the country since;

fixed-lines are 5 per 100 and mobile-cellular 15 per 100 persons (2019)

international: country code—850; satellite earth stations—2 (1 Intelsat—Indian Ocean, 1 Russian—Indian Ocean region); other international connections through Moscow and Beijing
note: the COVID-19 outbreak is negatively impacting telecommunications production and supply chains globally; consumer spending on telecom devices and services has also slowed due to the pandemic's effect on economies worldwide; overall progress towards improvements in all facets of the telecom industry—mobile, fixed-line, broadband, submarine cable and satellite—has moderated

Broadcast media: no independent media; radios and TVs are pre-tuned to government stations; 4 government-owned TV stations; the Korean Workers' Party owns and operates the Korean Central Broadcasting Station, and the state-run Voice of Korea operates an external broadcast service; the government prohibits listening to and jams foreign broadcasts (2019)

Internet country code: .kp

TRANSPORTATION

National air transport system: *number of registered air carriers:* 1 (2020)
inventory of registered aircraft operated by air carriers: 4
annual passenger traffic on registered air carriers: 103,560 (2018)
annual freight traffic on registered air carriers: 250,000 mt-km (2018)

Civil aircraft registration country code prefix: P (2016)

Airports: 82 (2013)
country comparison to the world: 67

Airports—with paved runways:
total: 39 (2017)
over 3,047 m: 3 (2017)
2,438 to 3,047 m: 22 (2017)
1,524 to 2,437 m: 8 (2017)
914 to 1,523 m: 2 (2017)
under 914 m: 4 (2017)

Airports—with unpaved runways:
total: 43 (2013)
2,438 to 3,047 m: 3 (2013)
1,524 to 2,437 m: 17 (2013)
914 to 1,523 m: 15 (2013)
under 914 m: 8 (2013)

Heliports: 23 (2013)

Pipelines: 6 km oil (2013)

Railways: *total:* 7,435 km (2014)
standard gauge: 7,435 km 1.435-m gauge (5,400 km electrified) (2014)
note: figures are approximate; some narrow-gauge railway also exists
country comparison to the world: 29

Roadways: *total:* 25,554 km (2006)
paved: 724 km (2006)
unpaved: 24,830 km (2006)
country comparison to the world: 104

Waterways: 2,250 km (most navigable only by small craft) (2011)

country comparison to the world: 37

Merchant marine: *total:* 264

by type: bulk carrier 9, container ship 5, general cargo 188, oil tanker 33, other 29 (2019) *country comparison to the world:* 58

Ports and terminals: *major seaport(s):* Ch'ongjin, Haeju, Hungnam, Namp'o, Songnim, Sonbong (formerly Unggi), Wonsan

MILITARY AND SECURITY

Military and security forces: *Korean People's Army (KPA):* KPA Ground Forces, KPA Navy, KPA Air Force (includes air defense), KPA Strategic Force (missile forces); Guard Command (protects the Kim family, other senior North Korean leadership figures, and government facilities in Pyongyang); Ministry of Public Security: Border Guards, civil security forces (2019) Military and security service personnel strengths: assessments of the size of the Korean People's Army (KPA) vary widely; approximately 1.1-1.2 million active troops (950,000-1.0 million Army; 110-120,000 Air Force; 60,000 Navy; 10,000 Strategic Missile Forces); est. 200,000 Public Security forces (2019)

Military equipment inventories and acquisitions: the KPA is equipped mostly with older weapon systems originally acquired from the former Soviet Union, Russia, and China; North Korea manufactures copies and provides some upgrades to these weapon systems; it also has a robust domestic ballistic missile program based largely on missiles acquired from the former Soviet Union; since 2010, there were no publicly-reported transfers of weapons to North Korea; between 2000 and 2010,

Russia was the only recorded provider of arms (2019 est.)

Military service age and obligation: 17 years of age for compulsory male and female military service; service obligation 10 years for men, to age 23 for women (2015)

TRANSNATIONAL ISSUES

Disputes—international: risking arrest, imprisonment, and deportation, tens of thousands of North Koreans cross into China to escape famine, economic privation, and political oppression; North Korea and China dispute the sovereignty of certain islands in Yalu and Tumen Rivers; Military Demarcation Line within the 4-km-wide Demilitarized Zone has separated North from South Korea since 1953; periodic incidents in the Yellow Sea with South Korea which claims the Northern Limiting Line as a maritime boundary; North Korea supports South Korea in rejecting Japan's claim to Liancourt Rocks (Tok-do/ Take-shima)

Refugees and internally displaced persons: *IDPs:* undetermined (periodic flooding and famine during mid-1990s) (2019)

Trafficking in persons: *current situation:* North Korea is a source country for men, women, and children who are subjected to forced labor and sex trafficking; many North Korean workers recruited to work abroad under bilateral contracts with foreign governments, most often Russia and China, are subjected to forced labor and do not have a choice in the work the government assigns them, are not free to change jobs, and face government reprisals if they try to escape

or complain to outsiders; tens of thousands of North Koreans, including children, held in prison camps are subjected to forced labor, including logging, mining, and farming; many North Korean women and girls, lured by promises of food, jobs, and freedom, have migrated to China illegally to escape poor social and economic conditions only to be forced into prostitution, domestic service, or agricultural work through forced marriages

tier rating: Tier 3—North Korea does not fully comply with minimum standards for the elimination of trafficking and is not making significant efforts to do so; the government continued to participate in human trafficking through its use of domestic forced labor camps and the provision of forced labor to foreign governments through bilateral contracts; officials did not demonstrate any efforts to address human trafficking through prosecution, protection, or prevention measures; no known investigations, prosecutions, or convictions of trafficking offenders or officials complicit in trafficking-related offenses were conducted; the government also made no efforts to identify or protect trafficking victims and did not permit NGOs to assist victims (2015)

Illicit drugs: at present there is insufficient information to determine the current level of involvement of government officials in the production or trafficking of illicit drugs, but for years, from the 1970s into the 2000s, citizens of the Democratic People's Republic of (North) Korea (DPRK), many of them diplomatic employees of the government, were apprehended abroad while trafficking in narcotics; police investigations in Taiwan and Japan in recent years have linked North Korea to large illicit shipments of heroin and methamphetamine

KOREA, SOUTH

INTRODUCTION

Background: An independent kingdom for much of its long history, Korea was occupied by Japan

beginning in 1905 following the Russo-Japanese War. In 1910, Tokyo formally annexed the entire Peninsula. Korea regained its independence following Japan's surrender to the US in 1945. After World War II, a democratic government (Republic of Korea, ROK) was set up in the southern half of the Korean Peninsula while a communist-style government was installed in the north (Democratic People's Republic of Korea, DPRK). During the Korean War (1950-53), US troops and UN forces fought alongside ROK soldiers to defend South Korea from a DPRK invasion supported by communist China and the Soviet Union. A 1953 armistice split the Peninsula along a demilitarized zone at about the 38th parallel. PARK Chung-hee took over leadership of the country in a 1961 coup. During his regime, from 1961 to 1979, South Korea achieved rapid economic growth, with per capita income rising to roughly 17 times the level of North Korea in 1979.

South Korea held its first free presidential election under a revised democratic constitution in 1987, with former ROK Army general ROH

Tae-woo winning a close race. In 1993, KIM Young-sam (1993-98) became the first civilian president of South Korea's new democratic era. President KIM Dae-jung (1998-2003) won the Nobel Peace Prize in 2000 for his contributions to South Korean democracy and his "Sunshine" policy of engagement with North Korea. President PARK Geun-hye, daughter of former ROK President PARK Chung-hee, took office in February 2013 as South Korea's first female leader. In December 2016, the National Assembly passed an impeachment motion against President PARK over her alleged involvement in a corruption and influence-peddling scandal, immediately suspending her presidential authorities. The impeachment was upheld in March 2017, triggering an early presidential election in May 2017 won by MOON Jae-in. South Korea hosted the Winter Olympic and Paralympic Games in February 2018, in which North Korea also participated. Discord with North Korea has permeated inter-Korean relations for much of the past decade, highlighted by the North's attacks on a South Korean ship and island in 2010, the exchange of artillery fire across the DMZ in 2015, and multiple nuclear and missile tests in 2016 and 2017. North Korea's participation in the Winter Olympics, dispatch of a senior delegation to Seoul, and three inter-Korean summits in 2018 appear to have ushered in a temporary period of respite, buoyed by the historic US-DPRK summits in 2018 and 2019.

GEOGRAPHY

Location: Eastern Asia, southern half of the Korean Peninsula bordering the Sea of Japan and the Yellow Sea

Geographic coordinates: 37 00 N, 127 30 E

Map references: Asia

Area: *total:* 99,720 sq km
land: 96,920 sq km
water: 2,800 sq km
country comparison to the world: 110

Area—comparative: slightly smaller than Pennsylvania; slightly larger than Indiana

Land boundaries: *total:* 237 km
border countries (1): North Korea 237 km

Coastline: 2,413 km

Maritime claims: *territorial sea:* 12 nm; between 3 nm and 12 nm in the Korea Strait
exclusive economic zone: 200 nm
contiguous zone: 24 nm
continental shelf: not specified

Climate: temperate, with rainfall heavier in summer than winter; cold winters

Terrain: mostly hills and mountains; wide coastal plains in west and south

Elevation: *mean elevation:* 282 m
lowest point: Sea of Japan 0 m
highest point: Halla-san 1,950 m

Natural resources: coal, tungsten, graphite, molybdenum, lead, hydropower potential

Land use: *agricultural land:* 18.1% (2011 est.)
arable land: 15.3% (2011 est.) / permanent crops: 2.2% (2011 est.) / permanent pasture: 0.6% (2011 est.)
forest: 63.9% (2011 est.)
other: 18% (2011 est.)

Irrigated land: 7,780 sq km (2012)

Population distribution: with approximately 70% of the country considered mountainous, the country's population is primarily concentrated in the lowland areas, where density is quite high; Gyeonggi Province in the northwest, which surrounds the capital of Seoul and contains the port of Incheon, is the most densely populated province; Gangwon in the northeast is the least populated

Natural hazards: occasional typhoons bring high winds and floods; low-level seismic activity common in southwest
volcanism: Halla (1,950 m) is considered historically active although it has not erupted in many centuries

Environment—current issues: air pollution in large cities; acid rain; water pollution from the

discharge of sewage and industrial effluents; drift net fishing; solid waste disposal; transboundary pollution

Environment—international agreements: *party to:* Antarctic-Environmental Protocol, Antarctic-Marine Living Resources, Antarctic Treaty, Biodiversity, Climate Change, Climate Change-Kyoto Protocol, Desertification, Endangered Species, Environmental Modification, Hazardous Wastes, Law of the Sea, Marine Dumping, Ozone Layer Protection, Ship Pollution, Tropical Timber 83, Tropical Timber 94, Wetlands, Whaling
signed, but not ratified: none of the selected agreements

Geography—note: strategic location on Korea Strait; about 3,000 mostly small and uninhabited islands lie off the western and southern coasts

PEOPLE AND SOCIETY

Population: 51,835,110 (July 2020 est.)
country comparison to the world: 28

Nationality: *noun:* Korean(s)
adjective: Korean

Ethnic groups: homogeneous

Languages: Korean, English (widely taught in elementary, junior high, and high school)

Religions: Protestant 19.7%, Buddhist 15.5%, Catholic 7.9%, none 56.9% (2015 est.)
note: many people also carry on at least some Confucian traditions and practices

Age structure: *0-14 years:* 12.77% (male 3,401,815/female 3,219,589)
15-24 years: 11.18% (male 3,030,027/female 2,764,860)
25-54 years: 44.66% (male 12,043,626/female 11,106,927)
55-64 years: 15.47% (male 3,927,496/female 4,089,033)
65 years and over: 15.92% (male 3,572,855/female 4,678,882) (2020 est.)

Dependency ratios: *total dependency ratio:* 39.5
youth dependency ratio: 17.5
elderly dependency ratio: 22
potential support ratio: 4.5 (2020 est.)

Median age: *total:* 43.2 years
male: 41.6 years
female: 45 years (2020 est.)
country comparison to the world: 29

Population growth rate: 0.39% (2020 est.)
country comparison to the world: 162

Birth rate: 8.2 births/1,000 population (2020 est.)
country comparison to the world: 220

Death rate: 6.8 deaths/1,000 population (2020 est.)
country comparison to the world: 135

Net migration rate: 2.3 migrant(s)/1,000 population (2020 est.)
country comparison to the world: 45

Population distribution: with approximately 70% of the country considered mountainous, the country's population is primarily concentrated in the lowland areas, where density is quite high;

Gyeonggi Province in the northwest, which surrounds the capital of Seoul and contains the port of Incheon, is the most densely populated province; Gangwon in the northeast is the least populated

Urbanization: *urban population:* 81.4% of total population (2020)
rate of urbanization: 0.3% annual rate of change (2015-20 est.)
total population growth rate v. urban population growth rate, 2000-2030: Major urban areas—population: 9.963 million SEOUL (capital), 3.465 million Busan, 2.801 million Incheon, 2.199 million Daegu (Taegu), 1.566 million Daejon (Taejon), 1.522 million Gwangju (Kwangju) (2020)

Sex ratio: *at birth:* 1.05 male(s)/female
0-14 years: 1.06 male(s)/female
15-24 years: 1.1 male(s)/female
25-54 years: 1.08 male(s)/female
55-64 years: 0.96 male(s)/female
65 years and over: 0.76 male(s)/female
total population: 1.01 male(s)/female (2020 est.)

Mother's mean age at first birth: 31 years (2014 est.)

Maternal mortality rate: 11 deaths/100,000 live births (2017 est.)
country comparison to the world: 142

Infant mortality rate: *total:* 3 deaths/1,000 live births
male: 3.2 deaths/1,000 live births
female: 2.8 deaths/1,000 live births (2020 est.)
country comparison to the world: 216

Life expectancy at birth: *total population:* 82.6 years
male: 79.4 years
female: 85.9 years (2020 est.)
country comparison to the world: 15

Total fertility rate: 1.29 children born/woman (2020 est.)
country comparison to the world: 223

Contraceptive prevalence rate: 82.3% (2018)
note: percent of women aged 20-49

Drinking water source:

improved: *total:* 100% of population

unimproved: *total:* 0% of population (2017 est.)

Current Health Expenditure: 7.6% (2017)

Physicians density: 2.36 physicians/1,000 population (2017)

Hospital bed density: 12.3 beds/1,000 population (2017)

Sanitation facility access: *improved:* total: 100% of population
unimproved: total: 0% of population (2017 est.)

HIV/AIDS—adult prevalence rate: NA

HIV/AIDS—people living with HIV/AIDS: NA

HIV/AIDS—deaths: NA

Major infectious diseases: Covid-19 (see note) (2020)
note: a novel coronavirus is causing an outbreak of respiratory illness (COVID-19) in South Korea; illness with this virus has ranged from mild to

severe with fatalities reported; the US Department of State has issued a Travel Advisory recommending avoiding all international travel due to COVID-19; the Centers for Disease Control and Prevention has recommended against all international travel and published additional guidance at https://wwwnc.cdc.gov/travel/notices/warning/; as of 10 November 2020, South Korea has reported 27,427 confirmed cases of COVID19 with 478 deaths

Obesity—adult prevalence rate: 4.7% (2016)
country comparison to the world: 184

Children under the age of 5 years underweight: 0.7% (2010)
country comparison to the world: 128

Education expenditures: 5.3% of GDP (2015)
country comparison to the world: 48

Literacy: *definition:* age 15 and over can read and write (2019)
total population: 98% (2019)
male: 99.2%
female: 96.6%

School life expectancy (primary to tertiary education): *total:* 17 years
male: 17 years
female: 16 years (2018)

Unemployment, youth ages 15-24: *total:* 10.2%
male: 10.6%
female: 10% (2018 est.)
country comparison to the world: 124

GOVERNMENT

Country name: *conventional long form:* Republic of Korea
conventional short form: South Korea
local long form: Taehan-min'guk
local short form: Han'guk
abbreviation: ROK
etymology: derived from the Chinese name for Goryeo, which was the Korean dynasty that united the peninsula in the 10th century A.D.; the South Korean name "Han'guk" derives from the long form, "Taehan-min'guk," which is itself a derivation from "Daehan-je'guk," which means "the Great Empire of the Han"; "Han" refers to the "Sam'han" or the "Three Han Kingdoms" (Goguryeo, Baekje, and Silla from the Three Kingdoms Era, 1st-7th centuries A.D.)

Government type: presidential republic

Capital: *name:* Seoul; note—Sejong, located some 120 km (75 mi) south of Seoul, is serving as an administrative capital for segments of the South Korean Government
geographic coordinates: 37 33 N, 126 59 E
time difference: UTC+9 (14 hours ahead of Washington, DC, during Standard Time)
etymology: the name originates from the Korean word meaning "capital city" and which is believed to be derived from Seorabeol, the name of the capital of the ancient Korean Kingdom of Silla

Administrative divisions: 9 provinces (do, singular and plural), 6 metropolitan cities (gwangyeoksi,

singular and plural), 1 special city (teugbyeolsi), and 1 special self-governing city (teukbyeoljachisi)
provinces: Chungbuk (North Chungcheong), Chungnam (South Chungcheong), Gangwon, Gyeongbuk (North Gyeongsang), Gyeonggi, Gyeongnam (South Gyeongsang), Jeju, Jeonbuk (North Jeolla), Jeonnam (South Jeolla)
metropolitan cities: Busan (Pusan), Daegu (Taegu), Daejeon (Taejon), Gwangju (Kwangju), Incheon (Inch'on), Ulsan
special city: Seoul
special self-governing city: Sejong

Independence: 15 August 1945 (from Japan)

National holiday: Liberation Day, 15 August (1945)

Constitution: *history:* several previous; latest passed by National Assembly 12 October 1987, approved in referendum 28 October 1987, effective 25 February 1988
amendments: proposed by the president or by majority support of the National Assembly membership; passage requires at least two-thirds majority vote by the Assembly membership, approval in a referendum by more than one half of the votes by more than one half of eligible voters, and promulgation by the president; amended several times, last in 1987

Legal system: mixed legal system combining European civil law, Anglo-American law, and Chinese classical thought

International law organization participation: has not submitted an ICJ jurisdiction declaration; accepts ICCt jurisdiction

Citizenship: *citizenship by birth:* no
citizenship by descent only: at least one parent must be a citizen of South Korea
dual citizenship recognized: no
residency requirement for naturalization: 5 years

Suffrage: 18 years of age; universal; note—the voting age was lowered from 19 to 18 beginning with the 2020 national election

Executive branch: *chief of state:* President MOON Jae-in (since 10 May 2017); the president is both chief of state and head of government; Prime Minister CHUNG Sye-kyun (since 14 January 2020) serves as the principal executive assistant to the president, similar to the role of a vice president
head of government: President MOON Jae-in (since 10 May 2017)
cabinet: State Council appointed by the president on the prime minister's recommendation
elections/appointments: president directly elected by simple majority popular vote for a single 5-year term; election last held on 9 May 2017 (next to be held in March 2022); prime minister appointed by president with consent of National Assembly
election results: MOON Jae-in elected president; percent of vote—MOON Jae-in (DP) 41.1%, HONG Joon-pyo (LKP) 25.5%, AHN Cheol-soo (PP) 21.4%, other 12%

Legislative branch: *description:* unicameral National Assembly or Kuk Hoe (300 seats

statutory); 253 members directly elected in single-seat constituencies by simple majority vote and 47 directly elected in a single national constituency by proportional representation vote; members serve 4-year terms)
elections: last held on 15 April 2020 (next to be held in April 2024)
election results: percent of vote by party—NA; seats by party—DP/TCP 180, UFP/FKP 103, JP 6, ODP 3, PP 3, independent 5; composition—men 249, women 51, percent of women 17%

Judicial branch: *highest courts:* Supreme Court (consists of a chief justice and 13 justices); Constitutional Court (consists of a court head and 8 justices)
judge selection and term of office: Supreme Court chief justice appointed by the president with the consent of the National Assembly; other justices appointed by the president upon the recommendation of the chief justice and consent of the National Assembly; position of the chief justice is a 6-year nonrenewable term; other justices serve 6-year renewable terms; Constitutional Court justices appointed—3 by the president, 3 by the National Assembly, and 3 by the Supreme Court chief justice; court head serves until retirement at age 70, while other justices serve 6-year renewable terms with mandatory retirement at age 65
subordinate courts: High Courts; District Courts; Branch Courts (organized under the District Courts); specialized courts for family and administrative issues

Political parties and leaders: Bareun Mirae Party or BMP [SOHN Hak-kyu] (merger of Bareun Party and People's Party)
Democratic Party or DP [LEE Hae-chan] (renamed from Minjoo Party of Korea or MPK in October 2016; formerly New Politics Alliance for Democracy or NPAD, which was a merger of the Democratic Party or DP (formerly DUP) [KIM Han-gil] and the New Political Vision Party or NPVP [AHN Cheol-soo] in March 2014)
Justice Party or JP [SIM Sang-jung]
Minjung Party or MP (formed from the merger of the New People's Party (formerly the New People's Political Party or NPP) and the People's United Party or PUP)
Open Democratic Pary or ODP [LEE Keun-shik] (formed in early 2020)
Our Republic Party [CHO Won-jin and HONG Moon-jong] (formerly Korean Patriots' Party or KPP)
Party for Democracy and Peace or PDP [CHUNG Dong-young]
People Party or PP [AHN Cheol-soo] (formed in February 2020)
Together Citizens' Party [WOO Hee-jong, ChOI Bae-geun] (formed in early 2020 in alliance with the Democratic Party) United Future Party or UFP (formed in early 2020 by the merger of Liberty Korea Party, New Conservative Party, Onward for Future 4.0, and several other minor parties; it has a sister relationship with the Future Korea Party

International organization participation: ADB, AfDB (nonregional member), APEC, Arctic

Council (observer), ARF, ASEAN (dialogue partner), Australia Group, BIS, CD, CICA, CP, EAS, EBRD, FAO, FATF, G-20, IADB, IAEA, IBRD, ICAO, ICC (national committees), ICCt, ICRM, IDA, IEA, IFAD, IFC, IFRCS, IHO, ILO, IMF, IMO, IMSO, Interpol, IOC, IOM, IPU, ISO, ITSO, ITU, ITUC (NGOs), LAIA (observer), MIGA, MINURSO, MINUSTAH, NEA, NSG, OAS (observer), OECD, OPCW, OSCE (partner), Pacific Alliance (observer), Paris Club (associate), PCA, PIF (partner), SAARC (observer), SICA (observer), UN, UNAMID, UNCTAD, UNESCO, UNHCR, UNIDO, UNIFIL, UNMIL, UNMISS, UNMOGIP, UNOCI, UNWTO, UPU, WCO, WHO, WIPO, WMO, WTO, ZC

Diplomatic representation in the US: *chief of mission:* Ambassador LEE Soo-hyuck (since 6 January 2020)
chancery: 2450 Massachusetts Avenue NW, Washington, DC 20008
telephone: [1] (202) 939-5600
FAX: [1] (202) 797-0595
consulate(s) general: Agana (Guam), Anchorage (AK), Atlanta, Boston, Chicago, Honolulu, Houston, Los Angeles, New York, San Francisco, Seattle

Diplomatic representation from the US: *chief of mission:* Ambassador Harry HARRIS (since 10 July 2018)
telephone: [82] (2) 397-4114
embassy: 188 Sejong-daero, Jongno-gu, Seoul 03141
mailing address: US Embassy Seoul, 9600 Seoul Place Washington, D.C., 20521-9600
FAX: [82] (2) 725-0152

Flag description: white with a red (top) and blue yin-yang symbol in the center; there is a different black trigram from the ancient I Ching (Book of Changes) in each corner of the white field; the South Korean national flag is called Taegukki; white is a traditional Korean color and represents peace and purity; the blue section represents the negative cosmic forces of the yin, while the red symbolizes the opposite positive forces of the yang; each trigram (kwae) denotes one of the four universal elements, which together express the principle of movement and harmony

National symbol(s): *taegeuk (yin yang symbol), Hibiscus syriacus (Rose of Sharon), Siberian tiger; national colors:* red, white, blue, black

National anthem: *name:* "Aegukga" (Patriotic Song)
lyrics/music: YUN Ch'i-Ho or AN Ch'ang-Ho/ AHN Eaktay
note: adopted 1948, well-known by 1910; both North Korea's and South Korea's anthems share the same name and have a vaguely similar melody but have different lyrics

ECONOMY

Economy—overview: After emerging from the 1950-53 war with North Korea, South Korea emerged as one of the 20th century's most remarkable economic success stories, becoming a developed, globally connected, high-technology society within decades. In the 1960s, GDP per capita was comparable with levels in the poorest countries in the world. In 2004, South Korea's GDP surpassed one trillion dollars.

Beginning in the 1960s under President PARK Chung-hee, the government promoted the import of raw materials and technology, encouraged saving and investment over consumption, kept wages low, and directed resources to export- oriented industries that remain important to the economy to this day. Growth surged under these policies, and frequently reached double-digits in the 1960s and 1970s. Growth gradually moderated in the 1990s as the economy matured, but remained strong enough to propel South Korea into the ranks of the advanced economies of the OECD by 1997. These policies also led to the emergence of family-owned chaebol conglomerates such as Daewoo, Hyundai, and Samsung, which retained their dominant positions even as the government loosened its grip on the economy amid the political changes of the 1980s and 1990s.

The Asian financial crisis of 1997-98 hit South Korea's companies hard because of their excessive reliance on short-term borrowing, and GDP ultimately plunged by 7% in 1998. South Korea tackled difficult economic reforms following the crisis, including restructuring some chaebols, increasing labor market flexibility, and opening up to more foreign investment and imports. These steps lead to a relatively rapid economic recovery. South Korea also began expanding its network of free trade agreements to help bolster exports, and has since implemented 16 free trade agreements covering 58 countries— including the United State and China—that collectively cover more than three-quarters of global GDP.

In 2017, the election of President MOON Jae-in brought a surge in consumer confidence, in part, because of his successful efforts to increase wages and government spending. These factors combined with an uptick in export growth to drive real GDP growth to more than 3%, despite disruptions in South Korea's trade with China over the deployment of a US missile defense system in South Korea.

In 2018 and beyond, South Korea will contend with gradually slowing economic growth—in the 2-3% range—not uncommon for advanced economies. This could be partially offset by efforts to address challenges arising from its rapidly aging population, inflexible labor market, continued dominance of the chaebols, and heavy reliance on exports rather than domestic consumption. Socioeconomic problems also persist, and include rising inequality, poverty among the elderly, high youth unemployment, long working hours, low worker productivity, and corruption.

GDP (purchasing power parity): $2.035 trillion (2017 est.)
$1.974 trillion (2016 est.)
$1.918 trillion (2015 est.)
note: data are in 2017 dollars
country comparison to the world: 14

GDP (official exchange rate): $1.54 trillion (2017 est.)

GDP—real growth rate: 2.04% (2019 est.)
2.91% (2018 est.)
3.16% (2017 est.)
country comparison to the world: 134

GDP—per capita (PPP): $39,500 (2017 est.)
$38,500 (2016 est.)
$37,600 (2015 est.)
note: data are in 2017 dollars
country comparison to the world: 46

Gross national saving: 36.6% of GDP (2017 est.)
36.3% of GDP (2016 est.)
36.6% of GDP (2015 est.)
country comparison to the world: 15

GDP—composition, by end use: *household consumption:* 48.1% (2017 est.)
government consumption: 15.3% (2017 est.)
investment in fixed capital: 31.1% (2017 est.)
investment in inventories: 0% (2017 est.)
exports of goods and services: 43.1% (2017 est.)
imports of goods and services: -37.7% (2017 est.)

GDP—composition, by sector of origin: *agriculture:* 2.2% (2017 est.)
industry: 39.3% (2017 est.)
services: 58.3% (2017 est.)

Agriculture—products: rice, root crops, barley, vegetables, fruit, cattle, pigs, chickens, milk, eggs, fish
Industries: electronics, telecommunications, automobile production, chemicals, shipbuilding, steel

Industrial production growth rate: 4.6% (2017 est.)
country comparison to the world: 63

Labor force: 26.839 million (2020 est.)
country comparison to the world: 19

Labor force—by occupation: *agriculture:* 4.8%
industry: 24.6%
services: 70.6% (2017 est.)

Unemployment rate: 3.76% (2019 est.)
3.85% (2018 est.)
country comparison to the world: 55

Population below poverty line: 14.4% (2016 est.)

Household income or consumption by percentage share: *lowest 10%:* 6.8%
highest 10%: 48.5% (2015 est.)

Budget: *revenues:* 357.1 billion (2017 est.)
expenditures: 335.8 billion (2017 est.)

Taxes and other revenues: 23.2% (of GDP) (2017 est.)
country comparison to the world: 129

Budget surplus (+) or deficit (-): 1.4% (of GDP) (2017 est.)
country comparison to the world: 24

Public debt: 39.5% of GDP (2017 est.)
39.9% of GDP (2016 est.)
country comparison to the world: 133

Fiscal year: calendar year

Inflation rate (consumer prices): 1.9% (2017 est.)
1% (2016 est.)
country comparison to the world: 97

Current account balance: $59.971 billion (2019 est.)

$77.467 billion (2018 est.)
country comparison to the world: 9

Exports: $577.4 billion (2017 est.)
$512 billion (2016 est.)
country comparison to the world: 5

Exports—partners: China 25.1%, US 12.2%, Vietnam 8.2%, Hong Kong 6.9%, Japan 4.7% (2017)

Exports—commodities: semiconductors, petrochemicals, automobile/auto parts, ships, wireless communication equipment, flat displays, steel, electronics, plastics, computers

Imports: $457.5 billion (2017 est.)
$393.1 billion (2016 est.)
country comparison to the world: 9

Imports—commodities: crude oil/petroleum products, semiconductors, natural gas, coal, steel, computers, wireless communication equipment, automobiles, fine chemicals, textiles

Imports—partners: China 20.5%, Japan 11.5%, US 10.5%, Germany 4.2%, Saudi Arabia 4.1% (2017)

Reserves of foreign exchange and gold: $389.2 billion (31 December 2017 est.)
$371.1 billion (31 December 2016 est.)
country comparison to the world: 9

Debt—external: $384.6 billion (31 December 2017 est.)
$384.1 billion (31 December 2016 est.)
country comparison to the world: 29

Exchange rates: South Korean won (KRW) per US dollar -
1,130.48 (2017 est.)
1,160.41 (2016 est.)
1,160.77 (2015 est.)
1,130.95 (2014 est.)
1,052.96 (2013 est.)

ENERGY

Electricity access: *electrification—total population:* 100% (2020)

Electricity—production: 526 billion kWh (2016 est.)
country comparison to the world: 10

Electricity—consumption: 507.6 billion kWh (2016 est.)
country comparison to the world: 9

Electricity—exports: 0 kWh (2016 est.)
country comparison to the world: 155

Electricity—imports: 0 kWh (2016 est.)
country comparison to the world: 166

Electricity—installed generating capacity: 111. 2 million kW (2016 est.)
country comparison to the world: 11

Electricity—from fossil fuels: 70% of total installed capacity (2016 est.)
country comparison to the world: 109

Electricity—from nuclear fuels: 21% of total installed capacity (2017 est.)
country comparison to the world: 7

Electricity—from hydroelectric plants: 2% of total installed capacity (2017 est.)
country comparison to the world: 139

Electricity—from other renewable sources: 8% of total installed capacity (2017 est.)
country comparison to the world: 86

Crude oil—production: 0 bbl/day (2018 est.)
country comparison to the world: 157

Crude oil—exports: 0 bbl/day (2017 est.)
country comparison to the world: 148

Crude oil—imports: 3.057 million bbl/day (2017 est.)
country comparison to the world: 5

Crude oil—proved reserves: NA (1 January 2017 est.)

Refined petroleum products—production: 3.302 million bbl/day (2017 est.)
country comparison to the world: 6

Refined petroleum products—consumption: 2.584 million bbl/day (2017 est.)
country comparison to the world: 8

Refined petroleum products—exports: 1.396 million bbl/day (2017 est.)
country comparison to the world: 6

Refined petroleum products—imports: 908,800 bbl/day (2017 est.)
country comparison to the world: 6

Natural gas—production: 339.8 million cu m (2017 est.)
country comparison to the world: 76

Natural gas—consumption: 45.28 billion cu m (2017 est.)
country comparison to the world: 18

Natural gas—exports: 0 cu m (2017 est.)
country comparison to the world: 131

Natural gas—imports: 48.65 billion cu m (2017 est.)
country comparison to the world: 9

Natural gas—proved reserves: 7.079 billion cu m (1 January 2018 est.)
country comparison to the world: 82

Carbon dioxide emissions from consumption of energy: 778.4 million Mt (2017 est.)
country comparison to the world: 7

COMMUNICATIONS

Telephones—fixed lines: *total subscriptions:* 24,924,607
subscriptions per 100 inhabitants: 48.27 (2019 est.)
country comparison to the world: 10

Telephones—mobile cellular: *total subscriptions:* 69,445,005
subscriptions per 100 inhabitants: 134.49 (2019 est.)
country comparison to the world: 23

Telecommunication systems: *general assessment:* excellent domestic and international services featuring rapid incorporation of new technologies; ranked 2nd out of 34 Asian telecom companies; exceedingly high mobile and mobile broadband

penetration and very high fixed broadband penetration; highest number of broadband per capita; strong support from govt., savvy population has catapulted the nation into one of the world's most active telecommunication markets; 5G services live for enterprise customers in 2019, all 3 mobile operators offer 5G networks; slower growth predicted over the next five years to 2023 due to saturation and maturity of market; Chinese telecommunications company Huawei has partnered with other MNOs in South Korea (2020)

domestic: fixed-line 48 per 100 and mobile-cellular services 135 per 100 persons; rapid assimilation of a full range of telecommunications technologies leading to a boom in e-commerce (2019)

international: country code—82; landing points for EAC-C2C, FEA, SeaMeWe-3, TPE, APCN-2, APG, FLAG North Asia Loop/REACH North Asia Loop, KJCN, NCP, and SJC2 submarine cables providing links throughout Asia, Australia, the Middle East, Africa, Europe, Southeast Asia and US; satellite earth stations—66 (2019)

note: the COVID-19 outbreak is negatively impacting telecommunications production and supply chains globally; consumer spending on telecom devices and services has also slowed due to the pandemic's effect on economies worldwide; overall progress towards improvements in all facets of the telecom industry—mobile, fixed-line, broadband, submarine cable and satellite—has moderated

Broadcast media: multiple national TV networks with 2 of the 3 largest networks publicly operated; the largest privately owned network, Seoul Broadcasting Service (SBS), has ties with other commercial TV networks; cable and satellite TV subscription services available; publicly operated radio broadcast networks and many privately owned radio broadcasting networks, each with multiple affiliates, and independent local stations

Internet country code: .kr

Internet users: *total:* 49,309,955
percent of population: 95.9% (July 2018 est.)
country comparison to the world: 17

Broadband—fixed subscriptions: *total:* 21,285,858
subscriptions per 100 inhabitants: 41 (2018 est.)
country comparison to the world: 9

TRANSPORTATION

National air transport system: *number of registered air carriers:* 14 (2020)
inventory of registered aircraft operated by air carriers: 424
annual passenger traffic on registered air carriers: 88,157,579 (2018)
annual freight traffic on registered air carriers: 11,929,560,000 mt-km (2018)

Civil aircraft registration country code prefix: HL (2016)

Airports: 111 (2013)
country comparison to the world: 52

Airports—with paved runways: *total:* 71 (2017)
over 3,047 m: 4 (2017)

531

2,438 to 3,047 m: 19 (2017)
1,524 to 2,437 m: 12 (2017)
914 to 1,523 m: 13 (2017)
under 914 m: 23 (2017)

Airports—with unpaved runways: *total:* 40 (2013)
914 to 1,523 m: 2 (2013)
under 914 m: 38 (2013)

Heliports: 466 (2013)

Pipelines: 3790 km gas, 16 km oil, 889 km refined products (2017)

Railways: *total:* 3,979 km (2016)
standard gauge: 3,979 km 1.435-m gauge (2,727 km electrified) (2016)
country comparison to the world: 49

Roadways: *total:* 100,428 km (2016)
paved: 92,795 km (includes 4,193 km of expressways) (2016)
unpaved: 7,633 km (2016)
country comparison to the world: 47

Waterways: 1,600 km (most navigable only by small craft) (2011)
country comparison to the world: 49

Merchant marine: *total:* 1,880
by type: bulk carrier 83, container ship 86, general cargo 368, oil tanker 187, other 1,156 (2019)
country comparison to the world: 12

Ports and terminals: *major seaport(s):* Busan, Incheon, Gunsan, Kwangyang, Mokpo, Pohang, Ulsan, Yeosu

container port(s) (TEUs): Busan (20,493,000), Incheon (3,050,000), Kwangyang (2,230,000) (2017)

LNG terminal(s) (import): Incheon, Kwangyang, Pyeongtaek, Samcheok, Tongyeong, Yeosu

MILITARY AND SECURITY

Military and security forces: *Armed Forces of the Republic of Korea:* Republic of Korea Army (ROKA), Navy (ROKN, includes Marine Corps, ROKMC), Air Force (ROKAF); Military reserves include Mobilization Reserve Forces (First Combat Forces) and Homeland Defense Forces (Regional Combat Forces); Ministry of Maritime Affairs and Fisheries: Korea Coast Guard (2019)

Military expenditures: 2.7% of GDP (2019)
2.6% of GDP (2018)
2.4% of GDP (2017)
2.5% of GDP (2016)
2.5% of GDP (2015)
country comparison to the world: 33

Military and security service personnel strengths: the Republic of Korea Armed Forces have approximately 600,000 active duty personnel (465,000 Army; 70,000 Navy/Marines; 65,000 Air Force) (2019)

Military equipment inventories and acquisitions: the Republic of Korea Armed Forces are equipped with a mix of domestically-produced and imported weapons systems; domestic production includes armored fighting vehicles, artillery, aircraft, and

naval ships; the top foreign weapons supplier is the US and some domestically-produced systems are built under US license; Germany is the second largest supplier of armaments since 2010 (2019 est.)

Military deployments: 280 Lebanon (UNIFIL); 270 South Sudan (UNMISS); 170 United Arab Emirates; note—since 2009, the ROK has kept a naval flotilla with approximately 300 personnel in the waters off of the Horn of Africa and the Arabian Peninsula (2020)

Military service age and obligation: 18-28 years of age for compulsory military service; minimum conscript service obligation varies by service- 21 months (Army, Marines), 23 months (Navy), 24 months (Air Force); 18-26 years of age for voluntary military service; women, in service since 1950, are able to serve in all branches (2019)
note: South Korea intends to reduce the length of military service to 18 – 22 months by 2022

TRANSNATIONAL ISSUES

Disputes—international: Military Demarcation Line within the 4-km-wide Demilitarized Zone has separated North from South Korea since 1953; periodic incidents with North Korea in the Yellow Sea over the Northern Limit Line, which South Korea claims as a maritime boundary; South Korea and Japan claim Liancourt Rocks (Tok-do/Takeshima), occupied by South Korea since 1954

Refugees and internally displaced persons: *stateless persons:* 197 (2019)

KOSOVO

INTRODUCTION

Background: The central Balkans were part of the Roman and Byzantine Empires before ethnic Serbs migrated to the territories of modern Kosovo in the 7th century. During the medieval period, Kosovo became the center of a Serbian Empire and

saw the construction of many important Serb religious sites, including many architecturally significant Serbian Orthodox monasteries. The defeat of Serbian forces at the Battle of Kosovo in 1389 led to five centuries of Ottoman rule during which large numbers of Turks and Albanians moved to Kosovo. By the end of the 19th century, Albanians replaced Serbs as the dominant ethnic group in Kosovo. Serbia reacquired control over the region from the Ottoman Empire during the First Balkan War of 1912. After World War II, Kosovo's present-day boundaries were established when Kosovo became an autonomous province of Serbia in the Socialist Federal Republic of Yugoslavia (S.F.R.Y.). Despite legislative concessions, Albanian nationalism increased in the 1980s, which led to riots and calls for Kosovo's independence. The Serbs—many of whom viewed Kosovo as their cultural heartland—instituted a new constitution in 1989 revoking Kosovo's autonomous status. Kosovo's Albanian leaders responded in 1991 by organizing a referendum declaring Kosovo independent. Serbia undertook repressive measures against the Kosovar Albanians in the 1990s, provoking a Kosovar Albanian insurgency.

Beginning in 1998, Serbia conducted a brutal counterinsurgency campaign that resulted in massacres and massive expulsions of ethnic Albanians (some 800,000 ethnic Albanians were forced from their homes in Kosovo). After international attempts to mediate the conflict failed, a three-month NATO military operation against Serbia beginning in March 1999 forced the Serbs to agree to withdraw their military and police forces from Kosovo. UN Security Council Resolution 1244 (1999) placed Kosovo under a transitional administration, the UN Interim Administration Mission in Kosovo (UNMIK), pending a determination of Kosovo's future status. A UN-led process began in late 2005 to determine Kosovo's final status. The 2006-07 negotiations ended without agreement between Belgrade and Pristina, though the UN issued a comprehensive report on Kosovo's final status that endorsed independence. On 17 February 2008, the Kosovo Assembly declared Kosovo independent. Since then, over 100 countries have recognized Kosovo, and it has joined numerous international organizations. In October 2008, Serbia sought an advisory opinion from the International Court of Justice (ICJ) on the legality under international law of Kosovo's declaration of independence. The ICJ released the advisory opinion in July 2010 affirming that Kosovo's declaration of independence did not violate general principles of international law, UN Security Council Resolution 1244, or the Constitutive Framework. The opinion was closely tailored to Kosovo's unique history and circumstances.

Demonstrating Kosovo's development into a sovereign, multi-ethnic, democratic country the international community ended the period of Supervised Independence in 2012. Kosovo held its most recent national and municipal elections in 2017. Serbia continues to reject Kosovo's independence, but the two countries agreed in April 2013 to normalize their relations through EU-facilitated talks, which produced several subsequent agreements the parties are engaged in implementing, though they have not yet reached a comprehensive normalization of relations. Kosovo seeks full integration into the international community, and has pursued bilateral recognitions and memberships in international organizations. Kosovo signed a Stabilization and Association Agreement with the EU in 2015, and was named by a 2018 EU report as one of six Western Balkan countries that will be able to join the organization once it meets the criteria to accede. Kosovo also seeks memberships in the UN and in NATO.

GEOGRAPHY

Location: Southeast Europe, between Serbia and Macedonia

Geographic coordinates: 42 35 N, 21 00 E

Map references: Europe

Area: *total:* 10,887 sq km
land: 10,887 sq km
water: 0 sq km
country comparison to the world: 168

Area—comparative: slightly larger than Delaware

Land boundaries:
total: 714 km
border countries (4): Albania 112 km, Macedonia 160 km, Montenegro 76 km, Serbia 366 km

Coastline: 0 km (landlocked)

Maritime claims: none (landlocked)

Climate: influenced by continental air masses resulting in relatively cold winters with heavy snowfall and hot, dry summers and autumns; Mediterranean and alpine influences create regional variation; maximum rainfall between October and December

Terrain: flat fluvial basin at an elevation of 400-700 m above sea level surrounded by several high mountain ranges with elevations of 2,000 to 2,500 m

Elevation: *mean elevation:* 450 m
lowest point: Drini i Bardhe/Beli Drim (located on the border with Albania) 297 m
highest point: Gjeravica/Deravica 2,656 m

Natural resources: nickel, lead, zinc, magnesium, lignite, kaolin, chrome, bauxite

Land use: *agricultural land:* 52.8% (2001 est.)

arable land: 27.4% (2001 est.) / *permanent crops:* 1.9% (2001 est.) / *permanent pasture:* 23.5% (2001 est.)
forest: 41.7% (2001 est.)
other: 5.5% (2001 est.)

Irrigated land: NA

Population distribution: population clusters exist throughout the country, the largest being in the east in and around the capital of Pristina

Environment—current issues: air pollution (pollution from power plants and nearby lignite mines take a toll on people's health); water scarcity and pollution; land degradation

Geography—note: *the 41-km long Nerodimka River divides into two branches each of which flows into a different sea:* the northern branch flows into the Sitnica River, which via the Ibar, Morava, and Danube Rivers ultimately flows into the Black Sea; the southern branch flows via the Lepenac and Vardar Rivers into the Aegean Sea

PEOPLE AND SOCIETY

Population: 1,932,774 (July 2020 est.)
country comparison to the world: 150

Nationality: *noun:* Kosovar (Albanian)
adjective: Kosovo
note: Kosovo, a neutral term, is sometimes also used as a noun or adjective as in Kosovo Albanian, Kosovo Serb, Kosovo minority, or Kosovo citizen

Ethnic groups: Albanians 92.9%, Bosniaks 1.6%, Serbs 1.5%, Turk 1.1%, Ashkali 0.9%, Egyptian 0.7%, Gorani 0.6%, Romani 0.5%, other/unspecified 0.2% (2011 est.)
note: these estimates may under-represent Serb, Romani, and some other ethnic minorities because they are based on the 2011 Kosovo national census, which excluded northern Kosovo (a largely Serb-inhabited region) and was partially boycotted by Serb and Romani communities in southern Kosovo

Languages: Albanian (official) 94.5%, Bosnian 1.7%, Serbian (official) 1.6%, Turkish 1.1%, other 0.9% (includes Romani), unspecified 0.1% (2011 est.)
note: in municipalities where a community's mother tongue is not one of Kosovo's official languages, the language of that community may be given official status according to the 2006 Law on the Use of Languages

Religions: Muslim 95.6%, Roman Catholic 2.2%, Orthodox 1.5%, other 0.1%, none 0.1%, unspecified 0.6% (2011 est.)

Age structure: *0-14 years:* 24.07% (male 241,563/female 223,568)
15-24 years: 16.95% (male 170,566/female 157,063)
25-54 years: 42.56% (male 433,914/female 388,595)
55-64 years: 8.67% (male 85,840/female 81,782)
65 years and over: 7.75% (male 63,943/female 85,940) (2020 est.)

Median age:
total: 30.5 years

male: 30.2 years
female: 30.8 years (2020 est.)
country comparison to the world: 120

Population growth rate: 0.66% (2020 est.)
country comparison to the world: 144

Birth rate: 15.4 births/1,000 population (2020 est.)
country comparison to the world: 117

Death rate: 7 deaths/1,000 population (2020 est.)
country comparison to the world: 126

Net migration rate: -1.8 migrant(s)/1,000 population (2020 est.)
country comparison to the world: 162

Population distribution: population clusters exist throughout the country, the largest being in the east in and around the capital of Pristina

Major urban areas—population: 214,688 PRISTINA (capital) (2018)

Sex ratio: *at birth:* 1.08 male(s)/female
0-14 years: 1.08 male(s)/female
15-24 years: 1.09 male(s)/female
25-54 years: 1.12 male(s)/female
55-64 years: 1.05 male(s)/female
65 years and over: 0.74 male(s)/female
total population: 1.06 male(s)/female (2020 est.)

Infant mortality rate: *total:* 30.2 deaths/1,000 live births
male: 31.3 deaths/1,000 live births
female: 28.9 deaths/1,000 live births (2020 est.)
country comparison to the world: 56

Life expectancy at birth: *total population:* 72.7 years
male: 70.5 years
female: 75.1 years (2020 est.)
country comparison to the world: 151

Total fertility rate: 1.95 children born/woman (2020 est.)
country comparison to the world: 119

HIV/AIDS—adult prevalence rate: NA

HIV/AIDS—people living with HIV/AIDS: NA

HIV/AIDS—deaths: NA

Education expenditures: NA

Unemployment, youth ages 15-24:
total: 55.4%
male: 51.5%
female: 64.8% (2018 est.)
country comparison to the world: 2

GOVERNMENT

Country name: *conventional long form:* Republic of Kosovo
conventional short form: Kosovo
local long form: Republika e Kosoves (Republika Kosovo)
local short form: Kosove (Kosovo)
etymology: name derives from the Serbian "kos" meaning "blackbird," an ellipsis (linguistic omission) for "kosove polje" or "field of the blackbirds"

Government type: parliamentary republic

Capital: *name:* Pristina (Prishtine, Prishtina)
geographic coordinates: 42 40 N, 21 10 E

533

time difference: UTC+1 (6 hours ahead of Washington, DC, during Standard Time)

daylight saving time: +1hr, begins last Sunday in March; ends last Sunday in October

etymology: the name may derive from a Proto-Slavic word reconstructed as "pryshchina," meaning "spring (of water)"

Administrative divisions: 38 municipalities (komunat, singular—komuna (Albanian); opstine, singular—opstina (Serbian)); Decan (Decani), Dragash (Dragas), Ferizaj (Urosevac), Fushe Kosove (Kosovo Polje), Gjakove (Dakovica), Gjilan (Gnjilane), Gllogovc (Glogovac), Gracanice (Gracanica), Hani i Elezit (Deneral Jankovic), Istog (Istok), Junik, Kacanik, Kamenice (Kamenica), Kline (Klina), Kllokot (Klokot), Leposaviq (Leposavic), Lipjan (Lipljan), Malisheve (Malisevo), Mamushe (Mamusa), Mitrovice e Jugut (Juzna Mitrovica) [South Mitrovica], Mitrovice e Veriut (Severna Mitrovica) [North Mitrovica], Novoberde (Novo Brdo), Obiliq (Obilic), Partesh (Partes), Peje (Pec), Podujeve (Podujevo), Prishtine (Pristina), Prizren, Rahovec (Orahovac), Ranillug (Ranilug), Shterpce (Strpce), Shtime (Stimlje), Skenderaj (Srbica), Suhareke (Suva Reka), Viti (Vitina), Vushtrri (Vucitrn), Zubin Potok, Zvecan

Independence: 17 February 2008 (from Serbia)

National holiday: Independence Day, 17 February (2008)

Constitution: *history:* previous 1974, 1990; latest (postindependence) draft finalized 2 April 2008, signed 7 April 2008, ratified 9 April 2008, entered into force 15 June 2008; note—amendment 24, passed by the Assembly in August 2015, established the Kosovo Relocated Specialist Institution, referred to as the Kosovo Specialist Chamber or "Specialist Court," to try war crimes allegedly committed by members of the Kosovo Liberation Army in the late 1990s

amendments: proposed by the government, by the president of the republic, or by one fourth of Assembly deputies; passage requires two-thirds majority vote of the Assembly, including two-thirds majority vote of deputies representing non-majority communities, followed by a favorable Constitutional Court assessment; amended several times, last in 2016

Legal system: civil law system; note—the European Union Rule of Law Mission (EULEX) retained limited executive powers within the Kosovo judiciary for complex cases from 2008 to 2018

International law organization participation: has not submitted an ICJ jurisdiction declaration; non-party state to the ICCt

Citizenship: *citizenship by birth:* no

citizenship by descent only: at least one parent must be a citizen of Kosovo

dual citizenship recognized: yes

residency requirement for naturalization: 5 years

Suffrage: 18 years of age; universal

Executive branch: *chief of state:* Acting President Vjosa OSMANI (since 5 November 2020); note: President Hashim THACI (since 7 April 2016) resigned 5 November 2020

head of government: Prime Minister Avdullah HOTI (since 3 June 2020)

cabinet: Cabinet elected by the Assembly

elections/appointments: president indirectly elected by at least two-thirds majority vote of the Assembly for a 5-year term; if a candidate does not attain a two-thirds threshold in the first two ballots, the candidate winning a simple majority vote in the third ballot is elected (eligible for a second term); election last held on 26 February 2016 (next to be held in 2021); prime minister indirectly elected by the Assembly

election results: Hashim THACI elected president in the third ballot; Assembly vote—Hashim THACI (PDK) 71, Rafet RAMA (PDK) 0, invalid 10; Avdullah HOTI (LDK) elected prime minister; Assembly vote—61 of 85

Legislative branch: *description:* unicameral Assembly or Kuvendi i Kosoves/Skupstina Kosova (120 seats; 100 members directly elected by open-list proportional representation vote with 20 seats reserved for ethnic minorities—10 for Serbs and 10 for other ethnic minorities; members serve 4-year terms)

elections: last held on 6 October 2019 (next to be held in 2023); note—early elections were held on 6 October 2019 following the dissolution of parliament on 22 August 2019, as a result of political deadlock since the resignation of Prime Minister HARADINAJ on 19 July 2019

election results: percent of vote by party/coalition—VV 25.5%, LDK 24.8%, PDK 21.2%, AAK-PSD 11.6%, Serb List 6.6%, other 10.3%; seats by party/coalition—VV 31, LDK 30, PDK 25, AAK-PSD 14, Serb List 10, Vakat 2, KDTP 2, other 6; composition—men NA, women NA, percent of women NA%

Judicial branch: *highest courts:* Supreme Court (consists of the court president and 18 judges and organized into Appeals Panel of the Kosovo Property Agency and Special Chamber); Constitutional Court (consists of the court president, vice president, and 7 judges)

judge selection and term of office: Supreme Court judges nominated by the Kosovo Judicial Council, a 13-member independent body staffed by judges and lay members, and also responsible for overall administration of Kosovo's judicial system; judges appointed by the president of the Republic of Kosovo; judges appointed until mandatory retirement age; Constitutional Court judges nominated by the Kosovo Assembly and appointed by the president of the republic to serve single, 9-year terms

subordinate courts: Court of Appeals (organized into 4 departments: General, Serious Crime, Commercial Matters, and Administrative Matters); Basic Court (located in 7 municipalities, each with several branches)

note: in August 2015, the Kosovo Assembly approved a constitutional amendment that establishes the Kosovo Relocated Specialist

Judicial Institution, also referred to as the Kosovo Specialist Chambers or "Special Court"; the court, located at the Hague in the Netherlands, began operating in late 2016 and has jurisdiction to try crimes against humanity, war crimes, and other crimes under Kosovo law that occurred in the 1998-2000 period

Political parties and leaders: Alliance for the Future of Kosovo or AAK [Ramush HARADINAJ]

Alternativa [Mimoza KUSARI-LILA]

Democratic League of Kosovo or LDK [Isa MUSTAFA]

Democratic Party of Kosovo or PDK [Kadri VESELI]

Independent Liberal Party or SLS [Slobodan PETROVIC]

Initiative for Kosovo or NISMA [Fatmir LIMAJ]

Movement for Self-Determination (Vetevendosje) or VV [Albin KURTI]

New Kosovo Alliance or AKR [Behgjet PACOLLI]

Serb List [Goran RAKIC]

Social Democratic Party of Kosovo or PSD [Shpend AHMETI]

Turkish Democratic Party of Kosovo or KDTP [Mahir YAGCILAR]

Vakat Coalition or VAKAT [Rasim DEMIRI]

International organization participation: IBRD, IDA, IFC, IMF, ITUC (NGOs), MIGA, OIF (observer)

Diplomatic representation in the US: *chief of mission:* Ambassador Vlora CITAKU (since 17 September 2015)

chancery: 2175 K Street NW, Suite 300, Washington, DC 20037

telephone: [1] (202) 450-2130

FAX: [1] (202) 735-0609

consulate(s) general: New York

consulate(s): Des Moines (IA)

Diplomatic representation from the US: *chief of mission:* Ambassador Philip KOSNETT (since 3 December 2018)

telephone: [383] 38 59 59 3000

embassy: Arberia/Dragodan, Nazim Hikmet 30, Pristina

mailing address: use embassy street address

FAX: [383] 38 549 890

Flag description: *centered on a dark blue field is a gold-colored silhouette of Kosovo surmounted by six white, five-pointed stars arrayed in a slight arc; each star represents one of the major ethnic groups of Kosovo:* Albanians, Serbs, Turks, Gorani, Roma, and Bosniaks

note: one of only two national flags that uses a map as a design element; the flag of Cyprus is the other

National symbol(s): *six, five-pointed, white stars; national colors:* blue, gold, white

National anthem:

name: Europe

lyrics/music: no lyrics/Mendi MENGJIQI

note: adopted 2008; Kosovo chose to exclude lyrics in its anthem so as not to offend the country's minority ethnic groups

ECONOMY

Economy—overview: Kosovo's economy has shown progress in transitioning to a market-based system and maintaining macroeconomic stability, but it is still highly dependent on the international community and the diaspora for financial and technical assistance. Remittances from the diaspora—located mainly in Germany, Switzerland, and the Nordic countries—are estimated to account for about 17% of GDP and international donor assistance accounts for approximately 10% of GDP. With international assistance, Kosovo has been able to privatize a majority of its state-owned enterprises.

Kosovo's citizens are the second poorest in Europe, after Moldova, with a per capita GDP (PPP) of $10,400 in 2017. An unemployment rate of 33%, and a youth unemployment rate near 60%, in a country where the average age is 26, encourages emigration and fuels a significant informal, unreported economy. Most of Kosovo's population lives in rural towns outside of the capital, Pristina. Inefficient, near-subsistence farming is common—the result of small plots, limited mechanization, and a lack of technical expertise. Kosovo enjoys lower labor costs than the rest of the region. However, high levels of corruption, little contract enforcement, and unreliable electricity supply have discouraged potential investors. The official currency of Kosovo is the euro, but the Serbian dinar is also used illegally in Serb majority communities. Kosovo's tie to the euro has helped keep core inflation low.

Minerals and metals production—including lignite, lead, zinc, nickel, chrome, aluminum, magnesium, and a wide variety of construction materials—once the backbone of industry, has declined because of aging equipment and insufficient investment, problems exacerbated by competing and unresolved ownership claims of Kosovo's largest mines. A limited and unreliable electricity supply is a major impediment to economic development. The US Government is cooperating with the Ministry of Economic Development (MED) and the World Bank to conclude a commercial tender for the construction of Kosovo C, a new lignite-fired power plant that would leverage Kosovo's large lignite reserves. MED also has plans for the rehabilitation of an older bituminous-fired power plant, Kosovo B, and the development of a coal mine that could supply both plants.

In June 2009, Kosovo joined the World Bank and International Monetary Fund, the Central Europe Free Trade Area (CEFTA) in 2006, the European Bank for Reconstruction and Development in 2012, and the Council of Europe Development Bank in 2013. In 2016, Kosovo implemented the Stabilization and Association Agreement (SAA) negotiations with the EU, focused on trade liberalization. In 2014, nearly 60% of customs duty-eligible imports into Kosovo were EU goods. In August 2015, as part of its EU-facilitated normalization process with Serbia, Kosovo signed agreements on telecommunications and energy distribution, but disagreements over who owns economic assets, such as the Trepca mining conglomerate, within Kosovo continue.

Kosovo experienced its first federal budget deficit in 2012, when government expenditures climbed sharply. In May 2014, the government introduced a 25% salary increase for public sector employees and an equal increase in certain social benefits. Central revenues could not sustain these increases, and the government was forced to reduce its planned capital investments. The government, led by Prime Minister MUSTAFA—a trained economist—recently made several changes to its fiscal policy, expanding the list of duty-free imports, decreasing the Value Added Tax (VAT) for basic food items and public utilities, and increasing the VAT for all other goods.

While Kosovo's economy continued to make progress, unemployment has not been reduced, nor living standards raised, due to lack of economic reforms and investment.

GDP (purchasing power parity): $19.6 billion (2017 est.)
$18.89 billion (2016 est.)
$18.16 billion (2015 est.)
note: data are in 2017 dollars
country comparison to the world: 151

GDP (official exchange rate): $7.094 billion (2017 est.)

GDP—real growth rate: 3.7% (2017 est.)
4.1% (2016 est.)
4.1% (2015 est.)
country comparison to the world: 79

GDP—per capita (PPP): $10,900 (2017 est.)
$10,600 (2016 est.)
$10,200 (2015 est.)
note: data are in 2016 US dollars
country comparison to the world: 137

Gross national saving: 17.3% of GDP (2017 est.)
13.2% of GDP (2016 est.)
15.1% of GDP (2015 est.)
country comparison to the world: 118

GDP—composition, by end use:
household consumption: 84.3% (2017 est.)
government consumption: 13.6% (2017 est.)
investment in fixed capital: 29% (2017 est.)
investment in inventories: 0% (2016 est.)
exports of goods and services: 27% (2017 est.)
imports of goods and services: -53.8% (2017 est.)

GDP—composition, by sector of origin:
agriculture: 11.9% (2017 est.)
industry: 17.7% (2017 est.)
services: 70.4% (2017 est.)

Agriculture—products: wheat, corn, berries, potatoes, peppers, fruit; dairy, livestock; fish
Industries: mineral mining, construction materials, base metals, leather, machinery, appliances, foodstuffs and beverages, textiles

Industrial production growth rate: 1.2% (2016 est.)
country comparison to the world: 149

Labor force: 500,300 (2017 est.)
note: includes those estimated to be employed in the gray economy
country comparison to the world: 155

Labor force—by occupation: *agriculture:* 4.4%
industry: 17.4%
services: 78.2% (2017 est.)

Unemployment rate: 30.5% (2017 est.)
27.5% (2016 est.)
note: Kosovo has a large informal sector that may not be reflected in these data
country comparison to the world: 207

Population below poverty line: 17.6% (2015 est.)

Household income or consumption by percentage share: *lowest 10%:* 3.8%
highest 10%: 22% (2015 est.)

Budget: *revenues:* 2.054 billion (2017 est.)
expenditures: 2.203 billion (2017 est.)

Taxes and other revenues: 29% (of GDP) (2017 est.)
country comparison to the world: 87

Budget surplus (+) or deficit (-): -2.1% (of GDP) (2017 est.)
country comparison to the world: 108

Public debt: 21.2% of GDP (2017 est.)
19.4% of GDP (2016 est.)
country comparison to the world: 186

Inflation rate (consumer prices): 1.5% (2017 est.)
0.3% (2016 est.)
country comparison to the world: 82

Current account balance: -$467 million (2017 est.)
-$533 million (2016 est.)
country comparison to the world: 120

Exports: $428 million (2017 est.)
$340 million (2016 est.)
country comparison to the world: 179

Exports—partners: Albania 16%, India 14%, North Macedonia 12.1%, Serbia 10.6%, Switzerland 5.6%, Germany 5.4% (2017)

Exports—commodities: mining and processed metal products, scrap metals, leather products, machinery, appliances, prepared foodstuffs, beverages and tobacco, vegetable products, textiles and apparel

Imports: $3.223 billion (2017 est.)
$2.876 billion (2016 est.)
country comparison to the world: 145

Imports—commodities: foodstuffs, livestock, wood, petroleum, chemicals, machinery, minerals, textiles, stone, ceramic and glass products, electrical equipment

Imports—partners: Germany 12.4%, Serbia 12.3%, Turkey 9.6%, China 9.1%, Italy 6.4%, North Macedonia 5.1%, Albania 5%, Greece 4.4% (2017)

Reserves of foreign exchange and gold: $683.9 million (31 December 2016 est.)
$708.7 million (31 December 2015 est.)
country comparison to the world: 142

Debt—external: $506 million (31 December 2017 est.)
$448 million (31 December 2016 est.)
country comparison to the world: 178

Exchange rates: euros (EUR) per US dollar—
0.885 (2017 est.)

535

0.903 (2016 est.)
0.9214 (2015 est.)
0.885 (2014 est.)
0.7634 (2013 est.)

ENERGY

Electricity access: *electrification—total population:* 100% (2020)

Electricity—production: 5.638 billion kWh (2016 est.)
country comparison to the world: 117

Electricity—consumption: 3.957 billion kWh (2016 est.)
country comparison to the world: 127

Electricity—exports: 885.7 million kWh (2017 est.)
country comparison to the world: 60

Electricity—imports: 557 million kWh (2016 est.)
country comparison to the world: 78

Electricity—installed generating capacity: 1.573 million kW (2016 est.)
country comparison to the world: 121

Electricity—from fossil fuels: 97% of total installed capacity (2016 est.)
country comparison to the world: 35

Electricity—from nuclear fuels: 0% of total installed capacity (2017 est.)
country comparison to the world: 121

Electricity—from hydroelectric plants: 3% of total installed capacity (2017 est.)
country comparison to the world: 135

Electricity—from other renewable sources: 1% of total installed capacity (2017 est.)
country comparison to the world: 156

Crude oil—production: 0 bbl/day (2017 est.)
country comparison to the world: 158

Crude oil—exports: 0 bbl/day (2015 est.)
country comparison to the world: 149

Crude oil—imports: 0 bbl/day (2015 est.)
country comparison to the world: 147

Crude oil—proved reserves: 0 bbl NA (2017 est.)
country comparison to the world: 153

Refined petroleum products—production: 0 bbl/day (2015 est.)
country comparison to the world: 161

Refined petroleum products—consumption: 14,000 bbl/day (2016 est.)
country comparison to the world: 155

Refined petroleum products—exports: 192 bbl/day (2015 est.)
country comparison to the world: 117

Refined petroleum products—imports: 14,040 bbl/day (2015 est.)
country comparison to the world: 139

Natural gas—production: 0 cu m (2017 est.)
country comparison to the world: 153

Natural gas—consumption: 0 cu m (2017 est.)
country comparison to the world: 164

Natural gas—exports: 0 cu m (2017 est.)
country comparison to the world: 132

Natural gas—imports: 0 cu m (2017 est.)
country comparison to the world: 145

Natural gas—proved reserves: 0 cu m NA (2017 est.)
country comparison to the world: 154

Carbon dioxide emissions from consumption of energy: 10.05 million Mt (2017 est.)
country comparison to the world: 106

COMMUNICATIONS

Telephones—fixed lines:
total subscriptions: 117,317
subscriptions per 100 inhabitants: 6.11 (2019 est.)
country comparison to the world: 136

Telephones—mobile cellular:
total subscriptions: 620,186
subscriptions per 100 inhabitants: 32.3 (2019 est.)
country comparison to the world: 168

Telecommunication systems: *general assessment:* Kosovo being part of the EU pre-accession process has helped with their progress in the telecom industry, following a regulatory framework, European standards, and a market of new players encourages development in its telecommunications; 2 MNOs dominate the sector; poor telecom infrastructure means low fixed-line penetration; little expansion of fiber networks for broadband; expansion of LTE services (2020)
domestic: fixed-line stands at 6 per 100 and mobile-cellular 32 per 100 persons (2019)
international: country code—383
note: the COVID-19 outbreak is negatively impacting telecommunications production and supply chains globally; consumer spending on telecom devices and services has also slowed due to the pandemic's effect on economies worldwide; overall progress towards improvements in all facets of the telecom industry—mobile, fixed-line, broadband, submarine cable and satellite—has moderated

Internet country code: .xk
note: assigned as a temporary code under UN Security Council resolution 1244/99

Internet users:
total: 1,706,150
percent of population: 89.44% (July 2018 est.)
country comparison to the world: 125

TRANSPORTATION

National air transport system: *number of registered air carriers:* 0 (2020)

Civil aircraft registration country code prefix: Z6 (2016)

Airports: 6 (2013)
country comparison to the world: 173

Airports—with paved runways:
total: 3 (2019)
2,438 to 3,047 m: 1
1,524 to 2,437 m: 1
under 914 m: 1

Airports—with unpaved runways: *total:* 3 (2013)
under 914 m: 3 (2013)

Heliports: 2 (2013)

Railways: *total:* 333 km (2015)
standard gauge: 333 km 1.435-m gauge (2015)
country comparison to the world: 120

Roadways:
total: 2,012 km (2015)
paved: 1,921 km (includes 78 km of expressways) (2015)
unpaved: 91 km (2015)
country comparison to the world: 172

MILITARY AND SECURITY

Military and security forces: *Kosovo Security Force (KSF):* Land Force Command; National Guard Command; Logistic Command and Doctrine and Training Command (2020)

Military expenditures: 0.8% of GDP (2019)
0.8% of GDP (2018)
0.8% of GDP (2017)
0.8% of GDP (2016)
0.8% of GDP (2015)
country comparison to the world: 130

Military and security service personnel strengths: *the Kosovo Security Force (KSF) has approximately 3-4,000 personnel; note:* Kosovo plans for the KSF to eventually number around 5,000 troops (2019)

Military equipment inventories and acquisitions: the Kosovo Security Force is equipped with small arms and light vehicles only; its only recorded delivery since 2010 was light-armored patrol vehicles from Turkey (2019 est.)

TRANSNATIONAL ISSUES

Disputes—international: Serbia with several other states protest the US and other states' recognition of Kosovo's declaration of its status as a sovereign and independent state in February 2008; ethnic Serbian municipalities along Kosovo's northern border challenge final status of Kosovo-Serbia boundary; NATO-led Kosovo Force peacekeepers under UN Interim Administration Mission in Kosovo authority continue to ensure a safe and secure environment and freedom of movement for all Kosovo citizens; Kosovo and North Macedonia completed demarcation of their boundary in September 2008; Kosovo ratified the border demarcation agreement with Montenegro in March 2018, but the actual demarcation has not been completed

Refugees and internally displaced persons: *IDPs:* 16,000 (primarily ethnic Serbs displaced during the 1998-1999 war fearing reprisals from the majority ethnic- Albanian population; a smaller number of ethnic Serbs, Roma, Ashkali, and Egyptians fled their homes in 2,004 as a result of violence) (2019)
note: 5,639 estimated refugee and migrant arrivals (January 2015-October 2020)

KUWAIT

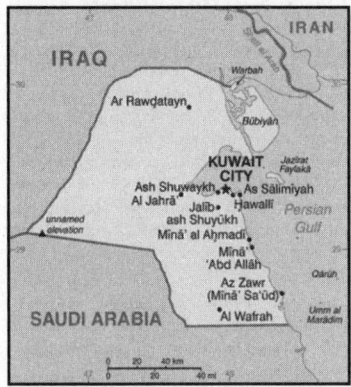

IRAN
IRAQ
Warbah
Ar Rawdatayn
Bübiyän
KUWAIT CITY   Jazirat Faylakä
Ash Shuwaykh
Al Jahrä   As Sälimiyah
Jalib   Hawalli
ash Shuyükh   Persian
Minä' al Ahmadi   Gulf
unnamed
elevation   Minä'
'Abd Alläh
Az Zawr
(Minä' Sa'üd)   Qärüh
SAUDI ARABIA   'Al Wafrah   Umm al
Marädim
0   20   40 km
0   20   40 mi

INTRODUCTION

Background: Kuwait has been ruled by the AL-SABAH dynasty since the 18th century. The threat of Ottoman invasion in 1899 prompted Amir Mubarak AL-SABAH to seek protection from Britain, ceding foreign and defense responsibility to Britain until 1961, when the country attained its independence. Kuwait was attacked and overrun by Iraq in August 1990. Following several weeks of aerial bombardment, a US-led UN coalition began a ground assault in February 1991 that liberated Kuwait in four days. In 1992, the Amir reconstituted the parliament that he had dissolved in 1986. Amid the 2010-11 uprisings and protests across the Arab world, stateless Arabs, known as Bidoon, staged small protests in early 2011 demanding citizenship, jobs, and other benefits available to Kuwaiti nationals. Other demographic groups, notably Islamists and Kuwaitis from tribal backgrounds, soon joined the growing protest movements, which culminated in late 2011 with the resignation of the prime minister amidst allegations of corruption. Demonstrations renewed in late 2012 in response to an amiri decree amending the electoral law that lessened the voting power of the tribal blocs.

An opposition coalition of Sunni Islamists, tribal populists, and some liberals, largely boycotted legislative elections in 2012 and 2013, which ushered in a legislature more amenable to the government's agenda. Faced with the prospect of painful subsidy cuts, oppositionists and independents actively participated in the November 2016 election, winning nearly half of the seats but a cohesive opposition alliance largely ceased to exist with the 2016 election and the opposition became increasingly factionalized. Since coming to power in 2006, the Amir has dissolved the National Assembly on seven occasions (the Constitutional Court annulled the Assembly elections in June 2012 and again in June 2013) and shuffled the cabinet over a dozen times, usually citing political stagnation and gridlock between the legislature and the government.

GEOGRAPHY

Location: Middle East, bordering the Persian Gulf, between Iraq and Saudi Arabia

Geographic coordinates: 29 30 N, 45 45 E

Map references: Middle East

Area: *total:* 17,818 sq km
land: 17,818 sq km
water: 0 sq km
country comparison to the world: 158

Area—comparative: slightly smaller than New Jersey

Land boundaries:
total: 475 km
border countries (2): Iraq 254 km, Saudi Arabia 221 km

Coastline: 499 km

Maritime claims: *territorial sea:* 12 nm

Climate: dry desert; intensely hot summers; short, cool winters

Terrain: flat to slightly undulating desert plain

Elevation: *mean elevation:* 108 m
lowest point: Persian Gulf 0 m
highest point: 3.6 km W. of Al-Salmi Border Post 300 m

Natural resources: petroleum, fish, shrimp, natural gas

Land use: *agricultural land:* 8.5% (2011 est.)
arable land: 0.6% (2011 est.) / *permanent crops:* 0.3% (2011 est.) / *permanent pasture:* 7.6% (2011 est.)
forest: 0.4% (2011 est.)
other: 91.1% (2011 est.)

Irrigated land: 105 sq km (2012)

Population distribution: densest settlement is along the Persian Gulf, particularly in Kuwait City and on Bubiyan Island; significant population threads extend south and west along highways that radiate from the capital, particularly in the southern half of the country

Natural hazards: sudden cloudbursts are common from October to April and bring heavy rain, which can damage roads and houses; sandstorms and dust storms occur throughout the year but are most common between March and August

Environment—current issues: limited natural freshwater resources; some of world's largest and most sophisticated desalination facilities provide much of the water; air and water pollution; desertification; loss of biodiversity

Environment—international agreements: *party to:* Biodiversity, Climate Change, Climate Change-Kyoto Protocol, Desertification, Endangered Species, Environmental Modification, Hazardous Wastes, Law of the Sea, Ozone Layer Protection

signed, but not ratified: Marine Dumping

Geography—note: strategic location at head of Persian Gulf

PEOPLE AND SOCIETY

Population: 2,993,706 (July 2020 est.)
note: Kuwait's Public Authority for Civil Information estimates the country's total population to be 4,420,110 for 2019, with non-Kuwaitis accounting for nearly 70% of the population
country comparison to the world: 138

Nationality: *noun:* Kuwaiti(s)
adjective: Kuwaiti

Ethnic groups: Kuwaiti 30.4%, other Arab 27.4%, Asian 40.3%, African 1%, other .9% (includes European, North American, South American, and Australian) (2018 est.)

Languages: Arabic (official), English widely spoken

Religions: Muslim (official) 74.6%, Christian 18.2%, other and unspecified 7.2% (2013 est.)
note: data represent the total population; about 69% of the population consists of immigrants

MENA religious affiliation: *Age structure:* 0-14 years: 24.29% (male 378,778/female 348,512)
15-24 years: 14.96% (male 245,354/female 202,642)
25-54 years: 52.39% (male 984,813/female 583,632)
55-64 years: 5.43% (male 90,583/female 72,026)
65 years and over: 2.92% (male 38,614/female 48,752) (2020 est.)

Dependency ratios: *total dependency ratio:* 32.4
youth dependency ratio: 28.4
elderly dependency ratio: 4
potential support ratio: 24.9 (2020 est.)

Median age: *total:* 29.7 years
male: 30.7 years
female: 27.9 years (2020 est.)
country comparison to the world: 127

Population growth rate: 1.27% (2020 est.)
country comparison to the world: 86

Birth rate: 18 births/1,000 population (2020 est.)
country comparison to the world: 90

Death rate: 2.3 deaths/1,000 population (2020 est.)
country comparison to the world: 227

Net migration rate: -3.3 migrant(s)/1,000 population (2020 est.)
country comparison to the world: 181

Population distribution: densest settlement is along the Persian Gulf, particularly in Kuwait City and on Bubiyan Island; significant population threads extend south and west along highways that radiate from the capital, particularly in the southern half of the country

Urbanization: *urban population:* 100% of total population (2020)

rate of urbanization: 1.78% annual rate of change (2015-20 est.)

total population growth rate v. urban population growth rate, 2000-2030: Major urban areas—population: 3.115 million KUWAIT (capital) (2020)

Sex ratio: *at birth:* 1.05 male(s)/female

0-14 years: 1.09 male(s)/female

15-24 years: 1.21 male(s)/female

25-54 years: 1.69 male(s)/female

55-64 years: 1.26 male(s)/female

65 years and over: 0.79 male(s)/female

total population: 1.38 male(s)/female (2020 est.)

Maternal mortality rate: 12 deaths/100,000 live births (2017 est.)

country comparison to the world: 140

Infant mortality rate: *total:* 6.5 deaths/1,000 live births

male: 6.4 deaths/1,000 live births

female: 6.7 deaths/1,000 live births (2020 est.)

country comparison to the world: 162

Life expectancy at birth: *total population:* 78.6 years

male: 77.2 years

female: 80.2 years (2020 est.)

country comparison to the world: 65

Total fertility rate: 2.26 children born/woman (2020 est.)

country comparison to the world: 86

Drinking water source:

improved: *total:* 100% of population

unimproved: *total:* 0% of population (2017 est.)

Current Health Expenditure: 5.3% (2017)

Physicians density: 2.65 physicians/1,000 population (2015)

Hospital bed density: 2 beds/1,000 population (2017)

Sanitation facility access:

improved: *total:* 100% of population

unimproved: *total:* 0% of population (2017 est.)

HIV/AIDS—adult prevalence rate: <.1% (2018 est.)

HIV/AIDS—people living with HIV/AIDS: <1000 (2018 est.)

HIV/AIDS—deaths: <100 (2018 est.)

Obesity—adult prevalence rate: 37.9% (2016)

country comparison to the world: 11

Children under the age of 5 years underweight: 3% (2014)

country comparison to the world: 101

Education expenditures: NA

Literacy: *definition:* age 15 and over can read and write

total population: 96.1%

male: 96.7%

female: 94.9% (2018)

School life expectancy (primary to tertiary education): *total:* 15 years

male: 14 years

female: 16 years (2015)

Unemployment, youth ages 15-24: *total:* 15.4%

male: 9.4% N/A

female: 30% N/A (2016 est.)

country comparison to the world: 88

GOVERNMENT

Country name: *conventional long form:* State of Kuwait

conventional short form: Kuwait

local long form: Dawlat al Kuwayt

local short form: Al Kuwayt

etymology: the name derives from the capital city, which is from Arabic "al-Kuwayt" a diminutive of "kut" meaning "fortress," possibly a reference to a small castle built on the current location of Kuwait City by the Beni Khaled tribe in the 17th century

Government type: constitutional monarchy (emirate)

Capital: *name:* Kuwait City

geographic coordinates: 29 22 N, 47 58 E

time difference: UTC+3 (8 hours ahead of Washington, DC, during Standard Time)

etymology: the name derives from Arabic "al-Kuwayt" a diminutive of "kut" meaning "fortress," possibly a reference to a small castle built on the current location of Kuwait City by the Beni Khaled tribe in the 17th century

Administrative divisions: 6 governorates (muhafazat, singular—muhafazah); Al Ahmadi, Al 'Asimah, Al Farwaniyah, Al Jahra', Hawalli, Mubarak al Kabir

Independence: 19 June 1961 (from the UK)

National holiday: National Day, 25 February (1950)

Constitution: *history:* approved and promulgated 11 November 1962

amendments: proposed by the amir or supported by at least one third of the National Assembly; passage requires twothirds consent of the Assembly membership and promulgation by the amir; constitutional articles on the initiation, approval, and promulgation of general legislation cannot be amended

Legal system: mixed legal system consisting of English common law, French civil law, and Islamic sharia law

International law organization participation: has not submitted an ICJ jurisdiction declaration; non-party state to the ICCt

Citizenship: *citizenship by birth:* no

citizenship by descent only: at least one parent must be a citizen of Kuwait

dual citizenship recognized: no

residency requirement for naturalization: not specified

Suffrage: 21 years of age and at least 20-year citizenship

Executive branch: *chief of state:* Amir NAWAF al-Ahmad al-Jabir al-Sabah (since 30 September 2020); Crown Prince Sheikh MESHAAL Al Ahmad Al Sabah, born in 1940, is the brother of Amir NAWAF al-Ahmad al-Jabir al-Sabah

head of government: Prime Minister JABIR AL-MUBARAK al-Hamad al-Sabah (since 30 November 2011); First Deputy Prime Minister NASIR Sabah al-Ahmad al-Sabah (since 11 December 2017); Deputy Prime Ministers SABAH KHALID al- Hamid al-Sabah (since 13 December 2011), KHALID al-Jarrah al-Sabah (since 4 August 2013), Anas Khalid al-SALEH (since 4 August 2013); note—on 14 November 2019, the government of Prime Minister JABIR AL-MUBARAK al-Hamad al-Sabah resigned

cabinet: Council of Ministers appointed by the prime minister, approved by the amir

elections/appointments: amir chosen from within the ruling family, confirmed by the National Assembly; prime minister and deputy prime ministers appointed by the amir; crown prince appointed by the amir and approved by the National Assembly

Legislative branch: *description:* unicameral National Assembly or Majlis al-Umma (65 seats; 50 members directly elected from 5 multi-seat constituencies by simple majority vote and 15 ex-officio members (cabinet ministers) appointed by the amir; members serve 4-year terms)

elections: last held on 26 November 2016 (next to be held on 5 December 2020)

election results: seats won—oppositionists and independents, including populists, Islamists, and liberals 26, progovernment loyalists 24; composition for elected members only—men 49, women 1, percent of women 1.5%

note— seats as of May 2019—oppositionists and independents, including populists, Islamists, and liberals 25, progovernment loyalists 25; composition as of May 2019 for elected members only—men 49, women 1, percent of women 2%

Judicial branch: *highest courts:* Constitutional Court (consists of 5 judges); Supreme Court or Court of Cassation (organized into several circuits, each with 5 judges)

judge selection and term of office: all Kuwaiti judges appointed by the Amir upon recommendation of the Supreme Judicial Council, a consultative body comprised of Kuwaiti judges and Ministry of Justice officials

subordinate courts: High Court of Appeal; Court of First Instance; Summary Court

Political parties and leaders: none; the government does not recognize any political parties or allow their formation, although no formal law bans political parties

International organization participation: ABEDA, AfDB (nonregional member), AFESD, AMF, BDEAC, CAEU, CD, FAO, G-77, GCC, IAEA, IBRD, ICAO, ICC (national committees), ICRM, IDA, IDB, IFAD, IFC, IFRCS, IHO, ILO, IMF, IMO, IMSO, Interpol, IOC, IPU, ISO, ITSO, ITU, ITUC (NGOs), LAS, MIGA, NAM, OAPEC, OIC, OPCW, OPEC, Paris Club (associate), PCA, UN, UNCTAD, UNESCO, UNIDO, UNRWA, UN Security Council (temporary), UNWTO, UPU, WCO, WFTU (NGOs), WHO, WIPO, WMO, WTO

Diplomatic representation in the US: *chief of mission:* Ambassador SALIM al-Abdallah al-Jabir al-Sabah (since 10 October 2001)

chancery: 2940 Tilden Street NW, Washington, DC 20008
telephone: [1] (202) 966-0702
FAX: [1] (202) 966-8468
consulate(s) general: New York City
consulate(s): Lost Angeles

Diplomatic representation from the US: *chief of mission:* Alina L. Romanowski (since 6 January 2020)
telephone: [965] 2259-1001
embassy: P.O. Box 77, Safat 13001
mailing address: P. O. Box 77 Safat 13001 Kuwait; or PSC 1280 APO AE 09880-9000
FAX: [965] 2538-6562

Flag description: three equal horizontal bands of green (top), white, and red with a black trapezoid based on the hoist side; colors and design are based on the Arab Revolt flag of World War I; green represents fertile fields, white stands for purity, red denotes blood on Kuwaiti swords, black signifies the defeat of the enemy

National symbol(s): *golden falcon; national colors:* green, white, red, black

National anthem: *name:* "Al-Nasheed Al-Watani" (National Anthem)
lyrics/music: Ahmad MUSHARI al-Adwani/ Ibrahim Nasir al-SOULA
note: adopted 1978; the anthem is only used on formal occasions

ECONOMY

Economy—overview: Kuwait has a geographically small, but wealthy, relatively open economy with crude oil reserves of about 102 billion barrels—more than 6% of world reserves. Kuwaiti officials plan to increase production to 4 million barrels of oil equivalent per day by 2020. Petroleum accounts for over half of GDP, 92% of export revenues, and 90% of government income.

With world oil prices declining, Kuwait realized a budget deficit in 2015 for the first time more than a decade; in 2016, the deficit grew to 16.5% of GDP. Kuwaiti authorities announced cuts to fuel subsidies in August 2016, provoking outrage among the public and National Assembly, and the Amir dissolved the government for the seventh time in ten years. In 2017 the deficit was reduced to 7.2% of GDP, and the government raised $8 billion by issuing international bonds. Despite Kuwait's dependence on oil, the government has cushioned itself against the impact of lower oil prices, by saving annually at least 10% of government revenue in the Fund for Future Generations.

Kuwait has failed to diversify its economy or bolster the private sector, because of a poor business climate, a large public sector that employs about 74% of citizens, and an acrimonious relationship between the National Assembly and the executive branch that has stymied most economic reforms. The Kuwaiti Government has made little progress on its longterm economic development plan first passed in 2010. While the government planned to spend up to $104 billion over four years to diversify the economy, attract more

investment, and boost private sector participation in the economy, many of the projects did not materialize because of an uncertain political situation or delays in awarding contracts. To increase non-oil revenues, the Kuwaiti Government in August 2017 approved draft bills supporting a Gulf Cooperation Council-wide value added tax scheduled to take effect in 2018.

GDP (purchasing power parity): $289.7 billion (2017 est.)
$299.7 billion (2016 est.)
$293.2 billion (2015 est.)
note: data are in 2017 dollars
country comparison to the world: 58

GDP (official exchange rate): $120.7 billion (2017 est.)

GDP—real growth rate: -3.3% (2017 est.)
2.2% (2016 est.)
-1% (2015 est.)
country comparison to the world: 214

GDP—per capita (PPP): $65,800 (2017 est.)
$69,900 (2016 est.)
$69,200 (2015 est.)
note: data are in 2017 dollars
country comparison to the world: 15

Gross national saving: 35.4% of GDP (2017 est.)
32.9% of GDP (2016 est.)
37.1% of GDP (2015 est.)
country comparison to the world: 16

GDP—composition, by end use: *household consumption:* 43.1% (2017 est.)
government consumption: 24.5% (2017 est.)
investment in fixed capital: 26.5% (2017 est.)
investment in inventories: 3.5% (2017 est.)
exports of goods and services: 49.4% (2017 est.)
imports of goods and services: -47% (2017 est.)

GDP—composition, by sector of origin: *agriculture:* 0.4% (2017 est.)
industry: 58.7% (2017 est.)
services: 40.9% (2017 est.)

Agriculture—products: fish
Industries: petroleum, petrochemicals, cement, shipbuilding and repair, water desalination, food processing, construction materials

Industrial production growth rate: 2.8% (2017 est.)
country comparison to the world: 109

Labor force: 2.695 million (2017 est.)
note: non-Kuwaitis represent about 60% of the labor force
country comparison to the world: 111

Labor force—by occupation: *agriculture:* NA
industry: NA
services: NA

Unemployment rate: 1.1% (2017 est.)
1.1% (2016 est.)
country comparison to the world: 12

Population below poverty line: NA

Household income or consumption by percentage share: *lowest 10%:* NA
highest 10%: NA

Budget: *revenues:* 50.5 billion (2017 est.)
expenditures: 62.6 billion (2017 est.)

Taxes and other revenues: 41.8% (of GDP) (2017 est.)
country comparison to the world: 32

Budget surplus (+) or deficit (-): -10% (of GDP) (2017 est.)
country comparison to the world: 210

Public debt: 20.6% of GDP (2017 est.)
9.9% of GDP (2016 est.)
country comparison to the world: 188

Fiscal year: 1 April—31 March

Inflation rate (consumer prices): 1.5% (2017 est.)
3.5% (2016 est.)
country comparison to the world: 83

Current account balance: $7.127 billion (2017 est.)
-$5.056 billion (2016 est.)
country comparison to the world: 28

Exports: $55.17 billion (2017 est.)
$46.26 billion (2016 est.)
country comparison to the world: 50

Exports—partners: South Korea 18.3%, China 17.4%, Japan 11.5%, India 11.2%, Singapore 6.3%, US 5.7% (2017)

Exports—commodities: oil and refined products, fertilizers

Imports: $29.53 billion (2017 est.)
$26.56 billion (2016 est.)
country comparison to the world: 69

Imports—commodities: food, construction materials, vehicles and parts, clothing

Imports—partners: China 13.5%, US 13.3%, UAE 9.5%, Saudi Arabia 5.8%, Germany 5.4%, Japan 5%, India 4.7%, Italy 4.5% (2017)

Reserves of foreign exchange and gold: $33.7 billion (31 December 2017 est.)
$31.13 billion (31 December 2016 est.)
country comparison to the world: 48

Debt—external: $47.24 billion (31 December 2017 est.)
$38.34 billion (31 December 2016 est.)
country comparison to the world: 68

Exchange rates: Kuwaiti dinars (KD) per US dollar -
0.3041 (2017 est.)
0.3022 (2016 est.)
0.3022 (2015 est.)
0.3009 (2014 est.)
0.2845 (2013 est.)

ENERGY

Electricity access: *electrification—total population:* 100% (2020)

Electricity—production: 65.95 billion kWh (2016 est.)
country comparison to the world: 44

Electricity—consumption: 57.78 billion kWh (2016 est.)
country comparison to the world: 44

Electricity—exports: 0 kWh (2016 est.)
country comparison to the world: 156

Electricity—imports: 0 kWh (2016 est.)

country comparison to the world: 167

Electricity—installed generating capacity: 18.89 million kW (2016 est.)
country comparison to the world: 47

Electricity—from fossil fuels: 100% of total installed capacity (2016 est.)
country comparison to the world: 10

Electricity—from nuclear fuels: 0% of total installed capacity (2017 est.)
country comparison to the world: 122

Electricity—from hydroelectric plants: 0% of total installed capacity (2017 est.)
country comparison to the world: 182

Electricity—from other renewable sources: 0% of total installed capacity (2017 est.)
country comparison to the world: 196

Crude oil—production: 2.807 million bbl/day (2018 est.)
country comparison to the world: 9

Crude oil—exports: 479,700 bbl/day (2015 est.)
country comparison to the world: 21

Crude oil—imports: 0 bbl/day (2015 est.)
country comparison to the world: 148

Crude oil—proved reserves: 101.5 billion bbl (1 January 2018 est.)
country comparison to the world: 6

Refined petroleum products—production: 915,800 bbl/day (2015 est.)
country comparison to the world: 22

Refined petroleum products—consumption: 446,000 bbl/day (2016 est.)
country comparison to the world: 34

Refined petroleum products—exports: 705,500 bbl/day (2015 est.)
country comparison to the world: 11

Refined petroleum products—imports: 0 bbl/day (2015 est.)
country comparison to the world: 212

Natural gas—production: 17.1 billion cu m (2017 est.)
country comparison to the world: 34

Natural gas—consumption: 21.72 billion cu m (2017 est.)
country comparison to the world: 36

Natural gas—exports: 0 cu m (2017 est.)
country comparison to the world: 133

Natural gas—imports: 5.125 billion cu m (2017 est.)
country comparison to the world: 35

Natural gas—proved reserves: 1.784 trillion cu m (1 January 2018 est.)
country comparison to the world: 19

Carbon dioxide emissions from consumption of energy: 106.5 million Mt (2017 est.)
country comparison to the world: 41

COMMUNICATIONS

Telephones—fixed lines: total subscriptions: 368,305
subscriptions per 100 inhabitants: 12.46 (2019 est.)

country comparison to the world: 106

Telephones—mobile cellular: *total subscriptions:* 5,147,990
subscriptions per 100 inhabitants: 174.16 (2019 est.)
country comparison to the world: 118

Telecommunication systems: *general assessment:* the quality of service is excellent; new telephone exchanges provide a large capacity for new subscribers; trunk traffic is carried by microwave radio relay, coaxial cable, and open-wire and fiber-optic cable; a 4G LTE mobile-cellular telephone system operates throughout Kuwait; Internet access is available via 4G LTE connections for fixed and mobile users; high ownership of smart phone in Kuwait; one of the highest mobile penetration rates in the world; exploring 5G opportunities; improvements to fiber-broadband underway (2020)
domestic: fixed-line subscriptions are 12 per 100 and mobile-cellular stands at 174 per 100 subscriptions (2019)
international: country code—965; landing points for the FOG, GBICS, MENA, Kuwait-Iran, and FALCON submarine cables linking Africa, the Middle East, and Asia; microwave radio relay to Saudi Arabia; satellite earth stations—6 (3 Intelsat—1 Atlantic Ocean and 2 Indian Ocean, 1 Inmarsat—Atlantic Ocean, and 2 Arabsat) (2019)
note: the COVID-19 outbreak is negatively impacting telecommunications production and supply chains globally; consumer spending on tele-com devices and services has also slowed due to the pandemic's effect on economies worldwide; overall progress towards improvements in all facets of the telecom industry—mobile, fixed line, broadband, submarine cable and satellite—has moderated

Broadcast media: state-owned TV broadcaster operates 4 networks and a satellite channel; several private TV broadcasters have emerged; satellite TV available and pan-Arab TV stations are especially popular; state-owned Radio Kuwait broadcasts on a number of channels in Arabic and English; first private radio station emerged in 2005; transmissions of at least 2 international radio broadcasters are available (2019)

Internet country code: .kw

Internet users: *total:* 2,904,801
percent of population: 99.6% (July 2018 est.)
country comparison to the world: 99

Broadband—fixed subscriptions: *total:* 103,821
subscriptions per 100 inhabitants: 4 (2018 est.)
country comparison to the world: 121

TRANSPORTATION

National air transport system: *number of registered air carriers:* 2 (2020)
inventory of registered aircraft operated by air carriers: 44
annual passenger traffic on registered air carriers: 6,464,847 (2018)
annual freight traffic on registered air carriers: 392.36 million mt-km (2018)

Civil aircraft registration country code prefix: 9K (2016)

Airports: 7 (2013)
country comparison to the world: 169

Airports—with paved runways: *total:* 4 (2019)
over 3,047 m: 1
2,438 to 3,047 m: 2
914 to 1,523 m: 1

Airports—with unpaved runways: *total:* 3 (2013)
1,524 to 2,437 m: 1 (2013)
under 914 m: 2 (2013)

Heliports: 4 (2013)

Pipelines: 261 km gas, 540 km oil, 57 km refined products (2013)

Roadways: *total:* 5,749 km (2018)
paved: 4,887 km (2018)
unpaved: 862 km (2018)
country comparison to the world: 145

Merchant marine: *total:* 154
by type: general cargo 15, oil tanker 24, other 115 (2019)
country comparison to the world: 71

Ports and terminals: *major seaport(s):* Ash Shu'aybah, Ash Shuwaykh, Az Zawr (Mina' Sa'ud), Mina' 'Abd Allah, Mina' al Ahmadi

MILITARY AND SECURITY

Military and security forces: *Kuwaiti Armed Forces:* Kuwaiti Land Forces (KLF), Kuwaiti Navy, Kuwaiti Air Force (Al-Quwwat al-Jawwiya al-Kuwaitiya; includes Kuwaiti Air Defense Force, KADF), 25th Commando Brigade, and the Kuwait Emiri Guard Brigade; Kuwaiti National Guard (KNG); Coast Guard (Ministry of Interior) (2019)
note: the Kuwait Emiri Guard Authority and the 25th Commando Brigade exercise independent command authority within the Kuwaiti Armed Forces, although activities such as training and equipment procurement are often coordinated with the other services; the KNG possesses an independent command structure, equipment inventory, and logistics corps separate from the Ministry of Defense, the regular armed services, and the Ministry of Interior

Military expenditures: 5.6% of GDP (2019)
5.1% of GDP (2018)
5.6% of GDP (2017)
5.8% of GDP (2016)
5% of GDP (2015)
country comparison to the world: 5

Military and security service personnel strengths: the Kuwaiti Armed Forces have approximately 17,000 active personnel (12,500 Army; 2,000 Navy; 2,500 Air Force); est. 6,500 National Guard; note – Army figures include the Kuwait Emiri Guard Authority (est. 500) and the 25th Commando Brigade (N/A) (2019 est.)

Military equipment inventories and acquisitions: the inventory of the Kuwaiti Armed Forces consists of a range of European- and US-sourced weapons systems; the US is the leading supplier of arms to Kuwait since 2010 (2019 est.)

Military service age and obligation: 17-21 years of age for voluntary military service; Kuwait reintroduced one-year mandatory service for men aged 18-35 in May 2017 after having suspended conscription in 2001; service is divided in two phases – four months for training and eight months for military service (2018)

TRANSNATIONAL ISSUES

Disputes—international: Kuwait and Saudi Arabia continue negotiating a joint maritime boundary with Iran; no maritime boundary exists with Iraq in the Persian Gulf

Refugees and internally displaced persons: *stateless persons:* 92,020 (2019); note—Kuwait's 1959 Nationality Law defined citizens as persons who settled in the country before 1920 and who had maintained normal residence since then; one-third of the population, descendants of Bedouin tribes, missed the window of opportunity to register for nationality rights after Kuwait became independent in 1961 and were classified as bidun (meaning "without"); since the 1980s Kuwait's bidun have progressively lost their rights, including opportunities for employment and education, amid official claims that they are nationals of other countries who have destroyed their identification documents in hopes of gaining Kuwaiti citizenship; Kuwaiti authorities have delayed processing citizenship applications and labeled biduns as "illegal residents," denying them access to civil documentation, such as birth and marriage certificates

Trafficking in persons: *current situation:* Kuwait is a destination country for men and women subjected to forced labor and, to a lesser degree, forced prostitution; men and women migrate from South and Southeast Asia, Egypt, the Middle East, and increasingly Africa to work in Kuwait, most of them in the domestic service, construction, and sanitation sectors; although most of these migrants enter Kuwait voluntarily, upon arrival some are subjected to conditions of forced labor by their sponsors and labor agents, including debt bondage; Kuwait's sponsorship law restricts workers' movements and penalizes them for running away from abusive workplaces, making domestic workers particularly vulnerable to forced labor in private homes

tier rating: Tier 3—Kuwait does not fully comply with the minimum standards for the elimination of trafficking and is not making sufficient efforts to do so; although investigations into visa fraud rings lead to the referral of hundreds of people for prosecution, including complicit officials, the government has not prosecuted or convicted any suspected traffickers; authorities made no effort to enforce the prohibition against withholding workers' passports, as mandated under Kuwaiti law; punishment of forced labor cases was limited to shutting down labor recruitment firms, assessing fines, and ordering the return of withheld passports and the paying of back-wages; the government made progress in victims' protection by opening a high-capacity shelter for runaway domestic workers but still lacks formal procedures to identify and refer victims to care services (2015)

KYRGYZSTAN

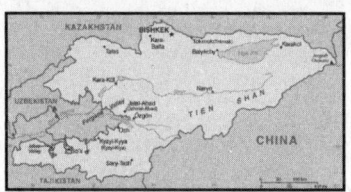

INTRODUCTION

Background: A Central Asian country of incredible natural beauty and proud nomadic traditions, most of the territory of the present-day Kyrgyz Republic was formally annexed to the Russian Empire in 1876. The Kyrgyz staged a major revolt against the Tsarist Empire in 1916 in which almost one-sixth of the Kyrgyz population was killed. The Kyrgyz Republic became a Soviet republic in 1936 and achieved independence in 1991 when the USSR dissolved. Nationwide demonstrations in 2005 and 2010 resulted in the ouster of the country's first two presidents, Askar AKAEV and Kurmanbek BAKIEV. Interim President Roza OTUNBAEVA led a transitional government and following a nation-wide election, President Almazbek ATAMBAEV was sworn in as president in 2011. In 2017, ATAMBAEV became the first Kyrgyzstani president to step down after serving one full six-year term as required in the country's constitution. Former prime minister and ruling Social- Democratic Party of Kyrgyzstan member Sooronbay JEENBEKOV replaced him after winning an October 2017 presidential election that was the most competitive in the country's history, although international and local election observers noted cases of vote buying and abuse of public resources. The president holds substantial powers as head of state even though the prime minister oversees the Kyrgyzstani Government and selects most cabinet members. The president represents the country internationally and can sign or veto laws, call for new elections, and nominate Supreme Court judges, cabinet members for posts related to security or defense, and numerous other high-level positions. Continuing concerns for the Kyrgyz Republic include the trajectory of democratization, endemic corruption, a history of tense, and at times violent, interethnic relations, border security vulnerabilities, and potential terrorist threats.

GEOGRAPHY

Location: Central Asia, west of China, south of Kazakhstan

Geographic coordinates: 41 00 N, 75 00 E

Map references: Asia

Area: *total:* 199,951 sq km
land: 191,801 sq km
water: 8,150 sq km
country comparison to the world: 88

Area—comparative: slightly smaller than South Dakota

Land boundaries: *total:* 4,573 km
border countries (4): China 1063 km, Kazakhstan 1212 km, Tajikistan 984 km, Uzbekistan 1314 km

Coastline: 0 km (landlocked)

Maritime claims: none (landlocked)

Climate: dry continental to polar in high Tien Shan Mountains; subtropical in southwest (Fergana Valley); temperate in northern foothill zone

Terrain: peaks of the Tien Shan mountain range and associated valleys and basins encompass the entire country

Elevation: *mean elevation:* 2,988 m
lowest point: Kara-Daryya (Karadar'ya) 132 m
highest point: Jengish Chokusu (Pik Pobedy) 7,439 m

Natural resources: abundant hydropower; gold; rare earth metals; locally exploitable coal, oil, and natural gas; other deposits of nepheline, mercury, bismuth, lead, and zinc

Land use: *agricultural land:* 55.4% (2011 est.)
arable land: 6.7% (2011 est.) / permanent crops: 0.4% (2011 est.) / permanent pasture: 48.3% (2011 est.)
forest: 5.1% (2011 est.)
other: 39.5% (2011 est.)

Irrigated land: 10,233 sq km (2012)

Population distribution: the vast majority of Kyrgyzstanis live in rural areas; densest population settlement is to the north in and around the capital, Bishkek, followed by Osh in the west; the least densely populated area is the east, southeast in the Tien Shan mountains

Natural hazards: major flooding during snow melt; prone to earthquakes

Environment—current issues: water pollution; many people get their water directly from contaminated streams and wells; as a result, water-borne diseases are prevalent; increasing soil salinity from faulty irrigation practices; air pollution due to rapid increase of traffic

Environment—international agreements: *party to:* Air Pollution, Biodiversity, Climate Change, Climate Change-Kyoto Protocol, Desertification, Hazardous Wastes, Ozone Layer Protection, Wetlands

signed, but not ratified: none of the selected agreements

Geography—note: landlocked; entirely mountainous, dominated by the Tien Shan range; 94% of the country is 1,000 m above sea level with an average elevation of 2,750 m; many tall peaks, glaciers, and high-altitude lakes

PEOPLE AND SOCIETY

Population: 5,964,897 (July 2020 est.)
country comparison to the world: 114

Nationality: *noun:* Kyrgyzstani(s)
adjective: Kyrgyzstani

Ethnic groups: Kyrgyz 73.5%, Uzbek 14.7%, Russian 5.5%, Dungan 1.1%, other 5.2% (includes Uyghur, Tajik, Turk, Kazakh, Tatar, Ukrainian, Korean, German) (2019 est.)

Languages: Kyrgyz (official) 71.4%, Uzbek 14.4%, Russian (official) 9%, other 5.2% (2009 est.)

Religions: Muslim 90% (majority Sunni), Christian 7% (Russian Orthodox 3%), other 3% (includes Jewish, Buddhist, Baha'i) (2017 est.)

Age structure: *0-14 years:* 30.39% (male 930,455/female 882,137)
15-24 years: 15.7% (male 475,915/female 460,604)
25-54 years: 40.02% (male 1,172,719/female 1,214,624)
55-64 years: 8.09% (male 210,994/female 271,480)
65 years and over: 5.8% (male 132,134/female 213,835) (2020 est.)

Dependency ratios: *total dependency ratio:* 59.7
youth dependency ratio: 52.1
elderly dependency ratio: 7.5
potential support ratio: 13.2 (2020 est.)

Median age: *total:* 27.3 years
male: 26.1 years
female: 28.5 years (2020 est.)
country comparison to the world: 147

Population growth rate: 0.96% (2020 est.)
country comparison to the world: 109

Birth rate: 20.6 births/1,000 population (2020 est.)
country comparison to the world: 74

Death rate: 6.3 deaths/1,000 population (2020 est.)
country comparison to the world: 151

Net migration rate: -5 migrant(s)/1,000 population (2020 est.)
country comparison to the world: 197

Population distribution: the vast majority of Kyrgyzstanis live in rural areas; densest population settlement is to the north in and around the capital, Bishkek, followed by Osh in the west; the least densely populated area is the east, southeast in the Tien Shan mountains

Urbanization: *urban population:* 36.9% of total population (2020)
rate of urbanization: 2.03% annual rate of change (2015-20 est.)

total population growth rate v. urban population growth rate, 2000-2030: Major urban areas - population: 1.038 million BISHKEK (capital) (2020)

Sex ratio: *at birth:* 1.07 male(s)/female
0-14 years: 1.05 male(s)/female
15-24 years: 1.03 male(s)/female
25-54 years: 0.97 male(s)/female
55-64 years: 0.78 male(s)/female
65 years and over: 0.62 male(s)/female
total population: 0.96 male(s)/female (2020 est.)

Mother's mean age at first birth: 22.9 years (2017 est.)

Maternal mortality rate: 60 deaths/100,000 live births (2017 est.)
country comparison to the world: 89

Infant mortality rate: *total:* 23.3 deaths/1,000 live births
male: 27.2 deaths/1,000 live births
female: 19.2 deaths/1,000 live births (2020 est.)
country comparison to the world: 69

Life expectancy at birth: *total population:* 71.8 years
male: 67.7 years
female: 76.2 years (2020 est.)
country comparison to the world: 159

Total fertility rate: 2.54 children born/woman (2020 est.)
country comparison to the world: 69

Contraceptive prevalence rate: 39.4% (2018)

Drinking water source:

improved: *urban:* 97.1% of population
rural: 84.4% of population
total: 89.3% of population

unimproved: *urban:* 2.9% of population
rural: 15.6% of population
total: 10.7% of population (2017 est.)

Current Health Expenditure: 6.2% (2017)

Physicians density: 2.21 physicians/1,000 population (2014)

Hospital bed density: 4.4 beds/1,000 population (2014)

Sanitation facility access:

improved: *urban:* 99.6% of population
rural: 100% of population
total: 99.3% of population

unimproved: *urban:* 0.4% of population
rural: 0% of population
total: 0.1% of population (2017 est.)

HIV/AIDS—adult prevalence rate: 0.2% (2019 est.)
country comparison to the world: 102

HIV/AIDS—people living with HIV/AIDS: 10,000 (2019 est.)
country comparison to the world: 103

HIV/AIDS—deaths: <500 (2019 est.)

Obesity—adult prevalence rate: 16.6% (2016)
country comparison to the world: 121

Children under the age of 5 years underweight: 1.8% (2018)
country comparison to the world: 115

Education expenditures: 7.2% of GDP (2017)

country comparison to the world: 12

Literacy: *definition:* age 15 and over can read and write
total population: 99.6%
male: 99.7%
female: 99.5% (2018)

School life expectancy (primary to tertiary education): *total:* 13 years
male: 13 years
female: 13 years (2019)

Unemployment, youth ages 15-24: *total:* 14.2%
male: 10.1%
female: 22.3% (2018 est.)
country comparison to the world: 96

GOVERNMENT

Country name: *conventional long form:* Kyrgyz Republic
conventional short form: Kyrgyzstan
local long form: Kyrgyz Respublikasy
local short form: Kyrgyzstan
former: Kirghiz Soviet Socialist Republic
etymology: a combination of the Turkic words "kyrg" (forty) and "-yz" (tribes) with the Persian suffix "-stan" (country) creating the meaning "Land of the Forty Tribes"; the name refers to the 40 clans united by the legendary Kyrgyz hero, MANAS

Government type: parliamentary republic

Capital: *name:* Bishkek
geographic coordinates: 42 52 N, 74 36 E
time difference: UTC+6 (11 hours ahead of Washington, DC, during Standard Time)
etymology: founded in 1868 as a Russian settlement on the site of a previously destroyed fortress named "Pishpek"; the name was retained and over-time became "Bishkek"

Administrative divisions: 7 provinces (oblustar, singular—oblus) and 2 cities* (shaarlar, singular - shaar); Batken Oblusu, Bishkek Shaary*, Chuy Oblusu (Bishkek), Jalal-Abad Oblusu, Naryn Oblusu, Osh Oblusu, Osh Shaary*, Talas Oblusu, Ysyk-Kol Oblusu (Karakol)
note: administrative divisions have the same names as their administrative centers (exceptions have the administrative center name following in parentheses)

Independence: 31 August 1991 (from the Soviet Union)

National holiday: Independence Day, 31 August (1991)

Constitution: *history:* previous 1993; latest adopted by referendum 27 June 2010, effective 2 July 2010; note—constitutional amendments that bolstered some presidential powers and transferred others from the president to the prime minister passed in a referendum in December 2016, effective December 2017
amendments: proposed as a draft law by the majority of the Supreme Council membership or by petition of 300,000 voters; passage requires at least two-thirds majority vote of the Council membership in each of at least three readings of the draft two months apart; the draft may be submitted

to a referendum if approved by two thirds of the Council membership; adoption requires the signature of the president; amended 2017

Legal system: civil law system, which includes features of French civil law and Russian Federation laws

International law organization participation: has not submitted an ICJ jurisdiction declaration; non-party state to the ICCt

Citizenship: *citizenship by birth:* no
citizenship by descent only: at least one parent must be a citizen of Kyrgyzstan
dual citizenship recognized: yes, but only if a mutual treaty on dual citizenship is in force
residency requirement for naturalization: 5 years

Suffrage: 18 years of age; universal

Executive branch: *chief of state:* Acting President Talant MAMYTOV (since 14 November 2020); President Sooronbay JEENBEKOV resigned on 16 October 2020 following massive protests brought on by disputed legislative election results of 4 October 2020
head of government: Acting Prime Minister Artem NOVIKOV (since 14 November 2020); note—Prime Minister Kubatbek BORONOV resigned on 9 October 2020 following massive protests brought on by disputed legislative election results of 4 October 2020
cabinet: Cabinet of Ministers proposed by the prime minister, appointed by the president upon approval by the Supreme Council; defense and security committee chairs appointed by the president
elections/appointments: president directly elected by absolute majority popular vote in 2 rounds if needed for a single 6-year term; election last held on 15 October 2017 (next to be held 10 January 2021); prime minister nominated by the majority party or majority coalition in the Supreme Council, appointed by the president upon approval by the Supreme Council
election results: Sooronbay JEENBEKOV elected president in first round; percent of vote - Sooronbay JEENBEKOV (SDPK) 54.2%, Omurbek BABANOV (Respublika) 33.5%, Adakhan MADUMAROV (Butun Kyrgyzstan) 6.6%, Temir SARIYEV (Akshumar) 2.5%, other 3.2%; note—JEENBEKOV resigned as president on 16 October 2020; BORONOV resigned as prime minister on 9 October 2020

Legislative branch: *description:* unicameral Supreme Council or Jogorku Kengesh (120 seats; parties directly elected in a single nationwide constituency by closed party-list proportional representation vote; members selected from party lists to serve 5-year terms)
elections: last held on 4 October 2020 (next to be held NA); note—the results of the 2020 election were annulled on 6 October 2020 following mass protests
election results: percent of vote by party—NA; seats by party—NA

Judicial branch: *highest courts:* Supreme Court (consists of 25 judges); Constitutional Chamber of the Supreme Court (consists of the chairperson, deputy chairperson, and 9 judges)
judge selection and term of office: Supreme Court and Constitutional Court judges appointed by the Supreme Council on the recommendation of the president; Supreme Court judges serve for 10 years, Constitutional Court judges serve for 15 years; mandatory retirement at age 70 for judges of both courts
subordinate courts: Higher Court of Arbitration; oblast (provincial) and city courts

Political parties and leaders: Ata-Meken (Fatherland) [Almambet SHYKMAMATOV]
Bir Bol (Stay United) [Altynbek SULAYMANOV]
Kyrgyzstan Party [Almazbek BAATYRBEKOV]
Onuguu-Progress (Development-Progress) [Bakyt TOROBAEV]
Respublika-Ata-Jurt (Republic-Homeland) [Jyrgalbek TURUSKULOV] (parliamentary faction)
Social-Democratic Party of Kyrgyzstan or SDPK [Almazbek ATAMBAEV, Isa OMURKULOV]

International organization participation: ADB, CICA, CIS, CSTO, EAEC, EAEU, EAPC, EBRD, ECO, EITI (compliant country), FAO, GCTU, IAEA, IBRD, ICAO, ICC (NGOs), ICRM, IDA, IDB, IFAD, IFC, IFRCS, ILO, IMF, Interpol, IOC, IOM, IPU, ISO (correspondent), ITSO, ITU, MIGA, NAM (observer), OIC, OPCW, OSCE, PCA, PFP, SCO, UN, UNAMID, UNCTAD, UNESCO, UNIDO, UNISFA, UNMIL, UNMISS, UNWTO, UPU, WCO, WFTU (NGOs), WHO, WIPO, WMO, WTO

Diplomatic representation in the US: *chief of mission:* Ambassador Bolot I. OTUNBAEV (since 8 April 2018)
chancery: 2360 Massachusetts Avenue NW, Washington, DC 20008
telephone: [1] (202) 449-9822
FAX: [1] (202) 449-8275
honorary consulate(s): Maple Valley (WA)

Diplomatic representation from the US: *chief of mission:* Ambassador Donald LU (since 18 September 2018)
telephone: [996] (312) 597-000
embassy: 171 Prospect Mira, Bishkek 720016
mailing address: use embassy street address
FAX: [996] (312) 597-744

Flag description: red field with a yellow sun in the center having 40 rays representing the 40 Kyrgyz tribes; on the obverse side the rays run counter-clockwise, on the reverse, clockwise; in the center of the sun is a red ring crossed by two sets of three lines, a stylized representation of a "tunduk"—the crown of a traditional Kyrgyz yurt; red symbolizes bravery and valor, the sun evinces peace and wealth

National symbol(s): *white falcon; national colors:* red, yellow

National anthem: *name:* "Kyrgyz Respublikasynyn Mamlekettik Gimni" (National Anthem of the Kyrgyz Republic)
lyrics/music: Djamil SADYKOV and Eshmambet KULUEV/Nasyr DAVLESOV and Kalyi MOLDOBASANOV

note: adopted 1992

ECONOMY

Economy—overview: Kyrgyzstan is a landlocked, mountainous, lower middle income country with an economy dominated by minerals extraction, agriculture, and reliance on remittances from citizens working abroad. Cotton, wool, and meat are the main agricultural products, although only cotton is exported in any quantity. Other exports include gold, mercury, uranium, natural gas, and—in some years—electricity. The country has sought to attract foreign investment to expand its export base, including construction of hydroelectric dams, but a difficult investment climate and an ongoing legal battle with a Canadian firm over the joint ownership structure of the nation's largest gold mine deter potential investors. Remittances from Kyrgyz migrant workers, predominantly in Russia and Kazakhstan, are equivalent to more than one-quarter of Kyrgyzstan's GDP.

Following independence, Kyrgyzstan rapidly implemented market reforms, such as improving the regulatory system and instituting land reform. In 1998, Kyrgyzstan was the first Commonwealth of Independent States country to be accepted into the World Trade Organization. The government has privatized much of its ownership shares in public enterprises. Despite these reforms, the country suffered a severe drop in production in the early 1990s and has again faced slow growth in recent years as the global financial crisis and declining oil prices have dampened economies across Central Asia. The Kyrgyz government remains dependent on foreign donor support to finance its annual budget deficit of approximately 3 to 5% of GDP.

Kyrgyz leaders hope the country's August 2015 accession to the Eurasian Economic Union (EAEU) will bolster trade and investment, but slowing economies in Russia and China and low commodity prices continue to hamper economic growth. Large-scale trade and investment pledged by Kyrgyz leaders has been slow to develop. Many Kyrgyz entrepreneurs and politicians complain that non-tariff measures imposed by other EAEU member states are hurting certain sectors of the Kyrgyz economy, such as meat and dairy production, in which they have comparative advantage. Since acceding to the EAEU, the Kyrgyz Republic has continued harmonizing its laws and regulations to meet EAEU standards, though many local entrepreneurs believe this process as disjointed and incomplete. Kyrgyzstan's economic development continues to be hampered by corruption, lack of administrative transparency, lack of diversity in domestic industries, and difficulty attracting foreign aid and investment.

GDP (purchasing power parity): $23.15 billion (2017 est.)
$22.14 billion (2016 est.)
$21.22 billion (2015 est.)
note: data are in 2017 dollars
country comparison to the world: 144

GDP (official exchange rate): $7.565 billion (2017 est.)

GDP—real growth rate: 4.6% (2017 est.)
4.3% (2016 est.)
3.9% (2015 est.)
country comparison to the world: 59

GDP—per capita (PPP): $3,700 (2017 est.)
$3,600 (2016 est.)
$3,500 (2015 est.)
note: data are in 2017 dollars
country comparison to the world: 183

Gross national saving: 27.3% of GDP (2017 est.)
20.1% of GDP (2016 est.)
18.3% of GDP (2015 est.)
country comparison to the world: 42

GDP—composition, by end use: *household consumption:* 85.4% (2017 est.)
government consumption: 18.9% (2017 est.)
investment in fixed capital: 33.2% (2017 est.)
investment in inventories: 1.8% (2017 est.)
exports of goods and services: 39.7% (2017 est.)
imports of goods and services: -79% (2017 est.)

GDP—composition, by sector of origin:
agriculture: 14.6% (2017 est.)
industry: 31.2% (2017 est.)
services: 54.2% (2017 est.)

Agriculture—products: cotton, potatoes, vegetables, grapes, fruits and berries; sheep, goats, cattle, wool
Industries: small machinery, textiles, food processing, cement, shoes, lumber, refrigerators, furniture, electric motors, gold, rare earth metals

Industrial production growth rate: 10.9% (2017 est.)
country comparison to the world: 10

Labor force: 2,841 million (2017 est.)
country comparison to the world: 106

Labor force—by occupation: *agriculture:* 48%
industry: 12.5%
services: 39.5% (2005 est.)

Unemployment rate: 3.18% (2019 est.)
2.59% (2018 est.)
country comparison to the world: 43

Population below poverty line: 32.1% (2015 est.)

Household income or consumption by percentage share: *lowest 10%:* 4.4%
highest 10%: 22.9% (2014 est.)

Budget: *revenues:* 2.169 billion (2017 est.)
expenditures: 2.409 billion (2017 est.)

Taxes and other revenues: 28.7% (of GDP) (2017 est.)
country comparison to the world: 92

Budget surplus (+) or deficit (-): -3.2% (of GDP) (2017 est.)
country comparison to the world: 140

Public debt: 56% of GDP (2017 est.)
55.9% of GDP (2016 est.)
country comparison to the world: 79

Fiscal year: calendar year

Inflation rate (consumer prices): 3.2% (2017 est.)
0.4% (2016 est.)
country comparison to the world: 135

Current account balance: -$306 million (2017 est.)

-$792 million (2016 est.)
country comparison to the world: 109

Exports: $1.84 billion (2017 est.)
$1.544 billion (2016 est.)
country comparison to the world: 144

Exports—partners: Switzerland 59.1%, Uzbekistan 9.4%, Kazakhstan 5.1%, Russia 4.9%, UK 4% (2017)

Exports—commodities: gold, cotton, wool, garments, meat; mercury, uranium, electricity; machinery; shoes

Imports: $4.187 billion (2017 est.)
$3.709 billion (2016 est.)
country comparison to the world: 137

Imports—commodities: oil and gas, machinery and equipment, chemicals, foodstuffs

Imports—partners: China 32.6%, Russia 24.8%, Kazakhstan 16.4%, Turkey 4.8%, US 4.2% (2017)

Reserves of foreign exchange and gold: $2.177 billion (31 December 2017 est.)
$1.97 billion (31 December 2016 est.)
country comparison to the world: 120

Debt—external: $8.164 billion (31 December 2017 est.)
$8.182 billion (31 December 2016 est.)
country comparison to the world: 121

Exchange rates: soms (KGS) per US dollar -
68.35 (2017 est.)
69.914 (2016 est.)
69.914 (2015 est.)
64.462 (2014 est.)
53.654 (2013 est.)

ENERGY

Electricity access: *electrification—total population:* 100% (2020)

Electricity—production: 13.04 billion kWh (2016 est.)
country comparison to the world: 93

Electricity—consumption: 10.52 billion kWh (2016 est.)
country comparison to the world: 94

Electricity—exports: 184 million kWh (2015 est.)
country comparison to the world: 77

Electricity—imports: 331 million kWh (2016 est.)
country comparison to the world: 87

Electricity—installed generating capacity: 4.046 million kW (2016 est.)
country comparison to the world: 87

Electricity—from fossil fuels: 24% of total installed capacity (2016 est.)
country comparison to the world: 190

Electricity—from nuclear fuels: 0% of total installed capacity (2017 est.)
country comparison to the world: 123

Electricity—from hydroelectric plants: 76% of total installed capacity (2017 est.)
country comparison to the world: 13

Electricity—from other renewable sources: 0% of total installed capacity (2017 est.)
country comparison to the world: 197

Crude oil—production: 1,000 bbl/day (2018 est.)
country comparison to the world: 92

Crude oil—exports: 0 bbl/day (2015 est.)
country comparison to the world: 150

Crude oil—imports: 4,480 bbl/day (2015 est.)
country comparison to the world: 77

Crude oil—proved reserves: 40 million bbl (1 January 2018 est.)
country comparison to the world: 79

Refined petroleum products—production: 6,996 bbl/day (2015 est.)
country comparison to the world: 102

Refined petroleum products—consumption: 37,000 bbl/day (2016 est.)
country comparison to the world: 114

Refined petroleum products—exports: 2,290 bbl/day (2015 est.)
country comparison to the world: 103

Refined petroleum products—imports: 34,280 bbl/day (2015 est.)
country comparison to the world: 96

Natural gas—production: 28.32 million cu m (2017 est.)
country comparison to the world: 88

Natural gas—consumption: 186.9 million cu m (2017 est.)
country comparison to the world: 106

Natural gas—exports: 0 cu m (2017 est.)
country comparison to the world: 134

Natural gas—imports: 169.9 million cu m (2017 est.)
country comparison to the world: 74

Natural gas—proved reserves: 5.663 billion cu m (1 January 2018 est.)
country comparison to the world: 89

Carbon dioxide emissions from consumption of energy: 10.02 million Mt (2017 est.)
country comparison to the world: 108

COMMUNICATIONS

Telephones—fixed lines: *total subscriptions:* 275,311
subscriptions per 100 inhabitants: 4.66 (2019 est.)
country comparison to the world: 111

Telephones—mobile cellular: *total subscriptions:* 7,940,306
subscriptions per 100 inhabitants: 134.4 (2019 est.)
country comparison to the world: 98

Telecommunication systems: *general assessment:* fixed-line phones declining quickly by roll-out of 4G LTE mobile networks; digital radio-relay stations, and fiber-optic links; low fixed-line and fixed-broadband penetration and moderate mobile broadband penetration; international connectivity continues to grow; 4 mobile networks in operation; 4G networks cover over 50% of the nation, eventually 5G networks will be available (2020)
domestic: fixed-line penetration 5 per 100 persons remains low and concentrated in urban areas;

mobile-cellular subscribership up to over 134 per 100 persons (2019)

international: country code—996; connections with other CIS (Commonwealth of Independent States, 9 members postSoviet Republics in EU) countries by landline or microwave radio relay and with other countries by leased connections with Moscow international gateway switch and by satellite; satellite earth stations—2 (1 Intersputnik, 1 Intelsat) (2019)

note: the COVID-19 outbreak is negatively impacting telecommunications production and supply chains globally; consumer spending on telecom devices and services has also slowed due to the pandemic's effect on economies worldwide; overall progress towards improvements in all facets of the telecom industry—mobile, fixed-line, broadband, submarine cable and satellite—has moderated

Broadcast media: state-funded public TV broadcaster KTRK has nationwide coverage; also operates Ala-Too 24 news channel which broadcasts 24/7 and 4 other educational, cultural, and sports channels; ELTR and Channel 5 are state-owned stations with national reach; the switchover to digital TV in 2017 resulted in private TV station growth; approximately 20 stations are struggling to increase their own content up to 50% of airtime, as required by law, instead of rebroadcasting primarily programs from Russian channels or airing unlicensed movies and music; 3 Russian TV stations also broadcast; state-funded radio stations and about 10 significant private radio stations also exist (2019)

Internet country code: .kg

Internet users: *total:* 2,222,732
percent of population: 38% (July 2018 est.)
country comparison to the world: 117

Broadband—fixed subscriptions: *total:* 355,640
subscriptions per 100 inhabitants: 6 (2018 est.)
country comparison to the world: 95

National air transport system: *number of registered air carriers:* 5 (2020)
inventory of registered aircraft operated by air carriers: 17
annual passenger traffic on registered air carriers: 709,198 (2018)

Civil aircraft registration country code prefix: EX (2016)

Airports: 28 (2013)
country comparison to the world: 122

Airports—with paved runways: *total:* 18 (2017)
over 3,047 m: 1 (2017)
2,438 to 3,047 m: 3 (2017)
1,524 to 2,437 m: 11 (2017)
under 914 m: 3 (2017)

Airports—with unpaved runways: *total:* 10 (2013)
1,524 to 2,437 m: 1 (2013)
914 to 1,523 m: 1 (2013)
under 914 m: 8 (2013)

Pipelines: 3566 km gas (2018), 16 km oil (2013)

Railways: *total:* 424 km (2018)
broad gauge: 424 km 1.520-m gauge (2018)
country comparison to the world: 118

Roadways: *total:* 34,000 km (2018)
country comparison to the world: 94

Waterways: 600 km (2010)
country comparison to the world: 78

Ports and terminals: *lake port(s):* Balykchy (Ysyk-Kol or Rybach'ye)(Lake Ysyk-Kol)

Military and security forces: *Kyrgyz Armed Forces:* Land Forces, Air Defense Forces, National Guard; State Border Service; Internal Troops (2019)

Military expenditures: 1.5% of GDP (2019)
1.6% of GDP (2018)

1.6% of GDP (2017)
1.7% of GDP (2016)
1.8% of GDP (2015)
country comparison to the world: 81

Military and security service personnel strengths: the Kyrgyz Armed Forces have approximately 11,000 active duty troops (8,500 Land Forces; 2,500 Air Force/Air Defense) (2019 est.)

Military equipment inventories and acquisitions: the Kyrgyz Armed Forces' inventory is comprised of older Russian and Soviet-era equipment; outside of a small delivery by China in 2019, Russia continues to be the only supplier of weapons systems to Kyrgyzstan (2020)

Military deployments: contributes a battalion-sized unit to CSTO's Rapid Reaction Force (2019 est.)

Military service age and obligation: 18-27 years of age for compulsory or voluntary male military service in the Armed Forces or Interior Ministry; 1-year service obligation (9 months for university graduates), with optional fee-based 3-year service in the call-up mobilization reserve; women may volunteer at age 19; 16-17 years of age for military cadets, who cannot take part in military operations (2016)

Disputes—international: disputes in Isfara Valley delay completion of delimitation with Tajikistan; delimitation of approximately 15% or 200 km of border with Uzbekistan is hampered by serious disputes over enclaves and other areas

Refugees and internally displaced persons: *stateless persons:* 58 (2019)

Illicit drugs: limited illicit cultivation of cannabis and opium poppy for CIS markets; limited government eradication of illicit crops; transit point for Southwest Asian narcotics bound for Russia and the rest of Europe; major consumer of opiates

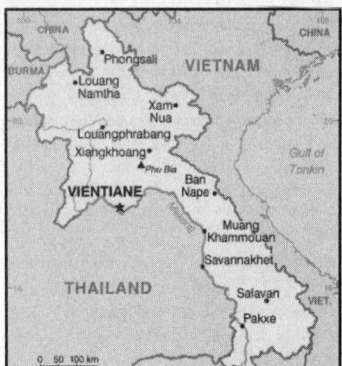

INTRODUCTION

Background: Modern-day Laos has its roots in the ancient Lao kingdom of Lan Xang, established in the 14th century under King FA NGUM. For 300 years Lan Xang had influence reaching into present-day Cambodia and Thailand, as well as over all of what is now Laos. After centuries of gradual decline, Laos came under the domination of Siam (Thailand) from the late 18th century until the late 19th century, when it became part of French Indochina. The Franco-Siamese Treaty of 1907 defined the current Lao border with Thailand. In 1975, the communist Pathet Lao took control of the government, ending a six-century-old monarchy and instituting a strict socialist regime closely aligned to Vietnam. A gradual, limited return to private enterprise and the liberalization of foreign investment laws began in 1988. Laos became a member of ASEAN in 1997 and the WTO in 2013.

GEOGRAPHY

Location: Southeastern Asia, northeast of Thailand, west of Vietnam

Geographic coordinates: 18 00 N, 105 00 E

Map references: Southeast Asia

Area: *total:* 236,800 sq km
land: 230,800 sq km
water: 6,000 sq km
country comparison to the world: 85

Area—comparative: about twice the size of Pennsylvania; slightly larger than Utah

Land boundaries: *total:* 5,274 km
border countries (5): Burma 238 km, Cambodia 555 km, China 475 km, Thailand 1845 km, Vietnam 2161 km

Coastline: 0 km (landlocked)

Maritime claims: none (landlocked)

Climate: tropical monsoon; rainy season (May to November); dry season (December to April)

Terrain: *mostly rugged mountains; some plains and plateaus*

Elevation: *mean elevation:* 710 m
lowest point: Mekong River 70 m
highest point: Phu Bia 2,817 m

Natural resources: timber, hydropower, gypsum, tin, gold, gemstones

Land use: *agricultural land:* 10.6% (2011 est.)
arable land: 6.2% (2011 est.) / permanent crops: 0.7% (2011 est.) / permanent pasture: 3.7% (2011 est.)
forest: 67.9% (2011 est.)
other: 21.5% (2011 est.)
Irrigated land: 3,100 sq km (2012)

Population distribution: most densely populated area is in and around the capital city of Vientiane; large communities are primarily found along the Mekong River along the southwestern border; overall density is considered one of the lowest in Southeast Asia

Natural hazards: floods, droughts

Environment—current issues: unexploded ordnance; deforestation; soil erosion; loss of biodiversity; water pollution, most of the population does not have access to potable water

Environment—international agreements: *party to:* Biodiversity, Climate Change, Climate Change-Kyoto Protocol, Desertification, Endangered Species, Environmental Modification, Hazardous Wastes, Law of the Sea, Ozone Layer Protection
signed, but not ratified: none of the selected agreements

Geography—note: landlocked; most of the country is mountainous and thickly forested; the Mekong River forms a large part of the western boundary with Thailand

PEOPLE AND SOCIETY

Population: 7,447,396 (July 2020 est.)
country comparison to the world: 101

Nationality: *noun:* Lao(s) or Laotian(s)
adjective: Lao or Laotian

Ethnic groups: Lao 53.2%, Khmou 11%, Hmong 9.2%, Phouthay 3.4%, Tai 3.1%, Makong 2.5%, Katong 2.2%, Lue 2%, Akha 1.8%, other 11.6% (2015 est.)
note: the Laos Government officially recognizes 49 ethnic groups, but the total number of ethnic groups is estimated to be well over 200

Languages: Lao (official), French, English, various ethnic languages

Religions: Buddhist 64.7%, Christian 1.7%, none 31.4%, other/not stated 2.1% (2015 est.)

Age structure: *0-14 years:* 31.25% (male 1,177,297/female 1,149,727)
15-24 years: 20.6% (male 763,757/female 770,497)
25-54 years: 38.29% (male 1,407,823/female 1,443,774)

55-64 years: 5.73% (male 206,977/female 219,833)
65 years and over: 4.13% (male 139,665/female 168,046) (2020 est.)

Dependency ratios: *total dependency ratio:* 56.8
youth dependency ratio: 50.1
elderly dependency ratio: 6.7
potential support ratio: 15 (2020 est.)

Median age: *total:* 24 years
male: 23.7 years
female: 24.4 years (2020 est.)
country comparison to the world: 170

Population growth rate: 1.44% (2020 est.)
country comparison to the world: 76

Birth rate: 22.4 births/1,000 population (2020 est.)
country comparison to the world: 64

Death rate: 7.2 deaths/1,000 population (2020 est.)
country comparison to the world: 119

Net migration rate: -1 migrant(s)/1,000 population (2020 est.)
country comparison to the world: 144

Population distribution: most densely populated area is in and around the capital city of Vientiane; large communities are primarily found along the Mekong River along the southwestern border; overall density is considered one of the lowest in Southeast Asia

Urbanization: *urban population:* 36.3% of total population (2020)
rate of urbanization: 3.28% annual rate of change (2015-20 est.)
total population growth rate v. urban population growth rate, 2000-2030:

Major urban areas—population: 683,000 VIENTIANE (capital) (2020)

Sex ratio: *at birth:* 1.04 male(s)/female
0-14 years: 1.02 male(s)/female
15-24 years: 0.99 male(s)/female
25-54 years: 0.98 male(s)/female
55-64 years: 0.94 male(s)/female
65 years and over: 0.83 male(s)/female
total population: 0.99 male(s)/female (2020 est.)

Maternal mortality rate: 185 deaths/100,000 live births (2017 est.)
country comparison to the world: 50

Infant mortality rate: *total:* 45.6 deaths/1,000 live births
male: 50.5 deaths/1,000 live births
female: 40.4 deaths/1,000 live births (2020 est.)
country comparison to the world: 29

Life expectancy at birth: *total population:* 65.7 years
male: 63.6 years
female: 67.9 years (2020 est.)
country comparison to the world: 193

Total fertility rate: 2.53 children born/woman (2020 est.)
country comparison to the world: 70

Contraceptive prevalence rate: 54.1% (2017)

Drinking water source:
improved:
urban: 94.4% of population
rural: 76.8% of population
total: 82.1% of population
unimproved:
urban: 5.6% of population
rural: 23.2% of population
total: 17.9% of population (2017 est.)

Current Health Expenditure: 2.5% (2017)

Physicians density: 0.49 physicians/1,000 population (2014)

Hospital bed density: 1.5 beds/1,000 population (2012)

Sanitation facility access:
improved:
urban: 98% of population
rural: 66.3% of population
total: 77.2% of population
unimproved:
urban: 2% of population
rural: 33.7% of population
total: 22.8% of population (2017 est.)

HIV/AIDS—adult prevalence rate: 0.3% (2019 est.)
country comparison to the world: 90

HIV/AIDS—people living with HIV/AIDS: 13,000 (2019 est.)
country comparison to the world: 96

HIV/AIDS—deaths: <500 (2019 est.)
Major infectious diseases: degree of risk: very high (2020)
food or waterborne diseases: bacterial and protozoal diarrhea, hepatitis A, and typhoid fever
vectorborne diseases: dengue fever and malaria

Obesity—adult prevalence rate: 5.3% (2016)
country comparison to the world: 179

Children under the age of 5 years underweight: 21.1% (2017)
country comparison to the world: 21

Education expenditures: 2.9% of GDP (2014)
country comparison to the world: 141

Literacy: *definition:* age 15 and over can read and write
total population: 84.7%
male: 90%
female: 79.4% (2015)

School life expectancy (primary to tertiary education): *total:* 11 years
male: 11 years
female: 10 years (2019)

Unemployment, youth ages 15-24: *total:* 18.2%
male: 20.8%
female: 15.5% (2017 est.)
country comparison to the world: 72

GOVERNMENT

Country name: *conventional long form:* Lao People's Democratic Republic
conventional short form: Laos
local long form: Sathalanalat Paxathipatai Paxaxon Lao

local short form: Mueang Lao (unofficial)
etymology: name means "Land of the Lao [people]"

Government type: communist state

Capital: *name:* Vientiane (Viangchan)

geographic coordinates: 17 58 N, 102 36 E
time difference: UTC+7 (12 hours ahead of Washington, DC, during Standard Time)
etymology: the meaning in Pali, a Buddhist liturgical language, is "city of sandalwood"

Administrative divisions: 17 provinces (khoueng, singular and plural) and 1 prefecture* (kampheng nakhon); Attapu, Bokeo, Bolikhamxai, Champasak, Houaphan, Khammouan, Louangnamtha, Louangphabang, Oudomxai, Phongsali, Salavan, Savannakhet, Viangchan (Vientiane)*, Viangchan, Xaignabouli, Xaisomboun, Xekong, Xiangkhouang

Independence: 19 July 1949 (from France by the Franco-Lao General Convention); 22 October 1953 (Franco-Lao Treaty recognizes full independence)

National holiday: Republic Day (National Day), 2 December (1975)

Constitution: *history:* previous 1947 (preindependence); latest promulgated 13-15 August 1991
amendments: proposed by the National Assembly; passage requires at least two-thirds majority vote of the Assembly membership and promulgation by the president of the republic; amended 2003, 2015

Legal system: civil law system similar in form to the French system

International law organization participation: has not submitted an ICJ jurisdiction declaration; non-party state to the ICCt

Citizenship: *citizenship by birth:* no
citizenship by descent only: at least one parent must be a citizen of Laos
dual citizenship recognized: no
residency requirement for naturalization: 10 years

Suffrage: 18 years of age; universal

Executive branch: *chief of state:* President BOUNNYANG Vorachit (since 20 April 2016); Vice President PHANKHAM Viphavan (since 20 April 2016)

head of government: Prime Minister THONGLOUN Sisoulit (since 20 April 2016); Deputy Prime Ministers BOUNTHONG Chitmani, SONXAI Siphandon, SOMDI Douangdi (since 20 April 2016)
cabinet: Council of Ministers appointed by the president, approved by the National Assembly
elections/appointments: president and vice president indirectly elected by the National Assembly for a 5-year term (no term limits); election last held on 20 April 2016 (next to be held in 2021); prime minister nominated by the president, elected by the National Assembly for 5-year term
election results: BOUNNYANG Vorachit (LPRP) elected president; PHANKHAM Viphavan (LPRP) elected vice president; percent of

National Assembly vote—NA; THONGLOUN Sisoulit (LPRP) elected prime minister; percent of National Assembly vote—NA

Legislative branch: *description:* unicameral National Assembly or Sapha Heng Xat (149 seats; members directly elected in multi-seat constituencies by simple majority vote from candidate lists provided by the Lao People's Revolutionary Party; members serve 5-year terms)
elections: last held on 20 March 2016 (next to be held in 2021)
election results: percent of vote by party—NA; seats by party—LPRP 144, independent 5; composition—men 108, women 41, percent of women 27.5%

Judicial branch: *highest courts:* People's Supreme Court (consists of the court president and organized into criminal, civil, administrative, commercial, family, and juvenile chambers, each with a vice president and several judges)
judge selection and term of office: president of People's Supreme Court appointed by the National Assembly upon the recommendation of the president of the republic for a 5-year term; vice presidents of the People's Supreme Court appointed by the president of the republic upon the recommendation of the National Assembly; appointment of chamber judges NA; tenure of court vice presidents and chamber judges NA
subordinate courts: appellate courts; provincial, municipal, district, and military courts

Political parties and leaders: Lao People's Revolutionary Party or LPRP [BOUNNYANG Vorachit]
note: other parties proscribed

International organization participation: ADB, ARF, ASEAN, CP, EAS, FAO, G-77, IAEA, IBRD, ICAO, ICRM, IDA, IFAD, IFC, IFRCS, ILO, IMF, Interpol, IOC, IPU, ISO (subscriber), ITU, MIGA, NAM, OIF, OPCW, PCA, UN, UNCTAD, UNESCO, UNIDO, UNWTO, UPU, WCO, WFTU (NGOs), WHO, WIPO, WMO, WTO

Diplomatic representation in the US: *chief of mission:* Ambassador KHAMPHAN Anlavan (since January 2019)
chancery: 2222 S Street NW, Washington, DC 20008
telephone: [1] (202) 332-6416
FAX: [1] (202) 332-4923
consulate(s): New York

Diplomatic representation from the US: *chief of mission:* Ambassador Peter HAYMOND (since 7 February 2020)
telephone: [856] 21-48-7000
embassy: Thadeua Road, Kilometer 9, Ban Somvang Tai, Hatsayfong District, Vientiane
mailing address: American Embassy Vientiane, Unit 46222, APO AP 96546-6222
FAX: [856] 21-48-7190

Flag description: three horizontal bands of red (top), blue (double width), and red with a large white disk centered in the blue band; the red bands recall the blood shed for liberation; the blue band represents the Mekong River and prosperity; the white disk symbolizes the full moon against the Mekong River, but also signifies the unity of the people under the Lao People's Revolutionary Party, as well as the country's bright future

National symbol(s): *elephant; national colors:* red, white, blue

National anthem: *name:* "Pheng Xat Lao" (Hymn of the Lao People)
lyrics/music: SISANA Sisane/THONGDY Sounthonevichit
note: music adopted 1945, lyrics adopted 1975; the anthem's lyrics were changed following the 1975 Communist revolution that overthrew the monarchy
0:00 / 0:50

ECONOMY

Economy—overview: The government of Laos, one of the few remaining one-party communist states, began decentralizing control and encouraging private enterprise in 1986. Economic growth averaged more than 6% per year in the period 1988-2008, and Laos' growth has more recently been amongst the fastest in Asia, averaging more than 7% per year for most of the last decade.

Nevertheless, Laos remains a country with an underdeveloped infrastructure, particularly in rural areas. It has a basic, but improving, road system, and limited external and internal land-line telecommunications. Electricity is available to 83% of the population. Agriculture, dominated by rice cultivation in lowland areas, accounts for about 20% of GDP and 73% of total employment. Recently, the country has faced a persistent current account deficit, falling foreign currency reserves, and growing public debt.

Laos' economy is heavily dependent on capital-intensive natural resource exports. The economy has benefited from high-profile foreign direct investment in hydropower dams along the Mekong River, copper and gold mining, logging, and construction, although some projects in these industries have drawn criticism for their environmental impacts.

Laos gained Normal Trade Relations status with the US in 2004 and applied for Generalized System of Preferences trade benefits in 2013 after being admitted to the World Trade Organization earlier in the year. Laos held the chairmanship of ASEAN in 2016. Laos is in the process of implementing a value-added tax system. The government appears committed to raising the country's profile among foreign investors and has developed special economic zones replete with generous tax incentives, but a limited labor pool, a small domestic market, and corruption remain impediments to investment. Laos also has ongoing problems with the business environment, including onerous registration requirements, a gap between legislation and implementation, and unclear or conflicting regulations.

GDP (purchasing power parity):
$49.34 billion (2017 est.)
$46.16 billion (2016 est.)
$43.13 billion (2015 est.)
note: data are in 2017 dollars
country comparison to the world: 111

GDP (official exchange rate):
$16.97 billion (2017 est.)

GDP—real growth rate: 6.9% (2017 est.)
7% (2016 est.)
7.3% (2015 est.)
country comparison to the world: 20

GDP—per capita (PPP): $7,400 (2017 est.)
$7,000 (2016 est.)
$6,600 (2015 est.)
note: data are in 2017 dollars
country comparison to the world: 155

Gross national saving:
22.7% of GDP (2017 est.)
21.3% of GDP (2016 est.)
15.8% of GDP (2015 est.)
country comparison to the world: 79

GDP—composition, by end use:
household consumption: 63.7% (2017 est.)
government consumption: 14.1% (2017 est.)
investment in fixed capital: 30.9% (2017 est.)
investment in inventories: 3.1% (2017 est.)
exports of goods and services: 34.6% (2017 est.)
imports of goods and services: -43.2% (2017 est.)

GDP—composition, by sector of origin:
agriculture: 20.9% (2017 est.)
industry: 33.2% (2017 est.)
services: 45.9% (2017 est.)

Agriculture—products: sweet potatoes, vegetables, corn, coffee, sugarcane, tobacco, cotton, tea, peanuts, rice; cassava (manioc, tapioca), water buffalo, pigs, cattle, poultry

Industries: mining (copper, tin, gold, gypsum); timber, electric power, agricultural processing, rubber, construction, garments, cement, tourism

Industrial production growth rate: 8% (2017 est.)
country comparison to the world: 23

Labor force: 3.582 million (2017 est.)
country comparison to the world: 99

Labor force—by occupation: *agriculture:* 73.1%
industry: 6.1%
services: 20.6% (2012 est.)

Unemployment rate: 0.7% (2017 est.)
0.7% (2016 est.)
country comparison to the world: 4

Population below poverty line: 22% (2013 est.)

Household income or consumption by percentage share: *lowest 10%:* 3.3%
highest 10%: 30.3% (2008)

Budget: *revenues:* 3.099 billion (2017 est.)
expenditures: 4.038 billion (2017 est.)

Taxes and other revenues: 18.3% (of GDP) (2017 est.)
country comparison to the world: 161

Budget surplus (+) or deficit (-): -5.5% (of GDP) (2017 est.)
country comparison to the world: 172

Public debt: 63.6% of GDP (2017 est.)
58.4% of GDP (2016 est.)
country comparison to the world: 64

Fiscal year: 1 October—30 September

Inflation rate (consumer prices): 0.8% (2017 est.)
1.6% (2016 est.)
country comparison to the world: 42

Current account balance: -$2.057 billion (2017 est.)
-$2.07 billion (2016 est.)
country comparison to the world: 167

Exports: $3.654 billion (2017 est.)
$2.705 billion (2016 est.)
country comparison to the world: 120

Exports—partners: Thailand 42.6%, China 28.7%, Vietnam 10.4%, India 4.4% (2017)

Exports—commodities: wood products, coffee, electricity, tin, copper, gold, cassava

Imports: $4.976 billion (2017 est.)
$4.739 billion (2016 est.)
country comparison to the world: 131

Imports—commodities: machinery and equipment, vehicles, fuel, consumer goods

Imports—partners: Thailand 59.1%, China 21.5%, Vietnam 9.8% (2017)

Reserves of foreign exchange and gold: $1.27 billion (31 December 2017 est.)
$940.1 million (31 December 2016 est.)
country comparison to the world: 128

Debt—external: $14.9 billion (31 December 2017 est.)
$12.9 billion (31 December 2016 est.)
country comparison to the world: 104

Exchange rates: kips (LAK) per US dollar—
8,231.1 (2017 est.)
8,129.1 (2016 est.)
8,129.1 (2015 est.)
8,147.9 (2014 est.)
8,049 (2013 est.)

ENERGY

Electricity access: *electrification—total population:* 95% (2019)
electrification—urban areas: 98% (2019)
electrification—rural areas: 93% (2019)

Electricity—production: 29.74 billion kWh (2016 est.)
country comparison to the world: 66

Electricity—consumption: 5.471 billion kWh (2016 est.)
country comparison to the world: 120

Electricity—exports: 8.469 billion kWh (2015 est.)
country comparison to the world: 24

Electricity—imports: 2.5 billion kWh (2016 est.)
country comparison to the world: 53

Electricity—installed generating capacity: 6.94 million kW (2016 est.)
country comparison to the world: 75

Electricity—from fossil fuels: 28% of total installed capacity (2016 est.)
country comparison to the world: 186

Electricity—from nuclear fuels: 0% of total installed capacity (2017 est.)

country comparison to the world: 124

Electricity—from hydroelectric plants: 72% of total installed capacity (2017 est.)
country comparison to the world: 15

Electricity—from other renewable sources: 1% of total installed capacity (2017 est.)
country comparison to the world: 157

Crude oil—production: 0 bbl/day (2018 est.)
country comparison to the world: 159

Crude oil—exports: 0 bbl/day (2015 est.)
country comparison to the world: 151

Crude oil—imports: 0 bbl/day (2015 est.)
country comparison to the world: 149

Crude oil—proved reserves: 0 bbl (1 January 2018 est.)
country comparison to the world: 154

Refined petroleum products—production: 0 bbl/day (2015 est.)
country comparison to the world: 162

Refined petroleum products—consumption: 18,000 bbl/day (2016 est.)
country comparison to the world: 146

Refined petroleum products—exports: 0 bbl/day (2015 est.)
country comparison to the world: 169

Refined petroleum products—imports: 17,460 bbl/day (2015 est.)
country comparison to the world: 132

Natural gas—production: 0 cu m (2017 est.)
country comparison to the world: 154

Natural gas—consumption: 0 cu m (2017 est.)
country comparison to the world: 165

Natural gas—exports: 0 cu m (2017 est.)
country comparison to the world: 135

Natural gas—imports: 0 cu m (2017 est.)
country comparison to the world: 146

Natural gas—proved reserves: 0 cu m (1 January 2014 est.)
country comparison to the world: 155

Carbon dioxide emissions from consumption of energy: 10.42 million Mt (2017 est.)
country comparison to the world: 105

COMMUNICATIONS

Telephones—fixed lines: *total subscriptions:* 1,526,232
subscriptions per 100 inhabitants: 20.79 (2019 est.)
country comparison to the world: 63

Telephones—mobile cellular: *total subscriptions:* 4,466,375
subscriptions per 100 inhabitants: 60.84 (2019 est.)
country comparison to the world: 125

Telecommunication systems: *general assessment:* the government relies on a radiotelephone network to communicate with remote areas; the regulatory reform is below industry standards but is trying to strengthen its telecommunication infrastructure and subsequently attract foreign investment; low fixed-broadband penetration due to dominance of mobile platforms; strong boost in mobile broadband penetration but still low compared to other Asian markets; mobile sector growth held back by regulators trying to keep hold on pricing and open competition; development of mobile broadband Internet services given the expansion of 4G services (2020)

domestic: fixed-line 21 per 100 and 61 per 100 for mobile-cellular subscriptions (2019)

international: country code—856; satellite earth station—1 Intersputnik (Indian Ocean region) and a second to be developed by China
note: the COVID-19 outbreak is negatively impacting telecommunications production and supply chains globally; consumer spending on telecom devices and services has also slowed due to the pandemic's effect on economies worldwide; overall progress towards improvements in all facets of the telecom industry—mobile, fixed-line, broadband, submarine cable and satellite—has moderated

Broadcast media: 6 TV stations operating out of Vientiane—3 government-operated and the others commercial; 17 provincial stations operating with nearly all programming relayed via satellite from the government-operated stations in Vientiane; Chinese and Vietnamese programming relayed via satellite from Lao National TV; broadcasts available from stations in Thailand and Vietnam in border areas; multi-channel satellite and cable TV systems provide access to a wide range of foreign stations; state-controlled radio with state-operated Lao National Radio (LNR) broadcasting on 5 frequencies—1 AM, 1 SW, and 3 FM; LNR's AM and FM programs are relayed via satellite constituting a large part of the programming schedules of the provincial radio stations; Thai radio broadcasts available in border areas and transmissions of multiple international broadcasters are also accessible

Internet country code: .la

Internet users: *total:* 1,845,437

percent of population: 25.51% (July 2018 est.)
country comparison to the world: 124

Broadband—fixed subscriptions:
total: 45,379
subscriptions per 100 inhabitants: 1 less than 1 (2018 est.)
country comparison to the world: 136

TRANSPORTATION

National air transport system: *number of registered air carriers:* 1 (2020)
inventory of registered aircraft operated by air carriers: 12
annual passenger traffic on registered air carriers: 1,251,961 (2018)
annual freight traffic on registered air carriers: 1.53 million mt-km (2018)

Civil aircraft registration country code prefix: RDPL (2016)

Airports: 41 (2013)
country comparison to the world: 104

Airports—with paved runways:
total: 8 (2017)
2,438 to 3,047 m: 3 (2017)
1,524 to 2,437 m: 4 (2017)
914 to 1,523 m: 1 (2017)

Airports—with unpaved runways: *total:* 33 (2013)
1,524 to 2,437 m: 2 (2013)
914 to 1,523 m: 9 (2013)
under 914 m: 22 (2013)

Pipelines: 540 km refined products (2013)

Roadways: *total:* 39,586 km (2009)
paved: 5,415 km (2009)
unpaved: 34,171 km (2009)
country comparison to the world: 90

Waterways: 4,600 km (primarily on the Mekong River and its tributaries; 2,900 additional km are intermittently navigable by craft drawing less than 0.5 m) (2012)
country comparison to the world: 23

Merchant marine: *total:* 1
by type: general cargo 1 (2019)
country comparison to the world: 176

MILITARY AND SECURITY

Military and security forces: *Lao People's Armed Forces (LPAF):* Lao People's Army (LPA, includes Riverine Force), Air Force, Self-Defense Militia Forces (2019)

Military expenditures: 0.2% of GDP (2013)
0.2% of GDP (2012)
0.2% of GDP (2011)
note: no public figures available for 2014-2019
country comparison to the world: 157

Military and security service personnel strengths: information is limited and estimates for the size of the Lao People's Armed Forces (LPAF) vary; approximately 29,000 active duty troops (26,000 Army; 3500 Air Force); approximately 100,000 Self-Defense Militia Forces (2019)

Military equipment inventories and acquisitions: the LPAF is armed largely with weapons from the former Soviet Union with a smaller mix of more modern weapons from China, Russia, and Ukraine; since 2010, China and Russia are the top suppliers of military hardware to Laos (2019 est.)

Military service age and obligation: 18 years of age for compulsory or voluntary military service; conscript service obligation—minimum 18 months (2019)

TRANSNATIONAL ISSUES

Disputes—international: southeast Asian states have enhanced border surveillance to check the spread of avian flu; talks continue on completion of demarcation with Thailand but disputes remain over islands in the Mekong River; Cambodia and Laos have a longstanding border demarcation dispute; concern among Mekong River Commission members that China's construction of eight dams on the Upper Mekong River and construction of more dams on its tributaries will affect water levels, sediment flows, and fisheries; Cambodia and Vietnam are concerned about Laos' extensive plans for upstream dam construction for the same reasons

Trafficking in persons: current situation: Laos is a source and, to a lesser extent, transit and destination country for men, women, and children subjected to forced labor and sex trafficking; Lao economic migrants may encounter conditions of forced labor or sexual exploitation in destination countries, most often Thailand; Lao women and girls are exploited in Thailand's commercial sex trade, domestic service, factories, and agriculture; a small, possibly growing, number of Lao women and girls are sold as brides in China and South Korea and subsequently sex trafficked; Lao men and boys are victims of forced labor in the Thai fishing, construction, and agriculture industries; some Lao children, as well as Vietnamese and Chinese women and girls, are subjected to sex trafficking in Laos; other Vietnamese and Chinese, and possibly Burmese, adults and girls transit Laos for sexual and labor exploitation in neighboring countries, particularly Thailand

tier rating: Tier 2 Watch List – Laos does not fully comply with the minimum standards for the elimination of trafficking; however, it is making significant efforts to do so; authorities sustained moderate efforts to investigate, prosecute, and convict trafficking offenders; the government failed to make progress in proactively identifying victims exploited within the country or among those deported from abroad; the government continues to rely almost entirely on local and international

organizations to provide and fund services to trafficking victims; although Lao men and boys are trafficked, most protective services are only available to women and girls, and long-term support is lacking; modest prevention efforts include the promotion of anti-trafficking awareness on state-controlled media (2015)

Illicit drugs: estimated opium poppy cultivation in 2015 was estimated to be 5,700 hectares, compared with 6,200 hectares in 2014; estimated potential production of between 84 and 176 mt of raw opium; unsubstantiated reports of domestic methamphetamine production; growing domestic methamphetamine problem

LATVIA

INTRODUCTION

Background: Several eastern Baltic tribes merged in medieval times to form the ethnic core of the Latvian people (ca. 8th-12th centuries A.D.). The region subsequently came under the control of Germans, Poles, Swedes, and finally, Russians. A Latvian republic emerged following World War I, but it was annexed by the USSR in 1940—an action never recognized by the US and many other countries. Latvia reestablished its independence in 1991 following the breakup of the Soviet Union. Although the last Russian troops left in 1994, the status of the Russian minority (some 26% of the population) remains of concern to Moscow. Latvia acceded to both NATO and the EU in the spring of 2004; it joined the euro zone in 2014 and the OECD in 2016. A dual citizenship law was adopted in 2013, easing naturalization for non-citizen children.

GEOGRAPHY

Location: Eastern Europe, bordering the Baltic Sea, between Estonia and Lithuania

Geographic coordinates: 57 00 N, 25 00 E

Map references: Europe

Area: *total:* 64,589 sq km
land: 62,249 sq km
water: 2,340 sq km
country comparison to the world: 125

Area—comparative: slightly larger than West Virginia

Land boundaries: *total:* 1,370 km
border countries (4): Belarus 161 km, Estonia 333 km, Lithuania 544 km, Russia 332 km

Coastline: 498 km

Maritime claims: *territorial sea:* 12 nm
exclusive economic zone: limits as agreed by Estonia, Finland, Latvia, Sweden, and Russia
continental shelf: 200 m depth or to the depth of exploitation

Climate: maritime; wet, moderate winters

Terrain: low plain

Elevation: *mean elevation:* 87 m
lowest point: Baltic Sea 0 m
highest point: Gaizina Kalns 312 m

Natural resources: peat, limestone, dolomite, amber, hydropower, timber, arable land

Land use: *agricultural land:* 29.2% (2011 est.)
arable land: 18.6% (2011 est.) / permanent crops: 0.1% (2011 est.) / permanent pasture: 10.5% (2011 est.)
forest: 54.1% (2011 est.)
other: 16.7% (2011 est.)
Irrigated land: 12 sq km (2012)
note: land in Latvia is often too wet and in need of drainage not irrigation; approximately 16,000 sq km or 85% of agricultural land has been improved by drainage

Population distribution: largest concentration of people is found in and around the port and capital city of Riga; small agglomerations are scattered throughout the country

Natural hazards: large percentage of agricultural fields can become waterlogged and require drainage

Environment—current issues: while land, water, and air pollution are evident, Latvia's environment has benefited from a shift to service industries after the country regained independence; improvements have occurred in drinking water quality, sewage treatment, household and hazardous waste management, as well as reduction of air pollution; concerns include nature protection and the management of water resources and the protection of the Baltic Sea

Environment—international agreements: *party to:* Air Pollution, Air Pollution-Persistent Organic Pollutants, Biodiversity, Climate Change, Climate Change-Kyoto Protocol, Desertification, Endangered Species, Hazardous Wastes, Law of the Sea, Ozone Layer Protection, Ship Pollution, Wetlands
signed, but not ratified: none of the selected agreements

Geography—note: most of the country is composed of fertile low-lying plains with some hills in the east

PEOPLE AND SOCIETY

Population: 1,881,232 (July 2020 est.)
country comparison to the world: 153

Nationality: *noun:* Latvian(s)
adjective: Latvian

Ethnic groups: Latvian 62.2%, Russian 25.2%, Belarusian 3.2%, Ukrainian 2.2%, Polish 2.1%, Lithuanian 1.2%, other 1.5%, unspecified 2.3% (2018 est.)

Languages: Latvian (official) 56.3%, Russian 33.8%, other 0.6% (includes Polish, Ukrainian, and Belarusian), unspecified 9.4% (2011 est.)
note: data represent language usually spoken at home

Religions: Lutheran 36.2%, Roman Catholic 19.5%, Orthodox 19.1%, other Christian 1.6%, other 0.1%, unspecified/none 23.5% (2017 est.)

Age structure: *0-14 years:* 15.32% (male 148,120/ female 140,028)

15-24 years: 9% (male 87,372/female 81,965)

25-54 years: 40.41% (male 380,817/female 379,359)

55-64 years: 14.77% (male 125,401/female 152,548)

65 years and over: 20.5% (male 128,151/female 257,471) (2020 est.)

Dependency ratios: *total dependency ratio:* 59
youth dependency ratio: 26.1
elderly dependency ratio: 32.9
potential support ratio: 3 (2020 est.)

Median age: *total:* 44.4 years
male: 40.5 years
female: 48 years (2020 est.)
country comparison to the world: 16

Population growth rate: -1.12% (2020 est.)
country comparison to the world: 231

Birth rate: 9.2 births/1,000 population (2020 est.)
country comparison to the world: 204

Death rate: 14.6 deaths/1,000 population (2020 est.)
country comparison to the world: 4

Net migration rate: -5.9 migrant(s)/1,000 population (2020 est.)
country comparison to the world: 202

Population distribution: largest concentration of people is found in and around the port and capital city of Riga; small agglomerations are scattered throughout the country

Urbanization: *urban population:* 68.3% of total population (2020)
rate of urbanization: -0.93% annual rate of change (2015-20 est.)
total population growth rate v. urban population growth rate, 2000-2030:

Major urban areas—population: 631,000 RIGA (capital) (2020)

Sex ratio: *at birth:* 1.05 male(s)/female
0-14 years: 1.06 male(s)/female
15-24 years: 1.07 male(s)/female
25-54 years: 1 male(s)/female
55-64 years: 0.82 male(s)/female
65 years and over: 0.5 male(s)/female
total population: 0.86 male(s)/female (2020 est.)

Mother's mean age at first birth: 27.6 years (2017 est.)

Maternal mortality rate: 19 deaths/100,000 live births (2017 est.)
country comparison to the world: 124

Infant mortality rate: *total:* 5 deaths/1,000 live births
male: 5.4 deaths/1,000 live births
female: 4.6 deaths/1,000 live births (2020 est.)
country comparison to the world: 177

Life expectancy at birth: *total population:* 75.4 years
male: 70.9 years
female: 80.1 years (2020 est.)
country comparison to the world: 116

Total fertility rate: 1.53 children born/woman (2020 est.)
country comparison to the world: 198

Drinking water source:
improved:
urban: 98.8% of population
rural: 98.2% of population
total: 98.6% of population
unimproved:
urban: 1.2% of population
rural: 1.8% of population
total: 1.4% of population (2017 est.)

Current Health Expenditure: 6% (2017)

Physicians density: 3.19 physicians/1,000 population (2017)

Hospital bed density: 5.6 beds/1,000 population (2017)

Sanitation facility access:
improved:
urban: 98.9% of population
rural: 84.6% of population
total: 94.3% of population
unimproved:
urban: 1.1% of population
rural: 15.4% of population
total: 5.7% of population (2017 est.)

HIV/AIDS—adult prevalence rate: 0.3% (2019 est.)
country comparison to the world: 91

HIV/AIDS—people living with HIV/AIDS: 5,600 (2019 est.)
country comparison to the world: 122

HIV/AIDS—deaths: <100 (2019 est.)

Major infectious diseases: *degree of risk:* intermediate (2020)
vectorborne diseases: tickborne encephalitis

Obesity—adult prevalence rate: 23.6% (2016)
country comparison to the world: 65

Education expenditures: 5.3% of GDP (2015)
country comparison to the world: 49

Literacy: *definition:* age 15 and over can read and write
total population: 99.9%
male: 99.9%
female: 99.9% (2015)

School life expectancy (primary to tertiary education): *total:* 16 years
male: 16 years
female: 17 years (2018)

Unemployment, youth ages 15-24: *total:* 12.2%
male: 12.5%
female: 11.8% (2018 est.)
country comparison to the world: 109

GOVERNMENT

Country name: *conventional long form:* Republic of Latvia
conventional short form: Latvia
local long form: Latvijas Republika
local short form: Latvija
former: Latvian Soviet Socialist Republic
etymology: the name "Latvia" originates from the ancient Latgalians, one of four eastern Baltic tribes that formed the ethnic core of the Latvian people (ca. 8th-12th centuries A.D.)

Government type: parliamentary republic

Capital: *name:* Riga

geographic coordinates: 56 57 N, 24 06 E
time difference: UTC+2 (7 hours ahead of Washington, DC, during Standard Time)
daylight saving time: +1hr, begins last Sunday in March; ends last Sunday in October
etymology: of the several theories explaining the name's origin, the one relating to the city's role in Baltic and North Sea commerce is the most probable; the name is likely related to the Latvian word "rija," meaning "warehouse," where the 'j' became a 'g' under the heavy German influence in the city from the late Middle Ages to the early 20th century

Administrative divisions: 110 municipalities (novadi, singular—novads) and 9 cities

municipalities: Adazi, Aglona, Aizkraukle, Aizpute, Akniste, Aloja, Alsunga, Aluksne, Amata, Ape, Auce, Babite, Baldone, Baltinava, Balvi, Bauska, Beverina, Broceni, Burtnieki, Carnikava, Cesis, Cesvaine, Cibla, Dagda, Daugavpils, Dobele, Dundaga, Durbe, Engure, Ergli, Garkalne, Grobina, Gulbene, Iecava, Ikskile, Ilukste, Incukalns, Jaunjelgava, Jaunpiebalga, Jaunpils, Jekabpils, Jelgava, Kandava, Karsava, Kegums, Kekava, Koceni, Koknese, Kraslava, Krimulda, Krustpils, Kuldiga, Lielvarde, Ligatne, Limbazi, Livani, Lubanas, Ludza, Madona, Malpils, Marupe, Mazsalaca, Mersrags, Naukseni, Nereta, Nica, Ogre, Olaine, Ozolnieki, Pargauja, Pavilosta, Plavinas, Preili, Priekule, Priekuli, Rauna, Rezekne, Riebini, Roja, Ropazi, Rucava, Rugaji, Rujiena, Rundale, Salacgriva, Sala, Salaspils, Saldus, Saulkrasti, Seja, Sigulda, Skriveri, Skrunda, Smiltene, Stopini, Strenci, Talsi, Tervete, Tukums, Vainode, Valka, Varaklani, Varkava, Vecpiebalga, Vecumnieki, Ventspils, Viesites, Vilaka, Vilani, Zilupe

cities: Daugavpils, Jekabpils, Jelgava, Jurmala, Liepaja, Rezekne, Riga, Valmiera, Ventspils

Independence: 18 November 1918 (from Soviet Russia); 4 May 1990 (declared from the Soviet Union); 6 September 1991 (recognized by the Soviet Union)

National holiday: Independence Day (Republic of Latvia Proclamation Day), 18 November (1918); note—18 November 1918 was the date Latvia established its statehood and its concomitant independence from Soviet Russia; 4 May 1990 was the date it declared the restoration of Latvian statehood and its concomitant independence from the Soviet Union

Constitution: *history:* several previous (pre-1991 independence); note—following the restoration of independence in 1991, parts of the 1922 constitution were reintroduced 4 May 1990 and fully reintroduced 6 July 1993
amendments: proposed by two thirds of Parliament members or by petition of one tenth of qualified voters submitted through the president; passage requires at least two-thirds majority vote of Parliament in each of three readings; amendment of constitutional articles, including national

sovereignty, language, the parliamentary electoral system, and constitutional amendment procedures, requires passage in a referendum by majority vote of at least one half of the electorate; amended several times, last in 2019

Legal system: civil law system with traces of socialist legal traditions and practices

International law organization participation: has not submitted an ICJ jurisdiction declaration; accepts ICCt jurisdiction

Citizenship: *citizenship by birth:* no
citizenship by descent only: at least one parent must be a citizen of Latvia
dual citizenship recognized: no
residency requirement for naturalization: 5 years

Suffrage: 18 years of age; universal

Executive branch: *chief of state:* President Egils LEVITS (since 8 July 2019)

head of government: Prime Minister Krisjanis KARINS (since 23 January 2019)
cabinet: Cabinet of Ministers nominated by the prime minister, appointed by Parliament
elections/appointments: president indirectly elected by Parliament for a 4-year term (eligible for a second term); election last held on 29 May 2019 (next to be held in 2023); prime minister appointed by the president, confirmed by Parliament
election results: Egils LEVITS elected president; Parliament vote—Egils LEVITS 61 votes, Didzis SMITS 24, Juris JANSONS 8; Krisjanis KARINS confirmed prime minister 61-39

Legislative branch: *description:* unicameral Parliament or Saeima (100 seats; members directly elected in multi-seat constituencies by party list proportional representation vote; members serve 4-year terms)
elections: last held on 6 October 2018 (next to be held in October 2022)
election results: percent of vote by party—SDPS 19.8%, KPV LV 14.3%, JKP 13.6%, AP! 12%, NA 11%, ZZS 9.9%, V 6.7%, other 12.7%; seats by party—SDPS 23, KPV LV 16, JKP 16, AP! 13, NA 13, ZZS 11, V 8; composition—men 69, women 31, percent of women 31%

Judicial branch: *highest courts:* Supreme Court (consists of the Senate with 36 judges); Constitutional Court (consists of 7 judges)
judge selection and term of office: Supreme Court judges nominated by chief justice and confirmed by the Saeima; judges serve until age 70, but term can be extended 2 years; Constitutional Court judges—3 nominated by Saeima members, 2 by Cabinet ministers, and 2 by plenum of Supreme Court; all judges confirmed by Saeima majority vote; Constitutional Court president and vice president serve in their positions for 3 years; all judges serve 10-year terms; mandatory retirement at age 70
subordinate courts: district (city) and regional courts

Political parties and leaders: Development/For! or AP! [Daniels PAVLUTS, Juris PUCE]

National Alliance "All For Latvia!"-"For Fatherland and Freedom/LNNK" or NA [Raivis DZINTARS] New Conservative Party or JKP [Janis BORDANS]
Social Democratic Party "Harmony" or SDPS [Nils USAKOVS] Union of Greens and Farmers or ZZS [Armands KRAUZE] Unity or V [Arvils ASERADENS]
Who Owns the State? or KPV LV [Artuss KAIMINS]

International organization participation: Australia Group, BA, BIS, CBSS, CD, CE, EAPC, EBRD, ECB, EIB, EMU, ESA (cooperating state), EU, FAO, IAEA, IBRD, ICAO, ICC (NGOs), ICCt, ICRM, IDA, IFC, IFRCS, IHO, ILO, IMF, IMO, IMSO, Interpol, IOC, IOM, IPU, ISO (correspondent), ITU, ITUC (NGOs), MIGA, NATO, NIB, NSG, OAS (observer), OIF (observer), OPCW, OSCE, PCA, Schengen Convention, UN, UNCTAD, UNESCO, UNHCR, UNWTO, UPU, WCO, WHO, WIPO, WMO, WTO

Diplomatic representation in the US: *chief of mission:* Ambassador Andris TEIKMANIS (since 16 September 2016)
chancery: 2306 Massachusetts Avenue NW, Washington, DC 20008
telephone: [1] (202) 328-2840
FAX: [1] (202) 328-2860

Diplomatic representation from the US: *chief of mission:* Ambassador John Leslie CARWILE (since 5 November 2019)
telephone: [371] 6710-7000
embassy: 1 Samnera Velsa St, Riga LV-1510
mailing address: Embassy of the United States of America, 1 Samnera Velsa St, Riga, LV-1510
FAX: [371] 6710-7050

Flag description: three horizontal bands of maroon (top), white (half-width), and maroon; the flag is one of the older banners in the world; a medieval chronicle mentions a red standard with a white stripe being used by Latvian tribes in about 1280

National symbol(s): *white wagtail (bird); national colors:* maroon, white

National anthem: *name:* "Dievs, sveti Latviju!" (God Bless Latvia)
lyrics/music: Karlis BAUMANIS
note: adopted 1920, restored 1990; first performed in 1873 while Latvia was a part of Russia; banned during the Soviet occupation from 1940 to 1990
0:00 / 0:00

ECONOMY

Economy—overview: Latvia is a small, open economy with exports contributing more than half of GDP. Due to its geographical location, transit services are highly-developed, along with timber and wood-processing, agriculture and food products, and manufacturing of machinery and electronics industries. Corruption continues to be an impediment to attracting foreign direct investment and Latvia's low birth rate and decreasing population are major challenges to its long-term economic vitality.

Latvia's economy experienced GDP growth of more than 10% per year during 2006-07, but entered a severe recession in 2008 as a result of an unsustainable current account deficit and large debt exposure amid the slowing world economy. Triggered by the collapse of the second largest bank, GDP plunged by more than 14% in 2009 and, despite strong growth since 2011, the economy took until 2017 return to pre-crisis levels in real terms. Strong investment and consumption, the latter stoked by rising wages, helped the economy grow by more than 4% in 2017, while inflation rose to 3%. Continued gains in competitiveness and investment will be key to maintaining economic growth, especially in light of unfavorable demographic trends, including the emigration of skilled workers, and one of the highest levels of income inequality in the EU.

In the wake of the 2008-09 crisis, the IMF, EU, and other international donors provided substantial financial assistance to Latvia as part of an agreement to defend the currency's peg to the euro in exchange for the government's commitment to stringent austerity measures. The IMF/EU program successfully concluded in December 2011, although, the austerity measures imposed large social costs. The majority of companies, banks, and real estate have been privatized, although the state still holds sizable stakes in a few large enterprises, including 80% ownership of the Latvian national airline. Latvia officially joined the World Trade Organization in February 1999 and the EU in May 2004. Latvia also joined the euro zone in 2014 and the OECD in 2016.

GDP (purchasing power parity):
$54.02 billion (2017 est.)
$51.67 billion (2016 est.)
$50.55 billion (2015 est.)
note: data are in 2017 dollars
country comparison to the world: 108

GDP (official exchange rate):
$30.33 billion (2017 est.)

GDP—real growth rate: 2.08% (2019 est.)
4.2% (2018 est.)
3.23% (2017 est.)
country comparison to the world: 133

GDP—per capita (PPP): $27,700 (2017 est.)
$26,200 (2016 est.)
$25,500 (2015 est.)
note: data are in 2017 dollars
country comparison to the world: 76

Gross national saving:
20.7% of GDP (2017 est.)
21% of GDP (2016 est.)
21.8% of GDP (2015 est.)
country comparison to the world: 91

GDP—composition, by end use:
household consumption: 61.8% (2017 est.)
government consumption: 18.2% (2017 est.)
investment in fixed capital: 19.9% (2017 est.)
investment in inventories: 1.5% (2017 est.)
exports of goods and services: 60.6% (2017 est.)
imports of goods and services: -61.9% (2017 est.)

GDP—composition, by sector of origin:
agriculture: 3.9% (2017 est.)

industry: 22.4% (2017 est.)
services: 73.7% (2017 est.)

Agriculture—products: grain, rapeseed, potatoes, vegetables; pork, poultry, milk, eggs; fish

Industries: processed foods, processed wood products, textiles, processed metals, pharmaceuticals, railroad cars, synthetic fibers, electronics

Industrial production growth rate: 10.6% (2017 est.)
country comparison to the world: 12

Labor force: 885,000 (2020 est.)
country comparison to the world: 142

Labor force—by occupation: *agriculture:* 7.7%
industry: 24.1%
services: 68.1% (2016 est.)

Unemployment rate: 6.14% (2019 est.)
6.51% (2018 est.)
country comparison to the world: 99

Population below poverty line: 25.5% (2015)

Household income or consumption by percentage share: *lowest 10%:* 2.2%
highest 10%: 26.3% (2015)

Budget: *revenues:* 11.39 billion (2017 est.)
expenditures: 11.53 billion (2017 est.)

Taxes and other revenues: 37.5% (of GDP) (2017 est.)
country comparison to the world: 54

Budget surplus (+) or deficit (-): -0.5% (of GDP) (2017 est.)
country comparison to the world: 62

Public debt: 36.3% of GDP (2017 est.)
37.4% of GDP (2016 est.)
note: data cover general government debt, and includes debt instruments issued (or owned) by government entities, including sub-sectors of central government, state government, local government, and social security funds
country comparison to the world: 147

Fiscal year: calendar year

Inflation rate (consumer prices): 2.9% (2017 est.)
0.1% (2016 est.)
country comparison to the world: 130

Current account balance: -$222 million (2019 est.)
-$99 million (2018 est.)
country comparison to the world: 102

Exports: $12.84 billion (2017 est.)
$11.35 billion (2016 est.)
country comparison to the world: 81

Exports—partners: Lithuania 15.8%, Russia 14%, Estonia 10.9%, Germany 6.9%, Sweden 5.7%, UK 4.9%, Poland 4.3%, Denmark 4.1% (2017)

Exports—commodities: foodstuffs, wood and wood products, metals, machinery and equipment, textiles

Imports: $15.79 billion (2017 est.)
$13.61 billion (2016 est.)
country comparison to the world: 86

Imports—commodities: machinery and equipment, consumer goods, chemicals, fuels, vehicles

Imports—partners: Lithuania 17.6%, Germany 11.7%, Poland 8.7%, Estonia 7.6%, Russia 7.1%, Netherlands 4.2%, Finland 4.2%, Italy 4% (2017)

Reserves of foreign exchange and gold: $4.614 billion (31 December 2017 est.)
$3.514 billion (31 December 2016 est.)
country comparison to the world: 97

Debt—external: $40.02 billion (31 March 2016 est.)
$38.19 billion (31 March 2015 est.)
country comparison to the world: 74

Exchange rates: euros (EUR) per US dollar—
0.885 (2017 est.)
0.903 (2016 est.)
0.9214 (2015 est.)
0.885 (2014 est.)
0.7634 (2013 est.)

ENERGY

Electricity access: *electrification—total population:* 100% (2020)

Electricity—production: 6.241 billion kWh (2016 est.)
country comparison to the world: 115

Electricity—consumption: 6.798 billion kWh (2016 est.)
country comparison to the world: 109

Electricity—exports: 3.795 billion kWh (2016 est.)
country comparison to the world: 38

Electricity—imports: 4.828 billion kWh (2016 est.)
country comparison to the world: 39

Electricity—installed generating capacity: 2.932 million kW (2016 est.)
country comparison to the world: 98

Electricity—from fossil fuels: 39% of total installed capacity (2016 est.)
country comparison to the world: 173

Electricity—from nuclear fuels: 0% of total installed capacity (2017 est.)
country comparison to the world: 125

Electricity—from hydroelectric plants: 53% of total installed capacity (2017 est.)
country comparison to the world: 33

Electricity—from other renewable sources:
8% of total installed capacity (2017 est.)
country comparison to the world: 87

Crude oil—production: 0 bbl/day (2018 est.)
country comparison to the world: 160

Crude oil—exports: 0 bbl/day (2017 est.)
country comparison to the world: 152

Crude oil—imports: 0 bbl/day (2017 est.)
country comparison to the world: 150

Crude oil—proved reserves: 0 bbl (1 January 2018 est.)
country comparison to the world: 155

Refined petroleum products—production: 0 bbl/day (2017 est.)
country comparison to the world: 163

Refined petroleum products—consumption: 44,600 bbl/day (2017 est.)
country comparison to the world: 111

Refined petroleum products—exports: 16,180 bbl/day (2017 est.)
country comparison to the world: 72

Refined petroleum products—imports: 54,370 bbl/day (2017 est.)
country comparison to the world: 77

Natural gas—production: 0 cu m (2017 est.)
country comparison to the world: 155

Natural gas—consumption: 1.218 billion cu m (2017 est.)
country comparison to the world: 88

Natural gas—exports: 0 cu m (2017 est.)
country comparison to the world: 136

Natural gas—imports: 1.246 billion cu m (2017 est.)
country comparison to the world: 58

Natural gas—proved reserves: 0 cu m (2014 est.)
country comparison to the world: 156

Carbon dioxide emissions from consumption of energy: 8.632 million Mt (2017 est.)
country comparison to the world: 114

COMMUNICATIONS

Telephones—fixed lines:
total subscriptions: 227,149
subscriptions per 100 inhabitants: 11.94 (2019 est.)
country comparison to the world: 120

Telephones—mobile cellular:
total subscriptions: 2,067,174
subscriptions per 100 inhabitants: 108.66 (2019 est.)
country comparison to the world: 151

Telecommunication systems: *general assessment:* recent efforts focused on bringing competition to the telecommunications sector; the number of fixed-line is decreasing as mobile-cellular telephone service expands; EU regulatory policies, and framework provide guidelines for growth; govt. adopted measures to build a national fiber broadband network, part-funded by European Commission; new competition in mobile markets with extensive LTE-A technologies and 5G service growth (2020)

domestic: fixed-line 12 per 100 and mobile-cellular 109 per 100 subscriptions (2019)

international: country code—371; the Latvian network is now connected via fiber-optic cable to Estonia, Finland, and Sweden
note: the COVID-19 outbreak is negatively impacting telecommunications production and supply chains globally; consumer spending on telecom devices and services has also slowed due to the pandemic's effect on economies worldwide; overall progress towards improvements in all facets of the telecom industry—mobile, fixed-line, broadband, submarine cable and satellite—has moderated

Broadcast media: several national and regional commercial TV stations are foreign-owned, 2 national TV stations are publicly owned; system supplemented by privately owned regional and local TV stations; cable and satellite multi-channel TV services with domestic and foreign broadcasts available; publicly owned broadcaster operates 4 radio networks with dozens of stations throughout the country; dozens of private broadcasters also operate radio stations

Internet country code: .lv

Internet users: total: 1,607,711

percent of population: 83.58% (July 2018 est.)
country comparison to the world: 129

Broadband—fixed subscriptions:
total: 525,995
subscriptions per 100 inhabitants: 27 (2018 est.)
country comparison to the world: 83

TRANSPORTATION

National air transport system:
number of registered air carriers: 3 (2020)
inventory of registered aircraft operated by air carriers: 53
annual passenger traffic on registered air carriers: 4,058,762 (2018)
annual freight traffic on registered air carriers: 4.01 million mt-km (2018)

Civil aircraft registration country code prefix: YL (2016)

Airports: 42 (2013)
country comparison to the world: 101

Airports—with paved runways:
total: 18 (2017)
over 3,047 m: 1 (2017)
2,438 to 3,047 m: 3 (2017)
1,524 to 2,437 m: 4 (2017)
914 to 1,523 m: 3 (2017)
under 914 m: 7 (2017)

Airports—with unpaved runways: total: 24 (2013)
under 914 m: 24 (2013)
Heliports: 1 (2013)

Pipelines: 1,213 km gas, 417 km refined products (2018)

Railways: total: 1,860 km (2018)
narrow gauge: 34 km 0.750-m gauge (2018)
broad gauge: 1,826 km 1.520-m gauge (2018)

country comparison to the world: 75

Roadways: total: 70,244 km (2018)
paved: 15,158 km (2018)
unpaved: 55,086 km (2018)
country comparison to the world: 70

Waterways: 300 km (navigable year-round) (2010)
country comparison to the world: 92

Merchant marine: total: 58
by type: general cargo 11, oil tanker 8, other 39 (2019)
country comparison to the world: 110

Ports and terminals: major seaport(s): Riga, Ventspils

MILITARY AND SECURITY

Military and security forces: National Armed Forces (Nacionalie Brunotie Speki): Land Forces (Latvijas Sauszemes Speki), Naval Force (Latvijas Juras Speki, includes Coast Guard (Latvijas Kara Flote)), Air Force (Latvijas Gaisa Speki), National Guard (2019)

Military expenditures:
2.01% of GDP (2019 est.)
2.08% of GDP (2018)
1.59% of GDP (2017)
1.45% of GDP (2016)
1.04% of GDP (2015)
country comparison to the world: 49

Military and security service personnel strengths: the National Armed Forces of Latvia have approximately 6,000 active duty troops (5,000 Land Forces, inc. joint service personnel and active duty National Guard; 500 Naval Force/Coast Guard; 500 Air Force) (2019 est.)

Military equipment inventories and acquisitions: the Latvian military's inventory is limited and consists of a European, Israeli, and US weapons systems; since 2010, it has received mostly second-hand equipment from Austria, Denmark, Germany, the Netherlands, Sweden, the UK, and the US (2019 est.)

Military service age and obligation: 18 years of age for voluntary male and female military service;

no conscription; under current law, every citizen is entitled to serve in the armed forces for life (2017)

TRANSNATIONAL ISSUES

Disputes—international: Russia demands better Latvian treatment of ethnic Russians in Latvia; boundary demarcated with Latvia and Lithuania; the Latvian parliament has not ratified its 1998 maritime boundary treaty with Lithuania, primarily due to concerns over oil exploration rights; as a member state that forms part of the EU's external border, Latvia has implemented the strict Schengen border rules with Russia

Refugees and internally displaced persons: stateless persons: 216,851 (2019); note—individuals who were Latvian citizens prior to the 1940 Soviet occupation and their descendants were recognized as Latvian citizens when the country's independence was restored in 1991; citizens of the former Soviet Union residing in Latvia who have neither Latvian nor other citizenship are considered non-citizens (officially there is no statelessness in Latvia) and are entitled to non-citizen passports; children born after Latvian independence to stateless parents are entitled to Latvian citizenship upon their parents' request; non-citizens cannot vote or hold certain government jobs and are exempt from military service but can travel visa-free in the EU under the Schengen accord like Latvian citizens; non-citizens can obtain naturalization if they have been permanent residents of Latvia for at least five years, pass tests in Latvian language and history, and know the words of the Latvian national anthem

Illicit drugs: transshipment and destination point for cocaine, synthetic drugs, opiates, and cannabis from Southwest Asia, Western Europe, Latin America, and neighboring Baltic countries; despite improved legislation, vulnerable to money laundering due to nascent enforcement capabilities and comparatively weak regulation of offshore companies and the gaming industry; CIS organized crime (including counterfeiting, corruption, extortion, stolen cars, and prostitution) accounts for most laundered proceeds

LEBANON

INTRODUCTION

Background: Following World War I, France acquired a mandate over the northern portion of the former Ottoman Empire province of Syria. The French demarcated the region of Lebanon in 1920 and granted this area independence in 1943. Since independence, the country has been marked by periods of political turmoil interspersed with prosperity built on its position as a regional center for finance and trade. The country's 1975-90 civil war, which resulted in an estimated 120,000 fatalities, was followed by years of social and

political instability. Sectarianism is a key element of Lebanese political life. Neighboring Syria has historically influenced Lebanon's foreign policy and internal policies, and its military occupied Lebanon from 1976 until 2005. The Lebanon-based Hizballah militia and Israel continued attacks and counterattacks against each other after Syria's withdrawal, and fought a brief war in 2006. Lebanon's borders with Syria and Israel remain unresolved.

GEOGRAPHY

Location: Middle East, bordering the Mediterranean Sea, between Israel and Syria

Geographic coordinates: 33 50 N, 35 50 E

Map references: Middle East

Area: total: 10,400 sq km
land: 10,230 sq km
water: 170 sq km
country comparison to the world: 169

Area—comparative: about one-third the size of Maryland

Land boundaries: *total:* 484 km
border countries (2): Israel 81 km, Syria 403 km

Coastline: 225 km

Maritime claims: *territorial sea:* 12 nm

Climate: Mediterranean; mild to cool, wet winters with hot, dry summers; the Lebanon Mountains experience heavy winter snows

Terrain: narrow coastal plain; El Beqaa (Bekaa Valley) separates Lebanon and Anti-Lebanon Mountains

Elevation: *mean elevation:* 1,250 m
lowest point: Mediterranean Sea 0 m
highest point: Qornet es Saouda 3,088 m

Natural resources: limestone, iron ore, salt, water-surplus state in a water-deficit region, arable land

Land use: *agricultural land:* 63.3% (2011 est.)
arable land: 11.9% (2011 est.) / permanent crops: 12.3% (2011 est.) / permanent pasture: 39.1% (2011 est.)
forest: 13.4% (2011 est.)
other: 23.3% (2011 est.)
Irrigated land: 1,040 sq km (2012)

Population distribution: the majority of the people live on or near the Mediterranean coast, and of these most live in and around the capital, Beirut; favorable growing conditions in the Bekaa Valley, on the southeastern side of the Lebanon Mountains, have attracted farmers and thus the area exhibits a smaller population density

Natural hazards: earthquakes; dust storms, sandstorms

Environment—current issues: deforestation; soil deterioration, erosion; desertification; species loss; air pollution in Beirut from vehicular traffic and the burning of industrial wastes; pollution of coastal waters from raw sewage and oil spills; waste-water management

Environment—international agreements: *party to:* Biodiversity, Climate Change, Climate Change-Kyoto Protocol, Desertification, Hazardous Wastes, Law of the Sea, Ozone Layer Protection, Ship Pollution, Wetlands
signed, but not ratified: Environmental Modification, Marine Life Conservation

Geography—note: smallest country in continental Asia; Nahr el Litani is the only major river in Near East not crossing an international boundary; rugged terrain historically helped isolate, protect, and develop numerous factional groups based on religion, clan, and ethnicity

PEOPLE AND SOCIETY

Population: 5,469,612 (July 2020 est.)
country comparison to the world: 118

Nationality: *noun:* Lebanese (singular and plural)
adjective: Lebanese

Ethnic groups: Arab 95%, Armenian 4%, other 1%
note: many Christian Lebanese do not identify themselves as Arab but rather as descendants of the ancient Canaanites and prefer to be called Phoenicians

Languages: Arabic (official), French, English, Armenian

Religions: Muslim 61.1% (30.6% Sunni, 30.5% Shia, smaller percentages of Alawites and Ismailis), Christian 33.7% (Maronite Catholics are the largest Christian group), Druze 5.2%, very small numbers of Jews, Baha'is, Buddhists, and Hindus (2018 est.)
note: data represent the religious affiliation of the citizen population (data do not include Lebanon's sizable Syrian and Palestinian refugee populations); 18 religious sects recognized

MENA religious affiliation:
Age structure: 0-14 years: 20.75% (male 581,015/female 554,175)
15-24 years: 14.98% (male 417,739/female 401,357)
25-54 years: 46.69% (male 1,296,250/female 1,257,273)
55-64 years: 9.62% (male 250,653/female 275,670)
65 years and over: 7.96% (male 187,001/female 248,479) (2020 est.)

Dependency ratios: *total dependency ratio:* 48.4
youth dependency ratio: 37.2
elderly dependency ratio: 11.2
potential support ratio: 8.9 (2020 est.)

Median age: *total:* 33.7 years
male: 33.1 years
female: 34.4 years (2020 est.)
country comparison to the world: 95

Population growth rate: -6.68% (2020 est.)
country comparison to the world: 237

Birth rate: 13.6 births/1,000 population (2020 est.)
country comparison to the world: 138

Death rate: 5.4 deaths/1,000 population (2020 est.)
country comparison to the world: 185

Net migration rate: -88.7 migrant(s)/1,000 population (2020 est.)
country comparison to the world: 228

Population distribution: the majority of the people live on or near the Mediterranean coast, and of these most live in and around the capital, Beirut; favorable growing conditions in the Bekaa

Valley, on the southeastern side of the Lebanon Mountains, have attracted farmers and thus the area exhibits a smaller population density

Urbanization: *urban population:* 88.9% of total population (2020)
rate of urbanization: 0.75% annual rate of change (2015-20 est.)
total population growth rate v. urban population growth rate, 2000-2030:

Major urban areas—population: 2.424 million BEIRUT (capital) (2020)

Sex ratio: *at birth:* 1.05 male(s)/female
0-14 years: 1.05 male(s)/female
15-24 years: 1.04 male(s)/female
25-54 years: 1.03 male(s)/female
55-64 years: 0.91 male(s)/female
65 years and over: 0.75 male(s)/female
total population: 1 male(s)/female (2020 est.)

Maternal mortality rate: 29 deaths/100,000 live births (2017 est.)
country comparison to the world: 112

Infant mortality rate: *total:* 6.8 deaths/1,000 live births
male: 7.2 deaths/1,000 live births
female: 6.4 deaths/1,000 live births (2020 est.)
country comparison to the world: 161

Life expectancy at birth: *total population:* 78.3 years
male: 76.9 years
female: 79.8 years (2020 est.)
country comparison to the world: 68

Total fertility rate: 1.71 children born/woman (2020 est.)
country comparison to the world: 170

Drinking water source: *improved:* total: 100% of population
unimproved:
total: 0% of population (2017 est.)

Current Health Expenditure: 8.2% (2017)

Physicians density: 2.03 physicians/1,000 population (2017)

Hospital bed density: 2.7 beds/1,000 population (2017)

Sanitation facility access:
improved:
total: 99% of population
unimproved:
total: 1% of population (2017 est.)

HIV/AIDS—adult prevalence rate: <.1% (2019 est.)

HIV/AIDS—people living with HIV/AIDS: 2,700 (2019 est.)
country comparison to the world: 134

HIV/AIDS—deaths: <100 (2019 est.)

Obesity—adult prevalence rate: 32% (2016)
country comparison to the world: 19

Education expenditures: 2.5% of GDP (2013)
country comparison to the world: 161

Literacy: *definition:* age 15 and over can read and write
total population: 95.1%
male: 96.9%

female: 93.3% (2018)

School life expectancy (primary to tertiary education): *total:* 11 years
male: 12 years
female: 11 years (2014)

GOVERNMENT

Country name: *conventional long form:* Lebanese Republic
conventional short form: Lebanon
local long form: Al Jumhuriyah al Lubnaniyah
local short form: Lubnan
former: Greater Lebanon
etymology: derives from the Semitic root "lbn" meaning "white" and refers to snow-capped Mount Lebanon

Government type: parliamentary republic

Capital: name: Beirut

geographic coordinates: 33 52 N, 35 30 E
time difference: UTC+2 (7 hours ahead of Washington, DC, during Standard Time)
daylight saving time: +1hr, begins last Sunday in March; ends last Sunday in October
etymology: derived from the Canaanite or Phoenician word "ber'ot," meaning "the wells" or "fountain," which referred to the site's accessible water table

Administrative divisions: 8 governorates (mohafazat, singular—mohafazah); Aakkar, Baalbek-Hermel, Beqaa (Bekaa), Beyrouth (Beirut), Liban-Nord (North Lebanon), Liban-Sud (South Lebanon), Mont-Liban (Mount Lebanon), Nabatiye

Independence: 22 November 1943 (from League of Nations mandate under French administration)

National holiday: Independence Day, 22 November (1943)

Constitution: *history:* drafted 15 May 1926, adopted 23 May 1926
amendments: proposed by the president of the republic and introduced as a government bill to the National Assembly or proposed by at least 10 members of the Assembly and agreed upon by two thirds of its members; if proposed by the National Assembly, review and approval by two-thirds majority of the Cabinet is required; if approved, the proposal is next submitted to the Cabinet for drafting as an amendment; Cabinet approval requires at least two-thirds majority, followed by submission to the National Assembly for discussion and vote; passage requires at least two-thirds majority vote of a required two-thirds quorum of the Assembly membership and promulgation by the president; amended several times, last in 1989

Legal system: mixed legal system of civil law based on the French civil code, Ottoman legal tradition, and religious laws covering personal status, marriage, divorce, and other family relations of the Jewish, Islamic, and Christian communities

International law organization participation: has not submitted an ICJ jurisdiction declaration; non-party state to the ICCt

Citizenship: *citizenship by birth:* no

citizenship by descent only: the father must be a citizen of Lebanon
dual citizenship recognized: yes
residency requirement for naturalization: unknown

Suffrage: 21 years of age; authorized for all men and women regardless of religion; excludes persons convicted of felonies and other crimes or those imprisoned; excludes all military and security service personnel regardless of rank

Executive branch: *chief of state:* President Michel AWN (since 31 October 2016)

head of government: Prime Minister Saad HARIRI (since 22 October 2020)
cabinet: Cabinet chosen by the prime minister in consultation with the president and National Assembly
elections/appointments: president indirectly elected by the National Assembly with two-thirds majority vote in the first round and if needed absolute majority vote in a second round for a 6-year term (eligible for non-consecutive terms); last held on 31 October 2016 (next to be held in 2022); prime minister appointed by the president in consultation with the National Assembly; deputy prime minister determined during cabinet formation
election results: Michel AWN elected president in second round; National Assembly vote—Michel AWN (FPM) 83; note—in the initial election held on 23 April 2014, no candidate received the required two-thirds vote, and subsequent attempts failed because the Assembly lacked the necessary quorum to hold a vote; the president was finally elected in its 46th attempt on 31 October 2016

Legislative branch: *description:* unicameral National Assembly or Majlis al-Nuwab in Arabic or Assemblee Nationale in French (128 seats; members directly elected by listed-based proportional representation vote; members serve 4-year terms); prior to 2017, the electoral system was by majoritarian vote
elections: last held on 6 May 2018 (next to be held in 2022)
election results: percent of vote by coalition—NA; seats by coalition – Strong Lebanon Bloc (Free Patriotic Movementled) 25; Future Bloc (Future Movement-led) 20; Development and Liberation Bloc (Amal Movement-led) 16; Loyalty to the Resistance Bloc (Hizballah-led) 15; Strong Republic Bloc (Lebanese Forces-led) 15; Democratic Gathering (Progressive Socialist Party-led) 9; Independent Centre Bloc 4; National Bloc (Marada Movement-led) 3; Syrian Social Nationalist Party 3; Tashnaq 3; Kata'ib 3; other 8; independent 4; composition—men 122, women 6, percent of women 4.6%
note: Lebanon's constitution states the National Assembly cannot conduct regular business until it elects a president when the position is vacant

Judicial branch: *highest courts:* Court of Cassation or Supreme Court (organized into 8 chambers, each with a presiding judge and 2 associate judges); Constitutional Council (consists of 10 members)

judge selection and term of office: Court of Cassation judges appointed by Supreme Judicial Council, a 10-member body headed by the chief justice, and includes other judicial officials; judge tenure NA; Constitutional Council members appointed—5 by the Council of Ministers and 5 by parliament; members serve 5-year terms
subordinate courts: Courts of Appeal; Courts of First Instance; specialized tribunals, religious courts; military courts

Political parties and leaders:
Al-Ahbash or Association of Islamic Charitable Projects [Adnan TARABULSI]
Amal Movement [Nabih BERRI]
Azm Movement [Najib MIQATI]
Ba'th Arab Socialist Party of Lebanon [Fayiz SHUKR]
Free Patriotic Movement or FPM [Gibran BASSIL]
Future Movement Bloc [Sa'ad al-HARIRI]
Hizballah [Hassan NASRALLAH]
Islamic Actions Front [Sheikh Zuhayr al-JU'AYD]
Kata'ib Party [Sami GEMAYEL]
Lebanese Democratic Party [Talal ARSLAN]
Lebanese Forces or LF [Samir JA'JA]
Marada Movement [Sulayman FRANJIEH]
Progressive Socialist Party or PSP [Walid JUNBLATT]
Social Democrat Hunshaqian Party [Sabuh KALPAKIAN]Syrian Social Nationalist Party [Ali QANSO]
Syrian Social Nationalist Party [Hanna al-NASHIF]
Tashnaq or Armenian Revolutionary Federation [Hagop PAKRADOUNIAN]

International organization participation: ABEDA, AFESD, AMF, CAEU, FAO, G-24, G-77, IAEA, IBRD, ICAO, ICC (national committees), ICRM, IDA, IDB, IFAD, IFC, IFRCS, ILO, IMF, IMO, IMSO, Interpol, IOC, IPU, ISO, ITSO, ITU, LAS, MIGA, NAM, OAS (observer), OIC, OIF, OPCW, PCA, UN, UNCTAD, UNESCO, UNHCR, UNIDO, UNRWA, UNWTO, UPU, WCO, WFTU (NGOs), WHO, WIPO, WMO, WTO (observer)

Diplomatic representation in the US: *chief of mission:* Ambassador Gabriel ISSA (since 24 January 2018)
chancery: 2560 28th Street NW, Washington, DC 20008
telephone: [1] (202) 939-6300
FAX: [1] (202) 939-6324
consulate(s) general: Detroit, New York, Los Angeles

Diplomatic representation from the US: *chief of mission:* Ambassador Dorothy SHEA (since 11 March 2020)
telephone: [961] (04) 543 600
embassy: Awkar-Facing the Municipality, Main Street, Beirut
mailing address: P. O. Box 70-840, Antelias, Lebanon; from US: US Embassy Beirut, 6070 Beirut Place, Washington, DC 20521-6070
FAX: [961] (4) 544136

Flag description: three horizontal bands consisting of red (top), white (middle, double width), and red (bottom) with a green cedar tree centered in the white band; the red bands symbolize blood shed for liberation, the white band denotes peace, the snow of the mountains, and purity; the green cedar tree is the symbol of Lebanon and represents eternity, steadiness, happiness, and prosperity

National symbol(s): *cedar tree; national colors:* red, white, green

National anthem: *name:* "Kulluna lil-watan" (All Of Us, For Our Country!)
lyrics/music: Rachid NAKHLE/Wadih SABRA
note: adopted 1927; chosen following a nation-wide competition
0:00 / 0:50

ECONOMY

Economy—overview: Lebanon has a free-market economy and a strong laissez-faire commercial tradition. The government does not restrict foreign investment; however, the investment climate suffers from red tape, corruption, arbitrary licensing decisions, complex customs procedures, high taxes, tariffs, and fees, archaic legislation, and inadequate intellectual property rights protection. The Lebanese economy is service-oriented; main growth sectors include banking and tourism.

The 1975-90 civil war seriously damaged Lebanon's economic infrastructure, cut national output by half, and derailed Lebanon's position as a Middle Eastern banking hub. Following the civil war, Lebanon rebuilt much of its war-torn physical and financial infrastructure by borrowing heavily, mostly from domestic banks, which saddled the government with a huge debt burden. Pledges of economic and financial reforms made at separate international donor conferences during the 2000s have mostly gone unfulfilled, including those made during the Paris III Donor Conference in 2007, following the July 2006 war. The "CEDRE" investment event hosted by France in April 2018 again rallied the international community to assist Lebanon with concessional financing and some grants for capital infrastructure improvements, conditioned upon long-delayed structural economic reforms in fiscal management, electricity tariffs, and transparent public procurement, among many others.

The Syria conflict cut off one of Lebanon's major markets and a transport corridor through the Levant. The influx of nearly one million registered and an estimated 300,000 unregistered Syrian refugees has increased social tensions and heightened competition for low-skill jobs and public services. Lebanon continues to face several long-term structural weaknesses that predate the Syria crisis, notably, weak infrastructure, poor service delivery, institutionalized corruption, and bureaucratic over-regulation. Chronic fiscal deficits have increased Lebanon's debt-to-GDP ratio, the third highest in the world; most of the debt is held internally by Lebanese banks. These factors combined to slow economic growth to the 1-2% range in 2011-17, after four years of averaging 8% growth.

Weak economic growth limits tax revenues, while the largest government expenditures remain debt servicing, salaries for government workers, and transfers to the electricity sector. These limitations constrain other government spending, limiting its ability to invest in necessary infrastructure improvements, such as water, electricity, and transportation. In early 2018, the Lebanese government signed long-awaited contract agreements with an international consortium for petroleum exploration and production as part of the country's first offshore licensing round. Exploration is expected to begin in 2019.

GDP (purchasing power parity):
$88.25 billion (2017 est.)
$86.94 billion (2016 est.)
$85.45 billion (2015 est.)
note: data are in 2017 dollars
country comparison to the world: 92

GDP (official exchange rate):
$54.18 billion (2017 est.)

GDP—real growth rate: 1.5% (2017 est.)
1.7% (2016 est.)
0.2% (2015 est.)
country comparison to the world: 154

GDP—per capita (PPP): $19,600 (2017 est.)
$19,500 (2016 est.)
$19,300 (2015 est.)
note: data are in 2017 dollars
country comparison to the world: 91

Gross national saving:
-0.7% of GDP (2017 est.)
0.7% of GDP (2016 est.)
4.5% of GDP (2015 est.)
country comparison to the world: 181

GDP—composition, by end use:
household consumption: 87.6% (2017 est.)
government consumption: 13.3% (2017 est.)
investment in fixed capital: 21.8% (2017 est.)
investment in inventories: 0.5% (2017 est.)
exports of goods and services: 23.6% (2017 est.)
imports of goods and services: -46.4% (2017 est.)

GDP—composition, by sector of origin:
agriculture: 3.9% (2017 est.)
industry: 13.1% (2017 est.)
services: 83% (2017 est.)

Agriculture—products: citrus, grapes, tomatoes, apples, vegetables, potatoes, olives, tobacco; sheep, goats

Industries: banking, tourism, real estate and construction, food processing, wine, jewelry, cement, textiles, mineral and chemical products, wood and furniture products, oil refining, metal fabricating

Industrial production growth rate: -21.1% (2017 est.)
country comparison to the world: 201

Labor force: 2.166 million (2016 est.)
note: excludes as many as 1 million foreign workers and refugees
country comparison to the world: 120

Labor force—by occupation: *agriculture:* 39% NA (2009 est.)
industry: NA

services: NA

Unemployment rate: 9.7% (2007)
country comparison to the world: 143

Population below poverty line: 28.6% (2004 est.)

Household income or consumption by percentage share: *lowest 10%:* NA
highest 10%: NA

Budget: *revenues:* 11.62 billion (2017 est.)
expenditures: 15.38 billion (2017 est.)

Taxes and other revenues: 21.5% (of GDP) (2017 est.)
country comparison to the world: 136

Budget surplus (+) or deficit (-): -6.9% (of GDP) (2017 est.)
country comparison to the world: 193

Public debt: 146.8% of GDP (2017 est.)
145.5% of GDP (2016 est.)
note: data cover central government debt and exclude debt instruments issued (or owned) by government entities other than the treasury; the data include treasury debt held by foreign entities; the data include debt issued by subnational entities, as well as intragovernmental debt; intra-governmental debt consists of treasury borrowings from surpluses in the social funds, such as for retirement, medical care, and unemployment
country comparison to the world: 4

Fiscal year: calendar year

Inflation rate (consumer prices): 4.5% (2017 est.)
-0.8% (2016 est.)
country comparison to the world: 167

Current account balance: -$12.37 billion (2017 est.)
-$11.18 billion (2016 est.)
country comparison to the world: 195

Exports: $3.524 billion (2017 est.)
$3.689 billion (2016 est.)
country comparison to the world: 122

Exports—partners: China 13%, UAE 9.9%, South Africa 7.5%, Saudi Arabia 6.5%, Syria 6.5%, Iraq 5.8%, Turkey 4.6% (2017)

Exports—commodities: jewelry, base metals, chemicals, consumer goods, fruit and vegetables, tobacco, construction minerals, electric power machinery and switchgear, textile fibers, paper

Imports: $18.34 billion (2017 est.)
$17.71 billion (2016 est.)
country comparison to the world: 80

Imports—commodities: petroleum products, cars, medicinal products, clothing, meat and live animals, consumer goods, paper, textile fabrics, tobacco, electrical machinery and equipment, chemicals

Imports—partners: China 10.2%, Italy 8.9%, Greece 7%, Germany 6.6%, US 6.3%, Turkey 4.5%, Egypt 4.2% (2017)

Reserves of foreign exchange and gold: $55.42 billion (31 December 2017 est.)
$54.04 billion (31 December 2016 est.)
country comparison to the world: 37

Debt—external: $39.3 billion (31 December 2017 est.)

$36.6 billion (31 December 2016 est.)
country comparison to the world: 76

Exchange rates: Lebanese pounds (LBP) per US dollar—
1,507.5 (2017 est.)
1,507.5 (2016 est.)
1,507.5 (2015 est.)
1,507.5 (2014 est.)
1,507.5 (2013 est.)

ENERGY

Electricity access: *electrification—total population:* 100% (2020)

Electricity—production: 17.59 billion kWh (2016 est.)
country comparison to the world: 82

Electricity—consumption: 15.71 billion kWh (2016 est.)
country comparison to the world: 77

Electricity—exports: 0 kWh (2016 est.)
country comparison to the world: 157

Electricity—imports: 69 million kWh (2016 est.)
country comparison to the world: 104

Electricity—installed generating capacity: 2.346 million kW (2016 est.)
country comparison to the world: 110

Electricity—from fossil fuels: 88% of total installed capacity (2016 est.)
country comparison to the world: 59

Electricity—from nuclear fuels: 0% of total installed capacity (2017 est.)
country comparison to the world: 126

Electricity—from hydroelectric plants: 11% of total installed capacity (2017 est.)
country comparison to the world: 114

Electricity—from other renewable sources: 1% of total installed capacity (2017 est.)
country comparison to the world: 158

Crude oil—production: 0 bbl/day (2018 est.)
country comparison to the world: 161

Crude oil—exports: 0 bbl/day (2015 est.)
country comparison to the world: 153

Crude oil—imports: 0 bbl/day (2015 est.)
country comparison to the world: 151

Crude oil—proved reserves: 0 bbl (1 January 2018 est.)
country comparison to the world: 156

Refined petroleum products—production: 0 bbl/day (2015 est.)
country comparison to the world: 164

Refined petroleum products—consumption: 154,000 bbl/day (2016 est.)
country comparison to the world: 65

Refined petroleum products—exports: 0 bbl/day (2015 est.)
country comparison to the world: 170

Refined petroleum products—imports: 151,100 bbl/day (2015 est.)
country comparison to the world: 41

Natural gas—production: 0 cu m (2017 est.)
country comparison to the world: 156

Natural gas—consumption: 0 cu m (2017 est.)
country comparison to the world: 166

Natural gas—exports: 0 cu m (2017 est.)
country comparison to the world: 137

Natural gas—imports: 0 cu m (2017 est.)
country comparison to the world: 147

Natural gas—proved reserves: 0 cu m (1 January 2014 est.)
country comparison to the world: 157

Carbon dioxide emissions from consumption of energy: 23.36 million Mt (2017 est.)
country comparison to the world: 83

COMMUNICATIONS

Telephones—fixed lines: *total subscriptions:* 752,547
subscriptions per 100 inhabitants: 12.87 (2019 est.)
country comparison to the world: 82

Telephones—mobile cellular: *total subscriptions:* 3,614,797
subscriptions per 100 inhabitants: 61.82 (2019 est.)
country comparison to the world: 135

Telecommunication systems: *general assessment:* two mobile-cellular networks provide good service, with 4G LTE services; future improvements to fiber-optic infrastructure for total nation coverage proposed by 2020; in 2018 first successful 5G trial conducted and in 2019 first live mobile 5G site launched, unfortunately, the COVID-19 pandemic has impacted telecoms industry and pricing has been raised (2020)

domestic: fixed-line 13 per 100 and 62 per 100 for mobile-cellular subscriptions (2019)

international: country code—961; landing points for the IMEWE, BERYTAR AND CADMOS submarine cable links to Europe, Africa, the Middle East and Asia; satellite earth stations—2 Intelsat (1 Indian Ocean and 1 Atlantic Ocean) (2019)

note: the COVID-19 outbreak is negatively impacting telecommunications production and supply chains globally; consumer spending on telecom devices and services has also slowed due to the pandemic's effect on economies worldwide; overall progress towards improvements in all facets of the telecom industry—mobile, fixed-line, broadband, submarine cable and satellite—has moderated

Broadcast media: 7 TV stations, 1 of which is state owned; more than 30 radio stations, 1 of which is state owned; satellite and cable TV services available; transmissions of at least 2 international broadcasters are accessible through partner stations (2019)

Internet country code: .lb

Internet users: *total:* 4,769,039

percent of population: 78.18% (July 2018 est.)
country comparison to the world: 85

Broadband—fixed subscriptions: *total:* 9,395
subscriptions per 100 inhabitants: less than 1 (2018 est.)
country comparison to the world: 171

TRANSPORTATION

National air transport system: *number of registered air carriers:* 1 (2020)
inventory of registered aircraft operated by air carriers: 21
annual passenger traffic on registered air carriers: 2,981,937 (2018)
annual freight traffic on registered air carriers: 56.57 million mt-km (2018)

Civil aircraft registration country code prefix: OD (2016)

Airports: 8 (2013)
country comparison to the world: 160

Airports—with paved runways:
total: 5 (2019)
over 3,047 m: 1
2,438 to 3,047 m: 2
1,524 to 2,437 m: 1
under 914 m: 1

Airports—with unpaved runways:
total: 3 (2013)
914 to 1,523 m: 2 (2013)
under 914 m: 1 (2013)

Heliports: 1 (2013)

Pipelines: 88 km gas (2013)

Railways: *total:* 401 km (2017)
standard gauge: 319 km 1.435-m gauge (2017)
narrow gauge: 82 km 1.050-m gauge (2017)
note: rail system is still unusable due to damage sustained from fighting in the 1980s and in 2006
country comparison to the world: 119

Roadways: *total:* 21,705 km (2017)
country comparison to the world: 111

Merchant marine: *total:* 55
by type: bulk carrier 2, container ship 1, general cargo 39, oil tanker 1, other 12 (2019)
country comparison to the world: 114

Ports and terminals: *major seaport(s):* Beirut, Tripoli
container port(s) (TEUs): Beirut (1,305,038) (2017)

MILITARY AND SECURITY

Military and security forces: *Lebanese Armed Forces (LAF):* Army Command (includes Presidential Guard Brigade, Land Border Regiments), Naval Forces, Air Forces; Lebanese Internal Security Forces Directorate (includes Mobile Gendarmerie); Directorate for General Security (DGS); Directorate General for State Security (2019)

Military expenditures: 4.2% of GDP (2019)
4.9% of GDP (2018)
4.5% of GDP (2017)
5.1% of GDP (2016)
4.5% of GDP (2015)
country comparison to the world: 9

Military and security service personnel strengths: the Lebanese Armed Forces (LAF) have approximately 58,000 active troops (55,000 Army; 1,500

Navy; 1,500 AF); est. 20,000 Internal Security Forces (2019 est.)

Military equipment inventories and acquisitions: the LAF inventory includes a wide mix of mostly older equipment, largely from the US and European countries, particularly France and Germany; since 2010, the US is the leading supplier of armaments (mostly second hand equipment) to Lebanon (2019 est.)

Military service age and obligation: 17-25 years of age for voluntary military service (including women); no conscription (2019)

Military—note: the United Nations Interim Force In Lebanon (UNIFIL) has operated in the country since 1978, originally under UNSCRs 425 and 426 to confirm Israeli withdrawal from Lebanon, restore international peace and security and assist the Lebanese Government in restoring its effective authority in the area; following the July-August 2006 war, the UN Security Council adopted resolution 1701 enhancing UNIFIL and deciding that in addition to the original mandate, it would, among other things, monitor the cessation of hostilities; accompany and support the Lebanese Armed Forces (LAF) as they deploy throughout the south of Lebanon; and extend its assistance to help ensure humanitarian access to civilian populations and the voluntary and safe return of displaced persons; UNIFIL had about 10,200 personnel deployed in the country as of March 2020 (2020)

TERRORISM

Terrorist group(s): Abdallah Azzam Brigades; al-Aqsa Martyrs Brigade; Asbat al-Ansar; Islamic Revolutionary Guard Corps/Qods Force; Hizballah; al-Nusrah Front (Hay'at Tahrir al-Sham); Palestine Liberation Front;

PFLP-General Command; Popular Front for the Liberation of Palestine (2019)

note: details about the history, aims, leadership, organization, areas of operation, tactics, targets, weapons, size, and

sources of support of the group(s) appear(s) in Appendix-T

TRANSNATIONAL ISSUES

Disputes—international: lacking a treaty or other documentation describing the boundary, portions of the Lebanon-Syria boundary are unclear with several sections in dispute; since 2000, Lebanon has claimed Shab'a Farms area in the Israeli-controlled Golan Heights; the roughly 2,000-strong UN Interim Force in Lebanon has been in place since 1978

Refugees and internally displaced persons: *refugees (country of origin):* 879,529 (Syria), 476,033 (Palestinian refugees) (2020)
IDPs: 11,000 (2007 Lebanese security forces' destruction of Palestinian refugee camp) (2019)
stateless persons: undetermined (2016); note— tens of thousands of persons are stateless in Lebanon, including many Palestinian refugees and their descendants, Syrian Kurds denaturalized in Syria in 1962, children born to Lebanese women married to foreign or stateless men; most babies born to Syrian refugees, and Lebanese children whose births are unregistered

Trafficking in persons: *current situation:* Lebanon is a source and destination country for women and children subjected to forced labor and sex trafficking and a transit point for Eastern European women and children subjected to sex trafficking in other Middle Eastern countries; women and girls from South and Southeast Asia and an increasing number from East and West Africa are recruited by agencies to work in

domestic service but are subject to conditions of forced labor; under Lebanon's artiste visa program, women from Eastern Europe, North Africa, and the Dominican Republic enter Lebanon to work in the adult entertainment industry but are often forced into the sex trade; Lebanese children are reportedly forced into street begging and commercial sexual exploitation, with small numbers of Lebanese girls sex trafficked in other Arab countries; Syrian refugees are vulnerable to forced labor and prostitution

tier rating: Tier 2 Watch List – Lebanon does not fully comply with the minimum standards for the elimination of trafficking; however, it is making significant efforts to do so; in 2014, Lebanon was granted a waiver from an otherwise required downgrade to Tier 3 because its government has a written plan that, if implemented would constitute making significant efforts to bring itself into compliance with the minimum standards for the elimination of trafficking; law enforcement efforts in 2014 were uneven; the number of convicted traffickers increased, but judges lack of familiarity with anti-trafficking law meant that many offenders were not brought to justice; the government relied heavily on an NGO to identify and provide service to trafficking victims; and its lack of thoroughly implemented victim identification procedures resulted in victims continuing to be arrested, detained, and deported for crimes committed as a direct result of being trafficked (2015)

Illicit drugs: Lebanon is a transit country for hashish, cocaine, heroin, and fenethylene; fenethylene, cannabis, hashish, and some opium are produced in the Bekaa Valley; small amounts of Latin American cocaine and Southwest Asian heroin transit country on way to European markets and for Middle Eastern consumption; money laundering of drug proceeds fuels concern that extremists are benefiting from drug trafficking

LESOTHO

INTRODUCTION

SOUTH AFRICA

Maputsoe · Butha-Buthe
Leribe
Teyateyaneng · Katse Reservoir
MASERU
Thaba-Tseka · Mokhotlong
Thabana Ntlenyana
Maleteng
Orange
Mohale's Hoek · Qacha's Nek
Quthing
SOUTH AFRICA

0 20 40 km
0 20 40 mi

Background: Paramount chief MOSHOESHOE I consolidated what would become Basutoland in the early 19th century and made himself king in 1822. Continuing encroachments by Dutch settlers from the neighboring Orange Free State caused the king to enter into an 1868 agreement with the UK by which Basutoland became a British protectorate, and after 1884, a crown colony. Upon independence in 1966, the country was renamed the Kingdom of Lesotho. The Basotho National Party ruled the country during its first two decades. King MOSHOESHOE II was exiled in 1990, but returned to Lesotho in 1992 and was reinstated in 1995 and subsequently succeeded by his son, King LETSIE III, in 1996. Constitutional government was restored in 1993 after seven years of military rule. In 1998, violent protests and a

military mutiny following a contentious election prompted a brief but bloody intervention by South African and Botswana military forces under the aegis of the Southern African Development Community. Subsequent constitutional reforms restored relative political stability. Peaceful parliamentary elections were held in 2002, but the National Assembly elections in 2007 were hotly contested and aggrieved parties disputed how the electoral law was applied to award proportional seats in the Assembly. In 2012, competitive elections involving 18 parties saw Prime Minister Motsoahae Thomas THABANE form a coalition government—the first in the country's history— that ousted the 14-year incumbent, Pakalitha MOSISILI, who peacefully transferred power the following month. MOSISILI returned to power in snap elections in February 2015 after the collapse of THABANE's coalition government and an alleged attempted military coup. In June 2017, THABANE returned to become prime minister.

GEOGRAPHY

Location: Southern Africa, an enclave of South Africa

Geographic coordinates: 29 30 S, 28 30 E

Map references: Africa

Area: *total:* 30,355 sq km
land: 30,355 sq km
water: 0 sq km
country comparison to the world: 142

Area—comparative: slightly smaller than Maryland

Land boundaries: *total:* 1,106 km
border countries (l): South Africa 1106 km

Coastline: 0 km (landlocked)

Maritime claims: none (landlocked)

Climate: temperate; cool to cold, dry winters; hot, wet summers

Terrain: mostly highland with plateaus, hills, and mountains

Elevation: *mean elevation:* 2,161 m
lowest point: junction of the Orange and Makhaleng Rivers 1,400 m
highest point: Thabana Ntlenyana 3,482 m

Natural resources: water, agricultural and grazing land, diamonds, sand, clay, building stone

Land use: *agricultural land:* 76.1% (2011 est.)
arable land: 10.1% (2011 est.) / *permanent crops:* 0.1% (2011 est.) / *permanent pasture:* 65.9% (2011 est.)
forest: 1.5% (2011 est.)
other: 22.4% (2011 est.)

Irrigated land: 30 sq km (2012)

Population distribution: relatively higher population density in the western half of the nation, with the capital of Maseru, and the smaller cities of Mafeteng, Teyateyaneng, and Leribe attracting the most people as shown in this population distribution map

Natural hazards: periodic droughts

Environment—current issues: population pressure forcing settlement in marginal areas results in overgrazing, severe soil erosion, and soil exhaustion; desertification; Highlands Water Project controls, stores, and redirects water to South Africa

Environment—international agreements: *party to:* Biodiversity, Climate Change, Climate Change-Kyoto Protocol, Desertification, Endangered Species, Hazardous Wastes, Law of the Sea, Marine Life Conservation, Ozone Layer Protection, Wetlands
signed, but not ratified: none of the selected agreements

Geography—note: landlocked, an enclave of (completely surrounded by) South Africa; mountainous, more than 80% of the country is 1,800 m above sea level

PEOPLE AND SOCIETY

Population: 1,969,334 (July 2020 est.)
note: estimates for this country explicitly take into account the effects of excess mortality due to AIDS; this can result in lower life expectancy, higher infant mortality, higher death rates, lower population growth rates, and changes in the distribution of population by age and sex than would otherwise be expected
country comparison to the world: 149

Nationality: *noun:* Mosotho (singular), Basotho (plural)
adjective: Basotho

Ethnic groups: Sotho 99.7%, Europeans, Asians, and other 0.3%

Languages: Sesotho (official) (southern Sotho), English (official), Zulu, Xhosa

Religions: Protestant 47.8% (Pentecostal 23.1%, Lesotho Evangelical 17.3%, Anglican 7.4%), Roman Catholic 39.3%, other Christian 9.1%, non-Christian 1.4%, none 2.3% (2014 est.)

Demographic profile: Lesotho faces great socioeconomic challenges. More than half of its population lives below the property line, and the country's HIV/AIDS prevalence rate is the second highest in the world. In addition, Lesotho is a small, mountainous, landlocked country with little arable land, leaving its population vulnerable to food shortages and reliant on remittances. Lesotho's persistently high infant, child, and maternal mortality rates have been increasing during the last decade, according to the last two Demographic and Health Surveys. Despite these significant shortcomings, Lesotho has made good progress in education; it is on-track to achieve universal primary education and has one of the highest adult literacy rates in Africa.

Lesotho's migration history is linked to its unique geography; it is surrounded by South Africa with which it shares linguistic and cultural traits. Lesotho at one time had more of its workforce employed outside its borders than any other country. Today remittances equal about 17% of its GDP. With few job options at home, a high rate of poverty, and higher wages available across the border, labor migration to South Africa replaced agriculture as the prevailing Basotho source of income decades ago. The majority of Basotho migrants were single men contracted to work as gold miners in South Africa. However, migration trends changed in the 1990s, and fewer men found mining jobs in South Africa because of declining gold prices, stricter immigration policies, and a preference for South African workers.

Although men still dominate cross-border labor migration, more women are working in South Africa, mostly as domestics, because they are widows or their husbands are unemployed. Internal rural-urban flows have also become more frequent, with more women migrating within the country to take up jobs in the garment industry or moving to care for loved ones with HIV/AIDS. Lesotho's small population of immigrants is increasingly composed of Taiwanese and Chinese migrants who are involved in the textile industry and small retail businesses.

Age structure: *0-14 years:* 31.3% (male 309,991/female 306,321)
15-24 years: 19.26% (male 181,874/female 197,452)
25-54 years: 38.86% (male 373,323/female 391,901)
55-64 years: 4.98% (male 52,441/female 45,726)
65 years and over: 5.6% (male 57,030/female 53,275) (2020 est.)

Dependency ratios: *total dependency ratio:* 59.2
youth dependency ratio: 51.3
elderly dependency ratio: 7.9
potential support ratio: 12.7 (2020 est.)

Median age: *total:* 24.7 years
male: 24.7 years
female: 24.7 years (2020 est.)
country comparison to the world: 164

Population growth rate: 0.16% (2020 est.)
country comparison to the world: 182

Birth rate: 23.2 births/1,000 population (2020 est.)
country comparison to the world: 56

Death rate: 15.4 deaths/1,000 population (2020 est.)
country comparison to the world: 1

Net migration rate: -6.1 migrant(s)/1,000 population (2020 est.)
country comparison to the world: 206

Population distribution: relatively higher population density in the western half of the nation, with the capital of Maseru, and the smaller cities of Mafeteng, Teyateyaneng, and Leribe attracting the most people as shown in this population distribution map

Urbanization: *urban population:* 29% of total population (2020)
rate of urbanization: 2.83% annual rate of change (2015-20 est.)
total population growth rate v. urban population growth rate, 2000-2030:

Major urban areas—population: 202,000 MASERU (capital) (2018)

Sex ratio: *at birth:* 1.03 male(s)/female
0-14 years: 1.01 male(s)/female
15-24 years: 0.92 male(s)/female
25-54 years: 0.95 male(s)/female
55-64 years: 1.15 male(s)/female
65 years and over: 1.07 male(s)/female
total population: 0.98 male(s)/female (2020 est.)

Mother's mean age at first birth: 21 years (2014 est.)
note: median age at first birth among women 25-29

Maternal mortality rate: 544 deaths/100,000 live births (2017 est.)
country comparison to the world: 17

Infant mortality rate: *total:* 41.5 deaths/1,000 live births
male: 44.8 deaths/1,000 live births
female: 38.1 deaths/1,000 live births (2020 est.)
country comparison to the world: 38

Life expectancy at birth: *total population:* 53 years
male: 53.1 years
female: 53 years (2020 est.)
country comparison to the world: 227

Total fertility rate: 2.5 children born/woman (2020 est.)
country comparison to the world: 74

Contraceptive prevalence rate: 64.9% (2018)

Drinking water source:
improved:
urban: 93% of population
rural: 72.4% of population
total: 78.2% of population
unimproved:
urban: 7% of population
rural: 27.6% of population
total: 21.8% of population (2017 est.)

Current Health Expenditure: 8.8% (2017)

Physicians density: 0.07 physicians/1,000 population (2010)

Sanitation facility access:
improved:
urban: 88.6% of population
rural: 52.3% of population
total: 62.4% of population
unimproved:
urban: 11.4% of population
rural: 47.7% of population
total: 37.6% of population (2017 est.)

HIV/AIDS—adult prevalence rate: 23.1% (2019 est.)
country comparison to the world: 2

HIV/AIDS—people living with HIV/AIDS: 340,000 (2019 est.)
country comparison to the world: 22

HIV/AIDS—deaths: 4,800 (2019 est.)
country comparison to the world: 27

Major infectious diseases: *degree of risk:* intermediate (2020)
food or waterborne diseases: bacterial diarrhea, hepatitis A, and typhoid fever

Obesity—adult prevalence rate: 16.6% (2016)
country comparison to the world: 122

Children under the age of 5 years underweight: 10.5% (2018)
country comparison to the world: 61

Education expenditures: 6.4% of GDP (2018)
country comparison to the world: 23

Literacy: *definition:* age 15 and over can read and write
total population: 79.4%
male: 70.1%
female: 88.3% (2015)

School life expectancy (primary to tertiary education): *total:* 12 years
male: 12 years
female: 13 years (2017)

Unemployment, youth ages 15-24: *total:* 34.4%
male: NA
female: NA (2013 est.)
country comparison to the world: 23

GOVERNMENT

Country name: *conventional long form:* Kingdom of Lesotho
conventional short form: Lesotho
local long form: Kingdom of Lesotho
local short form: Lesotho
former: Basutoland
etymology: the name translates as "Land of the Sesotho Speakers"

Government type: parliamentary constitutional monarchy

Capital: *name:* Maseru
geographic coordinates: 29 19 S, 27 29 E
time difference: UTC+ 2 (7 hours ahead of Washington, DC, during Standard Time)
etymology: in the Sesotho language the name means "[place of] red sandstones"

Administrative divisions: 10 districts; Berea, Butha-Buthe, Leribe, Mafeteng, Maseru, Mohale's Hoek, Mokhotlong, Qacha's Nek, Quthing, Thaba-Tseka

Independence: 4 October 1966 (from the UK)

National holiday: Independence Day, 4 October (1966)

Constitution: *history:* previous 1959, 1967; latest adopted 2 April 1993 (effectively restoring the 1967 version)
amendments: proposed by Parliament; passage of amendments affecting constitutional provisions, including fundamental rights and freedoms, sovereignty of the kingdom, the office of the king, and powers of Parliament, requires a majority vote by the National Assembly, approval by the Senate, approval in a referendum by a majority of qualified voters, and assent of the king; passage of amendments other than those specified provisions requires at least a two- thirds majority vote in both houses of Parliament; amended several times, last in 2011

Legal system: mixed legal system of English common law and Roman-Dutch law; judicial review of legislative acts in High Court and Court of Appeal

International law organization participation: accepts compulsory ICJ jurisdiction with reservations; accepts ICCt jurisdiction

Citizenship: *citizenship by birth:* yes
citizenship by descent only: yes
dual citizenship recognized: no
residency requirement for naturalization: 5 years

Suffrage: 18 years of age; universal

Executive branch: *chief of state:* King LETSIE III (since 7 February 1996); note—King LETSIE III formerly occupied the throne from November 1990 to February 1995 while his father was in exile
head of government: Prime Minister Moeketsi MAJORO (since 20 May 2020); note—Prime Minister Thomas THABANE resigned on 19 May 2020
cabinet: consists of the prime minister, appointed by the King on the advice of the Council of State, the deputy prime minister, and 26 other ministers

elections/appointments: the monarchy is hereditary, but under the terms of the constitution that came into effect after the March 1993 election, the monarch is a "living symbol of national unity" with no executive or legislative powers; under traditional law, the college of chiefs has the power to depose the monarch, to determine next in line of succession, or to serve as regent in the event that a successor is not of mature age; following legislative elections, the leader of the majority party or majority coalition in the Assembly automatically becomes prime minister

Legislative branch: *description:* bicameral Parliament consists of:
Senate (33 seats; 22 principal chiefs and 11 other senators nominated by the king with the advice of the Council of State, a 13-member body of key government and non-government officials; members serve 5-year terms)
National Assembly (120 seats; 80 members directly elected in single-seat constituencies by simple majority vote and 40 elected through proportional representation; members serve 5-year terms)
elections: Senate—last nominated by the king 11 July 2017 (next NA) National Assembly—last held on 3 June 2017 (next to be held in 2022)
election results: Senate—percent of votes by party—NA, seats by party—NA; composition—men 25, women 8, percent of women 24.2%
National Assembly—percent of votes by party—ABC 40.5%, DC 25.8%, LCD 9%, AD 7.3%, MEC 5.1%, BNP 4.1, PFD 2.3%, other 5.9%; seats by party—ABC 51, DC 30, LCD 11, AD 9, MEC 6, BNP 5, PFD 3, other 5; composition—men 95, women 27, percent of women 22.5%; note—total Parliament percent of women 22.9%

Judicial branch: *highest courts:* Court of Appeal (consists of the court president, such number of justices of appeal as set by Parliament, and the Chief Justice and the puisne judges of the High Court ex officio); High Court (consists of the chief justice and such number of puisne judges as set by Parliament); note—both the Court of Appeal and the High Court have jurisdiction in constitutional issues
judge selection and term of office: Court of Appeal president and High Court chief justice appointed by the monarch on the advice of the prime minister; puisne judges appointed by the monarch on advice of the Judicial Service Commission, an independent body of judicial officers and officials designated by the monarch; judges of both courts can serve until age 75
subordinate courts: Magistrate Courts; customary or traditional courts; military courts

Political parties and leaders:
All Basotho Convention or ABC [Thomas Motsoahae THABANE]
Alliance of Democrats or AD [Monyane MOLELEKI]
Basotho Congress Party or BCP [Thulo MAHLAKENG]
Basotho National Party or BNP [Thesele MASERIBANE]

Democratic Congress or DC [Pakalitha MOSISILI]
Democratic Party of Lesotho or DPL [Limpho TAU]
Lesotho Congress for Democracy or LCD [Mothetjoa METSING]
Movement of Economic Change or MEC [Selibe MOCHOBOROANE]
National Independent Party or NIP [Kimetso MATHABA]
Popular Front for Democracy of PFD [Lekhetho RAKUOANE]
Reformed Congress of Lesotho or RCL [Keketso RANTSO]

International organization participation: ACP, AfDB, AU, C, CD, FAO, G-77, IAEA, IBRD, ICAO, ICCt, ICRM, IDA, IFAD, IFC, IFRCS, ILO, IMF, Interpol, IOC, IOM, IPU, ISO (correspondent), ITU, MIGA, NAM, OPCW, SACU, SADC, UN, UNAMID, UNCTAD, UNESCO, UNHCR, UNIDO, UNWTO, UPU, WCO, WFTU (NGOs), WHO, WIPO, WMO, WTO

Diplomatic representation in the US: *chief of mission:* Ambassador Sankatana Gabriel MAJA (since 22 June 2018)
chancery: 2511 Massachusetts Avenue NW, Washington, DC 20008
telephone: [1] (202) 797-5533
FAX: [1] (202) 234-6815

Diplomatic representation from the US: chief of mission: Ambassador Rebecca E. GONZALES (since 8 February 2018)
telephone: [266] 22 312 666
embassy: 254 Kingsway Road, Maseru West
mailing address: P. O. Box 333, Maseru 100, Lesotho
FAX: [266] 22 310 116

Flag description: three horizontal stripes of blue (top), white, and green in the proportions of 3 : 4 : 3; the colors represent rain, peace, and prosperity respectively; centered in the white stripe is a black Basotho hat representing the indigenous people; the flag was unfurled in October 2006 to celebrate 40 years of independence

National symbol(s): mokorotio (Basotho hat); national colors: blue, white, green, black

National anthem: *name:* "Lesotho fatse la bo ntat'a rona" (Lesotho, Land of Our Fathers)
lyrics/music: Francois COILLARD/Ferdinand-Samuel LAUR
note: adopted 1967; music derives from an 1823 Swiss songbook

ECONOMY

Economy—overview: Small, mountainous, and completely landlocked by South Africa, Lesotho depends on a narrow economic base of textile manufacturing, agriculture, remittances, and regional customs revenue. About three-fourths of the people live in rural areas and engage in animal herding and subsistence agriculture, although Lesotho produces less than 20% of the nation's demand for food. Agriculture is vulnerable to weather and climate variability.

Lesotho relies on South Africa for much of its economic activity; Lesotho imports 85% of the goods it consumes from South Africa, including most agricultural inputs. Households depend heavily on remittances from family members working in South Africa in mines, on farms, and as domestic workers, though mining employment has declined substantially since the 1990s. Lesotho is a member of the Southern Africa Customs Union (SACU), and revenues from SACU accounted for roughly 26% of total GDP in 2016; however, SACU revenues are volatile and expected to decline over the next 5 years. Lesotho also gains royalties from the South African Government for water transferred to South Africa from a dam and reservoir system in Lesotho. However, the government continues to strengthen its tax system to reduce dependency on customs duties and other transfers.

The government maintains a large presence in the economy—government consumption accounted for about 26% of GDP in 2017. The government remains Lesotho's largest employer; in 2016, the government wage bill rose to 23% of GDP – the largest in Sub-Saharan Africa. Lesotho's largest private employer is the textile and garment industry—approximately 36,000 Basotho, mainly women, work in factories producing garments for export to South Africa and the US. Diamond mining in Lesotho has grown in recent years and accounted for nearly 35% of total exports in 2015. Lesotho managed steady GDP growth at an average of 4.5% from 2010 to 2014, dropping to about 2.5% in 2015-16, but poverty remains widespread around 57% of the total population.

GDP (purchasing power parity):
$6.656 billion (2017 est.)
$6.762 billion (2016 est.)
$6.561 billion (2015 est.)
note: data are in 2017 dollars
country comparison to the world: 170

GDP (official exchange rate):
$2.749 billion (2017 est.)

GDP—real growth rate: -1.6% (2017 est.)
3.1% (2016 est.)
2.5% (2015 est.)
country comparison to the world: 204

GDP—per capita (PPP): $3,300 (2017 est.)
$3,400 (2016 est.)
$3,300 (2015 est.)
note: data are in 2017 dollars
country comparison to the world: 190

Gross national saving:
20.3% of GDP (2017 est.)
19.7% of GDP (2016 est.)
24.7% of GDP (2015 est.)
country comparison to the world: 96

GDP—composition, by end use:
household consumption: 69.2% (2017 est.)
government consumption: 26.4% (2017 est.)
investment in fixed capital: 31.4% (2017 est.)
investment in inventories:—13.4% (2017 est.)
exports of goods and services: 40.8% (2017 est.)
imports of goods and services:—54.4% (2017 est.)

GDP—composition, by sector of origin:
agriculture: 5.8% (2016 est.)
industry: 39.2% (2016 est.)
services: 54.9% (2017 est.)

Agriculture—products: corn, wheat, pulses, sorghum, barley; livestock

Industries: food, beverages, textiles, apparel assembly, handicrafts, construction, tourism

Industrial production growth rate: 12.5% (2017 est.)
country comparison to the world: 6

Labor force: 930,800 (2017 est.)
country comparison to the world: 141

Labor force—by occupation: *agriculture:* 86%
industry and services: 14% (2002 est.)
note: most of the resident population is engaged in subsistence agriculture; roughly 35% of the active male wage earners work in South Africa

Unemployment rate:
28.1% (2014 est.)
25% (2008 est.)
country comparison to the world: 203

Population below poverty line: 57% (2016 est.)

Household income or consumption by percentage share: *lowest 10%:* 1%
highest 10%: 39.4% (2003)

Budget:
revenues: 1.09 billion (2017 est.)
expenditures: 1.255 billion (2017 est.)

Taxes and other revenues: 39.7% (of GDP) (2017 est.)
country comparison to the world: 43

Budget surplus (+) or deficit (-): -6% (of GDP) (2017 est.)
country comparison to the world: 184

Public debt: 33.7% of GDP (2017 est.)
36.2% of GDP (2016 est.)
country comparison to the world: 155

Fiscal year: 1 April–31 March

Inflation rate (consumer prices):
5.3% (2017 est.)
6.2% (2016 est.)
country comparison to the world: 173

Current account balance:
-$102 million (2017 est.)
-$201 million (2016 est.)
country comparison to the world: 87

Exports: $1.028 billion (2017 est.)
$894 million (2016 est.)
country comparison to the world: 159

Exports—partners: South Africa 57%, US 33.5% (2017)

Exports—commodities: manufactures (clothing, footwear), wool and mohair, food and live animals, electricity, water, diamonds

Imports: $1.826 billion (2017 est.)
$1.613 billion (2016 est.)
country comparison to the world: 172

Imports—commodities: food; building materials, vehicles, machinery, medicines, petroleum products

Imports—partners: South Africa 87.2% (2017)

Reserves of foreign exchange and gold: $657.7 million (31 December 2017 est.)
$925.2 million (31 December 2016 est.)
country comparison to the world: 143

Debt—external: $934.6 million (31 December 2017 est.)
$921.3 million (31 December 2016 est.)
country comparison to the world: 166

Exchange rates: maloti (LSL) per US dollar—
14.48 (2017 est.)
14.71 (2016 est.)
14.71 (2015 est.)
12.76 (2014 est.)
10.85 (2013 est.)

ENERGY

Electricity access: *population without electricity:* 1 million (2019)
electrification—total population: 36% (2019)
electrification—urban areas: 63% (2019)
electrification—rural areas: 26% (2019)

Electricity—production: 510 million kWh (2016 est.)
country comparison to the world: 165

Electricity—consumption: 847.3 million kWh (2016 est.)
country comparison to the world: 160

Electricity—exports: 0 kWh (2016 est.)
country comparison to the world: 158

Electricity—imports: 373 million kWh (2016 est.)
country comparison to the world: 84

Electricity—installed generating capacity: 80,400 kW (2016 est.)
country comparison to the world: 184

Electricity—from fossil fuels: 0% of total installed capacity (2016 est.)
country comparison to the world: 213

Electricity—from nuclear fuels: 0% of total installed capacity (2017 est.)
country comparison to the world: 127

Electricity—from hydroelectric plants: 100% of total installed capacity (2017 est.)
country comparison to the world: 1

Electricity—from other renewable sources: 1% of total installed capacity (2017 est.)
country comparison to the world: 159

Crude oil—production: 0 bbl/day (2018 est.)
country comparison to the world: 162

Crude oil—exports: 0 bbl/day (2015 est.)
country comparison to the world: 154

Crude oil—imports: 0 bbl/day (2015 est.)
country comparison to the world: 152

Crude oil—proved reserves: 0 bbl (1 January 2018 est.)
country comparison to the world: 157

Refined petroleum products—production: 0 bbl/day (2015 est.)
country comparison to the world: 165

Refined petroleum products—consumption: 5,000 bbl/day (2016 est.)
country comparison to the world: 179

Refined petroleum products—exports: bbl/day (2015 est.)
country comparison to the world: 171

Refined petroleum products—imports: 5,118 bbl/day (2015 est.)
country comparison to the world: 170

Natural gas—production: 0 cu m (2017 est.)
country comparison to the world: 157

Natural gas—consumption: 0 cu m (2017 est.)
country comparison to the world: 167

Natural gas—exports: 0 cu m (2017 est.)
country comparison to the world: 138

Natural gas—imports: 0 cu m (2017 est.)
country comparison to the world: 148

Natural gas—proved reserves: 0 cu m (1 January 2014 est.)
country comparison to the world: 158

Carbon dioxide emissions from consumption of energy: 711,100 Mt (2017 est.)
country comparison to the world: 177

COMMUNICATIONS

Telephones—fixed lines: *total subscriptions:* 7,865
subscriptions per 100 inhabitants: less than 1 (2019 est.)
country comparison to the world: 194
Telephones—mobile cellular:
total subscriptions: 2,238,186
subscriptions per 100 inhabitants: 113.83 (2019 est.)
country comparison to the world: 147

Telecommunication systems: *general assessment:* mobile penetration remains below regional average; introduction of mobile broadband in the country & LTE technology, with 5G trials in early 2019; fixed-line teledensity is low; mobile-cellular telephone system is growth sector; regulator considering improving SIM card registration (2020)
domestic: fixed-line is 1 per 100 subscriptions; mobile-cellular service dominates the market with a subscribership now over 114 per 100 persons; rudimentary system consisting of a modest number of landlines, a small microwave radio relay system, and a small radiotelephone communication system (2019)
international: country code—266; Internet accessibility has improved with several submarine fiber optic cables that land on African east and west coasts, but the country's land locked position makes access prices expensive; satellite earth station—1 Intelsat (Atlantic Ocean) (2019)
note: the COVID-19 outbreak is negatively impacting telecommunications production and supply chains globally; consumer spending on telecom devices and services has also slowed due to the pandemic's effect on economies worldwide; overall progress towards improvements in all facets of the telecom industry—mobile, fixed-line, broadband, submarine cable and satellite—has moderated

Broadcast media: 1 state-owned TV station and 2 state-owned radio stations; government controls most private broadcast media; satellite TV subscription service available; transmissions of multiple international broadcasters obtainable (2019)

Internet country code: .ls

Internet users: *total:* 569,114

percent of population: 29% (July 2018 est.)
country comparison to the world: 149

Broadband—fixed subscriptions: *total:* 5,763
subscriptions per 100 inhabitants: less than 1 (2018 est.)
country comparison to the world: 178

TRANSPORTATION

Civil aircraft registration country code prefix: 7P (2016)

Airports: 24 (2013)
country comparison to the world: 130

Airports—with paved runways: *total:* 3 (2019)
over 3,047 m: 1
914 to 1,523 m: 1
under 914 m: 1

Airports—with unpaved runways:
total: 21 (2013)
914 to 1,523 m: 5 (2013)
under 914 m: 16 (2013)

Roadways: *total:* 5,940 km (2011)
paved: 1,069 km (2011)
unpaved: 4,871 km (2011)
country comparison to the world: 144

MILITARY AND SECURITY

Military and security forces:

Lesotho Defense Force (LDF): Army (includes Air Wing) (2019)

Military expenditures:
1.5% of GDP (2019)
1.8% of GDP (2018)
2% of GDP (2017)
1.8% of GDP (2016)
1.9% of GDP (2015)
country comparison to the world: 82

Military and security service personnel strengths: the Lesotho Defense Force (LDF) has approximately 2,000 personnel, including 150 for its air wing (2019 est.)

Military equipment inventories and acquisitions: the LDF's inventory consists of older equipment from a variety of countries; the only reported delivery to the LDF since 2007 was two helicopters from France in 2017 (2019 est.)

Military service age and obligation: 18-24 years of age for voluntary military service; no conscription; women serve as commissioned officers (2019)

Military—note: Lesotho's declared policy for its military is the maintenance of the country's sovereignty and the preservation of internal security; in practice, external security is guaranteed by South Africa

TRANSNATIONAL ISSUES

Disputes—international: South Africa has placed military units to assist police operations along the border of Lesotho, Zimbabwe, and Mozambique to control smuggling, poaching, and illegal migration

Trafficking in persons: current situation: Lesotho is a source, transit, and destination country for women and children subjected to forced labor and sex trafficking and for men subjected to forced labor; in Lesotho and South Africa, Basotho women and children are subjected to domestic servitude, and Basotho children increasingly endure commercial sexual exploitation; some Basotho men who voluntarily migrate to South Africa for work become victims of forced labor in agriculture and mining or are coerced into committing crimes; foreign nationals continue to traffic fellow citizens in Lesotho

tier rating: Tier 2 Watch List – Lesotho does not fully comply with the minimum standards for the elimination of trafficking; however, it is making significant efforts to do so; in 2014, Lesotho was granted a waiver from an otherwise required downgrade to Tier 3 because its government has a written plan that, if implemented would constitute making significant efforts to bring itself into compliance with the minimum standards for the elimination of trafficking; the government failed to initiate any prosecutions against alleged traffickers and has not convicted any offenders under the 2011 anti-trafficking act, which remains unimplemented for a fifth year; authorities did not develop formal victim identification and referral procedures, did not establish victim care centers, as required under the 2011 anti-trafficking act, and did not support NGOs offering victims protective services (2015)

LIBERIA

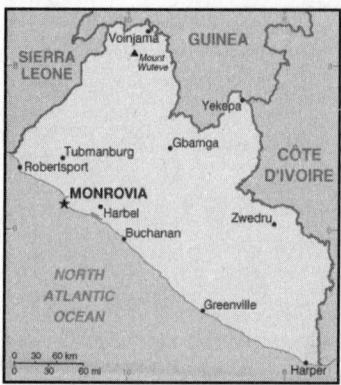

INTRODUCTION

Background: Settlement of freed slaves from the US in what is today Liberia began in 1822; by 1847, the Americo-Liberians were able to establish a republic. William TUBMAN, president from 1944-71, did much to promote foreign investment and to bridge the economic, social, and political gaps between the descendants of the original settlers and the inhabitants of the interior. In 1980, a military coup led by Samuel DOE ushered in a decade of authoritarian rule. In December 1989, Charles TAYLOR launched a rebellion against DOE's regime that led to a prolonged civil war in which DOE was killed. A period of relative peace in 1997 allowed for an election that brought TAYLOR to power, but major fighting resumed in 2000. An August 2003 peace agreement ended the war and prompted the resignation of former president Charles TAYLOR, who was convicted by the UN- backed Special Court for Sierra Leone in The Hague for his involvement in Sierra Leone's civil war. After two years of rule by a transitional government, democratic elections in late 2005 brought President Ellen JOHNSON SIRLEAF to power. She subsequently won reelection in 2011 but was challenged to rebuild Liberia's economy, particularly following the 2014-15 Ebola epidemic, and to reconcile a nation still recovering from 14 years of fighting. Constitutional term limits barred President JOHNSON SIRLEAF from running for re-election. Legal challenges delayed the 2017 presidential runoff election, which was eventually won by George WEAH. In March 2018, the UN completed its 15-year peacekeeping mission in Liberia.

GEOGRAPHY

Location: Western Africa, bordering the North Atlantic Ocean, between Cote d'Ivoire and Sierra Leone

Geographic coordinates: 6 30 N, 9 30 W

Map references: Africa

Area: *total:* 111,369 sq km
land: 96,320 sq km
water: 15,049 sq km
country comparison to the world: 105

Area—comparative: slightly larger than Virginia

Land boundaries: *total:* 1,667 km
border countries (3): Guinea 590 km, Cote d'Ivoire 778 km, Sierra Leone 299 km

Coastline: 579 km

Maritime claims: *territorial sea:* 200 nm

Climate: tropical; hot, humid; dry winters with hot days and cool to cold nights; wet, cloudy summers with frequent heavy showers

Terrain: mostly flat to rolling coastal plains rising to rolling plateau and low mountains in northeast

Elevation: mean elevation: 243 m
lowest point: Atlantic Ocean 0 m
highest point: Mount Wuteve 1,447 m

Natural resources: iron ore, timber, diamonds, gold, hydropower

Land use: *agricultural land:* 28.1% (2011 est.)
arable land: 5.2% (2011 est.) / permanent crops: 2.1% (2011 est.) / permanent pasture: 20.8% (2011 est.)
forest: 44.6% (2011 est.)
other: 27.3% (2011 est.)

Irrigated land: 30 sq km (2012)

Population distribution: more than half of the population lives in urban areas, with approximately one-third living within an 80-km radius of Monrovia as shown in this population distribution map

Natural hazards: dust-laden harmattan winds blow from the Sahara (December to March)

Environment—current issues: tropical rain forest deforestation; soil erosion; loss of biodiversity; hunting of endangered species for bushmeat; pollution of coastal waters from oil residue and raw sewage; pollution of rivers from industrial run-off; burning and dumping of household waste

Environment—international agreements: *party to:* Biodiversity, Climate Change, Climate Change-Kyoto Protocol, Desertification, Endangered Species, Hazardous Wastes, Law of the Sea, Ozone Layer Protection, Ship Pollution, Tropical Timber 83, Tropical Timber 94, Wetlands
signed, but not ratified: Environmental Modification, Marine Life Conservation

Geography—note: facing the Atlantic Ocean, the coastline is characterized by lagoons, mangrove swamps, and river-deposited sandbars; the inland grassy plateau supports limited agriculture

PEOPLE AND SOCIETY

Population: 5,073,296 (July 2020 est.)
country comparison to the world: 124

Nationality: noun: Liberian(s)
adjective: Liberian

Ethnic groups: Kpelle 20.3%, Bassa 13.4%, Grebo 10%, Gio 8%, Mano 7.9%, Kru 6%, Lorma 5.1%, Kissi 4.8%, Gola 4.4%, Krahn 4%, Vai 4%, Mandingo 3.2%, Gbandi 3%, Mende 1.3%, Sapo 1.3%, other Liberian 1.7%, other African 1.4%, non-African 1% (2008 est.)

Languages: English 20% (official), some 20 ethnic group languages few of which can be written or used in correspondence

Religions: Christian 85.6%, Muslim 12.2%, Traditional 0.6%, other 0.2%, none 1.5% (2008 est.)

Demographic profile: Liberia's high fertility rate of nearly 5 children per woman and large youth cohort – more than 60% of the population is under the age of 25 – will sustain a high dependency ratio for many years to come. Significant progress has been made in preventing child deaths, despite a lack of health care workers and infrastructure. Infant and child mortality have dropped nearly 70% since 1990; the annual reduction rate of about 5.4% is the highest in Africa.

Nevertheless, Liberia's high maternal mortality rate remains among the world's worst; it reflects a high unmet need for family planning services, frequency of early childbearing, lack of quality obstetric care, high adolescent fertility, and a low proportion of births attended by a medical professional. Female mortality is also increased by the prevalence of female genital cutting (FGC), which is practiced by 10 of Liberia's 16 tribes and affects more than two- thirds of women and girls. FGC is an initiation ritual performed in rural bush schools, which teach traditional beliefs on marriage and motherhood and are an obstacle to formal classroom education for Liberian girls.

Liberia has been both a source and a destination for refugees. During Liberia's 14-year civil war (1989-2003), more than 250,000 people became refugees and another half million were internally displaced. Between 2004 and the cessation of refugee status for Liberians in June 2012, the UNHCR helped more than 155,000 Liberians to voluntarily repatriate, while others returned home on their own. Some Liberian refugees spent more than two decades living in other West African countries. Liberia hosted more than 125,000 Ivoirian refugees escaping post-election violence in 2010-11; as of mid- 2017, about 12,000 Ivoirian refugees were still living in Liberia as of October 2017 because of instability.

Age structure: *0-14 years:* 43.35% (male 1,111,479/female 1,087,871)
15-24 years: 20.35% (male 516,136/female 516,137)
25-54 years: 30.01% (male 747,983/female 774,615)
55-64 years: 3.46% (male 89,150/female 86,231)
65 years and over: 2.83% (male 70,252/female 73,442) (2020 est.)

Dependency ratios: *total dependency ratio:* 77.6
youth dependency ratio: 71.7
elderly dependency ratio: 5.9
potential support ratio: 17 (2020 est.)

Median age: *total:* 18 years
male: 17.7 years
female: 18.2 years (2020 est.)
country comparison to the world: 216

Population growth rate: 2.71% (2020 est.)
country comparison to the world: 14

Birth rate: 37.3 births/1,000 population (2020 est.)
country comparison to the world: 13

Death rate: 7 deaths/1,000 population (2020 est.)
country comparison to the world: 127

Net migration rate: -2.9 migrant(s)/1,000 population (2020 est.)

country comparison to the world: 174

Population distribution: more than half of the population lives in urban areas, with approximately one-third living within an 80-km radius of Monrovia as shown in this population distribution map

Urbanization: *urban population:* 52.1% of total population (2020)
rate of urbanization: 3.41% annual rate of change (2015-20 est.)
total population growth rate v. urban population growth rate, 2000-2030:

Major urban areas—population: 1.517 million MONROVIA (capital) (2020)

Sex ratio: *at birth:* 1.03 male(s)/female
0-14 years: 1.02 male(s)/female
15-24 years: 1 male(s)/female
25-54 years: 0.97 male(s)/female
55-64 years: 1.03 male(s)/female
65 years and over: 0.96 male(s)/female
total population: 1 male(s)/female (2020 est.)

Mother's mean age at first birth: 19.2 years (2013 est.)
note: median age at first birth among women 25-29

Maternal mortality rate: 661 deaths/100,000 live births (2017 est.)
country comparison to the world: 10

Infant mortality rate:
total: 47.4 deaths/1,000 live births
male: 51.7 deaths/1,000 live births
female: 43.1 deaths/1,000 live births (2020 est.)
country comparison to the world: 27

Life expectancy at birth:
total population: 64.7 years
male: 62.5 years
female: 67 years (2020 est.)
country comparison to the world: 200

Total fertility rate: 4.9 children born/woman (2020 est.)
country comparison to the world: 15

Contraceptive prevalence rate: 31.2% (2016)

Drinking water source:
improved:
urban: 93.8% of population
rural: 67.9% of population
total: 81% of population
unimproved:
urban: 6.2% of population
rural: 32.1% of population
total: 19% of population (2017 est.)

Current Health Expenditure: 8.2% (2017)

Physicians density: 0.04 physicians/1,000 population (2015)

Hospital bed density: 0.8 beds/1,000 population (2010)

Sanitation facility access:
improved:
urban: 64.1% of population
rural: 23.5% of population
total: 44.1% of population
unimproved:

urban: 35.9% of population
rural: 76.5% of population
total: 55.9% of population (2017 est.)

HIV/AIDS—adult prevalence rate: 1.5% (2019 est.)
country comparison to the world: 29

HIV/AIDS—people living with HIV/AIDS: 47,000 (2019 est.)
country comparison to the world: 63

HIV/AIDS—deaths: 1,900 (2019 est.)
country comparison to the world: 46

Major infectious diseases: *degree of risk:* very high (2020)
food or waterborne diseases: bacterial and protozoal diarrhea, hepatitis A, and typhoid fever
vectorborne diseases: malaria, dengue fever, and yellow fever
water contact diseases: schistosomiasis
animal contact diseases: rabies
aerosolized dust or soil contact diseases: Lassa fever

Obesity—adult prevalence rate: 9.9% (2016)
country comparison to the world: 141

Children under the age of 5 years underweight: 13.6% (2016)
country comparison to the world: 43

Education expenditures: 3.8% of GDP (2017)
country comparison to the world: 114

Literacy: *definition:* age 15 and over can read and write
total population: 48.3%
male: 62.7%
female: 34.1% (2017)

Unemployment, youth ages 15-24: total: 2.3%
male: 2.4%
female: 2.2% (2016 est.)
country comparison to the world: 174

GOVERNMENT

Country name: *conventional long form:* Republic of Liberia
conventional short form: Liberia
etymology: name derives from the Latin word "liber" meaning "free"; so named because the nation was created as a homeland for liberated African-American slaves

Government type: presidential republic

Capital: *name:* Monrovia

geographic coordinates: 6 18 N, 10 48 W
time difference: UTC 0 (5 hours ahead of Washington, DC, during Standard Time)
etymology: named after James Monroe (1758-1831), the fifth president of the United States and supporter of the colonization of Liberia by freed slaves; one of two national capitals named for a US president, the other is Washington, D. C.

Administrative divisions: 15 counties; Bomi, Bong, Gbarpolu, Grand Bassa, Grand Cape Mount, Grand Gedeh, Grand Kru, Lofa, Margibi, Maryland, Montserrado, Nimba, River Cess, River Gee, Sinoe

Independence: 26 July 1847

National holiday: Independence Day, 26 July (1847)

Constitution: *history:* previous 1847 (at independence); latest drafted 19 October 1983, revised version adopted by referendum 3 July 1984, effective 6 January 1986

amendments: proposed by agreement of at least two thirds of both National Assembly houses or by petition of at least 10,000 citizens; passage requires at least two- thirds majority approval of both houses and approval in a referendum by at least two- thirds majority of registered voters; amended 2011

Legal system: mixed legal system of common law, based on Anglo-American law, and customary law

International law organization participation: accepts compulsory ICJ jurisdiction with reservations; accepts ICCt jurisdiction

Citizenship: *citizenship by birth:* no

citizenship by descent only: at least one parent must be a citizen of Liberia

dual citizenship recognized: no

residency requirement for naturalization: 2 years

Suffrage: 18 years of age; universal

Executive branch: *chief of state:* President George WEAH (since 22 January 2018); Vice President Jewel HOWARD- TAYLOR (since 22 January 2018); note—the president is both chief of state and head of government

head of government: President George WEAH (since 22 January 2018); Vice President Jewel HOWARD- TAYLOR (since 22 January 2018)

cabinet: Cabinet appointed by the president, confirmed by the Senate

elections/appointments: president directly elected by absolute majority popular vote in 2 rounds if needed for a 6-year term (eligible for a second term); election last held on 10 October 2017 with a run-off on 26 December 2017) (next to be held on 10 October 2023); the runoff originally scheduled for 7 November 2017 was delayed due to allegations of fraud in the first round, which the Supreme Court dismissed

election results: George WEAH elected president in second round; percent of vote in first round—George WEAH (Coalition for Democratic Change) 38.4%, Joseph BOAKAI (UP) 28.8%, Charles BRUMSKINE (LP) 9.6%, Prince JOHNSON (MDR) 8.2%, Alexander B. CUMMINGS (ANC) 7.2%, other 7.8%; percentage of vote in second round—George WEAH 61.5%, Joseph BOAKAI 38.5%

Legislative branch: *description:* bicameral National Assembly consists of:

The Liberian Senate (30 seats; members directly elected in 15 2-seat districts by simple majority vote to serve 9-year staggered terms; each district elects 1 senator and elects the second senator 3 years later, followed by a 6-year hiatus, after which the first Senate seat is up for election)

House of Representatives (73 seats; members directly elected in single- seat districts by simple majority vote to serve 6-year terms; eligible for a second term)

elections: Senate—last held on 20 December 2014; by election to fill the senate seats vacated by WEAH and HOWARD-TAYLOR was held on 31 July 2018 (next general election to be held on 31 December 2020)

House of Representatives—last held on 10 October 2017 (next to be held in October 2023)

election results: Senate—percent of vote by party—CDC 29.8%, UP 10.3%, LP 11.5%, NPP 6.1%, PUP 4.9%, ANC 4.2%, NDC 1.3%, other 7.6%, independent 24.3%; seats by party—UP 4, CDC 2, LP 2, ANC 1, NDC 1, NPP 1, PUP 1, independent 3; composition—men 27, women 3, percent of women 10%

House of Representatives—percent of vote by party/coalition—Coalition for Democratic Change 15.6%, UP 14%, LP 8.7%, ANC 6.1%, PUP 5.9%, ALP 5.1%, MDR 3.4%, other 41.2%; seats by coalition/party—Coalition for Democratic Change 21, UP 20, PUP 5, LP 3, ALP 3, MDR 2, independent 13, other 6; composition—men 64, women 9, percent of women 12.3%; total Parliament percent of women 11.7%

Judicial branch: *highest courts:* Supreme Court (consists of a chief justice and 4 associate justices); note—the Supreme Court has jurisdiction for all constitutional cases

judge selection and term of office: chief justice and associate justices appointed by the president of Liberia with consent of the Senate; judges can serve until age 70

subordinate courts: judicial circuit courts; special courts, including criminal, civil, labor, traffic; magistrate and traditional or customary courts

Political parties and leaders:

Alliance for Peace and Democracy or APD [Marcus S. G. DAHN]

All Liberian Party or ALP [Benoi UREY]

Alternative National Congress or ANC [Orishil GOULD]

Coalition for Democratic Change [George WEAH] (includes CDC, NPP, and LPDP)

Congress for Democratic Change or CDC [George WEAH]

Liberia Destiny Party or LDP [Nathaniel BARNES]

Liberia National Union or LINU [Nathaniel BLAMA]

Liberia Transformation Party or LTP [Julius SUKU]

Liberian People Democratic Party or LPDP [Alex J. TYLER]

Liberian People's Party or LPP

Liberty Party or LP [J. Fonati KOFFA]

Movement for Democracy and Reconstruction or MDR [Prince Y. JOHNSON]

Movement for Economic Empowerment [J. Mill JONES, Dr.]

Movement for Progressive Change or MPC [Simeon FREEMAN]

National Democratic Coalition or NDC [Dew MAYSON]

National Democratic Party of Liberia or NDPL [D. Nyandeh SIEH]

National Patriotic Party or NPP [Jewel HOWARD TAYLOR]

National Reformist Party or NRP [Maximillian T. W. DIABE]

National Union for Democratic Progress or NUDP [Victor BARNEY]

People's Unification Party or PUP [Isobe GBORKORKOLLIE]

Unity Party or UP [Varney SHERMAN]

United People's Party [MacDonald WENTO]

Victory for Change Party [Marcus R. JONES]

International organization participation: ACP, AfDB, AU, ECOWAS, EITI (compliant country), FAO, G- 77, IAEA, IBRD, ICAO, ICC (NGOs), ICCt, ICRM, IDA, IFAD, IFC, IFRCS, ILO, IMF, IMO, IMSO, Interpol, IOC, IOM, ISO (correspondent), ITU, ITUC (NGOs), MIGA, MINUSMA, NAM, OPCW, UN, UNCTAD, UNESCO, UNIDO, UNWTO, UPU, WCO, WFTU (NGOs), WHO, WIPO, WMO, WTO (observer)

Diplomatic representation in the US: *chief of mission:* Ambassador George PATTEN (since 11 January 2019)

chancery: 5201 16th Street NW, Washington, DC 20011

telephone: [1] (202) 723-0437

FAX: [1] (202) 723-0436

consulate(s) general: New York

Diplomatic representation from the US: *chief of mission:* Charge d'Affaires Alyson GRUNDER (since 21 March 2020)

telephone: [231] 77-677-7000

embassy: U. S. Embassy, 502 Benson Street, Monrovia

mailing address: P. O. Box 98, Monrovia

FAX: [231] 77-677-7370

Flag description: 11 equal horizontal stripes of red (top and bottom) alternating with white; a white five-pointed star appears on a blue square in the upper hoist- side corner; the stripes symbolize the signatories of the Liberian Declaration of Independence; the blue square represents the African mainland, and the star represents the freedom granted to the ex- slaves; according to the constitution, the blue color signifies liberty, justice, and fidelity, the white color purity, cleanliness, and guilelessness, and the red color steadfastness, valor, and fervor

note: the design is based on the US flag

National symbol(s): white star; national colors: red, white, blue

National anthem: *name:* All Hail, Liberia Hail!

lyrics/music: Daniel Bashiel WARNER/Olmstead LUCA

note: lyrics adopted 1847, music adopted 1860; the anthem's author later became the third president of Liberia

0:00 / 1:28

ECONOMY

Economy—overview: Liberia is a low-income country that relies heavily on foreign assistance and remittances from the diaspora. It is richly endowed with water, mineral resources, forests, and a climate favorable to agriculture. Its principal exports are iron ore, rubber, diamonds, and gold. Palm oil and cocoa are emerging as new export products. The government has attempted to revive raw timber extraction and is encouraging oil exploration.

In the 1990s and early 2000s, civil war and government mismanagement destroyed much of Liberia's economy, especially infrastructure in and around the capital. Much of the conflict was fueled by control over Liberia's natural resources. With the conclusion of fighting and the installation of a democratically elected government in 2006, businesses that had fled the country began to return. The country achieved high growth during the period 2010-13 due to favorable world prices for its commodities. However, during the 2014-2015 Ebola crisis, the economy declined and many foreign-owned businesses departed with their capital and expertise. The epidemic forced the government to divert scarce resources to combat the spread of the virus, reducing funds available for needed public investment. The cost of addressing the Ebola epidemic coincided with decreased economic activity reducing government revenue, although higher donor support significantly offset this loss. During the same period, global commodities prices for key exports fell and have yet to recover to pre-Ebola levels.

In 2017, gold was a key driver of growth, as a new mining project began its first full year of production; iron ore exports are also increased as Arcelor Mittal opened new mines at Mount Gangra. The completion of the rehabilitation of the Mount Coffee Hydroelectric Dam increased electricity production to support ongoing and future economic activity, although electricity tariffs remain high relative to other countries in the region and transmission infrastructure is limited. Presidential and legislative elections in October 2017 generated election-related spending pressures.

Revitalizing the economy in the future will depend on economic diversification, increasing investment and trade, higher global commodity prices, sustained foreign aid and remittances, development of infrastructure and institutions, combating corruption, and maintaining political stability and security.

GDP (purchasing power parity):
$6.112 billion (2017 est.)
$5.965 billion (2016 est.)
$6.064 billion (2015 est.)
note: data are in 2017 dollars
country comparison to the world: 173

GDP (official exchange rate):
$3.285 billion (2017 est.)

GDP—real growth rate: 2.5% (2017 est.)
-1.6% (2016 est.)
0% (2015 est.)
country comparison to the world: 111

GDP—per capita (PPP): $1,300 (2017 est.)
$1,300 (2016 est.)
$1,300 (2015 est.)
note: data are in 2017 dollars

country comparison to the world: 221

Gross national saving:
NA% (2017)
-21.9% of GDP (2016 est.)
1.9% of GDP (2016 est.)
country comparison to the world: 184

GDP—composition, by end use:
household consumption: 128.8% (2016 est.)
government consumption: 16.7% (2016 est.)
investment in fixed capital: 19.5% (2016 est.)
investment in inventories: 6.7% (2016 est.)
exports of goods and services: 17.5% (2016 est.)
imports of goods and services: -89.2% (2016 est.)

GDP—composition, by sector of origin:
agriculture: 34% (2017 est.)
industry: 13.8% (2017 est.)
services: 52.2% (2017 est.)

Agriculture—products: rubber, coffee, cocoa, rice, cassava (manioc, tapioca), palm oil, sugarcane, bananas; sheep, goats; timber

Industries: mining (iron ore and gold), rubber processing, palm oil processing, diamonds

Industrial production growth rate:
9% (2017 est.)
country comparison to the world: 19

Labor force: 1.677 million (2017 est.)
country comparison to the world: 125

Labor force—by occupation: agriculture: 70%
industry: 8%
services: 22% (2000 est.)

Unemployment rate: 2.8% (2014 est.)
country comparison to the world: 33

Population below poverty line: 54.1% (2014 est.)

Household income or consumption by percentage share: *lowest 10%:* 2.4%
highest 10%: 30.1% (2007)

Budget: *revenues:* 553.6 million (2017 est.)
expenditures: 693.8 million (2017 est.)

Taxes and other revenues: 16.9% (of GDP) (2017 est.)
country comparison to the world: 174

Budget surplus (+) or deficit (-): -4.3% (of GDP) (2017 est.)
country comparison to the world: 161

Public debt: 34.4% of GDP (2017 est.)
28.3% of GDP (2016 est.)
country comparison to the world: 154

Fiscal year: calendar year

Inflation rate (consumer prices): 12.4% (2017 est.)
8.8% (2016 est.)
country comparison to the world: 207

Current account balance: -$627 million (2017 est.)
-$464 million (2016 est.)
country comparison to the world: 129

Exports: $260.6 million (2017 est.)
$169.8 million (2016 est.)
country comparison to the world: 185

Exports—partners: Germany 36.2%, Switzerland 14.2%, UAE 8.8%, US 6.8%, Indonesia 4.7% (2017)

Exports—commodities: rubber, timber, iron, diamonds, cocoa, coffee

Imports: $1.166 billion (2017 est.)
$1.296 billion (2016 est.)
country comparison to the world: 179

Imports—commodities: fuels, chemicals, machinery, transportation equipment, manufactured goods; foodstuffs

Imports—partners: Singapore 29.8%, China 24.4%, South Korea 17.5%, Japan 9.4% (2017)

Reserves of foreign exchange and gold:
$459.8 million (31 December 2017 est.)
$528.7 million (31 December 2016 est.)
country comparison to the world: 154

Debt—external: $1.036 billion (31 December 2017 est.)
$938.9 million (31 December 2016 est.)
country comparison to the world: 164

Exchange rates: Liberian dollars (LRD) per US dollar—
109.4 (2017 est.)
93.4 (2016 est.)
93.4 (2015 est.)
85.3 (2014 est.)
83.893 (2013 est.)

ENERGY

Electricity access: *population without electricity:* 4 million (2019)
electrification—total population: 12% (2019)
electrification—urban areas: 18% (2019)
electrification—rural areas: 6% (2019)

Electricity—production: 300 million kWh (2016 est.)
note: according to a 2014 household survey, only 4.5% of Liberians use Liberia Electricity Corporation (LEC) power, 4.9% use a community generator, 4.4% have their own generator, 3.9% use vehicle batteries, and 0.8% use other sources of electricity, and 81.3% have no access to electricity; LEC accounts for roughly 70 million kWh of ouput.
country comparison to the world: 184

Electricity—consumption: 279 million kWh (2016 est.)
country comparison to the world: 187

Electricity—exports: 0 kWh (2016 est.)
country comparison to the world: 159

Electricity—imports: 0 kWh (2016 est.)
country comparison to the world: 168

Electricity—installed generating capacity: 151,000 kW (2016 est.)
country comparison to the world: 173

Electricity—from fossil fuels: 57% of total installed capacity (2016 est.)
country comparison to the world: 138

Electricity—from nuclear fuels: 0% of total installed capacity (2017 est.)

country comparison to the world: 128

Electricity—from hydroelectric plants: 43% of total installed capacity (2017 est.)
country comparison to the world: 46

Electricity—from other renewable sources: 0% of total installed capacity (2017 est.)
country comparison to the world: 198

Crude oil—production: 0 bbl/day (2018 est.)
country comparison to the world: 163

Crude oil—exports: 0 bbl/day (2015 est.)
country comparison to the world: 155

Crude oil—imports: 0 bbl/day (2015 est.)
country comparison to the world: 153

Crude oil—proved reserves: 0 bbl (1 January 2018 est.)
country comparison to the world: 158

Refined petroleum products—production: 0 bbl/day (2017 est.)
country comparison to the world: 166

Refined petroleum products—consumption: 8,000 bbl/day (2016 est.)
country comparison to the world: 164

Refined petroleum products—exports: 0 bbl/day (2015 est.)
country comparison to the world: 172

Refined petroleum products—imports: 8,181 bbl/day (2015 est.)
country comparison to the world: 152

Natural gas—production: 0 cu m (2017 est.)
country comparison to the world: 158

Natural gas—consumption: 0 cu m (2017 est.)
country comparison to the world: 168

Natural gas—exports: 0 cu m (2017 est.)
country comparison to the world: 139

Natural gas—imports: 0 cu m (2017 est.)
country comparison to the world: 149

Natural gas—proved reserves: 0 cu m (1 January 2014 est.)
country comparison to the world: 159

Carbon dioxide emissions from consumption of energy: 1.163 million Mt (2017 est.)
country comparison to the world: 164

COMMUNICATIONS

Telephones—fixed lines: *total subscriptions:* 8,394
subscriptions per 100 inhabitants: less than 1 (2019 est.)
country comparison to the world: 193

Telephones—mobile cellular: *total subscriptions:* 2,793,316
subscriptions per 100 inhabitants: 56.57 (2019 est.)
country comparison to the world: 145

Telecommunication systems: *general assessment:* the limited services available are found almost exclusively in the capital, Monrovia; fixed-line service is stagnant and extremely limited; telephone coverage recently extended to a number

of other towns and rural areas by four mobile-cellular network operators; Liberia is almost entirely a wireless telecommunications market; a number of operators avoid paying dues and operate despite regulations; govt. regulatory impose SIM card registration in an attempt to reduce crime, but makes mobile penetration seem low; the high cost and limited bandwidth of connections means that Internet access is expensive and data rates are very low (2020)
domestic: fixed-line less than 1 per 100; mobile-cellular subscription base growing and teledensity approached 57 per 100 persons (2019)
international: country code—231; landing point for the ACE submarine cable linking 20 West African countries and Europe; satellite earth station—1 Intelsat (Atlantic Ocean) (2019)
note: the COVID-19 outbreak is negatively impacting telecommunications production and supply chains globally; consumer spending on telecom devices and services has also slowed due to the pandemic's effect on economies worldwide; overall progress shows improvements in all facets of the telecom industry—mobile, fixed-line, broadband, submarine cable and satellite—has moderated

Broadcast media: 8 private and 1 government-owned TV station; satellite TV service available; 1 state-owned radio station; approximately 20 independent radio stations broadcasting in Monrovia, with approximately 80 more local stations operating in other areas; transmissions of 4 international (including the British Broadcasting Corporation and Radio France Internationale) broadcasters are available (2019)

Internet country code: .lr

Internet users: *total:* 383,819
percent of population: 7.98% (July 2018 est.)
country comparison to the world: 159

Broadband—fixed subscriptions: *total:* 8,000
subscriptions per 100 inhabitants: less than 1 (2017 est.)
country comparison to the world: 174

TRANSPORTATION

Civil aircraft registration country code prefix: A8 (2016)

Airports: 29 (2013)
country comparison to the world: 116

Airports—with paved runways: *total:* 2 (2019)
over 3,047 m: 1
1,524 to 2,437 m: 1

Airports—with unpaved runways: *total:* 27 (2013)
1,524 to 2,437 m: 5 (2013)
914 to 1,523 m: 8 (2013)
under 914 m: 14 (2013)

Pipelines: 4 km oil (2013)

Railways: *total:* 429 km (2008)
standard gauge: 345 km 1.435-m gauge (2008)
narrow gauge: 84 km 1.067-m gauge (2008)
note: most sections of the railways inoperable due to damage sustained during the civil wars from 1980 to 2003, but many are being rebuilt

country comparison to the world: 117

Roadways: *total:* 10,600 km (2018)
paved: 657 km (2018)
unpaved: 9,943 km (2018)
country comparison to the world: 135

Merchant marine: *total:* 3,496
by type: bulk carrier 1,161, container ship 854, general cargo 145, oil tanker 761, other 575 (2019)
country comparison to the world: 7

Ports and terminals: *major seaport(s):* Buchanan, Monrovia
Military and security forces:

MILITARY AND SECURITY

Armed Forces of Liberia (AFL): Army, Liberia Air Wing, Liberian Coast Guard (2019)

Military expenditures:
0.5% of GDP (2019)
0.4% of GDP (2018)
0.4% of GDP (2017)
0.4% of GDP (2016)
0.5% of GDP (2015)
country comparison to the world: 147

Military and security service personnel strengths: the Armed Forces of Liberia (AFL) have approximately 2,000 personnel (2019)

Military equipment inventories and acquisitions: the AFL has almost no significant combat hardware as nearly all aircraft, equipment, materiel, and facilities were damaged or destroyed during the country's civil war; it has received little new equipment outside of ammunition, small arms, and trucks from China in 2008 and boats donated to the Coast Guard by the US in 2011 and 2016 (2019)

Military deployments: 150 Mali (MINUSMA) (2020)

Military service age and obligation: 18 years of age for voluntary military service; no conscription (2012)

TRANSNATIONAL ISSUES

Disputes—international: as the UN Mission in Liberia (UNMIL) continues to drawdown prior to the 1 March 2018 closure date, the peacekeeping force is being reduced to 434 soldiers and two police units; some Liberian refugees still remain in Guinea, Cote d'Ivoire, Sierra Leone, and Ghana; Liberia shelters 8,804 Ivoirian refugees, as of 2019

Refugees and internally displaced persons: *refugees (country of origin):* 8,098 (Cote d'Ivoire) (2020)

Illicit drugs: transshipment point for Southeast and Southwest Asian heroin and South American cocaine for the European and US markets; corruption, criminal activity, arms-dealing, and diamond trade provide significant potential for money laundering, but the lack of well-developed financial system limits the country's utility as a major money-laundering center

LIBYA

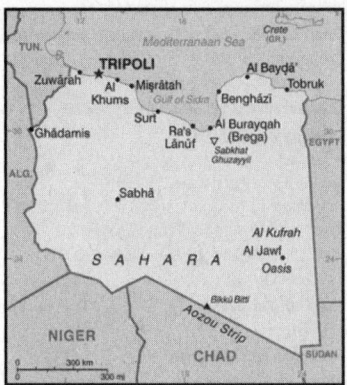

INTRODUCTION

Background: Berbers have inhabited central north Africa since ancient times, but the region has been settled and ruled by Phoenicians, Greeks, Carthaginians, Persians, Egyptians, Greeks, Romans, and Vandals. In the 7th century, Islam spread through the region; in the mid-16th century, Ottoman rule began. The Italians supplanted the Ottoman Turks in the area around Tripoli in 1911 and did not relinquish their hold until 1943 when they were defeated in World War II. Libya then passed to UN administration and achieved independence in 1951. Following a 1969 military coup, Col. Muammar al-QADHAFI assumed leadership and began to espouse his political system at home, which was a combination of socialism and Islam. During the 1970s, QADHAFI used oil revenues to promote his ideology outside Libya, supporting subversive and terrorist activities that included the downing of two airliners—one over Scotland, another in Northern Africa—and a discotheque bombing in Berlin. UN sanctions in 1992 isolated QADHAFI politically and economically following the attacks; sanctions were lifted in 2003 following Libyan acceptance of responsibility for the bombings and agreement to claimant compensation. QADHAFI also agreed to end Libya's program to develop weapons of mass destruction, and he made significant strides in normalizing relations with Western nations.

Unrest that began in several Middle Eastern and North African countries in late 2010 erupted in Libyan cities in early 2011. QADHAFI's brutal crackdown on protesters spawned a civil war that triggered UN authorization of air and naval intervention by the international community. After months of seesaw fighting between government and opposition forces, the QADHAFI regime was toppled in mid- 2011 and replaced by a transitional government known as the National Transitional Council (NTC). In 2012, the NTC handed power to an elected parliament, the General National

Congress (GNC). Voters chose a new parliament to replace the GNC in June 2014—the House of Representatives (HoR), which relocated to the eastern city of Tobruk after fighting broke out in Tripoli and Benghazi in July 2014.

In December 2015, the UN brokered an agreement among a broad array of Libyan political parties and social groups—known as the Libyan Political Agreement (LPA). Members of the Libyan Political Dialogue, including representatives of the HoR and GNC, signed the LPA in December 2015. The LPA called for the formation of an interim Government of National Accord or GNA, with a nine-member Presidency Council, the HoR, and an advisory High Council of State that most ex-GNC members joined. The LPA's roadmap for a transition to a new constitution and elected government was subsequently endorsed by UN Security Council Resolution 2259, which also called upon member states to cease official contact with parallel institutions. In January 2016, the HoR voted to approve the LPA, including the Presidency Council, while voting against a controversial provision on security leadership positions and the Presidency Council's proposed cabinet of ministers. In March 2016, the GNA Presidency Council seated itself in Tripoli. In 2016, the GNA twice announced a slate of ministers who operate in an acting capacity, but the HoR did not endorse the ministerial list. The HoR and defunct-GNC- affiliated political hardliners continued to oppose the GNA and hamper the LPA's implementation. In September 2017, UN Special Representative Ghassan SALAME announced a new roadmap for national political reconciliation. SALAME's plan called for amendments to the LPA, a national conference of Libyan leaders, and a constitutional referendum and general elections. In November 2018, the international partners supported SALAME's recalibrated Action Plan for Libya that aimed to break the political deadlock by holding a National Conference in Libya in 2019 on a timeline for political transition. The National Conference was delayed following a failure of the parties to implement an agreement mediated by SALAME in Abu Dhabi on February 27, and the subsequent military action by Khalifa HAFTAR's Libyan National Army against GNA forces in Tripoli that began in April 2019.

GEOGRAPHY

Location: Northern Africa, bordering the Mediterranean Sea, between Egypt, Tunisia, and Algeria

Geographic coordinates: 25 00 N, 17 00 E

Map references: Africa

Area: *total:* 1,759,540 sq km
land: 1,759,540 sq km
water: 0 sq km
country comparison to the world: 18

Area—comparative: about 2.5 times the size of Texas; slightly larger than Alaska

Land boundaries: *total:* 4,339 km
border countries (6): Algeria 989 km, Chad 1050 km, Egypt 1115 km, Niger 342 km, Sudan 382 km, Tunisia 461 km

Coastline: 1,770 km

Maritime claims: *territorial sea:* 12 nm
exclusive fishing zone: 62 nm
note: Gulf of Sidra closing line—32 degrees, 30 minutes north

Climate: Mediterranean along coast; dry, extreme desert interior

Terrain: mostly barren, flat to undulating plains, plateaus, depressions

Elevation: *mean elevation:* 423 m
lowest point: Sabkhat Ghuzayyil -47 m
highest point: Bikku Bitti 2,267 m

Natural resources: petroleum, natural gas, gypsum

Land use: *agricultural land:* 8.8% (2011 est.)
arable land: 1% (2011 est.) / permanent crops:
0.2% (2011 est.) / permanent pasture: 7.6% (2011 est.)
forest: 0.1% (2011 est.)
other: 91.1% (2011 est.)
Irrigated land: 4,700 sq km (2012)

Population distribution: well over 90% of the population lives along the Mediterranean coast in and between Tripoli to the west and Al Bayda to the east; the interior remains vastly underpopulated due to the Sahara and lack of surface water as shown in this population distribution map

Natural hazards: hot, dry, dust-laden ghibli is a southern wind lasting one to four days in spring and fall; dust storms, sandstorms

Environment—current issues: desertification; limited natural freshwater resources; the Great Manmade River Project, the largest water development scheme in the world, brings water from large aquifers under the Sahara to coastal cities; water pollution is a significant problem; the combined impact of sewage, oil byproducts, and industrial waste threatens Libya's coast and the Mediterranean Sea

Environment—international agreements: *party to:* Biodiversity, Climate Change, Climate Change-Kyoto Protocol, Desertification, Endangered Species, Hazardous Wastes, Marine Dumping, Ozone Layer Protection, Ship Pollution, Wetlands
signed, but not ratified: Law of the Sea
Geography—note:
note 1: more than 90% of the country is desert or semidesert
note 2: the volcano Waw an Namus lies in south central Libya in the middle of the Sahara; the caldera is an oasis—the name means "oasis of mosquitoes"—containing several small lakes surrounded by vegetation and hosting various insects and a large diversity of birds

PEOPLE AND SOCIETY

Population: 6,890,535 (July 2020 est.)
note: immigrants make up just over 12% of the total population, according to UN data (2017)
country comparison to the world: 107

Nationality: *noun:* Libyan(s)
adjective: Libyan

Ethnic groups: Berber and Arab 97%, other 3% (includes Egyptian, Greek, Indian, Italian, Maltese, Pakistani, Tunisian, and Turkish)

Languages: Arabic (official), Italian, English (all widely understood in the major cities); Berber (Nafusi, Ghadamis, Suknah, Awjilah, Tamasheq)

Religions: Muslim (official; virtually all Sunni) 96.6%, Christian 2.7%, Buddhist 0.3%, Hindu < 0.1, Jewish < 0.1, folk religion < 0.1, unaffiliated 0.2%, other < 0.1 (2010 est.)
note: non-Sunni Muslims include native Ibadhi Muslims (< 1% of the population) and foreign Muslims

MENA religious affiliation: *Demographic profile:* Despite continuing unrest, Libya remains a destination country for economic migrants. It is also a hub for transit migration to Europe because of its proximity to southern Europe and its lax border controls. Labor migrants have been drawn to Libya since the development of its oil sector in the 1960s. Until the latter part of the 1990s, most migrants to Libya were Arab (primarily Egyptians and Sudanese). However, international isolation stemming from Libya's involvement in international terrorism and a perceived lack of support from Arab countries led QADHAFI in 1998 to adopt a decade-long pan-African policy that enabled large numbers of Sub-Saharan migrants to enter Libya without visas to work in the construction and agricultural industries. Although Sub-Saharan Africans provided a cheap labor source, they were poorly treated and were subjected to periodic mass expulsions.

By the mid-2000s, domestic animosity toward African migrants and a desire to reintegrate into the international community motivated QADHAFI to impose entry visas on Arab and African immigrants and to agree to joint maritime patrols and migrant repatriations with Italy, the main recipient of illegal migrants departing Libya. As his regime neared collapse in 2011, QADHAFI reversed his policy of cooperating with Italy to curb illegal migration and sent boats loaded with migrants and asylum seekers to strain European resources. Libya's 2011 revolution decreased immigration drastically and prompted nearly 800,000 migrants to flee to third countries, mainly Tunisia and Egypt, or to their countries of origin. The inflow of migrants declined in 2012 but returned to normal levels by 2013, despite continued hostility toward Sub-Saharan Africans and a less-inviting job market.

While Libya is not an appealing destination for migrants, since 2014, transiting migrants – primarily from East and West Africa – continue to exploit its political instability and weak border controls and use it as a primary departure area to migrate across the central Mediterranean to Europe in growing numbers. In addition, more than 200,000 people were displaced internally as of August 2017 by fighting between armed groups in eastern and western Libya and, to a lesser extent, by inter-tribal clashes in the country's south.

Age structure: *0-14 years:* 33.65% (male 1,184,755/female 1,134,084)
15-24 years: 15.21% (male 534,245/female 513,728)
25-54 years: 41.57% (male 1,491,461/female 1,373,086)
55-64 years: 5.52% (male 186,913/female 193,560)
65 years and over: 4.04% (male 129,177/female 149,526) (2020 est.)

Dependency ratios: *total dependency ratio:* 47.7
youth dependency ratio: 41
elderly dependency ratio: 6.7
potential support ratio: 15 (2020 est.)

Median age: *total:* 25.8 years
male: 25.9 years
female: 25.7 years (2020 est.)
country comparison to the world: 156

Population growth rate: 1.94% (2020 est.)
country comparison to the world: 50

Birth rate: 23 births/1,000 population (2020 est.)
country comparison to the world: 59

Death rate: 3.5 deaths/1,000 population (2020 est.)
country comparison to the world: 219

Net migration rate: -0.7 migrant(s)/1,000 population (2020 est.)
country comparison to the world: 132

Population distribution: well over 90% of the population lives along the Mediterranean coast in and between Tripoli to the west and Al Bayda to the east; the interior remains vastly underpopulated due to the Sahara and lack of surface water as shown in this population distribution map

Urbanization: *urban population:* 80.7% of total population (2020)
rate of urbanization: 1.68% annual rate of change (2015-20 est.)
total population growth rate v. urban population growth rate, 2000-2030:

Major urban areas—population: 1.165 million TRIPOLI (capital), 881,000 Misratah, 824,000 Benghazi (2020)

Sex ratio: *at birth:* 1.05 male(s)/female
0-14 years: 1.04 male(s)/female
15-24 years: 1.04 male(s)/female
25-54 years: 1.09 male(s)/female
55-64 years: 0.97 male(s)/female
65 years and over: 0.86 male(s)/female
total population: 1.05 male(s)/female (2020 est.)

Maternal mortality rate: 72 deaths/100,000 live births (2017 est.)
country comparison to the world: 81

Infant mortality rate: *total:* 11.5 deaths/1,000 live births
male: 12.9 deaths/1,000 live births
female: 10 deaths/1,000 live births (2020 est.)
country comparison to the world: 113

Life expectancy at birth: *total population:* 76.7 years
male: 74.4 years
female: 79.1 years (2020 est.)
country comparison to the world: 90

Total fertility rate: 3.17 children born/woman (2020 est.)
country comparison to the world: 48

Contraceptive prevalence rate: 27.7% (2014)

Drinking water source:
improved:
total: 98.5% of population
unimproved:
total: 1.5% of population (2017 est.)

Physicians density: 2.09 physicians/1,000 population (2017)

Hospital bed density: 3.2 beds/1,000 population (2017)

Sanitation facility access:
improved:
total: 100% of population
unimproved:
total: 0% of population (2017 est.)

HIV/AIDS—adult prevalence rate: 0.2% (2019)
country comparison to the world: 103

HIV/AIDS—people living with HIV/AIDS: 9,500 (2019)
country comparison to the world: 106

HIV/AIDS—deaths: <500 (2019)

Obesity—adult prevalence rate: 32.5% (2016)
country comparison to the world: 16

Children under the age of 5 years underweight: 11.7% (2014)
country comparison to the world: 56

Education expenditures: NA

Literacy: *definition:* age 15 and over can read and write
total population: 91%
male: 96.7%
female: 85.6% (2015)

Unemployment, youth ages 15-24:
total: 48.7%
male: 40.8%
female: 67.8% (2012 est.)
country comparison to the world: 4

GOVERNMENT

Country name: *conventional long form:* State of Libya
conventional short form: Libya
local long form: Dawiat Libiya
local short form: Libiya
etymology: name derives from the Libu, an ancient Libyan tribe first mentioned in texts from the 13th century B. C.

Government type: in transition

Capital: *name:* Tripoli (Tarabulus)

geographic coordinates: 32 53 N, 13 10 E
time difference: UTC+ 2 (7 hours ahead of Washington, DC, during Standard Time)

etymology: originally founded by the Phoenicians as Oea in the 7th century B. C., the city changed rulers many times over the successive centuries; by the beginning of the 3rd century A.D. the region around the city was referred to as Regio Tripolitana by the Romans, meaning "region of the three cities"—namely Oea (i. e., modern Tripoli), Sabratha (to the west), and Leptis Magna (to the east); over time, the shortened name of "Tripoli" came to refer to just Oea, which derives from the Greek words "tria" and "polis" meaning "three cities"

Administrative divisions: 22 governorates (muhafazah, singular—muhafazat); Al Butnan, Al Jabal al Akhdar, Al Jabal al Gharbi, Al Jafarah, Al Jufrah, Al Kufrah, Al Marj, Al Marqab, Al Wahat, An Nuqat al Khams, Az Zawiyah, Banghazi (Benghazi), Darnah, Ghat, Misratah, Murzuq, Nalut, Sabha, Surt, Tarabulus (Tripoli), Wadi al Hayat, Wadi ash Shati

Independence: 24 December 1951 (from UN trusteeship)

National holiday: Liberation Day, 23 October (2011)

Constitution: *history:* previous 1951, 1977; in July 2017, the Constitutional Assembly completed and approved a draft of a new permanent constitution; in September 2018, the House of Representatives passed a constitutional referendum law in a session with contested reports of the quorum needed to pass the vote, and submitted it to the High National Elections Commission in December to begin preparations for a constitutional referendum

Legal system: Libya's post-revolution legal system is in flux and driven by state and non-state entities

International law organization participation: has not submitted an ICJ jurisdiction declaration; non-party state to the ICC

Citizenship: *citizenship by birth:* no
citizenship by descent only: at least one parent or grandparent must be a citizen of Libya
dual citizenship recognized: no
residency requirement for naturalization: varies from 3 to 5 years

Suffrage: 18 years of age, universal

Executive branch: *chief of state:* Chairman, Presidential Council, Fayiz al-SARAJ (since December 2015)

head of government: Prime Minister Fayiz al-SARAJ (since December 2015)
cabinet: GNA Presidency Council (pending approval by the House of Representatives—as of December 2018)

elections/appointments: direct presidential election to be held pending election-related legislation and constitutional referendum law

election results: NA

Legislative branch: *description:* unicameral House of Representatives (Majlis Al Nuwab) or HoR (200 seats including 32 reserved for women; members directly elected by majority vote; member term NA); note—the High Council of State serves as an advisory group for the HoR

elections: last held on 25 June 2014 (parliamentary election to be held pending election-related legislation); note—the Libyan Supreme Court in November 2014 declared the HoR election unconstitutional, but the HoR and the international community rejected the ruling

election results: percent of vote by party—NA; seats by party—NA; composition—men 158, women 30, percent of women 16%; note—only 188 of the 200 seats were filled in the June 2014 election because of boycotts and lack of security at some polling stations; some elected members of the HoR also boycotted the election

Judicial branch: NA; note—government is in transition

Political parties and leaders: NA

International organization participation: ABEDA, AfDB, AFESD, AMF, AMU, AU, BDEAC, CAEU, COMESA, FAO, G- 77, IAEA, IBRD, ICAO, ICC (NGOs), ICRM, IDA, IDB, IFAD, IFC, IFRCS, ILO, IMF, IMO, IMSO, Interpol, IOC, IOM, IPU, ISO, ITSO, ITU, LAS, MIGA, NAM, OAPEC, OIC, OPCW, OPEC, PCA, UN, UNCTAD, UNESCO, UNIDO, UNWTO, UPU, WCO, WFTU (NGOs), WHO, WIPO, WMO, WTO (observer)

Diplomatic representation in the US: chief of mission:
Ambassador Wafa M. T. BUGHAIGHIS (since 29 November 2017)
chancery: 1460 Dahlia Street NW, Washington, DC
telephone: [1] (202) 944-9601
FAX: [1] (202) 944-9606

Diplomatic representation from the US: *chief of mission:* Ambassador Richard B. NORLAND (since 22 August 2019)
telephone: [218] (0) 91-220-3239
embassy: Sidi Slim Area/Walie Al-Ahed Road, Tripoli (temporarily closed); please direct inquiries regarding US citizens in Libya to Libya Emergency USC@ state. gov
mailing address: US Embassy, 8850 Tripoli Place, Washington, DC 20521-8850
note: the US Embassy in Tripoli closed in July 2014 due to fighting near the embassy related to Libyan civil unrest; embassy staff and operations temporarily first relocated to Valetta, Malta and currently are temporarily relocated to Tunis, Tunisia

Flag description: three horizontal bands of red (top), black (double width), and green with a white crescent and star centered on the black stripe; the National Transitional Council reintroduced this flag design of the former Kingdom of Libya (1951-1969) on 27 February 2011; it replaced the former all-green banner promulgated by the QADHAFI regime in 1977; the colors represent the three major regions of the country: red stands for Fezzan, black symbolizes Cyrenaica, and green denotes Tripolitania; the crescent and star represent Islam, the main religion of the country

National symbol(s): star and crescent, hawk; national colors: red, black, green

National anthem: *name:* Libya, Libya, Libya
lyrics/music: Al Bashir AL AREBI/Mohamad Abdel WAHAB
note: also known as "Ya Beladi" or "Oh, My Country!"; adopted 1951; readopted 2011 with some modification to the lyrics; during the QADHAFI years between 1969 and 2011, the anthem was "Allahu Akbar," (God is Great) a marching song of the Egyptian Army in the 1956 Suez War

ECONOMY

Economy—overview: Libya's economy, almost entirely dependent on oil and gas exports, has struggled since 2014 given security and political instability, disruptions in oil production, and decline in global oil prices. The Libyan dinar has lost much of its value since 2014 and the resulting gap between official and black market exchange rates has spurred the growth of a shadow economy and contributed to inflation. The country suffers from widespread power outages, caused by shortages of fuel for power generation. Living conditions, including access to clean drinking water, medical services, and safe housing have all declined since 2011. Oil production in 2017 reached a five-year high, driving GDP growth, with daily average production rising to 879,000 barrels per day. However, oil production levels remain below the average pre- Revolution highs of 1.6 million barrels per day.

The Central Bank of Libya continued to pay government salaries to a majority of the Libyan workforce and to fund subsidies for fuel and food, resulting in an estimated budget deficit of about 17% of GDP in 2017. Low consumer confidence in the banking sector and the economy as a whole has driven a severe liquidity shortage.

GDP (purchasing power parity):
$61.97 billion (2017 est.)
$37.78 billion (2016 est.)
$40.8 billion (2015 est.)
note: data are in 2017 dollars
country comparison to the world: 106

GDP (official exchange rate):
$30.57 billion (2017 est.)

GDP—real growth rate: 64% (2017 est.)
-7.4% (2016 est.)
-13% (2015 est.)
country comparison to the world: 1

GDP—per capita (PPP): $9,600 (2017 est.)
$5,900 (2016 est.)
$6,500 (2015 est.)
note: data are in 2017 dollars
country comparison to the world: 141

Gross national saving:
5% of GDP (2017 est.)
-9% of GDP (2016 est.)
-25.1% of GDP (2015 est.)
country comparison to the world: 177

GDP—composition, by end use:
household consumption: 71.6% (2017 est.)
government consumption: 19.4% (2017 est.)
investment in fixed capital: 2.7% (2017 est.)

571

investment in inventories: 1.3% (2016 est.)
exports of goods and services: 38.8% (2017 est.)
imports of goods and services: -33.8% (2017 est.)

GDP—composition, by sector of origin:
agriculture: 1.3% (2017 est.)
industry: 52.3% (2017 est.)
services: 46.4% (2017 est.)

Agriculture—products: wheat, barley, olives, dates, citrus, vegetables, peanuts, soybeans; cattle

Industries: petroleum, petrochemicals, aluminum, iron and steel, food processing, textiles, handicrafts, cement

Industrial production growth rate: 60.3% (2017 est.)
country comparison to the world: 1

Labor force: 1.114 million (2017 est.)
country comparison to the world: 137

Labor force—by occupation: *agriculture:* 17%
industry: 23%
services: 59% (2004 est.)

Unemployment rate: 30% (2004 est.)
country comparison to the world: 206

Population below poverty line:
note: about one-third of Libyans live at or below the national poverty line

Household income or consumption by percentage share: *lowest 10%:* NA
highest 10%: NA

Budget: *revenues:* 15.78 billion (2017 est.)
expenditures: 23.46 billion (2017 est.)

Taxes and other revenues: 51.6% (of GDP) (2017 est.)
country comparison to the world: 14

Budget surplus (+) or deficit (-): -25.1% (of GDP) (2017 est.)
country comparison to the world: 219

Public debt: 4.7% of GDP (2017 est.)
7.5% of GDP (2016 est.)
country comparison to the world: 205

Fiscal year: calendar year

Inflation rate (consumer prices): 28.5% (2017 est.)
25.9% (2016 est.)
country comparison to the world: 221

Current account balance: $2.574 billion (2017 est.)
-$4.575 billion (2016 est.)
country comparison to the world: 36

Exports: $18.38 billion (2017 est.)
$11.99 billion (2016 est.)
country comparison to the world: 71

Exports—partners: Italy 19%, Spain 12.5%, France 11%, Egypt 8.6%, Germany 8.6%, China 8.3%, US 4.9%, UK 4.6%, Netherlands 4.5% (2017)

Exports—commodities: crude oil, refined petroleum products, natural gas, chemicals

Imports: $11.36 billion (2017 est.)
$8.667 billion (2016 est.)
country comparison to the world: 94

Imports—commodities: machinery, semi-finished goods, food, transport equipment, consumer products

Imports—partners: China 13.5%, Turkey 11.3%, Italy 6.9%, South Korea 5.9%, Spain 4.8% (2017)

Reserves of foreign exchange and gold:
$74.71 billion (31 December 2017 est.)
$66.05 billion (31 December 2016 est.)
country comparison to the world: 31

Debt—external:
$3.02 billion (31 December 2017 est.)
$3.116 billion (31 December 2016 est.)
country comparison to the world: 143

Exchange rates: Libyan dinars (LYD) per US dollar—
1.413 (2017 est.)
1.3904 (2016 est.)
1.3904 (2015 est.)
1.379 (2014 est.)
1.2724 (2013 est.)

ENERGY

Electricity access: electrification—total population: 98.5% (2016)
electrification—urban areas: 99.1% (2016)
electrification—rural areas: 96.4% (2016)

Electricity—production: 34.24 billion kWh (2016 est.)
note: persistent electricity shortages have contributed to the ongoing instability throughout the country
country comparison to the world: 61

Electricity—consumption: 27.3 billion kWh (2016 est.)
country comparison to the world: 65

Electricity—exports: 0 kWh (2015 est.)
country comparison to the world: 160

Electricity—imports: 376 million kWh (2016 est.)
country comparison to the world: 83

Electricity—installed generating capacity: 9.46 million kW (2016 est.)
country comparison to the world: 62

Electricity—from fossil fuels: 100% of total installed capacity (2016 est.)
country comparison to the world: 11

Electricity—from nuclear fuels: 0% of total installed capacity (2017 est.)
country comparison to the world: 129

Electricity—from hydroelectric plants: 0% of total installed capacity (2017 est.)
country comparison to the world: 183

Electricity—from other renewable sources: 0% of total installed capacity (2017 est.)
country comparison to the world: 199

Crude oil—production: 1.039 million bbl/day (2018 est.)
country comparison to the world: 19

Crude oil—exports: 337,800 bbl/day (2015 est.)
note: Libyan crude oil export values are highly volatile because of continuing protests and other disruptions across the country
country comparison to the world: 23

Crude oil—imports: 0 bbl/day (2015 est.)
country comparison to the world: 154

Crude oil—proved reserves: 48.36 billion bbl (1 January 2018 est.)
country comparison to the world: 9

Refined petroleum products—production: 89,620 bbl/day (2015 est.)
country comparison to the world: 69

Refined petroleum products—consumption: 260,000 bbl/day (2016 est.)
country comparison to the world: 49

Refined petroleum products—exports: 16,880 bbl/day (2015 est.)
country comparison to the world: 71

Refined petroleum products—imports: 168,200 bbl/day (2015 est.)
country comparison to the world: 36

Natural gas—production: 9.089 billion cu m (2017 est.)
country comparison to the world: 43

Natural gas—consumption: 4.446 billion cu m (2017 est.)
country comparison to the world: 64

Natural gas—exports: 4.644 billion cu m (2017 est.)
country comparison to the world: 31

Natural gas—imports: 0 cu m (2017 est.)
country comparison to the world: 150

Natural gas—proved reserves: 1.505 trillion cu m (1 January 2018 est.)
country comparison to the world: 21

Carbon dioxide emissions from consumption of energy: 46.48 million Mt (2017 est.)
country comparison to the world: 62

COMMUNICATIONS

Telephones—fixed lines: *total subscriptions:* 1,618,511
subscriptions per 100 inhabitants: 23.95 (2019 est.)
country comparison to the world: 62

Telephones—mobile cellular: *total subscriptions:* 6,182,105
subscriptions per 100 inhabitants: 91.48 (2019 est.)
country comparison to the world: 111

Telecommunication systems: *general assessment:* political and security instability in Libya has disrupted its telecommunications sector, but much of its infrastructure remains superior to that in most other African countries; registering a SIM card now requires proof of ID; govt. established new independent regulatory authority; LTE- based fixed broadband network launched; highest market penetration rates in Africa; growth opportunity in broadband sector (2020)
domestic: 24 per 100 fixed-line and 91 per 100 mobile-cellular subscriptions; service generally adequate (2019)
international: country code—218; landing points for LFON, EIG, Italy- Libya, Silphium and Tobrok-Emasaed submarine cable system

connecting Europe, Africa, the Middle East and Asia; satellite earth stations—4 Intelsat, Arabsat, and Intersputnik; microwave radio relay to Tunisia and Egypt; tropospheric scatter to Greece; participant in Medarabtel (2019)

note: the COVID-19 outbreak is negatively impacting telecommunications production and supply chains globally; consumer spending on telecom devices and services has also slowed due to the pandemic's effect on economies worldwide; overall progress towards improvements in all facets of the telecom industry—mobile, fixed-line, broadband, submarine cable and satellite—has moderated

Broadcast media: state-funded and private TV stations; some provinces operate local TV stations; pan-Arab satellite TV stations are available; state-funded radio (2019)

Internet country code: .ly

Internet users: *total:* 1,440,859
percent of population: 21.76% (July 2018 est.)
country comparison to the world: 131

Broadband—fixed subscriptions: *total:* 168,920
subscriptions per 100 inhabitants: 3 (2017 est.)
country comparison to the world: 113

TRANSPORTATION

National air transport system: number of registered air carriers: 9 (2020)
inventory of registered aircraft operated by air carriers: 55
annual passenger traffic on registered air carriers: 927,153 (2018)

Civil aircraft registration country code prefix: 5A (2016)

Airports: 146 (2013)
country comparison to the world: 39

Airports—with paved runways: *total:* 68 (2017)
over 3,047 m: 23 (2017)
2,438 to 3,047 m: 7 (2017)
1,524 to 2,437 m: 30 (2017)
914 to 1,523 m: 7 (2017)
under 914 m: 1 (2017)

Airports—with unpaved runways: *total:* 78 (2013)
over 3,047 m: 2 (2013)
2,438 to 3,047 m: 5 (2013)
1,524 to 2,437 m: 14 (2013)
914 to 1,523 m: 37 (2013)
under 914 m: 20 (2013)

Heliports: 2 (2013)

Pipelines: 882 km condensate, 3743 km gas, 7005 km oil (2013)

Roadways: *total:* 37,000 km (2010)
paved: 34,000 km (2010)
unpaved: 3,000 km (2010)
country comparison to the world: 92

Merchant marine: *total:* 94
by type: general cargo 2, oil tanker 12, other 80 (2019)
country comparison to the world: 94

Ports and terminals: *major seaport(s):* Marsa al Burayqah (Marsa el Brega), Tripoli
oil terminal(s): Az Zawiyah, Ra's Lanuf
LNG terminal(s) (export): Marsa el Brega

MILITARY AND SECURITY

Military and security forces: note—in transition; the Government of National Accord (GNA) has various ground, air, naval, and coast guard forces under its command; the forces are comprised of a mix of semi- regular military units, tribal militias, civilian volunteers, and foreign troops and mercenaries

forces under Khalifa HAFTER, known as the Libyan National Army (LNA), also include various ground, air, and naval units comprised of semi-regular military personnel, tribal militias, and foreign troops and mercenaries (2019)

Military and security service personnel strengths: the sizes of the forces of both the Government of National Accord and the Libyan National Army are unknown (2020 est.)

Military equipment inventories and acquisitions: both the forces of the Government of National Accord and the Libyan National Army are largely equipped with weapons of Russian or Soviet origin (2020 est.)

TERRORISM

Terrorist group(s): Ansar al-Sharia groups; Islamic State of Iraq and ash-Sham – Libya; al-Mulathamun Battalion (al-Mourabitoun); al-Qa'ida in the Islamic Maghreb (2019)

note: details about the history, aims, leadership, organization, areas of operation, tactics, targets, weapons, size, and sources of support of the group(s) appear(s) in Appendix- T

TRANSNATIONAL ISSUES

Disputes—international: dormant disputes include Libyan claims of about 32,000 sq km still reflected on its maps of southeastern Algeria and the FLN's assertions of a claim to Chirac Pastures in southeastern Morocco; various Chadian rebels from the Aozou region reside in southern Libya

Refugees and internally displaced persons: *refugees (country of origin):* 16,820 (Syria) (refugees and asylum seekers), 12,220 (Sudan) (refugees and asylum seekers), 5, 899 (Eritrea) (refugees and asylum seekers) (2019)

IDPs: 392,241 (conflict between pro- QADHAFI and anti-QADHAFI forces in 2011; post-QADHAFI tribal clashes 2014) (2020)

Trafficking in persons: *current situation:* Libya is a destination and transit country for men and women from Sub-Saharan Africa and Asia subjected to forced labor and forced prostitution; migrants who seek employment in Libya as laborers and domestic workers or who transit Libya en route to Europe are vulnerable to forced labor; private employers also exploit migrants from detention centers as forced laborers on farms and construction sites, returning them to detention when they are no longer needed; some Sub-Saharan women are reportedly forced to work in Libyan brothels, particularly in the country's south; since 2013, militia groups and other informal armed groups, including some affiliated with the government, are reported to conscript Libyan children under the age of 18; large-scale violence driven by militias, civil unrest, and increased lawlessness increased in 2014, making it more difficult to obtain information on human trafficking

tier rating: Tier 3—the Libyan Government does not fully comply with the minimum standards for the elimination of trafficking and is not making significant efforts to do so; in 2014, the government's capacity to address human trafficking was hampered by the ongoing power struggle and violence; the judicial system was not functioning, preventing any efforts to investigate, prosecute, or convict traffickers, complicit detention camp guards or government officials, or militias or armed groups that used child soldiers; the government failed to identify or provide protection to trafficking victims, including child conscripts, and continued to punish victims for unlawful acts committed as a direct result of being trafficked; no public anti- trafficking awareness campaigns were conducted (2015)

LIECHTENSTEIN

INTRODUCTION

Background: The Principality of Liechtenstein was established within the Holy Roman Empire in 1719. Occupied by both French and Russian troops during the Napoleonic Wars, it became a sovereign state in 1806 and joined the German Confederation in 1815. Liechtenstein became fully independent in 1866 when the Confederation dissolved. Until the end of World War I, it was closely tied to Austria, but the economic devastation caused by that conflict forced Liechtenstein to enter into a customs and monetary union with Switzerland. Since World War II (in which Liechtenstein remained neutral), the country's low taxes have spurred outstanding economic growth.

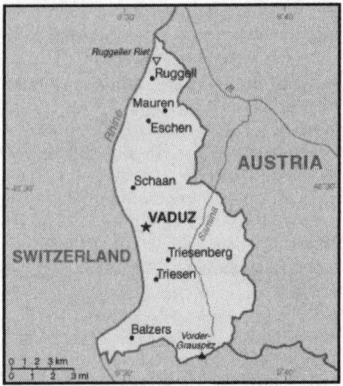

In 2000, shortcomings in banking regulatory oversight resulted in concerns about the use of financial institutions for money laundering. However, Liechtenstein implemented anti-money laundering legislation and a Mutual Legal Assistance Treaty with the US that went into effect in 2003.

GEOGRAPHY

Location: Central Europe, between Austria and Switzerland

Geographic coordinates: 47 16 N, 9 32 E

Map references: Europe

Area: *total:* 160 sq km
land: 160 sq km
water: 0 sq km
country comparison to the world: 219

Area—comparative: about 0.9 times the size of Washington, DC

Land boundaries: *total:* 75 km
border countries (2): Austria 34 km, Switzerland 41 km

Coastline: 0 km (doubly landlocked)

Maritime claims: none (landlocked)

Climate: continental; cold, cloudy winters with frequent snow or rain; cool to moderately warm, cloudy, humid summers

Terrain: mostly mountainous (Alps) with Rhine Valley in western third

Elevation: *lowest point:* Ruggeller Riet 430 m
highest point: Vorder-Grauspitz 2,599 m

Natural resources: hydroelectric potential, arable land

Land use: *agricultural land:* 37.6% (2011 est.)
arable land: 18.8% (2011 est.) / permanent crops: 0% (2011 est.) / permanent pasture: 18.8% (2011 est.)
forest: 43.1% (2011 est.)
other: 19.3% (2011 est.)
Irrigated land: 0 sq km (2012)

Population distribution: most of the population is found in the western half of the country along the Rhine River

Natural hazards: avalanches, landslides

Environment—current issues: some air pollution generated locally, some transfered from surrounding countries

Environment—international agreements: *party to:* Air Pollution, Air Pollution-Nitrogen Oxides, Air Pollution-Persistent Organic Pollutants, Air Pollution-Sulfur 85, Air Pollution-Sulfur 94, Air Pollution-Volatile Organic Compounds, Biodiversity, Climate Change, Climate Change-Kyoto Protocol, Desertification, Endangered Species, Hazardous Wastes, Ozone Layer Protection, Wetlands
signed, but not ratified: Law of the Sea

Geography—note: along with Uzbekistan, one of only two doubly landlocked countries in the world; variety of microclimatic variations based on elevation

PEOPLE AND SOCIETY

Population: 39,137 (July 2020 est.)
note: immigrants make up 65% of the total population, according to UN data (2017)
country comparison to the world: 213

Nationality: *noun:* Liechtensteiner(s)
adjective: Liechtenstein

Ethnic groups: Liechtensteiner 66%, Swiss 9.6%, Austrian 5.8%, German 4.3%, Italian 3.1%, other 11.2% (2017 est.)
note: data represent population by nationality

Languages: German 91.5% (official) (Alemannic is the main dialect), Italian 1.5%, Turkish 1.3%, Portuguese 1.1%, other 4.6% (2015 est.)

Religions: Roman Catholic (official) 73.4%, Protestant Reformed 6.3%, Muslim 5.9%, Christian Orthodox 1.3%, Lutheran 1.2%, other Protestant .7%, other Christian .3%, other .8%, none 7%, unspecified 3.3% (2015 est.)

Age structure: *0-14 years:* 15.2% (male 3,259/female 2,688)
15-24 years: 11.29% (male 2,238/female 2,181)
25-54 years: 40.22% (male 7,869/female 7,872)
55-64 years: 14.41% (male 2,711/female 2,930)
65 years and over: 18.88% (male 3,403/female 3,986) (2020 est.)

Median age: *total:* 43.7 years
male: 42 years
female: 45.3 years (2020 est.)
country comparison to the world: 22

Population growth rate: 0.75% (2020 est.)
country comparison to the world: 134

Birth rate: 10.4 births/1,000 population (2020 est.)
country comparison to the world: 189

Death rate: 7.8 deaths/1,000 population (2020 est.)
country comparison to the world: 98

Net migration rate: 4.9 migrant(s)/1,000 population (2020 est.)
country comparison to the world: 24

Population distribution: most of the population is found in the western half of the country along the Rhine River

Urbanization: *urban population:* 14.4% of total population (2020)
rate of urbanization: 0.81% annual rate of change (2015-20 est.)
total population growth rate v. urban population growth rate, 2000-2030:

Major urban areas—population: 5,000 VADUZ (capital) (2018)

Sex ratio: *at birth:* 1.26 male(s)/female
0-14 years: 1.21 male(s)/female
15-24 years: 1.03 male(s)/female
25-54 years: 1 male(s)/female
55-64 years: 0.93 male(s)/female
65 years and over: 0.85 male(s)/female
total population: 0.99 male(s)/female (2020 est.)

Mother's mean age at first birth: 31.3 years (2017)

Infant mortality rate: *total:* 4.2 deaths/1,000 live births
male: 4.5 deaths/1,000 live births
female: 3.8 deaths/1,000 live births (2020 est.)
country comparison to the world: 186

Life expectancy at birth: *total population:* 82.2 years
male: 79.9 years
female: 85 years (2020 est.)
country comparison to the world: 21

Total fertility rate: 1.69 children born/woman (2020 est.)
country comparison to the world: 177

Drinking water source:
improved:
total: 100% of population
unimproved:
total: 0% of population (2017 est.)

Sanitation facility access:
improved:
total: 98% of population
unimproved:
total: 0% of population (2017)

HIV/AIDS—adult prevalence rate: NA

HIV/AIDS—people living with HIV/AIDS: NA

HIV/AIDS—deaths: NA

Education expenditures: 2.6% of GDP (2011)
country comparison to the world: 154

School life expectancy (primary to tertiary education): *total:* 15 years
male: 16 years
female: 14 years (2018)

GOVERNMENT

Country name: *conventional long form:* Principality of Liechtenstein
conventional short form: Liechtenstein
local long form: Fuerstentum Liechtenstein
local short form: Liechtenstein
etymology: named after the Liechtenstein dynasty that purchased and united the counties of Schellenburg and Vaduz and that was allowed by the Holy Roman Emperor in 1719 to rename the new property after their family; the name in German means "light (bright) stone"

Government type: constitutional monarchy

Capital: *name:* Vaduz

geographic coordinates: 47 08 N, 9 31 E

time difference: UTC+1 (6 hours ahead of Washington, DC, during Standard Time)

daylight saving time: +1hr, begins last Sunday in March; ends last Sunday in October

etymology: may be a conflation from the Latin "vallis" (valley) and the High German "diutisk" (meaning "German") to produce "Valdutsch" (German valley), which over time simplified and came to refer specifically to Vaduz, the town

Administrative divisions: 11 communes (Gemeinden, singular—Gemeinde); Balzers, Eschen, Gamprin, Mauren, Planken, Ruggell, Schaan, Schellenberg, Triesen, Triesenberg, Vaduz

Independence: 23 January 1719 (Principality of Liechtenstein established); 12 July 1806 (independence from the Holy Roman Empire); 24 August 1866 (independence from the German Confederation)

National holiday: National Day, 15 August (1940); note—a National Day was originally established in 1940 to combine celebrations for the Feast of the Assumption (15 August) with those honoring the birthday of former Prince FRANZ JOSEF II (1906-1989) whose birth fell on 16 August; after the prince's death, National Day became the official national holiday by law in 1990

Constitution: *history:* previous 1862; latest adopted 5 October 1921

amendments: proposed by Parliament, by the reigning prince (in the form of "Government" proposals), by petition of at least 1,500 qualified voters, or by at least four communes; passage requires unanimous approval of Parliament members in one sitting or three-quarters majority vote in two successive sittings; referendum required only if petitioned by at least 1,500 voters or by at least four communes; passage by referendum requires absolute majority of votes cast; amended many times, last in 2018(2019)

Legal system: civil law system influenced by Swiss, Austrian, and German law

International law organization participation: accepts compulsory ICJ jurisdiction with reservations; accepts ICCt jurisdiction

Citizenship: *citizenship by birth:* no

citizenship by descent only: the father must be a citizen of Liechtenstein; in the case of a child born out of wedlock, the mother must be a citizen

dual citizenship recognized: no

residency requirement for naturalization: 5 years

Suffrage: 18 years of age; universal

Executive branch: *chief of state:* Prince HANS-ADAM II (since 13 November 1989, assumed executive powers on 26 August 1984); Heir Apparent and Regent of Liechtenstein Prince ALOIS (son of the monarch, born 11 June 1968); note—15 August 2004, HANS-ADAM II transferred the official duties of the ruling prince to ALOIS, but HANS-ADAM II retains status of chief of state

head of government: Prime Minister Adrian HASLER (since 27 March 2013)

cabinet: Cabinet elected by the Parliament, confirmed by the monarch

elections/appointments: the monarchy is hereditary; following legislative elections, the leader of the majority party in the Parliament usually appointed the head of government by the monarch, and the leader of the largest minority party in the Landtag usually appointed the deputy head of government by the monarch if there is a coalition government

Legislative branch: *description:* unicameral Parliament or Landtag (25 seats; members directly elected in 2 multi-seat constituencies by proportional representation vote to serve 4-year terms)

elections: last held on 5 February 2017 (next to be held on 7 February 2021)

election results: percent of vote by party—FBP 35.2%, VU 33.7%, DU 18.4% FL 12.6%; seats by party—FBP 9, VU 8, DU 5, FL 3; composition—men 22, women 3, percent of women 12%

Judicial branch: *highest courts:* Supreme Court or Oberster Gerichtshof (consists of 5 judges); Constitutional Court or Verfassungsgericht (consists of 5 judges and 5 alternates)

judge selection and term of office: judges of both courts elected by the Landtag and appointed by the monarch; Supreme Court judges serve 4-year renewable terms; Constitutional Court judges appointed for renewable 5-year terms

subordinate courts: Court of Appeal or Obergericht (second instance), Court of Justice (first instance), Administrative Court, county courts

Political parties and leaders: Fatherland Union (Vaterlaendische Union) or VU [Guenther FRITZ]

Progressive Citizens' Party (Fortschrittliche Buergerpartei) or FBP [Thomas BANZER]

The Free List (Die Freie Liste) or FL [Pepo FRICK and Conny BUECHEL BRUEHWILER]

The Independents (Die Unabhaengigen) or DU [Harry QUADERER]

International organization participation: CD, CE, EBRD, EFTA, IAEA, ICCt, ICRM, IFRCS, Interpol, IOC, IPU, ITSO, ITU, ITUC (NGOs), OAS (observer), OPCW, OSCE, PCA, Schengen Convention, UN, UNCTAD, UPU, WIPO, WTO

Diplomatic representation in the US: *chief of mission:* Ambassador Kurt JAEGER (since 16 September 2016)

chancery: 2900 K Street NW, Suite 602B, Washington, DC 20007

telephone: [1] (202) 331-0590

FAX: [1] (202) 331-3221

Diplomatic representation from the US: the US does not have an embassy in Liechtenstein; the US Ambassador to Switzerland is accredited to Liechtenstein

Flag description: two equal horizontal bands of blue (top) and red with a gold crown on the hoist side of the blue band; the colors may derive from the blue and red livery design used in the principality's household in the 18th century; the prince's crown was introduced in 1937 to distinguish the flag from that of Haiti

National symbol(s): *princely hat (crown); national colors:* blue, red

National anthem: *name:* "Oben am jungen Rhein" (High Above the Young Rhine)

lyrics/music: Jakob Joseph JAUCH/Josef FROMMELT

note: adopted 1850, revised 1963; uses the tune of "God Save the Queen"

0:00 / 1:24

ECONOMY

Economy—overview: Despite its small size and lack of natural resources, Liechtenstein has developed into a prosperous, highly industrialized, free-enterprise economy with a vital financial services sector and one of the highest per capita income levels in the world. The Liechtenstein economy is widely diversified with a large number of small and medium-sized businesses, particularly in the services sector. Low business taxes—a flat tax of 12.5% on income is applied—and easy incorporation rules have induced many holding companies to establish nominal offices in Liechtenstein, providing 30% of state revenues.

The country participates in a customs union with Switzerland and uses the Swiss franc as its national currency. It imports more than 90% of its energy requirements. Liechtenstein has been a member of the European Economic Area (an organization serving as a bridge between the European Free Trade Association and the EU) since May 1995. The government is working to harmonize its economic policies with those of an integrated EU. As of 2015, 54% of Liechtenstein's workforce consisted of cross-border commuters, largely from Austria, Germany, and Switzerland.

Since 2008, Liechtenstein has faced renewed international pressure—particularly from Germany and the US—to improve transparency in its banking and tax systems. In December 2008, Liechtenstein signed a Tax Information Exchange Agreement with the US. Upon Liechtenstein's conclusion of 12 bilateral information-sharing agreements, the OECD in October 2009 removed the principality from its "grey list" of countries that had yet to implement the organization's Model Tax Convention. By the end of 2010, Liechtenstein had signed 25 Tax Information Exchange Agreements or Double Tax Agreements. In 2011, Liechtenstein joined the Schengen area, which allows passport-free travel across 26 European countries. In 2015, Liechtenstein and the EU agreed to clamp down on tax fraud and evasion and in 2018 will start automatically exchanging information on the bank accounts of each other's residents.

GDP (purchasing power parity):
$4.978 billion (2014 est.)
$3.2 billion (2009 est.)
$3.216 billion (2008 est.)

country comparison to the world: 179

GDP (official exchange rate):
$6.672 billion (2014 est.)

GDP—real growth rate: 1.8% (2012 est.)
-0.5% (2011 est.)
3.1% (2007 est.)
country comparison to the world: 147

GDP—per capita (PPP): $139,100 (2009 est.)
$90,100 (2008 est.)
$91,300 (2007 est.)
country comparison to the world: 1

GDP—composition, by sector of origin:
agriculture: 7% (2014)
industry: 41% (2014)
services: 52% (2014)

Agriculture—products: wheat, barley, corn, potatoes; livestock, dairy products

Industries: electronics, metal manufacturing, dental products, ceramics, pharmaceuticals, food products, precision instruments, tourism, optical instruments

Industrial production growth rate: NA

Labor force: 38,520 (2012) (2015 est.)
note: 51% of the labor force in Liechtenstein commute daily from Austria, Switzerland, and Germany
country comparison to the world: 199

Labor force—by occupation: *agriculture:* 0.8%
industry: 36.9%
services: 62.3% (2015)

Unemployment rate: 2.4% (2015)
2.4% (2014)
country comparison to the world: 26

Population below poverty line: NA

Household income or consumption by percentage share: *lowest 10%:* NA
highest 10%: NA

Budget: *revenues:* 995.3 million (2012 est.)
expenditures: 890.4 million (2011 est.)

Taxes and other revenues: 14.9% (of GDP) (2012 est.)
country comparison to the world: 195

Budget surplus (+) or deficit (-): 1.6% (of GDP) (2012 est.)
country comparison to the world: 20

Fiscal year: calendar year

Inflation rate (consumer prices): -0.4% (2016 est.)
-0.2% (2013)
country comparison to the world: 7

Exports: $3.217 billion (2015 est.)
$3.774 billion (2014 est.)
note: trade data exclude trade with Switzerland
country comparison to the world: 127

Exports—commodities: small specialty machinery, connectors for audio and video, parts for motor vehicles, dental products, hardware, prepared foodstuffs, electronic equipment, optical products

Imports: NA (2015 est.)
$2.23 billion (2014 est.)
note: trade data exclude trade with Switzerland
country comparison to the world: 163

Imports—commodities: agricultural products, raw materials, energy products, machinery, metal goods, textiles, foodstuffs, motor vehicles

Debt—external: $0 (2015 est.)
note: public external debt only; private external debt unavailable
country comparison to the world: 206

Exchange rates: Swiss francs (CHF) per US dollar—
0.9875 (2017 est.)
0.9852 (2016 est.)
0.9852 (2015 est.)
0.9627 (2014 est.)
0.9152 (2013 est.)

ENERGY

Electricity access: *electrification—total population:* 100% (2020)

Electricity—production: 68.43 million kWh (2015 est.)
country comparison to the world: 202

Electricity—consumption: 393.6 million kWh (2015 est.)
country comparison to the world: 174

Electricity—exports: 0 kWh (2015 est.) (2015 est.)
country comparison to the world: 161

Electricity—imports: 325.2 million kWh (2015 est.)
country comparison to the world: 88

COMMUNICATIONS

Telephones—fixed lines: *total subscriptions:* 14,337
subscriptions per 100 inhabitants: 36.91 (2019 est.)
country comparison to the world: 185

Telephones—mobile cellular: *total subscriptions:* 49,355
subscriptions per 100 inhabitants: 127.06 (2019 est.)
country comparison to the world: 206

Telecommunication systems: *general assessment:* automatic telephone system; 44 Internet service providers in Liechtenstein and Switzerland combined; FttP (fiber to the home) penetration marketed 3rd highest in EU; fiber network reaches 3/4 of the population (2020)

domestic: fixed-line 37 per 100 and mobile-cellular services 127 per 100 (2019)

international: country code—423; linked to Swiss networks by cable and microwave radio relay
note: the COVID-19 outbreak is negatively impacting telecommunications production and supply chains globally; consumer spending on telecom devices and services has also slowed due to the pandemic's effect on economies worldwide; overall progress towards improvements in all facets of the telecom industry—mobile, fixed-line, broadband, submarine cable and satellite—has moderated

Broadcast media: relies on foreign terrestrial and satellite broadcasters for most broadcast media services; first Liechtenstein-based TV station established August 2008; Radio Liechtenstein operates multiple radio stations; a Swiss-based broadcaster operates one radio station in Liechtenstein

Internet country code: .li

Internet users: *total:* 37,815

percent of population: 98.1% (July 2018 est.)
country comparison to the world: 202

Broadband—fixed subscriptions: *total:* 16,712
subscriptions per 100 inhabitants: 43 (2018 est.)
country comparison to the world: 158

TRANSPORTATION

Civil aircraft registration country code prefix: HB (2016)

Pipelines:
434.5 km gas (2018)

Railways: *total:* 9 km (2018)
standard gauge: 9 km 1.435-m gauge (electrified) (2018)
note: belongs to the Austrian Railway System connecting Austria and Switzerland
country comparison to the world: 136

Roadways: *total:* 630 km (2019)
country comparison to the world: 191

Waterways: 28 km (2010)
country comparison to the world: 105

MILITARY AND SECURITY

Military and security forces: no regular military forces; National Police maintain close relations with neighboring forces (2019)

TRANSNATIONAL ISSUES

Disputes—international: none

Illicit drugs: has strengthened money laundering controls, but money laundering remains a concern due to Liechtenstein's sophisticated offshore financial services sector

LITHUANIA

INTRODUCTION

Background: Lithuanian lands were united under MINDAUGAS in 1236; over the next century, through alliances and conquest, Lithuania extended its territory to include most of present-day Belarus and Ukraine. By the end of the 14th century Lithuania was the largest state in Europe. An alliance with Poland in 1386 led the two countries into a union through the person of a common ruler. In 1569, Lithuania and Poland formally united into a single dual state, the Polish-Lithuanian Commonwealth. This entity survived until 1795 when its remnants were partitioned by surrounding countries. Lithuania regained its independence following World War I but was annexed by the USSR in 1940—an action never recognized by the US and many other countries. On 11 March 1990, Lithuania became the first of the Soviet republics to declare its independence, but Moscow did not recognize this proclamation until September of 1991 (following the abortive coup in Moscow). The last Russian troops withdrew in 1993. Lithuania subsequently restructured its economy for integration into Western European institutions; it joined both NATO and the EU in the spring of 2004. In 2015, Lithuania joined the euro zone, and it joined the Organization for Economic Cooperation and Development in 2018.

GEOGRAPHY

Location: Eastern Europe, bordering the Baltic Sea, between Latvia and Russia, west of Belarus

Geographic coordinates: 56 00 N, 24 00 E

Map references: Europe

Area: *total:* 65,300 sq km
land: 62,680 sq km
water: 2,620 sq km
country comparison to the world: 124

Area—comparative: slightly larger than West Virginia

Land boundaries: *total:* 1,549 km
border countries (4): Belarus 640 km, Latvia 544 km, Poland 104 km, Russia (Kaliningrad) 261 km

Coastline: 90 km

Maritime claims: *territorial sea:* 12 nm

Climate: transitional, between maritime and continental; wet, moderate winters and summers

Terrain: lowland, many scattered small lakes, fertile soil

Elevation: *mean elevation:* 110 m
lowest point: Baltic Sea 0 m
highest point: Aukstojas 294 m

Natural resources: peat, arable land, amber

Land use: *agricultural land:* 44.8% (2011 est.)
arable land: 34.9% (2011 est.) / permanent crops: 0.5% (2011 est.) / permanent pasture: 9.4% (2011 est.)
forest: 34.6% (2011 est.)
other: 20.6% (2011 est.)
Irrigated land: 44 sq km (2012)

Population distribution: fairly even population distribution throughout the country, but somewhat greater concentrations in the southern cities of Vilnius and Kaunas, and the western port of Klaipeda

Natural hazards: occasional floods, droughts

Environment—current issues: water pollution; air pollution; deforestation; threatened animal and plant species; chemicals and waste materials released into the environment contaminate soil and groundwater; soil degradation and erosion

Environment—international agreements: *party to:* Air Pollution, Air Pollution-Nitrogen Oxides, Air Pollution-Persistent Organic Pollutants, Air Pollution-Sulphur 85, Air Pollution-Sulphur 94, Air Pollution-Volatile Organic Compounds, Biodiversity, Climate Change, Climate Change-Kyoto Protocol, Desertification, Endangered Species, Environmental Modification, Hazardous Wastes, Law of the Sea, Ozone Layer Protection, Ship Pollution, Wetlands
signed, but not ratified: none of the selected agreements

Geography—note: fertile central plains are separated by hilly uplands that are ancient glacial deposits

PEOPLE AND SOCIETY

Population: 2,731,464 (July 2020 est.)
country comparison to the world: 141

Nationality: *noun:* Lithuanian(s)
adjective: Lithuanian

Ethnic groups: Lithuanian 84.1%, Polish 6.6%, Russian 5.8%, Belarusian 1.2%, other 1.1%, unspecified 1.2% (2011 est.)

Languages: Lithuanian (official) 82%, Russian 8%, Polish 5.6%, other 0.9%, unspecified 3.5% (2011 est.)

Religions: Roman Catholic 77.2%, Russian Orthodox 4.1%, Old Believer 0.8%, Evangelical Lutheran 0.6%, Evangelical Reformist 0.2%, other (including Sunni Muslim, Jewish, Greek Catholic, and Karaite) 0.8%, none 6.1%, unspecified 10.1% (2011 est.)

Age structure: *0-14 years:* 15.26% (male 213,802/female 202,948)
15-24 years: 10.23% (male 144,679/female 134,822)
25-54 years: 38.96% (male 528,706/female 535,485)
55-64 years: 15.1% (male 183,854/female 228,585)
65 years and over: 20.45% (male 190,025/female 368,558) (2020 est.)

Dependency ratios: *total dependency ratio:* 56.5
youth dependency ratio: 24.2
elderly dependency ratio: 32.3
potential support ratio: 3.1 (2020 est.)

Median age: *total:* 44.5 years
male: 40.2 years
female: 48.2 years (2020 est.)
country comparison to the world: 15

Population growth rate: -1.13% (2020 est.)
country comparison to the world: 232

Birth rate: 9.5 births/1,000 population (2020 est.)
country comparison to the world: 197

Death rate: 15 deaths/1,000 population (2020 est.)
country comparison to the world: 2

Net migration rate: -5.9 migrant(s)/1,000 population (2020 est.)
country comparison to the world: 203

Population distribution: fairly even population distribution throughout the country, but somewhat greater concentrations in the southern cities of Vilnius and Kaunas, and the western port of Klaipeda

Urbanization: *urban population:* 68% of total population (2020)
rate of urbanization: -0.31% annual rate of change (2015-20 est.)
total population growth rate v. urban population growth rate, 2000-2030:

Major urban areas—population: 539,000 VILNIUS (capital) (2020)

Sex ratio: *at birth:* 1.06 male(s)/female
0-14 years: 1.05 male(s)/female
15-24 years: 1.07 male(s)/female
25-54 years: 0.99 male(s)/female
55-64 years: 0.8 male(s)/female
65 years and over: 0.52 male(s)/female
total population: 0.86 male(s)/female (2020 est.)

Mother's mean age at first birth: 27.5 years (2017 est.)

Maternal mortality rate: 5 deaths/100,000 live births (2017 est.)
country comparison to the world: 167

Infant mortality rate: *total:* 3.8 deaths/1,000 live births
male: 4.2 deaths/1,000 live births
female: 3.3 deaths/1,000 live births (2020 est.)
country comparison to the world: 192

Life expectancy at birth: *total population:* 75.5 years
male: 70.3 years
female: 81.1 years (2020 est.)
country comparison to the world: 115

Total fertility rate: 1.6 children born/woman (2020 est.)
country comparison to the world: 185

Drinking water source:
improved:
urban: 100% of population
rural: 92.8% of population
total: 97.5% of population
unimproved:
urban: 0% of population
rural: 7.2% of population
total: 2.5% of population (2017 est.)

Current Health Expenditure: 6.5% (2017)

Physicians density: 4.83 physicians/1,000 population (2017)

Hospital bed density: 6.6 beds/1,000 population (2017)

Sanitation facility access:
improved:
urban: 99.3% of population (2015 est.)
rural: 87.5% of population
total: 95.5% of population
unimproved:
urban: 0.7% of population
rural: 12.5% of population
total: 4.5% of population (2017 est.)

HIV/AIDS—adult prevalence rate: 0.1% (2019 est.)
country comparison to the world: 125

HIV/AIDS—people living with HIV/AIDS: 3,400 (2019 est.)
country comparison to the world: 130

HIV/AIDS—deaths: <100 (2019 est.)

Major infectious diseases: *degree of risk:* intermediate (2020)

vectorborne diseases: tickborne encephalitis

Obesity—adult prevalence rate: 26.3% (2016)
country comparison to the world: 43

Education expenditures: 4.2% of GDP (2015)
country comparison to the world: 93

Literacy: *definition:* age 15 and over can read and write

total population: 99.8%
male: 99.8%
female: 99.8% (2015)

School life expectancy (primary to tertiary education): *total:* 17 years
male: 16 years
female: 17 years (2018)

Unemployment, youth ages 15-24: *total:* 11.1 %
male: 12%
female: 10.1% (2018 est.)
country comparison to the world: 117

GOVERNMENT

Country name: *conventional long form:* Republic of Lithuania
conventional short form: Lithuania
local long form: Lietuvos Respublika
local short form: Lietuva
former: Lithuanian Soviet Socialist Republic
etymology: meaning of the name "Lietuva" remains unclear; it may derive from the Lietava, a stream in east central Lithuania

Government type: semi-presidential republic

Capital: *name:* Vilnius

geographic coordinates: 54 41 N, 25 19 E
time difference: UTC+2 (7 hours ahead of Washington, DC, during Standard Time)
daylight saving time: +1hr, begins last Sunday in March; ends last Sunday in October
etymology: named after the Vilnia River, which flows into the Neris River at Vilnius; the river name derives from the Lithuanian word "vilnis" meaning "a surge"

Administrative divisions: 60 municipalities (savivaldybe, singular—savivaldybe); Akmene, Alytaus Miestas, Alytus, Anksciai, Birstono, Birzai, Druskininkai, Elektrenai, Ignalina, Jonava, Joniskis, Jurbarkas, Kaisiadorys, Kalvarijos, Kauno Miestas, Kaunas, Kazlu Rudos, Kedainiai, Kelme, Klaipedos Miestas, Klaipeda, Kretinga, Kupiskis, Lazdijai, Marijampole, Mazeikiai, Moletai, Neringa, Pagegiai, Pakruojis, Palangos Miestas, Panevezio Miestas, Panevezys, Pasvalys, Plunge, Prienai, Radviliskis, Raseiniai, Rietavo, Rokiskis, Sakiai, Salcininkai, Siauliu Miestas, Siauliai, Silale, Silute, Sirvintos, Skuodas, Svencionys, Taurage, Telsiai, Trakai, Ukmerge, Utena, Varena, Vilkaviskis, Vilniaus Miestas, Vilnius, Visaginas, Zarasai

Independence: 16 February 1918 (from Soviet Russia and Germany); 11 March 1990 (declared from the Soviet Union); 6 September 1991 (recognized by the Soviet Union); notable earlier dates: 6 July 1253 (coronation of MINDAUGAS, traditional founding date); 1 July 1569 (Polish-Lithuanian Commonwealth created)

National holiday: Independence Day (or National Day), 16 February (1918); note—16 February 1918 was the date Lithuania established its statehood and its concomitant independence from Soviet Russia and Germany; 11 March 1990 was the date it declared the restoration of Lithuanian statehood and its concomitant independence from the Soviet Union

Constitution: *history:* several previous; latest adopted by referendum 25 October 1992, entered into force 2 November 1992
amendments: proposed by at least one fourth of all Parliament members or by petition of at least 300,000 voters; passage requires two-thirds majority vote of Parliament in each of two readings

three months apart and a presidential signature; amendments to constitutional articles on national sovereignty and constitutional amendment procedure also require three-fourths voter approval in a referendum; amended 1996, 2003, 2006

Legal system: civil law system; legislative acts can be appealed to the Constitutional Court

International law organization participation: accepts compulsory ICJ jurisdiction with reservations; accepts ICCt jurisdiction

Citizenship: *citizenship by birth:* no
citizenship by descent only: at least one parent must be a citizen of Lithuania
dual citizenship recognized: no
residency requirement for naturalization: 10 years

Suffrage: 18 years of age; universal

Executive branch: *chief of state:* President Gitanas NAUSEDA (since 12 July 2019)

head of government: Prime Minister Ingrida SIMONYTE (since 24 November 2020)
cabinet: Council of Ministers nominated by the prime minister, appointed by the president, and approved by Parliament
elections/appointments: president directly elected by absolute majority popular vote in 2 rounds if needed for a 5-year term (eligible for a second term); election last held on 12 and 26 May 2019 (next to be held in May 2024); prime minister appointed by the president, approved by Parliament
election results: Gitanas NAUSEDA elected president in second round; percent of vote—Gitanas NAUSEDA (independent) 66.7%, Ingrida SIMONYTE (independent) 33.3%; Saulius SKVERNELIS (LVZS) approved as prime minister by Parliament vote—90 to 4
Legislative branch: description: unicameral Parliament or Seimas (141 seats; 71 members directly elected in single-seat constituencies by absolute majority vote and 70 directly elected in a single nationwide constituency by proportional representation vote; members serve 4-year terms)
elections: last held on11 and 25 October 2020 (next to be held in October 2024)
election results: percent of vote by party—NA; seats by party—TS-LKD 50, LVZS 32, LSDP 13, LRLS 13, Freedom 11, DP 10, AWPL 3, LSDDP 3, LT 1, Greens 1, independent 4

Judicial branch: *highest courts:* Supreme Court (consists of 37 judges); Constitutional Court (consists of 9 judges)
judge selection and term of office: Supreme Court judges nominated by the president and appointed by the Seimas; judges serve 5-year renewable terms; Constitutional Court judges appointed by the Seimas from nominations—3 each by the president of the republic, the Seimas chairperson, and the Supreme Court president; judges serve 9-year, nonrenewable terms; one-third of membership reconstituted every 3 years
subordinate courts: Court of Appeals; district and local courts

Political parties and leaders: Electoral Action of Lithuanian Poles or LLRA [Valdemar TOMASEVSKI]

Farmers and Greens Union or LVZS [Ramunas KARBAUSKIS]

Freedom Party or LP [Ausrine ARMONAITE]

Homeland Union-Lithuanian Christian Democrats or TS-LKD [Gabrielius LANDSBERGIS]

Labor Party or DP [Viktor USPASKICH]

Lithuanian Center Party or LCP [Naglis PUTEIKIS]

Lithuanian Green Party or LZP [Remigijus LAPINSKAS]]

Lithuanian Liberal Movement or LS or LRLS [Viktorija CMILYTE]

Lithuanian List or LL [Darius KUOLYS]

Lithuanian Social Democratic Party or LSDP [Gintautas PALUCKAS]

Lithuanian Social Democratic Labor Party or LSDDP [Gediminas KIRKILAS]

Freedom and Justice Party or LT [Remigijus ZEMAITAITIS]

International organization participation: Australia Group, BA, BIS, CBSS, CD, CE, EAPC, EBRD, ECB, EIB, EU, FAO, IAEA, IBRD, ICAO, ICC (national committees), ICCt, ICRM, IDA, IFC, IFRCS, ILO, IMF, IMO, Interpol, IOC, IOM, IPU, ISO, ITU, ITUC (NGOs), MIGA, NATO, NIB, NSG, OAS (observer), OECD, OIF (observer), OPCW, OSCE, PCA, Schengen Convention, UN, UNCTAD, UNESCO, UNIDO, UNWTO, UPU, WCO, WHO, WIPO, WMO, WTO

Diplomatic representation in the US: *chief of mission:* Ambassador Rolandas KRISCIUNAS (since 17 September 2015)

chancery: 2622 16th Street NW, Washington, DC 20009

telephone: [1] (202) 234-5860

FAX: [1] (202) 328-0466

consulate(s) general: Chicago, Los Angeles, New York

Diplomatic representation from the US: *chief of mission:* Ambassador Robert S. GILCHRIST (since 4 February 2010)

telephone: [370] (5) 266-5500

embassy: Akmenu gatve 6, Vilnius, LT-03106

mailing address: American Embassy, Akmenu Gatve 6, Vilnius LT-03106

FAX: [370] (5) 266-5510

Flag description: three equal horizontal bands of yellow (top), green, and red; yellow symbolizes golden fields, as well as the sun, light, and goodness; green represents the forests of the countryside, in addition to nature, freedom, and hope; red stands for courage and the blood spilled in defense of the homeland

National symbol(s): mounted knight known as Vytis (the Chaser); white stork; national colors: yellow, green, red

National anthem: *name:* "Tautiska giesme" (The National Song)

lyrics/music: Vincas KUDIRKA

note: adopted 1918, restored 1990; written in 1898 while Lithuania was a part of Russia; banned during the Soviet occupation from 1940 to 1990

0:00 / 1:46

ECONOMY

Economy—overview: After the country declared independence from the Soviet Union in 1990, Lithuania faced an initial dislocation that is typical during transitions from a planned economy to a free-market economy. Macroeconomic stabilization policies, including privatization of most state-owned enterprises, and a strong commitment to a currency board arrangement led to an open and rapidly growing economy and rising consumer demand. Foreign investment and EU funding aided in the transition. Lithuania joined the WTO in May 2001, the EU in May 2004, and the euro zone in January 2015, and is now working to complete the OECD accession roadmap it received in July 2015. In 2017, joined the OECD Working Group on Bribery, an important step in the OECD accession process.

The Lithuanian economy was severely hit by the 2008-09 global financial crisis, but it has rebounded and become one of the fastest growing in the EU. Increases in exports, investment, and wage growth that supported consumption helped the economy grow by 3.6% in 2017. In 2015, Russia was Lithuania's largest trading partner, followed by Poland, Germany, and Latvia; goods and services trade between the US and Lithuania totaled $2.2 billion. Lithuania opened a self-financed liquefied natural gas terminal in January 2015, providing the first non-Russian supply of natural gas to the Baltic States and reducing Lithuania's dependence on Russian gas from 100% to approximately 30% in 2016.

Lithuania's ongoing recovery hinges on improving the business environment, especially by liberalizing labor laws, and improving competitiveness and export growth, the latter hampered by economic slowdowns in the EU and Russia. In addition, a steady outflow of young and highly educated people is causing a shortage of skilled labor, which, combined with a rapidly aging population, could stress public finances and constrain long-term growth.

GDP (purchasing power parity):
$91.47 billion (2017 est.)
$88.07 billion (2016 est.)
$86.05 billion (2015 est.)
note: data are in 2017 dollars
country comparison to the world: 87

GDP (official exchange rate):
$47.26 billion (2017 est.)

GDP—real growth rate: 4.33% (2019 est.)
3.99% (2018 est.)
4.37% (2017 est.)
country comparison to the world: 64

GDP—per capita (PPP): $32,400 (2017 est.)
$30,700 (2016 est.)
$29,600 (2015 est.)
note: data are in 2017 dollars
country comparison to the world: 63

Gross national saving:
18% of GDP (2017 est.)

16.2% of GDP (2016 est.)
17.8% of GDP (2015 est.)
country comparison to the world: 111

GDP—composition, by end use:
household consumption: 63.9% (2017 est.)
government consumption: 16.6% (2017 est.)
investment in fixed capital: 18.8% (2017 est.)
investment in inventories: -1.3% (2017 est.)
exports of goods and services: 81.6% (2017 est.)
imports of goods and services: -79.3% (2017 est.)

GDP—composition, by sector of origin:
agriculture: 3.5% (2017 est.)
industry: 29.4% (2017 est.)
services: 67.2% (2017 est.)

Agriculture—products: grain, potatoes, sugar beets, flax, vegetables; beef, milk, eggs, pork, cheese; fish

Industries: metal-cutting machine tools, electric motors, televisions, refrigerators and freezers, petroleum refining, shipbuilding (small ships), furniture, textiles, food processing, fertilizer, agricultural machinery, optical equipment, lasers, electronic components, computers, amber jewelry, information technology, video game development, app/software development, biotechnology

Industrial production growth rate: 5.9% (2017 est.)
country comparison to the world: 43

Labor force: 1.333 million (2020 est.)
country comparison to the world: 130

Labor force—by occupation: *agriculture:* 9.1%
industry: 25.2%
services: 65.8% (2015 est.)

Unemployment rate: 8.4% (2019 est.)
8.5% (2018 est.)
country comparison to the world: 129

Population below poverty line: 22.2% (2015 est.)

Household income or consumption by percentage share: *lowest 10%:* 2.2%
highest 10%: 28.8% (2015)

Budget: *revenues:* 15.92 billion (2017 est.)
expenditures: 15.7 billion (2017 est.)

Taxes and other revenues: 33.7% (of GDP) (2017 est.)
country comparison to the world: 65

Budget surplus (+) or deficit (-): 0.5% (of GDP) (2017 est.)
country comparison to the world: 38

Public debt: 39.7% of GDP (2017 est.)
40.1% of GDP (2016 est.)
note: official data; data cover general government debt and include debt instruments issued (or owned) by government entities other than the treasury; the data include treasury debt held by foreign entities, debt issued by subnational entities, as well as intragovernmental debt; intragovernmental debt consists of treasury borrowings from surpluses in the social funds, such as for retirement, medical care, and unemployment; debt instruments for the social funds are sold at public auctions
country comparison to the world: 129

Fiscal year: calendar year

Inflation rate (consumer prices): 3.7% (2017 est.)
0.7% (2016 est.)

579

country comparison to the world: 147

Current account balance: $1.817 billion (2019 est.)

$131 million (2018 est.)

country comparison to the world: 42

Exports: $29.12 billion (2017 est.)

$24.23 billion (2016 est.)

country comparison to the world: 64

Exports—partners: Russia 15%, Latvia 9.9%, Poland 8.1%, Germany 7.3%, US 5.2%, Estonia 5%, Sweden 4.8% (2017)

Exports—commodities: refined fuel, machinery and equipment, chemicals, textiles, foodstuffs, plastics

Imports: $31.56 billion (2017 est.)

$26.21 billion (2016 est.)

country comparison to the world: 65

Imports—commodities: oil, natural gas, machinery and equipment, transport equipment, chemicals, textiles and clothing, metals

Imports—partners: Russia 13%, Germany 12.3%, Poland 10.6%, Latvia 7.1%, Italy 5.2%, Netherlands 5.1%, Sweden 4% (2017)

Reserves of foreign exchange and gold: $4.45 billion (31 December 2017 est.)

$1.697 billion (31 December 2015 est.)

country comparison to the world: 98

Debt—external: $34.48 billion (31 March 2016 est.)

$31.6 billion (31 March 2015 est.)

country comparison to the world: 78

Exchange rates: litai (LTL) per US dollar—

0.884 (2017 est.)

0.9037 (2016 est.)

0.9037 (2015 est.)

0.9012 (2014 est.)

0.7525 (2013 est.)

ENERGY

Electricity access: *electrification—total population:* 100% (2020)

Electricity—production: 3.131 billion kWh (2016 est.)

country comparison to the world: 131

Electricity—consumption: 10.5 billion kWh (2016 est.)

country comparison to the world: 95

Electricity—exports: 730 million kWh (2015 est.)

country comparison to the world: 62

Electricity—imports: 11.11 billion kWh (2016 est.)

country comparison to the world: 22

Electricity—installed generating capacity: 3.71 million kW (2016 est.)

country comparison to the world: 93

Electricity—from fossil fuels: 73% of total installed capacity (2016 est.)

country comparison to the world: 100

Electricity—from nuclear fuels: 0% of total installed capacity (2017 est.)

country comparison to the world: 130

Electricity—from hydroelectric plants: 4% of total installed capacity (2017 est.)

country comparison to the world: 132

Electricity—from other renewable sources: 23% of total installed capacity (2017 est.)

country comparison to the world: 31

Crude oil—production: 2,000 bbl/day (2018 est.)

country comparison to the world: 87

Crude oil—exports: 1,002 bbl/day (2015 est.)

country comparison to the world: 75

Crude oil—imports: 182,900 bbl/day (2015 est.)

country comparison to the world: 32

Crude oil—proved reserves: 12 million bbl (1 January 2018 est.)

country comparison to the world: 88

Refined petroleum products—production: 196,500 bbl/day (2015 est.)

country comparison to the world: 51

Refined petroleum products—consumption: 58,000 bbl/day (2016 est.)

country comparison to the world: 98

Refined petroleum products—exports: 174,800 bbl/day (2015 est.)

country comparison to the world: 32

Refined petroleum products—imports: 42,490 bbl/day (2015 est.)

country comparison to the world: 87

Natural gas—production: 0 cu m (2017 est.)

country comparison to the world: 159

Natural gas—consumption: 2.492 billion cu m (2017 est.)

country comparison to the world: 79

Natural gas—exports: 0 cu m (2017 est.)

country comparison to the world: 110

Natural gas—imports: 2.492 billion cu m (2017 est.)

country comparison to the world: 47

Natural gas—proved reserves: 0 cu m (2016 est.)

country comparison to the world: 160

Carbon dioxide emissions from consumption of energy: 13.49 million Mt (2017 est.)

country comparison to the world: 97

COMMUNICATIONS

Telephones—fixed lines: *total subscriptions:* 368,515

subscriptions per 100 inhabitants: 13.34 (2019 est.)

country comparison to the world: 105

Telephones—mobile cellular: *total subscriptions:* 4,663,627

subscriptions per 100 inhabitants: 168.82 (2019 est.)

country comparison to the world: 123

Telecommunication systems: *general assessment:* adequate; improved international capability and better residential access; SIM card penetration is high for the region; prepaid sector accounts for most subscribers; postpaid subscribers is increasing; LTE networks available to more than 99% of the population; Lithuanian FttP (fiber to the home

cable connections for Internet) penetration ranked third highest in Europe; govt. and telecoms invest in fiber, fiber accounts for most new broadband connections; effective competition with 3 network operators in mobile sector and all investing in LTE and mobile data services (2020)

domestic: 13 per 100 for fixed-line subscriptions; rapid expansion of mobile-cellular services has resulted in a steady decline in the number of fixed-line connections; mobile-cellular teledensity stands at about 169 per 100 persons (2019)

international: country code—370; landing points for the BCS East, BCS East-West Interlink and NordBalt connecting Lithuania to Sweden, and Latvia ; further transmission by satellite; landline connections to Latvia and Poland (2019)

note: the COVID-19 outbreak is negatively impacting telecommunications production and supply chains globally; consumer spending on telecom devices and services has also slowed due to the pandemic's effect on economies worldwide; overall progress towards improvements in all facets of the telecom industry—mobile, fixed-line, broadband, submarine cable and satellite—has moderated

Broadcast media: public broadcaster operates 3 channels with the third channel—a satellite channel—introduced in 2007; various privately owned commercial TV broadcasters operate national and multiple regional channels; many privately owned local TV stations; multi-channel cable and satellite TV services available; publicly owned broadcaster operates 3 radio networks; many privately owned commercial broadcasters, with repeater stations in various regions throughout the country

Internet country code: .lt

Internet users: *total:* 2,226,806

percent of population: 79.72% (July 2018 est.)

country comparison to the world: 116

Broadband—fixed subscriptions: *total:* 788,743

subscriptions per 100 inhabitants: 28 (2018 est.)

country comparison to the world: 76

TRANSPORTATION

National air transport system: *number of registered air carriers:* 3 (2020)

inventory of registered aircraft operated by air carriers: 50

annual passenger traffic on registered air carriers: 26,031 (2018)

Civil aircraft registration country code prefix: LY (2016)

Airports: 61 (2013)

country comparison to the world: 79

Airports—with paved runways: *total:* 22 (2017)

over 3,047 m: 3 (2017)

2,438 to 3,047 m: 1 (2017)

1,524 to 2,437 m: 7 (2017)

914 to 1,523 m: 2 (2017)

under 914 m: 9 (2017)

Airports—with unpaved runways: *total:* 39 (2013)

over 3,047 m: 1 (2013)

914 to 1,523 m: 2 (2013)

under 914 m: 36 (2013)

Pipelines: 1921 km gas, 121 km refined products (2013)

Railways: *total:* 1,768 km (2014)
standard gauge: 22 km 1.435-m gauge (2014)
broad gauge: 1,746 km 1.520-m gauge (122 km electrified) (2014)
country comparison to the world: 79

Roadways: *total:* 84,166 km (2012)
paved: 72,297 km (includes 312 km of expressways) (2012)
unpaved: 11,869 km (2012)
country comparison to the world: 60

Waterways: 441 km (navigable year-round) (2007)
country comparison to the world: 86

Merchant marine: *total:* 58
by type: bulk carrier 3, general cargo 21, oil tanker 2, other 32 (2019)
country comparison to the world: 111

Ports and terminals: *major seaport(s):* Klaipeda
oil terminal(s): Butinge oil terminal
LNG terminal(s) (import): Klaipeda

MILITARY AND SECURITY

Military and security forces: *Lithuanian Armed Forces (Lietuvos Ginkluotosios Pajegos):* Land Forces (Sausumos Pajegos), Naval Forces (Karines Juru Pajegos), Air Forces (Karines Oro Pajegos), Special Operations Forces (Specialiuju Operaciju Pajegos); National Defense Volunteer Forces (Savanoriu Pajegos); National Riflemen's Union (paramilitary force that acts as an additional reserve force) (2020)

Military expenditures: 2.03% of GDP (2019 est.)
1.98% of GDP (2018)
1.72% of GDP (2017)
1.48% of GDP (2016)
1.14% of GDP (2015)
country comparison to the world: 48

Military and security service personnel strengths: estimates for the Lithuanian Armed Forces vary; approximately 17,000 active duty personnel (12,500 Army, including about 5,000 National Defense Voluntary Forces; 700 Navy; 1,000 Air Force; 3,000 other, including special operations forces, logistics support, training, etc); est. 11,000 Riflemen Union (2020)

Military equipment inventories and acquisitions: the Lithuanian Armed Forces' inventory is mostly a mix of Western weapons systems and Soviet-era equipment (primarily aircraft and helicopters); Germany and the UK are the leading suppliers of armaments to Lithuania since 2010 (2019 est.)

Military deployments: contributes about 350 troops to the Lithuania, Poland, and Ukraine joint military brigade (LITPOLUKRBRIG), which was established in 2014; the brigade is headquartered in Warsaw and is comprised of an international staff, three battalions, and specialized units (2019)

Military service age and obligation: 19-26 years of age for conscripted military service (males); 9-month service obligation; in 2015, Lithuania reinstated conscription after having converted to a professional military in 2008; 18-38 for voluntary service (male and female) (2019)

TRANSNATIONAL ISSUES

Disputes—international: Lithuania and Russia committed to demarcating their boundary in 2006 in accordance with the land and maritime treaty ratified by Russia in May 2003 and by Lithuania in 1999; Lithuania operates a simplified transit regime for Russian nationals traveling from the Kaliningrad coastal exclave into Russia, while still conforming, as a EU member state having an external border with a non-EU member, to strict Schengen border rules; boundary demarcated with Latvia and Lithuania; as of January 2007, ground demarcation of the boundary with Belarus was complete and mapped with final ratification documents in preparation

Refugees and internally displaced persons: *stateless persons:* 2,904 (2019)

Illicit drugs: transshipment and destination point for cannabis, cocaine, ecstasy, and opiates from Southwest Asia, Latin America, Western Europe, and neighboring Baltic countries; growing production of high-quality amphetamines, but limited production of cannabis, methamphetamines; susceptible to money laundering despite changes to banking legislation

LUXEMBOURG

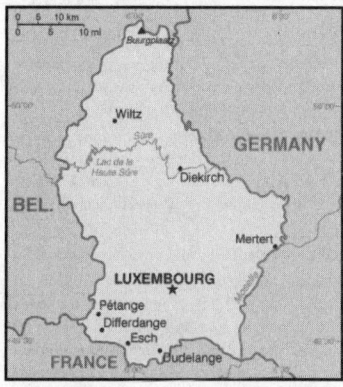

INTRODUCTION

Background: Founded in 963, Luxembourg became a grand duchy in 1815 and an independent state under the Netherlands. It lost more than half of its territory to Belgium in 1839 but gained a larger measure of autonomy. In 1867, Luxembourg attained full independence under the condition that it promise perpetual neutrality. Overrun by Germany in both world wars, it ended its neutrality in 1948 when it entered into the Benelux Customs Union and when it joined NATO the following year. In 1957, Luxembourg became one of the six founding countries of the EEC (later the EU), and in 1999 it joined the euro currency zone.

GEOGRAPHY

Location: Western Europe, between France and Germany

Geographic coordinates: 49 45 N, 6 10 E

Map references: Europe

Area: *total:* 2,586 sq km
land: 2,586 sq km
water: 0 sq km
country comparison to the world: 179

Area—comparative: slightly smaller than Rhode Island; about half the size of Delaware

Land boundaries: *total:* 327 km
border countries (3): Belgium 130 km, France 69 km, Germany 128 km

Coastline: 0 km (landlocked)

Maritime claims: none (landlocked)

Climate: modified continental with mild winters, cool summers

Terrain: mostly gently rolling uplands with broad, shallow valleys; uplands to slightly mountainous in the north; steep slope down to Moselle flood plain in the southeast

Elevation: *mean elevation:* 325 m
lowest point: Moselle River 133 m
highest point: Buurgplaatz 559 m

Natural resources: iron ore (no longer exploited), arable land

Land use: *agricultural land:* 50.7% (2011 est.)
arable land: 24% (2011 est.) / *permanent crops:* 0.6% (2011 est.) / *permanent pasture:* 26.1% (2011 est.)
forest: 33.5% (2011 est.)
other: 15.8% (2011 est.)
Irrigated land: 0 sq km (2012)

Population distribution: most people live in the south, on or near the border with France

Natural hazards: occasional flooding

Environment—current issues: air and water pollution in urban areas, soil pollution of farmland; unsustainable patterns of consumption (transport, energy, recreation, space) threaten biodiversity and landscapes

Environment—international agreements: *party to:* Air Pollution, Air Pollution-Nitrogen Oxides,

Air Pollution-Persistent Organic Pollutants, Air Pollution-Sulfur 85, Air Pollution-Sulfur 94, Air Pollution-Volatile Organic Compounds, Biodiversity, Climate Change, Climate Change-Kyoto Protocol, Desertification, Endangered Species, Hazardous Wastes, Law of the Sea, Marine Dumping, Ozone Layer Protection, Ship Pollution, Tropical Timber 83, Tropical Timber 94, Wetlands *signed, but not ratified:* Environmental Modification

Geography—note: landlocked; the only grand duchy in the world

PEOPLE AND SOCIETY

Population: 628,381 (July 2020 est.)
country comparison to the world: 169

Nationality: *noun:* Luxembourger(s)
adjective: Luxembourg

Ethnic groups: Luxembourger 51.1%, Portuguese 15.7%, French 7.5%, Italian 3.6%, Belgian 3.3%, German 2.1%, Spanish 1.1%, British 1%, other 14.6% (2019 est.)
note: data represent population by nationality

Languages: Luxembourgish (official administrative and judicial language and national language (spoken vernacular)) 55.8%, Portuguese 15.7%, French (official administrative, judicial, and legislative language) 12.1%, German (official administrative and judicial language) 3.1%, Italian 2.9%, English 2.1%, other 8.4% (2011 est.)

Religions: Christian (predominantly Roman Catholic) 70.4%, Muslim 2.3%, other (includes Buddhist, folk religions, Hindu, Jewish) 0.5%, none 26.8% (2010 est.)

Age structure: *0-14 years:* 16.73% (male 54,099/female 51,004)
15-24 years: 11.78% (male 37,946/female 36,061)
25-54 years: 43.93% (male 141,535/female 134,531)
55-64 years: 12.19% (male 39,289/female 37,337)
65 years and over: 15.37% (male 43,595/female 52,984) (2020 est.)

Dependency ratios: *total dependency ratio:* 42.8
youth dependency ratio: 22.2
elderly dependency ratio: 20.5
potential support ratio: 4.9 (2020 est.)

Median age: *total:* 39.5 years
male: 38.9 years
female: 40 years (2020 est.)
country comparison to the world: 56

Population growth rate: 1.8% (2020 est.)
country comparison to the world: 57

Birth rate: 11.6 births/1,000 population (2020 est.)
country comparison to the world: 170

Death rate: 7.3 deaths/1,000 population (2020 est.)
country comparison to the world: 114

Net migration rate: 13.3 migrant(s)/1,000 population (2020 est.)
country comparison to the world: 3

Population distribution: most people live in the south, on or near the border with France

Urbanization: *urban population:* 91.5% of total population (2020)
rate of urbanization: 1.55% annual rate of change (2015-20 est.)
total population growth rate v. urban population growth rate, 2000-2030:

Major urban areas—population: 120,000 LUXEMBOURG (capital) (2018)

Sex ratio: *at birth:* 1.06 male(s)/female
0-14 years: 1.06 male(s)/female
15-24 years: 1.05 male(s)/female
25-54 years: 1.05 male(s)/female
55-64 years: 1.05 male(s)/female
65 years and over: 0.82 male(s)/female
total population: 1.02 male(s)/female (2020 est.)

Mother's mean age at first birth: 30.7 years (2017 est.)

Maternal mortality rate: 5 deaths/100,000 live births (2017 est.)
country comparison to the world: 168

Infant mortality rate: *total:* 3.3 deaths/1,000 live births
male: 3.7 deaths/1,000 live births
female: 3 deaths/1,000 live births (2020 est.)
country comparison to the world: 209

Life expectancy at birth: *total population:* 82.6 years
male: 80.1 years
female: 85.2 years (2020 est.)
country comparison to the world: 16

Total fertility rate: 1.62 children born/woman (2020 est.)
country comparison to the world: 182

Drinking water source:
improved:
urban: 100% of population
rural: 98.8% of population
total: 99% of population
unimproved:
urban: 0% of population
rural: 1.2% of population
total: 1% of population (2017 est.)

Current Health Expenditure: 5.5% (2017)

Physicians density: 3.01 physicians/1,000 population (2017)

Hospital bed density: 4.7 beds/1,000 population (2017)

Sanitation facility access:
improved:
urban: 100% of population (2015 est.)
rural: 99.9% of population
total: 100% of population
unimproved:
urban: 0% of population
rural: 0.1% of population
total: 0% of population (2017 est.)

HIV/AIDS—adult prevalence rate: 0.3% (2018 est.)
country comparison to the world: 92

HIV/AIDS—people living with HIV/AIDS: 1,200 (2018 est.)
country comparison to the world: 141

HIV/AIDS—deaths: <100 (2018 est.)

Obesity—adult prevalence rate: 22.6% (2016)
country comparison to the world: 74

Education expenditures: 3.9% of GDP (2015)
country comparison to the world: 107

School life expectancy (primary to tertiary education): *total:* 14 years
male: 14 years
female: 14 years (2018)

Unemployment, youth ages 15-24: *total:* 14.2%
male: 16.3%
female: 11.9% (2018 est.)
country comparison to the world: 97

GOVERNMENT

Country name: *conventional long form:* Grand Duchy of Luxembourg
conventional short form: Luxembourg
local long form: Grand Duche de Luxembourg
local short form: Luxembourg
etymology: the name derives from the Celtic "lucilem" (little) and the German "burg" (castle or fortress) to produce the meaning of the "little castle"; the name is actually ironic, since for centuries the Fortress of Luxembourg was one of Europe's most formidable fortifications; the name passed to the surrounding city and then to the country itself

Government type: constitutional monarchy

Capital: *name:* Luxembourg
geographic coordinates: 49 36 N, 6 07 E
time difference: UTC+1 (6 hours ahead of Washington, DC, during Standard Time)
daylight saving time: +1hr, begins last Sunday in March; ends last Sunday in October
etymology: the name derives from the Celtic "lucilem" (little) and the German "burg" (castle or fortress) to produce the meaning of the "little castle"; the name is actually ironic, since for centuries the Fortress of Luxembourg was one of Europe's most formidable fortifications; the name passed to the city that grew around the fortress

Administrative divisions: 12 cantons (cantons, singular—canton); Capellen, Clervaux, Diekirch, Echternach, Esch-sur-Alzette, Grevenmacher, Luxembourg, Mersch, Redange, Remich, Vianden, Wiltz

Independence: 1839 (from the Netherlands)

National holiday: National Day (birthday of Grand Duke HENRI), 23 June; note—this date of birth is not the true date of birth for any of the Royals, but the national festivities were shifted in 1962 to allow observance during a more favorable time of year

Constitution: *history:* previous 1842 (heavily amended 1848, 1856); latest effective 17 October 1868
amendments: proposed by the Chamber of Deputies or by the monarch to the Chamber; passage requires at least two thirds majority vote by the Chamber in two successive readings three months apart; a referendum can be substituted for the second reading if approved by more than a quarter of the Chamber members or by 25,000

valid voters; adoption by referendum requires a majority of all valid voters; amended many times, last in 2009

Legal system: civil law system

International law organization participation: accepts compulsory ICJ jurisdiction; accepts ICCt jurisdiction

Citizenship: *citizenship by birth:* limited to situations where the parents are either unknown, stateless, or when the nationality law of the parents' state of origin does not permit acquisition of citizenship by descent when the birth occurs outside of national territory

citizenship by descent only: at least one parent must be a citizen of Luxembourg

dual citizenship recognized: yes

residency requirement for naturalization: 7 years

Suffrage: 18 years of age; universal and compulsory

Executive branch: *chief of state:* Grand Duke HENRI (since 7 October 2000); Heir Apparent Prince GUILLAUME (son of the monarch, born 11 November 1981)

head of government: Prime Minister Xavier BETTEL (since 4 December 2013); Deputy Prime Minister Etienne SCHNEIDER (since 4 December 2013); Deputy Prime Minister Felix BRAZ (since 5 December 2018)

cabinet: Council of Ministers recommended by the prime minister, appointed by the monarch

elections/appointments: the monarchy is hereditary; following elections to the Chamber of Deputies, the leader of the majority party or majority coalition usually appointed prime minister by the monarch; deputy prime minister appointed by the monarch; prime minister and deputy prime minister are responsible to the Chamber of Deputies

Legislative branch: *description:* unicameral Chamber of Deputies or Chambre des Deputes (60 seats; members directly elected in multi-seat constituencies by party-list proportional representation vote; members serve 5-year terms); note—a 21-member Council of State appointed by the Grand Duke on the advice of the prime minister serves as an advisory body to the Chamber of Deputies

elections: last held on 14 October 2018 (next to be held by October 2023)

election results: percent of vote by party—CSV 28.3%, LSAP 17.6%, DP 16.9%, Green Party 15.1%, ADR 8.3%, Pirate Party 6.4%, The Left 5.5%, other 1.9%; seats by party—CSV 21, DP 12, LSAP 10, Green Party 9, ADR 4, Pirate Party 2, The Left 2; composition—men 46, women 14, percent of women 23.3%

Judicial branch: *highest courts:* Supreme Court of Justice includes Court of Appeal and Court of Cassation (consists of 27 judges on 9 benches); Constitutional Court (consists of 9 members)

judge selection and term of office: judges of both courts appointed by the monarch for life

subordinate courts: Court of Accounts; district and local tribunals and courts

Political parties and leaders: Alternative Democratic Reform Party or ADR [Jean SCHOOS]
Christian Social People's Party or CSV [Marc SPAUTZ]
Democratic Party or DP [Corinne CAHEN]
Green Party [Francoise FOLMER, Christian KMIOTEK]
Luxembourg Socialist Workers' Party or LSAP [Claude HAAGEN]
The Left (dei Lenk/la Gauche) [collective leadership, Central Committee] other minor parties

International organization participation: ADB (nonregional member), Australia Group, Benelux, BIS, CD, CE, EAPC, EBRD, ECB, EIB, EMU, ESA, EU, FAO, FATF, IAEA, IBRD, ICAO, ICC (national committees), ICCt, ICRM, IDA, IEA, IFAD, IFC, IFRCS, ILO, IMF, IMO, Interpol, IOC, IOM, IPU, ISO, ITSO, ITU, ITUC (NGOs), MIGA, NATO, NEA, NSG, OAS (observer), OECD, OIF, OPCW, OSCE, PCA, Schengen Convention, UN, UNCTAD, UNESCO, UNHCR, UNIDO, UNRWA, UPU, WCO, WHO, WIPO, WMO, WTO, ZC

Diplomatic representation in the US: *chief of mission:* Ambassador Sylvie LUCAS (since 16 September 2016)

chancery: 2200 Massachusetts Avenue NW, Washington, DC 20008

telephone: [1] (202) 265-4171

FAX: [1] (202) 328-8270

consulate(s) general: New York, San Francisco

Diplomatic representation from the US: *chief of mission:* Ambassador James Randolph "Randy" EVANS (since 19 June 2018)

telephone: [352] 46-01-23 00

embassy: 22 Boulevard Emmanuel Servais, L-2535 Luxembourg City

mailing address: Unit 3560, APO AE 09126-3560 (official mail)

FAX: [352] 46-14-01

Flag description: three equal horizontal bands of red (top), white, and light blue; similar to the flag of the Netherlands, which uses a darker blue and is shorter; the coloring is derived from the Grand Duke's coat of arms (a red lion on a white and blue striped field)

National symbol(s): *red, rampant lion; national colors:* red, white, light blue

National anthem: *name:* "Ons Heemecht" (Our Motherland); "De Wilhelmus" (The William)

lyrics/music: Michel LENTZ/Jean-Antoine ZINNEN; Nikolaus WELTER/unknown

note: "Ons Heemecht," adopted 1864, is the national anthem, while "De Wilhelmus," adopted 1919, serves as a royal anthem for use when members of the grand ducal family enter or exit a ceremony in Luxembourg

0:00 / 1:23

ECONOMY

Economy—overview: This small, stable, high-income economy has historically featured solid growth, low inflation, and low unemployment.

Luxembourg, the only Grand Duchy in the world, is a landlocked country in northwestern Europe surrounded by Belgium, France, and Germany. Despite its small landmass and small population, Luxembourg is the fifth-wealthiest country in the world when measured on a gross domestic product (PPP) per capita basis. Luxembourg has one of the highest current account surpluses as a share of GDP in the euro zone, and it maintains a healthy budgetary position, with a 2017 surplus of 0.5% of GDP, and the lowest public debt level in the region.

Since 2002, Luxembourg's government has proactively implemented policies and programs to support economic diversification and to attract foreign direct investment. The government focused on key innovative industries that showed promise for supporting economic growth: logistics, information and communications technology (ICT); health technologies, including biotechnology and biomedical research; clean energy technologies, and more recently, space technology and financial services technologies. The economy has evolved and flourished, posting strong GDP growth of 3.4% in 2017, far outpacing the European average of 1.8%.

Luxembourg remains a financial powerhouse – the financial sector accounts for more than 35% of GDP—because of the exponential growth of the investment fund sector through the launch and development of cross-border funds (UCITS) in the 1990s. Luxembourg is the world's second-largest investment fund asset domicile, after the US, with $4 trillion of assets in custody in financial institutions.

Luxembourg has lost some of its advantage as a favorable tax location because of OECD and EU pressure, as well as the "LuxLeaks" scandal, which revealed advantageous tax treatments offered to foreign corporations. In 2015, the government's compliance with EU requirements to implement automatic exchange of tax information on savings accounts—thus ending banking secrecy—has constricted banking activity. Likewise, changes to the way EU members collect taxes from e-commerce has cut Luxembourg's sales tax revenues, requiring the government to raise additional levies and to reduce some direct social benefits as part of the tax reform package of 2017. The tax reform package also included reductions in the corporate tax rate and increases in deductions for families, both intended to increase purchasing power and increase competitiveness.

GDP (purchasing power parity):
$62.11 billion (2017 est.)
$60.71 billion (2016 est.)
$58.9 billion (2015 est.)
note: data are in 2017 dollars
country comparison to the world: 105

GDP (official exchange rate):
$62.53 billion (2017 est.)

GDP—real growth rate: 2.31% (2019 est.)
3.14% (2018 est.)
1.81% (2017 est.)
country comparison to the world: 120

GDP—per capita (PPP): $105,100 (2017 est.)
$105,400 (2016 est.)
$104,600 (2015 est.)
note: data are in 2017 dollars
country comparison to the world: 5

Gross national saving: 22.3% of GDP (2017 est.)
23% of GDP (2016 est.)
23.2% of GDP (2015 est.)
country comparison to the world: 82

GDP—composition, by end use:
household consumption: 30.2% (2017 est.)
government consumption: 16.5% (2017 est.)
investment in fixed capital: 16.2% (2017 est.)
investment in inventories: 1.1% (2017 est.)
exports of goods and services: 230% (2017 est.)
imports of goods and services: -194% (2017 est.)

GDP—composition, by sector of origin:
agriculture: 0.3% (2017 est.)
industry: 12.8% (2017 est.)
services: 86.9% (2017 est.)

Agriculture—products: grapes, barley, oats, potatoes, wheat, fruits; dairy and livestock products

Industries: banking and financial services, construction, real estate services, iron, metals, and steel, information technology, telecommunications, cargo transportation and logistics, chemicals, engineering, tires, glass, aluminum, tourism, biotechnology

Industrial production growth rate:
1.9% (2017 est.)
country comparison to the world: 134

Labor force: 476,000 (2020 est.)
note: data exclude foreign workers; in addition to the figure for domestic labor force, about 150,000 workers commute daily from France, Belgium, and Germany
country comparison to the world: 156

Labor force—by occupation: *agriculture:* 1.1%
industry: 20%
services: 78.9% (2013 est.)

Unemployment rate: 5.36% (2019 est.)
5.46% (2018 est.)
country comparison to the world: 86

Population below poverty line: NA

Household income or consumption by percentage share: *lowest 10%:* 3.5%
highest 10%: 23.8% (2000)

Budget: *revenues:* 27.75 billion (2017 est.)
expenditures: 26.8 billion (2017 est.)

Taxes and other revenues: 44.4% (of GDP) (2017 est.)
country comparison to the world: 24

Budget surplus (+) or deficit (-): 1.5% (of GDP) (2017 est.)
country comparison to the world: 23

Public debt: 23% of GDP (2017 est.)
20.8% of GDP (2016 est.)
note: data cover general government debt and include debt instruments issued (or owned) by government entities other than the treasury; the data include treasury debt held by foreign entities; the data include debt issued by subnational entities, as well as intragovernmental debt;

intragovernmental debt consists of treasury borrowings from surpluses in the social funds, such as for retirement, medical care, and unemployment; debt instruments for the social funds are not sold at public auctions
country comparison to the world: 183

Fiscal year: calendar year

Inflation rate (consumer prices): 2.1% (2017 est.)
0% (2016 est.)
country comparison to the world: 109

Current account balance: $3.254 billion (2019 est.)
$3.296 billion (2018 est.)
country comparison to the world: 33

Exports: $15.99 billion (2017 est.)
$16.37 billion (2016 est.)
country comparison to the world: 72

Exports—partners: Germany 25.6%, Belgium 17.6%, France 14%, Netherlands 5.1%, Italy 4.1%, UK 4.1% (2017)

Exports—commodities: machinery and equipment, steel products, chemicals, rubber products, glass

Imports: $20.66 billion (2017 est.)
$20.41 billion (2016 est.)
country comparison to the world: 75

Imports—commodities: commercial aircraft, minerals, chemicals, metals, foodstuffs, luxury consumer goods

Imports—partners: Belgium 32%, Germany 24.9%, France 11.1%, US 5.7%, Netherlands 4.9% (2017)

Reserves of foreign exchange and gold: $878 million (31 December 2017 est.)
$974 million (31 December 2016 est.)
country comparison to the world: 137

Debt—external: $3.781 trillion (31 March 2016 est.)
$3.806 trillion (31 March 2015 est.)
country comparison to the world: 6

Exchange rates: euros (EUR) per US dollar—
0.885 (2017 est.)
0.903 (2016 est.)
0.9214 (2015 est.)
0.885 (2014 est.)
0.7634 (2013 est.)

ENERGY

Electricity access: *electrification—total population:* 100% (2020)

Electricity—production: 334.5 million kWh (2016 est.)
country comparison to the world: 178

Electricity—consumption: 6.475 billion kWh (2016 est.)
country comparison to the world: 111

Electricity—exports: 1.42 billion kWh (2016 est.)
country comparison to the world: 51

Electricity—imports: 7.718 billion kWh (2016 est.)
country comparison to the world: 30

Electricity—installed generating capacity: 1.709 million kW (2016 est.)
country comparison to the world: 118

Electricity—from fossil fuels: 25% of total installed capacity (2016 est.)
country comparison to the world: 189

Electricity—from nuclear fuels: 0% of total installed capacity (2017 est.)
country comparison to the world: 131

Electricity—from hydroelectric plants: 8% of total installed capacity (2017 est.)
country comparison to the world: 122

Electricity—from other renewable sources: 67% of total installed capacity (2017 est.)
country comparison to the world: 2

Crude oil—production: 0 bbl/day (2018 est.)
country comparison to the world: 164

Crude oil—exports: 0 bbl/day (2017 est.)
country comparison to the world: 156

Crude oil—imports: 0 bbl/day (2017 est.)
country comparison to the world: 155

Crude oil—proved reserves: 0 bbl (1 January 2018 est.)
country comparison to the world: 159

Refined petroleum products—production: 0 bbl/day (2017 est.)
country comparison to the world: 167

Refined petroleum products—consumption: 59,850 bbl/day (2017 est.)
country comparison to the world: 96

Refined petroleum products—exports: 0 bbl/day (2017 est.)
country comparison to the world: 173

Refined petroleum products—imports: 59,020 bbl/day (2017 est.)
country comparison to the world: 73

Natural gas—production: 0 cu m (2017 est.)
country comparison to the world: 160

Natural gas—consumption: 792.8 million cu m (2017 est.)
country comparison to the world: 96

Natural gas—exports: 0 cu m (2017 est.)
country comparison to the world: 141

Natural gas—imports: 792.8 million cu m (2017 est.)
country comparison to the world: 63

Natural gas—proved reserves: 0 cu m (1 January 2014 est.)
country comparison to the world: 161

Carbon dioxide emissions from consumption of energy: 10.72 million Mt (2017 est.)
country comparison to the world: 103

COMMUNICATIONS

Telephones—fixed lines: *total subscriptions:* 268,043
subscriptions per 100 inhabitants: 43.43 (2019 est.)
country comparison to the world: 114

Telephones—mobile cellular: *total subscriptions:* 837,890

subscriptions per 100 inhabitants: 135.76 (2019 est.)
country comparison to the world: 164

Telecommunication systems: *general assessment:* highly developed; by 2020 the government is to provide a 1Gb/s service to all citizens, and to make Luxembourg the first fully fibered country in Europe; new law requiring SIM cards be registered has slowed down growth for mobile subscribers; regulator planning a multi-spectrum auction for 5G use by mid-2020 (2020)

domestic: fixed-line teledensity about 43 per 100 persons; nationwide mobile-cellular telephone system with market for mobile-cellular phones virtually saturated with 136 per 100 mobile-cellular (2019)

international: country code—352
note: the COVID-19 outbreak is negatively impacting telecommunications production and supply chains globally; consumer spending on telecom devices and services has also slowed due to the pandemic's effect on economies worldwide; overall progress towards improvements in all facets of the telecom industry—mobile, fixed-line, broadband, submarine cable and satellite—has moderated

Broadcast media: Luxembourg has a long tradition of operating radio and TV services for pan-European audiences and is home to Europe's largest privately owned broadcast media group, the RTL Group, which operates 46 TV stations and 29 radio stations in Europe; also home to Europe's largest satellite operator, Societe Europeenne des Satellites (SES); domestically, the RTL Group operates TV and radio networks; other domestic private radio and TV operators and French and German stations available; satellite and cable TV services available

Internet country code: .lu

Internet users: *total:* 587,955
percent of population: 97.06% (July 2018 est.)
country comparison to the world: 148

Broadband—fixed subscriptions: *total:* 224,300
subscriptions per 100 inhabitants: 37 (2018 est.)
country comparison to the world: 106

TRANSPORTATION

National air transport system: *number of registered air carriers:* 4 (2020)
inventory of registered aircraft operated by air carriers: 66
annual passenger traffic on registered air carriers: 2,099,102 (2018)
annual freight traffic on registered air carriers: 7,323,040,000 mt-km (2018)

Civil aircraft registration country code prefix: LX (2016)

Airports: 2 (2013)
country comparison to the world: 201

Airports—with paved runways:
total: 1 (2019)
over 3,047 m: 1

Airports—with unpaved runways:
total: 1 (2013)
under 914 m: 1 (2013)

Heliports: 1 (2013)

Pipelines: 142 km gas, 27 km refined products (2013)

Railways: *total:* 275 km (2014)
standard gauge: 275 km 1.435-m gauge (275 km electrified) (2014)
country comparison to the world: 124

Roadways: *total:* 2,875 km (2019)
country comparison to the world: 165

Waterways: 37 km (on Moselle River) (2010)
country comparison to the world: 104

Merchant marine: *total:* 143
by type: bulk carrier 4, container ship 4, general cargo 14, oil tanker 3, other 118 (2019)
country comparison to the world: 74

Ports and terminals: *river port(s):* Mertert (Moselle)

MILITARY AND SECURITY

Military and security forces: Luxembourg Army (l'Armée Luxembourgeoise) (2019)

Military expenditures:
0.56% of GDP (2019 est.)
0.51% of GDP (2018)
0.52% of GDP (2017)
0.4% of GDP (2016)
0.44% of GDP (2015)
country comparison to the world: 145

Military and security service personnel strengths: the Luxembourg Army has approximately 900 active personnel (2019 est.)

Military equipment inventories and acquisitions: the inventory of Luxembourg's Army is a small mix of European and US equipment; since 2010, it has received small quantities of equipment from Germany, Norway, and Sweden (2019 est.)

Military service age and obligation: 18-26 years of age for male and female voluntary military service; no conscription; Luxembourg citizen or EU citizen with 3-year residence in Luxembourg (2019)

TRANSNATIONAL ISSUES

Disputes—international: none

Refugees and internally displaced persons: *stateless persons:* 83 (2019)

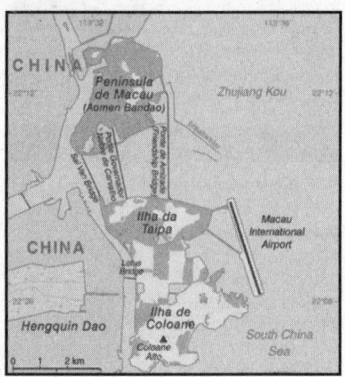

INTRODUCTION

Background: Colonized by the Portuguese in the 16th century, Macau was the first European settlement in the Far East. Pursuant to an agreement signed by China and Portugal on 13 April 1987, Macau became the Macau Special Administrative Region of the People's Republic of China on 20 December 1999. In this agreement, China promised that, under its "one country, two systems" formula, China's political and economic system would not be imposed on Macau, and that Macau would enjoy a "high degree of autonomy" in all matters except foreign affairs and defense for the subsequent 50 years.

GEOGRAPHY

Location: Eastern Asia, bordering the South China Sea and China

Geographic coordinates: 22 10 N, 113 33 E

Map references: Southeast Asia

Area: *total:* 28 sq km
land: 28.2 sq km
water: 0 sq km
country comparison to the world: 237

Area—comparative: less than one-sixth the size of Washington, DC

Land boundaries: *total:* 3 km
regional borders (1): China 3 km

Coastline: 41 km

Maritime claims: not specified

Climate: subtropical; marine with cool winters, warm summers

Terrain: generally flat

Elevation: *lowest point:* South China Sea 0 m
highest point: Alto Coloane 172 m

Natural resources: NEGL

Land use: *agricultural land:* 0% (2011 est.)
arable land: 0% (2011 est.) / *permanent crops:* 0% (2011 est.) / *permanent pasture:* 0% (2011 est.)

forest: 0% (2011 est.)
other: 100% (2011 est.)
Irrigated land: 0 sq km (2012)

Population distribution: population fairly equally distributed

Natural hazards: typhoons

Environment—current issues: air pollution; coastal waters pollution; insufficient policies in reducing and recycling solid wastes; increasing population density worsening noise pollution

Environment—international agreements: *party to:* Marine Dumping (associate member), Ship Pollution (associate member)

Geography—note: essentially urban; an area of land reclaimed from the sea measuring 5.2 sq km and known as Cotai now connects the islands of Coloane and Taipa; the island area is connected to the mainland peninsula by three bridges

PEOPLE AND SOCIETY

Population: 614,458 (July 2020 est.)
country comparison to the world: 170

Nationality: *noun:* Chinese
adjective: Chinese

Ethnic groups: Chinese 88.7%, Portuguese 1.1%, mixed 1.1%, other 9.2% (includes Macanese—mixed Portuguese and Asian ancestry) (2016 est.)

Languages: Cantonese 80.1%, Mandarin 5.5%, other Chinese dialects 5.3%, Tagalog 3%, English 2.8%, Portuguese 0.6%, other 2.8% (2016 est.)
note: Chinese and Portuguese are official languages

Religions: folk religionist 58.9%, Buddhist 17.3%, Christian 7.2%, other 1.2%, none 15.4% (2010 est.)

Age structure: *0-14 years:* 13.43% (male 42,449/female 40,051)
15-24 years: 10.45% (male 33,845/female 30,354)
25-54 years: 49% (male 134,302/female 166,762)
55-64 years: 14.57% (male 44,512/female 45,007)
65 years and over: 12.56% (male 36,223/female 40,953) (2020 est.)

Dependency ratios: *total dependency ratio:* 35.7
youth dependency ratio: 19.5
elderly dependency ratio: 16.2
potential support ratio: 6.2 (2020 est.)

Median age: *total:* 40.8 years
male: 40.7 years
female: 40.9 years (2020 est.)
country comparison to the world: 49

Population growth rate: 0.64% (2020 est.)
country comparison to the world: 147

Birth rate: 7.9 births/1,000 population (2020 est.)
country comparison to the world: 224

Death rate: 4.9 deaths/1,000 population (2020 est.)
country comparison to the world: 201

Net migration rate: 3.3 migrant(s)/1,000 population (2020 est.)
country comparison to the world: 35

Population distribution: population fairly equally distributed

Urbanization: *urban population:* 100% of total population (2020)
rate of urbanization: 1.63% annual rate of change (2015-20 est.)
total population growth rate v. urban population growth rate, 2000-2030:

Sex ratio: *at birth:* 1.05 male(s)/female
0-14 years: 1.06 male(s)/female
15-24 years: 1.12 male(s)/female
25-54 years: 0.81 male(s)/female
55-64 years: 0.99 male(s)/female
65 years and over: 0.88 male(s)/female
total population: 0.9 male(s)/female (2020 est.)

Infant mortality rate: *total:* 3.1 deaths/1,000 live births
male: 3.2 deaths/1,000 live births
female: 2.9 deaths/1,000 live births (2020 est.)
country comparison to the world: 215

Life expectancy at birth: *total population:* 84.6 years
male: 81.7 years
female: 87.7 years (2020 est.)
country comparison to the world: 4

Total fertility rate: 0.96 children born/woman (2020 est.)
country comparison to the world: 227

Drinking water source:
improved:
urban: 100% of population
total: 100% of population
unimproved:
urban: 0% of population
total: 0% of population (2017 est.)

Physicians density: 2.41 physicians/1,000 population (2010)

HIV/AIDS—adult prevalence rate: NA

HIV/AIDS—people living with HIV/AIDS: NA

HIV/AIDS—deaths: NA

Education expenditures: 2.7% of GDP (2017)
country comparison to the world: 152

Literacy: *definition:* age 15 and over can read and write
total population: 96.5%
male: 98.2%
female: 95% (2016)

School life expectancy (primary to tertiary education): *total:* 16 years
male: 16 years
female: 17 years (2019)

Unemployment, youth ages 15-24: *total:* 5.3%
male: 6.7%
female: 3.9% (2017 est.)
country comparison to the world: 165

GOVERNMENT

Country name: *conventional long form:* Macau Special Administrative Region
conventional short form: Macau
official long form: Aomen Tebie Xingzhengqu (Chinese); Regiao Administrativa Especial de Macau (Portuguese)
official short form: Aomen (Chinese); Macau (Portuguese)
etymology: name is thought to derive from the A-Ma Temple—built in 1488 and dedicated to Mazu, the goddess of seafarers and fishermen—which is referred to locally as "Maa Gok"—and in Portuguese became "Macau"; the Chinese name Aomen means "inlet gates"

Dependency status: special administrative region of the People's Republic of China

Government type: executive-led limited democracy; a special administrative region of the People's Republic of China

Administrative divisions: none (special administrative region of the People's Republic of China)

Independence: none (special administrative region of China)

National holiday: National Day (anniversary of the Founding of the People's Republic of China), 1 October (1949); note—20 December (1999) is celebrated as Macau Special Administrative Region Establishment Day

Constitution: *history:* previous 1976 (Organic Statute of Macau, under Portuguese authority); latest adopted 31 March 1993, effective 20 December 1999 (Basic Law of the Macau Special Administrative Region of the People's Republic of China serves as Macau's constitution)
amendments: proposed by the Standing Committee of the National People's Congress (NPC), the People's Republic of China State Council, and the Macau Special Administrative Region; submittal of proposals to the NPC requires two-thirds majority vote by the Legislative Assembly of Macau, approval by two thirds of Macau's deputies to the NPC, and consent of the Macau chief executive; final passage requires approval by the NPC; amended 2005, 2012

Legal system: civil law system based on the Portuguese model

Citizenship: see China

Suffrage: 18 years of age in direct elections for some legislative positions, universal for permanent residents living in Macau for the past 7 years; note—indirect elections are limited to organizations registered as "corporate voters" and an election committee for the chief executive drawn from broad regional groupings, municipal organizations, central government bodies, and elected Macau officials

Executive branch: *chief of state:* President of China XI Jinping (since 14 March 2013)
head of government: Chief Executive HO Iat Seng (since 20 December 2019)

cabinet: Executive Council appointed by the chief executive
elections/appointments: president indirectly elected by National People's Congress for a 5-year term (eligible for a second term); election last held on 17 March 2018 (next to be held in March 2023);chief executive chosen by a 400- member Election Committee for a 5-year term (eligible for a second term); election last held on 24 August 2019 (next to be held in 2024)
election results: Fernando CHUI Sai On reelected chief executive; Election Committee vote—380 of 396; note—HO Iat Seng was elected chief executive (receiving 392 out of 400 votes) on 24 August 2019 and will take office on 20 December 2019

Legislative branch: *description:* unicameral Legislative Assembly or Regiao Administrativa Especial de Macau (33 seats; 14 members directly elected by proportional representation vote, 12 indirectly elected by an electoral college of professional and commercial interest groups, and 7 appointed by the chief executive; members serve 4-year terms)
elections: last held on 17 September 2017 (next to be held in 2021)
election results: percent of vote—UMG 10%, UPD 9.7%, ACUM 8.6%, NE 8.3%, UPP 7.2%, ANMD 6.6%, NUDM 6.1%, ACDM 5.9%, APMD 5.8%, Civic Watch 5.6%, ABL 5.5%, ANPM 5.3%, other 15.4%; seats by political group—UMG 2, UPD 2, ABL 1, ACDM 1, ACUM 1, ANMD 1, ANPM 1, APMD 1, Civic Watch 1, NE 1, NUDM 1, UPP 1; 12 seats filled by professional and business groups; 7 members appointed by the chief executive; composition—men 27, women 6, percent of women 18.6%

Judicial branch: *highest courts:* Court of Final Appeal of Macau Special Administrative Region (consists of the court president and 2 associate justices)
judge selection and term of office: justices appointed by the Macau chief executive upon the recommendation of an independent commission of judges, lawyers, and "eminent" persons; judge tenure NA
subordinate courts: Court of Second Instance; Court of First instance; Lower Court; Administrative Court

Political parties and leaders: Alliance for Change or APM [Melinda CHAN Mei-yi]
Alliance for a Happy Home or ABL [WONG Kit-cheng] (an electoral list of UPP)
Civic Watch or Civico [Agnes LAM Iok-fong]
Macau-Guangdong Union or UMG [MAK Soi-kun]
Macau Citizens' Development Association or ACDM [Becky SONG Pek-kei] (an electoral list of ACUM)New Democratic
Macau Association or ANMD [AU Kam-san]
New Hope or NE [Jose Maria Pereira COUTINHO]
New Macau Association (New Macau Progressives) or AMN or ANPM [Sulu SOU Ka-hou]

New Union for Macau's Development or NUDM [Angela LEONG On-kei]
Prosperous Democratic Macau Association or APMD (an electoral list of AMN)
Union for Development or UPD [Ella LEI Cheng-I]
Union for Promoting Progress or UPP [HO Ion-sang]
United Citizens Association of Macau or ACUM [CHAN Meng-kam]
note: there is no political party ordinance, so there are no registered political parties; politically active groups register as societies or companies

International organization participation: ICC (national committees), IHO, IMF, IMO (associate), Interpol (subbureau), ISO (correspondent), UNESCO (associate), UNWTO (associate), UPU, WCO, WMO, WTO

Diplomatic representation in the US: none (Special Administrative Region of China)

Diplomatic representation from the US: the US has no offices in Macau; US Consulate General in Hong Kong is accredited to Macau

Flag description: *green with a lotus flower above a stylized bridge and water in white, beneath an arc of five gold, five-pointed stars:* one large in the center of the arc and two smaller on either side; the lotus is the floral emblem of Macau, the three petals represent the peninsula and two islands that make up Macau; the five stars echo those on the flag of China

National symbol(s): *lotus blossom; national colors:* green, white, yellow

National anthem: *note:* as a Special Administrative Region of China, "Yiyongjun Jinxingqu" is the official anthem (see China)
0:00 / 0:43

ECONOMY

Economy—overview: Since opening up its locally-controlled casino industry to foreign competition in 2001, Macau has attracted tens of billions of dollars in foreign investment, transforming the territory into one of the world's largest gaming centers. Macau's gaming and tourism businesses were fueled by China's decision to relax travel restrictions on Chinese citizens wishing to visit Macau. In 2016, Macau's gaming-related taxes accounted for more than 76% of total government revenue.

Macau's economy slowed dramatically in 2009 as a result of the global economic slowdown, but strong growth resumed in the 2010-13 period, largely on the back of tourism from mainland China and the gaming sectors. In 2015, this city of 646,800 hosted nearly 30.7 million visitors. Almost 67% came from mainland China. Macau's traditional manufacturing industry has slowed greatly since the termination of the Multi-Fiber Agreement in 2005. Services export — primarily gaming — increasingly has driven Macau's economic performance. Mainland China's anti-corruption campaign brought Macau's gambling boom to a halt in 2014, with spending in casinos contracting 34.3% in 2015. As a result, Macau's inflation-adjusted GDP contracted 21.5% in 2015 and another 2.1% in 2016—down from double-digit expansion rates in the period 2010-13—but the economy recovered handsomely in 2017. Macau continues to face the challenges of managing its growing casino industry, risks from money-laundering activities, and the need to diversify the economy away from heavy dependence on gaming revenues. Macau's currency, the pataca, is closely tied to the Hong Kong dollar, which is also freely accepted in the territory.

GDP (purchasing power parity):
$77.33 billion (2018)
$71.82 billion (2017 est.)
$65.84 billion (2016 est.)
note: data are in 2017 dollars
country comparison to the world: 97

GDP (official exchange rate):
$50.36 billion (2017 est.)

GDP—real growth rate: 9.1% (2017 est.)
-0.9% (2016 est.)
-21.6% (2015 est.)
country comparison to the world: 6

GDP—per capita (PPP): $122,000 (2018)
$110,000 (2017 est.)
$102,100 (2016 est.)
country comparison to the world: 3

GDP—composition, by end use:
household consumption: 24.2% (2017 est.)
government consumption: 9.9% (2017 est.)
investment in fixed capital: 18.5% (2017 est.)
investment in inventories: 0.8% (2017 est.)
exports of goods and services: 79.4% (2017 est.)
imports of goods and services: -32% (2017 est.)

GDP—composition, by sector of origin:
agriculture: 0% (2016 est.)
industry: 6.3% (2017 est.)
services: 93.7% (2017 est.)

Agriculture—products: only 2% of land area is cultivated, mainly by vegetable growers; fishing, mostly for crustaceans, is important; some of the catch is exported to Hong Kong

Industries: tourism, gambling, clothing, textiles, electronics, footwear, toys

Industrial production growth rate: 2% (2017 est.)
country comparison to the world: 131

Labor force: 392,000 (2020 est.)
country comparison to the world: 160

Labor force—by occupation: *agriculture:* 2.5%
industry: 9.8%
services: 4.4%
industry and services: 12.4%
agriculture/fishing/forestry/mining: 15%
manufacturing: 25.9%
construction: 7.1%
transportation and utilities: 2.6%
commerce: 20.3% (2013 est.)

Unemployment rate: 2% (2017 est.)
1.9% (2016 est.)
country comparison to the world: 19

Population below poverty line: NA

Household income or consumption by percentage share: *lowest 10%:* NA

highest 10%: NA

Budget: *revenues:* 14.71 billion (2017 est.)
expenditures: 9.684 billion (2017 est.)

Taxes and other revenues: 29.2% (of GDP) (2017 est.)
country comparison to the world: 85

Budget surplus (+) or deficit (-): 10% (of GDP) (2017 est.)
country comparison to the world: 2

Public debt: 0% of GDP (2017 est.)
0% of GDP (2016 est.)
country comparison to the world: 210

Fiscal year: calendar year

Inflation rate (consumer prices): 1.2% (2017 est.)
2.4% (2016 est.)
country comparison to the world: 65

Current account balance: $16.75 billion (2017 est.)
$12.22 billion (2016 est.)
country comparison to the world: 17

Exports: $1.45 billion (2018)
note: includes reexports
country comparison to the world: 148

Exports—partners: Hong Kong 62.1%, China 16.5%, US 1% (2018)

Exports—commodities: clothing, textiles, footwear, toys, electronics, machinery and parts

Imports: $11.1 billion (2018)
$9.7 billion (2017 est.)
country comparison to the world: 97

Imports—commodities: raw materials and semi-manufactured goods, consumer goods (foodstuffs, beverages, tobacco, garments and footwear, motor vehicles), capital goods, mineral fuels and oils

Imports—partners: China 35%, Italy 8.6%, Hong Kong 7.8%, France 8.4%, Switzerland 7.7%, Japan 8.1%, US 4.1% (2018)

Reserves of foreign exchange and gold:
$20.17 billion (31 December 2017 est.)
$18.89 billion (31 December 2015 est.)
note: the Fiscal Reserves Act that came into force on 1 January 2012 requires the fiscal reserves to be separated from the foreign exchange reserves and to be managed separately; the transfer of assets took place in February 2012
country comparison to the world: 59

Debt—external: $0 (31 December 2013)
$0 (31 December 2012)
country comparison to the world: 207

Exchange rates: patacas (MOP) per US dollar—8 (2017 est.)
7.9951 (2016 est.)
7.9951 (2015 est.)
7.985 (2014 est.)
7.9871 (2013 est.)

ENERGY

Electricity access: *electrification—total population:* 100% (2020)

Electricity—production: 929 million kWh (2016 est.)
country comparison to the world: 154

Electricity—consumption: 5.077 billion kWh (2016 est.)
country comparison to the world: 123

Electricity—exports: 0 kWh (2016 est.)
country comparison to the world: 162

Electricity—imports: 4.306 billion kWh (2016 est.)
country comparison to the world: 43

Electricity—installed generating capacity: 472,000 kW (2016 est.)
country comparison to the world: 150

Electricity—from fossil fuels: 100% of total installed capacity (2016 est.)
country comparison to the world: 12

Electricity—from nuclear fuels: 0% of total installed capacity (2017 est.)
country comparison to the world: 132

Electricity—from hydroelectric plants: 0% of total installed capacity (2017 est.)
country comparison to the world: 184

Electricity—from other renewable sources: 0% of total installed capacity (2017 est.)
country comparison to the world: 200

Crude oil—production: 0 bbl/day (2018 est.)
country comparison to the world: 165

Crude oil—exports: 0 bbl/day (2015 est.)
country comparison to the world: 157

Crude oil—imports: 0 bbl/day (2015 est.)
country comparison to the world: 156

Crude oil—proved reserves: 0 bbl (1 January 2018 est.)
country comparison to the world: 160

Refined petroleum products—production: 0 bbl/day (2015 est.)
country comparison to the world: 168

Refined petroleum products—consumption: 12,700 bbl/day (2016 est.)
country comparison to the world: 158

Refined petroleum products—exports: 0 bbl/day (2015 est.)
country comparison to the world: 174

Refined petroleum products—imports: 14,180 bbl/day (2015 est.)
country comparison to the world: 137

Natural gas—production: 0 cu m (2017 est.)
country comparison to the world: 161

Natural gas—consumption: 178.2 million cu m (2017 est.)
country comparison to the world: 107

Natural gas—exports: 0 cu m (2017 est.)
country comparison to the world: 142

Natural gas—imports: 175.5 million cu m (2017 est.)
country comparison to the world: 73

Natural gas—proved reserves: 0 cu m (1 January 2014 est.)
country comparison to the world: 162

Carbon dioxide emissions from consumption of energy: 2.563 million Mt (2017 est.)
country comparison to the world: 153

COMMUNICATIONS

Telephones—fixed lines: *total subscriptions:* 119,355
subscriptions per 100 inhabitants: 19.55 (2019 est.)
country comparison to the world: 135

Telephones—mobile cellular: *total subscriptions:* 2,108,274
subscriptions per 100 inhabitants: 345.33 (2019 est.)
country comparison to the world: 148

Telecommunication systems: *general assessment:* modern communication facilities maintained for domestic and international services; high mobile subscriber numbers and mobile penetration with 4 network operators; offering 4G, LTE services and 1st phase of 5G network rollout; possible synchronizing with neighboring regions; Macau's smart city project spans areas of transportation, medical services, tourism and 3-government (2020)

domestic: fixed-line 20 per 100 and mobile-cellular 345 per 100 persons (2019)

international: country code—853; landing point for the SEA-ME-WE-3 submarine cable network that provides links to Asia,. Africa, Australia, the Middle East, and Europe; HF radiotelephone communication facility; satellite earth station—1 Intelsat (Indian Ocean) (2019)

note: the COVID-19 outbreak is negatively impacting telecommunications production and supply chains globally; consumer spending on telecom devices and services has also slowed due to the pandemic's effect on economies worldwide; overall progress towards improvements in all facets of the telecom industry—mobile, fixed-line, broadband, submarine cable and satellite—has moderated

Broadcast media: local government dominates broadcast media; 2 television stations operated by the government with one broadcasting in Portuguese and the other in Cantonese and Mandarin; 1 cable TV and 4 satellite TV services available; 3 radio stations broadcasting, of which 2 are government-operated (2019)

Internet country code: .mo

Internet users: *total:* 508,052

percent of population: 83.79% (July 2018 est.)
country comparison to the world: 152

Broadband—fixed subscriptions: *total:* 193,057
subscriptions per 100 inhabitants: 32 (2018 est.)
country comparison to the world: 108

TRANSPORTATION

National air transport system: *number of registered air carriers:* 1 (registered in China) (2020)
inventory of registered aircraft operated by air carriers: 21 (registered in China)
annual passenger traffic on registered air carriers: 3,157,724 (2018)

annual freight traffic on registered air carriers: 31.84 million mt-km (2018)

Civil aircraft registration country code prefix: B-M (2016)

Airports: 1 (2013)
country comparison to the world: 227

Airports—with paved runways: *total:* 1 (2019)
over 3,047 m: 1

Heliports: 2 (2013)

Roadways: *total:* 428 km (2017)
paved: 428 km (2017)
country comparison to the world: 199

Merchant marine: *total:* 1
by type: other 1 (2019)
country comparison to the world: 177

Ports and terminals: *major seaport(s):* Macau

MILITARY AND SECURITY

Military and security forces: no regular indigenous military forces

Military—note: defense is the responsibility of China and the Chinese People's Liberation Army (PLA) maintains a garrison in Macau

TRANSNATIONAL ISSUES

Disputes—international: none

Illicit drugs: transshipment point for drugs going into mainland China; consumer of opiates and amphetamines

MADAGASCAR

INTRODUCTION

Background: Madagascar was one of the last major habitable landmasses on earth settled by humans.

While there is some evidence of human presence on the island in the millennia B. C., large-scale settlement began between A. D. 350 and 550 with settlers from present- day Indonesia. The island attracted Arab and Persian traders as early as the 7th century, and migrants from Africa arrived around A. D. 1000.Madagascar was a pirate stronghold during the late 17th and early 18th centuries, and served as a slave trading center into the 19th century. From the 16th to the late 19th century, a native Merina Kingdom dominated much of Madagascar. The island was conquered by the French in 1896 who made it a colony; independence was regained in 1960.

During 1992-93, free presidential and National Assembly elections were held ending 17 years of single- party rule. In 1997, in the second presidential race, Didier RATSIRAKA, the leader during the 1970s and 1980s, returned to the presidency. The 2001 presidential election was contested between the followers of Didier RATSIRAKA and Marc RAVALOMANANA, nearly causing secession of half of the country. In 2002, the High Constitutional Court announced RAVALOMANANA the winner. RAVALOMANANA won a second term in 2006 but, following protests in 2009, handed over power to the military, which then conferred the presidency on the mayor of Antananarivo, Andry RAJOELINA, in what amounted to a coup d'etat. Following a lengthy mediation process led by the Southern African Development Community, Madagascar held UN-supported presidential and parliamentary elections in 2013.Former de facto finance minister Hery RAJAONARIMAMPIANINA won a runoff election in December 2013 and was inaugurated in January 2014. In January 2019, RAJOELINA was declared the winner of a runoff election against RAVALOMANANA; both RATSIRAKA and RAJAONARIMAMPIANINA also ran in the first round of the election, which took place in November 2018.

GEOGRAPHY

Location: Southern Africa, island in the Indian Ocean, east of Mozambique

Geographic coordinates: 20 00 S, 47 00 E

Map references: Africa

Area: *total:* 587,041 sq km
land: 581,540 sq km
water: 5,501 sq km
country comparison to the world: 48

Area—comparative: almost four times the size of Georgia; slightly less than twice the size of Arizona

Land boundaries: 0 km

Coastline: 4,828 km

Maritime claims: territorial sea: 12 nm
exclusive economic zone: 200 nm
contiguous zone: 24 nm
continental shelf: 200 nm or 100 nm from the 2,500—m isobath

Climate: tropical along coast, temperate inland, arid in south

Terrain: narrow coastal plain, high plateau and mountains in center

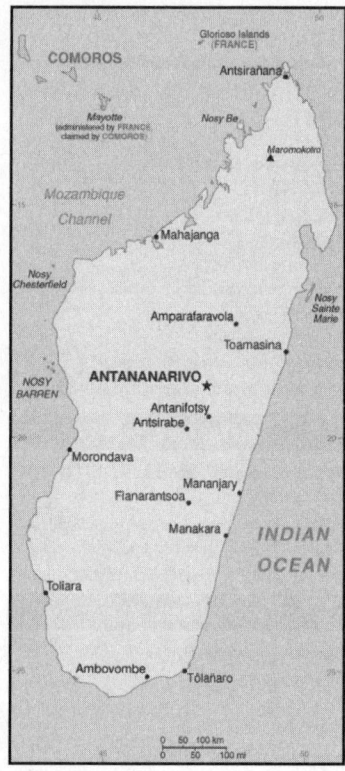

Elevation: *mean elevation:* 615 m
lowest point: Indian Ocean 0 m
highest point: Maromokotro 2,876 m

Natural resources: graphite, chromite, coal, bauxite, rare earth elements, salt, quartz, tar sands, semiprecious stones, mica, fish, hydropower

Land use: *agricultural land:* 71.1% (2011 est.)
arable land: 6% (2011 est.) / permanent crops: 1% (2011 est.) / permanent pasture: 64.1% (2011 est.)
forest: 21.5% (2011 est.)
other: 7.4% (2011 est.)
Irrigated land: 10,860 sq km (2012)

Population distribution: most of population lives on the eastern half of the island; significant clustering is found in the central highlands and eastern coastline as shown in this population distribution map

Natural hazards: periodic cyclones; drought; and locust infestation
volcanism: Madagascar' s volcanoes have not erupted in historical times

Environment—current issues: erosion and soil degradation results from deforestation and overgrazing; desertification; agricultural fires; surface water contaminated with raw sewage and other organic wastes; wildlife preservation (endangered species of flora and fauna unique to the island)

Environment—international agreements: *party to:* Biodiversity, Climate Change, Climate Change-Kyoto Protocol, Desertification, Endangered Species, Hazardous Wastes, Law of the Sea, Marine Life Conservation, Ozone Layer Protection, Ship Pollution, Wetlands
signed, but not ratified: none of the selected agreements

Geography—note: world's fourth- largest island; strategic location along Mozambique Channel; despite Madagascar's close proximity to the African continent, ocean currents isolate the island resulting in high rates of endemic plant and animal species; approximately 90% of the flora and fauna on the island are found nowhere else

PEOPLE AND SOCIETY

Population: 26,955,737 (July 2020 est.)
country comparison to the world: 53

Nationality: *noun:* Malagasy (singular and plural)
adjective: Malagasy

Ethnic groups: Malayo- Indonesian (Merina and related Betsileo), Cotiers (mixed African, Malayo-Indonesian, and Arab ancestry—Betsimisaraka, Tsimihety, Antaisaka, Sakalava), French, Indian, Creole, Comoran

Languages: French (official), Malagasy (official), English

Religions: Christian, indigenous, Muslim

Demographic profile: Madagascar's youthful population—just over 60% are under the age of 25—and high total fertility rate of more than 4 children per women ensures that the Malagasy population will continue its rapid growth trajectory for the foreseeable future. The population is predominantly rural and poor; chronic malnutrition is prevalent, and large families are the norm. Many young Malagasy girls are withdrawn from school, marry early (often pressured to do so by their parents), and soon begin having children. Early childbearing, coupled with Madagascar's widespread poverty and lack of access to skilled health care providers during delivery, increases the risk of death and serious health problems for young mothers and their babies.

Child marriage perpetuates gender inequality and is prevalent among the poor, the uneducated, and rural households—as of 2013, of Malagasy women aged 20 to 24, more than 40% were married and more than a third had given birth by the age of 18.Although the legal age for marriage is 18, parental consent is often given for earlier marriages or the law is flouted, especially in rural areas that make up nearly 65% of the country. Forms of arranged marriage whereby young girls are married to older men in exchange for oxen or money are traditional. If a union does not work out, a girl can be placed in another marriage, but the dowry paid to her family diminishes with each unsuccessful marriage.

Madagascar's population consists of 18 main ethnic groups, all of whom speak the same Malagasy language. Most Malagasy are multi- ethnic, however, reflecting the island's diversity of settlers and historical contacts (see Background). Madagascar's legacy of hierarchical societies practicing domestic slavery (most notably the Merina

Kingdom of the 16th to the 19th century) is evident today in persistent class tension, with some ethnic groups maintaining a caste system. Slave descendants are vulnerable to unequal access to education and jobs, despite Madagascar's constitutional guarantee of free compulsory primary education and its being party to several international conventions on human rights. Historical distinctions also remain between central highlanders and coastal people.

Age structure: *0-14 years:* 38.86% (male 5,278,838/female 5,196,036)
15-24 years: 20.06% (male 2,717,399/female 2,689,874)
25-54 years: 33.02% (male 4,443,147/female 4,456,691)
55-64 years: 4.6% (male 611,364/female 627,315)
65 years and over: 3.47% (male 425,122/female 509,951) (2020 est.)

Dependency ratios:
total dependency ratio: 75.9
youth dependency ratio: 70.5
elderly dependency ratio: 5.5
potential support ratio: 18.3 (2020 est.)

Median age: *total:* 20.3 years
male: 20.1 years
female: 20.5 years (2020 est.)
country comparison to the world: 194

Population growth rate: 2.39% (2020 est.)
country comparison to the world: 27

Birth rate: 29.9 births/ 1,000 population (2020 est.)
country comparison to the world: 32

Death rate: 6.2 deaths/ 1,000 population (2020 est.)
country comparison to the world: 155

Net migration rate: 0 migrant(s)/ 1,000 population (2020 est.)
country comparison to the world: 88

Population distribution: most of population lives on the eastern half of the island; significant clustering is found in the central highlands and eastern coastline as shown in this population distribution map

Urbanization: *urban population:* 38.5% of total population (2020)
rate of urbanization: 4.48% annual rate of change (2015-20 est.)
total population growth rate v. urban population growth rate, 2000-2030:

Major urban areas—population: 3.369 million ANTANANARIVO (capital) (2020)

Sex ratio: *at birth:* 1.03 male(s)/female
0-14 years: 1.02 male(s)/female
15-24 years: 1.01 male(s)/female
25-54 years: 1 male(s)/female
55-64 years: 0.97 male(s)/female
65 years and over: 0.83 male(s)/female
total population: 1 male(s)/female (2020 est.)

Mother's mean age at first birth: 19.5 years (2008/09 est.)
note: median age at first birth among women 25-29

Maternal mortality rate: 335 deaths/100,000 live births (2017 est.)
country comparison to the world: 33

Infant mortality rate: *total:* 37.8 deaths/1,000 live births
male: 41.5 deaths/1,000 live births
female: 34.1 deaths/1,000 live births (2020 est.)
country comparison to the world: 42

Life expectancy at birth: *total population:* 67.3 years
male: 65.7 years
female: 68.9 years (2020 est.)
country comparison to the world: 182

Total fertility rate: 3.78 children born/woman (2020 est.)
country comparison to the world: 34

Contraceptive prevalence rate: 44.3% (2018)

Drinking water source:
improved:
urban: 87.9% of population
rural: 36.3% of population
total: 55.5% of population
unimproved:
urban: 12.1% of population
rural: 63.7% of population
total: 44.5% of population (2017 est.)

Current Health Expenditure: 5.5% (2017)

Physicians density: 0.18 physicians/1,000 population (2014)

Hospital bed density: 0.2 beds/1,000 population (2010)

Sanitation facility access:
improved:
urban: 42.5% of population
rural: 16.6% of population
total: 26.1% of population
unimproved:
urban: 57.5% of population
rural: 83.4% of population
total: 73.9% of population (2017 est.)

HIV/AIDS—adult prevalence rate: 0.2% (2019 est.)
country comparison to the world: 104

HIV/AIDS—people living with HIV/AIDS: 39,000 (2019 est.)
country comparison to the world: 68

HIV/AIDS—deaths: 1,400 (2019 est.)
country comparison to the world: 52

Major infectious diseases: *degree of risk:* very high (2020)
food or waterborne diseases: bacterial diarrhea, hepatitis A, and typhoid fever
vectorborne diseases: malaria and dengue fever
water contact diseases: schistosomiasis
animal contact diseases: rabies

Obesity—adult prevalence rate: 5.3% (2016)
country comparison to the world: 180

Children under the age of 5 years underweight: 26.4% (2018)
country comparison to the world: 12

Education expenditures: 2.8% of GDP (2014)
country comparison to the world: 147

Literacy: *definition:* age 15 and over can read and write
total population: 74.8%
male: 77.3%
female: 72.4% (2018)

School life expectancy (primary to tertiary education):
total: 10 years
male: 10 years
female: 10 years (2018)

Unemployment, youth ages 15-24:
total: 1%
male: 1%
female: 1% (2012 est.)
country comparison to the world: 179

GOVERNMENT

Country name: *conventional long form:* Republic of Madagascar
conventional short form: Madagascar
local long form: Republique de Madagascar/ Repoblikan'i Madagasikara
local short form: Madagascar/Madagasikara
former: Malagasy Republic
etymology: the name " Madageiscar" was first used by the 13th- century Venetian explorer Marco POLO, as a corrupted transliteration of Mogadishu, the Somali port with which POLO confused the island

Government type: semi- presidential republic

Capital: *name:* Antananarivo
geographic coordinates: 18 55 S, 47 31 E
time difference: UTC+ 3 (8 hours ahead of Washington, DC, during Standard Time)
etymology: the name, which means " City of the Thousand," was bestowed by 17th century King Adrianjakaking to honor the soldiers assigned to guard the city

Administrative divisions: 6 provinces (faritany); Antananarivo, Antsiranana, Fianarantsoa, Mahajanga, Toamasina, Toliara

Independence: 26 June 1960 (from France)

National holiday: Independence Day, 26 June (1960)

Constitution: *history:* previous 1992; latest passed by referendum 17 November 2010, promulgated 11 December 2010
amendments: proposed by the president of the republic in consultation with the cabinet or supported by a least two thirds of both the Senate and National Assembly membership; passage requires at least three- fourths approval of both the Senate and National Assembly and approval in a referendum; constitutional articles, including the form and powers of government, the sovereignty of the state, and the autonomy of Madagascar's collectivities, cannot be amended
Legal system: civil law system based on the old French civil code and customary law in matters of marriage, family, and obligation

International law organization participation: accepts compulsory ICJ jurisdiction with reservations; accepts ICCt jurisdiction

Citizenship: citizenship by birth: no
citizenship by descent only: the father must be a citizen of Madagascar; in the case of a child born out of wedlock, the mother must be a citizen
dual citizenship recognized: no
residency requirement for naturalization: unknown

Suffrage: 18 years of age; universal

Executive branch: *chief of state:* President Andry RAJOELINA (since 21 January 2019) (2019)

head of government: Prime Minister Christian NTSAY (since 6 June 2018 and re- appointed 19 July 2019)
cabinet: Council of Ministers appointed by the prime minister
elections/appointments: president directly elected by absolute majority popular vote in 2 rounds if needed for a 5-year term (eligible for a second term); election last held on 7 November and 19 December 2018 (next to be held in 2023); prime minister nominated by the National Assembly, appointed by the president
election results: Andry RAJOELINA elected President in second round; percent of vote— Andry RAJOELINA (TGV) 55.7%, Marc RAVALOMANANA 44.3% (TIM)

Legislative branch: *description:* bicameral Parliament consists of: Senate or Antenimieran-Doholona (reestablished on 22 January 2016, following the December 2015 senatorial election) (63 seats; 42 members indirectly elected by an electoral college of municipal, communal, regional, and provincial leaders and 21 appointed by the president of the republic; members serve 5-year terms)
National Assembly or Antenimierampirenena (151 seats; 87 members directly elected in single-seat constituencies by simple majority vote and 64 directly elected in multi- seat constituencies by closed-list proportional representation vote; members serve 5-year terms)
elections: Senate—last held 29 December 2015 (next to be held in 2021)
National Assembly—last held on 27 May 2019 (next to be held in 2024)
election results: Senate—percent of vote by party—NA; seats by party—HVM 34, TIM 3, MAPAR 2, LEADER- Fanilo 1, independent 2, appointed by the president 21; composition—men 51, women 12, percent of women 19%
National Assembly—percent of vote by party—Independent Pro- HVM 18%, MAPAR 17%, MAPAR pro- HVM 16%, VPM- MMM 10%, VERTS 3%, LEADER FANILO 3%, HIARAKA ISIKA 3%, GPS/ARD 7%, INDEPENDENT 9%, TAMBATRA 1%, TIM 13%; composition—men 120, women 31, percent of women 20.5%; note— total National Assembly percent of women 20.1%

Judicial branch: *highest courts:* Supreme Court or Cour Supreme (consists of 11 members; addresses judicial administration issues only); High Constitutional Court or Haute Cour Constitutionnelle (consists of 9 members); note—the judiciary includes a High Court of Justice responsible for adjudicating crimes and misdemeanors by government officials, including the president

judge selection and term of office: Supreme Court heads elected by the president and judiciary officials to serve 3-year, single renewable terms; High Constitutional Court members appointed—3 each by the president, by both legislative bodies, and by the Council of Magistrates; members serve single, 7-year terms

subordinate courts: Courts of Appeal; Courts of First Instance

Political parties and leaders: Economic liberalism and democratic action for national recovery or LEADER FANILO [Jean Max RAKOTOMAMONJY] FOMBA [Ny Rado RAFALIMANANA]

Gideons fighting against poverty in Madagascar (Gedeona Miady amin'ny Fahantrana eto Madagasikara) or GFFM [Andre Christian Dieu Donne MAILHOL]

Green party or VERTS (Antoko Maintso) [Alexandre GEORGET]

I Love Madagascar (Tiako I Madagasikara) or TIM [Marc RAVALOMANANA] Malagasy aware (Malagasy Tonga Saina) or MTS [Roland RATSIRAKA]

Malagasy raising together (Malagasy Miara-Miainga) or MMM [Hajo ANDRIANAINARIVELO]

New Force for Madagascar (Hery Vaovao ho an'ny Madagasikara) or HVM [Hery Martial RAJAONARIMAMPIANINA Rakotoarimanana]

Total Refoundation of Madagascar (Refondation Totale de Madagascar) or RTM [Joseph Martin RANDRIAMAMPIONONA] Vanguard for the renovation of Madagascar (Avant- Garde pour la renovation de Madagascar) or AREMA [DidierRATSIRAKA]

Young Malagasies Determined (Malagasy: Tanora malaGasy Vonona) or TGV [Andry RAJOELINA] and MAPAR [Andry RAJOELINA], and IRD (We are all with Andy Rajoelina) [Andry RAJOELINA]

International organization participation: ACP, AfDB, AU, CD, COMESA, EITI (candidate country), FAO, G- 77, IAEA, IBRD, ICAO, ICC (NGOs), ICCt, ICRM, IDA, IFAD, IFC, IFRCS, ILO, IMF, IMO, InOC, Interpol, IOC, IOM, IPU, ISO (correspondent), ITSO, ITU, ITUC (NGOs), MIGA, NAM, OIF, OPCW, PCA, SADC, UN, UNCTAD, UNESCO, UNHCR, UNIDO, UNWTO, UPU, WCO, WFTU (NGOs), WHO, WIPO, WMO, WTO

Diplomatic representation in the US: *chief of mission:* Ambassador Eric ANDRIAMIHAJA Robson (since March 2018)

chancery: 2374 Massachusetts Avenue NW, Washington, DC 20008

telephone: [1] (202) 265-5525

FAX: [1] (202) 265-3034

consulate(s) general: New York

Diplomatic representation from the US: *chief of mission:* Ambassador Michael PELLETIER (since 14 February 2019)

telephone: [261] 20 23 480 00

embassy: Lot 207A, Point Liberty, Andranoro, Antehiroka, 105 Antananarivo

mailing address: B. P. 620, Antsahavola, Antananarivo

FAX: [261] 20 23 480 35 or [261] 33 44 328 17

Flag description: two equal horizontal bands of red (top) and green with a vertical white band of the same width on hoist side; by tradition, red stands for sovereignty, green for hope, white for purity

National symbol(s): traveller's palm, zebu; national colors: red, green, white

National anthem: name: " Ry Tanindraza nay malala o" (Oh, Our Beloved Fatherland)

lyrics/music: Pasteur RAHAJASON/Norbert RAHARISOA

note: adopted 1959

0:00 / 1:02

ECONOMY

Economy—overview: Madagascar is a mostly unregulated economy with many untapped natural resources, but no capital markets, a weak judicial system, poorly enforced contracts, and rampant government corruption. The country faces challenges to improve education, healthcare, and the environment to boost long- term economic growth. Agriculture, including fishing and forestry, is a mainstay of the economy, accounting for more than one- fourth of GDP and employing roughly 80% of the population. Deforestation and erosion, aggravated by bushfires, slash-and-burn clearing techniques, and the use of firewood as the primary source of fuel, are serious concerns for the agriculture dependent economy.

After discarding socialist economic policies in the mid- 1990s, Madagascar followed a World Bank- and IMF- led policy of privatization and liberalization until a 2009 coup d'état led many nations, including the United States, to suspend non-humanitarian aid until a democratically-elected president was inaugurated in 2014.The pre- coup strategy had placed the country on a slow and steady growth path from an extremely low starting point. Exports of apparel boomed after gaining duty-free access to the US market in 2000 under the African Growth and Opportunity Act (AGOA); however, Madagascar' s failure to comply with the requirements of the AGOA led to the termination of the country' s duty- free access in January 2010, a sharp fall in textile production, a loss of more than 100,000 jobs, and a GDP drop of nearly 11%.

Madagascar regained AGOA access in January 2015 and ensuing growth has been slow and fragile. Madagascar produces around 80% of the world's vanilla and its reliance on this commodity for most of its foreign exchange is a significant source of vulnerability. Economic reforms have been modest and the country's financial sector remains weak, limiting the use of monetary policy to control inflation. An ongoing IMF program aims to strengthen financial and investment management capacity.

GDP (purchasing power parity):
$ 39.85 billion (2017 est.)
$ 38.25 billion (2016 est.)
$ 36.72 billion (2015 est.)
note: data are in 2017 dollars
country comparison to the world: 120

GDP (official exchange rate):
$ 11.5 billion (2017 est.)

GDP—real growth rate:
4.2% (2017 est.)
4.2% (2016 est.)
3.1% (2015 est.)
country comparison to the world: 69

GDP—per capita (PPP):
$ 1,600 (2017 est.)
$ 1,500 (2016 est.)
$ 1,500 (2015 est.)
note: data are in 2017 dollars
country comparison to the world: 218

Gross national saving:
14.8% of GDP (2017 est.)
15.4% of GDP (2016 est.)
11.2% of GDP (2015 est.)
country comparison to the world: 138

GDP—composition, by end use:
household consumption: 67.1% (2017 est.)
government consumption: 11.2% (2017 est.)
investment in fixed capital: 15.1% (2017 est.)
investment in inventories: 8.8% (2017 est.)
exports of goods and services: 31.5% (2017 est.)
imports of goods and services:—33.7% (2017 est.)

GDP—composition, by sector of origin:
agriculture: 24% (2017 est.)
industry: 19.5% (2017 est.)
services: 56.4% (2017 est.)

Agriculture—products: coffee, vanilla, sugarcane, cloves, cocoa, rice, cassava (manioc, tapioca), beans, bananas, peanuts; livestock products

Industries: meat processing, seafood, soap, beer, leather, sugar, textiles, glassware, cement, automobile assembly plant, paper, petroleum, tourism, mining

Industrial production growth rate:
5.2% (2017 est.)
country comparison to the world: 54

Labor force: 13.4 million (2017 est.)
country comparison to the world: 40

Unemployment rate:
1.8% (2017 est.)
1.8% (2016 est.)
country comparison to the world: 18

Population below poverty line: 70.7% (2012 est.)

Household income or consumption by percentage share: *lowest 10%:* 2.2%
highest 10%: 34.7% (2010 est.)

Budget: *revenues:* 1.828 billion (2017 est.)
expenditures: 2.136 billion (2017 est.)

Taxes and other revenues:
15.9% (of GDP) (2017 est.)
country comparison to the world: 185

Budget surplus (+) or deficit (-):

- 2.7% (of GDP) (2017 est.)
country comparison to the world: 123

Public debt:
36% of GDP (2017 est.)
38.4% of GDP (2016 est.)
country comparison to the world: 148

Fiscal year: calendar year

Inflation rate (consumer prices):
8.3% (2017 est.)
6.7% (2016 est.)
country comparison to the world: 199

Current account balance:
-$ 35 million (2017 est.)
$ 57 million (2016 est.)
country comparison to the world: 77

Exports: $ 2.29 billion (2017 est.)
$ 2.26 billion (2016 est.)
country comparison to the world: 136

Exports—partners: France 24.8%, US 16.5%, China 6.7%, Germany 6.5%, Japan 6%, Netherlands 4.7% (2017)

Exports—commodities: coffee, vanilla, shellfish, sugar, cotton cloth, clothing, chromite, petroleum products, gems, ilmenite, cobalt, nickel

Imports: $ 2.738 billion (2017 est.)
$ 2.427 billion (2016 est.)
country comparison to the world: 152

Imports—commodities: capital goods, petroleum, consumer goods, food

Imports—partners: China 18.7%, India 9.3%, France 6.4%, South Africa 5.6%, UAE 5.3% (2017)

Reserves of foreign exchange and gold:
$ 1.6 billion (31 December 2017 est.)
$ 1.076 billion (31 December 2016 est.)
country comparison to the world: 124

Debt—external:
$ 4.089 billion (31 December 2017 est.)
$ 3.425 billion (31 December 2016 est.)
country comparison to the world: 139

Exchange rates: Malagasy ariary (MGA) per US dollar—
3,116.1 (2017 est.)
3,176.5 (2016 est.)
3,176.5 (2015 est.)
2,933.5 (2014 est.)
2,414.8 (2013 est.)

ENERGY

Electricity access: population without electricity: 17 million (2019)
electrification—total population: 39% (2019)
electrification—urban areas: 64% (2019)
electrification—rural areas: 23% (2019)

Electricity—production: 1.706 billion kWh (2016 est.)
country comparison to the world: 142

Electricity—consumption: 1.587 billion kWh (2016 est.)
country comparison to the world: 147

Electricity—exports: 0 kWh (2016 est.)
country comparison to the world: 163

Electricity—imports: 0 kWh (2016 est.)
country comparison to the world: 169

Electricity—installed generating capacity: 675,400 kW (2016 est.)
country comparison to the world: 136

Electricity—from fossil fuels: 74% of total installed capacity (2016 est.)
country comparison to the world: 97

Electricity—from nuclear fuels: 0% of total installed capacity (2017 est.)
country comparison to the world: 134

Electricity—from hydroelectric plants: 24% of total installed capacity (2017 est.)
country comparison to the world: 80

Electricity—from other renewable sources: 2% of total installed capacity (2017 est.)
country comparison to the world: 141

Crude oil—production: 0 bbl/day (2018 est.)
country comparison to the world: 167

Crude oil—exports: 0 bbl/day (2015 est.)
country comparison to the world: 158

Crude oil—proved reserves: 0 bbl (1 January 2018 est.)
country comparison to the world: 162

Refined petroleum products—production: 0 bbl/day (2015 est.)
country comparison to the world: 170

Refined petroleum products—consumption: 18,000 bbl/ day (2016 est.)
country comparison to the world: 147

Refined petroleum products—exports: 0 bbl/ day (2015 est.)
country comparison to the world: 175

Refined petroleum products—imports: 18,880 bbl/ day (2015 est.)
country comparison to the world: 125

Natural gas—production: 0 cu m (2017 est.)
country comparison to the world: 163

Natural gas—consumption: 0 cu m (2017 est.)
country comparison to the world: 169

Natural gas—exports: 0 cu m (2017 est.)
country comparison to the world: 144

Natural gas—imports: 0 cu m (2017 est.)
country comparison to the world: 151

Natural gas—proved reserves: 0 cu m (1 January 2012 est.)
country comparison to the world: 164

Carbon dioxide emissions from consumption of energy: 4.021 million Mt (2017 est.)
country comparison to the world: 138

COMMUNICATIONS

Telephones—fixed lines: *total subscriptions:* 68,426
subscriptions per 100 inhabitants: less than 1 (2019 est.)
country comparison to the world: 150

Telephones—mobile cellular:
total subscriptions: 10,677,153
subscriptions per 100 inhabitants: 40.57 (2019 est.)

country comparison to the world: 83

Telecommunication systems: *general assessment:* system is above average for the region; competition among the four mobile service providers has spurred recent growth in the mobile market and helped the service to be less expensive for the consumer; 3G and LTE services available; Telecom service tax raised to 10% (2020)
domestic: less than 1 per 100 for fixed- line and mobile- cellular teledensity about 41 per 100 persons (2019)
international: country code—261; landing points for the EASSy, METISS, and LION fiber- optic submarine cable systems connecting to numerous Indian Ocean Islands, South Africa, and Eastern African countries; satellite earth stations—2 (1 Intelsat—Indian Ocean, 1 Intersputnik—Atlantic Ocean region) (2019)
note: the COVID- 19 outbreak is negatively impacting telecommunications production and supply chains globally; consumer spending on telecom devices and services has also slowed due to the pandemic's effect on economies worldwide; overall progress towards improvements in all facets of the telecom industry—mobile, fixed-line, broadband, submarine cable and satellite—has moderated

Broadcast media: state-owned Radio Nationale Malagasy (RNM) and Television Malagasy (TVM) have an extensive national network reach; privately owned radio and TV broadcasters in cities and major towns; state-run radio dominates in rural areas; relays of 2 international broadcasters are available in Antananarivo (2019)

Internet country code: . mg

Internet users: *total:* 2,516,994

percent of population: 9.8% (July 2018 est.)
country comparison to the world: 109

Broadband—fixed subscriptions:
total: 27,211
subscriptions per 100 inhabitants: less than 1 (2018 est.)
country comparison to the world: 144

TRANSPORTATION

National air transport system: number of registered air carriers: 4 (2020)
inventory of registered aircraft operated by air carriers: 18
annual passenger traffic on registered air carriers: 541,290 (2018)
annual freight traffic on registered air carriers: 16.25 million mt-km (2018)

Civil aircraft registration country code prefix: 5R (2016)

Airports: 83 (2013)
country comparison to the world: 66

Airports—with paved runways:
total: 26 (2017)
over 3,047 m: 1 (2017)
2,438 to 3,047 m: 2 (2017)
1,524 to 2,437 m: 6 (2017)
914 to 1,523 m: 16 (2017)

under 914 m: 1 (2017)

Airports—with unpaved runways:
total: 57 (2013)
1,524 to 2,437 m: 1 (2013)
914 to 1,523 m: 38 (2013)
under 914 m: 18 (2013)

Railways: *total:* 836 km (2018)
narrow gauge: 836 km 1.000-m gauge (2018)
country comparison to the world: 96

Roadways: *total:* 31,640 km (2018)
country comparison to the world: 96

Waterways: 600 km (432 km navigable) (2011)
country comparison to the world: 79

Merchant marine: total: 28
by type: general cargo 15, oil tanker 2, other 11
(2019)
country comparison to the world: 136

Ports and terminals: *major seaport(s):*
Antsiranana (Diego Suarez), Mahajanga,
Toamasina, Toliara (Tulear)

MILITARY AND SECURITY

Military and security forces: *People's Armed
Forces:* Army, Navy, Air Force; National
Gendarmerie (operates under the Ministry of
Defense) (2019)

Military expenditures:
0.6% of GDP (2019)
0.6% of GDP (2018)
0.6% of GDP (2017)
0.6% of GDP (2016)
0.6% of GDP (2015)
country comparison to the world: 143

Military and security service personnel strengths:
the Peoples Armed Forces (PAF) have approximately 21,500 personnel (12,500 Army; 500 Navy;
500 Air Force; 8,000 Gendarmerie) (2019 est.)

Military equipment inventories and acquisitions:
the PAF's inventory consists mostly of ageing
Soviet- era equipment; since 2010, it has received

limited amounts of second- hand equipment from
South Africa and France (2019)

Military service age and obligation: Madagascar
has an all-volunteer military; 18-25 years of age for
males; service obligation 18 months; women are
permitted to serve in all branches (2018)

TRANSNATIONAL ISSUES

Disputes—international: claims Bassas da India,
Europa Island, Glorioso Islands, and Juan de Nova
Island (all administered by France); the vegetated
drying cays of Banc du Geyser, which were claimed
by Madagascar in 1976, also fall within the EEZ
claims of the Comoros and France (Glorioso
Islands, part of the French Southern and Antarctic
Lands)

Illicit drugs: illicit producer of cannabis (cultivated and wild varieties) used mostly for domestic
consumption; transshipment point for heroin

MALAWI

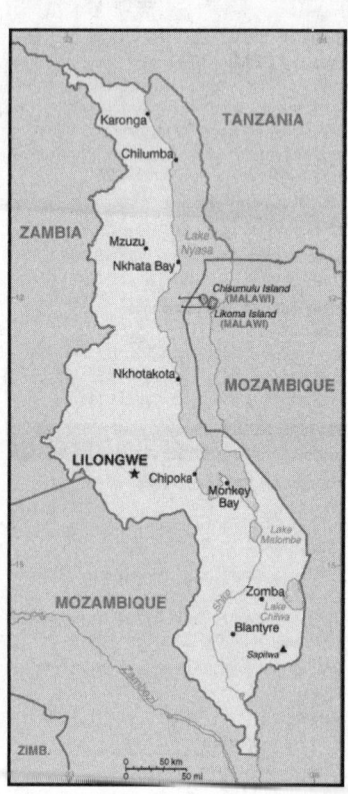

INTRODUCTION

Background: From the late 15th to the 18th centuries, a prosperous Kingdom of Maravi—from
which the name Malawi derives—extended its
reach into what are now areas of Zambia and
Mozambique. British missionary and trading activity increased in the area around Lake Malawi in
the second half of the 19th century. In 1889, a
British Central African Protectorate was established, which was renamed Nyasaland in 1907,
and which became the independent nation of
Malawi in 1964. After three decades of one-party
rule under President Hastings Kamuzu BANDA,
the country held multiparty presidential and parliamentary elections in 1994, under a provisional
constitution that came into full effect the following year. Bakili MULUZI became the first freely
elected president of Malawi when he won the
presidency in 1994; he won re-election in 1999.
President Bingu wa MUTHARIKA, elected in
2004 after a failed attempt by the previous president to amend the constitution to permit another
term, struggled to assert his authority against his
predecessor and subsequently started his own
party, the Democratic Progressive Party in 2005.
MUTHARIKA was reelected to a second term in
2009. He oversaw some economic improvement in
his first term, but was accused of economic mismanagement and poor governance in his second
term. He died abruptly in 2012 and was succeeded
by vice president, Joyce BANDA, who had earlier started her own party, the People's Party.
MUTHARIKA's brother, Peter MUTHARIKA,
defeated BANDA in the 2014 election. Peter
MUTHARIKA was reelected in a disputed
2019 election that resulted in countrywide protests. Population growth, increasing pressure on

agricultural lands, corruption, and the scourge of
HIV/AIDS pose major problems for Malawi.

GEOGRAPHY

Location: Southern Africa, east of Zambia, west
and north of Mozambique

Geographic coordinates:
13 30 S, 34 00 E

Map references: Africa

Area: *total:* 118,484 sq km
land: 94,080 sq km
water: 24,404 sq km
country comparison to the world: 101

Area—comparative: slightly smaller than
Pennsylvania

Land boundaries: *total:* 2,857 km
border countries (3): Mozambique 1498 km,
Tanzania 512 km, Zambia 847 km

Coastline: 0 km (landlocked)

Maritime claims: *none (landlocked)*

Climate: sub-tropical; rainy season (November to
May); dry season (May to November)

Terrain: narrow elongated plateau with rolling
plains, rounded hills, some mountains

Elevation: *mean elevation:* 779 m
lowest point: junction of the Shire River and
international boundary with Mozambique 37 m
highest point: Sapitwa (Mount Mlanje) 3,002 m

Natural resources: limestone, arable land, hydropower, unexploited deposits of uranium, coal, and
bauxite

Land use: *agricultural land:* 59.2% (2011 est.)
arable land: 38.2% (2011 est.) / permanent crops:
1.4% (2011 est.) / permanent pasture: 19.6%
(2011 est.)

forest: 34% (2011 est.)
other: 6.8% (2011 est.)
Irrigated land: 740 sq km (2012)

Population distribution: population density is highest south of Lake Nyasa as shown in this population distribution map

Natural hazards: flooding; droughts; earthquakes

Environment—current issues: deforestation; land degradation; water pollution from agricultural run-off, sewage, industrial wastes; siltation of spawning grounds endangers fish populations; negative effects of climate change (extreme high temperatures, changing precipatation pattens)

Environment—international agreements: *party to:* Biodiversity, Climate Change, Climate Change-Kyoto Protocol, Desertification, Endangered Species, Environmental Modification, Hazardous Wastes, Marine Life Conservation, Ozone Layer Protection, Ship Pollution, Wetlands
signed, but not ratified: Law of the Sea

Geography—note: landlocked; Lake Nyasa, some 580 km long, is the country's most prominent physical feature; it contains more fish species than any other lake on earth

PEOPLE AND SOCIETY

Population: 21,196,629 (July 2020 est.)
note: estimates for this country explicitly take into account the effects of excess mortality due to AIDS; this can result in lower life expectancy, higher infant mortality, higher death rates, lower population growth rates, and changes in the distribution of population by age and sex than would otherwise be expected
country comparison to the world: 60

Nationality: *noun:* Malawian(s)
adjective: Malawian

Ethnic groups: Chewa 34.3%, Lomwe 18.8%, Yao 13.2%, Ngoni 10.4%, Tumbuka 9.2%, Sena 3.8%, Mang'anja 3.2%, Tonga 1.8%, Nyanja 1.8%, Nkhonde 1%, other 2.2%, foreign .3% (2018 est.)

Languages: English (official), Chewa (common), Lambya, Lomwe, Ngoni, Nkhonde, Nyakyusa, Nyanja, Sena, Tonga, Tumbuka, Yao
note: Chewa and Nyanja are mutually intelligible dialects; Nkhonde and Nyakyusa are mutually intelligible dialects

Religions: Protestant 33.5% (includes Church of Central Africa Presbyterian 14.2%, Seventh Day Adventist/Baptist 9.4%, Pentecostal 7.6%, Anglican 2.3%), Roman Catholic 17.2%, other Christian 26.6%, Muslim 13.8%, traditionalist 1.1%, other 5.6%, none 2.1% (2018 est.)

Demographic profile: Malawi has made great improvements in maternal and child health, but has made less progress in reducing its high fertility rate. In both rural and urban areas, very high proportions of mothers are receiving prenatal care and skilled birth assistance, and most children are being vaccinated. Malawi's fertility rate, however, has only declined slowly, decreasing from more than 7 children per woman in the 1980s to about 5.5 today. Nonetheless, Malawians prefer

smaller families than in the past, and women are increasingly using contraceptives to prevent or space pregnancies. Rapid population growth and high population density is putting pressure on Malawi's land, water, and forest resources. Reduced plot sizes and increasing vulnerability to climate change, further threaten the sustainability of Malawi's agriculturally based economy and will worsen food shortages. About 80% of the population is employed in agriculture.

Historically, Malawians migrated abroad in search of work, primarily to South Africa and present-day Zimbabwe, but international migration became uncommon after the 1970s, and most migration in recent years has been internal. During the colonial period, Malawians regularly migrated to southern Africa as contract farm laborers, miners, and domestic servants. In the decade and a half after independence in 1964, the Malawian Government sought to transform its economy from one dependent on small-scale farms to one based on estate agriculture. The resulting demand for wage labor induced more than 300,000 Malawians to return home between the mid-1960s and the mid-1970s. In recent times, internal migration has generally been local, motivated more by marriage than economic reasons.

Age structure: 0-14 years: 45.87% (male 4,843,107/female 4,878,983)
15-24 years: 20.51% (male 2,151,417/female 2,195,939)
25-54 years: 27.96% (male 2,944,936/female 2,982,195)
55-64 years: 2.98% (male 303,803/female 328,092)
65 years and over: 2.68% (male 249,219/female 318,938) (2020 est.)

Dependency ratios: *total dependency ratio:* 83.9
youth dependency ratio: 79.1
elderly dependency ratio: 4.9
potential support ratio: 20.6 (2020 est.)

Median age: *total:* 16.8 years
male: 16.7 years
female: 16.9 years (2020 est.)
country comparison to the world: 222

Population growth rate: 3.3% (2020 est.)
country comparison to the world: 6

Birth rate: 40.1 births/1,000 population (2020 est.)
country comparison to the world: 9

Death rate: 7.2 deaths/1,000 population (2020 est.)
country comparison to the world: 120

Net migration rate: 0 migrant(s)/1,000 population (2020 est.)
country comparison to the world: 89

Population distribution: population density is highest south of Lake Nyasa as shown in this population distribution map

Urbanization: *urban population:* 17.4% of total population (2020)
rate of urbanization: 4.19% annual rate of change (2015-20 est.)

Major urban areas—population: 1.122 million LILONGWE (capital), 932,000 Blantyre-Limbe (2020)

Sex ratio: *at birth:* 1.02 male(s)/female
0-14 years: 0.99 male(s)/female
15-24 years: 0.98 male(s)/female
25-54 years: 0.99 male(s)/female
55-64 years: 0.93 male(s)/female
65 years and over: 0.78 male(s)/female
total population: 0.98 male(s)/female (2020 est.)

Mother's mean age at first birth: 18.9 years (2015/16 est.)
note: median age at first birth among women 25-29

Maternal mortality rate: 349 deaths/100,000 live births (2017 est.)
country comparison to the world: 31

Infant mortality rate: *total:* 39.5 deaths/1,000 live births
male: 45.8 deaths/1,000 live births
female: 33.2 deaths/1,000 live births (2020 est.)
country comparison to the world: 40

Life expectancy at birth: *total population:* 63.2 years
male: 61.2 years
female: 65.3 years (2020 est.)
country comparison to the world: 205

Total fertility rate: 5.31 children born/woman (2020 est.)
country comparison to the world: 11

Contraceptive prevalence rate: 59.2% (2015/16)

Drinking water source:
improved:
urban: 95.9% of population
rural: 87.3% of population
total: 88.7% of population
unimproved:
urban: 4.1% of population
rural: 12.7% of population
total: 11.3% of population (2017 est.)

Current Health Expenditure: 9.6% (2017)

Physicians density: 0.02 physicians/1,000 population (2016)

Hospital bed density: 1.3 beds/1,000 population (2011)

Sanitation facility access:
improved:
urban: 58.2% of population
rural: 35.9% of population
total: 39.6% of population
unimproved:
urban: 41.8% of population
rural: 64.1% of population
total: 60.4% of population (2017 est.)

HIV/AIDS—adult prevalence rate: 9.5% (2019 est.)
country comparison to the world: 9

HIV/AIDS—people living with HIV/AIDS: 1.1 million (2019 est.)
country comparison to the world: 10

HIV/AIDS—deaths: 13,000 (2019 est.)
country comparison to the world: 18

Major infectious diseases: *degree of risk:* very high (2020)
food or waterborne diseases: bacterial and protozoal diarrhea, hepatitis A, and typhoid fever

595

vectorborne diseases: malaria and dengue fever
water contact diseases: schistosomiasis
animal contact diseases: rabies

Obesity—adult prevalence rate: 5.8% (2016)
country comparison to the world: 173

Children under the age of 5 years underweight: 11.8% (2018)
country comparison to the world: 54

Education expenditures: 4% of GDP (2017)
country comparison to the world: 104

Literacy: *definition:* age 15 and over can read and write
total population: 62.1 %
male: 69.8%
female: 55.2% (2015)

School life expectancy (primary to tertiary education):
total: 11 years
male: 11 years
female: 11 years (2011)

Unemployment, youth ages 15-24:
total: 40.5%
male: 33.1%
female: 47.7% (2017 est.)
country comparison to the world: 10

GOVERNMENT

Country name: conventional long form: Republic of Malawi
conventional short form: Malawi
local long form: Dziko la Malawi
local short form: Malawi
former: British Central African Protectorate, Nyasaland Protectorate, Nyasaland
etymology: named for the East African Maravi Kingdom of the 16th century; the word "maravi" means "fire flames"

Government type: presidential republic

Capital: *name:* Lilongwe
geographic coordinates: 13 58 S, 33 47 E
time difference: UTC+2 (7 hours ahead of Washington, DC, during Standard Time)
etymology: named after the Lilongwe River that flows through the city

Administrative divisions: 28 districts; Balaka, Blantyre, Chikwawa, Chiradzulu, Chitipa, Dedza, Dowa, Karonga, Kasungu, Likoma, Lilongwe, Machinga, Mangochi, Mchinji, Mulanje, Mwanza, Mzimba, Neno, Ntcheu, Nkhata Bay, Nkhotakota, Nsanje, Ntchisi, Phalombe, Rumphi, Salima, Thyolo, Zomba

Independence: 6 July 1964 (from the UK)

National holiday: Independence Day, 6 July (1964); note—also called Republic Day since 6 July 1966

Constitution: *history:* previous 1953 (preindependence), 1966; latest drafted January to May 1994, approved 16 May 1994, entered into force 18 May 1995
amendments: proposed by the National Assembly; passage of amendments affecting constitutional articles, including the sovereignty and territory of the state, fundamental constitutional principles, human rights, voting rights, and the judiciary, requires majority approval in a referendum and majority approval by the Assembly; passage of other amendments requires at least two-thirds majority vote of the Assembly; amended several times, last in 2017

Legal system: mixed legal system of English common law and customary law; judicial review of legislative acts in the Supreme Court of Appeal

International law organization participation: accepts compulsory ICJ jurisdiction with reservations; accepts ICCt jurisdiction

Citizenship: *citizenship by birth:* no
citizenship by descent only: at least one parent must be a citizen of Malawi
dual citizenship recognized: no
residency requirement for naturalization: 7 years

Suffrage: 18 years of age; universal

Executive branch: *chief of state:* President Lazarus CHAKWERA (since 28 June 2020); Vice President Saulos CHILIMA (since 3 February 2020); note—the president is both chief of state and head of government
head of government: President Lazarus CHAKWERA (since 28 June 2020); Vice President Saulos CHILIMA (since 3 February 2020)
cabinet: Cabinet named by the president
elections/appointments: president directly elected by simple majority popular vote for a 5-year term (eligible for a second term); election last held on 23 June 2020 (next to be held in 2025)
election results: Lazarus CHAKWERA elected president; Lazarus CHAKWERA (MCP) 59.3%, Peter Mutharika (DPP) 39.9%, other 0.7%

Legislative branch: *description:* unicameral National Assembly (193 seats; members directly elected in single-seat constituencies by simple majority vote to serve 5-year terms)
elections: last held on 21 May 2019 (next to be held in May 2024)
election results: percent of vote by party—n/a; seats by party—DPP 62, MCP 55, UDF 10, PP 5, other 5, independent 55, vacant 1; composition—men 161, women 32, percent of women 16.6%

Judicial branch: *highest courts:* Supreme Court of Appeal (consists of the chief justice and at least 3 judges)
judge selection and term of office: Supreme Court chief justice appointed by the president and confirmed by the National Assembly; other judges appointed by the president upon the recommendation of the Judicial Service Commission, which regulates judicial officers; judges serve until age 65
subordinate courts: High Court; magistrate courts; Industrial Relations Court; district and city traditional or local courts

Political parties and leaders:
Democratic Progressive Party or DPP [Peter MUTHARIKA]
Malawi Congress Party or MCP [Lazarus CHAKWERA]
Peoples Party or PP [Joyce BANDA]
United Democratic Front or UDF [Atupele MULUZI]
United Transformation Movement or UTM [Saulos CHILIMA]

International organization participation: ACP, AfDB, AU, C, CD, COMESA, FAO, G-77, IAEA, IBRD, ICAO, ICCt, ICRM, IDA, IFAD, IFC, IFRCS, ILO, IMF, IMO, Interpol, IOC, IOM, IPU, ISO (correspondent), ITSO, ITU, ITUC (NGOs), MIGA, MINURSO, MONUSCO, NAM, OPCW, SADC, UN, UNCTAD, UNESCO, UNIDO, UNISFA, UNOCI, UNWTO, UPU, WCO, WFTU (NGOs), WHO, WIPO, WMO, WTO

Diplomatic representation in the US: *chief of mission:* Ambassador Edward Yakobe SAWERENGERA (since 16 September 2016)
chancery: 2408 Massachusetts Avenue NW, Washington, DC 20008
telephone: [1] (202) 721-0270
FAX: [1] (202) 721-0288

Diplomatic representation from the US: *chief of mission:* Ambassador Robert SCOTT (since 6 August 2019)
telephone: +(265) 1-773-166, 1-773-342 and 1-773-367 (Dial "0" before the "1" within Malawi); EMER: +(265) (0) 999-591-024 or +(265) (0) 888-734-826
embassy: 16 Jomo Kenyatta Road, Lilongwe 3
mailing address: P.O. Box 30016, Lilongwe 3, Malawi
FAX: 265 (0) 1770471

Flag description: three equal horizontal bands of black (top), red, and green with a radiant, rising, red sun centered on the black band; black represents the native peoples, red the blood shed in their struggle for freedom, and green the color of nature; the rising sun represents the hope of freedom for the continent of Africa

National symbol(s): lion; national colors: black, red, green

National anthem: *name:* "Mulungu dalitsa Malawi" (Oh God Bless Our Land of Malawi)
lyrics/music: Michael-Fredrick Paul SAUKA
note: adopted 1964

ECONOMY

Economy—overview: Landlocked Malawi ranks among the world's least developed countries. The country's economic performance has historically been constrained by policy inconsistency, macroeconomic instability, poor infrastructure, rampant corruption, high population growth, and poor health and education outcomes that limit labor productivity. The economy is predominately agricultural with about 80% of the population living in rural areas. Agriculture accounts for about one-third of GDP and 80% of export revenues. The performance of the tobacco sector is key to short-term growth as tobacco accounts for more than half of exports, although Malawi is looking to diversify away from tobacco to other cash crops.

The economy depends on substantial inflows of economic assistance from the IMF, the World

Bank, and individual donor nations. Donors halted direct budget support from 2013 to 2016 because of concerns about corruption and fiscal carelessness, but the World Bank resumed budget support in May 2017. In 2006, Malawi was approved for relief under the Heavily Indebted Poor Countries (HIPC) program but recent increases in domestic borrowing mean that debt servicing in 2016 exceeded the levels prior to HIPC debt relief.

Heavily dependent on rain-fed agriculture, with corn being the staple crop, Malawi's economy was hit hard by the El Ninodriven drought in 2015 and 2016, and now faces threat from the fall armyworm. The drought also slowed economic activity, led to two consecutive years of declining economic growth, and contributed to high inflation rates. Depressed food prices over 2017 led to a significant drop in inflation (from an average of 21.7% in 2016 to 12.3% in 2017), with a similar drop in interest rates.

GDP (purchasing power parity):
$22.42 billion (2017 est.)
$21.56 billion (2016 est.)
$21.08 billion (2015 est.)
note: data are in 2017 dollars
country comparison to the world: 145

GDP (official exchange rate):
$6.24 billion (2017 est.)

GDP—real growth rate: 4% (2017 est.)
2.3% (2016 est.)
3% (2015 est.)
country comparison to the world: 74

GDP—per capita (PPP): $1,200 (2017 est.)
$1,200 (2016 est.)
$1,200 (2015 est.)
note: data are in 2017 dollars
country comparison to the world: 223

Gross national saving:
3.9% of GDP (2017 est.)
-2.8% of GDP (2016 est.)
2.8% of GDP (2015 est.)
country comparison to the world: 178

GDP—composition, by end use:
household consumption: 84.3% (2017 est.)
government consumption: 16.3% (2017 est.)
investment in fixed capital: 15.3% (2017 est.)
investment in inventories: 0% (2017 est.)
exports of goods and services: 27.9% (2017 est.)
imports of goods and services: -43.8% (2017 est.)

GDP—composition, by sector of origin:
agriculture: 28.6% (2017 est.)
industry: 15.4% (2017 est.)
services: 56% (2017 est.)

Agriculture—products: tobacco, sugarcane, tea, corn, potatoes, sweet potatoes, cassava (manioc, tapioca), sorghum, pulses, cotton, groundnuts, macadamia nuts, coffee; cattle, goats

Industries: tobacco, tea, sugar, sawmill products, cement, consumer goods

Industrial production growth rate: 1.2% (2017 est.)
country comparison to the world: 150

Labor force: 7 million (2013 est.)
country comparison to the world: 64

Labor force—by occupation:
agriculture: 76.9%
industry: 4.1%
services: 19% (2013 est.)

Unemployment rate: 20.4% (2013 est.)
country comparison to the world: 190

Population below poverty line: 50.7% (2010 est.)

Household income or consumption by percentage share: *lowest 10%:* 2.2%
highest 10%: 37.5% (2010 est.)

Budget: *revenues:* 1.356 billion (2017 est.)
expenditures: 1.567 billion (2017 est.)

Taxes and other revenues:
21.7% (of GDP) (2017 est.)
country comparison to the world: 135

Budget surplus (+) or deficit (-): -3.4% (of GDP) (2017 est.)
country comparison to the world: 144

Public debt:
59.2% of GDP (2017 est.)
60.3% of GDP (2016 est.)
country comparison to the world: 75

Fiscal year: 1 July—30 June

Inflation rate (consumer prices):
12.2% (2017 est.)
21.7% (2016 est.)
country comparison to the world: 205

Current account balance:
-$591 million (2017 est.)
-$744 million (2016 est.)
country comparison to the world: 125

Exports: $1.42 billion (2017 est.)
$1.361 billion (2016 est.)
country comparison to the world: 150

Exports—partners: Zimbabwe 13.1%, Mozambique 11.8%, Belgium 10.7%, South Africa 6.3%, Netherlands 5%, UK 4.7%, Germany 4.3%, US 4.2% (2017)

Exports—commodities: tobacco (55%), dried legumes (8.8%), sugar (6.7%), tea (5.7%), cotton (2%), peanuts, coffee, soy (2015 est.)

Imports:
$2.312 billion (2017 est.)
$2.277 billion (2016 est.)
country comparison to the world: 161

Imports—commodities: food, petroleum products, semi-manufactures, consumer goods, transportation equipment

Imports—partners: South Africa 20.7%, China 14.2%, India 11.6%, UAE 7%, Netherlands 4.4% (2017)

Reserves of foreign exchange and gold:
$780.2 million (31 December 2017 est.)
$585.7 million (31 December 2016 est.)
country comparison to the world: 140

Debt—external:
$2.102 billion (31 December 2017 est.)
$1.5 billion (31 December 2016 est.)
country comparison to the world: 151

Exchange rates: Malawian kwachas (MWK) per US dollar—

731.69 (2017 est.)
720.1 (2016 est.)
713.85 (2015 est.)
499.6 (2014 est.)
424.9 (2013 est.)

ENERGY

Electricity access: *population without electricity:* 16 million (2019)
electrification—total population: 13% (2019)
electrification—urban areas: 55% (2019)
electrification—rural areas: 5% (2019)

Electricity—production: 1.42 billion kWh (2016 est.)
country comparison to the world: 144

Electricity—consumption: 1.321 billion kWh (2016 est.)
country comparison to the world: 150

Electricity—exports: 0 kWh (2016 est.)
country comparison to the world: 164

Electricity—imports: 0 kWh (2016 est.)
country comparison to the world: 170

Electricity—installed generating capacity: 375.000 kW (2016 est.)
country comparison to the world: 152

Electricity—from fossil fuels: 1% of total installed capacity (2016 est.)
country comparison to the world: 212

Electricity—from nuclear fuels: 0% of total installed capacity (2017 est.)
country comparison to the world: 135

Electricity—from hydroelectric plants: 93% of total installed capacity (2017 est.)
country comparison to the world: 7

Electricity—from other renewable sources: 6% of total installed capacity (2017 est.)
country comparison to the world: 100

Crude oil—production: 0 bbl/day (2018 est.)
country comparison to the world: 168

Crude oil—exports: 0 bbl/day (2015 est.)
country comparison to the world: 159

Crude oil—imports: 0 bbl/day (2015 est.)
country comparison to the world: 159

Crude oil—proved reserves: 0 bbl (1 January 2018 est.)
country comparison to the world: 163

Refined petroleum products—production: 0 bbl/day (2015 est.)
country comparison to the world: 171

Refined petroleum products—consumption: 6.000 bbl/day (2016 est.)
country comparison to the world: 171

Refined petroleum products—exports: 0 bbl/day (2015 est.)
country comparison to the world: 176

Refined petroleum products—imports: 4,769 bbl/day (2015 est.)
country comparison to the world: 173

Natural gas—production: 0 cu m (2017 est.)
country comparison to the world: 164

597

Natural gas—consumption: 0 cu m (2017 est.)
country comparison to the world: 170

Natural gas—exports: 0 cu m (2017 est.)
country comparison to the world: 145

Natural gas—imports: 0 cu m (2017 est.)
country comparison to the world: 152

Natural gas—proved reserves: 0 cu m (2017 est.)
country comparison to the world: 165

Carbon dioxide emissions from consumption of energy: 1.082 million Mt (2017 est.)
country comparison to the world: 167

COMMUNICATIONS

Telephones—fixed lines: *total subscriptions:* 14,357
subscriptions per 100 inhabitants: less than 1 (2019 est.)
country comparison to the world: 184

Telephones—mobile cellular: *total subscriptions:* 9,799,352
subscriptions per 100 inhabitants: 47.78 (2019 est.)
country comparison to the world: 87

Telecommunication systems: *general assessment:* rudimentary; 2 fixed-line and 3 mobile-cellular operators govern the market; some mobile services to rural areas; in a resolution to discourage crime the regulatory has imposed SIM card registration since 2018; 50 licensed ISPs; DSL services are available; LTE services are available; mobile penetration low in comparison to the region average; potential for growth; national fiber backbone nearing completion; prospect of gaining access to international submarine fiber optic cables from neighboring countries (2020)
domestic: limited fixed-line subscribership less than 1 per 100 households; mobile-cellular services are expanding but network coverage is limited and is based around the main urban areas; mobile-cellular subscribership 48 per 100 *households (2019)*
international: country code—265; satellite earth stations—2 Intelsat (1 Indian Ocean, 1 Atlantic Ocean) (2019)
note: the COVID-19 outbreak is negatively impacting telecommunications production and supply chains globally; consumer spending on telecom devices and services has also slowed due to the pandemic's effect on economies worldwide; overall progress towards improvements in all facets of the telecom industry—mobile, fixed-line, broadband, submarine cable and satellite—has moderated

Broadcast media: radio is the main broadcast medium; privately owned Zodiak radio has the widest national broadcasting reach, followed by state-run radio; numerous private and community radio stations broadcast in cities and towns around the country; the largest TV network is government-owned, but at least 4 private TV networks broadcast in urban areas; relays of multiple international broadcasters are available (2019)

Internet country code: .mw

Internet users: *total:* 2,734,305
percent of population: 13.78% (July 2018 est.)
country comparison to the world: 103

Broadband—fixed subscriptions: *total:* 11,358
subscriptions per 100 inhabitants: less than 1 (2018 est.)
country comparison to the world: 168

TRANSPORTATION

National air transport system:
number of registered air carriers: 2 (2020)
inventory of registered aircraft operated by air carriers: 9
annual passenger traffic on registered air carriers: 10,545 (2018)
annual freight traffic on registered air carriers: 10,000 mt-km (2018)

Civil aircraft registration country code prefix: 7Q (2016)

Airports: 32 (2013)
country comparison to the world: 112

Airports—with paved runways:
total: 7 (2019)
over 3,047 m: 1
1,524 to 2,437 m: 2
914 to 1,523 m: 4

Airports—with unpaved runways:
total: 25 (2013)
1,524 to 2,437 m: 1 (2013)
914 to 1,523 m: 11 (2013)
under 914 m: 13 (2013)

Railways: *total:* 767 km (2014)
narrow gauge: 767 km 1.067-m gauge (2014)
country comparison to the world: 99

Roadways: *total:* 15,452 km (2015)
paved: 4,074 km (2015)
unpaved: 11,378 km (2015)
country comparison to the world: 123

Waterways: 700 km (on Lake Nyasa [Lake Malawi] and Shire River) (2010)
country comparison to the world: 75

Ports and terminals: *lake port(s):* Chipoka, Monkey Bay, Nkhata Bay, Nkhotakota, Chilumba (Lake Nyasa)

MILITARY AND SECURITY

Military and security forces: *Malawi Defense Force (MDF):* Army (includes Air Wing, Marine Unit); note—a 2017 amendment to Malawi's Defense Force Act established a separate Army, Air Force, and Maritime Force within the MDF, but these services have yet to develop independent budgets, chains of command, and training institutions (2019)

Military expenditures:
0.9% of GDP (2019)
0.9% of GDP (2018)
0.8% of GDP (2017)
0.6% of GDP (2016)
0.6% of GDP (2015)
country comparison to the world: 128

Military and security service personnel strengths: size estimates for the Malawi Defense Force vary; approximately 8,000 personnel (including about 200 in the Air Wing and 200 in the Marine Unit) (2019 est.)

Military equipment inventories and acquisitions: the Malawi Defense Force inventory is comprised of mostly obsolescent or second-hand equipment from France, Germany, South Africa, and the UK; since 2010, it has taken deliveries of additional second-hand equipment from South Africa (2012-15) and the UK (2015), as well as new patrol boats from China (2019) and non-lethal equipment donated by the US (2019) (2019)

Military deployments: 730 Democratic Republic of the Congo (MONUSCO) (2020)

Military service age and obligation: 18 years of age for voluntary military service; high school equivalent required for enlisted recruits and college equivalent for officer recruits; initial engagement is 7 years for enlisted personnel and 10 years for officers (2014)

TRANSNATIONAL ISSUES

Disputes—international: dispute with Tanzania over the boundary in Lake Nyasa (Lake Malawi) and the meandering Songwe River; Malawi contends that the entire lake up to the Tanzanian shoreline is its territory, while Tanzania claims the border is in the center of the lake; the conflict was reignited in 2012 when Malawi awarded a license to a British company for oil exploration in the lake

Refugees and internally displaced persons: *refugees (country of origin):* 29,416 (Democratic Republic of the Congo) (refugees and asylum seekers), 10,838 (Burundi) (refugees and asylum seekers), 6,696 (Rwanda) (refugees and asylum seekers) (2020)

MALAYSIA

INTRODUCTION

Background: The adoption of Islam in the 14th century saw the rise of a number of powerful sultanates on the Malay Peninsula and island of Borneo. The Portuguese in the 16th century and the Dutch in the 17th century were the first European colonial powers to establish themselves on the Malay Peninsula and Southeast Asia. However, it was the British who ultimately secured their hegemony across the territory and during the late 18th and 19th centuries established colonies and protectorates in the area that is now Malaysia. These holdings were occupied by Japan from 1942 to 1945. In 1948, the British-ruled territories on the Malay Peninsula except Singapore formed the Federation of Malaya, which became independent in 1957. Malaysia was formed in 1963 when the former British colonies of Singapore, as well as Sabah and Sarawak on the northern coast of Borneo, joined the Federation. The first several years of the country's independence were marred by a communist insurgency, Indonesian confrontation with Malaysia, Philippine claims to Sabah, and Singapore's withdrawal in 1965. During the 22-year term of Prime Minister MAHATHIR Mohamad (1981-2003), Malaysia was successful in diversifying its economy from dependence on exports of raw materials to the development of manufacturing, services, and tourism. Prime Minister MAHATHIR and a newly-formed coalition of opposition parties defeated Prime Minister Mohamed NAJIB bin Abdul Razak's United Malays National Organization (UMNO) in May 2018, ending over 60 years of uninterrupted rule by UMNO. MAHATHIR resigned in February 2020 amid a political dispute. King ABDULLAH then selected Tan Sri MUHYIDDIN Yassin as the new prime minister.

GEOGRAPHY

Location: Southeastern Asia, peninsula bordering Thailand and northern one-third of the island of Borneo, bordering Indonesia, Brunei, and the South China Sea, south of Vietnam

Geographic coordinates: 2 30 N, 112 30 E

Map references: Southeast Asia

Area: *total:* 329,847 sq km
land: 328,657 sq km
water: 1,190 sq km
country comparison to the world: 68

Area—comparative: slightly larger than New Mexico

Land boundaries: *total:* 2,742 km
border countries (3): Brunei 266 km, Indonesia 1881 km, Thailand 595 km

Coastline: 4,675 km (Peninsular Malaysia 2,068 km, East Malaysia 2,607 km)

Maritime claims: *territorial sea:* 12 nm
exclusive economic zone: 200 nm
continental shelf: 200-m depth or to the depth of exploitation; specified boundary in the South China Sea

Climate: tropical; annual southwest (April to October) and northeast (October to February) monsoons

Terrain: coastal plains rising to hills and mountains

Elevation: *mean elevation:* 419 m
lowest point: Indian Ocean 0 m
highest point: Gunung Kinabalu 4,095 m

Natural resources: tin, petroleum, timber, copper, iron ore, natural gas, bauxite

Land use: *agricultural land:* 23.2% (2011 est.)
arable land: 2.9% (2011 est.) / permanent crops: 19.4% (2011 est.) / permanent pasture: 0.9% (2011 est.)
forest: 62% (2011 est.)
other: 14.8% (2011 est.)
Irrigated land: 3,800 sq km (2012)

Population distribution: a highly uneven distribution with over 80% of the population residing on the Malay Peninsula

Natural hazards: flooding; landslides; forest fires

Environment—current issues: air pollution from industrial and vehicular emissions; water pollution from raw sewage; deforestation; smoke/haze from Indonesian forest fires; endangered species; coastal reclamation damaging mangroves and turtle nesting sites

Environment—international agreements: *party to:* Biodiversity, Climate Change, Climate Change-Kyoto Protocol, Desertification, Endangered Species, Hazardous Wastes, Law of the Sea, Marine Life Conservation, Ozone Layer Protection, Ship Pollution, Tropical Timber 83, Tropical Timber 94, Wetlands
signed, but not ratified: none of the selected agreements

Geography—note: strategic location along Strait of Malacca and southern South China Sea

PEOPLE AND SOCIETY

Population: 32,652,083 (July 2020 est.)
country comparison to the world: 42

Nationality: *noun:* Malaysian(s)
adjective: Malaysian

Ethnic groups: Bumiputera 62% (Malays and indigenous peoples, including Orang Asli, Dayak, Anak Negeri), Chinese 20.6%, Indian 6.2%, other 0.9%, non-citizens 10.3% (2017 est.)

Languages: Bahasa Malaysia (official), English, Chinese (Cantonese, Mandarin, Hokkien, Hakka, Hainan, Foochow), Tamil, Telugu, Malayalam, Panjabi, Thai
note: Malaysia has 134 living languages—112 indigenous languages and 22 non-indigenous languages; in East Malaysia, there are several indigenous languages; the most widely spoken are Iban and Kadazan

Religions: Muslim (official) 61.3%, Buddhist 19.8%, Christian 9.2%, Hindu 6.3%, Confucianism, Taoism, other traditional Chinese religions 1.3%, other 0.4%, none 0.8%, unspecified 1% (2010 est.)

Age structure: *0-14 years:* 26.8% (male 4,504,562/female 4,246,681)
15-24 years: 16.63% (male 2,760,244/female 2,670,186)
25-54 years: 40.86% (male 6,737,826/female 6,604,776)
55-64 years: 8.81% (male 1,458,038/female 1,418,280)
65 years and over: 6.9% (male 1,066,627/female 1,184,863) (2020 est.)

Dependency ratios: *total dependency ratio:* 44.2
youth dependency ratio: 33.8
elderly dependency ratio: 10.4
potential support ratio: 9.7 (2020 est.)

Median age: *total:* 29.2 years
male: 28.9 years
female: 29.6 years (2020 est.)
country comparison to the world: 134

Population growth rate: 1.29% (2020 est.)
country comparison to the world: 83

Birth rate: 18.3 births/1,000 population (2020 est.)
country comparison to the world: 85

Death rate: 5.3 deaths/1,000 population (2020 est.)
country comparison to the world: 191

Net migration rate: -0.3 migrant(s)/1,000 population (2020 est.)
country comparison to the world: 119

Population distribution: a highly uneven distribution with over 80% of the population residing on the Malay Peninsula

Urbanization: *urban population:* 77.2% of total population (2020)
rate of urbanization: 2.13% annual rate of change (2015-20 est.)
total population growth rate v. urban population growth rate, 2000-2030:

Major urban areas—population: 7.997 million KUALA LUMPUR (capital), 1.024 million Johor Bahru, 814,000 Ipoh (2020)

Sex ratio: *at birth:* 1.07 male(s)/female
0-14 years: 1.06 male(s)/female
15-24 years: 1.03 male(s)/female

25-54 years: 1.02 male(s)/female
55-64 years: 1.03 male(s)/female
65 years and over: 0.9 male(s)/female
total population: 1.03 male(s)/female (2020 est.)

Maternal mortality rate: 29 deaths/100,000 live births (2017 est.)
country comparison to the world: 113

Infant mortality rate: total: 11.4 deaths/1,000 live births
male: 13.2 deaths/1,000 live births
female: 9.6 deaths/1,000 live births (2020 est.)
country comparison to the world: 116

Life expectancy at birth: total population: 75.9 years
male: 73 years
female: 78.9 years (2020 est.)
country comparison to the world: 108

Total fertility rate: 2.43 children born/woman (2020 est.)
country comparison to the world: 78

Contraceptive prevalence rate: 52.2% (2014)

Drinking water source:
improved:
urban: 100% of population
rural: 89.3% of population
total: 96.7% of population
unimproved:
urban: 0% of population
rural: 11.7% of population
total: 3.3% of population (2017 est.)

Current Health Expenditure: 3.9% (2017)

Physicians density: 1.54 physicians/1,000 population (2015)

Hospital bed density: 1.9 beds/1,000 population (2017)

Sanitation facility access:
improved:
urban: 100% of population
rural: 98.7% of population
total: 100% of population
unimproved:
urban: 0% of population
rural: 1.3% of population
total: 0% of population (2017 est.)

HIV/AIDS—adult prevalence rate: 0.4% (2019 est.)
country comparison to the world: 79

HIV/AIDS—people living with HIV/AIDS: 88,000 (2019 est.)
country comparison to the world: 48

HIV/AIDS—deaths: 2,700 (2019 est.)
country comparison to the world: 39

Major infectious diseases: degree of risk: intermediate (2020)
food or waterborne diseases: bacterial diarrhea
vectorborne diseases: dengue fever
water contact diseases: leptospirosis

Obesity—adult prevalence rate: 15.6% (2016)
country comparison to the world: 125

Children under the age of 5 years underweight: 13.7% (2016)
country comparison to the world: 42

Education expenditures: 4.7% of GDP (2017)

country comparison to the world: 80

Literacy: definition: age 15 and over can read and write
total population: 93.7%
male: 96.3%
female: 91.1% (2016)

School life expectancy (primary to tertiary education): total: 14 years
male: 13 years
female: 14 years (2017)

Unemployment, youth ages 15-24: total: 10.5%
male: 9.8%
female: 11.4% (2016 est.)
country comparison to the world: 122

GOVERNMENT

Country name: conventional long form: none
conventional short form: Malaysia
local long form: none
local short form: Malaysia
former: Federation of Malaya
etymology: the name means "Land of the Malays"

Government type: federal parliamentary constitutional monarchy
note: all Peninsular Malaysian states have hereditary rulers (commonly referred to as sultans) except Melaka (Malacca) and Pulau Pinang (Penang); those two states along with Sabah and Sarawak in East Malaysia have governors appointed by government; powers of state governments are limited by the federal constitution; under terms of federation, Sabah and Sarawak retain certain constitutional prerogatives (e.g., right to maintain their own immigration controls)

Capital: name: Kuala Lumpur; note—nearby Putrajaya is referred to as a federal government administrative center but not the capital; Parliament meets in Kuala Lumpur

geographic coordinates: 3 10 N, 101 42 E
time difference: UTC+8 (13 hours ahead of Washington, DC, during Standard Time)
etymology: the Malay word for "river junction or estuary" is "kuala" and "lumpur" means "mud"; together the words render the meaning of "muddy confluence"

Administrative divisions: 13 states (negeri-negeri, singular—negeri); Johor, Kedah, Kelantan, Melaka, Negeri Sembilan, Pahang, Perak, Perlis, Pulau Pinang, Sabah, Sarawak, Selangor, Terengganu; and 1 federal territory (Wilayah Persekutuan) with 3 components, Kuala Lumpur, Labuan, and Putrajaya

Independence: 31 August 1957 (from the UK)

National holiday: Independence Day (or Merdeka Day), 31 August (1957) (independence of Malaya); Malaysia Day, 16 September (1963) (formation of Malaysia)

Constitution: history: previous 1948; latest drafted 21 February 1957, effective 27 August 1957
amendments: proposed as a bill by Parliament; passage requires at least two-thirds majority vote by the Parliament membership in the bill's second and third readings; a number of constitutional

sections are excluded from amendment or repeal; amended many times, last in 2010

Legal system: mixed legal system of English common law, Islamic (sharia) law, and customary law; judicial review of legislative acts in the Federal Court at request of supreme head of the federation

International law organization participation: has not submitted an ICJ jurisdiction declaration; non-party state to the ICCt

Citizenship: citizenship by birth: no
citizenship by descent only: at least one parent must be a citizen of Malaysia
dual citizenship recognized: no
residency requirement for naturalization: 10 out 12 years preceding application

Suffrage: 18 years of age; universal (2019)

Executive branch: chief of state: King Sultan ABDULLAH Sultan Ahmad Shah (since 24 January 2019); note—King MUHAMMAD V (formerly known as Tuanku Muhammad Faris Petra) (selected on 14 October 2016; installed on 13 December 2016) resigned on 6 January 2019; the position of the king is primarily ceremonial, but he is the final arbiter on the appointment of the prime minister
head of government: Prime Minister Tan Sri MUHYIDDIN Yassin (since 1 March 2020); note—Prime Minister MAHATHIR resigned on 24 February 2020 but King ABDULLAH asked that he stay on as interim prime minister until Malaysian's King ABDULLAH picked MUHYIDDIN to step in as Prime Minister; note—previous Deputy Prime Minister WAN AZIZAH Wan Ismail (21 May 2018—24 February 2020) was the first female in this position (2019)
cabinet: Cabinet appointed by the prime minister from among members of Parliament with the consent of the king; note—cabinet dissolved 24 February 2020 with Prime Minister MAHATHIR resignation
elections/appointments: king elected by and from the hereditary rulers of 9 states for a 5-year term; election is on a rotational basis among rulers of the 9 states; election last held on 24 January 2019 (next to be held in 2024); prime minister designated from among members of the House of Representatives; following legislative elections, the leader who commands support of the majority of members in the House becomes prime minister

Legislative branch: description: bicameral Parliament of Malaysia or Parlimen Malaysia consists of:
Senate or Dewan Negara (70 seats; 44 members appointed by the king and 26 indirectly elected by 13 state legislatures; members serve 3-year terms)
House of Representatives or Dewan Rakyat (222 seats; members directly elected in single-seat constituencies by simple majority vote to serve 5-year terms) (2019)
elections: Senate—appointed
House of Representatives—last held on 9 May 2018 (next to be held no later than May 2023)
election results: Senate—appointed; composition—men 54, women 14, percent of women 20.6%

House of Representatives—percent of vote by party/coalition—PH 45.6%, BN 33.8%, PAS 16.9%, WARISAN 2.3%, other 1.4%; seats by party/coalition—PH 113, BN 79, PAS 18, WARISAN 8, USA 1, independent 3; composition—men 199, women 23, percent of women 10.4%; note—total Parliament percent of women 12.8%

note: as of 16 November 2019, seats by party—PH 129, BN 41, GS 18, GPS 18, WARISAN 9, GBS 3, UPKO 1, PSB 1, independent 1, vacant 1

Judicial branch: *highest courts:* Federal Court (consists of the chief justice, president of the Court of Appeal, chief justice of the High Court of Malaya, chief judge of the High Court of Sabah and Sarawak, 8 judges, and 1 "additional" judge); note—Malaysia has a dual judicial hierarchy of civil and religious (sharia) courts

judge selection and term of office: Federal Court justices appointed by the monarch on advice of the prime minister; judges serve until mandatory retirement at age 66 with the possibility of a single 6-month extension

subordinate courts: Court of Appeal; High Court; Sessions Court; Magistrates' Court

Political parties and leaders: *National Front (Barisan Nasional) or BN:* Malaysian Chinese Association (Persatuan China Malaysia) or MCA [LIOW Tiong Lai]

Malaysian Indian Congress (Kongres India Malaysia) or MIC [S. SUBRAMANIAM]

United Malays National Organization or UMNO [MOHAMAD Hasan, acting]

(Formerly—Coalition of Hope (Pakatan Harapan) or PH (formerly the People's Alliance, before former PM MAHATHIR resigns 24 February 2020): Democratic Action Party (Parti Tindakan Demokratik) or DAP [TAN Kok Wai]

Malaysian United Indigenous Party (Parti Pribumi Bersatu Malaysia) or PPBM [Tan Sri MUHYIDDIN Yassin; note—former PM MAHATHIR steps down 24 Feb 2020]

National Trust Party (Parti Amanah Negara) or AMANAH [Mohamad SABU]

People's Justice Party (Parti Keadilan Rakyat) or PKR [ANWAR Ibrahim]

New—Fighters of the Nation Party (Parti Pejuang Tanah Air) or Pejuang [former PM MAHATHIR bin Mohamad; interim president Datuk Seri Mukhriz Mahathir note—started August 2020]

Other: Pan-Malaysian Islamic Party (Parti Islam se Malaysia) or PAS [Abdul HADI Awang]

Progressive Democratic Party or PDP [TIONG King Sing]

Sabah Heritage Party (Parti Warisan Sabah) or WARISAN [SHAFIE Apdal]

Sarawak Parties Alliance (Gabungan Parti Sarawak) or GPS [ABANG JOHARI Openg] (includes PBB, SUPP, PRS, PDP) Sarawak People's Party (Parti Rakyat Sarawak) or PRS [James MASING]

Sarawak United People's Party (Parti Bersatu Rakyat Sarawak) or SUPP [Dr. SIM Kui Hian]

United Pasokmomogun Kadazandusun Murut Organization (Pertubuhan Pasko Momogun Kadazan Dusun Bersatu) or

UPKO [Wilfred Madius TANGAU]

United Sabah Alliance or USA (Gabungan Sabah)

United Sabah Party (Parti Bersatu Sabah) or PBS [Maximus ONGKILI]

United Sabah People's (Party Parti Bersatu Rakyat Sabah) or PBRS [Joseph KURUP]

United Traditional Bumiputera Party (Parti Pesaka Bumiputera Bersata) or PBB; note—PBB is listed under GPS above

International organization participation: ADB, APEC, ARF, ASEAN, BIS, C, CICA (observer), CP, D-8, EAS, FAO, G-15, G-77, IAEA, IBRD, ICAO, ICC (national committees), ICRM, IDA, IDB, IFAD, IFC, IFRCS, IHO, ILO, IMF, IMO, IMSO, Interpol, IOC, IPU, ISO, ITSO, ITU, ITUC (NGOs), MIGA, MINURSO, MONUSCO, NAM, OIC, OPCW, PCA, PIF (partner), UN, UNAMID, UNCTAD, UNESCO, UNIDO, UNIFIL, UNISFA, UNMIL, UNWTO, UPU, WCO, WFTU (NGOs), WHO, WIPO, WMO, WTO

Diplomatic representation in the US: *chief of mission:* Ambassador Dato' AZMIL Zabidi (since February 2019)

chancery: 3516 International Court NW, Washington, DC 20008

telephone: [1] (202) 572-9700

FAX: [1] (202) 572-9882

consulate(s) general: Los Angeles, New York

Diplomatic representation from the US: *chief of mission:* Ambassador Kamala Shirin LAKHDHIR (since 21 February 2017)

telephone: [60] (3) 2168-5000

embassy: 376 Jalan Tun Razak, 50400 Kuala Lumpur

mailing address: US Embassy Kuala Lumpur, APO AP 96535-8152

FAX: [60] (3) 2142-2207

Flag description: 14 equal horizontal stripes of red (top) alternating with white (bottom); there is a dark blue rectangle in the upper hoistside corner bearing a yellow crescent and a yellow 14-pointed star; the flag is often referred to as Jalur Gemilang (Stripes of Glory); the 14 stripes stand for the equal status in the federation of the 13 member states and the federal government; the 14 points on the star represent the unity between these entities; the crescent is a traditional symbol of Islam; blue symbolizes the unity of the Malay people and yellow is the royal color of Malay rulers

note: the design is based on the flag of the US

National symbol(s): *tiger, hibiscus; national colors:* gold, black

National anthem: *name:* "Negaraku" (My Country)

lyrics/music: collective, led by Tunku ABDUL RAHMAN/Pierre Jean DE BERANGER

note: adopted 1957; full version only performed in the presence of the king; the tune, which was adopted from a popular French melody titled "La Rosalie," was originally the anthem of Perak, one of Malaysia's 13 states

0:00 / 0:42

ECONOMY

Economy—overview: Malaysia, an upper middle-income country, has transformed itself since the 1970s from a producer of raw materials into a multi-sector economy. Under current Prime Minister NAJIB, Malaysia is attempting to achieve high-income status by 2020 and to move further up the value-added production chain by attracting investments in high technology, knowledge-based industries and services. NAJIB's Economic Transformation Program is a series of projects and policy measures intended to accelerate the country's economic growth. The government has also taken steps to liberalize some services subsectors. Malaysia is vulnerable to a fall in world commodity prices or a general slowdown in global economic activity.

The NAJIB administration is continuing efforts to boost domestic demand and reduce the economy's dependence on exports. Domestic demand continues to anchor economic growth, supported mainly by private consumption, which accounts for 53% of GDP. Nevertheless, exports—particularly of electronics, oil and gas, and palm oil—remain a significant driver of the economy. In 2015, gross exports of goods and services were equivalent to 73% of GDP. The oil and gas sector supplied about 22% of government revenue in 2015, down significantly from prior years amid a decline in commodity prices and diversification of government revenues. Malaysia has embarked on a fiscal reform program aimed at achieving a balanced budget by 2020, including rationalization of subsidies and the 2015 introduction of a 6% value added tax. Sustained low commodity prices throughout the period not only strained government finances, but also shrunk Malaysia's current account surplus and weighed heavily on the Malaysian ringgit, which was among the region's worst performing currencies during 2013-17. The ringgit hit new lows following the US presidential election amid a broader selloff of emerging market assets.

Bank Negara Malaysia (the central bank) maintains adequate foreign exchange reserves; a well-developed regulatory regime has limited Malaysia's exposure to riskier financial instruments, although it remains vulnerable to volatile global capital flows. In order to increase Malaysia's competitiveness, Prime Minister NAJIB raised possible revisions to the special economic and social preferences accorded to ethnic Malays under the New Economic Policy of 1970, but retreated in 2013 after he encountered significant opposition from Malay nationalists and other vested interests. In September 2013 NAJIB launched the new Bumiputra Economic Empowerment Program, policies that favor and advance the economic condition of ethnic Malays.

Malaysia signed the 12-nation Trans-Pacific Partnership (TPP) free trade agreement in February 2016, although the future of the TPP remains unclear following the US withdrawal from the agreement. Along with nine other ASEAN members, Malaysia established the ASEAN

601

Economic Community in 2015, which aims to advance regional economic integration.

GDP (purchasing power parity):
$933.3 billion (2017 est.)
$881.3 billion (2016 est.)
$845.6 billion (2015 est.)
note: data are in 2017 dollars
country comparison to the world: 26

GDP (official exchange rate):
$312.4 billion (2017 est.)

GDP—real growth rate: 4.31% (2019 est.)
4.77% (2018 est.)
5.81% (2017 est.)
country comparison to the world: 65

GDP—per capita (PPP): $29,100 (2017 est.)
$27,900 (2016 est.)
$27,100 (2015 est.)
note: data are in 2017 dollars
country comparison to the world: 71

Gross national saving:
28.5% of GDP (2017 est.)
28.3% of GDP (2016 est.)
28.2% of GDP (2015 est.)
country comparison to the world: 38

GDP—composition, by end use:
household consumption: 55.3% (2017 est.)
government consumption: 12.2% (2017 est.)
investment in fixed capital: 25.3% (2017 est.)
investment in inventories: 0.3% (2017 est.)
exports of goods and services: 71.4% (2017 est.)
imports of goods and services: -64.4% (2017 est.)

GDP—composition, by sector of origin:
agriculture: 8.8% (2017 est.)
industry: 37.6% (2017 est.)
services: 53.6% (2017 est.)

Agriculture—products: Peninsular Malaysia—palm oil, rubber, cocoa, rice;Sabah—palm oil, subsistence crops; rubber, timber;Sarawak—palm oil, rubber, timber; pepper

Industries: Peninsular Malaysia—rubber and oil palm processing and manufacturing, petroleum and natural gas, light manufacturing, pharmaceuticals, medical technology, electronics and semiconductors, timber processing;Sabah—logging, petroleum and natural gas production;Sarawak—agriculture processing, petroleum and natural gas production, logging

Industrial production growth rate:
5% (2017 est.)
country comparison to the world: 55

Labor force: 15.139 million (2020 est.)
country comparison to the world: 35

Labor force—by occupation: *agriculture:* 11%
industry: 36%
services: 53% (2012 est.)

Unemployment rate: 3.3% (2019 est.)
3.33% (2018 est.)
country comparison to the world: 45

Population below poverty line: 3.8% (2009 est.)

Household income or consumption by percentage share: *lowest 10%:* 1.8%
highest 10%: 34.7% (2009 est.)

Budget: *revenues:* 51.25 billion (2017 est.)

expenditures: 60.63 billion (2017 est.)

Taxes and other revenues: 16.4% (of GDP) (2017 est.)
country comparison to the world: 180

Budget surplus (+) or deficit (-): -3% (of GDP) (2017 est.)
country comparison to the world: 133

Public debt:
54.1% of GDP (2017 est.)
56.2% of GDP (2016 est.)
note: this figure is based on the amount of federal government debt, RM501.6 billion ($167.2 billion) in 2012; this includes Malaysian Treasury bills and other government securities, as well as loans raised externally and bonds and notes issued overseas; this figure excludes debt issued by non-financial public enterprises and guaranteed by the federal government, which was an additional $47.7 billion in 2012
country comparison to the world: 86

Fiscal year: calendar year

Inflation rate (consumer prices): 3.8% (2017 est.)
2.1% (2016 est.)
note: approximately 30% of goods are price-controlled
country comparison to the world: 151

Current account balance: $12.295 billion (2019 est.)
$8.027 billion (2018 est.)
country comparison to the world: 23

Exports: $187.9 billion (2017 est.)
$165.3 billion (2016 est.)
country comparison to the world: 28

Exports—partners: Singapore 15.1%, China 12.6%, US 9.4%, Japan 8.2%, Thailand 5.7%, Hong Kong 4.5% (2017)

Exports—commodities: semiconductors and electronic equipment, palm oil, petroleum and liquefied natural gas, wood and wood products, palm oil, rubber, textiles, chemicals, solar panels

Imports: $160.7 billion (2017 est.)
$141 billion (2016 est.)
country comparison to the world: 27

Imports—commodities: electronics, machinery, petroleum products, plastics, vehicles, iron and steel products, chemicals

Imports—partners: China 19.9%, Singapore 10.8%, US 8.4%, Japan 7.6%, Thailand 5.8%, South Korea 4.5%, Indonesia 4.4% (2017)

Reserves of foreign exchange and gold: $102.4 billion (31 December 2017 est.)
$94.5 billion (31 December 2016 est.)
country comparison to the world: 25

Debt—external: $217.2 billion (31 December 2017 est.)
$195.3 billion (31 December 2016 est.)
country comparison to the world: 33

Exchange rates: ringgits (MYR) per US dollar—
4.343 (2017 est.)
4.15 (2016 est.)
4.15 (2015 est.)
3.91 (2014 est.)
3.27 (2013 est.)

Electricity access: *electrification—total population:* 100% (2020)

Electricity—production: 148.3 billion kWh (2016 est.)
country comparison to the world: 28

Electricity—consumption: 136.9 billion kWh (2016 est.)
country comparison to the world: 26

Electricity—exports: 3 million kWh (2015 est.)
country comparison to the world: 93

Electricity—imports: 33 million kWh (2016 est.)
country comparison to the world: 109

Electricity—installed generating capacity: 33 million kW (2016 est.)
country comparison to the world: 31

Electricity—from fossil fuels: 78% of total installed capacity (2016 est.)
country comparison to the world: 90

Electricity—from nuclear fuels: 0% of total installed capacity (2017 est.)
country comparison to the world: 136

Electricity—from hydroelectric plants: 18% of total installed capacity (2017 est.)
country comparison to the world: 95

Electricity—from other renewable sources: 4% of total installed capacity (2017 est.)
country comparison to the world: 113

Crude oil—production: 647,000 bbl/day (2018 est.)
country comparison to the world: 26

Crude oil—exports: 326,200 bbl/day (2015 est.)
country comparison to the world: 24

Crude oil—imports: 166,000 bbl/day (2015 est.)
country comparison to the world: 35

Crude oil—proved reserves: 3.6 billion bbl (1 January 2018 est.)
country comparison to the world: 27

Refined petroleum products—production: 528,300 bbl/day (2015 est.)
country comparison to the world: 32

Refined petroleum products—consumption: 704,000 bbl/day (2016 est.)
country comparison to the world: 28

Refined petroleum products—exports: 208,400 bbl/day (2015 est.)
country comparison to the world: 31

Refined petroleum products—imports: 304,600 bbl/day (2015 est.)
country comparison to the world: 24

Natural gas—production: 69.49 billion cu m (2017 est.)
country comparison to the world: 13

Natural gas—consumption: 30.44 billion cu m (2017 est.)
country comparison to the world: 31

Natural gas—exports: 38.23 billion cu m (2017 est.)
country comparison to the world: 9

Natural gas—imports: 2.803 billion cu m (2017 est.)
country comparison to the world: 45

Natural gas—proved reserves: 1.183 trillion cu m (1 January 2018 est.)
country comparison to the world: 23

Carbon dioxide emissions from consumption of energy: 226.8 million Mt (2017 est.)
country comparison to the world: 31

COMMUNICATIONS

Telephones—fixed lines: *total subscriptions:* 6,530,410
subscriptions per 100 inhabitants: 20.26 (2019 est.)
country comparison to the world: 24

Telephones—mobile cellular: *total subscriptions:* 44,997,299
subscriptions per 100 inhabitants: 139.6 (2019 est.)
country comparison to the world: 34

Telecommunication systems: *general assessment:* one of the most advanced telecom networks in the developing world; strong commitment to developing a technological society; Malaysia is promoting itself as an information tech hub in the Asian region; closing the urban rural divide; 4G and 5G networks with strong competition, mobile dominance over fixed-broadband; roll-out of a national broadband network (2020)

domestic: fixed-line 20 per 100 and mobile-cellular teledensity exceeds 140 per 100 persons; domestic satellite system with 2 earth stations (2019)

international: country code—60; landing points for BBG, FEA, SAFE, SeaMeWe-3 & 4 & 5, AAE-1, JASUKA, BDM, Dumai-Melaka Cable System, BRCS, ACE, AAG, East-West Submarine Cable System, SEAX-1, SKR1M, APCN-2, APG, BtoBe, BaSICS, and Labuan-Brunei Submarine and MCT submarine cables providing connectivity to Asia, the Middle East, Southeast Asia, Australia and Europe; satellite earth stations—2 Intelsat (1 Indian Ocean, 1 Pacific Ocean); launch of Kacific-1 satellite in 2019 (2019)

note: the COVID-19 outbreak is negatively impacting telecommunications production and supply chains globally; consumer spending on telecom devices and services has also slowed due to the pandemic's effect on economies worldwide; overall progress towards improvements in all facets of the telecom industry—mobile, fixed-line, broadband, submarine cable, and satellite—has moderated

Broadcast media: state-owned TV broadcaster operates 2 TV networks with relays throughout the country, and the leading private commercial media group operates 4 TV stations with numerous relays throughout the country; satellite TV subscription service is available; state-owned radio broadcaster operates multiple national networks, as well as regional and local stations; many private commercial radio broadcasters and some subscription satellite radio services are available; about 55 radio stations overall (2019)

Internet country code: .my
Internet users: *total:* 25,829,444
percent of population: 81.2% (July 2018 est.)
country comparison to the world: 30

Broadband—fixed subscriptions: *total:* 2.696 million
subscriptions per 100 inhabitants: 8 (2018 est.)
country comparison to the world: 46

TRANSPORTATION

National air transport system: *number of registered air carriers:* 13 (2020)
inventory of registered aircraft operated by air carriers: 270
annual passenger traffic on registered air carriers: 60,481,772 (2018)
annual freight traffic on registered air carriers: 1,404,410,000 mt-km (2018)

Civil aircraft registration country code prefix: 9M (2016)

Airports: 114 (2013)
country comparison to the world: 50

Airports—with paved runways:
total: 39 (2017)
over 3,047 m: 8 (2017)
2,438 to 3,047 m: 8 (2017)
1,524 to 2,437 m: 7 (2017)
914 to 1,523 m: 8 (2017)
under 914 m: 8 (2017)

Airports—with unpaved runways:
total: 75 (2013)
914 to 1,523 m: 6 (2013)
under 914 m: 69 (2013)

Heliports: 4 (2013)

Pipelines: 354 km condensate, 6439 km gas, 155 km liquid petroleum gas, 1937 km oil, 43 km oil/gas/water, 114 km refined products, 26 km water (2013)

Railways: *total:* 1,851 km (2014)
standard gauge: 59 km 1.435-m gauge (59 km electrified) (2014)
narrow gauge: 1,792 km 1.000-m gauge (339 km electrified) (2014)
country comparison to the world: 77

Roadways: *total:* 144,403 km (excludes local roads) (2010)
paved: 116,169 km (includes 1,821 km of expressways) (2010)
unpaved: 28,234 km (2010)
country comparison to the world: 35

Waterways: 7,200 km (Peninsular Malaysia 3,200 km; Sabah 1,500 km; Sarawak 2,500 km) (2011)
country comparison to the world: 19

Merchant marine: *total:* 1,748
by type: bulk carrier 15, container ship 22, general cargo 176, oil tanker 140, other 1,395 (2019)
country comparison to the world: 15

Ports and terminals: *major seaport(s):* Bintulu, Johor Bahru, George Town (Penang), Port Kelang (Port Klang), Tanjung Pelepas
container port(s) (TEUs): Port Kelang (Port Klang) (11,978,000), Tanjung Pelepas (8,260,000) (2017)

LNG terminal(s) (export): Bintulu (Sarawak)
LNG terminal(s) (import): Sungei Udang

MILITARY AND SECURITY

Military and security forces: *Malaysian Armed Forces (Angkatan Tentera Malaysia, ATM):* Malaysian Army (Tentera Darat Malaysia), Royal Malaysian Navy (Tentera Laut Diraja Malaysia, TLDM), Royal Malaysian Air Force (Tentera Udara Diraja Malaysia, TUDM); Ministry of Home Affairs: the Royal Malaysian Police (PRMD, includes the General Operations Force, a paramilitary force with a variety of roles, including patrolling borders, counter-terrorism, maritime security, and counterinsurgency) (2019)
note: Malaysia created a National Special Operations Force in 2016 for combating terrorism threats; the force is comprised of personnel from the Armed Forces, the Royal Malaysian Police, and the Malaysian Maritime Enforcement Agency (Malaysian Coast Guard, MMEA)

Military expenditures: 1% of GDP (2019)
1% of GDP (2018)
1.1% of GDP (2017)
1.4% of GDP (2016)
1.5% of GDP (2015)
country comparison to the world: 116

Military and security service personnel strengths: the Malaysian Armed Forces have approximately 115,000 active duty troops (80,000 Army; 18,000 Navy; 17,000 Air Force); approximately 18,000 General Operations Force (2019 est.)

Military equipment inventories and acquisitions: the Malaysian Armed Forces field a diverse mix of imported weapons systems; the chief suppliers since 2010 are Germany, South Korea, Spain, and Turkey (2019)

Military deployments: 820 Lebanon (UNIFIL) (2020)

Military service age and obligation: 17 years 6 months of age for voluntary military service (younger with parental consent and proof of age); mandatory retirement age 60; women serve in the Malaysian Armed Forces; no conscription (2017)

Maritime threats: the International Maritime Bureau reports that the territorial and offshore waters in the Strait of Malacca and South China Sea remain high risk for piracy and armed robbery against ships; in the past, commercial vessels have been attacked and hijacked both at anchor and while underway; hijacked vessels are often disguised and cargo diverted to ports in East Asia; crews have been murdered or cast adrift; 11 attacks were reported in 2018 including eight ships boarded and seven crew taken hostage

TERRORISM

Terrorist group(s): Islamic State of Iraq and ash-Sham (ISIS); Jemaah Islamiyah (2019)
note: details about the history, aims, leadership, organization, areas of operation, tactics, targets, weapons, size, and sources of support of the group(s) appear(s) in Appendix-T

TRANSNATIONAL ISSUES

Disputes—international: while the 2002 "Declaration on the Conduct of Parties in the South China Sea" has eased tensions over the Spratly Islands, it is not the legally binding "code of conduct" sought by some parties, which is currently being negotiated between China and ASEAN; Malaysia was not party to the March 2005 joint accord among the national oil companies of China, the Philippines, and Vietnam on conducting marine seismic activities in the Spratly Islands; disputes continue over deliveries of fresh water to Singapore, Singapore's land reclamation, bridge construction, and maritime boundaries in the Johor and Singapore Straits; in 2008, ICJ awarded sovereignty of Pedra Branca (Pulau Batu Puteh/Horsburgh Island) to Singapore, and Middle Rocks to Malaysia, but did not rule on maritime regimes, boundaries, or disposition of South Ledge; land and maritime negotiations with Indonesia are ongoing, and disputed areas include the controversial Tanjung Datu and Camar Wulan border area in Borneo and the maritime boundary in the Ambalat oil block in the Celebes Sea; separatist violence in Thailand's predominantly Muslim southern provinces prompts measures to close and monitor border with Malaysia to stem terrorist activities; Philippines retains a dormant claim to Malaysia's Sabah State in northern Borneo; per Letters of Exchange signed in 2009, Malaysia in 2010 ceded two hydrocarbon concession blocks to Brunei in exchange for Brunei's sultan dropping claims to the Limbang corridor, which divides Brunei; piracy remains a problem in the Malacca Strait

Refugees and internally displaced persons: *refugees (country of origin):* 119,230 (Burma) (2019)

stateless persons: 108,332 (2019); note— Malaysia's stateless population consists of Rohingya refugees from Burma, ethnic Indians, and the children of Filipino and Indonesian illegal migrants; Burma stripped the Rohingya of their nationality in 1982; Filipino and Indonesian children who have not been registered for birth certificates by their parents or who received birth certificates stamped "foreigner" are not eligible to attend government schools; these children are vulnerable to statelessness should they not be able to apply to their parents' country of origin for passports

Trafficking in persons: *current situation:* Malaysia is a destination and, to a lesser extent, a source and transit country for men, women, and children subjected to forced labor and women and children subjected to sex trafficking; Malaysia is mainly a destination country for foreign workers who migrate willingly from countries, including Indonesia, Bangladesh, the Philippines, Nepal, Burma, and other Southeast Asian countries, but subsequently encounter forced labor or debt bondage in agriculture, construction, factories, and domestic service at the hands of employers, employment agents, and labor recruiters; women from Southeast Asia and, to a much lesser extent, Africa, are recruited for legal work in restaurants, hotels, and salons but are forced into prostitution; refugees, including Rohingya adults and children, are not legally permitted to work and are vulnerable to trafficking; a small number of Malaysians are trafficked internally and subjected to sex trafficking abroad

tier rating: Tier 2 Watch list—Malaysia does not fully comply with the minimum standards for the elimination of trafficking; however, it is making significant efforts to do so; in 2014, amendments to strengthen existing anti-trafficking laws, including enabling victims to move freely and to work and for NGOs to run protective facilities, were drafted by the government and are pending approval from Parliament; authorities more than doubled investigations and prosecutions but convicted only three traffickers for forced labor and none for sex trafficking, a decline from 2013 and a disproportionately small number compared to the scale of the country's trafficking problem; NGOs provided the majority of victim rehabilitation and counseling services with no financial support from the government (2015)

Illicit drugs: drug trafficking prosecuted vigorously, including enforcement of the death penalty; heroin still primary drug of abuse, but synthetic drug demand remains strong; continued ecstasy and methamphetamine producer for domestic users and, to a lesser extent, the regional drug market

MALDIVES

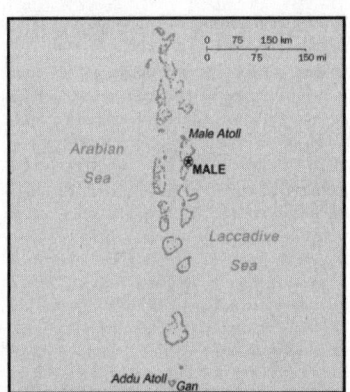

INTRODUCTION

Background: A sultanate since the 12th century, the Maldives became a British protectorate in 1887. The islands became a republic in 1968, three years after independence. President Maumoon Abdul GAYOOM dominated Maldives' political scene for 30 years, elected to six successive terms by single-party referendums. Following political demonstrations in the capital Male in August 2003, GAYOOM and his government pledged to embark upon a process of liberalization and democratic reforms, including a more representative political system and expanded political freedoms. Political parties were legalized in 2005.

In June 2008, a constituent assembly—termed the "Special Majlis"—finalized a new constitution ratified by GAYOOM in August 2008. The first-ever presidential elections under a multi-candidate, multi-party system were held in October 2008. GAYOOM was defeated in a runoff poll by Mohamed NASHEED, a political activist who had been jailed several years earlier by the GAYOOM regime. In early February 2012, after several weeks of street protests in response to his ordering the arrest of a top judge, NASHEED purportedly resigned the presidency and handed over power to Vice President Mohammed WAHEED Hassan Maniku. A government-appointed Commission of National Inquiry concluded there was no evidence of a coup, but NASHEED contends that police and military personnel forced him to resign. NASHEED, WAHEED, and Abdulla YAMEEN Abdul Gayoom ran in the 2013 elections with YAMEEN ultimately winning the presidency after three rounds of voting. As president, YAMEEN weakened democratic institutions, curtailed civil liberties, jailed his political opponents, restricted the press, and exerted control over the judiciary to strengthen his hold on power and limit dissent. In September 2018, YAMEEN lost his reelection bid to Ibrahim Mohamed SOLIH, a parliamentarian of the Maldivian Democratic Party (MDP), who had the support of a coalition of four parties that came together to defeat YAMEEN and restore democratic norms to Maldives. In April 2019, SOLIH's MDP won 65 of 87 seats in parliament.

GEOGRAPHY

Location: Southern Asia, group of atolls in the Indian Ocean, south-southwest of India

Geographic coordinates: 3 15 N, 73 00 E

Map references: Asia

Area: *total:* 298 sq km
land: 298 sq km
water: 0 sq km
country comparison to the world: 210

Area—comparative: about 1.7 times the size of Washington, DC

Land boundaries: 0 km

Coastline: 644 km

Maritime claims: *territorial sea:* 12 nm
exclusive economic zone: 200 nm
contiguous zone: 24 nm
measured from claimed archipelagic straight baselines

Climate: tropical; hot, humid; dry, northeast monsoon (November to March); rainy, southwest monsoon (June to August)

Terrain: flat, with white sandy beaches

Elevation: *mean elevation:* 2 m
lowest point: Indian Ocean 0 m
highest point: 8th tee, golf course, Villingi Island 5 m

Natural resources: fish

Land use: *agricultural land:* 23.3% (2011 est.)
arable land: 10% (2011 est.) / *permanent crops:* 10% (2011 est.) / *permanent pasture:* 3.3% (2011 est.)
forest: 3% (2011 est.)
other: 73.7% (2011 est.)
Irrigated land: 0 sq km (2012)

Population distribution: about a third of the population lives in the centrally located capital city of Male and almost a tenth in southern Addu City; the remainder of the populace is spread over the 200 or so populated islands of the archipelago

Natural hazards: tsunamis; low elevation of islands makes them sensitive to sea level rise

Environment—current issues: depletion of freshwater aquifers threatens water supplies; inadequate sewage treatment; coral reef bleaching

Environment—international agreements: *party to:* Biodiversity, Climate Change, Climate Change-Kyoto Protocol, Desertification, Hazardous Wastes, Law of the Sea, Ozone Layer Protection, Ship Pollution
signed, but not ratified: none of the selected agreements

Geography—note: smallest Asian country; archipelago of 1,190 coral islands grouped into 26 atolls (200 inhabited islands, plus 80 islands with tourist resorts); strategic location astride and along major sea lanes in Indian Ocean

PEOPLE AND SOCIETY

Population: 391,904 (July 2020 est.)
country comparison to the world: 177

Nationality: *noun:* Maldivian(s)
adjective: Maldivian

Ethnic groups: homogeneous mixture of Sinhalese, Dravidian, Arab, Australasian, and African resulting from historical changes in regional hegemony over marine trade routes

Languages: Dhivehi (official, dialect of Sinhala, script derived from Arabic), English (spoken by most government officials)

Religions: Sunni Muslim (official)

Age structure: *0-14 years:* 22.13% (male 44,260/female 42,477)
15-24 years: 17.24% (male 37,826/female 29,745)
25-54 years: 48.91% (male 104,217/female 87,465)
55-64 years: 6.91% (male 12,942/female 14,123)

65 years and over: 4.81% (male 8,417/female 10,432) (2020 est.)

Dependency ratios: *total dependency ratio:* 30.2
youth dependency ratio: 25.5
elderly dependency ratio: 4.7
potential support ratio: 21.4 (2020 est.)

Median age: *total:* 29.5 years
male: 29.2 years
female: 30 years (2020 est.)
country comparison to the world: 129

Population growth rate: -0.08% (2020 est.)
country comparison to the world: 203

Birth rate: 16 births/1,000 population (2020 est.)
country comparison to the world: 111

Death rate: 4.1 deaths/1,000 population (2020 est.)
country comparison to the world: 213

Net migration rate: -12.7 migrant(s)/1,000 population (2020 est.)
country comparison to the world: 221

Population distribution: about a third of the population lives in the centrally located capital city of Male and almost a tenth in southern Addu City; the remainder of the populace is spread over the 200 or so populated islands of the archipelago

Urbanization: *urban population:* 40.7% of total population (2020)
rate of urbanization: 2.93% annual rate of change (2015-20 est.)
total population growth rate v. urban population growth rate, 2000-2030:

Major urban areas—population: 177,000 MALE (capital) (2018)

Sex ratio: *at birth:* 1.05 male(s)/female
0-14 years: 1.04 male(s)/female
15-24 years: 1.27 male(s)/female
25-54 years: 1.19 male(s)/female
55-64 years: 0.92 male(s)/female
65 years and over: 0.81 male(s)/female
total population: 1.13 male(s)/female (2020 est.)

Mother's mean age at first birth: 24.5 years (2009 est.)
note: median age at first birth among women 25-29

Maternal mortality rate: 53 deaths/100,000 live births (2017 est.)
country comparison to the world: 92

Infant mortality rate: *total:* 19.8 deaths/1,000 live births
male: 22 deaths/1,000 live births
female: 17.5 deaths/1,000 live births (2020 est.)
country comparison to the world: 78

Life expectancy at birth: *total population:* 76.4 years
male: 74 years
female: 78.9 years (2020 est.)
country comparison to the world: 96

Total fertility rate: 1.71 children born/woman (2020 est.)
country comparison to the world: 171

Contraceptive prevalence rate: 18.8% (2016/17)

Drinking water source:

improved:
urban: 98.3% of population
rural: 100% of population
total: 100% of population
unimproved:
urban: 1.7% of population
rural: 0% of population
total: 0% of population (2017 est.)

Current Health Expenditure: 9% (2017)

Physicians density: 3.72 physicians/1,000 population (2017)

Hospital bed density: 4.3 beds/1,000 population (2009)

Sanitation facility access:
improved:
urban: 100% of population
rural: 100% of population
total: 100% of population
unimproved:
urban: 0% of population
rural: 0% of population
total: 0% of population (2017 est.)

HIV/AIDS—adult prevalence rate: NA

HIV/AIDS—people living with HIV/AIDS: NA

HIV/AIDS—deaths: NA

Obesity—adult prevalence rate: 8.6% (2016)
Maldives — The World Factbook—Central Intelligence Agency
country comparison to the world: 148

Children under the age of 5 years underweight: 17.7% (2009)
country comparison to the world: 32

Education expenditures: 4.1% of GDP (2016)
country comparison to the world: 97

Literacy: *definition:* age 15 and over can read and write
total population: 97.7%
male: 97.3%
female: 98.1% (2016)

Unemployment, youth ages 15-24: *total:* 15.9%
male: 19.1%
female: 12.1% (2016 est.)
country comparison to the world: 85

GOVERNMENT

Country name: *conventional long form:* Republic of Maldives
conventional short form: Maldives
local long form: Dhivehi Raajjeyge Jumhooriyyaa
local short form: Dhivehi Raajje
etymology: archipelago apparently named after the main island (and capital) of Male; the word "Maldives" means "the islands (dives) of Male"; alternatively, the name may derive from the Sanskrit word "maladvipa" meaning "garland of islands"; Dhivehi Raajje in Dhivehi means "Kingdom of the Dhivehi people"

Government type: presidential republic

Capital: *name:* Male

geographic coordinates: 4 10 N, 73 30 E
time difference: UTC+5 (10 hours ahead of Washington, DC, during Standard Time)

etymology: derived from the Sanskrit word "mahaalay" meaning "big house"

Administrative divisions: 21 administrative atolls (atholhuthah, singular—atholhu); Addu (Addu City), Ariatholhu Dhekunuburi (South Ari Atoll), Ariatholhu Uthuruburi (North Ari Atoll), Faadhippolhu, Felidhuatholhu (Felidhu Atoll), Fuvammulah, Hahdhunmathi, Huvadhuatholhu Dhekunuburi (South Huvadhu Atoll), Huvadhuatholhu Uthuruburi (North Huvadhu Atoll), Kolhumadulu, Maale (Male), Maaleatholhu (Male Atoll), Maalhosmadulu Dhekunuburi (South Maalhosmadulu), Maalhosmadulu Uthuruburi (North Maalhosmadulu), Miladhunmadulu Dhekunuburi (South Miladhunmadulu), Miladhunmadulu Uthuruburi (North Miladhunmadulu), Mulakatholhu (Mulaku Atoll), Nilandheatholhu Dhekunuburi (South Nilandhe Atoll), Nilandheatholhu Uthuruburi (North Nilandhe Atoll), Thiladhunmathee Dhekunuburi (South Thiladhunmathi), Thiladhunmathee Uthuruburi (North Thiladhunmathi)

Independence: 26 July 1965 (from the UK)

National holiday: Independence Day, 26 July (1965)

Constitution: *history:* many previous; latest ratified 7 August 2008
amendments: proposed by Parliament; passage requires at least three-quarters majority vote by its membership and the signature of the president of the republic; passage of amendments to constitutional articles on rights and freedoms and the terms of office of Parliament and of the president also requires a majority vote in a referendum; amended 2015

Legal system: Islamic (sharia) legal system with English common law influences, primarily in commercial matters

International law organization participation: has not submitted an ICJ jurisdiction declaration; accepts ICCt jurisdiction

Citizenship: *citizenship by birth:* no
citizenship by descent only: at least one parent must be a citizen of Maldives
dual citizenship recognized: yes
residency requirement for naturalization: unknown

Suffrage: 18 years of age; universal

Executive branch: *chief of state:* President Ibrahim "Ibu" Mohamed SOLIH (since 17 November 2018); Vice President Faisal NASEEM (since 17 November 2018); the president is both chief of state and head of government
head of government: President Ibrahim Mohamed SOLIH (since 17 November 2018); Vice President Faisal NASEEM (since 17 November 2018)
cabinet: Cabinet of Ministers appointed by the president, approved by Parliament
elections/appointments: president directly elected by absolute majority popular vote in 2 rounds if needed for a 5-year term (eligible for a second term); election last held on 23 September 2018 (next to be held in 2023)

election results: Ibrahim Mohamed SOLIH elected president (in 1 round); Ibrahim Mohamed SOLIH (MDP) 58.3%, Abdulla YAMEEN Abdul Gayoom (PPM) 41.7%

Legislative branch: *description:* unicameral Parliament or People's Majlis (87 seats—includes 2 seats added by the Elections Commission in late 2018; members directly elected in single-seat constituencies by simple majority vote to serve 5-year terms)
elections: last held on 6 April 2019 (next to be held in 2023)
election results: percent of vote—MDP 44.7%, JP 10.8%, PPM 8.7%, PNC 6.4%, MDA 2.8%, other 5.6%, independent 21%; seats by party—MDP 65, JP 5, PPM 5, PNC 3, MDA 2, independent 7; composition—men 83, women 4, percent of women 4.6%

Judicial branch: *highest courts:* Supreme Court (consists of the chief justice and 4-6 justices; note—3 justices as of late 2019)
judge selection and term of office: Supreme Court judges appointed by the president in consultation with the Judicial Service Commission—a 10-member body of selected high government officials and the public—and upon confirmation by voting members of the People's Majlis; judges serve until mandatory retirement at age 70
subordinate courts: High Court; Criminal, Civil, Family, Juvenile, and Drug Courts; Magistrate Courts (on each of the inhabited islands)

Political parties and leaders: Adhaalath (Justice) Party or AP [Sheikh Imran ABDULLA]
Dhivehi Rayyithunge Party or DRP [Ahmed Thasmeen ALI]
Maldives Development Alliance or MDA [Ahmed Shiyam MOHAMED]
Maldivian Democratic Party or MDP [Mohamed NASHEED]
Maldives Labor and Social Democratic Party or MLSDP [Ahmed SHIHAM]
Maldives Thirdway Democrats or MTD [Ahmed ADEEB]
Maumoon/Maldives Reform Movement or MRM [Maumoon Abdul GAYOOM]
National Democratic Congress [Yousuf Maaniu] (formed in 2020)
People's National Congress or PNC [Abdul Raheem ABDULLA] (formed in early 2019)
Progressive Party of Maldives or PPM
Republican (Jumhooree) Party or JP [Qasim IBRAHIM] (2020)

International organization participation: ADB, AOSIS, C, CP, FAO, G-77, IBRD, ICAO, ICC (NGOs), ICCt, IDA, IDB, IFAD, IFC, IFRCS, ILO, IMF, IMO, Interpol, IOC, IOM, IPU, ITU, MIGA, NAM, OIC, OPCW, SAARC, SACEP, UN, UNCTAD, UNESCO, UNIDO, UNWTO, UPU, WCO, WHO, WIPO, WMO, WTO

Diplomatic representation in the US: *chief of mission:* Ambassador THILMEEZA Hussain (since 8 July 2019)
chancery: 801 Second Avenue, Suite 400E, New York, NY 10017
telephone: [1] (212) 599-6194 and 599-6195

FAX: [1] (212) 661-6405

Diplomatic representation from the US: the US does not have an embassy in Maldives; US Ambassador to Sri Lanka and Maldives, Alaina TEPLITZ (since 1 November 2018), is accredited to both countries; note: Secretary of State Mike Pompeo spoke of establishing an embassy on his trip to Maldives in October of 2020

Flag description: red with a large green rectangle in the center bearing a vertical white crescent moon; the closed side of the crescent is on the hoist side of the flag; red recalls those who have sacrificed their lives in defense of their country, the green rectangle represents peace and prosperity, and the white crescent signifies Islam

National symbol(s): *coconut palm, yellowfin tuna; national colors:* red, green, white

National anthem: *name:* "Gaumee Salaam" (National Salute)
lyrics/music: Mohamed Jameel DIDI/ Wannakuwattawaduge DON AMARADEVA
note: lyrics adopted 1948, music adopted 1972; between 1948 and 1972, the lyrics were sung to the tune of "Auld Lang Syne"
0:00 / 1:08

ECONOMY

Economy—overview: Maldives has quickly become a middle-income country, driven by the rapid growth of its tourism and fisheries sectors, but the country still contends with a large and growing fiscal deficit. Infrastructure projects, largely funded by China, could add significantly to debt levels. Political turmoil and the declaration of a state of emergency in February 2018 led to the issuance of travel warnings by several countries whose citizens visit Maldives in significant numbers, but the overall impact on tourism revenue was unclear.

In 2015, Maldives' Parliament passed a constitutional amendment legalizing foreign ownership of land; foreign landbuyers must reclaim at least 70% of the desired land from the ocean and invest at least $1 billion in a construction project approved by Parliament.

Diversifying the economy beyond tourism and fishing, reforming public finance, increasing employment opportunities, and combating corruption, cronyism, and a growing drug problem are near-term challenges facing the government. Over the longer term, Maldivian authorities worry about the impact of erosion and possible global warming on their low-lying country; 80% of the area is 1 meter or less above sea level.

GDP (purchasing power parity):
$6.901 billion (2017 est.)
$6.583 billion (2016 est.)
$6.3 billion (2015 est.)
note: data are in 2017 dollars
country comparison to the world: 168

GDP (official exchange rate):
$4.505 billion (2017 est.)

GDP—real growth rate: 4.8% (2017 est.)
4.5% (2016 est.)

2.2% (2015 est.)
country comparison to the world: 54

GDP—per capita (PPP): $19,200 (2017 est.)
$18,600 (2016 est.)
$18,100 (2015 est.)
note: data are in 2017 dollars
country comparison to the world: 93

Gross national saving:
0.5% of GDP (2017 est.)
-4.5% of GDP (2016 est.)
12.6% of GDP (2015 est.)
country comparison to the world: 180

GDP—composition, by end use:
household consumption: NA (2016 est.)
government consumption: NA (2016 est.)
investment in fixed capital: NA (2016 est.)
investment in inventories: NA (2016 est.)
exports of goods and services: 93.6% (2016 est.)
imports of goods and services: 89% (2016 est.)

GDP—composition, by sector of origin:
agriculture: 3% (2015 est.)
industry: 16% (2015 est.)
services: 81% (2015 est.)

Agriculture—products: coconuts, corn, sweet potatoes; fish

Industries: tourism, fish processing, shipping, boat building, coconut processing, woven mats, rope, handicrafts, coral and sand mining

Industrial production growth rate: 14% (2012 est.)
country comparison to the world: 4

Labor force: 222,200 (2017 est.)
country comparison to the world: 167

Labor force—by occupation: *agriculture:* 7.7%
industry: 22.8%
services: 69.5% (2017 est.)

Unemployment rate: 2.9% (2017 est.)
3.2% (2016 est.)
country comparison to the world: 35

Population below poverty line: 15% (2009 est.)

Household income or consumption by percentage share: *lowest 10%:* 1.2%
highest 10%: 33.3% (FY09/10)

Budget: *revenues:* 1.19 billion (2016 est.)
expenditures: 1.643 billion (2016 est.)

Taxes and other revenues: 26.4% (of GDP) (2016 est.)
country comparison to the world: 111

Budget surplus (+) or deficit (-): -10.1% (of GDP) (2016 est.)
country comparison to the world: 212

Public debt: 63.9% of GDP (2017 est.)
61.7% of GDP (2016 est.)
country comparison to the world: 62

Fiscal year: calendar year

Inflation rate (consumer prices): 2.3% (2017 est.)
0.8% (2016 est.)
country comparison to the world: 116

Current account balance: -$876 million (2017 est.)
-$1.033 billion (2016 est.)
country comparison to the world: 140

Exports: $256.2 million (2016 est.)

$239.8 million (2015 est.)
country comparison to the world: 187

Exports—partners: Thailand 42.8%, Sri Lanka 8.7%, Bangladesh 6.4%, France 6.2%, US 6.1%, Germany 5%, Ireland 4.6% (2017)

Exports—commodities: fish

Imports: $2.125 billion (2016 est.)
$1.896 billion (2015 est.)
country comparison to the world: 165

Imports—commodities: petroleum products, clothing, intermediate and capital goods

Imports—partners: UAE 17.1%, India 13.5%, Singapore 13.3%, China 10.8%, Sri Lanka 6.7%, Malaysia 6%, Thailand 4.5% (2017)

Reserves of foreign exchange and gold: $477.9 million (31 December 2016 est.)
$575.8 million (31 December 2015 est.)
country comparison to the world: 153

Debt—external: $848.8 million (31 December 2016 est.)
$696.2 million (31 December 2015 est.)
country comparison to the world: 167

Exchange rates: rufiyaa (MVR) per US dollar—15.42 (2017 est.)
15.35 (2016 est.)

ENERGY

Electricity access: *electrification—total population:* 100% (2020)

Electricity—production: 402 million kWh (2016 est.)
country comparison to the world: 171

Electricity—consumption: 373.9 million kWh (2016 est.)
country comparison to the world: 178

Electricity—exports: 0 kWh (2016 est.)
country comparison to the world: 165

Electricity—imports: 0 kWh (2016 est.)
country comparison to the world: 171

Electricity—installed generating capacity: 278,000 kW (2016 est.)
country comparison to the world: 161

Electricity—from fossil fuels: 96% of total installed capacity (2016 est.)
country comparison to the world: 40

Electricity—from nuclear fuels: 0% of total installed capacity (2017 est.)
country comparison to the world: 137

Electricity—from hydroelectric plants: 0% of total installed capacity (2017 est.)
country comparison to the world: 185

Electricity—from other renewable sources: 4% of total installed capacity (2017 est.)
country comparison to the world: 114

Crude oil—production: 0 bbl/day (2018 est.)
country comparison to the world: 169

Crude oil—exports: 0 bbl/day (2015 est.)
country comparison to the world: 160

Crude oil—imports: 0 bbl/day (2015 est.)
country comparison to the world: 160

Crude oil—proved reserves: 0 bbl (1 January 2018 est.)
country comparison to the world: 164

Refined petroleum products—production: 0 bbl/day (2015 est.)
country comparison to the world: 172

Refined petroleum products—consumption: 11,000 bbl/day (2016 est.)
country comparison to the world: 160

Refined petroleum products—exports: 0 bbl/day (2015 est.)
country comparison to the world: 177

Refined petroleum products—imports: 10,840 bbl/day (2015 est.)
country comparison to the world: 144

Natural gas—production: 0 cu m (2017 est.)
country comparison to the world: 165

Natural gas—consumption: 0 cu m (2017 est.)
country comparison to the world: 171

Natural gas—exports: 0 cu m (2017 est.)
country comparison to the world: 146

Natural gas—imports: 0 cu m (2017 est.)
country comparison to the world: 153

Natural gas—proved reserves: 0 cu m (1 January 2016 est.)
country comparison to the world: 166

Carbon dioxide emissions from consumption of energy: 1.648 million Mt (2017 est.)
country comparison to the world: 161

COMMUNICATIONS

Telephones—fixed lines: *total subscriptions:* 12,316
subscriptions per 100 inhabitants: 3.14 (2019 est.)
country comparison to the world: 186

Telephones—mobile cellular: *total subscriptions:* 611,662
subscriptions per 100 inhabitants: 155.95 (2019 est.)
country comparison to the world: 170

Telecommunication systems: *general assessment:* upgrades to telecom infrastructure extended to outer islands; two mobile operators extend LTE coverage; tourism has strengthened the telecom market with investment and accounts for the high mobile penetration rate; mobile penetration passes 250%; launches 5G trials (2020)

domestic: fixed-line is at 3 per 100 persons and high mobile-cellular subscriptions stands at 156 per 100 persons (2019)

international: country code—960; landing points for Dhiraagu Cable Network, NaSCOM, Dhiraagu-SLT Submarine Cable Networks and WARF submarine cables providing connections to 8 points in Maldives, India, and Sri Lanka; satellite earth station—3 Intelsat (Indian Ocean) (2019)
note: the COVID-19 outbreak is negatively impacting telecommunications production and supply chains globally; consumer spending on telecom devices and services has also slowed due to the pandemic's effect on economies worldwide; overall

607

progress towards improvements in all facets of the telecom industry—mobile, fixed-line, broadband, submarine cable and satellite—has moderated

Broadcast media: state-owned radio and TV monopoly until recently; 4 state-operated and 7 privately owned TV stations and 4 state- operated and 7 privately owned radio stations (2019)

Internet country code: .mv

Internet users: total: 248,004

percent of population: 63.19% (July 2018 est.)
country comparison to the world: 170

Broadband—fixed subscriptions: total: 53,470
subscriptions per 100 inhabitants: 14 (2018 est.)
country comparison to the world: 133

TRANSPORTATION

National air transport system: number of registered air carriers: 3 (2020)
inventory of registered aircraft operated by air carriers: 36
annual passenger traffic on registered air carriers: 1,147,247 (2018)
annual freight traffic on registered air carriers: 7.75 million (2018)

Civil aircraft registration country code prefix: 8Q (2016)

Airports: 9 (2013)
country comparison to the world: 157

Airports—with paved runways: total: 7 (2017)
over 3,047 m: 1 (2017)
2,438 to 3,047 m: 1 (2017)
1,524 to 2,437 m: 1 (2017)
914 to 1,523 m: 4 (2017)

Airports—with unpaved runways: total: 2 (2013)
914 to 1,523 m: 2 (2013)

Roadways: total: 93 km (2018)
paved: 93 km—60 km in Male; 16 km on Addu Atolis; 17 km on Laamu (2018)
note: island roads are mainly compacted coral

country comparison to the world: 214

Merchant marine: total: 62
by type: bulk carrier 1, general cargo 20, oil tanker 16, other 25 (2019)
country comparison to the world: 107

Ports and terminals: major seaport(s): Male

MILITARY AND SECURITY

Military and security forces: the Republic of Maldives has no distinct army, navy, or air force but a single security unit called the Maldives National Defence Force (MNDF) comprised of ground forces, an air element, a coastguard, a presidential security division, and a special protection group (2020)
note: the MNDF is primarily tasked to reinforce the Maldives Police Service (MPS) and ensure security in the country's exclusive economic zone

Military and security service personnel strengths: the Maldives National Defense Force (MNDF) has approximately 2,500 personnel (2019 est.)

Military equipment inventories and acquisitions: India has provided most of the equipment in the MNDF's inventory (2020)

Military service age and obligation: 18-28 years of age for voluntary service; no conscription; 10th grade or equivalent education required; must not be a member of a political party

TERRORISM

Terrorist group(s): Islamic State of Iraq and ash-Sham (ISIS) (2019)
note: details about the history, aims, leadership, organization, areas of operation, tactics, targets, weapons, size, and sources of support of the group(s) appear(s) in Appendix-T

TRANSNATIONAL ISSUES

Disputes—international: none

Trafficking in persons: current situation: Maldives is a destination country for men, women, and children subjected to forced labor and sex trafficking and a source country for women and children subjected to labor and sex trafficking; primarily Bangladeshi and Indian migrants working both legally and illegally in the construction and service sectors face conditions of forced labor, including fraudulent recruitment, confiscation of identity and travel documents, nonpayment and withholding of wages, and debt bondage; a small number of women from Asia, Eastern Europe, and former Soviet states are trafficked to Maldives for sexual exploitation; Maldivian women may be subjected to sex trafficking domestically or in Sri Lanka; some Maldivian children are transported to the capital for domestic service, where they may also be victims of sexual abuse and forced labor

tier rating: Tier 2 Watch List—Maldives does not fully comply with the minimum standards for the elimination of trafficking; however, it is making significant efforts to do so; the government adopted a national action plan for 2015-19 and is continuing to develop victim identification, protection, and referral procedures, but overall its anti-trafficking efforts did not increase; only five trafficking investigations were conducted, no new prosecutions were initiated for the second consecutive year, and no convictions were made, down from one in 2013; some officials warned businesses in advance of planned raids for suspected trafficking offenses; victim protection deteriorated when the state-run shelter for female victims barred access to victims shortly after opening in January 2014, in part because of bureaucratic disputes, which dissuaded victims from pursuing charges against perpetrators; the government did not prosecute or hold accountable any employers or government officials for withholding passports (2015)

MALI

INTRODUCTION

Background: Present-day Mali is named for the Mali Empire that at its peak in the 14th century covered an area about twice the size of modern-day France and stretched to the west coast of Africa. In the late 19th century, France seized control of Mali. The Sudanese Republic and Senegal became independent of France in 1960 as the Mali Federation. When Senegal withdrew after only a few months, what formerly made up the Sudanese Republic was renamed Mali. Rule by dictatorship was brought to a close in 1991 by a military coup that ushered in a period of democratic rule. President Alpha Oumar KONARE won Mali's first two democratic presidential elections in 1992 and 1997. In keeping with Mali's two-term

constitutional limit, he stepped down in 2002 and was succeeded by Amadou Toumani TOURE, who was elected to a second term in a 2007 election that was widely judged to be free and fair. Malian returnees from Libya in 2011 exacerbated tensions in northern Mali, and Tuareg ethnic militias rebelled in January 2012. Low- and mid-level soldiers, frustrated with the poor handling of the rebellion, overthrew TOURE on 22 March. Intensive mediation efforts led by the Economic Community of West African States (ECOWAS) returned power to a civilian administration in April with the appointment of Interim President Dioncounda TRAORE.

The post-coup chaos led to rebels expelling the Malian military from the country's three northern regions and allowed Islamic militants to set up strongholds. Hundreds of thousands of northern Malians fled the violence to southern Mali and neighboring countries, exacerbating regional food shortages in host communities. A French-led international military intervention to retake the three northern regions began in January 2013 and within a month, most of the north had been retaken. In a democratic presidential election conducted in July and August of 2013, Ibrahim Boubacar KEITA was elected president. The Malian Government and northern armed groups signed an internationally mediated peace accord in June 2015, however, the parties to the peace accord have made little progress in the accord's implementation, despite a June 2017 target for its completion. Furthermore, extremist groups outside the peace process made steady inroads into rural areas of central Mali following the consolidation of three major terrorist organizations in March 2017. In central and northern Mali, terrorist groups have exploited age-old ethnic rivalries between pastoralists and sedentary communities and inflicted serious losses on the Malian military. Intercommunal violence incidents such as targeted killings occur with increasing regularity. KEITA was reelected president in 2018 in an election that was deemed credible by international observers, despite some security and logistic shortfalls.

GEOGRAPHY

Location: interior Western Africa, southwest of Algeria, north of Guinea, Cote d'Ivoire, and Burkina Faso, west of Niger

Geographic coordinates: 17 00 N, 4 00 W

Map references: Africa

Area: *total:* 1,240,192 sq km
land: 1,220,190 sq km
water: 20,002 sq km
country comparison to the world: 25

Area—comparative: slightly less than twice the size of Texas

Land boundaries: *total:* 7,908 km
border countries (7): Algeria 1359 km, Burkina Faso 1325 km, Cote d'Ivoire 599 km, Guinea 1062 km, Mauritania 2236 km, Niger 838 km, Senegal 489 km

Coastline: 0 km (landlocked)

Maritime claims: none (landlocked)

Climate: subtropical to arid; hot and dry (February to June); rainy, humid, and mild (June to November); cool and dry (November to February)

Terrain: mostly flat to rolling northern plains covered by sand; savanna in south, rugged hills in northeast

Elevation: *mean elevation:* 343 m
lowest point: Senegal River 23 m
highest point: Hombori Tondo 1,155 m

Natural resources: gold, phosphates, kaolin, salt, limestone, uranium, gypsum, granite, hydropower, note, bauxite, iron ore, manganese, tin, and copper deposits are known but not exploited

Land use: *agricultural land:* 34.1% (2011 est.)

arable land: 5.6% (2011 est.) / permanent crops: 0.1% (2011 est.) / permanent pasture: 28.4% (2011 est.)
forest: 10.2% (2011 est.)
other: 55.7% (2011 est.)
Irrigated land: 3,780 sq km (2012)

Population distribution: the overwhelming majority of the population lives in the southern half of the country, with greater density along the border with Burkina Faso as shown in this population distribution map

Natural hazards: hot, dust-laden harmattan haze common during dry seasons; recurring droughts; occasional Niger River flooding

Environment—current issues: deforestation; soil erosion; desertification; loss of pasture land; inadequate supplies of potable water

Environment—international agreements: *party to:* Biodiversity, Climate Change, Climate Change-Kyoto Protocol, Desertification, Endangered Species, Hazardous Wastes, Law of the Sea, Ozone Layer Protection, Wetlands, Whaling
signed, but not ratified: none of the selected agreements

Geography—note: *landlocked; divided into three natural zones:* the southern, cultivated Sudanese; the central, semiarid Sahelian; and the northern, arid Saharan

PEOPLE AND SOCIETY

Population: 19,553,397 (July 2020 est.)
country comparison to the world: 62
Nationality: noun: Malian(s)
adjective: Malian

Ethnic groups: Bambara 33.3%, Fulani (Peuhl) 13.3%, Sarakole/Soninke/Marka 9.8%, Senufo/Manianka 9.6%, Malinke 8.8%, Dogon 8.7%, Sonrai 5.9%, Bobo 2.1 %, Tuareg/Bella 1.7%, other Malian 6%, from members of Economic Community of West Africa .4%, other .3% (2018 est.)

Languages: French (official), Bambara 46.3%, Peuhl/Foulfoulbe 9.4%, Dogon 7.2%, Maraka/Soninke 6.4%, Malinke 5.6%, Sonrhai/Djerma 5.6%, Minianka 4.3%, Tamacheq 3.5%, Senoufo 2.6%, Bobo 2.1%, unspecified 0.7%, other 6.3% (2009 est.)
note: Mali has 13 national languages in addition to its official language

Religions: Muslim 93.9%, Christian 2.8%, animist .7%, none 2.5% (2018 est.)

Demographic profile: Mali's total population is expected to double by 2035; its capital Bamako is one of the fastest-growing cities in Africa. A young age structure, a declining mortality rate, and a sustained high total fertility rate of 6 children per woman — the third highest in the world — ensure continued rapid population growth for the foreseeable future. Significant outmigration only marginally tempers this growth. Despite decreases, Mali's infant, child, and maternal mortality rates remain among the highest in Sub-Saharan Africa because of limited access to and adoption of family

planning, early childbearing, short birth intervals, the prevalence of female genital cutting, infrequent use of skilled birth attendants, and a lack of emergency obstetrical and neonatal care.

Mali's high total fertility rate has been virtually unchanged for decades, as a result of the ongoing preference for large families, early childbearing, the lack of female education and empowerment, poverty, and extremely low contraceptive use. Slowing Mali's population growth by lowering its birth rate will be essential for poverty reduction, improving food security, and developing human capital and the economy.

Mali has a long history of seasonal migration and emigration driven by poverty, conflict, demographic pressure, unemployment, food insecurity, and droughts. Many Malians from rural areas migrate during the dry period to nearby villages and towns to do odd jobs or to adjoining countries to work in agriculture or mining. Pastoralists and nomads move seasonally to southern Mali or nearby coastal states. Others migrate long term to Mali's urban areas, Cote d'Ivoire, other neighboring countries, and in smaller numbers to France, Mali's former colonial ruler. Since the early 1990s, Mali's role has grown as a transit country for regional migration flows and illegal migration to Europe. Human smugglers and traffickers exploit the same regional routes used for moving contraband drugs, arms, and cigarettes.

Between early 2012 and 2013, renewed fighting in northern Mali between government forces and Tuareg secessionists and their Islamist allies, a French-led international military intervention, as well as chronic food shortages, caused the displacement of hundreds of thousands of Malians. Most of those displaced domestically sought shelter in urban areas of southern Mali, except for pastoralist and nomadic groups, who abandoned their traditional routes, gave away or sold their livestock, and dispersed into the deserts of northern Mali or crossed into neighboring countries. Almost all Malians who took refuge abroad (mostly Tuareg and Maure pastoralists) stayed in the region, largely in Mauritania, Niger, and Burkina Faso.

Age structure: *0-14 years:* 47.69% (male 4,689,121/female 4,636,685)
15-24 years: 19% (male 1,768,772/female 1,945,582)
25-54 years: 26.61% (male 2,395,566/female 2,806,830)
55-64 years: 3.68% (male 367,710/female 352,170)
65 years and over: 3.02% (male 293,560/female 297,401) (2020 est.)

Dependency ratios: *total dependency ratio:* 98
youth dependency ratio: 93.1
elderly dependency ratio: 4.9
potential support ratio: 20.4 (2020 est.)

Median age: *total:* 16 years
male: 15.3 years
female: 16.7 years (2020 est.)
country comparison to the world: 225

Population growth rate: 2.95% (2020 est.)
country comparison to the world: 9

Birth rate: 42.2 births/1,000 population (2020 est.)

country comparison to the world: 4

Death rate: 9 deaths/1,000 population (2020 est.)
country comparison to the world: 63

Net migration rate: -3.9 migrant(s)/1,000 population (2020 est.)
country comparison to the world: 188

Population distribution: the overwhelming majority of the population lives in the southern half of the country, with greater density along the border with Burkina Faso as shown in this population distribution map

Urbanization: *urban population:* 43.9% of total population (2020)
rate of urbanization: 4.86% annual rate of change (2015-20 est.)
total population growth rate v. urban population growth rate, 2000-2030:

Major urban areas—population: 2.618 million BAMAKO (capital) (2020)

Sex ratio: *at birth:* 1.03 male(s)/female
0-14 years: 1.01 male(s)/female
15-24 years: 0.91 male(s)/female
25-54 years: 0.85 male(s)/female
55-64 years: 1.04 male(s)/female
65 years and over: 0.99 male(s)/female
total population: 0.95 male(s)/female (2020 est.)

Mother's mean age at first birth: 18.9 years (2018 est.)
note: median age at first birth among women 25-29

Maternal mortality rate: 562 deaths/100,000 live births (2017 est.)
country comparison to the world: 15

Infant mortality rate: *total:* 64 deaths/1,000 live births
male: 69.6 deaths/1,000 live births
female: 58.3 deaths/1,000 live births (2020 est.)
country comparison to the world: 9

Life expectancy at birth: *total population:* 61.6 years
male: 59.4 years
female: 63.9 years (2020 est.)
country comparison to the world: 211

Total fertility rate: 5.72 children born/woman (2020 est.)
country comparison to the world: 4

Contraceptive prevalence rate: 17.2% (2018)

Drinking water source:
improved:
urban: 97.1% of population
rural: 72.8% of population
total: 82.9% of population
unimproved:
urban: 2.9% of population
rural: 27.2% of population
total: 17.1% of population (2017 est.)

Current Health Expenditure: 3.8% (2017)

Physicians density: 0.14 physicians/1,000 population (2016)

Hospital bed density: 0.1 beds/1,000 population (2010)

Sanitation facility access:

improved:
urban: 82.5% of population
rural: 34.1% of population
total: 54.2% of population
unimproved:
urban: 17.5% of population
rural: 65.9% of population
total: 45.8% of population (2017 est.)

HIV/AIDS—adult prevalence rate: 1 .2% (2019 est.)
country comparison to the world: 39

HIV/AIDS—people living with HIV/AIDS: 140,000 (2019 est.)
country comparison to the world: 38

HIV/AIDS—deaths: 5,800 (2019 est.)
country comparison to the world: 24

Major infectious diseases: *degree of risk:* very high (2020)
food or waterborne diseases: bacterial and protozoal diarrhea, hepatitis A, and typhoid fever
vectorborne diseases: malaria and dengue fever
water contact diseases: schistosomiasis
animal contact diseases: rabies
respiratory diseases: meningococcal meningitis

Obesity—adult prevalence rate: 8.6% (2016)
country comparison to the world: 149

Children under the age of 5 years underweight: 18.6% (2018)
country comparison to the world: 28

Education expenditures: 3.1% of GDP (2016)
country comparison to the world: 134

Literacy: *definition:* age 15 and over can read and write
total population: 35.5%
male: 46.2%
female: 25.7% (2018)

School life expectancy (primary to tertiary education):
total: 8 years
male: 8 years
female: 7 years (2017)

Unemployment, youth ages 15-24:
total: 16.9%
male: 15.3%
female: 18.8% (2018 est.)
country comparison to the world: 79

GOVERNMENT

Country name: *conventional long form:* Republic of Mali
conventional short form: Mali
local long form: Republique de Mali
local short form: Mali
former: French Sudan and Sudanese Republic
etymology: name derives from the West African Mali Empire of the 13th to 16th centuries A.D.

Government type: semi-presidential republic

Capital: *name:* Bamako
geographic coordinates: 12 39 N, 8 00 W
time difference: UTC 0 (5 hours ahead of Washington, DC, during Standard Time)

etymology: the name in the Bambara language can mean either "crocodile tail" or "crocodile river" and three crocodiles appear on the city seal

Administrative divisions: 10 regions (regions, singular—region), 1 district*; District de Bamako*, Gao, Kayes, Kidal, Koulikoro, Menaka, Mopti, Segou, Sikasso, Taoudenni, Tombouctou (Timbuktu); note—Menaka and Taoudenni were legislated in 2016, but implementation has not been confirmed by the US Board on Geographic Names

Independence: 22 September 1960 (from France)

National holiday: Independence Day, 22 September (1960)

Constitution: *history:* several previous; latest drafted August 1991, approved by referendum 12 January 1992, effective 25 February 1992, suspended briefly in 2012
amendments: proposed by the president of the republic or by members of the National Assembly; passage requires two-thirds majority vote by the Assembly and approval in a referendum; constitutional sections on the integrity of the state, its republican and secular form of government, and its multiparty system cannot be amended; amended 1999

Legal system: civil law system based on the French civil law model and influenced by customary law; judicial review of legislative acts in the Constitutional Court

International law organization participation: has not submitted an ICJ jurisdiction declaration; accepts ICC jurisdiction

Citizenship: *citizenship by birth:* no
citizenship by descent only: at least one parent must be a citizen of Mali
dual citizenship recognized: yes
residency requirement for naturalization: 5 years

Suffrage: 18 years of age; universal

Executive branch: *chief of state:* President of transitional government, Bah NDAW (since 25 September 2020); vice president of the transitional government, Assimi GOITA (since 25 September 2020); former president Ibrahim Boubacar KEITA was deposed by the Malian military on 18 August 2020; on 21 September, a group of 17 electors chosen by the Malian military junta, called the National Committee for the Salvation of the People (NCSP), selected former Malian defense minister and chairman of the NCSP, Bah NDAW, as transitional president, and retired Malian Army Colonel Assimi GOITA as transitional vice president; the transitional government was inaugurated on 25 September 2020
head of government: Prime Minister Moctar OUANE (appointed by the transitional government on 27 Sep 2020; former PM Boubou CISSE was removed on 18 August 2020 following the military coup)
cabinet: Council of Ministers appointed by the prime minister
elections/appointments: president directly elected by absolute majority popular vote in 2

rounds if needed for a 5-year term (eligible for a second term); election last held on 29 July 2018 with a runoff on 12 August 2018; prime minister appointed by the president

election results: Ibrahim Boubacar KEITA elected president in second round; percent of vote—Ibrahim Boubacar KEITA (RPM) 77.6%, Soumaila CISSE (URD) 22.4%

Legislative branch: *description:* unicameral National Assembly or Assemblee Nationale (147 seats; members directly elected in single and multi-seat constituencies by absolute majority vote in 2 rounds if needed; 13 seats reserved for citizens living abroad; members serve 5-year terms)

note—the National Assembly was dissolved on 18 August 2020 following a military coup and the resignation of President KEITA

elections: last held on 30 March and 19 April 2020 (prior to the August 2020 coup, the next election was scheduled to be held in 2025)

election results: percent of vote by party—NA; seats by party—NA composition—NA

Judicial branch: *highest courts:* Supreme Court or Cour Supreme (consists of 19 judges organized into judicial, administrative, and accounting sectons); Constitutional Court (consists of 9 judges)

judge selection and term of office: Supreme Court judges appointed by the Ministry of Justice to serve 5-year terms; Constitutional Court judges selected—3 each by the president, the National Assembly, and the Supreme Council of the Magistracy; members serve single renewable 7-year terms

subordinate courts: Court of Appeal; High Court of Justice (jurisdiction limited to cases of high treason or criminal offenses by the president or ministers while in office); administrative courts (first instance and appeal); commercial courts; magistrate courts; labor courts; juvenile courts; special court of state security

Political parties and leaders:
African Solidarity for Democracy and Independence or SADI [Oumar MARIKO]
Alliance for Democracy in Mali-Pan-African Party for Liberty, Solidarity, and Justice or ADEMA-PASJ [Tiemoko SANGARE] Alliance for Democracy and Progress or ADP-Maliba [Amadou THIAM]
Alliance for the Solidarity of Mali-Convergence of Patriotic Forces or ASMA-CFP [Soumeylou Boubeye MAIGA]
Alternative Forces for Renewal and Emergence or FARE [Modibo SIDIBE]
Convergence for the Development of Mali or CODEM [Housseyni Amion GUINDO]
Democratic Alliance for Peace or ADP-Maliba [Aliou Boubacar DIALLO]
Economic and Social Development Party or PDES [Jamille BITTAR]
Front for Democracy and the Republic or FDR (coalition of smaller opposition parties)
National Congress for Democratic Initiative or CNID [Mountaga TALL]
Party for National Renewal or PARENA [Tiebile DRAME]

Patriotic Movement for Renewal or MPR [Choguel Kokalla MAIGA]
Rally for Mali or RPM [Boucary TRETA]
Union for Republic and Democracy or URD [Younoussi TOURE]

International organization participation: ACP, AfDB, AU, CD, ECOWAS, EITI (compliant country), FAO, FZ, G-77, IAEA, IBRD, ICAO, ICCt, ICRM, IDA, IDB, IFAD, IFC, IFRCS, ILO, IMF, Interpol, IOC, IOM, IPU, ISO, ITSO, ITU, ITUC (NGOs), MIGA, MONUSCO, NAM, OIC, OIF, OPCW, UN, UNAMID, UNCTAD, UNESCO, UNIDO, UNISFA, UNMISS, UNWTO, UPU, WADB (regional), WAEMU, WCO, WFTU (NGOs), WHO, WIPO, WMO, WTO

Diplomatic representation in the US: *chief of mission:* Ambassador Mahamadou NIMAGA (since 22 June 2018)
chancery: 2130 R Street NW, Washington, DC 20008
telephone: [1] (202) 332-2249, 939-8950
FAX: [1] (202) 332-6603

Diplomatic representation from the US: *chief of mission:* Ambassador Dennis B. HANKINS (since 15 March 2019)
telephone: [223] 2070-2300
embassy: ACI 2000, Rue 243, (located off the Roi Bin Fahad Aziz Bridge west of the Bamako central district), Porte 297, Bamako
mailing address: ACI 2000, Rue 243, Porte 297, Bamako
FAX: [223] 2070-2479

Flag description: three equal vertical bands of green (hoist side), yellow, and red
note: uses the popular Pan-African colors of Ethiopia; the colors from left to right are the same as those of neighboring Senegal (which has an additional green central star) and the reverse of those on the flag of neighboring Guinea

National symbol(s): Great Mosque of Djenne; national colors: green, yellow, red

National anthem: name: "Le Mali" (Mali)
lyrics/music: Seydou Badian KOUYATE/ Banzoumana SISSOKO
note: adopted 1962; also known as "Pour L'Afrique et pour toi, Mali" (For Africa and for You, Mali) and "A ton appel Mali" (At Your Call, Mali)

ECONOMY

Economy—overview: Among the 25 poorest countries in the world, landlocked Mali depends on gold mining and agricultural exports for revenue. The country's fiscal status fluctuates with gold and agricultural commodity prices and the harvest; cotton and gold exports make up around 80% of export earnings. Mali remains dependent on foreign aid.

Economic activity is largely confined to the riverine area irrigated by the Niger River; about 65% of Mali's land area is desert or semidesert. About 10% of the population is nomadic and about 80% of the labor force is engaged in farming and fishing. Industrial activity is concentrated

on processing farm commodities. The government subsidizes the production of cereals to decrease the country's dependence on imported foodstuffs and to reduce its vulnerability to food price shocks.

Mali is developing its iron ore extraction industry to diversify foreign exchange earnings away from gold, but the pace will depend on global price trends. Although the political coup in 2012 slowed Mali's growth, the economy has since bounced back, with GDP growth above 5% in 2014-17, although physical insecurity, high population growth, corruption, weak infrastructure, and low levels of human capital continue to constrain economic development. Higher rainfall helped to boost cotton output in 2017, and the country's 2017 budget increased spending more than 10%, much of which was devoted to infrastructure and agriculture. Corruption and political turmoil are strong downside risks in 2018 and beyond.

GDP (purchasing power parity):
$41.22 billion (2017 est.)
$39.1 billion (2016 est.)
$36.97 billion (2015 est.)
note: data are in 2017 dollars
country comparison to the world: 117

GDP (official exchange rate):
$15.37 billion (2017 est.)

GDP—real growth rate: 5.4% (2017 est.)
5.8% (2016 est.)
6.2% (2015 est.)
country comparison to the world: 35

GDP—per capita (PPP): $2,200 (2017 est.)
$2,100 (2016 est.)
$2,100 (2015 est.)
note: data are in 2017 dollars
country comparison to the world: 206

Gross national saving:
16.5% of GDP (2017 est.)
15.5% of GDP (2016 est.)
15.4% of GDP (2015 est.)
country comparison to the world: 126

GDP—composition, by end use:
household consumption: 82.9% (2017 est.)
government consumption: 17.4% (2017 est.)
investment in fixed capital: 19.3% (2017 est.)
investment in inventories: -0.7% (2017 est.)
exports of goods and services: 22.1% (2017 est.)
imports of goods and services: -41.1% (2017 est.)

GDP—composition, by sector of origin:
agriculture: 41.8% (2017 est.)
industry: 18.1% (2017 est.)
services: 40.5% (2017 est.)

Agriculture—products: cotton, millet, rice, corn, vegetables, peanuts; cattle, sheep, goats

Industries: food processing; construction; phosphate and gold mining

Industrial production growth rate: 6.3% (2017 est.)
country comparison to the world: 37

Labor force: 6.447 million (2017 est.)
country comparison to the world: 69

Labor force—by occupation: agriculture: 80%

industry and services: 20% (2005 est.)

Unemployment rate: 7.9% (2017 est.)

7.8% (2016 est.)
country comparison to the world: 122

Population below poverty line: 36.1% (2005 est.)

Household income or consumption by percentage share: *lowest 10%:* 3.5%
highest 10%: 25.8% (2010 est.)

Budget: *revenues:* 3.075 billion (2017 est.)
expenditures: 3.513 billion (2017 est.)

Taxes and other revenues: 20% (of GDP) (2017 est.)
country comparison to the world: 153

Budget surplus (+) or deficit (-): -2.9% (of GDP) (2017 est.)
country comparison to the world: 128

Public debt: 35.4% of GDP (2017 est.)
36% of GDP (2016 est.)
country comparison to the world: 150

Fiscal year: calendar year

Inflation rate (consumer prices): 1.8% (2017 est.)
-1.8% (2016 est.)
country comparison to the world: 94

Current account balance: -$886 million (2017 est.)
-$1.015 billion (2016 est.)
country comparison to the world: 141

Exports: $3.06 billion (2017 est.)
$2.803 billion (2016 est.)
country comparison to the world: 129

Exports—partners: Switzerland 31.8%, UAE 15.4%, Burkina Faso 7.8%, Cote d'Ivoire 7.3%, South Africa 5%, Bangladesh 4.6% (2017)

Exports—commodities: cotton, gold, livestock

Imports: $3.644 billion (2017 est.)
$3.403 billion (2016 est.)
country comparison to the world: 142

Imports—commodities: petroleum, machinery and equipment, construction materials, foodstuffs, textiles

Imports—partners: Senegal 24.4%, China 13.2%, Cote d'Ivoire 9%, France 7.3% (2017)

Reserves of foreign exchange and gold: $647.8 million (31 December 2017 est.)
$395.7 million (31 December 2016 est.)
country comparison to the world: 144

Debt—external: $4.192 billion (31 December 2017 est.)
$3.981 billion (31 December 2016 est.)
country comparison to the world: 137

Exchange rates: Communaute Financiere Africaine francs (XOF) per US dollar—
605.3 (2017 est.)
593.1 (2016 est.)
593.1 (2015 est.)
591.45 (2014 est.)
494.42 (2013 est.)

ENERGY

Electricity access: *population without electricity:* 10 million (2019)
electrification—total population: 50% (2019)
electrification—urban areas: 78% (2019)
electrification—rural areas: 28% (2019)

Electricity—production: 2.489 billion kWh (2016 est.)
country comparison to the world: 136

Electricity—consumption: 2.982 billion kWh (2016 est.)
country comparison to the world: 136

Electricity—exports: 0 kWh (2016 est.)
country comparison to the world: 166

Electricity—imports: 800 million kWh (2016 est.)
country comparison to the world: 73

Electricity—installed generating capacity: 590,000 kW (2016 est.)
country comparison to the world: 140

Electricity—from fossil fuels: 68% of total installed capacity (2016 est.)
country comparison to the world: 113

Electricity—from nuclear fuels: 0% of total installed capacity (2017 est.)
country comparison to the world: 138

Electricity—from hydroelectric plants: 31% of total installed capacity (2017 est.)
country comparison to the world: 68

Electricity—from other renewable sources: 1% of total installed capacity (2017 est.)
country comparison to the world: 160

Crude oil—production: 0 bbl/day (2018 est.)
country comparison to the world: 170

Crude oil—exports: 0 bbl/day (2015 est.)
country comparison to the world: 161

Crude oil—imports: 0 bbl/day (2015 est.)
country comparison to the world: 161

Crude oil—proved reserves: 0 bbl (1 January 2018 est.)
country comparison to the world: 165

Refined petroleum products—production: 0 bbl/day (2015 est.)
country comparison to the world: 173

Refined petroleum products—consumption: 22,000 bbl/day (2016 est.)
country comparison to the world: 134

Refined petroleum products—exports: 0 bbl/day (2015 est.)
country comparison to the world: 178

Refined petroleum products—imports: 20,610 bbl/day (2015 est.)
country comparison to the world: 119

Natural gas—production: 0 cu m (2017 est.)
country comparison to the world: 166

Natural gas—consumption: 0 cu m (2017 est.)
country comparison to the world: 172

Natural gas—exports: 0 cu m (2017 est.)
country comparison to the world: 147

Natural gas—imports: 0 cu m (2017 est.)
country comparison to the world: 154

Natural gas—proved reserves: 0 cu m (1 January 2014 est.)
country comparison to the world: 167

Carbon dioxide emissions from consumption of energy: 3.388 million Mt (2017 est.)
country comparison to the world: 143

COMMUNICATIONS

Telephones—fixed lines: total subscriptions: 227,831
subscriptions per 100 inhabitants: 1.2 (2019 est.)
country comparison to the world: 119

Telephones—mobile cellular: total subscriptions: 21,850,850
subscriptions per 100 inhabitants: 115.09 (2019 est.)
country comparison to the world: 55

Telecommunication systems: *general assessment:* telecoms infrastructure is barely adequate in most town and not available in many areas of the country; geography is a challenge for telecommunications; poverty, security, high illiteracy and low PC use has taken its toll; 4 mobile operators in market; mobile penetration high and potential for mobile broadband service; local plans for Internet Exchange Point; as Mali is landlocked there is hope that neighboring countries will allow use of international bandwidth; G5 Sahel countries adopt free roaming measures; Chinese company Huawei attempts to build a national backbone network but security issues make this difficult (2020)
domestic: fixed-line subscribership 1 per 100 persons; mobile-cellular subscribership has increased sharply to over 115 per 100 persons; increasing use of local radio loops to extend network coverage to remote areas (2019)
international: country code—223; satellite communications center and fiber-optic links to neighboring countries; satellite earth stations—2 Intelsat (1 Atlantic Ocean, 1 Indian Ocean)
note: the COVID-19 outbreak is negatively impacting telecommunications production and supply chains globally; consumer spending on telecom devices and services has also slowed due to the pandemic's effect on economies worldwide; overall progress towards improvements in all facets of the telecom industry—mobile, fixed-line, broadband, submarine cable and satellite—has moderated

Broadcast media: national public TV broadcaster; 2 privately owned companies provide subscription services to foreign multi-channel TV packages; national public radio broadcaster supplemented by a large number of privately owned and community broadcast stations; transmissions of multiple international broadcasters are available (2019)

Internet country code: .ml

Internet users: total: 2,395,886
percent of population: 13% (July 2018 est.)
country comparison to the world: 111

Broadband—fixed subscriptions:
total: 120,934
subscriptions per 100 inhabitants: 1 less than 1 (2018 est.)
country comparison to the world: 119

TRANSPORTATION

National air transport system: *number of registered air carriers:* 0 (2020)

Civil aircraft registration country code prefix: TZ, TT (2016)

Airports: 25 (2013)
country comparison to the world: 127

Airports—with paved runways: *total:* 8 (2019)
over 3,047 m: 1
2,438 to 3,047 m: 4
1,524 to 2,437 m: 2
914 to 1,523 m: 1

Airports—with unpaved runways: *total:* 17 (2013)
1,524 to 2,437 m: 3 (2013)
914 to 1,523 m: 9 (2013)
under 914 m: 5 (2013)

Heliports: 2 (2013)

Railways: *total:* 593 km (2014)
narrow gauge: 593 km 1.000-m gauge (2014)
country comparison to the world: 110

Roadways: *total:* 139,107 km (2018)
country comparison to the world: 38

Waterways: 1,800 km (downstream of Koulikoro; low water levels on the River Niger cause problems in dry years; in the months before the rainy season the river is not navigable by commercial vessels) (2011)
country comparison to the world: 43

Ports and terminals: *river port(s):* Koulikoro (Niger)

MILITARY AND SECURITY

Military and security forces: Malian Armed Forces (FAMa): Army (Armee de Terre), Republic of Mali Air Force (Force Aerienne de la Republique du Mali, FARM); National Gendarmerie; National Guard (Garde National du Mali) (2019)
note(s): the Gendarmerie and the National Guard are under the authority of the Ministry of Defense and Veterans Affairs (Ministere De La Defense Et Des Anciens Combattants, MDAC), but operational control is shared between the MDAC and the Ministry of Internal Security and Civil Protection the Gendarmerie's primary mission is internal security and public order; its duties also include territorial defense, humanitarian operations, intelligence gathering, and protecting private property, mainly in rural areas the National Guard is a military force responsible for providing security to government facilities and institutions, prison service, public order, humanitarian operations, some border security, and intelligence gathering; it has special units on camels (the Camel Corps) for patrolling the deserts and borders of northern Mali

Military expenditures:
2.7% of GDP (2019)
2.9% of GDP (2018)
3% of GDP (2017)
2.6% of GDP (2016)
2.4% of GDP (2015)
country comparison to the world: 34

Military and security service personnel strengths: estimates for the size of the Malian Armed Forces (FAMa) vary; approximately 19,000 total troops

(13,000 Army; 800 Air Force; 3,000 Gendarmerie; 2,000 National Guard) (2019 est.)

Military equipment inventories and acquisitions: the FAMa's inventory consists primarily of Soviet-era equipment, although in recent years it has received limited quantities of mostly second-hand armaments from a variety of countries; since 2010, the leading suppliers have been Brazil, Bulgaria, France, Russia, South Africa, Spain, and the United Arab Emirates (2019 est.)

Military service age and obligation: 18 years of age for selective compulsory and voluntary military service (men and women); 2-year conscript service obligation (2014)

Military—note: prior to the August 2020 coup, the Malian military had intervened in the political arena at least five times since the country gained independence in 1960; two attempts failed (1976 and 1978), while three succeeded (1968, 1991, and 2012); the military collapsed in 2012 during the fighting against Tuareg rebels and Islamic militants

since 2017, the FAMa, along with other government security and paramilitary forces, has conducted multiple major operations against militants in the eastern, central, and northern parts of the country; up to 4,000 troops reportedly have been deployed; the stated objectives for the most recent operation (Operation Maliko in early 2020) was to end terrorist activity and restore government authority in seven of the country's 10 regions, including Mopti, Ségou, Gao, Kidal, Ménaka, Taoudénit, and Timbuktu

Mali is part of a five-nation anti-jihadist task force known as the G5 Sahel Group, set up in 2014 with Burkina Faso, Chad, Mauritania, and Niger; it has committed 1,100 troops and 200 gendarmes to the force; in early 2020, G5 Sahel military chiefs of staff agreed to allow defense forces from each of the states to pursue terrorist fighters up to 100 km into neighboring countries; the G5 force is backed by the UN, US, and France; G5 troops periodically conduct joint operations with French forces deployed to the Sahel under Operation Barkhane the United Nations Multidimensional Integrated Stabilization Mission in Mali (MINUSMA) has operated in the country since 2013; the Mission's responsibilities include providing security, rebuilding Malian security forces, supporting national political dialogue, and assisting in the reestablishment of Malian government authority; as of March 2020, MINUSMA had around 15,500 military, police, and civilian personnel deployed

the European Union Training Mission in Mali (EUTM-M) also has operated in the country since 2013; the EUTM-M provides advice and training to the Malian Armed Forces and military assistance to the G5 Sahel Joint Force; as of August 2020, the mission included more than 600 personnel from 28 European countries (2020)

TERRORISM

Terrorist group(s): Ansar al-Dine; Islamic State of Iraq and ash-Sham in the Greater Sahara; Jama'at

Nusrat al-Islam wal-Muslimin; al- Mulathamin Battalion (al-Mourabitoun) (2020)
note: details about the history, aims, leadership, organization, areas of operation, tactics, targets, weapons, size, and sources of support of the group(s) appear(s) in Appendix-T

TRANSNATIONAL ISSUES

Disputes—international: demarcation is underway with Burkina Faso

Refugees and internally displaced persons: *refugees (country of origin):* 16,938 (Niger), 15,316 (Mauritania), 12,890 (Burkina Faso) (2020)
IDPs: 287,496 (Tuareg rebellion since 2012) (2020)

Trafficking in persons: *current situation:* Mali is a source, transit, and destination country for men, women, and children subjected to forced labor and sex trafficking; internal trafficking is more prevalent than transnational trafficking, but foreign women and girls are forced into domestic servitude, agricultural labor, and support roles in gold mines, as well as subjected to sex trafficking; Malian boys are forced to work in agricultural settings, gold mines, the informal commercial sector and to beg within Mali and neighboring countries; Malians and other Africans who travel through Mali to Mauritania, Algeria, or Libya in hopes of reaching Europe are particularly at risk of becoming victims of human trafficking; men and boys, primarily of Songhai ethnicity, are subjected to debt bondage in the salt mines of Taoudenni in northern Mali; some members of Mali's Tamachek community are subjected to hereditary slavery-related practices; Malian women and girls are victims of sex trafficking in Gabon, Libya, Lebanon, and Tunisia; the recruitment of child soldiers by armed groups in northern Mali decreased

tier rating: Tier 2 Watch List—Mali does not fully comply with the minimum standards for the elimination of trafficking; however, it is making significant efforts to do so; in 2014, Mali was granted a waiver from an otherwise required downgrade to Tier 3 because its government has a written plan that, if implemented would constitute making significant efforts to bring itself into compliance with the minimum standards for the elimination of trafficking; officials failed to distribute the 2012 anti-trafficking law to judicial and law enforcement personnel, perpetuating a lack of understanding and awareness of the legislation; anti-trafficking law enforcement efforts decreased in 2014, with only one case investigated and no prosecutions or convictions; fewer victims were identified, and the government did not support the privately funded NGOs and international organizations it relied upon to provide victims with services; the government did not conduct any awareness-raising campaigns, workshops, or training sessions (2015)

MALTA

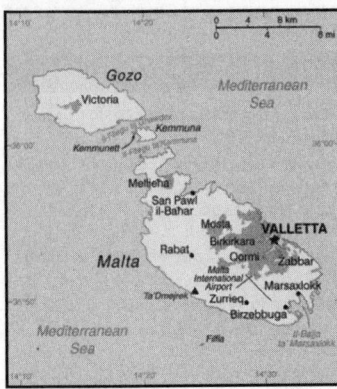

INTRODUCTION

Background: With a civilization that dates back thousands of years, Malta boasts some of the oldest megalithic sites in the world. Situated in the center of the Mediterranean, Malta's islands have long served as a strategic military asset, with the islands at various times having come under control of the Phoenicians, Carthaginians, Greeks, Romans, Byzantines, Moors, Normans, Sicilians, Spanish, Knights of St. John, and the French. Most recently a British colony (since 1814), Malta gained its independence in 1964 and declared itself a republic ten years later. While under British rule, the island staunchly supported the UK through both world wars. Since about the mid-1980s, the island has transformed itself into a freight transshipment point, a financial center, and a tourist destination while its key industries moved toward more service-oriented activities. Malta became an EU member in May 2004 and began using the euro as currency in 2008.

GEOGRAPHY

Location: Southern Europe, islands in the Mediterranean Sea, south of Sicily (Italy)

Geographic coordinates: 35 50 N, 14 35 E

Map references: Europe

Area: *total:* 316 sq km
land: 316 sq km
water: 0 sq km
country comparison to the world: 209

Area—comparative: slightly less than twice the size of Washington, DC

Land boundaries: 0 km

Coastline: 196.8 km (excludes 56 km for the island of Gozo)

Maritime claims: *territorial sea:* 12 nm
contiguous zone: 24 nm

continental shelf: 200-m depth or to the depth of exploitation
exclusive fishing zone: 25 nm

Climate: Mediterranean; mild, rainy winters; hot, dry summers

Terrain: mostly low, rocky, flat to dissected plains; many coastal cliffs

Elevation: *lowest point:* Mediterranean Sea 0 m
highest point: Ta'Dmejrek on Dingli Cliffs 253 m

Natural resources: limestone, salt, arable land

Land use: *agricultural land:* 32.3% (2011 est.)
arable land: 28.4% (2011 est.) / permanent crops: 3.9% (2011 est.) / permanent pasture: 0% (2011 est.)
forest: 0.9% (2011 est.)
other: 66.8% (2011 est.)
Irrigated land: 35 sq km (2012)

Population distribution: most of the population lives on the eastern half of Malta, the largest of the three inhabited islands

Natural hazards: occasional droughts

Environment—current issues: limited natural freshwater resources; increasing reliance on desalination; deforestation; wildlife preservation

Environment—international agreements: *party to:* Air Pollution, Biodiversity, Climate Change, Climate Change-Kyoto Protocol, Desertification, Endangered Species, Hazardous Wastes, Law of the Sea, Marine Dumping, Ozone Layer Protection, Ship Pollution, Wetlands
signed, but not ratified: none of the selected agreements

Geography—note: the country comprises an archipelago, with only the three largest islands (Malta, Ghawdex or Gozo, and Kemmuna or Comino) inhabited; numerous bays provide good harbors; Malta and Tunisia are discussing oil exploration on the continental shelf between their countries, although no commercially viable reserves have been found as of 2017

PEOPLE AND SOCIETY

Population: 457,267 (July 2020 est.)
country comparison to the world: 175

Nationality: *noun:* Maltese (singular and plural)
adjective: Maltese

Ethnic groups: Maltese (descendants of ancient Carthaginians and Phoenicians with strong elements of Italian and other Mediterranean stock)

Languages: Maltese (official) 90.1%, English (official) 6%, multilingual 3%, other 0.9% (2005 est.)

Religions: Roman Catholic (official) more than 90% (2006 est.)

Age structure: *0-14 years:* 14.38% (male 33,934/female 31,823)
15-24 years: 10.33% (male 24,445/female 22,811)
25-54 years: 41.1% (male 97,685/female 90,264)

55-64 years: 12.88% (male 29,533/female 29,353)
65 years and over: 21.3% (male 44,644/female 52,775) (2020 est.)

Dependency ratios: *total dependency ratio:* 55.5
youth dependency ratio: 22.4
elderly dependency ratio: 33.2
potential support ratio: 3 (2020 est.)

Median age: *total:* 42.3 years
male: 41.2 years
female: 43.5 years (2020 est.)
country comparison to the world: 35

Population growth rate: 0.87% (2020 est.)
country comparison to the world: 119

Birth rate: 9.9 births/1,000 population (2020 est.)
country comparison to the world: 192

Death rate: 8.3 deaths/1,000 population (2020 est.)
country comparison to the world: 79

Net migration rate: 6.6 migrant(s)/1,000 population (2020 est.)
country comparison to the world: 16

Population distribution: most of the population lives on the eastern half of Malta, the largest of the three inhabited islands

Urbanization: *urban population:* 94.7% of total population (2020)
rate of urbanization: 0.38% annual rate of change (2015-20 est.)
total population growth rate v. urban population growth rate, 2000-2030:

Major urban areas—population: 213,000 VALLETTA (capital) (2018)

Sex ratio: *at birth:* 1.04 male(s)/female
0-14 years: 1.07 male(s)/female
15-24 years: 1.07 male(s)/female
25-54 years: 1.08 male(s)/female
55-64 years: 1.01 male(s)/female
65 years and over: 0.85 male(s)/female
total population: 1.01 male(s)/female (2020 est.)

Mother's mean age at first birth: 28.9 years (2017 est.)
note: data refer to the average of the different childbearing ages of first-order births

Maternal mortality rate: 6 deaths/100,000 live births (2017 est.)
country comparison to the world: 161

Infant mortality rate: *total:* 4.6 deaths/1,000 live births
male: 4.5 deaths/1,000 live births
female: 4.7 deaths/1,000 live births (2020 est.)
country comparison to the world: 181

Life expectancy at birth: *total population:* 82.8 years
male: 80.7 years
female: 85 years (2020 est.)
country comparison to the world: 12

Total fertility rate: 1.49 children born/woman (2020 est.)

country comparison to the world: 205

Drinking water source:
improved:
urban: 100% of population
rural: 100% of population
total: 100% of population
unimproved:
urban: 0% of population
rural: 0% of population
total: 0% of population (2017 est.)

Current Health Expenditure: 9.3% (2017)

Physicians density: 2.86 physicians/1,000 population (2015)

Hospital bed density: 4.5 beds/1,000 population (2017)

Sanitation facility access:
improved:
urban: 100% of population
rural: 100% of population
total: 100% of population
unimproved:
urban: 0% of population
rural: 0% of population
total: 0% of population (2017 est.)

HIV/AIDS—adult prevalence rate: 0.1% (2016 est.)
country comparison to the world: 126

HIV/AIDS—people living with HIV/AIDS: <500 (2016 est.)

HIV/AIDS—deaths: <100 (2016 est.)

Obesity—adult prevalence rate: 28.9% (2016)
country comparison to the world: 28

Education expenditures: 5.3% of GDP (2015)
country comparison to the world: 50

Literacy: *definition:* age 15 and over can read and write
total population: 94.5%
male: 93%
female: 96% (2018)

School life expectancy (primary to tertiary education): *total:* 17 years
male: 16 years
female: 17 years (2018)

Unemployment, youth ages 15-24: *total:* 9.1%
male: 11.2%
female: 6.8% (2018 est.)
country comparison to the world: 133

GOVERNMENT

Country name: *conventional long form:* Republic of Malta
conventional short form: Malta
local long form: Repubblika ta' Malta
local short form: Malta
etymology: the ancient Greeks called the island "Melite" meaning "honey-sweet" from the Greek word "meli" meaning "honey" and referring to the island's honey production

Government type: parliamentary republic

Capital: *name:* Valletta

geographic coordinates: 35 53 N, 14 30 E

time difference: UTC+1 (6 hours ahead of Washington, DC, during Standard Time)
daylight saving time: +1hr, begins last Sunday in March; ends last Sunday in October
etymology: named in honor of Jean de Valette, the Grand Master of the Order of Saint John (crusader knights), who successfully led a defense of the island from an Ottoman invasion in 1565

Administrative divisions: 68 localities (Il-lokalita); Attard, Balzan, Birgu, Birkirkara, Birzebbuga, Bormla, Dingli, Fgura, Floriana, Fontana, Ghajnsielem, Gharb, Gharghur, Ghasri, Ghaxaq, Gudja, Gzira, Hamrun, Iklin, Imdina, Imgarr, Imqabba, Imsida, Imtarfa, Isla, Kalkara, Kercem, Kirkop, Lija, Luqa, Marsa, Marsaskala, Marsaxlokk, Mellieha, Mosta, Munxar, Nadur, Naxxar, Paola, Pembroke, Pieta, Qala, Qormi, Qrendi, Rabat, Rabat (Ghawdex), Safi, San Giljan/Saint Julian, San Gwann/Saint John, San Lawrenz/Saint Lawrence, Sannat, San Pawl il-Bahar/Saint Paul's Bay, Santa Lucija/Saint Lucia, Santa Venera/Saint Venera, Siggiewi, Sliema, Swieqi, Tarxien, Ta' Xbiex, Valletta, Xaghra, Xewkija, Xghajra, Zabbar, Zebbug, Zebbug (Ghawdex), Zejtun, Zurrieq

Independence: 21 September 1964 (from the UK)

National holiday: Independence Day, 21 September (1964); Republic Day, 13 December (1974)

Constitution: *history:* many previous; latest adopted 21 September 1964
amendments: proposals (Acts of Parliament) require at least two-thirds majority vote by the House of Representatives; passage of Acts requires majority vote by referendum, followed by final majority vote by the House and assent of the president of the republic; amended many times, last in 2016

Legal system: mixed legal system of English common law and civil law based on the Roman and Napoleonic civil codes; subject to European Union law

International law organization participation: accepts compulsory ICJ jurisdiction with reservations; accepts ICCt jurisdiction

Citizenship: *citizenship by birth:* no
citizenship by descent only: at least one parent must be a citizen of Malta
dual citizenship recognized: no
residency requirement for naturalization: 5 years

Suffrage: 18 years of age (16 in local council elections); universal

Executive branch: *chief of state:* President George VELLA (since 4 April 2019)
head of government: Prime Minister Robert ABELA (13 January 2020)
cabinet: Cabinet appointed by the president on the advice of the prime minister
elections/appointments: president indirectly elected by the House of Representatives for a single 5-year term; election last held on 2 April 2019 (next to be held by April 2024); following legislative elections, the leader of the majority party or majority coalition usually appointed prime

minister by the president for a 5-year term; deputy prime minister appointed by the president on the advice of the prime minister
election results: George VELLA (PL) elected president; House of Representatives vote—unanimous; Joseph MUSCAT (PL) reappointed prime minister

Legislative branch: *description:* unicameral House of Representatives or Il-Kamra Tad-Deputati, a component of the Parliament of Malta (normally 65 seats but can include at-large members; members directly elected in 5 multi-seat constituencies by proportional representation vote; members serve 5-year terms); note—the parliament elected in 2013 had 69 seats; an additional two seats were added in 2016 by the Constitutional Court to correct for mistakes made in the 2013 votecounting process
elections: last held on 3 June 2017 (next to be held in 2022); note—Prime Minister MUSCAT called for early elections amid corruption allegations
election results: percent of vote by party—PL 55%, PN 43.7%, other 1.3%; seats by party—PL 37 PN 30; note—PN was awarded two additional seats for a total of 30 in accordance with the proportionality provisions specified in the constitution; PD candidates ran under the PN list; composition—men 57, women 10, percent of women 14.9%

Judicial branch: *highest courts:* Court of Appeal (consists of either 1 or 3 judges); Constitutional Court (consists of 3 judges); Court of Criminal Appeal (consists of either 1 or 3 judges)
judge selection and term of office: Court of Appeal and Constitutional Court judges appointed by the president, usually upon the advice of the prime minister; judges of both courts serve until age 65
subordinate courts: Civil Court (divided into the General Jurisdiction Section, Family Section, and Voluntary Section); Criminal Court; Court of Magistrates; Gozo Courts (for the islands of Gozo and Comino)

Political parties and leaders: Democratic Party (Partit Demokratiku) or PD [Godfrey FARRUGIA]
Labor Party (Partit Laburista) or PL [Joseph MUSCAT]
Nationalist Party (Partit Nazzjonalista) or PN [Adrian DELIA]

International organization participation: Australia Group, C, CD, CE, EAPC, EBRD, ECB, EIB, EMU, EU, FAO, IAEA, IBRD, ICAO, ICC (NGOs), ICCt, ICRM, IDA, IFAD, IFC, IFRCS, ILO, IMF, IMO, IMSO, Interpol, IOC, IOM, IPU, ISO, ITSO, ITU, ITUC (NGOs), MIGA, NSG, OAS (observer), OPCW, OSCE, PCA, PFP, Schengen Convention, UN, UNCTAD, UNESCO, UNIDO, Union Latina (observer), UNWTO, UPU, WCO, WHO, WIPO, WMO, WTO

Diplomatic representation in the US: *chief of mission:* Ambassador Keith AZZOPARDI (since 17 September 2018)
chancery: 2017 Connecticut Avenue NW, Washington, DC 20008

telephone: [1] (202) 462-3611 through 3612
FAX: [1] (202) 387-5470

Diplomatic representation from the US: *chief of mission:* Ambassador (vacant); Charge d'Affaires Mark A. SCHAPIRO (since 29 September 2018)
telephone: [356] 2561-4000
embassy: Ta' Qali National Park, Attard, ATD 4000
mailing address: 5800 Valletta Place, Dulles, VA 20189
FAX: [356] 2561-4183

Flag description: two equal vertical bands of white (hoist side) and red; in the upper hoist-side corner is a representation of the George Cross, edged in red; according to legend, the colors are taken from the red and white checkered banner of Count Roger of Sicily who removed a bi-colored corner and granted it to Malta in 1091; an uncontested explanation is that the colors are those of the Knights of Saint John who ruled Malta from 1530 to 1798; in 1942, King George VI of the UK awarded the George Cross to the islanders for their exceptional bravery and gallantry in World War II; since independence in 1964, the George Cross bordered in red has appeared directly on the white field

National symbol(s): *Maltese eight-pointed cross; national colors:* red, white

National anthem: *name:* "L-Innu Malti" (The Maltese Anthem)
lyrics/music: Dun Karm PSAILA/Robert SAMMUT
note: adopted 1945; written in the form of a prayer
0:00 / 0:00

ECONOMY

Economy—overview: Malta's free market economy—the smallest economy in the euro-zone—relies heavily on trade in both goods and services, principally with Europe. Malta produces less than a quarter of its food needs, has limited fresh water supplies, and has few domestic energy sources. Malta's economy is dependent on foreign trade, manufacturing, and tourism. Malta joined the EU in 2004 and adopted the euro on 1 January 2008.

Malta has weathered the euro-zone crisis better than most EU member states due to a low debt-to-GDP ratio and financially sound banking sector. It maintains one of the lowest unemployment rates in Europe, and growth has fully recovered since the 2009 recession. In 2014 through 2016, Malta led the euro zone in growth, expanding more than 4.5% per year.

Malta's services sector continues to grow, with sustained growth in the financial services and online gaming sectors. Advantageous tax schemes remained attractive to foreign investors, though EU discussions of anti-tax avoidance measures have raised concerns among Malta's financial services and insurance providers, as the measures could have a significant impact on those sectors. The tourism sector also continued to grow, with 2016 showing record-breaking numbers of both air and cruise passenger arrivals.

Malta's GDP growth remains strong and is supported by a strong labor market. The government has implemented new programs, including free childcare, to encourage increased labor participation. The high cost of borrowing and small labor market remain potential constraints to future economic growth. Increasingly, other EU and European migrants are relocating to Malta for employment, though wages have remained low compared to other European countries. Inflation remains low.

GDP (purchasing power parity):
$19.26 billion (2017 est.)
$18.05 billion (2016 est.)
$17.16 billion (2015 est.)
note: data are in 2017 dollars
country comparison to the world: 152

GDP (official exchange rate):
$12.58 billion (2017 est.)

GDP—real growth rate: 4.94% (2019 est.)
5.17% (2018 est.)
8.03% (2017 est.)
country comparison to the world: 49

GDP—per capita (PPP): $41,900 (2017 est.)
$40,100 (2016 est.)
$39,000 (2015 est.)
note: data are in 2017 dollars
country comparison to the world: 43

Gross national saving:
33.5% of GDP (2017 est.)
31.8% of GDP (2016 est.)
31.2% of GDP (2015 est.)
country comparison to the world: 22

GDP—composition, by end use:
household consumption: 45.2% (2017 est.)
government consumption: 15.3% (2017 est.)
investment in fixed capital: 21.1% (2017 est.)
investment in inventories: 0.3% (2017 est.)
exports of goods and services: 136.1% (2017 est.)
imports of goods and services: -117.9% (2017 est.)

GDP—composition, by sector of origin:
agriculture: 1.1% (2017 est.)
industry: 10.2% (2017 est.)
services: 88.7% (2017 est.)

Agriculture—products: potatoes, cauliflower, grapes, wheat, barley, tomatoes, citrus, cut flowers, green peppers; pork, milk, poultry, eggs

Industries: tourism, electronics, ship building and repair, construction, food and beverages, pharmaceuticals, footwear, clothing, tobacco, aviation services, financial services, information technology services

Industrial production growth rate: -3.3% (2016 est.)
country comparison to the world: 189

Labor force: 223,000 (2019 est.)
country comparison to the world: 166

Labor force—by occupation: *agriculture:* 1.6%
industry: 20.7%
services: 77.7% (2016 est.)

Unemployment rate: 0.78% (2019 est.)
0.89% (2018 est.)

country comparison to the world: 5

Population below poverty line: 16.3% (2015 est.)

Household income or consumption by percentage share: *lowest 10%:* NA
highest 10%: NA

Budget: *revenues:* 5.076 billion (2017 est.)
expenditures: 4.583 billion (2017 est.)

Taxes and other revenues: 40.4% (of GDP) (2017 est.)
country comparison to the world: 38

Budget surplus (+) or deficit (-): 3.9% (of GDP) (2017 est.)
country comparison to the world: 9

Public debt: 50.7% of GDP (2017 est.)
56.3% of GDP (2016 est.)
note: Malta reports public debt at nominal value outstanding at the end of the year, according to guidelines set out in the Maastricht Treaty for general government gross debt; the data include the following categories of government liabilities (as defined in ESA95): currency and deposits (AF.2), securities other than shares excluding financial derivatives (AF.3, excluding AF.34), and loans (AF.4); general government comprises the central, state, and local governments, and social security funds
country comparison to the world: 99

Fiscal year: calendar year

Inflation rate (consumer prices): 1.3% (2017 est.)
0.9% (2016 est.)
country comparison to the world: 69

Current account balance: $1.561 billion (2019 est.)
$1.55 billion (2018 est.)
country comparison to the world: 46

Exports: $3.272 billion (2017 est.)
$2.493 billion (2016 est.)
country comparison to the world: 125

Exports—partners: Germany 17.3%, France 10.2%, Italy 9.4%, Singapore 5.9%, Hong Kong 5.8%, US 5.7%, Japan 4.9%, Libya 4.5% (2017)

Exports—commodities: machinery and mechanical appliances; mineral fuels, oils and petroleum products; pharmaceutical products; books and newspapers; aircraft/spacecraft and parts; toys, games, and sports equipment

Imports: $4.996 billion (2017 est.)
$4.965 billion (2016 est.)
country comparison to the world: 129

Imports—commodities: mineral fuels, oils and products; electrical machinery; aircraft/spacecraft and parts thereof; machinery and mechanical appliances; plastic and other semi-manufactured goods; vehicles and parts

Imports—partners: Italy 23%, Germany 7.9%, UK 7.7%, Spain 5%, Canada 4.5%, US 4.3%, France 4.2% (2017)

Reserves of foreign exchange and gold: $833 million (31 December 2017 est.)
$677.1 million (31 December 2016 est.)
country comparison to the world: 139

Debt—external: $90.98 billion (September 2016 est.)
$99.02 billion (31 December 2015 est.)
country comparison to the world: 53

Exchange rates: euros (EUR) per US dollar—
0.885 (2017 est.)
0.903 (2016 est.)
0.9214 (2015 est.)
0.885 (2014 est.)
0.7634 (2013 est.)

ENERGY

Electricity access: *electrification—total population:* 100% (2020)

Electricity—production: 813 million kWh (2016 est.)
country comparison to the world: 156

Electricity—consumption: 2.122 billion kWh (2016 est.)
country comparison to the world: 142

Electricity—exports: 0 kWh (2016 est.)
country comparison to the world: 167

Electricity—imports: 1.525 billion kWh (2016 est.)
country comparison to the world: 61

Electricity—installed generating capacity: 575,100 kW (2016 est.)
country comparison to the world: 142

Electricity—from fossil fuels: 81% of total installed capacity (2016 est.)
country comparison to the world: 80

Electricity—from nuclear fuels: 0% of total installed capacity (2017 est.)
country comparison to the world: 139

Electricity—from hydroelectric plants: 0% of total installed capacity (2017 est.)
country comparison to the world: 186

Electricity—from other renewable sources: 19% of total installed capacity (2017 est.)
country comparison to the world: 43

Crude oil—production: 0 bbl/day (2018 est.)
country comparison to the world: 171

Crude oil—exports: 0 bbl/day (2015 est.)
country comparison to the world: 162

Crude oil—imports: 0 bbl/day (2015 est.)
country comparison to the world: 162

Crude oil—proved reserves: 0 bbl (1 January 2018 est.)
country comparison to the world: 166

Refined petroleum products—production: 0 bbl/day (2017 est.)
country comparison to the world: 174

Refined petroleum products—consumption: 45,000 bbl/day (2016 est.)
country comparison to the world: 110

Refined petroleum products—exports: 10,400 bbl/day (2015 est.)
country comparison to the world: 81

Refined petroleum products—imports: 52,290 bbl/day (2015 est.)
country comparison to the world: 79

Natural gas—production: 0 cu m (2017 est.)
country comparison to the world: 167

Natural gas—consumption: 283.2 million cu m (2017 est.)
country comparison to the world: 102

Natural gas—exports: 0 cu m (2017 est.)
country comparison to the world: 148

Natural gas—imports: 311.5 million cu m (2017 est.)
country comparison to the world: 69

Natural gas—proved reserves: 0 cu m (1 January 2014 est.)
country comparison to the world: 168

Carbon dioxide emissions from consumption of energy: 8.141 million Mt (2017 est.)
country comparison to the world: 116

COMMUNICATIONS

Telephones—fixed lines: *total subscriptions:* 264,557
subscriptions per 100 inhabitants: 58.36 (2019 est.)
country comparison to the world: 115

Telephones—mobile cellular: *total subscriptions:* 653,414
subscriptions per 100 inhabitants: 144.14 (2019 est.)
country comparison to the world: 166

Telecommunication systems: *general assessment:* one of the most advanced telecoms in Europe, high penetration of mobile and broadband, and a way forward to expand e-commerce opportunities; stimulated by regulator measures to reduce consumer prices; extensive FttP network and investment in LTE and fiber thru 2023; launches 5G ready network (2020)

domestic: fixed-line 58 per 100 persons and mobile-cellular subscribership 144 per 100 persons; automatic system featuring submarine cable and microwave radio relay between islands (2019)

international: country code—356; landing points for the Malta-Gozo Cable, VMSCS, GO-1 Mediterranean Cable System, Malta Italy Interconnector, Melita-1, and the Italy-Malta submarine cable connections to Italy; satellite earth station—1 Intelsat (Atlantic Ocean) (2019)
note: the COVID-19 outbreak is negatively impacting telecommunications production and supply chains globally; consumer spending on telecom devices and services has also slowed due to the pandemic's effect on economies worldwide; overall progress towards improvements in all facets of the telecom industry—mobile, fixed-line, broadband, submarine cable and satellite—has moderated

Broadcast media: 2 publicly owned TV stations, Television Malta broadcasting nationally plus an educational channel; several privately owned national television stations, 2 of which are owned by political parties; Italian and British broadcast programs are available; multi-channel cable and satellite TV services are available; publicly owned radio broadcaster operates 3 stations; roughly 20 commercial radio stations (2019)

Internet country code: .mt

Internet users: *total:* 365,521

percent of population: 81.4% (July 2018 est.)
country comparison to the world: 161

Broadband—fixed subscriptions: *total:* 191,833
subscriptions per 100 inhabitants: 43 (2018 est.)
country comparison to the world: 110

TRANSPORTATION

National air transport system: *number of registered air carriers:* 13 (2020)
inventory of registered aircraft operated by air carriers: 180
annual passenger traffic on registered air carriers: 2,576,898 (2018)
annual freight traffic on registered air carriers: 5.14 million mt-km (2018)

Civil aircraft registration country code prefix: 9H (2016)

Airports: 1 (2013)
country comparison to the world: 228

Airports—with paved runways: *total:* 1 (2019)
over 3,047 m: 1

Heliports: 2 (2013)

Roadways: *total:* 2,254 km (2001)
paved: 1,973 km (2001)
unpaved: 281 km (2001)
urban: 1,422 km (2001)

non-urban: 832 km (2001)
country comparison to the world: 169

Merchant marine: *total:* 2,172
by type: bulk carrier 637, container ship 268, general cargo 253, oil tanker 412, other 602 (2019)
country comparison to the world: 11

Ports and terminals: *major seaport(s):* Marsaxlokk (Malta Freeport), Valletta
container port(s) (TEUs): Marsaxlokk (3,150,000) (2017)

MILITARY AND SECURITY

Military and security forces: Armed Forces of Malta (AFM, includes land, maritime, and air elements, plus a Volunteer Reserve Force) (2020)

Military expenditures: 0.6% of GDP (2019)
0.5% of GDP (2018)
0.5% of GDP (2017)
0.5% of GDP (2016)
0.5% of GDP (2015)
country comparison to the world: 144

Military and security service personnel strengths: the Armed Forces of Malta have approximately 2,000 active duty personnel (2019 est.)

Military equipment inventories and acquisitions: the small inventory of the Armed Forces of Malta consists of equipment from a mix of European countries, particularly Italy, and the US; since 2010, Italy and the US are the only providers of military equipment to Malta (2019 est.)

Military service age and obligation: 18-30 years of age for voluntary military service; no conscription (2019)

TRANSNATIONAL ISSUES

Disputes—international: none

Refugees and internally displaced persons: *stateless persons:* 11 (2018)
note: 7,256 estimated refugee and migrant arrivals by sea (January 2015-October 2020)

Illicit drugs: minor transshipment point for hashish from North Africa to Western Europe

MARSHALL ISLANDS

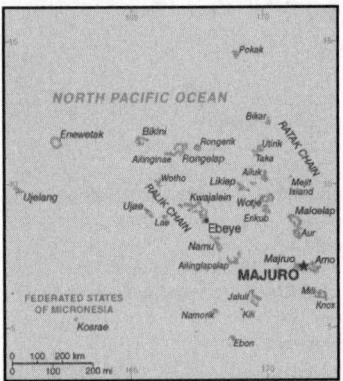

INTRODUCTION

Background: After almost four decades under US administration as the easternmost part of the UN Trust Territory of the Pacific Islands, the Marshall Islands attained independence in 1986 under a Compact of Free Association. Compensation claims continue as a result of US nuclear testing conducted on some of the atolls between 1947 and 1962 (67 tests total). The Marshall Islands hosts the US Army Kwajalein Atoll Reagan Missile Test Site, a key installation in the US missile defense network. Kwajalein also hosts one of four dedicated ground antennas that assist in the operation of the Global Positioning System (GPS) navigation system (the others are at Cape Canaveral, Florida (US), on Ascension (Saint Helena, Ascension, and Tristan da Cunha), and at Diego Garcia (British Indian Ocean Territory)).

GEOGRAPHY

Location: Oceania, consists of 29 atolls and five isolated islands in the North Pacific Ocean, about halfway between Hawaii and Australia; the atolls and islands are situated in two, almost-parallel island chains—the Ratak (Sunrise) group and the Ralik (Sunset) group; the total number of islands and islets is about 1,225; 22 of the atolls and four of the islands are uninhabited

Geographic coordinates: 9 00 N, 168 00 E

Map references: Oceania

Area: *total:* 181 sq km
land: 181 sq km
water: 0 sq km

note: the archipelago includes 11,673 sq km of lagoon waters and encompasses the atolls of Bikini, Enewetak, Kwajalein, Majuro, Rongelap, and Utirik
country comparison to the world: 217

Area—comparative: about the size of Washington, DC

Land boundaries: 0 km

Coastline: 370.4 km

Maritime claims: *territorial sea:* 12 nm
exclusive economic zone: 200 nm
contiguous zone: 24 nm

Climate: tropical; hot and humid; wet season May to November; islands border typhoon belt

Terrain: low coral limestone and sand islands

Elevation: *mean elevation:* 2 m
lowest point: Pacific Ocean 0 m
highest point: East-central Airik Island, Maloelap Atoll 14 m

Natural resources: coconut products, marine products, deep seabed minerals

Land use: *agricultural land:* 50.7% (2011 est.)
arable land: 7.8% (2011 est.) / *permanent crops:* 31.2% (2011 est.) / *permanent pasture:* 11.7% (2011 est.)
forest: 49.3% (2011 est.)
other: 0% (2011 est.)
Irrigated land: 0 sq km (2012)

Population distribution: most people live in urban clusters found on many of the country's islands; more than two-thirds of the population lives on the atolls of Majuro and Ebeye

Natural hazards: infrequent typhoons

Environment—current issues: inadequate supplies of potable water; pollution of Majuro lagoon from household waste and discharges from fishing vessels; sea level rise

Environment—international agreements: *party to:* Biodiversity, Climate Change, Climate Change-Kyoto Protocol, Desertification, Hazardous Wastes, Law of the Sea, Ozone Layer Protection, Ship Pollution, Wetlands, Whaling
signed, but not ratified: none of the selected agreements

Geography—note: the islands of Bikini and Enewetak are former US nuclear test sites; Kwajalein atoll, famous as a World War II battleground, surrounds the world's largest lagoon and is used as a US missile test range; the island city of Ebeye is the second largest settlement in the Marshall Islands, after the capital of Majuro, and

one of the most densely populated locations in the Pacific

PEOPLE AND SOCIETY

Population: 77,917 (July 2020 est.)
country comparison to the world: 200

Nationality: *noun:* Marshallese (singular and plural)
adjective: Marshallese

Ethnic groups: Marshallese 92.1%, mixed Marshallese 5.9%, other 2% (2006)

Languages: Marshallese (official) 98.2%, other languages 1.8% (1999 census)
note: English (official), widely spoken as a second language

Religions: Protestant 80.5% (United Church of Christ 47%, Assembly of God 16.2%, Bukot Nan Jesus 5.4%, Full Gospel 3.3%, Reformed Congressional Church 3%, Salvation Army 1.9%, Seventh Day Adventist 1.4%, Meram in Jesus 1.2%, other Protestant 1.1%), Roman Catholic 8.5%, Mormon 7%, Jehovah's Witness 1.7%, other 1.2%, none 1.1% (2011 est.)

Age structure: *0-14 years:* 32.94% (male 13,090/female 12,575)
15-24 years: 19.09% (male 7,568/female 7,308)
25-54 years: 37.35% (male 14,834/female 14,270)
55-64 years: 5.92% (male 2,269/female 2,341)
65 years and over: 4.7% (male 1,805/female 1,857) (2020 est.)

Median age: *total:* 23.8 years
male: 23.6 years
female: 23.9 years (2020 est.)
country comparison to the world: 173

Population growth rate: 1.43% (2020 est.)
country comparison to the world: 78

Birth rate: 22.8 births/1,000 population (2020 est.)
country comparison to the world: 61

Death rate: 4.3 deaths/1,000 population (2020 est.)
country comparison to the world: 210

Net migration rate: -4.5 migrant(s)/1,000 population (2020 est.)
country comparison to the world: 193

Population distribution: most people live in urban clusters found on many of the country's islands; more than two-thirds of the population lives on the atolls of Majuro and Ebeye

Urbanization: *urban population:* 77.8% of total population (2020)

rate of urbanization: 0.61% annual rate of change (2015-20 est.)

total population growth rate v. urban population growth rate, 2000-2030:

Major urban areas—population: 31,000 MAJURO (capital) (2018)

Sex ratio: *at birth:* 1.05 male(s)/female
0-14 years: 1.04 male(s)/female
15-24 years: 1.04 male(s)/female
25-54 years: 1.04 male(s)/female
55-64 years: 0.97 male(s)/female
65 years and over: 0.97 male(s)/female
total population: 1.03 male(s)/female (2020 est.)

Infant mortality rate: *total:* 17.4 deaths/1,000 live births
male: 19.7 deaths/1,000 live births
female: 15.1 deaths/1,000 live births (2020 est.)
country comparison to the world: 86

Life expectancy at birth: *total population:* 74.1 years
male: 71.8 years
female: 76.5 years (2020 est.)
country comparison to the world: 135

Total fertility rate: 2.86 children born/woman (2020 est.)
country comparison to the world: 58

Drinking water source:
improved:
urban: 99.8% of population
rural: 99.7% of population
total: 99.8% of population
unimproved:
urban: 0.2% of population
rural: 0.3% of population
total: 0.2% of population (2017 est.)

Current Health Expenditure: 16.4% (2017)

Physicians density: 0.42 physicians/1,000 population (2012)

Hospital bed density: 2.7 beds/1,000 population (2010)

Sanitation facility access:
improved:
urban: 96.3% of population
rural: 65.4% of population
total: 89.1% of population
unimproved:
urban: 15.5% of population
rural: 34.6% of population
total: 10.9% of population (2017 est.)

HIV/AIDS—adult prevalence rate: NA

HIV/AIDS—people living with HIV/AIDS: NA

HIV/AIDS—deaths: NA

Major infectious diseases: *degree of risk:* high (2020)
food or waterborne diseases: bacterial diarrhea
vectorborne diseases: malaria

Obesity—adult prevalence rate: 52.9% (2016)
country comparison to the world: 4

Children under the age of 5 years underweight: 11.9% (2017)
country comparison to the world: 53

Education expenditures: NA

Literacy: *definition:* age 15 and over can read and write
total population: 98.3%
male: 98.3%
female: 98.2% (2011)

School life expectancy (primary to tertiary education): *total:* 10 years
male: 10 years
female: 10 years (2019)

Unemployment, youth ages 15-24: *total:* 11%
male: 12.2%
female: 8.7% (2010 est.)
country comparison to the world: 118

GOVERNMENT

Country name: *conventional long form:* Republic of the Marshall Islands
conventional short form: Marshall Islands
local long form: Republic of the Marshall Islands
local short form: Marshall Islands
former: Trust Territory of the Pacific Islands, Marshall Islands District
abbreviation: RMI
etymology: named after British Captain John MARSHALL, who charted many of the islands in 1788

Government type: mixed presidential-parliamentary system in free association with the US

Capital: *name:* Majuro; note—the capital is an atoll of 64 islands; governmental buildings are housed on three fused islands on the eastern side of the atoll: Djarrit, Uliga, and Delap
geographic coordinates: 7 06 N, 171 23 E
time difference: UTC+12 (17 hours ahead of Washington, DC, during Standard Time)
etymology: Majuro means "two openings" or "two eyes" and refers to the two major northern passages through the atoll into the Majuro lagoon

Administrative divisions: 24 municipalities; Ailinglaplap, Ailuk, Arno, Aur, Bikini & Kili, Ebon, Enewetak & Ujelang, Jabat, Jaluit, Kwajalein, Lae, Lib, Likiep, Majuro, Maloelap, Mejit, Mili, Namorik, Namu, Rongelap, Ujae, Utrik, Wotho, Wotje

Independence: 21 October 1986 (from the US-administered UN trusteeship)

National holiday: Constitution Day, 1 May (1979)

Constitution: *history:* effective 1 May 1979
amendments: proposed by the National Parliament or by a constitutional convention; passage by Parliament requires at least two-thirds majority vote of the total membership in each of two readings and approval by a majority of votes in a referendum; amendments submitted by a constitutional convention require approval of at least two thirds of votes in a referendum; amended several times, last in 1995

Legal system: mixed legal system of US and English common law, customary law, and local statutes

International law organization participation: accepts compulsory ICJ jurisdiction with reservations; accepts ICCt jurisdiction

Citizenship: *citizenship by birth:* no
citizenship by descent only: at least one parent must be a citizen of the Marshall Islands
dual citizenship recognized: no
residency requirement for naturalization: 5 years

Suffrage: 18 years of age; universal

Executive branch: *chief of state:* President David KABUA (since 13 January 2020); note—the president is both chief of state and head of government

head of government: President David KABUA (since 13 January 2020)
cabinet: Cabinet nominated by the president from among members of the Nitijela, appointed by Nitijela speaker
elections/appointments: president indirectly elected by the Nitijela from among its members for a 4-year term (no term limits); election last held on 6 January 2020 (next to be held in 2024)
election results: David KABUA elected president; Parliament vote—David KABUA 20, Hilda C. HEINE 12

Legislative branch: *description:* bicameral National Parliament consists of: Council of Iroij, a 12-member group of tribal leaders advises the Presidential Cabinet and reviews legislation affecting customary law or any traditional practice); members appointed to serve 1-year terms
Nitijela (33 seats; members in 19 single- and 5 multi-seat constituencies directly elected by simple majority vote to serve 4-year terms); note—legislative power resides in the Nitijela
elections: last held on 18 November 2019 (next to be held by November 2023)
election results: percent of vote by party—NA; seats by party—independent 33

Judicial branch: *highest courts:* Supreme Court (consists of the chief justice and 2 associate justices)
judge selection and term of office: judges appointed by the Cabinet upon the recommendation of the Judicial Service Commission (consists of the chief justice of the High Court, the attorney general and a private citizen selected by the Cabinet) and upon approval of the Nitijela; the current chief justice, appointed in 2013, serves for 10 years; Marshallese citizens appointed as justices serve until retirement at age 72
subordinate courts: High Court; District Courts; Traditional Rights Court; Community Courts

Political parties and leaders: traditionally there have been no formally organized political parties; what has existed more closely resembles factions or interest groups because they do not have party headquarters, formal platforms, or party structures; the following two "groupings" have competed in legislative balloting in recent years—Aelon Kein Ad Party [Imata KABUA] and United Democratic Party or UDP [Litokwa TOMEING]

International organization participation: ACP, ADB, AOSIS, FAO, G-77, IAEA, IBRD, ICAO, ICCt, IDA, IFAD, IFC, ILO, IMF, IMO, IMSO,

Interpol, IOC, IOM, ITU, OPCW, PIF, Sparteca, SPC, UN, UNCTAD, UNESCO, WHO

Diplomatic representation in the US: *chief of mission:* Ambassador Gerald M. ZACKIOS (since 16 September 2016)
chancery: 2433 Massachusetts Avenue NW, 1st Floor, Washington, DC 20008
telephone: [1] (202) 234-5414
FAX: [1] (202) 232-3236
consulate(s) general: Honolulu, Springdale (AR)
consulate(s): Agana (Guam)

Diplomatic representation from the US: *chief of mission:* Ambassador Karen Brevard STEWART (since 25 July 2016)
telephone: [692] 247-4011
embassy: Oceanside, Mejen Weto, Long Island, Majuro
mailing address: P. O. Box 1379, Majuro, Republic of the Marshall Islands 96960-1379
FAX: [692] 247-4012

Flag description: blue with two stripes radiating from the lower hoist-side corner—orange (top) and white; a white star with four large rays and 20 small rays appears on the hoist side above the two stripes; blue represents the Pacific Ocean, the orange stripe signifies the Ralik Chain or sunset and courage, while the white stripe signifies the Ratak Chain or sunrise and peace; the star symbolizes the cross of Christianity, each of the 24 rays designates one of the electoral districts in the country and the four larger rays highlight the principal cultural centers of Majuro, Jaluit, Wotje, and Ebeye; the rising diagonal band can also be interpreted as representing the equator, with the star showing the archipelago's position just to the north

National symbol(s): *a 24-rayed star; national colors:* blue, white, orange

National anthem: *name:* Forever Marshall Islands
lyrics/music: Amata KABUA
note: adopted 1981
0:00/ 2:18

ECONOMY

Economy—overview: US assistance and lease payments for the use of Kwajalein Atoll as a US military base are the mainstay of this small island country. Agricultural production, primarily subsistence, is concentrated on small farms; the most important commercial crops are coconuts and breadfruit. Industry is limited to handicrafts, tuna processing, and copra. Tourism holds some potential. The islands and atolls have few natural resources, and imports exceed exports.

The Marshall Islands received roughly $1 billion in aid from the US during the period 1986-2001 under the original Compact of Free Association (Compact). In 2002 and 2003, the US and the Marshall Islands renegotiated the Compact's financial package for a 20-year period, 2004 to 2024. Under the amended Compact, the Marshall Islands will receive roughly $1.5 billion in direct US assistance. Under the amended Compact, the US and Marshall Islands are also

jointly funding a Trust Fund for the people of the Marshall Islands that will provide an income stream beyond 2024, when direct Compact aid ends.

GDP (purchasing power parity):
$196 million (2017 est.)
$191.3 million (2016 est.)
$184.6 million (2015 est.)
note: data are in 2017 dollars
country comparison to the world: 221

GDP (official exchange rate):
$222 million (2017 est.)

GDP—real growth rate: 2.5% (2017 est.)
3.6% (2016 est.)
2% (2015 est.)
country comparison to the world: 130

GDP—per capita (PPP): $3,600 (2017 est.)
$3,500 (2016 est.)
$3,400 (2015 est.)
note: data are in 2017 dollars
country comparison to the world: 186

GDP—composition, by end use:
government consumption: 50% (2016 est.)
investment in fixed capital: 17.8% (2016 est.)
investment in inventories: 0.2% (2016 est.)
exports of goods and services: 52.9% (2016 est.)
imports of goods and services: -102.3% (2016 est.)

GDP—composition, by sector of origin:
agriculture: 4.4% (2013 est.)
industry: 9.9% (2013 est.)
services: 85.7% (2013 est.)

Agriculture—products: coconuts, tomatoes, melons, taro, breadfruit, fruits; pigs, chickens

Industries: copra, tuna processing, tourism, craft items (from seashells, wood, and pearls)

Industrial production growth rate: NA

Labor force: 10,670 (2013 est.)
country comparison to the world: 217

Labor force—by occupation: *agriculture:* 11%
industry: 16.3%
services: 72.7% (2011 est.)

Unemployment rate: 36% (2006 est.)
30.9% (2000 est.)
country comparison to the world: 212

Population below poverty line: NA

Household income or consumption by percentage share: *lowest 10%:* NA
highest 10%: NA

Budget: *revenues:* 116.7 million (2013 est.)
expenditures: 113.9 million (2013 est.)

Taxes and other revenues: 52.6% (of GDP) (2013 est.)
country comparison to the world: 13

Budget surplus (+) or deficit (-): 1.3% (of GDP) (2013 est.)
country comparison to the world: 26

Public debt:
25.5% of GDP (2017 est.)
30% of GDP (2016 est.)
country comparison to the world: 173

Fiscal year: 1 October—30 September

Inflation rate (consumer prices): 0% (2017 est.)
-1.5% (2016 est.)
country comparison to the world: 11

Current account balance: -$1 million (2017 est.)
$15 million (2016 est.)
country comparison to the world: 68

Exports: $0 (2013 est.)
country comparison to the world: 223

Exports—commodities: copra cake, coconut oil, handicrafts, fish

Imports: $103.8 million (2016 est.)
$133.7 million (2013 est.)
country comparison to the world: 216

Imports—commodities: foodstuffs, machinery and equipment, fuels, beverages, tobacco

Debt—external: $97.96 million (2013 est.)
$87 million (2008 est.)
country comparison to the world: 193

Exchange rates: the US dollar is used

ENERGY

Electricity access: *electrification—total population:* 93.1% (2016)
electrification—urban areas: 94.6% (2016)
electrification—rural areas: 89.1% (2016)

Electricity—production: 650 million kWh (2016 est.)
country comparison to the world: 161

Electricity—consumption: 604.5 million kWh (2016 est.)
country comparison to the world: 167

Electricity—exports: 0 kWh (2016 est.)
country comparison to the world: 168

Electricity—imports: 0 kWh (2016 est.)
country comparison to the world: 172

Electricity—installed generating capacity: 52,000 kW (2016 est.)
country comparison to the world: 190

Electricity—from fossil fuels: 81% of total installed capacity (2016 est.)
country comparison to the world: 81

Electricity—from nuclear fuels: 0% of total installed capacity (2017 est.)
country comparison to the world: 140

Electricity—from hydroelectric plants: 19% of total installed capacity (2017 est.)
country comparison to the world: 90

Electricity—from other renewable sources: 0% of total installed capacity (2017 est.)
country comparison to the world: 201

Crude oil—production: 0 bbl/day (2017 est.)
country comparison to the world: 172

Crude oil—exports: 0 bbl/day (2015 est.)
country comparison to the world: 163

Crude oil—imports: 0 bbl/day (2015 est.)
country comparison to the world: 163

Crude oil—proved reserves: 0 bbl (1 January 2018 est.)
country comparison to the world: 167

Refined petroleum products—production: 0 bbl/day (2015 est.)
country comparison to the world: 175

Refined petroleum products—consumption: 2,000 bbl/day (2016 est.)
country comparison to the world: 195

Refined petroleum products—exports: 0 bbl/day (2015 est.)
country comparison to the world: 179

Refined petroleum products—imports: 2,060 bbl/day (2015 est.)
country comparison to the world: 190

Natural gas—production: 0 cu m (2017 est.)
country comparison to the world: 168

Natural gas—consumption: 0 cu m (2017 est.)
country comparison to the world: 173

Natural gas—exports: 0 cu m (2017 est.)
country comparison to the world: 149

Natural gas—imports: 0 cu m (2017 est.)
country comparison to the world: 155

Carbon dioxide emissions from consumption of energy: 293,700 Mt (2017 est.)
country comparison to the world: 192

COMMUNICATIONS

Telephones—fixed lines: *total subscriptions:* 3,172
subscriptions per 100 inhabitants: 4.13 (2019 est.)
country comparison to the world: 212

Telephones—mobile cellular: *total subscriptions:* 21,169
subscriptions per 100 inhabitants: 27.56 (2019 est.)
country comparison to the world: 213

Telecommunication systems: *general assessment:* some telecom infrastructure improvements made in recent years; modern services include fiber optic cable service, cellular, Internet, international calling, caller ID, and leased data circuits; the US Government, World Bank, UN and International Telecommunication Union (ITU), have aided in improvements and monetary aid to the islands telecom; mobile penetrations is around 30%; radio communication is especially vital to remote islands (2018)

domestic: Majuro Atoll and Ebeye and Kwajalein islands have regular, seven-digit, direct-dial telephones; other islands interconnected by high frequency radiotelephone (used mostly for government purposes) and mini-satellite telephones;

fixed-line 4 per 100 persons and mobile-cellular is 28 per 100 persons (2019)

international: country code—692; satellite earth stations—2 Intelsat (Pacific Ocean); US Government satellite communications system on Kwajalein
note: the COVID-19 outbreak is negatively impacting telecommunications production and supply chains globally; consumer spending on telecom devices and services has also slowed due to the pandemic's effect on economies worldwide; overall progress towards improvements in all facets of the telecom industry—mobile, fixed-line, broadband, submarine cable and satellite—has moderated

Broadcast media: no TV broadcast station; a cable network is available on Majuro with programming via videotape replay and satellite relays; 4 radio broadcast stations; American Armed Forces Radio and Television Service (AFRTS) provides satellite radio and television service to Kwajalein Atoll (2019)

Internet country code: .mh

Internet users: *total:* 29,290

percent of population: 38.7% (July 2018 est.)
country comparison to the world: 205

Broadband—fixed subscriptions: *total:* 1,000
subscriptions per 100 inhabitants: 1 (2017 est.)
country comparison to the world: 196

Communications—note: Kwajalein hosts one of four dedicated ground antennas that assist in the operation of the Global Positioning System (GPS) navigation system (the others are at Cape Canaveral, Florida (US), on Ascension (Saint Helena, Ascension, and Tristan da Cunha), and at Diego Garcia (British Indian Ocean Territory))

TRANSPORTATION

National air transport system: *number of registered air carriers:* 1 (2020)
inventory of registered aircraft operated by air carriers: 3
annual passenger traffic on registered air carriers: 24,313 (2018)
annual freight traffic on registered air carriers: 130,000 mt-km (2018)

Civil aircraft registration country code prefix: V7 (2016)

Airports: 15 (2013)
country comparison to the world: 147

Airports—with paved runways:
total: 4 (2017)
1,524 to 2,437 m: 3 (2017)
914 to 1,523 m: 1 (2017)

Airports—with unpaved runways:
total: 11 (2013)
914 to 1,523 m: 10 (2013)
under 914 m: 1 (2013)

Roadways: *total:* 2,028 km (2007)
paved: 75 km (2007)
unpaved: 1,953 km
country comparison to the world: 171

Merchant marine: *total:* 3,537
by type: bulk carrier 1,537, container ship 257, general cargo 66, oil tanker 856, other 821 (2019)
country comparison to the world: 6

Ports and terminals: *major seaport(s):* Enitwetak Island, Kwajalein, Majuro

MILITARY AND SECURITY

Military and security forces: no regular military forces; Marshall Islands Police Department (2019)

Military—note: defense is the responsibility of the US

TRANSNATIONAL ISSUES

Disputes—international: claims US territory of Wake Island

Trafficking in persons: *current situation:* The Marshall Islands is a source and destination country for Marshallese women and girls and women from East Asia subjected to sex trafficking; Marshallese and foreign women are forced into prostitution in businesses frequented by crew members of fishing and transshipping vessels that dock in Majuro; some Chinese women are recruited to the Marshall Islands with promises of legitimate work and are subsequently forced into prostitution

tier rating: Tier 3—The Marshall Islands do not fully comply with the minimum standards for the elimination of trafficking and is not making significant efforts to do so; the government made no anti-trafficking law enforcement efforts, including developing a written plan to combat trafficking; no new trafficking investigations were opened in 2014, and no prosecutions or convictions were made for the fourth consecutive year; no efforts were made to identify trafficking victims, especially among women in prostitution or men working on foreign fishing vessels in Marshallese waters, and no attempt was made to ensure their access to protective services; limited awareness-raising events were conducted by an international organization (2015)

MAURITANIA

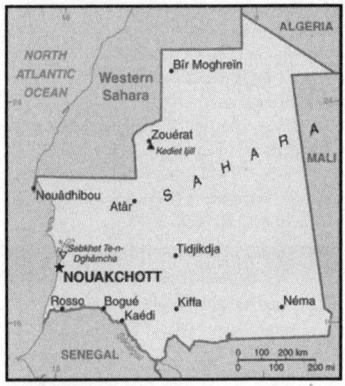

INTRODUCTION

Background: Berbers moved south into the area of today's Mauritania beginning in the 3rd century. Beginning in the 8th century, Mauritania experienced a slow but constant infiltration of Arabs and Arab influence from the north, pressing the Berbers, who resisted assimilation, to move farther south. One particular Arab group, the Bani Hassan, continued to migrate southward until, by the end of the 17th century, they dominated the entire country. Having finally been defeated, Berber groups turned to clericalism to regain a degree of ascendancy. At the bottom of the social structure were the slaves, subservient to both the Arabic warriors and Islamic Berber holy men. All of the social rivalries were fully exploited by the French as they colonized Mauritania in the late 19th century. Independent from France in 1960, Mauritania annexed the southern third of the former Spanish Sahara (now Western Sahara) in 1976 but relinquished it after three years of raids by the Polisario guerrilla front seeking independence for the territory. Maaouya Ould Sid Ahmed TAYA seized power in a coup in 1984 and ruled Mauritania with a heavy hand for more than two decades. A series of presidential elections that he held were widely seen as flawed. A bloodless coup in August 2005 deposed President TAYA and ushered in a military council that oversaw a transition to democratic rule. Independent candidate Sidi Ould Cheikh ABDALLAHI was inaugurated in April 2007 as Mauritania's first freely and fairly elected president. His term ended prematurely in August 2008 when a military junta led by General Mohamed Ould Abdel AZIZ deposed him and installed a military council government. AZIZ was subsequently elected president in 2009 and reelected in 2014 to a second and final term. He was replaced in 2019 by Mohamed Cheikh El GHAZOUANI. The country continues to experience ethnic tensions among three major groups: Arabic-speaking descendants of slaves

(Haratines), Arabic-speaking "White Moors" (Beydane), and members of Sub-Saharan ethnic groups mostly originating in the Senegal River valley (Halpulaar, Soninke, and Wolof).

Al-Qaeda in the Islamic Maghreb (AQIM) launched a series of attacks in Mauritania between 2005 and 2011, murdering American and foreign tourists and aid workers, attacking diplomatic and government facilities, and ambushing Mauritanian soldiers and gendarmes. A successful strategy against terrorism that combines dialogue with the terrorists and military actions has prevented the country from further terrorist attacks since 2011. However, AQIM and similar groups remain active in neighboring Mali and elsewhere in the Sahel region and continue to pose a threat to Mauritanians and foreign visitors.

GEOGRAPHY

Location: Western Africa, bordering the North Atlantic Ocean, between Senegal and Western Sahara

Geographic coordinates: 20 00 N, 12 00 W

Map references: Africa

Area: *total:* 1,030,700 sq km
land: 1,030,700 sq km
water: 0 sq km
country comparison to the world: 30

Area—comparative: slightly larger than three times the size of New Mexico; about six times the size of Florida

Land boundaries: *total:* 5,002 km
border countries (4): Algeria 460 km, Mali 2236 km, Senegal 742 km, Western Sahara 1564 km

Coastline: 754 km

Maritime claims: *territorial sea:* 12 nm
exclusive economic zone: 200 nm
contiguous zone: 24 nm
continental shelf: 200 nm or to the edge of the continental margin

Climate: desert; constantly hot, dry, dusty

Terrain: mostly barren, flat plains of the Sahara; some central hills

Elevation: *mean elevation:* 276 m
lowest point: Sebkhet Te-n-Dghamcha -5 m
highest point: Kediet Ijill 915 m

Natural resources: iron ore, gypsum, copper, phosphate, diamonds, gold, oil, fish

Land use: *agricultural land:* 38.5% (2011 est.)
arable land: 0.4% (2011 est.) / *permanent crops:* 0% (2011 est.) / *permanent pasture:* 38.1% (2011 est.)
forest: 0.2% (2011 est.)
other: 61.3% (2011 est.)
Irrigated land: 450 sq km (2012)

Population distribution: with most of the country being a desert, vast areas of the country, particularly in the central, northern, and eastern areas,

are without sizeable population clusters; half the population lives in or around the coastal capital of Nouakchott; smaller clusters are found near the southern border with Mali and Senegal as shown in this population distribution map

Natural hazards: hot, dry, dust/sand-laden sirocco wind primarily in March and April; periodic droughts

Environment—current issues: overgrazing, deforestation, and soil erosion aggravated by drought are contributing to desertification; limited natural freshwater resources away from the Senegal, which is the only perennial river; locust infestation

Environment—international agreements: *party to:* Biodiversity, Climate Change, Climate Change-Kyoto Protocol, Desertification, Endangered Species, Hazardous Wastes, Law of the Sea, Ozone Layer Protection, Ship Pollution, Wetlands, Whaling
signed, but not ratified: none of the selected agreements

Geography—note: Mauritania is considered both a part of North Africa's Maghreb region and West Africa's Sahel region; most of the population is concentrated in the cities of Nouakchott and Nouadhibou and along the Senegal River in the southern part of the country

PEOPLE AND SOCIETY

Population: 4,005,475 (July 2020 est.)
country comparison to the world: 128

Nationality: *noun:* Mauritanian(s)
adjective: Mauritanian

Ethnic groups: black Moors (Haratines—Arab-speaking slaves, former slaves, and their descendants of African origin, enslaved by white Moors) 40%, white Moors (of Arab-Berber descent, known as Beydane) 30%, Sub-Saharan Mauritanians (non-Arabic speaking, largely resident in or originating from the Senegal River Valley, including Halpulaar, Fulani, Soninke, Wolof, and Bambara ethnic groups) 30%

Languages: Arabic (official and national), Pular, Soninke, Wolof (all national languages), French
note: the spoken Arabic in Mauritania differs considerably from the modern standard Arabic used for official written purposes or in the media; the Mauritanian dialect, which incorporates many Berber words, is referred to as Hassaniya

Religions: Muslim (official) 100 %

Demographic profile: With a sustained total fertility rate of about 4 children per woman and almost 60% of the population under the age of 25, Mauritania's population is likely to continue growing for the foreseeable future. Mauritania's large youth cohort is vital to its development prospects, but available schooling does not adequately prepare students for the workplace. Girls continue to be underrepresented in the classroom, educational quality remains poor, and the dropout rate is high. The literacy rate is only about 50%, even though access to primary education has improved since the mid-2000s. Women's restricted access to education and discriminatory laws maintain gender inequality—worsened by early and forced marriages and female genital cutting.

The denial of education to black Moors also helps to perpetuate slavery. Although Mauritania abolished slavery in 1981 (the last country in the world to do so) and made it a criminal offense in 2007, the millenniums-old practice persists largely because anti-slavery laws are rarely enforced and the custom is so ingrained. According to a 2018 nongovernmental organization's report, a little more than 2% of Mauritania's population is enslaved, which includes individuals subjected to forced labor and forced marriage, although many thousands of individuals who are legally free contend with discrimination, poor education, and a lack of identity papers and, therefore, live in de facto slavery. The UN and international press outlets have claimed that up to 20% of Mauritania's population is enslaved, which would be the highest rate worldwide.

Drought, poverty, and unemployment have driven outmigration from Mauritania since the 1970s. Early flows were directed toward other West African countries, including Senegal, Mali, Cote d'Ivoire, and Gambia. The 1989 Mauritania-Senegal conflict forced thousands of black Mauritanians to take refuge in Senegal and pushed labor migrants toward the Gulf, Libya, and Europe in the late 1980s and early 1990s. Mauritania has accepted migrants from neighboring countries to fill labor shortages since its independence in 1960 and more recently has received refugees escaping civil wars, including tens of thousands of Tuaregs who fled Mali in 2012.

Mauritania was an important transit point for Sub-Saharan migrants moving illegally to North Africa and Europe. In the mid-2000s, as border patrols increased in the Strait of Gibraltar, security increased around Spain's North African enclaves (Ceuta and Melilla), and Moroccan border controls intensified, illegal migration flows shifted from the Western Mediterranean to Spain's Canary Islands. In 2006, departure points moved southward along the West African coast from Morocco and Western Sahara to Mauritania's two key ports (Nouadhibou and the capital Nouakchott), and illegal migration to the Canaries peaked at almost 32,000. The numbers fell dramatically in the following years because of joint patrolling off the West African coast by Frontex (the EU's border protection agency), Spain, Mauritania, and Senegal; the expansion of Spain's border surveillance system; and the 2008 European economic downturn.

Age structure: *0-14 years:* 37.56% (male 755,788/female 748,671)
15-24 years: 19.71% (male 387,140/female 402,462)
25-54 years: 33.91% (male 630,693/female 727,518)
55-64 years: 4.9% (male 88,888/female 107,201)

65 years and over: 3.92% (male 66,407/female 90,707) (2020 est.)

Dependency ratios: *total dependency ratio:* 75
youth dependency ratio: 69.5
elderly dependency ratio: 5.6
potential support ratio: 18 (2020 est.)

Median age: *total:* 21 years
male: 20.1 years
female: 22 years (2020 est.)
country comparison to the world: 188

Population growth rate: 2.09% (2020 est.)
country comparison to the world: 42

Birth rate: 29 births/1,000 population (2020 est.)
country comparison to the world: 36

Death rate: 7.5 deaths/1,000 population (2020 est.)
country comparison to the world: 105

Net migration rate: -0.8 migrant(s)/1,000 population (2020 est.)
country comparison to the world: 135

Population distribution: with most of the country being a desert, vast areas of the country, particularly in the central, northern, and eastern areas, are without sizeable population clusters; half the population lives in or around the coastal capital of Nouakchott; smaller clusters are found near the southern border with Mali and Senegal as shown in this population distribution map

Urbanization: *urban population:* 55.3% of total population (2020)
rate of urbanization: 4.28% annual rate of change (2015-20 est.)
total population growth rate v. urban population growth rate, 2000-2030:

Major urban areas—population:
1.315 million NOUAKCHOTT (capital) (2020)

Sex ratio: *at birth:* 1.03 male(s)/female
0-14 years: 1.01 male(s)/female
15-24 years: 0.96 male(s)/female
25-54 years: 0.87 male(s)/female
55-64 years: 0.83 male(s)/female
65 years and over: 0.73 male(s)/female
total population: 0.93 male(s)/female (2020 est.)

Maternal mortality rate: 766 deaths/100,000 live births (2017 est.)
country comparison to the world: 7

Infant mortality rate:
total: 47.9 deaths/1,000 live births
male: 52.5 deaths/1,000 live births
female: 43.1 deaths/1,000 live births (2020 est.)
country comparison to the world: 25

Life expectancy at birth: *total population:* 64.5 years
male: 62.1 years
female: 67 years (2020 est.)
country comparison to the world: 201

Total fertility rate: 3.65 children born/woman (2020 est.)
country comparison to the world: 37

Contraceptive prevalence rate: 17.8% (2015)

Drinking water source:
improved:

urban: 98.7% of population
rural: 68.4% of population
total: 84.4% of population
unimproved:
urban: 1.3% of population
rural: 31.6% of population
total: 15.6% of population (2017 est.)

Current Health Expenditure: 4.4% (2017)

Physicians density: 0.18 physicians/1,000 population (2017)

Sanitation facility access:
improved:
urban: 83.5% of population
rural: 25.2% of population
total: 56% of population
unimproved:
urban: 16.5% of population
rural: 74.8% of population
total: 44% of population (2017 est.)

HIV/AIDS—adult prevalence rate: 0.2% (2019 est.)
country comparison to the world: 105

HIV/AIDS—people living with HIV/AIDS: 5,700 (2019 est.)
country comparison to the world: 121

HIV/AIDS—deaths: <500 (2019 est.)

Major infectious diseases: *degree of risk:* very high (2020)
food or waterborne diseases: bacterial and protozoal diarrhea, hepatitis A, and typhoid fever
vectorborne diseases: malaria and dengue fever
animal contact diseases: rabies
respiratory diseases: meningococcal meningitis

Obesity—adult prevalence rate: 12.7% (2016)
country comparison to the world: 132

Children under the age of 5 years underweight: 19.2% (2018)
country comparison to the world: 24

Education expenditures: 2.6% of GDP (2016)
country comparison to the world: 155

Literacy: *definition:* age 15 and over can read and write
total population: 53.5%
male: 63.7%
female: 43.4% (2017)

School life expectancy (primary to tertiary education): *total:* 9 years
male: 9 years
female: 10 years (2019)

Unemployment, youth ages 15-24: *total:* 15.2%
male: 14.1%
female: 17% (2012 est.)
country comparison to the world: 90

GOVERNMENT

Country name: *conventional long form:* Islamic Republic of Mauritania
conventional short form: Mauritania
local long form: Al Jumhuriyah al Islamiyah al Muritaniyah
local short form: Muritaniyah

etymology: named for the ancient kingdom of Mauretania (3rd century B.C. to 1st century A.D.) and the subsequent Roman province (1st-7th centuries A.D.), which existed further north in present-day Morocco; the name derives from the Mauri (Moors), the Berber-speaking peoples of northwest Africa

Government type: presidential republic

Capital: *name:* Nouakchott

geographic coordinates: 18 04 N, 15 58 W

time difference: UTC 0 (5 hours ahead of Washington, DC, during Standard Time)

etymology: may derive from the Berber "nawak-shut" meaning "place of the winds"

Administrative divisions: 15 regions (wilayas, singular—wilaya); Adrar, Assaba, Brakna, Dakhlet Nouadhibou, Gorgol, Guidimaka, Hodh ech Chargui, Hodh El Gharbi, Inchiri, Nouakchott Nord, Nouakchott Ouest, Nouakchott Sud, Tagant, Tiris Zemmour, Trarza

Independence: 28 November 1960 (from France)

National holiday: Independence Day, 28 November (1960)

Constitution: *history:* previous 1964; latest adopted 12 July 1991

amendments: proposed by the president of the republic or by Parliament; consideration of amendments by Parliament requires approval of at least one third of the membership; a referendum is held only if the amendment is approved by two-thirds majority vote; passage by referendum requires simple majority vote by eligible voters; passage of amendments proposed by the president can bypass a referendum if approved by at least three-fifths majority vote by Parliament; amended many times, last in 2017 (by referendum)

Legal system: mixed legal system of Islamic and French civil law

International law organization participation: has not submitted an ICJ jurisdiction declaration; non-party state to the ICCt

Citizenship: *citizenship by birth:* no

citizenship by descent only: at least one parent must be a citizen of Mauritania

dual citizenship recognized: no

residency requirement for naturalization: 5 years

Suffrage: 18 years of age; universal

Executive branch: *chief of state:* President Mohamed Cheikh El GHAZOUANI (since 1 August 2019)

head of government: Prime Minister Mohamed Ould BILAL (since 6 August 2020)

cabinet: Council of Ministers—nominees suggested by the prime minister, appointed by the president

elections/appointments: president directly elected by absolute majority popular vote in 2 rounds if needed for a 5-year term (eligible for a second term); election last held on 22 June 2019 (next scheduled for 22 June 2024); prime minister appointed by the president

election results: Mohamed Cheikh El GHAZOUANI elected president in first round; percent of vote—Mahamed Cheikh El GHAZOUANI (UPR) 52%, Biram Dah Ould ABEID (independent) 18.6%, Sidi Mohamed Ould BOUBACAR (independent) 17.9%, other 11.55%

Legislative branch: *description:* unicameral Parliament or Barlamane consists of the National Assembly or Al Jamiya Al Wataniya (157 seats; 113 members in single- and multi-seat constituencies directly elected by a combination of plurality and proportional representation voting systems, 40 members in a single, nationwide constituency directly elected by proportional representation vote, and 4 members directly elected by the diaspora; all members serve 5-year terms)

elections: first held as the unicameral National Assembly in 2 rounds on 1 and 15 September 2018 (next to be held in 2023)

election results: National Assembly—percent of vote by party—NA; seats by party—NA; composition—NA

note: a referendum held in August 2017 approved a constitutional amendment to change the Parliament structure from bicameral to unicameral by abolishing the Senate and creating Regional Councils for local development

Judicial branch: *highest courts:* Supreme Court or Cour Supreme (subdivided into 7 chambers: 2 civil, 2 labor, 1 commercial, 1 administrative, and 1 criminal, each with a chamber president and 2 councilors); Constitutional Council (consists of 6 members)

judge selection and term of office: Supreme Court president appointed by the president of the republic to serve a 5- year renewable term; Constitutional Council members appointed—3 by the president of the republic, 2 by the president of the National Assembly, and 1 by the president of the Senate; members serve single, 9-year terms with one-third of membership renewed every 3 years

subordinate courts: Courts of Appeal; courts of first instance or wilya courts are established in the regions' headquarters and include commercial and labor courts, criminal courts, Moughataa (district) Courts, and informal/customary courts

Political parties and leaders:
Alliance for Justice and Democracy/Movement for Renewal or AJD/MR [Ibrahima Moctar SARR]
Burst of Youth for the Nation [Lalla Mint CHERIF]
Coalition of Majority Parties or CPM (includes UPR, UDP)
El Karama Party [Cheikhna Ould Mohamed Ould HAJBOU]
El Vadila Party [Ethmane Ould Ahmed ABOULMAALY]
National Forum for Democracy and Unity or FNDU [Mohamed Ould MAOLOUD] (coalition of hard-line opposition parties, includes RNRD-TAWASSOUL)
National Rally for Reform and Development or RNRD-TAWASSOUL [Mohamed Mahmoud Ould SEYIDI]
Party of Unity and Development or PUD [Mohamed BARO]
Popular Progressive Alliance or APP [Messaoud Ould BOULKHEIR]
Rally of Democratic Forces or RFD [Ahmed Ould DADDAH]
Ravah Party [Mohamed Ould VALL]
Republican Party for Democracy and Renewal or PRDR [Mintata Mint HEDEID]
Union for Democracy and Progressor UDP [Naha Mint MOUKNASS]
Union of Progress Forces [Mohamed Ould MAOULOUD]
Union for the Republic or UPR [Seyidna Ali Ould MOHAMED KHOUNA]

International organization participation: ABEDA, ACP, AfDB, AFESD, AMF, AMU, AU, CAEU (candidate), EITI (compliant country), FAO, G-77, IAEA, IBRD, ICAO, ICC (NGOs), ICRM, IDA, IDB, IFAD, IFC, IFRCS, IHO (pending member), ILO, IMF, IMO, Interpol, IOC, IOM, IPU, ISO (correspondent), ITSO, ITU, ITUC (NGOs), LAS, MIGA, MIUSMA, NAM, OIC, OIF, OPCW, UN, UNCTAD, UNESCO, UNIDO, UNWTO, UPU, WCO, WHO, WIPO, WMO, WTO

Diplomatic representation in the US: *chief of mission:* Ambassador Mohamedoun DADDAH (since 27 June 2016)

chancery: 2129 Leroy Place NW, Washington, DC 20008

telephone: [1] (202) 232-5700 through 5701

FAX: [1] (202) 319-2623

Diplomatic representation from the US: *chief of mission:* Ambassador Michael J. DODMAN (since 5 January 2018)

telephone: [222] 4525-2660 or [222] 2660-2663

embassy: Avenue Al Quds, Nouadhibou, Nouadhibou Road, Nouakchott

mailing address: use embassy street address

FAX: [222] 4525-1592

Flag description: green with a yellow, five-pointed star between the horns of a yellow, upward-pointing crescent moon; red stripes along the top and bottom edges; the crescent, star, and color green are traditional symbols of Islam; green also represents hope for a bright future; the yellow color stands for the sands of the Sahara; red symbolizes the blood shed in the struggle for independence

National symbol(s): five-pointed star between the horns of a horizontal crescent moon; national colors: green, yellow

National anthem: *name:* «Hymne National de la Republique Islamique de Mauritanie» (National Anthem of the Islamic Republic of Mauritania)

lyrics/music: Baba Ould CHEIKH/traditional, arranged by Tolia NIKIPROWETZKY

note: adopted 1960; the unique rhythm of the Mauritanian anthem makes it particularly challenging to sing; Mauritania in November 2017 adopted a new national anthem, "Bilada-l ubati-l hudati-l kiram" (The Country of Fatherhood is the Honorable Gift) composed by Rageh Daoud (sound file of the new anthem is forthcoming)

ECONOMY

Economy—overview: Mauritania's economy is dominated by extractive industries (oil and mines), fisheries, livestock, agriculture, and services. Half the population still depends on farming and raising livestock, even though many nomads and subsistence farmers were forced into the cities by recurrent droughts in the 1970s, 1980s, 2000s, and 2017. Recently, GDP growth has been driven largely by foreign investment in the mining and oil sectors.

Mauritania's extensive mineral resources include iron ore, gold, copper, gypsum, and phosphate rock, and exploration is ongoing for tantalum, uranium, crude oil, and natural gas. Extractive commodities make up about three-quarters of Mauritania's total exports, subjecting the economy to price swings in world commodity markets. Mining is also a growing source of government revenue, rising from 13% to 30% of total revenue from 2006 to 2014. The nation's coastal waters are among the richest fishing areas in the world, and fishing accounts for about 15% of budget revenues, 45% of foreign currency earnings. Mauritania processes a total of 1,800,000 tons of fish per year, but overexploitation by foreign and national fleets threaten the sustainability of this key source of revenue.

The economy is highly sensitive to international food and extractive commodity prices. Other risks to Mauritania's economy include its recurring droughts, dependence on foreign aid and investment, and insecurity in neighboring Mali, as well as significant shortages of infrastructure, institutional capacity, and human capital. In December 2017, Mauritania and the IMF agreed to a three year agreement under the Extended Credit Facility to foster economic growth, maintain macroeconomic stability, and reduce poverty. Investment in agriculture and infrastructure are the largest components of the country's public expenditures.

GDP (purchasing power parity):
$17.28 billion (2017 est.)
$16.7 billion (2016 est.)
$16.4 billion (2015 est.)
note: data are in 2017 dollars
country comparison to the world: 154

GDP (official exchange rate):
$4.935 billion (2017 est.)

GDP—real growth rate: 3.5% (2017 est.)
1.8% (2016 est.)
0.4% (2015 est.)
country comparison to the world: 85

GDP—per capita (PPP): $4,500 (2017 est.)
$4,400 (2016 est.)
$4,400 (2015 est.)
note: data are in 2017 dollars
country comparison to the world: 173

Gross national saving:
24.2% of GDP (2017 est.)
24.8% of GDP (2016 est.)
19% of GDP (2015 est.)
country comparison to the world: 68

GDP—composition, by end use:
household consumption: 64.9% (2017 est.)
government consumption: 21.8% (2017 est.)
investment in fixed capital: 56.1% (2017 est.)
investment in inventories: -3.2% (2017 est.)
exports of goods and services: 39% (2017 est.)
imports of goods and services: -78.6% (2017 est.)

GDP—composition, by sector of origin:
agriculture: 27.8% (2017 est.)
industry: 29.3% (2017 est.)
services: 42.9% (2017 est.)

Agriculture—products: dates, millet, sorghum, rice, corn; cattle, camel and sheep

Industries: fish processing, oil production, mining (iron ore, gold, copper)
note: gypsum deposits have never been exploited

Industrial production growth rate: 1% (2017 est.)
country comparison to the world: 157

Labor force: 1.437 million (2017 est.)
country comparison to the world: 129

Labor force—by occupation: *agriculture:* 50%
industry: 1.9%
services: 48.1% (2014 est.)

Unemployment rate: 10.2% (2017 est.)
10.1% (2016 est.)
country comparison to the world: 149

Population below poverty line: 31% (2014 est.)

Household income or consumption by percentage share: *lowest 10%:* 2.5%
highest 10%: 29.5% (2000)

Budget: *revenues:* 1.354 billion (2017 est.)
expenditures: 1.396 billion (2017 est.)

Taxes and other revenues: 27.4% (of GDP) (2017 est.)
country comparison to the world: 100

Budget surplus (+) or deficit (-): -0.8% (of GDP) (2017 est.)
country comparison to the world: 69

Public debt: 96.6% of GDP (2017 est.)
100% of GDP (2016 est.)
country comparison to the world: 21

Fiscal year: calendar year

Inflation rate (consumer prices): 2.3% (2017 est.)
1.5% (2016 est.)
country comparison to the world: 117

Current account balance: -$711 million (2017 est.)
-$707 million (2016 est.)
country comparison to the world: 133

Exports: $1.722 billion (2017 est.)
$1.401 billion (2016 est.)
country comparison to the world: 147

Exports—partners: China 31.2%, Switzerland 14.4%, Spain 10.1%, Germany 8.2%, Japan 8.1% (2017)

Exports—commodities: iron ore, fish and fish products, livestock, gold, copper, crude oil

Imports: $2.094 billion (2017 est.)
$1.9 billion (2016 est.)
country comparison to the world: 166

Imports—commodities: machinery and equipment, petroleum products, capital goods, foodstuffs, consumer goods

Imports—partners: Belgium 11.5%, UAE 11.3%, US 9.2%, China 7.5%, France 7.4%, Netherlands 6.1%, Morocco 6%, Slovenia 4.8%, Vanuatu 4.7%, Spain 4.7% (2017)

Reserves of foreign exchange and gold:
$875 million (31 December 2017 est.)
$849.3 million (31 December 2016 est.)
country comparison to the world: 138

Debt—external:
$4.15 billion (31 December 2017 est.)
$3.899 billion (31 December 2016 est.)
country comparison to the world: 138

Exchange rates: ouguiyas (MRO) per US dollar—
363.6 (2017 est.)
352.37 (2016 est.)
352.37 (2015 est.)
319.7 (2014 est.)
299.5 (2013 est.)

ENERGY

Electricity access: *population without electricity:* 3 million (2019)
electrification—total population: 32% (2019)
electrification—urban areas: 56% (2019)
electrification—rural areas: 4% (2019)

Electricity—production: 1.139 billion kWh (2016 est.)
country comparison to the world: 147

Electricity—consumption: 1.059 billion kWh (2016 est.)
country comparison to the world: 154

Electricity—exports: 0 kWh (2016 est.)
country comparison to the world: 169

Electricity—imports: 0 kWh (2016 est.)
country comparison to the world: 173

Electricity—installed generating capacity: 558.000 kW (2016 est.)
country comparison to the world: 144

Electricity—from fossil fuels: 65% of total installed capacity (2016 est.)
country comparison to the world: 118

Electricity—from nuclear fuels: 0% of total installed capacity (2017 est.)
country comparison to the world: 141

Electricity—from hydroelectric plants: 16% of total installed capacity (2017 est.)
country comparison to the world: 100

Electricity—from other renewable sources: 20% of total installed capacity (2017 est.)
country comparison to the world: 39

Crude oil—production: 4.000 bbl/day (2018 est.)
country comparison to the world: 81

Crude oil—exports: 5,333 bbl/day (2015 est.)
country comparison to the world: 65

Crude oil—imports: 0 bbl/day (2015 est.)
country comparison to the world: 164

Crude oil—proved reserves: 20 million bbl (1 January 2018 est.)

country comparison to the world: 83

Refined petroleum products—production: 0 bbl/day (2015 est.)
country comparison to the world: 176

Refined petroleum products—consumption: 17.000 bbl/day (2016 est.)
country comparison to the world: 150

Refined petroleum products—exports: 0 bbl/day (2015 est.)
country comparison to the world: 180

Refined petroleum products—imports: 17,290 bbl/day (2015 est.)
country comparison to the world: 133

Natural gas—production: 0 cu m (2017 est.)
country comparison to the world: 169

Natural gas—consumption: 0 cu m (2017 est.)
country comparison to the world: 174

Natural gas—exports: 0 cu m (2017 est.)
country comparison to the world: 150

Natural gas—imports: 0 cu m (2017 est.)
country comparison to the world: 156

Natural gas—proved reserves: 28.32 billion cu m (1 January 2018 est.)
country comparison to the world: 70

Carbon dioxide emissions from consumption of energy: 2.615 million Mt (2017 est.)
country comparison to the world: 152

COMMUNICATIONS

Telephones—fixed lines: total subscriptions: 53,742
subscriptions per 100 inhabitants: 1.37 (2019 est.)
country comparison to the world: 158

Telephones—mobile cellular: total subscriptions: 4,083,199
subscriptions per 100 inhabitants: 104.09 (2019 est.)
country comparison to the world: 130

Telecommunication systems: *general assessment:* limited system of cable and open-wire lines, minor microwave radio relay links, and radiotelephone communications stations; mobile-cellular services expanding; 3 mobile network operators; monopolies and little stimulus for competition; 3G penetration high yet little development in LTE and consequently mobile broadband access speeds are low; World Bank and European Investment Bank support attempts to improve telecom and improve regulatory measures; regulator struggles to enforce good quality of service; efforts to improve backbone of network (2020)
domestic: fixed-line teledensity 1 per 100 persons; mobile-cellular network coverage extends mainly to urban areas with a teledensity of roughly 104 per 100 persons; mostly cable and open-wire lines; a domestic satellite telecommunications system links Nouakchott with regional capitals (2019)
international: country code—222; landing point for the ACE submarine cable for connectivity to 19 West African countries and 2 European

countries; satellite earth stations—3 (1 Intelsat—Atlantic Ocean, 2 Arabsat) (2019)
note: the COVID-19 outbreak is negatively impacting telecommunications production and supply chains globally; consumer spending on telecom devices and services has also slowed due to the pandemic's effect on economies worldwide; overall progress towards improvements in all facets of the telecom industry—mobile, fixed-line, broadband, submarine cable and satellite—has moderated

Broadcast media: 10 TV stations: 5 government-owned and 5 private; in October 2017, the government suspended all private TV stations due to non-payment of broadcasting fees; as of April 2018, only one private TV station was broadcasting, Al Mourabitoune, the official TV of the Mauritanian Islamist party, Tewassoul; the other stations are negotiating payment options with the government and hope to be back on the air soon; 18 radio broadcasters: 15 government-owned, 3 (Radio Nouakchott Libre, Radio Tenwir, Radio Kobeni) private; all 3 private radio stations broadcast from Nouakchott; of the 15 government stations, 3 broadcast from Nouakchott (Radio Mauritanie, Radio Jeunesse, Radio Koran) and the other 12 broadcast from each of the 12 regions outside Nouakchott; Radio Jeunesse and Radio Koran are now also being rebroadcast in the regions (2019)

Internet country code: .mr

Internet users: *total:* 798,809

percent of population: 20.8% (July 2018 est.)
country comparison to the world: 145

Broadband—fixed subscriptions: *total:* 13,222
subscriptions per 100 inhabitants: less than 1 (2018 est.)
country comparison to the world: 163

TRANSPORTATION

National air transport system: *number of registered air carriers:* 1 (2020)
inventory of registered aircraft operated by air carriers: 6
annual passenger traffic on registered air carriers: 454,435 (2018)

Civil aircraft registration country code prefix: 5T (2016)

Airports: 30 (2013)
country comparison to the world: 114

Airports—with paved runways: *total:* 9 (2017)
2,438 to 3,047 m: 5 (2017)
1,524 to 2,437 m: 4 (2017)

Airports—with unpaved runways: *total:* 21 (2013)
2,438 to 3,047 m: 1 (2013)
1,524 to 2,437 m: 10 (2013)
914 to 1,523 m: 8 (2013)
under 914 m: 2 (2013)

Railways: *total:* 728 km (2014)
standard gauge: 728 km 1.435-m gauge (2014)
country comparison to the world: 100

Roadways:
total: 12,253 km (2018)
paved: 3,988 km (2018)

unpaved: 8,265 km (2018)
country comparison to the world: 131

Waterways: (some navigation possible on the Senegal River) (2011)

Merchant marine: *total:* 5
by type: general cargo 2, other 3 (2019)
country comparison to the world: 165

Ports and terminals: *major seaport(s):* Nouadhibou, Nouakchott

MILITARY AND SECURITY

Military and security forces: *Mauritanian Armed Forces:* Army, Mauritanian Navy (Marine Mauritanienne), Islamic Republic of Mauritania Air Group (Groupement Aerienne Islamique de Mauritanie, GAIM); Ministry of Interior: Gendarmerie, National Guard (2019)

Military expenditures:
2.8% of GDP (2019)
3% of GDP (2018)
2.9% of GDP (2017)
2.9% of GDP (2016)
2.8% of GDP (2015)
country comparison to the world: 31

Military and security service personnel strengths: the Mauritanian Armed Forces have approximately 16,000 active personnel (15,000 Army; 700 Navy; 300 Air Force); est. 3,000 Gendarmerie; est. 2,000 National Guard) (2019)

Military equipment inventories and acquisitions: the Mauritanian Armed Forces' inventory is limited and made up largely of older French and Soviet-era equipment; since 2010, Mauritania has received mostly secondhand military equipment from a variety of suppliers, including Brazil, China, France, and Turkey (2019 est.)

Military deployments: 450 Central African Republic (MINUSCA) (2020)

Military service age and obligation: 18 is the legal minimum age for voluntary military service; no conscription (2012)

Military—note: since a spate of terrorist attacks in the 2000s, including a 2008 attack on a military base in the country's north that resulted in the deaths of 12 soldiers, the Mauritanian Government has increased the defense budget and military equipment acquisitions, enhanced military training, heightened security cooperation with its neighbors and the international community, and built up the military's special operations and civil-military affairs forces

Mauritania is part of a five-nation anti-jihadist task force known as the G5 Sahel Group, set up in 2014 with Burkina Faso, Chad, Mali, and Niger; it has committed 550 troops and 100 gendarmes to the force; in early 2020, G5 Sahel military chiefs of staff agreed to allow defense forces from each of the states to pursue terrorist fighters up to 100 km into neighboring countries; the G5 force is backed by the UN, US, and France; G5 troops periodically conduct joint operations with French forces deployed to the Sahel under Operation Barkhane (2020)

Disputes—international: Mauritanian claims to Western Sahara remain dormant

Refugees and internally displaced persons: *refugees (country of origin):* 26,001 (Western Saharan Sahrawis) (2019); 60,455 (Mali) (2020)

Trafficking in persons: current situation: Mauritania is a source and destination country for men, women, and children subjected to forced labor and sex trafficking; adults and children from traditional slave castes are subjected to slavery-related practices rooted in ancestral master-slave relationships; Mauritanian boy students called talibes are trafficked within the country by religious teachers for forced begging; Mauritanian girls, as well as girls from Mali, Senegal, The Gambia, and other West African countries, are forced into domestic servitude; Mauritanian women and girls are forced into prostitution domestically or transported to countries in the Middle East for the same purpose, sometimes through forced marriages

tier rating: Tier 3—Mauritania does not fully comply with the minimum standards for the elimination of trafficking and is not making significant efforts to do so; anti-trafficking law enforcement efforts were negligible; one slavery case identified by an NGO was investigated, but no prosecutions or convictions were made, including among the 4,000 child labor cases NGOs referred to the police; the 2007 anti-slavery law remains ineffective because it requires slaves, most of whom are illiterate, to file their own legal complaint, and the government agency that can submit claims on them did not file any in 2014; authorities arrested, prosecuted, and convicted several anti-slavery activists; NGOs continued to provide the majority of protective services to trafficking victims without support from the government; some steps were taken to raise public awareness about human trafficking (2015)

MAURITIUS

Location: Southern Africa , island in the Indian Ocean , about 800 km (500 mi) east of Madagascar

Geographic coordinates: 20 17 S, 57 33 E

Map references: Africa

Area: *total:* 2,040 sq km
land: 2,030 sq km
water: 10 sq km
note: includes Agalega Islands, Cargados Carajos Shoals (Saint Brandon), and Rodrigues
country comparison to the world: 181

Area—comparative: almost 11 times the size of Washington, DC

Land boundaries: 0 km

Coastline: 177 km

Maritime claims: *territorial sea:* 12 nm
exclusive economic zone: 200 nm
continental shelf: 200 nm or to the edge of the continental margin measured from claimed archipelagic straight baselines

Climate: tropical, modified by southeast trade winds; warm, dry winter (May to November); hot, wet, humid summer (November to May)

Terrain: small coastal plain rising to discontinuous mountains encircling central plateau

Elevation: *lowest point:* Indian Ocean 0 m
highest point: Mont Piton 828 m

Natural resources: arable land, fish

Land use: *agricultural land:* 43.8% (2011 est.)
arable land: 38.4% (2011 est.) / permanent crops: 2% (2011 est.) / permanent pasture: 3.4% (2011 est.)
forest: 17.3% (2011 est.)
other: 38.9% (2011 est.)
Irrigated land: 190 sq km (2012)

Population distribution: population density is one of the highest in the world; urban cluster are found throught the main island, with a greater density in and around Port Luis; population on Rodrigues Island is spread across the island with a slightly denser cluster on the north coast as shown in this population distribution map

Natural hazards: cyclones (November to April); almost completely surrounded by reefs that may pose maritime hazards

Environment—current issues: water pollution, degradation of coral reefs; soil erosion; wildlife preservation; solid waste disposal

Environment—international agreements: party to: Antarctic-Marine Living Resources, Biodiversity, Climate Change, Climate Change-Kyoto Protocol, Desertification, Endangered Species, Environmental Modification, Hazardous Wastes, Law of the Sea, Marine Life Conservation, Ozone Layer Protection, Ship Pollution, Wetlands
signed, but not ratified: none of the selected agreements

Geography—note: the main island, from which the country derives its name, is of volcanic origin and is almost entirely surrounded by coral reefs; former home of the dodo, a large flightless bird related to pigeons, driven to extinction by the end of the 17th century through a combination of hunting and the introduction of predatory species

Population: 1,379,365 (July 2020 est.)
country comparison to the world: 156

Nationality: *noun:* Mauritian(s)
adjective: Mauritian

Ethnic groups: Indo-Mauritian (compose approximately two thirds of the total population), Creole, Sino-Mauritian, Franco-Mauritian
note: Mauritius has not had a question on ethnicity on its national census since 1972

Languages: Creole 86.5%, Bhojpuri 5.3%, French 4.1%, two languages 1.4%, other 2.6% (includes English, the official language of the National Assembly, which is spoken by less than 1% of the population), unspecified 0.1% (2011 est.)

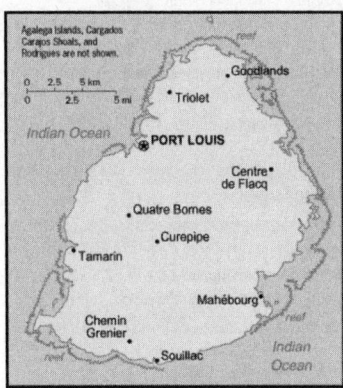

Agalega Islands, Cargados Carajos Shoals, and Rodrigues are not shown.

Background: Although known to Arab and Malay sailors as early as the 10th century, the uninhabited island of Mauritius was first explored by the Portuguese in the 16th century and subsequently settled by the Dutch—who named it in honor of Prince Maurits van NASSAU—in the 17th century. The French assumed control in 1715, developing the island into an important naval base overseeing Indian Ocean trade, and establishing a plantation economy of sugar cane. The British captured the island in 1810, during the Napoleonic Wars. Mauritius remained a strategically important British naval base, and later an air station, playing an important role during World War II for anti-submarine and convoy operations, as well as the collection of signals intelligence. Independence from the UK was attained in 1968. A stable democracy with regular free elections and a positive human rights record, the country has attracted considerable foreign investment and has one of Africa's highest per capita incomes. Mauritius claims the French island of Tromelin and the British Chagos Archipelago (British Indian Ocean Territory).

Religions: Hindu 48.5%, Roman Catholic 26.3%, Muslim 17.3%, other Christian 6.4%, other 0.6%, none 0.7%, unspecified 0.1% (2011 est.)

Demographic profile: Mauritius has transitioned from a country of high fertility and high mortality rates in the 1950s and mid-1960s to one with among the lowest population growth rates in the developing world today. After World War II, Mauritius' population began to expand quickly due to increased fertility and a dramatic drop in mortality rates as a result of improved health care and the eradication of malaria. This period of heightened population growth — reaching about 3% a year — was followed by one of the world's most rapid birth rate declines.

The total fertility rate fell from 6.2 children per women in 1963 to 3.2 in 1972 — largely the result of improved educational attainment, especially among young women, accompanied by later marriage and the adoption of family planning methods. The family planning programs' success was due to support from the government and eventually the traditionally pronatalist religious communities, which both recognized that controlling population growth was necessary because of Mauritius' small size and limited resources. Mauritius' fertility rate has consistently been below replacement level since the late 1990s, a rate that is substantially lower than nearby countries in southern Africa.

With no indigenous population, Mauritius' ethnic mix is a product of more than two centuries of European colonialism and continued international labor migration. Sugar production relied on slave labor mainly from Madagascar, Mozambique, and East Africa from the early 18th century until its abolition in 1835, when slaves were replaced with indentured Indians.

Most of the influx of indentured labor — peaking between the late 1830s and early 1 860—settled permanently creating massive population growth of more than 7% a year and reshaping the island's social and cultural composition. While Indians represented about 12% of Mauritius' population in 1837, and their descendants accounted for roughly two-thirds by the end of the 19th century. Most were Hindus, but the majority of the free Indian traders were Muslims.

Mauritius again turned to overseas labor when its success in clothing and textile exports led to a labor shortage in the mid-1980s. Clothing manufacturers brought in contract workers (increasingly women) from China, India, and, to a lesser extent Bangladesh and Madagascar, who worked longer hours for lower wages under poor conditions and were viewed as more productive than locals. Downturns in the sugar and textile industries in the mid-2000s and a lack of highly qualified domestic workers for Mauritius' growing services sector led to the emigration of low-skilled workers and a reliance on skilled foreign labor. Since 2007, Mauritius has pursued a circular migration program to enable citizens to acquire new skills and savings abroad and then return home to start businesses and to invest in the country's development.

Age structure: *0-14 years:* 19.44% (male 137,010/female 131,113)

15-24 years: 14.06% (male 98,480/female 95,472)
25-54 years: 43.11% (male 297,527/female 297,158)
55-64 years: 12.31% (male 80,952/female 88,785)
65 years and over: 11.08% (male 63,230/female 89,638) (2020 est.)

Dependency ratios: *total dependency ratio:* 41.5
youth dependency ratio: 23.7
elderly dependency ratio: 17.7
potential support ratio: 5.6 (2020 est.)

Median age: *total:* 36.3 years
male: 35 years
female: 37.6 years (2020 est.)
country comparison to the world: 80

Population growth rate: 0.54% (2020 est.)
country comparison to the world: 153

Birth rate: 12.6 births/1,000 population (2020 est.)
country comparison to the world: 152

Death rate: 7.3 deaths/1,000 population (2020 est.)
country comparison to the world: 115

Net migration rate: 0 migrant(s)/1,000 population (2020 est.)
country comparison to the world: 90

Population distribution: population density is one of the highest in the world; urban cluster are found throught the main island, with a greater density in and around Port Luis; population on Rodrigues Island is spread across the island with a slightly denser cluster on the north coast as shown in this population distribution map

Urbanization: *urban population:* 40.8% of total population (2020)
rate of urbanization: 0.11% annual rate of change (2015-20 est.)
total population growth rate v. urban population growth rate, 2000-2030:

Major urban areas—population: 149,000 PORT LOUIS (capital) (2018)

Sex ratio: *at birth:* 1.05 male(s)/female
0-14 years: 1.04 male(s)/female
15-24 years: 1.03 male(s)/female
25-54 years: 1 male(s)/female
55-64 years: 0.91 male(s)/female
65 years and over: 0.71 male(s)/female
total population: 0.96 male(s)/female (2020 est.)

Maternal mortality rate: 61 deaths/100,000 live births (2017 est.)
country comparison to the world: 87

Infant mortality rate: *total:* 9 deaths/1,000 live births
male: 10.7 deaths/1,000 live births
female: 7.3 deaths/1,000 live births (2020 est.)
country comparison to the world: 140

Life expectancy at birth: *total population:* 76.5 years
male: 73 years
female: 80.1 years (2020 est.)
country comparison to the world: 95

Total fertility rate: 1.73 children born/woman (2020 est.)
country comparison to the world: 166

Contraceptive prevalence rate: 63.8% (2014)

Drinking water source:
improved:
urban: 100% of population
rural: 100% of population
total: 100% of population
unimproved:
urban: 0% of population
rural: 0% of population
total: 0% of population (2017 est.)

Current Health Expenditure: 5.7% (2017)

Physicians density: 2.6 physicians/1,000 population (2019)

Hospital bed density: 3.4 beds/1,000 population (2019)

Sanitation facility access:
improved:
urban: 99.9% of population
rural: 99.2% of population
total: 99.5% of population
unimproved:
urban: 0.1% of population
rural: 0.8% of population
total: 0.5% of population (2017 est.)

HIV/AIDS—adult prevalence rate: 1% (2019)
country comparison to the world: 45

HIV/AIDS—people living with HIV/AIDS: 11,000 (2019)
country comparison to the world: 98

HIV/AIDS—deaths: <1000 (2018)

Obesity—adult prevalence rate: 10.8% (2016)
country comparison to the world: 137

Education expenditures: 4.8% of GDP (2018)
country comparison to the world: 73

Literacy: *definition:* age 15 and over can read and write
total population: 91.3%
male: 93.4%
female: 89.4% (2018)

School life expectancy (primary to tertiary education):
total: 15 years
male: 14 years
female: 16 years (2017)

Unemployment, youth ages 15-24:
total: 23.9%
male: 20.6%
female: 28% (2018 est.)
country comparison to the world: 53

GOVERNMENT

Country name: *conventional long form:* Republic of Mauritius
conventional short form: Mauritius
local long form: Republic of Mauritius
local short form: Mauritius
etymology: island named after Prince Maurice VAN NASSAU, stadtholder of the Dutch Republic, in 1598
note: pronounced mah-rish-us

Government type: parliamentary republic

Capital: *name:* Port Louis
geographic coordinates: 20 09 S, 57 29 E

time difference: UTC+4 (9 hours ahead of Washington, DC, during Standard Time)

etymology: named after Louis XV, who was king of France in 1736 when the port became the administrative center of Mauritius and a major reprovisioning stop for French ships traveling between Europe and Asia

Administrative divisions: 9 districts and 3 dependencies*; Agalega Islands*, Black River, Cargados Carajos Shoals*, Flacq, Grand Port, Moka, Pamplemousses, Plaines Wilhems, Port Louis, Riviere du Rempart, Rodrigues*, Savanne

Independence: 12 March 1968 (from the UK)

National holiday: Independence and Republic Day, 12 March (1968 & 1992); note—became independent and a republic on the same date in 1968 and 1992 respectively

Constitution: *history:* several previous; latest adopted 12 March 1968
amendments: proposed by the National Assembly; passage of amendments affecting constitutional articles, including the sovereignty of the state, fundamental rights and freedoms, citizenship, or the branches of government, requires approval in a referendum by at least three-fourths majority of voters followed by a unanimous vote by the Assembly; passage of other amendments requires only two-thirds majority vote by the Assembly; amended many times, last in 2016

Legal system: civil legal system based on French civil law with some elements of English common law

International law organization participation: accepts compulsory ICJ jurisdiction with reservations; accepts ICCt jurisdiction

Citizenship: *citizenship by birth:* yes
citizenship by descent only: yes
dual citizenship recognized: yes
residency requirement for naturalization: 5 out of the previous 7 years including the last 12 months

Suffrage: 18 years of age; universal

Executive branch: *chief of state:* President Pritivirajsing ROOPUN (since December 2019); Vice President Marie Cyril Eddy Boissezon (2 December 2019) note—President Ameenah GURIB-FAKIM (since 5 June 2015) resigned on 23 March 2018 amid a credit card scandal
head of government: Prime Minister Pravind JUGNAUTH (since 23 January 2017, remains PM after parliamentary election 7 Nov 2019); note—Prime Minister Sir Anerood JUGNAUTH (since 17 December 2014) stepped down on 23 January 2017 in favor of his son, Pravind Kumar JUGNAUTH, who was then appointed prime minister; 7 Nov 2019 Pravind Jugnauth remains prime minister and home affairs minister and also becomes defense minister (2019)
cabinet: Cabinet of Ministers (Council of Ministers) appointed by the president on the recommendation of the prime minister
elections/appointments: president and vice president indirectly elected by the National Assembly for 5-year renewable terms; election last held on 7 Nov 2019 (next to be held in 2024); prime minister and deputy prime minister appointed by the president, responsible to the National Assembly (2019)
election results: seats by party as of 7/11/2019- (MSM) 38, (PTR) 14, (MMM) 8, (OPR) 2; note—GURIB-FAKIM, Mauritius'- first female president, resigned on 23 March 2018 (2018)

Legislative branch: *description:* unicameral National Assembly or Assemblee Nationale (70 seats maximum; 62 members directly elected multiseat constituencies by simple majority vote and up to 8 seats allocated to non-elected party candidates by the Office of Electoral Commissioner; members serve a 5-year term)
elections: last held on 7 November 2019 (next to be held by late 2024)
election results: percent of vote by party—MSM 61%, Labour Party 23%, MMM 13%, OPR 3%; elected seats by party as of—the Militant Socialist Movement (MSM) wins 38 seats, the Labour Party (PTR) or (MLP) 14, Mauritian Militant Movement (MMM) 8 and the Rodrigues People's Organization (OPR) 2; composition—men 49, women 13; percent of women 20% (2019)

Judicial branch: *highest courts:* Supreme Court of Mauritius (consists of the chief justice, a senior puisne judge, and 18 puisne judges); note—the Judicial Committee of the Privy Council (in London) serves as the final court of appeal
judge selection and term of office: chief justice appointed by the president after consultation with the prime minister; senior puisne judge appointed by the president with the advice of the chief justice; other puisne judges appointed by the president with the advice of the Judicial and Legal Commission, a 4-member body of judicial officials including the chief justice; all judges serve until retirement at age 67
subordinate courts: lower regional courts known as District Courts, Court of Civil Appeal; Court of Criminal Appeal; Public Bodies Appeal Tribunal

Political parties and leaders:
Alliance Lepep (Alliance of the People) [Pravind JUGNAUTH] (coalition includes MSM and ML)
Labor Party (Parti Travailliste) or PTR or MLP [Navinchandra RAMGOOLAM]
Mauritian Militant Movement (Mouvement Militant Mauricien) or MMM [Paul BERENGER]
Mauritian Social Democratic Party (Parti Mauricien Social Democrate) or PMSD [Xavier Luc DUVAL]
Mauritian Solidarity Front (Front Solidarite Mauricienne) or FSM [Cehl FAKEERMEEAH, aka Cehl MEEAH]
Militant Socialist Movement (Mouvement Socialist Mauricien) or MSM [Pravind JUGNAUTH]
Muvman Liberater or ML [Ivan COLLENDAVELLOO]
Patriotic Movement (Mouvement Patriotic) [Alan GANOO]
Rodrigues Peoples Organization (Organisationdu Peuple Rodriguais) or OPR [Serge CLAIR]

International organization participation: ACP, AfDB, AOSIS, AU, C, CD, COMESA, CPLP (associate), FAO, G-77, IAEA, IBRD, ICAO, ICC (NGOs), ICCt, ICRM, IDA, IFAD, IFC, IFRCS, IHO, ILO, IMF, IMO, IMSO, InOC, Interpol, IOC, IOM, IPU, ISO, ITSO, ITU, ITUC (NGOs), MIGA, NAM, OIF, OPCW, PCA, SAARC (observer), SADC, UN, UNCTAD, UNESCO, UNIDO, UNWTO, UPU, WCO, WFTU (NGOs), WHO, WIPO, WMO, WTO

Diplomatic representation in the US: *chief of mission:* Ambassador Sooroojdev PHOKEER (since 3 August 2015)
chancery: 1709 N Street NW, Washington, DC 20036; administrative offices at 3201 Connecticut Avenue NW, Suite 441, Washington, DC 20036
telephone: [1] (202) 244-1491 through 1492
FAX: [1] (202) 966-0983

Diplomatic representation from the US: *chief of mission:* Ambassador David D. REIMER (since 10 January 2018); note—also accredited to Seychelles
telephone: [230] 202-4400
embassy: 4th Floor, Rogers House, John Kennedy Avenue, Port Louis
mailing address: international mail: P.O. Box 544, Port Louis; US mail: American Embassy, Port Louis, US Department of State, Washington, DC 20521-2450
FAX: [230] 208-9534

Flag description: four equal horizontal bands of red (top), blue, yellow, and green; red represents self-determination and independence, blue the Indian Ocean surrounding the island, yellow has been interpreted as the new light of independence, golden sunshine, or the bright future, and green can symbolize either agriculture or the lush vegetation of the island
note: while many national flags consist of three—and in some cases five—horizontal bands of color, the flag of Mauritius is the world's only national flag to consist of four horizontal color bands

National symbol(s): dodo bird, Trochetia Boutoniana flower; national colors: red, blue, yellow, green

National anthem: *name:* Motherland
lyrics/music: Jean Georges PROSPER/Philippe GENTIL
note: adopted 1968

ECONOMY

Economy—overview: Since independence in 1968, Mauritius has undergone a remarkable economic transformation from a low-income, agriculturally based economy to a diversified, upper middle-income economy with growing industrial, financial, and tourist sectors. Mauritius has achieved steady growth over the last several decades, resulting in more equitable income distribution, increased life expectancy, lowered infant mortality, and a much-improved infrastructure.

The economy currently depends on sugar, tourism, textiles and apparel, and financial services, but is expanding into fish processing, information and communications technology, education, and hospitality and property development. Sugarcane is grown on about 90% of the cultivated land area but sugar makes up only around 3-4% of national GDP. Authorities plan to emphasize services and innovation in the coming years. After several years of slow growth, government policies now seek to stimulate economic growth in five areas: serving as a gateway for international investment into Africa; increasing the use of renewable energy; developing smart cities; growing the ocean economy; and upgrading and modernizing infrastructure, including public transportation, the port, and the airport.

Mauritius has attracted more than 32,000 offshore entities, many aimed at commerce in India, South Africa, and China. The Mauritius International Financial Center is under scrutiny by international bodies promoting fair tax competition and Mauritius has been cooperating with the European Union and the United states in the automatic exchange of account information. Mauritius is also a member of the OECD/G20's Inclusive Framework on Base Erosion and Profit Shifting and is under pressure to review its Double Taxation Avoidance Agreements. The offshore sector is vulnerable to changes in the tax framework and authorities have been working on a Financial Services Sector Blueprint to enable Mauritius to transition to a jurisdiction of higher value added. Mauritius' textile sector has taken advantage of the Africa Growth and Opportunity Act, a preferential trade program that allows duty free access to the US market, with Mauritian exports to the US growing by 35.6 % from 2000 to 2014. However, lack of local labor as well as rising labor costs eroding the competitiveness of textile firms in Mauritius.

Mauritius' sound economic policies and prudent banking practices helped mitigate negative effects of the global financial crisis in 2008-09. GDP grew in the 3-4% per year range in 2010-17, and the country continues to expand its trade and investment outreach around the globe. Growth in the US and Europe fostered services and services exports, including tourism, while lower oil prices kept inflation low. Mauritius continues to rank as one of the most business-friendly environments on the continent and passed a Business Facilitation Act to improve competitiveness and long-term growth prospects. A new National Economic Development Board was set up in 2017-2018 to spearhead efforts to promote exports and attract inward investment.

GDP (purchasing power parity):
$28.27 billion (2017 est.)
$27.23 billion (2016 est.)
$26.23 billion (2015 est.)
note: data are in 2017 dollars
country comparison to the world: 137

GDP (official exchange rate):
$13.33 billion (2017 est.)

GDP—real growth rate: 3.8% (2017 est.)
3.8% (2016 est.)
3.6% (2015 est.)
country comparison to the world: 77

GDP—per capita (PPP): $22,300 (2017 est.)
$21,500 (2016 est.)
$20,800 (2015 est.)
note: data are in 2017 dollars
country comparison to the world: 86

Gross national saving:
16.9% of GDP (2017 est.)
15.8% of GDP (2016 est.)
15.2% of GDP (2015 est.)
country comparison to the world: 122

GDP—composition, by end use:
household consumption: 81% (2017 est.)
government consumption: 15.1% (2017 est.)
investment in fixed capital: 17.3% (2017 est.)
investment in inventories: -0.4% (2017 est.)
exports of goods and services: 42.1% (2017 est.)
imports of goods and services: -55.1% (2017 est.)

GDP—composition, by sector of origin:
agriculture: 4% (2017 est.)
industry: 21.8% (2017 est.)
services: 74.1% (2017 est.)

Agriculture—products: sugarcane, tea, corn, potatoes, bananas, pulses; cattle, goats; fish

Industries: food processing (largely sugar milling), textiles, clothing, mining, chemicals, metal products, transport equipment, nonelectrical machinery, tourism

Industrial production growth rate: 3.2% (2017 est.)
country comparison to the world: 98

Labor force: 554,0 (2020 est.)
country comparison to the world: 154

Labor force—by occupation: agriculture: 8%
industry: 29.8%
services: 62.2% (2014 est.)

Unemployment rate: 6.65% (2019 est.)
6.84% (2018 est.)
country comparison to the world: 106

Population below poverty line: 8% (2006 est.)

Household income or consumption by percentage share: *lowest 10%:* NA
highest 10%: NA

Budget: *revenues:* 2.994 billion (2017 est.)
expenditures: 3.038 billion (2017 est.)

Taxes and other revenues: 22.5% (of GDP) (2017 est.)
country comparison to the world: 133

Budget surplus (+) or deficit (-): -0.3% (of GDP) (2017 est.)
country comparison to the world: 54

Public debt: 64% of GDP (2017 est.)
66.1% of GDP (2016 est.)
country comparison to the world: 60

Fiscal year: 1 July–30 June

Inflation rate (consumer prices): 3.7% (2017 est.)
1% (2016 est.)
country comparison to the world: 148

Current account balance: -$875 million (2017 est.)
-$531 million (2016 est.)
country comparison to the world: 139

Exports: $2.36 billion (2017 est.)
$2.359 billion (2016 est.)

country comparison to the world: 135

Exports—partners: France 16.7%, US 12.5%, UK 12%, South Africa 9%, Madagascar 6.7%, Italy 6.6%, Spain 5.2% (2017)

Exports—commodities: clothing and textiles, sugar, cut flowers, molasses, fish, primates (for research)

Imports: $4.986 billion (2017 est.)
$4.406 billion (2016 est.)
country comparison to the world: 130

Imports—commodities: manufactured goods, capital equipment, foodstuffs, petroleum products, chemicals

Imports—partners: India 17.9%, China 15.7%, France 11.1%, South Africa 9.7% (2017)

Reserves of foreign exchange and gold:
$5.984 billion (31 December 2017 est.)
$4.967 billion (31 December 2016 est.)
country comparison to the world: 92

Debt—external: $19.99 billion (31 December 2017 est.)
$14.34 billion (31 December 2016 est.)
country comparison to the world: 92

Exchange rates: Mauritian rupees (MUR) per US dollar—
35.17 (2017 est.)
35.542 (2016 est.)
35.542 (2015 est.)
35.057 (2014 est.)
30.622 (2013 est.)

ENERGY

Electricity access: electrification—total population: 100% (2020)

Electricity—production: 2.898 billion kWh (2016 est.)
country comparison to the world: 134

Electricity—consumption: 2.726 billion kWh (2016 est.)
country comparison to the world: 140

Electricity—exports: 0 kWh (2016 est.)
country comparison to the world: 170

Electricity—imports: 0 kWh (2016 est.)
country comparison to the world: 174

Electricity—installed generating capacity:
894,0 kW (2016 est.)
country comparison to the world: 132

Electricity—from fossil fuels: 79% of total installed capacity (2016 est.)
country comparison to the world: 86

Electricity—from nuclear fuels: 0% of total installed capacity (2017 est.)
country comparison to the world: 142

Electricity—from hydroelectric plants: 7% of total installed capacity (2017 est.)
country comparison to the world: 126

Electricity—from other renewable sources: 14% of total installed capacity (2017 est.)
country comparison to the world: 62

Crude oil—production: 0 bbl/day (2018 est.)
country comparison to the world: 173

Crude oil—exports: 0 bbl/day (2015 est.)
country comparison to the world: 164

Crude oil—imports: 0 bbl/day (2015 est.)
country comparison to the world: 165

Crude oil—proved reserves: 0 bbl (1 January 2018 est.)
country comparison to the world: 168

Refined petroleum products—production: 0 bbl/day (2017 est.)
country comparison to the world: 177

Refined petroleum products—consumption: 27,000 bbl/day (2016 est.)
country comparison to the world: 123

Refined petroleum products—exports: 0 bbl/day (2015 est.)
country comparison to the world: 181

Refined petroleum products—imports: 26,960 bbl/day (2015 est.)
country comparison to the world: 102

Natural gas—production: 0 cu m (2017 est.)
country comparison to the world: 170

Natural gas—consumption: 0 cu m (2017 est.)
country comparison to the world: 175

Natural gas—exports: 0 cu m (2017 est.)
country comparison to the world: 151

Natural gas—imports: 0 cu m (2017 est.)
country comparison to the world: 157

Natural gas—proved reserves: 0 cu m (1 January 2014 est.)
country comparison to the world: 169

Carbon dioxide emissions from consumption of energy: 6.429 million Mt (2017 est.)
country comparison to the world: 125

COMMUNICATIONS

Telephones—fixed lines: *total subscriptions:* 470,166
subscriptions per 100 inhabitants: 34.27 (2019 est.)
country comparison to the world: 95

Telephones—mobile cellular:
total subscriptions: 2,076,577
subscriptions per 100 inhabitants: 151.36 (2019 est.)
country comparison to the world: 150

Telecommunication systems: *general assessment:* small system with good service; LTE and fiber broadband service are available; government supports building a national Wi-Fi network; partial privatization of biggest telecommunications company, open to competition; 3 mobile network operators; the country is a hub for submarine cables providing international connectivity; successfully pursuing a policy to make telecommunications a pillar of economic growth and to have a fully digital based infrastructure (2020)
domestic: fixed-line teledensity 34 per 100 persons and mobile-cellular services teledensity approaching 151 per 100 persons (2019)
international: country code—230; landing points for the SAFE, MARS, IOX Cable System, METISS and LION submarine cable system that

provides links to Asia, Africa, Southeast Asia, Indian Ocean Islands of Reunion, Madagascar, and Mauritius; satellite earth station—1 Intelsat (Indian Ocean); new microwave link to Reunion; HF radiotelephone links to several countries (2019)
note: the COVID-19 outbreak is negatively impacting telecommunications production and supply chains globally; consumer spending on telecom devices and services has also slowed due to the pandemic's effect on economies worldwide; overall progress towards improvements in all facets of the telecom industry—mobile, fixed-line, broadband, submarine cable and satellite—has moderated

Broadcast media: the government maintains control over TV broadcasting through the Mauritius Broadcasting Corporation (MBC), which only operates digital TV stations since June 2015; MBC is a shareholder in a local company that operates 2 pay-TV stations; the state retains the largest radio broadcast network with multiple stations; several private radio broadcasters have entered the market since 2001; transmissions of at least 2 international broadcasters are available (2019)

Internet country code: .mu

Internet users: *total:* 799,470
percent of population: 58.6% (July 2018 est.)
country comparison to the world: 144

Broadband—fixed subscriptions: total: 274,200
subscriptions per 100 inhabitants: 20 (2018 est.)
country comparison to the world: 104

TRANSPORTATION

National air transport system: *number of registered air carriers:* 1 (2020)
inventory of registered aircraft operated by air carriers: 13
annual passenger traffic on registered air carriers: 1,745,291 (2018)
annual freight traffic on registered air carriers: 233.72 million mt-km (2018)

Civil aircraft registration country code prefix: 3B (2016)

Airports: 5 (2013)
country comparison to the world: 181

Airports—with paved runways:
total: 2 (2019)
over 3,047 m: 1
914 to 1,523 m: 1

Airports—with unpaved runways:
total: 3 (2013)
914 to 1,523 m: 2 (2013)
under 914 m: 1 (2013)

Roadways:
total: 2,428 km (2015)
paved: 2,379 km (includes 99 km of expressways) (2015)
unpaved: 49 km (2015)
country comparison to the world: 168

Merchant marine:
total: 28
by type: general cargo 1, oil tanker 4, other 23 (2019)

country comparison to the world: 137

Ports and terminals: *major seaport(s):* Port Louis

MILITARY AND SECURITY

Military and security forces: no regular military forces; Mauritius Police Force includes a Special Mobile Force (a paramilitary force formed as a mobile infantry battalion) and the National Coast Guard (2019)

Military expenditures:
0.2% of GDP (2018)
0.2% of GDP (2017)
0.2% of GDP (2016)
0.2% of GDP (2015)
0.2% of GDP (2014)
country comparison to the world: 158

Military and security service personnel strengths: police paramilitary forces for Mauritius number about 2,500 (est. 1.700 Special Mobile Force; 800 National Coast Guard) (2019 est.)

Military equipment inventories and acquisitions: the Special Mobile Force's inventory includes mostly second-hand equipment from France and the UK; since 2014, India has provided the majority of the Coast Guard's equipment, including patrol boats and aircraft (2019 est.)

TRANSNATIONAL ISSUES

Disputes—international: Mauritius and Seychelles claim the Chagos Islands; claims French-administered Tromelin Island

Trafficking in persons: *current situation:* Mauritius is a source, transit, and destination country for men, women, and children subjected to forced labor and sex trafficking; Mauritian girls are induced or sold into prostitution, often by peers, family members, or businessmen offering other forms of employment; Mauritian adults have been identified as labor trafficking victims in the UK, Belgium, and Canada, while Mauritian women from Rodrigues Island are also subject to domestic servitude in Mauritius; Malagasy women transit Mauritius en route to the Middle East for jobs as domestic servants and subsequently are subjected to forced labor; Cambodian men are victims of forced labor on foreign fishing vessels in Mauritius' territorial waters; other migrant workers from East and South Asia and Madagascar are also subject to forced labor in Mauritius' manufacturing and construction sectors

tier rating: Tier 2 Watch List — Mauritius does not fully comply with the minimum standards for the elimination of trafficking; however, it is making significant efforts to do so; in 2014, the government made modest efforts to address child sex trafficking but none related to adult forced labor; law enforcement lacks an understanding of trafficking crimes outside of child sex trafficking, despite increasing evidence of other forms of human trafficking; authorities made no trafficking prosecutions or convictions and made modest efforts to assist a couple of child sex trafficking victims; officials sustained an extensive public awareness campaign to prevent child sex trafficking, but no efforts were made to raise awareness or reduce demand for forced adult or child labor (2015)

Illicit drugs: consumer and transshipment point for heroin from South Asia; small amounts of cannabis produced and consumed locally; significant offshore financial industry creates potential for

money laundering, but corruption levels are relatively low and the government appears generally to be committed to regulating its banking industry

MEXICO

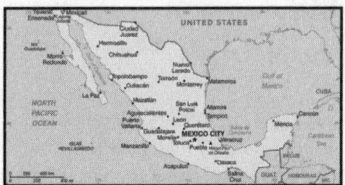

INTRODUCTION

Background: The site of several advanced Amerindian civilizations—including the Olmec, Toltec, Teotihuacan, Zapotec, Maya, and Aztec—Mexico was conquered and colonized by Spain in the early 16th century. Administered as the Viceroyalty of New Spain for three centuries, it achieved independence early in the 19th century. Elections held in 2000 marked the first time since the 1910 Mexican Revolution that an opposition candidate—Vicente FOX of the National Action Party (PAN)—defeated the party in government, the Institutional Revolutionary Party (PRI). He was succeeded in 2006 by another PAN candidate Felipe CALDERON, but Enrique PENA NIETO regained the presidency for the PRI in 2012. Left-leaning antiestablishment politician and former mayor of Mexico City (2000-05) Andres Manuel LOPEZ OBRADOR, from the National Regeneration Movement (MORENA), became president in December 2018.

The global financial crisis in late 2008 caused a massive economic downturn in Mexico the following year, although growth returned quickly in 2010. Ongoing economic and social concerns include low real wages, high underemployment, inequitable income distribution, and few advancement opportunities for the largely indigenous population in the impoverished southern states. Since 2007, Mexico's powerful drug-trafficking organizations have engaged in bloody feuding, resulting in tens of thousands of drug-related homicides.

GEOGRAPHY

Location: North America, bordering the Caribbean Sea and the Gulf of Mexico, between Belize and the United States and bordering the North Pacific Ocean, between Guatemala and the United States

Geographic coordinates: 23 00 N, 102 00 W

Map references: North America

Area: *total:* 1,964,375 sq km
land: 1,943,945 sq km
water: 20,430 sq km

country comparison to the world: 15

Area—comparative: slightly less than three times the size of Texas

Land boundaries: *total:* 4,389 km
border countries (3): Belize 276 km, Guatemala 958 km, US 3155 km

Coastline: 9,330 km

Maritime claims: *territorial sea:* 12 nm
exclusive economic zone: 200 nm
contiguous zone: 24 nm
continental shelf: 200 nm or to the edge of the continental margin

Climate: varies from tropical to desert

Terrain: high, rugged mountains; low coastal plains; high plateaus; desert

Elevation: *mean elevation:* 1,111 m
lowest point: Laguna Salada -10 m
highest point: Volcan Pico de Orizaba 5,636 m

Natural resources: petroleum, silver, antimony, copper, gold, lead, zinc, natural gas, timber

Land use: *agricultural land:* 54.9% (2011 est.)
arable land: 11.8% (2011 est.) / permanent crops: 1.4% (2011 est.) / permanent pasture: 41.7% (2011 est.)
forest: 33.3% (2011 est.)
other: 11.8% (2011 est.)
Irrigated land: 65,000 sq km (2012)

Population distribution: most of the population is found in the middle of the country between the states of Jalisco and Veracruz; approximately a quarter of the population lives in and around Mexico City

Natural hazards: tsunamis along the Pacific coast, volcanoes and destructive earthquakes in the center and south, and hurricanes on the Pacific, Gulf of Mexico, and Caribbean coasts

volcanism: volcanic activity in the central-southern part of the country; the volcanoes in Baja California are mostly dormant; Colima (3,850 m), which erupted in 2010, is Mexico's most active volcano and is responsible for causing periodic evacuations of nearby villagers; it has been deemed a Decade Volcano by the International Association of Volcanology and Chemistry of the Earth's Interior, worthy of study due to its explosive history and close proximity to human populations; Popocatepetl (5,426 m) poses a threat to Mexico City; other historically active volcanoes include Barcena, Ceboruco, El Chichon, Michoacan-Guanajuato, Pico de Orizaba, San Martin, Socorro, and Tacana; see note 2 under "Geography—note"

Environment—current issues: scarcity of hazardous waste disposal facilities; rural to urban migration; natural freshwater resources scarce and polluted in north, inaccessible and poor quality in center and extreme southeast; raw sewage and industrial effluents polluting rivers in urban areas; deforestation; widespread erosion; desertification; deteriorating agricultural lands; serious air and water pollution in the national capital and urban centers along US-Mexico border; land subsidence in Valley of Mexico caused by groundwater depletion

note: the government considers the lack of clean water and deforestation national security issues

Environment—international agreements: *party to:* Biodiversity, Climate Change, Climate Change-Kyoto Protocol, Desertification, Endangered Species, Hazardous Wastes, Law of the Sea, Marine Dumping, Marine Life Conservation, Ozone Layer Protection, Ship Pollution, Wetlands, Whaling
signed, but not ratified: none of the selected agreements

Geography—note: *note 1:* strategic location on southern border of the US; Mexico is one of the countries along the Ring of Fire, a belt of active volcanoes and earthquake epicenters bordering the Pacific Ocean; up to 90% of the world's earthquakes and some 75% of the world's volcanoes occur within the Ring of Fire
note 2: the "Three Sisters" companion plants—winter squash, maize (corn), and climbing beans—served as the main agricultural crops for various North American Indian groups; all three apparently originated in Mexico but then were widely disseminated through much of North America; vanilla, the world's most popular aroma and flavor spice, also emanates from Mexico
note 3: the Sac Actun cave system at 348 km (216 mi) is the longest underwater cave in the world and the second longest cave worldwide, after Mammoth Cave in the United States (see "Geography—note" under United States)

note 4: the prominent Yucatan Peninsula that divides the Gulf of Mexico from the Caribbean Sea is shared by Mexico, Guatemala, and Belize; just on the northern coast of Yucatan, near the town of Chicxulub (pronounce cheek-sha-loob), lie the remnants of a massive crater (some 150 km in diameter and extending well out into the Gulf of Mexico); formed by an asteroid or comet when it struck the earth 66 million years ago, the impact is now widely accepted as initiating a worldwide climate disruption that caused a mass extinction of 75% of all the earth's plant and animal species—including the non-avian dinosaurs

PEOPLE AND SOCIETY

Population: 128,649,565 (July 2020 est.)
country comparison to the world: 10

Nationality: *noun:* Mexican(s)
adjective: Mexican

Ethnic groups: mestizo (Amerindian-Spanish) 62%, predominantly Amerindian 21%, Amerindian 7%, other 10% (mostly European) (2012 est.)
note: Mexico does not collect census data on ethnicity

Languages: Spanish only 92.7%, Spanish and indigenous languages 5.7%, indigenous only 0.8%, unspecified 0.8% (2005)
note: indigenous languages include various Mayan, Nahuatl, and other regional languages

Religions: Roman Catholic 82.7%, Pentecostal 1.6%, Jehovah's Witness 1.4%, other Evangelical Churches 5%, other 1.9%, none 4.7%, unspecified 2.7% (2010 est.)

Age structure: *0-14 years:* 26.01% (male 17,111,199/female 16,349,767)
15-24 years: 16.97% (male 11,069,260/female 10,762,784)
25-54 years: 41.06% (male 25,604,223/female 27,223,720)
55-64 years: 8.29% (male 4,879,048/female 5,784,176)
65 years and over: 7.67% (male 4,373,807/female 5,491,581) (2020 est.)

Dependency ratios: *total dependency ratio:* 50.3
youth dependency ratio: 38.8
elderly dependency ratio: 11.4
potential support ratio: 8.7 (2020 est.)

Median age: *total:* 29.3 years
male: 28.2 years
female: 30.4 years (2020 est.)
country comparison to the world: 132

Population growth rate: 1.04% (2020 est.)
country comparison to the world: 102

Birth rate: 17.6 births/1,000 population (2020 est.)
country comparison to the world: 95

Death rate: 5.4 deaths/1,000 population (2020 est.)
country comparison to the world: 186

Net migration rate: -1.9 migrant(s)/1,000 population (2020 est.)
country comparison to the world: 166

Population distribution: most of the population is found in the middle of the country between the states of Jalisco and Veracruz; approximately a quarter of the population lives in and around Mexico City

Urbanization: *urban population:* 80.7% of total population (2020)
rate of urbanization: 1.59% annual rate of change (2015-20 est.)
total population growth rate v. urban population growth rate, 2000-2030:

Major urban areas—population: 21.782 million MEXICO CITY (capital), 5.179 million Guadalajara, 4.874 million Monterrey, 3.195 million Puebla, 2.467 million Toluca de Lerdo, 2.140 million Tijuana (2020)

Sex ratio: *at birth:* 1.05 male(s)/female
0-14 years: 1.05 male(s)/female
15-24 years: 1.03 male(s)/female
25-54 years: 0.94 male(s)/female
55-64 years: 0.84 male(s)/female
65 years and over: 0.8 male(s)/female
total population: 0.96 male(s)/female (2020 est.)

Mother's mean age at first birth: 21.3 years (2008 est.)

Maternal mortality rate: 33 deaths/100,000 live births (2017 est.)
country comparison to the world: 108

Infant mortality rate: *total:* 10.7 deaths/1,000 live births
male: 12 deaths/1,000 live births
female: 9.2 deaths/1,000 live births (2020 est.)
country comparison to the world: 127

Life expectancy at birth: *total population:* 76.7 years
male: 73.9 years
female: 79.6 years (2020 est.)
country comparison to the world: 91

Total fertility rate: 2.19 children born/woman (2020 est.)
country comparison to the world: 94

Contraceptive prevalence rate: 73.1% (2018)

Drinking water source:
improved:
urban: 100% of population
rural: 96.6% of population
total: 100% of population
unimproved:
urban: 0% of population
rural: 3.4% of population
total: 0% of population (2017 est.)

Current Health Expenditure: 5.5% (2017)

Physicians density: 2.38 physicians/1,000 population (2017)

Hospital bed density: 1.5 beds/1,000 population (2015)

Sanitation facility access:
improved:
urban: 99.3% of population
rural: 91.9% of population
total: 97.8% of population
unimproved:
urban: 0.7% of population
rural: 8.1% of population
total: 2.2% of population (2017 est.)

HIV/AIDS—adult prevalence rate: 0.2% (2018 est.)
country comparison to the world: 106

HIV/AIDS—people living with HIV/AIDS: 230,000 (2018 est.)
country comparison to the world: 25

HIV/AIDS—deaths: 4,000 (2017 est.)
country comparison to the world: 30

Major infectious diseases: *degree of risk:* intermediate (2020)

food or waterborne diseases: bacterial diarrhea and hepatitis A

vectorborne diseases: dengue fever

note: a new coronavirus is causing sustained community spread of respiratory illness (COVID-19) in Mexico; sustained community spread means that people have been infected with the virus, but how or where they became infected is not known, and the spread is ongoing; illness with this virus has ranged from mild to severe with fatalities reported; as of 5 August 2020, Mexico has reported 443,813 confirmed cases of COVID19 with 48,012 deaths

Obesity—adult prevalence rate: 28.9% (2016)
country comparison to the world: 29

Children under the age of 5 years underweight: 4.2% (2016)
country comparison to the world: 87

Education expenditures: 4.9% of GDP (2016)
country comparison to the world: 66

Literacy: *definition:* age 15 and over can read and write
total population: 95.4%
male: 95.8%
female: 94.6% (2018)

School life expectancy (primary to tertiary education): *total:* 15 years
male: 15 years
female: 15 years (2018)

Unemployment, youth ages 15-24: *total:* 6.9%
male: 6.5%
female: 7.6% (2018 est.)
country comparison to the world: 152

GOVERNMENT

Country name: *conventional long form:* United Mexican States
conventional short form: Mexico
local long form: Estados Unidos Mexicanos
local short form: Mexico
etymology: named after the capital city, whose name stems from the Mexica, the largest and most powerful branch of the Aztecs; the meaning of the name is uncertain

Government type: federal presidential republic

Capital: *name:* Mexico City (Ciudad de Mexico)
geographic coordinates: 19 26 N, 99 08 W
time difference: UTC-6 (1 hour behind Washington, DC, during Standard Time)
daylight saving time: +1hr, begins first Sunday in April; ends last Sunday in October
note: Mexico has four time zones
etymology: named after the Mexica, the largest and most powerful branch of the Aztecs; the meaning of the name is uncertain

Administrative divisions: 32 states (estados, singular—estado); Aguascalientes, Baja California, Baja California Sur, Campeche, Chiapas, Chihuahua, Coahuila, Colima, Cuidad de Mexico, Durango, Guanajuato, Guerrero, Hidalgo, Jalisco, Mexico, Michoacan, Morelos, Nayarit, Nuevo Leon, Oaxaca, Puebla, Queretaro, Quintana Roo, San

Luis Potosi, Sinaloa, Sonora, Tabasco, Tamaulipas, Tlaxcala, Veracruz, Yucatan, Zacatecas

Independence: 16 September 1810 (declared independence from Spain); 27 September 1821 (recognized by Spain)

National holiday: Independence Day, 16 September (1810)

Constitution: *history:* several previous; latest approved 5 February 1917
amendments: proposed by the Congress of the Union; passage requires approval by at least two thirds of the members present and approval by a majority of the state legislatures; amended many times, last in 2020

Legal system: civil law system with US constitutional law influence; judicial review of legislative acts

International law organization participation: accepts compulsory ICJ jurisdiction with reservations; accepts ICCt jurisdiction

Citizenship: *citizenship by birth:* yes
citizenship by descent only: yes
dual citizenship recognized: not specified
residency requirement for naturalization: 5 years

Suffrage: 18 years of age; universal and compulsory

Executive branch: *chief of state:* President Andres Manuel LOPEZ OBRADOR (since 1 December 2018); note—the president is both chief of state and head of government

head of government: President Andres Manuel LOPEZ OBRADOR (since 1 December 2018)
cabinet: Cabinet appointed by the president; note—appointment of attorney general, the head of the Bank of Mexico, and senior treasury officials require consent of the Senate
elections/appointments: president directly elected by simple majority popular vote for a single 6-year term; election last held on 1 July 2018 (next to be held in July 2024)
election results: Andres Manuel LOPEZ OBRADOR elected president; percent of vote—Andres Manuel LOPEZ OBRADOR (MORENA) 53.2%, Ricardo ANAYA (PAN) 22.3%, Jose Antonio MEADE Kuribrena (PRI) 16.4%, Jaime RODRIGUEZ Calderon 5.2% (independent), other 2.9%

Legislative branch: *description:* bicameral National Congress or Congreso de la Union consists of: Senate or Camara de Senadores (128 seats; 96 members directly elected in multiseat constituencies by simple majority vote and 32 directly elected in a single, nationwide constituency by proportional representation vote; members serve 6 year terms)
Chamber of Deputies or Camara de Diputados (500 seats; 300 members directly elected in single-seat constituencies by simple majority vote and 200 directly elected in a single, nationwide constituency by proportional representation vote; members serve 3-year terms)
elections: Senate—last held on 1 July 2018 (next to be held on 1 July 2024)

Chamber of Deputies—last held on 1 July 2018 (next to be held on 1 July 2021)
election results: Senate—percent of vote by party—percent of vote by party—NA; seats by party—MORENA 58, PAN 22, PRI 14, PRD 9, MC 7, PT 7, PES 5, PVEM 5, PNA/PANAL 1; composition—men 65, women 63, percent of women 49.3%
Chamber of Deputies—percent of vote by party—NA; seats by party—MORENA 193, PAN 79, PT 61, PES 58, PRI 42, MC 26, PRD 23, PVEM 17, PNA/PANAL 1; composition—men 259, women 241, percent of women 48.2%; note—total National Congress percent of women 48.4%
note: for the 2018 election, senators will be eligible for a second term and deputies up to 4 consecutive terms

Judicial branch: *highest courts:* Supreme Court of Justice or Suprema Corte de Justicia de la Nacion (consists of the chief justice and 11 justices and organized into civil, criminal, administrative, and labor panels) and the Electoral Tribunal of the Federal Judiciary (organized into the superior court, with 7 judges including the court president, and 5 regional courts, each with 3 judges)
judge selection and term of office: Supreme Court justices nominated by the president of the republic and approved by two-thirds vote of the members present in the Senate; justices serve 15-year terms; Electoral Tribunal superior and regional court judges nominated by the Supreme Court and elected by two-thirds vote of members present in the Senate; superior court president elected from among its members to hold office for a 4-year term; other judges of the superior and regional courts serve staggered, 9-year terms
subordinate courts: federal level includes circuit, collegiate, and unitary courts; state and district level courts
Note: in mid-February 2020, the Mexican president endorsed a bill on judicial reform, which proposes changes to 7 articles of the constitution and the issuance of a new Organic Law on the Judicial Branch of the Federation

Political parties and leaders: Citizen's Movement (Movimiento Ciudadano) or MC [Clemente CASTANEDA]
Institutional Revolutionary Party (Partido Revolucionario Institucional) or PRI [Claudia RUIZ Massieu]
Labor Party (Partido del Trabajo) or PT [Alberto ANAYA Gutierrez]
Mexican Green Ecological Party (Partido Verde Ecologista de Mexico) or PVEM [Carlos Alberto PUENTE Salas]
Movement for National Regeneration (Movimiento Regeneracion Nacional) or MORENA [Andres Manuel LOPEZ Obrador]
National Action Party (Partido Accion Nacional) or PAN [Damian ZEPEDA Vidales]
Party of the Democratic Revolution (Partido de la Revolucion Democratica) or PRD [Manuel GRANADOS]

International organization participation: APEC, Australia Group, BCIE, BIS, CAN (observer),

Caricom (observer), CD, CDB, CE (observer), CELAC, CSN (observer), EBRD, FAO, FATF, G-3, G-15, G-20, G-24, G-5, IADB, IAEA, IBRD, ICAO, ICC (national committees), ICCt, ICRM, IDA, IFAD, IFC, IFRCS, IHO, ILO, IMF, IMO, IMSO, Interpol, IOC, IOM, IPU, ISO, ITSO, ITU, ITUC (NGOs), LAES, LAIA, MIGA, NAFTA, NAM (observer), NEA, NSG, OAS, OECD, OPANAL, OPCW, Pacific Alliance, Paris Club (associate), PCA, SICA (observer), UN, UNASUR (observer), UNCTAD, UNESCO, UNHCR, UNIDO, Union Latina (observer), UNWTO, UPU, WCO, WFTU (NGOs), WHO, WIPO, WMO, WTO

Diplomatic representation in the US: *chief of mission:* Ambassador Martha BARCENA Coqui (since 11 January 2019); note—Ambassador BARCENA Coqui is Mexico'a first-ever female ambassador to the US

chancery: 1911 Pennsylvania Avenue NW, Washington, DC 20006
telephone: [1] (202) 728-1600
FAX: [1] (202) 728-1698

consulate(s) general: Atlanta, Austin, Boston, Chicago, Dallas, Denver, El Paso (TX), Houston, Laredo (TX), Los Angeles, Miami, New York, Nogales (AZ), Phoenix, Sacramento (CA), San Antonio (TX), San Diego, San Francisco, San Jose (CA), San Juan (Puerto Rico), Saint Paul (MN)

consulate(s): Albuquerque (NM), Anchorage (AK), Boise (ID), Brownsville (TX), Calexico (CA), Del Rio (TX), Detroit, Douglas (AZ), Eagle Pass (TX), Fresno (CA), Indianapolis (IN), Kansas City (MO), Las Vegas, Little Rock (AR), McAllen (TX), Minneapolis (MN), New Orleans, Omaha (NE), Orlando (FL), Oxnard (CA), Philadelphia, Portland (OR), Presidio (TX), Raleigh (NC), Salt Lake City, San Bernardino (CA), Santa Ana (CA), Seattle, Tucson (AZ), Yuma (AZ); note—Washington DC Consular Section is located in a separate building from the Mexican Embassy and has jurisdiction over DC, parts of Virginia, Maryland, and West Virginia

Diplomatic representation from the US: *chief of mission:* Ambassador Christopher LANDAU (since 26 August 2019)
telephone: (011) [52]-55-5080-2000
embassy: Paseo de la Reforma 305, Colonia Cuauhtemoc, 06500 Mexico, Distrito Federal
mailing address: P. O. Box 9000, Brownsville, TX 78520-9000
FAX: (011) 52-55-5080-2005
consulate(s) general: Ciudad Juarez, Guadalajara, Hermosillo, Matamoros, Merida, Monterrey, Nogales, Nuevo Laredo, Tijuana

Flag description: three equal vertical bands of green (hoist side), white, and red; Mexico's coat of arms (an eagle with a snake in its beak perched on a cactus) is centered in the white band; green signifies hope, joy, and love; white represents peace and honesty; red stands for hardiness, bravery, strength, and valor; the coat of arms is derived from a legend that the wandering Aztec people were to settle at a location where they would see an eagle on a cactus eating a snake; the city they founded, Tenochtitlan, is now Mexico City

note: similar to the flag of Italy, which is shorter, uses lighter shades of green and red, and does not display anything in its white band

National symbol(s): *golden eagle; national colors:* green, white, red

National anthem: *name:* "Himno Nacional Mexicano" (National Anthem of Mexico)

lyrics/music: Francisco Gonzalez BOCANEGRA/ Jaime Nuno ROCA

note: adopted 1943, in use since 1854; also known as "Mexicanos, al grito de Guerra" (Mexicans, to the War Cry); according to tradition, Francisco Gonzalez BOCANEGRA, an accomplished poet, was uninterested in submitting lyrics to a national anthem contest; his fiancee locked him in a room and refused to release him until the lyrics were completed

0:00 /1:41

ECONOMY

Economy—overview: Mexico's $2.4 trillion economy—11th largest in the world—has become increasingly oriented toward manufacturing since the North American Free Trade Agreement (NAFTA) entered into force in 1994. Per capita income is roughly one-third that of the US; income distribution remains highly unequal.

Mexico has become the US' second-largest export market and third-largest source of imports. In 2017, two-way trade in goods and services exceeded $623 billion. Mexico has free trade agreements with 46 countries, putting more than 90% of its trade under free trade agreements. In 2012, Mexico formed the Pacific Alliance with Peru, Colombia, and Chile.

Mexico's current government, led by President Enrique PENA NIETO, has emphasized economic reforms, passing and implementing sweeping energy, financial, fiscal, and telecommunications reform legislation, among others, with the longterm aim to improve competitiveness and economic growth across the Mexican economy. Since 2015, Mexico has held public auctions of oil and gas exploration and development rights and for long-term electric power generation contracts. Mexico has also issued permits for private sector import, distribution, and retail sales of refined petroleum products in an effort to attract private investment into the energy sector and boost production.

Since 2013, Mexico's economic growth has averaged 2% annually, falling short of private-sector expectations that President PENA NIETO's sweeping reforms would bolster economic prospects. Growth is predicted to remain below potential given falling oil production, weak oil prices, structural issues such as low productivity, high inequality, a large informal sector employing over half of the workforce, weak rule of law, and corruption. Mexico's economy remains vulnerable to uncertainty surrounding the future of NAFTA — because the United States is its top trading partner and the two countries share integrated supply chains — and to potential shifts in domestic policies following the inauguration of a new a president in December 2018.

GDP (purchasing power parity):
$2.463 trillion (2017 est.)
$2.413 trillion (2016 est.)
$2.346 trillion (2015 est.)
note: data are in 2017 dollars
country comparison to the world: 11

GDP (official exchange rate):
$1.151 trillion (2017 est.)

GDP—real growth rate: -0.3% (2019 est.)
2.19% (2018 est.)
2.34% (2017 est.)
country comparison to the world: 197

GDP—per capita (PPP): $19,900 (2017 est.)
$19,700 (2016 est.)
$19,400 (2015 est.)
note: data are in 2017 dollars
country comparison to the world: 90

Gross national saving:
21.4% of GDP (2017 est.)
21.6% of GDP (2016 est.)
20.7% of GDP (2015 est.)
country comparison to the world: 85

GDP—composition, by end use:
household consumption: 67% (2017 est.)
government consumption: 11.8% (2017 est.)
investment in fixed capital: 22.3% (2017 est.)
investment in inventories: 0.8% (2017 est.)
exports of goods and services: 37.8% (2017 est.)
imports of goods and services: -39.7% (2017 est.)

GDP—composition, by sector of origin:
agriculture: 3.6% (2017 est.)
industry: 31.9% (2017 est.)
services: 64.5% (2017 est.)

Agriculture—products: corn, wheat, soybeans, rice, beans, cotton, coffee, fruit, tomatoes; beef, poultry, dairy products; wood products

Industries: food and beverages, tobacco, chemicals, iron and steel, petroleum, mining, textiles, clothing, motor vehicles, consumer durables, tourism

Industrial production growth rate: -0.6% (2017 est.)
country comparison to the world: 174

Labor force: 50.914 million (2020 est.)
country comparison to the world: 12

Labor force—by occupation: *agriculture:* 13.4%
industry: 24.1%
services: 61.9% (2011)

Unemployment rate: 3.49% (2019 est.)
3.33% (2018 est.)
note: underemployment may be as high as 25%
country comparison to the world: 48

Population below poverty line: 46.2% (2014 est.)
note: from a food-based definition of poverty; asset-based poverty amounted to more than 47%

Household income or consumption by percentage share: *lowest 10%:* 2%
highest 10%: 40% (2014)

Budget: *revenues:* 261.4 billion (2017 est.)

expenditures: 273.8 billion (2017 est.)

Taxes and other revenues: 22.7% (of GDP) (2017 est.)
country comparison to the world: 131

Budget surplus (+) or deficit (-): -1.1% (of GDP) (2017 est.)
country comparison to the world: 83

Public debt: 54.3% of GDP (2017 est.)
56.8% of GDP (2016 est.)
country comparison to the world: 82

Fiscal year: calendar year

Inflation rate (consumer prices): 6% (2017 est.)
2.8% (2016 est.)
country comparison to the world: 186

Current account balance: -$4.351 billion (2019 est.)
-$25.415 billion (2018 est.)
country comparison to the world: 181

Exports: $409.8 billion (2017 est.)
$374.3 billion (2016 est.)
country comparison to the world: 12

Exports—partners: US 79.9% (2017)

Exports—commodities: manufactured goods, electronics, vehicles and auto parts, oil and oil products, silver, plastics, fruits, vegetables, coffee, cotton; Mexico is the world's leading producer of silver

Imports: $420.8 billion (2017 est.)
$387.4 billion (2016 est.)
country comparison to the world: 14

Imports—commodities: metalworking machines, steel mill products, agricultural machinery, electrical equipment, automobile parts for assembly and repair, aircraft, aircraft parts, plastics, natural gas and oil products

Imports—partners: US 46.4%, China 17.7%, Japan 4.3% (2017)

Reserves of foreign exchange and gold: $175.3 billion (31 December 2017 est.)
$178.4 billion (31 December 2016 est.)
note: Mexico also maintains access to an $88 million Flexible Credit Line with the IMF
country comparison to the world: 14

Debt—external: $445.8 billion (31 December 2017 est.)
$450.2 billion (31 December 2016 est.)
country comparison to the world: 28

Exchange rates: Mexican pesos (MXN) per US dollar—
18.26 (2017 est.)
18.664 (2016 est.)
18.664 (2015 est.)
15.848 (2014 est.)
13.292 (2013 est.)

ENERGY

Electricity access: *electrification—total population:* 100% (2020)

Electricity—production: 302.7 billion kWh (2016 est.)
country comparison to the world: 13

635

Electricity—consumption: 258.7 billion kWh (2016 est.)
country comparison to the world: 14

Electricity—exports: 7.308 billion kWh (2016 est.)
country comparison to the world: 27

Electricity—imports: 3.532 billion kWh (2016 est.)
country comparison to the world: 47

Electricity—installed generating capacity: 72.56 million kW (2016 est.)
country comparison to the world: 17

Electricity—from fossil fuels: 71% of total installed capacity (2016 est.)
country comparison to the world: 106

Electricity—from nuclear fuels: 2% of total installed capacity (2017 est.)
country comparison to the world: 27

Electricity—from hydroelectric plants: 17% of total installed capacity (2017 est.)
country comparison to the world: 96

Electricity—from other renewable sources: 9% of total installed capacity (2017 est.)
country comparison to the world: 82

Crude oil—production: 1.852 million bbl/day (2018 est.)
country comparison to the world: 13

Crude oil—exports: 1.214 million bbl/day (2017 est.)
country comparison to the world: 11

Crude oil—imports: 0 bbl/day (2017 est.)
country comparison to the world: 166

Crude oil—proved reserves: 6.63 billion bbl (1 January 2018 est.)
country comparison to the world: 19

Refined petroleum products—production: 844,600 bbl/day (2017 est.)
country comparison to the world: 23

Refined petroleum products—consumption: 1.984 million bbl/day (2017 est.)
country comparison to the world: 11

Refined petroleum products—exports: 155,800 bbl/day (2017 est.)
country comparison to the world: 35

Refined petroleum products—imports: 867,500 bbl/day (2017 est.)
country comparison to the world: 10

Natural gas—production: 31.57 billion cu m (2017 est.)
country comparison to the world: 24

Natural gas—consumption: 81.61 billion cu m (2017 est.)
country comparison to the world: 9

Natural gas—exports: 36.81 million cu m (2017 est.)
country comparison to the world: 51

Natural gas—imports: 50.12 billion cu m (2017 est.)
country comparison to the world: 8

Natural gas—proved reserves: 279.8 billion cu m (1 January 2018 est.)

country comparison to the world: 38

Carbon dioxide emissions from consumption of energy: 454.1 million Mt (2017 est.)
country comparison to the world: 14

COMMUNICATIONS

Telephones—fixed lines: *total subscriptions:* 22,471,647
subscriptions per 100 inhabitants: 17.65 (2019 est.)
country comparison to the world: 11

Telephones—mobile cellular: *total subscriptions:* 121,117,720
subscriptions per 100 inhabitants: 95.13 (2019 est.)
country comparison to the world: 14

Telecommunication systems: *general assessment:* adequate telephone service for business and government; improving quality and increasing mobile cellular availability, with mobile subscribers far outnumbering fixed-line subscribers; relatively low broadband and mobile penetration, potential for growth; extensive microwave radio relay network; considerable use of fiber-optic cable and coaxial cable; two main MNOs despite efforts for competition; 5G development slow given the existing capabilities of LTE; Mexico's first local Internet Exchange Point opens in Mexico City; regulator strives to bring competition and foreign investment to Mexico; regulator brings back SIM card registration program (2020)

domestic: competition has spurred the mobile-cellular market; fixed-line teledensity exceeds 18 per 100 persons; mobile- cellular teledensity is about 95 per 100 persons; domestic satellite system with 120 earth stations (2019)

international: country code—52; Columbus-2 fiber-optic submarine cable with access to the US, Virgin Islands, Canary Islands, Spain, and Italy; the ARCOS-1 and the MAYA-1 submarine cable system together provide access to Central America, parts of South America and the Caribbean, and the US; satellite earth stations—120 (32 Intelsat, 2 Solidaridad (giving Mexico improved access to South America, Central America, and much of the US as well as enhancing domestic communications), 1 Panamsat, numerous Inmarsat mobile earth stations); linked to Central American Microwave System of trunk connections (2016)
note: the COVID-19 outbreak is negatively impacting telecommunications production and supply chains globally; consumer spending on telecom devices and services has also slowed due to the pandemic's effect on economies worldwide; overall progress towards improvements in all facets of the telecom industry—mobile, fixed-line, broadband, submarine cable, and satellite—has moderated

Broadcast media: telecom reform in 2013 enabled the creation of new broadcast television channels after decades of a quasi-monopoly; Mexico has 821 TV stations and 1,745 radio stations and most are privately owned; the Televisa group once had a virtual monopoly in TV broadcasting, but new broadcasting groups and foreign satellite

and cable operators are now available; in 2016, Mexico became the first country in Latin America to complete the transition from analog to digital transmissions, allowing for better image and audio quality and a wider selection of programming from networks

Internet country code: .mx

Internet users: *total:* 82,843,369
percent of population: 65.77% (July 2018 est.)
country comparison to the world: 9

Broadband—fixed subscriptions: *total:* 18,359,028
subscriptions per 100 inhabitants: 15 (2018 est.)
country comparison to the world: 10

TRANSPORTATION

National air transport system: *number of registered air carriers:* 16 (2020)
inventory of registered aircraft operated by air carriers: 370
annual passenger traffic on registered air carriers: 64,569,640 (2018)
annual freight traffic on registered air carriers: 1,090,380,000 mt-km (2018)

Civil aircraft registration country code prefix: XA (2016)

Airports: 1,714 (2013)
country comparison to the world: 3

Airports—with paved runways: *total:* 243 (2017)
over 3,047 m: 12 (2017)
2,438 to 3,047 m: 32 (2017)
1,524 to 2,437 m: 80 (2017)
914 to 1,523 m: 86 (2017)
under 914 m: 33 (2017)

Airports—with unpaved runways: *total:* 1,471 (2013)
over 3,047 m: 1 (2013)
2,438 to 3,047 m: 1 (2013)
1,524 to 2,437 m: 42 (2013)
914 to 1,523 m: 281 (2013)
under 914 m: 1,146 (2013)

Heliports: 1 (2013)

Pipelines: 15,986 km natural gas (2019), 10,365 km oil (2017), 8,946 km refined products (2016)

Railways: *total:* 20,825 km (2017)
standard gauge: 20,825 km 1.435-m gauge (27 km electrified) (2017)
country comparison to the world: 14

Roadways: *total:* 398,148 km (2017)
paved: 174,911 km (includes 10,362 km of expressways) (2017)
unpaved: 223,237 km (2017)
country comparison to the world: 18

Waterways: 2,900 km (navigable rivers and coastal canals mostly connected with ports on the country's east coast) (2012)
country comparison to the world: 33

Merchant marine: *total:* 637
by type: bulk carrier 6, general cargo 10, oil tanker 35, other 586 (2019)
country comparison to the world: 35

Ports and terminals: *major seaport(s):* Altamira, Coatzacoalcos, Lazaro Cardenas, Manzanillo, Veracruz

oil terminal(s): Cayo Arcas terminal, Dos Bocas terminal

cruise port(s): Cancun, Cozumel, Ensenada

container port(s) (TEUs): Manzanillo (2,830,370), Lazaro Cardenas (1,149,079) (2017)

LNG terminal(s) (import): Altamira, Ensenada

MILITARY AND SECURITY

Military and security forces: *Secretariat of National Defense (Secretaria de Defensa Nacional, SEDENA):* Army (Ejercito), Mexican Air Force (Fuerza Aerea Mexicana, FAM); Secretariat of the Navy (Secretaria de Marina, SEMAR): Mexican Navy (Armada de Mexico (ARM), includes Naval Air Force (FAN), Mexican Naval Infantry Corps (Cuerpo de Infanteria de Marina, Mexmar or CIM)); Ministry of Security and Citizen Protection: Federal Police (includes Gendarmerie), National Guard (2019)

note: the National Guard was formed in 2019 and consists of personnel from the Federal Police and military police units of the Army and Navy

Military expenditures:
0.5% of GDP (2019)
0.5% of GDP (2018)
0.5% of GDP (2017)
0.6% of GDP (2016)
0.7% of GDP (2015)
country comparison to the world: 148

Military and security service personnel strengths: the Mexican armed forces have approximately 270,000 active personnel (200,000 Army; 60,000 Navy; 8,000 Air Force); approximately 60-80,000 National Guard (2019 est.)

Military equipment inventories and acquisitions: the Mexican military inventory includes a mix of domestically-produced and imported equipment from a variety of mostly Western suppliers; since 2010, France, Spain, and the US are the leading suppliers of military hardware to Mexico; Mexico's defense industry produces naval vessels and light armored vehicles (2019 est.)

Military service age and obligation: 18 years of age for compulsory military service (selection for service determined by lottery), conscript service obligation is 12 months; 16 years of age with consent for voluntary enlistment; cadets enrolled in military schools from the age of 15 are considered members of the armed forces; women are eligible for voluntary military service (2012)

TRANSNATIONAL ISSUES

Disputes—international: abundant rainfall in recent years along much of the Mexico-US border region has ameliorated periodically strained watersharing arrangements; the US has intensified security measures to monitor and control legal and illegal personnel, transport, and commodities across its border with Mexico; Mexico must deal with thousands of impoverished Guatemalans and other Central Americans who cross the porous border looking for work in Mexico and the US; Belize and Mexico are working to solve minor border demarcation discrepancies arising from inaccuracies in the 1898 border treaty

Refugees and internally displaced persons: *refugees (country of origin):* 9,257 (Honduras) (2019); 73,494 (Venezuela) (economic and political crisis; includes Venezuelans who have claimed asylum, are recognized as refugees, or have received alternative legal stay) (2020)

IDPs: 345,000 (government's quashing of Zapatista uprising in 1994 in eastern Chiapas Region; drug cartel violence and government's military response since 2007; violence between and within indigenous groups) (2019)

stateless persons: 13 (2019)

Illicit drugs: major drug-producing and transit nation; Mexico is estimated to be the world's third largest producer of opium with poppy cultivation in 2015 estimated to be 28,000 hectares yielding a potential production of 475 metric tons of raw opium; government conducts the largest independent illicit-crop eradication program in the world; continues as the primary transshipment country for US-bound cocaine from South America, with an estimated 95% of annual cocaine movements toward the US stopping in Mexico; major drug syndicates control the majority of drug trafficking throughout the country; producer and distributor of ecstasy; significant money-laundering center; major supplier of heroin and largest foreign supplier of marijuana and methamphetamine to the US market

MICRONESIA, FEDERATED STATES OF

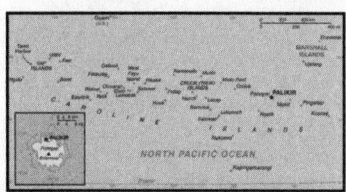

INTRODUCTION

Background: The Caroline Islands are a widely scattered archipelago in the western Pacific Ocean; they became part of a UN Trust Territory under US administration following World War II. The eastern four island groups adopted a constitution in 1979 and chose to become the Federated States of Micronesia (FSM). (The westernmost island group became Palau.) Independence came in 1986 under a Compact of Free Association (COFA) with the US, which was amended in 2004. The COFA has been a force for stability and democracy in the FSM since it came into force in 1986. Present concerns include economic uncertainty after 2023 when direct US economic assistance is scheduled to end, large-scale unemployment, overfishing, overdependence on US foreign aid, and state perceptions of inequitable allocation of US aid.

As a signatory to the COFA with the US, eligible Micronesians can live, work, and study in any part of the US and its territories without a visa—this privilege reduces stresses on the island economy and the environment. Micronesians serve in the US armed forces and military recruiting from the Federated States of Micronesia, per capita, is higher than many US states.

GEOGRAPHY

Location: Oceania, island group in the North Pacific Ocean, about three-quarters of the way from Hawaii to Indonesia

Geographic coordinates: 6 55 N, 158 15 E

Map references: Oceania

Area: *total:* 702 sq km
land: 702 sq km
water: 0 sq km (fresh water only)
note: includes Pohnpei (Ponape), Chuuk (Truk) Islands, Yap Islands, and Kosrae (Kosaie)
country comparison to the world: 192

Area—comparative: four times the size of Washington, DC (land area only)

Land boundaries: 0 km

Coastline: 6,112 km

Maritime claims: *territorial sea:* 12 nm
exclusive economic zone: 200 nm

Climate: tropical; heavy year-round rainfall, especially in the eastern islands; located on southern edge of the typhoon belt with occasionally severe damage

Terrain: islands vary geologically from high mountainous islands to low, coral atolls; volcanic outcroppings on Pohnpei, Kosrae, and Chuuk

Elevation: *lowest point:* Pacific Ocean 0 m
highest point: Nanlaud on Pohnpei 782 m

Natural resources: timber, marine products, deep-seabed minerals, phosphate

Land use: *agricultural land:* 25.5% (2011 est.)
arable land: 2.3% (2011 est.) / permanent crops: 19.7% (2011 est.) / permanent pasture: 3.5% (2011 est.)
forest: 74.5% (2011 est.)
other: 0% (2011 est.)
Irrigated land: 0 sq km NA (2012)

637

Population distribution: the majority of the population lives in the coastal areas of the high islands; the mountainous interior is largely uninhabited; less than half of the population lives in urban areas

Natural hazards: typhoons (June to December)

Environment—current issues: overfishing; climate change; water pollution, toxic pollution from mining; solid waste disposal

Environment—international agreements: *party to:* Biodiversity, Climate Change, Climate Change-Kyoto Protocol, Desertification, Hazardous Wastes, Law of the Sea, Ozone Layer Protection *signed, but not ratified:* none of the selected agreements

Geography—note: composed of four major island groups totaling 607 islands

PEOPLE AND SOCIETY

Population: 102,436 (July 2020 est.)
country comparison to the world: 194

Nationality: *noun:* Micronesian(s)
adjective: Micronesian; Chuukese, Kosraen(s), Pohnpeian(s), Yapese

Ethnic groups: Chuukese/Mortlockese 49.3%, Pohnpeian 29.8%, Kosraean 6.3%, Yapese 5.7%, Yap outer islanders 5.1%, Polynesian 1.6%, Asian 1.4%, other 0.8% (2010 est.)

Languages: English (official and common language), Chuukese, Kosrean, Pohnpeian, Yapese, Ulithian, Woleaian, Nukuoro, Kapingamarangi

Religions: Roman Catholic 54.7%, Protestant 41.1% (includes Congregational 38.5%, Baptist 1.1%, Seventh Day Adventist 0.8%, Assembly of God 0.7%), Mormon 1.5%, other 1.9%, none 0.7%, unspecified 0.1% (2010 est.)

Age structure: *0-14 years:* 28.88% (male 15,046/female 14,542)
15-24 years: 18.94% (male 9,710/female 9,696)
25-54 years: 40.32% (male 19,903/female 21,395)
55-64 years: 7.24% (male 3,572/female 3,842)
65 years and over: 4.62% (male 2,130/female 2,600) (2020 est.)

Dependency ratios: *total dependency ratio:* 55.2
youth dependency ratio: 48.4
elderly dependency ratio: 6.8
potential support ratio: 14.7 (2020 est.)

Median age: *total:* 26.3 years
male: 25.5 years
female: 27.1 years (2020 est.)
country comparison to the world: 154

Population growth rate: -0.6% (2020 est.)
country comparison to the world: 227

Birth rate: 18.9 births/1,000 population (2020 est.)
country comparison to the world: 81

Death rate: 4.3 deaths/1,000 population (2020 est.)
country comparison to the world: 211

Net migration rate: -20.9 migrant(s)/1,000 population (2020 est.)
country comparison to the world: 225

Population distribution: the majority of the population lives in the coastal areas of the high islands; the mountainous interior is largely uninhabited; less than half of the population lives in urban areas

Urbanization: *urban population:* 22.9% of total population (2020)
rate of urbanization: 1.05% annual rate of change (2015-20 est.)
total population growth rate v. urban population growth rate, 2000-2030:

Major urban areas—population: 7,000 PALIKIR (capital) (2018)

Sex ratio: *at birth:* 1.05 male(s)/female
0-14 years: 1.03 male(s)/female
15-24 years: 1 male(s)/female
25-54 years: 0.93 male(s)/female
55-64 years: 0.93 male(s)/female
65 years and over: 0.82 male(s)/female
total population: 0.97 male(s)/female (2020 est.)

Maternal mortality rate: 88 deaths/100,000 live births (2017 est.)
country comparison to the world: 75

Infant mortality rate: *total:* 17.8 deaths/1,000 live births
male: 19.8 deaths/1,000 live births
female: 15.8 deaths/1,000 live births (2020 est.)
country comparison to the world: 84

Life expectancy at birth: *total population:* 73.9 years
male: 71.8 years
female: 76.1 years (2020 est.)
country comparison to the world: 137

Total fertility rate: 2.29 children born/woman (2020 est.)
country comparison to the world: 85

Drinking water source:
improved:
total: 78.6% of population
unimproved:
total: 21.4% of population (2017 est.)

Current Health Expenditure: 12.4% (2017)

Hospital bed density: 3.2 beds/1,000 population (2009)

Sanitation facility access:
improved:
total: 88.3% of population
unimproved:
total: 11.7% of population (2017 est.)

HIV/AIDS—adult prevalence rate: NA

HIV/AIDS—people living with HIV/AIDS: NA

HIV/AIDS—deaths: NA

Major infectious diseases: *degree of risk:* high (2020)
food or waterborne diseases: bacterial diarrhea
vectorborne diseases: malaria

Obesity—adult prevalence rate: 45.8% (2016)
country comparison to the world: 10

Education expenditures: 12.5% of GDP (2015)
country comparison to the world: 2

Unemployment, youth ages 15-24: *total:* 18.9%
male: 10.4%
female: 29.9% (2014)
country comparison to the world: 70

GOVERNMENT

Country name: *conventional long form:* Federated States of Micronesia
conventional short form: none
local long form: Federated States of Micronesia
local short form: none
former: New Philippines; Caroline Islands; Trust Territory of the Pacific Islands, Ponape, Truk, and Yap Districts
abbreviation: FSM
etymology: the term "Micronesia" is a 19th-century construct of two Greek words, "micro" (small) and "nesoi" (islands), and refers to thousands of small islands in the western Pacific Ocean

Government type: federal republic in free association with the US

Capital: *name:* Palikir
geographic coordinates: 6 55 N, 158 09 E
time difference: UTC+11 (16 hours ahead of Washington, DC, during Standard Time)
note 1: Micronesia has two time zones
note 2: Palikir became the new capital of the country in 1989, three years after independence; Kolonia, the former capital, remains the site for many foreign embassies; it also serves as the Pohnpei state capital

Administrative divisions: 4 states; Chuuk (Truk), Kosrae (Kosaie), Pohnpei (Ponape), Yap

Independence: 3 November 1986 (from the US-administered UN trusteeship)

National holiday: Constitution Day, 10 May (1979)

Constitution: *history:* drafted June 1975, ratified 1 October 1978, entered into force 10 May 1979
amendments: proposed by Congress, by a constitutional convention, or by public petition; passage requires approval by at least three-fourths majority vote in at least three fourths of the states; amended 1990; note—at least every 10 years as part of a general or special election, voters are asked whether to hold a constitution convention; a majority of affirmative votes is required to proceed

Legal system: mixed legal system of common and customary law

International law organization participation: has not submitted an ICJ jurisdiction declaration; non-party state to the ICCt

Citizenship: *citizenship by birth:* no
citizenship by descent only: at least one parent must be a citizen of FSM
dual citizenship recognized: no
residency requirement for naturalization: 5 years

Suffrage: 18 years of age; universal

Executive branch: *chief of state:* President David W. PANUELO (since 11 May 2019); Vice President Yosiwo P. GEORGE (since 11 May 2015); note—the president is both chief of state and head of government

head of government: President David W. PANUELO (since 11 May 2019); Vice President Yosiwo P. GEORGE (since 11 May 2015) *cabinet:* Cabinet includes the vice president and the heads of the 8 executive departments

elections/appointments: president and vice president indirectly elected by Congress from among the 4 'at large' senators for a 4-year term (eligible for a second term); election last held on 11 May 2019 (next to be held in 2023)

election results: David W. PANUELO elected president by Congress; Yosiwo P. GEORGE reelected vice president

Legislative branch: *description:* unicameral Congress (14 seats; 10 members directly elected in single-seat constituencies by simple majority vote to serve 2-year terms and 4 at- large members directly elected from each of the 4 states by proportional representation vote to serve 4-year terms)

elections: last held on 5 March 2019 (next to be held in March 2021)

election results: percent of vote—NA; seats—independent 14; composition—men 14, women 0

Judicial branch: *highest courts:* Federated States of Micronesia (FSM) Supreme Court (consists of the chief justice and not more than 5 associate justices and organized into appellate and criminal divisions)

judge selection and term of office: justices appointed by the FSM president with the approval of two-thirds of Congress; justices appointed for life

subordinate courts: the highest state-level courts are: Chuuk Supreme Court; Korsae State Court; Pohnpei State Court; Yap State Court

Political parties and leaders: no formal parties

International organization participation: ACP, ADB, AOSIS, FAO, G-77, IBRD, ICAO, ICRM, IDA, IFC, IFRCS, IMF, IOC, IOM, IPU, ITSO, ITU, MIGA, OPCW, PIF, Sparteca, SPC, UN, UNCTAD, UNESCO, WHO, WMO

Diplomatic representation in the US: *chief of mission:* Ambassador Akillino Harris SUSAIA (since 24 April 2017)

chancery: 1725 N Street NW, Washington, DC 20036

telephone: [1] (202) 223-4383

FAX: [1] (202) 223-4391

consulate(s) general: Honolulu, Tamuning (Guam)

Diplomatic representation from the US: *chief of mission:* Ambassador Carmen G. CANTOR (since 31 January 2020)

telephone: [691] 320-2187

embassy: US Embassy in Kolonia, PO Box 1286, Kolonia, Pohnpei, FSM 96941

mailing address: P. O. Box 1286, Kolonia, Pohnpei, 96941; U.S. Embassy in Micronesia, 4120 Kolonia Place, Washington, D.C. 20521-4120

FAX: [691] 320-2186

Flag description: light blue with four white five-pointed stars centered; the stars are arranged in a diamond pattern; blue symbolizes the Pacific Ocean, the stars represent the four island groups of Chuuk, Kosrae, Pohnpei, and Yap

National symbol(s): *four, five-pointed, white stars on a light blue field, hibiscus flower;* national colors: light blue, white

National anthem: *name:* Patriots of Micronesia
lyrics/music: unknown
note: adopted 1991; also known as "Across All Micronesia"; the music is based on the 1820 German patriotic song "Ich hab mich ergeben", which was the West German national anthem from 1949-1950; variants of this tune are used in Johannes Brahms' "Festival Overture" and Gustav Mahler's "Third Symphony"
0:00/ 0:00

ECONOMY

Economy—overview: Economic activity consists largely of subsistence farming and fishing, and government, which employs two-thirds of the adult working population and receives funding largely—58% in 2013—from Compact of Free Association assistance provided by the US. The islands have few commercially valuable mineral deposits. The potential for tourism is limited by isolation, lack of adequate facilities, and limited internal air and water transportation.

Under the terms of the original Compact, the US provided $1.3 billion in grants and aid from 1986 to 2001. The US and the Federated States of Micronesia (FSM) negotiated a second (amended) Compact agreement in 2002-03 that took effect in 2004. The amended Compact runs for a 20-year period to 2023; during which the US will provide roughly $2.1 billion to the FSM. The amended Compact also develops a trust fund for the FSM that will provide a comparable income stream beyond 2024 when Compact grants end.

The country's medium-term economic outlook appears fragile because of dependence on US assistance and lackluster performance of its small and stagnant private sector.

GDP (purchasing power parity):
$348 million (2017 est.)
$341.1 million (2016 est.)
$331.4 million (2015 est.)
note: data are in 2017 dollars
country comparison to the world: 215

GDP (official exchange rate):
$328 million (2017 est.)

GDP—real growth rate: 2% (2017 est.)
2.9% (2016 est.)
3.9% (2015 est.)
country comparison to the world: 138

GDP—per capita (PPP): $3,400 (2017 est.)
$3,300 (2016 est.)
$3,200 (2015 est.)
note: data are in 2017 dollars
country comparison to the world: 189

GDP—composition, by end use:
household consumption: 83.5% (2013 est.)
government consumption: 48.4% (2016 est.)
investment in fixed capital: 29.5% (2016 est.)
investment in inventories: 1.9% (2016 est.)
exports of goods and services: 27.5% (2016 est.)
imports of goods and services: -77% (2016 est.)

GDP—composition, by sector of origin:
agriculture: 26.3% (2013 est.)
industry: 18.9% (2013 est.)
services: 54.8% (2013 est.)

Agriculture—products: taro, yams, coconuts, bananas, cassava (manioc, tapioca), sakau (kava), Kosraen citrus, betel nuts, black pepper, fish, pigs, chickens

Industries: tourism, construction; specialized aquaculture, craft items (shell and wood)

Industrial production growth rate: NA

Labor force: 37,920 (2010 est.)
country comparison to the world: 200

Labor force—by occupation: *agriculture:* 0.9%
industry: 5.2%
services: 93.9% (2013 est.)
note: two-thirds of the labor force are government employees

Unemployment rate: 16.2% (2010 est.)
country comparison to the world: 179

Population below poverty line: 26.7% (2000 est.)

Household income or consumption by percentage share: *lowest 10%:* NA

highest 10%: NA

Budget: *revenues:* 213.8 million (FY12/13 est.)
expenditures: 192.1 million (FY12/13 est.)

Taxes and other revenues: 65.2% (of GDP) (FY12/13 est.)
country comparison to the world: 7

Budget surplus (+) or deficit (-): 6.6% (of GDP) (FY12/13 est.)
country comparison to the world: 4

Public debt: 24.5% of GDP (2017 est.)

25.3% of GDP (2016 est.)
country comparison to the world: 176

Fiscal year: 1 October—30 September

Inflation rate (consumer prices): 0.5% (2017 est.)
0.5% (2016 est.)
country comparison to the world: 28

Current account balance: $12 million (2017 est.)
$11 million (2016 est.)
country comparison to the world: 62

Exports: $88.3 million (2013 est.)
country comparison to the world: 197

Exports—commodities: fish, sakau (kava), betel nuts, black pepper

Imports: $167.8 million (2015 est.)
$258.5 million (2013 est.)
country comparison to the world: 212

Imports—commodities: food, beverages, clothing, computers, household electronics, appliances, manufactured goods, automobiles, machinery and equipment, furniture, tools

Reserves of foreign exchange and gold: $203.7 million (31 December 2017 est.)
$135.1 million (31 December 2015 est.)
country comparison to the world: 174

Debt—external: $93.6 million (2013 est.)
$93.5 million (2012 est.)
country comparison to the world: 194

Exchange rates: the US dollar is used

ENERGY

Electricity access: *electrification—total population:* 75.4% (2016)
electrification—urban areas: 91.9% (2016)
electrification—rural areas: 70.7% (2016)

Electricity—production: 192 million kWh (2002)
country comparison to the world: 193

Electricity—consumption: 178.6 million kWh (2002)
country comparison to the world: 195

Electricity—exports: 0 kWh (2013 est.)
country comparison to the world: 171

Electricity—imports: 0 kWh (2013 est.)
country comparison to the world: 175

Electricity—installed generating capacity: 18,000 kW (2015 est.)
country comparison to the world: 206

Electricity—from fossil fuels: 96% of total installed capacity (2015 est.)
country comparison to the world: 41

Electricity—from nuclear fuels: 0% of total installed capacity (2015 est.)
country comparison to the world: 143

Electricity—from hydroelectric plants: 1% of total installed capacity (2013 est.)
country comparison to the world: 149

Electricity—from other renewable sources: 3% of total installed capacity (2013 est.)
country comparison to the world: 127

Crude oil—production: 0 bbl/day (2014)
country comparison to the world: 174

Crude oil—exports: 0 bbl/day (2014)
country comparison to the world: 165

Crude oil—imports: 0 bbl/day (2014)
country comparison to the world: 167

Crude oil—proved reserves: 0 bbl (1 January 2014)
country comparison to the world: 169

Refined petroleum products—production: 0 bbl/day (2014)
country comparison to the world: 178

Refined petroleum products—exports: 0 bbl/day
country comparison to the world: 182

Natural gas—production: 0 cu m (2014)
country comparison to the world: 171

Natural gas—proved reserves: 0 cu m
country comparison to the world: 170

Carbon dioxide emissions from consumption of energy: 105 Mt (2010 est.)
country comparison to the world: 214

COMMUNICATIONS

Telephones—fixed lines: *total subscriptions:* 6,420
subscriptions per 100 inhabitants: 6.23 (2019 est.)
country comparison to the world: 202

Telephones—mobile cellular: *total subscriptions:* 21,374
subscriptions per 100 inhabitants: 20.74 (2019 est.)
country comparison to the world: 212

Telecommunication systems: *general assessment:* adequate system, the demand for mobile broadband is increasing due to mobile services being the primary and most wide-spread source for Internet access across the region (2020)

domestic: islands interconnected by shortwave radiotelephone, satellite (Intelsat) ground stations, and some coaxial and fiber-optic cable; mobile-cellular service available on the major islands; fixed line teledensity 6 per 100 and mobile- cellular 21 per 100 (2019)

international: country code—691; landing points for the Chuukk-Pohnpei Cable and HANTRU-1 submarine cable system linking the Federated States of Micronesia and the US; satellite earth stations—5 Intelsat (Pacific Ocean) (2019)
note: the COVID-19 outbreak is negatively impacting telecommunications production and supply chains globally; consumer spending on telecom devices and services has also slowed due to the pandemic's effect on economies worldwide; overall progress towards improvements in all facets of the telecom industry—mobile, fixed-line, broadband, submarine cable and satellite—has moderated

Broadcast media: no TV broadcast stations; each state has a multi-channel cable service with TV transmissions carrying roughly 95% imported programming and 5% local programming; about a half-dozen radio stations (2009)

Internet country code: .fm

Internet users: *total:* 36,586

percent of population: 35.3% (July 2018 est.)
country comparison to the world: 203

Broadband—fixed subscriptions: *total:* 3,776
subscriptions per 100 inhabitants: 4 (2017 est.)
country comparison to the world: 185

TRANSPORTATION

Civil aircraft registration country code prefix: V6 (2016)

Airports: 6 (2013)
country comparison to the world: 174

Airports—with paved runways:
total: 6 (2017)
1,524 to 2,437 m: 4 (2017)
914 to 1,523 m: 2 (2017)

Roadways: note—paved and unpaved circumferential roads, most interior roads are unpaved

Merchant marine: *total:* 39
by type: general cargo 19, oil tanker 4, other 16 (2019)
country comparison to the world: 123

Ports and terminals: *major seaport(s):* Colonia (Tamil Harbor), Molsron Lele Harbor, Pohnepi Harbor

MILITARY AND SECURITY

Military and security forces: no military forces; Federated States of Micronesia National Police (2019)

Military—note: defense is the responsibility of the US

TRANSNATIONAL ISSUES

Disputes—international: none

Illicit drugs: major consumer of cannabis

MOLDOVA

INTRODUCTION

Background: A large portion of present day Moldovan territory became a province of the Russian Empire in 1812 and then unified with Romania in 1918 in the aftermath of World War I. This territory was then incorporated into the Soviet Union at the close of World War II Although Moldova has been independent from the Soviet Union since 1991, Russian forces have remained on Moldovan territory east of the Nistru River in the breakaway region of Transnistria, whose population is roughly equally composed of ethnic Ukrainians, Russians, and Moldovans.

Years of Communist Party rule in Moldova from 2001-2009 ultimately ended with election-related violent protests and a rerun of parliamentary elections in 2009. Since then, a series of pro-European ruling coalitions have governed Moldova As a result of the country's most recent legislative election in February 2019, parliamentary seats are split among the left-leaning Socialist Party (35 seats), the former ruling Democratic Party (30 seats), and the center-right ACUM bloc (26 seats). Parliament voted in Prime Minister Ion CHICU and his cabinet on 14 November 2019, two days after voting to remove his predecessor, ACUM co-leader Maia SANDU, who had been in office since June 2019.

GEOGRAPHY

Location: Eastern Europe, northeast of Romania
Geographic coordinates: 47 00 N, 29 00 E
Map references: Europe
Area: *total:* 33,851 sq km

land: 32,891 sq km
water: 960 sq km
country comparison to the world: 140

Area—comparative: slightly larger than Maryland

Land boundaries: *total:* 1,885 km
border countries (2): Romania 683 km, Ukraine 1202 km

Coastline: 0 km (landlocked)

Maritime claims: none (landlocked)

Climate: moderate winters, warm summers

Terrain: rolling steppe, gradual slope south to Black Sea

Elevation: *mean elevation:* 139 m
lowest point: Dniester (Nistru) 2 m
highest point: Dealul Balanesti 430 m

Natural resources: lignite, phosphorites, gypsum, limestone, arable land

Land use: *agricultural land:* 74.9% (2011 est.)
arable land: 55.1% (2011 est.) / permanent crops: 9.1% (2011 est.) / permanent pasture: 10.7% (2011 est.)
forest: 11.9% (2011 est.)
other: 13.2% (2011 est.)
Irrigated land: 2,283 sq km (2012)

Population distribution: pockets of agglomeration exist throughout the country, the largest being in the center of the country around the capital of Chisinau, followed by Tiraspol and Balti

Natural hazards: landslides

Environment—current issues: heavy use of agricultural chemicals, has contaminated soil and groundwater; extensive soil erosion and declining soil fertility from poor farming methods

Environment—international agreements: *party to:* Air Pollution, Air Pollution-Persistent Organic Pollutants, Biodiversity, Climate Change, Climate Change-Kyoto Protocol, Desertification, Endangered Species, Hazardous Wastes, Ozone Layer Protection, Ship Pollution, Wetlands
signed, but not ratified: none of the selected agreements

Geography—note: landlocked; well endowed with various sedimentary rocks and minerals including sand, gravel, gypsum, and limestone

PEOPLE AND SOCIETY

Population: 3,364,496 (July 2020 est.)
country comparison to the world: 133

Nationality: *noun:* Moldovan(s)
adjective: Moldovan

Ethnic groups: Moldovan 75.1%, Romanian 7%, Ukrainian 6.6%, Gagauz 4.6%, Russian 4.1%, Bulgarian 1.9%, other 0.8% (2014 est.)

Languages: Moldovan/Romanian 80.2% (official) (56.7% identify their mother tongue as Moldovan, which is virtually the same as Romanian; 23.5% identify Romanian as their mother tongue), Russian 9.7%, Gagauz 4.2% (a Turkish language), Ukrainian 3.9%, Bulgarian 1.5%, Romani 0.3%, other 0.2% (2014 est.)
note: data represent mother tongue

Religions: Orthodox 90.1%, other Christian 2.6%, other 0.1%, agnostic (2014 est.)

Age structure: *0-14 years:* 18.31% (male 317,243/female 298,673)
15-24 years: 11.27% (male 196,874/female 182,456)
25-54 years: 43.13% (male 738,103/female 712,892)
55-64 years: 13.26% (male 205,693/female 240,555)
65 years and over: 14.03% (male 186,949/female 285,058) (2020 est.)

Dependency ratios: *total dependency ratio:* 39.6
youth dependency ratio: 22.2
elderly dependency ratio: 17.4
potential support ratio: 5.7 (2020 est.)

Median age: *total:* 37.7 years
male: 36.2 years
female: 39.5 years (2020 est.)
country comparison to the world: 68

Population growth rate: -1.08% (2020 est.)
country comparison to the world: 230

Birth rate: 10.7 births/1,000 population (2020 est.)
country comparison to the world: 184

Death rate: 12.6 deaths/1,000 population (2020 est.)
country comparison to the world: 13

Net migration rate: -9 migrant(s)/1,000 population (2020 est.)
country comparison to the world: 215

Population distribution: pockets of agglomeration exist throughout the country, the largest being in the center of the country around the capital of Chisinau, followed by Tiraspol and Balti

Urbanization: *urban population:* 42.8% of total population (2020)
rate of urbanization: -0.07% annual rate of change (2015-20 est.)
total population growth rate v. urban population growth rate, 2000-2030:

Major urban areas—population: 499,000 CHISINAU (capital) (2020)

Sex ratio: *at birth:* 1.06 male(s)/female
0-14 years: 1.06 male(s)/female
15-24 years: 1.08 male(s)/female
25-54 years: 1.04 male(s)/female
55-64 years: 0.86 male(s)/female
65 years and over: 0.66 male(s)/female
total population: 0.96 male(s)/female (2020 est.)

Mother's mean age at first birth: 24.8 years (2017 est.)

Maternal mortality rate: 19 deaths/100,000 live births (2017 est.)
country comparison to the world: 125

Infant mortality rate: *total:* 11.1 deaths/1,000 live births
male: 12.8 deaths/1,000 live births
female: 9.3 deaths/1,000 live births (2020 est.)
country comparison to the world: 120

Life expectancy at birth: *total population:* 71.9 years
male: 68 years
female: 76 years (2020 est.)
country comparison to the world: 157

Total fertility rate: 1.58 children born/woman (2020 est.)
country comparison to the world: 189

Contraceptive prevalence rate: 59.5% (2012)

Drinking water source:
improved:
urban: 98.5% of population
rural: 84.6% of population
total: 90.5% of population
unimproved:
urban: 1.5% of population
rural: 15.4% of population
total: 9.5% of population (2017 est.)

Current Health Expenditure: 7% (2017)

Physicians density: 3.21 physicians/1,000 population (2017)

Hospital bed density: 5.8 beds/1,000 population (2013)

Sanitation facility access:
improved:
urban: 98.3% of population
rural: 78.9% of population
total: 87.2% of population
unimproved:
urban: 1.7% of population
rural: 21.1% of population
total: 12.8% of population (2017 est.)

HIV/AIDS—adult prevalence rate: 0.6% (2019 est.)

country comparison to the world: 63

HIV/AIDS—people living with HIV/AIDS: 15,000 (2019 est.)
country comparison to the world: 89

HIV/AIDS—deaths: <500 (2019 est.)

Obesity—adult prevalence rate: 18.9% (2016)
country comparison to the world: 115

Children under the age of 5 years underweight: 2.2% (2012)
country comparison to the world: 110

Education expenditures: 6.7% of GDP (2017)
country comparison to the world: 17

Literacy: *definition:* age 15 and over can read and write
total population: 99.4%
male: 99.7%
female: 99.1% (2015)

School life expectancy (primary to tertiary education): *total:* 11 years
male: 11 years
female: 12 years (2019)

Unemployment, youth ages 15-24: *total:* 7.4%
male: 7.5%
female: 7.2% (2018 est.)
country comparison to the world: 147

GOVERNMENT

Country name: *conventional long form:* Republic of Moldova
conventional short form: Moldova
local long form: Republica Moldova
local short form: Moldova

former: Moldavian Soviet Socialist Republic, Moldova Soviet Socialist Republic
etymology: named for the Moldova River in neighboring eastern Romania

Government type: parliamentary republic

Capital: *name:* Chisinau in Moldovan (Kishinev in Russian)

geographic coordinates: 47 00 N, 28 51 E
time difference: UTC+2 (7 hours ahead of Washington, DC, during Standard Time)
daylight saving time: +1hr, begins last Sunday in March; ends last Sunday in October
note: pronounced KEE-shee-now (KIH-shi-nyov)
etymology: origin unclear but may derive from the archaic Romanian word "chisla" ("spring" or "water source") and "noua" ("new") because the original settlement was built at the site of a small spring

Administrative divisions: 32 raions (raioane, singular—raion), 3 municipalities (municipii, singular—municipiul), 1 autonomous territorial unit (unitatea teritoriala autonoma), and 1 territorial unit (unitatea teritoriala)

raions: Anenii Noi, Basarabeasca, Briceni, Cahul, Cantemir, Calarasi, Causeni, Cimislia, Criuleni, Donduseni, Drochia, Dubasari, Edinet, Falesti, Floresti, Glodeni, Hincesti, Ialoveni, Leova, Nisporeni, Ocnita, Orhei, Rezina, Riscani, Singerei, Soldanesti, Soroca, Stefan Voda, Straseni, Taraclia, Telenesti, Ungheni

municipalities: Balti, Bender, Chisinau

autonomous territorial unit: Gagauzia

territorial unit: Stinga Nistrului (Transnistria)

Independence: 27 August 1991 (from the Soviet Union)

National holiday: Independence Day, 27 August (1991)

Constitution: *history:* previous 1978; latest adopted 29 July 1994, effective 27 August 1994
amendments: proposed by voter petition (at least 200,000 eligible voters), by at least one third of Parliament members, or by the government; passage requires two-thirds majority vote of Parliament within one year of initial proposal; revisions to constitutional articles on sovereignty, independence, and neutrality require majority vote by referendum; articles on fundamental rights and freedoms cannot be amended; amended several times, last in 2010; note — in early 2016, the Moldovan Constitutional Court decision returned the country to direct presidential elections, reversing a 2000 constitutional amendment that allowed Parliament to select the president

Legal system: civil law system with Germanic law influences; Constitutional Court review of legislative acts

International law organization participation: has not submitted an ICJ jurisdiction declaration; accepts ICCt jurisdiction

Citizenship: *citizenship by birth:* no
citizenship by descent only: at least one parent must be a citizen of Moldova
dual citizenship recognized: no
residency requirement for naturalization: 10 years

Suffrage: 18 years of age; universal

Executive branch: *chief of state:* President Igor DODON (since 23 December 2016); note — in 2017-19, DODON was temporarily suspended several times by the Moldovan Constitutional Court for rejecting ministerial appointments and for refusing to sign legislation

head of government: Prime Minister Ion CHICU (since 14 November 2019)
cabinet: Cabinet proposed by the prime minister-designate, nominated by the president, approved through a vote of confidence in Parliament
elections/appointments: president directly elected for a 4-year term (eligible for a second term); election last held on 15 November 2020 (next to be held in fall 2024); prime minister designated by the president upon consultation with Parliament; within 15 days from designation, the prime minister-designate must request a vote of confidence for his/her proposed work program from the Parliament
election results: Maia SANDU elected president; percent of vote—Maia SANDU (PAS) 57.8%, Igor DODON (PSRM) 42.2%; Ion CHICU designated prime minister; Parliament vote—62 of 101

Legislative branch: *description:* unicameral Parliament (101 seats; 51 members directly elected in single-seat constituencies by simple majority vote and 50 members directly elected in a single, nationwide constituency by closed party-list proportional representation vote; all members serve 4-year terms
elections: last held on 24 February 2019 (next scheduled for February 2023)
election results: percent of vote by party—PSRM 31.2%, ACUM (PPDA + PAS) 26.8%, PDM 23.6%, PS 8.3%, other 10.1%; seats by party—PSRM 35, ACUM (PPDA + PAS) 26, PDM 30, PS 7, independent 3; composition—men 78, women 23, percent of women 22.8%

Judicial branch: *highest courts:* Supreme Court of Justice (consists of the chief judge, 3 deputy-chief judges, 45 judges, and 7 assistant judges); Constitutional Court (consists of the court president and 6 judges); note—the Constitutional Court is autonomous to the other branches of government; the Court interprets the Constitution and reviews the constitutionality of parliamentary laws and decisions, decrees of the president, and acts of the government
judge selection and term of office: Supreme Court of Justice judges appointed by the president upon the recommendation of the Superior Council of Magistracy, an 11-member body of judicial officials; all judges serve 4-year renewable terms; Constitutional Court judges appointed 2 each by Parliament, the president, and the Higher Council of Magistracy for 6-year terms; court president elected by other court judges for a 3-year term
subordinate courts: Courts of Appeal; Court of Business Audit; municipal courts

Political parties and leaders: *represented in Parliament:* Action and Solidarity Party or PAS [Maia SANDU]
Democratic Party of Moldova or PDM [Vladimir PLAHOTNIUC]
Dignity and Truth Platform or PPDA [Andrei NASTASE]
NOW Platform or ACUM (PPDA + PAS)
Shor Party or PS [Ilan SHOR]
Socialist Party of the Republic of Moldova or PSRM [Zinaida GRECEANII]
not represented in Parliament, participated in recent elections (2014-2019): Anti-Mafia Movement or MPA [Sergiu MOCANU]
Centrist Union of Moldova or UCM [Mihai PETRACHE]
Christian Democratic People's Party or PPCD [Victor CIOBANU]
Communist Party of the Republic of Moldova or PCRM [Vladimir VORONIN]
Conservative Party or PC [Natalia NIRCA]
Democracy at Home Party or PDA [Vasile COSTIUC]
Democratic Action Party or PAD [Mihai GODEA]
Ecologist Green Party or PVE [Anatolie PROHNITCHI]
European People's Party of Moldova or EPPM [Iurie LEANCA]
Law and Justice Party or PLD [Nicolae ALEXEI]

Liberal Democratic Party of Moldova or PLDM [Tudor DELIU]

Liberal Party or PL [Dorin CHIRTOACA]

"Motherland" Party or PP [Sergiu BIRIUCOV]

National Liberal Party or PNL [Vitalia PAVLICENKO]

Our Home Moldova or PCNM [Grigore PETRENCO]

Our Party or PN [Renato USATII]

Party of National Unity [Anatol SALARU]

People's Party of Moldova or PPRM [Alexandru OLEINIC]

Regions Party of Moldova or PRM [Alexandr KALININ]

Socialist People's Party of Moldova or PPSM [Victor STEPANIUC]

International organization participation: BSEC, CD, CE, CEI, CIS, EAEC (observer), EAPC, EBRD, FAO, GCTU, GUAM, IAEA, IBRD, ICAO, ICC (NGOs), ICCt, ICRM, IDA, IFAD, IFC, IFRCS, ILO, IMF, IMO, Interpol, IOC, IOM, IPU, ISO (correspondent), ITU, ITUC (NGOs), MIGA, OIF, OPCW, OSCE, PFP, SELEC, UN, UNCTAD, UNESCO, UNHCR, UNIDO, Union Latina, UNMIL, UNMISS, UNOCI, UNWTO, UPU, WCO, WHO, WIPO, WMO, WTO

Diplomatic representation in the US: *chief of mission:* Ambassador Eugen CARAS (since 17 July 2020)

chancery: 2101 S Street NW, Washington, DC 20008

telephone: [1] (202) 667-1130

FAX: [1] (202) 667-1204

Diplomatic representation from the US: *chief of mission:* Ambassador Dereck J. HOGAN (since 15 October 2018)

telephone: [373] (22) 40-8300

embassy: 103 Mateevici Street, Chisinau MD-2009

mailing address: use embassy street address

FAX: [373] (22) 23-3044

Flag description: three equal vertical bands of Prussian blue (hoist side), chrome yellow, and vermilion red; emblem in center of flag is of a Roman eagle of dark gold (brown) outlined in black with a red beak and talons carrying a yellow cross in its beak and a green olive branch in its right talons and a yellow scepter in its left talons; on its breast is a shield divided horizontally red over blue with a stylized aurochs head, star, rose, and crescent all in black-outlined yellow; based on the color scheme of the flag of Romania—with which Moldova shares a history and culture—but Moldova's blue band is lighter; the reverse of the flag displays a mirrored image of the coat of arms

note: one of only three national flags that differ on their obverse and reverse sides—the others are Paraguay and Saudi Arabia

National symbol(s): *aurochs (a type of wild cattle); national colors: blue, yellow, red*

National anthem: *name:* "Limba noastra" (Our Language)

lyrics/music: Alexei MATEEVICI/Alexandru CRISTEA

note: adopted 1994

0:00/ 1:10

ECONOMY

Economy—overview: Despite recent progress, Moldova remains one of the poorest countries in Europe. With a moderate climate and productive farmland, Moldova's economy relies heavily on its agriculture sector, featuring fruits, vegetables, wine, wheat, and tobacco. Moldova also depends on annual remittances of about $1.2 billion—almost 15% of GDP—from the roughly one million Moldovans working in Europe, Israel, Russia, and elsewhere.

With few natural energy resources, Moldova imports almost all of its energy supplies from Russia and Ukraine. Moldova's dependence on Russian energy is underscored by a more than $6 billion debt to Russian natural gas supplier Gazprom, largely the result of unreimbursed natural gas consumption in the breakaway region of Transnistria. Moldova and Romania inaugurated the Ungheni-Iasi natural gas interconnector project in August 2014. The 43-kilometer pipeline between Moldova and Romania, allows for both the import and export of natural gas. Several technical and regulatory delays kept gas from flowing into Moldova until March 2015. Romanian gas exports to Moldova are largely symbolic. In 2018, Moldova awarded a tender to Romanian Transgaz to construct a pipeline connecting Ungheni to Chisinau, bringing the gas to Moldovan population centers. Moldova also seeks to connect with the European power grid by 2022.

The government's stated goal of EU integration has resulted in some market-oriented progress. Moldova experienced better than expected economic growth in 2017, largely driven by increased consumption, increased revenue from agricultural exports, and improved tax collection. During fall 2014, Moldova signed an Association Agreement and a Deep and Comprehensive Free Trade Agreement with the EU (AA/DCFTA), connecting Moldovan products to the world's largest market. The EU AA/DCFTA has contributed to significant growth in Moldova's exports to the EU. In 2017, the EU purchased over 65% of Moldova's exports, a major change from 20 years previously when the Commonwealth of Independent States (CIS) received over 69% of Moldova's exports. A $1 billion asset-stripping heist of Moldovan banks in late 2014 delivered a significant shock to the economy in 2015; the subsequent bank bailout increased inflationary pressures and contributed to the depreciation of the leu and a minor recession. Moldova's growth has also been hampered by endemic corruption, which limits business growth and deters foreign investment, and Russian restrictions on imports of Moldova's agricultural products. The government's push to restore stability and implement meaningful reform led to the approval in 2016 of a $179 million three-year IMF program focused on improving the banking and fiscal environments, along with additional assistance programs from the EU, World Bank, and Romania.

Moldova received two IMF tranches in 2017, totaling over $42.5 million.

Over the longer term, Moldova's economy remains vulnerable to corruption, political uncertainty, weak administrative capacity, vested bureaucratic interests, energy import dependence, Russian political and economic pressure, heavy dependence on agricultural exports, and unresolved separatism in Moldova's Transnistria region.

GDP (purchasing power parity):
$23.72 billion (2017 est.)
$22.69 billion (2016 est.)
$21.75 billion (2015 est.)
note: data are in 2017 dollars
country comparison to the world: 143

GDP (official exchange rate):
$9.556 billion (2017 est.)

GDP—real growth rate: 4.5% (2017 est.)
4.3% (2016 est.)
-0.4% (2015 est.)
country comparison to the world: 62

GDP—per capita (PPP): $6,700 (2017 est.)
$6,400 (2016 est.)
$6,100 (2015 est.)
note: data are in 2017 dollars
country comparison to the world: 162

Gross national saving: 13.5% of GDP (2017 est.)
15.9% of GDP (2016 est.)
14.5% of GDP (2015 est.)
country comparison to the world: 143

GDP—composition, by end use:
household consumption: 85.8% (2017 est.)
government consumption: 19% (2017 est.)
investment in fixed capital: 21.9% (2017 est.)
investment in inventories: 1.4% (2017 est.)
exports of goods and services: 42.5% (2017 est.)
imports of goods and services: -70.7% (2017 est.)

GDP—composition, by sector of origin:
agriculture: 17.7% (2017 est.)
industry: 20.3% (2017 est.)
services: 62% (2017 est.)

Agriculture—products: vegetables, fruits, grapes, grain, sugar beets, sunflower seeds, tobacco; beef, milk; wine

Industries: sugar processing, vegetable oil, food processing, agricultural machinery; foundry equipment, refrigerators and freezers, washing machines; hosiery, shoes, textiles

Industrial production growth rate: 3% (2017 est.)
country comparison to the world: 104

Labor force: 1.295 million (2017 est.)
country comparison to the world: 131

Labor force—by occupation: *agriculture:* 32.3%
industry: 12%
services: 55.7% (2017 est.)

Unemployment rate: 4.99% (2019 est.)
3.16% (2018 est.)
country comparison to the world: 76

Population below poverty line: 9.6% (2015 est.)

Household income or consumption by percentage share: *lowest 10%:* 4.2%
highest 10%: 22.1% (2014 est.)

Budget: *revenues:* 2.886 billion (2017 est.)
expenditures: 2.947 billion (2017 est.)
note: National Public Budget

Taxes and other revenues: 30.2% (of GDP) (2017 est.)
country comparison to the world: 77

Budget surplus (+) or deficit (-): -0.6% (of GDP) (2017 est.)
country comparison to the world: 65

Public debt: 31.5% of GDP (2017 est.)
35.8% of GDP (2016 est.)
country comparison to the world: 163

Fiscal year: calendar year

Inflation rate (consumer prices): 6.6% (2017 est.)
6.4% (2016 est.)
country comparison to the world: 191

Current account balance: -$602 million (2017 est.)
-$268 million (2016 est.)
country comparison to the world: 126

Exports: $1.858 billion (2017 est.)
$2.045 billion (2016 est.)
country comparison to the world: 142

Exports—partners: Romania 24.6%, Russia 13.7%, Italy 9.1%, Germany 6.2%, Ukraine 5.3%, UK 4.6%, Poland 4.6% (2017)

Exports—commodities: foodstuffs, textiles, machinery

Imports: $4.427 billion (2017 est.)
$3.635 billion (2016 est.)
country comparison to the world: 136

Imports—commodities: mineral products and fuel, machinery and equipment, chemicals, textiles

Imports—partners: Romania 15.5%, Ukraine 11.4%, Russia 10.6%, China 10.4%, Germany 8.9%, Italy 6.9%, Turkey 6.1% (2017)

Reserves of foreign exchange and gold: $2.803 billion (31 December 2017 est.)
$2.206 billion (31 December 2016 est.)
country comparison to the world: 111

Debt—external: $6.549 billion (31 December 2017 est.)
$6.138 billion (31 December 2016 est.)
country comparison to the world: 126

Exchange rates: Moldovan lei (MDL) per US dollar—
18.49 (2017 est.)
19.924 (2016 est.)
19.924 (2015 est.)
19.83 (2014 est.)
14.036 (2013 est.)

ENERGY

Electricity access: *electrification—total population:* 100% (2020)

Electricity—production: 5.49 billion kWh (2016 est.)
country comparison to the world: 118

Electricity—consumption: 4.4 billion kWh (2016 est.)
country comparison to the world: 125

Electricity—exports: 0 kWh (2016 est.)
country comparison to the world: 172

Electricity—imports: 4 million kWh (2016 est.)
country comparison to the world: 116

Electricity—installed generating capacity: 515,000 kW (2016 est.)
note: excludes Transnistria
country comparison to the world: 148

Electricity—from fossil fuels: 86% of total installed capacity (2016 est.)
country comparison to the world: 67

Electricity—from nuclear fuels: 0% of total installed capacity (2017 est.)
country comparison to the world: 144

Electricity—from hydroelectric plants: 12% of total installed capacity (2017 est.)
country comparison to the world: 112

Electricity—from other renewable sources: 2% of total installed capacity (2017 est.)
country comparison to the world: 142

Crude oil—production: 0 bbl/day (2018 est.)
country comparison to the world: 175

Crude oil—exports: 0 bbl/day (2015 est.)
country comparison to the world: 166

Crude oil—imports: 20 bbl/day (2015 est.)
country comparison to the world: 83

Crude oil—proved reserves: 0 bbl (1 January 2018 est.)
country comparison to the world: 170

Refined petroleum products—production: 232 bbl/day (2015 est.)
country comparison to the world: 107

Refined petroleum products—consumption: 18,000 bbl/day (2016 est.)
country comparison to the world: 148

Refined petroleum products—exports: 275 bbl/day (2015 est.)
country comparison to the world: 116

Refined petroleum products—imports: 18,160 bbl/day (2015 est.)
country comparison to the world: 130

Natural gas—production: 11.33 million cu m (2017 est.)
country comparison to the world: 92

Natural gas—consumption: 2.52 billion cu m (2017 est.)
note: excludes breakaway Transnistria
country comparison to the world: 78

Natural gas—exports: 0 cu m (2017 est.)
country comparison to the world: 152

Natural gas—imports: 2.52 billion cu m (2017 est.)
note: excludes breakaway Transnistria
country comparison to the world: 46

Natural gas—proved reserves: NA cu m (1 January 2017 est.)

Carbon dioxide emissions from consumption of energy: 7.653 million Mt (2017 est.)
country comparison to the world: 121

COMMUNICATIONS

Telephones—fixed lines: *total subscriptions:* 901,317
subscriptions per 100 inhabitants: 26.5 (2019 est.)
country comparison to the world: 78

Telephones—mobile cellular: *total subscriptions:* 3,039,990
subscriptions per 100 inhabitants: 89.38 (2019 est.)
country comparison to the world: 139

Telecommunication systems: *general assessment:* the mobile market has extended the reach of services to outside the cities and across most of the country; endeavors to join the EU have promoted regulatory issues to be in line with EU principles and standards; LTE services available; market is competitive with 94 ISPs active; by mid-2019 fiber accounted for about 62% of all fixed broadband connections; most telecom revenue is from the mobile market (2020)

domestic: competition among mobile telephone providers has spurred subscriptions; little interest in expanding fixed-line service 27 per 100; mobile-cellular teledensity sits at 89 per 100 persons (2019)

international: country code—373; service through Romania and Russia via landline; satellite earth stations—at least 3—Intelsat, Eutelsat, and Intersputnik

note: the COVID-19 outbreak is negatively impacting telecommunications production and supply chains globally; consumer spending on telecom devices and services has also slowed due to the pandemic's effect on economies worldwide; overall progress towards improvements in all facets of the telecom industry—mobile, fixed-line, broadband, submarine cable and satellite—has moderated

Broadcast media: state-owned national radio-TV broadcaster operates 1 TV and 1 radio station; a total of nearly 70 terrestrial TV channels and some 50 radio stations are in operation; Russian and Romanian channels also are available (2019)

Internet country code: .md

Internet users: *total:* 2,616,792

percent of population: 76.12% (July 2018 est.)
country comparison to the world: 108

Broadband—fixed subscriptions: *total:* 623,135
subscriptions per 100 inhabitants: 18 (2018 est.)
country comparison to the world: 78

TRANSPORTATION

National air transport system: *number of registered air carriers:* 6 (2020)
inventory of registered aircraft operated by air carriers: 21
annual passenger traffic on registered air carriers: 1,135,999 (2018)
annual freight traffic on registered air carriers: 640,000 mt-km (2018)

Civil aircraft registration country code prefix: ER (2016)

Airports: 7 (2013)
country comparison to the world: 170

Airports—with paved runways: *total:* 5 (2017)
over 3,047 m: 1 (2017)
2,438 to 3,047 m: 2 (2017)
1,524 to 2,437 m: 2 (2017)

Airports—with unpaved runways: *total:* 2 (2013)
1,524 to 2,437 m: 1 (2013)
under 914 m: 1 (2013)

Pipelines: 1916 km gas (2014)

Railways: *total:* 1,171 km (2014)
standard gauge: 14 km 1.435-m gauge (2014)
broad gauge: 1,157 km 1.520-m gauge (2014)
country comparison to the world: 87

Roadways: *total:* 9,352 km (2012)
paved: 8,835 km (2012)
unpaved: 517 km (2012)
country comparison to the world: 1 36

Waterways: 558 km (in public use on Danube, Dniester and Prut Rivers) (2011)
country comparison to the world: 82

Merchant marine: *total:* 142
by type: bulk carrier 4, container ship 4, general cargo 98, oil tanker 8, other 28 (2019)
country comparison to the world: 75

MILITARY AND SECURITY

Military and security forces: *National Army:* Land Forces Command, Air Forces Command (includes air defense unit); Carabinieri Troops (a component of the Ministry of Internal Affairs that also has official status as a service of the Armed Forces during wartime) (2020)

Military expenditures: 0.4% of GDP (2019)
0.4% of GDP (2018)
0.4% of GDP (2017)
0.45% of GDP (2016)
0.4% of GDP (2015)
country comparison to the world: 153

Military and security service personnel strengths: estimates of the size of the Moldovan National Army vary; approximately 6,000 active troops (5,000 Land Forces; 1,000 Air Force) (2019 est.)

Military equipment inventories and acquisitions: the Moldovan military's inventory is limited and almost entirely comprised of older Russian and Soviet-era equipment; since 2000, it has received small amounts of donated material from other nations, including the US (2019 est.)

Military service age and obligation: 18-27 years of age for compulsory or voluntary military service; male registration required at age 16; 1-year service obligation (2019)

note: Moldova intends to abolish military conscription by 2021

TRANSNATIONAL ISSUES

Disputes—international: Moldova and Ukraine operate joint customs posts to monitor the transit of people and commodities through Moldova's break-away Transnistria region, which remains under the auspices of an Organization for Security and Cooperation in Europe-mandated peacekeeping mission comprised of Moldovan, Transnistrian, Russian, and Ukrainian troops

Refugees and internally displaced persons: *refugees (country of origin):* 6,779 applicants for forms of legal stay other than asylum (Ukraine) (2015)

stateless persons: 3,500 (2019)

Illicit drugs: limited cultivation of opium poppy and cannabis, mostly for CIS consumption; transshipment point for illicit drugs from Southwest Asia via Central Asia to Russia, Western Europe, and possibly the US; widespread crime and underground economic activity

MONACO

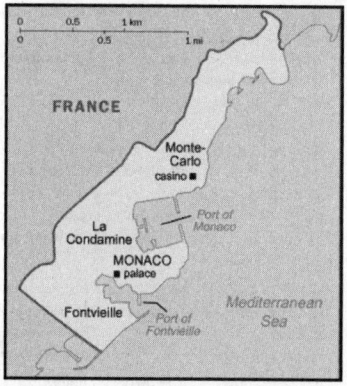

INTRODUCTION

Background: The Genoese built a fortress on the site of present day Monaco in 1215. The current ruling GRIMALDI family first seized control in 1297 but was not able to permanently secure its holding until 1419. Economic development was spurred in the late 19th century with a railroad linkup to France and the opening of a casino. Since then, the principality's mild climate, splendid scenery, and gambling facilities have made Monaco world famous as a tourist and recreation center.

GEOGRAPHY

Location: Western Europe, bordering the Mediterranean Sea on the southern coast of France, near the border with Italy

Geographic coordinates: 43 44 N, 7 24 E

Map references: Europe

Area: *total:* 2 sq km
land: 2 sq km
water: 0 sq km
country comparison to the world: 255

Area—comparative: about three times the size of the National Mall in Washington, DC

Land boundaries: *total:* 6 km
border countries (1): France 6 km

Coastline: 4.1 km

Maritime claims: *territorial sea:* 12 nm
exclusive economic zone: 12 nm

Climate: Mediterranean with mild, wet winters and hot, dry summers

Terrain: hilly, rugged, rocky

Elevation: *lowest point:* Mediterranean Sea 0 m
highest point: Chemin des Revoires on Mont Agel 162 m

Natural resources: none

Land use: *agricultural land:* 1% (2011 est.)
arable land: 0% (2011 est.) / *permanent crops:* 1% (2011 est.) / *permanent pasture:* 0% (2011 est.)
forest: 0% (2011 est.)
other: 99% (2011 est.)

Irrigated land: 0 sq km (2012)

Population distribution: the second most densely populated country in the world (after Macau); its entire population living on 2 square km

Natural hazards: none

Environment—current issues: no serious issues; actively monitors pollution levels in air and water

Environment—international agreements: *party to:* Air Pollution, Air Pollution-Sulfur 94, Air Pollution-Volatile Organic Compounds, Biodiversity, Climate Change, Climate Change-Kyoto Protocol, Desertification, Endangered Species, Hazardous Wastes, Law of the Sea, Marine Dumping, Ozone Layer Protection, Ship Pollution, Wetlands, Whaling
signed, but not ratified: none of the selected agreements

Geography—note: second-smallest independent state in the world (after the Holy See); smallest country with a coastline; almost entirely urban

PEOPLE AND SOCIETY

Population: 39,000 (2019 est.)
note: immigrants make up almost 68% of the total population, according to UN data (2019)
country comparison to the world: 214

Nationality: *noun:* Monegasque(s) or Monacan(s)
adjective: Monegasque or Monacan

Ethnic groups: Monegasque 32.1%, French 19.9%, Italian 15.3%, British 5%, Belgian 2.3%, Swiss

2%, German 1.9%, Russian 1.8%, American 1.1%, Dutch 1.1%, Moroccan 1%, other 16.6%

note: data represent population by country of birth
French 24.9%, Monegasque 22.5%, Italian 21.9%, British 7.5%, Swiss 3.2%, Belgian 2.9%, German 2.4%, Russian 2%, Dutch 1.5%, Portuguese 1.4%, Greek 1.1%, American 1%, other 7.7%

note: data represent population by nationality (2016 est.)

Languages: French (official), English, Italian, Monegasque

Religions: Roman Catholic 90% (official), other 10%

Age structure: *0-14 years:* 9.41% (male 1,497/female 1,415)
15-24 years: 9.52% (male 1,538/female 1,406)
25-54 years: 30.46% (male 4,779/female 4,644)
55-64 years: 15.47% (male 2,370/female 2,417)
65 years and over: 35.15% (male 4,817/female 6,057) (2020 est.)

Median age: *total:* 55.4 years
male: 53.7 years
female: 57 years (2020 est.)
country comparison to the world: 1

Population growth rate: 0.37% (2020 est.)
country comparison to the world: 165

Birth rate: 6.4 births/1,000 population (2020 est.)
country comparison to the world: 229

Death rate: 10.8 deaths/1,000 population (2020 est.)
country comparison to the world: 24

Net migration rate: 8.3 migrant(s)/1,000 population (2020 est.)
country comparison to the world: 10

Population distribution: the second most densely populated country in the world (after Macau); its entire population living on 2 square km

Urbanization: *urban population:* 100% of total population (2020)
rate of urbanization: 0.51% annual rate of change (2015-20 est.)
total population growth rate v. urban population growth rate, 2000-2030:

Major urban areas—population: 39,000 MONACO (capital) (2018)

Sex ratio: *at birth:* 1.03 male(s)/female
0-14 years: 1.06 male(s)/female
15-24 years: 1.09 male(s)/female
25-54 years: 1.03 male(s)/female
55-64 years: 0.98 male(s)/female
65 years and over: 0.8 male(s)/female
total population: 0.94 male(s)/female (2020 est.)

Infant mortality rate: *total:* 1.9 deaths/1,000 live births
male: 2.1 deaths/1,000 live births
female: 1.6 deaths/1,000 live births (2020 est.)
country comparison to the world: 227

Life expectancy at birth: *total population:* 89.3 years
male: 85.4 years
female: 93.3 years (2020 est.)
country comparison to the world: 1

Total fertility rate: 1.55 children born/woman (2020 est.)
country comparison to the world: 195

Drinking water source:
improved:
urban: 100% of population
total: 100% of population
unimproved:
urban: 0% of population
total: 0% of population (2017 est.)

Current Health Expenditure: 1.8% (2017)

Physicians density: 7.51 physicians/1,000 population (2014)

Hospital bed density: 13.8 beds/1,000 population (2012)

Sanitation facility access:
improved:
urban: 100% of population
total: 100% of population
unimproved:
urban: 0% of population
total: 0% of population (2017 est.)

HIV/AIDS—adult prevalence rate: NA

HIV/AIDS—people living with HIV/AIDS: NA

HIV/AIDS—deaths: NA

Education expenditures: 1.5% of GDP (2017)
country comparison to the world: 173

Unemployment, youth ages 15-24: *total:* 26.6%
male: 25.7%
female: 27.9% (2016 est.)
country comparison to the world: 43

GOVERNMENT

Country name: *conventional long form:* Principality of Monaco
conventional short form: Monaco
local long form: Principaute de Monaco
local short form: Monaco
etymology: founded as a Greek colony in the 6th century B.C., the name derives from two Greek words "monos" (single, alone) and "oikos" (house) to convey the sense of a people "living apart" or in a "single habitation"

Government type: constitutional monarchy

Capital: *name:* Monaco

geographic coordinates: 43 44 N, 7 25 E
time difference: UTC + 1 (6 hours ahead of Washington, DC, during Standard Time)
daylight saving time: +1hr, begins last Sunday in March; ends last Sunday in October

Administrative divisions: none; there are no first-order administrative divisions as defined by the US Government, but there are 4 quarters (quartiers, singular—quartier); Fontvieille, La Condamine, Monaco-Ville, Monte-Carlo; note—Moneghetti, a part of La Condamine, is sometimes called the 5th quarter of Monaco

Independence: 1419 (beginning of permanent rule by the House of GRIMALDI)

National holiday: National Day (Saint Rainier's Day), 19 November (1857)

Constitution: *history:* previous 1911 (suspended 1959); latest adopted 17 December 1962
amendments: proposed by joint agreement of the chief of state (the prince) and the National Council; passage requires two-thirds majority vote of National Council members; amended 2002

Legal system: civil law system influenced by French legal tradition

International law organization participation: has not submitted an ICJ jurisdiction declaration; non-party state to the ICCt

Citizenship: *citizenship by birth:* no
citizenship by descent only: the father must be a citizen of Monaco; in the case of a child born out of wedlock, the mother must be a citizen and father unknown
dual citizenship recognized: no
residency requirement for naturalization: 10 years

Suffrage: 18 years of age; universal

Executive branch: *chief of state:* Prince ALBERT II (since 6 April 2005)
head of government: Minister of State Serge TELLE (since 1 February 2016)
cabinet: Council of Government under the authority of the monarch
elections/appointments: the monarchy is hereditary; minister of state appointed by the monarch from a list of three French national candidates presented by the French Government

Legislative branch: *description:* unicameral National Council or Conseil National (24 seats; 16 members directly elected in multi-seat constituencies by simple majority vote and 8 directly elected by proportional representation vote; members serve 5-year terms)
elections: last held on 11 February 2018 (next to be held in February 2023)
election results: percent of vote by party—Priorite Monaco 57.7%, Horizon Monaco 26.1%, Union Monegasque 16.2%; seats by party—Priorite Monaco 21 , Horizon Monaco 2, Union Monegasque 1; composition—men 16, women 8, percent of women 33.3%

Judicial branch: *highest courts:* Supreme Court (consists of 5 permanent members and 2 substitutes)
judge selection and term of office: Supreme Court members appointed by the monarch upon the proposals of the National Council, State Council, Crown Council, Court of Appeal, and Trial Court
subordinate courts: Court of Appeal; Civil Court of First Instance

Political parties and leaders: Horizon Monaco [Laurent NOUVION]
Priorite Monaco [Stephane VALERI]
Renaissance [SBM (public corporation)]
Union Monegasque [Jean-Francois ROBILLON]

International organization participation: CD, CE, FAO, IAEA, ICAO, ICC (national committees), ICRM, IFRCS, IHO, IMO, IMSO, Interpol, IOC, IPU, ITSO, ITU, OAS (observer), OIF, OPCW, OSCE, Schengen Convention (de facto member),

UN, UNCTAD, UNESCO, UNIDO, Union Latina, UNWTO, UPU, WHO, WIPO, WMO

Diplomatic representation in the US: *chief of mission:* Ambassador Maguy MACCARIO-DOYLE (since 3 December 2013)

chancery: 3400 International Drive NW, Suite 2K-100, Washington, DC 20008
telephone: (202) 234-1530
FAX: (202) 244-7656
consulate(s) general: New York

Diplomatic representation from the US: US does not have an embassy in Monaco; the US Ambassador to France is accredited to Monaco; the US Consul General in Marseille (France), under the authority of the US Ambassador to France, handles diplomatic and consular matters concerning Monaco; +(33)(1) 43-12-22-22, enter zero "0" after the automated greeting; US Embassy Paris, 2 Avenue Gabriel, 75008 Paris, France

Flag description: two equal horizontal bands of red (top) and white; the colors are those of the ruling House of Grimaldi and have been in use since 1339, making the flag one of the world's oldest national banners
note: similar to the flag of Indonesia which is longer and the flag of Poland which is white (top) and red

National symbol(s): *red and white lozenges (diamond shapes); national colors:* red, white

National anthem: *name:* "A Marcia de Muneghu" (The March of Monaco)
lyrics/music: Louis NOTARI/Charles ALBRECHT
note: music adopted 1867, lyrics adopted 1931; although French is commonly spoken, only the Monegasque lyrics are official; the French version is known as «Hymne Monegasque" (Monegasque Anthem); the words are generally only sung on official occasions
0:00/ 1:27

ECONOMY

Economy—overview: Monaco, bordering France on the Mediterranean coast, is a popular resort, attracting tourists to its casino and pleasant climate. The principality also is a banking center and has successfully sought to diversify into services and small, high- value-added, nonpolluting industries. The state retains monopolies in a number of sectors, including tobacco, the telephone network, and the postal service. Living standards are high, roughly comparable to those in prosperous French metropolitan areas.

The state has no income tax and low business taxes and thrives as a tax haven both for individuals who have established residence and for foreign companies that have set up businesses and offices. Monaco, however, is not a tax-free shelter; it charges nearly 20% value-added tax, collects stamp duties, and companies face a 33% tax on profits unless they can show that three-quarters of profits are generated within the principality. Monaco was formally removed from the OECD's "grey list" of uncooperative tax jurisdictions in

late 2009, but continues to face international pressure to abandon its banking secrecy laws and help combat tax evasion. In October 2014, Monaco officially became the 84th jurisdiction participating in the OECD's Multilateral Convention on Mutual Administrative Assistance in Tax Matters, an effort to combat offshore tax avoidance and evasion.

Monaco's reliance on tourism and banking for its economic growth has left it vulnerable to downturns in France and other European economies which are the principality's main trade partners. In 2009, Monaco's GDP fell by 11.5% as the eurozone crisis precipitated a sharp drop in tourism and retail activity and home sales. A modest recovery ensued in 2010 and intensified in 2013, with GDP growth of more than 9%, but Monaco's economic prospects remain uncertain.

GDP (purchasing power parity):
$7.672 billion (2015 est.)
$7.279 billion (2014 est.)
$6.79 billion (2013 est.)
note: data are in 2015 US dollars
country comparison to the world: 165

GDP (official exchange rate):
$6.006 billion (2015 est.)

GDP—real growth rate: 5.4% (2015 est.)
7.2% (2014 est.)
9.6% (2013 est.)
country comparison to the world: 36

GDP—per capita (PPP): $115,700 (2015 est.)
$109,200 (2014 est.)
$101,900 (2013 est.)
country comparison to the world: 4

GDP—composition, by sector of origin:
agriculture: 0% (2013)
industry: 14% (2013)
services: 86% (2013)

Agriculture—products: none

Industries: banking, insurance, tourism, construction, small-scale industrial and consumer products

Industrial production growth rate: 6.8% (2015)
country comparison to the world: 33

Labor force: 52,000 (2014 est.)
note: includes all foreign workers
country comparison to the world: 190

Labor force—by occupation: *agriculture:* 0%
industry: 16.1%
services: 83.9% (2012 est.)

Unemployment rate: 2% (2012)
country comparison to the world: 20

Population below poverty line: NA

Household income or consumption by percentage share: *lowest 10%:* NA
highest 10%: NA

Budget: *revenues:* 896.3 million (2011 est.)
expenditures: 953.6 million (2011 est.)

Taxes and other revenues: 14.9% (of GDP) (2011 est.)
country comparison to the world: 196

Budget surplus (+) or deficit (-): -1% (of GDP) (2011 est.)

country comparison to the world: 80

Fiscal year: calendar year

Inflation rate (consumer prices): 1.5% (2010)
country comparison to the world: 84

Exports: $964.6 million (2017 est.)
$1.115 billion (2011)
note: full customs integration with France, which collects and rebates Monegasque trade duties; also participates in EU market system through customs union with France
country comparison to the world: 161

Imports: $1.371 billion (2017 est.)
$1.162 billion (2011 est.)
note: full customs integration with France, which collects and rebates Monegasque trade duties; also participates in EU market system through customs union with France
country comparison to the world: 176

Debt—external: NA

Exchange rates: euros (EUR) per US dollar—0.885 (2017 est.)
0.903 (2016 est.)
0.9214 (2015 est.)
0.885 (2014 est.)
0.7634 (2013 est.)

ENERGY

Electricity access: *electrification—total population:* 100% (2020)

COMMUNICATIONS

Telephones—fixed lines: *total subscriptions:* 34,903
subscriptions per 100 inhabitants: 113.23 (2019 est.)
country comparison to the world: 166

Telephones—mobile cellular: *total subscriptions:* 26,725
subscriptions per 100 inhabitants: 86.7 (2019 est.)
country comparison to the world: 211

Telecommunication systems: *general assessment:* modern automatic telephone system; the country's sole fixed-line operator offers a full range of services to residential and business customers; competitive mobile telephony market; 4G LTE widely available (2020)

domestic: fixed-line 113 per 100 and mobile-cellular teledensity exceeds 87 per 100 persons (2019)

international: country code—377; landing points for the EIG and Italy-Monaco submarine cables connecting Monaco to Europe, Africa, the Middle East and Asia; no satellite earth stations; connected by cable into the French communications system (2019)

note: the COVID-19 outbreak is negatively impacting telecommunications production and supply chains globally; consumer spending on telecom devices and services has also slowed due to the pandemic's effect on economies worldwide; overall progress towards improvements in all facets of the

telecom industry—mobile, fixed-line, broadband, submarine cable and satellite—has moderated

Broadcast media: TV Monte-Carlo operates a TV network; cable TV available; Radio Monte-Carlo has extensive radio networks in France and Italy with French-language broadcasts to France beginning in the 1960s and Italian-language broadcasts to Italy beginning in the 1970s; other radio stations include Riviera Radio and Radio Monaco

Internet country code: .mc

Internet users: *total:* 29,821

percent of population: 97.05% (July 2018 est.)
country comparison to the world: 204

Broadband—fixed subscriptions: *total:* 19,822

subscriptions per 100 inhabitants: 65 (2018 est.)
country comparison to the world: 152

TRANSPORTATION

Civil aircraft registration country code prefix: 3A (2016)

Heliports: 1 (2012)

Railways: *note:* Monaco has a single railway station but does not operate its own train service; the French operator SNCF operates rail services in Monaco

Ports and terminals: *major seaport(s):* Hercules Port

MILITARY AND SECURITY

Military and security forces: *no regular military forces; Ministry of Interior:* Compagnie des Carabiniers du Prince (Prince's Company of Carabiniers (Palace Guard)), Corps des Sapeurs-pompiers de Monaco (Fire and Emergency), Police Department (2019)

Military—note: defense is the responsibility of France

TRANSNATIONAL ISSUES

Disputes—international: none

MONGOLIA

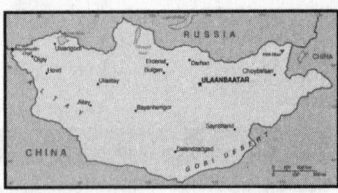

INTRODUCTION

Background: The Mongols gained fame in the 13th century when under Chinggis KHAAN they established a huge Eurasian empire through conquest. After his death the empire was divided into several powerful Mongol states, but these broke apart in the 14th century. The Mongols eventually retired to their original steppe homelands and in the late 17th century came under Chinese rule. Mongolia declared its independence from the Manchu-led Qing Empire in 1911 and achieved limited autonomy until 1919, when it again came under Chinese control. The Mongolian Revolution of 1921 ended Chinese dominance, and a communist regime, the Mongolian People's Republic, took power in 1924.

The modern country of Mongolia, represents only part of the Mongols' historical homeland; today, more ethnic Mongolians live in the Inner Mongolia Autonomous Region in the People's Republic of China than in Mongolia. Since the country's peaceful democratic revolution in 1990, the ex-communist Mongolian People's Revolutionary Party (MPRP)—which took the name Mongolian People's Party (MPP) in 2010—has competed for political power with the Democratic Party (DP) and several other smaller parties, including a new party formed by former President ENKHBAYAR, which confusingly adopted for itself the MPRP name. In the country's most recent parliamentary elections in June 2016, Mongolians handed the MPP overwhelming control of Parliament, largely pushing out the DP, which had overseen a sharp decline in Mongolia's economy during its control of Parliament in the

preceding years. Mongolians elected a DP member, Khaltmaa BATTULGA, as president in 2017.

GEOGRAPHY

Location: Northern Asia, between China and Russia

Geographic coordinates: 46 00 N, 105 00 E

Map references: Asia

Area: *total:* 1,564,116 sq km
land: 1,553,556 sq km
water: 10,560 sq km
country comparison to the world: 20

Area—comparative: slightly smaller than Alaska; more than twice the size of Texas

Land boundaries: *total:* 8,082 km
border countries (2): China 4630 km, Russia 3452 km

Coastline: 0 km (landlocked)

Maritime claims: none (landlocked)

Climate: desert; continental (large daily and seasonal temperature ranges)

Terrain: vast semidesert and desert plains, grassy steppe, mountains in west and southwest; Gobi Desert in south-central

Elevation: *mean elevation:* 1,528 m
lowest point: Hoh Nuur 560 m
highest point: Nayramadlin Orgil (Khuiten Peak) 4,374 m

Natural resources: oil, coal, copper, molybdenum, tungsten, phosphates, tin, nickel, zinc, fluorspar, gold, silver, iron

Land use: *agricultural land:* 73% (2011 est.)
arable land: 0.4% (2011 est.) / permanent crops: 0% (2011 est.) / permanent pasture: 72.6% (2011 est.)
forest: 7% (2011 est.)
other: 20% (2011 est.)
Irrigated land: 840 sq km (2012)

Population distribution: sparsely distributed population throughout the country; the capital of

Ulaanbaatar and the northern city of Darhan support the highest population densities

Natural hazards: dust storms; grassland and forest fires; drought; "zud," which is harsh winter conditions

Environment—current issues: limited natural freshwater resources in some areas; the burning of soft coal in power plants and the lack of enforcement of environmental laws leads to air pollution in Ulaanbaatar; deforestation and overgrazing increase soil erosion from wind and rain; water pollution; desertification and mining activities have a deleterious effect on the environment

Environment—international agreements: *party to:* Biodiversity, Climate Change, Climate Change-Kyoto Protocol, Desertification, Endangered Species, Environmental Modification, Hazardous Wastes, Law of the Sea, Ozone Layer Protection, Ship Pollution, Wetlands, Whaling
signed, but not ratified: none of the selected agreements

Geography—note: landlocked; strategic location between China and Russia

PEOPLE AND SOCIETY

Population: 3,168,026 (July 2020 est.)
note: Mongolia is one of the least densely populated countries in the world (2 people per sq km); twice as many ethnic Mongols (some 6 million) live in Inner Mongolia (Nei Mongol) in neighboring China
country comparison to the world: 135

Nationality: *noun:* Mongolian(s)
adjective: Mongolian

Ethnic groups: Khalkh 84.5%, Kazak 3.9%, Dorvod 2.4%, Bayad 1.7%, Buryat-Bouriates 1.3%, Zakhchin 1%, other 5.2% (2015 est.)

Languages: Mongolian 90% (official) (Khalkha dialect is predominant), Turkic, Russian (1999)

Religions: Buddhist 53%, Muslim 3%, Shamanist 2.9%, Christian 2.2%, other 0.4%, none 38.6% (2010 est.)

Age structure: *0-14 years:* 26.96% (male 435,596/female 418,524)

15-24 years: 14.93% (male 239,495/female 233,459)

25-54 years: 45.29% (male 694,481/female 740,334)

55-64 years: 8.04% (male 115,560/female 139,129)

65 years and over: 4.78% (male 60,966/female 90,482) (2020 est.)

Dependency ratios: *total dependency ratio:* 54.8
youth dependency ratio: 48.1
elderly dependency ratio: 6.7
potential support ratio: 15 (2020 est.)

Median age: *total:* 29.8 years
male: 28.8 years
female: 30.7 years (2020 est.)
country comparison to the world: 126

Population growth rate: 0.99% (2020 est.)
country comparison to the world: 105

Birth rate: 16.6 births/1,000 population (2020 est.)
country comparison to the world: 102

Death rate: 6.3 deaths/1,000 population (2020 est.)
country comparison to the world: 152

Net migration rate: -0.8 migrant(s)/1,000 population (2020 est.)
country comparison to the world: 136

Population distribution: sparsely distributed population throughout the country; the capital of Ulaanbaatar and the northern city of Darhan support the highest population densities

Urbanization: *urban population:* 68.7% of total population (2020)
rate of urbanization: 1.63% annual rate of change (2015-20 est.)
total population growth rate v. urban population growth rate, 2000-2030:

Major urban areas—population: 1.584 million ULAANBAATAR (capital) (2020)

Sex ratio: *at birth:* 1.05 male(s)/female
0-14 years: 1.04 male(s)/female
15-24 years: 1.03 male(s)/female
25-54 years: 0.94 male(s)/female
55-64 years: 0.83 male(s)/female
65 years and over: 0.67 male(s)/female
total population: 0.95 male(s)/female (2020 est.)

Mother's mean age at first birth: 20.5 years (2008 est.)

note: median age at first birth among women 20-24

Maternal mortality rate: 45 deaths/100,000 live births (2017 est.)
country comparison to the world: 97

Infant mortality rate: *total:* 19.2 deaths/1,000 live births
male: 22.2 deaths/1,000 live births
female: 16.2 deaths/1,000 live births (2020 est.)
country comparison to the world: 81

Life expectancy at birth: *total population:* 70.8 years
male: 66.6 years
female: 75.2 years (2020 est.)
country comparison to the world: 164

Total fertility rate: 1.95 children born/woman (2020 est.)
country comparison to the world: 120

Contraceptive prevalence rate: 48.1% (2018)

Drinking water source:
improved:
urban: 97.6% of population
rural: 59.2% of population
total: 85.4% of population
unimproved:
urban: 2.4% of population
rural: 40.8% of population
total: 14.6% of population (2017 est.)

Current Health Expenditure: 4% (2017)

Physicians density: 2.86 physicians/1,000 population (2016)

Hospital bed density: 8 beds/1,000 population (2017)

Sanitation facility access:
improved:
urban: 96.4% of population
rural: 65.1% of population
total: 86.5% of population
unimproved:
urban: 3.6% of population
rural: 34.9% of population
total: 13.5% of population (2017 est.)

HIV/AIDS—adult prevalence rate: <.1% (2019 est.)

HIV/AIDS—people living with HIV/AIDS: <1000 (2019 est.)

HIV/AIDS—deaths: <100 (2019 est.)

Obesity—adult prevalence rate: 20.6% (2016)
country comparison to the world: 96

Children under the age of 5 years underweight: 1.9% (2018)
country comparison to the world: 113

Education expenditures: 4.1% of GDP (2017)
country comparison to the world: 98

Literacy: *definition:* age 15 and over can read and write
total population: 98.4%
male: 98.2%
female: 98.6% (2018)

School life expectancy (primary to tertiary education): *total:* 15 years
male: 14 years
female: 16 years (2015)

Unemployment, youth ages 15-24: *total:* 16.8%
male: 15.8%
female: 18.4% (2018 est.)
country comparison to the world: 80

GOVERNMENT

Country name: *conventional long form:* none
conventional short form: Mongolia
local long form: none
local short form: Mongol Uls

former: Outer Mongolia, Mongolian People's Republic

etymology: the name means "Land of the Mongols" in Latin; the Mongolian name Mongol Uls translates as "Mongol State"

Government type: semi-presidential republic

Capital: *name:* Ulaanbaatar

geographic coordinates: 47 55 N, 106 55 E
time difference: UTC+8 (13 hours ahead of Washington, DC, during Standard Time)
daylight saving time: +1hr, begins last Saturday in March; ends last Saturday in September
note: Mongolia has two time zones—Ulaanbaatar Time (8 hours in advance of UTC) and Hovd Time (7 hours in advance of UTC)
etymology: the name means "red hero" in Mongolian and honors national hero Damdin Sukhbaatar, leader of the partisan army that with Soviet Red Army help, liberated Mongolia from Chinese occupation in the early 1920s

Administrative divisions: 21 provinces (aymguud, singular—aymag) and 1 municipality* (singular—hot); Arhangay, Bayanhongor, Bayan-Olgiy, Bulgan, Darhan-Uul, Dornod, Dornogovi, Dundgovi, Dzavhan (Zavkhan), Govi-Altay, Govisumber, Hentiy, Hovd, Hovsgol, Omnogovi, Orhon, Ovorhangay, Selenge, Suhbaatar, Tov, Ulaanbaatar*, Uvs

Independence: 29 December 1911 (independence declared from China; in actuality, autonomy attained); 11 July 1921 (from China)

National holiday: Naadam (games) holiday (commemorates independence from China in the 1921 Revolution), 11-15 July; Constitution Day (marks the date that the Mongolian People's Republic was created under a new constitution), 26 November (1924)

Constitution: *history:* several previous; latest adopted 13 January 1992, effective 12 February 1992
amendments: proposed by the State Great Hural, by the president of the republic, by the government, or by petition submitted to the State Great Hural by the Constitutional Court; conducting referenda on proposed amendments requires at least two-thirds majority vote of the State Great Hural; passage of amendments by the State Great Hural requires at least three-quarters majority vote; passage by referendum requires majority participation of qualified voters and a majority of votes; amended 1999, 2000, 2019

Legal system: civil law system influenced by Soviet and Romano-Germanic legal systems; constitution ambiguous on judicial review of legislative acts

International law organization participation: has not submitted an ICJ jurisdiction declaration; accepts ICCt jurisdiction

Citizenship: *citizenship by birth:* no
citizenship by descent only: both parents must be citizens of Mongolia; one parent if born within Mongolia
dual citizenship recognized: no
residency requirement for naturalization: 5 years

649

Suffrage: 18 years of age; universal

Executive branch: *chief of state:* President Khaltmaa BATTULGA (since 10 July 2017)

head of government: Prime Minister Ukhnaa KHURELSUKH (since 4 October 2017; re-elected by the Parliament 2 July 2020); Deputy Prime Minister Ulziisaikhan ENKHTUVSHIN (since 18 October 2017); note—Prime Minister Jargaltulga ERDENEBAT (since 8 July 2016) was voted out of office by the Parliament on 7 September 2017

cabinet: directly appointed by the prime minister following a constitutional amendment ratified in November 2019; prior to the amendment, the cabinet was nominated by the prime minister in consultation with the president and confirmed by the State Great Hural (parliament)

elections/appointments: presidential candidates nominated by political parties represented in the State Great Hural and directly elected by simple majority popular vote for a 4-year term (eligible for a second term); election last held on 26 June 2017 with a runoff held 7 July 2017 (next to be held in 2021); following legislative elections, the leader of the majority party or majority coalition is usually elected prime minister by the State Great Hural

election results: Khaltmaa BATTULGA elected president in second round; percent of vote in first round—Khaltmaa BATTULGA (DP) 38.1%, Miyegombo ENKHBOLD (MPP) 30.3%, Sainkhuu GANBAATAR (MPRP) 30.2%, invalid 1.4%; percent of vote in second round—Khaltmaa BATTULGA 55.2%, Miyegombo ENKHBOLD 44.8%; on 2 July 2020, Prime Minister Ukhnaa KHURELSUKH was reelected prime minister by the State Great Hural

Legislative branch: *description:* unicameral State Great Hural or Ulsyn Ikh Khural (76 seats; members directly elected in single-seat constituencies by simple majority vote; each constituency requires at least 50% voter participation for the poll to be valid; members serve 4-year terms)

elections: last held on 24 June 2020 (next to be held in 2024)

election results: percent of vote by party—MPP 44.9%, DP 24.5%, Our Coalition 8.1%, independent 8.7%, Right Person Electorate Coalition 5.2%, other 8.5%; seats by party—MPP 62, DP 11, Our Coalition 1, Right Person Electorate Coalition 1; independent 1; composition—63 men, 13 women; percent of women 17.1%; note—the MPRP, Civil Will-Green Party, and Mongolian Traditionally United Party formed Our Coalition for the 2020 election; the Right Person Electorate Coalition was established in 2020 by the National Labor Party, Mongolian Social Democratic Party, and Justice Party

Judicial branch: *highest courts:* Supreme Court (consists of the Chief Justice and 24 judges organized into civil, criminal, and administrative chambers); Constitutional Court or Tsets (consists of the chairman and 8 members)

judge selection and term of office: Supreme Court chief justice and judges appointed by the president upon recommendation by the General Council of

Courts—a 14-member body of judges and judicial officials—to the State Great Hural; appointment is for life; chairman of the Constitutional Court elected from among its members; members appointed from nominations by the State Great Hural—3 each by the president, the State Great Hural, and the Supreme Court; appointment is 6 years; chairmanship limited to a single renewable 3-year term

subordinate courts: aimag (provincial) and capital city appellate courts; soum, inter-soum, and district courts; Administrative Cases Courts

Political parties and leaders: Democratic Party or DP [Sodnomzundui ERDENE; resigned June 2020]
Mongolian National Democratic Party or MNDP [Bayanjargal TSOGTGEREL]
Mongolian People's Party or MPP [Ukhnaa KHURELSUKH]
Mongolian People's Revolutionary Party or MPRP [Nambar ENKHBAYAR]
Civil Will-Green Party or CWGP [Tserendorjiin GANKHUYAG]
Mongolian Traditionally United Party or MTUP [Batdelgeriin BATBOLD]
National Labor Party or HUN [B. NAIDALAA]
Mongolian Social Democratic Party or MSDP [A. GANBAATAR]
Justice Party [B. NASANBILEG]
note—there are 36 total registered parties as of March 2020

International organization participation: ADB, ARF, CD, CICA, CP, EBRD, EITI (compliant country), FAO, G-77, IAEA, IBRD, ICAO, ICC (NGOs), ICCt, ICRM, IDA, IFAD, IFC, IFRCS, ILO, IMF, IMO, IMSO, Interpol, IOC, IOM, IPU, ISO, ITSO, ITU, ITUC, MIGA, MINURSO, MONUSCO, NAM, OPCW, OSCE, SCO (observer), UN, UNAMID, UNCTAD, UNESCO, UNIDO, UNISFA, UNMISS, UNWTO, UPU, WCO, WHO, WIPO, WMO, WTO

Diplomatic representation in the US: *chief of mission:* Ambassador Yondon OTGONBAYAR (since 28 March 2018)

chancery: 2833 M Street NW, Washington, DC 20007

telephone: [1] (202) 333-7117

FAX: [1] (202) 298-9227

consulate(s) general: New York, San Francisco

Diplomatic representation from the US: *chief of mission:* Ambassador Michael S. KLECHESKI (since 22 February 2019)

telephone: [976] 7007-6001

embassy: Denver Street #3, 11th Micro-District, Ulaanbaatar 14190

mailing address: P.O. Box 341, Ulaanbaatar 14192

FAX: [976] 7007-6016

Flag description: three, equal vertical bands of red (hoist side), blue, and red; centered on the hoist-side red band in yellow is the national emblem ("soyombo"—a columnar arrangement of abstract and geometric representation for fire, sun, moon, earth, water, and the yin-yang symbol);

blue represents the sky, red symbolizes progress and prosperity

National symbol(s): *soyombo emblem; national colors:* red, blue, yellow

National anthem: *name:* "Mongol ulsyn toriin duulal" (National Anthem of Mongolia)

lyrics/music: Tsendiin DAMDINSUREN/Bilegiin DAMDINSUREN and Luvsanjamts MURJORJ

note: music adopted 1950, lyrics adopted 2006; lyrics altered on numerous occasions

ECONOMY

Economy—overview: Foreign direct investment in Mongolia's extractive industries—which are based on extensive deposits of copper, gold, coal, molybdenum, fluorspar, uranium, tin, and tungsten—has transformed Mongolia's landlocked economy from its traditional dependence on herding and agriculture. Exports now account for more than 40% of GDP. Mongolia depends on China for more than 60% of its external trade—China receives some 90% of Mongolia's exports and supplies Mongolia with more than one-third of its imports. Mongolia also relies on Russia for 90% of its energy supplies, leaving it vulnerable to price increases. Remittances from Mongolians working abroad, particularly in South Korea, are significant.

Soviet assistance, at its height one-third of GDP, disappeared almost overnight in 1990 and 1991 at the time of the dismantlement of the USSR. The following decade saw Mongolia endure both deep recession, because of political inaction, and natural disasters, as well as strong economic growth, because of market reforms and extensive privatization of the formerly state-run economy. The country opened a fledgling stock exchange in 1991. Mongolia joined the WTO in 1997 and seeks to expand its participation in regional economic and trade regimes.

Growth averaged nearly 9% per year in 2004-08 largely because of high copper prices globally and new gold production. By late 2008, Mongolia was hit by the global financial crisis and Mongolia's real economy contracted 1.3% in 2009. In early 2009, the IMF reached a $236 million Stand-by Arrangement with Mongolia and it emerged from the crisis with a stronger banking sector and better fiscal management. In October 2009, Mongolia passed long-awaited legislation on an investment agreement to develop the Oyu Tolgoi (OT) mine, among the world's largest untapped copper-gold deposits. However, a dispute with foreign investors developing OT called into question the attractiveness of Mongolia as a destination for foreign investment. This caused a severe drop in FDI, and a slowing economy, leading to the dismissal of Prime Minister Norovyn ALTANKHUYAG in November 2014. The economy had grown more than 10% per year between 2011 and 2013—largely on the strength of commodity exports and high government spending—before slowing to 7.8% in 2014, and falling to the 2% level in 2015. Growth rebounded from a brief 1.6% contraction in the third quarter of 2016 to 5.8% during the first three quarters of 2017, largely due to rising commodity prices.

The May 2015 agreement with Rio Tinto to restart the OT mine and the subsequent $4.4 billion finance package signing in December 2015 stemmed the loss of investor confidence. The current government has made restoring investor trust and reviving the economy its top priority, but has failed to invigorate the economy in the face of the large drop-off in foreign direct investment, mounting external debt, and a sizeable budget deficit. Mongolia secured a $5.5 billion financial assistance package from the IMF and a host of international creditors in May 2017, which is expected to improve Mongolia's long-term fiscal and economic stability as long as Ulaanbaatar can advance the agreement's difficult contingent reforms, such as consolidating the government's off-balance sheet liabilities and rehabilitating the Mongolian banking sector.

GDP (purchasing power parity):
$43.54 billion (2018)
$39.73 billion (2017 est.)
$37.81 billion (2016 est.)
note: data are in 2017 dollars
country comparison to the world: 114

GDP (official exchange rate):
$11.14 billion (2017 est.)

GDP—real growth rate: 5.1% (2017 est.)
1.2% (2016 est.)
2.4% (2015 est.)
country comparison to the world: 44

GDP—per capita (PPP): $13,700 (2018)
$13,000 (2017 est.)
$12,500 (2016 est.)
note: data are in 2017 dollars
country comparison to the world: 118

Gross national saving:
26.9% of GDP (2017 est.)
23.1% of GDP (2016 est.)
22.4% of GDP (2015 est.)
country comparison to the world: 46

GDP—composition, by end use:
household consumption: 49.2% (2017 est.)
government consumption: 12.3% (2017 est.)
investment in fixed capital: 23.8% (2017 est.)
investment in inventories: 12.4% (2017 est.)
exports of goods and services: 59.5% (2017 est.)
imports of goods and services: -57.1% (2017 est.)

GDP—composition, by sector of origin:
agriculture: 12.1% (2017 est.)
industry: 38.2% (2017 est.)
services: 49.7% (2017 est.)

Agriculture—products: wheat, barley, vegetables, forage crops; sheep, goats, cattle, camels, horses

Industries: construction and construction materials; mining (coal, copper, molybdenum, fluorspar, tin, tungsten, gold); oil; food and beverages; processing of animal products, cashmere and natural fiber manufacturing

Industrial production growth rate: -1% (2017 est.)
country comparison to the world: 176

Labor force: 1.241 million (2017 est.)

country comparison to the world: 132

Labor force—by occupation: *agriculture:* 31.1%
industry: 18.5%
services: 50.5% (2016)

Unemployment rate: 8% (2017 est.)
7.9% (2016 est.)
country comparison to the world: 124

Population below poverty line: 29.6% (2016 est.)

Household income or consumption by percentage share: *lowest 10%:* 13.7%
highest 10%: 5.7% (2011)

Budget: *revenues:* 2.967 billion (2017 est.)
expenditures: 3.681 billion (2017 est.)

Taxes and other revenues: 26.6% (of GDP) (2017 est.)
country comparison to the world: 106

Budget surplus (+) or deficit (-): -6.4% (of GDP) (2017 est.)
country comparison to the world: 187

Public debt: 91.4% of GDP (2017 est.)
90% of GDP (2016 est.)
country comparison to the world: 24

Fiscal year: calendar year

Inflation rate (consumer prices): 4.6% (2017 est.)
0.5% (2016 est.)
country comparison to the world: 169

Current account balance: -$1.155 billion (2017 est.)

-$700 million (2016 est.)
country comparison to the world: 149

Exports: $7.012 billion (2018)
$5.834 billion (2017 est.)
$4.916 billion (2016 est.)
country comparison to the world: 100

Exports—partners: China 93.3%, UK 2.5% (2017)

Exports—commodities: copper, apparel, livestock, animal products, cashmere, wool, hides, fluorspar, other nonferrous metals, coal, crude oil

Imports: $5.875 billion (2018)
$4.345 billion (2017 est.)
$3.466 billion (2016 est.)
country comparison to the world: 121

Imports—commodities: machinery and equipment, fuel, cars, food products, industrial consumer goods, chemicals, building materials, cigarettes and tobacco, appliances, soap and detergent

Imports—partners: China 32.6%, Russia 28.1%, Japan 8.4%, US 4.8%, South Korea 4.6% (2017)

Reserves of foreign exchange and gold: $3.016 billion (31 December 2017 est.)
$1.296 billion (31 December 2016 est.)
country comparison to the world: 109

Debt—external: $25.33 billion (31 December 2017 est.)
$24.63 billion (31 December 2016 est.)
country comparison to the world: 88

Exchange rates: togrog/tugriks (MNT) per US dollar—
2,378.1 (2017 est.)
2,140.3 (2016 est.)

2,140.3 (2015 est.)
1,970.3 (2014 est.)
1,817.9 (2013 est.)

ENERGY

Electricity access: *electrification—total population:* 91% (2019)
electrification—urban areas: 99% (2019)
electrification—rural areas: 73% (2019)

Electricity—production: 5.339 billion kWh (2016 est.)
country comparison to the world: 120

Electricity—consumption: 5.932 billion kWh (2016 est.)
country comparison to the world: 115

Electricity—exports: 51 million kWh (2015 est.)
country comparison to the world: 87

Electricity—imports: 1.446 billion kWh (2016 est.)
country comparison to the world: 62

Electricity—installed generating capacity: 1.134 million kW (2016 est.)
country comparison to the world: 125

Electricity—from fossil fuels: 87% of total installed capacity (2016 est.)
country comparison to the world: 63

Electricity—from nuclear fuels: 0% of total installed capacity (2017 est.)
country comparison to the world: 145

Electricity—from hydroelectric plants: 2% of total installed capacity (2017 est.)
country comparison to the world: 140

Electricity—from other renewable sources: 11% of total installed capacity (2017 est.)
country comparison to the world: 79

Crude oil—production: 20,000 bbl/day (2018 est.)
country comparison to the world: 66

Crude oil—exports: 14,360 bbl/day (2015 est.)
country comparison to the world: 56

Crude oil—imports: 0 bbl/day (2015 est.)
country comparison to the world: 168

Crude oil—proved reserves: NA bbl (1 January 2017)

Refined petroleum products—production: 0 bbl/day (2015 est.)
country comparison to the world: 179

Refined petroleum products—consumption: 27,000 bbl/day (2016 est.)
country comparison to the world: 124

Refined petroleum products—exports: 0 bbl/day (2015 est.)
country comparison to the world: 183

Refined petroleum products—imports: 24,190 bbl/day (2015 est.)
country comparison to the world: 109

Natural gas—production: 0 cu m (2017 est.)
country comparison to the world: 172

Natural gas—consumption: 0 cu m (2017 est.)
country comparison to the world: 176

Natural gas—exports: 0 cu m (2017 est.)
country comparison to the world: 153

Natural gas—imports: 0 cu m (2017 est.)
country comparison to the world: 158

Natural gas—proved reserves: 0 cu m (1 January 2014 est.)
country comparison to the world: 171

Carbon dioxide emissions from consumption of energy: 19.86 million Mt (2017 est.)
country comparison to the world: 86

COMMUNICATIONS

Telephones—fixed lines: *total subscriptions:* 385,191
subscriptions per 100 inhabitants: 12.28 (2019 est.)
country comparison to the world: 102

Telephones—mobile cellular: *total subscriptions:* 4,297,643
subscriptions per 100 inhabitants: 137.01 (2019 est.)
country comparison to the world: 127

Telecommunication systems: *general assessment:* liberalized and competitive telecoms market; mobile broadband seen steady growth, but fixed-line broadband is an attractive option; a fiber-optic network has been installed that is improving broadband and communication services between major urban centers with multiple companies providing inter-city fiber-optic cable services; compared to other Asian countries, Mongolia's growth in telecommunications is moderate; mobile broadband is growing with 4 competitive MNOs (mobile network operators) along with better tariffs; 3G mobile broadband products are very popular, launch of 4G LTE services by all major operators; in May 2018 a South Korean company completed the sale of 40% stake back to Mongolian government (2020)

domestic: very low fixed-line teledensity 12 per 100; there are four mobile-cellular providers and subscribership is increasing with 137 per 100 persons (2019)

international: country code—976; satellite earth stations—7 (2016)
note: the COVID-19 outbreak is negatively impacting telecommunications production and supply chains globally; consumer spending on telecom devices and services has also slowed due to the pandemic's effect on economies worldwide; overall progress towards improvements in all facets of the telecom industry—mobile, fixed-line, broadband, submarine cable and satellite—has moderated

Broadcast media: following a law passed in 2005, Mongolia's state-run radio and TV provider converted to a public service provider; also available are 68 radio and 160 TV stations, including multi-channel satellite and cable TV providers; transmissions of multiple international broadcasters are available (2019)

Internet country code: .mn

Internet users: *total:* 735,823

percent of population: 23.71% (July 2018 est.)
country comparison to the world: 147

Broadband—fixed subscriptions: *total:* 306,150
subscriptions per 100 inhabitants: 10 (2018 est.)
country comparison to the world: 102

TRANSPORTATION

National air transport system: *number of registered air carriers: 4 (2020)*
inventory of registered aircraft operated by air carriers: 12
annual passenger traffic on registered air carriers: 670,360 (2018)
annual freight traffic on registered air carriers: 7.82 million mt-km (2018)

Civil aircraft registration country code prefix: JU (2016)

Airports: 44 (2013)
country comparison to the world: 98

Airports—with paved runways: *total:* 15 (2017)
over 3,047 m: 2 (2017)
2,438 to 3,047 m: 10 (2017)
1,524 to 2,437 m: 3 (2017)

Airports—with unpaved runways: *total:* 29 (2013)
over 3,047 m: 2 (2013)
2,438 to 3,047 m: 2 (2013)
1,524 to 2,437 m: 24 (2013)
under 914 m: 1 (2013)

Heliports: 1 (2013)

Railways: *total:* 1,815 km (2017)
broad gauge: 1,815 km 1.520-m gauge (2017)
note: national operator Ulaanbaatar Railway is jointly owned by the Mongolian Government and by the Russian State Railway
country comparison to the world: 78

Roadways: *total:* 113,200 km (2017)
paved: 10,600 km (2017)
unpaved: 102,600 km (2017)
country comparison to the world: 44

Waterways: 580 km (the only waterway in operation is Lake Hovsgol) (135 km); Selenge River

(270 km) and Orhon River (175 km) are navigable but carry little traffic; lakes and rivers ice free from May to September) (2010)
country comparison to the world: 81

Merchant marine: *total:* 271
by type: bulk carrier 2, container ship 3, general cargo 100, oil tanker 72, other 94 (2019)
country comparison to the world: 55

MILITARY AND SECURITY

Military and security forces: *Mongolian Armed Forces (Mongol ulsyn zevsegt huchin):* Mongolian Army (includes Border Troops), Mongolian Air Force, National Center for Emergency and Disaster Relief (coordinates the military's efforts as first-responders for earthquakes, wildfires, and forest fires; contagious diseases; and snow and dust storms as well as severe winters (known as zu d)); paramilitary forces: Internal Security Troops (2019)

Military expenditures: 0.7% of GDP (2019)
0.7% of GDP (2018)
0.8% of GDP (2017)
0.9% of GDP (2016)
0.9% of GDP (2015)
country comparison to the world: 138

Military and security service personnel strengths: size estimates for the Mongolian Armed Forces (MAF) vary; approximately 8,000 active duty troops (7,000 Army; 800 Air Force); est. 6,000 Border Guard; est. 1,200 Internal Security Troops (2019)

Military equipment inventories and acquisitions: the MAF are armed with Soviet-era equipment supplemented by deliveries of second-hand Russian weapons; since 2010, Russia is the sole provider of armaments to Mongolia (2019 est.)

Military deployments: 850 South Sudan (UNMISS); 230 Afghanistan (NATO) (2020)

Military service age and obligation: 18-27 years of age for compulsory and voluntary military service; 1-year conscript service obligation in army or air forces or police for males only; after conscription, soldiers can contract into military service for 2 or 4 years; citizens can also voluntarily join the armed forces (2017)

TRANSNATIONAL ISSUES

Disputes—international: none

Refugees and internally displaced persons: *stateless persons:* 17 (2019)

MONTENEGRO

INTRODUCTION

Background: The use of the name Crna Gora or Black Mountain (Montenegro) began in the 13th century in reference to a highland region in the Serbian province of Zeta. The later medieval state of Zeta maintained its existence until 1496 when Montenegro finally fell under Ottoman rule. Over subsequent centuries, Montenegro managed to maintain a level of autonomy within the Ottoman Empire. From the 16th to 19th centuries, Montenegro was a theocracy ruled by a series of bishop princes; in 1852, it transformed into a secular principality. Montenegro was recognized as an independent sovereign principality at the Congress of Berlin in 1878. After World War I,

during which Montenegro fought on the side of the Allies, Montenegro was absorbed by the Kingdom of Serbs, Croats, and Slovenes, which became the Kingdom of Yugoslavia in 1929. At the conclusion of World War II, it became a constituent republic of the Socialist Federal Republic of Yugoslavia. When the latter dissolved in 1992, Montenegro joined with Serbia, creating the Federal Republic of Yugoslavia and, after 2003, shifting to a looser State Union of Serbia and Montenegro. In May 2006, Montenegro invoked its right under the Constitutional Charter of Serbia and Montenegro to hold a referendum on independence from the two-state union. The vote for severing ties with Serbia barely exceeded 55%—the threshold set by the EU—allowing Montenegro to formally restore its independence on 3 June 2006. In 2017, Montenegro joined NATO and is currently completing its EU accession process, having officially applied to join the EU in December 2008.

GEOGRAPHY

Location: Southeastern Europe, between the Adriatic Sea and Serbia

Geographic coordinates: 42 30 N, 19 18 E

Map references: Europe

Area: *total:* 13,812 sq km
land: 13,452 sq km
water: 360 sq km
country comparison to the world: 162

Area—comparative: slightly smaller than Connecticut; slightly larger than twice the size of Delaware

Land boundaries: *total:* 680 km
border countries (5): Albania 186 km, Bosnia and Herzegovina 242 km, Croatia 19 km, Kosovo 76 km, Serbia 157 km

Coastline: 293.5 km

Maritime claims: *territorial sea:* 12 nm
continental shelf: defined by treaty

Climate: Mediterranean climate, hot dry summers and autumns and relatively cold winters with heavy snowfalls inland

Terrain: highly indented coastline with narrow coastal plain backed by rugged high limestone mountains and plateaus

Elevation: *mean elevation:* 1,086 m
lowest point: Adriatic Sea 0 m
highest point: Bobotov Kuk 2,522 m

Natural resources: bauxite, hydroelectricity

Land use: *agricultural land:* 38.2% (2011 est.)
arable land: 12.9% (2011 est.) / *permanent crops:* 1.2% (2011 est.) / *permanent pasture:* 24.1% (2011 est.)
forest: 40.4% (2011 est.)
other: 21.4% (2011 est.)
Irrigated land: 24 sq km (2012)

Population distribution: highest population density is concentrated in the south, southwest; the extreme eastern border is the least populated area

Natural hazards: destructive earthquakes

Environment—current issues: pollution of coastal waters from sewage outlets, especially in tourist-related areas such as Kotor; serious air pollution in Podgorica, Pljevlja and Niksie; air pollution in Pljevlja is caused by the nearby lignite power plant and the domestic use of coal and wood for household heating

Environment—international agreements: *party to:* Air Pollution, Biodiversity, Climate Change, Climate Change-Kyoto Protocol, Desertification, Hazardous Wastes, Law of the Sea, Marine Dumping, Marine Life Conservation, Ozone Layer Protection, Ship Pollution
signed, but not ratified: none of the selected agreements

Geography—note: strategic location along the Adriatic coast

PEOPLE AND SOCIETY

Population: 609,859 (July 2020 est.)
country comparison to the world: 171

Nationality: *noun:* Montenegrin(s)
adjective: Montenegrin

Ethnic groups: Montenegrin 45%, Serbian 28.7%, Bosniak 8.7%, Albanian 4.9%, Muslim 3.3%, Romani 1%, Croat 1%, other 2.6%, unspecified 4.9% (2011 est.)

Languages: Serbian 42.9%, Montenegrin (official) 37%, Bosnian 5.3%, Albanian 5.3%, Serbo-Croat 2%, other 3.5%, unspecified 4% (2011 est.)

Religions: Orthodox 72.1%, Muslim 19.1%, Catholic 3.4%, atheist 1.2%, other 1.5%, unspecified 2.6% (2011 est.)

Age structure: *0-14 years:* 18.14% (male 57,402/female 53,217)
15-24 years: 12.78% (male 40,220/female 37,720)
25-54 years: 39.65% (male 120,374/female 121,461)
55-64 years: 13.41% (male 40,099/female 41,670)
65 years and over: 16.02% (male 42,345/female 55,351) (2020 est.)

Dependency ratios: *total dependency ratio:* 51.1
youth dependency ratio: 27.3
elderly dependency ratio: 23.8
potential support ratio: 4.2 (2020 est.)

Median age: *total:* 39.6 years
male: 38.1 years

female: 41.1 years (2020 est.)
country comparison to the world: 54

Population growth rate: -0.37% (2020 est.)
country comparison to the world: 221

Birth rate: 11.5 births/1,000 population (2020 est.)
country comparison to the world: 171

Death rate: 10.4 deaths/1,000 population (2020 est.)
country comparison to the world: 30

Net migration rate: -4.9 migrant(s)/1,000 population (2020 est.)
country comparison to the world: 196

Population distribution: highest population density is concentrated in the south, southwest; the extreme eastern border is the least populated area

Urbanization: *urban population:* 67.5% of total population (2020)
rate of urbanization: 0.54% annual rate of change (2015-20 est.)
total population growth rate v. urban population growth rate, 2000-2030:

Major urban areas—population: 177,000 PODGORICA (capital) (2018)

Sex ratio: *at birth:* 1.04 male(s)/female
0-14 years: 1.08 male(s)/female
15-24 years: 1.07 male(s)/female
25-54 years: 0.99 male(s)/female
55-64 years: 0.96 male(s)/female
65 years and over: 0.77 male(s)/female
total population: 0.97 male(s)/female (2020 est.)

Mother's mean age at first birth: 26.3 years (2010 est.)

Maternal mortality rate: 6 deaths/100,000 live births (2017 est.)
country comparison to the world: 162

Infant mortality rate: *total:* 3.4 deaths/1,000 live births
male: 2.8 deaths/1,000 live births
female: 4 deaths/1,000 live births (2020 est.)
country comparison to the world: 202

Life expectancy at birth: *total population:* 77.3 years
male: 74.8 years
female: 79.8 years (2020 est.)
country comparison to the world: 84

Total fertility rate: 1.82 children born/woman (2020 est.)
country comparison to the world: 148

Contraceptive prevalence rate: 20.7% (2018)

Drinking water source:
improved:
urban: 100% of population
rural: 100% of population
total: 99.8% of population
unimproved:
urban: 0% of population
rural: 0% of population
total: 0.2% of population (2017 est.)

Current Health Expenditure: 7.6% (2016)

Physicians density: 2.38 physicians/1,000 population (2015)

Hospital bed density: 3.9 beds/1,000 population (2017)

Sanitation facility access:
improved:
urban: 100% of population
rural: 93.9% of population
total: 97.8% of population
unimproved:
urban: 0% of population
rural: 6.1% of population
total: 2.2% of population (2017 est.)

HIV/AIDS—adult prevalence rate: <.1% (2019 est.)

HIV/AIDS—people living with HIV/AIDS: <500 (2019 est.)

HIV/AIDS—deaths: <100 (2019 est.)

Major infectious diseases: *degree of risk:* intermediate (2020)

food or waterborne diseases: bacterial diarrhea

vectorborne diseases: Crimean-Congo hemorrhagic fever

Obesity—adult prevalence rate: 23.3% (2016)
country comparison to the world: 66

Children under the age of 5 years underweight: 1% (2013)
country comparison to the world: 127

Education expenditures: NA

Literacy: *definition:* age 15 and over can read and write
total population: 98.8%
male: 99.5%
female: 98.3% (2018)

School life expectancy (primary to tertiary education): *total:* 15 years
male: 15 years
female: 15 years (2019)

Unemployment, youth ages 15-24: *total:* 29.4%
male: 33.3%
female: 23.6% (2018 est.)
country comparison to the world: 34

GOVERNMENT

Country name: *conventional long form:* none
conventional short form: Montenegro
local long form: none
local short form: Crna Gora

former: People's Republic of Montenegro, Socialist Republic of Montenegro, Republic of Montenegro
etymology: the country's name locally as well as in most Western European languages means "black mountain" and refers to the dark coniferous forests on Mount Lovcen and the surrounding area

Government type: parliamentary republic

Capital: *name:* Podgorica; note—Cetinje retains the status of "Old Royal Capital"

geographic coordinates: 42 26 N, 19 16 E
time difference: UTC + 1 (6 hours ahead of Washington, DC, during Standard Time)
daylight saving time: +1 hr, begins last Sunday in March; ends last Sunday in October

etymology: the name translates as "beneath Gorica"; the meaning of Gorica is "hillock"; the reference is to the small hill named Gorica that the city is built around

Administrative divisions: 24 municipalities (opstine, singular—opstina); Andrijevica, Bar, Berane, Bijelo Polje, Budva, Cetinje, Danilovgrad, Gusinje, Herceg Novi, Kolasin, Kotor, Mojkovac, Niksic, Petnijica, Plav, Pljevlja, Pluzine, Podgorica, Rozaje, Savnik, Tivat, Tuzi, Ulcinj, Zabljak

Independence: *3 June 2006 (from the State Union of Serbia and Montenegro); notable earlier dates:* 13 March 1852 (Principality of Montenegro established); 13 July 1878 (Congress of Berlin recognizes Montenegrin independence); 28 August 1910 (Kingdom of Montenegro established)

National holiday: National Day, 13 July (1878, the day the Berlin Congress recognized Montenegro as the 27th independent state in the world, and 1941, the day the Montenegrins staged an uprising against fascist occupiers and sided with the partisan communist movement)

Constitution: *history:* several previous; latest adopted 22 October 2007
amendments: proposed by the president of Montenegro, by the government, or by at least 25 members of the Assembly; passage of draft proposals requires two-thirds majority vote of the Assembly, followed by a public hearing; passage of draft amendments requires two-thirds majority vote of the Assembly; changes to certain constitutional articles, such as sovereignty, state symbols, citizenship, and constitutional change procedures, require three-fifths majority vote in a referendum; amended 2013, 2014

Legal system: civil law

International law organization participation: has not submitted an ICJ jurisdiction declaration; accepts ICCt jurisdiction

Citizenship: *citizenship by birth:* no
citizenship by descent only: at least one parent must be a citizen of Montenegro
dual citizenship recognized: no
residency requirement for naturalization: 10 years

Suffrage: 18 years of age; universal

Executive branch: *chief of state:* President Milo DJUKANOVIC (since 20 May 2018)

head of government: Prime Minister Dusko MARKOVIC (since 28 November 2016)
cabinet: Ministers act as cabinet
elections/appointments: president directly elected by absolute majority popular vote in 2 rounds if needed for a 5-year term (eligible for a second term); election last held on 15 April 2018 (next to be held in 2023); prime minister nominated by the president, approved by the Assembly
election results: Milo DJUKANOVIC elected president in the first round; percent of vote—Milo DJUKANOVIC (DPS) 53.9%, Mladen BOJANIC (independent) 33.4%, Draginja VUKSANOVIC (SDP) 8.2%, Marko MILACIC (PRAVA) 2.8%, other 1.7%

Legislative branch: *description:* unicameral Assembly or Skupstina (81 seats; members directly elected in a single nationwide constituency by proportional representation vote; members serve 4-year terms)
elections: last held on 30 August 2020 (next to be held in 2024)
election results: percent of vote by party/coalition—DPS 35.1%, ZBCG 32.6%, MNIM 12.5%, URA 5.5%, SD 4.1%, BS 3.9%, SDP 3.1%, AL 1.6%, AK 1.1%, other 0.4%; seats by party/coalition—DPS 30, ZBCG 27, MNIM 10, URA 4, BS 3, SD 3, SDP 2, AL 1, AK 1.; composition—men 57, women 24, percent of women 29.6%

Judicial branch: *highest courts:* Supreme Court or Vrhovni Sud (consists of the court president, deputy president, and 15 judges); Constitutional Court or Ustavni Sud (consists of the court president and 7 judges)
judge selection and term of office: Supreme Court president proposed by general session of the Supreme Court and elected by the Judicial Council, a 9-member body consisting of judges, lawyers designated by the Assembly, and the minister of judicial affairs; Supreme Court president elected for a single renewable, 5-year term; other judges elected by the Judicial Council for life; Constitutional Court judges—2 proposed by the president of Montenegro and 5 by the Assembly, and elected by the Assembly; court president elected from among the court members; court president elected for a 3-year term, other judges serve 9-year terms

subordinate courts: Administrative Courts; Appellate Court; Commercial Courts; High Courts; basic courts

Political parties and leaders: Albanian Alternative or AA [Nik DJELOSAJ]
Albanian Coalition (includes DP, DSCG, DUA)
Albanian Coalition Perspective or AKP
Albanian List (coalition includes AA, Forca, AKP, DSA)
Bosniak Party or BS [Rafet HUSOVIC]
Croatian Civic Initiative or HGI [Marija VUCINOVIC]
Croatian Reform Party [Marija VUCINOVIC]
Democratic Alliance or DEMOS [Miodrag LEKIC]
Democratic Front or DF [collective leadership] (coalition includes NOVA, PZP, DNP, RP)
Democratic League in Montenegro or DSCG [Mehmet BARDHI]
Democratic League of Albanians or DSA
Democratic Montenegro or DCG [Alexsa BECIC]
Democratic Party or DP [Fatmir GJEKA]
Democratic Party of Socialists or DPS [Milo DJUKANOVIC]
Democratic Party of Unity or DSJ [Nebojsa JUSKOVIC]
Democratic People's Party or DNP [Milan KNEZEVIC]
Democratic Serb Party or DSS [Dragica PEROVIC]
Democratic Union of Albanians or DUA [Mehmet ZENKA]
For the Future of Montenegro or ZBCG [Zdravko KRIVOKAPIC] (electoral coalition includes SNP

and 2 alliances—DF, NP) Liberal Party or LP [Andrija POPOVIC]

Movement for Change or PZP [Nebojsa MEDOJEVIC]

New Democratic Power or FORCA [Nazif CUNGU]

New Serb Democracy or NOVA [Andrija MANDIC]

Party of Pensioners, Disabled, and Restitution or PUPI [Momir JOKSIMOVIC]

Peace is Our Nation or MNIM [Alexa BECIC] (coalition includes Democrats, DEMOS, New Left, PUPI]

Popular Movement or NP [Miodrag DAVIDOVIC] (coalition includes DEMOS, RP, UCG, and several minor parties)

Social Democratic Party or SDP [Ranko KRIVOKAPIC]

Social Democrats or SD [Ivan BRAJOVIC]

Socialist People's Party or SNP [Vladimir JOKOVIC]

True Montenegro or PRAVA [Marko MILACIC]

United Montenegro or UCG [Goran DANILOVIC] (split from DEMOS)

United Reform Action or URA [Dritan ABAZOVIC]

Workers' Party or RP [Janko VUCINIC]

International organization participation: CE, CEI, EAPC, EBRD, FAO, IAEA, IBRD, ICAO, ICC (NGOs), ICCt, ICRM, IDA, IFC, IFRCS, IHO, ILO, IMF, IMO, IMSO, Interpol, IOC, IOM, IPU, ISO (correspondent), ITSO, ITU, ITUC (NGOs), MIGA, OAS (observer), OIF (obsérver), OPCW, OSCE, PCA, PFP, SELEC, UN, UNCTAD, UNESCO, UNHCR, UNIDO, UNWTO, UPU, WCO, WHO, WIPO, WMO, WTO

Diplomatic representation in the US: *chief of mission:* Ambassador Nebojsa KALUDEROVIC (since 18 January 2017)

chancery: 1610 New Hampshire Avenue NW, Washington, DC, 20009
telephone: [1] (202) 234-6108
FAX: [1] (202) 234-6109
consulate(s) general: New York

Diplomatic representation from the US: *chief of mission:* Ambassador Judy Rising REINKE (since 20 December 2018)
telephone: +382 (0)20 410 500
embassy: Dzona Dzeksona 2, 81000 Podgorica
mailing address: use embassy street address
FAX: [382] 20-241-358

Flag description: a red field bordered by a narrow golden-yellow stripe with the Montenegrin coat of arms centered; the arms consist of a double-headed golden eagle—symbolizing the unity of church and state—surmounted by a crown; the eagle holds a golden scepter in its right claw and a blue orb in its left; the breast shield over the eagle shows a golden lion passant on a green field in front of a blue sky; the lion is a symbol of episcopal authority and harkens back to the three and a half centuries when Montenegro was ruled as a theocracy

National symbol(s): *double-headed eagle; national colors:* red, gold

National anthem: *name:* "Oj, svijetla majska zoro" (Oh, Bright Dawn of May)
lyrics/music: Sekula DRLJEVIC/unknown, arranged by Zarko MIKOVIC
note: adopted 2004; music based on a Montenegrin folk song
0:00/ 2:05

ECONOMY

Economy — overview: Montenegro's economy is transitioning to a market system. Around 90% of Montenegrin state-owned companies have been privatized, including 100% of banking, telecommunications, and oil distribution. Tourism, which accounts for more than 20% of Montenegro's GDP, brings in three times as many visitors as Montenegro's total population every year. Several new luxury tourism complexes are in various stages of development along the coast, and a number are being offered in connection with nearby boating and yachting facilities. In addition to tourism, energy and agriculture are considered two distinct pillars of the economy. Only 20% of Montenegro's hydropower potential is utilized. Montenegro plans to become a net energy exporter, and the construction of an underwater cable to Italy, which will be completed by the end of 2018, will help meet its goal.

Montenegro uses the euro as its domestic currency, though it is not an official member of the euro zone. In January 2007, Montenegro joined the World Bank and IMF, and in December 2011, the WTO. Montenegro began negotiations to join the EU in 2012, having met the conditions set down by the European Council, which called on Montenegro to take steps to fight corruption and organized crime.

The government recognizes the need to remove impediments in order to remain competitive and open the economy to foreign investors. Net foreign direct investment in 2017 reached $848 million and investment per capita is one of the highest in Europe, due to a low corporate tax rate. The biggest foreign investors in Montenegro in 2017 were Norway, Russia, Italy, Azerbaijan and Hungary.

Montenegro is currently planning major overhauls of its road and rail networks, and possible expansions of its air transportation system. In 2014, the Government of Montenegro selected two Chinese companies to construct a 41 km-long section of the country's highway system, which will become part of China's Belt and Road Initiative. Cheaper borrowing costs have stimulated Montenegro's growing debt, which currently sits at 65.9% of GDP, with a forecast, absent fiscal consolidation, to increase to 80% once the repayment to China's Ex/Im Bank of a €800 million highway loan begins in 2019. Montenegro first instituted a value-added tax (VAT) in April 2003, and introduced differentiated VAT rates of 17% and 7% (for tourism) in January 2006. The Montenegrin Government increased the non-tourism Value Added Tax (VAT) rate to 21% as of January 2018, with the goal of reducing its public debt.

GDP (purchasing power parity):
$11.08 billion (2017 est.)
$10.63 billion (2016 est.)
$10.32 billion (2015 est.)
note: data are in 2017 dollars
country comparison to the world: 160

GDP (official exchange rate):
$4.784 billion (2017 est.)

GDP—real growth rate: 4.3% (2017 est.)
2.9% (2016 est.)
3.4% (2015 est.)
country comparison to the world: 67

GDP—per capita (PPP): $17,800 (2017 est.)
$17,100 (2016 est.)
$16,600 (2015 est.)
note: data are in 2017 dollars
country comparison to the world: 100

Gross national saving:
13.2% of GDP (2017 est.)
9.9% of GDP (2016 est.)
9.1% of GDP (2015 est.)
country comparison to the world: 144

GDP—composition, by end use:
household consumption: 76.8% (2016 est.)
government consumption: 19.6% (2016 est.)
investment in fixed capital: 23.2% (2016 est.)
investment in inventories: 2.9% (2016 est.)
exports of goods and services: 40.5% (2016 est.)
imports of goods and services: -63% (2016 est.)

GDP—composition, by sector of origin:
agriculture: 7.5% (2016 est.)
industry: 15.9% (2016 est.)
services: 76.6% (2016 est.)

Agriculture—products: tobacco, potatoes, citrus fruits, olives and related products, grapes; sheep, wine

Industries: steelmaking, aluminum, agricultural processing, consumer goods, tourism

Industrial production growth rate: -4.2% (2017 est.)
country comparison to the world: 194

Labor force: 167,000 (2020 est.)
country comparison to the world: 174

Labor force—by occupation: *agriculture:* 7.9%
industry: 17.1%
services: 75% (2017 est.)

Unemployment rate: 15.82% (2019 est.)
18.8% (2018 est.)
country comparison to the world: 177

Population below poverty line: 8.6% (2013 est.)

Household income or consumption by percentage share: *lowest 10%:* 3.5%
highest 10%: 25.7% (2014 est.)

Budget: *revenues:* 1.78 billion (2017 est.)
expenditures: 2.05 billion (2017 est.)

Taxes and other revenues: 37.2% (of GDP) (2017 est.)
country comparison to the world: 55

Budget surplus (+) or deficit (-): -5.6% (of GDP) (2017 est.)
country comparison to the world: 175

Public debt: 67.2% of GDP (2017 est.)

66.4% of GDP (2016 est.)

note: data cover general government debt, and includes debt instruments issued (or owned) by government entities other than the treasury; the data include treasury debt held by foreign entities; the data include debt issued by subnational entities, as well as intragovernmental debt; intragovernmental debt consists of treasury borrowings from surpluses in the social funds, such as for retirement, medical care, and unemployment; debt instruments for the social funds are not sold at public auctions
country comparison to the world: 55

Fiscal year: calendar year

Inflation rate (consumer prices): 2.4% (2017 est.)
-0.3% (2016 est.)
country comparison to the world: 120

Current account balance: -$780 million (2017 est.)
-$710 million (2016 est.)
country comparison to the world: 137

Exports: $422.2 million (2017 est.)
$362 million (2016 est.)
country comparison to the world: 180

Imports: $2.618 billion (2017 est.)
$2.29 billion (2016 est.)
country comparison to the world: 154

Reserves of foreign exchange and gold: $1.077 billion (31 December 2017 est.)
$846.5 million (31 December 2016 est.)
country comparison to the world: 131

Debt—external: $2.516 billion (31 December 2017 est.)
$2.224 billion (31 December 2016 est.)
country comparison to the world: 148

Exchange rates: euros (EUR) per US dollar
0.885 (2017 est.)
0.903 (2016 est.)
0.9214 (2015 est.)
0.885 (2014 est.)
0.7634 (2013 est.)

ENERGY

Electricity access: *electrification—total population:* 100% (2020)

Electricity—production: 3.045 billion kWh (2016 est.)
country comparison to the world: 132

Electricity—consumption: 2.808 billion kWh (2016 est.)
country comparison to the world: 138

Electricity—exports: 914 million kWh (2016 est.)
country comparison to the world: 59

Electricity—imports: 1.21 billion kWh (2016 est.)
country comparison to the world: 65

Electricity—installed generating capacity: 890,000 kW (2016 est.)
country comparison to the world: 133

Electricity—from fossil fuels: 23% of total installed capacity (2016 est.)
country comparison to the world: 192

Electricity—from nuclear fuels: 0% of total installed capacity (2017 est.)
country comparison to the world: 146

Electricity—from hydroelectric plants: 69% of total installed capacity (2017 est.)
country comparison to the world: 18

Electricity—from other renewable sources: 8% of total installed capacity (2017 est.)
country comparison to the world: 88

Crude oil—production: 0 bbl/day (2018 est.)
country comparison to the world: 176

Crude oil—exports: 0 bbl/day (2015 est.)
country comparison to the world: 167

Crude oil—imports: 0 bbl/day (2015 est.)
country comparison to the world: 169

Crude oil—proved reserves: 0 bbl (1 January 2018 est.)
country comparison to the world: 171

Refined petroleum products—production: 0 bbl/day (2015 est.)
country comparison to the world: 180

Refined petroleum products—consumption: 6,000 bbl/day (2016 est.)
country comparison to the world: 172

Refined petroleum products—exports: 357 bbl/day (2015 est.)
country comparison to the world: 114

Refined petroleum products—imports: 6,448 bbl/day (2015 est.)
country comparison to the world: 163

Natural gas—production: 0 cu m (2017 est.)
country comparison to the world: 173

Natural gas—consumption: 0 cu m (2017 est.)
country comparison to the world: 177

Natural gas—exports: 0 cu m (2017 est.)
country comparison to the world: 154

Natural gas—imports: 0 cu m (2017 est.)
country comparison to the world: 159

Natural gas—proved reserves: 0 cu m (2016 est.)
country comparison to the world: 172

Carbon dioxide emissions from consumption of energy: 2.287 million Mt (2017 est.)
country comparison to the world: 156

COMMUNICATIONS

Telephones—fixed lines: *total subscriptions:* 183,387
subscriptions per 100 inhabitants: 29.96 (2019 est.)
country comparison to the world: 124

Telephones—mobile cellular: *total subscriptions:* 1,121,870
subscriptions per 100 inhabitants: 183.28 (2019 est.)
country comparison to the world: 160

Telecommunication systems: *general assessment:* modern telecommunications system with access to European satellites; telecom sector in-line with EU norms which means competition, access and tariff structures; DSL, cable, leased line, fiber and wireless available; seasonal tourist have boosted

mobile penetration; wide availability of LTE technologies has made mobile broadband a viable alternative to fixed-line broadband in rural areas; 5G services anticipated in the future (2020)

domestic: GSM mobile-cellular service, available through multiple providers with national coverage growing; fixed-line 30 per 100 and mobile-cellular 183 per 100 persons (2019)

international: country code—382; 2 international switches connect the national system
note: the COVID-19 outbreak is negatively impacting telecommunications production and supply chains globally; consumer spending on telecom devices and services has also slowed due to the pandemic's effect on economies worldwide; overall progress towards improvements in all facets of the telecom industry—mobile, fixed-line, broadband, submarine cable and satellite—has moderated

Broadcast media: state-funded national radio-TV broadcaster operates 2 terrestrial TV networks, 1 satellite TV channel, and 2 radio networks; 4 local public TV stations and 14 private TV stations; 14 local public radio stations, 35 private radio stations, and several on-line media (2019)

Internet country code: .me

Internet users: *total:* 439,311

percent of population: 71.52% (July 2018 est.)
country comparison to the world: 155

Broadband—fixed subscriptions: *total:* 159,029
subscriptions per 100 inhabitants: 26 (2018 est.)
country comparison to the world: 115

TRANSPORTATION

National air transport system: *number of registered air carriers:* 1 (2020)
inventory of registered aircraft operated by air carriers: 4
annual passenger traffic on registered air carriers: 565,522 (2018)
annual freight traffic on registered air carriers: 130,000 mt-km (2018)

Civil aircraft registration country code prefix: 4O (2016)

Airports: 5 (2013)
country comparison to the world: 182

Airports—with paved runways: *total:* 5 (2019)
2,438 to 3,047 m: 2
1,524 to 2,437 m: 1
914 to 1,523 m: 1
under 914 m:1

Heliports: 1 (2012)

Railways: *total:* 250 km (2017)
standard gauge: 250 km 1.435-m gauge (224 km electrified) (2017)
country comparison to the world: 125

Roadways: *total:* 7,762 km (2010)
paved: 7,141 km (2010)
unpaved: 621 km (2010)
country comparison to the world: 139

Merchant marine: *total:* 12
by type: bulk carrier 4, other 8 (2019)
country comparison to the world: 150

Ports and terminals: *major seaport(s):* Bar

MILITARY AND SECURITY

Military and security forces: *Armed Forces of the Republic of Montenegro:* Army of Montenegro (includes Ground Troops (Kopnena Vojska), Montenegrin Navy (Mornarica Crne Gore, MCG)), Air Force (2019)

Military expenditures:
1.66% of GDP (2019 est.)
1.39% of GDP (2018)
1.35% of GDP (2017)
1.42% of GDP (2016)
1.4% of GDP (2015)

country comparison to the world: 68

Military and security service personnel strengths: the Armed Forces of the Republic of Montenegro have approximately 2,400 total active duty troops (1,400 Army; 400 Navy; 200 Air Force; 400 other) (2019 est.)

Military equipment inventories and acquisitions: the inventory of the Armed Forces of Montenegro is small and consists mostly of equipment inherited from the former Yugoslavia military, with a limited mix of other imported systems, such as French-made helicopters; since 2010, it has received small quantities of equipment from Austria, Turkey, and the US (2019 est.)

Military service age and obligation: 18 is the legal minimum age for voluntary military service; no conscription (2012)

Terrorism

TRANSNATIONAL ISSUES

Disputes—international: Kosovo ratified the border demarcation agreement with Montenegro in March 2018, but the actual demarcation has not been completed

Refugees and internally displaced persons: *stateless persons:* 142 (2019)
note: 17,051 estimated refugee and migrant arrivals (January 2015-November 2020)

MONTSERRAT

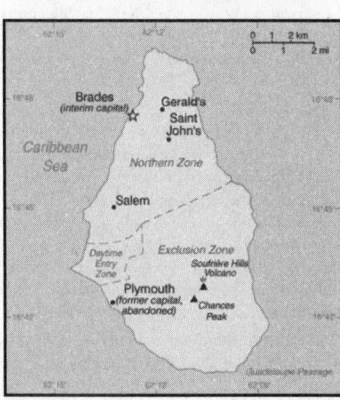

INTRODUCTION

Background: English and Irish colonists from St. Kitts first settled on Montserrat in 1632; the first African slaves arrived three decades later. The British and French fought for possession of the island for most of the 18th century, but it finally was confirmed as a British possession in 1783. The island's sugar plantation economy was converted to small farm landholdings in the mid-19th century. Much of this island was devastated and two-thirds of the population fled abroad because of the eruption of the Soufriere Hills Volcano that began on 18 July 1995. Montserrat has endured volcanic activity since, with the last eruption occurring in 2013.

GEOGRAPHY

Location: Caribbean, island in the Caribbean Sea, southeast of Puerto Rico

Geographic coordinates: 16 45 N, 62 12 W

Map references: Central America and the Caribbean

Area: *total:* 102 sq km

land: 102 sq km
water: 0 sq km
country comparison to the world: 226

Area—comparative: about 0.6 times the size of Washington, DC

Land boundaries: 0 km

Coastline: 40 km

Maritime claims: *territorial sea:* 12 nm
exclusive fishing zone: 200 nm

Climate: tropical; little daily or seasonal temperature variation

Terrain: volcanic island, mostly mountainous, with small coastal lowland

Elevation: *lowest point:* Caribbean Sea 0 m

highest point: Soufriere Hills volcano pre-eruption height was 915 m; current lava dome is subject to periodic build up and collapse; estimated dome height was 1,050 m in 2015

Natural resources: NEGL

Land use: *agricultural land:* 30% (2011 est.)
arable land: 20% (2011 est.) / permanent crops: 0% (2011 est.) / permanent pasture: 10% (2011 est.)
forest: 25% (2011 est.)
other: 45% (2011 est.)
Irrigated land: 0 sq km (2012)

Population distribution: only the northern half of the island is populated, the southern portion is uninhabitable due to volcanic activity

Natural hazards: volcanic eruptions; severe hurricanes (June to November)
volcanism: Soufriere Hills volcano (915 m), has erupted continuously since 1995; a massive eruption in 1997 destroyed most of the capital, Plymouth, and resulted in approximately half of the island becoming uninhabitable; the island of Montserrat is part of the volcanic island arc of the Lesser Antilles that extends from Saba in the north to Grenada in the south

Environment—current issues: land erosion occurs on slopes that have been cleared for cultivation

Geography—note: the island is entirely volcanic in origin and comprised of three major volcanic centers of differing ages

PEOPLE AND SOCIETY

Population: 5,373 (July 2020 est.)
note: an estimated 8,000 refugees left the island following the resumption of volcanic activity in July 1995; some have returned
country comparison to the world: 227

Nationality: *noun:* Montserratian(s)
adjective: Montserratian

Ethnic groups: African/black 88.4%, mixed 3.7%, hispanic/Spanish 3%, caucasian/white 2.7%, East Indian/Indian 1.5%, other 0.7% (2011 est.)

Languages: English

Religions: Protestant 67.1% (includes Anglican 21.8%, Methodist 17%, Pentecostal 14.1%, Seventh Day Adventist 10.5%, and Church of God 3.7%), Roman Catholic 11.6%, Rastafarian 1.4%, other 6.5%, none 2.6%, unspecified 10.8% (2001 est.)

Age structure: *0-14 years:* 16.02% (male 443/ female 418)
15-24 years: 20.55% (male 579/female 525)
25-54 years: 47.09% (male 1,217/female 1,313)
55-64 years: 9.79% (male 246/female 280)
65 years and over: 6.55% (male 196/female 156) (2020 est.)

Median age: *total:* 34.8 years
male: 34.1 years
female: 35.6 years (2020 est.)
country comparison to the world: 88

Population growth rate: 0.58% (2020 est.)
country comparison to the world: 152

Birth rate: 11.7 births/1,000 population (2020 est.)
country comparison to the world: 167

Death rate: 6 deaths/1,000 population (2020 est.)
country comparison to the world: 164

Net migration rate: 0 migrant(s)/1,000 population (2020 est.)

country comparison to the world: 91

Population distribution: only the northern half of the island is populated, the southern portion is uninhabitable due to volcanic activity

Urbanization: *urban population:* 9.1% of total population (2020)
rate of urbanization: 0.64% annual rate of change (2015-20 est.)
total population growth rate v. urban population growth rate, 2000-2030:

Sex ratio: *at birth:* 1.03 male(s)/female
0-14 years: 1.06 male(s)/female
15-24 years: 1.1 male(s)/female
25-54 years: 0.93 male(s)/female
55-64 years: 0.88 male(s)/female
65 years and over: 1.26 male(s)/female
total population: 1 male(s)/female (2020 est.)

Infant mortality rate: *total:* 11.1 deaths/1,000 live births
male: 8.9 deaths/1,000 live births
female: 13.4 deaths/1,000 live births (2020 est.)
country comparison to the world: 121

Life expectancy at birth: *total population:* 75.3 years
male: 76.4 years
female: 74.1 years (2020 est.)
country comparison to the world: 118

Total fertility rate: 1.36 children born/woman (2020 est.)
country comparison to the world: 220

Drinking water source:
improved:
urban: 99% of population
rural: 99% of population
total: 99% of population
unimproved:
urban: 1% of population
rural: 1% of population
total: 1% of population (2015 est.)

HIV/AIDS—adult prevalence rate: NA

HIV/AIDS—people living with HIV/AIDS: NA

HIV/AIDS—deaths: NA

Education expenditures: 5.1% of GDP (2009)
country comparison to the world: 59

School life expectancy (primary to tertiary education): *total:* 14 years
male: 13 years
female: 15 years (2019)

GOVERNMENT

Country name: *conventional long form:* none
conventional short form: Montserrat
etymology: island named by explorer Christopher COLUMBUS in 1493 after the Benedictine abbey Santa Maria de Montserrat, near Barcelona, Spain

Dependency status: overseas territory of the UK

Government type: parliamentary democracy; self-governing overseas territory of the UK

Capital: *name:* Plymouth; note—Plymouth was abandoned in 1997 because of volcanic activity; interim government buildings have been built at Brades Estate, the de facto capital, in the Carr's

Bay/Little Bay vicinity at the northwest end of Montserrat

geographic coordinates: 16 42 N, 62 13 W
time difference: UTC-4 (1 hour ahead of Washington, DC, during Standard Time)
etymology and note: now entirely deserted because of volcanic activity, the city was originally named after Plymouth, England; de jure, Plymouth remains the capital city of Montserrat; it is therefore the only ghost town that serves as the capital of a political entity

Administrative divisions: 3 parishes; Saint Anthony, Saint Georges, Saint Peter

Independence: none (overseas territory of the UK)

National holiday: Birthday of Queen ELIZABETH II, usually celebrated the Monday after the second Saturday in June (1926)

Constitution: *history:* previous 1960; latest effective 1 September 2010 (The Montserrat Constitution Order 2010)
amendments: amended 2011

Legal system: English common law

Citizenship: *see United Kingdom*

Suffrage: 18 years of age; universal

Executive branch: *chief of state:* Queen ELIZABETH II (since 6 February 1952); represented by Governor Andrew PEARCE (since 1 February 2018)
head of government: Premier Easton TAYLOR-FARRELL (since 19 November 2019); note—effective with The Constitution Order 2010, effective October 2010, the office of premier replaced the office of chief minister
cabinet: Executive Council consists of the governor, the premier, 3 other ministers, the attorney general, and the finance secretary
elections/appointments: the monarchy is hereditary; governor appointed by the monarch; following legislative elections, the leader of the majority party usually becomes premier

Legislative branch: *description:* unicameral Legislative Assembly (11 seats; 9 members directly elected in a single constituency by absolute majority vote in 2 rounds to serve 5-year terms and 2 ex-officio members—the attorney general and financial secretary)
elections: last held on 18 November 2019 (next scheduled for 2024)
election results: percent of vote by party—MCAP 42.7%, PDM 29.9%, other 17.1%; seats by party—MCAP 5, PDM 3, independent 1

Judicial branch: *highest courts:* the Eastern Caribbean Supreme Court (ECSC) is the superior court of the Organization of Eastern Caribbean States; the ECSC—headquartered on St. Lucia—consists of the Court of Appeal—headed by the chief justice and 4 judges—and the High Court with 18 judges; the Court of Appeal is itinerant, traveling to member states on a schedule to hear appeals from the High Court and subordinate courts; High Court judges reside in the member states, with 1 assigned to Montserrat; Montserrat is also a member of the Caribbean Court of Justice

judge selection and term of office: chief justice of Eastern Caribbean Supreme Court appointed by the Her Majesty, Queen ELIZABETH II; other justices and judges appointed by the Judicial and Legal Services Commission, and independent body of judicial officials; Court of Appeal justices appointed for life with mandatory retirement at age 65; High Court judges appointed for life with mandatory retirement at age 62
subordinate courts: magistrate's court

Political parties and leaders: Movement for Change and Prosperity or MCAP [Easton Taylor FARRELL]
People's Democratic Movement or PDM [Donaldson ROMERO]

International organization participation: Caricom, CDB, Interpol (subbureau), OECS, UPU

Diplomatic representation in the US: none (overseas territory of the UK)

Diplomatic representation from the US: none (overseas territory of the UK); alternate contact is the US Embassy in Barbados [1] (246) 227-4000; US Embassy in Bridgetown, Wildey Business Park, St. Michael BB 14006, Barbados, WI

Flag description: blue with the flag of the UK in the upper hoist-side quadrant and the Montserratian coat of arms centered in the outer half of the flag; the arms feature a woman in green dress, Erin, the female personification of Ireland, standing beside a yellow harp and embracing a large dark cross with her right arm; Erin and the harp are symbols of Ireland reflecting the territory's Irish ancestry; blue represents awareness, trustworthiness, determination, and righteousness

National anthem: *note:* as a territory of the UK, "God Save the Queen" is official (see United Kingdom)
0:00/ 1:02

ECONOMY

Economy—overview: Severe volcanic activity, which began in July 1995, has put a damper on this small, open economy. A catastrophic eruption in June 1997 closed the airport and seaports, causing further economic and social dislocation. Two-thirds of the 12,000 inhabitants fled the island. Some began to return in 1998 but lack of housing limited the number. The agriculture sector continued to be affected by the lack of suitable land for farming and the destruction of crops.

Prospects for the economy depend largely on developments in relation to the volcanic activity and on public sector construction activity. Half of the island remains uninhabitable. In January 2013, the EU announced the disbursement of a $55.2 million aid package to Montserrat in order to boost the country's economic recovery, with a specific focus on public finance management, public sector reform, and prudent economic management. Montserrat is tied to the EU through the UK. Although the UK is leaving the EU, Montserrat's aid will not be affected as Montserrat maintains a direct agreement with the EU regarding aid.

GDP (purchasing power parity):

$167.4 million (2011 est.)
$155.9 million (2010 est.)
$162.7 million (2009 est.)
country comparison to the world: 223

GDP (official exchange rate):
$167.4 million (2011 est.)

GDP—real growth rate: 7.4% (2011 est.)
-4.2% (2010 est.)
country comparison to the world: 15

GDP—per capita (PPP): $34,000 (2011 est.)
$31,100 (2010 est.)
$32,300 (2009 est.)
country comparison to the world: 60

GDP—composition, by end use:
household consumption: 90.8% (2017 est.)
government consumption: 50.4% (2017 est.)
investment in fixed capital: 17.9% (2017 est.)
investment in inventories: -0.1% (2017 est.)
exports of goods and services: 29.5% (2017 est.)
imports of goods and services: -88.6% (2017 est.)

GDP—composition, by sector of origin:
agriculture: 1.9% (2017 est.)
industry: 7.8% (2017 est.)
services: 90.3% (2017 est.)

Agriculture—products: cabbages, carrots, cucumbers, tomatoes, onions, peppers; livestock products

Industries: tourism, rum, textiles, electronic appliances

Industrial production growth rate: -21% (2017 est.)
country comparison to the world: 200

Labor force: 4,521 (2012)
country comparison to the world: 222

Labor force—by occupation: *agriculture:* 1.4%
industry: 12.7%
services: 85.9% (2017 est.)

Unemployment rate: 5.6% (2017 est.)
6% (1998 est.)
country comparison to the world: 89

Population below poverty line: NA

Household income or consumption by percentage share: *lowest 10%:* NA
highest 10%: NA

Budget: *revenues:* 66.67 million (2017 est.)
expenditures: 47.04 million (2017 est.)

Fiscal year: 1 April—31 March

Inflation rate (consumer prices): 1.2% (2017 est.)
-0.2% (2016 est.)
country comparison to the world: 66

Current account balance:
-$15.4 million (2017 est.)
-$12.2 million (2016 est.)
country comparison to the world: 71

Exports: $4.4 million (2017 est.)
$5.2 million (2016 est.)
country comparison to the world: 218

Exports—partners: US 29%, France 23%, Saint Kitts and Nevis 22.2% (2017)

Exports—commodities: electronic components, plastic bags, apparel; hot peppers, limes, live plants; cattle

Imports: $39.44 million (2017 est.)
$36.1 million (2016 est.)
country comparison to the world: 221

Imports—commodities: machinery and transportation equipment, foodstuffs, manufactured goods, fuels, lubricants

Imports—partners: US 72.8%, Trinidad and Tobago 6%, UK 4.1% (2017)

Reserves of foreign exchange and gold:
$47.58 million (31 December 2017 est.)
$51.47 million (31 December 2015 est.)
country comparison to the world: 187

Debt—external: $8.9 million (1997)
country comparison to the world: 200

Exchange rates: East Caribbean dollars (XCD) per US dollar -
2.7 (2017 est.)
2.7 (2016 est.)
2.7 (2015 est.)
2.7 (2014 est.)
2.7 (2013 est.)

ENERGY

Electricity—production: 24 million kWh (2016 est.)
country comparison to the world: 211

Electricity—consumption: 22.32 million kWh (2016 est.)
country comparison to the world: 211

Electricity—exports: 0 kWh (2016 est.)
country comparison to the world: 173

Electricity—imports: 0 kWh (2016 est.)
country comparison to the world: 176

Electricity—installed generating capacity: 5,000 kW (2016 est.)
country comparison to the world: 213

Electricity—from fossil fuels: 100% of total installed capacity (2016 est.)
country comparison to the world: 13

Electricity—from nuclear fuels: 0% of total installed capacity (2017 est.)
country comparison to the world: 147

Electricity—from hydroelectric plants: 0% of total installed capacity (2017 est.)
country comparison to the world: 187

Electricity—from other renewable sources: 0% of total installed capacity (2017 est.)
country comparison to the world: 202

Crude oil—production: 0 bbl/day (2018 est.)
country comparison to the world: 177

Crude oil—exports: 0 bbl/day (2015 est.)
country comparison to the world: 168

Crude oil—imports: 0 bbl/day (2015 est.)
country comparison to the world: 170

Crude oil—proved reserves: 0 bbl (1 January 2018 est.)
country comparison to the world: 172

Refined petroleum products—production: 0 bbl/day (2015 est.)
country comparison to the world: 181

Refined petroleum products—consumption: 400 bbl/day (2016 est.)
country comparison to the world: 212

Refined petroleum products—exports: 0 bbl/day (2015 est.)
country comparison to the world: 184

Refined petroleum products—imports: 406 bbl/day (2015 est.)
country comparison to the world: 208

Natural gas—production: 0 cu m (2017 est.)
country comparison to the world: 174

Natural gas—consumption: 0 cu m (2017 est.)
country comparison to the world: 178

Natural gas—exports: 0 cu m (2017 est.)
country comparison to the world: 155

Natural gas—imports: 0 cu m (2017 est.)
country comparison to the world: 160

Natural gas—proved reserves: 0 cu m (1 January 2014 est.)
country comparison to the world: 173

Carbon dioxide emissions from consumption of energy: 57,180 Mt (2017 est.)
country comparison to the world: 210

COMMUNICATIONS

Telephones—fixed lines: *total subscriptions:* 3,227
subscriptions per 100 inhabitants: 60.4 (2019 est.)
country comparison to the world: 211

Telephones—mobile cellular: *total subscriptions:* 5,377
subscriptions per 100 inhabitants: 100.66 (2019 est.)
country comparison to the world: 217

Telecommunication systems: *general assessment:* telecom market one of growth in Caribbean and fully digitalized; high dependency on tourism and offshore financial services; operators expand FttP (Fiber to Home) services; LTE launches and operators invest in mobile networks; effective competition in all sectors (2020)

domestic: fixed-line 60 per 100 and mobile-cellular teledensity 101 per 100 persons (2019)

international: country code—1-664; landing point for the ECFS optic submarine cable with links to 14 other islands in the eastern Caribbean extending from the British Virgin Islands to Trinidad (2019)

note: the COVID-19 outbreak is negatively impacting telecommunications production and supply chains globally; consumer spending on telecom devices and services has also slowed due to the pandemic's effect on economies worldwide; overall progress towards improvements in all facets of the telecom industry—mobile, fixed-line, broadband, submarine cable and satellite—has moderated

Broadcast media: Radio Montserrat, a public radio broadcaster, transmits on 1 station and has a repeater transmission to a second station; repeater transmissions from the GEM Radio Network of

Trinidad and Tobago provide another 2 radio stations; cable and satellite TV available (2007)

Internet country code: .ms

Internet users: total: 2,860
percent of population: 54.6% (July 2016 est.)
country comparison to the world: 222

TRANSPORTATION

National air transport system: *number of registered air carriers:* 1 (2020)
inventory of registered aircraft operated by air carriers: 3

Civil aircraft registration country code prefix: VP-M (2016)

Airports: 1 (2013)
country comparison to the world: 229

Airports—with paved runways:
total: 1 (2019)
*under 914 m:*1

Roadways: *note:* volcanic eruptions that began in 1995 destroyed most of the 227 km road system; a new road infrastructure has been built on the north end of the island

Ports and terminals: *major seaport(s):* Little Bay, Plymouth

MILITARY AND SECURITY

Military and security forces: no regular military forces; Royal Montserrat Defence Force (ceremonial, civil defense duties), Montserrat Police Force (2019)

Military—note: defense is the responsibility of the UK

TRANSNATIONAL ISSUES

Disputes—international: none

Illicit drugs: transshipment point for South American narcotics destined for the US and Europe

MOROCCO

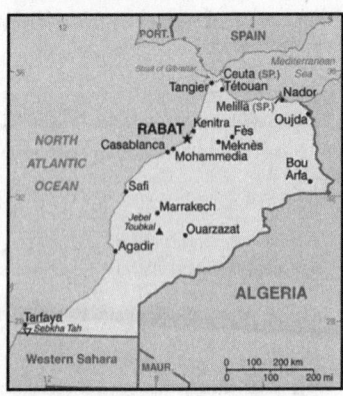

INTRODUCTION

Background: In 788, about a century after the Arab conquest of North Africa, a series of Moroccan Muslim dynasties began to rule in Morocco. In the 16th century, the Sa'adi monarchy, particularly under Ahmad al-MANSUR (1578-1603), repelled foreign invaders and inaugurated a golden age. The Alaouite Dynasty, to which the current Moroccan royal family belongs, dates from the 17th century. In 1860, Spain occupied northern Morocco and ushered in a half-century of trade rivalry among European powers that saw Morocco's sovereignty steadily erode; in 1912, the French imposed a protectorate over the country. A protracted independence struggle with France ended successfully in 1956. The internationalized city of Tangier and most Spanish possessions were turned over to the new country that same year. Sultan MOHAMMED V, the current monarch's grandfather, organized the new state as a constitutional monarchy and in 1957 assumed the title of king. Since Spain's 1976 withdrawal from what is today called Western Sahara, Morocco has extended its de facto administrative control to

roughly 75% of this territory; however, the UN does not recognize Morocco as the administering power for Western Sahara. The UN since 1991 has monitored a cease-fire between Morocco and the Polisario Front—an organization advocating the territory's independence—and restarted negotiations over the status of the territory in December 2018.

King MOHAMMED VI in early 2011 responded to the spread of pro-democracy protests in the region by implementing a reform program that included a new constitution, passed by popular referendum in July 2011, under which some new powers were extended to parliament and the prime minister, but ultimate authority remains in the hands of the monarch. In November 2011, the Justice and Development Party (PJD)—a moderate Islamist party—won the largest number of seats in parliamentary elections, becoming the first Islamist party to lead the Moroccan Government. In September 2015, Morocco held its first direct elections for regional councils, one of the reforms included in the 2011 constitution. The PJD again won the largest number of seats in nationwide parliamentary elections in October 2016.

GEOGRAPHY

Location: Northern Africa, bordering the North Atlantic Ocean and the Mediterranean Sea, between Algeria and Western Sahara

Geographic coordinates: 32 00 N, 5 00 W

Map references: Africa

Area: *total:* 446,550 sq km
land: 446,300 sq km
water: 250 sq km
country comparison to the world: 59

Area—comparative: slightly more than three times the size of New York; slightly larger than California

Land boundaries: *total:* 2,362.5 km

border countries (4): Algeria 1900 km, Western Sahara 444 km, Spain (Ceuta) 8 km, Spain (Melilla) 10.5 km

note: an additional 75-meter border segment exists between Morocco and the Spanish exclave of Penon de Velez de la Gomera

Coastline: 1,835 km

Maritime claims: *territorial sea:* 12 nm
exclusive economic zone: 200 nm
contiguous zone: 24 nm
continental shelf: 200-m depth or to the depth of exploitation

Climate: Mediterranean, becoming more extreme in the interior

Terrain: mountainous northern coast (Rif Mountains) and interior (Atlas Mountains) bordered by large plateaus with intermontane valleys, and fertile coastal plains

Elevation: *mean elevation:* 909 m
lowest point: Sebkha Tah -59 m
highest point: Jebel Toubkal 4,165 m

Natural resources: phosphates, iron ore, manganese, lead, zinc, fish, salt

Land use: *agricultural land:* 67.5% (2011 est.)
arable land: 17.5% (2011 est.) / permanent crops: 2.9% (2011 est.) / permanent pasture: 47.1% (2011 est.)
forest: 11.5% (2011 est.)
other: 21% (2011 est.)
Irrigated land: 14,850 sq km (2012)

Population distribution: the highest population density is found along the Atlantic and Mediterranean coasts; a number of densely populated agglomerations are found scattered through the Atlas Mountains as shown in this population distribution map

Natural hazards: northern mountains geologically unstable and subject to earthquakes; periodic droughts; windstorms, flash floods; landslides

Environment—current issues: land degradation/desertification (soil erosion resulting from farming of marginal areas, overgrazing, destruction of vegetation); water and soil pollution due to dumping of industrial wastes into the ocean and inland water sources, and onto the land

Environment—international agreements: *party to:* Biodiversity, Climate Change, Climate Change-Kyoto Protocol, Desertification, Endangered Species, Hazardous Wastes, Law of the Sea, Marine Dumping, Ozone Layer Protection, Ship Pollution, Wetlands, Whaling
signed, but not ratified: Environmental Modification

Geography—note: strategic location along Strait of Gibraltar; the only African nation to have both Atlantic and Mediterranean coastlines

PEOPLE AND SOCIETY

Population: 35,561,654 (July 2020 est.)
country comparison to the world: 40

Nationality: *noun:* Moroccan(s)
adjective: Moroccan

Ethnic groups: Arab-Berber 99%, other 1%

Languages: Arabic (official), Berber languages (Tamazight (official), Tachelhit, Tarifit), French (often the language of business, government, and diplomacy)
note: the proportion of Berber speakers is disputed

Religions: Muslim 99% (official; virtually all Sunni, <0.1% Shia), other 1% (includes Christian, Jewish, and Baha'i); note—Jewish about 6,000 (2010 est.)

MENA religious affiliation: *Demographic profile:* Morocco is undergoing a demographic transition. Its population is growing but at a declining rate, as people live longer and women have fewer children. Infant, child, and maternal mortality rates have been reduced through better health care, nutrition, hygiene, and vaccination coverage, although disparities between urban and rural and rich and poor households persist. Morocco's shrinking child cohort reflects the decline of its total fertility rate from 5 in mid-1980s to 2.2 in 2010, which is a result of increased female educational attainment, higher contraceptive use, delayed marriage, and the desire for smaller families. Young adults (persons aged 15-29) make up almost 26% of the total population and represent a potential economic asset if they can be gainfully employed. Currently, however, many youths are unemployed because Morocco's job creation rate has not kept pace with the growth of its working-age population. Most youths who have jobs work in the informal sector with little security or benefits.

During the second half of the 20th century, Morocco became one of the world's top emigration countries, creating large, widely dispersed migrant communities in Western Europe. The Moroccan Government has encouraged emigration since its independence in 1956, both to secure remittances for funding national development and as an outlet to prevent unrest in rebellious (often Berber) areas. Although Moroccan labor migrants earlier targeted Algeria and France, the flood of

Moroccan "guest workers" from the mid-1960s to the early 1970s spread widely across northwestern Europe to fill unskilled jobs in the booming manufacturing, mining, construction, and agriculture industries. Host societies and most Moroccan migrants expected this migration to be temporary, but deteriorating economic conditions in Morocco related to the 1973 oil crisis and tighter European immigration policies resulted in these stays becoming permanent.

A wave of family migration followed in the 1970s and 1980s, with a growing number of second generation Moroccans opting to become naturalized citizens of their host countries. Spain and Italy emerged as new destination countries in the mid-1980s, but their introduction of visa restrictions in the early 1990s pushed Moroccans increasingly to migrate either legally by marrying Moroccans already in Europe or illegally to work in the underground economy. Women began to make up a growing share of these labor migrants. At the same time, some higher-skilled Moroccans went to the US and Quebec, Canada.

In the mid-1990s, Morocco developed into a transit country for asylum seekers from Sub-Saharan Africa and illegal labor migrants from Sub-Saharan Africa and South Asia trying to reach Europe via southern Spain, Spain's Canary Islands, or Spain's North African enclaves, Ceuta and Melilla. Forcible expulsions by Moroccan and Spanish security forces have not deterred these illegal migrants or calmed Europe's security concerns. Rabat remains unlikely to adopt an EU agreement to take back third-country nationals who have entered the EU illegally via Morocco. Thousands of other illegal migrants have chosen to stay in Morocco until they earn enough money for further travel or permanently as a "second-best" option. The launching of a regularization program in 2014 legalized the status of some migrants and granted them equal access to education, health care, and work, but xenophobia and racism remain obstacles.

Age structure: *0-14 years:* 27.04% (male 4,905,626/female 4,709,333)
15-24 years: 16.55% (male 2,953,523/female 2,930,708)
25-54 years: 40.64% (male 7,126,781/female 7,325,709)
55-64 years: 8.67% (male 1,533,771/female 1,548,315)
65 years and over: 7.11% (male 1,225,307/female 1,302,581) (2020 est.)

Dependency ratios: *total dependency ratio:* 52.4
youth dependency ratio: 40.8
elderly dependency ratio: 11.6

potential support ratio: 8.6 (2020 est.)

Median age: *total:* 29.1 years
male: 28.7 years
female: 29.6 years (2020 est.)
country comparison to the world: 137

Population growth rate: 0.96% (2020 est.)
country comparison to the world: 110

Birth rate: 17.9 births/1,000 population (2020 est.)
country comparison to the world: 91

Death rate: 6.6 deaths/1,000 population (2020 est.)
country comparison to the world: 141

Net migration rate: -1.9 migrant(s)/1,000 population (2020 est.)
country comparison to the world: 167

Population distribution: the highest population density is found along the Atlantic and Mediterranean coasts; a number of densely populated agglomerations are found scattered through the Atlas Mountains as shown in this population distribution map

Urbanization: *urban population:* 63.5% of total population (2020)
rate of urbanization: 2.14% annual rate of change (2015-20 est.)
total population growth rate v. urban population growth rate, 2000-2030:

Major urban areas—population: 3.752 million Casablanca, 1.885 million RABAT (capital), 1.224 million Fes, 1.198 million Tangier, 1.003 million Marrakech, 924,000 Agadir (2020)

Sex ratio: *at birth:* 1.05 male(s)/female
0-14 years: 1.04 male(s)/female
15-24 years: 1.01 male(s)/female
25-54 years: 0.97 male(s)/female
55-64 years: 0.99 male(s)/female
65 years and over: 0.94 male(s)/female
total population: 1 male(s)/female (2020 est.)

Maternal mortality rate: 70 deaths/100,000 live births (2017 est.)
country comparison to the world: 84

Infant mortality rate: *total:* 1 8.2 deaths/1,000 live births
male: 20.2 deaths/1,000 live births
female: 1 6.1 deaths/1,000 live births (2020 est.)
country comparison to the world: 82

Life expectancy at birth: *total population:* 73.3 years
male: 71.6 years
female: 75.1 years (2020 est.)
country comparison to the world: 147

Total fertility rate: 2.31 children born/woman (2020 est.)
country comparison to the world: 84

Contraceptive prevalence rate: 70.8% (2018)

Drinking water source:
improved:
urban: 98.3% of population
rural: 79.1% of population
total: 91% of population
unimproved:
urban: 1.7% of population
rural: 20.9% of population
total: 9% of population (2017 est.)

Current Health Expenditure: 5.2% (2017)

Physicians density: 0.73 physicians/1,000 population (2017)

Hospital bed density: 1 beds/1,000 population (2017)

Sanitation facility access:
improved:
urban: 99.1% of population

661

rural: 81.1% of population
total: 92.2% of population
unimproved:
urban: 0.9% of population
rural: 18.9% of population
total: 7.3% of population (2017 est.)

HIV/AIDS—adult prevalence rate: <.1% (2019 est.)

HIV/AIDS—people living with HIV/AIDS: 21,000 (2019 est.)

country comparison to the world: 86

HIV/AIDS—deaths: <500 (2019 est.)

Major infectious diseases: note: clusters of cases of a respiratory illness caused by the novel coronavirus (COVID-19) are occurring in Morocco; as of 10 November 2020, Morocco has reported a total of 252,185 cases of COVID-19 or 6,832 cumulative cases of COVID- 19 per 1 million population with 114 cumulative deaths per 1 million population

Obesity—adult prevalence rate: 26.1% (2016)
country comparison to the world: 45 Children under the age of 5 years underweight: 2.6% (2017/18)

country comparison to the world: 106 Education expenditures:
5.3% of GDP (2009)
country comparison to the world: 51

Literacy: definition: age 15 and over can read and write
total population: 73.8%
male: 83.3%
female: 64.6% (2018)

School life expectancy (primary to tertiary education): *total:* 14 years
male: 14 years
female: 14 years (2019)

Unemployment, youth ages 15-24: *total:* 22.2%
male: 22%
female: 22.8% (2016 est.)
country comparison to the world: 57

GOVERNMENT

Country name: *conventional long form:* Kingdom of Morocco
conventional short form: Morocco
local long form: Al Mamlakah al Maghribiyah
local short form: Al Maghrib
former: French Protectorate in Morocco, Spanish Protectorate in Morocco
etymology: the English name "Morocco" derives from, respectively, the Spanish and Portuguese names "Marruecos" and "Marrocos," which stem from "Marrakesh" the Latin name for the former capital of ancient Morocco; the Arabic name "Al Maghrib" translates as "The West"

Government type: parliamentary constitutional monarchy

Capital: *name:* Rabat

geographic coordinates: 34 01 N, 6 49 W
time difference: UTC 0 (5 hours ahead of Washington, DC, during Standard Time)

daylight saving time: +1 hr, begins last Sunday in March; ends last Sunday in October
etymology: name derives from the Arabic title "Ribat el-Fath," meaning "stronghold of victory," applied to the newly constructed citadel in 1170

Administrative divisions: 11 regions (recognized); Beni Mellal-Khenifra, Casablanca-Settat, Draa-Tafilalet, Fes-Meknes, Guelmim-Oued Noun, Laayoune-Sakia al Hamra, Oriental, Marrakech-Safi, Rabat-Sale-Kenitra, Souss-Massa, Tanger-Tetouan-Al Hoceima
note: Morocco claims the territory of Western Sahara, the political status of which is considered undetermined by the US Government; portions of the regions Guelmim-Oued Noun and Laayoune-Sakia al Hamra as claimed by Morocco lie within Western Sahara; Morocco also claims a 12th region, Dakhla-Oued ed Dahab, that falls entirely within Western Sahara

Independence: 2 March 1956 (from France)

National holiday: Throne Day (accession of King MOHAMMED VI to the throne), 30 July (1999)

Constitution: *history:* several previous; latest drafted 17 June 2011, approved by referendum 1 July 2011; note—sources disagree on whether the 2011 referendum was for a new constitution or for reforms to the previous constitution
amendments: proposed by the king, by the prime minister, or by members in either chamber of Parliament; passage requires at least two-thirds majority vote by both chambers and approval in a referendum; the king can opt to submit selfinitiated proposals directly to a referendum

Legal system: mixed legal system of civil law based on French civil law and Islamic (sharia) law; judicial review of legislative acts by Constitutional Court

International law organization participation: has not submitted an ICJ jurisdiction declaration; non-party state to the ICCt

Citizenship: *citizenship by birth:* no
citizenship by descent only: the father must be a citizen of Morocco; if the father is unknown or stateless, the mother must be a citizen
dual citizenship recognized: yes
residency requirement for naturalization: 5 years

Suffrage: 18 years of age; universal Executive branch: chief of state: King MOHAMMED VI (since 30 July 1999)

head of government: Prime Minister Saad-Eddine al-OTHMANI (since 17 March 2017)
cabinet: Council of Ministers chosen by the prime minister in consultation with Parliament and appointed by the monarch
elections/appointments: the monarchy is hereditary; prime minister appointed by the monarch from the majority party following legislative elections

Legislative branch: *description:* bicameral Parliament consists of:
Chamber of Advisors (120 seats; members indirectly elected by an electoral college of local councils, professional organizations, and labor

unions; members serve 6-year terms) Chamber of Representatives (395 seats; 305 members directly elected in multi-seat constituencies by proportional representation vote and 90 directly elected in a single nationwide constituency by proportional representation vote; members serve 5-year terms); note—in the national constituency, 60 seats are reserved for women and 30 reserved for those under age 40
elections: Chamber of Advisors—last held on 2 October 2015 (next to be held in fall 2021)
Chamber of Representatives—last held on 7 October 2016 (next to be held in fall 2021)
election results: Chamber of Advisors—percent of vote by party—NA; seats by party—NA; composition—men 106, women 14, percent of women 11.7%
Chamber of Representatives—percent of vote by party NA; seats by party—PJD 125, PAM 102, PI 46, RNI 37, MP 27, USFP 20, UC 19, PPS 12, MDS 3, other 4; composition—men 314, women 81, percent of women 20.5%; note—total Parliament percent of women 18.4%

Judicial branch: *highest courts:* Supreme Court or Court of Cassation (consists of 5-judge panels organized into civil, family matters, commercial, administrative, social, and criminal sections); Constitutional Court (consists of 12 members)
judge selection and term of office: Supreme Court judges appointed by the Superior Council of Judicial Power, a 20- member body presided by the monarch, which includes the Supreme Court president, the prosecutor general, representatives of the appeals and first instance courts (among them 1 woman magistrate), the president of the National Council of the Rights of Man, and 5 "notable persons" appointed by the monarch; judges appointed for life; Constitutional Court members—6 designated by the monarch and 6 elected by Parliament; court president appointed by the monarch from among the court members; members serve 9-year nonrenewable terms
subordinate courts: courts of appeal; High Court of Justice; administrative and commercial courts; regional and sadad courts (for religious, civil and administrative, and penal adjudication); first instance courts

Political parties and leaders: Action Party or PA [Mohammed EL IDRISSI]
Amal (hope) Party [Mohamed BANI]
An-Nahj Ad-Dimocrati or An-Nahj [Mustapha BRAHMA]
Authenticity and Modernity Party or PAM [Ilyas al-OMARI]
Constitutional Union Party or UC [Mohamed SAJID]
Democratic and Social Movement or MDS [Abdessamad ARCHANE]
Democratic Forces Front or FFD [Mustapha BENALI]
Democratic Oath Party or SD
Democratic Socialist Vanguard Party or PADS [Abderrahman BENAMROU]
Democratic Society Party [Zhour CHAKKAFI]
Environment and Development Party or PED [Karim HRITAN]

Green Left Party [Mohamed FARES]

Istiqlal (Independence) Party or PI [Nizar BARAKA]

Ittihadi National Congress or CNI [Abdesalam EL AZIZ]

Labor Party or P.T

Moroccan Liberal Party or PML [Mohammed ZIANE]

Moroccan Union for Democracy or UMD [Jamal MANDRI]

National Rally of Independents or RNI [Aziz AKHANNOUCH]

Neo-Democrats Party [Mohamed DARIF]

Party of Development Reform or PRD [Abderrahmane EL KOHEN]

Party of Justice and Development or PJD [Saad Eddine al-OTHMANI]

Party of Liberty and Social Justice [Miloud MOUSSAOUI]

Popular Movement or MP [Mohand LAENSER]

Progress and Socialism Party or PPS [Nabil BENABDELLAH]

Renaissance and Virtue Party [Mohamed KHALIDI]

Renaissance Party [Said EL GHENNIOUI]

Renewal and Equity Party or PRE [Chakir ACHEHABAR]

Shoura (consultation) and Istiqlal Party [Ahmed BELGHAZI]

Social Center Party or PCS [Lahcen MADIH]

Socialist Party [Abdelmajid BOUZOUBAA]

Socialist Union of Popular Forces or USFP [Driss LACHGAR]

Unified Socialist Party or GSU [Nabila MOUNIB]

Unity and Democracy Party [Ahmed FITRI]

International organization participation: ABEDA, AfDB, AFESD, AMF, AMU, CAEU, CD, EBRD, FAO, G-11, G-77, IAEA, IBRD, ICAO, ICC (national committees), ICRM, IDA, IDB, IFAD, IFC, IFRCS, IHO, ILO, IMF, IMO, IMSO, Interpol, IOC, IOM, IPU, ISO, ITSO, ITU, ITUC (NGOs), LAS, MIGA, MONUSCO, NAM, OAS (observer), OIC, OIF, OPCW, OSCE (partner), Pacific Alliance (observer), Paris Club (associate), PCA, SICA (observer), UN, UNCTAD, UNESCO, UNHCR, UNIDO, UNOCI, UNSC (temporary), UNWTO, UPU, WCO, WHO, WIPO, WMO, WTO

Diplomatic representation in the US: *chief of mission:* Ambassador Lalla Joumala ALAOUI (since 24 April 2017)

chancery: 3508 International Drive NW, Washington, DC 20008

telephone: [1] (202) 462-7979

FAX: [1] (202) 462-7643

consulate(s) general: New York

Diplomatic representation from the US: *chief of mission:* Ambassador David T. FISCHER (since 22 January 2020)

telephone: [212] 537 637 200

embassy: Km 5.7 Avenue Mohammed VI, Souissi, Rabat 10170

mailing address: Unit 9400, Box Front Office, DPO AE 09718

FAX: [212] 537 637 201

consulate(s) general: Casablanca

Flag description: red with a green pentacle (five-pointed, linear star) known as Sulayman's (Solomon's) seal in the center of the flag; red and green are traditional colors in Arab flags, although the use of red is more commonly associated with the Arab states of the Persian Gulf; the pentacle represents the five pillars of Islam and signifies the association between God and the nation; design dates to 1912

National symbol(s): pentacle symbol, lion; national colors: red, green

National anthem: name: "Hymne Cherifien" (Hymn of the Sharif)

lyrics/music: Ali Squalli HOUSSAINI/Leo MORGAN

note: music adopted 1956, lyrics adopted 1970

ECONOMY

Economy—overview: Morocco has capitalized on its proximity to Europe and relatively low labor costs to work towards building a diverse, open, market-oriented economy. Key sectors of the economy include agriculture, tourism, aerospace, automotive, phosphates, textiles, apparel, and subcomponents. Morocco has increased investment in its port, transportation, and industrial infrastructure to position itself as a center and broker for business throughout Africa. Industrial development strategies and infrastructure improvements—most visibly illustrated by a new port and free trade zone near Tangier—are improving Morocco's competitiveness.

In the 1980s, Morocco was a heavily indebted country before pursuing austerity measures and pro-market reforms, overseen by the IMF. Since taking the throne in 1999, King MOHAMMED VI has presided over a stable economy marked by steady growth, low inflation, and gradually falling unemployment, although poor harvests and economic difficulties in Europe contributed to an economic slowdown. To boost exports, Morocco entered into a bilateral Free Trade Agreement with the US in 2006 and an Advanced Status agreement with the EU in 2008. In late 2014, Morocco eliminated subsidies for gasoline, diesel, and fuel oil, dramatically reducing outlays that weighed on the country's budget and current account. Subsidies on butane gas and certain food products remain in place. Morocco also seeks to expand its renewable energy capacity with a goal of making renewable more than 50% of installed electricity generation capacity by 2030.

Despite Morocco's economic progress, the country suffers from high unemployment, poverty, and illiteracy, particularly in rural areas. Key economic challenges for Morocco include reforming the education system and the judiciary.

GDP (purchasing power parity):
$298.6 billion (2017 est.)
$286.8 billion (2016 est.)
$283.6 billion (2015 est.)
note: data are in 2017 dollars
country comparison to the world: 57

GDP (official exchange rate):
$109.3 billion (2017 est.)

GDP—real growth rate: 2.5% (2019 est.)
2.96% (2018 est.)
3.98% (2017 est.)
country comparison to the world: 110

GDP—per capita (PPP): $8,600 (2017 est.)
$8,300 (2016 est.)
$8,300 (2015 est.)
note: data are in 2017 dollars
country comparison to the world: 147

Gross national saving:
30.1% of GDP (2017 est.)
28.9% of GDP (2016 est.)
28.8% of GDP (2015 est.)
country comparison to the world: 31

GDP—composition, by end use:
household consumption: 58% (2017 est.)
government consumption: 18.9% (2017 est.)
investment in fixed capital: 28.4% (2017 est.)
investment in inventories: 4.2% (2017 est.)
exports of goods and services: 37.1% (2017 est.)
imports of goods and services: -46.6% (2017 est.)

GDP—composition, by sector of origin:
agriculture: 14% (2017 est.)
industry: 29.5% (2017 est.)
services: 56.5% (2017 est.)

Agriculture—products: barley, wheat, citrus fruits, grapes, vegetables, olives; livestock; wine

Industries: automotive parts, phosphate mining and processing, aerospace, food processing, leather goods, textiles, construction, energy, tourism

Industrial production growth rate: 2.8% (2017 est.)
country comparison to the world: 110

Labor force: 10.399 million (2020 est.)
country comparison to the world: 49

Labor force—by occupation: *agriculture:* 39.1%
industry: 20.3%
services: 40.5% (2014 est.)

Unemployment rate: 9.23% (2019 est.)
9.65% (2018 est.)
country comparison to the world: 140

Population below poverty line: 15% (2007 est.)

Household income or consumption by percentage share: *lowest 10%:* 2.7%
highest 10%: 33.2% (2007)

Budget: *revenues:* 22.81 billion (2017 est.
expenditures: 26.75 billion (2017 est.)

Taxes and other revenues: 20.9% (of GDP) (2017 est.)
country comparison to the world: 144

Budget surplus (+) or deficit (-): -3.6% (of GDP) (2017 est.)
country comparison to the world: 150

Public debt: 65.1% of GDP (2017 est.)
64.9% of GDP (2016 est.)
country comparison to the world: 58

Fiscal year: calendar year

Inflation rate (consumer prices): 0.8% (2017 est.)
1.6% (2016 est.)
country comparison to the world: 43

Current account balance: -$5.075 billion (2019 est.)
-$6.758 billion (2018 est.)
country comparison to the world: 184

Exports: $21.48 billion (2017 est.)
$22.66 billion (2016 est.)
country comparison to the world: 69

Exports—partners: Spain 23.2%, France 22.6%, Italy 4.5%, US 4.2% (2017)

Exports—commodities: clothing and textiles, automobiles, electric components, inorganic chemicals, transistors, crude minerals, fertilizers (including phosphates), petroleum products, citrus fruits, vegetables, fish

Imports: $39.64 billion (2017 est.)
$36.59 billion (2016 est.)
country comparison to the world: 59

Imports—commodities: crude petroleum, textile fabric, telecommunications equipment, wheat, gas and electricity, transistors, plastics

Imports—partners: Spain 16.7%, France 12.2%, China 9.2%, US 6.9%, Germany 6%, Italy 5.9%, Turkey 4.5% (2017)

Reserves of foreign exchange and gold: $26.27 billion (31 December 2017 est.)
$25.37 billion (31 December 2016 est.)
country comparison to the world: 53

Debt—external: $51.48 billion (31 December 2017 est.)
$44.65 billion (31 December 2016 est.)
country comparison to the world: 65

Exchange rates: Moroccan dirhams (MAD) per US dollar—9.639 (2017 est.)
9.7787 (2016 est.)
9.7787 (2015 est.)
9.7351 (2014 est.)
8.3798 (2013 est.)

ENERGY

Electricity access: *electrification—total population:* 100% (2020)

Electricity—production: 28.75 billion kWh (2016 est.)
country comparison to the world: 68

Electricity—consumption: 28.25 billion kWh (2016 est.)
country comparison to the world: 64

Electricity—exports: 165 million kWh (2015 est.)
country comparison to the world: 79

Electricity—imports: 5.289 billion kWh (2016 est.)
country comparison to the world: 37

Electricity—installed generating capacity: 8.303 million kW (2016 est.)
country comparison to the world: 68

Electricity—from fossil fuels: 68% of total installed capacity (2016 est.)
country comparison to the world: 114

Electricity—from nuclear fuels: 0% of total installed capacity (2017 est.)
country comparison to the world: 148

Electricity—from hydroelectric plants: 16% of total installed capacity (2017 est.)
country comparison to the world: 101

Electricity—from other renewable sources: 15% of total installed capacity (2017 est.)
country comparison to the world: 60

Crude oil—production: 160 bbl/day (2018 est.)
country comparison to the world: 98

Crude oil—exports: 0 bbl/day (2015 est.)
country comparison to the world: 169

Crude oil—imports: 61,160 bbl/day (2015 est.)
country comparison to the world: 53

Crude oil—proved reserves: 684,000 bbl (1 January 2018 est.)
country comparison to the world: 97

Refined petroleum products—production: 66,230 bbl/day (2017 est.)
country comparison to the world: 74

Refined petroleum products—consumption: 278,000 bbl/day (2016 est.)
country comparison to the world: 44

Refined petroleum products—exports: 9,504 bbl/day (2015 est.)
country comparison to the world: 83

Refined petroleum products—imports: 229,300 bbl/day (2015 est.)
country comparison to the world: 30

Natural gas—production: 87.78 million cu m (2017 est.)
country comparison to the world: 82

Natural gas—consumption: 1.218 billion cu m (2017 est.)
country comparison to the world: 89

Natural gas—exports: 0 cu m (2017 est.)
country comparison to the world: 156

Natural gas—imports: 1.133 billion cu m (2017 est.)
country comparison to the world: 61

Natural gas—proved reserves: 1.444 billion cu m (1 January 2018 est.)
country comparison to the world: 97

Carbon dioxide emissions from consumption of energy: 55.4 million Mt (2017 est.)
country comparison to the world: 56

COMMUNICATIONS

Telephones—fixed lines: total subscriptions: 1,982,934
subscriptions per 100 inhabitants: 5.63 (2019 est.)
country comparison to the world: 55

Telephones—mobile cellular: total subscriptions: 45,065,083
subscriptions per 100 inhabitants: 127.95 (2019 est.)
country comparison to the world: 33

Telecommunication systems: *general assessment:* national network nearly 100% digital using fiber-optic links; improved rural service employs microwave radio relay; one of the most state-of-the-art markets in Africa; high mobile penetration rates in the region with low cost for broadband Internet access; improvement in LTE and VoD (Video on Demand) reach and capabilities; some market limitations with lack of competition; mobile internet accounts for 93.2% of all Internet connections (2020)

domestic: fixed-line teledensity is 6 per 100 persons and mobile-cellular subscribership exceeds 128 per 100 persons; good system composed of open-wire lines, cables, and microwave radio relay links; principal switching centers are Casablanca and Rabat (2019)

international: country code—212; landing point for the Atlas Offshore, Estepona-Tetouan, Canalink and SEA-ME-WE-3 fiber-optic telecommunications undersea cables that provide connectivity to Asia, Africa, the Middle East, Europe and Australia; satellite earth stations—2 Intelsat (Atlantic Ocean) and 1 Arabsat; microwave radio relay to Gibraltar, Spain, and Western Sahara (2019)

note: the COVID-19 outbreak is negatively impacting telecommunications production and supply chains globally; consumer spending on telecom devices and services has also slowed due to the pandemic's effect on economies worldwide; overall progress towards improvements in all facets of the telecom industry—mobile, fixed-line, broadband, submarine cable and satellite—has moderated

Broadcast media: 2 TV broadcast networks with state-run Radio-Television Marocaine (RTM) operating one network and the state partially owning the other; foreign TV broadcasts are available via satellite dish; 3 radio broadcast networks with RTM operating one; the government-owned network includes 10 regional radio channels in addition to its national service (2019)

Internet country code: .ma

Internet users: total: 22,596,729

percent of population: 64.8% (July 2018 est.)
country comparison to the world: 33

Broadband—fixed subscriptions:
total: 1,552,599
subscriptions per 100 inhabitants: 4 (2018 est.)
country comparison to the world: 61

Communications—note: the University of al-Quarawiyyin Library in Fez is recognized as the oldest existing, continually operating library in the world, dating back to A.D. 859; among its holdings are approximately 4,000 ancient Islamic manuscripts (2018)

TRANSPORTATION :: MOROCCO

National air transport system: number of registered air carriers: 3 (2020)
inventory of registered aircraft operated by air carriers: 76
annual passenger traffic on registered air carriers: 8,132,917 (2018)
annual freight traffic on registered air carriers: 97.71 million mt-km (2018)

Civil aircraft registration country code prefix: CN (2016)

Airports: 55 (2013)
country comparison to the world: 83

Airports—with paved runways: *total:* 31 (2017)
over 3,047 m: 11 (2017)
2,438 to 3,047 m: 9 (2017)
1,524 to 2,437 m: 7 (2017)
914 to 1,523 m: 4 (2017)

Airports—with unpaved runways:
total: 24 (2013)
2.438 to 3,047 m: 1 (2013)
1.524 to 2,437 m: 7 (2013)
914 to 1,523 m: 11 (2013)
under 914 m: 5 (2013)

Heliports: 1 (2013)

Pipelines: 944 km gas, 270 km oil, 175 km refined products (2013)

Railways: *total:* 2,067 km (2014)
standard gauge: 2,067 km 1.435-m gauge (1,022 km electrified) (2014)
country comparison to the world: 73

Roadways: *total:* 57,300 km (2018)
country comparison to the world: 80

Merchant marine: *total:* 86
by type: bulk carrier 4, general cargo 4, oil tanker 3, other 75 (2019)
country comparison to the world: 97

Ports and terminals: *major seaport(s):* Casablanca, Jorf Lasfar, Mohammedia, Safi, Tangier
container port(s) (TEUs): Tangier (3,312,409) (2017)
LNG terminal(s) (import): Jorf Lasfar

MILITARY AND SECURITY

Military and security forces: Royal Armed Forces: Royal Moroccan Army, Royal Moroccan Navy (includes Coast Guard, marines), Royal Moroccan Air Force, Royal Morroccan Gendarmerie, Morroccan Royal Guard (provides security for the royal family; officially part of the Royal Army); Force Auxiliaire (a paramilitary force under the Ministry of Interior that supplements the military and the police as needed) (2019)

Military expenditures:
3.1% of GDP (2019)
3.1% of GDP (2018)
3.2% of GDP (2017)
3.2% of GDP (2016)
3.2% of GDP (2015)
country comparison to the world: 25

Military and security service personnel strengths: the Royal Armed Forces have approximately 197,000 active personnel (175,000 Army; 9,000 Navy; 13,000 Air Force); est. 25,000 Gendarmerie (2019 est.)

Military equipment inventories and acquisitions: the Moroccan military's inventory is comprised of mostly older French and US equipment; since 2010, France and the US are the leading suppliers of weapons to Morocco, followed by China and the Netherlands (2019 est.)

Military deployments: 750 Central African Republic (MINUSCA); 960 Democratic Republic of the Congo (MONUSCO) (2020)

Military service age and obligation: 19 years of age for compulsory military service (reintroduced in 2019); both sexes are obligated to military service; conscript service obligation—12 months (2019)

TRANSNATIONAL ISSUES

Disputes—international: claims and administers Western Sahara whose sovereignty remains unresolved; Morocco protests Spain's control over the coastal enclaves of Ceuta, Melilla, and Penon de Velez de la Gomera, the islands of Penon de Alhucemas and Islas Chafarinas, and surrounding waters; both countries claim Isla Pereiil (Leila Island); discussions have not progressed on a comprehensive maritime delimitation, setting limits on resource exploration and refugee interdiction, since Morocco's 2002 rejection of Spain's unilateral designation of a median line from the Canary Islands; Morocco serves as one of the primary launching areas of illegal migration into Spain from North Africa; Algeria's border with Morocco remains an irritant to bilateral relations, each nation accusing the other of harboring militants and arms smuggling; the National Liberation Front's assertions of a claim to Chirac Pastures in southeastern Morocco is a dormant dispute

Illicit drugs: the world's largest producer and exporter of cannabis; total production for 2015-2016 growing season estimated to be 700 metric tons; shipments of hashish mostly directed to Western Europe; transit point for cocaine from South America destined for Western Europe; significant consumer of cannabis

MOZAMBIQUE

INTRODUCTION

Background: In the first half of the second millennium A.D., northern Mozambican port towns were frequented by traders from Somalia, Ethiopia, Egypt, Arabia, Persia, and India. The Portuguese were able to wrest much of the coastal trade from Arab Muslims in the centuries after 1500 and to set up their own colonies. Portugal did not relinquish Mozambique until 1975. Large-scale emigration, economic dependence on South Africa, a severe drought, and a prolonged civil war hindered the country's development until the mid-1990s. The ruling Front for the Liberation of Mozambique (FRELIMO) party formally abandoned Marxism in 1989, and a new constitution the following year provided for multiparty elections and a free market economy. A UN-negotiated peace agreement between FRELIMO and rebel Mozambique National Resistance (RENAMO) forces ended the fighting in 1992. In 2004, Mozambique underwent a delicate transition as Joaquim CHISSANO stepped down after 18 years in office. His elected successor, Armando GUEBUZA, served two terms and then passed executive power to Filipe NYUSI in 2015. RENAMO's residual armed forces intermittently engaged in a low-level insurgency after 2012, but a late December 2016 ceasefire eventually led to the two sides signing a comprehensive peace deal in August 2019. Elections in October 2019, challenged by Western observers and civil society as being problematic, resulted in resounding wins for NYUSI and FRELIMO across the country. Since October 2017, violent extremists—who an official ISIS media outlet recognized as ISIS's network in Mozambique for the first time in June 2019—have been conducting attacks against civilians and security services in the northern province of Cabo Delgado.

GEOGRAPHY

Location: Southeastern Africa, bordering the Mozambique Channel, between South Africa and Tanzania

Geographic coordinates: 18 15 S, 35 00 E

Map references: Africa

Area: *total:* 799,380 sq km
land: 786,380 sq km
water: 13,000 sq km
country comparison to the world: 36

Area—comparative: slightly more than five times the size of Georgia; slightly less than twice the size of California

Land boundaries: *total:* 4,783 km
border countries (6): Malawi 1498 km, South Africa 496 km, Eswatini 108 km, Tanzania 840 km, Zambia 439 km, Zimbabwe 1402 km

Coastline: 2,470 km

Maritime claims: *territorial sea:* 12 nm
exclusive economic zone: 200 nm

Climate: tropical to subtropical

Terrain: mostly coastal lowlands, uplands in center, high plateaus in northwest, mountains in west

Elevation: *mean elevation:* 345 m
lowest point: Indian Ocean 0 m
highest point: Monte Binga 2,436 m

Natural resources: coal, titanium, natural gas, hydropower, tantalum, graphite

Land use: *agricultural land:* 56.3% (2011 est.)
arable land: 6.4% (2011 est.) / *permanent crops:* 0.3% (2011 est.) / *permanent pasture:* 49.6% (2011 est.)
forest: 43.7% (2011 est.)
other: 0% (2011 est.)

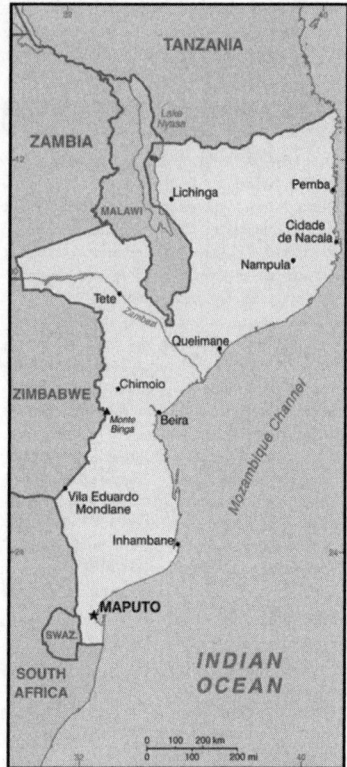

Irrigated land: 1,180 sq km (2012)

Population distribution: three large populations clusters are found along the southern coast between Maputo and Inhambane, in the central area between Beira and Chimoio along the Zambezi River, and in and around the northern cities of Nampula, Cidade de Nacala, and Pemba; the northwest and southwest are the least populated areas as shown in this population distribution map

Natural hazards: severe droughts; devastating cyclones and floods in central and southern provinces Environment—current issues: increased migration of the population to urban and coastal areas with adverse environmental consequences; desertification; soil erosion; deforestation; water pollution caused by artisanal mining; pollution of surface and coastal waters; wildlife preservation (elephant poaching for ivory)

Environment—international agreements: *party to:* Biodiversity, Climate Change, Climate Change-Kyoto Protocol, Desertification, Endangered Species, Hazardous Wastes, Law of the Sea, Ozone Layer Protection, Ship Pollution, Wetlands
signed, but not ratified: none of the selected agreements

Geography—note: the Zambezi River flows through the north-central and most fertile part of the country

PEOPLE AND SOCIETY

Population: 30,098,197 (July 2020 est.)
note: estimates for this country explicitly take into account the effects of excess mortality due to AIDS; this can result in lower life expectancy, higher infant mortality, higher death rates, lower population growth rates, and changes in the distribution of population by age and sex than would otherwise be expected
country comparison to the world: 47

Nationality: *noun:* Mozambican(s)
adjective: Mozambican

Ethnic groups: African 99% (Makhuwa, Tsonga, Lomwe, Sena, and others), mestizo 0.8%, other (includes European, Indian, Pakistani, Chinese) .2% (2017 est.)

Languages: Makhuwa 26.1%, Portuguese (official) 16.6%, Tsonga 8.6%, Nyanja 8.1, Sena 7.1%, Lomwe 7.1%, Chuwabo 4.7%, Ndau 3.8%, Tswa 3.8%, other Mozambican languages 11.8%, other 0.5%, unspecified 1.8% (2017 est.)

Religions: Roman Catholic 27.2%, Muslim 18.9%, Zionist Christian 15.6%, Evangelical/Pentecostal 15.3%, Anglican 1.7%, other 4.8%, none 13.9%, unspecified 2.5% (2017 est.)

Demographic profile: Mozambique is a poor, sparsely populated country with high fertility and mortality rates and a rapidly growing youthful population — 45% of the population is younger than 15. Mozambique's high poverty rate is sustained by natural disasters, disease, high population growth, low agricultural productivity, and the unequal distribution of wealth. The country's birth rate is among the world's highest, averaging around more than 5 children per woman (and higher in rural areas) for at least the last three decades. The sustained high level of fertility reflects gender inequality, low contraceptive use, early marriages and childbearing, and a lack of education, particularly among women. The high population growth rate is somewhat restrained by the country's high HIV/AIDS and overall mortality rates. Mozambique ranks among the worst in the world for HIV/AIDS prevalence, HIV/AIDS deaths, and life expectancy at birth.

Mozambique is predominantly a country of emigration, but internal, rural-urban migration has begun to grow.

Mozambicans, primarily from the country's southern region, have been migrating to South Africa for work for more than a century. Additionally, approximately 1.7 million Mozambicans fled to Malawi, South Africa, and other neighboring countries between 1979 and 1992 to escape from civil war. Labor migrants have usually been men from rural areas whose crops have failed or who are unemployed and have headed to South Africa to work as miners; multiple generations of the same family often become miners. Since the abolition of apartheid in South Africa in 1991, other job opportunities have opened to Mozambicans, including in the informal and manufacturing sectors, but mining remains their main source of employment.

Age structure: *0-14 years:* 45.57% (male 6,950,800/female 6,766,373)
15-24 years: 19.91% (male 2,997,529/female 2,994,927)
25-54 years: 28.28% (male 3,949,085/female 4,564,031)
55-64 years: 3.31% (male 485,454/female 509,430)
65 years and over: 2.93% (male 430,797/female 449,771) (2020 est.)

Dependency ratios: *total dependency ratio:* 88.4
youth dependency ratio: 83
elderly dependency ratio: 5.4
potential support ratio: 18.5 (2020 est.)

Median age: *total:* 17 years
male: 16.3 years
female: 17.6 years (2020 est.)
country comparison to the world: 220

Population growth rate: 2.62% (2020 est.)
country comparison to the world: 19

Birth rate: 38.6 births/1,000 population (2020 est.)
country comparison to the world: 12

Death rate: 11 deaths/1,000 population (2020 est.)
country comparison to the world: 22

Net migration rate: -1.7 migrant(s)/1,000 population (2020 est.)
country comparison to the world: 159

Population distribution: three large populations clusters are found along the southern coast between Maputo and Inhambane, in the central area between Beira and Chimoio along the Zambezi River, and in and around the northern cities of Nampula, Cidade de Nacala, and Pemba; the northwest and southwest are the least populated areas as shown in this population distribution map

Urbanization: urban population: 37.1% of total population (2020)
rate of urbanization: 4.35% annual rate of change (2015-20 est.)
total population growth rate v. urban population growth rate, 2000-2030:

Major urban areas—population: 1.706 million Matola, 1.11 million MAPUTO (capital), 848,000 Nampula (2020)

Sex ratio: *at birth:* 1.03 male(s)/female
0-14 years: 1.03 male(s)/female
15-24 years: 1 male(s)/female
25-54 years: 0.87 male(s)/female
55-64 years: 0.95 male(s)/female
65 years and over: 0.96 male(s)/female
total population: 0.97 male(s)/female (2020 est.)

Mother's mean age at first birth: 18.9 years (2011 est.)
median age at first birth among women 25-29

Maternal mortality rate: 289 deaths/100,000 live births (2017 est.)
country comparison to the world: 39

Infant mortality rate: *total:* 64.7 deaths/1,000 live births
male: 66.8 deaths/1,000 live births
female: 62.6 deaths/1,000 live births (2020 est.)
country comparison to the world: 7

Life expectancy at birth: total population: 55.9 years
male: 54.4 years
female: 57.4 years (2020 est.)
country comparison to the world: 222

Total fertility rate: 4.97 children born/woman (2020 est.)
country comparison to the world: 13

Contraceptive prevalence rate: 27.1% (2015)

Drinking water source:
improved:
urban: 93.2% of population
rural: 58.3% of population
total: 70.7% of population
unimproved:
urban: 6.8% of population
rural: 41.7% of population
total: 29.3% of population (2017 est.)

Current Health Expenditure: 4.9% (2017)

Physicians density: 0.08 physicians/1,000 population (2017)

Hospital bed density: 0.7 beds/1,000 population (2011)

Sanitation facility access:
improved:
urban: 61.8% of population (2015 est.)
rural: 18.8% of population
total: 34.1% of population
unimproved:
urban: 38.2% of population
rural: 81.2% of population
total: 65.9% of population (2017 est.)

HIV/AIDS—adult prevalence rate: 12.1% (2019 est.)
country comparison to the world: 7

HIV/AIDS—people living with HIV/AIDS: 2.2 million (2019 est.)
country comparison to the world: 2

HIV/AIDS—deaths: 51,000 (2019 est.)
country comparison to the world: 3

Major infectious diseases: *degree of risk:* very high (2020)
food or waterborne diseases: bacterial and protozoal diarrhea, hepatitis A, and typhoid fever
vectorborne diseases: malaria and dengue fever
water contact diseases: schistosomiasis
animal contact diseases: rabies

Obesity—adult prevalence rate: 7.2% (2016)
country comparison to the world: 160

Children under the age of 5 years underweight: 15.6% (2014/15)
country comparison to the world: 38

Education expenditures: 6.5% of GDP (2013)
country comparison to the world: 21

Literacy: *definition:* age 15 and over can read and write
total population: 60.7%
male: 72.6%
female: 50.3% (2017)

School life expectancy (primary to tertiary education): *total:* 10 years
male: 11 years
female: 10 years (2017)

Unemployment, youth ages 15-24: *total:* 7.4%
male: 7.7%
female: 7.1% (2015 est.)
country comparison to the world: 148

GOVERNMENT

Country name: *conventional long form:* Republic of Mozambique
conventional short form: Mozambique
local long form: Republica de Mocambique
local short form: Mocambique
former: Portuguese East Africa, People's Republic of Mozambique
etymology: named for the offshore island of Mozambique; the island was apparently named after Mussa al-BIK, an influential Arab slave trader who set himself up as sultan on the island in the 15th century

Government type: presidential republic

Capital: name: Maputo
geographic coordinates: 25 57 S, 32 35 E
time difference: UTC+2 (7 hours ahead of Washington, DC, during Standard Time)
etymology: reputedly named after the Maputo River, which drains into Maputo Bay south of the city

Administrative divisions: 10 provinces (provincias, singular—provincia), 1 city (cidade)*; Cabo Delgado, Gaza, Inhambane, Manica, Maputo, Cidade de Maputo*, Nampula, Niassa, Sofala, Tete, Zambezia

Independence: 25 June 1975 (from Portugal)

National holiday: Independence Day, 25 June (1975)

Constitution: *history:* previous 1975, 1990; latest adopted 16 November 2004, effective 21 December 2004
amendments: proposed by the president of the republic or supported by at least one third of the Assembly of the Republic membership; passage of amendments affecting constitutional provisions, including the independence and sovereignty of the state, the republican form of government, basic rights and freedoms, and universal suffrage, requires at least a two-thirds majority vote by the Assembly and approval in a referendum; referenda not required for passage of other amendments; amended 2007, 2018

Legal system: mixed legal system of Portuguese civil law and customary law; note—in rural, apply where applicable predominantly Muslim villages with no formal legal system, Islamic law may be applied

International law organization participation: has not submitted an ICJ jurisdiction declaration; non-party state to the ICCt

Citizenship: *citizenship by birth:* no
citizenship by descent only: at least one parent must be a citizen of Mozambique
dual citizenship recognized: no
residency requirement for naturalization: 5 years

Suffrage: 18 years of age; universal

Executive branch: *chief of state:* President Filipe Jacinto NYUSI (since 15 January 2015, re-elected 15 Oct 2019) (2019)
head of government: President Filipe Jacinto NYUSI (since 15 January 2015); Prime Minister Carlos Agostinho DO ROSARIO (since 17 January 2015; reconfirmed DO ROSARIO 17 January 2020) (2020)
cabinet: Cabinet appointed by the president
elections/appointments: president elected directly by absolute majority popular vote (in 2 rounds, if needed) for a 5- year term (eligible for 2 consecutive terms); election last held on 15 October 2019 (next to be held on 15 October 2024); prime minister appointed by the president (2019)
election results: Filipe NYUSI elected president in first round; percent of vote—Filipe NYUSI (FRELIMO) 73.0%, Ossufo MOMADE (RENAMO) 21.9%, Daviz SIMANGO (MDM) 5.1% (2019)

Legislative branch: *description:* unicameral Assembly of the Republic or Assembleia da Republica (250 seats; 248 members elected in multiseat constituencies by party-list proportional representation vote and 2 single members representing Mozambicans abroad directly elected by simple majority vote; members serve 5-year terms) (2019)
elections: last held on 15 October 2019 (next to be held on 15 October 2024) (2019)
election results: percent of vote by party—FRELIMO 71%, RENAMO 23%, MDM 4%; seats by party—FRELIMO 184, RENAMO 60, MDM 6; composition—men 151, women 99, percent of women 39.6% (2019)

Judicial branch: *highest courts:* Supreme Court (consists of the court president, vice president, and 5 judges); Constitutional Council (consists of 7 judges); note—the Higher Council of the Judiciary Magistracy is responsible for judiciary management and discipline
judge selection and term of office: Supreme Court president appointed by the president of the republic; vice president appointed by the president in consultation with the Higher Council of the Judiciary (CSMJ) and ratified by the Assembly of the Republic; other judges elected by the Assembly; judges serve 5-year renewable terms; Constitutional Council judges appointed—1 by the president, 5 by the Assembly, and 1 by the CSMJ; judges serve 5-year nonrenewable terms
subordinate courts: Administrative Court (capital city only); provincial courts or Tribunais Judicias de Provincia; District Courts or Tribunais Judicias de Districto; customs courts; maritime courts; courts marshal; labor courts; community courts

Political parties and leaders:
Democratic Movement of Mozambique (Movimento Democratico de Mocambique) or MDM [Daviz SIMANGO]

Front for the Liberation of Mozambique (Frente de Liberatacao de Mocambique) or FRELIMO [Filipe NYUSI]

Mozambican National Resistance (Resistencia Nacional Mocambicana) or RENAMO [Ossufo MOMADE]

Optimistic Party for the Development of Mozambique or Podemos [Helder Mendonca]

International organization participation: ACP, AfDB, AU, C, CD, CPLP, EITI (compliant country), FAO, G-77, IAEA, IBRD, ICAO, ICC (NGOs), ICRM, IDA, IDB, IFAD, IFC, IFRCS, IHO, ILO, IMF, IMO, IMSO, Interpol, IOC, IOM, IPU, ISO (correspondent), ITSO, ITU, ITUC (NGOs), MIGA, NAM, OIC, OIF (observer), OPCW, SADC, UN, UNCTAD, UNESCO, UNHCR, UNIDO, Union Latina, UNISFA, UNWTO, UPU, WCO, WFTU (NGOs), WHO, WIPO, WMO, WTO

Diplomatic representation in the US: *chief of mission:* Ambassador Carlos DOS SANTOS (since 28 January 2016)
chancery: 1525 New Hampshire Avenue NW, Washington, DC 20036
telephone: [1] (202) 293-7146
FAX: [1] (202) 835-0245

Diplomatic representation from the US: *chief of mission:* Ambassador Dennis W. HEARNE (since 22 February 2019)
telephone: [258] (21) 49 2797
embassy: Avenida Kenneth Kuanda 193, Caixa Postal, 783, Maputo
mailing address: P.O. Box 783, Maputo
FAX: [258] (21) 49 0114

Flag description: three equal horizontal bands of green (top), black, and yellow with a red isosceles triangle based on the hoist side; the black band is edged in white; centered in the triangle is a yellow five-pointed star bearing a crossed rifle and hoe in black superimposed on an open white book; green represents the riches of the land, white peace, black the African continent, yellow the country's minerals, and red the struggle for independence; the rifle symbolizes defense and vigilance, the hoe refers to the country's agriculture, the open book stresses the importance of education, and the star represents Marxism and internationalism
note: one of only two national flags featuring a firearm, the other is Guatemala

National symbol(s): *national colors:* green, black, yellow, white, red

National anthem: *name:* "Patria Amada" (Lovely Fatherland)
lyrics/music: Salomao J. MANHICA/unknown
note: adopted 2002

ECONOMY

Economy—overview: At independence in 1975, Mozambique was one of the world's poorest countries. Socialist policies, economic mismanagement, and a brutal civil war from 1977 to 1992 further impoverished the country. In 1987, the government embarked on a series of macroeconomic reforms designed to stabilize the economy.

These steps, combined with donor assistance and with political stability since the multi-party elections in 1994, propelled the country's GDP, in purchasing power parity terms, from $4 billion in 1993 to about $37 billion in 2017. Fiscal reforms, including the introduction of a value-added tax and reform of the customs service, have improved the government's revenue collection abilities. In spite of these gains, about half the population remains below the poverty line and subsistence agriculture continues to employ the vast majority of the country's work force.

Mozambique's once substantial foreign debt was reduced through forgiveness and rescheduling under the IMF's Heavily Indebted Poor Countries (HIPC) and Enhanced HIPC initiatives. However, in 2016, information surfaced revealing that the Mozambican Government was responsible for over $2 billion in government-backed loans secured between 2012-14 by state-owned defense and security companies without parliamentary approval or national budget inclusion; this prompted the IMF and international donors to halt direct budget support to the Government of Mozambique. An international audit was performed on Mozambique's debt in 201 6-17, but debt restructuring and resumption of donor support have yet to occur.

Mozambique grew at an average annual rate of 6%-8% in the decade leading up to 2015, one of Africa's strongest performances, but the sizable external debt burden, donor withdrawal, elevated inflation, and currency depreciation contributed to slower growth in 2016-17.

Two major International consortiums, led by American companies ExxonMobil and Anadarko, are seeking approval to develop massive natural gas deposits off the coast of Cabo Delgado province, in what has the potential to become the largest infrastructure project in Africa. . The government predicts sales of liquefied natural gas from these projects could generate several billion dollars in revenues annually sometime after 2022.

GDP (purchasing power parity):
$37.09 billion (2017 est.)
$35.76 billion (2016 est.)
$34.46 billion (2015 est.)
note: data are in 2017 dollars
country comparison to the world: 122

GDP (official exchange rate):
$12.59 billion (2017 est.)

GDP—real growth rate: 3.11% (2018 est.)
3.7% (2017 est.)
4.07% (2017 est.)
country comparison to the world: 95

GDP—per capita (PPP): $1,300 (2017 est.)
$1,200 (2016 est.)
$1,200 (2015 est.)
note: data are in 2017 dollars
country comparison to the world: 222

Gross national saving:
16.8% of GDP (2017 est.)
-1.2% of GDP (2016 est.)
5% of GDP (2015 est.)
country comparison to the world: 123

GDP—composition, by end use:

household consumption: 69.7% (2017 est.)
government consumption: 27.2% (2017 est.)
investment in fixed capital: 21.7% (2017 est.)
investment in inventories: 13.9% (2017 est.)
exports of goods and services: 38.3% (2017 est.)
imports of goods and services: -70.6% (2017 est.)

GDP—composition, by sector of origin:
agriculture: 23.9% (2017 est.)
industry: 19.3% (2017 est.)
services: 56.8% (2017 est.)

Agriculture—products: cotton, cashew nuts, sugarcane, tea, cassava (manioc, tapioca), corn, coconuts, sisal, citrus and tropical fruits, potatoes, sunflowers; beef, poultry

Industries: aluminum, petroleum products, chemicals (fertilizer, soap, paints), textiles, cement, glass, asbestos, tobacco, food, beverages

Industrial production growth rate: 4.9% (2017 est.)
country comparison to the world: 61

Labor force: 12.9 million (2017 est.)
country comparison to the world: 42

Labor force—by occupation: *agriculture:* 74.4%
industry: 3.9%
services: 21.7% (2015 est.)

Unemployment rate: 24.5% (2017 est.)
25% (2016 est.)
country comparison to the world: 196

Population below poverty line: 46.1% (2015 est.)

Household income or consumption by percentage share: *lowest 10%:* 1.9%
highest 10%: 36.7% (2008)

Budget: *revenues:* 3.356 billion (2017 est.)
expenditures: 4.054 billion (2017 est.)

Taxes and other revenues: 26.7% (of GDP) (2017 est.)
country comparison to the world: 105

Budget surplus (+) or deficit (-): -5.6% (of GDP) (2017 est.)
country comparison to the world: 176

Public debt: 102.1% of GDP (2017 est.)
121.6% of GDP (2016 est.)
country comparison to the world: 15

Fiscal year: calendar year

Inflation rate (consumer prices): 15.3% (2017 est.)
19.2% (2016 est.)
country comparison to the world: 212

Current account balance: -$3.025 billion (2019 est.)
-$4.499 billion (2018 est.)
country comparison to the world: 172

Exports: $4.725 billion (2017 est.)
$3.328 billion (2016 est.)
country comparison to the world: 110

Exports—partners: India 28.1%, Netherlands 24.4%, South Africa 16.7% (2017)

Exports—commodities: aluminum, prawns, cashews, cotton, sugar, citrus, timber; bulk electricity

Imports: $5.223 billion (2017 est.)
$4.733 billion (2016 est.)

country comparison to the world: 124

Imports—commodities: machinery and equipment, vehicles, fuel, chemicals, metal products, foodstuffs, textiles

Imports—partners: South Africa 36.8%, China 7%, UAE 6.8%, India 6.2%, Portugal 4.4% (2017)

Reserves of foreign exchange and gold: $3.361 billion (31 December 2017 est.) $2.081 billion (31 December 2016 est.) *country comparison to the world:* 106

Debt—external: $10.91 billion (31 December 2017 est.) $10.48 billion (31 December 2016 est.) *country comparison to the world:* 110

Exchange rates: meticais (MZM) per US dollar—64.4 (2017 est.) 63.067 (2016 est.) 63.067 (2015 est.) 39.983 (2014 est.) 31.367 (2013 est.)

ENERGY

Electricity access: *population without electricity:* 20 million (2019)
electrification—total population: 35% (2019)
electrification—urban areas: 57% (2019)
electrification—rural areas: 22% (2019)

Electricity—production: 18.39 billion kWh (2016 est.)
country comparison to the world: 79

Electricity—consumption: 11.57 billion kWh (2016 est.)
country comparison to the world: 90

Electricity—exports: 12.88 billion kWh (2015 est.)
country comparison to the world: 16

Electricity—imports: 9.928 billion kWh (2016 est.)
country comparison to the world: 25

Electricity—installed generating capacity: 2.626 million kW (2016 est.)
country comparison to the world: 102

Electricity—from fossil fuels: 16% of total installed capacity (2016 est.)
country comparison to the world: 199

Electricity—from nuclear fuels: 0% of total installed capacity (2017 est.)
country comparison to the world: 149

Electricity—from hydroelectric plants: 83% of total installed capacity (2017 est.)
country comparison to the world: 12

Electricity—from other renewable sources: 1% of total installed capacity (2017 est.)
country comparison to the world: 161

Crude oil—production: 0 bbl/day (2018 est.)
country comparison to the world: 178

Crude oil—exports: 0 bbl/day (2015 est.)
country comparison to the world: 170

Crude oil—imports: 0 bbl/day (2015 est.)
country comparison to the world: 171

Crude oil—proved reserves: 0 bbl (1 January 2018 est.)
country comparison to the world: 173

Refined petroleum products—production: 0 bbl/day (2015 est.)
country comparison to the world: 182

Refined petroleum products—consumption: 26,000 bbl/day (2016 est.)
country comparison to the world: 128

Refined petroleum products—exports: 0 bbl/day (2015 est.)
country comparison to the world: 185

Refined petroleum products—imports: 25,130 bbl/day (2015 est.)
country comparison to the world: 107

Natural gas—production: 6.003 billion cu m (2017 est.)
country comparison to the world: 47

Natural gas—consumption: 1841 billion cu m (2017 est.)
country comparison to the world: 84

Natural gas—exports: 4.162 billion cu m (2017 est.)
country comparison to the world: 32

Natural gas—imports: 0 cu m (2017 est.)
country comparison to the world: 161

Natural gas—proved reserves: 2.832 trillion cu m (1 January 2018 est.)
country comparison to the world: 13

Carbon dioxide emissions from consumption of energy: 11.12 million Mt (2017 est.)
country comparison to the world: 102

COMMUNICATIONS

Telephones—fixed lines: *total subscriptions:* 61,575
subscriptions per 100 inhabitants: less than 1 (2019 est.)
country comparison to the world: 153

Telephones—mobile cellular: *total subscriptions:* 13,992,090
subscriptions per 100 inhabitants: 47.72 (2019 est.)
country comparison to the world: 70

Telecommunication systems: *general assessment:* the mobile segment has shown strong growth; poor fixed-line infrastructure means most Internet access is through mobile accounts; DSL, cable broadband, WiMAX (broadband over long distances), 3G and some fiber broadband available; first LTE services launched in 2018; govt. implemented legislation to enforce the registration of SIM cards; submarine cables reduced the cost of bandwidth (2020)
domestic: extremely low fixed-line teledensity contrasts with rapid growth in the mobile-cellular network; operators provide coverage that includes all the main cities and key roads; fixed-line less than 1 per 100 and 48 per 100 mobilecellular teledensity (2019)
international: country code—258; landing points for the EASSy and SEACOM/ Tata TGN-Eurasia fiber-optic submarine cable systems linking numerous east African countries, the Middle East and Asia ; satellite earth stations—5 Intelsat (2 Atlantic Ocean and 3 Indian Ocean); TdM contracts for Itelsat for satellite broadband and bulk haul services (2020)
note: the COVID-19 outbreak is negatively impacting telecommunications production and supply chains globally; consumer spending on telecom devices and services has also slowed due to the pandemic's effect on economies worldwide; overall progress towards improvements in all facets of the telecom industry—mobile, fixed-line, broadband, submarine cable and satellite—has moderated

Broadcast media: 1 state-run TV station supplemented by private TV station; Portuguese state TV's African service, RTP Africa, and Brazilian-owned TV Miramar are available; state-run radio provides nearly 100% territorial coverage and broadcasts in multiple languages; a number of privately owned and community-operated stations; transmissions of multiple international broadcasters are available (2019)

Internet country code: .mz

Internet users: *total:* 2,855,670
percent of population: 10% (July 2018 est.)
country comparison to the world: 100

Broadband—fixed subscriptions: total: 70,142
subscriptions per 100 inhabitants: less than 1 (2018 est.)
country comparison to the world: 128

TRANSPORTATION

National air transport system: *number of registered air carriers:* 2 (2020)
inventory of registered aircraft operated by air carriers: 11
annual passenger traffic on registered air carriers: 540,124 (2018)
annual freight traffic on registered air carriers: 4.78 million mt-km (2018)

Civil aircraft registration country code prefix: C9 (2016)

Airports: 98 (2013)
country comparison to the world: 57

Airports—with paved runways:
total: 21 (2017)
over 3,047 m: 1 (2017)
2,438 to 3,047 m: 2 (2017)
1,524 to 2,437 m: 9 (2017)
914 to 1,523 m: 5 (2017)
under 914 m: 4 (2017)

Airports—with unpaved runways:
total: 77 (2013)
2,438 to 3,047 m: 1 (2013)
1,524 to 2,437 m: 9 (2013)
914 to 1,523 m: 29 (2013)
under 914 m: 38 (2013)

Pipelines: 972 km gas, 278 km refined products (2013)

Railways: *total:* 4,787 km (2014)
narrow gauge: 4,787 km 1.067-m gauge (2014)
country comparison to the world: 41

669

Roadways:

total: 31,083 km (2015)

paved: 7,365 km (2015)

unpaved: 23,718 km (2015)

country comparison to the world: 97

Waterways: 460 km (Zambezi River navigable to Tete and along Cahora Bassa Lake) (2010)

country comparison to the world: 85

Merchant marine: *total:* 29

by type: general cargo 9, other 20 (2019)

country comparison to the world: 134

Ports and terminals: *major seaport(s):* Beira, Maputo, Nacala

MILITARY AND SECURITY

Military and security forces: Armed Defense Forces of Mozambique (Forcas Armadas de Defesa de Mocambique, FADM): Mozambique Army, Mozambique Navy (Marinha de Guerra de Mocambique, MGM), Mozambique Air Force (Forca Aerea de Mocambique, FAM)

Ministry of Interior: National Police (PRM), the National Criminal Investigation Service (SERNIC), Rapid Intervention Unit (UIR; police special forces), Border Security Force (2019)

note: the FADM and Ministry of Interior forces are referred to collectively as the Defense and Security Forces (DFS)

Military expenditures:

0.99% of GDP (2018)

1.02% of GDP (2017)

1.03% of GDP (2016)

0.81% of GDP (2015)

1.02% of GDP (2014)

country comparison to the world: 122

Military and security service personnel strengths: the Armed Defense Forces of Mozambique (FADM) are comprised of approximately 11,000 personnel (10,000 Army; 200 Navy; 1,000 Air Force) (2019 est.)

Military equipment inventories and acquisitions: the FADM's inventory consists primarily of Soviet-era equipment, although in recent years it has received limited quantities of newer equipment, particularly aircraft and maritime patrol craft (mostly as aid/donations); India is the leading supplier since 2010 (2019)

Military service age and obligation: registration for military service is mandatory for all males and females at 18 years of age; 18-35 years of age for selective compulsory military service; 18 years of age for voluntary service; 2-year service obligation; women may serve as officers or enlisted (2019)

Military—note: the Mozambique Defense and Security Forces are facing a growing insurgency involving terrorist/militant groups with ties to the Islamic State in Central Africa in the northern province of Cabo Delgado, an area known for rich liquid natural gas deposits; attacks in the province began around 2017 and as of November 2020, the fighting had left an estimated 2,000 dead and over 200-400,000 displaced; Mozambique has brought in private military companies based in Russia and South Africa to provide assistance to its security forces (2020)

TERRORISM

Terrorist group(s): Islamic State of Iraq and ash-Sham—Central Africa/Mozambique (2020)

note: details about the history, aims, leadership, organization, areas of operation, tactics, targets, weapons, size, and sources of support of the group(s) appear(s) in Appendix-T

TRANSNATIONAL ISSUES

Disputes—international: South Africa has placed military units to assist police operations along the border of Lesotho, Zimbabwe, and Mozambique to control smuggling, poaching, and illegal migration

Refugees and internally displaced persons: *refugees (country of origin):* 9,953 (Democratic Republic of Congo) (refugees and asylum seekers), 8,658 (Burundi) (refugees and asylum seekers) (2020)

IDPs: 369,220 (violence between the government and an opposition group, violence associated with extremists groups in 2018, political violence 2019) (2020)

Illicit drugs: southern African transit point for South Asian hashish and heroin, and South American cocaine probably destined for the European and South African markets; producer of cannabis (for local consumption) and methaqualone (for export to South Africa); corruption and poor regulatory capability make the banking system vulnerable to money laundering, but the lack of a well-developed financial infrastructure limits the country's utility as a money-laundering center

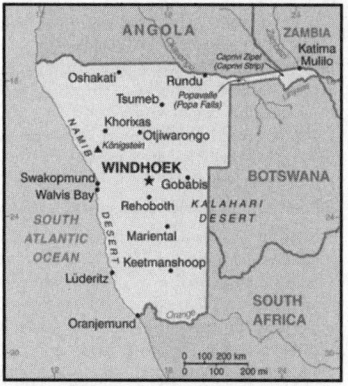

INTRODUCTION

Background: Various ethnic groups occupied south western Africa prior to Germany establishing a colony over most of the territory in 1884. South Africa occupied the colony, then known as German South West Africa, in 1915 during World War I and administered it as a mandate until after World War II, when it annexed the territory. In 1966, the Marxist South-West Africa People's Organization (SWAPO) guerrilla group launched a war of independence for the area that became Namibia, but it was not until 1988 that South Africa agreed to end its administration in accordance with a UN peace plan for the entire region. Namibia gained independence in 1990 and has been governed by SWAPO since, though the party has dropped much of its Marxist ideology. President Hage GEINGOB was elected in 2014 in a landslide victory, replacing Hifikepunye POHAMBA who stepped down after serving two terms. SWAPO retained its parliamentary super majority in the 2014 elections. In 2019 elections, GEINGOB was reelected but by a substantially reduced majority and SWAPO narrowly lost its super majority in parliament. Namibia gained independence in 1990.

GEOGRAPHY

Location: Southern Africa, bordering the South Atlantic Ocean, between Angola and South Africa

Geographic coordinates: 22 00 S, 17 00 E

Map references: Africa

Area: *total:* 824,292 sq km
land: 823,290 sq km
water: 1,002 sq km
country comparison to the world: 35

Area—comparative: almost seven times the size of Pennsylvania; slightly more than half the size of Alaska

Land boundaries: *total:* 4,220 km

border countries (4): Angola 1427 km, Botswana 1544 km, South Africa 1005 km, Zambia 244 km

Coastline: 1,572 km

Maritime claims: *territorial sea:* 12 nm
exclusive economic zone: 200 nm
contiguous zone: 24 nm

Climate: desert; hot, dry; rainfall sparse and erratic

Terrain: mostly high plateau; Namib Desert along coast; Kalahari Desert in east

Elevation: *mean elevation:* 1,141 m
lowest point: Atlantic Ocean 0 m
highest point: Konigstein on Brandberg 2,573 m

Natural resources: diamonds, copper, uranium, gold, silver, lead, tin, lithium, cadmium, tungsten, zinc, salt, hydropower, fish, note, suspected deposits of oil, coal, and iron ore

Land use: *agricultural land:* 47.2% (2011 est.)
arable land: 1% (2011 est.) / *permanent crops:* 0% (2011 est.) / *permanent pasture:* 46.2% (2011 est.)
forest: 8.8% (2011 est.)
other: 44% (2011 est.)
Irrigated land: 80 sq km (2012)

Population distribution: population density is very low, with the largest clustering found in the extreme north-central area along the border with Angola as shown in this population distribution map

Natural hazards: prolonged periods of drought
Environment—current issues: depletion and degradation of water and aquatic resources; desertification; land degradation; loss of biodiversity and biotic resources; wildlife poaching

Environment—international agreements: *party to:* Antarctic-Marine Living Resources, Biodiversity, Climate Change, Climate Change-Kyoto Protocol, Desertification, Endangered Species, Hazardous Wastes, Law of the Sea, Ozone Layer Protection, Wetlands
signed, but not ratified: none of the selected agreements

Geography—note: the Namib Desert, after which the country is named, is considered to be the oldest desert in the world; Namibia is the first country in the world to incorporate the protection of the environment into its constitution; some 14% of the land is protected, including virtually the entire Namib Desert coastal strip; Namib-Naukluft National Park (49,768 sq km), is the largest game park in Africa and one of the largest in the world

PEOPLE AND SOCIETY

Population: 2,630,073 (July 2020 est.)
note: estimates for this country explicitly take into account the effects of excess mortality due to AIDS; this can result in lower life expectancy, higher infant mortality, higher death rates, lower population growth rates, and changes in the distribution of population by age and sex than would otherwise be expected

country comparison to the world: 142

Nationality: *noun:* Namibian(s)
adjective: Namibian

Ethnic groups: Ovambo 50%, Kavangos 9%, Herero 7%, Damara 7%, mixed European and African ancestry 6.5%, European 6%, Nama 5%, Caprivian 4%, San 3%, Baster 2%, Tswana .5%

Languages: Oshiwambo languages 49.7%, Nama/Damara 11%, Kavango languages 10.4%, Afrikaans 9.4% (also a common language), Herero languages 9.2%, Zambezi languages 4.9%, English (official) 2.3%, other African languages 1.5%, other European languages .7%, other 1% (2016 est.)
note: Namibia has 13 recognized national languages, including 10 indigenous African languages and 3 European languages

Religions: Christian 80% to 90% (at least 50% Lutheran), indigenous beliefs 10% to 20%

Demographic profile: Planning officials view Namibia's reduced population growth rate as sustainable based on the country's economic growth over the past decade. Prior to independence in 1990, Namibia's relatively small population grew at about 3% annually, but declining fertility and the impact of HIV/AIDS slowed this growth to 1.4% by 2011, rebounding to close to 2% by 2016. Namibia's fertility rate has fallen over the last two decades — from about 4.5 children per woman in 1996 to 3.4 in 2016 — due to increased contraceptive use, higher educational attainment among women, and greater female participation in the labor force. The average age at first birth has stayed fairly constant, but the age at first marriage continues to increase, indicating a rising incidence of premarital childbearing.

The majority of Namibians are rural dwellers (about 55%) and live in the better-watered north and northeast parts of the country. Migration, historically male-dominated, generally flows from northern communal areas — non-agricultural lands where blacks were sequestered under the apartheid system—to agricultural, mining, and manufacturing centers in the center and south. After independence from South Africa, restrictions on internal movement eased, and rural-urban migration increased, bolstering urban growth.

Some Namibians — usually persons who are better-educated, more affluent, and from urban areas —continue to legally migrate to South Africa temporarily to visit family and friends and, much less frequently, to pursue tertiary education or better economic opportunities. Namibians concentrated along the country's other borders make unauthorized visits to Angola, Zambia, Zimbabwe, or Botswana, to visit family and to trade agricultural goods. Few Namibians express interest in permanently settling in other countries; they prefer the safety of their homeland, have a strong national identity, and enjoy a well-supplied retail sector. Although Namibia is receptive to foreign investment and cross-border trade, intolerance toward non-citizens is widespread.

Age structure: *0-14 years:* 35.68% (male 473,937/ female 464,453)

15-24 years: 20.27% (male 267,106/female 265,882)

25-54 years: 35.47% (male 449,132/female 483,811)

55-64 years: 4.68% (male 54,589/female 68,619)

65 years and over: 3.9% (male 43,596/female 58,948) (2020 est.)

Dependency ratios: *total dependency ratio:* 67.9 *youth dependency ratio:* 61.8 *elderly dependency ratio:* 6 *potential support ratio:* 16.6 (2020 est.)

Median age: *total:* 21.8 years *male:* 21.1 years *female:* 22.6 years (2020 est.) *country comparison to the world:* 183

Population growth rate: 1.86% (2020 est.) *country comparison to the world:* 55

Birth rate: 25.7 births/1,000 population (2020 est.) *country comparison to the world:* 48

Death rate: 7.3 deaths/1,000 population (2020 est.) *country comparison to the world:* 116

Net migration rate: 0 migrant(s)/1,000 population (2020 est.) *country comparison to the world:* 92

Population distribution: population density is very low, with the largest clustering found in the extreme north-central area along the border with Angola as shown in this population distribution map

Urbanization: *urban population:* 52% of total population (2020) *rate of urbanization:* 4.2% annual rate of change (2015-20 est.) total population growth rate v. urban population growth rate, 2000-2030:

Major urban areas—population: 431,000 WINDHOEK (capital) (2020)

Sex ratio: *at birth:* 1.03 male(s)/female 0-14 years: 1.02 male(s)/female 15-24 years: 1 male(s)/female 25-54 years: 0.93 male(s)/female 55-64 years: 0.8 male(s)/female 65 years and over: 0.74 male(s)/female *total population:* 0.96 male(s)/female (2020 est.)

Mother's mean age at first birth: 21.5 years (2013 est.) *note:* median age at first birth among women 25-29

Maternal mortality rate: 195 deaths/100,000 live births (2017 est.) *country comparison to the world:* 48

Infant mortality rate: *total:* 31 .4 deaths/1,000 live births *male:* 33.5 deaths/1,000 live births *female:* 29.2 deaths/1,000 live births (2020 est.) *country comparison to the world:* 52

Life expectancy at birth: *total population:* 65.3 years

male: 63.3 years *female:* 67.3 years (2020 est.) *country comparison to the world:* 195

Total fertility rate: 3.07 children born/woman (2020 est.) *country comparison to the world:* 49

Contraceptive prevalence rate: 56.1% (2013)

Drinking water source: improved: *urban:* 98.9% of population *rural:* 80.8% of population *total:* 89.7% of population unimproved: *urban:* 1.1% of population *rural:* 19.2% of population *total:* 10.3% of population (2017 est.)

Current Health Expenditure: 8.6% (2017)

Physicians density: 0.59 physicians/1,000 population (2017)

Hospital bed density: 2.7 beds/1,000 population (2009)

Sanitation facility access: improved: *urban:* 72.9% of population *rural:* 22% of population *total:* 46.9% of population unimproved: *urban:* 27.1% of population *rural:* 78% of population *total:* 53.1% of population (2017 est.)

HIV/AIDS—adult prevalence rate: 12.7% (2019 est.) *country comparison to the world:* 6

HIV/AIDS—people living with HIV/AIDS: 210.000 (2019 est.) *country comparison to the world:* 28

HIV/AIDS—deaths: 3.000 (2019 est.) *country comparison to the world:* 135

Major infectious diseases: *degree of risk:* high (2020) *food or waterborne diseases:* bacterial diarrhea, hepatitis A, and typhoid fever *vectorborne diseases:* malaria *water contact diseases:* schistosomiasis

Obesity—adult prevalence rate: 17.2% (2016) *country comparison to the world:* 119

Children under the age of 5 years underweight: 13.2% (2013) *country comparison to the world:* 47

Education expenditures: 3.1% of GDP (2014) *country comparison to the world:* 135

Literacy: *definition:* age 15 and over can read and write *total population:* 91.5% *male:* 91.6% *female:* 91.4% (2018)

Unemployment, youth ages 15-24: *total:* 38% *male:* 37.5% *female:* 38.5% (2016 est.) *country comparison to the world:* 16

Country name: *conventional long form:* Republic of Namibia *conventional short form:* Namibia *local long form:* Republic of Namibia *local short form:* Namibia *former:* German South-West Africa (Deutsch-Suedwestafrika), South-West Africa *etymology:* named for the coastal Namib Desert; the name "namib" means "vast place" in the Nama/Damara language

Government type: presidential republic

Capital: *name:* Windhoek *geographic coordinates:* 22 34 S, 17 05 E *time difference:* UTC + 1 (6 hours ahead of Washington, DC, during Standard Time) *daylight saving time:* +1hr, begins first Sunday in September; ends first Sunday in April *etymology:* may derive from the Afrikaans word "wind-hoek" meaning "windy corner"

Administrative divisions: 14 regions; Erongo, Hardap, //Karas, Kavango East, Kavango West, Khomas, Kunene, Ohangwena, Omaheke, Omusati, Oshana, Oshikoto, Otjozondjupa, Zambezi; note—the Karas Region was renamed // Karas in September 2013 to include the alveolar lateral click of the Khoekhoegowab language

Independence: 21 March 1990 (from South African mandate)

National holiday: Independence Day, 21 March (1990)

Constitution: *history:* adopted 9 February 1990, entered into force 21 March 1990 *amendments:* initiated by the Cabinet; passage requires two-thirds majority vote of the National Assembly membership and of the National Council of Parliament and assent of the president of the republic; if the National Council fails to pass an amendment, the president can call for a referendum; passage by referendum requires two-thirds majority of votes cast; amendments that detract from or repeal constitutional articles on fundamental rights and freedoms cannot be amended, and the requisite majorities needed by Parliament to amend the constitution cannot be changed; amended 1998, 2010, 2014

Legal system: mixed legal system of uncodified civil law based on Roman-Dutch law and customary law

International law organization participation: has not submitted an ICJ jurisdiction declaration; accepts ICCt jurisdiction

Citizenship: *citizenship by birth:* no *citizenship by descent only:* at least one parent must be a citizen of Namibia *dual citizenship recognized:* no *residency requirement for naturalization:* 5 years

Suffrage: 18 years of age; universal

Executive branch: *chief of state:* President Hage GEINGOB (since 21 March 2015); Vice President Nangola MBUMBA (since 8 February 2018);

note—the president is both chief of state and head of government

head of government: President Hage GEINGOB (since 21 March 2015); Vice President Nangola MBUMBA (since 8 February 2018); Prime Minister Saara KUUGONGELWA-AMADHILA (since 21 March 2015)

cabinet: Cabinet appointed by the president from among members of the National Assembly

elections/appointments: president elected by absolute majority popular vote in 2 rounds if needed for a 5-year term (eligible for a second term); election last held on 28 November 2019 (next to be held in 2024)

election results: Hage GEINGOB elected president in the first round; percent of vote—Hage GEINGOB (SWAPO) 56.3%, Panduleni ITULA (Independent) 29.4%, McHenry VENAANI (PDM) 5.3%, Bernadus SWARTBOOI (LPM) 2.7%, Apius AUCHAB (UDF) 2.7%, Esther MUINJANGUE (NUDO) 1.5%, other 2%

Legislative branch: *description:* bicameral Parliament consists of:

National Council (42 seats); members indirectly elected 3 each by the 14 regional councils to serve 5-year terms); note—the Council primarily reviews legislation passed and referred by the National Assembly

National Assembly (104 seats; 96 members directly elected in multi-seat constituencies by closed list, proportional representation vote to serve 5-year terms and 8 nonvoting members appointed by the president)

elections: National Council—elections for regional councils to determine members of the National Council held on 27 November 2015 (next to be held on 27 November 2020

National Assembly—last held on 27 November 2019 (next to be held in 2024)

election results: National Council—percent of vote by party—NA; seats by party—SWAPO 40, NUDO 1, DPM 1; composition—men 32, women 10, percent of women 23.8%

National Assembly—percent of vote by party— SWAPO 65.5%, PDM 16.6%, LPM 4.7%, NUDO 1.9%, APP 1.8%, UDF 1.8%, RP 1.8%, NEFF 1.7%, RDP 1.1%, CDV .7%, SWANU .6%, other 1.8%; seats by party—SWAPO 63, PDM 16, LPM 4, NUDO 2, APP 2, UDF 2, RP 2, NEFF 2, RDP 1, CDV 1, SWANU 1; composition—NA

Judicial branch: *highest courts:* Supreme Court (consists of the chief justice and at least 3 judges in quorum sessions)

judge selection and term of office: judges appointed by the president of Namibia upon the recommendation of the Judicial Service Commission; judges serve until age 65, but terms can be extended by the president until age 70

subordinate courts: High Court; Electoral Court, Labor Court; regional and district magistrates' courts; community courts

Political parties and leaders:

All People's Party or APP [Ignatius SHIXWAMENI]

Christian Democratic Voice or CDV [Gothard KANDUME]

Landless People's Movement or LPM [Bernadus SWARTBOOI]

National Unity Democratic Organization or NUDO [Estes MUINJANGUE]

Namibian Economic Freedom Fighters or NEFF [Epafras MUKWIILONGO]

Popular Democratic Movement or PDM (formerly DTA) [McHenry VENAANI]

Rally for Democracy and Progress or RDP [Mike KAVEKOTORA]

Republican Party or RP [Henk MUDGE]

South West Africa National Union or SWANU [Tangeni IIYAMBO]

South West Africa People's Organization or SWAPO [Hage GEINGOB]

United Democratic Front or UDF [Apius AUCHAB]

United People's Movement or UPM [Jan J. VAN WYK]

Workers' Revolutionary Party or WRP (formerly CPN) [MPs Salmon FLEERMUYS and Benson KAAPALA]

International organization participation: ACP, AfDB, AU, C, CD, CPLP (associate observer), FAO, G-77, IAEA, IBRD, ICAO, ICCt, ICRM, IDA, IFAD, IFC, IFRCS, ILO, IMF, IMO, Interpol, IOC, IOM, IPU, ISO, ITSO, ITU, ITUC (NGOs), MIGA, NAM, OPCW, SACU, SADC, UN, UNAMID, UNCTAD, UNESCO, UNHCR, UNIDO, UNISFA, UNMIL, UNMISS, UNOCI, UNWTO, UPU, WCO, WHO, WIPO, WMO, WTO

Diplomatic representation in the US: *chief of mission:* Charge d'Affaires Jerome Mutamba MUTAMBA (since 3 August 2020)

chancery: 1605 New Hampshire Avenue NW, Washington, DC 20009

telephone: [1] (202) 986-0540

FAX: [1] (202) 986-0443

Diplomatic representation from the US: *chief of mission:* Ambassador Lisa A. JOHNSON (since 3 February 2018)

telephone: [264] (061) 295-8500

embassy: 14 Lossen Street, Windhoek

mailing address: Private Bag 12029 Ausspannplatz, Windhoek

FAX: [264] (061) 295-8603

Flag description: a wide red stripe edged by narrow white stripes divides the flag diagonally from lower hoist corner to upper fly corner; the upper hoist-side triangle is blue and charged with a golden-yellow, 12-rayed sunburst; the lower fly-side triangle is green; red signifies the heroism of the people and their determination to build a future of equal opportunity for all; white stands for peace, unity, tranquility, and harmony; blue represents the Namibian sky and the Atlantic Ocean, the country's precious water resources and rain; the golden-yellow sun denotes power and existence; green symbolizes vegetation and agricultural resources

National symbol(s): oryx (antelope); national colors: blue, red, green, white, yellow

National anthem: *name:* Namibia, Land of the Brave

lyrics/music: Axali DOESEB

note: adopted 1991

0:00/ 1:28

ECONOMY

Economy—overview: Namibia's economy is heavily dependent on the extraction and processing of minerals for export. Mining accounts for about 12.5% of GDP, but provides more than 50% of foreign exchange earnings. Rich alluvial diamond deposits make Namibia a primary source for gem-quality diamonds. Marine diamond mining is increasingly important as the terrestrial diamond supply has dwindled. The rising cost of mining diamonds, especially from the sea, combined with increased diamond production in Russia and China, has reduced profit margins. Namibian authorities have emphasized the need to add value to raw materials, do more in-country manufacturing, and exploit the services market, especially in the logistics and transportation sectors.

Namibia is one of the world's largest producers of uranium. The Chinese-owned Husab uranium mine began producing uranium ore in 2017, and is expected to reach full production in August 2018 and produce 15 million pounds of uranium a year. Namibia also produces large quantities of zinc and is a smaller producer of gold and copper. Namibia's economy remains vulnerable to world commodity price fluctuations and drought.

Namibia normally imports about 50% of its cereal requirements; in drought years, food shortages are problematic in rural areas. A high per capita GDP, relative to the region, obscures one of the world's most unequal income distributions; the current government has prioritized exploring wealth redistribution schemes while trying to maintain a pro-business environment. GDP growth in 2017 slowed to about 1%, however, due to contractions in both the construction and mining sectors, as well as an ongoing drought. Growth is expected to recover modestly in 2018.

A five-year Millennium Challenge Corporation compact ended in September 2014. As an upper middle income country, Namibia is ineligible for a second compact. The Namibian economy is closely linked to South Africa with the Namibian dollar pegged one-to-one to the South African rand. Namibia receives 30%-40% of its revenues from the Southern African Customs Union (SACU); volatility in the size of Namibia's annual SACU allotment and global mineral prices complicates budget planning.

GDP (purchasing power parity):

$26.6 billion (2017 est.)

$26.81 billion (2016 est.)

$26.62 billion (2015 est.)

note: data are in 2017 dollars

country comparison to the world: 139

GDP (official exchange rate):

$13.24 billion (2017 est.)

GDP—real growth rate: -1.56% (2019 est.)

1.13% (2018 est.)

-1.02% (2017 est.)
country comparison to the world: 203

GDP—per capita (PPP):
$11,200 (2017 est.)
$11,500 (2016 est.)
$11,700 (2015 est.)
note: data are in 2017 dollars
country comparison to the world: 135

Gross national saving:
16.7% of GDP (2017 est.)
9.6% of GDP (2016 est.)
19.1% of GDP (2015 est.)
country comparison to the world: 125

GDP—composition, by end use:
household consumption: 68.7% (2017 est.)
government consumption: 24.5% (2017 est.)
investment in fixed capital: 16% (2017 est.)
investment in inventories: 1.6% (2017 est.)
exports of goods and services: 36.7% (2017 est.)
imports of goods and services: -47.5% (2017 est.)

GDP—composition, by sector of origin:
agriculture: 6.7% (2016 est.)
industry: 26.3% (2016 est.)
services: 67% (2017 est.)

Agriculture—products: millet, sorghum, peanuts, grapes; livestock; fish

Industries: meatpacking, fish processing, dairy products, pasta, beverages; mining (diamonds, lead, zinc, tin, silver, tungsten, uranium, copper)

Industrial production growth rate: -0.4% (2017 est.)
country comparison to the world: 171

Labor force: 956,800 (2017 est.)
country comparison to the world: 140

Labor force—by occupation:
agriculture: 31%
industry: 14%
services: 54% (2013 est.)
note: about half of Namibia's people are unemployed while about two-thirds live in rural areas; roughly two-thirds of rural dwellers rely on subsistence agriculture

Unemployment rate:
34% (2016 est.)
28.1% (2014 est.)
country comparison to the world: 210

Population below poverty line: 28.7% (2010 est.)

Household income or consumption by percentage share: *lowest 10%:* 2.4%
highest 10%: 42% (2010)

Budget: *revenues:* 4.268 billion (2017 est.)
expenditures: 5 billion (2017 est.)

Taxes and other revenues:
32.2% (of GDP) (2017 est.)
country comparison to the world: 68

Budget surplus (+) or deficit (-):
-5.5% (of GDP) (2017 est.)
country comparison to the world: 173

Public debt:
41.3% of GDP (2017 est.)
39.5% of GDP (2016 est.)
country comparison to the world: 121

Fiscal year: 1 April—31 March

Inflation rate (consumer prices): 6.1% (2017 est.)
6.7% (2016 est.)
country comparison to the world: 187

Current account balance:
-$216 million (2019 est.)
-$465 million (2018 est.)
country comparison to the world: 101

Exports: $3.995 billion (2017 est.)
$4.003 billion (2016 est.)
country comparison to the world: 117

Exports—partners: South Africa 27.1%, Botswana 14.9%, Switzerland 12%, Zambia 5.7%, China 4.6%, Italy 4.4% (2017)

Exports—commodities: diamonds, copper, gold, zinc, lead, uranium; cattle, white fish and mollusks

Imports: $5.384 billion (2017 est.)
$5.625 billion (2016 est.)
country comparison to the world: 123

Imports—commodities: foodstuffs; petroleum products and fuel, machinery and equipment, chemicals

Imports—partners: South Africa 61.4% (2017)

Reserves of foreign exchange and gold:
$2.432 billion (31 December 2017 est.)
$1.834 billion (31 December 2016 est.)
country comparison to the world: 115

Debt—external:
$7.969 billion (31 December 2017 est.)
$6.904 billion (31 December 2016 est.)
country comparison to the world: 123

Exchange rates: Namibian dollars (NAD) per US dollar—
13.67 (2017 est.)
14.7096 (2016 est.)
14.7096 (2015 est.)
12.7589 (2014 est.)
10.8526 (2013 est.)

ENERGY

Electricity access: *population without electricity:* 1 million (2019)
electrification—total population: 57% (2019)
electrification—urban areas: 78% (2019)
electrification—rural areas: 36% (2019)

Electricity—production: 1.403 billion kWh (2016 est.)
country comparison to the world: 145

Electricity—consumption: 3.891 billion kWh (2016 est.)
country comparison to the world: 128

Electricity—exports: 88 million kWh (2015 est.)
country comparison to the world: 83

Electricity—imports: 3.073 billion kWh (2016 est.)
country comparison to the world: 50

Electricity—installed generating capacity: 535,500 kW (2016 est.)
country comparison to the world: 146

Electricity—from fossil fuels: 28% of total installed capacity (2016 est.)

Electricity—from nuclear fuels: 0% of total installed capacity (2017 est.)
country comparison to the world: 150

Electricity—from hydroelectric plants: 64% of total installed capacity (2017 est.)
country comparison to the world: 26

Electricity—from other renewable sources: 8% of total installed capacity (2017 est.)
country comparison to the world: 89

Crude oil—production: 0 bbl/day (2018 est.)
country comparison to the world: 179

Crude oil—exports: 0 bbl/day (2015 est.)
country comparison to the world: 171

Crude oil—imports: 0 bbl/day (2015 est.)
country comparison to the world: 172

Crude oil—proved reserves: 0 bbl (1 January 2018 est.)
country comparison to the world: 174

Refined petroleum products—production: 0 bbl/day (2015 est.)
country comparison to the world: 183

Refined petroleum products—consumption: 27,000 bbl/day (2016 est.)
country comparison to the world: 125

Refined petroleum products—exports: 80 bbl/day (2015 est.)
country comparison to the world: 120

Refined petroleum products—imports: 26,270 bbl/day (2015 est.)
country comparison to the world: 105

Natural gas—production: 0 cu m (2017 est.)
country comparison to the world: 175

Natural gas—consumption: 0 cu m (2017 est.)
country comparison to the world: 179

Natural gas—exports: 0 cu m (2017 est.)
country comparison to the world: 157

Natural gas—imports: 0 cu m (2017 est.)
country comparison to the world: 162

Natural gas—proved reserves: 62.29 billion cu m (1 January 2018 est.)
country comparison to the world: 60

Carbon dioxide emissions from consumption of energy: 3.958 million Mt (2017 est.)
country comparison to the world: 139

COMMUNICATIONS

Telephones—fixed lines: *total subscriptions:* 144,575
subscriptions per 100 inhabitants: 5.6 (2019 est.)
country comparison to the world: 126

Telephones—mobile cellular: *total subscriptions:* 2,921,697
subscriptions per 100 inhabitants: 113.17 (2019 est.)
country comparison to the world: 142

Telecommunication systems: *general assessment:* fixed-line still a govt. monopoly; penetration rates rise above regional average with the rise of competition in the mobile market; 3G and LTE-A services; Internet and broadband sector fairly competitive; infrastructure investment through 2021; working on implementing 5G (2020)

domestic: fixed-line subscribership is 6 per 100 and mobile-cellular 113 per 100 persons (2019)

international: country code—264; landing points for the ACE and WACS fiber-optic submarine cable linking southern and western African countries to Europe; satellite earth stations—4 Intelsat (2019)

note: the COVID-19 outbreak is negatively impacting telecommunications production and supply chains globally; consumer spending on telecom devices and services has also slowed due to the pandemic's effect on economies worldwide; overall progress towards improvements in all facets of the telecom industry—mobile, fixed-line, broadband, submarine cable and satellite—has moderated

Broadcast media: 1 private and 1 state-run TV station; satellite and cable TV service available; state-run radio service broadcasts in multiple languages; about a dozen private radio stations; transmissions of multiple international broadcasters available

Internet country code: .na

Internet users: *total:* 1,291,944
percent of population: 51% (July 2018 est.)
country comparison to the world: 134

Broadband—fixed subscriptions: *total:* 61,968
subscriptions per 100 inhabitants: 2 (2018 est.)
country comparison to the world: 131

TRANSPORTATION

National air transport system: *number of registered air carriers:* 2 (2020)
inventory of registered aircraft operated by air carriers: 21
annual passenger traffic on registered air carriers: 602,893 (2018)
annual freight traffic on registered air carriers: 26.29 million mt-km (2018)

Civil aircraft registration country code prefix: V5 (2016)

Airports: 112 (2013)
country comparison to the world: 51

Airports—with paved runways:
total: 19 (2017)
over 3,047 m: 4 (2017)
2,438 to 3,047 m: 2 (2017)
1,524 to 2,437 m: 12 (2017)
914 to 1,523 m: 1 (2017)

Airports—with unpaved runways:
total: 93 (2013)
1,524 to 2,437 m: 25 (2013)
914 to 1,523 m: 52 (2013)
under 914 m: 16 (2013)

Railways: *total:* 2,628 km (2014)
narrow gauge: 2,628 km 1.067-m gauge (2014)
country comparison to the world: 65

Roadways:
total: 48,875 km (2018)
paved: 7,893 km (2018)
unpaved: 40,982 km (2018)
country comparison to the world: 83

Merchant marine: *total:* 12
by type: general cargo 1, other 11 (2019)
country comparison to the world: 151

Ports and terminals: *major seaport(s):* Luderitz, Walvis Bay

MILITARY AND SECURITY

Military and security forces: *Namibian Defense Force (NDF):* Army, Navy, Air Force; Namibian Police Force: Special Field Force (paramilitary unit responsible for protecting borders and government installations) (2019)

Military expenditures:
3% of GDP (2019)
3.4% of GDP (2018)
3.6% of GDP (2017)
3.9% of GDP (2016)
4.5% of GDP (2015)
country comparison to the world: 27

Military and security service personnel strengths: size assessments for the Namibian Defense Force (NDF) vary; approximately 13,000 personnel (11,000 Army; 1,000 Navy; 700 Air Force) (2019 est.)

Military equipment inventories and acquisitions: the inventory of the Namibian Defense Force consists mostly of Soviet-era equipment; China is the leading supplier of weapons to Namibia since 2010 (2019 est.) ·

Military service age and obligation: 18-25 years of age for voluntary military service; no conscription (2019)

TRANSNATIONAL ISSUES

Disputes—international: concerns from international experts and local populations over the Okavango Delta ecology in Botswana and human displacement scuttled Namibian plans to construct a hydroelectric dam on Popa Falls along the Angola-Namibia border; the governments of South Africa and Namibia have not signed or ratified the text of the 1994 Surveyor's General agreement placing the boundary in the middle of the Orange River; Namibia has supported, and in 2004 Zimbabwe dropped objections to, plans between Botswana and Zambia to build a bridge over the Zambezi River, thereby de facto recognizing a short, but not clearly delimited, Botswana-Zambia boundary in the river

Refugees and internally displaced persons: *refugees (country of origin):* 6,595 (Democratic Republic of Congo) (2020)

Trafficking in persons: *current situation:* Namibia is a country of origin and destination for children and, to a lesser extent, women subjected to forced labor and sex trafficking; victims, lured by promises of legitimate jobs, are forced to work in urban centers and on commercial farms; traffickers exploit Namibian children, as well as children from Angola, Zambia, and Zimbabwe, for forced labor in agriculture, cattle herding, domestic service, fishing, and street vending; children are also forced into prostitution, often catering to tourists from southern Africa and Europe; San and Zemba children are particularly vulnerable; foreign adults and Namibian adults and children are reportedly subjected to forced labor in Chinese-owned retail, construction, and fishing operations

tier rating: Tier 2 Watch List — Namibia does not fully comply with the minimum standards for the elimination of trafficking; however, it is making significant efforts to do so; Namibia was granted a waiver from an otherwise required downgrade to Tier 3 because its government has a written plan that, if implemented would constitute making significant efforts to bring itself into compliance with the minimum standards for the elimination of trafficking; in 2015, the Child Care and Protection Bill passed, criminalizing child trafficking; the government's first sex trafficking prosecution remained pending; no new prosecutions were initiated and no trafficking offenders have ever been convicted; accusations of forced labor at Chinese construction and mining companies continue to go uninvestigated; authorities failed to fully implement victim identification and referral processes, which led to the deportation of possible victims (2015)

NAURU

INTRODUCTION

Background: The exact origins of the Nauruans are unclear since their language does not resemble any other in the Pacific region. Germany annexed the island in 1888. A German-British consortium began mining the island's phosphate deposits early in the 20th century. Australian forces occupied Nauru in World War I; it subsequently became a League of Nations mandate. After the Second World War—and a brutal occupation by Japan—Nauru became a UN trust territory. It achieved independence in 1968 and became one of the richest countries in the world because of its extensive phosphate stocks; however, the phosphate was depleted in the early 1980s and the quality of life began to decline. In 2001, an Australian offshore refugee processing center was opened in Nauru, providing an economic lifeline. Nauru is one of Taiwan's few remaining diplomatic partners, and in 2008, Nauru recognized the breakaway Georgian republics of Abkhazia and South Ossetia.

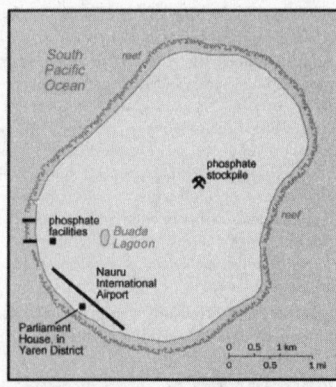

GEOGRAPHY

Location: Oceania, island in the South Pacific Ocean, south of the Marshall Islands

Geographic coordinates: 0 32 S, 166 55 E

Map references: Oceania

Area: *total:* 21 sq km
land: 21 sq km
water: 0 sq km
country comparison to the world: 240

Area—comparative: about 0.1 times the size of Washington, DC

Land boundaries: 0 km

Coastline: 30 km

Maritime claims: *territorial sea:* 12 nm
exclusive economic zone: 200 nm
contiguous zone: 24 nm

Climate: tropical with a monsoonal pattern; rainy season (November to February)

Terrain: sandy beach rises to fertile ring around raised coral reefs with phosphate plateau in center

Elevation: *lowest point:* Pacific Ocean 0 m
highest point: Command Ridge 70 m

Natural resources: phosphates, fish

Land use: *agricultural land:* 20% (2011 est.)
arable land: 0% (2011 est.) / *permanent crops:* 20% (2011 est.) / *permanent pasture:* 0% (2011 est.)
forest: 0% (2011 est.)
other: 80% (2011 est.)
Irrigated land: 0 sq km (2012)

Population distribution: extensive phosphate mining made approximately 90% of the island unsuitable for farming; most people live in the fertile coastal areas, especially along the southwest coast

Natural hazards: periodic droughts

Environment—current issues: limited natural freshwater resources, roof storage tanks that collect rainwater and desalination plants provide water; a century of intensive phosphate mining beginning in 1906 left the central 90% of Nauru a wasteland; cadmium residue, phosphate dust, and other contaminants have caused air and water pollution with negative impacts on health; climate change has brought on rising sea levels and inland water shortages

Environment—international agreements: *party to:* Biodiversity, Climate Change, Climate Change-Kyoto Protocol, Desertification, Hazardous Wastes, Law of the Sea, Marine Dumping, Ozone Layer Protection, Whaling
signed, but not ratified: none of the selected agreements

Geography—note: Nauru is the third-smallest country in the world behind the Holy See (Vatican City) and Monaco; it is the smallest country in the Pacific Ocean, the smallest country outside Europe, the world's smallest island country, and the the world's smallest independent republic; situated just 53 km south of the Equator, Nauru is one of the three great phosphate rock islands in the Pacific Ocean—the others are Banaba (Ocean Island) in Kiribati and Makatea in French Polynesia

PEOPLE AND SOCIETY

Population: 11,000 (2019 est.)
country comparison to the world: 223

Nationality: *noun:* Nauruan(s)
adjective: Nauruan

Ethnic groups: Nauruan 88.9%, part Nauruan 6.6%, I-Kiribati 2%, other 2.5% (2007 est.)

Languages: Nauruan 93% (official, a distinct Pacific Island language), English 2% (widely understood, spoken, and used for most government and commercial purposes), other 5% (includes I-Kiribati 2% and Chinese 2%) (2011 est.)
note: percentages represent main language spoken at home; Nauruan is spoken by 95% of the population, English by 66%, and other languages by 12%

Religions: Protestant 60.4% (includes Nauru Congregational 35.7%, Assembly of God 13%, Nauru Independent Church 9.5%, Baptist 1.5%, and Seventh Day Adventist 0.7%), Roman Catholic 33%, other 3.7%, none 1.8%, unspecified 1.1% (2011 est.)

Age structure: *0-14 years:* 30.87% (male 1,337/ female 1,684)
15-24 years: 16.35% (male 734/female 866)
25-54 years: 42.57% (male 2,115/female 2,050)
55-64 years: 6.72% (male 262/female 396)
65 years and over: 3.48% (male 122/female 219) (2020 est.)

Median age: *total:* 27 years
male: 28.2 years
female: 25.9 years (2020 est.)
country comparison to the world: 150

Population growth rate: 0.46% (2020 est.)
country comparison to the world: 158

Birth rate: 21.9 births/1,000 population (2020 est.)
country comparison to the world: 67

Death rate: 6 deaths/1,000 population (2020 est.)
country comparison to the world: 165

Net migration rate: -11.3 migrant(s)/1,000 population (2020 est.)
country comparison to the world: 219

Population distribution: extensive phosphate mining made approximately 90% of the island unsuitable for farming; most people live in the fertile coastal areas, especially along the southwest coast

Urbanization: *urban population:* 100% of total population (2020)
rate of urbanization: -0.06% annual rate of change (2015-20 est.)
total population growth rate v. urban population growth rate, 2000-2030:

Sex ratio: *at birth:* 0.84 male(s)/female
0-14 years: 0.79 male(s)/female
15-24 years: 0.85 male(s)/female
25-54 years: 1.03 male(s)/female
55-64 years: 0.66 male(s)/female
65 years and over: 0.56 male(s)/female
total population: 0.88 male(s)/female (2020 est.)

Infant mortality rate: *total:* 7.4 deaths/1,000 live births
male: 9.4 deaths/1,000 live births
female: 5.7 deaths/1,000 live births (2020 est.)
country comparison to the world: 157

Life expectancy at birth: *total population:* 68.4 years
male: 64.3 years
female: 71.9 years (2020 est.)
country comparison to the world: 175

Total fertility rate: 2.68 children born/woman (2020 est.)
country comparison to the world: 66

Drinking water source:
improved:
urban: 100% of population
total: 100% of population
unimproved:
urban: 0% of population
total: 0% of population (2017 est.)

Current Health Expenditure: 11% (2017)

Physicians density: 1.35 physicians/1,000 population (2015)

Hospital bed density: 5 beds/1,000 population (2010)

Sanitation facility access:
improved:
urban: 96.3% of population
total: 96.3% of population
unimproved:
urban: 3.7% of population
total: 3.7% of population (2017 est.)

HIV/AIDS—adult prevalence rate: NA

HIV/AIDS—people living with HIV/AIDS: NA

HIV/AIDS—deaths: NA

Major infectious diseases: *degree of risk:* high (2020)

food or waterborne diseases: bacterial diarrhea

vectorborne diseases: malaria

Obesity—adult prevalence rate: 61% (2016)
country comparison to the world: 1

Education expenditures: NA

School life expectancy (primary to tertiary education): *total:* 9 years
male: 9 years

female: 10 years (2008)

Unemployment, youth ages 15-24: *total:* 26.6%
male: 20.9%
female: 37.5% (2013)
country comparison to the world: 44

GOVERNMENT

Country name: *conventional long form:* Republic of Nauru
conventional short form: Nauru
local long form: Republic of Nauru
local short form: Nauru
former: Pleasant Island
etymology: the island name may derive from the Nauruan word "anaoero" meaning "I go to the beach"

Government type: parliamentary republic

Capital: *name:* no official capital; government offices in the Yaren District
time difference: UTC+12 (17 hours ahead of Washington, DC, during Standard Time)

Administrative divisions: 14 districts; Aiwo, Anabar, Anetan, Anibare, Baitsi, Boe, Buada, Denigomodu, Ewa, Ijuw, Meneng, Nibok, Uaboe, Yaren

Independence: 31 January 1968 (from the Australia-, NZ-, and UK-administered UN trusteeship)

National holiday: Independence Day, 31 January (1968)

Constitution: *history:* effective 29 January 1968
amendments: proposed by Parliament; passage requires two-thirds majority vote of Parliament; amendments to constitutional articles, such as the republican form of government, protection of fundamental rights and freedoms, the structure and authorities of the executive and legislative branches, also requires two-thirds majority of votes in a referendum; amended 1968, 2009, 2014

Legal system: mixed legal system of common law based on the English model and customary law

International law organization participation: has not submitted an ICJ jurisdiction declaration; accepts ICCt jurisdiction

Suffrage: 20 years of age; universal and compulsory

Executive branch: *chief of state:* President Lionel AINGIMEA (since 27 August 2019); note—the president is both chief of state and head of government

head of government: President Lionel AINGIMEA (since 27 August 2019)
cabinet: Cabinet appointed by the president from among members of Parliament
elections/appointments: president indirectly elected by Parliament (eligible for a second term); election last held on 27 August 2019 (next to be held in 2022)
election results: Lionel AINGIMEA elected president; Parliament vote—Lionel AINGIMEA (independent) 12, David ADEANG (Nauru First) 6

Legislative branch: *description:* unicameral parliament (19 seats; members directly elected in multi-seat constituencies by majority vote using the "Dowdall" counting system by which voters rank candidates on their ballots; members serve 3-year terms)
elections: last held on 24 August 2019 (next to be held in 2022)
election results: percent of vote—NA; seats—independent 19; composition—men 17, women 2, percent of women 10.5%

Judicial branch: *highest courts:* Supreme Court (consists of the chief justice and several justices); note—in late 2017, the Nauruan Government revoked the 1976 High Court Appeals Act, which had allowed appeals beyond the Nauruan Supreme Court, and in early 2018, the government formed its own appeals court
judge selection and term of office: judges appointed by the president to serve until age 65
subordinate courts: District Court, Family Court

Political parties and leaders: Democratic Party [Kennan ADEANG]
Nauru First (Naoero Amo) Party
Nauru Party (informal)
note: loose multiparty system

International organization participation: ACP, ADB, AOSIS, C, FAO, G-77, ICAO, ICCt, IFAD, Interpol, IOC, IOM, ITU, OPCW, PIF, Sparteca, SPC, UN, UNCTAD, UNESCO, UPU, WHO

Diplomatic representation in the US: *chief of mission:* Ambassador Marlene Inemwin MOSES (since 13 March 2006)
chancery: 800 2nd Avenue, Suite 400 D, New York, NY 10017
telephone: [1] (212) 937-0074
FAX: [1] (212) 937-0079

Diplomatic representation from the US: the US does not have an embassy in Nauru; the US Ambassador to Fiji is accredited to Nauru

Flag description: blue with a narrow, horizontal, gold stripe across the center and a large white 12-pointed star below the stripe on the hoist side; blue stands for the Pacific Ocean, the star indicates the country's location in relation to the Equator (the gold stripe) and the 12 points symbolize the 12 original tribes of Nauru; the star's white color represents phosphate, the basis of the island's wealth

National symbol(s): *frigatebird, calophyllum flower;* *national colors:* blue, yellow, white

National anthem: *name:* "Nauru Bwiema" (Song of Nauru)
lyrics/music: Margaret HENDRIE/Laurence Henry HICKS
note: adopted 1968

ECONOMY

Economy—overview: Revenues of this tiny island—a coral atoll with a land area of 21 square kilometers—traditionally have come from exports of phosphates. Few other resources exist, with most necessities being imported, mainly from Australia, its former occupier and later major source of support. Primary reserves of phosphates were exhausted and mining ceased in 2006, but mining of a deeper layer of "secondary phosphate" in the interior of the island began the following year. The secondary phosphate deposits may last another 30 years. Earnings from Nauru's export of phosphate remains an important source of income. Few comprehensive statistics on the Nauru economy exist; estimates of Nauru's GDP vary widely.

The rehabilitation of mined land and the replacement of income from phosphates are serious long-term problems. In anticipation of the exhaustion of Nauru's phosphate deposits, substantial amounts of phosphate income were invested in trust funds to help cushion the transition and provide for Nauru's economic future.

Although revenue sources for government are limited, the opening of the Australian Regional Processing Center for asylum seekers since 2012 has sparked growth in the economy. Revenue derived from fishing licenses under the "vessel day scheme" has also boosted government income. Housing, hospitals, and other capital plant are deteriorating. The cost to Australia of keeping the Nauruan government and economy afloat continues to climb.

GDP (purchasing power parity):
$160 million (2017 est.)
$153.9 million (2016 est.)
$139.4 million (2015 est.)
note: data are in 2015 dollars
country comparison to the world: 224

GDP (official exchange rate):
$114 million (2017 est.)

GDP—real growth rate: 4% (2017 est.)
10.4% (2016 est.)
2.8% (2015 est.)
country comparison to the world: 75

GDP—per capita (PPP): $12,300 (2017 est.)
$11,800 (2016 est.)
$11,600 (2015 est.)
note: data are in 2015 US dollars
country comparison to the world: 129

GDP—composition, by end use:
household consumption: 98% (2016 est.)
government consumption: 37.6% (2016 est.)
investment in fixed capital: 42.2% (2016 est.)
exports of goods and services: 11.2% (2016 est.)
imports of goods and services: -89.1% (2016 est.)

GDP—composition, by sector of origin:
agriculture: 6.1% (2009 est.)
industry: 33% (2009 est.)
services: 60.8% (2009 est.)

Agriculture—products: coconuts

Industries: phosphate mining, offshore banking, coconut products

Industrial production growth rate: NA

Labor force: NA

Labor force—by occupation: *note:* most of the labor force is employed in phosphate mining, public administration, education, and transportation

Unemployment rate: 23% (2011 est.)
90% (2004 est.)

country comparison to the world: 193

Population below poverty line: NA

Household income or consumption by percentage share: *lowest 10%:* NA
highest 10%: NA

Budget: *revenues:* 103 million (2017 est.)
expenditures: 113.4 million (2017 est.)

Taxes and other revenues: 90.3% (of GDP) (2017 est.)
country comparison to the world: 2

Budget surplus (+) or deficit (-): -9.2% (of GDP) (2017 est.)
country comparison to the world: 206

Public debt: 62% of GDP (2017 est.)
65% of GDP (2016 est.)
country comparison to the world: 71

Fiscal year: 1 July—30 June

Inflation rate (consumer prices): 5.1% (2017 est.)
8.2% (2016 est.)
country comparison to the world: 172

Current account balance: $5 million (2017 est.)
$2 million (2016 est.)
country comparison to the world: 63

Exports: $125 million (2013 est.)
$110.3 million (2012 est.)
country comparison to the world: 194

Exports—partners: Nigeria 38.6%, Japan 16.6%, Australia 15.9%, South Korea 13.7%, NZ 5.7% (2017)

Exports—commodities: phosphates

Imports: $64.9 million (2016 est.)
$143.1 million (2013 est.)
country comparison to the world: 219

Imports—commodities: food, fuel, manufactures, building materials, machinery

Imports—partners: Australia 67.5%, Fiji 9.2%, India 8.1%, Singapore 5.4% (2017)

Debt—external: $33.3 million (2004 est.)
country comparison to the world: 199

Exchange rates: Australian dollars (AUD) per US dollar—
1.311 (2017 est.)
1.3452 (2016 est.)
1.3452 (2015 est.)
1.3291 (2014 est.)
1.1094 (2013 est.)

ENERGY

Electricity—production: 24 million kWh (2016 est.)
country comparison to the world: 212

Electricity—consumption: 22.32 million kWh (2016 est.)
country comparison to the world: 212

Electricity—exports: 0 kWh (2016 est.)
country comparison to the world: 174

Electricity—imports: 0 kWh (2016 est.)
country comparison to the world: 177

Electricity—installed generating capacity: 7,000 kW (2016 est.)
country comparison to the world: 211

Electricity—from fossil fuels: 86% of total installed capacity (2016 est.)
country comparison to the world: 68

Electricity—from nuclear fuels: 0% of total installed capacity (2017 est.)
country comparison to the world: 151

Electricity—from hydroelectric plants: 0% of total installed capacity (2017 est.)
country comparison to the world: 188

Electricity—from other renewable sources: 14% of total installed capacity (2017 est.)
country comparison to the world: 63

Crude oil—production: 0 bbl/day (2018 est.)
country comparison to the world: 180

Crude oil—exports: 0 bbl/day (2015 est.)
country comparison to the world: 172

Crude oil—imports: 0 bbl/day (2015 est.)
country comparison to the world: 173

Crude oil—proved reserves: 0 bbl (1 January 2018 est.)
country comparison to the world: 175

Refined petroleum products—production: 0 bbl/day (2015 est.)
country comparison to the world: 184

Refined petroleum products—consumption: 470 bbl/day (2016 est.)
country comparison to the world: 210

Refined petroleum products—exports: 0 bbl/day (2015 est.)
country comparison to the world: 186

Refined petroleum products—imports: 449 bbl/day (2015 est.)
country comparison to the world: 206

Natural gas—production: 0 cu m (2017 est.)
country comparison to the world: 176

Natural gas—consumption: 0 cu m (2017 est.)
country comparison to the world: 180

Natural gas—exports: 0 cu m (2017 est.)
country comparison to the world: 158

Natural gas—imports: 0 cu m (2017 est.)
country comparison to the world: 163

Natural gas—proved reserves: 0 cu m (1 January 2014 est.)
country comparison to the world: 174

Carbon dioxide emissions from consumption of energy: 76,540 Mt (2017 est.)
country comparison to the world: 208

COMMUNICATIONS

Telephones—fixed lines: *total subscriptions:* 1,900
subscriptions per 100 inhabitants: 14 (July 2016 est.)
country comparison to the world: 219

Telephones—mobile cellular: *total subscriptions:* 9,212
subscriptions per 100 inhabitants: 94.58 (2019 est.)
country comparison to the world: 214

Telecommunication systems: *general assessment:* adequate local and international radiotelephone communication provided via Australian facilities; geography is a challenge for the islands; there is a need to service the tourism sector and the South Pacific Islands economy; mobile technology is booming (2018)
domestic: fixed-line 14 per 100 and mobile-cellular 95 per 100 (2019)
international: country code—674; satellite earth station—1 Intelsat (Pacific Ocean)
note: the COVID-19 outbreak is negatively impacting telecommunications production and supply chains globally; consumer spending on telecom devices and services has also slowed due to the pandemic's effect on economies worldwide; overall progress towards improvements in all facets of the telecom industry—mobile, fixed-line, broadband, submarine cable and satellite—has moderated

Broadcast media: 1 government-owned TV station broadcasting programs from New Zealand sent via satellite or on videotape; 1 government-owned radio station, broadcasting on AM and FM, utilizes Australian and British programs (2019)

Internet country code: .nr

Internet users: *total:* 5,524

percent of population: 57% (July 2018 est.)
country comparison to the world: 215

TRANSPORTATION

National air transport system: *number of registered air carriers:* 1 (2020)
inventory of registered aircraft operated by air carriers: 5
annual passenger traffic on registered air carriers: 45,457 (2018)
annual freight traffic on registered air carriers: 7,94 million mt-km (2018)

Civil aircraft registration country code prefix: C2 (2016)

Airports: 1 (2013)
country comparison to the world: 230

Airports—with paved runways: *total:* 1 (2019)
1,524 to 2,437 m: 1

Roadways: *total:* 30 km (2002)
paved: 24 km (2002)
unpaved: 6 km (2002)
country comparison to the world: 219

Merchant marine: *total:* 2
by type: oil tanker 1, other 1 (2019)
country comparison to the world: 175

Ports and terminals: *major seaport(s):* Nauru

MILITARY AND SECURITY

Military and security forces: no regular military forces (2019)

Military—note: Nauru maintains no defense forces; under an informal agreement, defense is the responsibility of Australia

TRANSNATIONAL ISSUES

Disputes—international: none

NAVASSA ISLAND

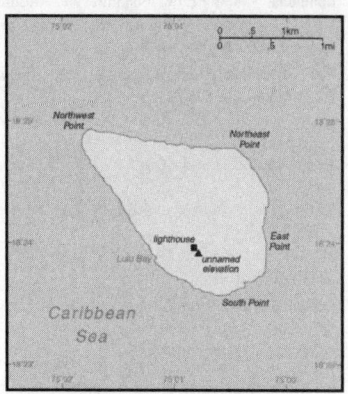

INTRODUCTION

Background: This uninhabited island was claimed by the US in 1857 for its guano. Mining took place between 1865 and 1898. The lighthouse, built in 1917, was shut down in 1996 and administration of Navassa Island transferred from the US Coast Guard to the Department of the Interior, Office of Insular Affairs. A 1998 scientific expedition to the island described it as a "unique preserve of Caribbean biodiversity." The following year it became a National Wildlife Refuge and annual scientific expeditions have continued.

GEOGRAPHY

Location: Caribbean, island in the Caribbean Sea, 30 nm west of Tiburon Peninsula of Haiti

Geographic coordinates: 18 25 N, 75 02 W

Map references: Central America and the Caribbean

Area: *total:* 5 sq km
land: 5.4 sq km
water: 0 sq km

country comparison to the world: 249

Area—comparative: about nine times the size of the National Mall in Washington, DC

Land boundaries: 0 km

Coastline: 8 km

Maritime claims: *territorial sea:* 12 nm
exclusive economic zone: 200 nm

Climate: marine, tropical

Terrain: raised flat to undulating coral and limestone plateau; ringed by vertical white cliffs (9 to 15 m high)

Elevation: *lowest point:* Caribbean Sea 0 m
highest point: 200 m NNW of lighthouse 85 m

Natural resources: guano (mining discontinued in 1898)

Land use: *agricultural land:* 0% (2011 est.)
arable land: 0% (2011 est.) / permanent crops: 0% (2011 est.) / permanent pasture: 0% (2011 est.)
forest: 0% (2011 est.)
other: 100% (2011 est.)

Natural hazards: hurricanes

Environment—current issues: some coral bleaching

Geography—note: strategic location 160 km south of the US Naval Base at Guantanamo Bay, Cuba; mostly exposed rock with numerous solution holes (limestone sinkholes) but with enough grassland to support goat herds; dense stands of fig trees, scattered cactus

PEOPLE AND SOCIETY

Population: uninhabited; transient Haitian fishermen and others camp on the island

Education expenditures: NA

GOVERNMENT

Country name: *conventional long form:* none

conventional short form: Navassa Island
etymology: the flat island was named "Navaza" by some of Christopher COLUMBUS' sailors in 1504; the name derives from the Spanish term "nava" meaning "flat land, plain, or field"

Dependency status: unorganized, unincorporated territory of the US; administered by the Fish and Wildlife Service, US Department of the Interior from the Caribbean Islands National Wildlife Refuge in Boqueron, Puerto Rico; in September 1996, the Coast Guard ceased operations and maintenance of the Navassa Island Light, a 46-meter-tall lighthouse on the southern side of the island; Haiti has claimed the island since the 19th century

Legal system: the laws of the US apply where applicable

Diplomatic representation from the US: none (territory of the US)

Flag description: the flag of the US is used

ECONOMY

Economy—overview: Subsistence fishing and commercial trawling occur within refuge waters.

TRANSPORTATION

Ports and terminals: none; offshore anchorage only

MILITARY AND SECURITY

Military—note: defense is the responsibility of the US

TRANSNATIONAL ISSUES

Disputes—international: claimed by Haiti, source of subsistence fishing

NEPAL

INTRODUCTION

Background: During the late 18th-early 19th centuries, the principality of Gorkha united many of the other principalities and states of the sub-Himalayan region into a Nepali Kingdom. Nepal retained its independence following the Anglo-Nepalese War of 1814-16 and the subsequent peace treaty laid the foundations for two centuries of amicable relations between Britain and Nepal. (The Brigade of Gurkhas continues to serve in the British Army to the present day.) In 1951, the Nepali monarch ended the century-old system of

rule by hereditary premiers and instituted a cabinet system that brought political parties into the government. That arrangement lasted until 1960, when political parties were again banned, but was reinstated in 1990 with the establishment of a multiparty democracy within the framework of a constitutional monarchy.

An insurgency led by Maoists broke out in 1996. During the ensuing 10-year civil war between Maoist and government forces, the monarchy dissolved the cabinet and parliament and re-assumed absolute power in 2002, after the crown prince massacred the royal family in 2001.

A peace accord in 2006 led to the promulgation of an interim constitution in 2007. Following a nationwide Constituent Assembly (CA) election in 2008, the newly formed CA declared Nepal a federal democratic republic, abolished the monarchy, and elected the country's first president. After the CA failed to draft a constitution by a 2012 deadline set by the Supreme Court, then-Prime Minister Baburam BHATTARAI dissolved the CA. Months of negotiations ensued until 2013 when the major political parties agreed to create an interim government headed by then-Chief Justice Khil Raj REGMI with a mandate to hold

elections for a new CA. Elections were held in 2013, in which the Nepali Congress (NC) won the largest share of seats in the CA and in 2014 formed a coalition government with the second-place Communist Party of Nepal-Unified Marxist-Leninist (UML) with NC President Sushil KOIRALA serving as prime minister. Nepal's new constitution came into effect in 2015, at which point the CA became the Parliament. Khagda Prasad Sharma OLI served as the first post-constitution prime minister from 2015 to 2016. OLI resigned ahead of a noconfidence motion against him, and Parliament elected Communist Party of Nepal-Maoist (CPN-M) leader Pushpa Kamal DAHAL (aka "Prachanda") prime minister. The constitution provided for a transitional period during which three sets of elections – local, provincial, and national – needed to take place. The first local elections in 20 years occurred in three phases between May and September 2017, and state and federal elections proceeded in two phases in November and December 2017. The parties headed by OLI and DAHAL ran in coalition and swept the parliamentary elections, and OLI, who led the larger of the two parties, was sworn in as prime minister in February 2018. In May 2018, OLI and DAHAL announced the merger of their parties—the UML and CPN-M—to establish the Nepal Communist Party (NCP), which is now the ruling party in Parliament.

GEOGRAPHY

Location: Southern Asia, between China and India

Geographic coordinates: 28 00 N, 84 00 E

Map references: Asia

Area: *total:* 147,181 sq km
land: 143,351 sq km
water: 3,830 sq km
country comparison to the world: 96

Area—comparative: slightly larger than New York state

Land boundaries: *total:* 3,159 km
border countries (2): China 1389 km, India 1770 km

Coastline: 0 km (landlocked)

Maritime claims: none (landlocked)

Climate: varies from cool summers and severe winters in north to subtropical summers and mild winters in south

Terrain: Tarai or flat river plain of the Ganges in south; central hill region with rugged Himalayas in north

Elevation: *mean elevation:* 2,565 m
lowest point: Kanchan Kalan 70 m

highest point: Mount Everest (highest peak in Asia and highest point on earth above sea level) 8,848 m

Natural resources: quartz, water, timber, hydropower, scenic beauty, small deposits of lignite, copper, cobalt, iron ore

Land use: *agricultural land:* 28.8% (2011 est.)
arable land: 15.1% (2011 est.) / *permanent crops:* 1.2% (2011 est.) / *permanent pasture:* 12.5% (2011 est.)
forest: 25.4% (2011 est.)
other: 45.8% (2011 est.)
Irrigated land: 13,320 sq km (2012)

Population distribution: most of the population is divided nearly equally between a concentration in the southern-most plains of the Tarai region and the central hilly region; overall density is quite low

Natural hazards: severe thunderstorms; flooding; landslides; drought and famine depending on the timing, intensity, and duration of the summer monsoons

Environment—current issues: deforestation (overuse of wood for fuel and lack of alternatives); forest degradation; soil erosion; contaminated water (with human and animal wastes, agricultural runoff, and industrial effluents); unmanaged solid-waste; wildlife conservation; vehicular emissions

Environment—international agreements: *party to:* Biodiversity, Climate Change, Climate Change-Kyoto Protocol, Desertification, Endangered Species, Hazardous Wastes, Law of the Sea, Ozone Layer Protection, Tropical Timber 83, Tropical Timber 94, Wetlands
signed, but not ratified: Marine Life Conservation

Geography—note: landlocked; strategic location between China and India; contains eight of world's 10 highest peaks, including Mount Everest and Kanchenjunga—the world's tallest and third tallest mountains—on the borders with China and India respectively

PEOPLE AND SOCIETY

Population: 30,327,877 (July 2020 est.)
country comparison to the world: 46

Nationality: *noun:* Nepali (singular and plural)
adjective: Nepali

Ethnic groups: Chhettri 16.6%, Brahman-Hill 12.2%, Magar 7.1%, Tharu 6.6%, Tamang 5.8%, Newar 5%, Kami 4.8%, Muslim 4.4%, Yadav 4%, Rai 2.3%, Gurung 2%, Damai/Dholii 1.8%, Thakuri 1.6%, Limbu 1.5%, Sarki 1.4%, Teli 1.4%, Chamar/Harijan/Ram 1.3%, Koiri/Kushwaha 1.2%, other 19% (2011 est.)
note: 125 caste/ethnic groups were reported in the 2011 national census

Languages: Nepali (official) 44.6%, Maithali 11.7%, Bhojpuri 6%, Tharu 5.8%, Tamang 5.1%, Newar 3.2%, Bajjika 3%, Magar 3%, Doteli 3%, Urdu 2.6%, Avadhi 1.9%, Limbu 1.3%, Gurung 1.2%, Baitadeli 1%, other 6.4%, unspecified 0.2% (2011 est.)

note: 123 languages reported as mother tongue in 2011 national census; many in government and business also speak English

Religions: Hindu 81.3%, Buddhist 9%, Muslim 4.4%, Kirant 3.1%, Christian 1.4%, other 0.5%, unspecified 0.2% (2011 est.)

Age structure: *0-14 years:* 28.36% (male 4,526,786/female 4,073,642)
15-24 years: 20.93% (male 3,276,431/female 3,070,843)
25-54 years: 38.38% (male 5,251,553/female 6,387,365)
55-64 years: 6.64% (male 954,836/female 1,059,360)
65 years and over: 5.69% (male 852,969/female 874,092) (2020 est.)

Dependency ratios: *total dependency ratio:* 53
youth dependency ratio: 44.1
elderly dependency ratio: 8.9
potential support ratio: 11.2 (2020 est.)

Median age: *total:* 25.3 years
male: 23.9 years
female: 26.9 years (2020 est.)
country comparison to the world: 161

Population growth rate: 0.98% (2020 est.)
country comparison to the world: 107

Birth rate: 18.1 births/1,000 population (2020 est.)
country comparison to the world: 89

Death rate: 5.7 deaths/1,000 population (2020 est.)
country comparison to the world: 177

Net migration rate: -3.1 migrant(s)/1,000 population (2020 est.)
country comparison to the world: 178

Population distribution: most of the population is divided nearly equally between a concentration in the southern-most plains of the Tarai region and the central hilly region; overall density is quite low

Urbanization: *urban population:* 20.6% of total population (2020)
rate of urbanization: 3.15% annual rate of change (2015-20 est.)
total population growth rate v. urban population growth rate, 2000-2030:

Major urban areas—population: 1.424 million KATHMANDU (capital) (2020)

Sex ratio: *at birth:* 1.06 male(s)/female
0-14 years: 1.11 male(s)/female
15-24 years: 1.07 male(s)/female
25-54 years: 0.82 male(s)/female
55-64 years: 0.9 male(s)/female
65 years and over: 0.98 male(s)/female
total population: 0.96 male(s)/female (2020 est.)

Mother's mean age at first birth: 20.8 years (2016 est.)
note: median age at first birth among women 25-29

Maternal mortality rate: 186 deaths/100,000 live births (2017 est.)
country comparison to the world: 49

Infant mortality rate: *total:* 25.1 deaths/1,000 live births

male: 26.3 deaths/1,000 live births
female: 23.8 deaths/1,000 live births (2020 est.)
country comparison to the world: 68

Life expectancy at birth: *total population:* 71.8 years
male: 71.1 years
female: 72.6 years (2020 est.)
country comparison to the world: 160

Total fertility rate: 1.96 children born/woman (2020 est.)
country comparison to the world: 117

Contraceptive prevalence rate: 52.6% (2016/17)

Drinking water source:
improved:
urban: 91.7% of population
rural: 91.4% of population
total: 91.5% of population
unimproved:
urban: 8.3% of population
rural: 8.6% of population
total: 8.5% of population (2017 est.)

Current Health Expenditure: 5.6% (2017)

Physicians density: 0.91 physicians/1,000 population (2017)

Hospital bed density: 0.3 beds/1,000 population (2012)

Sanitation facility access:
improved:
urban: 91.7% of population
rural: 71.9% of population
total: 75.7% of population
unimproved:
urban: 7.3% of population
rural: 28.1% of population
total: 24.3% of population (2017 est.)

HIV/AIDS—adult prevalence rate: 0.1% (2019 est.)
country comparison to the world: 127

HIV/AIDS—people living with HIV/AIDS: 30,000 (2019 est.)
country comparison to the world: 75

HIV/AIDS—deaths: <1000 (2019 est.)

Major infectious diseases: *degree of risk:* high (2020)

food or waterborne diseases: bacterial diarrhea, hepatitis A and E, and typhoid fever

vectorborne diseases: Japanese encephalitis, malaria, and dengue fever

Obesity—adult prevalence rate: 4.1% (2016)
country comparison to the world: 187

Children under the age of 5 years underweight: 27.2% (2016)
country comparison to the world: 11

Education expenditures: 5.2% of GDP (2018)
country comparison to the world: 56

Literacy: *definition:* age 15 and over can read and write
total population: 67.9%
male: 78.6%
female: 59.7% (2018)

School life expectancy (primary to tertiary education): *total:* 13 years

male: 13 years
female: 13 years (2019)

Unemployment, youth ages 15-24: *total:* 21.4%
male: 19.7%
female: 23.9% (2017 est.)
country comparison to the world: 59

GOVERNMENT

Country name: *conventional long form:* Federal Democratic Republic of Nepal
conventional short form: Nepal
local long form: Sanghiya Loktantrik Ganatantra Nepal
local short form: Nepal
etymology: the Newar people of the Kathmandu Valley and surrounding areas apparently gave their name to the country; the terms "Nepal," "Newar," "Nepar," and "Newal" are phonetically different forms of the same word

Government type: federal parliamentary republic

Capital: *name:* Kathmandu

geographic coordinates: 27 43 N, 85 19 E
time difference: UTC+5.75 (10.75 hours ahead of Washington, DC, during Standard Time)
etymology: name derives from the Kasthamandap temple that stood in Durbar Square; in Sanskrit, "kastha" means "wood" and "mandapa" means "pavilion"; the three-story structure was made entirely of wood, without iron nails or supports, and dated to the late 16th century; it collapsed during a 2015 earthquake

Administrative divisions: 7 provinces; Gandaki Pradesh, Karnali Pradesh, Province No. One, Province No. Two, Province No. Three, Province No. Five, Sudurpashchim Pradesh

Independence: 1768 (unified by Prithvi Narayan SHAH)

National holiday: Constitution Day, 20 September (2015); note—marks the promulgation of Nepal's constitution in 2015 and replaces the previous 28 May Republic Day as the official national day in Nepal; the Gregorian day fluctuates based on Nepal's Hindu calendar

Constitution: *history:* several previous; latest approved by the Second Constituent Assembly 16 September 2015, signed by the president and effective 20 September 2015
amendments: proposed as a bill by either house of the Federal Parliament; bills affecting a state border or powers delegated to a state must be submitted to the affected state assembly; passage of such bills requires a majority vote of that state assembly membership; bills not requiring state assembly consent require at least two-thirds majority vote by the membership of both houses of the Federal Parliament; parts of the constitution on the sovereignty, territorial integrity, independence, and sovereignty vested in the people cannot be amended; last amended 2016

Legal system: English common law and Hindu legal concepts; note—new criminal and civil codes came into effect on 17 August 2018

International law organization participation: has not submitted an ICJ jurisdiction declaration; non-party state to the ICCt

Citizenship: *citizenship by birth:* yes
citizenship by descent only: yes
dual citizenship recognized: no
residency requirement for naturalization: 15 years

Suffrage: 18 years of age; universal

Executive branch: *chief of state:* President Bidhya Devi BHANDARI (since October 2015)

head of government: Prime Minister Khadga Prasad (KP) Sharma OLI (since 15 February 2018); deputy prime ministers Ishwar POKHREL, Upendra YADAV (since 1 June 2018) (an)
cabinet: Council of Ministers appointed by the prime minister; cabinet dominated by the Nepal Communist Party
elections/appointments: president indirectly elected by an electoral college of the Federal Parliament and of the state assemblies for a 5-year term (eligible for a second term); election last held 13 March 2018 (next to be held in 2023); prime minister indirectly elected by the Federal Parliament
election results: Bidhya Devi BHANDARI reelected president; electoral vote—Bidhya Devi BHANDARI (CPN-UML) 39,275, Kumari Laxmi RAI (NC) 11,730

head of state: President Bidhya Devi BHANDARI (since 29 October 2015); Vice President Nanda Bahadur PUN (since 31 October 2015)

Legislative branch: *description:* bicameral Federal Parliament consists of: National Assembly (59 seats; 56 members, including at least 3 women, 1 Dalit, 1 member with disabilities, or 1 minority indirectly elected by an electoral college of state and municipal government leaders, and 3 members, including 1 woman, nominated by the president of Nepal on the recommendation of the government; members serve 6-year terms with renewal of one-third of the membership every 2 years)

House of Representatives (275 seats; 165 members directly elected in single-seat constituencies by simple majority vote and 110 members directly elected in a single nationwide constituency by party-list proportional representation vote; members serve 5-year terms)
elections: first election for the National Assembly held on 7 February 2018 (next to be held in 2024)
first election for House of Representatives held on 26 November and 7 December 2017 (next to be held in 2022)
election results: National Assembly—percent of vote by party—NA; seats by party—NCP 42, NC 13, FSFN 2, RJPN 2; composition—men 37, women 22, percent of women 37.3%

House of Representatives—percent of vote by party—NA; seats by party—NCP 174, NC 63, RJPN 17, FSFN 16, other 4, independent 1; composition—men 185, women 90, percent of women 32.7%; note—total Federal Parliament percent of women 33.5%

Judicial branch: *highest courts:* Supreme Court (consists of the chief justice and up to 20 judges) *judge selection and term of office:* Supreme Court chief justice appointed by the president upon the recommendation of the Constitutional Council, a 5-member, high-level advisory body headed by the prime minister; other judges appointed by the president upon the recommendation of the Judicial Council, a 5-member advisory body headed by the chief justice; the chief justice serves a 6-year term; judges serve until age 65

subordinate courts: High Court; district courts

Political parties and leaders: *the Election Commission of Nepal granted ballot access under the proportional system to 88 political parties for the November-December 2017 House of Representatives election to the Federal Parliament; of these, the following 8 parties won seats:* Federal Socialist Forum, Nepal or FSFN [Upendra YADAV]
Naya Shakti Party, Nepal [Baburam BHATTARAI]
Nepal Communist Party or NCP [Khadga Prasad OLI, Pushpa Kamal DAHAL] Nepali Congress or NC [Sher Bahadur DEUBA]
Nepal Mazdoor Kisan Party [Narayan Man BIJUKCHHE] Rastriya Janamorcha [Chitra Bahadur K.C.]
Rastriya Janata Party or RJPN [Mahanta THAKUR] Rastriya Prajatantra party or RPP [Kamal THAPA]

International organization participation: ADB, BIMSTEC, CD, CP, FAO, G-77, IAEA, IBRD, ICAO, ICC (NGOs), ICRM, IDA, IFAD, IFC, IFRCS, ILO, IMF, IMO, Interpol, IOC, IOM, IPU, ISO, ITSO, ITU, ITUC (NGOs), MIGA, MINURSO, MINUSMA, MINUSTAH, MONUSCO, NAM, OPCW, SAARC, SACEP, UN, UNAMID, UNCTAD, UNDOF, UNESCO, UNIDO, UNIFIL, UNMIL, UNMISS, UNOCI, UNTSO, UNWTO, UPU, WCO, WFTU (NGOs), WHO, WIPO, WMO, WTO

Diplomatic representation in the US: *chief of mission:* Ambassador Arjun Kumar KARKI (since 18 May 2015)
chancery: 2730 34th Place NW, Washington, DC 20007
telephone: [1] (202) 667-4550
FAX: [1] (202) 667-5534
consulate(s) general: Chicago (IL), New York

Diplomatic representation from the US: *chief of mission:* Ambassador Randy BERRY (since 25 October 2018)
telephone: [977] (1) 423-4000
embassy: Maharajgunj, Kathmandu
mailing address: US Embassy, Maharajgunj Chakrapath, Kathmandu, Nepal 44600
FAX: [977] (1) 400-7272

Flag description: crimson red with a blue border around the unique shape of two overlapping right triangles; the smaller, upper triangle bears a white stylized moon and the larger, lower triangle displays a white 12-pointed sun; the color red represents the rhododendron (Nepal's national flower) and is a sign of victory and bravery, the blue border signifies peace and harmony; the two right triangles are a combination of two single pennons (pennants) that originally symbolized the Himalaya Mountains while their charges represented the families of the king (upper) and the prime minister, but today they are understood to denote Hinduism and Buddhism, the country's two main religions; the moon represents the serenity of the Nepalese people and the shade and cool weather in the Himalayas, while the sun depicts the heat and higher temperatures of the lower parts of Nepal; the moon and the sun are also said to express the hope that the nation will endure as long as these heavenly bodies

note: Nepal is the only country in the world whose flag is not rectangular or square

National symbol(s): *rhododendron blossom; national color:* red

National anthem: *name:* "Sayaun Thunga Phool Ka" (Hundreds of Flowers)
lyrics/music: Pradeep Kumar RAI/Ambar GURUNG
note: adopted 2007; after the abolition of the monarchy in 2006, a new anthem was required because of the previous anthem's praise for the king
0:00 / 0:00

ECONOMY

Economy—overview: Nepal is among the least developed countries in the world, with about one-quarter of its population living below the poverty line. Nepal is heavily dependent on remittances, which amount to as much as 30% of GDP. Agriculture is the mainstay of the economy, providing a livelihood for almost two-thirds of the population but accounting for less than a third of GDP. Industrial activity mainly involves the processing of agricultural products, including pulses, jute, sugarcane, tobacco, and grain.

Nepal has considerable scope for exploiting its potential in hydropower, with an estimated 42,000 MW of commercially feasible capacity. Nepal has signed trade and investment agreements with India, China, and other countries, but political uncertainty and a difficult business climate have hampered foreign investment. The United States and Nepal signed a $500 million Millennium Challenge Corporation Compact in September 2017 which will expand Nepal's electricity infrastructure and help maintain transportation infrastructure.

Massive earthquakes struck Nepal in early 2015, which damaged or destroyed infrastructure and homes and set back economic development. Although political gridlock and lack of capacity have hindered post-earthquake recovery, government-led reconstruction efforts have progressively picked up speed, although many hard hit areas still have seen little assistance. Additional challenges to Nepal's growth include its landlocked geographic location, inconsistent electricity supply, and underdeveloped transportation infrastructure.

GDP (purchasing power parity):
$79.19 billion (2017 est.)
$73.39 billion (2016 est.)
$72.96 billion (2015 est.)

note: data are in 2017 dollars
country comparison to the world: 95

GDP (official exchange rate):
$24.88 billion (2017 est.)

GDP—real growth rate: 7.9% (2017 est.)
0.6% (2016 est.)
3.3% (2015 est.)
country comparison to the world: 9

GDP—per capita (PPP): $2,700 (2017 est.)
$2,500 (2016 est.)
$2,500 (2015 est.)
note: data are in 2017 dollars
country comparison to the world: 195

Gross national saving:
45.4% of GDP (2017 est.)
40.2% of GDP (2016 est.)
44% of GDP (2015 est.)
country comparison to the world: 7

GDP—composition, by end use:
household consumption: 78% (2017 est.)
government consumption: 11.7% (2017 est.)
investment in fixed capital: 33.8% (2017 est.)
investment in inventories: 8.7% (2017 est.)
exports of goods and services: 9.8% (2017 est.)
imports of goods and services: -42% (2017 est.)

GDP—composition, by sector of origin:
agriculture: 27% (2017 est.)
industry: 13.5% (2017 est.)
services: 59.5% (2017 est.)

Agriculture—products: pulses, rice, corn, wheat, sugarcane, jute, root crops; milk, water buffalo meat

Industries: tourism, carpets, textiles; small rice, jute, sugar, and oilseed mills; cigarettes, cement and brick production

Industrial production growth rate: 12.4% (2017 est.)
country comparison to the world: 7

Labor force: 16.81 million (2017 est.)
note: severe lack of skilled labor
country comparison to the world: 32

Labor force—by occupation: *agriculture:* 69%
industry: 12%
services: 19% (2015 est.)

Unemployment rate: 3% (2017 est.)
3.2% (2016 est.)
country comparison to the world: 37

Population below poverty line: 25.2% (2011 est.)

Household income or consumption by percentage share: *lowest 10%:* 3.2%
highest 10%: 29.5% (2011)

Budget: *revenues:* 5.925 billion (2017 est.)
expenditures: 5.945 billion (2017 est.)

Taxes and other revenues: 23.8% (of GDP) (2017 est.)
country comparison to the world: 122

Budget surplus (+) or deficit (-): -0.1% (of GDP) (2017 est.)
country comparison to the world: 48

Public debt: 26.4% of GDP (2017 est.)
27.9% of GDP (2016 est.)
country comparison to the world: 171

Fiscal year: 16 July—15 July

Inflation rate (consumer prices): 4.5% (2017 est.)
9.9% (2016 est.)
country comparison to the world: 168

Current account balance: -$93 million (2017 est.)
$1.339 billion (2016 est.)
country comparison to the world: 85

Exports: $818.7 million (2017 est.)
$761.6 million (2016 est.)
country comparison to the world: 168

Exports—partners: India 53.1%, US 11.8%, Turkey 7.2% (2017)

Exports—commodities: clothing, pulses, carpets, textiles, juice, jute goods

Imports: $10 billion (2017 est.)
$8.764 billion (2016 est.)
country comparison to the world: 101

Imports—commodities: petroleum products, machinery and equipment, gold, electrical goods, medicine

Imports—partners: India 70.2%, China 7.5% (2017)

Reserves of foreign exchange and gold: $9.091 billion (31 December 2017 est.)
$8.506 billion (31 December 2016 est.)
country comparison to the world: 76

Debt—external: $5.849 billion (31 December 2017 est.)
$4.321 billion (31 December 2016 est.)
country comparison to the world: 129

Exchange rates: Nepalese rupees (NPR) per US dollar—
104 (2017 est.)
107.38 (2016 est.)
107.38 (2015 est.)
102.41 (2014 est.)
99.53 (2013 est.)

ENERGY

Electricity access: *population without electricity:* 2 million (2019)
electrification—total population: 93% (2019)
electrification—urban areas: 94% (2019)
electrification—rural areas: 93% (2019)

Electricity—production: 4.244 billion kWh (2016 est.)
country comparison to the world: 125

Electricity—consumption: 4.983 billion kWh (2016 est.)
country comparison to the world: 124

Electricity—exports: 2.69 million kWh (FY 2017 est.)
country comparison to the world: 94

Electricity—imports: 2.175 billion kWh (2016 est.)
country comparison to the world: 57

Electricity—installed generating capacity: 943,100 kW (2016 est.)
country comparison to the world: 130

Electricity—from fossil fuels: 5% of total installed capacity (2016 est.)

country comparison to the world: 203

Electricity—from nuclear fuels: 0% of total installed capacity (2017 est.)
country comparison to the world: 152

Electricity—from hydroelectric plants: 92% of total installed capacity (2017 est.)
country comparison to the world: 10

Electricity—from other renewable sources: 3% of total installed capacity (2017 est.)
country comparison to the world: 128

Crude oil—production: 0 bbl/day (2018 est.)
country comparison to the world: 181

Crude oil—exports: 0 bbl/day (2015 est.)
country comparison to the world: 173

Crude oil—imports: 0 bbl/day (2015 est.)
country comparison to the world: 174

Crude oil—proved reserves: 0 bbl (1 January 2018 est.)
country comparison to the world: 176

Refined petroleum products—production: 0 bbl/day (2015 est.)
country comparison to the world: 185

Refined petroleum products—consumption: 27,000 bbl/day (2016 est.)
country comparison to the world: 126

Refined petroleum products—exports: 0 bbl/day (2015 est.)
country comparison to the world: 187

Refined petroleum products—imports: 26,120 bbl/day (2015 est.)
country comparison to the world: 106

Natural gas—production: 0 cu m (2017 est.)
country comparison to the world: 177

Natural gas—consumption: 0 cu m (2017 est.)
country comparison to the world: 181

Natural gas—exports: 0 cu m (2017 est.)
country comparison to the world: 159

Natural gas—imports: 0 cu m (2017 est.)
country comparison to the world: 164

Natural gas—proved reserves: 0 cu m (1 January 2014 est.)
country comparison to the world: 175

Carbon dioxide emissions from consumption of energy: 8.396 million Mt (2017 est.)
country comparison to the world: 115

COMMUNICATIONS

Telephones—fixed lines: *total subscriptions:* 855,926
subscriptions per 100 inhabitants: 2.85 (2019 est.)
country comparison to the world: 80

Telephones—mobile cellular: *total subscriptions:* 41,880,311
subscriptions per 100 inhabitants: 139.45 (2019 est.)
country comparison to the world: 36

Telecommunication systems: *general assessment:* mountainous topography hinders development of telecom infrastructure; mobile service has been extended to all 75 districts covering 90% of

Nepal's land area; fixed broadband is low due to limited number of fixed lines and preeminence of the mobile platform, with overall penetration 2.8%; 3G and 4G subscribers, early stages for mobile broadband market; first launch of a Nepalese satellite (2020)
domestic: 3G coverage is available in 20 major cities (2019); disparity between high coverage in cities and coverage available in underdeveloped rural regions; fixed-line 3 per 100 persons and mobile-cellular 139 per 100 persons; fair radiotelephone communication service; 20% of the market share is fixed (wired) broadband, 2% is fixed (wireless) broadband, and 78% is mobile broadband (2019)
international: country code—977; Nepal, China and Tibet connected across borders with underground and all-dielectric self-supporting (ADSS) fiber-optic cables; radiotelephone communications; microwave and fiber landlines to India; satellite earth station—1 Intelsat (Indian Ocean) (2019)
note: the COVID-19 outbreak is negatively impacting telecommunications production and supply chains globally; consumer spending on telecom devices and services has also slowed due to the pandemic's effect on economies worldwide; overall progress towards improvements in all facets of the telecom industry—mobile, fixed-line, broadband, submarine cable and satellite—has moderated

Broadcast media: state operates 3 TV stations, as well as national and regional radio stations; 117 television channels are licensed, among those 71 are cable television channels, three are distributed through Direct-To-Home (DTH) system, and four are digital terrestrial; 736 FM radio stations are licensed and at least 314 of those radio stations are community radio stations (2019)

Internet country code: .np

Internet users: *total:* 10,103,980

percent of population: 34% (July 2018 est.)
country comparison to the world: 52

Broadband—fixed subscriptions: *total:* 791,961
subscriptions per 100 inhabitants: 3 (2018 est.)
country comparison to the world: 75

TRANSPORTATION

National air transport system: *number of registered air carriers:* 6 (2020)
inventory of registered aircraft operated by air carriers: 39
annual passenger traffic on registered air carriers: 3,296,953 (2018)
annual freight traffic on registered air carriers: 4.66 million mt-km (2018)

Civil aircraft registration country code prefix: 9N (2016)

Airports: 47 (2013)
country comparison to the world: 92

Airports—with paved runways: *total:* 11 (2017)
over 3,047 m: 1 (2017)
1,524 to 2,437 m: 3 (2017)
914 to 1,523 m: 6 (2017)

under 914 m: 1 (2017)

Airports—with unpaved runways: *total:* 36 (2013)
1,524 to 2,437 m: 1 (2013)
914 to 1,523 m: 6 (2013)
under 914 m: 29 (2013)

Railways: *total:* 59 km (2018)
narrow gauge: 59 km 0.762-m gauge (2018)
country comparison to the world: 131

Roadways: *total:* 27,990 km (2016)
paved: 11,890 km (2016)
unpaved: 16,100 km (2016)
country comparison to the world: 100

MILITARY AND SECURITY

Military and security forces: Nepal Army (includes Air Wing); Nepal Armed Police Force (under the Ministry of Home Affairs; paramilitary force responsible for border and internal security, including counter-insurgency, and assisting the Army in the event of an external invasion) (2019)

Military expenditures: 1.6% of GDP (2019)
1.6% of GDP (2018)
1.7% of GDP (2017)
1.7% of GDP (2016)
1.6% of GDP (2015)
country comparison to the world: 75

Military and security service personnel strengths: the Nepal Army has approximately 95,000 active troops (including a small air wing of about 500 personnel); approximately 15,000 Nepal Armed Police (2019 est.)

Military equipment inventories and acquisitions: the Army's inventory includes a mix of older equipment largely of British, Chinese, Indian, Russian, and South African origin; since 2010, China, Italy, and Russia are the top suppliers of military hardware to Nepal (2019 est.)

Military deployments: 720 Central African Republic (MINUSCA); 880 Democratic Republic of the Congo (MONUSCO); 400 Golan Heights (UNDOF); 870 Lebanon (UNIFIL); 230 Liberia (UNSMIL); 150 Mali (MINUSMA); 1,700 South Sudan (UNMISS) (2020)

Military service age and obligation: 18 years of age for voluntary military service (including women); no conscription (2019)

TERRORISM

Terrorist group(s): Indian Mujahedeen (2019)
note: details about the history, aims, leadership, organization, areas of operation, tactics, targets, weapons, size, and sources of support of the group(s) appear(s) in Appendix-T

TRANSNATIONAL ISSUES

Disputes—international: joint border commission continues to work on contested sections of boundary with India, including the 400 sq km dispute over the source of the Kalapani River; India has instituted a stricter border regime to restrict transit of illegal cross-border activities

Refugees and internally displaced persons: *refugees (country of origin):* 12,540 (Tibet/China), 6,396 (Bhutan) (2019)

stateless persons: undetermined (2016); note—the UNHCR is working with the Nepali Government to address the large number of individuals lacking citizenship certificates in Nepal; smaller numbers of Bhutanese Hindu refugees of Nepali origin (the Lhotshampa) who were stripped of Bhutanese nationality and forced to flee their country in the late 1980s and early 1990s—and undocumented Tibetan refugees who arrived in Nepal prior to the 1990s—are considered stateless

Illicit drugs: illicit producer of cannabis and hashish for the domestic and international drug markets; transit point for opiates from Southeast Asia to the West

NETHERLANDS

A modern, industrialized nation, the Netherlands is also a large exporter of agricultural products. The country was a founding member of NATO and the EEC (now the EU) and participated in the introduction of the euro in 1999. In October 2010, the former Netherlands Antilles was dissolved and the three smallest islands—Bonaire, Sint Eustatius, and Saba—became special municipalities in the Netherlands administrative structure. The larger islands of Sint Maarten and Curacao joined the Netherlands and Aruba as constituent countries forming the Kingdom of the Netherlands.

In February 2018, the Sint Eustatius island council (governing body) was dissolved and replaced by a government commissioner to restore the integrity of public administration. According to the Dutch Government, the intervention will be as "short as possible and as long as needed."

INTRODUCTION

Background: The Dutch United Provinces declared their independence from Spain in 1579; during the 17th century, they became a leading seafaring and commercial power, with settlements and colonies around the world. After a 20-year French occupation, a Kingdom of the Netherlands was formed in 1815. In 1830, Belgium seceded and formed a separate kingdom. The Netherlands remained neutral in World War I, but suffered German invasion and occupation in World War II.

GEOGRAPHY

Location: Western Europe, bordering the North Sea, between Belgium and Germany

Geographic coordinates: 52 30 N, 5 45 E

Map references: Europe

Area: *total:* 41,543 sq km
land: 33,893 sq km
water: 7,650 sq km
country comparison to the world: 135

Area—comparative: slightly less than twice the size of New Jersey

Land boundaries: *total:* 1,053 km
border countries (2): Belgium 478 km, Germany 575 km

Coastline: 451 km

Maritime claims: *territorial sea:* 12 nm
contiguous zone: 24 nm

exclusive fishing zone: 200 nm

Climate: temperate; marine; cool summers and mild winters

Terrain: mostly coastal lowland and reclaimed land (polders); some hills in southeast

Elevation: *mean elevation:* 30 m
lowest point: Zuidplaspolder -7 m
highest point: Mount Scenery (on the island of Saba in the Caribbean, now considered an integral part of the Netherlands following the dissolution of the Netherlands Antilles) 862 m
note: the highest point on continental Netherlands is Vaalserberg at 322 m

Natural resources: natural gas, petroleum, peat, limestone, salt, sand and gravel, arable land

Land use: *agricultural land:* 55.1% (2011 est.)
arable land: 29.8% (2011 est.) / permanent crops: 1.1% (2011 est.) / permanent pasture: 74.2% (2011 est.)
forest: 10.8% (2011 est.)
other: 34.1% (2011 est.)

Irrigated land: 4,860 sq km (2012)

Population distribution: an area known as the Randstad, anchored by the cities of Amsterdam, Rotterdam, the Hague, and Utrecht, is the most densely populated region; the north tends to be less dense, though sizeable communities can be found throughout the entire country

Natural hazards: flooding

volcanism: Mount Scenery (887 m), located on the island of Saba in the Caribbean, last erupted in 1640;; Round Hill (601 m), a dormant volcano also known as The Quill, is located on the island of St. Eustatius in the Caribbean;; these islands are at the northern end of the volcanic island arc of the Lesser Antilles that extends south to Grenada

Environment—current issues: water and air pollution are significant environmental problems; pollution of the country's rivers from industrial and agricultural chemicals, including heavy metals, organic compounds, nitrates, and phosphates; air pollution from vehicles and refining activities

Environment—international agreements: *party to:* Air Pollution, Air Pollution-Nitrogen Oxides, Air Pollution-Persistent Organic Pollutants, Air Pollution-Sulfur 85, Air Pollution-Sulfur 94, Air Pollution-Volatile Organic Compounds, Antarctic-Environmental Protocol, Antarctic-Marine Living Resources, Antarctic Treaty, Biodiversity, Climate Change, Climate Change-Kyoto Protocol, Desertification, Endangered Species, Environmental Modification, Hazardous Wastes, Law of the Sea, Marine Dumping, Marine Life Conservation, Ozone Layer Protection, Ship Pollution, Tropical Timber 83, Tropical Timber 94, Wetlands, Whaling
signed, but not ratified: none of the selected agreements

Geography—note: located at mouths of three major European rivers (Rhine, Maas or Meuse, and Schelde); about a quarter of the country lies below sea level and only about half of the land exceeds one meter above sea level

PEOPLE AND SOCIETY

Population: 17,280,397 (July 2020 est.)
country comparison to the world: 67

Nationality: *noun:* Dutchman(men), Dutchwoman(women)
adjective: Dutch

Ethnic groups: Dutch 76.9%, EU 6.4%, Turkish 2.4%, Moroccan 2.3%, Indonesian 2.1%, German 2.1%, Surinamese 2%, Polish 1%, other 4.8% (2018 est.)

Languages: Dutch (official)
note: Frisian is an official language in Fryslan province; Frisian, Low Saxon, Limburgish, Romani, and Yiddish have protected status under the European Charter for Regional or Minority Languages; Dutch is the official language of the three special municipalities of the Caribbean Netherlands; English is a recognized regional language on Sint Eustatius and Saba; Papiamento is a recognized regional language on Bonaire

Religions: Roman Catholic 23.6%, Protestant 14.9% (includes Dutch Reformed 6.4%, Protestant Church of The Netherlands 5.6%, Calvinist 2.9%), Muslim 5.1%, other 5.6% (includes Hindu, Buddhist, Jewish), none 50.7% (2017 est.)

Age structure: *0-14 years:* 16.11% (male 1,425,547/female 1,358,894)
15-24 years: 11.91% (male 1,049,000/female 1,008,763)
25-54 years: 38.47% (male 3,334,064/female 3,313,238)
55-64 years: 13.69% (male 1,177,657/female 1,188,613)
65 years and over: 19.82% (male 1,558,241/female 1,866,380) (2020 est.)

Dependency ratios: *total dependency ratio:* 55.6
youth dependency ratio: 24.4
elderly dependency ratio: 31.2
potential support ratio: 3.2 (2020 est.)

Median age: *total:* 42.8 years
male: 41.6 years
female: 44 years (2020 est.)
country comparison to the world: 32

Population growth rate: 0.37% (2020 est.)
country comparison to the world: 166

Birth rate: 11 births/1,000 population (2020 est.)
country comparison to the world: 179

Death rate: 9.2 deaths/1,000 population (2020 est.)
country comparison to the world: 56

Net migration rate: 1.9 migrant(s)/1,000 population (2020 est.)
country comparison to the world: 50

Population distribution: an area known as the Randstad, anchored by the cities of Amsterdam, Rotterdam, the Hague, and Utrecht, is the most densely populated region; the north tends to be less dense, though sizeable communities can be found throughout the entire country

Urbanization: *urban population:* 92.2% of total population (2020)
rate of urbanization: 0.74% annual rate of change (2015-20 est.)
total population growth rate v. urban population growth rate, 2000-2030:

Major urban areas—population: 1.149 million AMSTERDAM (capital), 1.010 million Rotterdam (2020)

Sex ratio: *at birth:* 1.05 male(s)/female
0-14 years: 1.05 male(s)/female
15-24 years: 1.04 male(s)/female
25-54 years: 1.01 male(s)/female
55-64 years: 0.99 male(s)/female
65 years and over: 0.83 male(s)/female
total population: 0.98 male(s)/female (2020 est.)

Mother's mean age at first birth: 29.8 years (2017 est.)

Maternal mortality rate: 5 deaths/100,000 live births (2017 est.)
country comparison to the world: 169

Infant mortality rate: *total:* 3.5 deaths/1,000 live births
male: 3.7 deaths/1,000 live births

female: 3.2 deaths/1,000 live births (2020 est.)
country comparison to the world: 199

Life expectancy at birth: *total population:* 81.7 years
male: 79.5 years
female: 84.1 years (2020 est.)
country comparison to the world: 28

Total fertility rate: 1.77 children born/woman (2020 est.)
country comparison to the world: 154

Contraceptive prevalence rate: 73% (2013)
note: percent of women aged 18-45

Drinking water source:
improved:
urban: 100% of population
rural: 100% of population
total: 100% of population
unimproved:
urban: 0% of population
rural: 0% of population
total: 0% of population (2017 est.)

Current Health Expenditure: 10.1% (2017)

Physicians density: 3.61 physicians/1,000 population (2017)

Hospital bed density: 3.3 beds/1,000 population (2017)

Sanitation facility access:
improved:
urban: 100% of population
rural: 100% of population
total: 100% of population
unimproved:
urban: 0% of population
rural: 0% of population
total: 0% of population (2017 est.)

HIV/AIDS—adult prevalence rate: 0.2% (2019 est.)
country comparison to the world: 107

HIV/AIDS—people living with HIV/AIDS: 24,000 (2019 est.)
country comparison to the world: 83

HIV/AIDS—deaths: <200 (2019 est.)

Obesity—adult prevalence rate: 20.4% (2016)
country comparison to the world: 99

Education expenditures: 5.5% of GDP (2016)
country comparison to the world: 39

School life expectancy (primary to tertiary education): *total:* 19 years
male: 18 years
female: 19 years (2018)

Unemployment, youth ages 15-24: *total:* 7.2%
male: 7.7%
female: 6.6% (2018 est.)
country comparison to the world: 150

GOVERNMENT

Country name: *conventional long form:* Kingdom of the Netherlands
conventional short form: Netherlands
local long form: Koninkrijk der Nederlanden
local short form: Nederland
abbreviation: NL

etymology: the country name literally means "the lowlands" and refers to the geographic features of the land being both flat and down river from higher areas (i.e., at the estuaries of the Scheldt, Meuse, and Rhine Rivers; only about half of the Netherlands is more than 1 meter above sea level)

Government type: parliamentary constitutional monarchy; part of the Kingdom of the Netherlands

Capital: *name:* Amsterdam; note—The Hague is the seat of government

geographic coordinates: 52 21 N, 4 55 E

time difference: UTC + 1 (6 hours ahead of Washington, DC, during Standard Time)

daylight saving time: +1hr, begins last Sunday in March; ends last Sunday in October

note: time descriptions apply to the continental Netherlands only, for the constituent countries in the Caribbean, the time difference is UTC-4

etymology: the original Dutch name, Amstellerdam, meaning "a dam on the Amstel River," dates to the 13th century; over time the name simplified to Amsterdam

Administrative divisions: 12 provinces (provincies, singular—provincie), 3 public entities* (openbare lichamen, singular—openbaar lichaam (Dutch); entidatnan publiko, singular—entidat publiko (Papiamento)); Bonaire*, Drenthe, Flevoland, Fryslan (Friesland), Gelderland, Groningen, Limburg, Noord-Brabant (North Brabant), Noord-Holland (North Holland), Overijssel, Saba*, Sint Eustatius*, Utrecht, Zeeland (Zealand), Zuid-Holland (South Holland)

note l: the Netherlands is one of four constituent countries of the Kingdom of the Netherlands; the other three, Aruba, Curacao, and Sint Maarten, are all islands in the Caribbean; while all four parts are considered equal partners, in practice, most of the Kingdom's affairs are administered by the Netherlands, which makes up about 98% of the Kingdom's total land area and population

note 2: although Bonaire, Saba, and Sint Eustatius are officially incorporated into the country of the Netherlands under the broad designation of "public entities," Dutch Government sources regularly apply to them the more descriptive term of "special municipalities"; Bonaire, Saba, and Sint Eustatius are collectively referred to as the Caribbean Netherlands

Dependent areas: Aruba, Curacao, Sint Maarten

Independence: 23 January 1579 (the northern provinces of the Low Countries conclude the Union of Utrecht breaking with Spain; on 26 July 1581, they formally declared their independence with an Act of Abjuration; however, it was not until 30 January 1648 and the Peace of Westphalia that Spain recognized this independence)

National holiday: King's Day (birthday of King WILLEM-ALEXANDER), 27 April (1967); note—King's or Queen's Day are observed on the ruling monarch's birthday, currently celebrated on 26 April if 27 April is a Sunday

Constitution: *history:* previous 1597, 1798; latest adopted 24 August 1815 (substantially revised in 1848)

amendments: proposed as an Act of Parliament by or on behalf of the king or by the Second Chamber of the States General; the Second Chamber is dissolved after its first reading of the Act; passage requires a second reading by both the First Chamber and the newly elected Second Chamber, followed by at least two-thirds majority vote of both chambers, and ratification by the king; amended many times, last in 2010

Legal system: civil law system based on the French system; constitution does not permit judicial review of acts of the States General

International law organization participation: accepts compulsory ICJ jurisdiction with reservations; accepts ICCt jurisdiction

Citizenship: *citizenship by birth:* no

citizenship by descent only: at least one parent must be a citizen of the Netherlands

dual citizenship recognized: no

residency requirement for naturalization: 5 years

Suffrage: 18 years of age; universal

Executive branch: *chief of state:* King WILLEM-ALEXANDER (since 30 April 2013); Heir Apparent Princess Catharina-Amalia (daughter of King WILLEM-ALEXANDER, born 7 December 2003)

head of government: Prime Minister Mark RUTTE (since 14 October 2010; Deputy Prime Ministers (since 26 October 2017) Hugo DE JONGE, Karin Kajsa OLLONGREN, and Carola SCHOUTEN (since 26 October 2017); note—Mark RUTTE heads his third cabinet put in place since 26 October 2017

cabinet: Council of Ministers appointed by the monarch

elections/appointments: the monarchy is hereditary; following Second Chamber elections, the leader of the majority party or majority coalition is usually appointed prime minister by the monarch; deputy prime ministers are appointed by the monarch

Legislative branch: *description:* bicameral States General or Staten Generaal consists of: First Chamber or Eerste Kamer (75 seats; members indirectly elected by the country's 12 provincial council members by proportional representation vote; members serve 4-year terms)

Second Chamber or Tweede Kamer (150 seats; members directly elected in multi-seat constituencies by proportional representation vote to serve up to 4-year terms)

elections: First Chamber—last held on 27 May 2019 (next to be held on NA May 2023)

Second Chamber—last held on 15 March 2017 (next to be held 15 March 2021)

election results: First Chamber—percent of vote by party—NA; seats by party—FvD 12, VVD 12, CDA 9, GL 8, D66 7, MvdA 6, PVV 5, SP 4, CU 4, other 8; composition—men 49, women 26, percent of women 34.7%

Second Chamber—percent of vote by party— VVD 21.3%, PVV 13.1%, CDA 12.4%, D66 12.2%, GL 9.1%, SP 9.1%, PvdA 5.7%, CU 3.4%,

PvdD 3.2%, 50 Plus 3.1%, other 7.4%; seats by party—VVD 33, PVV 20, CDA 19, D66 19, GL 14, SP 14, PvdA 9, CU 5, PvdD 5, 50 Plus 4, other 8; composition—men 96, women 54, percent of women 36%; note—total States General percent of women 35.6%

Judicial branch: *highest courts:* Supreme Court or Hoge Raad (consists of 41 judges: the president, 6 vice presidents, 31 justices or raadsheren, and 3 justices in exceptional service, referred to as buitengewone dienst); the court is divided into criminal, civil, tax, and ombuds chambers

judge selection and term of office: justices appointed by the monarch from a list provided by the Second Chamber of the States General; justices appointed for life or until mandatory retirement at age 70

subordinate courts: courts of appeal; district courts, each with up to 5 subdistrict courts; Netherlands Commercial Court

Political parties and leaders: Christian Democratic Appeal or CDA [Sybrand VAN HAERSMA BUMA]

Christian Union or CU [Gert-Jan SEGERS]

Democrats 66 or D66 [Rob JETTEN]

Denk [Tunahan KUZU]

50 Plus [Henk KROL]

Forum for Democracy or FvD [Thierry BAUDET]

Green Left or GL [Jesse KL AVER]

Labor Party or PvdA [Lodewijk ASSCHER]

Party for Freedom or PVV [Geert WILDERS]

Party for the Animals or PvdD [Marianne THIEME]

People's Party for Freedom and Democracy or VVD [Mark RUTTE]

Reformed Political Party or SGP [Kees VAN DER STAAIJ]

Socialist Party or SP [Emile ROEMER]

plus a few minor parties

International organization participation: ADB (nonregional member), AfDB (nonregional member), Arctic Council (observer), Australia Group, Benelux, BIS, CBSS (observer), CD, CE, CERN, EAPC, EBRD, ECB, EIB, EITI (implementing country), EMU, ESA, EU, FAO, FATF, G-10, IADB, IAEA, IBRD, ICAO, ICC (national committees), ICCt, ICRM, IDA, IEA, IFAD, IFC, IFRCS, IGAD (partners), IHO, ILO, IMF, IMO, IMSO, Interpol, IOC, IOM, IPU, ISO, ITSO, ITU, ITUC (NGOs), MIGA, MINUSMA, NATO, NEA, NSG, OAS (observer), OECD, OPCW, OSCE, Pacific Alliance (observer), Paris Club, PCA, Schengen Convention, SELEC (observer), UN, UNCTAD, UNDOF, UNESCO, UNHCR, UNIDO, UNMISS, UNRWA, UN Security Council (temporary), UNTSO, UNWTO, UPU, WCO, WHO, WIPO, WMO, WTO, ZC

Diplomatic representation in the US: *chief of mission:* Ambassador Andre HASPELS (since 16 September 2019)

chancery: 4200 Linnean Avenue NW, Washington, DC 20008

telephone: [1] (202) 244-5300,

[1] 877-388-2443

FAX: [1] (202) 362-3430

consulate(s) general: Chicago, Miami, New York, San Francisco

Diplomatic representation from the US: *chief of mission:* Ambassador Peter HOEKSTRA (since 10 January 2018)
telephone: [31] (70) 310-2209
embassy: John Adams Park 1, 2244 BZ Wassenaar
mailing address: PSC 71, Box 1000, APO AE 09715
FAX: [31] (70) 310-2207

consulate(s) general: Amsterdam

Flag description: three equal horizontal bands of red (bright vermilion; top), white, and blue (cobalt); similar to the flag of Luxembourg, which uses a lighter blue and is longer; the colors were derived from those of WILLIAM I, Prince of Orange, who led the Dutch Revolt against Spanish sovereignty in the latter half of the 16th century; originally the upper band was orange, but because its dye tended to turn red over time, the red shade was eventually made the permanent color; the banner is perhaps the oldest tricolor in continuous use

National symbol(s): *lion, tulip; national color:* orange

National anthem: *name:* "Het Wilhelmus" (The William)
lyrics/music: Philips VAN MARNIX van Sint Aldegonde (presumed)/unknown
note: adopted 1932, in use since the 17th century, making it the oldest national anthem in the world; also known as "Wilhelmus van Nassouwe" (William of Nassau), it is in the form of an acrostic, where the first letter of each stanza spells the name of the leader of the Dutch Revolt
0:00/ 0:48

ECONOMY

Economy—overview: The Netherlands, the sixth-largest economy in the European Union, plays an important role as a European transportation hub, with a consistently high trade surplus, stable industrial relations, and low unemployment. Industry focuses on food processing, chemicals, petroleum refining, and electrical machinery. A highly mechanized agricultural sector employs only 2% of the labor force but provides large surpluses for food-processing and underpins the country's status as the world's second largest agricultural exporter.

The Netherlands is part of the euro zone, and as such, its monetary policy is controlled by the European Central Bank. The Dutch financial sector is highly concentrated, with four commercial banks possessing over 80% of banking assets, and is four times the size of Dutch GDP.

In 2008, during the financial crisis, the government budget deficit hit 5.3% of GDP. Following a protracted recession from 2009 to 2013, during which unemployment doubled to 7.4% and household consumption contracted for four consecutive years, economic growth began inching forward in 2014. Since 2010, Prime Minister Mark RUTTE's government has implemented significant austerity

measures to improve public finances and has instituted broad structural reforms in key policy areas, including the labor market, the housing sector, the energy market, and the pension system. In 2017, the government budget returned to a surplus of 0.7% of GDP, with economic growth of 3.2%, and GDP per capita finally surpassed pre-crisis levels. The fiscal policy announced by the new government in the 2018-2021 coalition plans for increases in government consumption and public investment, fueling domestic demand and household consumption and investment. The new government's policy also plans to increase demand for workers in the public and private sector, forecasting a further decline in the unemployment rate, which hit 4.8% in 2017.

GDP (purchasing power parity):
$924.4 billion (2017 est.)
$898.6 billion (2016 est.)
$879.4 billion (2015 est.)
note: data are in 2017 dollars
country comparison to the world: 27

GDP (official exchange rate):
$832.2 billion (2017 est.)

GDP—real growth rate: 1.63% (2019 est.)
2.32% (2018 est.)
3.02% (2017 est.)
country comparison to the world: 150

GDP—per capita (PPP): $53,900 (2017 est.)
$52,800 (2016 est.)
$51,900 (2015 est.)
note: data are in 2017 dollars
country comparison to the world: 23

Gross national saving:
31.2% of GDP (2017 est.)
28.5% of GDP (2016 est.)
28.8% of GDP (2015 est.)
country comparison to the world: 28

GDP—composition, by end use:
household consumption: 44.3% (2017 est.)
government consumption: 24.2% (2017 est.)
investment in fixed capital: 20.5% (2017 est.)
investment in inventories: 0.2% (2017 est.)
exports of goods and services: 83% (2017 est.)
imports of goods and services: -72.3% (2017 est.)

GDP—composition, by sector of origin:
agriculture: 1.6% (2017 est.)
industry: 17.9% (2017 est.)
services: 70.2% (2017 est.)

Agriculture—products: vegetables, ornamentals, dairy, poultry and livestock products; propagation materials

Industries: agroindustries, metal and engineering products, electrical machinery and equipment, chemicals, petroleum, construction, microelectronics, fishing

Industrial production growth rate: 3.3% (2017 est.)
country comparison to the world: 96

Labor force: 8.907 million (2020 est.)
country comparison to the world: 53

Labor force—by occupation: *agriculture:* 1.2%
industry: 17.2%
services: 81.6% (2015 est.)

Unemployment rate: 3.41% (2019 est.)
3.84% (2018 est.)
country comparison to the world: 46

Population below poverty line: 8.8% (2015 est.)

Household income or consumption by percentage share: *lowest 10%:* 2.3%
highest 10%: 24.9% (2014 est.)

Budget: *revenues:* 361.4 billion (2017 est.)
expenditures: 352.4 billion (2017 est.)

Taxes and other revenues: 43.4% (of GDP) (2017 est.)
country comparison to the world: 27

Budget surplus (+) or deficit (-): 1.1% (of GDP) (2017 est.)
country comparison to the world: 32

Public debt: 56.5% of GDP (2017 est.)
61.3% of GDP (2016 est.)
note: data cover general government debt and include debt instruments issued (or owned) by government entities other than the treasury; the data include treasury debt held by foreign entities; the data include debt issued by subnational entities, as well as intragovernmental debt; intragovernmental debt consists of treasury borrowings from surpluses in the social funds, such as for retirement, medical care, and unemployment, debt instruments for the social funds are not sold at public auctions
country comparison to the world: 78 Fiscal year: calendar year

Inflation rate (consumer prices): 1.3% (2017 est.)
0.1% (2016 est.)
country comparison to the world: 70

Current account balance: $90.207 billion (2019 est.)
$98.981 billion (2018 est.)
country comparison to the world: 4

Exports: $555.6 billion (2017 est.)
$495.4 billion (2016 est.)
country comparison to the world: 6

Exports—partners: Germany 24.2%, Belgium 10.7%, UK 8.8%, France 8.8%, Italy 4.2% (2017)

Exports—commodities: machinery and transport equipment, chemicals, mineral fuels; food and livestock, manufactured goods

Imports: $453.8 billion (2017 est.)
$402.9 billion (2016 est.)
country comparison to the world: 10

Imports—commodities: machinery and transport equipment, chemicals, fuels, foodstuffs, clothing

Imports—partners: China 16.4%, Germany 15.3%, Belgium 8.5%, US 6.9%, UK 5.1%, Russia 4.3% (2017)

Reserves of foreign exchange and gold: $38.44 billion (31 December 2017 est.)
$38.21 billion (31 December 2015 est.)
country comparison to the world: 46

Debt—external: $4.063 trillion (31 December 2016 est.)
$4.054 trillion (31 December 2015 est.)
country comparison to the world: 5

Exchange rates: euros (EUR) per US dollar—

0.885 (2017 est.)
0.903 (2016 est.)
0.9214 (2015 est.)
0.885 (2014 est.)
0.7634 (2013 est.)

ENERGY

Electricity access: *electrification—total population:* 100% (2020)

Electricity—production: 109.3 billion kWh (2016 est.)
country comparison to the world: 33

Electricity—consumption: 108.8 billion kWh (2016 est.)
country comparison to the world: 32

Electricity—exports: 19.34 billion kWh (2016 est.)
country comparison to the world: 8

Electricity—imports: 24.26 billion kWh (2016 est.)
country comparison to the world: 7

Electricity—installed generating capacity: 34.17 million kW (2016 est.)
country comparison to the world: 29

Electricity—from fossil fuels: 75% of total installed capacity (2016 est.)
country comparison to the world: 95

Electricity—from nuclear fuels: 1% of total installed capacity (2017 est.)
country comparison to the world: 31

Electricity—from hydroelectric plants: 0% of total installed capacity (2017 est.)
country comparison to the world: 189

Electricity—from other renewable sources: 23% of total installed capacity (2017 est.)
country comparison to the world: 32

Crude oil—production: 18,000 bbl/day (2018 est.)
country comparison to the world: 67

Crude oil—exports: 7,984 bbl/day (2017 est.)
country comparison to the world: 62

Crude oil—imports: 1.094 million bbl/day (2017 est.)
country comparison to the world: 10

Crude oil—proved reserves: 81.13 million bbl (1 January 2018 est.)
country comparison to the world: 72

Refined petroleum products—production: 1.282 million bbl/day (2017 est.)
country comparison to the world: 17

Refined petroleum products—consumption: 954,500 bbl/day (2017 est.)
country comparison to the world: 23

Refined petroleum products—exports: 2.406 million bbl/day (2017 est.)
country comparison to the world: 3

Refined petroleum products—imports: 2.148 million bbl/day (2017 est.)
country comparison to the world: 3

Natural gas—production: 45.33 billion cu m (2017 est.)

note: the Netherlands has curbed gas production due to seismic activity in the province of Groningen, largest source of gas reserves
country comparison to the world: 17

Natural gas—consumption: 43.38 billion cu m (2017 est.)
country comparison to the world: 21

Natural gas—exports: 51.25 billion cu m (2017 est.)
country comparison to the world: 8

Natural gas—imports: 51 billion cu m (2017 est.)
country comparison to the world: 7

Natural gas—proved reserves: 801.4 billion cu m (1 January 2018 est.)
country comparison to the world: 26

Carbon dioxide emissions from consumption of energy: 250.2 million Mt (2017 est.)
country comparison to the world: 26

COMMUNICATIONS

Telephones—fixed lines: *total subscriptions:* 5,598,798
subscriptions per 100 inhabitants: 32.52 (2019 est.)
country comparison to the world: 25

Telephones—mobile cellular: *total subscriptions:* 21,914,852
subscriptions per 100 inhabitants: 127.29 (2019 est.)
country comparison to the world: 54

Telecommunication systems: *general assessment:* highly developed and well maintained; while fixed-line voice market is in decline the VoIP (voice over Internet protocol) and mobile platforms advance; one of the highest fixed broadband penetration rates in the world, due to government investments; plans for 3G network shutdown in 2022; operators are concentrating investment on LTE-A and 5G services; MNOs and banks launch m-payments system (2020)
domestic: extensive fixed-line, fiber-optic network; large cellular telephone system with five major operators utilizing the third generation of the Global System for Mobile Communications technology; one in five households now use Voice over the Internet Protocol services; fixed-line 33 per 100 and mobile-cellular 127 per 100 persons (2019)
international: country code—31; landing points for Farland North, TAT-14, Circe North, Concerto, Ulysses 2, AC-1, UK- Netherlands 14, and COBRAcable submarine cables which provide links to the US and Europe; satellite earth stations—5 (3 Intelsat—1 Indian Ocean and 2 Atlantic Ocean, 1 Eutelsat, and 1 Inmarsat) (2019)
note: the COVID-19 outbreak is negatively impacting telecommunications production and supply chains globally; consumer spending on telecom devices and services has also slowed due to the pandemic's effect on economies worldwide; overall progress towards improvements in all facets of the telecom industry—mobile, fixed-line, broadband, submarine cable and satellite—has moderated

Broadcast media: more than 90% of households are connected to cable or satellite TV systems that provide a wide range of domestic and foreign channels; public service broadcast system includes multiple broadcasters, 3 with a national reach and the remainder operating in regional and local markets; 2 major nationwide commercial television companies, each with 3 or more stations, and many commercial TV stations in regional and local markets; nearly 600 radio stations with a mix of public and private stations providing national or regional coverage

Internet country code: .nl

Internet users: *total:* 16,243,928

percent of population: 94.71% (July 2018 est.)
country comparison to the world: 41

Broadband—fixed subscriptions: *total:* 7,406,700
subscriptions per 100 inhabitants: 43 (2018 est.)
country comparison to the world: 23

TRANSPORTATION

National air transport system: *number of registered air carriers:* 8 (2020)
inventory of registered aircraft operated by air carriers: 238
annual passenger traffic on registered air carriers: 43,996,044 (2018)
annual freight traffic on registered air carriers: 5,886,510,000 mt-km (2018)

Civil aircraft registration country code prefix: PH (2016)

Airports: 29 (2013)
country comparison to the world: 117

Airports—with paved runways: *total:* 23 (2017)
over 3,047 m: 3 (2017)
2,438 to 3,047 m: 11 (2017)
1,524 to 2,437 m: 1 (2017)
914 to 1,523 m: 6 (2017)
under 914 m: 2 (2017)

Airports—with unpaved runways: *total:* 6 (2013)
914 to 1,523 m: 4 (2013)
under 914 m: 2 (2013)

Heliports: 1 (2013)

Pipelines: 14000 km gas, 2500 km oil and refined products, 3000 km chemicals (2016)

Railways: *total:* 3,058 km (2016)
standard gauge: 3,058 km 1.435-m gauge (2,314 km electrified) (2016)
country comparison to the world: 61

Roadways: *total:* 139,124 km (includes 3,654 km of expressways) (2016)
country comparison to the world: 37

Waterways: 6,237 km (navigable by ships up to 50 tons) (2012)
country comparison to the world: 21

Merchant marine: *total:* 1,217
by type: bulk carrier 13, container ship 43, general cargo 568, oil tanker 22, other 571 (2019)
country comparison to the world: 24

Ports and terminals: *major seaport(s):* IJmuiden, Vlissingen

container port(s) (TEUs): Rotterdam (13,734,000) (2017)
LNG terminal(s) (import): Rotterdam
river port(s): Amsterdam (Nordsee Kanaal); Moerdijk (Hollands Diep River); Rotterdam (Rhine River); Terneuzen (Western Scheldt River)

MILITARY AND SECURITY

Military and security forces: Royal Netherlands Army, Royal Netherlands Navy (includes Naval Air Service and Marine Corps), Royal Netherlands Air Force (Koninklijke Luchtmacht, KLu), Royal Netherlands Marechaussee (Military Constabulary) (2019)

note: the Netherlands Coast Guard and the Dutch Caribbean Coast Guard are civilian in nature, but managed by the Royal Netherlands Navy

Military expenditures:
1.36% of GDP (2019 est.)
1.21% of GDP (2018)
1.15% of GDP (2017)
1.16% of GDP (2016)
1.13% of GDP (2015)
country comparison to the world: 88

Military and security service personnel strengths: the Netherlands Armed Forces have approximately 41,000 active duty personnel (19,000 Army; 8,500 Navy; 8,000 Air Force; 5,800 Constabulary) (2019 est.)

Military equipment inventories and acquisitions: the inventory of the Netherlands Armed Forces consists of a mix of domestically-produced and modern European- and US-sourced equipment; since 2010, the US is the leading supplier of weapons systems to the Netherlands, followed by Germany, Italy, and Sweden; the Netherlands has an advanced domestic defense industry that focuses on armored vehicles, naval ships, and air defense systems; it also participates with the US and other European countries on joint development and production of advanced weapons systems (2019)

Military deployments: 160 Afghanistan (NATO); 270 Lithuania (NATO) (2020)

Military service age and obligation: 1 7 years of age for an all-volunteer force (201 6)

Military—note: in 2018, the Defense Ministers of Belgium, Denmark and the Netherlands signed a Memorandum of Understanding (MOU) for the creation of a Composite Special Operations Component Command (C-SOCC); C-SOCC is scheduled to be fully operational in 2021 (2020)

TERRORISM

Terrorist group(s): Islamic State of Iraq and ash-Sham (ISIS) (2019)
note: details about the history, aims, leadership, organization, areas of operation, tactics, targets, weapons, size, and sources of support of the group(s) appear(s) in Appendix-T

TRANSNATIONAL ISSUES

Disputes—international: none

Refugees and internally displaced persons: *refugees (country of origin):* 31,694 (Syria), 14,809 (Eritrea), 13,007 (Somalia), 8,423 (Iraq), 5,815 (Afghanistan) (2019)

stateless persons: 1,951 (2019)

Illicit drugs: major European producer of synthetic drugs, including ecstasy, and cannabis cultivator; important gateway for cocaine, heroin, and hashish entering Europe; major source of US-bound ecstasy and a significant consumer of ecstasy; a large financial sector vulnerable to money laundering

NEW CALEDONIA

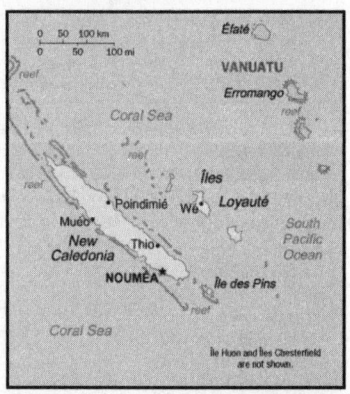

INTRODUCTION

Background: Settled by both Britain and France during the first half of the 19th century, the island became a French possession in 1853. It served as a penal colony for four decades after 1864. Agitation for independence during the 1980s and early 1990s ended in the 1998 Noumea Accord, which over two decades transferred an increasing amount of governing responsibility from France to New Caledonia. In a referendum held in November 2018, residents rejected independence and decided to retain their territorial status, although two additional referendums may occur in 2020 and 2022, per the Noumea Accord.

GEOGRAPHY

Location: Oceania, islands in the South Pacific Ocean, east of Australia

Geographic coordinates: 21 30 S, 165 30 E

Map references: Oceania

Area: *total:* 18,575 sq km
land: 18,275 sq km
water: 300 sq km
country comparison to the world: 156

Area—comparative: slightly smaller than New Jersey

Land boundaries: 0 km

Coastline: 2,254 km

Maritime claims: *territorial sea:* 12 nm
exclusive economic zone: 200 nm

Climate: tropical; modified by southeast trade winds; hot, humid

Terrain: coastal plains with interior mountains

Elevation: *lowest point:* Pacific Ocean 0 m
highest point: Mont Panie 1,628 m

Natural resources: nickel, chrome, iron, cobalt, manganese, silver, gold, lead, copper

Land use: *agricultural land:* 10.4% (2011 est.)

arable land: 0.4% (2011 est.) / permanent crops: 0.2% (2011 est.) / permanent pasture: 9.8% (2011 est.)
forest: 45.9% (2011 est.)
other: 43.7% (2011 est.)
Irrigated land: 100 sq km (2012)

Population distribution: most of the populace lives in the southern part of the main island, in and around the capital of Noumea

Natural hazards: cyclones, most frequent from November to March

volcanism: Matthew and Hunter Islands are historically active

Environment—current issues: preservation of coral reefs; prevention of invasive species; limiting erosion caused by nickel mining and forest fires

Geography—note: consists of the main island of New Caledonia (one of the largest in the Pacific Ocean), the archipelago of Iles Loyaute, and numerous small, sparsely populated islands and atolls

PEOPLE AND SOCIETY

Population: 290,009 (July 2020 est.)
country comparison to the world: 183

Nationality: *noun:* New Caledonian(s)
adjective: New Caledonian

Ethnic groups: Kanak 39.1%, European 27.1%, Wallisian, Futunian 8.2%, Tahitian 2.1%, Indonesian 1.4%, Ni-Vanuatu 1%, Vietnamese 0.9%, other 17.7%, unspecified 2.5% (2014 est.)

Languages: French (official), 33 Melanesian-Polynesian dialects

Religions: Roman Catholic 60%, Protestant 30%, other 10%

Age structure: 0-14 years: 21.74% (male 32,227/female 30,819)
15-24 years: 15.63% (male 23,164/female 22,163)
25-54 years: 43.73% (male 63,968/female 62,856)
55-64 years: 9.06% (male 12,700/female 13,568)
65 years and over: 9.84% (male 12,552/female 15,992) (2020 est.)

Dependency ratios: *total dependency ratio:* 46.6
youth dependency ratio: 32.4
elderly dependency ratio: 14.2
potential support ratio: 7 (2020 est.)

Median age: *total:* 32.9 years
male: 32.1 years
female: 33.7 years (2020 est.)
country comparison to the world: 103

Population growth rate: 1.25% (2020 est.)
country comparison to the world: 88

Birth rate: 14.5 births/1,000 population (2020 est.)
country comparison to the world: 129

Death rate: 5.9 deaths/1,000 population (2020 est.)
country comparison to the world: 173

Net migration rate: 3.8 migrant(s)/1,000 population (2020 est.)
note: there has been steady emigration from Wallis and Futuna to New Caledonia
country comparison to the world: 32

Population distribution: most of the populace lives in the southern part of the main island, in and around the capital of Noumea

Urbanization: *urban population:* 71.5% of total population (2020)
rate of urbanization: 1.89% annual rate of change (2015-20 est.)
total population growth rate v. urban population growth rate, 2000-2030:

Major urban areas—population: 198,000 NOUMEA (capital) (2018)

Sex ratio: *at birth:* 1.05 male(s)/female
0-14 years: 1.05 male(s)/female
15-24 years: 1.05 male(s)/female
25-54 years: 1.02 male(s)/female
55-64 years: 0.94 male(s)/female
65 years and over: 0.78 male(s)/female
total population: 1 male(s)/female (2020 est.)

Infant mortality rate: *total:* 5 deaths/1,000 live births
male: 5.9 deaths/1,000 live births
female: 4.1 deaths/1,000 live births (2020 est.)
country comparison to the world: 178

Life expectancy at birth: *total population:* 78.4 years
male: 74.4 years
female: 82.5 years (2020 est.)
country comparison to the world: 67

Total fertility rate: 1.88 children born/woman (2020 est.)
country comparison to the world: 134

Drinking water source:
improved:
total: 100% of population
unimproved:
total: 0% of population (2017 est.)

Physicians density: 2.22 physicians/1,000 population (2009)

Sanitation facility access:
improved:
total: 100% of population
unimproved:
total: 0% of population (2017 est.)

HIV/AIDS—adult prevalence rate: NA

HIV/AIDS—people living with HIV/AIDS: NA

HIV/AIDS—deaths: NA

Major infectious diseases: *degree of risk:* high (2020)

food or waterborne diseases: bacterial diarrhea

vectorborne diseases: malaria

Education expenditures: NA

Literacy: *definition:* age 15 and over can read and write
total population: 96.9%
male: 97.3%
female: 96.5% (2015)

Unemployment, youth ages 15-24: *total:* 38.4%
male: 37.1%
female: 40% (2014 est.)
country comparison to the world: 15

GOVERNMENT

Country name: *conventional long form:* Territory of New Caledonia and Dependencies
conventional short form: New Caledonia
local long form: Territoire des Nouvelle-Caledonie et Dependances
local short form: Nouvelle-Caledonie
etymology: British explorer Captain James COOK discovered and named New Caledonia in 1774; he used the appellation because the northeast of the island reminded him of Scotland (Caledonia is the Latin designation for Scotland)

Dependency status: special collectivity (or a sui generis collectivity) of France since 1998; note—independence referenda took place on 4 November 2018 and 4 October 2020 with a majority voting to reject independence in favor of maintaining the status quo; an additional referenda, still unscheduled, may occur in 2022

Government type: parliamentary democracy (Territorial Congress); an overseas collectivity of France

Capital: *name:* Noumea

geographic coordinates: 22 16 S, 166 27 E
time difference: UTC+11 (16 hours ahead of Washington, DC, during Standard Time)
etymology: established in 1854 as Port-de-France, the settlement was renamed Noumea in 1866, in order to avoid any confusion with Fort-de-France

in Martinique; the New Caledonian language of Ndrumbea (also spelled Ndumbea, Dubea, and Drubea) spoken in the area gave its name to the capital city, Noumea, as well as to the neighboring town (suburb) of Dumbea

Administrative divisions: 3 provinces; Province Iles (Islands Province), Province Nord (North Province), and Province Sud (South Province)

Independence: none (overseas collectivity of France); note—in two independence referenda, on 4 November 2018 and 4 October 2020, the majority voted to reject independence in favor of maintaining the status quo

National holiday: Fete de la Federation, 14 July (1790); note—the local holiday is New Caledonia Day, 24 September (1853)

Constitution: *history:* 4 October 1958 (French Constitution with changes as reflected in the Noumea Accord of 5 May 1998)
amendments: French constitution amendment procedures apply

Legal system: civil law system based on French civil law

Citizenship: see France

Suffrage: 18 years of age; universal

Executive branch: *chief of state:* President Emmanuel MACRON (since 14 May 2017); represented by High Commissioner Laurent PREVOST (since 5 August 2019)
head of government: President of the Government Thierry SANTA (since 9 July 2019); Temporary Vice President Gilbert TUIENON (since 9 July 2019); note—Temporary Vice President Gilbert TUIENON was elected so that the new government could take over; Philippe GERMAIN's government remained caretaker government until the new government was settled
cabinet: Cabinet elected from and by the Territorial Congress
elections/appointments: French president directly elected by absolute majority popular vote in 2 rounds if needed for a 5-year term (eligible for a second term); high commissioner appointed by the French president on the advice of the French Ministry of Interior; president of New Caledonia elected by Territorial Congress for a 5-year term (no term limits); election last held on 13 June 2017 (next to be held in 2022)
election results: Thierry SANTA elected president by Territorial Congress with 6 votes out of 11

Legislative branch: *description:* unicameral Territorial Congress or Congrès du Territoire (54 seats; members indirectly selected proportionally by the partisan makeup of the 3 Provincial Assemblies or Assemblés Provinciales; members of the 3 Provincial Assemblies directly elected by proportional representation vote; members serve 5-year terms); note—the Customary Senate is the assembly of the various traditional councils of the Kanaks, the indigenous population, which rules on laws affecting the indigenous population

New Caledonia indirectly elects 2 members to the French Senate by an electoral colleges for a 6-year term with one seat renewed every 3 years and directly elects 2 members to the French National Assembly by absolute majority vote in 2 rounds if needed for a 5-year term

elections: Territorial Congress—last held on 12 May 2019 (next to be held in May 2024) French Senate—election last held on 24 September 2017 (next to be held not later than 2019)

French National Assembly—election last held on 11 and 18 June 2017 (next to be held by June 2022)

election results: Territorial Congress—percent of vote by party—N/A; seats by party—Future With Confidence 18, UNI 9, UC 9, CE 7, FLNKS 6, Oceanic Awakening 3, PT 1, LKS 1 (Anti-Independence 28, Pro-Independence 26); composition—men 30, women 24, percent of women 44.4%

French Senate—percent of vote by party—NA; seats by party—UMP 2

French National Assembly—percent of vote by party—NA; seats by party—CE 2

Judicial branch: *highest courts:* Court of Appeal in Noumea or Cour d'Appel; organized into civil, commercial, social, and pre-trial investigation chambers; court bench normally includes the court president and 2 counselors); Administrative Court (number of judges NA); note—final appeals beyond the Court of Appeal are referred to the Court of Cassation or Cour de Cassation (in Paris); final appeals beyond the Administrative Court are referred to the Administrative Court of Appeal (in Paris)

judge selection and term of office: judge appointment and tenure based on France's judicial system

subordinate courts: Courts of First Instance include: civil, juvenile, commercial, labor, police, criminal, assizes, and also a pre-trial investigation chamber; Joint Commerce Tribunal; administrative courts

Political parties and leaders: Build Our Rainbow Nation
Caledonia Together or CE [Philippe GERMAIN]
Caledonian Union or UC [Daniel GOA]
Future Together (l'Avenir Ensemble) [Harold MARTIN]
Kanak Socialist Front for National Liberation or FLNKS (alliance includes PALIKA, UNI, UC, and UPM) [Victor TUTUGORO]
Labor Party (Parti Travailliste) or PT [Louis Kotra UREGEI]
National Union for Independence (Union Nationale pour l'Independance) or UNI
Party of Kanak Liberation (Parti de Liberation Kanak) or PALIKA [Paul NEAOUTYINE]
Socialist Kanak Liberation or LKS [Nidoish NAISSELINE]
The Republicans (formerly The Rally or UMP) [interim leader Thierry SANTA]
Union for Caledonia in France

International organization participation: ITUC (NGOs), PIF (associate member), SPC, UPU, WFTU (NGOs), WMO

Diplomatic representation in the US: none (overseas territory of France)

Diplomatic representation from the US: none (overseas territory of France)

Flag description: New Caledonia has two official flags; alongside the flag of France, the Kanak (indigenous Melanesian) flag has equal status; the latter consists of three equal horizontal bands of blue (top), red, and green; a large yellow disk—diameter two-thirds the height of the flag—shifted slightly to the hoist side is edged in black and displays a black fleche faitiere symbol, a native rooftop adornment

National symbol(s): *fleche faitiere (native rooftop adornment), kagu bird; national colors:* gray, red

National anthem: *name:* "Soyons unis, devenons freres" (Let Us Be United, Let Us Become Brothers)
lyrics/music: Chorale Melodia (a local choir)
note: adopted 2008; contains a mixture of lyrics in both French and Nengone (an indigenous language); as a selfgoverning territory of France, in addition to the local anthem, "La Marseillaise" is official (see France)

ECONOMY

Economy—overview: New Caledonia has 11% of the world's nickel reserves, representing the second largest reserves on the planet. Only a small amount of the land is suitable for cultivation, and food accounts for about 20% of imports. In addition to nickel, substantial financial support from France—equal to more than 15% of GDP—and tourism are keys to the health of the economy.

With the gradual increase in the production of two new nickel plants in 2015, average production of metallurgical goods stood at a record level of 94 thousand tons. However, the sector is exposed to the high volatility of nickel prices, which have been in decline since 2016. In 2017, one of the three major mining firms on the island, Vale, put its operations up for sale, triggering concerns of layoffs ahead of the 2018 independence referendum.

GDP (purchasing power parity):
$11.11 billion (2017 est.)
$10.89 billion (2016 est.)
$10.77 billion (2015 est.)
note: data are in 2015 dollars
country comparison to the world: 159

GDP (official exchange rate):
$9.77 billion (2017 est.)

GDP—real growth rate: 2% (2017 est.)
1.1% (2016 est.)
3.2% (2015 est.)
country comparison to the world: 139

GDP—per capita (PPP): $31,100 (2015 est.)
$32,100 (2014 est.)
$29,800 (2012 est.)
country comparison to the world: 66

GDP—composition, by end use:
household consumption: 64.3% (2017 est.)
government consumption: 24% (2017 est.)
investment in fixed capital: 38.4% (2017 est.)

investment in inventories: 0% (2017 est.)
exports of goods and services: 18.7% (2017 est.)
imports of goods and services: -45.5% (2017 est.)

GDP—composition, by sector of origin:
agriculture: 1.4% (2017 est.)
industry: 26.4% (2017 est.)
services: 72.1% (2017 est.)

Agriculture—products: vegetables; beef, venison, other livestock products; fish

Industries: nickel mining and smelting

Industrial production growth rate: 3.5% (2017 est.)
country comparison to the world: 87

Labor force: 119,500 (2016 est.)
country comparison to the world: 181

Labor force—by occupation: *agriculture:* 2.7%
industry: 22.4%
services: 74.9% (2010)

Unemployment rate: 14.7% (2014)
14% (2009)
country comparison to the world: 173

Population below poverty line: 17% (2008)

Household income or consumption by percentage share: *lowest 10%:* NA
highest 10%: NA

Budget: *revenues:* 1.995 billion (2015 est.)
expenditures: 1.993 billion (2015 est.)

Taxes and other revenues: 20.4% (of GDP) (2015 est.)
country comparison to the world: 147

Budget surplus (+) or deficit (-): 0% (of GDP) (2015 est.)
country comparison to the world: 45

Public debt: 6.5% of GDP (2015 est.)
6.5% of GDP (2014 est.)
country comparison to the world: 203

Fiscal year: calendar year

Inflation rate (consumer prices): 1.4% (2017 est.)
0.6% (2016 est.)
country comparison to the world: 79

Current account balance: -$1.469 billion (2014 est.)
-$1.861 billion (2013 est.)
country comparison to the world: 159

Exports: $2.207 billion (2014 est.)
country comparison to the world: 137

Exports—partners: China 25.4%, Japan 16.6%, South Korea 14.8%, France 8.2%, Belgium 5%, US 4.6% (2017)

Exports—commodities: ferronickels, nickel ore, fish

Imports: $2.715 billion (2015 est.)
$4.4 billion (2014 est.)
country comparison to the world: 153

Imports—commodities: machinery and equipment, fuels, chemicals, foodstuffs

Imports—partners: France 24.2%, Singapore 13.1%, China 9.2%, Australia 7.1%, South Korea 5.2%, Malaysia 4.7%, NZ 4.4%, US 4.4% (2017)

Debt—external: $112 million (31 December 2013 est.)
$79 million (31 December 1998 est.)

country comparison to the world: 192

Exchange rates: Comptoirs Francais du Pacifique francs (XPF) per US dollar—
110.2 (2017 est.)
107.84 (2016 est.)
107.84 (2015 est.)
89.85 (2013 est.)
90.56 (2012 est.)

ENERGY

Electricity access: *electrification—total population:* 100% (2020)

Electricity—production: 2.945 billion kWh (2016 est.)
country comparison to the world: 133

Electricity—consumption: 2.739 billion kWh (2016 est.)
country comparison to the world: 139

Electricity—exports: 0 kWh (2016 est.)
country comparison to the world: 175

Electricity—imports: 0 kWh (2016 est.)
country comparison to the world: 178

Electricity—installed generating capacity: 996,200 kW (2016 est.)
country comparison to the world: 128

Electricity—from fossil fuels: 87% of total installed capacity (2016 est.)
country comparison to the world: 64

Electricity—from nuclear fuels: 0% of total installed capacity (2017 est.)
country comparison to the world: 153

Electricity—from hydroelectric plants: 8% of total installed capacity (2017 est.)
country comparison to the world: 123

Electricity—from other renewable sources: 6% of total installed capacity (2017 est.)
country comparison to the world: 101

Crude oil—production: 0 bbl/day (2018 est.)
country comparison to the world: 182

Crude oil—exports: 0 bbl/day (2015 est.)
country comparison to the world: 174

Crude oil—imports: 0 bbl/day (2015 est.)
country comparison to the world: 175

Crude oil—proved reserves: 0 bbl (1 January 2018 est.)
country comparison to the world: 177

Refined petroleum products—production: 0 bbl/day (2015 est.)
country comparison to the world: 186

Refined petroleum products—consumption: 20,000 bbl/day (2016 est.)

country comparison to the world: 142

Refined petroleum products—exports: 0 bbl/day (2015 est.)
country comparison to the world: 188

Refined petroleum products—imports: 19,100 bbl/day (2015 est.)
country comparison to the world: 124

Natural gas—production: 0 cu m (2017 est.)
country comparison to the world: 178

Natural gas—consumption: 0 cu m (2017 est.)
country comparison to the world: 182

Natural gas—exports: 0 cu m (2017 est.)
country comparison to the world: 160

Natural gas—imports: 0 cu m (2017 est.)
country comparison to the world: 165

Natural gas—proved reserves: 0 cu m (1 January 2014 est.)
country comparison to the world: 176

Carbon dioxide emissions from consumption of energy: 6.165 million Mt (2017 est.)
country comparison to the world: 128

COMMUNICATIONS

Telephones—fixed lines: *total subscriptions:* 82,111
subscriptions per 100 inhabitants: 28.67 (2019 est.)
country comparison to the world: 142

Telephones—mobile cellular: *total subscriptions:* 275,002
subscriptions per 100 inhabitants: 96.02 (2019 est.)
country comparison to the world: 180

Telecommunication systems: *general assessment:* well advanced telecoms sector; 3G & 4G network services; one of the highest smart phone adoption rates in the region; telecommunications sector is dominated by govt. owned company with a monopoly on fixed and mobile services, Internet and broadband access (2020)
domestic: fixed-line 29 per 100 and mobile-cellular telephone subscribership 96 per 100 persons (2019)
international: country code—687; landing points for the Gondwana-1 and Picot-1 providing connectivity via submarine cables around New Caledonia and to Australia; satellite earth station—1 Intelsat (Pacific Ocean) (2019)
note: the COVID-19 outbreak is negatively impacting telecommunications production and supply chains globally; consumer spending on telecom devices and services has also slowed due to the pandemic's effect on economies worldwide; overall

progress towards improvements in all facets of the telecom industry—mobile, fixed line, broadband, submarine cable and satellite—has moderated

Broadcast media: the publicly owned French Overseas Network (RFO), which operates in France's overseas departments and territories, broadcasts over the RFO Nouvelle-Calédonie TV and radio stations; a small number of privately owned radio stations also broadcast

Internet country code: .nc

Internet users: *total:* 231,887

percent of population: 82.01% (July 2018 est.)
country comparison to the world: 172

TRANSPORTATION

National air transport system: *number of registered air carriers:* 3 (registered in France) (2020)
inventory of registered aircraft operated by air carriers: 15 (registered in France)

Airports: 25 (2013)
country comparison to the world: 128

Airports—with paved runways: *total:* 12 (2019)
over 3,047 m: 1
914 to 1,523 m: 10
under 914 m: 1

Airports—with unpaved runways: *total:* 13 (2013)
914 to 1,523 m: 5 (2013)
under 914 m: 8 (2013)

Heliports: 8 (2013)

Roadways: *total:* 5,622 km (2006)
country comparison to the world: 146

Merchant marine: *total:* 19
by type: general cargo 5, oil tanker 1, other 13 (2019)
country comparison to the world: 144

Ports and terminals: *major seaport(s):* Noumea

MILITARY AND SECURITY

Military and security forces: no regular military forces; France bases land, air, and naval forces on New Caledonia (Forces Armées de la Nouvelle-Calédonie, FANC) (2019)

Military—note: defense is the responsibility of France

TRANSNATIONAL ISSUES

Disputes—international: Matthew and Hunter Islands east of New Caledonia claimed by France and Vanuatu

NEW ZEALAND

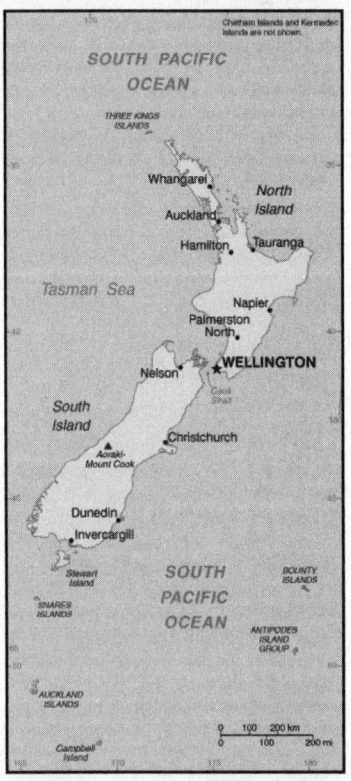

land: 264,537 sq km

water: 4,301 sq km

note: includes Antipodes Islands, Auckland Islands, Bounty Islands, Campbell Island, Chatham Islands, and Kermadec Islands

country comparison to the world: 77

Area—comparative: almost twice the size of North Carolina; about the size of Colorado

Land boundaries: 0 km

Coastline: 15,134 km

Maritime claims: *territorial sea:* 12 nm

exclusive economic zone: 200 nm

contiguous zone: 24 nm

continental shelf: 200 nm or to the edge of the continental margin

Climate: temperate with sharp regional contrasts

Terrain: predominately mountainous with large coastal plains

Elevation: *mean elevation:* 388 m

lowest point: Pacific Ocean 0 m

highest point: Aoraki/Mount Cook 3,724 m; *note*—the mountain's height was 3,764 m until 14 December 1991 when it lost about 10 m in an avalanche of rock and ice; erosion of the ice cap since then has brought the height down another 30 m

Natural resources: natural gas, iron ore, sand, coal, timber, hydropower, gold, limestone

Land use: *agricultural land:* 43.2% (2011 est.)

arable land: 1.8% (2011 est.) / permanent crops: 0.3% (2011 est.) / permanent pasture: 41.1% (2011 est.)

forest: 31.4% (2011 est.)

other: 25.4% (2011 est.)

Irrigated land: 7,210 sq km (2012)

Population distribution: over three-quarters of New Zealanders, including the indigenous Maori, live on the North Island, primarily in urban areas

Natural hazards: earthquakes are common, though usually not severe; volcanic activity

volcanism: significant volcanism on North Island; Ruapehu (2,797 m), which last erupted in 2007, has a history of large eruptions in the past century; Taranaki has the potential to produce dangerous avalanches and lahars; other historically active volcanoes include Okataina, Raoul Island, Tongariro, and White Island; see note 2 under "Geography—note"

Environment—current issues: water quality and availability; rapid urbanisation; deforestation; soil erosion and degradation; native flora and fauna hard-hit by invasive species; negative effects of climate change

Environment—international agreements: *party to:* Antarctic-Environmental Protocol, Antarctic-Marine Living Resources, Antarctic Treaty, Biodiversity, Climate Change, Climate Change-Kyoto Protocol, Desertification, Endangered Species, Environmental Modification, Hazardous Wastes, Law of the Sea, Marine Dumping, Ozone Layer Protection, Ship Pollution, Tropical Timber 83, Tropical Timber 94, Wetlands, Whaling

signed, but not ratified: Antarctic Seals, Marine Life Conservation

Geography—note: *note 1:* consists of two main islands and a number of smaller islands; South Island, the larger main island, is the 12th largest island in the world and is divided along its length by the Southern Alps; North Island is the 14th largest island in the world and is not as mountainous, but it is marked by volcanism

note 2: New Zealand lies along the Ring of Fire, a belt of active volcanoes and earthquake epicenters bordering the Pacific Ocean; up to 90% of the world's earthquakes and some 75% of the world's volcanoes occur within the Ring of Fire

note 3: almost 90% of the population lives in cities and over three-quarters on North Island; Wellington is the southernmost national capital in the world

PEOPLE AND SOCIETY

Population: 4,925,477 (July 2020 est.)

country comparison to the world: 125

Nationality: *noun:* New Zealander(s)

adjective: New Zealand

Ethnic groups: European 64.1%, Maori 16.5%, Chinese 4.9%, Indian 4.7%, Samoan 3.9%, Tongan 1.8%, Cook Islands Maori 1.7%, English 1.5%, Filipino 1.5%, New Zealander 1%, other 13.7% (2018 est.)

note: based on the 2018 census of the usually resident population; percentages add up to more than 100% because respondents were able to identify more than one ethnic group

Languages: English (de facto official) 95.4%, Maori (de jure official) 4%, Samoan 2.2%, Northern Chinese 2%, Hindi 1.5%, French 1.2%, Yue 1.1%, New Zealand Sign Language (de jure official) .5%, other or not stated 17.2% (2018 est.)

note: shares sum to 124.1% due to multiple responses on the 2018 census

Religions: Christian 37.3% (Catholic 10.1%, Anglican 6.8%, Presbyterian and Congregational 5.2%, Pentecostal 1.8%, Methodist 1.6%, Mormon 1.2%, other 10.7%), Hindu 2.7%, Maori 1.3%, Muslim, 1.3%, Buddhist 1.1%, other religion 1.6% (includes Judaism, Spiritualism and New Age religions, Baha'i, Asian religions other than Buddhism), no religion 48.6%, objected to answering 6.7% (2018 est.)

note: based on the 2018 census of the usually resident population; percentages add up to more than 100% because respondents were able to identify more than one religion

Age structure: *0-14 years:* 19.63% (male 496,802/ female 469,853)

INTRODUCTION

Background: The Polynesian Maori reached New Zealand sometime between A.D. 1250 and 1300. In 1840, their chieftains entered into a compact with Great Britain, the Treaty of Waitangi, in which they ceded sovereignty to Queen Victoria while retaining territorial rights. That same year, the British began the first organized colonial settlement. A series of land wars between 1843 and 1872 ended with the defeat of the native peoples. The British colony of New Zealand became an independent dominion in 1907 and supported the UK militarily in both world wars. New Zealand's full participation in a number of defense alliances lapsed by the 1980s. In recent years, the government has sought to address longstanding Maori grievances.

GEOGRAPHY

Location: Oceania, islands in the South Pacific Ocean, southeast of Australia

Geographic coordinates: 41 00 S, 174 00 E

Map references: Oceania

Area: *total:* 268,838 sq km

15-24 years: 12.92% (male 328,327/female 308,132)

25-54 years: 39.98% (male 996,857/female 972,566)

55-64 years: 11.93% (male 285,989/female 301,692)

65 years and over: 15.54% (male 358,228/female 407,031) (2020 est.)

Dependency ratios: *total dependency ratio:* 55.8
youth dependency ratio: 30.3
elderly dependency ratio: 25.5
potential support ratio: 3.9 (2020 est.)

Median age: *total:* 37.2 years
male: 36.4 years
female: 37.9 years (2020 est.)
country comparison to the world: 73

Population growth rate: 1.44% (2020 est.)
country comparison to the world: 77

Birth rate: 12.8 births/1,000 population (2020 est.)
country comparison to the world: 147

Death rate: 6.9 deaths/1,000 population (2020 est.)
country comparison to the world: 133

Net migration rate: 8 migrant(s)/1,000 population (2020 est.)
country comparison to the world: 12

Population distribution: over three-quarters of New Zealanders, including the indigenous Maori, live on the North Island, primarily in urban areas

Urbanization: *urban population:* 86.7% of total population (2020)
rate of urbanization: 1.01% annual rate of change (2015-20 est.)
total population growth rate v. urban population growth rate, 2000-2030:

Major urban areas—population: 1.607 million Auckland, 415,000 WELLINGTON (capital) (2020)

Sex ratio: *at birth:* 1.05 male(s)/female
0-14 years: 1.06 male(s)/female
15-24 years: 1.07 male(s)/female
25-54 years: 1.02 male(s)/female
55-64 years: 0.95 male(s)/female
65 years and over: 0.88 male(s)/female
total population: 1 male(s)/female (2020 est.)

Mother's mean age at first birth: 27.8 years (2009 est.)
note: median age at first birth

Maternal mortality rate: 9 deaths/100,000 live births (2017 est.)
country comparison to the world: 148

Infant mortality rate: *total:* 3.5 deaths/1,000 live births
male: 3.7 deaths/1,000 live births
female: 3.4 deaths/1,000 live births (2020 est.)
country comparison to the world: 200

Life expectancy at birth: *total population:* 82.1 years
male: 80.4 years
female: 84 years (2020 est.)
country comparison to the world: 22

Total fertility rate: 1.87 children born/woman (2020 est.)

country comparison to the world: 137

Contraceptive prevalence rate: 79.9% (2014/15)
note: percent of women aged 18-45

Drinking water source:
improved:
urban: 100% of population
rural: 100% of population
total: 100% of population
unimproved:
urban: 0% of population
rural: 0% of population
total: 0% of population (2017 est.)

Current Health Expenditure: 9.2% (2017)

Physicians density: 3.47 physicians/1,000 population (2017)

Hospital bed density: 2.7 beds/1,000 population (2017)

Sanitation facility access:
improved:
urban: 100% of population
rural: 100% of population
total: 100% of population
unimproved:
urban: 0% of population
rural: 0% of population
total: 0% of population (2017)

HIV/AIDS—adult prevalence rate: <.1% (2019 est.)

HIV/AIDS—people living with HIV/AIDS: 3,500 (2019 est.)
country comparison to the world: 129

HIV/AIDS—deaths: <100 (2019 est.)

Obesity—adult prevalence rate: 30.8% (2016)
country comparison to the world: 22

Education expenditures: 6.4% of GDP (2016)
country comparison to the world: 24

School life expectancy (primary to tertiary education): *total:* 19 years
male: 18 years
female: 20 years (2018)

Unemployment, youth ages 15-24: *total:* 11.5%
male: 12.3%
female: 10.7% (2018 est.)
country comparison to the world: 114

GOVERNMENT

Country name: *conventional long form:* none
conventional short form: New Zealand
abbreviation: NZ
etymology: Dutch explorer Abel TASMAN was the first European to reach New Zealand in 1642; he named it Staten Landt, but Dutch cartographers renamed it Nova Zeelandia in 1645 after the Dutch province of Zeeland; British explorer Captain James COOK subsequently anglicized the name to New Zealand when he mapped the islands in 1769

Government type: parliamentary democracy under a constitutional monarchy; a Commonwealth realm

Capital: *name:* Wellington

geographic coordinates: 41 18 S, 174 47 E
time difference: UTC+12 (17 hours ahead of Washington, DC, during Standard Time)

daylight saving time: +1hr, begins last Sunday in September; ends first Sunday in April
note: New Zealand has two time zones: New Zealand standard time (UTC+12) and Chatham Islands time (45 minutes in advance of New Zealand standard time; UTC+12:45)
etymology: named in 1840 after Arthur Wellesley, the first Duke of Wellington and victorious general at the Battle of Waterloo

Administrative divisions: 16 regions and 1 territory*; Auckland, Bay of Plenty, Canterbury, Chatham Islands*, Gisborne, Hawke's Bay, Manawatu- Wanganui, Marlborough, Nelson, Northland, Otago, Southland, Taranaki, Tasman, Waikato, Wellington, West Coast

Dependent areas: Cook Islands, Niue, Tokelau

Independence: 26 September 1907 (from the UK)

National holiday: Waitangi Day (Treaty of Waitangi established British sovereignty over New Zealand), 6 February (1840); Anzac Day (commemorated as the anniversary of the landing of troops of the Australian and New Zealand Army Corps during World War I at Gallipoli, Turkey), 25 April (1915)

Constitution: *history:* New Zealand has no single constitution document; the Constitution Act 1986, effective 1 January 1987, includes only part of the uncodified constitution; others include a collection of statutes or "acts of Parliament," the Treaty of Waitangi, Orders in Council, letters patent, court decisions, and unwritten conventions
amendments: proposed as bill by Parliament or by referendum called either by the government or by citizens; passage of a bill as an act normally requires two separate readings with committee reviews in between to make changes and corrections, a third reading approved by the House of Representatives membership or by the majority of votes in a referendum, and assent of the governor-general; passage of amendments to reserved constitutional provisions affecting the term of Parliament, electoral districts, and voting restrictions requires approval by 75% of the House membership or the majority of votes in a referendum; amended many times, last in 2014

Legal system: common law system, based on English model, with special legislation and land courts for the Maori

International law organization participation: accepts compulsory ICJ jurisdiction with reservations; accepts ICCt jurisdiction

Citizenship: *citizenship by birth:* no
citizenship by descent only: at least one parent must be a citizen of New Zealand
dual citizenship recognized: yes
residency requirement for naturalization: 3 years

Suffrage: 18 years of age; universal

Executive branch: *chief of state:* Queen ELIZABETH II (since 6 February 1952);

represented by Governor-General Dame Patricia Lee REDDY (since 28 September 2016)

head of government: Prime Minister Jacinda ARDERN (since 26 October 2017); Deputy Prime Minister Grant ROBERTSON (since 2 November 2020)

cabinet: Executive Council appointed by the governor-general on the recommendation of the prime minister

elections/appointments: the monarchy is hereditary; governor-general appointed by the monarch on the advice of the prime minister; following legislative elections, the leader of the majority party or majority coalition usually appointed prime minister by the governor-general; deputy prime minister appointed by the governor-general; note—Prime Minister ARDERN heads up a minority coalition government consisting of the Labor and New Zealand First parties with confidence and supply support from the Green Party

Legislative branch: *description:* unicameral House of Representatives—commonly called Parliament (120 seats for 2020-23 term); 72 members directly elected in 65 single-seat constituencies and 7 Maori constituencies by simple majority vote and 48 directly elected by closed party-list proportional representation vote; members serve 3-year terms)

elections: last held on 17 October 2020 (next scheduled for 2023)

election results: percent of vote by party—Labor Party 49.1%, National Party 26.8%, ACT Party 8%, Green Party 6.3%, Maori Party 1%; seats by party—Labor Party 64, National Party 35, Green Party 10, ACT Party 10, Maori Party 1; composition—men 63, women 57, percent of women 47.5%

Judicial branch: *highest courts:* Supreme Court (consists of 5 justices, including the chief justice); note—the Supreme Court in 2004 replaced the Judicial Committee of the Privy Council (in London) as the final appeals court

judge selection and term of office: justices appointed by the governor-general upon the recommendation of the attorney- general; justices appointed until compulsory retirement at age 70

subordinate courts: Court of Appeal; High Court; tribunals and authorities; district courts; specialized courts for issues related to employment, environment, family, Maori lands, youth, military; tribunals

Political parties and leaders: ACT New Zealand [David SEYMOUR]

Green Party [James SHAW]

Mana Movement [Hone HARAWIRA] (formerly Mana Party)

Maori Party [Che WILSON and Kaapua SMITH]

New Zealand First Party or NZ First [Winston PETERS]

New Zealand Labor Party [Jacinda ARDERN]

New Zealand National Party [Judith COLLINS]

United Future New Zealand [Damian LIGHT]

International organization participation: ADB, ANZUS, APEC, ARF, ASEAN (dialogue partner), Australia Group, BIS, C, CD, CP, EAS, EBRD, FAO, FATF, IAEA, IBRD, ICAO, ICC (national committees), ICCt, ICRM, IDA, IEA, IFAD, IFC, IFRCS, IHO, ILO, IMF, IMO, IMSO, Interpol, IOC, IOM, IPU, ISO, ITSO, ITU, ITUC (NGOs), MIGA, NSG, OECD, OPCW, Pacific Alliance (observer), Paris Club (associate), PCA, PIF, SICA (observer), Sparteca, SPC, UN, UNCTAD, UNESCO, UNHCR, UNIDO, UNMISS, UNTSO, UPU, WCO, WFTU (NGOs), WHO, WIPO, WMO, WTO

Diplomatic representation in the US: *chief of mission:* Ambassador Rosemary BANKS (since 11 January 2019)

chancery: 37 Observatory Circle NW, Washington, DC 20008

telephone: [1] (202) 328-4800

FAX: [1] (202) 667-5227

consulate(s) general: Honolulu (HI), Los Angeles, New York

Diplomatic representation from the US: *chief of mission:* Ambassador Scott P. BROWN (since 27 June 2017) note—also accredited to Samoa

telephone: [64] (4) 462-6000

embassy: 29 Fitzherbert Terrace, Thorndon, Wellington

mailing address: P. O. Box 1190, Wellington; PSC 467, Box 1, APO AP 96531-1034

FAX: [64] (4) 499-0490

consulate(s) general: Auckland

Flag description: blue with the flag of the UK in the upper hoist-side quadrant with four red five-pointed stars edged in white centered in the outer half of the flag; the stars represent the Southern Cross constellation

National symbol(s): *Southern Cross constellation (four, five-pointed stars), kiwi (bird), silver fern; national colors: black, white, red (ochre)*

National anthem: *name:* God Defend New Zealand

lyrics/music: Thomas BRACKEN [English], Thomas Henry SMITH [Maori]/John Joseph WOODS

note: adopted 1940 as national song, adopted 1977 as co-national anthem; New Zealand has two national anthems with equal status; as a commonwealth realm, in addition to "God Defend New Zealand," "God Save the Queen" serves as a national anthem (see United Kingdom); "God Save the Queen" normally played only when a member of the royal family or the governor-general is present; in all other cases, "God Defend New Zealand" is played

0:00/ 1:02

ECONOMY

Economy—overview: Over the past 40 years, the government has transformed New Zealand from an agrarian economy, dependent on concessionary British market access, to a more industrialized, free market economy that can compete globally. This dynamic growth has boosted real incomes, but left behind some at the bottom of the ladder and broadened and deepened the technological capabilities of the industrial sector.

Per capita income rose for 10 consecutive years until 2007 in purchasing power parity terms, but fell in 2008-09. Debt- driven consumer spending drove robust growth in the first half of the decade, fueling a large balance of payments deficit that posed a challenge for policymakers. Inflationary pressures caused the central bank to raise its key rate steadily from January 2004 until it was among the highest in the OECD in 2007 and 2008. The higher rate attracted international capital inflows, which strengthened the currency and housing market while aggravating the current account deficit. Rising house prices, especially in Auckland, have become a political issue in recent years, as well as a policy challenge in 2016 and 2017, as the ability to afford housing has declined for many.

Expanding New Zealand's network of free trade agreements remains a top foreign policy priority. New Zealand was an early promoter of the Trans-Pacific Partnership (TPP) and was the second country to ratify the agreement in May 2017. Following the United States' withdrawal from the TPP in January 2017, on 10 November 2017 the remaining 11 countries agreed on the core elements of a modified agreement, which they renamed the Comprehensive and Progressive Agreement for Trans-Pacific Partnership (CPTPP). In November 2016, New Zealand opened negotiations to upgrade its FTA with China; China is one of New Zealand's most important trading partners.

GDP (purchasing power parity):
$189 billion (2017 est.)
$183.4 billion (2016 est.)
$176.1 billion (2015 est.)
note: data are in 2017 dollars
country comparison to the world: 68

GDP (official exchange rate):
$201.4 billion (2017 est.)

GDP—real growth rate: 2.22% (2019 est.)
3.22% (2018 est.)
3.8% (2017 est.)
country comparison to the world: 127

GDP—per capita (PPP): $39,000 (2017 est.)
$38,600 (2016 est.)
$37,900 (2015 est.)
note: data are in 2017 dollars
country comparison to the world: 48

Gross national saving:
21% of GDP (2017 est.)
21.5% of GDP (2016 est.)
20.2% of GDP (2015 est.)
country comparison to the world: 89

GDP—composition, by end use:
household consumption: 57.2% (2017 est.)
government consumption: 18.2% (2017 est.)
investment in fixed capital: 23.4% (2017 est.)
investment in inventories: 0.3% (2017 est.)
exports of goods and services: 27% (2017 est.)
imports of goods and services: -26.1% (2017 est.)

GDP—composition, by sector of origin:
agriculture: 5.7% (2017 est.)
industry: 21.5% (2017 est.)
services: 72.8% (2017 est.)

Agriculture—products: dairy products, sheep, beef, poultry, fruit, vegetables, wine, seafood, wheat and barley

Industries: agriculture, forestry, fishing, logs and wood articles, manufacturing, mining, construction, financial services, real estate services, tourism

Industrial production growth rate: 1.8% (2017 est.)
country comparison to the world: 137

Labor force: 2.709 million (2020 est.)
country comparison to the world: 109

Labor force—by occupation: *agriculture:* 6.6%
industry: 20.7%
services: 72.7% (2017 est.)

Unemployment rate: 4.13% (2019 est.)
4.32% (2018 est.)
country comparison to the world: 62

Population below poverty line: NA

Household income or consumption by percentage share: *lowest 10%:* NA
highest 10%: NA

Budget: *revenues:* 74.11 billion (2017 est.)
expenditures: 70.97 billion (2017 est.)

Taxes and other revenues: 36.8% (of GDP) (2017 est.)
country comparison to the world: 56

Budget surplus (+) or deficit (-): 1.6% (of GDP) (2017 est.)
country comparison to the world: 21

Public debt: 31.7% of GDP (2017 est.)
33.5% of GDP (2016 est.)
country comparison to the world: 162

Fiscal year: 1 April—31 March
note: data to the fiscal year for tax purposes

Inflation rate (consumer prices): 1.9% (2017 est.)
0.6% (2016 est.)
country comparison to the world: 98

Current account balance: -$6.962 billion (2019 est.)
-$8.742 billion (2018 est.)
country comparison to the world: 187

Exports: $37.35 billion (2017 est.)
$33.61 billion (2016 est.)
country comparison to the world: 56

Exports—partners: China 22.4%, Australia 16.4%, US 9.9%, Japan 6.1% (2017)

Exports—commodities: dairy products, meat and edible offal, logs and wood articles, fruit, crude oil, wine

Imports: $39.74 billion (2017 est.)
$35.53 billion (2016 est.)
country comparison to the world: 58

Imports—commodities: petroleum and products, mechanical machinery, vehicles and parts, electrical machinery, textiles

Imports—partners: China 19%, Australia 12.1%, US 10.5%, Japan 7.3%, Germany 5.3%, Thailand 4.6% (2017)

Reserves of foreign exchange and gold: $20.68 billion (31 December 2017 est.)
$17.81 billion (31 December 2016 est.)
country comparison to the world: 58

Debt—external: $91.62 billion (31 December 2017 est.)
$84.03 billion (31 December 2016 est.)
country comparison to the world: 51

Exchange rates: New Zealand dollars (NZD) per US dollar—
1.416 (2017 est.)
1.4341 (2016 est.)
1.4341 (2015 est.)
1.4279 (2014 est.)
1.2039 (2013 est.)

ENERGY

Electricity access: *electrification—total population:* 100% (2020)

Electricity—production: 42.53 billion kWh (2016 est.)
country comparison to the world: 56

Electricity—consumption: 39.5 billion kWh (2016 est.)
country comparison to the world: 55

Electricity—exports: 0 kWh (2016 est.)
country comparison to the world: 176

Electricity—imports: 0 kWh (2016 est.)
country comparison to the world: 179

Electricity—installed generating capacity: 9.301 million kW (2016 est.)
country comparison to the world: 63

Electricity—from fossil fuels: 23% of total installed capacity (2016 est.)
country comparison to the world: 193

Electricity—from nuclear fuels: 0% of total installed capacity (2017 est.)
country comparison to the world: 154

Electricity—from hydroelectric plants: 58% of total installed capacity (2017 est.)
country comparison to the world: 29

Electricity—from other renewable sources: 20% of total installed capacity (2017 est.)
country comparison to the world: 40

Crude oil—production: 24,000 bbl/day (2018 est.)
country comparison to the world: 64

Crude oil—exports: 26,440 bbl/day (2017 est.)
country comparison to the world: 48

Crude oil—imports: 108,900 bbl/day (2017 est.)
country comparison to the world: 43

Crude oil—proved reserves: 51.8 million bbl (1 January 2018 est.)
country comparison to the world: 76

Refined petroleum products—production: 115,100 bbl/day (2017 est.)
country comparison to the world: 65

Refined petroleum products—consumption: 169,100 bbl/day (2017 est.)
country comparison to the world: 61

Refined petroleum products—exports: 1,782 bbl/day (2017 est.)
country comparison to the world: 106

Refined petroleum products—imports: 56,000 bbl/day (2017 est.)
country comparison to the world: 76

Natural gas—production: 5.097 billion cu m (2017 est.)
country comparison to the world: 51

Natural gas—consumption: 5.182 billion cu m (2017 est.)
country comparison to the world: 57

Natural gas—exports: 0 cu m (2017 est.)
country comparison to the world: 161

Natural gas—imports: 0 cu m (2017 est.)
country comparison to the world: 166

Natural gas—proved reserves: 33.7 billion cu m (1 January 2018 est.)
country comparison to the world: 67

Carbon dioxide emissions from consumption of energy: 37.75 million Mt (2017 est.)
country comparison to the world: 68

COMMUNICATIONS

Telephones—fixed lines: *total subscriptions:* 1,801,645
subscriptions per 100 inhabitants: 37.11 (2019 est.)
country comparison to the world: 60

Telephones—mobile cellular: *total subscriptions:* 6,550,687
subscriptions per 100 inhabitants: 134.93 (2019 est.)
country comparison to the world: 108

Telecommunication systems: *general assessment:* excellent domestic and international systems; mobile and P2P services soar; LTE rates some of the fastest in the world; growth in mobile broadband and fiber sectors; roll out of 5G; investment and development of infrastructure enable network capabilities to propel the digital economy; digital media sector along with e-government, ecommerce across the country; newest and most powerful commercial satellite, Kacific-1 satellite, launched in 2019 to improve telecommunications in the Asia Pacific region (2020)
domestic: fixed-line 37 per 100 and mobile-cellular telephone subscribership 135 per 100 persons (2019)
international: country code—64; landing points for the Southern Cross NEXT, Aqualink, Nelson-Levin, SCCN and Hawaiki submarine cable system providing links to Australia, Fiji, American Samoa, Kiribati, Samo, Tokelau, US and around New Zealand; satellite earth stations—8 (1 Inmarsat—Pacific Ocean, 7 other) (2019)
note: the COVID-19 outbreak is negatively impacting telecommunications production and supply chains globally; consumer spending on telecom devices and services has also slowed due to the pandemic's effect on economies worldwide; overall progress towards improvements in all facets of the telecom industry—mobile, fixed-line, broadband, submarine cable and satellite—has moderated

Broadcast media: state-owned Television New Zealand operates multiple TV networks and state-owned Radio New Zealand operates 3 radio networks and an external shortwave radio service to the South Pacific region; a small number of national commercial TV and radio stations and many regional commercial television and radio stations are available; cable and satellite TV systems are available, as are a range of streaming services (2019)

Internet country code: .nz

Internet users: total: 4,340,672

percent of population: 90.81% (July 2018 est.)
country comparison to the world: 91

Broadband—fixed subscriptions: total: 1.647 million
subscriptions per 100 inhabitants: 34 (2018 est.)
country comparison to the world: 59

TRANSPORTATION

National air transport system: *number of registered air carriers:* 15 (2020)
inventory of registered aircraft operated by air carriers: 199
annual passenger traffic on registered air carriers: 17,249,049 (2018)
annual freight traffic on registered air carriers: 1,349,300,000 mt-km (2018)

Civil aircraft registration country code prefix: ZK (2016)

Airports: 123 (2013)
country comparison to the world: 47

Airports—with paved runways: total: 39 (2017)
over 3,047 m: 2 (2017)
2,438 to 3,047 m: 1 (2017)

1,524 to 2,437 m: 12 (2017)
914 to 1,523 m: 23 (2017)
under 914 m: 1 (2017)

Airports—with unpaved runways: *total:* 84 (2013)
1,524 to 2,437 m: 3 (2013)
914 to 1,523 m: 33 (2013)
under 914 m: 48 (2013)

Pipelines: 331 km condensate, 2500 km gas, 172 km liquid petroleum gas, 288 km oil, 198 km refined products (2018)

Railways: total: 4,128 km (2018)
narrow gauge: 4,128 km 1.067-m gauge (506 km electrified) (2018)
country comparison to the world: 46

Roadways: total: 94,000 km (2017)
paved: 61,600 km (includes 199 km of expressways) (2017)
unpaved: 32,400 km (2017)
country comparison to the world: 53

Merchant marine: total: 113
by type: general cargo 12, oil tanker 4, other 97 (2019)
country comparison to the world: 84

Ports and terminals: *major seaport(s):* Auckland, Lyttelton, Manukau Harbor, Marsden Point, Tauranga, Wellington

---

MILITARY AND SECURITY

Military and security forces: *New Zealand Defence Force (NZDF):* New Zealand Army,

Royal New Zealand Navy, Royal New Zealand Air Force (2020)

Military expenditures: 1.5% of GDP (2019)
1.3% of GDP (2018)
1.2% of GDP (2017)
1.2% of GDP (2016)
1.2% of GDP (2015)
country comparison to the world: 83

Military and security service personnel strengths: the New Zealand Defense Force (NZDF) has about 9,600 active duty troops (4,700 Army; 2,300 Navy; 2,600 Air Force) (2020)

Military equipment inventories and acquisitions: NZDF is equipped mostly with imported weapons and equipment from Western suppliers; Australia, France, and the US are the leading suppliers since 2010 (2019 est.)

Military deployments: up to 220 Antarctica (summer season only) (2020)

Military service age and obligation: 17 years of age for voluntary military service; soldiers cannot be deployed until the age of 18; no conscription (2019)

TRANSNATIONAL ISSUES

Disputes—international: asserts a territorial claim in Antarctica (Ross Dependency)

Illicit drugs: significant consumer of amphetamines

NICARAGUA

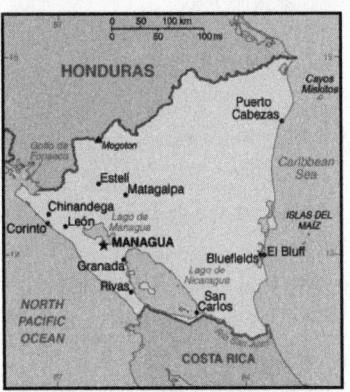

INTRODUCTION

Background: The Pacific coast of Nicaragua was settled as a Spanish colony from Panama in the early 16th century. Independence from Spain was declared in 1821 and the country became an independent republic in 1838. Britain occupied the Caribbean Coast in the first half of the 19th century, but gradually ceded control of the region in

subsequent decades. Violent opposition to governmental manipulation and corruption spread to all classes by 1978 and resulted in a short-lived civil war that brought a civic-military coalition, spearheaded by the Marxist Sandinista guerrillas led by Daniel ORTEGA Saavedra to power in 1979. Nicaraguan aid to leftist rebels in El Salvador prompted the US to sponsor anti-Sandinista contra guerrillas through much of the 1980s. After losing free and fair elections in 1990, 1996, and 2001, former Sandinista President Daniel ORTEGA was elected president in 2006, 2011, and most recently in 2016. Municipal, regional, and national-level elections since 2008 have been marred by widespread irregularities. Democratic institutions have weakened under the ORTEGA administration as the president has garnered full control over all branches of government, especially after cracking down on a nationwide antigovernment protest movement in 2018.

GEOGRAPHY

Location: Central America, bordering both the Caribbean Sea and the North Pacific Ocean, between Costa Rica and Honduras

Geographic coordinates: 13 00 N, 85 00 W

Map references: Central America and the Caribbean

Area: *total:* 130,370 sq km
land: 119,990 sq km
water: 10,380 sq km
country comparison to the world: 99

Area—comparative: slightly larger than Pennsylvania; slightly smaller than New York state

Land boundaries: *total:* 1,253 km
border countries (2): Costa Rica 313 km, Honduras 940 km

Coastline: 910 km

Maritime claims: *territorial sea:* 12 nm
contiguous zone: 24 nm
continental shelf: natural prolongation

Climate: tropical in lowlands, cooler in highlands

Terrain: extensive Atlantic coastal plains rising to central interior mountains; narrow Pacific coastal plain interrupted by volcanoes

Elevation: *mean elevation:* 298 m
lowest point: Pacific Ocean 0 m
highest point: Mogoton 2,085 m

Natural resources: gold, silver, copper, tungsten, lead, zinc, timber, fish

Land use: *agricultural land:* 42.2% (2011 est.)

arable land: 12.5% (2011 est.) / permanent crops: 2.5% (2011 est.) / permanent pasture: 27.2% (2011 est.)

forest: 25.3% (2011 est.)

other: 32.5% (2011 est.)

Irrigated land: 1,990 sq km (2012)

Population distribution: the overwhelming majority of the population resides in the western half of the country, with much of the urban growth centered in the capital city of Managua; coastal areas also show large population clusters

Natural hazards: destructive earthquakes; volcanoes; landslides; extremely susceptible to hurricanes

volcanism: significant volcanic activity; Cerro Negro (728 m), which last erupted in 1999, is one of Nicaragua's most active volcanoes; its lava flows and ash have been known to cause significant damage to farmland and buildings; other historically active volcanoes include Concepcion, Cosiguina, Las Pilas, Masaya, Momotombo, San Cristobal, and Telica

Environment—current issues: deforestation; soil erosion; water pollution; drought

Environment—international agreements: *party to:* Biodiversity, Climate Change, Climate Change-Kyoto Protocol, Desertification, Endangered Species, Environmental Modification, Hazardous Wastes, Law of the Sea, Ozone Layer Protection, Ship Pollution, Wetlands, Whaling

signed, but not ratified: none of the selected agreements

Geography—note: largest country in Central America; contains the largest freshwater body in Central America, Lago de Nicaragua

PEOPLE AND SOCIETY

Population: 6,203,441 (July 2020 est.)
country comparison to the world: 111

Nationality: *noun:* Nicaraguan(s)
adjective: Nicaraguan

Ethnic groups: mestizo (mixed Amerindian and white) 69%, white 17%, black 9%, Amerindian 5%

Languages: Spanish (official) 95.3%, Miskito 2.2%, Mestizo of the Caribbean coast 2%, other 0.5% (2005 est.)
note: English and indigenous languages found on the Caribbean coast

Religions: Roman Catholic 50%, Evangelical 33.2%, other 2.9%, unspecified 13.2%, none 0.7% (2017 est.)

Demographic profile: Despite being one of the poorest countries in Latin America, Nicaragua has improved its access to potable water and sanitation and has ameliorated its life expectancy, infant and child mortality, and immunization rates. However, income distribution is very uneven, and the poor, agriculturalists, and indigenous people continue to have less access to healthcare services. Nicaragua's total fertility rate has fallen from around 6 children per woman in 1980 to below replacement level today, but the high birth rate among adolescents

perpetuates a cycle of poverty and low educational attainment.

Nicaraguans emigrate primarily to Costa Rica and to a lesser extent the United States. Nicaraguan men have been migrating seasonally to Costa Rica to harvest bananas and coffee since the early 20th century. Political turmoil, civil war, and natural disasters from the 1970s through the 1990s dramatically increased the flow of refugees and permanent migrants seeking jobs, higher wages, and better social and healthcare benefits. Since 2000, Nicaraguan emigration to Costa Rica has slowed and stabilized. Today roughly 300,000 Nicaraguans are permanent residents of Costa Rica—about 75% of the foreign population—and thousands more migrate seasonally for work, many illegally.

Age structure: *0-14 years:* 25.63% (male 811,731/female 777,984)

15-24 years: 19.51% (male 609,962/female 600,567)

25-54 years: 42.41% (male 1,254,683/female 1,376,052)

55-64 years: 6.63% (male 188,591/female 222,766)

65 years and over: 5.82% (male 159,140/female 201,965) (2020 est.)

Dependency ratios: *total dependency ratio:* 54.3
youth dependency ratio: 45.5
elderly dependency ratio: 8.8
potential support ratio: 11.4 (2020 est.)

Median age: *total:* 27.3 years
male: 26.4 years
female: 28.2 years (2020 est.)
country comparison to the world: 148

Population growth rate: 0.96% (2020 est.)
country comparison to the world: 111

Birth rate: 17.1 births/1,000 population (2020 est.)
country comparison to the world: 97

Death rate: 5.2 deaths/1,000 population (2020 est.)
country comparison to the world: 196

Net migration rate: -2.4 migrant(s)/1,000 population (2020 est.)
country comparison to the world: 170

Population distribution: the overwhelming majority of the population resides in the western half of the country, with much of the urban growth centered in the capital city of Managua; coastal areas also show large population clusters

Urbanization: *urban population:* 59% of total population (2020)
rate of urbanization: 1.45% annual rate of change (2015-20 est.)
total population growth rate v. urban population growth rate, 2000-2030: Major urban areas—population: 1.064 million MANAGUA (capital) (2020)

Sex ratio: *at birth:* 1.05 male(s)/female
0-14 years: 1.04 male(s)/female
15-24 years: 1.02 male(s)/female
25-54 years: 0.91 male(s)/female
55-64 years: 0.85 male(s)/female
65 years and over: 0.79 male(s)/female
total population: 0.95 male(s)/female (2020 est.)

Mother's mean age at first birth: 19.7 years (2011/12 est.)
note: median age at first birth among women 25-29

Maternal mortality rate: 198 deaths/100,000 live births (2017 est.)
country comparison to the world: 47

Infant mortality rate: *total:* 16.5 deaths/1,000 live births
male: 19 deaths/1,000 live births
female: 13.9 deaths/1,000 live births (2020 est.)
country comparison to the world: 91

Life expectancy at birth: *total population:* 74.2 years
male: 72 years
female: 76.6 years (2020 est.)
country comparison to the world: 134

Total fertility rate: 1.82 children born/woman (2020 est.)
country comparison to the world: 149

Contraceptive prevalence rate: 80.4% (2011/12)

Drinking water source:
improved:
urban: 97.6% of population
rural: 62.6% of population
total: 83.1% of population
unimproved:
urban: 2.4% of population
rural: 37.4% of population
total: 16.9% of population (2017 est.)

Current Health Expenditure: 8.6% (2017)

Physicians density: 1.01 physicians/1,000 population (2018)

Hospital bed density: 0.9 beds/1,000 population (2017)

Sanitation facility access:
improved:
urban: 89.8% of population
rural: 66.5% of population
total: 80.1% of population
unimproved:
urban: 10.2% of population
rural: 33.5% of population
total: 19.9% of population (2017 est.)

HIV/AIDS—adult prevalence rate: 0.2% (2019 est.)
country comparison to the world: 108

HIV/AIDS—people living with HIV/AIDS: 9,600 (2019 est.)
country comparison to the world: 105

HIV/AIDS—deaths: <200 (2019 est.)

Major infectious diseases: *degree of risk:* high (2020)
food or waterborne diseases: bacterial diarrhea, hepatitis A, and typhoid fever
vectorborne diseases: dengue fever and malaria

Obesity—adult prevalence rate: 23.7% (2016)
country comparison to the world: 63

Children under the age of 5 years underweight: 4.6% (2012)
country comparison to the world: 86

Education expenditures: 4.3% of GDP (2017)

country comparison to the world: 90

Literacy: *definition:* age 15 and over can read and write

total population: 82.6%

male: 82.4%

female: 82.8% (2015)

Unemployment, youth ages 15-24: *total:* 8.5%

male: 6.4%

female: 12.9% (2014 est.)

country comparison to the world: 141

GOVERNMENT

Country name: *conventional long form:* Republic of Nicaragua

conventional short form: Nicaragua

local long form: Republica de Nicaragua

local short form: Nicaragua

etymology: Nicarao was the name of the largest indigenous settlement at the time of Spanish arrival; conquistador Gil GONZALEZ Davila, who explored the area (1622-23), combined the name of the community with the Spanish word "agua" (water), referring to the two large lakes in the west of the country (Lake Managua and Lake Nicaragua)

Government type: presidential republic

Capital: *name:* Managua

geographic coordinates: 12 08 N, 86 15 W

time difference: UTC-6 (1 hour behind Washington, DC, during Standard Time)

etymology: may derive from the indigenous Nahuatl term "mana-ahuac," which translates as "adjacent to the water" or a site "surrounded by water"; the city is situated on the southwestern shore of Lake Managua

Administrative divisions: 15 departments (departamentos, singular—departamento) and 2 autonomous regions* (regiones autonomistas, singular—region autonoma); Boaco, Carazo, Chinandega, Chontales, Costa Caribe Norte*, Costa Caribe Sur*, Esteli, Granada, Jinotega, Leon, Madriz, Managua, Masaya, Matagalpa, Nueva Segovia, Rio San Juan, Rivas

Independence: 15 September 1821 (from Spain)

National holiday: Independence Day, 15 September (1821)

Constitution: *history:* several previous; latest adopted 19 November 1986, effective 9 January 1987

amendments: proposed by the president of the republic or assent of at least half of the National Assembly membership; passage requires approval by 60% of the membership of the next elected Assembly and promulgation by the president of the republic; amended several times, last in 2014

Legal system: civil law system; Supreme Court may review administrative acts

International law organization participation: accepts compulsory ICJ jurisdiction with reservations; non-party state to the ICCt

Citizenship: *citizenship by birth:* yes

citizenship by descent only: yes

dual citizenship recognized: no, except in cases where bilateral agreements exist

residency requirement for naturalization: 4 years

Suffrage: 16 years of age; universal

Executive branch: *chief of state:* President Jose Daniel ORTEGA Saavedra (since 10 January 2007); Vice President Rosario MURILLO Zambrana (since 10 January 2017); note—the president is both chief of state and head of government

head of government: President Jose Daniel ORTEGA Saavedra (since 10 January 2007); Vice President Rosario MURILLO Zambrana (since 10 January 2017)

cabinet: Council of Ministers appointed by the president

elections/appointments: president and vice president directly elected on the same ballot by qualified plurality vote for a 5-year term (no term limits as of 2014); election last held on 6 November 2016 (next to be held by November 2021)

election results: Jose Daniel ORTEGA Saavedra reelected president; percent of vote—Jose Daniel ORTEGA Saavedra (FSLN) 72.4%, Maximino RODRIGUEZ (PLC) 15%, Jose del Carmen ALVARADO (PLI) 4.5%, Saturnino CERRATO Hodgson (ALN) 4.3%, other 3.7%

Legislative branch: *description:* unicameral National Assembly or Asamblea Nacional (92 seats; 70 members in multi-seat constituencies and 20 members in a single nationwide constituency directly elected by proportional representation vote; 2 seats reserved for the previous president and the runner-up candidate in the previous presidential election; members serve 5-year terms)

elections: last held on 6 November 2016 (next to be held by November 2021)

election results: percent of vote by party—NA; seats by party—FSLN 71, PLC 14, ALN 2, PLI 2, APRE 1, PC 1, YATAMA 1; composition—men 50, women 42, percent of women 45.7%

Judicial branch: *highest courts:* Supreme Court or Corte Suprema de Justicia (consists of 16 judges organized into administrative, civil, criminal, and constitutional chambers)

judge selection and term of office: Supreme Court judges elected by the National Assembly to serve 5-year staggered terms

subordinate courts: Appeals Court; first instance civil, criminal, and labor courts; military courts are independent of the Supreme Court

Political parties and leaders: Alliance for the Republic or APRE [Carlos CANALES]

Conservative Party or PC [Alfredo CESAR]

Independent Liberal Party or PLI [Jose del Carmen ALVARADO]

Liberal Constitutionalist Party or PLC [Maria Haydee OSUNA]

Nicaraguan Liberal Alliance or ALN [Alejandro MEJIA Ferreti]

Sandinista National Liberation Front or FSLN [Jose Daniel ORTEGA Saavedra]

Sandinista Renovation Movement or MRS [Suyen BARAHONA]

Sons of Mother Earth or YATAMA [Brooklyn RIVERA]

International organization participation: BCIE, CACM, CD, CELAC, FAO, G-77, IADB, IAEA, IBRD, ICAO, ICRM, IDA, IFAD, IFC, IFRCS, ILO, IMF, IMO, Interpol, IOC, IOM, IPU, ISO (correspondent), ITSO, ITU, ITUC (NGOs), LAES, LAIA (observer), MIGA, NAM, OAS, OPANAL, OPCW, PCA, Petrocaribe, SICA, UN, UNCTAD, UNESCO, UNHCR, UNIDO, Union Latina, UNWTO, UPU, WCO, WHO, WIPO, WMO, WTO

Diplomatic representation in the US: *chief of mission:* Ambassador Francisco Obadiah CAMPBELL Hooker (since 28 June 2010)

chancery: 1627 New Hampshire Avenue NW, Washington, DC 20009

telephone: [1] (202) 939-6570, 6573

FAX: [1] (202) 939-6545

consulate(s) general: Houston, Los Angeles, Miami, New York, San Francisco

Diplomatic representation from the US: *chief of mission:* Ambassador Kevin K. SULLIVAN (since 18 December 2018)

telephone: [505] 2252-7100, 2252-7888; 2252-7100 or 8767-7100 (after hours)

embassy: Kilometer 5.5 Carretera Sur, Managua

mailing address: American Embassy Managua, APO AA 34021

FAX: [505] 2252-7250

Flag description: three equal horizontal bands of blue (top), white, and blue with the national coat of arms centered in the white band; the coat of arms features a triangle encircled by the words REPUBLICA DE NICARAGUA on the top and AMERICA CENTRAL on the bottom; the banner is based on the former blue-white-blue flag of the Federal Republic of Central America; the blue bands symbolize the Pacific Ocean and the Caribbean Sea, while the white band represents the land between the two bodies of water

note: similar to the flag of El Salvador, which features a round emblem encircled by the words REPUBLICA DE EL SALVADOR EN LA AMERICA CENTRAL centered in the white band; also similar to the flag of Honduras, which has five blue stars arranged in an X pattern centered in the white band

National symbol(s): *turquoise-browed motmot (bird); national colors:* blue, white

National anthem: *name:* "Salve a ti, Nicaragua" (Hail to Thee, Nicaragua)

lyrics/music: Salomon Ibarra MAYORGA/traditional, arranged by Luis Abraham DELGADILLO

note: although only officially adopted in 1971, the music was approved in 1918 and the lyrics in 1939; the tune, originally from Spain, was used as an anthem for Nicaragua from the 1830s until 1876

ECONOMY

Economy—overview: Nicaragua, the poorest country in Central America and the second poorest in the Western Hemisphere, has widespread underemployment and poverty. GDP growth of

4.5% in 2017 was insufficient to make a significant difference. Textiles and agriculture combined account for nearly 50% of Nicaragua's exports. Beef, coffee, and gold are Nicaragua's top three export commodities.

The Dominican Republic-Central America-United States Free Trade Agreement has been in effect since April 2006 and has expanded export opportunities for many Nicaraguan agricultural and manufactured goods.

In 2013, the government granted a 50-year concession with the option for an additional 50 years to a newly formed Chinese-run company to finance and build an inter-oceanic canal and related projects, at an estimated cost of $50 billion. The canal construction has not started.

GDP (purchasing power parity):
$36.4 billion (2017 est.)
$34.71 billion (2016 est.)
$33.17 billion (2015 est.)
note: data are in 2017 dollars
country comparison to the world: 124

GDP (official exchange rate):
$13.81 billion (2017 est.)

GDP—real growth rate: 4.9% (2017 est.)
4.7% (2016 est.)
4.8% (2015 est.)
country comparison to the world: 54

GDP—per capita (PPP): $5,900 (2017 est.)
$5,600 (2016 est.)
$5,500 (2015 est.)
note: data are in 2017 dollars
country comparison to the world: 165

Gross national saving:
24% of GDP (2017 est.)
23.2% of GDP (2016 est.)
23.6% of GDP (2015 est.)
country comparison to the world: 69

GDP—composition, by end use:
household consumption: 69.9% (2017 est.)
government consumption: 15.3% (2017 est.)
investment in fixed capital: 28.1% (2017 est.)
investment in inventories: 1.7% (2017 est.)
exports of goods and services: 41.2% (2017 est.)
imports of goods and services: -55.4% (2017 est.)

GDP—composition, by sector of origin:
agriculture: 15.5% (2017 est.)
industry: 24.4% (2017 est.)
services: 60% (2017 est.)

Agriculture—products: coffee, bananas, sugarcane, rice, corn, tobacco, cotton, sesame, soya, beans, beef, veal, pork, poultry, dairy products, shrimp, lobsters, peanuts

Industries: food processing, chemicals, machinery and metal products, knit and woven apparel, petroleum refining and distribution, beverages, footwear, wood, electric wire harness manufacturing, mining

Industrial production growth rate: 3.5% (2017 est.)
country comparison to the world: 88

Labor force: 3.046 million (2017 est.)
country comparison to the world: 103

Labor force—by occupation: *agriculture:* 31%

industry: 18%
services: 50% (2011 est.)

Unemployment rate: 6.4% (2017 est.)
6.2% (2016 est.)
note: underemployment was 46.5% in 2008
country comparison to the world: 95

Population below poverty line: 29.6% (2015 est.)

Household income or consumption by percentage share: *lowest 10%:* 1.8%
highest 10%: 47.1% (2014)

Budget: *revenues:* 3.871 billion (2017 est.)
expenditures: 4.15 billion (2017 est.)

Taxes and other revenues: 28% (of GDP) (2017 est.)
country comparison to the world: 97

Budget surplus (+) or deficit (-): -2% (of GDP) (2017 est.)
country comparison to the world: 106

Public debt: 33.3% of GDP (2017 est.)
31.2% of GDP (2016 est.)
note: official data; data cover general government debt and include debt instruments issued (or owned) by Government entities other than the treasury; the data include treasury debt held by foreign entities, as well as intragovernmental debt; intragovernmental debt consists of treasury borrowings from surpluses in the social funds, such as retirement, medical care, and unemployment, debt instruments for the social funds are not sold at public auctions; Nicaragua rebased its GDP figures in 2012, which reduced the figures for debt as a percentage of GDP
country comparison to the world: 157

Fiscal year: calendar year

Inflation rate (consumer prices): 3.9% (2017 est.)
3.5% (2016 est.)
country comparison to the world: 153

Current account balance: -$694 million (2017 est.)
-$989 million (2016 est.)
country comparison to the world: 127

Exports: $3.819 billion (2017 est.)
$3.772 billion (2016 est.)
country comparison to the world: 119

Exports—partners: US 44.2%, El Salvador 6.4%, Venezuela 5.5%, Costa Rica 5.5% (2017)

Exports—commodities: coffee, beef, gold, sugar, peanuts, shrimp and lobster, tobacco, cigars, automobile wiring harnesses, textiles, apparel

Imports: $6.613 billion (2017 est.)
$6.384 billion (2016 est.)
country comparison to the world: 116

Imports—commodities: consumer goods, machinery and equipment, raw materials, petroleum products

Imports—partners: US 20.8%, China 14.3%, Mexico 11.1%, Costa Rica 7.9%, Guatemala 7%, El Salvador 5.6% (2017)

Reserves of foreign exchange and gold: $2.758 billion (31 December 2017 est.)
$2.448 billion (31 December 2016 est.)
country comparison to the world: 113

Debt—external: $11.31 billion (31 December 2017 est.)
$10.87 billion (31 December 2016 est.)
country comparison to the world: 109

Exchange rates: cordobas (NIO) per US dollar—
30.11 (2017 est.)
28.678 (2016 est.)
28.678 (2015 est.)
27.257 (2014 est.)
26.01 (2013 est.)

ENERGY

Electricity access: *electrification—total population:* 81.8% (2016)
electrification—urban areas: 99.2% (2016)
electrification—rural areas: 56.6% (2016)

Electricity—production: 4.454 billion kWh (2016 est.)
country comparison to the world: 124

Electricity—consumption: 3.59 billion kWh (2016 est.)
country comparison to the world: 132

Electricity—exports: 17.87 million kWh (2016 est.)
country comparison to the world: 91

Electricity—imports: 205 million kWh (2016 est.)
country comparison to the world: 93

Electricity—installed generating capacity: 1.551 million kW (2016 est.)
country comparison to the world: 123

Electricity—from fossil fuels: 56% of total installed capacity (2016 est.)
country comparison to the world: 139

Electricity—from nuclear fuels: 0% of total installed capacity (2017 est.)
country comparison to the world: 155

Electricity—from hydroelectric plants: 9% of total installed capacity (2017 est.)
country comparison to the world: 120

Electricity—from other renewable sources: 35% of total installed capacity (2017 est.)
country comparison to the world: 9

Crude oil—production: 0 bbl/day (2018 est.)
country comparison to the world: 183

Crude oil—exports: 0 bbl/day (2015 est.)
country comparison to the world: 175

Crude oil—imports: 16,180 bbl/day (2015 est.)
country comparison to the world: 69

Crude oil—proved reserves: 0 bbl (1 January 2018 est.)
country comparison to the world: 178

Refined petroleum products—production: 14,720 bbl/day (2015 est.)
country comparison to the world: 95

Refined petroleum products—consumption: 37,000 bbl/day (2016 est.)
country comparison to the world: 115

Refined petroleum products—exports: 460 bbl/day (2015 est.)
country comparison to the world: 111

Refined petroleum products—imports: 20,120 bbl/day (2015 est.)
country comparison to the world: 121

Natural gas—production: 0 cu m (2017 est.)
country comparison to the world: 179

Natural gas—consumption: 0 cu m (2017 est.)
country comparison to the world: 183

Natural gas—exports: 0 cu m (2017 est.)
country comparison to the world: 162

Natural gas—imports: 0 cu m (2017 est.)
country comparison to the world: 167

Natural gas—proved reserves: 0 cu m (1 January 2015 est.)
country comparison to the world: 177

Carbon dioxide emissions from consumption of energy: 5.405 million Mt (2017 est.)
country comparison to the world: 132

COMMUNICATIONS

Telephones—fixed lines: *total subscriptions:* 215,055
subscriptions per 100 inhabitants: 3.5 (2019 est.)
country comparison to the world: 121

Telephones—mobile cellular: *total subscriptions:* 5,433,530
subscriptions per 100 inhabitants: 88.43 (2019 est.)
country comparison to the world: 116

Telecommunication systems: *general assessment:* system being upgraded by foreign investment; new canal being built between Pacific and Caribbean with Chinese funding; nearly all installed telecommunications capacity now uses digital technology, owing to investments since privatization of the formerly state-owned telecommunications company; lowest fixed-line teledensity and mobile penetration in Central America; Internet cafe's provide access to Internet and email services; telecom is bigger in the cities and marginal in rural area; liberalization slow; a Russian state corporation is operating in the area; LTE service in 60 towns and cities (2020)
domestic: since privatization, access to fixed-line and mobile-cellular services has improved; fixed-line teledensity roughly 4 per 100 persons; mobile-cellular telephone subscribership has increased to 88 per 100 persons (2019)
international: country code—505; landing point for the ARCOS fiber-optic submarine cable which provides connectivity to South and Central America, parts of the Caribbean, and the US; satellite earth stations—1 Intersputnik (Atlantic Ocean region) and 1 Intelsat (Atlantic Ocean) (2019)
note: the COVID-19 outbreak is negatively impacting telecommunications production and supply chains globally; consumer spending on telecom devices and services has also slowed due to the pandemic's effect on economies worldwide; overall progress towards improvements in all facets of the telecom industry—mobile, fixed-line, broadband, submarine cable and satellite—has moderated

Broadcast media: multiple terrestrial TV stations, supplemented by cable TV in most urban areas; nearly all are government-owned or affiliated; more than 300 radio stations, both government-affiliated and privately owned (2019)

Internet country code: .ni

Internet users: *total:* 1,695,340
percent of population: 27.86% (July 2018 est.)
country comparison to the world: 126

Broadband—fixed subscriptions: *total:* 192,413
subscriptions per 100 inhabitants: 3 (2018 est.)
country comparison to the world: 109

TRANSPORTATION

National air transport system: *number of registered air carriers:* 1 (2020)
inventory of registered aircraft operated by air carriers: 7

Civil aircraft registration country code prefix: YN (2016)

Airports: 147 (2013)
country comparison to the world: 38

Airports—with paved runways: *total:* 12 (2017)
2,438 to 3,047 m: 3 (2017)
1,524 to 2,437 m: 2 (2017)
914 to 1,523 m: 3 (2017)
under 914 m: 4 (2017)

Airports—with unpaved runways: *total:* 135 (2013)
1,524 to 2,437 m: 1 (2013)
914 to 1,523 m: 15 (2013)
under 914 m: 119 (2013)

Pipelines: 54 km oil (2013)

Roadways: *total:* 23,897 km (2014)
paved: 3,346 km (2014)
unpaved: 20,551 km (2014)
country comparison to the world: 107

Waterways: 2,220 km (navigable waterways as well as the use of the large Lake Managua and Lake Nicaragua; rivers serve only the sparsely populated eastern part of the country) (2011)
country comparison to the world: 39

Merchant marine: *total:* 6
by type: general cargo 2, oil tanker 1, other 3 (2019)
country comparison to the world: 164

Ports and terminals: *major seaport(s):* Bluefields, Corinto

MILITARY AND SECURITY

Military and security forces: Army of Nicaragua (Ejercito de Nicaragua, EN; includes Navy, Air Force) (2020)

Military expenditures:
0.7% of GDP (2019)
0.6% of GDP (2018)
0.6% of GDP (2017)
0.6% of GDP (2016)
0.8% of GDP (2015)
country comparison to the world: 139

Military and security service personnel strengths: the Army of Nicaragua has approximately 12,000 active personnel (10,000 Army; 800 Navy; 1,200 Air Force) (2019 est.)

Military equipment inventories and acquisitions: the Nicaraguan military's inventory includes mostly Russian/Soviet-era equipment; since 2010, Russia is the leading arms supplier to Nicaragua (2019 est.)

Military service age and obligation: 18-30 years of age for voluntary military service; no conscription; tour of duty 18-36 months; requires Nicaraguan nationality and 6th-grade education (2017)

TRANSNATIONAL ISSUES

Disputes—international: the 1992 ICJ ruling for El Salvador and Honduras advised a tripartite resolution to establish a maritime boundary in the Gulf of Fonseca, which considers Honduran access to the Pacific; Nicaragua and Costa Rica regularly file border dispute cases over the delimitations of the San Juan River and the northern tip of Calero Island to the ICJ; there is an ongoing case in the ICJ to determine Pacific and Atlantic ocean maritime borders as well as land borders; in 2009, the ICJ ruled that Costa Rican vessels carrying out police activities could not use the river, but official Costa Rican vessels providing essential services to riverside inhabitants and Costa Rican tourists could travel freely on the river; in 2011, the ICJ provisionally ruled that both countries must remove personnel from the disputed area; in 2013, the ICJ rejected Nicaragua's 2012 suit to halt Costa Rica's construction of a highway paralleling the river on the grounds of irreparable environmental damage; in 2013, the ICJ, regarding the disputed territory, ordered that Nicaragua should refrain from dredging or canal construction and refill and repair damage caused by trenches connecting the river to the Caribbean and upheld its 2010 ruling that Nicaragua must remove all personnel; in early 2014, Costa Rica brought Nicaragua to the ICJ over offshore oil concessions in the disputed region; Nicaragua filed a case against Colombia in 2013 over the delimitation of the Continental shelf beyond the 200 nautical miles from the Nicaraguan coast, as well as over the alleged violation by Colombia of Nicaraguan maritime space in the Caribbean Sea

Illicit drugs: transshipment point for cocaine destined for the US and transshipment point for arms-for-drugs dealing

NIGER

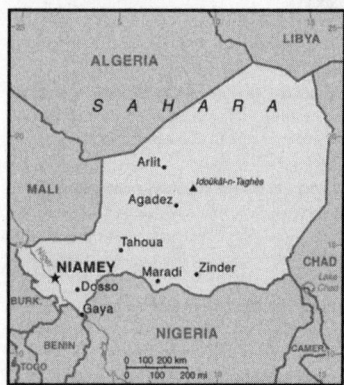

INTRODUCTION

Background: In the late 19th century, the British and French agreed to partition the middle regions of the Niger River into British Nigeria and French Niger. In subsequent decades French administration spread until in 1922 Niger officially became a colony. Following independence from France in 1960, the country experienced single-party and military rule until 1991, when Gen. Ali SAIBOU was forced by public pressure to allow multiparty elections, which resulted in a democratic government in 1993. Political infighting brought the government to a standstill and in 1996 led to a coup by Col. Ibrahim BARE. In 1999, BARE was killed in a counter coup by military officers who restored democratic rule and held elections that brought Mamadou TANDJA to power in December of that year. TANDJA was reelected in 2004 and in 2009 spearheaded a constitutional amendment allowing him to extend his term as president. In February 2010, military officers led a coup that deposed TANDJA and suspended the constitution. ISSOUFOU Mahamadou was elected in April 2011 following the coup and reelected to a second term in early 2016. Niger is one of the poorest countries in the world with minimal government services and insufficient funds to develop its resource base, and is ranked last in the world on the United Nations Development Programme's Human Development Index. The largely agrarian and subsistence-based economy is frequently disrupted by extended droughts common to the Sahel region of Africa. The Nigerien Government continues its attempts to diversify the economy through increased oil production and mining projects. A Tuareg rebellion emerged in 2007 and ended in 2009. Niger is facing increased security concerns on its borders from various external threats including insecurity in Libya, spillover from the conflict in Mali, and violent extremism in northeastern Nigeria.

GEOGRAPHY

Location: Western Africa, southeast of Algeria

Geographic coordinates: 16 00 N, 8 00 E

Map references: Africa

Area: *total:* 1.267 million sq km
land: 1,266,700 sq km
water: 300 sq km
country comparison to the world: 23

Area—comparative: slightly less than twice the size of Texas

Land boundaries: *total:* 5,834 km
border countries (7): Algeria 951 km, Benin 277 km, Burkina Faso 622 km, Chad 1196 km, Libya 342 km, Mali 838 km, Nigeria 1608 km

Coastline: 0 km (landlocked)

Maritime claims: none (landlocked)

Climate: desert; mostly hot, dry, dusty; tropical in extreme south

Terrain: predominately desert plains and sand dunes; flat to rolling plains in south; hills in north

Elevation: *mean elevation:* 474 m
lowest point: Niger River 200 m
highest point: Idoukal-n-Taghes 2,022 m

Natural resources: uranium, coal, iron ore, tin, phosphates, gold, molybdenum, gypsum, salt, petroleum

Land use: *agricultural land:* 35.1% (2011 est.)
arable land: 12.3% (2011 est.) / permanent crops: 0.1% (2011 est.) / permanent pasture: 22.7% (2011 est.)
forest: 1% (2011 est.)
other: 63.9% (2011 est.)
Irrigated land: 1,000 sq km (2012)

Population distribution: majority of the populace is located in the southernmost extreme of the country along the border with Nigeria and Benin as shown in this population distribution map

Natural hazards: recurring droughts

Environment—current issues: overgrazing; soil erosion; deforestation; desertification; contaminated water; inadequate potable water; wildlife populations (such as elephant, hippopotamus, giraffe, and lion) threatened because of poaching and habitat destruction

Environment—international agreements: *party to:* Biodiversity, Climate Change, Climate Change-Kyoto Protocol, Desertification, Endangered Species, Environmental Modification, Hazardous Wastes, Ozone Layer Protection, Wetlands
signed, but not ratified: Law of the Sea

Geography—note: landlocked; one of the hottest countries in the world; northern four-fifths is desert, southern one-fifth is savanna, suitable for livestock and limited agriculture

PEOPLE AND SOCIETY

Population: 22,772,361 (July 2020 est.)
country comparison to the world: 58

Nationality: *noun:* Nigerien(s)
adjective: Nigerien

Ethnic groups: Hausa 53.1%, Zarma/Songhai 21.2%, Tuareg 11%, Fulani (Peuhl) 6.5%, Kanuri 5.9%, Gurma 0.8%, Arab 0.4%, Tubu 0.4%, other/unavailable 0.9% (2006 est.)

Languages: French (official), Hausa, Djerma

Religions: Muslim 99.3%, Christian 0.3%, animist 0.2%, none 0.1% (2012 est.)

Demographic profile: Niger has the highest total fertility rate (TFR) of any country in the world, averaging close to 7 children per woman in 2016. A slight decline in fertility over the last few decades has stalled. This leveling off of the high fertility rate is in large part a product of the continued desire for large families. In Niger, the TFR is lower than the desired fertility rate, which makes it unlikely that contraceptive use will increase. The high TFR sustains rapid population growth and a large youth population – almost 70% of the populace is under the age of 25. Gender inequality, including a lack of educational opportunities for women and early marriage and childbirth, also contributes to high population growth.

Because of large family sizes, children are inheriting smaller and smaller parcels of land. The dependence of most Nigeriens on subsistence farming on increasingly small landholdings, coupled with declining rainfall and the resultant shrinkage of arable land, are all preventing food production from keeping up with population growth.

For more than half a century, Niger's lack of economic development has led to steady net outmigration. In the 1960s, Nigeriens mainly migrated to coastal West African countries to work on a seasonal basis. Some headed to Libya and Algeria in the 1970s to work in the booming oil industry until its decline in the 1980s. Since the 1990s, the principal destinations for Nigerien labor migrants have been West African countries, especially Burkina Faso and Cote d'Ivoire, while emigration to Europe and North America has remained modest. During the same period, Niger's desert trade route town Agadez became a hub for West African and other Sub-Saharan migrants crossing the Sahara to North Africa and sometimes onward to Europe.

More than 60,000 Malian refugees have fled to Niger since violence between Malian government troops and armed rebels began in early 2012. Ongoing attacks by the Boko Haram Islamist insurgency, dating to 2013 in northern Nigeria and February 2015 in southeastern Niger, have pushed tens of thousands of Nigerian refugees and Nigerien returnees across the border to Niger and to displace thousands of locals in Niger's already impoverished Diffa region.

Age structure: *0-14 years:* 50.58% (male 5,805,102/female 5,713,815)

15-24 years: 19.99% (male 2,246,670/female 2,306,285)

25-54 years: 23.57% (male 2,582,123/female 2,784,464)

55-64 years: 3.17% (male 357,832/female 364,774)

65 years and over: 2.68% (male 293,430/female 317,866) (2020 est.)

Dependency ratios: *total dependency ratio:* 109.5
youth dependency ratio: 104.1
elderly dependency ratio: 5.4
potential support ratio: 18.4 (2020 est.)

Median age: *total:* 14.8 years
male: 14.5 years
female: 15.1 years (2020 est.)
country comparison to the world: 228

Population growth rate: 3.66% (2020 est.)
country comparison to the world: 2

Birth rate: 47.5 births/1,000 population (2020 est.)
country comparison to the world: 1

Death rate: 10.2 deaths/1,000 population (2020 est.)
country comparison to the world: 35

Net migration rate: -0.7 migrant(s)/1,000 population (2020 est.)
country comparison to the world: 133

Population distribution: majority of the populace is located in the southernmost extreme of the country along the border with Nigeria and Benin as shown in this population distribution map

Urbanization: *urban population:* 16.6% of total population (2020)
rate of urbanization: 4.27% annual rate of change (2015-20 est.)
total population growth rate v. urban population growth rate, 2000-2030:

Major urban areas—population: 1.292 million NIAMEY (capital) (2020)

Sex ratio: *at birth:* 1.03 male(s)/female
0-14 years: 1.02 male(s)/female
15-24 years: 0.97 male(s)/female
25-54 years: 0.93 male(s)/female
55-64 years: 0.98 male(s)/female
65 years and over: 0.92 male(s)/female
total population: 0.98 male(s)/female (2020 est.)

Mother's mean age at first birth: 18.1 years (2012 est.)
note: median age at first birth among women 25-29

Maternal mortality rate: 509 deaths/100,000 live births (2017 est.)
country comparison to the world: 20

Infant mortality rate: *total:* 67.7 deaths/1,000 live births
male: 72 deaths/1,000 live births
female: 63.3 deaths/1,000 live births (2020 est.)
country comparison to the world: 6

Life expectancy at birth: *total population:* 59.3 years
male: 57.8 years
female: 60.8 years (2020 est.)
country comparison to the world: 219

Total fertility rate: 7 children born/woman (2020 est.)
country comparison to the world: 1

Contraceptive prevalence rate: 11% (2017/18)

Drinking water source:
improved:
urban: 95.7% of population
rural: 59.2% of population
total: 65.2% of population
unimproved:
urban: 4.3% of population
rural: 40.8% of population
total: 34.8% of population (2017 est.)

Current Health Expenditure: 7.7% (2017)

Physicians density: 0.04 physicians/1,000 population (2016)

Hospital bed density: 0.4 beds/1,000 population (2017)

Sanitation facility access:
improved:
urban: 76.6% of population
rural: 12.9% of population
total: 23.3% of population
unimproved:
urban: 23.4% of population
rural: 87.1% of population
total: 76.7% of population (2017 est.)

HIV/AIDS—adult prevalence rate: 0.3% (2019 est.)
country comparison to the world: 93

HIV/AIDS—people living with HIV/AIDS: 33,000 (2019 est.)
country comparison to the world: 71

HIV/AIDS—deaths: 1,100 (2019 est.)
country comparison to the world: 58

Major infectious diseases: *degree of risk:* very high (2020)
food or waterborne diseases: bacterial and protozoal diarrhea, hepatitis A, and typhoid fever
vectorborne diseases: malaria and dengue fever
water contact diseases: schistosomiasis
animal contact diseases: rabies
respiratory diseases: meningococcal meningitis

Obesity—adult prevalence rate: 5.5% (2016)
country comparison to the world: 177

Children under the age of 5 years underweight: 21.8% (2018)
country comparison to the world: 18

Education expenditures: 3.5% of GDP (2017)
country comparison to the world: 123

Literacy: *definition:* age 15 and over can read and write
total population: 19.1%
male: 27.3%
female: 11% (2015)

School life expectancy (primary to tertiary education): *total:* 6 years
male: 7 years
female: 6 years (2017)

Unemployment, youth ages 15-24: *total:* 0.7%
male: 0.9%
female: 0.4% (2014 est.)

country comparison to the world: 180

Country name: *conventional long form:* Republic of Niger
conventional short form: Niger
local long form: Republique du Niger
local short form: Niger
etymology: named for the Niger River that passes through the southwest of the country; from a native term "Ni Gir" meaning "River Gir"
note: pronounced nee-zher

Government type: semi-presidential republic

Capital: *name:* Niamey
geographic coordinates: 13 31 N, 2 07 E
time difference: UTC+1 (6 hours ahead of Washington, DC, during Standard Time)
etymology: according to tradition, the site was originally a fishing village named after a prominent local tree referred to as "nia niam"

Administrative divisions: 7 regions (regions, singular—region) and 1 capital district* (communaute urbaine); Agadez, Diffa, Dosso, Maradi, Niamey*, Tahoua, Tillaberi, Zinder

Independence: 3 August 1960 (from France)

National holiday: Republic Day, 18 December (1958); note—commemorates the founding of the Republic of Niger which predated independence from France in 1960

Constitution: *history:* several previous; passed by referendum 31 October 2010, entered into force 25 November 2010
amendments: proposed by the president of the republic or by the National Assembly; consideration of amendments requires at least three-fourths majority vote by the Assembly; passage requires at least four-fifths majority vote; if disapproved, the proposed amendment is dropped or submitted to a referendum; constitutional articles on the form of government, the multiparty system, the separation of state and religion, disqualification of Assembly members, amendment procedures, and amnesty of participants in the 2010 coup cannot be amended; amended 2011

Legal system: mixed legal system of civil law, based on French civil law, Islamic law, and customary law

International law organization participation: has not submitted an ICJ jurisdiction declaration; accepts ICCt jurisdiction

Citizenship: *citizenship by birth:* no
citizenship by descent only: at least one parent must be a citizen of Niger
dual citizenship recognized: yes
residency requirement for naturalization: unknown

Suffrage: 18 years of age; universal

Executive branch: *chief of state:* President ISSOUFOU Mahamadou (since 7 April 2011)
head of government: Prime Minister Brigi RAFINI (since 7 April 2011)
cabinet: Cabinet appointed by the president

elections/appointments: president directly elected by absolute majority popular vote in 2 rounds if needed for a 5-year term (eligible for a second term); election last held on 21 February 2016 with a runoff on 20 March 2016 (next to be held in 2021); prime minister appointed by the president, authorized by the National Assembly
election results: ISSOUFOU Mahamadou reelected president in second round; percent of vote in first round—ISSOUFOU Mahamadou (PNDS-Tarrayya) 48.6%, Hama AMADOU (MODEN/FA Lumana Africa) 17.8%, Seini OUMAROU (MNSD-Nassara) 11.3%, other 22.3%; percent of vote in second round—ISSOUFOU Mahamadou 92%, Hama AMADOU 8%

Legislative branch: *description:* unicameral National Assembly or Assemblee Nationale (171 seats; 158 members directly elected from 8 multi-member constituencies in 7 regions and Niamey by party-list proportional representation, 8 reserved for minorities elected in special single-seat constituencies by simple majority vote, 5 seats reserved for Nigeriens living abroad—1 seat per continent—elected in single-seat constituencies by simple majority vote; members serve 5-year terms)
elections: last held on 21 February 2016 (next to be held in 2021)
election results: percent of vote by party—PNDS-Tarrayya 44.1%, MODEN/FA Lumana 14.7%, MNSD-Nassara 11.8%, MPR-Jamhuriya 7.1%, MNRD Hankuri-PSDN Alheri 3.5%, MPN-Kishin Kassa 2.9%, ANDP-Zaman Lahiya 2.4%, RSD-Gaskiya 2.4%, CDS-Rahama 1.8%, CPR-Inganci 1.8%, RDP-Jama'a 1.8%, AMEN AMIN 1.8%, other 3.9%; seats by party—PNDS-Tarrayya 75, MODEN/FA Lumana 25, MNSD-Nassara 20, MPR-Jamhuriya 12, MNRD Hankuri-PSDN Alheri 6, MPN-Kishin Kassa 5, ANDP-Zaman Lahiya 4, RSD-Gaskiya 4, CDS-Rahama 3, CPR-Inganci 3, RDP-Jama'a 3, RDP-Jama'a 3, AMEN AMIN 3, other 8; composition—men 146, women 24 percent of women 14.6%

Judicial branch: *highest courts:* Constitutional Court (consists of 7 judges); High Court of Justice (consists of 7 members)
judge selection and term of office: Constitutional Court judges nominated/elected—1 by the president of the Republic, 1 by the president of the National Assembly, 2 by peer judges, 2 by peer lawyers, 1 law professor by peers, and 1 from within Nigerien society; all appointed by the president; judges serve 6-year nonrenewable terms with one-third of membership renewed every 2 years; High Judicial Court members selected from among the legislature and judiciary; members serve 5-year terms
subordinate courts: Court of Cassation; Council of State; Court of Finances; various specialized tribunals and customary courts

Political parties and leaders: Alliance of Movements for the Emergence of Niger or AMEN AMIN [Omar Hamidou TCHIANA]
Congress for the Republic or CPR-Inganci [Kassoum MOCTAR]
Democratic Alliance for Niger or ADN-Fusaha [Habi Mahamadou SALISSOU]
Democratic and Social Convention-Rahama or CDS-Rahama [Abdou LABO]
National Movement for the Development of Society-Nassara or MNSD-Nassara [Seini OUMAROU]
Nigerien Alliance for Democracy and Progress-Zaman Lahiya or ANDP-Zaman Lahiya [Moussa Moumouni DJERMAKOYE]
Nigerien Democratic Movement for an African Federation or MODEN/FA Lumana [Hama AMADOU]
Nigerien Movement for Democratic Renewal or MNRD-Hankuri [Mahamane OUSMANE]
Nigerien Party for Democracy and Socialism or PNDS- Tarrayya [Mahamadou ISSOUFOU]
Nigerien Patriotic Movement or MPN-Kishin Kassa [Ibrahim YACOUBA]
Party for Socialism and Democracy in Niger or PSDN-Alheri
Patriotic Movement for the Republic or MPR-Jamhuriya [Albade ABOUBA]
Rally for Democracy and Progress-Jama'a or RDP-Jama'a [Hamid ALGABID]
Social and Democratic Rally or RSD-Gaskiyya [Amadou CHEIFFOU]
Social Democratic Party or PSD-Bassira [Mohamed BEN OMAR]
Union for Democracy and the Republic-Tabbat or UDR-Tabbat [Amadou Boubacar CISSE]
note: the SPLM and SPLM-DC are banned political parties

International organization participation: ACP, AfDB, AU, CD, ECOWAS, EITI (compliant country), Entente, FAO, FZ, G-77, IAEA, IBRD, ICAO, ICCt, ICRM, IDA, IDB, IFAD, IFC, IFRCS, ILO, IMF, Interpol, IOC, IOM, IPU, ISO (correspondent), ITSO, ITU, ITUC (NGOs), MIGA, MINUSMA, MONUSCO, NAM, OIC, OIF, OPCW, UN, UNCTAD, UNESCO, UNIDO, UNMIL, UNOCI, UNWTO, UPU, WADB (regional), WAEMU, WCO, WFTU (NGOs), WHO, WIPO, WMO, WTO

Diplomatic representation in the US: *chief of mission:* Ambassador Hassana ALIDOU (since 23 February 2015)
chancery: 2204 R Street NW, Washington, DC 20008
telephone: [1] (202) 483-4224 through 4227
FAX: [1] (202) 483-3169

Diplomatic representation from the US: *chief of mission:* Ambassador Eric P. WHITAKER (since 26 January 2018) telephone: [227] 20-72-26-61
embassy: BP 11201, Rue Des Ambassades, Niamey
mailing address: 2420 Niamey Place, Washington DC 20521-2420
FAX: [227] 20-73-55-60

Flag description: three equal horizontal bands of orange (top), white, and green with a small orange disk centered in the white band; the orange band denotes the drier northern regions of the Sahara; white stands for purity and innocence; green symbolizes hope and the fertile and productive southern and western areas, as well as the Niger River;
the orange disc represents the sun and the sacrifices made by the people
note: similar to the flag of India, which has a blue spoked wheel centered in the white band

National symbol(s): zebu; national colors: orange, white, green

National anthem: *name:* "La Nigerienne" (The Nigerien)
lyrics/music: Maurice Albert THIRIET/Robert JACQUET and Nicolas Abel Francois FRIONNET
note: adopted 1961

ECONOMY

Economy—overview: Niger is a landlocked, Sub-Saharan nation, whose economy centers on subsistence crops, livestock, and some of the world's largest uranium deposits. Agriculture contributes approximately 40% of GDP and provides livelihood for over 80% of the population. The UN ranked Niger as the second least developed country in the world in 2016 due to multiple factors such as food insecurity, lack of industry, high population growth, a weak educational sector, and few prospects for work outside of subsistence farming and herding.

Since 2011 public debt has increased due to efforts to scale-up public investment, particularly that related to infrastructure, as well as due to increased security spending. The government relies on foreign donor resources for a large portion of its fiscal budget. The economy in recent years has been hurt by terrorist activity near its uranium mines and by instability in Mali and in the Diffa region of the country; concerns about security have resulted in increased support from regional and international partners on defense. Low uranium prices, demographics, and security expenditures may continue to put pressure on the government's finances.

The Government of Niger plans to exploit oil, gold, coal, and other mineral resources to sustain future growth. Although Niger has sizable reserves of oil, the prolonged drop in oil prices has reduced profitability. Food insecurity and drought remain perennial problems for Niger, and the government plans to invest more in irrigation. Niger's three-year $131 million IMF Extended Credit Facility (ECF) agreement for the years 2012-15 was extended until the end of 2016. In February 2017, the IMF approved a new 3-year $134 million ECF. In June 2017, The World Bank's International Development Association (IDA) granted Niger $1 billion over three years for IDA18, a program to boost the country's development and alleviate poverty. A $437 million Millennium Challenge Account compact for Niger, commencing in FY18, will focus on large-scale irrigation infrastructure development and community-based, climate-resilient agriculture, while promoting sustainable increases in agricultural productivity and sales.

Formal private sector investment needed for economic diversification and growth remains a challenge, given the country's limited domestic markets, access to credit, and competitiveness. Although President ISSOUFOU is courting foreign investors, including those from the US, as of April 2017, there were no US firms operating in Niger. In November 2017, the National Assembly passed the 2018 Finance Law that was geared towards raising government revenues and moving away from international support.

GDP (purchasing power parity):
$21.86 billion (2017 est.)
$20.84 billion (2016 est.)
$19.87 billion (2015 est.)
note: data are in 2017 dollars
country comparison to the world: 146

GDP (official exchange rate):
$8.224 billion (2017 est.)

GDP—real growth rate: 4.9% (2017 est.)
4.9% (2016 est.)
4.3% (2015 est.)
country comparison to the world: 51

GDP—per capita (PPP): $1,200 (2017 est.)
$1,100 (2016 est.)
$1,100 (2015 est.)
note: data are in 2017 dollars
country comparison to the world: 224

Gross national saving:
22.4% of GDP (2017 est.)
20.6% of GDP (2016 est.)
21.2% of GDP (2015 est.)
country comparison to the world: 80

GDP—composition, by end use:
household consumption: 70.2% (2017 est.)
government consumption: 9.4% (2017 est.)
investment in fixed capital: 38.6% (2017 est.)
investment in inventories: 0% (2017 est.)
exports of goods and services: 16.4% (2017 est.)
imports of goods and services: -34.6% (2017 est.)

GDP—composition, by sector of origin:
agriculture: 41.6% (2017 est.)
industry: 19.5% (2017 est.)
services: 38.7% (2017 est.)

Agriculture—products: cowpeas, cotton, peanuts, millet, sorghum, cassava (manioc, tapioca), rice; cattle, sheep, goats, camels, donkeys, horses, poultry

Industries: uranium mining, petroleum, cement, brick, soap, textiles, food processing, chemicals, slaughterhouses

Industrial production growth rate: 6% (2017 est.)
country comparison to the world: 41

Labor force: 6.5 million (2017 est.)
country comparison to the world: 68

Labor force—by occupation:
agriculture: 79.2%
industry: 3.3%
services: 17.5% (2012 est.)

Unemployment rate:
0.3% (2017 est.)
0.3% (2016 est.)
country comparison to the world: 3

Population below poverty line:
45.4% (2014 est.)

Household income or consumption by percentage share:
lowest 10%: 3.2%

highest 10%: 26.8% (2014)

Budget: *revenues:* 1.757 billion (2017 est.)
expenditures: 2.171 billion (2017 est.)

Taxes and other revenues: 21.4% (of GDP) (2017 est.)
country comparison to the world: 140

Budget surplus (+) or deficit (-): -5% (of GDP) (2017 est.)
country comparison to the world: 169

Public debt: 45.3% of GDP (2017 est.)
45.2% of GDP (2016 est.)
country comparison to the world: 115

Fiscal year: calendar year

Inflation rate (consumer prices):
2.4% (2017 est.)
0.2% (2016 est.)
country comparison to the world: 121

Current account balance: -$1.16 billion (2017 est.)
-$1.181 billion (2016 est.)
country comparison to the world: 150

Exports: $4.143 billion (2017 est.)
$1.101 billion (2016 est.)
country comparison to the world: 115

Exports—partners: France 30.2%, Thailand 18.3%, Malaysia 9.9%, Nigeria 8.3%, Mali 5%, Switzerland 4.9% (2017)

Exports—commodities: uranium ore, livestock, cowpeas, onions

Imports: $1.829 billion (2017 est.)
$1.715 billion (2016 est.)
country comparison to the world: 171

Imports—commodities: foodstuffs, machinery, vehicles and parts, petroleum, cereals

Imports—partners: France 28.8%, China 14.4%, Malaysia 5.7%, Nigeria 5.4%, Thailand 5.3%, US 5.1%, India 4.9% (2017)

Reserves of foreign exchange and gold:
$1.314 billion (31 December 2017 est.)
$1.186 billion (31 December 2016 est.)
country comparison to the world: 126

Debt—external: $3.728 billion (31 December 2017 est.)
$2.926 billion (31 December 2016 est.)
country comparison to the world: 140

Exchange rates: Communaute Financiere Africaine francs (XOF) per US dollar—
605.3 (2017 est.)
593.01 (2016 est.)
593.01 (2015 est.)
591.45 (2014 est.)
494.42 (2013 est.)

ENERGY

Electricity access: *population without electricity:* 20 million (2019)
electrification—total population: 14% (2019)
electrification—urban areas: 71% (2019)
electrification—rural areas: 2% (2019)

Electricity—production: 494.7 million kWh (2016 est.)

country comparison to the world: 167

Electricity—consumption: 1.065 billion kWh (2016 est.)
country comparison to the world: 153

Electricity—exports: 0 kWh (2016 est.)
country comparison to the world: 177

Electricity—imports: 779 million kWh (2016 est.)
country comparison to the world: 74

Electricity—installed generating capacity: 184,000 kW (2016 est.)
country comparison to the world: 168

Electricity—from fossil fuels: 95% of total installed capacity (2016 est.)
country comparison to the world: 45

Electricity—from nuclear fuels: 0% of total installed capacity (2017 est.)
country comparison to the world: 156

Electricity—from hydroelectric plants: 0% of total installed capacity (2017 est.)
country comparison to the world: 190

Electricity—from other renewable sources: 5% of total installed capacity (2017 est.)
country comparison to the world: 108

Crude oil—production: 9.000 bbl/day (2018 est.)
country comparison to the world: 79

Crude oil—exports: 0 bbl/day (2015 est.)
country comparison to the world: 176

Crude oil—imports: 0 bbl/day (2015 est.)
country comparison to the world: 176

Crude oil—proved reserves: 150 million bbl (1 January 2018 est.)
country comparison to the world: 61

Refined petroleum products—production: 15,280 bbl/day (2015 est.)
country comparison to the world: 94

Refined petroleum products—consumption: 14.000 bbl/day (2016 est.)
country comparison to the world: 156

Refined petroleum products—exports: 5,422 bbl/day (2015 est.)
country comparison to the world: 90

Refined petroleum products—imports: 3,799 bbl/day (2015 est.)
country comparison to the world: 180

Natural gas—production: 0 cu m (2017 est.)
country comparison to the world: 180

Natural gas—consumption: 0 cu m (2017 est.)
country comparison to the world: 184

Natural gas—exports: 0 cu m (2017 est.)
country comparison to the world: 163

Natural gas—imports: 0 cu m (2017 est.)
country comparison to the world: 168

Natural gas—proved reserves: 0 cu m (1 January 2016 est.)
country comparison to the world: 178

Carbon dioxide emissions from consumption of energy: 2.534 million Mt (2017 est.)
country comparison to the world: 154

COMMUNICATIONS

Telephones—fixed lines: *total subscriptions:* 116,352
subscriptions per 100 inhabitants: less than 1 (2019 est.)
country comparison to the world: 138

Telephones—mobile cellular: *total subscriptions:* 8,921,769
subscriptions per 100 inhabitants: 40.64 (2019 est.)
country comparison to the world: 92

Telecommunication systems: *general assessment:* mobile services stronger than fixed telecoms; broadband penetration inconsequential; adopts free mobile roaming with other G5 Sahel countries; govt. contributes to Trans-Sahara Backbone network; LTE license awarded; govt. tax of telecom sector (2020)
domestic: fixed-line 1 per 100 persons and mobile-cellular teledensity remains 41 per 100 persons despite a rapidly increasing cellular subscribership base; small system of wire, radio telephone communications, and microwave radio relay links concentrated in southwestern Niger; domestic satellite system with 3 earth stations and 1 planned (2019)
international: country code—227; satellite earth stations—2 Intelsat (1 Atlantic Ocean and 1 Indian Ocean)
note: the COVID-19 outbreak is negatively impacting telecommunications production and supply chains globally; consumer spending on telecom devices and services has also slowed due to the pandemic's effect on economies worldwide; overall progress towards improvements in all facets of the telecom industry—mobile, fixed-line, broadband, submarine cable and satellite—has moderated

Broadcast media: state-run TV station; 3 private TV stations provide a mix of local and foreign programming; state-run radio has only radio station with national coverage; about 30 private radio stations operate locally; as many as 100 community radio stations broadcast; transmissions of multiple international broadcasters are available

Internet country code: .ne

Internet users: *total:* 1,110,778
percent of population: 5.25% (July 2018 est.)
country comparison to the world: 137

Broadband—fixed subscriptions: *total:* 8,650
subscriptions per 100 inhabitants: less than 1 (2017 est.)
country comparison to the world: 173

TRANSPORTATION

National air transport system: *number of registered air carriers:* 2 (2020)
inventory of registered aircraft operated by air carriers: 3

Civil aircraft registration country code prefix: 5U (2016)

Airports: 30 (2013)

country comparison to the world: 115
Airports—with paved runways: *total:* 10 (2017)
2,438 to 3,047 m: 3 (2017)
1,524 to 2,437 m: 6 (2017)
914 to 1,523 m: 1 (2017)

Airports—with unpaved runways:
total: 20 (2013)
1,524 to 2,437 m: 3 (2013)
914 to 1,523 m: 15 (2013)
under 914 m: 2 (2013)

Heliports: 1 (2013)

Pipelines: 464 km oil

Roadways:
total: 18,949 km (2010)
paved: 3,912 km (2010)
unpaved: 15,037 km (2010)
country comparison to the world: 118

Waterways: 300 km (the Niger, the only major river, is navigable to Gaya between September and March) (2012)
country comparison to the world: 93

Merchant marine: *total:* 1
by type: general cargo 1 (2019)
country comparison to the world: 178

MILITARY AND SECURITY

Military and security forces: *Nigerien Armed Forces (Forces Armees Nigeriennes, FAN):* Army, Nigerien Air Force, Niger Gendarmerie (GN); Ministry of Interior: Niger National Guard (GNN), National Police (2019)
note: the Gendarmerie is subordinate to the Ministry of Defense and has primary responsibility for rural security; the National Guard is responsible for domestic security and the protection of high-level officials and government buildings

Military expenditures:
1.8% of GDP (2019)
2.5% of GDP (2018)
2.5% of GDP (2017)
2.2% of GDP (2016)
country comparison to the world: 62

Military and security service personnel strengths: size estimates for the Nigerien Armed Forces (FAN) vary; approximately 7,000-9,500 active troops (est. 5,500-6,000 Army; 200 Air Force; 1,500-3,500 Gendarmerie); est. 2,500-3,000 National Guard (2019)

Military equipment inventories and acquisitions: the FAN's inventory consists of a wide variety of foreign-supplied weapons, including Chinese, French, German, Russian, and US; since 2015, the FAN has received limited amounts of equipment from China, France, Russia, Sweden, and the US, some of which were donations (2019 est.)

Military deployments: 860 Mali (MINUSMA) (2020)

Military service age and obligation: 18 is the legal minimum age for compulsory or voluntary military service; enlistees must be Nigerien citizens and unmarried; 2-year service term; women may serve in health care (2017)

Military—note: as of September 2020, the FAN was conducting counterinsurgency and counter-terrorism operations against Islamic militants on two fronts; in the Diffa region, the Nigeria-based Boko Haram terrorist group has conducted dozens of attacks on security forces, army bases, and civilians; on Niger's western border with Mali, the Islamic State-West Africa (ISWA) has conducted numerous attacks on security personnel; a series of ISWA attacks on FAN forces near the Malian border in December of 2019 and January of 2020 resulted in the deaths of more than 170 soldiers

Niger is part of a five-nation anti-jihadist task force known as the G5 Sahel Group, set up in 2014 with Burkina Faso, Mali, Mauritania, and Chad; it has committed 1,100 troops and 200 gendarmes to the force; in early 2020, G5 Sahel military chiefs of staff agreed to allow defense forces from each of the states to pursue terrorist fighters up to 100 km into neighboring countries; the G5 force is backed by the UN, US, and France; G5 troops periodically conduct joint operations with French forces deployed to the Sahel under Operation Barkhane

Niger also has about 1,000 troops committed to the Multinational Joint Task Force (MNJTF) against Boko Haram; national MNJTF troop contingents are deployed within their own country territories, although cross-border operations are conducted periodically (2020)

TERRORISM

Terrorist group(s): Boko Haram; Islamic State of Iraq and ash-Sham in the Greater Sahara; Islamic State of Iraq and ash-Sham – West Africa; Jama'at Nusrat al-Islam wal-Muslimin; al-Mulathamun Battalion (al-Mourabitoun) (2020)
note: details about the history, aims, leadership, organization, areas of operation, tactics, targets, weapons, size, and sources of support of the group(s) appear(s) in Appendix-T

TRANSNATIONAL ISSUES

Disputes—international: Libya claims about 25,000 sq km in a currently dormant dispute in the Tommo region; location of Benin-Niger-Nigeria tripoint is unresolved; only Nigeria and Cameroon have heeded the Lake Chad Commission's admonition to ratify the delimitation treaty that also includes the Chad-Niger and Niger-Nigeria boundaries; the dispute with Burkina Faso was referred to the ICJ in 2010

Refugees and internally displaced persons: *refugees (country of origin):* 168,081 (Nigeria), 58,702 (Mali) (2020)

IDPs: 257,095 (includes the regions of Diffa, Tillaberi, and Tahoua; unknown how many of the 11,000 people displaced by clashes between government forces and the Tuareg militant group, Niger Movement for Justice, in 2007 are still displaced; inter-communal violence; Boko Haram attacks in southern Niger, 2015) (2020)

NIGERIA

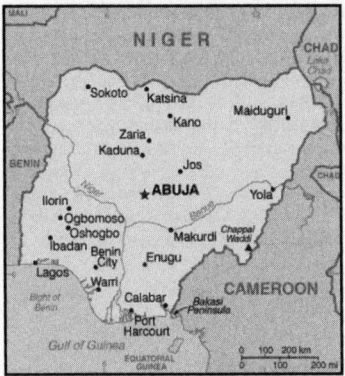

INTRODUCTION

Background: In ancient and pre-colonial times, the area of present-day Nigeria was occupied by a great diversity of ethnic groups with very different languages and traditions. British influence and control over what would become Nigeria and Africa's most populous country grew through the 19th century. A series of constitutions after World War II granted Nigeria greater autonomy. After independence in 1960, politics were marked by coups and mostly military rule, until the death of a military head of state in 1998 allowed for a political transition. In 1999, a new constitution was adopted and a peaceful transition to civilian government was completed. The government continues to face the daunting task of institutionalizing democracy and reforming a petroleum-based economy, whose revenues have been squandered through decades of corruption and mismanagement. In addition, Nigeria continues to experience longstanding ethnic and religious tensions. Although both the 2003 and 2007 presidential elections were marred by significant irregularities and violence, Nigeria is currently experiencing its longest period of civilian rule since independence. The general elections of 2007 marked the first civilian-to-civilian transfer of power in the country's history. National and state elections in 2011 and 2015 were generally regarded as credible. The 2015 election was also heralded for the fact that the then-umbrella opposition party, the All Progressives Congress, defeated the long-ruling People's Democratic Party that had governed since 1999, and assumed the presidency, marking the first peaceful transfer of power from one party to another. Presidential and legislative elections were held in early 2019 and deemed broadly free and fair despite voting irregularities, intimidation, and violence.

GEOGRAPHY

Location: Western Africa, bordering the Gulf of Guinea, between Benin and Cameroon

Geographic coordinates: 10 00 N, 8 00 E

Map references: Africa

Area: *total:* 923,768 sq km
land: 910,768 sq km
water: 13,000 sq km
country comparison to the world: 33

Area—comparative: about six times the size of Georgia; slightly more than twice the size of California

Land boundaries: *total:* 4,477 km
border countries (4): Benin 809 km, Cameroon 1975 km, Chad 85 km, Niger 1608 km

Coastline: 853 km

Maritime claims: *territorial sea:* 12 nm
exclusive economic zone: 200 nm
continental shelf: 200-m depth or to the depth of exploitation

Climate: varies; equatorial in south, tropical in center, arid in north

Terrain: southern lowlands merge into central hills and plateaus; mountains in southeast, plains in north

Elevation: *mean elevation:* 380 m
lowest point: Atlantic Ocean 0 m
highest point: Chappal Waddi 2,419 m

Natural resources: natural gas, petroleum, tin, iron ore, coal, limestone, niobium, lead, zinc, arable land

Land use: *agricultural land:* 78% (2011 est.)
arable land: 37.3% (2011 est.) / permanent crops: 7.4% (2011 est.) / permanent pasture: 33.3% (2011 est.)
forest: 9.5% (2011 est.)
other: 12.5% (2011 est.)
Irrigated land: 2,930 sq km (2012)

Population distribution: largest population of any African nation; significant population clusters are scattered throughout the country, with the highest density areas being in the south and southwest as shown in this population distribution map

Natural hazards: periodic droughts; flooding

Environment—current issues: serious overpopulation and rapid urbanization have led to numerous environmental problems; urban air and water pollution; rapid deforestation; soil degradation; loss of arable land; oil pollution—water, air, and soil have suffered serious damage from oil spills

Environment—international agreements: *party to:* Biodiversity, Climate Change, Climate Change-Kyoto Protocol, Desertification, Endangered Species, Hazardous Wastes, Law of the Sea, Marine Dumping, Marine Life Conservation, Ozone Layer Protection, Ship Pollution, Wetlands

signed, but not ratified: none of the selected agreements

Geography—note: the Niger River enters the country in the northwest and flows southward through tropical rain forests and swamps to its delta in the Gulf of Guinea

PEOPLE AND SOCIETY

Population: 214,028,302 (July 2020 est.)
note: estimates for this country explicitly take into account the effects of excess mortality due to AIDS; this can result in lower life expectancy, higher infant mortality, higher death rates, lower population growth rates, and changes in the distribution of population by age and sex than would otherwise be expected
country comparison to the world: 6

Nationality: *noun:* Nigerian(s)
adjective: Nigerian

Ethnic groups: Hausa 30%, Yoruba 15.5%, Igbo (Ibo) 15.2%, Fulani 6%, Tiv 2.4%, Kanuri/Beriberi 2.4%, Ibibio 1.8%, Ijaw/Izon 1.8%, other 24.7% (2018 est.)
note: Nigeria, Africa's most populous country, is composed of more than 250 ethnic groups

Languages: English (official), Hausa, Yoruba, Igbo (Ibo), Fulani, over 500 additional indigenous languages

Religions: Muslim 53.5%, Roman Catholic 10.6%, other Christian 35.3%, other .6% (2018 est.)

Demographic profile: Nigeria's population is projected to grow from more than 186 million people in 2016 to 392 million in 2050, becoming the world's fourth most populous country. Nigeria's sustained high population growth rate will continue for the foreseeable future because of population momentum and its high birth rate. Abuja has not successfully implemented family planning programs to reduce and space births because of a lack of political will, government financing, and the availability and affordability of services and products, as well as a cultural preference for large families. Increased educational attainment, especially among women, and improvements in health care are needed to encourage and to better enable parents to opt for smaller families.

Nigeria needs to harness the potential of its burgeoning youth population in order to boost economic development, reduce widespread poverty, and channel large numbers of unemployed youth into productive activities and away from ongoing religious and ethnic violence. While most movement of Nigerians is internal, significant emigration regionally and to the West provides an outlet for Nigerians looking for economic opportunities, seeking asylum, and increasingly pursuing higher education. Immigration largely of West Africans continues to be insufficient to offset emigration and the loss of highly skilled workers. Nigeria also is a major source, transit, and destination country for forced labor and sex trafficking.

Age structure: *0-14 years:* 41.7% (male 45,571,738/female 43,674,769)

15-24 years: 20.27% (male 22,022,660/female 21,358,753)

25-54 years: 30.6% (male 32,808,913/female 32,686,474)

55-64 years: 4.13% (male 4,327,847/female 4,514,264)

65 years and over: 3.3% (male 3,329,083/female 3,733,801) (2020 est.)

Dependency ratios: *total dependency ratio:* 86
youth dependency ratio: 80.9
elderly dependency ratio: 5.1
potential support ratio: 19.6 (2020 est.)

Median age: *total:* 18.6 years
male: 18.4 years
female: 18.9 years (2020 est.)
country comparison to the world: 208

Population growth rate: 2.53% (2020 est.)
country comparison to the world: 23

Birth rate: 34.6 births/1,000 population (2020 est.)
country comparison to the world: 21

Death rate: 9.1 deaths/1,000 population (2020 est.)
country comparison to the world: 59

Net migration rate: -0.2 migrant(s)/1,000 population (2020 est.)
country comparison to the world: 108

Population distribution: largest population of any African nation; significant population clusters are scattered throughout the country, with the highest density areas being in the south and southwest as shown in this population distribution map

Urbanization: *urban population:* 52% of total population (2020)
rate of urbanization: 4.23% annual rate of change (2015-20 est.)
total population growth rate v. urban population growth rate, 2000-2030:

Major urban areas—population: 14.368 million Lagos, 3.999 million Kano, 3.552 million Ibadan, 3.278 million ABUJA (capital), 3.020 million Port Harcourt, 1.727 million Benin City (2020)

Sex ratio: *at birth:* 1.06 male(s)/female
0-14 years: 1.04 male(s)/female
15-24 years: 1.03 male(s)/female
25-54 years: 1 male(s)/female
55-64 years: 0.96 male(s)/female
65 years and over: 0.89 male(s)/female
total population: 1.02 male(s)/female (2020 est.)

Mother's mean age at first birth: 20.3 years (2013 est.)
note: median age at first birth among women 25-29

Maternal mortality rate: 917 deaths/100,000 live births (2017 est.)
country comparison to the world: 4

Infant mortality rate: *total:* 59.8 deaths/1,000 live births
male: 65.4 deaths/1,000 live births
female: 54 deaths/1,000 live births (2020 est.)

country comparison to the world: 12

Life expectancy at birth: *total population:* 60.4 years
male: 58.6 years
female: 62.3 years (2020 est.)
country comparison to the world: 217

Total fertility rate: 4.72 children born/woman (2020 est.)
country comparison to the world: 18

Contraceptive prevalence rate: 16.6% (2018)

Drinking water source:
improved:
urban: 92.6% of population
rural: 63.6% of population
total: 77.9% of population
unimproved:
urban: 7.4% of population
rural: 36.4% of population
total: 22.1% of population (2017 est.)

Current Health Expenditure: 3.8% (2017)

Physicians density: 0.45 physicians/1,000 population (2016)

Sanitation facility access:
improved:
urban: 80.2% of population
rural: 39.5% of population
total: 59.7% of population
unimproved:
urban: 19.8% of population
rural: 60.5% of population
total: 40.3% of population (2017 est.)

HIV/AIDS—adult prevalence rate: 1.3% (2019 est.)
country comparison to the world: 35

HIV/AIDS—people living with HIV/AIDS: 1.8% million (2019 est.)
country comparison to the world: 4

HIV/AIDS—deaths: 45,000 (2019 est.)
country comparison to the world: 4

Major infectious diseases: *degree of risk:* very high (2020)
food or waterborne diseases: bacterial and protozoal diarrhea, hepatitis A and E, and typhoid fever
vectorborne diseases: malaria, dengue fever, and yellow fever
water contact diseases: leptospirosis and schistosomiasis
animal contact diseases: rabies
respiratory diseases: meningococcal meningitis
aerosolized dust or soil contact diseases: Lassa fever
note: on 7 October 2019, the Centers for Disease Control and Prevention issued a Travel Health Notice for a Yellow Fever outbreak in Nigeria; a large, ongoing outbreak of yellow fever in Nigeria began in September 2017; the outbreak is now spread throughout the country with the Nigerian Ministry of Health reporting cases of the disease in all 36 states and the Federal Capital Territory; the CDC recommends travelers going to Nigeria should receive vaccination against yellow fever at least 10 days before travel and should take steps to prevent mosquito bites while there; those never

vaccinated against yellow fever should avoid travel to Nigeria during the outbreak
note: widespread ongoing transmission of a respiratory illness caused by the novel coronavirus (COVID-19) is occurring throughout Nigeria; as of 10 November 2020, Nigeria has reported a total of 63,790 cases of COVID-19 or 309 cumulative cases of COVID-19 per 1 million population with 6 cumulative deaths per 1 million population; as of 19 March 2020, the Government of Nigeria has restricted entry into Nigeria for travelers from the following high incidence countries: China, Italy, Iran, Norway, South Korea, Spain, Japan, France, Germany, US, UK, Netherlands, and Switzerland

Obesity—adult prevalence rate: 8.9% (2016)
country comparison to the world: 145

Children under the age of 5 years underweight: 21.8% (2018)
country comparison to the world: 19

Education expenditures: NA

Literacy: *definition:* age 15 and over can read and write
total population: 62%
male: 71.3%
female: 52.7% (2018)

School life expectancy (primary to tertiary education): *total:* 9 years
male: 9 years
female: 8 years (2011)

Unemployment, youth ages 15-24: *total:* 13.8%
male: NA
female: NA (2016 est.)
country comparison to the world: 100

GOVERNMENT

Country name: *conventional long form;* Federal Republic of Nigeria
conventional short form: Nigeria
etymology: named for the Niger River that flows through the west of the country to the Atlantic Ocean; from a native term "Ni Gir" meaning "River Gir"

Government type: federal presidential republic

Capital: *name:* Abuja

geographic coordinates: 9 05 N, 7 32 E
time difference: UTC+1 (6 hours ahead of Washington, DC, during Standard Time)
etymology: Abuja is a planned capital city, it replaced Lagos in 1991; situated in the center of the country, Abuja takes its name from a nearby town, now renamed Suleja

Administrative divisions: 36 states and 1 territory*; Abia, Adamawa, Akwa Ibom, Anambra, Bauchi, Bayelsa, Benue, Borno, Cross River, Delta, Ebonyi, Edo, Ekiti, Enugu, Federal Capital Territory*, Gombe, Imo, Jigawa, Kaduna, Kano, Katsina, Kebbi, Kogi, Kwara, Lagos, Nasarawa, Niger, Ogun, Ondo, Osun, Oyo, Plateau, Rivers, Sokoto, Taraba, Yobe, Zamfara

Independence: 1 October 1960 (from the UK)

National holiday: Independence Day (National Day), 1 October (1960)

Constitution: *history:* several previous; latest adopted 5 May 1999, effective 29 May 1999
amendments: proposed by the National Assembly; passage requires at least two-thirds majority vote of both houses and approval by the Houses of Assembly of at least two thirds of the states; amendments to constitutional articles on the creation of a new state, fundamental constitutional rights, or constitution-amending procedures requires at least four-fifths majority vote by both houses of the National Assembly and approval by the Houses of Assembly in at least two thirds of the states; passage of amendments limited to the creation of a new state require at least two-thirds majority vote by the proposing National Assembly house and approval by the Houses of Assembly in two thirds of the states; amended several times, last in 2018

Legal system: mixed legal system of English common law, Islamic law (in 12 northern states), and traditional law

International law organization participation: accepts compulsory ICJ jurisdiction with reservations; accepts ICCt jurisdiction

Citizenship: *citizenship by birth:* no
citizenship by descent only: at least one parent must be a citizen of Nigeria
dual citizenship recognized: yes
residency requirement for naturalization: 15 years

Suffrage: 18 years of age; universal

Executive branch: *chief of state:* President Maj. Gen. (ret.) Muhammadu BUHARI (since 29 May 2015); Vice President Oluyemi "Yemi" OSINBAJO (since 29 May 2015); note—the president is both chief of state, head of government, and commander-in-chief of the armed forces
head of government: President Maj.Gen. (ret.) Muhammadu BUHARI (since 29 May 2015); Vice President Oluyemi "Yemi" OSINBAJO (since 29 May 2015)
cabinet: Federal Executive Council appointed by the president but constrained constitutionally to include at least one member from each of the 36 states
elections/appointments: president directly elected by qualified majority popular vote and at least 25% of the votes cast in 24 of Nigeria's 36 states; president elected for a 4-year term (eligible for a second term); election last held on 23 February 2019 (next to be held in February 2023); note: the election was scheduled for 16 February 2019, but postponed on 16 February 2019
election results: Muhammadu BUHARI elected president; percent of vote—Muhammadu BUHARI (APC) 53%, Atiku ABUBAKER (PDP) 39%, other 8%

Legislative branch: *description:* bicameral National Assembly consists of:
Senate (109 seats—3 each for the 36 states and 1 for Abuja-Federal Capital Territory; members directly elected in singleseat constituencies by simple majority vote to serve 4-year terms)

House of Representatives (360 seats; members directly elected in single-seat constituencies by simple majority vote to serve 4-year terms)
elections: Senate—last held on 23 February 2019 (next to be held on 23 February 2023); note: election was scheduled for 16 February 2019 but was postponed on 15 February 2019
House of Representatives—last held on 23 February 2019 (next to be held on 23 February 2023); note: election was scheduled for 16 February 2019 but was postponed on 15 February 2019
election results: Senate—percent of vote by party—NA; seats by party—APC 65, PDP 39, YPP 1, TBD 3; composition—men 103, women 6, percent of women 5.5%
House of Representatives—percent of vote by party—NA; seats by party—APC 217, PDP 115, other 20, TBD 8; composition—men 346, women 14, percent of women 3.9%; note—total National Assembly percent of women 4.3%

Judicial branch: *highest courts:* Supreme Court (consists of the chief justice and 15 justices)
judge selection and term of office: judges appointed by the president upon the recommendation of the National Judicial Council, a 23-member independent body of federal and state judicial officials; judge appointments confirmed by the Senate; judges serve until age 70
subordinate courts: Court of Appeal; Federal High Court; High Court of the Federal Capital Territory; Sharia Court of Appeal of the Federal Capital Territory; Customary Court of Appeal of the Federal Capital Territory; state court system similar in structure to federal system

Political parties and leaders:
Accord Party or ACC [Mohammad Lawal MALADO]
All Progressives Congress or APC [Adams OSHIOMHOLE]
All Progressives Grand Alliance or APGA [Victor Ike OYE]
Democratic Peoples Party or DPP [Biodun OGUNBIYI]
Labor Party or LP [Alhai Abdulkadir ABDULSALAM]
Peoples Democratic Party or PDP [Uche SECONDUS]
Young Progressive Party or YPP [Kingsley MOGHALU]

International organization participation: ACP, AfDB, AU, C, CD, D-8, ECOWAS, EITI (compliant country), FAO, G-15, G-24, G-77, IAEA, IBRD, ICAO, ICC (national committees), ICCt, ICRM, IDA, IDB, IFAD, IFC, IFRCS, IHO, ILO, IMF, IMO, IMSO, Interpol, IOC, IOM, IPU, ISO, ITSO, ITU, ITUC (NGOs), MIGA, MINURSO, MINUSMA, MONUSCO, NAM, OAS (observer), OIC, OPCW, OPEC, PCA, UN, UNAMID, UNCTAD, UNESCO, UNHCR, UNIDO, UNIFIL, UNISFA, UNITAR, UNMIL, UNMISS, UNOCI, UNWTO, UPU, WCO, WFTU (NGOs), WHO, WIPO, WMO, WTO

Diplomatic representation in the US: chief of mission: Ambassador Sylvanus Adiewere NSOFOR (since 29 November 2017)
chancery: 3519 International Court NW, Washington, DC 20008
telephone: [1] (202) 516-4277
FAX: [1] (202) 362-6541
consulate(s) general: Atlanta, New York

Diplomatic representation from the US: *chief of mission:* Ambassador Mary Beth LEONARD (since 24 December 2019)
telephone: [234] (9) 461-4000
embassy: Plot 1075 Diplomatic Drive, Central District Area, Abuja
mailing address: P. O. Box 5760, Garki, Abuja
FAX: [234] (9) 461-4036
consulate(s): Lagos

Flag description: three equal vertical bands of green (hoist side), white, and green; the color green represents the forests and abundant natural wealth of the country, white stands for peace and unity

National symbol(s): eagle; national colors: green, white

National anthem: *name:* Arise Oh Compatriots, Nigeria's Call Obey
lyrics/music: John A. ILECHUKWU, Eme Etim AKPAN, B.A. OGUNNAIKE, Sotu OMOIGUI and P.O. ADERIBIGBE/Benedict Elide ODIASE
note: adopted 1978; lyrics are a mixture of the five top entries in a national contest
0:00 / 0:48

ECONOMY

Economy—overview: Nigeria is Sub Saharan Africa's largest economy and relies heavily on oil as its main source of foreign exchange earnings and government revenues. Following the 2008-09 global financial crises, the banking sector was effectively recapitalized and regulation enhanced. Since then, Nigeria's economic growth has been driven by growth in agriculture, telecommunications, and services. Economic diversification and strong growth have not translated into a significant decline in poverty levels; over 62% of Nigeria's over 180 million people still live in extreme poverty.

Despite its strong fundamentals, oil-rich Nigeria has been hobbled by inadequate power supply, lack of infrastructure, delays in the passage of legislative reforms, an inefficient property registration system, restrictive trade policies, an inconsistent regulatory environment, a slow and ineffective judicial system, unreliable dispute resolution mechanisms, insecurity, and pervasive corruption. Regulatory constraints and security risks have limited new investment in oil and natural gas, and Nigeria's oil production had been contracting every year since 2012 until a slight rebound in 2017.

President BUHARI, elected in March 2015, has established a cabinet of economic ministers that includes several technocrats, and he has announced plans to increase transparency, diversify the economy away from oil, and improve fiscal management, but has taken a primarily protectionist approach that favors domestic producers at the expense of consumers. President BUHARI ran on an anti-corruption platform, and has made some headway in alleviating corruption, such as implementation of a Treasury Single Account that allows the government to better manage its resources and a more transparent government payroll and personnel system that eliminated duplicate and "ghost workers." The government also is working to develop stronger public-private partnerships for roads, agriculture, and power.

Nigeria entered recession in 2016 as a result of lower oil prices and production, exacerbated by militant attacks on oil and gas infrastructure in the Niger Delta region, coupled with detrimental economic policies, including foreign exchange restrictions. GDP growth turned positive in 2017 as oil prices recovered and output stabilized.

GDP (purchasing power parity):
$1.121 trillion (2017 est.)
$1.112 trillion (2016 est.)
$1.13 trillion (2015 est.)
note: data are in 2017 dollars
country comparison to the world: 24

GDP (official exchange rate):
$376.4 billion (2017 est.)

GDP—real growth rate: 0.8% (2017 est.)
-1.6% (2016 est.)
2.7% (2015 est.)
country comparison to the world: 178

GDP—per capita (PPP): $5,900 (2017 est.)
$6,100 (2016 est.)
$6,300 (2015 est.)
note: data are in 2017 dollars
country comparison to the world: 166

Gross national saving:
18.2% of GDP (2017 est.)
16% of GDP (2016 est.)
12.3% of GDP (2015 est.)
country comparison to the world: 110

GDP—composition, by end use:
household consumption: 80% (2017 est.)
government consumption: 5.8% (2017 est.)
investment in fixed capital: 14.8% (2017 est.)
investment in inventories: 0.7% (2017 est.)
exports of goods and services: 11.9% (2017 est.)
imports of goods and services: -13.2% (2017 est.)

GDP—composition, by sector of origin:
agriculture: 21.1% (2016 est.)
industry: 22.5% (2016 est.)
services: 56.4% (2017 est.)

Agriculture—products: cocoa, peanuts, cotton, palm oil, corn, rice, sorghum, millet, cassava (manioc, tapioca), yams, rubber; cattle, sheep, goats, pigs; timber; fish

Industries: crude oil, coal, tin, columbite; rubber products, wood; hides and skins, textiles, cement and other construction materials, food products, footwear, chemicals, fertilizer, printing, ceramics, steel

Industrial production growth rate: 2.2% (2017 est.)
country comparison to the world: 126

Labor force: 60.08 million (2017 est.)
country comparison to the world: 9

Labor force—by occupation:
agriculture: 70%
industry: 10%
services: 20% (1999 est.)

Unemployment rate:
16.5% (2017 est.)
13.9% (2016 est.)
country comparison to the world: 180

Population below poverty line: 70% (2010 est.)

Household income or consumption by percentage share: *lowest 10%:* 1.8%
highest 10%: 38.2% (2010 est.)

Budget: *revenues:* 12.92 billion (2017 est.)
expenditures 19.54 billion (2017 est.)

Taxes and other revenues: 3.4% (of GDP) (2017 est.)
country comparison to the world: 220

Budget surplus (+) or deficit (-): -1.8% (of GDP) (2017 est.)
country comparison to the world: 99

Public debt: 21.8% of GDP (2017 est.)
19.6% of GDP (2016 est.)
country comparison to the world: 185

Fiscal year: calendar year

Inflation rate (consumer prices): 16.5% (2017 est.)
15.7% (2016 est.)
country comparison to the world: 213

Current account balance: $10.38 billion (2017 est.)
$2.714 billion (2016 est.)
country comparison to the world: 24

Exports: $1.146 billion (2017 est.)
$34.7 billion (2016 est.)
country comparison to the world: 154

Exports—partners: India 30.6%, US 12.1%, Spain 6.6%, China 5.6%, France 5.5%, Netherlands 4.4%, Indonesia 4.4% (2017)

Exports—commodities: petroleum and petroleum products 95%, cocoa, rubber (2012 est.)

Imports: $32.67 billion (2017 est.)
$35.24 billion (2016 est.)
country comparison to the world: 62

Imports—commodities: machinery, chemicals, transport equipment, manufactured goods, food and live animals

Imports—partners: China 21.1%, Belgium 8.7%, US 8.4%, South Korea 7.5%, UK 4.4% (2017)

Reserves of foreign exchange and gold:
$38.77 billion (31 December 2017 est.)
$25.84 billion (31 December 2016 est.)
country comparison to the world: 45

Debt—external:
$40.96 billion (31 December 2017 est.)
$31.41 billion (31 December 2016 est.)
country comparison to the world: 73

Exchange rates: nairas (NGN) per US dollar—
323.5 (2017 est.)
253 (2016 est.)

253 (2015 est.)
192.73 (2014 est.)
158.55 (2013 est.)

ENERGY

Electricity access: *population without electricity:* 77 million (2019)
electrification—total population: 62% (2019)
electrification—urban areas: 91% (2019)
electrification—rural areas: 30% (2019)

Electricity—production: 29.35 billion kWh (2016 est.)
country comparison to the world: 67

Electricity—consumption: 24.72 billion kWh (2016 est.)
country comparison to the world: 69

Electricity—exports: 0 kWh (2016 est.)
country comparison to the world: 178

Electricity—imports: 0 kWh (2016 est.)
country comparison to the world: 180

Electricity—installed generating capacity: 10.52 million kW (2016 est.)
country comparison to the world: 58

Electricity—from fossil fuels: 80% of total installed capacity (2016 est.)
country comparison to the world: 83

Electricity—from nuclear fuels: 0% of total installed capacity (2017 est.)
country comparison to the world: 157

Electricity—from hydroelectric plants: 19% of total installed capacity (2017 est.)
country comparison to the world: 91

Electricity—from other renewable sources: 0% of total installed capacity (2017 est.)
country comparison to the world: 203

Crude oil—production: 1.989 million bbl/day (2018 est.)
country comparison to the world: 11

Crude oil—exports: 2.096 million bbl/day (2015 est.)
country comparison to the world: 6

Crude oil—imports: 0 bbl/day (2015 est.)
country comparison to the world: 177

Crude oil—proved reserves: 37.45 billion bbl (1 January 2018 est.)
country comparison to the world: 10

Refined petroleum products—production: 35,010 bbl/day (2017 est.)
country comparison to the world: 83

Refined petroleum products—consumption: 325,000 bbl/day (2016 est.)
country comparison to the world: 41

Refined petroleum products—exports: 2,332 bbl/day (2015 est.)
country comparison to the world: 102

Refined petroleum products—imports: 223,400 bbl/day (2015 est.)
country comparison to the world: 31

Natural gas—production: 44.48 billion cu m (2017 est.)
country comparison to the world: 18

Natural gas—consumption: 17.24 billion cu m (2017 est.)
country comparison to the world: 41

Natural gas—exports: 27.21 billion cu m (2017 est.)
country comparison to the world: 13

Natural gas—imports: 0 cu m (2017 est.)
country comparison to the world: 169

Natural gas—proved reserves: 5.475 trillion cu m (1 January 2018 est.)
country comparison to the world: 8

Carbon dioxide emissions from consumption of energy: 104 million Mt (2017 est.)
country comparison to the world: 42

COMMUNICATIONS

Telephones—fixed lines: total subscriptions: 146,075
subscriptions per 100 inhabitants: less than 1 (2019 est.)
country comparison to the world: 125

Telephones—mobile cellular: *total subscriptions:* 184,013,243
subscriptions per 100 inhabitants: 88.18 (2019 est.)
country comparison to the world: 7

Telecommunication systems: *general assessment:* one of the larger telecom markets in Africa; most Internet connections are via mobile networks; foreign investment presence, particularly China; market competition; LTE-A technologies available but GSM technology dominate; mobile penetration rate of 123% and 173 million subscribers; unified licensing regime; government committed to expanding broadband penetration; in Q1 2018, the Nigerian Communications Commission approved seven licenses to telecom companies to deploy fiber optic cable in the six geopolitical zones and Lagos; operators invest in base stations to take care of network congestion (2020)
domestic: fixed-line subscribership remains less than 1 per 100 persons; mobile-cellular services growing rapidly, in part responding to the short-comings of the fixed-line network; multiple cellular providers operate nationally with subscribership base over 88 per 100 persons (2019)
international: country code—234; landing point for the SAT-3/WASC, NCSCS, MainOne, Glo-1 & 2, ACE, and Equiano fiber-optic submarine cable that provides connectivity to Europe and South and West Africa; satellite earth stations—3 Intelsat (2 Atlantic Ocean and 1 Indian Ocean) (2019)
note: the COVID-19 outbreak is negatively impacting telecommunications production and supply chains globally; consumer spending on telecom devices and services has also slowed due to the pandemic's effect on economies worldwide; overall progress towards improvements in all facets of the telecom industry—mobile, fixed-line, broadband, submarine cable and satellite—has moderated

Broadcast media: nearly 70 federal government-controlled national and regional TV stations; all 36 states operate TV stations; several private TV stations operational; cable and satellite TV subscription services are available; network of federal governmentcontrolled national, regional, and state radio stations; roughly 40 state government-owned radio stations typically carry their own programs except for news broadcasts; about 20 private radio stations; transmissions of international broadcasters are available; digital broadcasting migration process completed in three states in 2018 (2019)

Internet country code: .ng

Internet users: total: 85,450,052
percent of population: 42% (July 2018 est.)
country comparison to the world: 8

Broadband—fixed subscriptions: total: 73,965
subscriptions per 100 inhabitants: less than 1 (2018 est.)
country comparison to the world: 126

TRANSPORTATION

National air transport system: *number of registered air carriers:* 13 (2020)
inventory of registered aircraft operated by air carriers: 104
annual passenger traffic on registered air carriers: 8,169,192 (2018)
annual freight traffic on registered air carriers: 19.42 million mt-km (2018)

Civil aircraft registration country code prefix: 5N (2016)

Airports: 54 (2013)
country comparison to the world: 87

Airports—with paved runways:
total: 40 (2017)
over 3,047 m: 10 (2017)
2,438 to 3,047 m: 12 (2017)
1,524 to 2,437 m: 9 (2017)
914 to 1,523 m: 6 (2017)
under 914 m: 3 (2017)

Airports—with unpaved runways:
total: 14 (2013)
1,524 to 2,437 m: 2 (2013)
914 to 1,523 m: 9 (2013)
under 914 m: 3 (2013)

Heliports: 5 (2013)

Pipelines: 124 km condensate, 4045 km gas, 164 km liquid petroleum gas, 4441 km oil, 3940 km refined products (2013)

Railways: *total:* 3,798 km (2014)
standard gauge: 293 km 1.435-m gauge (2014)
narrow gauge: 3,505 km 1.067-m gauge (2014)
note: as of the end of 2018, there were only six operational locomotives in Nigeria primarily used for passenger service; the majority of the rail lines are in a severe state of disrepair and need to be replaced
country comparison to the world: 54

Roadways:
total: 195,000 km (2017)
paved: 60,000 km (2017)
unpaved: 135,000 km (2017)
country comparison to the world: 29

Waterways: 8,600 km (Niger and Benue Rivers and smaller rivers and creeks) (2011)
country comparison to the world: 15

Merchant marine: *total:* 677
by type: general cargo 15, oil tanker 105, other 557 (2019)
country comparison to the world: 33

Ports and terminals: *major seaport(s):* Bonny Inshore Terminal, Calabar, Lagos
LNG terminal(s) (export): Bonny Island

MILITARY AND SECURITY

Military and security forces: *Nigerian Armed Forces:* Army, Navy (includes Coast Guard), Air Force; Ministry of Interior: Nigeria Security and Civil Defence Corps (NSCDC, a paramilitary agency commissioned to assist the military in the management of threats to internal security, including attacks and natural disasters) (2020)

Military expenditures:
0.5% of GDP (2019)
0.5% of GDP (2018)
0.4% of GDP (2017)
0.4% of GDP (2016)
0.4% of GDP (2015)
country comparison to the world: 149

Military and security service personnel strengths: size estimates for the Nigerian Armed Forces vary; approximately 135,000 active personnel (100,000 Army; 20,000 Navy/Coast Guard; 15,000 Air Force); est. 80,000 Security and Civil Defense Corps (2019 est.)

Military equipment inventories and acquisitions: the Nigerian Armed Forces' inventory consists of a wide variety of imported weapons systems of Chinese, European, Middle Eastern, Russian (including Soviet-era), and US origin; since 2010, the leading suppliers include China, France, Italy, Russia, South Korea, Ukraine, and the US; Nigeria has been the largest arms importer in sub-Saharan Africa since 2014; Nigeria is also developing a defense-industry capacity, including small arms, armored personnel vehicle, and small-scale naval production (2019)

Military deployments: 200 Ghana (ECOMIG); MNJTF (1 brigade or approximately 3,000 troops committed; note—the national MNJTF troop contingents are deployed within their own country territories, although cross-border operations are conducted periodically) (2020)

Military service age and obligation: 18 years of age for voluntary military service; no conscription (2012)

Maritime threats: the International Maritime Bureau reports the territorial and offshore waters in the Niger Delta and Gulf of Guinea as very high risk for piracy and armed robbery of ships; in 2018, 48 commercial vessels were boarded or attacked compared with 33 attacks in 2017; in 2018, 29 ships were boarded eight of which were underway, 12 were fired upon, and 78 crew members were abducted; Nigerian pirates have extended the range of their attacks to as far away as Cote d'Ivoire and as far as 170 nm offshore; the Maritime Administration of the US Department of Transportation has issued a Maritime Advisory (2019-010-Gulf of Guinea-Piracy/Armed Robbery/Kidnapping for Ransom) effective 19 July 2019, which states in part "Piracy, armed robbery, and kidnapping for ransom (KFR) continue to serve as significant threats to U.S. flagged vessels transiting or operating in the Gulf of Guinea (GoG). ...According to the Office of Naval Intelligence's "Weekly Piracy Reports" 72 reported incidents of piracy and armed robbery at sea occurred in the GoG region this year as of July 9, 2019. Attacks, kidnappings for ransom (KFR), and boardings to steal valuables from the ships and crews are the most common types of incidents with approximately 75 percent of all incidents taking place off Nigeria. During the first six months of 2019, there were 15 kidnapping and 3 hijackings in the GoG."

Military—note: the Nigerian Armed Forces are used primarily for internal security operations; in the northeast, the military is conducting counterinsurgency/counter-terrorist operations against the Boko Haram (BH) and Islamic State in West Africa (ISWA) terrorist groups, where up to 70,000 troops have been deployed at times; in the northwest, it faces threats from criminal gangs, bandits, and militants associated with ongoing herder-farmer violence, as well as BH and ISWA terrorists; the military also focuses on the Niger Delta region to protect the oil industry against militants and criminal activity, although the levels of violence there have decreased in recent years (2020)

TERRORISM

Terrorist group(s): Boko Haram; Islamic State of Iraq and ash-Sham – West Africa; Jama'atu Ansarul Muslimina Fi Biladis-Sudan (Ansaru) (2020)
note: details about the history, aims, leadership, organization, areas of operation, tactics, targets, weapons, size, and sources of support of the group(s) appear(s) in Appendix-T

TRANSNATIONAL ISSUES

Disputes—international: Joint Border Commission with Cameroon reviewed 2002 ICJ ruling on the entire boundary and bilaterally resolved differences, including June 2006 Greentree Agreement that immediately cedes sovereignty of the Bakassi Peninsula to Cameroon with a phaseout of Nigerian control within two years while resolving patriation issues; the ICJ ruled on an equidistance settlement of Cameroon-Equatorial Guinea-Nigeria maritime boundary in the Gulf of Guinea, but imprecisely defined coordinates in the ICJ decision and a sovereignty dispute between Equatorial Guinea and Cameroon over an island at the mouth of the Ntem River all contribute to the delay in implementation; only Nigeria and Cameroon have heeded the Lake Chad Commission's admonition to ratify the delimitation treaty which also includes the Chad-Niger and Niger-Nigeria boundaries; location of Benin-Niger-Nigeria tripoint is unresolved

Refugees and internally displaced persons: *refugees (country of origin):* 61,774 (Cameroon) (2020)
IDPs: 3,214,506 (northeast Nigeria; Boko Haram attacks and counterinsurgency efforts in northern Nigeria; communal violence between Christians and Muslims in the middle belt region, political violence; flooding; forced evictions; cattle rustling; competition for resources) (2020)

Illicit drugs: a transit point for heroin and cocaine intended for European, East Asian, and North American markets; consumer of amphetamines; safe haven for Nigerian narcotraffickers operating worldwide; major money-laundering center; massive corruption and criminal activity; Nigeria has improved some anti-money-laundering controls, resulting in its removal from the Financial Action Task Force's (FATF's) Noncooperative Countries and Territories List in June 2006; Nigeria's antimoney-laundering regime continues to be monitored by FATF

NIUE

GEOGRAPHY

Location: Oceania, island in the South Pacific Ocean, east of Tonga

Geographic coordinates: 19 02 S, 169 52 W

Map references: Oceania

Area: *total:* 260 sq km
land: 260 sq km
water: 0 sq km
country comparison to the world: 213

Area—comparative: 1.5 times the size of Washington, DC

Land boundaries: 0 km

Coastline: 64 km

Maritime claims: *territorial sea:* 12 nm
exclusive economic zone: 200 nm

Climate: tropical; modified by southeast trade winds

Terrain: steep limestone cliffs along coast, central plateau

Elevation: *lowest point:* Pacific Ocean 0 m
highest point: unnamed elevation 1.4 km east of Hikutavake 80 m

Natural resources: arable land, fish

Land use: *agricultural land:* 19.1% (2011 est.)

arable land: 3.8% (2011 est.) / permanent crops: 11.5% (2011 est.) / permanent pasture: 3.8% (2011 est.)
forest: 71.2% (2011 est.)
other: 9.7% (2011 est.)
Irrigated land: 0 sq km (2012)

Population distribution: population distributed around the peripheral coastal areas of the island

Natural hazards: tropical cyclones

Environment—current issues: increasing attention to conservationist practices to counter loss of soil fertility from traditional slash and burn agriculture

Environment—international agreements: *party to:* Biodiversity, Climate Change, Climate Change-Kyoto Protocol, Desertification, Law of the Sea, Ozone Layer Protection

Geography—note: one of world's largest coral islands; the only major break in the surrounding coral reef occurs in the central western part of the coast

PEOPLE AND SOCIETY

Population: 2,000 (2019)
note: because of the island's limited economic and educational opportunities, Niueans have emigrated for decades—primarily to New Zealand, but

INTRODUCTION

Background: Niue's remoteness, as well as cultural and linguistic differences between its Polynesian inhabitants and those of the adjacent Cook Islands, has caused it to be separately administered by New Zealand. The population of the island has trended downwards over recent decades (from a peak of 5,200 in 1966 to 1,618 in 2017) with substantial emigration to New Zealand 2,400 km to the southwest.

also to Australia and other Pacific island states; Niue's population peaked in 1966 at 5,194, but by 2005 had fallen to 1,508; since then it has rebounded slightly; as of 2013, 23,883 people of Niuean ancestry lived in New Zealand—with more than 20% Niue-born; this means that there are about 15 times as many persons of Niuean living in New Zealand as in Niue, possibly the most eccentric population distribution in the world *country comparison to the world:* 232

Nationality: *noun:* Niuean(s)
adjective: Niuean

Ethnic groups: Niuean 66.5%, part-Niuean 13.4%, non-Niuean 20.1% (includes 12% European and Asian and 8% other Pacific Islanders) (2011 est.)

Languages: Niuean (official) 46% (a Polynesian language closely related to Tongan and Samoan), Niuean and English 32%, English (official) 11%, Niuean and others 5%, other 6% (2011 est.)

Religions: Ekalesia Niue (Congregational Christian Church of Niue—a Protestant church founded by missionaries from the London Missionary Society) 67%, other Protestant 3% (includes Seventh Day Adventist 1%, Presbyterian 1%, and Methodist 1%), Mormon 10%, Roman Catholic 10%, Jehovah's Witnesses 2%, other 6%, none 2% (2011 est.)

Population growth rate: -0.03% (2014 est.)
country comparison to the world: 198

Population distribution: population distributed around the peripheral coastal areas of the island

Urbanization: *urban population:* 46.2% of total population (2020)
rate of urbanization: 1.69% annual rate of change (2015-20 est.)
total population growth rate v. urban population growth rate, 2000-2030:

Major urban areas—population: 1,000 ALOFI (capital) (2018)

Sex ratio: NA

Infant mortality rate: *total:* NA (2018)
male: NA
female: NA

Life expectancy at birth: *total population:* NA (2017 est.)
male: NA
female: NA

Total fertility rate: NA

Drinking water source: *improved:* total: 98.2% of population
unimproved: *total:* 1.8% of population (2017 est.)

Current Health Expenditure: 8.6% (2017)

Sanitation facility access:
improved:
total: 96.8% of population
unimproved:
total: 3.2% of population (2017 est.)

HIV/AIDS—adult prevalence rate: NA

HIV/AIDS—people living with HIV/AIDS: NA

HIV/AIDS—deaths: NA

Major infectious diseases: *degree of risk:* high (2020)

food or waterborne diseases: bacterial diarrhea

vectorborne diseases: malaria

Obesity—adult prevalence rate: 50% (2016)
country comparison to the world: 6

Education expenditures: NA

GOVERNMENT

Country name: *conventional long form:* none
conventional short form: Niue
former: Savage Island
etymology: the origin of the name is obscure; in Niuean, the word supposedly translates as "behold the coconut"
note: pronunciation falls between nyu-way and new-way, but not like new-wee

Dependency status: self-governing in free association with New Zealand since 1974; Niue is fully responsible for internal affairs; New Zealand retains responsibility for external affairs and defense; however, these responsibilities confer no rights of control and are only exercised at the request of the Government of Niue

Government type: parliamentary democracy

Capital: *name:* Alofi

geographic coordinates: 19 01 S, 169 55 W
time difference: UTC-11 (6 hours behind Washington, DC, during Standard Time)

Administrative divisions: none; note—there are no first-order administrative divisions as defined by the US Government, but there are 14 villages at the second order

Independence: 19 October 1974 (Niue became a self-governing state in free association with New Zealand)

National holiday: Waitangi Day (Treaty of Waitangi established British sovereignty over New Zealand), 6 February (1840)

Constitution: *history:* several previous (New Zealand colonial statutes); latest 19 October 1974 (Niue Constitution Act 1974)
amendments: proposed by the Assembly; passage requires at least two-thirds majority vote of the Assembly membership in each of three readings and approval by the majority of votes in a referendum; passage of amendments to a number of sections, including Niue's self-governing status, British nationality and New Zealand citizenship, external affairs and defense, economic and administrative assistance by New Zealand, and amendment procedures, requires at least two- thirds majority vote by the Assembly and at least two thirds of votes in a referendum; amended 1992, 2007

Legal system: English common law

Suffrage: 18 years of age; universal

Executive branch: *chief of state:* Queen ELIZABETH II (since 6 February 1952); represented by Governor-General of New Zealand Dame Patricia Lee REDDY (since 28 September 2016); the UK and New Zealand are represented by New Zealand High Commissioner Kirk YATES (since May 2018)

head of government: Premier Dalton TAGELAGI (since 10 June 2020)
cabinet: Cabinet chosen by the premier
elections/appointments: the monarchy is hereditary; premier indirectly elected by the Legislative Assembly for a 3-year term; election last held on 10 June 2020 (next to be held in 2023)
election results: Dalton TAGELAGI elected premier; Legislative Assembly vote—Dalton TAGELAGI (independent) 13, O'Love JACOBSEN (independent) 7; Toke TALAGI lost his seat in election

Legislative branch: *description:* unicameral Assembly or Fono Ekepule (20 seats; 14 members directly elected in single-seat constituencies by simple majority vote and 6 directly elected from the National Register or "common roll" by majority vote; members serve 3-year terms)
elections: last held on 30 May 2020 (next to be held in 2023)
election results: percent of vote by party—NA; seats by party—independent 20

Judicial branch: *highest courts:* Court of Appeal (consists of the chief justice and up to 3 judges); note—the Judicial Committee of the Privy Council (in London) is the final appeal court beyond the Niue Court of Appeal
judge selection and term of office: Niue chief justice appointed by the governor general on the advice of the Cabinet and tendered by the premier; other judges appointed by the governor general on the advice of the Cabinet and tendered by the chief justice and the minister of justice; judges serve until age 68
subordinate courts: High Court
note: Niue is a participant in the Pacific Judicial Development Program, which is designed to build governance and the rule of law in 15 Pacific island countries

Political parties and leaders: Alliance of Independents or AI
Niue People's Action Party or NPP [Young VIVIAN]

International organization participation: ACP, AOSIS, FAO, IFAD, OPCW, PIF, Sparteca, SPC, UNESCO, UPU, WHO, WIPO, WMO

Diplomatic representation in the US: none (self-governing territory in free association with New Zealand)

Diplomatic representation from the US: none (self-governing territory in free association with New Zealand)

Flag description: yellow with the flag of the UK in the upper hoist-side quadrant; the flag of the UK bears five yellow five-pointed stars—a large star on a blue disk in the center and a smaller star on each arm of the bold red cross; the larger star stands for Niue, the smaller stars recall the Southern Cross constellation on the New Zealand flag and symbolize links with that country; yellow represents the bright sunshine of Niue and the warmth and friendship between Niue and New Zealand

National symbol(s): *yellow, five-pointed star; national color:* yellow

National anthem: *name:* "Ko e Iki he Lagi" (The Lord in Heaven)

lyrics/music: unknown/unknown, prepared by Sioeli FUSIKATA

note: adopted 1974

ECONOMY

Economy—overview: The economy suffers from the typical Pacific island problems of geographic isolation, few resources, and a small population. The agricultural sector consists mainly of subsistence gardening, although some cash crops are grown for export. Industry consists primarily of small factories for processing passion fruit, lime oil, honey, and coconut cream. The sale of postage stamps to foreign collectors is an important source of revenue.

Government expenditures regularly exceed revenues, and the shortfall is made up by critically needed grants from New Zealand that are used to pay wages to public employees. Economic aid allocation from New Zealand in FY13/14 was US$10.1 million. Niue has cut government expenditures by reducing the public service by almost half.

The island in recent years has suffered a serious loss of population because of emigration to New Zealand. Efforts to increase GDP include the promotion of tourism and financial services, although the International Banking Repeal Act of 2002 resulted in the termination of all offshore banking licenses.

GDP (purchasing power parity):
$10.01 million (2003 est.)
country comparison to the world: 228

GDP (official exchange rate):
$10.01 million (2003) (2003)

GDP—real growth rate: 6.2% (2003 est.)
country comparison to the world: 26

GDP—per capita (PPP): $5,800 (2003 est.)
country comparison to the world: 168

GDP—composition, by sector of origin:
agriculture: 23.5% (2003)
industry: 26.9% (2003)
services: 49.5% (2003)

Agriculture—products: coconuts, passion fruit, honey, limes, taro, yams, cassava (manioc, tapioca), sweet potatoes; pigs, poultry, beef cattle

Industries: handicrafts, food processing

Industrial production growth rate: NA

Labor force: 663 (2001)
country comparison to the world: 230

Labor force—by occupation: *note:* most work on family plantations; paid work exists only in government service, small industry, and the Niue Development Board

Unemployment rate: 12% (2001)
country comparison to the world: 164

Population below poverty line: NA

Household income or consumption by percentage share: *lowest 10%:* NA

highest 10%: NA

Budget: *revenues:* 15.07 million (FY04/05)
expenditures: 16.33 million (FY04/05)

Budget surplus (+) or deficit (-): -12.6% (of GDP) (FY04/05)
country comparison to the world: 215

Fiscal year: 1 April—31 March

Inflation rate (consumer prices): 4% (2005)
country comparison to the world: 155

Exports: $201,400 (2004 est.)
country comparison to the world: 221

Exports—commodities: canned coconut cream, copra, honey, vanilla, passion fruit products, pawpaws, root crops, limes, footballs, stamps, handicrafts

Imports: $9.038 million (2004 est.)
country comparison to the world: 223

Imports—commodities: food, live animals, manufactured goods, machinery, fuels, lubricants, chemicals, drugs

Debt—external: $418,000 (2002 est.)
country comparison to the world: 202

Exchange rates: New Zealand dollars (NZD) per US dollar—
1.416 (2017 est.)
1.4279 (2016 est.)
1.4279 (2015)
1.4279 (2014 est.)
1.2039 (2013 est.)

ENERGY

Electricity—production: 3 million kWh (2016 est.)
country comparison to the world: 216

Electricity—consumption: 2.79 million kWh (2016 est.)
country comparison to the world: 215

Electricity—exports: 0 kWh (2016 est.)
country comparison to the world: 179

Electricity—imports: 0 kWh (2016 est.)
country comparison to the world: 181

Electricity—installed generating capacity: 2,300 kW (2016 est.)
country comparison to the world: 214

Electricity—from fossil fuels: 87% of total installed capacity (2016 est.)
country comparison to the world: 65

Electricity—from nuclear fuels: 0% of total installed capacity (2017 est.)
country comparison to the world: 158

Electricity—from hydroelectric plants: 0% of total installed capacity (2017 est.)
country comparison to the world: 191

Electricity—from other renewable sources: 13% of total installed capacity (2017 est.)
country comparison to the world: 68

Crude oil—production: 0 bbl/day (2018 est.)
country comparison to the world: 184

Crude oil—exports: 0 bbl/day (2015 est.)
country comparison to the world: 177

Crude oil—imports: 0 bbl/day (2015 est.)

country comparison to the world: 178

Crude oil—proved reserves: 0 bbl (1 January 2018 est.)
country comparison to the world: 179

Refined petroleum products—production: 0 bbl/day (2017 est.)
country comparison to the world: 187

Refined petroleum products—consumption: 50 bbl/day (2016 est.)
country comparison to the world: 215

Refined petroleum products—exports: 0 bbl/day (2015 est.)
country comparison to the world: 189

Refined petroleum products—imports: 54 bbl/day (2015 est.)
country comparison to the world: 211

Natural gas—production: 0 cu m (2017 est.)
country comparison to the world: 181

Natural gas—consumption: 0 cu m (2017 est.)
country comparison to the world: 185

Natural gas—exports: 0 cu m (2017 est.)
country comparison to the world: 164

Natural gas—imports: 0 cu m (2017 est.)
country comparison to the world: 170

Natural gas—proved reserves: 0 cu m (1 January 2014 est.)
country comparison to the world: 179

Carbon dioxide emissions from consumption of energy: 7,252 Mt (2017 est.)
country comparison to the world: 213

COMMUNICATIONS

Telecommunication systems: *general assessment:* sole provider service for over 1000 landlines and fixed wireless lines; cellular telephone service operates on AMPS and GSM platforms; difficult geography presents challenges for rural areas; mobile is primary source of Internet access; mobile broadband demand is growing due to mobile services (2020)

domestic: single-line (fixed line) telephone system connects all villages (and virtually all households) on island (2018)

international: country code—683; landing point for the Manatua submarine cable linking Niue to several South Pacific Ocean Islands; expansion of satellite services (2019)

note: the COVID-19 outbreak is negatively impacting telecommunications production and supply chains globally; consumer spending on telecom devices and services has also slowed due to the pandemic's effect on economies worldwide; overall progress towards improvements in all facets of the telecom industry—mobile, fixed-line, broadband, submarine cable and satellite—has moderated

Broadcast media: 1 government-owned TV station with many of the programs supplied by Television New Zealand; 1 government-owned radio station broadcasting in AM and FM (2019)

Internet country code: .nu

Internet users: *total:* 1,090

percent of population: 91.6% (July 2016 est.)

country comparison to the world: 225

TRANSPORTATION

Airports: 1 (2013)
country comparison to the world: 231
Airports—with paved runways: *total:* 1 (2017)
1,524 to 2,437 m: 1 (2017)
Airports—with unpaved runways: *total:* 1 (2013)
1,524 to 2,437 m: 1 (2013)

Roadways: *total:* 234 km (2017)
paved: 210 km (2017)
unpaved: 24 km
country comparison to the world: 206
Merchant marine: *total:* 61
by type: bulk carrier 4, container ship 1, general cargo 29, oil tanker 2, other 25 (2019)
country comparison to the world: 108
Ports and terminals: *major seaport(s):* Alofi

MILITARY AND SECURITY

Military and security forces: no regular indigenous military forces; Police Force (2019)
Military—note: defense is the responsibility of New Zealand

TRANSNATIONAL ISSUES

Disputes—international: none

NORFOLK ISLAND

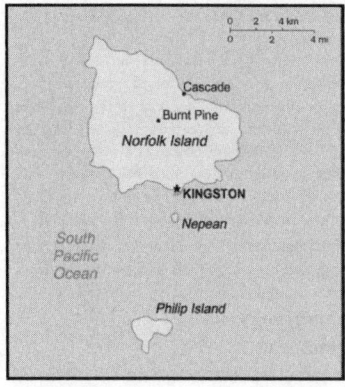

INTRODUCTION ISLAND

Background: Two British attempts at establishing the island as a penal colony (1788-1814 and 1825-55) were ultimately abandoned. In 1856, the island was resettled by Pitcairn Islanders, descendants of the Bounty mutineers and their Tahitian companions.

GEOGRAPHY ISLAND

Location: Oceania, island in the South Pacific Ocean, east of Australia
Geographic coordinates: 29 02 S, 167 57 E
Map references: Oceania
Area: *total:* 36 sq km
land: 36 sq km
water: 0 sq km
country comparison to the world: 235
Area—comparative: about 0.2 times the size of Washington, DC
Land boundaries: 0 km
Coastline: 32 km
Maritime claims: *territorial sea:* 12 nm
exclusive fishing zone: 200 nm
Climate: subtropical; mild, little seasonal temperature variation
Terrain: volcanic island with mostly rolling plains

Elevation: *lowest point:* Pacific Ocean 0 m
highest point: Mount Bates 319 m
Natural resources: fish
Land use: *agricultural land:* 25% (2011 est.)
arable land: 0% (2011 est.) / permanent crops: 0% (2011 est.) / permanent pasture: 25% (2011 est.)
forest: 11.5% (2011 est.)
other: 63.5% (2011 est.)
Irrigated land: 0 sq km (2012)
Population distribution: population concentrated around the capital of Kingston
Natural hazards: tropical cyclones (especially May to July)
Environment—current issues: inadequate solid waste management; most freshwater obtained through rainwater catchment; preservation of unique ecosystem
Geography—note: most of the 32 km coastline consists of almost inaccessible cliffs, but the land slopes down to the sea in one small southern area on Sydney Bay, where the capital of Kingston is situated

PEOPLE AND SOCIETY

Population: 1,748 (2016 est.)
country comparison to the world: 233
Nationality: *noun:* Norfolk Islander(s)
adjective: Norfolk Islander(s)
Ethnic groups: Australian 22.8%, English 22.4%, Pitcairn 20%, Scottish 6%, Irish 5.2% (2011 est.)
note: respondents were able to identify up to two ancestries; percentages represent a proportion of all responses from people in Norfolk Island, including those who did not identify an ancestry; only top responses are shown
Languages: English (official) 44.9%, Norfolk (also known as Norfuk or Norf'k, which is a mixture of 18th century English and ancient Tahitian) 40.3%, Fijian 1.8%, other 6.8%, unspecified 6.2% (2016 est.)
note: data represent language spoken at home
Religions: Protestant 46.8% (Anglican 29.2%, Uniting Church in Australia 9.8%, Presbyterian 2.9%, Seventh Day Adventist 2.7%, other 2.2%), Roman Catholic 12.6%, other Christian 2.9%,

other 1.4%, none 26.7%, unspecified 9.5% (2016 est.)
Population growth rate: 0.01% (2014 est.)
country comparison to the world: 192
Population distribution: population concentrated around the capital of Kingston
Sex ratio: NA
Infant mortality rate: *total:* NA (2018)
male: NA
female: NA
Life expectancy at birth: *total population:* NA (2017 est.)
male: NA
female: NA
Total fertility rate: NA
HIV/AIDS—adult prevalence rate: NA
HIV/AIDS—people living with HIV/AIDS: NA
HIV/AIDS—deaths: NA
Education expenditures: NA

GOVERNMENT ISLAND

Country name: *conventional long form:* Territory of Norfolk Island
conventional short form: Norfolk Island
etymology: named by British explorer Captain James COOK after Mary HOWARD, Duchess of Norfolk, in 1774
Dependency status: self-governing territory of Australia; administered from Canberra by the Department of Regional Australia, Local Government, Arts, and Sport
Government type: non-self-governing overseas territory of Australia; note—the Norfolk Island Regional Council, which began operations 1 July 2016, is responsible for planning and managing a variety of public services, including those funded by the Government of Australia
Capital: *name:* Kingston
geographic coordinates: 29 03 S, 167 58 E
time difference: UTC+11 (16 hours ahead of Washington, DC, during Standard Time)
etymology: the name is a blending of the words "king's" and "town"; the British king at the time of the town's settlement in the late 18th century was George III

715

Administrative divisions: none (territory of Australia)

Independence: none (territory of Australia)

National holiday: Bounty Day (commemorates the arrival of Pitcairn Islanders), 8 June (1856)

Constitution: *history:* previous 1913, 1957; latest effective 7 August 1979
amendments: amended many times, last in 2015

Legal system: English common law and the laws of Australia

Citizenship: see Australia

Suffrage: 18 years of age; universal

Executive branch: *chief of state:* Queen ELIZABETH II (since 6 February 1952); represented by Governor General of the Commonwealth of Australia General Sir Peter COSGROVE (since 28 March 2014)

head of government: Administrator Eric HUTCHINSON (since 1 April 2017)
cabinet: Executive Council consists of 4 Legislative Assembly members
elections/appointments: the monarchy is hereditary; governor general appointed by the monarch; administrator appointed by the governor general of Australia for a 2-year term and represents the monarch and Australia

Legislative branch: *description:* unicameral Norfolk Island Regional Council (5 seats; councillors directly elected by simple majority vote to serve 4-year terms); mayor elected annually by the councillors
elections: elections last held 28 May 2016 (next to be held in 2020)
election results: seats by party—independent 5; composition—men 4, women 1, percent of women 20%
note: following an administrative restructuring of local government, the Legislative Assembly was dissolved on 18 June 2015 and replaced by an interim Norfolk Island Advisory Council effective 1 July 2015; the Advisory Council consisted of 5 members appointed by the Norfolk Island administrator based on nominations from the community; following elections on 28 May 2016, the new Norfolk Island Regional Council commenced operations on 1 July 2016

Judicial branch: *highest courts:* Supreme Court of Norfolk Island (consists of the chief justice and several justices); note—appeals beyond the Supreme Court of Norfolk Island are heard by the Federal Court and the High Court of Australia
judge selection and term of office: justices appointed by the governor general of Australia from among justices of the Federal Court of Australia; justices serve until mandatory retirement at age 70
subordinate courts: Petty Court of Sessions; specialized courts, including a Coroner's Court and the Employment Tribunal

Political parties and leaders:
Norfolk Island Labor Party [Mike KELLY]
Norfolk Liberals [John BROWN]

International organization participation: UPU

Diplomatic representation in the US: none (territory of Australia)

Diplomatic representation from the US: none (territory of Australia)

Flag description: three vertical bands of green (hoist side), white, and green with a large green Norfolk Island pine tree centered in the slightly wider white band; green stands for the rich vegetation on the island, and the pine tree—endemic to the island—is a symbol of Norfolk Island
note: somewhat reminiscent of the flag of Canada with its use of only two colors and depiction of a prominent local floral symbol in the central white band; also resembles the green and white triband of Nigeria

National symbol(s): Norfolk Island pine

National anthem: *name:* Come Ye Blessed
lyrics/music: New Testament/John Prindle SCOTT
note: the local anthem, whose lyrics consist of the words from Matthew 25:34-36, 40, is also known as "The Pitcairn Anthem;" the island does not recognize "Advance Australia Fair" (which other Australian territories use); instead "God Save the Queen" is official (see United Kingdom)

ECONOMY ISLAND

Economy—overview: Norfolk Island is suffering from a severe economic downturn. Tourism, the primary economic activity, is the main driver of economic growth. The agricultural sector has become self-sufficient in the production of beef, poultry, and eggs.

GDP (purchasing power parity): NA

Agriculture—products: Norfolk Island pine seed, Kentia palm seed, cereals, vegetables, fruit; cattle, poultry

Industries: tourism, light industry, ready mixed concrete

Labor force: 978 (2006)
country comparison to the world: 229

Labor force—by occupation: *agriculture:* 6%
industry: 14%
services: 80% (2006 est.)

Budget: *revenues:* 4.6 million (FY99/00)
expenditures: 4.8 million (FY99/00)

Fiscal year: 1 July–30 June

Exports: NA

Exports—commodities: postage stamps, seeds of the Norfolk Island pine and Kentia palm, small quantities of avocados

Imports: $NA

Imports—commodities: NA

Debt—external: NA

Exchange rates: Australian dollars (AUD) per US dollar—
1.311 (2017 est.)
1.3291 (2016 est.)
1.3291 (2015)
1.3291 (2014 est.)
1.1094 (2013 est.)

COMMUNICATIONS ISLAND

Telecommunication systems: *general assessment:* adequate, 4G mobile telecommunication network (2020)
domestic: free local calls
international: country code—672; submarine cable links with Australia and New Zealand; satellite earth station—1
note: the COVID-19 outbreak is negatively impacting telecommunications production and supply chains globally; consumer spending on telecom devices and services has also slowed due to the pandemic's effect on economies worldwide; overall progress towards improvements in all facets of the telecom industry—mobile, fixed-line, broadband, submarine cable and satellite—has moderated

Broadcast media: 1 local radio station; broadcasts of several Australian radio and TV stations available via satellite (2009)

Internet country code: .nf

Internet users: *total:* 765

percent of population: 34.6% (July 2016 est.)
country comparison to the world: 228

TRANSPORTATION ISLAND

Airports: 1 (2013)
country comparison to the world: 232

Airports—with paved runways: *total:* 1 (2019)
1,524 to 2,437 m: 1

Roadways: *total:* 80 km (2008)
paved: 53 km (2008)
unpaved: 27 km (2008)
country comparison to the world: 215

Ports and terminals: *major seaport(s):* Kingston

MILITARY AND SECURITY

Military—note: defense is the responsibility of Australia

TRANSNATIONAL ISSUES

Disputes—international: none

NORTH MACEDONIA

INTRODUCTION

Background: North Macedonia gained its independence peacefully from Yugoslavia in 1991 under the name of "Macedonia." Greek objection to the new country's name, insisting it implied territorial pretensions to the northern Greek province of Macedonia, and democratic backsliding for several years stalled the country's movement toward Euro-Atlantic integration. Immediately after Macedonia declared independence, Greece sought to block Macedonian efforts to gain UN membership if the name "Macedonia" was used. The country was eventually admitted to the UN in 1993 as "The former Yugoslav Republic of Macedonia," and at the same time it agreed to UN-sponsored negotiations on the name dispute. In 1995, Greece lifted a 20-month trade embargo and the two countries agreed to normalize relations, but the issue of the name remained unresolved and negotiations for a solution continued. Over time, the US and over 130 other nations recognized Macedonia by its constitutional name, Republic of Macedonia. Ethnic Albanian grievances over perceived political and economic inequities escalated into a conflict in 2001 that eventually led to the internationally brokered Ohrid Framework Agreement, which ended the fighting and established guidelines for constitutional amendments and the creation of new laws that enhanced the rights of minorities. In January 2018, the government adopted a new law on languages, which elevated the Albanian language to an official language at the national level, with the Macedonian language remaining the sole official language in international relations. Relations between ethnic Macedonians and ethnic Albanians remain complicated, however.

North Macedonia's pro-Western government has used its time in office since 2017 to sign a historic deal with Greece in June 2018 to end the name dispute and revive Skopje's NATO and EU membership prospects. This followed a nearly three-year political crisis that engulfed the country but ended in June 2017 following a six-month-long government formation period after a closely contested election in December 2016. The crisis began after the 2014 legislative and presidential election, and escalated in 2015 when the opposition party began releasing wiretapped material that revealed alleged widespread government corruption and abuse. Although an EU candidate since 2005, North Macedonia has yet to open EU accession negotiations. The country still faces challenges, including fully implementing reforms to overcome years of democratic backsliding and stimulating economic growth and development. In June 2018, Macedonia and Greece signed the Prespa Accord whereby the Republic of Macedonia agreed to change its name to the Republic of North Macedonia. Following ratification by both countries, the agreement went in to force on 12 February 2019. North Macedonia signed an accession protocol to become a NATO member state in February 2019.

GEOGRAPHY

Location: Southeastern Europe, north of Greece

Geographic coordinates: 41 50 N, 22 00 E

Map references: Europe

Area: *total:* 25,713 sq km
land: 25,433 sq km
water: 280 sq km
country comparison to the world: 150

Area—comparative: slightly larger than Vermont; almost four times the size of Delaware

Land boundaries: *total:* 838 km
border countries (5): Albania 181 km, Bulgaria 162 km, Greece 234 km, Kosovo 160 km, Serbia 101 km

Coastline: 0 km (landlocked)

Maritime claims: none (landlocked)

Climate: warm, dry summers and autumns; relatively cold winters with heavy snowfall

Terrain: mountainous with deep basins and valleys; three large lakes, each divided by a frontier line; country bisected by the Vardar River

Elevation: *mean elevation:* 741 m
lowest point: Vardar River 50 m
highest point: Golem Korab (Maja e Korabit) 2,764 m

Natural resources: low-grade iron ore, copper, lead, zinc, chromite, manganese, nickel, tungsten, gold, silver, asbestos, gypsum, timber, arable land

Land use: *agricultural land:* 44.3% (2011 est.)
arable land: 16.4% (2011 est.) / permanent crops: 1.4% (2011 est.) / permanent pasture: 26.5% (2011 est.)
forest: 39.8% (2011 est.)
other: 15.9% (2011 est.)
Irrigated land: 1,280 sq km (2012)

Population distribution: a fairly even distribution throughout most of the country, with urban areas attracting larger and denser populations

Natural hazards: high seismic risks

Environment—current issues: air pollution from metallurgical plants; Skopje has severe air pollution problems every winter as a result of industrial emissions, smoke from wood-buring stoves, and exhaust fumes from old cars

Environment—international agreements: *party to:* Air Pollution, Biodiversity, Climate Change, Climate Change-Kyoto Protocol, Desertification, Endangered Species, Hazardous Wastes, Law of the Sea, Ozone Layer Protection, Wetlands
signed, but not ratified: none of the selected agreements

Geography—note: landlocked; major transportation corridor from Western and Central Europe to Aegean Sea and Southern Europe to Western Europe

PEOPLE AND SOCIETY

Population: 2,125,971 (July 2020 est.)
country comparison to the world: 147

Nationality: *noun:* Macedonian(s)
adjective: Macedonian

Ethnic groups: Macedonian 64.2%, Albanian 25.2%, Turkish 3.9%, Romani 2.7%, Serb 1.8%, other 2.2% (2002 est.)
note: North Macedonia has not conducted a census since 2002; Romani populations are usually underestimated in official statistics and may represent 6.5-13% of North Macedonia's population

Languages: Macedonian (official) 66.5%, Albanian 25.1%, Turkish 3.5%, Romani 1.9%, Serbian 1.2%, other (includes Aromanian (Vlach) and Bosnian) 1.8% (2002 est.)
note: minority languages are co-official with Macedonian in municipalities where they are spoken by at least 20% of the population; Albanian is co-official in Tetovo, Brvenica, Vrapciste, and other municipalities; Turkish is co-official in Centar Zupa and Plasnica; Romani is co-official in Suto Orizari; Aromanian is co-official in Krusevo; Serbian is co-official in Cucer Sandevo

Religions: Macedonian Orthodox 64.8%, Muslim 33.3%, other Christian 0.4%, other and unspecified 1.5% (2002 est.)

Age structure: *0-14 years:* 16.16% (male 177,553/female 165,992)
15-24 years: 12.65% (male 139,250/female 129,770)
25-54 years: 44.47% (male 480,191/female 465,145)
55-64 years: 12.55% (male 131,380/female 135,407)
65 years and over: 14.17% (male 131,674/female 169,609) (2020 est.)

Dependency ratios: *total dependency ratio:* 44.5
youth dependency ratio: 23.6
elderly dependency ratio: 20.9
potential support ratio: 4.8 (2020 est.)

Median age: *total:* 39 years
male: 38 years
female: 40 years (2020 est.)
country comparison to the world: 58

Population growth rate: 0.15% (2020 est.)
country comparison to the world: 183

Birth rate: 10.7 births/1,000 population (2020 est.)
country comparison to the world: 183

Death rate: 9.6 deaths/1,000 population (2020 est.)
country comparison to the world: 43

Net migration rate: 0.4 migrant(s)/1,000 population (2020 est.)
country comparison to the world: 68

Population distribution: a fairly even distribution throughout most of the country, with urban areas attracting larger and denser populations

Urbanization: *urban population:* 58.5% of total population (2020)
rate of urbanization: 0.45% annual rate of change (2015-20 est.)
total population growth rate v. urban population growth rate, 2000-2030:

Major urban areas—population: 595,000 SKOPJE (capital) (2020)

Sex ratio: *at birth:* 1.07 male(s)/female
0-14 years: 1.07 male(s)/female
15-24 years: 1.07 male(s)/female
25-54 years: 1.03 male(s)/female
55-64 years: 0.97 male(s)/female
65 years and over: 0.78 male(s)/female
total population: 0.99 male(s)/female (2020 est.)

Mother's mean age at first birth: 27.2 years (2017 est.)

Maternal mortality rate: 7 deaths/100,000 live births (2017 est.)
country comparison to the world: 155

Infant mortality rate: *total:* 7.4 deaths/1,000 live births
male: 8.2 deaths/1,000 live births
female: 6.5 deaths/1,000 live births (2020 est.)
country comparison to the world: 156

Life expectancy at birth: *total population:* 76.3 years
male: 74.2 years
female: 78.6 years (2020 est.)
country comparison to the world: 97

Total fertility rate: 1.5 children born/woman (2020 est.)
country comparison to the world: 202

Contraceptive prevalence rate: 40.2% (2011)

Drinking water source:
improved:
urban: 99.8% of population
rural: 98.9% of population
total: 99.4% of population
unimproved:
urban: 0.2% of population

rural: 1.1% of population
total: 0.6% of population (2017 est.)

Current Health Expenditure: 6.1% (2017)

Physicians density: 2.87 physicians/1,000 population (2015)

Hospital bed density: 4.3 beds/1,000 population (2017)

Sanitation facility access:
improved:
urban: 100% of population
rural: 97.9% of population
total: 100% of population
unimproved:
urban: 2.8% of population
rural: 17.4% of population
total: 9.1% of population (2017 est.)

HIV/AIDS—adult prevalence rate: <.1% (2018 est.)

HIV/AIDS—people living with HIV/AIDS: <500 (2018 est.)

HIV/AIDS—deaths: 300 (2018 est.)
country comparison to the world: 60

Obesity—adult prevalence rate: 22.4% (2016)
country comparison to the world: 77

Children under the age of 5 years underweight: 1.3% (2011)
country comparison to the world: 124

Education expenditures: NA

Literacy: *definition:* age 15 and over can read and write
total population: 97.8%
male: 98.8%
female: 96.8% (2015)

School life expectancy (primary to tertiary education): *total:* 14 years
male: 13 years
female: 14 years (2018)

Unemployment, youth ages 15-24: *total:* 45.4%
male: 46.6%
female: 43.2% (2018 est.)
country comparison to the world: 7

GOVERNMENT

Country name: *conventional long form:* Republic of North Macedonia
conventional short form: North Macedonia
local long form: Republika Severna Makedonija
local short form: Severna Makedonija
former: Democratic Federal Macedonia, People's Republic of Macedonia, Socialist Republic of Macedonia, Republic of Macedonia
etymology: the country name derives from the ancient kingdom of Macedon (7th to 2nd centuries B.C.)

Government type: parliamentary republic

Capital: *name:* Skopje

geographic coordinates: 42 00 N, 21 26 E
time difference: UTC+1 (6 hours ahead of Washington, DC, during Standard Time)
daylight saving time: +1hr, begins last Sunday in March; ends last Sunday in October

etymology: Skopje derives from its ancient name Scupi, the Latin designation of a classical era Greco-Roman frontier fortress town; the name may go back even further to a pre-Greek, Illyrian name

Administrative divisions: 70 municipalities (opstini, singular—opstina) and 1 city* (grad); Aracinovo, Berovo, Bitola, Bogdanci, Bogovinje, Bosilovo, Brvenica, Caska, Centar Zupa, Cesinovo-Oblesevo, Cucer Sandevo, Debar, Debarca, Delcevo, Demir Hisar, Demir Kapija, Dojran, Dolneni, Gevgelija, Gostivar, Gradsko, Ilinden, Jegunovce, Karbinci, Kavadarci, Kicevo, Kocani, Konce, Kratovo, Kriva Palanka, Krivogastani, Krusevo, Kumanovo, Lipkovo, Lozovo, Makedonska Kamenica, Makedonski Brod, Mavrovo i Rostusa, Mogila, Negotino, Novaci, Novo Selo, Ohrid, Pehcevo, Petrovec, Plasnica, Prilep, Probistip, Radovis, Rankovce, Resen, Rosoman, Skopje*, Sopiste, Staro Nagoricane, Stip, Struga, Strumica, Studenicani, Sveti Nikole, Tearce, Tetovo, Valandovo, Vasilevo, Veles, Vevcani, Vinica, Vrapciste, Zelenikovo, Zelino, Zrnovci

Independence: 8 September 1991 (referendum by registered voters endorsed independence from Yugoslavia)

National holiday: Independence Day, 8 September (1991), also known as National Day

Constitution: *history:* several previous; latest adopted 17 November 1991, effective 20 November 1991
amendments: proposed by the president of the republic, by the government, by at least 30 members of the Assembly, or by petition by at least 150,000 citizens; final approval requires a two-thirds majority vote by the Assembly; amended several times, last in 2019

Legal system: civil law system; judicial review of legislative acts

International law organization participation: has not submitted an ICJ jurisdiction declaration; accepts ICCt jurisdiction

Citizenship: *citizenship by birth:* no
citizenship by descent only: at least one parent must be a citizen of North Macedonia
dual citizenship recognized: no
residency requirement for naturalization: 8 years

Suffrage: 18 years of age; universal

Executive branch: *chief of state:* President Stevo PENDAROVSKI (since 12 May 2019)

head of government: Prime Minister Zoran ZAEV (since 31 August 2020); note—Prime Minister ZAEV resigned on 3 January 2019 but was reelected by the Assembly on 31 August 2020 (62-51) following the delayed Assembly general election on 15 July 2020
cabinet: Council of Ministers elected by the Assembly by simple majority vote
elections/appointments: president directly elected using a modified 2-round system; a candidate can only be elected in the first round with an absolute majority from all registered voters; in the

second round, voter turnout must be at least 40% for the result to be deemed valid; president elected for a 5-year term (eligible for a second term); election last held on 21 April and 5 May 2019 (next to be held in 2024); following legislative elections, the leader of the majority party or majority coalition is usually elected prime minister by the Assembly; Zoran ZAEV reelected prime minister by the Assembly on 31 August 2020; Assembly vote—62 for, 51 against

election results: Stevo PENDAROVSKI elected president in second round; percent of vote in first round—Stevo PENDAROVSKI (SDSM) 44.8%, Gordana SILJANOVSKA-DAVKOVA (VMRO-DPMNE) 44.2%, Blenim REKA (independent) 11.1%; percent of vote in second round—Stevo PENDAROVSKI 53.6%, Gordana SILJANOVSKA-DAVKOVA 46.4%

Legislative branch: *description:* unicameral Assembly—Sobraine in Macedonian, Kuvend in Albanian (between 120 and 140 seats, currently 120; members directly elected in multiseat constituencies by closed-list proportional representation vote; possibility of 3 directly elected in diaspora constituencies by simple majority vote provided there is sufficient voter turnout; members serve 4-year terms)

elections: last election was to be held on 12 April 2020 but was postponed until 15 July 2020 due to the COVID-19 pandemic (next to be held in 2024)

election results: percent of vote by party/coalition—We Can 35.9%, Renewal 34.6%, BDI 11.5%, AfA-Alternative 9%, The Left 4.1%, PDSh 1.5%, other 3.4%; seats by party/coalition—We Can 46, Renewal 44, BDI 15, AfA-Alternative 12, The Left 2, PDSh 1

Judicial branch: *highest courts:* Supreme Court (consists of 22 judges); Constitutional Court (consists of 9 judges)

judge selection and term of office: Supreme Court judges nominated by the Judicial Council, a 7-member body of legal professionals, and appointed by the Assembly; judge tenure NA; Constitutional Court judges appointed by the Assembly for nonrenewable, 9-year terms

subordinate courts: Courts of Appeal; Basic Courts

Political parties and leaders: Alliance for Albanians or AfA [Ziadin SELA]
Alternative (Alternativa) [Afrim GASHI]
Besa Movement [Bilal KASAMI]
Democratic Party of Albanians or PDSh [Menduh THACI]
Democratic Union for Integration or BDI [Ali AHMETI]
Internal Macedonian Revolutionary Organization—Democratic Party for Macedonian National Unity or VMRO-DPMNE [Hristijan MICKOSKI]
Internal Macedonian Revolutionary Organization—People's Party or VMRO-NP [Ljubco GEORGIEVSKI]
Liberal Democratic Party or LDP [Goran MILEVSKI]

Renewal (VMRO-DPMNE coalition)
Social Democratic Union of Macedonia or SDSM [Zoran ZAEV]
The Left (Levica) [Dimitar APASIEV]
Turkish Democratic Party of DPT [Beycan ILYAS]
We Can (coalition includes SDSM/Besa/VMRO-NP, DPT, LDP)

International organization participation: BIS, CD, CE, CEI, EAPC, EBRD, EU (candidate country), FAO, IAEA, IBRD, ICAO, ICC (NGOs), ICCt, ICRM, IDA, IFAD, IFC, IFRCS, ILO, IMF, IMO, Interpol, IOC, IOM, IPU, ISO, ITU, ITUC (NGOs), MIGA, NATO, OAS (observer), OIF, OPCW, OSCE, PCA, PFP, SELEC, UN, UNCTAD, UNESCO, UNHCR, UNIDO, UNIFIL, UNWTO, UPU, WCO, WHO, WIPO, WMO, WTO

Diplomatic representation in the US: *chief of mission:* Ambassador Vasko NAUMOVSKI (since 18 November 2014)
chancery: 2129 Wyoming Avenue NW, Washington, DC 20008
telephone: [1] (202) 667-0501
FAX: [1] (202) 667-2131
consulate(s) general: Chicago, Detroit, New York

Diplomatic representation from the US: *chief of mission:* Ambassador Kate Marie BYRNES (since 12 July 2019)
telephone: [389] (2) 310-2000
embassy: Str. Samoilova, Nr. 21, 1000 Skopje
mailing address: American Embassy Skopje, US Department of State, 7120 Skopje Place, Washington, DC 20521-7120 (pouch)
FAX: [389] (2) 310-2499

Flag description: a yellow sun (the Sun of Liberty) with eight broadening rays extending to the edges of the red field; the red and yellow colors have long been associated with Macedonia

National symbol(s): *eight-rayed sun; national colors:* red, yellow

National anthem: *name:* "Denes nad Makedonija" (Today Over Macedonia)
lyrics/music: Vlado MALESKI/Todor SKALOVSKI
note: written in 1943 and adopted in 1991, the song previously served as the anthem of the Socialist Republic of Macedonia while part of Yugoslavia
0:00/ 0:49

ECONOMY

Economy—overview: Since its independence in 1991, Macedonia has made progress in liberalizing its economy and improving its business environment. Its low tax rates and free economic zones have helped to attract foreign investment, which is still low relative to the rest of Europe. Corruption and weak rule of law remain significant problems. Some businesses complain of opaque regulations and unequal enforcement of the law.

Macedonia's economy is closely linked to Europe as a customer for exports and source of investment, and has suffered as a result of prolonged weakness in the euro zone. Unemployment

has remained consistently high at about 23% but may be overstated based on the existence of an extensive gray market, estimated to be between 20% and 45% of GDP, which is not captured by official statistics.

Macedonia is working to build a country-wide natural gas pipeline and distribution network. Currently, Macedonia receives its small natural gas supplies from Russia via Bulgaria. In 2016, Macedonia signed a memorandum of understanding with Greece to build an interconnector that could connect to the Trans Adriatic Pipeline that will traverse the region once complete, or to an LNG import terminal in Greece.

Macedonia maintained macroeconomic stability through the global financial crisis by conducting prudent monetary policy, which keeps the domestic currency pegged to the euro, and inflation at a low level. However, in the last two years, the internal political crisis has hampered economic performance, with GDP growth slowing in 2016 and 2017, and both domestic private and public investments declining. Fiscal policies were lax, with unproductive public expenditures, including subsidies and pension increases, and rising guarantees for the debt of state owned enterprises, and fiscal targets were consistently missed. In 2017, public debt stabilized at about 47% of GDP, still relatively low compared to its Western Balkan neighbors and the rest of Europe.

GDP (purchasing power parity):
$31.03 billion (2017 est.)
$31.02 billion (2016 est.)
$30.15 billion (2015 est.)
note: data are in 2017 dollars; Macedonia has a large informal sector that may not be reflected in these data
country comparison to the world: 131

GDP (official exchange rate):
$11.37 billion (2017 est.)

GDP—real growth rate: 0% (2017 est.)
2.9% (2016 est.)
3.9% (2015 est.)
country comparison to the world: 194

GDP—per capita (PPP): $14,900 (2017 est.)
$15,000 (2016 est.)
$14,600 (2015 est.)
note: data are in 2017 dollars
country comparison to the world: 113

Gross national saving: '
30.3% of GDP (2017 est.)
29.9% of GDP (2016 est.)
28.5% of GDP (2015 est.)
country comparison to the world: 29

GDP—composition, by end use:
household consumption: 65.6% (2017 est.)
government consumption: 15.6% (2017 est.)
investment in fixed capital: 13.6% (2017 est.)
investment in inventories: 20.2% (2017 est.)
exports of goods and services: 54% (2017 est.)
imports of goods and services: -69% (2017 est.)

GDP—composition, by sector of origin:
agriculture: 10.9% (2017 est.)
industry: 26.6% (2017 est.)

services: 62.5% (2017 est.)

Agriculture—products: grapes, tobacco, vegetables, fruits; milk, eggs

Industries: food processing, beverages, textiles, chemicals, iron, steel, cement, energy, pharmaceuticals, automotive parts

Industrial production growth rate: -7.8% (2017 est.)
country comparison to the world: 198

Labor force: 793,000 (2020 est.)
country comparison to the world: 146

Labor force—by occupation: *agriculture:* 16.2%
industry: 29.2%
services: 54.5% (2017 est.)

Unemployment rate: 17.29% (2019 est.)
20.7% (2018 est.)
country comparison to the world: 181

Population below poverty line: 21.5% (2015 est.)

Household income or consumption by percentage share: *lowest 10%:* 1.7%
highest 10%: 25% (2015 est.)

Budget: *revenues:* 3.295 billion (2017 est.)
expenditures: 3.605 billion (2017 est.)

Taxes and other revenues: 29% (of GDP) (2017 est.)
country comparison to the world: 88

Budget surplus (+) or deficit (-): -2.7% (of GDP) (2017 est.)
country comparison to the world: 122

Public debt: 39.3% of GDP (2017 est.)
39.5% of GDP (2016 est.)
note: official data from Ministry of Finance; data cover central government debt; this data excludes debt instruments issued (or owned) by government entities other than the treasury; includes treasury debt held by foreign entitites; excludes debt issued by sub-national entities; there are no debt instruments sold for social funds
country comparison to the world: 134

Fiscal year: calendar year

Inflation rate (consumer prices): 1.4% (2017 est.)
-0.2% (2016 est.)
country comparison to the world: 78

Current account balance: -$151 million (2017 est.)
-$293 million (2016 est.)
country comparison to the world: 93

Exports: $4.601 billion (2017 est.)
$3.75 billion (2016 est.)
country comparison to the world: 112

Exports—partners: Germany 46.7%, Bulgaria 6.1%, Serbia 4.4%, Belgium 4.1% (2017)

Exports—commodities: foodstuffs, beverages, tobacco; textiles, miscellaneous manufactures, iron, steel; automotive parts

Imports: $6.63 billion (2017 est.)
$5.805 billion (2016 est.)
country comparison to the world: 115

Imports—commodities: machinery and equipment, automobiles, chemicals, fuels, food products

Imports—partners: Germany 11.9%, UK 10%, Greece 8%, Serbia 7.1%, China 5.9%, Italy 5.5%, Turkey 4.5%, Bulgaria 4.3% (2017)

Reserves of foreign exchange and gold: $2.802 billion (31 December 2017 est.)
$2.755 billion (31 December 2016 est.)
country comparison to the world: 112

Debt—external: $8.79 billion (31 December 2017 est.)
$7.685 billion (31 December 2016 est.)
country comparison to the world: 117

Exchange rates: Macedonian denars (MKD) per US dollar—55.8 (2017 est.)
55.733 (2016 est.)
55.733 (2015 est.)
55.537 (2014 est.)
46.437 (2013 est.)

ENERGY

Electricity access: *electrification—total population:* 100% (2020)

Electricity—production: 5.396 billion kWh (2016 est.)
country comparison to the world: 119

Electricity—consumption: 6.42 billion kWh (2016 est.)
country comparison to the world: 112

Electricity—exports: 58.5 million kWh (2016 est.)
country comparison to the world: 84

Electricity—imports: 2.191 billion kWh (2016 est.)
country comparison to the world: 55

Electricity—installed generating capacity: 1.828 million kW (2016 est.)
country comparison to the world: 115

Electricity—from fossil fuels: 60% of total installed capacity (2016 est.)
country comparison to the world: 132

Electricity—from nuclear fuels: 0% of total installed capacity (2017 est.)
country comparison to the world: 133

Electricity—from hydroelectric plants: 37% of total installed capacity (2017 est.)
country comparison to the world: 57

Electricity—from other renewable sources: 3% of total installed capacity (2017 est.)
country comparison to the world: 126

Crude oil—production: 0 bbl/day (2018 est.)
country comparison to the world: 166

Crude oil—exports: 142 bbl/day (2015 est.)
country comparison to the world: 80

Crude oil—imports: 0 bbl/day (2015 est.)
country comparison to the world: 157

Crude oil—proved reserves: 0 bbl (1 January 2018 est.)
country comparison to the world: 161

Refined petroleum products—production: 0 bbl/day (2015 est.)
country comparison to the world: 169

Refined petroleum products—consumption: 21,000 bbl/day (2016 est.)

country comparison to the world: 138

Refined petroleum products—exports: 3,065 bbl/day (2015 est.)
country comparison to the world: 99

Refined petroleum products—imports: 23,560 bbl/day (2015 est.)
country comparison to the world: 111

Natural gas—production: 0 cu m (2017 est.)
country comparison to the world: 162

Natural gas—consumption: 198.2 million cu m (2017 est.)
country comparison to the world: 105

Natural gas—exports: 0 cu m (2017 est.)
country comparison to the world: 143

Natural gas—imports: 198.2 million cu m (2017 est.)
country comparison to the world: 72

Natural gas—proved reserves: 0 cu m (31 December 2016 est.)
country comparison to the world: 163

Carbon dioxide emissions from consumption of energy: 7.459 million Mt (2017 est.)
country comparison to the world: 123

COMMUNICATIONS

Telephones—fixed lines: *total subscriptions:* 402,250
subscriptions per 100 inhabitants: 18.95 (2019 est.)
country comparison to the world: 100

Telephones—mobile cellular: *total subscriptions:* 2,094,037
subscriptions per 100 inhabitants: 98.65 (2019 est.)
country comparison to the world: 149

Telecommunication systems: *general assessment:* being part of the EU pre-accession process has led to stronger teledensity with a closer regulatory framework and independent regulators; administrative ties with the European Union have led to progress; broadband services are widely available; more customers moving to fiber networks; 2 mobile network operators; end of roaming tariffs (2020)
domestic: fixed-line 19 per 100 and mobile-cellular 99 per 100 subscriptions (2019)
international: country code—389
note: the COVID-19 outbreak is negatively impacting telecommunications production and supply chains globally; consumer spending on telecom devices and services has also slowed due to the pandemic's effect on economies worldwide; overall progress towards improvements in all facets of the telecom industry—mobile, fixed-line, broadband, submarine cable and satellite—has moderated

Broadcast media: public service TV broadcaster Macedonian Radio and Television operates 3 national terrestrial TV channels and 2 satellite TV channels; additionally, there are 10 regional TV stations that broadcast nationally using terrestrial transmitters, 54 TV channels with concession for cable TV, 9 regional TV stations with concessions for cable TV; 4 satellite TV channels broadcasting on a national level, 21 local commercial TV channels, and a large number of cable operators that offer domestic and international programming; the public radio broadcaster operates over 3 stations; there are 4 privately owned radio stations that broadcast nationally; 17 regional radio stations, and 49 local commercial radio stations (2019)

Internet country code: .mk

Internet users: *total:* 1,677,569

percent of population: 79.17% (July 2018 est.)
country comparison to the world: 127

Broadband—fixed subscriptions: *total:* 427,964
subscriptions per 100 inhabitants: 20 (2018 est.)
country comparison to the world: 88

TRANSPORTATION

Civil aircraft registration country code prefix: Z3 (2016)

Airports: 10 (2013)
country comparison to the world: 155

Airports—with paved runways: *total:* 8 (2017)
2,438 to 3,047 m: 2 (2017)
under 914 m: 6 (2017)

Airports—with unpaved runways: *total:* 2 (2013)
914 to 1,523 m: 1 (2013)
under 914 m: 1 (2013)

Pipelines: 262 km gas, 120 km oil (2017)

Railways: *total:* 925 km (2017)
standard gauge: 925 km 1.435-m gauge (313 km electrified) (2017)
country comparison to the world: 93

Roadways: *total:* 14,182 km (includes 290 km of expressways) (2017)
paved: 9,633 km (2017)
unpaved: 4,549 km (2017)
country comparison to the world: 128

MILITARY AND SECURITY

Military and security forces: Army of the Republic of North Macedonia (ARSM; includes a General Staff and subordinate Operations Command, Logistic Support Command, Training and Doctrine Command, and Center for Electronic Reconnaissance) (2020)
note: the Operations Command includes air, ground, special operations, support, and reserve forces

Military expenditures: 1.2% of GDP (2019)
0.9% of GDP (2018)
0.9% of GDP (2017)
1% of GDP (2016)
1% of GDP (2015)
country comparison to the world: 104

Military and security service personnel strengths: the Army of the Republic of North Macedonia

(ARSM) has approximately 8,000 active duty personnel (2019 est.)

Military equipment inventories and acquisitions: the inventory of North Macedonia's Army consists mostly of Soviet-era equipment; since 2010, it has received small amounts of equipment from Ireland and Turkey (2019 est.)

Military service age and obligation: 18 years of age for voluntary military service; conscription abolished in 2008 (2013)

TRANSNATIONAL ISSUES

Disputes—international: Kosovo and North Macedonia completed demarcation of their boundary in September 2008

Refugees and internally displaced persons: *stateless persons:* 567 (2019)
note: 506,774 estimated refugee and migrant arrivals (January 2015-August 2020); North Macedonia is predominantly a transit country and hosts fewer than 50 refugees and asylum seekers as of October 2017; 3,132 migrant arrivals in 2018

Illicit drugs: major transshipment point for Southwest Asian heroin and hashish; minor transit point for South American cocaine destined for Europe; although not a financial center and most criminal activity is thought to be domestic, money laundering is a problem due to a mostly cash-based economy and weak enforcement

NORTHERN MARIANA ISLANDS

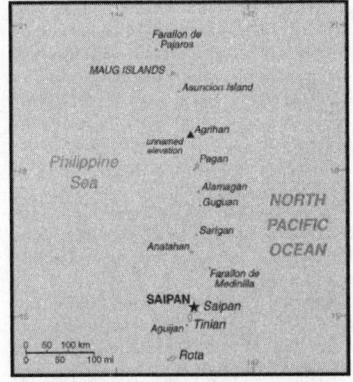

INTRODUCTION

Background: Under US administration as part of the UN Trust Territory of the Pacific, the people of the Northern Mariana Islands decided in the 1970s not to seek independence but instead to forge closer links with the US. Negotiations for territorial status began in 1972. A covenant to establish a commonwealth in political union with the US was approved in 1975, and came into force

on 24 March 1976. A new government and constitution went into effect in 1978.

GEOGRAPHY

Location: Oceania, islands in the North Pacific Ocean, about three-quarters of the way from Hawaii to the Philippines

Geographic coordinates: 15 12 N, 145 45 E

Map references: Oceania

Area: *total:* 464 sq km
land: 464 sq km
water: 0 sq km
note: consists of 14 islands including Saipan, Rota, and Tinian
country comparison to the world: 197

Area—comparative: 2.5 times the size of Washington, DC

Land boundaries: 0 km

Coastline: 1,482 km

Maritime claims: *territorial sea:* 12 nm
exclusive economic zone: 200 nm

Climate: tropical marine; moderated by northeast trade winds, little seasonal temperature variation; dry season December to June, rainy season July to October

Terrain: the southern islands in this north-south trending archipelago are limestone, with fringing coral reefs; the northern islands are volcanic, with active volcanoes on several islands

Elevation: *lowest point:* Pacific Ocean 0 m
highest point: unnamed elevation on Agrihan 965 m

Natural resources: arable land, fish

Land use: *agricultural land:* 6.6% (2011 est.)
arable land: 2.2% (2011 est.) / permanent crops: 2.2% (2011 est.) / permanent pasture: 2.2% (2011 est.)
forest: 65.5% (2011 est.)
other: 27.9% (2011 est.)
Irrigated land: 1 sq km (2012)

Population distribution: approximately 90% of the population lives on the island of Saipan

Natural hazards: active volcanoes on Pagan and Agrihan; typhoons (especially August to November)

Environment—current issues: contamination of groundwater on Saipan may contribute to disease; clean-up of landfill; protection of endangered species conflicts with development

Geography—note: strategic location in the North Pacific Ocean

PEOPLE AND SOCIETY

Population: 51,433 (July 2020 est.)
country comparison to the world: 210

Nationality: *noun:* NA (US citizens)
adjective: NA

Ethnic groups: Asian 50% (includes Filipino 35.3%, Chinese 6.8%, Korean 4.2%, and other Asian 3.7%), Native Hawaiian or other Pacific Islander 34.9% (includes Chamorro 23.9%, Carolinian 4.6%, and other Native Hawaiian or Pacific Islander 6.4%), other 2.5%, two or more ethnicities or races 12.7% (2010 est.)

Languages: Philippine languages 32.8%, Chamorro (official) 24.1%, English (official) 17%, other Pacific island languages 10.1%, Chinese 6.8%, other Asian languages 7.3%, other 1.9% (2010 est.)

Religions: Christian (Roman Catholic majority, although traditional beliefs and taboos may still be found)

Age structure: *0-14 years:* 25.02% (male 6,937/female 5,934)
15-24 years: 16.28% (male 4,518/female 3,857)
25-54 years: 37.44% (male 9,934/female 9,325)
55-64 years: 14.01% (male 3,921/female 3,286)
65 years and over: 7.23% (male 1,988/female 1,733) (2020 est.)

Median age: *total:* 32.8 years
male: 31.8 years
female: 34.1 years (2020 est.)
country comparison to the world: 105

Population growth rate: -0.55% (2020 est.)
country comparison to the world: 226

Birth rate: 15.1 births/1,000 population (2020 est.)
country comparison to the world: 118

Death rate: 5.3 deaths/1,000 population (2020 est.)
country comparison to the world: 192

Net migration rate: -15.4 migrant(s)/1,000 population (2020 est.)
country comparison to the world: 223

Population distribution: approximately 90% of the population lives on the island of Saipan

Urbanization: *urban population:* 91.8% of total population (2020)
rate of urbanization: 0.29% annual rate of change (2015-20 est.)
total population growth rate v. urban population growth rate, 2000-2030:

Major urban areas—population: 51,000 SAIPAN (capital) (2018)

Sex ratio: *at birth:* 1.16 male(s)/female
0-14 years: 1.17 male(s)/female
15-24 years: 1.17 male(s)/female
25-54 years: 1.07 male(s)/female
55-64 years: 1.19 male(s)/female
65 years and over: 1.15 male(s)/female
total population: 1.13 male(s)/female (2020 est.)

Infant mortality rate: *total:* 11.5 deaths/1,000 live births
male: 13.7 deaths/1,000 live births

female: 9 deaths/1,000 live births (2020 est.)
country comparison to the world: 114

Life expectancy at birth: *total population:* 76.1 years
male: 74 years
female: 78.5 years (2020 est.)
country comparison to the world: 105

Total fertility rate: 2.7 children born/woman (2020 est.)
country comparison to the world: 65

Drinking water source:
improved:
total: 100% of population
unimproved:
total: 0% of population (2017 est.)

Sanitation facility access:
improved:
total: 97.7% of population
unimproved:
total: 2.2% of population (2017 est.)

HIV/AIDS—adult prevalence rate: NA

HIV/AIDS—people living with HIV/AIDS: NA

HIV/AIDS—deaths: NA

Education expenditures: NA

GOVERNMENT

Country name: *conventional long form:* Commonwealth of the Northern Mariana Islands
conventional short form: Northern Mariana Islands
former: Trust Territory of the Pacific Islands, Mariana Islands District
abbreviation: CNMI
etymology: formally claimed and named by Spain in 1667 in honor of the Spanish Queen MARIANA of Austria

Dependency status: commonwealth in political union with and under the sovereignty of the US; federal funds to the Commonwealth administered by the US Department of the Interior, Office of Insular Affairs

Government type: republican form of government with separate executive, legislative, and judicial branches; a commonwealth in political union with and under the sovereignty of the US

Capital: *name:* Saipan

geographic coordinates: 15 12 N, 145 45 E
time difference: UTC+10 (15 hours ahead of Washington, DC, during Standard Time)
etymology: the entire island of Saipan is organized as a single municipality and serves as the capital; according to legend, when the first native voyagers arrived in their outrigger canoes they found an uninhabited island; to them it was like an empty voyage, so they named the island "saay" meaning "a voyage," and "peel" meaning "empty"; over time Saaypeel—"island of the empty voyage"—became Saipan

Administrative divisions: none (*commonwealth in political union with the US*); there are no first-order administrative divisions as defined by the US Government, but there are

4 municipalities at the second order: Northern Islands, Rota, Saipan, Tinian

Independence: none (commonwealth in political union with the US)

National holiday: Commonwealth Day, 8 January (1978)

Constitution: *history:* partially effective 9 January 1978 (Constitution of the Commonwealth of the Northern Mariana Islands); fully effective 4 November 1986 (Covenant Agreement)
amendments: proposed by constitutional convention, by public petition, or by the Legislature; ratification of proposed amendments requires approval by voters at the next general election or special election; amendments proposed by constitutional convention or by petition become effective if approved by a majority of voters and at least two-thirds majority of voters in each of two senatorial districts; amendments proposed by the Legislature are effective if approved by majority vote; amended several times, last in 2012

Legal system: the laws of the US apply, except for customs and some aspects of taxation

Citizenship: see United States

Suffrage: 18 years of age; universal; note—indigenous inhabitants are US citizens but do not vote in US presidential elections

Executive branch: *chief of state:* President Donald J. TRUMP (since 20 January 2017); Vice President Michael R. PENCE (since 20 January 2017)

head of government: Governor Ralph TORRES (since 29 December 2015); Lieutenant Governor Victor HOCOG (since 29 December 2015)
cabinet: Cabinet appointed by the governor with the advice and consent of the Senate
elections/appointments: president and vice president indirectly elected on the same ballot by an Electoral College of 'electors' chosen from each state; president and vice president serve a 4-year term (eligible for a second term); under the US Constitution, residents of the Northern Mariana Islands do not vote in elections for US president and vice president; however, they may vote in Democratic and Republican party presidential primary elections; governor directly elected by absolute majority vote in 2 rounds if needed; election last held on 13 November 2018 (next to be held in 2022)
election results: Ralph TORRES elected governor; percent of vote—Ralph TORRES (Republican) 62.2%, Juan BABAUTA (Independent) 37.8%; Arnold PALACIOS elected Lieutenant Governor

Legislative branch: *description:* bicameral Northern Marianas Commonwealth Legislature consists of: Senate (9 seats; members directly elected in single-seat constituencies by simple majority vote to serve 4-year terms) House of Representatives (20 seats; members directly elected in single-seat constituencies by simple majority vote to serve 2-year terms)
the Northern Mariana Islands directly elects 1 delegate to the US House of Representatives by simple majority vote to serve a 2-year term

elections: CNMI Senate—last held on 8 November 2016 (next to be held in November 2020)

CNMI House of Representatives—last held on 13 November 2018 (next to be held in November 2020)

Commonwealth of Northern Mariana Islands delegate to the US House of Representatives—last held on 13 November 2018 (next to be held in November 2020)

election results: CNMI Senate—percent of vote by party—NA; seats by party—Republican Party 6, independent 3; composition—men 8, women 1, percent of women 11.1%

CNMI House of Representatives—percent of vote by party—NA; seats by party—Republican Party 13, independent 7; composition—men 17, women 3, percent of women 15%; note—total CNMI Legislature percent of women 13.8%

delegate to US House of Representatives—seat won by Democratic Party; composition—1 man
note: the Northern Mariana Islands delegate to the US House of Representatives can vote when serving on a committee and when the House meets as the "Committee of the Whole House" but not when legislation is submitted for a "full floor" House vote

Judicial branch: *highest courts:* Supreme Court of the Commonwealth of the Northern Mariana Islands (CNMI) (consists of the chief justice and 2 associate justices); US Federal District Court (consists of 1 judge); note—US Federal District Court jurisdiction limited to US federal laws; appeals beyond the CNMI Supreme Court are referred to the US Supreme Court
judge selection and term of office: CNMI Supreme Court judges appointed by the governor and confirmed by the CNMI Senate; judges appointed for 8-year terms and another term if directly elected in a popular election; US Federal District Court judges appointed by the US president and confirmed by the US Senate; judges appointed for renewable 10-year terms
subordinate courts: Superior Court

Political parties and leaders: Democratic Party [Daniel QUITUGUA]
Republican Party [James ADA]

International organization participation: PIF (observer), SPC, UPU

Diplomatic representation from the US: none (commonwealth in political union with the US)

Flag description: blue with a white, five-pointed star superimposed on a gray latte stone (the traditional foundation stone used in building) in the center, surrounded by a wreath; blue symbolizes the Pacific Ocean, the star represents the Commonwealth; the latte stone and the floral head wreath display elements of the native Chamorro culture

National symbol(s): *latte stone; national colors:* blue, white

National anthem: *name:* "Gi Talo Gi Halom Tasi" (In the Middle of the Sea)

lyrics/music: Jose S. PANGELINAN [Chamoru], David PETER [Carolinian]/Wilhelm GANZHORN
note: adopted 1996; the Carolinian version of the song is known as "Satil Matawal Pacifico;" as a commonwealth of the US, in addition to the local anthem, "The Star-Spangled Banner" is official (see United States)
0:00/ 1:27

ECONOMY

Economy—overview: The economy of the Commonwealth of the Northern Mariana Islands(CNMI) has been on the rebound in the last few years, mainly on the strength of its tourism industry. In 2016, the CNMI's real GDP increased 28.6% over the previous year, following two years of relatively rapid growth in 2014 and 2015. Chinese and Korean tourists have supplanted Japanese tourists in the last few years. The Commonwealth is making a concerted effort to broaden its tourism by extending casino gambling from the small Islands of Tinian and Rota to the main Island of Saipan, its political and commercial center. Investment is concentrated on hotels and casinos in Saipan, the CNMI's largest island and home to about 90% of its population.

Federal grants have also contributed to economic growth and stability. In 2016, federal grants amounted to $101.4 billion which made up 26% of the CNMI government's total revenues. A small agriculture sector consists of cattle ranches and small farms producing coconuts, breadfruit, tomatoes, and melons.

Legislation is pending in the US Congress to extend the transition period to allow foreign workers to work in the CNMI on temporary visas.

GDP (purchasing power parity):
$1.242 billion (2016 est.)
$933 million (2015 est.)
$845 million (2014 est.)
note: GDP estimate includes US subsidy; data are in 2013 dollars
country comparison to the world: 203

GDP (official exchange rate):
$1.242 billion (2016 est.)

GDP—real growth rate: 28.6% (2016 est.)
3.8% (2015 est.)
3.5% (2014 est.)
country comparison to the world: 2

GDP—per capita (PPP): $24,500 (2016 est.)
$18,400 (2015 est.)
$16,600 (2014 est.)
country comparison to the world: 84

GDP—composition, by end use:
household consumption: 43.1% (2016 est.)
government consumption: 28.9% (2016 est.)
investment in fixed capital: 26.3% (2016 est.)
investment in inventories: NA (2016 est.)
exports of goods and services: 73.6% (2016 est.)
imports of goods and services: -71.9% (2016 est.)

GDP—composition, by sector of origin:
agriculture: 1.7% (2016)
industry: 58.1% (2016 est.)

services: 40.2% (2016)

Agriculture—products: vegetables and melons, fruits and nuts; ornamental plants; livestock, poultry, eggs; fish and aquaculture products

Industries: tourism, banking, construction, fishing, handicrafts, other services

Industrial production growth rate: NA

Labor force: 27,970 (2010 est.)
note: includes foreign workers
country comparison to the world: 205

Labor force—by occupation: *agriculture:* 1.9%
industry: 10%
services: 88.1% (2010 est.)

Unemployment rate: 11.2% (2010 est.)
8% (2005 est.)
country comparison to the world: 157

Population below poverty line: NA

Household income or consumption by percentage share: *lowest 10%:* NA
highest 10%: NA

Budget: *revenues:* 389.6 million (2016 est.)
expenditures: 344 million (2015 est.)

Taxes and other revenues: 31.4% (of GDP) (2016 est.)
country comparison to the world: 72

Budget surplus (+) or deficit (-): 3.7% (of GDP) (2016 est.)
country comparison to the world: 11

Public debt: 7.1% of GDP (2017 est.)
country comparison to the world: 201

Fiscal year: 1 October—30 September

Inflation rate (consumer prices): 0.3% (2016 est.)
0.1% (2015 est.)
country comparison to the world: 21

Exports: $914 million (2016 est.)
$520 million (2015 est.)
country comparison to the world: 162

Exports—commodities: garments

Imports: $893 million (2016 est.)
$638 million (2015 est.)
country comparison to the world: 187

Imports—commodities: food, construction equipment and materials, petroleum products

Debt—external: NA

Exchange rates: the US dollar is used

ENERGY

Electricity access: *electrification—total population:* 100% (2020)

Electricity—production: 60,600 kWh (2009)
country comparison to the world: 217

Electricity—consumption: 48,300 kWh (2009)
country comparison to the world: 217

Electricity—exports: 0 kWh (2009 est.)
country comparison to the world: 180

Electricity—imports: 0 kWh (January 2009 est.)
country comparison to the world: 182

COMMUNICATIONS

Telephones—fixed lines: *total subscriptions:* 20,398
subscriptions per 100 inhabitants: 39.44 (2019 est.)
country comparison to the world: 176

Telecommunication systems: *general assessment:* digital fiber-optic cables and satellites connect the islands to worldwide networks; demand for broadband growing given that mobile services are the source for Internet across region; future launch of 5G (2020)
domestic: wide variety of services available including dial-up and broadband Internet, mobile cellular, international private lines, payphones, phone cards, voicemail, and automatic call distribution systems; fixed-line teledensity 39 per 100 persons (2019)
international: country code—1-670; landing points for the Atisa and Mariana-Guam submarine cables linking Mariana islands to Guam; satellite earth stations—2 Intelsat (Pacific Ocean) (2019)

note: the COVID-19 outbreak is negatively impacting telecommunications production and supply chains globally; consumer spending on telecom devices and services has also slowed due to the pandemic's effect on economies worldwide; overall progress towards improvements in all facets of the telecom industry—mobile, fixed-line, broadband, submarine cable and satellite—has moderated

Broadcast media: 1 TV broadcast station on Saipan; multi-channel cable TV services are available on Saipan; 9 licensed radio broadcast stations (2009)

Internet country code: .mp

Internet users: *total:* 16,000

percent of population: 30.6% (July 2016 est.)
country comparison to the world: 211

TRANSPORTATION

Airports: 5 (2013)
country comparison to the world: 183

Airports—with paved runways: *total:* 3 (2019)
2,438 to 3,047 m: 2

1,524 to 2,437 m: 1

Airports—with unpaved runways: *total:* 2 (2013)
2,438 to 3,047 m: 1 (2013)
under 914 m: 1 (2013)

Heliports: 1 (2013)

Roadways: *total:* 536 km (2008)
country comparison to the world: 194

Merchant marine: *total:* 1
by type: other 1 (2019)
country comparison to the world: 179

Ports and terminals: *major seaport(s):* Saipan, Tinian, Rota

MILITARY AND SECURITY

Military—note: defense is the responsibility of the US

TRANSNATIONAL ISSUES

Disputes—international: none

NORWAY

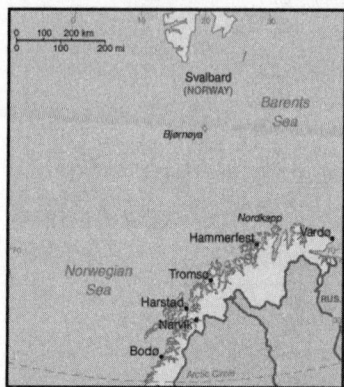

INTRODUCTION

Background: Two centuries of Viking raids into Europe tapered off following the adoption of Christianity by King Olav TRYGGVASON in 994; conversion of the Norwegian kingdom occurred over the next several decades. In 1397, Norway was absorbed into a union with Denmark that lasted more than four centuries. In 1814, Norwegians resisted the cession of their country to Sweden and adopted a new constitution. Sweden then invaded Norway but agreed to let Norway keep its constitution in return for accepting the union under a Swedish king. Rising nationalism throughout the 19th century led to a 1905 referendum granting Norway independence. Although Norway remained neutral in World War I, it

suffered heavy losses to its shipping. Norway proclaimed its neutrality at the outset of World War II, but was nonetheless occupied for five years by Nazi Germany (1940-45). In 1949, Norway abandoned neutrality and became a member of NATO. Discovery of oil and gas in adjacent waters in the late 1960s boosted Norway's economic fortunes. In referenda held in 1972 and 1994, Norway rejected joining the EU. Key domestic issues include immigration and integration of ethnic minorities, maintaining the country's extensive social safety net with an aging population, and preserving economic competitiveness.

GEOGRAPHY

Location: Northern Europe, bordering the North Sea and the North Atlantic Ocean, west of Sweden

Geographic coordinates: 62 00 N, 10 00 E

Map references: Europe

Area: *total:* 323,802 sq km
land: 304,282 sq km
water: 19,520 sq km
country comparison to the world: 69

Area—comparative: slightly larger than twice the size of Georgia; slightly larger than New Mexico

Land boundaries: *total:* 2,566 km
border countries (3): Finland 709 km, Sweden 1666 km, Russia 191 km

Coastline: 25,148 km (includes mainland 2,650 km, as well as long fjords, numerous small islands, and minor indentations 22,498 km; length of island coastlines 58,133 km)

Maritime claims: *territorial sea:* 12 nm
exclusive economic zone: 200 nm
contiguous zone: 10 nm
continental shelf: 200 nm

Climate: temperate along coast, modified by North Atlantic Current; colder interior with increased precipitation and colder summers; rainy year-round on west coast

Terrain: glaciated; mostly high plateaus and rugged mountains broken by fertile valleys; small, scattered plains; coastline deeply indented by fjords; arctic tundra in north

Elevation: *mean elevation:* 460 m
lowest point: Norwegian Sea 0 m
highest point: Galdhopiggen 2,469 m

Natural resources: petroleum, natural gas, iron ore, copper, lead, zinc, titanium, pyrites, nickel, fish, timber, hydropower

Land use: *agricultural land:* 2.7% (2011 est.)
arable land: 2.2% (2011 est.) / permanent crops: 0% (2011 est.) / permanent pasture: 0.5% (2011 est.)
forest: 27.8% (2011 est.)
other: 69.5% (2011 est.)
Irrigated land: 900 sq km (2012)

Population distribution: most Norweigans live in the south where the climate is milder and there is better connectivity to mainland Europe; population clusters are found all along the North Sea coast in the southwest, and Skaggerak in the southeast; the interior areas of the north remain sparsely populated

Natural hazards: rockslides, avalanches

volcanism: Beerenberg (2,227 m) on Jan Mayen Island in the Norwegian Sea is the country's only active volcano

Environment—current issues: water pollution; acid rain damaging forests and adversely affecting lakes, threatening fish stocks; air pollution from vehicle emissions

Environment—international agreements: *party to:* Air Pollution, Air Pollution-Nitrogen Oxides, Air Pollution-Persistent Organic Pollutants, Air Pollution-Sulfur 85, Air Pollution-Sulfur 94, Air Pollution-Volatile Organic Compounds, Antarctic-Environmental Protocol, Antarctic-Marine Living Resources, Antarctic Seals, Antarctic Treaty, Biodiversity, Climate Change, Climate Change-Kyoto Protocol, Desertification, Endangered Species, Environmental Modification, Hazardous Wastes, Law of the Sea, Marine Dumping, Ozone Layer Protection, Ship Pollution, Tropical Timber 83, Tropical Timber 94, Wetlands, Whaling
signed, but not ratified: none of the selected agreements

Geography—note: about two-thirds mountains; some 50,000 islands off its much-indented coastline; strategic location adjacent to sea lanes and air routes in North Atlantic; one of the most rugged and longest coastlines in the world

PEOPLE AND SOCIETY

Population: 5,467,439 (July 2020 est.)
country comparison to the world: 119

Nationality: *noun:* Norwegian(s)
adjective: Norwegian

Ethnic groups: Norwegian 83.2% (includes about 60,000 Sami), other European 8.3%, other 8.5% (2017 est.)

Languages: Bokmal Norwegian (official), Nynorsk Norwegian (official), small Sami- and Finnish-speaking minorities
note: Sami has three dialects: Lule, North Sami, and South Sami; Sami is an official language in nine municipalities in Norway's three northernmost counties: Finnmark, Nordland, and Troms

Religions: Church of Norway (Evangelical Lutheran—official) 70.6%, Muslim 3.2%, Roman Catholic 3%, other Christian 3.7%, other 2.5%, unspecified 17% (2016 est.)

Age structure: *0-14 years:* 17.96% (male 503,013/female 478,901)
15-24 years: 12.02% (male 336,597/female 320,720)
25-54 years: 40.75% (male 1,150,762/female 1,077,357)
55-64 years: 11.84% (male 328,865/female 318,398)
65 years and over: 17.43% (male 442,232/female 510,594) (2020 est.)

Dependency ratios: *total dependency ratio:* 53.3
youth dependency ratio: 26.5
elderly dependency ratio: 26.9
potential support ratio: 3.7 (2020 est.)
note: data include Svalbard and Jan Mayen Islands

Median age: *total:* 39.5 years
male: 38.8 years
female: 40.2 years (2020 est.)
country comparison to the world: 57

Population growth rate: 0.85% (2020 est.)
country comparison to the world: 123

Birth rate: 12.2 births/1,000 population (2020 est.)
country comparison to the world: 159

Death rate: 8.1 deaths/1,000 population (2020 est.)
country comparison to the world: 89

Net migration rate: 4 migrant(s)/1,000 population (2020 est.)
country comparison to the world: 29

Population distribution: most Norwegians live in the south where the climate is milder and there is better connectivity to mainland Europe; population clusters are found all along the North Sea coast in the southwest, and Skaggerak in the southeast; the interior areas of the north remain sparsely populated

Urbanization: *urban population:* 83% of total population (2020)
rate of urbanization: 1.4% annual rate of change (2015-20 est.)
note: data include Svalbard and Jan Mayen Islands
total population growth rate v. urban population growth rate, 2000-2030:

Major urban areas—population: 1.041 million OSLO (capital) (2020)

Sex ratio: *at birth:* 1.05 male(s)/female
0-14 years: 1.05 male(s)/female
15-24 years: 1.05 male(s)/female
25-54 years: 1.07 male(s)/female
55-64 years: 1.03 male(s)/female
65 years and over: 0.87 male(s)/female
total population: 1.02 male(s)/female (2020 est.)

Mother's mean age at first birth: 29.3 years (2017 est.)
note: data is calculated based on actual age at first births

Maternal mortality rate: 2 deaths/100,000 live births (2017 est.)
country comparison to the world: 183

Infant mortality rate: *total:* 2.5 deaths/1,000 live births
male: 2.8 deaths/1,000 live births
female: 2.2 deaths/1,000 live births (2020 est.)
country comparison to the world: 223

Life expectancy at birth: *total population:* 82.1 years
male: 80 years
female: 84.4 years (2020 est.)
country comparison to the world: 23

Total fertility rate: 1.84 children born/woman (2020 est.)
country comparison to the world: 141

Drinking water source:
improved:
urban: 100% of population
rural: 100% of population
total: 100% of population
unimproved:

urban: 0% of population
rural: 0% of population
total: 0% of population (2017 est.)

Current Health Expenditure: 10.4% (2017)

Physicians density: 2.83 physicians/1,000 population (2017)

Hospital bed density: 3.6 beds/1,000 population (2017)

Sanitation facility access:
improved:
urban: 100% of population
rural: 100% of population
total: 100% of population
unimproved:
urban: 0% of population
rural: 0% of population
total: 0% of population (2017 est.)

HIV/AIDS—adult prevalence rate: 0.1% (2018 est.)
country comparison to the world: 128

HIV/AIDS—people living with HIV/AIDS: 5,800 (2018 est.)
country comparison to the world: 119

HIV/AIDS—deaths: <100 (2018 est.)

Obesity—adult prevalence rate: 23.1% (2016)
country comparison to the world: 68

Education expenditures: 8% of GDP (2016)
country comparison to the world: 5

School life expectancy (primary to tertiary education): *total:* 18 years
male: 18 years
female: 19 years (2018)

Unemployment, youth ages 15-24: *total:* 9.7%
male: 10.7%
female: 8.6% (2018 est.)
country comparison to the world: 127

GOVERNMENT

Country name: *conventional long form:* Kingdom of Norway
conventional short form: Norway
local long form: Kongeriket Norge
local short form: Norge
etymology: derives from the Old Norse words "nordr" and "vegr" meaning "northern way" and refers to the long coastline of western Norway

Government type: parliamentary constitutional monarchy

Capital: *name:* Oslo

geographic coordinates: 59 55 N, 10 45 E
time difference: UTC+ 1 (6 hours ahead of Washington, DC, during Standard Time)
daylight saving time: +1hr, begins last Sunday in March; ends last Sunday in October
etymology: the medieval name was spelt "Aslo"; the "as" component refered either to the Ekeberg ridge southeast of the town ("as" in modern Norwegian), or to the Aesir (Norse gods); "lo" refered to "meadow," so the most likely interpretations would have been either "the meadow beneath the ridge" or "the meadow of the gods"; both explanations are considered equally plausible

Administrative divisions: 18 counties (fylker, singular—fylke); Akershus, Aust-Agder, Buskerud, Finnmark, Hedmark, Hordaland, More og Romsdal, Nordland, Oppland, Oslo, Ostfold, Rogaland, Sogn og Fjordane, Telemark, Troms, Trondelag, Vest-Agder, Vestfold

Dependent areas: Bouvet Island, Jan Mayen, Svalbard

Independence: 7 June 1905 (declared the union with Sweden dissolved); 26 October 1905 (Sweden agreed to the repeal of the union); notable earlier dates: ca. 872 (traditional unification of petty Norwegian kingdoms by HARALD Fairhair); 1397 (Kalmar Union of Denmark, Norway, and Sweden); 1524 (Denmark-Norway); 17 May 1814 (Norwegian constitution adopted); 4 November 1814 (Sweden-Norway union confirmed)

National holiday: Constitution Day, 17 May (1814)

Constitution: history: drafted spring 1814, adopted 16 May 1814, signed by Constituent Assembly 17 May 1814
amendments: proposals submitted by members of Parliament or by the government within the first three years of Parliament's four-year term; passage requires two-thirds majority vote of a two-thirds quorum in the next elected Parliament; amended over 400 times, last in 2020 (2020)

Legal system: mixed legal system of civil, common, and customary law; Supreme Court can advise on legislative acts

International law organization participation: accepts compulsory ICJ jurisdiction with reservations; accepts ICCt jurisdiction

Citizenship: citizenship by birth: no
citizenship by descent only: at least one parent must be a citizen of Norway
dual citizenship recognized: no
residency requirement for naturalization: 7 years

Suffrage: 18 years of age; universal

Executive branch: chief of state: King HARALD V (since 17 January 1991); Heir Apparent Crown Prince HAAKON MAGNUS (son of the monarch, born 20 July 1973)

head of government: Prime Minister Erna SOLBERG (since 16 October 2013)
cabinet: Council of State appointed by the monarch, approved by Parliament
elections/appointments: the monarchy is hereditary; following parliamentary elections, the leader of the majority party or majority coalition usually appointed prime minister by the monarch with the approval of the parliament

Legislative branch: description: unicameral Parliament or Storting (169 seats; members directly elected in multi-seat constituencies by proportional representation vote; members serve 4-year terms)
elections: last held on 11 September 2017 (next to be held in September 2021)
election results: percent of vote by party—Ap 27.4%, H 25%, FrP 15.2%, SP 10.3%, SV 6%, V 4.4%, KrF 4.2%, MDG 3.2%, R 2.4%, other/invalid 1.9%; seats by party—Ap 49, H 45, FrP 27, SP 19, SV 11, V 8, KrF 8, MDG 1, R 1; composition—men 99, women 70, percent of women 41.4%

Judicial branch: highest courts: Supreme Court or Hoyesterett (consists of the chief justice and 18 associate justices)
judge selection and term of office: justices appointed by the monarch (King in Council) upon the recommendation of the Judicial Appointments Board; justices can serve until mandatory retirement at age 70
subordinate courts: Courts of Appeal or Lagmennsrett; regional and district courts; Conciliation Boards; ordinary and special courts; note—in addition to professionally trained judges, elected lay judges sit on the bench with professional judges in the Courts of Appeal and district courts

Political parties and leaders: Center Party or Sp [Trygve Slagsvold VEDUM]
Christian Democratic Party or KrF [Kjell Ingolf ROPSTADT]
Conservative Party or H [Erna SOLBERG]
Green Party or MDG [Rasmus HANSSON and Une Aina BASTHOLM]
Labor Party or Ap [Jonas Gahr STORE]
Liberal Party or V [Trine SKEI GRANDE]
Progress Party or FrP [Siv JENSEN]
Red Party or R [Bionar MOXNES]
Socialist Left Party or SV [Audun LYSBAKKEN]

International organization participation: ADB (nonregional member), AfDB (nonregional member), Arctic Council, Australia Group, BIS, CBSS, CD, CE, CERN, EAPC, EBRD, EFTA, EITI (implementing country), ESA, FAO, FATF, IADB, IAEA, IBRD, ICAO, ICC (national committees), ICCt, ICRM, IDA, IEA, IFAD, IFC, IFRCS, IGAD (partners), IHO, ILO, IMF, IMO, IMSO, Interpol, IOC, IOM, IPU, ISO, ITSO, ITU, ITUC (NGOs), MIGA, MINUSMA, NATO, NC, NEA, NIB, NSG, OAS (observer), OECD, OPCW, OSCE, Paris Club, PCA, Schengen Convention, UN, UNCTAD, UNESCO, UNHCR, UNIDO, UNITAR, UNMISS, UNRWA, UNTSO, UNWTO, UPU, WCO, WHO, WIPO, WMO, WTO, ZC

Diplomatic representation in the US: chief of mission: Ambassador Anniken Ramberg KRUTNES (since 17 September 2020)
chancery: 2720 34th Street NW, Washington, DC 20008
telephone: [1] (202) 333-6000
FAX: [1] (202) 469-3990
consulate(s) general: Houston, New York, San Francisco

Diplomatic representation from the US: chief of mission: Ambassador Kenneth BRAITHWAITE (since 8 February 2018)
telephone: [47] 21-30-85-40
embassy: Morgedalsvegen 36, 0378 Oslo
mailing address: PO Box 4075 AMB 0244 Oslo
FAX: [47] 22-44-33-63, 22-56-27-51

Flag description: red with a blue cross outlined in white that extends to the edges of the flag; the vertical part of the cross is shifted to the hoist side in the style of the Dannebrog (Danish flag); the colors recall Norway's past political unions with Denmark (red and white) and Sweden (blue)

National symbol(s): lion; national colors: red, white, blue

National anthem: name: "Ja, vi elsker dette landet" (Yes, We Love This Country) lyrics/music: lyrics/music: Bjornstjerne BJORNSON/Rikard NORDRAAK
note: adopted 1864; in addition to the national anthem, "Kongesangen" (Song of the King), which uses the tune of "God Save the Queen," serves as the royal anthem
0:00/ 1:02

ECONOMY

Economy—overview: Norway has a stable economy with a vibrant private sector, a large state sector, and an extensive social safety net. Norway opted out of the EU during a referendum in November 1994. However, as a member of the European Economic Area, Norway partially participates in the EU's single market and contributes sizably to the EU budget.

The country is richly endowed with natural resources such as oil and gas, fish, forests, and minerals. Norway is a leading producer and the world's second largest exporter of seafood, after China. The government manages the country's petroleum resources through extensive regulation. The petroleum sector provides about 9% of jobs, 12% of GDP, 13% of the state's revenue, and 37% of exports, according to official national estimates. Norway is one of the world's leading petroleum exporters, although oil production is close to 50% below its peak in 2000. Gas production, conversely, has more than doubled since 2000. Although oil production is historically low, it rose in 2016 for the third consecutive year due to the higher production of existing oil fields and to new fields coming on stream. Norway's domestic electricity production relies almost entirely on hydropower.

In anticipation of eventual declines in oil and gas production, Norway saves state revenue from petroleum sector activities in the world's largest sovereign wealth fund, valued at over $1 trillion at the end of 2017. To help balance the federal budget each year, the government follows a "fiscal rule," which states that spending of revenues from petroleum and fund investments shall correspond to the expected real rate of return on the fund, an amount it estimates is sustainable over time. In February 2017, the government revised the expected rate of return for the fund downward from 4% to 3%.

After solid GDP growth in the 2004-07 period, the economy slowed in 2008, and contracted in 2009, before returning to modest, positive growth from 2010 to 2017. The Norwegian economy has been adjusting to lower energy prices, as demonstrated by growth in labor force participation and employment in 2017. GDP growth was about 1.5% in 2017, driven largely by domestic demand, which has been boosted by the rebound in the labor market and supportive fiscal policies. Economic growth is expected to remain constant or improve slightly in the next few years.

GDP (purchasing power parity):
$381.2 billion (2017 est.)
$374 billion (2016 est.)
$370 billion (2015 est.)
note: data are in 2017 dollars
country comparison to the world: 48

GDP (official exchange rate):
$398.8 billion (2017 est.)

GDP—real growth rate: 0.86% (2019 est.)
1.36% (2018 est.)
2.75% (2017 est.)
country comparison to the world: 176

GDP—per capita (PPP): $72,100 (2017 est.)
$71,200 (2016 est.)
$71,100 (2015 est.)
note: data are in 2017 dollars
country comparison to the world: 11

Gross national saving:
34.3% of GDP (2017 est.)
33.1% of GDP (2016 est.)
35.5% of GDP (2015 est.)
country comparison to the world: 18

GDP—composition, by end use:
household consumption: 44.8% (2017 est.)
government consumption: 24% (2017 est.)
investment in fixed capital: 24.1% (2017 est.)
investment in inventories: 4.8% (2017 est.)
exports of goods and services: 35.5% (2017 est.)
imports of goods and services: -33.2% (2017 est.)

GDP—composition, by sector of origin:
agriculture: 2.3% (2017 est.)
industry: 33.7% (2017 est.)
services: 64% (2017 est.)

Agriculture—products: barley, wheat, potatoes; pork, beef, veal, milk; fish

Industries: petroleum and gas, shipping, fishing, aquaculture, food processing, shipbuilding, pulp and paper products, metals, chemicals, timber, mining, textiles

Industrial production growth rate: 1.5% (2017 est.)
country comparison to the world: 143

Labor force: 2.699 million (2020 est.)
country comparison to the world: 110

Labor force—by occupation: *agriculture:* 2.1%
industry: 19.3%
services: 78.6% (2016 est.)

Unemployment rate: 3.72% (2019 est.)
3.89% (2018 est.)
country comparison to the world: 53

Population below poverty line: NA

Household income or consumption by percentage share: *lowest 10%:* 3.8%
highest 10%: 21.2% (2014)

Budget: *revenues:* 217.1 billion (2017 est.)
expenditures: 199.5 billion (2017 est.)

Taxes and other revenues: 54.4% (of GDP) (2017 est.)
country comparison to the world: 9

Budget surplus (+) or deficit (-): 4.4% (of GDP) (2017 est.)
country comparison to the world: 8

Public debt: 36.5% of GDP (2017 est.)
36.4% of GDP (2016 est.)
note: data cover general government debt and include debt instruments issued (or owned) by government entities other than the treasury; the data exclude treasury debt held by foreign entities; the data exclude debt issued by subnational entities, as well as intragovernmental debt; intragovernmental debt consists of treasury borrowings from surpluses in the social funds, such as for retirement, medical care, and unemployment; debt instruments for the social funds are not sold at public auctions
country comparison to the world: 146

Fiscal year: calendar year

Inflation rate (consumer prices): 1.9% (2017 est.)
3.6% (2016 est.)
country comparison to the world: 99

Current account balance: $16.656 billion (2019 est.)
$31.111 billion (2018 est.)
country comparison to the world: 18

Exports: $102.8 billion (2017 est.)
$88.88 billion (2016 est.)
country comparison to the world: 36

Exports—partners: UK 21.1%, Germany 15.5%, Netherlands 9.9%, Sweden 6.6%, France 6.4%, Belgium 4.8%, Denmark 4.7%, US 4.6% (2017)

Exports—commodities: petroleum and petroleum products, machinery and equipment, metals, chemicals, ships, fish

Imports: $95.06 billion (2017 est.)
$74.94 billion (2016 est.)
country comparison to the world: 36

Imports—commodities: machinery and equipment, chemicals, metals, foodstuffs

Imports—partners: Sweden 11.4%, Germany 11%, China 9.8%, US 6.8%, South Korea 6.7%, Denmark 5.4%, UK 4.7% (2017)

Reserves of foreign exchange and gold: $65.92 billion (31 December 2017 est.)
$57.46 billion (31 December 2015 est.)
country comparison to the world: 34

Debt—external: $642.3 billion (31 March 2016 est.)
$640.1 billion (31 March 2015 est.)
note: Norway is a net external creditor
country comparison to the world: 17

Exchange rates: Norwegian kroner (NOK) per US dollar—8.308 (2017 est.)
8.3978 (2016 est.)
8.3978 (2015 est.)
8.0646 (2014 est.)
6.3021 (2013 est.)

ENERGY

Electricity access: *electrification—total population:* 100% (2020)

Electricity—production: 147.7 billion kWh (2016 est.)
country comparison to the world: 29

Electricity—consumption: 122.2 billion kWh (2016 est.)
country comparison to the world: 29

Electricity—exports: 15.53 billion kWh (2016 est.)
country comparison to the world: 12

Electricity—imports: 5.741 billion kWh (2016 est.)
country comparison to the world: 34

Electricity—installed generating capacity: 33.86 million kW (2016 est.)
country comparison to the world: 30

Electricity—from fossil fuels: 3% of total installed capacity (2016 est.)
country comparison to the world: 208

Electricity—from nuclear fuels: 0% of total installed capacity (2017 est.)
country comparison to the world: 159

Electricity—from hydroelectric plants: 93% of total installed capacity (2017 est.)
country comparison to the world: 8

Electricity—from other renewable sources: 4% of total installed capacity (2017 est.)
country comparison to the world: 115

Crude oil—production: 1.517 million bbl/day (2018 est.)
country comparison to the world: 15

Crude oil—exports: 1.383 million bbl/day (2017 est.)
country comparison to the world: 10

Crude oil—imports: 36,550 bbl/day (2017 est.)
country comparison to the world: 58

Crude oil—proved reserves: 6.376 billion bbl (1 January 2018)
country comparison to the world: 20

Refined petroleum products—production: 371,600 bbl/day (2017 est.)
country comparison to the world: 38

Refined petroleum products—consumption: 205.300 bbl/day (2017 est.)
country comparison to the world: 57

Refined petroleum products—exports: 432,800 bbl/day (2017 est.)
country comparison to the world: 20

Refined petroleum products—imports: 135.300 bbl/day (2017 est.)
country comparison to the world: 43

Natural gas—production: 123.9 billion cu m (2017 est.)
country comparison to the world: 7

Natural gas—consumption: 4.049 billion cu m (2017 est.)
country comparison to the world: 65

Natural gas—exports: 120.2 billion cu m (2017 est.)
country comparison to the world: 3

Natural gas—imports: 5.663 million cu m (2017 est.)
country comparison to the world: 77

Natural gas—proved reserves: 1.782 trillion cu m (1 January 2018 est.)
country comparison to the world: 20

Carbon dioxide emissions from consumption of energy: 39.8 million Mt (2017 est.)
country comparison to the world: 65

COMMUNICATIONS

Telephones—fixed lines: *total subscriptions:* 571,958
subscriptions per 100 inhabitants: 10.55 (2019 est.)
country comparison to the world: 90

Telephones—mobile cellular: *total subscriptions:* 5,810,113
subscriptions per 100 inhabitants: 107.17 (2019 est.)
country comparison to the world: 113

Telecommunication systems: *general assessment:* one of the most advanced telecommunications networks in Europe; high mobile and broadband penetration rates and highly developed digital media sector; forward leaning in LTE-A developments; migrate all DSL subscribers to fiber by 2023; looking to close 2G and 3G networks by 2025; regulator competes 700 MHz auction and assigns spectrum for 5G, partners with Chinese company Huawei (2020)
domestic: Norway has a domestic satellite system; the prevalence of rural areas encourages the wide use of mobilecellular systems; fixed-line 11 per 100 and mobile-cellular 107 per 100 (2019)
international: country code—47; landing points for the Svalbard Undersea Cable System, Polar Circle Cable, Bodo-Rost Cable, NOR5KE Viking, Celtic Norse, Tempnet Offshore FOC Network, England Cable, Denmark-Norway6, Havfrue/AEC- 2, Skagerrak 4, and the Skagenfiber West & East submarine cables providing links to other Nordic countries, Europe and the US; satellite earth stations—Eutelsat, Intelsat (Atlantic Ocean), and 1 Inmarsat (Atlantic and Indian Ocean regions); note—Norway shares the Inmarsat earth station with the other Nordic countries (Denmark, Finland, Iceland, and Sweden) (2019)
note: the COVID-19 outbreak is negatively impacting telecommunications production and supply chains globally; consumer spending on telecom devices and services has also slowed due to the pandemic's effect on economies worldwide; overall progress towards improvements in all facets of the telecom industry—mobile, fixed-line, broadband, submarine cable and satellite—has moderated

Broadcast media: state-owned public radio-TV broadcaster operates 3 nationwide TV stations, 3 nationwide radio stations, and 16 regional radio stations; roughly a dozen privately owned TV stations broadcast nationally and roughly another 25 local TV stations broadcasting; nearly 75% of households have access to multi-channel cable or satellite TV; 2 privately owned radio stations broadcast nationwide and another 240 stations operate locally; Norway is the first country in the world to phase out FM radio in favor of Digital Audio Broadcasting (DAB), a process scheduled for completion in late 2017 (2019)

Internet country code: .no

Internet users: *total:* 5,183,627

percent of population: 96.49% (July 2018 est.)
country comparison to the world: 82

Broadband—fixed subscriptions: *total:* 2,206,519
subscriptions per 100 inhabitants: 41 (2018 est.)
country comparison to the world: 53

TRANSPORTATION

National air transport system: *number of registered air carriers:* 8 (2020)
inventory of registered aircraft operated by air carriers: 125

Civil aircraft registration country code prefix: LN (2016)

Airports: 95 (2013)
country comparison to the world: 61

Airports—with paved runways: *total:* 67 (2017)
2,438 to 3,047 m: 14 (2017)
1,524 to 2,437 m: 10 (2017)
914 to 1,523 m: 22 (2017)
under 914 m: 21 (2017)

Airports—with unpaved runways: *total:* 28 (2013)
914 to 1,523 m: 6 (2013)
under 914 m: 22 (2013)

Heliports: 1 (2013)

Pipelines: 8520 km gas, 1304 km oil/condensate (2017)

Railways: *total:* 4,200 km (2019)
standard gauge: 4,200 km 1.435-m gauge (2,480 km electrified) (2019)
country comparison to the world: 45

Roadways: *total:* 94,902 km (includes 455 km of expressways) (2018)
country comparison to the world: 52

Waterways: 1,577 km (2010)
country comparison to the world: 51

Merchant marine: *total:* 1,576
by type: bulk carrier 98, general cargo 216, oil tanker 87, other 1,175 (2019)
country comparison to the world: 18

Ports and terminals: *major seaport(s):* Bergen, Haugesund, Maaloy, Mongstad, Narvik, Sture
LNG terminal(s) (export): Kamoy, Kollsnes, Melkoya Islan

LNG terminal(s) (import): Fredrikstad, Mosjoen

MILITARY AND SECURITY

Military and security forces: *Norwegian Armed Forces:* Norwegian Army (Haeren), Royal Norwegian Navy (Kongelige Norske Sjoeforsvaret; includes Coastal Rangers and Coast Guard (Kystvakt)), Royal Norwegian Air Force (Kongelige Norske Luftforsvaret), Home Guard (Heimevernet, HV) (2020)

Military expenditures: 1.8% of GDP (2019 est.)
1.73% of GDP (2018)
1.71% of GDP (2017)
1.73% of GDP (2016)
1.59% of GDP (2015)
country comparison to the world: 63

Military and security service personnel strengths: the Norwegian Armed Forces have approximately 23,000 active personnel (8,400 Army; 3,500 Navy; 3,500 Air Force; 600 active Home Guard; 7,000 other, including special operations, cyber, joint staff, intelligence, logistics support, etc.) (2020)
note: the Home Guard has approximately 40,000 total personnel

Military equipment inventories and acquisitions: the Norwegian Armed Forces inventory includes mostly imported European and US weapons systems, as well as a limited mix of domestically-produced equipment, particularly small naval craft; since 2010, the US is the leading supplier of weapons systems to Norway, followed by France, Italy, South Korea, and Spain (2019 est.)

Military deployments: 120 Lithuania (NATO) (2020)

Military service age and obligation: 19-35 years of age for male and female selective compulsory military service; 17 years of age for male volunteers (16 in wartime); 18 years of age for women; 19-month service obligation; conscripts first serve 12 months from 19-28, and then up to 4-5 refresher training periods until age 35, 44, 55, or 60 depending on rank and function. (2019)

TRANSNATIONAL ISSUES

Disputes—international: Norway asserts a territorial claim in Antarctica (Queen Maud Land and its continental shelf); Denmark (Greenland) and Norway have made submissions to the Commission on the Limits of the Continental Shelf (CLCS) and Russia is collecting additional data to augment its 2001 CLCS submission; Norway and Russia signed a comprehensive maritime boundary agreement in 2010

Refugees and internally displaced persons: *refugees (country of origin):* 14,359 (Syria), 14,038 (Eritrea), 6,518 (Somalia), 5,108 (Afghanistan) (2019)

stateless persons: 2,272 (2019)

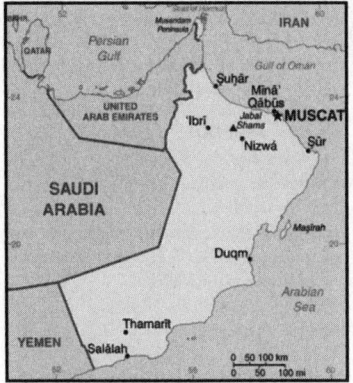

INTRODUCTION

Background: The inhabitants of the area of Oman have long prospered from Indian Ocean trade. In the late 18th century, the nascent sultanate in Muscat signed the first in a series of friendship treaties with Britain. Over time, Oman's dependence on British political and military advisors increased, although the sultanate never became a British colony. In 1970, QABOOS bin Said Al-Said overthrew his father, and has since ruled as sultan. Sultan QABOOS has no children and has not designated a successor publicly; the Basic Law of 1996 outlines Oman's succession procedure. Sultan QABOOS' extensive modernization program opened the country to the outside world, and the sultan has prioritized strategic ties with the UK and US. Oman's moderate, independent foreign policy has sought to maintain good relations with its neighbors and to avoid external entanglements.

Inspired by the popular uprisings that swept the Middle East and North Africa beginning in January 2011, some Omanis staged demonstrations, calling for more jobs and economic benefits and an end to corruption. In response to those protester demands, QABOOS in 2011 pledged to implement economic and political reforms, such as granting Oman's bicameral legislative body more power and authorizing direct elections for its lower house, which took place in November 2011. Additionally, the Sultan increased unemployment benefits, and, in August 2012, issued a royal directive mandating the speedy implementation of a national job creation plan for thousands of public and private sector Omani jobs. As part of the government's efforts to decentralize authority and allow greater citizen participation in local governance, Oman successfully conducted its first municipal council elections in December 2012. Announced by the sultan in 2011, the municipal councils have the power to advise the Royal Court on the needs of local districts across Oman's 11 governorates. Sultan QABOOS, Oman's longest reigning monarch, died on 11 January 2020. His cousin, HAYTHAM bin Tariq Al-Said, former Minister of Heritage and Culture, was sworn in as Oman's new sultan the same day.

GEOGRAPHY

Location: Middle East, bordering the Arabian Sea, Gulf of Oman, and Persian Gulf, between Yemen and the UAE

Geographic coordinates: 21 00 N, 57 00 E

Map references: Middle East

Area: *total:* 309,500 sq km
land: 309,500 sq km
water: 0 sq km
country comparison to the world: 72

Area—comparative: twice the size of Georgia

Land boundaries: *total:* 1,561 km
border countries (3): Saudi Arabia 658 km, UAE 609 km, Yemen 294 km

Coastline: 2,092 km

Maritime claims: *territorial sea:* 12 nm
exclusive economic zone: 200 nm
contiguous zone: 24 nm

Climate: dry desert; hot, humid along coast; hot, dry interior; strong southwest summer monsoon (May to September) in far south

Terrain: central desert plain, rugged mountains in north and south

Elevation: *mean elevation:* 310 m
lowest point: Arabian Sea 0 m
highest point: Jabal Shams 3,004 m

Natural resources: petroleum, copper, asbestos, some marble, limestone, chromium, gypsum, natural gas

Land use: *agricultural land:* 4.7% (2011 est.)
arable land: 0.1% (2011 est.) / permanent crops: 0.1% (2011 est.) / permanent pasture: 4.5% (2011 est.)
forest: 0% (2011 est.)
other: 95.3% (2011 est.)
Irrigated land: 590 sq km (2012)

Population distribution: the vast majority of the population is located in and around the Al Hagar Mountains in the north of the country; another smaller cluster is found around the city of Salalah in the far south; most of the country remains sparsely poplulated

Natural hazards: summer winds often raise large sandstorms and dust storms in interior; periodic droughts

Environment—current issues: limited natural freshwater resources; high levels of soil and water salinity in the coastal plains; beach pollution from oil spills; industrial effluents seeping into the water tables and aquifers; desertificaiton due to high winds driving desert sand into arable lands

Environment—international agreements: *party to:* Biodiversity, Climate Change, Climate Change-Kyoto Protocol, Desertification, Hazardous Wastes, Law of the Sea, Marine Dumping, Ozone Layer Protection, Ship Pollution, Whaling
signed, but not ratified: none of the selected agreements

Geography—note: consists of Oman proper and two northern exclaves, Musandam and Al Madhah; the former is a peninsula that occupies a strategic location adjacent to the Strait of Hormuz, a vital transit point for world crude oil

PEOPLE AND SOCIETY

Population: 4,664,844 (December 2019 est.)
note: immigrants make up approximately 46% of the total population (2019)
country comparison to the world: 126

Nationality: *noun:* Omani(s)
adjective: Omani

Ethnic groups: Arab, Baluchi, South Asian (Indian, Pakistani, Sri Lankan, Bangladeshi), African

Languages: Arabic (official), English, Baluchi, Swahili, Urdu, Indian dialects

Religions: Muslim 85.9%, Christian 6.5%, Hindu 5.5%, Buddhist 0.8%, Jewish <0.1%, other 1%, unaffiliated 0.2% (2010 est.)
note: Omani citizens represent approximately 56.4% of the population and are overwhelming Muslim (Ibadhi and Sunni sects each constitute about 45% and Shia about 5%); Christians, Hindus, and Buddhists account for roughly 5% of Omani citizens

MENA religious affiliation: *Age structure:* 0-14 years: 30.15% (male 561,791/female 533,949)
15-24 years: 17.35% (male 331,000/female 299,516)
25-54 years: 44.81% (male 928,812/female 699,821)
55-64 years: 4.02% (male 77,558/female 68,427)
65 years and over: 3.68% (male 64,152/female 69,663) (2020 est.)

Dependency ratios: *total dependency ratio:* 33.3
youth dependency ratio: 30
elderly dependency ratio: 3.3
potential support ratio: 29.9 (2020 est.)

Median age: *total:* 26.2 years
male: 27.2 years
female: 25.1 years (2020 est.)
country comparison to the world: 155

Population growth rate: 1.96% (2020 est.)
country comparison to the world: 49

Birth rate: 23.1 births/1,000 population (2020 est.)
country comparison to the world: 57

Death rate: 3.3 deaths/1,000 population (2020 est.)
country comparison to the world: 224

Net migration rate: -0.4 migrant(s)/1,000 population (2020 est.)

country comparison to the world: 124

Population distribution: the vast majority of the population is located in and around the Al Hagar Mountains in the north of the country; another smaller cluster is found around the city of Salalah in the far south; most of the country remains sparsely poplulated

Urbanization: *urban population:* 86.3% of total population (2020)
rate of urbanization: 5.25% annual rate of change (2015-20 est.)
total population growth rate v. urban population growth rate, 2000-2030:

Major urban areas—population: 1.550 million MUSCAT (capital) (2020)

Sex ratio: *at birth:* 1.05 male(s)/female
0-14 years: 1.05 male(s)/female
15-24 years: 1.11 male(s)/female
25-54 years: 1.33 male(s)/female
55-64 years: 1.13 male(s)/female
65 years and over: 0.92 male(s)/female
total population: 1.18 male(s)/female (2020 est.)

Maternal mortality rate: 19 deaths/100,000 live births (2017 est.)
country comparison to the world: 126

Infant mortality rate: *total:* 11.7 deaths/1,000 live births
male: 12 deaths/1,000 live births
female: 11.4 deaths/1,000 live births (2020 est.)
country comparison to the world: 110

Life expectancy at birth: *total population:* 76.3 years
male: 74.4 years
female: 78.4 years (2020 est.)
country comparison to the world: 98

Total fertility rate: 2.76 children born/woman (2020 est.)
country comparison to the world: 62

Contraceptive prevalence rate: 29.7% (2014)

Drinking water source:
improved:
urban: 100% of population
rural: 100% of population
total: 100% of population
unimproved:
urban: 0% of population
rural: 0% of population
total: 0% of population (2017 est.)

Current Health Expenditure: 3.8% (2017)

Physicians density: 1.96 physicians/1,000 population (2017)

Hospital bed density: 1.5 beds/1,000 population (2017)

Sanitation facility access:
improved:
urban: 100% of population
rural: 100% of population
total: 100% of population
unimproved:
urban: 0% of population
rural: 0% of population
total: 0% of population (2017 est.)

HIV/AIDS—adult prevalence rate: 0.1% (2019)
country comparison to the world: 129

HIV/AIDS—people living with HIV/AIDS: 2,500 (2019)
country comparison to the world: 136

HIV/AIDS—deaths: <100 (2019)

Obesity—adult prevalence rate: 27% (2016)
country comparison to the world: 39

Children under the age of 5 years underweight: 11.2% (2017)
country comparison to the world: 59

Education expenditures: 6.8% of GDP (2017)
country comparison to the world: 16

Literacy: *definition:* age 15 and over can read and write
total population: 95.7%
male: 97%
female: 92.7% (2018)

School life expectancy (primary to tertiary education): *total:* 14 years
male: 14 years
female: 15 years (2019)

Unemployment, youth ages 15-24: *total:* 13.7%
male: 10.3%
female: 33.9% (2016)
country comparison to the world: 101

GOVERNMENT

Country name: *conventional long form:* Sultanate of Oman
conventional short form: Oman
local long form: Saltanat Uman
local short form: Uman
former: Sultanate of Muscat and Oman
etymology: the origin of the name is uncertain, but it apparently dates back at least 2,000 years since an "Omana" is mentioned by Pliny the Elder (1st century A.D.) and an "Omanon" by Ptolemy (2nd century A.D.)

Government type: absolute monarchy

Capital: *name:* Muscat
geographic coordinates: 23 37 N, 58 35 E
time difference: UTC+4 (9 hours ahead of Washington, DC, during Standard Time)
etymology: the name, whose meaning is uncertain, traces back almost two millennia; two 2nd century A.D. scholars, the geographer Ptolemy and the historian Arrian, both mention an Arabian Sea coastal town of Moscha, which most likely referred to Muscat

Administrative divisions: 11 governorates (muhafazat, singular—muhafaza); Ad Dakhiliyah, Al Buraymi, Al Wusta, Az Zahirah, Janub al Batinah (Al Batinah South), Janub ash Sharqiyah (Ash Sharqiyah South), Masqat (Muscat), Musandam, Shamal al Batinah (Al Batinah North), Shamal ash Sharqiyah (Ash Sharqiyah North), Zufar (Dhofar)

Independence: 1650 (expulsion of the Portuguese)

National holiday: National Day, 18 November; note—celebrates Oman's independence from Portugal in 1650 and the birthday of Sultan

QABOOS bin Said al Said, who reigned from 1970 to 2020

Constitution: *history:* promulgated by royal decree 6 November 1996 (the Basic Law of the Sultanate of Oman serves as the constitution)amended by royal decree in 2011
amendments: promulgated by the sultan or proposed by the Council of Oman and drafted by a technical committee as stipulated by royal decree and then promulgated through royal decree; amended by royal decree in 2011

Legal system: mixed legal system of Anglo-Saxon law and Islamic law

International law organization participation: has not submitted an ICJ jurisdiction declaration; non-party state to the ICCt

Citizenship: *citizenship by birth:* no
citizenship by descent only: the father must be a citizen of Oman
dual citizenship recognized: no
residency requirement for naturalization: unknown

Suffrage: 21 years of age; universal; note—members of the military and security forces by law cannot vote

Executive branch: *chief of state:* Sultan and Prime Minister HAYTHAM bin Tariq bin Taimur Al-Said (since 11 January 2020); note—the monarch is both chief of state and head of government

head of government: Sultan and Prime Minister HAYTHAM bin Tariq bin Taimur Al-Said (since 11 January 2020)
cabinet: Cabinet appointed by the monarch
elections/appointments: members of the Ruling Family Council determine a successor from the sultan's extended family; if the Council cannot form a consensus within 3 days of the sultan's death or incapacitation, the Defense Council will relay a predetermined heir as chosen by the sultan

Legislative branch: *description:* bicameral Council of Oman or Majlis Oman consists of: Council of State or Majlis al-Dawla (85 seats including the chairman; members appointed by the sultan from among former government officials and prominent educators, businessmen, and citizens)

Consultative Council or Majlis al-Shura (86 seats; members directly elected in single- and 2-seat constituencies by simple majority popular vote to serve renewable 4-year terms); note—since political reforms in 2011, legislation from the Consultative Council is submitted to the Council of State for review by the Royal Court
elections: Council of State—last appointments on 11 July 2019 (next—NA)

Consultative Assembly—last held on 27 October 2019 (next to be held in October 2023)
election results: Council of State—composition—men 70, women 15, percent of women 17.6%

Consultative Council percent of vote by party—NA, seats by party—NA (organized political parties in Oman are legally banned); composition men 84, women 2, percent of women 2.3%;

note—total Council of Oman percent of women 9.9%

Judicial branch: *highest courts:* Supreme Court (consists of 5 judges)

judge selection and term of office: judges nominated by the 9-member Supreme Judicial Council (chaired by the monarch) and appointed by the monarch; judges appointed for life

subordinate courts: Courts of Appeal; Administrative Court; Courts of First Instance; sharia courts; magistrates' courts; military courts

Political parties and leaders: none; note—organized political parties are legally banned in Oman, and loyalties tend to form around tribal affiliations

International organization participation: ABEDA, AFESD, AMF, CAEU, FAO, G-77, GCC, IAEA, IBRD, ICAO, ICC (NGOs), IDA, IDB, IFAD, IFC, IHO, ILO, IMF, IMO, IMSO, Interpol, IOC, IPU, ISO, ITSO, ITU, LAS, MIGA, NAM, OIC, OPCW, UN, UNCTAD, UNESCO, UNIDO, UNWTO, UPU, WCO, WFTU (NGOs), WHO, WIPO, WMO, WTO

Diplomatic representation in the US: *chief of mission:* Ambassador Hunaina bint Sultan bin Ahmad al-MUGHAIRI (since 2 December 2005)

chancery: 2535 Belmont Road, NW, Washington, DC 20008

telephone: [1] (202) 387-1980

FAX: [1] (202) 745-4933

Diplomatic representation from the US: *chief of mission:* Ambassador Leslie M. TSOU (since 19 January 2020)

telephone: [968] 24-643-400

embassy: P.C. 115, Madinat Al Sultan Qaboos, Muscat

mailing address: P.O. Box 202, P.C. 115, Madinat Al Sultan Qaboos, Muscat

FAX: [968] 24-643-740

Flag description: three horizontal bands of white (top), red, and green of equal width with a broad, vertical, red band on the hoist side; the national emblem (a khanjar dagger in its sheath superimposed on two crossed swords in scabbards) in white is centered near the top of the vertical band; white represents peace and prosperity, red recalls battles against foreign invaders, and green symbolizes the Jebel al Akhdar (Green Mountains) and fertility

National symbol(s): *khanjar dagger superimposed on two crossed swords; national colors:* red, white, green

National anthem: *name:* "Nashid as-Salaam as-Sultani" (The Sultan's Anthem)

lyrics/music: Rashid bin Uzayyiz al KHUSAIDI/ James Frederick MILLS, arranged by Bernard EBBINGHAUS

note: adopted 1932; new lyrics written after QABOOS bin Said al Said gained power in 1970; first performed by the band of a British ship as a salute to the Sultan during a 1932 visit to Muscat; the bandmaster of the HMS Hawkins was asked to write a salutation to the Sultan on the occasion of his ship visit

0:00 / 1:53

ECONOMY

Economy—overview: Oman is heavily dependent on oil and gas resources, which can generate between and 68% and 85% of government revenue, depending on fluctuations in commodity prices. In 2016, low global oil prices drove Oman's budget deficit to $13.8 billion, or approximately 20% of GDP, but the budget deficit is estimated to have reduced to 12% of GDP in 2017 as Oman reduced government subsidies. As of January 2018, Oman has sufficient foreign assets to support its currency's fixed exchange rates. It is issuing debt to cover its deficit.

Oman is using enhanced oil recovery techniques to boost production, but it has simultaneously pursued a development plan that focuses on diversification, industrialization, and privatization, with the objective of reducing the oil sector's contribution to GDP. The key components of the government's diversification strategy are tourism, shipping and logistics, mining, manufacturing, and aquaculture.

Muscat also has notably focused on creating more Omani jobs to employ the rising number of nationals entering the workforce. However, high social welfare benefits—that had increased in the wake of the 2011 Arab Spring—have made it impossible for the government to balance its budget in light of current oil prices. In response, Omani officials imposed austerity measures on its gasoline and diesel subsidies in 2016. These spending cuts have had only a moderate effect on the government's budget, which is projected to again face a deficit of $7.8 billion in 2018.

GDP (purchasing power parity):
$190.1 billion (2017 est.)
$191.9 billion (2016 est.)
$182.8 billion (2015 est.)
note: data are in 2017 dollars
country comparison to the world: 67

GDP (official exchange rate):
$70.78 billion (2017 est.)

GDP—real growth rate: -0.9% (2017 est.)
5% (2016 est.)
4.7% (2015 est.)
country comparison to the world: 201

GDP—per capita (PPP): $46,000 (2017 est.)
$47,900 (2016 est.)
$48,400 (2015 est.)
note: data are in 2017 dollars
country comparison to the world: 37

Gross national saving:
16.1% of GDP (2017 est.)
10.5% of GDP (2016 est.)
14.3% of GDP (2015 est.)
country comparison to the world: 127

GDP—composition, by end use:
household consumption: 36.8% (2017 est.)
government consumption: 26.2% (2017 est.)
investment in fixed capital: 27.8% (2017 est.)

investment in inventories: 3% (2017 est.)
exports of goods and services: 51.5% (2017 est.)
imports of goods and services: -46.6% (2017 est.)

GDP—composition, by sector of origin:
agriculture: 1.8% (2017 est.)
industry: 46.4% (2017 est.)
services: 51.8% (2017 est.)

Agriculture—products: dates, limes, bananas, alfalfa, vegetables; camels, cattle; fish

Industries: crude oil production and refining, natural and liquefied natural gas production; construction, cement, copper, steel, chemicals, optic fiber

Industrial production growth rate: -3% (2017 est.)
country comparison to the world: 188

Labor force: 2.255 million (2016 est.)
note: about 60% of the labor force is non-national
country comparison to the world: 119

Labor force—by occupation: *agriculture:* 4.7% NA
industry: 49.6% NA
services: 45% NA (2016 est.)

Unemployment rate: NA

Population below poverty line: NA

Household income or consumption by percentage share: *lowest 10%:* NA
highest 10%: NA

Budget: *revenues:* 22.14 billion (2017 est.)
expenditures: 31.92 billion (2017 est.)

Taxes and other revenues: 31.3% (of GDP) (2017 est.)
country comparison to the world: 73

Budget surplus (+) or deficit (-): -13.8% (of GDP) (2017 est.)
country comparison to the world: 216

Public debt: 46.9% of GDP (2017 est.)
32.5% of GDP (2016 est.)
note: excludes indebtedness of state-owned enterprises
country comparison to the world: 113

Fiscal year: calendar year

Inflation rate (consumer prices): 1.6% (2017 est.)
1.1% (2016 est.)
country comparison to the world: 90

Current account balance:
-$10.76 billion (2017 est.)
-$12.32 billion (2016 est.)
country comparison to the world: 192

Exports: $103.3 billion (2017 est.)
$27.54 billion (2016 est.)
country comparison to the world: 35

Exports—partners: China 43.7%, UAE 11%, South Korea 7.9%, Saudi Arabia 4.2% (2017)

Exports—commodities: petroleum, reexports, fish, metals, textiles

Imports: $24.12 billion (2017 est.)
$21.29 billion (2016 est.)
country comparison to the world: 70

Imports—commodities: machinery and transport equipment, manufactured goods, food, livestock, lubricants

Imports—partners: UAE 35.5%, US 27.8%, Brazil 4% (2017)

Reserves of foreign exchange and gold: $16.09 billion (31 December 2017 est.)
$20.26 billion (31 December 2016 est.)
country comparison to the world: 64

Debt—external: $46.27 billion (31 December 2017 est.)
$27.05 billion (31 December 2016 est.)
country comparison to the world: 70

Exchange rates: Omani rials (OMR) per US dollar—
0.3845 (2017 est.)
0.3845 (2016 est.)
0.3845 (2015 est.)
0.3845 (2014 est.)
0.3845 (2013 est.)

ENERGY

Electricity access: *electrification—total population:* 99% (2016)
electrification—urban areas: 100% (2016)
electrification—rural areas: 93% (2016)

Electricity—production: 32.16 billion kWh (2016 est.)
country comparison to the world: 62

Electricity—consumption: 28.92 billion kWh (2016 est.)
country comparison to the world: 63

Electricity—exports: 0 kWh (2016 est.)
country comparison to the world: 181

Electricity—imports: 0 kWh (2016 est.)
country comparison to the world: 183

Electricity—installed generating capacity: 8.167 million kW (2016 est.)
country comparison to the world: 70

Electricity—from fossil fuels: 100% of total installed capacity (2016 est.)
country comparison to the world: 14

Electricity—from nuclear fuels: 0% of total installed capacity (2017 est.)
country comparison to the world: 160

Electricity—from hydroelectric plants: 0% of total installed capacity (2017 est.)
country comparison to the world: 192

Electricity—from other renewable sources: 0% of total installed capacity (2017 est.)
country comparison to the world: 204

Crude oil—production: 979,000 bbl/day (2018 est.)
country comparison to the world: 21

Crude oil—exports: 844,100 bbl/day (2015 est.)
country comparison to the world: 14

Crude oil—imports: 0 bbl/day (2015 est.)
country comparison to the world: 179

Crude oil—proved reserves: 5.373 billion bbl (1 January 2018 est.)
country comparison to the world: 21

Refined petroleum products—production: 229,600 bbl/day (2015 est.)
country comparison to the world: 48

Refined petroleum products—consumption: 188,000 bbl/day (2016 est.)

country comparison to the world: 59

Refined petroleum products—exports: 33,700 bbl/day (2015 est.)
country comparison to the world: 60

Refined petroleum products—imports: 6,041 bbl/day (2015 est.)
country comparison to the world: 165

Natural gas—production: 31.23 billion cu m (2017 est.)
country comparison to the world: 26

Natural gas—consumption: 21.94 billion cu m (2017 est.)
country comparison to the world: 35

Natural gas—exports: 11.16 billion cu m (2017 est.)
country comparison to the world: 20

Natural gas—imports: 1.982 billion cu m (2017 est.)
country comparison to the world: 53

Natural gas—proved reserves: 651.3 billion cu m (1 January 2018 est.)
country comparison to the world: 28

Carbon dioxide emissions from consumption of energy: 68.94 million Mt (2017 est.)
country comparison to the world: 52

COMMUNICATIONS

Telephones—fixed lines: *total subscriptions:* 456,940
subscriptions per 100 inhabitants: 12.82 (2019 est.)
country comparison to the world: 99

Telephones—mobile cellular: *total subscriptions:* 4,926,099
subscriptions per 100 inhabitants: 138.23 (2019 est.)
country comparison to the world: 121

Telecommunication systems: *general assessment:* modern system consisting of open-wire, microwave, and radiotelephone communication stations; coaxial cable; domestic satellite system with 8 earth stations; progressive mobile sector with both 3G and 4G LTE networks and reediness for 5G launch; competition among 3 (mobile network operators) MNO (2020)
domestic: fixed-line 13 per 100 and mobile-cellular 138 per 100, subscribership both increasing with fixed-line phone service gradually being introduced to remote villages using wireless local loop systems (2019)
international: country code—968; landing points for GSA, AAE-1, SeaMeWe-5, Tata TGN-Gulf, FALCON, GBICS/MENA, MENA/Guld Bridge International, TW1, BBG, EIG, OMRAN/EPEG, and POI submarine cables providing connectivity to Asia, Africa, the Middle East, Southeast Asia and Europe; satellite earth stations—2 Intelsat (Indian Ocean) (2019)
note: the COVID-19 outbreak is negatively impacting telecommunications production and supply chains globally; consumer spending on telecom devices and services has also slowed due

to the pandemic's effect on economies worldwide; overall progress towards improvements in all facets of the telecom industry—mobile, fixed-line, broadband, submarine cable and satellite—has moderated

Broadcast media: 1 state-run TV broadcaster; TV stations transmitting from Saudi Arabia, the UAE, Iran, and Yemen available via satellite TV; state-run radio operates multiple stations; first private radio station began operating in 2007 and several additional stations now operating (2019)

Internet country code: .om

Internet users: *total:* 2,801,932

percent of population: 80.19% (July 2018 est.)
country comparison to the world: 102

Broadband—fixed subscriptions: *total:* 422,035
subscriptions per 100 inhabitants: 12 (2018 est.)
country comparison to the world: 89

TRANSPORTATION

National air transport system: *number of registered air carriers:* 2 (2020)
inventory of registered aircraft operated by air carriers: 57
annual passenger traffic on registered air carriers: 10,438,241 (2018)
annual freight traffic on registered air carriers: 510.43 million mt-km (2018)

Civil aircraft registration country code prefix: A4O (2016)

Airports: 132 (2013)
country comparison to the world: 43

Airports—with paved runways:
total: 13 (2017)
over 3,047 m: 7 (2017)
2,438 to 3,047 m: 5 (2017)
914 to 1,523 m: 1 (2017)

Airports—with unpaved runways:
total: 119 (2013)
over 3,047 m: 2 (2013)
2,438 to 3,047 m: 7 (2013)
1,524 to 2,437 m: 51 (2013)
914 to 1,523 m: 33 (2013)
under 914 m: 26 (2013)

Heliports: 3 (2013)

Pipelines: 106 km condensate, 4224 km gas, 3558 km oil, 33 km oil/gas/water, 264 km refined products (2013)

Roadways: *total:* 60,230 km (2012)
paved: 29,685 km (includes 1,943 km of expressways) (2012)
unpaved: 30,545 km (2012)
country comparison to the world: 75

Merchant marine: *total:* 51
by type: general cargo 10, other 41 (2019)
country comparison to the world: 117

Ports and terminals: *major seaport(s):* Mina' Qabus, Salalah, Suhar
container port(s) (TEUs): Salalah (3,946,421) (2017)
LNG terminal(s) (export): Qalhat

MILITARY AND SECURITY

Military and security forces: *Sultan's Armed Forces (SAF):* Royal Army of Oman (RAO), Royal Navy of Oman (RNO), Royal Air Force of Oman (RAFO), Royal Guard of Oman (RGO); Royal Oman Police Coast Guard; Tribal Home Guard (2020)

Military expenditures: 8.8% of GDP (2019)
8.2% of GDP (2018)
9.6% of GDP (2017)
12% of GDP (2016)
10.9% of GDP (2015)
country comparison to the world: 1

Military and security service personnel strengths: the Sultan's Armed Forces (SAF) have approximately 40,000 total active troops (25,000 Army, 4,200 Navy; 4,500 Air Force; 6,400 Royal Guard); 400 Coast Guard; 4,000 Tribal Home Guard (2019)

Military equipment inventories and acquisitions: the SAF's inventory includes mostly a mix of older and some more modern British and US weapons systems, with smaller quantities of equipment from South Africa and a variety of European countries; since 2010, the UK and the US are the leading suppliers of armaments to Oman (2019 est.)

Military service age and obligation: 18-30 years of age for voluntary military service; no conscription (2012)

Maritime threats: the Maritime Administration of the US Department of Transportation has issued a Maritime Advisory (2019-012-Persian Gulf, Strait of Hormuz, Gulf of Oman, Arabian Sea, Red Sea-Threats to US and International Shipping from Iran) effective 7 August 2019, which states in part that "heightened military activities and increased political tensions in this region continue to present risk to commercial shipping...there is a continued possibility that Iran and/or its regional proxies could take actions against US and partner interests in the region;" at present, Iran has seized two foreign-flagged tankers in the Persian Gulf; the US and UK navies have established Operation Sentinel to provide escorts for commercial shipping transiting the Persian Gulf, Strait of Hormuz, and Gulf of Oman

TRANSNATIONAL ISSUES

Disputes—international: boundary agreement reportedly signed and ratified with UAE in 2003 for entire border, including Oman's Musandam Peninsula and Al Madhah exclave, but details of the alignment have not been made public

Refugees and internally displaced persons: *refugees (country of origin):* 5,000 (Yemen) (2017)

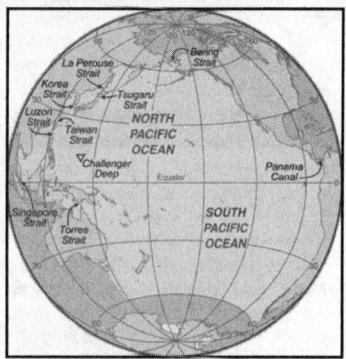

INTRODUCTION

Background: The Pacific Ocean is the largest of the world's five oceans (followed by the Atlantic Ocean, Indian Ocean, Southern Ocean, and Arctic Ocean). Strategically important access waterways include the La Perouse, Tsugaru, Tsushima, Taiwan, Singapore, and Torres Straits. The decision by the International Hydrographic Organization in the spring of 2000 to delimit a fifth ocean, the Southern Ocean, removed the portion of the Pacific Ocean south of 60 degrees south.

GEOGRAPHY

Location: body of water between the Southern Ocean, Asia, Australia, and the Western Hemisphere

Geographic coordinates: 0 00 N, 160 00 W

Map references: Political Map of the World

Area: *total:* 155.557 million sq km
note: includes Bali Sea, Bering Sea, Bering Strait, Coral Sea, East China Sea, Gulf of Alaska, Gulf of Tonkin, Philippine Sea, Sea of Japan, Sea of Okhotsk, South China Sea, Tasman Sea, and other tributary water bodies

Area—comparative: about 15 times the size of the US; covers about 28% of the global surface; almost equal to the total land area of the world

Coastline: 135,663 km

Climate: planetary air pressure systems and resultant wind patterns exhibit remarkable uniformity in the south and east; trade winds and westerly winds are well-developed patterns, modified by seasonal fluctuations; tropical cyclones (hurricanes) may form south of Mexico from June to October and affect Mexico and Central America; continental influences cause climatic uniformity to be much less pronounced in the eastern and western regions at the same latitude in the North Pacific Ocean; the western Pacific is monsoonal - a rainy season occurs during the summer months, when moisture-laden winds blow from the ocean

over the land, and a dry season during the winter months, when dry winds blow from the Asian landmass back to the ocean; tropical cyclones (typhoons) may strike southeast and east Asia from May to December

Terrain: surface dominated by two large gyres (broad, circular systems of currents), one in the northern Pacific and another in the southern Pacific; in the northern Pacific, sea ice forms in the Bering Sea and Sea of Okhotsk in winter; in the southern Pacific, sea ice from Antarctica reaches its northernmost extent in October; the ocean floor in the eastern Pacific is dominated by the East Pacific Rise, while the western Pacific is dissected by deep trenches, including the Mariana Trench, which is the world's deepest at 10,924 m
major surface currents: clockwise North Pacific Gyre formed by the warm northward flowing Kuroshio Current in the west, the eastward flowing North Pacific Current in the north, the southward flowing cold California Current in the east, and the westward flowing North Equatorial Current in the south; the counterclockwise South Pacific Gyre composed of the southward flowing warm East Australian Current in the west, the eastward flowing South Pacific Current in the south, the northward flowing cold Peru (Humbolt) Current in the east, and the westward flowing South Equatorial Current in the north

Elevation: mean depth: -2,970 m
lowest point: Challenger Deep in the Mariana Trench -10,924 m
highest point: sea level

Natural resources: oil and gas fields, polymetallic nodules, sand and gravel aggregates, placer deposits, fish

Natural hazards: surrounded by a zone of violent volcanic and earthquake activity sometimes referred to as the "Pacific Ring of Fire"; subject to tropical cyclones (typhoons) in southeast and east Asia from May to December (most frequent from July to October); tropical cyclones (hurricanes) may form south of Mexico and strike Central America and Mexico from June to October (most common in August and September); cyclical El Nino/La Nina phenomenon occurs in the equatorial Pacific, influencing weather in the Western Hemisphere and the western Pacific; ships subject to superstructure icing in extreme north from October to May; persistent fog in the northern Pacific can be a maritime hazard from June to December

Environment—current issues: pollution (such as sewage, runoff from land and toxic waste); habitat destruction; over-fishing; climate change leading to sea level rise, ocean acidification, and warming; endangered marine species include the dugong, sea lion, sea otter, seals, turtles, and whales; oil pollution in Philippine Sea and South China Sea

Geography—note: the major chokepoints are the Bering Strait, Panama Canal, Luzon Strait,

and the Singapore Strait; the Equator divides the Pacific Ocean into the North Pacific Ocean and the South Pacific Ocean; dotted with low coral islands and rugged volcanic islands in the southwestern Pacific Ocean; much of the Pacific Ocean's rim lies along the Ring of Fire, a belt of active volcanoes and earthquake epicenters that accounts for up to 90% of the world's earthquakes and some 75% of the world's volcanoes

GOVERNMENT

Country name: etymology: named by Portuguese explorer Ferdinand MAGELLAN during the Spanish circumnavigation of the world in 1521; encountering favorable winds upon reaching the ocean, he called it "Mar Pacifico," which means "peaceful sea" in both Portuguese and Spanish

ECONOMY

Economy—overview: The Pacific Ocean is a major contributor to the world economy and particularly to those nations its waters directly touch. It provides low-cost sea transportation between East and West, extensive fishing grounds, offshore oil and gas fields, minerals, and sand and gravel for the construction industry. In 1996, over 60% of the world's fish catch came from the Pacific Ocean. Exploitation of offshore oil and gas reserves is playing an ever-increasing role in the energy supplies of the US, Australia, NZ, China, and Peru. The high cost of recovering offshore oil and gas, combined with the wide swings in world prices for oil since 1985, has led to fluctuations in new drillings.

Marine fisheries: the Pacific Ocean fisheries are the most important in the world accounting for 56.6%, or 45,580,140 mt, of the global marine capture in 2017; of the six regions delineated by the Food and Agriculture Organization in the Pacific Ocean, the following are the most important:
Northwest Pacific region (Region 61) is the world's most important fishery producing 25% of the global catch or 20,234,899 mt in 2017; it encompasses the waters north of 20º north latitude and west of 175º west longitude with the major producers including China (12,589,877 mt), Japan (2,917,663 mt), South Korea (948,670 mt), and Taiwan (341,260 mt); the principal catches include Alaska Pollock, Japanese anchovy, chub mackerel, and scads
Western Central Pacific region (Region 71) is the world's second most important fishing region producing 15%, or 12,530,652 mt, of the global catch in 2017; tuna is the most important species in this region; the region includes the waters between 20º North and 25º South latitude and west of 175º West longitude with the major producers including Indonesia (4,281,018 mt), Vietnam (3,118,696 mt), Philippines (1,724,272 mt), Thailand (912,863 mt), and Malaysia (741,561 mt); the principal catches include Skipjack and Yellowfin tuna, sardinellas, and cephalopods

Southeast Pacific region (Region 87) is the third major Pacific fishery and fourth largest in the world producing 9%, or 7,223,740 mt, of the global catch in 2017; this region includes the nutrient rich upwelling waters off the west coast of South America between 5º North and 60º South latitude and east of 120º West longitude with the major producers including Peru (4,128,760 mt), Chile (1,918,611 mt), and Ecuador (554,961 mt); the principal catches include Peruvian anchovy (50% of the catch), Jumbo flying squid, and Chilean jack mackerel

Pacific Northeast region (Region 67) is the fourth largest Pacific Ocean fishery and eighth largest in the world producing 4% of the global catch or 3,379,432 mt in 2017; this region encompasses the waters north of 40º North latitude and east of 175º West longitude including the Gulf of Alaska and Bering Sea with the major producers including the US (3,186,515 mt), Canada (180,929 mt), and Russia (11,988 mt); the principal catches include Alaska pollock, Pacific cod, and North Pacific hake

FAO map of world fishing regions; used with permission.:

TRANSPORTATION

Ports and terminals: *major seaport(s):* Bangkok (Thailand), Hong Kong (China), Kao-hsiung (Taiwan), Los Angeles (US), Manila (Philippines), Pusan (South Korea), San Francisco (US), Seattle (US), Shanghai (China), Singapore, Sydney (Australia), Vladivostok (Russia), Wellington (NZ), Yokohama (Japan)

MILITARY AND SECURITY

Maritime threats: the International Maritime Bureau reports the territorial waters of littoral states and offshore waters in the South China Sea as high risk for piracy and armed robbery against ships; an emerging threat area lies in the Celebes and Sulu Seas between the Philippines and Malaysia where three crew were kidnapped or taken hostage in

2018; numerous commercial vessels have been attacked and hijacked both at anchor and while underway; hijacked vessels are often disguised and cargoes stolen; crew and passengers are often held for ransom, murdered, or cast adrift; the Maritime Administration (MARAD) of the US Department of Transportation has issued a Maritime Advisory (2019-011-Sulu and Celebes Seas- Piracy/Armed Robbery/Terrorism) which states in part "In 2018, there were at least 12 reported boardings, attempted boardings, attacks, hijackings, and kidnappings in the Sulu and Celebes Seas. Recent kidnapping incidents in this area were reportedly linked to the Abu Sayyaf Group (ASG), a violent Islamic separatist group operating in the southern Philippines..." and advises ships to adhere to counter-piracy practices to minimize risk

TRANSNATIONAL ISSUES

Disputes—International: some maritime disputes (see littoral states)

PAKISTAN

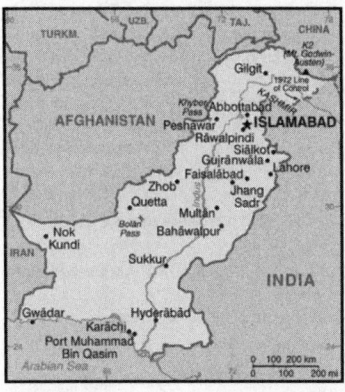

INTRODUCTION

Background: The Indus Valley civilization, one of the oldest in the world and dating back at least 5,000 years, spread over much of what is presently Pakistan. During the second millennium B.C., remnants of this culture fused with the migrating Indo-Aryan peoples. The area underwent successive invasions in subsequent centuries from the Persians, Greeks, Scythians, Arabs (who brought Islam), Afghans, and Turks. The Mughal Empire flourished in the 16th and 17th centuries; the British came to dominate the region in the 18th century. The separation in 1947 of British India into the Muslim state of Pakistan (with West and East sections) and largely Hindu India was never satisfactorily resolved, and India and Pakistan fought two wars and a limited conflict - in 1947-48, 1965, and 1999 respectively - over the disputed

Kashmir territory. A third war between these countries in 1971 - in which India assisted an indigenous movement reacting to the marginalization of Bengalis in Pakistani politics - resulted in East Pakistan becoming the separate nation of Bangladesh.

In response to Indian nuclear weapons testing, Pakistan conducted its own tests in mid-1998. India-Pakistan relations improved in the mid-2000s but have been rocky since the November 2008 Mumbai attacks and have been further strained by attacks in India by militants believed to be based in Pakistan. Imran KHAN took office as prime minister in 2018 after the Pakistan Tehreek-e-Insaaf (PTI) party won a plurality of seats in the July 2018 general elections. Pakistan has been engaged in a decades-long armed conflict with militant groups that target government institutions and civilians, including the Tehreek-e-Taliban Pakistan (TTP) and other militant networks.

GEOGRAPHY

Location: Southern Asia, bordering the Arabian Sea, between India on the east and Iran and Afghanistan on the west and China in the north

Geographic coordinates: 30 00 N, 70 00 E

Map references: Asia

Area: *total:* 796,095 sq km
land: 770,875 sq km
water: 25,220 sq km
country comparison to the world: 37

Area - comparative: slightly more than five times the size of Georgia; slightly less than twice the size of California

Land boundaries: *total:* 7,257 km

border countries (4): Afghanistan 2670 km, China 438 km, India 3190 km, Iran 959 km

Coastline: 1,046 km

Maritime claims: *territorial sea:* 12 nm
exclusive economic zone: 200 nm
contiguous zone: 24 nm
continental shelf: 200 nm or to the edge of the continental margin

Climate: mostly hot, dry desert; temperate in northwest; arctic in north

Terrain: divided into three major geographic areas: the northern highlands, the Indus River plain in the center and east, and the Balochistan Plateau in the south and west

Elevation: *mean elevation:* 900 m
lowest point: Arabian Sea 0 m
highest point: K2 (Mt. Godwin-Austen) 8,611 m

Natural resources: arable land, extensive natural gas reserves, limited petroleum, poor quality coal, iron ore, copper, salt, limestone

Land use: *agricultural land:* 35.2% (2011 est.)
arable land: 27.6% (2011 est.) / permanent crops: 1.1% (2011 est.) / permanent pasture: 6.5% (2011 est.)
forest: 2.1% (2011 est.)
other: 62.7% (2011 est.)
Irrigated land: 202,000 sq km (2012)

Population distribution: the Indus River and its tributaries attract most of the settlement, with Punjab province the most densely populated

Natural hazards: frequent earthquakes, occasionally severe especially in north and west; flooding along the Indus after heavy rains (July and August)

Environment - current issues: water pollution from raw sewage, industrial wastes, and agricultural runoff; limited natural freshwater resources; most of the population does not have access to potable water; deforestation; soil erosion; desertification; air pollution and noise pollution in urban areas

Environment - international agreements: *party to:* Biodiversity, Climate Change, Climate Change-Kyoto Protocol, Desertification, Endangered Species, Environmental Modification, Hazardous Wastes, Law of the Sea, Marine Dumping, Ozone Layer Protection, Ship Pollution, Wetlands
signed, but not ratified: Marine Life Conservation

Geography - note: controls Khyber Pass and Bolan Pass, traditional invasion routes between Central Asia and the Indian Subcontinent

PEOPLE AND SOCIETY

Population: 233,500,636 (July 2020 est.)
note: provisional results of Pakistan's 2017 national census estimate the country's total population to be 207,774,000
country comparison to the world: 5

Nationality: *noun:* Pakistani(s)
adjective: Pakistani

Ethnic groups: Punjabi 44.7%, Pashtun (Pathan) 15.4%, Sindhi 14.1%, Saraiki 8.4%, Muhajirs 7.6%, Balochi 3.6%, other 6.3%

Languages: Punjabi 48%, Sindhi 12%, Saraiki (a Punjabi variant) 10%, Pashto (alternate name, Pashtu) 8%, Urdu (official) 8%, Balochi 3%, Hindko 2%, Brahui 1%, English (official; lingua franca of Pakistani elite and most government ministries), Burushaski, and other 8%

Religions: Muslim (official) 96.4% (Sunni 85-90%, Shia 10-15%), other (includes Christian and Hindu) 3.6% (2010 est.)

Age structure: *0-14 years:* 36.01% (male 42,923,925/female 41,149,694)
15-24 years: 19.3% (male 23,119,205/female 21,952,976)
25-54 years: 34.7% (male 41,589,381/female 39,442,046)
55-64 years: 5.55% (male 6,526,656/female 6,423,993)
65 years and over: 4.44% (male 4,802,165/female 5,570,595) (2020 est.)

Dependency ratios: *total dependency ratio:* 64.4
youth dependency ratio: 57.2
elderly dependency ratio: 7.1
potential support ratio: 14 (2020 est.)

Median age: *total:* 22 years
male: 21.9 years
female: 22.1 years (2020 est.)
country comparison to the world: 180

Population growth rate: 2.07% (2020 est.)
country comparison to the world: 45

Birth rate: 27.4 births/1,000 population (2020 est.)
country comparison to the world: 41

Death rate: 6.2 deaths/1,000 population (2020 est.)
country comparison to the world: 156

Net migration rate: -0.9 migrant(s)/1,000 population (2020 est.)
country comparison to the world: 140

Population distribution: the Indus River and its tributaries attract most of the settlement, with Punjab province the most densely populated

Urbanization: *urban population:* 37.2% of total population (2020)
rate of urbanization: 2.53% annual rate of change (2015-20 est.)

total population growth rate v. urban population growth rate, 2000-2030:

Major urban areas—population: 16.094 million Karachi, 12.642 million Lahore, 3.462 million Faisalabad, 2.237 million Rawalpindi, 2.229 million Gujranwala, 1.129 million ISLAMABAD (capital) (2020)

Sex ratio: *at birth:* 1.05 male(s)/female
0-14 years: 1.04 male(s)/female
15-24 years: 1.05 male(s)/female
25-54 years: 1.05 male(s)/female
55-64 years: 1.02 male(s)/female
65 years and over: 0.86 male(s)/female
total population: 1.04 male(s)/female (2020 est.)

Mother's mean age at first birth: 23.6 years (2017/18 est.)
note: median age at first birth among women 25-29

Maternal mortality rate: 140 deaths/100,000 live births (2017 est.)
country comparison to the world: 61

Infant mortality rate: *total:* 52.3 deaths/1,000 live births
male: 55.9 deaths/1,000 live births
female: 48.5 deaths/1,000 live births (2020 est.)
country comparison to the world: 20

Life expectancy at birth: *total population:* 69.2 years
male: 67.2 years
female: 71.3 years (2020 est.)
country comparison to the world: 171

Total fertility rate: 3.6 children born/woman (2020 est.)
country comparison to the world: 40

Contraceptive prevalence rate: 34.2% (2017/18)

Drinking water source:
improved:
urban: 94.2% of population
rural: 89.9% of population
total: 91.5% of population
unimproved:
urban: 5.8% of population
rural: 10.1% of population
total: 8.5% of population (2017 est.)

Current Health Expenditure: 2.9% (2017)

Physicians density: 1 physicians/1,000 population (2017)

Hospital bed density: 0.6 beds/1,000 population (2017)

Sanitation facility access:
improved:
urban: 82.5% of population
rural: 62.9% of population
total: 70.1% of population
unimproved:
urban: 17.5% of population

rural: 37.1% of population
total: 29.9% of population (2017 est.)

HIV/AIDS—adult prevalence rate: 0.1% (2019 est.)
country comparison to the world: 130

HIV/AIDS—people living with HIV/AIDS: 190,000 (2019 est.)
country comparison to the world: 32

HIV/AIDS—deaths: 6,800 (2019 est.)
country comparison to the world: 22

Major infectious diseases: *degree of risk:* high (2020)
food or waterborne diseases: bacterial diarrhea, hepatitis A and E, and typhoid fever
vectorborne diseases: dengue fever and malaria
animal contact diseases: rabies
note: widespread ongoing transmission of a respiratory illness caused by the novel coronavirus (COVID-19) is occurring throughout Pakistan; as of 10 November 2020, Pakistan has reported a total of 341,753 cases of COVID-19 or 1,547 cumulative cases of COVID-19 per 1 million population with 31 cumulative deaths per 1 million population; the Government of Pakistan will permit commercial outbound passenger flights from all international airports except Gwadar and Turbat effective 30 May 2020, but inbound passenger flights remain suspended; limited domestic flight operations from five major airports – Islamabad, Karachi, Lahore, Peshawar, and Quetta are available; on 7 May 2020, the Government of Pakistan announced an ease in some of the nationwide lockdown restrictions; additionally, the Islamabad Capital Territory and Sindh, Punjab, Balochistan, and Khyber Pakhtunkhwa provinces have varying degrees of lockdowns

Obesity—adult prevalence rate: 8.6% (2016)
country comparison to the world: 150

Children under the age of 5 years underweight: 23.1% (2018)
country comparison to the world: 15

Education expenditures: 2.9% of GDP (2017)
country comparison to the world: 142

Literacy: *definition:* age 15 and over can read and write
total population: 59.1%
male: 71.1%
female: 46.5% (2015)

School life expectancy (primary to tertiary education): *total:* 8 years
male: 9 years
female: 8 years (2018)

Unemployment, youth ages 15-24: *total:* 7.8%
male: 8.2%
female: 6.8% (2018 est.)
country comparison to the world: 145

GOVERNMENT

Country name: *conventional long form:* Islamic Republic of Pakistan
conventional short form: Pakistan
local long form: Jamhuryat Islami Pakistan
local short form: Pakistan

former: West Pakistan

etymology: the word "pak" means "pure" in Persian or Pashto, while the Persian suffix "-stan" means "place of" or "country," so the word Pakistan literally means "Land of the Pure"

Government type: federal parliamentary republic

Capital: *name:* Islamabad

geographic coordinates: 33 41 N, 73 03 E

time difference: UTC+5 (10 hours ahead of Washington, DC, during Standard Time)

etymology: derived from two words: "Islam," an Urdu word referring to the religion of Islam, and "-abad," a Persian suffix indicating an "inhabited place" or "city," to render the meaning "City of Islam"

Administrative divisions: 4 provinces, 2 Pakistan-administered areas*, and 1 capital territory**; Azad Kashmir*, Balochistan, Gilgit-Baltistan*, Islamabad Capital Territory**, Khyber Pakhtunkhwa, Punjab, Sindh

Independence: 14 August 1947 (from British India)

National holiday: Pakistan Day (also referred to as Pakistan Resolution Day or Republic Day), 23 March (1940); note—commemorates both the adoption of the Lahore Resolution by the All-India Muslim League during its 22-24 March 1940 session, which called for the creation of independent Muslim states, and the adoption of the first constitution of Pakistan on 23 March 1956 during the transition to the Islamic Republic of Pakistan

Constitution: *history:* several previous; latest endorsed 12 April 1973, passed 19 April 1973, entered into force 14 August 1973 (suspended and restored several times)

amendments: proposed by the Senate or by the National Assembly; passage requires at least two-thirds majority vote of both houses; amended many times, last in 2018

Legal system: common law system with Islamic law influence

International law organization participation: accepts compulsory ICJ jurisdiction with reservations; non-party state to the ICCt

Citizenship: *citizenship by birth:* yes

citizenship by descent only: at least one parent must be a citizen of Pakistan

dual citizenship recognized: yes, but limited to select countries

residency requirement for naturalization: 4 out of the previous 7 years and including the 12 months preceding application

Suffrage: 18 years of age; universal; note—there are joint electorates and reserved parliamentary seats for women and non-Muslims

Executive branch: *chief of state:* President Arif ALVI (since 9 September 2018)

head of government: Prime Minister Imran KHAN (since 18 August 2018)

cabinet: Cabinet appointed by the president upon the advice of the prime minister

elections/appointments: president indirectly elected by the Electoral College consisting of members of the Senate, National Assembly, and provincial assemblies for a 5-year term (limited to 2 consecutive terms); election last held on 4 September 2018 (next to be held in 2023); prime minister elected by the National Assembly on 17 August 2018

election results: Arif ALVI elected president; Electoral College vote—Arif ALVI (PTI) 352, Fazl-ur-REHMAN (MMA) 184, Aitzaz AHSAN (PPP) 124; Imran KHAN elected prime minister; National Assembly vote—Imran KHAN (PTI) 176, Shehbaz SHARIF (PML-N) 96

Legislative branch: *description:* bicameral Parliament or Majlis-e-Shoora consists of:

Senate (104 seats; members indirectly elected by the 4 provincial assemblies and the territories' representatives by proportional representation vote; members serve 6-year terms with one-half of the membership renewed every 3 years); note—the byelection scheduled for 15 April 2020 has been postponed due to the COVID-19 pandemic

National Assembly (342 seats; 272 members directly elected in single-seat constituencies by simple majority vote and 70 members—60 women and 10 non-Muslims—directly elected by proportional representation vote; all members serve 5-year terms)

elections: Senate—last held on 3 March 2018 (next to be held in March 2021)

National Assembly—last held on 25 July 2018 (next to be held on 25 July 2023)

election results: Senate—percent of vote by party—NA; seats by party as of December 2019—PPP 19, PML-N 16, PTI 14, MQM-P 5, JUI- F 4, BAP 2, JI 2, PkMAP 2, ANP 1, BNP 1, PML-F 1, other 7, independent 30

National Assembly—percent of votes by party NA; seats by party as of December 2019—PTI 156, PML-N 84, PPP 55, MMA 16, MQM-P 7, BAP 5, PML-Q 5, BNP 4, GDA 3, AML 1, ANP 1, JWP 1, independent 4

Judicial branch: *highest courts:* Supreme Court of Pakistan (consists of the chief justice and 16 judges)

judge selection and term of office: justices nominated by an 8-member parliamentary committee upon the recommendation of the Judicial Commission, a 9-member body of judges and other judicial professionals, and appointed by the president; justices can serve until age 65

subordinate courts: High Courts; Federal Shariat Court; provincial and district civil and criminal courts; specialized courts for issues, such as taxation, banking, and customs

Political parties and leaders: Awami National Party or ANP [Asfandyar Wali KHAN]

Awami Muslim League or AML [Sheikh Rashid AHMED]

Balochistan National Party-Awami or BNP-A [Mir Israr Ullah ZEHRI]

Balochistan National Party-Mengal or BNP-M [Sardar Akhtar Jan MENGAL]

Grand Democratic Alliance or GDA (alliance of several parties)

Jamhoori Wattan Party or JWP [Shahzain BUGTI]

Jamaat-i Islami or JI [Sirajul HAQ]

Jamiat-i Ulema-i Islam Fazl-ur Rehman or JUI-F [Fazlur REHMAN]

Muttahida Quami Movement-London or MQM-L [Altaf HUSSAIN] (MQM split into two factions in 2016)

Muttahida Quami Movement-Pakistan or MQM-P [Dr. Khalid Maqbool SIDDIQUI] (MQM split into two factions in 2016) Muttahida Majlis-e-Amal or MMA [Fazl-ur- REHMAN] (alliance of several parties)

National Party or NP [Mir Hasil Khan BIZENJO]

Pakhtunkhwa Milli Awami Party or PMAP or PkMAP [Mahmood Khan ACHAKZAI]

Pakistan Muslim League-Functional or PML-F [Pir PAGARO or Syed Shah Mardan SHAH-II]

Pakistan Muslim League-Nawaz or PML-N [Shehbaz SHARIF]

Pakistan Muslim League – Quaid-e-Azam Group or PML-Q [Chaudhry Shujaat HUSSAIN]

Pakistan Peoples Party or PPP [Bilawal BHUTTO ZARDARI, Asif Ali ZARDARI]

Pakistan Tehrik-e Insaaf or PTI (Pakistan Movement for Justice) [Imran KHAN]Pak Sarzameen Party or PSP [Mustafa KAMAL]

Quami Watan Party or QWP [Aftab Ahmed Khan SHERPAO]

note: political alliances in Pakistan shift frequently

International organization participation: ADB, ARF, ASEAN (dialogue partner), C, CICA, CP, D-8, ECO, FAO, G-11, G-24, G-77, IAEA, IBRD, ICAO, ICC (national committees), ICRM, IDA, IDB, IFAD, IFC, IFRCS, IHO, ILO, IMF, IMO, IMSO, Interpol, IOC, IOM, IPU, ISO, ITSO, ITU, ITUC (NGOs), MIGA, MINURSO, MONUSCO, NAM, OAS (observer), OIC, OPCW, PCA, SAARC, SACEP, SCO (observer), UN, UNAMID, UNCTAD, UNESCO, UNHCR, UNIDO, UNMIL, UNOCI, UNWTO, UPU, WCO, WFTU (NGOs), WHO, WIPO, WMO, WTO

Diplomatic representation in the US: *chief of mission:* Ambassador Asad Majeed KHAN (since 11 January 2019)

chancery: 3517 International Court NW, Washington, DC 20008

telephone: [1] (202) 243-6500

FAX: [1] (202) 686-1534

consulate(s) general: Chicago, Houston, Los Angeles, New York

consulate(s): Louisville (KY), San Francisco

Diplomatic representation from the US: *chief of mission:* Ambassador (vacant); Charge d'Affaires Ambassador Paul W. JONES (since 24 September 2018)

telephone: [92] 51-201-4000

embassy: Diplomatic Enclave, Ramna 5, Islamabad

mailing address: 8100 Islamabad Place, Washington, DC 20521-8100

FAX: [92] 51-227-6427

consulate(s) general: Karachi, Lahore, Peshawar

Flag description: green with a vertical white band (symbolizing the role of religious minorities) on the hoist side; a large white crescent and star are centered in the green field; the crescent, star, and color green are traditional symbols of Islam

National symbol(s): *five-pointed star between the horns of a waxing crescent moon, jasmine; national colors: green, white*

National anthem: *name:* "Qaumi Tarana" (National Anthem)

lyrics/music: Abu-Al-Asar Hafeez JULLANDHURI/Ahmed Ghulamali CHAGLA

note: adopted 1954; also known as "Pak sarzamin shad bad" (Blessed Be the Sacred Land)

0:00 / 0:00

ECONOMY

Economy—overview: Decades of internal political disputes and low levels of foreign investment have led to underdevelopment in Pakistan. Pakistan has a large English-speaking population, with English-language skills less prevalent outside urban centers. Despite some progress in recent years in both security and energy, a challenging security environment, electricity shortages, and a burdensome investment climate have traditionally deterred investors. Agriculture accounts for one-fifth of output and two-fifths of employment. Textiles and apparel account for more than half of Pakistan's export earnings; Pakistan's failure to diversify its exports has left the country vulnerable to shifts in world demand. Pakistan's GDP growth has gradually increased since 2012, and was 5.3% in 2017. Official unemployment was 6% in 2017, but this fails to capture the true picture, because much of the economy is informal and underemployment remains high. Human development continues to lag behind most of the region.

In 2013, Pakistan embarked on a $6.3 billion IMF Extended Fund Facility, which focused on reducing energy shortages, stabilizing public finances, increasing revenue collection, and improving its balance of payments position. The program concluded in September 2016. Although Pakistan missed several structural reform criteria, it restored macroeconomic stability, improved its credit rating, and boosted growth. The Pakistani rupee has remained relatively stable against the US dollar since 2015, though it declined about 10% between November 2017 and March 2018. Balance of payments concerns have reemerged, however, as a result of a significant increase in imports and weak export and remittance growth.

Pakistan must continue to address several long-standing issues, including expanding investment in education, healthcare, and sanitation; adapting to the effects of climate change and natural disasters; improving the country's business environment; and widening the country's tax base. Given demographic challenges, Pakistan's leadership will be pressed to implement economic reforms, promote further development of the energy sector, and attract foreign investment to support sufficient economic growth necessary to employ its growing and rapidly urbanizing population, much of which is under the age of 25.

In an effort to boost development, Pakistan and China are implementing the "China-Pakistan Economic Corridor" (CPEC) with $60 billion in investments targeted towards energy and other infrastructure projects. Pakistan believes CPEC investments will enable growth rates of over 6% of GDP by laying the groundwork for increased exports. CPEC-related obligations, however, have raised IMF concern about Pakistan's capital outflows and external financing needs over the medium term.

GDP (purchasing power parity): $1.061 trillion (2017 est.)
$1.007 trillion (2016 est.)
$962.8 billion (2015 est.)
note: data are in 2017 dollars data are for fiscal years
country comparison to the world: 25

GDP (official exchange rate): $305 billion (2017 est.)

GDP—real growth rate: 5.4% (2017 est.)
4.6% (2016 est.)
4.1% (2015 est.)
note: data are for fiscal years
country comparison to the world: 37

GDP—per capita (PPP): $5,400 (2017 est.)
$5,200 (2016 est.)
$5,100 (2015 est.)
note: data are in 2017 dollars data are for fiscal years
country comparison to the world: 171

Gross national saving:
12% of GDP (2017 est.)
13.9% of GDP (2016 est.)
14.7% of GDP (2015 est.)
note: data are for fiscal years
country comparison to the world: 151

GDP—composition, by end use:
household consumption: 82% (2017 est.)
government consumption: 11.3% (2017 est.)
investment in fixed capital: 14.5% (2017 est.)
investment in inventories: 1.6% (2017 est.)
exports of goods and services: 8.2% (2017 est.)
imports of goods and services: -17.6% (2017 est.)

GDP—composition, by sector of origin:
agriculture: 24.4% (2016 est.)
industry: 19.1% (2016 est.)
services: 56.5% (2017 est.)

Agriculture—products: cotton, wheat, rice, sugarcane, fruits, vegetables; milk, beef, mutton, eggs

Industries: textiles and apparel, food processing, pharmaceuticals, surgical instruments, construction materials, paper products, fertilizer, shrimp

Industrial production growth rate: 5.4% (2017 est.)
country comparison to the world: 53

Labor force: 61.71 million (2017 est.)
note: extensive export of labor, mostly to the Middle East, and use of child labor
country comparison to the world: 8

Labor force—by occupation: *agriculture:* 42.3%
industry: 22.6%
services: 35.1% (FY2015 est.)

Unemployment rate: 6% (2017 est.)
6% (2016 est.)
note: Pakistan has substantial underemployment

country comparison to the world: 96

Population below poverty line: 29.5% (FY2013 est.)

Household income or consumption by percentage share: *lowest 10%:* 4%
highest 10%: 26.1% (FY2013)

Budget: *revenues:* 46.81 billion (2017 est.)
expenditures: 64.49 billion (2017 est.)
note: data are for fiscal years

Taxes and other revenues: 15.4% (of GDP) (2017 est.)
country comparison to the world: 190

Budget surplus (+) or deficit (-): -5.8% (of GDP) (2017 est.)
country comparison to the world: 178

Public debt: 67% of GDP (2017 est.)
67.6% of GDP (2016 est.)
country comparison to the world: 56

Fiscal year: 1 July—30 June

Inflation rate (consumer prices): 4.1% (2017 est.)
2.9% (2016 est.)
country comparison to the world: 161

Current account balance:
-$7.143 billion (2019 est.)
-$19.482 billion (2018 est.)
country comparison to the world: 188

Exports: $32.88 billion (2017 est.)
$21.97 billion (2016 est.)
country comparison to the world: 60

Exports—partners: US 17.7%, UK 7.7%, China 6%, Germany 5.8%, Afghanistan 5.2%, UAE 4.5%, Spain 4.1% (2017)

Exports—commodities: textiles (garments, bed linen, cotton cloth, yarn), rice, leather goods, sporting goods, chemicals, manufactures, surgical instruments, carpets and rugs

Imports:
$53.11 billion (2017 est.)
$42.69 billion (2016 est.)
country comparison to the world: 51

Imports—commodities: petroleum, petroleum products, machinery, plastics, transportation equipment, edible oils, paper and paperboard, iron and steel, tea

Imports—partners: China 27.4%, UAE 13.7%, US 4.9%, Indonesia 4.3%, Saudi Arabia 4.2% (2017)

Reserves of foreign exchange and gold:
$18.46 billion (31 December 2017 est.)
$22.05 billion (31 December 2016 est.)
country comparison to the world: 62

Debt—external:
$82.19 billion (31 December 2017 est.)
$70.45 billion (31 December 2016 est.)
country comparison to the world: 55

Exchange rates: Pakistani rupees (PKR) per US dollar -
105.1 (2017 est.)
104.769 (2016 est.)
104.769 (2015 est.)
102.769 (2014 est.)
101.1 (2013 est.)

ENERGY

Electricity access: *population without electricity:* 45 million (2019)
electrification—total population: 79% (2019)
electrification—urban areas: 91% (2019)
electrification—rural areas: 72% (2019)

Electricity—production: 109.7 billion kWh (2016 est.)
country comparison to the world: 32

Electricity—consumption: 92.33 billion kWh (2016 est.)
country comparison to the world: 34

Electricity—exports: 0 kWh (2016 est.)
country comparison to the world: 182

Electricity—imports: 490 million kWh (2016 est.)
country comparison to the world: 80

Electricity—installed generating capacity: 26.9 million kW (2016 est.)
country comparison to the world: 35

Electricity—from fossil fuels: 62% of total installed capacity (2016 est.)
country comparison to the world: 125

Electricity—from nuclear fuels: 5% of total installed capacity (2017 est.)
country comparison to the world: 22

Electricity—from hydroelectric plants: 27% of total installed capacity (2017 est.)
country comparison to the world: 74

Electricity—from other renewable sources: 7% of total installed capacity (2017 est.)
country comparison to the world: 94

Crude oil—production: 90,000 bbl/day (2018 est.)
country comparison to the world: 45

Crude oil—exports: 13,150 bbl/day (2015 est.)
country comparison to the world: 58

Crude oil—imports: 168,200 bbl/day (2015 est.)
country comparison to the world: 34

Crude oil—proved reserves: 332.2 million bbl (1 January 2018 est.)
country comparison to the world: 52

Refined petroleum products—production: 291,200 bbl/day (2015 est.)
country comparison to the world: 43

Refined petroleum products—consumption: 557,000 bbl/day (2016 est.)
country comparison to the world: 33

Refined petroleum products—exports: 25,510 bbl/day (2015 est.)
country comparison to the world: 68

Refined petroleum products—imports: 264,500 bbl/day (2015 est.)
country comparison to the world: 27

Natural gas—production: 39.05 billion cu m (2017 est.)
country comparison to the world: 21

Natural gas—consumption: 45.05 billion cu m (2017 est.)
country comparison to the world: 20

Natural gas—exports: 0 cu m (2017 est.)
country comparison to the world: 165

Natural gas—imports: 6.003 billion cu m (2017 est.)
country comparison to the world: 32

Natural gas—proved reserves: 588.8 billion cu m (1 January 2018 est.)
country comparison to the world: 30

Carbon dioxide emissions from consumption of energy: 179.5 million Mt (2017 est.)
country comparison to the world: 33

COMMUNICATIONS

Telephones—fixed lines: *total subscriptions:* 2,607,495
subscriptions per 100 inhabitants: 1.14 (2019 est.)
country comparison to the world: 49

Telephones—mobile cellular:
total subscriptions: 174,702,132
subscriptions per 100 inhabitants: 76.38 (2019 est.)
country comparison to the world: 9

Telecommunication systems: *general assessment:* the telecommunications infrastructure is improving, with investments in mobile-cellular networks increasing, fixed-line subscriptions declining; system consists of microwave radio relay, coaxial cable, fiber-optic cable, cellular, and satellite networks; 4G mobile services broadly available; 5G not before 2030; mobile platform and mobile broadband doing well and dominate over fixed broadband sector (2020)
domestic: mobile-cellular subscribership has skyrocketed; more than 90% of Pakistanis live within areas that have cell phone coverage; fiber-optic networks are being constructed throughout the country to increase broadband access, though broadband penetration in Pakistan is still relatively low; fixed-line 1 per 100 and mobile-cellular 76 per 100 persons (2019)
international: country code—92; landing points for the SEA-ME-WE-3, -4, -5, AAE-1, IMEWE, Orient Express, PEACE Cable, and TW1 submarine cable systems that provide links to Europe, Africa, the Middle East, Asia, Southeast Asia, and Australia; satellite earth stations—3 Intelsat (1 Atlantic Ocean and 2 Indian Ocean); 3 operational international gateway exchanges (1 at Karachi and 2 at Islamabad); microwave radio relay to neighboring countries (2019)
note: the COVID-19 outbreak is negatively impacting telecommunications production and supply chains globally; consumer spending on telecom devices and services has also slowed due to the pandemic's effect on economies worldwide; overall progress towards improvements in all facets of the telecom industry—mobile, fixed-line, broadband, submarine cable and satellite—has moderated

Broadcast media: media is government regulated; 1 dominant state-owned TV broadcaster, Pakistan Television Corporation (PTV), operates a network consisting of 8 channels; private TV broadcasters are permitted; to date 69 foreign satellite channels are operational; the state-owned radio network operates more than 30 stations; nearly 200 commercially licensed, privately owned radio stations provide programming mostly limited to music and talk shows (2019)

Internet country code: .pk

Internet users: *total:* 34,734,689
percent of population: 15.51% (July 2018 est.)
country comparison to the world: 22

Broadband—fixed subscriptions: *total:* 1,811,365
subscriptions per 100 inhabitants: 1 (2018 est.)
country comparison to the world: 57

TRANSPORTATION

National air transport system: *number of registered air carriers:* 5 (2020)
inventory of registered aircraft operated by air carriers: 52
annual passenger traffic on registered air carriers: 6,880,637 (2018)
annual freight traffic on registered air carriers: 217.53 million mt-km (2018)

Civil aircraft registration country code prefix: AP (2016)

Airports: 151 (2013)
country comparison to the world: 35

Airports—with paved runways:
total: 108 (2017)
over 3,047 m: 15 (2017)
2,438 to 3,047 m: 20 (2017)
1,524 to 2,437 m: 43 (2017)
914 to 1,523 m: 20 (2017)
under 914 m: 10 (2017)

Airports—with unpaved runways:
total: 43 (2013)
2,438 to 3,047 m: 1 (2013)
1,524 to 2,437 m: 9 (2013)
914 to 1,523 m: 9 (2013)
under 914 m: 24 (2013)

Heliports: 23 (2013)

Pipelines: 12,984 km gas, 3,470 km oil, 1,170 km refined products (2019)

Railways: *total:* 11,881 km (2019)
narrow gauge: 389 km 1.000-m gauge (2019)
broad gauge: 11,492 km 1.676-m gauge (293 km electrified) (2019)
country comparison to the world: 22

Roadways: *total:* 263,775 km (2019)
paved: 185,063 km (includes 708 km of expressways) (2019)
unpaved: 78,712 km (2019)
country comparison to the world: 22

Merchant marine: *total:* 54
by type: bulk carrier 5, oil tanker 5, other 44 (2019)
country comparison to the world: 115

Ports and terminals: *major seaport(s):* Karachi, Port Muhammad Bin Qasim
container port(s) (TEUs): Karachi (2,224,000) (2017)
LNG terminal(s) (import): Port Qasim

MILITARY AND SECURITY

Military and security forces: *Pakistan Army (includes National Guard), Pakistan Navy (includes marines, Maritime Security Agency), Pakistan Air Force (Pakistan Fizaia); Ministry*

739

of Interior paramilitary forces: Frontier Corps, Pakistan Rangers (2019)

note: the National Guard is a paramilitary force and one of the Army's reserve forces, along with the Pakistan Army Reserve, the Frontier Corps, and the Pakistan Rangers

Military expenditures:
4% of GDP (2019)
4.1% of GDP (2018)
3.8% of GDP (2017)
3.6% of GDP (2016)
3.6% of GDP (2015)
country comparison to the world: 11

Military and security service personnel strengths: estimates of the size of the Pakistan military's active force vary; approximately 650,000 active personnel (560,000 Army; 30,000 Navy; 60,000 Air Force); est. 70,000 Frontier Corps; est. 25,000 Pakistan Rangers (2019)

Military equipment inventories and acquisitions: the Pakistan military inventory includes a broad mix of equipment, primarily from China, France, Ukraine, the UK, and the US; since 2010, China and the US are the leading suppliers of arms to Pakistan; Pakistan also has a large domestic defense industry capable of upgrading existing air, land, and sea weapons systems (2019 est.)

Military deployments: 1,230 Central African Republic (MINUSCA); 1,950 Democratic Republic of the Congo (MONUSCO); 140 Mali (MINUSMA); 900 Sudan (UNAMID) (2020)

Military service age and obligation: 16-23 years of age for voluntary military service; soldiers cannot be deployed for combat until age 18; women serve in all three armed forces; reserve obligation to age 45 for enlisted men, age 50 for officers (2019)

TERRORISM

Terrorist group(s): Haqqani Network; Harakat ul-Jihad-i-Islami; Harakat ul-Mujahidin; Hizbul Mujahideen; Indian Mujahedeen; Islamic State of Iraq and ash-Sham-Khorasan; Islamic State of ash-Sham – India; Islamic State of ash-Sham – Pakistan; Islamic Movement of Uzbekistan; Jaish-e-Mohammed; Jaysh al Adl (Jundallah); Lashkar i Jhangvi; Lashkar-e Tayyiba; Tehrik-e-Taliban Pakistan; al-Qa'ida; al-Qa'ida in the Indian Subcontinent (2019)

note: details about the history, aims, leadership, organization, areas of operation, tactics, targets, weapons, size, and sources of support of the group(s) appear(s) in Appendix-T

TRANSNATIONAL ISSUES

Disputes—international: various talks and confidence-building measures cautiously have begun to defuse tensions over Kashmir, particularly since the October 2005 earthquake in the region; Kashmir nevertheless remains the site of the world's largest and most militarized territorial dispute with portions under the de facto administration of China (Aksai Chin), India (Jammu and Kashmir), and Pakistan (Azad Kashmir and Northern Areas); UN Military Observer Group in India and Pakistan has maintained a small group of peacekeepers since 1949; India does not recognize Pakistan's ceding historic Kashmir lands to China in 1964; India and Pakistan have maintained their 2004 cease-fire in Kashmir and initiated discussions on defusing the armed standoff in the Siachen glacier region; Pakistan protests India's fencing the highly militarized Line of Control and construction of the Baglihar Dam on the Chenab River in Jammu and Kashmir, which is part of the larger dispute on water sharing of the Indus River and its tributaries; to defuse tensions and prepare for discussions on a maritime boundary, India and Pakistan seek technical resolution of the disputed boundary in Sir Creek estuary at the mouth of the Rann of Kutch in the Arabian Sea; Pakistani maps continue to show the Junagadh claim in India's Gujarat State; since 2002, with UN assistance, Pakistan has repatriated 3.8 million Afghan refugees, leaving about 2.6 million; Pakistan has sent troops across and built fences along some remote tribal areas of its treaty-defined Durand Line border with Afghanistan, which serve as bases for foreign terrorists and other illegal activities; Afghan, Coalition, and Pakistan military meet periodically to clarify the alignment of the boundary on the ground and on maps

Refugees and internally displaced persons: *refugees (country of origin):* 2.58-2.68 million (1.4 million registered, 1.18-1.28 million undocumented) (Afghanistan) (2017)
IDPs: 106,000 (primarily those who remain displaced by counter-terrorism and counter-insurgency operations and violent conflict between armed non-state groups in the Federally Administered Tribal Areas and Khyber-Paktunkwa Province; more than 1 million displaced in northern Waziristan in 2014; individuals also have been displaced by repeated monsoon floods) (2019)

Trafficking in persons: *current situation:* Pakistan is a source, transit, and destination country for men, women, and children subjected to forced labor and sex trafficking; the largest human trafficking problem is bonded labor in agriculture, brickmaking and, to a lesser extent, fishing, mining and carpet-making; children are bought, sold, rented, and placed in forced begging rings, domestic service, small shops, brick-making factories, or prostitution; militant groups also force children to spy, fight, or die as suicide bombers, kidnapping the children or getting them from poor parents through sale or coercion; women and girls are forced into prostitution or marriages; Pakistani adults migrate to the Gulf States and African and European states for low- skilled jobs and sometimes become victims of forced labor, debt bondage, or prostitution; foreign adults and children, particularly from Afghanistan, Bangladesh, and Sri Lanka, may be subject to forced labor, and foreign women may be sex trafficked in Pakistan, with refugees and ethnic minorities being most vulnerable

tier rating: Tier 2 Watch List – Pakistan does not fully comply with the minimum standards for the elimination of trafficking; however, it is making significant efforts to do so; the government lacks political will and capacity to fully address human trafficking, as evidenced by ineffective law enforcement efforts, official complicity, penalization of victims, and the continued conflation of migrant smuggling and human trafficking by many officials; not all forms of trafficking are prohibited; an anti-trafficking bill drafted in 2013 to address gaps in existing legislation remains pending, and a national action plan drafted in 2014 is not finalized; feudal landlords and brick kiln owners use their political influence to protect their involvement in bonded labor, while some police personnel have taken bribes to ignore prostitution that may have included sex trafficking; authorities began to use standard procedures for the identification and referral of trafficking victims, but it is not clear how widely these methods were practiced; in other instances, police were reluctant to assist NGOs with rescues and even punished victims for crimes committed as a direct result of being trafficked (2015)

Illicit drugs: significant transit area for Afghan drugs, including heroin, opium, morphine, and hashish, bound for Iran, Western markets, the Gulf States, Africa, and Asia; financial crimes related to drug trafficking, terrorism, corruption, and smuggling remain problems; opium poppy cultivation estimated to be 930 hectares in 2015; federal and provincial authorities continue to conduct anti-poppy campaigns that utilizes forced eradication, fines, and arrests

PALAU

INTRODUCTION

Background: After three decades as part of the UN Trust Territory of the Pacific under US administration, this westernmost cluster of the Caroline Islands opted for independence in 1978 rather than join the Federated States of Micronesia. A Compact of Free Association with the US was approved in 1986 but not ratified until 1993. It entered into force the following year when the islands gained independence.

GEOGRAPHY

Location: Oceania, group of islands in the North Pacific Ocean, southeast of the Philippines
Geographic coordinates: 7 30 N, 134 30 E
Map references: Oceania

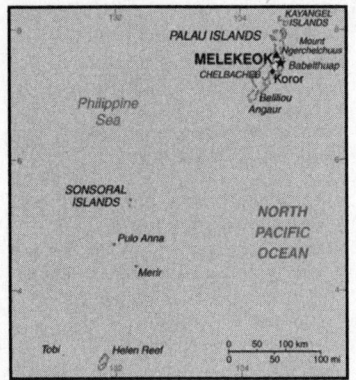

Area: *total:* 459 sq km
land: 459 sq km
water: 0 sq km
country comparison to the world: 198

Area—comparative: slightly more than 2.5 times the size of Washington, DC

Land boundaries: 0 km

Coastline: 1,519 km

Maritime claims: *territorial sea:* 12 nm
exclusive economic zone: 200 nm
contiguous zone: 24 nm
continental shelf: 200 nm

Climate: tropical; hot and humid; wet season May to November

Terrain: varying topography from the high, mountainous main island of Babelthuap to low, coral islands usually fringed by large barrier reefs

Elevation: *lowest point:* Pacific Ocean 0 m
highest point: Mount Ngerchelchuus 242 m

Natural resources: forests, minerals (especially gold), marine products, deep-seabed minerals

Land use: *agricultural land:* 10.8% (2011 est.)
arable land: 2.2% (2011 est.) / *permanent crops:* 4.3% (2011 est.) / *permanent pasture:* 4.3% (2011 est.)
forest: 87.6% (2011 est.)
other: 1.6% (2011 est.)

Irrigated land: 0 sq km (2012)

Population distribution: most of the population is located on the southern end of the main island of Babelthuap

Natural hazards: typhoons (June to December)

Environment—current issues: inadequate facilities for disposal of solid waste; threats to the marine ecosystem from sand and coral dredging, illegal and destructive fishing practices, and overfishing; climate change contributes to rising sea level and coral bleaching; drought

Environment—international agreements: *party to:* Biodiversity, Climate Change, Climate Change-Kyoto Protocol, Desertification, Hazardous Wastes, Law of the Sea, Ozone Layer Protection, Wetlands, Whaling
signed, but not ratified: none of the selected agreements

Geography—note: westernmost archipelago in the Caroline chain, consists of six island groups totaling more than 300 islands; includes World War II battleground of Beliliou (Peleliu) and world-famous Rock Islands

PEOPLE AND SOCIETY

Population: 21,685 (July 2020 est.)
country comparison to the world: 219

Nationality: *noun:* Palauan(s)
adjective: Palauan

Ethnic groups: Palauan (Micronesian with Malayan and Melanesian admixtures) 73%, Carolinian 2%, Asian 21.7%, caucasian 1.2%, other 2.1% (2015 est.)

Languages: Palauan (official on most islands) 65.2%, other Micronesian 1.9%, English (official) 19.1%, Filipino 9.9%, Chinese 1.2%, other 2.8% (2015 est.)
note: Sonsoralese is official in Sonsoral; Tobian is official in Tobi; Angaur and Japanese are official in Angaur

Religions: Roman Catholic 45.3%, Protestant 34.9% (includes Evangelical 26.4%, Seventh Day Adventist 6.9%, Assembly of God .9%, Baptist .7%), Modekngei 5.7% (indigenous to Palau), Muslim 3%, Mormon 1.5%, other 9.7% (2015 est.)

Age structure: *0-14 years:* 18.68% (male 2,090/female 1,961)
15-24 years: 15.86% (male 1,723/female 1,716)
25-54 years: 45.33% (male 6,026/female 3,804)
55-64 years: 10.68% (male 853/female 1,463)
65 years and over: 9.45% (male 501/female 1,548) (2020 est.)

Median age: *total:* 33.9 years
male: 32.9 years
female: 35.9 years (2020 est.)
country comparison to the world: 94

Population growth rate: 0.39% (2020 est.)
country comparison to the world: 163

Birth rate: 11.3 births/1,000 population (2020 est.)
country comparison to the world: 174

Death rate: 8.3 deaths/1,000 population (2020 est.)
country comparison to the world: 80

Net migration rate: 0.9 migrant(s)/1,000 population (2020 est.)
country comparison to the world: 62

Population distribution: most of the population is located on the southern end of the main island of Babelthuap

Urbanization: *urban population:* 81% of total population (2020)
rate of urbanization: 1.77% annual rate of change (2015-20 est.)

total population growth rate v. urban population growth rate, 2000-2030:

Major urban areas—population: 277 NGERULMUD (capital) (2018)

Sex ratio: *at birth:* 1.07 male(s)/female
0-14 years: 1.07 male(s)/female
15-24 years: 1 male(s)/female

25-54 years: 1.58 male(s)/female
55-64 years: 0.58 male(s)/female
65 years and over: 0.32 male(s)/female
total population: 1.07 male(s)/female (2020 est.)

Infant mortality rate: *total:* 9.8 deaths/1,000 live births
male: 11.2 deaths/1,000 live births
female: 8.3 deaths/1,000 live births (2020 est.)
country comparison to the world: 133

Life expectancy at birth: *total population:* 74.1 years
male: 70.9 years
female: 77.5 years (2020 est.)
country comparison to the world: 136

Total fertility rate: 1.7 children born/woman (2020 est.)
country comparison to the world: 173

Drinking water source:
improved:
urban: 100% of population
rural: 100% of population
total: 100% of population
unimproved:
urban: 0% of population
rural: 0% of population
total: 0% of population (2017 est.)

Current Health Expenditure: 12% (2017)

Physicians density: 1.42 physicians/1,000 population (2014)

Hospital bed density: 4.8 beds/1,000 population (2010)

Sanitation facility access:
improved:
urban: 100% of population
rural: 100% of population
total: 100% of population
unimproved:
urban: 0% of population
rural: 0% of population
total: 0% of population (2017 est.)

HIV/AIDS—adult prevalence rate: NA

HIV/AIDS—people living with HIV/AIDS: NA

HIV/AIDS—deaths: NA

Major infectious diseases: *degree of risk:* high (2020)
food or waterborne diseases: bacterial diarrhea
vectorborne diseases: malaria

Obesity—adult prevalence rate: 55.3% (2016)
country comparison to the world: 3

Education expenditures: NA

Literacy: *definition:* age 15 and over can read and write
total population: 96.6%
male: 96.8%
female: 96.3% (2015)

School life expectancy (primary to tertiary education): *total:* 17 years
male: 16 years
female: 17 years (2013)

Unemployment, youth ages 15-24: 5.6%
country comparison to the world: 162

GOVERNMENT

Country name: *conventional long form:* Republic of Palau

conventional short form: Palau

local long form: Beluu er a Belau

local short form: Belau

former: Trust Territory of the Pacific Islands, Palau District

etymology: from the Palauan name for the islands, Belau, which likely derives from the Palauan word "beluu" meaning "village"

Government type: presidential republic in free association with the US

Capital: *name:* Ngerulmud

geographic coordinates: 7 30 N, 134 37 E

time difference: UTC+9 (14 hours ahead of Washington, DC, during Standard Time)

etymology: the Palauan meaning is "place of fermented 'mud'" ('mud' being the native name for the keyhole angelfish); the site of the new capitol (established in 2006) had been a large hill overlooking the ocean, Ngerulmud, on which women would communally gather to offer fermented angelfish to the gods

note: Ngerulmud, on Babeldaob Island, is the smallest national capital on earth by population, with only a few hundred people; the name is pronounced en-jer-al-mud; Koror, on Koror Island, with over 11,000 residents is by far the largest settlement in Palau; it served as the country's capital from independence in 1994 to 2006

Administrative divisions: 16 states; Aimeliik, Airai, Angaur, Hatohobei, Kayangel, Koror, Melekeok, Ngaraard, Ngarchelong, Ngardmau, Ngatpang, Ngchesar, Ngeremlengui, Ngiwal, Peleliu, Sonsorol

Independence: 1 October 1994 (from the US-administered UN trusteeship)

National holiday: Constitution Day, 9 July (1981), day of a national referendum to pass the new constitution; Independence Day, 1 October (1994)

Constitution: *history:* ratified 9 July 1980, effective 1 January 1981

amendments: proposed by a constitutional convention (held at least once every 15 years with voter approval), by public petition of at least 25% of eligible voters, or by a resolution adopted by at least three fourths of National Congress members; passage requires approval by a majority of votes in at least three fourths of the states in the next regular general election; amended 1992, 2004, 2008

Legal system: mixed legal system of civil, common, and customary law

International law organization participation: has not submitted an ICJ jurisdiction declaration; non-party state to the ICCt

Citizenship: *citizenship by birth:* no

citizenship by descent only: at least one parent must be a citizen of Palau

dual citizenship recognized: no

residency requirement for naturalization: note—no procedure for naturalization

Suffrage: 18 years of age; universal

Executive branch: *chief of state:* President Tommy REMENGESAU (since 17 January 2013); Vice President Raynold OILUCH (since 19 January 2017); note—the president is both chief of state and head of government

head of government: President Tommy REMENGESAU (since 17 January 2013); Vice President Raynold OILUCH (since 19 January 2017)

cabinet: Cabinet appointed by the president with the advice and consent of the Senate; also includes the vice president; the Council of Chiefs consists of chiefs from each of the states who advise the president on issues concerning traditional laws, customs, and their relationship to the constitution and laws of Palau

elections/appointments: president and vice president directly elected on separate ballots by absolute majority popular vote in 2 rounds if needed for a 4-year term (eligible for a second term); election last held on 3 November 2020 (next to be held on November 2024)

election results: Surangel WHIPPS, Jr. elected president; percent of vote—Surangel WHIPPS, Jr.(independent) 57.4%, Raynold OILUCH (independent) 42.6%

Legislative branch: *description:* bicameral National Congress or Olbiil Era Kelulau consists of:

Senate (13 seats; members directly elected in single-seat constituencies by majority vote to serve 4-year terms)

House of Delegates (16 seats; members directly elected in single-seat constituencies by simple majority vote to serve 4year terms)

elections:

Senate—last held on 1 November 2016 (next to be held on 3 November 2020)

House of Delegates—last held on 1 November 2016 (next to be held on 3 November 2020)

election results:

Senate—percent of vote—NA; seats—independent 13; composition—men 11, women 2, percent of women 15.4%

House of Delegates—percent of vote—NA; seats—independent 16; composition—men 14, women 2, percent of women 12.5%; note—total National Congress percent of women 13.8%

Judicial branch: *highest courts:* Supreme Court (consists of the chief justice and 3 associate justices organized into appellate trial divisions; the Supreme Court organization also includes the Common Pleas and Land Courts)

judge selection and term of office: justices nominated by a 7-member independent body consisting of judges, presidential appointees, and lawyers and appointed by the president; judges can serve until mandatory retirement at age 65

subordinate courts: National Court and other 'inferior' courts

Political parties and leaders: none

International organization participation: ACP, ADB, AOSIS, FAO, IAEA, IBRD, ICAO, ICRM, IDA, IFC, IFRCS, ILO, IMF, IMO, IMSO, IOC, IPU, MIGA, OPCW, PIF, Sparteca, SPC, UN, UNAMID, UNCTAD, UNESCO, WHO

Diplomatic representation in the US: *chief of mission:* Ambassador Hersey KYOTA (since 12 November 1997)

chancery: 1701 Pennsylvania Avenue NW, Suite 300, Washington, DC 20036

telephone: [1] (202) 452-6814

FAX: [1] (202) 452-6281

consulate(s): Tamuning (Guam)

Diplomatic representation from the US: *chief of mission:* Ambassador Amy HYATT (since 9 March 2015)

telephone: [680] 587-2920

embassy: Omsangel/Beklelachieb, Airai 96940

mailing address: P. O. Box 6028, Koror, Republic of Palau 96940

FAX: [680] 587-2911

Flag description: light blue with a large yellow disk shifted slightly to the hoist side; the blue color represents the ocean, the disk represents the moon; Palauans consider the full moon to be the optimum time for human activity; it is also considered a symbol of peace, love, and tranquility

National symbol(s): *bai (native meeting house);*

national colors: blue, yellow

National anthem: *name:* "Belau rekid" (Our Palau)

lyrics/music: multiple/Ymesei O. EZEKIEL

note: adopted 1980

0:00/ 0:51

ECONOMY

Economy—overview: The economy is dominated by tourism, fishing, and subsistence agriculture. Government is a major employer of the work force relying on financial assistance from the US under the Compact of Free Association (Compact) with the US that took effect after the end of the UN trusteeship on 1 October 1994. The US provided Palau with roughly $700 million in aid for the first 15 years following commencement of the Compact in 1994 in return for unrestricted access to its land and waterways for strategic purposes. The population enjoys a per capita income roughly double that of the Philippines and much of Micronesia.

Business and leisure tourist arrivals reached a record 167,966 in 2015, a 14.4% increase over the previous year, but fell to 138,408 in 2016. Long-run prospects for tourism have been bolstered by the expansion of air travel in the Pacific, the rising prosperity of industrial East Asia, and the willingness of foreigners to finance infrastructure development. Proximity to Guam, the region's major destination for tourists from East Asia, and a regionally competitive tourist infrastructure enhance Palau's advantage as a destination.

GDP (purchasing power parity): $264 million (2017 est.)

$274.2 million (2016 est.)

$274.1 million (2015 est.)

note: data are in 2017 dollars

country comparison to the world: 217

GDP (official exchange rate):

$292 million (2017 est.)

GDP—real growth rate: -3.7% (2017 est.)
0% (2016 est.)
10.1% (2015 est.)
country comparison to the world: 215

GDP—per capita (PPP): $14,700 (2017 est.)
$15,200 (2016 est.)
$15,200 (2015 est.)
note: data are in 2017 dollars
country comparison to the world: 115

Gross national saving:
48.7% of GDP (2016 est.)
50.1% of GDP (2015 est.)
country comparison to the world: 2

GDP—composition, by end use:
household consumption: 60.5% (2016 est.)
government consumption: 27.2% (2016 est.)
investment in fixed capital: 22.7% (2016 est.)
investment in inventories: 1.9% (2016 est.)
exports of goods and services: 55.2% (2016 est.)
imports of goods and services: -67.6% (2016 est.)

GDP—composition, by sector of origin:
agriculture: 3% (2016 est.)
industry: 19% (2016 est.)
services: 78% (2016 est.)

Agriculture—products: coconuts, cassava (manioc, tapioca), sweet potatoes; fish, pigs, chickens, eggs, bananas, papaya, breadfruit, calamansi, soursop, Polynesian chestnuts, Polynesian almonds, mangoes, taro, guava, beans, cucumbers, squash/pumpkins (various), eggplant, green onions, kangkong (watercress), cabbages (various), radishes, betel nuts, melons, peppers, noni, okra

Industries: tourism, fishing, subsistence agriculture

Industrial production growth rate: NA

Labor force: 11,610 (2016)
country comparison to the world: 216

Labor force—by occupation: *agriculture:* 1.2%
industry: 12.4%
services: 86.4% (2016)

Unemployment rate: 1.7% (2015 est.)
4.1% (2012)
country comparison to the world: 16

Population below poverty line: 24.9% NA (2006)

Household income or consumption by percentage share: *lowest 10%:* NA
highest 10%: NA

Budget: *revenues:* 193 million (2012 est.)
expenditures: 167.3 million (2012 est.)

Taxes and other revenues: 66.1% (of GDP) (2016 est.)
country comparison to the world: 6

Budget surplus (+) or deficit (-): 8.8% (of GDP) (2016 est.)
country comparison to the world: 3

Public debt:
24.1% of GDP (2016 est.)
21.6% of GDP (2015)
country comparison to the world: 179

Fiscal year: 1 October—30 September

Inflation rate (consumer prices): 0.9% (2017 est.)
-1% (2016 est.)
country comparison to the world: 46

Current account balance:
-$53 million (2017 est.)
-$36 million (2016 est.)
country comparison to the world: 80

Exports: $23.17 billion (2017 est.)
$14.8 million (2015 est.)
country comparison to the world: 68

Exports—partners: Japan 51.3%, US 15.8%, India 13.8%, Guam 8% (2017)

Exports—commodities: shellfish, tuna, other fish (many species)

Imports: $4.715 billion (2018 est.)
$4.079 billion (2017 est.)
country comparison to the world: 134

Imports—commodities: machinery and equipment, fuels, metals; foodstuffs

Imports—partners: US 33.4%, Guam 15.8%, Japan 15.7%, China 13.5%, South Korea 5.3% (2017)

Reserves of foreign exchange and gold: $0 (31 December 2017 est.)
$580.9 million (31 December 2015 est.)
country comparison to the world: 193

Debt—external:
$18.38 billion (31 December 2014 est.)
$16.47 billion (31 December 2013 est.)
country comparison to the world: 94

Exchange rates: the US dollar is used

ENERGY

Electricity access: *electrification—total population:* 99.3% (2016)
electrification—urban areas: 99.6% (2016)
electrification—rural areas: 97.2% (2016)

COMMUNICATIONS

Telephones—fixed lines: *total subscriptions:* 8,808
subscriptions per 100 inhabitants: 40.78 (2019 est.)
country comparison to the world: 191

Telephones—mobile cellular: *total subscriptions:* 29,033
subscriptions per 100 inhabitants: 134.41 (2019 est.)
country comparison to the world: 210

Telecommunication systems: *general assessment:* well-developed mobile sector, recently boosted by satellite network capacity upgrades; 3G services available with satellite; lack of telecom regulations; newest and most powerful commercial satellite, Kacific-1 satellite, launched in 2019 to improve telecommunications in the Asia Pacific region (2020)
domestic: fixed-line 41 per 100 and mobile-cellular services 134 per 100 persons (2019)
international: country code—680; landing point for the SEA-US submarine cable linking Palau, Philippines, Micronesia, Indonesia, Hawaii (US), Guam (US) and California (US); satellite earth station—1 Intelsat (Pacific Ocean) (2019)

note: the COVID-19 outbreak is negatively impacting telecommunications production and supply chains globally; consumer spending on telecom devices and services has also slowed due to the pandemic's effect on economies worldwide; overall progress towards improvements in all facets of the telecom industry—mobile, fixed-line, broadband, submarine cable and satellite—has moderated

Broadcast media: no broadcast TV stations; a cable TV network covers the major islands and provides access to 4 local cable stations, rebroadcasts (on a delayed basis) of a number of US stations, as well as access to a number of real-time satellite TV channels; about a half dozen radio stations (1 government-owned) (2019)

Internet country code: .pw

Internet users: *total:* 7,650
percent of population: 36% (July 2016 est.)
country comparison to the world: 214

TRANSPORTATION

National air transport system: *number of registered air carriers:* 1 (2020)
inventory of registered aircraft operated by air carriers: 1

Airports: 3 (2013)
country comparison to the world: 196

Airports—with paved runways: *total:* 1 (2019)
1,524 to 2,437 m: 1

Airports—with unpaved runways: *total:* 2 (2013)
1,524 to 2,437 m: 2 (2013)

Roadways: *total:* 125 km (2018)
paved: 89 km (2018)
unpaved: 36 km (2018)
country comparison to the world: 211

Merchant marine: *total:* 203
by type: bulk carrier 9, container ship 11, general cargo 88, oil tanker 31, other 64 (2019)
country comparison to the world: 66

Ports and terminals: *major seaport(s):* Koror

MILITARY AND SECURITY

Military and security forces: no regular military forces; the Ministry of Justice includes divisions/bureaus for public security, police functions, and maritime law enforcement. (2019)

Military equipment inventories and acquisitions: since 2018, Australia and Japan have provided patrol boats to the Palau's Division of Marine Law Enforcement (2020)

Military—note: Under a 1994 Compact of Free Association between Palau and the US, the US until 2044 is responsible for the defense of Palaus and the US military is granted access to the islands, but it has not stationed any military forces there. (2020)

TRANSNATIONAL ISSUES

Disputes—international: maritime delineation negotiations continue with Philippines, Indonesia

PANAMA

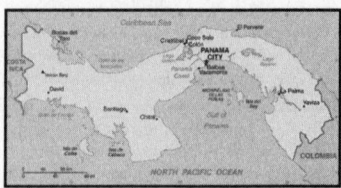

INTRODUCTION

Background: Explored and settled by the Spanish in the 16th century, Panama broke with Spain in 1821 and joined a union of Colombia, Ecuador, and Venezuela—named the Republic of Gran Colombia. When the latter dissolved in 1830, Panama remained part of Colombia. With US backing, Panama seceded from Colombia in 1903 and promptly signed a treaty with the US allowing for the construction of a canal and US sovereignty over a strip of land on either side of the structure (the Panama Canal Zone). The Panama Canal was built by the US Army Corps of Engineers between 1904 and 1914. In 1977, an agreement was signed for the complete transfer of the Canal from the US to Panama by the end of the century. Certain portions of the Zone and increasing responsibility over the Canal were turned over in the subsequent decades. With US help, dictator Manuel NORIEGA was deposed in 1989. The entire Panama Canal, the area supporting the Canal, and remaining US military bases were transferred to Panama by the end of 1999. An ambitious expansion project to more than double the Canal's capacity—by allowing for more Canal transits and larger ships—was carried out between 2007 and 2016.

GEOGRAPHY

Location: Central America, bordering both the Caribbean Sea and the North Pacific Ocean, between Colombia and Costa Rica

Geographic coordinates: 9 00 N, 80 00 W

Map references: Central America and the Caribbean

Area: *total:* 75,420 sq km
land: 74,340 sq km
water: 1,080 sq km
country comparison to the world: 119

Area—comparative: slightly smaller than South Carolina

Land boundaries: *total:* 687 km
border countries (2): Colombia 339 km, Costa Rica 348 km

Coastline: 2,490 km

Maritime claims: *territorial sea:* 12 nm
exclusive economic zone: 200 nm or edge of continental margin
contiguous zone: 24 nm

Climate: tropical maritime; hot, humid, cloudy; prolonged rainy season (May to January), short dry season (January to May)

Terrain: interior mostly steep, rugged mountains with dissected, upland plains; coastal plains with rolling hills

Elevation: *mean elevation:* 360 m
lowest point: Pacific Ocean 0 m
highest point: Volcan Baru 3,475 m

Natural resources: copper, mahogany forests, shrimp, hydropower

Land use: *agricultural land:* 30.5% (2011 est.)
arable land: 7.3% (2011 est.) / permanent crops: 2.5% (2011 est.) / permanent pasture: 20.7% (2011 est.)
forest: 43.6% (2011 est.)
other: 25.9% (2011 est.)
Irrigated land: 321 sq km (2012)

Population distribution: population is concentrated towards the center of the country, particularly around the Canal, but a sizeable segment of the populace also lives in the far west around David; the eastern third of the country is sparsely inhabited

Natural hazards: occasional severe storms and forest fires in the Darien area

Environment—current issues: water pollution from agricultural runoff threatens fishery resources; deforestation of tropical rain forest; land degradation and soil erosion threatens siltation of Panama Canal; air pollution in urban areas; mining threatens natural resources

Environment—international agreements: *party to:* Biodiversity, Climate Change, Climate Change-Kyoto Protocol, Desertification, Endangered Species, Environmental Modification, Hazardous Wastes, Law of the Sea, Marine Dumping, Ozone Layer Protection, Ship Pollution, Tropical Timber 83, Tropical Timber 94, Wetlands, Whaling
signed, but not ratified: Marine Life Conservation

Geography—note: strategic location on eastern end of isthmus forming land bridge connecting North and South America; controls Panama Canal that links North Atlantic Ocean via Caribbean Sea with North Pacific Ocean

PEOPLE AND SOCIETY

Population: 3,894,082 (July 2020 est.)
country comparison to the world: 130

Nationality: *noun:* Panamanian(s)
adjective: Panamanian

Ethnic groups: mestizo (mixed Amerindian and white) 65%, Native American 12.3% (Ngabe 7.6%, Kuna 2.4%, Embera 0.9%, Bugle 0.8%, other 0.4%, unspecified 0.2%), black or African descent 9.2%, mulatto 6.8%, white 6.7% (2010 est.)

Languages: Spanish (official), indigenous languages (including Ngabere (or Guaymi), Buglere, Kuna, Embera, Wounaan, Naso (or Teribe), and Bri Bri), Panamanian English Creole (similar to Jamaican English Creole; a mixture of English and Spanish with elements of Ngabere; also known as Guari Guari and Colon Creole), English, Chinese (Yue and Hakka), Arabic, French Creole, other (Yiddish, Hebrew, Korean, Japanese)
note: many Panamanians are bilingual

Religions: Roman Catholic 85%, Protestant 15%

Demographic profile: Panama is a country of demographic and economic contrasts. It is in the midst of a demographic transition, characterized by steadily declining rates of fertility, mortality, and population growth, but disparities persist based on wealth, geography, and ethnicity. Panama has one of the fastest growing economies in Latin America and dedicates substantial funding to social programs, yet poverty and inequality remain prevalent. The indigenous population accounts for a growing share of Panama's poor and extreme poor, while the non-indigenous rural poor have been more successful at rising out of poverty through rural-to-urban labor migration. The government's large expenditures on untargeted, indirect subsidies for water, electricity, and fuel have been ineffective, but its conditional cash transfer program has shown some promise in helping to decrease extreme poverty among the indigenous population.

Panama has expanded access to education and clean water, but the availability of sanitation and, to a lesser extent, electricity remains poor. The increase in secondary schooling—led by female enrollment—is spreading to rural and indigenous areas, which probably will help to alleviate poverty if educational quality and the availability of skilled jobs improve. Inadequate access to sanitation contributes to a high incidence of diarrhea in Panama's children, which is one of the main causes of Panama's elevated chronic malnutrition rate, especially among indigenous communities.

Age structure: *0-14 years:* 25.56% (male 508,131/female 487,205)
15-24 years: 16.59% (male 329,250/female 316,796)
25-54 years: 40.31% (male 794,662/female 774,905)
55-64 years: 8.54% (male 165,129/female 167,317)
65 years and over: 9.01% (male 160,516/female 190,171) (2020 est.)

Dependency ratios: *total dependency ratio:* 53.9
youth dependency ratio: 40.8
elderly dependency ratio: 13.1
potential support ratio: 7.6 (2020 est.)

Median age: *total:* 30.1 years
male: 29.6 years
female: 30.5 years (2020 est.)

country comparison to the world: 122

Population growth rate: 1.2% (2020 est.)
country comparison to the world: 90

Birth rate: 17.1 births/1,000 population (2020 est.)
country comparison to the world: 98

Death rate: 5.1 deaths/1,000 population (2020 est.)
country comparison to the world: 197

Net migration rate: -0.1 migrant(s)/1,000 population (2020 est.)
country comparison to the world: 103

Population distribution: population is concentrated towards the center of the country, particularly around the Canal, but a sizeable segment of the populace also lives in the far west around David; the eastern third of the country is sparsely inhabited

Urbanization: *urban population:* 68.4% of total population (2020)
rate of urbanization: 2.06% annual rate of change (2015-20 est.)

total population growth rate v. urban population growth rate, 2000-2030:

Major urban areas—population: *1.860 million* PANAMA CITY *(capital) (2020)*
Sex ratio: at birth: 1.04 male(s)/female
0-14 years: 1.04 male(s)/female
15-24 years: 1.04 male(s)/female
25-54 years: 1.03 male(s)/female
55-64 years: 0.99 male(s)/female
65 years and over: 0.84 male(s)/female
total population: 1.01 male(s)/female (2020 est.)

Maternal mortality rate: 52 deaths/100,000 live births (2017 est.)
country comparison to the world: 93

Infant mortality rate: *total:* 9.1 deaths/1,000 live births
male: 9.8 deaths/1,000 live births
female: 8.4 deaths/1,000 live births (2020 est.)
country comparison to the world: 138

Life expectancy at birth: *total population:* 79.2 years
male: 76.4 years
female: 82.2 years (2020 est.)
country comparison to the world: 60

Total fertility rate: 2.23 children born/woman (2020 est.)
country comparison to the world: 90

Contraceptive prevalence rate: 50.8% (2014/15)

Drinking water source:
improved:
urban: 100% of population
rural: 94.8% of population
total: 98.3% of population
unimproved:
urban: 0% of population
rural: 5.2% of population
total: 1.7% of population (2017 est.)

Current Health Expenditure: 7.3% (2017)

Physicians density: 1.57 physicians/1,000 population (2016)

Hospital bed density: 2.3 beds/1,000 population (2016)

Sanitation facility access:
improved:
urban: 97.2% of population
rural: 72.4% of population
total: 89.1% of population
unimproved:
urban: 2.8% of population
rural: 27.6% of population
total: 10.9% of population (2017 est.)

HIV/AIDS—adult prevalence rate: 0.9% (2018 est.)
country comparison to the world: 49

HIV/AIDS—people living with HIV/AIDS: 26,000 (2018 est.)
country comparison to the world: 81

HIV/AIDS—deaths: <500 (2018 est.)

Major infectious diseases: *degree of risk:* intermediate
food or waterborne diseases: bacterial diarrhea
vectorborne diseases: dengue fever

Obesity—adult prevalence rate: 22.7% (2016)
country comparison to the world: 73

Education expenditures: 3.2% of GDP (2011)
country comparison to the world: 132

Literacy: *definition:* age 15 and over can read and write
total population: 95.4%
male: 96%
female: 94.9% (2018)

School life expectancy (primary to tertiary education): *total:* 13 years
male: 12 years
female: 14 years (2016)

Unemployment, youth ages 15-24: *total:* 10.2%
male: 7.4%
female: 15.3% (2018 est.)
country comparison to the world: 125

GOVERNMENT

Country name: *conventional long form:* Republic of Panama
conventional short form: Panama
local long form: Republica de Panama
local short form: Panama
etymology: named after the capital city which was itself named after a former indigenous fishing village

Government type: presidential republic

Capital: *name:* Panama City
geographic coordinates: 8 58 N, 79 32 W
time difference: UTC-5 (same time as Washington, DC, during Standard Time)
etymology: according to tradition, the name derives from a former fishing area near the present capital—an indigenous village and its adjacent beach—that were called "Panama" meaning "an abundance of fish"

Administrative divisions: 10 provinces (provincias, singular—provincia) and 3 indigenous regions* (comarcas); Bocas del Toro, Chiriqui,

Cocle, Colon, Darien, Embera-Wounaan*, Herrera, Guna Yala*, Los Santos, Ngobe-Bugle*, Panama, Panama Oeste, Veraguas

Independence: 3 November 1903 (from Colombia; became independent from Spain on 28 November 1821)

National holiday: Independence Day (Separation Day), 3 November (1903)

Constitution: *history:* several previous; latest effective 11 October 1972

amendments: *proposed by the National Assembly, by the Cabinet, or by the Supreme Court of Justice; passage requires approval by one of two procedures:* 1) absolute majority vote of the Assembly membership in each of three readings and by absolute majority vote of the next elected Assembly in a single reading without textual modifications; 2) absolute majority vote of the Assembly membership in each of three readings, followed by absolute majority vote of the next elected Assembly in each of three readings with textual modifications, and approval in a referendum; amended several times, last in 2004

Legal system: civil law system; judicial review of legislative acts in the Supreme Court of Justice

International law organization participation: accepts compulsory ICJ jurisdiction with reservations; accepts ICCt jurisdiction

Citizenship: *citizenship by birth:* yes
citizenship by descent only: yes
dual citizenship recognized: no
residency requirement for naturalization: 5 years

Suffrage: 18 years of age; universal

Executive branch: *chief of state:* President Laurentino "Nito" CORTIZO Cohen (since 1 July 2019); Vice President Jose Gabriel CARRIZO Jaen (since 1 July 2019); note—the president is both chief of state and head of government

head of government: President Laurentino "Nito" CORTIZO Cohen (since 1 July 2019); Vice President Jose Gabriel CARRIZO Jaen (since 1 July 2019)
cabinet: Cabinet appointed by the president
elections/appointments: president and vice president directly elected on the same ballot by simple majority popular vote for a 5-year term; president eligible for a single non-consecutive term); election last held on 5 May 2019 (next to be held in 2024)
election results: Laurentino "Nito" CORTIZO Cohen elected president; percent of vote—Laurentino CORTIZO Cohen (PRD) 33.3%, Romulo ROUX (CD) 31%, Ricardo LOMBANA (independent) 18.8%, Jose BLANDON (Panamenista Party) 10.8%, Ana Matilde GOMEZ Ruiloba (independent) 4.8%, other 1.3%

Legislative branch: *description:* unicameral National Assembly or Asamblea Nacional (71 seats; 45 members directly elected in multi-seat constituencies—populous towns and cities—by proportional representation vote and 26 directly elected in single-seat constituencies—outlying rural districts—by plurality vote; members serve 5-year terms)

elections: last held on 5 May 2019 (next to be held in May 2024)

election results: percent of vote by party—NA; seats by party—PRD 35, CD 18, Panamenista 8, MOLIRENA 5, independent 5; composition—men 55, women 16, percent of women 22.5%

Judicial branch: *highest courts:* Supreme Court of Justice or Corte Suprema de Justicia (consists of 9 magistrates and 9 alternates and divided into civil, criminal, administrative, and general business chambers)

judge selection and term of office: magistrates appointed by the president for staggered 10-year terms

subordinate courts: appellate courts or Tribunal Superior; Labor Supreme Courts; Court of Audit; circuit courts or Tribunal Circuital (2 each in 9 of the 10 provinces); municipal courts; electoral, family, maritime, and adolescent courts

Political parties and leaders: Democratic Change or CD [Romulo ROUX]

Democratic Revolutionary Party or PRD [Benicio ROBINSON]

Nationalist Republican Liberal Movement or MOLIRENA [Francisco "Pancho" ALEMAN]

Panamenista Party [Jose Luis "Popi" VARELA Rodriguez] (formerly the Arnulfista Party)

Popular Party or PP [Juan Carlos ARANGO Reese] (formerly Christian Democratic Party or PDC)

International organization participation: BCIE, CAN (observer), CD, CELAC, FAO, G-77, IADB, IAEA, IBRD, ICAO, ICC (national committees), ICCt, ICRM, IDA, IFAD, IFC, IFRCS, ILO, IMF, IMO, IMSO, Interpol, IOC, IOM, IPU, ISO, ITSO, ITU, ITUC (NGOs), LAES, LAIA, MIGA, NAM, OAS, OPANAL, OPCW, Pacific Alliance (observer), PCA, SICA, UN, UNASUR (observer), UNCTAD, UNESCO, UNIDO, Union Latina, UNWTO, UPU, WCO, WFTU (NGOs), WHO, WIPO, WMO, WTO

Diplomatic representation in the US: *chief of mission:* Ambassador Juan Ricardo DE DIANOUS HENRIQUEZ (since 16 September 2019)

chancery: 2862 McGill Terrace NW, Washington, DC 20007

telephone: [1] (202) 483-1407

FAX: [1] (202) 483-8413

consulate(s) general: Houston, Miami, Los Angeles, New Orleans, New York, Philadelphia, Tampa, Washington DC

Diplomatic representation from the US: *chief of mission:* Ambassador (vacant), Charge d'Affairs Roxanne CABRAL (since 9 March 2018)

telephone: [507] 317-5000

embassy: Edificio 783, Avenida Demetrio Basilio Lakas Avenue, Clayton

mailing address: American Embassy Panama, Unit 0945, APO AA 34002; American Embassy Panama, 9100 Panama City PL, Washington, DC 20521-9100

FAX: [507] 317-5445 (2018)

Flag description: divided into four, equal rectangles; the top quadrants are white (hoist side) with a blue five-pointed star in the center and plain red; the bottom quadrants are plain blue (hoist side) and white with a red five-pointed star in the center; the blue and red colors are those of the main political parties (Conservatives and Liberals respectively) and the white denotes peace between them; the blue star stands for the civic virtues of purity and honesty, the red star signifies authority and law

National symbol(s): *harpy eagle; national colors:* blue, white, red

National anthem: *name:* "Himno Istmeno" (Isthmus Hymn)

lyrics/music: Jeronimo DE LA OSSA/Santos A. JORGE

note: adopted 1925

0:00 / 0:00

ECONOMY

Economy—overview: Panama's dollar-based economy rests primarily on a well-developed services sector that accounts for more than threequarters of GDP. Services include operating the Panama Canal, logistics, banking, the Colon Free Trade Zone, insurance, container ports, flagship registry, and tourism and Panama is a center for offshore banking. Panama's transportation and logistics services sectors, along with infrastructure development projects, have boosted economic growth; however, public debt surpassed $37 billion in 2016 because of excessive government spending and public works projects. The US-Panama Trade Promotion Agreement was approved by Congress and signed into law in October 2011, and entered into force in October 2012.

Future growth will be bolstered by the Panama Canal expansion project that began in 2007 and was completed in 2016 at a cost of $5.3 billion—about 10-15% of current GDP. The expansion project more than doubled the Canal's capacity, enabling it to accommodate high-capacity vessels such as tankers and neopanamax vessels that are too large to traverse the existing canal. The US and China are the top users of the Canal.

Strong economic performance has not translated into broadly shared prosperity, as Panama has the second worst income distribution in Latin America. About one-fourth of the population lives in poverty; however, from 2006 to 2012 poverty was reduced by 10 percentage points.

GDP (purchasing power parity):
$104.1 billion (2017 est.)
$98.82 billion (2016 est.)
$94.12 billion (2015 est.)
note: data are in 2017 dollars
country comparison to the world: 83

GDP (official exchange rate):
$61.84 billion (2017 est.)

GDP—real growth rate: 5.4% (2017 est.)
5% (2016 est.)
5.8% (2015 est.)
country comparison to the world: 38

GDP—per capita (PPP): $25,400 (2017 est.)
$24,500 (2016 est.)
$23,700 (2015 est.)
note: data are in 2017 dollars
country comparison to the world: 80

Gross national saving:
38.9% of GDP (2017 est.)
39.2% of GDP (2016 est.)
36.8% of GDP (2015 est.)
country comparison to the world: 10

GDP—composition, by end use:
household consumption: 45.6% (2017 est.)
government consumption: 10.7% (2017 est.)
investment in fixed capital: 42.9% (2017 est.)
investment in inventories: 3% (2017 est.)
exports of goods and services: 41.9% (2017 est.)
imports of goods and services: -44.2% (2017 est.)

GDP—composition, by sector of origin:
agriculture: 2.4% (2017 est.)
industry: 15.7% (2017 est.)
services: 82% (2017 est.)

Agriculture—products: bananas, rice, corn, coffee, sugarcane, vegetables; livestock; shrimp

Industries: construction, brewing, cement and other construction materials, sugar milling

Industrial production growth rate: 6.3% (2017 est.)
country comparison to the world: 38

Labor force: 1.633 million (2017 est.)
note: shortage of skilled labor, but an oversupply of unskilled labor
country comparison to the world: 127

Labor force—by occupation: *agriculture:* 17%
industry: 18.6%
services: 64.4% (2009 est.)

Unemployment rate: 6.14% (2018 est.)
6% (2017 est.)
country comparison to the world: 98

Population below poverty line: 23% (2015 est.)

Household income or consumption by percentage share: *lowest 10%:* 1.1%
highest 10%: 38.9% (2014 est.)

Budget: *revenues:* 12.43 billion (2017 est.)
expenditures: 13.44 billion (2017 est.)

Taxes and other revenues: 20.1% (of GDP) (2017 est.)
country comparison to the world: 152

Budget surplus (+) or deficit (-): -1.6% (of GDP) (2017 est.)
country comparison to the world: 94

Public debt: 3
7.8% of GDP (2017 est.)
37.4% of GDP (2016 est.)
country comparison to the world: 138

Fiscal year: calendar year

Inflation rate (consumer prices):
0.9% (2017 est.)
0.7% (2016 est.)
country comparison to the world: 47

Current account balance:
-$3.036 billion (2017 est.)
-$3.16 billion (2016 est.)
country comparison to the world: 174

Exports: $15.5 billion (2017 est.)
$14.7 billion (2016 est.)
note: includes the Colon Free Zone
country comparison to the world: 74

Exports—partners: US 18.9%, Netherlands 16.6%, China 6.5%, Costa Rica 5.4%, India 5.1%, Vietnam 5% (2017)

Exports—commodities: fruit and nuts, fish, iron and steel waste, wood

Imports: $21.91 billion (2017 est.)
$20.51 billion (2016 est.)
note: includes the Colon Free Zone
country comparison to the world: 72

Imports—commodities: fuels, machinery, vehicles, iron and steel rods, pharmaceuticals

Imports—partners: US 24.4%, China 9.8%, Mexico 4.9% (2017)

Reserves of foreign exchange and gold:
$2.703 billion (31 December 2017 est.)
$3.878 billion (31 December 2016 est.)
country comparison to the world: 114

Debt—external: $91.53 billion (31 December 2017 est.)

$83.81 billion (31 December 2016 est.)
country comparison to the world: 52

Exchange rates: balboas (PAB) per US dollar -
1 (2017 est.)
1 (2016 est.)
1 (2015 est.)
1 (2014 est.)
1 (2013 est.)

ENERGY

Electricity access: *electrification—total population:* 93.4% (2016)
electrification—urban areas: 99.4% (2016)
electrification—rural areas: 81.3% (2016)

Electricity—production: 10.6 billion kWh (2016 est.)
country comparison to the world: 101

Electricity—consumption: 8.708 billion kWh (2016 est.)
country comparison to the world: 103

Electricity—exports: 139 million kWh (2015 est.)
country comparison to the world: 80

Electricity—imports: 30 million kWh (2016 est.)
country comparison to the world: 110

Electricity—installed generating capacity: 3.4 million kW (2016 est.)
country comparison to the world: 97

Electricity—from fossil fuels: 36% of total installed capacity (2016 est.)
country comparison to the world: 175

Electricity—from nuclear fuels: 0% of total installed capacity (2017 est.)
country comparison to the world: 161

Electricity—from hydroelectric plants: 51% of total installed capacity (2017 est.)
country comparison to the world: 36

Electricity—from other renewable sources: 13% of total installed capacity (2017 est.)
country comparison to the world: 69

Crude oil—production: 0 bbl/day (2018 est.)
country comparison to the world: 185

Crude oil—exports: 0 bbl/day (2015 est.)
country comparison to the world: 178

Crude oil—imports: 0 bbl/day (2015 est.)
country comparison to the world: 180

Crude oil—proved reserves: 0 bbl (1 January 2018)
country comparison to the world: 180

Refined petroleum products—production: 0 bbl/day (2015 est.)
country comparison to the world: 188

Refined petroleum products—consumption: 146,000 bbl/day (2016 est.)
country comparison to the world: 67

Refined petroleum products—exports: 66 bbl/day (2015 est.)
country comparison to the world: 121

Refined petroleum products—imports: 129,200 bbl/day (2015 est.)
country comparison to the world: 45

Natural gas—production: 0 cu m (2017 est.)
country comparison to the world: 182

Natural gas—consumption: 0 cu m (2017 est.)
country comparison to the world: 186

Natural gas—exports: 0 cu m (2017 est.)
country comparison to the world: 166

Natural gas—imports: 0 cu m (2017 est.)
country comparison to the world: 171

Natural gas—proved reserves: 0 cu m (1 January 2014 est.)
country comparison to the world: 180

Carbon dioxide emissions from consumption of energy: 26.08 million Mt (2017 est.)
country comparison to the world: 79

COMMUNICATIONS

Telephones—fixed lines: *total subscriptions:* 671,799
subscriptions per 100 inhabitants: 17.46 (2019 est.)
country comparison to the world: 86

Telephones—mobile cellular: *total subscriptions:* 5,073,123
subscriptions per 100 inhabitants: 131.85 (2019 est.)
country comparison to the world: 119

Telecommunication systems: *general assessment:* domestic and international facilities well-developed; investment from international operators; competition among operators helps reduce price of services; launch of LTE services; govt. fixed-line projects and popularity of mobile broadband connectivity see growth; Chinese company Huawei helps with G-fast technologies (2020)
domestic: fixed-line 17 per 100 and rapid subscribership of mobile-cellular telephone 132 per 100 (2019)
international: country code—507; landing points for the PAN-AM, ARCOS, SAC, AURORA, PCCS, PAC, and the MAYA-1 submarine cable systems that together provide links to the US and parts of the Caribbean, Central America, and South America; satellite earth stations—2 Intelsat (Atlantic Ocean); connected to the Central American Microwave System (2019)
note: the COVID-19 outbreak is negatively impacting telecommunications production and supply chains globally; consumer spending on telecom devices and services has also slowed due to the pandemic's effect on economies worldwide; overall progress towards improvements in all facets of the telecom industry—mobile, fixed-line, broadband, submarine cable and satellite—has moderated

Broadcast media: multiple privately owned TV networks and a government-owned educational TV station; multi-channel cable and satellite TV subscription services are available; more than 100 commercial radio stations (2019)

Internet country code: .pa

Internet users: *total:* 2,199,433
percent of population: 57.87% (July 2018 est.)
country comparison to the world: 118

Broadband—fixed subscriptions: *total:* 540,220
subscriptions per 100 inhabitants: 14 (2018 est.)
country comparison to the world: 82

TRANSPORTATION

National air transport system: *number of registered air carriers:* 4 (2020)
inventory of registered aircraft operated by air carriers: 122
annual passenger traffic on registered air carriers: 12,939,350 (2018)
annual freight traffic on registered air carriers: 47.63 million mt-km (2018)

Civil aircraft registration country code prefix: HP (2016)

Airports: 117 (2013)
country comparison to the world: 49

Airports—with paved runways:
total: 57 (2017)
over 3,047 m: 1 (2017)
2,438 to 3,047 m: 3 (2017)
1,524 to 2,437 m: 3 (2017)
914 to 1,523 m: 20 (2017)
under 914 m: 30 (2017)

Airports—with unpaved runways:
total: 60 (2013)
1,524 to 2,437 m: 1 (2013)
914 to 1,523 m: 8 (2013)
under 914 m: 51 (2013)

Heliports: 3 (2013)

Pipelines: 128 km oil (2013)

Railways: *total:* 77 km (2014)
standard gauge: 77 km 1.435-m gauge (2014)
country comparison to the world: 128

Waterways: 800 km (includes the 82-km Panama Canal that is being widened) (2011)
country comparison to the world: 71

Merchant marine: *total:* 7,860

747

by type: bulk carrier 2,567, container ship 609, general cargo 1,325, oil tanker 798, other 2,561 (2019)

country comparison to the world: 2

Ports and terminals: *major seaport(s):* Balboa, Colon, Cristobal

container port(s) (TEUs): Balboa (2,905,049), Colon (3,891,209) (2017)

MILITARY AND SECURITY

Military and security forces: no regular military forces; Panamanian Public Security Forces (subordinate to the Ministry of Public Security), comprising the National Police (PNP), National Air-Naval Service (SENAN), National Border Service (SENAFRONT) (2020)

note: on 10 February 1990, the government of then President Guillermo ENDARA abolished Panama's military and reformed the security apparatus by creating the Panamanian Public Forces; in October 1994, Panama's National Assembly approved a constitutional amendment prohibiting the creation of a standing military force but allowing the temporary establishment of special police units to counter acts of "external aggression"

Military and security service personnel strengths: the Panamanian Public Security Forces are comprised of approximately 26,000 personnel (20,000 National Police Force; 4,000 National Border Service; 2,000 National Air-Naval Service) (2019 est.)

Military equipment inventories and acquisitions: Panama's security forces do not maintain heavy military equipment, instead focusing on light air transport, patrol, and surveillance capabilities; since 2010, Italy and the US have been the leading suppliers to the security forces (2019 est.)

TRANSNATIONAL ISSUES

Disputes—international: organized illegal narcotics operations in Colombia operate within the remote border region with Panama

Refugees and internally displaced persons: *refugees (country of origin):* 79,155 (Venezuela) (economic and political crisis; includes Venezuelans who have claimed asylum or have received alternative legal stay) (2020)

Illicit drugs: major cocaine transshipment point and primary money-laundering center for narcotics revenue; money-laundering activity is especially heavy in the Colon Free Zone; offshore financial center; negligible signs of coca cultivation; monitoring of financial transactions is improving; official corruption remains a major problem

PAPUA NEW GUINEA

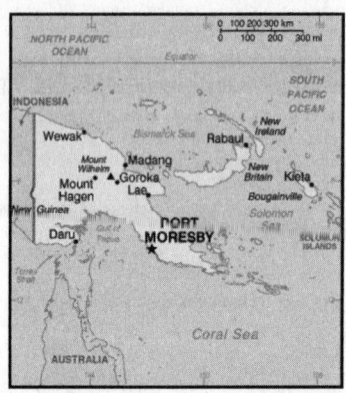

INTRODUCTION

Background: The eastern half of the island of New Guinea—second largest in the world—was divided between Germany (north) and the UK (south) in 1885. The latter area was transferred to Australia in 1902, which occupied the northern portion during World War I and continued to administer the combined areas until independence in 1975. A nine-year secessionist revolt on the island of Bougainville ended in 1997 after claiming some 20,000 lives. Since 2001, Bougainville has experienced autonomy; a referendum asking the population if they would like independence or greater self rule occurred in November 2019, with almost 98% of voters choosing independence.

GEOGRAPHY

Location: Oceania, group of islands including the eastern half of the island of New Guinea between the Coral Sea and the South Pacific Ocean, east of Indonesia

Geographic coordinates: 6 00 S, 147 00 E

Map references: Oceania

Area: *total:* 462,840 sq km
land: 452,860 sq km
water: 9,980 sq km
country comparison to the world: 56

Area—comparative: slightly larger than California

Land boundaries: *total:* 874 km
border countries (1): Indonesia 824 km

Coastline: 5,152 km

Maritime claims: *territorial sea:* 12 nm
continental shelf: 200-m depth or to the depth of exploitation
exclusive fishing zone: 200 nm
measured from claimed archipelagic baselines

Climate: tropical; northwest monsoon (December to March), southeast monsoon (May to October); slight seasonal temperature variation

Terrain: mostly mountains with coastal lowlands and rolling foothills

Elevation: *mean elevation:* 667 m
lowest point: Pacific Ocean 0 m
highest point: Mount Wilhelm 4,509 m

Natural resources: gold, copper, silver, natural gas, timber, oil, fisheries

Land use: *agricultural land:* 2.6% (2011 est.)
arable land: 0.7% (2011 est.) / permanent crops: 1.5% (2011 est.) / permanent pasture: 0.4% (2011 est.)
forest: 63.1% (2011 est.)
other: 34.3% (2011 est.)
Irrigated land: 0 sq km (2012)

Population distribution: population concentrated in the highlands and eastern coastal areas on the island of New Guinea; predominantly a rural distribution with only about one-fifth of the population residing in urban areas

Natural hazards: active volcanism; the country is subject to frequent and sometimes severe earthquakes; mud slides; tsunamis

volcanism: severe volcanic activity; Ulawun (2,334 m), one of Papua New Guinea's potentially most dangerous volcanoes, has been deemed a Decade Volcano by the International Association of Volcanology and Chemistry of the Earth's Interior, worthy of study due to its explosive history and close proximity to human populations; Rabaul (688 m) destroyed the city of Rabaul in 1937 and 1994; Lamington erupted in 1951 killing 3,000 people; Manam's 2004 eruption forced the island's abandonment; other historically active volcanoes include Bam, Bagana, Garbuna, Karkar, Langila, Lolobau, Long Island, Pago, St. Andrew Strait, Victory, and Waiowa; see note 2 under "Geography—note"

Environment—current issues: rain forest loss as a result of growing commercial demand for tropical timber; unsustainable logging practices result in soil erosion, water quality degradation, and loss of habitat and biodiversity; large-scale mining projects cause adverse impacts on forests and water quality (discharge of heavy metals, cyanide, and acids into rivers); severe drought; inappropriate farming practices accelerate land degradation (soil erosion, siltation, loss of soil fertility); destructive fishing practices and coastal pollution due to run-off from land-based activities and oil spills

Environment—international agreements: *party to:* Antarctic Treaty, Biodiversity, Climate Change, Climate Change-Kyoto Protocol, Desertification, Endangered Species, Environmental Modification, Hazardous Wastes, Law of the Sea, Marine Dumping, Ozone Layer Protection, Ship Pollution, Tropical Timber 83, Tropical Timber 94, Wetlands

signed, but not ratified: none of the selected agreements

Geography—note: *note 1:* shares island of New Guinea with Indonesia; generally east-west trending highlands break up New Guinea into diverse ecoregions; one of world's largest swamps along southwest coast

note 2: Papua New Guinea is one of the countries along the Ring of Fire, a belt of active volcanoes and earthquake epicenters bordering the Pacific Ocean; up to 90% of the world's earthquakes and some 75% of the world's volcanoes occur within the Ring of Fire

PEOPLE AND SOCIETY

Population: 7,259,456 (July 2020 est.)
country comparison to the world: 102

Nationality: *noun:* Papua New Guinean(s)
adjective: Papua New Guinean

Ethnic groups: Melanesian, Papuan, Negrito, Micronesian, Polynesian

Languages: Tok Pisin (official), English (official), Hiri Motu (official), some 839 indigenous languages spoken (about 12% of the world's total); many languages have fewer than 1,000 speakers
note: Tok Pisin, a creole language, is widely used and understood; English is spoken by 1%-2%; Hiri Motu is spoken by less than 2%

Religions: Protestant 64.3% (Evangelical Lutheran 18.4%, Seventh Day Adventist 12.9%, Pentecostal 10.4%, United Church 10.3%, Evangelical Alliance 5.9%, Anglican 3.2%, Baptist 2.8%, Salvation Army .4%), Roman Catholic 26%, other Christian 5.3%, non-Christian 1.4%, unspecified 3.1% (2011 est.)
note: data represent only the citizen population; roughly .3% of the population are non-citizens, consisting of Christian 52% (predominantly Roman Catholic), other 10.7% , none 37.3%

Age structure: *0-14 years:* 31.98% (male 1,182,539/female 1,139,358)
15-24 years: 19.87% (male 731,453/female 711,164)
25-54 years: 37.68% (male 1,397,903/female 1,337,143)
55-64 years: 5.83% (male 218,529/female 204,717)
65 years and over: 4.64% (male 164,734/female 171,916) (2020 est.)

Dependency ratios: *total dependency ratio:* 63.2
youth dependency ratio: 57.4
elderly dependency ratio: 5.8
potential support ratio: 17.2 (2020 est.)

Median age: *total:* 24 years
male: 24 years
female: 24 years (2020 est.)
country comparison to the world: 171

Population growth rate: 1.6% (2020 est.)
country comparison to the world: 63

Birth rate: 22.5 births/1,000 population (2020 est.)
country comparison to the world: 63

Death rate: 6.7 deaths/1,000 population (2020 est.)
country comparison to the world: 137

Net migration rate: 0 migrant(s)/1,000 population (2020 est.)
country comparison to the world: 93

Population distribution: population concentrated in the highlands and eastern coastal areas on the island of New Guinea; predominantly a rural distribution with only about one-fifth of the population residing in urban areas

Urbanization: *urban population:* 13.3% of total population (2020)
rate of urbanization: 2.51% annual rate of change (2015-20 est.)

total population growth rate v. urban population growth rate, 2000-2030:

Major urban areas—population: 383,000 PORT MORESBY (capital) (2020)

Sex ratio: *at birth:* 1.05 male(s)/female
0-14 years: 1.04 male(s)/female
15-24 years: 1.03 male(s)/female
25-54 years: 1.05 male(s)/female
55-64 years: 1.07 male(s)/female
65 years and over: 0.96 male(s)/female
total population: 1.04 male(s)/female (2020 est.)

Maternal mortality rate: 145 deaths/100,000 live births (2017 est.)
country comparison to the world: 58

Infant mortality rate: *total:* 33.2 deaths/1,000 live births
male: 36.4 deaths/1,000 live births
female: 29.8 deaths/1,000 live births (2020 est.)
country comparison to the world: 46

Life expectancy at birth: *total population:* 67.8 years
male: 65.6 years
female: 70 years (2020 est.)
country comparison to the world: 179

Total fertility rate: 2.84 children born/woman (2020 est.)
country comparison to the world: 60

Contraceptive prevalence rate: 36.7% (2016/18)

Drinking water source:
improved:
urban: 89.4% of population
rural: 36.1% of population
total: 43% of population
unimproved:
urban: 10.6% of population
rural: 63.9% of population
total: 57% of population (2017 est.)

Current Health Expenditure: 2.5% (2017)

Physicians density: 0.05 physicians/1,000 population (2010)

Sanitation facility access:
improved:
urban: 55.5% of population
rural: 9.1% of population
total: 15.2% of population
unimproved:
urban: 44.5% of population
rural: 90.9% of population
total: 84.8% of population (2017 est.)

HIV/AIDS—adult prevalence rate: 0.8% (2019 est.)
country comparison to the world: 53

HIV/AIDS—people living with HIV/AIDS: 52,000 (2019 est.)
country comparison to the world: 59

HIV/AIDS—deaths: <1000 (2019 est.)

Major infectious diseases: *degree of risk:* very high (2020)
food or waterborne diseases: bacterial diarrhea, hepatitis A, and typhoid fever
vectorborne diseases: dengue fever and malaria

Obesity—adult prevalence rate: 21.3% (2016)
country comparison to the world: 91

Children under the age of 5 years underweight: 27.8% (2010)
country comparison to the world: 8

Education expenditures: NA

Literacy: *definition:* age 15 and over can read and write
total population: 64.2%
male: 65.6%
female: 62.8% (2015)

Unemployment, youth ages 15-24: *total:* 3.6%
male: 4.3%
female: 3% (2010 est.)
country comparison to the world: 172

People—note: the indigenous population of Papua New Guinea (PNG) is one of the most heterogeneous in the world; PNG has several thousand separate communities, most with only a few hundred people; divided by language, customs, and tradition, some of these communities have engaged in low-scale tribal conflict with their neighbors for millennia; the advent of modern weapons and modern migrants into urban areas has greatly magnified the impact of this lawlessness

GOVERNMENT

Country name: *conventional long form:* Independent State of Papua New Guinea
conventional short form: Papua New Guinea
local short form: Papuaniugini
former: Territory of Papua and New Guinea
abbreviation: PNG
etymology: the word "papua" derives from the Malay "papuah" describing the frizzy hair of the Melanesians; Spanish explorer Ynigo ORTIZ de RETEZ applied the term "Nueva Guinea" to the island of New Guinea in 1545 after noting the resemblance of the locals to the peoples of the Guinea coast of Africa

Government type: parliamentary democracy under a constitutional monarchy; a Commonwealth realm

Capital: *name:* Port Moresby
geographic coordinates: 9 27 S, 147 11 E
time difference: UTC+10 (15 hours ahead of Washington, DC, during Standard Time)
note: Papua New Guinea has two time zones, including Bougainville (UTC+11)
etymology: named in 1873 by Captain John Moresby (1830-1922) in honor of his father, British Admiral Sir Fairfax Moresby (1786-1877)

Administrative divisions: 20 provinces, 1 autonomous region*, and 1 district**; Bougainville*, Central, Chimbu, Eastern Highlands, East New Britain, East Sepik, Enga, Gulf, Hela, Jiwaka, Madang, Manus, Milne Bay, Morobe, National Capital**, New Ireland, Northern, Southern Highlands, Western, Western Highlands, West New Britain, West Sepik

Independence: 16 September 1975 (from the Australia-administered UN trusteeship)

National holiday: Independence Day, 16 September (1975)

Constitution: *history:* adopted 15 August 1975, effective at independence 16 September 1975
amendments: proposed by the National Parliament; passage has prescribed majority vote requirements depending on the constitutional sections being amended – absolute majority, two-thirds majority, or three-fourths majority; amended many times, last in 2014

Legal system: mixed legal system of English common law and customary law

International law organization participation: has not submitted an ICJ jurisdiction declaration; non-party state to the ICCt

Citizenship: *citizenship by birth:* no
citizenship by descent only: at least one parent must be a citizen of Papua New Guinea
dual citizenship recognized: no
residency requirement for naturalization: 8 years

Suffrage: 18 years of age; universal

Executive branch: *chief of state:* Queen ELIZABETH II (since 6 February 1952); represented by Governor General Grand Chief Sir Bob DADAE (since 28 February 2017)

head of government: Prime Minister James MARAPE (since 30 May 2019); Deputy Prime Minister Charles ABEL (since 4 August 2017)
cabinet: National Executive Council appointed by the governor general on the recommendation of the prime minister
elections/appointments: the monarchy is hereditary; governor general nominated by the National Parliament and appointed by the chief of state; following legislative elections, the leader of the majority party or majority coalition usually appointed prime minister by the governor general pending the outcome of a National Parliament vote
election results: Peter Paire O'NEILL (PNC) reelected prime minister; National Parliament vote—60 to 46

Legislative branch: *description:* unicameral National Parliament (111 seats; members directly elected in single-seat constituencies—89 local, 20 provincial, the autonomous province of Bouganville, and the National Capital District—by majority preferential vote; members serve 5-year terms); note—the constitution allows up to 126 seats
elections: last held from 24 June 2017 to 8 July 2017 (next to be held in June 2022)

election results: percent of vote by party—PNC 37%; NA 13%; Pangu 14%; URP 11%; PPP 4%; SDP 4%; Independents 3%; and smaller parties 14%; seats by party—NA; composition—men 108, women 3, percent of women 3%

Judicial branch: *highest courts:* Supreme Court (consists of the chief justice, deputy chief justice, 35 justices, and 5 acting justices); National Courts (consists of 13 courts located in the provincial capitals, with a total of 19 resident judges)
judge selection and term of office: Supreme Court chief justice appointed by the governor general upon advice of the National Executive Council (cabinet) after consultation with the National Justice Administration minister; deputy chief justice and other justices appointed by the Judicial and Legal Services Commission, a 5-member body that includes the Supreme Court chief and deputy chief justices, the chief ombudsman, and a member of the National Parliament; full-time citizen judges appointed for 10-year renewable terms; non-citizen judges initially appointed for 3-year renewable terms and after first renewal can serve until age 70; appointment and tenure of National Court resident judges NA
subordinate courts: district, village, and juvenile courts, military courts, taxation courts, coronial courts, mining warden courts, land courts, traffic courts, committal courts, grade five courts

Political parties and leaders:
National Alliance Party or NAP [Patrick PRUAITCH]
Papua and Niugini Union Party or PANGU [Sam BASIL]
Papua New Guinea Party or PNGP [Belden NAMAH]
People's National Congress Party or PNC [Peter Paire O'NEILL]
People's Party or PP [Peter IPATAS]
People's Progress Party or PPP [Sir Julius CHAN]
Social Democratic Party or SDP [Powes PARKOP]
Triumph Heritage Empowerment Party or THE [Don POLYE]
United Resources Party or URP [William DUMA]
note: as of 8 July 2017, 45 political parties were registered

International organization participation: ACP, ADB, AOSIS, APEC, ARF, ASEAN (observer), C, CD, CP, EITI (candidate country), FAO, G-77, IAEA, IBRD, ICAO, ICRM, IDA, IFAD, IFC, IFRCS, IHO, ILO, IMF, IMO, Interpol, IOC, IOM, IPU, ISO (correspondent), ITSO, ITU, MIGA, NAM, OPCW, PIF, Sparteca, SPC, UN, UNCTAD, UNESCO, UNIDO, UNMISS, UNWTO, UPU, WCO, WFTU (NGOs), WHO, WIPO, WMO, WTO

Diplomatic representation in the US: *chief of mission:* Ambassador (vacant); Charge D'Affaires Cephas KAYO (since 31 January 2018)
chancery: 1779 Massachusetts Avenue NW, Suite 805, Washington, DC 20036
telephone: [1] (202) 745-3680
FAX: [1] (202) 745-3679

Diplomatic representation from the US: *chief of mission:* Ambassador Erin Elizabeth MCKEE

(since 27 November 2019); note—also accredited to the Solomon Islands and Vanuatu
telephone: [675] 321-1455
embassy: P.O. Box 1492, Port Moresby
mailing address: 4240 Port Moresby Place, US Department of State, Washington DC 20521-4240
FAX: [675] 321-3423

Flag description: divided diagonally from upper hoist-side corner; the upper triangle is red with a soaring yellow bird of paradise centered; the lower triangle is black with five, white, five-pointed stars of the Southern Cross constellation centered; red, black, and yellow are traditional colors of Papua New Guinea; the bird of paradise—endemic to the island of New Guinea—is an emblem of regional tribal culture and represents the emergence of Papua New Guinea as a nation; the Southern Cross, visible in the night sky, symbolizes Papua New Guinea's connection with Australia and several other countries in the South Pacific

National symbol(s): *bird of paradise; national colors:* red, black

National anthem: *name:* O Arise All You Sons
lyrics/music: Thomas SHACKLADY
note: adopted 1975
0:00 / 0:00

ECONOMY

Economy—overview: Papua New Guinea (PNG) is richly endowed with natural resources, but exploitation has been hampered by rugged terrain, land tenure issues, and the high cost of developing infrastructure. The economy has a small formal sector, focused mainly on the export of those natural resources, and an informal sector, employing the majority of the population. Agriculture provides a subsistence livelihood for 85% of the people. The global financial crisis had little impact because of continued foreign demand for PNG's commodities.

Mineral deposits, including copper, gold, and oil, account for nearly two-thirds of export earnings. Natural gas reserves amount to an estimated 155 billion cubic meters. Following construction of a $19 billion liquefied natural gas (LNG) project, PNG LNG, a consortium led by ExxonMobil, began exporting liquefied natural gas to Asian markets in May 2014. The project was delivered on time and only slightly above budget. The success of the project has encouraged other companies to look at similar LNG projects. French supermajor Total is hopes to begin construction on the Papua LNG project by 2020. Due to lower global commodity prices, resource revenues of all types have fallen dramatically. PNG's government has recently been forced to adjust spending levels downward.

Numerous challenges still face the government of Peter O'NEILL, including providing physical security for foreign investors, regaining investor confidence, restoring integrity to state institutions, promoting economic efficiency by privatizing moribund state institutions, and maintaining good relations with Australia, its former colonial ruler. Other sociocultural challenges could upend the economy including chronic law and order and land tenure issues. In August, 2017, PNG launched its first-ever national trade policy, PNG Trade Policy 2017-2032. The policy goal is to maximize trade and investment by increasing exports, to reduce imports, and to increase foreign direct investment (FDI).

GDP (purchasing power parity):
$30.19 billion (2017 est.)
$29.44 billion (2016 est.)
$28.98 billion (2015 est.)
note: data are in 2017 dollars
country comparison to the world: 132

GDP (official exchange rate):
$19.82 billion (2017 est.)

GDP—real growth rate: 2.5% (2017 est.)
1.6% (2016 est.)
5.3% (2015 est.)
country comparison to the world: 113

GDP—per capita (PPP): $3,700 (2017 est.)
$3,600 (2016 est.)
$3,700 (2015 est.)
note: data are in 2017 dollars
country comparison to the world: 184

Gross national saving:
36.8% of GDP (2017 est.)
38% of GDP (2016 est.)
33.7% of GDP (2015 est.)
country comparison to the world: 14

GDP—composition, by end use:
household consumption: 43.7% (2017 est.)
government consumption: 19.7% (2017 est.)
investment in fixed capital: 10% (2017 est.)
investment in inventories: 0.4% (2017 est.)
exports of goods and services: 49.3% (2017 est.)
imports of goods and services: -22.3% (2017 est.)

GDP—composition, by sector of origin:
agriculture: 22.1% (2017 est.)
industry: 42.9% (2017 est.)
services: 35% (2017 est.)

Agriculture—products: coffee, cocoa, copra, palm kernels, tea, sugar, rubber, sweet potatoes, fruit, vegetables, vanilla; poultry, pork; shellfish

Industries: copra crushing, palm oil processing, plywood production, wood chip production; mining (gold, silver, copper); crude oil and petroleum products; construction, tourism, livestock (pork, poultry, cattle), dairy products, spice products (turmeric, vanilla, ginger, cardamom, chili, pepper, citronella, and nutmeg), fisheries products

Industrial production growth rate: 3.3% (2017 est.)
country comparison to the world: 97

Labor force: 3.681 million (2017 est.)
country comparison to the world: 96

Labor force—by occupation: *agriculture:* 85%
industry: NA
services: NA

Unemployment rate: 2.5% (2017 est.)
2.5% (2016 est.)
country comparison to the world: 28

Population below poverty line: 37% (2002 est.)

Household income or consumption by percentage share: *lowest 10%:* 1.7%
highest 10%: 40.5% (1996)

Budget: *revenues:* 3.638 billion (2017 est.)
expenditures: 4.591 billion (2017 est.)

Taxes and other revenues: 18.4% (of GDP) (2017 est.)
country comparison to the world: 160

Budget surplus (+) or deficit (-): -4.8% (of GDP) (2017 est.)
country comparison to the world: 168

Public debt:
36.9% of GDP (2017 est.)
36.9% of GDP (2016 est.)
country comparison to the world: 144

Fiscal year: calendar year

Inflation rate (consumer prices): 5.4% (2017 est.)
6.7% (2016 est.)
country comparison to the world: 177

Current account balance:
$4.859 billion (2017 est.)
$4.569 billion (2016 est.)
country comparison to the world: 30

Exports: $8.522 billion (2017 est.)
$9.224 billion (2016 est.)
country comparison to the world: 95

Exports—partners: Australia 18.9%, Singapore 17.5%, Japan 13.8%, China 12.7%, Philippines 4.7%, Netherlands 4.2%, India 4.2% (2017)

Exports—commodities: liquefied natural gas, oil, gold, copper ore, nickel, cobalt logs, palm oil, coffee, cocoa, copra, spice (turmeric, vanilla, ginger, and cardamom), crayfish, prawns, tuna, sea cucumber

Imports: $1.876 billion (2017 est.)
$2.077 billion (2016 est.)
country comparison to the world: 170

Imports—commodities: machinery and transport equipment, manufactured goods, food, fuels, chemicals

Imports—partners: Australia 30.1%, China 17.3%, Singapore 10.2%, Malaysia 8.2%, Indonesia 4% (2017)

Reserves of foreign exchange and gold:
$1.735 billion (31 December 2017 est.)
$1.656 billion (31 December 2016 est.)
country comparison to the world: 123

Debt—external:
$17.94 billion (31 December 2017 est.)
$18.28 billion (31 December 2016 est.)
country comparison to the world: 95

Exchange rates: kina (PGK) per US dollar—
3.179 (2017 est.)
3.133 (2016 est.)
3.133 (2015 est.)
2.7684 (2014 est.)
2.4614 (2013 est.)

ENERGY

Electricity access: *electrification—total population:* 22.9% (2016)

electrification—urban areas: 72.7% (2016)
electrification—rural areas: 15.5% (2016)

Electricity—production: 3.481 billion kWh (2016 est.)
country comparison to the world: 129

Electricity—consumption: 3.237 billion kWh (2016 est.)
country comparison to the world: 134

Electricity—exports: 0 kWh (2017 est.)
country comparison to the world: 183

Electricity—imports: 0 kWh (2016 est.)
country comparison to the world: 184

Electricity—installed generating capacity: 900,900 kW (2016 est.)
country comparison to the world: 131

Electricity—from fossil fuels: 63% of total installed capacity (2016 est.)
country comparison to the world: 123

Electricity—from nuclear fuels: 0% of total installed capacity (2017 est.)
country comparison to the world: 162

Electricity—from hydroelectric plants: 30% of total installed capacity (2017 est.)
country comparison to the world: 69

Electricity—from other renewable sources: 7% of total installed capacity (2017 est.)
country comparison to the world: 95

Crude oil—production: 45,000 bbl/day (2018 est.)
country comparison to the world: 55

Crude oil—exports: 55,600 bbl/day (2015 est.)
country comparison to the world: 41

Crude oil—imports: 22,220 bbl/day (2015 est.)
country comparison to the world: 62

Crude oil—proved reserves: 183.8 million bbl (1 January 2018 est.)
country comparison to the world: 57

Refined petroleum products—production: 22,170 bbl/day (2015 est.)
country comparison to the world: 88

Refined petroleum products—consumption: 37,000 bbl/day (2016 est.)
country comparison to the world: 116

Refined petroleum products—exports: 0 bbl/day (2015 est.)
country comparison to the world: 190

Refined petroleum products—imports: 17,110 bbl/day (2015 est.)
country comparison to the world: 134

Natural gas—production: 11.18 billion cu m (2017 est.)
country comparison to the world: 39

Natural gas—consumption: 99.11 million cu m (2017 est.)
country comparison to the world: 109

Natural gas—exports: 11.1 billion cu m (2017 est.)
country comparison to the world: 21

Natural gas—imports: 0 cu m (2017 est.)
country comparison to the world: 172

Natural gas—proved reserves: 210.5 billion cu m (1 January 2018 est.)

country comparison to the world: 41

Carbon dioxide emissions from consumption of energy: 6.082 million Mt (2017 est.)
country comparison to the world: 129

COMMUNICATIONS

Telephones—fixed lines: *total subscriptions:* 133,593
subscriptions per 100 inhabitants: 1.87 (2019 est.)
country comparison to the world: 130

Telephones—mobile cellular: *total subscriptions:* 3,401,971
subscriptions per 100 inhabitants: 47.62 (2019 est.)
country comparison to the world: 137

Telecommunication systems: *general assessment:* services are minimal; Internet slow and expensive; facilities provide radiotelephone and telegraph, coastal radio, aeronautical radio, and international radio communication services; a great deal of the population is under served in telecommunications; terrain, living conditions and economic stability is not high; 2G still exists in rural areas, 3G and 4G LTE in urban areas; the launch of the Kacific-1 satellite in 2019, will improve most services in the region (2020)
domestic: access to telephone services is not widely available; fixed-line 2 per 100 and mobile-cellular 48 per 100 person, teledensity has increased (2019)
international: country code—675; landing points for the Kumul Domestic Submarine Cable System, PNG-LNG, APNG-2, CSCS and the PPC-1 submarine cables to Australia, Guam, PNG and Solomon Islands, satellite earth station 1 Intelsar (Pacific Ocean) (2019)
note: the COVID-19 outbreak is negatively impacting telecommunications production and supply chains globally; consumer spending on tele-com devices and services has also slowed due to the pandemic's effect on economies worldwide; overall progress towards improvements in all facets of the telecom industry—mobile, fixed-line, broadband, submarine cable and satellite—has moderated

Broadcast media: *4 TV stations:* 1 commercial station operating since 1987, 1 state-run station launched in 2008, 1 digital free-to-view network launched in 2014, and 1 satellite network Click TV (PNGTV) launched in 2015; the state-run National Broadcasting Corporation operates 3 radio networks with multiple repeaters and about 20 provincial stations; several commercial radio stations with multiple transmission points as well as several community stations; transmissions of several international broadcasters are accessible (2018)

Internet country code: .pg

Internet users: *total:* 787,764
percent of population: 11.21% (July 2018 est.)
country comparison to the world: 146

Broadband—fixed subscriptions: *total:* 17,000

subscriptions per 100 inhabitants: less than 1 (2017 est.)
country comparison to the world: 157

TRANSPORTATION

National air transport system: *number of registered air carriers:* 6 (2020)
inventory of registered aircraft operated by air carriers: 48
annual passenger traffic on registered air carriers: 964,713 (2018)
annual freight traffic on registered air carriers: 30.93 million mt-km (2018)

Civil aircraft registration country code prefix: P2 (2016)

Airports: 561 (2013)
country comparison to the world: 11

Airports—with paved runways: *total:* 21 (2017)
over 3,047 m: 1 (2017)
2,438 to 3,047 m: 2 (2017)
1,524 to 2,437 m: 12 (2017)
914 to 1,523 m: 5 (2017)
under 914 m: 1 (2017)

Airports—with unpaved runways: *total:* 540 (2013)
1,524 to 2,437 m: 11 (2013)
914 to 1,523 m: 53 (2013)
under 914 m: 476 (2013)

Heliports: 2 (2013)

Pipelines: 264 km oil (2013)

Roadways: *total:* 9,349 km (2011)
paved: 3,000 km (2011)
unpaved: 6,349 km (2011)
country comparison to the world: 137

Waterways: 11,000 km (2011)
country comparison to the world: 12

Merchant marine: *total:* 171
by type: bulk carrier 7, general cargo 76, oil tanker 3, other 85 (2019)
country comparison to the world: 69

Ports and terminals: *major seaport(s):* Kimbe, Lae, Madang, Rabaul, Wewak
LNG terminal(s) (export): Port Moresby

MILITARY AND SECURITY

Military and security forces: Papua New Guinea Defense Force (PNGDF, includes land, maritime, and air elements) (2019)

Military expenditures:
0.4% of GDP (2019)
0.4% of GDP (2018)
0.4% of GDP (2017)
0.4% of GDP (2016)
0.5% of GDP (2015)
country comparison to the world: 154

Military and security service personnel strengths: the Papau New Guinea Defense Force has approximately 3,000 active duty troops (2,700 Ground; 200 Maritime; 100 Air) (2019)

Military equipment inventories and acquisitions: the PNGDF has a limited inventory consisting

of a diverse mix of foreign-supplied weapons and equipment; Papau New Guinea receives most of its military assistance from Australia; since 2010, it has also received equipment from China and New Zealand (2019 est.)

Military service age and obligation: 16 years of age for voluntary military service (with parental consent); no conscription; graduation from grade 12 required (2013)

TRANSNATIONAL ISSUES

Disputes—international: relies on assistance from Australia to keep out illegal cross-border activities from primarily Indonesia, including goods smuggling, illegal narcotics trafficking, and squatters and secessionists

Refugees and internally displaced persons: *refugees (country of origin):* 9,368 (Indonesia) (2019)
IDPs: 14,000 (natural disasters, tribal conflict, inter-communal violence, development projects) (2019)

Trafficking in persons: *current situation:* Papua New Guinea is a source and destination country for men, women, and children subjected to sex trafficking and forced labor; foreign and Papua New Guinean women and children are subjected to sex trafficking, domestic servitude, forced begging, and street vending; parents may sell girls into forced marriages to settle debts or as peace offerings or trade them to another tribe to forge a political alliance, leaving them vulnerable to forced domestic service, or, in urban areas, they may prostitute their children for income or to pay school fees; Chinese, Malaysian, and local men are forced to labor in logging and mining camps through debt bondage schemes; migrant women from Indonesia, Malaysia, Thailand, China, and the Philippines are subjected to sex trafficking and domestic servitude at logging and mining camps, fisheries, and entertainment sites
tier rating: Tier 2 Watch List—Papua New Guinea does not fully comply with the minimum standards for the elimination of trafficking; however, it is making significant efforts to do so; the Criminal Code Amendment of 2013, which prohibits all forms of trafficking was brought into force in 2014; the government also formed an anti-trafficking committee, which drafted a national action plan; despite corruption problems, trafficking-related crimes were prosecuted in village courts rather than criminal courts, resulting in restitution to the victim but no prison time for offenders; the government did not investigate, prosecute, or convict any officials or law enforcement personnel complicit in trafficking offenses; the government made no efforts to proactively identify trafficking victims, has no formal victim identification and referral mechanism, and does not provide care facilities to victims or funding to shelters run by NGOs or international organizations (2015)

Illicit drugs: major consumer of cannabis

PARACEL ISLANDS

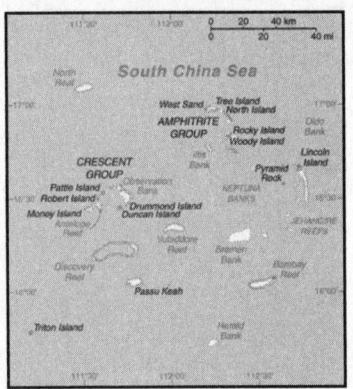

Land boundaries: 0 km

Coastline: 518 km

Maritime claims: NA

Climate: tropical

Terrain: mostly low and flat

Elevation: *lowest point:* South China Sea 0 m
highest point: unnamed location on Rocky Island 14 m

Natural resources: none

Land use: *agricultural land:* 0% (2011 est.)
arable land: 0% (2011 est.) / permanent crops: 0% (2011 est.) / permanent pasture: 0% (2011 est.)
forest: 0% (2011 est.)
other: 100% (2011 est.)

Irrigated land: 0 sq km (2012)

Population distribution: a population of over 1,000 Chinese resides on Woody Island, the largest of the Paracels; there are scattered Chinese garrisons on some other islands

Natural hazards: typhoons

Environment—current issues: China's use of dredged sand and coral to build artificial islands harms reef systems; ongoing human activities, including military operations, infrastructure construction, and tourism endangers local ecosystem including birds, fish, marine mammals, and marine reptiles

Geography—note: composed of 130 small coral islands and reefs divided into the northeast Amphitrite Group and the western Crescent Group

PEOPLE AND SOCIETY

Population: 1,440 (2014 est.)
note: Chinese activity has increased in recent years, particularly on Woody Island, where the population exceeds 1,000; there are scattered Chinese garrisons on some other islands
country comparison to the world: 235

Population distribution: a population of over 1,000 Chinese resides on Woody Island, the largest of the Paracels; there are scattered Chinese garrisons on some other islands

INTRODUCTION

Background: The Paracel Islands are surrounded by productive fishing grounds and by potential oil and gas reserves. In 1932, French Indochina annexed the islands and set up a weather station on Pattle Island; maintenance was continued by its successor, Vietnam. China has occupied all the Paracel Islands since 1974, when its troops seized a South Vietnamese garrison occupying the western islands. China built a military installation on Woody Island with an airfield and artificial harbor. The islands also are claimed by Taiwan and Vietnam.

GEOGRAPHY

Location: Southeastern Asia, group of small islands and reefs in the South China Sea, about one-third of the way from central Vietnam to the northern Philippines

Geographic coordinates: 16 30 N, 112 00 E

Map references: Southeast Asia

Area: *total:* 8 sq km ca.
land: 7.75 sq km ca.
water: 0 sq km
country comparison to the world: 244

Area—comparative: land area is about 13 times the size of the National Mall in Washington, DC

GOVERNMENT

Country name: *conventional long form:* none
conventional short form: Paracel Islands
etymology: Portuguese navigators began to refer to the "Ilhas do Pracel" in the 16th century as a designation of low lying islets, sandbanks, and reefs scattered over a wide area; over time the name changed to "parcel" and then "paracel"

ECONOMY

Economy—overview: The islands have the potential for oil and gas development. Waters around the islands support commercial fishing, but the islands themselves are not populated on a permanent basis.

TRANSPORTATION

Airports: 1 (2013)
country comparison to the world: 233

Airports—with paved runways: *total:* 1 (2019)
1,524 to 2,437 m: 1

Ports and terminals: small Chinese port facilities on Woody Island and Duncan Island

MILITARY AND SECURITY

Military—note: occupied by China, which is assessed to maintain 20 outposts in the Paracels (Antelope, Bombay, and North reefs; Drummond, Duncan, Lincoln, Middle, Money, North, Pattle, Quanfu, Robert, South, Tree, Triton, Woody, and Yagong islands; South Sand and West Sand; Observation Bank); the outposts range in size from one or two buildings to bases with significant military infrastructure; Woody Island is the main base in the Paracels and includes an airstrip with fighter aircraft hangers, naval facilities, surveillance radars, and defenses such as surface-to-air missiles and anti-ship cruise missiles; fighter aircraft have deployed to the island (2020)

TRANSNATIONAL ISSUES

Disputes—international: occupied by China, also claimed by Taiwan and Vietnam

PARAGUAY

INTRODUCTION

Background: Paraguay achieved its independence from Spain in 1811. In the disastrous War of the Triple Alliance (1865-70)—between Paraguay and Argentina, Brazil, and Uruguay—Paraguay lost two-thirds of its adult males and much of its territory. The country stagnated economically for the next half century. Following the Chaco War of 1932-35 with Bolivia, Paraguay gained a large part of the Chaco lowland region. The 35-year military dictatorship of Alfredo STROESSNER ended in 1989, and Paraguay has held relatively free and regular presidential elections since the country's return to democracy.

GEOGRAPHY

Location: Central South America, northeast of Argentina, southwest of Brazil

Geographic coordinates: 23 00 S, 58 00 W

Map references: South America

Area: *total:* 406,752 sq km
land: 397,302 sq km
water: 9,450 sq km
country comparison to the world: 61

Area—comparative: about three times the size of New York state; slightly smaller than California

Land boundaries: *total:* 4,655 km
border countries (3): Argentina 2531 km, Bolivia 753 km, Brazil 1371 km

Coastline: 0 km (landlocked)

Maritime claims: none (landlocked)

Climate: subtropical to temperate; substantial rainfall in the eastern portions, becoming semiarid in the far west

Terrain: grassy plains and wooded hills east of Rio Paraguay; Gran Chaco region west of Rio Paraguay mostly low, marshy plain near the river, and dry forest and thorny scrub elsewhere

Elevation: *mean elevation:* 178 m
lowest point: junction of Rio Paraguay and Rio Parana 46 m
highest point: Cerro Pero 842 m

Natural resources: hydropower, timber, iron ore, manganese, limestone

Land use: *agricultural land:* 53.8% (2011 est.)
arable land: 10.8% (2011 est.) / *permanent crops:* 0.2% (2011 est.) / *permanent pasture:* 42.8% (2011 est.)
forest: 43.8% (2011 est.)
other: 2.4% (2011 est.)
Irrigated land: 1,362 sq km (2012)

Population distribution: most of the population resides in the eastern half of the country; to the west lies the Gran Chaco (a semi-arid lowland plain), which accounts for 60% of the land territory, but only 2% of the overall population

Natural hazards: local flooding in southeast (early September to June); poorly drained plains may become boggy (early October to June)

Environment—current issues: deforestation; water pollution; rivers suffer from toxic dumping; tanneries release mercury and chromium into rivers and streams; loss of wetlands; inadequate means

for waste disposal pose health risks for many urban residents

Environment—international agreements: *party to:* Biodiversity, Climate Change, Climate Change-Kyoto Protocol, Desertification, Endangered Species, Hazardous Wastes, Law of the Sea, Ozone Layer Protection, Wetlands
signed, but not ratified: none of the selected agreements

Geography—note: landlocked; lies between Argentina, Bolivia, and Brazil; population concentrated in eastern and southern part of country

PEOPLE AND SOCIETY

Population: 7,191,685 (July 2020 est.)
country comparison to the world: 104

Nationality: *noun:* Paraguayan(s)
adjective: Paraguayan

Ethnic groups: mestizo (mixed Spanish and Amerindian) 95%, other 5%

Languages: Spanish (official) and Guarani (official) 46.3%, only Guarani 34%, only Spanish 15.2%, other (includes Portuguese, German, other indigenous languages) 4.1% , no response .4% (2012 est.)
note: data represent predominant household language

Religions: Roman Catholic 89.6%, Protestant 6.2%, other Christian 1.1%, other or unspecified 1.9%, none 1.1% (2002 census)

Demographic profile: Paraguay falls below the Latin American average in several socioeconomic categories, including immunization rates, potable water, sanitation, and secondary school enrollment, and has greater rates of income inequality and child and maternal mortality. Paraguay's poverty rate has declined in recent years but remains high, especially in rural areas, with more than a third of the population below the poverty line. However, the well-being of the poor in many regions has improved in terms of housing quality and access to clean water, telephone service, and electricity. The fertility rate continues to drop, declining sharply from an average 4.3 births per woman in the late 1990s to about 2 in 2013, as a result of the greater educational attainment of women, increased use of contraception, and a desire for smaller families among young women.

Paraguay is a country of emigration; it has not attracted large numbers of immigrants because of political instability, civil wars, years of dictatorship, and the greater appeal of neighboring countries. Paraguay first tried to encourage immigration in 1870 in order to rebound from the heavy death toll it suffered during the War of the Triple Alliance, but it received few European and Middle Eastern immigrants. In the 20th century, limited numbers of immigrants arrived from Lebanon, Japan, South Korea, and China, as well as Mennonites from Canada, Russia, and Mexico. Large flows of Brazilian immigrants have been arriving since the 1960s, mainly to work in agriculture. Paraguayans continue to emigrate to

Argentina, Brazil, Uruguay, the United States, Italy, Spain, and France.

Age structure: *0-14 years:* 23.41% (male 857,303/female 826,470)
15-24 years: 17.71% (male 640,400/female 633,525)
25-54 years: 42.63% (male 1,532,692/female 1,532,851)
55-64 years: 8.37% (male 306,100/female 295,890)
65 years and over: 7.88% (male 267,351/female 299,103) (2020 est.)

Dependency ratios: *total dependency ratio:* 55.5
youth dependency ratio: 49.9
elderly dependency ratio: 10.6
potential support ratio: 9.4 (2020 est.)

Median age: *total:* 29.7 years
male: 29.5 years
female: 29.9 years (2020 est.)
country comparison to the world: 128

Population growth rate: 1.16% (2020 est.)
country comparison to the world: 93

Birth rate: 16.6 births/1,000 population (2020 est.)
country comparison to the world: 103

Death rate: 4.9 deaths/1,000 population (2020 est.)
country comparison to the world: 202

Net migration rate: -0.1 migrant(s)/1,000 population (2020 est.)
country comparison to the world: 104

Population distribution: most of the population resides in the eastern half of the country; to the west lies the Gran Chaco (a semi-arid lowland plain), which accounts for 60% of the land territory, but only 2% of the overall population

Urbanization: *urban population:* 62.2% of total population (2020)
rate of urbanization: 1.71% annual rate of change (2015-20 est.)

total population growth rate v. urban population growth rate, 2000-2030: *Major urban areas—population:* 3.337 million ASUNCION (capital) (2020)

Sex ratio: *at birth:* 1.05 male(s)/female
0-14 years: 1.04 male(s)/female
15-24 years: 1.01 male(s)/female
25-54 years: 1 male(s)/female
55-64 years: 1.03 male(s)/female
65 years and over: 0.89 male(s)/female
total population: 1 male(s)/female (2020 est.)

Mother's mean age at first birth: 22.9 years (2008 est.)
note: median age at first birth among women 25-29

Maternal mortality rate: 84 deaths/100,000 live births (2017 est.)
country comparison to the world: 77

Infant mortality rate: *total:* 16.9 deaths/1,000 live births
male: 20 deaths/1,000 live births
female: 13.7 deaths/1,000 live births (2020 est.)
country comparison to the world: 89

Life expectancy at birth: *total population:* 77.9 years

male: 75.2 years

female: 80.7 years (2020 est.)

country comparison to the world: 72

Total fertility rate: 1.89 children born/woman (2020 est.)

country comparison to the world: 132

Contraceptive prevalence rate: 68.4% (2016)

Drinking water source:

improved:

urban: 100% of population

rural: 100% of population

total: 100% of population

unimproved:

urban: 0% of population

rural: 0% of population

total: 0% of population (2017 est.)

Current Health Expenditure: 6.7% (2017)

Physicians density: 1.37 physicians/1,000 population (2018)

Hospital bed density: 0.8 beds/1,000 population (2016)

Sanitation facility access:

improved:

urban: 98.4% of population

rural: 84.8% of population

total: 93.1% of population

unimproved:

urban: 1.6% of population

rural: 15.2% of population

total: 6.8% of population (2017 est.)

HIV/AIDS—adult prevalence rate: 0.4% (2019 est.)

country comparison to the world: 80

HIV/AIDS—people living with HIV/AIDS: 22,000 (2019 est.)

country comparison to the world: 85

HIV/AIDS—deaths: <1000 (2019 est.)

Major infectious diseases: *degree of risk:* intermediate (2020)

food or waterborne diseases: bacterial diarrhea, hepatitis A, and typhoid fever

vectorborne diseases: dengue fever

Obesity—adult prevalence rate: 20.3% (2016)

country comparison to the world: 100

Children under the age of 5 years underweight: 1.3% (2016)

country comparison to the world: 125

Education expenditures: 3.4% of GDP (2016)

country comparison to the world: 126

Literacy: *definition:* age 15 and over can read and write

total population: 94%

male: 94.5%

female: 93.5% (2018)

School life expectancy (primary to tertiary education): *total:* 12 years

male: 12 years

female: 13 years (2010)

Unemployment, youth ages 15-24: *total:* 14.5%

male: 11.8%

female: 18.7% (2018 est.)

country comparison to the world: 95

GOVERNMENT

Country name: *conventional long form:* Republic of Paraguay

conventional short form: Paraguay

local long form: Republica del Paraguay

local short form: Paraguay

etymology: the precise meaning of the name Paraguay is unclear, but it seems to derive from the river of the same name; one explanation has the name meaning "water of the Payagua" (an indigenous tribe that lived along the river)

Government type: presidential republic

Capital: *name:* Asuncion

geographic coordinates: 25 16 S, 57 40 W

time difference: UTC-4 (1 hour ahead of Washington, DC, during Standard Time)

daylight saving time: +1hr, begins first Sunday in October; ends last Sunday in March

etymology: the name means "assumption" and derives from the original name given to the city at its founding in 1537, Nuestra Senora Santa Maria de la Asuncion (Our Lady Saint Mary of the Assumption)

Administrative divisions: 17 departments (departamentos, singular—departamento) and 1 capital city*; Alto Paraguay, Alto Parana, Amambay, Asuncion*, Boqueron, Caaguazu, Caazapa, Canindeyu, Central, Concepcion, Cordillera, Guaira, Itapua, Misiones, Neembucu, Paraguari, Presidente Hayes, San Pedro

Independence: 14-15 May 1811 (from Spain); note—the uprising against Spanish authorities took place during the night of 14-15 May 1811 and both days are celebrated in Paraguay

National holiday: Independence Day, 14-15 May (1811) (observed 15 May); 14 May is celebrated as Flag Day

Constitution: *history:* several previous; latest approved and promulgated 20 June 1992

amendments: proposed at the initiative of at least one quarter of either chamber of the National Congress, by the president of the republic, or by petition of at least 30,000 voters; passage requires absolute majority vote by both chambers and approval in a referendum; amended 2011, 2014; note—in April 2017, a proposed amendment to extend presidential term limits was defeated by the lower house of the National Congress

Legal system: civil law system with influences from Argentine, Spanish, Roman, and French civil law models; judicial review of legislative acts in Supreme Court of Justice

International law organization participation: accepts compulsory ICJ jurisdiction; accepts ICCt jurisdiction

Citizenship: *citizenship by birth:* yes

citizenship by descent only: at least one parent must be a native-born citizen of Paraguay

dual citizenship recognized: yes

residency requirement for naturalization: 3 years

Suffrage: 18 years of age; universal and compulsory until the age of 75

Executive branch: *chief of state:* President Mario Abdo BENITEZ (since 15 August 2018); Vice President Hugo Adalberto VELAZQUEZ Moreno (since 15 August 2018); note—the president is both chief of state and head of government

head of government: President Mario Abdo BENITEZ (since 15 August 2018); Vice President Hugo Adalberto VELAZQUEZ Moreno (since 15 August 2018)

cabinet: Council of Ministers appointed by the president

elections/appointments: president and vice president directly elected on the same ballot by simple majority popular vote for a single 5-year term; election last held on 22 April 2018 (next to be held in April 2023)

election results: Mario Abdo BENITEZ elected president; percent of vote—Mario Abdo BENITEZ (ANR) 46.4%, Efrain ALEGRE (PLRA) 42.7%, Juan Bautista YBANEZ 3.3%, other 7.6%

Legislative branch: *description:* bicameral National Congress or Congreso Nacional consists of:

Chamber of Senators or Camara de Senadores (45 seats; members directly elected in a single nationwide constituency by proportional representation vote to serve 5-year terms)

Chamber of Deputies or Camara de Diputados (80 seats; members directly elected in 18 multiseat constituencies corresponding to the country's 17 departments and capital city—by proportional representation vote to serve 5-year terms)

elections: Chamber of Senators—last held on 22 April 2018 (next to be held in April 2023)

Chamber of Deputies—last held on 22 April 2018 (next to be held in April 2023)

election results: Chamber of Senators—percent of vote by party/coalition—ANR 32.52%, PLRA 24.18%, FG 11.83%, PPQ 6.77%, MH 4.47%, PDP 3.66%, MCN 2.48%, UNACE 2.12%, other 11.97%; seats by party/coalition—ANR 17, PLRA 13, FG 6, PPQ 3, MH 2, PDP 2, MCN 1, UNACE 1; composition—men 36, women 9, percent of women 20%

Chamber of Deputies—percent of vote by party/coalition—ANR 39.1%, PLRA 17.74%, Ganar Alliance 12.08%, PPQ 4.46%, MH 3.19%; other 23.43%; seats by party/coalition—ANR 42, PLRA 17, Ganar Alliance 13, PPQ 3, MH 2, other 3; composition—men 66, women 14, percent of women 17.5%; note—total National Congress percent of women 18.4%

Judicial branch: *highest courts:* Supreme Court of Justice or Corte Suprema de Justicia (consists of 9 justices divided 3 each into the Constitutional Court, Civil and Commercial Chamber, and Criminal Division)

judge selection and term of office: justices proposed by the Council of Magistrates or Consejo de la Magistratura, a 6-member independent body, and appointed by the Chamber of Senators with presidential concurrence; judges can serve until mandatory retirement at age 75

subordinate courts: appellate courts; first instance courts; minor courts, including justices of the peace

Political parties and leaders: Asociacion Nacional Republicana—Colorado Party or ANR [Pedro ALLIANA]

Avanza Pais coalition or AP [Adolfo FERREIRO]

Broad Front coalition (Frente Guasu) or FG [Esperanza MARTINEZ]

Ganar Alliance (alliance between PLRA and Guasu Front)

Movimiento Cruzada Nacional or MCN

Movimiento Hagamos or MH [Antonio "Tony" APURIL]

Movimiento Union Nacional de Ciudadanos Eticos or UNACE [Jorge OVIEDO MATTO]

Partido del Movimiento al Socialismo or P-MAS [Camilo Ernesto SOARES Machado]

Partido Democratica Progresista or PDP [Rafael FILIZZOLA]

Partido Encuentro Nacional or PEN [Hermann RATZLAFFIN Klippemstein]

Partido Liberal Radical Autentico or PLRA [Efrain ALEGRE]

Partido Pais Solidario or PPS [Carlos Alberto FILIZZOLA Pallares]

Partido Popular Tekojoja or PPT [Sixto PEREIRA Galeano]

Patria Querida (Beloved Fatherland Party) or PPQ [Miguel CARRIZOSA]

International organization participation: CAN (associate), CD, CELAC, FAO, G-11, G-77, IADB, IAEA, IBRD, ICAO, ICC (national committees), ICCt, ICRM, IDA,IFAD, IFC, IFRCS, ILO, IMF, IMO, Interpol, IOC, IOM, IPU, ISO (correspondent), ITSO, ITU, ITUC (NGOs), LAES, LAIA, Mercosur, MIGA, MINURSO, MINUSTAH, MONUSCO, NAM (observer), OAS, OPANAL, OPCW, Pacific Alliance (observer),PCA, UN, UNASUR, UNCTAD, UNESCO, UNFICYP, UNIDO, Union Latina, UNISFA, UNMIL, UNMISS, UNOCI, UNWTO,UPU, WCO, WHO, WIPO, WMO, WTO

Diplomatic representation in the US: *chief of mission:* Ambassador Manuel Maria CACERES (since 11 January 2019)

chancery: 2400 Massachusetts Avenue NW, Washington, DC 20008

telephone: [1] (202) 483-6960 through 6962

FAX: [1] (202) 234-4508

consulate(s) general: Los Angeles, Miami, New York

Diplomatic representation from the US: *chief of mission:* Ambassador Lee MCCLENNY (since 20 February 2018)

telephone: [595] (21) 213-715

embassy: 1776 Avenida Mariscal Lopez, Casilla Postal 402, Asuncion

mailing address: Unit 4711, DPO AA 34036-0001

FAX: [595] (21) 213-728

Flag description: three equal, horizontal bands of red (top), white, and blue with an emblem centered in the white band; unusual flag in that the emblem is different on each side; the obverse (hoist side at the left) bears the national coat of arms (a yellow five- pointed star within a green

wreath capped by the words REPUBLICA DEL PARAGUAY, all within two circles); the reverse (hoist side at the right) bears a circular seal of the treasury (a yellow lion below a red Cap of Liberty and the words PAZ Y JUSTICIA (Peace and Justice)); red symbolizes bravery and patriotism, white represents integrity and peace, and blue denotes liberty and generosity

note: the three color bands resemble those on the flag of the Netherlands; one of only three national flags that differ on their obverse and reverse sides—the others are Moldova and Saudi Arabia

National symbol(s): *lion; national colors:* red, white, blue

National anthem: *name:* "Paraguayos, Republica o muerte!" (Paraguayans, The Republic or Death!)

lyrics/music: Francisco Esteban ACUNA de Figueroa/disputed

note: adopted 1934, in use since 1846; officially adopted following its re-arrangement in 1934

0:00 /3:15

ECONOMY

Economy—overview: Landlocked Paraguay has a market economy distinguished by a large informal sector, featuring re-export of imported consumer goods to neighboring countries, as well as the activities of thousands of microenterprises and urban street vendors. A large percentage of the population, especially in rural areas, derives its living from agricultural activity, often on a subsistence basis. Because of the importance of the informal sector, accurate economic measures are difficult to obtain.

On a per capita basis, real income has grown steadily over the past five years as strong world demand for commodities, combined with high prices and favorable weather, supported Paraguay's commodity-based export expansion. Paraguay is the fifth largest soy producer in the world. Drought hit in 2008, reducing agricultural exports and slowing the economy even before the onset of the global recession. The economy fell 3.8% in 2009, as lower world demand and world prices caused exports to contract. Severe drought and outbreaks of hoof-and-mouth disease in 2012 led to a brief drop in beef and other agricultural exports. Since 2014, however, Paraguay's economy has grown at a 4% average annual rate due to strong production and high global prices, at a time when other countries in the region have contracted.

The Paraguayan Government recognizes the need to diversify its economy and has taken steps in recent years to do so. In addition to looking for new commodity markets in the Middle East and Europe, Paraguayan officials have promoted the country's low labor costs, cheap energy from its massive Itaipu Hydroelectric Dam, and single-digit tax rate on foreign firms. As a result, the number of factories operating in the country – mostly transplants from Brazil—has tripled since 2014.

Corruption, limited progress on structural reform, and deficient infrastructure are the main obstacles to long-term growth. Judicial corruption

is endemic and is seen as the greatest barrier to attracting more foreign investment. Paraguay has been adverse to public debt throughout its history, but has recently sought to finance infrastructure improvements to attract foreign investment.

GDP (purchasing power parity): $88.91 billion (2017 est.)

$84.87 billion (2016 est.)

$81.36 billion (2015 est.)

note: data are in 2017 dollars

country comparison to the world: 91

GDP (official exchange rate): $38.94 billion (2017 est.)

GDP—real growth rate: 4.8% (2017 est.)

4.3% (2016 est.)

3.1% (2015 est.)

country comparison to the world: 55

GDP—per capita (PPP): $12,800 (2017 est.)

$12,400 (2016 est.)

$12,000 (2015 est.)

note: data are in 2017 dollars

country comparison to the world: 123

Gross national saving: 18.6% of GDP (2017 est.)

20.9% of GDP (2016 est.)

20% of GDP (2015 est.)

country comparison to the world: 108

GDP—composition, by end use:

household consumption: 66.7% (2017 est.)

government consumption: 11.3% (2017 est.)

investment in fixed capital: 17.3% (2017 est.)

investment in inventories: 0.3% (2017 est.)

exports of goods and services: 46.6% (2017 est.)

imports of goods and services: -42.2% (2017 est.)

GDP—composition, by sector of origin:

agriculture: 17.9% (2017 est.)

industry: 27.7% (2017 est.)

services: 54.5% (2017 est.)

Agriculture—products: cotton, sugarcane, soybeans, corn, wheat, tobacco, cassava (manioc, tapioca), fruits, vegetables; beef, pork, eggs, milk; timber

Industries: sugar processing, cement, textiles, beverages, wood products, steel, base metals, electric power

Industrial production growth rate: 2% (2017 est.)

country comparison to the world: 132

Labor force: 3.428 million (2017 est.)

country comparison to the world: 100

Labor force—by occupation: *agriculture:* 26.5%

industry: 18.5%

services: 55% (2008)

Unemployment rate: 5.7% (2017 est.)

6% (2016 est.)

country comparison to the world: 92

Population below poverty line: 22.2% (2015 est.)

Household income or consumption by percentage share: *lowest 10%:* 1.5%

highest 10%: 37.6% (2013 est.)

Budget: *revenues:* 5.524 billion (2017 est.)

expenditures: 5.968 billion (2017 est.)

Taxes and other revenues: 14.2% (of GDP) (2017 est.)
country comparison to the world: 202

Budget surplus (+) or deficit (-): -1.1% (of GDP) (2017 est.)
country comparison to the world: 84

Public debt: 19.5% of GDP (2017 est.)
18.9% of GDP (2016 est.)
country comparison to the world: 191

Fiscal year: calendar year

Inflation rate (consumer prices):
3.6% (2017 est.)
4.1% (2016 est.)
country comparison to the world: 144

Current account balance:
-$298 million (2017 est.)
$416 million (2016 est.)
country comparison to the world: 108

Exports: $11.73 billion (2017 est.)
$10.86 billion (2016 est.)
country comparison to the world: 83

Exports—partners: Brazil 31.9%, Argentina 15.9%, Chile 6.9%, Russia 5.9% (2017)

Exports—commodities: soybeans, livestock feed, cotton, meat, edible oils, wood, leather, gold

Imports: $11.35 billion (2017 est.)
$9.617 billion (2016 est.)
country comparison to the world: 95

Imports—commodities: road vehicles, consumer goods, tobacco, petroleum products, electrical machinery, tractors, chemicals, vehicle parts

Imports—partners: China 31.3%, Brazil 23.4%, Argentina 12.9%, US 7.4% (2017)

Reserves of foreign exchange and gold: $7.877 billion (31 December 2017 est.)
$6.881 billion (31 December 2016 est.)
country comparison to the world: 79

Debt—external:
$17.7 billion (31 December 2017 est.)
$16.48 billion (31 December 2016 est.)
country comparison to the world: 96

Exchange rates: guarani (PYG) per US dollar -
5,628.1 (2017 est.)
5,680.7 (2016 est.)
5,680.7 (2015 est.)
5,160.4 (2014 est.)
4,462.2 (2013 est.)

ENERGY

Electricity access: electrification—total population: 100% (2020)

Electricity—production: 63.13 billion kWh (2016 est.)
country comparison to the world: 45

Electricity—consumption: 10.9 billion kWh (2016 est.)
country comparison to the world: 92

Electricity—exports: 41.13 billion kWh (2015 est.)
country comparison to the world: 4

Electricity—imports: 0 kWh (2016 est.)

country comparison to the world: 185

Electricity—installed generating capacity: 8.87 million kW (2016 est.)
country comparison to the world: 65

Electricity—from fossil fuels: 0% of total installed capacity (2016 est.)
country comparison to the world: 214

Electricity—from nuclear fuels: 0% of total installed capacity (2017 est.)
country comparison to the world: 163

Electricity—from hydroelectric plants: 99% of total installed capacity (2017 est.)
country comparison to the world: 3

Electricity—from other renewable sources: 1% of total installed capacity (2017 est.)
country comparison to the world: 162

Crude oil—production: 0 bbl/day (2018 est.)
country comparison to the world: 186

Crude oil—exports: 0 bbl/day (2015 est.)
country comparison to the world: 179

Crude oil—imports: 0 bbl/day (2015 est.)
country comparison to the world: 181

Crude oil—proved reserves: 0 bbl (1 January 2018 est.)
country comparison to the world: 181

Refined petroleum products—production: 0 bbl/day (2015 est.)
country comparison to the world: 189

Refined petroleum products—consumption: 43,000 bbl/day (2016 est.)
country comparison to the world: 112

Refined petroleum products—exports: 0 bbl/day (2015 est.)
country comparison to the world: 191

Refined petroleum products—imports: 40,760 bbl/day (2015 est.)
country comparison to the world: 89

Natural gas—production: 0 cu m (2017 est.)
country comparison to the world: 183

Natural gas—consumption: 0 cu m (2017 est.)
country comparison to the world: 187

Natural gas—exports: 0 cu m (2017 est.)
country comparison to the world: 167

Natural gas—imports: 0 cu m (2017 est.)
country comparison to the world: 173

Natural gas—proved reserves: 0 cu m (1 January 2014 est.)
country comparison to the world: 181

Carbon dioxide emissions from consumption of energy: 7.74 million Mt (2017 est.)
country comparison to the world: 117

COMMUNICATIONS

Telephones—fixed lines: *total subscriptions:* 309,221
subscriptions per 100 inhabitants: 4.35 (2019 est.)
country comparison to the world: 107

Telephones—mobile cellular: *total subscriptions:* 7,602,566

subscriptions per 100 inhabitants: 106.95 (2019 est.)
country comparison to the world: 100

Telecommunication systems: *general assessment:* the fixed-line market is a state monopoly and fixed-line telephone service is meager; principal switching center is in Asuncion; DSL, cable modem, FttP (fiber to the home) and WiMAX technologies available; competition in mobile market among 4 operators; 18 mobile phones for every fixed-line service phone; mobile and Internet market operators bring new investment and working towards LTE (2020)
domestic: deficiencies in provision of fixed-line service have resulted in a rapid expansion of mobile-cellular services fostered by competition among multiple providers; Internet market also open to competition; fixed-line 4 per 100 and mobile-cellular 107 per 100 (2019)
international: country code—595; Paraguay's landlocked position means they must depend on neighbors for interconnection with submarine cable networks, making it cost more for broadband services; satellite earth station—1 Intelsat (Atlantic Ocean) (2019)
note: the COVID-19 outbreak is negatively impacting telecommunications production and supply chains globally; consumer spending on telecom devices and services has also slowed due to the pandemic's effect on economies worldwide; overall progress towards improvements in all facets of the telecom industry—mobile, fixed-line, broadband, submarine cable and satellite—has moderated

Broadcast media: 6 privately owned TV stations; about 75 commercial and community radio stations; 1 state-owned radio network (2019)

Internet country code: .py

Internet users: *total:* 4,566,043
percent of population: 64.99% (July 2018 est.)
country comparison to the world: 88

Broadband—fixed subscriptions: *total:* 320,700
subscriptions per 100 inhabitants: 5 (2018 est.)
country comparison to the world: 99

TRANSPORTATION

National air transport system: *number of registered air carriers:* 2 (2020)
inventory of registered aircraft operated by air carriers: 8
annual passenger traffic on registered air carriers: 560,631 (2018)
annual freight traffic on registered air carriers: 1.97 million mt-km (2018)

Civil aircraft registration country code prefix: ZP (2016)

Airports: 799 (2013)
country comparison to the world: 9

Airports—with paved runways: *total:* 15 (2017)
over 3,047 m: 3 (2017)
1,524 to 2,437 m: 7 (2017)
914 to 1,523 m: 5 (2017)

Airports—with unpaved runways: *total:* 784 (2013)

1,524 to 2,437 m: 23 (2013)
914 to 1,523 m: 290 (2013)
under 914 m: 471 (2013)

Railways: total: 30 km (2014)
standard gauge: 30 km 1.435-m gauge (2014)
country comparison to the world: 133

Roadways: total: 74,676 km (2017)
paved: 6,167 km (2017)
unpaved: 68,509 km (2017)
country comparison to the world: 66

Waterways: 3,100 km (primarily on the Paraguay and Paraná River systems) (2012)
country comparison to the world: 32

Merchant marine: total: 106
by type: bulk carrier 3, general cargo 24, oil tanker 5, other 74 (2019)
note: as of 2017, Paraguay registered 2,012 fluvial vessels of which 1,741 were commercial barges
country comparison to the world: 87

Ports and terminals: river port(s): Asuncion, Villeta, San Antonio, Encarnacion (Parana)

Military and security forces: Armed Forces Command (Commando de las Fuerzas Militares): Army, National Navy (Armada Nacional, includes marines), Paraguayan Air Force (Fuerza Aerea Paraguay, FAP) (2020)

Military expenditures:
1% of GDP (2019)
0.9% of GDP (2018)
0.9% of GDP (2017)
1% of GDP (2016)
1.1% of GDP (2015)
country comparison to the world: 117

Military and security service personnel strengths: the Armed Forces of Paraguay have approximately 14,000 active personnel (8,500 Army; 3,000 Navy; 2,500 Air Force) (2019 est.)

Military equipment inventories and acquisitions: the Paraguayan military forces inventory is comprised of mostly older equipment from a variety of foreign suppliers, particularly Brazil and the US; since 2010, Paraguay has acquired limited quantities of mostly second-hand military equipment

from Argentina, Brazil, Israel, Spain, Taiwan, and the US (2019 est.)

Military service age and obligation: 18 years of age for compulsory and voluntary military service; conscript service obligation is 12 months for Army, 24 months for Navy; volunteers for the Air Force must be younger than 22 years of age with a secondary school diploma (2016)

TRANSNATIONAL ISSUES

Disputes—international: unruly region at convergence of Argentina-Brazil-Paraguay borders is locus of money laundering, smuggling, arms and illegal narcotics trafficking, and fundraising for violent extremist organizations

Illicit drugs: major illicit producer of cannabis, most or all of which is consumed in Brazil, Argentina, and Chile; transshipment country for Andean cocaine headed for Brazil, other Southern Cone markets, and Europe; weak border controls, extensive corruption and money-laundering activity, especially in the Tri-Border Area; weak anti-money-laundering laws and enforcement

PERU

INTRODUCTION

Background: Ancient Peru was the seat of several prominent Andean civilizations, most notably that of the Incas whose empire was captured by Spanish conquistadors in 1533. Peru declared its independence in 1821, and remaining Spanish forces were defeated in 1824. After a dozen years of military rule, Peru returned to democratic leadership in 1980, but experienced economic problems and the growth of a violent insurgency. President Alberto FUJIMORI's election in 1990 ushered in a decade that saw a dramatic turnaround in the economy and significant progress in curtailing guerrilla activity. Nevertheless, the president's increasing reliance on authoritarian measures and an economic slump in the late 1990s

generated mounting dissatisfaction with his regime, which led to his resignation in 2000. A caretaker government oversaw a new election in the spring of 2001, which installed Alejandro TOLEDO Manrique as the new head of government—Peru's first democratically elected president of indigenous ethnicity. The presidential election of 2006 saw the return of Alan GARCIA Perez who, after a disappointing presidential term from 1985 to 1990, oversaw a robust economic rebound. Former army officer Ollanta HUMALA Tasso was elected president in June 2011, and carried on the sound, market- oriented economic policies of the three preceding administrations. Poverty and unemployment levels have fallen dramatically in the last decade, and today Peru boasts one of the best performing economies in Latin America. Pedro Pablo KUCZYNSKI Godard won a very narrow presidential runoff election in June 2016. Facing impeachment after evidence surfaced of his involvement in a vote-buying scandal, President KUCZYNSKI offered his resignation on 21 March 2018. Two days later, First Vice President Martin Alberto VIZCARRA Cornejo was sworn in as president. On 30 September 2019, President VIZCARRA invoked his constitutional authority to dissolve Peru's Congress after months of battling with the body over anticorruption reforms. New congressional elections took place on 26 January 2020 resulting in the return of an opposition-led legislature. President VIZCARRA was impeached by Congress on 9 November 2020 for a second time and removed from office after being accused of corruption and mishandling of the COVID-19 pandemic. Because of vacancies in the vice-presidential positions, constitutional succession led to

the President of the Peruvian Congress, Manuel MERINO, becoming the next president of Peru. His ascension to office was not well received by the population, and large protests forced his resignation on 15 November 2020. On 17 November, Francisco SAGASTI assumed the position of President of Peru after being appointed President of the Congress the previous day.

GEOGRAPHY

Location: Western South America, bordering the South Pacific Ocean, between Chile and Ecuador

Geographic coordinates: 10 00 S, 76 00 W

Map references: South America

Area: total: 1,285,216 sq km
land: 1,279,996 sq km
water: 5,220 sq km
country comparison to the world: 21

Area—comparative: almost twice the size of Texas; slightly smaller than Alaska

Land boundaries: total: 7,062 km

border countries (5): Bolivia 1212 km, Brazil 2659 km, Chile 168 km, Colombia 1494 km, Ecuador 1529 km

Coastline: 2,414 km

Maritime claims: territorial sea: 200 nm
continental shelf: 200 nm

Climate: varies from tropical in east to dry desert in west; temperate to frigid in Andes

Terrain: western coastal plain (costa), high and rugged Andes in center (sierra), eastern lowland jungle of Amazon Basin (selva)

Elevation: *mean elevation:* 1,555 m
lowest point: Pacific Ocean 0 m
highest point: Nevado Huascaran 6,746 m

Natural resources: copper, silver, gold, petroleum, timber, fish, iron ore, coal, phosphate, potash, hydropower, natural gas

Land use: *agricultural land:* 18.8% (2011 est.)
arable land: 3.1% (2011 est.) / permanent crops: 1.1% (2011 est.) / permanent pasture: 14.6% (2011 est.)
forest: 53% (2011 est.)
other: 28.2% (2011 est.)
Irrigated land: 25,800 sq km (2012)

Population distribution: approximately one-third of the population resides along the desert coastal belt in the west, with a strong focus on the capital city of Lima; the Andean highlands, or sierra, which is strongly identified with the country's Amerindian population, contains roughly half of the overall population; the eastern slopes of the Andes, and adjoining rainforest, are sparsely populated

Natural hazards: earthquakes, tsunamis, flooding, landslides, mild volcanic activity
volcanism: volcanic activity in the Andes Mountains; Ubinas (5,672 m), which last erupted in 2009, is the country's most active volcano; other historically active volcanoes include El Misti, Huaynaputina, Sabancaya, and Yucamane; see note 2 under "Geography—note"

Environment—current issues: deforestation (some the result of illegal logging); overgrazing of the slopes of the costa and sierra leading to soil erosion; desertification; air pollution in Lima; pollution of rivers and coastal waters from municipal and mining wastes; overfishing

Environment—international agreements: *party to:* Antarctic-Environmental Protocol, Antarctic-Marine Living Resources, Antarctic Treaty, Biodiversity, Climate Change, Climate Change-Kyoto Protocol, Desertification, Endangered Species, Hazardous Wastes, Marine Dumping, Ozone Layer Protection, Ship Pollution, Tropical Timber 83, Tropical Timber 94, Wetlands, Whaling
signed, but not ratified: none of the selected agreements

Geography—note: *note 1:* shares control of Lago Titicaca, world's highest navigable lake, with Bolivia; a remote slope of Nevado Mismi, a 5,316 m peak, is the ultimate source of the Amazon River
note 2: Peru is one of the countries along the Ring of Fire, a belt of active volcanoes and earthquake epicenters bordering the Pacific Ocean; up to 90% of the world's earthquakes and some 75% of the world's volcanoes occur within the Ring of Fire
note 3: on 19 February 1600, Mount Huaynaputina in the southern Peruvian Andes erupted in the largest volcanic explosion in South America in historical times; intermittent eruptions lasted until 5 March 1600 and pumped an estimated 16 to 32 million metric tons of particulates into the atmosphere reducing the amount of sunlight reaching the earth's surface and affecting weather

worldwide; over the next two and a half years, millions died around the globe in famines from bitterly cold winters, cool summers, and the loss of crops and animals
note 4: the southern regions of Peru and the extreme northwestern part of Bolivia are considered to be the place of origin for the common potato

PEOPLE AND SOCIETY

Population: 31,914,989 (July 2020 est.)
country comparison to the world: 44

Nationality: *noun:* Peruvian(s)
adjective: Peruvian

Ethnic groups: mestizo (mixed Amerindian and white) 60.2%, Amerindian 25.8%, white 5.9%, African descent 3.6%, other (includes Chinese and Japanese descent) 1.2%, unspecified 3.3% (2017 est.)

Languages: Spanish (official) 82.9%, Quechua (official) 13.6%, Aymara (official) 1.6%, Ashaninka 0.3%, other native languages (includes a large number of minor Amazonian languages) 0.8%, other (includes foreign languages and sign language) 0.2%, none .1%, unspecified .7% (2017 est.)

Religions: Roman Catholic 60%, Christian 14.6% (includes evangelical 11.1%, other 3.5%), other .3%, none 4%, unspecified 21.1% (2017 est.)

Demographic profile: Peru's urban and coastal communities have benefited much more from recent economic growth than rural, Afro-Peruvian, indigenous, and poor populations of the Amazon and mountain regions. The poverty rate has dropped substantially during the last decade but remains stubbornly high at about 30% (more than 55% in rural areas). After remaining almost static for about a decade, Peru's malnutrition rate began falling in 2005, when the government introduced a coordinated strategy focusing on hygiene, sanitation, and clean water. School enrollment has improved, but achievement scores reflect ongoing problems with educational quality. Many poor children temporarily or permanently drop out of school to help support their families. About a quarter to a third of Peruvian children aged 6 to 14 work, often putting in long hours at hazardous mining or construction sites.

Peru was a country of immigration in the 19th and early 20th centuries, but has become a country of emigration in the last few decades. Beginning in the 19th century, Peru brought in Asian contract laborers mainly to work on coastal plantations. Populations of Chinese and Japanese descent—among the largest in Latin America—are economically and culturally influential in Peru today. Peruvian emigration began rising in the 1980s due to an economic crisis and a violent internal conflict, but outflows have stabilized in the last few years as economic conditions have improved. Nonetheless, more than 2 million Peruvians have emigrated in the last decade, principally to the US, Spain, and Argentina.

Age structure: *0-14 years:* 25.43% (male 4,131,985/female 3,984,546)
15-24 years: 17.21% (male 2,756,024/female 2,736,394)
25-54 years: 41.03% (male 6,279,595/female 6,815,159)
55-64 years: 8.28% (male 1,266,595/female 1,375,708)
65 years and over: 8.05% (male 1,207,707/female 1,361,276) (2020 est.)

Dependency ratios: *total dependency ratio:* 50.2
youth dependency ratio: 37.1
elderly dependency ratio: 13.1
potential support ratio: 7.6 (2020 est.)

Median age: *total:* 29.1 years
male: 28.3 years
female: 29.9 years (2020 est.)
country comparison to the world: 138

Population growth rate: 0.92% (2020 est.)
country comparison to the world: 116

Birth rate: 17 births/1,000 population (2020 est.)
country comparison to the world: 101

Death rate: 6.2 deaths/1,000 population (2020 est.)
country comparison to the world: 157

Net migration rate: -1.8 migrant(s)/1,000 population (2020 est.)
country comparison to the world: 163

Population distribution: approximately one-third of the population resides along the desert coastal belt in the west, with a strong focus on the capital city of Lima; the Andean highlands, or sierra, which is strongly identified with the country's Amerindian population, contains roughly half of the overall population; the eastern slopes of the Andes, and adjoining rainforest, are sparsely populated

Urbanization: *urban population:* 78.3% of total population (2020)
rate of urbanization: 1.44% annual rate of change (2015-20 est.)

total population growth rate v. urban population growth rate, 2000-2030:

Major urban areas—population: *10.719 million LIMA (capital), 923,000 Arequipa, 865,000 Trujillo (2020)*

Sex ratio: *at birth:* 1.05 male(s)/female
0-14 years: 1.04 male(s)/female
15-24 years: 1.01 male(s)/female
25-54 years: 0.92 male(s)/female
55-64 years: 0.92 male(s)/female
65 years and over: 0.89 male(s)/female
total population: 0.96 male(s)/female (2020 est.)

Mother's mean age at first birth: 22.2 years (2013 est.)
note: median age at first birth among women 25-29

Maternal mortality rate: 88 deaths/100,000 live births (2017 est.)
country comparison to the world: 76

Infant mortality rate: *total:* 16.7 deaths/1,000 live births
male: 18.7 deaths/1,000 live births

female: 14.6 deaths/1,000 live births (2020 est.)
country comparison to the world: 90

Life expectancy at birth: *total population:* 74.7 years
male: 72.6 years
female: 76.9 years (2020 est.)
country comparison to the world: 127

Total fertility rate: 2.04 children born/woman (2020 est.)
country comparison to the world: 107

Contraceptive prevalence rate: 76.3% (2018)

Drinking water source:
improved:
urban: 95.6% of population
rural: 77.4% of population
total: 92.1% of population
unimproved:
urban: 4.4% of population
rural: 22.6% of population
total: 7.9% of population (2017 est.)

Current Health Expenditure: 5% (2017)

Physicians density: 1.3 physicians/1,000 population (2016)

Hospital bed density: 1.6 beds/1,000 population (2017)

Sanitation facility access:
improved:
urban: 92.2% of population
rural: 60.8% of population
total: 85.2% of population
unimproved:
urban: 7.8% of population
rural: 14.8% of population (2017 est.)
total: 14.8% of population (2015 est.)

HIV/AIDS—adult prevalence rate: 0.4% (2019 est.)
country comparison to the world: 81

HIV/AIDS—people living with HIV/AIDS: 87,000 (2019 est.)
country comparison to the world: 50

HIV/AIDS—deaths: <1000 (2019 est.)

Major infectious diseases: *degree of risk:* very high (2020)
food or waterborne diseases: bacterial diarrhea, hepatitis A, and typhoid fever
vectorborne diseases: dengue fever, malaria, and Bartonellosis (Oroya fever)
note: widespread ongoing transmission of a respiratory illness caused by the novel coronavirus (COVID-19) is occurring throughout Peru; as of 10 November 2020, Peru has reported a total of 917,503 cases of COVID-19 or 27,827 cumulative cases of COVID-19 per 1 million population with 1,055 cumulative deaths per 1 million population; at this time, there are no specific limitations or quarantine requirements for US citizens and Lawful Permanent Residents entering the US from Peru; on 3 June 2020, Peruvian President Martín VIZCARRA signed a supreme decree extending Peru's Health State of Emergency for 90 days beginning Wednesday, 10 June 2020; this is not an extension of the national quarantine, although social distancing and the use of facemasks will be required for the foreseeable future

Obesity—adult prevalence rate: 19.7% (2016)
country comparison to the world: 110

Children under the age of 5 years underweight: 2.6% (2018)
country comparison to the world: 107

Education expenditures: 3.9% of GDP (2017)
country comparison to the world: 108

Literacy: *definition:* age 15 and over can read and write
total population: 94.4%
male: 97.1%
female: 91.7% (2018)

School life expectancy (primary to tertiary education): *total:* 15 years
male: 14 years
female: 15 years (2017)

Unemployment, youth ages 15-24: *total:* 14.7%
male: 14.3%
female: 15% (2018 est.)
country comparison to the world: 93

GOVERNMENT

Country name: *conventional long form:* Republic of Peru
conventional short form: Peru
local long form: Republica del Peru
local short form: Peru
etymology: exact meaning is obscure, but the name may derive from a native word "biru" meaning "river"

Government type: presidential republic

Capital: *name:* Lima
geographic coordinates: 12 03 S, 77 03 W
time difference: UTC-5 (same time as Washington, DC, during Standard Time)
etymology: the word "Lima" derives from the Spanish pronunciation of "Limaq," the native name for the valley in which the city was founded in 1535; "limaq" means "talker" in coastal Quechua and referred to an oracle that was situated in the valley but which was eventually destroyed by the Spanish and replaced with a church

Administrative divisions: 25 regions (regiones, singular—region) and 1 province* (provincia); Amazonas, Ancash, Apurimac, Arequipa, Ayacucho, Cajamarca, Callao, Cusco, Huancavelica, Huanuco, Ica, Junin, La Libertad, Lambayeque, Lima, Lima*, Loreto, Madre de Dios, Moquegua, Pasco, Piura, Puno, San Martin, Tacna, Tumbes, Ucayali
note: Callao, the largest port in Peru, is also referred to as a constitutional province, the only province of the Callao region

Independence: 28 July 1821 (from Spain)

National holiday: Independence Day, 28-29 July (1821)

Constitution: *history:* several previous; latest promulgated 29 December 1993, enacted 31 December 1993
amendments: proposed by Congress, by the president of the republic with the approval of the "Cabinet, " or by petition of at least 0.3% of voters; passage requires absolute majority approval by the Congress membership, followed by approval in a referendum; a referendum is not required if Congress approves the amendment by greater than two-thirds majority vote in each of two successive sessions; amended many times, last in 2018

Legal system: civil law system

International law organization participation: accepts compulsory ICJ jurisdiction with reservations; accepts ICCt jurisdiction

Citizenship: *citizenship by birth:* yes
citizenship by descent only: yes
dual citizenship recognized: yes
residency requirement for naturalization: 2 years

Suffrage: 18 years of age; universal and compulsory until the age of 70

Executive branch: *chief of state:* President Francisco Rafael SAGASTI Hochhausler (since 17 November 2020); First Vice President (vacant); Second Vice President (vacant); note—President Martin Alberto VIZCARRA was impeached and removed from office on 9 November 2020; after the resignation of his successor, Manuel Arturo MERINO, President SAGASTI assumed the office and will serve as president until 28 July 2021; new elections are slated for April 2021; the president is both chief of state and head of government

head of government: President Francisco Rafael SAGASTI Hochhausler (since 17 November 2020); First Vice President (vacant); Second Vice President (vacant)
cabinet: Council of Ministers appointed by the president
elections/appointments: president directly elected by absolute majority popular vote in 2 rounds if needed for a 5-year term (eligible for nonconsecutive terms); election last held on 10 April 2016 with a runoff on 5 June 2016 (next to be held in April 2021)
election results: Pedro Pablo KUCZYNSKI Godard elected president in second round; percent of vote in first round—Keiko FUJIMORI Higuchi (Fuerza Popular) 39.9%, Pedro Pablo KUCZYNSKI Godard (PeruanosPor el Kambio) 21.1%, Veronika MENDOZA (Broad Front) 18.7%, Alfredo BARNECHEA (Popular Action) 7%, Alan GARCIA (APRA) 5.8%, other 7.5%; percent of vote in second round—Pedro Pablo KUCZYNSKI Godard 50.1%, Keiko FUJIMORI Higuchi 49.9%
note: President Martin Alberto VIZCARRA Cornejo assumed office after President Pedro Pablo KUCZYNSKI Godard resigned from office on 21 March 2018; after VIZCARRA was impeached on 9 November 2020, the constitutional line of succession led to the inauguration of the President of the Peruvian Congress, Manuel Arturo MERINO, as President of Peru on 10 November 2020; following his resignation only days later on 15 November 2020, Francisco Rafael SAGASTI Hochhausler—who had been elected by the legislature to be the new President of Congress on 16 November 2020—was then sworn in as President of Peru on 17 November 2020 by line of succession

note: Prime Minister Violeta BERMUDEZ (since 18 November 2020) does not exercise executive power; this power rests with the president

Legislative branch: *description:* unicameral Congress of the Republic of Peru or Congreso de la Republica del Peru (130 seats; members directly elected in multi-seat constituencies by closed party-list proportional representation vote to serve single 5-year terms); note—a referendum held in December 2018 banned congressional reelection, holding members to a single consecutive term

elections: last held on 10 April 2016 with run-off election on 6 June 2016 (next to be held in April 2021); note—President VIZCARRA dissolved the Congress on 30 September 2019 and called new congressional elections for 26 January 2020; the new Congress will serve an abbreviated term, with the next regular election to be held in April 2021

election results: percent of vote by party/coalition—Fuerza Popular 36.3%, PPK 16.5%, Frente Amplio 13.9%, APP 9.2%; APRA 8.3%; AP 7.2%, other 8.6%; seats by party/coalition—Fuerza Popular 73, Frente Amplio 20, PPK 18, APP 9; APRA 5; AP 5; composition—men 94, women 36, percent of women 27.7%

Judicial branch: *highest courts:* Supreme Court (consists of 16 judges and divided into civil, criminal, and constitutional-social sectors)

judge selection and term of office: justices proposed by the National Board of Justice (a 7-member independent body), nominated by the president, and confirmed by the Congress; justices can serve until mandatory retirement at age 70

subordinate courts: Court of Constitutional Guarantees; Superior Courts or Cortes Superiores; specialized civil, criminal, and mixed courts; 2 types of peace courts in which professional judges and selected members of the local communities preside

Political parties and leaders:
Alliance for Progress (Alianzapara el Progreso) or APP [Cesar ACUNA Peralta] American Popular Revolutionary Alliance or APRA
Broad Front (Frente Amplio; also known as El Frente Amplio por Justicia, Vida y Libertad) (coalition includes Nuevo Peru [Veronika Mendoza], Tierra y Libertad [Marco ARANA Zegarra], and Fuerza Social [Susana VILLARAN de la Puente]
Fuerza Popular (formerly Fuerza 2011) [Keiko FUJIMORI Higuchi]
National Solidarity (Solidaridad Nacional) or SN [Luis CASTANEDA Lossio]
Peru Posible or PP (coalition includes Accion Popular and Somos Peru) [Alejandro TOLEDO Manrique]
Peruvian Aprista Party (Partido Aprista Peruano) or PAP [Javier VELASQUEZ Quesquen] (also referred to by its original name Alianza Popular Revolucionaria Americana or APRA)
Peruvian Nationalist Party [Ollanta HUMALA]
Peruvians for Change (Peruanos Por el Kambio) or PPK [Pedro Pablo KUCZYNSKI]
Popular Action (Accion Popular) or AP [Mesias GUEVARA Amasifuen]

Popular Christian Party (Partido Popular Cristiano) or PPC [Lourdes FLORES Nano]

International organization participation: APEC, BIS, CAN, CD, CELAC, EITI (compliant country), FAO, G-24, G-77, IADB, IAEA, IBRD, ICAO, ICC (NGOs), ICCt, ICRM, IDA, IFAD, IFC, IFRCS, IHO, ILO, IMF, IMO, IMSO, Interpol, IOC, IOM, IPU, ISO, ITSO, ITU, ITUC (NGOs), LAES, LAIA, Mercosur (associate), MIGA, MINUSTAH, MONUSCO, NAM, OAS, OPANAL, OPCW, Pacific Alliance, PCA, SICA (observer), UN, UNAMID, UNASUR, UNCTAD, UNESCO, UNHCR, UNIDO, Union Latina, UNISFA, UNMISS, UNOCI, UN Security Council (temporary), UNWTO, UPU, WCO, WFTU (NGOs), WHO, WIPO, WMO, WTO

Diplomatic representation in the US: *chief of mission:* Ambassador Hugo DE ZELA Martínez (since 8 July 2019)

chancery: 1700 Massachusetts Avenue NW, Washington, DC 20036

telephone: [1] (202) 833-9860 through 9869

FAX: [1] (202) 659-8124

consulate(s) general: Atlanta, Boston, Chicago, Dallas, Denver, Hartford (CT), Houston, Los Angeles, Miami, New York, Paterson (NJ), San Francisco, Washington DC

Diplomatic representation from the US: *chief of mission:* Ambassador Krishna R. URS (since 18 October 2017)

telephone: [51] (1) 618-2000

embassy: Avenida La Encalada, Cuadra 17 s/n, Surco, Lima 33

mailing address: P. O. Box 1995, Lima 1; American Embassy (Lima), APO AA 34031-5000

FAX: [51] (1) 618-2397

Flag description: three equal, vertical bands of red (hoist side), white, and red with the coat of arms centered in the white band; the coat of arms features a shield bearing a vicuna (representing fauna), a cinchona tree (the source of quinine, signifying flora), and a yellow cornucopia spilling out coins (denoting mineral wealth); red recalls blood shed for independence, white symbolizes peace

National symbol(s): *vicuna (a camelid related to the llama); national colors:* red, white

National anthem: *name:* "Himno Nacional del Peru" (National Anthem of Peru)

lyrics/music: Jose DE LA TORRE Ugarte/Jose Bernardo ALZEDO

note: adopted 1822; the song won a national anthem contest

0:00 **/3:**09

ECONOMY

Economy—overview: Peru's economy reflects its varied topography—an arid lowland coastal region, the central high sierra of the Andes, and the dense forest of the Amazon. A wide range of important mineral resources are found in the mountainous and coastal areas, and Peru's coastal waters provide excellent fishing grounds. Peru is

the world's second largest producer of silver and copper.

The Peruvian economy grew by an average of 5.6% per year from 2009-13 with a stable exchange rate and low inflation. This growth was due partly to high international prices for Peru's metals and minerals exports, which account for 55% of the country's total exports. Growth slipped from 2014 to 2017, due to weaker world prices for these resources. Despite Peru's strong macroeconomic performance, dependence on minerals and metals exports and imported foodstuffs makes the economy vulnerable to fluctuations in world prices.

Peru's rapid expansion coupled with cash transfers and other programs have helped to reduce the national poverty rate by over 35 percentage points since 2004, but inequality persists and continued to pose a challenge for the Ollanta HUMALA administration, which championed a policy of social inclusion and a more equitable distribution of income. Poor infrastructure hinders the spread of growth to Peru's non-coastal areas. The HUMALA administration passed several economic stimulus packages in 2014 to bolster growth, including reforms to environmental regulations in order to spur investment in Peru's lucrative mining sector, a move that was opposed by some environmental groups. However, in 2015, mining investment fell as global commodity prices remained low and social conflicts plagued the sector.

Peru's free trade policy continued under the HUMALA administration; since 2006, Peru has signed trade deals with the US, Canada, Singapore, China, Korea, Mexico, Japan, the EU, the European Free Trade Association, Chile, Thailand, Costa Rica, Panama, Venezuela, Honduras, concluded negotiations with Guatemala and the Trans-Pacific Partnership, and begun trade talks with El Salvador, India, and Turkey. Peru also has signed a trade pact with Chile, Colombia, and Mexico, called the Pacific Alliance, that seeks integration of services, capital, investment and movement of people. Since the US-Peru Trade Promotion Agreement entered into force in February 2009, total trade between Peru and the US has doubled. President Pedro Pablo KUCZYNSKI succeeded HUMALA in July 2016 and is focusing on economic reforms and free market policies aimed at boosting investment in Peru. Mining output increased significantly in 2016-17, which helped Peru attain one of the highest GDP growth rates in Latin America, and Peru should maintain strong growth in 2018. However, economic performance was depressed by delays in infrastructure mega-projects and the start of a corruption scandal associated with a Brazilian firm. Massive flooding in early 2017 also was a drag on growth, offset somewhat by additional public spending aimed at recovery efforts.

GDP (purchasing power parity):
$430.3 billion (2017 est.)
$420 billion (2016 est.)
$403.7 billion (2015 est.)
note: data are in 2017 dollars
country comparison to the world: 46

GDP (official exchange rate): $214.2 billion (2017 est.)

GDP—real growth rate:
2.18% (2019 est.)
3.97% (2018 est.)
2.48% (2017 est.)
country comparison to the world: 129

GDP—per capita (PPP):
$13,500 (2017 est.)
$13,300 (2016 est.)
$13,000 (2015 est.)
note: data are in 2017 dollars
country comparison to the world: 120

Gross national saving:
19.8% of GDP (2017 est.)
19.5% of GDP (2016 est.)
19% of GDP (2015 est.)
country comparison to the world: 100

GDP—composition, by end use:
household consumption: 64.9% (2017 est.)
government consumption: 11.7% (2017 est.)
investment in fixed capital: 21.7% (2017 est.)
investment in inventories: -0.2% (2017 est.)
exports of goods and services: 24% (2017 est.)
imports of goods and services: -22% (2017 est.)

GDP—composition, by sector of origin:
agriculture: 7.6% (2017 est.)
industry: 32.7% (2017 est.)
services: 59.9% (2017 est.)

Agriculture—products: artichokes, asparagus, avocados, blueberries, coffee, cocoa, cotton, sugarcane, rice, potatoes, corn, plantains, grapes, oranges, pineapples, guavas, bananas, apples, lemons, pears, coca, tomatoes, mangoes, barley, medicinal plants, quinoa, palm oil, marigolds, onions, wheat, dry beans; poultry, beef, pork, dairy products; guinea pigs; fish

Industries: mining and refining of minerals; steel, metal fabrication; petroleum extraction and refining, natural gas and natural gas liquefaction; fishing and fish processing, cement, glass, textiles, clothing, food processing, beer, soft drinks, rubber, machinery, electrical machinery, chemicals, furniture

Industrial production growth rate: 2.7% (2017 est.)
country comparison to the world: 113

Labor force: 3.421 million (2020 est.)
note: individuals older than 14 years of age
country comparison to the world: 101

Labor force—by occupation: *agriculture:* 25.8%
industry: 17.4%
services: 56.8% (2011)

Unemployment rate: 6.58% (2019 est.)
6.73% (2018 est.)
note: data are for metropolitan Lima; widespread underemployment
country comparison to the world: 103

Population below poverty line: 22.7% (2014 est.)

Household income or consumption by percentage share: *lowest 10%:* 1.4%
highest 10%: 36.1% (2010 est.)

Budget: *revenues:* 58.06 billion (2017 est.)
expenditures: 64.81 billion (2017 est.)

Taxes and other revenues: 27.1% (of GDP) (2017 est.)
country comparison to the world: 103

Budget surplus (+) or deficit (-): -3.1% (of GDP) (2017 est.)
country comparison to the world: 135

Public debt:
25.4% of GDP (2017 est.)
24.5% of GDP (2016 est.)
note: data cover general government debt, and includes debt instruments issued by government entities other than the treasury; the data exclude treasury debt held by foreign entities; the data include debt issued by subnational entities
country comparison to the world: 174

Fiscal year: calendar year

Inflation rate (consumer prices):
2.8% (2017 est.)
3.6% (2016 est.)
note: data are for metropolitan Lima, annual average
country comparison to the world: 128

Current account balance:
-$3.531 billion (2019 est.)
-$3.821 billion (2018 est.)
country comparison to the world: 177

Exports:
$44.92 billion (2017 est.)
$37.02 billion (2016 est.)
country comparison to the world: 53

Exports—partners: China 26.5%, US 15.2%, Switzerland 5.2%, South Korea 4.4%, Spain 4.1%, India 4.1% (2017)

Exports—commodities: copper, gold, lead, zinc, tin, iron ore, molybdenum, silver; crude petroleum and petroleum products, natural gas; coffee, asparagus and other vegetables, fruit, apparel and textiles, fishmeal, fish, chemicals, fabricated metal products and machinery, alloys

Imports:
$38.65 billion (2017 est.)
$35.13 billion (2016 est.)
country comparison to the world: 61

Imports—commodities: petroleum and petroleum products, chemicals, plastics, machinery, vehicles, TV sets, power shovels, front-end loaders, telephones and telecommunication equipment, iron and steel, wheat, corn, soybean products, paper, cotton, vaccines and medicines

Imports—partners: China 22.3%, US 20.1%, Brazil 6%, Mexico 4.4% (2017)

Reserves of foreign exchange and gold: $63.83 billion (31 December 2017 est.)

$61.81 billion (31 December 2016 est.)
country comparison to the world: 35

Debt—external: $66.25 billion (31 December 2017 est.)

$66.76 billion (31 December 2016 est.)
country comparison to the world: 60

Exchange rates: nuevo sol (PEN) per US dollar -
3.265 (2017 est.)
3.3751 (2016 est.)

3.3751 (2015 est.)
3.185 (2014 est.)
2.8383 (2013 est.)

ENERGY

Electricity access: *population without electricity:* 2 million (2017)
electrification—total population: 95% (2017)
electrification—urban areas: 97% (2017)
electrification—rural areas: 89% (2017)

Electricity—production: 50.13 billion kWh (2016 est.)
country comparison to the world: 54

Electricity—consumption: 44.61 billion kWh (2016 est.)
country comparison to the world: 53

Electricity—exports: 55 million kWh (2015 est.)
country comparison to the world: 86

Electricity—imports: 22 million kWh (2016 est.)
country comparison to the world: 112

Electricity—installed generating capacity: 14.73 million kW (2016 est.)
country comparison to the world: 51

Electricity—from fossil fuels: 61% of total installed capacity (2016 est.)
country comparison to the world: 128

Electricity—from nuclear fuels: 0% of total installed capacity (2017 est.)
country comparison to the world: 164

Electricity—from hydroelectric plants: 35% of total installed capacity (2017 est.)
country comparison to the world: 60

Electricity—from other renewable sources: 4% of total installed capacity (2017 est.)
country comparison to the world: 116

Crude oil—production: 49,000 bbl/day (2018 est.)
country comparison to the world: 54

Crude oil—exports: 7,995 bbl/day (2015 est.)
country comparison to the world: 61

Crude oil—imports: 86,060 bbl/day (2015 est.)
country comparison to the world: 46

Crude oil—proved reserves: 434.9 million bbl (1 January 2018 est.)
country comparison to the world: 47

Refined petroleum products—production: 166,600 bbl/day (2015 est.)
country comparison to the world: 57

Refined petroleum products—consumption: 250,000 bbl/day (2016 est.)
country comparison to the world: 50

Refined petroleum products—exports: 62,640 bbl/day (2015 est.)
country comparison to the world: 49

Refined petroleum products—imports: 65,400 bbl/day (2015 est.)
country comparison to the world: 71

Natural gas—production: 12.99 billion cu m (2017 est.)
country comparison to the world: 37

Natural gas—consumption: 7.483 billion cu m (2017 est.)

country comparison to the world: 53

Natural gas—exports: 5.505 billion cu m (2017 est.)
country comparison to the world: 28

Natural gas—imports: 0 cu m (2017 est.)
country comparison to the world: 174

Natural gas—proved reserves: 455.9 billion cu m (1 January 2018 est.)
country comparison to the world: 32

Carbon dioxide emissions from consumption of energy: 55.94 million Mt (2017 est.)
country comparison to the world: 55

COMMUNICATIONS

Telephones—fixed lines: *total subscriptions:* 3,099,172
subscriptions per 100 inhabitants: 9.8 (2019 est.)
country comparison to the world: 41

Telephones—mobile cellular: *total subscriptions:* 39,138,119
subscriptions per 100 inhabitants: 123.76 (2019 est.)
country comparison to the world: 38

Telecommunication systems: *general assessment:* good mobile operator competition with LTE services; broadband subscriber penetration low compared to other Latin American countries; 3G network and new LTE services expanded providing mobile broadband to rural communities, regulator auctions of 700 MHz spectrum for LTE services; Peru is seen as a potential market for growth in broadband, with government work to install fiber-optic backbone to remote areas (2020)
domestic: fixed-line teledensity is only about 10 per 100 persons; mobile-cellular teledensity, spurred by competition among multiple providers, now 124 telephones per 100 persons; nationwide microwave radio relay system and a domestic satellite system with 12 earth stations (2019)
international: country code—51; landing points for the SAM-1, IGW, American Movil-Telxius, SAC and PAN-AM submarine cable systems that provide links to parts of Central and South America, the Caribbean, and US; satellite earth stations—2 Intelsat (Atlantic Ocean) (2019)
note: the COVID-19 outbreak is negatively impacting telecommunications production and supply chains globally; consumer spending on telecom devices and services has also slowed due to the pandemic's effect on economies worldwide; overall progress towards improvements in all facets of the telecom industry—mobile, fixed-line, broadband, submarine cable and satellite—has moderated

Broadcast media: 10 major TV networks of which only one, Television Nacional de Peru, is state owned; multi-channel cable TV services are available; in excess of 2,000 radio stations including a substantial number of indigenous language stations (2019)

Internet country code: .pe

Internet users: *total:* 16,461,427
percent of population: 52.54% (July 2018 est.)
country comparison to the world: 40

Broadband—fixed subscriptions: *total:* 2,310,217
subscriptions per 100 inhabitants: 7 (2017 est.)

country comparison to the world: 52

TRANSPORTATION

National air transport system: *number of registered air carriers:* 6 (2020)
inventory of registered aircraft operated by air carriers: 62
annual passenger traffic on registered air carriers: 17,758,527 (2018)
annual freight traffic on registered air carriers: 313.26 million mt-km (2018)

Civil aircraft registration country code prefix: OB (2016)

Airports: 191 (2013)
country comparison to the world: 30

Airports—with paved runways: *total:* 59 (2017)
over 3,047 m: 5 (2017)
2,438 to 3,047 m: 21 (2017)
1,524 to 2,437 m: 16 (2017)
914 to 1,523 m: 12 (2017)
under 914 m: 5 (2017)

Airports—with unpaved runways: *total:* 132 (2013)
2,438 to 3,047 m: 1 (2013)
1,524 to 2,437 m: 19 (2013)
914 to 1,523 m: 30 (2013)
under 914 m: 82 (2013)

Heliports: 5 (2013)

Pipelines: 786 km extra heavy crude, 1526 km gas, 679 km liquid petroleum gas, 1033 km oil, 15 km refined products (2013)

Railways: *total:* 1,854 km (2014)
standard gauge: 1,730.4 km 1.435-m gauge (34 km electrified) (2014)
narrow gauge: 124 km 0.914-m gauge (2014)
country comparison to the world: 76

Roadways: *total:* 140,672 km (18,699 km paved) (2012)
note: includes 24,593 km of national roads (14,748 km paved), 24,235 km of departmental roads (2,340 km paved), and 91,844 km of local roads (1,611 km paved)
country comparison to the world: 36

Waterways: 8,808 km (8,600 km of navigable tributaries on the Amazon River system and 208 km on Lago Titicaca) (2011)
country comparison to the world: 14

Merchant marine: *total:* 98
by type: bulk carrier 1, oil tanker 10, other 87 (2019)
country comparison to the world: 91

Ports and terminals: *major seaport(s):* Callao, Matarani, Paita
oil terminal(s): Conchan oil terminal, La Pampilla oil terminal
container port(s) (TEUs): Callao (2,250,200) (2017)
river port(s): Iquitos, Pucallpa, Yurimaguas (Amazon)

MILITARY AND SECURITY

Military and security forces: *Joint Command of the Armed Forces of Peru:* Peruvian Army

(Ejercito del Peru), Peruvian Navy (Marina de Guerra del Peru, MGP, includes naval air, naval infantry, and Coast Guard), Air Force of Peru (Fuerza Aerea del Peru, FAP); Ministry of the Interior (Ministerio del Interior): Peruvian National Police (Policía Nacional del Perú, PNP) (2020)

Military expenditures:
1.2% of GDP (2019)
1.2% of GDP (2018)
1.2% of GDP (2017)
1.3% of GDP (2016)
1.7% of GDP (2015)
country comparison to the world: 105

Military and security service personnel strengths: Peruvian military size estimates vary widely; approximately 95,000 active personnel (55,000 Army; 25,000 Navy; 15,000 Air Force) (2019 est.)

Military equipment inventories and acquisitions: the Peruvian military's inventory is a mix of mostly older equipment from a wide variety of suppliers, including Brazil, Europe, the former Soviet Union, and the US; the leading suppliers of military equipment since 2010 are Italy, Russia, and South Korea (2019 est.)

Military deployments: 200 Central African Republic (MINUSCA) (2020)

Military service age and obligation: 18-50 years of age for male and 18-45 years of age for female voluntary military service; no conscription (2013)

Maritime threats: the International Maritime Bureau reports the territorial waters of Peru are a risk for armed robbery against ships; in 2018, four attacks against commercial vessels were reported, a slight increase from the two reported in 2017; most of these occured in the main port of Callao

TERRORISM

Terrorist group(s): Shining Path (Sendero Luminoso) (2019)
note: details about the history, aims, leadership, organization, areas of operation, tactics, targets, weapons, size, and sources of support of the group(s) appear(s) in Appendix-T

TRANSNATIONAL ISSUES

Disputes—international: Chile and Ecuador rejected Peru's November 2005 unilateral legislation to shift the axis of their joint treaty-defined maritime boundaries along the parallels of latitude to equidistance lines which favor Peru; organized illegal narcotics operations in Colombia have penetrated Peru's shared border; Peru rejects Bolivia's claim to restore maritime access through a sovereign corridor through Chile along the Peruvian border

Refugees and internally displaced persons: *refugees (country of origin):* 959,631 (Venezuela) (economic and political crisis; includes Venezuelans who have claimed asylum, are recognized as refugees, or have received alternative legal stay) (2020)

IDPs: 60,000 (civil war from 1980-2000; most IDPs are indigenous peasants in Andean and Amazonian regions; as of 2011, no new information on the situation of these IDPs) (2019)

Illicit drugs: until 1996 the world's largest coca leaf producer, Peru is now the world's second largest producer of coca leaf, though it lags far behind Colombia; cultivation of coca in Peru was estimated at 44,000 hectares in 2016, a decrease of 16 per cent over 2015; second largest producer of cocaine, estimated at 410 metric tons of potential pure cocaine in 2016; finished cocaine is shipped out from Pacific ports to the international drug market; increasing amounts of base and finished cocaine, however, are being moved to Brazil, Chile, Argentina, and Bolivia for use in the Southern Cone or transshipment to Europe and Africa; increasing domestic drug consumption

PHILIPPINES

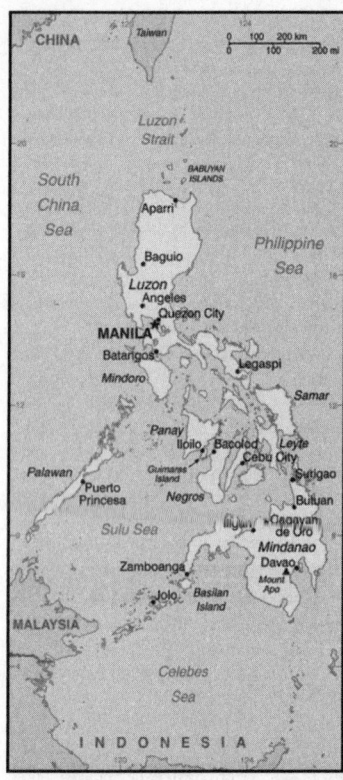

INTRODUCTION

Background: The Philippine Islands became a Spanish colony during the 16th century; they were ceded to the US in 1898 following the Spanish-American War. In 1935 the Philippines became a self-governing commonwealth. Manuel QUEZON was elected president and was tasked with preparing the country for independence after a 10-year transition. In 1942 the islands fell under Japanese occupation during World War II, and US forces and Filipinos fought together during 1944-45 to regain control. On 4 July 1946 the Republic of the Philippines attained its independence. A 21-year rule by Ferdinand MARCOS ended in 1986, when a "people power" movement in Manila ("EDSA 1") forced him into exile and installed Corazon AQUINO as president. Her presidency was hampered by several coup attempts that prevented a return to full political stability and economic development. Fidel RAMOS was elected president in 1992. His administration was marked by increased stability and by progress on economic reforms. In 1992, the US closed its last military bases on the islands. Joseph ESTRADA was elected president in 1998. He was succeeded by his vice-president, Gloria MACAPAGAL-ARROYO, in January 2001 after ESTRADA's stormy impeachment trial on corruption charges broke down and another "people power" movement ("EDSA 2") demanded his resignation. MACAPAGAL-ARROYO was elected to a six-year term as president in May 2004. Her presidency was marred by several corruption allegations but the Philippine economy was one of the few to avoid contraction following the 2008 global financial crisis, expanding each year of her administration. Benigno AQUINO III was elected to a six-year term as president in May 2010 and was succeeded by Rodrigo DUTERTE in May 2016.

The Philippine Government faces threats from several groups, some of which are on the US Government's Foreign Terrorist Organization list. Manila has waged a decades-long struggle against ethnic Moro insurgencies in the southern Philippines, which led to a peace accord with the Moro National Liberation Front and a separate agreement with a break away faction, the Moro Islamic Liberation Front. The decades-long Maoist-inspired New People's Army insurgency also operates through much of the country. In 2017, Philippine armed forces battled an ISIS-Philippines siege in Marawi City, driving DUTERTE to declare martial law in the region. The Philippines faces increased tension with China over disputed territorial and maritime claims in the South China Sea.

GEOGRAPHY

Location: Southeastern Asia, archipelago between the Philippine Sea and the South China Sea, east of Vietnam

Geographic coordinates: 13 00 N, 122 00 E

Map references: Southeast Asia

Area: *total:* 300,000 sq km
land: 298,170 sq km
water: 1,830 sq km
country comparison to the world: 74

Area—comparative: slightly less than twice the size of Georgia; slightly larger than Arizona

Land boundaries: 0 km

Coastline: 36,289 km

Maritime claims: *territorial sea:* irregular polygon extending up to 100 nm from coastline as defined by 1898 treaty; since late 1970s has also claimed polygonal-shaped area in South China Sea as wide as 285 nm
exclusive economic zone: 200 nm
continental shelf: to the depth of exploitation

Climate: tropical marine; northeast monsoon (November to April); southwest monsoon (May to October)

Terrain: mostly mountains with narrow to extensive coastal lowlands

Elevation: *mean elevation:* 442 m
lowest point: Philippine Sea 0 m
highest point: Mount Apo 2,954 m

Natural resources: timber, petroleum, nickel, cobalt, silver, gold, salt, copper

Land use: *agricultural land:* 41% (2011 est.)
arable land: 18.2% (2011 est.) / *permanent crops:* 17.8% (2011 est.) / permanent pasture: 5% (2011 est.)
forest: 25.9% (2011 est.)
other: 33.1% (2011 est.)

Irrigated land: 16,270 sq km (2012)

Population distribution: population concentrated where good farmlands lie; highest concentrations are northwest and south-central Luzon, the southeastern extension of Luzon, and the islands of the Visayan Sea, particularly Cebu and Negros; Manila is home to one-eighth of the entire national population

Natural hazards: astride typhoon belt, usually affected by 15 and struck by five to six cyclonic storms each year; landslides; active volcanoes; destructive earthquakes; tsunamis
volcanism: significant volcanic activity; Taal (311 m), which has shown recent unrest and may erupt in the near future, has been deemed a Decade Volcano by the International Association of Volcanology and Chemistry of the Earth's Interior, worthy of study due to its explosive history and close proximity to human populations; Mayon (2,462 m), the country's most active volcano, erupted in 2009 forcing over 33,000 to be evacuated; other historically active volcanoes include

Biliran, Babuyan Claro, Bulusan, Camiguin, Camiguin de Babuyanes, Didicas, Iraya, Jolo, Kanlaon, Makaturing, Musuan, Parker, Pinatubo, and Ragang; see note 2 under "Geography—note"

Environment—current issues: uncontrolled deforestation especially in watershed areas; illegal mining and logging; soil erosion; air and water pollution in major urban centers; coral reef degradation; increasing pollution of coastal mangrove swamps that are important fish breeding grounds; coastal erosion; dynamite fishing; wildlife extinction

Environment—international agreements: *party to:* Biodiversity, Climate Change, Climate Change-Kyoto Protocol, Desertification, Endangered Species, Hazardous Wastes, Law of the Sea, Marine Dumping, Ozone Layer Protection, Ship Pollution, Tropical Timber 83, Tropical Timber 94, Wetlands, Whaling
signed, but not ratified: Air Pollution-Persistent Organic Pollutants

Geography—note: *note 1:* for decades, the Philippine archipelago was reported as having 7,107 islands; in 2016, the national mapping authority reported that hundreds of new islands had been discovered and increased the number of islands to 7,641—though not all of the new islands have been verified; the country is favorably located in relation to many of Southeast Asia's main water bodies: the South China Sea, Philippine Sea, Sulu Sea, Celebes Sea, and Luzon Strait
note 2: Philippines is one of the countries along the Ring of Fire, a belt of active volcanoes and earthquake epicenters bordering the Pacific Ocean; up to 90% of the world's earthquakes and some 75% of the world's volcanoes occur within the Ring of Fire
note 3: the Philippines sits astride the Pacific typhoon belt and an average of 9 typhoons make landfall on the islands each year—with about 5 of these being destructive; the country is the most exposed in the world to tropical storms

PEOPLE AND SOCIETY

Population: 109,180,815 (July 2020 est.)
country comparison to the world: 12

Nationality: *noun:* Filipino(s)
adjective: Philippine

Ethnic groups: Tagalog 24.4%, Bisaya/Binisaya 11.4%, Cebuano 9.9%, Ilocano 8.8%, Hiligaynon/Ilonggo 8.4%, Bikol/Bicol 6.8%, Waray 4%, other local ethnicity 26.1%, other foreign ethnicity .1% (2010 est.)

Languages: unspecified Filipino (official; based on Tagalog) and English (official); eight major dialects—Tagalog, Cebuano, Ilocano, Hiligaynon or Ilonggo, Bicol, Waray, Pampango, and Pangasinan

Religions: Roman Catholic 80.6%, Protestant 8.2% (includes Philippine Council of Evangelical Churches 2.7%, National Council of Churches in the Philippines 1.2%, other Protestant 4.3%), other Christian 3.4%, Muslim 5.6%, tribal religions .2%, other 1.9%, none .1% (2010 est.)

Age structure: *0-14 years:* 32.42% (male 18,060,976/female 17,331,781)
15-24 years: 19.16% (male 10,680,325/female 10,243,047)
25-54 years: 37.37% (male 20,777,741/female 20,027,153)
55-64 years: 6.18% (male 3,116,485/female 3,633,301)
65 years and over: 4.86% (male 2,155,840/female 3,154,166) (2020 est.)

Dependency ratios: *total dependency ratio:* 55.2
youth dependency ratio: 46.6
elderly dependency ratio: 8.6
potential support ratio: 11.7 (2020 est.)

Median age: *total:* 24.1 years
male: 23.6 years
female: 24.6 years (2020 est.)
country comparison to the world: 168

Population growth rate: 1.52% (2020 est.)
country comparison to the world: 68

Birth rate: 22.9 births/1,000 population (2020 est.)
country comparison to the world: 60

Death rate: 6 deaths/1,000 population (2020 est.)
country comparison to the world: 166

Net migration rate: -1.8 migrant(s)/1,000 population (2020 est.)
country comparison to the world: 164

Population distribution: population concentrated where good farmlands lie; highest concentrations are northwest and south-central Luzon, the southeastern extension of Luzon, and the islands of the Visayan Sea, particularly Cebu and Negros; Manila is home to one-eighth of the entire national population

Urbanization: *urban population:* 47.4% of total population (2020)
rate of urbanization: 1.99% annual rate of change (2015-20 est.)

total population growth rate v. urban population growth rate, 2000-2030: *Major urban areas—population:* 13.923 million MANILA (capital), 1.825 million Davao, 980,000 Cebu City, 917,000 Zamboanga, 881,000 Antipolo, 753,000 Cagayan de Oro City (2020)

Sex ratio: *at birth:* 1.05 male(s)/female
0-14 years: 1.04 male(s)/female
15-24 years: 1.04 male(s)/female
25-54 years: 1.04 male(s)/female
55-64 years: 0.86 male(s)/female
65 years and over: 0.68 male(s)/female
total population: 1.01 male(s)/female (2020 est.)

Mother's mean age at first birth: 22.8 years (2017 est.)
note: median age at first birth among women 25-29

Maternal mortality rate: 121 deaths/100,000 live births (2017 est.)
country comparison to the world: 64

Infant mortality rate: *total:* 20 deaths/1,000 live births
male: 22.9 deaths/1,000 live births
female: 17 deaths/1,000 live births (2020 est.)
country comparison to the world: 77

Life expectancy at birth: *total population:* 70 years
male: 66.5 years
female: 73.8 years (2020 est.)
country comparison to the world: 166

Total fertility rate: 2.92 children born/woman (2020 est.)
country comparison to the world: 54

Contraceptive prevalence rate: 54.1% (2017)

Drinking water source:
improved:
urban: 97.7% of population
rural: 92.7% of population
total: 95.4% of population
unimproved:
urban: 2.3% of population
rural: 7.3% of population
total: 4.6% of population (2017 est.)

Current Health Expenditure: 4.4% (2017)

Physicians density: 0.6 physicians/1,000 population (2017)

Hospital bed density: 1 beds/1,000 population (2014)

Sanitation facility access:
improved:
urban: 95% of population
rural: 88.2% of population
total: 91.4% of population
unimproved:
urban: 5% of population
rural: 11.8% of population
total: 8.6% of population (2017 est.)

HIV/AIDS—adult prevalence rate: 0.1% (2019 est.)
country comparison to the world: 131

HIV/AIDS—people living with HIV/AIDS: 97,000 (2019 est.)
country comparison to the world: 47

HIV/AIDS—deaths: 1,600 (2019 est.)
country comparison to the world: 49

Major infectious diseases: *degree of risk:* high (2020)
food or waterborne diseases: bacterial diarrhea, hepatitis A, and typhoid fever
vectorborne diseases: dengue fever and malaria
water contact diseases: leptospirosis
note—on 8 October 2019, the Centers for Disease Control and Prevention issued a Travel Health Notice regarding a polio outbreak in the Philippines; CDC recommends that all travelers to the Philippines be vaccinated fully against polio; before traveling to the Philippines, adults who completed their routine polio vaccine series as children should receive a single, lifetime adult booster dose of polio vaccine

Obesity—adult prevalence rate: 6.4% (2016)
country comparison to the world: 168

Children under the age of 5 years underweight: 19.1% (2018)
country comparison to the world: 26

Education expenditures: 2.7% of GDP (2009)
country comparison to the world: 153

Literacy: *definition:* age 15 and over can read and write

total population: 98.2%
male: 98.1%
female: 98.2% (2015)

School life expectancy (primary to tertiary education): *total:* 13 years
male: 13 years
female: 15 years (2017)

Unemployment, youth ages 15-24: *total:* 6.7%
male: 5.8%
female: 8.2% (2018 est.)
country comparison to the world: 156

People—note: one of only two predominantly Christian nations in Southeast Asia, the other being Timor-Leste

GOVERNMENT

Country name: *conventional long form:* Republic of the Philippines
conventional short form: Philippines
local long form: Republika ng Pilipinas
local short form: Pilipinas
etymology: named in honor of King PHILLIP II of Spain by Spanish explorer Ruy LOPEZ de VILLALOBOS, who visited some of the islands in 1543

Government type: presidential republic

Capital: *name:* Manila
geographic coordinates: 14 36 N, 120 58 E
time difference: UTC+8 (13 hours ahead of Washington, DC, during Standard Time)
etymology: derives from the Tagalog "may-nila" meaning "where there is indigo" and refers to the presence of indigoyielding plants growing in the area surrounding the original settlement

Administrative divisions: 81 provinces and 38 chartered cities
provinces: Abra, Agusan del Norte, Agusan del Sur, Aklan, Albay, Antique, Apayao, Aurora, Basilan, Bataan, Batanes, Batangas, Biliran, Benguet, Bohol, Bukidnon, Bulacan, Cagayan, Camarines Norte, Camarines Sur, Camiguin, Capiz, Catanduanes, Cavite, Cebu, Cotabato, Davao del Norte, Davao del Sur, Davao de Oro, Davao Occidental, Davao Oriental, Dinagat Islands, Eastern Samar, Guimaras, Ifugao, Ilocos Norte, Ilocos Sur, Iloilo, Isabela, Kalinga, Laguna, Lanao del Norte, Lanao del Sur, La Union, Leyte, Maguindanao, Marinduque, Masbate, Mindoro Occidental, Mindoro Oriental, Misamis Occidental, Misamis Oriental, Mountain, Negros Occidental, Negros Oriental, Northern Samar, Nueva Ecija, Nueva Vizcaya, Palawan, Pampanga, Pangasinan, Quezon, Quirino, Rizal, Romblon, Samar, Sarangani, Siquijor, Sorsogon, South Cotabato, Southern Leyte, Sultan Kudarat, Sulu, Surigao del Norte, Surigao del Sur, Tarlac, Tawi-Tawi, Zambales, Zamboanga del Norte, Zamboanga del Sur, Zamboanga Sibugay;
chartered cities: Angeles, Bacolod, Baguio, Butuan, Cagayan de Oro, Caloocan, Cebu, Cotabato, Dagupan, Davao, General Santos, Iligan, Iloilo, Lapu-Lapu, Las Pinas, Lucena, Makati, Malabon, Mandaluyong, Mandaue, Manila, Marikina, Muntinlupa, Naga, Navotas,

Olongapo, Ormoc, Paranaque, Pasay, Pasig, Puerto Princesa, Quezon, San Juan, Santiago, Tacloban, Taguig, Valenzuela, Zamboanga

Independence: 4 July 1946 (from the US)

National holiday: Independence Day, 12 June (1898); note—12 June 1898 was date of declaration of independence from Spain; 4 July 1946 was date of independence from the US

Constitution: *history:* several previous; latest ratified 2 February 1987, effective 11 February 1987
amendments: proposed by Congress if supported by three fourths of the membership, by a constitutional convention called by Congress, or by public petition; passage by either of the three proposal methods requires a majority vote in a national referendum; note—the constitution has not been amended since its enactment in 1987

Legal system: mixed legal system of civil, common, Islamic (sharia), and customary law

International law organization participation: accepts compulsory ICJ jurisdiction with reservations; withdrew from the ICCt in March 2019

Citizenship: *citizenship by birth:* no
citizenship by descent only: at least one parent must be a citizen of the Philippines
dual citizenship recognized: no
residency requirement for naturalization: 10 years

Suffrage: 18 years of age; universal

Executive branch: *chief of state:* President Rodrigo DUTERTE (since 30 June 2016); Vice President Leni ROBREDO (since 30 June 2016); note—the president is both chief of state and head of government

head of government: President Rodrigo DUTERTE (since 30 June 2016); Vice President Leni ROBREDO (since 30 June 2016)
cabinet: Cabinet appointed by the president with the consent of the Commission of Appointments, an independent body of 25 Congressional members including the Senate president (ex officio chairman), appointed by the president
elections/appointments: president and vice president directly elected on separate ballots by simple majority popular vote for a single 6-year term; election last held on 9 May 2016 (next to be held in May 2022)
election results: Rodrigo DUTERTE elected president; percent of vote—Rodrigo DUTERTE (PDP-Laban) 39%, Manuel "Mar" ROXAS (LP) 23.5%, Grace POE (independent) 21.4%, Jejomar BINAY (UNA) 12.7%, Miriam Defensor SANTIAGO (PRP) 3.4%; Leni ROBREDO elected vice president; percent of vote Leni ROBREDO (LP) 35.1%, Bongbong MARCOS (independent) 34.5%, Alan CAYETANO 14.4%, Francis ESCUDERO (independent) 12%, Antonio TRILLANES (independent) 2.1%, Gregorio HONASAN (UNA) 1.9%

Legislative branch: *description:* bicameral Congress or Kongreso consists of:
Senate or Senado (24 seats; members directly elected in multi-seat constituencies by majority

vote; members serve 6year terms with one-half of the membership renewed every 3 years)

House of Representatives or Kapulungan Ng Mga Kinatawan (297 seats; 238 members directly elected in single-seat constituencies by simple majority vote and 59 representing minorities directly elected by party-list proportional representation vote; members serve 3-year terms)
elections: Senate—elections last held on 9 May 2016 (next to be held on 13 May 2019) House of Representatives—elections last held on 9 May 2016 (next to be held on 13 May 2019)
election results: Senate—percent of vote by party—LP 31.3%, NPC 10.1%, UNA 7.6%, Akbayan 5.0%, other 30.9%, independent 15.1%; seats by party—LP 6, NPC 3, UNA 4, Akbayan 1, other 10; composition—men 18, women 6, percent of women 25% House of Representatives—percent of vote by party—LP 41.7%, NPC 17.0%, UNA 6.6%, NUP 9.7%, NP 9.4%, independent 6.0%, others 10.1%; seats by party—LP 115, NPC 42, NUP 23, NP 24, UNA 11, other 19, independent 4, party-list 59; composition—men 210, women 87, percent of women 29.8%; note—total Congress percent of women 29.4%

Judicial branch: *highest courts:* Supreme Court (consists of a chief justice and 14 associate justices)
judge selection and term of office: justices are appointed by the president on the recommendation of the Judicial and Bar Council, a constitutionally created, 6-member body that recommends Supreme Court nominees; justices serve until age 70
subordinate courts: Court of Appeals; Sandiganbayan (special court for corruption cases of government officials); Court of Tax Appeals; regional, metropolitan, and municipal trial courts; sharia courts

Political parties and leaders:
Akbayan [Machris CABREROS]
Laban ng Demokratikong Pilipino (Struggle of Filipino Democrats) or LDP [Edgardo ANGARA]
Lakas ng EDSA-Christian Muslim Democrats or Lakas-CMD [Ferdinand Martin ROMUALDEZ]
Liberal Party or LP [Francis PANGILINAN]
Nacionalista Party or NP [Manuel "Manny" VILLAR]
Nationalist People's Coalition or NPC [Eduardo COJUNGCO, Jr.]
National Unity Party or NUP [Albert GARCIA]
PDP-Laban [Aquilino PIMENTEL III]
People's Reform Party or PRP [Narciso SANTIAGO]
Puwersa ng Masang Pilipino (Force of the Philippine Masses) or PMP [Joseph ESTRADA]
United Nationalist Alliance or UNA

International organization participation: ADB, APEC, ARF, ASEAN, BIS, CD, CICA (observer), CP, EAS, FAO, G-24, G-77, IAEA, IBRD, ICAO, ICC (national committees), ICCt, ICRM, IDA, IFAD, IFC, IFRCS, IHO, ILO, IMF, IMO, IMSO, Interpol, IOC, IOM, IPU, ISO, ITSO, ITU, ITUC (NGOs), MIGA, MINUSTAH, NAM, OAS (observer), OPCW, PCA, PIF (partner), UN, UNCTAD, UNESCO, UNHCR, UNIDO, Union Latina, UNMIL, UNMOGIP, UNOCI, UNWTO,

UPU, WCO, WFTU (NGOs), WHO, WIPO, WMO, WTO

Diplomatic representation in the US: *chief of mission:* Ambassador Jose Manuel del Gallego ROMUALDEZ (since 29 November 2017)
chancery: 1600 Massachusetts Avenue NW, Washington, DC 20036
telephone: [1] (202) 467-9300
FAX: [1] (202) 328-7614
consulate(s) general: Chicago, Honolulu, Los Angeles, New York, Saipan (Northern Mariana Islands), San Francisco, Tamuning (Guam)

Diplomatic representation from the US: *chief of mission:* Ambassador Sung KIM (since 6 December 2016)
telephone: [63] (2) 301-2000
embassy: 1201 Roxas Boulevard, Manila 1000
mailing address: PSC 500, FPO AP 96515-1000
FAX: [63] (2) 301-2017

Flag description: two equal horizontal bands of blue (top) and red; a white equilateral triangle is based on the hoist side; the center of the triangle displays a yellow sun with eight primary rays; each corner of the triangle contains a small, yellow, five-pointed star; blue stands for peace and justice, red symbolizes courage, the white equal-sided triangle represents equality; the rays recall the first eight provinces that sought independence from Spain, while the stars represent the three major geographical divisions of the country: Luzon, Visayas, and Mindanao; the design of the flag dates to 1897
note: in wartime the flag is flown upside down with the red band at the top

National symbol(s): *three stars and sun, Philippine eagle;* national colors: red, white, blue, yellow

National anthem: *name:* "Lupang Hinirang" (Chosen Land)
lyrics/music: Jose PALMA (revised by Felipe PADILLA de Leon)/Julian FELIPE
note: music adopted 1898, original Spanish lyrics adopted 1899, Filipino (Tagalog) lyrics adopted 1956; although the original lyrics were written in Spanish, later English and Filipino versions were created; today, only the Filipino version is used
0:00 / 1:01

ECONOMY

Economy—overview: The economy has been relatively resilient to global economic shocks due to less exposure to troubled international securities, lower dependence on exports, relatively resilient domestic consumption, large remittances from about 10 million overseas Filipino workers and migrants, and a rapidly expanding services industry. During 2017, the current account balance fell into the negative range, the first time since the 2008 global financial crisis, in part due to an ambitious new infrastructure spending program announced this year. However, international reserves remain at comfortable levels and the banking system is stable.

Efforts to improve tax administration and expenditures management have helped ease the Philippines' debt burden and tight fiscal situation. The Philippines received investment-grade credit ratings on its sovereign debt under the former AQUINO administration and has had little difficulty financing its budget deficits. However, weak absorptive capacity and implementation bottlenecks have prevented the government from maximizing its expenditure plans. Although it has improved, the low tax-to-GDP ratio remains a constraint to supporting increasingly higher spending levels and sustaining high and inclusive growth over the longer term.

Economic growth has accelerated, averaging over 6% per year from 2011 to 2017, compared with 4.5% under the MACAPAGAL-ARROYO government; and competitiveness rankings have improved. Although 2017 saw a new record year for net foreign direct investment inflows, FDI to the Philippines has continued to lag regional peers, in part because the Philippine constitution and other laws limit foreign investment and restrict foreign ownership in important activities/sectors—such as land ownership and public utilities.

Although the economy grew at a rapid pace under the AQUINO government, challenges to achieving more inclusive growth remain. Wealth is concentrated in the hands of the rich. The unemployment rate declined from 7.3% to 5.7% between 2010 and 2017; while there has been some improvement, underemployment remains high at around 17% to 18% of the employed population. At least 40% of the employed work in the informal sector. Poverty afflicts more than a fifth of the total population but is as high as 75% in some areas of the southern Philippines. More than 60% of the poor reside in rural areas, where the incidence of poverty (about 30%) is more severe—a challenge to raising rural farm and non-farm incomes. Continued efforts are needed to improve governance, the judicial system, the regulatory environment, the infrastructure, and the overall ease of doing business.

2016 saw the election of President Rodrigo DUTERTE, who has pledged to make inclusive growth and poverty reduction his top priority. DUTERTE believes that illegal drug use, crime and corruption are key barriers to economic development. The administration wants to reduce the poverty rate to 17% and graduate the economy to upper-middle income status by the end of President DUTERTE's term in 2022. Key themes under the government's Ten-Point Socioeconomic Agenda include continuity of macroeconomic policy, tax reform, higher investments in infrastructure and human capital development, and improving competitiveness and the overall ease of doing business. The administration sees infrastructure shortcomings as a key barrier to sustained economic growth and has pledged to spend $165 billion on infrastructure by 2022. Although the final outcome has yet to be seen, the current administration is shepherding legislation for a comprehensive tax reform program to raise revenues for its ambitious infrastructure spending plan and to promote a more equitable and efficient tax system. However, the need to finance rehabilitation and reconstruction efforts in the southern region of Mindanao following the 2017 Marawi City siege may compete with other spending on infrastructure.

GDP (purchasing power parity):
$877.2 billion (2017 est.)
$822.2 billion (2016 est.)
$769.3 billion (2015 est.)
note: data are in 2017 dollars
country comparison to the world: 29

GDP (official exchange rate): $313.6 billion (2017 est.)

GDP—real growth rate:
6.04% (2019 est.)
6.34% (2018 est.)
6.94% (2017 est.)
country comparison to the world: 30

GDP—per capita (PPP):
$8,400 (2017 est.)
$8,000 (2016 est.)
$7,600 (2015 est.)
note: data are in 2017 dollars
country comparison to the world: 148

Gross national saving:
24.3% of GDP (2017 est.)
24% of GDP (2016 est.)
23.7% of GDP (2015 est.)
country comparison to the world: 67

GDP—composition, by end use:
household consumption: 73.5% (2017 est.)
government consumption: 11.3% (2017 est.)
investment in fixed capital: 25.1% (2017 est.)
investment in inventories: 0.1% (2017 est.)
exports of goods and services: 31% (2017 est.)
imports of goods and services: -40.9% (2017 est.)

GDP—composition, by sector of origin:
agriculture: 9.6% (2017 est.)
industry: 30.6% (2017 est.)
services: 59.8% (2017 est.)

Agriculture—products: rice, fish, livestock, poultry, bananas, coconut/copra, corn, sugarcane, mangoes, pineapple, cassava

Industries: semiconductors and electronics assembly, business process outsourcing, food and beverage manufacturing, construction, electric/gas/water supply, chemical products, radio/television/communications equipment and apparatus, petroleum and fuel, textile and garments, non-metallic minerals, basic metal industries, transport equipment

Industrial production growth rate: 7.2% (2017 est.)
country comparison to the world: 30

Labor force: 41.533 million (2020 est.)
country comparison to the world: 14

Labor force—by occupation: *agriculture:* 25.4%
industry: 18.3%
services: 56.3% (2017 est.)

Unemployment rate:
5.11% (2019 est.)
5.29% (2018 est.)
country comparison to the world: 81

Population below poverty line: 21.6% (2017 est.)

Household income or consumption by percentage share: *lowest 10%:* 3.2%
highest 10%: 29.5% (2015 est.)

Budget: *revenues:* 49.07 billion (2017 est.)
expenditures: 56.02 billion (2017 est.)

Taxes and other revenues: 15.6% (of GDP) (2017 est.)
country comparison to the world: 188

Budget surplus (+) or deficit (-): -2.2% (of GDP) (2017 est.)
country comparison to the world: 109

Public debt: 39.9% of GDP (2017 est.)
39% of GDP (2016 est.)
country comparison to the world: 128

Fiscal year: calendar year

Inflation rate (consumer prices): 2.9% (2017 est.)
1.3% (2016 est.)
country comparison to the world: 131

Current account balance:
-$3.386 billion (2019 est.)
-$8.877 billion (2018 est.)
country comparison to the world: 176

Exports: $48.2 billion (2017 est.)
$57.41 billion (2016 est.)
country comparison to the world: 52

Exports—partners: Japan 16.4%, US 14.6%, Hong Kong 13.7%, China 11%, Singapore 6.1%, Thailand 4.3%, Germany 4.1%, South Korea 4% (2017)

Exports—commodities: semiconductors and electronic products, machinery and transport equipment, wood manufactures, chemicals, processed food and beverages, garments, coconut oil, copper concentrates, seafood, bananas/fruits

Imports:
$89.39 billion (2017 est.)
$78.28 billion (2016 est.)
country comparison to the world: 39

Imports—commodities: electronic products, mineral fuels, machinery and transport equipment, iron and steel, textile fabrics, grains, chemicals, plastic

Imports—partners: China 18.1%, Japan 11.4%, South Korea 8.8%, US 7.4%, Thailand 7.1%, Indonesia 6.7%, Singapore 5.9% (2017)

Reserves of foreign exchange and gold:
$81.57 billion (31 December 2017 est.)
$80.69 billion (31 December 2016 est.)
country comparison to the world: 29

Debt—external:
$76.18 billion (31 December 2017 est.)
$74.76 billion (31 December 2016 est.)
country comparison to the world: 57

Exchange rates: Philippine pesos (PHP) per US dollar -
50.4 (2017 est.)
47.493 (2016 est.)
47.493 (2015 est.)
45.503 (2014 est.)
44.395 (2013 est.)

ENERGY

Electricity access: *population without electricity:* 4 million (2019)
electrification—total population: 96% (2019)
electrification—urban areas: 100% (2019)
electrification—rural areas: 93% (2019)

Electricity—production: 86.59 billion kWh (2016 est.)
country comparison to the world: 36

Electricity—consumption: 78.3 billion kWh (2016 est.)
country comparison to the world: 37

Electricity—exports: 0 kWh (2017 est.)
country comparison to the world: 184

Electricity—imports: 0 kWh (2016 est.)
country comparison to the world: 186

Electricity—installed generating capacity: 22.13 million kW (2016 est.)
country comparison to the world: 39

Electricity—from fossil fuels: 67% of total installed capacity (2016 est.)
country comparison to the world: 116

Electricity—from nuclear fuels: 0% of total installed capacity (2017 est.)
country comparison to the world: 165

Electricity—from hydroelectric plants: 17% of total installed capacity (2017 est.)
country comparison to the world: 97

Electricity—from other renewable sources: 16% of total installed capacity (2017 est.)
country comparison to the world: 54

Crude oil—production: 13,000 bbl/day (2018 est.)
country comparison to the world: 76

Crude oil—exports: 16,450 bbl/day (2015 est.)
country comparison to the world: 52

Crude oil—imports: 211,400 bbl/day (2015 est.)
country comparison to the world: 30

Crude oil—proved reserves: 138.5 million bbl (1 January 2018 est.)
country comparison to the world: 64

Refined petroleum products—production: 215,500 bbl/day (2015 est.)
country comparison to the world: 50

Refined petroleum products—consumption: 424,000 bbl/day (2016 est.)
country comparison to the world: 36

Refined petroleum products—exports: 26,710 bbl/day (2015 est.)
country comparison to the world: 65

Refined petroleum products—imports: 211,400 bbl/day (2015 est.)
country comparison to the world: 33

Natural gas—production: 3.058 billion cu m (2017 est.)
country comparison to the world: 58

Natural gas—consumption: 3.143 billion cu m (2017 est.)
country comparison to the world: 72

Natural gas—exports: 0 cu m (2017 est.)
country comparison to the world: 168

Natural gas—imports: 0 cu m (2017 est.)
country comparison to the world: 175

Natural gas—proved reserves: 98.54 billion cu m (1 January 2018 est.)
country comparison to the world: 51

Carbon dioxide emissions from consumption of energy: 117.2 million Mt (2017 est.)
country comparison to the world: 38

COMMUNICATIONS

Telephones—fixed lines: *total subscriptions:* 4,140,108
subscriptions per 100 inhabitants: 3.85 (2019 est.)
country comparison to the world: 33

Telephones—mobile cellular: *total subscriptions:* 166,421,595
subscriptions per 100 inhabitants: 154.76 (2019 est.)
country comparison to the world: 10

Telecommunication systems: *general assessment:* good international radiotelephone and submarine cable services; domestic and interisland service adequate; National Broadband Plan to improve connectivity in rural areas underway; dominance of mobile platform and mobile broadband over fixed broadband penetration; 4G available now in most areas with 5G roll outs soon; smart city pilot has begun; with more mobile services there is demand for data center services and iCloud; launch of the Kacific-1 satellite in 2019 will improve telecommunication for the region (2020)

domestic: telecommunications infrastructure includes the following platforms: fixed line, mobile cellular, cable TV, over- the-air TV, radio and (very small aperture terminal) VSAT, fiber-optic cable, and satellite for redundant international connectivity; fixed-line 4 per 100 and mobile-cellular 155 per 100 (2019)

international: country code—63; landing points for the NDTN, TGN-IA, AAG, PLCN, EAC-02C, DFON, SJC, APCN-2, SeaMeWe, Boracay-Palawan Submarine Cable System, Palawa-Illoilo Cable System, NDTN, SEA-US, SSSFOIP, ASE and JUPITAR submarine cables that together provide connectivity to the US, Southeast Asia, Asia, Europe, Africa, the Middle East, and Australia (2019)

note: the COVID-19 outbreak is negatively impacting telecommunications production and supply chains globally; consumer spending on telecom devices and services has also slowed due to the pandemic's effect on economies worldwide; overall progress towards improvements in all facets of the telecom industry—mobile, fixed-line, broadband, submarine cable and satellite—has moderated

Broadcast media: multiple national private TV and radio networks; multi-channel satellite and cable TV systems available; more than 400 TV stations; about 1,500 cable TV providers with more than 2 million subscribers, and some 1,400 radio stations; the Philippines adopted Japan's Integrated Service Digital Broadcast – Terrestrial standard for digital terrestrial television in November 2013 and is scheduled to complete the switch from analog to digital broadcasting by the end of 2023 (2019)

Internet country code: .ph

Internet users: *total:* 63,588,975
percent of population: 60.05% (July 2018 est.)
country comparison to the world: 12

Broadband—fixed subscriptions: *total:* 3,919,713
subscriptions per 100 inhabitants: 4 (2018 est.)
country comparison to the world: 37

TRANSPORTATION

National air transport system: *number of registered air carriers:* 13 (2020)
inventory of registered aircraft operated by air carriers: 200
annual passenger traffic on registered air carriers: 43,080,118 (2018)
annual freight traffic on registered air carriers: 835.9 million mt-km (2018)

Civil aircraft registration country code prefix: RP (2016)

Airports: 247 (2013)
country comparison to the world: 24

Airports—with paved runways: *total:* 89 (2019)
over 3,047 m: 4
2,438 to 3,047 m: 8
1,524 to 2,437 m: 33
914 to 1,523 m: 34
under 914 m: 10

Airports—with unpaved runways: *total:* 158 (2013)
1,524 to 2,437 m: 3 (2013)
914 to 1,523 m: 56 (2013)
under 914 m: 99 (2013)

Heliports: 2 (2013)

Pipelines: 530 km gas, 138 km oil (non-operational), 185 km refined products (2017)

Railways: *total:* 77 km (2017)
standard gauge: 49 km 1.435-m gauge (2017)
narrow gauge: 28 km 1.067-m gauge (2017)
country comparison to the world: 129

Roadways: *total:* 216,387 km (2014)
paved: 61,093 km (2014)
unpaved: 155,294 km (2014)
country comparison to the world: 25

Waterways: 3,219 km (limited to vessels with draft less than 1.5 m) (2011)
country comparison to the world: 30

Merchant marine: *total:* 1,706
by type: bulk carrier 54, container ship 46, general cargo 685, oil tanker 197, other 724 (2019)
country comparison to the world: 17

Ports and terminals: *major seaport(s):* Batangas, Cagayan de Oro, Cebu, Davao, Liman, Manila
container port(s) (TEUs): Manila (4,782,240) (2017)

MILITARY AND SECURITY

Military and security forces: *Armed Forces of the Philippines (AFP):* Army, Navy (includes Marine Corps), Air Force (2020)
note: the Philippine Coast Guard is an armed and uniformed service under the Department of Transportation; it would be attached to the AFP in wartime; the Philippine National Police Force (PNP) falls under the Ministry of Interior and Local Government

Military expenditures:
1% of GDP (2019)
0.9% of GDP (2018)
1.3% of GDP (2017)
1.1% of GDP (2016)
1.1% of GDP (2015)
country comparison to the world: 118

Military and security service personnel strengths: the Armed Forces of the Philippines (AFP) have approximately 130,000 active duty personnel (90,000 Army; 24,000 Navy; 16,000 Air Force); note—the Navy includes about 8,500 marines) (2019 est.)

Military equipment inventories and acquisitions: the AFP is equipped with a mix of imported weapons systems, particularly second-hand equipment from the US; since 2014, its top weapons suppliers are Brazil, Indonesia, South Korea, and the US (2019)

Military service age and obligation: 18-23 years of age (officers 21-29) for voluntary military service; no conscription (2019)

Maritime threats: the International Maritime Bureau reports the territorial and offshore waters in the South China Sea as high risk for piracy and armed robbery against ships; during 2018, 10 attacks were reported in and around the Philippines including six ships that were boarded, one fired upon, and three crewman kidnapped for ransom; an emerging threat area lies in the Celebes and Sulu Seas between the Philippines and Malaysia where it is believed the pirates involved are associated with the Abu Sayyaf Group (ASG) terrorist organization; numerous commercial vessels have been attacked and hijacked both at anchor and while underway; hijacked vessels are often disguised and cargo diverted to ports in East Asia; crews have been murdered or cast adrift; the Maritime Administration (MARAD) of the US Department of Transportation has issued a Maritime Advisory (2019-011-Sulu and Celebes Seas-Piracy/Armed Robbery/Terrorism) which states in part "In 2018, there were at least 12 reported boardings, attempted boardings, attacks, hijackings, and kidnappings in the Sulu and Celebes Seas. Recent kidnapping incidents in this area were reportedly linked to the Abu

Sayyaf Group (ASG), a violent Islamic separatist group operating in the southern Philippines..." and advises ships to adhere to counter-piracy practices to minimize risk

Military—note: the AFP's primary operational focus is on internal security duties, particularly in the south, where several insurgent and terrorist groups operate and an estimated 60% of the armed forces were deployed as of 2019; the Philippines National Police (PNP) also has an active role in counterinsurgency and counter-terrorism operations alongside the AFP (2019)

TERRORISM

Terrorist group(s): Abu Sayyaf Group; Communist Party of the Philippines/New People's Army; Islamic State of Iraq and ash-Sham – East Asia (ISIS-EA) in the Philippines (2020)
note: details about the history, aims, leadership, organization, areas of operation, tactics, targets, weapons, size, and sources of support of the group(s) appear(s) in Appendix-T

TRANSNATIONAL ISSUES

Disputes—international: Philippines claims sovereignty over Scarborough Reef (also claimed by China together with Taiwan) and over certain of the Spratly Islands, known locally as the Kalayaan (Freedom) Islands, also claimed by China, Malaysia, Taiwan, and Vietnam; the 2002 "Declaration on the Conduct of Parties in the South China Sea," has eased tensions in the Spratly Islands but falls short of a legally binding "code of conduct" desired by several of the disputants; in March 2005, the national oil companies of China, the Philippines, and Vietnam signed a joint accord to conduct marine seismic activities in the Spratly Islands; Philippines retains a dormant claim to Malaysia's Sabah State in northern Borneo based on the Sultanate of Sulu's granting the Philippines Government power of attorney to pursue a sovereignty claim on his behalf; maritime delimitation negotiations continue with Palau

Refugees and internally displaced persons: *IDPs:* 182,000 (government troops fighting the Moro Islamic Liberation Front, the Abu Sayyaf Group, and the New People's Army; clan feuds; armed attacks, political violence, and communal tensions in Mindanao) (2019)
stateless persons: 383 (2019); note—stateless persons are descendants of Indonesian migrants

Illicit drugs: domestic methamphetamine production has been a growing problem in recent years despite government crackdowns; major consumer of amphetamines; longstanding marijuana producer mainly in rural areas where Manila's control is limited

PITCAIRN ISLANDS

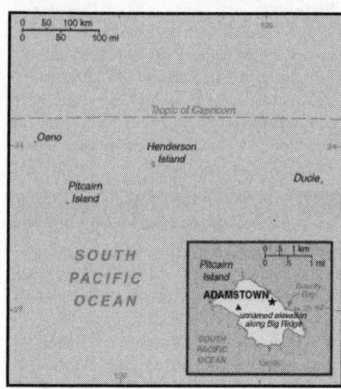

INTRODUCTIONS

Background: Pitcairn Island was discovered in 1767 by the British and settled in 1790 by the Bounty mutineers and their Tahitian companions. Pitcairn was the first Pacific island to become a British colony (in 1838) and today remains the last vestige of that empire in the South Pacific. Outmigration, primarily to New Zealand, has thinned the population from a peak of 233 in 1937 to less than 50 today.

GEOGRAPHYS

Location: Oceania, islands in the South Pacific Ocean, about midway between Peru and New Zealand

Geographic coordinates: 25 04 S, 130 06 W

Map references: Oceania

Area: *total:* 47 sq km
land: 47 sq km
water: 0 sq km
country comparison to the world: 234

Area—comparative: about three-tenths the size of Washington, DC

Land boundaries: 0 km

Coastline: 51 km

Maritime claims: *territorial sea:* 12 nm
exclusive economic zone: 200 nm

Climate: tropical; hot and humid; modified by southeast trade winds; rainy season (November to March)

Terrain: rugged volcanic formation; rocky coastline with cliffs

Elevation: *lowest point:* Pacific Ocean 0 m
highest point: Palwala Valley Point on Big Ridge 347 m

Natural resources: miro trees (used for handicrafts), fish, note, manganese, iron, copper, gold, silver, and zinc have been discovered offshore

Land use: *agricultural land:* 0% (2011 est.)
arable land: 0% (2011 est.) / *permanent crops:* 0% (2011 est.) / permanent pasture: 0% (2011 est.)
forest: 74.5% (2011 est.)
other: 25.5% (2011 est.)

Irrigated land: 0 sq km (2012)

Population distribution: less than 50 inhabitants on Pitcairn Island, most reside near the village of Adamstown

Natural hazards: occasional tropical cyclones (especially November to March), but generally only heavy tropical storms; landslides

Environment—current issues: deforestation (only a small portion of the original forest remains because of burning and clearing for settlement)

Geography—note: Britain's most isolated dependency; only the larger island of Pitcairn is inhabited but it has no port or natural harbor; supplies must be transported by rowed longboat from larger ships stationed offshore

PEOPLE AND SOCIETYS

Population: 50 (2020 est. est.)
country comparison to the world: 238

Nationality: *noun:* Pitcairn Islander(s)
adjective: Pitcairn Islander

Ethnic groups: other descendants of the Bounty mutineers and their Tahitian wives

Languages: English (official), Pitkern (mixture of an 18th century English dialect and a Tahitian dialect)

Religions: Seventh-Day Adventist 100%

Population growth rate: 0% (2014 est.)
country comparison to the world: 195

Population distribution: less than 50 inhabitants on Pitcairn Island, most reside near the village of Adamstown

Urbanization: *urban population:* 0% of total population (2012)
rate of urbanization: NA

Sex ratio: NA

Infant mortality rate: *total:* NA (2018)
male: NA
female: NA

Life expectancy at birth: *total population:* NA (2017 est.)
male: NA
female: NA

Total fertility rate: NA

HIV/AIDS—adult prevalence rate: NA

HIV/AIDS—people living with HIV/AIDS: NA

HIV/AIDS—deaths: NA

Major infectious diseases: *degree of risk:* high (2020)
food or waterborne diseases: bacterial diarrhea
vectorborne diseases: malaria

GOVERNMENTS

Country name: *conventional long form:* Pitcairn, Henderson, Ducie, and Oeno Islands
conventional short form: Pitcairn Islands
etymology: named after Midshipman Robert PITCAIRN who first sighted the island in 1767

Dependency status: overseas territory of the UK

Government type: parliamentary democracy

Capital: *name:* Adamstown
geographic coordinates: 25 04 S, 130 05 W
time difference: UTC-9 (4 hours behind Washington, DC, during Standard Time)
etymology: named after John Adams (1767–1829), the last survivor of the Bounty mutineers who settled on Pitcairn Island in January 1790

Administrative divisions: none (overseas territory of the UK)

Independence: none (overseas territory of the UK)

National holiday: Birthday of Queen ELIZABETH II, second Saturday in June (1926); Discovery Day (Pitcairn Day), 2 July (1767)

Constitution: *history:* several previous; latest drafted 10 February 2010, presented 17 February 2010, effective 4 March 2010

Legal system: local island by-laws

Citizenship: see United Kingdom

Suffrage: 18 years of age; universal with three years residency

Executive branch: *chief of state:* Queen ELIZABETH II (since 6 February 1952); represented by UK High Commissioner to New Zealand and Governor (nonresident) of the Pitcairn Islands Laura CLARK (since 25 January 2018)
head of government: Mayor and Chairman of the Island Council Charlene WARREN-PEU (since 1 January 2020)
cabinet: none
elections/appointments: the monarchy is hereditary; governor or commissioner appointed by the monarch; island mayor directly elected by majority popular vote for a 3-year term; election last held on 6 November 2019 (next to be held not later than December 2022)
election results: Charlene WARREN-PEU elected mayor and chairman of the Island Council; Island Council vote—NA

Legislative branch: *description:* unicameral Island Council (10 seats; 4 members directly elected by proportional representation vote, 1 nominated by the elected Council members, 2 appointed by the governor, and 3 ex-officio members—the governor, deputy governor, and commissioner; elected members serve 1-year terms)
elections: last held in November 2017 (next to be held not later than December 2019)
election results: percent of vote—NA; seats— 5 independent; composition—men 5, women 5, percent of women 50%

Judicial branch: *highest courts:* Pitcairn Court of Appeal (consists of the court president, 2 judges, and the Supreme Court chief justice, an ex-officio member); Pitcairn Supreme Court (consists of the chief justice and 2 judges); note—appeals beyond the Pitcairn Court of Appeal are referred to the Judicial Committee of the Privy Council (in London)

judge selection and term of office: all judges of both courts appointed by the governor of the Pitcairn Islands on the instructions of the Queen of England through the Secretary of State; all judges can serve until retirement, normally at age 75

subordinate courts: Magistrate's Court

Political parties and leaders: none

International organization participation: SPC, UPU

Diplomatic representation in the US: none (overseas territory of the UK)

Diplomatic representation from the US: none (overseas territory of the UK)

Flag description: blue with the flag of the UK in the upper hoist-side quadrant and the Pitcairn Islander coat of arms centered on the outer half of the flag; the green, yellow, and blue of the shield represents the island rising from the ocean; the green field features a yellow anchor surmounted by a bible (both the anchor and the bible were items found on the HMS Bounty); sitting on the crest is a Pitcairn Island wheelbarrow from which springs a flowering twig of miro (a local plant)

National anthem: *name:* We From Pitcairn Island *lyrics/music:* unknown/Frederick M. LEHMAN *note:* serves as a local anthem; as a territory of the UK, "God Save the Queen" is official (see United Kingdom)

Economy—overview: The inhabitants of this tiny isolated economy exist on fishing, subsistence farming, handicrafts, and postage stamps. The fertile soil of the valleys produces a wide variety of fruits and vegetables, including citrus, sugarcane, watermelons, bananas, yams, and beans. Bartering is an important part of the economy. The major sources of revenue are the sale of postage stamps to collectors and the sale of handicrafts to passing ships.

GDP (purchasing power parity): NA

Agriculture—products: honey; wide variety of fruits and vegetables; goats, chickens; fish

Industries: postage stamps, handicrafts, beekeeping, honey

Labor force: 15 (2004) *country comparison to the world:* 232

Labor force—by occupation: *note:* no business community in the usual sense; some public works; subsistence farming and fishing

Budget: *revenues:* 746,000 (FY04/05) *expenditures:* 1.028 million (FY04/05)

Fiscal year: 1 April—31 March

Exports: NA

Exports—commodities: honey, fruits, vegetables, curios, postage stamps

Imports: NA

Imports—commodities: fuel oil, machinery, building materials, flour, sugar, other foodstuffs

Exchange rates: New Zealand dollars (NZD) per US dollar -
1.416 (2017 est.)
1.4279 (2016 est.)
1.4279 (2015)
1.4279 (2014 est.)
1.2039 (2013 est.)

Telecommunication systems: *general assessment:* satellite-based phone services; rural connectivity a challenge; 2G services widespread; demand for mobile broadband due to mobile services providing Internet source; the launch of the Kacific-1 satellite in 2019 will improve telecommunications in the region (2020)

domestic: local phone service with international connections via Internet (2018)

international: country code—872; satellite earth station—1 Inmarsat

note: the COVID-19 outbreak is negatively impacting telecommunications production and supply chains globally; consumer spending on telecom devices and services has also slowed due to the pandemic's effect on economies worldwide; overall progress towards improvements in all facets of the telecom industry—mobile, fixed-line, broadband, submarine cable and satellite—has moderated

Broadcast media: satellite TV from Fiji-based Sky Pacific offering a wide range of international channels

Internet country code: .pn

Internet users: *total:* 54 *percent of population:* 100% (July 2016 est.) *country comparison to the world:* 229

Communications—note: satellite-based local phone service and broadband Internet connections available in all homes

Roadways: *total:* 0 km *country comparison to the world:* 224

Ports and terminals: *major seaport(s):* Adamstown (on Bounty Bay)

Military—note: defense is the responsibility of the UK

Disputes—international: none

POLAND

Background: Poland's history as a state began near the middle of the 10th century. By the mid-16th century, the Polish-Lithuanian Commonwealth ruled a vast tract of land in Central and Eastern Europe. During the 18th century, internal disorders weakened the nation, and in a series of agreements between 1772 and 1795, Russia, Prussia, and Austria partitioned Poland among themselves. Poland regained its independence in 1918 only to be overrun by Germany and the Soviet Union in World War II. It became a Soviet satellite state following the war. Labor turmoil in 1980 led to the formation of the independent trade union "Solidarity" that over time became a political force with over 10 million members. Free elections in 1989 and 1990 won Solidarity control of the parliament and the presidency, bringing the communist era to a close. A "shock therapy" program during the early 1990s enabled the country to transform its economy into one of the most robust in Central Europe. Poland joined NATO in 1999 and the EU in 2004. With its transformation to a democratic, marketoriented country largely completed and with large investments in defense, energy, and other infrastructure, Poland is an increasingly active member of Euro-Atlantic organizations.

Location: Central Europe, east of Germany

Geographic coordinates: 52 00 N, 20 00 E

Map references: Europe

Area: *total:* 312,685 sq km
land: 304,255 sq km
water: 8,430 sq km
country comparison to the world: 71

Area—comparative: about twice the size of Georgia; slightly smaller than New Mexico
Land boundaries:
total: 3,071 km
border countries (7): Belarus 418 km, Czech Republic 796 km, Germany 467 km, Lithuania 104 km, Russia (Kaliningrad Oblast) 210 km, Slovakia 541 km, Ukraine 535 km

Coastline: 440 km

Maritime claims: *territorial sea:* 12 nm
exclusive economic zone: defined by international treaties

Climate: temperate with cold, cloudy, moderately severe winters with frequent precipitation; mild summers with frequent showers and thundershowers

Terrain: mostly flat plain; mountains along southern border

Elevation: *mean elevation:* 173 m
lowest point: near Raczki Elblaskie -2 m
highest point: Rysy 2,499 m

Natural resources: coal, sulfur, copper, natural gas, silver, lead, salt, amber, arable land

Land use: *agricultural land:* 48.2% (2011 est.)
arable land: 36.2% (2011 est.) / *permanent crops:* 1.3% (2011 est.) / *permanent pasture:* 10.7% (2011 est.)
forest: 30.6% (2011 est.)
other: 21.2% (2011 est.)

Irrigated land: 970 sq km (2012)

Population distribution: population concentrated in the southern area around Krakow and the central area around Warsaw and Lodz, with an extension to the northern coastal city of Gdansk

Natural hazards: flooding

Environment—current issues: decreased emphasis on heavy industry and increased environmental concern by post-communist governments has improved environment; air pollution remains serious because of emissions from burning low-quality coals in homes and from coal-fired power plants; the resulting acid rain causes forest damage; water pollution from industrial and municipal sources is a problem, as is disposal of hazardous wastes

Environment—international agreements: *party to:* Air Pollution, Antarctic-Environmental Protocol, Antarctic-Marine Living Resources, Antarctic Seals, Antarctic Treaty, Biodiversity, Climate Change, Climate Change-Kyoto Protocol, Desertification, Endangered Species, Environmental Modification, Hazardous Wastes, Law of the Sea, Marine Dumping, Ozone Layer Protection, Ship Pollution, Wetlands signed, but not ratified: Air Pollution-Nitrogen Oxides, Air Pollution-Persistent Organic Pollutants, Air Pollution-Sulfur 94

Geography—note: historically, an area of conflict because of flat terrain and the lack of natural barriers on the North European Plain

PEOPLE AND SOCIETY

Population: 38,282,325 (July 2020 est.)
country comparison to the world: 37

Nationality: *noun:* Pole(s)
adjective: Polish

Ethnic groups: Polish 96.9%, Silesian 1.1%, German 0.2%, Ukrainian 0.1%, other and unspecified 1.7% (2011 est.)
note: represents ethnicity declared first

Languages: Polish (official) 98.2%, Silesian 1.4%, other 1.1%, unspecified 1.3% (2011 est.)
note: data represents the language spoken at home; shares sum to more than 100% because some respondents gave more than one answer on the census; Poland ratified the European Charter for Regional or Minority Languages in 2009 recognizing Kashub as a regional language, Czech, Hebrew, Yiddish, Belarusian, Lithuanian, German, Armenian, Russian, Slovak, and Ukrainian as national minority languages, and Karaim, Lemko, Romani (Polska Roma and Bergitka Roma), and Tatar as ethnic minority languages

Religions: Catholic 85.9% (includes Roman Catholic 85.6% and Greek Catholic, Armenian Catholic, and Byzantine-Slavic Catholic .3%), Orthodox 1.3% (almost all are Polish Autocephalous Orthodox), Protestant 0.4% (mainly Augsburg Evangelical and Pentacostal), other 0.4% (includes Jehovah's Witness, Buddhist, Hare Krishna, Gaudiya Vaishnavism, Muslim, Jewish, Mormon), unspecified 12.1% (2017 est.)

Age structure: *0-14 years:* 14.83% (male 2,918,518/female 2,756,968)
15-24 years: 9.8% (male 1,928,637/female 1,823,691)
25-54 years: 43.33% (male 8,384,017/female 8,203,646)
55-64 years: 13.32% (male 2,424,638/female 2,675,351)
65 years and over: 18.72% (male 2,867,315/female 4,299,341) (2020 est.)

Dependency ratios: *total dependency ratio:* 51.4
youth dependency ratio: 23
elderly dependency ratio: 28.4
potential support ratio: 3.5 (2020 est.)
Median age:
total: 41.9 years
male: 40.3 years
female: 43.6 years (2020 est.)
country comparison to the world: 39

Population growth rate: -0.19% (2020 est.)
country comparison to the world: 210

Birth rate: 8.9 births/1,000 population (2020 est.)
country comparison to the world: 206

Death rate: 10.6 deaths/1,000 population (2020 est.)
country comparison to the world: 28

Net migration rate: -0.3 migrant(s)/1,000 population (2020 est.)
country comparison to the world: 120

Population distribution: population concentrated in the southern area around Krakow and

the central area around Warsaw and Lodz, with an extension to the northern coastal city of Gdansk

Urbanization: *urban population:* 60% of total population (2020)
rate of urbanization: -0.25% annual rate of change (2015-20 est.)
total population growth rate v. urban population growth rate, 2000-2030:

Major urban areas—population: 1.783 million WARSAW (capital), 769,000 Krakow (2020)

Sex ratio: *at birth:* 1.06 male(s)/female
0-14 years: 1.06 male(s)/female
15-24 years: 1.06 male(s)/female
25-54 years: 1.02 male(s)/female
55-64 years: 0.91 male(s)/female
65 years and over: 0.67 male(s)/female
total population: 0.94 male(s)/female (2020 est.)

Mother's mean age at first birth: 27.8 years (2017 est.)

Maternal mortality rate: 2 deaths/100,000 live births (2017 est.)
country comparison to the world: 184

Infant mortality rate: *total:* 4.3 deaths/1,000 live births
male: 4.6 deaths/1,000 live births
female: 3.9 deaths/1,000 live births (2020 est.)
country comparison to the world: 185

Life expectancy at birth: *total population:* 78.3 years
male: 74.5 years
female: 82.3 years (2020 est.)
country comparison to the world: 69

Total fertility rate: 1.38 children born/woman (2020 est.)
country comparison to the world: 218

Contraceptive prevalence rate: 62.3% (2014)
Drinking water source:
improved:
urban: 100% of population
rural: 100% of population
total: 100% of population
unimproved:
urban: 0% of population
rural: 0% of population
total: 0% of population (2017 est.)

Current Health Expenditure: 6.5% (2017)

Physicians density: 2.38 physicians/1,000 population (2017)

Hospital bed density: 6.6 beds/1,000 population (2017)
Sanitation facility access:
improved:
urban: 99.7% of population
rural: 100% of population
total: 99.8% of population
unimproved:
urban: 0.3% of population
rural: 0% of population
total: 0.2% of population (2017 est.)

HIV/AIDS—adult prevalence rate: NA

HIV/AIDS—people living with HIV/AIDS: NA

HIV/AIDS—deaths: NA

Major infectious diseases: *degree of risk:* intermediate (2016)
vectorborne diseases: tickborne encephalitis (2016)

Obesity—adult prevalence rate: 23.1% (2016)
country comparison to the world: 69

Education expenditures: 4.6% of GDP (2016)
country comparison to the world: 84

Literacy: *definition:* age 15 and over can read and write
total population: 99.8%
male: 99.9%
female: 99.7% (2015)

School life expectancy (primary to tertiary education): *total:* 16 years
male: 15 years
female: 17 years (2018)

Unemployment, youth ages 15-24: *total:* 11.7%
male: 11.5%
female: 12.1% (2018 est.)
country comparison to the world: 112

GOVERNMENT

Country name: *conventional long form:* Republic of Poland
conventional short form: Poland
local long form: Rzeczpospolita Polska
local short form: Polska
former: Polish People's Republic
etymology: name derives from the Polanians, a west Slavic tribe that united several surrounding Slavic groups (9th-10th centuries A.D.) and who passed on their name to the country; the name of the tribe likely comes from the Slavic "pole" (field or plain), indicating the flat nature of their country

Government type: parliamentary republic

Capital: *name:* Warsaw
geographic coordinates: 52 15 N, 21 00 E
time difference: UTC+1 (6 hours ahead of Washington, DC, during Standard Time)
daylight saving time: +1hr, begins last Sunday in March; ends last Sunday in October
etymology: the origin of the name is unknown; the Polish designation "Warszawa" was the name of a fishing village and several legends/traditions link the city's founding to a man named Wars or Warsz

Administrative divisions: 16 voivodships [provinces] (wojewodztwa, singular—wojewodztwo); Dolnoslaskie (Lower Silesia), Kujawsko-Pomorskie (Kuyavia-Pomerania), Lodzkie (Lodz), Lubelskie (Lublin), Lubuskie (Lubusz), Malopolskie (Lesser Poland), Mazowieckie (Masovia), Opolskie (Opole), Podkarpackie (Subcarpathia), Podlaskie, Pomorskie (Pomerania), Slaskie (Silesia), Swietokrzyskie (Holy Cross), Warminsko-Mazurskie (Warmia-Masuria), Wielkopolskie (Greater Poland), Zachodniopomorskie (West Pomerania)

Independence: 11 November 1918 (republic proclaimed); *notable earlier dates:* 14 April 966 (adoption of Christianity, traditional founding date), 1 July 1569 (Polish-Lithuanian Commonwealth created)

National holiday: Constitution Day, 3 May (1791)

Constitution: *history:* several previous; latest adopted 2 April 1997, approved by referendum 25 May 1997, effective 17 October 1997
amendments: proposed by at least one fifth of Sejm deputies, by the Senate, or by the president of the republic; passage requires at least two-thirds majority vote in the Sejm and absolute majority vote in the Senate; amendments to articles relating to sovereignty, personal freedoms, and constitutional amendment procedures also require passage by majority vote in a referendum; amended 2006, 2009

Legal system: civil law system; judicial review of legislative, administrative, and other governmental acts; constitutional law rulings of the Constitutional Tribunal are final

International law organization participation: accepts compulsory ICJ jurisdiction with reservations; accepts ICCt jurisdiction

Citizenship: *citizenship by birth:* no
citizenship by descent only: both parents must be citizens of Poland
dual citizenship recognized: no
residency requirement for naturalization: 5 years

Suffrage: 18 years of age; universal

Executive branch: *chief of state:* President Andrzej DUDA (since 6 August 2015)
head of government: Prime Minister Mateusz MORAWIECKI (since 11 December 2017); Deputy Prime Ministers Piotr GLINSKI and Jaroslaw GOWIN (since 16 November 2015), Jacek SASIN (since 4 June 2019)
cabinet: Council of Ministers proposed by the prime minister, appointed by the president, and approved by the Sejm
elections/appointments: president directly elected by absolute majority popular vote in 2 rounds if needed for a 5-year term (eligible for a second term); election last held on 28 June 2020 with a second round on 12 July 2020 (next to be held in 2025); prime minister, deputy prime ministers, and Council of Ministers appointed by the president and confirmed by the Sejm
election results: Andrzej DUDA reelected president in runoff; percent of vote—Andrzej DUDA (independent) 51%, Rafal TRZASKOWSKI (KO) 49%

Legislative branch: *description:* bicameral legislature consists of:
Senate or Senat (100 seats; members directly elected in single-seat constituencies by simple majority vote to serve 4- year terms)
Sejm (460 seats; members elected in multi-seat constituencies by party-list proportional representation vote with 5% threshold of total votes needed for parties and 8% for coalitions to gain seats; minorities exempt from threshold; members serve 4-year terms)
elections: Senate—last held on 13 October 2019 (next to be held in October 2023)
Sejm—last held on 13 October 2019 (next to be held in October 2023)

election results: Senate—percent of vote by party—NA; seats by party—PiS 48, KO 43, PSL 3, SLD 2, independent 4; composition—men 87, women 13, percent of women 13%
Sejm—percent of vote by party—PiS 43.6%, KO 27.4%, SLD 12.6%, PSL 8.5% Confederation 6.8%, other 1.1%; seats by party—PiS 235, KO 134, SLD 49, PSL 30, KWiN 11, MN 1; men 334, women 126, percent of women 27.4%; note—total legislature percent of women 24.8%
note: the designation National Assembly or Zgromadzenie Narodowe is only used on those rare occasions when the 2 houses meet jointly

Judicial branch: *highest courts:* Supreme Court or Sad Najwyzszy (consists of the first president of the Supreme Court and 120 justices organized in criminal, civil, labor and social insurance, and extraordinary appeals and public affairs and disciplinary chambers); Constitutional Tribunal (consists of 15 judges, including the court president and vice president)
judge selection and term of office: president of the Supreme Court nominated by the General Assembly of the Supreme Court and selected by the president of Poland; other judges nominated by the 25-member National Judicial Council and appointed by the president of Poland; judges serve until retirement, usually at age 65, but tenure can be extended; Constitutional Tribunal judges chosen by the Sejm for 9-year terms
subordinate courts: administrative courts; military courts; local, regional and appellate courts subdivided into military, civil, criminal, labor, and family courts

Political parties and leaders:
Civic Coalition or KO [Grzegorz SCHETYNA]
Confederation Liberty and Independence or KWiN [Janusz KORWIN-MIKKE, Robert WINNICKI, Grzegorz BRAUN] Democratic Left Alliance or SLD [Wlodzimierz CZARZASTY]
German Minority or MN [Ryszard GALLA]
Kukiz 15 or K15 [Pawel KUKIZ]
Law and Justice or PiS [Jaroslaw KACZYNSKI]
TERAZ! (NOW!) [Ryszard PETRU]
Nowoczesna (Modern) or N [Katarzyna LUBNAUER]
Polish People's Party or PSL [Wladyslaw KOSINIAK-KAMYSZ]
Razem (Together) [collective leadership]
Wiosna (Spring) [Robert BIEDRON]

International organization participation: Arctic Council (observer), Australia Group, BIS, BSEC (observer), CBSS, CD, CE, CEI, CERN, EAPC, EBRD, ECB, EIB, ESA, EU, FAO, IAEA, IBRD, ICAO, ICC (national committees), ICCt, ICRM, IDA, IEA, IFC, IFRCS, IHO, ILO, IMF, IMO, IMSO, Interpol, IOC, IOM, IPU, ISO, ITSO, ITU, ITUC (NGOs), MIGA, MONUSCO, NATO, NEA, NSG, OAS (observer), OECD, OIF (observer), OPCW, OSCE, PCA, Schengen Convention, UN, UNCTAD, UNESCO, UNHCR, UNIDO, UNMIL, UNMISS, UNOCI, UN Security Council (temporary), UNWTO, UPU, WCO, WFTU (NGOs), WHO, WIPO, WMO, WTO, ZC

Diplomatic representation in the US: *chief of mission:* Ambassador Piotr Antoni WILCZEK (since 18 January 2017)
chancery: 2640 16th Street NW, Washington, DC 20009
telephone: [1] (202) 499-1700
FAX: [1] (202) 328-6271
consulate(s) general: Chicago, Los Angeles, New York

Diplomatic representation from the US: *chief of mission:* Ambassador Georgette MOSBACHER (since 6 September 2018)
telephone: [48] (22) 504-2000
embassy: Aleje Ujazdowskie 29/31 00-540 Warsaw
mailing address: American Embassy Warsaw, US Department of State, Washington, DC 20521-5010 (pouch)
FAX: [48] (22) 504-2226
consulate(s) general: Krakow

Flag description: two equal horizontal bands of white (top) and red; colors derive from the Polish emblem—a white eagle on a red field
note: similar to the flags of Indonesia and Monaco which are red (top) and white

National symbol(s): *white crowned eagle; national colors:* white, red

National anthem: *name:* "Mazurek Dabrowskiego" (Dabrowski's Mazurka)
lyrics/music: Jozef WYBICKI/traditional
note: adopted 1927; the anthem, commonly known as written in 1 797; the lyrics resonate strongly with Poles lands have been occupied "Jeszcze Polska nie zginela" (Poland Has Not Yet Perished), was because they reflect the numerous occupied in which the nation's
0:00 / 0:42

ECONOMY

Economy—overview: Poland has the sixth-largest economy in the EU and has long had a reputation as a business-friendly country with largely sound macroeconomic policies. Since 1990, Poland has pursued a policy of economic liberalization. During the 2008-09 economic slowdown Poland was the only EU country to avoid a recession, in part because of the government's loose fiscal policy combined with a commitment to rein in spending in the medium-term Poland is the largest recipient of EU development funds and their cyclical allocation can significantly impact the rate of economic growth.

The Polish economy performed well during the 2014-17 period, with the real GDP growth rate generally exceeding 3%, in part because of increases in government social spending that have helped to accelerate consumer-driven growth.However, since 2015, Poland has implemented new business restrictions and taxes on foreign-dominated economic sectors, including banking and insurance, energy, and healthcare, that have dampened investor sentiment and has increased the government's ownership of some firms. The government reduced the retirement age in 2016 and has had mixed success in introducing new taxes and boosting tax compliance to offset the increased costs of social spending programs and relieve upward pressure on the budget deficit. Some credit ratings agencies estimate that Poland during the next few years is at risk of exceeding the EU's 3%-of-GDP limit on budget deficits, possibly impacting its access to future EU funds. Poland's economy is projected to perform well in the next few years in part because of an anticipated cyclical increase in the use of its EU development funds and continued, robust household spending.

Poland faces several systemic challenges, which include addressing some of the remaining deficiencies in its road and rail infrastructure, business environment, rigid labor code, commercial court system, government red tape, and burdensome tax system, especially for entrepreneurs. Additional long-term challenges include diversifying Poland's energy mix, strengthening investments in innovation, research, and development, as well as stemming the outflow of educated young Poles to other EU member states, especially in light of a coming demographic contraction due to emigration, persistently low fertility rates, and the aging of the Solidarity-era baby boom generation.

GDP (purchasing power parity):
$1.126 trillion (2017 est.)
$1.076 trillion (2016 est.)
$1.045 trillion (2015 est.)
note: data are in 2017 dollars
country comparison to the world: 23

GDP (official exchange rate): $524.8 billion (2017 est.)

GDP—real growth rate:
4.55% (2019 est.)
5.36% (2018 est.)
4.83% (2017 est.)
country comparison to the world: 61

GDP—per capita (PPP):
$29,600 (2017 est.)
$28,300 (2016 est.)
$27,500 (2015 est.)
note: data are in 2017 dollars
country comparison to the world: 69

Gross national saving:
20% of GDP (2017 est.)
19.2% of GDP (2016 est.)
19.9% of GDP (2015 est.)
country comparison to the world: 97

GDP—composition, by end use:
household consumption: 58.6% (2017 est.)
government consumption: 17.7% (2017 est.)
investment in fixed capital: 17.7% (2017 est.)
investment in inventories: 2% (2017 est.)
exports of goods and services: 54% (2017 est.)
imports of goods and services: -49.9% (2017 est.)

GDP—composition, by sector of origin:
agriculture: 2.4% (2017 est.)
industry: 40.2% (2017 est.)
services: 57.4% (2017 est.)

Agriculture—products: potatoes, fruits, vegetables, wheat; poultry, eggs, pork, dairy

Industries: machine building, iron and steel, coal mining, chemicals, shipbuilding, food processing, glass, beverages, textiles

Industrial production growth rate: 7.5% (2017 est.)

country comparison to the world: 28

Labor force: 9.561 million (2020 est.)
country comparison to the world: 51

Labor force—by occupation: *agriculture:* 11.5%
industry: 30.4%
services: 57.6% (2015)

Unemployment rate: 5.43% (2019 est.)
6.08% (2018 est.)
country comparison to the world: 87

Population below poverty line: 17.6% (2015 est.)

Household income or consumption by percentage share: *lowest 10%:* 3%
highest 10%: 23.9% (2015 est.)

Budget: *revenues:* 207.5 billion (2017 est.)
expenditures: 216.2 billion (2017 est.)

Taxes and other revenues: 39.5% (of GDP) (2017 est.)
country comparison to the world: 45

Budget surplus (+) or deficit (-): -1.7% (of GDP) (2017 est.)
country comparison to the world: 96

Public debt:
50.6% of GDP (2017 est.)
54.2% of GDP (2016 est.)
note: data cover general government debt and include debt instruments issued (or owned) by government entities other than the treasury; the data include treasury debt held by foreign entities, the data include subnational entities, as well as intragovernmental debt; intragovernmental debt consists of treasury borrowings from surpluses in the social funds, such as for retirement, medical care, and unemployment; debt instruments for the social funds are not sold at public auctions
country comparison to the world: 100

Fiscal year: calendar year

Inflation rate (consumer prices): 2% (2017 est.)
-0.6% (2016 est.)
country comparison to the world: 106

Current account balance: $2.92 billion (2019 est.)
-$7.52 billion (2018 est.)
country comparison to the world: 35

Exports: $224.6 billion (2017 est.)
$195.7 billion (2016 est.)
country comparison to the world: 23

Exports—partners: Germany 27.4%, Czech Republic 6.4%, UK 6.4%, France 5.6%, Italy 4.9%, Netherlands 4.4% (2017)

Exports—commodities: machinery and transport equipment 37.8%, intermediate manufactured goods 23.7%, miscellaneous manufactured goods 17.1%, food and live animals 7.6% (2012 est.)

Imports: $223.8 billion (2017 est.)
$193.2 billion (2016 est.)
country comparison to the world: 23

Imports—commodities: machinery and transport equipment 38%, intermediate manufactured goods 21%, chemicals 15%, minerals, fuels, lubricants, and related materials 9% (2011 est.)

Imports—partners: Germany 27.9%, China 8%, Russia 6.4%, Netherlands 6%, Italy 5.3%, France 4.2%, Czech Republic 4% (2017)

Reserves of foreign exchange and gold: $113.3 billion (31 December 2017 est.)

$114.4 billion (31 December 2016 est.)
country comparison to the world: 22

Debt—external:
$241 billion (31 December 2017 est.)
$347.8 billion (31 December 2016 est.)
country comparison to the world: 31

Exchange rates: zlotych (PLN) per US dollar –
3.748 (2017 est.)
3.9459 (2016 est.)
3.9459 (2015 est.)
3.7721 (2014 est.)
3.1538 (2013 est.)

ENERGY

Electricity access: *electrification—total population:* 100% (2020)

Electricity—production: 156.9 billion kWh (2016 est.)
country comparison to the world: 25

Electricity—consumption: 149.4 billion kWh (2016 est.)
country comparison to the world: 24

Electricity—exports: 12.02 billion kWh (2016)
country comparison to the world: 17

Electricity—imports: 14.02 billion kWh (2016 est.)
country comparison to the world: 17
Electricity—installed generating capacity: 38.11 million kW (2016 est.)
country comparison to the world: 28

Electricity—from fossil fuels: 79% of total installed capacity (2016 est.)
country comparison to the world: 87

Electricity—from nuclear fuels: 0% of total installed capacity (2017 est.)
country comparison to the world: 166

Electricity—from hydroelectric plants: 2% of total installed capacity (2017 est.)
country comparison to the world: 141

Electricity—from other renewable *sources:* 19% of total installed capacity (2017 est.)
country comparison to the world: 44

Crude oil—production: 21,000 bbl/day (2018 est.)
country comparison to the world: 65

Crude oil—exports: 4,451 bbl/day (2017 est.)
country comparison to the world: 66

Crude oil—imports: 493,100 bbl/day (2017 est.)
country comparison to the world: 19

Crude oil—proved reserves: 126 million bbl (1 January 2018)
country comparison to the world: 66

Refined petroleum products—production: 554,200 bbl/day (2017 est.)
country comparison to the world: 30

Refined petroleum products—consumption: 649,600 bbl/day (2017 est.)
country comparison to the world: 30

Refined petroleum products—exports: 104,800 bbl/day (2017 est.)
country comparison to the world: 43

Refined petroleum products—imports: 222,300 bbl/day (2017 est.)
country comparison to the world: 32

Natural gas—production: 5.748 billion cu m (2017 est.)
country comparison to the world: 49

Natural gas—consumption: 20.1 billion cu m (2017 est.)
country comparison to the world: 38

Natural gas—exports: 1.246 billion cu m (2017 est.)
country comparison to the world: 39

Natural gas—imports: 15.72 billion cu m (2017 est.)
country comparison to the world: 20

Natural gas—proved reserves: 79.79 billion cu m (1 January 2018 est.)
country comparison to the world: 56
Carbon dioxide emissions from consumption of energy:
359 million Mt (2017 est.)
country comparison to the world: 18

COMMUNICATIONS

Telephones—fixed lines: *total subscriptions:* 6,907,937
subscriptions per 100 inhabitants: 18.01 (2019 est.)
country comparison to the world: 22
Telephones—mobile cellular:
total subscriptions: 52,916,105
subscriptions per 100 inhabitants: 137.96 (2019 est.)
country comparison to the world: 30

Telecommunication systems: *general assessment:* fixed-line service is dominated by the former state-owned company, yet it is dwarfed by the growth in mobile-cellular services; regulatory is framed by EU principles of competition; regulator measures have improved wholesale market access; rapid extension of LTE networks and development of mobile data service; mobile penetration is above European average; regulator to auction 700MHz spectrum of 5G services; good market competition (2020)
domestic: several nation-wide networks provide mobile-cellular service; coverage is generally good; fixed-line 18 per 100 service lags in rural areas, mobile-cellular 138 per 100 persons (2019)
international: country code—48; landing points for the Baltica and the Denmark-Poland2 submarine cables connecting Poland, Denmark and Sweden; international direct dialing with automated exchanges; satellite earth station—1 with access to Intelsat, Eutelsat, Inmarsat, and Intersputnik (2019)
note: the COVID-19 outbreak is negatively impacting telecommunications production and supply chains globally; consumer spending on telecom devices and services has also slowed due to the pandemic's effect on economies worldwide; overall progress towards improvements in all facets of the telecom industry—mobile, fixed-line, broadband, submarine cable and satellite—has moderated

Broadcast media: state-run public TV operates 2 national channels supplemented by 16 regional channels and several niche channels; privately owned entities operate several national TV networks and a number of special interest channels; many privately owned channels broadcasting locally; roughly half of all households are linked to either satellite or cable TV systems providing access to foreign television networks; state-run public radio operates 5 national networks and 17 regional radio stations; 2 privately owned national radio networks, several commercial stations broadcasting to multiple cities, and many privately owned local radio stations (2019)

Internet country code: .pl
Internet users:
total: 29,791,401
percent of population: 77.54% (July 2018 est.)
country comparison to the world: 28
Broadband—fixed subscriptions:
total: 6,114,926
subscriptions per 100 inhabitants: 16 (2018 est.)
country comparison to the world: 27

TRANSPORTATION

National air transport system: *number of registered air carriers:* 6 (2020)
inventory of registered aircraft operated by air carriers: 169
annual passenger traffic on registered air carriers: 9,277,538 (2018)
annual freight traffic on registered air carriers: 271.49 million mt-km (2018)

Civil aircraft registration country code prefix: SP (2016)

Airports: 126 (2013)
country comparison to the world: 46

Airports—with paved runways: *total:* 87 (2017)
over 3,047 m: 5 (2017)
2,438 to 3,047 m: 30 (2017)
1,524 to 2,437 m: 36 (2017)
914 to 1,523 m: 10 (2017)
under 914 m: 6 (2017)

Airports—with unpaved runways: *total:* 39 (2013)
1,524 to 2,437 m: 1 (2013)
914 to 1,523 m: 17 (2013)
under 914 m: 21 (2013)

Heliports: 6 (2013)

Pipelines: 14198 km gas, 1374 km oil, 2483 km refined products (2016)

Railways: *total:* 19,231 km (2016)
standard gauge: 18,836 km 1.435-m gauge (11,874 km electrified) (2016)
broad gauge: 395 km 1.524-m gauge (2016)
country comparison to the world: 16

Roadways: *total:* 420,000 km (2016)
paved: 291,000 km (includes 1,492 km of expressways, 1,559 of motorways) (2016)
unpaved: 129,000 km (2016)
country comparison to the world: 17

Waterways: 3,997 km (navigable rivers and canals) (2009)
country comparison to the world: 27

Merchant marine:
total: 144
by type: general cargo 12, oil tanker 7, other 125 (2019)
country comparison to the world: 73

Ports and terminals: *major seaport(s):* Gdansk, Gdynia, Swinoujscie
container port(s) (TEUs): Gdansk (1,593,761) (2017)
LNG terminal(s) (import): Swinoujscie
river port(s): Szczecin (River Oder)

MILITARY AND SECURITY

Military and security forces: *Polish Armed Forces:* Land Forces (Wojska Ladowe), Navy (Marynarka Wojenna), Air Force (Sily Powietrzne), Special Forces (Wojska Specjalne), Territorial Defense Force (Wojska Obrony Terytorialnej); Ministry of the Interior: Border Guard (includes coast guard duties) (2019)

Military expenditures:
2% of GDP (2019 est.)
2.02% of GDP (2018)
1.89% of GDP (2017)
1 .99% of GDP (2016)
2.22% of GDP (2015)

country comparison to the world: 52

Military and security service personnel strengths: the Polish Armed Forces have approximately 105,000 total active duty personnel (60,000 Army; 7,000 Navy; 17,000 Air Force; 3,500 Special Forces; 3,000 Territorial Defense Forces; 14,000 other); approximately 20,000 total Territorial Defense Forces including reservists (2019 est.)

note—in June 2019, the Polish Government approved a plan to increase the size of the military by 50,000 troops over the coming decade

Military equipment inventories and acquisitions: the inventory of the Polish Armed Forces consists of a mix of Soviet-era and more modern Western weapons systems; since 2010, the leading suppliers of armaments to Poland are Finland, Germany, Italy, and the US (2019 est.)

Military deployments: 360 Afghanistan (NATO); 230 Kosovo (NATO); up to 200 Latvia (NATO); 220 Lebanon (UNIFIL); contributes about 3,500 troops to the Lithuania, Poland, and Ukraine joint military brigade (LITPOLUKRBRIG), which was established in 2014; the brigade is headquartered in Warsaw and is comprised of an international staff, three battalions, and specialized units (2020)

Military service age and obligation: 18-28 years of age for male and female voluntary military service; conscription phased out in 2009-12; professional soldiers serve on a permanent basis (for an unspecified period of time) or on a contract basis (for a specified period of time); initial contract period is 24 months; women serve in the military on the same terms as men (201 9)

TRANSNATIONAL ISSUES

Disputes—international: as a member state that forms part of the EU's external border, Poland has implemented the strict Schengen border rules to restrict illegal immigration and trade along its eastern borders with Belarus and Ukraine

Refugees and internally displaced persons: *refugees (country of origin):* 9,870 (Russia) (2019)
stateless persons: 1,328 (2019)

Illicit drugs: despite diligent counternarcotics measures and international information sharing on cross-border crimes, a major illicit producer of synthetic drugs for the international market; minor transshipment point for Southwest Asian heroin and Latin American cocaine to Western Europe

PORTUGAL

INTRODUCTION

Background: Following its heyday as a global maritime power during the 15th and 16th centuries, Portugal lost much of its wealth and status with the destruction of Lisbon in a 1755 earthquake, occupation during the Napoleonic Wars, and the independence of Brazil, its wealthiest colony, in 1822. A 1910 revolution deposed the monarchy, and for most of the next six decades, repressive governments ran the country. In 1974, a left-wing military coup installed broad democratic reforms. The following year, Portugal granted independence to all of its African colonies. Portugal is a founding member of NATO and entered the EC (now the EU) in 1986.

GEOGRAPHY

Location: Southwestern Europe, bordering the North Atlantic Ocean, west of Spain

Geographic coordinates: 39 30 N, 8 00 W

Map references: Europe

Area: *total:* 92,090 sq km
land: 91,470 sq km
water: 620 sq km
note: includes Azores and Madeira Islands
country comparison to the world: 112

Area—comparative: slightly smaller than Virginia

Land boundaries: *total:* 1,224 km
border countries (1): Spain 1224 km

Coastline: 1,793 km

Maritime claims:
territorial sea: 12 nm
exclusive economic zone: 200 nm
contiguous zone: 24 nm
continental shelf: 200-m depth or to the depth of exploitation

Climate: maritime temperate; cool and rainy in north, warmer and drier in south

Terrain: *the west-flowing Tagus River divides the country:* the north is mountainous toward the interior, while the south is characterized by rolling plains

Elevation: *mean elevation:* 372 m
lowest point: Atlantic Ocean 0 m
highest point: Ponta do Pico (Pico or Pico Alto) on Ilha do Pico in the Azores 2,351 m

Natural resources: fish, forests (cork), iron ore, copper, zinc, tin, tungsten, silver, gold, uranium, marble, clay, gypsum, salt, arable land, hydropower

Land use: *agricultural land:* 39.7% (2011 est.)
arable land: *11.9% (2011 est.)* / *permanent crops:* 7.8% (2011 est.) / *permanent pasture:* 20% (2011 est.)
forest: 37.8% (2011 est.)
other: 22.5% (2011 est.)

Irrigated land: 5,400 sq km (2012)

Population distribution: concentrations are primarily along or near the Atlantic coast; both Lisbon and the second largest city, Porto, are coastal cities

Natural hazards: Azores subject to severe earthquakes

volcanism: limited volcanic activity in the Azores Islands; Fayal or Faial (1,043 m) last erupted in 1958; most volcanoes have not erupted in centuries; historically active volcanoes include Agua de Pau, Furnas, Pico, Picos Volcanic System, San Jorge, Sete Cidades, and Terceira

Environment—current issues: soil erosion; air pollution caused by industrial and vehicle emissions; water pollution, especially in urban centers and coastal areas

Environment—international agreements: *party to:* Air Pollution, Biodiversity, Climate Change, Climate Change-Kyoto Protocol, Desertification, Endangered Species, Hazardous Wastes, Law of the Sea, Marine Dumping, Marine Life Conservation, Ozone Layer Protection, Ship Pollution, Tropical Timber 83, Tropical Timber 94, Wetlands, Whaling

signed, but not ratified: Air Pollution-Persistent Organic Pollutants, Air Pollution-Volatile Organic Compounds, Environmental Modification

Geography—note: Azores and Madeira Islands occupy strategic locations along western sea approaches to Strait of Gibraltar

PEOPLE AND SOCIETY

Population: 10,302,674 (July 2020 est.)
country comparison to the world: 89

Nationality: *noun:* Portuguese (singular and plural)
adjective: Portuguese

Ethnic groups: white homogeneous Mediterranean population; citizens of black African descent who immigrated to mainland during decolonization number less than 100,000; since 1990, Eastern Europeans have migrated to Portugal

Languages: Portuguese (official), Mirandese (official, but locally used)

Religions: Roman Catholic 81%, other Christian 3.3%, other (includes Jewish, Muslim) 0.6%, none 6.8%, unspecified 8.3% (2011 est.)
note: represents population 15 years of age and older

Age structure: *0-14 years:* 13.58% (male 716,102/female 682,582)
15-24 years: 10.94% (male 580,074/female 547,122)
25-54 years: 41.49% (male 2,109,693/female 2,164,745)
55-64 years: 13.08% (male 615,925/female 731,334)
65 years and over: 20.92% (male 860,198/female 1,294,899) (2020 est.)

Dependency ratios:
total dependency ratio: 55.8
youth dependency ratio: 20.3
elderly dependency ratio: 35.5
potential support ratio: 2.8 (2020 est.)
Median age:
total: 44.6 years
male: 42.7 years

female: 46.5 years (2020 est.)
country comparison to the world: 13

Population growth rate: -0.25% (2020 est.)
country comparison to the world: 213

Birth rate: 8.1 births/1,000 population (2020 est.)
country comparison to the world: 221

Death rate: 10.8 deaths/1,000 population (2020 est.)
country comparison to the world: 25

Net migration rate: 0.3 migrant(s)/1,000 population (2020 est.)
country comparison to the world: 71

Population distribution: concentrations are primarily along or near the Atlantic coast; both Lisbon and the second largest city, Porto, are coastal cities

Urbanization: *urban population:* 66.3% of total population (2020)
rate of urbanization: 0.47% annual rate of change (2015-20 est.)

total population growth rate v. urban population growth rate, 2000-2030:

Major urban areas—population: *2.957 million LISBON (capital), 1.313 million Porto (2020)*

Sex ratio: *at birth:* 1.05 male(s)/female
0-14 years: 1.05 male(s)/female
15-24 years: 1.06 male(s)/female
25-54 years: 0.97 male(s)/female
55-64 years: 0.84 male(s)/female
65 years and over: 0.66 male(s)/female
total population: 0.9 male(s)/female (2020 est.)

Mother's mean age at first birth: 29.6 years (2017 est.)

Maternal mortality rate: 8 deaths/100,000 live births (2017 est.)
country comparison to the world: 152

Infant mortality rate: *total:* 2.6 deaths/1,000 live births
male: 3 deaths/1,000 live births
female: 2.3 deaths/1,000 live births (2020 est.)
country comparison to the world: 219

Life expectancy at birth: *total population:* 81.1 years
male: 77.9 years
female: 84.4 years (2020 est.)
country comparison to the world: 39

Total fertility rate: 1.41 children born/woman (2020 est.)
country comparison to the world: 216

Contraceptive prevalence rate: 73.9% (2014)

Drinking water source:
improved:
urban: 100% of population
rural: 100% of population
total: 100% of population
unimproved:
urban: 0% of population
rural: 0% of population
total: 0% of population (2017 est.)

Current Health Expenditure: 9% (2017)

Physicians density: 5.12 physicians/1,000 population (2017)

Hospital bed density: 3.4 beds/1,000 population (2017)

Sanitation facility access:
improved:
urban: 100% of population
rural: 100% of population
total: 100% of population
unimproved:
urban: 0% of population
rural: 0% of population
total: 0% of population (2017 est.)

HIV/AIDS—adult prevalence rate: 0.5% (2018 est.)
country comparison to the world: 70

HIV/AIDS—people living with HIV/AIDS: 41,000 (2018 est.)
country comparison to the world: 65

HIV/AIDS—deaths: <500 (2018 est.)

Obesity—adult prevalence rate: 20.8% (2016)
country comparison to the world: 95

Education expenditures: 4.9% of GDP (2015)
country comparison to the world: 67

Literacy: *definition:* age 15 and over can read and write
total population: 96.1%
male: 97.4%
female: 95.1% (2018)

School life expectancy (primary to tertiary education): *total:* 17 years
male: 17 years
female: 17 years (2018)

Unemployment, youth ages 15-24: *total:* 20.3%
male: 19.8%
female: 20.9% (2018 est.)
country comparison to the world: 66

GOVERNMENT

Country name: *conventional long form:* Portuguese Republic
conventional short form: Portugal
local long form: Republica Portuguesa
local short form: Portugal
etymology: name derives from the Roman designation "Portus Cale" meaning "Port of Cale"; Cale was an ancient Celtic town and port in present-day northern Portugal

Government type: semi-presidential republic

Capital: *name:* Lisbon
geographic coordinates: 38 43 N, 9 08 W
time difference: UTC 0 (5 hours ahead of Washington, DC, during Standard Time)
daylight saving time: +1hr, begins last Sunday in March; ends last Sunday in October
note: Portugal has two time zones, including the Azores (UTC-1)
etymology: Lisbon is one of Europe's oldest cities (the second oldest capital city after Athens) and the origin of the name is lost in time; it may have been founded as an ancient Celtic settlement that subsequently maintained close commercial relations with the Phoenicians (beginning about 1200 B.C.); the name of the settlement may have been derived from the pre-Roman appellation for

the Tagus River that runs through the city, Lisso or Lucio; the Romans named the city "Olisippo" when they took it from the Carthaginians in 205 B.C.; under the Visigoths the city name became "Ulixbona," under the Arabs it was "al-Ushbuna"; the medieval version of "Lissabona" became today's Lisboa

Administrative divisions: 18 districts (distritos, singular—distrito) and 2 autonomous regions* (regioes autonomas, singular—regiao autonoma); Aveiro, Acores (Azores)*, Beja, Braga, Braganca, Castelo Branco, Coimbra, Evora, Faro, Guarda, Leiria, Lisboa (Lisbon), Madeira*, Portalegre, Porto, Santarem, Setubal, Viana do Castelo, Vila Real, Viseu

Independence: 1143 (Kingdom of Portugal recognized); 1 December 1640 (independence reestablished following 60 years of Spanish rule); 5 October 1910 (republic proclaimed)

National holiday: Portugal Day (Dia de Portugal), 10 June (1580); note—also called Camoes Day, the day that revered national poet Luis DE CAMOES (1524-80) died

Constitution: *history:* several previous; latest adopted 2 April 1976, effective 25 April 1976
amendments: proposed by the Assembly of the Republic; adoption requires two-thirds majority vote of Assembly members; amended several times, last in 2005

Legal system: civil law system; Constitutional Court review of legislative acts

International law organization participation: accepts compulsory ICJ jurisdiction with reservations; accepts ICCt jurisdiction

Citizenship: citizenship by birth: no
citizenship by descent only: at least one parent must be a citizen of Portugal
dual citizenship recognized: yes
residency requirement for naturalization: 10 years; 6 years if from a Portuguese-speaking country

Suffrage: 18 years of age; universal

Executive branch: *chief of state:* President Marcelo REBELO DE SOUSA (since 9 March 2016)
head of government: Prime Minister Antonio Luis Santos da COSTA (since 24 November 2015)
cabinet: Council of Ministers appointed by the president on the recommendation of the prime minister
elections/appointments: president directly elected by absolute majority popular vote in 2 rounds if needed for a 5-year term (eligible for a second term); election last held on 24 January 2016 (next to be held on 31 January 2021); following legislative elections the leader of the majority party or majority coalition is usually appointed prime minister by the president
election results: Marcelo REBELO DE SOUSA elected president in the first round; percent of vote—Marcelo REBELO DE SOUSA (PSD) 52%, Antonio Sampaio da NOVOA (independent) 22.9%, Marisa MATIAS (BE) 10.1%, Maria de BELEM (independent) 4.2%, other 10.8%

note: there is also a Council of State that acts as a consultative body to the president

Legislative branch: *description:* unicameral Assembly of the Republic or Assembleia da Republica (230 seats; 226 members directly elected in multi-seat constituencies by closed-list proportional representation vote and 4 members—2 each in 2 constituencies representing Portuguese living abroad—directly elected by proportional representation vote; members serve 4-year terms) (e.g. 2019)
elections: last held on 6 October 2019 (next to be held 2023) (e.g. 2019)
election results: percent of vote by party—PS 36.4%, PSD 27.8%, B.E. 9.5%, CDU 6.5%, other 20.8%; seats by party—PS 108, PSD 79, B.E. 19, CDU 12, other 12; composition—men 158, women 72, percent of women 31.3% (e.g. 2019)

Judicial branch: *highest courts:* Supreme Court or Supremo Tribunal de Justica (consists of 12 justices); Constitutional Court or Tribunal Constitucional (consists of 13 judges)
judge selection and term of office: Supreme Court justices nominated by the president and appointed by the Assembly of the Republic; judges can serve for life; Constitutional Court judges—10 elected by the Assembly and 3 elected by the other Constitutional Court judges; judges elected for 6-year nonrenewable terms
subordinate courts: Supreme Administrative Court (Supremo Tribunal Administrativo); Audit Court (Tribunal de Contas); appellate, district, and municipal courts

Political parties and leaders:
Democratic and Social Center/Popular Party (Partido do Centro Democratico Social-Partido Popular) or CDS-PP [Manuel CRISTAS]
Ecologist Party "The Greens" or "Os Verdes" (Partido Ecologista-Os Verdes) or PEV [Heloisa APOLONIA] People-Animals-Nature Party (Pessoas-Animais-Natureza) or PAN [Andre SILVA]
Portuguese Communist Party (Partido Comunista Portugues) or PCP [Jeronimo DE SOUSA]
Social Democratic Party (Partido Social Democrata) or PSD (original name Partido Popular Democratico) or PPD [Rui RIO] Socialist Party (Partido Socialista) or PS [Antonio COSTA]
The Left Bloc (Bloco de Esquerda) or BE or O Bloco [Catarina MARTINS]
Unitary Democratic Coalition (Coligacao Democratica Unitaria) or CDU [Jeronimo DE SOUSA] (includes PCP and PEV)

International organization participation: ADB (nonregional member), AfDB (nonregional member), Australia Group, BIS, CD, CE, CERN, CPLP, EAPC, EBRD, ECB, EIB, EMU, ESA, EU, FAO, FATF, IADB, IAEA, IBRD, ICAO, ICC (national committees), ICCt, ICRM, IDA, IEA, IFAD, IFC, IFRCS, IHO, ILO, IMF, IMO, IMSO, Interpol, IOC, IOM, IPU, ISO, ITSO, ITU, ITUC (NGOs), LAIA (observer), MIGA, MINUSMA, NATO, NEA, NSG, OAS (observer), OECD, OPCW, OSCE, Pacific Alliance (observer), Paris Club (associate), PCA, Schengen Convention,

SELEC (observer), UN, UNCTAD, UNESCO, UNHCR, UNIDO, Union Latina, UNWTO, UPU, WCO, WFTU (NGOs), WHO, WIPO, WMO, WTO, ZC

Diplomatic representation in the US: *chief of mission:* Ambassador Domingos Teixeira de Abreu FEZAS VITAL (since 28 January 2016)
chancery: 2012 Massachusetts Avenue NW, Washington, DC 20036
telephone: [1] (202) 332-3007
FAX: [1] (202) 223-3926
consulate(s) general: Boston, New York, San Francisco
consulate(s): New Bedford (MA), Newark (NJ), Providence (RI)

Diplomatic representation from the US: *chief of mission:* Ambassador George E. GLASS (since 25 August 2017)
telephone: [351] (21) 727-3300
embassy: Avenida das Forcas Armadas, 1600-081 Lisbon
mailing address: Apartado 43033, 1601-301 Lisboa; PSC 83, APO AE 09726
FAX: [351] (21) 726-9109
consulate(s): Ponta Delgada (Azores)

Flag description: two vertical bands of green (hoist side, two-fifths) and red (three-fifths) with the national coat of arms (armillary sphere and Portuguese shield) centered on the dividing line; explanations for the color meanings are ambiguous, but a popular interpretation has green symbolizing hope and red the blood of those defending the nation

National symbol(s): *armillary sphere (a spherical astrolabe modeling objects in the sky and representing the Republic); national colors:* red, green

National anthem: *name:* "A Portuguesa" (The Song of the Portuguese)
lyrics/music: Henrique LOPES DE MENDOCA/ Alfredo KEIL
note: adopted 1910; "A Portuguesa" was originally written to protest the Portuguese monarchy's acquiescence to the 1890 British ultimatum forcing Portugal to give up areas of Africa; the lyrics refer to the "insult" that resulted from the event 0:00/ 1:10

ECONOMY

Economy — overview: Portugal has become a diversified and increasingly service-based economy since joining the European Community—the EU's predecessor—in 1986. Over the following two decades, successive governments privatized many state-controlled firms and liberalized key areas of the economy, including the financial and telecommunications sectors. The country joined the Economic and Monetary Union in 1999 and began circulating the euro on 1 January 2002 along with 11 other EU members.

The economy grew by more than the EU average for much of the 1990s, but the rate of growth slowed in 2001-08. After the global financial crisis in 2008, Portugal's economy contracted in 2009 and fell into recession from 2011 to 2013, as the government implemented spending cuts and tax increases to comply with conditions of an EU-IMF financial rescue package, signed in May 2011. Portugal successfully exited its EU-IMF program in May 2014, and its economic recovery gained traction in 2015 because of strong exports and a rebound in private consumption. GDP growth accelerated in 2016, and probably reached 2.5 % in 2017. Unemployment remained high, at 9.7% in 2017, but has improved steadily since peaking at 18% in 2013.

The center-left minority Socialist government has unwound some unpopular austerity measures while managing to remain within most EU fiscal targets. The budget deficit fell from 11.2% of GDP in 2010 to 1.8% in 2017, the country's lowest since democracy was restored in 1974, and surpassing the EU and IMF projections of 3%. Portugal exited the EU's excessive deficit procedure in mid-2017.

GDP (purchasing power parity):
$314.1 billion (2017 est.)
$305.9 billion (2016 est.)
$301 billion (2015 est.)
note: data are in 2017 dollars
country comparison to the world: 55

GDP (official exchange rate): $218 billion (2017 est.)

GDP—real growth rate:
2.24% (2019 est.)
2.85% (2018 est.)
3.51% (2017 est.)
country comparison to the world: 126

GDP—per capita (PPP):
$30,500 (2017 est.)
$29,600 (2016 est.)
$29,100 (2015 est.)
note: data are in 2017 dollars
country comparison to the world: 67

Gross national saving:
16.8% of GDP (2017 est.)
16.1% of GDP (2016 est.)
15.9% of GDP (2015 est.)
country comparison to the world: 124

GDP—composition, by end use:
household consumption: 65.1% (2017 est.)
government consumption: 17.6% (2017 est.)
investment in fixed capital: 16.2% (2017 est.)
investment in inventories: 0.1% (2017 est.)
exports of goods and services: 43.1% (2017 est.)
imports of goods and services: -42.1% (2017 est.)

GDP—composition, by sector of origin:
agriculture: 2.2% (2017 est.)
industry: 22.1% (2017 est.)
services: 75.7% (2017 est.)

Agriculture—products: grain, potatoes, tomatoes, olives, grapes; sheep, cattle, goats, pigs, poultry, dairy products; fish

Industries: textiles, clothing, footwear, wood and cork, paper and pulp, chemicals, fuels and lubricants, automobiles and auto parts, base metals, minerals, porcelain and ceramics, glassware, technology, telecommunications; dairy products, wine, other foodstuffs; ship construction and refurbishment; tourism, plastics, financial services, optics

Industrial production growth rate: 3.5% (2017 est.)
country comparison to the world: 89

Labor force: 4.717 million (2020 est.)
country comparison to the world: 82

Labor force—by occupation: *agriculture:* 8.6%
industry: 23.9%
services: 67.5% (2014 est.)

Unemployment rate:
6.55% (2019 est.)
7.05% (2018 est.)
country comparison to the world: 102

Population below poverty line: 19% (2015 est.)

Household income or consumption by percentage share: *lowest 10%:* 2.6%
highest 10%: 25.9% (2015 est.)

Budget: *revenues:* 93.55 billion (2017 est.)
expenditures: 100 billion (2017 est.)

Taxes and other revenues: 42.9% (of GDP) (2017 est.)
country comparison to the world: 29

Budget surplus (+) or deficit (-): -3% (of GDP) (2017 est.)
country comparison to the world: 134

Public debt: 125.7% of GDP (2017 est.)

129.9% of GDP (2016 est.)
note: data cover general government debt and include debt instruments issued (or owned) by government entities other than the treasury; the data include treasury debt held by foreign entities; the data include debt issued by subnational entities, as well as intragovernmental debt; intragovernmental debt consists of treasury borrowings from surpluses in the social funds, such as for retirement, medical care, and unemployment; debt instruments for the social funds are not sold at public auctions
country comparison to the world: 9

Fiscal year: calendar year

Inflation rate (consumer prices): 1.6% (2017 est.)
0.6% (2016 est.)
country comparison to the world: 91

Current account balance: -$203 million (2019 est.)

$988 million (2018 est.)
country comparison to the world: 100

Exports: $61 billion (2017 est.)
$54.76 billion (2016 est.)
country comparison to the world: 47

Exports—partners: Spain 25.2%, France 12.5%, Germany 11.3%, UK 6.6%, US 5.2%, Netherlands 4% (2017)

Exports—commodities: agricultural products, foodstuffs, wine, oil products, chemical products, plastics and rubber, hides, leather, wood and cork, wood pulp and paper, textile materials, clothing, footwear, machinery and tools, base metals

Imports:
$74.73 billion (2017 est.)
$64.98 billion (2016 est.)
country comparison to the world: 45

Imports—commodities: agricultural products, chemical products, vehicles and other transport material, optical and precision instruments, computer accessories and parts, semiconductors and related devices, oil products, base metals, food products, textile materials

Imports—partners: Spain 32%, Germany 13.7%, France 7.4%, Italy 5.5%, Netherlands 5.4% (2017)

Reserves of foreign exchange and gold:
$26.11 billion (31 December 2017 est.)
$19.4 billion (31 December 2015 est.)
country comparison to the world: 55

Debt—external:
$449 billion (31 March 2016 est.)
$447 billion (31 March 2015 est.)
country comparison to the world: 27

Exchange rates: euros (EUR) per US dollar –
0.885 (2017 est.)
0.903 (2016 est.)
0.9214 (2015 est.)
0.7525 (2014 est.)
0.7634 (2013 est.)

ENERGY

Electricity access: *electrification—total population:* 100% (2020)

Electricity—production: 56.9 billion kWh (2016 est.)
country comparison to the world: 51

Electricity—consumption: 46.94 billion kWh (2016 est.)
country comparison to the world: 52

Electricity—exports: 9.701 billion kWh (2016 est.)
country comparison to the world: 21

Electricity—imports: 4.616 billion kWh (2016 est.)
country comparison to the world: 40

Electricity—installed generating capacity: 20.56 million kW (2016 est.)
country comparison to the world: 43

Electricity—from fossil fuels: 41% of total installed capacity (2016 est.)
country comparison to the world: 168

Electricity—from nuclear fuels: 0% of total installed capacity (2017 est.)
country comparison to the world: 167

Electricity—from hydroelectric plants: 25% of total installed capacity (2017 est.)
country comparison to the world: 78

Electricity—from other renewable sources: 35% of total installed capacity (2017 est.)
country comparison to the world: 10

Crude oil—production: 0 bbl/day (2018 est.)
country comparison to the world: 187

Crude oil—exports: 0 bbl/day (2017 est.)
country comparison to the world: 180

Crude oil—imports: 285,200 bbl/day (2017 est.)
country comparison to the world: 26

Crude oil—proved reserves: 0 bbl (1 January 2018 est.)
country comparison to the world: 182

Refined petroleum products—production: 323,000 bbl/day (2017 est.)
country comparison to the world: 39

Refined petroleum products—consumption: 247,200 bbl/day (2017 est.)
country comparison to the world: 51

Refined petroleum products—exports: 143,500 bbl/day (2017 est.)
country comparison to the world: 36

Refined petroleum products—imports: 78,700 bbl/day (2017 est.)
country comparison to the world: 64

Natural gas—production: 0 cu m (2017 est.)
country comparison to the world: 184

Natural gas—consumption: 6.258 billion cu m (2017 est.)
country comparison to the world: 54

Natural gas—exports: 0 cu m (2017 est.)
country comparison to the world: 169

Natural gas—imports: 6.541 billion cu m (2017 est.)
country comparison to the world: 30

Natural gas—proved reserves: 0 cu m (1 January 2014 est.)
country comparison to the world: 182

Carbon dioxide emissions from consumption of energy: 54.97 million Mt (2017 est.)
country comparison to the world: 57

COMMUNICATIONS

Telephones—fixed lines: *total subscriptions:* 5,179,685
subscriptions per 100 inhabitants: 50.15 (2019 est.)
country comparison to the world: 28

Telephones—mobile cellular: *total subscriptions:* 12,028,436
subscriptions per 100 inhabitants: 116.46 (2019 est.)
country comparison to the world: 76

Telecommunication systems: *general assessment:* telephone system has a state-of-the-art network with broadband, high-speed capabilities; FttP in 2020; 3G universal and 4G upgrades; regulator release 700MHz spectrum for 5G use; DSL moves to fiber services; FttP for over 5 million customers in 2020 providing national coverage; fiber subscriber base grows 24% in 2018; development in M-payment solutions (2020)
domestic: integrated network of coaxial cables, open-wire, microwave radio relay, and domestic satellite earth stations; fixed-line 50 per 100 persons and mobile-cellular 116 per 100 persons (2019)
international: country code—351; landing points for the Ella Link, BUGIO, EIG, SAT-3/WASC, SeaMeWe-3, Equino, MainOne, Tat TGN-Western Europe, WACS, ACE, Atlantis2 and Columbus-III submarine cables provide connectivity to Europe, Africa, the Middle East, Asia, Southeast Asia, Australia, South America and the US; satellite earth stations—3 Intelsat (2 Atlantic Ocean and 1 Indian Ocean), NA Eutelsat; tropospheric scatter to Azores (2019)
note: the COVID-19 outbreak is negatively impacting telecommunications production and supply chains globally; consumer spending on telecom devices and services has also slowed due to the pandemic's effect on economies worldwide; overall progress towards improvements in all facets of the telecom industry—mobile, fixed-line, broadband, submarine cable and satellite—has moderated

Broadcast media: Radio e Televisao de Portugal (RTP), the publicly owned TV broadcaster, operates 4 domestic channels and external service channels to Africa; overall, roughly 40 domestic TV stations; viewers have widespread access to international broadcasters with more than half of all households connected to multi-channel cable or satellite TV systems; publicly owned radio operates 3 national networks and provides regional and external services; several privately owned national radio stations and some 300 regional and local commercial radio stations

Internet country code: pt
Internet users:
total: 7,731,411
percent of population: 74.66% (July 2018 est.)
country comparison to the world: 62

Broadband—fixed subscriptions: *total:* 3,784,684
subscriptions per 100 inhabitants: 37 (2018 est.)
country comparison to the world: 38

TRANSPORTATION

National air transport system: *number of registered air carriers:* 10 (2020)
inventory of registered aircraft operated by air carriers: 168
annual passenger traffic on registered air carriers: 17,367,956 (2018)
annual freight traffic on registered air carriers: 454.21 million mt-km (2018)

Civil aircraft registration country code prefix: CR, CS (2016)

Airports: 64 (2013)
country comparison to the world: 77

Airports—with paved runways: *total:* 43 (2017)
over 3,047 m: 5 (2017)
2,438 to 3,047 m: 7 (2017)
1,524 to 2,437 m: 8 (2017)
914 to 1,523 m: 15 (2017)
under 914 m: 8 (2017)

Airports—with unpaved runways: *total:* 21 (2013)
914 to 1,523 m: 1 (2013)
under 914 m: 20 (2013)

Pipelines: 1344 km gas, 11 km oil, 188 km refined products (2013)

Railways: *total:* 3,075 km (2014)
narrow gauge: 108.1 km 1.000 m gauge (2014)
broad gauge: 2,439 km 1.668-m gauge (1,633.4 km electrified) (2014)
other: 528 km (gauge unspecified) (2014)

country comparison to the world: 60

Roadways: *total:* 82,900 km (2008)
paved: 71,294 km (includes 2,613 km of expressways) (2008)
unpaved: 11,606 km (2008)
country comparison to the world: 61

Waterways: 210 km (on Douro River from Porto) (2011)
country comparison to the world: 95

Merchant marine: *total:* 624
by type: bulk carrier 65, container ship 249, general cargo 99, oil tanker 21, other 190 (2019)
country comparison to the world: 37

Ports and terminals: *major seaport(s):* Leixoes, Lisbon, Setubal, Sines
container port(s) (TEUs): Sines (1,669,057) (2017)
LNG terminal(s) (import): Sines

MILITARY AND SECURITY

Military and security forces: *Portuguese Armed Forces:* Portuguese Army (Exercito Portuguesa), Portuguese Navy (Marinha Portuguesa; includes Marine Corps), Portuguese Air Force (Forca Aerea Portuguesa, FAP); Portuguese National Republican Guard (Guarda Nacional Republicana, GNR) (2019)
note: the GNR is a national gendarmerie force comprised of military personnel with law enforcement, internal security, civil defense, disaster response, and coast guard duties; it is responsible to the Minister of Internal Administration and to the Minister of National Defense; in the event of war or crisis, it may be placed under the Chief of the General Staff of the Armed Forces

Military expenditures:
1.52% of GDP (2019 est.)
1.43% of GDP (2018)
1.25% of GDP (2017)
1.27% of GDP (2016)
1.33% of GDP (2015)
country comparison to the world: 77

Military and security service personnel strengths: the Portuguese Armed Forces have approximately 26,500 active duty personnel (13,000 Army; 7,500 Navy; 6,000 Air Force); 24,700 National Republican Guard (military personnel) (2019 est.)

Military equipment inventories and acquisitions: the Portuguese Armed Forces inventory includes mostly European and US-origin weapons systems along with a smaller mix of domestically-produced equipment; since 2010, Germany and the US are the leading suppliers of armaments to Portugal; Portugal's defense industry is primarily focused on shipbuilding (2019 est.)

Military deployments: 190 Afghanistan (NATO); 200 Central African Republic (MINUSCA/EUTM); up to 120 Baltic States (NATO) (2020)

Military service age and obligation: 18-30 years of age for voluntary or contract military service; no compulsory military service, but conscription possible if insufficient volunteers available; women serve in the armed forces, on naval ships since 1992, but are prohibited from serving in some combatant specialties; contract service lasts for an initial period from two to six years, and can be extended to a maximum of 20 years of service. Voluntary military service lasts 12 months; reserve obligation to age 35 (2017)

Disputes—international: Portugal does not recognize Spanish sovereignty over the territory of Olivenza based on a difference of interpretation of the 1815 Congress of Vienna and the 1801 Treaty of Badajoz

Refugees and internally displaced persons: *stateless persons:* 14 (2019)

Illicit drugs: seizing record amounts of Latin American cocaine destined for Europe; a European gateway for Southwest Asian heroin; transshipment point for hashish from North Africa to Europe; consumer of Southwest Asian heroin

PUERTO RICO

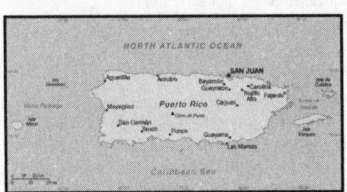

INTRODUCTION

Background: Populated for centuries by aboriginal peoples, the island was claimed by the Spanish Crown in 1493 following Christopher COLUMBUS' second voyage to the Americas. In 1898, after 400 years of colonial rule that saw the indigenous population nearly exterminated and African slave labor introduced, Puerto Rico was ceded to the US as a result of the Spanish-American War. Puerto Ricans were granted US citizenship in 1917. Popularly elected governors have served since 1948. In 1952, a constitution was enacted providing for internal self-government. In plebiscites held in 1967, 1993, and 1998, voters chose not to alter the existing political status with the US, but the results of a 2012 vote left open the possibility of American statehood. Economic recession on the island has led to a net population loss since about 2005, as large numbers of residents moved to the US mainland. The trend has accelerated since 2010; in 2014, Puerto Rico experienced a net population loss to the mainland of 64,000, more than double the net loss of 26,000 in 2010. Hurricane Maria struck the island on 20 September 2017 causing catastrophic damage, including destruction of the electrical grid that had been cripled by Hurricane Irma just two weeks before. It was the worst storm to hit the island in eight decades, and damage is estimated in the tens of billions of dollars.

GEOGRAPHY

Location: Caribbean, island between the Caribbean Sea and the North Atlantic Ocean, east of the Dominican Republic

Geographic coordinates: 18 15 N, 66 30 W

Map references: Central America and the Caribbean

Area: *total:* 9,104 sq km
land: 8,959 sq km
water: 145 sq km
country comparison to the world: 171

Area—comparative: slightly less than three times the size of Rhode Island

Land boundaries: 0 km

Coastline: 501 km

Maritime claims: *territorial sea:* 12 nm
exclusive economic zone: 200 nm

Climate: tropical marine, mild; little seasonal temperature variation

Terrain: mostly mountains with coastal plain in north; precipitous mountains to the sea on west coast; sandy beaches along most coastal areas

Elevation: *mean elevation:* 261 m
lowest point: Caribbean Sea 0 m
highest point: Cerro de Punta 1,338 m

Natural resources: some copper and nickel; potential for onshore and offshore oil

Land use: *agricultural land:* 22% (2011 est.)
arable land: 6.6% (2011 est.) / *permanent crops:* 5.6% (2011 est.) / *permanent pasture:* 9.8% (2011 est.)
forest: 63.2% (2011 est.)
other: 14.8% (2011 est.)
Irrigated land: 220 sq km (2012)

Population distribution: population clusters tend to be found along the coast, the largest of these is found in and around San Juan; an exception to this is a sizeable population located in the interior of the island immediately south of the capital around Caguas; most of the interior, particularly in the western half of the island, is dominated by the Cordillera Central mountains, where population density is low

Natural hazards: periodic droughts; hurricanes

Environment—current issues: soil erosion; occasional droughts cause water shortages; industrial pollution

Geography—note: important location along the Mona Passage—a key shipping lane to the Panama Canal; San Juan is one of the biggest and best natural harbors in the Caribbean; many small rivers and high central mountains ensure land is well watered; south coast relatively dry; fertile coastal plain belt in north

PEOPLE AND SOCIETY

Population: 3,189,068 (July 2020 est.)
country comparison to the world: 134

Nationality: *noun:* Puerto Rican(s) (US citizens)
adjective: Puerto Rican

Ethnic groups: white 75.8%, black/African American 12.4%, other 8.5% (includes American Indian, Alaskan Native, Native Hawaiian, other Pacific Islander, and others), mixed 3.3% (2010 est.)
note: 99% of the population is Latino

Languages: Spanish, English

Religions: Roman Catholic 85%, Protestant and other 15%

Age structure:
0-14 years: 14.22% (male 231,406/female 222,061)
15-24 years: 12.78% (male 207,169/female 200,373)
25-54 years: 37.73% (male 573,114/female 630,276)
55-64 years: 13.5% (male 197,438/female 232,931)
65 years and over: 21.77% (male 297,749/female 396,551) (2020 est.)

Dependency ratios: *total dependency ratio:* 57.7
youth dependency ratio: 24.8
elderly dependency ratio: 32.8
potential support ratio: 3 (2020 est.)

Median age: *total:* 43.6 years
male: 41.6 years
female: 45.3 years (2020 est.)
country comparison to the world: 25

Population growth rate: -1.59% (2020 est.)
country comparison to the world: 235

Birth rate: 8 births/1,000 population (2020 est.)
country comparison to the world: 222

Death rate: 9.5 deaths/1,000 population (2020 est.)
country comparison to the world: 46

Net migration rate: -14.1 migrant(s)/1,000 population (2020 est.)
country comparison to the world: 222

Population distribution: population clusters tend to be found along the coast, the largest of these is found in and around San Juan; an exception to this is a sizeable population located in the interior of the island immediately south of the capital around Caguas; most of the interior, particularly in the western half of the island, is dominated by the Cordillera Central mountains, where population density is low

Urbanization: *urban population:* 93.6% of total population (2020)
rate of urbanization: -0.14% annual rate of change (2015-20 est.)

total population growth rate v. urban population growth rate, 2000-2030:

Major urban areas—population: *2.448 million SAN JUAN (capital) (2020)*

Sex ratio: *at birth:* 1.06 male(s)/female
0-14 years: 1.04 male(s)/female
15-24 years: 1.03 male(s)/female
25-54 years: 0.91 male(s)/female
55-64 years: 0.85 male(s)/female
65 years and over: 0.75 male(s)/female
total population: 0.9 male(s)/female (2020 est.)

Maternal mortality rate: 21 deaths/100,000 live births (2017 est.)
country comparison to the world: 123

Infant mortality rate: *total:* 6 deaths/1,000 live births
male: 6.6 deaths/1,000 live births
female: 5.4 deaths/1,000 live births (2020 est.)
country comparison to the world: 166

Life expectancy at birth: *total population:* 81.3 years
male: 78 years
female: 84.7 years (2020 est.)
country comparison to the world: 34

Total fertility rate: 1.24 children born/woman (2020 est.)
country comparison to the world: 224

Drinking water source: *improved:* total: 97% of population
unimproved: *total:* 0% of population (2017 est.)

Physicians density: 3.06 physicians/1,000 population (2018)

Sanitation facility access: *improved:* total: 97.2% of population
unimproved: *total:* 2.8% of population (2017 est.)

HIV/AIDS—adult prevalence rate: NA

HIV/AIDS—people living with HIV/AIDS: NA

HIV/AIDS—deaths: NA

Education expenditures: 6.1% of GDP (2014)
country comparison to the world: 30

Literacy: *definition:* age 15 and over can read and write
total population: 92.4%
male: 92.4%
female: 92.4% (2017)

School life expectancy (primary to tertiary education): *total:* 16 years
male: 15 years
female: 18 years (2018)

Unemployment, youth ages 15-24: *total:* 26.6%
male: 28.9%
female: 23.1% (2012 est.)

country comparison to the world: 45

GOVERNMENT

Country name: *conventional long form:* Commonwealth of Puerto Rico
conventional short form: Puerto Rico
abbreviation: PR
etymology: Christopher COLUMBUS named the island San Juan Bautista (Saint John the Baptist) and the capital city and main port Cuidad de Puerto Rico (Rich Port City); over time, however, the names were shortened and transposed and the island came to be called Puerto Rico and its capital San Juan

Dependency status: unincorporated organized territory of the US; policy relations between Puerto Rico and the US conducted under the jurisdiction of the Office of the President

Government type: republican form of government with separate executive, legislative, and judicial branches; unincorporated organized territory of the US with local self-government
Note: reference Puerto Rican Federal Relations Act, 2 March 1917, as amended by Public Law 600, 3 July 1950

Capital: *name:* San Juan
geographic coordinates: 18 28 N, 66 07 W
time difference: UTC-4 (1 hour ahead of Washington, DC, during Standard Time)
etymology: the name dates to 1521 and the founding of the city under the name "Ciudad de San Juan Bautista de Puerto Rico" (City of Saint John the Baptist of Puerto Rico)

Administrative divisions: none (territory of the US); there are no first-order administrative divisions as defined by the US Government, but there are 78 municipalities (municipios, singular—municipio) at the second order; Adjuntas, Aguada, Aguadilla, Aguas Buenas, Aibonito, Anasco, Arecibo, Arroyo, Barceloneta, Barranquitas, Bayamon, Cabo Rojo, Caguas, Camuy, Canovanas, Carolina, Catano, Cayey, Ceiba, Ciales, Cidra, Coamo, Comerio, Corozal, Culebra, Dorado, Fajardo, Florida, Guanica, Guayama, Guayanilla, Guaynabo, Gurabo, Hatillo, Hormigueros, Humacao, Isabela, Jayuya, Juana Diaz, Juncos, Lajas, Lares, Las Marias, Las Piedras, Loiza, Luquillo, Manati, Maricao, Maunabo, Mayaguez, Moca, Morovis, Naguabo, Naranjito, Orocovis, Patillas, Penuelas, Ponce, Quebradillas, Rincon, Rio Grande, Sabana Grande, Salinas, San German, San Juan, San Lorenzo, San Sebastian, Santa Isabel, Toa Alta, Toa Baja, Trujillo Alto, Utuado, Vega Alta, Vega Baja, Vieques, Villalba, Yabucoa, Yauco

Independence: none (territory of the US with commonwealth status)

National holiday: US Independence Day, 4 July (1776); Puerto Rico Constitution Day, 25 July (1952)

Constitution: *history:* previous 1900 (Organic Act, or Foraker Act); latest ratified by referendum 3 March 1952, approved 3 July 1952, effective 25 July 1952

amendments: proposed by a concurrent resolution of at least two-thirds majority by the total Legislative Assembly membership; approval requires at least two-thirds majority vote by the membership of both houses and approval by a majority of voters in a special referendum; if passed by at least three-fourths Assembly vote, the referendum can be held concurrently with the next general election; constitutional articles such as the republican form of government or the bill of rights cannot be amended; amended 1952

Legal system: civil law system based on the Spanish civil code and within the framework of the US federal system

Citizenship: see United States

Suffrage: 18 years of age; universal; note—island residents are US citizens but do not vote in US presidential elections

Executive branch: *chief of state:* President Donald J. TRUMP (since 20 January 2017); Vice President Michael R. PENCE (since 20 January 2017)
head of government: Governor Wanda VAZQUEZ (since 7 August 2019)
cabinet: Cabinet appointed by governor with the consent of the Legislative Assembly
elections/appointments: president and vice president indirectly elected on the same ballot by an Electoral College of 'electors' chosen from each state; president and vice president serve a 4-year term (eligible for a second term); under the US Constitution, residents of Puerto Rico do not vote in elections for US president and vice president; however, they may vote in Democratic and Republican party presidential primary elections; governor directly elected by simple majority popular vote for a 4-year term (no term limits); election last held on 8 November 2016 (next to be held on 3 November 2020)
election results: Ricardo ROSSELLO elected governor; percent of vote—Ricardo ROSSELLO (PNP) 41.8%, David BERNIER (PPD) 38.9%, Alexandra LUGARO (independent) 11.1%, Manuel CIDRE (independent) 5.7%
note: on 24 July 2019, Governor Ricardo ROSSELLO announced his resignation effective 2 August 2019; as Secretary of State, Pedro PIERLUISI succeeded Governor Ricardo ROSSELLO; on 7 August 2019 the Supreme Court of Puerto Rico ruled Pedro PIERLUISI accession was unconstitutional and Wanda VAZQUEZ is sworn in as governor

Legislative branch: *description:* bicameral Legislative Assembly or Asamblea Legislativa consists of:
Senate or Senado (30 seats; 16 members directly elected in 8 2-seat constituencies by simple majority vote and 14 atlarge members directly elected by simple majority vote to serve 4-year terms)
House of Representatives or Camara de Representantes (51 seats; members directly elected in single-seat constituencies by simple majority vote to serve 4-year terms)
elections: Senate—last held on 8 November 2016 (next to be held on 3 November 2020)

House of Representatives—last held on 8 November 2016 (next to be held on 3 November 2020)

election results: Senate—percent of vote by party—NA; seats by party—PNP 21, PPD 7, PIP 1, Independent 1; composition—men 23, women 7, percent of women 23.3%

House of Representatives—percent of vote by party—NA; seats by party—PNP 34, PPD 16, PIP 1; composition—men 11, women 4, percent of women 26.7%; total Legislative Assembly percent of women 16%

note: Puerto Rico directly elects 1 member by simple majority vote to serve a 4-year term as a commissioner to the US House of Representatives; the commissioner can vote when serving on a committee and when the House meets as the Committee of the Whole House but not when legislation is submitted for a 'full floor' House vote; election of commissioner last held on 6 November 2018 (next to be held in November 2022)

Judicial branch: *highest courts:* Supreme Court (consists of the chief justice and 8 associate justices)

judge selection and term of office: justices appointed by the governor and confirmed by majority Senate vote; judges serve until compulsory retirement at age 70

subordinate courts: Court of Appeals; First Instance Court comprised of superior and municipal courts

Political parties and leaders: National Democratic Party [Charlie RODRIGUEZ]

National Republican Party of Puerto Rico [Jenniffer GONZALEZ]

New Progressive Party or PNP [Ricardo ROSSELLO] (pro-US statehood)

Popular Democratic Party or PPD [Alejandro GARCIA Padillo] (pro-commonwealth)

Puerto Rican Independence Party or PIP [Ruben BERRIOS Martinez] (pro-independence)

International organization participation: AOSIS (observer), Caricom (observer), Interpol (subbureau), IOC, UNWTO (associate), UPU, WFTU (NGOs)

Diplomatic representation in the US: none (territory of the US)

Diplomatic representation from the US: none (territory of the US with commonwealth status)

Flag description: five equal horizontal bands of red (top, center, and bottom) alternating with white; a blue isosceles triangle based on the hoist side bears a large, white, five-pointed star in the center; the white star symbolizes Puerto Rico; the three sides of the triangle signify the executive, legislative and judicial parts of the government; blue stands for the sky and the coastal waters; red symbolizes the blood shed by warriors, while white represents liberty, victory, and peace

note: design initially influenced by the US flag, but similar to the Cuban flag, with the colors of the bands and triangle reversed

National symbol(s): *Puerto Rican spindalis (bird), coqui (frog); national colors:* red, white, blue

National anthem: *name:* "La Borinquena" (The Puerto Rican)

lyrics/music: Manuel Fernandez JUNCOS/Felix Astol ARTES

note: music adopted 1952, lyrics adopted 1977; the local anthem's name is a reference to the indigenous name of the island, Borinquen; the music was originally composed as a dance in 1867 and gained popularity in the early 20th century; there is some evidence that the music was written by Francisco RAMIREZ; as a commonwealth of the US, "The Star-Spangled Banner" is official (see United States)

0:00 / 1:34

ECONOMY

Economy—overview: Puerto Rico had one of the most dynamic economies in the Caribbean region until 2006; however, growth has been

negative for each of the last 11 years. The downturn coincided with the phaseout of tax preferences that had led US firms to invest heavily in the Commonwealth since the 1950s, and a steep rise in the price of oil, which generates most of the island's electricity.

Diminished job opportunities prompted a sharp rise in outmigration, as many Puerto Ricans sought jobs on the US mainland. Unemployment reached 16% in 2011, but declined to 11.5% in December 2017. US minimum wage laws apply in Puerto Rico, hampering job expansion. Per capita income is about two-thirds that of the US mainland.

The industrial sector greatly exceeds agriculture as the locus of economic activity and income. Tourism has traditionally been an important source of income with estimated arrivals of more than 3.6 million tourists in 2008. Puerto Rico's merchandise trade surplus is exceptionally strong, with exports nearly 50% greater than imports, and its current account surplus about 10% of GDP.

Closing the budget deficit while restoring economic growth and employment remain the central concerns of the government. The gap between revenues and expenditures amounted to 0.6% of GDP in 2016, although analysts believe that not all expenditures have been accounted for in the budget and a better accounting of costs would yield an overall deficit of roughly 5% of GDP. Public debt remained steady at 92.5% of GDP in 2017, about $17,000 per person, or nearly three times the per capita debt of the State of Connecticut, the highest in the US. Much of that debt was issued by staterun schools and public corporations, including water and electric utilities. In June 2015, Governor Alejandro GARCIA Padilla announced that the island could not pay back at least $73 billion in debt and that it would seek a deal with its creditors.

Hurricane Maria hit Puerto Rico square on in September 2017, causing electrical power outages to 90% of the territory, as well as extensive loss of housing and infrastructure and contamination of

potable water. Despite massive efforts, more than 40% of the territory remained without electricity as of yearend 2017. As a result of the destruction, many Puerto Ricans have emigrated to the US mainland.

GDP (purchasing power parity):
$130 billion (2017 est.)
$133.1 billion (2016 est.)
$134.9 billion (2015 est.)
note: data are in 2017 dollars
country comparison to the world: 81

GDP (official exchange rate): $104.2 billion (2017 est.)

GDP—real growth rate:
-2.4% (2017 est.)
-1.3% (2016 est.)
-1% (2015 est.)
country comparison to the world: 207

GDP—per capita (PPP):
$39,400 (2017 est.)
$39,000 (2016 est.)
$38,800 (2015 est.)
note: data are in 2017 dollars
country comparison to the world: 47

GDP—composition, by end use:
household consumption: 87.7% (2017 est.)
government consumption: 12.2% (2017 est.)
investment in fixed capital: 11.7% (2017 est.)
investment in inventories: 0.5% (2017 est.)
exports of goods and services: 117.8% (2017 est.)
imports of goods and services: -129.8% (2017 est.)

GDP—composition, by sector of origin:
agriculture: 0.8% (2017 est.)
industry: 50.1% (2017 est.)
services: 49.1% (2017 est.)

Agriculture—products: sugarcane, coffee, pineapples, plantains, bananas; livestock products, chickens

Industries: pharmaceuticals, electronics, apparel, food products, tourism

Industrial production growth rate: -2.1% (2017 est.)
country comparison to the world: 184

Labor force: 1.139 million (December 2014 est.)
country comparison to the world: 136

Labor force—by occupation: *agriculture:* 2.1%
industry: 19%
services: 79% (2005 est.)

Unemployment rate: 10.8% (2017 est.)
11.8% (2016 est.)
country comparison to the world: 153

Population below poverty line: NA

Household income or consumption by percentage share: *lowest 10%:* NA
highest 10%: NA

Budget: *revenues:* 9.268 billion (2017 est.)
expenditures: 9.974 billion (2017 est.)

Taxes and other revenues: 8.9% (of GDP) (2017 est.)
country comparison to the world: 217

783

Budget surplus (+) or deficit (-): -0.7% (of GDP) (2017 est.)
country comparison to the world: 68

Public debt: 51.6% of GDP (2017 est.)
50.1% of GDP (2016 est.)
country comparison to the world: 97

Fiscal year: 1 July—30 June

Inflation rate (consumer prices): 1.8% (2017 est.)
-0.3% (2016 est.)
country comparison to the world: 95

Current account balance: $0 (2017 est.)
$0 (2016 est.)
country comparison to the world: 65

Exports: $73.17 billion (2017 est.)
$73.2 billion (2016 est.)
country comparison to the world: 41

Exports—commodities: chemicals, electronics, apparel, canned tuna, rum, beverage concentrates, medical equipment

Imports: $49.01 billion (2017 est.)
$48.86 billion (2016 est.)
country comparison to the world: 54

Imports—commodities: chemicals, machinery and equipment, clothing, food, fish, petroleum products

Debt—external: $56.82 billion (31 December 2010 est.)
$52.98 billion (31 December 2009 est.)
country comparison to the world: 61

Exchange rates: the US dollar is used

ENERGY

Electricity access: *electrification—total population:* 100% (2020)

Electricity—production: 20.95 billion kWh (2016 est.)
country comparison to the world: 75

Electricity—consumption: 19.48 billion kWh (2016 est.)
country comparison to the world: 72

Electricity—exports: 0 kWh (2016 est.)
country comparison to the world: 185

Electricity—imports: 0 kWh (2016 est.)
country comparison to the world: 187

Electricity—installed generating capacity: 6.294 million kW (2016 est.)
country comparison to the world: 76

Electricity—from fossil fuels: 94% of total installed capacity (2016 est.)
country comparison to the world: 47

Electricity—from nuclear fuels: 0% of total installed capacity (2017 est.)
country comparison to the world: 168

Electricity—from hydroelectric plants: 2% of total installed capacity (2017 est.)
country comparison to the world: 142

Electricity—from other renewable sources: 4% of total installed capacity (2017 est.)
country comparison to the world: 117

Crude oil—production: 0 bbl/day (2018 est.)
country comparison to the world: 188

Crude oil—exports: 0 bbl/day (2015 est.)
country comparison to the world: 181

Crude oil—imports: 0 bbl/day (2015 est.)
country comparison to the world: 182

Crude oil—proved reserves: 0 bbl (1 January 2018 est.)
country comparison to the world: 183

Refined petroleum products—production: 0 bbl/day (2015 est.)
country comparison to the world: 190

Refined petroleum products—consumption: 98,000 bbl/day (2016 est.)
country comparison to the world: 81

Refined petroleum products—exports: 18,420 bbl/day (2015 est.)
country comparison to the world: 70

Refined petroleum products—imports: 127,100 bbl/day (2015 est.)
country comparison to the world: 46

Natural gas—production: 0 cu m (2017 est.)
country comparison to the world: 185

Natural gas—consumption: 1.303 billion cu m (2017 est.)
country comparison to the world: 86

Natural gas—exports: 0 cu m (2017 est.)
country comparison to the world: 170

Natural gas—imports: 1.303 billion cu m (2017 est.)
country comparison to the world: 57

Natural gas—proved reserves: 0 cu m (1 January 2014 est.)
country comparison to the world: 183

Carbon dioxide emissions from consumption of energy: 19.85 million Mt (2017 est.)
country comparison to the world: 87

COMMUNICATIONS

Telephones—fixed lines: *total subscriptions:* 758,869
subscriptions per 100 inhabitants: 23.42 (2019 est.)
country comparison to the world: 81

Telephones—mobile cellular: *total subscriptions:* 3,724,680
subscriptions per 100 inhabitants: 114.95 (2019 est.)
country comparison to the world: 133

Telecommunication systems: *general assessment:* modern system integrated with that of the US by high-capacity submarine cable and Intelsat with high-speed data capability; havoc caused by hurricane Maria in 2017 and earthquake in 2020, has left the island lagging behind the mainland US both economically and technologically; competition among network operators helps with growth; availability of LTE coverage increasing to 90%; operators expanding and securing 600 MHz spectrum, LTE reach and launching services based on 5G to majority of the population (2020)
domestic: digital telephone system; mobile-cellular services; fixed-line 23 per 100 and mobile-cellular 115 per 100 persons (2019)

international: country code—1-787, 939; landing points for the GTMO-PR, AMX-1, BRUSA, GCN, PCCS, SAm-1, Southern Caribbean Fiber, Americas-II, Antillas, ARCOS, SMPR-1, and Taino-Carib submarine cables providing connectivity to the mainland US, Caribbean, Central and South America; satellite earth station—1 Intelsat (2019)
note: the COVID-19 outbreak is negatively impacting telecommunications production and supply chains globally; consumer spending on telecom devices and services has also slowed due to the pandemic's effect on economies worldwide; overall progress towards improvements in all facets of the telecom industry—mobile, fixed-line, broadband, submarine cable and satellite—has moderated

Broadcast media: more than 30 TV stations operating; cable TV subscription services are available; roughly 125 radio stations

Internet country code: .pr

Internet users: *total:* 2,326,006
percent of population: 70.6% (July 2018 est.)
country comparison to the world: 114

Broadband—fixed subscriptions: *total:* 609,027
subscriptions per 100 inhabitants: 18 (2018 est.)
country comparison to the world: 80

TRANSPORTATION

Airports: 29 (2013)
country comparison to the world: 118

Airports—with paved runways: *total:* 17 (2017)
over 3,047 m: 2 (2017)
2,438 to 3,047 m: 1 (2017)
1,524 to 2,437 m: 2 (2017)
914 to 1,523 m: 7 (2017)
under 914 m: 5 (2017)

Airports—with unpaved runways: *total:* 12 (2013)
1,524 to 2,437 m: 1 (2013)
914 to 1,523 m: 1 (2013)
under 914 m: 10 (2013)

Roadways: *total:* 26,862 km (includes 454 km of expressways) (2012)
country comparison to the world: 102

Ports and terminals: *major seaport(s):* Ensenada Honda, Mayaguez, Playa de Guayanilla, Playa de Ponce, San Juan
container port(s) (TEUs): San Juan (1,210,503) (2015)
LNG terminal(s) (import): Guayanilla Bay

MILITARY AND SECURITY

Military and security forces: no regular indigenous military forces; National Guard, State Guard, Police Force

Military—note: defense is the responsibility of the US

TRANSNATIONAL ISSUES

Disputes—international: increasing numbers of illegal migrants from the Dominican Republic cross the Mona Passage to Puerto Rico each year looking for work

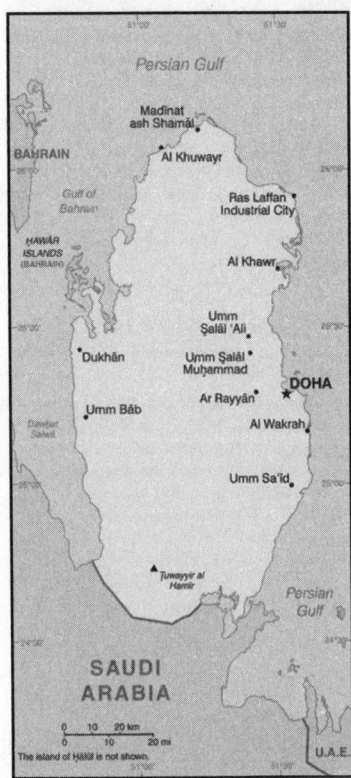

INTRODUCTION

Background: Ruled by the Al Thani family since the mid-1800s, Qatar within the last 60 years transformed itself from a poor British protectorate noted mainly for pearling into an independent state with significant oil and natural gas revenues. Former Amir HAMAD bin Khalifa Al Thani, who overthrew his father in a bloodless coup in 1995, ushered in wide-sweeping political and media reforms, unprecedented economic investment, and a growing Qatari regional leadership role, in part through the creation of the pan-Arab satellite news network Al-Jazeera and Qatar's mediation of some regional conflicts. In the 2000s, Qatar resolved its longstanding border disputes with both Bahrain and Saudi Arabia and by 2007 had attained the highest per capita income in the world. Qatar did not experience domestic unrest or violence like that seen in other Near Eastern and North African countries in 2011, due in part to its immense wealth and patronage network. In mid-2013, HAMAD peacefully abdicated, transferring power to his son, the current Amir TAMIM bin Hamad. TAMIM is popular with the Qatari public, for his role in shepherding the country through

an economic embargo by some other regional countries, for his efforts to improve the country's healthcare and education systems, and for his expansion of the country's infrastructure in anticipation of Doha's hosting of the 2022 World Cup.

Recently, Qatar's relationships with its neighbors have been tense, although since the fall of 2019 there have been signs of improved prospects for a thaw. Following the outbreak of regional unrest in 2011, Doha prided itself on its support for many popular revolutions, particularly in Libya and Syria. This stance was to the detriment of Qatar's relations with Bahrain, Egypt, Saudi Arabia, and the United Arab Emirates (UAE), which temporarily recalled their respective ambassadors from Doha in March 2014. TAMIM later oversaw a warming of Qatar's relations with Bahrain, Egypt, Saudi Arabia, and the UAE in November 2014 following Kuwaiti mediation and signing of the Riyadh Agreement. This reconciliation, however, was short-lived. In June 2017, Bahrain, Egypt, Saudi Arabia, and the UAE (the "Quartet") cut diplomatic and economic ties with Qatar in response to alleged violations of the agreement, among other complaints.

GEOGRAPHY

Location: Middle East, peninsula bordering the Persian Gulf and Saudi Arabia

Geographic coordinates: 25 30 N, 51 15 E

Map references: Middle East

Area: *total:* 11,586 sq km
land: 11,586 sq km
water: 0 sq km
country comparison to the world: 165

Area—comparative: almost twice the size of Delaware; slightly smaller than Connecticut

Land boundaries: *total:* 87 km
border countries (1): Saudi Arabia 87 km

Coastline: 563 km

Maritime claims: *territorial sea:* 12 nm
exclusive economic zone: as determined by bilateral agreements or the median line
contiguous zone: 24 nm

Climate: arid; mild, pleasant winters; very hot, humid summers

Terrain: mostly flat and barren desert

Elevation: *mean elevation:* 28 m
lowest point: Persian Gulf 0 m
highest point: Tuwayyir al Hamir 103 m

Natural resources: petroleum, fish, natural gas

Land use: *agricultural land:* 5.6% (2011 est.)
arable land: *1.1% (2011 est.) / permanent crops:* 0.2% (2011 est.) / permanent pasture: 4.3% (2011 est.)
forest: 0% (2011 est.)
other: 94.4% (2011 est.)

Irrigated land: 130 sq km (2012)

Population distribution: most of the population is clustered in or around the capital of Doha on the eastern side of the peninsula

Natural hazards: haze, dust storms, sandstorms common

Environment—current issues: air, land, and water pollution are significant environmental issues; limited natural freshwater resources are increasing dependence on large-scale desalination facilities; other issues include conservation of oil supplies and preservation of the natural wildlife heritage

Environment—international agreements: *party to:* Biodiversity, Climate Change, Climate Change-Kyoto Protocol, Desertification, Endangered Species, Hazardous Wastes, Law of the Sea, Ozone Layer Protection, Ship Pollution
signed, but not ratified: none of the selected agreements

Geography—note: the peninsula occupies a strategic location in the central Persian Gulf near major petroleum deposits

PEOPLE AND SOCIETY

Population: 2,444,174 (July 2020 est.)
country comparison to the world: 143

Nationality: *noun:* Qatari(s)
adjective: Qatari

Ethnic groups: non-Qatari 88.4%, Qatari 11.6% (2015 est.)

Languages: Arabic (official), English commonly used as a second language

Religions: Muslim 67.7%, Christian 13.8%, Hindu 13.8%, Buddhist 3.1%, folk religion <.1%, Jewish <.1%, other 0.7%, unaffiliated 0.9% (2010 est.)

MENA religious affiliation: *Age structure:* 0-14 years: 12.84% (male 158,702/female 155,211)
15-24 years: 11.78% (male 203,703/female 84,323)
25-54 years: 70.66% (male 1,439,364/female 287,575)
55-64 years: 3.53% (male 66,561/female 19,600)
65 years and over: 1.19% (male 19,067/female 10,068) (2020 est.)

Dependency ratios: *total dependency ratio:* 18.1
youth dependency ratio: 16.1
elderly dependency ratio: 2
potential support ratio: 50.1 (2020 est.)

Median age: *total:* 33.7 years
male: 35 years
female: 28.2 years (2020 est.)
country comparison to the world: 96

Population growth rate: 1.55% (2020 est.)
country comparison to the world: 66

Birth rate: 9.3 births/1,000 population (2020 est.)
country comparison to the world: 201

Death rate: 1.6 deaths/1,000 population (2020 est.)
country comparison to the world: 229

Net migration rate: 6.5 migrant(s)/1,000 population (2020 est.)
country comparison to the world: 18

Population distribution: most of the population is clustered in or around the capital of Doha on the eastern side of the peninsula

Urbanization: *urban population:* 99.2% of total population (2020)
rate of urbanization: 2.41% annual rate of change (2015-20 est.)

total population growth rate v. urban population growth rate, 2000-2030:

Major urban areas—population: *641,000 DOHA (capital) (2020)*

Sex ratio: *at birth:* 1.02 male(s)/female
0-14 years: 1.02 male(s)/female
15-24 years: 2.42 male(s)/female
25-54 years: 5.01 male(s)/female
55-64 years: 3.4 male(s)/female
65 years and over: 1.89 male(s)/female
total population: 3.39 male(s)/female (2020 est.)

Maternal mortality rate: 9 deaths/100,000 live births (2017 est.)
country comparison to the world: 149

Infant mortality rate: *total:* 5.7 deaths/1,000 live births
male: 6 deaths/1,000 live births
female: 5.4 deaths/1,000 live births (2020 est.)
country comparison to the world: 167

Life expectancy at birth: *total population:* 79.4 years
male: 77.2 years
female: 81.6 years (2020 est.)
country comparison to the world: 54

Total fertility rate: 1.88 children born/woman (2020 est.)
country comparison to the world: 135

Contraceptive prevalence rate: 37.5% (2012)

Drinking water source: *improved:* total: 100% of population
unimproved: total: 0% of population (2017 est.)

Current Health Expenditure: 2.6% (2017)

Physicians density: 2.69 physicians/1,000 population (2016)

Hospital bed density: 1.3 beds/1,000 population (2017)

Sanitation facility access: *improved:* total: 100% of population (2015 est.)
unimproved: total: 0% of population (2017 est.)

HIV/AIDS—adult prevalence rate: 0.1% (2017 est.)
country comparison to the world: 132

HIV/AIDS—people living with HIV/AIDS: <500 (2017 est.)

HIV/AIDS—deaths: <100 (2017 est.)

Obesity—adult prevalence rate: 35.1% (2016)
country comparison to the world: 15

Education expenditures: 2.9% of GDP (2017)
country comparison to the world: 143

Literacy: *definition:* age 15 and over can read and write

total population: 93.5%
male: 92.4%
female: 94.7% (2017)

School life expectancy (primary to tertiary education): *total:* 12 years
male: 12 years
female: 14 years (2019)

Unemployment, youth ages 15-24: *total:* 0.4%
male: 0.2%
female: 1.5% (2018 est.)
country comparison to the world: 181

GOVERNMENT

Country name: *conventional long form:* State of Qatar
conventional short form: Qatar
local long form: Dawlat Qatar
local short form: Qatar
etymology: the origin of the name is uncertain, but it dates back at least 2,000 years since a term "Catharrei" was used to describe the inhabitants of the peninsula by Pliny the Elder (1st century A.D.), and a "Catara" peninsula is depicted on a map by Ptolemy (2nd century A.D.)
note: closest approximation of the native pronunciation is gattar or cottar

Government type: absolute monarchy

Capital: *name:* Doha
geographic coordinates: 25 17 N, 51 32 E
time difference: UTC+3 (8 hours ahead of Washington, DC, during Standard Time)
etymology: derives from the Arabic term "dohat," meaning "roundness," and refers to the small rounded bays along the area's coastline

Administrative divisions: 8 municipalities (baladiyat, singular—baladiyah); Ad Dawhah, Al Khawr wa adh Dhakhirah, Al Wakrah, Ar Rayyan, Ash Shamal, Ash Shihaniyah, Az Za'ayin, Umm Salal

Independence: 3 September 1971 (from the UK)

National holiday: National Day, 18 December (1878), anniversary of Al Thani family accession to the throne; Independence Day, 3 September (1971)

Constitution: *history:* previous 1972 (provisional); latest drafted 2 July 2002, approved by referendum 29 April 2003, endorsed 8 June 2004, effective 9 June 2005
amendments: proposed by the Amir or by one third of Advisory Council members; passage requires two-thirds majority vote of Advisory Council members and approval and promulgation by the emir; articles pertaining to the rule of state and its inheritance, functions of the emir, and citizen rights and liberties cannot be amended

Legal system: mixed legal system of civil law and Islamic (sharia) law (in family and personal matters)

International law organization participation: has not submitted an ICJ jurisdiction declaration; non-party state to the ICCt

Citizenship: *citizenship by birth:* no
citizenship by descent only: the father must be a citizen of Qatar

dual citizenship recognized: no
residency requirement for naturalization: 20 years; 15 years if an Arab national

Suffrage: 18 years of age; universal

Executive branch: *chief of state:* Amir TAMIM bin Hamad Al Thani (since 25 June 2013)
head of government: Prime Minister and Minister of Interior Sheikh KHALID ibn Khalifa ibn Abdul Aziz Al Thani (since 28 January 2020); Deputy Prime Minister and Minister of State for Defense Affairs KHALID bin Mohamed AL Attiyah (since 14 November 2017); Deputy Prime Minister and Minister of Foreign Affairs MOHAMED bin Abdulrahman Al Thani (since 14 November 2017)
cabinet: Council of Ministers appointed by the amir
elections/appointments: the monarchy is hereditary; prime minister and deputy prime minister appointed by the amir

Legislative branch: *description:* unicameral Advisory Council or Majlis al-Shura (45 seats; 30 members directly elected by popular vote for 4-year re-electable terms; 15 members appointed by the monarch to serve until resignation or until relieved; note—legislative drafting authority rests with the Council of Ministers and is reviewed by the Advisory Council or Majlis al-Shura
elections: last on 17 June 2016 (next in 2019); note—in late 2019, the amir announced the formation of a committee to oversee preparations for the first elected council, although Doha has not selected a date for elections
election results: NA; composition—men 41, women 4, percent of women 8.9%

Judicial branch: *highest courts:* Supreme Court or Court of Cassation (consists of the court president and several judges); Supreme Constitutional Court (consists of the chief justice and 6 members)
judge selection and term of office: Supreme Court judges nominated by the Supreme Judiciary Council, a 9-member independent body consisting of judiciary heads appointed by the Amir; judges appointed for 3-year renewable terms; Supreme Constitutional Court members nominated by the Supreme Judiciary Council and appointed by the monarch; term of appointment NA
subordinate courts: Courts of Appeal; Administrative Court; Courts of First Instance; sharia courts; Courts of Justice; Qatar International Court and Dispute Resolution Center, established in 2009, provides dispute resolution services for institutions and bodies in Qatar, as well as internationally

Political parties and leaders: political parties are banned

International organization participation: ABEDA, AFESD, AMF, CAEU, CD, CICA (observer), EITI (implementing country), FAO, G-77, GCC, IAEA, IBRD, ICAO, ICC (national committees), ICRM, IDA, IDB, IFAD, IFC, IFRCS, IHO, ILO, IMF, IMO, IMSO, Interpol, IOC, IOM (observer), IPU, ISO, ITSO, ITU, LAS, MIGA, NAM, OAPEC, OAS (observer), OIC, OIF, OPCW, OPEC, PCA, UN, UNCTAD, UNESCO,

UNIDO, UNIFIL, UNWTO, UPU, WCO, WHO, WIPO, WMO, WTO

Diplomatic representation in the US: *chief of mission:* Ambassador MISHAL bin Hamad bin Muhammad Al Thani (since 24 April 2017)
chancery: 2555 M Street NW, Washington, DC 20037
telephone: [1] (202) 274-1600
FAX: [1] (202) 237-0682
consulate(s) general: Houston, Los Angeles

Diplomatic representation from the US: *chief of mission:* Ambassador (vacant); Charge d'Affaires Phillip NELSON (since 2 March 2020)
telephone: [974] 4496-6000
embassy: 22 February Street, Al Luqta District, P. O. Box 2399, Doha
mailing address: P. O. Box 2399, Doha
FAX: [974] 4488-4298

Flag description: maroon with a broad white serrated band (nine white points) on the hoist side; maroon represents the blood shed in Qatari wars, white stands for peace; the nine-pointed serrated edge signifies Qatar as the ninth member of the "reconciled emirates" in the wake of the Qatari-British treaty of 1916
note: the other eight emirates are the seven that compose the UAE and Bahrain; according to some sources, the dominant color was formerly red, but this darkened to maroon upon exposure to the sun and the new shade was eventually adopted

National symbol(s): *a maroon field surmounted by a white serrated band with nine white points; national colors:* maroon, white

National anthem: *name:* "Al-Salam Al-Amiri" (The Amiri Salute)
lyrics/music: Sheikh MUBARAK bin Saif al-Thani/Abdul Aziz Nasser OBAIDAN
note: adopted 1996; anthem first performed that year at a meeting of the Gulf Cooperative Council hosted by Qatar

0:00 / 1:26

ECONOMY

Economy—overview: Qatar's oil and natural gas resources are the country's main economic engine and government revenue source, driving Qatar's high economic growth and per capita income levels, robust state spending on public entitlements, and booming construction spending, particularly as Qatar prepares to host the World Cup in 2022. Although the government has maintained high capital spending levels for ongoing infrastructure projects, low oil and natural gas prices in recent years have led the Qatari Government to tighten some spending to help stem its budget deficit.

Qatar's reliance on oil and natural gas is likely to persist for the foreseeable future. Proved natural gas reserves exceed 25 trillion cubic meters—13% of the world total and, among countries, third largest in the world. Proved oil reserves exceed 25 billion barrels, allowing production to continue at current levels for about 56 years. Despite the dominance of oil and natural gas, Qatar has made significant gains in strengthening non-oil sectors, such as manufacturing, construction, and financial services, leading non-oil GDP to steadily rise in recent years to just over half the total.

Following trade restriction imposed by Saudi Arabia, the UAE, Bahrain, and Egypt in 2017, Qatar established new trade routes with other countries to maintain access to imports.

GDP (purchasing power parity):
$339.5 billion (2017 est.)
$334.2 billion (2016 est.)
$327.3 billion (2015 est.)
note: data are in 2017 dollars
country comparison to the world: 52

GDP (official exchange rate): $166.9 billion (2017 est.)

GDP—real growth rate:
1.6% (2017 est.)
2.1% (2016 est.)
3.7% (2015 est.)
country comparison to the world: 153

GDP—per capita (PPP):
$124,100 (2017 est.)
$127,700 (2016 est.)
$134,200 (2015 est.)
note: data are in 2017 dollars
country comparison to the world: 2

Gross national saving:
50.2% of GDP (2017 est.)
42.4% of GDP (2016 est.)
47.4% of GDP (2015 est.)
country comparison to the world: 1

GDP—composition, by end use:
household consumption: 24.6% (2017 est.)
government consumption: 17% (2017 est.)
investment in fixed capital: 43.1% (2017 est.)
investment in inventories: 1.5% (2017 est.)
exports of goods and services: 51% (2017 est.)
imports of goods and services: -37.3% (2017 est.)

GDP—composition, by sector of origin:
agriculture: 0.2% (2017 est.)
industry: 50.3% (2017 est.)
services: 49.5% (2017 est.)

Agriculture—products: fruits, vegetables; poultry, dairy products, beef; fish

Industries: liquefied natural gas, crude oil production and refining, ammonia, fertilizer, petrochemicals, steel reinforcing bars, cement, commercial ship repair

Industrial production growth rate: 3% (2017 est.)
country comparison to the world: 105

Labor force: 1.953 million (2017 est.)
country comparison to the world: 122

Unemployment rate: 8.9% (2017 est.)
11.1% (2016 est.)
country comparison to the world: 135

Population below poverty line: NA

Household income or consumption by percentage share: *lowest 10%:* 1.3%
highest 10%: 35.9% (2007)

Budget: *revenues:* 44.1 billion (2017 est.)
expenditures: 53.82 billion (2017 est.)

Taxes and other revenues: 26.4% (of GDP) (2017 est.)
country comparison to the world: 112

Budget surplus (+) or deficit (-): -5.8% (of GDP) (2017 est.)
country comparison to the world: 179

Public debt: 53.8% of GDP (2017 est.)
46.7% of GDP (2016 est.)
country comparison to the world: 88

Fiscal year: 1 April—31 March

Inflation rate (consumer prices): 0.4% (2017 est.)
2.7% (2016 est.)
country comparison to the world: 25

Current account balance: $6.426 billion (2017 est.)
-$8.27 billion (2016 est.)
country comparison to the world: 29

Exports: $67.5 billion (2017 est.)
$57.25 billion (2016 est.)
country comparison to the world: 44

Exports—partners: Japan 17.3%, South Korea 16%, India 12.6%, China 11.2%, Singapore 8.2%, UAE 6.4% (2017)

Exports—commodities: liquefied natural gas (LNG), petroleum products, fertilizers, steel

Imports: $30.77 billion (2017 est.)
$31.93 billion (2016 est.)
country comparison to the world: 67

Imports—commodities: machinery and transport equipment, food, chemicals

Imports—partners: China 10.9%, US 8.9%, UAE 8.5%, Germany 8.1%, UK 5.5%, India 5.4%, Japan 5.3%, Italy 4.3% (2017)

Reserves of foreign exchange and gold: $15.01 billion (31 December 2017 est.)
$31.89 billion (31 December 2016 est.)
country comparison to the world: 68

Debt—external: $167.8 billion (31 December 2017 est.)

$157.9 billion (31 December 2016 est.)
country comparison to the world: 39

Exchange rates: Qatari rials (QAR) per US dollar—
3.64 (2017 est.)
3.64 (2016 est.)
3.64 (2015 est.)
3.64 (2014 est.)
3.64 (2013 est.)

ENERGY

Electricity access: *electrification—total population:* 100% (2020)

Electricity—production: 39.78 billion kWh (2016 est.)
country comparison to the world: 58

Electricity—consumption: 37.24 billion kWh (2016 est.)
country comparison to the world: 58

Electricity—exports: 0 kWh (2016 est.)
country comparison to the world: 186

Electricity—imports: 0 kWh (2016 est.)

country comparison to the world: 188

Electricity—installed generating capacity: 8.796 million kW (2016 est.)
country comparison to the world: 66

Electricity—from fossil fuels: 100% of total installed capacity (2016 est.)
country comparison to the world: 15

Electricity—from nuclear fuels: 0% of total installed capacity (2017 est.)
country comparison to the world: 169

Electricity—from hydroelectric plants: 0% of total installed capacity (2017 est.)
country comparison to the world: 193

Electricity—from other renewable sources: 1% of total installed capacity (2017 est.)
country comparison to the world: 163

Crude oil—production: 1.464 million bbl/day (2018 est.)
country comparison to the world: 17

Crude oil—exports: 1.15 million bbl/day (2015 est.)
country comparison to the world: 13

Crude oil—imports: 0 bbl/day (2015 est.)
country comparison to the world: 183

Crude oil—proved reserves: 25.24 billion bbl (1 January 2018 est.)
country comparison to the world: 13

Refined petroleum products—production: 273,800 bbl/day (2015 est.)
country comparison to the world: 46

Refined petroleum products—consumption: 277,000 bbl/day (2016 est.)
country comparison to the world: 45

Refined petroleum products—exports: 485,000 bbl/day (2015 est.)
country comparison to the world: 18

Refined petroleum products—imports: 12,300 bbl/day (2015 est.)
country comparison to the world: 143

Natural gas—production: 166.4 billion cu m (2017 est.)
country comparison to the world: 4

Natural gas—consumption: 39.9 billion cu m (2017 est.)
country comparison to the world: 26

Natural gas—exports: 126.5 billion cu m (2017 est.)
country comparison to the world: 2

Natural gas—imports: 0 cu m (2017 est.)
country comparison to the world: 176

Natural gas—proved reserves: 24.07 trillion cu m (1 January 2018 est.)
country comparison to the world: 3

Carbon dioxide emissions from consumption of energy: 114.2 million Mt (2017 est.)
country comparison to the world: 40

COMMUNICATIONS

Telephones—fixed lines: total subscriptions: 392,048
subscriptions per 100 inhabitants: 16.29 (2019 est.)
country comparison to the world: 101

Telephones—mobile cellular: total subscriptions: 3,329,155
subscriptions per 100 inhabitants: 138.33 (2019 est.)
country comparison to the world: 138

Telecommunication systems: general assessment: regional leaders in telecom; highest fixed-line and mobile penetrations in Middle East; deployed over 90 5G base stations for 5G launch, claiming 1st commercial launch of 5G in the world May 2018; telecom system centered in Doha; steady LTE networks; good broadband penetration, ADSL, (Fiber-to-the-Home/Premises) FttP, wireless and mobile services; largest users of the Internet and use of OTT services and bundled services (2020)
domestic: fixed-line 16 per 100 and mobile-cellular telephone subscribership 138 telephones per 100 persons (209)
international: country code—974; landing points for the Qatar-UAE Submarine Cable System, AAE-1, FOG, GBICS/East North Africa MENA and the FALCON submarine cable network that provides links to Asia, Africa, the Middle East, Europe and Southeast Asia; tropospheric scatter to Bahrain; microwave radio relay to Saudi Arabia and the UAE; satellite earth stations—2 Intelsat (1 Atlantic Ocean and 1 Indian Ocean) and 1 Arabsat; retains full ownership of two commercial satellites, Es'hailSat 1 and 2 (2019)
note: the COVID-19 outbreak is negatively impacting telecommunications production and supply chains globally; consumer spending on telecom devices and services has also slowed due to the pandemic's effect on economies worldwide; overall progress towards improvements in all facets of the telecom industry—mobile, fixed-line, broadband, submarine cable and satellite—has moderated

Broadcast media: TV and radio broadcast licensing and access to local media markets are state controlled; home of the satellite TV channel Al-Jazeera, which was originally owned and financed by the Qatari government but has evolved to independent corporate status; Al-Jazeera claims editorial independence in broadcasting; local radio transmissions include state, private, and international broadcasters on FM frequencies in Doha; in August 2013, Qatar's satellite company Es'hailSat launched its first communications satellite Es'hail 1 (manufactured in the US), which entered commercial service in December 2013 to provide improved television broadcasting capability and expand availability of voice and Internet; Es'hailSat launched its second commercial satellite in 2018 with aid of SpaceX (2019)

Internet country code: .qa

Internet users: total: 2,355,297
percent of population: 99.65% (July 2018 est.)
country comparison to the world: 112

Broadband—fixed subscriptions: total: 267,906
subscriptions per 100 inhabitants: 11 (2018 est.)
country comparison to the world: 105

TRANSPORTATION

National air transport system: number of registered air carriers: 3 (2020)

inventory of registered aircraft operated by air carriers: 251
annual passenger traffic on registered air carriers: 29,178,923 (2018)
annual freight traffic on registered air carriers: 12,666,710,000 mt-km (2018)

Civil aircraft registration country code prefix: A7 (2016)

Airports: 6 (2013)
country comparison to the world: 175

Airports—with paved runways: total: 4 (2017)
over 3,047 m: 3 (2017)
1,524 to 2,437 m: 1 (2017)

Airports—with unpaved runways: total: 2 (2013)
914 to 1,523 m: 1 (2013)
under 914 m: 1 (2013)

Heliports: 1 (2013)

Pipelines: 288 km condensate, 221 km condensate/gas, 2383 km gas, 90 km liquid petroleum gas, 745 km oil, 103 km refined products (2013)

Roadways: total: 7,039 km (2016)
country comparison to the world: 142

Merchant marine: total: 136
by type: bulk carrier 9, container ship 6, general cargo 5, oil tanker 7, other 109 (2019)
country comparison to the world: 77

Ports and terminals: major seaport(s): Doha, Musay'id, Ra's Laffan
LNG terminal(s) (export): Ras Laffan

MILITARY AND SECURITY

Military and security forces: Qatari Amiri Land Force (QALF, includes Emiri Guard), Qatari Amiri Navy (QAN, includes Coast Guard), Qatari Amiri Air Force (QAAF); Internal Security Forces: Mobile Gendarmerie (2019)

Military and security service personnel strengths: size assessments for the Qatari Amiri military vary; approximately 14,000 active personnel (10,000 Land Force, including Emiri Guard; 2,000 Navy, including Coast Guard; 2,000 Air Force); est. 5,000 Internal Security Forces (2019)

Military equipment inventories and acquisitions: the Qatari military's inventory includes a mix of older and modern weapons systems, mostly from the US and Europe, particularly France, Germany, and the UK; the leading providers of armaments to Qatar since 2010 are France, Germany, and the US; Qatar is scheduled to receive several ships from Italy beginning in 2021 and a large shipment of fighter aircraft from the UK in 2022 (2019 est.)

Military service age and obligation: conscription for males aged 18-35; compulsory service times range from 4 months to up to a year, depending on the cadets educational and professional circumstances; women are permitted to serve in the armed forces, including as uniformed officers and pilots (2019)

TERRORISM

TRANSNATIONAL ISSUES

Disputes—international: none

Refugees and internally displaced persons: *stateless persons:* 1,200 (2019)

Trafficking in persons: *current situation:* Qatar is a destination country for men, women, and children subjected to forced labor, and, to a much lesser extent, forced prostitution; the predominantly foreign workforce migrates to Qatar legally for low- and semi-skilled work but often experiences situations of forced labor, including debt bondage, delayed or nonpayment of salaries, confiscation of passports, abuse, hazardous working conditions, and squalid living arrangements; foreign female domestic workers are particularly vulnerable to trafficking because of their isolation in private homes and lack of protection under Qatari labor laws; some women who migrate for work are also forced into prostitution

tier rating: Tier 2 Watch List—Qatar does not fully comply with the minimum standards for the elimination of trafficking; however, it is making significant efforts to do so; the government investigated 11 trafficking cases but did not prosecute or convict any offenders, including exploitative employers and recruitment agencies; the primary solution for resolving labor violations was to transfer a worker's sponsorship to a new employer with minimal effort to investigate whether a forced labor violation had occurred; authorities increased their efforts to protect some trafficking victims, although many victims of forced labor, particularly domestic workers, remained unidentified and unprotected and were sometimes punished for immigration violations or running away from an employer or sponsor; authorities visited worksites throughout the country to meet and educate workers and employers on trafficking regulations, but the government failed to abolish or reform the sponsorship system, perpetuating Qatar's forced labor problem (2015)

INTRODUCTION

Background: The principalities of Wallachia and Moldavia—for centuries under the suzerainty of the Turkish Ottoman Empire—secured their autonomy in 1856; they were de facto linked in 1859 and formally united in 1862 under the new name of Romania. The country gained recognition of its independence in 1878. It joined the Allied Powers in World War I and acquired new territories—most notably Transylvania—following the conflict. In 1940, Romania allied with the Axis powers and participated in the 1941 German invasion of the USSR. Three years later, overrun by the Soviets, Romania signed an armistice. The post-war Soviet occupation led to the formation of a communist "people's republic" in 1947 and the abdication of the king. The decades-long rule of dictator Nicolae CEAUSESCU, who took power in 1965, and his Securitate police state became increasingly oppressive and draconian through the 1980s. CEAUSESCU was overthrown and executed in late 1989. Former communists dominated the government until 1996 when they were swept from power. Romania joined NATO in 2004 and the EU in 2007.

GEOGRAPHY

Location: Southeastern Europe, bordering the Black Sea, between Bulgaria and Ukraine

Geographic coordinates: 46 00 N, 25 00 E

Map references: Europe

Area: *total:* 238,391 sq km
land: 229,891 sq km
water: 8,500 sq km
country comparison to the world: 84

Area—comparative: twice the size of Pennsylvania; slightly smaller than Oregon

Land boundaries: *total:* 2,844 km
border countries (5): Bulgaria 605 km, Hungary 424 km, Moldova 683 km, Serbia 531 km, Ukraine 601 km

Coastline: 225 km

Maritime claims: *territorial sea:* 12 nm
exclusive economic zone: 200 nm
contiguous zone: 24 nm
continental shelf: 200-m depth or to the depth of exploitation

Climate: temperate; cold, cloudy winters with frequent snow and fog; sunny summers with frequent showers and thunderstorms

Terrain: central Transylvanian Basin is separated from the Moldavian Plateau on the east by the Eastern Carpathian Mountains and separated from the Walachian Plain on the south by the Transylvanian Alps

Elevation: *mean elevation:* 414 m
lowest point: Black Sea 0 m
highest point: Moldoveanu 2,544 m

Natural resources: petroleum (reserves declining), timber, natural gas, coal, iron ore, salt, arable land, hydropower

Land use: *agricultural land:* 60.7% (2011 est.)

arable land: 39.1% (2011 est.) / *permanent crops:* 1.9% (2011 est.) / permanent pasture: 19.7% (2011 est.)
forest: 28.7% (2011 est.)
other: 10.6% (2011 est.)

Irrigated land: 31,490 sq km (2012)

Population distribution: urbanization is not particularly high, and a fairly even population distribution can be found throughout most of the country, with urban areas attracting larger and denser populations; Hungarians, the country's largest minority, have a particularly strong presence in eastern Transylvania

Natural hazards: earthquakes, most severe in south and southwest; geologic structure and climate promote landslides

Environment—current issues: soil erosion, degradation, and desertification; water pollution; air pollution in south from industrial effluents; contamination of Danube delta wetlands

Environment—international agreements: *party to:* Air Pollution, Air Pollution-Persistent Organic Pollutants, Antarctic-Environmental Protocol, Antarctic Treaty, Biodiversity, Climate Change, Climate Change-Kyoto Protocol, Desertification, Endangered Species, Environmental Modification, Hazardous Wastes, Law of the Sea, Ozone Layer Protection, Ship Pollution, Wetlands
signed, but not ratified: none of the selected agreements

Geography—note: controls the most easily traversable land route between the Balkans, Moldova, and Ukraine; the Carpathian Mountains dominate the center of the country, while the Danube River forms much of the southern boundary with Serbia and Bulgaria

PEOPLE AND SOCIETY

Population: 21,302,893 (July 2020 est.)

country comparison to the world: 59

Nationality: *noun:* Romanian(s)
adjective: Romanian

Ethnic groups: Romanian 83.4%, Hungarian 6.1%, Romani 3.1%, Ukrainian 0.3%, German 0.2%, other 0.7%, unspecified 6.1% (2011 est.)
note: Romani populations are usually underestimated in official statistics and may represent 5-11% of Romania's population

Languages: Romanian (official) 85.4%, Hungarian 6.3%, Romani 1.2%, other 1%, unspecified 6.1% (2011 est.)

Religions: Eastern Orthodox (including all sub-denominations) 81.9%, Protestant (various denominations including Reformed and Pentecostal) 6.4%, Roman Catholic 4.3%, other (includes Muslim) 0.9%, none or atheist 0.2%, unspecified 6.3% (2011 est.)

Age structure: *0-14 years:* 14.12% (male 1,545,196/female 1,463,700)
15-24 years: 10.31% (male 1,126,997/female 1,068,817)
25-54 years: 46.26% (male 4,993,886/female 4,860,408)
55-64 years: 11.73% (male 1,176,814/female 1,322,048)
65 years and over: 17.58% (male 1,516,472/female 2,228,555) (2020 est.)

Dependency ratios: *total dependency ratio:* 53.3
youth dependency ratio: 23.8
elderly dependency ratio: 29.5
potential support ratio: 3.4 (2020 est.)

Median age: *total:* 42.5 years
male: 41 years
female: 44 years (2020 est.)
country comparison to the world: 34

Population growth rate: -0.37% (2020 est.)
country comparison to the world: 222

Birth rate: 8.5 births/1,000 population (2020 est.)
country comparison to the world: 216

Death rate: 12 deaths/1,000 population (2020 est.)
country comparison to the world: 18

Net migration rate: -0.2 migrant(s)/1,000 population (2020 est.)
country comparison to the world: 109

Population distribution: urbanization is not particularly high, and a fairly even population distribution can be found throughout most of the country, with urban areas attracting larger and denser populations; Hungarians, the country's largest minority, have a particularly strong presence in eastern Transylvania

Urbanization: *urban population:* 54.2% of total population (2020)
rate of urbanization: -0.38% annual rate of change (2015-20 est.)

total population growth rate v. urban population growth rate, 2000-2030:

Major urban areas—population: *1.803 million* BUCHAREST (capital) (2020)

Sex ratio: *at birth:* 1.06 male(s)/female
0-14 years: 1.06 male(s)/female
15-24 years: 1.05 male(s)/female
25-54 years: 1.03 male(s)/female
55-64 years: 0.89 male(s)/female
65 years and over: 0.68 male(s)/female
total population: 0.95 male(s)/female (2020 est.)

Mother's mean age at first birth: 27.1 years (2017 est.)

Maternal mortality rate: 19 deaths/100,000 live births (2017 est.)
country comparison to the world: 127

Infant mortality rate: *total:* 8.7 deaths/1,000 live births
male: 9.9 deaths/1,000 live births
female: 7.4 deaths/1,000 live births (2020 est.)
country comparison to the world: 144

Life expectancy at birth: *total population:* 76 years
male: 72.6 years
female: 79.7 years (2020 est.)
country comparison to the world: 107

Total fertility rate: 1.38 children born/woman (2020 est.)
country comparison to the world: 219

Drinking water source:
improved:
urban: 100% of population
rural: 100% of population
total: 100% of population
unimproved:
urban: 0% of population
rural: 0% of population
total: 0% of population (2017 est.)

Current Health Expenditure: 5.2% (2017)

Physicians density: 2.98 physicians/1,000 population (2017)

Hospital bed density: 6.9 beds/1,000 population (2017)

Sanitation facility access:
improved:
urban: 95.3% of population
rural: 71.5% of population
total: 84.3% of population
unimproved:
urban: 4.7% of population
rural: 28.5% of population
total: 15.7% of population (2017 est.)

HIV/AIDS—adult prevalence rate: 0.1% (2019 est.)
country comparison to the world: 133

HIV/AIDS—people living with HIV/AIDS: 190,000 (2019 est.)
country comparison to the world: 33

HIV/AIDS—deaths: <500 (2019 est.)

Obesity—adult prevalence rate: 22.5% (2016)
country comparison to the world: 75

Education expenditures: 3.1% of GDP (2015)
country comparison to the world: 136

Literacy: *definition:* age 15 and over can read and write
total population: 98.8%

male: 99.1%
female: 98.6% (2018)

School life expectancy (primary to tertiary education): *total:* 14 years
male: 14 years
female: 15 years (2018)

Unemployment, youth ages 15-24: *total:* 16.2%
male: 16.3%
female: 16.2% (2018 est.)
country comparison to the world: 84

GOVERNMENT

Country name: *conventional long form:* none
conventional short form: Romania
local long form: none
local short form: Romania
former: Kingdom of Romania, Romanian People's Republic, Socialist Republic of Romania
etymology: the name derives from the Latin "Romanus" meaning "citizen of Rome" and was used to stress the common ancient heritage of Romania's three main regions—Moldavia, Transylvania, and Wallachia—during their gradual unification between the mid-19th century and early 20th century

Government type: semi-presidential republic

Capital: *name:* Bucharest
geographic coordinates: 44 26 N, 26 06 E
time difference: UTC+2 (7 hours ahead of Washington, DC, during Standard Time)
daylight saving time: +1hr, begins last Sunday in March; ends last Sunday in October
etymology: related to the Romanian word "bucura" that is believed to be of Dacian origin and whose meaning is "to be glad (happy)"; Bucharest's meaning is thus akin to "city of joy"

Administrative divisions: 41 counties (judete, singular—judet) and 1 municipality* (municipiu); Alba, Arad, Arges, Bacau, Bihor, Bistrita-Nasaud, Botosani, Braila, Brasov, Bucuresti (Bucharest)*, Buzau, Calarasi, Caras-Severin, Cluj, Constanta, Covasna, Dambovita, Dolj, Galati, Gorj, Giurgiu, Harghita, Hunedoara, Ialomita, Iasi, Ilfov, Maramures, Mehedinti,Mures, Neamt, Olt, Prahova, Salaj, Satu Mare, Sibiu, Suceava, Teleorman, Timis, Tulcea, Vaslui, Valcea, Vrancea

Independence: 9 May 1877 (independence proclaimed from the Ottoman Empire; 13 July 1878 (independence recognized by the Treaty of Berlin); 26 March 1881 (kingdom proclaimed); 30 December 1947 (republic proclaimed)

National holiday: Unification Day (unification of Romania and Transylvania), 1 December (1918)

Constitution: *history:* several previous; latest adopted 21 November 1991, approved by referendum and effective 8 December 1991
amendments: initiated by the president of Romania through a proposal by the government, by at least one fourth of deputies or senators in Parliament, or by petition of eligible voters representing at least half of Romania's counties; passage requires at least two-thirds majority vote by both chambers or — if mediation is required—by

three-fourths majority vote in a joint session, followed by approval in a referendum; articles, including those on national sovereignty, form of government, political pluralism, and fundamental rights and freedoms, cannot be amended; amended 2003

Legal system: civil law system

International law organization participation: accepts compulsory ICJ jurisdiction with reservations; accepts ICCt jurisdiction

Citizenship: *citizenship by birth:* no
citizenship by descent only: at least one parent must be a citizen of Romania
dual citizenship recognized: yes
residency requirement for naturalization: 5 years

Suffrage: 18 years of age; universal

Executive branch: *chief of state:* President Klaus Werner IOHANNIS (since 21 December 2014)
head of government: Prime Minister Ludovic ORBAN (since 4 November 2019); Deputy Prime Minister Raluca TURCAN (since 4 November 2019); note—Prime Minister ORBAN lost a no-confidence vote on 5 February 2020; President IOHANNIS asked ORBAN to form a new government on 6 February 2020; Prime Minister ORBAN announced an unchanged government on 10 February 2020; on 24 February, the Constitutional Court rules that the president must nominate for Prime Minister someone who can get enough support in parliament to assume office, not a Prime Minister-designate who has been previously ousted in a no-confidence vote; on 13 March President IOHANNIS again asked ORBAN to form a new government; Prime Minister ORBAN's unchanged cabinet was approved by parliament on 14 March 2020
cabinet: Council of Ministers appointed by the prime minister
elections/appointments: president directly elected by absolute majority popular vote in 2 rounds if needed for a 5-year term (eligible for a second term); election last held on 10 November 2019 with a runoff on 24 November 2019 (next to be held in November 2024); prime minister appointed by the president with consent of Parliament
election results: Klaus IOHANNIS reelected president in second round; percent of vote—Klaus IOHANNIS (PNL) 66.1%, Viorica DANCILA (PSD) 33.9%; Ludovic ORBAN approved as prime minister with 240 votes

Legislative branch: *description:* bicameral Parliament or Parlament consists of:

Senate or Senat (136 seats; members directly elected in single- and multi-seat constituencies—including 2 seats for diaspora—by party-list, proportional representation vote; members serve 4-year terms)

Chamber of Deputies or Camera Deputatilor (329 seats; members directly elected in single- and multi-seat constituencies—including 4 seats for diaspora—by party-list, proportional representation vote; members serve 4-year terms)

elections: Senate—last held on 11 December 2016 (next to be held on 6 December 2020)

791

Chamber of Deputies—last held on 11 December 2016 (next to be held on 6 December 2020)

election results: Senate—percent of vote by party—PSD 45.7%, PNL 20.4%, USR 8.9%, UDMR 6.2%, ALDE 6%, PMP 5.7%, other 7.1%; seats by party—PSD 67, PNL 30, USR 13, UDMR 9, ALDE 9, PMP 8; composition—men 116, women 20, percent of women 14.7%

Chamber of Deputies—percent of vote by party—PSD 45.5%, PNL 20%, USR 8.9%, UDMR 6.2%, ALDE 5.6%, PMP 5.4%, other 8.4%; seats by party—PSD 154, PNL 69, USR 30, UDMR 21, ALDE 20, PMP 18, minorities 17; composition men 261, women 68, percent of women 20.7%; note—total Parliament percent of women 20.7%

Judicial branch: *highest courts:* High Court of Cassation and Justice (consists of 111 judges organized into civil, penal, commercial, contentious administrative and fiscal business, and joint sections); Supreme Constitutional Court (consists of 9 members)

judge selection and term of office: High Court of Cassation and Justice judges appointed by the president upon nomination by the Superior Council of Magistracy, a 19-member body of judges, prosecutors, and law specialists; judges appointed for 6-year renewable terms; Constitutional Court members—6 elected by Parliament and 3 appointed by the president; members serve 9-year, nonrenewable terms

subordinate courts: Courts of Appeal; regional tribunals; first instance courts; military and arbitration courts

Political parties and leaders: Christian-Democratic National Peasants' Party or PNT CD [Aurelian PAVELESCU]

Democratic Union of Hungarians in Romania or UDMR [Hunor KELEMEN]

Civic Hungarian Party [Zsolt BIRO]

Ecologist Party of Romania or PER [Danut POP]

Greater Romania Party or PRM [Adrian POPESCU]

M10 Party [Ioana CONSTANTIN]

National Liberal Party or PNL [Ludovic ORBAN]

New Romania Party or PNR [Sebastian POPESCU]

Our Romania Alliance [Marian MUNTEANU]

Party of the Alliance of Liberals and Democrats or ALDE [Calin POPESCU TARICEANU]

Popular Movement Party or PMP [Traian BASESCU]

Romanian Social Party or PSRo [Mircea GEOANA]

Save Romania Union Party or Partidul USR [Dan BARNA]

Social Democratic Party or PSD [Marcel CIOLACU, interim leader]

United Romania Party or PRU [Robert BUGA]

International organization participation: Australia Group, BIS, BSEC, CBSS (observer), CD, CE, CEI, EAPC, EBRD, ECB, EIB, ESA, EU, FAO, G-9, IAEA, IBRD, ICAO, ICC (national committees), ICCt, ICRM, IDA, IFAD, IFC, IFRCS, IHO, ILO, IMF, IMO, IMSO, Interpol, IOC, IOM, IPU, ISO, ITSO, ITU, ITUC (NGOs), LAIA (observer), MIGA, MONUSCO, NATO, NSG, OAS (observer), OIF, OPCW, OSCE, PCA, SELEC, UN, UNCTAD, UNESCO, UNHCR, UNIDO, Union Latina, UNMIL, UNMISS, UNOCI, UNWTO, UPU, WCO, WFTU (NGOs), WHO, WIPO, WMO, WTO, ZC

Diplomatic representation in the US: *chief of mission:* Ambassador George Cristian MAIOR (since 17 September 2015)

chancery: 1607 23rd Street NW, Washington, DC 20008

telephone: [1] (202) 332-4846, 4848, 4851, 4852

FAX: [1] (202) 232-4748

consulate(s) general: Chicago, Los Angeles, New York

Diplomatic representation from the US: *chief of mission:* Ambassador Adrian ZUCKERMAN (since 17 December 2019)

telephone: [40] (21) 200-3300

embassy: 4-6, Dr. Liviu Librescu Blvd., District 1, Bucharest, 015118

mailing address: American Embassy Bucharest, US Department of State, 5260 Bucharest Place, Washington, DC 20521-5260 (pouch)

FAX: [40] (21) 200-3442

Flag description: three equal vertical bands of cobalt blue (hoist side), chrome yellow, and vermilion red; modeled after the flag of France, the colors are those of the principalities of Walachia (red and yellow) and Moldavia (red and blue), which united in 1862 to form Romania; the national coat of arms that used to be centered in the yellow band has been removed

note: now similar to the flag of Chad, whose blue band is darker; also resembles the flags of Andorra and Moldova

National symbol(s): *golden eagle; national colors:* blue, yellow, red

National anthem: *name:* "Desteapta-te romane!" (Wake up, Romanian!)

lyrics/music: Andrei MURESIANU/Anton PANN

note: adopted 1990; the anthem was written during the 1848 Revolution

0:00/ 2:18

ECONOMY

Economy—overview: Romania, which joined the EU on 1 January 2007, began the transition from communism in 1989 with a largely obsolete industrial base and a pattern of output unsuited to the country's needs. Romania's macroeconomic gains have only recently started to spur creation of a middle class and to address Romania's widespread poverty. Corruption and red tape continue to permeate the business environment.

In the aftermath of the global financial crisis, Romania signed a $26 billion emergency assistance package from the IMF, the EU, and other international lenders, but GDP contracted until 2011. In March 2011, Romania and the IMF/EU/World Bank signed a 24-month precautionary standby agreement, worth $6.6 billion, to promote fiscal discipline, encourage progress on structural reforms, and strengthen financial sector stability; no funds were drawn. In September 2013, Romanian authorities and the IMF/EU agreed to a follow-on standby agreement, worth $5.4 billion, to continue with reforms. This agreement expired in September 2015, and no funds were drawn. Progress on structural reforms has been uneven, and the economy still is vulnerable to external shocks.

Economic growth rebounded in the 2013-17 period, driven by strong industrial exports, excellent agricultural harvests, and, more recently, expansionary fiscal policies in 2016-2017 that nearly quadrupled Bucharest's annual fiscal deficit, from +0.8% of GDP in 2015 to -3% of GDP in 2016 and an estimated -3.4% in 2017. Industry outperformed other sectors of the economy in 2017. Exports remained an engine of economic growth, led by trade with the EU, which accounts for roughly 70% of Romania trade. Domestic demand was the major driver, due to tax cuts and large wage increases that began last year and are set to continue in 2018.

An aging population, emigration of skilled labor, significant tax evasion, insufficient health care, and an aggressive loosening of the fiscal package compromise Romania's long-term growth and economic stability and are the economy's top vulnerabilities.

GDP (purchasing power parity): $483.4 billion (2017 est.)

$452 billion (2016 est.)

$431.2 billion (2015 est.)

note: data are in 2017 dollars

country comparison to the world: 41

GDP (official exchange rate): $211.9 billion (2017 est.)

GDP—real growth rate: 4.2% (2019 est.)

4.54% (2018 est.)

7.11% (2017 est.)

country comparison to the world: 71

GDP—per capita (PPP): $24,600 (2017 est.)

$22,900 (2016 est.)

$21,700 (2015 est.)

note: data are in 2017 dollars

country comparison to the world: 83

Gross national saving: 21.1% of GDP (2017 est.)

21.7% of GDP (2016 est.)

23.9% of GDP (2015 est.)

country comparison to the world: 87

GDP—composition, by end use: *household consumption:* 70% (2017 est.)

government consumption: 7.7% (2017 est.)

investment in fixed capital: 22.6% (2017 est.)

investment in inventories: 1.9% (2017 est.)

exports of goods and services: 41.4% (2017 est.)

imports of goods and services: -43.6% (2017 est.)

GDP—composition, by sector of origin: *agriculture:* 4.2% (2017 est.)

industry: 33.2% (2017 est.)

services: 62.6% (2017 est.)

Agriculture—products: wheat, corn, barley, sugar beets, sunflower seed, potatoes, grapes; eggs, sheep

Industries: electric machinery and equipment, auto assembly, textiles and footwear, light machinery, metallurgy, chemicals, food processing, petroleum refining, mining, timber, construction materials

Industrial production growth rate: 5.5% (2017 est.)
country comparison to the world: 50

Labor force: 4.889 million (2020 est.)
country comparison to the world: 80

Labor force—by occupation: *agriculture:* 28.3%
industry: 28.9%
services: 42.8% (2014)

Unemployment rate: 3.06% (2019 est.)
3.56% (2018 est.)
country comparison to the world: 40

Population below poverty line: 22.4% (2012 est.)

Household income or consumption by percentage share: *lowest 10%:* 15.3%
highest 10%: 7.6% (2014 est.)

Budget: *revenues:* 62.14 billion (2017 est.)
expenditures: 68.13 billion (2017 est.)

Taxes and other revenues: 29.3% (of GDP) (2017 est.)
country comparison to the world: 84

Budget surplus (+) or deficit (-): -2.8% (of GDP) (2017 est.)
country comparison to the world: 125

Public debt: 36.8% of GDP (2017 est.)
38.8% of GDP (2016 est.)

note: *defined by the EU's Maastricht Treaty as consolidated general government gross debt at nominal value, outstanding at the end of the year in the following categories of government liabilities: currency and deposits, securities other than shares excluding financial derivatives, and loans; general government sector comprises the subsectors: central government, state government, local government, and social security funds*
country comparison to the world: 145

Fiscal year: calendar year

Inflation rate (consumer prices): 1.3% (2017 est.)
-1.6% (2016 est.)
country comparison to the world: 71

Current account balance: -$11.389 billion (2019 est.)
-$10.78 billion (2018 est.)
country comparison to the world: 194

Exports: $64.58 billion (2017 est.)
$57.72 billion (2016 est.)
country comparison to the world: 45

Exports—partners: Germany 23%, Italy 11.2%, France 6.8%, Hungary 4.7%, UK 4.1% (2017)

Exports—commodities: machinery and equipment, other manufactured goods, agricultural products and foodstuffs, metals and metal products, chemicals, minerals and fuels, raw materials

Imports: $78.12 billion (2017 est.)
$68 billion (2016 est.)
country comparison to the world: 43

Imports—commodities: machinery and equipment, other manufactured goods, chemicals, agricultural products and foodstuffs, fuels and minerals, metals and metal products, raw materials

Imports—partners: Germany 20%, Italy 10%, Hungary 7.5%, Poland 5.5%, France 5.3%, China 5%, Netherlands 4% (2017)

Reserves of foreign exchange and gold: $44.43 billion (31 December 2017 est.)
$40 billion (31 December 2016 est.)
country comparison to the world: 43

Debt—external: $95.97 billion (31 December 2017 est.)

$93.71 billion (31 December 2016 est.)
country comparison to the world: 49

Exchange rates: lei (RON) per US dollar –
4.077 (2017 est.)
4.0592 (2016 est.)
4.0592 (2015 est.)
4.0057 (2014 est.)
3.3492 (2013 est.)

ENERGY

Electricity access: *electrification—total population:* 100% (2020)

Electricity—production: 61.78 billion kWh (2016 est.)
country comparison to the world: 47

Electricity—consumption: 49.64 billion kWh (2016 est.)
country comparison to the world: 49

Electricity—exports: 11.22 billion kWh (2015 est.)
country comparison to the world: 18

Electricity—imports: 4.177 billion kWh (2016 est.)
country comparison to the world: 45

Electricity—installed generating capacity: 23.94 million kW (2016 est.)
country comparison to the world: 38

Electricity—from fossil fuels: 47% of total installed capacity (2016 est.)
country comparison to the world: 156

Electricity—from nuclear fuels: 6% of total installed capacity (2017 est.)
country comparison to the world: 20

Electricity—from hydroelectric plants: 29% of total installed capacity (2017 est.)
country comparison to the world: 70

Electricity—from other renewable sources: 19% of total installed capacity (2017 est.)
country comparison to the world: 45

Crude oil—production: 70,000 bbl/day (2018 est.)
country comparison to the world: 46

Crude oil—exports: 2,076 bbl/day (2015 est.)
country comparison to the world: 70

Crude oil—imports: 145,300 bbl/day (2015 est.)
country comparison to the world: 38

Crude oil—proved reserves: 600 million bbl (1 January 2018 est.)
country comparison to the world: 42

Refined petroleum products—production: 232,600 bbl/day (2015 est.)
country comparison to the world: 47

Refined petroleum products—consumption: 198.000 bbl/day (2016 est.)
country comparison to the world: 58

Refined petroleum products—exports: 103.000 bbl/day (2015 est.)
country comparison to the world: 44

Refined petroleum products—imports: 49,420 bbl/day (2015 est.)
country comparison to the world: 81

Natural gas—production: 10.87 billion cu m (2017 est.)
country comparison to the world: 40

Natural gas—consumption: 11.58 billion cu m (2017 est.)
country comparison to the world: 45

Natural gas—exports: 22.65 million cu m (2017 est.)
country comparison to the world: 53

Natural gas—imports: 1.218 billion cu m (2017 est.)
country comparison to the world: 59

Natural gas—proved reserves: 105.5 billion cu m (1 January 2018 est.)
country comparison to the world: 50

Carbon dioxide emissions from consumption of energy: 72.07 million Mt (2017 est.)
country comparison to the world: 50

COMMUNICATIONS

Telephones—fixed lines: *total subscriptions:* 3,731,047
subscriptions per 100 inhabitants: 17.45 (2019 est.)
country comparison to the world: 36

Telephones—mobile cellular: *total subscriptions:* 25,033,292
subscriptions per 100 inhabitants: 117.08 (2019 est.)
country comparison to the world: 50

Telecommunication systems: *general assessment: the telecommunications sector is being expanded; domestic and international service improving rapidly, especially mobile-cellular services; competition among a number of telecoms; LTE and 5G services; 1Gb/FttP offering; govt. secures EU funding to extend broadband to areas of the country not yet connected and does away with SIM card registration; operators invest in networks capacity upgrades (2020)*
domestic: fixed-line teledensity is about 17 telephones per 100 persons; mobile market served by four mobile network operators; mobile-cellular teledensity over 117 telephones per 100 persons (2019)
international: country code—40; landing point for the Diamond Link Global submarine cable linking Romania with Georgia; satellite earth stations—10; digital, international, direct-dial exchanges operate in Bucharest (2019)

note: the COVID-19 outbreak is negatively impacting telecommunications production and supply chains globally; consumer spending on telecom devices and services has also slowed due to the pandemic's effect on economies worldwide; overall progress towards improvements in all facets of the telecom industry—mobile, fixed-line, broadband, submarine cable and satellite—has moderated

Broadcast media: a mixture of public and private TV stations; there are 7 public TV stations (2 national, 5 regional) using terrestrial broadcasting and 187 private TV stations (out of which 171 offer local coverage) using terrestrial broadcasting, plus 11 public TV stations using satellite broadcasting and 86 private TV stations using satellite broadcasting; state-owned public radio broadcaster operates 4 national networks and regional and local stations, having in total 20 public radio stations by terrestrial broadcasting plus 4 public radio stations by satellite broadcasting; there are 502 operational private radio stations using terrestrial broadcasting and 26 private radio stations using satellite broadcasting

Internet country code: .ro

Internet users: *total:* 15,165,890
percent of population: 70.68% (July 2018 est.)
country comparison to the world: 43

Broadband—fixed subscriptions: *total:* 5.083 million
subscriptions per 100 inhabitants: 24 (2018 est.)
country comparison to the world: 30

TRANSPORTATION

National air transport system: *number of registered air carriers: 8 (2020)*
inventory of registered aircraft operated by air carriers: 60
annual passenger traffic on registered air carriers: 4,908,235 (2018)
annual freight traffic on registered air carriers: 2.71 million mt-km (2018)

Civil aircraft registration country code prefix: YR (2016)

Airports: 45 (2013)
country comparison to the world: 95

Airports—with paved runways: *total:* 26 (2017)

over 3,047 m: 4 (2017)
2,438 to 3,047 m: 10 (2017)
1,524 to 2,437 m: 11 (2017)
under 914 m: 1 (2017)

Airports—with unpaved runways: *total:* 19 (2013)
914 to 1,523 m: 5 (2013)
under 914 m: 14 (2013)

Heliports: 2 (2013)

Pipelines: 3726 km gas, 2451 km oil (2013)

Railways: *total:* 11,268 km (2014)
standard gauge: 10,781 km 1.435-m gauge (3,292 km electrified) (2014)
narrow gauge: 427 km 0.760-m gauge (2014)
broad gauge: 60 km 1.524-m gauge (2014)
country comparison to the world: 23

Roadways: *total:* 84,185 km (2012)
paved: 49,873 km (includes 337 km of expressways) (2012)
unpaved: 34,312 km (2012)
country comparison to the world: 59

Waterways: 1,731 km (includes 1,075 km on the Danube River, 524 km on secondary branches, and 132 km on canals) (2010)
country comparison to the world: 45

Merchant marine: *total:* 120
by type: general cargo 11, oil tanker 7, other 102 (2019)
country comparison to the world: 80

Ports and terminals: *major seaport(s):* Constanta, Midia
river port(s): Braila, Galati (Galatz), Mancanului (Giurgiu), Tulcea (Danube River)

MILITARY AND SECURITY

Military and security forces: *Romanian Armed Forces:* Land Forces, Naval Forces, Air Force; Ministry of Internal Affairs: Romanian Gendarmerie (2019)

Military expenditures: 2.04% of GDP (2019 est.)
1.82% of GDP (2018)
1.72% of GDP (2017)
1.4% of GDP (2016)
1.45% of GDP (2015)

country comparison to the world: 47

Military and security service personnel strengths: the Romanian Armed Forces have approximately 72,000 active duty personnel (40,000 Land Forces; 7,000 Naval Forces; 10,000 Air Force; 15,000 joint) (2019 est.)

Military equipment inventories and acquisitions: the inventory of the Romanian Armed Forces is comprised mostly of Soviet-era and older domestically-produced weapons systems; there is also a smaller mix of Western-origin equipment; Italy, Portugal (second-hand fighter aircraft), and the US are the leading suppliers of armaments to Romania since 2010 (2019 est.)

Military deployments: 740 Afghanistan (NATO); 240 Mali (MINUSMA/EUTM); up to 120 Poland (NATO) (2020)

Military service age and obligation: conscription ended 2006; 18 years of age for male and female voluntary service; all military inductees (including women) contract for an initial 5-year term of service, with subsequent successive 3-year terms until age 36 (2015)

TRANSNATIONAL ISSUES

Disputes—international: the ICJ ruled largely in favor of Romania in its dispute submitted in 2004 over Ukrainian-administered Zmiyinyy/Serpilor (Snake) Island and Black Sea maritime boundary delimitation; Romania opposes Ukraine's reopening of a navigation canal from the Danube border through Ukraine to the Black Sea

Refugees and internally displaced persons: *stateless persons:* 192 (2019)
note: 6,036 estimated refugee and migrant arrivals (January 2015-November 2020)

Illicit drugs: major transshipment point for Southwest Asian heroin transiting the Balkan route and small amounts of Latin American cocaine bound for Western Europe; although not a significant financial center, role as a narcotics conduit leaves it vulnerable to laundering, which occurs via the banking system, currency exchange houses, and casinos

RUSSIA

INTRODUCTION

Background: Founded in the 12th century, the Principality of Muscovy was able to emerge from over 200 years of Mongol domination (13th-15th centuries) and to gradually conquer and absorb surrounding principalities. In the early 17th century, a new ROMANOV Dynasty continued this policy of expansion across Siberia to the Pacific. Under PETER I (ruled 1682-1725), hegemony was extended to the Baltic Sea and the country was renamed the Russian Empire. During the 19th century, more territorial acquisitions were made in Europe and Asia. Defeat in the Russo-Japanese War of 1904-05 contributed to the Revolution of 1905, which resulted in the formation of a parliament and other reforms. Devastating defeats and food shortages in World War I led to widespread rioting in the major cities of the Russian Empire and to the overthrow in 1917 of the ROMANOV Dynasty. The communists under Vladimir LENIN seized power soon after and formed the USSR. The brutal rule of Iosif STALIN (1928-53) strengthened communist rule and Russian dominance of the Soviet Union at a cost of tens of millions of lives. After defeating Germany in World War II as part of an alliance with the US (1939-1945), the USSR expanded its territory and influence in Eastern Europe and emerged as a global power. The USSR was the principal adversary of the US during the Cold War (1947-1991). The Soviet economy and society stagnated in the decades following Stalin's rule, until General Secretary Mikhail GORBACHEV (1985-91) introduced glasnost (openness) and perestroika (restructuring) in an attempt to modernize communism, but his initiatives inadvertently released forces that by December 1991 led to the dissolution of the USSR into Russia and 14 other independent states.

Following economic and political turmoil during President Boris YELTSIN's term (1991-99), Russia shifted toward a centralized authoritarian state under President Vladimir PUTIN (2000-2008, 2012-present) in which the regime seeks to legitimize its rule through managed elections, populist appeals, a foreign policy focused on enhancing the country's geopolitical influence, and commodity-based economic growth. Russia faces a largely subdued rebel movement in Chechnya and some other surrounding regions, although violence still occurs throughout the North Caucasus.

GEOGRAPHY

Location: North Asia bordering the Arctic Ocean, extending from Europe (the portion west of the Urals) to the North Pacific Ocean

Geographic coordinates: 60 00 N, 100 00 E

Map references: Asia

Area: *total:* 17,098,242 sq km
land: 16,377,742 sq km
water: 720,500 sq km
country comparison to the world: 1

Area—comparative: approximately 1.8 times the size of the US

Land boundaries: *total:* 22,408 km
border countries (15): Azerbaijan 338 km, Belarus 1312 km, China (southeast) 4133 km, China (south) 46 km, Estonia 324 km, Finland 1309 km, Georgia 894 km, Kazakhstan 7644 km, North Korea 18 km, Latvia 332 km, Lithuania (Kaliningrad Oblast) 261 km, Mongolia 3452 km, Norway 191 km, Poland (Kaliningrad Oblast) 210 km, Ukraine 1944 km

Coastline: 37,653 km

Maritime claims: *territorial sea:* 12 nm
exclusive economic zone: 200 nm
contiguous zone: 24 nm
continental shelf: 200-m depth or to the depth of exploitation

Climate: ranges from steppes in the south through humid continental in much of European Russia; subarctic in Siberia to tundra climate in the polar north; winters vary from cool along Black Sea coast to frigid in Siberia; summers vary from warm in the steppes to cool along Arctic coast

Terrain: broad plain with low hills west of Urals; vast coniferous forest and tundra in Siberia; uplands and mountains along southern border regions

Elevation: *mean elevation:* 600 m
lowest point: Caspian Sea -28 m
highest point: Gora El'brus (highest point in Europe) 5,642 m

Natural resources: wide natural resource base including major deposits of oil, natural gas, coal, and many strategic minerals, bauxite, reserves of rare earth elements, timber, note, formidable obstacles of climate, terrain, and distance hinder exploitation of natural resources

Land use: *agricultural land:* 13.1% (2011 est.)
arable land: 7.3% (2011 est.) / *permanent crops:* 0.1% (2011 est.) / *permanent pasture:* 5.7% (2011 est.)
forest: 49.4% (2011 est.)
other: 37.5% (2011 est.)

Irrigated land: 43,000 sq km (2012)

Population distribution: population is heavily concentrated in the westernmost fifth of the country extending from the Baltic Sea, south to the Caspian Sea, and eastward parallel to the Kazakh border; elsewhere, sizeable pockets are isolated and generally found in the south

Natural hazards: permafrost over much of Siberia is a major impediment to development; volcanic activity in the Kuril Islands; volcanoes and earthquakes on the Kamchatka Peninsula; spring floods and summer/autumn forest fires throughout Siberia and parts of European Russia
volcanism: significant volcanic activity on the Kamchatka Peninsula and Kuril Islands; the peninsula alone is home to some 29 historically active volcanoes, with dozens more in the Kuril Islands; Kliuchevskoi (4,835 m), which erupted in 2007 and 2010, is Kamchatka's most active volcano; Avachinsky and Koryaksky volcanoes, which pose a threat to the city of Petropavlovsk-Kamchatsky, have been deemed Decade Volcanoes by the International Association of Volcanology and Chemistry of the Earth's Interior, worthy of study due to their explosive history and close proximity to human populations; other notable historically active volcanoes include Bezymianny, Chikurachki, Ebeko, Gorely, Grozny, Karymsky, Ketoi, Kronotsky, Ksudach, Medvezhia, Mutnovsky, Sarychev Peak, Shiveluch, Tiatia, Tolbachik, and Zheltovsky; see note 2 under "Geography—note"

Environment—current issues: air pollution from heavy industry, emissions of coal-fired electric plants, and transportation in major cities; industrial, municipal, and agricultural pollution of inland waterways and seacoasts; deforestation; soil erosion; soil contamination from improper application of agricultural chemicals; nuclear waste disposal; scattered areas of sometimes intense radioactive contamination; groundwater contamination from toxic waste; urban solid waste management; abandoned stocks of obsolete pesticides

Environment—international agreements: *party to:* Air Pollution, Air Pollution-Nitrogen Oxides, Air Pollution-Sulfur 85, Antarctic-Environmental Protocol, Antarctic- Marine Living Resources, Antarctic Seals, Antarctic Treaty, Biodiversity, Climate Change, Climate Change-Kyoto Protocol, Desertification, Endangered Species, Environmental Modification, Hazardous Wastes, Law of the Sea, Marine Dumping, Ozone Layer Protection, Ship Pollution, Tropical Timber 83, Wetlands, Whaling

signed, but not ratified: Air Pollution-Sulfur 94

Geography—note: *note 1:* largest country in the world in terms of area but unfavorably located in relation to major sea lanes of the world; despite its size, much of the country lacks proper soils and climates (either too cold or too dry) for agriculture
note 2: Russia's far east, particularly the Kamchatka Peninsula, lies along the Ring of Fire, a belt of active volcanoes and earthquake epicenters bordering the Pacific Ocean; up to 90% of the world's earthquakes and some 75% of the world's volcanoes occur within the Ring of Fire
note 3: Mount El'brus is Europe's tallest peak; Lake Baikal, the deepest lake in the world, is estimated to hold one fifth of the world's fresh surface water
note 4: Kaliningrad oblast is an exclave annexed from Germany following World War II (it was formerly part of East Prussia); its capital city of Kaliningrad—formerly Koenigsberg—is the only Baltic port in Russia that remains ice free in the winter

PEOPLE AND SOCIETY

Population: 141,722,205 (July 2020 est.)
country comparison to the world: 9

Nationality: *noun:* Russian(s)
adjective: Russian

Ethnic groups: Russian 77.7%, Tatar 3.7%, Ukrainian 1.4%, Bashkir 1.1%, Chuvash 1%, Chechen 1%, other 10.2%, unspecified 3.9% (2010 est.)
note: nearly 200 national and/or ethnic groups are represented in Russia's 2010 census

Languages: Russian (official) 85.7%, Tatar 3.2%, Chechen 1%, other 10.1% (2010 est.)
note: data represent native language spoken

Religions: Russian Orthodox 15-20%, Muslim 10-15%, other Christian 2% (2006 est.)
note: estimates are of practicing worshipers; Russia has large populations of non-practicing believers and non-believers, a legacy of over seven decades of official atheism under Soviet rule; Russia officially recognizes Orthodox Christianity, Islam, Judaism, and Buddhism as the country's traditional religions

Age structure: *0-14 years:* 17.24% (male 12,551,611/female 11,881,297)
15-24 years: 9.54% (male 6,920,070/female 6,602,776)
25-54 years: 43.38% (male 30,240,260/female 31,245,104)
55-64 years: 14.31% (male 8,808,330/female 11,467,697)
65 years and over: 15.53% (male 7,033,381/female 14,971,679) (2020 est.)

Dependency ratios: *total dependency ratio:* 51.2
youth dependency ratio: 27.8
elderly dependency ratio: 23.5
potential support ratio: 4.3 (2020 est.)

Median age: *total:* 40.3 years
male: 37.5 years
female: 43.2 years (2020 est.)
country comparison to the world: 52

Population growth rate: -0.16% (2020 est.)
country comparison to the world: 205

Birth rate: 10 births/1,000 population (2020 est.)
country comparison to the world: 191

Death rate: 13.4 deaths/1,000 population (2020 est.)
country comparison to the world: 7

Net migration rate: 1.7 migrant(s)/1,000 population (2020 est.)
country comparison to the world: 52

Population distribution: population is heavily concentrated in the westernmost fifth of the country extending from the Baltic Sea, south to the Caspian Sea, and eastward parallel to the Kazakh border; elsewhere, sizeable pockets are isolated and generally found in the south

Urbanization: *urban population:* 74.8% of total population (2020)
rate of urbanization: 0.18% annual rate of change (2015-20 est.)

total population growth rate v. urban population growth rate, 2000-2030:

Major urban areas—population: *12.538 million MOSCOW (capital), 5.468 million Saint Petersburg, 1.664 million Novosibirsk, 1.504 million Yekaterinburg, 1.272 million Kazan, 1.258 million Nizhniy Novgorod (2020)*

Sex ratio: *at birth:* 1.06 male(s)/female
0-14 years: 1.06 male(s)/female
15-24 years: 1.05 male(s)/female
25-54 years: 0.97 male(s)/female
55-64 years: 0.77 male(s)/female
65 years and over: 0.47 male(s)/female
total population: 0.86 male(s)/female (2020 est.)

Mother's mean age at first birth: 25.2 years (2013 est.)

Maternal mortality rate: 17 deaths/100,000 live births (2017 est.)
country comparison to the world: 130

Infant mortality rate: *total:* 6.5 deaths/1,000 live births
male: 7.3 deaths/1,000 live births
female: 5.6 deaths/1,000 live births (2020 est.)
country comparison to the world: 163

Life expectancy at birth: *total population:* 71.9 years
male: 66.3 years
female: 77.8 years (2020 est.)
country comparison to the world: 158

Total fertility rate: 1.6 children born/woman (2020 est.)
country comparison to the world: 186

Contraceptive prevalence rate: 68% (2011)
note: percent of women aged 15-44

Drinking water source:
improved:
urban: 98.6% of population
rural: 94.2% of population
total: 97.1% of population
unimproved:
urban: 1.4% of population
rural: 5.8% of population
total: 2.9% of population (2017 est.)

Current Health Expenditure: 5.3% (2017)

Physicians density: 4.01 physicians/1,000 population (2016)

Hospital bed density: 8.1 beds/1,000 population (2017)

Sanitation facility access:
improved:
urban: 94.8% of population
rural: 78.1% of population
total: 90.5% of population
unimproved:
urban: 5.2% of population
rural: 21.9% of population
total: 9.5% of population (2017 est.)

HIV/AIDS—adult prevalence rate: 1.2% (2017 est.)
country comparison to the world: 40

HIV/AIDS—people living with HIV/AIDS: 1 million (2017 est.)
country comparison to the world: 11

HIV/AIDS—deaths: NA

Major infectious diseases: *degree of risk:* intermediate (2020)
food or waterborne diseases: bacterial diarrhea
vectorborne diseases: Crimean-Congo hemorrhagic fever, tickborne encephalitis
note: widespread ongoing transmission of a respiratory illness caused by the novel coronavirus (COVID-19) is occurring throughout the Russia; as of 10 November 2020, Russia has reported a total of 1,774,334 cases of COVID-19 or 12,158 cumulative cases of COVID-19 per 1 million population with 209 cumulative deaths per 1 million population

Obesity—adult prevalence rate: 23.1% (2016)
country comparison to the world: 70

Education expenditures: 3.7% of GDP (2016)
country comparison to the world: 118

Literacy: *definition:* age 15 and over can read and write
total population: 99.7%
male: 99.7%
female: 99.7% (2018)

School life expectancy (primary to tertiary education): *total:* 16 years
male: 16 years
female: 16 years (2018)

Unemployment, youth ages 15-24: *total:* 17%
male: 16.2%
female: 17.9% (2018 est.)
country comparison to the world: 78

GOVERNMENT

Country name: *conventional long form:* Russian Federation
conventional short form: Russia
local long form: Rossiyskaya Federatsiya
local short form: Rossiya
former: Russian Empire, Russian Soviet Federative Socialist Republic
etymology: Russian lands were generally referred to as Muscovy until PETER I officially declared the Russian Empire in 1721; the new name

sought to invoke the patrimony of the medieval eastern European Rus state centered on Kyiv in present-day Ukraine; the Rus were a Varangian (eastern Viking) elite that imposed their rule and eventually their name on their Slavic subjects

Government type: semi-presidential federation

Capital: *name:* Moscow
geographic coordinates: 55 45 N, 37 36 E
time difference: UTC+3 (8 hours ahead of Washington, DC, during Standard Time)
daylight saving time: does not observe daylight savings time
note: Russia has 11 time zones, the largest number of contiguous time zones of any country in the world; in 2014, two time zones were added and DST dropped
etymology: named after the Moskva River; the origin of the river's name is obscure but may derive from the appellation "Mustajoki" given to the river by the Finno-Ugric people who originally inhabited the area and whose meaning may have been "dark" or "turbid"

Administrative divisions: 46 provinces (oblasti, singular—oblast), 21 republics (respubliki, singular—respublika), 4 autonomous okrugs (avtonomnyye okrugi, singular—avtonomnyy okrug), 9 krays (kraya, singular—kray), 2 federal cities (goroda, singular—gorod), and 1 autonomous oblast (avtonomnaya oblast')
oblasts: Amur (Blagoveshchensk), Arkhangel'sk, Astrakhan', Belgorod, Bryansk, Chelyabinsk, Irkutsk, Ivanovo, Kaliningrad, Kaluga, Kemerovo, Kirov, Kostroma, Kurgan, Kursk, Leningrad, Lipetsk, Magadan, Moscow, Murmansk, Nizhniy Novgorod, Novgorod, Novosibirsk, Omsk, Orenburg, Orel, Penza, Pskov, Rostov, Ryazan', Sakhalin (Yuzhno- Sakhalinsk), Samara, Saratov, Smolensk, Sverdlovsk (Yekaterinburg), Tambov, Tomsk, Tula, Tver', Tyumen', Ul'yanovsk, Vladimir, Volgograd, Vologda, Voronezh, Yaroslavl'
republics: Adygeya (Maykop), Altay (Gorno-Altaysk), Bashkortostan (Ufa), Buryatiya (Ulan-Ude), Chechnya (Groznyy), Chuvashiya (Cheboksary), Dagestan (Makhachkala), Ingushetiya (Magas), Kabardino-Balkariya (Nal'chik), Kalmykiya (Elista), Karachayevo-Cherkesiya (Cherkessk), Kareliya (Petrozavodsk), Khakasiya (Abakan), Komi (Syktyvkar), Mariy-El (Yoshkar-Ola), Mordoviya (Saransk), North Ossetia (Vladikavkaz), Sakha [Yakutiya] (Yakutsk), Tatarstan (Kazan'), Tyva (Kyzyl), Udmurtiya (Izhevsk)
autonomous okrugs: Chukotka (Anadyr'), Khanty-Mansi-Yugra (Khanty-Mansiysk), Nenets (Nar'yan-Mar), Yamalo-Nenets (Salekhard)
krays: Altay (Barnaul), Kamchatka (Petropavlovsk-Kamchatskiy), Khabarovsk, Krasnodar, Krasnoyarsk, Perm', Primorskiy [Maritime] (Vladivostok), Stavropol', Zabaykal'sk [Transbaikal] (Chita)
federal cities: Moscow [Moskva], Saint Petersburg [Sankt-Peterburg]
autonomous oblast: Yevreyskaya [Jewish] (Birobidzhan)

note: administrative divisions have the same names as their administrative centers (exceptions have the administrative center name following in parentheses)

note: the United States does not recognize Russia's annexation of Ukraine's Autonomous Republic of Crimea and the municipality of Sevastopol, nor their redesignation as the "Republic of Crimea" and the "Federal City of Sevastopol"

Independence: *25 December 1991 (from the Soviet Union; Russian SFSR renamed Russian Federation); notable earlier dates:* 1157 (Principality of Vladimir-Suzdal created); 16 January 1547 (Tsardom of Muscovy established); 22 October 1721 (Russian Empire proclaimed); 30 December 1922 (Soviet Union established)

National holiday: Russia Day, 12 June (1990); note—commemorates the adoption of the Declaration of State Sovereignty of the Russian Soviet Federative Socialist Republic (RSFSR)

Constitution: *history:* several previous (during Russian Empire and Soviet era); latest drafted 12 July 1993, adopted by referendum 12 December 1993, effective 25 December 1993

amendments: proposed by the president of the Russian Federation, by either house of the Federal Assembly, by the government of the Russian Federation, or by legislative (representative) bodies of the Federation's constituent entities; proposals to amend the government's constitutional system, human and civil rights and freedoms, and procedures for amending or drafting a new constitution require formation of a Constitutional Assembly; passage of such amendments requires two-thirds majority vote of its total membership; passage in a referendum requires participation of an absolute majority of eligible voters and an absolute majority of valid votes; approval of proposed amendments to the government structure, authorities, and procedures requires approval by the legislative bodies of at least two thirds of the Russian Federation's constituent entities; amended 2008, 2014, 2020

Legal system: civil law system; judicial review of legislative acts

International law organization participation: has not submitted an ICJ jurisdiction declaration; non-party state to the ICCt

Citizenship: *citizenship by birth:* no

citizenship by descent only: at least one parent must be a citizen of Russia

dual citizenship recognized: yes

residency requirement for naturalization: 3-5 years

Suffrage: 18 years of age; universal

Executive branch: *chief of state:* President Vladimir Vladimirovich PUTIN (since 7 May 2012)

head of government: Premier Mikhail MISHUSTIN (since 16 January 2020); First Deputy Premier Andrey Removich BELOUSOV (since 21 January 2020); Deputy Premiers Yuriy TRUTNEV (since 31 August 2013), Yuriy Ivanovich BORISOV, Tatiana

Alekseyevna GOLIKOVA (since 18 May 2018), Dmitriy Yuriyevich GRIGORENKO, Viktoriya Valeriyevna ABRAMCHENKO, Aleksey Logvinovich OVERCHUK, Marat Shakirzyanovich KHUSNULLIN, Dmitriy Nikolayevich CHERNYSHENKO (since 21 January 2020), Aleksandr NOVAK (since 10 November 2020)

cabinet: the "Government" is composed of the premier, his deputies, and ministers, all appointed by the president; the premier is also confirmed by the Duma

elections/appointments: president directly elected by absolute majority popular vote in 2 rounds if needed for a 6-year term (2020 constitutional amendments allow a second consecutive term); election last held on 18 March 2018 (next to be held in March 2024); note—for the 2024 presidential election, previous presidential terms are discounted; there is no vice president; premier appointed by the president with the approval of the Duma

election results: Vladimir PUTIN reelected president; percent of vote—Vladimir PUTIN (independent) 77.5%, Pavel GRUDININ (CPRF) 11.9%, Vladimir ZHIRINOVSKIY (LDPR) 5.7%, other 5.8%; Mikhail MISHUSTIN (independent) approved as premier by Duma; vote—383 to 0

note: there is also a Presidential Administration that provides staff and policy support to the president, drafts presidential decrees, and coordinates policy among government agencies; a Security Council also reports directly to the president

Legislative branch: *description:* bicameral Federal Assembly or Federalnoye Sobraniye consists of:

Federation Council or Sovet Federatsii (170 seats; 2 members in each of the 83 federal administrative units (see note below)—oblasts, krays, republics, autonomous okrugs and oblasts, and federal cities of Moscow and Saint Petersburg—appointed by the top executive and legislative officials; members serve 4-year terms)

State Duma or Gosudarstvennaya Duma (450 seats (see note below); as of February 2014, the electoral system reverted to a mixed electoral system for the 2016 election, in which one-half of the members are directly elected by simple majority vote and one-half directly elected by proportional representation vote; members serve 5-year terms)

elections: State Duma—last held on 18 September 2016 (next to be held in fall 2021)

election results: Federation Council (members appointed); composition—men 145, women 25, percent of women 14.7%

State Duma—United Russia 54.2%, CPRF 13.3%, LDPR 13.1%, A Just Russia 6.2%, Rodina 1.5%, CP 0.2%, other minor parties 11.5%; seats by party—United Russia 343, CPRF 42, LDPR 39, A Just Russia 23, Rodina 1, CP 1, independent 1

note 1: the State Duma now includes 3 representatives from the "Republic of Crimea," while the Federation Council includes 2 each from the "Republic of Crimea" and the "Federal City of Sevastopol," both regions that Russia occupied

and attempted to annex from Ukraine and that the US does not recognize as part of Russia

note 2: seats by party as of December 2018—United Russia 341, CPRF 43, LDPR 39, A Just Russia 23, independent 2, vacant 2; composition as of October 2018—men 393, women 57, percent of women 12.7%; note—total Federal Assembly percent of women 13.2%

Judicial branch: *highest courts:* Supreme Court of the Russian Federation (consists of 170 members organized into the Judicial Panel for Civil Affairs, the Judicial Panel for Criminal Affairs, and the Military Panel); Constitutional Court (consists of 11 members, including the chairperson and deputy); note—in February 2014, Russia's Higher Court of Arbitration was abolished and its former authorities transferred to the Supreme Court, which in addition is the country's highest judicial authority for appeals, civil, criminal, administrative, and military cases, and the disciplinary judicial board, which has jurisdiction over economic disputes

judge selection and term of office: all members of Russia's 3 highest courts nominated by the president and appointed by the Federation Council (the upper house of the legislature); members of all 3 courts appointed for life

subordinate courts: regional (kray) and provincial (oblast) courts; Moscow and St. Petersburg city courts; autonomous province and district courts; note—the 21 Russian Republics have court systems specified by their own constitutions

Political parties and leaders: A Just Russia [Sergey MIRONOV]

Civic Platform or CP [Rifat SHAYKHUTDINOV]

Communist Party of the Russian Federation or CPRF [Gennadiy ZYUGANOV]

Liberal Democratic Party of Russia or LDPR [Vladimir ZHIRINOVSKIY]

Rodina [Aleksei ZHURAVLYOV]

United Russia [Dmitriy MEDVEDEV]

note: 64 political parties are registered with Russia's Ministry of Justice (as of September 2018), but only four parties maintain representation in Russia's national legislature

International organization participation: APEC, Arctic Council, ARF, ASEAN (dialogue partner), BIS, BRICS, BSEC, CBSS, CD, CE, CERN (observer), CICA, CIS, CSTO, EAEC, EAEU, EAPC, EAS, EBRD, FAO, FATF, G-20, GCTU, IAEA, IBRD, ICAO, ICC (national committees), ICRM, IDA, IFAD, IFC, IFRCS, IHO, ILO, IMF, IMO, IMSO, Interpol, IOC, IOM (observer), IPU, ISO, ITSO, ITU, ITUC (NGOs), LAIA (observer), MIGA, MINURSO, MONUSCO, NEA, NSG, OAS (observer), OIC (observer), OPCW, OSCE, Paris Club, PCA, PFP, SCO, UN, UNCTAD, UNESCO, UNHCR, UNIDO, UNISFA, UNMIL, UNMISS, UNOCI, UN Security Council (permanent), UNTSO, UNWTO, UPU, WCO, WFTU (NGOs), WHO, WIPO, WMO, WTO, ZC

Diplomatic representation in the US: *chief of mission:* Ambassador Anatoliy Ivanovich ANTONOV (since 8 September 2017)

chancery: 2650 Wisconsin Avenue NW, Washington, DC 20007

telephone: [1] (202) 298-5700, 5701, 5704, 5708

FAX: [1] (202) 298-5735

consulate(s) general: Houston, New York, Seattle

Diplomatic representation from the US: *chief of mission:* Ambassador Jon M. HUNTSMAN, Jr. (since 3 October 2017)

telephone: [7] (495) 728-5000

embassy: Bolshoy Deviatinskiy Pereulok No. 8, 121099 Moscow

mailing address: PSC-77, APO AE 09721

FAX: [7] (495) 728-5090

consulate(s) general: Saint Petersburg, Vladivostok, Yekaterinburg

Flag description: three equal horizontal bands of white (top), blue, and red

note: the colors may have been based on those of the Dutch flag; despite many popular interpretations, there is no official meaning assigned to the colors of the Russian flag; this flag inspired several other Slav countries to adopt horizontal tricolors of the same colors but in different arrangements, and so red, blue, and white became the Pan-Slav colors

National symbol(s): *bear, double-headed eagle; national colors:* white, blue, red

National anthem: *name:* "Gimn Rossiyskoy Federatsii" (National Anthem of the Russian Federation)

lyrics/music: Sergey Vladimirovich MIKHALKOV/ Aleksandr Vasilyevich ALEKSANDROV

note: in 2000, Russia adopted the tune of the anthem of the former Soviet Union (composed in 1939); the lyrics, also adopted in 2000, were written by the same person who authored the Soviet lyrics in 1943

0:00 / 1:07

ECONOMY

Economy—overview: Russia has undergone significant changes since the collapse of the Soviet Union, moving from a centrally planned economy towards a more market-based system. Both economic growth and reform have stalled in recent years, however, and Russia remains a predominantly statist economy with a high concentration of wealth in officials' hands. Economic reforms in the 1990s privatized most industry, with notable exceptions in the energy, transportation, banking, and defense-related sectors. The protection of property rights is still weak, and the state continues to interfere in the free operation of the private sector.

Russia is one of the world's leading producers of oil and natural gas, and is also a top exporter of metals such as steel and primary aluminum. Russia is heavily dependent on the movement of world commodity prices as reliance on commodity exports makes it vulnerable to boom and bust cycles that follow the volatile swings in global prices. The economy, which had averaged 7% growth during the 1998-2008 period as oil prices rose rapidly, has seen diminishing growth rates

since then due to the exhaustion of Russia's commodity-based growth model.

A combination of falling oil prices, international sanctions, and structural limitations pushed Russia into a deep recession in 2015, with GDP falling by close to 2.8%. The downturn continued through 2016, with GDP contracting another 0.2%, but was reversed in 2017 as world demand picked up. Government support for import substitution has increased recently in an effort to diversify the economy away from extractive industries.

GDP (purchasing power parity):
$4.016 trillion (2017 est.)
$3.955 trillion (2016 est.)
$3.963 trillion (2015 est.)
note: data are in 2017 dollars
country comparison to the world: 6

GDP (official exchange rate):
$1.578 trillion (2017 est.)

GDP—real growth rate:
1.34% (2019 est.)
2.54% (2018 est.)
1.83% (2017 est.)
country comparison to the world: 161

GDP—per capita (PPP):
$27,900 (2017 est.)
$27,500 (2016 est.)
$27,500 (2015 est.)
note: data are in 2017 dollars
country comparison to the world: 74

Gross national saving:
26.5% of GDP (2017 est.)
25.9% of GDP (2016 est.)
26.8% of GDP (2015 est.)
country comparison to the world: 48

GDP—composition, by end use:
household consumption: 52.4% (2017 est.)
government consumption: 18% (2017 est.)
investment in fixed capital: 21.6% (2017 est.)
investment in inventories: 2.3% (2017 est.)
exports of goods and services: 26.2% (2017 est.)
imports of goods and services: -20.6% (2017 est.)

GDP—composition, by sector of origin:
agriculture: 4.7% (2017 est.)
industry: 32.4% (2017 est.)
services: 62.3% (2017 est.)

Agriculture—products: grain, sugar beets, sunflower seeds, vegetables, fruits; beef, milk

Industries: complete range of mining and extractive industries producing coal, oil, gas, chemicals, and metals; all forms of machine building from rolling mills to high-performance aircraft and space vehicles; defense industries (including radar, missile production, advanced electronic components), shipbuilding; road and rail transportation equipment; communications equipment; agricultural machinery, tractors, and construction equipment; electric power generating and transmitting equipment; medical and scientific instruments; consumer durables, textiles, foodstuffs, handicrafts

Industrial production growth rate: -1% (2017 est.)
country comparison to the world: 177

Labor force: 69.923 million (2020 est.)
country comparison to the world: 5

Labor force—by occupation: *agriculture:* 9.4%
industry: 27.6%
services: 63% (2016 est.)

Unemployment rate: 4.6% (2019 est.)
4.8% (2018 est.)
country comparison to the world: 69

Population below poverty line: 13.3% (2015 est.)

Household income or consumption by percentage share: *lowest 10%:* 2.3%
highest 10%: 32.2% (2012 est.)

Budget: *revenues:* 258.6 billion (2017 est.)
expenditures: 281.4 billion (2017 est.)

Taxes and other revenues: 16.4% (of GDP) (2017 est.)
country comparison to the world: 181

Budget surplus (+) or deficit (-): -1.4% (of GDP) (2017 est.)
country comparison to the world: 88

Public debt: 15.5% of GDP (2017 est.)
16.1% of GDP (2016 est.)
note: data cover general government debt and include debt instruments issued (or owned) by government entities other than the treasury; the data include treasury debt held by foreign entities; the data include debt issued by subnational entities, as well as intragovernmental debt; intragovernmental debt consists of treasury borrowings from surpluses in the social funds, such as for retirement, medical care, and unemployment, debt instruments for the social funds are not sold at public auctions
country comparison to the world: 194

Fiscal year: calendar year

Inflation rate (consumer prices): 3.7% (2017 est.)
7.1% (2016 est.)
country comparison to the world: 149

Current account balance: $65.311 billion (2019 est.)
$115.68 billion (2018 est.)
country comparison to the world: 6

Exports: $353 billion (2017 est.)
$281.9 billion (2016 est.)
country comparison to the world: 14

Exports—partners: China 10.9%, Netherlands 10%, Germany 7.1%, Belarus 5.1%, Turkey 4.9% (2017)

Exports—commodities: petroleum and petroleum products, natural gas, metals, wood and wood products, chemicals, and a wide variety of civilian and military manufactures

Imports: $238 billion (2017 est.)
$191.6 billion (2016 est.)
country comparison to the world: 20

Imports—commodities: machinery, vehicles, pharmaceutical products, plastic, semi-finished metal products, meat, fruits and nuts, optical and medical instruments, iron, steel

Imports—partners: China 21.2%, Germany 10.7%, US 5.6%, Belarus 5%, Italy 4.5%, France 4.2% (2017)

Reserves of foreign exchange and gold: $432.7 billion (31 December 2017 est.)
$377.7 billion (31 December 2016 est.)
country comparison to the world: 6

Debt—external: $539.6 billion (31 December 2017 est.)
$434.8 billion (31 December 2016 est.)
country comparison to the world: 22

Exchange rates: Russian rubles (RUB) per US dollar -
58.39 (2017 est.)
67.056 (2016 est.)
67.056 (2015 est.)
60.938 (2014 est.)
38.378 (2013 est.)

ENERGY

Electricity access: *electrification—total population:* 100% (2020)

Electricity—production: 1.031 trillion kWh (2016 est.)
country comparison to the world: 4

Electricity—consumption: 909.6 billion kWh (2016 est.)
country comparison to the world: 5

Electricity—exports: 13.13 billion kWh (2016 est.)
country comparison to the world: 14

Electricity—imports: 3.194 billion kWh (2016 est.)
country comparison to the world: 48

Electricity—installed generating capacity: 244.9 million kW (2016 est.)
country comparison to the world: 5

Electricity—from fossil fuels: 68% of total installed capacity (2016 est.)
country comparison to the world: 115

Electricity—from nuclear fuels: 11% of total installed capacity (2017 est.)
country comparison to the world: 13

Electricity—from hydroelectric plants: 21% of total installed capacity (2017 est.)
country comparison to the world: 86

Electricity—from other renewable sources: 1% of total installed capacity (2017 est.)
country comparison to the world: 164

Crude oil—production: 10.759 million bbl/day (2018 est.)
country comparison to the world: 2

Crude oil—exports: 4.921 million bbl/day (2015 est.)
country comparison to the world: 2

Crude oil—imports: 76,220 bbl/day (2015 est.)
country comparison to the world: 48

Crude oil—proved reserves: 80 billion bbl (1 January 2018 est.)
country comparison to the world: 8

Refined petroleum products—production: 6.076 million bbl/day (2015 est.)
country comparison to the world: 3

Refined petroleum products—consumption: 3.65 million bbl/day (2016 est.)
country comparison to the world: 5

Refined petroleum products—exports: 2.671 million bbl/day (2015 est.)
country comparison to the world: 2

Refined petroleum products—imports: 41,920 bbl/day (2015 est.)
country comparison to the world: 88

Natural gas—production: 665.6 billion cu m (2017 est.)
country comparison to the world: 2

Natural gas—consumption: 467.5 billion cu m (2017 est.)
country comparison to the world: 2

Natural gas—exports: 210.2 billion cu m (2017 est.)
country comparison to the world: 1

Natural gas—imports: 15.77 billion cu m (2017 est.)
country comparison to the world: 19

Natural gas—proved reserves: 47.8 trillion cu m (1 January 2018 est.)
country comparison to the world: 1

Carbon dioxide emissions from consumption of energy: 1.847 billion Mt (2017 est.)
country comparison to the world: 4

COMMUNICATIONS

Telephones—fixed lines: *total subscriptions:* 31,171,043
subscriptions per 100 inhabitants: 21.96 (2019 est.)
country comparison to the world: 7

Telephones—mobile cellular: *total subscriptions:* 233,342,795
subscriptions per 100 inhabitants: 164.39 (2019 est.)
country comparison to the world: 5

Telecommunication systems: *general assessment:* telecom sector impacted by sanctions related to the annexations in Ukraine; the estimated number of mobile subscribers jumped from fewer than 1 million in 1998 to 255 million in 2016; fixed-line service has improved but a large demand remains; Russia with low broadband penetration is one of Europe's fastest growing markets for fiberbased broadband and moving from DSL to fiber; use by the population of multiple SIM cards; regulator ended roaming charges and works to bring down prices; 4 major operators in the mobile market; deployment of LTE support mobile broadband and data services, mobile on the cusp of 5G (2020)
domestic: cross-country digital trunk lines run from Saint Petersburg to Khabarovsk, and from Moscow to Novorossiysk; the telephone systems in 60 regional capitals have modern digital infrastructures; cellular services, both analog and digital, are available in many areas; in rural areas, telephone services are still outdated, inadequate, and low-density; 22 per 100 for fixed-line and mobile-cellular 164 per 100 persons (2019)

international: country code—7; landing points for the Far East Submarine Cable System, HSCS, Sakhalin-Kuril Island Cable, RSCN, BCS North-Phase 2, Kerch Strait Cable and the Georgia-Russian submarine cable system connecting Russia, Japan, Finland, Georgia and Ukraine; satellite earth stations provide access to Intelsat, Intersputnik, Eutelsat, Inmarsat, and Orbita systems (2019)
note: the COVID-19 outbreak is negatively impacting telecommunications production and supply chains globally; consumer spending on telecom devices and services has also slowed due to the pandemic's effect on economies worldwide; overall progress towards improvements in all facets of the telecom industry—mobile, fixed-line, broadband, submarine cable and satellite—has moderated

Broadcast media: 13 national TV stations with the federal government owning 1 and holding a controlling interest in a second; state-owned Gazprom maintains a controlling interest in 2 of the national channels; government-affiliated Bank Rossiya owns controlling interest in a fourth and fifth, while a sixth national channel is owned by the Moscow city administration; the Russian Orthodox Church and the Russian military, respectively, own 2 additional national channels; roughly 3,300 national, regional, and local TV stations with over two-thirds completely or partially controlled by the federal or local governments; satellite TV services are available; 2 state-run national radio networks with a third majority-owned by Gazprom; roughly 2,400 public and commercial radio stations

Internet country code: .ru; note—Russia also has responsibility for a legacy domain ".su" that was allocated to the Soviet Union and is being phased out

Internet users: *total:* 114,920,477
percent of population: 80.86% (July 2018 est.)
country comparison to the world: 5

Broadband—fixed subscriptions: *total:* 32,062,780
subscriptions per 100 inhabitants: 23 (2018 est.)
country comparison to the world: 5

TRANSPORTATION

National air transport system: *number of registered air carriers:* 32 (2020)
inventory of registered aircraft operated by air carriers: 958
annual passenger traffic on registered air carriers: 99,327,311 (2018)
annual freight traffic on registered air carriers: 6,810,610,000 mt-km (2018)

Civil aircraft registration country code prefix: RA (2016)

Airports: 1,218 (2013)
country comparison to the world: 5

Airports—with paved runways: *total:* 594 (2017)
over 3,047 m: 54 (2017)
2,438 to 3,047 m: 197 (2017)
1,524 to 2,437 m: 123 (2017)
914 to 1,523 m: 95 (2017)

under 914 m: 125 (2017)

Airports—with unpaved runways: *total:* 624 (2013)
over 3,047 m: 4 (2013)
2,438 to 3,047 m: 13 (2013)
1,524 to 2,437 m: 69 (2013)
914 to 1,523 m: 81 (2013)
under 914 m: 457 (2013)

Heliports: 49 (2013)

Pipelines: 177700 km gas, 54800 km oil, 19300 km refined products (2016)

Railways: *total:* 87,157 km (2014)
narrow gauge: 957 km 1.067-m gauge (on Sakhalin Island) (2014)
broad gauge: 86,200 km 1.520-m gauge (40,300 km electrified) (2014)
note: an additional 30,000 km of non-common carrier lines serve industries
country comparison to the world: 3

Roadways: *total:* 1,283,387 km (2012)
paved: 927,721 km (includes 39,143 km of expressways) (2012)
unpaved: 355,666 km (2012)
country comparison to the world: 5

Waterways: 102,000 km (including 48,000 km with guaranteed depth; the 72,000-km system in European Russia links Baltic Sea, White Sea, Caspian Sea, Sea of Azov, and Black Sea) (2009)
country comparison to the world: 2

Merchant marine: *total:* 2,739
by type: bulk carrier 16, container ship 13, general cargo 899, oil tanker 404, other 1,407 (2019)
country comparison to the world: 9

Ports and terminals: *major seaport(s):* Kaliningrad, Nakhodka, Novorossiysk, Primorsk, Vostochnyy
oil terminal(s): Kavkaz oil terminal
container port(s) (TEUs): Saint Petersburg (1,848,700) (2017)
LNG terminal(s) (export): Sakhalin Island
river port(s): Saint Petersburg (Neva River)

MILITARY AND SECURITY

Military and security forces: *Armed Forces of the Russian Federation:* Ground Troops (Sukhoputnyye Voyskia, SV), Navy (Voyenno-Morskoy Flot, VMF), Aerospace Forces (Vozdushno-Kosmicheskiye Sily, VKS); Airborne Troops (Vozdushno-Desantnyye Voyska, VDV), and Missile Troops of Strategic Purpose (Raketnyye Voyska Strategicheskogo Naznacheniya, RVSN) referred to commonly as Strategic Rocket Forces, are independent "combat arms," not subordinate to any of the three branches
Federal National Guard Troops Service of the Russian Federation (National Guard, Russian Guard, or Rosgvardiya): created in 2016 as an independent agency for internal/regime security, combating terrorism and narcotics trafficking, protecting important state facilities and government personnel, and supporting border security; forces include Interior Troops that formerly belong to the Interior Ministry, special police units, rapid response units, and other air, ground, maritime, and police forces

Federal Security Services Border Troops (includes land and maritime forces) (2019)

Military expenditures:
3.9% of GDP (2019)
3.8% of GDP (2018)
4.2% of GDP (2017)
5.5% of GDP (2016)
4.9% of GDP (2015)
country comparison to the world: 14

Military and security service personnel strengths: size estimates for the Armed Forces of the Russian Federation vary; approximately 900,000 total active duty troops (400,000 Ground Troops, including 40,000 Airborne Troops; 150,000 Navy; 200,000 Aerospace Forces; 60,000 Strategic Rocket Forces; 90,000 other uniformed personnel (special forces, command and control, support, etc.); est. 200,000 Federal National Guard Troops (2019 est.)

Military equipment inventories and acquisitions: the Russian Federation's military and paramilitary services are equipped with domestically-produced weapons systems, although since 2010 Russia has imported limited amounts of military hardware from Czechia, France, Israel, Italy, Turkey, and Ukraine; the Russian defense industry is capable of designing, developing, and producing a full range of advanced air, land, missile, and naval systems (2019)

Military deployments: est. 3,000-5,000 Armenia; est. 7,000-10,000 Georgia; est. 500 Kyrgyzstan; est. 1,500 Moldova; est. 4,000-5,000 Syria; est. 5,000-7,000 Tajikistan; est. 25,000-30,000 Ukraine; contributes approximately 8,000 personnel to CSTO's Rapid Reaction Force (2019)
it is assessed that as many as 2,500 personnel from a Russian Government-backed private military company are present in Libya supporting Libyan National Army forces (June 2020)

Military service age and obligation: 18-27 years of age for compulsory or voluntary military service; males are registered for the draft at 17 years of age; one- year service obligation (Russia offers the option of serving on a two-year contract instead of completing a one-year conscription period); reserve obligation for non-officers to age 50; enrollment in military schools from the age of 16, cadets classified as members of the armed forces (2019)
note: in April of 2019, the Russian government pledged its intent to end conscription

TERRORISM

Terrorist group(s): Aum Shimrikyo (AUM/Aleph); Islamic State of Iraq and ash-Sham; Islamic State of Iraq and ash-Sham – Caucasus Province (2019)
note: details about the history, aims, leadership, organization, areas of operation, tactics, targets, weapons, size, and sources of support of the group(s) appear(s) in Appendix-T

TRANSNATIONAL ISSUES

Disputes—international: Russia remains concerned about the smuggling of poppy derivatives from Afghanistan through Central Asian countries; China and Russia have demarcated the once disputed islands at the Amur and Ussuri confluence and in the Argun River in accordance with the 2004 Agreement, ending their centuries-long border disputes; the sovereignty dispute over the islands of Etorofu, Kunashiri, Shikotan, and the Habomai group, known in Japan as the "Northern Territories" and in Russia as the "Southern Kurils," occupied by the Soviet Union in 1945, now administered by Russia, and claimed by Japan, remains the primary sticking point to signing a peace treaty formally ending World War II hostilities; Russia's military support and subsequent recognition of Abkhazia and South Ossetia independence in 2008 continue to sour relations with Georgia; Azerbaijan, Kazakhstan, and Russia ratified Caspian seabed delimitation treaties based on equidistance, while Iran continues to insist on a one-fifth slice of the sea; Norway and Russia signed a comprehensive maritime boundary agreement in 2010; various groups in Finland advocate restoration of Karelia (Kareliya) and other areas ceded to the Soviet Union following World War II but the Finnish Government asserts no territorial demands; Russia and Estonia signed a technical border agreement in May 2005, but Russia recalled its signature in June 2005 after the Estonian parliament added to its domestic ratification act a historical preamble referencing the Soviet occupation and Estonia's pre-war borders under the 1920 Treaty of Tartu; Russia contends that the preamble allows Estonia to make territorial claims on Russia in the future, while Estonian officials deny that the preamble has any legal impact on the treaty text; Russia demands better treatment of the Russian-speaking population in Estonia and Latvia; Russia remains involved in the conflict in eastern Ukraine while also occupying Ukraine's territory of Crimea; Lithuania and Russia committed to demarcating their boundary in 2006 in accordance with the land and maritime treaty ratified by Russia in May 2003 and by Lithuania in 1999; Lithuania operates a simplified transit regime for Russian nationals traveling from the Kaliningrad coastal exclave into Russia, while still conforming, as an EU member state with an EU external border, where strict Schengen border rules apply; preparations for the demarcation delimitation of land boundary with Ukraine have commenced; the dispute over the boundary between Russia and Ukraine through the Kerch Strait and Sea of Azov is suspended due to the occupation of Crimea by Russia; Kazakhstan and Russia boundary delimitation was ratified on November 2005 and field demarcation should commence in 2007; Russian Duma has not yet ratified 1990 Bering Sea Maritime Boundary Agreement with the US; Denmark (Greenland) and Norway have made submissions to the Commission on the Limits of the Continental Shelf (CLCS) and Russia is collecting additional data to augment its 2001 CLCS submission

Refugees and internally displaced persons: *refugees (country of origin):* 41,251 (Ukraine) (2019)

stateless persons: 68,209 (2019); note—Russia's stateless population consists of Roma, Meskhetian Turks, and exSoviet citizens from the former republics; between 2003 and 2010 more than 600,000 stateless people were naturalized; most Meskhetian Turks, followers of Islam with origins in Georgia, fled or were evacuated from Uzbekistan after a 1989 pogrom and have lived in Russia for more than the required five-year residency period; they continue to be denied registration for citizenship and basic rights by local Krasnodar Krai authorities on the grounds that they are temporary illegal migrants

Trafficking in persons: *current situation:* Russia is a source, transit, and destination country for men, women, and children who are subjected to forced labor and sex trafficking; with millions of foreign workers, forced labor is Russia's predominant

human trafficking problem and sometimes involves organized crime syndicates; workers from Russia, other European countries, Central Asia, and East and Southeast Asia, including North Korea and Vietnam, are subjected to forced labor in the construction, manufacturing, agricultural, textile, grocery store, maritime, and domestic service industries, as well as in forced begging, waste sorting, and street sweeping; women and children from Europe, Southeast Asia, Africa, and Central Asia are subject to sex trafficking in Russia; Russian women and children are victims of sex trafficking domestically and in Northeast Asia, Europe, Central Asia, Africa, the US, and the Middle East

tier rating: Tier 3—Russia does not fully comply with the minimum standards for the elimination of trafficking and is not making a significant effort to do so; prosecutions of trafficking offenders remained low in comparison to the scope of Russia's trafficking problem; the government did not develop or employ a formal system for identifying trafficking victims or referring them to protective services, although authorities reportedly

assisted a limited number of victims on an ad hoc basis; foreign victims, the largest group in Russia, were not entitled to state-provided rehabilitative services and were routinely detained and deported; the government has not reported investigating reports of slave-like conditions among North Korean workers in Russia; authorities have made no effort to reduce the demand for forced labor or to develop public awareness of forced labor or sex trafficking (2015)

Illicit drugs: limited cultivation of illicit cannabis and opium poppy and producer of methamphetamine, mostly for domestic consumption; government has active illicit crop eradication program; used as transshipment point for Asian opiates, cannabis, and Latin American cocaine bound for growing domestic markets, to a lesser extent Western and Central Europe, and occasionally to the US; major source of heroin precursor chemicals; corruption and organized crime are key concerns; major consumer of opiates

RWANDA

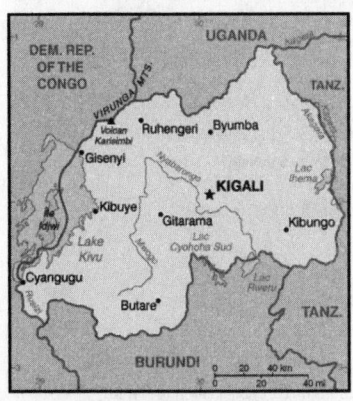

INTRODUCTION

Background: A Rwandan kingdom dominated the region from the mid-18th century onward, with the Tutsi rulers conquering others militarily, centralizing power, and increasingly enacting anti-Hutu policies. German colonial rule began in 1898, but Belgian forces captured Rwanda in 1916 during World War I. Both European nations ruled through the kings and pursued a pro-Tutsi policy. In 1959, three years before independence from Belgium, the majority ethnic group, the Hutus, overthrew the ruling Tutsi king. Over the next several years, thousands of Tutsis were killed, and some 150,000 driven into exile in neighboring countries. The children of these

exiles later formed a rebel group, the Rwandan Patriotic Front (RPF), and began a civil war in 1990. The war, along with several political and economic upheavals, exacerbated ethnic tensions, culminating in April 1994 in a state-orchestrated genocide, in which Rwandans killed approximately 800,000 of their fellow citizens, including approximately three-quarters of the Tutsi population. The genocide ended later that same year when the predominantly Tutsi RPF, operating out of Uganda and northern Rwanda, defeated the national army and Hutu militias, and established an RPF-led government of national unity. Rwanda held its first local elections in 1999 and its first post-genocide presidential and legislative elections in 2003. Rwanda joined the Commonwealth in late 2009. President Paul KAGAME won the presidential election in August 2017 after changing the constitution in 2016 to allow him to run for a third term.

GEOGRAPHY

Location: Central Africa, east of the Democratic Republic of the Congo, north of Burundi

Geographic coordinates: 2 00 S, 30 00 E

Map references: Africa

Area: *total:* 26,338 sq km
land: 24,668 sq km
water: 1,670 sq km
country comparison to the world: 149

Area—comparative: slightly smaller than Maryland

Land boundaries: *total:* 930 km

border countries (4): Burundi 315 km, Democratic Republic of the Congo 221 km, Tanzania 222 km, Uganda 172 km

Coastline: 0 km (landlocked)

Maritime claims: none (landlocked)

Climate: temperate; two rainy seasons (February to April, November to January); mild in mountains with frost and snow possible

Terrain: mostly grassy uplands and hills; relief is mountainous with altitude declining from west to east

Elevation: *mean elevation:* 1,598 m
lowest point: Rusizi River 950 m
highest point: Volcan Karisimbi 4,519 m

Natural resources: gold, cassiterite (tin ore), wolframite (tungsten ore), methane, hydropower, arable land

Land use: *agricultural land:* 74.5% (2011 est.)
arable land: 47% (2011 est.) / *permanent crops:* 10.1% (2011 est.) / permanent pasture: 17.4% (2011 est.)
forest: 18% (2011 est.)
other: 7.5% (2011 est.)

Irrigated land: 96 sq km (2012)

Population distribution: one of Africa's most densely populated countries; large concentrations tend to be in the central regions and along the shore of Lake Kivu in the west as shown in this population distribution map

Natural hazards: periodic droughts; the volcanic Virunga Mountains are in the northwest along the border with Democratic Republic of the Congo *volcanism:* Visoke (3,711 m), located on the border with the Democratic Republic of the Congo, is the country's only historically active volcano

Environment—current issues: deforestation results from uncontrolled cutting of trees for fuel; overgrazing; land degradation; soil erosion; a decline in soil fertility (soil exhaustion); wetland degradation and loss of biodiversity; widespread poaching

Environment—international agreements: *party to:* Biodiversity, Climate Change, Climate Change-Kyoto Protocol, Desertification, Endangered Species, Hazardous Wastes, Ozone Layer Protection, Wetlands
signed, but not ratified: Law of the Sea

Geography—note: landlocked; most of the country is intensively cultivated and rugged with the population predominantly rural

PEOPLE AND SOCIETY

Population: 12,712,431 (July 2020 est.)
note: estimates for this country explicitly take into account the effects of excess mortality due to AIDS; this can result in lower life expectancy, higher infant mortality, higher death rates, lower population growth rates, and changes in the distribution of population by age and sex than would otherwise be expected
country comparison to the world: 75

Nationality: *noun:* Rwandan(s)
adjective: Rwandan

Ethnic groups: Hutu, Tutsi, Twa (Pygmy)

Languages: Kinyarwanda (official, universal Bantu vernacular) 93.2%, French (official) <.1, English (official) <.1, Swahili/Kiswahili (official, used in commercial centers) <.1, more than one language, other 6.3%, unspecified 0.3% (2002 est.)

Religions: Protestant 49.5% (includes Adventist 11.8% and other Protestant 37.7%), Roman Catholic 43.7%, Muslim 2%, other 0.9% (includes Jehovah's Witness), none 2.5%, unspecified 1.3% (2012 est.)

Demographic profile: Rwanda's fertility rate declined sharply during the last decade, as a result of the government's commitment to family planning, the increased use of contraceptives, and a downward trend in ideal family size. Increases in educational attainment, particularly among girls, and exposure to social media also contributed to the reduction in the birth rate. The average number of births per woman decreased from a 5.6 in 2005 to 4.5 in 2016. Despite these significant strides in reducing fertility, Rwanda's birth rate remains very high and will continue to for an extended period of time because of its large population entering reproductive age. Because Rwanda is one of the most densely populated countries in Africa, its persistent high population growth and increasingly small agricultural landholdings will put additional strain on families' ability to raise foodstuffs and access potable water. These conditions will also hinder the government's efforts to reduce poverty and prevent environmental degradation.

The UNHCR recommended that effective 30 June 2013 countries invoke a cessation of refugee status for those Rwandans who fled their homeland between 1959 and 1998, including the 1994 genocide, on the grounds that the conditions that drove them to seek protection abroad no longer exist. The UNHCR's decision is controversial because many Rwandan refugees still fear persecution if they return home, concerns that are supported by the number of Rwandans granted asylum since 1998 and by the number exempted from the cessation. Rwandan refugees can still seek an exemption or local integration, but host countries are anxious to send the refugees back to Rwanda and are likely to avoid options that enable them to stay. Conversely, Rwanda itself hosts almost 160,000 refugees as of 2017; virtually all of them fleeing conflict in neighboring Burundi and the Democratic Republic of the Congo.

Age structure: *0-14 years:* 39.95% (male 2,564,893/female 2,513,993)
15-24 years: 20.1% (male 1,280,948/female 1,273,853)
25-54 years: 33.06% (male 2,001,629/female 2,201,132)
55-64 years: 4.24% (male 241,462/female 298,163)
65 years and over: 2.65% (male 134,648/female 201,710) (2020 est.)

Dependency ratios: *total dependency ratio:* 74.2
youth dependency ratio: 68.8
elderly dependency ratio: 5.4
potential support ratio: 18.4 (2020 est.)

Median age: *total:* 19.7 years
male: 18.9 years
female: 20.4 years (2020 est.)
country comparison to the world: 200

Population growth rate: 2% (2020 est.)
country comparison to the world: 47

Birth rate: 27.9 births/1,000 population (2020 est.)
country comparison to the world: 40

Death rate: 6.1 deaths/1,000 population (2020 est.)
country comparison to the world: 160

Net migration rate: -3.3 migrant(s)/1,000 population (2020 est.)
country comparison to the world: 182

Population distribution: one of Africa's most densely populated countries; large concentrations tend to be in the central regions and along the shore of Lake Kivu in the west as shown in this population distribution map

Urbanization: *urban population:* 17.4% of total population (2020)
rate of urbanization: 2.86% annual rate of change (2015-20 est.)

total population growth rate v. urban population growth rate, 2000-2030: *Major urban areas—population:* 1.132 million KIGALI (capital) (2020)

Sex ratio: *at birth:* 1.03 male(s)/female
0-14 years: 1.02 male(s)/female
15-24 years: 1.01 male(s)/female

25-54 years: 0.91 male(s)/female
55-64 years: 0.81 male(s)/female
65 years and over: 0.67 male(s)/female
total population: 0.96 male(s)/female (2020 est.)

Mother's mean age at first birth: 23 years (2014/15 est.)
note: median age at first birth among women 25-29

Maternal mortality rate: 248 deaths/100,000 live births (2017 est.)
country comparison to the world: 44

Infant mortality rate: *total:* 28 deaths/1,000 live births
male: 30.6 deaths/1,000 live births
female: 25.3 deaths/1,000 live births (2020 est.)
country comparison to the world: 61

Life expectancy at birth: *total population:* 65.1 years
male: 63.2 years
female: 67.1 years (2020 est.)
country comparison to the world: 196

Total fertility rate: 3.52 children born/woman (2020 est.)
country comparison to the world: 41

Contraceptive prevalence rate: 53.2% (2014/15)

Drinking water source:
improved:
urban: 92% of population
rural: 76.9% of population
total: 79.5% of population
unimproved:
urban: 8% of population
rural: 23.1% of population
total: 20.5% of population (2017 est.)

Current Health Expenditure: 6.6% (2017)

Physicians density: 0.14 physicians/1,000 population (2017)

Sanitation facility access:
improved:
urban: 88.4% of population
rural: 79.4% of population
total: 80.9% of population
unimproved:
urban: 11.6% of population
rural: 20.6% of population
total: 19.1% of population (2017 est.)

HIV/AIDS—adult prevalence rate: 2.9% (2019 est.)
country comparison to the world: 19

HIV/AIDS—people living with HIV/AIDS: 230,000 (2019 est.)
country comparison to the world: 26

HIV/AIDS—deaths: 2,800 (2019 est.)
country comparison to the world: 37

Major infectious diseases: *degree of risk:* very high (2020)
food or waterborne diseases: bacterial diarrhea, hepatitis A, and typhoid fever
vectorborne diseases: malaria and dengue fever
animal contact diseases: rabies

Obesity—adult prevalence rate: 5.8% (2016)
country comparison to the world: 174

Children under the age of 5 years underweight: 9.6% (2015)
country comparison to the world: 66

Education expenditures: 3.1% of GDP (2018)
country comparison to the world: 137

Literacy: *definition:* age 15 and over can read and write
total population: 73.2%
male: 77.6%
female: 69.4% (2018)

School life expectancy (primary to tertiary education): *total:* 11 years
male: 11 years
female: 11 years (2019)

Unemployment, youth ages 15-24: *total:* 20.6%
male: 18.8%
female: 22.6% (2018 est.)
country comparison to the world: 64

GOVERNMENT

Country name: *conventional long form:* Republic of Rwanda
conventional short form: Rwanda
local long form: Republika y'u Rwanda
local short form: Rwanda
former: Ruanda, German East Africa
etymology: the name translates as "domain" in the native Kinyarwanda language

Government type: presidential republic

Capital: *name:* Kigali
geographic coordinates: 1 57 S, 30 03 E
time difference: UTC+2 (7 hours ahead of Washington, DC, during Standard Time)
etymology: the city takes its name from nearby Mount Kigali; the name "Kigali" is composed of the Bantu prefix "ki" and the Rwandan "gali" meaning "broad" and likely refers to the broad, sprawling hill that has been dignified with the title of "mount"

Administrative divisions: 4 provinces (in French—provinces, singular—province; in Kinyarwanda—intara for singular and plural) and 1 city* (in French—ville; in Kinyarwanda—umujyi); Est (Eastern), Kigali*, Nord (Northern), Ouest (Western), Sud (Southern)

Independence: 1 July 1962 (from Belgium-administered UN trusteeship)

National holiday: Independence Day, 1 July (1962)

Constitution: *history:* several previous; latest adopted by referendum 26 May 2003, effective 4 June 2003
amendments: proposed by the president of the republic (with Council of Ministers approval) or by two-thirds majority vote of both houses of Parliament; passage requires at least three-quarters majority vote in both houses; changes to constitutional articles on national sovereignty, the presidential term, the form and system of government, and political pluralism also require approval in a referendum; amended 2008, 2010, 2015

Legal system: mixed legal system of civil law, based on German and Belgian models, and customary law; judicial review of legislative acts in the Supreme Court

International law organization participation: has not submitted an ICJ jurisdiction declaration; non-party state to the ICCt

Citizenship: *citizenship by birth:* no
citizenship by descent only: the father must be a citizen of Rwanda; if the father is stateless or unknown, the mother must be a citizen
dual citizenship recognized: no
residency requirement for naturalization: 10 years

Suffrage: 18 years of age; universal

Executive branch: *chief of state:* President Paul KAGAME (since 22 April 2000)
head of government: Prime Minister Edouard NGIRENTE (since 30 August 2017)
cabinet: Council of Ministers appointed by the president
elections/appointments: president directly elected by simple majority vote for a 5-year term (eligible for a second term); note—a constitutional amendment approved in December 2016 reduced the presidential term from 7 to 5 years but included an exception that allowed President KAGAME to serve another 7-year term in 2017, potentially followed by two additional 5-year terms; election last held on 4 August 2017 (next to be held in August 2024); prime minister appointed by the president
election results: Paul KAGAME reelected president; Paul KAGAME (RPF) 98.8%, Philippe MPAYIMANA (independent) 0.7%, Frank HABINEZA (DGPR) 0.5%

Legislative branch: *description:* bicameral Parliament consists of:
Senate or Senat (26 seats; 12 members indirectly elected by local councils, 8 appointed by the president, 4 appointed by the Political Organizations Forum—a body of registered political parties, and 2 selected by institutions of higher learning; members serve 8-year terms)
Chamber of Deputies or Chambre des Deputes (80 seats; 53 members directly elected by proportional representation vote, 24 women selected by special interest groups, and 3 selected by youth and disability organizations; members serve 5-year terms)
elections: Senate—last held on 16-18 September 2019 (next to be held in 2027)
Chamber of Deputies—last held on 3 September 2018 (next to be held in September 2023)
election results: Senate—percent of vote by party—NA; seats by party—NA; composition—men 16, women 10, percent of women 38.5%
Chamber of Deputies—percent of vote by party—NA; seats by party—Rwandan Patriotic Front Coalition 40, PSD 5, PL 4, other 4 indirectly elected 27; composition—men 26, women 54, percent of women 67.5%; note—total Parliament percent of women 60.4%

Judicial branch: *highest courts:* Supreme Court (consists of the chief and deputy chief justices and 15 judges; normally organized into 3-judge panels); High Court (consists of the court president, vice president, and a minimum of 24 judges and organized into 5 chambers)
judge selection and term of office: Supreme Court judges nominated by the president after consultation with the Cabinet and the Superior Council of the Judiciary (SCJ), a 27-member body of judges, other judicial officials, and legal professionals) and approved by the Senate; chief and deputy chief justices appointed for 8-year nonrenewable terms; tenure of judges NA; High Court president and vice president appointed by the president of the republic upon approval by the Senate; judges appointed by the Supreme Court chief justice upon approval of the SCJ; judge tenure NA
subordinate courts: High Court of the Republic; commercial courts including the High Commercial Court; intermediate courts; primary courts; and military specialized courts

Political parties and leaders: Democratic Green Party of Rwanda or DGPR [Frank HABINEZA]
Liberal Party or PL [Donatille MUKABALISA]
Party for Progress and Concord or PPC [Dr. Alivera MUKABARAMBA]
Party Imberakuri or PS-Imberakuri [Christine MUKABUNANI]
Rwandan Patriotic Front or RPF [Paul KAGAME]
Rwandan Patriotic Front Coalition (includes RPF, PPC) [Paul KAGAME]
Social Democratic Party or PSD [Vincent BIRUTA]

International organization participation: ACP, AfDB, AU, C, CEPGL, COMESA, EAC, EADB, FAO, G-77, IAEA, IBRD, ICAO, ICRM, IDA, IFAD, IFC, IFRCS, ILO, IMF, Interpol, IOC, IOM, IPU, ISO, ITSO, ITU, ITUC (NGOs), MIGA, MINUSMA, NAM, OIF, OPCW, PCA, UN, UNAMID, UNCTAD, UNESCO, UNHCR, UNIDO, UNISFA, UNMISS, UNWTO, UPU, WCO, WHO, WIPO, WMO, WTO

Diplomatic representation in the US: *chief of mission:* Ambassador Mathilde MUKANTABANA (since 18 July 2013)
chancery: 1875 Connecticut Avenue NW, Suite 418, Washington, DC 20009
telephone: [1] (202) 232-2882
FAX: [1] (202) 232-4544

Diplomatic representation from the US: *chief of mission:* Ambassador Peter H. VROOMAN (since 5 April 2018)
telephone: [250] 252 596-400
embassy: 2657 Avenue de la Gendarmerie, P. O. Box 28, Kigali
mailing address: B.P. 28, Kigali
FAX: [250] 252 580 325

Flag description: three horizontal bands of sky blue (top, double width), yellow, and green, with a golden sun with 24 rays near the fly end of the blue band; blue represents happiness and peace, yellow economic development and mineral wealth, green hope of prosperity and natural resources; the sun symbolizes unity, as well as enlightenment and transparency from ignorance

National symbol(s): *traditional woven basket with peaked lid; national colors:* blue, yellow, green

National anthem: *name:* "Rwanda nziza" (Rwanda, Our Beautiful Country)
lyrics/music: Faustin MURIGO/Jean-Bosco HASHAKAIMANA
note: adopted 2001
0:00 / 2:18

ECONOMY

Economy—overview: Rwanda is a rural, agrarian country with agriculture accounting for about 63% of export earnings, and with some mineral and agro-processing. Population density is high but, with the exception of the capital Kigali, is not concentrated in large cities – its 12 million people are spread out on a small amount of land (smaller than the state of Maryland). Tourism, minerals, coffee, and tea are Rwanda's main sources of foreign exchange. Despite Rwanda's fertile ecosystem, food production often does not keep pace with demand, requiring food imports. Energy shortages, instability in neighboring states, and lack of adequate transportation linkages to other countries continue to handicap private sector growth.

The 1994 genocide decimated Rwanda's fragile economic base, severely impoverished the population, particularly women, and temporarily stalled the country's ability to attract private and external investment. However, Rwanda has made substantial progress in stabilizing and rehabilitating its economy well beyond pre-1994 levels. GDP has rebounded with an average annual growth of 6%-8% since 2003 and inflation has been reduced to single digits. In 2015, 39% of the population lived below the poverty line, according to government statistics, compared to 57% in 2006.

The government has embraced an expansionary fiscal policy to reduce poverty by improving education, infrastructure, and foreign and domestic investment. Rwanda consistently ranks well for ease of doing business and transparency.

The Rwandan Government is seeking to become a regional leader in information and communication technologies and aims to reach middle-income status by 2020 by leveraging the service industry. In 2012, Rwanda completed the first modern Special Economic Zone (SEZ) in Kigali. The SEZ seeks to attract investment in all sectors, but specifically in agribusiness, information and communications, trade and logistics, mining, and construction. In 2016, the government launched an online system to give investors information about public land and its suitability for agricultural development.

GDP (purchasing power parity): $24.68 billion (2017 est.)
$23.26 billion (2016 est.)
$21.94 billion (2015 est.)
note: data are in 2017 dollars
country comparison to the world: 142

GDP (official exchange rate): $9.136 billion (2017 est.)

GDP—real growth rate: 6.1% (2017 est.)
6% (2016 est.)
8.9% (2015 est.)
country comparison to the world: 29

GDP—per capita (PPP): $2,100 (2017 est.)
$2,000 (2016 est.)
$1,900 (2015 est.)
note: data are in 2017 dollars
country comparison to the world: 208

Gross national saving: 12.5% of GDP (2017 est.)
6.1% of GDP (2016 est.)
7.5% of GDP (2015 est.)
country comparison to the world: 147

GDP—composition, by end use:
household consumption: 75.9% (2017 est.)
government consumption: 15.2% (2017 est.)
investment in fixed capital: 22.9% (2017 est.)
investment in inventories: 0.5% (2017 est.)
exports of goods and services: 18.2% (2017 est.)
imports of goods and services: -32.8% (2017 est.)

GDP—composition, by sector of origin:
agriculture: 30.9% (2017 est.)
industry: 17.6% (2017 est.)
services: 51.5% (2017 est.)

Agriculture—products: coffee, tea, pyrethrum (insecticide made from chrysanthemums), bananas, beans, sorghum, potatoes; livestock

Industries: cement, agricultural products, small-scale beverages, soap, furniture, shoes, plastic goods, textiles, cigarettes

Industrial production growth rate: 4.2% (2017 est.)
country comparison to the world: 72

Labor force: 6.227 million (2017 est.)
country comparison to the world: 70

Labor force—by occupation: *agriculture:* 75.3%
industry: 6.7%
services: 18% (2012 est.)

Unemployment rate: 2.7% (2014 est.)
country comparison to the world: 31

Population below poverty line: 39.1% (2015 est.)

Household income or consumption by percentage share: *lowest 10%:* 2.1%
highest 10%: 43.2% (2011 est.)

Budget: *revenues:* 1.943 billion (2017 est.)
expenditures: 2.337 billion (2017 est.)

Taxes and other revenues: 21.3% (of GDP) (2017 est.)
country comparison to the world: 142

Budget surplus (+) or deficit (-): -4.3% (of GDP) (2017 est.)
country comparison to the world: 162

Public debt: 40.5% of GDP (2017 est.)
37.3% of GDP (2016 est.)
country comparison to the world: 125

Fiscal year: calendar year

Inflation rate (consumer prices): 4.8% (2017 est.)
5.7% (2016 est.)
country comparison to the world: 170

Current account balance: -$622 million (2017 est.)
-$1.336 billion (2016 est.)
country comparison to the world: 128

Exports: $1.05 billion (2017 est.)
$745 million (2016 est.)
country comparison to the world: 157

Exports—partners: UAE 38.3%, Kenya 15.1%, Switzerland 9.9%, Democratic Republic of the Congo 9.5%, US 4.9%, Singapore 4.5% (2017)

Exports—commodities: coffee, tea, hides, tin ore

Imports: $1.922 billion (2017 est.)
$2.036 billion (2016 est.)
country comparison to the world: 168

Imports—commodities: foodstuffs, machinery and equipment, steel, petroleum products, cement and construction material

Imports—partners: China 20.4%, Uganda 11%, India 7.2%, Kenya 7.1%, Tanzania 5.3%, UAE 5.1% (2017)

Reserves of foreign exchange and gold: $997.6 million (31 December 2017 est.)
$1.104 billion (31 December 2016 est.)
country comparison to the world: 132

Debt—external: $3.258 billion (31 December 2017 est.)
$2.611 billion (31 December 2016 est.)
country comparison to the world: 141

Exchange rates: Rwandan francs (RWF) per US dollar -
839.1 (2017 est.)
787.25 (2016 est.)
787.25 (2015 est.)
720.54 (2014 est.)
680.95 (2013 est.)

ENERGY

Electricity access: *population without electricity:* 6 million (2019)
electrification—total population: 53% (2019)
electrification—urban areas: 76% (2019)
electrification—rural areas: 48% (2019)

Electricity—production: 525 million kWh (2016 est.)
country comparison to the world: 164

Electricity—consumption: 527.3 million kWh (2016 est.)
country comparison to the world: 169

Electricity—exports: 4 million kWh (2015 est.)
country comparison to the world: 92

Electricity—imports: 42 million kWh (2016 est.)
country comparison to the world: 108

Electricity—installed generating capacity: 191,000 kW (2016 est.)
country comparison to the world: 166

Electricity—from fossil fuels: 42% of total installed capacity (2016 est.)
country comparison to the world: 164

Electricity—from nuclear fuels: 0% of total installed capacity (2017 est.)
country comparison to the world: 170

Electricity—from hydroelectric plants: 51% of total installed capacity (2017 est.)
country comparison to the world: 37

Electricity—from other renewable sources: 7% of total installed capacity (2017 est.)
country comparison to the world: 96

Crude oil—production: 0 bbl/day (2018 est.)

country comparison to the world: 189

Crude oil—exports: 0 bbl/day (2015 est.)
country comparison to the world: 182

Crude oil—imports: 0 bbl/day (2015 est.)
country comparison to the world: 184

Crude oil—proved reserves: 0 bbl (1 January 2018 est.)
country comparison to the world: 184

Refined petroleum products—production: 0 bbl/day (2015 est.)
country comparison to the world: 191

Refined petroleum products—consumption: 6,700 bbl/day (2016 est.)
country comparison to the world: 167

Refined petroleum products—exports: 0 bbl/day (2015 est.)
country comparison to the world: 192

Refined petroleum products—imports: 6,628 bbl/day (2015 est.)
country comparison to the world: 162

Natural gas—production: 0 cu m (2017 est.)
country comparison to the world: 186

Natural gas—consumption: 0 cu m (2017 est.)
country comparison to the world: 188

Natural gas—exports: 0 cu m (2017 est.)
country comparison to the world: 171

Natural gas—imports: 0 cu m (2017 est.)
country comparison to the world: 177

Natural gas—proved reserves: 56.63 billion cu m (1 January 2018 est.)
country comparison to the world: 61

Carbon dioxide emissions from consumption of energy: 985,600 Mt (2017 est.)
country comparison to the world: 169

COMMUNICATIONS

Telephones—fixed lines: *total subscriptions:* 11,215
subscriptions per 100 inhabitants: less than 1 (2019 est.)
country comparison to the world: 188

Telephones—mobile cellular: *total subscriptions:* 9,531,609
subscriptions per 100 inhabitants: 76.49 (2019 est.)
country comparison to the world: 90

Telecommunication systems: *general assessment:* govt. invests in smart city infrastructure; expanding wholesale LTE services; govt. launches SIM card registration; growing economy and foreign aid help launch telecom sector, despite widespread poverty; slow to liberalize mobile sector; competing operators roll out national fiber optic backbone that connects to submarine cables of neighboring countries ending expensive dependence on satellite (2020)
domestic: the capital, Kigali, is connected to provincial centers by microwave radio relay, and recently by cellular telephone service; much of the network depends on wire and HF radiotelephone; fixed-line less than 1 per 100 and mobile-cellular telephone density has increased to 76 telephones per 100 persons (2019)
international: country code—250; international connections employ microwave radio relay to neighboring countries and satellite communications to more distant countries; satellite earth stations—1 Intelsat (Indian Ocean) in Kigali (includes telex and telefax service); international submarine fiber-optic cables on the African east coast has brought international bandwidth and lessened the dependency on satellites
note: the COVID-19 outbreak is negatively impacting telecommunications production and supply chains globally; consumer spending on telecom devices and services has also slowed due to the pandemic's effect on economies worldwide; overall progress towards improvements in all facets of the telecom industry—mobile, fixed-line, broadband, submarine cable and satellite—has moderated

Broadcast media: 13 TV stations; 35 radio stations registered, including international broadcasters, government owns most popular TV and radio stations; regional satellite-based TV services available

Internet country code: .rw

Internet users: *total:* 2,653,197
percent of population: 21.77% (July 2018 est.)
country comparison to the world: 107

Broadband—fixed subscriptions: *total:* 7,501
subscriptions per 100 inhabitants: less than 1 (2018 est.)
country comparison to the world: 175

TRANSPORTATION

National air transport system: *number of registered air carriers:* 1 (2020)
inventory of registered aircraft operated by air carriers: 12
annual passenger traffic on registered air carriers: 1,073,528 (2018)

Civil aircraft registration country code prefix: 9XR (2016)

Airports: 7 (2013)
country comparison to the world: 171

Airports—with paved runways: *total:* 4 (2019)
over 3,047 m: 1
914 to 1,523 m: 2

under 914 m: 1

Airports—with unpaved runways: *total:* 3 (2013)
914 to 1,523 m: 2 (2013)
under 914 m: 1 (2013)

Roadways: *total:* 4,700 km (2012)
paved: 1,207 km (2012)
unpaved: 3,493 km (2012)
country comparison to the world: 148

Waterways: (Lac Kivu navigable by shallow-draft barges and native craft) (2011)

Ports and terminals: *lake port(s):* Cyangugu, Gisenyi, Kibuye (Lake Kivu)

MILITARY AND SECURITY

Military and security forces: *Rwanda Defense Force (RDF):* Rwanda Army (Rwanda Land Force), Rwanda Air Force (Force Aerienne Rwandaise, FAR), Rwanda Reserve Force (2020)

Military expenditures: 1.2% of GDP (2019)
1.2% of GDP (2018)
1.3% of GDP (2017)
1.3% of GDP (2016)
1.3% of GDP (2015)
country comparison to the world: 106

Military and security service personnel strengths: the Rwanda Defense Force (RDF) has approximately 32,500 active personnel (32,000 Army; 500 Air Force) (2019 est.)

Military equipment inventories and acquisitions: the RDF's inventory includes mostly Soviet-era and older Western—mostly French and South African—equipment; Russia is the largest supplier of equipment to the RDF since 2010 (2019 est.)

Military deployments: 1,370 Central African Republic (MINUSCA); 1,090 Sudan (UNAMID); 2,750 South Sudan (UNMISS) (2020)

Military service age and obligation: 18 years of age for voluntary military service; no conscription; Rwandan citizenship is required, as is a 9th-grade education for enlisted recruits and an A-level certificate for officer candidates; enlistment is either as contract (5-years, renewable twice) or career; retirement (for officers and senior NCOs) after 20 years of service or at 40-60 years of age (2013)

TRANSNATIONAL ISSUES

Disputes—international: Burundi and Rwanda dispute two sq km (0.8 sq mi) of Sabanerwa, a farmed area in the Rukurazi Valley where the Akanyaru/Kanyaru River shifted its course southward after heavy rains in 1965; fighting among ethnic groups—loosely associated political rebels, armed gangs, and various government forces in Great Lakes region transcending the boundaries of Burundi, Democratic Republic of the Congo (DROC), Rwanda, and Uganda—abated substantially from a decade ago due largely to UN peacekeeping, international mediation, and efforts by local governments to create civil societies; nonetheless, 57,000 Rwandan refugees still reside in 21 African states, including Zambia, Gabon, and 20,000 who fled to Burundi in 2005 and 2006 to escape drought and recriminations from traditional courts investigating the 1994 massacres; the 2005 DROC and Rwanda border verification mechanism to stem rebel actions on both sides of the border remains in place

Refugees and internally displaced persons: *refugees (country of origin):* 77,017 (Democratic Republic of the Congo), 72,007 (Burundi) (2020)

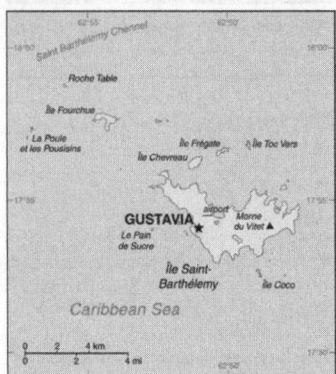

Background: Discovered in 1493 by Christopher COLUMBUS who named it for his brother Bartolomeo, Saint Barthelemy was first settled by the French in 1648. In 1784, the French sold the island to Sweden, which renamed the largest town Gustavia, after the Swedish King GUSTAV III, and made it a free port; the island prospered as a trade and supply center during the colonial wars of the 18th century. France repurchased the island in 1877 and took control the following year. It was placed under the administration of Guadeloupe. Saint Barthelemy retained its free port status along with various Swedish appellations such as Swedish street and town names, and the three-crown symbol on the coat of arms. In 2003, the islanders voted to secede from Guadeloupe, and in 2007, the island became a French overseas collectivity. In 2012, it became an overseas territory of the EU, allowing it to exert local control over the permanent and temporary immigration of foreign workers including non-French European citizens.

GEOGRAPHY

Location: Caribbean, island between the Caribbean Sea and the North Atlantic Ocean; located in the Leeward Islands (northern) group; Saint Barthelemy lies east of the US Virgin Islands

Geographic coordinates: 17 90 N, 62 85 W

Map references: Central America and the Caribbean

Area: *total:* 25 sq km
land: 25 sq km
water: negligible
country comparison to the world: 239

Area—comparative: less than one-eighth the size of Washington, DC

Land boundaries: 0 km

Climate: tropical, with practically no variation in temperature; has two seasons (dry and humid)

Terrain: hilly, almost completely surrounded by shallow-water reefs, with plentiful beaches

Elevation: *lowest point:* Caribbean Ocean 0 m
highest point: Morne du Vitet 286 m

Natural resources: few natural resources; beaches foster tourism

Population distribution: most of the populace concentrated in and around the capital of Gustavia, but scattered settlements exist around the island periphery

Environment—current issues: land-based pollution; urbanization; with no natural rivers or streams, fresh water is in short supply, especially in summer, and is provided by the desalination of sea water, the collection of rain water, or imported via water tanker; overfishing

Geography—note: a 1,200-hectare marine nature reserve, the Reserve Naturelle, is made up of five zones around the island that form a network to protect the island's coral reefs, seagrass, and endangered marine species

PEOPLE AND SOCIETY

Population: 7,122 (July 2020 est.)
country comparison to the world: 226

Ethnic groups: other white, Creole (mulatto), black, Guadeloupe Mestizo (French-East Asia)

Languages: French (primary), English

Religions: Roman Catholic, Protestant, Jehovah's Witnesses

Age structure: *0-14 years:* 15.16% (male 555/female 525)
15-24 years: 7.34% (male 275/female 248)
25-54 years: 41.86% (male 1,618/female 1,363)
55-64 years: 16.29% (male 630/female 530)
65 years and over: 19.35% (male 690/female 688) (2020 est.)

Median age: *total:* 45.6 years
male: 45.5 years
female: 45.8 years (2020 est.)
country comparison to the world: 8

Population growth rate: -0.25% (2020 est.)
country comparison to the world: 214

Birth rate: 9.3 births/1,000 population (2020 est.)
country comparison to the world: 202

Death rate: 8.7 deaths/1,000 population (2020 est.)
country comparison to the world: 70

Net migration rate: -3 migrant(s)/1,000 population (2020 est.)
country comparison to the world: 176

Population distribution: most of the populace concentrated in and around the capital of Gustavia, but scattered settlements exist around the island periphery

Sex ratio: *at birth:* 1.06 male(s)/female
0-14 years: 1.06 male(s)/female
15-24 years: 1.11 male(s)/female

25-54 years: 1.19 male(s)/female
55-64 years: 1.19 male(s)/female
65 years and over: 1 male(s)/female
total population: 1.12 male(s)/female (2020 est.)

Infant mortality rate: *total:* 5.6 deaths/1,000 live births
male: 6.4 deaths/1,000 live births
female: 4.8 deaths/1,000 live births (2020 est.)
country comparison to the world: 169

Life expectancy at birth: *total population:* 80.2 years
male: 77 years
female: 83.4 years (2020 est.)
country comparison to the world: 46

Total fertility rate: 1.64 children born/woman (2020 est.)
country comparison to the world: 181

HIV/AIDS—people living with HIV/AIDS: NA

Education expenditures: NA

GOVERNMENT

Country name: *conventional long form:* Overseas Collectivity of Saint Barthelemy
conventional short form: Saint Barthelemy
local long form: Collectivite d'outre mer de Saint-Barthelemy
local short form: Saint-Barthelemy
abbreviation: Saint-Barth (French); St. Barts or St. Barths (English)
etymology: explorer Christopher COLUMBUS named the island in honor of his brother Bartolomeo's namesake saint in 1493

Dependency status: overseas collectivity of France

Government type: parliamentary democracy (Territorial Council); overseas collectivity of France

Capital: *name:* Gustavia

geographic coordinates: 17 53 N, 62 51 W
time difference: UTC-4 (1 hour ahead of Washington, DC, during Standard Time)
etymology: named in honor of King Gustav III (1746-1792) of Sweden during whose reign the island was obtained from France in 1784; the name was retained when in 1878 the island was sold back to France

Independence: none (overseas collectivity of France)

National holiday: Fete de la Federation, 14 July (1790); note—local holiday is St. Barthelemy Day, 24 August (1572)

Constitution: *history:* 4 October 1958 (French Constitution)
amendments: amendment procedures of France's constitution apply

Legal system: French civil law

Citizenship: see France

Suffrage: 18 years of age, universal

Executive branch: *chief of state:* President Emmanuel MACRON (since 14 May 2017), represented by Prefect Anne LAUBIES (since 8 June 2015)

head of government: President of Territorial Council Bruno MAGRAS (since 16 July 2007)

cabinet: Executive Council elected by the Territorial Council; note—there is also an advisory, economic, social, and cultural council

elections/appointments: French president directly elected by absolute majority popular vote in 2 rounds if needed for a 5-year term (eligible for a second term); prefect appointed by the French president on the advice of French Ministry of Interior; president of Territorial Council indirectly elected by its members for a 5-year term; election last held on 2 April 2017 (next to be held in 2022)

election results: Bruno MAGRAS (SBA) reelected president; Territorial Council vote—NA

Legislative branch: *description:* unicameral Territorial Council (19 seats; members elected by absolute majority vote in the first round vote and proportional representation vote in the second round; members serve 5-year terms); Saint Barthelemy indirectly elects 1 senator to the French Senate by an electoral college for a 6-year term and directly elects 1 deputy (shared with Saint Martin) to the French National Assembly

elections: Territorial Council—last held on 19 March 2017 (next to be held in September 2022) French Senate—election last held 24 September 2017 (next to be held in September 2020) French National Assembly—election last held on 11 and 18 June 2017 (next to be held by June 2022)

election results: Territorial Council—percent of vote by party—SBA 53.7%, United for Saint Barth 20.6%, Saint Barth Essential 18.1%, All for Saint Barth 7.7%; seats by party—SBA 14, United for Saint Barth 2, Saint Barth Essential 2, All for Saint Barth 1; composition—men 9, women 10, percent of women 52.6%; French Senate—percent of vote by party NA; seats by party UMP 1 French National Assembly—percent of vote by party NA; seats by party UMP 1

Political parties and leaders: All for Saint Barth (Tous pour Saint-Barth) [Bettina COINTRE]

Saint Barth Essential (Saint-Barth Autrement) [Marie-Helene BERNIER]

Saint Barth First! (Saint-Barth d'Abord!) or SBA [Bruno MAGRAS]

Saint Barth United (Unis pour Saint-Barthelemy) [Xavier LEDEE]

International organization participation: UPU

Diplomatic representation in the US: none (overseas collectivity of France)

Diplomatic representation from the US: none (overseas collectivity of France)

Flag description: the flag of France is used

National symbol(s): pelican

National anthem: *name:* "L'Hymne a St. Barthelemy" (Hymn to St. Barthelemy)

lyrics/music: Isabelle Massart DERAVIN/Michael VALENTI

note: local anthem in use since 1999; as a collectivity of France, "La Marseillaise" is official (see France)

Economy—overview: The economy of Saint Barthelemy is based upon high-end tourism and duty-free luxury commerce, serving visitors primarily from North America. The luxury hotels and villas host 70,000 visitors each year with another 130,000 arriving by boat. The relative isolation and high cost of living inhibits mass tourism. The construction and public sectors also enjoy significant investment in support of tourism. With limited fresh water resources, all food must be imported, as must all energy resources and most manufactured goods. The tourism sector creates a strong employment demand and attracts labor from Brazil and Portugal. The country's currency is the euro.

Exchange rates: 2013 est.)
0.885 (2017 est.)
0.903 (2016 est.)
0.9214 (2015 est.)
0.885 (2014 est.)

Telecommunication systems: *general assessment:* fully integrated access; 4G and LTE services (2020)

domestic: direct dial capability with both fixed and wireless systems, 3 FM channels, no broadcasting (2018)

international: country code—590; landing points for the SSCS and the Southern Caribbean Fiber submarine cables providing voice and data connectivity to numerous Caribbean Islands (2019)

note: the COVID-19 outbreak is negatively impacting telecommunications production and supply chains globally; consumer spending on telecom devices and services has also slowed due to the pandemic's effect on economies worldwide; overall progress towards improvements in all facets of the telecom industry—mobile, fixed-line, broadband, submarine cable and satellite—has moderated

Broadcast media: no local TV broadcasters; 3 FM radio channels (2019)

Internet country code: .bl; note—.gp, the Internet country code for Guadeloupe, and .fr, the Internet country code for France, might also be encountered

Airports: 1 (2013)

country comparison to the world: 234

Airports—with paved runways: *total:* 1 (2019)

under 914 m: 1

Roadways: *total:* 40 km

country comparison to the world: 217

Ports and terminals: *major seaport(s):* Gustavia

Transportation—note: nearest airport for international flights is Princess Juliana International Airport (SXM) located on Sint Maarten

Military—note: defense is the responsibility of France

SAINT HELENA, ASCENSION, AND TRISTAN DA CUNHA

Background: Saint Helena is a British Overseas Territory consisting of Saint Helena and Ascension Islands, and the island group of Tristan da Cunha. *Saint Helena:* Uninhabited when first discovered by the Portuguese in 1502, Saint Helena was garrisoned by the British during the 17th century. It acquired fame as the place of Napoleon BONAPARTE's exile from 1815 until his death in 1821, but its importance as a port of call declined after the opening of the Suez Canal in 1869. During the Anglo-Boer War in South Africa, several thousand Boer prisoners were confined on the island between 1900 and 1903.;

Saint Helena is one of the most remote populated places in the world. The British Government committed to building an airport on Saint Helena in 2005. After more than a decade of delays and construction, a commercial air service to South Africa via Namibia was inaugurated in October of 2017. The weekly service to Saint Helena from Johannesburg via Windhoek in Namibia takes just over six hours (including the refueling stop in Windhoek) and replaces the mail ship that had made a five-day journey to the island every three weeks.;

Ascension Island: This barren and uninhabited island was discovered and named by the Portuguese in 1503. The British garrisoned the island in 1815 to prevent a rescue of Napoleon from Saint Helena. It served as a provisioning station for the Royal Navy's West Africa Squadron on anti-slavery patrol. The island remained under Admiralty control until 1922, when it became a dependency of Saint Helena. During World War II, the UK permitted the US to construct an airfield on Ascension in support of transatlantic flights to Africa and anti-submarine operations in the South Atlantic. In the 1960s the island became an important space tracking station for the US. In 1982, Ascension was an essential staging area for British forces during the Falklands War. It remains a critical refueling point in the air-bridge from the UK to the South Atlantic.;

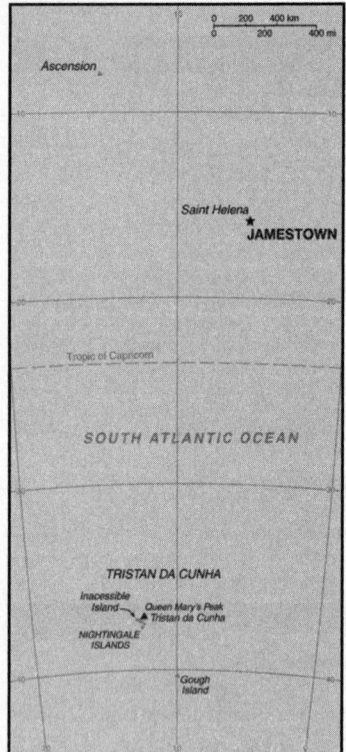

The island hosts one of four dedicated ground antennas that assist in the operation of the Global Positioning System (GPS) navigation system (the others are on Diego Garcia (British Indian Ocean Territory), Kwajalein (Marshall Islands), and at Cape Canaveral, Florida (US)). NASA and the US Air Force also operate a Meter-Class Autonomous Telescope (MCAT) on Ascension as part of the deep space surveillance system for tracking orbital debris, which can be a hazard to spacecraft and astronauts.

Tristan da Cunha: The island group consists of Tristan da Cunha, Nightingale, Inaccessible, and Gough Islands. Tristan da Cunha, named after its Portuguese discoverer (1506), was garrisoned by the British in 1816 to prevent any attempt to rescue Napoleon from Saint Helena. Gough and Inaccessible Islands have been designated World Heritage Sites. South Africa leases a site for a meteorological station on Gough Island.

GEOGRAPHY

Location: islands in the South Atlantic Ocean, about midway between South America and Africa; Ascension Island lies 1,300 km (800 mi) northwest of Saint Helena; Tristan da Cunha lies 4,300 km (2,700 mi) southwest of Saint Helena

Geographic coordinates: *Saint Helena:* 15 57 S, 5 42 W;
Ascension Island: 7 57 S, 14 22 W;

Tristan da Cunha island group: 37 15 S, 12 30 W

Map references: Africa

Area: *total:* 394 sq km
land: 122 sq km Saint Helena Island
water: 0 sq km
88 sq km Ascension Island, 184 sq km Tristan da Cunha island group (includes Tristan (98 sq km), Inaccessible, Nightingale, and Gough islands)
country comparison to the world: 204

Area—comparative: slightly more than twice the size of Washington, DC

Land boundaries: 0 km

Coastline: *Saint Helena:* 60 km
Ascension Island: NA
Tristan da Cunha (island only): 34 km

Maritime claims: *territorial sea:* 12 nm
exclusive fishing zone: 200 nm

Climate: *Saint Helena:* tropical marine; mild, tempered by trade winds;
Ascension Island: tropical marine; mild, semi-arid;
Tristan da Cunha: temperate marine; mild, tempered by trade winds (tends to be cooler than Saint Helena)

Terrain: the islands of this group are of volcanic origin associated with the Atlantic Mid-Ocean Ridge
Saint Helena: rugged, volcanic; small scattered plateaus and plains;
Ascension: surface covered by lava flows and cinder cones of 44 dormant volcanoes; terrain rises to the east;
Tristan da Cunha: sheer cliffs line the coastline of the nearly circular island; the flanks of the central volcanic peak are deeply dissected; narrow coastal plain lies between The Peak and the coastal cliffs

Elevation: *lowest point:* Atlantic Ocean 0 m
highest point: Queen Mary's Peak on Tristan da Cunha 859 m; Green Mountain on Ascension Island 818 m; Mount Actaeon on Saint Helena Island 2,060 m

Natural resources: fish, lobster

Land use: *agricultural land:* 30.8% (2011 est.)
arable land: 10.3% (2011 est.) / permanent crops: 0% (2011 est.) / permanent pasture: 20.5% (2011 est.)
forest: 5.1% (2011 est.)
other: 64.1% (2011 est.)
Irrigated land: 0 sq km (2012)

Population distribution: Saint Helena—population is concentrated in and around the capital Jamestown in the northwest, with another significant cluster in the interior Longwood area; Ascension—largest settlement, and location of most of the population, is Georgetown; Tristan da Cunha—most of the nearly 300 inhabitants live in the northern coastal town of Edinburgh of the Seven Seas

Natural hazards: active volcanism on Tristan da Cunha
volcanism: the island volcanoes of Tristan da Cunha (2,060 m) and Nightingale Island (365 m)

experience volcanic activity; Tristan da Cunha erupted in 1962 and Nightingale in 2004

Environment—current issues: development threatens unique biota on Saint Helena

Geography—note: Saint Helena harbors at least 40 species of plants unknown elsewhere in the world; Ascension is a breeding ground for sea turtles and sooty terns; Queen Mary's Peak on Tristan da Cunha is the highest island mountain in the South Atlantic and a prominent landmark on the sea lanes around southern Africa

PEOPLE AND SOCIETY

Population: 7,862 (July 2020 est.)
note: Saint Helena's Statistical Office estimated the de facto population to be 4,577 in 2019; only Saint Helena, Ascension, and Tristan da Cunha islands are inhabited, none of the other nearby islands/islets are
country comparison to the world: 225

Nationality: *noun:* Saint Helenian(s)
adjective: Saint Helenian
note: referred to locally as "Saints"

Ethnic groups: African descent 50%, white 25%, Chinese 25%

Languages: English

Religions: Protestant 75.9% (includes Anglican 68.9, Baptist 2.1%, Seventh Day Adventist 1.8%, Salvation Army 1.7%, New Apostolic 1.4%), Jehovah's Witness 4.1%, Roman Catholic 1.2%, other 2.5% (includes Baha'i), unspecified 0.8%, none 6.1%, no response 9.4% (2016 est.)
note: data represent Saint Helena only

Demographic profile: The vast majority of the population of Saint Helena, Ascension, and Tristan da Cunha live on Saint Helena. Ascension has no indigenous or permanent residents and is inhabited only by persons contracted to work on the island (mainly with the UK and US military or in the space and communications industries) or their dependents, while Tristan da Cunha – the main island in a small archipelago – has fewer than 300 residents. The population of Saint Helena consists of the descendants of 17th century British sailors and settlers from the East India Company, African slaves, and indentured servants and laborers from India, Indonesia, and China. Most of the population of Ascension are Saint Helenians, Britons, and Americans, while that of Tristan da Cunha descends from shipwrecked sailors and Saint Helenians.

Change in Saint Helena's population size is driven by net outward migration. Since the 1980s, Saint Helena's population steadily has shrunk and aged as the birth rate has decreased and many working-age residents left for better opportunities elsewhere. The restoration of British citizenship in 2002 accelerated family emigration; from 1998 to 2008 alone, population declined by about 20%.

In the last few years, population has experienced some temporary growth, as foreigners and returning Saint Helenians, have come to build an international airport, but numbers are beginning to fade as the project reaches completion and workers depart. In the long term, once the airport is fully operational, increased access to the remote island has the potential to boost tourism and fishing, provide more jobs for Saint Helenians domestically, and could encourage some expatriots to return home. In the meantime, however, Saint Helena, Ascension, and Tristan da Cunha have to contend with the needs of an aging population. The elderly population of the islands has risen from an estimated 9.4% in 1998 to 20.4% in 2016.

Age structure: *0-14 years:* 15.15% (male 607/female 584)
15-24 years: 12.12% (male 486/female 467)
25-54 years: 43.06% (male 1,685/female 1,700)
55-64 years: 12.96% (male 503/female 516)
65 years and over: 16.71% (male 670/female 644) (2020 est.)

Median age: *total:* 43.2 years
male: 43.2 years
female: 43.3 years (2020 est.)
country comparison to the world: 30

Population growth rate: 0.13% (2020 est.)
country comparison to the world: 185

Birth rate: 9.4 births/1,000 population (2020 est.)
country comparison to the world: 199

Death rate: 8.3 deaths/1,000 population (2020 est.)
country comparison to the world: 81

Net migration rate: 0 migrant(s)/1,000 population (2020 est.)
country comparison to the world: 94

Population distribution: Saint Helena—population is concentrated in and around the capital Jamestown in the northwest, with another significant cluster in the interior Longwood area; Ascension—largest settlement, and location of most of the population, is Georgetown; Tristan da Cunha—most of the nearly 300 inhabitants live in the northern coastal town of Edinburgh of the Seven Seas

Urbanization: *urban population:* 40.1% of total population (2020)
rate of urbanization: 0.73% annual rate of change (2015-20 est.)

total population growth rate v. urban population growth rate, 2000-2030:

Major urban areas—population: 1,000 JAMESTOWN (capital) (2018)

Sex ratio: *at birth:* 1.06 male(s)/female
0-14 years: 1.04 male(s)/female
15-24 years: 1.04 male(s)/female
25-54 years: 0.99 male(s)/female
55-64 years: 0.97 male(s)/female
65 years and over: 1.04 male(s)/female
total population: 1.01 male(s)/female (2020 est.)

Infant mortality rate: *total:* 12 deaths/1,000 live births
male: 14.1 deaths/1,000 live births
female: 9.7 deaths/1,000 live births (2020 est.)
country comparison to the world: 107

Life expectancy at birth: *total population:* 80 years
male: 77.1 years

female: 83.1 years (2020 est.)
country comparison to the world: 50

Total fertility rate: 1.6 children born/woman (2020 est.)
country comparison to the world: 187

Drinking water source: *improved:* total: 100% of population

unimproved: *total:* 0% of population (2017 est.)

Sanitation facility access: *improved:* total: 100% of population

unimproved: *total:* 0% of population (2017)

HIV/AIDS—adult prevalence rate: NA

HIV/AIDS—people living with HIV/AIDS: NA

HIV/AIDS—deaths: NA

Education expenditures: NA

GOVERNMENT

Country name: *conventional long form:* Saint Helena, Ascension, and Tristan da Cunha
conventional short form: none
etymology: Saint Helena was discovered in 1502 by Galician navigator Joao da NOVA, sailing in the service of the Kingdom of Portugal, who named it "Santa Helena"; Ascension was named in 1503 by Portuguese navigator Afonso de ALBUQUERQUE who sighted the island on the Feast Day of the Ascension; Tristan da Cunha was discovered in 1506 by Portuguese explorer Tristao da CUNHA who christened the main island after himself (the name was subsequently anglicized)

Dependency status: overseas territory of the UK

Government type: parliamentary democracy

Capital: *name:* Jamestown
geographic coordinates: 15 56 S, 5 43 W
time difference: UTC 0 (5 hours ahead of Washington, DC, during Standard Time)
etymology: founded in 1659 and named after James, Duke of York, who would become King James II of England (r. 1785-1788)

Administrative divisions: 3 administrative areas; Ascension, Saint Helena, Tristan da Cunha

Independence: none (overseas territory of the UK)

National holiday: Birthday of Queen ELIZABETH II, third Monday in April (1926)

Constitution: *history:* several previous; latest effective 1 September 2009 (St Helena, Ascension and Tristan da Cunha Constitution Order, 2009)

Legal system: English common law and local statutes

Citizenship: see United Kingdom

Suffrage: 18 years of age

Executive branch: *chief of state:* Queen ELIZABETH II (since 6 February 1952)
head of government: Governor Philip RUSHBROOK (since 11 May 2019)
cabinet: Executive Council consists of the governor, 3 ex-officio officers, and 5 elected members of the Legislative Council
elections/appointments: none; the monarchy is hereditary; governor appointed by the monarch

note: the constitution order provides for an administrator for Ascension and Tristan da Cunha appointed by the governor

Legislative branch: *description:* unicameral Legislative Council (17 seats including the speaker and deputy speaker; 12 members directly elected in a single countrywide constituency by simple majority vote and 3 ex-officio members—the chief secretary, financial secretary, and attorney general; members serve 4-year terms)
elections: last held on 26 July 2017 (next to be held in 2021)
election results: percent of vote—NA; seats by party—independent 12; composition—men 14, women 3, percent women 17.6%
note: the Constitution Order provides for separate Island Councils for both Ascension and Tristan da Cunha

Judicial branch: *highest courts:* Court of Appeal (consists of the court president and 2 justices); Supreme Court (consists of the chief justice—a nonresident—and NA judges); note—appeals beyond the Court of Appeal are heard by the Judicial Committee of the Privy Council (in London)
judge selection and term of office: Court of Appeal and Supreme Court justices appointed by the governor acting upon the instructions from a secretary of state acting on behalf of Queen ELIZABETH II; justices of both courts serve until retirement at age 70, but terms can be extended
subordinate courts: Magistrates' Court; Small Claims Court; Juvenile Court

Political parties and leaders: none

International organization participation: UPU

Diplomatic representation in the US: none (overseas territory of the UK)

Diplomatic representation from the US: none (overseas territory of the UK)

Flag description: blue with the flag of the UK in the upper hoist-side quadrant and the Saint Helenian shield centered on the outer half of the flag; the upper third of the shield depicts a white plover (wire bird) on a yellow field; the remainder of the shield depicts a rocky coastline on the left, offshore is a three-masted sailing ship with sails furled but flying an English flag

National symbol(s): Saint Helena plover (bird)

National anthem: *note:* as a territory of the UK, "God Save the Queen" is official (see United Kingdom)
0:00 / 1:02

ECONOMY

Economy—overview: The economy depends largely on financial assistance from the UK, which amounted to about $27 million in FY06/07 or more than twice the level of annual budgetary revenues. The local population earns income from fishing, raising livestock, and sales of handicrafts. Because there are few jobs, 25% of the work force has left to seek employment on Ascension Island, on the Falklands, and in the UK.

GDP (purchasing power parity): $31.1 million (FY09/10 est.)
country comparison to the world: 227

GDP (official exchange rate): NA

GDP—real growth rate: NA

GDP—per capita (PPP): $7,800 (FY09/10 est.)
country comparison to the world: 153

GDP—composition, by sector of origin:
agriculture: NA
industry: NA
services: NA

Agriculture—products: coffee, corn, potatoes, vegetables; fish, lobster; livestock; timber

Industries: construction, crafts (furniture, lacework, fancy woodwork), fishing, collectible postage stamps

Industrial production growth rate: NA

Labor force: 2,486 (1998 est.)
country comparison to the world: 226

Labor force—by occupation: *agriculture:* 6%
industry: 48%
services: 46% (1987 est.)

Unemployment rate: 14% (1998 est.)
country comparison to the world: 170

Population below poverty line: NA

Household income or consumption by percentage share: *lowest 10%:* NA
highest 10%: NA

Budget: *revenues:* 8.427 million (FY06/07 est.)
expenditures: 20.7 million (FY06/07 est.)
note: revenue data reflect only locally raised revenues; the budget deficit is resolved by grant aid from the UK

Fiscal year: 1 April—31 March

Inflation rate (consumer prices): 4% (2012 est.)
country comparison to the world: 156

Exports: $19 million (2004 est.)
country comparison to the world: 211

Exports—commodities: fish (frozen, canned, and salt-dried skipjack, tuna), coffee, handicrafts

Imports: $20.53 million (2010 est.)
country comparison to the world: 222

Imports—commodities: food, beverages, tobacco, fuel oils, animal feed, building materials, motor vehicles and parts, machinery and parts

Debt—external: NA

Exchange rates: Saint Helenian pounds (SHP) per US dollar -
0.7836 (2017 est.)
0.6542 (2016 est.)
0.6542 (2015)
0.607 (2014 est.)
0.6391 (2013 est.)

ENERGY

Electricity—production: 7 million kWh (2016 est.)
country comparison to the world: 215

Electricity—consumption: 6.51 million kWh (2016 est.)
country comparison to the world: 214

Electricity—exports: 0 kWh (2016 est.)
country comparison to the world: 187

Electricity—imports: 0 kWh (2016 est.)
country comparison to the world: 189

Electricity—installed generating capacity: 8,000 kW (2016 est.)
country comparison to the world: 210

Electricity—from fossil fuels: 100% of total installed capacity (2016 est.)
country comparison to the world: 16

Electricity—from nuclear fuels: 0% of total installed capacity (2017 est.)
country comparison to the world: 171

Electricity—from hydroelectric plants: 0% of total installed capacity (2017 est.)
country comparison to the world: 194

Electricity—from other renewable sources: 0% of total installed capacity (2017 est.)
country comparison to the world: 205

Crude oil—production: 0 bbl/day (2018 est.)
country comparison to the world: 190

Crude oil—exports: 0 bbl/day (2015 est.)
country comparison to the world: 183

Crude oil—imports: 0 bbl/day (2015 est.)
country comparison to the world: 185

Crude oil—proved reserves: 0 bbl (1 January 2018 est.)
country comparison to the world: 185

Refined petroleum products—production: 0 bbl/day (2015 est.)
country comparison to the world: 192

Refined petroleum products—consumption: 70 bbl/day (2016 est.)
country comparison to the world: 214

Refined petroleum products—exports: 0 bbl/day (2015 est.)
country comparison to the world: 193

Refined petroleum products—imports: 65 bbl/day (2015 est.)
country comparison to the world: 210

Natural gas—production: 0 cu m (2017 est.)
country comparison to the world: 187

Natural gas—consumption: 0 cu m (2017 est.)
country comparison to the world: 189

Natural gas—exports: 0 cu m (2017 est.)
country comparison to the world: 172

Natural gas—imports: 0 cu m (2017 est.)
country comparison to the world: 178

Natural gas—proved reserves: 0 cu m (1 January 2014 est.)
country comparison to the world: 184

Carbon dioxide emissions from consumption of energy: 10,650 Mt (2017 est.)
country comparison to the world: 212

COMMUNICATIONS

Telephones—fixed lines: *total subscriptions:* 3,921
subscriptions per 100 inhabitants: 49.93 (2019 est.)
country comparison to the world: 208

Telephones—mobile cellular: *total subscriptions:* 5,228
subscriptions per 100 inhabitants: 66.58 (2019 est.)
country comparison to the world: 218

Telecommunication systems: *general assessment:* capability to communicate worldwide; ADSL-broadband service; LTE coverage of 95% of population, includes voice calls, text messages, mobile data as well as inbound and outbound roaming; Wi-Fi hotspots in Jamestown, 1 ISP, many services are not offered locally but made available for visitors; some sun outages due to the reliance of international telephone and Internet communication relying on single satellite link (2020)
domestic: automatic digital network; fixed-line 50 per 100 and mobile-cellular 67 per 100 persons (2019)
international: country code (Saint Helena)—290, (Ascension Island)—247; landing point for the SaEx1 submarine cable providing connectivity to South Africa, Brazil, Virginia Beach (US) and islands in Saint Helena, Ascension and Tristan de Cunha; international direct dialing; satellite voice and data communications; satellite earth stations—5 (Ascension Island—4, Saint Helena—1)
note: the COVID-19 outbreak is negatively impacting telecommunications production and supply chains globally; consumer spending on telecom devices and services has also slowed due to the pandemic's effect on economies worldwide; overall progress towards improvements in all facets of the telecom industry—mobile, fixed-line, broadband, submarine cable and satellite—has moderated

Broadcast media: Saint Helena has no local TV station; 2 local radio stations, one of which is relayed to Ascension Island; satellite TV stations rebroadcast terrestrially; Ascension Island has no local TV station but has 1 local radio station and receives relays of broadcasts from 1 radio station on Saint Helena; broadcasts from the British Forces Broadcasting Service (BFBS) are available, as well as TV services for the US military; Tristan da Cunha has 1 local radio station and receives BFBS TV and radio broadcasts

Internet country code: .sh; note—Ascension Island assigned .ac

Internet users: *total:* 1,800
percent of population: 23.1% (July 2016 est.)
country comparison to the world: 223

Broadband—fixed subscriptions: *total:* 1,347
subscriptions per 100 inhabitants: 17 (2017 est.)
country comparison to the world: 193

Communications—note: Ascension Island hosts one of four dedicated ground antennas that assist in the operation of the Global Positioning System (GPS) navigation system (the others are on Diego Garcia (British Indian Ocean Territory), Kwajalein (Marshall Islands), and at Cape Canaveral, Florida (US)); South Africa maintains a meteorological station on Gough Island in the Tristan da Cunha archipelago

TRANSPORTATION

Civil aircraft registration country code prefix: VQ-H (2016)

Airports: 2 (2015)
country comparison to the world: 202

Airports—with paved runways: *total:* 2 (2019)
over 3,047 m: 1 Ascension Island—Wideawake Field (ASI)
1,524 to 2,437 m: 1 Saint Helena (HLE);
note—weekly commercial air service to South Africa via Namibia commenced on 14 October 2017

Roadways: *total:* 198 km (Saint Helena 138 km, Ascension 40 km, Tristan da Cunha 20 km) (2002)

paved: 168 km (Saint Helena 118 km, Ascension 40 km, Tristan da Cunha 10 km) (2002)
unpaved: 30 km (Saint Helena 20 km, Tristan da Cunha 10 km) (2002)
country comparison to the world: 208

Ports and terminals: *major seaport(s):* Saint Helena
Saint Helena: Jamestown
Ascension Island: Georgetown
Tristan da Cunha: Calshot Harbor (Edinburgh)

Transportation—note: the new airport on Saint Helena opened for limited operations in July 2016, and the first commercial flight took place on 14 October 2017, marking the start of weekly air service between Saint Helena and South Africa

via Namibia; the military airport on Ascension Island is closed to civilian traffic; there is no air connection to Tristan da Cunha and very limited sea connections making it one of the most isolated communities on the planet

MILITARY AND SECURITY

Military—note: defense is the responsibility of the UK

TRANSNATIONAL ISSUES

Disputes—international: none

SAINT KITTS AND NEVIS

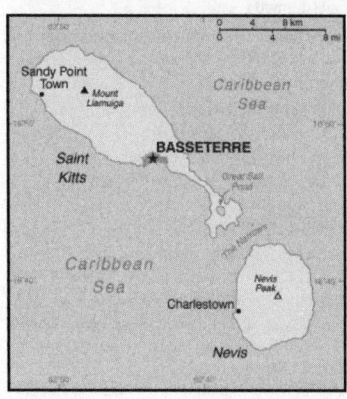

INTRODUCTION

Background: Carib Indians occupied the islands of the West Indies for hundreds of years before the British and French began settlement in 1623. During the course of 17th century, Saint Kitts became the premier base for English and French expansion into the Caribbean. The French ceded the territory to the UK in 1713. At the turn of the 18th century, Saint Kitts was the richest British Crown Colony per capita in the Caribbean, a result of the sugar trade. Although small in size and separated by only 3 km (2 mi) of water, Saint Kitts and Nevis were viewed and governed as different states until the late-19th century, when the British forcibly unified them along with the island of Anguilla. In 1967, the island territory of Saint Christopher-Nevis-Anguilla became an associated state of the UK with full internal autonomy. The island of Anguilla rebelled and was allowed to secede in 1971. The remaining islands achieved independence in 1983 as Saint Kitts and Nevis. In 1998, a referendum on Nevis to separate from Saint Kitts fell short of the two-thirds majority vote needed.

GEOGRAPHY

Location: Caribbean, islands in the Caribbean Sea, about one-third of the way from Puerto Rico to Trinidad and Tobago

Geographic coordinates: 17 20 N, 62 45 W

Map references: Central America and the Caribbean

Area: *total:* 261 sq km (Saint Kitts 168 sq km; Nevis 93 sq km)
land: 261 sq km
water: 0 sq km
country comparison to the world: 212

Area—comparative: 1.5 times the size of Washington, DC

Land boundaries: 0 km

Coastline: 135 km

Maritime claims: *territorial sea:* 12 nm
exclusive economic zone: 200 nm
contiguous zone: 24 nm
continental shelf: 200 nm or to the edge of the continental margin

Climate: tropical, tempered by constant sea breezes; little seasonal temperature variation; rainy season (May to November)

Terrain: volcanic with mountainous interiors

Elevation: *lowest point:* Caribbean Sea 0 m
highest point: Mount Liamuiga 1,156 m

Natural resources: arable land

Land use: *agricultural land:* 23.1% (2011 est.)
arable land: 19.2% (2011 est.) / permanent crops: 0.4% (2011 est.) / permanent pasture: 3.5% (2011 est.)
forest: 42.3% (2011 est.)
other: 34.6% (2011 est.)
Irrigated land: 8 sq km (2012)

Population distribution: population clusters are found in the small towns located on the periphery of both islands

Natural hazards: hurricanes (July to October)

volcanism: Mount Liamuiga (1,156 m) on Saint Kitts, and Nevis Peak (985 m) on Nevis, are both volcanoes that are part of the volcanic island arc of the Lesser Antilles, which extends from Saba in the north to Grenada in the south

Environment—current issues: deforestation; soil erosion and silting affects marine life on coral reefs; water pollution from uncontrolled dumping of sewage

Environment—international agreements: *party to:* Biodiversity, Climate Change, Climate Change-Kyoto Protocol, Desertification, Endangered Species, Hazardous Wastes, Law of the Sea, Marine Dumping, Ozone Layer Protection, Ship Pollution, Whaling
signed, but not ratified: none of the selected agreements

Geography—note: smallest country in the Western Hemisphere both in terms of area and population; with coastlines in the shape of a baseball bat and ball, the two volcanic islands are separated by a 3-km-wide channel called The Narrows; on the southern tip of long, baseball bat-shaped Saint Kitts lies the Great Salt Pond; Nevis Peak sits in the center of its almost circular namesake island and its ball shape complements that of its sister island

PEOPLE AND SOCIETY

Population: 53,821 (July 2020 est.)
country comparison to the world: 208

Nationality: *noun:* Kittitian(s), Nevisian(s)
adjective: Kittitian, Nevisian

Ethnic groups: African descent 92.5%, mixed 3%, white 2.1%, East Indian 1.5%, other .6%, unspecified .3% (2001 est.)

Languages: English (official)

Religions: Protestant 74.4% (includes Anglican 20.6%, Methodist 19.1%, Pentecostal 8.2%, Church of God 6.8%, Moravian 5.5%,

Baptist 4.8%, Seventh Day Adventist 4.7%, Evangelical 2.6%, Bretheren 1.8%, other .3%), Roman Catholic 6.7%,
Rastafarian 1.7%, Jehovah's Witness 1.3%, other 7.6%, none 5.2%, unspecified 3.2% (2001 est.)

Age structure: 0-14 years: 19.87% (male 5,357/female 5,336)
15-24 years: 13.46% (male 3,504/female 3,741)
25-54 years: 43.64% (male 12,010/female 11,477)
55-64 years: 13.03% (male 3,527/female 3,485)
65 years and over: 10% (male 2,540/female 2,844) (2020 est.)

Median age: total: 36.5 years
male: 36.7 years
female: 36.3 years (2020 est.)
country comparison to the world: 79

Population growth rate: 0.67% (2020 est.)
country comparison to the world: 141

Birth rate: 12.6 births/1,000 population (2020 est.)
country comparison to the world: 153

Death rate: 7.3 deaths/1,000 population (2020 est.)
country comparison to the world: 117

Net migration rate: 1.2 migrant(s)/1,000 population (2020 est.)
country comparison to the world: 58

Population distribution: population clusters are found in the small towns located on the periphery of both islands

Urbanization: urban population: 30.8% of total population (2020)
rate of urbanization: 0.92% annual rate of change (2015-20 est.)

total population growth rate v urban population growth rate, 2000-2030: Major urban areas—population: 14,000 BASSETERRE (capital) (2018)

Sex ratio: at birth: 1.02 male(s)/female
0-14 years: 1 male(s)/female
15-24 years: 0.94 male(s)/female
25-54 years: 1.05 male(s)/female
55-64 years: 1.01 male(s)/female
65 years and over: 0.89 male(s)/female
total population: 1 male(s)/female (2020 est.)

Infant mortality rate: total: 7.8 deaths/1,000 live births
male: 5.7 deaths/1,000 live births
female: 10 deaths/1,000 live births (2020 est.)
country comparison to the world: 150

Life expectancy at birth: total population: 76.6 years
male: 74.1 years
female: 79.1 years (2020 est.)
country comparison to the world: 94

Total fertility rate: 1.77 children born/woman (2020 est.)
country comparison to the world: 155

Drinking water source:
improved:
urban: 98.3% of population
rural: 98.3% of population
total: 98.3% of population
unimproved:

urban: 1.7% of population
rural: 1.7% of population
total: 1.7% of population (2015 est.)

Current Health Expenditure: 5% (2017)

Physicians density: 2.68 physicians/1,000 population (2015)

Hospital bed density: 4.8 beds/1,000 population (2012)

Sanitation facility access:
improved:
urban: 87.3% of population (2007 est.)
rural: 87.3% of population (2007 est.)
total: 87.3% of population (2007 est.)
unimproved:
urban: 12.7% of population (2007 est.)
rural: 12.7% of population (2007 est.)
total: 12.7% of population (2007 est.)

HIV/AIDS—adult prevalence rate: 0.5% (2018)
country comparison to the world: 71

HIV/AIDS—people living with HIV/AIDS: <200 (2018)

HIV/AIDS—deaths: <100 (2018)

Obesity—adult prevalence rate: 22.9% (2016)
country comparison to the world: 71

Education expenditures: 2.6% of GDP (2015)
country comparison to the world: 156

School life expectancy (primary to tertiary education): total: 18 years
male: 16 years
female: 1,619 years (2015)

GOVERNMENT

Country name: conventional long form: Federation of Saint Kitts and Nevis
conventional short form: Saint Kitts and Nevis
former: Federation of Saint Christopher and Nevis
etymology: Saint Kitts was, and still is, referred to as Saint Christopher and this name was well established by the 17th century (although who first applied the name is unclear); in the 17th century a common nickname for Christopher was Kit or Kitt, so the island began to be referred to as "Saint Kitt's Island" or just "Saint Kitts"; Nevis is derived from the original Spanish name "Nuestra Senora de las Nieves" (Our Lady of the Snows) and refers to the white halo of clouds that generally wreathes Nevis Peak
note: Nevis is pronounced neevis

Government type: federal parliamentary democracy under a constitutional monarchy; a Commonwealth realm

Capital: name: Basseterre

geographic coordinates: 17 18 N, 62 43 W
time difference: UTC-4 (1 hour ahead of Washington, DC, during Standard Time)
etymology: the French name translates as "low land" in English; the reference is to the city's low-lying location within a valley, as well as to the fact that the city is on the leeward (downwind) part of the island, and is thus a safe anchorage

Administrative divisions: 14 parishes; Christ Church Nichola Town, Saint Anne Sandy Point,

Saint George Basseterre, Saint George Gingerland, Saint James Windward, Saint John Capesterre, Saint John Figtree, Saint Mary Cayon, Saint Paul Capesterre, Saint Paul Charlestown, Saint Peter Basseterre, Saint Thomas Lowland, Saint Thomas Middle Island, Trinity Palmetto Point

Independence: 19 September 1983 (from the UK)

National holiday: Independence Day, 19 September (1983)

Constitution: history: several previous (preindependence); latest presented 22 June 1983, effective 23 June 1983
amendments: proposed by the National Assembly; passage requires approval by at least two-thirds majority vote of the total Assembly membership and assent of the governor general; amendments to constitutional provisions such as the sovereignty of the federation, fundamental rights and freedoms, the judiciary, and the Nevis Island Assembly also require approval in a referendum by at least two thirds of the votes cast in Saint Kitts and in Nevis

Legal system: English common law

International law organization participation: has not submitted an ICJ jurisdiction declaration; accepts ICCt jurisdiction

Citizenship: citizenship by birth: yes
citizenship by descent only: yes
dual citizenship recognized: yes
residency requirement for naturalization: 14 years

Suffrage: 18 years of age; universal

Executive branch: chief of state: Queen ELIZABETH II (since 6 February 1952); represented by Governor General Samuel W.T. SEATON (since 2 September 2015); note—SEATON was acting Governor General from 20 May to 2 September 2015
head of government: Prime Minister Timothy HARRIS (since 18 February 2015); Deputy Prime Minister Shawn RICHARDS (since 22 February 2015)
cabinet: Cabinet appointed by governor general in consultation with prime minister
elections/appointments: the monarchy is hereditary; governor general appointed by the monarch; following legislative elections, the leader of the majority party or majority coalition usually appointed prime minister by governor general; deputy prime minister appointed by governor general

Legislative branch: description: unicameral National Assembly (14 or 15 seats, depending on inclusion of attorney general; 11 members directly elected in single-seat constituencies by simple majority vote and 3 appointed by the governor general—2 on the advice of the prime minister and the third on the advice of the opposition leader; members serve 5-year terms)
elections: last held on 5 June 2020 (next to be held on 2025)
election results: percent of vote by party—Team Unity (PAM, CCM,PLP) 56.4%, SKNLP 34.5%, NRP 9%; seats by party—PAM 4, SKNLP 2, CCM 3, PLP 2

Judicial branch: *highest courts:* the Eastern Caribbean Supreme Court (ECSC) is the superior court of the Organization of Eastern Caribbean States; the ECSC—headquartered on St. Lucia—consists of the Court of Appeal—headed by the chief justice and 4 judges—and the High Court with 18 judges; the Court of Appeal is itinerant, traveling to member states on a schedule to hear appeals from the High Court and subordinate courts; High Court judges reside in the member states, with 2 assigned to Saint Kitts and Nevis; note—the ECSC in 2003 replaced the Judicial Committee of the Privy Council (in London) as the final court of appeal on Saint Kitts and Nevis; Saint Kitts and Nevis is also a member of the Caribbean Court of Justice

judge selection and term of office: chief justice of Eastern Caribbean Supreme Court appointed by Her Majesty, Queen ELIZABETH II; other justices and judges appointed by the Judicial and Legal Services Commission, an independent body of judicial officials; Court of Appeal justices appointed for life with mandatory retirement at age 65; High Court judges appointed for life with mandatory retirement at age 62

subordinate courts: magistrates' courts

Political parties and leaders: Concerned Citizens Movement or CCM [Mark BRANTLEY]
Nevis Reformation Party or NRP [Joseph PARRY]
People's Action Movement or PAM [Shawn RICHARDS]
People's Labour Party or PLP [Dr. Timothy HARRIS]
Saint Kitts and Nevis Labor Party or SKNLP [Dr. Denzil DOUGLAS]

International organization participation: ACP, AOSIS, C, Caricom, CDB, CELAC, FAO, G-77, IBRD, ICAO, ICCt, ICRM, IDA, IFAD, IFC, IFRCS, ILO, IMF, IMO, Interpol, IOC, ITU, MIGA, OAS, OECS, OPANAL, OPCW, Petrocaribe, UN, UNCTAD, UNESCO, UNIDO, UPU, WHO, WIPO, WTO

Diplomatic representation in the US: *chief of mission:* Ambassador Dr. Thelma Patricia PHILLIP-BROWNE (since 28 January 2016)
chancery: 3216 New Mexico Avenue NW, Washington, DC 20016
telephone: [1] (202) 686-2636
FAX: [1] (202) 686-5740
consulate(s) general: Los Angeles, New York

Diplomatic representation from the US: the US does not have an embassy in Saint Kitts and Nevis; the US Ambassador to Barbados is accredited to Saint Kitts and Nevis

Flag description: divided diagonally from the lower hoist side by a broad black band bearing two white, five-pointed stars; the black band is edged in yellow; the upper triangle is green, the lower triangle is red; green signifies the island's fertility, red symbolizes the struggles of the people from slavery, yellow denotes year-round sunshine, and black represents the African heritage of the people; the white stars stand for the islands of Saint Kitts and Nevis, but can also express hope and liberty, or independence and optimism

National symbol(s): *brown pelican, royal poinciana (flamboyant) tree; national colors:* green, yellow, red, black, white

National anthem: *name:* Oh Land of Beauty!
lyrics/music: Kenrick Anderson GEORGES
note: adopted 1983
0:00 / 1:02

ECONOMY

Economy—overview: The economy of Saint Kitts and Nevis depends on tourism; since the 1970s, tourism has replaced sugar as the economy's traditional mainstay. Roughly 200,000 tourists visited the islands in 2009, but reduced tourism arrivals and foreign investment led to an economic contraction in the 2009-2013 period, and the economy returned to growth only in 2014. Like other tourist destinations in the Caribbean, Saint Kitts and Nevis is vulnerable to damage from natural disasters and shifts in tourism demand.

Following the 2005 harvest, the government closed the sugar industry after several decades of losses. To compensate for lost jobs, the government has embarked on a program to diversify the agricultural sector and to stimulate other sectors of the economy, such as export-oriented manufacturing and offshore banking. The government has made notable progress in reducing its public debt, from 154% of GDP in 2011 to 83% in 2013, although it still faces one of the highest levels in the world, largely attributable to public enterprise losses. Saint Kitts and Nevis is among other countries in the Caribbean that supplement their economic activity through economic citizenship programs, whereby foreigners can obtain citizenship from Saint Kitts and Nevis by investing there.

GDP (purchasing power parity): $1.55 billion (2017 est.)
$1.518 billion (2016 est.)
$1.476 billion (2015 est.)
note: data are in 2017 dollars
country comparison to the world: 199

GDP (official exchange rate): $964 million (2017 est.)

GDP—real growth rate: 2.1% (2017 est.)
2.9% (2016 est.)
2.7% (2015 est.)
country comparison to the world: 132

GDP—per capita (PPP): $28,200 (2017 est.)
$27,600 (2016 est.)
$27,300 (2015 est.)
note: data are in 2017 dollars
country comparison to the world: 73

Gross national saving: 19.9% of GDP (2017 est.)
19.3% of GDP (2016 est.)
15.4% of GDP (2015 est.)
country comparison to the world: 98

GDP—composition, by end use:
household consumption: 41.4% (2017 est.)
government consumption: 25.9% (2017 est.)
investment in fixed capital: 30.8% (2017 est.)
investment in inventories: 0% (2017 est.)
exports of goods and services: 62.5% (2017 est.)
imports of goods and services: -60.4% (2017 est.)

GDP—composition, by sector of origin:
agriculture: 1.1% (2017 est.)
industry: 30% (2017 est.)
services: 68.9% (2017 est.)

Agriculture—products: sugarcane, rice, yams, vegetables, bananas; fish

Industries: tourism, cotton, salt, copra, clothing, footwear, beverages

Industrial production growth rate: 5% (2017 est.)
country comparison to the world: 56

Labor force: 18,170 (June 1995 est.)
country comparison to the world: 212

Unemployment rate: 4.5% (1997)
country comparison to the world: 68

Population below poverty line: NA

Household income or consumption by percentage share: *lowest 10%:* NA
highest 10%: NA

Budget: *revenues:* 307 million (2017 est.)
expenditures: 291.1 million (2017 est.)

Taxes and other revenues: 31.9% (of GDP) (2017 est.)
country comparison to the world: 70

Budget surplus (+) or deficit (-): 1.7% (of GDP) (2017 est.)
country comparison to the world: 18

Public debt: 62.9% of GDP (2017 est.)
61.5% of GDP (2016 est.)
country comparison to the world: 67

Fiscal year: calendar year

Inflation rate (consumer prices): 0% (2017 est.)
-0.3% (2016 est.)
country comparison to the world: 12

Current account balance: -$97 million (2017 est.)
-$102 million (2016 est.)
country comparison to the world: 86

Exports: $57.4 million (2017 est.)
$53.9 million (2016 est.)
country comparison to the world: 202

Exports—partners: US 49.6%, Poland 15.2%, Turkey 11.6% (2016)

Exports—commodities: machinery, food, electronics, beverages, tobacco

Imports: $335.3 million (2017 est.)
$307.9 million (2016 est.)
country comparison to the world: 202

Imports—commodities: machinery, manufactures, food, fuels

Imports—partners: US 56.8%, Trinidad and Tobago 6.8%, Cyprus 6.2%, Japan 4% (2016)

Reserves of foreign exchange and gold: $365.1 million (31 December 2017 est.)
$320.5 million (31 December 2016 est.)
country comparison to the world: 162

Debt—external: $201.8 million (31 December 2017 est.)
$187.9 million (31 December 2016 est.)
country comparison to the world: 187

Exchange rates: East Caribbean dollars (XCD) per US dollar -
2.7 (2017 est.)
2.7 (2016 est.)
2.7 (2015 est.)
2.7 (2014 est.)
2.7 (2013 est.)

ENERGY

Electricity access: *electrification—total population:* 100% (2020)

Electricity—production: 208 million kWh (2016 est.)
country comparison to the world: 191

Electricity—consumption: 193.4 million kWh (2016 est.)
country comparison to the world: 193

Electricity—exports: 0 kWh (2016 est.)
country comparison to the world: 188

Electricity—imports: 0 kWh (2016 est.)
country comparison to the world: 190

Electricity—installed generating capacity: 64,200 kW (2016 est.)
country comparison to the world: 187

Electricity—from fossil fuels: 94% of total installed capacity (2016 est.)
country comparison to the world: 48

Electricity—from nuclear fuels: 0% of total installed capacity (2017 est.)
country comparison to the world: 172

Electricity—from hydroelectric plants: 0% of total installed capacity (2017 est.)
country comparison to the world: 195

Electricity—from other renewable sources: 6% of total installed capacity (2017 est.)
country comparison to the world: 102

Crude oil—production: 0 bbl/day (2018 est.)
country comparison to the world: 191

Crude oil—exports: 0 bbl/day (2015 est.)
country comparison to the world: 184

Crude oil—imports: 0 bbl/day (2015 est.)
country comparison to the world: 186

Crude oil—proved reserves: 0 bbl (1 January 2018 est.)
country comparison to the world: 186

Refined petroleum products—production: 0 bbl/day (2015 est.)
country comparison to the world: 193

Refined petroleum products—consumption: 1,700 bbl/day (2016 est.)
country comparison to the world: 196

Refined petroleum products—exports: 0 bbl/day (2015 est.)
country comparison to the world: 194

Refined petroleum products—imports: 1,743 bbl/day (2015 est.)
country comparison to the world: 192

Natural gas—production: 0 cu m (2017 est.)
country comparison to the world: 188

Natural gas—consumption: 0 cu m (2017 est.)
country comparison to the world: 190

Natural gas—exports: 0 cu m (2017 est.)
country comparison to the world: 173

Natural gas—imports: 0 cu m (2017 est.)
country comparison to the world: 179

Natural gas—proved reserves: 0 cu m (1 January 2014 est.)
country comparison to the world: 185

Carbon dioxide emissions from consumption of energy: 248,100 Mt (2017 est.)
country comparison to the world: 195

COMMUNICATIONS

Telephones—fixed lines: *total subscriptions:* 17,766
subscriptions per 100 inhabitants: 33.23 (2019 est.)
country comparison to the world: 180

Telephones—mobile cellular: *total subscriptions:* 78,970
subscriptions per 100 inhabitants: 147.71 (2019 est.)
country comparison to the world: 197

Telecommunication systems: *general assessment:* good interisland and international connections, broadband access; expanded FttP (Fiber to the Home) and LTE markets; regulatory development; telecom sector contributes greatly to the overall GDP; telecom sector is a growth area (2020)
domestic: interisland links via ECFS; fixed-line teledensity about 33 per 100 persons; mobile-cellular teledensity is roughly 148 per 100 persons (2019)
international: country code—1-869; landing points for the ECFS, Southern Caribbean Fiber and the SSCS submarine cables providing connectivity for numerous Caribbean Islands (2019)
note: the COVID-19 outbreak is negatively impacting telecommunications production and supply chains globally; consumer spending on telecom devices and services has also slowed due to the pandemic's effect on economies worldwide; overall progress towards improvements in all facets of the telecom industry—mobile, fixed-line, broadband, submarine cable and satellite—has moderated

Broadcast media: the government operates a national TV network that broadcasts on 2 channels; cable subscription services provide access to local and international channels; the government operates a national radio network; a mix of government-owned and privately owned broadcasters operate roughly 15 radio stations

Internet country code: .kn

Internet users: *total:* 42,852
percent of population: 80.71% (July 2018 est.)
country comparison to the world: 200

Broadband—fixed subscriptions: *total:* 16,400
subscriptions per 100 inhabitants: 31 (2017 est.)
country comparison to the world: 159

TRANSPORTATION

Civil aircraft registration country code prefix: V4 (2016)

Airports: 2 (2013)
country comparison to the world: 203

Airports—with paved runways: *total:* 2 (2019)
1,524 to 2,437 m: 1
914 to 1,523 m: 1

Railways: *total:* 50 km (2008)
narrow gauge: 50 km 0.762-m gauge on Saint Kitts for tourists (2008)
country comparison to the world: 132

Roadways: *total:* 383 km (2002)
paved: 163 km (2002)
unpaved: 220 km (2002)
country comparison to the world: 200

Merchant marine: *total:* 218
by type: bulk carrier 3, container ship 3, general cargo 31, oil tanker 51, other 130 (2019)
country comparison to the world: 64

Ports and terminals: *major seaport(s):* Basseterre, Charlestown

MILITARY AND SECURITY

Military and security forces: *Ministry of Foreign Affairs, National Security, Labour, Immigration, and Social Security:* Royal Saint Kitts and Nevis Defense Force (includes Coast Guard), Royal Saint Kitts and Nevis Police Force (2013)

Military service age and obligation: 18 years of age for voluntary military service; no conscription (2012)

TRANSNATIONAL ISSUES

Disputes—international: joins other Caribbean states to counter Venezuela's claim that Aves Island sustains human habitation, a criterion under UN Convention on the Law of the Sea, which permits Venezuela to extend its EEZ/continental shelf over a large portion of the eastern Caribbean Sea

Illicit drugs: transshipment point for South American drugs destined for the US and Europe; some money-laundering activity

SAINT LUCIA

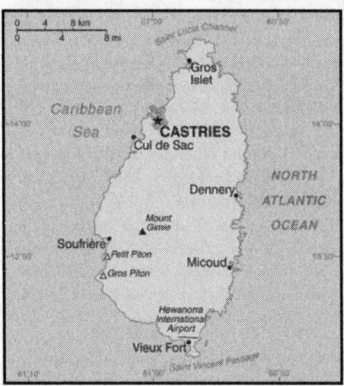

INTRODUCTION

Background: The island, with its fine natural harbor at Castries and burgeoning sugar industry, was contested between England and France throughout the 17th and early 18th centuries (changing possession 14 times); it was finally ceded to the UK in 1814 and became part of the British Windward Islands colony. Even after the abolition of slavery on its plantations in 1834, Saint Lucia remained an agricultural island, dedicated to producing tropical commodity crops. In the mid-20th century, Saint Lucia joined the West Indies Federation (1958–1962) and in 1967 became one of the six members of the West Indies Associated States, with internal self-government. In 1979, Saint Lucia gained full independence.

GEOGRAPHY

Location: Caribbean, island between the Caribbean Sea and North Atlantic Ocean, north of Trinidad and Tobago

Geographic coordinates: 13 53 N, 60 58 W

Map references: Central America and the Caribbean

Area: *total:* 616 sq km
land: 606 sq km
water: 10 sq km
country comparison to the world: 193

Area—comparative: three and a half times the size of Washington, DC

Land boundaries: 0 km

Coastline: 158 km

Maritime claims: *territorial sea:* 12 nm
exclusive economic zone: 200 nm
contiguous zone: 24 nm
continental shelf: 200 nm or to the edge of the continental margin

Climate: tropical, moderated by northeast trade winds; dry season January to April, rainy season May to August

Terrain: volcanic and mountainous with broad, fertile valleys

Elevation: *lowest point:* Caribbean Sea 0 m
highest point: Mount Gimie 948 m

Natural resources: forests, sandy beaches, minerals (pumice), mineral springs, geothermal potential

Land use: *agricultural land:* 17.4% (2011 est.)
arable land: 4.9% (2011 est.) / *permanent crops:* 11.5% (2011 est.) / *permanent pasture:* 1% (2011 est.)
forest: 77% (2011 est.)
other: 5.6% (2011 est.)
Irrigated land: 30 sq km (2012)

Population distribution: most of the population is found on the periphery of the island, with a larger concentration in the north around the capital of Castries

Natural hazards: hurricanes
volcanism: Mount Gimie (948 m), also known as Qualibou, is a caldera on the west of the island; the iconic twin pyramidal peaks of Gros Piton (771 m) and Petit Piton (743 m) are lava dome remnants associated with the Soufriere volcano; there have been no historical magmatic eruptions, but a minor steam eruption in 1766 spread a thin layer of ash over a wide area; Saint Lucia is part of the volcanic island arc of the Lesser Antilles that extends from Saba in the north to Grenada in the south

Environment—current issues: deforestation; soil erosion, particularly in the northern region

Environment—international agreements: *party to:* Biodiversity, Climate Change, Climate Change-Kyoto Protocol, Desertification, Endangered Species, Environmental Modification, Hazardous Wastes, Law of the Sea, Marine Dumping, Ozone Layer Protection, Ship Pollution, Wetlands, Whaling
signed, but not ratified: none of the selected agreements

Geography—note: the twin Pitons (Gros Piton and Petit Piton), striking cone-shaped peaks south of Soufriere, are one of the scenic natural highlights of the Caribbean

PEOPLE AND SOCIETY

Population: 166,487 (July 2020 est.)
country comparison to the world: 187

Nationality: *noun:* Saint Lucian(s)
adjective: Saint Lucian

Ethnic groups: black/African descent 85.3%, mixed 10.9%, East Indian 2.2%, other 1.6%, unspecified 0.1% (2010 est.)

Languages: English (official), French patois

Religions: Roman Catholic 61.5%, Protestant 25.5% (includes Seventh Day Adventist 10.4%, Pentecostal 8.9%, Baptist 2.2%, Anglican 1.6%, Church of God 1.5%, other Protestant 0.9%), other Christian 3.4% (includes Evangelical 2.3% and Jehovah's Witness 1.1%), Rastafarian 1.9%, other 0.4%, none 5.9%, unspecified 1.4% (2010 est.)

Age structure: *0-14 years:* 19.24% (male 16,484/female 15,546)
15-24 years: 13.6% (male 11,475/female 11,165)
25-54 years: 42.83% (male 34,436/female 36,868)
55-64 years: 11.23% (male 8,624/female 10,075)
65 years and over: 13.1% (male 9,894/female 11,920) (2020 est.)

Dependency ratios: *total dependency ratio:* 39.4
youth dependency ratio: 25
elderly dependency ratio: 14.4
potential support ratio: 7 (2020 est.)

Median age: *total:* 36.9 years
male: 35.7 years
female: 38 years (2020 est.)
country comparison to the world: 75

Population growth rate: 0.29% (2020 est.)
country comparison to the world: 172

Birth rate: 12.5 births/1,000 population (2020 est.)
country comparison to the world: 155

Death rate: 8.1 deaths/1,000 population (2020 est.)
country comparison to the world: 90

Net migration rate: -1.7 migrant(s)/1,000 population (2020 est.)
country comparison to the world: 160

Population distribution: most of the population is found on the periphery of the island, with a larger concentration in the north around the capital of Castries

Urbanization: *urban population:* 18.8% of total population (2020)
rate of urbanization: 0.8% annual rate of change (2015-20 est.)

total population growth rate v. urban population growth rate, 2000-2030: *Major urban areas—population:* 22,000 CASTRIES (capital) (2018)

Sex ratio: *at birth:* 1.06 male(s)/female
0-14 years: 1.06 male(s)/female
15-24 years: 1.03 male(s)/female
25-54 years: 0.93 male(s)/female
55-64 years: 0.86 male(s)/female
65 years and over: 0.83 male(s)/female
total population: 0.95 male(s)/female (2020 est.)

Maternal mortality rate: 117 deaths/100,000 live births (2017 est.)
country comparison to the world: 67

Infant mortality rate: *total:* 10.1 deaths/1,000 live births
male: 9.8 deaths/1,000 live births
female: 10.4 deaths/1,000 live births (2020 est.)
country comparison to the world: 130

Life expectancy at birth: *total population:* 78.5 years
male: 75.7 years
female: 81.4 years (2020 est.)
country comparison to the world: 66

Total fertility rate: 1.73 children born/woman (2020 est.)
country comparison to the world: 167

Contraceptive prevalence rate: 55.5% (2011/12)

Drinking water source:
improved:
urban: 100% of population
rural: 100% of population
total: 100% of population
unimproved:
urban: 0% of population
rural: 0% of population
total: 0% of population (2017 est.)

Current Health Expenditure: 4.5% (2017)

Physicians density: 0.64 physicians/1,000 population (2017)

Hospital bed density: 1.3 beds/1,000 population (2017)

Sanitation facility access:
improved:
urban: 95.6% of population
rural: 100% of population
total: 99.2% of population
unimproved:
urban: 4.4% of population
rural: 0% of population
total: 0.8% of population (2017 est.)

HIV/AIDS—adult prevalence rate: 0.6% (2018)
country comparison to the world: 64

HIV/AIDS—people living with HIV/AIDS: <1000 (2018)

HIV/AIDS—deaths: <100 (2018)

Obesity—adult prevalence rate: 19.7% (2016)
country comparison to the world: 111

Children under the age of 5 years underweight: 2.8% (2012)
country comparison to the world: 104

Education expenditures: 3.8% of GDP (2018)
country comparison to the world: 115

School life expectancy (primary to tertiary education): *total:* 13 years
male: 13 years
female: 13 years (2019)

Unemployment, youth ages 15-24: *total:* 46.2%
male: 42.6%
female: 51% (2016 est.)
country comparison to the world: 6

GOVERNMENT

Country name: *conventional long form:* none
conventional short form: Saint Lucia
etymology: named after Saint LUCY of Syracuse by French sailors who were shipwrecked on the island on 13 December 1502, the saint's feast day; Saint Lucia is the only country named specifically after a woman
note: pronounced saynt-looshya

Government type: parliamentary democracy under a constitutional monarchy; a Commonwealth realm

Capital: *name:* Castries

geographic coordinates: 14 00 N, 61 00 W

time difference: UTC-4 (1 hour ahead of Washington, DC, during Standard Time)
etymology: in 1785, the village of Carenage was renamed Castries, after Charles Eugene Gabriel de La Croix de Castries (1727-1801), who was then the French Minister of the Navy and Colonies

Administrative divisions: 10 districts; Anse-la-Raye, Canaries, Castries, Choiseul, Dennery, Gros-Islet, Laborie, Micoud, Soufriere, Vieux-Fort

Independence: 22 February 1979 (from the UK)

National holiday: Independence Day, 22 February (1979)

Constitution: *history:* previous 1958, 1960 (preindependence); latest presented 20 December 1978, effective 22 February 1979
amendments: proposed by Parliament; passage requires at least two-thirds majority vote by the House of Assembly membership in the final reading and assent of the governor general; passage of amendments to various constitutional sections, such as those on fundamental rights and freedoms, government finances, the judiciary, and procedures for amending the constitution, require at least three-quarters majority vote by the House and assent of the governor general; passage of amendments approved by the House but rejected by the Senate require a majority of votes cast in a referendum

Legal system: English common law

International law organization participation: has not submitted an ICJ jurisdiction declaration; accepts ICCt jurisdiction

Citizenship: *citizenship by birth:* yes
citizenship by descent only: at least one parent must be a citizen of Saint Lucia
dual citizenship recognized: yes
residency requirement for naturalization: 8 years

Suffrage: 18 years of age; universal

Executive branch: *chief of state:* Queen ELIZABETH II (since 6 February 1952); represented by Governor General Neville CENAC (since 12 January 2018)
head of government: Prime Minister Allen CHASTANET (since 7 June 2016)
cabinet: Cabinet appointed by the governor general on the advice of the prime minister
elections/appointments: the monarchy is hereditary; governor general appointed by the monarch; following legislative elections, the leader of the majority party or majority coalition usually appointed prime minister by governor general; deputy prime minister appointed by governor general

Legislative branch: *description:* bicameral Parliament consists of:
Senate (11 seats; 6 members appointed on the advice of the prime minister, 3 on the advice of the leader of the opposition, and 2 upon consultation with religious, economic, and social groups; members serve 5-year terms)
House of Assembly (17 seats; members directly elected in single-seat constituencies by simple majority vote to serve 5-year terms)

elections: Senate—last appointments on 12 July 2016 (next in 2021)
House of Assembly—last held on 6 June 2016 (next to be held in 2021)
election results: Senate—percent of vote by party—NA; seats by party—NA; composition—men 8, women 3, percent of women 27.3%
House of Assembly—percent of vote by party—UWP 54.8%, SLP 44.1%, other 1.1%; seats by party—UWP 11, SLP 6; composition—men 14, women 3, percent of women 17.6%; note—total Parliament percent of women 21.4%

Judicial branch: *highest courts:* the Eastern Caribbean Supreme Court (ECSC) is the superior court of the Organization of Eastern Caribbean States; the ECSC—headquartered on St. Lucia—consists of the Court of Appeal—headed by the chief justice and 4 judges—and the High Court with 18 judges; the Court of Appeal is itinerant, traveling to member states on a schedule to hear appeals from the High Court and subordinate courts; High Court judges reside in the member states with 4 on Saint Lucia; Saint Lucia is a member of the Caribbean Court of Justice
judge selection and term of office: chief justice of Eastern Caribbean Supreme Court appointed by Her Majesty, Queen ELIZABETH II; other justices and judges appointed by the Judicial and Legal Services Commission, an independent body of judicial officials; Court of Appeal justices appointed for life with mandatory retirement at age 65; High Court judges appointed for life with mandatory retirement at age 62
subordinate courts: magistrate's court

Political parties and leaders: Lucian People's Movement or LPM [Therold PRUDENT]
Saint Lucia Labor Party or SLP [Philip J. PIERRE]
United Workers Party or UWP [Allen CHASTANET]

International organization participation: ACP, AOSIS, C, Caricom, CD, CDB, CELAC, FAO, G-77, IBRD, ICAO, ICCt, ICRM, IDA, IFAD, IFC, IFRCS, ILO, IMF, IMO, Interpol, IOC, ISO, ITU, ITUC (NGOs), MIGA, NAM, OAS, OECS, OIF, OPANAL, OPCW, Petrocaribe, UN, UNCTAD, UNESCO, UNIDO, UPU, WCO, WFTU (NGOs), WHO, WIPO, WMO, WTO

Diplomatic representation in the US: *chief of mission:* Ambassador Anton Edsel EDMUNDS (since 8 September 2017)
chancery: 1628 K Street NW, Suite 1250, Washington, DC 20006
telephone: [1] (202) 364-6792 through 6795
FAX: [1] (202) 364-6723
consulate(s) general: New York

Diplomatic representation from the US: the US does not have an embassy in Saint Lucia; the US Ambassador to Barbados is accredited to Saint Lucia

Flag description: cerulean blue with a gold isosceles triangle below a black arrowhead; the upper edges of the arrowhead have a white border; the blue color represents the sky and sea, gold stands for sunshine and prosperity, and white and black the racial composition of the island (with the latter being dominant); the two major triangles invoke the twin Pitons (Gros Piton and Petit Piton), cone-shaped volcanic plugs that are a symbol of the island

National symbol(s): *twin pitons (volcanic peaks), Saint Lucia parrot; national colors:* cerulean blue, gold, black, white

National anthem: *name:* Sons and Daughters of St. Lucia
lyrics/music: Charles JESSE/Leton Felix THOMAS
note: adopted 1967

ECONOMY

Economy—overview: The island nation has been able to attract foreign business and investment, especially in its offshore banking and tourism industries. Tourism is Saint Lucia's main source of jobs and income—accounting for 65% of GDP—and the island's main source of foreign exchange earnings. The manufacturing sector is the most diverse in the Eastern Caribbean area. Crops such as bananas, mangos, and avocados continue to be grown for export, but St. Lucia's once solid banana industry has been devastated by strong competition.

Saint Lucia is vulnerable to a variety of external shocks, including volatile tourism receipts, natural disasters, and dependence on foreign oil. Furthermore, high public debt—77% of GDP in 2012—and high debt servicing obligations constrain the CHASTANET administration's ability to respond to adverse external shocks.

St. Lucia has experienced anemic growth since the onset of the global financial crisis in 2008, largely because of a slowdown in tourism—airlines cut back on their routes to St. Lucia in 2012. Also, St. Lucia introduced a value added tax in 2012 of 15%, becoming the last country in the Eastern Caribbean to do so. In 2013, the government introduced a National Competitiveness and Productivity Council to address St. Lucia's high public wages and lack of productivity.

GDP (purchasing power parity): $2.542 billion (2017 est.)
$2.469 billion (2016 est.)
$2.388 billion (2015 est.)
note: data are in 2017 dollars
country comparison to the world: 191

GDP (official exchange rate): $1.686 billion (2017 est.)

GDP—real growth rate: 3% (2017 est.)
3.4% (2016 est.)
-0.9% (2015 est.)
country comparison to the world: 115

GDP—per capita (PPP): $14,400 (2017 est.)
$14,200 (2016 est.)
$13,800 (2015 est.)
note: data are in 2017 dollars
country comparison to the world: 117

Gross national saving: 19.4% of GDP (2017 est.)
15.5% of GDP (2016 est.)

24.3% of GDP (2015 est.)
country comparison to the world: 102

GDP—composition, by end use:
household consumption: 66.1% (2017 est.)
government consumption: 11.2% (2017 est.)
investment in fixed capital: 16.9% (2017 est.)
investment in inventories: 0.1% (2017 est.)
exports of goods and services: 62.7% (2017 est.)
imports of goods and services: -56.9% (2017 est.)

GDP—composition, by sector of origin:
agriculture: 2.9% (2017 est.)
industry: 14.2% (2017 est.)
services: 82.8% (2017 est.)

Agriculture—products: bananas, coconuts, vegetables, citrus, root crops, cocoa

Industries: tourism; clothing, assembly of electronic components, beverages, corrugated cardboard boxes, lime processing, coconut processing

Industrial production growth rate: 6% (2017 est.)
country comparison to the world: 42

Labor force: 79,700 (2012 est.)
country comparison to the world: 183

Labor force—by occupation: *agriculture:* 21.7%
industry: 24.7%
services: 53.6% (2002 est.)

Unemployment rate: 20% (2003 est.)
country comparison to the world: 186

Population below poverty line: NA

Household income or consumption by percentage share: *lowest 10%:* NA
highest 10%: NA

Budget: *revenues:* 398.2 million (2017 est.)
expenditures: 392.8 million (2017 est.)

Taxes and other revenues: 23.6% (of GDP) (2017 est.)
country comparison to the world: 125

Budget surplus (+) or deficit (-): 0.3% (of GDP) (2017 est.)
country comparison to the world: 41

Public debt: 70.7% of GDP (2017 est.)
69.2% of GDP (2016 est.)
country comparison to the world: 49

Fiscal year: 1 April—31 March

Inflation rate (consumer prices): 0.1% (2017 est.)
-3.1% (2016 est.)
country comparison to the world: 15

Current account balance: $21 million (2017 est.)
-$31 million (2016 est.)
country comparison to the world: 61

Exports: $185.1 million (2017 est.)
$188.2 million (2016 est.)
country comparison to the world: 191

Exports—partners: US 67.6%, UK 5.9%, Trinidad and Tobago 5.5% (2017)

Exports—commodities: bananas 41%, clothing, cocoa, avocados, mangoes, coconut oil (2010 est.)

Imports: $600 million (2017 est.)
$575.9 million (2016 est.)
country comparison to the world: 196

Imports—commodities: food, manufactured goods, machinery and transportation equipment, chemicals, fuels

Imports—partners: US 53.3%, Trinidad and Tobago 10.8% (2017)

Reserves of foreign exchange and gold: $321.8 million (31 December 2017 est.)
$320.7 million (31 December 2016 est.)
country comparison to the world: 166

Debt—external: $570.6 million (31 December 2017 est.)
$529 million (31 December 2015 est.)
country comparison to the world: 175

Exchange rates: East Caribbean dollars (XCD) per US dollar -
2.7 (2017 est.)
2.7 (2016 est.)
2.7 (2015 est.)
2.7 (2014 est.)
2.7 (2013 est.)

ENERGY

Electricity access: *electrification—total population:* 97.8% (2016)
electrification—urban areas: 94.9% (2016)
electrification—rural areas: 98.4% (2016)

Electricity—production: 369 million kWh (2016 est.)
country comparison to the world: 174

Electricity—consumption: 343.2 million kWh (2016 est.)
country comparison to the world: 181

Electricity—exports: 0 kWh (2016 est.)
country comparison to the world: 189

Electricity—imports:
0 kWh (2016 est.)
country comparison to the world: 191

Electricity—installed generating capacity: 89,000 kW (2016 est.)
country comparison to the world: 180

Electricity—from fossil fuels: 99% of total installed capacity (2016 est.)
country comparison to the world: 25

Electricity—from nuclear fuels: 0% of total installed capacity (2017 est.)
country comparison to the world: 173

Electricity—from hydroelectric plants: 0% of total installed capacity (2017 est.)
country comparison to the world: 196

Electricity—from other renewable sources: 1% of total installed capacity (2017 est.)
country comparison to the world: 165

Crude oil—production: 0 bbl/day (2018 est.)
country comparison to the world: 192

Crude oil—exports: 0 bbl/day (2015 est.)
country comparison to the world: 185

Crude oil—imports: 0 bbl/day (2015 est.)
country comparison to the world: 187

Crude oil—proved reserves: 0 bbl (1 January 2018 est.)

country comparison to the world: 187

Refined petroleum products—production: 0 bbl/day (2015 est.)
country comparison to the world: 194

Refined petroleum products—consumption: 3,100 bbl/day (2016 est.)
country comparison to the world: 187

Refined petroleum products—exports: 0 bbl/day (2015 est.)
country comparison to the world: 195

Refined petroleum products—imports: 3,113 bbl/day (2015 est.)
country comparison to the world: 184

Natural gas—production: 0 cu m (2017 est.)
country comparison to the world: 189

Natural gas—consumption: 0 cu m (2017 est.)
country comparison to the world: 191

Natural gas—exports: 0 cu m (2017 est.)
country comparison to the world: 174

Natural gas—imports: 0 cu m (2017 est.)
country comparison to the world: 180

Natural gas—proved reserves: 0 cu m (1 January 2014 est.)
country comparison to the world: 186

Carbon dioxide emissions from consumption of energy: 437,900 Mt (2017 est.)
country comparison to the world: 186

COMMUNICATIONS

Telephones—fixed lines: *total subscriptions:* 33,285
subscriptions per 100 inhabitants: 20.05 (2019 est.)
country comparison to the world: 169

Telephones—mobile cellular: *total subscriptions:* 168,797

subscriptions per 100 inhabitants: 101.68 (2019 est.)
country comparison to the world: 187

Telecommunication systems: *general assessment:* an adequate system that is automatically switched; good interisland and international connections; broadband access; expanded FttP (Fiber to the Home) and LTE markets; regulatory development; telecom sector contributes to the overall GDP; telecom sector is a growth area (2020)
domestic: fixed-line teledensity is 20 per 100 persons and mobile-cellular teledensity is roughly 102 per 100 persons (2019)
international: country code—1-758; landing points for the ECFS and Southern Caribbean Fiber submarine cables providing connectivity to numerous Caribbean islands; direct microwave radio relay link with Martinique and Saint Vincent and the Grenadines; tropospheric scatter to Barbados (2019)
note: the COVID-19 outbreak is negatively impacting telecommunications production and supply chains globally; consumer spending on telecom devices and services has also slowed due to the pandemic's effect on economies worldwide; overall progress towards improvements in all facets of the telecom industry—mobile, fixed-line, broadband, submarine cable and satellite—has moderated

Broadcast media: 3 privately owned TV stations; 1 public TV station operating on a cable network; multi-channel cable TV service available; a mix of state-owned and privately owned broadcasters operate nearly 25 radio stations including repeater transmission stations

Internet country code: .lc

Internet users: *total:* 94,112
percent of population: 50.82% (July 2018 est.)
country comparison to the world: 180

Broadband—fixed subscriptions: *total:* 32,265

subscriptions per 100 inhabitants: 19 (2018 est.)
country comparison to the world: 140

TRANSPORTATION

Civil aircraft registration country code prefix: J6 (2016)

Airports: 2 (2013)
country comparison to the world: 204

Airports—with paved runways: *total:* 2 (2019)
2,438 to 3,047 m: 1
1,524 to 2,437 m: 1

Roadways: *total:* 1,210 km (2011)
paved: 847 km (2011)
unpaved: 363 km (2011)
country comparison to the world: 179

Ports and terminals: *major seaport(s):* Castries, Cul-de-Sac, Vieux-Fort

MILITARY AND SECURITY

Military and security forces: no regular military forces; Royal Saint Lucia Police Force (includes Special Service Unit, Marine Unit) (2018)

Military—note: St. Lucia is a member of the Regional Security System (RSS), an international agreement for the defense and security of the eastern Caribbean region.

TRANSNATIONAL ISSUES

Disputes—international: joins other Caribbean states to counter Venezuela's claim that Aves Island sustains human habitation, a criterion under UN Convention on the Law of the Sea, which permits Venezuela to extend its EEZ/continental shelf over a large portion of the eastern Caribbean Sea

Illicit drugs: transit point for South American drugs destined for the US and Europe

SAINT MARTIN

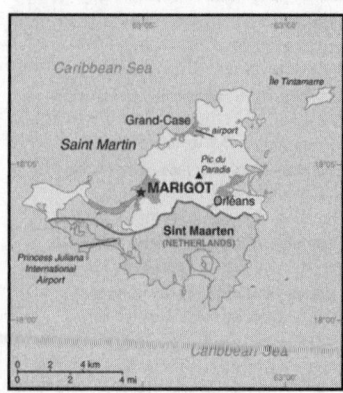

INTRODUCTION

Background: Although sighted by Christopher COLUMBUS in 1493 and claimed for Spain, it was the Dutch who occupied the island in 1631 to exploit its salt deposits. The Spanish retook the island in 1633, but continued to be harassed by the Dutch. The Spanish finally relinquished Saint Martin to the French and Dutch, who divided it between themselves in 1648. Friction between the two sides caused the border to frequently fluctuate over the next two centuries, with the French eventually holding the greater portion of the island (about 61%). The cultivation of sugar cane introduced African slavery to the island in the late 18th century; the practice was not abolished until 1848. The island became a free port in 1939; the tourism industry was dramatically expanded during the

1970s and 1980s. In 2003, the populace of Saint Martin voted to secede from Guadeloupe and in 2007, the northern portion of the island became a French overseas collectivity. In 2010, the southern Dutch portion of the island became the independent nation of Sint Maarten within the Kingdom of the Netherlands. On 6 September 2017, Hurricane Irma passed over the island of Saint Martin causing extensive damage to roads, communications, electrical power, and housing; the UN estimated that 90% of the buildings were damaged or destroyed.

GEOGRAPHY

Location: Caribbean, located in the Leeward Islands (northern) group; French part of the island of Saint Martin in the Caribbean Sea; Saint Martin lies east of the US Virgin Islands

Geographic coordinates: 18 05 N, 63 57 W

Map references: Central America and the Caribbean

Area: *total:* 54 sq km
land: 54.4 sq km
water: negligible
country comparison to the world: 231

Area—comparative: more than one-third the size of Washington, DC

Land boundaries: *total:* 16 km
border countries (1): Sint Maarten 16 km

Coastline: 58.9 km (for entire island)

Climate: temperature averages 27-29 degrees Celsius all year long; low humidity, gentle trade winds, brief, intense rain showers; hurricane season stretches from July to November

Elevation: *lowest point:* Caribbean Ocean 0 m
highest point: Pic du Paradis 424 m

Natural resources: salt

Population distribution: most of the population is found along the coast, with a largest concentrations around the capital Marigot, Orleans, and Grand-Case

Natural hazards: subject to hurricanes from July to November

Environment—current issues: excessive population pressure (increasing settlement); waste management; salinity intrusions into the main land of the island; fresh water supply is dependent on desalination of sea water; over exploitation of marine resources (reef fisheries, coral and shell); indiscriminate anchoring of boats damages coral reefs, causing underwater pollution and changes the sediment dynamics of Saint Martin's Island

Geography—note: the southern border is shared with Sint Maarten, a country within the Kingdom of the Netherlands; together, these two entities make up the smallest landmass in the world shared by two self-governing states

PEOPLE AND SOCIETY

Population: 32,556 (July 2020 est.)
country comparison to the world: 217

Ethnic groups: other Creole (mulatto), black, Guadeloupe Mestizo (French-East Asia), white, East Indian

Languages: French (official), English, Dutch, French Patois, Spanish, Papiamento (dialect of Netherlands Antilles)

Religions: Roman Catholic, Jehovah's Witness, Protestant, Hindu

Age structure: *0-14 years:* 25.63% (male 4,148/female 4,197)
15-24 years: 10.28% (male 1,647/female 1,701)
25-54 years: 46.2% (male 7,201/female 7,841)
55-64 years: 8.71% (male 1,328/female 1,508)
65 years and over: 9.17% (male 1,305/female 1,680) (2020 est.)

Median age: *total:* 33.3 years
male: 32.5 years
female: 34.1 years (2020 est.)

country comparison to the world: 100

Population growth rate: 0.4% (2020 est.)
country comparison to the world: 160

Birth rate: 14.3 births/1,000 population (2020 est.)
country comparison to the world: 131

Death rate: 4.6 deaths/1,000 population (2020 est.)
country comparison to the world: 206

Net migration rate: -6 migrant(s)/1,000 population (2020 est.)
country comparison to the world: 205

Population distribution: most of the population is found along the coast, with a largest concentrations around the capital Marigot, Orleans, and Grand-Case

Sex ratio: *at birth:* 1.04 male(s)/female
0-14 years: 0.99 male(s)/female
15-24 years: 0.97 male(s)/female
25-54 years: 0.92 male(s)/female
55-64 years: 0.88 male(s)/female
65 years and over: 0.78 male(s)/female
total population: 0.92 male(s)/female (2020 est.)

Infant mortality rate: *total:* 5.6 deaths/1,000 live births
male: 6.4 deaths/1,000 live births
female: 4.8 deaths/1,000 live births (2020 est.)
country comparison to the world: 170

Life expectancy at birth: *total population:* 80.2 years
male: 77 years
female: 83.4 years (2020 est.)
country comparison to the world: 47

Total fertility rate: 1.81 children born/woman (2020 est.)
country comparison to the world: 150

HIV/AIDS—adult prevalence rate: NA

HIV/AIDS—people living with HIV/AIDS: NA

Education expenditures: NA

GOVERNMENT

Country name: *conventional long form:* Overseas Collectivity of Saint Martin
conventional short form: Saint Martin
local long form: Collectivite d'outre mer de Saint-Martin
local short form: Saint-Martin
etymology: explorer Christopher COLUMBUS named the island after Saint MARTIN of Tours because the 11 November 1493 day of discovery was the saint's feast day

Dependency status: overseas collectivity of France
note: the only French overseas collectivity that is part of the EU

Government type: parliamentary democracy (Territorial Council); overseas collectivity of France

Capital: *name:* Marigot

geographic coordinates: 18 04 N, 63 05 W
time difference: UTC-4 (1 hour ahead of Washington, DC, during Standard Time)
etymology: marigot is a French term referring to a body of water, a watercourse, a side-stream, or a

tributary rivulet; the name likely refers to a stream at the site of the city's original founding

Independence: none (overseas collectivity of France)

National holiday: Fete de la Federation, 14 July (1790); note—local holiday is Schoelcher Day (Slavery Abolition Day) 12 July (1848), as well as St. Martin's Day, 11 November (1985), which commemorates the discovery of the island by COLUMBUS on Saint Martin's Day, 11 November 1493; the latter holiday celebrated on both halves of the island

Constitution: *history:* 4 October 1958 (French Constitution)
amendments: amendment procedures of France's constitution apply

Legal system: French civil law

Citizenship: see France

Suffrage: 18 years of age, universal

Executive branch: *chief of state:* President Emmanuel MACRON (since 14 May 2017); represented by Prefect Anne LAUBIES (since 8 June 2015)
head of government: President of Territorial Council Daniel GIBBS (since 2 April 2017); First Vice President Valerie DAMASEAU (since 2 April 2017)
cabinet: Executive Council; note—there is also an advisory economic, social, and cultural council
elections/appointments: French president directly elected by absolute majority popular vote in 2 rounds if needed for a 5-year term (eligible for a second term); prefect appointed by French president on the advice of French Ministry of Interior; president of Territorial Council elected by its members for a 5-year term; election last held on 26 March 2017
election results: Daniel GIBBS (TDG) elected president; Territorial Council vote—18 votes, 4 blank, 1 invalid

Legislative branch: *description:* unicameral Territorial Council (23 seats; members directly elected by absolute majority vote in 2 rounds if needed to serve 5-year terms); Saint Martin elects 1 member to the French Senate and one member (shared with Saint Barthelemy) to the French National Assembly
elections: Territorial Council—last held on 18 and 25 March 2017 (next to be held in March 2022)
election results: Territorial Council—percent of vote by party (first round)—TDG 49.1%, MJP 13.7%, MVP 12.3%, HOPE 8.7%, Continuons pour Saint-Martin 6.5%, other 9.7%; seats by party—NA; percent of vote by party (second round)—TDG 64.3%, MJP 24.2%, MVP 11.5.5%; seats by party—TDG 18, MJP 4, MVP 1; composition—men 13, women 10, percent of women 43.5%
French Senate—held on 28 September 2014 (next to be held not later than September 2020) French National Assembly—last held on 11 and 18 June 2017 (next to be held by June 2022) French Senate—1 seat: UMP 1 French National Assembly—1 seat: UMP 1

Political parties and leaders: Continuons pour St. Martin [Aline HANSON]
En marche vers le progres or MVP [Alain RICHARDSON]
Gereration Hope or HOPE [Jules CHARVILLE]
Movement for Justice and Prosperity or MJP [Louis MUSSINGTON]
New Direction [Jeanne VANTERPOOL]
Rally Responsibility Success (Rassemblement Responsabilite Reussite or RRR [Alain RICHARDSON]
Team Daniel Gibbs 2017 or TDG [Daniel GIBBS]
Union for Progress (Union Pour le Progres or UPP) [Louis-Constant FLEMING]; affiliated with UMP

International organization participation: UPU

Diplomatic representation in the US: none (overseas collectivity of France)

Diplomatic representation from the US: none (overseas collectivity of France)

Flag description: the flag of France is used

National symbol(s): brown pelican

National anthem: *name:* O Sweet Saint Martin's Land
lyrics/music: Gerard KEMPS
note: the song, written in 1958, is used as an unofficial anthem for the entire island (both French and Dutch sides); as a collectivity of France, in addition to the local anthem, "La Marseillaise" remains official on the French side (see France); as a constituent part of the Kingdom of the Netherlands, in addition to the local anthem, "Het Wilhelmus" remains official on the Dutch side (see Netherlands)

ECONOMY

Economy—overview: The economy of Saint Martin centers on tourism with 85% of the labor force engaged in this sector. Over one million visitors come to the island each year with most arriving through the Princess Juliana International Airport in Sint Maarten. The financial sector is also important to Saint Martin's economy as it facilitates financial mediation for its thriving tourism sector. No significant agriculture and limited local fishing means that almost all food must be imported.

Energy resources and manufactured goods are also imported, primarily from Mexico and the US. Saint Martin is reported to have one of the highest per capita income in the Caribbean. As with the rest of the Caribbean, Saint Martin's financial sector is having to deal with losing correspondent banking relationships.

In September 2017, Hurricane Irma destroyed 95% of the French side of Saint Martin. Along the coastline of Marigot, the nerve center of the economy, the storm wiped out restaurants, shops, banks and open-air markets impacting more than 36,000 inhabitants.

GDP (purchasing power parity):
$561.5 million (2005 est.)
country comparison to the world: 212

GDP (official exchange rate): $561.5 million (2005 est.)

GDP—per capita (PPP):
$19,300 (2005 est.)
country comparison to the world: 92

GDP—composition, by sector of origin:
agriculture: 1% (2000)
industry: 15% (2000)
services: 84% (2000)

Industries: tourism, light industry and manufacturing, heavy industry

Labor force: 17,300 (2008 est.)
country comparison to the world: 214

Labor force—by occupation: 85 directly or indirectly employed in tourist industry

Imports—commodities: crude petroleum, food, manufactured items

Exchange rates: euros (EUR) per US dollar -
0.885 (2017 est.)
0.903 (2016 est.)
0.9214 (2015 est.)
0.885 (2014 est.)
0.7634 (2013 est.)

ENERGY

Electricity access: *electrification—total population:* 72% (2016)
electrification—urban areas: 89.8% (2016)

COMMUNICATIONS

Telecommunication systems: *general assessment:* fully integrated access; good interisland and international connections; broadband access; expanded FttP (Fiber to the Home) and LTE markets; regulatory development; telecom sector contributes greatly to the overall GDP; telecom sector is a growth area (2020)
domestic: direct dial capability with both fixed and wireless systems (2018)
international: country code—590; landing points for the SMPR-1, Southern Caribbean Fiber and the SSCS submarine cables providing connectivity to numerous Caribbean islands (2019)
note: the COVID-19 outbreak is negatively impacting telecommunications production and supply chains globally; consumer spending on telecom devices and services has also slowed due to the pandemic's effect on economies worldwide; overall progress towards improvements in all facets of the telecom industry—mobile, fixed-line, broadband, submarine cable and satellite—has moderated

Broadcast media: 1 local TV station; access to about 20 radio stations, including RFO Guadeloupe radio broadcasts via repeater

Internet country code: .mf; note—.gp, the Internet country code for Guadeloupe, and .fr, the Internet country code for France, might also be encountered

Internet users: *total:* 1,100
percent of population: 3.5% (July 2016 est.)
country comparison to the world: 224

TRANSPORTATION

Airports: 1 (2013)
country comparison to the world: 235

Airports—with paved runways: *total:* 1 (2019)
914 to 1,523 m: 1

Transportation—note: nearest airport for international flights is Princess Juliana International Airport (SXM) located on Sint Maarten

MILITARY AND SECURITY

Military—note: defense is the responsibility of France

SAINT PIERRE AND MIQUELON

INTRODUCTION

Background: First settled by the French in the early 17th century, the islands represent the sole remaining vestige of France's once vast North American possessions. They attained the status of an overseas collectivity in 2003.

GEOGRAPHY

Location: Northern North America, islands in the North Atlantic Ocean, south of Newfoundland (Canada)

Geographic coordinates: 46 50 N, 56 20 W

Map references: North America

Area: *total:* 242 sq km
land: 242 sq km
water: 0 sq km
note: includes eight small islands in the Saint Pierre and the Miquelon groups
country comparison to the world: 214

Area—comparative: one and half times the size of Washington, DC

Land boundaries: 0 km

Coastline: 120 km

Maritime claims: *territorial sea:* 12 nm
exclusive economic zone: 200 nm

Climate: cold and wet, with considerable mist and fog; spring and autumn are often windy

Terrain: mostly barren rock

Elevation: *lowest point:* Atlantic Ocean 0 m
highest point: Morne de la Grande Montagne 240 m

Natural resources: fish, deepwater ports

Land use: *agricultural land:* 8.7% (2011 est.)

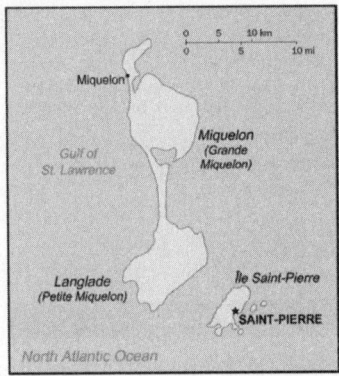

arable land: 8.7% (2011 est.) / permanent crops: 0% (2011 est.) / permanent pasture: 0% (2011 est.)
forest: 12.5% (2011 est.)
other: 78.8% (2011 est.)
Irrigated land: 0 sq km (2012)

Population distribution: most of the population is found on Saint Pierre Island; a small settlement is located on the north end of Miquelon Island

Natural hazards: persistent fog throughout the year can be a maritime hazard

Environment—current issues: overfishing; recent test drilling for oil in waters around Saint Pierre and Miquelon may bring future development that would impact the environment

Geography—note: vegetation scanty; the islands are actually part of the northern Appalachians along with Newfoundland

PEOPLE AND SOCIETY

Population: 5,347 (July 2020 est.)
country comparison to the world: 228

Nationality: noun: Frenchman(men), Frenchwoman(women)
adjective: French

Ethnic groups: Basques and Bretons (French fishermen)

Languages: French (official)

Religions: Roman Catholic 99%, other 1%

Age structure: 0-14 years: 14.31% (male 395/female 370)
15-24 years: 8.83% (male 245/female 227)
25-54 years: 40% (male 1,039/female 1,100)
55-64 years: 14.49% (male 400/female 375)
65 years and over: 22.37% (male 513/female 683) (2020 est.)

Median age: total: 48.5 years
male: 47.9 years
female: 49 years (2020 est.)
country comparison to the world: 3

Population growth rate: -1.15% (2020 est.)
country comparison to the world: 233

Birth rate: 6.7 births/1,000 population (2020 est.)
country comparison to the world: 228

Death rate: 10.9 deaths/1,000 population (2020 est.)
country comparison to the world: 23

Net migration rate: -7.7 migrant(s)/1,000 population (2020 est.)
country comparison to the world: 212

Population distribution: most of the population is found on Saint Pierre Island; a small settlement is located on the north end of Miquelon Island

Urbanization: urban population: 90% of total population (2020)
rate of urbanization: 0.36% annual rate of change (2015-20 est.)

total population growth rate v. urban population growth rate, 2000-2030: Major urban areas—population: 6,000 SAINT-PIERRE (capital) (2018)

Sex ratio: at birth: 1.06 male(s)/female
0-14 years: 1.07 male(s)/female
15-24 years: 1.08 male(s)/female
25-54 years: 0.94 male(s)/female
55-64 years: 1.07 male(s)/female
65 years and over: 0.75 male(s)/female
total population: 0.94 male(s)/female (2020 est.)

Infant mortality rate: total: 6.1 deaths/1,000 live births
male: 7 deaths/1,000 live births
female: 5.1 deaths/1,000 live births (2020 est.)
country comparison to the world: 165

Life expectancy at birth: total population: 81 years
male: 78.6 years
female: 83.5 years (2020 est.)
country comparison to the world: 41

Total fertility rate: 1.58 children born/woman (2020 est.)
country comparison to the world: 190

Drinking water source: improved: total: 91.4% of population
unimproved: total: 8.6% of population (2017 est.)

HIV/AIDS—adult prevalence rate: NA

HIV/AIDS—people living with HIV/AIDS: NA

HIV/AIDS—deaths: NA

Education expenditures: NA

GOVERNMENT

Country name: conventional long form: Territorial Collectivity of Saint Pierre and Miquelon
conventional short form: Saint Pierre and Miquelon
local long form: Departement de Saint-Pierre et Miquelon
local short form: Saint-Pierre et Miquelon
etymology: Saint-Pierre is named after Saint PETER, the patron saint of fishermen; Miquelon may be a corruption of the Basque name Mikelon

Dependency status: overseas collectivity of France

Government type: parliamentary democracy (Territorial Council); overseas collectivity of France

Capital: name: Saint-Pierre
geographic coordinates: 46 46 N, 56 11 W
time difference: UTC-3 (2 hours ahead of Washington, DC, during Standard Time)
daylight saving time: +1hr, begins second Sunday in March; ends first Sunday in November
etymology: named after Saint Peter, the patron saint of fisherman

Administrative divisions: none (territorial overseas collectivity of France); note—there are no first-order administrative divisions as defined by the US Government, but there are 2 communes at the second order—Saint Pierre, Miquelon

Independence: none (overseas collectivity collectivity of France; has been under French control since 1763)

National holiday: Fete de la Federation, 14 July (1790)

Constitution: history: 4 October 1958 (French Constitution)
amendments: amendment procedures of France's constitution apply

Legal system: French civil law

Citizenship: see France

Suffrage: 18 years of age; universal

Executive branch: chief of state: President Emmanuel MACRON (since 14 May 2017); represented by Prefect Thierry DEVIMEUX (since 17 January 2018)
head of government: President of Territorial Council Stephane LENORMAND (since 24 October 2017)
cabinet: Le Cabinet du Prefet
elections/appointments: French president directly elected by absolute majority popular vote in 2 rounds if needed for a 5-year term (eligible for a second term); election last held on 23 April and 6 May 2017 (next to be held in 2022); prefect appointed by French president on the advice of French Ministry of Interior

Legislative branch: description: unicameral Territorial Council or Conseil Territorial (19 seats—Saint Pierre 15, Miquelon 4; members directly elected in single-seat constituencies by absolute majority vote in 2 rounds if needed to serve 6-year terms);
Saint Pierre and Miquelon indirectly elects 1 senator to the French Senate by an electoral college to serve a 6-year term and directly elects 1 deputy to the French National Assembly by absolute majority vote to serve a 5-year term
elections: Territorial Council—last held on 19 March 2017 (next to be held in March 2023)
French Senate—last held on 24 September 2017 (next to be held no later than September 2020)
French National Assembly—last held on 11 and 18 June 2017 (next to be held by June 2022)
election results: Territorial Council—percent of vote by party—AD 70.2%, Cap surl'Avenir 29.8%; seats by party—AD 17, Cap surl'Avenir 2; composition—men 10, women 9, percent of women 47.4%

French Senate—percent of vote by party—NA; seats by party—PS 1 (affiliated with UMP)

French National Assembly—percent of vote by party—NA; seats by party—Ensemble pour l'Avenir 1 (affiliated with PRG); the Republicans (LR) 1

Judicial branch: *highest courts:* Superior Tribunal of Appeals or Tribunal Superieurd'Appel (composition NA)

judge selection and term of office: judge selection and tenure NA

subordinate courts: NA

Political parties and leaders: Archipelago Tomorrow or AD (affiliated with UMP)

Cap surl'Avenir [Annick GIRARDIN] (affiliated with Left Radical Party)

Togerther for the Future (Ensemble pour l'Avenir) (affiliated with PRG) SPM ensemble

International organization participation: UPU, WFTU (NGOs)

Diplomatic representation in the US: none (territorial overseas collectivity of France)

Diplomatic representation from the US: none (territorial overseas collectivity of France)

Flag description: *a yellow three-masted sailing ship facing the hoist side rides on a blue background with scattered, white, wavy lines under the ship; a continuous black-over-white wavy line divides the ship from the white wavy lines; on the hoist side, a vertical band is divided into three parts:* the top part (called ikkurina) is red with a green diagonal cross extending to the corners overlaid by a white cross dividing the rectangle into four sections; the middle part has a white background with an ermine pattern; the third part has a red background with two stylized yellow lions outlined in black, one above the other; these three heraldic arms represent settlement by colonists from the Basque Country (top), Brittany, and Normandy; the blue on the main portion of the flag symbolizes the Atlantic Ocean and the stylized ship represents the Grande Hermine in which Jacques Cartier "discovered" the islands in 1536

note: the flag of France used for official occasions

National symbol(s): 16th-century sailing ship

National anthem: *note:* as a collectivity of France, "La Marseillaise" is official (see France)

0:00 /1:19

ECONOMY

Economy—overview: The inhabitants have traditionally earned their livelihood by fishing and by servicing fishing fleets operating off the coast of Newfoundland. The economy has been declining, however, because of disputes with Canada over fishing quotas and a steady decline in the number of ships stopping at Saint Pierre. The services sector accounted for 86% of GDP in 2010, the last year data is available for. Government employment accounts for than 46% of the GDP, and 78% of the population is working age.

The government hopes an expansion of tourism will boost economic prospects. Fish farming,

crab fishing, and agriculture are being developed to diversify the local economy. Recent test drilling for oil may pave the way for development of the energy sector. Trade is the second largest sector in terms of value added created, where it contributes significantly to economic activity. The extractive industries and energy sector is the third largest sector of activity in the archipelago, attributable in part to the construction of a new thermal power plant in 2015.

GDP (purchasing power parity): $261.3 million (2015 est.)

$215.3 million (2006 est.)

note: supplemented by annual payments from France of about $60 million

country comparison to the world: 218

GDP (official exchange rate): $261.3 million (2015 est.)

GDP—real growth rate: NA

GDP—per capita (PPP): $46,200 (2006 est.)

$34,900 (2005)

country comparison to the world: 36

GDP—composition, by sector of origin:

agriculture: 2% (2006 est.)

industry: 15% (2006 est.)

services: 83% (2006 est.)

Agriculture—products: vegetables; poultry, cattle, sheep, pigs; fish

Industries: fish processing and supply base for fishing fleets; tourism

Industrial production growth rate: NA

Labor force: 4,429 (2015)

country comparison to the world: 224

Labor force—by occupation: *agriculture:* 18%

industry: 41%

services: 41% (1996 est.)

Unemployment rate: 8.7% (2015 est.)

9.9% (2008 est.)

country comparison to the world: 132

Population below poverty line: NA

Household income or consumption by percentage share: *lowest 10%:* NA

highest 10%: NA

Budget: *revenues:* 70 million (1996 est.)

expenditures: 60 million (1996 est.)

Taxes and other revenues: 26.8% (of GDP) (1996 est.)

country comparison to the world: 104

Budget surplus (+) or deficit (-): 3.8% (of GDP) (1996 est.)

country comparison to the world: 10

Fiscal year: calendar year

Inflation rate (consumer prices): 1.5% (2015)

4.5% (2010)

country comparison to the world: 85

Exports: $6.641 million (2010 est.)

$5.5 million (2005 est.)

country comparison to the world: 217

Exports—commodities: fish and fish products, soybeans, animal feed, mollusks and crustaceans, fox and mink pelts

Imports: $95.35 million (2010 est.)

$68.2 million (2005 est.)

country comparison to the world: 217

Imports—commodities: meat, clothing, fuel, electrical equipment, machinery, building materials

Debt—external: NA

Exchange rates: euros (EUR) per US dollar - 0.885 (2017 est.)

0.903 (2016 est.)

0.9214 (2015 est.)

0.885 (2014 est.)

0.7634 (2013 est.)

ENERGY

Electricity—production: 46 million kWh (2016 est.)

country comparison to the world: 206

Electricity—consumption: 42.78 million kWh (2016 est.)

country comparison to the world: 206

Electricity—exports: 0 kWh (2016 est.)

country comparison to the world: 190

Electricity—imports: 0 kWh (2016 est.)

country comparison to the world: 192

Electricity—installed generating capacity: 27,600 kW (2016 est.)

country comparison to the world: 202

Electricity—from fossil fuels: 96% of total installed capacity (2016 est.)

country comparison to the world: 42

Electricity—from nuclear fuels: 0% of total installed capacity (2017 est.)

country comparison to the world: 174

Electricity—from hydroelectric plants: 0% of total installed capacity (2017 est.)

country comparison to the world: 197

Electricity—from other renewable sources: 4% of total installed capacity (2017 est.)

country comparison to the world: 118

Crude oil—production: 0 bbl/day (2018 est.)

country comparison to the world: 193

Crude oil—exports: 0 bbl/day (2015 est.)

country comparison to the world: 186

Crude oil—imports: 0 bbl/day (2015 est.)

country comparison to the world: 188

Crude oil—proved reserves: 0 bbl (1 January 2018 est.)

country comparison to the world: 188

Refined petroleum products—production: 0 bbl/day (2015 est.)

country comparison to the world: 195

Refined petroleum products—consumption: 660 bbl/day (2016 est.)

country comparison to the world: 208

Refined petroleum products—exports: 0 bbl/day (2015 est.)

country comparison to the world: 196

Refined petroleum products—imports: 650 bbl/day (2015 est.)

country comparison to the world: 204

Natural gas—production: 0 cu m (2017 est.)

country comparison to the world: 190

Natural gas—consumption: 0 cu m (2017 est.)
country comparison to the world: 192

Natural gas—exports: 0 cu m (2017 est.)
country comparison to the world: 175

Natural gas—imports: 0 cu m (2017 est.)
country comparison to the world: 181

Natural gas—proved reserves: 0 cu m (1 January 2014 est.)
country comparison to the world: 187

Carbon dioxide emissions from consumption of energy: 100,200 Mt (2017 est.)
country comparison to the world: 206

Telephones—fixed lines: *total subscriptions:* 4,086
subscriptions per 100 inhabitants: 75.55 (2019 est.)
country comparison to the world: 206

Telecommunication systems: *general assessment:* adequate (2018)

domestic: fixed-line teledensity 76 per 100 persons (2019)

international: country code—508; landing point for the St Pierre and Miquelon Cable connecting Saint Pierre & Miquelon and Canada; radiotelephone communication with most countries in the world; satellite earth station—1 in French domestic satellite system (2019)

note: the COVID-19 outbreak is negatively impacting telecommunications production and supply chains globally; consumer spending on telecom devices and services has also slowed due to the pandemic's effect on economies worldwide; overall progress towards improvements in all facets of the telecom industry—mobile, fixed-line, broadband, submarine cable and satellite—has moderated

Broadcast media: 2 TV stations with a third repeater station, all part of the French Overseas Network; radio stations on St. Pierre and on Miquelon are part of the French Overseas Network

Internet country code: .pm

Internet users: *total:* 4,500
percent of population: 79.5% (July 2016 est.)

country comparison to the world: 218

Airports: 2 (2013)
country comparison to the world: 205

Airports—with paved runways: *total:* 2 (2019)
1,524 to 2,437 m: 1
914 to 1,523 m: 1

Roadways: *total:* 117 km (2009)
paved: 80 km (2009)
unpaved: 37 km (2009)
country comparison to the world: 213

Ports and terminals: *major seaport(s):* Saint-Pierre

Military—note: defense is the responsibility of France

Disputes—international: none

SAINT VINCENT AND THE GRENADINES

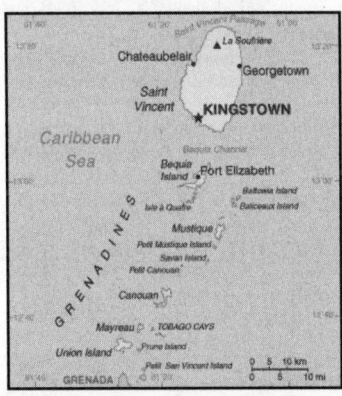

Background: Resistance by native Caribs prevented colonization on Saint Vincent until 1719. Disputed between France and the UK for most of the 18th century, the island was ceded to the latter in 1783. The British prized Saint Vincent due to its fertile soil, which allowed for thriving slave-run plantations of sugar, coffee, indigo, tobacco, cotton, and cocoa. In 1834, the British abolished slavery. Immigration of indentured servants eased the ensuing labor shortage, as did subsequent Portuguese immigrants from Madeira and East Indian laborers. Conditions remained harsh for both former slaves and immigrant agricultural workers, however, as depressed world sugar prices kept the economy stagnant until the early 1900s. The economy then went into a period of decline

with many landowners abandoning their estates and leaving the land to be cultivated by liberated slaves. Between 1960 and 1962, Saint Vincent and the Grenadines was a separate administrative unit of the Federation of the West Indies. Autonomy was granted in 1969 and independence in 1979.

Location: Caribbean, islands between the Caribbean Sea and North Atlantic Ocean, north of Trinidad and Tobago

Geographic coordinates: 13 15 N, 61 12 W

Map references: Central America and the Caribbean

Area: *total:* 389 sq km (Saint Vincent 344 sq km)
land: 389 sq km
water: 0 sq km
country comparison to the world: 205

Area—comparative: twice the size of Washington, DC

Area comparison map: *[INSERT IMAGE: SAINT VINCENT AND THE GRENADINES—Area comparison map]*

Land boundaries: 0 km

Coastline: 84 km

Maritime claims: *territorial sea:* 12 nm
exclusive economic zone: 200 nm
contiguous zone: 24 nm
continental shelf: 200 nm

Climate: tropical; little seasonal temperature variation; rainy season (May to November)

Terrain: volcanic, mountainous

Elevation: *lowest point:* Caribbean Sea 0 m
highest point: La Soufriere 1,234 m

Natural resources: hydropower, arable land

Land use: *agricultural land:* 25.6% (2011 est.)
arable land: 12.8% (2011 est.) / permanent crops: 7.7% (2011 est.) / permanent pasture: 5.1% (2011 est.)
forest: 68.7% (2011 est.)
other: 5.7% (2011 est.)
Irrigated land: 10 sq km (2012)

Population distribution: most of the population is concentrated in and around the capital of Kingstown

Natural hazards: hurricanes; La Soufriere volcano on the island of Saint Vincent is a constant threat
volcanism: La Soufriere (1,234 m) on the island of Saint Vincent last erupted in 1979; the island of Saint Vincent is part of the volcanic island arc of the Lesser Antilles that extends from Saba in the north to Grenada in the south

Environment—current issues: pollution of coastal waters and shorelines from discharges by pleasure yachts and other effluents; in some areas, pollution is severe enough to make swimming prohibitive; poor land use planning; deforestation; watershed management and squatter settlement control

Environment—international agreements: *party to:* Biodiversity, Climate Change, Climate Change-Kyoto Protocol, Desertification, Endangered Species,
Environmental Modification, Hazardous Wastes, Law of the Sea, Marine Dumping, Ozone Layer Protection, Ship Pollution, Whaling

signed, but not ratified: none of the selected agreements

Geography—note: the administration of the islands of the Grenadines group is divided between Saint Vincent and the Grenadines and Grenada; Saint Vincent and the Grenadines is comprised of 32 islands and cays

PEOPLE AND SOCIETY

Population: 101,390 (July 2020 est.)
country comparison to the world: 195

Nationality: *noun:* Saint Vincentian(s) or Vincentian(s)
adjective: Saint Vincentian or Vincentian

Ethnic groups: African descent 71.2%, mixed 23%, indigenous 3%, East Indian/Indian 1.1%, European 1.5%, other .2% (2012 est.)

Languages: English, Vincentian Creole English, French patois

Religions: Protestant 75% (Pentecostal 27.6%, Anglican 13.9%, Seventh Day Adventist 11.6%, Baptist 8.9%, Methodist 8.7%, Evangelical 3.8%, Salvation Army .3%, Presbyterian/Congregational .3%), Roman Catholic 6.3%, Rastafarian 1.1%, Jehovah's Witness 0.8%, other 4.7%, none 7.5%, unspecified 4.7% (2012 est.)

Age structure: *0-14 years:* 20.15% (male 10,309/ female 10,121)
15-24 years: 14.83% (male 7,582/female 7,451)
25-54 years: 42.63% (male 22,395/female 20,824)
55-64 years: 11.68% (male 6,136/female 5,703)
65 years and over: 10.72% (male 5,167/female 5,702) (2020 est.)

Dependency ratios: *total dependency ratio:* 46.7
youth dependency ratio: 32.1
elderly dependency ratio: 14.5
potential support ratio: 6.9 (2020 est.)

Median age: *total:* 35.3 years
male: 35.4 years
female: 35.1 years (2020 est.)
country comparison to the world: 86

Population growth rate: -0.22% (2020 est.)
country comparison to the world: 211

Birth rate: 12.6 births/1,000 population (2020 est.)
country comparison to the world: 154

Death rate: 7.6 deaths/1,000 population (2020 est.)
country comparison to the world: 101

Net migration rate: -7.2 migrant(s)/1,000 population (2020 est.)
country comparison to the world: 210

Population distribution: most of the population is concentrated in and around the capital of Kingstown

Urbanization: *urban population:* 53% of total population (2020)
rate of urbanization: 1.03% annual rate of change (2015-20 est.)

total population growth rate v. urban population growth rate, 2000-2030: *Major urban areas—population:* 27,000 KINGSTOWN (capital) (2018)

Sex ratio: *at birth:* 1.03 male(s)/female
0-14 years: 1.02 male(s)/female
15-24 years: 1.02 male(s)/female
25-54 years: 1.08 male(s)/female
55-64 years: 1.08 male(s)/female
65 years and over: 0.91 male(s)/female
total population: 1.04 male(s)/female (2020 est.)

Maternal mortality rate: 98 deaths/100,000 live births (2017 est.)
country comparison to the world: 70

Infant mortality rate: *total:* 11 deaths/1,000 live births
male: 12 deaths/1,000 live births
female: 10 deaths/1,000 live births (2020 est.)
country comparison to the world: 123

Life expectancy at birth: *total population:* 76.2 years
male: 74.1 years
female: 78.3 years (2020 est.)
country comparison to the world: 101

Total fertility rate: 1.76 children born/woman (2020 est.)
country comparison to the world: 158

Drinking water source: *improved:* total: 95.1% of population
unimproved: *total:* 4.9% of population (2017 est.)

Current Health Expenditure: 4.5% (2017)

Physicians density: 0.66 physicians/1,000 population (2010)

Hospital bed density: 4.3 beds/1,000 population (2016)

Sanitation facility access: *improved:* total: 90.2% of population
unimproved: *total:* 9.8% of population (2017 est.)

HIV/AIDS—adult prevalence rate: 1.5% (2018)
country comparison to the world: 30

HIV/AIDS—people living with HIV/AIDS: 1,200 (2018)
country comparison to the world: 142

HIV/AIDS—deaths: <100 (2018)

Obesity—adult prevalence rate: 23.7% (2016)
country comparison to the world: 64

Education expenditures: 5.8% of GDP (2017)
country comparison to the world: 34

School life expectancy (primary to tertiary education): *total:* 14 years
male: 14 years
female: 15 years (2015)

GOVERNMENT

Country name: *conventional long form:* none
conventional short form: Saint Vincent and the Grenadines
etymology: Saint Vincent was named by explorer Christopher COLUMBUS after Saint VINCENT of Saragossa because the 22 January 1498 day of discovery was the saint's feast day

Government type: parliamentary democracy under a constitutional monarchy; a Commonwealth realm

Capital: *name:* Kingstown

geographic coordinates: 13 08 N, 61 13 W

time difference: UTC-4 (1 hour ahead of Washington, DC, during Standard Time)
etymology: an earlier French settlement was renamed Kingstown by the British in 1763 when they assumed control of the island; the king referred to in the name is George III (r. 1760-1820)

Administrative divisions: 6 parishes; Charlotte, Grenadines, Saint Andrew, Saint David, Saint George, Saint Patrick

Independence: 27 October 1979 (from the UK)

National holiday: Independence Day, 27 October (1979)

Constitution: *history:* previous 1969, 1975; latest drafted 26 July 1979, effective 27 October 1979 (The Saint Vincent Constitution Order 1979)
amendments: proposed by the House of Assembly; passage requires at least two-thirds majority vote of the Assembly membership and assent of the governor general; passage of amendments to constitutional sections on fundamental rights and freedoms, citizen protections, various government functions and authorities, and constitutional amendment procedures requires approval by the Assembly membership, approval in a referendum of at least two thirds of the votes cast, and assent of the governor general

Legal system: English common law

International law organization participation: has not submitted an ICJ jurisdiction declaration; accepts ICCt jurisdiction

Citizenship: *citizenship by birth:* yes
citizenship by descent only: at least one parent must be a citizen of Saint Vincent and the Grenadines
dual citizenship recognized: yes
residency requirement for naturalization: 7 years

Suffrage: 18 years of age; universal

Executive branch: *chief of state:* Queen ELIZABETH II (since 6 February 1952); represented by Governor General Susan DOUGAN (since 1 August 2019)
head of government: Prime Minister Ralph E. GONSALVES (since 29 March 2001)
cabinet: Cabinet appointed by the governor general on the advice of the prime minister
elections/appointments: the monarchy is hereditary; governor general appointed by the monarch; following legislative elections, the leader of the majority party usually appointed prime minister by the governor general; deputy prime minister appointed by the governor general on the advice of the prime minister

Legislative branch: *description:* unicameral House of Assembly (23 seats; 15 representatives directly elected in single-seat constituencies by simple majority vote, 6 senators appointed by the governor general, and 2 ex officio members—the speaker of the house and the attorney general; members serve 5-year terms)
elections: last held on 5 November 2020 (next to be held in 2025)
election results: percent of vote by party—ULP 49.58%, NDP 50.34%, other 0.8%; seats by party—ULP 9, NDP 6

Judicial branch: *highest courts:* the Eastern Caribbean Supreme Court (ECSC) is the superior court of the Organization of Eastern Caribbean States; the ECSC—headquartered on St. Lucia—consists of the Court of Appeal—headed by the chief justice and 4 judges—and the High Court with 18 judges; the Court of Appeal is itinerant, traveling to member states on a schedule to hear appeals from the High Court and subordinate courts; High Court judges reside in the member states, with 2 assigned to Saint Vincent and the Grenadines; note—Saint Vincent and the Grenadines is also a member of the Caribbean Court of Justice

judge selection and term of office: chief justice of Eastern Caribbean Supreme Court appointed by Her Majesty, Queen ELIZABETH II; other justices and judges appointed by the Judicial and Legal Services Commission, an independent body of judicial officials; Court of Appeal justices appointed for life with mandatory retirement at age 65; High Court judges appointed for life with mandatory retirement at age 62

subordinate courts: magistrates' courts

Political parties and leaders: Democratic Republican Party or DRP [Anesia BAPTISTE] New Democratic Party or NDP [Godwin L. FRIDAY] Unity Labor Party or ULP [Dr. Ralph GONSALVES] (formed in 1994 by the coalition of Saint Vincent Labor Party or SVLP and the Movement for National Unity or MNU) SVG Green Party or SVGP [Ivan O'NEAL]

International organization participation: ACP, AOSIS, C, Caricom, CDB, CELAC, FAO, G-77, IBRD, ICAO, ICCt, ICRM, IDA, IFAD, IFRCS, ILO, IMF, IMO, Interpol, IOC, IOM, ISO (subscriber), ITU, MIGA, NAM, OAS, OECS, OPANAL, OPCW, Petrocaribe, UN, UNCTAD, UNESCO, UNIDO, UPU, WFTU (NGOs), WHO, WIPO, WTO

Diplomatic representation in the US: *chief of mission:* Ambassador Lou-Anne Gaylene GILCHRIST (since 18 January 2017)
chancery: 1627 K Street, NW, Suite 1202, Washington, DC 20006
telephone: [1] (202) 364-6730
FAX: [1] (202) 364-6730
consulate(s) general: New York

Diplomatic representation from the US: the US does not have an embassy in Saint Vincent and the Grenadines; the US Ambassador to Barbados is accredited to Saint Vincent and the Grenadines

Flag description: three vertical bands of blue (hoist side), gold (double width), and green; the gold band bears three green diamonds arranged in a V pattern, which stands for Vincent; the diamonds recall the islands as "the Gems of the Antilles" and are set slightly lowered in the gold band to reflect the nation's position in the Antilles; blue conveys the colors of a tropical sky and crystal waters, yellow signifies the golden Grenadine sands, and green represents lush vegetation

National symbol(s): *Saint Vincent parrot; national colors:* blue, gold, green

National anthem: *name:* St. Vincent! Land So Beautiful!

lyrics/music: Phyllis Joyce MCCLEAN PUNNETT/Joel Bertram MIGUEL
note: adopted 1967

ECONOMY

Economy—overview: Success of the economy hinges upon seasonal variations in agriculture, tourism, and construction activity, as well as remittances. Much of the workforce is employed in banana production and tourism. Saint Vincent and the Grenadines is home to a small offshore banking sector and continues to fully adopt international regulatory standards.

This lower-middle-income country remains vulnerable to natural and external shocks. The economy has shown some signs of recovery due to increased tourist arrivals, falling oil prices and renewed growth in the construction sector. The much anticipated international airport opened in early 2017 with hopes for increased airlift and tourism activity. The government's ability to invest in social programs and respond to external shocks is constrained by its high public debt burden, which was 67% of GDP at the end of 2013.

GDP (purchasing power parity): $1.265 billion (2017 est.)
$1.256 billion (2016 est.)
$1.246 billion (2015 est.)
note: data are in 2017 dollars
country comparison to the world: 202

GDP (official exchange rate): $785 million (2017 est.)

GDP—real growth rate: 0.7% (2017 est.)
0.8% (2016 est.)
0.8% (2015 est.)
country comparison to the world: 182

GDP—per capita (PPP): $11,500 (2017 est.)
$11,400 (2016 est.)
$11,300 (2015 est.)
note: data are in 2017 dollars
country comparison to the world: 133

Gross national saving: 12.1% of GDP (2017 est.)
10.3% of GDP (2016 est.)
10.4% of GDP (2015 est.)
country comparison to the world: 148

GDP—composition, by end use:
household consumption: 87.3% (2017 est.)
government consumption: 16.6% (2017 est.)
investment in fixed capital: 10.8% (2017 est.)
investment in inventories: -0.2% (2017 est.)
exports of goods and services: 37.1% (2017 est.)
imports of goods and services: -51.7% (2017 est.)

GDP—composition, by sector of origin:
agriculture: 7.1% (2017 est.)
industry: 17.4% (2017 est.)
services: 75.5% (2017 est.)

Agriculture—products: bananas, coconuts, sweet potatoes, spices; small numbers of cattle, sheep, pigs, goats; fish

Industries: tourism; food processing, cement, furniture, clothing, starch

Industrial production growth rate: 2.5% (2017 est.)
country comparison to the world: 118

Labor force: 57,520 (2007 est.)
country comparison to the world: 188

Labor force—by occupation: *agriculture:* 26%
industry: 17%
services: 57% (1980 est.)

Unemployment rate: 18.8% (2008 est.)
country comparison to the world: 184

Population below poverty line: NA

Household income or consumption by percentage share: *lowest 10%:* NA
highest 10%: NA

Budget: *revenues:* 225.2 million (2017 est.)
expenditures: 230 million (2017 est.)

Taxes and other revenues: 28.7% (of GDP) (2017 est.)
country comparison to the world: 93

Budget surplus (+) or deficit (-): -0.6% (of GDP) (2017 est.)
country comparison to the world: 66

Public debt: 73.8% of GDP (2017 est.)
82.8% of GDP (2016 est.)
country comparison to the world: 42

Fiscal year: calendar year

Inflation rate (consumer prices): 2.2% (2017 est.)
-0.2% (2016 est.)
country comparison to the world: 115

Current account balance: -$116 million (2017 est.)
-$122 million (2016 est.)
country comparison to the world: 90

Exports: $48.6 million (2017 est.)
$47.3 million (2016 est.)
country comparison to the world: 203

Exports—partners: Jordan 40.7%, France 12.5%, Barbados 7%, St. Lucia 6.8%, Antigua and Barbuda 5.7%, US 5.5%, Trinidad and Tobago 4.7% (2017)

Exports—commodities: bananas, eddoes and dasheen (taro), arrowroot starch; tennis racquets

Imports: $295.9 million (2017 est.)
$294.6 million (2016 est.)
country comparison to the world: 205

Imports—commodities: foodstuffs, machinery and equipment, chemicals and fertilizers, minerals and fuels

Imports—partners: US 36.8%, Trinidad and Tobago 19.1%, UK 7%, China 5.8% (2017)

Reserves of foreign exchange and gold: $182.1 million (31 December 2017 est.)
$192.3 million (31 December 2016 est.)
country comparison to the world: 178

Debt—external: $362.2 million (31 December 2017 est.)
$330.8 million (31 December 2016 est.)
country comparison to the world: 182

Exchange rates: East Caribbean dollars (XCD) per US dollar -
2.7 (2017 est.)
2.7 (2016 est.)
2.7 (2015 est.)
2.7 (2014 est.)
2.7 (2013 est.)

ENERGY

Electricity access: *electrification—total population:* 100% (2020)

Electricity—production: 157 million kWh (2016 est.)
country comparison to the world: 196

Electricity—consumption: 146 million kWh (2016 est.)
country comparison to the world: 198

Electricity—exports: 0 kWh (2016 est.)
country comparison to the world: 191

Electricity—imports: 0 kWh (2016 est.)
country comparison to the world: 193

Electricity—installed generating capacity: 54,000 kW (2016 est.)
country comparison to the world: 189

Electricity—from fossil fuels: 85% of total installed capacity (2016 est.)
country comparison to the world: 72

Electricity—from nuclear fuels: 0% of total installed capacity (2017 est.)
country comparison to the world: 175

Electricity—from hydroelectric plants: 13% of total installed capacity (2017 est.)
country comparison to the world: 110

Electricity—from other renewable sources: 2% of total installed capacity (2017 est.)
country comparison to the world: 143

Crude oil—production: 0 bbl/day (2018 est.)
country comparison to the world: 194

Crude oil—exports: 0 bbl/day (2015 est.)
country comparison to the world: 107

Crude oil—imports: 0 bbl/day (2015 est.)
country comparison to the world: 189

Crude oil—proved reserves: 0 bbl (1 January 2018 est.)
country comparison to the world: 189

Refined petroleum products—production: 0 bbl/day (2015 est.)
country comparison to the world: 196

Refined petroleum products—consumption: 1,620 bbl/day (2016 est.)
country comparison to the world: 198

Refined petroleum products—exports: 0 bbl/day (2015 est.)
country comparison to the world: 197

Refined petroleum products—imports: 1,621 bbl/day (2015 est.)
country comparison to the world: 194

Natural gas—production: 0 cu m (2017 est.)
country comparison to the world: 191

Natural gas—consumption: 0 cu m (2017 est.)
country comparison to the world: 193

Natural gas—exports: 0 cu m (2017 est.)
country comparison to the world: 176

Natural gas—imports: 0 cu m (2017 est.)
country comparison to the world: 182

Natural gas—proved reserves: 0 cu m (1 January 2014 est.)

country comparison to the world: 188

Carbon dioxide emissions from consumption of energy: 226,800 Mt (2017 est.)
country comparison to the world: 197

COMMUNICATIONS

Telephones—fixed lines: *total subscriptions:* 11,889
subscriptions per 100 inhabitants: 11.7 (2019 est.)
country comparison to the world: 187

Telephones—mobile cellular: *total subscriptions:* 94,367
subscriptions per 100 inhabitants: 92.87 (2019 est.)
country comparison to the world: 194

Telecommunication systems: *general assessment:* adequate island-wide, fully automatic telephone system; broadband access; expanded FttP (Fiber to the Home) markets; LTE launches; regulatory development; telecom sector contributes greatly to the overall GDP; telecom sector is a growth area (2020)
domestic: fixed-line teledensity exceeds 12 per 100 persons and mobile-cellular teledensity is about 93 per 100 persons (2019)
international: country code—1-784; landing points for the ECFS, CARCIP and Southern Caribbean Fiber submarine cables providing connectivity to US and Caribbean Islands; connectivity also provided by VHF/UHF radiotelephone from Saint Vincent to Barbados; SHF radiotelephone to Grenada and Saint Lucia; access to Intelsat earth station in Martinique through Saint Lucia (2019)
note: the COVID-19 outbreak is negatively impacting telecommunications production and supply chains globally; consumer spending on telecom devices and services has also slowed due to the pandemic's effect on economies worldwide; overall progress towards improvements in all facets of the telecom industry—mobile, fixed-line, broadband, submarine cable and satellite—has moderated

Broadcast media: St. Vincent and the Grenadines Broadcasting Corporation operates 1 TV station and 5 repeater stations that provide near total coverage to the multi-island state; multi-channel cable TV service available; a partially government-funded national radio service broadcasts on 1 station and has 2 repeater stations; about a dozen privately owned radio stations and repeater stations

Internet country code: .vc

Internet users: *total:* 22,803
percent of population: 22.39% (July 2018 est.)
country comparison to the world: 208

Broadband—fixed subscriptions: *total:* 24,613
subscriptions per 100 inhabitants: 24 (2018 est.)
country comparison to the world: 149

TRANSPORTATION

National air transport system: *number of registered air carriers:* 2 (2020)

inventory of registered aircraft operated by air carriers: 11

Civil aircraft registration country code prefix: J8 (2016)

Airports: 6 (2013)
country comparison to the world: 176

Airports—with paved runways: *total:* 5 (2017)
1,524 to 2,437 m: 1 (2017)
914 to 1,523 m: 3 (2017)
under 914 m: 1 (2017)

Airports—with unpaved runways: *total:* 1 (2013)
under 914 m: 1 (2013)

Merchant marine: *total:* 810
by type: bulk carrier 19, container ship 12, general cargo 172, oil tanker 16, other 591 (2019)
country comparison to the world: 28

Ports and terminals: *major seaport(s):* Kingstown

MILITARY AND SECURITY

Military and security forces: no regular military forces; the Special Services Unit (SSU) is the paramilitary arm of the Royal Saint Vincent and the Grenadines Police Force (RSVPF) (2019)

TRANSNATIONAL ISSUES

Disputes—international: joins other Caribbean states to counter Venezuela's claim that Aves Island sustains human habitation, a criterion under UN Convention on the Law of the Sea, which permits Venezuela to extend its EEZ/continental shelf over a large portion of the eastern Caribbean Sea

Trafficking in persons: *current situation:* Saint Vincent and the Grenadines is a source, transit, and destination country for men, women, and children subjected to forced labor and sex trafficking; some children under 18 are pressured to engage in sex acts in exchange for money or gifts; foreign workers may experience forced labor and are particularly vulnerable when employed by small, foreign-owned companies; adults and children are vulnerable to forced labor domestically, especially in the agriculture sector
tier rating: Tier 2 Watch List – Saint Vincent and the Grenadines does not fully comply with the minimum standards for the elimination of trafficking; however, it is making significant efforts to do so; the government for the first time acknowledged a trafficking problem, launched an anti-trafficking public awareness campaign, and conducted antitrafficking training for law enforcement, immigration, and labor officials; in 2014, authorities initiated three trafficking investigations, two of which were ultimately determined not to be trafficking cases, and did not prosecute or convict any trafficking offenders; the government did not identify or refer any potential trafficking victims to care (2015)

Illicit drugs: transshipment point for South American drugs destined for the US and Europe; small-scale cannabis cultivation

SAMOA

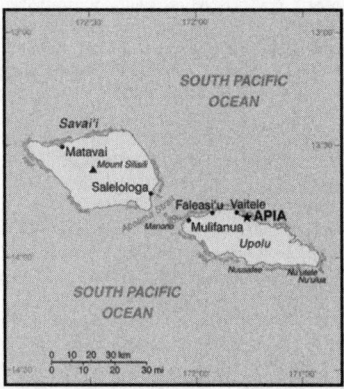

INTRODUCTION

Background: New Zealand occupied the German protectorate of Western Samoa at the outbreak of World War I in 1914. It continued to administer the islands as a mandate and then as a trust territory until 1962, when the islands became the first Polynesian nation to reestablish independence in the 20th century. The country dropped the "Western" from its name in 1997.

In the late 2000s, Samoa began making efforts to more closely align with Australia and New Zealand. In 2009, Samoa changed its driving orientation to the left side of the road, in line with other Commonwealth countries. In 2011, Samoa jumped forward one day—skipping December 30—by moving to the west of the International Date Line so that it was one hour ahead of New Zealand and three hours ahead of the east coast of Australia, rather than 23 and 21 hours behind, respectively.

GEOGRAPHY

Location: Oceania, group of islands in the South Pacific Ocean, about halfway between Hawaii and New Zealand

Geographic coordinates: 13 35 S, 172 20 W

Map references: Oceania

Area: *total:* 2,831 sq km
land: 2,821 sq km
water: 10 sq km
country comparison to the world: 178

Area—comparative: slightly smaller than Rhode Island

Land boundaries: 0 km

Coastline: 403 km

Maritime claims: *territorial sea:* 12 nm
exclusive economic zone: 200 nm
contiguous zone: 24 nm

Climate: tropical; rainy season (November to April), dry season (May to October)

Terrain: two main islands (Savaii, Upolu) and several smaller islands and uninhabited islets; narrow coastal plain with volcanic, rugged mountains in interior

Elevation: *lowest point:* Pacific Ocean 0 m
highest point: Mount Silisili 1,857 m

Natural resources: hardwood forests, fish, hydropower

Land use: *agricultural land:* 12.4% (2011 est.)
arable land: 2.8% (2011 est.) / permanent crops: 7.8% (2011 est.) / permanent pasture: 1.8% (2011 est.)
forest: 60.4% (2011 est.)
other: 27.2% (2011 est.)
Irrigated land: 0 sq km (2012)

Population distribution: about three-quarters of the population lives on the island of Upolu

Natural hazards: occasional cyclones; active volcanism
volcanism: Savai'I Island (1,858 m), which last erupted in 1911, is historically active

Environment—current issues: soil erosion, deforestation, invasive species, overfishing

Environment—international agreements: *party to:* Biodiversity, Climate Change, Climate Change-Kyoto Protocol, Desertification, Hazardous Wastes, Law of the Sea, Ozone Layer Protection, Ship Pollution, Wetlands
signed, but not ratified: none of the selected agreements

Geography—note: occupies an almost central position within Polynesia

PEOPLE AND SOCIETY

Population: 203,774 (July 2020 est.)
country comparison to the world: 185

Nationality: *noun:* Samoan(s)
adjective: Samoan

Ethnic groups: Samoan 96%, Samoan/New Zealander 2%, other 1.9% (2011 est.)
note: data represent the population by country of citizenship

Languages: Samoan (Polynesian) (official) 91.1%, Somoan/English 6.7%, English (official) 0.5%, other 0.2%, unspecified 1.6% (2006 est.)

Religions: Protestant 54.9% (Congregationalist 29%, Methodist 12.4%, Assembly of God 6.8%, Seventh Day Adventist 4.4%, other Protestant 2.3%), Roman Catholic 18.8%, Mormon 16.9%, Worship Centre 2.8%, other Christian 3.6%, other 2.9% (includes Baha'i, Muslim), none 0.2% (2016 est.)

Age structure: *0-14 years:* 29.31% (male 30,825/female 28,900)
15-24 years: 19.61% (male 20,519/female 19,439)
25-54 years: 37.4% (male 39,011/female 37,200)
55-64 years: 7.5% (male 7,780/female 7,505)

65 years and over: 6.18% (male 5,513/female 7,082) (2020 est.)

Dependency ratios: *total dependency ratio:* 73.3
youth dependency ratio: 64.5
elderly dependency ratio: 8.8
potential support ratio: 11.4 (2020 est.)

Median age: *total:* 25.6 years
male: 25.3 years
female: 26 years (2020 est.)
country comparison to the world: 159

Population growth rate: 0.61% (2020 est.)
country comparison to the world: 149

Birth rate: 19.6 births/1,000 population (2020 est.)
country comparison to the world: 77

Death rate: 5.4 deaths/1,000 population (2020 est.)
country comparison to the world: 187

Net migration rate: -8.1 migrant(s)/1,000 population (2020 est.)
country comparison to the world: 214

Population distribution: about three-quarters of the population lives on the island of Upolu

Urbanization: *urban population:* 17.9% of total population (2020)
rate of urbanization: -0.47% annual rate of change (2015-20 est.)

total population growth rate v. urban population growth rate, 2000-2030: *Major urban areas—population:* 36,000 APIA (capital) (2018)

Sex ratio: *at birth:* 1.05 male(s)/female
0-14 years: 1.07 male(s)/female
15-24 years: 1.06 male(s)/female
25-54 years: 1.05 male(s)/female
55-64 years: 1.04 male(s)/female
65 years and over: 0.78 male(s)/female
total population: 1.04 male(s)/female (2020 est.)

Mother's mean age at first birth: 23.6 years (2009 est.)
note: median age at first birth among women 25-29

Maternal mortality rate: 43 deaths/100,000 live births (2017 est.)
country comparison to the world: 98

Infant mortality rate: *total:* 17 deaths/1,000 live births
male: 20 deaths/1,000 live births
female: 13.9 deaths/1,000 live births (2020 est.)
country comparison to the world: 88

Life expectancy at birth: *total population:* 74.7 years
male: 71.7 years
female: 77.7 years (2020 est.)
country comparison to the world: 128

Total fertility rate: 2.5 children born/woman (2020 est.)
country comparison to the world: 75

Contraceptive prevalence rate: 26.9% (2014)

Drinking water source:

improved:
urban: 100% of population
rural: 96.8% of population
total: 97.4% of population
unimproved:
urban: 0% of population
rural: 3.2% of population
total: 2.6% of population (2017 est.)

Current Health Expenditure: 5.5% (2017)

Physicians density: 0.34 physicians/1,000 population (2016)

Sanitation facility access:
improved:
urban: 98.5% of population
rural: 98.1% of population
total: 98.2% of population
unimproved:
urban: 1.5% of population
rural: 1.9% of population
total: 1.8% of population (2017 est.)

HIV/AIDS—adult prevalence rate: NA

HIV/AIDS—people living with HIV/AIDS: NA

HIV/AIDS—deaths: NA

Major infectious diseases: *degree of risk:* high (2020)
food or waterborne diseases: bacterial diarrhea
vectorborne diseases: malaria

Obesity—adult prevalence rate: 47.3% (2016)
country comparison to the world: 8

Children under the age of 5 years underweight: 3.2% (2014)
country comparison to the world: 97

Education expenditures: 4.1% of GDP (2016)
country comparison to the world: 99

Literacy: *definition:* age 15 and over can read and write
total population: 99.1%
male: 99%
female: 99.2% (2018)

Unemployment, youth ages 15-24: *total:* 31.9%
male: 24.6%
female: 43.4% (2017 est.)
country comparison to the world: 28

GOVERNMENT

Country name: *conventional long form:* Independent State of Samoa
conventional short form: Samoa
local long form: Malo Sa'oloto Tuto'atasi o Samoa
local short form: Samoa
former: Western Samoa
etymology: the meaning of Samoa is disputed; some modern explanations are that the "sa" connotes "sacred" and "moa" indicates "center," so the name can mean "Holy Center"; alternatively, some assertions state that it can mean "place of the sacred moa bird" of Polynesian mythology; the name, however, may go back to Proto Polynesian (PPn) times (before 1000 B.C.); a plausible PPn reconstruction has the first syllable as "sa'a" meaning "tribe or people" and "moa" meaning "deep sea or ocean" to convey the meaning "people of the deep sea"

Government type: parliamentary republic

Capital: *name:* Apia

geographic coordinates: 13 49 S, 171 46 W
time difference: UTC+13 (18 hours ahead of Washington, DC, during Standard Time)
daylight saving time: +1hr, begins last Sunday in September; ends first Sunday in April
etymology: name derives from the native village around which the capital was constructed in the 1850s; the village still exists within the larger modern capital

Administrative divisions: 11 districts; A'ana, Aiga-i-le-Tai, Atua, Fa'asaleleaga, Gaga'emauga, Gagaifomauga, Palauli, Satupa'itea, Tuamasaga, Va'a-o-Fonoti, Vaisigano

Independence: 1 January 1962 (from New Zealand-administered UN trusteeship)

National holiday: Independence Day Celebration, 1 June (1962); note—1 January 1962 is the date of independence from the New Zealand- administered UN trusteeship, but it is observed in June

Constitution: *history:* several previous (preindependence); latest 1 January 1962
amendments: proposed as an act by the Legislative Assembly; passage requires at least two-thirds majority vote by the Assembly membership in the third reading—provided at least 90 days have elapsed since the second reading, and assent of the chief of state; passage of amendments affecting constitutional articles on customary land or constitutional amendment procedures also requires at least two-thirds majority approval in a referendum; amended several times, last in 2015

Legal system: mixed legal system of English common law and customary law; judicial review of legislative acts with respect to fundamental rights of the citizen

International law organization participation: has not submitted an ICJ jurisdiction declaration; accepts ICCt jurisdiction

Citizenship: *citizenship by birth:* no
citizenship by descent only: at least one parent must be a citizen of Samoa
dual citizenship recognized: no
residency requirement for naturalization: 5 years

Suffrage: 21 years of age; universal

Executive branch: *chief of state:* TUIMALEALI'IFANO Va'aletoa Sualauvi II (since 21 July 2017)
head of government: Prime Minister TUILA'EPA Lupesoliai Sailele Malielegaoi (since 23 November 1998); Deputy Prime Minister FIAME Naomi Mata'afa (since 2016)
cabinet: Cabinet appointed by the chief of state on the prime minister's advice
elections/appointments: chief of state indirectly elected by the Legislative Assembly to serve a 5-year term (2- term limit); election last held on 4 July 2017 (next to be held in 2022); following legislative elections, the leader of the majority party is usually appointed prime minister by the chief of state, approved by the Legislative Assembly

election results: TUIMALEALI'IFANO Va'aletoa Sualauvi unanimously elected by the Legislative Assembly on 5 July 2017

Legislative branch: *description:* unicameral Legislative Assembly or Fono (50 seats for 2016-2021 term); members from 49 single-seat constituencies directly elected by simple majority vote and 1 seat for a woman, added for the 2016 election to meet the mandated 10% representation of women in the Assembly; members serve 5-year terms)
elections: election last held on 4 March 2016 (next election to be held no later than March 2021)
election results: percent of vote by party—HRPP 89.8%, Tautua Samoa 4.1%, independent 6.1%; seats by party – initial election results—HRPP 44, Tautua Samoa 2, independents 3; post-election party affiliation – HRPP 47, (informal) opposition 3; composition—men 45, women 5, percent of women 10%

Judicial branch: *highest courts:* Court of Appeal (consists of the chief justice and 2 Supreme Court judges and meets once or twice a year); Supreme Court (consists of the chief justice and several judges)
judge selection and term of office: chief justice appointed by the chief of state upon the advice of the prime minister; other Supreme Court judges appointed by the Judicial Service Commission, a 3-member body chaired by the chief justice and includes the attorney general and an appointee of the Minister of Justice; judges normally serve until retirement at age 68
subordinate courts: District Court; Magistrates' Courts; Land and Titles Courts; village fono or village chief councils

Political parties and leaders: Human Rights Protection Party or HRPP [TUILA'EPA Sailele Malielegaoi]

International organization participation: ACP, ADB, AOSIS, C, FAO, G-77, IBRD, ICAO, ICCt, ICRM, IDA, IFAD, IFC, IFRCS, ILO, IMF, IMO, Interpol, IOC, IPU, ITU, ITUC (NGOs), MIGA, OPCW, PIF, Sparteca, SPC, UN, UNCTAD, UNESCO, UNIDO, UPU, WCO, WHO, WIPO, WMO, WTO

Diplomatic representation in the US: *chief of mission:* Ambassador Aliioaiga Feturi ELISAIA (since 4 December 2003)
chancery: 800 Second Avenue, Suite 400J, New York, NY 10017
telephone: [1] (212) 599-6196 through 6197
FAX: [1] (212) 599-0797
consulate(s) general: Pago Pago (American Samoa)

Diplomatic representation from the US: *chief of mission:* the US Ambassador to New Zealand is accredited to Samoa
telephone: [685] 21-631 (2018)
embassy: Accident Corporation Building, 5th Floor, Matafele, Apia
mailing address: P. O. Box 3430, Matafele, Apia
FAX: [685] 22-030 (2018)

Flag description: red with a blue rectangle in the upper hoist-side quadrant bearing five white, five-pointed stars representing the Southern Cross constellation; red stands for courage, blue represents freedom, and white signifies purity
note: similar to the flag of Taiwan

National symbol(s): *Southern Cross constellation (five, five-pointed stars); national colors:* red, white, blue

National anthem: *name:* "O le Fu'a o le Sa'olotoga o Samoa" (The Banner of Freedom)
lyrics/music: Sauni Liga KURESA
note: adopted 1962; also known as "Samoa Tula'i" (Samoa Arise)

ECONOMY

Economy—overview: The economy of Samoa has traditionally been dependent on development aid, family remittances from overseas, tourism, agriculture, and fishing. It has a nominal GDP of $844 million. Agriculture, including fishing, furnishes 90% of exports, featuring fish, coconut oil, nonu products, and taro. The manufacturing sector mainly processes agricultural products. Industry accounts for nearly 22% of GDP while employing less than 6% of the work force. The service sector accounts for nearly two-thirds of GDP and employs approximately 50% of the labor force. Tourism is an expanding sector accounting for 25% of GDP; 132,000 tourists visited the islands in 2013.

The country is vulnerable to devastating storms. In September 2009, an earthquake and the resulting tsunami severely damaged Samoa and nearby American Samoa, disrupting transportation and power generation, and resulting in about 200 deaths. In December 2012, extensive flooding and wind damage from Tropical Cyclone Evan killed four people, displaced over 6,000, and damaged or destroyed an estimated 1,500 homes on Samoa's Upolu Island.

The Samoan Government has called for deregulation of the country's financial sector, encouragement of investment, and continued fiscal discipline, while at the same time protecting the environment. Foreign reserves are relatively healthy and

inflation is low, but external debt is approximately 45% of GDP. Samoa became the 155th member of the WTO in May 2012, and graduated from least developed country status in January 2014.

GDP (purchasing power parity): $1.137 billion (2017 est.)
$1.11 billion (2016 est.)
$1.036 billion (2015 est.)
note: data are in 2017 dollars
country comparison to the world: 204

GDP (official exchange rate): $841 million (2017 est.)

GDP—real growth rate: 2.5% (2017 est.)
7.1% (2016 est.)
1.6% (2015 est.)
country comparison to the world: 114

GDP—per capita (PPP): $5,700 (2017 est.)

$5,700 (2016 est.)
$5,300 (2015 est.)
note: data are in 2017 dollars
country comparison to the world: 169

GDP—composition, by end use:
household consumption: NA
government consumption: NA
investment in fixed capital: NA
investment in inventories: NA
exports of goods and services: 27.2% (2015 est.)
imports of goods and services: -50.5% (2015 est.)

GDP—composition, by sector of origin:
agriculture: 10.4% (2017 est.)
industry: 23.6% (2017 est.)
services: 66% (2017 est.)

Agriculture—products: coconuts, nonu, bananas, taro, yams, coffee, cocoa

Industries: food processing, building materials, auto parts

Industrial production growth rate: -1.8% (2017 est.)
country comparison to the world: 180

Labor force: 50,700 (2016 est.)
country comparison to the world: 193

Labor force—by occupation: *agriculture:* 65%
industry: 6%
services: 29% (2015 est.)

Unemployment rate: 5.2% (2017 est.)
5.5% (2016 est.)

NA
country comparison to the world: 83

Population below poverty line: NA

Household income or consumption by percentage share: *lowest 10%:* NA
highest 10%: NA

Budget: *revenues:* 237.3 million (2017 est.)
expenditures: 276.8 million (2017 est.)

Taxes and other revenues: 28.2% (of GDP) (2017 est.)
country comparison to the world: 95

Budget surplus (+) or deficit (-): -4.7% (of GDP) (2017 est.)
country comparison to the world: 166

Public debt: 49.1% of GDP (2017 est.)
52.6% of GDP (2016 est.)
country comparison to the world: 103

Fiscal year: June 1—May 31

Inflation rate (consumer prices): 1.3% (2017 est.)
0.1% (2016 est.)
country comparison to the world: 72

Current account balance: -$19 million (2017 est.)
-$37 million (2016 est.)
country comparison to the world: 72

Exports: $27.5 million (2014 est.)
country comparison to the world: 207

Exports—partners: Australia 22.9%, NZ 22.8%, American Samoa 22.1%, Afghanistan 14.9%, US 5.9% (2017)

Exports—commodities: fish, coconut oil and cream, nonu, copra, taro, automotive parts, garments, beer

Imports: $89.29 billion (2018 est.)
$312.6 million (2016 est.)
country comparison to the world: 41

Imports—commodities: machinery and equipment, industrial supplies, foodstuffs

Imports—partners: NZ 22%, Singapore 20.7%, US 12.5%, China 10.1%, Australia 8.6%, Fiji 5.2% (2017)

Reserves of foreign exchange and gold: $133 million (31 December 2017 est.)
$122.5 million (31 December 2015 est.)
country comparison to the world: 180

Debt—external: $447.2 million (31 December 2013 est.)
country comparison to the world: 179

Exchange rates: tala (SAT) per US dollar -
2.566 (2017 est.)
2.565 (2016 est.)
2.565 (2015 est.)
2.5609 (2014 est.)
2.3318 (2013 est.)

ENERGY

Electricity access: *electrification—total population:* 100% (2020)

Electricity—production: 132 million kWh (2016 est.)
country comparison to the world: 197

Electricity—consumption: 122.8 million kWh (2016 est.)
country comparison to the world: 199

Electricity—exports: 0 kWh (2016 est.)
country comparison to the world: 192

Electricity—imports: 0 kWh (2016 est.)
country comparison to the world: 194

Electricity—installed generating capacity: 45,000 kW (2016 est.)
country comparison to the world: 194

Electricity—from fossil fuels: 48% of total installed capacity (2016 est.)
country comparison to the world: 155

Electricity—from nuclear fuels: 0% of total installed capacity (2017 est.)
country comparison to the world: 176

Electricity—from hydroelectric plants: 23% of total installed capacity (2017 est.)
country comparison to the world: 85

Electricity—from other renewable sources: 29% of total installed capacity (2017 est.)
country comparison to the world: 20

Crude oil—production: 0 bbl/day (2018 est.)
country comparison to the world: 195

Crude oil—exports: 0 bbl/day (2015 est.)
country comparison to the world: 188

Crude oil—imports: 0 bbl/day (2015 est.)
country comparison to the world: 190

Crude oil—proved reserves: 0 bbl (1 January 2018 est.)
country comparison to the world: 190

Refined petroleum products—production: 0 bbl/day (2017 est.)

country comparison to the world: 197

Refined petroleum products—consumption: 2,400 bbl/day (2016 est.)
country comparison to the world: 191

Refined petroleum products—exports: 0 bbl/day (2015 est.)
country comparison to the world: 198

Refined petroleum products—imports: 2,363 bbl/day (2015 est.)
country comparison to the world: 187

Natural gas—production: 0 cu m (2017 est.)
country comparison to the world: 192

Natural gas—consumption: 0 cu m (2017 est.)
country comparison to the world: 194

Natural gas—exports: 0 cu m (2017 est.)
country comparison to the world: 177

Natural gas—imports: 0 cu m (2017 est.)
country comparison to the world: 183

Natural gas—proved reserves: 0 cu m (1 January 2014 est.)
country comparison to the world: 189

Carbon dioxide emissions from consumption of energy: 341,100 Mt (2017 est.)
country comparison to the world: 191

COMMUNICATIONS

Telephones—fixed lines: *total subscriptions:* 8,770
subscriptions per 100 inhabitants: 4.33 (2019 est.)
country comparison to the world: 192

Telephones—mobile cellular: *total subscriptions:* 128,776
subscriptions per 100 inhabitants: 63.58 (2019 est.)
country comparison to the world: 190

Telecommunication systems: *general assessment:* most households have at least one mobile phone; all businesses in the greater Apia area have access to broadband and Wi-Fi, which is reasonably reliable and fast; in rural Upolu and on Savaii Island there is now readily available high-speed Internet and Wi-Fi; due to the establishment of a regulatory infrastructure, liberalization and competition of the mobile market the telecom market has increased coverage and reduced cost; 4G LTE services accessible to about 95% of residents; working to increase speed, reliability and connectivity (2020)
domestic: fixed-line 4 per 100 and mobile-cellular teledensity 64 telephones per 100 persons (2019)
international: country code—685; landing points for the Tui-Samo, Manatua, SAS, and Southern Cross NEXT submarine cables providing connectivity to Samoa, Fiji, Wallis & Futuna, Cook Islands, Niue, French Polynesia, American Samoa, Australia, New Zealand, Kiribati, Los Angeles (US), and Tokelau; satellite earth station—1 Intelsat (Pacific Ocean) (2019)
note: the COVID-19 outbreak is negatively impacting telecommunications production and supply chains globally; consumer spending on telecom devices and services has also slowed due to the pandemic's effect on economies worldwide; overall progress towards improvements in all facets of the telecom industry—mobile, fixed-line, broadband, submarine cable and satellite—has moderated

Broadcast media: state-owned TV station privatized in 2008; 4 privately owned television broadcast stations; about a half-dozen privately owned radio stations and one state-owned radio station; TV and radio broadcasts of several stations from American Samoa are available (2019)

Internet country code: .ws

Internet users: *total:* 67,662
percent of population: 33.61% (July 2018 est.)
country comparison to the world: 190

Broadband—fixed subscriptions: *total:* 1,692

subscriptions per 100 inhabitants: 1 (2017 est.)
country comparison to the world: 187

TRANSPORTATION

National air transport system: *number of registered air carriers:* 1 (2020)
inventory of registered aircraft operated by air carriers: 4
annual passenger traffic on registered air carriers: 137,770 (2018)

Civil aircraft registration country code prefix: 5W (2016)

Airports: 4 (2013)
country comparison to the world: 188

Airports—with paved runways: *total:* 1 (2019)
2,438 to 3,047 m: 1

Airports—with unpaved runways: *total:* 3 (2013)
under 914 m: 3 (2013)

Roadways: *total:* 1,150 km (2018)
country comparison to the world: 181

Merchant marine: *total:* 13
by type: general cargo 5, oil tanker 1, other 7 (2019)
country comparison to the world: 149

Ports and terminals: *major seaport(s):* Apia

MILITARY AND SECURITY

Military and security forces: no regular military forces; Samoa Police Force (2019)

Military—note: Samoa has no formal defense structure or regular armed forces; informal defense ties exist with NZ, which is required to consider any Samoan request for assistance under the 1962 Treaty of Friendship

TRANSNATIONAL ISSUES

Disputes—international: none

SAN MARINO

INTRODUCTION

Background: Geographically the third smallest state in Europe (after the Holy See and Monaco), San Marino also claims to be the world's oldest republic. According to tradition, it was founded by a Christian stonemason named MARINUS in A.D. 301. San Marino's foreign policy is aligned with that of the EU, although it is not a member; social and political trends in the republic track closely with those of its larger neighbor, Italy.

GEOGRAPHY

Location: Southern Europe, an enclave in central Italy

Geographic coordinates: 43 46 N, 12 25 E

Map references: Europe

Area: *total:* 61 sq km
land: 61 sq km
water: 0 sq km
country comparison to the world: 229

Area—comparative: about one-third the size of Washington, DC

Land boundaries: *total:* 37 km
border countries (1): Italy 37 km

Coastline: 0 km (landlocked)

Maritime claims: none (landlocked)

Climate: Mediterranean; mild to cool winters; warm, sunny summers

Terrain: rugged mountains

Elevation: *lowest point:* Torrente Ausa 55 m
highest point: Monte Titano 739 m

Natural resources: building stone

Land use: *agricultural land:* 16.7% (2011 est.) *arable land:* 16.7% (2011 est.) / *permanent crops:* 0% (2011 est.) / *permanent pasture:* 0% (2011 est.)
forest: 0% (2011 est.)
other: 83.3% (2011 est.)
Irrigated land: 0 sq km (2012)
Natural hazards: occasional earthquakes

Environment—current issues: air pollution; urbanization decreasing rural farmlands; water shortage

Environment—international agreements: *party to:* Biodiversity, Climate Change, Desertification, Whaling
signed, but not ratified: Air Pollution

Geography—note: landlocked; an enclave of (completely surrounded by) Italy; smallest independent state in Europe after the Holy See and Monaco; dominated by the Apennine Mountains

PEOPLE AND SOCIETY

Population: 34,232 (July 2020 est.)
country comparison to the world: 216

Nationality: *noun:* Sammarinese (singular and plural)
adjective: Sammarinese

Ethnic groups: Sammarinese, Italian

Languages: Italian

Religions: Roman Catholic

Age structure: *0-14 years:* 14.73% (male 2,662/female 2,379)
15-24 years: 11.64% (male 2,091/female 1,894)
25-54 years: 39.12% (male 6,310/female 7,081)
55-64 years: 14.28% (male 2,367/female 2,520)
65 years and over: 20.24% (male 3,123/female 3,805) (2020 est.)

Median age: *total:* 45.2 years
male: 43.9 years
female: 46.3 years (2020 est.)
country comparison to the world: 10

Population growth rate: 0.65% (2020 est.)
country comparison to the world: 146

Birth rate: 8.8 births/1,000 population (2020 est.)
country comparison to the world: 209

Death rate: 9 deaths/1,000 population (2020 est.)
country comparison to the world: 64

Net migration rate: 6.6 migrant(s)/1,000 population (2020 est.)
country comparison to the world: 17

Urbanization: *urban population:* 97.5% of total population (2020)
rate of urbanization: 0.67% annual rate of change (2015-20 est.)

total population growth rate v. urban population growth rate, 2000-2030:]

Major urban areas—population: 4,000 SAN MARINO (2018)

Sex ratio: *at birth:* 1.09 male(s)/female
0-14 years: 1.12 male(s)/female
15-24 years: 1.1 male(s)/female
25-54 years: 0.89 male(s)/female

55-64 years: 0.94 male(s)/female
65 years and over: 0.82 male(s)/female
total population: 0.94 male(s)/female (2020 est.)

Mother's mean age at first birth: 32 years (2017)

Infant mortality rate: *total:* 4.2 deaths/1,000 live births
male: 4.4 deaths/1,000 live births
female: 4 deaths/1,000 live births (2020 est.)
country comparison to the world: 187

Life expectancy at birth: *total population:* 83.5 years
male: 80.9 years
female: 86.3 years (2020 est.)
country comparison to the world: 5

Total fertility rate: 1.52 children born/woman (2020 est.)
country comparison to the world: 199

Drinking water source: *improved:* total: 100% of population
unimproved: *total:* 0% of population (2017 est.)

Current Health Expenditure: 7.4% (2017)

Physicians density: 6.11 physicians/1,000 population (2014)

Hospital bed density: 3.8 beds/1,000 population (2012)

Sanitation facility access: *improved:* total: 100% of population
unimproved: *total:* 0% of population (2017)

HIV/AIDS—adult prevalence rate: NA

HIV/AIDS—people living with HIV/AIDS: NA

HIV/AIDS—deaths: NA

Education expenditures: 3% of GDP (2017)
country comparison to the world: 139

Literacy: *total population:* 99.9%
male: 99.9%

School life expectancy (primary to tertiary education): *total:* 16 years
male: 15 years
female: 16 years (2012)

Unemployment, youth ages 15-24: *total:* 27.4%
male: 21.4%
female: 36% (2016 est.)
country comparison to the world: 42

GOVERNMENT

Country name: *conventional long form:* Republic of San Marino
conventional short form: San Marino
local long form: Repubblica di San Marino
local short form: San Marino
etymology: named after Saint MARINUS, who in A.D. 301 founded the monastic settlement around which the city and later the state of San Marino coalesced

Government type: parliamentary republic

Capital: *name:* San Marino (city)

geographic coordinates: 43 56 N, 12 25 E
time difference: UTC+ 1 (6 hours ahead of Washington, DC, during Standard Time)

daylight saving time: +1hr, begins last Sunday in March; ends last Sunday in October
etymology: named after Saint MARINUS, who in A.D. 301 founded a monastic settlement around which the city and later the state of San Marino coalesced

Administrative divisions: 9 municipalities (castelli, singular—castello); Acquaviva, Borgo Maggiore, Chiesanuova, Domagnano, Faetano, Fiorentino, Montegiardino, San Marino Citta, Serravalle

Independence: 3 September 301 (traditional founding date)

National holiday: Founding of the Republic (or Feast of Saint Marinus), 3 September (A. D. 301)

Constitution: *history:* San Marino's principal legislative instruments consist of old customs (antiche consuetudini), the Statutory Laws of San Marino (Leges Statutae Sancti Marini), old statutes (antichi statute) from the1600s, Brief Notes on the Constitutional Order and Institutional Organs of the Republic of San Marino (Brevi Cenni sull'Ordinamento Costituzionale e gli Organi Istituzionali della Repubblica di San Marino) and successive legislation, chief among them is the Declaration of the Rights of Citizens and Fundamental Principles of the San Marino Legal Order (Dichiarazione dei Diritti dei Cittadini e dei Principi Fondamentali dell'Ordinamento Sammarinese), approved 8 July 1974; Declaration last amended 2019
amendments: proposed by the Great and General Council; passage requires two-thirds majority Council vote; Council passage by absolute majority vote also requires passage in a referendum; Declaration of Civil Rights amended several times, last in 2019

Legal system: civil law system with Italian civil law influences

International law organization participation: has not submitted an ICJ jurisdiction declaration; accepts ICCt jurisdiction

Citizenship: *citizenship by birth:* no
citizenship by descent only: at least one parent must be a citizen of San Marino
dual citizenship recognized: no
residency requirement for naturalization: 30 years

Suffrage: 18 years of age; universal

Executive branch: *chief of state:* co-chiefs of state Captain Regent Alessandro CARDELLI and Captain Regent Mirko DOLCINI (for the period 1 October 2020—31 March 2021)
head of government: Secretary of State for Foreign and Political Affairs Luca BECCARI (since 8 January 2020)
cabinet: Congress of State elected by the Grand and General Council
elections/appointments: co-chiefs of state (captains regent) indirectly elected by the Grand and General Council for a single 6-month term; election last held in March 2020 (next to be held in September 2020); secretary of state for foreign and political affairs indirectly elected by the Grand and General Council for a single 5-year term; election

831

last held on 28 December 2019 (next to be held by November 2024)

election results: Alessandro MANCINI (PSD) and Grazia ZAFFERANI (RETE Movement) elected captains regent; percent of Grand and General Council vote—NA; Luca BECCARI (PDCS) elected secretary of state for foreign and political affairs; percent of Grand and General Council vote—NA

note: the captains regent preside over meetings of the Grand and General Council and its cabinet (Congress of State), which has 7 other members who are selected by the Grand and General Council; assisting the captains regent are 7 secretaries of state; the secretary of state for Foreign Affairs has some prime ministerial roles

Legislative branch: *description:* unicameral Grand and General Council or Consiglio Grande e Generale (60 seats; members directly elected in single- and multi-seat constituencies by proportional representation vote in 2 rounds if needed; members serve 5-year terms)

elections: last held on 8 December 2019 (next to be held by December 2024)

election results: percent of vote by coalition/party—PDCS 33.3%, Tomorrow in Movement coalition 24.7% (RETE Movement 18.2%, Domani Motus Liberi 6.2%, other 0.3%), Free San Marino 16.5%, We for the Republic 13.1%, Future Republic 10.3%, I Elect for a New Republic 2%; seats by coalition/party—PDCS 21, Tomorrow in Movement coalition 15 (RETE Movement 11, Domani Motus Liberi 4), Free San Marino 10, We for the Republic 8, Future Republic 6; composition—men 42, women 18, percent of women 30%

Judicial branch: *highest courts:* Council of Twelve or Consiglio dei XII (consists of 12 members); note—the College of Guarantors for the Constitutionality and General Norms functions as San Marino's constitutional court

judge selection and term of office: judges elected by the Grand and General Council from among its own to serve 5- year terms

subordinate courts: first instance and first appeal criminal, administrative, and civil courts; Court for the Trust and Trustee Relations; justices of the peace or conciliatory judges

Political parties and leaders: DOMANI—Modus Liberi or DML Free San Marino (Libera)

Future Republic or RF [Mario VENTURINI]

I Elect for a New Republic

Party of Socialists and Democrats or PSD [Paride ANDREOLI]

RETE Movement

Sammarinese Christian Democratic Party (PDCS) [Marco GATTI]

Socialist Party or PS [Alessandro BEVITORI]

Tomorrow in Movement coalition (includes RETE Movement, DML)

We for the Republic

International organization participation: CE, FAO, IAEA, IBRD, ICAO, ICC (NGOs), ICCt, ICRM, IDA, IFRCS, ILO, IMF, IMO, Interpol, IOC, IOM (observer), IPU, ITU, ITUC (NGOs), LAIA (observer), OPCW, OSCE, Schengen

Convention (de facto member), UN, UNCTAD, UNESCO, Union Latina, UNWTO, UPU, WHO, WIPO

Diplomatic representation in the US: *chief of mission:* Ambassador Damiano BELEFFI (since 21 July 2017)

chancery: 327 E 50th Street, New York, NY 10022

Embassy address: 1711 North Street, NW (2nd Floor)

Washington, DC 22036

telephone: [1] (212) 751-1234

[1] (202) 223-2418

[1] (202) 751-1436

FAX: [1] (212) 751-1436

Diplomatic representation from the US: the United States does not have an Embassy in San Marino; the US Ambassador to Italy is accredited to San Marino, and the US Consulate general in Florence maintains day-to-day ties

Flag description: *two equal horizontal bands of white (top) and light blue with the national coat of arms superimposed in the center; the main colors derive from the shield of the coat of arms, which features three white towers on three peaks on a blue field; the towers represent three castles built on San Marino's highest feature, Mount Titano; the coat of arms is flanked by a wreath, below a crown and above a scroll bearing the word LIBERTAS (Liberty); the white and blue colors are also said to stand for peace and liberty respectively*

National symbol(s): *three peaks each displaying a tower; national colors: white, blue*

National anthem: *name:* "Inno Nazionale della Repubblica" (National Anthem of the Republic)

lyrics/music: no lyrics/Federico CONSOLO

note: adopted 1894; the music for the lyric-less anthem is based on a 10th century chorale piece

Economy ::SAN MARINO

Economy—overview: San Marino's economy relies heavily on tourism, banking, and the manufacture and export of ceramics, clothing, fabrics, furniture, paints, spirits, tiles, and wine. The manufacturing and financial sectors account for more than half of San Marino's GDP. The per capita level of output and standard of living are comparable to those of the most prosperous regions of Italy.

San Marino's economy contracted considerably in the years since 2008, largely due to weakened demand from Italy—which accounts for nearly 90% of its export market—and financial sector consolidation. Difficulties in the banking sector, the global economic downturn, and the sizable decline in tax revenues all contributed to negative real GDP growth. The government adopted measures to counter the downturn, including subsidized credit to businesses and is seeking to shift its growth model away from a reliance on bank and tax secrecy. San Marino does not issue public debt securities; when necessary, it finances deficits by drawing down central bank deposits.

The economy benefits from foreign investment due to its relatively low corporate taxes and low taxes on interest earnings. The income tax rate

is also very low, about one-third the average EU level. San Marino continues to work towards harmonizing its fiscal laws with EU and international standards. In September 2009, the OECD removed San Marino from its list of tax havens that have yet to fully adopt global tax standards, and in 2010 San Marino signed Tax Information Exchange Agreements with most major countries. In 2013, the San Marino Government signed a Double Taxation Agreement with Italy, but a referendum on EU membership failed to reach the quorum needed to bring it to a vote.

GDP (purchasing power parity): $2.064 billion (2017 est.)

$2.026 billion (2016 est.)

$1.983 billion (2015 est.)

note: data are in 2017 dollars

country comparison to the world: 195

GDP (official exchange rate): $1.643 billion (2017 est.)

GDP—real growth rate: 1.9% (2017 est.)

2.2% (2016 est.)

0.6% (2015 est.)

country comparison to the world: 143

GDP—per capita (PPP): $59,000 (2017 est.)

$59,600 (2016 est.)

$58,300 (2015 est.)

note: data are in 2017 dollars

country comparison to the world: 20

GDP—composition, by end use:

household consumption: NA (2011 est.)

government consumption: NA (2011 est.)

investment in fixed capital: NA (2011 est.)

investment in inventories: NA (2011 est.)

exports of goods and services: 176.6% (2011)

imports of goods and services: -153.3% (2011)

GDP—composition, by sector of origin:

agriculture: 0.1% (2009)

industry: 39.2% (2009)

services: 60.7% (2009)

Agriculture—products: wheat, grapes, corn, olives; cattle, pigs, horses, beef, cheese, hides

Industries: tourism, banking, textiles, electronics, ceramics, cement, wine

Industrial production growth rate: -1.1% (2012 est.)

country comparison to the world: 178

Labor force: 21,960 (September 2013 est.)

country comparison to the world: 211

Labor force—by occupation: *agriculture:* 0.2%

industry: 33.5%

services: 66.3% (September 2013 est.)

Unemployment rate: 8.1% (2017 est.)

8.6% (2016 est.)

country comparison to the world: 127

Population below poverty line: NA

Household income or consumption by percentage share: *lowest 10%:* NA

highest 10%: NA

Budget: *revenues:* 667.7 million (2011 est.)

expenditures: 715.3 million (2011 est.)

Taxes and other revenues: 40.6% (of GDP) (2011 est.)
country comparison to the world: 36

Budget surplus (+) or deficit (-): -2.9% (of GDP) (2011 est.)
country comparison to the world: 129

Public debt: 24.1% of GDP (2017 est.)
22.5% of GDP (2016 est.)
country comparison to the world: 180

Fiscal year: calendar year

Inflation rate (consumer prices): 1% (2017 est.)
0.6% (2016 est.)
country comparison to the world: 53

Current account balance: $0 (2017 est.)
$0 (2016 est.)
country comparison to the world: 66

Exports: $3.827 billion (2011 est.)
$2.576 billion (2010 est.)
country comparison to the world: 118

Exports—commodities: building stone, lime, wood, chestnuts, wheat, wine, baked goods, hides, ceramics

Imports: $2.551 billion (2011 est.)
$2.132 billion (2010 est.)
country comparison to the world: 156

Imports—commodities: wide variety of consumer manufactures, food, energy

Reserves of foreign exchange and gold: $392 million (2014 est.)
$539.3 million (2013 est.)
country comparison to the world: 161

Debt—external: NA

Exchange rates: euros (EUR) per US dollar –
0.885 (2017 est.)
0.903 (2016 est.)
0.9214 (2015 est.)
0.885 (2014 est.)
0.7634 (2013 est.)

Electricity access: *electrification—total population:* 100% (2020)

Telephones—fixed lines: *total subscriptions:* 16,070
subscriptions per 100 inhabitants: 47.25 (2019 est.)
country comparison to the world: 182

Telephones—mobile cellular: *total subscriptions:* 38,921
subscriptions per 100 inhabitants: 114.44 (2019 est.)
country comparison to the world: 207

Telecommunication systems: *general assessment:* automatic telephone system completely integrated into Italian system (2018)
domestic: fixed-line 47 per 100 and mobile-cellular teledensity 114 telephones per 100 persons (2019)
international: country code—378; connected to Italian international network
note: the COVID-19 outbreak is negatively impacting telecommunications production and supply chains globally; consumer spending on telecom devices and services has also slowed due to the pandemic's effect on economies worldwide; overall progress towards improvements in all facets of the telecom industry—mobile, fixed-line, broadband, submarine cable and satellite—has moderated

Broadcast media: state-owned public broadcaster operates 1 TV station and 3 radio stations; receives radio and TV broadcasts from Italy (2019)

Internet country code: . sm

Internet users: *total:* 20,328
percent of population: 60.18% (July 2018 est.)
country comparison to the world: 209

Broadband—fixed subscriptions: *total:* 12,500
subscriptions per 100 inhabitants: 37 (2017 est.)
country comparison to the world: 166

Civil aircraft registration country code prefix: T7 (2016)

Roadways: *total:* 292 km (2006)
paved: 292 km (2006)
country comparison to the world: 203

Military and security forces: No regular military forces; Voluntary Military Corps (Corpi Militari), which includes a Uniformed Militia (performs ceremonial duties and limited police support functions) and Guard of the Great and General Council (defends the Captains Regent and the Great and General Council, participates in official ceremonies, cooperates with the maintenance of public order on special occasions, and performs guard duties during parliamentary sittings); the Police Corps includes the Gendarmerie, which is responsible for maintaining public order, protecting citizens and their property, and providing assistance during disasters (2019)

Military service age and obligation: 18 is the legal minimum age for voluntary military service; no conscription; government has the authority to call up all San Marino citizens from 16-60 years of age to service in the military (2012)

Military—note: defense is the responsibility of Italy

Disputes—international: none

SAO TOME AND PRINCIPE

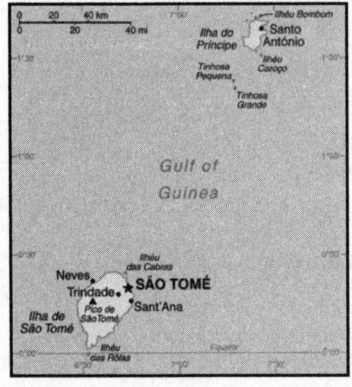

Background: Portugal discovered and colonized the uninhabited islands in the late 15th century, setting up a sugar-based economy that gave way to coffee and cocoa in the 19th century—all grown with African plantation slave labor, a form of which lingered into the 20th century. While independence was achieved in 1975, democratic reforms were not instituted until the late 1980s. The country held its first free elections in 1991, but frequent internal wrangling between the various political parties precipitated repeated changes in leadership and four failed, non-violent coup attempts in 1995, 1998, 2003, and 2009. In 2012, three opposition parties combined in a no confidence vote to bring down the majority government of former Prime Minister Patrice TROVOADA, but in 2014, legislative elections returned him to the office. President Evaristo CARVALHO, of the same political party as Prime Minister TROVOADA, was elected in September 2016, marking a rare instance in which the positions of president and prime minister are held by the same party. Prime Minister TROVOADA resigned at the end of 2018 and was replaced by Jorge BOM JESUS. New oil discoveries in the Gulf of Guinea may attract increased attention to the small island nation.

Location: Central Africa, islands in the Gulf of Guinea, just north of the Equator, west of Gabon

Geographic coordinates: 1 00 N, 7 00 E

Map references: Africa

Area: *total:* 964 sq km
land: 964 sq km
water: 0 sq km
country comparison to the world: 185

Area—comparative: more than five times the size of Washington, DC

Land boundaries: 0 km

Coastline: 209 km

Maritime claims: *territorial sea:* 12 nm
exclusive economic zone: 200 nm

measured from claimed archipelagic baselines

Climate: tropical; hot, humid; one rainy season (October to May)

Terrain: volcanic, mountainous

Elevation: *lowest point:* Atlantic Ocean 0 m
highest point: Pico de Sao Tome 2,024 m

Natural resources: fish, hydropower

Land use: *agricultural land:* 50.7% (2011 est.)
arable land: 9.1% (2011 est.) / *permanent crops:* 40.6% (2011 est.) / *permanent pasture:* 1% (2011 est.)
forest: 28.1% (2011 est.)
other: 21.2% (2011 est.)
Irrigated land: 100 sq km (2012)

Population distribution: Sao Tome, the capital city, has roughly a quarter of the nation's population; Santo Antonio is the largest town on Principe; the northern areas of both islands have the highest population densities as shown in this population distribution map

Natural hazards: flooding

Environment—current issues: deforestation and illegal logging; soil erosion and exhaustion; inadequate sewage treatment in cities; biodiversity preservation

Environment—international agreements: *party to:* Biodiversity, Climate Change, Climate Change-Kyoto Protocol, Desertification, Endangered Species, Environmental Modification, Hazardous Wastes, Law of the Sea, Ozone Layer Protection, Ship Pollution, Wetlands
signed, but not ratified: none of the selected agreements

Geography—note: the second-smallest African country (after the Seychelles); the two main islands form part of a chain of extinct volcanoes and both are mountainous

PEOPLE AND SOCIETY

Population: 211,122 (July 2020 est.)
country comparison to the world: 184

Nationality: *noun:* Sao Tomean(s)
adjective: Sao Tomean

Ethnic groups: mestico, angolares (descendants of Angolan slaves), forros (descendants of freed slaves), servicais (contract laborers from Angola, Mozambique, and Cabo Verde), tongas (children of servicais born on the islands), Europeans (primarily Portuguese), Asians (mostly Chinese)

Languages: Portuguese 98.4% (official), Forro 36.2%, Cabo Verdian 8.5%, French 6.8%, Angolar 6.6%, English 4.9%, Lunguie 1%, other (including sign language) 2.4% (2012 est.)
note: shares sum to more than 100% because some respondents gave more than one answer on the census

Religions: Catholic 55.7%, Adventist 4.1%, Assembly of God 3.4%, New Apostolic 2.9%, Mana 2.3%, Universal Kingdom of God 2%, Jehovah's Witness 1.2%, other 6.2%, none 21.2%, unspecified 1% (2012 est.)

Demographic profile: Sao Tome and Principe's youthful age structure – more than 60% of the population is under the age of 25 – and high fertility rate ensure future population growth. Although Sao Tome has a net negative international migration rate, emigration is not a sufficient safety valve to reduce already high levels of unemployment and poverty. While literacy and primary school attendance have improved in recent years, Sao Tome still struggles to improve its educational quality and to increase its secondary school completion rate. Despite some improvements in education and access to healthcare, Sao Tome and Principe has much to do to decrease its high poverty rate, create jobs, and increase its economic growth.

The population of Sao Tome and Principe descends primarily from the islands' colonial Portuguese settlers, who first arrived in the late 15th century, and the much larger number of African slaves brought in for sugar production and the slave trade. For about 100 years after the abolition of slavery in 1876, the population was further shaped by the widespread use of imported unskilled contract laborers from Portugal's other African colonies, who worked on coffee and cocoa plantations. In the first decades after abolition, most workers were brought from Angola under a system similar to slavery. While Angolan laborers were technically free, they were forced or coerced into long contracts that were automatically renewed and extended to their children. Other contract workers from Mozambique and famine-stricken Cape Verde first arrived in the early 20th century under short-term contracts and had the option of repatriation, although some chose to remain in Sao Tome and Principe.

Today's Sao Tomean population consists of mesticos (creole descendants of the European immigrants and African slaves that first inhabited the islands), forros (descendants of freed African slaves), angolares (descendants of runaway African slaves that formed a community in the south of Sao Tome Island and today are fishermen), servicais (contract laborers from Angola, Mozambique, and Cape Verde), tongas (locally born children of contract laborers), and lesser numbers of Europeans and Asians.

Age structure: *0-14 years:* 39.77% (male 42,690/female 41,277)
15-24 years: 21.59% (male 23,088/female 22,487)
25-54 years: 31.61% (male 32,900/female 33,834)
55-64 years: 4.17% (male 4,095/female 4,700)
65 years and over: 2.87% (male 2,631/female 3,420) (2020 est.)

Dependency ratios: *total dependency ratio:* 81
youth dependency ratio: 75.6
elderly dependency ratio: 5.4
potential support ratio: 18.4 (2020 est.)

Median age: *total:* 19.3 years
male: 18.9 years
female: 19.7 years (2020 est.)
country comparison to the world: 205

Population growth rate: 1.58% (2020 est.)
country comparison to the world: 65

Birth rate: 29.7 births/1,000 population (2020 est.)
country comparison to the world: 33

Death rate: 6.3 deaths/1,000 population (2020 est.)
country comparison to the world: 153

Net migration rate: -7.9 migrant(s)/1,000 population (2020 est.)
country comparison to the world: 213

Population distribution: Sao Tome, the capital city, has roughly a quarter of the nation's population; Santo Antonio is the largest town on Principe; the northern areas of both islands have the highest population densities as shown in this population distribution map

Urbanization: *urban population:* 74.4% of total population (2020)
rate of urbanization: 3.33% annual rate of change (2015-20 est.)

total population growth rate v. urban population growth rate, 2000-2030: *Major urban areas—population:* 80,000 SAO TOME (capital) (2018)

Sex ratio: *at birth:* 1.03 male(s)/female
0-14 years: 1.03 male(s)/female
15-24 years: 1.03 male(s)/female
25-54 years: 0.97 male(s)/female
55-64 years: 0.87 male(s)/female
65 years and over: 0.77 male(s)/female
total population: 1 male(s)/female (2020 est.)

Mother's mean age at first birth: 19.4 years (2008/09 est.)
note: median age at first birth among women 25-29

Maternal mortality rate: 130 deaths/100,000 live births (2017 est.)
country comparison to the world: 62

Infant mortality rate: *total:* 41.7 deaths/1,000 live births
male: 43.6 deaths/1,000 live births
female: 39.8 deaths/1,000 live births (2020 est.)
country comparison to the world: 36

Life expectancy at birth: *total population:* 66.3 years
male: 64.9 years
female: 67.8 years (2020 est.)
country comparison to the world: 187

Total fertility rate: 3.82 children born/woman (2020 est.)
country comparison to the world: 33

Contraceptive prevalence rate: 40.6% (2014)

Drinking water source:
improved:
urban: 100% of population

rural: 88.4% of population
total: 96.8% of population
unimproved:
urban: 0% of population
rural: 11.6% of population
total: 3.2% of population (2017 est.)

Current Health Expenditure: 6.2% (2017)

Physicians density: 0.05 physicians/1,000 population (2017)

Hospital bed density: 2.9 beds/1,000 population (2011)

Sanitation facility access:
improved:
urban: 54.4% of population
rural: 35.3% of population
total: 49.1% of population
unimproved:
urban: 45.6% of population
rural: 64.7% of population
total: 50.9% of population (2017 est.)

HIV/AIDS—adult prevalence rate: 0.7% (2018)
country comparison to the world: 56

HIV/AIDS—people living with HIV/AIDS: 1,100 (2018)
country comparison to the world: 144

HIV/AIDS—deaths: <100 (2018)

Major infectious diseases: *degree of risk:* high (2020)
food or waterborne diseases: bacterial diarrhea, hepatitis A, and typhoid fever
vectorborne diseases: malaria and dengue fever
water contact diseases: schistosomiasis

Obesity—adult prevalence rate: 12.4% (2016)
country comparison to the world: 133

Children under the age of 5 years underweight: 8.8% (2014)
country comparison to the world: 69

Education expenditures: 4.9% of GDP (2017)
country comparison to the world: 68

Literacy: *definition:* age 15 and over can read and write
total population: 92.8%
male: 96.2%
female: 89.5% (2018)

School life expectancy (primary to tertiary education): *total:* 12 years
male: 12 years
female: 13 years (2015)

Unemployment, youth ages 15-24: *total:* 20.8%
male: NA
female: NA (2012 est.)
country comparison to the world: 62

COUNTRY NAME: GOVERNMENT

conventional long form: Democratic Republic of Sao Tome and Principe
conventional short form: Sao Tome and Principe
local long form: Republica Democratica de Sao Tome e Principe
local short form: Sao Tome e Principe
etymology: Sao Tome was named after Saint THOMAS the Apostle by the Portuguese who

discovered the island on 21 December 1470 (or 1471), the saint's feast day; Principe is a shortening of the original Portuguese name of "Ilha do Principe" (Isle of the Prince) referring to the Prince of Portugal to whom duties on the island's sugar crop were paid

Government type: semi-presidential republic

Capital: *name:* Sao Tome
geographic coordinates: 0 20 N, 6 44 E
time difference: UTC 0 (5 hours ahead of Washington, DC, during Standard Time)
etymology: named after Saint Thomas the Apostle

Administrative divisions: 6 districts (distritos, singular—distrito), 1 autonomous region* (regiao autonoma); Agua Grande, Cantagalo, Caue, Lemba, Lobata, Me-Zochi, Principe*

Independence: 12 July 1975 (from Portugal)

National holiday: Independence Day, 12 July (1975)

Constitution: *history:* approved 5 November 1975
amendments: proposed by the National Assembly; passage requires two-thirds majority vote by the Assembly; the Assembly can propose to the president of the republic that an amendment be submitted to a referendum; revised several times, last in 2006

Legal system: mixed legal system of civil law based on the Portuguese model and customary law

International law organization participation: has not submitted an ICJ jurisdiction declaration; non-party state to the ICCt

Citizenship: *citizenship by birth:* no
citizenship by descent only: at least one parent must be a citizen of Sao Tome and Principe
dual citizenship recognized: no
residency requirement for naturalization: 5 years

Suffrage: 18 years of age; universal

Executive branch: *chief of state:* President Evaristo CARVALHO (since 3 September 2016)
head of government: Prime Minister Jorge Bom JESUS (since 3 December 2018)
cabinet: Council of Ministers proposed by the prime minister, appointed by the president
elections/appointments: president directly elected by absolute majority popular vote in 2 rounds if needed for a 5-year term (eligible for a second term); election last held on 7 July 2016 and 7 August 2016 (next to be held in July 2021); prime minister chosen by the National Assembly and approved by the president
election results: Evaristo CARVALHO elected president; percent of vote—Evaristo CARVALHO (ADI) 49.8%, Manuel Pinto DA COSTA (independent) 24.8%, Maria DAS NEVES (MLSTP-PSD) 24.1%; note—first round results for CARVALHO were revised downward from just over 50%, prompting the 7 August runoff; however, on 1 August 2016 DA COSTA withdrew from the runoff, citing voting irregularities, and CARVALHO was declared the winner

Legislative branch: *description:* unicameral National Assembly or Assembleia Nacional (55

seats; members directly elected in multi-seat constituencies by closed party-list proportional representation vote to serve 4-year terms)
elections: last held on 7 October 2018 (next to be held in October 2022)
election results: percent of vote by party—ADI 41.8%, MLSTP/PSD 40.3%, PCD-GR 9.5%, MCISTP 2.1%, other 6.3%; seats by party—ADI 25, MLSTP-PSD 23, PCD-MDFM-UDD 5, MCISTP 2; composition—men 45, women 10, percent of women 18.2%

Judicial branch: *highest courts:* Supreme Court or Supremo Tribunal Justica (consists of 5 judges); Constitutional Court or Tribunal Constitucional (consists of 5 judges, 3 of whom are from the Supreme Court)
judge selection and term of office: Supreme Court judges appointed by the National Assembly; judge tenure NA; Constitutional Court judges nominated by the president and elected by the National Assembly for 5-year terms
subordinate courts: Court of First Instance; Audit Court

Political parties and leaders: Force for Democratic Change Movement or MDFM [Fradique Bandeira Melo DE MENEZES]
Independent Democratic Action or ADI [vacant]
Movement for the Liberation of Sao Tome and Principe-Social Democratic Party or MLSTP-PSD [Aurelio MARTINS]
Party for Democratic Convergence-Reflection Group or PCD-GR [Leonel Mario D'ALVA]
other small parties

International organization participation: ACP, AfDB, AOSIS, AU, CD, CEMAC, CPLP, EITI (candidate country), FAO, G-77, IBRD, ICAO, ICRM, IDA, IFAD, IFC, IFRCS, ILO, IMF, IMO, Interpol, IOC, IOM (observer), IPU, ITU, ITUC (NGOs), MIGA, NAM, OIF, OPCW, PCA, UN, UNCTAD, UNESCO, UNIDO, Union Latina, UNWTO, UPU, WCO, WHO, WIPO, WMO, WTO (observer)

Diplomatic representation in the US: *chief of mission:* Ambassador Carlos Filomeno Azevedo Agostinho das NEVES (since 3 December 2013)
chancery: 675 Third Avenue, Suite 1807, New York, NY 10017
telephone: [1] (212) 651-8116
FAX: [1] (212) 651-8117

Diplomatic representation from the US: the US does not have an embassy in Sao Tome and Principe; the US Ambassador to Gabon is accredited to Sao Tome and Principe

Flag description: three horizontal bands of green (top), yellow (double width), and green with two black five-pointed stars placed side by side in the center of the yellow band and a red isosceles triangle based on the hoist side; green stands for the country's rich vegetation, red recalls the struggle for independence, and yellow represents cocoa, one of the country's main agricultural products; the two stars symbolize the two main islands
note: uses the popular Pan-African colors of Ethiopia

835

National symbol(s): *palm tree; national colors:* green, yellow, red, black

National anthem: *name:* "Independencia total" (Total Independence)
lyrics/music: Alda Neves DA GRACA do Espirito Santo/Manuel dos Santos Barreto de Sousa e ALMEIDA
note: adopted 1975
0:00 / 2:04

ECONOMY

Economy—overview: The economy of São Tomé and Príncipe is small, based mainly on agricultural production, and, since independence in 1975, increasingly dependent on the export of cocoa beans. Cocoa production has substantially declined in recent years because of drought and mismanagement. Sao Tome depends heavily on imports of food, fuels, most manufactured goods, and consumer goods, and changes in commodity prices affect the country's inflation rate. Maintaining control of inflation, fiscal discipline, and increasing flows of foreign direct investment into the nascent oil sector are major economic problems facing the country. In recent years the government has attempted to reduce price controls and subsidies. In 2017, several business-related laws were enacted that aim to improve the business climate.

São Tomé and Príncipe has had difficulty servicing its external debt and has relied heavily on concessional aid and debt rescheduling. In April 2011, the country completed a Threshold Country Program with The Millennium Challenge Corporation to help increase tax revenues, reform customs, and improve the business environment. In 2016, São Tomé and Portugal signed a five-year cooperation agreement worth approximately $64 million, some of which will be provided as loans. In 2017, China and Sao Tomé signed a mutual cooperation agreement in areas such as infrastructure, health, and agriculture worth approximately $146 million over five years.

Considerable potential exists for development of tourism, and the government has taken steps to expand tourist facilities in recent years. Potential also exists for the development of petroleum resources in São Tomé and Principe's territorial waters in the oil-rich Gulf of Guinea, some of which are being jointly developed in a 60-40 split with Nigeria, but production is at least several years off.

Volatile aid and investment inflows have limited growth, and poverty remains high. Restricteded capacity at the main port increases the periodic risk of shortages of consumer goods. Contract enforcement in the country's judicial system is difficult. The IMF in late 2016 expressed concern about vulnerabilities in the country's banking sector, although the country plans some austerity measures in line with IMF recommendations under their three year extended credit facility. Deforestation, coastal erosion, poor waste management, and misuse of natural resources also are challenging issues.

GDP (purchasing power parity): $686 million (2017 est.)
$660.4 million (2016 est.)
$633.9 million (2015 est.)
note: data are in 2017 dollars
country comparison to the world: 208

GDP (official exchange rate): $393 million (2017 est.)

GDP—real growth rate: 3.9% (2017 est.)
4.2% (2016 est.)
3.8% (2015 est.)
country comparison to the world: 76

GDP—per capita (PPP): $3,200 (2017 est.)
$3,200 (2016 est.)
$3,100 (2015 est.)
note: data are in 2017 dollars
country comparison to the world: 191

Gross national saving: 18.7% of GDP (2017 est.)
21% of GDP (2016 est.)
19.3% of GDP (2015 est.)
country comparison to the world: 107

GDP—composition, by end use:
household consumption: 81.4% (2017 est.)
government consumption: 17.6% (2017 est.)
investment in fixed capital: 33.4% (2017 est.)
investment in inventories: 0% (2017 est.)
exports of goods and services: 7.9% (2017 est.)
imports of goods and services: -40.4% (2017 est.)

GDP—composition, by sector of origin:
agriculture: 11.8% (2017 est.)
industry: 14.8% (2017 est.)
services: 73.4% (2017 est.)

Agriculture—products: cocoa, coconuts, palm kernels, copra, cinnamon, pepper, coffee, bananas, papayas, beans; poultry; fish

Industries: light construction, textiles, soap, beer, fish processing, timber

Industrial production growth rate: 5% (2017 est.)
country comparison to the world: 57

Labor force: 72,600 (2017 est.)
country comparison to the world: 186

Labor force—by occupation: *agriculture:* 26.1%
industry: 21.4%
services: 52.5% (2014 est.)

Unemployment rate: 12.2% (2017 est.)
12.6% (2016 est.)
country comparison to the world: 166

Population below poverty line: 66.2% (2009 est.)

Household income or consumption by percentage share: *lowest 10%:* NA
highest 10%: NA

Budget: *revenues:* 103 million (2017 est.)
expenditures: 112.4 million (2017 est.)

Taxes and other revenues: 26.2% (of GDP) (2017 est.)
country comparison to the world: 114

Budget surplus (+) or deficit (-): -2.4% (of GDP) (2017 est.)
country comparison to the world: 113

Public debt: 88.4% of GDP (2017 est.)
93.1% of GDP (2016 est.)
country comparison to the world: 27

Fiscal year: calendar year

Inflation rate (consumer prices): 5.7% (2017 est.)
5.4% (2016 est.)
country comparison to the world: 182

Current account balance: -$32 million (2017 est.)
-$23 million (2016 est.)
country comparison to the world: 76

Exports: $15.6 million (2017 est.)
$9.31 million (2016 est.)
country comparison to the world: 215

Exports—partners: Guyana 43.7%, Germany 23.6%, Portugal 6%, Netherlands 5.5%, Poland 4.4% (2017)

Exports—commodities: cocoa 68%, copra, coffee, palm oil (2010 est.)

Imports: $127.7 million (2017 est.)
$119.1 million (2016 est.)
country comparison to the world: 213

Imports—commodities: machinery and electrical equipment, food products, petroleum products

Imports—partners: Portugal 54.7%, Angola 16.5%, China 5.6% (2017)

Reserves of foreign exchange and gold: $58.95 million (31 December 2017 est.)
$61.5 million (31 December 2016 est.)
country comparison to the world: 185

Debt—external: $292.9 million (31 December 2017 est.)
$308.5 million (31 December 2016 est.)

country comparison to the world: *184 Exchange rates:* dobras (STD) per US dollar -
22,689 (2017 est.)
21,797 (2016 est.)
22,149 (2015 est.)
22,091 (2014 est.)
18,466 (2013 est.)

ENERGY

Electricity access: *electrification—total population:* 71% (2019)
electrification—urban areas: 87% (2019)
electrification—rural areas: 25% (2019)

Electricity—production: 66 million kWh (2016 est.)
country comparison to the world: 203

Electricity—consumption: 61.38 million kWh (2016 est.)
country comparison to the world: 203

Electricity—exports: 0 kWh (2016)
country comparison to the world: 193

Electricity—imports: 0 kWh (2016 est.)
country comparison to the world: 195

Electricity—installed generating capacity: 18,100 kW (2016 est.)
country comparison to the world: 205

Electricity—from fossil fuels: 88% of total installed capacity (2016 est.)
country comparison to the world: 60

Electricity—from nuclear fuels: 0% of total installed capacity (2017 est.)
country comparison to the world: 177

Electricity—from hydroelectric plants: 11% of total installed capacity (2017 est.)
country comparison to the world: 115

Electricity—from other renewable sources: 1% of total installed capacity (2017 est.)
country comparison to the world: 166

Crude oil—production: 0 bbl/day (2018 est.)
country comparison to the world: 196

Crude oil—exports: 0 bbl/day (2015 est.)
country comparison to the world: 189

Crude oil—imports: 0 bbl/day (2015 est.)
country comparison to the world: 191

Crude oil—proved reserves: 0 bbl (1 January 2018)
country comparison to the world: 191

Refined petroleum products—production: 0 bbl/day (2017 est.)
country comparison to the world: 198

Refined petroleum products—consumption: 1,000 bbl/day (2016 est.)
country comparison to the world: 206

Refined petroleum products—exports: 0 bbl/day (2015 est.)
country comparison to the world: 199

Refined petroleum products—imports: 1,027 bbl/day (2015 est.)
country comparison to the world: 202

Natural gas—production: 0 cu m (2017 est.)
country comparison to the world: 193

Natural gas—consumption: 0 cu m (2017 est.)
country comparison to the world: 195

Natural gas—exports: 0 cu m (2017 est.)
country comparison to the world: 178

Natural gas—imports: 0 cu m (2017 est.)
country comparison to the world: 184

Natural gas—proved reserves: 0 cu m (1 January 2014 est.)
country comparison to the world: 190

Carbon dioxide emissions from consumption of energy: 148,100 Mt (2017 est.)

country comparison to the world: 204

COMMUNICATIONS

Telephones—fixed lines: *total subscriptions:* 4,614
subscriptions per 100 inhabitants: 2.22 (2019 est.)
country comparison to the world: 205

Telephones—mobile cellular: *total subscriptions:* 160,189
subscriptions per 100 inhabitants: 77.08 (2019 est.)
country comparison to the world: 188

Telecommunication systems: *general assessment:* local telephone network of adequate quality with most lines connected to digital switches; mobile cellular superior choice to landland; dial-up quality low; broadband expensive (2018)
domestic: fixed-line 2 per 100 and mobile-cellular teledensity 77 telephones per 100 persons (2019)
international: country code—239; landing points for the Ultramar GE and ACE submarine cables from South Africa to over 20 West African countries and Europe; satellite earth station—1 Intelsat (Atlantic Ocean) (2019)
note: the COVID-19 outbreak is negatively impacting telecommunications production and supply chains globally; consumer spending on telecom devices and services has also slowed due to the pandemic's effect on economies worldwide; overall progress towards improvements in all facets of the telecom industry—mobile, fixed-line, broadband, submarine cable and satellite—has moderated

Broadcast media: 1 government-owned TV station; 1 government-owned radio station; 3 independent local radio stations authorized in 2005 with 2 operating at the end of 2006; transmissions of multiple international broadcasters are available

Internet country code: .st

Internet users: *total:* 61,193
percent of population: 29.93% (July 2018 est.)
country comparison to the world: 192

Broadband—fixed subscriptions: *total:* 1,557
subscriptions per 100 inhabitants: 1 (2018 est.)
country comparison to the world: 190

TRANSPORTATION

National air transport system: *number of registered air carriers:* 1 (2020)
inventory of registered aircraft operated by air carriers: 1

Civil aircraft registration country code prefix: S9 (2016)

Airports: 2 (2013)
country comparison to the world: 206

Airports—with paved runways: *total:* 2 (2019)
1,524 to 2,437 m: 1
914 to 1,523 m: 1

Roadways: *total:* 1,300 km (2018)
paved: 230 km (2018)
unpaved: 1,070 km (2018)
country comparison to the world: 177

Merchant marine: *total:* 15
by type: general cargo 11, other 4 (2019)
country comparison to the world: 148

Ports and terminals: *major seaport(s):* Sao Tome

MILITARY AND SECURITY

Military and security forces: *Armed Forces of Sao Tome and Principe (Forcas Armadas de Sao Tome e Principe, FASTP):* Army, Coast Guard of Sao Tome e Principe (Guarda Costeira de Sao Tome e Principe, GCSTP; also called "Navy"), Presidential Guard, National Guard (2019)

Military service age and obligation: 18 is the legal minimum age for compulsory military service; 17 is the legal minimum age for voluntary service (2012)

TRANSNATIONAL ISSUES

Disputes—international: none

SAUDI ARABIA

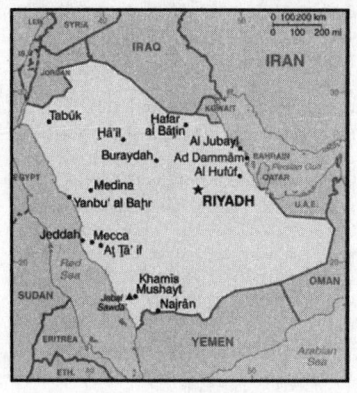

INTRODUCTION

Background: Saudi Arabia is the birthplace of Islam and home to Islam's two holiest shrines in Mecca and Medina. The king's official title is the Custodian of the Two Holy Mosques. The modern Saudi state was founded in 1932 by ABD AL-AZIZ bin Abd al-Rahman Al SAUD (Ibn Saud) after a 30-year campaign to unify most of the Arabian Peninsula. One of his male descendants rules the country today, as required by the country's 1992 Basic Law. Following Iraq's invasion of Kuwait in 1990, Saudi Arabia accepted the Kuwaiti royal family and 400,000 refugees while allowing Western and Arab troops to deploy on its soil for the liberation of Kuwait the following year. The continuing presence of foreign troops on Saudi soil after the liberation of Kuwait became a source of tension between the royal family and the public until all operational US troops left the country in 2003. Major terrorist attacks in May and November 2003 spurred a strong ongoing campaign against domestic terrorism and extremism. US troops returned to the Kingdom in October 2019 after attacks on Saudi oil infrastructure.

From 2005 to 2015, King ABDALLAH bin Abd al-Aziz Al Saud incrementally modernized the Kingdom. Driven by personal ideology and political pragmatism, he introduced a series of social and economic initiatives, including expanding employment and social opportunities for women, attracting foreign investment, increasing the role of the private sector in the economy, and discouraging businesses from hiring foreign workers. These reforms have accelerated under King SALMAN bin Abd al-Aziz, who ascended to the throne in 2015, and has since lifted the Kingdom's ban on women driving and allowed cinemas to operate for the first time in decades. Saudi Arabia saw some protests during the 2011 Arab Spring but not the level of bloodshed seen in protests elsewhere in the region. Shia Muslims in the Eastern Province protested primarily against the detention of political prisoners, endemic discrimination, and Bahraini and Saudi Government actions in Bahrain. Riyadh took a cautious but firm approach by arresting some protesters but releasing most of them quickly and by using its state-sponsored clerics to counter political and Islamist activism.

The government held its first-ever elections in 2005 and 2011, when Saudis went to the polls to elect municipal councilors. In December 2015, women were allowed to vote and stand as candidates for the first time in municipal council elections, with 19 women winning seats. After King SALMAN ascended to the throne in 2015, he placed the first next-generation prince, MUHAMMAD BIN NAYIF bin Abd al-Aziz Al Saud, in the line of succession as Crown Prince. He designated his son, MUHAMMAD BIN SALMAN bin Abd al-Aziz Al Saud, as the Deputy Crown Prince. In March 2015, Saudi Arabia led a coalition of 10 countries in a military campaign to restore the legitimate government of Yemen, which had been ousted by Huthi forces allied with former president ALI ABDULLAH al-Salih. The war in Yemen has drawn international criticism for civilian casualties and its effect on the country's dire humanitarian situation. In December 2015, then Deputy Crown Prince MUHAMMAD BIN SALMAN announced Saudi Arabia would lead a 34-nation Islamic Coalition to fight terrorism (it has since grown to 41 nations). In May 2017, Saudi Arabia inaugurated the Global Center for Combatting Extremist Ideology (also known as "Etidal") as part of its ongoing efforts to counter violent extremism. In June 2017, King SALMAN elevated MUHAMMAD BIN SALMAN to Crown Prince.

The country remains a leading producer of oil and natural gas and holds about 16% of the world's proven oil reserves as of 2015. The government continues to pursue economic reform and diversification, particularly since Saudi Arabia's accession to the WTO in 2005, and promotes foreign investment in the Kingdom. In April 2016, the Saudi Government announced a broad set of socio-economic reforms, known as Vision 2030. Low global oil prices throughout 2015 and 2016 significantly lowered Saudi Arabia's governmental revenue. In response, the government cut subsidies on water, electricity, and gasoline; reduced government employee compensation packages; and announced limited new land taxes. In coordination with OPEC and some key non-OPEC countries, Saudi Arabia agreed cut oil output in early 2017 to regulate supply and help elevate global prices.

GEOGRAPHY

Location: Middle East, bordering the Persian Gulf and the Red Sea, north of Yemen

Geographic coordinates: 25 00 N, 45 00 E

Map references: Middle East

Area: *total:* 2,149,690 sq km
land: 2,149,690 sq km
water: 0 sq km
country comparison to the world: 14

Area—comparative: slightly more than one-fifth the size of the US

Land boundaries: *total:* 4,272 km
border countries (7): Iraq 811 km, Jordan 731 km, Kuwait 221 km, Oman 658 km, Qatar 87 km, UAE 457 km, Yemen 1307 km

Coastline: 2,640 km

Maritime claims: *territorial sea:* 12 nm
contiguous zone: 18 nm
continental shelf: not specified

Climate: harsh, dry desert with great temperature extremes

Terrain: mostly sandy desert

Elevation: *mean elevation:* 665 m
lowest point: Persian Gulf 0 m
highest point: Jabal Sawda' 3,133 m

Natural resources: petroleum, natural gas, iron ore, gold, copper

Land use: *agricultural land:* 80.7% (2011 est.)
arable land: 1.5% (2011 est.) / *permanent crops:* 0.1% (2011 est.) / *permanent pasture:* 79.1% (2011 est.)
forest: 0.5% (2011 est.)
other: 18.8% (2011 est.)
Irrigated land: 16,200 sq km (2012)

Population distribution: historically a population that was mostly nomadic or semi-nomadic, the Saudi population has become more settled since petroleum was discovered in the 1930s; most of the economic activities—and with it the country's population—is concentrated in a wide area across the middle of the peninsula, from Ad Dammam in the east, through Riyadh in the interior, to Mecca-Medina in the west near the Red Sea

Natural hazards: frequent sand and dust storms
volcanism: despite many volcanic formations, there has been little activity in the past few centuries; volcanoes include Harrat Rahat, Harrat Khaybar, Harrat Lunayyir, and Jabal Yar

Environment—current issues: desertification; depletion of underground water resources; the lack of perennial rivers or permanent water bodies has prompted the development of extensive seawater desalination facilities; coastal pollution from oil spills; air pollution; waste management

Environment—international agreements: *party to:* Biodiversity, Climate Change, Climate Change-Kyoto Protocol, Desertification, Endangered Species, Hazardous Wastes, Law of the Sea, Marine Dumping, Ozone Layer Protection, Ship Pollution
signed, but not ratified: none of the selected agreements

Geography—note: Saudi Arabia is the largest country in the world without a river; extensive coastlines on the Persian Gulf and Red Sea allow for considerable shipping (especially of crude oil) through the Persian Gulf and Suez Canal

PEOPLE AND SOCIETY

Population: 34,173,498 (July 2020 est.)
note: immigrants make up 38.3% of the total population, according to UN data (2019)
country comparison to the world: 41

Nationality: *noun:* Saudi(s)
adjective: Saudi or Saudi Arabian

Ethnic groups: Arab 90%, Afro-Asian 10%

Languages: Arabic (official)

Religions: Muslim (official; citizens are 85-90% Sunni and 10-15% Shia), other (includes Eastern Orthodox, Protestant, Roman Catholic, Jewish, Hindu, Buddhist, and Sikh) (2012 est.)
note: despite having a large expatriate community of various faiths (more than 30% of the population), most forms of public religious expression inconsistent with the government-sanctioned interpretation of Sunni Islam are restricted; non-Muslims are not allowed to have Saudi citizenship and non-Muslim places of worship are not permitted (2013)

MENA religious affiliation: *Age structure:* 0-14 years: 24.84% (male 4,327,830/female 4,159,242)
15-24 years: 15.38% (male 2,741,371/female 2,515,188)
25-54 years: 50.2% (male 10,350,028/female 6,804,479)
55-64 years: 5.95% (male 1,254,921/female 778,467)
65 years and over: 3.63% (male 657,393/female 584,577) (2020 est.)

Dependency ratios: *total dependency ratio:* 39.3
youth dependency ratio: 34.4
elderly dependency ratio: 4.9
potential support ratio: 20.5 (2020 est.)

Median age: *total:* 30.8 years
male: 33 years
female: 27.9 years (2020 est.)
country comparison to the world: 119

Population growth rate: 1.6% (2020 est.)
country comparison to the world: 64

Birth rate: 14.7 births/1,000 population (2020 est.)
country comparison to the world: 124

Death rate: 3.4 deaths/1,000 population (2020 est.)
country comparison to the world: 221

Net migration rate: 4.7 migrant(s)/1,000 population (2020 est.)
country comparison to the world: 27

Population distribution: historically a population that was mostly nomadic or semi-nomadic, the Saudi population has become more settled since petroleum was discovered in the 1930s; most of the economic activities—and with it the country's population—is concentrated in a wide area across the middle of the peninsula, from Ad Dammam in the east, through Riyadh in the interior, to Mecca-Medina in the west near the Red Sea

Urbanization: *urban population:* 84.3% of total population (2020)
rate of urbanization: 2.17% annual rate of change (2015-20 est.)

total population growth rate v. urban population growth rate, 2000-2030: *Major urban areas—population:* 7.231 million RIYADH (capital), 4.610 million Jeddah, 2.042 million Mecca, 1.489 million Medina, 1.253 million Ad Dammam (2020)

Sex ratio: *at birth:* 1.05 male(s)/female
0-14 years: 1.04 male(s)/female
15-24 years: 1.09 male(s)/female
25-54 years: 1.52 male(s)/female
55-64 years: 1.61 male(s)/female
65 years and over: 1.12 male(s)/female
total population: 1.3 male/female (2020 est.)

Maternal mortality rate: 17 deaths/100,000 live births (2017 est.)
country comparison to the world: 131

Infant mortality rate: *total:* 11.3 deaths/1,000 live births
male: 12.2 deaths/1,000 live births
female: 10.4 deaths/1,000 live births (2020 est.)
country comparison to the world: 117

Life expectancy at birth: *total population:* 76.2 years
male: 74.6 years
female: 77.8 years (2020 est.)
country comparison to the world: 102

Total fertility rate: 1.95 children born/woman (2020 est.)
country comparison to the world: 121

Contraceptive prevalence rate: 24.6% (2016)

Drinking water source: *improved:* total: 100% of population
unimproved: total: 0% of population (2017 est.)

Current Health Expenditure: 5.2% (2017)

Physicians density: 2.54 physicians/1,000 population (2017)

Hospital bed density: 2.2 beds/1,000 population (2017)

Sanitation facility access: *improved:* total: 100% of population
unimproved: total: 0% of population (2017 est.)

HIV/AIDS—adult prevalence rate: <.1% (2016 est.)

HIV/AIDS—people living with HIV/AIDS: 8,200 (2016 est.)
country comparison to the world: 110

HIV/AIDS—deaths: <500 (2016 est.)

Major infectious diseases: *note:* sporadic cases of a respiratory illness caused by the novel coronavirus (COVID-19) are occurring throughout Saudi Arabia; as of 10 November 2020, Saudi Arabia has reported a total of 350,229 cases of COVID-19 or 10,060 cumulative cases of COVID-19 per 1 million population with 159 cumulative deaths per 1 million population

Obesity—adult prevalence rate: 35.4% (2016)
country comparison to the world: 14

Education expenditures: NA

Literacy: *definition:* age 15 and over can read and write
total population: 95.3%
male: 97.1%
female: 92.7% (2017)

School life expectancy (primary to tertiary education): *total:* 17 years
male: 16 years
female: 16 years (2019)

Unemployment, youth ages 15-24: *total:* 28.8%
male: 19.9%
female: 62.6% (2018 est.)
country comparison to the world: 38

GOVERNMENT

Country name: *conventional long form:* Kingdom of Saudi Arabia
conventional short form: Saudi Arabia
local long form: Al Mamlakah al Arabiyah as Suudiyah
local short form: Al Arabiyah as Suudiyah
etymology: named after the ruling dynasty of the country, the House of Saud; the name "Arabia" can be traced back many centuries B.C., the ancient Egyptians referred to the region as "Ar Rabi"

Government type: absolute monarchy

Capital: *name:* Riyadh
geographic coordinates: 24 39 N, 46 42 E
time difference: UTC+3 (8 hours ahead of Washington, DC, during Standard Time)
etymology: the name derives from the Arabic word «riyadh,» meaning «gardens,» and refers to various oasis towns in the area that merged to form the city

Administrative divisions: 13 regions (manatiq, singular—mintaqah); Al Bahah, Al Hudud ash Shamaliyah (Northern Border), Al Jawf, Al Madinah al Munawwarah (Medina), Al Qasim, Ar Riyad (Riyadh), Ash Sharqiyah (Eastern), 'Asir, Ha'il, Jazan, Makkah al Mukarramah (Mecca), Najran, Tabuk

Independence: 23 September 1932 (unification of the kingdom)

National holiday: Saudi National Day (Unification of the Kingdom), 23 September (1932)

Constitution: *history:* 1 March 1992—Basic Law of Government, issued by royal decree, serves as the constitutional framework and is based on the Qur'an and the life and traditions of the Prophet Muhammad
amendments: proposed by the king directly or proposed to the king by the Consultative Assembly or by the Council of Ministers; passage by the king through royal decree; Basic Law amended many times, last in 2017

Legal system: Islamic (sharia) legal system with some elements of Egyptian, French, and customary law; *note*—several secular codes have been

introduced; commercial disputes handled by special committees

International law organization participation: has not submitted an ICJ jurisdiction declaration; non-party state to the ICCt

Citizenship: *citizenship by birth:* no
citizenship by descent only: the father must be a citizen of Saudi Arabia; a child born out of wedlock in Saudi Arabia to a Saudi mother and unknown father
dual citizenship recognized: no
residency requirement for naturalization: 5 years

Suffrage: 18 years of age; restricted to males; universal for municipal elections

Executive branch: *chief of state:* King and Prime Minister SALMAN bin Abd al-Aziz Al Saud (since 23 January 2015); Crown Prince MUHAMMAD BIN SALMAN bin Abd al-Aziz Al Saud (born 31 August 1985); note—the monarch is both chief of state and head of government
head of government: King and Prime Minister SALMAN bin Abd al-Aziz Al Saud (since 23 January 2015); Crown Prince MUHAMMAD BIN SALMAN bin Abd al-Aziz Al Saud (born 31 August 1985)
cabinet: Council of Ministers appointed by the monarch every 4 years and includes many royal family members
elections/appointments: none; the monarchy is hereditary; an Allegiance Council created by royal decree in October 2006 established a committee of Saudi princes for a voice in selecting future Saudi kings

Legislative branch: *description:* unicameral Consultative Council or Majlis al-Shura (150 seats; members appointed by the monarch to serve 4-year terms); note—in early 2013, the monarch granted women 30 seats on the Council
note: composition as of 2013—men 121, women 30, percent of women 19.9%

Judicial branch: *highest courts:* High Court (consists of the court chief and organized into circuits with 3-judge panels, except for the criminal circuit, which has a 5-judge panel for cases involving major punishments)
judge selection and term of office: High Court chief and chiefs of the High Court Circuits appointed by royal decree upon the recommendation of the Supreme Judiciary Council, a 10-member body of high-level judges and other judicial heads; new judges and assistant judges serve 1- and 2-year probations, respectively, before permanent assignment
subordinate courts: Court of Appeals; Specialized Criminal Court, first-degree courts composed of general, criminal, personal status, and commercial courts; Labor Court; a hierarchy of administrative courts

Political parties and leaders: none

International organization participation: ABEDA, AfDB (nonregional member), AFESD, AMF, BIS, CAEU, CP, FAO, G-20, G-77, GCC, IAEA, IBRD, ICAO, ICC (national committees), ICRM,

IDA, IDB, IFAD, IFC, IFRCS, IHO, ILO, IMF, IMO, IMSO, Interpol, IOC, IOM (observer), IPU, ISO, ITSO, ITU, LAS, MIGA, NAM, OAPEC, OAS (observer), OIC, OPCW, OPEC, PCA, UN, UNCTAD, UNESCO, UNIDO, UNRWA, UNWTO, UPU, WCO, WFTU (NGOs), WHO, WIPO, WMO, WTO

Diplomatic representation in the US: *chief of mission:* Ambassador Princess REEMA bint Bandar Al Saud (since 8 July 2019)
chancery: 601 New Hampshire Avenue NW, Washington, DC 20037
telephone: [1] (202) 342-3800
FAX: [1] (202) 944-5983
consulate(s) general: Houston, Los Angeles, New York

Diplomatic representation from the US: *chief of mission:* Ambassador John P. ABIZAID (since 8 May 2019)
telephone: [966] (11) 488-3800
embassy: P.O. Box 94309, Riyadh 11693
mailing address: *American Embassy, Unit 61307, APO AE 09803-1307; International Mail:* P. O. Box 94309, Riyadh 11693
FAX: [966] (11) 488-7360
consulate(s) general: Dhahran, Jiddah (Jeddah)

Flag description: green, a traditional color in Islamic flags, with the Shahada or Muslim creed in large white Arabic script (translated as "There is no god but God; Muhammad is the Messenger of God") above a white horizontal saber (the tip points to the hoist side); design dates to the early twentieth century and is closely associated with the Al Saud family, which established the kingdom in 1932; the flag is manufactured with differing obverse and reverse sides so that the Shahada reads—and the sword points—correctly from right to left on both sides
note: the only national flag to display an inscription as its principal design; one of only three national flags that differ on their obverse and reverse sides—the others are Moldova and Paraguay

National symbol(s): *palm tree surmounting two crossed swords; national colors: green, white*

National anthem: *name:* "Aash Al Maleek" (Long Live Our Beloved King)
lyrics/music: Ibrahim KHAFAJI/Abdul Rahman al-KHATEEB
note: music adopted 1947, lyrics adopted 1984
0:00 / 0:35

ECONOMY

Economy—overview: Saudi Arabia has an oil-based economy with strong government controls over major economic activities. It possesses about 16% of the world's proven petroleum reserves, ranks as the largest exporter of petroleum, and plays a leading role in OPEC. The petroleum sector accounts for roughly 87% of budget revenues, 42% of GDP, and 90% of export earnings.

Saudi Arabia is encouraging the growth of the private sector in order to diversify its economy and to employ more Saudi nationals. Approximately 6 million foreign workers play an important role in the Saudi economy, particularly in the oil and service sectors; at the same time, however, Riyadh is struggling to reduce unemployment among its own nationals. Saudi officials are particularly focused on employing its large youth population.

In 2017, the Kingdom incurred a budget deficit estimated at 8.3% of GDP, which was financed by bond sales and drawing down reserves. Although the Kingdom can finance high deficits for several years by drawing down its considerable foreign assets or by borrowing, it has cut capital spending and reduced subsidies on electricity, water, and petroleum products and recently introduced a value-added tax of 5%. In January 2016, Crown Prince and Deputy Prime Minister MUHAMMAD BIN SALMAN announced that Saudi Arabia intends to list shares of its state-owned petroleum company, ARAMCO—another move to increase revenue and outside investment. The government has also looked at privatization and diversification of the economy more closely in the wake of a diminished oil market. Historically, Saudi Arabia has focused diversification efforts on power generation, telecommunications, natural gas exploration, and petrochemical sectors. More recently, the government has approached investors about expanding the role of the private sector in the health care, education and tourism industries. While Saudi Arabia has emphasized their goals of diversification for some time, current low oil prices may force the government to make more drastic changes ahead of their long-run timeline.

GDP (purchasing power parity): $1.775 trillion (2017 est.)
$1.79 trillion (2016 est.)
$1.761 trillion (2015 est.)
note: data are in 2017 dollars
country comparison to the world: 16

GDP (official exchange rate): $686.7 billion (2017 est.)

GDP—real growth rate: -0.9% (2017 est.)
1.7% (2016 est.)
4.1% (2015 est.)
country comparison to the world: 199

GDP—per capita (PPP): $54,500 (2017 est.)
$56,400 (2016 est.)
$56,800 (2015 est.)
note: data are in 2017 dollars
country comparison to the world: 22

Gross national saving: 30.1% of GDP (2017 est.)
27.2% of GDP (2016 est.)
26.5% of GDP (2015 est.)
country comparison to the world: 32

GDP—composition, by end use:
household consumption: 41.3% (2017 est.)
government consumption: 24.5% (2017 est.)
investment in fixed capital: 23.2% (2017 est.)
investment in inventories: 4.7% (2017 est.)
exports of goods and services: 34.8% (2017 est.)
imports of goods and services: -28.6% (2017 est.)

GDP—composition, by sector of origin:
agriculture: 2.6% (2017 est.)
industry: 44.2% (2017 est.)

services: 53.2% (2017 est.)

Agriculture—products: wheat, barley, tomatoes, melons, dates, citrus; mutton, chickens, eggs, milk

Industries: crude oil production, petroleum refining, basic petrochemicals, ammonia, industrial gases, sodium hydroxide (caustic soda), cement, fertilizer, plastics, metals, commercial ship repair, commercial aircraft repair, construction

Industrial production growth rate: -2.4% (2017 est.)
country comparison to the world: 186

Labor force: 13.8 million (2017 est.)
note: comprised of 3.1 million Saudis and 10.7 million non-Saudis
country comparison to the world: 39

Labor force—by occupation: *agriculture:* 6.7%
industry: 21.4%
services: 71.9% (2005 est.)

Unemployment rate: 6% (2017 est.)
5.6% (2016 est.)
note: data are for total population; unemployment among Saudi nationals is more than double
country comparison to the world: 97

Population below poverty line: NA

Household income or consumption by percentage share: *lowest 10%:* NA
highest 10%: NA

Budget: *revenues:* 181 billion (2017 est.)
expenditures: 241.8 billion (2017 est.)

Taxes and other revenues: 26.4% (of GDP) (2017 est.)
country comparison to the world: 113

Budget surplus (+) or deficit (-): -8.9% (of GDP) (2017 est.)
country comparison to the world: 204

Public debt: 17.2% of GDP (2017 est.)
13.1% of GDP (2016 est.)
country comparison to the world: 193

Fiscal year: calendar year

Inflation rate (consumer prices): -0.9% (2017 est.)
2% (2016 est.)
country comparison to the world: 3

Current account balance: $15.23 billion (2017 est.)
-$23.87 billion (2016 est.)
country comparison to the world: 19

Exports: $221.1 billion (2017 est.)
$183.6 billion (2016 est.)
country comparison to the world: 24

Exports—partners: Japan 12.2%, China 11.7%, South Korea 9%, India 8.9%, US 8.3%, UAE 6.7%, Singapore 4.2% (2017)

Exports—commodities: petroleum and petroleum products 90% (2012 est.)

Imports: $119.3 billion (2017 est.)
$127.8 billion (2016 est.)
country comparison to the world: 33

Imports—commodities: machinery and equipment, foodstuffs, chemicals, motor vehicles, textiles

Imports—partners: China 15.4%, US 13.6%, UAE 6.5%, Germany 5.8%, Japan 4.1%, India 4.1%, South Korea 4% (2017)

Reserves of foreign exchange and gold: $496.4 billion (31 December 2017 est.)
$535.8 billion (31 December 2016 est.)
country comparison to the world: 4

Debt—external: $205.1 billion (31 December 2017 est.)
$189.3 billion (31 December 2016 est.)
country comparison to the world: 36

Exchange rates: Saudi riyals (SAR) per US dollar -
3.75 (2017 est.)
3.75 (2016 est.)
3.75 (2015 est.)
3.75 (2014 est.)
3.75 (2013 est.)

ENERGY

Electricity access: *electrification—total population:* 100% (2020)

Electricity—production: 324.1 billion kWh (2016 est.)
country comparison to the world: 11

Electricity—consumption: 296.2 billion kWh (2016 est.)
country comparison to the world: 12

Electricity—exports: 0 kWh (2016 est.)
country comparison to the world: 194

Electricity—imports: 0 kWh (2016 est.)
country comparison to the world: 196

Electricity—installed generating capacity: 82.94 million kW (2016 est.)
country comparison to the world: 14

Electricity—from fossil fuels: 100% of total installed capacity (2016 est.)
country comparison to the world: 17

Electricity—from nuclear fuels: 0% of total installed capacity (2017 est.)
country comparison to the world: 178

Electricity—from hydroelectric plants: 0% of total installed capacity (2017 est.)
country comparison to the world: 198

Electricity—from other renewable sources: 0% of total installed capacity (2017 est.)
country comparison to the world: 206

Crude oil—production: 10.425 million bbl/day (2018 est.)
country comparison to the world: 3

Crude oil—exports: 7.341 million bbl/day (2015 est.)
country comparison to the world: 1

Crude oil—imports: 0 bbl/day (2015 est.)
country comparison to the world: 192

Crude oil—proved reserves: 266.2 billion bbl (1 January 2018 est.)
country comparison to the world: 2

Refined petroleum products—production: 2.476 million bbl/day (2015 est.)
country comparison to the world: 8

Refined petroleum products—consumption: 3.287 million bbl/day (2016 est.)
country comparison to the world: 6

Refined petroleum products—exports: 1.784 million bbl/day (2015 est.)
country comparison to the world: 5

Refined petroleum products—imports: 609,600 bbl/day (2015 est.)
country comparison to the world: 13

Natural gas—production: 109.3 billion cu m (2017 est.)
country comparison to the world: 8

Natural gas—consumption: 109.3 billion cu m (2017 est.)
country comparison to the world: 7

Natural gas—exports: 0 cu m (2017 est.)
country comparison to the world: 179

Natural gas—imports: 0 cu m (2017 est.)
country comparison to the world: 185

Natural gas—proved reserves: 8.619 trillion cu m (1 January 2018 est.)
country comparison to the world: 4

Carbon dioxide emissions from consumption of energy: 657.1 million Mt (2017 est.)
country comparison to the world: 8

COMMUNICATIONS

Telephones—fixed lines: *total subscriptions:* 5,276,773
subscriptions per 100 inhabitants: 15.69 (2019 est.)
country comparison to the world: 27

Telephones—mobile cellular: *total subscriptions:* 40,532,610
subscriptions per 100 inhabitants: 120.52 (2019 est.)
country comparison to the world: 37

Telecommunication systems: *general assessment:* one of the most progressive telecom markets in the Middle East; mobile penetration high, with a saturated market; mobile operators competitive and meeting the demand for workers, students and citizens working from home; 5G launched, partners include Chinese company Huawei; broadband is available with DSL, fiber, and wireless; mobile penetration is steep in Saudi Arabia (2020)
domestic: fixed-line 16 per 100 and mobile-cellular subscribership has been increasing rapidly to 121 per 100 persons (2019)
international: country code—966; landing points for the SeaMeWe-3, -4, -5, AAE-1, EIG, FALCON, FEA, IMEWE, MENA/Gulf Bridge International, SEACOM, SAS-1, -2, GBICS/MENA, and the Tata TGN-Gulf submarine cables providing connectivity to Europe, Africa, the Middle East, Asia, Southeast Asia and Australia; microwave radio relay to Bahrain,
    Jordan, Kuwait, Qatar, UAE, Yemen, and Sudan; coaxial cable to Kuwait and Jordan; satellite earth stations—5 Intelsat (3 Atlantic Ocean and 2 Indian Ocean), 1 Arabsat, and 1 Inmarsat (Indian Ocean region) (2019)

note: the COVID-19 outbreak is negatively impacting telecommunications production and supply chains globally; consumer spending on telecom devices and services has also slowed due to the pandemic's effect on economies worldwide; overall progress towards improvements in all facets of the telecom industry—mobile, fixed-line, broadband, submarine cable and satellite—has moderated

Broadcast media: broadcast media are state-controlled; state-run TV operates 4 networks; Saudi Arabia is a major market for pan-Arab satellite TV broadcasters; state-run radio operates several networks; multiple international broadcasters are available

Internet country code: .sa

Internet users: *total:* 30,877,318
percent of population: 93.31% (July 2018 est.)
country comparison to the world: 26

Broadband—fixed subscriptions: *total:* 6,821,873
subscriptions per 100 inhabitants: 21 (2018 est.)
country comparison to the world: 24

Communications—note: the innovative King Abdulaziz Center for World Culture (informally known as Ithra, meaning "enrichment") opened on 1 December 2017 in Dhahran, Eastern Region; its facilities include a grand library, several museums, an archive, an Idea Lab, a theater, a cinema, and an Energy Exhibit, all which are meant to provide visitors an immersive and transformative experience

TRANSPORTATION

National air transport system: *number of registered air carriers:* 12 (2020)
inventory of registered aircraft operated by air carriers: 230
annual passenger traffic on registered air carriers: 39,141,660 (2018)
annual freight traffic on registered air carriers: 1,085,470,000 mt-km (2018)

Civil aircraft registration country code prefix: HZ (2016)

Airports: 214 (2013)
country comparison to the world: 26

Airports—with paved runways: *total:* 82 (2017)
over 3,047 m: 33 (2017)
2,438 to 3,047 m: 16 (2017)
1,524 to 2,437 m: 27 (2017)
914 to 1,523 m: 2 (2017)
under 914 m: 4 (2017)

Airports—with unpaved runways: *total:* 132 (2013)
2,438 to 3,047 m: 7 (2013)
1,524 to 2,437 m: 72 (2013)
914 to 1,523 m: 37 (2013)
under 914 m: 16 (2013)

Heliports: 10 (2013)

Pipelines: 209 km condensate, 2940 km gas, 1183 km liquid petroleum gas, 5117 km oil, 1151 km refined products (2013)

Railways: *total:* 5,410 km (2016)

standard gauge: 5,410 km 1.435-m gauge (with branch lines and sidings) (2016)
country comparison to the world: 36

Roadways: *total:* 221,372 km (2006)
paved: 47,529 km (includes 3,891 km of expressways) (2006)
unpaved: 173,843 km (2006)
country comparison to the world: 24

Merchant marine: *total:* 374
by type: bulk carrier 5, container ship 1, general cargo 20, oil tanker 57, other 291 (2019)
country comparison to the world: 48

Ports and terminals: *major seaport(s):* Ad Dammam, Al Jubayl, Jeddah, King Abdulla, Yanbu'
container port(s) (TEUs): Ad Dammam (1,582,388), Jeddah (4,150,000), King Abdulla (1,695,322) (2017)

MILITARY AND SECURITY

Military and security forces: *Ministry of Defense:* Royal Saudi Land Forces, Royal Saudi Naval Forces (includes marines, special forces, naval aviation), Royal Saudi Air Force, Royal Saudi Air Defense Forces, Royal Saudi Strategic Missiles Force; Ministry of the National Guard (SANG); Ministry of Interior: Border Guard, Facilities Security Force (2020)
note: SANG (also known as the White Army) is a land force separate from the Ministry of Defense that is responsible for internal security, protecting the royal family, and external defense

Military expenditures: 8% of GDP (2019)
9.5% of GDP (2018)
10.2% of GDP (2017)
10% of GDP (2016)
13% of GDP (2015)
country comparison to the world: 2

Military and security service personnel strengths: the Saudi military forces have about 225,000 active troops; approximately 125,000 under the Ministry of Defense (75,000 Land Forces; 13,500 Naval Forces; 35,000 Air Force/Air Defense; 2,500 Strategic Missile Forces) and approximately 100,000 in the Saudi Arabia National Guard (SANG) (2019)
note: SANG also has an irregular force (Fowj), primarily Bedouin tribal volunteers, with a total strength of approximately 25,000 men

Military equipment inventories and acquisitions: the inventory of the Saudi military forces, including the SANG, includes a mix of mostly modern weapons systems from the US and Europe, particularly France and the UK; since 2010, France, the UK, and the US are the leading suppliers of armaments, followed by Germany, Spain, and Canada; the Saudi Navy is in the midst of a major modernization/procurement program (2020)

Military deployments: est. 2,500-10,000 Yemen (probably varies depending on operations) (April 2020)

Military service age and obligation: 17 is the legal minimum age for voluntary military service; no conscription; in 2018, women were allowed to serve as soldiers in the internal security services under certain requirements (2018)

TERRORISM

Terrorist group(s): Islamic State of Iraq and ash-Sham; al-Qa'ida; al-Qa'ida in the Arabian Peninsula (2019)
note: details about the history, aims, leadership, organization, areas of operation, tactics, targets, weapons, size, and sources of support of the group(s) appear(s) in Appendix-T

TRANSNATIONAL ISSUES

Disputes—international: Saudi Arabia has reinforced its concrete-filled security barrier along sections of the now fully demarcated border with Yemen to stem illegal cross-border activities; Kuwait and Saudi Arabia continue discussions on a maritime boundary with Iran; Saudi Arabia claims Egyptian-administered islands of Tiran and Sanafir

Refugees and internally displaced persons: *stateless persons:* 70,000 (2019); note—thousands of biduns (stateless Arabs) are descendants of nomadic tribes who were not officially registered when national borders were established, while others migrated to Saudi Arabia in search of jobs; some have temporary identification cards that must be renewed every five years, but their rights remain restricted; most Palestinians have only legal resident status; some naturalized Yemenis were made stateless after being stripped of their passports when Yemen backed Iraq in its invasion of Kuwait in 1990; Saudi women cannot pass their

citizenship on to their children, so if they marry a non-national, their children risk statelessness

Trafficking in persons: *current situation:* Saudi Arabia is a destination country for men and women subjected to forced labor and, to a lesser extent, forced prostitution; men and women from South and East Asia, the Middle East, and Africa who voluntarily travel to Saudi Arabia as domestic servants or low-skilled laborers subsequently face conditions of involuntary servitude, including nonpayment and withholding of passports; some migrant workers are forced to work indefinitely beyond the term of their contract because their employers will not grant them a required exit visa; female domestic workers are particularly vulnerable because of their isolation in private homes; women, primarily from Asian and African countries, are believed to be forced into prostitution in Saudi Arabia, while other foreign women were reportedly kidnapped and forced into prostitution after running away from abusive employers; children from South Asia, East Africa, and Yemen are subjected to forced labor as beggars and street vendors in Saudi Arabia, facilitated by criminal gangs
tier rating: Tier 2 Watch List—Saudi Arabia does not fully comply with the minimum standards for the elimination of trafficking; however, it is making significant efforts to do so; government officials and high-level religious leaders demonstrated greater political will to combat trafficking and publically acknowledged the problem – specifically forced labor; the government reported increased numbers of prosecutions and convictions of trafficking offenders; however, it did not proactively investigate and prosecute employers for potential labor trafficking crimes following their withholding of workers' wages and passports, which are illegal; authorities did not systematically use formal criteria to proactively identify victims, resulting in some unidentified victims being arrested, detained, deported, and sometimes prosecuted; more victims were identified and referred to protective services in 2014 than the previous year, but victims of sex trafficking and male trafficking victims were not provided with shelter and remained vulnerable to punishment (2015)

Illicit drugs: regularly enforces the death penalty for drug traffickers, with foreigners being convicted and executed disproportionately; improving anti-money-laundering legislation and enforcement

SENEGAL

INTRODUCTION

Background: A Jolof Empire ruled parts of Senegal from 1350 to 1549. Various European powers, including Portugal, the Netherlands, France, and Great Britain, competed for trade in the area from the 15th century onward. A slave station on the island of Goree, next to modern Dakar, was used

as a base to purchase slaves from the warring chiefdoms on the mainland. Having abolished slavery in 1815, the French began to expand onto the Senegalese mainland in the second half of the 19th century and made it a French colony. The French colonies of Senegal and French Sudan were merged in 1959 and granted independence in 1960 as the Mali Federation. The union broke

up after only a few months. Senegal joined with The Gambia to form the nominal confederation of Senegambia in 1982. The envisaged integration of the two countries was never implemented, and the union was dissolved in 1989. The Movement of Democratic Forces in the Casamance has led a low-level separatist insurgency in southern Senegal since the 1980s. Several attempts at reaching a

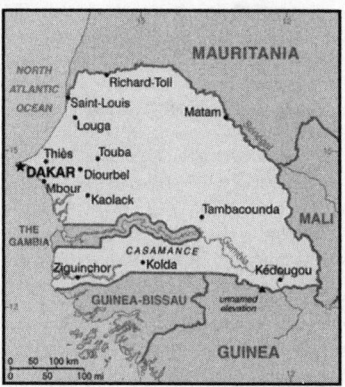

comprehensive peace agreement have failed to resolve the conflict but, despite sporadic incidents of violence, an unofficial cease-fire has remained largely in effect since 2012. Senegal remains one of the most stable democracies in Africa and has a long history of participating in international peacekeeping and regional mediation. Senegal was ruled by the Socialist Party of Senegal, first under President Léopold Sédar SENGHOR, and then President Abdou DIOUF, for 40 years until Abdoulaye WADE was elected president in 2000. He was re-elected in 2007 and during his two terms amended Senegal's constitution over a dozen times to increase executive power and weaken the opposition. His decision to run for a third presidential term sparked a large public backlash that led to his defeat in a March 2012 runoff with Macky SALL. A 2016 constitutional referendum reduced the term to five years with a maximum of two consecutive terms for future presidents—the change did not apply to SALL's first term. SALL won his bid for re-election in February 2019; his term will end in 2024. A month after the election, the National Assembly voted to abolish the office of the prime minister. Opposition organizations and civil society have criticized the decision as a further concentration of power in the executive branch at the expense of the legislative and judicial branches.

GEOGRAPHY

Location: Western Africa, bordering the North Atlantic Ocean, between Guinea-Bissau and Mauritania

Geographic coordinates: 14 00 N, 14 00 W

Map references: Africa

Area: *total:* 196,722 sq km
land: 192,530 sq km
water: 4,192 sq km
country comparison to the world: 89

Area—comparative: slightly smaller than South Dakota; slightly larger than twice the size of Indiana

Land boundaries: *total:* 2,684 km
border countries (5): The Gambia 749 km, Guinea 363 km, Guinea-Bissau 341 km, Mali 489 km, Mauritania 742 km

Coastline: 531 km

Maritime claims: *territorial sea:* 12 nm
exclusive economic zone: 200 nm
contiguous zone: 24 nm
continental shelf: 200 nm or to the edge of the continental margin

Climate: tropical; hot, humid; rainy season (May to November) has strong southeast winds; dry season (December to April) dominated by hot, dry, harmattan wind

Terrain: generally low, rolling, plains rising to foothills in southeast

Elevation: *mean elevation:* 69 m
lowest point: Atlantic Ocean 0 m
highest point: unnamed elevation 2.8 km southeast of Nepen Diaka 648 m

Natural resources: fish, phosphates, iron ore

Land use: *agricultural land:* 46.8% (2011 est.)
arable land: 17.4% (2011 est.) / permanent crops: 0.3% (2011 est.) / permanent pasture: 29.1% (2011 est.)
forest: 43.8% (2011 est.)
other: 9.4% (2011 est.)
Irrigated land: 1,200 sq km (2012)

Population distribution: the population is concentrated in the west, with Dakar anchoring a well-defined core area; approximately 70% of the population is rural as shown in this population distribution map

Natural hazards: lowlands seasonally flooded; periodic droughts

Environment—current issues: deforestation; overgrazing; soil erosion; desertification; periodic droughts; seasonal flooding; overfishing; weak environmental protective laws; wildlife populations threatened by poaching

Environment—international agreements: *party to:* Biodiversity, Climate Change, Climate Change-Kyoto Protocol, Desertification, Endangered Species, Hazardous Wastes, Law of the Sea, Marine Life Conservation, Ozone Layer Protection, Ship Pollution, Wetlands, Whaling
signed, but not ratified: none of the selected agreements

Geography—note: westernmost country on the African continent; The Gambia is almost an enclave within Senegal

PEOPLE AND SOCIETY

Population: 15,736,368 (July 2020 est.)
country comparison to the world: 72

Nationality: *noun:* Senegalese (singular and plural)
adjective: Senegalese

Ethnic groups: Wolof 37.1%, Pular 26.2%, Serer 17%, Mandinka 5.6%, Jola 4.5%, Soninke 1.4%, other 8.3% (includes Europeans and persons of Lebanese descent) (2017 est.)

Languages: French (official), Wolof, Pular, Jola, Mandinka, Serer, Soninke

Religions: Muslim 95.9% (most adhere to one of the four main Sufi brotherhoods), Christian 4.1% (mostly Roman Catholic) (2016 est.)

Demographic profile: Senegal has a large and growing youth population but has not been successful in developing its potential human capital. Senegal's high total fertility rate of almost 4.5 children per woman continues to bolster the country's large youth cohort – more than 60% of the population is under the age of 25. Fertility remains high because of the continued desire for large families, the low use of family planning, and early childbearing. Because of the country's high illiteracy rate (more than 40%), high unemployment (even among university graduates), and widespread poverty, Senegalese youths face dim prospects; women are especially disadvantaged.

Senegal historically was a destination country for economic migrants, but in recent years West African migrants more often use Senegal as a transit point to North Africa – and sometimes illegally onward to Europe. The country also has been host to several thousand black Mauritanian refugees since they were expelled from their homeland during its 1989 border conflict with Senegal. The country's economic crisis in the 1970s stimulated emigration; departures accelerated in the 1990s. Destinations shifted from neighboring countries, which were experiencing economic decline, civil wars, and increasing xenophobia, to Libya and Mauritania because of their booming oil industries and to developed countries (most notably former colonial ruler France, as well as Italy and Spain). The latter became attractive in the 1990s because of job opportunities and their periodic regularization programs (legalizing the status of illegal migrants).

Additionally, about 16,000 Senegalese refugees still remain in The Gambia and Guinea-Bissau as a result of more than 30 years of fighting between government forces and rebel separatists in southern Senegal's Casamance region.

Age structure: *0-14 years:* 40.38% (male 3,194,454/female 3,160,111)
15-24 years: 20.35% (male 1,596,896/female 1,606,084)
25-54 years: 31.95% (male 2,327,424/female 2,700,698)
55-64 years: 4.21% (male 283,480/female 378,932)
65 years and over: 3.1% (male 212,332/female 275,957) (2020 est.)

Dependency ratios: *total dependency ratio:* 84.2
youth dependency ratio: 78.4
elderly dependency ratio: 5.7
potential support ratio: 17.5 (2020 est.)

Median age: *total:* 19.4 years
male: 18.5 years
female: 20.3 years (2020 est.)
country comparison to the world: 204

Population growth rate: 2.31% (2020 est.)
country comparison to the world: 30

Birth rate: 31.8 births/1,000 population (2020 est.)
country comparison to the world: 29

Death rate: 7.6 deaths/1,000 population (2020 est.)
country comparison to the world: 102

Net migration rate: -1.3 migrant(s)/1,000 population (2020 est.)
country comparison to the world: 149

Population distribution: the population is concentrated in the west, with Dakar anchoring a well-defined core area; approximately 70% of the population is rural as shown in this population distribution map

Urbanization: *urban population:* 48.1% of total population (2020)
rate of urbanization: 3.73% annual rate of change (2015-20 est.)

total population growth rate v. urban population growth rate, 2000-2030: *Major urban areas—population:* 3.140 million DAKAR (capital) (2020)

Sex ratio: *at birth:* 1.03 male(s)/female
0-14 years: 1.01 male(s)/female
15-24 years: 0.99 male(s)/female
25-54 years: 0.86 male(s)/female
55-64 years: 0.75 male(s)/female
65 years and over: 0.77 male(s)/female
total population: 0.94 male(s)/female (2020 est.)

Mother's mean age at first birth: 21.9 years (2018 est.)
note: median age at first birth among women 25-29

Maternal mortality rate: 315 deaths/100,000 live births (2017 est.)
country comparison to the world: 35

Infant mortality rate: *total:* 45.7 deaths/1,000 live births
male: 51.3 deaths/1,000 live births
female: 40 deaths/1,000 live births (2020 est.)
country comparison to the world: 28

Life expectancy at birth: *total population:* 63.2 years
male: 61.1 years
female: 65.4 years (2020 est.)
country comparison to the world: 206

Total fertility rate: 4.04 children born/woman (2020 est.)
country comparison to the world: 30

Contraceptive prevalence rate: 27.8% (2017)

Drinking water source:
improved:
urban: 92.3% of population
rural: 74.5% of population
total: 83.3% of population
unimproved:
urban: 6.7% of population
rural: 25.5% of population
total: 16.7% of population (2017 est.)

Current Health Expenditure: 4.1% (2017)

Physicians density: 0.07 physicians/1,000 population (2017)

Hospital bed density: 0.3 beds/1,000 population (2000)

Sanitation facility access:
improved:
urban: 91.2% of population
rural: 48.5% of population
total: 68.4% of population

unimproved:
urban: 8.8% of population
rural: 51.5% of population
total: 31.6% of population (2017 est.)

HIV/AIDS—adult prevalence rate: 0.4% (2019 est.)
country comparison to the world: 82

HIV/AIDS—people living with HIV/AIDS: 41,000 (2019 est.)
country comparison to the world: 66

HIV/AIDS—deaths: 1,200 (2019 est.)
country comparison to the world: 55

Major infectious diseases: *degree of risk:* very high (2020)
food or waterborne diseases: bacterial and protozoal diarrhea, hepatitis A, and typhoid fever
vectorborne diseases: malaria and dengue fever
water contact diseases: schistosomiasis
animal contact diseases: rabies
respiratory diseases: meningococcal meningitis

Obesity—adult prevalence rate: 8.8% (2016)
country comparison to the world: 146

Children under the age of 5 years underweight: 13.3% (2019)
country comparison to the world: 46

Education expenditures: 4.8% of GDP (2017)
country comparison to the world: 74

Literacy: *definition:* age 15 and over can read and write
total population: 51.9%
male: 64.8%
female: 39.8% (2017)

School life expectancy (primary to tertiary education): *total:* 9 years
male: 8 years
female: 9 years (2019)

Unemployment, youth ages 15-24: *total:* 8.1%
male: 7.4%
female: 8.9% (2015 est.)
country comparison to the world: 142

GOVERNMENT

Country name: *conventional long form:* Republic of Senegal
conventional short form: Senegal
local long form: Republique du Senegal
local short form: Senegal
former: Senegambia (along with The Gambia), Mali Federation
etymology: named for the Senegal River that forms the northern border of the country; many theories exist for the origin of the river name; perhaps the most widely cited derives the name from "Azenegue," the Portuguese appellation for the Berber Zenaga people who lived north of the river

Government type: presidential republic

Capital: *name:* Dakar
geographic coordinates: 14 44 N, 17 38 W
time difference: UTC 0 (5 hours ahead of Washington, DC, during Standard Time)
etymology: the Atlantic coast trading settlement of Ndakaaru came to be called "Dakar" by French colonialists

Administrative divisions: 14 regions (regions, singular—region); Dakar, Diourbel, Fatick, Kaffrine, Kaolack, Kedougou, Kolda, Louga, Matam, Saint-Louis, Sedhiou, Tambacounda, Thies, Ziguinchor

Independence: 4 April 1960 (from France); note—complete independence achieved upon dissolution of federation with Mali on 20 August 1960

National holiday: Independence Day, 4 April (1960)

Constitution: *history:* previous 1959 (preindependence), 1963; latest adopted by referendum 7 January 2001, promulgated 22 January 2001
amendments: proposed by the president of the republic or by the National Assembly; passage requires Assembly approval and approval in a referendum; the president can bypass a referendum and submit an amendment directly to the Assembly, which requires at least three-fifths majority vote; the republican form of government is not amendable; amended several times, last in 2019

Legal system: civil law system based on French law; judicial review of legislative acts in Constitutional Court

International law organization participation: accepts compulsory ICJ jurisdiction with reservations; accepts ICCt jurisdiction

Citizenship: *citizenship by birth:* no
citizenship by descent only: at least one parent must be a citizen of Senegal
dual citizenship recognized: no, but Senegalese citizens do not automatically lose their citizenship if they acquire citizenship in another state
residency requirement for naturalization: 5 years

Suffrage: 18 years of age; universal

Executive branch: *chief of state:* President Macky SALL (since 2 April 2012)
head of government: President Macky SALL (since 2 April 2012)
cabinet: Council of Ministers appointed by the president
elections/appointments: president directly elected by absolute majority popular vote in 2 rounds if needed for a single renewable 5-year term; election last held on 24 February 2019 (next to be held in February 2024)
election results: Macky SALL elected president in first round; percent of vote—Macky SALL (APR) 58.3%, Idrissa SECK (Rewmi) 20.5%, Ousmane SONKO (PASTEF) 15.7%

Legislative branch: *description:* unicameral National Assembly or Assemblée Nationale (165 seats; 105 members including 15 representing Senegalese diaspora directly elected by plurality vote in single- and multi-seat constituencies and 60 members directly elected by proportional representation vote in single- and multi-seat constituencies)
elections: National Assembly—last held on 2 July 2017 (next to be held in July 2022)
election results: National Assembly results—percent of vote by party/coalition—BBK 49.5%, CGWS 16.7%, MTS 11.7%, PUR 4.7%,

CP-Kaddu Askan Wi 2%, other 15.4%; seats by party/coalition—BBY 125, CGWS 19, MTS 7, PUR 3, CP-Kaddu Askan Wi 2, other 9; composition—men 96, women 69, percent of women 41.8%

Judicial branch: *highest courts:* Supreme Court or Cour Supreme (consists of the court president and 12 judges and organized into civil and commercial, criminal, administrative, and social chambers); Constitutional Council or Conseil Constitutionel (consists of 7 members, including the court president, vice president, and 5 judges)
judge selection and term of office: Supreme Court judges appointed by the president of the republic upon recommendation of the Superior Council of the Magistrates, a body chaired by the president and minister of justice; judge tenure varies, with mandatory retirement either at 65 or 68 years; Constitutional Council members appointed—5 by the president and 2 by the National Assembly speaker; judges serve 6-year terms, with renewal of 2 members every 2 years
subordinate courts: High Court of Justice (for crimes of high treason by the president); Courts of Appeal; Court of Auditors; assize courts; regional and district courts; Labor Court

Political parties and leaders: Alliance for the Republic-Yakaar or APR-Yakaar [Macky SALL]
Alliance of Forces of Progress or AFP [Moustapha NIASSE]
Alliance for Citizenship and Labor or ACT [Abdoul MBAYE]
And-Jef/African Party for Democracy and Socialism or AJ/PADS [Mamadou DIOP Decriox]
Benno Bokk Yakaar or BBY (United in Hope) [Macky SALL] (coalition includes AFP, APR, BGC, LD-MPT, PIT, PS, and UNP)
Bokk Gis Gis coalition [Pape DIOP]
Citizen Movement for National Reform or MCRN-Bes Du Nakk [Mansour Sy DJAMIL]
Democratic League-Labor Party Movement or LD-MPT [Abdoulaye BATHILY]
Dare the Future movement [Aissata Tall SALL]
Front for Socialism and Democracy/Benno Jubel or FSD/BJ [Cheikh Abdoulaye Bamba DIEYE]
Gainde Centrist Bloc or BGC [Jean-Paul DIAS]
General Alliance for the Interests of the Republic or AGIR [Thierno BOCOUM]
Grand Party or GP [Malick GAKOU]
Independence and Labor Party or PIT [Magatte THIAM]
Madicke 2019 coalition [Madicke NIANG]
National Union for the People or UNP [Souleymane Ndene NDIAYE]
Only Senegal movement [Pierre Goudiaby ATEPA]
Party for Truth and Development or PVD [Cheikh Ahmadou Kara MBAKE]
Party of Unity and Rally or PUR [El Hadji SALL]
Patriotic Convergence Kaddu Askan Wi or CP-Kaddu Askan Wi [Abdoulaye BALDE]
Patriots of Senegal for Ethics, Work and Fraternity or (PASTEF) [Ousmane SONKO]
Rewmi Party [Idrissa SECK]
Senegalese Democratic Party or PDS [Abdoulaye WADE]
Socialist Party or PS [Ousmane Tanor DIENG]
Tekki Movement [Mamadou Lamine DIALLO]

International organization participation: ACP, AfDB, AU, CD, CPLP (associate), ECOWAS, EITI (candidate country), FAO, FZ, G-15, G-77, IAEA, IBRD, ICAO, ICC (national committees), ICCr, ICRM, IDA, IDB, IFAD, IFC, IFRCS, ILO, IMF, IMO, IMSO, Interpol, IOC, IOM, IPU, ISO, ITSO, ITU, ITUC (NGOs), MIGA, MINUSMA, MONUSCO, NAM, OIC, OIF, OPCW, PCA, UN, UNAMID, UNCTAD, UNESCO, UNHCR, UNIDO, UNMIL, UNMISS, UNOCI, UNWTO, UPU, WADB (regional), WAEMU, WCO, WFTU (NGOs), WHO, WIPO, WMO, WTO

Diplomatic representation in the US: *chief of mission:* Ambassador Mansour KANE (since 6 January 2020)
chancery: 2215 M Street NW, Washington, DC 20007
telephone: [1] (202) 234-0540
FAX: [1] (202) 629-2961
consulate(s) general: Houston, New York

Diplomatic representation from the US: *chief of mission:* Ambassador Tulinabo S. MUSHINGI (since August 2017); note—also accredited to Guinea-Bissau
telephone: [221] 33-879-4000
embassy: Route des Almadies, Dakar
mailing address: B.P. 49, Dakar
FAX: [221] 33-822-2991

Flag description: three equal vertical bands of green (hoist side), yellow, and red with a small green five-pointed star centered in the yellow band; green represents Islam, progress, and hope; yellow signifies natural wealth and progress; red symbolizes sacrifice and determination; the star denotes unity and hope
note: uses the popular Pan-African colors of Ethiopia; the colors from left to right are the same as those of neighboring Mali and the reverse of those on the flag of neighboring Guinea

National symbol(s): *lion; national colors:* green, yellow, red

National anthem: *name:* "Pincez Tous vos Koras, Frappez les Balafons" (Pluck Your Koras, Strike the Balafons)
lyrics/music: Leopold Sedar SENGHOR/Herbert PEPPER
note: adopted 1960; lyrics written by Leopold Sedar SENGHOR, Senegal's first president; the anthem sometimes played incorporating the Koras (harp-like stringed instruments) and Balafons (types of xylophones) mentioned in the title
0:00 / 0:55

ECONOMY

Economy—overview: Senegal's economy is driven by mining, construction, tourism, fisheries and agriculture, which are the primary sources of employment in rural areas. The country's key export industries include phosphate mining, fertilizer production, agricultural products and commercial fishing and Senegal is also working on oil exploration projects. It relies heavily on donor assistance, remittances and foreign direct investment. Senegal reached a growth rate of 7% in 2017, due in part to strong performance in agriculture despite erratic rainfall.

President Macky SALL, who was elected in March 2012 under a reformist policy agenda, inherited an economy with high energy costs, a challenging business environment, and a culture of overspending. President SALL unveiled an ambitious economic plan, the Emerging Senegal Plan (ESP), which aims to implement priority economic reforms and investment projects to increase economic growth while preserving macroeconomic stability and debt sustainability. Bureaucratic bottlenecks and a challenging business climate are among the perennial challenges that may slow the implementation of this plan.

Senegal receives technical support from the IMF under a Policy Support Instrument (PSI) to assist with implementation of the ESP. The PSI implementation continues to be satisfactory as concluded by the IMF's fifth review in December 2017. Financial markets have signaled confidence in Senegal through successful Eurobond issuances in 2014, 2017, and 2018.

The government is focusing on 19 projects under the ESP to continue The government's goal under the ESP is structural transformation of the economy. Key projects include the Thiès-Touba Highway, the new international airport opened in December 2017, and upgrades to energy infrastructure. The cost of electricity is a chief constraint for Senegal's development. Electricity prices in Senegal are among the highest in the world. Power Africa, a US presidential initiative led by USAID, supports Senegal's plans to improve reliability and increase generating capacity.

GDP (purchasing power parity): $54.8 billion (2017 est.)
$51.15 billion (2016 est.)
$48.15 billion (2015 est.)
note: data are in 2017 dollars
country comparison to the world: 107

GDP (official exchange rate): $21.11 billion (2017 est.)

GDP—real growth rate: 7.2% (2017 est.)
6.2% (2016 est.)
6.4% (2015 est.)
country comparison to the world: 16

GDP—per capita (PPP): $3,500 (2017 est.)
$3,300 (2016 est.)
$3,200 (2015 est.)
note: data are in 2017 dollars
country comparison to the world: 188

Gross national saving: 21.2% of GDP (2017 est.)
21.3% of GDP (2016 est.)
20.4% of GDP (2015 est.)
country comparison to the world: 86

GDP—composition, by end use:
household consumption: 71.9% (2017 est.)
government consumption: 15.2% (2017 est.)
investment in fixed capital: 25.1% (2017 est.)
investment in inventories: 3.4% (2017 est.)
exports of goods and services: 27% (2017 est.)
imports of goods and services: -42.8% (2017 est.)

GDP—composition, by sector of origin:
agriculture: 16.9% (2017 est.)
industry: 24.3% (2017 est.)
services: 58.8% (2017 est.)

Agriculture—products: peanuts, millet, corn, sorghum, rice, cotton, tomatoes, green vegetables; cattle, poultry, pigs; fish

Industries: agricultural and fish processing, phosphate mining, fertilizer production, petroleum refining, zircon, and gold mining, construction materials, ship construction and repair

Industrial production growth rate: 7.7% (2017 est.)
country comparison to the world: 26

Labor force: 6.966 million (2017 est.)
country comparison to the world: 65

Labor force—by occupation: *agriculture:* 77.5%
industry: 22.5%
industry and services: 22.5% (2007 est.)

Unemployment rate: 48% (2007 est.)
country comparison to the world: 216

Population below poverty line: 46.7% (2011 est.)

Household income or consumption by percentage share: *lowest 10%:* 2.5%
highest 10%: 31.1% (2011)

Budget: *revenues:* 4.139 billion (2017 est.)
expenditures: 4.9 billion (2017 est.)

Taxes and other revenues: 19.6% (of GDP) (2017 est.)
country comparison to the world: 155

Budget surplus (+) or deficit (-): -3.6% (of GDP) (2017 est.)
country comparison to the world: 151

Public debt: 48.3% of GDP (2017 est.)
47.8% of GDP (2016 est.)
country comparison to the world: 108

Fiscal year: calendar year

Inflation rate (consumer prices): 1.3% (2017 est.)
0.8% (2016 est.)
country comparison to the world: 73

Current account balance: -$1.547 billion (2017 est.)
-$769 million (2016 est.)
country comparison to the world: 160

Exports: $2.362 billion (2017 est.)
$2.498 billion (2016 est.)
country comparison to the world: 133

Exports—partners: Mali 14.8%, Switzerland 11.4%, India 6%, Cote dIvoire 5.3%, UAE 5.1%, Gambia, The 4.2%, Spain 4.1% (2017)

Exports—commodities: fish, groundnuts (peanuts), petroleum products, phosphates, cotton

Imports: $5.217 billion (2017 est.)
$4.966 billion (2016 est.)
country comparison to the world: 125

Imports—commodities: food and beverages, capital goods, fuels

Imports—partners: France 16.3%, China 10.4%, Nigeria 8%, India 7.2%, Netherlands 4.8%, Spain 4.2% (2017)

Reserves of foreign exchange and gold: $1.827 billion (31 December 2017 est.)

$116.9 million (31 December 2016 est.)
country comparison to the world: 122

Debt—external: $8.571 billion (31 December 2017 est.)
$6.327 billion (31 December 2016 est.)
country comparison to the world: 119

Exchange rates: Communaute Financiere Africaine francs (XOF) per US dollar –

617.4 (2017 est.)
593.01 (2016 est.)
593.01 (2015 est.)
591.45 (2014 est.)
494.42 (2013 est.)

ENERGY

Electricity access: *population without electricity:* 5 million (2019)
electrification—total population: 71% (2019)
electrification—urban areas: 94% (2019)
electrification—rural areas: 50% (2019)

Electricity—production: 4.167 billion kWh (2016 est.)
country comparison to the world: 126

Electricity—consumption: 3.497 billion kWh (2016 est.)
country comparison to the world: 133

Electricity—exports: 0 kWh (2016 est.)
country comparison to the world: 195

Electricity—imports: 0 kWh (2016 est.)
country comparison to the world: 197

Electricity—installed generating capacity: 977,000 kW (2016 est.)
country comparison to the world: 129

Electricity—from fossil fuels: 82% of total installed capacity (2016 est.)
country comparison to the world: 79

Electricity—from nuclear fuels: 0% of total installed capacity (2017 est.)
country comparison to the world: 179

Electricity—from hydroelectric plants: 7% of total installed capacity (2017 est.)
country comparison to the world: 127

Electricity—from other renewable sources: 11% of total installed capacity (2017 est.)
country comparison to the world: 80

Crude oil—production: 0 bbl/day (2018 est.)
country comparison to the world: 197

Crude oil—exports: 0 bbl/day (2015 est.)
country comparison to the world: 190

Crude oil—imports: 17,880 bbl/day (2015 est.)
country comparison to the world: 65

Crude oil—proved reserves: 0 bbl (1 January 2018 est.)
country comparison to the world: 192

Refined petroleum products—production: 17,590 bbl/day (2015 est.)
country comparison to the world: 90

Refined petroleum products—consumption: 48,000 bbl/day (2016 est.)
country comparison to the world: 107

Refined petroleum products—exports: 4,063 bbl/day (2015 est.)
country comparison to the world: 94

Refined petroleum products—imports: 32,050 bbl/day (2015 est.)
country comparison to the world: 98

Natural gas—production: 59.46 million cu m (2017 est.)
country comparison to the world: 85

Natural gas—consumption: 59.46 million cu m (2017 est.)
country comparison to the world: 111

Natural gas—exports: 0 cu m (2017 est.)
country comparison to the world: 180

Natural gas—imports: 0 cu m (2017 est.)
country comparison to the world: 186

Natural gas—proved reserves: 0 cu m (1 January 2012 est.)
country comparison to the world: 191

Carbon dioxide emissions from consumption of energy: 8.644 million Mt (2017 est.)
country comparison to the world: 113

COMMUNICATIONS

Telephones—fixed lines: *total subscriptions:* 195,288
subscriptions per 100 inhabitants: 1.27 (2019 est.)
country comparison to the world: 122

Telephones—mobile cellular: *total subscriptions:* 16,871,654
subscriptions per 100 inhabitants: 109.72 (2019 est.)
country comparison to the world: 63

Telecommunication systems: *general assessment:* mobile penetration reached 108% in March 2019; mobile broadband accounts for close to 100% (97.2%) Internet accesses; 3G and LTE services for 50% of population; growth in the intel market along with economic growth for the country; regulator awards more MVNO licenses, deactivated some 5 million unregistered SIM cards (2020)
domestic: generally reliable urban system with a fiber-optic network; about two-thirds of all fixed-line connections are in Dakar; mobile-cellular service is steadily displacing fixed-line service, even in urban areas; fixed-line 1 per 100 and mobile-cellular 110 per 100 persons (2019)
international: country code—221; landing points for the ACE, Atlantis-2, MainOne and SAT-3/WASC submarine cables providing connectivity from South Africa, numerous western African countries, Europe and South America; satellite earth station—1 Intelsat (Atlantic Ocean) (2019)
note: the COVID-19 outbreak is negatively impacting telecommunications production and supply chains globally; consumer spending on telecom devices and services has also slowed due to the pandemic's effect on economies worldwide; overall progress towards improvements in all facets of the telecom industry—mobile, fixed-line, broadband, submarine cable and satellite—has moderated

Broadcast media: state-run Radiodiffusion Television Senegalaise (RTS) broadcasts TV programs from five cities in Senegal; in most regions of the country, viewers can receive TV programming from at least 7 private broadcasters; a wide range of independent TV programming is available via satellite; RTS operates a national radio network and a number of regional FM stations; at least 7 community radio stations and 18 private-broadcast radio stations are available; transmissions of at least 5 international broadcasters are accessible on FM in Dakar (2019)

Internet country code: .sn

Internet users: *total:* 6,909,635
percent of population: 46% (July 2018 est.)
country comparison to the world: 72

Broadband—fixed subscriptions: *total:* 129,820
subscriptions per 100 inhabitants: 1 (2018 est.)
country comparison to the world: 117

TRANSPORTATION

National air transport system: *number of registered air carriers:* 2 (2020)
inventory of registered aircraft operated by air carriers: 11
annual passenger traffic on registered air carriers: 21,038 (2018)
annual freight traffic on registered air carriers: 40,000 mt-km (2018)

Civil aircraft registration country code prefix: 6V (2016)

Airports: 20 (2013)
country comparison to the world: 135

Airports—with paved runways: *total:* 9 (2017)
over 3,047 m: 2 (2017)
1,524 to 2,437 m: 6 (2017)
914 to 1,523 m: 1 (2017)

Airports—with unpaved runways: *total:* 11 (2013)
1,524 to 2,437 m: 7 (2013)
914 to 1,523 m: 3 (2013)
under 914 m: 1 (2013)

Pipelines: 43 km gas, 8 km refined products (2017)

Railways: *total:* 906 km (713 km operational in 2017) (2017)
narrow gauge: 906 km 1.000-m gauge (2017)
country comparison to the world: 94

Roadways: *total:* 16,665 km (2017)
paved: 6,126 km (includes 241 km of expressways) (2017)
unpaved: 10,539 km (2017)
country comparison to the world: 120

Waterways: 1,000 km (primarily on the Senegal, Saloum, and Casamance Rivers) (2012)
country comparison to the world: 63

Merchant marine: *total:* 32
by type: general cargo 4, oil tanker 1, other 27 (2019)
country comparison to the world: 130

Ports and terminals: *major seaport(s):* Dakar

MILITARY AND SECURITY

Military and security forces: *Senegalese Armed Forces:* Army, Senegalese National Navy (Marine Senegalaise, MNS), Senegalese Air Force (Armee de l'Air du Senegal), National Gendarmerie (includes Territorial and Mobile components) (2020)

Military expenditures: 1.5% of GDP (2019 est.)
1.6% of GDP (2018)
1.5% of GDP (2017)
1.6% of GDP (2016)
1.2% of GDP (2015)
country comparison to the world: 84

Military and security service personnel strengths: the Senegalese Armed Forces (SAF) consist of approximately 19,000 active personnel (12,000 Army; 1,000 Navy/Coast Guard; 800 Air Force; 5,000 National Gendarmerie) (2019 est.)

Military equipment inventories and acquisitions: the SAF inventory includes mostly older or second-hand equipment from a variety of countries, including France, South Africa, and Russia/former Soviet Union; in recent years, the SAF has attempted to modernize, particularly its air force; China and France are the leading suppliers of newer military hardware to the SAF since 2010 (2019 est.)

Military deployments: 1,000 Mali (MINUSMA) (2020)

Military service age and obligation: 18 years of age for voluntary military service; 20 years of age for selective conscript service; 2-year service obligation; women have been accepted into military service since 2008 (2016)

TRANSNATIONAL ISSUES

Disputes—international: cross-border trafficking in persons, timber, wildlife, and cannabis; rebels from the Movement of Democratic Forces in the Casamance find refuge in Guinea-Bissau

Refugees and internally displaced persons: *refugees (country of origin):* 14,114 (Mauritania) (2020)
IDPs: 8,400 (clashes between government troops and separatists in Casamance region in the 1990s and early 2000s) (2019)

Illicit drugs: transshipment point for Southwest and Southeast Asian heroin and South American cocaine moving to Europe and North America; illicit cultivator of cannabis

SERBIA

INTRODUCTION

Background: The Kingdom of Serbs, Croats, and Slovenes was formed in 1918; its name was changed to Yugoslavia in 1929. Communist Partisans resisted the Axis occupation and division of Yugoslavia from 1941 to 1945 and fought nationalist opponents and collaborators as well. The military and political movement headed by Josip Broz "TITO" (Partisans) took full control of Yugoslavia when their domestic rivals and the occupiers were defeated in 1945. Although communists, TITO and his successors (Tito died in 1980) managed to steer their own path between the Warsaw Pact nations and the West for the next four and a half decades. In 1989, Slobodan MILOSEVIC became president of the Republic of Serbia and his ultranationalist calls for Serbian domination led to the violent breakup of Yugoslavia along ethnic lines. In 1991, Croatia, Slovenia, and Macedonia declared independence, followed by Bosnia in 1992. The remaining republics of Serbia and Montenegro declared a new Federal Republic of Yugoslavia (FRY) in April 1992 and under MILOSEVIC's leadership, Serbia led various military campaigns to unite ethnic

Serbs in neighboring republics into a "Greater Serbia." These actions ultimately failed and, after international intervention, led to the signing of the Dayton Peace Accords in 1995.

MILOSEVIC retained control over Serbia and eventually became president of the FRY in 1997. In 1998, an ethnic Albanian insurgency in the formerly autonomous Serbian province of Kosovo provoked a Serbian counterinsurgency campaign that resulted in massacres and massive expulsions of ethnic Albanians living in Kosovo. The MILOSEVIC government's rejection of a proposed international settlement led to NATO's bombing of Serbia in the spring of 1999. Serbian military and police forces withdrew from Kosovo in June 1999, and the UN Security Council authorized an interim UN administration and a NATO-led security force in Kosovo. FRY elections in late 2000 led to the ouster of MILOSEVIC and the installation of democratic government. In 2003, the FRY became the State Union of Serbia and Montenegro, a loose federation of the two republics. Widespread violence predominantly targeting ethnic Serbs in Kosovo in March 2004 led to more intense calls to address Kosovo's status, and the UN began facilitating status talks in 2006. In June 2006, Montenegro seceded from the federation and declared itself an independent nation. Serbia subsequently gave notice that it was the successor state to the union of Serbia and Montenegro.

In February 2008, after nearly two years of inconclusive negotiations, Kosovo declared itself independent of Serbia—an action Serbia refuses to recognize. At Serbia's request, the UN General Assembly (UNGA) in October 2008 sought an advisory opinion from the International Court of Justice (ICJ) on whether Kosovo's unilateral declaration of independence was in accordance with international law. In a ruling considered unfavorable to Serbia, the ICJ issued an advisory opinion in July 2010 stating that international law did not prohibit declarations of independence. In late 2010, Serbia agreed to an EU-drafted UNGA Resolution acknowledging the ICJ's decision and calling for a new round of talks between Serbia and Kosovo, this time on practical issues rather than Kosovo's status. Serbia and Kosovo signed the first agreement of principles governing the normalization of relations between the two countries in April 2013 and are in the process of implementing its provisions. In 2015, Serbia and Kosovo reached four additional agreements within the EU-led Brussels Dialogue framework. These included agreements on the Community of Serb-Majority Municipalities; telecommunications; energy production and distribution; and freedom of movement. President Aleksandar VUCIC has promoted an ambitious goal of Serbia joining the EU by 2025. Under his leadership as prime minister, in 2014 Serbia opened formal negotiations for accession.

GEOGRAPHY

Location: Southeastern Europe, between Macedonia and Hungary

Geographic coordinates: 44 00 N, 21 00 E

Map references: Europe

Area: *total:* 77,474 sq km
land: 77,474 sq km
water: 0 sq km
country comparison to the world: 118

Area—comparative: slightly smaller than South Carolina

Land boundaries: *total:* 2,322 km
border countries (8): Bosnia and Herzegovina 345 km, Bulgaria 344 km, Croatia 314 km, Hungary 164 km, Kosovo 366 km, Macedonia 101 km, Montenegro 157 km, Romania 531 km

Coastline: 0 km (landlocked)
Maritime claims: none (landlocked)

Climate: in the north, continental climate (cold winters and hot, humid summers with well-distributed rainfall); in other parts, continental and Mediterranean climate (relatively cold winters with heavy snowfall and hot, dry summers and autumns)

Terrain: extremely varied; to the north, rich fertile plains; to the east, limestone ranges and basins; to the southeast, ancient mountains and hills

Elevation: *mean elevation:* 442 m
lowest point: Danube and Timok Rivers 35 m
highest point: Midzor 2,169 m

Natural resources: oil, gas, coal, iron ore, copper, zinc, antimony, chromite, gold, silver, magnesium, pyrite, limestone, marble, salt, arable land

Land use: *agricultural land:* 57.9% (2011 est.)
arable land: 37.7% (2011 est.) / permanent crops: 3.4% (2011 est.) / permanent pasture: 16.8% (2011 est.)
forest: 31.6% (2011 est.)
other: 10.5% (2011 est.)
Irrigated land: 950 sq km (2012)

Population distribution: a fairly even distribution throughout most of the country, with urban areas attracting larger and denser populations

Natural hazards: destructive earthquakes

Environment—current issues: air pollution around Belgrade and other industrial cities; water pollution from industrial wastes dumped into the Sava which flows into the Danube; inadequate management of domestic, industrial, and hazardous waste

Environment—international agreements: *party to:* Air Pollution, Biodiversity, Climate Change, Climate Change-Kyoto Protocol, Desertification, Endangered Species, Hazardous Wastes, Law of the Sea, Marine Dumping, Marine Life Conservation, Ozone Layer Protection, Ship Pollution, Wetlands
signed, but not ratified: none of the selected agreements

Geography—note: landlocked; controls one of the major land routes from Western Europe to Turkey and the Near East

PEOPLE AND SOCIETY

Population: 7,012,165 (July 2020 est.)
note: does not include the population of Kosovo
country comparison to the world: 105

Nationality: *noun:* Serb(s)
adjective: Serbian

Ethnic groups: Serb 83.3%, Hungarian 3.5%, Romani 2.1%, Bosniak 2%, other 5.7%, undeclared or unknown 3.4% (2011 est.)
note: most ethnic Albanians boycotted the 2011 census; Romani populations are usually underestimated in official statistics and may represent 5-11% of Serbia's population

Languages: Serbian (official) 88.1%, Hungarian 3.4%, Bosnian 1.9%, Romani 1.4%, other 3.4%, undeclared or unknown 1.8% (2011 est.)
note: Serbian, Hungarian, Slovak, Romanian, Croatian, and Ruthenian (Rusyn) are official in the Autonomous Province of Vojvodina; most ethnic Albanians boycotted the 2011 census

Religions: Orthodox 84.6%, Catholic 5%, Muslim 3.1%, Protestant 1%, atheist 1.1%, other 0.8% (includes agnostics, other Christians, Eastern, Jewish), undeclared or unknown 4.5% (2011 est.)

note: most ethnic Albanians boycotted the 2011 census

Age structure: *0-14 years:* 14.07% (male 508,242/ female 478,247)
15-24 years: 11.04% (male 399,435/female 374,718)
25-54 years: 41.19% (male 1,459,413/female 1,429,176)
55-64 years: 13.7% (male 464,881/female 495,663)
65 years and over: 20% (male 585,705/female 816,685) (2020 est.)

Dependency ratios: *total dependency ratio:* 52.5
youth dependency ratio: 23.4
elderly dependency ratio: 29.1
potential support ratio: 3.4 (2020 est.)
note: data include Kosovo

Median age: *total:* 43.4 years
male: 41.7 years
female: 45 years (2020 est.)
country comparison to the world: 26

Population growth rate: -0.47% (2020 est.)
country comparison to the world: 224

Birth rate: 8.8 births/1,000 population (2020 est.)
country comparison to the world: 210

Death rate: 13.5 deaths/1,000 population (2020 est.)
comparison to the world: 6

Net migration rate: 0 migrant(s)/1,000 population (2020 est.)
country comparison to the world: 95

Population distribution: a fairly even distribution throughout most of the country, with urban areas attracting larger and denser populations

Urbanization: *urban population:* 56.4% of total population (2020)
rate of urbanization: -0.07% annual rate of change (2015-20 est.)
note: data include Kosovo

total population growth rate v. urban population growth rate, 2000-2030: *Major urban areas—population:* 1.398 million BELGRADE (capital) (2020)

Sex ratio: *at birth:* 1.07 male(s)/female
0-14 years: 1.06 male(s)/female
15-24 years: 1.07 male(s)/female
25-54 years: 1.02 male(s)/female
55-64 years: 0.94 male(s)/female
65 years and over: 0.72 male(s)/female
total population: 0.95 male(s)/female (2020 est.)

Mother's mean age at first birth: 28.4 years (2017 est.)
note: data do not cover Kosovo or Metohija

Maternal mortality rate: 12 deaths/100,000 live births (2017 est.)
country comparison to the world: 141

Infant mortality rate: *total:* 5.6 deaths/1,000 live births
male: 6.4 deaths/1,000 live births
female: 4.7 deaths/1,000 live births (2020 est.)
country comparison to the world: 171

Life expectancy at birth: *total population:* 76.3 years

male: 73.4 years

female: 79.4 years (2020 est.)

country comparison to the world: 99

Total fertility rate: 1.46 children born/woman (2020 est.)

country comparison to the world: 211

Contraceptive prevalence rate: 58.4% (2014)

Drinking water source:

improved:

urban: 99.4% of population

rural: 99% of population

total: 99.2% of population

unimproved:

urban: 0.6% of population

rural: 1% of population

total: 0.8% of population (2017 est.)

Current Health Expenditure: 8.4% (2017)

Physicians density: 3.11 physicians/1,000 population (2016)

Hospital bed density: 5.6 beds/1,000 population (2017)

Sanitation facility access:

improved:

urban: 100% of population

rural: 95.1% of population

total: 97.6% of population

unimproved:

urban: 0% of population

rural: 4.9% of population

total: 2.4% of population (2017 est.)

HIV/AIDS—adult prevalence rate: <.1% (2019 est.)

HIV/AIDS—people living with HIV/AIDS: 3,200 (2019 est.)

country comparison to the world: 132

HIV/AIDS—deaths: <100 (2019 est.)

Major infectious diseases: *degree of risk:* intermediate (2020)

food or waterborne diseases: bacterial diarrhea

Obesity—adult prevalence rate: 21.5% (2016)

country comparison to the world: 88

Children under the age of 5 years underweight: 1.8% (2014)

country comparison to the world: 116

Education expenditures: 4% of GDP (2017)

country comparison to the world: 105

Literacy: *definition:* age 15 and over can read and write

total population: 98.3%

male: 99.1%

female: 97.5% (2016)

School life expectancy (primary to tertiary education): *total:* 15 years

male: 14 years

female: 15 years (2019)

Unemployment, youth ages 15-24: *total:* 29.7%

male: 28.3%

female: 32% (2018 est.)

country comparison to the world: 30

GOVERNMENT

Country name: *conventional long form:* Republic of Serbia

conventional short form: Serbia

local long form: Republika Srbija

local short form: Srbija

former: People's Republic of Serbia, Socialist Republic of Serbia

etymology: the origin of the name is uncertain, but seems to be related to the name of the West Slavic Sorbs who reside in the Lusatian region in present-day eastern Germany; by tradition, the Serbs migrated from that region to the Balkans in about the 6th century A.D.

Government type: parliamentary republic

Capital: *name:* Belgrade (Beograd)

geographic coordinates: 44 50 N, 20 30 E

time difference: UTC+1 (6 hours ahead of Washington, DC, during Standard Time)

daylight saving time: +1hr, begins last Sunday in March; ends last Sunday in October

etymology: the Serbian "Beograd" means "white fortress" or "white city" and dates back to the 9th century; the name derives from the white fortress wall that once enclosed the city

Administrative divisions: 119 municipalities (opstine, singular—opstina) and 26 cities (gradovi, singular—grad)

municipalities: Ada*, Aleksandrovac, Aleksinac, Alibunar*, Apatin*, Arandelovac, Arilje, Babusnica, Bac*, Backa Palanka*, Backa Topola*, Backi Petrovac*, Bajina Basta, Batocina, Becej*, Bela Crkva*, Bela Palanka*, Beocin*, Blace, Bogatic, Bojnik, Boljevac, Bor, Bosilegrad, Brus, Bujanovac, Cajetina, Cicevac, Coka*, Crna Trava, Cuprija, Despotovac, Dimitrov, Doljevac, Gadzin Han, Golubac, Gornji Milanovac, Indija*, Irig*, Ivanjica, Kanjiza*, Kladovo, Knic, Knjazevac, Koceljeva, Kosjeric, Kovacica*, Kovin*, Krupanj, Kucevo, Kula*, Kursumlija, Lajkovac, Lapovo, Lebane, Ljig, Ljubovija, Lucani, Majdanpek, Mali Idos*, Mali Zvornik, Malo Crnice, Medveda, Merosina, Mionica, Negotin, Nova Crnja*, Nova Varos, Novi Becej*, Novi Knezevac*, Odzaci*, Opovo*, Osecina, Paracin, Pecinci*, Petrovac na Mlavi, Plandiste*, Pozega, Presevo, Priboj, Prijepolje, Prokuplje, Raca, Raska, Razanj, Rekovac, Ruma*, Secanj*, Senta*, Sid*, Sjenica, Smederevska Palanka, Sokobanja, Srbobran*, Sremski Karlovci*, Stara Pazova*, Surdulica, Svilajnac, Svrljig, Temerin*, Titel*, Topola, Trgoviste, Trstenik, Tutin, Ub, Varvarin, Velika Plana, Veliko Gradiste, Vladicin Han, Vladimirci, Vlasotince, Vrbas*, Vrnjacka Banja, Zabalj*, Zabari, Zagubica, Zititste*, Zitorada;

cities: Beograd, Cacak, Jagodina, Kikinda*, Kragujevac, Kraljevo, Krusevac, Leskovac, Loznica, Nis, Novi Pazar, Novi Sad*, Pancevo*, Pirot, Pozarevac, Sabac, Smederevo, Sombor*, Sremska Mitrovica*, Subotica*, Uzice, Valjevo, Vranje, Vrsac*, Zajecar, Zrenjanin*

note: the northern 37 municipalities and 8 cities—about 28% of Serbia's area—compose the

Autonomous Province of Vojvodina and are indicated with *

Independence: 5 June 2006 (from the State Union of Serbia and Montenegro); notable earlier dates: 1217 (Serbian Kingdom established); 16 April 1346 (Serbian Empire established); 13 July 1878 (Congress of Berlin recognizes Serbian independence); 1 December 1918 (Kingdom of Serbs, Croats, and Slovenes (Yugoslavia) established)

National holiday: National Day (Statehood Day), 15 February (1835), the day the first constitution of the country was adopted

Constitution: *history:* many previous; latest adopted 30 September 2006, approved by referendum 28-29 October 2006, effective 8 November 2006

amendments: proposed by at least one third of deputies in the National Assembly, by the president of the republic, by the government, or by petition of at least 150,000 voters; passage of proposals and draft amendments each requires at least two-thirds majority vote in the Assembly; amendments to constitutional articles including the preamble, constitutional principles, and human and minority rights and freedoms also require passage by simple majority vote in a referendum

Legal system: civil law system

International law organization participation: has not submitted an ICJ jurisdiction declaration; accepts ICCt jurisdiction

Citizenship: *citizenship by birth:* no

citizenship by descent only: at least one parent must be a citizen of Serbia

dual citizenship recognized: yes

residency requirement for naturalization: 3 years

Suffrage: 18 years of age, 16 if employed; universal

Executive branch: *chief of state:* President Aleksandar VUCIC (since 31 May 2017)

head of government: Prime Minister Ana BRNABIC (since 29 June 2017)

cabinet: Cabinet elected by the National Assembly

elections/appointments: president directly elected by absolute majority popular vote in 2 rounds if needed for a 5-year term (eligible for a second term); election last held on 2 April 2017 (next to be held in 2022); prime minister elected by the National Assembly

election results: Aleksandar VUCIC elected president in the first round; percent of vote— Aleksandar VUCIC (SNS) 55.1%, Sasa JANKOVIC (independent) 16.4%, Luka MAKSIMOVIC (independent) 9.4%, Vuk JEREMIC (independent) 5.7%, Vojislav SESELJ (SRS) 4.5%, Bosko OBRADOVIC (Dveri) 2.3%, other 5.0%, invalid/blank 1.6%; Prime Minister Ana BRNABIC reelected by the National Assembly on 5 October 2020

Legislative branch: *description:* unicameral National Assembly or Narodna Skupstina (250 seats; members directly elected by party list proportional representation vote in a single nationwide constituency to serve 4-year terms)

elections: last held on 21 June 2020 (originally scheduled for 26 April 2020 but postponed due to the COVID-19 pandemic) (next to be held in 2024)

election results: percent of vote by party/coalition—For Our Children 60.7%, SPS-JS 10.4%, SPAS 3.8%, SVM 2.2%, Straight Ahead 1%, Albanian Democratic Alternative .8%, SDA .8%, other 20.3%; seats by party/coalition For Our Children 188, SPS-JS 32, SPAS 11, SVM 9, Straight Ahead 4, Albanian Democratic Alternative 3, SDA 3; composition (preliminary)-men 165, women 85, percent of women 30%

note: seats by party as of May 2019—SNS 91, SRS 22, SPS 20, DS 13, SDPS 10, PUPS 9, Dveri 6, JS 6, LDP 4, SDS 4, SVM 4, other 36, independent 25; composition—men 157, women 93, percent of women 37.2%

Judicial branch: *highest courts:* Supreme Court of Cassation (consists of 36 judges, including the court president); Constitutional Court (consists of 15 judges, including the court president and vice president)

judge selection and term of office: Supreme Court justices proposed by the High Judicial Council (HJC), an 11 -member independent body consisting of 8 judges elected by the National Assembly and 3 ex-officio members; justices appointed by the National Assembly; Constitutional Court judges elected—5 each by the National Assembly, the president, and the Supreme Court of Cassation; initial appointment of Supreme Court judges by the HJC is 3 years and beyond that period tenure is permanent; Constitutional Court judges elected for 9-year terms

subordinate courts: basic courts, higher courts, appellate courts; courts of special jurisdiction include the Administrative Court, commercial courts, and misdemeanor courts

Political parties and leaders: Albanian Democratic Alternative (coalition of ethnic Albanian parties) Shaip KAMBERI Alliance of Vojvodina Hungarians or SVM [Istvan PASZTOR] Democratic Party or DS [Zoran LUTOVAC] Democratic Party of Macedonians or DPM [Nenad KRSTESKI] Democratic Party of Serbia or DSS [Milos JOVANOVIC] Dveri [Bosko OBRADOVIC] For Our Children (electoral alliance includes SNS, SDP, PS, PUPS, PSS, SNP, SPO, NSS) [Aleksandar VUCIC] Justice and Reconciliation Party or SPP [Muamer ZUKORLIC] (formerly Bosniak Democratic Union of Sandzak or BDZS) Movement of Socialists or PS [Aleksandar VULIN] Party of Democratic Action of the Sandzak or SDA [Sulejman UGLJANIN] Party of United Pensioners of Serbia or PUPS [Milan KRKOBABIC] People's Party or NARODNA [Vuk JEREMIC] People's Peasant Party or NSS [Marijan RISTICEVIC]

Serbian Patriotic Alliance or SPAS [Aleksandar SAPIC] Serbian People's Party or SNP [Nenad POPOVIC] Serbian Progressive Party or SNS [Aleksandar VUCIC] Serbian Radical Party or SRS [Vojislav SESELJ] Serbian Renewal Movement or SPO [Vuk DRASKOVIC] Social Democratic Party or SDS [Boris TADIC] Social Democratic Party of Serbia or SDPS [Rasim LJAJIC] Socialist Party of Serbia or SPS [Ivica DACIC] Straight Ahead (electoral coalition includes SPP, DPM) Strength of Serbia or PSS [Bogoljub KARIC] Together for Serbia or ZZS [Nebojsa ZELENOVIC] United Serbia or JS [Dragan MARKOVIC]

note: Serbia has more than 110 registered political parties and citizens' associations

International organization participation: BIS, BSEC, CD, CE, CEI, EAPC, EBRD, EU (candidate country), FAO, G-9, IAEA, IBRD, ICAO, ICC (national committees), ICCt, ICRM, I DA , I FC, IFRCS, IHO, ILO, IMF, IMO, IMSO, Interpol, IOC, IOM, IPU, ISO, ITSO, ITU, ITUC (NGOs), MIGA, MONUSCO, NAM (observer), NSG, OAS (observer), OIF (observer), OPCW, OSCE, PCA, PFP, SELEC, UN, UNCTAD, UNESCO, UNFICYP, UNHCR, UNIDO, UNIFIL, UNMIL, UNOCI, UNTSO, UNWTO, UPU, WCO, WHO, WIPO, WMO, WTO (observer)

Diplomatic representation in the US: *chief of mission:* Ambassador Djerdj MATKOVIC (since 23 February 2015)

chancery: 2233 Wisconsin Ave NW, #410, Washington, DC 20007

telephone: [1] (202) 332-0333

FAX: [1] (202) 332-3933

consulate(s) general: Chicago, New York

Diplomatic representation from the US: *chief of mission:* Ambassador Anthony GODFREY (since 24 October 2019)

telephone: [381] (11) 706-4000

embassy: 92 Bulevar kneza Aleksandra Karadjordjevica, 11040 Belgrade

mailing address: 5070 Belgrade Place, Washington, DC 20521-5070

FAX: [381] (11) 706-4005

Flag description: three equal horizontal stripes of red (top), blue, and white—the Pan-Slav colors representing freedom and revolutionary ideals; charged with the coat of arms of Serbia shifted slightly to the hoist side; the principal field of the coat of arms represents the Serbian state and displays a white two-headed eagle on a red shield; a smaller red shield on the eagle represents the Serbian nation, and is divided into four quarters by a white cross; interpretations vary as to the meaning and origin of the white, curved symbols resembling firesteels (fire strikers) or Cyrillic "C's" in each quarter; a royal crown surmounts the coat of arms

note: the Pan-Slav colors were inspired by the 19th-century flag of Russia

National symbol(s): *white double-headed eagle; national colors:* red, blue, white

National anthem: *name:* "Boze pravde" (God of Justice)

lyrics/music: Jovan DORDEVIC/Davorin JENKO

note: adopted 1904; song originally written as part of a play in 1872 and has been used as an anthem by the Serbian people throughout the 20th and 21st centuries

0:00/ 1:46

ECONOMY

Economy—overview: Serbia has a transitional economy largely dominated by market forces, but the state sector remains significant in certain areas. The economy relies on manufacturing and exports, driven largely by foreign investment. MILOSEVIC-era mismanagement of the economy, an extended period of international economic sanctions, civil war, and the damage to Yugoslavia's infrastructure and industry during the NATO airstrikes in 1999 left the economy worse off than it was in 1990. In 2015, Serbia's GDP was 27.5% below where it was in 1989.

After former Federal Yugoslav President MILOSEVIC was ousted in September 2000, the Democratic Opposition of Serbia (DOS) coalition government implemented stabilization measures and embarked on a market reform program. Serbia renewed its membership in the IMF in December 2000 and rejoined the World Bank and the European Bank for Reconstruction and Development. Serbia has made progress in trade liberalization and enterprise restructuring and privatization, but many large enterprises—including the power utilities, telecommunications company, natural gas company, and others—remain state-owned. Serbia has made some progress towards EU membership, gaining candidate status in March 2012. In January 2014, Serbia's EU accession talks officially opened and, as of December 2017, Serbia had opened 12 negotiating chapters including one on foreign trade. Serbia's negotiations with the WTO are advanced, with the country's complete ban on the trade and cultivation of agricultural biotechnology products representing the primary remaining obstacle to accession. Serbia maintains a three-year Stand-by Arrangement with the IMF worth approximately $1.3 billion that is scheduled to end in February 2018. The government has shown progress implementing economic reforms, such as fiscal consolidation, privatization, and reducing public spending.

Unemployment in Serbia, while relatively low (16% in 2017) compared with its Balkan neighbors, remains significantly above the European average. Serbia is slowly implementing structural economic reforms needed to ensure the country's long-term prosperity. Serbia reduced its budget deficit to 1.7% of GDP and its public debt to 71% of GDP in 2017. Public debt had more than doubled between 2008 and 2015. Serbia's concerns about inflation and exchange-rate stability preclude the use of expansionary monetary policy.

Major economic challenges ahead include: stagnant household incomes; the need for private sector job creation; structural reforms of state-owned companies; strategic public sector reforms; and the need for new foreign direct investment. Other serious longer-term challenges include an inefficient judicial system, high levels of corruption, and an aging population. Factors favorable to Serbia's economic growth include the economic reforms it is undergoing as part of its EU accession process and IMF agreement, its strategic location, a relatively inexpensive and skilled labor force, and free trade agreements with the EU, Russia, Turkey, and countries that are members of the Central European Free Trade Agreement.

GDP (purchasing power parity): $105.7 billion (2017 est.)
$103.8 billion (2016 est.)
$101 billion (2015 est.)
note: data are in 2017 dollars
country comparison to the world: 82

GDP (official exchange rate): $41.43 billion (2017 est.)

GDP—real growth rate: 4.18% (2019 est.)
4.4% (2018 est.)
2.05% (2017 est.)
country comparison to the world: 72

GDP—per capita (PPP): $15,100 (2017 est.)
$14,700 (2016 est.)
$14,200 (2015 est.)
note: data are in 2017 dollars
country comparison to the world: 111

Gross national saving: 15.3% of GDP (2017 est.)
16% of GDP (2016 est.)
14.1% of GDP (2015 est.)
country comparison to the world: 134

GDP—composition, by end use:
household consumption: 78.2% (2017 est.)
government consumption: 10.1% (2017 est.)
investment in fixed capital: 18.5% (2017 est.)
investment in inventories: 2% (2017 est.)
exports of goods and services: 52.5% (2017 est.)
imports of goods and services: -61.3% (2017 est.)

GDP—composition, by sector of origin:
agriculture: 9.8% (2017 est.)
industry: 41.1% (2017 est.)
services: 49.1% (2017 est.)

Agriculture—products: wheat, maize, sunflower, sugar beets, grapes/wine, fruits (raspberries, apples, sour cherries), vegetables (tomatoes, peppers, potatoes), beef, pork, and meat products, milk and dairy products

Industries: automobiles, base metals, furniture, food processing, machinery, chemicals, sugar, tires, clothes, pharmaceuticals

Industrial production growth rate: 3.9% (2017 est.)
country comparison to the world: 78

Labor force: 3 million (2020 est.)
country comparison to the world: 104

Labor force—by occupation: *agriculture:* 19.4%
industry: 24.5%
services: 56.1% (2017 est.)

Unemployment rate: 14.1% (2017 est.)

15.9% (2016 est.)
country comparison to the world: 171

Population below poverty line: 8.9% (2014 est.)

Household income or consumption by percentage share: *lowest 10%:* 2.2%
highest 10%: 23.8% (2011)

Budget: *revenues:* 17.69 billion (2017 est.)
expenditures: 17.59 billion (2017 est.)
note: data include both central government and local goverment budgets

Taxes and other revenues: 42.7% (of GDP) (2017 est.)
country comparison to the world: 30

Budget surplus (+) or deficit (-): 0.2% (of GDP) (2017 est.)
country comparison to the world: 43

Public debt: 62.5% of GDP (2017 est.)
73.1% of GDP (2016 est.)
country comparison to the world: 70

Inflation rate (consumer prices): 3.1% (2017 est.)
1.1% (2016 est.)
country comparison to the world: 133

Current account balance: -$2.354 billion (2017 est.)
-$1.189 billion (2016 est.)
country comparison to the world: 170

Exports: $15.92 billion (2017 est.)
$13.99 billion (2016 est.)
country comparison to the world: 73

Exports—partners: Italy 13.5%, Germany 12.8%, Bosnia and Herzegovina 8.2%, Russia 6%, Romania 4.9% (2017)

Exports—commodities: automobiles, iron and steel, rubber, clothes, wheat, fruit and vegetables, nonferrous metals, electric appliances, metal products, weapons and ammunition

Imports: $20.44 billion (2017 est.)
$17.63 billion (2016 est.)
country comparison to the world: 76

Imports—commodities: machinery and transport equipment, fuels and lubricants, manufactured goods, chemicals, food and live animals, raw materials

Imports—partners: Germany 12.7%, Italy 10%, China 8.2%, Russia 7.3%, Hungary 4.9%, Poland 4.1% (2017)

Reserves of foreign exchange and gold: $11.91 billion (31 December 2017 est.)
$10.76 billion (31 December 2016 est.)
country comparison to the world: 70

Debt—external: $29.5 billion (31 December 2017 est.)
$30.38 billion (31 December 2016 est.)
country comparison to the world: 81

Exchange rates: Serbian dinars (RSD) per US dollar –
112.4 (2017 est.)
111.278 (2016 est.)
111.278 (2015 est.)
108.811 (2014 est.)
88.405 (2013 est.)

Electricity **access:** *electrification—total population:* 100% (2020)

Electricity—production: 36.54 billion kWh (2016 est.)
country comparison to the world: 59

Electricity—consumption: 29.81 billion kWh (2016 est.)
country comparison to the world: 62

Electricity—exports: 6.428 billion kWh (2016 est.)
country comparison to the world: 29

Electricity—imports: 5.068 billion kWh (2016 est.)
country comparison to the world: 38

Electricity—installed generating capacity: 7.342 million kW (2016 est.)
country comparison to the world: 73

Electricity—from fossil fuels: 65% of total installed capacity (2016 est.)
country comparison to the world: 119

Electricity—from nuclear fuels: 0% of total installed capacity (2017 est.)
country comparison to the world: 180

Electricity—from hydroelectric plants: 35% of total installed capacity (2017 est.)
country comparison to the world: 61

Electricity—from other renewable sources: 1% of total installed capacity (2017 est.)
country comparison to the world: 167

Crude oil—production: 17,000 bbl/day (2018 est.)
country comparison to the world: 69

Crude oil—exports: 123 bbl/day (2015 est.)
country comparison to the world: 81

Crude oil—imports: 40,980 bbl/day (2015 est.)
country comparison to the world: 56

Crude oil—proved reserves: 77.5 million bbl (1 January 2018 est.)
country comparison to the world: 73

Refined petroleum products—production: 74,350 bbl/day (2015 est.)
country comparison to the world: 71

Refined **petroleum** **products—consumption:** 74,000 bbl/day (2016 est.)
country comparison to the world: 90

Refined petroleum products—exports: 15,750 bbl/day (2015 est.)
country comparison to the world: 73

Refined petroleum products—imports: 18,720 bbl/day (2015 est.)
country comparison to the world: 126

Natural gas—production: 509.7 million cu m (2017 est.)
country comparison to the world: 71

Natural gas—consumption: 2.718 billion cu m (2017 est.)
country comparison to the world: 75

Natural gas—exports: 0 cu m (2017 est.)
country comparison to the world: 181

Natural gas—imports: 2.01 billion cu m (2017 est.)
country comparison to the world: 52

Natural gas—proved reserves: 48.14 billion cu m (1 January 2018 est.)
country comparison to the world: 63

Carbon dioxide emissions from consumption of energy: 50.21 million Mt (2017 est.)
country comparison to the world: 60

COMMUNICATIONS

Telephones—fixed lines: *total subscriptions:* 2,060,005
subscriptions per 100 inhabitants: 29.24 (2019 est.)
country comparison to the world: 53

Telephones—mobile cellular: *total subscriptions:* 6,789,423
subscriptions per 100 inhabitants: 96.37 (2019 est.)
country comparison to the world: 104

Telecommunication systems: *general assessment:* Serbia's integration with the EU has helped regulator reforms and promotion of telecoms; wireless service is available through multiple providers; national coverage is growing very rapidly; best telecommunications services are centered in urban centers; 4G/LTE mobile network launched; 5G trials; high mobile penetration the result of multiple SIM cards (2020)
domestic: fixed-line 29 per 100 and mobile-cellular 96 per 100 persons (2019)
international: country code—381
note: the COVID-19 outbreak is negatively impacting telecommunications production and supply chains globally; consumer spending on telecom devices and services has also slowed due to the pandemic's effect on economies worldwide; overall progress towards improvements in all facets of the telecom industry—mobile, fixed-line, broadband, submarine cable and satellite—has moderated

Internet country code: .rs

Internet users: *total:* 5,192,501
percent of population: 73.36% (July 2018 est.)
country comparison to the world: 81

Broadband—fixed subscriptions: *total:* 1,552,160
subscriptions per 100 inhabitants: 22 (2018 est.)
country comparison to the world: 62

TRANSPORTATION

National air transport system: *number of registered air carriers:* 4 (2020)

inventory of registered aircraft operated by air carriers: 43
annual passenger traffic on registered air carriers: 2,262,703 (2018)
annual freight traffic on registered air carriers: 17.71 million mt-km (2018)

Civil aircraft registration country code prefix: YU (2016)

Airports: 26 (2013)
country comparison to the world: 125

Airports—with paved runways: *total:* 10 (2017)
over 3,047 m: 2 (2017)
2,438 to 3,047 m: 3 (2017)
1,524 to 2,437 m: 3 (2017)
914 to 1,523 m: 2 (2017)

Airports—with unpaved runways: *total:* 16 (2013)
1,524 to 2,437 m: 1 (2013)
914 to 1,523 m: 10 (2013)
under 914 m: 5 (2013)

Heliports: 2 (2012)

Pipelines: 1936 km gas, 413 km oil

Railways: *total:* 3,809 km (2015)
standard gauge: 3,809 km 1.435-m gauge (3,526 km one-track lines and 283 km double-track lines) out of which 1,279 km electrified (1,000 km one-track lines and 279 km double-track lines) (2015)
country comparison to the world: 53

Roadways: *total:* 44,248 km (2016)
paved: 28,000 km (16,162 km state roads, out of which 741 km highways) (2016)
unpaved: 16,248 km (2016)
country comparison to the world: 86

Waterways: 587 km (primarily on the Danube and Sava Rivers) (2009)
country comparison to the world: 80

Ports and terminals: *river port(s):* Belgrade (Danube)

MILITARY AND SECURITY

Military and security forces: Serbian Armed Forces (Vojska Srbije, VS): Land Forces (includes Riverine Component, consisting of a river flotilla on the Danube), Air and Air Defense Forces, Serbian Guard; Ministry of Interior: Gendarmerie (2019)
note: the Guard is a brigade-sized unit that is directly subordinate to the Serbian Armed Forces Chief of General Staff

Military expenditures: 2.2% of GDP (2019)
1.6% of GDP (2018 est.)
1.8% of GDP (2017 est.)
1.7% of GDP (2016)
1.8% of GDP (2015)

country comparison to the world: 43

Military and security service personnel strengths: size estimates for the Serbian Armed Forces vary; approximately 25,000 active duty troops (13,500 Land Forces; 5,000 Air/Air Defense; 1,500 Guards Brigade; 5,000 other, including training, logistics, intelligence, medical, and other support staff) (2019 est.)

Military equipment inventories and acquisitions: the inventory of the Serbian Armed Forces consists of Russian and Soviet-era weapons systems; since 2010, most of its weapons imports have come from Russia, but it has also received equipment from Belarus (second-hand aircraft), Germany, Montenegro (second-hand aircraft), and the US (2019 est.)

Military deployments: 175 Lebanon (UNIFIL) (2020)

Military service age and obligation: 18 years of age for voluntary military service; conscription abolished December 2010 (2019)

TRANSNATIONAL ISSUES

Disputes—international: Serbia with several other states protest the US and other states' recognition of Kosovo's declaration of its status as a sovereign and independent state in February 2008; ethnic Serbian municipalities along Kosovo's northern border challenge final status of Kosovo-Serbia boundary; several thousand NATO-led Kosovo Force peacekeepers under UN Interim Administration Mission in Kosovo authority continue to keep the peace within Kosovo between the ethnic Albanian majority and the Serb minority in Kosovo; Serbia delimited about half of the boundary with Bosnia and Herzegovina, but sections along the Drina River remain in dispute

Refugees and internally displaced persons: *refugees (country of origin):* 17,972 (Croatia), 8,198 (Bosnia and Herzegovina) (2019)
IDPs: 199,584 (most are Kosovar Serbs, some are Roma, Ashkalis, and Egyptian (RAE); some RAE IDPs are unregistered) (2019)
stateless persons: 2,052 (includes stateless persons in Kosovo) (2018)
note: 741,752 estimated refugee and migrant arrivals (January 2015-November 2020); Serbia is predominantly a transit country and hosts an estimated 6,304 migrants and asylum seekers as of the end of May 2020; 8,827 migrant arrivals in 2018

Illicit drugs: transshipment point for Southwest Asian heroin moving to Western Europe on the Balkan route; economy vulnerable to money laundering

SEYCHELLES

INTRODUCTION

Background: Seychelles was uninhabited prior to being discovered by Europeans early in the 16th century. A lengthy struggle between France and Great Britain for the islands ended in 1814, when they were ceded to the latter. During colonial rule, a plantation-based economy developed that relied on imported labor, primarily from European colonies in Africa. Independence came in 1976. Following a coup d'etat in 1977, the country was a socialist one-party state until adopting a new

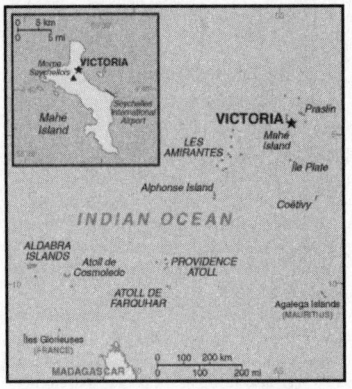

constitution and holding free elections in 1993. President France-Albert RENE, who had served since 1977, was reelected in 2001, but stepped down in 2004. Vice President James Alix MICHEL took over the presidency and in 2006 was elected to a new five-year term; he was reelected in 2011 and again in 2015. In 2016, James MICHEL resigned and handed over the presidency to his vice-president, Danny FAURE.

GEOGRAPHY

Location: archipelago in the Indian Ocean, northeast of Madagascar

Geographic coordinates: 4 35 S, 55 40 E

Map references: Africa

Area: *total:* 455 sq km
land: 455 sq km
water: 0 sq km
country comparison to the world: 199

Area—comparative: 2.5 times the size of Washington, DC

Land boundaries: 0 km

Coastline: 491 km

Maritime claims: *territorial sea:* 12 nm
exclusive economic zone: 200 nm
contiguous zone: 24 nm
continental shelf: 200 nm or to the edge of the continental margin

Climate: tropical marine; humid; cooler season during southeast monsoon (late May to September); warmer season during northwest monsoon (March to May)

Terrain: Mahe Group is volcanic with a narrow coastal strip and rocky, hilly interior; others are coral, flat, elevated reefs

Elevation: *lowest point:* Indian Ocean 0 m
highest point: Morne Seychellois 905 m

Natural resources: fish, coconuts (copra), cinnamon trees

Land use: *agricultural land:* 6.5% (2011 est.)
arable land: 2.2% (2011 est.) / permanent crops: 4.3% (2011 est.) / permanent pasture: 0% (2011 est.)
forest: 88.5% (2011 est.)

other: 5% (2011 est.)
Irrigated land: 3 sq km (2012)

Population distribution: more than three-quarters of the population lives on the main island of Mahe; Praslin contains less than 10%; a smaller percent on La Digue and the outer islands as shown in this population distribution map

Natural hazards: lies outside the cyclone belt, so severe storms are rare; occasional short droughts

Environment—current issues: water supply depends on catchments to collect rainwater; water pollution; biodiversity maintainance

Environment—international agreements: *party to:* Biodiversity, Climate Change, Climate Change-Kyoto Protocol, Desertification, Endangered Species, Hazardous Wastes, Law of the Sea, Marine Dumping, Ozone Layer Protection, Ship Pollution, Wetlands
signed, but not ratified: none of the selected agreements

Geography—note: *the smallest African country in terms of both area and population; the constitution of the Republic of Seychelles lists 155 islands: 42 granitic and 113 coralline; by far the largest island is Mahe, which is home to about 90% of the population and the site of the capital city of Victoria*

PEOPLE AND SOCIETY

Population: 95,981 (July 2020 est.)
country comparison to the world: 198

Nationality: *noun:* Seychellois (singular and plural)
adjective: Seychellois

Ethnic groups: predominantly creole (mainly of East African and Malagasy heritage); also French, Indian, Chinese, and Arab populations

Languages: Seychellois Creole (official) 89.1%, English (official) 5.1%, French (official) 0.7%, other 3.8%, unspecified 1.4% (2010 est.)

Religions: Roman Catholic 76.2%, Protestant 10.5% (Anglican 6.1%, Pentecostal Assembly 1.5%, Seventh Day Adventist 1.2%, other Protestant 1.7%), other Christian 2.4%, Hindu 2.4%, Muslim 1.6%, other non-Christian 1.1%, unspecified 4.8%, none 0.9% (2010 est.)

Demographic profile: Seychelles has no indigenous population and was first permanently settled by a small group of French planters, African slaves, and South Indians in 1770. Seychelles' modern population is composed of the descendants of French and later British settlers, Africans, and Indian, Chinese, and Middle Eastern traders and is concentrated on three of its 155 islands – the vast majority on Mahe and lesser numbers on Praslin and La Digue. Seychelles' population grew rapidly during the second half of the 20th century, largely due to natural increase, but the pace has slowed because of fertility decline. The total fertility rate dropped sharply from 4.0 children per woman in 1980 to 1.9 in 2015, mainly as a result of a family planning program, free education and health care, and increased female labor force participation.

Life expectancy has increased steadily, but women on average live 9 years longer than men, a difference that is higher than that typical of developed countries.

The combination of reduced fertility and increased longevity has resulted in an aging population, which will put pressure on the government's provision of pensions and health care. Seychelles' sustained investment in social welfare services, such as free primary health care and education up to the post-secondary level, have enabled the country to achieve a high human development index score – among the highest in Africa. Despite some of its health and education indicators being nearly on par with Western countries, Seychelles has a high level of income inequality.

An increasing number of migrant workers – mainly young men – have been coming to Seychelles in recent years to work in the construction and tourism industries. As of 2011, foreign workers made up nearly a quarter of the workforce. Indians are the largest non-Seychellois population – representing half of the country's foreigners – followed by Malagasy.

Age structure: *0-14 years:* 18.85% (male 9,297/female 8,798)
15-24 years: 12.39% (male 6,283/female 5,607)
25-54 years: 49.03% (male 25,209/female 21,851)
55-64 years: 11.46% (male 5,545/female 5,455)
65 years and over: 8.27% (male 3,272/female 4,664) (2020 est.)

Dependency ratios: *total dependency ratio:* 46.7
youth dependency ratio: 34.9
elderly dependency ratio: 11.8
potential support ratio: 8.5 (2020 est.)

Median age: *total:* 36.8 years
male: 36.3 years
female: 37.4 years (2020 est.)
country comparison to the world: 76

Population growth rate: 0.69% (2020 est.)
country comparison to the world: 139

Birth rate: 12.8 births/1,000 population (2020 est.)
country comparison to the world: 148

Death rate: 7.1 deaths/1,000 population (2020 est.)
country comparison to the world: 122

Net migration rate: 1 migrant(s)/1,000 population (2020 est.)
country comparison to the world: 60

Population distribution: more than three-quarters of the population lives on the main island of Mahe; Praslin contains less than 10%; a smaller percent on La Digue and the outer islands as shown in this population distribution map

Urbanization: *urban population:* 57.5% of total population (2020)
rate of urbanization: 1.26% annual rate of change (2015-20 est.)

total population growth rate v. urban population growth rate, 2000-2030: *Major urban areas—population:* 28,000 VICTORIA (capital) (2018)

Sex ratio: *at birth:* 1.03 male(s)/female
0-14 years: 1.06 male(s)/female

15-24 years: 1.12 male(s)/female
25-54 years: 1.15 male(s)/female
55-64 years: 1.02 male(s)/female
65 years and over: 0.7 male(s)/female
total population: 1.07 male(s)/female (2020 est.)

Infant mortality rate: *total:* 9.3 deaths/1,000 live births
male: 11.5 deaths/1,000 live births
female: 6.9 deaths/1,000 live births (2020 est.)
country comparison to the world: 136

Life expectancy at birth: *total population:* 75.6 years
male: 71.1 years
female: 80.2 years (2020 est.)
country comparison to the world: 112

Total fertility rate: 1.83 children born/woman (2020 est.)
country comparison to the world: 146

Drinking water source: *improved:* total: 96.2% of population
unimproved: *total:* 3.8% of population (2017 est.)

Current Health Expenditure: 5% (2017)

Physicians density: 2.12 physicians/1,000 population (2016)

Hospital bed density: 3.6 beds/1,000 population (2011)

Sanitation facility access: *improved:* total: 100% of population
unimproved: *total:* 0% of population (2017 est.)

HIV/AIDS—adult prevalence rate: NA

HIV/AIDS—people living with HIV/AIDS: NA

HIV/AIDS—deaths: NA

Obesity—adult prevalence rate: 14% (2016)
country comparison to the world: 130

Children under the age of 5 years underweight: 3.6% (2012)
country comparison to the world: 93

Education expenditures: 4.4% of GDP (2016)
country comparison to the world: 89

Literacy: *definition:* age 15 and over can read and write
total population: 95.9%
male: 95.4%
female: 96.4% (2018)

School life expectancy (primary to tertiary education): *total:* 14 years
male: 13 years
female: 16 years (2019)

Unemployment, youth ages 15-24: *total:* 11.6%
male: 12.6%
female: 10.4% (2018 est.)
country comparison to the world: 113

GOVERNMENT

Country name: *conventional long form:* Republic of Seychelles
conventional short form: Seychelles
local long form: Republic of Seychelles
local short form: Seychelles
etymology: named by French Captain Corneille Nicholas MORPHEY after Jean Moreau de

SECHELLES, the finance minister of France, in 1756

Government type: presidential republic

Capital: *name:* Victoria

geographic coordinates: 4 37 S, 55 27 E
time difference: UTC+4 (9 hours ahead of Washington, DC, during Standard Time)
etymology: founded as L'etablissement in 1778 by French colonists, the town was renamed in 1841 by the British after Queen Victoria (1819-1901); "victoria" is the Latin word for "victory"

Administrative divisions: 27 administrative districts; Anse aux Pins, Anse Boileau, Anse Etoile, Anse Royale, Au Cap, Baie Lazare, Baie Sainte Anne, Beau Vallon, Bel Air, Bel Ombre, Cascade, Glacis, Grand Anse Mahe, Grand Anse Praslin, Ile Perseverance I, Ile Perseverance II, La Digue, La Riviere Anglaise, Les Mamelles, Mont Buxton, Mont Fleuri, Plaisance, Pointe Larue, Port Glaud, Roche Caiman, Saint Louis, Takamaka

Independence: 29 June 1976 (from the UK)

National holiday: Constitution Day, 18 June (1993); Independence Day (National Day), 29 June (1976)

Constitution: *history:* previous 1970, 1979; latest drafted May 1993, approved by referendum 18 June 1993, effective 23 June 1993
amendments: proposed by the National Assembly; passage requires at least two-thirds majority vote by the National Assembly; passage of amendments affecting the country's sovereignty, symbols and languages, the supremacy of the constitution, fundamental rights and freedoms, amendment procedures, and dissolution of the Assembly also requires approval by at least 60% of voters in a referendum; amended several times, last in 2017

Legal system: mixed legal system of English common law, French civil law, and customary law

International law organization participation: has not submitted an ICJ jurisdiction declaration; accepts ICCt jurisdiction

Citizenship: *citizenship by birth:* no
citizenship by descent only: at least one parent must be a citizen of the Seychelles
dual citizenship recognized: no
residency requirement for naturalization: 5 years

Suffrage: 18 years of age; universal

Executive branch: *chief of state:* President Wavel RAMKALAWAN (since 26 October 2020); Vice President Ahmed AFIF (since 27 October 2020); the president is both chief of state and head of government
head of government: President Wavel RAMKALAWAN (since 26 October 2020); Vice President Ahmed AFIF (since 27 October 2020)
cabinet: Council of Ministers appointed by the president
elections/appointments: president directly elected by absolute majority popular vote in 2 rounds if needed for a 5-year term (eligible for 1 additional term); election last held on 22-24 Oct 2020 (originally scheduled for December 2020 but

moved up to coincide with the 22-24 October National Assembly election in order to cut election costs)
election results: Wavel RAMKALAWAN elected president; Wavel RAMKALAWAN (LDS) 54.9%, Danny FAURE (US) 43.5%

Legislative branch: *description:* unicameral National Assembly or Assemblee Nationale (35 seats in the 2020 -25 term; 26 members directly elected in single-seat constituencies by simple majority vote and up to 9 members elected by proportional representation vote; members serve 5-year terms)
elections: last held on 22-24 Oct 2020 (next to be held October 2025); note—the election was originally scheduled for 2021 but was moved up a year and will be held alongside the presidential election in order to cut election costs
election results: percent of vote by party—LDS 54.8%, US 42.3% , other 2.9%; seats by party—LDS 25, US10; composition—men 25, women 10, percent of women 29%

Judicial branch: *highest courts:* Seychelles Court of Appeal (consists of the court president and 4 justices); Supreme Court of Seychelles (consists of the chief justice and 9 puisne judges); Constitutional Court (consists of 3 Supreme Court judges)
judge selection and term of office: all judges appointed by the president of the republic upon the recommendation of the Constitutional Appointments Authority, a 3-member body, with 1 member appointed by the president of the republic, 1 by the opposition leader in the National Assembly, and 1 by the other 2 appointees; judges serve until retirement at age 70
subordinate courts: Magistrates' Courts of Seychelles; Family Tribunal for issues such as domestic violence, child custody, and maintenance; Employment Tribunal for labor-related disputes

Political parties and leaders: Lafors Seselwa Demokratik or LSD [Martin AGLAE]
One Seychelles [Alain St. ANGE]
Seselwa (Seychelles) United Party or SUP [Robert ERNESTA] (formerly the New Democratic Party or NDP)
Seychelles National Party or SNP [Wavel RAMKALAWAN] (formerly the United Opposition or UO)
Seychelles Party for Social Justice and Democracy or SPSD [Alexia AMESBURY]
Seychelles Patriotic Movement or SPM [Vincent LARUER]
Seychelloise Alliance (Lalyans Seselwa) [Patrick PILLAY]
Seychellois Democratic Alliance (Linyon Demokratik Seselwa) or LDS [Roger MANCIENNE] (includes SNP, SPSD, and SUP)
United Seychelles or US [Vincent MERITON] (formerly People's Party (Parti Lepep) or PL; (formerly SPPF)

International organization participation: ACP, AfDB, AOSIS, AU, C, CD, COMESA, EITI (candidate country), FAO, G-77, IAEA, IBRD, ICAO,

ICC (NGOs), ICCt, ICRM, IDA, IFAD, IFC, IFRCS, ILO, IMF, IMO, IMO, InOC, Interpol, IOC, IOM, IPU, ISO (correspondent), ITU, MIGA, NAM, OIF, OPCW, SADC, UN, UNCTAD, UNESCO, UNIDO, UNWTO, UPU, WCO, WHO, WIPO, WMO, WTO (observer)

Diplomatic representation in the US: *chief of mission:* Ambassador Ronald Jean JUMEAU (since 8 September 2017)
chancery: 800 Second Avenue, Suite 400C, New York, NY 10017
telephone: [1] (212) 972-1785
FAX: [1] (212) 972-1786
consulate(s) general: New York

Diplomatic representation from the US: the US does not have an embassy in Seychelles; the US Ambassador to Mauritius is accredited to Seychelles

Flag description: five oblique bands of blue (hoist side), yellow, red, white, and green (bottom) radiating from the bottom of the hoist side; the oblique bands are meant to symbolize a dynamic new country moving into the future; blue represents sky and sea, yellow the sun giving light and life, red the peoples' determination to work for the future in unity and love, white social justice and harmony, and green the land and natural environment

National symbol(s): *coco de mer (sea coconut); national colors:* blue, yellow, red, white, green

National anthem: *name:* "Koste Seselwa" (Seychellois Unite)
lyrics/music: David Francois Marc ANDRE and George Charles Robert PAYET
note: adopted 1996
0:00 / 1:07

ECONOMY

Economy—overview: Since independence in 1976, per capita output in this Indian Ocean archipelago has expanded to roughly seven times the pre-independence, near-subsistence level, moving the island into the high income group of countries. Growth has been led by the tourism sector, which directly employs about 26% of the labor force and directly and indirectly accounts for more than 55% of GDP, and by tuna fishing. In recent years, the government has encouraged foreign investment to upgrade hotels and tourism industry services. At the same time, the government has moved to reduce the dependence on tourism by promoting the development of the offshore financial, information, and communication sectors and renewable energy.

In 2008, having depleted its foreign exchange reserves, Seychelles defaulted on interest payments due on a $230 million Eurobond, requested assistance from the IMF, and immediately enacted a number of significant structural reforms, including liberalization of the exchange rate, reform of the public sector to include layoffs, and the sale of some state assets. In December 2013, the IMF declared that Seychelles had successfully transitioned to a market-based economy with full employment and a fiscal surplus. However,

state-owned enterprises still play a prominent role in the economy. Effective 1 January 2017, Seychelles was no longer eligible for trade benefits under the US African Growth and Opportunities Act after having gained developed country status. Seychelles grew at 5% in 2017 because of a strong tourism sector and low commodity prices. The Seychellois Government met the IMF's performance criteria for 2017 but recognizes a need to make additional progress to reduce high income inequality, represented by a Gini coefficient of 46.8.

As a very small open economy dependent on tourism, Seychelles remains vulnerable to developments such as economic downturns in countries that supply tourists, natural disasters, and changes in local climatic conditions and ocean temperature. One of the main challenges facing the government is implementing strategies that will increase Seychelles' long-term resilience to climate change without weakening economic growth.

GDP (purchasing power parity): $2.75 billion (2017 est.)
$2.612 billion (2016 est.)
$2.499 billion (2015 est.)
note: data are in 2017 dollars
country comparison to the world: 190

GDP (official exchange rate): $1.498 billion (2017 est.)

GDP—real growth rate: 5.3% (2017 est.)
4.5% (2016 est.)
4.9% (2015 est.)
country comparison to the world: 40

GDP—per capita (PPP): $29,300 (2017 est.)
$27,800 (2016 est.)
$26,900 (2015 est.)
note: data are in 2017 dollars
country comparison to the world: 70

Gross national saving: 8.1% of GDP (2017 est.)
10.2% of GDP (2016 est.)
15.2% of GDP (2015 est.)
country comparison to the world: 169

GDP—composition, by end use:
household consumption: 52.7% (2017 est.)
government consumption: 34.4% (2017 est.)
investment in fixed capital: 26.7% (2017 est.)
investment in inventories: 0% (2017 est.)
exports of goods and services: 79.4% (2017 est.)
imports of goods and services: -93.2% (2017 est.)

GDP—composition, by sector of origin:
agriculture: 2.5% (2017 est.)
industry: 13.8% (2017 est.)
services: 83.7% (2017 est.)

Agriculture—products: coconuts, cinnamon, vanilla, sweet potatoes, cassava (manioc, tapioca), copra, bananas; tuna

Industries: fishing, tourism, beverages

Industrial production growth rate: 2.3% (2017 est.)
country comparison to the world: 121

Labor force: 51,000 (2018 est.)
country comparison to the world: 192

Labor force—by occupation: *agriculture:* 3%
industry: 23%

services: 74% (2006)

Unemployment rate: 3% (2017 est.)
2.7% (2016 est.)
country comparison to the world: 38

Population below poverty line: 39.3% (2013 est.)

Household income or consumption by percentage share: *lowest 10%:* 4.7%
highest 10%: 15.4% (2007)

Budget: *revenues:* 593.4 million (2017 est.)
expenditures: 600.7 million (2017 est.)

Taxes and other revenues: 39.6% (of GDP) (2017 est.)
country comparison to the world: 44

Budget surplus (+) or deficit (-): -0.5% (of GDP) (2017 est.)
country comparison to the world: 63

Public debt: 63.6% of GDP (2017 est.)
69.1% of GDP (2016 est.)
country comparison to the world: 65

Fiscal year: calendar year

Inflation rate (consumer prices): 2.9% (2017 est.)
-1% (2016 est.)
country comparison to the world: 132

Current account balance: -$307 million (2017 est.)
-$286 million (2016 est.)
country comparison to the world: 110

Exports: $564.8 million (2017 est.)
$477.6 million (2016 est.)
country comparison to the world: 172

Exports—partners: UAE 28.5%, France 24%, UK 13.8%, Italy 8.9%, Germany 4.6% (2017)

Exports—commodities: canned tuna, frozen fish, petroleum products (reexports)

Imports: $1.155 billion (2017 est.)
$991 million (2016 est.)
country comparison to the world: 180

Imports—commodities: machinery and equipment, foodstuffs, petroleum products, chemicals, other manufactured goods

Imports—partners: UAE 13.4%, France 9.4%, Spain 5.7%, South Africa 5% (2017)

Reserves of foreign exchange and gold: $545.2 million (31 December 2017 est.)
$523.5 million (31 December 2016 est.)
country comparison to the world: 149

Debt—external: $2.559 billion (31 December 2017 est.)
$2.651 billion (31 December 2016 est.)
country comparison to the world: 147

Exchange rates: Seychelles rupees (SCR) per US dollar -
13.64 (2017 est.)
13.319 (2016 est.)
13.319 (2015 est.)
13.314 (2014 est.)
12.747 (2013 est.)

ENERGY

Electricity access: *electrification—total population:* 100% (2020)

Electricity—production: 350 million kWh (2016 est.)
country comparison to the world: 175

Electricity—consumption: 325.5 million kWh (2016 est.)
country comparison to the world: 182

Electricity—exports: 0 kWh (2016 est.)
country comparison to the world: 196

Electricity—imports: 0 kWh (2016 est.)
country comparison to the world: 198

Electricity—installed generating capacity: 88,000 kW (2016 est.)
country comparison to the world: 181

Electricity—from fossil fuels: 91% of total installed capacity (2016 est.)
country comparison to the world: 56

Electricity—from nuclear fuels: 0% of total installed capacity (2017 est.)
country comparison to the world: 181

Electricity—from hydroelectric plants: 0% of total installed capacity (2017 est.)
country comparison to the world: 199

Electricity—from other renewable sources: 9% of total installed capacity (2017 est.)
country comparison to the world: 83

Crude oil—production: 0 bbl/day (2018 est.)
country comparison to the world: 198

Crude oil—exports: 0 bbl/day (2015 est.)
country comparison to the world: 191

Crude oil—imports: 0 bbl/day (2015 est.)
country comparison to the world: 193

Crude oil—proved reserves: 0 bbl (1 January 2018 est.)
country comparison to the world: 193

Refined petroleum products—production: 0 bbl/day (2015 est.)
country comparison to the world: 199

Refined petroleum products—consumption: 7,300 bbl/day (2016 est.)
country comparison to the world: 166

Refined petroleum products—exports: 0 bbl/day (2015 est.)
country comparison to the world: 200

Refined petroleum products—imports: 7,225 bbl/day (2015 est.)
country comparison to the world: 155

Natural gas—production: 0 cu m (2017 est.)
country comparison to the world: 194

Natural gas—consumption: 0 cu m (2017 est.)
country comparison to the world: 196

Natural gas—exports: 0 cu m (2017 est.)
country comparison to the world: 182

Natural gas—imports: 0 cu m (2017 est.)
country comparison to the world: 187

Natural gas—proved reserves: 0 cu m (1 January 2014 est.)
country comparison to the world: 192

Carbon dioxide emissions from consumption of energy: 1.15 million Mt (2017 est.)
country comparison to the world: 165

COMMUNICATIONS

Telephones—fixed lines: *total subscriptions:* 19,627
subscriptions per 100 inhabitants: 20.59 (2019 est.)
country comparison to the world: 177

Telephones—mobile cellular: *total subscriptions:* 188,879
subscriptions per 100 inhabitants: 198.15 (2019 est.)
country comparison to the world: 183

Telecommunication systems: *general assessment:* effective system; direct international calls to over 100 countries; radiotelephone communications between islands in the archipelago; 3 ISPs; use of Internet cafes' for access to Internet; 4G services and 5G pending (2020)
domestic: fixed-line 21 per 100 and mobile-cellular teledensity is 198 telephones per 100 persons (2019)
international: country code—248; landing points for the PEACE and the SEAS submarine cables providing connectivity to Europe, the Middle East, Africa and Asia; direct radiotelephone communications with adjacent island countries and African coastal countries; satellite earth station—1 Intelsat (Indian Ocean) (2019)
note: the COVID-19 outbreak is negatively impacting telecommunications production and supply chains globally; consumer spending on telecom devices and services has also slowed due to the pandemic's effect on economies worldwide; overall progress towards improvements in all facets of the telecom industry—mobile, fixed-line, broadband, submarine cable and satellite—has moderated

Broadcast media: the national broadcaster, Seychelles Broadcasting Corporation (SBC), which is funded by taxpayer money, operates the only terrestrial TV station, which provides local programming and airs broadcasts from international services; a privately owned Internet Protocol Television (IPTV) channel also provides local programming multi-channel cable and satellite TV are available through 2 providers; the national broadcaster operates 1 AM and 1 FM radio station; there are 2 privately operated radio stations; transmissions of 2 international broadcasters are accessible in Victoria (2019)

Internet country code: .sc

Internet users: *total:* 55,616
percent of population: 58.77% (July 2018 est.)
country comparison to the world: 194

Broadband—fixed subscriptions: *total:* 19,696
subscriptions per 100 inhabitants: 21 (2018 est.)
country comparison to the world: 153

TRANSPORTATION

National air transport system: *number of registered air carriers:* 1 (2020)

inventory of registered aircraft operated by air carriers: 7
annual passenger traffic on registered air carriers: 455,201 (2018)
annual freight traffic on registered air carriers: 7.79 million mt-km (2018)

Civil aircraft registration country code prefix: S7 (2016)

Airports: 14 (2013)
country comparison to the world: 150

Airports—with paved runways: *total:* 7 (2019)
2,438 to 3,047 m: 1
914 to 1,523 m: 5
under 914 m:1

Airports—with unpaved runways: *total:* 7 (2013)
914 to 1,523 m: 2 (2013)
under 914 m: 5 (2013)

Heliports: 1 (2013)

Roadways: *total:* 526 km (2015)
paved: 514 km (2015)
unpaved: 12 km (2015)
country comparison to the world: 195

Merchant marine: *total:* 25
by type: general cargo 4, oil tanker 6, other 15 (2019)
country comparison to the world: 138

Ports and terminals: *major seaport(s):* Victoria

MILITARY AND SECURITY

Military and security forces: *Seychelles People's Defence Forces (SPDF):* Army (includes infantry, Special Forces (Tazar), and Presidential Security Unit), Coast Guard, and Air Force (2019)

Military expenditures: 1.3% of GDP (2019)
1.44% of GDP (2018)
1.57% of GDP (2017)
1.29% of GDP (2016)
1.21% of GDP (2015)
country comparison to the world: 94

Military and security service personnel strengths: the Seychelles People's Defence Forces (SPDF) is comprised of about 500 personnel (200 Land Forces; 200 Coast Guard; 100 Air Force) (2019)

Military equipment inventories and acquisitions: the SPDF's inventory primarily consists of Soviet-era equipment delivered in the 1970s and 1980s; since 2010, China and India are the leading suppliers of newer equipment (mostly donations of patrol boats and aircraft) (2019 est.)

Military service age and obligation: 18-28 years of age for voluntary military service (18-25 for officers); 6-year initial commitment; no conscription (2019)

TRANSNATIONAL ISSUES

Disputes—international: Mauritius and Seychelles claim the Chagos Islands (UK-administered British Indian Ocean Territory)

SIERRA LEONE

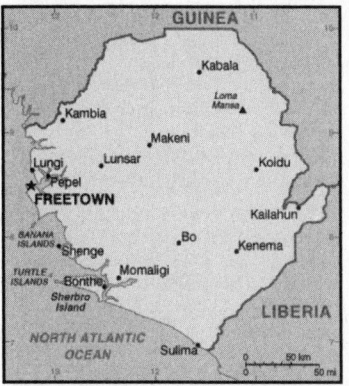

INTRODUCTION

Background: The British set up a trading post near present-day Freetown in the 17th century. Originally, the trade involved timber and ivory, but later it expanded to slaves. Following the American Revolution, a colony was established in 1787 and Sierra Leone became a destination for resettling black loyalists who had originally been resettled in Nova Scotia. After the abolition of the slave trade in 1807, British crews delivered thousands of Africans liberated from illegal slave ships to Sierra Leone, particularly Freetown. The colony gradually expanded inland during the course of the 19th century; independence was attained in 1961. Democracy is slowly being reestablished after the civil war (1991-2002) that resulted in tens of thousands of deaths and the displacement of more than 2 million people (about one-third of the population). The military, which took over full responsibility for security following the departure of UN peacekeepers at the end of 2005, has developed as a guarantor of the country's stability; the armed forces remained on the sideline during the 2007, 2012, and 2018 national elections. In March 2014, the closure of the UN Integrated Peacebuilding Office in Sierra Leone marked the end of more than 15 years of peacekeeping and political operations in Sierra Leone. The government's stated priorities include free primary and secondary education, economic growth, accountable governance, health, and infrastructure.

GEOGRAPHY

Location: Western Africa, bordering the North Atlantic Ocean, between Guinea and Liberia

Geographic coordinates: 8 30 N, 11 30 W

Map references: Africa

Area: *total:* 71,740 sq km
land: 71,620 sq km
water: 120 sq km

country comparison to the world: 120

Area—comparative: slightly smaller than South Carolina

Land boundaries: *total:* 1,093 km
border countries (2): Guinea 794 km, Liberia 299 km

Coastline: 402 km

Maritime claims: *territorial sea:* 12 nm
exclusive economic zone: 200 nm
contiguous zone: 24 nm
continental shelf: 200 nm

Climate: tropical; hot, humid; summer rainy season (May to December); winter dry season (December to April)

Terrain: coastal belt of mangrove swamps, wooded hill country, upland plateau, mountains in east

Elevation: *mean elevation:* 279 m
lowest point: Atlantic Ocean 0 m
highest point: Loma Mansa (Bintimani) 1,948 m

Natural resources: diamonds, titanium ore, bauxite, iron ore, gold, chromite

Land use: *agricultural land:* 56.2% (2011 est.)
arable land: 23.4% (2011 est.) / permanent crops: 2.3% (2011 est.) / permanent pasture: 30.5% (2011 est.)
forest: 37.5% (2011 est.)
other: 6.3% (2011 est.)
Irrigated land: 300 sq km (2012)

Population distribution: population clusters are found in the lower elevations of the south and west; the northern third of the country is less populated as shown on this population distribution map

Natural hazards: dry, sand-laden harmattan winds blow from the Sahara (December to February); sandstorms, dust storms

Environment—current issues: rapid population growth pressuring the environment; overharvesting of timber, expansion of cattle grazing, and slash-and-burn agriculture have resulted in deforestation, soil exhaustion, and flooding; loss of biodiversity; air pollution; water pollution; overfishing

Environment—international agreements: *party to:* Biodiversity, Climate Change, Climate Change-Kyoto Protocol, Desertification, Endangered Species, Hazardous Wastes, Law of the Sea, Marine Life Conservation, Ozone Layer Protection, Ship Pollution, Wetlands
signed, but not ratified: Environmental Modification

Geography—note: rainfall along the coast can reach 495 cm (195 inches) a year, making it one of the wettest places along coastal, western Africa

PEOPLE AND SOCIETY

Population: 6,624,933 (July 2020 est.)
country comparison to the world: 108

Nationality: *noun:* Sierra Leonean(s)
adjective: Sierra Leonean

Ethnic groups: Temne 35.5%, Mende 33.2%, Limba 6.4%, Kono 4.4%, Fullah 3.4%, Loko 2.9%, Koranko 2.8%, Sherbro 2.6%, Mandingo 2.4%, Creole 1.2% (descendants of freed Jamaican slaves who were settled in the Freetown area in the late-18th century; also known as Krio), other Sierra Leone 4.7%, other foreign 0.3% (includes refugees from Liberia's civil war, and small numbers of Europeans, Lebanese, Pakistanis, and Indians), unspecified 0.2% (2013 est.)

Languages: English (official, regular use limited to literate minority), Mende (principal vernacular in the south), Temne (principal vernacular in the north), Krio (English-based Creole, spoken by the descendants of freed Jamaican slaves who were settled in the Freetown area, a lingua franca and a first language for 10% of the population but understood by 95%)

Religions: Muslim 78.6%, Christian 20.8%, other 0.3%, unspecified 0.2% (2013 est.)

Demographic profile: Sierra Leone's youthful and growing population is driven by its high total fertility rate (TFR) of almost 5 children per woman, which has declined little over the last two decades. Its elevated TFR is sustained by the continued desire for large families, the low level of contraceptive use, and the early start of childbearing. Despite its high TFR, Sierra Leone's population growth is somewhat tempered by high infant, child, and maternal mortality rates that are among the world's highest and are a result of poverty, a lack of potable water and sanitation, poor nutrition, limited access to quality health care services, and the prevalence of female genital cutting.

Sierra Leone's large youth cohort – about 60% of the population is under the age of 25 – continues to struggle with high levels of unemployment, which was one of the major causes of the country's 1991-2002 civil war and remains a threat to stability today. Its estimated 60% youth unemployment rate is attributed to high levels of illiteracy and unskilled labor, a lack of private sector jobs, and low pay.

Sierra Leone has been a source of and destination for refugees. Sierra Leone's civil war internally displaced as many as 2 million people, or almost half the population, and forced almost another half million to seek refuge in neighboring countries (370,000 Sierra Leoneans fled to Guinea and 120,000 to Liberia). The UNHCR has helped almost 180,000 Sierra Leoneans to return home, while more than 90,000 others have repatriated on their own. Of the more than 65,000 Liberians who took refuge in Sierra Leone during their country's civil war (1989-2003), about 50,000 have been voluntarily repatriated by the UNHCR and others have returned home independently. As of 2015, less than 1,000 Liberians still reside in Sierra Leone.

Age structure: *0-14 years:* 41.38% (male 1,369,942/female 1,371,537)

15-24 years: 18.83% (male 610,396/female 636,880)

25-54 years: 32.21% (male 1,020,741/female 1,112,946)

55-64 years: 3.89% (male 121,733/female 135,664)

65 years and over: 3.7% (male 100,712/female 144,382) (2020 est.)

Dependency ratios: *total dependency ratio:* 76.3
youth dependency ratio: 71.1
elderly dependency ratio: 5.2
potential support ratio: 19.4 (2020 est.)

Median age: *total:* 19.1 years
male: 18.5 years
female: 19.7 years (2020 est.)
country comparison to the world: 207

Population growth rate: 2.43% (2020 est.)
country comparison to the world: 26

Birth rate: 35.4 births/1,000 population (2020 est.)
country comparison to the world: 19

Death rate: 9.8 deaths/1,000 population (2020 est.)
country comparison to the world: 41

Net migration rate: -1.2 migrant(s)/1,000 population (2020 est.)
country comparison to the world: 147

Population distribution: population clusters are found in the lower elevations of the south and west; the northern third of the country is less populated as shown on this population distribution map

Urbanization: *urban population:* 42.9% of total population (2020)
rate of urbanization: 3.12% annual rate of change (2015-20 est.)

total population growth rate v. urban population growth rate, 2000-2030: *Major urban areas—population:* 1.202 million FREETOWN (capital) (2020)

Sex ratio: *at birth:* 1.03 male(s)/female
0-14 years: 1 male(s)/female
15-24 years: 0.96 male(s)/female
25-54 years: 0.92 male(s)/female
55-64 years: 0.9 male(s)/female
65 years and over: 0.7 male(s)/female
total population: 0.95 male(s)/female (2020 est.)

Mother's mean age at first birth: 19.2 years (2013 est.)
note: median age at first birth among women 25-29

Maternal mortality rate: 1,120 deaths/100,000 live births (2017 est.)
country comparison to the world: 3

Infant mortality rate: *total:* 63.6 deaths/1,000 live births
male: 71.6 deaths/1,000 live births
female: 55.4 deaths/1,000 live births (2020 est.)
country comparison to the world: 10

Life expectancy at birth: *total population:* 59.8 years
male: 57.1 years

female: 62.6 years (2020 est.)
country comparison to the world: 218

Total fertility rate: 4.62 children born/woman (2020 est.)
country comparison to the world: 21

Contraceptive prevalence rate: 21.2% (2019)

Drinking water source:
improved:
urban: 89.5% of population
rural: 55.7% of population
total: 69.8% of population
unimproved:
urban: 10.5% of population
rural: 44.3% of population
total: 30.2% of population (2017 est.)

Current Health Expenditure: 13.4% (2017)

Physicians density: 0.03 physicians/1,000 population (2011)

Sanitation facility access:
improved:
urban: 74.3% of population
rural: 31.9% of population
total: 49.6% of population
unimproved:
urban: 25.7% of population
rural: 68.1% of population
total: 50.4% of population (2017 est.)

HIV/AIDS—adult prevalence rate: 1.5% (2019 est.)
country comparison to the world: 31

HIV/AIDS—people living with HIV/AIDS: 78,000 (2019 est.)
country comparison to the world: 52

HIV/AIDS—deaths: 2,600 (2019 est.)
country comparison to the world: 40

Major infectious diseases: *degree of risk:* very high (2020)
food or waterborne diseases: bacterial and protozoal diarrhea, hepatitis A, and typhoid fever
vectorborne diseases: malaria and dengue fever
water contact diseases: schistosomiasis
animal contact diseases: rabies
aerosolized dust or soil contact diseases: Lassa fever

Obesity—adult prevalence rate: 8.7% (2016)
country comparison to the world: 147

Children under the age of 5 years underweight: 13.6% (2019)
country comparison to the world: 44

Education expenditures: 4.6% of GDP (2017)
country comparison to the world: 85

Literacy: *definition:* age 15 and over can read and write English, Mende, Temne, or Arabic
total population: 43.2%
male: 51.6%
female: 39.8% (2018)

Unemployment, youth ages 15-24: *total:* 9.4%
male: 14.8%
female: 6.1% (2014 est.)
country comparison to the world: 131

GOVERNMENT

Country name: *conventional long form:* Republic of Sierra Leone
conventional short form: Sierra Leone
local long form: Republic of Sierra Leone
local short form: Sierra Leone
etymology: the Portuguese explorer Pedro de SINTRA named the country "Serra Leoa" (Lion Mountains) for the impressive mountains he saw while sailing the West African coast in 1462

Government type: presidential republic

Capital: *name:* Freetown

geographic coordinates: 8 29 N, 13 14 W
time difference: UTC 0 (5 hours ahead of Washington, DC, during Standard Time)
etymology: name derived from the fact that the original settlement served as a haven for free-born and freed African Americans, as well as for liberated Africans rescued from slave ships

Administrative divisions: 4 provinces and 1 area*; Eastern, Northern, North Western, Southern, Western*

Independence: 27 April 1961 (from the UK)

National holiday: Independence Day, 27 April (1961)

Constitution: *history:* several previous; latest effective 1 October 1991
amendments: proposed by Parliament; passage of amendments requires at least two-thirds majority vote of Parliament in two successive readings and assent of the president of the republic; passage of amendments affecting fundamental rights and freedoms and many other constitutional sections also requires approval in a referendum with participation of at least one half of qualified voters and at least two thirds of votes cast; amended several times, last in 2013

Legal system: mixed legal system of English common law and customary law

International law organization participation: has not submitted an ICJ jurisdiction declaration; accepts ICCt jurisdiction

Citizenship: *citizenship by birth:* no
citizenship by descent only: at least one parent or grandparent must be a citizen of Sierra Leone
dual citizenship recognized: yes
residency requirement for naturalization: 5 years

Suffrage: 18 years of age; universal

Executive branch: *chief of state:* President Julius Maada BIO (since 4 April 2018); Vice President Mohamed Juldeh JALLOH (since 4 April 2018) ; note—the president is both chief of state, head of government, and minister of defense
head of government: President Julius Maada BIO (since 4 April 2018); Vice President Mohamed Juldeh JALLOH (since 4 April 2018)
cabinet: Ministers of State appointed by the president, approved by Parliament; the cabinet is responsible to the president
elections/appointments: president directly elected by absolute majority popular vote in 2

rounds if needed for a 5-year term (eligible for a second term); election last held on 4 April 2018 (next to be in 2023)

election results: Julius Maada BIO elected president in second round; percent of vote—Julius Maada BIO (SLPP) 51.8%, Samura KAMARA (APC) 48.2%

Legislative branch: *description:* unicameral Parliament (146 seats; 132 members directly elected in single-seat constituencies by simple majority vote and 14 seats filled in separate elections by non-partisan members of Parliament called "paramount chiefs;" members serve 5-year terms)

elections: last held on 7 March 2018 (next to be held in March 2023)

election results: percent of vote by party—n/a; seats by party—APC 68, SLPP 49, C4C 8, other 7; composition—men 131, women 15, percent of women 10.3%

Judicial branch: *highest courts:* Superior Court of Judicature (consists of the Supreme Court—at the apex—with the chief justice and 4 other judges, the Court of Appeal with the chief justice and 7 other judges, and the High Court of Justice with the chief justice and 9 other judges); note – the Judicature has jurisdiction in all civil, criminal, and constitutional matters

judge selection and term of office: Supreme Court chief justice and other judges of the Judicature appointed by the president on the advice of the Judicial and Legal Service Commission, a 7-member independent body of judges, presidential appointees, and the Commission chairman, and are subject to approval by Parliament; all Judicature judges serve until retirement at age 65

subordinate courts: magistrates' courts; District Appeals Court; local courts

Political parties and leaders: All People's Congress or APC [Ernest Bai KOROMA] Coalition for Change or C4C [Tamba R. SANDY] National Grand Coalition or NGC [Dr. Dennis BRIGHT] Sierra Leone People's Party or SLPP [Dr. Prince HARDING] numerous other parties

International organization participation: ACP, AfDB, AU, C, ECOWAS, EITI (compliant country), FAO, G-77, IAEA, IBRD, ICAO, ICCt, ICRM, IDA, IDB, IFAD, IFC, IFRCS, IHO (pending member), ILO, IMF, IMO, Interpol, IOC, IOM, IPU, ISO (correspondent), ITU, ITUC (NGOs), MIGA, MINUSMA, NAM, OIC, OPCW, UN, UNAMID, UNCTAD, UNESCO, UNIDO, UNIFIL, UNISFA, UNWTO, UPU, WCO, WFTU (NGOs), WHO, WIPO, WMO, WTO

Diplomatic representation in the US: *chief of mission:* Ambassador Sidique Abou-Bakarr WAI (since 4 April 2008)

chancery: 1701 19th Street NW, Washington, DC 20009

telephone: [1] (202) 939-9261 through 9263

FAX: [1] (202) 483-1793

Diplomatic representation from the US: *chief of mission:* Ambassador Maria E. BREWER (since 20 December 2017)

telephone: [232] 99 105 000

embassy: Southridge-Hill Station, Freetown

mailing address: use embassy street address

FAX: [232] 99 515 355

Flag description: three equal horizontal bands of light green (top), white, and light blue; green symbolizes agriculture, mountains, and natural resources, white represents unity and justice, and blue the sea and the natural harbor in Freetown

National symbol(s): *lion; national colors:* green, white, blue

National anthem: *name:* High We Exalt Thee, Realm of the Free

lyrics/music: Clifford Nelson FYLE/John Joseph AKA

note: adopted 1961

0:00/ 0:46

ECONOMY

Economy—overview: Sierra Leone is extremely poor and nearly half of the working-age population engages in subsistence agriculture. The country possesses substantial mineral, agricultural, and fishery resources, but it is still recovering from a civil war that destroyed most institutions before ending in the early 2000s.

In recent years, economic growth has been driven by mining—particularly iron ore. The country's principal exports are iron ore, diamonds, and rutile, and the economy is vulnerable to fluctuations in international prices. Until 2014, the government had relied on external assistance to support its budget, but it was gradually becoming more independent. The Ebola outbreak of 2014 and 2015, combined with falling global commodities prices, caused a significant contraction of economic activity in all areas. While the World Health Organization declared an end to the Ebola outbreak in Sierra Leone in November 2015, low commodity prices in 2015-2016 contributed to the country's biggest fiscal shortfall since 2001. In 2017, increased iron ore exports, together with the end of the Ebola epidemic, supported a resumption of economic growth.

Continued economic growth will depend on rising commodities prices and increased efforts to diversify the sources of growth. Non-mining activities will remain constrained by inadequate infrastructure, such as power and roads, even though power sector projects may provide some additional electricity capacity in the near term. Pervasive corruption and undeveloped human capital will continue to deter foreign investors. Sustained international donor support in the near future will partially offset these fiscal constraints.

GDP (purchasing power parity): $11.55 billion (2017 est.)
$11.14 billion (2016 est.)
$10.48 billion (2015 est.)
note: data are in 2017 dollars
country comparison to the world: 158

GDP (official exchange rate): $3.612 billion (2017 est.)

GDP—real growth rate: 3.7% (2017 est.)
6.3% (2016 est.)
-20.5% (2015 est.)
country comparison to the world: 80

GDP—per capita (PPP): $1,600 (2017 est.)
$1,500 (2016 est.)
$1,500 (2015 est.)
note: data are in 2017 dollars
country comparison to the world: 219

Gross national saving: 10% of GDP (2017 est.)
7.9% of GDP (2016 est.)
-5.9% of GDP (2015 est.)
country comparison to the world: 163

GDP—composition, by end use:
household consumption: 97.9% (2017 est.)
government consumption: 12.1% (2017 est.)
investment in fixed capital: 18.1% (2017 est.)
investment in inventories: 0.4% (2017 est.)
exports of goods and services: 26.8% (2017 est.)
imports of goods and services: -55.3% (2017 est.)

GDP—composition, by sector of origin:
agriculture: 60.7% (2017 est.)
industry: 6.5% (2017 est.)
services: 32.9% (2017 est.)

Agriculture—products: rice, coffee, cocoa, palm kernels, palm oil, peanuts, cashews; poultry, cattle, sheep, pigs; fish

Industries: diamond mining; iron ore, rutile and bauxite mining; small-scale manufacturing (beverages, textiles, footwear)

Industrial production growth rate: 15.5% (2017 est.)
country comparison to the world: 3

Labor force: 132,000 (2013 est.)
country comparison to the world: 178

Labor force—by occupation: *agriculture:* 61.1%
industry: 5.5%
services: 33.4% (2014 est.)

Unemployment rate: 15% (2017 est.)
17.2% (2016 est.)
country comparison to the world: 174

Population below poverty line: 70.2% (2004 est.)

Household income or consumption by percentage share: *lowest 10%:* 2.6%
highest 10%: 33.6% (2003)

Budget: *revenues:* 562 million (2017 est.)
expenditures: 846.4 million (2017 est.)

Taxes and other revenues: 15.6% (of GDP) (2017 est.)
country comparison to the world: 189

Budget surplus (+) or deficit (-): -7.9% (of GDP) (2017 est.)
country comparison to the world: 199

Public debt: 63.9% of GDP (2017 est.)
54.9% of GDP (2016 est.)
country comparison to the world: 63

Fiscal year: calendar year

Inflation rate (consumer prices): 18.2% (2017 est.)

10.9% (2016 est.)
country comparison to the world: 215

Current account balance: -$407 million (2017 est.)
-$88 million (2016 est.)
country comparison to the world: 117

Exports: $808.4 million (2017 est.)
$670 million (2016 est.)
country comparison to the world: 169

Exports—partners: Cote dIvoire 37.7%, Belgium 20.5%, US 15.7%, China 10.2%, Netherlands 6.1% (2017)

Exports—commodities: iron ore, diamonds, rutile, cocoa, coffee, fish

Imports: $1.107 billion (2017 est.)
$972.8 million (2016 est.)
country comparison to the world: 183

Imports—commodities: foodstuffs, machinery and equipment, fuels and lubricants, chemicals

Imports—partners: China 11.5%, US 9.2%, Belgium 8.8%, UAE 7.7%, India 7.4%, Turkey 5.2%, Senegal 5.1%, Netherlands 4.3% (2017)

Reserves of foreign exchange and gold: $478 million (31 December 2017 est.)
$497.2 million (31 December 2016 est.)
country comparison to the world: 152

Debt—external: $1.615 billion (31 December 2017 est.)
$1.503 billion (31 December 2016 est.)
country comparison to the world: 158

Exchange rates: leones (SLL) per US dollar -
7,396.3 (2017 est.)
6,289.9 (2016 est.)
6,289.9 (2015 est.)
5,080.8 (2014 est.)
4,524.2 (2013 est.)

ENERGY

Electricity access: *population without electricity:* 6 million (2019)
electrification—total population: 26% (2019)
electrification—urban areas: 52% (2019)
electrification—rural areas: 6% (2019)

Electricity—production: 300 million kWh (2016 est.)
country comparison to the world: 185

Electricity—consumption: 279 million kWh (2016 est.)
country comparison to the world: 188

Electricity—exports: 0 kWh (2016 est.)
country comparison to the world: 197

Electricity—imports: 0 kWh (2016 est.)
country comparison to the world: 199

Electricity—installed generating capacity: 113,300 kW (2016 est.)
country comparison to the world: 179

Electricity—from fossil fuels: 23% of total installed capacity (2016 est.)
country comparison to the world: 194

Electricity—from nuclear fuels: 0% of total installed capacity (2017 est.)

country comparison to the world: 182

Electricity—from hydroelectric plants: 51% of total installed capacity (2017 est.)
country comparison to the world: 38

Electricity—from other renewable sources: 26% of total installed capacity (2017 est.)
country comparison to the world: 27

Crude oil—production: 0 bbl/day (2018 est.)
country comparison to the world: 199

Crude oil—exports: 0 bbl/day (2015 est.)
country comparison to the world: 192

Crude oil—imports: 0 bbl/day (2015 est.)
country comparison to the world: 194

Crude oil—proved reserves: 0 bbl (1 January 2018 est.)
country comparison to the world: 194

Refined petroleum products—production: 0 bbl/day (2017 est.)
country comparison to the world: 200

Refined petroleum products—consumption: 6,500 bbl/day (2016 est.)
country comparison to the world: 169

Refined petroleum products—exports: 0 bbl/day (2015 est.)
country comparison to the world: 201

Refined petroleum products—imports: 6,439 bbl/day (2015 est.)
country comparison to the world: 164

Natural gas—production: 0 cu m (2017 est.)
country comparison to the world: 195

Natural gas—consumption: 0 cu m (2017 est.)
country comparison to the world: 197

Natural gas—exports: 0 cu m (2017 est.)
country comparison to the world: 183

Natural gas—imports: 0 cu m (2017 est.)
country comparison to the world: 188

Natural gas—proved reserves: 0 cu m (1 January 2014 est.)
country comparison to the world: 193

Carbon dioxide emissions from consumption of energy: 984,800 Mt (2017 est.)
country comparison to the world: 170

COMMUNICATIONS

Telephones—fixed lines: *total subscriptions:* 2,586
subscriptions per 100 inhabitants: less than 1 (2019 est.)
country comparison to the world: 215

Telephones—mobile cellular: *total subscriptions:* 5,569,221
subscriptions per 100 inhabitants: 86.13 (2019 est.)
country comparison to the world: 115

Telecommunication systems: *general assessment:* the stability in the country has led to international investment; telecom regulator continues to improve the market; telephone service improving with the expansion of the mobile sector; mobile-cellular service has grown rapidly from a small base, overcoming the deficiencies of

the fixed-line sector; mobile sector has a high penetration; regulator approves 27% price increase for mobile voice calls; LTE launched in 2018 to compete with state owned almost monopoly on fixed-line (2020)
domestic: fixed-line less than 1 per 100 and mobile-cellular 86 per 100 (2019)
international: country code—232; landing point for the ACE submarine cable linking to South Africa, over 20 western African countries and Europe; satellite earth station—1 Intelsat (Atlantic Ocean) (2019)
note: the COVID-19 outbreak is negatively impacting telecommunications production and supply chains globally; consumer spending on telecom devices and services has also slowed due to the pandemic's effect on economies worldwide; overall progress towards improvements in all facets of the telecom industry—mobile, fixed-line, broadband, submarine cable and satellite—has moderated

Broadcast media: 1 government-owned TV station; 3 private TV stations; a pay-TV service began operations in late 2007; 1 government-owned national radio station; about two-dozen private radio stations primarily clustered in major cities; transmissions of several international broadcasters are available (2019)

Internet country code: sl

Internet users: *total:* 568,099
percent of population: 9% (July 2018 est.)
country comparison to the world: 150

TRANSPORTATION

National air transport system: *annual passenger traffic on registered air carriers:* 50,193 (2015)
annual freight traffic on registered air carriers: 0 mt-km (2015)

Civil aircraft registration country code prefix: 9L (2016)

Airports: 8 (2013)
country comparison to the world: 161

Airports—with paved runways: *total:* 1 (2019)
over 3,047 m: 1

Airports—with unpaved runways: *total:* 7 (2013)
914 to 1,523 m: 7 (2013)

Heliports: 2 (2013)

Roadways: *total:* 11,700 km (2015)
paved: 1,051 km (2015)
unpaved: 10,650 km (2015)
urban: 3,000 km (2015)
non-urban: 8,700 km (2015)
country comparison to the world: 134

Waterways: 800 km (600 km navigable year-round) (2011)
country comparison to the world: 72

Merchant marine: *total:* 518
by type: bulk carrier 30, container ship 10, general cargo 263, oil tanker 95, other 120 (2019)
country comparison to the world: 42

Ports and terminals: *major seaport(s):* Freetown, Pepel, Sherbro Islands

Military and security forces: *Republic of Sierra Leone Armed Forces (RSLAF):* Army (includes Maritime Wing and Air Wing) (2019)

Military expenditures: 0.7% of GDP (2019)
0.8% of GDP (2018)
1.1% of GDP (2017)
1.1% of GDP (2016)
0.9% of GDP (2015)
country comparison to the world: 140

Military and security service personnel strengths: the Republic of Sierra Leone Armed Forces (RSLAF) is comprised of about 8,500 personnel, including an estimated 300 in the air and maritime wings (2019 est.)

Military equipment inventories and acquisitions: the RSLAF's small inventory includes a mix of Soviet-origin and other older foreign-supplied equipment; since 2010, it has received limited quantities of material from China and South Africa (2019 est.)

Military service age and obligation: 18-29 for voluntary military service; women are eligible to serve; no conscription (2019)

TRANSNATIONAL ISSUES

Disputes—international: Sierra Leone opposes Guinean troops' continued occupation of Yenga, a small village on the Makona River that serves as a border with Guinea; Guinea's forces came to Yenga in the mid-1990s to help the Sierra Leonean

military to suppress rebels and to secure their common border but have remained there even after both countries signed a 2005 agreement acknowledging that Yenga belonged to Sierra Leone; in 2012, the two sides signed a declaration to demilitarize the area

Refugees and internally displaced persons: *IDPs:* 5,500 (displacement caused by post-electoral violence in 2018 and clashes in the Pujehun region in 2019) (2019)

SINGAPORE

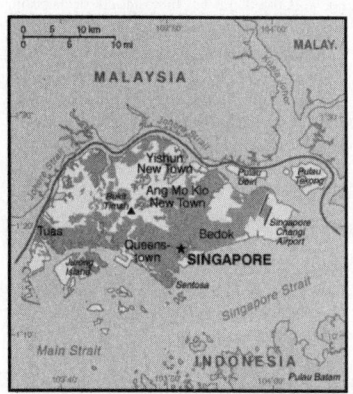

INTRODUCTION

Background: A Malay trading port known as Temasek existed on the island of Singapore by the 14th century. The settlement changed hands several times in the ensuing centuries and was eventually burned in the 17th century and fell into obscurity. The British founded modern Singapore as a trading colony on the site in 1819. It joined the Malaysian Federation in 1963 but was ousted two years later and became independent. Singapore subsequently became one of the world's most prosperous countries with strong international trading links (its port is one of the world's busiest in terms of tonnage handled) and with per capita GDP equal to that of the leading nations of Western Europe.

GEOGRAPHY

Location: Southeastern Asia, islands between Malaysia and Indonesia

Geographic coordinates: 1 22 N, 103 48 E

Map references: Southeast Asia

Area: *total:* 719 sq km
land: 709.2 sq km
water: 10 sq km
country comparison to the world: 191

Area—comparative: slightly more than 3.5 times the size of Washington, DC

Land boundaries: 0 km

Coastline: 193 km

Maritime claims: *territorial sea:* 3 nm
exclusive fishing zone: within and beyond territorial sea, as defined in treaties and practice

Climate: tropical; hot, humid, rainy; two distinct monsoon seasons—northeastern monsoon (December to March) and southwestern monsoon (June to September); inter-monsoon—frequent afternoon and early evening thunderstorms

Terrain: lowlying, gently undulating central plateau

Elevation: *lowest point:* Singapore Strait 0 m
highest point: Bukit Timah 166 m

Natural resources: fish, deepwater ports

Land use: *agricultural land:* 1% (2011 est.)
arable land: 0.9% (2011 est.) / *permanent crops:* 0.1% (2011 est.) / *permanent pasture:* 0% (2011 est.)
forest: 3.3% (2011 est.)
other: 95.7% (2011 est.)
Irrigated land: 0 sq km (2012)

Population distribution: most of the urbanization is along the southern coast, with relatively dense population clusters found in the central areas

Natural hazards: flash floods

Environment—current issues: water pollution; industrial pollution; limited natural freshwater resources; limited land availability presents waste disposal problems; air pollution; deforestation; seasonal smoke/haze resulting from forest fires in Indonesia

Environment—international agreements: *party to:* Biodiversity, Climate Change, Climate Change-Kyoto Protocol, Desertification, Endangered Species, Hazardous Wastes, Law of the Sea, Ozone Layer Protection, Ship Pollution
signed, but not ratified: none of the selected agreements

Geography—note: focal point for Southeast Asian sea routes; consists of about 60 islands, by far the largest of which is Pulau Ujong; land reclamation has removed many former islands and created a number of new ones

PEOPLE AND SOCIETY

Population: 6,209,660 (July 2020 est.)
country comparison to the world: 110

Nationality: *noun:* Singaporean(s)
adjective: Singapore

Ethnic groups: Chinese 74.3%, Malay 13.4%, Indian 9%, other 3.2% (2018 est.)

note: *individuals self-identify; the population is divided into four categories:* Chinese, Malay (includes indigenous Malays and Indonesians), Indian (includes Indian, Pakistani, Bangladeshi, or Sri Lankan), and other ethnic groups (includes Eurasians, Caucasians, Japanese, Filipino, Vietnamese)

Languages: English (official) 36.9%, Mandarin (official) 34.9%, other Chinese dialects (includes Hokkien, Cantonese, Teochew, Hakka) 12.2%, Malay (official) 10.7%, Tamil (official) 3.3%, other 2% (2015 est.)

note: data represent language most frequently spoken at home

Religions: Buddhist 33.2%, Christian 18.8%, Muslim 14%, Taoist 10%, Hindu 5%, other 0.6%, none 18.5% (2015 est.)

Age structure: *0-14 years:* 12.8% (male 406,983/female 387,665)

15-24 years: 15.01% (male 457,190/female 474,676)

25-54 years: 50.73% (male 1,531,088/female 1,618,844)

55-64 years: 10.58% (male 328,024/female 328,808)

65 years and over: 10.89% (male 310,123/female 366,259) (2020 est.)

Dependency ratios: *total dependency ratio:* 34.5
youth dependency ratio: 16.5
elderly dependency ratio: 18
potential support ratio: 5.6 (2020 est.)

Median age: *total:* 35.6 years
male: 35.4 years
female: 35.7 years (2020 est.)
country comparison to the world: 82

Population growth rate: 1.73% (2020 est.)
country comparison to the world: 59

Birth rate: 8.9 births/1,000 population (2020 est.)
country comparison to the world: 207

Death rate: 3.6 deaths/1,000 population (2020 est.)
country comparison to the world: 218

Net migration rate: 11.8 migrant(s)/1,000 population (2020 est.)
country comparison to the world: 5

Population distribution: most of the urbanization is along the southern coast, with relatively dense population clusters found in the central areas

Urbanization: *urban population:* 100% of total population (2020)
rate of urbanization: 1.39% annual rate of change (2015-20 est.)

total population growth rate v. urban population growth rate, 2000-2030: *Major urban areas—population:* 5.935 million SINGAPORE (capital) (2020)

Sex ratio: *at birth:* 1.07 male(s)/female
0-14 years: 1.05 male(s)/female
15-24 years: 0.96 male(s)/female
25-54 years: 0.95 male(s)/female
55-64 years: 1 male(s)/female
65 years and over: 0.85 male(s)/female
total population: 0.96 male(s)/female (2020 est.)

Mother's mean age at first birth: 30.5 years (2015 est.)

median age

Maternal mortality rate: 8 deaths/100,000 live births (2017 est.)
country comparison to the world: 153

Infant mortality rate: *total:* 2.3 deaths/1,000 live births
male: 2.4 deaths/1,000 live births
female: 2 deaths/1,000 live births (2020 est.)
country comparison to the world: 224

Life expectancy at birth: *total population:* 86 years
male: 83.3 years
female: 88.9 years (2020 est.)
country comparison to the world: 3

Total fertility rate: 0.87 children born/woman (2020 est.)

country comparison to the world: 228

Drinking water source:
improved:
urban: 100% of population
total: 100% of population
unimproved:
urban: 0% of population
total: 0% of population (2017 est.)

Current Health Expenditure: 4.4% (2017)

Physicians density: 2.29 physicians/1,000 population (2016)

Hospital bed density: 2.5 beds/1,000 population (2017)

Sanitation facility access:
improved:
urban: 100% of population (2015 est.)
total: 100% of population
unimproved:
urban: 0% of population
total: 0% of population (2017 est.)

HIV/AIDS—adult prevalence rate: 0.2% (2019 est.)
country comparison to the world: 109

HIV/AIDS—people living with HIV/AIDS: 7,900 (2019 est.)
country comparison to the world: 111

HIV/AIDS—deaths: <100 (2019 est.)

Obesity—adult prevalence rate: 6.1% (2016)
country comparison to the world: 171

Education expenditures: 2.9% of GDP (2013)
country comparison to the world: 144

Literacy: *definition:* age 15 and over can read and write
total population: 97.3%
male: 98.9%
female: 95.9% (2018)

School life expectancy (primary to tertiary education): *total:* 17 years
male: 16 years
female: 17 years (2018)

Unemployment, youth ages 15-24: *total:* 9.1%
male: 6.2%
female: 12.5% (2016 est.)
country comparison to the world: 134

GOVERNMENT

Country name: *conventional long form:* Republic of Singapore
conventional short form: Singapore
local long form: Republic of Singapore
local short form: Singapore
etymology: name derives from the Sanskrit words "simha" (lion) and "pura" (city) to describe the city-state's leonine symbol

Government type: parliamentary republic

Capital: *name:* Singapore

geographic coordinates: 1 17 N, 103 51 E
time difference: UTC+8 (13 hours ahead of Washington, DC, during Standard Time)
etymology: name derives from the Sanskrit words "simha" (lion) and "pura" (city), thus creating the city's epithet "lion city"

Administrative divisions: *no first order administrative divisions; there are five community development councils:* Central Singapore Development Council, North East Development Council, North West Development Council, South East Development Council, South West Development Council (2019)

Independence: 9 August 1965 (from Malaysian Federation)

National holiday: National Day, 9 August (1965)

Constitution: *history:* several previous; latest adopted 22 December 1965
amendments: proposed by Parliament; passage requires two-thirds majority vote in the second and third readings by the elected Parliament membership and assent of the president of the republic; passage of amendments affecting sovereignty or control of the Police Force or the Armed Forces requires at least two-thirds majority vote in a referendum; amended many times, last in 2016

Legal system: English common law

International law organization participation: has not submitted an ICJ jurisdiction declaration; non-party state to the ICC (2019)

Citizenship: *citizenship by birth:* no
citizenship by descent only: at least one parent must be a citizen of Singapore
dual citizenship recognized: no
residency requirement for naturalization: 10 years

Suffrage: 21 years of age; universal and compulsory

Executive branch: *chief of state:* President HALIMAH Yacob (since 14 September 2017); note—President TAN's term ended on 31 August 2017; HALIMAH is Singapore's first female president; the head of the Council of Presidential Advisors, J.Y. PILLAY, served as acting president until HALIMAH was sworn in as president on 14 September 2017
head of government: Prime Minister LEE Hsien Loong (since 12 August 2004, reelected 10 July 2020); Deputy Prime Ministers HENG Swee Keat (since 1 May 2019) (2019)
cabinet: Cabinet appointed by the president on the advice of the prime minister; Cabinet responsible to Parliament
elections/appointments: president directly elected by simple majority popular vote for a fixed term of 6-years (there are no term limits); election last held on 13 September 2017 (next to be held in 2023); following legislative elections, leader of majority party or majority coalition appointed prime minister by president; deputy prime ministers appointed by the president
election results: HALIMAH Yacob was declared president on 13 September 2017, being the only eligible candidate; Tony TAN Keng Yam elected president in the previous contested election on 27 August 2011; percent of vote—Tony TAN Keng Yam (independent) 35.2% , TAN Cheng Bock (independent) 34.9%, TAN Jee Say (independent) 25%, TAN Kin Lian (independent) 4.9%

Legislative branch: *description:* unicameral Parliament (104 seats; 93 members directly elected by popular vote, up to 9 nominated by a parliamentary selection committee and appointed by the president, and up to 12 non-constituency members from opposition parties to ensure political diversity; members serve 5-year terms); note—the number of nominated members will increase to 12 for the 2020 election for the first time (2020)

elections: last held on 10 July 2020 (next must be held by 2025)

election results: percent of vote by party—PAP 61.2%, WP 11.2%, PSP 10.2%; seats by party—PAP 83, WP 10, PSP 2; composition—men 79, women 25, percent of women 24%

Judicial branch: *highest courts:* Supreme Court (although the number of judges varies—as of April 2019, the court totaled 20 judges, 7 judicial commissioners, 4 judges of appeal, and 16 international judges); the court is organized into an upper tier Appeal Court and a lower tier High Court

judge selection and term of office: judges appointed by the president from candidates recommended by the prime minister after consultation with the chief justice; judges usually serve until retirement at age 65, but terms can be extended

subordinate courts: district, magistrates', juvenile, family, community, and coroners' courts; small claims tribunals; employment claims tribunals

Political parties and leaders: National Solidarity Party or NSP [Reno FONG]
People's Action Party or PAP [LEE Hsien Loong]
People's Power Party or (PPP) [Goh Meng SENG]
People's Voice or PV [Lim TEAN]
Progress Singapore Party or PSP [Tan Cheng Bock]
Red Dot United or RDU [Ravi PHILEMON]
Reform Party or RP [Kenneth JEYARETNAM]
Singapore Democratic Alliance or SDA [Abu MOHAMED]
Singapore Democratic Party or SDP [Dr. CHEE Soon Juan]
Singapore People's Party or SPP [Steve Chia]
Workers' Party or WP [Pritam SINGH] (2020)

International organization participation: ADB, AOSIS, APEC, Arctic Council (observer), ARF, ASEAN, BIS, C, CP, EAS, FAO, FATF, G-77, IAEA, IBRD, ICAO, ICC (national committees), ICCt, ICRM, IDA, IFC, IFRCS, IHO, ILO, IMF, IMO, IMSO, Interpol, IOC, IPU, ISO, ITSO, ITU, ITUC (NGOs), MIGA, NAM, OPCW, Pacific Alliance (observer), PCA, UN, UNCTAD, UNESCO, UNHCR, UPU, WCO, WHO, WIPO, WMO, WTO

Diplomatic representation in the US: *chief of mission:* Ambassador Ashok KUMAR Mirpuri since 30 July 2012)

chancery: 3501 International Place NW, Washington, DC 20008

telephone: [1] (202) 537-3100

FAX: [1] (202) 537-0876

consulate(s) general: San Francisco

consulate(s): New York

Diplomatic representation from the US: *chief of mission:* Ambassador (vacant); Charge d'Affaires Rafik MANSOUR (since July 2019)

telephone: [65] 6476-9100

embassy: 27 Napier Road, Singapore 258508

mailing address: FPO AP 96507-0001

FAX: [65] 6476-9340

Flag description: two equal horizontal bands of red (top) and white; near the hoist side of the red band, there is a vertical, white crescent (closed portion is toward the hoist side) partially enclosing five white five-pointed stars arranged in a circle; red denotes brotherhood and equality; white signifies purity and virtue; the waxing crescent moon symbolizes a young nation on the ascendancy; the five stars represent the nation's ideals of democracy, peace, progress, justice, and equality

National symbol(s): *lion, merlion (mythical half lion-half fish creature), orchid; national colors:* red, white

National anthem: *name:* "Majulah Singapura" (Onward Singapore)

lyrics/music: ZUBIR Said

note: adopted 1965; first performed in 1958 at the Victoria Theatre, the anthem is sung only in Malay

0:00 / 1:19

ECONOMY

Economy—overview: Singapore has a highly developed and successful free-market economy. It enjoys an open and corruption-free environment, stable prices, and a per capita GDP higher than that of most developed countries. Unemployment is very low. The economy depends heavily on exports, particularly of electronics, petroleum products, chemicals, medical and optical devices, pharmaceuticals, and on Singapore's vibrant transportation, business, and financial services sectors.

The economy contracted 0.6% in 2009 as a result of the global financial crisis, but has continued to grow since 2010. Growth from 2012-2017 was slower than during the previous decade, a result of slowing structural growth—as Singapore reached high-income levels—and soft global demand for exports. Growth recovered to 3.6% in 2017 with a strengthening global economy.

The government is attempting to restructure Singapore's economy to reduce its dependence on foreign labor, raise productivity growth, and increase wages amid slowing labor force growth and an aging population. Singapore has attracted major investments in advanced manufacturing, pharmaceuticals, and medical technology production and will continue efforts to strengthen its position as Southeast Asia's leading financial and technology hub. Singapore is a signatory of the Comprehensive and Progressive Agreement for Trans-Pacific Partnership (CPTPP), and a party to the Regional Comprehensive Economic Partnership (RCEP) negotiations with nine other ASEAN members plus Australia, China, India, Japan, South Korea, and New Zealand. In 2015, Singapore formed, with the other ASEAN members, the ASEAN Economic Community.

GDP (purchasing power parity): $528.1 billion (2017 est.)

$509.7 billion (2016 est.)

$497.8 billion (2015 est.)

note: data are in 2017 dollars

country comparison to the world: 38

GDP (official exchange rate): $323.9 billion (2017 est.)

GDP—real growth rate: 0.73% (2019 est.)

3.48% (2018 est.)

4.34% (2017 est.)

country comparison to the world: 180

GDP—per capita (PPP): $94,100 (2017 est.)

$90,900 (2016 est.)

$89,900 (2015 est.)

note: data are in 2017 dollars

country comparison to the world: 7

Gross national saving: 46.5% of GDP (2017 est.)

46% of GDP (2016 est.)

45.7% of GDP (2015 est.)

country comparison to the world: 5

GDP—composition, by end use:

household consumption: 35.6% (2017 est.)

government consumption: 10.9% (2017 est.)

investment in fixed capital: 24.8% (2017 est.)

investment in inventories: 2.8% (2017 est.)

exports of goods and services: 173.3% (2017 est.)

imports of goods and services: -149.1% (2017 est.)

GDP—composition, by sector of origin:

agriculture: 0% (2017 est.)

industry: 24.8% (2017 est.)

services: 75.2% (2017 est.)

Agriculture—products: vegetables; poultry, eggs; fish, ornamental fish, orchids

Industries: electronics, chemicals, financial services, oil drilling equipment, petroleum refining, biomedical products, scientific instruments, telecommunication equipment, processed food and beverages, ship repair, offshore platform construction, entrepot trade

Industrial production growth rate: 5.7% (2017 est.)

country comparison to the world: 46

Labor force: 3.778 million (2019 est.)

note: excludes non-residents

country comparison to the world: 92

Labor force—by occupation: *agriculture:* 0.7%

industry: 25.6%

services: 73.7% (2017)

note: excludes non-residents

Unemployment rate: 2.25% (2019 est.)

2.1% (2018 est.)

country comparison to the world: 22

Population below poverty line: NA

Household income or consumption by percentage share: *lowest 10%:* 1.6%

highest 10%: 27.5% (2017)

Budget: *revenues:* 50.85 billion (2017 est.)

expenditures: 51.87 billion (2017 est.)

note: expenditures include both operational and development expenditures

Taxes and other revenues: 15.7% (of GDP) (2017 est.)

country comparison to the world: 187

863

Budget surplus (+) or deficit (-): -0.3% (of GDP) (2017 est.)
country comparison to the world: 55

Public debt: 111. 1% of GDP (2017 est.) 106.8% of GDP (2016 est.)
note: Singapore's public debt consists largely of Singapore Government Securities (SGS) issued to assist the Central Provident Fund (CPF), which administers Singapore's defined contribution pension fund; special issues of SGS are held by the CPF, and are non-tradable; the government has not borrowed to finance deficit expenditures since the 1980s; Singapore has no external public debt
country comparison to the world: 11

Fiscal year: 1 April—31 March

Inflation rate (consumer prices): 0.6% (2017 est.) -0.5% (2016 est.)
country comparison to the world: 33

Current account balance: $63.109 billion (2019 est.)
$64.042 billion (2018 est.)
country comparison to the world: 8

Exports: $396.8 billion (2017 est.)
$338 billion (2016 est.)
country comparison to the world: 13

Exports—partners: China 14.7%, Hong Kong 12.6%, Malaysia 10.8%, US 6.6%, Indonesia 5.8%, Japan 4.7%, South Korea 4.6%, Thailand 4% (2017)

Exports—commodities: machinery and equipment (including electronics and telecommunications), pharmaceuticals and other chemicals, refined petroleum products, foodstuffs and beverages

Imports: $312.1 billion (2017 est.)
$277.6 billion (2016 est.)
country comparison to the world: 16

Imports—commodities: machinery and equipment, mineral fuels, chemicals, foodstuffs, consumer goods

Imports—partners: China 13.9%, Malaysia 12%, US 10.7%, Japan 6.3%, South Korea 5% (2017)

Reserves of foreign exchange and gold: $279.9 billion (31 December 2017 est.)
$271.8 billion (31 December 2016 est.)
country comparison to the world: 11

Debt—external: $566.1 billion (31 December 2017 est.)
$464.1 billion (30 September 2017 est.)
country comparison to the world: 20

Exchange rates: Singapore dollars (SGD) per US dollar -
1.3 (2017 est.)
1.35 (2016 est.)
1.3815 (2015 est.)
1.3748 (2014 est.)
1.2671 (2013 est.)

ENERGY

Electricity access: *electrification—total population:* 100% (2020)

Electricity—production: 48.66 billion kWh (2016 est.)

country comparison to the world: 55

Electricity—consumption: 47.69 billion kWh (2016 est.)
country comparison to the world: 51

Electricity—exports: 0 kWh (2016 est.)
country comparison to the world: 198

Electricity—imports: 0 kWh (2016 est.)
country comparison to the world: 200

Electricity—installed generating capacity: 13.35 million kW (2016 est.)
country comparison to the world: 53

Electricity—from fossil fuels: 98% of total installed capacity (2016 est.)
country comparison to the world: 29

Electricity—from nuclear fuels: 0% of total installed capacity (2017 est.)
country comparison to the world: 183

Electricity—from hydroelectric plants: 0% of total installed capacity (2017 est.)
country comparison to the world: 200

Electricity—from other renewable sources: 2% of total installed capacity (2017 est.)
country comparison to the world: 144

Crude oil—production: 0 bbl/day (2018 est.)
country comparison to the world: 200

Crude oil—exports: 14,780 bbl/day (2015 est.)
country comparison to the world: 54

Crude oil—imports: 783,300 bbl/day (2015 est.)
country comparison to the world: 15

Crude oil—proved reserves: 0 bbl (1 January 2018 est.)
country comparison to the world: 195

Refined petroleum products—production: 755,000 bbl/day (2015 est.)
country comparison to the world: 24

Refined petroleum products—consumption: 1.322 million bbl/day (2016 est.)
country comparison to the world: 17

Refined petroleum products—exports: 1.82 million bbl/day (2015 est.)
country comparison to the world: 4

Refined petroleum products—imports: 2.335 million bbl/day (2015 est.)
country comparison to the world: 1

Natural gas—production: 0 cu m (2017 est.)
country comparison to the world: 196

Natural gas—consumption: 12.97 billion cu m (2017 est.)
country comparison to the world: 44

Natural gas—exports: 622.9 million cu m (2017 est.)
country comparison to the world: 41

Natural gas—imports: 13.48 billion cu m (2017 est.)
country comparison to the world: 23

Natural gas—proved reserves: 0 cu m (1 January 2017 est.)
country comparison to the world: 194

Carbon dioxide emissions from consumption of energy: 249.5 million Mt (2017 est.)
country comparison to the world: 27

COMMUNICATIONS

Telephones—fixed lines: *total subscriptions:* 2,003,594
subscriptions per 100 inhabitants: 32.83 (2019 est.)
country comparison to the world: 54

Telephones—mobile cellular: *total subscriptions:* 9,543,773
subscriptions per 100 inhabitants: 156.38 (2019 est.)
country comparison to the world: 89

Telecommunication systems: *general assessment:* excellent service; world leader in telecommunications and perhaps the first 'Smart Nation' where a sensor network is implemented, for water and air, smart logistics and smart sensor in the home of elderly or chronically ill; roll out of 4G and 5G networks to ensure faster speeds; wireless and fiber broadband growing segments of telecommunications; roll out of 'Next Generation Network' (NGNBN) almost complete with FttH and wireless network fiber based services; mobile sector saturated, but with mobile operators competing to offer more to the consumer such as value-added services; 4 MNVO; demand for data storage in Singapore (2020)
domestic: excellent domestic facilities; fixed-line 33 per 100 and mobile-cellular 156 per 100 teledensity; multiple providers of high-speed Internet connectivity (2019)
international: country code—65; landing points for INDIGO-West, SeaMeWe -3,-4,-5, SIGMAR, SJC, i2icn, PGASCOM, BSCS, IGG, B3JS, SAEx2, APCN-2, APG, ASC, SEAX-1, ASE, EAC-C2C, Matrix Cable System and SJC2 submarine cables providing links throughout Asia, Southeast Asia, Africa, Australia, the Middle East, and Europe; satellite earth stations—3, Bukit Timah, Seletar, and Sentosa; supplemented by VSAT coverage (2019)
note: the COVID-19 outbreak is negatively impacting telecommunications production and supply chains globally; consumer spending on telecom devices and services has also slowed due to the pandemic's effect on economies worldwide; overall progress towards improvements in all facets of the telecom industry—mobile, fixed-line, broadband, submarine cable and satellite—has moderated

Broadcast media: state controls broadcast media; 6 domestic TV stations operated by MediaCorp which is wholly owned by a state investment company; broadcasts from Malaysian and Indonesian stations available; satellite dishes banned; multi-channel cable TV services available; a total of 19 domestic radio stations broadcasting, with MediaCorp operating 11, Singapore Press Holdings, also government-linked, another 5, 2 controlled by the Singapore Armed Forces Reservists Association and one owned by BBC Radio; Malaysian and Indonesian radio stations are available as is BBC; a number of Internet service radio stations are also available (2019)

Internet country code: .sg

Internet users: *total:* 5,286,665
percent of population: 88.17% (July 2018 est.)

country comparison to the world: 80

Broadband—fixed subscriptions: *total:* 1,610,500
subscriptions per 100 inhabitants: 27 (2018 est.)
country comparison to the world: 60

National air transport system: *number of registered air carriers:* 4 (2020)
inventory of registered aircraft operated by air carriers: 230
annual passenger traffic on registered air carriers: 40,401,515 (2018)
annual freight traffic on registered air carriers: 5,194,900,000 mt-km (2018)

Civil aircraft registration country code prefix: 9V (2016)

Airports: 9 (2013)
country comparison to the world: 158

Airports—with paved runways: *total:* 9 (2017)
over 3,047 m: 2 (2017)
2,438 to 3,047 m: 2 (2017)
1,524 to 2,437 m: 3 (2017)
914 to 1,523 m: 1 (2017)
under 914 m: 1 (2017)

Pipelines: 3220 km domestic gas (2014), 1122 km cross-border pipelines (2017), 8 km refined products (2013)

Roadways: *total:* 3,500 km (2017)
paved: 3,500 km (includes 164 km of expressways) (2017)
country comparison to the world: 158

Merchant marine: *total:* 3,433
by type: bulk carrier 585, container ship 492, general cargo 130, oil tanker 724, other 1,502 (2019)
country comparison to the world: 8

Ports and terminals: *major seaport(s):* Singapore

container port(s) (TEUs): Singapore (33,666,000) (2017)
LNG terminal(s) (import): Singapore

MILITARY AND SECURITY

Military and security forces: *Singapore Armed Forces:* Singapore Army, Republic of Singapore Navy, Republic of Singapore Air Force (includes air defense); Police Coast Guard (subordinate to the Singapore Police Force) (2019)

Military expenditures: 3.2% of GDP (2019)
3.1% of GDP (2018)
3.1% of GDP (2017)
3.2% of GDP (2016)
3.1% of GDP (2015)
country comparison to the world: 24

Military and security service personnel strengths: the Singapore Armed Forces (SAF) have approximately 62,000 active duty troops (45,000 Army; 7,000 Navy; 10,000 Air Force) (2019)

Military equipment inventories and acquisitions: the SAF has a diverse and largely modern mix of domestically-produced and imported weapons; Singapore has the most developed arms industry in Southeast Asia and is also the largest importer of weapons; the chief suppliers since 2010 are France, Germany, Spain, and the US (2019 est.)

Military deployments: maintains permanent training bases and detachments of military personnel in Australia, France, and the US (June 2020)

Military service age and obligation: 18-21 years of age for male compulsory military service; 16 1/2 years of age for voluntary enlistment (with parental consent); 2-year conscript service obligation, with a reserve obligation to age 40 (enlisted) or age 50 (officers) (2019)

Maritime threats: the International Maritime Bureau reports the territorial and offshore waters in the South China Sea as high risk for piracy and armed robbery against ships; numerous commercial vessels have been attacked and hijacked both at anchor and while underway; hijacked vessels are often disguised and cargo diverted to ports in East Asia; crews have been murdered or cast adrift; the Singapore Straits saw three attacks against commercial vessels in 2018, a slight decrease from the four attacks in 2017 (2018)

TERRORISM

TRANSNATIONAL ISSUES

Disputes—international: disputes with Malaysia over territorial waters, airspace, the price of fresh water delivered to Singapore from Malaysia, Singapore's extensive land reclamation works, bridge construction, and maritime boundaries in the Johor and Singapore Straits; in 2008, ICJ awarded sovereignty of Pedra Branca (Pulau Batu Puteh/Horsburgh Island) to Singapore, and Middle Rocks to Malaysia, but did not rule on maritime regimes, boundaries, or disposition of South Ledge; Indonesia and Singapore continue to work on finalization of their 1973 maritime boundary agreement by defining unresolved areas north of Indonesia's Batam Island; piracy remains a problem in the Malacca Strait

Refugees and internally displaced persons: *stateless persons:* 1,303 (2019)

Illicit drugs: drug abuse limited because of aggressive law enforcement efforts, including carrying out death sentences; as a transportation and financial services hub, Singapore is vulnerable, despite strict laws and enforcement, as a venue for money laundering

SINT MAARTEN

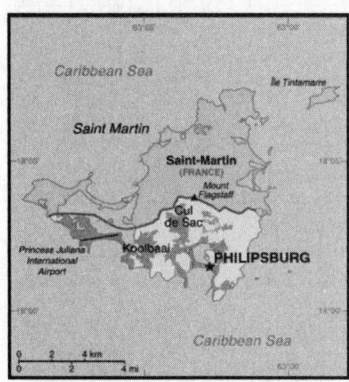

INTRODUCTION

Background: Although sighted by Christopher COLUMBUS in 1493 and claimed for Spain, it was the Dutch who occupied the island in 1631 and began exploiting its salt deposits. The Spanish retook the island in 1633, but the Dutch continued to assert their claims. The Spanish finally relinquished the island of Saint Martin to the French and Dutch, who divided it between themselves in 1648. The establishment of cotton, tobacco, and sugar plantations dramatically expanded African slavery on the island in the 18th and 19th centuries; the practice was not abolished in the Dutch half until 1863. The island's economy declined until 1939 when it became a free port; the tourism industry was dramatically expanded beginning in the 1950s. In 1954, Sint Maarten and several other Dutch Caribbean possessions became part of the Kingdom of the Netherlands as the Netherlands Antilles. In a 2000 referendum, the citizens of Sint Maarten voted to become a self-governing country within the Kingdom of the Netherlands, effective October 2010. On 6 September 2017, Hurricane Irma hit Saint Martin/Sint Maarten, causing extensive damage to roads, communications, electrical power, and housing. The UN estimated the storm destroyed or damaged 90% of the buildings, and Princess Juliana International Airport was heavily damaged and closed to commercial air traffic for five weeks.

GEOGRAPHY

Location: Caribbean, located in the Leeward Islands (northern) group; Dutch part of the island of Saint Martin in the Caribbean Sea; Sint Maarten lies east of the US Virgin Islands

Geographic coordinates: 18 4 N, 63 4 W

Map references: Central America and the Caribbean

Area: *total:* 34 sq km
land: 34 sq km
water: 0 sq km
note: Dutch part of the island of Saint Martin
country comparison to the world: 236

Area—comparative: one-fifth the size of Washington, DC

Land boundaries: *total:* 16 km
border countries (1): Saint Martin (France) 16 km

Coastline: 58.9 km (for entire island)

Maritime claims: *territorial sea:* 12 nm
exclusive economic zone: 200 nm

Climate: tropical marine climate, ameliorated by northeast trade winds, results in moderate temperatures; average rainfall of 150 cm/year; hurricane season stretches from July to November

Terrain: low, hilly terrain, volcanic origin

Elevation: *lowest point:* Caribbean Sea 0 m
highest point: Mount Flagstaff 383 m

Natural resources: fish, salt

Population distribution: most populous areas are Lower Prince's Quarter (north of Philipsburg), followed closely by Cul de Sac

Natural hazards: subject to hurricanes from July to November

Environment—current issues: scarcity of potable water (increasing percentage provided by desalination); inadequate solid waste management; pollution from construction, chemical runoff, and sewage harms reefs

Geography—note: the northern border is shared with the French overseas collectivity of Saint Martin; together, these two entities make up the smallest landmass in the world shared by two self-governing states

PEOPLE AND SOCIETY

Population: 45,847 (July 2020 est.)
country comparison to the world: 212

Ethnic groups: Saint Maarten 29.9%, Dominican Republic 10.2%, Haiti 7.8%, Jamaica 6.6%, Saint Martin 5.9%, Guyana 5%, Dominica 4.4%, Curacao 4.1%, Aruba 3.4%, Saint Kitts and Nevis 2.8%, India 2.6%, Netherlands 2.2%, US 1.6%, Suriname 1.4%, Saint Lucia 1.3%, Anguilla 1.1%, other 8%, unspecified 1.7% (2011 est.)
note: data represent population by country of birth

Languages: English (official) 67.5%, Spanish 12.9%, Creole 8.2%, Dutch (official) 4.2%, Papiamento (a Spanish-Portuguese-Dutch-English dialect) 2.2%, French 1.5%, other 3.5% (2001 est.)

Religions: Protestant 41.9% (Pentecostal 14.7%, Methodist 10.0%, Seventh Day Adventist 6.6%, Baptist 4.7%, Anglican 3.1%, other Protestant 2.8%), Roman Catholic 33.1%, Hindu 5.2%, Christian 4.1%, Jehovah's Witness 1.7%, Evangelical 1.4%, Muslim/Jewish 1.1%, other 1.3% (includes Buddhist, Sikh, Rastafarian), none 7.9%, no response 2.4% (2011 est.)

Age structure: *0-14 years:* 18.64% (male 4,242/female 3,932)
15-24 years: 13.26% (male 2,967/female 2,849)
25-54 years: 39.08% (male 8,417/female 8,717)
55-64 years: 17.47% (male 3,638/female 4,020)
65 years and over: 11.55% (male 2,385/female 2,680) (2020 est.)

Median age: *total:* 41.1 years
male: 39.6 years
female: 42.7 years (2020 est.)
country comparison to the world: 46

Population growth rate: 1.34% (2020 est.)
country comparison to the world: 82

Birth rate: 12.9 births/1,000 population (2020 est.)
country comparison to the world: 145

Death rate: 5.8 deaths/1,000 population (2020 est.)
country comparison to the world: 175

Net migration rate: 6 migrant(s)/1,000 population (2020 est.)
country comparison to the world: 19

Population distribution: most populous areas are Lower Prince's Quarter (north of Philipsburg), followed closely by Cul de Sac

Urbanization: *urban population:* 100% of total population (2020)
rate of urbanization: 1.56% annual rate of change (2015-20 est.)

total population growth rate v. urban population growth rate, 2000-2030: *Major urban areas—population:* 1,327 PHILIPSBURG (capital) (2011)

Sex ratio: *at birth:* 1.05 male(s)/female
0-14 years: 1.08 male(s)/female
15-24 years: 1.04 male(s)/female
25-54 years: 0.97 male(s)/female
55-64 years: 0.9 male(s)/female
65 years and over: 0.89 male(s)/female
total population: 0.98 male(s)/female (2020 est.)

Infant mortality rate: *total:* 7.5 deaths/1,000 live births
male: 8.2 deaths/1,000 live births
female: 6.8 deaths/1,000 live births (2020 est.)
country comparison to the world: 154

Life expectancy at birth: *total population:* 78.8 years
male: 76.4 years
female: 81.3 years (2020 est.)
country comparison to the world: 64

Total fertility rate: 2.02 children born/woman (2020 est.)
country comparison to the world: 111

Drinking water source: *improved:* total: 95.1% of population
unimproved: total: 4.9% of population (2017 est.)

Sanitation facility access: *improved:* total: 98.8% of population
unimproved: total: 1.2% of population (2017)

HIV/AIDS—adult prevalence rate: NA

HIV/AIDS—people living with HIV/AIDS: NA

HIV/AIDS—deaths: NA

Education expenditures: NA

School life expectancy (primary to tertiary education):
total: 12 years
male: 12 years
female: 12 years (2014)

GOVERNMENT

Country name: *conventional long form:* Country of Sint Maarten
conventional short form: Sint Maarten
local long form: Land Sint Maarten (Dutch); Country of Sint Maarten (English)
local short form: Sint Maarten (Dutch and English)
former: Netherlands Antilles; Curacao and Dependencies
etymology: explorer Christopher COLUMBUS named the island after Saint MARTIN of Tours because the 11 November 1493 day of discovery was the saint's feast day

Dependency status: constituent country within the Kingdom of the Netherlands; full autonomy in internal affairs granted in 2010; Dutch Government responsible for defense and foreign affairs

Government type: parliamentary democracy under a constitutional monarchy

Capital: *name:* Philipsburg
geographic coordinates: 18 1 N, 63 2 W
time difference: UTC-4 (1 hour ahead of Washington, DC, during Standard Time)
etymology: founded and named in 1763 by John PHILIPS, a Scottish captain in the Dutch navy

Administrative divisions: none (part of the Kingdom of the Netherlands)
note: Sint Maarten is one of four constituent countries of the Kingdom of the Netherlands; the other three are the Netherlands, Aruba, and Curacao

Independence: none (part of the Kingdom of the Netherlands)

National holiday: King's Day (birthday of King WILLEM-ALEXANDER), 27 April (1967); note—King's or Queen's Day are observed on the ruling monarch's birthday; celebrated on 26 April if 27 April is a Sunday; local holiday Sint Maarten's Day, 11 November (1985), commemorates the discovery of the island by COLUMBUS on Saint Martin's Day, 11 November 1493; celebrated on both halves of the island

Constitution: *history:* previous 1947, 1955; latest adopted 21 July 2010, entered into force 10 October 2010 (regulates governance of Sint Maarten but is subordinate to the Charter for the Kingdom of the Netherlands)

Legal system: based on Dutch civil law system with some English common law influence

Citizenship: see the Netherlands

Suffrage: 18 years of age; universal

Executive branch: *chief of state:* King WILLEM-ALEXANDER of the Netherlands (since 30 April 2013); represented by Governor General Eugene HOLIDAY (since 10 October 2010)
head of government: Interim Prime Minister Silveria JACOBS (since 16 January 2020)
cabinet: Cabinet nominated by the prime minister and appointed by the governor-general
elections/appointments: the monarch is hereditary; governor general appointed by the monarch

for a 6-year term; following parliamentary elections, the leader of the majority party usually elected prime minister by Parliament

note—on 16 January 2020, Governor Eugene HOLIDAY appoints Silveria JACOBS as formateur of a new government

Legislative branch: *description:* unicameral Parliament of Sint Maarten (15 seats; members directly elected by proportional representation vote to serve 4-year terms)

elections: last held 9 January 2020 (next to be held in 2024)

election results: percent of vote by party—NA 35.2%, UP 24.2%, US Party 13.2%, PFP 10.6%, UD 8.7%, other 8.1%; seats by party—NA 6, UP 4, PFP 2, US Party 2, UD 1

Judicial branch: *highest courts:* Joint Court of Justice of Aruba, Curacao, Sint Maarten, and of Bonaire, Sint Eustatius and Saba or "Joint Court of Justice" (consists of the presiding judge, other members, and their substitutes); final appeals heard by the Supreme Court (in The Hague, Netherlands); note—prior to 2010, the Joint Court of Justice was the Common Court of Justice of the Netherlands Antilles and Aruba

judge selection and term of office: Joint Court judges appointed by the monarch serve for life

subordinate courts: Courts in First Instance

Political parties and leaders: National Alliance or NA [William MARLIN]

Party for Progress or PFP [Melissa GUMBS]

Sint Maarten Christian Party or SMCP [Wycliffe SMITH]

United Democrats Party or UD [Theodore HEYLIGER]

United Peoples Party or UP [NA]

United Sint Maarten Party or US Party [Frans RICHARDSON]

International organization participation: Caricom (observer), ILO, Interpol, UNESCO (associate), UPU, WMO

Diplomatic representation in the US: none (represented by the Kingdom of the Netherlands)

Diplomatic representation from the US: the US does not have an embassy in Sint Maarten; the Consul General to Curacao is accredited to Sint Maarten

Flag description: two equal horizontal bands of red (top) and blue with a white isosceles triangle based on the hoist side; the center of the triangle displays the Sint Maarten coat of arms; the arms consist of an orange-bordered blue shield prominently displaying the white court house in Philipsburg, as well as a bouquet of yellow sage (the national flower) in the upper left, and the silhouette of a Dutch-French friendship monument in the upper right; the shield is surmounted by a yellow rising sun in front of which is a brown pelican in flight; a yellow scroll below the shield bears the motto: SEMPER PROGREDIENS (Always Progressing); the three main colors are identical to those on the Dutch flag

note: the flag somewhat resembles that of the Philippines but with the main red and blue bands reversed; the banner more closely evokes the wartime Philippine flag

National symbol(s): *brown pelican, yellow sage (flower); national colors:* red, white, blue

National anthem: *name:* O Sweet Saint Martin's Land

lyrics/music: Gerard KEMPS

note: the song, written in 1958, is used as an unofficial anthem for the entire island (both French and Dutch sides); as a collectivity of France, in addition to the local anthem, "La Marseillaise" is official on the French side (see France); as a constituent part of the Kingdom of the Netherlands, in addition to the local anthem, "Het Wilhelmus" is official on the Dutch side (see Netherlands)

ECONOMY

Economy—overview: The economy of Sint Maarten centers around tourism with nearly four-fifths of the labor force engaged in this sector. Nearly 1.8 million visitors came to the island by cruise ship and roughly 500,000 visitors arrived through Princess Juliana International Airport in 2013. Cruise ships and yachts also call on Sint Maarten's numerous ports and harbors. Limited agriculture and local fishing means that almost all food must be imported. Energy resources and manufactured goods are also imported. Sint Maarten had the highest per capita income among the five islands that formerly comprised the Netherlands Antilles.

GDP (purchasing power parity): $365.8 million (2014 est.)
$353.5 million (2013 est.)
$339.6 million (2012 est.)
note: data are in 2014 US dollars
country comparison to the world: 214

GDP (official exchange rate): $304.1 million (2014 est.)

GDP—real growth rate: 3.6% (2014 est.)
4.1% (2013 est.)
1.9% (2012 est.)
country comparison to the world: 83

GDP—per capita (PPP): $66,800 (2014 est.)
$65,500 (2013 est.)
$63,900 (2012 est.)
note: data are in 2015 US dollars
country comparison to the world: 14

GDP—composition, by sector of origin:
agriculture: 0.4% (2008 est.)
industry: 18.3% (2008 est.)
services: 81.3% (2008 est.)

Agriculture—products: sugar

Industries: tourism, light industry

Labor force: 23,200 (2008 est.)
country comparison to the world: 210

Labor force—by occupation: *agriculture:* 1.1%
industry: 15.2%
services: 83.7% (2008 est.)

Unemployment rate: 12% (2012 est.)
10.6% (2008 est.)
country comparison to the world: 165

Inflation rate (consumer prices): 4% (2012 est.)
0.7% (2009 est.)
country comparison to the world: 157

Exports—commodities: sugar

Exchange rates: Netherlands Antillean guilders (ANG) per US dollar -
1.79 (2017 est.)
1.79 (2016 est.)
1.79 (2015 est.)
1.79 (2014 est.)
1.79 (2013 est.)

ENERGY

Electricity access: *electrification—total population:* 100% (2020)

Electricity—production: 304.3 million kWh (2008 est.)
country comparison to the world: 181

Crude oil—exports: 0 bbl/day (2015 est.)
country comparison to the world: 193

Crude oil—imports: 0 bbl/day (2015 est.)
country comparison to the world: 195

Refined petroleum products—production: 0 bbl/day (2015 est.)
country comparison to the world: 201

Refined petroleum products—consumption: 10,600 bbl/day (2016 est.)
country comparison to the world: 161

Refined petroleum products—exports: 0 bbl/day (2015 est.)
country comparison to the world: 202

Refined petroleum products—imports: 10,440 bbl/day (2015 est.)
country comparison to the world: 148

COMMUNICATIONS

Telephones—mobile cellular: *total subscriptions:* 84,773
subscriptions per 100 inhabitants: 195.94 (2019 est.)
country comparison to the world: 196

Telecommunication systems: *general assessment:* generally adequate facilities; growth sectors include mobile telephone and data segments; effective competition; LTE expansion; tourism and telecom sector contribute greatly to the GDP (2018)

domestic: extensive interisland microwave radio relay links; 196 per 100 mobile-cellular teledensity (2019)

international: country code—1-721; landing points for SMPR-1 and the ECFS submarine cables providing connectivity to the Caribbean; satellite earth stations—2 Intelsat (Atlantic Ocean) (2019)

note: the COVID-19 outbreak is negatively impacting telecommunications production and supply chains globally; consumer spending on telecom devices and services has also slowed due to the pandemic's effect on economies worldwide; overall progress towards improvements in all facets of the

telecom industry—mobile, fixed-line, broadband, submarine cable and satellite—has moderated

Internet country code: .sx; note—IANA has designated .sx for Sint Maarten, but has not yet assigned it to a sponsoring organization

Airports: 1 (2013)
country comparison to the world: 236

Airports—with paved runways: *total:* 1 (2019)
1,524 to 2,437 m: 1

note: Princess Juliana International Airport (SXM) was severely damaged on 6 September 2017 by hurricane Irma, but resumed commercial operations on 10 October 2017

Roadways: *total:* 53 km
country comparison to the world: 216

Ports and terminals: *major seaport(s):* Philipsburg
oil terminal(s): Coles Bay oil terminal

MILITARY AND SECURITY

Military and security forces: no regular military forces; Police Department for local law enforcement, supported by the Royal Netherlands Marechaussee (Gendarmerie), the Dutch Caribbean Police Force (Korps Politie Caribisch Nederland or KPCN), and the Dutch Caribbean Coast Guard (Kustwacht Caribisch Gebied or KWCARIB0)) (2019)

Military—note: defense is the responsibility of the Kingdom of the Netherlands

SLOVAKIA

INTRODUCTION

Background: Slovakia traces its roots to the 9th century state of Great Moravia. Subsequently, the Slovaks became part of the Hungarian Kingdom, where they remained for the next 1,000 years. After the formation of the dual Austro-Hungarian monarchy in 1867, backlash to language and education policies favoring the use of Hungarian (Magyarization) encouraged the strengthening of Slovak nationalism and a cultivation of cultural ties with the closely related Czechs, who fell administratively under the Austrian half of the empire. After the dissolution of the Austro-Hungarian Empire at the close of World War I, the Slovaks joined the Czechs to form Czechoslovakia. The new state was envisioned as a nation with Czech and Slovak branches. During the interwar period, Slovak nationalist leaders pushed for autonomy within Czechoslovakia, and in 1939 Slovakia became an independent state created by and allied with Nazi Germany. Following World War II, Czechoslovakia was reconstituted and came under communist rule within Soviet-dominated Eastern Europe. In 1968, an invasion by Warsaw Pact troops ended the efforts of Czechoslovakia's leaders to liberalize communist rule and create "socialism with a human face," ushering in a period of repression known as "normalization." The peaceful "Velvet Revolution" swept the Communist Party from power at the end of 1989 and inaugurated a return to democratic rule and a market economy. On 1 January 1993, Czechoslovakia underwent a nonviolent "velvet divorce" into its two national components, Slovakia and the Czech Republic. Slovakia joined both NATO and the EU in the spring of 2004 and the euro zone on 1 January 2009.

GEOGRAPHY

Location: Central Europe, south of Poland

Geographic coordinates: 48 40 N, 19 30 E

Map references: Europe

Area: *total:* 49,035 sq km
land: 48,105 sq km
water: 930 sq km
country comparison to the world: 131

Area—comparative: about one and a half times the size of Maryland; about twice the size of New Hampshire

Land boundaries: *total:* 1,611 km
border countries (5): Austria 105 km, Czech Republic 241 km, Hungary 627 km, Poland 541 km, Ukraine 97 km

Coastline: 0 km (landlocked)

Maritime claims: none (landlocked)

Climate: temperate; cool summers; cold, cloudy, humid winters

Terrain: rugged mountains in the central and northern part and lowlands in the south

Elevation: *mean elevation:* 458 m
lowest point: Bodrok River 94 m
highest point: Gerlachovsky Stit 2,655 m

Natural resources: lignite, small amounts of iron ore, copper and manganese ore; salt; arable land

Land use: *agricultural land:* 40.1% (2011 est.)
arable land: 28.9% (2011 est.) / permanent crops: 0.4% (2011 est.) / permanent pasture: 10.8% (2011 est.)
forest: 40.2% (2011 est.)
other: 19.7% (2011 est.)
Irrigated land: 869 sq km (2012)

Population distribution: a fairly even distribution throughout most of the country; slightly larger concentration in the west in proximity to the Czech border

Natural hazards: flooding

Environment—current issues: air pollution and acid rain present human health risks and damage forests; land erosion caused by agricultural and mining practices; water pollution

Environment—international agreements: *party to:* Air Pollution, Air Pollution-Nitrogen Oxides, Air Pollution-Persistent Organic Pollutants, Air Pollution-Sulfur 85, Air Pollution-Sulfur 94, Air Pollution-Volatile Organic Compounds, Antarctic Treaty, Biodiversity, Climate Change, Climate Change-Kyoto Protocol, Desertification, Endangered Species, Environmental Modification, Hazardous Wastes, Law of the Sea, Ozone Layer Protection, Ship Pollution, Wetlands, Whaling
signed, but not ratified: none of the selected agreements

Geography—note: landlocked; most of the country is rugged and mountainous; the Tatra Mountains in the north are interspersed with many scenic lakes and valleys

Population: *People and Society*
5,440,602 (July 2020 est.)
country comparison to the world: 120

Nationality: *noun:* Slovak(s)
adjective: Slovak

Ethnic groups: Slovak 80.7%, Hungarian 8.5%, Romani 2%, other 1.8% (includes Czech, Ruthenian, Ukrainian, Russian, German, Polish), unspecified 7% (2011 est.)
note: data represent population by nationality; Romani populations are usually underestimated in official statistics and may represent 7–11% of Slovakia's population

Languages: Slovak (official) 78.6%, Hungarian 9.4%, Roma 2.3%, Ruthenian 1%, other or unspecified 8.8% (2011 est.)

Religions: Roman Catholic 62%, Protestant 8.2%, Greek Catholic 3.8%, other or unspecified 12.5%, none 13.4% (2011 est.)

Age structure: *0-14 years:* 15.13% (male 423,180/female 400,128)
15-24 years: 10.06% (male 280,284/female 266,838)

25-54 years: 44.61% (male 1,228,462/female 1,198,747)
55-64 years: 13.15% (male 342,124/female 373,452)
65 years and over: 17.05% (male 366,267/female 561,120) (2020 est.)

Dependency ratios: total dependency ratio: 47.6
youth dependency ratio: 23
elderly dependency ratio: 24.6
potential support ratio: 4.1 (2020 est.)

Median age: total: 41.8 years
male: 40.1 years
female: 43.6 years (2020 est.)
country comparison to the world: 41

Population growth rate: -0.05% (2020 est.)
country comparison to the world: 199

Birth rate: 9.3 births/1,000 population (2020 est.)
country comparison to the world: 203

Death rate: 10.1 deaths/1,000 population (2020 est.)
country comparison to the world: 37

Net migration rate: 0.2 migrant(s)/1,000 population (2020 est.)
country comparison to the world: 72

Population distribution: a fairly even distribution throughout most of the country; slightly larger concentration in the west in proximity to the Czech border

Urbanization: urban population: 53.8% of total population (2020)
rate of urbanization: 0% annual rate of change (2015-20 est.)

total population growth rate v. urban population growth rate, 2000-2030: Major urban areas—population: 435,000 BRATISLAVA (capital) (2020)

Sex ratio: at birth: 1.07 male(s)/female
0-14 years: 1.06 male(s)/female
15-24 years: 1.05 male(s)/female
25-54 years: 1.02 male(s)/female
55-64 years: 0.92 male(s)/female
65 years and over: 0.65 male(s)/female
total population: 0.94 male(s)/female (2020 est.)

Mother's mean age at first birth: 27.8 years (2014 est.)

Maternal mortality rate: 5 deaths/100,000 live births (2017 est.)
country comparison to the world: 170

Infant mortality rate: total: 4.9 deaths/1,000 live births
male: 5.5 deaths/1,000 live births
female: 4.3 deaths/1,000 live births (2020 est.)
country comparison to the world: 179

Life expectancy at birth: total population: 77.8 years
male: 74.3 years
female: 81.6 years (2020 est.)
country comparison to the world: 75

Total fertility rate: 1.44 children born/woman (2020 est.)
country comparison to the world: 212

Drinking water source:

improved:
urban: 100% of population
rural: 100% of population
total: 100% of population
unimproved:
urban: 0% of population
rural: 0% of population
total: 0% of population (2017 est.)

Current Health Expenditure: 6.7% (2017)

Physicians density: 3.42 physicians/1,000 population (2017)

Hospital bed density: 5.8 beds/1,000 population (2017)

Sanitation facility access:
improved:
urban: 99.9% of population
rural: 100% of population
total: 100% of population
unimproved:
urban: 0.1% of population
rural: 0% of population
total: 0% of population (2017 est.)

HIV/AIDS—adult prevalence rate: <.1% (2018 est.)

HIV/AIDS—people living with HIV/AIDS: 1,200 (2018 est.)
country comparison to the world: 143

HIV/AIDS—deaths: <100 (2018 est.)

Obesity—adult prevalence rate: 20.5% (2016)
country comparison to the world: 98

Education expenditures:

+.
3.9% of GDP (2016)
country comparison to the world: 109

School life expectancy (primary to tertiary education): total: 15 years
male: 14 years
female: 15 years (2018)

Unemployment, youth ages 15-24: total: 14.9%
male: 14.3%
female: 16.1% (2018 est.)
country comparison to the world: 91

GOVERNMENT

Country name: conventional long form: Slovak Republic
conventional short form: Slovakia
local long form: Slovenska republika
local short form: Slovensko
etymology: may derive from the medieval Latin word "Slavus" (Slav), which had the local form "Sloven", used since the 13th century to refer to the territory of Slovakia and its inhabitants

Government type: parliamentary republic

Capital: name: Bratislava

geographic coordinates: 48 09 N, 17 07 E
time difference: UTC+1 (6 hours ahead of Washington, DC, during Standard Time)
daylight saving time: +1hr, begins last Sunday in March; ends last Sunday in October
etymology: the name was adopted in 1919 after Czechoslovakia gained its independence and may

derive from later transliterations of the 9th century military commander, Braslav, or the 11th century Bohemian Duke Bretislav I; alternatively, the name may derive from the Slovak words "brat" (brother) and "slava" (glory)

Administrative divisions: 8 regions (kraje, singular—kraj); Banskobystricky, Bratislavsky, Kosicky, Nitriansky, Presovsky, Trenciansky, Trnavsky, Zilinsky

Independence: 1 January 1993 (Czechoslovakia split into the Czech Republic and Slovakia)

National holiday: Constitution Day, 1 September (1992)

Constitution: history: several previous (preindependence); latest passed by the National Council 1 September 1992, signed 3 September 1992, effective 1 October 1992
amendments: proposed by the National Council; passage requires at least three-fifths majority vote of Council members; amended many times, last in 2017

Legal system: civil law system based on Austro-Hungarian codes; note—legal code modified to comply with the obligations of Organization on Security and Cooperation in Europe

International law organization participation: accepts compulsory ICJ jurisdiction with reservations; accepts ICCt jurisdiction

Citizenship: citizenship by birth: no
citizenship by descent only: at least one parent must be a citizen of Slovakia
dual citizenship recognized: no
residency requirement for naturalization: 5 years

Suffrage: 18 years of age; universal

Executive branch: chief of state: President Zuzana CAPUTOVA (since 15 June 2014)
head of government: Prime Minister Peter PELLIGRINI (since 22 March 2018); Deputy Prime Ministers Richard RASI (since 22 March 2018), Laszlo SOLYMOS (since 22 March 2018), Gabriela MATECNA (since 29 November 2017)
cabinet: Cabinet appointed by the president on the recommendation of the prime minister
elections/appointments: president directly elected by absolute majority popular vote in 2 rounds if needed for a 5-year term (eligible for a second term); election last held on 16 March and 30 March 2019 (next to be held March 2024); following National Council elections (every 4 years), the president designates a prime minister candidate, usually the leader of the party or coalition that wins the most votes, who must win a vote of confidence in the National Council
election results: Zuzana CAPUTOVA elected president in second round; percent of vote—Zuzana CAPUTOVA (PS) 58.4%, Maros SEFCOVIC (independent) 41.6%

Legislative branch: description: unicameral National Council or Narodna Rada (150 seats; members directly elected in a single- and multiseat constituencies by closed, party-list proportional representation vote; members serve 4-year terms)

elections: last held on 29 February 2020 (next to be held March 2024)

election results: percent of vote by party—OLaNO-NOVA 25%, Smer-SD 18.3%, Sme-Rodina 8.2%, LSNS 8%, PS-SPOLU 7%, SaS 6.2%, Za Ludi 5.8%, other 21.5%; seats by party—OLaNO-NOVA 53, Smer-SD 38, Sme-Rodina 17, LSNS 17, SaS 13, Za Ludi 12, PS-SPOLU 0; composition—men 120, women 30, percent of women 20%

Judicial branch: *highest courts:* Supreme Court of the Slovak Republic (consists of the court president, vice president, and approximately 80 judges organized into criminal, civil, commercial, and administrative divisions with 3- and 5-judge panels); Constitutional Court of the Slovak Republic (consists of 13 judges organized into 3-judge panels)

judge selection and term of office: Supreme Court judge candidates nominated by the Judicial Council of the Slovak Republic, an 18-member self-governing body that includes the Supreme Court chief justice and presidential, governmental, parliamentary, and judiciary appointees; judges appointed by the president serve for life subject to removal by the president at age 65; Constitutional Court judges nominated by the National Council of the Republic and appointed by the president; judges serve 12-year terms

subordinate courts: regional and district civil courts; Special Criminal Court; Higher Military Court; military district courts; Court of Audit;

Political parties and leaders: Christian Democratic Movement or KDH [Alojz HLINA] Bridge or Most-Hid [Bela BUGAR] Direction-Social Democracy or Smer-SD [Robert FICO] For the People or Za Ludi [Andrej KISKA] Freedom and Solidarity or SaS [Richard SULIK] Kotleba-People's Party Our Slovakia or LSNS [Marian KOTLEBA] Ordinary People and Independent Personalities—New Majority or OLaNO-NOVA [Igor MATOVIC] Party of the Hungarian Community or SMK [Jozsef MENYHART] Progressive Slovakia or PS [Michal TRUBAN] Slovak National Party or SNS [Andrej DANKO] Together or SPOLU [Miroslav BEBLAVY] We Are Family or Sme-Rodina [Boris KOLLAR]

International organization participation: Australia Group, BIS, BSEC (observer), CBSS (observer), CD, CE, CEI, CERN, EAPC, EBRD, ECB, EIB, EMU, EU, FAO, IAEA, IBRD, ICAO, ICC (national committees), ICRM, IDA, IEA, IFC, IFRCS, ILO, IMF, IMO, IMSO, Interpol, IOC, IOM, IPU, ISO, ITU, ITUC (NGOs), MIGA, NATO, NEA, NSG, OAS (observer), OECD, OIF (observer), OPCW, OSCE, PCA, Schengen Convention, SELEC (observer), UN, UNCTAD, UNESCO, UNFICYP, UNIDO, UNTSO, UNWTO, UPU, WCO, WFTU (NGOs), WHO, WIPO, WMO, WTO, ZC

Diplomatic representation in the US: *chief of mission:* Charge d'Affaires Josef POLAKOVIC (since 7 April 2020)

chancery: 3523 International Court NW, Washington, DC 20008
telephone: [1] (202) 237-1054
FAX: [1] (202) 237-6438
consulate(s) general: Los Angeles, New York

Diplomatic representation from the US: *chief of mission:* Ambassador Bridget A. BRINK (since 20 August 2019)
telephone: [421] (2) 5443-3338
embassy: P.O. Box 309, 814 99 Bratislava
mailing address: P.O. Box 309, 814 99 Bratislava
FAX: [421] (2) 5441-5148

Flag description: three equal horizontal bands of white (top), blue, and red derive from the Pan-Slav colors; the Slovakian coat of arms (consisting of a red shield bordered in white and bearing a white double-barred cross of St. Cyril and St. Methodius surmounting three blue hills) is centered over the bands but offset slightly to the hoist side
note: the Pan-Slav colors were inspired by the 19th-century flag of Russia

National symbol(s): *double-barred cross (Cross of St. Cyril and St. Methodius) surmounting three peaks; national colors:* white, blue, red

National anthem: *name:* "Nad Tatrou sa blyska" (Lightning Over the Tatras)
lyrics/music: Janko MATUSKA/traditional
note: adopted 1993, in use since 1844; music based on the Slovak folk song "Kopala studienku"
0:00 / 0:33

ECONOMY

Economy—overview: Slovakia's economy suffered from a slow start in the first years after its separation from the Czech Republic in 1993, due to the country's authoritarian leadership and high levels of corruption, but economic reforms implemented after 1998 have placed Slovakia on a path of strong growth. With a population of 5.4 million, the Slovak Republic has a small, open economy driven mainly by automobile and electronics exports, which account for more than 80% of GDP. Slovakia joined the EU in 2004 and the euro zone in 2009. The country's banking sector is sound and predominantly foreign owned.

Slovakia has been a regional FDI champion for several years, attractive due to a relatively low-cost yet skilled labor force, and a favorable geographic location in the heart of Central Europe. Exports and investment have been key drivers of Slovakia's robust growth in recent years. The unemployment rate fell to historical lows in 2017, and rising wages fueled increased consumption, which played a more prominent role in 2017 GDP growth. A favorable outlook for the Eurozone suggests continued strong growth prospects for Slovakia during the next few years, although inflation is also expected to pick up.

Among the most pressing domestic issues potentially threatening the attractiveness of the Slovak market are shortages in the qualified labor force, persistent corruption issues, and an inadequate judiciary, as well as a slow transition to an innovation-based economy. The energy sector in particular is characterized by unpredictable regulatory oversight and high costs, in part driven by government interference in regulated tariffs. Moreover, the government's attempts to maintain low household energy prices could harm the profitability of domestic energy firms while undercutting energy efficiency initiatives.

GDP (purchasing power parity): $179.7 billion (2017 est.)
$173.8 billion (2016 est.)
$168.2 billion (2015 est.)
note: data are in 2017 dollars
country comparison to the world: 69

GDP (official exchange rate): $95.96 billion (2017 est.)

GDP—real growth rate: 2.4% (2019 est.)
3.9% (2018 est.)
3.04% (2017 est.)
country comparison to the world: 117

GDP—per capita (PPP): $33,100 (2017 est.)
$32,000 (2016 est.)
$31,000 (2015 est.)
note: data are in 2017 dollars
country comparison to the world: 61

Gross national saving: 20.6% of GDP (2017 est.)
21.1% of GDP (2016 est.)
22.5% of GDP (2015 est.)
country comparison to the world: 92

GDP—composition, by end use:
household consumption: 54.7% (2017 est.)
government consumption: 19.2% (2017 est.)
investment in fixed capital: 21.2% (2017 est.)
investment in inventories: 1.2% (2017 est.)
exports of goods and services: 96.3% (2017 est.)
imports of goods and services: -92.9% (2017 est.)

GDP—composition, by sector of origin:
agriculture: 3.8% (2017 est.)
industry: 35% (2017 est.)
services: 61.2% (2017 est.)

Agriculture—products: grains, potatoes, sugar beets, hops, fruit; pigs, cattle, poultry; forest products

Industries: automobiles; metal and metal products; electricity, gas, coke, oil, nuclear fuel; chemicals, synthetic fibers, wood and paper products; machinery; earthenware and ceramics; textiles; electrical and optical apparatus; rubber products; food and beverages; pharmaceutical

Industrial production growth rate: 2.7% (2017 est.)
country comparison to the world: 114

Labor force: 2.511 million (2020 est.)
country comparison to the world: 114

Labor force—by occupation: *agriculture:* 3.9%
industry: 22.7%
services: 73.4% (2015)

Unemployment rate: 5% (2019 est.)
5.42% (2018 est.)
country comparison to the world: 77

Population below poverty line: 12.3% (2015 est.)

Household income or consumption by percentage share: *lowest 10%:* 3.3%
highest 10%: 19.3% (2015 est.)

Budget: *revenues:* 37.79 billion (2017 est.) *expenditures:* 38.79 billion (2017 est.)

Taxes and other revenues: 39.4% (of GDP) (2017 est.)
country comparison to the world: 47

Budget surplus (+) or deficit (-): -1% (of GDP) (2017 est.)
country comparison to the world: 81

Public debt: 50.9% of GDP (2017 est.)
51.8% of GDP (2016 est.)
note: data cover general Government Gross Debt and include debt instruments issued (or owned) by Government entities, including sub-sectors of central, state, local government, and social security funds
country comparison to the world: 98

Fiscal year: calendar year

Inflation rate (consumer prices): 1.3% (2017 est.)
-0.5% (2016 est.)
country comparison to the world: 74

Current account balance: -$3.026 billion (2019 est.)
-$2.635 billion (2018 est.)
country comparison to the world: 173

Exports: $80.8 billion (2017 est.)
$75.53 billion (2016 est.)
country comparison to the world: 40

Exports—partners: Germany 20.7%, Czech Republic 11.6%, Poland 7.7%, France 6.3%, Italy 6.1%, UK 6%, Hungary 6%, Austria 6% (2017)

Exports—commodities: vehicles and related parts 27%, machinery and electrical equipment 20%, nuclear reactors and furnaces 12%, iron and steel 4%, mineral oils and fuels 5% (2015 est.)

Imports: $80.07 billion (2017 est.)
$72.51 billion (2016 est.)
country comparison to the world: 42

Imports—commodities: machinery and electrical equipment 20%, vehicles and related parts 14%, nuclear reactors and furnaces 12%, fuel and mineral oils 9% (2015 est.)

Imports—partners: Germany 19.1%, Czech Republic 16.3%, Austria 10.3%, Poland 6.5%, Hungary 6.4%, South Korea 4.5%, Russia 4.5%, France 4.3%, China 4.2% (2017)

Reserves of foreign exchange and gold: $3.622 billion (31 December 2017 est.)
$2.892 billion (31 December 2016 est.)
country comparison to the world: 102

Debt—external: $75.04 billion (31 March 2016 est.)
$74.19 billion (31 March 2015 est.)
country comparison to the world: 58

Exchange rates: euros (EUR) per US dollar -
0.885 (2017 est.)
0.903 (2016 est.)
0.9214 (2015 est.)
0.885 (2014 est.)
0.7634 (2013 est.)

ENERGY

Electricity access: *electrification—total population:* 100% (2020)

Electricity—production: 25.32 billion kWh (2016 est.)
country comparison to the world: 72

Electricity—consumption: 26.64 billion kWh (2016 est.)
country comparison to the world: 66

Electricity—exports: 10.6 billion kWh (2016 est.)
country comparison to the world: 19

Electricity—imports: 13.25 billion kWh (2016 est.)
country comparison to the world: 19

Electricity—installed generating capacity: 7.644 million kW (2016 est.)
country comparison to the world: 72

Electricity—from fossil fuels: 36% of total installed capacity (2016 est.)
country comparison to the world: 176

Electricity—from nuclear fuels: 27% of total installed capacity (2017 est.)
country comparison to the world: 3

Electricity—from hydroelectric plants: 24% of total installed capacity (2017 est.)
country comparison to the world: 81

Electricity—from other renewable sources: 13% of total installed capacity (2017 est.)
country comparison to the world: 70

Crude oil—production: 200 bbl/day (2018 est.)
country comparison to the world: 95

Crude oil—exports: 1,022 bbl/day (2017 est.)
country comparison to the world: 74

Crude oil—imports: 111,200 bbl/day (2017 est.)
country comparison to the world: 42

Crude oil—proved reserves: 9 million bbl (1 January 2018 est.)
country comparison to the world: 91

Refined petroleum products—production: 131,300 bbl/day (2017 est.)
country comparison to the world: 64

Refined petroleum products—consumption: 85,880 bbl/day (2017 est.)
country comparison to the world: 85

Refined petroleum products—exports: 81,100 bbl/day (2017 est.)
country comparison to the world: 46

Refined petroleum products—imports: 38,340 bbl/day (2017 est.)
country comparison to the world: 91

Natural gas—production: 104.8 million cu m (2017 est.)
country comparison to the world: 81

Natural gas—consumption: 4.672 billion cu m (2017 est.)
country comparison to the world: 62

Natural gas—exports: 0 cu m (2017 est.)
country comparison to the world: 184

Natural gas—imports: 4.984 billion cu m (2017 est.)
country comparison to the world: 37

Natural gas—proved reserves: 14.16 billion cu m (1 January 2018 est.)
country comparison to the world: 75

Carbon dioxide emissions from consumption of energy: 34.86 million Mt (2017 est.)
country comparison to the world: 73

COMMUNICATIONS

Telephones—fixed lines: *total subscriptions:* 673,341
subscriptions per 100 inhabitants: 12.37 (2019 est.)
country comparison to the world: 85

Telephones—mobile cellular: *total subscriptions:* 7,381,164
subscriptions per 100 inhabitants: 135.6 (2019 est.)
country comparison to the world: 101

Telecommunication systems: *general assessment:* a modern telecommunications system; near monopoly of fixed-line market; competition in mobile and fixed broadband market; broadband growth in recent years; competition among DSL, cable and fiber platforms; FttP growth in cities; mid-2019 launched 1G cable broadband service in 3 cities and 200,000 premises; EU funds development and improvement of e-govt. and online services; regulator prepares groundwork for 5G services (2020)
domestic: four companies have a license to operate cellular networks and provide nationwide cellular services; a few other companies provide services but do not have their own networks; fixed-line 12 per 100 and mobile-cellular 136 per 100 teledensity (2019)
international: country code—421; 3 international exchanges (1 in Bratislava and 2 in Banska Bystrica) are available; Slovakia is participating in several international telecommunications projects that will increase the availability of external services; connects to DREAM cable (2017)
note: the COVID-19 outbreak is negatively impacting telecommunications production and supply chains globally; consumer spending on telecom devices and services has also slowed due to the pandemic's effect on economies worldwide; overall progress towards improvements in all facets of the telecom industry—mobile, fixed-line, broadband, submarine cable and satellite—has moderated

Broadcast media: state-owned public broadcaster, Radio and Television of Slovakia (RTVS), operates 2 national TV stations and multiple national and regional radio networks; roughly 50 privately owned TV stations operating nationally, regionally, and locally; about 40% of households are connected to multi-channel cable or satellite TV; 32 privately owned radio stations

Internet country code: .sk

Internet users: *total:* 4,391,969
percent of population: 80.66% (July 2018 est.)
country comparison to the world: 89

Broadband—fixed subscriptions: *total:* 1,507,998
subscriptions per 100 inhabitants: 28 (2018 est.)
country comparison to the world: 64

TRANSPORTATION

National air transport system: *number of registered air carriers:* 4 (2020)

inventory of registered aircraft operated by air carriers: 45

Civil aircraft registration country code prefix: OM (2016)

Airports: 35 (2013)
country comparison to the world: 110

Airports—with paved runways: *total:* 19 (2019)
over 3,047 m: 2
2,438 to 3,047 m: 2
1,524 to 2,437 m: 3
914 to 1,523 m: 3
under 914 m: 9

Airports—with unpaved runways: *total:* 15 (2019)
914 to 1,523 m: 10
under 914 m: 5

Heliports: 1 (2019)

Pipelines: 2270 km gas transmission pipelines, 6278 km high-pressure gas distribution pipelines, 27023 km mid- and low-pressure gas distribution pipelines (2016), 510 km oil (2015)

Railways: *total:* 3,580 km (2016)
standard gauge: 3,435 km 1.435-m gauge (1,587 km electrified) (2016)
narrow gauge: 46 km 1.000-m or 0.750-m gauge (2016)
broad gauge: 99 km 1.520-m gauge (2016)
country comparison to the world: 56

Roadways: *total:* 56,926 km (includes local roads, national roads, and 464 km of highways) (2016)
country comparison to the world: 81

Waterways: 172 km (on Danube River) (2012)
country comparison to the world: 99

Ports and terminals: *river port(s):* Bratislava, Komarno (Danube)

MILITARY AND SECURITY

Military and security forces: *Armed Forces of the Slovak Republic (Ozbrojene Sily Slovenskej Republiky):* Land Forces, Air and Air Defense Forces, and a Joint Training and Support Command (2019)

Military expenditures: 1.74% of GDP (2019 est.)
1.22% of GDP (2018)
1.1% of GDP (2017)
1.12% of GDP (2016)
1.12% of GDP (2015)
country comparison to the world: 64

Military and security service personnel strengths: the Armed Forces of the Slovak Republic have approximately 15,000 active duty personnel (6,000 Land Forces; 4,000 Air and Air Defense; 5,000 other, including central staff, support, and training duties) (2019 est.)

Military equipment inventories and acquisitions: the inventory of the Slovakian military consists mostly of Soviet-era platforms; since 2010, it has imported limited quantities of equipment from China, Czechia, Italy, Russia, and the US (2019 est.)

Military deployments: 240 Cyprus (UNFICYP); up to 150 Latvia (NATO) (2020)

Military service age and obligation: 18-30 years of age for voluntary military service; conscription in peacetime suspended in 2006; women are eligible to serve (2012)

TRANSNATIONAL ISSUES

Disputes—international: bilateral government, legal, technical and economic working group negotiations continued between Slovakia and Hungary over Hungary's completion of its portion of the Gabcikovo-Nagymaros hydroelectric dam project along the Danube; as a member state that forms part of the EU's external border, Slovakia has implemented strict Schengen border rules

Refugees and internally displaced persons: *stateless persons:* 1,523 (2019)

Illicit drugs: transshipment point for Southwest Asian heroin bound for Western Europe; producer of synthetic drugs for regional market; consumer of ecstasy

SLOVENIA

INTRODUCTION

Background: The Slovene lands were part of the Austro-Hungarian Empire until the latter's dissolution at the end of World War I. In 1918, the Slovenes joined the Serbs and Croats in forming a new multinational state, which was named Yugoslavia in 1929. After World War II, Slovenia was one of the republics in the restored Yugoslavia, which, though communist, soon distanced itself from the Soviet Union and spearheaded the Non-Aligned Movement. Dissatisfied with the exercise

of power by the majority Serbs, the Slovenes succeeded in establishing their independence in 1991 after a short 10-day war. Historical ties to Western Europe, a growing economy, and a stable democracy have assisted in Slovenia's postcommunist transition. Slovenia acceded to both NATO and the EU in the spring of 2004; it joined the euro zone and the Schengen zone in 2007.

GEOGRAPHY

Location: south Central Europe, Julian Alps between Austria and Croatia

Geographic coordinates: 46 07 N, 14 49 E

Map references: Europe

Area: *total:* 20,273 sq km
land: 20,151 sq km
water: 122 sq km
country comparison to the world: 155

Area—comparative: slightly smaller than New Jersey

Land boundaries: *total:* 1,211 km
border countries (4): Austria 299 km, Croatia 600 km, Hungary 94 km, Italy 218 km

Coastline: 46.6 km

Maritime claims: *territorial sea:* 12 nm

Climate: Mediterranean climate on the coast, continental climate with mild to hot summers and cold winters in the plateaus and valleys to the east

Terrain: a short southwestern coastal strip of Karst topography on the Adriatic; an alpine mountain region lies adjacent to Italy and Austria in the north; mixed mountains and valleys with numerous rivers to the east

Elevation: *mean elevation:* 492 m
lowest point: Adriatic Sea 0 m
highest point: Triglav 2,864 m

Natural resources: lignite, lead, zinc, building stone, hydropower, forests

Land use: *agricultural land:* 22.8% (2011 est.)
arable land: 8.4% (2011 est.) / *permanent crops:* 1.3% (2011 est.) / *permanent pasture:* 13.1% (2011 est.)
forest: 62.3% (2011 est.)
other: 14.9% (2011 est.)
Irrigated land: 60 sq km (2012)

Population distribution: a fairly even distribution throughout most of the country, with urban areas attracting larger and denser populations; pockets in the mountainous northwest exhibit less density than elsewhere

Natural hazards: flooding; earthquakes

Environment—current issues: air pollution from road traffic, domestic heating (wood buring), power generation, and industry; water pollution; biodiversity protection

Environment—international agreements: *party to:* Air Pollution, Air Pollution-Nitrogen Oxides,

Air Pollution-Persistent Organic Pollutants, Air Pollution-Sulfur 94, Biodiversity, Climate Change, Climate Change-Kyoto Protocol, Desertification, Endangered Species, Environmental Modification, Hazardous Wastes, Law of the Sea, Marine Dumping, Ozone Layer Protection, Ship Pollution, Wetlands, Whaling

signed, but not ratified: none of the selected agreements

Geography—note: despite its small size, this eastern Alpine country controls some of Europe's major transit routes

PEOPLE AND SOCIETY

Population: 2,102,678 (July 2020 est.)
country comparison to the world: 148

Nationality: *noun:* Slovene(s)
adjective: Slovenian

Ethnic groups: Slovene 83.1%, Serb 2%, Croat 1.8%, Bosniak 1.1%, other or unspecified 12% (2002 est.)

Languages: Slovene (official) 91.1%, Serbo-Croatian 4.5%, other or unspecified 4.4%, Italian (official, only in municipalities where Italian national communities reside), Hungarian (official, only in municipalities where Hungarian national communities reside) (2002 census)

Religions: Catholic 57.8%, Muslim 2.4%, Orthodox 2.3%, other Christian 0.9%, unaffiliated 3.5%, other or unspecified 23%, none 10.1% (2002 est.)

Age structure: *0-14 years:* 14.84% (male 160,134/female 151,960)
15-24 years: 9.01% (male 98,205/female 91,318)
25-54 years: 40.73% (male 449,930/female 406,395)
55-64 years: 14.19% (male 148,785/female 149,635)
65 years and over: 21.23% (male 192,420/female 253,896) (2020 est.)

Dependency ratios: *total dependency ratio:* 55.9
youth dependency ratio: 23.6
elderly dependency ratio: 32.3
potential support ratio: 3.1 (2020 est.)

Median age: *total:* 44.9 years
male: 43.4 years
female: 46.6 years (2020 est.)
country comparison to the world: 11

Population growth rate: 0.01% (2020 est.)
country comparison to the world: 191

Birth rate: 8.7 births/1,000 population (2020 est.)
country comparison to the world: 212

Death rate: 10.3 deaths/1,000 population (2020 est.)
country comparison to the world: 32

Net migration rate: 1.5 migrant(s)/1,000 population (2020 est.)
country comparison to the world: 55

Population distribution: a fairly even distribution throughout most of the country, with urban areas attracting larger and denser populations; pockets

in the mountainous northwest exhibit less density than elsewhere

Urbanization: *urban population:* 55.1% of total population (2020)
rate of urbanization: 0.56% annual rate of change (2015-20 est.)

total population growth rate v. urban population growth rate, 2000-2030: *Major urban areas—population:* 286,000 LJUBLJANA (capital) (2018)

Sex ratio: *at birth:* 1.04 male(s)/female
0-14 years: 1.05 male(s)/female
15-24 years: 1.08 male(s)/female
25-54 years: 1.11 male(s)/female
55-64 years: 0.99 male(s)/female
65 years and over: 0.76 male(s)/female
total population: 1 male(s)/female (2020 est.)

Mother's mean age at first birth: 29.4 years (2017 est.)

Maternal mortality rate: 7 deaths/100,000 live births (2017 est.)
country comparison to the world: 156

Infant mortality rate: *total:* 1.7 deaths/1,000 live births
male: 1.8 deaths/1,000 live births
female: 1.6 deaths/1,000 live births (2020 est.)
country comparison to the world: 228

Life expectancy at birth: *total population:* 81.4 years
male: 78.5 years
female: 84.4 years (2020 est.)
country comparison to the world: 32

Total fertility rate: 1.59 children born/woman (2020 est.)
country comparison to the world: 188

Drinking water source:
improved:
urban: 100% of population
rural: 100% of population
total: 100% of population
unimproved:
urban: 0% of population
rural: 0% of population
total: 0% of population (2017 est.)

Current Health Expenditure: 8.2% (2017)

Physicians density: 3.09 physicians/1,000 population (2017)

Hospital bed density: 4.5 beds/1,000 population (2017)

Sanitation facility access:
improved:
urban: 100% of population
rural: 100% of population
total: 100% of population
unimproved:
urban: 0% of population
rural: 0% of population
total: 0% of population (2017 est.)

HIV/AIDS—adult prevalence rate: <.1% (2018 est.)

HIV/AIDS—people living with HIV/AIDS: <1000 (2017 est.)

HIV/AIDS—deaths: <100 (2018 est.)

Obesity—adult prevalence rate: 20.2% (2016)
country comparison to the world: 104

Education expenditures: 4.8% of GDP (2016)
country comparison to the world: 75

Literacy: *definition:* NA
total population: 99.7%
male: 99.7%
female: 99.7% (2015)

School life expectancy (primary to tertiary education): *total:* 18 years
male: 17 years
female: 18 years (2018)

Unemployment, youth ages 15-24: *total:* 8.8%
male: 8.3%
female: 9.6% (2018 est.)
country comparison to the world: 135

GOVERNMENT

Country name: *conventional long form:* Republic of Slovenia
conventional short form: Slovenia
local long form: Republika Slovenija
local short form: Slovenija
former: People's Republic of Slovenia, Socialist Republic of Slovenia
etymology: the country's name means "Land of the Slavs" in Slovene

Government type: parliamentary republic

Capital: *name:* Ljubljana

geographic coordinates: 46 03 N, 14 31 E
time difference: UTC+1 (6 hours ahead of Washington, DC, during Standard Time)
daylight saving time: +1hr, begins last Sunday in March; ends last Sunday in October
etymology: likely related to the Slavic root "ljub", meaning "to like" or "to love"; by tradition, the name is related to the Slovene word "ljubljena" meaning "beloved"

Administrative divisions: 201 municipalities (obcine, singular—obcina) and 11 urban municipalities (mestne obcine, singular—mestna obcina)
municipalities: Ajdovscina, Ankaran, Apace, Beltinci, Benedikt, Bistrica ob Sotli, Bled, Bloke, Bohinj, Borovnica, Bovec, Braslovce, Brda, Brezice, Brezovica, Cankova, Cerklje na Gorenjskem, Cerknica, Cerkno, Cerkvenjak, Cirkulane, Crensovci, Crna na Koroskem, Crnomelj, Destrnik, Divaca, Dobje, Dobrepolje, Dobrna, Dobrova-Polhov Gradec, Dobrovnik/Dobronak, Dolenjske Toplice, Dol pri Ljubljani, Domzale, Dornava, Dravograd, Duplek, Gorenja Vas-Poljane, Gorisnica, Gorje, Gornja Radgona, Gornji Grad, Gornji Petrovci, Grad, Grosuplje, Hajdina, Hoce-Slivnica, Hodos, Horjul, Hrastnik, Hrpelje-Kozina, Idrija, Ig, Ilirska Bistrica, Ivancna Gorica, Izola/Isola, Jesenice, Jezersko, Jursinci, Kamnik, Kanal, Kidricevo, Kobarid, Kobilje, Kocevje, Komen, Komenda, Kosanjevica na Krki, Kostel, Kozje, Kranjska Gora, Krizevci, Krsko, Kungota, Kuzma, Lasko, Lenart, Lendava/Lendva, Litija, Ljubno, Ljutomer, Log-Dragomer, Logatec,

Loska Dolina, Loski Potok, Lovrenc na Pohorju, Luce, Lukovica,

Majsperk, Makole, Markovci, Medvode, Menges, Metlika, Mezica, Miklavz na Dravskem Polju, Miren-Kostanjevica, Mirna, Mirna Pec, Mislinja, Mokronog-Trebelno, Moravce, Moravske Toplice, Mozirje, Muta, Naklo, Nazarje, Odranci, Oplotnica, Ormoz, Osilnica, Pesnica, Piran/Pirano, Pivka, Podcetrtek, Podlehnik, Podvelka, Poljcane, Polzela, Postojna, Prebold, Preddvor, Prevalje, Puconci, Race-Fram, Radece, Radenci, Radlje ob Dravi, Radovljica, Ravne na Koroskem, Razkrizje, Recica ob Savinji, Rence-Vogrsko, Ribnica, Ribnica na Pohorju, Rogaska Slatina, Rogasovci, Rogatec, Ruse, Selnica ob

Dravi, Semic, Sevnica, Sezana, Slovenska Bistrica, Slovenske Konjice, Sodrazica, Solcava, Sredisce ob Dravi, Starse, Straza, Sveta Ana, Sveta Trojica v Slovenskih Goricah, Sveti Andraz v Slovenskih Goricah, Sveti Jurij ob Scavnici, Sveti Jurij v Slovenskih Goricah, Sveti Tomaz, Salovci, Sempeter-Vrtojba, Sencur, Sentilj, Sentjernej, Sentjur, Sentrupert, Skocjan, Skofja Loka, Skofljica, Smarje pri Jelsah, Smarjeske Toplice, Smartno ob Paki, Smartno pri Litiji, Sostanj, Store, Tabor, Tisina, Tolmin, Trbovlje, Trebnje, Trnovska Vas, Trzic, Trzin, Turnisce, Velika Polana, Velike Lasce, Verzej, Videm, Vipava, Vitanje, Vodice, Vojnik, Vransko, Vrhnika, Vuzenica, Zagorje ob Savi, Zalec, Zavrc, Zelezniki, Zetale, Ziri, Zirovnica, Zrece, Zuzemberk

urban municipalities: Celje, Koper-Capodistria, Kranj, Ljubljana, Maribor, Murska Sobota, Nova Gorica, Novo Mesto, Ptuj, Slovenj Gradec, Velenje

Independence: 25 June 1991 (from Yugoslavia)

National holiday: Independence Day/Statehood Day, 25 June (1991)

Constitution: *history:* previous 1974 (preindependence); latest passed by Parliament 23 December 1991

amendments: proposed by at least 20 National Assembly members, by the government, or by petition of at least 30,000 voters; passage requires at least two-thirds majority vote by the Assembly; referendum required if agreed upon by at least 30 Assembly members; passage in a referendum requires participation of a majority of eligible voters and a simple majority of votes cast; amended several times, last in 2015

Legal system: civil law system

International law organization participation: has not submitted an ICJ jurisdiction declaration; accepts ICCt jurisdiction

Citizenship: *citizenship by birth:* no

citizenship by descent only: at least one parent must be a citizen of Slovenia; both parents if the child is born outside of Slovenia

dual citizenship recognized: yes, for select cases

residency requirement for naturalization: 10 years, the last 5 of which have been continuous

Suffrage: 18 years of age, 16 if employed; universal

Executive branch: *chief of state:* President Borut PAHOR (since 22 December 2012)

head of government: Prime Minister Janez JANSA (since 13 March 2020)

cabinet: Council of Ministers nominated by the prime minister, elected by the National Assembly

elections/appointments: president directly elected by absolute majority popular vote in 2 rounds if needed for a 5-year term (eligible for a second consecutive term); election last held on 22 October with a runoff on 12 November 2017 (next election to be held by November 2022); following National Assembly elections, the leader of the majority party or majority coalition usually nominated prime minister by the president and elected by the National Assembly

election results: Borut PAHOR is reelected president in second round; percent of vote in first round—Borut PAHOR (independent) 47.1%, Marjan SAREC (Marjan Sarec List) 25%, Romana TOMC (SDS) 13.7%, Ljudmila NOVAK (NSi) 7.2%, other 7%; percent of vote in second round—Borut PAHOR 52.9%, Marjan SAREC 47.1%; Janez JANSA (SDS) elected prime minister on 3 March 2020, National Assembly vote—52-31

Legislative branch: *description:* bicameral Parliament consists of:

National Council or Drzavni Svet (40 seats; members indirectly elected by an electoral college to serve 5-year terms);

note—the Council is primarily an advisory body with limited legislative powers

National Assembly or Drzavni Zbor (90 seats; 88 members directly elected in single-seat constituencies by proportional representation vote and 2 directly elected in special constituencies for Italian and Hungarian minorities by simple majority vote; members serve 4-year terms)

elections:

National Council—last held on 22 November 2017 (next to be held in 2022)

National Assembly—last held on 3 June 2018 (next to be held no later than 2022)

election results:

National Council—percent of vote by party—NA; seats by party—NA; composition—men 36, women 4, percent of women 10%

National Assembly—percent of vote by party—SDS 24.9%, LMS 12.7%, SD 9.9%, SMC 9.8%, Levica 9.3%, NSi 7.1%, Stranka AB 5.1%, DeSUS 4.9%, SNS 4.2%, other 12.1%; seats by party—SDS 25, LMS 13, SD 10, SMC 10, Levica 9, NSi 7, Stranka AB 5, DeSUS 5, SNS 4, Italian and Hungarian minorities 2; composition—men 68, women 22, percent of women 24.4%; note—total Parliament percent of women 20%

Judicial branch: *highest courts:* Supreme Court (consists of the court president and 37 judges organized into civil, criminal, commercial, labor and social security, administrative, and registry departments); Constitutional Court (consists of the court president, vice president, and 7 judges)

judge selection and term of office: Supreme Court president and vice president appointed by the National Assembly

upon the proposal of the Minister of Justice based on the opinions of the Judicial Council, an 11-member independent body elected by

the National Assembly from proposals submitted by the president, attorneys, law universities, and sitting judges; other Supreme Court judges elected by the National Assembly from candidates proposed by the Judicial Council; Supreme Court judges serve for life; Constitutional Court judges appointed by the National Assembly from nominations by the president of the republic; Constitutional Court president selected from among its own membership for a 3-year term; other judges elected for single 9-year terms

subordinate courts: county, district, regional, and high courts; specialized labor-related and social courts; Court of Audit; Administrative Court

Political parties and leaders: Democratic Party of Pensioners of Slovenia or DeSUS [Aleksandra PIVEC]

List of Marjan Sarec or LMS [Marjan SAREC]

Modern Center Party or SMC [Miro CERAR]

New Slovenia or NSi [Matej TONIN]

Party of Alenka Bratusek or Stranka AB [Alenka BRATUSEK] (formerly Alliance of Social Liberal Democrats or ZSD and before that Alliance of Alenka Bratusek or ZaAB)

Slovenian Democratic Party or SDS [Janez JANSA]

Slovenian National Party or SNS [Zmago JELINCIC Plemeniti]

Social Democrats or SD [Dejan ZIDAN]

The Left or Levica [Luka MESEC] (successor to United Left or ZL)

International organization participation: Australia Group, BIS, CD, CE, CEI, EAPC, EBRD, ECB, EIB, EMU, ESA (cooperating state), EU, FAO, IADB, IAEA, IBRD, ICAO, ICC (national committees), ICCt, ICRM, IDA, IFC, IFRCS, IHO, ILO, IMF, IMO, Interpol, IOC, IOM, IPU, ISO, ITU, MIGA, NATO, NEA, NSG, OAS (observer), OECD, OIF (observer), OPCW, OSCE, PCA, Schengen Convention, SELEC, UN, UNCTAD, UNESCO, UNHCR, UNIDO, UNIFIL, UNTSO, UNWTO, UPU, WCO, WHO, WIPO, WMO, WTO, ZC

Diplomatic representation in the US: *chief of mission:* Ambassador Stanislav VIDOVIC (since 21 July 2017)

chancery: 2410 California Street NW, Washington, DC 20008

telephone: [1] (202) 386-6601

FAX: [1] (202) 386-6633

consulate(s) general: Cleveland (OH)

Diplomatic representation from the US: *chief of mission:* Ambassador Lynda C. BLANCHARD (since 29 August 2019)

telephone: [386] (1) 200-5500

embassy: Presernova 31, 1000 Ljubljana

mailing address: American Embassy Ljubljana, US Department of State, 7140 Ljubljana Place, Washington, DC 20521-7140

FAX: [386] (1) 200-5555

Flag description: three equal horizontal bands of white (top), blue, and red, derive from the medieval coat of arms of the Duchy of Carniola; the Slovenian seal (a shield with the image of Triglav, Slovenia's highest peak, in white against a blue background at the center; beneath it are two wavy blue lines depicting seas and rivers, and above it are three six-pointed stars arranged in an inverted triangle, which are taken from the coat of arms of the Counts of Celje, the prominent Slovene dynastic house of the late 14th and early 15th centuries) appears in the upper hoist side of the flag centered on the white and blue bands

National symbol(s): *Mount Triglav; national colors:* white, blue, red

National anthem: *name:* "Zdravljica" (A Toast) *lyrics/music:* France PRESEREN/Stanko PREMRL *note:* adopted in 1989 while still part of Yugoslavia; originally written in 1848; the full poem, whose seventh verse is used as the anthem, speaks of pan-Slavic nationalism 0:00 / 0:56

ECONOMY

Economy—overview: With excellent infrastructure, a well-educated work force, and a strategic location between the Balkans and Western Europe, Slovenia has one of the highest per capita GDPs in Central Europe, despite having suffered a protracted recession in the 2008-09 period in the wake of the global financial crisis. Slovenia became the first 2004 EU entrant to adopt the euro (on 1 January 2007) and has experienced a stable political and economic transition.

In March 2004, Slovenia became the first transition country to graduate from borrower status to donor partner at the World Bank. In 2007, Slovenia was invited to begin the process for joining the OECD; it became a member in 2012. From 2014 to 2016, export-led growth, fueled by demand in larger European markets, pushed annual GDP growth above 2.3%. Growth reached 5.0% in 2017 and is projected to near or reach 5% in 2018. What used to be stubbornly high unemployment fell below 5.5% in early 2018, driven by strong exports and increasing consumption that boosted labor demand. Continued fiscal consolidation through increased tax collection and social security contributions will likely result in a balanced government budget in 2019.

Prime Minister CERAR's government took office in September 2014, pledging to press ahead with commitments to privatize a select group of state-run companies, rationalize public spending, and further stabilize the banking sector. Efforts to privatize Slovenia's largely state-owned banking sector have largely stalled, however, amid concerns about an ongoing dispute over Yugoslav-era foreign currency deposits.

GDP (purchasing power parity): $71.23 billion (2017 est.) $67.84 billion (2016 est.) $65.77 billion (2015 est.) *note:* data are in 2017 dollars *country comparison to the world:* 99

GDP (official exchange rate): $48.87 billion (2017 est.)

GDP—real growth rate: 2.4% (2019 est.)

4.24% (2018 est.) 5.14% (2017 est.) *country comparison to the world:* 118

GDP—per capita (PPP): $34,500 (2017 est.) $32,900 (2016 est.) $31,900 (2015 est.) *note:* data are in 2017 dollars *country comparison to the world:* 58

Gross national saving: 26.4% of GDP (2017 est.) 24.2% of GDP (2016 est.) 23.9% of GDP (2015 est.) *country comparison to the world:* 49

GDP—composition, by end use: *household consumption:* 52.6% (2017 est.) *government consumption:* 18.2% (2017 est.) *investment in fixed capital:* 18.4% (2017 est.) *investment in inventories:* 1.1% (2017 est.) *exports of goods and services:* 82.3% (2017 est.) *imports of goods and services:* -72.6% (2017 est.)

GDP—composition, by sector of origin: *agriculture:* 1.8% (2017 est.) *industry:* 32.2% (2017 est.) *services:* 65.9% (2017 est.)

Agriculture—products: hops, wheat, coffee, corn, apples, pears; cattle, sheep, poultry

Industries: ferrous metallurgy and aluminum products, lead and zinc smelting; electronics (including military electronics), trucks, automobiles, electric power equipment, wood products, textiles, chemicals, machine tools

Industrial production growth rate: 8.6% (2017 est.) *country comparison to the world:* 22

Labor force: 885,000 (2020 est.) *country comparison to the world:* 143

Labor force—by occupation: *agriculture:* 5.5% *industry:* 31.2% *services:* 63.3% (2017 est.)

Unemployment rate: 7.64% (2019 est.) 8.25% (2018 est.) *country comparison to the world:* 118

Population below poverty line: 13.9% (2016 est.)

Household income or consumption by percentage share: *lowest 10%:* 3.8% *highest 10%:* 20.1% (2016)

Budget: *revenues:* 21.07 billion (2017 est.) *expenditures:* 21.06 billion (2017 est.)

Taxes and other revenues: 43.1% (of GDP) (2017 est.) *country comparison to the world:* 28

Budget surplus (+) or deficit (-): 0% (of GDP) (2017 est.) *country comparison to the world:* 46

Public debt: 73.6% of GDP (2017 est.) 78.6% of GDP (2016 est.)

note: *defined by the EU's Maastricht Treaty as consolidated general government gross debt at nominal value, outstanding at the end of the year in the following categories of government liabilities:* currency and deposits, securities other than shares excluding financial derivatives, and loans; general government sector comprises

the central, state, local government, and social security funds *country comparison to the world:* 44

Fiscal year: calendar year

Inflation rate (consumer prices): 1.4% (2017 est.) -0.1% (2016 est.) *country comparison to the world:* 80

Current account balance: $3.05 billion (2019 est.) $3.17 billion (2018 est.) *country comparison to the world:* 34

Exports: $32.14 billion (2017 est.) $27.65 billion (2016 est.) *country comparison to the world:* 61

Exports—partners: Germany 18.9%, Italy 10.7%, Austria 7.4%, Croatia 7.1%, France 4.8%, Poland 4.2%, Hungary 4.2% (2017)

Exports—commodities: manufactured goods, machinery and transport equipment, chemicals, food

Imports: $30.38 billion (2017 est.) $25.95 billion (2016 est.) *country comparison to the world:* 68

Imports—commodities: machinery and transport equipment, manufactured goods, chemicals, fuels and lubricants, food

Imports—partners: Germany 16.5%, Italy 13.5%, Austria 9.3%, Turkey 5.8%, Croatia 4.8%, China 4.5% (2017)

Reserves of foreign exchange and gold: $889.9 million (31 December 2017 est.) $853 million (31 December 2016 est.) *country comparison to the world:* 135

Debt—external: $46.3 billion (31 January 2017 est.) $48.2 billion (31 January 2016 est.) *country comparison to the world:* 69

Exchange rates: euros (EUR) per US dollar - 0.885 (2017 est.) 0.903 (2016 est.) 0.9214 (2015 est.) 0.885 (2014 est.) 0.7634 (2013 est.)

ENERGY

Electricity access: *electrification—total population:* 100% (2020)

Electricity—production: 15.46 billion kWh (2016 est.) *country comparison to the world:* 88

Electricity—consumption: 13.4 billion kWh (2016 est.) *country comparison to the world:* 84

Electricity—exports: 7.972 billion kWh (2017 est.) *country comparison to the world:* 26

Electricity—imports: 8.359 billion kWh (2016 est.) *country comparison to the world:* 29

Electricity—installed generating capacity: 3.536 million kW (2016 est.) *country comparison to the world:* 95

875

Electricity—from fossil fuels: 37% of total installed capacity (2016 est.)
country comparison to the world: 174

Electricity—from nuclear fuels: 20% of total installed capacity (2017 est.)
country comparison to the world: 9

Electricity—from hydroelectric plants: 34% of total installed capacity (2017 est.)
country comparison to the world: 63

Electricity—from other renewable sources: 9% of total installed capacity (2017 est.)
country comparison to the world: 84

Crude oil—production: 5 bbl/day (2018 est.)
country comparison to the world: 100

Crude oil—exports: 0 bbl/day (2017 est.)
country comparison to the world: 194

Crude oil—imports: 0 bbl/day (2017 est.)
country comparison to the world: 196

Crude oil—proved reserves: 0 bbl (1 January 2018 est.)
country comparison to the world: 196

Refined petroleum products—production: 0 bbl/day (2017 est.)
country comparison to the world: 202

Refined petroleum products—consumption: 52,140 bbl/day (2017 est.)
country comparison to the world: 102

Refined petroleum products—exports: 29,350 bbl/day (2017 est.)
country comparison to the world: 63

Refined petroleum products—imports: 93,060 bbl/day (2017 est.)
country comparison to the world: 56

Natural gas—production: 8 million cu m (2017 est.)
country comparison to the world: 94

Natural gas—consumption: 906.1 million cu m (2017 est.)
country comparison to the world: 94

Natural gas—exports: 2.832 million cu m (2017 est.)
country comparison to the world: 55

Natural gas—imports: 906.1 million cu m (2017 est.)
country comparison to the world: 62

Natural gas—proved reserves: NA cu m (2017 est.)

Carbon dioxide emissions from consumption of energy: 14.37 million Mt (2017 est.)
country comparison to the world: 94

COMMUNICATIONS

Telephones—fixed lines: *total subscriptions:* 715,283
subscriptions per 100 inhabitants: 34.02 (2019 est.)
country comparison to the world: 84

Telephones—mobile cellular: *total subscriptions:* 2,540,917
subscriptions per 100 inhabitants: 120.85 (2019 est.)

country comparison to the world: 146

Telecommunication systems: *general assessment:* well-developed telecommunications infrastructure; four mobile network operators; increase in Internet community demanding e-govt., e-commerce and e-health; govt. funds to improve broadband to more municipalities; high mobile penetration rate and therefore retaining customers with bundled products; regulatory intervention has improved telecommunications; trials for use of 5G; FttP to 90% of population by 2020 (2020)
domestic: fixed-line 34 per 100 and mobile-cellular 121 per 100 teledensity (2019)
international: country code—386 (2016)
note: the COVID-19 outbreak is negatively impacting telecommunications production and supply chains globally; consumer spending on telecom devices and services has also slowed due to the pandemic's effect on economies worldwide; overall progress towards improvements in all facets of the telecom industry—mobile, fixed-line, broadband, submarine cable and satellite—has moderated

Broadcast media: public TV broadcaster, Radiotelevizija Slovenija (RTV), operates a system of national and regional TV stations; 35 domestic commercial TV stations operating nationally, regionally, and locally; about 60% of households are connected to multi-channel cable TV; public radio broadcaster operates 3 national and 4 regional stations; more than 75 regional and local commercial and non-commercial radio stations

Internet country code: .si

Internet users: *total:* 1,676,445
percent of population: 79.75% (July 2018 est.)
country comparison to the world: 128

Broadband—fixed subscriptions: *total:* 612,737
subscriptions per 100 inhabitants: 29 (2018 est.)
country comparison to the world: 79

TRANSPORTATION

National air transport system: *number of registered air carriers:* 2 (2020)
inventory of registered aircraft operated by air carriers: 21
annual passenger traffic on registered air carriers: 1,094,762 (2018)
annual freight traffic on registered air carriers: 540,000 mt-km (2018)

Civil aircraft registration country code prefix: S5 (2016)

Airports: 16 (2020)
country comparison to the world: 144

Airports—with paved runways: *total:* 9 (2020)
over 3,047 m: 1
2,438 to 3,047 m: 2
914 to 1,523 m: 3
under 914 m: 3

Airports—with unpaved runways: *total:* 7 (2020)
914 to 1,523 m: 4
under 914 m: 3

Pipelines: 1155 km gas, 5 km oil (2017)

Railways: *total:* 1,229 km (2014)

standard gauge: 1,229 km 1.435-m gauge (503 km electrified) (2014)
country comparison to the world: 86

Roadways: *total:* 38,985 km (2012)
paved: 38,985 km (includes 769 km of expressways) (2012)
country comparison to the world: 91

Waterways: (some transport on the Drava River) (2012)

Merchant marine: *total:* 8
by type: other 8 (2019)
country comparison to the world: 161

Ports and terminals: *major seaport(s):* Koper

MILITARY AND SECURITY

Military and security forces: *Slovenian Armed Forces (Slovenska Vojska, SV):* structured as a combined Force Command with air, land, logistical, maritime, support, and training components (2019)

Military expenditures: 1.04% of GDP (2019 est.)
1.01% of GDP (2018)
0.98% of GDP (2017)
1.01% of GDP (2016)
0.93% of GDP (2015)
country comparison to the world: 115

Military and security service personnel strengths: the Slovenian Armed Forces have approximately 7,000 active duty troops, including ground, air, and maritime elements (June 2020)

Military equipment inventories and acquisitions: the inventory of the Slovenian Armed Forces is a mix of Soviet-era and limited quantities of more modern Western equipment; since 2010, it has received weapons systems from Finland, Russia, and the US (2019 est.)

Military deployments: 230 Kosovo (NATO) (2020)

Military service age and obligation: 18-25 years of age for voluntary military service; conscription abolished in 2003 (2013)

TRANSNATIONAL ISSUES

Disputes—international: since the breakup of Yugoslavia in the early 1990s, Croatia and Slovenia have each claimed sovereignty over Piran Bay and four villages, and Slovenia has objected to Croatia's claim of an exclusive economic zone in the Adriatic Sea; in 2009, however Croatia and Slovenia signed a binding international arbitration agreement to define their disputed land and maritime borders, which led Slovenia to lift its objections to Croatia joining the EU; in June 2017 the arbitration panel issued a ruling on the border that Croatia has not implemented; as a member state that forms part of the EU's external border, Slovenia has implemented the strict Schengen border rules to curb illegal migration and commerce through southeastern Europe while encouraging close cross-border ties with Croatia; Slovenia continues to impose a hard border Schengen regime with Croatia, which joined the

EU in 2013 but has not yet fulfilled Schengen requirements

Refugees and internally displaced persons: *stateless persons:* 5 (2019)

note: 516,394 estimated refugee and migrant arrivals (January 2015-September 2020); migration through the Western Balkans has decreased significantly since March 2016; Slovenia is predominantly a transit country and hosts

approximately 300 asylum seekers as of the end of June 2018

Illicit drugs: minor transit point for cocaine and Southwest Asian heroin bound for Western Europe, and for precursor chemicals

SOLOMON ISLANDS

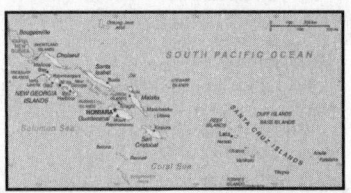

INTRODUCTION

Background: The UK established a protectorate over the Solomon Islands in the 1890s. Some of the bitterest fighting of World War II occurred on this archipelago and the Guadalcanal Campaign (August 1942-February 1943) proved a turning point in the Pacific War, since after the operation the Japanese lost their strategic initiative and remained on the defensive until thier final defeat in 1945. Self-government for the Solomon Islands came in 1976 and independence two years later. Ethnic violence, government malfeasance, endemic crime, and a narrow economic base have undermined stability and civil society. In June 2003, then Prime Minister Sir Allan KEMAKEZA sought the assistance of Australia in reestablishing law and order; the following month, an Australian-led multinational force arrived to restore peace and disarm ethnic militias. The Regional Assistance Mission to the Solomon Islands (RAMSI), which ended in June 2017, was generally effective in restoring law and order and rebuilding government institutions.

GEOGRAPHY

Location: Oceania, group of islands in the South Pacific Ocean, east of Papua New Guinea

Geographic coordinates: 8 00 S, 159 00 E

Map references: Oceania

Area: *total:* 28,896 sq km
land: 27,986 sq km
water: 910 sq km
country comparison to the world: 144

Area—comparative: slightly smaller than Maryland

Land boundaries: 0 km

Coastline: 5,313 km

Maritime claims: *territorial sea:* 12 nm
exclusive economic zone: 200 nm
continental shelf: 200 nm

measured from claimed archipelagic baselines

Climate: tropical monsoon; few temperature and weather extremes

Terrain: mostly rugged mountains with some low coral atolls

Elevation: *lowest point:* Pacific Ocean 0 m
highest point: Mount Popomanaseu 2,335 m

Natural resources: fish, forests, gold, bauxite, phosphates, lead, zinc, nickel

Land use: *agricultural land:* 3.9% (2011 est.)
arable land: 0.7% (2011 est.) / permanent crops: 2.9% (2011 est.) / permanent pasture: 0.3% (2011 est.)
forest: 78.9% (2011 est.)
other: 17.2% (2011 est.)
Irrigated land: 0 sq km NA (2012)

Population distribution: most of the population lives along the coastal regions; about one in five live in urban areas, and of these some two-thirds reside in Honiara, the largest town and chief port

Natural hazards: tropical cyclones, but rarely destructive; geologically active region with frequent earthquakes, tremors, and volcanic activity; tsunamis
volcanism: Tinakula (851 m) has frequent eruption activity, while an eruption of Savo (485 m) could affect the capital Honiara on nearby Guadalcanal

Environment—current issues: deforestation; soil erosion; many of the surrounding coral reefs are dead or dying; effects of climate change and rising sea levels

Environment—international agreements: *party to:* Biodiversity, Climate Change, Climate Change-Kyoto Protocol, Desertification, Environmental Modification, Law of the Sea, Marine Dumping, Marine Life Conservation, Ozone Layer Protection, Whaling
signed, but not ratified: none of the selected agreements

Geography—note: strategic location on sea routes between the South Pacific Ocean, the Solomon Sea, and the Coral Sea

PEOPLE AND SOCIETY

Population: 685,097 (July 2020 est.)
country comparison to the world: 167

Nationality: *noun:* Solomon Islander(s)
adjective: Solomon Islander

Ethnic groups: Melanesian 95.3%, Polynesian 3.1%, Micronesian 1.2%, other 0.3% (2009 est.)

Languages: Melanesian pidgin (in much of the country is lingua franca), English (official but spoken by only 1%-2% of the population), 120 indigenous languages

Religions: Protestant 73.4% (Church of Melanesia 31.9%, South Sea Evangelical 17.1%, Seventh Day Adventist 11.7%, United Church 10.1%, Christian Fellowship Church 2.5%), Roman Catholic 19.6%, other Christian 2.9%, other 4%, unspecified 0.1% (2009 est.)

Age structure: *0-14 years:* 32.99% (male 116,397/female 109,604)
15-24 years: 19.82% (male 69,914/female 65,874)
25-54 years: 37.64% (male 131,201/female 126,681)
55-64 years: 5.04% (male 17,844/female 16,704)
65 years and over: 4.51% (male 14,461/female 16,417) (2020 est.)

population pyramid: *[INSERT IMAGE:* SOLOMON ISLANDS-population pyramid]

Dependency ratios: *total dependency ratio:* 77.6
youth dependency ratio: 71.1
elderly dependency ratio: 6.5
potential support ratio: 15.3 (2020 est.)

Median age: *total:* 23.5 years
male: 23.2 years
female: 23.7 years (2020 est.)
country comparison to the world: 176

Population growth rate: 1.84% (2020 est.)
country comparison to the world: 56

Birth rate: 23.6 births/1,000 population (2020 est.)
country comparison to the world: 53

Death rate: 3.8 deaths/1,000 population (2020 est.)
country comparison to the world: 217

Net migration rate: -1.6 migrant(s)/1,000 population (2020 est.)
country comparison to the world: 157

Population distribution: most of the population lives along the coastal regions; about one in five live in urban areas, and of these some two-thirds reside in Honiara, the largest town and chief port

Urbanization: *urban population:* 24.7% of total population (2020)
rate of urbanization: 3.91% annual rate of change (2015-20 est.)

total population growth rate v. urban population growth rate, 2000-2030: *Major urban areas—population:* 82,000 HONIARA (capital) (2018)

Sex ratio: *at birth:* 1.05 male(s)/female
0-14 years: 1.06 male(s)/female
15-24 years: 1.06 male(s)/female

25-54 years: 1.04 male(s)/female
55-64 years: 1.07 male(s)/female
65 years and over: 0.88 male(s)/female
total population: 1.04 male(s)/female (2020 est.)

Mother's mean age at first birth: 22.6 years (2015 est.)

note: median age at first birth among women 25-29

Maternal mortality rate: 104 deaths/100,000 live births (2017 est.)
country comparison to the world: 69

Infant mortality rate: *total:* 13.4 deaths/1,000 live births
male: 15.3 deaths/1,000 live births
female: 11.4 deaths/1,000 live births (2020 est.)
country comparison to the world: 102

Life expectancy at birth: *total population:* 76.2 years
male: 73.5 years
female: 79 years (2020 est.)
country comparison to the world: 103

Total fertility rate: 2.97 children born/woman (2020 est.)
country comparison to the world: 52

Contraceptive prevalence rate: 29.3% (2015)

Drinking water source:
improved:
urban: 95% of population
rural: 67.1% of population
total: 73.6% of population
unimproved:
urban: 5% of population
rural: 32.9% of population
total: 26.4% of population (2017 est.)

Current Health Expenditure: 4.7% (2017)

Physicians density: 0.19 physicians/1,000 population (2016)

Hospital bed density: 1.4 beds/1,000 population (2012)

Sanitation facility access:
improved:
urban: 95.6% of population
rural: 22% of population
total: 39.1% of population
unimproved:
urban: 4.4% of population
rural: 78% of population
total: 60.9% of population (2017 est.)

HIV/AIDS—adult prevalence rate: NA

HIV/AIDS—people living with HIV/AIDS: NA

HIV/AIDS—deaths: NA

Major infectious diseases: *degree of risk:* high (2020)
food or waterborne diseases: bacterial diarrhea
vectorborne diseases: malaria

Obesity—adult prevalence rate: 22.5% (2016)
country comparison to the world: 76

Children under the age of 5 years underweight: 16.2% (2015)
country comparison to the world: 37

Education expenditures: 9.9% of GDP (2010)
country comparison to the world: 3

Unemployment, youth ages 15-24: *total:* 1.3%
male: 1%
female: 1.6% (2013)
country comparison to the world: 176

GOVERNMENT

Country name: *conventional long form:* none
conventional short form: Solomon Islands
local long form: none
local short form: Solomon Islands
former: British Solomon Islands
etymology: Spanish explorer Alvaro de MENDANA named the isles in 1568 after the wealthy biblical King SOLOMON in the mistaken belief that the islands contained great riches

Government type: parliamentary democracy under a constitutional monarchy; a Commonwealth realm

Capital: *name:* Honiara

geographic coordinates: 9 26 S, 159 57 E
time difference: UTC+11 (16 hours ahead of Washington, DC, during Standard Time)
etymology: the name derives from "nagho ni ara," which in one of the Guadalcanal languages roughly translates as "facing the eastern wind"

Administrative divisions: 9 provinces and 1 city*; Central, Choiseul, Guadalcanal, Honiara*, Isabel, Makira and Ulawa, Malaita, Rennell and Bellona, Temotu, Western

Independence: 7 July 1978 (from the UK)

National holiday: Independence Day, 7 July (1978)

Constitution: *history:* adopted 31 May 1978, effective 7 July 1978; note—in late 2017, provincial leaders agreed to adopt a new federal constitution, with passage expected in 2018, but it has been postponed indefinitely
amendments: proposed by the National Parliament; passage of constitutional sections, including those on fundamental rights and freedoms, the legal system, Parliament, alteration of the constitution and the ombudsman, requires three-fourths majority vote by Parliament and assent of the governor general; passage of other amendments requires two-thirds majority vote and assent of the governor general; amended several times, last in 2014

Legal system: mixed legal system of English common law and customary law

International law organization participation: has not submitted an ICJ jurisdiction declaration; non-party state to the ICCt

Citizenship: *citizenship by birth:* no
citizenship by descent only: at least one parent must be a citizen of the Solomon Islands
dual citizenship recognized: no
residency requirement for naturalization: 7 years

Suffrage: 21 years of age; universal

Executive branch: *chief of state:* Queen ELIZABETH II (since 6 February 1952); represented by Governor General David VUNAGI (since 8 July 2019)

head of government: Prime Minister Rick HOU (since 16 November 2017)
cabinet: Cabinet appointed by the governor general on the advice of the prime minister
elections/appointments: the monarchy is hereditary; governor general appointed by the monarch on the advice of the National Parliament for up to 5 years (eligible for a second term); following legislative elections, the leader of the majority party or majority coalition usually elected prime minister by the National Parliament; deputy prime minister appointed by the governor general on the advice of the prime minister from among members of the National Parliament
election results: Manasseh SOGAVARE (independent) defeated in no-confidence vote on 6 November 2017; Rick HOU elected prime minister on 15 November 2017

Legislative branch: *description:* unicameral National Parliament (50 seats; members directly elected in single-seat constituencies by simple majority vote to serve 4-year terms)
elections: last held on 19 November 2014 (next to be held 3 April 2019)
election results: percent of vote by party—UDP 10.7%, DAP 7.8%, PAP 4.4%, other 20.8%, independent 56.3%; seats by party—DAP 7, UDP 5, PAP 3, KPSI 1, SIPFP 1, SIPRA 1, independent 32; composition—men 49, women 1, percent of women 2%

Judicial branch: *highest courts:* Court of Appeal (consists of the court president and ex officio members including the High Court chief justice and its puisne judges); High Court (consists of the chief justice and puisne judges, as prescribed by the National Parliament)
judge selection and term of office: Court of Appeal and High Court president, chief justices, and puisne judges appointed by the governor general upon recommendation of the Judicial and Legal Service Commission, chaired by the chief justice and includes 5 members, mostly judicial officials and legal professionals; all judges serve until retirement at age 60
subordinate courts: Magistrates' Courts; Customary Land Appeal Court; local courts

Political parties and leaders: Democratic Alliance Party or DAP [Steve ABANA]
Kadere Party of Solomon Islands or KPSI [Peter BOYERS]
People's Alliance Party or PAP [Nathaniel WAENA]
Solomon Islands People First Party or SIPFP [Dr. Jimmie RODGERS]
Solomon Islands Party for Rural Advancement or SIPRA [Manasseh MAELANGA]
United Democratic Party or UDP [Sir Thomas Ko CHAN]
note: in general, Solomon Islands politics is characterized by fluid coalitions

International organization participation: ACP, ADB, AOSIS, C, EITI (candidate country), ESCAP, FAO, G-77, IBRD, ICAO, ICRM, IDA, IFAD, IFC, IFRCS, ILO, IMF, IMO, IOC, ITU, MIGA, OPCW, PIF, Sparteca, SPC, UN,

UNCTAD, UNESCO, UPU, WFTU, WHO, WMO, WTO

Diplomatic representation in the US: *chief of mission:* Ambassador (vacant); Charge d'Affaires Janice MOSE
chancery: 800 Second Avenue, Suite 400L, New York, NY 10017
telephone: [1] (212) 599-6192, 6193
FAX: [1] (212) 661-8925

Diplomatic representation from the US: the US does not have an embassy in the Solomon Islands; the US Ambassador to Papua New Guinea is accredited to the Solomon Islands

Flag description: divided diagonally by a thin yellow stripe from the lower hoist-side corner; the upper triangle (hoist side) is blue with five white five-pointed stars arranged in an X pattern; the lower triangle is green; blue represents the ocean, green the land, and yellow sunshine; the five stars stand for the five main island groups of the Solomon Islands

National symbol(s): *national colors:* blue, yellow, green, white

National anthem: *name:* God Save Our Solomon Islands
lyrics/music: Panapasa BALEKANA and Matila BALEKANA/Panapasa BALEKANA
note: adopted 1978

ECONOMY

Economy—overview: The bulk of the population depends on agriculture, fishing, and forestry for at least part of its livelihood. Most manufactured goods and petroleum products must be imported. The islands are rich in undeveloped mineral resources such as lead, zinc, nickel, and gold. Prior to the arrival of The Regional Assistance Mission to the Solomon Islands (RAMSI), severe ethnic violence, the closure of key businesses, and an empty government treasury culminated in economic collapse. RAMSI's efforts, which concluded in Jun 2017, to restore law and order and economic stability have led to modest growth as the economy rebuilds.

GDP (purchasing power parity): $1.33 billion (2017 est.)
$1.285 billion (2016 est.)
$1.242 billion (2015 est.)
note: data are in 2017 dollars
country comparison to the world: 200

GDP (official exchange rate): $1.298 billion (2017 est.)

GDP—real growth rate: 3.5% (2017 est.)
3.5% (2016 est.)
2.5% (2015 est.)
country comparison to the world: 100

GDP—per capita (PPP): $2,200 (2017 est.)
$2,100 (2016 est.)
$2,100 (2015 est.)
note: data are in 2017 dollars
country comparison to the world: 207

Gross national saving: 13.1% of GDP (2017 est.)
15.2% of GDP (2016 est.)

14.5% of GDP (2015 est.)
country comparison to the world: 145

GDP—composition, by end use:
household consumption: NA
government consumption: NA
investment in fixed capital: NA
investment in inventories: NA
exports of goods and services: 25.8% (2011 est.)
imports of goods and services: -49.6% (2011 est.)

GDP—composition, by sector of origin:
agriculture: 34.3% (2017 est.)
industry: 7.6% (2017 est.)
services: 58.1% (2017 est.)

Agriculture—products: cocoa, coconuts, palm kernels, rice, fruit; cattle, pigs; fish; timber

Industries: fish (tuna), mining, timber

Industrial production growth rate: 3.6% (2017 est.)
country comparison to the world: 83

Labor force: 202,500 (2007 est.)
country comparison to the world: 171

Labor force—by occupation: *agriculture:* 75%
industry: 5%
services: 20% (2000 est.)

Unemployment rate: NA

Population below poverty line: NA

Household income or consumption by percentage share: *lowest 10%:* NA
highest 10%: NA

Budget: *revenues:* 532.5 million (2017 est.)
expenditures: 570.5 million (2017 est.)

Taxes and other revenues: 41% (of GDP) (2017 est.)
country comparison to the world: 33

Budget surplus (+) or deficit (-): -2.9% (of GDP) (2017 est.)
country comparison to the world: 130

Public debt: 9.4% of GDP (2017 est.)
7.9% of GDP (2016 est.)
country comparison to the world: 198

Fiscal year: calendar year

Inflation rate (consumer prices): 0.5% (2017 est.)
0.5% (2016 est.)
country comparison to the world: 29

Current account balance: -$54 million (2017 est.)
-$49 million (2016 est.)
country comparison to the world: 81

Exports: $468.6 million (2017 est.)
$419.9 million (2016 est.)
country comparison to the world: 176

Exports—partners: China 64.5%, Italy 6.2%, Switzerland 4.6%, Philippines 4.4% (2017)

Exports—commodities: timber, fish, copra, palm oil, cocoa, coconut oil

Imports: $462.1 million (2017 est.)
$419.3 million (2016 est.)
country comparison to the world: 199

Imports—commodities: food, plant and equipment, manufactured goods, fuels, chemicals

Imports—partners: China 21.9%, Australia 19.6%, Singapore 10.7%, Vietnam 7.5%, NZ

6.2%, Papua New Guinea 5%, South Korea 4.7% (2017)

Reserves of foreign exchange and gold: $0 (31 December 2017 est.)
$421 million (31 December 2016 est.)
country comparison to the world: 194

Debt—external: $757 million (31 December 2017 est.)
$643 million (31 December 2016 est.)
country comparison to the world: 171

Exchange rates: Solomon Islands dollars (SBD) per US dollar -
7.9 (2017 est.)
7.94 (2016 est.)
7.94 (2015 est.)
7.9147 (2014 est.)
7.3754 (2013 est.)

ENERGY

Electricity access: *electrification—total population:* 47.9% (2016)
electrification—urban areas: 69.6% (2016)
electrification—rural areas: 41.5% (2016)

Electricity—production: 103 million kWh (2016 est.)
country comparison to the world: 200

Electricity—consumption: 95.79 million kWh (2016 est.)
country comparison to the world: 202

Electricity—exports: 0 kWh (2016 est.)
country comparison to the world: 199

Electricity—imports: 0 kWh (2016 est.)
country comparison to the world: 201

Electricity—installed generating capacity: 38,000 kW (2016 est.)
country comparison to the world: 198

Electricity—from fossil fuels: 92% of total installed capacity (2016 est.)
country comparison to the world: 52

Electricity—from nuclear fuels: 0% of total installed capacity (2017 est.)
country comparison to the world: 184

Electricity—from hydroelectric plants: 0% of total installed capacity (2017 est.)
country comparison to the world: 201

Electricity—from other renewable sources: 8% of total installed capacity (2017 est.)
country comparison to the world: 90

Crude oil—production: 0 bbl/day (2018 est.)
country comparison to the world: 201

Crude oil—exports: 0 bbl/day (2015 est.)
country comparison to the world: 195

Crude oil—imports: 0 bbl/day (2015 est.)
country comparison to the world: 197

Crude oil—proved reserves: 0 bbl (1 January 2018 est.)
country comparison to the world: 197

Refined petroleum products—production: 0 bbl/day (2015 est.)
country comparison to the world: 203

Refined petroleum products—consumption: 1,600 bbl/day (2016 est.)
country comparison to the world: 199

Refined petroleum products—exports: 0 bbl/day (2015 est.)
country comparison to the world: 203

Refined petroleum products—imports: 1,577 bbl/day (2015 est.)
country comparison to the world: 195

Natural gas—production: 0 cu m (2017 est.)
country comparison to the world: 197

Natural gas—consumption: 0 cu m (2017 est.)
country comparison to the world: 198

Natural gas—exports: 0 cu m (2017 est.)
country comparison to the world: 185

Natural gas—imports: 0 cu m (2017 est.)
country comparison to the world: 189

Natural gas—proved reserves: 0 cu m (1 January 2014 est.)
country comparison to the world: 195

Carbon dioxide emissions from consumption of energy: 233,500 Mt (2017 est.)
country comparison to the world: 196

COMMUNICATIONS

Telephones—fixed lines: *total subscriptions:* 7,130
subscriptions per 100 inhabitants: 1.06 (2019 est.)
country comparison to the world: 198

Telephones—mobile cellular: *total subscriptions:* 480,124
subscriptions per 100 inhabitants: 71.38 (2019 est.)
country comparison to the world: 173

Telecommunication systems: *general assessment:* Internet penetration has reached 20%; 3G and 4G LTE mobile network expansions, investment in mobile services in the region; otherwise 3G and satellite services for communication and Internet access; increase in broadband subscriptions; the launch of the Kacific-1 satellite in 2019 and the Coral Sea Cable System have vastly improved the telecom sector (2020)
domestic: fixed-line is 1 per 100 persons and mobile-cellular telephone density is about 71 per 100 persons; domestic cable system to extend to key major islands (2019)
international: country code—677; landing points for the CSCS and ICNS2 submarine cables providing connectivity from Solomon Islands, to PNG, Vanuatu and Australia; satellite earth station—1 Intelsat (Pacific Ocean) (2019)
note: the COVID-19 outbreak is negatively impacting telecommunications production and supply chains globally; consumer spending on telecom devices and services has also slowed due to the pandemic's effect on economies worldwide; overall progress towards improvements in all facets of the telecom industry—mobile, fixed-line, broadband, submarine cable and satellite—has moderated

Broadcast media: Solomon Islands Broadcasting Corporation (SIBC) does not broadcast television; multi-channel pay-TV is available; SIBC operates 2 national radio stations and 2 provincial stations; there are 2 local commercial radio stations; Radio Australia is **available via satellite feed (since 2009) (2019)**

Internet country code: .sb

Internet users: *total:* 78,686
percent of population: 11.92% (July 2018 est.)
country comparison to the world: 181

Broadband—fixed subscriptions: *total:* 1,488
subscriptions per 100 inhabitants: less than 1 (2018 est.)
country comparison to the world: 192

TRANSPORTATION

National air transport system: *number of registered air carriers:* 1 (2020)
inventory of registered aircraft operated by air carriers: 6
annual passenger traffic on registered air carriers: 427,806 (2018)
annual freight traffic on registered air carriers: 3.84 million mt-km (2018)

Civil aircraft registration country code prefix: H4 (2016)

Airports: 36 (2013)
country comparison to the world: 109

Airports—with paved runways: *total:* 1 (2019)
1,524 to 2,437 m: 1

Airports—with unpaved runways: *total:* 35 (2013)
1,524 to 2,437 m: 1 (2013)
914 to 1,523 m: 10 (2013)
under 914 m: 24 (2013)

Heliports: 3 (2013)

Roadways: *total:* 1,390 km (2011)
paved: 34 km (2011)
unpaved: 1,356 km (2011)
note: includes 920 km of private plantation roads
country comparison to the world: 175

Merchant marine: *total:* 23
by type: general cargo 7, oil tanker 1, other 15 (2019)
country comparison to the world: 141

Ports and terminals: *major seaport(s):* Honiara, Malloco Bay, Viru Harbor, Tulaghi

MILITARY AND SECURITY

Military and security forces: no regular military forces; Royal Solomon Islands Police Force (2019)

TRANSNATIONAL ISSUES

Disputes—international: since 2003, the Regional Assistance Mission to Solomon Islands, consisting of police, military, and civilian advisors drawn from 15 countries, has assisted in reestablishing and maintaining civil and political order while reinforcing regional stability and security

Trafficking in persons: *current situation:* the Solomon Islands is a source and destination country for local adults and children and Southeast Asian men and women subjected to forced labor and forced prostitution; women from China, Indonesia, Malaysia, and the Philippines are recruited for legitimate work and upon arrival are forced into prostitution; men from Indonesia and Malaysia recruited to work in the Solomon Islands' mining and logging industries may be subjected to forced labor; local children are forced into prostitution near foreign logging camps, on fishing vessels, at hotels, and entertainment venues; some local children are also sold by their parents for marriage to foreign workers or put up for "informal adoption" to pay off debts and then find themselves forced into domestic servitude or forced prostitution
tier rating: Tier 2 Watch List – the Solomon Islands does not fully comply with the minimum standards for the elimination of trafficking; however, it is making significant efforts to do so; in 2014, the Solomon Islands was granted a waiver from an otherwise required downgrade to Tier 3 because its government has a written plan that, if implemented, would constitute making significant efforts to bring itself into compliance with the minimum standards for the elimination of trafficking; the government gazetted implementing regulations for the 2012 immigration act prohibiting transnational trafficking, but the penalties are not sufficiently stringent because they allow the option of paying a fine; a new draft law to address these weaknesses awaits parliamentary review; no new trafficking investigations were conducted, even after labor inspections at logging and fishing companies, no existing cases led to prosecutions or convictions, and no funding was allocated for national anti-trafficking efforts; authorities did not identify or protect any victims and lack any procedures or shelters to do so; civil society and religious organizations provide most of the limited services available; a lack of understanding of the crime of trafficking remains a serious challenge (2015)

SOMALIA

INTRODUCTION

Background: Several powerful Somali states dominated the Indian Ocean trade from the 13th century onward. In the late 19th century, the area that would become Somalia was colonized by Britain in the north and Italy in the south. Britain withdrew from British Somaliland in 1960 to allow its protectorate to join with Italian Somaliland and form the new nation of Somalia. In 1969, a coup headed by Mohamed SIAD Barre ushered in an authoritarian socialist rule characterized by the

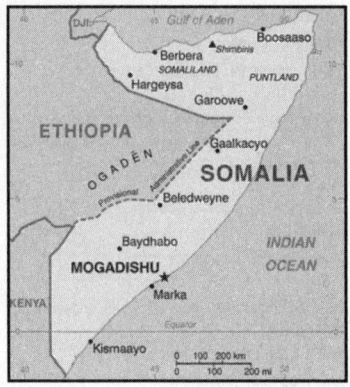

persecution, jailing, and torture of political opponents and dissidents. After the regime's collapse early in 1991, Somalia descended into turmoil, factional fighting, and anarchy. In May 1991, northern clans declared an independent Republic of Somaliland that now includes the administrative regions of Awdal, Woqooyi Galbeed, Togdheer, Sanaag, and Sool. Although not recognized by any government, this entity has maintained a stable existence and continues efforts to establish a constitutional democracy, including holding municipal, parliamentary, and presidential elections. The regions of Bari, Nugaal, and northern Mudug comprise a neighboring semi-autonomous state of Puntland, which has been self-governing since 1998 but does not aim at independence; it has also made strides toward reconstructing a legitimate, representative government but has suffered some civil strife. Puntland disputes its border with Somaliland as it also claims the regions of Sool and Sanaag, and portions of Togdheer. Beginning in 1993, a two-year UN humanitarian effort (primarily in southcentral Somalia) was able to alleviate famine conditions, but when the UN withdrew in 1995, having suffered significant casualties, order still had not been restored.

In 2000, the Somalia National Peace Conference (SNPC) held in Djibouti resulted in the formation of an interim government, known as the Transitional National Government (TNG). When the TNG failed to establish adequate security or governing institutions, the Government of Kenya, under the auspices of the Intergovernmental Authority on Development (IGAD), led a subsequent peace process that concluded in October 2004 with the election of Abdullahi YUSUF Ahmed as President of a second interim government, known as the Transitional Federal Government (TFG) of the Somali Republic. The TFG included a 275-member parliamentary body, known as the Transitional Federal Parliament (TFP). President YUSUF resigned late in 2008 while UN-sponsored talks between the TFG and the opposition Alliance for the Re-Liberation of Somalia (ARS) were underway in Djibouti. In January 2009, following the creation of a TFG-ARS unity government, Ethiopian military forces, which had entered

Somalia in December 2006 to support the TFG in the face of advances by the opposition Islamic Courts Union (ICU), withdrew from the country. The TFP was doubled in size to 550 seats with the addition of 200 ARS and 75 civil society members of parliament. The expanded parliament elected Sheikh SHARIF Sheikh Ahmed, the former ICU and ARS chairman as president in January 2009. The creation of the TFG was based on the Transitional Federal Charter (TFC), which outlined a five-year mandate leading to the establishment of a new Somali constitution and a transition to a representative government following national elections. In 2009, the TFP amended the TFC to extend TFG's mandate until 2011 and in 2011 Somali principals agreed to institute political transition by August 2012. The transition process ended in September 2012 when clan elders replaced the TFP by appointing 275 members to a new parliament who subsequently elected a new president.

GEOGRAPHY

Location: Eastern Africa, bordering the Gulf of Aden and the Indian Ocean, east of Ethiopia

Geographic coordinates: 10 00 N, 49 00 E

Map references: Africa

Area: *total:* 637,657 sq km
land: 627,337 sq km
water: 10,320 sq km
country comparison to the world: 45

Area—comparative: almost five times the size of Alabama; slightly smaller than Texas

Land boundaries: *total:* 2,385 km
border countries (3): Djibouti 61 km, Ethiopia 1640 km, Kenya 684 km

Coastline: 3,025 km

Maritime claims: *territorial sea:* 200 nm

Climate: principally desert; northeast monsoon (December to February), moderate temperatures in north and hot in south; southwest monsoon (May to October), torrid in the north and hot in the south, irregular rainfall, hot and humid periods (tangambili) between monsoons

Terrain: mostly flat to undulating plateau rising to hills in north

Elevation: *mean elevation:* 410 m
lowest point: Indian Ocean 0 m
highest point: Shimbiris 2,416 m

Natural resources: uranium and largely unexploited reserves of iron ore, tin, gypsum, bauxite, copper, salt, natural gas, likely oil reserves

Land use: *agricultural land:* 70.3% (2011 est.)
arable land: 1.8% (2011 est.) / permanent crops: 0% (2011 est.) / permanent pasture: 68.5% (2011 est.)
forest: 10.6% (2011 est.)
other: 19.1% (2011 est.)
Irrigated land: 2,000 sq km (2012)

Population distribution: distribution varies greatly throughout the country; least densely populated areas are in the northeast and central regions,

as well as areas along the Kenyan border; most populated areas are in and around the cities of Mogadishu, Marka, Boorama, Hargeysa, and Baidoa as shown on this population distribution map

Natural hazards: recurring droughts; frequent dust storms over eastern plains in summer; floods during rainy season

Environment—current issues: water scarcity; contaminated water contributes to human health problems; improper waste disposal; deforestation; land degradation; overgrazing; soil erosion; desertification

Environment—international agreements: *party to:* Biodiversity, Desertification, Endangered Species, Hazardous Wastes, Law of the Sea, Ozone Layer Protection
signed, but not ratified: none of the selected agreements

Geography—note: strategic location on Horn of Africa along southern approaches to Bab el Mandeb and route through Red Sea and Suez Canal

PEOPLE AND SOCIETY

Population: 11,757,124 (July 2020 est.)
note: this estimate was derived from an official census taken in 1975 by the Somali Government; population counting in Somalia is complicated by the large number of nomads and by refugee movements in response to famine and clan warfare
country comparison to the world: 78

Nationality: *noun:* Somali(s)
adjective: Somali

Ethnic groups: Somali 85%, Bantu and other non-Somali 15% (including 30,000 Arabs)

Languages: Somali (official, according to the 2012 Transitional Federal Charter), Arabic (official, according to the 2012 Transitional Federal Charter), Italian, English

Religions: Sunni Muslim (Islam) (official, according to the 2012 Transitional Federal Charter)

Demographic profile: Somalia scores very low for most humanitarian indicators, suffering from poor governance, protracted internal conflict, underdevelopment, economic decline, poverty, social and gender inequality, and environmental degradation. Despite civil war and famine raising its mortality rate, Somalia's high fertility rate and large proportion of people of reproductive age maintain rapid population growth, with each generation being larger than the prior one. More than 60% of Somalia's population is younger than 25, and the fertility rate is among the world's highest at almost 6 children per woman – a rate that has decreased little since the 1970s.

A lack of educational and job opportunities is a major source of tension for Somalia's large youth cohort, making them vulnerable to recruitment by extremist and pirate groups. Somalia has one of the world's lowest primary school enrollment rates – just over 40% of children are in school – and one of world's highest youth unemployment rates.

Life expectancy is low as a result of high infant and maternal mortality rates, the spread of preventable diseases, poor sanitation, chronic malnutrition, and inadequate health services.

During the two decades of conflict that followed the fall of the SIAD regime in 1991, hundreds of thousands of Somalis fled their homes. Today Somalia is the world's third highest source country for refugees, after Syria and Afghanistan. Insecurity, drought, floods, food shortages, and a lack of economic opportunities are the driving factors.

As of 2016, more than 1.1 million Somali refugees were hosted in the region, mainly in Kenya, Yemen, Egypt, Ethiopia, Djibouti, and Uganda, while more than 1.1 million Somalis were internally displaced. Since the implementation of a tripartite voluntary repatriation agreement among Kenya, Somalia, and the UNHCR in 2013, nearly 40,000 Somali refugees have returned home from Kenya's Dadaab refugee camp – still houses to approximately 260,000 Somalis. The flow sped up rapidly after the Kenyan Government in May 2016 announced its intention to close the camp, worsening security and humanitarian conditions in receiving communities in south-central Somalia. Despite the conflict in Yemen, thousands of Somalis and other refugees and asylum seekers from the Horn of Africa risk their lives crossing the Gulf of Aden to reach Yemen and beyond (often Saudi Arabia). Bossaso in Puntland overtook Obock, Djibouti, as the primary departure point in mid-2014.

Age structure: *0-14 years:* 42.38% (male 2,488,604/female 2,493,527)
15-24 years: 19.81% (male 1,167,807/female 1,161,040)
25-54 years: 30.93% (male 1,881,094/female 1,755,166)
55-64 years: 4.61% (male 278,132/female 264,325)
65 years and over: 2.27% (male 106,187/female 161,242) (2020 est.)

Dependency ratios: *total dependency ratio:* 96.3
youth dependency ratio: 90.6
elderly dependency ratio: 5.7
potential support ratio: 17.6 (2020 est.)

Median age: *total:* 18.5 years
male: 18.7 years
female: 18.3 years (2020 est.)
country comparison to the world: 211

Population growth rate: 2.21% (2020 est.)
country comparison to the world: 35

Birth rate: 38.7 births/1,000 population (2020 est.)
country comparison to the world: 11

Death rate: 12.4 deaths/1,000 population (2020 est.)
country comparison to the world: 14

Net migration rate: -3.8 migrant(s)/1,000 population (2020 est.)
country comparison to the world: 187

Population distribution: distribution varies greatly throughout the country; least densely populated areas are in the northeast and central regions,

as well as areas along the Kenyan border; most populated areas are in and around the cities of Mogadishu, Marka, Boorama, Hargeysa, and Baidoa as shown on this population distribution map

Urbanization: *urban population:* 46.1% of total population (2020)
rate of urbanization: 4.23% annual rate of change (2015-20 est.)

total population growth rate v. urban population growth rate, 2000-2030: *Major urban areas—population:* 2.282 million MOGADISHU (capital), 989,000 Hargeysa (2020)

Sex ratio: *at birth:* 1.03 male(s)/female
0-14 years: 1 male(s)/female
15-24 years: 1.01 male(s)/female
25-54 years: 1.07 male(s)/female
55-64 years: 1.05 male(s)/female
65 years and over: 0.66 male(s)/female
total population: 1.02 male(s)/female (2020 est.)

Maternal mortality rate: 829 deaths/100,000 live births (2017 est.)
country comparison to the world: 6

Infant mortality rate: *total:* 89.5 deaths/1,000 live births
male: 97.8 deaths/1,000 live births
female: 81 deaths/1,000 live births (2020 est.)
country comparison to the world: 2

Life expectancy at birth: *total population:* 54 years
male: 51.8 years
female: 56.2 years (2020 est.)
country comparison to the world: 225

Total fertility rate: 5.51 children born/woman (2020 est.)
country comparison to the world: 9

Drinking water source:
improved:
urban: 98.1% of population
rural: 72.5% of population
total: 83.8% of population
unimproved:
urban: 1.9% of population
rural: 27.5% of population
total: 16.2% of population (2017 est.)

Physicians density: 0.02 physicians/1,000 population (2014)

Hospital bed density: 0.9 beds/1,000 population (2017)

Sanitation facility access:
improved:
urban: 86.2% of population
rural: 27.1% of population
total: 53.3% of population
unimproved:
urban: 13.8% of population
rural: 72.9% of population
total: 46.7% of population (2017 est.)

HIV/AIDS—adult prevalence rate: 0.1% (2019 est.)
country comparison to the world: 134

HIV/AIDS—people living with HIV/AIDS: 11,000 (2019 est.)

country comparison to the world: 99

HIV/AIDS—deaths: <1000 (2019 est.)

Major infectious diseases: *degree of risk:* very high (2020)
food or waterborne diseases: bacterial and protozoal diarrhea, hepatitis A and E, and typhoid fever
vectorborne diseases: dengue fever, malaria, and Rift Valley fever
water contact diseases: schistosomiasis
animal contact diseases: rabies

Obesity—adult prevalence rate: 8.3% (2016)
country comparison to the world: 153

Children under the age of 5 years underweight: 23% (2009)
country comparison to the world: 16

Education expenditures: NA

GOVERNMENT

Country name: *conventional long form:* Federal Republic of Somalia
conventional short form: Somalia
local long form: Jamhuuriyadda Federaalkaa Soomaaliya
local short form: Soomaaliya
former: Somali Republic, Somali Democratic Republic
etymology: "Land of the Somali" (ethnic group)

Government type: federal parliamentary republic

Capital: *name:* Mogadishu
geographic coordinates: 2 04 N, 45 20 E
time difference: UTC+3 (8 hours ahead of Washington, DC, during Standard Time)
etymology: several theories attempt to explain the city's name; one of the more plausible is that it derives from "maq'ad- i-shah" meaning "the seat of the shah," reflecting the city's links with Persia

Administrative divisions: 18 regions (plural—NA, singular—gobolka); Awdal, Bakool, Banaadir, Bari, Bay, Galguduud, Gedo, Hiiraan, Jubbada Dhexe (Middle Jubba), Jubbada Hoose (Lower Jubba), Mudug, Nugaal, Sanaag, Shabeellaha Dhexe (Middle Shabeelle), Shabeellaha Hoose (Lower Shabeelle), Sool, Togdheer, Woqooyi Galbeed

Independence: 1 July 1960 (from a merger of British Somaliland, which became independent from the UK on 26 June 1960, and Italian Somaliland, which became independent from the Italian-administered UN trusteeship on 1 July 1960 to form the Somali Republic)

National holiday: Foundation of the Somali Republic, 1 July (1960); note—26 June (1960) in Somaliland

Constitution: *history:* previous 1961, 1979; latest drafted 12 June 2012, approved 1 August 2012 (provisional)
amendments: proposed by the federal government, by members of the state governments, the Federal Parliament, or by public petition; proposals require review by a joint committee of Parliament with inclusion of public comments and state legislatures' comments; passage requires

at least two-thirds majority vote in both houses of Parliament and approval by a majority of votes cast in a referendum; constitutional clauses on Islamic principles, the federal system, human rights and freedoms, powers and authorities of the government branches, and inclusion of women in national institutions cannot be amended

Legal system: mixed legal system of civil law, Islamic (sharia) law, and customary law (referred to as Xeer)

International law organization participation: accepts compulsory ICJ jurisdiction with reservations; non-party state to the ICCt

Citizenship: *citizenship by birth:* no
citizenship by descent only: the father must be a citizen of Somalia
dual citizenship recognized: no
residency requirement for naturalization: 7 years

Suffrage: 18 years of age; universal

Executive branch: *chief of state:* President Mohamed ABDULLAHI Mohamed "Farmaajo" (since 8 February 2017)
head of government: Prime Minister Mohamed Hussein ROBLE (since 27 September 2020)
cabinet: Cabinet appointed by the prime minister, approved by the House of the People
elections/appointments: president indirectly elected by the Federal Parliament by two-thirds majority vote in 2 rounds if needed for a single 4-year term; election last held on 8 February 2017 (previously scheduled for 30 September 2016 but postponed repeatedly); prime minister appointed by the president, approved by the House of People
election results: Mohamed ABDULLAHI Mohamed "Farmaajo" elected president in second round; Federal Parliament second round vote—Mohamed ABDULLAHI Mohamed "Farmaajo" (TPP) 184, HASSAN SHEIKH Mohamud (PDP) 97, Sheikh SHARIF Sheikh Ahmed (ARS) 46

Legislative branch: *description:* bicameral Federal Parliament to consist of:
Upper House (54 seats; senators indirectly elected by state assemblies to serve 4-year terms) House of the People (275 seats; members indirectly elected by electoral colleges, each consisting of 51 delegates selected by the 136 Traditional Elders in consultation with sub-clan elders; members serve 4-year terms)
elections:
Upper House—first held on 10 October 2016 (next to be held in November 2020)
House of the People—first held 23 October—10 November 2016 (next to be held in November 2020)
election results:
Upper House—percent of vote by party—NA; seats by party—NA; composition—men 41, women 13, percent of women 24.1%
House of the People—percent of vote by party—NA; seats by party—NA; composition—men 208, women 67, percent of women 24.4%; note—total Parliament percent of women 24.3%

note: the inaugural House of the People was appointed in September 2012 by clan elders; in 2016 and 2017, the Federal Parliament became bicameral with elections scheduled for 10 October 2016 for the Upper House and 23 October to 10 November 2016 for the House of the People; while the elections were delayed, they were eventually held in most regions despite voting irregularities; on 27 December 2016, 41 Upper House senators and 242 House of the People members were sworn in

Judicial branch: *highest courts:* the provisional constitution stipulates the establishment of the Constitutional Court (consists of 5 judges, including the chief judge and deputy chief judge); note—under the terms of the 2004 Transitional National Charter, a Supreme Court based in Mogadishu and the Appeal Court were established; yet most regions have reverted to local forms of conflict resolution, either secular, traditional Somali customary law, or Islamic law
judge selection and term of office: judges appointed by the president upon proposal of the Judicial Service Commission, a 9-member judicial and administrative body; judge tenure NA
subordinate courts: federal courts; federal member state-level courts; military courts; sharia courts

Political parties and leaders: Cosmopolitan Democratic Party [Yarow Sharef ADEN]
Daljir Party or DP [Hassan MOALIM]
Democratic Green Party of Somalia or DGPS [Abdullahi Y. MAHAMOUD]
Democratic Party of Somalia or DPS [Maslah Mohamed SIAD]
Green Leaf for Democracy or GLED
Hiil Qaran
Justice and Communist Party [Mohamed NUR]
Justice and Development of Democracy and Self-Respectfulness Party or CAHDI [Abdirahman Abdigani IBRAHIM Bile]
Justice Party [SAKARIYE Haji]
Liberal Party of Somalia
National Democratic Party [Abdirashid ALI]
National Unity Party (Xisbiga MIdnimo-Quaran) [Abdurahman BAADIYOW]
Peace and Development Party or PDP
Somali Green Party (local chapter of Federation of Green Parties of Africa)
Somali National Party or SNP [Mohammed Ameen Saeed AHMED]
Somali People's Party [Salad JEELE]
Somali Society Unity Party [Yasin MAALIM]
Tayo or TPP [Mohamed Abdullahi MOHAMED]
Tiir Party [Fadhil Sheik MOHAMUD]
Union for Peace and Development or UPD [HASSAN SHEIKH Mohamud]
United and Democratic Party [FAUZIA Haji]
United Somali Parliamentarians
United Somali Republican Party [Ali TIMA-JLIC]
inactive: Alliance for the Reliberation of Somalia; reportedly inactive since 2009

International organization participation: ACP, AfDB, AFESD, AMF, AU, CAEU (candidate), FAO, G-77, IBRD, ICAO, ICRM, IDA, IDB, IFAD, IFC, IFRCS, IGAD, ILO, IMF, IMO,

Interpol, IOC, IOM, IPU, ITSO, ITU, LAS, NAM, OIC, OPCW, OPCW (signatory), UN, UNCTAD, UNESCO, UNHCR, UNIDO, UPU, WFTU (NGOs), WHO, WIPO, WMO

Diplomatic representation in the US: *chief of mission:* Ambassador Ali Sharif AHMED (since 16 September 2019)
chancery: 1705 DeSales Street NW, Suite 300, Washington, DC 20036
telephone: [1] (202) 296-0570, [1] (202) 833-1523

Diplomatic representation from the US: *chief of mission:* Ambassador Donald YAMAMOTO (since 17 Nov 2018)
telephone: [254] 20 363-6000
embassy: Mogadishu, (reopened October 2019 on the grounds of the Mogadishu Airport)
mailing address: P.O. Box 606 Village Market 00621 Nairobi, Kenya
FAX: 254 20 363-6157

Flag description: *light blue with a large white five-pointed star in the center; the blue field was originally influenced by the flag of the UN but today is said to denote the sky and the neighboring Indian Ocean; the five points of the star represent the five regions in the horn of Africa that are inhabited by Somali people:* the former British Somaliland and Italian Somaliland (which together make up Somalia), Djibouti, Ogaden (Ethiopia), and the North East Province (Kenya)

National symbol(s): *leopard; national colors:* blue, white

National anthem: *name:* "Qolobaa Calankeed" (Every Nation Has its own Flag)
lyrics/music: *lyrics/music:* Abdullahi QARSHE
note: adopted 2012; written in 1959

Government—note: regional and local governing bodies continue to exist and control various areas of the country, including the self-declared Republic of Somaliland in northwestern Somalia

ECONOMY

Economy—overview: Despite the lack of effective national governance, Somalia maintains an informal economy largely based on livestock, remittance/money transfer companies, and telecommunications. Somalia's government lacks the ability to collect domestic revenue and external debt – mostly in arrears – was estimated at about 77% of GDP in 2017.

Agriculture is the most important sector, with livestock normally accounting for about 40% of GDP and more than 50% of export earnings. Nomads and semi-pastoralists, who are dependent upon livestock for their livelihood, make up a large portion of the population. Economic activity is estimated to have increased by 2.4% in 2017 because of growth in the agriculture, construction and telecommunications sector. Somalia's small industrial sector, based on the processing of agricultural products, has largely been looted and the machinery sold as scrap metal.

In recent years, Somalia's capital city, Mogadishu, has witnessed the development of the city's first gas stations, supermarkets, and air-line flights to Turkey since the collapse of central authority in 1991. Mogadishu's main market offers a variety of goods from food to electronic gadgets. Hotels continue to operate and are supported with private-security militias. Formalized economic growth has yet to expand outside of Mogadishu and a few regional capitals, and within the city, security concerns dominate business. Telecommunication firms provide wireless services in most major cities and offer the lowest international call rates on the continent. In the absence of a formal banking sector, money transfer/remittance services have sprouted throughout the country, handling up to $1.6 billion in remittances annually, although international concerns over the money transfers into Somalia continues to threaten these services' ability to operate in Western nations. In 2017, Somalia elected a new president and collected a record amount of foreign aid and investment, a positive sign for economic recovery.

GDP (purchasing power parity): $20.44 billion (2017 est.)
$19.98 billion (2016 est.)
$19.14 billion (2015 est.)
note: data are in 2016 US dollars
country comparison to the world: 148

GDP (official exchange rate): $7.052 billion (2017 est.)

GDP—real growth rate: 2.3% (2017 est.)
4.4% (2016 est.)
3.9% (2015 est.)
country comparison to the world: 122

GDP—per capita (PPP): $NA (2017)
$NA (2016)
$NA (2015)

GDP—composition, by end use:
household consumption: 72.6% (2015 est.)
government consumption: 8.7% (2015 est.)
investment in fixed capital: 20% (2015 est.)
investment in inventories: 0.8% (2016 est.)
exports of goods and services: 0.3% (2015 est.)
imports of goods and services: -1.6% (2015 est.)

GDP—composition, by sector of origin:
agriculture: 60.2% (2013 est.)
industry: 7.4% (2013 est.)
services: 32.5% (2013 est.)

Agriculture—products: bananas, sorghum, corn, coconuts, rice, sugarcane, mangoes, sesame seeds, beans; cattle, sheep, goats; fish

Industries: light industries, including sugar refining, textiles, wireless communication

Industrial production growth rate: 3.5% (2014 est.)
country comparison to the world: 90

Labor force: 4.154 million (2016 est.)
country comparison to the world: 87

Labor force—by occupation: *agriculture:* 71%
industry: 29%
industry and services: 29% (1975)

Unemployment rate: NA

Population below poverty line: NA

Household income or consumption by percentage share: *lowest 10%:* NA

highest 10%: NA

Budget: *revenues:* 145.3 million (2014 est.)
expenditures: 151.1 million (2014 est.)

Taxes and other revenues: 2.1% (of GDP) (2014 est.)
country comparison to the world: 221

Budget surplus (+) or deficit (-): -0.1% (of GDP) (2014 est.)
country comparison to the world: 49

Public debt: 76.7% of GDP (2017 est.)
93% of GDP (2014 est.)
country comparison to the world: 39

Fiscal year: NA

Inflation rate (consumer prices): 1.5% (2017 est.)
-71.1% (2016 est.)
country comparison to the world: 86

Current account balance: -$464 million (2017 est.)
-$427 million (2016 est.)
country comparison to the world: 119

Exports: $819 million (2014 est.)
$779 million (2013 est.)
country comparison to the world: 167

Exports—partners: Oman 31.7%, Saudi Arabia 18.7%, UAE 16.3%, Nigeria 5.1%, Yemen 4.8%, Pakistan 4% (2017)

Exports—commodities: livestock, bananas, hides, fish, charcoal, scrap metal

Imports: $94.43 billion (2018 est.)
$80.07 billion (2017 est.)
country comparison to the world: 38

Imports—commodities: manufactures, petroleum products, foodstuffs, construction materials, qat

Imports—partners: China 17.6%, India 17.2%, Ethiopia 10.5%, Oman 10.3%, Kenya 6.9%, Turkey 5.3%, Malaysia 4.1% (2017)

Reserves of foreign exchange and gold: $30.45 million (2014 est.)
country comparison to the world: 189

Debt—external: $5.3 billion (31 December 2014 est.)
country comparison to the world: 131

Exchange rates: Somali shillings (SOS) per US dollar -
23,960 (2016 est.)

ENERGY

Electricity access: *population without electricity:* 13 million (2019)
electrification—total population: 18% (2019)
electrification—urban areas: 34% (2019)
electrification—rural areas: 4% (2019)

Electricity—production: 339 million kWh (2016 est.)
country comparison to the world: 176

Electricity—consumption: 315.3 million kWh (2016 est.)
country comparison to the world: 183

Electricity—exports: 0 kWh (2016 est.)
country comparison to the world: 200

Electricity—imports: 0 kWh (2016 est.)
country comparison to the world: 202

Electricity—installed generating capacity: 85,000 kW (2016 est.)
country comparison to the world: 182

Electricity—from fossil fuels: 93% of total installed capacity (2016 est.)
country comparison to the world: 51

Electricity—from nuclear fuels: 0% of total installed capacity (2017 est.)
country comparison to the world: 185

Electricity—from hydroelectric plants: 0% of total installed capacity (2017 est.)
country comparison to the world: 202

Electricity—from other renewable sources: 7% of total installed capacity (2017 est.)
country comparison to the world: 97

Crude oil—production: 0 bbl/day (2018 est.)
country comparison to the world: 202

Crude oil—exports: 0 bbl/day (2015 est.)
country comparison to the world: 196

Crude oil—imports: 0 bbl/day (2015 est.)
country comparison to the world: 198

Crude oil—proved reserves: 0 bbl (1 January 2018 est.)
country comparison to the world: 198

Refined petroleum products—production: 0 bbl/day (2015 est.)
country comparison to the world: 204

Refined petroleum products—consumption: 5,600 bbl/day (2016 est.)
country comparison to the world: 174

Refined petroleum products—exports: 0 bbl/day (2015 est.)
country comparison to the world: 204

Refined petroleum products—imports: 5,590 bbl/day (2015 est.)
country comparison to the world: 107

Natural gas—production: 0 cu m (2017 est.)
country comparison to the world: 198

Natural gas—consumption: 0 cu m (2017 est.)
country comparison to the world: 199

Natural gas—exports: 0 cu m (2017 est.)
country comparison to the world: 186

Natural gas—imports: 0 cu m (2017 est.)
country comparison to the world: 190

Natural gas—proved reserves: 5.663 billion cu m (1 January 2018 est.)
country comparison to the world: 90

Carbon dioxide emissions from consumption of energy: 852,500 Mt (2017 est.)
country comparison to the world: 173

COMMUNICATIONS

Telephones—fixed lines: *total subscriptions:* 74,800
subscriptions per 100 inhabitants: 1 less than 1 (2018 est.)
country comparison to the world: 147

Telephones—mobile cellular: *total subscriptions:* 5,612,338
subscriptions per 100 inhabitants: 48.8 (2019 est.)
country comparison to the world: 114

Telecommunication systems: *general assessment:* the public telecom system was almost completely destroyed or dismantled during the civil war; private companies offer limited local fixed-line service, and private wireless companies offer service in most major cities; mobile sector has 7 networks improving the telecom sector along with submarine cables ending the expensive satellite dependency for Internet access; Al Shabaab Islamic militant group has forced closure of Internet services in some parts of the country; new telecom regulatory sector in place (2020)
domestic: seven networks compete for customers in the mobile sector; some of these mobile-service providers offer fixed-lines and Internet services; fixed-line less than 1 per 100 and mobile-cellular 49 per 100 (2019)
international: country code—252; landing points for the G2A, DARE1, PEACE, and EASSy fiber-optic submarine cable system linking East Africa, Indian Ocean Islands, the Middle East, North Africa and Europe (2019)
note: the COVID-19 outbreak is negatively impacting telecommunications production and supply chains globally; consumer spending on telecom devices and services has also slowed due to the pandemic's effect on economies worldwide; overall progress towards improvements in all facets of the telecom industry—mobile, fixed-line, broadband, submarine cable and satellite—has moderated

Broadcast media: 2 private TV stations rebroadcast Al-Jazeera and CNN; Somaliland has 1 government-operated TV station and Puntland has 1 private TV station; the transitional government operates Radio Mogadishu; 1 SW and roughly 10 private FM radio stations broadcast in Mogadishu; several radio stations operate in central and southern regions; Somaliland has 1 government-operated radio station; Puntland has roughly a half-dozen private radio stations; transmissions of at least 2 international broadcasters are available (2019)

Internet country code: .so

Internet users: *total:* 225,181
percent of population: 2% (July 2018 est.)
country comparison to the world: 173

Broadband—fixed subscriptions: *total:* 92,000
subscriptions per 100 inhabitants: 1 (2017 est.)
country comparison to the world: 123

TRANSPORTATION

National air transport system: *number of registered air carriers:* 6 (2020)
inventory of registered aircraft operated by air carriers: 7
annual passenger traffic on registered air carriers: 4,486 (2018)

Civil aircraft registration country code prefix: 6O (2016)

Airports: 52 (2020)
country comparison to the world: 89

Airports—with paved runways: *total:* 8 (2020)
over 3,047 m: 5
2,438 to 3,047 m: 1
1,524 to 2,437 m: 2

Airports—with unpaved runways: *total:* 44 (2020)
2,438 to 3,047 m: 5
1,524 to 2,437 m: 16
914 to 1,523 m: 22
under 914 m: 1

Roadways: *total:* 15,000 km (2018)
country comparison to the world: 125

Merchant marine: *total:* 4
by type: general cargo 1, other 3 (2019)
country comparison to the world: 168

Ports and terminals: *major seaport(s):* Berbera, Kismaayo

MILITARY AND SECURITY

Military and security forces: *Somali National Security Forces:* Somali National Army (SNA), Somali National Police (SNP, includes a maritime unit), National Intelligence and Security Agency (NISA) (2019)

Military and security service personnel strengths: estimates of the size of Somali National Army (SNA) vary widely because of inconsistent and unreliable data, as well as the ongoing integration of various militias; as of January 2020, estimates ranged from approximately 10,500-20,000; note—in 2017, the Somali Government announced a plan for the SNA to eventually number 18,000 troops; the same plan called for 32,000 federal and regional police (2019 est.)
note: the US-trained Danab ("Lightning") Brigade numbers about 850 personnel as of April 2020; the unit intends to eventually have as many as 3,000 soldiers

Military equipment inventories and acquisitions: the SNA inventory includes a variety of older, second-hand equipment largely from Italy, Russia, South Africa, and the UK; since 2015, it has received limited quantities of second-hand equipment from China, France, Italy, Qatar, and the United Arab Emirates, usually as aid/donations (2019 est.)

Military service age and obligation: 18 is the legal minimum age for compulsory and voluntary military service (2012)

Maritime threats: the International Maritime Bureau continues to report the territorial and offshore waters in the Gulf of Aden and Indian Ocean as a region of significant risk for piracy and armed robbery against ships; during 2018, two vessels were attacked compared with five in 2017; Operation Ocean Shield, the NATO naval task force established in 2009 to combat Somali piracy, concluded its operations in December 2016 as a result of the drop in reported incidents over the last few years; additional anti-piracy measures on the part of ship operators, including the use of on-board armed security teams, have reduced piracy incidents in that body of water; Somali pirates tend to be heavily armed with automatic weapons and rocket propelled grenades; the use of "mother ships" from which skiffs can be launched to attack vessels allows these pirates to extend the range of their operations hundreds of nautical miles offshore

Military—note: Somali military forces are heavily engaged in operations against the al-Shabaab

terrorist organization, including joint operations with the African Union Mission in Somalia (AMISOM); AMISOM has operated in the country with the approval of the United Nations (UN) since 2007; AMISOM's peacekeeping mission includes assisting Somali forces in providing security for a stable political process, enabling the gradual handing over of security responsibilities from AMISOM to the Somali security forces, and reducing the threat posed by Al-Shabaab and other armed opposition groups; as of early 2020, AMISOM had about 19,000 military troops and about 1,000 police personnel from six African countries deployed in Somalia

UN Assistance Mission in Somalia (UNSOM) is mandated by the Security Council to work with the Federal Government of Somalia to support national reconciliation, provide advice on peace-building and state-building, monitor the human rights situation, and help coordinate the efforts of the international community

the UN Support Office in Somalia (UNSOS) is responsible for providing logistical field support to AMISOM, UNSOM, the Somali National Army, and the Somali Police Force on joint operations with AMISOM

the European Union Training Mission in Somalia (EUTM-S) has operated in the country since 2010; the EUTM provides advice and training to the Somali military

the US and Turkey maintain separate unilateral military training missions in Somalia (2020)

TERRORISM

Terrorist group(s): al-Shabaab; Islamic State of Iraq and ash-Sham – Somalia (2020)
note: details about the history, aims, leadership, organization, areas of operation, tactics, targets, weapons, size, and sources of support of the group(s) appear(s) in Appendix-T

TRANSNATIONAL ISSUES

Disputes—international: Ethiopian forces invaded southern Somalia and routed Islamist Courts from Mogadishu in January 2007; "Somaliland" secessionists provide port facilities in Berbera to landlocked Ethiopia and have established commercial ties with other regional states; "Puntland" and "Somaliland" "governments" seek international support in their secessionist aspirations and overlapping border claims; the undemarcated former British administrative line has little meaning as a political separation to rival clans within Ethiopia's Ogaden and southern Somalia's Oromo region; Kenya works hard to prevent the clan and militia fighting in Somalia from spreading south across the border, which has long been open to nomadic pastoralists

Refugees and internally displaced persons: *refugees (country of origin):* 13,235 (Yemen) (2019)
IDPs: 2.65 million (civil war since 1988, clan-based competition for resources; 2011 famine; insecurity because of fighting between al-Shabaab and the Transitional Federal Government's allied forces) (2019)

SOUTH AFRICA

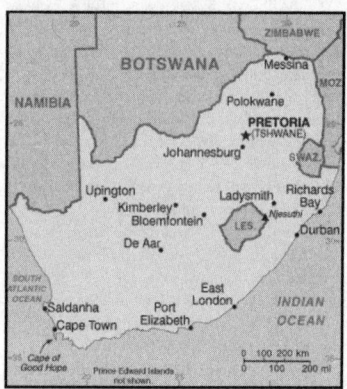

willingness to negotiate a peaceful transition to majority rule.

The first multi-racial elections in 1994 following the end of apartheid ushered in majority rule under an ANC-led government. South Africa has since struggled to address apartheid-era imbalances in wealth, housing, education, and health care. Jacob ZUMA became president in 2009 and was reelected in 2014, but resigned in February 2018 after numerous corruption scandals and gains by opposition parties in municipal elections in 2016. His successor, Cyril RAMAPHOSA, has made some progress in reigning in corruption, though many challenges persist. In May 2019 national elections, the country's sixth since the end of apartheid, the ANC won a majority of parliamentary seats, delivering RAMAPHOSA a five-year term.

INTRODUCTION

Background: Some of the earliest human remains in the fossil record are found in South Africa. By about A.D. 500, Bantu speaking groups began settling into what is now northeastern South Africa displacing Khoisan speaking groups to the southwest. Dutch traders landed at the southern tip of present-day South Africa in 1652 and established a stopover point on the spice route between the Netherlands and the Far East, founding the city of Cape Town. After the British seized the Cape of Good Hope area in 1806, many of the settlers of Dutch descent (Afrikaners, also called "Boers" (farmers) at the time) trekked north to found their own republics, Transvaal and Orange Free State. In the 1820s, several decades of wars began as the Zulus expanded their territory, moving out of what is today southeastern South Africa and clashing with other indigenous peoples and with expanding European settlements. The discovery of diamonds (1867) and gold (1886) spurred wealth and immigration from Europe.

The Anglo-Zulu War (1879) resulted in the incorporation of the Zulu kingdom's territory into the British Empire. Subsequently, the Afrikaner republics were incorporated into the British Empire after their defeat in the Second South African War (1899-1902). However, the British and the Afrikaners ruled together beginning in 1910 under the Union of South Africa, which became a republic in 1961 after a whites-only referendum. In 1948, the National Party was voted into power and instituted a policy of apartheid – billed as "separate development" of the races— which favored the white minority at the expense of the black majority and other non-white groups. The African National Congress (ANC) led the opposition to apartheid and many top ANC leaders, such as Nelson MANDELA, spent decades in South Africa's prisons. Internal protests and insurgency, as well as boycotts by some Western nations and institutions, led to the regime's eventual

GEOGRAPHY

Location: Southern Africa, at the southern tip of the continent of Africa

Geographic coordinates: 29 00 S, 24 00 E

Map references: Africa

Area: *total:* 1,219,090 sq km
land: 1,214,470 sq km
water: 4,620 sq km
note: includes Prince Edward Islands (Marion Island and Prince Edward Island)
country comparison to the world: 26

Area—comparative: slightly less than twice the size of Texas

Land boundaries: *total:* 5,244 km
border countries (6): Botswana 1969 km, Lesotho 1106 km, Mozambique 496 km, Namibia 1005 km, Eswatini 438 km, Zimbabwe 230 km

Coastline: 2,798 km

Maritime claims: *territorial sea:* 12 nm
exclusive economic zone: 200 nm
contiguous zone: 24 nm
continental shelf: 200 nm or to edge of the continental margin

Climate: mostly semiarid; subtropical along east coast; sunny days, cool nights

Terrain: vast interior plateau rimmed by rugged hills and narrow coastal plain

Elevation: *mean elevation:* 1,034 m
lowest point: Atlantic Ocean 0 m
highest point: Njesuthi 3,408 m

Natural resources: gold, chromium, antimony, coal, iron ore, manganese, nickel, phosphates, tin, rare earth elements, uranium, gem diamonds, platinum, copper, vanadium, salt, natural gas

Land use: *agricultural land:* 79.4% (2011 est.)
arable land: 9.9% (2011 est.) / permanent crops: 0.3% (2011 est.) / permanent pasture: 69.2% (2011 est.)
forest: 7.6% (2011 est.)

other: 13% (2011 est.)
Irrigated land: 16,700 sq km (2012)

Population distribution: the population concentrated along the southern and southeastern coast, and inland around Pretoria; the eastern half of the country is more densly populated than the west as shown in this population distribution map

Natural hazards: prolonged droughts
volcanism: the volcano forming Marion Island in the Prince Edward Islands, which last erupted in 2004, is South Africa's only active volcano

Environment—current issues: lack of important arterial rivers or lakes requires extensive water conservation and control measures; growth in water usage outpacing supply; pollution of rivers from agricultural runoff and urban discharge; air pollution resulting in acid rain; deforestation; soil erosion; land degradation; desertification; solid waste pollution; disruption of fragile ecosystem has resulted in significant floral extinctions

Environment—international agreements: *party to:* Antarctic-Environmental Protocol, Antarctic-Marine Living Resources, Antarctic Seals, Antarctic Treaty, Biodiversity, Climate Change, Climate Change-Kyoto Protocol, Desertification, Endangered Species, Hazardous Wastes, Law of the Sea, Marine Dumping, Marine Life Conservation, Ozone Layer Protection, Ship Pollution, Wetlands, Whaling
signed, but not ratified: none of the selected agreements

Geography—note: South Africa completely surrounds Lesotho and almost completely surrounds Eswatini

PEOPLE AND SOCIETY

Population: 56,463,617 (July 2020 est.)
note: estimates for this country explicitly take into account the effects of excess mortality due to AIDS; this can result in lower life expectancy, higher infant mortality, higher death rates, lower population growth rates, and changes in the distribution of population by age and sex than would otherwise be expected
country comparison to the world: 26

Nationality: *noun:* South African(s)
adjective: South African

Ethnic groups: black African 80.9%, colored 8.8%, white 7.8%, Indian/Asian 2.5% (2018 est.)
note: colored is a term used in South Africa, including on the national census, for persons of mixed race ancestry who developed a distinct cultural identity over several hundred years

Languages: isiZulu (official) 24.7%, isiXhosa (official) 15.6%, Afrikaans (official) 12.1%, Sepedi (official) 9.8%, Setswana (official) 8.9%, English (official) 8.4%, Sesotho (official) 8%, Xitsonga (official) 4%, siSwati (official) 2.6%, Tshivenda (official) 2.5%, isiNdebele (official) 1.6%, other (includes Khoi, Nama, and San languages) 1.9% (2017 est.)

note: data represent language spoken most often at home

Religions: Christian 86%, ancestral, tribal, animist, or other traditional African religions 5.4%, Muslim 1.9%, other 1.5%, nothing in particular 5.2% (2015 est.)

Demographic profile: South Africa's youthful population is gradually aging, as the country's total fertility rate (TFR) has declined dramatically from about 6 children per woman in the 1960s to roughly 2.2 in 2014. This pattern is similar to fertility trends in South Asia, the Middle East, and North Africa, and sets South Africa apart from the rest of Sub-Saharan Africa, where the average TFR remains higher than other regions of the world. Today, South Africa's decreasing number of reproductive age women is having fewer children, as women increase their educational attainment, workforce participation, and use of family planning methods; delay marriage; and opt for smaller families.

As the proportion of working-age South Africans has grown relative to children and the elderly, South Africa has been unable to achieve a demographic dividend because persistent high unemployment and the prevalence of HIV/AIDs have created a larger-than-normal dependent population. HIV/AIDS was also responsible for South Africa's average life expectancy plunging to less than 43 years in 2008; it has rebounded to 63 years as of 2017. HIV/AIDS continues to be a serious public health threat, although awareness-raising campaigns and the wider availability of anti-retroviral drugs is stabilizing the number of new cases, enabling infected individuals to live longer, healthier lives, and reducing mother-child transmissions.

Migration to South Africa began in the second half of the 17th century when traders from the Dutch East India Company settled in the Cape and started using slaves from South and southeast Asia (mainly from India but also from present-day Indonesia, Bangladesh, Sri Lanka, and Malaysia) and southeast Africa (Madagascar and Mozambique) as farm laborers and, to a lesser extent, as domestic servants. The Indian subcontinent remained the Cape Colony's main source of slaves in the early 18th century, while slaves were increasingly obtained from southeast Africa in the latter part of the 18th century and into the 19th century under British rule.

After slavery was completely abolished in the British Empire in 1838, South Africa's colonists turned to temporary African migrants and indentured labor through agreements with India and later China, countries that were anxious to export workers to alleviate domestic poverty and overpopulation. Of the more than 150,000 indentured Indian laborers hired to work in Natal's sugar plantations between 1860 and 1911, most exercised the right as British subjects to remain permanently (a small number of Indian immigrants came freely as merchants). Because of growing resentment toward Indian workers, the 63,000 indentured

Chinese workers who mined gold in Transvaal between 1904 and 1911 were under more restrictive contracts and generally were forced to return to their homeland.

In the late 19th century and nearly the entire 20th century, South Africa's then British colonies' and Dutch states' enforced selective immigration policies that welcomed "assimilable" white Europeans as permanent residents but excluded or restricted other immigrants. Following the Union of South Africa's passage of a law in 1913 prohibiting Asian and other non-white immigrants and its elimination of the indenture system in 1917, temporary African contract laborers from neighboring countries became the dominant source of labor in the burgeoning mining industries. Others worked in agriculture and smaller numbers in manufacturing, domestic service, transportation, and construction. Throughout the 20th century, at least 40% of South Africa's miners were foreigners; the numbers peaked at over 80% in the late 1960s. Mozambique, Lesotho, Botswana, and Eswatini were the primary sources of miners, and Malawi and Zimbabwe were periodic suppliers.

Under apartheid, a "two gates" migration policy focused on policing and deporting illegal migrants rather than on managing migration to meet South Africa's development needs. The exclusionary 1991 Aliens Control Act limited labor recruitment to the highly skilled as defined by the ruling white minority, while bilateral labor agreements provided exemptions that enabled the influential mining industry and, to a lesser extent, commercial farms, to hire temporary, low- paid workers from neighboring states. Illegal African migrants were often tacitly allowed to work for low pay in other sectors but were always under threat of deportation.

The abolishment of apartheid in 1994 led to the development of a new inclusive national identity and the strengthening of the country's restrictive immigration policy. Despite South Africa's protectionist approach to immigration, the downsizing and closing of mines, and rising unemployment, migrants from across the continent believed that the country held work opportunities. Fewer African labor migrants were issued temporary work permits and, instead, increasingly entered South Africa with visitors' permits or came illegally, which drove growth in cross-border trade and the informal job market. A new wave of Asian immigrants has also arrived over the last two decades, many operating small retail businesses.

In the post-apartheid period, increasing numbers of highly skilled white workers emigrated, citing dissatisfaction with the political situation, crime, poor services, and a reduced quality of life. The 2002 Immigration Act and later amendments were intended to facilitate the temporary migration of skilled foreign labor to fill labor shortages, but instead the legislation continues to create regulatory obstacles. Although the education system has improved and brain drain has slowed in the wake of the 2008 global financial crisis, South Africa continues to face skills shortages in several key sectors, such as health care and technology.

South Africa's stability and economic growth has acted as a magnet for refugees and asylum seekers from nearby countries, despite the prevalence of discrimination and xenophobic violence. Refugees have included an estimated 350,000 Mozambicans during its 1980s civil war and, more recently, several thousand Somalis, Congolese, and Ethiopians. Nearly all of the tens of thousands of Zimbabweans who have applied for asylum in South Africa have been categorized as economic migrants and denied refuge.

Age structure: *0-14 years:* 27.94% (male 7,894,742/female 7,883,266)

15-24 years: 16.8% (male 4,680,587/female 4,804,337)

25-54 years: 42.37% (male 12,099,441/female 11,825,193)

55-64 years: 6.8% (male 1,782,902/female 2,056,988)

65 years and over: 6.09% (male 1,443,956/female 1,992,205) (2020 est.)

Dependency ratios: *total dependency ratio:* 52.2
youth dependency ratio: 43.8
elderly dependency ratio: 8.4
potential support ratio: 11.9 (2020 est.)

Median age: *total:* 28 years
male: 27.9 years
female: 28.1 years (2020 est.)
country comparison to the world: 142

Population growth rate: 0.97% (2020 est.)
country comparison to the world: 108

Birth rate: 19.2 births/1,000 population (2020 est.)
country comparison to the world: 78

Death rate: 9.3 deaths/1,000 population (2020 est.)
country comparison to the world: 50

Net migration rate: -0.2 migrant(s)/1,000 population (2020 est.)
country comparison to the world: 110

Population distribution: the population concentrated along the southern and southeastern coast, and inland around Pretoria; the eastern half of the country is more densly populated than the west as shown in this population distribution map

Urbanization: *urban population:* 67.4% of total population (2020)
rate of urbanization: 1.97% annual rate of change (2015-20 est.)

total population growth rate v. urban population growth rate, 2000-2030: *Major urban areas— population:* 9.677 million Johannesburg (includes Ekurhuleni), 4.618 million Cape Town (legislative capital), 3.158 million Durban, 2.566 million PRETORIA (administrative capital), 1.254 million Port Elizabeth, 898,000 West Rand (2020)

Sex ratio: *at birth:* 1.02 male(s)/female
0-14 years: 1 male(s)/female
15-24 years: 0.97 male(s)/female
25-54 years: 1.02 male(s)/female
55-64 years: 0.87 male(s)/female
65 years and over: 0.72 male(s)/female
total population: 0.98 male(s)/female (2020 est.)

887

Maternal mortality rate: 119 deaths/100,000 live births (2017 est.)
country comparison to the world: 66

Infant mortality rate: *total:* 27.8 deaths/1,000 live births
male: 31 deaths/1,000 live births
female: 24.6 deaths/1,000 live births (2020 est.)
country comparison to the world: 63

Life expectancy at birth: *total population:* 64.8 years
male: 63.4 years
female: 66.2 years (2020 est.)
country comparison to the world: 198

Total fertility rate: 2.22 children born/woman (2020 est.)
country comparison to the world: 91

Contraceptive prevalence rate: 54.6% (2016)

Drinking water source:
improved:
urban: 98.9% of population
rural: 87.4% of population
total: 95.5% of population
unimproved:
urban: 1.1% of population
rural: 12.6% of population
total: 4.5% of population (2017 est.)

Current Health Expenditure: 8.1% (2017)

Physicians density: 0.91 physicians/1,000 population (2017)

Hospital bed density: 2.3 beds/1,000 population (2010)

Sanitation facility access:
improved:
urban: 95.6% of population
rural: 80.9% of population
total: 90.6% of population
unimproved:
urban: 4.4% of population
rural: 19.1% of population
total: 9.4% of population (2017 est.)

HIV/AIDS—adult prevalence rate: 17.3% (2019 est.)
country comparison to the world: 4

HIV/AIDS—people living with HIV/AIDS: 7.5 million (2019 est.)
country comparison to the world: 1

HIV/AIDS—deaths: 72,000 (2019 est.)
country comparison to the world: 1

Major infectious diseases: *degree of risk:* intermediate (2020)
food or waterborne diseases: bacterial diarrhea, hepatitis A, and typhoid fever
water contact diseases: schistosomiasis
note: widespread ongoing transmission of a respiratory illness caused by the novel coronavirus (COVID-19) is occurring throughout South Africa; as of 10 November 2020, South Africa has reported a total of 735,906 cases of COVID-19 or 12,408 cumulative cases of COVID-19 per 1 million population with 312 cumulative deaths per 1 million population; on 24 May 2020, the Government of South Africa announced the lockdown alert level for South Africa will be lowered

to level 3 with effect on 1 June 2020, except for some areas designated as "coronavirus hotspots"; per the lockdown, all airports in South Africa are closed to commercial traffic

Obesity—adult prevalence rate: 28.3% (2016)
country comparison to the world: 31

Children under the age of 5 years underweight: 5.9% (2016)
country comparison to the world: 76

Education expenditures: 6.2% of GDP (2018)
country comparison to the world: 29

Literacy: *definition:* age 15 and over can read and write
total population: 87%
male: 87.7%
female: 86.5% (2017)

School life expectancy (primary to tertiary education): *total:* 14 years
male: 13 years
female: 14 years (2018)

Unemployment, youth ages 15-24: *total:* 53.4%
male: 49.2%
female: 58.8% (2018 est.)
country comparison to the world: 3

GOVERNMENT

Country name: *conventional long form:* Republic of South Africa
conventional short form: South Africa
former: Union of South Africa
abbreviation: RSA
etymology: self-descriptive name from the country's location on the continent; "Africa" is derived from the Roman designation of the area corresponding to present-day Tunisia "Africa terra," which meant "Land of the Afri" (the tribe resident in that area), but which eventually came to mean the entire continent

Government type: parliamentary republic

Capital: *name:* Pretoria (administrative capital); Cape Town (legislative capital); Bloemfontein (judicial capital)
geographic coordinates: 25 42 S, 28 13 E
time difference: UTC+2 (7 hours ahead of Washington, DC, during Standard Time)
etymology: Pretoria is named in honor of Andries PRETORIUS, the father of voortrekker (pioneer) leader Marthinus PRETORIUS; Cape Town reflects its location on the Cape of Good Hope; Bloemfontein is a combination of the Dutch words "bloem" (flower) and "fontein" (fountain) meaning "fountain of flowers"

Administrative divisions: 9 provinces; Eastern Cape, Free State, Gauteng, KwaZulu-Natal, Limpopo, Mpumalanga, Northern Cape, North West, Western Cape

Independence: *31 May 1910 (Union of South Africa formed from four British colonies:* Cape Colony, Natal, Transvaal, and Orange Free State); 22 August 1934 (Status of the Union Act); 31 May 1961 (republic declared); 27 April 1994 (majority rule)

National holiday: Freedom Day, 27 April (1994)

Constitution: *history:* several previous; latest drafted 8 May 1996, approved by the Constitutional Court 4 December 1996, effective 4 February 1997
amendments: proposed by the National Assembly of Parliament; passage of amendments affecting constitutional sections on human rights and freedoms, non-racism and non-sexism, supremacy of the constitution, suffrage, the multi-party system of democratic government, and amendment procedures requires at least 75% majority vote of the Assembly, approval by at least six of the nine provinces represented in the National Council of Provinces, and assent of the president of the republic; passage of amendments affecting the Bill of Rights, and those related to provincial boundaries, powers, and authorities requires at least two-thirds majority vote of the Assembly, approval by at least six of the nine provinces represented in the National Council, and assent of the president; amended many times, last in 2013

Legal system: mixed legal system of Roman-Dutch civil law, English common law, and customary law

International law organization participation: has not submitted an ICJ jurisdiction declaration; accepts ICCt jurisdiction

Citizenship: *citizenship by birth:* no
citizenship by descent only: at least one parent must be a citizen of South Africa
dual citizenship recognized: yes, but requires prior permission of the government
residency requirement for naturalization: 1 year

Suffrage: 18 years of age; universal

Executive branch: *chief of state:* President Matamela Cyril RAMAPHOSA (since 15 February 2018); Deputy President David MABUZA (26 February 2018); note—the president is both chief of state and head of government; Jacob ZUMA resigned the presidency on 14 February 2018
head of government: President Matamela Cyril RAMAPHOSA (since 15 February 2018); deputy president David MABUZA (26 February 2018)
cabinet: Cabinet appointed by the president
elections/appointments: president indirectly elected by the National Assembly for a 5-year term (eligible for a second term); election last held on 22 May 2019 (next to be held in May 2024)
election results: Matamela Cyril RAMAPHOSA (ANC) elected president by the National Assembly unopposed

Legislative branch: *description:* bicameral Parliament consists of:
National Council of Provinces (90 seats; 10-member delegations appointed by each of the 9 provincial legislatures to serve 5-year terms; note—the Council has special powers to protect regional interests, including safeguarding cultural and linguistic traditions among ethnic minorities)
National Assembly (400 seats; members directly elected in multi-seat constituencies by proportional representation vote to serve 5-year terms)

elections: National Council of Provinces and National Assembly—last held on 8 May 2019 (next to be held in 2024)

election results: National Council of Provinces—percent of vote by party—NA; seats by party—ANC 29, DA 13, EFF 9, FF+ 2, IFP 1; note—36 appointed seats not filled

National Assembly—percent of vote by party—ANC 57.5%, DA 20.8%, EFF 10.8%, IFP 3.8%, FF+ 2.4%, other 4.7%; seats by party—ANC 230, DA 84, EFF 44, IFP 14, FF+ 10, other 18; composition—men 237, women 163, percent of women 40.8%

Judicial branch: *highest courts:* Supreme Court of Appeals (consists of the court president, deputy president, and 21 judges); Constitutional Court (consists of the chief and deputy chief justices and 9 judges)

judge selection and term of office: Supreme Court of Appeals president and vice president appointed by the national president after consultation with the Judicial Services Commission (JSC), a 23-member body chaired by the chief justice and includes other judges and judicial executives, members of parliament, practicing lawyers and advocates, a teacher of law, and several members designated by the president of South Africa; other Supreme Court judges appointed by the national president on the advice of the JSC and hold office until discharged from active service by an Act of Parliament; Constitutional Court chief and deputy chief justices appointed by the president of South Africa after consultation with the JSC and with heads of the National Assembly; other Constitutional Court judges appointed by the national president after consultation with the chief justice and leaders of the National Assembly; Constitutional Court judges serve 12-year nonrenewable terms or until age 70

subordinate courts: High Courts; Magistrates' Courts; labor courts; land claims courts

Political parties and leaders: African Christian Democratic Party or ACDP [Kenneth MESHOE]
African Independent Congress or AIC [Mandla GALO]
African National Congress or ANC [Cyril RAMAPHOSA]
African People's Convention or APC [Themba GODI]
Agang SA [Mike TSHISHONGA]
Congress of the People or COPE [Mosiuoa LEKOTA]
Democratic Alliance or DA [John STEENHUISEN]
Economic Freedom Fighters or EFF [Julius Sello MALEMA]
Freedom Front Plus or FF+ [Pieter GROENEWALD]
GOOD [Patricia de LILLE]
Inkatha Freedom Party or IFP [Mangosuthu BUTHELEZI]
National Freedom Party or NFP [Zanele kaMAGWAZA-MSIBI]
Pan-Africanist Congress of Azania or PAC [Luthanado MBINDA]

United Christian Democratic Party or UCDP [Isaac Sipho MFUNDISI]
United Democratic Movement or UDM [Bantu HOLOMISA]

International organization participation: ACP, AfDB, AU, BIS, BRICS, C, CD, FAO, FATF, G-20, G-24, G-5, G-77, IAEA, IBRD, ICAO, ICC (national committees), ICCt, ICRM, IDA, IFAD, IFC, IFRCS, IHO, ILO, IMF, IMO, IMSO, Interpol, IOC, IOM, IPU, ISO, ITSO, ITU, ITUC (NGOs), MIGA, MONUSCO, NAM, NSG, OECD (enhanced engagement), OPCW, Paris Club (associate), PCA, SACU, SADC, UN, UNAMID, UNCTAD, UNESCO, UNHCR, UNIDO, UNITAR, UNWTO, UPU, WCO, WFTU (NGOs), WHO, WIPO, WMO, WTO, ZC

Diplomatic representation in the US: *chief of mission:* Ambassador Nomaindiya MFEKETO (since 8 April 2020)
chancery: 3051 Massachusetts Avenue NW, Washington, DC 20008
telephone: [1] (202) 232-4400
FAX: [1] (202) 265-1607
consulate(s) general: Chicago, Los Angeles, New York

Diplomatic representation from the US: *chief of mission:* Ambassador Lana MARKS (since 28 January 2020)
telephone: [27] (12) 431-4000
embassy: 877 Pretorius Street, Arcadia, Pretoria
mailing address: P.O. Box 9536, Pretoria 0001
FAX: [27] (12) 342-2299
consulate(s) general: Cape Town, Durban, Johannesburg

Flag description: two equal width horizontal bands of red (top) and blue separated by a central green band that splits into a horizontal Y, the arms of which end at the corners of the hoist side; the Y embraces a black isosceles triangle from which the arms are separated by narrow yellow bands; the red and blue bands are separated from the green band and its arms by narrow white stripes; the flag colors do not have any official symbolism, but the Y stands for the "convergence of diverse elements within South African society, taking the road ahead in unity"; black, yellow, and green are found on the flag of the African National Congress, while red, white, and blue are the colors in the flags of the Netherlands and the UK, whose settlers ruled South Africa during the colonial era
note: the South African flag is one of only two national flags to display six colors as part of its primary design, the other is South Sudan's

National symbol(s): *springbok (antelope), king protea flower; national colors:* red, green, blue, yellow, black, white

National anthem: *name:* National Anthem of South Africa
lyrics/music: Enoch SONTONGA and Cornelius Jacob LANGENHOVEN/Enoch SONTONGA and Marthinus LOURENS de Villiers
note: adopted 1994; a combination of "N'kosi Sikelel' iAfrica" (God Bless Africa) and "Die Stem van Suid Afrika" (The Call of South Africa),

which were respectively the anthems of the non-white and white communities under apartheid; official lyrics contain a mixture of Xhosa, Zulu, Sesotho, Afrikaans, and English (i.e., the five most widely spoken of South Africa's 11 official languages); music incorporates the melody used in the Tanzanian and Zambian anthems
0:00 / 2:02

ECONOMY

Economy—overview: South Africa is a middle-income emerging market with an abundant supply of natural resources; well-developed financial, legal, communications, energy, and transport sectors; and a stock exchange that is Africa's largest and among the top 20 in the world.

Economic growth has decelerated in recent years, slowing to an estimated 0.7% in 2017. Unemployment, poverty, and inequality—among the highest in the world—remain a challenge. Official unemployment is roughly 27% of the workforce, and runs significantly higher among black youth. Even though the country's modern infrastructure supports a relatively efficient distribution of goods to major urban centers throughout the region, unstable electricity supplies retard growth. Eskom, the state-run power company, is building three new power stations and is installing new power demand management programs to improve power grid reliability but has been plagued with accusations of mismanagement and corruption and faces an increasingly high debt burden.

South Africa's economic policy has focused on controlling inflation while empowering a broader economic base; however, the country faces structural constraints that also limit economic growth, such as skills shortages, declining global competitiveness, and frequent work stoppages due to strike action. The government faces growing pressure from urban constituencies to improve the delivery of basic services to low-income areas, to increase job growth, and to provide university level-education at affordable prices. Political infighting among South Africa's ruling party and the volatility of the rand risks economic growth. International investors are concerned about the country's long-term economic stability; in late 2016, most major international credit ratings agencies downgraded South Africa's international debt to junk bond status.

GDP (purchasing power parity): $767.2 billion (2017 est.)
$757.2 billion (2016 est.)
$752.9 billion (2015 est.)
note: data are in 2017 dollars
country comparison to the world: 30

GDP (official exchange rate): $349.3 billion (2017 est.)

GDP—real growth rate: 0.06% (2019 est.)
0.7% (2018 est.)
1.4% (2017 est.)
country comparison to the world: 191

GDP—per capita (PPP): $13,600 (2017 est.)
$13,600 (2016 est.)

$13,800 (2015 est.)
note: data are in 2017 dollars
country comparison to the world: 119

Gross national saving: 16.1% of GDP (2017 est.)
16.6% of GDP (2016 est.)
16.4% of GDP (2015 est.)
country comparison to the world: 128

GDP—composition, by end use:
household consumption: 59.4% (2017 est.)
government consumption: 20.9% (2017 est.)
investment in fixed capital: 18.7% (2017 est.)
investment in inventories: -0.1% (2017 est.)
exports of goods and services: 29.8% (2017 est.)
imports of goods and services: -28.4% (2017 est.)

GDP—composition, by sector of origin:
agriculture: 2.8% (2017 est.)
industry: 29.7% (2017 est.)
services: 67.5% (2017 est.)

Agriculture—products: corn, wheat, sugarcane, fruits, vegetables; beef, poultry, mutton, wool, dairy products

Industries: mining (world's largest producer of platinum, gold, chromium), automobile assembly, metalworking, machinery, textiles, iron and steel, chemicals, fertilizer, foodstuffs, commercial ship repair

Industrial production growth rate: 1.2% (2017 est.)
country comparison to the world: 151

Labor force: 14.687 million (2020 est.)
country comparison to the world: 36

Labor force—by occupation: *agriculture*: 4.6%
industry: 23.5%
services: 71.9% (2014 est.)

Unemployment rate: 28.53% (2019 est.)
27.09% (2018 est.)
country comparison to the world: 204

Population below poverty line: 16.6% (2016 est.)

Household income or consumption by percentage share: *lowest 10%*: 1.2%
highest 10%: 51.3% (2011 est.)

Budget: *revenues*: 92.86 billion (2017 est.)
expenditures: 108.3 billion (2017 est.)

Taxes and other revenues: 26.6% (of GDP) (2017 est.)
country comparison to the world: 107

Budget surplus (+) or deficit (-): -4.4% (of GDP) (2017 est.)
country comparison to the world: 163

Public debt: 53% of GDP (2017 est.)
51.6% of GDP (2016 est.)
country comparison to the world: 92

Fiscal year: 1 April—31 March

Inflation rate (consumer prices): 5.3% (2017 est.)
6.3% (2016 est.)
country comparison to the world: 174

Current account balance: -$10.626 billion (2019 est.)
-$13.31 billion (2018 est.)
country comparison to the world: 191

Exports: $94.93 billion (2017 est.)
$75.16 billion (2016 est.)

country comparison to the world: 39

Exports—partners: China 9.5%, US 7.7%, Germany 7.1%, Japan 4.7%, India 4.6%, Botswana 4.3%, Namibia 4.1% (2017)

Exports—commodities: gold, diamonds, platinum, other metals and minerals, machinery and equipment

Imports: $89.36 billion (2017 est.)
$79.57 billion (2016 est.)
country comparison to the world: 40

Imports—commodities: machinery and equipment, chemicals, petroleum products, scientific instruments, foodstuffs

Imports—partners: China 18.3%, Germany 11.9%, US 6.6%, Saudi Arabia 4.7%, India 4.7% (2017)

Reserves of foreign exchange and gold: $50.72 billion (31 December 2017 est.)
$47.23 billion (31 December 2016 est.)
country comparison to the world: 39

Debt—external: $156.3 billion (31 December 2017 est.)
$144.6 billion (31 December 2016 est.)
country comparison to the world: 41

Exchange rates: rand (ZAR) per US dollar -
13.67 (2017 est.)
14.6924 (2016 est.)
14.6924 (2015 est.)
12.7581 (2014 est.)
10.8469 (2013 est.)

ENERGY

Electricity access: *population without electricity*: 3 million (2019)
electrification—total population: 94% (2019)
electrification—urban areas: 95% (2019)
electrification—rural areas: 92% (2019)

Electricity—production: 234.5 billion kWh (2016 est.)
country comparison to the world: 21

Electricity—consumption: 207.1 billion kWh (2016 est.)
country comparison to the world: 21

Electricity—exports: 16.55 billion kWh (2016 est.)
country comparison to the world: 11

Electricity—imports: 10.56 billion kWh (2016 est.)
country comparison to the world: 24

Electricity—installed generating capacity: 50.02 million kW (2016 est.)
country comparison to the world: 21

Electricity—from fossil fuels: 85% of total installed capacity (2016 est.)
country comparison to the world: 73

Electricity—from nuclear fuels: 4% of total installed capacity (2017 est.)
country comparison to the world: 24

Electricity—from hydroelectric plants: 1% of total installed capacity (2017 est.)
country comparison to the world: 150

Electricity—from other renewable sources: 10% of total installed capacity (2017 est.)
country comparison to the world: 81

Crude oil—production: 1,600 bbl/day (2018 est.)
country comparison to the world: 89

Crude oil—exports: 0 bbl/day (2015 est.)
country comparison to the world: 197

Crude oil—imports: 404,000 bbl/day (2015 est.)
country comparison to the world: 22

Crude oil—proved reserves: 15 million bbl (1 January 2018 est.)
country comparison to the world: 86

Refined petroleum products—production: 487,100 bbl/day (2015 est.)
country comparison to the world: 33

Refined petroleum products—consumption: 621,000 bbl/day (2016 est.)
country comparison to the world: 32

Refined petroleum products—exports: 105,600 bbl/day (2015 est.)
country comparison to the world: 42

Refined petroleum products—imports: 195,200 bbl/day (2015 est.)
country comparison to the world: 34

Natural gas—production: 906.1 million cu m (2017 est.)
country comparison to the world: 70

Natural gas—consumption: 5.069 billion cu m (2017 est.)
country comparison to the world: 60

Natural gas—exports: 0 cu m (2017 est.)
country comparison to the world: 187

Natural gas—imports: 4.162 billion cu m (2017 est.)
country comparison to the world: 39

Natural gas—proved reserves: 0 cu m (1 January 2012 est.)
country comparison to the world: 196

Carbon dioxide emissions from consumption of energy: 572.3 million Mt (2017 est.)
country comparison to the world: 11

COMMUNICATIONS

Telephones—fixed lines: *total subscriptions*: 1,934,778
subscriptions per 100 inhabitants: 3.46 (2019 est.)
country comparison to the world: 57

Telephones—mobile cellular: *total subscriptions*: 92,600,942
subscriptions per 100 inhabitants: 165.6 (2019 est.)
country comparison to the world: 18

Telecommunication systems: *general assessment*: the telecommunication system is the best-developed and most modern in Africa; mobile Internet accounts for about 95% of Internet connections; 94% with access to WiMAX/LTE services; LTE-A services launched for commercial use; the mobile sector for both voice and data service demand most investment; first region to launch commercial 5G services; regulator made provisions to anticipate spike in data traffic resulting from COVID-19 lockdown (2020)

domestic: fixed-line 3 per 100 persons and mobile-cellular 166 telephones per 100 persons; consists of carrier-equipped open-wire lines, coaxial cables, microwave radio relay links, fiber-optic cable, radiotelephone communication stations, and wireless local loops; key centers are Bloemfontein, Cape Town, Durban, Johannesburg, Port Elizabeth, and Pretoria (2019)

international: country code—27; landing points for the WACS, ACE, SAFE, SAT-3, Equiano, SABR, SAEx1, SAEx2, IOX Cable System, METISS, EASSy, and SEACOM/ Tata TGN-Eurasia fiber-optic submarine cable systems connecting South Africa, East Africa, West Africa, Europe, Southeast Asia, Asia, South America, Indian Ocean Islands, and the US; satellite earth stations—3 Intelsat (1 Indian Ocean and 2 Atlantic Ocean) (2019)

note: the COVID-19 outbreak is negatively impacting telecommunications production and supply chains globally; consumer spending on telecom devices and services has also slowed due to the pandemic's effect on economies worldwide; overall progress towards improvements in all facets of the telecom industry—mobile, fixed-line, broadband, submarine cable and satellite—has moderated

Broadcast media: the South African Broadcasting Corporation (SABC) operates 4 TV stations, 3 are free-to-air and 1 is pay TV; e.tv, a private station, is accessible to more than half the population; multiple subscription TV services provide a mix of local and international channels; well-developed mix of public and private radio stations at the national, regional, and local levels; the SABC radio network, state-owned and controlled but nominally independent, operates 18 stations, one for each of the 11 official languages, 4 community stations, and 3 commercial stations; more than 100 community-based stations extend coverage to rural areas

Internet country code: .za

Internet users: *total:* 31,107,064
percent of population: 56.17% (July 2018 est.)
country comparison to the world: 25

Broadband—fixed subscriptions: *total:* 1,107,013
subscriptions per 100 inhabitants: 2 (2018 est.)
country comparison to the world: 68

TRANSPORTATION

National air transport system: *number of registered air carriers:* 17 (2020)
inventory of registered aircraft operated by air carriers: 243

annual passenger traffic on registered air carriers: 23,921,748 (2018)
annual freight traffic on registered air carriers: 716.25 million mt-km (2018)

Civil aircraft registration country code prefix: ZS (2016)

Airports: 407 (2020)
country comparison to the world: 20

Airports—with paved runways: *total:* 130 (2020)
over 3,047 m: 11
2,438 to 3,047 m: 6
1,524 to 2,437 m: 46
914 to 1,523 m: 60
under 914 m: 7

Airports—with unpaved runways: *total:* 277 (2020)
2,438 to 3,047 m: 1
1,524 to 2,437 m: 19
914 to 1,523 m: 178
under 914 m: 79

Pipelines: 94 km condensate, 1293 km gas, 992 km oil, 1460 km refined products (2013)

Railways: *total:* 20,986 km (2014)
standard gauge: 80 km 1.435-m gauge (80 km electrified) (2014)
narrow gauge: 19,756 km 1.065-m gauge (8,271 km electrified) (2014)
other: 1,150 km (passenger rail, gauge unspecified, 1,115.5 km electrified) (2014)
country comparison to the world: 13

Roadways: *total:* 750,000 km (2016)
paved: 158,124 km (2016)
unpaved: 591,876 km (2016)
country comparison to the world: 10

Merchant marine: *total:* 103
by type: bulk carrier 2general cargo 1, oil tanker 6, other 94 (2019)
country comparison to the world: 90

Ports and terminals: *major seaport(s):* Cape Town, Durban, Port Elizabeth, Richards Bay, Saldanha Bay
container port(s) (TEUs): Durban (2,699,978) (2017)
LNG terminal(s) (import): Mossel Bay

MILITARY AND SECURITY

Military and security forces: *South African National Defence Force (SANDF):* South African Army (includes Reserve Force), South African Navy (SAN), South African Air Force (SAAF), South African Military Health Services (2019)

Military expenditures: 1% of GDP (2019)

1% of GDP (2018)
1% of GDP (2017)
1.1% of GDP (2016)
1.1% of GDP (2015)
country comparison to the world: 119

Military and security service personnel strengths: the South African National Defence Force (SANDF) is comprised of approximately 75,000 personnel (40,000 Army; 7,000 Navy; 10,000 Air Force; 8,000 Military Health Service; 10,000 other) (2020 est.)

Military equipment inventories and acquisitions: the SANDF's inventory consists of a mix of domestically-produced and foreign-supplied equipment; South Africa's domestic defense industry produced most of the Army's major weapons systems (some were jointly-produced with foreign companies), while the Air Force and Navy inventories include a mix of European, Israeli, and US-origin weapons systems; since 2010, Sweden was the largest supplier of weapons to the SANDF (2019 est.)

Military deployments: 1,050 Democratic Republic of the Congo (MONUSCO) (2020)

Military service age and obligation: 18-26 years of age for voluntary military service; women are eligible to serve in noncombat roles; 2-year service obligation (2019)

TRANSNATIONAL ISSUES

Disputes—international: South Africa has placed military units to assist police operations along the border of Lesotho, Zimbabwe, and Mozambique to control smuggling, poaching, and illegal migration; the governments of South Africa and Namibia have not signed or ratified the text of the 1994 Surveyor's General agreement placing the boundary in the middle of the Orange River

Refugees and internally displaced persons: *refugees (country of origin):* 27,113 (Somalia), 17,726 (Ethiopia), 5,273 (Republic of the Congo) (2019); 59,675 (Democratic Republic of the Congo) (refugees and asylum seekers) (2020)

Illicit drugs: transshipment center for heroin, hashish, and cocaine, as well as a major cultivator of marijuana in its own right; cocaine and heroin consumption on the rise; world's largest market for illicit methaqualone, usually imported illegally from India through various east African countries, but increasingly producing its own synthetic drugs for domestic consumption; attractive venue for money launderers given the increasing level of organized criminal and narcotics activity in the region and the size of the South African economy

SOUTH GEORGIA AND SOUTH SANDWICH ISLANDS

INTRODUCTION

Background: The islands, with large bird and seal populations, lie approximately 1,000 km east of the Falkland Islands and have been under British administration since 1908—except for a brief period in 1982 when Argentina occupied them. Grytviken, on South Georgia, was a 19th and early 20th century whaling station. Famed explorer Ernest SHACKLETON stopped there in 1914 en route to his ill-fated attempt to cross Antarctica on foot. He returned some 20 months later with

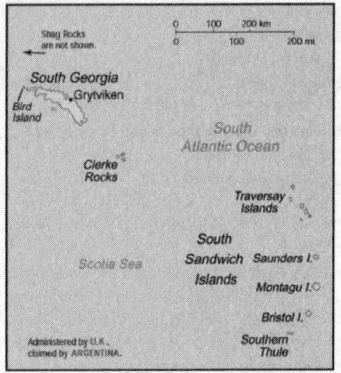

a few companions in a small boat and arranged a successful rescue for the rest of his crew, stranded off the Antarctic Peninsula. He died in 1922 on a subsequent expedition and is buried in Grytviken. Today, the station houses scientists from the British Antarctic Survey. Recognizing the importance of preserving the marine stocks in adjacent waters, the UK, in 1993, extended the exclusive fishing zone from 12 nm to 200 nm around each island.

GEOGRAPHY

Location: Southern South America, islands in the South Atlantic Ocean, east of the tip of South America

Geographic coordinates: 54 30 S, 37 00 W

Map references: Antarctic Region

Area: *total:* 3,903 sq km
land: 3,903 sq km
water: 0 sq km
note: includes Shag Rocks, Black Rock, Clerke Rocks, South Georgia Island, Bird Island, and the South Sandwich Islands, which consist of 11 islands
country comparison to the world: 177

Area—comparative: slightly larger than Rhode Island

Land boundaries: 0 km

Coastline: NA

Maritime claims: *territorial sea:* 12 nm
exclusive fishing zone: 200 nm

Climate: variable, with mostly westerly winds throughout the year interspersed with periods of calm; nearly all precipitation falls as snow

Terrain: most of the islands are rugged and mountainous rising steeply from the sea; South Georgia is largely barren with steep, glacier-covered

mountains; the South Sandwich Islands are of volcanic origin with some active volcanoes

Elevation: *lowest point:* Atlantic Ocean 0 m
highest point: Mount Paget (South Georgia) 2,934 m

Natural resources: fish

Land use: *agricultural land:* 0% (2011 est.)
arable land: 0% (2011 est.) / permanent crops: 0% (2011 est.) / permanent pasture: 0% (2011 est.)
forest: 0% (2011 est.)
other: 100% (2011 est.)
Irrigated land: 0 sq km (2011)

Natural hazards: the South Sandwich Islands have prevailing weather conditions that generally make them difficult to approach by ship; they are also subject to active volcanism

Environment—current issues: reindeer—introduced to the islands in the 20th century—devastated the native flora and bird species; some reindeer were translocated to the Falkland Islands in 2001, the rest were exterminated (2013-14); a parallel effort (2010-15) eradicated rats and mice that came to the islands as stowaways on ships as early as the late 18th century

Geography—note: the north coast of South Georgia has several large bays, which provide good anchorage

PEOPLE AND SOCIETY

Population: no indigenous inhabitants
note: the small military garrison on South Georgia withdrew in March 2001, replaced by a permanent group of scientists of the British Antarctic Survey, which also has a biological station on Bird Island; the South Sandwich Islands are uninhabited

GOVERNMENT

Country name: *conventional long form:* South Georgia and the South Sandwich Islands
conventional short form: South Georgia and South Sandwich Islands
abbreviation: SGSSI
etymology: South Georgia was named "the Isle of Georgia" in 1775 by Captain James COOK in honor of British King GEORGE III; the explorer also discovered the Sandwich Islands Group that year, which he named "Sandwich Land" after John MONTAGU, the Earl of Sandwich and First Lord of the Admiralty; the word "South" was later added to distinguish these islands from the other Sandwich Islands, now known as the Hawaiian Islands

Dependency status: overseas territory of the UK, also claimed by Argentina; administered from the

Falkland Islands by a commissioner, who is concurrently governor of the Falkland Islands, representing Queen ELIZABETH II

Legal system: the laws of the UK, where applicable, apply

International organization participation: UPU

Diplomatic representation in the US: none (overseas territory of the UK, also claimed by Argentina)

Diplomatic representation from the US: none (overseas territory of the UK, also claimed by Argentina)

Flag description: blue with the flag of the UK in the upper hoist-side quadrant and the South Georgia and South Sandwich Islands coat of arms centered on the outer half of the flag; the coat of arms features a shield with a golden lion rampant, holding a torch; the shield is supported by a fur seal on the left and a Macaroni penguin on the right; a reindeer appears above the crest, and below the shield on a scroll is the motto LEO TERRAM PROPRIAM PROTEGAT (Let the Lion Protect its Own Land); the lion with the torch represents the UK and discovery; the background of the shield, blue and white estoiles, are found in the coat of arms of James Cook, discoverer of the islands; all the outer supporting animals represented are native to the islands

ECONOMY

Economy—overview: Some fishing takes place in adjacent waters. Harvesting finfish and krill are potential sources of income. The islands receive income from postage stamps produced in the UK, the sale of fishing licenses, and harbor and landing fees from tourist vessels. Tourism from specialized cruise ships is increasing rapidly.

COMMUNICATIONS

TRANSPORTATION

Ports and terminals: *major seaport(s):* Grytviken

MILITARY AND SECURITY

Military—note: defense is the responsibility of the UK

TRANSNATIONAL ISSUES

Disputes—international: Argentina, which claims the islands in its constitution and briefly occupied them by force in 1982, agreed in 1995 to no longer seek settlement by force

SOUTH SUDAN

INTRODUCTION

Background: British explorer Samuel BAKER established the colony of Equatoria in 1870, in

the name of the Ottoman Khedive of Egypt who claimed the territory. Headquartered in Gondokoro (near modern day Juba), Equatoria in theory composed most of what is now South Sudan. After

being cut off from colonial administration during the Mahdist War from 1885-1898, Equatoria was made a state under the Anglo-Egyptian condominium in 1899. It was largely left to itself over the

South Sudan-Sudan boundary represents January 1, 1956 alignment; final alignment pending negotiations and demarcation.

following decades, but Christian missionaries converted much of the population and facilitated the spread of English, rather than Arabic. Equatoria was ruled by British colonial administrators separately from what is now Sudan until the two colonies were combined at the 1947 Juba Conference, as part of British plans to prepare the region for independence. When Sudan gained its independence in 1956, it was with the understanding that the southerners would be able to participate fully in the political system. When the Arab Khartoum government reneged on its promises, a mutiny began that led to two prolonged periods of conflict (1955-1972 and 1983-2005) in which perhaps 2.5 million people died—mostly civilians—due to starvation and drought. Ongoing peace talks finally resulted in a Comprehensive Peace Agreement, signed in January 2005. As part of this agreement, the south was granted a six-year period of autonomy to be followed by a referendum on final status. The result of this referendum, held in January 2011, was a vote of 98% in favor of secession.

Since independence on 9 July 2011, South Sudan has struggled with good governance and nation building and has attempted to control opposition forces operating in its territory. Economic conditions have deteriorated since January 2012 when the government decided to shut down oil production following bilateral disagreements with Sudan. In December 2013, conflict between government and opposition forces killed tens of thousands and led to a dire humanitarian crisis with millions of South Sudanese displaced and food insecure. The warring parties signed a peace agreement in August 2015 that created a transitional government of national unity in April 2016. However, in July 2016, fighting broke out in Juba between the two principal signatories, plunging the country back into conflict. A "revitalized" peace agreement was signed in September 2018 ending the fighting. Under the agreement, the government and various rebel groups agreed that the sides would form a unified national army and create a transitional government by May 2019. The agreement was extended until November 2019 and then subsequently to February 2020. However, implementation has been stalled, in part by a failure to agree on the country's internal political boundaries.

GEOGRAPHY

Location: East-Central Africa; south of Sudan, north of Uganda and Kenya, west of Ethiopia

Geographic coordinates: 8 00 N, 30 00 E

Map references: Africa

Area: *total:* 644,329 sq km
land: NA
water: NA
country comparison to the world: 43

Area—comparative: more than four times the size of Georgia; slightly smaller than Texas

Land boundaries: *total:* 6,018 km
border countries (6): Central African Republic 1055 km, Democratic Republic of the Congo 714 km, Ethiopia 1299 km, Kenya 317 km, Sudan 2158 km, Uganda 475 km
note: South Sudan-Sudan boundary represents 1 January 1956 alignment; final alignment pending negotiations and demarcation; final sovereignty status of Abyei Area pending negotiations between South Sudan and Sudan

Coastline: 0 km (landlocked)

Maritime claims: none (landlocked)

Climate: hot with seasonal rainfall influenced by the annual shift of the Inter-Tropical Convergence Zone; rainfall heaviest in upland areas of the south and diminishes to the north

Terrain: plains in the north and center rise to southern highlands along the border with Uganda and Kenya; the White Nile, flowing north out of the uplands of Central Africa, is the major geographic feature of the country; The Sudd (a name derived from floating vegetation that hinders navigation) is a large swampy area of more than 100,000 sq km fed by the waters of the White Nile that dominates the center of the country

Elevation: *lowest point:* White Nile 381 m
highest point: Kinyeti 3,187 m

Natural resources: hydropower, fertile agricultural land, gold, diamonds, petroleum, hardwoods, limestone, iron ore, copper, chromium ore, zinc, tungsten, mica, silver

Land use: *agricultural land:* 100%
arable land: 0% / permanent crops: 0% / permanent pasture: 100%
forest: 0%
other: 0%
Irrigated land: 1,000 sq km (2012)

Population distribution: clusters found in urban areas, particularly in the western interior and around the White Nile as shown in this population distribution map

Environment—current issues: water pollution; inadequate supplies of potable water; wildlife conservation and loss of biodiversity; deforestation; soil erosion; desertification; periodic drought

Geography—note: landlocked; The Sudd is a vast swamp in the north central region of South Sudan, formed by the White Nile, its size is variable but can reach some 15% of the country's total area during the rainy season; it is one of the world's largest wetlands

PEOPLE AND SOCIETY

Population: 10,561,244 (July 2020 est.)
country comparison to the world: 87

Nationality: *noun:* South Sudanese (singular and plural)
adjective: South Sudanese

Ethnic groups: Dinka (Jieng) 35.8%, Nuer (Naath) 15.6%, Shilluk (Chollo), Azande, Bari, Kakwa, Kuku, Murle, Mandari, Didinga, Ndogo, Bviri, Lndi, Anuak, Bongo, Lango, Dungotona, Acholi, Baka, Fertit (2011 est.)

Languages: English (official), Arabic (includes Juba and Sudanese variants), regional languages include Dinka, Nuer, Bari, Zande, Shilluk

Religions: animist, Christian, Muslim

Demographic profile: South Sudan, independent from Sudan since July 2011 after decades of civil war, is one of the world's poorest countries and ranks among the lowest in many socioeconomic categories. Problems are exacerbated by ongoing tensions with Sudan over oil revenues and land borders, fighting between government forces and rebel groups, and inter-communal violence. Most of the population lives off of farming, while smaller numbers rely on animal husbandry; more than 80% of the populace lives in rural areas. The maternal mortality rate is among the world's highest for a variety of reasons, including a shortage of health care workers, facilities, and supplies; poor roads and a lack of transport; and cultural beliefs that prevent women from seeking obstetric care. Most women marry and start having children early, giving birth at home with the assistance of traditional birth attendants, who are unable to handle complications.

Educational attainment is extremely poor due to the lack of schools, qualified teachers, and materials. Less than a third of the population is literate (the rate is even lower among women), and half live below the poverty line. Teachers and students are also struggling with the switch from Arabic to English as the language of instruction. Many adults missed out on schooling because of warfare and displacement.

Almost 2 million South Sudanese have sought refuge in neighboring countries since the current conflict began in December 2013. Another 1.96 million South Sudanese are internally displaced as of August 2017. Despite South Sudan's instability and lack of infrastructure and social services, more than 240,000 people have fled to South Sudan to escape fighting in Sudan.

Age structure: *0-14 years:* 41.58% (male 2,238,534/female 2,152,685)
15-24 years: 21.28% (male 1,153,108/female 1,094,568)
25-54 years: 30.67% (male 1,662,409/female 1,577,062)
55-64 years: 3.93% (male 228,875/female 186,571)
65 years and over: 2.53% (male 153,502/female 113,930) (2020 est.)

Dependency ratios: *total dependency ratio:* 80.8
youth dependency ratio: 74.7
elderly dependency ratio: 6.1
potential support ratio: 16.5 (2020 est.)

Median age: *total:* 18.6 years
male: 18.9 years
female: 18.3 years (2020 est.)
country comparison to the world: 209

Population growth rate: 2.7% (2020 est.)
country comparison to the world: 16

Birth rate: 38.8 births/1,000 population (2020 est.)
country comparison to the world: 10

Death rate: 11. 4 deaths/1,000 population (2020 est.)
country comparison to the world: 20

Net migration rate: 0.2 migrant(s)/1,000 population (2020 est.)
country comparison to the world: 73

Population distribution: clusters found in urban areas, particularly in the western interior and around the White Nile as shown in this population distribution map

Urbanization: *urban population:* 20.2% of total population (2020)
rate of urbanization: 4.1% annual rate of change (2015-20 est.)

total population growth rate v. urban population growth rate, 2000-2030: *Major urban areas— population:* 403,000 JUBA (capital) (2020)

Sex ratio: *at birth:* 1.05 male(s)/female
0-14 years: 1.04 male(s)/female
15-24 years: 1.05 male(s)/female
25-54 years: 1.05 male(s)/female
55-64 years: 1.23 male(s)/female
65 years and over: 1.35 male(s)/female
total population: 1.06 male(s)/female (2020 est.)

Maternal mortality rate: 1,150 deaths/100,000 live births (2017 est.)
country comparison to the world: 1

Infant mortality rate: *total:* 69.9 deaths/1,000 live births
male: 76 deaths/1,000 live births
female: 63.5 deaths/1,000 live births (2020 est.)
country comparison to the world: 4

Life expectancy at birth: *total population:* 55.5 years
male: 54.6 years
female: 56.5 years (2020 est.)
country comparison to the world: 223

Total fertility rate: 5.54 children born/woman (2020 est.)
country comparison to the world: 6

Contraceptive prevalence rate: 4% (2010)

Drinking water source:
improved:
urban: 85.2% of population
rural: 71.7% of population
total: 74.3% of population
unimproved:
urban: 14.8% of population
rural: 28.3% of population
total: 25.7% of population (2017 est.)

Current Health Expenditure: 9.8% (2017)

Sanitation facility access:
improved:
urban: 54.1% of population
rural: 10.7% of population
total: 19.1% of population
unimproved:
urban: 45.9% of population
rural: 89.3% of population
total: 80.9% of population (2017 est.)

HIV/AIDS—adult prevalence rate: 2.4% (2019 est.)
country comparison to the world: 21

HIV/AIDS—people living with HIV/AIDS: 190,000 (2019 est.)
country comparison to the world: 34

HIV/AIDS—deaths: 9,100 (2019 est.)
country comparison to the world: 20

Major infectious diseases: *degree of risk:* very high (2020)
food or waterborne diseases: bacterial and protozoal diarrhea, hepatitis A and E, and typhoid fever
vectorborne diseases: malaria, dengue fever, Trypanosomiasis-Gambiense (African sleeping sickness)
water contact diseases: schistosomiasis
animal contact diseases: rabies
respiratory diseases: meningococcal meningitis

Obesity—adult prevalence rate: 6.6% (2014)
country comparison to the world: 165

Children under the age of 5 years underweight: 27.7% (2010)
country comparison to the world: 9

Education expenditures: 1% of GDP (2017)
country comparison to the world: 175

Literacy: *definition:* age 15 and over can read and write
total population: 34.5%
male: 40.3%
female: 28.9% (2018)

Unemployment, youth ages 15-24: *total:* 38.6%
male: 39.5%
female: 37.4% (2017 est.)
country comparison to the world: 14

GOVERNMENT

Country name: *conventional long form:* Republic of South Sudan
conventional short form: South Sudan
etymology: self-descriptive name from the country's former position within Sudan prior to independence; the name "Sudan" derives from the Arabic "bilad-as-sudan" meaning "Land of the Black [peoples]"

Government type: presidential republic

Capital: *name:* Juba

geographic coordinates: 04 51 N, 31 37 E
time difference: UTC+3 (8 hours ahead of Washington, DC, during Standard Time)
etymology: the name derives from Djouba, another name for the Bari people of South Sudan

Administrative divisions: 10 states; Central Equatoria, Eastern Equatoria, Jonglei, Lakes, Northern Bahr el Ghazal, Unity, Upper Nile, Warrap, Western Bahr el Ghazal, Western Equatoria; note—in 2015, the creation of 28 new states was announced and in 2017 four additional; following the February 2020 peace agreement, the country was reportedly again reorganized into the 10 original states, plus 2 administrative areas, Pibor and Ruweng, and 1 special administrative status area, Abyei; this latest administrative revision has not yet been vetted by the US Board on Geographic Names

Independence: 9 July 2011 (from Sudan)

National holiday: Independence Day, 9 July (2011)

Constitution: *history:* previous 2005 (preindependence); latest signed 7 July 2011, effective 9 July 2011 (Transitional Constitution of the Republic of South Sudan, 2011)
amendments: proposed by the National Legislature or by the president of the republic; passage requires submission of the proposal to the Legislature at least one month prior to consideration, approval by at least two-thirds majority vote in both houses of the Legislature, and assent of the president; amended 2013, 2015, 2018

Citizenship: *citizenship by birth:* no
citizenship by descent only: at least one parent must be a citizen of South Sudan
dual citizenship recognized: yes
residency requirement for naturalization: 10 years

Suffrage: 18 years of age; universal

Executive branch: *chief of state:* President Salva KIIR Mayardit (since 9 July 2011); First Vice President Riek MACHAR Teny Dhurgon (since 22 February 2020); Vice President James Wani IGGA (since 26 April 2016); Vice President TABAN Deng Gai (since 22 February 2020); Vice President Rebecca Nyandeng Chol GARANG de Mabior (since 22 February 2020); Vice President Hussein ABDELBAGI Ayii (since 22 February 2020); note—the president is both chief of state and head of government
head of government: President Salva KIIR Mayardit (since 9 July 2011); First Vice President Taban Deng GAI (since 26 July 2016); Vice President James Wani IGGA (since 26 April 2016); Vice President TABAN Deng Gai (since 22 February 2020); Vice President Rebecca Nyandeng Chol GARANG de Mabior (since 22 February 2020); Vice President Hussein ABDELBAGI Ayii (since 22 February 2020); note—the president is both chief of state and head of government
cabinet: National Council of Ministers appointed by the president, approved by the Transitional National Legislative Assembly
elections/appointments: president directly elected by simple majority popular vote for a 4-year term (eligible for a second term); election last held on 11 15 April 2010 (next election scheduled for 2015 postponed to 2018 and again to 2021)
election results: Salva KIIR Mayardit elected president; percent of vote—Salva KIIR Mayardit (SPLM) 93%, Lam AKOL (SPLM-DC) 7%

Legislative branch: *description:* bicameral National Legislature consists of:
Council of States, established by presidential decree in August 2011 (50 seats; 20 former members of the Council of States and 30 appointed representatives)
Transitional National Legislative Assembly, established on 4 August 2016, in accordance with the August 2015 Agreement on the Resolution of the Conflict in the Republic of South Sudan (400 seats; 170 members elected in April 2010, 96 members of the former National Assembly, 66 members appointed after independence, and 68 members added as a result of the 2016 Agreement); the TNLA will be expanded to 550 members after the transitional government forms
elections:
Council of States—established and members appointed 1 August 2011
National Legislative Assembly—last held 11-15 April 2010 but did not take office until July 2011; current parliamentary term extended until 2021)
election results:
Council of States—percent of vote by party—NA; seats by party—SPLM 20, unknown 30; composition—men 44, women 6, percent of women 12%
National Legislative Assembly—percent of vote by party—NA; seats by party—SPLM 251, DCP 10, independent 6, unknown 133; composition—men 291, women 109, percent of women 27.3%; note—total National Legislature percent of women 25.6%

Judicial branch: *highest courts:* Supreme Court of South Sudan (consists of the chief and deputy chief justices, 9 other justices and normally organized into panels of 3 justices, except when sitting as a Constitutional panel of all 9 justices chaired by the chief justice)
judge selection and term of office: justices appointed by the president upon proposal of the Judicial Service Council, a 9-member judicial and administrative body; justice tenure set by the National Legislature
subordinate courts: national level—Courts of Appeal; High Courts; County Courts; state level—High Courts; County Courts; customary courts; other specialized courts and tribunals

Political parties and leaders: Democratic Change or DC [Onyoti Adigo NYIKWEC] (formerly Sudan People's Liberation Movement-Democratic Movement or SPLM-DC)
Sudan People's Liberation Movement or SPLM [Salva KIIR Mayardit]
Sudan People's Liberation Movement-In Opposition or SPLM-IO [Riek MACHAR Teny Dhurgon]

International organization participation: AU, FAO, G-77, IBRD, ICAO, ICRM, IDA, IFAD, IFC, IFRCS, ILO, IMF, Interpol, IOM, IPU, ITU, MIGA, UN, UNCTAD, UNESCO, UPU, WCO, WHO, WMO

Diplomatic representation in the US: *chief of mission:* Ambassador Philip Jada NATANA (since 17 September 2018)

chancery: 1015 31st Street NW, Third Floor, Washington, DC 20007
telephone: [1] (202) 293-7940
FAX: [1] (202) 293-7941

Diplomatic representation from the US: *chief of mission:* Ambassador Thomas HUSHEK (since 5 June 2018)
telephone: [211] 912-105-188
embassy: Kololo Road adjacent to the EU's compound, Juba

Flag description: three equal horizontal bands of black (top), red, and green; the red band is edged in white; a blue isosceles triangle based on the hoist side contains a gold, five-pointed star; black represents the people of South Sudan, red the blood shed in the struggle for freedom, green the verdant land, and blue the waters of the Nile; the gold star represents the unity of the states making up South Sudan
note: resembles the flag of Kenya; one of only two national flags to display six colors as part of its primary design, the other is South Africa's

National symbol(s): *African fish eagle; national colors:* red, green, blue, yellow, black, white

National anthem: *name:* South Sudan Oyee! (Hooray!)
lyrics/music: collective of 49 poets/Juba University students and teachers
note: adopted 2011; anthem selected in a national contest

ECONOMY

Economy—overview: Industry and infrastructure in landlocked South Sudan are severely underdeveloped and poverty is widespread, following several decades of civil war with Sudan. Continued fighting within the new nation is disrupting what remains of the economy. The vast majority of the population is dependent on subsistence agriculture and humanitarian assistance. Property rights are insecure and price signals are weak, because markets are not well-organized.

South Sudan has little infrastructure – about 10,000 kilometers of roads, but just 2% of them paved. Electricity is produced mostly by costly diesel generators, and indoor plumbing and potable water are scarce, so less than 2% of the population has access to electricity. About 90% of consumed goods, capital, and services are imported from neighboring countries – mainly Uganda, Kenya and Sudan. Chinese investment plays a growing role in the infrastructure and energy sectors.
Nevertheless, South Sudan does have abundant natural resources. South Sudan holds one of the richest agricultural areas in Africa, with fertile soils and abundant water supplies. Currently the region supports 10-20 million head of cattle. At independence in 2011, South Sudan produced nearly three-fourths of former Sudan's total oil output of nearly a half million barrels per day. The Government of South Sudan relies on oil for the vast majority of its budget revenues, although oil production has fallen sharply since independence. South Sudan is one of the most oil-dependent

countries in the world, with 98% of the government's annual operating budget and 80% of its gross domestic product (GDP) derived from oil. Oil is exported through a pipeline that runs to refineries and shipping facilities at Port Sudan on the Red Sea. The economy of South Sudan will remain linked to Sudan for some time, given the existing oil infrastructure. The outbreak of conflict in December 2013, combined with falling crude oil production and prices, meant that GDP fell significantly between 2014 and 2017. Since the second half of 2017 oil production has risen, and is currently about 130,000 barrels per day.
Poverty and food insecurity has risen due to displacement of people caused by the conflict. With famine spreading, 66% of the population in South Sudan is living on less than about $2 a day, up from 50.6% in 2009, according to the World Bank. About 80% of the population lives in rural areas, with agriculture, forestry and fishing providing the livelihood for a majority of the households. Much of rural sector activity is focused on low-input, low-output subsistence agriculture.
South Sudan is burdened by considerable debt because of increased military spending and high levels of government corruption. Economic mismanagement is prevalent. Civil servants, including police and the military, are not paid on time, creating incentives to engage in looting and banditry. South Sudan has received more than $11 billion in foreign aid since 2005, largely from the US, the UK, and the EU. Inflation peaked at over 800% per year in October 2016 but dropped to 118% in 2017. The government has funded its expenditures by borrowing from the central bank and foreign sources, using forward sales of oil as collateral. The central bank's decision to adopt a managed floating exchange rate regime in December 2015 triggered a 97% depreciation of the currency and spawned a growing black market.

Long-term challenges include rooting out public sector corruption, improving agricultural productivity, alleviating poverty and unemployment, improving fiscal transparency—particularly in regard to oil revenues, taming inflation, improving government revenues, and creating a rules-based business environment.

GDP (purchasing power parity): $20.01 billion (2017 est.)
$21.1 billion (2016 est.)
$24.52 billion (2015 est.)
note: data are in 2017 dollars
country comparison to the world: 149

GDP (official exchange rate): $3.06 billion (2017 est.)

GDP—real growth rate: -5.2% (2017 est.)
-13.9% (2016 est.)
-0.2% (2015 est.)
country comparison to the world: 218

GDP—per capita (PPP): $1,600 (2017 est.)
$1,700 (2016 est.)
$2,100 (2015 est.)
note: data are in 2017 dollars
country comparison to the world: 220

Gross national saving: 3.6% of GDP (2017 est.)

18.7% of GDP (2016 est.)
7.4% of GDP (2015 est.)
country comparison to the world: 179

GDP—composition, by end use:
household consumption: 34.9% (2011 est.)
government consumption: 17.1% (2011 est.)
investment in fixed capital: 10.4% (2011 est.)
exports of goods and services: 64.9% (2011 est.)
imports of goods and services: -27.2% (2011 est.)

Agriculture—products: sorghum, maize, rice, millet, wheat, gum arabic, sugarcane, mangoes, papayas, bananas, sweet potatoes, sunflower seeds, cotton, sesame seeds, cassava (manioc, tapioca), beans, peanuts; cattle, sheep

Population below poverty line: 66% (2015 est.)

Budget: *revenues:* 259.6 million (FY2017/18 est.)
expenditures: 298.6 million (FY2017/18 est.)

Taxes and other revenues: 8.5% (of GDP) (FY2017/18 est.)
country comparison to the world: 218

Budget surplus (+) or deficit (-): -1.3% (of GDP) (FY2017/18 est.)
country comparison to the world: 87

Public debt: 62.7% of GDP (2017 est.)
86.6% of GDP (2016 est.)
country comparison to the world: 69

Inflation rate (consumer prices): 187.9% (2017 est.)
379.8% (2016 est.)
country comparison to the world: 225

Current account balance: -$154 million (2017 est.)
$39 million (2016 est.)
country comparison to the world: 94

Exports: $1.13 billion (2016 est.)
country comparison to the world: 155

Imports: $3.795 billion (2016 est.)
country comparison to the world: 140

Reserves of foreign exchange and gold: $73 million (31 December 2016 est.)
country comparison to the world: 184

Exchange rates: South Sudanese pounds (SSP) per US dollar -
0.885 (2017 est.)
0.903 (2016 est.)
0.9214 (2015 est.)
0.885 (2014 est.)
0.7634 (2013 est.)

ENERGY

Electricity access: *population without electricity:* 11 million (2019)
electrification—total population: 1% (2019)
electrification—urban areas: 4% (2019)
electrification—rural areas: 1% (2019)

Electricity—production: 412.8 million kWh (2016 est.)
country comparison to the world: 169

Electricity—consumption: 391.8 million kWh (2016 est.)
country comparison to the world: 175

Electricity—exports: 0 kWh (2016 est.)

country comparison to the world: 201

Electricity—imports: 0 kWh (2016 est.)
country comparison to the world: 203

Electricity—installed generating capacity: 80,400 kW (2016 est.)
country comparison to the world: 185

Electricity—from fossil fuels: 100% of total installed capacity (2016 est.)
country comparison to the world: 18

Electricity—from nuclear fuels: 0% of total installed capacity (2017 est.)
country comparison to the world: 186

Electricity—from hydroelectric plants: 0% of total installed capacity (2017 est.)
country comparison to the world: 203

Electricity—from other renewable sources: 1% of total installed capacity (2017 est.)
country comparison to the world: 168

Crude oil—production: 150,200 bbl/day (2017 est.)
country comparison to the world: 39

Crude oil—exports: 147,300 bbl/day (2015 est.)
country comparison to the world: 32

Crude oil—imports: 0 bbl/day (2015 est.)
country comparison to the world: 199

Crude oil—proved reserves: 3.75 billion bbl (1 January 2017 est.)
country comparison to the world: 26

Refined petroleum products—production: 0 bbl/day (2017 est.)
country comparison to the world: 205

Refined petroleum products—consumption: 8,000 bbl/day (2016 est.)
country comparison to the world: 165

Refined petroleum products—exports: 0 bbl/day (2015 est.)
country comparison to the world: 205

Refined petroleum products—imports: 7,160 bbl/day (2015 est.)
country comparison to the world: 157

Natural gas—production: 0 cu m (2017 est.)
country comparison to the world: 199

Natural gas—consumption: 0 cu m (2017 est.)
country comparison to the world: 200

Natural gas—exports: 0 cu m (2017 est.)
country comparison to the world: 188

Natural gas—imports: 0 cu m (2017 est.)
country comparison to the world: 191

Natural gas—proved reserves: 63.71 billion cu m (1 January 2016 est.)
country comparison to the world: 59

Carbon dioxide emissions from consumption of energy: 1.224 million Mt (2017 est.)
country comparison to the world: 163

COMMUNICATIONS

Telephones—fixed lines: *total subscriptions:* 0
subscriptions per 100 inhabitants: less than 1 (2018 est.)
country comparison to the world: 225

Telephones—mobile cellular: *total subscriptions:* 3,439,784
subscriptions per 100 inhabitants: 33.46 (2019 est.)
country comparison to the world: 136

Telecommunication systems: *general assessment:* one of the least developed telecommunications and Internet systems in the world; the international community has provided billions in aid to help the young country, unfortunate instability, widespread poverty and low literacy rate all contribute to a struggle for their telecom sector; the few carriers in the market have reduced the areas in which they offer service, not expanded them; recently the government shut down the largest cellphone carrier isolating 1.4 million customers over a disputed service fee arrangement (2020)
domestic: fixed-line less than 1 per 100 subscriptions, mobile-cellular 33 per 100 persons (2019)
international: country code -211 (2017)
note: the COVID-19 outbreak is negatively impacting telecommunications production and supply chains globally; consumer spending on telecom devices and services has also slowed due to the pandemic's effect on economies worldwide; overall progress towards improvements in all facets of the telecom industry—mobile, fixed-line, broadband, submarine cable and satellite—has moderated

Broadcast media: a single TV channel and a radio station are controlled by the government; several community and commercial FM stations are operational, mostly sponsored by outside aid donors; some foreign radio broadcasts are available (2019)

Internet country code: .ss

Internet users: *total:* 814,326
percent of population: 7.98% (July 2018 est.)
country comparison to the world: 143

Broadband—fixed subscriptions: *total:* 200
subscriptions per 100 inhabitants: less than 1 (2018 est.)
country comparison to the world: 203

TRANSPORTATION

National air transport system: *number of registered air carriers:* 2 (2020)
inventory of registered aircraft operated by air carriers: 2
annual freight traffic on registered air carriers: 0 mt-km

Civil aircraft registration country code prefix: Z8 (2016)

Airports: 89 (2020)
country comparison to the world: 63

Airports—with paved runways: *total:* 4 (2020)
over 3,047 m: 1
2,438 to 3,047 m: 2
1,524 to 2,437 m: 1

Airports—with unpaved runways: *total:* 84 (2020)
2,438 to 3,047 m: 1
1,524 to 2,437 m: 12
914 to 1,523 m: 38
under 914 m: 33

Heliports: 3 (2020)

Railways: *total:* 248 km (2018)

note: a narrow gauge, single-track railroad between Babonosa (Sudan) and Wau, the only existing rail system, was repaired in 2010 with $250 million in UN funds, but is not currently operational

country comparison to the world: 126

Roadways: *total:* 90,200 km (2019)

paved: 300 km (2019)

unpaved: 89,900 km (2019)

note: most of the road network is unpaved and much of it is in disrepair

country comparison to the world: 55

Waterways: see entry for Sudan

MILITARY AND SECURITY

Military and security forces: *South Sudan People's Defence Force (SSPDF):* Ground Force, Air Force, Air Defense Forces, Presidential Guard (2019)

Military expenditures: 3.5% of GDP (2019)

3.7% of GDP (2018)

2.4% of GDP (2017)

4.6% of GDP (2016)

10% of GDP (2015)

country comparison to the world: 19

Military and security service personnel strengths: the South Sudan People's Defense Force (SSPDF) has an estimated 190,000 active personnel, including ground, air, and riverine forces (2019)

Military equipment inventories and acquisitions: the SSPDF inventory is primarily of Soviet origin; South Sudan was under a UN arms embargo through May 2020; from 2010 to 2015, Russian and the United Arab Emirates were the leading suppliers of arms and equipment (2020)

Military service age and obligation: 18 is the legal minimum age for compulsory and voluntary military service; the Government of South Sudan signed agreements in March 2012 and August 2015 that included the demobilization of all child soldiers within the armed forces and opposition, but the recruitment of child soldiers by the warring parties continues; as of the end of 2018, UNICEF estimated that more than 19,000 child soldiers had been used in the country's civil war since it began in December 2013 (2018)

Military—note: under the September 2018 peace agreement, all armed groups in South Sudan were to assemble at designated sites where fighters could be either disarmed and demobilized, or integrated into unified military and police forces; the unified forces were then to be retrained and deployed prior to the formation of a national unity government; all fighters were ordered to these sites in July 2019; some progress toward merging the various armed forces into a national army has been made; for example, in May 2020, South Sudan announced that it was graduating some unified forces at various training centers across the country, and in June the SSPDF incorporated some senior officers from the main opposition force, the Sudan People's Liberation Movement Army—in Opposition (SPLM/A-IO) into its rank structure; nevertheless, progress has been slow, and as of August 2020 armed clashes continued to occur between government forces and armed militant groups

the United Nations Mission in South Sudan (UNMISS) has operated in the country since 2011 with the objectives of consolidating peace and security and helping establish conditions for the successful economic and political development of South Sudan; UNMISS had more than 18,000 personnel deployed in the country as of May 2020

United Nations Interim Security Force for Abyei (UNISFA) has operated in the disputed Abyei region along the border between Sudan and South Sudan since 2011; UNISFA's mission includes ensuring security, protecting civilians, strengthening the capacity of the Abyei Police Service, de-mining, monitoring/verifying the redeployment of armed forces from the area, and facilitating the flow of humanitarian aid; UNISFA had about 4,000 personnel deployed as of March 2020 (2020)

TRANSNATIONAL ISSUES

Disputes—international: South Sudan-Sudan boundary represents 1 January 1956 alignment, final alignment pending negotiations and demarcation; final sovereignty status of Abyei Area pending negotiations between South Sudan and Sudan; periodic violent skirmishes with South Sudanese residents over water and grazing rights persist among related pastoral populations along the border with the Central African Republic; the boundary that separates Kenya and South Sudan's sovereignty is unclear in the "Ilemi Triangle," which Kenya has administered since colonial times

Refugees and internally displaced persons: *refugees (country of origin):* 729,530 (Sudan) (refugees and asylum seekers), 16,176 (Democratic Republic of the Congo) (refugees and asylum seekers) (2020)

IDPs: *1.66 million (alleged coup attempt and ethnic conflict beginning in December 2013; information is lacking on those displaced in earlier years by:* fighting in Abyei between the Sudanese Armed Forces and the Sudan People's Liberation Army (SPLA) in May 2011; clashes between the SPLA and dissident militia groups in South Sudan; inter-ethnic conflicts over resources and cattle; attacks from the Lord's Resistance Army; floods and drought) (2020)

Trafficking in persons: *current situation:* South Sudan is a source and destination country for men, women, and children subjected to forced labor and sex trafficking; South Sudanese women and girls, particularly those who are internally displaced, orphaned, refugees, or from rural areas, are vulnerable to forced labor and sexual exploitation, often in urban centers; children may be victims of forced labor in construction, market vending, shoe shining, car washing, rock breaking, brick making, delivery cart pulling, and begging; girls are also forced into marriages and subsequently subjected to sexual slavery or domestic servitude; women and girls migrate willingly from Uganda, Kenya, Ethiopia, Eritrea, and the Democratic Republic of the Congo to South Sudan with the promise of legitimate jobs and are forced into the sex trade; inter-ethnic abductions and abductions by criminal groups continue, with abductees subsequently forced into domestic servitude, herding, or sex trafficking; in 2014, the recruitment and use of child soldiers increased significantly within government security forces and was also prevalent among opposition forces

tier rating: Tier 3 – South Sudan does not fully comply with the minimum standards for the elimination of trafficking and is not making significant efforts to do so; despite the government's formal recommitment to an action plan to eliminate the recruitment and use of child soldiers by 2016, the practice expanded during 2014, and the government did not hold any officers criminally responsible; government officials reportedly are complicit in trafficking offenses but these activities continue to go uninvestigated; authorities reportedly identified five trafficking victims but did not transfer them to care facilities; law enforcement continued to arrest and imprison individuals for prostitution, including trafficking victims; no known steps were taken to address the exploitation of South Sudanese nationals working abroad or foreign workers in South Sudan (2015)

SOUTHERN OCEAN

INTRODUCTION

Background: A large body of recent oceanographic research has shown that the Antarctic Circumpolar Current (ACC), an ocean current that flows from west to east around Antarctica, plays a crucial role in global ocean circulation. The region where the cold waters of the ACC meet and mingle with the warmer waters of the north defines a distinct border—the Antarctic Convergence—which fluctuates with the seasons, but which encompasses a discrete body of water and a unique ecologic region. The Convergence concentrates nutrients, which promotes marine plant life, and which, in turn, allows for a greater abundance of animal life. In 2000, the International Hydrographic Organization delimited the waters within the Convergence as a fifth

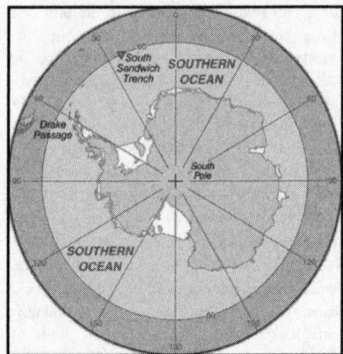

world ocean—the Southern Ocean—by combining the southern portions of the Atlantic Ocean, Indian Ocean, and Pacific Ocean. The Southern Ocean extends from the coast of Antarctica north to 60 degrees south latitude, which coincides with the Antarctic Treaty region and which approximates the extent of the Antarctic Convergence. As such, the Southern Ocean is now the fourth largest of the world's five oceans (after the Pacific Ocean, Atlantic Ocean, and Indian Ocean, but larger than the Arctic Ocean). It should be noted that inclusion of the Southern Ocean does not imply recognition of this feature as one of the world's primary oceans by the US Government.

GEOGRAPHY

Location: body of water between 60 degrees south latitude and Antarctica

Geographic coordinates: 60 00 S, 90 00 E (nominally), but the Southern Ocean has the unique distinction of being a large circumpolar body of water totally encircling the continent of Antarctica; this ring of water lies between 60 degrees south latitude and the coast of Antarctica and encompasses 360 degrees of longitude

Map references: Antarctic Region

Area: *total:* 20.327 million sq km
note: includes Amundsen Sea, Bellingshausen Sea, part of the Drake Passage, Ross Sea, a small part of the Scotia Sea, Weddell Sea, and other tributary water bodies

Area—comparative: slightly more than twice the size of the US

Coastline: 17,968 km

Climate: sea temperatures vary from about 10 degrees Celsius to -2 degrees Celsius; cyclonic storms travel eastward around the continent and frequently are intense because of the temperature contrast between ice and open ocean; the ocean area from about latitude 40 south to the Antarctic Circle has the strongest average winds found anywhere on Earth; in winter the ocean freezes outward to 65 degrees south latitude in the Pacific sector and 55 degrees south latitude in the Atlantic sector, lowering surface temperatures well below 0 degrees Celsius; at some coastal points intense persistent drainage winds from the interior keep the shoreline ice-free throughout the winter

Terrain: the Southern Ocean is 4,000 to 5,000-m deep over most of its extent with only limited areas of shallow water; the Antarctic continental shelf is generally narrow and unusually deep, its edge lying at depths of 400 to 800 m (the global mean is 133 m); the Antarctic icepack grows from an average minimum of 2.6 million sq km in March to about 18.8 million sq km in September, better than a six-fold increase in area

major surface currents: the cold, clockwise-flowing Antarctic Circumpolar Current (West Wind Drift; 21,000 km long) moves perpetually eastward around the continent and is the world's largest and strongest ocean current, transporting 130 million cubic meters of water per second—100 times the flow of all the world's rivers; it is also the only current that flows all the way around the planet and connects the Atlantic, Pacific, and Indian Oceans; the cold Antarctic Coastal Current (East Wind Drift) is the southernmost current in the world, flowing westward and parallel to the Antarctic coastline

Elevation: *mean depth:* -3,270 m
lowest point: southern end of the South Sandwich Trench -7,235 m
highest point: sea level

Natural resources: probable large oil and gas fields on the continental margin; manganese nodules, possible placer deposits, sand and gravel, fresh water as icebergs; squid, whales, and seals—none exploited; krill, fish

Natural hazards: huge icebergs with drafts up to several hundred meters; smaller bergs and iceberg fragments; sea ice (generally 0.5 to 1 m thick) with sometimes dynamic short-term variations and with large annual and interannual variations; deep continental shelf floored by glacial deposits varying widely over short distances; high winds and large waves much of the year; ship icing, especially May-October; most of region is remote from sources of search and rescue

Environment—current issues: changes to the ocean's physical, chemical, and biological systems have taken place because of climate change, ocean acidification, and commercial exploitation

Environment—international agreements: *the Southern Ocean is subject to all international agreements regarding the world's oceans; in addition, it is subject to these agreements specific to the Antarctic region:* International Whaling Commission (prohibits commercial whaling south of 40 degrees south [south of 60 degrees south between 50 degrees and 130 degrees west]); Convention on the Conservation of Antarctic Seals (limits sealing); Convention on the Conservation of Antarctic Marine Living Resources (regulates fishing)

note: many nations (including the US) prohibit mineral resource exploration and exploitation south of the fluctuating Polar Front (Antarctic Convergence), which is in the middle of the Antarctic Circumpolar Current and serves as the dividing line between the cold polar surface waters to the south and the warmer waters to the north

Geography—note: the major chokepoint is the Drake Passage between South America and Antarctica; the Polar Front (Antarctic Convergence) is the best natural definition of the northern extent of the Southern Ocean; it is a distinct region at the middle of the Antarctic Circumpolar Current that separates the cold polar surface waters to the south from the warmer waters to the north; the Front and the Current extend entirely around Antarctica, reaching south of 60 degrees south near New Zealand and near 48 degrees south in the far South Atlantic coinciding with the path of the maximum westerly winds

GOVERNMENT

Country name: *etymology:* the International Hydrographic Organization (IHO) included the ocean and its definition as the waters south of 60 degrees south in its year 2000 revision, but this has not formally been adopted; the 2000 IHO definition, however, was circulated in a draft edition in 2002 and has acquired de facto usage by many nations and organizations, including the CIA

ECONOMY

Economy—overview: Fisheries in 2013-14 landed 302,960 metric tons, of which 96% (291,370 tons-the highest reported catch since 1991) was krill and 4% (11,590 tons) Patagonian toothfish (also known as Chilean sea bass), compared to 15,330 tons in 2012-13 (estimated fishing from the area covered by the Convention of the Conservation of Antarctic Marine Living Resources, which extends slightly beyond the Southern Ocean area). International agreements were adopted in late 1999 to reduce illegal, unreported, and unregulated fishing, which in the 2000-01 season landed, by one estimate, 8,376 metric tons of Patagonian and Antarctic toothfish. In the 2014-15 Antarctic summer, 36,702 tourists visited the Southern Ocean, slightly lower than the 37,405 visitors in 2013-14 (estimates provided to the Antarctic Treaty by the International Association of Antarctica Tour Operators, and does not include passengers on overflights and those flying directly in and out of Antarctica).

Marine fisheries: the Southern Ocean fishery is relatively small with a total catch of 257,278 mt in 2017; the Food and Agriculture Organization has delineated three regions in the Southern Ocean (Regions 48, 58, 88) that generally encompass the waters south of 40° to 60° South latitude; the most important producers in these regions include Norway (156,884 mt), China (38,112 mt), and South Korea (34,506 mt); Antarctic Krill made up 92% of the total catch in 2017, while other important species include Patagonian and Antarctic toothfish

FAO map of world fishing regions; used with permission.: *Transportation*

Ports and terminals: *major seaport(s):* McMurdo, Palmer, and offshore anchorages in Antarctica *note:* few ports or harbors exist on the southern side of the Southern Ocean; ice conditions limit use of most to short periods in midsummer; even then some cannot be entered without icebreaker escort; most Antarctic ports are operated by government research stations and, except in an emergency, are not open to commercial or private vessels

Transportation—note: Drake Passage offers alternative to transit through the Panama Canal

TRANSNATIONAL ISSUES

Disputes—international: Antarctic Treaty defers claims (see Antarctica entry), but Argentina, Australia, Chile, France, NZ, Norway, and UK assert claims (some overlapping), including the continental shelf in the Southern Ocean; several states have expressed an interest in extending those continental shelf claims under the UN Convention on the Law of the Sea to include undersea ridges; the US and most other states do not recognize the land or maritime claims of other states and have made no claims themselves (the US and Russia have reserved the right to do so); no formal claims exist in the waters in the sector between 90 degrees west and 150 degrees west

SPAIN

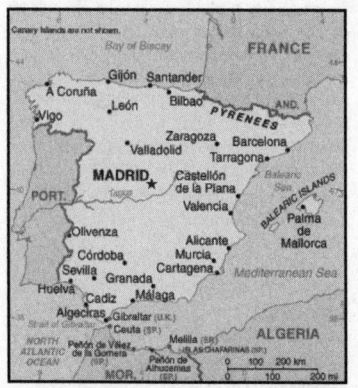

INTRODUCTION

Background: Spain's powerful world empire of the 16th and 17th centuries ultimately yielded command of the seas to England. Subsequent failure to embrace the mercantile and industrial revolutions caused the country to fall behind Britain, France, and Germany in economic and political power. Spain remained neutral in World War I and II, but suffered through a devastating civil war (1936-39). A peaceful transition to democracy following the death of dictator Francisco FRANCO in 1975, and rapid economic modernization (Spain joined the EU in 1986) gave Spain a dynamic and rapidly growing economy, and made it a global champion of freedom and human rights. More recently, Spain has emerged from a severe economic recession that began in mid-2008, posting four straight years of GDP growth above the EU average. Unemployment has fallen, but remains high, especially among youth. Spain is the Eurozone's fourth largest economy. The country has faced increased domestic turmoil in recent years due to the independence movement in its restive Catalonia region.

GEOGRAPHY

Location: Southwestern Europe, bordering the Mediterranean Sea, North Atlantic Ocean, Bay of Biscay, and Pyrenees Mountains; southwest of France

Geographic coordinates: 40 00 N, 4 00 W

Map references: Europe

Area: *total:* 505,370 sq km
land: 498,980 sq km
water: 6,390 sq km
note: there are two autonomous cities—Ceuta and Melilla—and 17 autonomous communities including Balearic Islands and Canary Islands, and three small Spanish possessions off the coast of Morocco—Islas Chafarinas, Penon de Alhucemas, and Penon de Velez de la Gomera
country comparison to the world: 53

Area—comparative: almost five times the size of Kentucky; slightly more than twice the size of Oregon

Land boundaries: *total:* 1,952.7 km
border countries (6): Andorra 63 km, France 646 km, Gibraltar 1.2 km, Portugal 1224 km, Morocco (Ceuta) 8 km, Morocco (Melilla) 10.5 km
note: an additional 75-meter border segment exists between Morocco and the Spanish exclave of Penon de Velez de la Gomera

Coastline: 4,964 km

Maritime claims: *territorial sea:* 12 nm
exclusive economic zone: 200 nm (applies only to the Atlantic Ocean)
contiguous zone: 24 nm

Climate: temperate; clear, hot summers in interior, more moderate and cloudy along coast; cloudy, cold winters in interior, partly cloudy and cool along coast

Terrain: large, flat to dissected plateau surrounded by rugged hills; Pyrenees Mountains in north

Elevation: *mean elevation:* 660 m
lowest point: Atlantic Ocean 0 m
highest point: Pico de Teide (Tenerife) on Canary Islands 3,718 m

Natural resources: coal, lignite, iron ore, copper, lead, zinc, uranium, tungsten, mercury, pyrites, magnesite, fluorspar, gypsum, sepiolite, kaolin, potash, hydropower, arable land

Land use: *agricultural land:* 54.1% (2011 est.)
arable land: 24.9% (2011 est.) / *permanent crops:* 9.1% (2011 est.) / *permanent pasture:* 20.1% (2011 est.)
forest: 36.8% (2011 est.)
other: 9.1% (2011 est.)

Irrigated land: 38,000 sq km (2012)

Population distribution: with the notable exception of Madrid, Sevilla, and Zaragoza, the largest urban agglomerations are found along the Mediterranean and Atlantic coasts; numerous smaller cities are spread throughout the interior reflecting Spain's agrarian heritage; very dense settlement around the capital of Madrid, as well as the port city of Barcelona

Natural hazards: periodic droughts, occasional flooding
volcanism: volcanic activity in the Canary Islands, located off Africa's northwest coast; Teide (3,715 m) has been deemed a Decade Volcano by the International Association of Volcanology and Chemistry of the Earth's Interior, worthy of study due to its explosive history and close proximity to human populations; La Palma (2,426 m), which last erupted in 1971, is the most active of the Canary Islands volcanoes; Lanzarote is the only other historically active volcano

Environment—current issues: pollution of the Mediterranean Sea from raw sewage and effluents from the offshore production of oil and gas; water quality and quantity nationwide; air pollution; deforestation; desertification

Environment—international agreements: *party to:* Air Pollution, Air Pollution-Nitrogen Oxides, Air Pollution-Sulfur 94, Air Pollution-Volatile Organic Compounds, Antarctic-Environmental Protocol, Antarctic-Marine Living Resources, Antarctic Treaty, Biodiversity, Climate Change, Climate Change-Kyoto Protocol, Desertification, Endangered Species, Environmental Modification, Hazardous Wastes, Law of the Sea, Marine Dumping, Marine Life Conservation, Ozone Layer Protection, Ship Pollution, Tropical Timber 83, Tropical Timber 94, Wetlands, Whaling
signed, but not ratified: Air Pollution-Persistent Organic Pollutants

Geography—note: strategic location along approaches to Strait of Gibraltar; Spain controls a number of territories in northern Morocco including the enclaves of Ceuta and Melilla, and the islands of Penon de Velez de la Gomera, Penon de Alhucemas, and Islas Chafarinas

899

PEOPLE AND SOCIETY

Population: 50,015,792 (July 2020 est.)
country comparison to the world: 29

Nationality: *noun:* Spaniard(s)
adjective: Spanish

Ethnic groups: Spanish 86.4%, Moroccan 1.8%, Romanian 1.3%, other 10.5% (2018 est.)
note: data represent population by country of birth

Languages: Castilian Spanish (official nationwide) 74%, Catalan (official in Catalonia, the Balearic Islands, and the Valencian Community (where it is known as Valencian)) 17%, Galician (official in Galicia) 7%, Basque (official in the Basque Country and in the Basque-speaking area of Navarre) 2%, Aranese (official in the northwest corner of Catalonia (Vall d'Aran) along with Catalan, <5,000 speakers)
note: Aragonese, Aranese Asturian, Basque, Calo, Catalan, Galician, and Valencian are recognized as regional languages under the European Charter for Regional or Minority Languages

Religions: Roman Catholic 68.9%, atheist 11.3%, agnostic 7.6%, other 2.8%, non-believer 8.2%, unspecified 1.1% (2019 est.)

Age structure: *0-14 years:* 15.02% (male 3,861,522/female 3,650,085)
15-24 years: 9.9% (male 2,557,504/female 2,392,498)
25-54 years: 43.61% (male 11,134,006/female 10,675,873)
55-64 years: 12.99% (male 3,177,080/female 3,319,823)
65 years and over: 18.49% (male 3,970,417/female 5,276,081) (2020 est.)

Dependency ratios: *total dependency ratio:* 52.4
youth dependency ratio: 21.9
elderly dependency ratio: 30.4
potential support ratio: 3.3 (2020 est.)

Median age: *total:* 43.9 years
male: 42.7 years
female: 45.1 years (2020 est.)
country comparison to the world: 19

Population growth rate: 0.67% (2020 est.)
country comparison to the world: 142

Birth rate: 8.7 births/1,000 population (2020 est.)
country comparison to the world: 213

Death rate: 9.3 deaths/1,000 population (2020 est.)
country comparison to the world: 51

Net migration rate: 7 migrant(s)/1,000 population (2020 est.)
country comparison to the world: 15

Population distribution: with the notable exception of Madrid, Sevilla, and Zaragoza, the largest urban agglomerations are found along the Mediterranean and Atlantic coasts; numerous smaller cities are spread throughout the interior reflecting Spain's agrarian heritage; very dense settlement around the capital of Madrid, as well as the port city of Barcelona

Urbanization: *urban population:* 80.8% of total population (2020)

rate of urbanization: 0.33% annual rate of change (2015-20 est.)
note: data include Canary Islands, Ceuta, and Melilla

total population growth rate v. urban population growth rate, 2000-2030: *Major urban areas—population:* 6.618 million MADRID (capital), 5.586 million Barcelona, 834,000 Valencia (2020)

Sex ratio: *at birth:* 1.07 male(s)/female
0-14 years: 1.06 male(s)/female
15-24 years: 1.07 male(s)/female
25-54 years: 1.04 male(s)/female
55-64 years: 0.96 male(s)/female
65 years and over: 0.75 male(s)/female
total population: 0.98 male(s)/female (2020 est.)

Mother's mean age at first birth: 30.9 years (2017 est.)

Maternal mortality rate: 4 deaths/100,000 live births (2017 est.)
country comparison to the world: 174

Infant mortality rate: *total:* 3.2 deaths/1,000 live births
male: 3.5 deaths/1,000 live births
female: 2.9 deaths/1,000 live births (2020 est.)
country comparison to the world: 213

Life expectancy at birth: *total population:* 82 years
male: 79 years
female: 85.2 years (2020 est.)
country comparison to the world: 24

Total fertility rate: 1.51 children born/woman (2020 est.)
country comparison to the world: 200

Contraceptive prevalence rate: 62.1% (2018)
note: percent of women aged 18-49

Drinking water source:
improved:
urban: 100% of population
rural: 100% of population
total: 100% of population
unimproved:
urban: 0% of population
rural: 0% of population
total: 0% of population (2017 est.)

Current Health Expenditure: 8.9% (2017)

Physicians density: 3.87 physicians/1,000 population (2017)

Hospital bed density: 3 beds/1,000 population (2017)

Sanitation facility access:
improved:
urban: 100% of population (5 est.)
rural: 100% of population
total: 100% of population
unimproved:
urban: 0% of population
rural: 0% of population
total: 0% of population (2017 est.)

HIV/AIDS—adult prevalence rate: 0.4% (2019 est.)
country comparison to the world: 83

HIV/AIDS—people living with HIV/AIDS: 150,000 (2019 est.)
country comparison to the world: 36

HIV/AIDS—deaths: <1000 (2019)

Major infectious diseases: Covid-19 (see note) (2020)
note: widespread ongoing transmission of a respiratory illness caused by the novel coronavirus (COVID-19) is occurring throughout Spain; as of 10 November 2020, Spain has reported a total of 1,328,832 cases of COVID-19 or 28,421 cumulative cases of COVID-19 per 1 million population with 831 cumulative deaths per 1 million population; the Spanish Government is gradually relaxing confinement measures in phases over the next several weeks; these measures will vary from region to region within Spain; the Department of Homeland Security has issued instructions requiring US passengers who have been in Spain to travel through select airports where the US Government has implemented enhanced screening procedures

Obesity—adult prevalence rate: 23.8% (2016)
country comparison to the world: 62

Education expenditures: 4.2% of GDP (2016)
country comparison to the world: 94

Literacy: *definition:* age 15 and over can read and write
total population: 98.4%
male: 98.9%
female: 98% (2018)

School life expectancy (primary to tertiary education): *total:* 18 years
male: 17 years
female: 18 years (2018)

Unemployment, youth ages 15-24: *total:* 34.3%
male: 35.2%
female: 33.3% (2018 est.)
country comparison to the world: 24

GOVERNMENT

Country name: *conventional long form:* Kingdom of Spain
conventional short form: Spain
local long form: Reino de Espana
local short form: Espana
etymology: derivation of the name "Espana" is uncertain, but may come from the Phoenician term "span," related to the word "spy," meaning "to forge metals," so, "i-spn-ya" would mean "place where metals are forged"; the ancient Phoenicians long exploited the Iberian Peninsula for its mineral wealth

Government type: parliamentary constitutional monarchy

Capital: *name:* Madrid

geographic coordinates: 40 24 N, 3 41 W
time difference: UTC+1 (6 hours ahead of Washington, DC, during Standard Time)
daylight saving time: +1hr, begins last Sunday in March; ends last Sunday in October
note: Spain has two time zones, including the Canary Islands (UTC 0)

etymology: the Romans named the original settlement "Matrice" after the river that ran through it; under Arab rule it became "Majerit," meaning "source of water"; in medieval Romance dialects (Mozarabic) it became "Matrit," which over time changed to "Madrid"

Administrative divisions: 17 autonomous communities (comunidades autonomas, singular—comunidad autonoma) and 2 autonomous cities* (ciudades autonomas, singular—ciudad autonoma); Andalucia; Aragon; Asturias; Canarias (Canary Islands); Cantabria; Castilla-La Mancha; Castilla-Leon; Cataluna (Castilian), Catalunya (Catalan), Catalonha (Aranese) [Catalonia]; Ceuta*; Comunidad Valenciana (Castilian), Comunitat Valenciana (Valencian) [Valencian Community]; Extremadura; Galicia; Illes Baleares (Balearic Islands); La Rioja; Madrid; Melilla*; Murcia; Navarra (Castilian), Nafarroa (Basque) [Navarre]; Pais Vasco (Castilian), Euskadi (Basque) [Basque Country]

note: the autonomous cities of Ceuta and Melilla plus three small islands of Islas Chafarinas, Penon de Alhucemas, and Penon de Velez de la Gomera, administered directly by the Spanish central government, are all along the coast of Morocco and are collectively referred to as Places of Sovereignty (Plazas de Soberania)

Independence: 1492; the Iberian peninsula was characterized by a variety of independent kingdoms prior to the Muslim occupation that began in the early 8th century A.D. and lasted nearly seven centuries; the small Christian redoubts of the north began the reconquest almost immediately, culminating in the seizure of Granada in 1492; this event completed the unification of several kingdoms and is traditionally considered the forging of present-day Spain

National holiday: National Day (Hispanic Day), 12 October (1492); note—commemorates the arrival of COLUMBUS in the Americas

Constitution: *history:* previous 1812; latest approved by the General Courts 31 October 1978, passed by referendum 6 December 1978, signed by the king 27 December 1978, effective 29 December 1978

amendments: proposed by the government, by the General Courts (the Congress or the Senate), or by the self-governing communities submitted through the government; passage requires three-fifths majority vote by both houses and passage by referendum if requested by one tenth of the members of either house; proposals disapproved by both houses are submitted to a joint committee, which submits an agreed upon text for another vote; passage requires two-thirds majority vote in Congress and simple majority vote in the Senate; amended 1992, 2007, 2011

Legal system: civil law system with regional variations

International law organization participation: accepts compulsory ICJ jurisdiction with reservations; accepts ICCt jurisdiction

Citizenship: *citizenship by birth:* no

citizenship by descent only: at least one parent must be a citizen of Spain

dual citizenship recognized: only with select Latin American countries

residency requirement for naturalization: 10 years for persons with no ties to Spain

Suffrage: 18 years of age; universal

Executive branch: *chief of state:* King FELIPE VI (since 19 June 2014); Heir Apparent Princess LEONOR, Princess of Asturias (daughter of the monarch, born 31 October 2005)

head of government: President of the Government (Prime Minister-equivalent) Pedro SANCHEZ Perez-Castejon (since 2 June 2018); Vice President (and Minister of the President's Office) Maria del Carmen CALVO Poyato (since 7 June 2018)

cabinet: Council of Ministers designated by the president

elections/appointments: the monarchy is hereditary; following legislative elections, the monarch usually proposes as president the leader of the party or coalition with the largest majority of seats, who is then indirectly elected by the Congress of Deputies; election last held on 10 November 2019 (next to be held November 2023); vice president and Council of Ministers appointed by the president

election results: percent of National Assembly vote—NA

note: there is also a Council of State that is the supreme consultative organ of the government, but its recommendations are non-binding

Legislative branch: *description:* bicameral General Courts or Las Cortes Generales consists of:
Senate or Senado (266 seats; 208 members directly elected in multi-seat constituencies by simple majority vote and 58 members indirectly elected by the legislatures of the autonomous communities; members serve 4-year terms)
Congress of Deputies or Congreso de los Diputados (350 seats; 348 members directly elected in 50 multi-seat constituencies by closed-list proportional representation vote, with a 3% threshold needed to gain a seat, and 2 directly elected from the North African Ceuta and Melilla enclaves by simple majority vote; members serve 4-year terms or until the government is dissolved)

elections:
Senate—last held on 10 November 2019 (next to be held no later than November 2023)
Congress of Deputies—last held on 10 November 2019 (next to be held no later than November 2023)

election results:
Senate—percent of vote by party—NA; seats by party—PSOE 113, PP 97, ERC 15, EAJ/PNV 10, C's 9, other 22; composition—men 163, women 103; percent of women 39%
Congress of Deputies—percent of vote by party—PSOE 28.7%, PP 20.8%, Vox 15.1%, Unidos Podemos 12.8%, C's 6.8%, ERC 3.6%, other 12.8%; seats by party—PSOE 120, PP 88, Vox 52, Unidos Podemos 35, C's 10, ERC 13, other 23;

composition—men 184, women 166; percent of women 47.4%; note—total General Courts percent of women 43.7%

Judicial branch: *highest courts:* Supreme Court or Tribunal Supremo (consists of the court president and organized into the Civil Room, with a president and 9 judges; the Penal Room, with a president and 14 judges; the Administrative Room, with a president and 32 judges; the Social Room, with a president and 12 judges; and the Military Room, with a president and 7 judges); Constitutional Court or Tribunal Constitucional de Espana (consists of 12 judges)

judge selection and term of office: Supreme Court judges appointed by the monarch from candidates proposed by the General Council of the Judiciary Power, a 20-member governing board chaired by the monarch that includes presidential appointees, lawyers, and jurists confirmed by the National Assembly; judges can serve until age 70; Constitutional Court judges nominated by the National Assembly, executive branch, and the General Council of the Judiciary, and appointed by the monarch for 9-year terms

subordinate courts: National High Court; High Courts of Justice (in each of the autonomous communities); provincial courts; courts of first instance

Political parties and leaders: Asturias Forum or FAC [Carmen MORIYON]
Basque Country Unite (Euskal Herria Bildu) or EH Bildu (coalition of 4 Basque pro-independence parties)
Basque Nationalist Party or PNV or EAJ [Andoni ORTUZAR]
Canarian Coalition or CC [Ana ORAMAS] (coalition of 5 parties)
Junts per Catalunia or JxCat [Carles PUIDGEMONT]
Ciudadanos Party or C's [Albert RIVERA]
Compromis—Communist Coalition [Joan BALDOVI]
New Canary or NCa [Pedro QUEVEDOS]
Unidas Podemos [Pablo IGLESIAS Turrion] (formerly Podemos IU; electoral coalition formed for May 2016 election) People's Party or PP [Pablo CASADO]
Republican Left of Catalonia or ERC [Oriol JUNQUERAS i Vies]
Spanish Socialist Workers Party or PSOE [Pedro SANCHEZ]
JxCat-Junts Together for Catalonia [Jordi SANCHEZ]
Union of People of Navarra or UPN [Javier ESPARZA]
Navarra Suma (electoral Coaltion formed by Navarrese People's Union (UPN), Ciudadanos (C's), and the Popular Partty (PP) ahead of the 2019 election)
Vox or Vox [Santiago ABASCAL]

International organization participation: ADB (nonregional member), AfDB (nonregional member), Arctic Council (observer), Australia Group, BCIE, BIS, CAN (observer), CBSS (observer), CD, CE, CERN, EAPC, EBRD, ECB, EIB, EITI (implementing country), EMU, ESA, EU,

FAO, FATF, IADB, IAEA, IBRD, ICAO, ICC (national committees), ICCt, ICRM, IDA, IEA, IFAD, IFC, IFRCS, IHO, ILO, IMF, IMO, IMSO, Interpol, IOC, IOM, IPU, ISO, ITSO, ITU, ITUC (NGOs), LAIA (observer), MIGA, NATO, NEA, NSG, OAS (observer), OECD, OPCW, OSCE, Pacific Alliance (observer), Paris Club, PCA, PIF (partner), Schengen Convention, SELEC (observer), SICA (observer), UN, UNCTAD, UNESCO, UNHCR, UNIDO, UNIFIL, Union Latina, UNOCI, UNRWA, UNWTO, UPU, WCO, WHO, WIPO, WMO, WTO, ZC

Diplomatic representation in the US: *chief of mission:* Ambassador Santiago CABANAS Ansorena (since 17 September 2018)
chancery: 2375 Pennsylvania Avenue NW, Washington, DC 20037
telephone: [1] (202) 452-0100, 728-2340
FAX: [1] (202) 833-5670
consulate(s) general: Boston, Chicago, Houston, Los Angeles, Miami, New York, San Francisco, San Juan (Puerto Rico)
consulate(s): Kansas City (MO)

Diplomatic representation from the US: *chief of mission:* Ambassador Richard Duke BUCHAN III (since 18 January 2018) note—also accredited to Andorra
telephone: [34] (91) 587-2200
embassy: Calle de Serrano 75, 28006 Madrid
mailing address: PSC 61, APO AE 09642
FAX: [34] (91) 587-2303
consulate(s) general: Barcelona

Flag description: three horizontal bands of red (top), yellow (double width), and red with the national coat of arms on the hoist side of the yellow band; the coat of arms is quartered to display the emblems of the traditional kingdoms of Spain (clockwise from upper left, Castile, Leon, Navarre, and Aragon) while Granada is represented by the stylized pomegranate at the bottom of the shield; the arms are framed by two columns representing the Pillars of Hercules, which are the two promontories (Gibraltar and Ceuta) on either side of the eastern end of the Strait of Gibraltar; the red scroll across the two columns bears the imperial motto of "Plus Ultra" (further beyond) referring to Spanish lands beyond Europe; the triband arrangement with the center stripe twice the width of the outer dates to the 18th century

note: *the red and yellow colors are related to those of the oldest Spanish kingdoms:* Aragon, Castile, Leon, and Navarre

National symbol(s): *Pillars of Hercules; national colors:* red, yellow

National anthem: *name:* "Himno Nacional Espanol" (National Anthem of Spain)
lyrics/music: no lyrics/unknown
note: officially in use between 1770 and 1931, restored in 1939; the Spanish anthem is the first anthem to be officially adopted, but it has no lyrics; in the years prior to 1931 it became known as "Marcha Real" (The Royal March); it first appeared in a 1761 military bugle call book and was replaced by "Himno de Riego" in the years between 1931 and 1939; the long version of the anthem is used for the king, while the short version is used for the prince, prime minister, and occasions such as sporting events 0:00 / 0:35

ECONOMY

Economy—overview: After a prolonged recession that began in 2008 in the wake of the global financial crisis, Spain marked the fourth full year of positive economic growth in 2017, with economic activity surpassing its pre-crisis peak, largely because of increased private consumption. The financial crisis of 2008 broke 16 consecutive years of economic growth for Spain, leading to an economic contraction that lasted until late 2013. In that year, the government successfully shored up its struggling banking sector—heavily exposed to the collapse of Spain's real estate boom—with the help of an EU-funded restructuring and recapitalization program.

Until 2014, contraction in bank lending, fiscal austerity, and high unemployment constrained domestic consumption and investment. The unemployment rate rose from a low of about 8% in 2007 to more than 26% in 2013, but labor reforms prompted a modest reduction to 16.4% in 2017. High unemployment strained Spain's public finances, as spending on social benefits increased while tax revenues fell. Spain's budget deficit peaked at 11.4% of GDP in 2010, but Spain gradually reduced the deficit to about 3.3% of GDP in 2017. Public debt has increased substantially – from 60.1% of GDP in 2010 to nearly 96.7% in 2017.

Strong export growth helped bring Spain's current account into surplus in 2013 for the first time since 1986 and sustain Spain's economic growth. Increasing labor productivity and an internal devaluation resulting from moderating labor costs and lower inflation have improved Spain's export competitiveness and generated foreign investor interest in the economy, restoring FDI flows.

In 2017, the Spanish Government's minority status constrained its ability to implement controversial labor, pension, health care, tax, and education reforms. The European Commission expects the government to meet its 2017 budget deficit target and anticipates that expected economic growth in 2018 will help the government meet its deficit target. Spain's borrowing costs are dramatically lower since their peak in mid-2012, and increased economic activity has generated a modest level of inflation, at 2% in 2017.

GDP (purchasing power parity): $1.778 trillion (2017 est.)
$1.727 trillion (2016 est.)
$1.674 trillion (2015 est.)
note: data are in 2017 dollars
country comparison to the world: 15

GDP (official exchange rate): $1.314 trillion (2017 est.)

GDP—real growth rate: 1.95% (2019 est.)
2.43% (2018 est.)
2.97% (2017 est.)
country comparison to the world: 141

GDP—per capita (PPP): $38,400 (2017 est.)
$37,200 (2016 est.)
$36,100 (2015 est.)
note: data are in 2017 dollars
country comparison to the world: 49

Gross national saving: 23% of GDP (2017 est.)
22.4% of GDP (2016 est.)
21.5% of GDP (2015 est.)
country comparison to the world: 76

GDP—composition, by end use:
household consumption: 57.7% (2017 est.)
government consumption: 18.5% (2017 est.)
investment in fixed capital: 20.6% (2017 est.)
investment in inventories: 0.6% (2017 est.)
exports of goods and services: 34.1% (2017 est.)
imports of goods and services: -31.4% (2017 est.)

GDP—composition, by sector of origin:
agriculture: 2.6% (2017 est.)
industry: 23.2% (2017 est.)
services: 74.2% (2017 est.)

Agriculture—products: grain, vegetables, olives, wine grapes, sugar beets, citrus; beef, pork, poultry, dairy products; fish

Industries: textiles and apparel (including footwear), food and beverages, metals and metal manufactures, chemicals, shipbuilding, automobiles, machine tools, tourism, clay and refractory products, footwear, pharmaceuticals, medical equipment

Industrial production growth rate: 4% (2017 est.)
country comparison to the world: 76

Labor force: 19.057 million (2020 est.)
country comparison to the world: 28

Labor force—by occupation: *agriculture:* 4.2%
industry: 24%
services: 71.7% (2009)

Unemployment rate: 14.13% (2019 est.)
15.25% (2018 est.)
country comparison to the world: 172

Population below poverty line: 21.1% (2012 est.)

Household income or consumption by percentage share: *lowest 10%:* 2.5%
highest 10%: 24% (2011)

Budget: *revenues:* 498.1 billion (2017 est.)
expenditures: 539 billion (2017 est.)

Taxes and other revenues: 37.9% (of GDP) (2017 est.)
country comparison to the world: 52

Budget surplus (+) or deficit (-): -3.1% (of GDP) (2017 est.)
country comparison to the world: 136

Public debt: 98.4% of GDP (2017 est.)
99% of GDP (2016 est.)
country comparison to the world: 18

Fiscal year: calendar year

Inflation rate (consumer prices): 2% (2017 est.)
-0.2% (2016 est.)
country comparison to the world: 107

Current account balance: $29.603 billion (2019 est.)
$27.206 billion (2018 est.)
country comparison to the world: 13

Exports: $313.7 billion (2017 est.)
$280.5 billion (2016 est.)
country comparison to the world: 16

Exports—partners: France 15.1%, Germany 11.3%, Italy 7.8%, Portugal 7.1%, UK 6.9%, US 4.4% (2017)

Exports—commodities: machinery, motor vehicles; foodstuffs, pharmaceuticals, medicines, other consumer goods

Imports: $338.6 billion (2017 est.)
$300.2 billion (2016 est.)
country comparison to the world: 15

Imports—commodities: machinery and equipment, fuels, chemicals, semi-finished goods, foodstuffs, consumer goods, measuring and medical control instruments

Imports—partners: Germany 14.2%, France 11.9%, China 6.9%, Italy 6.8%, Netherlands 5.1%, UK 4% (2017)

Reserves of foreign exchange and gold: $69.41 billion (31 December 2017 est.)
$63.14 billion (31 December 2016 est.)
country comparison to the world: 32

Debt—external: $2.094 trillion (31 December 2017 est.)
$1.963 trillion (31 March 2015 est.)
country comparison to the world: 10

Exchange rates: euros (EUR) per US dollar -
0.885 (2017 est.)
0.903 (2016 est.)
0.9214 (2015 est.)
0.7525 (2014 est.)
0.7634 (2013 est.)

ENERGY

Electricity access: *electrification—total population:* 100% (2020)

Electricity—production: 258.6 billion kWh (2016 est.)
country comparison to the world: 17

Electricity—consumption: 239.5 billion kWh (2016 est.)
country comparison to the world: 15

Electricity—exports: 14.18 billion kWh (2016 est.)
country comparison to the world: 13

Electricity—imports: 21.85 billion kWh (2016 est.)
country comparison to the world: 9

Electricity—installed generating capacity: 105.9 million kW (2016 est.)
country comparison to the world: 12

Electricity—from fossil fuels: 47% of total installed capacity (2016 est.)
country comparison to the world: 157

Electricity—from nuclear fuels: 7% of total installed capacity (2017 est.)
country comparison to the world: 19

Electricity—from hydroelectric plants: 14% of total installed capacity (2017 est.)
country comparison to the world: 108

Electricity—from other renewable sources: 32% of total installed capacity (2017 est.)
country comparison to the world: 15

Crude oil—production: 1,700 bbl/day (2018 est.)
country comparison to the world: 88

Crude oil—exports: 0 bbl/day (2017 est.)
country comparison to the world: 198

Crude oil—imports: 1.325 million bbl/day (2017 est.)
country comparison to the world: 8

Crude oil—proved reserves: 150 million bbl (1 January 2018 est.)
country comparison to the world: 62

Refined petroleum products—production: 1.361 million bbl/day (2017 est.)
country comparison to the world: 13

Refined petroleum products—consumption: 1.296 million bbl/day (2017 est.)
country comparison to the world: 18

Refined petroleum products—exports: 562,400 bbl/day (2017 est.)
country comparison to the world: 16

Refined petroleum products—imports: 464,800 bbl/day (2017 est.)
country comparison to the world: 18

Natural gas—production: 36.81 million cu m (2017 est.)
country comparison to the world: 87

Natural gas—consumption: 31.27 billion cu m (2017 est.)
country comparison to the world: 29

Natural gas—exports: 2.888 billion cu m (2017 est.)
country comparison to the world: 36

Natural gas—imports: 34.63 billion cu m (2017 est.)
country comparison to the world: 12

Natural gas—proved reserves: 2.548 billion cu m (1 January 2018 est.)
country comparison to the world: 96

Carbon dioxide emissions from consumption of energy: 286.7 million Mt (2017 est.)
country comparison to the world: 25

COMMUNICATIONS

Telephones—fixed lines: *total subscriptions:* 21,065,700
subscriptions per 100 inhabitants: 42.4 (2019 est.)
country comparison to the world: 12

Telephones—mobile cellular: *total subscriptions:* 58,750,448
subscriptions per 100 inhabitants: 118.25 (2019 est.)
country comparison to the world: 27

Telecommunication systems: *general assessment:* well-developed, one of the largest telecom markets in Europe, average mobile penetration for Europe; LTE universal; launch of 5G services; regulator has championed competition; Chinese company Huawei contributes to the telecom sector; fiber broadband accounts for 62% of all fixed-line broadband connections (2020)
domestic: fixed-line 42 per 100 and mobile-cellular 118 telephones per 100 persons (2019)
international: country code—34; landing points for the MAREA, Tata TGN-Western Europe, Pencan-9, SAT-3/WASC, Canalink, Atlantis-2, Columbus-111, Estepona-Tetouan, FEA, Balalink, ORVAL and PENBAL-5 submarine cables providing connectivity to Europe, the Middle East, Africa, South America, Asia, Southeast Asia and the US; satellite earth stations—2 Intelsat (1 Atlantic Ocean and 1 Indian Ocean), NA Eutelsat; tropospheric scatter to adjacent countries (2019)

note: the COVID-19 outbreak is negatively impacting telecommunications production and supply chains globally; consumer spending on telecom devices and services has also slowed due to the pandemic's effect on economies worldwide; overall progress towards improvements in all facets of the telecom industry—mobile, fixed-line, broadband, submarine cable and satellite—has moderated

Broadcast media: a mixture of both publicly operated and privately owned TV and radio stations; overall, hundreds of TV channels are available including national, regional, local, public, and international channels; satellite and cable TV systems available; multiple national radio networks, a large number of regional radio networks, and a larger number of local radio stations; overall, hundreds of radio stations

(2019)

Internet country code: .es

Internet users: *total:* 42,478,990
percent of population: 86.11% (July 2018 est.)
country comparison to the world: 20

Broadband—fixed subscriptions: *total:* 15,176,954
subscriptions per 100 inhabitants: 31 (2018 est.)
country comparison to the world: 13

TRANSPORTATION

National air transport system: *number of registered air carriers:* 21 (2020)
inventory of registered aircraft operated by air carriers: 552
annual passenger traffic on registered air carriers: 80,672,105 (2018)
annual freight traffic on registered air carriers: 1,117,070,000 mt-km (2018)

Civil aircraft registration country code prefix: EC (2016)

Airports: 135 (2020)
country comparison to the world: 40

Airports—with paved runways: *total:* 102 (2020)
over 3,047 m: 18
2,438 to 3,047 m: 16
1,524 to 2,437 m: 19
914 to 1,523 m: 26
under 914 m: 23

Airports—with unpaved runways: *total:* 33 (2020)
914 to 1,523 m: 14

under 914 m: 19

Heliports: 13 (2020)

Pipelines: 10481 km gas, 358 km oil, 4378 km refined products (2017)

Railways: *total:* 15,333 km (9,699 km electrified) (2017)
standard gauge: 2,571 km 1.435-m gauge (2,571 km electrified) (2017)
narrow gauge: 1,207 km 1.000-m gauge (400 km electrified) (2017)
broad gauge: 11,333 km 1.668-m gauge (6,538 km electrified) (2017)
mixed gauge: 190 km 1.668-m and 1.435m gage (190.1 km electrified); 28 km 0.914-m gauge (28 km electrified); 4 km 0.600-m gauge
country comparison to the world: 19

Roadways: *total:* 683,175 km (2011)
paved: 683,175 km (includes 16,205 km of expressways) (2011)
country comparison to the world: 11

Waterways: 1,000 km (2012)
country comparison to the world: 64

Merchant marine: *total:* 119
by type: container ship 2, general cargo 17, oil tanker 12, other 88 (2019)
country comparison to the world: 81

Ports and terminals: *major seaport(s):* Algeciras, Barcelona, Bilbao, Cartagena, Huelva, Tarragona, Valencia (all in Spain); Las Palmas, Santa Cruz de Tenerife (in the Canary Islands)
container port(s) (TEUs): Algeciras (4,389,836), Barcelona (2,968,757), Valencia (4,832,156) (2017)
LNG terminal(s) (import): Barcelona, Bilbao, Cartagena, Huelva, Mugardos, Sagunto

MILITARY AND SECURITY

Military and security forces: *Spanish Armed Forces:* Army (Ejercito de Tierra), Spanish Navy (Armada Espanola, AE, includes Marine Corps), Spanish Air Force (Ejercito del Aire Espanola, EdA); Civil Guard (Guardia Civil) (2019)
note: the Civil Guard is a military force with police duties (including coast guard) under both the Ministry of Defence and the Ministry of the Interior; it also responds to the needs of the Ministry of Finance

Military expenditures: 0.92% of GDP (2019 est.)
0.92% of GDP (2018)
0.9% of GDP (2017)
0.81% of GDP (2016)
0.92% of GDP (2015)
country comparison to the world: 125

Military and security service personnel strengths: the Spanish Armed Forces have approximately 120,000 active duty troops (70,000 Army; 20,000 Navy; 20,000 Air Force; 10,000 other/joint); 70-75,000 Guardia Civil (2019 est.)

Military equipment inventories and acquisitions: the inventory of the Spanish military is comprised of domestically-produced and imported Western weapons systems; France, Germany, and the US are the leading suppliers of military hardware since 2010; Spain's defense industry manufactures land, air, and sea weapons systems and is integrated within the European defense-industrial sector (2019 est.)

Military deployments: 350 Latvia (NATO); 630 Lebanon (UNIFIL); 180 Mali (EUTM); 150 Turkey (NATO) (2020)

Military service age and obligation: 18-26 years of age for voluntary military service by a Spanish citizen or legal immigrant, 2-3 year obligation; women allowed to serve in all SAF branches, including combat units; no conscription, but Spanish Government retains right to mobilize citizens 19-25 years of age in a national emergency; mandatory retirement of non-NCO enlisted personnel at age 45 or 58, depending on service length (2013)

TERRORISM

Terrorist group(s): Basque Fatherland and Liberty (disbanded); Islamic State of Iraq and ash-Sham; al-Qa'ida (2019)
note: details about the history, aims, leadership, organization, areas of operation, tactics, targets, weapons, size, and sources of support of the group(s) appear(s) in Appendix-T

TRANSNATIONAL ISSUES

Disputes—international: in 2002, Gibraltar residents voted overwhelmingly by referendum to reject any "shared sovereignty" arrangement; the Government of Gibraltar insists on equal participation in talks between the UK and Spain; Spain disapproves of UK plans to grant Gibraltar greater autonomy; after voters in the UK chose to leave the EU in a June 2016 referendum, Spain again proposed shared sovereignty of Gibraltar; UK officials rejected Spain's joint sovereignty proposal; Morocco protests Spain's control over the coastal enclaves of Ceuta, Melilla, and the islands of Penon de Velez de la Gomera, Penon de Alhucemas, and Islas Chafarinas, and surrounding waters; both countries claim Isla Perejil (Leila Island); Morocco serves as the primary launching site of illegal migration into Spain from North Africa; Portugal does not recognize Spanish sovereignty over the territory of Olivenza based on a difference of interpretation of the 1815 Congress of Vienna and the 1801 Treaty of Badajoz

Refugees and internally displaced persons: *refugees (country of origin):* 14,133 (Syria) (2019); 16,540 (Venezuela) (2020) (economic and political crisis; includes Venezuelans who have claimed asylum, are recognized as refugees, or have received alternative legal stay) (2020)
stateless persons: 4,246 (2019)
note: 184,338 estimated refugee and migrant arrivals (January 2015-October 2020); 65,325 migrant arrivals in 2018

Illicit drugs: despite rigorous law enforcement efforts, North African, Latin American, Galician, and other European traffickers take advantage of Spain's long coastline to land large shipments of cocaine and hashish for distribution to the European market; consumer for Latin American cocaine and North African hashish; destination and minor transshipment point for Southwest Asian heroin; money-laundering site for Colombian narcotics trafficking organizations and organized crime

SPRATLY ISLANDS

INTRODUCTION

Background: The Spratly Islands consist of more than 100 small islands or reefs surrounded by rich fishing grounds—and potentially by gas and oil deposits. They are claimed in their entirety by China, Taiwan, and Vietnam, while portions are claimed by Malaysia and the Philippines. About 45 islands are occupied by relatively small numbers of military forces from China, Malaysia, the Philippines, Taiwan, and Vietnam. Since 1985 Brunei has claimed a continental shelf that overlaps a southern reef but has not made any formal claim to the reef. Brunei claims an exclusive economic zone over this area.

GEOGRAPHY

Location: Southeastern Asia, group of reefs and islands in the South China Sea, about two-thirds of the way from southern Vietnam to the southern Philippines

Geographic coordinates: 8 38 N, 111 55 E

Map references: Southeast Asia

Area: *total:* 5 sq km less than
land: 5 sq km less than
water: 0 sq km
note: includes 100 or so islets, coral reefs, and sea mounts scattered over an area of nearly 410,000 sq km (158,000 sq mi) of the central South China Sea
country comparison to the world: 251

Area—comparative: land area is about seven times the size of the National Mall in Washington, DC

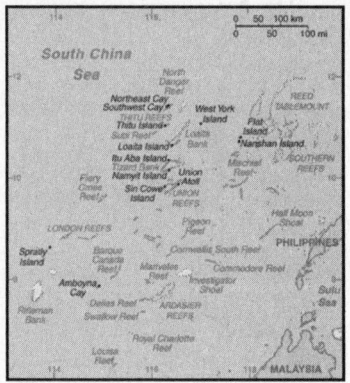

Land boundaries: 0 km

Coastline: 926 km

Maritime claims: NA

Climate: tropical

Terrain: small, flat islands, islets, cays, and reefs

Elevation: *lowest point:* South China Sea 0 m
highest point: unnamed location on Southwest Cay 6 m

Natural resources: fish, guano, undetermined oil and natural gas potential

Land use: *agricultural land:* 0% (2011 est.)
arable land: 0% (2011 est.) / permanent crops: 0% (2011 est.) / permanent pasture: 0% (2011 est.)
forest: 0% (2011 est.)
other: 100% (2011 est.)

Natural hazards: typhoons; numerous reefs and shoals pose a serious maritime hazard

Environment—current issues: China's use of dredged sand and coral to build artificial islands harms reef systems; illegal fishing practices indiscriminately harvest endangered species, including sea turtles and giant clams

Geography—note: strategically located near several primary shipping lanes in the central South China Sea; includes numerous small islands, atolls, shoals, and coral reefs

PEOPLE AND SOCIETY

Population: no indigenous inhabitants

---

note: there are scattered garrisons occupied by military personnel of several claimant states

GOVERNMENT

Country name: *conventional long form:* none
conventional short form: Spratly Islands
etymology: named after a British whaling captain Richard SPRATLY, who sighted Spratly Island in 1843; the name of the island eventually passed to the entire archipelago

ECONOMY

Economy—overview: Economic activity is limited to commercial fishing. The proximity to nearby oil- and gas-producing sedimentary basins indicate potential oil and gas deposits, but the region is largely unexplored. No reliable estimates of potential reserves are available. Commercial exploitation has yet to be developed.

TRANSPORTATION

Airports: 8 (2020)
country comparison to the world: 162
Airports—with paved runways: *total:* 6 (2020)
2,438 to 3,047 m: 3
914 to 1,523 m: 2
under 914 m: 1

Airports—with unpaved runways: *total:* 2 (2020)
914 to 1,523 m: 2

Heliports: 5 (2020)

Ports and terminals: none; offshore anchorage only

MILITARY AND SECURITY

Military—note: Spratly Islands consist of more than 100 small islands or reefs of which about 45 are claimed and occupied by China, Malaysia, the Philippines, Taiwan, and Vietnam
China: assessed to have 7 outposts (Fiery Cross, Mischief, Subi, Cuarteron, Gavin, Hughes, and Johnson reefs); the outposts on Fiery Cross, Mischief, and Subi include air bases with helipads and dozens of fighter jet hangers, naval port facilities, surveillance radars, air defense sites, anti-ship cruise missiles, and other military infrastructure such as communications, barracks, maintenance facilities, and ammunition and fuel bunkers

---

Malaysia: assessed to have 5 outposts in the southern portion of the archipelago, closest to the Malaysian state of Sabah (Ardasier Reef, Eric Reef, Mariveles Reef, Shallow Reef, and Investigator Shoal); all the outposts have helicopter landing pads, while Shallow Reef also has an airstrip
Philippines: assessed to occupy 9 features (Commodore Reef, Second Thomas Shoal, Flat Island, Loaita Cay, Loaita Island, Nanshan Island, Northeast Cay, Thitu Island, and West York Island); Thitu Island has the only Philippine airstrip in the Spratlys
Taiwan: maintains an outpost with an airstrip on Itu Aba Island
Vietnam: assessed to occupy about 49 outposts spread across 27 features, including facilities on 21 rocks and reefs in the Spratlys, plus 14 platforms known as "economic, scientific, and technological service stations," or Dich vu-Khoa (DK1), on six underwater banks to the southeast that Vietnam does not consider part of the disputed island chain, although China and Taiwan disagree; Spratly Islands outposts are on Alison Reef, Amboyna Cay, Barque Canada Reef, Central Reef, Collins Reef, Cornwallis South Reef, Discovery Great Reef, East Reef, Grierson Reef, Ladd Reef, Landsdowne Reef, Namyit Island, Pearson Reef, Petley Reef, Sand Cay, Sin Cowe Island, South Reef, Southwest Cay, Spratly Island, Tennent Reef, West Reef; Spratly Island includes an airstrip with aircraft hangers; the six underwater banks with outposts include Vanguard, Rifleman, Prince of Wales, Prince Consort, Grainger, and Alexandra

(2020)

TRANSNATIONAL ISSUES

Disputes—international: all of the Spratly Islands are claimed by China (including Taiwan) and Vietnam; parts of them are claimed by Brunei, Malaysia and the Philippines; despite no public territorial claim to Louisa Reef, Brunei implicitly lays claim by including it within the natural prolongation of its continental shelf and basis for a seabed median with Vietnam; claimants in November 2002 signed the "Declaration on the Conduct of Parties in the South China Sea," which has eased tensions but falls short of a legally binding "code of conduct"; in March 2005, the national oil companies of China, the Philippines, and Vietnam signed a joint accord to conduct marine seismic activities in the Spratly Islands

---

SRI LANKA

INTRODUCTION

Background: The first Sinhalese arrived in Sri Lanka late in the 6th century B.C., probably from northern India. Buddhism was introduced circa 250 B.C., and the first kingdoms developed at the cities of Anuradhapura (from circa 200 B.C. to circa A.D. 1000) and Polonnaruwa (from about 1070 to 1200). In the 14th century, a south Indian dynasty established a Tamil kingdom in northern Sri Lanka. The Portuguese controlled the coastal areas of the island in the 16th century followed by the Dutch in the 17th century. The island was ceded to the British in 1796, became a crown colony in 1802, and was formally united under British rule by 1815. As Ceylon, it became independent in 1948; its name was changed to Sri Lanka in 1972. Prevailing tensions between the Sinhalese majority and Tamil separatists erupted into war in July 1983. Fighting between the government and Liberation Tigers of Tamil Eelam (LTTE) continued for over a quarter century. Although Norway brokered peace negotiations that led to a ceasefire in 2002, the fighting slowly resumed and was again

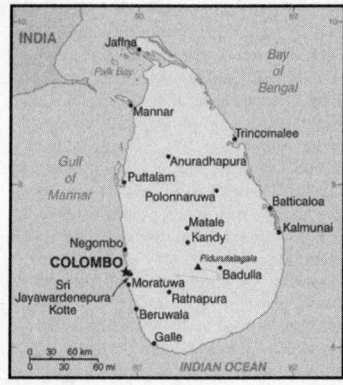

in full force by 2006. The government defeated the LTTE in May 2009.

During the post-conflict years under President Mahinda RAJAPAKSA, the government initiated infrastructure development projects, many of which were financed by loans from China. His regime faced significant allegations of human rights violations and a shrinking democratic space for civil society. In 2015, a new coalition government headed by President Maithripala SIRISENA of the Sri Lanka Freedom Party and Prime Minister Ranil WICKREMESINGHE of the United National Party came to power with pledges to advance economic, governance, anti-corruption, reconciliation, justice, and accountability reforms. However, implementation of these reforms has been uneven. In October 2018, President SIRISENA attempted to oust Prime Minister WICKREMESINGHE, swearing in former President RAJAPAKSA as the new prime minister and issuing an order to dissolve the parliament and hold elections. This sparked a seven-week constitutional crisis that ended when the Supreme Court ruled SIRISENA's actions unconstitutional, RAJAPAKSA resigned, and WICKREMESINGHE was reinstated. In November 2019, Gotabaya RAJAPAKSA won the presidential election and appointed his brother, Mahinda, prime minister.

GEOGRAPHY

Location: Southern Asia, island in the Indian Ocean, south of India

Geographic coordinates: 7 00 N, 81 00 E

Map references: Asia

Area: *total:* 65,610 sq km
land: 64,630 sq km
water: 980 sq km
country comparison to the world: 123

Area—comparative: slightly larger than West Virginia

Land boundaries: 0 km

Coastline: 1,340 km

Maritime claims: *territorial sea:* 12 nm
exclusive economic zone: 200 nm
contiguous zone: 24 nm

continental shelf: 200 nm or to the edge of the continental margin

Climate: tropical monsoon; northeast monsoon (December to March); southwest monsoon (June to October)

Terrain: mostly low, flat to rolling plain; mountains in south-central interior

Elevation: *mean elevation:* 228 m
lowest point: Indian Ocean 0 m
highest point: Pidurutalagala 2,524 m

Natural resources: limestone, graphite, mineral sands, gems, phosphates, clay, hydropower, arable land

Land use: *agricultural land:* 43.5% (2011 est.)
arable land: 20.7% (2011 est.) / permanent crops: 15.8% (2011 est.) / permanent pasture: 7% (2011 est.)
forest: 29.4% (2011 est.)
other: 27.1% (2011 est.)
Irrigated land: 5,700 sq km (2012)

Population distribution: the population is primarily concentrated within a broad wet zone in the southwest, urban centers along the eastern coast, and on the Jaffna Peninsula in the north

Natural hazards: occasional cyclones and tornadoes

Environment—current issues: deforestation; soil erosion; wildlife populations threatened by poaching and urbanization; coastal degradation from mining activities and increased pollution; coral reef destruction; freshwater resources being polluted by industrial wastes and sewage runoff; waste disposal; air pollution in Colombo

Environment—international agreements: *party to:* Biodiversity, Climate Change, Climate Change-Kyoto Protocol, Desertification, Endangered Species, Environmental Modification, Hazardous Wastes, Law of the Sea, Ozone Layer Protection, Ship Pollution, Wetlands
signed, but not ratified: Marine Life Conservation

Geography—note: strategic location near major Indian Ocean sea lanes; Adam's Bridge is a chain of limestone shoals between the southeastern coast of India and the northwestern coast of Sri Lanka; geological evidence suggests that this 50-km long Bridge once connected India and Sri Lanka; ancient records seem to indicate that a foot passage was possible between the two land masses until the 15th century when the land bridge broke up in a cyclone

PEOPLE AND SOCIETY

Population: 22,889,201 (July 2020 est.)
country comparison to the world: 57

Nationality: *noun:* Sri Lankan(s)
adjective: Sri Lankan

Ethnic groups: Sinhalese 74.9%, Sri Lankan Tamil 11.2%, Sri Lankan Moors 9.2%, Indian Tamil 4.2%, other 0.5% (2012 est.)

Languages: Sinhala (official and national language) 87%, Tamil (official and national language) 28.5%, English 23.8% (2012 est.)

note: data represent main languages spoken by the population aged 10 years and older; shares sum to more than 100% because some respondents gave more than one answer on the census; English is commonly used in government and is referred to as the "link language" in the constitution

Religions: Buddhist (official) 70.2%, Hindu 12.6%, Muslim 9.7%, Roman Catholic 6.1%, other Christian 1.3%, other 0.05% (2012 est.)

Age structure: *0-14 years:* 23.11% (male 2,696,379/female 2,592,450)
15-24 years: 14.58% (male 1,700,442/female 1,636,401)
25-54 years: 41.2% (male 4,641,842/female 4,789,101)
55-64 years: 10.48% (male 1,110,481/female 1,288,056)
65 years and over: 10.63% (male 1,023,315/female 1,410,734) (2020 est.)

Dependency ratios: *total dependency ratio:* 53.7
youth dependency ratio: 36.4
elderly dependency ratio: 17.3
potential support ratio: 5.8 (2020 est.)

Median age: *total:* 33.7 years
male: 32.3 years
female: 35.1 years (2020 est.)
country comparison to the world: 97

Population growth rate: 0.67% (2020 est.)
country comparison to the world: 143

Birth rate: 14.2 births/1,000 population (2020 est.)
country comparison to the world: 132

Death rate: 6.5 deaths/1,000 population (2020 est.)
country comparison to the world: 143

Net migration rate: -1.3 migrant(s)/1,000 population (2020 est.)
country comparison to the world: 150

Population distribution: the population is primarily concentrated within a broad wet zone in the southwest, urban centers along the eastern coast, and on the Jaffna Peninsula in the north

Urbanization: *urban population:* 18.7% of total population (2020)
rate of urbanization: 0.85% annual rate of change (2015-20 est.)

total population growth rate v. urban population growth rate, 2000-2030: *Major urban areas—population:* 103,000 Sri Jayewardenepura Kotte (legislative capital) (2018), 613,000 COLOMBO (capital) (2020)

Sex ratio: *at birth:* 1.04 male(s)/female
0-14 years: 1.04 male(s)/female
15-24 years: 1.04 male(s)/female
25-54 years: 0.97 male(s)/female
55-64 years: 0.86 male(s)/female
65 years and over: 0.73 male(s)/female
total population: 0.95 male(s)/female (2020 est.)

Mother's mean age at first birth: 25.6 years (2016 est.)
note: median age at first birth among women 30-34

Maternal mortality rate: 36 deaths/100,000 live births (2017 est.)

country comparison to the world: 106

Infant mortality rate: *total:* 7.8 deaths/1,000 live births
male: 8.8 deaths/1,000 live births
female: 6.9 deaths/1,000 live births (2020 est.)
country comparison to the world: 151

Life expectancy at birth: *total population:* 77.5 years
male: 74 years
female: 81.1 years (2020 est.)
country comparison to the world: 81

Total fertility rate: 2.01 children born/woman (2020 est.)
country comparison to the world: 112

Contraceptive prevalence rate: 61.7% (2016)

Drinking water source:
improved:
urban: 98.1% of population
rural: 91.4% of population
total: 92.6% of population
unimproved:
urban: 1.9% of population
rural: 8.6% of population
total: 7.4% of population (2017 est.)

Current Health Expenditure: 3.8% (2017)

Physicians density: 0.93 physicians/1,000 population (2017)

Hospital bed density: 4.2 beds/1,000 population (2017)

Sanitation facility access:
improved:
urban: 97.1% of population
rural: 99.3% of population
total: 98.9% of population
unimproved:
urban: 2.9% of population
rural: 0.7% of population
total: 1.1% of population (2017 est.)

HIV/AIDS—adult prevalence rate: <.1% (2019 est.)

HIV/AIDS—people living with HIV/AIDS: 3,600 (2019 est.)
country comparison to the world: 127

HIV/AIDS—deaths: <200 (2019 est.)

Major infectious diseases: *degree of risk:* intermediate
(2020)
vectorborne diseases: dengue fever
water contact diseases: leptospirosis
animal contact diseases: rabies

Obesity—adult prevalence rate: 5.2% (2016)
country comparison to the world: 182

Children under the age of 5 years underweight: 20.5% (2016)
country comparison to the world: 23

Education expenditures: 2.8% of GDP (2017)
country comparison to the world: 148

Literacy: *definition:* age 15 and over can read and write
total population: 91.9%
male: 93%
female: 91% (2017)

School life expectancy (primary to tertiary education): *total:* 14 years
male: 14 years
female: 15 years (2018)

Unemployment, youth ages 15-24: *total:* 21%
male: 16.8%
female: 28.4% (2016 est.)
country comparison to the world: 60

GOVERNMENT

Country name: *conventional long form:* Democratic Socialist Republic of Sri Lanka
conventional short form: Sri Lanka
local long form: Shri Lanka Prajatantrika Samajavadi Janarajaya/Ilankai Jananayaka Choshalichak Kutiyarachu
local short form: Shri Lanka/Ilankai
former: Serendib, Ceylon
etymology: the name means "resplendent island" in Sanskrit

Government type: presidential republic

Capital: *name:* Colombo (commercial capital); Sri Jayewardenepura Kotte (legislative capital)
geographic coordinates: 6 55 N, 79 50 E
time difference: UTC+5.5 (10.5 hours ahead of Washington, DC, during Standard Time)
etymology: Colombo may derive from the Sinhala "kolon thota," meaning "port on the river" (referring to the Kelani River that empties into the Indian Ocean at Colombo); alternatively, the name may derive from the Sinhala "kola amba thota" meaning "harbor with mango trees"; it is also possible that the Portuguese named the city after Christopher COLUMBUS, who lived in Portugal for many years (as Cristovao COLOMBO) before discovering the Americas for the Spanish crown in 1492—not long before the Portuguese made their way to Sri Lanka in 1505; Sri Jayewardenepura Kotte translates as "Resplendent City of Growing Victory" in Sinhala

Administrative divisions: 9 provinces; Central, Eastern, North Central, Northern, North Western, Sabaragamuwa, Southern, Uva, Western

Independence: 4 February 1948 (from the UK)

National holiday: Independence Day (National Day), 4 February (1948)

Constitution: *history:* several previous; latest adopted 16 August 1978, certified 31 August 1978
amendments: proposed by Parliament; passage requires at least two-thirds majority vote of its total membership, certification by the president of the republic or the Parliament speaker, and in some cases approval in a referendum by absolute majority of valid votes; amended many times, last in 2020

Legal system: mixed legal system of Roman-Dutch civil law, English common law, Jaffna Tamil customary law, and Muslim personal law

International law organization participation: has not submitted an ICJ jurisdiction declaration; non-party state to the ICCt

Citizenship: *citizenship by birth:* no

citizenship by descent only: at least one parent must be a citizen of Sri Lanka
dual citizenship recognized: no, except in cases where the government rules it is to the benefit of Sri Lanka
residency requirement for naturalization: 7 years

Suffrage: 18 years of age; universal

Executive branch: *chief of state:* President Gotabaya RAJAPAKSA (since 18 November 2019); note—the president is both chief of state and head of government; Prime Minister Mahinda RAJAPAKSA (since 21 November 2019)
head of government: President Gotabaya RAJAPAKSA (since 18 November 2019)
cabinet: Cabinet appointed by the president in consultation with the prime minister
elections/appointments: president directly elected by preferential majority popular vote for a 5-year term (eligible for a second term); election last held on 16 November 2019 (next to be held in 2024); prime minister appointed by the president from among members of Parliament for a 5-year term)
election results: Gotabaya RAJAPAKSA elected president; percent of vote—Gotabaya RAJAPAKSA (SLPP) 52.2%, Sajith PREMADASA (UNP) 42%, other 5.8%

Legislative branch: *description:* unicameral Parliament (225 seats; 196 members directly elected in multi-seat constituencies by proportional representation vote using a preferential method in which voters select 3 candidates in order of preference; remaining 29 seats allocated to other political parties and groups in proportion to share of national vote; members serve 5-year terms)
elections: last held on 17 August 2015 (next originally scheduled for 25 April 2020 but postponed to due to the COVID-19 pandemic)
election results: percent of vote by coalition/party—SLFPA 59.1%, SJB 23.9%, JVP 3.8%, TNA 2.8%, UNP 2.2%, TNPF 0.6%, EPDP 0.5%, other 7.1%; seats by coalition/party—SLFPA 145, SJB 54, TNA 10, JVP 3, other 13; composition—NA

Judicial branch: *highest courts:* Supreme Court of the Republic (consists of the chief justice and 9 justices); note—the court has exclusive jurisdiction to review legislation
judge selection and term of office: chief justice nominated by the Constitutional Council (CC), a 9-member high-level advisory body, and appointed by the president; other justices nominated by the CC and appointed by the president on the advice of the chief justice; all justices can serve until age 65
subordinate courts: Court of Appeals; High Courts; Magistrates' Courts; municipal and primary courts

Political parties and leaders: Crusaders for Democracy [Ganeshalingam CHANDRALINGAM]

Eelam People's Democratic Party or EPDP [Douglas DEVANANDA]
Eelam People's Revolutionary Liberation Front [Suresh PREMACHANDRAN]

Janatha Vimukthi Peramuna or JVP [Anura Kumara DISSANAYAKE]

Jathika Hela Urumaya or JHU [Karunarathna PARANAWITHANA, Ven. Hadigalle Wimalasara THERO]

National Peoples Power or JVP [Anura Kumara DISSANAYAKE]

Samagi Jana Balawegaya or SJB [Sajith PREMADASA]

Sri Lanka Freedom Party or SLFP [Maithripala SIRISENA]

Sri Lanka Muslim Congress or SLMC [Rauff HAKEEM]

Sri Lanka People's Freedom Alliance [Mahinda RAJAPAKSA]

Sri Lanka Podujana Peramuna or SLPP [G. L. PEIRIS]

Tamil National Alliance or TNA [Rajavarothiam SAMPANTHAN] (alliance includes Illankai Tamil Arasu Kachchi [Mavai SENATHIRAJAH], People's Liberation Organisation of Tamil Eelam [D. SIDDARTHAN], Tamil Eelam Liberation Organization [Selvam ADAIKALANATHAN])

Tamil National People's Front [Gajendrakumar PONNAMBALAM]

United National Front for Good Governance or UNFGG [Ranil WICKREMESINGHE] (coalition includes JHU, UNP)

United National Party or UNP [Ranil WICKREMESINGHE]

United People's Freedom Alliance or UPFA [Maithripala SIRISENA] (coalition includes SLFP)

International organization participation: ABEDA, ADB, ARF, BIMSTEC, C, CD, CICA (observer), CP, EMO, G 11, G-15, G-24, G-77, IAEA, IBRD, ICAO, ICC (national committees), ICRM, IDA, IFAD, IFC, IFRCS, IHO, ILO, IMF, IMO, IMSO, Interpol, IOC, IOM, IPU, ISO, ITSO, ITU, ITUC (NGOs), MIGA, MINURSO, MINUSTAH, MONUSCO, NAM, OAS (observer), OPCW, PCA, SAARC, SACEP, SCO (dialogue member), UN, UNCTAD, UNESCO, UNIDO, UNIFIL, UNISFA, UNMISS, UNWTO, UPU, WCO, WFTU (NGOs), WHO, WIPO, WMO, WTO

Diplomatic representation in the US: *chief of mission:* Ambassador E. Rodney M. PERERA (since 8 July 2019)
chancery: 3025 Whitehaven Street NW, Washington, DC 20008
telephone: [1] (202) 483-4025 through 4028
FAX: [1] (202) 232-7181
consulate(s) general: Los Angeles, New York

Diplomatic representation from the US: *chief of mission:* Ambassador Alaina B. TEPLITZ (since 1 November 2018); note—also accredited to Maldives
telephone: [94] (11) 249-8500
embassy: 210 Galle Road, Colombo 03
mailing address: P. O. Box 106, Colombo
FAX: [94] (11) 243-7345

Flag description: yellow with two panels; the smaller hoist-side panel has two equal vertical bands of green (hoist side) and orange; the other larger panel depicts a yellow lion holding a sword

on a maroon rectangular field that also displays a yellow bo leaf in each corner; the yellow field appears as a border around the entire flag and extends between the two panels; the lion represents Sinhalese ethnicity, the strength of the nation, and bravery; the sword demonstrates the sovereignty of the nation; the four bo leaves—symbolizing Buddhism and its influence on the country—stand for the four virtues of kindness, friendliness, happiness, and equanimity; orange signifies Sri Lankan Tamils, green Sri Lankan Moors, and maroon the Sinhalese majority; yellow denotes other ethnic groups; also referred to as the Lion Flag

National symbol(s): *lion, water lily; national colors:* maroon, yellow

National anthem: *name:* "Sri Lanka Matha" (Mother Sri Lanka)
lyrics/music: Ananda SAMARKONE
note: adopted 1951
0:00 / 2:31

ECONOMY

Economy—overview: Sri Lanka is attempting to sustain economic growth while maintaining macroeconomic stability under the IMF program it began in 2016. The government's high debt payments and bloated civil service, which have contributed to historically high budget deficits, remain a concern. Government debt is about 79% of GDP and remains among the highest of the emerging markets. In the coming years, Sri Lanka will need to balance its elevated debt repayment schedule with its need to maintain adequate foreign exchange reserves.

In May 2016, Sri Lanka regained its preferential trade status under the European Union's Generalized System of Preferences Plus, enabling many of its firms to export products, including its top export garments, tax free to the EU. In 2017, Parliament passed a new Inland Revenue Act in an effort to increase tax collection and broaden the tax base in response to recommendations made under its IMF program. In November 2017, the Financial Action Task Force on money laundering and terrorist financing listed Sri Lanka as non-compliant, but reported subsequently that Sri Lanka had made good progress in implementing an action plan to address deficiencies.

Tourism has experienced strong growth in the years since the resolution of the government's 26-year conflict with the Liberation Tigers of Tamil Eelam. In 2017, the government promulgated plans to transform the country into a knowledgebased, export-oriented Indian Ocean hub by 2025.

GDP (purchasing power parity): $275.8 billion (2017 est.)
$267 billion (2016 est.)
$255.6 billion (2015 est.)
note: data are in 2017 dollars
country comparison to the world: 61

GDP (official exchange rate): $87.35 billion (2017 est.)

GDP—real growth rate: 2.29% (2019 est.)
3.32% (2018 est.)
3.58% (2017 est.)
country comparison to the world: 123

GDP—per capita (PPP): $12,900 (2017 est.)
$12,600 (2016 est.)
$12,200 (2015 est.)
note: data are in 2017 dollars
country comparison to the world: 121

Gross national saving: 33.8% of GDP (2017 est.)
32.8% of GDP (2016 est.)
28.8% of GDP (2015 est.)
country comparison to the world: 20

GDP—composition, by end use:
household consumption: 62% (2017 est.)
government consumption: 8.5% (2017 est.)
investment in fixed capital: 26.3% (2017 est.)
investment in inventories: 10.2% (2017 est.)
exports of goods and services: 21.9% (2017 est.)
imports of goods and services: -29.1% (2017 est.)

GDP—composition, by sector of origin:
agriculture: 7.8% (2017 est.)
industry: 30.5% (2017 est.)
services: 61.7% (2017 est.)

Agriculture—products: rice, sugarcane, grains, pulses, oilseed, spices, vegetables, fruit, tea, rubber, coconuts; milk, eggs, hides, beef; fish

Industries: processing of rubber, tea, coconuts, tobacco and other agricultural commodities; telecommunications, insurance, banking; tourism, shipping; clothing, textiles; cement, petroleum refining, information technology services, construction

Industrial production growth rate: 4.6% (2017 est.)
country comparison to the world: 64

Labor force: 8 million (2020 est.)
country comparison to the world: 60

Labor force—by occupation: *agriculture:* 27%
industry: 26%
services: 47% (31 December 2016)

Unemployment rate: 4.83% (2019 est.)
4.44% (2018 est.)
country comparison to the world: 71

Population below poverty line: 6.7% (2012 est.)

Household income or consumption by percentage share: *lowest 10%:* 3%
highest 10%: 32.2% (2012 est.)

Budget: *revenues:* 12.07 billion (2017 est.)
expenditures: 16.88 billion (2017 est.)

Taxes and other revenues: 13.8% (of GDP) (2017 est.)
country comparison to the world: 204

Budget surplus (+) or deficit (-): -5.5% (of GDP) (2017 est.)
country comparison to the world: 174

Public debt: 79.1% of GDP (2017 est.)
79.6% of GDP (2016 est.)
note: covers central government debt and excludes debt instruments directly owned by government entities other than the treasury (e.g. commercial bank borrowings of a government corporation); the data includes treasury debt held by foreign

entities as well as intragovernmental debt; intragovernmental debt consists of treasury borrowings from surpluses in the social funds, such as for retirement; sub-national entities are usually not permitted to sell debt instruments
country comparison to the world: 35

Fiscal year: calendar year

Inflation rate (consumer prices): 6.5% (2017 est.)
4% (2016 est.)
country comparison to the world: 190

Current account balance: -$10 million (2019 est.)
-$17 million (2018 est.)
country comparison to the world: 69

Exports: $11.36 billion (2017 est.)
$10.31 billion (2016 est.)
country comparison to the world: 87

Exports—partners: US 24.6%, UK 9%, India 5.8%, Singapore 4.5%, Germany 4.3%, Italy 4.3% (2017)

Exports—commodities: textiles and apparel, tea and spices; rubber manufactures; precious stones; coconut products, fish

Imports: $20.98 billion (2017 est.)
$19.18 billion (2016 est.)
country comparison to the world: 73

Imports—commodities: petroleum, textiles, machinery and transportation equipment, building materials, mineral products, foodstuffs

Imports—partners: India 22%, China 19.9%, Singapore 6.9%, UAE 5.7%, Japan 4.9% (2017)

Reserves of foreign exchange and gold: $7.959 billion (31 December 2017 est.)
$6.019 billion (31 December 2016 est.)
country comparison to the world: 78

Debt—external: $51.72 billion (31 December 2017 est.)
$45.26 billion (31 December 2016 est.)
country comparison to the world: 64

Exchange rates: Sri Lankan rupees (LKR) per US dollar -
154.1 (2017 est.)
145.58 (2016 est.)
145.58 (2015 est.)
135.86 (2014 est.)
130.57 (2013 est.)

ENERGY

Electricity access: *electrification—total population:* 100% (2019)

Electricity—production: 13.66 billion kWh (2016 est.)
country comparison to the world: 90

Electricity—consumption: 12.67 billion kWh (2016 est.)
country comparison to the world: 86

Electricity—exports: 0 kWh (2016 est.)
country comparison to the world: 202

Electricity—imports: 0 kWh (2016 est.)
country comparison to the world: 204

Electricity—installed generating capacity: 3.998 million kW (2016 est.)

country comparison to the world: 89

Electricity—from fossil fuels: 52% of total installed capacity (2016 est.)
country comparison to the world: 146

Electricity—from nuclear fuels: 0% of total installed capacity (2017 est.)
country comparison to the world: 187

Electricity—from hydroelectric plants: 42% of total installed capacity (2017 est.)
country comparison to the world: 49

Electricity—from other renewable sources: 6% of total installed capacity (2017 est.)
country comparison to the world: 103

Crude oil—production: 0 bbl/day (2018 est.)
country comparison to the world: 203

Crude oil—exports: 0 bbl/day (2015 est.)
country comparison to the world: 199

Crude oil—imports: 33,540 bbl/day (2015 est.)
country comparison to the world: 60

Crude oil—proved reserves: 0 bbl (1 January 2018 est.)
country comparison to the world: 199

Refined petroleum products—production: 34,210 bbl/day (2017 est.)
country comparison to the world: 84

Refined petroleum products—consumption: 116,000 bbl/day (2016 est.)
country comparison to the world: 74

Refined petroleum products—exports: 3,871 bbl/day (2015 est.)
country comparison to the world: 96

Refined petroleum products—imports: 66,280 bbl/day (2015 est.)
country comparison to the world: 70

Natural gas—production: 0 cu m (2017 est.)
country comparison to the world: 200

Natural gas—consumption: 0 cu m (2017 est.)
country comparison to the world: 201

Natural gas—exports: 0 cu m (2017 est.)
country comparison to the world: 189

Natural gas—imports: 0 cu m (2017 est.)
country comparison to the world: 192

Natural gas—proved reserves: 0 cu m (1 January 2014 est.)
country comparison to the world: 197

Carbon dioxide emissions from consumption of energy: 25.19 million Mt (2017 est.)
country comparison to the world: 80

COMMUNICATIONS

Telephones—fixed lines: *total subscriptions:* 2,641,982
subscriptions per 100 inhabitants: 11.62 (2019 est.)
country comparison to the world: 47

Telephones—mobile cellular: *total subscriptions:* 26,160,623
subscriptions per 100 inhabitants: 115.06 (2019 est.)
country comparison to the world: 49

Telecommunication systems: *general assessment:* telephone services have improved significantly; strong growth anticipated as Sri Lanka is lagging behind other Asian telecoms; increase in mobile broadband penetration; govt. funds telecom sector to expand fiber and LTE networks and growing investment in 5G services (2020)
domestic: fixed-line 12 per 100 and mobile-cellular 115 per 100; national trunk network consists of digital microwave radio relay and fiber-optic links; fixed wireless local loops have been installed; competition is strong in mobile cellular systems and mobile cellular subscribership is increasing (2019)
international: country code—94; landing points for the SeaMeWe -3,-5, Dhiraagu-SLT Submarine Cable Network, WARF Submarine Cable, Bharat Lanka Cable System and the Bay of Bengal Gateway submarine cables providing connectivity to Asia, Africa, Southeast Asia, Australia, the Middle East, and Europe; satellite earth stations—2 Intelsat (Indian Ocean) (2019)
note: the COVID-19 outbreak is negatively impacting telecommunications production and supply chains globally; consumer spending on telecom devices and services has also slowed due to the pandemic's effect on economies worldwide; overall progress towards improvements in all facets of the telecom industry—mobile, fixed-line, broadband, submarine cable and satellite—has moderated

Broadcast media: government operates 5 TV channels and 19 radio channels; multi-channel satellite and cable TV subscription services available; 25 private TV stations and about 43 radio stations; 6 non-profit TV stations and 4 radio stations

Internet country code: .lk

Internet users: *total:* 7,700,876
percent of population: 34.11% (July 2018 est.)
country comparison to the world: 65

Broadband—fixed subscriptions: *total:* 1,544,313
subscriptions per 100 inhabitants: 7 (2018 est.)
country comparison to the world: 63

TRANSPORTATION

National air transport system: *number of registered air carriers:* 3 (2020)
inventory of registered aircraft operated by air carriers: 34
annual passenger traffic on registered air carriers: 5,882,376 (2018)
annual freight traffic on registered air carriers: 436.2 million mt-km (2018)

Civil aircraft registration country code prefix: 4R (2016)

Airports: 18 (2020)
country comparison to the world: 140

Airports—with paved runways: *total:* 11 (2020)
over 3,047 m: 2
1,524 to 2,437 m: 5
914 to 1,523 m: 4

Airports—with unpaved runways: *total:* 7 (2020)
1,524 to 2,437 m: 2
914 to 1,523 m: 3

under 914 m: 2

Heliports: 1 (2020)

Pipelines: 7 km refined products

Railways: *total:* 1,562 km (2016)
broad gauge: 1,562 km 1.676-m gauge (2016)
country comparison to the world: 82

Roadways: *total:* 114,093 km (2010)
paved: 16,977 km (2010)
unpaved: 97,116 km (2010)
country comparison to the world: 43

Waterways: 160 km (primarily on rivers in south-west) (2012)
country comparison to the world: 100

Merchant marine: *total:* 97
by type: bulk carrier 8, container ship 1, general cargo 17, oil tanker 13, other 58 (2019)
country comparison to the world: 92

Ports and terminals: *major seaport(s):* Colombo
container port(s) (TEUs): Colombo (6,209,000) (2017)

MILITARY AND SECURITY

Military and security forces: *Sri Lanka Army (includes National Guard and the Volunteer Force), Sri Lanka Navy (includes Marine Corps), Sri Lanka Air Force, Sri Lanka Coast Guard; Civil Security Department (Home Guard); Sri Lanka National Police:* Special Task Force (counter-terrorism and counter-insurgency) (2019)

Military expenditures: 1.9% of GDP (2019)
1.9% of GDP (2018)
2.1% of GDP (2017)
2.1% of GDP (2016)
2.6% of GDP (2015)
country comparison to the world: 56

Military and security service personnel strengths: the Sri Lankan military has approximately 250,000

total personnel (180,000 Army; 40,000 Navy; 30,000 Air Force) (2019)

Military equipment inventories and acquisitions: the Sri Lankan military inventory consists mostly of Chinese and Russian-origin equipment, as well as smaller amounts from Israel, the UK, and the US; since 2000, China, India, Israel, and the US have been the leading suppliers of arms to Sri Lanka (2019 est.)

Military deployments: 110 Central African Republic (MINUSCA); 140 Lebanon (UNIFIL); 240 Mali (MINUSMA); 170 South Sudan (UNMISS) (2019)

Military service age and obligation: 18-22 years of age for voluntary military service; no conscription (2019)

TERRORISM

Terrorist group(s): Islamic State of Iraq and ash-Sham; Liberation Tigers of Tamil Eelam (2019)
note: details about the history, aims, leadership, organization, areas of operation, tactics, targets, weapons, size, and sources of support of the group(s) appear(s) in Appendix-T

TRANSNATIONAL ISSUES

Disputes—international: none

Refugees and internally displaced persons: *IDPs:* 27,000 (civil war; more than half displaced prior to 2008; many of the more than 480,000 IDPs registered as returnees have not reached durable solutions) (2019)

Trafficking in persons: *current situation:* Sri Lanka is primarily a source and, to a lesser extent, a destination country for men, women, and children subjected to forced labor and sex trafficking; some Sri Lankan adults and children who migrate willingly to the Middle East, Southeast Asia, and Afghanistan to work in the construction, garment, and domestic service sectors are subsequently subjected to forced labor or debt bondage (incurred through high recruitment fees or money advances); some Sri Lankan women are forced into prostitution in Jordan, Maldives, Malaysia, Singapore, and other countries; within Sri Lanka, women and children are subjected to sex trafficking, and children are also forced to beg and work in the agriculture, fireworks, and fish-drying industries; a small number of women from Asia, Central Asia, Europe, and the Middle East have been forced into prostitution in Sri Lanka in recent years
tier rating: Tier 2 Watch List – Sri Lanka does not fully comply with the minimum standards for the elimination of trafficking; however, it is making significant efforts to do so; in 2014, Sri Lanka was granted a waiver from an otherwise required downgrade to Tier 3 because its government has a written plan that, if implemented, would constitute making significant efforts to bring itself into compliance with the minimum standards for the elimination of trafficking; law enforcement continues to demonstrate a lack of understanding of trafficking crimes and inadequate investigations, relying on trafficking cases to be prosecuted under the procurement statute rather than the trafficking statute, which carries more stringent penalties; authorities convicted only one offender under the procurement statute, a decrease from 2013; the government approved guidelines for the identification of victims and their referral to protective services but failed to ensure that victims were not jailed and charged for crimes committed as a direct result of being trafficked; no government employees were investigated or prosecuted, despite allegations of complicity (2015)

SUDAN

INTRODUCTION

Background: The region along the Nile River south of Egypt has long been referred to as Nubia. It was the site of the Kingdom of Kerma, which flourished for about a millennium (ca. 2500-1500 B.C.) until absorbed into the New Kingdom of Egypt. By the 11th century B.C., a Kingdom of Kush emerged and regained the region's independence from Egypt; it lasted in various forms until the middle of the fourth century A.D. After the fall of Kush, the Nubians formed three Christian kingdoms of Nobatia, Makuria, and Alodia, the latter two endured until around 1500. Between the 14th and 15th centuries much of Sudan was settled by Arab nomads, and between the 16th–19th centuries it underwent extensive Islamization.

Egyptian occupation early in the 19th century was overthrown by a native Mahdist Sudan state (1885-99) that was crushed by the British who then set up an Anglo-Egyptian Sudan—nominally a condominium, but in effect a British colony.

Following independence from Anglo-Egyptian co-rule in 1956, military regimes favoring Islamic-oriented governments have dominated national politics. Sudan was embroiled in two prolonged civil wars during most of the remainder of the 20th century. These conflicts were rooted in northern economic, political, and social domination of largely non-Muslim, nonArab southern Sudanese. The first civil war ended in 1972 but another broke out in 1983. Peace talks gained momentum in 2002-04 with the signing of several accords. The final North/South Comprehensive

Peace Agreement (CPA), signed in January 2005, granted the southern rebels autonomy for six years followed by a referendum on independence for Southern Sudan. The referendum was held in January 2011 and indicated overwhelming support for independence. South Sudan became independent on 9 July 2011. Sudan and South Sudan have yet to fully implement security and economic agreements signed in September 2012 relating to the normalization of relations between the two countries. The final disposition of the contested Abyei region has also to be decided. The 30-year reign of President Umar Hassan Ahmad alBASHIR ended in his ouster in April 2019, and a Sovereignty Council, a joint civilian-military-executive body, holds power as of November 2019.

Following South Sudan's independence, conflict broke out between the government and the Sudan People's Liberation Movement-North in Southern Kordofan and Blue Nile states (together known as the Two Areas), resulting in a humanitarian crisis affecting more than a million people. A earlier conflict that broke out in the western region of Darfur in 2003, displaced nearly 2 million people and caused thousands of deaths. While some repatriation has taken place, about 1.83 million IDPs remain in Sudan as of May 2019. Fighting in both the Two Areas and Darfur between government forces and opposition has largely subsided, however the civilian populations are affected by low-level violence including intertribal conflict and banditry, largely a result of weak rule of law. The UN and the African Union have jointly commanded a Darfur peacekeeping operation (UNAMID) since 2007, but are slowly drawing down as the situation in Darfur becomes more stable. Sudan also has faced refugee influxes from neighboring countries, primarily Ethiopia, Eritrea, Chad, Central African Republic, and South Sudan. Armed conflict, poor transport infrastructure, and denial of access by both the government and armed opposition have impeded the provision of humanitarian assistance to affected populations. However, Sudan's new transitional government has stated its priority to allow greater humanitarian access, as the food security and humanitarian situation in Sudan worsens and as it appeals to the West for greater engagement.

GEOGRAPHY

Location: north-eastern Africa, bordering the Red Sea, between Egypt and Eritrea

Geographic coordinates: 15 00 N, 30 00 E

Map references: Africa

Area: *total:* 1,861,484 sq km
land: 1,731,671 sq km
water: 129,813 sq km
country comparison to the world: 17

Area—comparative: slightly less than one-fifth the size of the US

Land boundaries: *total:* 6,819 km
border countries (7): Central African Republic 174 km, Chad 1403 km, Egypt 1276 km, Eritrea

682 km, Ethiopia 744 km, Libya 382 km, South Sudan 2158 km
note: Sudan-South Sudan boundary represents 1 January 1956 alignment; final alignment pending negotiations and demarcation; final sovereignty status of Abyei region pending negotiations between Sudan and South Sudan

Coastline: 853 km

Maritime claims: *territorial sea:* 12 nm
contiguous zone: 18 nm
continental shelf: 200-m depth or to the depth of exploitation

Climate: hot and dry; arid desert; rainy season varies by region (April to November)

Terrain: generally flat, featureless plain; desert dominates the north

Elevation: *mean elevation:* 568 m
lowest point: Red Sea 0 m
highest point: Jabal Marrah 3,042 m

Natural resources: petroleum; small reserves of iron ore, copper, chromium ore, zinc, tungsten, mica, silver, gold; hydropower

Land use: *agricultural land:* 100% (2011 est.)
arable land: 15.7% (2011 est.) / permanent crops: 0.2% (2011 est.) / permanent pasture: 84.2% (2011 est.)
forest: 0% (2011 est.)
other: 0% (2011 est.)
Irrigated land: 18,900 sq km (2012)

Population distribution: with the exception of a ribbon of settlement that corresponds to the banks of the Nile, northern Sudan, which extends into the dry Sahara, is sparsely populated; more abundant vegetation and broader access to water increases population distribution in the south extending habitable range along nearly the entire border with South Sudan; sizeable areas of population are found around Khartoum, southeast between the Blue and White Nile Rivers, and througout South Darfur as shown on this population distribution map

Natural hazards: dust storms and periodic persistent droughts

Environment—current issues: water pollution; inadequate supplies of potable water; water scarcity and periodic drought; wildlife populations threatened by excessive hunting; soil erosion; desertification; deforestation; loss of biodiversity

Environment—international agreements: *party to:* Biodiversity, Climate Change, Climate Change-Kyoto Protocol, Desertification, Endangered Species, Hazardous Wastes, Law of the Sea, Ozone Layer Protection, Wetlands
signed, but not ratified: none of the selected agreements

Geography—note: the Nile is Sudan's primary water source; its major tributaries, the White Nile and the Blue Nile, meet at Khartoum to form the River Nile which flows northward through Egypt to the Mediterranean Sea

PEOPLE AND SOCIETY

Population: 45,561,556 (July 2020 est.)

country comparison to the world: 31

Nationality: *noun:* Sudanese (singular and plural)
adjective: Sudanese

Ethnic groups: unspecified Sudanese Arab (approximately 70%), Fur, Beja, Nuba, Fallata

Languages: Arabic (official), English (official), Nubian, Ta Bedawie, Fur

Religions: Sunni Muslim, small Christian minority

Age structure: *0-14 years:* 42.01% (male 9,726,937/female 9,414,988)
15-24 years: 20.94% (male 4,852,903/female 4,687,664)
25-54 years: 29.89% (male 6,633,567/female 6,986,241)
55-64 years: 4.13% (male 956,633/female 923,688)
65 years and over: 3.03% (male 729,214/female 649,721) (2020 est.)

Dependency ratios: *total dependency ratio:* 76.9
youth dependency ratio: 70.4
elderly dependency ratio: 6.5
potential support ratio: 15.4 (2020 est.)

Median age: *total:* 18.3 years
male: 18.1 years
female: 18.5 years (2020 est.)
country comparison to the world: 212

Population growth rate: 2.69% (2020 est.)
country comparison to the world: 17

Birth rate: 33.8 births/1,000 population (2020 est.)
country comparison to the world: 23

Death rate: 6.5 deaths/1,000 population (2020 est.)
country comparison to the world: 144

Net migration rate: -0.4 migrant(s)/1,000 population (2020 est.)
country comparison to the world: 125

Population distribution: with the exception of a ribbon of settlement that corresponds to the banks of the Nile, northern Sudan, which extends into the dry Sahara, is sparsely populated; more abundant vegetation and broader access to water increases population distribution in the south extending habitable range along nearly the entire border with South Sudan; sizeable areas of population are found around Khartoum, southeast between the Blue and White Nile Rivers, and througout South Darfur as shown on this population distribution map

Urbanization: *urban population:* 35.3% of total population (2020)
rate of urbanization: 3.17% annual rate of change (2015-20 est.)

total population growth rate v. urban population growth rate, 2000-2030: *Major urban areas—population:* 5.829 million KHARTOUM (capital), 923,000 Nyala (2020)

Sex ratio: *at birth:* 1.05 male(s)/female
0-14 years: 1.03 male(s)/female
15-24 years: 1.04 male(s)/female
25-54 years: 0.95 male(s)/female
55-64 years: 1.04 male(s)/female
65 years and over: 1.12 male(s)/female

total population: 1.01 male(s)/female (2020 est.)

Maternal mortality rate: 295 deaths/100,000 live births (2017 est.)
country comparison to the world: 38

Infant mortality rate: *total:* 41.8 deaths/1,000 live births
male: 46.7 deaths/1,000 live births
female: 36.6 deaths/1,000 live births (2020 est.)
country comparison to the world: 35

Life expectancy at birth: *total population:* 66.5 years
male: 64.3 years
female: 68.8 years (2020 est.)
country comparison to the world: 186

Total fertility rate: 4.72 children born/woman (2020 est.)
country comparison to the world: 19

Contraceptive prevalence rate: 12.2% (2014)

Drinking water source:
improved:
urban: 99% of population
rural: 80.7% of population
total: 87% of population
unimproved:
urban: 1% of population
rural: 19.3% of population
total: 13% of population (2017 est.)

Current Health Expenditure: 6.3% (2017)

Physicians density: 0.26 physicians/1,000 population (2017)

Hospital bed density: 0.7 beds/1,000 population (2017)

Sanitation facility access
improved: urban: 72.1% of population
rural: 30.6% of population
total: 44.9% of population
unimproved:
urban: 27.9% of population
rural: 69.4% of population
total: 55.1% of population (2017 est.)

HIV/AIDS—adult prevalence rate: 0.2% (2019 est.)
country comparison to the world: 110

HIV/AIDS—people living with HIV/AIDS: 46,000 (2019 est.)
country comparison to the world: 64

HIV/AIDS—deaths: 2,300 (2019 est.)
country comparison to the world: 44

Major infectious diseases: *degree of risk:* very high (2020)
food or waterborne diseases: bacterial and protozoal diarrhea, hepatitis A and E, and typhoid fever
vectorborne diseases: malaria, dengue fever, and Rift Valley fever
water contact diseases: schistosomiasis
animal contact diseases: rabies
respiratory diseases: meningococcal meningitis

Obesity—adult prevalence rate: 6.6% (2014)
country comparison to the world: 166

Children under the age of 5 years underweight: 33.1% (2014)
country comparison to the world: 5

Education expenditures: 2.2% of GDP (2009)
country comparison to the world: 166

Literacy: *definition:* age 15 and over can read and write
total population: 60.7%
male: 65.4%
female: 56.1% (2018)

School life expectancy (primary to tertiary education): *total:* 8 years
male: 8 years
female: 7 years (2015)

GOVERNMENT

Country name: *conventional long form:* Republic of the Sudan
conventional short form: Sudan
local long form: Jumhuriyat as-Sudan
local short form: As-Sudan
former: Anglo-Egyptian Sudan, Democratic Republic of the Sudan
etymology: the name "Sudan" derives from the Arabic "bilad-as-sudan" meaning "Land of the Black [peoples]"

Government type: presidential republic

Capital: *name:* Khartoum
geographic coordinates: 15 36 N, 32 32 E
time difference: UTC+3 (8 hours ahead of Washington, DC, during Standard Time)
etymology: several explanations of the name exist; two of the more plausible are that it is derived from Arabic "al-jartum" meaning "elephant's trunk" or "hose," and likely referring to the narrow strip of land extending between the Blue and White Niles; alternatively, the name could derive from the Dinka words "khar-tuom," indicating a "place where rivers meet"

Administrative divisions: 18 states (wilayat, singular—wilayah); Blue Nile, Central Darfur, East Darfur, Gedaref, Gezira, Kassala, Khartoum, North Darfur, North Kordofan, Northern, Red Sea, River Nile, Sennar, South Darfur, South Kordofan, West Darfur, West Kordofan, White Nile
note: the peace accord signed in October 2020 included a protocol to restructure the country's current 18 provinces/states into eight regions

Independence: 1 January 1956 (from Egypt and the UK)

National holiday: Independence Day, 1 January (1956)

Constitution: *history:* previous 1973, 1998; 2005 (interim constitution, which was suspended in April 2019); latest initial draft completed by Transitional Military Council in May 2019; revised draft known as the "Draft Constitutional Charter for the 2019 Transitional Period," was signed by the Council and opposition coalition on 4 August 2019
amendments: NA

Legal system: mixed legal system of Islamic law and English common law; note—in mid-July 2020, Sudan amended 15 provisions of its 1991 penal code

International law organization participation: accepts compulsory ICJ jurisdiction with reservations; withdrew acceptance of ICCt jurisdiction in 2008

Citizenship: *citizenship by birth:* no
citizenship by descent only: the father must be a citizen of Sudan
dual citizenship recognized: no
residency requirement for naturalization: 10 years

Suffrage: 17 years of age; universal

Executive branch: *chief of state:* president (vacant); note—in August 2019, the ruling military council and civilian opposition alliance signed a power-sharing deal as the "Sovereignty Council," chaired by General Abd-al-Fatah al-BURHAN Abd-al-Rahman and consisting of 6 civilians and 5 generals; the Council is currently led by the military but is intended to transition to civilian leadership in May 2021 until elections can be held; General BURHAN serves as both chief of state and head of government
head of government: president (vacant); note—in August 2019, the ruling military council and civilian opposition alliance signed a power-sharing deal as the "Sovereignty Council," chaired by General Abd-al-Fatah al-BURHAN Abd-al-Rahman and consisting of 6 civilians and 5 generals; the Council is currently led by the military but is intended to transition to civilian leadership in May 2021 until elections can be held (Abd-al-Rahman)
cabinet: Council of Ministers appointed by the prime minister (2019)
elections/appointments: president directly elected by absolute majority popular vote in 2 rounds if needed; last held on 13-16 April 2015 (next to be held in 2022 at the end of the transitional period); prime minister typically appointed by the president; note—the position of prime minister was reinstated in December 2016 as a result of the 2015-16 national dialogue process, and President al-BASHIR appointed BAKRI Hassan Salih to the position on 2 March 2017; on 21 August 2019, the Forces for Freedom and Change, the civilian opposition alliance, named Abdallah HANDOUK as prime minister of Sudan for the transitional period
election results: Umar Hassan Ahmad al-BASHIR reelected president; percent of vote—Umar Hassan Ahmad al-BASHIR (NCP) 94.1%, other (15 candidates) 5.9%

Legislative branch: *description:* according to the August 2019 Constitutional Decree, which established Sudan's transitional government, the Transitional Legislative Council (TLC) will serve as the national legislature during the transitional period until elections can be held in 2022; as of early December 2019, the TLC had not been established
elections: Council of State—last held 1 June 2015 National Assembly—last held on 13-15 April 2015
note—elections for an as yet defined new legislature to be held in 2022 at the expiry of the Transnational Legislative Council

election results:
Council of State—percent of vote by party—NA; seats by party—NA; composition—men 35, women 19, percent of women 35.2%
National Assembly—percent of vote by party—NA; seats by party—NCP 323, DUP 25, Democratic Unionist Party 15, other 44, independent 19; composition—men 296 women 130, percent of women 30.5%; note—total National Legislature percent of women 31%

Judicial branch: *highest courts:* National Supreme Court (consists of 70 judges organized into panels of 3 judges and includes 4 circuits that operate outside the capital); Constitutional Court (consists of 9 justices including the court president); note—the Constitutional Court resides outside the national judiciary
judge selection and term of office: National Supreme Court and Constitutional Court judges selected by the Supreme Judicial Council, which replaced the National Judicial Service Commission upon enactment of the Draft Constitutional Charter for the 2019 Transitional Period
subordinate courts: Court of Appeal; other national courts; public courts; district, town, and rural courts

Political parties and leaders: Democratic Unionist Party or DUP [Jalal al-DIGAIR]
Democratic Unionist Party [Muhammad Uthman al-MIRGHANI]
Federal Umma Party [Dr. Ahmed Babikir NAHAR]
Muslim Brotherhood or MB
National Congress Party or NCP (in November 2019, Sudan's transitional government approved a law to "dismantle" the regime of former President Omar al-Bashir, including the dissolution of his political party, the NCP)
National Umma Party or NUP [Saddiq al-MAHDI]
Popular Congress Party or PCP [Hassan al-TURABI]
Reform Movement Now [Dr. Ghazi Salahuddin al-ATABANI] Sudan National Front [Ali Mahmud HASANAYN]
Sudanese Communist Party or SCP [Mohammed Moktar Al-KHATEEB]
Sudanese Congress Party or SCoP [Ibrahim Al-SHEIKH]
Umma Party for Reform and Development
Unionist Movement Party or UMP

International organization participation: ABEDA, ACP, AfDB, AFESD, AMF, AU, CAEU, COMESA, FAO, G-77, IAEA, IBRD, ICAO, ICC (NGOs), ICRM, IDA, IDB, IFAD, IFC, IFRCS, IGAD, ILO, IMF, IMO, Interpol, IOC, IOM, IPU, ISO, ITSO, ITU, LAS, MIGA, NAM, OIC, OPCW, PCA, UN, UNCTAD, UNESCO, UNHCR, UNIDO, UNWTO, UPU, WCO, WFTU (NGOs), WHO, WIPO, WMO, WTO (observer)

Diplomatic representation in the US: *chief of mission:* Ambassador Nureldin Mohamed Hamed SATTI (since 17 September 2020)

chancery: 2210 Massachusetts Avenue NW, Washington, DC 20008
telephone: [1] (202) 338-8565
FAX: [1] (202) 667-2406

Diplomatic representation from the US: *chief of mission:* Ambassador (vacant); Charge d'Affaires Brian SHUKAN (since September 2019)
telephone: [249] 18702-2000
embassy: Kilo 10, Soba, Khartoum
mailing address: P.O. Box 699, Kilo 10, Soba, Khartoum; APO AE 09829
FAX: [249] 18702-2547

Flag description: *three equal horizontal bands of red (top), white, and black with a green isosceles triangle based on the hoist side; colors and design based on the Arab Revolt flag of World War I, but the meanings of the colors are expressed as follows:* red signifies the struggle for freedom, white is the color of peace, light, and love, black represents the people of Sudan (in Arabic 'Sudan' means black), green is the color of Islam, agriculture, and prosperity

National symbol(s): *secretary bird; national colors:* red, white, black, green

National anthem: *name:* "Nahnu Djundulla Djundulwatan" (We Are the Army of God and of Our Land)
lyrics/music: Sayed Ahmad Muhammad SALIH/ Ahmad MURJAN
note: adopted 1956; originally served as the anthem of the Sudanese military
0:00/ 0:45

ECONOMY

Economy—overview: Sudan has experienced protracted social conflict and the loss of three quarters of its oil production due to the secession of South Sudan. The oil sector had driven much of Sudan's GDP growth since 1999. For nearly a decade, the economy boomed on the back of rising oil production, high oil prices, and significant inflows of foreign direct investment. Since the economic shock of South Sudan's secession, Sudan has struggled to stabilize its economy and make up for the loss of foreign exchange earnings. The interruption of oil production in South Sudan in 2012 for over a year and the consequent loss of oil transit fees further exacerbated the fragile state of Sudan's economy. Ongoing conflicts in Southern Kordofan, Darfur, and the Blue Nile states, lack of basic infrastructure in large areas, and reliance by much of the population on subsistence agriculture, keep close to half of the population at or below the poverty line.
Sudan was subject to comprehensive US sanctions, which were lifted in October 2017. Sudan is attempting to develop non-oil sources of revenues, such as gold mining and agriculture, while carrying out an austerity program to reduce expenditures. The world's largest exporter of gum Arabic, Sudan produces 75-80% of the world's total output. Agriculture continues to employ 80% of the work force.

Sudan introduced a new currency, still called the Sudanese pound, following South Sudan's secession, but the value of the currency has fallen since its introduction. Khartoum formally devalued the currency in June 2012, when it passed austerity measures that included gradually repealing fuel subsidies. Sudan also faces high inflation, which reached 47% on an annual basis in November 2012 but fell to about 35% per year in 2017.

(2017)

GDP (purchasing power parity): $177.4 billion (2017 est.)
$174.9 billion (2016 est.)
$169.8 billion (2015 est.)
note: data are in 2017 dollars
country comparison to the world: 71

GDP (official exchange rate): $45.82 billion (2017 est.)

GDP—real growth rate: 1.4% (2017 est.)
3% (2016 est.)
1.3% (2015 est.)
country comparison to the world: 160

GDP—per capita (PPP): $4,300 (2017 est.)
$4,400 (2016 est.)
$4,400 (2015 est.)
note: data are in 2017 dollars
country comparison to the world: 174

Gross national saving: 12.1% of GDP (2017 est.)
13.1% of GDP (2016 est.)
12.2% of GDP (2015 est.)
country comparison to the world: 149

GDP—composition, by end use:
household consumption: 77.3% (2017 est.)
government consumption: 5.8% (2017 est.)
investment in fixed capital: 18.4% (2017 est.)
investment in inventories: 0.6% (2017 est.)
exports of goods and services: 9.7% (2017 est.)
imports of goods and services: -11.8% (2017 est.)

GDP—composition, by sector of origin:
agriculture: 39.6% (2017 est.)
industry: 2.6% (2017 est.)
services: 57.8% (2017 est.)

Agriculture—products: cotton, groundnuts (peanuts), sorghum, millet, wheat, gum Arabic, sugarcane, cassava (manioc, tapioca), mangoes, papaya, bananas, sweet potatoes, sesame seeds; animal feed, sheep and other livestock

Industries: oil, cotton ginning, textiles, cement, edible oils, sugar, soap distilling, shoes, petroleum refining, pharmaceuticals, armaments, automobile/light truck assembly, milling

Industrial production growth rate: 4.5% (2017 est.)
country comparison to the world: 66

Labor force: 11. 92 million (2007 est.)
country comparison to the world: 46

Labor force—by occupation: *agriculture:* 80%
industry: 7%
services: 13% (1998 est.)

Unemployment rate: 19.6% (2017 est.)
20.6% (2016 est.)
country comparison to the world: 187

Population below poverty line: 46.5% (2009 est.)

Household income or consumption by percentage share: *lowest 10%:* 2.7%
highest 10%: 26.7% (2009 est.)

Budget: *revenues:* 8.48 billion (2017 est.)
expenditures: 13.36 billion (2017 est.)

Taxes and other revenues: 18.5% (of GDP) (2017 est.)
country comparison to the world: 159

Budget surplus (+) or deficit (-): -10.6% (of GDP) (2017 est.)
country comparison to the world: 213

Public debt: 121.6% of GDP (2017 est.)
99.5% of GDP (2016 est.)
country comparison to the world: 10

Fiscal year: calendar year

Inflation rate (consumer prices): 32.4% (2017 est.)
17.8% (2016 est.)
country comparison to the world: 223

Current account balance: -$4.811 billion (2017 est.)
-$4.213 billion (2016 est.)
country comparison to the world: 183

Exports: $4.1 billion (2017 est.)
$3.094 billion (2016 est.)
country comparison to the world: 116

Exports—partners: UAE 55.5%, Egypt 14.7%, Saudi Arabia 8.8% (2017)

Exports—commodities: gold; oil and petroleum products; cotton, sesame, livestock, peanuts, gum Arabic, sugar

Imports: $8.22 billion (2017 est.)
$7.48 billion (2016 est.)
country comparison to the world: 109

Imports—commodities: foodstuffs, manufactured goods, refinery and transport equipment, medicines, chemicals, textiles, wheat

Imports—partners: UAE 12.7%, Egypt 10.6%, India 10.5%, Turkey 10.2%, Japan 7.6%, Saudi Arabia 6%, Germany 4.6% (2017)

Reserves of foreign exchange and gold: $198 million (31 December 2017 est.)
$168.3 million (31 December 2016 est.)
country comparison to the world: 177

Debt—external: $56.05 billion (31 December 2017 est.)
$51.26 billion (31 December 2016 est.)
country comparison to the world: 62

Exchange rates: Sudanese pounds (SDG) per US dollar -
6.72 (2017 est.)
6.14 (2016 est.)
6.14 (2015 est.)
6.03 (2014 est.)
5.74 (2013 est.)

ENERGY

Electricity access: *population without electricity:* 23 million (2019)
electrification—total population: 47% (2019)
electrification—urban areas: 71% (2019)
electrification—rural areas: 35% (2019)

Electricity—production: 13.99 billion kWh (2016 est.)
country comparison to the world: 89

Electricity—consumption: 12.12 billion kWh (2016 est.)
country comparison to the world: 88

Electricity—exports: 0 kWh (2016 est.)
country comparison to the world: 203

Electricity—imports: 0 kWh (2016 est.)
country comparison to the world: 205

Electricity—installed generating capacity: 3.437 million kW (2016 est.)
country comparison to the world: 96

Electricity—from fossil fuels: 44% of total installed capacity (2016 est.)
country comparison to the world: 162

Electricity—from nuclear fuels: 0% of total installed capacity (2017 est.)
country comparison to the world: 188

Electricity—from hydroelectric plants: 51% of total installed capacity (2017 est.)
country comparison to the world: 39

Electricity—from other renewable sources: 6% of total installed capacity (2017 est.)
country comparison to the world: 104

Crude oil—production: 95,000 bbl/day (2018 est.)
country comparison to the world: 43

Crude oil—exports: 19,540 bbl/day (2015 est.)
country comparison to the world: 50

Crude oil—imports: 9,440 bbl/day (2015 est.)
country comparison to the world: 73

Crude oil—proved reserves: 5 billion bbl (1 January 2018 est.)
country comparison to the world: 22

Refined petroleum products—production: 94,830 bbl/day (2015 est.)
country comparison to the world: 68

Refined petroleum products—consumption: 112,000 bbl/day (2016 est.)
country comparison to the world: 75

Refined petroleum products—exports: 8,541 bbl/day (2015 est.)
country comparison to the world: 85

Refined petroleum products—imports: 24,340 bbl/day (2015 est.)
country comparison to the world: 108

Natural gas—production: 0 cu m (2017 est.)
country comparison to the world: 201

Natural gas—consumption: 0 cu m (2017 est.)
country comparison to the world: 202

Natural gas—exports: 0 cu m (2017 est.)
country comparison to the world: 190

Natural gas—imports: 0 cu m (2017 est.)
country comparison to the world: 193

Natural gas—proved reserves: 84.95 billion cu m (1 January 2018 est.)
country comparison to the world: 55

Carbon dioxide emissions from consumption of energy: 16.03 million Mt (2017 est.)
country comparison to the world: 92

COMMUNICATIONS

Telephones—fixed lines: *total subscriptions:* 141,922
subscriptions per 100 inhabitants: less than 1 (2019 est.)
country comparison to the world: 127

Telephones—mobile cellular: *total subscriptions:* 34,198,859
subscriptions per 100 inhabitants: 77.11 (2019 est.)
country comparison to the world: 44

Telecommunication systems: *general assessment:* well-equipped system by regional standards and being upgraded; despite economic hardships govt. boosts mobile infrastructure and builds fiber broadband network across country; economic climate has not encouraged growth in telecoms, but some investment has been made to build mobile towers and expand LTE services; launches its own Chinese built satellite in 2019 to develop space technology sector (2020)
domestic: consists of microwave radio relay, cable, fiber optic, radiotelephone communications, tropospheric scatter, and a domestic satellite system with 14 earth stations; teledensity fixed-line less than 1 per 100 and mobile-cellular 77 telephones per 100 persons (2019)
international: country code—249; landing points for the EASSy, FALCON and SAS-1,-2, fiber-optic submarine cable systems linking Africa, the Middle East, Indian Ocean Islands and Asia; satellite earth stations—1 Intelsat (Atlantic Ocean) (2019)
note: the COVID-19 outbreak is negatively impacting telecommunications production and supply chains globally; consumer spending on telecom devices and services has also slowed due to the pandemic's effect on economies worldwide; overall progress towards improvements in all facets of the telecom industry—mobile, fixed-line, broadband, submarine cable and satellite—has moderated

Broadcast media: the Sudanese Government directly controls TV and radio, requiring that both media reflect government policies; TV has a permanent military censor; a private radio station is in operation (2019)

Internet country code: .sd

Internet users: *total:* 13,311,404
percent of population: 30.87% (July 2018 est.)
country comparison to the world: 47

Broadband—fixed subscriptions: *total:* 31,352
subscriptions per 100 inhabitants: less than 1 (2018 est.)
country comparison to the world: 141

TRANSPORTATION

National air transport system: *number of registered air carriers:* 9 (2020)
inventory of registered aircraft operated by air carriers: 42
annual passenger traffic on registered air carriers: 269,958 (2018)

Civil aircraft registration country code prefix: ST (2016)

Airports: 67 (2020)
country comparison to the world: 74

Airports—with paved runways: *total:* 17 (2020)
over 3,047 m: 2
2,438 to 3,047 m: 11
1,524 to 2,437 m: 2
914 to 1,523 m: 1
under 914 m: 1

Airports—with unpaved runways: *total:* 50 (2020)
1,524 to 2,437 m: 17
914 to 1,523 m: 24
under 914 m: 9

Heliports: 7 (2020)

Pipelines: 156 km gas, 4070 km oil, 1613 km refined products (2013)

Railways: *total:* 7,251 km (2014)
narrow gauge: 5,851 km 1.067-m gauge (2014)
1,400 km 0.600-m gauge for cotton plantations
country comparison to the world: 31

Roadways: *total:* 31,000 km (2019)
paved: 8,000 km (2019)
unpaved: 23,000 km (2019)
urban: 1,000 km (2019)
country comparison to the world: 98

Waterways: 4,068 km (1,723 km open year-round on White and Blue Nile Rivers) (2011)
country comparison to the world: 24

Merchant marine: *total:* 17
by type: other 17 (2019)
country comparison to the world: 146

Ports and terminals: *major seaport(s):* Port Sudan

MILITARY AND SECURITY

Military and security forces: *Sudanese Armed Forces (SAF):* Ground Force, Navy, Sudanese Air Force; Rapid Support Forces (RSF, paramilitary); Reserve Department (formerly the paramilitary Popular Defense Forces) (2020)

the RSF is an autonomous paramilitary force formed in 2013 to fight armed rebel groups in Sudan, with Mohammed Hamdan DAGALLO (aka Hemeti) as its commander (he is also Deputy Chairman of the Sovereignty Council), from the remnants of the Janjaweed militia that participated in suppressing the Darfur rebellion; it was initially commanded by the National Intelligence and Security Service, then came under the direct command of former president Omar al-BASHIR, who boosted the RSF as his own personal security force; the RSF has been accused of committing rights abuses against civilians; it is also reportedly involved in business enterprises, such as gold mining; in late 2019, Sovereignty Council Chairman and SAF Commander-in-Chief General Abd-al-Fatah al-BURHAN said the RSF would be fully integrated into the SAF, but did not give a timeline

Military expenditures: 1.6% of GDP (2019)
2.3% of GDP (2018)
3.9% of GDP (2017)
3% of GDP (2016)

3% of GDP (2015)
country comparison to the world: 76

Military and security service personnel strengths: size assessments for the Sudanese Armed Forces (SAF) vary widely, ranging from about 100,000 to more than 200,000 active personnel, including approximately 1,500 Navy and 3,000 Air Force; est. 30-40,000 paramilitary Rapid Support Forces; est. 20,000 Reserve Department (formerly the paramilitary Popular Defense Forces) (2019)
note: in August 2020, Sudan and the major rebel group Sudan People's Liberation Movement-North (SPLM-N) signed an agreement to integrate the group's fighters into the Sudanese Army by the end of 2023

Military equipment inventories and acquisitions: the SAF's inventory includes a mix of Chinese, Russian, Soviet, Ukrainian, and domestically-produced weapons systems; since 2010, the leading arms providers to the SAF are Belarus, China, Russia, and Ukraine; Sudan has a domestic arms industry that manufactures ammunition, small arms, and armored vehicles, largely based on older Chinese and Russian systems (2019 est.)

Military deployments: estimates vary; approximately 1,000 Libya; approximately 1-3,000 Yemen (Dec 2019)

Military service age and obligation: 18-33 years of age for male and female compulsory or voluntary military service; 1-2 year service obligation (2013)

Military—note: United Nations Interim Security Force for Abyei (UNISFA) has operated in the disputed Abyei region along the border between Sudan and South Sudan since 2011; UNISFA's mission includes ensuring security, protecting civilians, strengthening the capacity of the Abyei Police Service, de-mining, monitoring/verifying the redeployment of armed forces from the area, and facilitating the flow of humanitarian aid; UNISFA had about 4,000 personnel deployed as of January 2020

in addition, the United Nations African Union Hybrid Operation in Darfur (UNAMID) has operated in the war-torn Darfur region since 2007; UNAMID is a joint African Union-UN peacekeeping force with the mission of bringing stability to Darfur, including protecting civilians, facilitating humanitarian assistance, and promoting mediation efforts, while peace talks on a final settlement continue; as of March 2020, UNAMID had about 6,500 personnel deployed (2020)

TRANSNATIONAL ISSUES

Disputes—international: the effects of Sudan's ethnic and rebel militia fighting since the mid-20th century have penetrated all of the neighboring states; Chad wishes to be a helpful mediator in resolving the Darfur conflict, and in 2010 established a joint border monitoring force with Sudan, which has helped to reduce cross-border banditry and violence; as of early 2019, more than 590,000 Sudanese refugees are being hosted in the Central African Republic, Chad, Egypt, Ethiopia, Kenya, and South Sudan; Sudan, in turn, is hosting more

than 975,000 refugees and asylum seekers, including more than 845,000 from South Sudan; Sudan accuses South Sudan of supporting Sudanese rebel groups; Sudan claims but Egypt de facto administers security and economic development of the Halaib region north of the 22nd parallel boundary; periodic violent skirmishes with Sudanese residents over water and grazing rights persist among related pastoral populations along the border with the Central African Republic; South Sudan-Sudan boundary represents 1 January 1956 alignment, final alignment pending negotiations and demarcation; final sovereignty status of Abyei Area pending negotiations between South Sudan and Sudan

Refugees and internally displaced persons: *refugees (country of origin):* 729,557 (South Sudan) (refugees and asylum seekers), 122,227 (Eritrea) (refugees and asylum seekers), 93,498 (Syria) (refugees and asylum seekers), 26,523 (Central African Republic), 13,130 (Ethiopia) (refugees and asylum seekers) (2020)
IDPs: 2.134 million (civil war 1983-2005; ongoing conflict in Darfur region; government and rebel fighting along South Sudan border; inter-tribal clashes) (2019)

Trafficking in persons: *current situation:* Sudan is a source, transit, and destination country for men, women, and children who are subjected to forced labor and sex trafficking; Sudanese women and girls, particularly those from rural areas or who are internally displaced, or refugees are vulnerable to domestic servitude in country, as well as domestic servitude and sex trafficking abroad; migrants from East and West Africa, South Sudan, Syria, and Nigeria smuggled into or through Sudan are vulnerable to exploitation; Ethiopian, Eritrean, and Filipino women are subjected to domestic servitude in Sudanese homes, and East African and possibly Thai women are forced into prostitution in Sudan; Sudanese children continue to be recruited and used as combatants by government forces and armed groups
tier rating: Tier 2 Watch List—Sudan does not fully comply with the minimum standards for the elimination of trafficking; however, it is making significant efforts to do so; the government increased its efforts to publically address and prevent trafficking, established a national anti-trafficking council, and began drafting a national action plan against trafficking; the government acknowledges cross-border trafficking but still denies the existence of forced labor, sex trafficking, and the recruitment of child soldiers domestically; law enforcement and judicial officials struggled to apply the national antitrafficking law, often relying on other statutes with lesser penalties; authorities did not use systematic procedure to identify victims or refer them to care and relied on international organizations and domestic groups to provide protective services; some foreign victims were penalized for unlawful acts committed as a direct result of being trafficked, such as immigration or prostitution violations (2015)

915

SURINAME

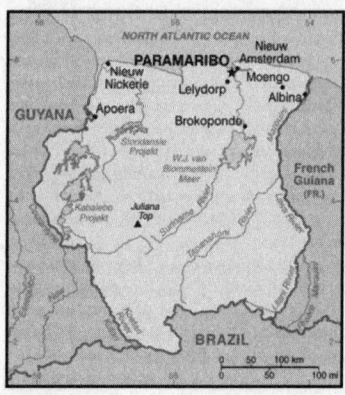

NORTH ATLANTIC OCEAN

INTRODUCTION

Background: First explored by the Spaniards in the 16th century and then settled by the English in the mid-17th century, Suriname became a Dutch colony in 1667. With the abolition of African slavery in 1863, workers were brought in from India and Java. The Netherlands granted the colony independence in 1975. Five years later the civilian government was replaced by a military regime that soon declared Suriname a socialist republic. It continued to exert control through a succession of nominally civilian administrations until 1987, when international pressure finally forced a democratic election. In 1990, the military overthrew the civilian leadership, but a democratically elected government—a four-party coalition—returned to power in 1991. The coalition expanded to eight parties in 2005 and ruled until August 2010, when voters returned former military leader Desire BOUTERSE and his opposition coalition to power. President BOUTERSE was reelected unopposed in 2015.

GEOGRAPHY

Location: Northern South America, bordering the North Atlantic Ocean, between French Guiana and Guyana

Geographic coordinates: 4 00 N, 56 00 W

Map references: South America

Area: *total:* 163,820 sq km
land: 156,000 sq km
water: 7,820 sq km
country comparison to the world: 93

Area—comparative: slightly larger than Georgia

Land boundaries: *total:* 1,907 km
border countries (3): Brazil 515 km, French Guiana 556 km, Guyana 836 km

Coastline: 386 km

Maritime claims: *territorial sea:* 12 nm
exclusive economic zone: 200 nm

Climate: tropical; moderated by trade winds

Terrain: mostly rolling hills; narrow coastal plain with swamps

Elevation: *mean elevation:* 246 m
lowest point: unnamed location in the coastal plain -2 m
highest point: Juliana Top 1,230 m

Natural resources: timber, hydropower, fish, kaolin, shrimp, bauxite, gold, and small amounts of nickel, copper, platinum, iron ore

Land use: *agricultural land:* 0.5% (2011 est.)
arable land: 0.4% (2011 est.) / permanent crops: 0% (2011 est.) / permanent pasture: 0.1% (2011 est.)
forest: 94.6% (2011 est.)
other: 4.9% (2011 est.)
Irrigated land: 570 sq km (2012)

Population distribution: population concentrated along the nothern coastal strip; the remainder of the country is sparsely populated

Natural hazards: flooding

Environment—current issues: deforestation as timber is cut for export; pollution of inland waterways by small-scale mining activities

Environment—international agreements: *party to:* Biodiversity, Climate Change, Climate Change-Kyoto Protocol, Desertification, Endangered Species, Hazardous Wastes, Law of the Sea, Marine Dumping, Ozone Layer Protection, Ship Pollution, Tropical Timber 94, Wetlands, Whaling *signed, but not ratified;* none of the selected agreements

Geography—note: smallest independent country on South American continent; mostly tropical rain forest; great diversity of flora and fauna that, for the most part, is increasingly threatened by new development; relatively small population, mostly along the coast

PEOPLE AND SOCIETY

Population: 609,569 (July 2020 est.)
country comparison to the world: 172

Nationality: *noun:* Surinamer(s)
adjective: Surinamese

Ethnic groups: Hindustani (also known locally as "East Indians"; their ancestors emigrated from northern India in the latter part of the 19th century) 27.4%, "Maroon" (their African ancestors were brought to the country in the 17th and 18th centuries as slaves and escaped to the interior) 21.7%, Creole (mixed white and black) 15.7%, Javanese 13.7%, mixed 13.4%, other 7.6%, unspecified 0.6% (2012 est.)

Languages: Dutch (official), English (widely spoken), Sranang Tongo (Surinamese, sometimes called Taki-Taki, is the native language of Creoles and much of the younger population and is lingua

franca among others), Caribbean Hindustani (a dialect of Hindi), Javanese

Religions: Protestant 23.6% (includes Evangelical 11.2%, Moravian 11.2%, Reformed .7%, Lutheran .5%), Hindu 22.3%, Roman Catholic 21.6%, Muslim 13.8%, other Christian 3.2%, Winti 1.8%, Jehovah's Witness 1.2%, other 1.7%, none 7.5%, unspecified 3.2% (2012 est.)

Demographic profile: Suriname is a pluralistic society consisting primarily of Creoles (persons of mixed African and European heritage), the descendants of escaped African slaves known as Maroons, and the descendants of Indian and Javanese (Indonesian) contract workers. The country overall is in full, post-industrial demographic transition, with a low fertility rate, a moderate mortality rate, and a rising life expectancy. However, the Maroon population of the rural interior lags behind because of lower educational attainment and contraceptive use, higher malnutrition, and significantly less access to electricity, potable water, sanitation, infrastructure, and health care.

Some 350,000 people of Surinamese descent live in the Netherlands, Suriname's former colonial ruler. In the 19th century, better-educated, largely Dutch-speaking Surinamese began emigrating to the Netherlands. World War II interrupted the outflow, but it resumed after the war when Dutch labor demands grew—emigrants included all segments of the Creole population. Suriname still is strongly influenced by the Netherlands because most Surinamese have relatives living there and it is the largest supplier of development aid. Other emigration destinations include French Guiana and the United States. Suriname's immigration rules are flexible, and the country is easy to enter illegally because rainforests obscure its borders. Since the mid-1980s, Brazilians have settled in Suriname's capital, Paramaribo, or eastern Suriname, where they mine gold. This immigration is likely to slowly re-orient Suriname toward its Latin American roots.

Age structure: *0-14 years:* 23.38% (male 72,642/ female 69,899)
15-24 years: 17.2% (male 53,427/female 51,438)
25-54 years: 44.09% (male 136,889/female 131,868)
55-64 years: 8.78% (male 26,435/female 27,066)
65 years and over: 6.55% (male 17,437/female 22,468) (2020 est.)

Dependency ratios: *total dependency ratio:* 51.1
youth dependency ratio: 40.3
elderly dependency ratio: 10.8
potential support ratio: 9.3 (2020 est.)

Median age: *total:* 31 years
male: 30.6 years
female: 31.4 years (2020 est.)
country comparison to the world: 118

Population growth rate: 0.95% (2020 est.)

country comparison to the world: 113

Birth rate: 14.9 births/1,000 population (2020 est.)
country comparison to the world: 120

Death rate: 6.2 deaths/1,000 population (2020 est.)
country comparison to the world: 158

Net migration rate: 0.5 migrant(s)/1,000 population (2020 est.)
country comparison to the world: 66

Population distribution: population concentrated along the nothern coastal strip; the remainder of the country is sparsely populated

Urbanization: *urban population:* 66.1% of total population (2020)
rate of urbanization: 0.9% annual rate of change (2015-20 est.)

total population growth rate v. urban population growth rate, 2000-2030: *Major urban areas—population:* 239,000 PARAMARIBO (capital) (2018)

Sex ratio: *at birth:* 1.05 male(s)/female
0-14 years: 1.04 male(s)/female
15-24 years: 1.04 male(s)/female
25-54 years: 1.04 male(s)/female
55-64 years: 0.98 male(s)/female
65 years and over: 0.78 male(s)/female
total population: 1.01 male(s)/female (2020 est.)

Maternal mortality rate: 120 deaths/100,000 live births (2017 est.)
country comparison to the world: 65

Infant mortality rate: *total:* 22.1 deaths/1,000 live births
male: 25.8 deaths/1,000 live births
female: 18.2 deaths/1,000 live births (2020 est.)
country comparison to the world: 70

Life expectancy at birth: *total population:* 73.3 years
male: 70.8 years
female: 75.9 years (2020 est.)
country comparison to the world: 148

Total fertility rate: 1.86 children born/woman (2020 est.)
country comparison to the world: 139

Contraceptive prevalence rate: 39.1% (2018)

Drinking water source:
improved:
urban: 98.2% of population
rural: 92% of population
total: 96.6% of population
unimproved:
urban: 1.8% of population
rural: 8% of population
total: 3.4% of population (2017 est.)

Current Health Expenditure: 6.2% (2017)

Physicians density: 1.23 physicians/1,000 population (2018)

Hospital bed density: 3 beds/1,000 population (2017)

Sanitation facility access:
improved:
urban: 98.5% of population
rural: 88.2% of population

total: 95% of population
unimproved:
urban: 1.5% of population
rural: 11.8% of population
total: 5% of population (2017 est.)

HIV/AIDS—adult prevalence rate: 1.3% (2019 est.)
country comparison to the world: 36

HIV/AIDS—people living with HIV/AIDS: 5,800 (2019 est.)
country comparison to the world: 120

HIV/AIDS—deaths: <200 (2019 est.)

Major infectious diseases: *degree of risk:* very high (2020)
food or waterborne diseases: bacterial and protozoal diarrhea, hepatitis A, and typhoid fever
vectorborne diseases: dengue fever and malaria

Obesity—adult prevalence rate: 26.4% (2016)
country comparison to the world: 42

Children under the age of 5 years underweight: 5.8% (2010)
country comparison to the world: 78

Education expenditures: NA

Literacy: *definition:* age 15 and over can read and write
total population: 94.4%
male: 96.1%
female: 92.7% (2018)

Unemployment, youth ages 15-24: *total:* 13.4%
male: 9%
female: 21.9% (2015 est.)
country comparison to the world: 104

GOVERNMENT

Country name: *conventional long form:* Republic of Suriname
conventional short form: Suriname
local long form: Republiek Suriname
local short form: Suriname
former: Netherlands Guiana, Dutch Guiana
etymology: name may derive from the indigenous "Surinen" people who inhabited the area at the time of European contact

Government type: presidential republic

Capital: *name:* Paramaribo

geographic coordinates: 5 50 N, 55 10 W
time difference: UTC-3 (2 hours ahead of Washington, DC, during Standard Time)
etymology: the name may be the corruption of a Carib (Kalina) village or tribe named Parmirbo

Administrative divisions: 10 districts (distrikten, singular—distrikt); Brokopondo, Commewijne, Coronie, Marowijne, Nickerie, Para, Paramaribo, **Saramacca, Sipaliwini, Wanica**

Independence: 25 November 1975 (from the Netherlands)

National holiday: Independence Day, 25 November (1975)

Constitution: *history:* previous 1975; latest ratified 30 September 1987, effective 30 October 1987

amendments: proposed by the National Assembly; passage requires at least two-thirds majority vote of the total membership; amended 1992

Legal system: civil law system influenced by Dutch civil law; note—a new criminal code was enacted in 2017

International law organization participation: accepts compulsory ICJ jurisdiction with reservations; accepts ICCt jurisdiction

Citizenship: *citizenship by birth:* no
citizenship by descent only: at least one parent must be a citizen of Suriname
dual citizenship recognized: no
residency requirement for naturalization: 5 years

Suffrage: 18 years of age; universal

Executive branch: *chief of state:* President Chandrikapersad SANTOKHI (since 16 July 2020); Vice President Ronnie BRUNSWIJK (since 16 July 2020); note—the president is both chief of state and head of government
head of government: President Chandrikapersad SANTOKHI (since 16 July 2020); Vice President Ronnie BRUNSWIJK (since 16 July 2020)
cabinet: Cabinet of Ministers appointed by the president
elections/appointments: president and vice president indirectly elected by the National Assembly; president and vice president serve a 5-year term (no term limits); election last held on 13 July 2020 (next to be held in May 2025)
election results: Chandrikapersad SANTOKHI elected president unopposed; National Assembly vote—NA

Legislative branch: *description:* unicameral National Assembly or Nationale Assemblee (51 seats; members directly elected in multi-seat constituencies by party-list proportional representation vote to serve 5-year terms)
elections: last held on 25 May 2020 (next to be held May 2025)
election results: percent of vote by party—VHP 41.1%, NDP 29.4%, ABOP 17.6%, NPS 7.8%, other 3.9%; seats by party—VHP 21, NDP 15, ABOP 9, NPS 4, other 2

Judicial branch: *highest courts:* High Court of Justice of Suriname (consists of the court president, vice president, and 4 judges); note—appeals beyond the High Court are referred to the Caribbean Court of Justice; human rights violations can be appealed to the Inter-American Commission on Human Rights with judgments issued by the Inter-American Court on Human Rights
judge selection and term of office: court judges appointed by the national president in consultation with the National Assembly, the State Advisory Council, and the Order of Private Attorneys; judges serve for life
subordinate courts: cantonal courts

Political parties and leaders: Alternative Combination or A-Com (coalition includes ABOP, KTPI, Party for Democracy and

Development) Brotherhood and Unity in Politics or BEP [Celsius WATERBERG]

Democratic Alternative '91 or DA91 [Angelique DEL CASTILLO]

General Liberation and Development Party or ABOP [Ronnie BRUNSWIJK]

National Democratic Party or NDP [Desire Delano BOUTERSE]

National Party of Suriname or NPS [Gregory RUSLAND]

Party for Democracy and Development in Unity or DOE [Carl BREEVELD]

Party for National Unity and Solidarity or KTPI [Willy SOEMITA]

People's Alliance (Pertjaja Luhur) or PL [Paul SOMOHARDJO]

Progressive Workers' and Farmers' Union or PALU [Jim HOK]

Progressive Reform Party or VHP [Chandrikapersad SANTOKHI]

Reform and Renewal Movement or HVB

Surinamese Labor Party or SPA [Guno CASTELEN]

International organization participation: ACP, AOSIS, Caricom, CD, CDB, CELAC, FAO, G-77, IADB, IBRD, ICAO, ICCt, ICRM, IDA, IDB, IFAD, IFC, IFRCS, IHO, ILO, IMF, IMO, Interpol, IOC, IOM, IPU, ISO (correspondent), ITU, ITUC (NGOs), LAES, MIGA, NAM, OAS, OIC, OPANAL, OPCW, PCA, Petrocaribe, UN, UNASUR, UNCTAD, UNESCO, UNIDO, UPU, WHO, WIPO, WMO, WTO

Diplomatic representation in the US: *chief of mission:* Ambassador Niermala Sakoentala BADRISING (since 21 July 2017)

chancery: 4301 Connecticut Avenue NW, Suite 460, Washington, DC 20008

telephone: [1] (202) 244-7488

FAX: [1] (202) 244-5878

consulate(s) general: Miami

Diplomatic representation from the US: *chief of mission:* Ambassador Karen Lynn WILLIAMS (since 20 November 2018)

telephone: [597] 472-900

embassy: 165 Kristalstraat, Paramaribo

mailing address: US Department of State, PO Box 1821, Paramaribo

FAX: [597] 410-972

Flag description: five horizontal bands of green (top, double width), white, red (quadruple width), white, and green (double width); a large, yellow, five-pointed star is centered in the red band; red stands for progress and love, green symbolizes hope and fertility, white signifies peace, justice, and freedom; the star represents the unity of all ethnic groups; from its yellow light the nation draws strength to bear sacrifices patiently while working toward a golden future

National symbol(s): *royal palm, faya lobi (flower); national colors:* green, white, red, yellow

National anthem: *name:* "God zij met ons Suriname!" (God Be With Our Suriname)

lyrics/music: Cornelis Atses HOEKSTRA and Henry DE ZIEL/Johannes Corstianus DE PUY

note: adopted 1959; originally adapted from a Sunday school song written in 1893 and contains lyrics in both Dutch and Sranang Tongo

ECONOMY

Economy—overview: Suriname's economy is dominated by the mining industry, with exports of oil and gold accounting for approximately 85% of exports and 27% of government revenues. This makes the economy highly vulnerable to mineral price volatility. The worldwide drop in international commodity prices and the cessation of alumina mining in Suriname significantly reduced government revenue and national income during the past few years. In November 2015, a major US aluminum company discontinued its mining activities in Suriname after 99 years of operation. Public sector revenues fell, together with exports, international reserves, employment, and private sector investment.

Economic growth declined annually from just under 5% in 2012 to -10.4% in 2016. In January 2011, the government devalued the currency by 20% and raised taxes to reduce the budget deficit. Suriname began instituting macro adjustments between September 2015 and 2016; these included another 20% currency devaluation in November 2015 and foreign currency interventions by the Central Bank until March 2016, after which time the Bank allowed the Surinamese dollar (SRD) to float. By December 2016, the SRD had lost 46% of its value against the dollar. Depreciation of the Surinamese dollar and increases in tariffs on electricity caused domestic prices in Suriname to rise 22.0% year-over-year by December 2017.

Suriname's economic prospects for the medium-term will depend on its commitment to responsible monetary and fiscal policies and on the introduction of structural reforms to liberalize markets and promote competition. The government's over-reliance on revenue from the extractive sector colors Suriname's economic outlook. Following two years of recession, the Fitch Credit Bureau reported a positive growth of 1.2% in 2017 and the World Bank predicted 2.2% growth in 2018. Inflation declined to 9%, down from 55% in 2016, and increased gold production helped lift exports. Yet continued budget imbalances and a heavy debt and interest burden resulted in a debt-to-GDP ratio of 83% in September 2017. Purchasing power has fallen rapidly due to the devalued local currency. The government has announced its intention to pass legislation to introduce a new value-added tax in 2018. Without this and other measures to strengthen the country's fiscal position, the government may face liquidity pressures.

GDP (purchasing power parity): $8.688 billion (2017 est.)

$8.526 billion (2016 est.)

$8.988 billion (2015 est.)

note: data are in 2017 dollars

country comparison to the world: 162

GDP (official exchange rate): $3.419 billion (2017 est.)

GDP—real growth rate: 1.9% (2017 est.)

-5.1% (2016 est.)

-2.6% (2015 est.)

country comparison to the world: 144

GDP—per capita (PPP): $14,900 (2017 est.)

$14,800 (2016 est.)

$15,900 (2015 est.)

note: data are in 2017 dollars

country comparison to the world: 114

Gross national saving: 46.6% of GDP (2017 est.)

55.6% of GDP (2016 est.) 53.6% of GDP (2015 est.)

country comparison to the world: 4

GDP—composition, by end use:

household consumption: 27.6% (2017 est.)

government consumption: 11.7% (2017 est.)

investment in fixed capital: 52.5% (2017 est.)

investment in inventories: 26.5% (2017 est.)

exports of goods and services: 68.9% (2017 est.)

imports of goods and services: -60.6% (2017 est.)

GDP—composition, by sector of origin:

agriculture: 11.6% (2017 est.)

industry: 31.1% (2017 est.)

services: 57.4% (2017 est.)

Agriculture—products: rice, bananas, seabob shrimp, yellow-fin tuna, vegetables

Industries: gold mining, oil, lumber, food processing, fishing

Industrial production growth rate: 1% (2017 est.)

country comparison to the world: 158

Labor force: 144,000 (2014 est.)

country comparison to the world: 176

Labor force—by occupation: *agriculture:* 11.2%

industry: 19.5%

services: 69.3% (2010)

Unemployment rate: 8.9% (2017 est.)

9.7% (2016 est.)

country comparison to the world: 136

Population below poverty line: 70% (2002 est.)

Household income or consumption by percentage share: *lowest 10%:* NA

highest 10%: NA

Budget: *revenues:* 560.7 million (2017 est.)

expenditures: 827.8 million (2017 est.)

Taxes and other revenues: 16.4% (of GDP) (2017 est.)

country comparison to the world: 182

Budget surplus (+) or deficit (-): -7.8% (of GDP) (2017 est.)

country comparison to the world: 197

Public debt: 69.3% of GDP (2017 est.)

75.8% of GDP (2016 est.)

country comparison to the world: 52

Fiscal year: calendar year

Inflation rate (consumer prices): 22% (2017 est.)

55.5% (2016 est.)

country comparison to the world: 216

Current account balance: -$2 million (2017 est.)

-$169 million (2016 est.)

country comparison to the world: 68

Exports: $2.028 billion (2017 est.)

$1.449 billion (2016 est.)

country comparison to the world: 139

Exports—partners: Switzerland 38%, Hong Kong 21.9%, Belgium 10.1%, UAE 7.2%, Guyana 6.1% (2017)

Exports—commodities: alumina, gold, crude oil, lumber, shrimp and fish, rice, bananas

Imports: $1.293 billion (2017 est.)
$1.203 billion (2016 est.)
country comparison to the world: 177

Imports—commodities: capital equipment, petroleum, foodstuffs, cotton, consumer goods

Imports—partners: US 30.6%, Netherlands 14.8%, Trinidad and Tobago 11.4%, China 7.6% (2017)

Reserves of foreign exchange and gold: $424.4 million (31 December 2017 est.)
$381.1 million (31 December 2016 est.)
country comparison to the world: 158

Debt—external: $1.7 billion (31 December 2017 est.)
$1.436 billion (31 December 2016 est.)
country comparison to the world: 155

Exchange rates: Surinamese dollars (SRD) per US dollar -
7.53 (2017 est.)
6.229 (2016 est.)
6.229 (2015 est.)
3.4167 (2014 est.)
3.3 (2013 est.)

ENERGY

Electricity access: *electrification—total population:* 87.2% (2016)
electrification—urban areas: 96.4% (2016)
electrification—rural areas: 69.3% (2016)

Electricity—production: 1.967 billion kWh (2016 est.)
country comparison to the world: 138

Electricity—consumption: 1.75 billion kWh (2016 est.)
country comparison to the world: 144

Electricity—exports: 0 kWh (2016 est.)
country comparison to the world: 204

Electricity—imports: 0 kWh (2016 est.)
country comparison to the world: 206

Electricity—installed generating capacity: 504,000 kW (2016 est.)
country comparison to the world: 149

Electricity—from fossil fuels: 61% of total installed capacity (2016 est.)
country comparison to the world: 129

Electricity—from nuclear fuels: 0% of total installed capacity (2017 est.)
country comparison to the world: 189

Electricity—from hydroelectric plants: 38% of total installed capacity (2017 est.)
country comparison to the world: 56

Electricity—from other renewable sources: 2% of total installed capacity (2017 est.)
country comparison to the world: 145

Crude oil—production: 17,000 bbl/day (2018 est.)

country comparison to the world: 70

Crude oil—exports: 0 bbl/day (2015 est.)
country comparison to the world: 200

Crude oil—imports: 820 bbl/day (2015 est.)
country comparison to the world: 80

Crude oil—proved reserves: 84.2 million bbl (1 January 2018 est.)
country comparison to the world: 70

Refined petroleum products—production: 7,571 bbl/day (2015 est.)
country comparison to the world: 101

Refined petroleum products—consumption: 13,000 bbl/day (2016 est.)
country comparison to the world: 157

Refined petroleum products—exports: 14,000 bbl/day (2015 est.)
country comparison to the world: 74

Refined petroleum products—imports: 10,700 bbl/day (2015 est.)
country comparison to the world: 145

Natural gas—production: 0 cu m (2017 est.)
country comparison to the world: 202

Natural gas—consumption: 0 cu m (2017 est.)
country comparison to the world: 203

Natural gas—exports: 0 cu m (2017 est.)
country comparison to the world: 191

Natural gas—imports: 0 cu m (2017 est.)
country comparison to the world: 194

Natural gas—proved reserves: 0 cu m (1 January 2011 est.)
country comparison to the world: 198

Carbon dioxide emissions from consumption of energy: 2.075 million Mt (2017 est.)
country comparison to the world: 159

COMMUNICATIONS

Telephones—fixed lines: *total subscriptions:* 96,310
subscriptions per 100 inhabitants: 15.95 (2019 est.)
country comparison to the world: 139

Telephones—mobile cellular: *total subscriptions:* 845,292
subscriptions per 100 inhabitants: 139.99 (2019 est.)
country comparison to the world: 163

Telecommunication systems: *general assessment:* international facilities are good; state-owned fixed-line teledensity and broadband services below regional average for Latin America and Caribbean, but mobile penetration is above regional average; fixed-line effective along the coastline and poor in the interior; competition in the mobile sector (2020)
domestic: fixed-line 16 per 100 and mobile-cellular teledensity 140 telephones per 100 persons; microwave radio relay network is in place (2019)
international: country code—597; landing point for the SG-SCS submarine cable linking South America with the Caribbean; satellite earth stations—2 Intelsat (Atlantic Ocean) (2019)

note: the COVID-19 outbreak is negatively impacting telecommunications production and supply chains globally; consumer spending on telecom devices and services has also slowed due to the pandemic's effect on economies worldwide; overall progress towards improvements in all facets of the telecom industry—mobile, fixed-line, broadband, submarine cable and satellite—has moderated

Broadcast media: 2 state-owned TV stations; 1 state-owned radio station; multiple private radio and TV stations (2019)

Internet country code: .sr

Internet users: *total:* 292,685
percent of population: 48.95% (July 2018 est.)
country comparison to the world: 166

Broadband—fixed subscriptions: *total:* 73,176
subscriptions per 100 inhabitants: 12 (2018 est.)
country comparison to the world: 127

TRANSPORTATION

National air transport system: *number of registered air carriers:* 4 (2020)
inventory of registered aircraft operated by air carriers: 20
annual passenger traffic on registered air carriers: 272,347 (2018)
annual freight traffic on registered air carriers: 33.2 million mt-km (2018)

Civil aircraft registration country code prefix: PZ (2016)

Airports: 55 (2013)
country comparison to the world: 84

Airports—with paved runways: *total:* 6 (2019)
over 3,047 m: 1
under 914 m: 5

Airports—with unpaved runways: *total:* 49 (2013)
914 to 1,523 m: 4 (2013)
under 914 m: 45 (2013)

Pipelines: 50 km oil (2013)

Roadways: *total:* 4,304 km (2003)
paved: 1,119 km (2003)
unpaved: 3,185 km (2003)
country comparison to the world: 151

Waterways: 1,200 km (most navigable by ships with drafts up to 7 m) (2011)
country comparison to the world: 59

Merchant marine: *total:* 10
by type: general cargo 5, oil tanker 3, other 2 (2019)
country comparison to the world: 155

Ports and terminals: *major seaport(s):* Paramaribo, Wageningen

MILITARY AND SECURITY

Military and security forces: *Suriname Army (National Leger, NL):* Army, Navy, Air Force, Military Police (2020)

Military and security service personnel strengths: the Suriname Army is comprised of approximately 1,800 active personnel (ground, air, naval, and military police) (2019 est.)

Military equipment inventories and acquisitions: the Suriname Army inventory includes a mix of equipment from several foreign suppliers, including Brazil, China, India, and the US; since 2010, Suriname has received small quantities of military hardware from Colombia, France, India, and the US (2019 est.)

Military service age and obligation: 18 is the legal minimum age for voluntary military service; no conscription (2019)

TRANSNATIONAL ISSUES

Disputes—international: area claimed by French Guiana between Riviere Litani and Riviere Marouini (both headwaters of the Lawa); Suriname claims a triangle of land between the New and Kutari/Koetari rivers in a historic dispute over the headwaters of the Courantyne; Guyana seeks UN Convention on the Law of the Sea arbitration to resolve the longstanding dispute with Suriname over the axis of the territorial sea boundary in potentially oil-rich waters

Trafficking in persons: *current situation:* Suriname is a source, transit, and destination country for women and children subjected to sex trafficking and men, women, and children subjected to forced labor; women and girls from Suriname, Guyana, Brazil, and the Dominican Republic are subjected to sex trafficking in the country, sometimes in interior mining camps; migrant workers in agriculture and on fishing boats and children working in informal urban sectors and gold mines are vulnerable to forced labor; traffickers from Suriname exploit victims in the Netherlands

tier rating: Tier 2 Watch List – Suriname does not fully comply with the minimum standards for the elimination of trafficking; however, it is making significant efforts to do so; in 2014, Suriname was granted a waiver from an otherwise required downgrade to Tier 3 because its government has a written plan that, if implemented, would constitute making significant efforts to bring itself into compliance with the minimum standards for the elimination of trafficking; authorities increased the number of trafficking investigations, prosecutions, and convictions as compared to 2013, but resources were insufficient to conduct investigations in the country's interior; more trafficking victims were identified in 2014 than in 2013, but protective services for adults and children were inadequate, with a proposed government shelter for women and child trafficking victims remaining unopened (2015)

Illicit drugs: growing transshipment point for South American drugs destined for Europe via the Netherlands and Brazil; transshipment point for arms-for-drugs dealing

SVALBARD

INTRODUCTION

Background: The archipelago may have been first discovered by Norse explorers in the 12th century; the islands served as an international whaling base during the 17th and 18th centuries. Norway's sovereignty was internationally recognized by treaty in 1920, and five years later it officially took over the territory. In the 20th century coal mining started and today a Norwegian and a Russian company are still functioning. Travel between the settlements is accomplished with snowmobiles, aircraft, and boats.

GEOGRAPHY

Location: Northern Europe, islands between the Arctic Ocean, Barents Sea, Greenland Sea, and Norwegian Sea, north of Norway

Geographic coordinates: 78 00 N, 20 00 E

Map references: Arctic Region

Area: *total:* 62,045 sq km
land: 62,045 sq km
water: 0 sq km
note: includes Spitsbergen and Bjornoya (Bear Island)
country comparison to the world: 126

Area—comparative: slightly smaller than West Virginia

Land boundaries: 0 km

Coastline: 3,587 km

Maritime claims: *territorial sea:* 12 nm
contiguous zone: 24 nm
continental shelf: extends to depth of exploitation
exclusive fishing zone: 200 nm

Climate: arctic, tempered by warm North Atlantic Current; cool summers, cold winters; North Atlantic Current flows along west and north coasts of Spitsbergen, keeping water open and navigable most of the year

Terrain: rugged mountains; much of the upland areas are ice covered; west coast clear of ice about half the year; fjords along west and north coasts

Elevation: *lowest point:* Arctic Ocean 0 m
highest point: Newtontoppen 1,717 m

Natural resources: coal, iron ore, copper, zinc, phosphate, wildlife, fish

Land use: *agricultural land:* 0% (2011 est.)
arable land: 0% (2011 est.) / permanent crops: 0% (2011 est.) / permanent pasture: 0% (2011 est.)
forest: 0% (2011 est.)
other: 100% (2011 est.)

Population distribution: the small population is primarily concentrated on the island of Spitsbergen in a handful of settlements on the south side of the Isfjorden, with Longyearbyen being the largest

Natural hazards: ice floes often block the entrance to Bellsund (a transit point for coal export) on the west coast and occasionally make parts of the northeastern coast inaccessible to maritime traffic

Environment—current issues: ice floes are a maritime hazard; past exploitation of mammal species (whale, seal, walrus, and polar bear) severely depleted the populations, but a gradual recovery seems to be occurring

Geography—note: northernmost part of the Kingdom of Norway; consists of nine main islands; glaciers and snowfields cover 60% of the total area; Spitsbergen Island is the site of the Svalbard Global Seed Vault, a seed repository established by the Global Crop Diversity Trust and the Norwegian Government

PEOPLE AND SOCIETY

Population: 2,926 (July 2019 est.)
country comparison to the world: 230

Ethnic groups: Norwegian 58%, foreign population 42% (consists primarily of Russians, Thais, Swedes, Filipinos, and Ukrainians) (2019 est.)
note: foreigners account for almost one third of the population of the Norwegian settlements, Longyearbyen and Ny-Alesund (where the majority of Svalbard's resident population lives), as of mid-2019

Languages: Norwegian, Russian

Population growth rate: -0.03% (2014 est.)
country comparison to the world: 197

Population distribution: the small population is primarily concentrated on the island of Spitsbergen in a handful of settlements on the south side of the Isfjorden, with Longyearbyen being the largest

Sex ratio: NA

Infant mortality rate: *total:* NA (2018)
male: NA
female: NA

Life expectancy at birth: *total population:* NA (2017 est.)
male: NA
female: NA

Total fertility rate: NA

HIV/AIDS—adult prevalence rate: NA

Education expenditures: NA

GOVERNMENT

Country name: *conventional long form:* none
conventional short form: Svalbard (sometimes referred to as Spitsbergen, the largest island in the archipelago)
etymology: 12th century Norse accounts speak of the discovery of a "Svalbard"—literally "cold shores"—but they may have referred to Jan Mayen Island or eastern Greenland; the archipelago was traditionally known as Spitsbergen, but Norway renamed it Svalbard in the 1920s when it assumed sovereignty of the islands

Dependency status: territory of Norway; administered by the Polar Department of the Ministry of Justice, through a governor (sysselmann) residing in Longyearbyen, Spitsbergen; by treaty (9 February 1920), sovereignty was awarded to Norway

Government type: non-self-governing territory of Norway

Capital: *name:* Longyearbyen

geographic coordinates: 78 13 N, 15 38 E
time difference: UTC+1 (6 hours ahead of Washington, DC, during Standard Time)
daylight saving time: +1hr, begins last Sunday in March; ends last Sunday in October
etymology: the name in Norwegian means Longyear Town; the site was established by and named after John LONGYEAR, whose Arctic Coal Company began mining operations there in 1906

Independence: none (territory of Norway)

Legal system: the laws of Norway where applicable apply; only the laws of Norway made explicitly applicable to Svalbard have effect there; the Svalbard Act and the Svalbard Environmental Protection Act, and certain regulations, apply only to Svalbard; the Spitsbergen Treaty and the Svalbard Treaty grant certain rights to citizens and corporations of signatory nations; as of June 2017, 45 nations had ratified the Svalbard Treaty

Citizenship: see Norway

Executive branch: *chief of state:* King HARALD V of Norway (since 17 January 1991); Heir Apparent Crown Prince Haakon MAGNUS (son of the king, born 20 July 1973)
head of government: Governor Kjerstin ASKHOLT (since 1 October 2015); Assistant Governor Berit SAGFOSSEN (since 1 April 2016)
elections/appointments: none; the monarchy is hereditary; governor and assistant governor responsible to the Polar Department of the Ministry of Justice

Legislative branch: *description:* unicameral Longyearbyen Community Council (15 seats; members directly elected by majority vote to serve 4-year-terms); note—the Council acts very much like a Norwegian municipality, responsible for infrastructure and utilities, including power, land-use and community planning, education, and child welfare; however, healthcare services are provided by the state
elections: last held on 7 October 2019 (next to be held in October 2023)
election results: seats by party—Conservatives 5, Labor Party 5, Liberals 3, Green Party 2

Judicial branch: none; note—Svalbard is subordinate to Norway's Nord-Troms District Court and Halogaland Court of Appeal, both located in Tromso

Political parties and leaders: Svalbard Conservative Party [Kjetil FIGENSCHOU] Svalbard Green Party [Helga Bardsdatter KRISTIANSEN, Espen Klungseth ROTEVATN] Svalbard Labor Party [Elise STROMSENG] Svalbard Liberal Party [Erik BERGER]
International organization participation: none

Flag description: the flag of Norway is used

National anthem: *note:* as a territory of Norway, "Ja, vi elsker dette landet" is official (see Norway) 0:00 / 1:02

ECONOMY

Economy—overview: Coal mining, tourism, and international research are Svalbard's major industries. Coal mining has historically been the dominant economic activity, and the Spitzbergen Treaty of 9 February 1920 gives the 45 countries that so far have ratified the treaty equal rights to exploit mineral deposits, subject to Norwegian regulation. Although US, UK, Dutch, and Swedish coal companies have mined in the past, the only companies still engaging in this are Norwegian and Russian. Low coal prices have forced the Norwegian coal company, Store Norske Spitsbergen Kulkompani, to close one of its two mines and to considerably reduce the activity of the other. Since the 1990s, the tourism and hospitality industry has grown rapidly, and Svalbard now receives 60,000 visitors annually.

The settlements on Svalbard were established as company towns, and at their height in the 1950s, the Norwegian stateowned coal company supported nearly 1,000 jobs. Today, only about 300 people work in the mining industry.

Goods such as alcohol, tobacco, and vehicles, normally highly taxed on mainland Norway, are considerably cheaper in Svalbard in an effort by the Norwegian Government to entice more people to live on the Arctic archipelago. By law, Norway collects only enough taxes to pay for the needs of the local government; none of tax proceeds go to the central government.

GDP—real growth rate: NA

Labor force: 1,590 (2013)
country comparison to the world: 228

Budget: *revenues:* NA

expenditures: NA

Taxes and other revenues: NA

Budget surplus (+) or deficit (-): NA

Exports: NA

Imports: $NA

Exchange rates: Norwegian kroner (NOK) per US dollar -
8.308 (2017 est.)
8.0646 (2016 est.)
8.0646 (2015)
8.0646 (2014 est.)
6.3021 (2013 est.)

ENERGY

Crude oil—production: 194,300 bbl/day (2014 est.)
country comparison to the world: 36

Crude oil—exports: 16,070 bbl/day (2012 est.)
country comparison to the world: 53

Crude oil—imports: 0 bbl/day (2012 est.)
country comparison to the world: 200

Refined petroleum products—consumption: 80,250 bbl/day (2013 est.)
country comparison to the world: 87

Refined petroleum products—exports: 4,488 bbl/day (2012 est.)
country comparison to the world: 93

Refined petroleum products—imports: 18,600 bbl/day (2012 est.)
country comparison to the world: 127

Natural gas—production: 0 cu m (2013 est.)
country comparison to the world: 203

Natural gas—consumption: 0 cu m (2013 est.)
country comparison to the world: 204

Natural gas—exports: 0 cu m (2013 est.)
country comparison to the world: 192

Natural gas—imports: 0 cu m (2013 est.)
country comparison to the world: 195

COMMUNICATIONS

Telecommunication systems: *general assessment:* modern, well-developed (2018)
domestic: the Svalbard Satellite Station—connected to the mainland via the Svalbard Undersea Cable System—is the only Arctic ground station that can see low-altitude, polar-orbiting satellites; it provides ground services to more satellites than any other facility in the world (2018)
international: country code—47-790; the Svalbard Undersea Cable System is a twin communications cable that connects Svalbard to mainland Norway; the system is the sole telecommunications link to the archipelago (2019)
note: the COVID-19 outbreak is negatively impacting telecommunications production and supply chains globally; consumer spending on telecom devices and services has also slowed due to the pandemic's effect on economies worldwide; overall progress towards improvements in all facets of the telecom industry—mobile, fixed-line, broadband, submarine cable and satellite—has moderated

Broadcast media: the Norwegian Broadcasting Corporation (NRK) began direct TV transmission to Svalbard via satellite in 1984; Longyearbyen households have access to 3 NRK radio and 2 TV stations

Internet country code: .sj

Airports: 4 (2013)
country comparison to the world: 189

Airports—with paved runways: *total:* 1 (2019)
2,438 to 3,047 m: 1

Airports—with unpaved runways: *total:* 3 (2013)
under 914 m: 3 (2013)

Heliports: 1 (2013)

Roadways: *total:* 40 km (2020)
country comparison to the world: 218

Ports and terminals: *major seaport(s):* Barentsburg, Longyearbyen, Ny-Alesund, Pyramiden

Military and security forces: no regular military forces; military installations prohibited by treaty

Military—note: Svalbard is a territory of Norway, demilitarized by treaty on 9 February 1920; Norwegian military activity is limited to fisheries surveillance by the Norwegian Coast Guard

Disputes—international: despite recent discussions, Russia and Norway dispute their maritime limits in the Barents Sea and Russia's fishing rights beyond Svalbard's territorial limits within the Svalbard Treaty zone

SWEDEN

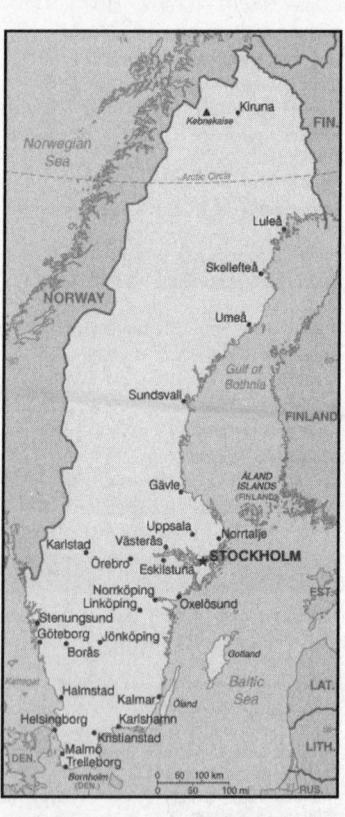

Background: A military power during the 17th century, Sweden has not participated in any war for two centuries. An armed neutrality was preserved in both World Wars. Since then, Sweden has pursued a successful economic formula consisting of a capitalist system intermixed with substantial welfare elements. Sweden joined the EU in 1995, but the public rejected the introduction of the euro in a 2003 referendum. The share of

Sweden's population born abroad increased from 11.3% in 2000 to 19.1% in 2018.

Location: Northern Europe, bordering the Baltic Sea, Gulf of Bothnia, Kattegat, and Skagerrak, between Finland and Norway

Geographic coordinates: 62 00 N, 15 00 E

Map references: Europe

Area: *total:* 450,295 sq km
land: 410,335 sq km
water: 39,960 sq km
country comparison to the world: 57

Area—comparative: almost three times the size of Georgia; slightly larger than California

Land boundaries: *total:* 2,211 km
border countries (2): Finland 545 km, Norway 1666 km

Coastline: 3,218 km

Maritime claims: *territorial sea:* 12 nm (adjustments made to return a portion of straits to high seas)
exclusive economic zone: agreed boundaries or midlines
continental shelf: 200-m depth or to the depth of exploitation

Climate: temperate in south with cold, cloudy winters and cool, partly cloudy summers; subarctic in north

Terrain: mostly flat or gently rolling lowlands; mountains in west

Elevation: *mean elevation:* 320 m
lowest point: reclaimed bay of Lake Hammarsjon, near Kristianstad -2.4 m
highest point: Kebnekaise 2,111 m

Natural resources: iron ore, copper, lead, zinc, gold, silver, tungsten, uranium, arsenic, feldspar, timber, hydropower

Land use: *agricultural land:* 7.5% (2011 est.)
arable land: 6.4% (2011 est.) / permanent crops: 0% (2011 est.) / permanent pasture: 1.1% (2011 est.)
forest: 68.7% (2011 est.)

other: 23.8% (2011 est.)
Irrigated land: 1,640 sq km (2012)

Population distribution: most Swedes live in the south where the climate is milder and there is better connectivity to mainland Europe; population clusters are found all along the Baltic coast in the east; the interior areas of the north remain sparsely populated

Natural hazards: ice floes in the surrounding waters, especially in the Gulf of Bothnia, can interfere with maritime traffic

Environment—current issues: marine pollution (Baltic Sea and North Sea); acid rain damage to soils and lakes; air pollution; inappropriate timber harvesting practices

Environment—international agreements: *party to:* Air Pollution, Air Pollution-Nitrogen Oxides, Air Pollution-Persistent Organic Pollutants, Air Pollution-Sulfur 85, Air Pollution-Sulfur 94, Air Pollution-Volatile Organic Compounds, Antarctic-Environmental Protocol, Antarctic-Marine Living Resources, Antarctic Treaty, Biodiversity, Climate Change, Climate Change-Kyoto Protocol, Desertification, Endangered Species, Environmental Modification, Hazardous Wastes, Law of the Sea, Marine Dumping, Ozone Layer Protection, Ship Pollution, Tropical Timber 83, Tropical Timber 94, Wetlands, Whaling
signed, but not ratified: none of the selected agreements

Geography—note: strategic location along Danish Straits linking Baltic and North Seas; Sweden has almost 100,000 lakes, the largest of which, Vanern, is the third largest in Europe

Population: 10,202,491 (July 2020 est.)
country comparison to the world: 91

Nationality: *noun:* Swede(s)
adjective: Swedish

Ethnic groups: Swedish 80.9%, Syrian 1.8%, Finnish 1.4%, Iraqi 1.4%, other 14.5% (2018 est.)
note: data represent the population by country of birth; the indigenous Sami people are estimated to number between 20,000 and 40,000

Languages: Swedish (official)

note: Finnish, Sami, Romani, Yiddish, and Meankieli are official minority languages

Religions: Church of Sweden (Lutheran) 60.2%, other (includes Roman Catholic, Orthodox, Baptist, Muslim, Jewish, and Buddhist) 8.5%, none or unspecified 31.3% (2017 est.)

note: estimates reflect registered members of faith communities eligible for state funding (not all religions are statefunded and not all people who identify with a particular religion are registered members); an estimated 57.7% of Sweden's population were members of the Church of Sweden in 2018

Age structure: *0-14 years:* 17.71% (male 928,413/female 878,028)

15-24 years: 10.8% (male 569,082/female 532,492)

25-54 years: 39.01% (male 2,016,991/female 1,962,617)

55-64 years: 11.9% (male 610,521/female 603,795)

65 years and over: 20.59% (male 974,410/female 1,126,142) (2020 est.)

Dependency ratios: *total dependency ratio:* 61.2

youth dependency ratio: 28.4

elderly dependency ratio: 32.8

potential support ratio: 3.1 (2020 est.)

Median age: *total:* 41.1 years

male: 40.1 years

female: 42.1 years (2020 est.)

country comparison to the world: 47

Population growth rate: 0.79% (2020 est.)

country comparison to the world: 130

Birth rate: 12.1 births/1,000 population (2020 est.)

country comparison to the world: 161

Death rate: 9.4 deaths/1,000 population (2020 est.)

country comparison to the world: 49

Net migration rate: 5.2 migrant(s)/1,000 population (2020 est.)

country comparison to the world: 22

Population distribution: most Swedes live in the south where the climate is milder and there is better connectivity to mainland Europe; population clusters are found all along the Baltic coast in the east; the interior areas of the north remain sparsely populated

Urbanization: *urban population:* 88% of total population (2020)

rate of urbanization: 1.05% annual rate of change (2015-20 est.)

total population growth rate v. urban population growth rate, 2000-2030: *Major urban areas—population:* 1.633 million STOCKHOLM (capital) (2020)

Sex ratio: *at birth:* 1.06 male(s)/female

0-14 years: 1.06 male(s)/female

15-24 years: 1.07 male(s)/female

25-54 years: 1.03 male(s)/female

55-64 years: 1.01 male(s)/female

65 years and over: 0.87 male(s)/female

total population: 1 male(s)/female (2020 est.)

Mother's mean age at first birth: 29.3 years (2015 est.)

Maternal mortality rate: 4 deaths/100,000 live births (2017 est.)

country comparison to the world: 175

Infant mortality rate: *total:* 2.6 deaths/1,000 live births

male: 2.9 deaths/1,000 live births

female: 2.3 deaths/1,000 live births (2020 est.)

country comparison to the world: 220

Life expectancy at birth: *total population:* 82.4 years

male: 80.4 years

female: 84.5 years (2020 est.)

country comparison to the world: 18

Total fertility rate: 1.87 children born/woman (2020 est.)

country comparison to the world: 138

Drinking water source:

improved:

urban: 100% of population

rural: 100% of population

total: 100% of population

unimproved:

urban: 0% of population

rural: 0% of population

total: 0% of population (2017 est.)

Current Health Expenditure: 11% (2017)

Physicians density: 3.98 physicians/1,000 population (2016)

Hospital bed density: 2.2 beds/1,000 population (2017)

Sanitation facility access:

improved:

urban: 100% of population

rural: 100% of population

total: 100% of population

unimproved:

urban: 0% of population

rural: 0% of population

total: 0% of population (2017 est.)

HIV/AIDS—adult prevalence rate: 0.2% (2016 est.)

country comparison to the world: 111

HIV/AIDS—people living with HIV/AIDS: 11,000 (2016 est.)

country comparison to the world: 100

HIV/AIDS—deaths: NA

Obesity—adult prevalence rate: 20.6% (2016)

country comparison to the world: 97

Education expenditures: 7.7% of GDP (2016)

country comparison to the world: 6

School life expectancy (primary to tertiary education): *total:* 20 years

male: 19 years

female: 21 years (2018)

Unemployment, youth ages 15-24: *total:* 16.8%

male: 18%

female: 15.5% (2018 est.)

country comparison to the world: 81

GOVERNMENT

Country name: *conventional long form:* Kingdom of Sweden

conventional short form: Sweden

local long form: Konungariket Sverige

local short form: Sverige

etymology: name ultimately derives from the North Germanic Svear tribe, which inhabited central Sweden and is first mentioned in the first centuries A.D.

Government type: parliamentary constitutional monarchy

Capital: *name:* Stockholm

geographic coordinates: 59 20 N, 18 03 E

time difference: UTC+1 (6 hours ahead of Washington, DC, during Standard Time)

daylight saving time: +1hr, begins last Sunday in March; ends last Sunday in October

etymology: "stock" and "holm" literally mean "log" and "islet" in Swedish, but there is no consensus as to what the words refer to

Administrative divisions: 21 counties (lan, singular and plural); Blekinge, Dalarna, Gavleborg, Gotland, Halland, Jamtland, Jonkoping, Kalmar, Kronoberg, Norrbotten, Orebro, Ostergotland, Skane, Sodermanland, Stockholm, Uppsala, Varmland, Vasterbotten, Vasternorrland, Vastmanland, Vastra Gotaland

Independence: 6 June 1523 (Gustav VASA elected king of Sweden, marking the abolishment of the Kalmar Union between Denmark, Norway, and Sweden)

National holiday: National Day, 6 June (1983); note—from 1916 to 1982 this date was celebrated as Swedish Flag Day

Constitution: *history:* several previous; latest adopted 1 January 1975

amendments: proposed by Parliament; passage requires simple majority vote in two consecutive parliamentary terms with an intervening general election; passage also requires approval by simple majority vote in a referendum if Parliament approves a motion for a referendum by one third of its members; amended several times, last in 2014 (changes to the "Instrument of Government")

Legal system: civil law system influenced by Roman-Germanic law and customary law

International law organization participation: accepts compulsory ICJ jurisdiction with reservations; accepts ICCt jurisdiction

Citizenship: *citizenship by birth:* no

citizenship by descent only: the father must be a citizen of Sweden; in the case of a child born out of wedlock, the mother must be a citizen of Sweden and the father unknown

dual citizenship recognized: no, unless the other citizenship was acquired involuntarily

residency requirement for naturalization: 5 years

Suffrage: 18 years of age; universal

Executive branch: *chief of state:* King CARL XVI GUSTAF (since 15 September 1973); Heir Apparent Princess VICTORIA Ingrid Alice Desiree (daughter of the monarch, born 14 July 1977) *head of government:* Prime Minister Stefan LOFVEN (since 3 October 2014); Deputy Prime Minister Isabella LOVIN (since 25 May 2016); note—Prime Minister Stefan LOFVEN was ousted in a no-confidence vote on 25 September 2018 and headed a caretaker government until the next government was formed; LOFVEN was reelected as Prime Minister and took office on 21 January 2019

cabinet: Cabinet appointed by the prime minister
elections/appointments: the monarchy is hereditary; following legislative elections, the leader of the majority party or majority coalition usually becomes the prime minister

Legislative branch: *description:* unicameral Parliament or Riksdag (349 seats; 310 members directly elected in multi-seat constituencies by closed, party-list proportional representation vote and 39 members in "at-large" seats directly elected by proportional representation vote; members serve 4-year terms)

elections: last held on 9 September 2018 (next to be held in 2022)

election results: percent of vote by party—SAP 28.3%, M 19.8%, SD 17.5%, C 8.6%, V 8%, KD 6.3%, L 5.5%, MP 4.4%, other 1.6%; seats by party—SAP 100, M 70, SD 62, C 31, V 28, KD 22, L 20, MP 16; composition—men 188, women 161, percent of women 46.1%

Judicial branch: *highest courts:* Supreme Court of Sweden (consists of 16 justices, including the court chairman); Supreme Administrative Court (consists of 18 justices, including the court president)

judge selection and term of office: Supreme Court and Supreme Administrative Court justices nominated by the Judges Proposal Board, a 9-member nominating body consisting of high-level judges, prosecutors, and members of Parliament; justices appointed by the Government; following a probationary period, justices' appointments are permanent

subordinate courts: first instance, appellate, general, and administrative courts; specialized courts that handle cases such as land and environment, immigration, labor, markets, and patents

Political parties and leaders: Center Party (Centerpartiet) or C [Annie LOOF]
Christian Democrats (Kristdemokraterna) or KD [Ebba Busch THOR]
Green Party (Miljopartiet de Grona) or MP [Isabella LOVIN and Per BOLUND]
Left Party (Vansterpartiet) or V [Jonas SJOSTEDT]
Liberal Party (Liberalerna) or L [Jan BJORKLUND]
Moderate Party (Moderaterna) or M [Ulf KRISTERSSON]
Swedish Social Democratic Party (Socialdemokraterna) or SAP [Stefan LOFVEN]
Sweden Democrats (Sverigedemokraterna) or SD [Jimmie AKESSON]

International organization participation: ADB (nonregional member), AfDB (nonregional member), Arctic Council, Australia Group, BIS, CBSS, CD, CE, CERN, EAPC, EBRD, ECB, EIB, EITI (implementing country), EMU, ESA, EU, FAO, FATF, G-9, G-10, IADB, IAEA, IBRD, ICAO, ICC (national committees), ICCt, ICRM, IDA, IEA, IFAD, IFC, IFRCS, IGAD (partners), IHO, ILO, IMF, IMO, IMSO, Interpol, IOC, IOM, IPU, ISO, ITSO, ITU, ITUC (NGOs), MIGA, MINUSMA, MONUSCO, NC, NEA, NIB, NSG, OAS (observer), OECD, OPCW, OSCE, Paris Club, PCA, PFP, Schengen Convention, UN, UNCTAD, UNESCO, UNHCR, UNIDO, UNMISS, UNMOGIP, UNRWA, UN Security Council (temporary), UNTSO, UPU, WCO, WFTU (NGOs), WHO, WIPO, WMO, WTO, ZC

Diplomatic representation in the US: *chief of mission:* Ambassador Karin Ulrika OLOFSDOTTER (since 17 September 2017)

chancery: The House of Sweden, 2900 K Street NW, Washington, DC 20007

telephone: [1] (202) 536-1500

FAX: [1] (202) 536-1501

consulate(s) general: New York, San Francisco

Diplomatic representation from the US: *chief of mission:* Ambassador Kenneth A. HOWERY (since 10 October 2019)

telephone: [46] (08) 783 53 00

embassy: Dag Hammarskjolds Vag 31, SE-11589 Stockholm

mailing address: American Embassy Stockholm, US Department of State, 5750 Stockholm Place, Washington, DC 205215750

FAX: [46] (08) 661 19 64

Flag description: blue with a golden yellow cross extending to the edges of the flag; the vertical part of the cross is shifted to the hoist side in the style of the Dannebrog (Danish flag); the colors reflect those of the Swedish coat of arms—three gold crowns on a blue field

National symbol(s): *three crowns, lion; national colors:* blue, yellow

National anthem: *name:* "Du Gamla, Du Fria" (Thou Ancient, Thou Free)

lyrics/music: Richard DYBECK/traditional

note: in use since 1844; also known as "Sang till Norden" (Song of the North), is based on a Swedish folk tune; it has never been officially adopted by the government; "Kungssangen" (The King's Song) serves as the royal anthem and is played in the presence of the royal family and during certain state ceremonies

0:00 / 1:27

ECONOMY

Economy—overview: Sweden's small, open, and competitive economy has been thriving and Sweden has achieved an enviable standard of living with its combination of free-market capitalism and extensive welfare benefits. Sweden remains outside the euro zone largely out of concern that joining the European Economic and Monetary Union would diminish the country's sovereignty over its welfare system.

Timber, hydropower, and iron ore constitute the resource base of a manufacturing economy that relies heavily on foreign trade. Exports, including engines and other machines, motor vehicles, and telecommunications equipment, account for more than 44% of GDP. Sweden enjoys a current account surplus of about 5% of GDP, which is one of the highest margins in Europe.

GDP grew an estimated 3.3% in 2016 and 2017 driven largely by investment in the construction sector. Swedish economists expect economic growth to ease slightly in the coming years as this investment subsides. Global economic growth boosted exports of Swedish manufactures further, helping drive domestic economic growth in 2017. The Central Bank is keeping an eye on deflationary pressures and bank observers expect it to maintain an expansionary monetary policy in 2018. Swedish prices and wages have grown only slightly over the past few years, helping to support the country's competitiveness.

In the short and medium term, Sweden's economic challenges include providing affordable housing and successfully integrating migrants into the labor market.

GDP (purchasing power parity): $518 billion (2017 est.)
$507.3 billion (2016 est.)
$494 billion (2015 est.)
note: data are in 2017 dollars
country comparison to the world: 40

GDP (official exchange rate): $535.6 billion (2017 est.)

GDP—real growth rate: 1.29% (2019 est.)
2.06% (2018 est.)
2.82% (2017 est.)
country comparison to the world: 163

GDP—per capita (PPP): $51,200 (2017 est.)
$50,800 (2016 est.)
$50,100 (2015 est.)
note: data are in 2017 dollars
country comparison to the world: 26

Gross national saving: 28.9% of GDP (2017 est.)
28.8% of GDP (2016 est.)
28.8% of GDP (2015 est.)
country comparison to the world: 34

GDP—composition, by end use:
household consumption: 44.1% (2017 est.)
government consumption: 26% (2017 est.)
investment in fixed capital: 24.9% (2017 est.)
investment in inventories: 0.8% (2017 est.)
exports of goods and services: 45.3% (2017 est.)
imports of goods and services: -41.1% (2017 est.)

GDP—composition, by sector of origin:
agriculture: 1.6% (2017 est.)
industry: 33% (2017 est.)
services: 65.4% (2017 est.)

Agriculture—products: barley, wheat, sugar beets; meat, milk

Industries: iron and steel, precision equipment (bearings, radio and telephone parts, armaments), wood pulp and paper products, processed foods, motor vehicles

Industrial production growth rate: 4.1% (2017 est.)
country comparison to the world: 74

Labor force: 5.029 million (2020 est.)
country comparison to the world: 77

Labor force—by occupation: *agriculture:* 2%
industry: 12%
services: 86% (2014 est.)

Unemployment rate: 6.78% (2019 est.)

6.33% (2018 est.)
country comparison to the world: 107

Population below poverty line: 15% (2014 est.)

Household income or consumption by percentage share: *lowest 10%:* 3.4%
highest 10%: 24% (2012)

Budget: *revenues:* 271.2 billion (2017 est.)
expenditures: 264.4 billion (2017 est.)

Taxes and other revenues: 50.6% (of GDP) (2017 est.)
country comparison to the world: 16

Budget surplus (+) or deficit (-): 1.3% (of GDP) (2017 est.)
country comparison to the world: 27

Public debt: 40.8% of GDP (2017 est.)
42.3% of GDP (2016 est.)
note: data cover general government debt and include debt instruments issued (or owned) by government entities other than the treasury; the data include treasury debt held by foreign entities; the data include debt issued by subnational entities, as well as intragovernmental debt; intragovernmental debt consists of treasury borrowings from surpluses in the social funds, such as for retirement, medical care, and unemployment; debt instruments for the social funds are not sold at public auctions
country comparison to the world: 124

Fiscal year: calendar year

Inflation rate (consumer prices): 1.9% (2017 est.)
1.1% (2016 est.)
country comparison to the world: 100

Current account balance: $22.339 billion (2019 est.)
$13.902 billion (2018 est.)
country comparison to the world: 16

Exports: $165.6 billion (2017 est.)
$151.4 billion (2016 est.)
country comparison to the world: 31

Exports—partners: Germany 11%, Norway 10.2%, Finland 6.9%, US 6.9%, Denmark 6.9%, UK 6.2%, Netherlands 5.5%, China 4.5%, Belgium 4.4%, France 4.2% (2017)

Exports—commodities: machinery (26%), motor vehicles, paper products, pulp and wood, iron and steel products, chemicals (2016 est.)

Imports: $153.2 billion (2017 est.)
$140.2 billion (2016 est.)
country comparison to the world: 30

Imports—commodities: machinery, petroleum and petroleum products, chemicals, motor vehicles, iron and steel; foodstuffs, clothing

Imports—partners: Germany 18.7%, Netherlands 8.9%, Norway 7.7%, Denmark 7.2%, China 5.5%, UK 5.1%, Finland 4.7%, Belgium 4.7% (2017)

Reserves of foreign exchange and gold: $62.22 billion (31 December 2017 est.)
$59.39 billion (31 December 2016 est.)
country comparison to the world: 36

Debt—external: $939.9 billion (31 March 2016 est.)
$929.4 billion (31 March 2015 est.)
country comparison to the world: 16

Exchange rates: Swedish kronor (SEK) per US dollar -
8.442 (2017 est.)
8.5605 (2016 est.)
8.5605 (2015 est.)
8.4335 (2014 est.)

6.8612 (2013 est.)

ENERGY

Electricity access: *electrification—total population:* 100% (2020)

Electricity—production: 152.9 billion kWh (2016 est.)
country comparison to the world: 27

Electricity—consumption: 133.5 billion kWh (2016 est.)
country comparison to the world: 27

Electricity—exports: 26.02 billion kWh (2016 est.)
country comparison to the world: 6

Electricity—imports: 14.29 billion kWh (2016 est.)
country comparison to the world: 16

Electricity—installed generating capacity: 40.29 million kW (2016 est.)
country comparison to the world: 26

Electricity—from fossil fuels: 5% of total installed capacity (2016 est.)
country comparison to the world: 204

Electricity—from nuclear fuels: 22% of total installed capacity (2017 est.)
country comparison to the world: 6

Electricity—from hydroelectric plants: 42% of total installed capacity (2017 est.)
country comparison to the world: 50

Electricity—from other renewable sources: 32% of total installed capacity (2017 est.)
country comparison to the world: 16

Crude oil—production: 0 bbl/day (2018 est.)
country comparison to the world: 204

Crude oil—exports: 14,570 bbl/day (2017 est.)
country comparison to the world: 55

Crude oil—imports: 400,200 bbl/day (2017 est.)
country comparison to the world: 23

Crude oil—proved reserves: 0 bbl (1 January 2018 est.)
country comparison to the world: 200

Refined petroleum products—production: 413,200 bbl/day (2017 est.)
country comparison to the world: 36

Refined petroleum products—consumption: 323,100 bbl/day (2017 est.)
country comparison to the world: 42

Refined petroleum products—exports: 371,100 bbl/day (2017 est.)
country comparison to the world: 23

Refined petroleum products—imports: 229,600 bbl/day (2017 est.)
country comparison to the world: 29

Natural gas—production: 0 cu m (2017 est.)
country comparison to the world: 204

Natural gas—consumption: 764.5 million cu m (2017 est.)
country comparison to the world: 97

Natural gas—exports: 0 cu m (2017 est.)
country comparison to the world: 193

Natural gas—imports: 764.5 million cu m (2017 est.)
country comparison to the world: 64

Natural gas—proved reserves: 0 cu m (1 January 2014 est.)
country comparison to the world: 199

Carbon dioxide emissions from consumption of energy: 52.31 million Mt (2017 est.)
country comparison to the world: 58

COMMUNICATIONS

Telephones—fixed lines: *total subscriptions:* 1,941,360
subscriptions per 100 inhabitants: 19.18 (2019 est.)
country comparison to the world: 56

Telephones—mobile cellular: *total subscriptions:* 12,785,850
subscriptions per 100 inhabitants: 126.32 (2019 est.)
country comparison to the world: 73

Telecommunication systems: *general assessment:* highly developed telecommunications infrastructure; ranked among leading countries for fixed-line, mobile-cellular, Internet, and broadband penetration; best developed LTE infrastructures in the region; first in the world to deliver 5G services (2020)
domestic: fixed-line 19 per 100 and mobile-cellular 126 per 100; coaxial and multiconductor cables carry most of the voice traffic; parallel microwave radio relay systems carry some additional telephone channels (2019)
international: country code—46; landing points for Botina, SFL, SFS-4, Baltic Sea Submarine Cable, Eastern Light, Sweden-Latvia, BCS North-Phase1, EE-S1, LV-SE1, BCS East-West Interlink, NordBalt, Baltica, Denmark-Sweden-15,-17,-18, Scandinavian Ring -North,-South, IP-Only Denmark-Sweden, Donica North, Kattegate-1,-2, Energinet Laeso-Varberg and GC2 submarine cables providing links to other Nordic countries and Europe; satellite earth stations—1 Intelsat (Atlantic Ocean), 1 Eutelsat, and 1 Inmarsat (Atlantic and Indian Ocean regions); note—Sweden shares the Inmarsat earth station with the other Nordic countries (Denmark, Finland, Iceland, and Norway) (2019)
note: the COVID-19 outbreak is negatively impacting telecommunications production and

supply chains globally; consumer spending on telecom devices and services has also slowed due to the pandemic's effect on economies worldwide; overall progress towards improvements in all facets of the telecom industry—mobile, fixed-line, broadband, submarine cable and satellite—has moderated

Broadcast media: publicly owned TV broadcaster operates 2 terrestrial networks plus regional stations; multiple privately owned TV broadcasters operating nationally, regionally, and locally; about 50 local TV stations; widespread access to pan-Nordic and international broadcasters through multi-channel cable and satellite TV; publicly owned radio broadcaster operates 3 national stations and a network of 25 regional channels; roughly 100 privately owned local radio stations with some consolidating into near national networks; an estimated 900 community and neighborhood radio stations broadcast intermittently

Internet country code: .se

Internet users: *total:* 9,251,773
percent of population: 92.14% (July 2018 est.)
country comparison to the world: 56

Broadband—fixed subscriptions: *total:* 3,973,622
subscriptions per 100 inhabitants: 40 (2018 est.)
country comparison to the world: 34

TRANSPORTATION

National air transport system: *number of registered air carriers:* 11 (2020)
inventory of registered aircraft operated by air carriers: 316

Civil aircraft registration country code prefix: SE (2016)

Airports: 231 (2013)
country comparison to the world: 25

Airports—with paved runways: *total:* 149 (2013)
over 3,047 m: 3 (2013)
2,438 to 3,047 m: 12 (2013)
1,524 to 2,437 m: 75 (2013)
914 to 1,523 m: 22 (2013)
under 914 m: 37 (2013)

Airports—with unpaved runways: *total:* 82 (2013)
914 to 1,523 m: 5 (2013)

under 914 m: 77 (2013)

Heliports: 2 (2013)

Pipelines: 1626 km gas (2013)

Railways: *total:* 14,127 km (2016)
standard gauge: 14,062 km 1.435-m gauge (12,322 km electrified) (2016)
narrow gauge: 65 km 0.891-m gauge (65 km electrified) (2016)
country comparison to the world: 20

Roadways: *total:* 573,134 km (includes 2,050 km of expressways) (2016)
paved: 140,100 km (2016)
unpaved: 433,034 km (2016)
note: includes 98,500 km of state roads, 433,034 km of private roads, and 41,600 km of municipal roads
country comparison to the world: 13

Waterways: 2,052 km (2010)
country comparison to the world: 40

Merchant marine: *total:* 360
by type: general cargo 51, oil tanker 20, other 289 (2019)
country comparison to the world: 50

Ports and terminals: *major seaport(s):* Brofjorden, Goteborg, Helsingborg, Karlshamn, Lulea, Malmo, Stockholm, Trelleborg, Visby
LNG terminal(s) (import): Brunnsviksholme, Lysekil

MILITARY AND SECURITY

Military and security forces: Swedish Armed Forces (Forsvarsmakten): Army, Navy, Air Force; Home Guard (2019)

Military expenditures: 1.1% of GDP (2019)
1% of GDP (2018)
1% of GDP (2017)
1.1% of GDP (2016)
1.1% of GDP (2015)
country comparison to the world: 114

Military and security service personnel strengths: the Swedish Armed Forces (Forsvarsmakten) have approximately 15,000 active duty troops (6,200 Army; 2,500 Navy; 2,800 Air Force; 3,500 other,

including staff, logistics, support, intelligence, etc); 22,000 Home Guard (2019 est.)

Military equipment inventories and acquisitions: the inventory of the Swedish Armed Forces is comprised of domestically-produced and imported Western weapons systems; since 2010, the US is the leading supplier of military hardware to Sweden, followed by France and Germany; Sweden's defense industry is capable of providing most of the military's equipment requirements, including advanced aircraft and submarines (2019 est.)

Military deployments: 200 Mali (MINUSMA and EUTM; plans to send an additional 150 personnel to the Sahel/Mali in early 2021) (2020)

Military service age and obligation: *18-47 years of age for male and female voluntary military service; service obligation:* 7.5 months (Army), 7-15 months (Navy), 8-12 months (Air Force); after completing initial service, soldiers have a reserve commitment until age 47; compulsory military service, abolished in 2010, was reinstated in January 2018; conscription is selective, includes both female and male (age 18), and requires 9-12 months of service (2018)

TERRORISM

Terrorist group(s): Islamic State of Iraq and ash-Sham (ISIS) (2019)
note: details about the history, aims, leadership, organization, areas of operation, tactics, targets, weapons, size, and

sources of support of the group(s) appear(s) in Appendix-T

TRANSNATIONAL ISSUES

Disputes—international: none

Refugees and internally displaced persons: *refugees (country of origin):* 113,418 (Syria), 27,933 (Eritrea), 30,546 (Afghanistan), 17,593 (Somalia), 12,460 (Iraq), 7,408 (Iran) (2019)
stateless persons: 30,305 (2019); note—the majority of stateless people are from the Middle East and Somalia

SWITZERLAND

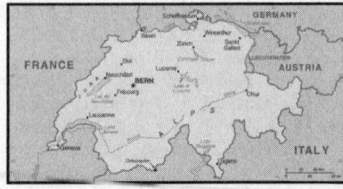

INTRODUCTION

Background: The Swiss Confederation was founded in 1291 as a defensive alliance among

three cantons. In succeeding years, other localities joined the original three. The Swiss Confederation secured its independence from the Holy Roman Empire in 1499. A constitution of 1848, subsequently modified in 1874 to allow voters to introduce referenda on proposed laws, replaced the confederation with a centralized federal government. Switzerland's sovereignty and neutrality have long been honored by the major European powers, and the country was not involved in either of the two world wars. The political and economic integration of Europe over the past half century, as well as Switzerland's role in many UN and international organizations, has strengthened

Switzerland's ties with its neighbors. However, the country did not officially become a UN member until 2002. Switzerland remains active in many UN and international organizations but retains a strong commitment to neutrality.

GEOGRAPHY

Location: Central Europe, east of France, north of Italy

Geographic coordinates: 47 00 N, 8 00 E

Map references: Europe

Area: *total:* 41,277 sq km

land: 39,997 sq km
water: 1,280 sq km
country comparison to the world: 136

Area—comparative: slightly less than twice the size of New Jersey

Land boundaries: *total:* 1,770 km
border countries (5): Austria 158 km, France 525 km, Italy 698 km, Liechtenstein 41 km, Germany 348 km

Coastline: 0 km (landlocked)

Maritime claims: none (landlocked)

Climate: temperate, but varies with altitude; cold, cloudy, rainy/snowy winters; cool to warm, cloudy, humid summers with occasional showers

Terrain: mostly mountains (Alps in south, Jura in northwest) with a central plateau of rolling hills, plains, and large lakes

Elevation: *mean elevation:* 1,350 m
lowest point: Lake Maggiore 195 m
highest point: Dufourspitze 4,634 m

Natural resources: hydropower potential, timber, salt

Land use: *agricultural land:* 38.7% (2011 est.)
arable land: 10.2% (2011 est.) / permanent crops: 0.6% (2011 est.) / permanent pasture: 27.9% (2011 est.)
forest: 31.5% (2011 est.)
other: 29.8% (2011 est.)
Irrigated land: 630 sq km (2012)

Population distribution: population distribution corresponds to elevation with the northern and western areas far more heavily populated; the higher Alps of the south limit settlement

Natural hazards: avalanches, landslides; flash floods

Environment—current issues: air pollution from vehicle emissions; water pollution from agricultural fertilizers; chemical contaminants and erosion damage the soil and limit productivity; loss of biodiversity

Environment—international agreements: *party to:* Air Pollution, Air Pollution-Nitrogen Oxides, Air Pollution-Persistent Organic Pollutants, Air Pollution-Sulfur 85, Air Pollution-Sulfur 94, Air Pollution-Volatile Organic Compounds, Antarctic Treaty, Biodiversity, Climate Change, Climate Change-Kyoto Protocol, Desertification, Endangered Species, Environmental Modification, Hazardous Wastes, Marine Dumping, Marine Life Conservation, Ozone Layer Protection, Ship Pollution, Tropical Timber 83, Tropical Timber 94, Wetlands, Whaling
signed, but not ratified: Law of the Sea

Geography—note: landlocked; crossroads of northern and southern Europe; along with southeastern France, northern Italy, and
southwestern Austria, has the highest elevations in the Alps

PEOPLE AND SOCIETY

Population: 8,403,994 (July 2020 est.)
country comparison to the world: 100

Nationality: *noun:* Swiss (singular and plural)
adjective: Swiss

Ethnic groups: Swiss 69.5%, German 4.2%, Italian 3.2%, Portuguese 2.6%, French 2%, Kosovo 1.1%, other 17.3%, unspecified .1% (2018 est.)
note: data represent permanent and non-permanent resident population by country of birth

Languages: German (or Swiss German) (official) 62.6%, French (official) 22.9%, Italian (official) 8.2%, English 5.4%, Portuguese 3.7%, Albanian 3.2%, Serbo-Croatian 2.5%, Spanish 2.4%, Romansh (official) 0.5%, other 7.7% (2017 est.)
note: German, French, Italian, and Romansh are all national and official languages; shares sum to more than 100% because some respondents gave more than one answer

Religions: Roman Catholic 35.9%, Protestant 23.8%, other Christian 5.9%, Muslim 5.4%, Jewish 0.3%, other 1.4%, none 26%, unspecified 1.4% (2017 est.)

Age structure: *0-14 years:* 15.34% (male 664,255/female 625,252)
15-24 years: 10.39% (male 446,196/female 426,708)
25-54 years: 42.05% (male 1,768,245/female 1,765,941)
55-64 years: 13.48% (male 569,717/female 563,482)
65 years and over: 18.73% (male 699,750/female 874,448) (2020 est.)

Dependency ratios: *total dependency ratio:* 51.6
youth dependency ratio: 22.7
elderly dependency ratio: 29
potential support ratio: 3.5 (2020 est.)

Median age: *total:* 42.7 years
male: 41.7 years
female: 43.7 years (2020 est.)
country comparison to the world: 33

Population growth rate: 0.66% (2020 est.)
country comparison to the world: 145

Birth rate: 10.5 births/1,000 population (2020 est.)
country comparison to the world: 187

Death rate: 8.5 deaths/1,000 population (2020 est.)
country comparison to the world: 73

Net migration rate: 4.6 migrant(s)/1,000 population (2020 est.)
country comparison to the world: 28

Population distribution: population distribution corresponds to elevation with the northern and western areas far more heavily populated; the higher Alps of the south limit settlement

Urbanization: *urban population:* 73.9% of total population (2020)
rate of urbanization: 0.88% annual rate of change (2015-20 est.)

total population growth rate v. urban population growth rate, 2000-2030: *Major urban areas—population:* 1.395 million Zurich, 430,000 BERN (capital) (2020)

Sex ratio: *at birth:* 1.06 male(s)/female
0-14 years: 1.06 male(s)/female
15-24 years: 1.05 male(s)/female

25-54 years: 1 male(s)/female
55-64 years: 1.01 male(s)/female
65 years and over: 0.8 male(s)/female
total population: 0.98 male(s)/female (2020 est.)

Mother's mean age at first birth: 30.7 years (2017 est.)

Maternal mortality rate: 5 deaths/100,000 live births (2017 est.)
country comparison to the world: 171

Infant mortality rate: *total:* 3.5 deaths/1,000 live births
male: 3.8 deaths/1,000 live births
female: 3.2 deaths/1,000 live births (2020 est.)
country comparison to the world: 201

Life expectancy at birth: *total population:* 82.8 years
male: 80.5 years
female: 85.3 years (2020 est.)
country comparison to the world: 13

Total fertility rate: 1.57 children born/woman (2020 est.)
country comparison to the world: 193

Contraceptive prevalence rate: 71.6% (2017)

Drinking water source:
improved:
urban: 100% of population
rural: 100% of population
total: 100% of population
unimproved:
urban: 0% of population
rural: 0% of population
total: 0% of population (2017 est.)

Current Health Expenditure: 12.3% (2017)

Physicians density: 4.3 physicians/1,000 population (2017)

Hospital bed density: 4.7 beds/1,000 population (2017)

Sanitation facility access:
improved:
urban: 100% of population
rural: 100% of population
total: 100% of population
unimproved:
urban: 0% of population
rural: 0% of population
total: 0% of population (2017 est.)

HIV/AIDS—adult prevalence rate: 0.2% (2019)
country comparison to the world: 112

HIV/AIDS—people living with HIV/AIDS: 17,000 (2019)
country comparison to the world: 88

HIV/AIDS—deaths: <100 (2019)

Obesity—adult prevalence rate: 19.5% (2016)
country comparison to the world: 112

Education expenditures: 5.1% of GDP (2016)
country comparison to the world: 60

School life expectancy (primary to tertiary education): *total:* 16 years
male: 17 years
female: 16 years (2018)

Unemployment, youth ages 15-24: *total:* 7.9%
male: 8.4%

female: 7.5% (2018 est.)
country comparison to the world: 144

GOVERNMENT

Country name: *conventional long form:* Swiss Confederation
conventional short form: Switzerland
local long form: Schweizerische Eidgenossenschaft (German); Confederation Suisse (French); Confederazione Svizzera (Italian); Confederaziun Svizra (Romansh)
local short form: Schweiz (German); Suisse (French); Svizzera (Italian); Svizra (Romansh)
abbreviation: CH
etymology: name derives from the canton of Schwyz, one of the founding cantons of the Old Swiss Confederacy that formed in the 14th century

Government type: federal republic (formally a confederation)

Capital: *name:* Bern

geographic coordinates: 46 55 N, 7 28 E
time difference: UTC+1 (6 hours ahead of Washington, DC, during Standard Time)
daylight saving time: +1hr, begins last Sunday in March; ends last Sunday in October
etymology: origin of the name is uncertain, but may derive from a 2nd century B.C. Celtic place name, possibly "berna" meaning "cleft," that was subsequently adopted by a Roman settlement

Administrative divisions: 26 cantons (cantons, singular—canton in French; cantoni, singular—cantone in Italian; Kantone, singular—Kanton in German); Aargau, Appenzell Ausserrhoden, Appenzell Innerrhoden, Basel-Landschaft, Basel-Stadt, Berne/Bern, Fribourg/Freiburg, Geneve (Geneva), Glarus, Graubuenden/Grigioni/Grischun, Jura, Luzern, Neuchatel, Nidwalden, Obwalden, Sankt Gallen, Schaffhausen, Schwyz, Solothurn, Thurgau, Ticino, Uri, Valais/Wallis, Vaud, Zug, Zuerich
note: 6 of the cantons—Appenzell Ausserrhoden, Appenzell Innerrhoden, Basel-Landschaft, Basel-Stadt, Nidwalden, Obwalden—are referred to as half cantons because they elect only one member (instead of two) to the Council of States and, in popular referendums where a majority of popular votes and a majority of cantonal votes are required, these 6 cantons only have a half vote

Independence: 1 August 1291 (founding of the Swiss Confederation)

National holiday: Founding of the Swiss Confederation in 1291; note—since 1 August 1891 celebrated as Swiss National Day

Constitution: *history:* previous 1848, 1874; latest adopted by referendum 18 April 1999, effective 1 January 2000
amendments: proposed by the two houses of the Federal Assembly or by petition of at least one hundred thousand voters (called the "federal popular initiative"); passage of proposals requires majority vote in a referendum; following drafting of an amendment by the Assembly, its passage requires approval by majority vote in a referendum and approval by the majority

of cantons; amended many times, last in 2018 (2020)

Legal system: civil law system; judicial review of legislative acts, except for federal decrees of a general obligatory character

International law organization participation: accepts compulsory ICJ jurisdiction with reservations; accepts ICCt jurisdiction

Citizenship: *citizenship by birth:* no
citizenship by descent only: at least one parent must be a citizen of Switzerland
dual citizenship recognized: yes
residency requirement for naturalization: 12 years including at least 3 of the last 5 years prior to application

Suffrage: 18 years of age; universal

Executive branch: *chief of state:* President of the Swiss Confederation Simonetta SOMMARUGA (since 1 January 2020; Vice President Guy PARMELIN (since 1 January 2020); note—the Federal Council, which is comprised of 7 federal councillors, constitutes the federal government of Switzerland; council members rotate the 1-year term of federal president (chief of state and head of government)

head of government: President of the Swiss Confederation Simonetta SOMMARUGA (since 1 January 2020; Vice President Guy

PARMELIN (since 1 January 2020)
cabinet: Federal Council or Bundesrat (in German), Conseil Federal (in French), Consiglio Federale (in Italian) indirectly elected by the Federal Assembly for a 4-year term
elections/appointments: president and vice president elected by the Federal Assembly from among members of the Federal Council for a 1-year, non-consecutive term; election last held on 11 December 2019 (next to be held in December 2020)
election results: Simonetta SOMMARUGA elected president; Federal Assembly vote—192 of 205; Guy PARMELIN elected vice president; Federal Assembly vote—191 of 204

Legislative branch: *description:* description: bicameral Federal Assembly or Bundesversammlung (in German), Assemblée Fédérale (in French), Assemblea Federale (in Italian) consists of:
    Council of States or Ständerat (in German), Conseil des États (in French), Consiglio degli Stati (in Italian) (46 seats; members in multi-seat constituencies representing cantons and single-seat constituencies representing half cantons directly elected by simple majority vote except Jura and Neuchatel cantons which use proportional representation vote; member term governed by cantonal law)
    National Council or Nationalrat (in German), Conseil National (in French), Consiglio Nazionale (in Italian) (200 seats; 195 members in cantons directly elected by proportional representation vote and 6 in half cantons directly elected by simple majority vote; members serve 4-year terms) (e.g. 2019)
elections: Council of States—last held in most cantons on 20 October 2019 (each canton determines when the next election will be held)

National Council—last held on 20 October 2019 (next to be held in 2023) (e.g. 2019)
election results: Council of States—percent of vote by party—NA; seats by party—CVP 13, FDP 12, SDP 9, Green Party 5, other 1; composition—NA
    National Council—percent of vote by party—SVP 25.6%, SP 16.8%, FDP 15.1%, Green Party 13.2%, CVP 11.4%, GLP 7.8%, other 10.1%; seats by party—SVP 53, SP 39, FDP 29, Green Party 28, CVP 25, GLP 16, other 10; composition—men 116, women 84, percent of women 42% (e.g. 2019)

Judicial branch: *highest courts:* Federal Supreme Court (consists of 38 justices and 19 deputy justices organized into 7 divisions)
judge selection and term of office: judges elected by the Federal Assembly for 6-year terms; note—judges are affiliated with political parties and are elected according to linguistic and regional criteria in approximate proportion to the level of party representation in the Federal Assembly
subordinate courts: Federal Criminal Court (established in 2004); Federal Administrative Court (established in 2007); note—each of Switzerland's 26 cantons has its own courts

Political parties and leaders: Christian Democratic People's Party (Christlichdemokratische Volkspartei der Schweiz or CVP, Parti Democrate-Chretien Suisse or PDC, Partito Popolare Democratico Svizzero or PPD, Partida Cristiandemocratica dalla Svizra or PCD) [Gerhard PFISTER]
Conservative Democratic Party (Buergerlich-Demokratische Partei Schweiz or BDP, Parti Bourgeois Democratique Suisse or PBD, Partito Borghese Democratico Svizzero or PBD, Partido burgais democratica Svizera or PBD) [Martin LANDOLT] Free Democratic Party or FDP. The Liberals (FDP.Die Liberalen, PLR.Les Liberaux-Radicaux, PLR.I Liberali, Ils Liberals) [Petra GOESSI]
Green Liberal Party (Gruenliberale Partei or GLP, Parti vert liberale or PVL, Partito Verde-Liberale or PVL, Partida Verde Liberale or PVL) [Juerg GROSSEN]
Green Party (Gruene Partei der Schweiz or Gruene, Parti Ecologiste Suisse or Les Verts, Partito Ecologista Svizzero or I Verdi, Partida Ecologica Svizra or La Verda) [Regula RYTZ]
Social Democratic Party (Sozialdemokratische Partei der Schweiz or SP, Parti Socialiste Suisse or PSS, Partito Socialista Svizzero or PSS, Partida Socialdemocratica de la Svizra or PSS) [Christian LEVRAT]
Swiss People's Party (Schweizerische Volkspartei or SVP, Union Democratique du Centre or UDC, Unione Democratica di Centro or UDC, Uniun Democratica dal Center or UDC) [Albert ROESTI]
other minor parties

International organization participation: ADB (nonregional member), AfDB (nonregional member), Australia Group, BIS, CD, CE, CERN, EAPC, EBRD, EFTA, EITI (implementing country), ESA, FAO, FATF, G-10, IADB, IAEA, IBRD, ICAO, ICC (national committees), ICCt, ICRM,

IDA, IEA, IFAD, IFC, IFRCS, IGAD (partners), ILO, IMF, IMO, IMSO, Interpol, IOC, IOM, IPU, ISO, ITSO, ITU, ITUC (NGOs), LAIA (observer), MIGA, MINUSMA, MONUSCO, NEA, NSG, OAS (observer), OECD, OIF, OPCW, OSCE, Pacific Alliance (observer), Paris Club, PCA, PFP, Schengen Convention, UN, UNCTAD, UNESCO, UNHCR, UNIDO, UNITAR, UNMISS, UNMOGIP, UNRWA, UNTSO, UNWTO, UPU, WCO, WHO, WIPO, WMO, WTO, ZC

Diplomatic representation in the US: *chief of mission:* Ambassador Jacques PITTELOUD (since 16 September 2019)
chancery: 2900 Cathedral Avenue NW, Washington, DC 20008
telephone: [1] (202) 745-7900
FAX: [1] (202) 387-2564
consulate(s) general: Atlanta, Chicago, Los Angeles, New York, San Francisco

Diplomatic representation from the US: *chief of mission:* Ambassador Edward "Ed" MCMULLEN, Jr. (since 21 November 2017) note—also accredited to

Liechtenstein
telephone: [41] (031) 357-70-11
embassy: Sulgeneckstrasse 19, CH-3007 Bern
mailing address: use embassy street address
FAX: [41] (031) 357-73-20

Flag description: red square with a bold, equilateral white cross in the center that does not extend to the edges of the flag; various medieval legends purport to describe the origin of the flag; a white cross used as identification for troops of the Swiss Confederation is first attested at the Battle of Laupen (1339)

National symbol(s): *Swiss cross (white cross on red field, arms equal length); national colors:* red, white

National anthem: Leonhard WIDMER [German], Charles CHATELANAT [French], Camillo VALSANGIACOMO [Italian], and Flurin CAMATHIAS [Romansch]/Alberik ZWYSSIG
the Swiss anthem has four names: "Schweizerpsalm" [German] "Cantique Suisse" [French] "Salmo svizzero," [Italian] "Psalm svizzer" [Romansch] (Swiss Psalm)
note: unofficially adopted 1961, officially 1981; the anthem has been popular in a number of Swiss cantons since its composition (in German) in 1841; translated into the other three official languages of the country (French, Italian, and Romansch), it is official in each of those languages
0:00 / 1:24

Economy—overview: Switzerland, a country that espouses neutrality, is a prosperous and modern market economy with low unemployment, a highly skilled labor force, and a per capita GDP among the highest in the world. Switzerland's economy benefits from a highly developed service sector, led by financial services, and a manufacturing industry that specializes in hightechnology,

knowledge-based production. Its economic and political stability, transparent legal system, exceptional infrastructure, efficient capital markets, and low corporate tax rates also make Switzerland one of the world's most competitive economies.

The Swiss have brought their economic practices largely into conformity with the EU's to gain access to the Union's Single Market and enhance the country's international competitiveness. Some trade protectionism remains, however, particularly for its small agricultural sector. The fate of the Swiss economy is tightly linked to that of its neighbors in the euro zone, which purchases half of Swiss exports. The global financial crisis of 2008 and resulting economic downturn in 2009 stalled demand for Swiss exports and put Switzerland into a recession. During this period, the Swiss National Bank (SNB) implemented a zero-interest rate policy to boost the economy, as well as to prevent appreciation of the franc, and Switzerland's economy began to recover in 2010.

The sovereign debt crises unfolding in neighboring euro-zone countries, however, coupled with economic instability in Russia and other Eastern European economies drove up demand for the Swiss franc by investors seeking a safehaven currency. In January 2015, the SNB abandoned the Swiss franc's peg to the euro, roiling global currency markets and making active SNB intervention a necessary hallmark of present-day Swiss monetary policy. The independent SNB has upheld its zero interest rate policy and conducted major market interventions to prevent further appreciation of the Swiss franc, but parliamentarians have urged it to do more to weaken the currency. The franc's strength has made Swiss exports less competitive and weakened the country's growth outlook; GDP growth fell below 2% per year from 2011 through 2017.

In recent years, Switzerland has responded to increasing pressure from neighboring countries and trading partners to reform its banking secrecy laws, by agreeing to conform to OECD regulations on administrative assistance in tax matters, including tax evasion. The Swiss Government has also renegotiated its double taxation agreements with numerous countries, including the US, to incorporate OECD standards.

GDP (purchasing power parity): $523.1 billion (2017 est.)
$514.5 billion (2016 est.)
$506.5 billion (2015 est.)
note: data are in 2017 dollars
country comparison to the world: 39

GDP (official exchange rate): $679 billion (2017 est.)

GDP—real growth rate: 1.11% (2019 est.)
3.04% (2018 est.)
1.65% (2017 est.)
country comparison to the world: 170

GDP—per capita (PPP): $62,100 (2017 est.)
$61,800 (2016 est.)
$61,500 (2015 est.)
note: data are in 2017 dollars
country comparison to the world: 17

Gross national saving: 33.8% of GDP (2017 est.)
32.3% of GDP (2016 est.)
33.9% of GDP (2015 est.)
country comparison to the world: 21

GDP—composition, by end use:
household consumption: 53.7% (2017 est.)
government consumption: 12% (2017 est.)
investment in fixed capital: 24.5% (2017 est.)
investment in inventories: -1.4% (2017 est.)
exports of goods and services: 65.1% (2017 est.)
imports of goods and services: -54% (2017 est.)

GDP—composition, by sector of origin:
agriculture: 0.7% (2017 est.)
industry: 25.6% (2017 est.)
services: 73.7% (2017 est.)

Agriculture—products: grains, fruits, vegetables; meat, eggs, dairy products

Industries: machinery, chemicals, watches, textiles, precision instruments, tourism, banking, insurance, pharmaceuticals

Industrial production growth rate: 3.4% (2017 est.)
country comparison to the world: 92

Labor force: 5.067 million (2020 est.)
country comparison to the world: 76

Labor force—by occupation: *agriculture:* 3.3%
industry: 19.8%
services: 76.9% (2015)

Unemployment rate: 2.31% (2019 est.)
2.55% (2018 est.)
country comparison to the world: 24

Population below poverty line: 6.6% (2014 est.)

Household income or consumption by percentage share: *lowest 10%:* 7.5%
highest 10%: 19% (2007)

Budget: *revenues:* 242.1 billion (2017 est.)
expenditures: 234.4 billion (2017 est.)
note: includes federal, cantonal, and municipal budgets

Taxes and other revenues: 35.7% (of GDP) (2017 est.)
country comparison to the world: 60

Budget surplus (+) or deficit (-): 1.1% (of GDP) (2017 est.)
country comparison to the world: 33

Public debt: 41.8% of GDP (2017 est.)
41.8% of GDP (2016 est.)
note: general government gross debt; gross debt consists of all liabilities that require payment or payments of interest and/or principal by the debtor to the creditor at a date or dates in the future; includes debt liabilities in the form of Special Drawing Rights (SDRs), currency and deposits, debt securities, loans, insurance, pensions and standardized guarantee schemes, and other accounts payable; all liabilities in the GFSM (Government Financial Systems Manual) 2001 system are debt, except for equity and investment fund shares and financial derivatives and employee stock options
country comparison to the world: 119

Fiscal year: calendar year

Inflation rate (consumer prices): 0.5% (2017 est.)

-0.4% (2016 est.)
country comparison to the world: 30

Current account balance: $79.937 billion (2019 est.)
$63.273 billion (2018 est.)
country comparison to the world: 5

Exports: $313.5 billion (2017 est.)
$318.1 billion (2016 est.)
note: trade data exclude trade with Switzerland
country comparison to the world: 17

Exports—partners: Germany 15.2%, US 12.3%, China 8.2%, India 6.7%, France 5.7%, UK 5.7%, Hong Kong 5.4%, Italy 5.3% (2017)

Exports—commodities: machinery, chemicals, metals, watches, agricultural products

Imports: $264.5 billion (2017 est.)
$266.3 billion (2016 est.)
country comparison to the world: 19

Imports—commodities: machinery, chemicals, vehicles, metals; agricultural products, textiles

Imports—partners: Germany 20.9%, US 7.9%, Italy 7.6%, UK 7.3%, France 6.8%, China 5% (2017)

Reserves of foreign exchange and gold: $811.2 billion (31 December 2017 est.)
$679.3 billion (31 December 2016 est.)
country comparison to the world: 3

Debt—external: $1.664 trillion (31 March 2016 est.)
$1.663 trillion (31 March 2015 est.)
country comparison to the world: 12

Exchange rates: Swiss francs (CHF) per US dollar -
0.9875 (2017 est.)
0.9852 (2016 est.)
0.9852 (2015 est.)
0.9627 (2014 est.)
0.9152 (2013 est.)

ENERGY

Electricity access: *electrification—total population:* 100% (2020)

Electricity—production: 59.01 billion kWh (2016 est.)
country comparison to the world: 50

Electricity—consumption: 58.46 billion kWh (2016 est.)
country comparison to the world: 43

Electricity—exports: 30.17 billion kWh (2016 est.)
country comparison to the world: 5

Electricity—imports: 34.1 billion kWh (2016 est.)
country comparison to the world: 4

Electricity—installed generating capacity: 20.84 million kW (2016 est.)
country comparison to the world: 42

Electricity—from fossil fuels. 3% of total installed capacity (2016 est.)
country comparison to the world: 209

Electricity—from nuclear fuels: 18% of total installed capacity (2017 est.)

country comparison to the world: 11

Electricity—from hydroelectric plants: 67% of total installed capacity (2017 est.)
country comparison to the world: 21

Electricity—from other renewable sources: 13% of total installed capacity (2017 est.)
country comparison to the world: 71

Crude oil—production: 0 bbl/day (2018 est.)
country comparison to the world: 205

Crude oil—exports: 0 bbl/day (2017 est.)
country comparison to the world: 201

Crude oil—imports: 57,400 bbl/day (2017 est.)
country comparison to the world: 54

Crude oil—proved reserves: 0 bbl (1 January 2018 est.)
country comparison to the world: 201

Refined petroleum products—production: 61,550 bbl/day (2017 est.)
country comparison to the world: 79

Refined petroleum products—consumption: 223,900 bbl/day (2017 est.)
country comparison to the world: 54

Refined petroleum products—exports: 7,345 bbl/day (2017 est.)
country comparison to the world: 88

Refined petroleum products—imports: 165,100 bbl/day (2017 est.)
country comparison to the world: 39

Natural gas—production: 0 cu m (2017 est.)
country comparison to the world: 205

Natural gas—consumption: 3.709 billion cu m (2017 est.)
country comparison to the world: 68

Natural gas—exports: 0 cu m (2017 est.)
country comparison to the world: 194

Natural gas—imports: 3.681 billion cu m (2017 est.)
country comparison to the world: 42

Natural gas—proved reserves: NA cu m (1 January 2011 est.)

Carbon dioxide emissions from consumption of energy: 38.95 million Mt (2017 est.)
country comparison to the world: 66

COMMUNICATIONS

Telephones—fixed lines: *total subscriptions:* 3,012,224
subscriptions per 100 inhabitants: 36.08 (2019 est.)
country comparison to the world: 45

Telephones—mobile cellular: *total subscriptions:* 10,618,759
subscriptions per 100 inhabitants: 127.19 (2019 est.)
country comparison to the world: 84

Telecommunication systems: *general assessment:* highly developed telecommunications infrastructure with extensive domestic and international services; one of the highest broadband penetration rates in Europe; although not a member of the EU, Switzerland follows

the EU's telecom framework, and regulations; expansive cable broadband network with effective cross-platform competition; despite the countries expansion of 5G services, and switching off 2G infrastructure, the Environmental Agency has raised concern regarding the 2,000 5G mobile antennas and asked the govt. to halt 5G transmissions, the developers of the 5G infrastructure are allowed to continue with future checks to be studied of the health implications of the radio frequency radiation; regulator auction of 5G spectrum (2020)
domestic: ranked among leading countries for fixed-line teledensity and infrastructure; fixed-line 36 per 100 and mobilecellular subscribership 127 per 100 persons; extensive cable and microwave radio relay networks (2019)
international: country code—41; satellite earth stations—2 Intelsat (Atlantic Ocean and Indian Ocean)
note: the COVID-19 outbreak is negatively impacting telecommunications production and supply chains globally; consumer spending on telecom devices and services has also slowed due to the pandemic's effect on economies worldwide; overall progress towards improvements in all facets of the telecom industry—mobile, fixed-line, broadband, submarine cable and satellite—has moderated

Broadcast media: the publicly owned radio and TV broadcaster, Swiss Broadcasting Corporation (SRG/SSR), operates 8 national TV networks, 3 broadcasting in German, 3 in French, and 2 in Italian; private commercial TV stations broadcast regionally and locally; TV broadcasts from stations in Germany, Italy, and France are widely available via multi-channel cable and satellite TV services; SRG/SSR operates 17 radio stations that, along with private broadcasters, provide national to local coverage) (2019)

Internet country code: .ch

Internet users: *total:* 7,437,820
percent of population: 89.69% (July 2018 est.)
country comparison to the world: 69

Broadband—fixed subscriptions: *total:* 3,957,669
subscriptions per 100 inhabitants: 48 (2018 est.)
country comparison to the world: 36

TRANSPORTATION

National air transport system: *number of registered air carriers:* 6 (2020)
inventory of registered aircraft operated by air carriers: 179
annual passenger traffic on registered air carriers: 28,857,994 (2018)
annual freight traffic on registered air carriers: 1,841,310,000 mt-km (2018)

Civil aircraft registration country code prefix: HB (2016)

Airports: 63 (2013)
country comparison to the world: 78

Airports—with paved runways: *total:* 40 (2013)
over 3,047 m: 3 (2013)

2,438 to 3,047 m: 2 (2013)
1,524 to 2,437 m: 12 (2013)
914 to 1,523 m: 6 (2013)
under 914 m: 17 (2013)

Airports—with unpaved runways: *total:* 23 (2013)
under 914 m: 23 (2013)

Heliports: 2 (2013)

Pipelines: 1,800 km gas, 94 km oil (of which 60 are inactive), 17 km refined products (2017)

Railways: *total:* 5,690 km (includes 19 km in neighboring countries) (2015)
standard gauge: 3,836 km 1.435-m gauge (3,634 km electrified) (2015)
narrow gauge: 1,630 km 1.200-m gauge (2 km electrified) (includes 19 km in neighboring countries) (2015)
1188 km 1.000-m gauge (1,167.3 km electrified)
36 km 0.800-m gauge (36.4 km electrified)
country comparison to the world: 34

Roadways: *total:* 71,557 km (2017)
paved: 71,557 km (includes 1,458 of expressways) (2017)
country comparison to the world: 68

Waterways: 1,292 km (there are 1,227 km of waterways on lakes and rivers for public transport and 65 km on the Rhine River between Basel-Rheinfelden and Schaffhausen-Bodensee for commercial goods transport) (2010)
country comparison to the world: 57

Merchant marine: *total:* 32 includes Liechtenstein
by type: bulk carrier 24, general cargo 4, oil tanker 1, other 3 (2019)
country comparison to the world: 131

Ports and terminals: *river port(s):* Basel (Rhine)

MILITARY AND SECURITY

Military and security forces: *Swiss Armed Forces:* Land Forces, Swiss Air Force (Schweizer Luftwaffe) (2019)

Military expenditures: 0.7% of GDP (2019)
0.7% of GDP (2018)
0.7% of GDP (2017)
0.7% of GDP (2016)
0.7% of GDP (2015)
country comparison to the world: 141

Military and security service personnel strengths: the Swiss Armed Forces maintain a full-time active duty cadre of about 3,000 Army and Air Force personnel along with approximately 18,500 conscripts brought in annually for 18-23 weeks of training (2019 est.)

Military equipment inventories and acquisitions: the Swiss Armed Forces inventory includes a mix of domestically-produced and imported weapons systems; the US is the leading supplier of military armaments to Switzerland since 2010; the Swiss defense industry produces a range of military land vehicles (2019 est.)

Military deployments: 165 Kosovo (NATO) (2020)

Military service age and obligation: 18-30 years of age generally for male compulsory military service; 18 years of age for voluntary male and female military service; every Swiss male has to serve at least 245 days in the armed forces; conscripts receive 18 weeks of mandatory training, followed by six 19-day intermittent recalls for training during the next 10 years (2019)

TRANSNATIONAL ISSUES

Disputes—international: none

Refugees and internally displaced persons: *refugees (country of origin):* 36,698 (Eritrea), 18,755 (Syria), 13,455 (Afghanistan), 5,819 (Sri Lanka) (2019)
stateless persons: 49 (2018)

Illicit drugs: a major international financial center vulnerable to the layering and integration stages of money laundering; despite significant legislation and reporting requirements, secrecy rules persist and nonresidents are permitted to conduct business through offshore entities and various intermediaries; transit country for and consumer of South American cocaine, Southwest Asian heroin, and Western European synthetics; domestic cannabis cultivation and limited ecstasy production

SYRIA

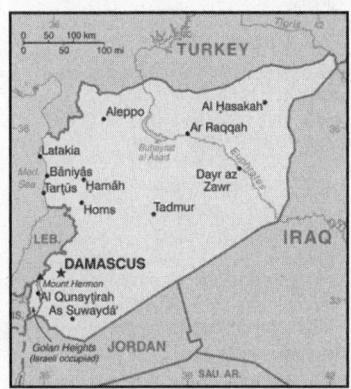

INTRODUCTION

Background: Following World War I, France acquired a mandate over the northern portion of the former Ottoman Empire province of Syria. The French administered the area as Syria until granting it independence in 1946. The new country lacked political stability and experienced a series of military coups. Syria united with Egypt in February 1958 to form the United Arab Republic.

In September 1961, the two entities separated, and the Syrian Arab Republic was reestablished. In the 1967 Arab-Israeli War, Syria lost the Golan Heights region to Israel. During the 1990s, Syria and Israel held occasional, albeit unsuccessful, peace talks over its return. In November 1970, Hafiz al-ASAD, a member of the socialist Ba'ath Party and the minority Alawi sect, seized power in a bloodless coup and brought political stability to the country. Following the death of President Hafiz al-ASAD, his son, Bashar al-ASAD, was approved as president by popular referendum in July 2000. Syrian troops—stationed in Lebanon since 1976 in an ostensible peacekeeping role—were withdrawn in April 2005. During the July-August 2006 conflict between Israel and Hizballah, Syria placed its military forces on alert but did not intervene directly on behalf of its ally Hizballah. In May 2007, Bashar al-ASAD's second term as president was approved by popular referendum.

Influenced by major uprisings that began elsewhere in the region, and compounded by additional social and economic factors, antigovernment protests broke out first in the southern province of Dar'a in March 2011 with protesters calling for the repeal of the restrictive Emergency Law allowing arrests without charge, the legalization of political parties, and the removal of corrupt local officials. Demonstrations and violent unrest

spread across Syria with the size and intensity of protests fluctuating. The government responded to unrest with a mix of concessions—including the repeal of the Emergency Law, new laws permitting new political parties, and liberalizing local and national elections—and with military force and detentions. The government's efforts to quell unrest and armed opposition activity led to extended clashes and eventually civil war between government forces, their allies, and oppositionists.

International pressure on the ASAD regime intensified after late 2011, as the Arab League, the EU, Turkey, and the US expanded economic sanctions against the regime and those entities that support it. In December 2012, the Syrian National Coalition, was recognized by more than 130 countries as the sole legitimate representative of the Syrian people. In September 2015, Russia launched a military intervention on behalf of the ASAD regime, and domestic and foreign government-aligned forces recaptured swaths of territory from opposition forces, and eventually the country's second largest city, Aleppo, in December 2016, shifting the conflict in the regime's favor. The regime, with this foreign support, also recaptured opposition strongholds in the Damascus suburbs and the southern province of Dar'a in 2018. The government lacks territorial control over much of the northeastern part of the country, which is dominated by the predominantly Kurdish Syrian Democratic Forces (SDF). The SDF has expanded its territorial hold over much of the northeast since 2014 as it has captured territory from the Islamic State of Iraq and Syria. Since 2016, Turkey has also conducted three large-scale military operations into Syria, capturing territory along Syria's northern border in the provinces of Aleppo, Ar Raqqah, and Al Hasakah. Political negotiations between the government and opposition delegations at UN-sponsored Geneva conferences since 2014 have failed to produce a resolution of the conflict. Since early 2017, Iran, Russia, and Turkey have held separate political negotiations outside of UN auspices to attempt to reduce violence in Syria. According to an April 2016 UN estimate, the death toll among Syrian Government forces, opposition forces, and civilians was over 400,000, though other estimates placed the number well over 500,000. As of December 2019, approximately 6 million Syrians were internally displaced. Approximately 11.1 million people were in need of humanitarian assistance across the country, and an additional 5.7 million Syrians were registered refugees in Turkey, Jordan, Iraq, Egypt, and North Africa. The conflict in Syria remains one of the largest humanitarian crises worldwide.

GEOGRAPHY

Location: Middle East, bordering the Mediterranean Sea, between Lebanon and Turkey

Geographic coordinates: 35 00 N, 38 00 E

Map references: Middle East

Area: *total:* 187,437 sq km
land: 185,887 sq km
water: 1,550 sq km
note: includes 1,295 sq km of Israeli-occupied territory
country comparison to the world: 90

Area—comparative: slightly more than 1.5 times the size of Pennsylvania

Land boundaries: *total:* 2,343 km
border countries (5): Iraq 599 km, Israel 79 km, Jordan 362 km, Lebanon 394 km, Turkey 909 km

Coastline: 193 km

Maritime claims: *territorial sea:* 12 nm
contiguous zone: 24 nm

Climate: mostly desert; hot, dry, sunny summers (June to August) and mild, rainy winters (December to February) along coast; cold weather with snow or sleet periodically in Damascus

Terrain: primarily semiarid and desert plateau; narrow coastal plain; mountains in west

Elevation: *mean elevation:* 514 m
lowest point: unnamed location near Lake Tiberias -208 m
highest point: Mount Hermon (Jabal a-Shayk) 2,814 m

Natural resources: petroleum, phosphates, chrome and manganese ores, asphalt, iron ore, rock salt, marble, gypsum, hydropower

Land use: *agricultural land:* 75.8% (2011 est.)

arable land: 25.4% (2011 est.) / permanent crops: 5.8% (2011 est.) / permanent pasture: 44.6% (2011 est.)
forest: 2.7% (2011 est.)
other: 21.5% (2011 est.)
Irrigated land: 14,280 sq km (2012)

Population distribution: significant population density along the Mediterranean coast; larger concentrations found in the major cities of Damascus, Aleppo (the country's largest city), and Hims (Homs); more than half of the population lives in the coastal plain, the province of Halab, and the Euphrates River valley
note: the ongoing civil war has altered the population distribution

Natural hazards: dust storms, sandstorms
volcanism: Syria's two historically active volcanoes, Es Safa and an unnamed volcano near the Turkish border have not erupted in centuries

Environment—current issues: deforestation; overgrazing; soil erosion; desertification; depletion of water resources; water pollution from raw sewage and petroleum refining wastes; inadequate potable water

Environment—international agreements: *party to:* Biodiversity, Climate Change, Climate Change-Kyoto Protocol, Desertification, Endangered Species, Hazardous Wastes, Ozone Layer Protection, Ship Pollution, Wetlands
signed, but not ratified: Environmental Modification

Geography—note: the capital of Damascus—located at an oasis fed by the Barada River—is thought to be one of the world's oldest continuously inhabited cities; there are 42 Israeli settlements and civilian land use sites in the Israeli-controlled Golan Heights (2017)

PEOPLE AND SOCIETY

Population: 19,398,448 (July 2020 est.)
note: approximately 22,000 Israeli settlers live in the Golan Heights (2016)
country comparison to the world: 63

Nationality: *noun:* Syrian(s)
adjective: Syrian

Ethnic groups: Arab ~50%, Alawite ~15%, Kurd ~10%, Levantine ~10%, other ~15% (includes Druze, Ismaili, Imami, Nusairi, Assyrian, Turkoman, Armenian)

Languages: Arabic (official), Kurdish, Armenian, Aramaic, Circassian, French, English

Religions: Muslim 87% (official; includes Sunni 74% and Alawi, Ismaili, and Shia 13%), Christian 10% (includes Orthodox, Uniate, and Nestorian), Druze 3%, Jewish (few remaining in Damascus and Aleppo)
note: the Christian population may be considerably smaller as a result of Christians fleeing the country during the ongoing civil war

MENA religious affiliation: *Age structure:* 0-14 years: 33.47% (male 3,323,072/female 3,170,444)
15-24 years: 19.34% (male 1,872,903/female 1,879,564)

25-54 years: 37.31% (male 3,558,241/female 3,679,596)
55-64 years: 5.41% (male 516,209/female 534,189)
65 years and over: 4.46% (male 404,813/female 459,417) (2020 est.)

Dependency ratios: *total dependency ratio:* 55.4
youth dependency ratio: 47.8
elderly dependency ratio: 7.6
potential support ratio: 13.2 (2020 est.)

Median age: *total:* 23.5 years
male: 23 years
female: 24 years (2020 est.)
country comparison to the world: 177

Population growth rate: 4.25% NA (2020 est.)
country comparison to the world: 1

Birth rate: 23.8 births/1,000 population (2020 est.)
country comparison to the world: 51

Death rate: 4.5 deaths/1,000 population (2020 est.)
country comparison to the world: 207

Net migration rate: 27.1 migrant(s)/1,000 population NA (2020 est.)
country comparison to the world: 1

Population distribution: significant population density along the Mediterranean coast; larger concentrations found in the major cities of Damascus, Aleppo (the country's largest city), and Hims (Homs); more than half of the population lives in the coastal plain, the province of Halab, and the Euphrates River valley
note: the ongoing civil war has altered the population distribution

Urbanization: *urban population:* 55.5% of total population (2020)
rate of urbanization: 1.43% annual rate of change (2015-20 est.)

total population growth rate v. urban population growth rate, 2000-2030: *Major urban areas—population:* 2.392 million DAMASCUS (capital), 1.917 million Aleppo, 1.336 million Hims (Homs), 922,000 Hamah (2020)

Sex ratio: *at birth:* 1.06 male(s)/female
0-14 years: 1.05 male(s)/female
15-24 years: 1 male(s)/female
25-54 years: 0.97 male(s)/female
55-64 years: 0.97 male(s)/female
65 years and over: 0.88 male(s)/female
total population: 1 male(s)/female (2020 est.)

Maternal mortality rate: 31 deaths/100,000 live births (2017 est.)
country comparison to the world: 110

Infant mortality rate: *total:* 16.5 deaths/1,000 live births
male: 18.1 deaths/1,000 live births
female: 14.7 deaths/1,000 live births (2020 est.)
country comparison to the world: 92

Life expectancy at birth: *total population:* 73.7 years
male: 72.3 years
female: 75.3 years (2020 est.)
country comparison to the world: 143

Total fertility rate: 2.9 children born/woman (2020 est.)
country comparison to the world: 55

Drinking water source:
improved:
urban: 99% of population
rural: 99.3% of population
total: 99.4% of population
unimproved:
urban: 1% of population
rural: 0.7% of population
total: 0.6% of population (2017 est.)

Physicians density: 1.29 physicians/1,000 population (2016)

Hospital bed density: 1.4 beds/1,000 population (2017)

Sanitation facility access:
improved:
urban: 99.6% of population
rural: 98.6% of population
total: 99.1% of population
unimproved:
urban: 0.4% of population
rural: 1.4% of population
total: 0.9% of population (2017 est.)

HIV/AIDS—adult prevalence rate: <.1% (2019)

HIV/AIDS—people living with HIV/AIDS: <1000 (2019)

HIV/AIDS—deaths: <100 (2019)

Obesity—adult prevalence rate: 27.8% (2016)
country comparison to the world: 35

Children under the age of 5 years underweight: 5.8% (2009/10)
country comparison to the world: 79

Education expenditures: 5.1% of GDP (2009)
country comparison to the world: 61

Literacy: *definition:* age 15 and over can read and write
total population: 86.4%
male: 91.7%
female: 81% (2015)

School life expectancy (primary to tertiary education): *total:* 9 years
male: 9 years
female: 9 years (2013)

Unemployment, youth ages 15-24: *total:* 35.8%
male: 26.6%
female: 71.1% (2011 est.)
country comparison to the world: 19

GOVERNMENT

Country name: *conventional long form:* Syrian Arab Republic
conventional short form: Syria
local long form: Al Jumhuriyah al Arabiyah as Suriyah
local short form: Suriyah
former: United Arab Republic (with Egypt)
etymology: name ultimately derived from the ancient Assyrians who dominated northern Mesopotamia, but whose reach also extended westward to the Levant; over time, the name

came to be associated more with the western area

Government type: presidential republic; highly authoritarian regime

Capital: *name:* Damascus

geographic coordinates: 33 30 N, 36 18 E
time difference: UTC+2 (7 hours ahead of Washington, DC, during Standard Time)
daylight saving time: +1hr, begins midnight on the last Friday in March; ends at midnight on the last Friday in October
etymology: Damascus is a very old city; its earliest name, Temeseq, first appears in an Egyptian geographical list of the 15th century B.C., but the meaning is uncertain

Administrative divisions: 14 provinces (muhafazat, singular—muhafazah); Al Hasakah, Al Ladhiqiyah (Latakia), Al Qunaytirah, Ar Raqqah, As Suwayda', Dar'a, Dayr az Zawr, Dimashq (Damascus), Halab (Aleppo), Hamah, Hims (Homs), Idlib, Rif Dimashq (Damascus Countryside), Tartus

Independence: 17 April 1946 (from League of Nations mandate under French administration)

National holiday: Independence Day (Evacuation Day), 17 April (1946); note—celebrates the leaving of the last French troops and the proclamation of full independence

Constitution: *history:* several previous; latest issued 15 February 2012, passed by referendum and effective 27 February 2012
amendments: proposed by the president of the republic or by one third of the People's Assembly members; following review by a special Assembly committee, passage requires at least three-quarters majority vote by the Assembly and approval by the president

Legal system: mixed legal system of civil and Islamic (sharia) law (for family courts)

International law organization participation: has not submitted an ICJ jurisdiction declaration; non-party state to the ICC

Citizenship: *citizenship by birth:* no
citizenship by descent only: the father must be a citizen of Syria; if the father is unknown or stateless, the mother must be a citizen of Syria
dual citizenship recognized: yes
residency requirement for naturalization: 10 years

Suffrage: 18 years of age; universal

Executive branch: *chief of state:* President Bashar al-ASAD (since 17 July 2000); Vice President Najah al-ATTAR (since 23 March 2006)
head of government: Prime Minister Hussein ARNOUS (since 30 August 2020); Deputy Prime Minister Ali Abdullah AYOUB (Gen.) (since 30 August 2020)
cabinet: Council of Ministers appointed by the president
elections/appointments: president directly elected by simple majority popular vote for a 7-year term (eligible for a second term); election last held on 3 June 2014 (next to be held in June

2021); the president appoints the vice presidents, prime minister, and deputy prime ministers
election results: Bashar al-ASAD elected president; percent of vote—Bashar al-ASAD (Ba'th Party) 88.7%, Hassan al- NOURI (independent) 4.3%, Maher HAJJER (independent) 3.2%, other/invalid 3.8%

Legislative branch: *description:* unicameral People's Assembly or Majlis al-Shaab (250 seats; members directly elected in multi-seat constituencies by simple majority preferential vote to serve 4-year terms)
elections: last held on 19 July 2020 (next to be held in 2024)
election results: percent of vote by party—NPF 80%, other 20%; seats by party—NPF 200, other 50; composition—men 217, women 33, percent of women 13.2%

Judicial branch: *highest courts:* Court of Cassation (organized into civil, criminal, religious, and military divisions, each with 3 judges); Supreme Constitutional Court (consists of 7 members)
judge selection and term of office: Court of Cassation judges appointed by the Supreme Judicial Council (SJC), a judicial management body headed by the minister of justice with 7 members, including the national president; judge tenure NA; Supreme Constitutional Court judges nominated by the president and appointed by the SJC; judges serve 4-year renewable terms
subordinate courts: courts of first instance; magistrates' courts; religious and military courts; Economic Security Court; Counterterrorism Court (established June 2012)

Political parties and leaders: *legal parties/alliances:*
Arab Socialist Ba'ath Party [Bashar al-ASAD, regional secretary]
Arab Socialist Renaissance (Ba'th) Party [President Bashar al-ASAD]
Arab Socialist Union of Syria or ASU [Safwan al-QUDSI]
National Progressive Front or NPF [Bashar al-ASAD, Suleiman QADDAH] (alliance includes Arab Socialist Renaissance (Ba'th) Party, Socialist Unionist Democratic Party)
Socialist Unionist Democratic Party [Fadlallah Nasr al-DIN]
Syrian Communist Party (two branches) [Wissal Farha BAKDASH, Yusuf Rashid FAYSAL]
Syrian Social Nationalist Party or SSNP [Ali HAIDAR]
Unionist Socialist Party [Fayez ISMAIL]
Major Kurdish parties
Kurdish Democratic Union Party or PYD [Shahoz HASAN and Aysha HISSO]
Kurdish National Council [Sa'ud MALA]
other: Syrian Democratic Party [Mustafa QALAAJI]

International organization participation: ABEDA, AFESD, AMF, CAEU, FAO, G-24, G-77, IAEA, IBRD, ICAO, ICC (national committees), ICRM, ICSID, IDA, IDB, IFAD, IFC, IFRCS, IHO, ILO, IMF, IMO, Interpol, IOC, IPU, ISO, ITSO, ITU, LAS, MIGA, NAM, OAPEC, OIC, OPCW,

UN, UNCTAD, UNESCO, UNIDO, UNRWA, UNWTO, UPU, WBG, WCO, WFTU (NGOs), WHO, WIPO, WMO, WTO (observer)

Diplomatic representation in the US: *chief of mission:* Ambassador (vacant)

chancery: 2215 Wyoming Avenue NW, Washington, DC 20008

telephone: [1] (202) 232-6313

FAX: [1] (202) 234-9548

note: Embassy ceased operations and closed on 18 March 2014

Diplomatic representation from the US: *chief of mission:* Ambassador (vacant); *note—*on 6 February 2012, the US closed its embassy in Damascus; Czechia serves as a protecting power for US interests in Syria

telephone: [963] (11) 3391-4444

embassy: Abou Roumaneh, 2 Al Mansour Street, Damascus

mailing address: P. O. Box 29, Damascus

FAX: [963] (11) 3391-3999

Flag description: three equal horizontal bands of red (top), white, and black; two small, green, five-pointed stars in a horizontal line centered in the white band; the band colors derive from the Arab Liberation flag and represent oppression (black), overcome through bloody struggle (red), to be replaced by a bright future (white); identical to the former flag of the United Arab Republic (1958-1961) where the two stars represented the constituent states of Syria and Egypt; the current design dates to 1980

note: similar to the flag of Yemen, which has a plain white band, Iraq, which has an Arabic inscription centered in the white band, and that of Egypt, which has a gold Eagle of Saladin centered in the white band

National symbol(s): *hawk; national colors:* red, white, black, green

National anthem: *name:* "Humat ad-Diyar" (Guardians of the Homeland)

lyrics/music: Khalil Mardam BEY/Mohammad Salim FLAYFEL and Ahmad Salim FLAYFEL

note: adopted 1936, restored 1961; between 1958 and 1961, while Syria was a member of the United Arab Republic with Egypt, the country had a different anthem

0:00 / 1:06

ECONOMY

Economy—overview: Syria's economy has deeply deteriorated amid the ongoing conflict that began in 2011, declining by more than 70% from 2010 to 2017. The government has struggled to fully address the effects of international sanctions, widespread infrastructure damage, diminished domestic consumption and production, reduced subsidies, and high inflation, which have caused dwindling foreign exchange reserves, rising budget and trade deficits, a decreasing value of the Syrian pound, and falling household purchasing power. In 2017, some economic indicators began to stabilize, including the exchange rate and inflation, but economic activity remains depressed and GDP almost certainly fell.

During 2017, the ongoing conflict and continued unrest and economic decline worsened the humanitarian crisis, necessitating high levels of international assistance, as more than 13 million people remain in need inside Syria, and the number of registered Syrian refugees increased from 4.8 million in 2016 to more than 5.4 million.

Prior to the turmoil, Damascus had begun liberalizing economic policies, including cutting lending interest rates, opening private banks, consolidating multiple exchange rates, raising prices on some subsidized items, and establishing the Damascus Stock Exchange, but the economy remains highly regulated. Long-run economic constraints include foreign trade barriers, declining oil production, high unemployment, rising budget deficits, increasing pressure on water supplies caused by heavy use in agriculture, industrial contaction, water pollution, and widespread infrastructure damage.

GDP (purchasing power parity): $50.28 billion (2015 est.)

$55.8 billion (2014 est.)

$61.9 billion (2013 est.)

note: data are in 2015 US dollars

the war-driven deterioration of the economy resulted in a disappearance of quality national level statistics in the 2012-13 period

country comparison to the world: 110

GDP (official exchange rate): $24.6 billion (2014 est.)

GDP—real growth rate: -36.5% (2014 est.)

-30.9% (2013 est.)

note: data are in 2015 dollars

country comparison to the world: 224

GDP—per capita (PPP): $2,900 (2015 est.)

$3,300 (2014 est.)

$2,800 (2013 est.)

note: data are in 2015 US dollars

country comparison to the world: 194

Gross national saving: 17% of GDP (2017 est.)

15.3% of GDP (2016 est.)

16.1% of GDP (2015 est.)

country comparison to the world: 121

GDP—composition, by end use:

household consumption: 73.1% (2017 est.)

government consumption: 26% (2017 est.)

investment in fixed capital: 18.6% (2017 est.)

investment in inventories: 12.3% (2017 est.)

exports of goods and services: 16.1% (2017 est.)

imports of goods and services: -46.1% (2017 est.)

GDP—composition, by sector of origin:

agriculture: 20% (2017 est.)

industry: 19.5% (2017 est.)

services: 60.8% (2017 est.)

Agriculture—products: wheat, barley, cotton, lentils, chickpeas, olives, sugar beets; beef, mutton, eggs, poultry, milk

Industries: petroleum, textiles, food processing, beverages, tobacco, phosphate rock mining, cement, oil seeds crushing, automobile assembly

Industrial production growth rate: 4.3% (2017 est.)

country comparison to the world: 70

Labor force: 3.767 million (2017 est.)

country comparison to the world: 93

Labor force—by occupation: *agriculture:* 17%

industry: 16%

services: 67% (2008 est.)

Unemployment rate: 50% (2017 est.)

50% (2016 est.)

country comparison to the world: 217

Population below poverty line: 82.5% (2014 est.)

Household income or consumption by percentage share: *lowest 10%:* NA

highest 10%: NA

Budget: *revenues:* 1.162 billion (2017 est.)

expenditures: 3.211 billion (2017 est.)

note: government projections for FY2016

Taxes and other revenues: 4.2% (of GDP) (2017 est.)

country comparison to the world: 219

Budget surplus (+) or deficit (-): -8.7% (of GDP) (2017 est.)

country comparison to the world: 203

Public debt: 94.8% of GDP (2017 est.)

91.3% of GDP (2016 est.)

country comparison to the world: 23

Fiscal year: calendar year

Inflation rate (consumer prices): 28.1% (2017 est.)

47.3% (2016 est.)

country comparison to the world: 220

Current account balance: -$2.123 billion (2017 est.)

-$2.077 billion (2016 est.)

country comparison to the world: 168

Exports: $1.85 billion (2017 est.)

$1.705 billion (2016 est.)

country comparison to the world: 143

Exports—partners: Lebanon 31.5%, Iraq 10.3%, Jordan 8.8%, China 7.8%, Turkey 7.5%, Spain 7.3% (2017)

Exports—commodities: crude oil, minerals, petroleum products, fruits and vegetables, cotton fiber, textiles, clothing, meat and live animals, wheat

Imports: $6.279 billion (2017 est.)

$5.496 billion (2016 est.)

country comparison to the world: 119

Imports—commodities: machinery and transport equipment, electric power machinery, food and livestock, metal and metal products, chemicals and chemical products, plastics, yarn, paper

Imports—partners: Russia 32.4%, Turkey 16.7%, China 9.5% (2017)

Reserves of foreign exchange and gold: $407.3 million (31 December 2017 est.)

$504.6 million (31 December 2016 est.)

country comparison to the world: 159

Debt—external: $4.989 billion (31 December 2017 est.)

$5.085 billion (31 December 2016 est.)

country comparison to the world: 133

Exchange rates: Syrian pounds (SYP) per US dollar -

514.6 (2017 est.)
459.2 (2016 est.)
459.2 (2015 est.)
236.41 (2014 est.)
153.695 (2013 est.)

ENERGY

Electricity access: *population without electricity:* 1 million (2017)
electrification—total population: 92% (2017)
electrification—urban areas: 100% (2017)
electrification—rural areas: 84% (2017)

Electricity—production: 17.07 billion kWh (2016 est.)
country comparison to the world: 84

Electricity—consumption: 14.16 billion kWh (2016 est.)
country comparison to the world: 82

Electricity—exports: 262 million kWh (2015 est.)
country comparison to the world: 72

Electricity—imports: 0 kWh (2016 est.)
country comparison to the world: 207

Electricity—installed generating capacity: 9.058 million kW (2016 est.)
country comparison to the world: 64

Electricity—from fossil fuels: 83% of total installed capacity (2016 est.)
country comparison to the world: 77

Electricity—from nuclear fuels: 0% of total installed capacity (2017 est.)
country comparison to the world: 190

Electricity—from hydroelectric plants: 17% of total installed capacity (2017 est.)
country comparison to the world: 98

Electricity—from other renewable sources: 0% of total installed capacity (2017 est.)
country comparison to the world: 207

Crude oil—production: 25,000 bbl/day (2018 est.)
country comparison to the world: 63

Crude oil—exports: 0 bbl/day (2015 est.)
country comparison to the world: 202

Crude oil—imports: 87,660 bbl/day (2015 est.)
country comparison to the world: 45

Crude oil—proved reserves: 2.5 billion bbl (1 January 2018 est.)
country comparison to the world: 30

Refined petroleum products—production: 111,600 bbl/day (2015 est.)
country comparison to the world: 66

Refined petroleum products—consumption: 134,000 bbl/day (2016 est.)
country comparison to the world: 71

Refined petroleum products—exports: 12,520 bbl/day (2015 est.)
country comparison to the world: 79

Refined petroleum products—imports: 38,080 bbl/day (2015 est.)
country comparison to the world: 92

Natural gas—production: 3.738 billion cu m (2017 est.)
country comparison to the world: 53

Natural gas—consumption: 3.738 billion cu m (2017 est.)
country comparison to the world: 67

Natural gas—exports: 0 cu m (2017 est.)
country comparison to the world: 195

Natural gas—imports: 0 cu m (2017 est.)
country comparison to the world: 196

Natural gas—proved reserves: 240.7 billion cu m (1 January 2018 est.)
country comparison to the world: 40

Carbon dioxide emissions from consumption of energy: 27.51 million Mt (2017 est.)
country comparison to the world: 75

COMMUNICATIONS

Telephones—fixed lines: *total subscriptions:* 3,097,164
subscriptions per 100 inhabitants: 16.66 (2019 est.)
country comparison to the world: 42

Telephones—mobile cellular: *total subscriptions:* 21.115 million
subscriptions per 100 inhabitants: 113.58 (2019 est.)
country comparison to the world: 58

Telecommunication systems: *general assessment:* the armed insurgency that began in 2011 has led to major disruptions to the network and has caused telephone and Internet outages throughout the country; 2018 saw some stabilizing; telecoms have become decentralized; fairly high mobile penetration of 98%; potential for growth given that subscription numbers are low; remote areas rely on expensive satellite communications; mobile broadband infrastructure is predominantly 3G for about 85% of the population; LTE launched in 2017; Syria has two mobile telephone operators (2020)
domestic: the number of fixed-line connections increased markedly prior to the civil war in 2011 and now stands at 17 per 100; mobile-cellular service stands at about 114 per 100 persons (2019)
international: country code—963; landing points for the Aletar, BERYTAR and UGART submarine cable connections to Egypt, Lebanon, and Cyprus; satellite earth stations—1 Intelsat (Indian Ocean) and 1 Intersputnik (Atlantic Ocean region); coaxial cable and microwave radio relay to Iraq, Jordan, Lebanon, and Turkey; participant in Medarabtel (2019)
note: the COVID-19 outbreak is negatively impacting telecommunications production and supply chains globally; consumer spending on telecom devices and services has also slowed due to the pandemic's effect on economies worldwide; overall progress towards improvements in all facets of the telecom industry—mobile, fixed-line, broadband, submarine cable and satellite—has moderated

Broadcast media: state-run TV and radio broadcast networks; state operates 2 TV networks and 5 satellite channels; roughly two-thirds of Syrian homes have a satellite dish providing access to foreign TV broadcasts; 3 state-run radio channels; first private radio station launched in 2005; private radio broadcasters prohibited from transmitting news or political content (2018)

Internet country code: .sy

Internet users: *total:* 6,077,510
percent of population: 34.25% (July 2018 est.)
country comparison to the world: 78

Broadband—fixed subscriptions: *total:* 1,328,688
subscriptions per 100 inhabitants: 7 (2018 est.)
country comparison to the world: 66

TRANSPORTATION

National air transport system: *number of registered air carriers:* 3 (2020)
inventory of registered aircraft operated by air carriers: 11
annual passenger traffic on registered air carriers: 17,896 (2018)
annual freight traffic on registered air carriers: 30,000 mt-km (2018)

Civil aircraft registration country code prefix: YK (2016)

Airports: 90 (2013)
country comparison to the world: 62

Airports—with paved runways: *total:* 29 (2013)
over 3,047 m: 5 (2013)
2,438 to 3,047 m: 16 (2013)
914 to 1,523 m: 3 (2013)
under 914 m: 5 (2013)

Airports—with unpaved runways: *total:* 61 (2013)
1,524 to 2,437 m: 1 (2013)
914 to 1,523 m: 12 (2013)
under 914 m: 48 (2013)

Heliports: 6 (2013)

Pipelines: 3170 km gas, 2029 km oil (2013)

Railways: *total:* 2,052 km (2014)
standard gauge: 1,801 km 1.435-m gauge (2014)
narrow gauge: 251 km 1.050-m gauge (2014)
country comparison to the world: 74

Roadways: *total:* 69,873 km (2010)
paved: 63,060 km (2010)
unpaved: 6,813 km (2010)
country comparison to the world: 71

Waterways: 900 km (navigable but not economically significant) (2011)
country comparison to the world: 68

Merchant marine: *total:* 25
by type: bulk carrier 1, general cargo 10, other 14 (2019)
country comparison to the world: 139

Ports and terminals: *major seaport(s):* Baniyas, Latakia, Tartus

MILITARY AND SECURITY

Military and security forces: *Syrian Armed Forces:* Syrian Arab Army, Syrian Naval Forces, Syrian Air Forces, Syrian Air Defense Forces, National Defense Forces (pro-government militia and auxiliary forces) (2019)
note: the Syrian government is working to demobilize militias or integrate them into its regular forces

Military and security service personnel strengths: N/A; the Syrian Armed Forces (SAF) are rebuilding and trying to integrate government-allied militias and auxiliary forces while continuing to engage in a civil war; prior to the start of the civil war in 2011, the SAF had approximately 300,000 active troops, including 200-225,000 Army; by 2018, its estimated size was reportedly less than 100,000 due to casualties and desertions (2019 est.)

Military equipment inventories and acquisitions: the SAF's inventory is comprised mostly of Russian and Soviet-era equipment; since 2010, Russia has supplied nearly all of Syria's imported weapons systems, although China and Iran have also provided military equipment (2019 est.)

Military service age and obligation: 18-42 years of age for compulsory and voluntary military service; conscript service obligation is 18 months; women are not conscripted but may volunteer to serve (2019)

Military—note: the United Nations Disengagement Observer Force (UNDOF) has operated in the Golan between Israel and Syria since 1974 to monitor the ceasefire following the 1973 Arab-Israeli War and supervise the areas of separation between the two countries; as of October 2019, UNDOF consisted of about 1,140 personnel

TERRORISM

Terrorist group(s): Abdallah Azzam Brigades; Ansar al-Islam; Asa'ib Ahl Al-Haq; Hizballah; Hurras al-Din; Islamic Jihad Union; Islamic Revolutionary Guard Corps -- Qods Force; Islamic State of Iraq and ash-Sham; Kata'ib Hizballah; Kurdistan Workers' Party; Mujahidin Shura Council in the Environs of Jerusalem; al-Nusrah Front (Hay'at Tahrir al-Sham); al-Qa'ida; Palestine Liberation Front; PFLP-General

Command; Popular Front for the Liberation of Palestine (2020)

note: details about the history, aims, leadership, organization, areas of operation, tactics, targets, weapons, size, and sources of support of the group(s) appear(s) in Appendix-T

TRANSNATIONAL ISSUES

Disputes—international: Golan Heights is Israeli-controlled with an almost 1,000-strong UN Disengagement Observer Force patrolling a buffer zone since 1964; lacking a treaty or other documentation describing the boundary, portions of the Lebanon-Syria boundary are unclear with several sections in dispute; since 2000, Lebanon has claimed Shab'a Farms in the Golan Heights; 2004 Agreement and pending demarcation would settle border dispute with Jordan

Refugees and internally displaced persons: *refugees (country of origin):* 13,311 (Iraq) (2019); 562,312 (Palestinian Refugees) (2020)
IDPs: 6.1 million (ongoing civil war since 2011) (2020)
stateless persons: 160,000 (2019); note—Syria's stateless population consists of Kurds and Palestinians; stateless persons are prevented from voting, owning land, holding certain jobs, receiving food subsidies or public healthcare, enrolling in public schools, or being legally married to Syrian citizens; in 1962, some 120,000 Syrian Kurds were stripped of their Syrian citizenship, rendering them and their descendants stateless; in 2011, the Syrian Government granted citizenship to thousands of Syrian Kurds as a means of appeasement; however, resolving the question of statelessness is not a priority given Syria's ongoing civil war
note: the ongoing civil war has resulted in more than 5.5 million registered Syrian

refugees—dispersed in Egypt, Iraq, Jordan, Lebanon, and Turkey—as of November 2020

Trafficking in persons: *current situation:* as conditions continue to deteriorate due to Syria's civil war, human trafficking has increased; Syrians remaining in the country and those that are refugees abroad are vulnerable to trafficking; Syria is a source and destination country for men, women and children subjected to forced labor and sex trafficking; Syrian children continue to be forcibly recruited by government forces, pro-regime militias, armed opposition groups, and terrorist organizations to serve as soldiers, human shields, and executioners; ISIL forces Syrian women and girls and Yazidi women and girls taken from Iraq to marry its fighters, where they experience domestic servitude and sexual violence; Syrian refugee women and girls are forced into exploitive marriages or prostitution in neighboring countries, while displaced children are forced into street begging domestically and abroad
tier rating: Tier 3—the government does not fully comply with the minimum standards for the elimination of trafficking and is not making significant efforts to do so; in 2014, Syria's violent conditions enabled human trafficking to flourish; the government made no effort to investigate, prosecute, or convict trafficking offenders or complicit government officials, including those who forcibly recruited child soldiers; authorities did not identify victims and failed to ensure victims, including child soldiers, were protected from arrest, detention, and severe abuse as a result of being trafficked (2015)

Illicit drugs: a transit point for opiates, hashish, and cocaine bound for regional and Western markets; weak anti-money-laundering controls and bank privatization may leave it vulnerable to money laundering

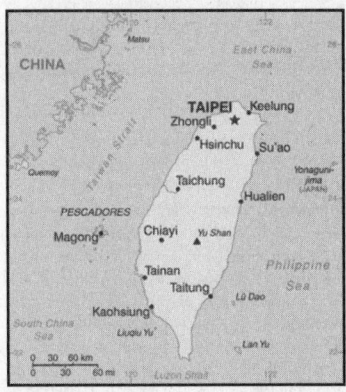

INTRODUCTION

Background: First inhabited by Austronesian people, Taiwan became home to Han immigrants beginning in the late Ming Dynasty (17th century). In 1895, military defeat forced China's Qing Dynasty to cede Taiwan to Japan, which then governed Taiwan for 50 years. Taiwan came under Chinese Nationalist (Kuomintang, KMT) control after World War II. With the communist victory in the Chinese civil war in 1949, the Nationalist-controlled Republic of China government and 2 million Nationalists fled to Taiwan and continued to claim to be the legitimate government for mainland China and Taiwan based on a 1947 Constitution drawn up for all of China. Until 1987, however, the Nationalist government ruled Taiwan under a civil war martial law declaration dating to 1948. Beginning in the 1970s, Nationalist authorities gradually began to incorporate the native population into the governing structure beyond the local level. The democratization process expanded rapidly in the 1980s, leading to the then illegal founding of Taiwan's first opposition party (the Democratic Progressive Party or DPP) in 1986 and the lifting of martial law the following year. Taiwan held legislative elections in 1992, the first in over forty years, and its first direct presidential election in 1996. In the 2000 presidential elections, Taiwan underwent its first peaceful transfer of power with the KMT loss to the DPP and afterwards experienced two additional democratic transfers of power in 2008 and 2016. Throughout this period, the island prospered, became one of East Asia's economic "Tigers," and after 2000 became a major investor in mainland China as cross-Strait ties matured. The dominant political issues continue to be economic reform and growth as well as management of sensitive relations between Taiwan and China.

GEOGRAPHY

Location: Eastern Asia, islands bordering the East China Sea, Philippine Sea, South China Sea, and Taiwan Strait, north of the Philippines, off the southeastern coast of China

Geographic coordinates: 23 30 N, 121 00 E

Map references: Southeast Asia

Area: *total:* 35,980 sq km
land: 32,260 sq km
water: 3,720 sq km
note: includes the Pescadores, Matsu, and Quemoy islands
country comparison to the world: 139

Area—comparative: slightly smaller than Maryland and Delaware combined

Land boundaries: 0 km

Coastline: 1,566.3 km

Maritime claims: *territorial sea:* 12 nm
exclusive economic zone: 200 nm

Climate: tropical; marine; rainy season during southwest monsoon (June to August); persistent and extensive cloudiness all year

Terrain: eastern two-thirds mostly rugged mountains; flat to gently rolling plains in west

Elevation: *mean elevation:* 1,150 m
lowest point: South China Sea 0 m
highest point: Yu Shan 3,952 m

Natural resources: small deposits of coal, natural gas, limestone, marble, asbestos, arable land

Land use: *agricultural land:* 22.7% (2011 est.)
arable land: 16.9% (2011 est.) / permanent crops: 5.8% (2011 est.)
other: 77.3% (2011 est.)

Irrigated land: 3,820 sq km (2012)

Population distribution: distribution exhibits a peripheral coastal settlement pattern, with the largest populations on the north and west coasts

Natural hazards: earthquakes; typhoons
volcanism: Kueishantao Island (401 m), east of Taiwan, is its only historically active volcano, although it has not erupted in centuries

Environment—current issues: air pollution; water pollution from industrial emissions, raw sewage; contamination of drinking water supplies; trade in endangered species; low-level radioactive waste disposal

Environment—international agreements: *party to:* none of the selected agreements because of Taiwan's international status

Geography—note: strategic location adjacent to both the Taiwan Strait and the Luzon Strait

PEOPLE AND SOCIETY

Population: 23,603,049 (July 2020 est.)
country comparison to the world: 56

Nationality: *noun:* Taiwan (singular and plural)
adjective: Taiwan (or Taiwanese)
note: example - he or she is from Taiwan; they are from Taiwan

Ethnic groups: Han Chinese (including Hoklo, who compose approximately 70% of Taiwan's population, Hakka, and other groups originating in mainland China) more than 95%, indigenous Malayo-Polynesian peoples 2.3%
note 1: there are 16 officially recognized indigenous groups: Amis, Atayal, Bunun, Hla'alua, Kanakaravu, Kavalan, Paiwan, Puyuma, Rukai, Saisiyat, Sakizaya, Seediq, Thao, Truku, Tsou, and Yami; Amis, Paiwan, and Atayal are the largest and account for roughly 70% of the indigenous population
note 2: although not definitive, the majority of current genetic, archeological, and linguistic data support the theory that Taiwan is the ultimate source for the spread of humans across the Pacific to Polynesia; the expansion (ca. 3000 B.C. to A.D. 1200) took place via the Philippines and eastern Indonesia and reached Fiji and Tonga by about 900 B.C.; from there voyagers spread across all of the rest of the Pacific islands over the next two millennia

Languages: Mandarin Chinese (official), Taiwanese (Min Nan), Hakka dialects, approximately 16 indigenous languages

Religions: Buddhist 35.3%, Taoist 33.2%, Christian 3.9%, folk (includes Confucian) approximately 10%, none or unspecified 18.2% (2005 est.)

Age structure: *0-14 years:* 12.42% (male 1,504,704/female 1,426,494)
15-24 years: 11.62% (male 1,403,117/female 1,339,535)
25-54 years: 45.51% (male 5,351,951/female 5,389,112)
55-64 years: 14.73% (male 1,698,555/female 1,778,529)
65 years and over: 15.72% (male 1,681,476/female 2,029,576) (2020 est.)

Dependency ratios: *total dependency ratio:* 40
youth dependency ratio: 17.8
elderly dependency ratio: 22.2
potential support ratio: 4.5 (2020 est.)

Median age: *total:* 42.3 years
male: 41.5 years
female: 43.1 years (2020 est.)
country comparison to the world: 36

Population growth rate: 0.11% (2020 est.)
country comparison to the world: 187

Birth rate: 8 births/1,000 population (2020 est.)
country comparison to the world: 223

Death rate: 7.9 deaths/1,000 population (2020 est.)
country comparison to the world: 97

Net migration rate: 0.8 migrant(s)/1,000 population (2020 est.)
country comparison to the world: 64

Population distribution: distribution exhibits a peripheral coastal settlement pattern, with the largest populations on the north and west coasts

Urbanization: *urban population:* 78.9% of total population (2020)
rate of urbanization: 0.8% annual rate of change (2015-20 est.)

Major urban areas - population: 4.398 million New Taipei City, 2.721 million TAIPEI (capital), 2.245 million Taoyuan, 1.538 million Kaohsiung, 1.321 million
Taichung, 850,000 Tainan (2020)

Sex ratio: *at birth:* 1.06 male(s)/female
0-14 years: 1.05 male(s)/female
15-24 years: 1.05 male(s)/female
25-54 years: 0.99 male(s)/female
55-64 years: 0.96 male(s)/female
65 years and over: 0.83 male(s)/female
total population: 0.97 male(s)/female (2020 est.)

Infant mortality rate: *total:* 4.2 deaths/1,000 live births
male: 4.6 deaths/1,000 live births
female: 3.8 deaths/1,000 live births (2020 est.)
country comparison to the world: 188

Life expectancy at birth: *total population:* 80.6 years
male: 77.5 years
female: 83.9 years (2020 est.)
country comparison to the world: 43

Total fertility rate: 1.14 children born/woman (2020 est.)
country comparison to the world: 226

HIV/AIDS—adult prevalence rate: NA

HIV/AIDS—people living with HIV/AIDS: NA

HIV/AIDS—deaths: NA

Education expenditures: NA

Literacy: *definition:* age 15 and over can read and write
total population: 98.5%
male: 99.7%
female: 97.3% (2014)

GOVERNMENT

Country name: *conventional long form:* none
conventional short form: Taiwan
local long form: none
local short form: Taiwan
former: Formosa
etymology: "Tayowan" was the name of the coastal sandbank where the Dutch erected their colonial headquarters on the island in the 17th century; the former name "Formosa" means "beautiful" in Portuguese

Government type: semi-presidential republic

Capital: *name:* Taipei
geographic coordinates: 25 02 N, 121 31 E
time difference: UTC+8 (13 hours ahead of Washington, DC, during Standard Time)

etymology: the Chinese meaning is "Northern Taiwan," reflecting the city's position in the far north of the island

Administrative divisions: includes main island of Taiwan plus smaller islands nearby and off coast of China's Fujian Province; Taiwan is divided into 13 counties (xian, singular and plural), 3 cities (shi, singular and plural), and 6 special municipalities directly under the jurisdiction of the Executive Yuan

countries: Changhua, Chiayi, Hsinchu, Hualien, Kinmen, Lienchiang, Miaoli, Nantou, Penghu, Pingtung, Taitung, Yilan, Yunlin

cities: Chiayi, Hsinchu, Keelung

special municipalities: Kaohsiung (city), New Taipei (city), Taichung (city), Tainan (city), Taipei (city), Taoyuan (city)
note: Taiwan uses a variety of romanization systems; while a modified Wade-Giles system still dominates, the city of Taipei has adopted a Pinyin romanization for street and place names within its boundaries; other local authorities use different romanization systems

National holiday: Republic Day (National Day), 10 October (1911); note - celebrates the anniversary of the Chinese Revolution, also known as Double Ten (10-10) Day

Constitution: *history:* previous 1912, 1931; latest adopted 25 December 1946, promulgated 1 January 1947, effective 25 December 1947

amendments: proposed by at least one fourth of the Legislative Yuan membership; passage requires approval by at least three-fourths majority vote of at least three fourths of the Legislative Yuan membership and approval in a referendum by more than half of eligible voters; revised several times, last in 2005

Legal system: civil law system

International law organization participation: has not submitted an ICJ jurisdiction declaration; non-party state to the ICCt

Citizenship: *citizenship by birth:* no
citizenship by descent only: at least one parent must be a citizen of Taiwan
dual citizenship recognized: yes, except that citizens of Taiwan are not recognized as dual citizens of the People's Republic of China
residency requirement for naturalization: 5 years

Suffrage: 20 years of age; universal; note - in mid-2016, the Legislative Yuan drafted a constitutional amendment to reduce the voting age to 18, but it has not passed as of December 2017

Executive branch: *chief of state:* President TSAI Ing-wen (since 20 May 2016; re-elected on 11 Jan 2020); Vice President CHEN Chien-jen (since 20 May 2016)

head of government: Premier SU Tseng-chang (President of the Executive Yuan) (since 11 January 2019); Vice Premier SHIH Jun-ji, Vice President of the Executive Yuan (since 8 September 2017)
cabinet: Executive Yuan - ministers appointed by president on recommendation of premier

elections/appointments: president and vice president directly elected on the same ballot by simple majority popular vote for a 4-year term (eligible for a second term); election last held on 11 January 2020 (next to be held on 11 January 2024); premier appointed by the president; vice premiers appointed by the president on the recommendation of the premier
election results: TSAI Ing-wen elected president; percent of vote - TSAI Ing-wen (DPP) 57.1%, HAN Kuo-yu (KMT) 38.6%; note - TSAI is the first woman elected president of Taiwan

Legislative branch: *description:* unicameral Legislative Yuan (113 seats; 73 members directly elected in single-seat constituencies by simple majority vote, 34 directly elected in a single island-wide constituency by proportional representation vote, and 6 directly elected in multi-seat aboriginal constituencies by proportional representation vote; members serve 4-year terms)
elections: last held on 11 January 2020 (next to be held on 11 January 2024)
election results: percent of vote by party - Democratic Progressive Party (DPP) 34.0%, Kuomintang (KMT) 33.4%, Taiwan People's Party (TPP) 11.2%; seats by party - DPP 61, KMT 38, TPP 5

Judicial branch: *highest courts:* Supreme Court (consists of the court president, vice president, and approximately 100 judges organized into 8 civil and 12 criminal divisions, each with a division chief justice and 4 associate justices); Constitutional Court (consists of the court president, vice president, and 13 justices)
judge selection and term of office: Supreme Court justices appointed by the president; Constitutional Court justices appointed by the president, with approval of the Legislative Yuan; Supreme Court justices serve for life; Constitutional Court justices appointed for 8-year terms, with half the membership renewed every 4 years
subordinate courts: high courts; district courts; hierarchy of administrative courts

Political parties and leaders: Democratic Progressive Party or DPP [CHO Jung-tai]
Kuomintang or KMT (Nationalist Party) [WU Den-yih]
New Power Party or NPP [CHIU Hsien-chih]
Non-Partisan Solidarity Union or NPSU [LIN Pin-kuan]
People First Party or PFP [James SOONG Chu-yu]
International organization participation: ADB (Taipei, China), APEC (Chinese Taipei), BCIE, IOC, ITUC (NGOs), SICA (observer), WTO (Taipei, China);
note - separate customs territory of Taiwan, Penghu, Kinmen, and Matsu

Diplomatic representation in the US: *chief of mission:* none; commercial and cultural relations with its citizens in the US are maintained through an unofficial instrumentality, the Taipei Economic and Cultural Representative Office in the United States (TECRO), a private nonprofit corporation that performs citizen and consular services similar to those at diplomatic posts, represented by

Stanley KAO (since 5 June 2016); office: 4201 Wisconsin Avenue NW, Washington, DC 20016; telephone: [1] 202 895-1800

Taipei Economic and Cultural Offices (branch offices): Atlanta, Boston, Chicago, Denver (CO), Houston, Honolulu, Los Angeles, Miami, New York, San Francisco, Seattle

Diplomatic representation from the US: *chief of mission:* the US does not have an embassy in Taiwan; commercial and cultural relations with the people of Taiwan are maintained through an unofficial instrumentality, the American Institute in Taiwan (AIT), a private nonprofit corporation that performs citizen and consular services similar to those at diplomatic posts; it is managed by Director William Brent CHRISTENSEN (since 11 August 2018); telephone [886] 7-335-5006; FAX [886] 7-338-0551
telephone: (+886) (02) 2162-2000
branch office(s): American Institute in Taiwan No. 100, Jinhu Road,
Neihu District 11461, Taipei City
other offices: Kaohsiung (Branch Office)

Flag description: red field with a dark blue rectangle in the upper hoist-side corner bearing a white sun with 12 triangular rays; the blue and white design of the canton (symbolizing the sun of progress) dates to 1895; it was later adopted as the flag of the Kuomintang Party; blue signifies liberty, justice, and democracy, red stands for fraternity, sacrifice, and nationalism, and white represents equality, frankness, and the people's livelihood; the 12 rays of the sun are those of the months and the twelve traditional Chinese hours (each ray equals two hours)
note: similar to the flag of Samoa

National symbol(s): *white, 12-rayed sun on blue field; national colors:* blue, white, red

National anthem: *name:* "Zhonghua Minguo guoge" (National Anthem of the Republic of China)
lyrics/music: HU Han-min, TAI Chi-t'ao, and LIAO Chung-k'ai/CHENG Mao-Yun
note: adopted 1930; also the song of the Kuomintang Party; it is informally known as "San Min Chu I" or "San Min Zhu Yi" (Three Principles of the People); because of political pressure from China, "Guo Qi Ge" (National Banner Song) is used at international events rather than the official anthem of Taiwan; the "National Banner Song" has gained popularity in Taiwan and is commonly used during flag raisings

ECONOMY

Economy—overview: Taiwan has a dynamic capitalist economy that is driven largely by industrial manufacturing, and especially exports of electronics, machinery, and petrochemicals. This heavy dependence on exports exposes the economy to fluctuations in global demand. Taiwan's diplomatic isolation, low birth rate, rapidly aging population, and increasing competition from China and other Asia Pacific markets are other major long-term challenges.

Following the landmark Economic Cooperation Framework Agreement (ECFA) signed with China in June 2010, Taiwan in July 2013 signed a free trade deal with New Zealand - Taipei's first-ever with a country with which it does not maintain diplomatic relations - and, in November of that year, inked a trade pact with Singapore. However, follow-on components of the ECFA, including a signed agreement on trade in services and negotiations on trade in goods and dispute resolution, have stalled. In early 2014, the government bowed to public demand and proposed a new law governing the oversight of cross-Strait agreements, before any additional deals with China are implemented; the legislature has yet to vote on such legislation, leaving the future of ECFA uncertain. President TSAI since taking office in May 2016 has promoted greater economic integration with South and Southeast Asia through the New Southbound Policy initiative and has also expressed interest in Taiwan joining the Trans-Pacific Partnership as well as bilateral trade deals with partners such as the US. These overtures have likely played a role in increasing Taiwan's total exports, which rose 11% during the first half of 2017, buoyed by strong demand for semiconductors.

Taiwan's total fertility rate of just over one child per woman is among the lowest in the world, raising the prospect of future labor shortages, falling domestic demand, and declining tax revenues. Taiwan's population is aging quickly, with the number of people over 65 expected to account for nearly 20% of the island's total population by 2025.

The island runs a trade surplus with many economies, including China and the US, and its foreign reserves are the world's fifth largest, behind those of China, Japan, Saudi Arabia, and Switzerland. In 2006, China overtook the US to become Taiwan's second-largest source of imports after Japan. China is also the island's number one destination for foreign direct investment. Taiwan since 2009 has gradually loosened rules governing Chinese investment and has also secured greater market access for its investors on the mainland. In August 2012, the Taiwan Central Bank signed a memorandum of understanding (MOU) on cross-Strait currency settlement with its Chinese counterpart. The MOU allows for the direct settlement of Chinese renminbi (RMB) and the New Taiwan dollar across the Strait, which has helped Taiwan develop into a local RMB hub.

Closer economic links with the mainland bring opportunities for Taiwan's economy but also pose challenges as political differences remain unresolved and China's economic growth is slowing. President TSAI's administration has made little progress on the domestic economic issues that loomed large when she was elected, including concerns about stagnant wages, high housing prices, youth unemployment, job security, and financial security in retirement. TSAI has made more progress on boosting trade with South and Southeast Asia, which may help insulate Taiwan's economy from a fall in mainland demand should China's growth slow in 2018.

GDP (purchasing power parity): $1.189 trillion (2017 est.)
$1.156 trillion (2016 est.)
$1.14 trillion (2015 est.)
note: data are in 2017 dollars
country comparison to the world: 22

GDP (official exchange rate): $572.6 billion (2017 est.)

GDP - real growth rate: 2.71% (2019 est.)
2.75% (2018 est.)
3.31% (2017 est.)
country comparison to the world: 105

GDP—per capita (PPP): $49,100 (2016 est.)
$48,500 (2015 est.)
note: data are in 2017 dollars
country comparison to the world: 28

Gross national saving: 34.9% of GDP (2017 est.)
35.5% of GDP (2016 est.)
36.3% of GDP (2015 est.)
country comparison to the world: 17

GDP—composition, by end use: *household consumption:* 53% (2017 est.)
government consumption: 14.1% (2017 est.)
investment in fixed capital: 20.5% (2017 est.)
investment in inventories: -0.2% (2017 est.)
exports of goods and services: 65.2% (2017 est.)
imports of goods and services: -52.6% (2017 est.)

GDP—composition, by sector of origin: *agriculture:* 1.8% (2017 est.)
industry: 36% (2017 est.)
services: 62.1% (2017 est.)

Agriculture—products: rice, vegetables, fruit, tea, flowers; pigs, poultry; fish

Industries: electronics, communications and information technology products, petroleum refining, chemicals, textiles, iron and steel, machinery, cement, food processing, vehicles, consumer products, pharmaceuticals

Industrial production growth rate: 3.9% (2017 est.)
country comparison to the world: 79

Labor force: 11.498 million (2020 est.)
country comparison to the world: 47

Labor force—by occupation: agriculture: 4.9%
industry: 35.9%
services: 59.2% (2016 est.)

Unemployment rate: 3.73% (2019 est.)
3.69% (2018 est.)
country comparison to the world: 54

Population below poverty line: 1.5% (2012 est.)

Household income or consumption by percentage share: *lowest 10%:* 6.4% (2010)
highest 10%: 40.3% (2010)

Budget: *revenues:* 91.62 billion (2017 est.)
expenditures: 92.03 billion (2017 est.)

Taxes and other revenues: 16% (of GDP) (2017 est.)
country comparison to the world: 184

Budget surplus (+) or deficit (-): -0.1% (of GDP) (2017 est.)
country comparison to the world: 50

Public debt: 35.7% of GDP (2017 est.)
36.2% of GDP (2016 est.)

note: data for central government
country comparison to the world: 149

Fiscal year: calendar year

Inflation rate (consumer prices): 1.1% (2017 est.)
1% (2016 est.)
country comparison to the world: 61

Current account balance: $65.173 billion (2019 est.)
$70.843 billion (2018 est.)
country comparison to the world: 7
Exports: $329.5 billion (2019)
country comparison to the world: 15

Exports—partners: China 27.9%, US 14.1%, Hong Kong 12.3%, Japan 7.1%, Singapore 5.5%, South Korea 5.1% (2019)

Exports—commodities: semiconductors, petrochemicals, automobile/auto parts, ships, wireless communication equipment, flat display displays, steel, electronics, plastics, computers
Imports: $285.9 billion (2019)
country comparison to the world: 18

Imports—commodities: oil/petroleum, semiconductors, natural gas, coal, steel, computers, wireless communication equipment, automobiles, fine chemicals, textiles

Imports—partners: China 20.1%, Japan 15.4%, US 12.3%, South Korea 6.2% (2019)
Reserves of foreign exchange and gold: $456.7 billion (31 December 2017 est.)
$439 billion (31 December 2016 est.)
country comparison to the world: 5

Debt—external: $181.9 billion (31 December 2017 est.)
$172.2 billion (31 December 2016 est.)
country comparison to the world: 38

Exchange rates: New Taiwan dollars (TWD) per US dollar—
30.68 (2017 est.)
32.325 (2016 est.)
32.325 (2015 est.)
31.911 (2014 est.)
30.363 (2013 est.)

ENERGY

Electricity—production: 246.1 billion kWh (2016 est.)
country comparison to the world: 18

Electricity—consumption: 237.4 billion kWh (2016 est.)
country comparison to the world: 16

Electricity—exports: 0 kWh (2016 est.)
country comparison to the world: 205

Electricity—imports: 0 kWh (2016 est.)
country comparison to the world: 208

Electricity—installed generating capacity: 49.52 million kW (2016 est.)
country comparison to the world: 22

Electricity—from fossil fuels: 79% of total installed capacity (2016 est.)
country comparison to the world: 88

Electricity—from nuclear fuels: 11% of total installed capacity (2017 est.)

Electricity—from hydroelectric plants: 4% of total installed capacity (2017 est.)
country comparison to the world: 133

Electricity—from other renewable sources: 6% of total installed capacity (2017 est.)
country comparison to the world: 105

Crude oil—production: 196 bbl/day (2018 est.)
country comparison to the world: 96

Crude oil—exports: 0 bbl/day (2015 est.)
country comparison to the world: 203

Crude oil—imports: 846,400 bbl/day (2015 est.)
country comparison to the world: 13

Crude oil—proved reserves: 2.38 million bbl (1 January 2018 est.)
country comparison to the world: 95

Refined petroleum products—production: 924,000 bbl/day (2015 est.)
country comparison to the world: 21

Refined petroleum products—consumption: 962,400 bbl/day (2016 est.)
country comparison to the world: 22

Refined petroleum products— exports: 349,600 bbl/day (2015 est.)
country comparison to the world: 26

Refined petroleum products—imports: 418,300 bbl/day (2015 est.)
country comparison to the world: 20

Natural gas—production: 237.9 million cu m (2017 est.)
country comparison to the world: 77

Natural gas—consumption: 22.45 billion cu m (2017 est.)
country comparison to the world: 34

Natural gas—exports: 0 cu m (2017 est.)
country comparison to the world: 196

Natural gas—imports: 22.14 billion cu m (2017 est.)
country comparison to the world: 15

Natural gas—proved reserves: 6.229 billion cu m (1 January 2018 est.)
country comparison to the world: 86

Carbon dioxide emissions from consumption of energy: 348.8 million Mt (2017 est.)
country comparison to the world: 21

COMMUNICATIONS

Telephones—fixed lines: *total subscriptions:* 12,863,860

subscriptions per 100 inhabitants: 54.56 (2019 est.)
country comparison to the world: 16

Telephones—mobile cellular: total subscriptions: 29,049,784

subscriptions per 100 inhabitants: 123.21 (2019 est.)
country comparison to the world: 46

Telecommunication systems: *general assessment.* good telecommunications infrastructure and competitive mobile market; Taiwan has a stable regulatory system and an educated workforce

country comparison to the world: 14

building on availability of fixed and mobile broadband networks; investors attracted to this excellent telecom infrastructure; fixed-line will decline in the next 5 years; 6 mobile network operators; 4G LTE service; regulator begins multi-spectrum auction for 5G services; govt. to release NT $20.5 billion to encourage development of 5G services (2020)

domestic: fixed-line 55 per 100 and mobile-cellular 123 per 100 (2019)

international: country code - 886; landing points for the EAC-C2C, APCN-2, FASTER, SJC2, TSE-1, TPE, APG, SeaMeWe-3, FLAG North Asia Loop/REACH North Asia Loop, HKA, NCP, and PLCN submarine fiber cables provide links throughout Asia, Australia, the Middle East, Europe, Africa and the US; satellite earth stations - 2 (2019)

note: the COVID-19 outbreak is negatively impacting telecommunications production and supply chains globally; consumer spending on telecom devices and services has also slowed due to the pandemic's effect on economies worldwide; overall progress towards improvements in all facets of the telecom industry - mobile, fixed-line, broadband, submarine cable and satellite - has moderated

Broadcast media: 5 nationwide television networks operating roughly 22 TV stations; more than 300 satellite TV channels are available; about 60% of households utilize multi-channel cable TV; 99.9% of households subscribe to digital cable TV; national and regional radio networks with about 171 radio stations (2019)

Internet country code: .tw

Internet users: total: 21,845,944

percent of population: 92.78% (July 2018 est.)
country comparison to the world: 34

Broadband—fixed subscriptions: total: 5,725,022

subscriptions per 100 inhabitants: 24 (2018 est.)
country comparison to the world: 28

TRANSPORTATION

National air transport system: *number of registered air carriers:* 7 (2020)

inventory of registered aircraft operated by air carriers: 216

Civil aircraft registration country code prefix: B (2016)

Airports: 37 (2013)
country comparison to the world: 107

Airports—with paved runways: total: 35 (2013)
over 3,047 m: 8 (2013)
2,438 to 3,047 m: 7 (2013)
1,524 to 2,437 m: 10 (2013)
914 to 1,523 m: 8 (2013)
under 914 m: 2 (2013)

Airports—with unpaved runways: total: 2 (2013)
1,524 to 2,437 m: 1 (2013)
under 914 m: 1 (2013)

Heliports: 31 (2013)

Pipelines: 25 km condensate, 2,200 km gas, 13,500 km oil (2018)

Railways: *total:* 1,613 km (2018)

standard gauge: 345 km 1.435-m gauge (345 km electrified) (2018)

narrow gauge: 1,118.1 km 1.067-m gauge (793.9 km electrified) (2018)

150 0.762-m gauge

note: the 0.762-gauge track belongs to three entities: the Forestry Bureau, Taiwan Cement, and TaiPower

country comparison to the world: 81

Roadways: *total:* 43,206 km (2017)

paved: 42,793 km (includes 1,348 km of highways and 737 km of expressways) (2017)

unpaved: 413 km (2017)

country comparison to the world: 88

Merchant marine: *total:* 389

by type: bulk carrier 30, container ship 47, general cargo 56, oil tanker 32, other 224 (2019)

country comparison to the world: 46

Ports and terminals: *major seaport(s):* Keelung (Chi-lung), Kaohsiung, Hualian, Taichung

container port(s) (TEUs): Kaohsiung (10,271,018), Taichung (1,660,663), Taipei (1,561,743) (2017)

LNG terminal(s) (import): Yung An (Kaohsiung), Taichung

MILITARY AND SECURITY

Military and security forces: *Taiwan Armed Forces:* Army, Navy (includes Marine Corps), Air Force, Military Police Command, Armed Forces Reserve Command; Taiwan Coast Guard Administration (a law enforcement organization with homeland security functions during peacetime and national defense missions during wartime) (2020)

Military expenditures: 1.7% of GDP (2019)

1.7% of GDP (2018)

1.8% of GDP (2017)

1.8% of GDP (2016)

1.9% of GDP (2015)

country comparison to the world: 66

Military and security service personnel strengths: the Taiwan military has approximately 170,000 active duty troops (90,000 Army; 40,000 Navy; 40,000 Air Force) (2019)

Military equipment inventories and acquisitions: the Taiwan military is armed mostly with second-hand weapons and equipment provided by the US; Taiwan also has a domestic defense industry capable of upgrading some weapons systems and building surface naval craft and submarines (2019)

Military service age and obligation: starting with those born in 1994, males 18-36 years of age may volunteer for military service or must complete 4 months of compulsory military training (or substitute civil service in some cases); men born before December 1993 are required to complete compulsory service for 1 year (military or civil); men are subject to training recalls up to four times for periods not to exceed 20 days for 8 years after discharge; women may enlist, but are restricted to noncombat roles in most cases; as part of its transition to an all-volunteer military in December 2018, the last cohort of one-year military conscripts completed their service obligations (2019)

TRANSPORTATION

Disputes—international: involved in complex dispute with Brunei, China, Malaysia, the Philippines, and Vietnam over the Spratly Islands, and with China and the Philippines over Scarborough Reef; the 2002 "Declaration on the Conduct of Parties in the South China Sea" has eased tensions but falls short of a legally binding "code of conduct" desired by several of the disputants; Paracel Islands are occupied by China, but claimed by Taiwan and Vietnam; in 2003, China and Taiwan became more vocal in

rejecting both Japan's claims to the uninhabited islands of the Senkaku-shoto (Diaoyu Tai) and Japan's unilaterally declared exclusive economic zone in the East China Sea where all parties engage in hydrocarbon prospecting

Illicit drugs: regional transit point for heroin, methamphetamine, and precursor chemicals; transshipment point for drugs to Japan; major problem with domestic consumption of methamphetamine and heroin; rising problems with use of ketamine and club drugs

TAJIKISTAN

INTRODUCTION

Background: The Tajik people came under Russian imperial rule in the 1860s and 1870s, but Russia's hold on Central Asia weakened following the Revolution of 1917. At that time, bands of indigenous guerrillas (called "basmachi") fiercely contested Bolshevik control of the area, which was not fully reestablished until 1925. Tajikistan was first created as an autonomous republic within Uzbekistan in 1924, but in 1929 the USSR designated Tajikistan a separate republic and transferred to it much of present-day Sughd province. Ethnic Uzbeks form a substantial minority in Tajikistan, and ethnic Tajiks an even larger minority in Uzbekistan. Tajikistan became independent in 1991 following the breakup of the Soviet Union, and experienced a civil war between political, regional, and religious factions from 1992 to 1997.

Though the country holds general elections for both the presidency (once every seven years) and parliament (once every five years), observers note an electoral system rife with irregularities and abuse, with results that are neither free nor fair. President Emomali RAHMON, who came to power in 1994 during the civil war, used an attack planned by a disaffected deputy defense minister in 2015 to ban the last major opposition political party in Tajikistan. In December 2015, RAHMON further strengthened his position by having himself declared "Founder of Peace and National Unity, Leader of the Nation," with limitless terms and lifelong immunity through constitutional amendments ratified in a referendum. The referendum also lowered the minimum age required to run for president from 35 to 30, which would make RAHMON's son Rustam EMOMALI, the current mayor of the capital city of Dushanbe, eligible to run for president in 2020.

The country remains the poorest in the former Soviet sphere. Tajikistan became a member of the WTO in March 2013. However, its economy continues to face major challenges, including dependence on remittances from Tajikistani migrant laborers working in Russia and Kazakhstan, pervasive corruption, and the opiate trade and other destabilizing violence emanating from neighboring Afghanistan. Tajikistan has endured several domestic security incidents since 2010, including armed conflict between government forces and local strongmen in the Rasht Valley and between government forces and criminal groups in Gorno-Badakhshan Autonomous Oblast. Tajikistan suffered its first ISIS-claimed attack in 2018, when assailants attacked a group of Western bicyclists with vehicles and knives, killing four.

GEOGRAPHY

Location: Central Asia, west of China, south of Kyrgyzstan

Geographic coordinates: 39 00 N, 71 00 E

Map references: Asia

Area: *total:* 144,100 sq km

land: 141,510 sq km

water: 2,590 sq km

country comparison to the world: 97

Area—comparative: slightly smaller than Wisconsin

Land boundaries: *total:* 4,130 km

941

border countries (4): Afghanistan 1357 km, China 477 km, Kyrgyzstan 984 km, Uzbekistan 1312 km

Coastline: 0 km (landlocked)

Maritime claims: none (landlocked)

Climate: mid-latitude continental, hot summers, mild winters; semiarid to polar in Pamir Mountains

Terrain: mountainous region dominated by the Trans-Alay Range in the north and the Pamirs in the southeast; western Fergana Valley in north, Kofarnihon and Vakhsh Valleys in southwest

Elevation: *mean elevation:* 3,186 m
lowest point: Syr Darya (Sirdaryo) 300 m
highest point: Qullai Ismoili Somoni 7,495 m

Natural resources: hydropower, some petroleum, uranium, mercury, brown coal, lead, zinc, antimony, tungsten, silver, gold

Land use: *agricultural land:* 34.7% (2011 est.)
arable land: 6.1% (2011 est.) / permanent crops: 0.9% (2011 est.) / permanent pasture: 27.7% (2011 est.)
forest: 2.9% (2011 est.)
other: 62.4% (2011 est.)

Irrigated land: 7,420 sq km (2012)

Population distribution: the country's population is concentrated at lower elevations, with perhaps as much as 90% of the people living in valleys; overall density increases from east to west

Natural hazards: earthquakes; floods

Environment—current issues: areas of high air pollution from motor vehicles and industry; water pollution from agricultural runoff and disposal of untreated industrial waste and sewage; poor management of water resources; soil erosion; increasing levels of soil salinity

Environment—international agreements: *party to:* Biodiversity, Climate Change, Climate Change-Kyoto Protocol, Desertification, Environmental Modification, Hazardous Wastes, Ozone Layer Protection, Wetlands
signed, but not ratified: none of the selected agreements

Geography—note: landlocked; highest point, Qullai Ismoili Somoni (formerly Communism Peak), was the tallest mountain in the former USSR

PEOPLE AND SOCIETY

Population: 8,873,669 (July 2020 est.)
country comparison to the world: 96

Nationality: *noun:* Tajikistani(s)
adjective: Tajikistani

Ethnic groups: Tajik 84.3% (includes Pamiri and Yagnobi), Uzbek 13.8%, other 2% (includes Kyrgyz, Russian, Turkmen, Tatar, Arab) (2014 est.)

Languages: Tajik (official) 84.4%, Uzbek 11.9%, Kyrgyz .8%, Russian .5%, other 2.4% (2010 est.)
note: Russian widely used in government and business

Religions: Muslim 98% (Sunni 95%, Shia 3%) other 2% (2014 est.)

Age structure: *0-14 years:* 31.43% (male 1,420,271/female 1,368,445)
15-24 years: 18.13% (male 816,658/female 792,231)
25-54 years: 40.58% (male 1,789,271/female 1,811,566)
55-64 years: 6.23% (male 253,862/female 299,378)
65 years and over: 3.63% (male 132,831/female 189,156) (2020 est.)

Dependency ratios: *total dependency ratio:* 67.9
youth dependency ratio: 62.6
elderly dependency ratio: 5.3
potential support ratio: 18.7 (2020 est.)

Median age: *total:* 25.3 years
male: 24.6 years
female: 26 years (2020 est.)
country comparison to the world: 162

Population growth rate: 1.52% (2020 est.)
country comparison to the world: 69

Birth rate: 21.8 births/1,000 population (2020 est.)
country comparison to the world: 68

Death rate: 5.8 deaths/1,000 population (2020 est.)
country comparison to the world: 176

Net migration rate: -1.1 migrant(s)/1,000 population (2020 est.)
country comparison to the world: 146

Population distribution: the country's population is concentrated at lower elevations, with perhaps as much as 90% of the people living in valleys; overall density increases from east to west

Urbanization: *urban population:* 27.5% of total population (2020)
rate of urbanization: 2.62% annual rate of change (2015-20 est.)
total population growth rate v. urban population growth rate, 2000-2030: Major urban areas - population: 916,000 DUSHANBE (capital) (2020)

Sex ratio: *at birth:* 1.05 male(s)/female
0-14 years: 1.04 male(s)/female
15-24 years: 1.03 male(s)/female
25-54 years: 0.99 male(s)/female
55-64 years: 0.85 male(s)/female
65 years and over: 0.7 male(s)/female
total population: 0.99 male(s)/female (2020 est.)

Mother's mean age at first birth: 23.2 years (2017 est.)
note: median age at first birth among women 25-29

Maternal mortality rate: 17 deaths/100,000 live births (2017 est.)
country comparison to the world: 132

Infant mortality rate: *total:* 28.8 deaths/1,000 live births
male: 32.7 deaths/1,000 live births
female: 24.8 deaths/1,000 live births (2020 est.)
country comparison to the world: 59

Life expectancy at birth: *total population:* 69 years

male: 65.9 years
female: 72.3 years (2020 est.)
country comparison to the world: 174

Total fertility rate: 2.51 children born/woman (2020 est.)
country comparison to the world: 73

Contraceptive prevalence rate: 29.3% (2017)

Drinking water source: *improved:* urban: 96.2% of population
rural: 78.6% of population
total: 83.5% of population
unimproved: urban: 3.8% of population
rural: 21.4% of population
total: 16.5% of population (2017 est.)

Current Health Expenditure: 7.2% (2017)

Physicians density: 2.1 physicians/1,000 population (2014)

Hospital bed density: 4.7 beds/1,000 population (2014)

Sanitation facility access: *improved:* urban: 99.7% of population
rural: 99.3% of population
total: 99.4% of population
unimproved: urban: 0.3% of population
rural: 0.7% of population
total: 2% of population (2017 est.)

HIV/AIDS—adult prevalence rate: 0.2% (2019 est.)
country comparison to the world: 113

HIV/AIDS—people living with HIV/AIDS: 14,000 (2019 est.)
country comparison to the world: 94

HIV/AIDS—deaths: <500 (2019 est.)

Major infectious diseases: *degree of risk:* high (2020)
food or waterborne diseases: bacterial diarrhea, hepatitis A, and typhoid fever
vectorborne diseases: malaria

Obesity—adult prevalence rate: 14.2% (2016)
country comparison to the world: 128

Children under the age of 5 years underweight: 7.6% (2017)
country comparison to the world: 71

Education expenditures: 5.2% of GDP (2015)
country comparison to the world: 57

Literacy: *definition:* age 15 and over can read and write
total population: 99.8%
male: 99.8%
female: 99.7% (2015)

School life expectancy (primary to tertiary education): *total:* 11 years
male: 12 years
female: 11 years (2013)

GOVERNMENT

Country name: *conventional long form:* Republic of Tajikistan
conventional short form: Tajikistan
local long form: Jumhurii Tojikiston
local short form: Tojikiston

former: Tajik Soviet Socialist Republic

etymology: the Persian suffix "-stan" means "place of" or "country," so the word Tajikistan literally means "Land of the Tajik [people]"

Government type: presidential republic

Capital: *name:* Dushanbe

geographic coordinates: 38 33 N, 68 46 E

time difference: UTC+5 (10 hours ahead of Washington, DC, during Standard Time)

etymology: today's city was originally at the crossroads where a large bazaar occurred on Mondays, hence the name

Dushanbe, which in Persian means Monday, i.e., the second day (du) after Saturday (shambe)

Administrative divisions: 2 provinces (viloyatho, singular - viloyat), 1 autonomous province* (viloyati mukhtor), 1 capital region** (viloyati poytakht), and 1 area referred to as Districts Under Republic Administration***; Dushanbe**, Khatlon (Bokhtar), Kuhistoni Badakhshon [Gorno-Badakhshan]* (Khorugh), Nohiyahoi Tobei Jumhuri*** (Sughd (Khujand)

note: the administrative center name follows in parentheses

Independence: 9 September 1991 (from the Soviet Union)

National holiday: Independence Day (or National Day), 9 September (1991)

Constitution: *history:* several previous; latest adopted 6 November 1994

amendments: proposed by the president of the republic or by at least one third of the total membership of both houses of the Supreme Assembly; adoption of any amendment requires a referendum, which includes approval of the president or approval by at least two-thirds majority of the Assembly of Representatives; passage in a referendum requires participation of an absolute majority of eligible voters and an absolute majority of votes; constitutional articles, including Tajikistan's form of government, its territory, and its democratic nature, cannot be amended; amended several times, last in 2016

Legal system: civil law system

International law organization participation: has not submitted an ICJ jurisdiction declaration; accepts ICCt jurisdiction

Citizenship: *citizenship by birth:* no

citizenship by descent only: at least one parent must be a citizen of Tajikistan

dual citizenship recognized: no

residency requirement for naturalization: 5 years or 3 years of continuous residence prior to application

Suffrage: 18 years of age; universal

Executive branch: *chief of state:* President Emomali RAHMON (since 6 November 1994; head of state and Supreme Assembly chairman since 19 November 1992)

head of government: Prime Minister Qohir RASULZODA (since 23 November 2013)

cabinet: Council of Ministers appointed by the president, approved by the Supreme Assembly

elections/appointments: president directly elected by simple majority popular vote for a 7-year term for a maximum of two terms; however, as the "Leader of the Nation" President RAHMON can run an unlimited number of times; election last held on 11 October 2020 (next to be held in 2027); prime minister appointed by the president

election results: Emomali RAHMON reelected president; percent of vote - Emomali RAHMON (PDPT) 92.1%, Rustam LATIFZODA (APT) 3.1%, Rustam RAHAMATZODA (PERT) 2.2%, Abduhalim GHAFFOROV (SPT) 1.5%, Miroj ABDULLOEV (CPT) 1.2%

Legislative branch: *description:* bicameral Supreme Assembly or Majlisi Oli consists of: National Assembly or Majlisi Milli (34 seats; 25 members indirectly elected by local representative assemblies or majlisi, 8 appointed by the president, and 1 reserved for each living former president; members serve 5-year terms) Assembly of Representatives or Majlisi Namoyandagon (63 seats; 41 members directly elected in single-seat constituencies by 2-round absolute majority vote and 22 directly elected in a single nationwide constituency by proportional representation vote; members serve 5-year terms)

elections: National Assembly - last held on 1 March 2020 (next to be held in 2025) Assembly of Representatives - last held on 1 March 2020 (next to be held in 2025)

election results: National Assembly - percent of vote by party - NA; seats by party - NA; composition - men 28, women 6, percent of women 17.6%

Assembly of Representatives—percent of vote by party—PDPT 50.4%, PERT 16.6%, APT 16.5%, SPT 5.2%, DPT 5.1%, CPT 3.1%, other 3.1%; seats by party—PDPT 47, APT 7, PERT 5, CPT 2, SPT 1, DPT 1; composition—men 50, women 13, percent of women 20.6%; note—total Supreme Assembly percent of women 19.6%

Judicial branch: *highest courts:* Supreme Court (consists of the chairman, deputy chairmen, and 34 judges organized into civil, family, criminal, administrative offense, and military chambers); Constitutional Court (consists of the court chairman, deputy chairman, and 5 judges); High Economic Court (consists of 16 judicial positions)

judge selection and term of office: Supreme Court, Constitutional Court, and High Economic Court judges nominated by the president and approved by the National Assembly; judges of all 3 courts appointed for 10-year renewable terms with no term limits, but the last appointment must occur before the age of 65

subordinate courts: regional and district courts; Dushanbe City Court; viloyat (province level) courts; Court of Gorno-Badakhshan Autonomous Region

Political parties and leaders: Agrarian Party of Tajikistan or APT [Rustam LATIFZODA]

Communist Party of Tajikistan or CPT [Miroj ABDULLOEV]

Democratic Party of Tajikistan or DPT [Saidjafar USMONZODA]

Party of Economic Reform of Tajikistan or PERT [Rustam OUDRATOV]

People's Democratic Party of Tajikistan or PDPT [Emomali RAHMON]

Social Democratic Party of Tajikistan or SDPT [Rahmatullo ZOIROV]

Socialist Party of Tajikistan or SPT [Abduhalim GHAFFOROV]

International organization participation: ADB, CICA, CIS, CSTO, EAEC, EAPC, EBRD, ECO, EITI (candidate country), FAO, G-77, GCTU, IAEA, IBRD, ICAO, ICC (NGOs), ICCt, ICRM, IDA, IDB, IFAD, IFC, IFRCS, ILO, IMF, Interpol, IOC, IOM, IPU, ISO (correspondent), ITSO, ITU, MIGA, NAM (observer), OIC, OPCW, OSCE, PFP, SCO, UN, UNCTAD, UNESCO, UNIDO, UNWTO, UPU, WCO, WFTU (NGOs), WHO, WIPO, WMO, WTO

Diplomatic representation in the US: *chief of mission:* Ambassador Farhod SALIM (since 21 May 2014)

chancery: 1005 New Hampshire Avenue NW, Washington, DC 20037

telephone: [1] (202) 223-6090

FAX: [1] (202) 223-6091

Diplomatic representation from the US: *chief of mission:* Ambassador John Mark POMMERSHEIM (since 15 March 2019)

telephone: [992] (37) 229-20-00, 992-37-229-2300 (consular direct line); EMER: 992-98-580-1032

embassy: 109-A Ismoili Somoni Avenue, Dushanbe 734019

mailing address: 7090 Dushanbe Place, Dulles, VA 20189

FAX: [992] (37) 229-20-50

Flag description: three horizontal stripes of red (top), a wider stripe of white, and green; a gold crown surmounted by seven gold, five-pointed stars is located in the center of the white stripe; red represents the sun, victory, and the unity of the nation, white stands for purity, cotton, and mountain snows, while green is the color of Islam and the bounty of nature; the crown symbolizes the Tajik people; the seven stars signify the Tajik magic number "seven" - a symbol of perfection and the embodiment of happiness

National symbol(s): *crown surmounted by an arc of seven, five-pointed stars; snow leopard; national colors:* red, white, green

National anthem: *name:* "Surudi milli" (National Anthem)

lyrics/music: Gulnazar KELDI/Sulaimon YUDAKOV

note: adopted 1991; after the fall of the Soviet Union, Tajikistan kept the music of the anthem from its time as a Soviet republic but adopted new lyrics

ECONOMY

Economy—overview: Tajikistan is a poor, mountainous country with an economy dominated by minerals extraction, metals processing, agriculture, and reliance on remittances from citizens working abroad. Mineral resources include silver, gold, uranium, antimony, tungsten, and coal. Industry consists mainly of small obsolete factories in food processing and light industry, substantial hydropower facilities, and a large aluminum plant - currently operating well below its capacity. The 1992-97 civil war severely damaged an already weak economic infrastructure and caused a sharp decline in industrial and agricultural production. Today, Tajikistan is the poorest among the former Soviet republics. Because less than 7% of the land area is arable and cotton is the predominant crop, Tajikistan imports approximately 70% of its food.

Since the end of the civil war, the country has pursued half-hearted reforms and privatizations in the economic sphere, but its poor business climate remains a hindrance to attracting foreign investment. Some experts estimate the value of narcotics transiting Tajikistan is equivalent to 30%-50% of GDP.

Because of a lack of employment opportunities in Tajikistan, more than one million Tajik citizens work abroad - roughly 90% in Russia - supporting families back home through remittances that in 2017 were equivalent to nearly 35% of GDP. Tajikistan's large remittances from migrant workers in Russia exposes it to monetary shocks. Tajikistan often delays devaluation of its currency for fear of inflationary pressures on food and other consumables. Recent slowdowns in the Russian and Chinese economies, low commodity prices, and currency fluctuations have hampered economic growth. The dollar value of remittances from Russia to Tajikistan dropped by almost 65% in 2015, and the government spent almost $500 million in 2016 to bail out the country's still troubled banking sector.

Tajikistan's growing public debt – currently about 50% of GDP – could result in financial difficulties. Remittances from Russia increased in 2017, however, bolstering the economy somewhat. China owns about 50% of Tajikistan's outstanding debt. Tajikistan has borrowed heavily to finance investment in the country's vast hydropower potential. In 2016, Tajikistan contracted with the Italian firm Salini Impregilo to build the Roghun dam over a 13-year period for $3.9 billion. A 2017 Eurobond has largely funded Roghun's first phase, after which sales from Roghun's output are expected to fund the rest of its construction. The government has not ruled out issuing another Eurobond to generate auxiliary funding for its second phase.

GDP (purchasing power parity): $28.43 billion (2017 est.)
$26.55 billion (2016 est.)
$24.83 billion (2015 est.)
note: data are in 2017 dollars
country comparison to the world: 135

GDP (official exchange rate): $7.144 billion (2017 est.)

GDP—real growth rate: 7.1% (2017 est.)
6.9% (2016 est.)
6% (2015 est.)
country comparison to the world: 17

GDP—per capita (PPP): $3,200 (2017 est.)
$3,000 (2016 est.)
$2,900 (2015 est.)
note: data are in 2017 dollars
country comparison to the world: 192

Gross national saving: 24.4% of GDP (2017 est.)
15.4% of GDP (2016 est.)
11.8% of GDP (2015 est.)
country comparison to the world: 66

GDP—composition, by end use: household consumption: 98.4% (2017 est.)
government consumption: 13.3% (2017 est.)
investment in fixed capital: 11.7% (2017 est.)
investment in inventories: 2.5% (2017 est.)
exports of goods and services: 10.7% (2017 est.)
imports of goods and services: -36.6% (2017 est.)

GDP—composition, by sector of origin: agriculture: 28.6% (2017 est.)
industry: 25.5% (2017 est.)
services: 45.9% (2017 est.)

Agriculture—products: cotton, grain, fruits, grapes, vegetables; cattle, sheep, goats

Industries: aluminum, cement, coal, gold, silver, antimony, textile, vegetable oil

Industrial production growth rate: 1% (2017 est.)
country comparison to the world: 159

Labor force: 2.295 million (2016 est.)
country comparison to the world: 116

Labor force—by occupation: agriculture: 43%
industry: 10.6%
services: 46.4% (2016 est.)

Unemployment rate: 2.4% (2016 est.)
2.5% (2015 est.)
note: official rate; actual unemployment is much higher
country comparison to the world: 27

Population below poverty line: 31.5% (2016 est.)

Household income or consumption by percentage share: lowest 10%: NA (2009 est.)
highest 10%: NA (2009 est.)

Budget: revenues: 2.269 billion (2017 est.)
expenditures: 2.374 billion (2017 est.)

Taxes and other revenues: 31.8% (of GDP) (2017 est.)
country comparison to the world: 71

Budget surplus (+) or deficit (-): -1.5% (of GDP) (2017 est.)
country comparison to the world: 90

Public debt: 50.4% of GDP (2017 est.)
42% of GDP (2016 est.)
country comparison to the world: 101

Fiscal year: calendar year

Inflation rate (consumer prices): 7.3% (2017 est.)
5.9% (2016 est.)
country comparison to the world: 193

Current account balance: -$35 million (2017 est.)
-$362 million (2016 est.)
country comparison to the world: 78

Exports: $873.1 million (2017 est.)
$691.1 million (2016 est.)
country comparison to the world: 165

Exports—partners: Turkey 27.5%, China 17.7%, Russia 13.4%, Switzerland 12.5%, Algeria 8.2%, Iran 7.1% (2017)

Exports—commodities: aluminum, electricity, cotton, fruits, vegetable oil, textiles

Imports: $2.39 billion (2017 est.)
$2.554 billion (2016 est.)
country comparison to the world: 159

Imports—commodities: petroleum products, aluminum oxide, machinery and equipment, foodstuffs

Imports—partners: Russia 38%, Kazakhstan 19%, China 8.7%, Iran 4.4% (2017)

Reserves of foreign exchange and gold: $1.292 billion (31 December 2017 est.)
$652.8 million (31 December 2016 est.)
country comparison to the world: 127

Debt—external: $5.75 billion (31 December 2017 est.)
$5.495 billion (31 December 2016 est.)
country comparison to the world: 130

Exchange rates: Tajikistani somoni (TJS) per US dollar -
8.764 (2017 est.)
7.8358 (2016 est.)
7.8358 (2015 est.)
6.1631 (2014 est.)
4.9348 (2013 est.)

ENERGY

Electricity access: electrification—total population: 100% (2020)

Electricity—production: 17.03 billion kWh (2016 est.)
country comparison to the world: 85

Electricity—consumption: 12.96 billion kWh (2016 est.)
country comparison to the world: 85

Electricity—exports: 1.4 billion kWh NA (2015 est.)
country comparison to the world: 52

Electricity—imports: 103 million kWh (2016 est.)
country comparison to the world: 98

Electricity—installed generating capacity: 5.508 million kW (2016 est.)
country comparison to the world: 78

Electricity—from fossil fuels: 6% of total installed capacity (2016 est.)
country comparison to the world: 201

Electricity—from nuclear fuels: 0% of total installed capacity (2017 est.)
country comparison to the world: 191

Electricity—from hydroelectric plants: 94% of total installed capacity (2017 est.)
country comparison to the world: 6

Electricity—from other renewable sources: 0% of total installed capacity (2017 est.)
country comparison to the world: 208

Crude oil—production: 180 bbl/day (2018 est.)
country comparison to the world: 97

Crude oil—exports: 0 bbl/day (2015 est.)

country comparison to the world: 204

Crude oil—imports: 0 bbl/day (2015 est.)
country comparison to the world: 201

Crude oil—proved reserves: 12 million bbl (1 January 2018 est.)
country comparison to the world: 89

Refined petroleum products—production: 172 bbl/day (2015 est.)
country comparison to the world: 108

Refined petroleum products—consumption: 24,000 bbl/day (2016 est.)
country comparison to the world: 130

Refined petroleum products—exports: 0 bbl/day (2015 est.)
country comparison to the world: 206

Refined petroleum products—imports: 22,460 bbl/day (2015 est.)
country comparison to the world: 114

Natural gas—production: 19.82 million cu m (2017 est.)
country comparison to the world: 89

Natural gas—consumption: 19.82 million cu m (2017 est.)
country comparison to the world: 114

Natural gas—exports: 0 cu m (2017 est.)
country comparison to the world: 197

Natural gas—imports: 0 cu m (2017 est.)
country comparison to the world: 197

Natural gas—proved reserves: 5.663 billion cu m (1 January 2018 est.)
country comparison to the world: 91

Carbon dioxide emissions from consumption of energy: 6.329 million Mt (2017 est.)
country comparison to the world: 126

COMMUNICATIONS

Telephones—fixed lines: *total subscriptions:* 471,090
subscriptions per 100 inhabitants: 5.39 (2019 est.)
country comparison to the world: 94

Telephones—mobile cellular: *total subscriptions:* 9,747,803
subscriptions per 100 inhabitants: 111.53 (2019 est.)
country comparison to the world: 88

Telecommunication systems: *general assessment:* foreign investment in the telephone system has resulted in major improvements; an increase in mobile broadband penetration, but still in the early stages and remains low compared to those in the region; the country has endeavored to launch 4G/LTE services with mixed results; 7 major cities have 4G coverage; 5 major operators in the market (2020)
domestic: fixed line availability has not changed significantly since 1998, while mobile cellular subscribership, aided by competition among multiple operators, has expanded rapidly; coverage now

extends to all major cities and towns; fixed-line 5 per 100 and mobile-cellular 112 per 100 (2019)
international: country code - 992; linked by cable and microwave radio relay to other CIS republics and by leased connections to the Moscow international gateway switch; Dushanbe linked by Intelsat to international gateway switch in Ankara (Turkey); 3 satellite earth stations - 2 Intelsat and 1 Orbita
note: the COVID-19 outbreak is negatively impacting telecommunications production and supply chains globally; consumer spending on telecom devices and services has also slowed due to the pandemic's effect on economies worldwide; overall progress towards improvements in all facets of the telecom industry - mobile, fixed-line, broadband, submarine cable and satellite - has moderated

Broadcast media: state-run TV broadcasters transmit nationally on 9 TV and 10 radio stations, and regionally on 4 stations; 31 independent TV and 20 radio stations broadcast locally and regionally; many households are able to receive Russian and other foreign stations via cable and satellite (2016)

Internet country code: .tj

Internet users: *total:* 1,889,632
percent of population: 21.96% (July 2018 est.)
country comparison to the world: 123

Broadband—fixed subscriptions: *total:* 6,000
subscriptions per 100 inhabitants: less than 1 (2017 est.)
country comparison to the world: 177

TRANSPORTATION

National air transport system: *number of registered air carriers:* 2 (2020)
inventory of registered aircraft operated by air carriers: 6
annual passenger traffic on registered air carriers: 492,320 (2018)
annual freight traffic on registered air carriers: 2.34 million mt-km (2018)

Civil aircraft registration country code prefix: EY (2016)

Airports: 24 (2013)
country comparison to the world: 131

Airports—with paved runways: *total:* 17 (2013)
over 3,047 m: 2 (2013)
2,438 to 3,047 m: 4 (2013)
1,524 to 2,437 m: 5 (2013)
914 to 1,523 m: 3 (2013)
under 914 m: 3 (2013)

Airports—with unpaved runways: *total:* 7 (2013)
1,524 to 2,437 m: 1 (2013)
914 to 1,523 m: 1 (2013)
under 914 m: 5 (2013)

Pipelines: 549 km gas, 38 km oil (2013)

Railways: *total:* 680 km (2014)
broad gauge: 680 km 1.520-m gauge (2014)
country comparison to the world: 102

Roadways: *total:* 30,000 km (2018)

country comparison to the world: 99

Waterways: 200 km (along Vakhsh River) (2011)
country comparison to the world: 98

MILITARY AND SECURITY

Military and security forces: *Armed Forces of the Republic of Tajikistan:* Land Forces, Mobile Forces, Air and Air Defense Forces; National Guard; Ministry of Internal Affairs: Internal Troops (reserves for Armed Forces in wartime); State Committee on National Security: Border Guard Forces (2019)

Military expenditures: 1.2% of GDP (2015)
1.1% of GDP (2014)
1% of GDP (2012)
1.1% of GDP (2011)
note: no public data available for 2013, 2016-2018
country comparison to the world: 107

Military and security service personnel strengths: the Armed Forces of the Republic of Tajikistan have approximately 9,500 active troops (8,000 Land and Mobile Forces; 1,500 Air and Air Defense Forces) (2019 est.)

Military equipment inventories and acquisitions: the Tajikistan Armed Forces' inventory is comprised of older Russian and Soviet-era equipment; it has received limited quantities of weapons systems since 2010, most of which was secondhand material from Russia, followed by Belarus and China (2019 est.)

Military deployments: contributes troops to CSTO's Rapid Reaction Force (2019)

Military service age and obligation: 18-27 years of age for compulsory or voluntary military service; 12-18 month conscript service obligation (2019)

TERRORISM

Terrorist group(s): Islamic State of Iraq and ash-Sham (ISIS) (2019)
note: details about the history, aims, leadership, organization, areas of operation, tactics, targets, weapons, size, and sources of support of the group(s) appear(s) in Appendix-T

TRANSNATIONAL ISSUES

Disputes—international: in 2006, China and Tajikistan pledged to commence demarcation of the revised boundary agreed to in the delimitation of 2002; talks continue with Uzbekistan to delimit border and remove minefields; disputes in Isfara Valley delay delimitation with Kyrgyzstan

Refugees and internally displaced persons: *stateless persons:* 7,151 (2019)

Illicit drugs: Tajikistan sits on one of the world's highest volume illicit drug trafficking routes, between Afghan opiate production to the south and the illicit drug markets of Russia and Eastern Europe to the north; limited illicit cultivation of opium poppy for domestic consumption; significant consumer of opiates

TANZANIA

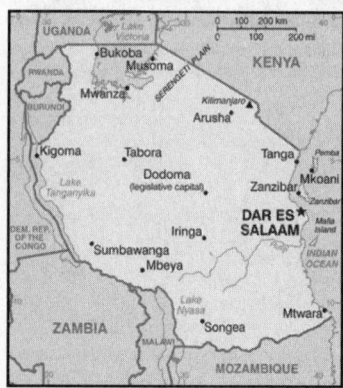

INTRODUCTION

Background: Mainland Tanzania fell under German rule during the late 19th century as part of German East Africa. After World War I, Britain governed the mainland as Tanganyika; the Zanzibar Archipelago remained a separate colonial jurisdiction. Shortly after achieving independence from Britain in the early 1960s, Tanganyika and Zanzibar merged to form the United Republic of Tanzania in 1964. In 1995, the country held its first democratic elections since the 1970s. Zanzibar maintains semi-autonomy and participates in national elections; popular political opposition on the isles led to four contentious elections since 1995, in which the ruling party claimed victory despite international observers' claims of voting irregularities.

GEOGRAPHY

Location: Eastern Africa, bordering the Indian Ocean, between Kenya and Mozambique

Geographic coordinates: 6 00 S, 35 00 E

Map references: Africa

Area: *total:* 947,300 sq km
land: 885,800 sq km
water: 61,500 sq km
note: includes the islands of Mafia, Pemba, and Zanzibar
country comparison to the world: 32

Area—comparative: more than six times the size of Georgia; slightly larger than twice the size of California

Land boundaries: *total:* 4,161 km
border countries (8): Burundi 589 km, Democratic Republic of the Congo 479 km, Kenya 775 km, Malawi 512 km, Mozambique 840 km, Rwanda 222 km, Uganda 391 km, Zambia 353 km

Coastline: 1,424 km

Maritime claims: *territorial sea:* 12 nm
exclusive economic zone: 200 nm

Climate: varies from tropical along coast to temperate in highlands

Terrain: plains along coast; central plateau; highlands in north, south

Elevation: *mean elevation:* 1,018 m
lowest point: Indian Ocean 0 m
highest point: Kilimanjaro (highest point in Africa) 5,895 m

Natural resources: hydropower, tin, phosphates, iron ore, coal, diamonds, gemstones, gold, natural gas, nickel

Land use: *agricultural land:* 43.7% (2011 est.)
arable land: 14.3% (2011 est.) / *permanent crops:* 2.3% (2011 est.) / *permanent pasture:* 27.1% (2011 est.)
forest: 37.3% (2011 est.)
other: 19% (2011 est.)

Irrigated land: 1,840 sq km (2012)

Population distribution: the largest and most populous East African country; population distribution is extremely uneven, but greater population clusters occur in the northern half of country and along the east coast as shown in this population distribution map

Natural hazards: flooding on the central plateau during the rainy season; drought
volcanism: limited volcanic activity; Ol Doinyo Lengai (2,962 m) has emitted lava in recent years; other historically active volcanoes include Kieyo and Meru

Environment—current issues: water polution; improper management of liquid waste; indoor air pollution caused by the burning of fuel wood or charcoal for cooking and heating is a large environmental health issue; soil degradation; deforestation; desertification; destruction of coral reefs threatens marine habitats; wildlife threatened by illegal hunting and trade, especially for ivory; loss of biodiversity; solid waste disposal

Environment—international agreements: *party to:* Biodiversity, Climate Change, Climate Change-Kyoto Protocol, Desertification, Endangered Species, Hazardous Wastes, Law of the Sea, Ozone Layer Protection, Wetlands *signed, but not ratified:* none of the selected agreements

Geography—note: Kilimanjaro is the highest point in Africa and one of only three mountain ranges on the continent that has glaciers (the others are Mount Kenya [in Kenya] and the Ruwenzori Mountains [on the Uganda-Democratic Republic of the Congo border]); Tanzania is bordered by three of the largest lakes on the continent: Lake Victoria (the world's second-largest freshwater lake) in the north, Lake Tanganyika (the world's second deepest) in the west, and Lake Nyasa (Lake Malawi) in the southwest

PEOPLE AND SOCIETY

Population: 58,552,845 (July 2020 est.)
note: estimates for this country explicitly take into account the effects of excess mortality due to AIDS; this can result in lower life expectancy, higher infant mortality, higher death rates, lower population growth rates, and changes in the distribution of population by age and sex than would otherwise be expected
country comparison to the world: 24

Nationality: *noun:* Tanzanian(s)
adjective: Tanzanian

Ethnic groups: mainland - African 99% (of which 95% are Bantu consisting of more than 130 tribes), other 1% (consisting of Asian, European, and Arab); Zanzibar - Arab, African, mixed Arab and African

Languages: Kiswahili or Swahili (official), Kiunguja (name for Swahili in Zanzibar), English (official, primary language of commerce, administration, and higher education), Arabic (widely spoken in Zanzibar), many local languages
note: Kiswahili (Swahili) is the mother tongue of the Bantu people living in Zanzibar and nearby coastal Tanzania; although Kiswahili is Bantu in structure and origin, its vocabulary draws on a variety of sources including Arabic and English; it has become the lingua franca of central and eastern Africa; the first language of most people is one of the local languages

Religions: Christian 61.4%, Muslim 35.2%, folk religion 1.8%, other 0.2%, unaffiliated 1.4% (2010 est.)
note: Zanzibar is almost entirely Muslim

Demographic profile: Tanzania has the largest population in East Africa and the lowest population density; almost a third of the population is urban. Tanzania's youthful population – about two-thirds of the population is under 25 – is growing rapidly because of the high total fertility rate of 4.8 children per woman. Progress in reducing the birth rate has stalled, sustaining the country's nearly 3% annual growth. The maternal mortality rate has improved since 2000, yet it remains very high because of early and frequent pregnancies, inadequate maternal health services, and a lack of skilled birth attendants – problems that are worse among poor and rural women. Tanzania has made strides in reducing under-5 and infant mortality rates, but a recent drop in immunization threatens to undermine gains in child health. Malaria is a leading killer of children under 5, while HIV is the main source of adult mortality

For Tanzania, most migration is internal, rural to urban movement, while some temporary labor migration from towns to plantations takes place seasonally for harvests. Tanzania was Africa's largest refugee-hosting country for decades, hosting hundreds of thousands of refugees from the Great Lakes region, primarily Burundi, over the last fifty years. However, the assisted repatriation and naturalization of tens of thousands of Burundian refugees between 2002 and 2014 dramatically reduced the refugee population. Tanzania is increasingly a transit country for illegal migrants from the Horn of Africa and the Great Lakes region who are heading to southern Africa for security reasons and/or economic opportunities. Some of these migrants choose to settle in Tanzania.

Age structure: *0-14 years:* 42.7% (male 12,632,772/female 12,369,115)
15-24 years: 20.39% (male 5,988,208/female 5,948,134)
25-54 years: 30.31% (male 8,903,629/female 8,844,180)
55-64 years: 3.52% (male 954,251/female 1,107,717)
65 years and over: 3.08% (male 747,934/female 1,056,905) (2020 est.)

population pyramid: [INSERT IMAGE: TANZANIA-population pyramid]

Dependency ratios: *total dependency ratio:* 85.9
youth dependency ratio: 81
elderly dependency ratio: 4.9
potential support ratio: 20.4 (2020 est.)

Median age: *total:* 18.2 years
male: 17.9 years
female: 18.4 years (2020 est.)
country comparison to the world: 213

Population growth rate: 2.71% (2020 est.)
country comparison to the world: 15

Birth rate: 34.6 births/1,000 population (2020 est.)
country comparison to the world: 22

Death rate: 7.1 deaths/1,000 population (2020 est.)
country comparison to the world: 123

Net migration rate: -0.4 migrant(s)/1,000 population (2020 est.)
country comparison to the world: 126

Population distribution: the largest and most populous East African country; population distribution is extremely uneven, but greater population clusters occur in the northern half of country and along the east coast as shown in this population distribution map

Urbanization: *urban population:* 35.2% of total population (2020)
rate of urbanization: 5.22% annual rate of change (2015-20 est.)

total population growth rate v. urban population growth rate, 2000-2030: *Major urban areas - population:* 262,000 Dodoma (legislative capital) (2018), 6.702 million DAR ES SALAAM (administrative capital), 1.120 million Mwanza (2020)

Sex ratio: *at birth:* 1.03 male(s)/female
0-14 years: 1.02 male(s)/female
15-24 years: 1.01 male(s)/female
25-54 years: 1.01 male(s)/female
55-64 years: 0.86 male(s)/female
65 years and over: 0.71 male(s)/female
total population: 1 male(s)/female (2020 est.)

Mother's mean age at first birth: 19.8 years (2015/16 est.)
note: median age at first birth among women 25-29

Maternal mortality rate: 524 deaths/100,000 live births (2017 est.)
country comparison to the world: 19

Infant mortality rate: *total:* 36.4 deaths/1,000 live births
male: 38.5 deaths/1,000 live births
female: 34.4 deaths/1,000 live births (2020 est.)
country comparison to the world: 43

Life expectancy at birth: *total population:* 63.9 years
male: 62.3 years
female: 65.5 years (2020 est.)
country comparison to the world: 203

Total fertility rate: 4.59 children born/woman (2020 est.)
country comparison to the world: 22

Contraceptive prevalence rate: 38.4% (2015/16)

Drinking water source:

improved: *urban:* 92.3% of population
rural: 56.2% of population
total: 68.2% of population

unimproved: *urban:* 7.7% of population
rural: 43.8% of population
total: 31.8% of population (2017 est.)

Current Health Expenditure: 3.6% (2017)

Physicians density: 0.01 physicians/1,000 population (2016)

Hospital bed density: 0.7 beds/1,000 population (2010)

Sanitation facility access:

improved: *urban:* 82.1% of population
rural: 29.5% of population
total: 46.9% of population

unimproved: *urban:* 17.9% of population
rural: 70.5% of population
total: 53.1% of population (2017 est.)

HIV/AIDS—adult prevalence rate: 5.1% (2019 est.)
country comparison to the world: 12

HIV/AIDS—people living with HIV/AIDS: 1.7 million (2019 est.)
country comparison to the world: 5

HIV/AIDS—deaths: 27,000 (2019 est.)
country comparison to the world: 6

Major infectious diseases: *degree of risk:* very high (2020)
food or waterborne diseases: bacterial diarrhea, hepatitis A, and typhoid fever

vectorborne diseases: malaria, dengue fever, and Rift Valley fever

water contact diseases: schistosomiasis

animal contact diseases: rabies

Obesity—adult prevalence rate: 8.4% (2016)
country comparison to the world: 151

Children under the age of 5 years underweight: 14.6% (2018)
country comparison to the world: 41

Education expenditures: 3.4% of GDP (2014)
country comparison to the world: 127

Literacy: *definition:* age 15 and over can read and write Kiswahili (Swahili), English, or Arabic
total population: 77.9%
male: 83.2%
female: 73.1% (2015)

School life expectancy (primary to tertiary education): *total:* 9 years
male: 9 years
female: 9 years (2019)

Unemployment, youth ages 15-24: *total:* 3.9%
male: 3.1%
female: 4.6% (2014 est.)
country comparison to the world: 167

GOVERNMENT

Country name: *conventional long form:* United Republic of Tanzania
conventional short form: Tanzania
local long form: Jamhuri ya Muungano wa Tanzania
local short form: Tanzania
former: German East Africa, Trust Territory of Tanganyika, United Republic of Tanganyika and Zanzibar
etymology: the country's name is a combination of the first letters of Tanganyika and Zanzibar, the two states that merged to form Tanzania in 1964

Government type: presidential republic

Capital: *name:* Dar es Salaam (administrative capital), Dodoma (legislative capital); note - Dodoma was designated the national capital in 1996 and serves as the meeting place for the National Assembly; Dar es Salaam remains the de facto capital, the country's largest city and commercial center, and the site of the executive branch offices and diplomatic representation; the government contends that it will complete the transfer of the executive branch to Dodoma by 2020

geographic coordinates: 6 48 S, 39 17 E
time difference: UTC+3 (8 hours ahead of Washington, DC, during Standard Time)
etymology: Dar es Salaam was the name given by Majid bin Said, the first sultan of Zanzibar, to the new city he founded on the Indian Ocean coast; the Arabic name is commonly translated as "abode/home of peace"; Dodoma, in the native Gogo language, means "it has sunk"; supposedly, one day during the rainy season, an elephant

drowned in the area; the villagers in that place were so struck by what had occurred, that ever since the locale has been referred to as the place where "it (the elephant) sunk"

Administrative divisions: 31 regions; Arusha, Dar es Salaam, Dodoma, Geita, Iringa, Kagera, Kaskazini Pemba (Pemba North), Kaskazini Unguja (Zanzibar North), Katavi, Kigoma, Kilimanjaro, Kusini Pemba (Pemba South), Kusini Unguja (Zanzibar Central/South), Lindi, Manyara, Mara, Mbeya, Mjini Magharibi (Zanzibar Urban/West), Morogoro, Mtwara, Mwanza, Njombe, Pwani (Coast), Rukwa, Ruvuma, Shinyanga, Simiyu, Singida, Songwe, Tabora, Tanga

Independence: 26 April 1964 (Tanganyika united with Zanzibar to form the United Republic of Tanganyika and Zanzibar); 29 October 1964 (renamed United Republic of Tanzania); notable earlier dates: 9 December 1961 (Tanganyika became independent from UK-administered UN trusteeship); 10 December 1963 (Zanzibar became independent from UK)

National holiday: Union Day (Tanganyika and Zanzibar), 26 April (1964)

Constitution: *history*: several previous; latest adopted 25 April 1977; note - progress enacting a new constitution drafted in 2014 by the Constituent Assembly stalled

amendments: proposed by the National Assembly; passage of amendments to constitutional articles including those on sovereignty of the United Republic, the authorities and powers of the government, the president, the Assembly, and the High Court requires two-thirds majority vote of the mainland Assembly membership and of the Zanzibar House of Representatives membership; House of Representatives approval of other amendments is not required; amended several times, last in 2017

Legal system: English common law; judicial review of legislative acts limited to matters of interpretation

International law organization participation: has not submitted an ICJ jurisdiction declaration; accepts ICCt jurisdiction

Citizenship: *citizenship by birth*: no
citizenship by descent only: at least one parent must be a citizen of Tanzania; if a child is born abroad, the father must be a citizen of Tanzania
dual citizenship recognized: no
residency requirement for naturalization: 5 years

Suffrage: 18 years of age; universal

Executive branch: *chief of state*: President John MAGUFULI, Dr. (since 5 November 2015; sworn in for second 5-year term on 5 November 2020); Vice President Samia Suluhu HASSAN (since 5 November 2015); note - the president is both chief of state and head of government

head of government: President John MAGUFULI, Dr. (since 5 November 2015; sworn in for second

5-year term on 5 November 2020); Vice President Samia Suluhu HASSAN (since 5 November 2015); note - Prime Minister Kassim Majaliwa MAJALIWA (since 20 November 2015; reappointed 13 November 2020) has authority over the day-to-day functions of the government, is the leader of government business in the National Assembly, and is head of the Cabinet
cabinet: Cabinet appointed by the president from among members of the National Assembly
elections/appointments: president and vice president directly elected on the same ballot by simple majority popular vote for a 5-year term (eligible for a second term); election last held on 25 October 2015 (next to be held 28 October 2020); prime minister appointed by the president
election results: John MAGUFULI elected president; percent of vote - John MAGUFULI (CCM) 58.5%, Edward LOWASSA (CHADEMA) 40%, other 1.5%
note: Zanzibar elects a president as head of government for internal matters; election held on 25 October 2015 was annulled by the Zanzibar Electoral Commission and rerun on 20 March 2016; President Ali Mohamed SHEIN reelected; percent of vote - Ali Mohamed SHEIN (CCM) 91.4%, Hamad Rashid MOHAMED (ADC) 3%, other 5.6%; the main opposition party in Zanzibar CUF boycotted the 20 March 2016 election rerun

Legislative branch: *description*: unicameral National Assembly or Parliament (Bunge) (393 seats; 264 members directly elected in single-seat constituencies by simple majority vote, 113 women indirectly elected by proportional representation vote, 5 indirectly elected by simple majority vote by the Zanzibar House of Representatives, 10 appointed by the president, and 1 seat reserved for the attorney general; members serve a 5-year term); note - in addition to enacting laws that apply to the entire United Republic of Tanzania, the National Assembly enacts laws that apply only to the mainland; Zanzibar has its own House of Representatives or Baraza La Wawakilishi (82 seats; 50 members directly elected in single-seat constituencies by simple majority vote, 20 women directly elected by proportional representation vote, 10 appointed by the Zanzibar president, 1 seat for the House speaker, and 1 ex-officio seat for the attorney general; elected members serve a 5-year term)
elections: Tanzania National Assembly and Zanzibar House of Representatives - elections last held on 25 October 2015 (next National Assembly election to be held in October 2020; next Zanzibar election either October 2020 or March 2021); note the Zanzibar Electoral Commission annulled the 2015 election; repoll held on 20 March 2016
election results: National Assembly - percent of vote by party - CCM 55%, Chadema 31.8%, CUF 8.6%, other 4.6%; seats by party - CCM 253, Chadema 70, CUF 42, other 2; composition as of September 2018 - men 245, women 145, percent of women 37.2%

Zanzibar House of Representatives - percent of vote by party - NA; seats by party - NA; composition - NA

Judicial branch: *highest courts*: Court of Appeal of the United Republic of Tanzania (consists of the chief justice and 14 justices); High Court of the United Republic for Mainland Tanzania (consists of the principal judge and 30 judges organized into commercial, land, and labor courts); High Court of Zanzibar (consists of the chief justice and 10 justices)
judge selection and term of office: Court of Appeal and High Court justices appointed by the national president after consultation with the Judicial Service Commission for Tanzania, a judicial body of high level judges and 2 members appointed by the national president; Court of Appeal and High Court judges serve until mandatory retirement at age 60, but terms can be extended; High Court of Zanzibar judges appointed by the national president after consultation with the Judicial Commission of Zanzibar; judges can serve until mandatory retirement at age 65
subordinate courts: Resident Magistrates Courts; Kadhi courts (for Islamic family matters); district and primary courts

Political parties and leaders: Alliance for Change and Transparency (Wazalendo) or ACT [Zitto KABWE]
Alliance for Democratic Change or ADC [Miraji ABDALLAH]
Civic United Front (Chama Cha Wananchi) or CUF [Ibrahim LIPUMBA]
National Convention for Construction and Reform-Mageuzi or NCCR-M [James Francis MBATIA]
National League for Democracy
Party of Democracy and Development (Chama Cha Demokrasia na Maendeleo) or Chadema [Freeman MBOWE]
Revolutionary Party (Chama Cha Mapinduzi) or CCM [John MAGUFULI]
Tanzania Labor Party or TLP [Augustine MREMA]
United Democratic Party or UDP [John Momose CHEYO]
note: in March 2014, four opposition parties (CUF, CHADEMA, NCCR-Mageuzi, and NLD) united to form Coalition for the People's Constitution (Umoja wa Katiba ya Wananchi) or UKAWA; during local elections held in October, 2014, UKAWA entered one candidate representing the three parties united in the coalition

International organization participation: ACP, AfDB, AU, C, CD, EAC, EADB, EITI, FAO, G-77, IAEA, IBRD, ICAO, ICC (NGOs), ICCt, ICRM, IDA, IFAD, IFC, IFRCS, ILO, IMF, IMO, IMSO, Interpol, IOC, IOM, IPU, ISO, ITSO, ITU, ITUC (NGOs), MIGA, MONUSCO, NAM, OPCW, SADC, UN, UNAMID, UNCTAD, UNESCO, UNHCR, UNIDO, UNIFIL, UNISFA, UNMISS, UNWTO, UPU, WCO, WFTU (NGOs), WHO, WIPO, WMO, WTO

Diplomatic representation in the US: *chief of mission:* Ambassador Wilson Mutagaywa MASILINGI (since 17 September 2015)
chancery: 1232 22nd Street NW, Washington, DC 20037
telephone: [1] (202) 939-6125
FAX: [1] (202) 797-7408

Diplomatic representation from the US: *chief of mission:* Ambassador Donald J. WRIGHT (since 2 April 2020)
telephone: (255) 22-229-4000, dial '1' for an emergency operator
embassy: 686 Old Bagamoyo Road, Msasani, Dar es Salaam
mailing address: P.O. Box 9123, Dar es Salaam
FAX: [255] (22) 229-4970 or 4971

Flag description: divided diagonally by a yellow-edged black band from the lower hoist-side corner; the upper triangle (hoist side) is green and the lower triangle is blue; the banner combines colors found on the flags of Tanganyika and Zanzibar; green represents the natural vegetation of the country, gold its rich mineral deposits, black the native Swahili people, and blue the country's many lakes and rivers, as well as the Indian Ocean

National symbol(s): *Uhuru (Freedom) torch, giraffe; national colors:* green, yellow, blue, black

National anthem: *name:* "Mungu ibariki Afrika" (God Bless Africa)

lyrics/music: collective/Enoch Mankayi SONTONGA
note: adopted 1961; the anthem, which is also a popular song in Africa, shares the same melody with that of Zambia but has different lyrics; the melody is also incorporated into South Africa's anthem
0:00 / 0:51

ECONOMY

Economy—overview: Tanzania has achieved high growth rates based on its vast natural resource wealth and tourism with GDP growth in 200917 averaging 6%-7% per year. Dar es Salaam used fiscal stimulus measures and easier monetary policies to lessen the impact of the global recession and in general, benefited from low oil prices. Tanzania has largely completed its transition to a market economy, though the government retains a presence in sectors such as telecommunications, banking, energy, and mining.

The economy depends on agriculture, which accounts for slightly less than one-quarter of GDP and employs about 65% of the work force, although gold production in recent years has increased to about 35% of exports. All land in Tanzania is owned by the government, which can lease land for up to 99 years. Proposed reforms to allow for land ownership, particularly foreign land ownership, remain unpopular.

The financial sector in Tanzania has expanded in recent years and foreign-owned banks account for about 48% of the banking industry's total assets. Competition among foreign commercial banks has resulted in significant improvements in the efficiency and quality of financial services, though interest rates are still relatively high, reflecting high fraud risk. Banking reforms have helped increase private-sector growth and investment.

The World Bank, the IMF, and bilateral donors have provided funds to rehabilitate Tanzania's aging infrastructure, including rail and port, which provide important trade links for inland countries. In 2013, Tanzania completed the world's largest Millennium Challenge Compact (MCC) grant, worth $698 million, but in late 2015, the MCC Board of Directors deferred a decision to renew Tanzania's eligibility because of irregularities in voting in Zanzibar and concerns over the government's use of a controversial cybercrime bill.

The new government elected in 2015 has developed an ambitious development agenda focused on creating a better business environment through improved infrastructure, access to financing, and education progress, but implementing budgets remains challenging for the government. Recent policy moves by President MAGUFULI are aimed at protecting domestic industry and have caused concern among foreign investors.

GDP (purchasing power parity): $162.5 billion (2017 est.)
$153.3 billion (2016 est.)
$143.3 billion (2015 est.)
note: data are in 2017 dollars
country comparison to the world: 75

GDP (official exchange rate): $51.76 billion (2017 est.)

GDP—real growth rate: 6.98% (2019 est.)
6.95% (2018 est.)
6.78% (2017 est.)
country comparison to the world: 18

GDP—per capita (PPP): $3,200 (2017 est.)
$3,100 (2016 est.)
$3,000 (2015 est.)
note: data are in 2017 dollars
country comparison to the world: 193

Gross national saving: 25% of GDP (2017 est.)
23.1% of GDP (2016 est.)
24.9% of GDP (2015 est.)
country comparison to the world: 60

GDP—composition, by end use: *household consumption:* 62.4% (2017 est.)
government consumption: 12.5% (2017 est.)
investment in fixed capital: 36.1% (2017 est.)
investment in inventories: -8.7% (2017 est.)
exports of goods and services: 18.1% (2017 est.)
imports of goods and services: -20.5% (2017 est.)

GDP—composition, by sector of origin: *agriculture:* 23.4% (2017 est.)
industry: 28.6% (2017 est.)
services: 47.6% (2017 est.)

Agriculture—products: coffee, sisal, tea, cotton, pyrethrum (insecticide made from chrysanthemums), cashew nuts, tobacco, cloves, corn, wheat, cassava (manioc, tapioca), bananas, fruits, vegetables; cattle, sheep, goats

Industries: agricultural processing (sugar, beer, cigarettes, sisal twine); mining (diamonds, gold, and iron), salt, soda ash; cement, oil refining, shoes, apparel, wood products, fertilizer

Industrial production growth rate: 12% (2017 est.)
country comparison to the world: 8

Labor force: 24.89 million (2017 est.)
country comparison to the world: 21

Labor force—by occupation: agriculture: 66.9%
industry: 6.4%
services: 26.6% (2014 est.)

Unemployment rate: 10.3% (2014 est.)
country comparison to the world: 150

Population below poverty line: 22.8% (2015 est.)

Household income or consumption by percentage share: *lowest 10%:* 2.8%
highest 10%: 29.6% (2007)

Budget: *revenues:* 7.873 billion (2017 est.)
expenditures: 8.818 billion (2017 est.)

Taxes and other revenues: 15.2% (of GDP) (2017 est.)
country comparison to the world: 192

Budget surplus (+) or deficit (-): -1.8% (of GDP) (2017 est.)
country comparison to the world: 100

Public debt: 37% of GDP (2017 est.)
38% of GDP (2016 est.)
country comparison to the world: 141

Fiscal year: 1 July - 30 June

Inflation rate (consumer prices): 5.3% (2017 est.)
5.2% (2016 est.)
country comparison to the world: 175

Current account balance: -$1.313 billion (2019 est.)
-$1.898 billion (2018 est.)
country comparison to the world: 155

Exports: $4.971 billion (2017 est.)
$5.697 billion (2016 est.)
country comparison to the world: 107

Exports—partners: India 21.8%, South Africa 17.9%, Kenya 8.8%, Switzerland 6.7%, Belgium 5.9%, Democratic Republic of the Congo 5.8%,

China 4.8% (2017)

Exports—commodities: gold, coffee, cashew nuts, manufactures, cotton

Imports: $7.869 billion (2017 est.)
$8.464 billion (2016 est.)
country comparison to the world: 111

Imports—commodities: consumer goods, machinery and transportation equipment, industrial raw materials, crude oil

Imports—partners: India 16.5%, China 15.8%, UAE 9.2%, Saudi Arabia 7.9%, South Africa 5.1%, Japan 4.9%, Switzerland 4.4% (2017)

Reserves of foreign exchange and gold: $5.301 billion (31 December 2017 est.)

$4.067 billion (31 December 2016 est.)
note: excludes gold
country comparison to the world: 94

Debt—external: $17.66 billion (31 December 2017 est.)
$15.21 billion (31 December 2016 est.)
country comparison to the world: 97

Exchange rates: Tanzanian shillings (TZS) per US dollar—
2,243.8 (2017 est.)
2,177.1 (2016 est.)
2,177.1 (2015 est.)
1,989.7 (2014 est.)
1,654 (2013 est.)

ENERGY

Electricity access: population without electricity: 35 million (2019)

electrification—total population: 40% (2019)

electrification—urban areas: 71% (2019)

electrification—rural areas: 23% (2019)

Electricity—production: 6.699 billion kWh (2016 est.)
country comparison to the world: 114

Electricity—consumption: 5.682 billion kWh (2016 est.)
country comparison to the world: 118

Electricity—exports: 0 kWh (2016 est.)
country comparison to the world: 206

Electricity—imports: 102 million kWh (2016 est.)
country comparison to the world: 99

Electricity—installed generating capacity: 1.457 million kW (2016 est.)
country comparison to the world: 124

Electricity—from fossil fuels: 55% of total installed capacity (2016 est.)
country comparison to the world: 141

Electricity—from nuclear fuels: 0% of total installed capacity (2017 est.)
country comparison to the world: 192

Electricity—from hydroelectric plants: 40% of total installed capacity (2017 est.)
country comparison to the world: 53

Electricity—from other renewable sources: 6% of total installed capacity (2017 est.)
country comparison to the world: 106

Crude oil—production: 0 bbl/day (2018 est.)
country comparison to the world: 206

Crude oil—exports: 0 bbl/day (2015 est.)
country comparison to the world: 205

Crude oil—imports: 0 bbl/day (2015 est.)
country comparison to the world: 202

Crude oil—proved reserves: 0 bbl (1 January 2018 est.)
country comparison to the world: 202

Refined petroleum products—production: 0 bbl/day (2015 est.)
country comparison to the world: 206

Refined petroleum products—consumption: 72,000 bbl/day (2016 est.)
country comparison to the world: 92

Refined petroleum products— exports: 0 bbl/day (2015 est.)
country comparison to the world: 207

Refined petroleum products—imports: 67,830 bbl/day (2015 est.)
country comparison to the world: 69

Natural gas—production: 3.115 billion cu m (2017 est.)
country comparison to the world: 56

Natural gas—consumption: 3.115 billion cu m (2017 est.)
country comparison to the world: 74

Natural gas—exports: 0 cu m (2017 est.)
country comparison to the world: 198

Natural gas—imports: 0 cu m (2017 est.)
country comparison to the world: 198

Natural gas—proved reserves: 6.513 billion cu m (1 January 2018 est.)
country comparison to the world: 85

Carbon dioxide emissions from consumption of energy: 14.57 million Mt (2017 est.)
country comparison to the world: 93

COMMUNICATIONS

Telephones—fixed lines: *total subscriptions:* 74,081

subscriptions per 100 inhabitants: less than 1 (2019 est.)
country comparison to the world: 148

Telephones—mobile cellular: total subscriptions: 46,847,405

subscriptions per 100 inhabitants: 82.21 (2019 est.)
country comparison to the world: 31

Telecommunication systems: *general assessment:* telecommunications services are marginal and operating below capacity; 1 fixed-line operator and 8 operational mobile networks; unfortunate high tariffs on telecoms; mobile use is growing at 85% penetration; 3G/LTE services; govt. allocates TZ $17.5 billion to improve rural telecom infrastructure and work on national fiber backbone network connecting population around country (2020)

domestic: fixed-line telephone network inadequate with less than 1 connection per 100 persons; mobile-cellular service, aided by multiple providers, is increasing rapidly and exceeds 82 telephones per 100 persons; trunk service provided by open-wire, microwave radio relay, tropospheric scatter, and fiber-optic cable; some links being made digital (2019)

international: country code - 255; landing points for the EASSy, SEACOM/Tata TGN-Eurasia, and SEAS fiber-optic submarine cable system linking East Africa with the Middle East; satellite earth stations - 2 Intelsat (1 Indian Ocean, 1 Atlantic Ocean) (2019)
note: the COVID-19 outbreak is negatively impacting telecommunications production and supply chains globally; consumer spending on telecom devices and services has also slowed due to the pandemic's effect on economies worldwide; overall progress towards improvements in all facets of the telecom industry - mobile, fixed-line, broadband, submarine cable and satellite - has moderated

Broadcast media: a state-owned TV station and multiple privately owned TV stations; state-owned national radio station supplemented by more than 40 privately owned radio stations; transmissions of several international broadcasters are available (2019)

Internet country code: .tz

Internet users: *total:* 13,862,836

percent of population: 25% (July 2018 est.)
country comparison to the world: 46

Broadband—fixed subscriptions: *total:* 861,234

subscriptions per 100 inhabitants: 2 (2018 est.)
country comparison to the world: 71

TRANSPORTATION

National air transport system: *number of registered air carriers:* 11 (2020)

inventory of registered aircraft operated by air carriers: 91

annual passenger traffic on registered air carriers: 1,481,557 (2018)

annual freight traffic on registered air carriers: 390,000 mt-km (2018)

Civil aircraft registration country code prefix: 5H (2016)

Airports: 166 (2013)
country comparison to the world: 33

Airports—with paved runways: *total:* 10 (2019)
over 3,047 m: 2
2,438 to 3,047 m: 2
1,524 to 2,437 m: 4
914 to 1,523 m: 2

Airports—with unpaved runways: *total:* 156 (2013)
over 3,047 m: 1 (2013)
1,524 to 2,437 m: 24 (2013)
914 to 1,523 m: 98 (2013)
under 914 m: 33 (2013)

Pipelines: 311 km gas, 891 km oil, 8 km refined products (2013)

Railways: *total:* 4,567 km (2014)
narrow gauge: 1,860 km 1.067-m gauge (2014)
2707 km 1.000-m gauge
country comparison to the world: 43

Roadways: *total:* 87,581 km (2015)
paved: 10,025 km (2015)
unpaved: 77,556 km (2015)
country comparison to the world: 56

Waterways: (Lake Tanganyika, Lake Victoria, and Lake Nyasa (Lake Malawi) are the principal avenues of commerce with neighboring countries; the rivers are not navigable) (2011)

Merchant marine: *total:* 337

by type: bulk carrier 4, container ship 8, general cargo 173, oil tanker 44, other 108 (2019)
country comparison to the world: 51

Ports and terminals: *major seaport(s):* Dar es Salaam, Zanzibar

MILITARY AND SECURITY

Military and security forces: Tanzania People's Defense Forces (TPDF or Jeshi la Wananchi la Tanzania, JWTZ): Land Forces Command, Naval Forces Command, Air Force Command, National Building Army (Jeshi la Kujenga Taifa, JKT), People's Militia (Reserves) (2019)
note: the National Building Army is a paramilitary organization under the Defense Forces that provides six months of military and vocational training to individuals as part of their two years of public service; after completion of training, some graduates join the regular Defense Forces while the remainder become part of the People's Militia

Military expenditures: 1.3% of GDP (2019)
1.3% of GDP (2018)
1.2% of GDP (2017)
1.1% of GDP (2016)
1.1% of GDP (2015)
country comparison to the world: 95

Military and security service personnel strengths: the Tanzania People's Defense Forces (TPDF) have an estimated 26,000 active personnel (22,000 Land Forces; 1,000 Naval Forces; 3,000 Air Force) (2019)

Military equipment inventories and acquisitions: the TPDF inventory includes mostly Soviet-era and older Chinese equipment; since 2010, China is the leading supplier of arms to the TPDF (2019 est.)

Military deployments: 450 Central African Republic (MINUSCA); 750 Democratic Republic of the Congo (MONUSCO); 120 Lebanon (UNIFIL); 330 Sudan (UNAMID) (2020)

Military service age and obligation: 18-25 years of age for voluntary military service; 6-year commitment (2019)

Maritime threats: The International Maritime Bureau reports that shipping in territorial and offshore waters in the Indian Ocean remain at risk for piracy and armed robbery against ships, especially as Somali-based pirates extend their activities south; numerous commercial vessels have been attacked and hijacked both at anchor and while underway; crews have been robbed and stores or cargoes stolen.

Military—note: the TPDF has deployed additional troops to its border with Mozambique to prevent a spillover of the growing violence in the northern Mozambican province of Cabo Delgado (2020)

TERRORISM

Terrorist group(s): Islamic State of Iraq and ash-Sham - Central Africa (2020)
note: details about the history, aims, leadership, organization, areas of operation, tactics, targets, weapons, size, and sources of support of the group(s) appear(s) in Appendix-T

TRANSPORTATION

Disputes—international: dispute with Tanzania over the boundary in Lake Nyasa (Lake Malawi) and the meandering Songwe River; Malawi contends that the entire lake up to the Tanzanian shoreline is its territory, while Tanzania claims the border is in the center of the lake; the conflict was reignited in 2012 when Malawi awarded a license to a British company for oil exploration in the lake

Refugees and internally displaced persons: *refugees (country of origin):* 154,163 (Burundi), 77,898 (Democratic Republic of the Congo) (2020)

Trafficking in persons: *current situation:* Tanzania is a source, transit, and destination country for men, women, and children subjected to forced labor and sex trafficking; the exploitation of young girls in domestic servitude continues to be Tanzania's largest human trafficking problem; Tanzanian boys are subject to forced labor mainly on farms but also in mines and quarries, in the informal commercial sector, in factories, in the sex trade, and possibly on small fishing boats; Tanzanian children and adults are subjected to domestic servitude, other forms of forced labor, and sex trafficking in other African countries, the Middle East, Europe, and the US; internal trafficking is more prevalent than transnational trafficking and is usually facilitated by friends, family members, or intermediaries with false offers of education or legitimate jobs; trafficking victims from Burundi, Kenya, South Asia, and Yemen are forced to work in Tanzania's agricultural, mining, and domestic service sectors or may be sex trafficked

tier rating: Tier 2 Watch List – Tanzania does not fully comply with the minimum standards for the elimination of trafficking; however, it is making significant efforts to do so; in 2014, Tanzania was granted a waiver from an otherwise required downgrade to Tier 3 because its government has a written plan that, if implemented, would constitute making significant efforts to bring itself into compliance with the minimum standards for the elimination of trafficking; the government adopted a three-year national action plan and implementing regulations for the 2008 anti-trafficking law; authorities somewhat increased their number of trafficking investigations and prosecutions and convicted one offender, but the penalty was a fine in lieu of prison, which was inadequate given the severity of the crime; the government did not operate any shelters for victims and relied on NGOs to provide protective services (2015)

Illicit drugs: targeted by traffickers moving hashish, Afghan heroin, and South American cocaine transported down the East African coastline, through airports, or overland through Central Africa; Zanzibar likely used by traffickers for drug smuggling; traffickers in the past have recruited Tanzanian couriers to move drugs through Iran into East Asia

THAILAND

INTRODUCTION

Background: A unified Thai kingdom was established in the mid-14th century. Known as Siam until 1939, Thailand is the only Southeast Asian country never to have been colonized by a European power. A bloodless revolution in 1932 led to the establishment of a constitutional monarchy. After the Japanese invaded Thailand in 1941, the government split into a pro-Japan faction and a pro-Ally faction backed by the King. Following the war, Thailand became a US treaty ally in 1954 after sending troops to Korea and later fighting alongside the US in Vietnam. Thailand since 2005 has experienced several rounds of political turmoil including a military coup in 2006 that ousted then Prime Minister THAKSIN Chinnawat, followed by large- scale street protests by competing political factions in 2008, 2009, and 2010. THAKSIN's youngest sister, YINGLAK Chinnawat, in 2011 led the Puea Thai Party to an electoral win and assumed control of the government.

In early May 2014, after months of large-scale anti-government protests in Bangkok beginning

in November 2013, YINGLAK was removed from office by the Constitutional Court and in late May 2014 the Royal Thai Army, led by Royal Thai Army Gen. PRAYUT Chan-ocha, staged a coup against the caretaker government. PRAYUT was appointed prime minister in August 2014. PRAYUT also serves as the head of the National Council for Peace and Order (NCPO), a military- affiliated body that oversees the interim government. This body created several interim institutions to promote reform and draft a new constitution, which was passed in a national referendum in August 2016. In late 2017, PRAYUT announced elections would be held by November 2018; he has subsequently suggested they might occur in February 2019. As of midDecember 2018, a previously held ban on campaigning and political activity has been lifted and per parliamentary laws, an election must be held within 150 days. King PHUMIPHON Adunyadet passed away in October 2016 after 70 years on the throne; his only son, WACHIRALONGKON Bodinthrathepphayawarangkun, ascended the throne in December 2016. He signed the new constitution in April 2017. Thailand has also experienced violence associated with the ethno-nationalist insurgency in its southern Malay-Muslim majority provinces. Since January 2004, thousands have been killed and wounded in the insurgency.

GEOGRAPHY

Location: Southeastern Asia, bordering the Andaman Sea and the Gulf of Thailand, southeast of Burma

Geographic coordinates: 15 00 N, 100 00 E

Map references: Southeast Asia

Area: *total:* 513,120 sq km
land: 510,890 sq km
water: 2,230 sq km
country comparison to the world: 52

Area—comparative: about three times the size of Florida; slightly more than twice the size of Wyoming

Area comparison map: [INSERT IMAGE: THAILAND-Area comparison map]

Land boundaries: *total:* 5,673 km
border countries (4): Burma 2416 km, Cambodia 817 km, Laos 1845 km, Malaysia 595 km

Coastline: 3,219 km

Maritime claims: *territorial sea:* 12 nm
exclusive economic zone: 200 nm
continental shelf: 200-m depth or to the depth of exploitation

Climate: tropical; rainy, warm, cloudy southwest monsoon (mid-May to September); dry, cool

northeast monsoon (November to midMarch); southern isthmus always hot and humid

Terrain: central plain; Khorat Plateau in the east; mountains elsewhere

Elevation: *mean elevation:* 287 m
lowest point: Gulf of Thailand 0 m
highest point: Doi Inthanon 2,565 m

Natural resources: tin, rubber, natural gas, tungsten, tantalum, timber, lead, fish, gypsum, lignite, fluorite, arable land

Land use: *agricultural land:* 41.2% (2011 est.)
arable land: 30.8% (2011 est.) / permanent crops: 8.8% (2011 est.) / permanent pasture: 1.6% (2011 est.)
forest: 37.2% (2011 est.)
other: 21.6% (2011 est.)

Irrigated land: 64,150 sq km (2012)

Population distribution: highest population density is found in and around Bangkok; significant population clusters found througout large parts of the country, particularly north and northeast of Bangkok and in the extreme southern region of the country

Natural hazards: land subsidence in Bangkok area resulting from the depletion of the water table; droughts

Environment—current issues: air pollution from vehicle emissions; water pollution from organic and factory wastes; water scarcity; deforestation; soil erosion; wildlife populations threatened by illegal hunting; hazardous waste disposal

Environment—international agreements: *party to:* Biodiversity, Climate Change, Climate Change-Kyoto Protocol, Desertification, Endangered Species, Hazardous Wastes, Marine Life Conservation, Ozone Layer Protection, Tropical Timber 83, Tropical Timber 94, Wetlands
signed, but not ratified: Law of the Sea

Geography—note: controls only land route from Asia to Malaysia and Singapore; ideas for the construction of a canal across the Kra Isthmus that would create a bypass to the Strait of Malacca and shorten shipping times around Asia continue to be discussed

PEOPLE AND SOCIETY

Population: 68,977,400 (July 2020 est.)
country comparison to the world: 20

Nationality: *noun:* Thai (singular and plural)
adjective: Thai

Ethnic groups: Thai 97.5%, Burmese 1.3%, other 1.1%, unspecified <.1% (2015 est.)
note: data represent population by nationality

Languages: Thai (official) only 90.7%, Thai and other languages 6.4%, only other languages 2.9% (includes Malay, Burmese) (2010 est.)

note: data represent population by language(s) spoken at home; English is a secondary language of the elite

Religions: Buddhist 94.6%, Muslim 4.3%, Christian 1%, other (2015 est.)

Age structure: 0-14 years: 16.45% (male 5,812,803/female 5,533,772)
15-24 years: 13.02% (male 4,581,622/female 4,400,997)
25-54 years: 45.69% (male 15,643,583/female 15,875,353)
55-64 years: 13.01% (male 4,200,077/female 4,774,801)
65 years and over: 11.82% (male 3,553,273/female 4,601,119) (2020 est.)

population pyramid: [INSERT IMAGE: THAILAND-population pyramid]

Dependency ratios: total dependency ratio: 41.9
youth dependency ratio: 23.5
elderly dependency ratio: 18.4
potential support ratio: 5.4 (2020 est.)

Median age: total: 39 years
male: 37.8 years
female: 40.1 years (2020 est.)
country comparison to the world: 59

Population growth rate: 0.25% (2020 est.)
country comparison to the world: 177

Birth rate: 10.7 births/1,000 population (2020 est.)
country comparison to the world: 185

Death rate: 8.3 deaths/1,000 population (2020 est.)
country comparison to the world: 82
Net migration rate: 0 migrant(s)/1,000 population (2020 est.)
country comparison to the world: 96

Population distribution: highest population density is found in and around Bangkok; significant population clusters found througout large parts of the country, particularly north and northeast of Bangkok and in the extreme southern region of the country

Urbanization: urban population: 51.4% of total population (2020)
rate of urbanization: 1.73% annual rate of change (2015-20 est.)

total population growth rate v. urban population growth rate, 2000-2030: Major urban areas - population: 10.539 million BANGKOK (capital), 1.399 Chon Buri, 1.307 million Samut Prakan, 1.167 million Chiang Mai, 967,000 Songkla, 963,000 Nothaburi (2020)

Sex ratio: at birth: 1.05 male(s)/female
0-14 years: 1.05 male(s)/female
15-24 years: 1.04 male(s)/female
25-54 years: 0.99 male(s)/female
55-64 years: 0.88 male(s)/female
65 years and over: 0.77 male(s)/female
total population: 0.96 male(s)/female (2020 est.)

Mother's mean age at first birth: 23.3 years (2009 est.)

Maternal mortality rate: 37 deaths/100,000 live births (2017 est.)
country comparison to the world: 103

Infant mortality rate: total: 8.6 deaths/1,000 live births
male: 9.5 deaths/1,000 live births
female: 7.6 deaths/1,000 live births (2020 est.)
country comparison to the world: 146

Life expectancy at birth: total population: 75.6 years
male: 72.4 years
female: 78.9 years (2020 est.)
country comparison to the world: 113

Total fertility rate: 1.54 children born/woman (2020 est.)
country comparison to the world: 196

Contraceptive prevalence rate: 78.4% (2015/16)

Drinking water source:

improved: urban: 100% of population
rural: 100% of population
total: 100% of population

unimproved: urban: 0% of population
rural: 0% of population
total: 0% of population (2017 est.)

Current Health Expenditure: 3.7% (2017)

Physicians density: 0.81 physicians/1,000 population (2017)

Hospital bed density: 2.1 beds/1,000 population (2010)

Sanitation facility access:

improved: urban: 100% of population
rural: 100% of population
total: 99.9% of population

unimproved: urban: 0% of population
rural: 0% of population
total: 0.1% of population (2017 est.)

HIV/AIDS—adult prevalence rate: 0.8% (2019 est.)
country comparison to the world: 54

HIV/AIDS—people living with HIV/AIDS: 480,000 (2018 est.)
country comparison to the world: 17

HIV/AIDS—deaths: 14,000 (2019 est.)
country comparison to the world: 15

Major infectious diseases: degree of risk: very high (2020)
food or waterborne diseases: bacterial diarrhea
vectorborne diseases: dengue fever, Japanese encephalitis, and malaria

Obesity—adult prevalence rate: 10% (2016)
country comparison to the world: 140

Children under the age of 5 years underweight: 6.7% (2016)
country comparison to the world: 74

Education expenditures: 4.1% of GDP (2013)
country comparison to the world: 100

Literacy: definition: age 15 and over can read and write
total population: 92.9%
male: 94.7%

female: 91.2% (2015)
School life expectancy (primary to tertiary education): total: 15 years
male: 15 years
female: 16 years (2016)
Unemployment, youth ages 15-24: total: 3.7%
male: 3%
female: 4.7% (2016 est.)
country comparison to the world: 170

GOVERNMENT

Country name: conventional long form: Kingdom of Thailand
conventional short form: Thailand
local long form: Ratcha Anachak Thai
local short form: Prathet Thai
former: Siam
etymology: Land of the Tai [People]"; the meaning of "tai" is uncertain, but may originally have meant "human beings," "people," or "free people

Government type: constitutional monarchy

Capital: name: Bangkok

geographic coordinates: 13 45 N, 100 31 E
time difference: UTC+7 (12 hours ahead of Washington, DC, during Standard Time)
etymology: Bangkok was likely originally a colloquial name, but one that was widely adopted by foreign visitors; the name may derive from "bang ko," where "bang" is the Thai word for "village on a stream" and "ko" means "island," both referencing the area's landscape, which was carved by rivers and canals; alternatively, the name may come from "bang makok," where "makok" is the name of the Java plum, a plant bearing olive-like fruit; this possibility is supported by the former name of Wat Arun, a historic temple in the area, that used to be called Wat Makok;

Krung Thep, the city's Thai name, means "City of the Deity" and is a shortening of the full ceremonial name: Krungthepmahanakhon Amonrattanakosin Mahintharayutthaya Mahadilokphop Noppharatratchathaniburirom Udomratchaniwetmahasathan Amonphimanawatansathit Sakkathattiyawitsanukamprasit; translated the meaning is: City of angels, great city of immortals, magnificent city of the nine gems, seat of the king, city of royal palaces, home of gods incarnate, erected by Vishvakarman at Indra's behest; it holds the world's record as the longest place name (169 letters)

Administrative divisions: 76 provinces (changwat, singular and plural) and 1 municipality* (maha nakhon); Amnat Charoen, Ang Thong, Bueng Kan, Buri Ram, Chachoengsao, Chai Nat, Chaiyaphum, Chanthaburi, Chiang Mai, Chiang Rai, Chon Buri, Chumphon, Kalasin, Kamphaeng Phet, Kanchanaburi, Khon Kaen, Krabi, Krung Thep* (Bangkok), Lampang, Lamphun, Loei, Lop Buri, Mae Hong Son, Maha Sarakham, Mukdahan, Nakhon Nayok, Nakhon Pathom, Nakhon Phanom, Nakhon Ratchasima, Nakhon Sawan, Nakhon Si Thammarat, Nan, Narathiwat, Nong

Bua Lamphu, Nong Khai, Nonthaburi, Pathum Thani, Pattani, Phangnga, Phatthalung, Phayao, Phetchabun, Phetchaburi, Phichit, Phitsanulok, Phra Nakhon Si Ayutthaya, Phrae, Phuket, Prachin Buri, Prachuap Khiri Khan, Ranong, Ratchaburi, Rayong, Roi Et, Sa Kaeo, Sakon Nakhon, Samut Prakan, Samut Sakhon, Samut Songkhram, Saraburi, Satun, Sing Buri, Si Sa Ket, Songkhla, Sukhothai, Suphan Buri, Surat Thani, Surin, Tak, Trang, Trat, Ubon Ratchathani, Udon Thani, Uthai Thani, Uttaradit, Yala, Yasothon

Independence: 1238 (traditional founding date; never colonized)

National holiday: Birthday of King WACHIRALONGKON, 28 July (1952)

Constitution: *history:* many previous; latest drafted and presented 29 March 2016, approved by referendum 7 August 2016, signed into law by the king 6 April 2017

amendments: proposed as a joint resolution by the Council of Ministers and the National Council for Peace and Order (the junta that has ruled Thailand since the 2014 coup) and submitted as a draft to the National Legislative Assembly; passage requires majority vote of the existing Assembly members and presentation to the monarch for assent and countersignature of the prime minister

Legal system: civil law system with common law influences

International law organization participation: has not submitted an ICJ jurisdiction declaration; non-party state to the ICCt

Citizenship: *citizenship by birth:* no
citizenship by descent only: at least one parent must be a citizen of Thailand
dual citizenship recognized: no
residency requirement for naturalization: 5 years

Suffrage: 18 years of age; universal and compulsory

Executive branch: *chief of state:* King WACHIRALONGKON, also spelled Vajiralongkorn, (since 1 December 2016); note - King PHUMIPHON Adunyadet, also spelled BHUMIBOL Adulyadej (since 9 June 1946) died 13 October 2016

head of government: Prime Minister PRAYUT Chan-ocha (since 25 August 2014); Deputy Prime Ministers PRAWIT Wongsuwan (since 31 August 2014), WITSANU Kruea-ngam (since 31 August 2014), SUPHATTHANAPHONG Phanmichao (since August 2020), CHURIN Laksanawisit (since November 2019), ANUTHIN Chanwirakun (since November 2019), DON Pramudwinai (since August 2020)

cabinet: Council of Ministers nominated by the prime minister, appointed by the king; a Privy Council advises the king
elections/appointments: the monarchy is hereditary; the House of Representatives and Senate approves a person for Prime Minister who must then be appointed by the King (as stated in the transitory provision of the 2017 constitution); the

office of prime minister can be held for up to a total of 8 years
note: PRAYUT Chan-ocha was appointed interim prime minister in August 2014, three months after he staged the coup that removed the previously elected government of Prime Minister YINGLAK Chinnawat; on 5 June 2019 PRAYUT (independent) was approved as prime minister by the parliament - 498 votes to 244 for THANATHON Chuengrungrueangkit (FFP)

Legislative branch: *description:* bicameral National Assembly or Rathhasapha consists of:
Senate or Wuthissapha (250 seats; members appointed by the Royal Thai Army to serve 5-year terms)
House of Representatives or Saphaphuthan Ratsadon (500 seats; 375 members directly elected in single-seat constituencies by simple majority vote and 150 members elected in a single nationwide constituency by party-list proportional representation vote; members serve 4-year terms)
elections: Senate - last held on 14 May 2019 (next to be held in 2024)
House of Representatives - last held on 24 March 2019 (next to be held in 2023)
election results: Senate - percent of vote by party - NA; seats by party - NA; composition - men 224, women 26, percent of women 10.4%
House of Representatives - percent of vote by party - PPRP 23.7%, PTP 22.2%, FFP 17.8%, DP 11.1%, PJT 10.5%, TLP 2.3%, CTP 2.2%, NEP 1.4%, PCC 1.4%, ACT 1.2%, PCP 1.2%, other 5.1%; seats by party - PTP 136, PPRP 116, FFP 81, DP 53, PJT 51, CTP 10, TLP 10, PCC 7, PCP 5, NEP 6, ACT 5, other 20; composition - men 421, women 79, percent of women 15.8%; note - total National Assembly percent of women 14%

Judicial branch: *highest courts:* Supreme Court of Justice (consists of the court president, 6 vice presidents, 60-70 judges, and organized into 10 divisions); Constitutional Court (consists of the court president and 8 judges); Supreme Administrative Court (number of judges determined by Judicial Commission of the Administrative Courts)
judge selection and term of office: Supreme Court judges selected by the Judicial Commission of the Courts of Justice and approved by the monarch; judge term determined by the monarch; Constitutional Court justices - 3 judges drawn from the Supreme Court, 2 judges drawn from the Administrative Court, and 4 judge candidates selected by the Selective Committee for Judges of the Constitutional Court, and confirmed by the Senate; judges appointed by the monarch serve single 9-year terms; Supreme Administrative Court judges selected by the Judicial Commission of the Administrative Courts and appointed by the monarch; judges serve for life
subordinate courts: courts of first instance and appeals courts within both the judicial and administrative systems; military courts

Political parties and leaders: Action Coalition of Thailand Party or ACT [TAWEESAK Na

Takuathung (acting); CHATUMONGKHON Sonakun resigned June 2020]
Anakhot Mai Party (Future Forward Party) or FFP [THANATHON Chuengrungrueangkit] (dissolved, February 2020)
Chat Phatthana Party (National Development Party) [THEWAN Liptaphanlop]
Chat Thai Phatthana Party (Thai Nation Development Party) or CTP [KANCHANA Sinlapa-acha]
New Economics Party or NEP [MINGKHWAN Sangsuwan]
Phalang Pracharat Party or PPP [UTTAMA Sawanayon]
Phumchai Thai Party (Thai Pride Party) or PJT [ANUTHIN Chanwirakun]
Prachachat Party of PCC [WAN Muhamad NOOR Matha]
Prachathipat Party (Democrat Party) or DP [CHURIN Laksanawisit]
Puea Chat Party (For Nation Party) or PCP [SONGKHRAM Kitletpairot]
Puea Thai Party (For Thais Party) or PTP [WIROT Paoin]
Puea Tham Party (For Dharma Party) [NALINI Thawisin]
Seri Ruam Thai Party (Thai Liberal Party) or TLP [SERIPHISUT Temiyawet]
Thai Forest Conservation Party or TFCP [DAMRONG Phidet]
Thai Local Power Party or TLP [collective leadership]
Thai Raksa Chat Party (Thai National Preservation Party) [PRICHAPHON Phongpanit]
note: as of 5 April 2018, 98 new parties applied to be registered with the Election Commission in accordance with the provisions of the new organic law on political parties

International organization participation: ADB, APEC, ARF, ASEAN, BIMSTEC, BIS, CD, CICA, CP, EAS, FAO, G-77, IAEA, IBRD, ICAO, ICC (national committees), ICRM, IDA, IFAD, IFC, IFRCS, IHO, ILO, IMF, IMO, IMSO, Interpol, IOC, IOM, IPU, ISO, ITSO, ITU, ITUC (NGOs), MIGA, NAM, OAS (observer), OIC (observer), OIF (observer), OPCW, OSCE (partner), PCA, PIF (partner), UN, UNAMID, UNCTAD, UNESCO, UNHCR, UNIDO, UNMOGIP, UNOCI, UNWTO, UPU, WCO, WFTU (NGOs), WHO, WIPO, WMO, WTO

Diplomatic representation in the US: *chief of mission:* Ambassador THANI Thongphakdi (since 6 January 2020)
chancery: 1024 Wisconsin Avenue NW, Suite 401, Washington, DC 20007
telephone: [1] (202) 944-3600
FAX: [1] (202) 944-3611

consulate(s) general: Chicago, Los Angeles, New York

Diplomatic representation from the US: *chief of mission:* Ambassador (vacant); Charge d'Affaires Michael HEATH (since August 2019)
telephone: [66] 2-205-4000
embassy: 95 Wireless Road, Bangkok 10330

mailing address: APO AP 96546
FAX: [66] 2-205-4306

consulate(s) general: Chiang Mai

Flag description: five horizontal bands of red (top), white, blue (double width), white, and red; the red color symbolizes the nation and the blood of life, white represents religion and the purity of Buddhism, and blue stands for the monarchy
note: similar to the flag of Costa Rica but with the blue and red colors reversed

National symbol(s): garuda (mythical half-man, half-bird figure), elephant; national colors: red, white, blue

National anthem: *name:* "Phleng Chat Thai" (National Anthem of Thailand)

lyrics/music: Luang SARANUPRAPAN/Phra JENDURIYANG
note: music adopted 1932, lyrics adopted 1939; by law, people are required to stand for the national anthem at 0800 and 1800 every day; the anthem is played in schools, offices, theaters, and on television and radio during this time; "Phleng Sanlasoen Phra Barami" (A Salute to the Monarch) serves as the royal anthem and is played in the presence of the royal family and during certain state ceremonies
0:00 / 0:43

ECONOMY

Economy—overview: With a relatively well-developed infrastructure, a free-enterprise economy, and generally pro-investment policies, Thailand is highly dependent on international trade, with exports accounting for about two thirds of GDP. Thailand's exports include electronics, agricultural commodities, automobiles and parts, and processed foods. The industry and service sectors produce about 90% of GDP. The agricultural sector, comprised mostly of small-scale farms, contributes only 10% of GDP but employs about one third of the labor force. Thailand has attracted an estimated 3.0-4.5 million migrant workers, mostly from neighboring countries.

Over the last few decades, Thailand has reduced poverty substantially. In 2013, the Thai Government implemented a nationwide 300 baht (roughly $10) per day minimum wage policy and deployed new tax reforms designed to lower rates on middle-income earners.

Thailand's economy is recovering from slow growth during the years since the 2014 coup. Thailand's economic fundamentals are sound, with low inflation, low unemployment, and reasonable public and external debt levels. Tourism and government spending - mostly on infrastructure and short-term stimulus measures – have helped to boost the economy, and The Bank of Thailand has been supportive, with several interest rate reductions.

Over the longer-term, household debt levels, political uncertainty, and an aging population pose risks to growth.

GDP (purchasing power parity): $1.236 trillion (2017 est.)
$1.19 trillion (2016 est.)
$1.152 trillion (2015 est.)
note: data are in 2017 dollars
country comparison to the world: 20

GDP (official exchange rate): $455.4 billion (2017 est.)

GDP—real growth rate: 2.62% (2019 est.)
4.31% (2018 est.)
4.26% (2017 est.)
country comparison to the world: 109

GDP—per capita (PPP): $17,900 (2017 est.)
$17,200 (2016 est.)
$16,700 (2015 est.)
note: data are in 2017 dollars
country comparison to the world: 99

Gross national saving: 34.1% of GDP (2017 est.)
32.8% of GDP (2016 est.)
30.3% of GDP (2015 est.)
country comparison to the world: 19

GDP—composition, by end use: *household consumption:* 48.8% (2017 est.)
government consumption: 16.4% (2017 est.)
investment in fixed capital: 23.2% (2017 est.)
investment in inventories: -0.4% (2017 est.)
exports of goods and services: 68.2% (2017 est.)
imports of goods and services: -54.6% (2017 est.)

GDP—composition, by sector of origin:
agriculture: 8.2% (2017 est.)
industry: 36.2% (2017 est.)
services: 55.6% (2017 est.)

Agriculture—products: rice, cassava (manioc, tapioca), rubber, corn, sugarcane, coconuts, palm oil, pineapple, livestock, fish products

Industries: tourism, textiles and garments, agricultural processing, beverages, tobacco, cement, light manufacturing such as jewelry and electric appliances, computers and parts, integrated circuits, furniture, plastics, automobiles and automotive parts, agricultural machinery, air conditioning and refrigeration, ceramics, aluminum, chemical, environmental management, glass, granite and marble, leather, machinery and metal work, petrochemical, petroleum refining, pharmaceuticals, printing, pulp and paper, rubber, sugar, rice, fishing, cassava, world's second-largest tungsten producer and third-largest tin producer

Industrial production growth rate: 1.6% (2017 est.)
country comparison to the world: 141

Labor force: 37.546 million (2020 est.)
country comparison to the world: 15

Labor force—by occupation: agriculture: 31.8%
industry: 16.7%
services: 51.5% (2015 est.)

Unemployment rate: 0.99% (2019 est.)
1.06% (2018 est.)
country comparison to the world: 7

Population below poverty line: 7.2% (2015 est.)

Household income or consumption by percentage share: *lowest 10%:* 2.8%
highest 10%: 31.5% (2009 est.)

Budget: *revenues:* 69.23 billion (2017 est.)
expenditures: 85.12 billion (2017 est.)

Taxes and other revenues: 15.2% (of GDP) (2017 est.)
country comparison to the world: 193

Budget surplus (+) or deficit (-): -3.5% (of GDP) (2017 est.)
country comparison to the world: 148

Public debt: 41.9% of GDP (2017 est.)
41.8% of GDP (2016 est.)
note: data cover general government debt and include debt instruments issued (or owned) by government entities other than the treasury; the data include treasury debt held by foreign entities; the data include debt issued by subnational entities, as well as intragovernmental debt; intragovernmental debt consists of treasury borrowings from surpluses in the social funds, such as for retirement, medical care, and unemployment; debt instruments for the social funds are sold at public auctions
country comparison to the world: 118

Fiscal year: 1 October - 30 September

Inflation rate (consumer prices): 0.7% (2017 est.)
0.2% (2016 est.)
country comparison to the world: 38

Current account balance: $37.033 billion (2019 est.)
$28.423 billion (2018 est.)
country comparison to the world: 11

Exports: $235.1 billion (2017 est.)
$214.3 billion (2016 est.)
country comparison to the world: 21

Exports—partners: China 12.4%, US 11.2%, Japan 9.5%, Hong Kong 5.2%, Vietnam 4.9%, Australia 4.5%, Malaysia 4.4% (2017)

Exports—commodities: automobiles and parts, computer and parts, jewelry and precious stones, polymers of ethylene in primary forms, refine fuels, electronic integrated circuits, chemical products, rice, fish products, rubber products, sugar, cassava, poultry, machinery and parts, iron and steel and their products

Imports: $203.2 billion (2017 est.)
$177.7 billion (2016 est.)
country comparison to the world: 25

Imports—commodities: machinery and parts, crude oil, electrical machinery and parts, chemicals, iron & steel and product, electronic integrated circuit, automobile's parts, jewelry including silver bars and gold, computers and parts, electrical household appliances, soybean, soybean meal, wheat, cotton, dairy products

Imports—partners: China 20%, Japan 14.5%, US 6.8%, Malaysia 5.4% (2017)

Reserves of foreign exchange and gold: $202.6 billion (31 December 2017 est.)

955

$171.9 billion (31 December 2016 est.)
country comparison to the world: 12

Debt—external: $132 billion (31 December 2017 est.)
$130.6 billion (31 December 2016 est.)
country comparison to the world: 44

Exchange rates: baht per US dollar—
34.34 (2017 est.)
35.296 (2016 est.)
35.296 (2015 est.)
34.248 (2014 est.)
32.48 (2013 est.)

ENERGY

Electricity access: *electrification—total population:* 100% (2020)

Electricity—production: 181.5 billion kWh (2016 est.)
country comparison to the world: 23

Electricity—consumption: 187.7 billion kWh (2016 est.)
country comparison to the world: 22

Electricity—exports: 2.267 billion kWh (2015 est.)
country comparison to the world: 44

Electricity—imports: 19.83 billion kWh (2016 est.)
country comparison to the world: 11

Electricity—installed generating capacity: 44.89 million kW (2016 est.)
country comparison to the world: 24

Electricity—from fossil fuels: 76% of total installed capacity (2016 est.)
country comparison to the world: 94

Electricity—from nuclear fuels: 0% of total installed capacity (2017 est.)
country comparison to the world: 193

Electricity—from hydroelectric plants: 8% of total installed capacity (2017 est.)
country comparison to the world: 124

Electricity—from other renewable sources: 16% of total installed capacity (2017 est.)
country comparison to the world: 55

Crude oil—production: 228,000 bbl/day (2018 est.)
country comparison to the world: 34

Crude oil—exports: 790 bbl/day (2015 est.)
country comparison to the world: 76

Crude oil—imports: 875,400 bbl/day (2015 est.)
country comparison to the world: 12

Crude oil—proved reserves: 349.4 million bbl (1 January 2018 est.)
country comparison to the world: 50

Refined petroleum products—production: 1.328 million bbl/day (2015 est.)
country comparison to the world: 14

Refined petroleum products—consumption: 1.326 million bbl/day (2016 est.)
country comparison to the world: 16

Refined petroleum products— exports: 278,300 bbl/day (2015 est.)
country comparison to the world: 29

Refined petroleum products—imports: 134,200 bbl/day (2015 est.)
country comparison to the world: 44

Natural gas—production: 38.59 billion cu m (2017 est.)
country comparison to the world: 22

Natural gas—consumption: 52.64 billion cu m (2017 est.)
country comparison to the world: 16

Natural gas—exports: 0 cu m (2017 est.)
country comparison to the world: 199

Natural gas—imports: 14.41 billion cu m (2017 est.)
country comparison to the world: 21

Natural gas—proved reserves: 193.4 billion cu m (1 January 2018 est.)
country comparison to the world: 43

Carbon dioxide emissions from consumption of energy: 355 million Mt (2017 est.)
country comparison to the world: 19

COMMUNICATIONS

Telephones—fixed lines: *total subscriptions:* 2,580,166

subscriptions per 100 inhabitants: 3.75 (2019 est.)
country comparison to the world: 50

Telephones—mobile cellular: total subscriptions: 128,086,321

subscriptions per 100 inhabitants: 186.16 (2019 est.)
country comparison to the world: 13

Telecommunication systems: *general assessment:* high quality system, especially in urban areas like Bangkok, mobile and mobile broadband penetration are on the increase; Fiber-to-the-home (FttH) has seen strong growth in the major cities; 4G TD-LTE available and moving to 5G services; seven smart cities with the hope of 100 smart cities within its borders in the next two decades; one of the biggest e-commerce markets in Southeast Asia; fixed broadband remains relative compared to other developed Asian telecom markets and with the dominance of the mobile platform (2020)

domestic: fixed-line system provided by both a government-owned and commercial provider; wireless service expanding rapidly; fixed-line 4 per 100 and mobile-cellular 186 per 100 (2019)

international: country code - 66; landing points for the AAE-1, FEA, SeaMeWe-3,-4, APG, SJC2, TIS, MCT and AAG submarine cable systems providing links throughout Asia, Australia, Africa, Middle East, Europe, and US; satellite earth stations - 2 Intelsat (1 Indian Ocean, 1 Pacific Ocean) (2019)

note: the COVID-19 outbreak is negatively impacting telecommunications production and supply chains globally; consumer spending on telecom devices and services has also slowed due to the pandemic's effect on economics worldwide; overall progress towards improvements in all facets of the telecom industry - mobile, fixed-line, broadband, submarine cable and satellite - has moderated

Broadcast media: 26 digital TV stations in Bangkok broadcast nationally, 6 terrestrial TV stations in Bangkok broadcast nationally via relay stations - 2 of the stations are owned by the military, the other 4 are government-owned or controlled, leased to private enterprise, and all are required to broadcast government-produced news programs twice a day; multi-channel satellite and cable TV subscription services are available; radio frequencies have been allotted for more than 500 government and commercial radio stations; many small community radio stations operate with low-power transmitters (2017)

Internet country code: .th

Internet users: total: 38,987,531

percent of population: 56.82% (July 2018 est.)
country comparison to the world: 21

Broadband—fixed subscriptions: *total:* 9.189 million

subscriptions per 100 inhabitants: 13 (2018 est.)
country comparison to the world: 19

TRANSPORTATION

National air transport system: *number of registered air carriers:* 15 (2020)

inventory of registered aircraft operated by air carriers: 283

annual passenger traffic on registered air carriers: 76,053,042 (2018)

annual freight traffic on registered air carriers: 2,666,260,000 mt-km (2018)

Civil aircraft registration country code prefix: HS (2016)

Airports: 101 (2013)
country comparison to the world: 56

Airports—with paved runways: *total:* 63 (2013)
over 3,047 m: 8 (2013)
2,438 to 3,047 m: 12 (2013)
1,524 to 2,437 m: 23 (2013)
914 to 1,523 m: 14 (2013)
under 914 m: 6 (2013)

Airports—with unpaved runways: *total:* 38 (2013)
2,438 to 3,047 m: 1 (2013)
1,524 to 2,437 m: 1 (2013)
914 to 1,523 m: 10 (2013)
under 914 m: 26 (2013)

Heliports: 7 (2013)

Pipelines: 2 km condensate, 5900 km gas, 85 km liquid petroleum gas, 1 km oil, 1097 km refined products (2013)

Railways: *total:* 4,127 km (2017)

standard gauge: 84 km 1.435-m gauge (84 km electrified) (2017)

narrow gauge: 4,043 km 1.000-m gauge (2017)
country comparison to the world: 47

Roadways: *total:* 180,053 km (includes 450 km of expressways) (2006)
country comparison to the world: 30

Waterways: 4,000 km (3,701 km navigable by boats with drafts up to 0.9 m) (2011)

country comparison to the world: 26

Merchant marine: *total:* 825

by type: bulk carrier 27, container ship 27, general cargo 89, oil tanker 243, other 439 (2019)
country comparison to the world: 27

Ports and terminals: *major seaport(s):* Bangkok, Laem Chabang, Map Ta Phut, Prachuap Port, Si Racha

container port(s) (TEUs): Laem Chabang (7,227,431) (2017)

LNG terminal(s) (import): Map Ta Phut

MILITARY AND SECURITY

Military and security forces: *Royal Thai Armed Forces (Kongthap Thai, RTARF):* Royal Thai Army (Kongthap Bok Thai, RTA; includes Thai Rangers (Thahan Phrahan)), Royal Thai Navy (Kongthap Ruea Thai, RTN; includes Royal Thai Marine Corps), Royal Thai Air Force (Kongthap Akaat Thai, RTAF); Interior Ministry paramilitary forces: Volunteer Defense Corps (2019)

note: the Thai Rangers (aka Thahan Phrahan or 'Hunter Soldiers') is a paramilitary force formed in 1978 to clear Communist Party of Thailand guerrillas from mountain strongholds in the country's northeast; it is a light infantry force led by regular officers and non-commissioned officers and comprised of both full - and part - time personnel; it conducts counterinsurgency operations in the southern, predominantly Muslim, region; on the eastern border with Laos and Cambodia, the Rangers have primary responsibility for border surveillance and protection

Military expenditures: 1.3% of GDP (2019)
1.4% of GDP (2018)
1.6% of GDP (2017)
1.6% of GDP (2016)
1.4% of GDP (2015)
country comparison to the world: 96

Military and security service personnel strengths: estimates for the size of the Royal Thai Armed Forces (RTARF) vary; approximately 360,000 active duty personnel (245,000 Army; 70,000 Navy; 45,000 Air Force); est. 20,000 Thai Rangers (2019)

Military equipment inventories and acquisitions: the RTARF has a diverse array of foreign-supplied weapons systems, including a large amount of obsolescent or secondhand US equipment; since 2015, the top suppliers are China, South Korea, Ukraine, and the US (2019 est.)

Military deployments: 270 South Sudan (UNMISS) (2020)

Military service age and obligation: 21 years of age for compulsory military service; 18 years of age for voluntary military service; males register at 18 years of age; 2-year conscript service obligation based on lottery (2018)

Military—note: including the most recent in 2014, the military has conducted 12 successful coups and attempted an additional seven since the fall of absolute monarchy in 1932; since 2004, the military has fought against separatist insurgents in the southern provinces of Pattani, Yala, and Narathiwat, as well as parts of Songkhla; as of 2019, approximately 60,000 security forces, including large numbers of paramilitary troops such as the Thai Rangers, were stationed in the south (2019)

TRANSPORTATION

Disputes—international: separatist violence in Thailand's predominantly Malay-Muslim southern provinces prompt border closures and controls with Malaysia to stem insurgent activities; Southeast Asian states have enhanced border surveillance to check the spread of avian flu; talks continue on completion of demarcation with Laos but disputes remain over several islands in the Mekong River; despite continuing border committee talks, Thailand must deal with Karen and other ethnic rebels, refugees, and illegal cross-border activities; Cambodia and Thailand dispute sections of boundary; in 2011, Thailand and Cambodia resorted to arms in the dispute over the location of the boundary on the precipice surmounted by Preah Vihear temple ruins, awarded to Cambodia by ICJ decision in 1962 and part of a planned UN World Heritage site; Thailand is studying the feasibility of jointly constructing the Hatgyi Dam on the Salween river near the border with Burma; in 2004, international environmentalist pressure prompted China to halt construction of 13 dams on the Salween River that flows through China, Burma, and Thailand; approximately 100,000 mostly Karen refugees fleeing civil strife, political upheaval and economic stagnation in Burma live in remote camps in Thailand near the border

Refugees and internally displaced persons:

refugees (country of origin): 91,806 (Burma) (2020)

stateless persons: 475,009 (2019) (estimate represents stateless persons registered with the Thai Government; actual number may be as high as 3.5 million); note - about half of Thailand's northern hill tribe people do not have citizenship and make up the bulk of Thailand's stateless population; most lack documentation showing they or one of their parents were born in Thailand; children born to Burmese refugees are not eligible for Burmese or Thai citizenship and are stateless; most Chao Lay, maritime nomadic peoples, who travel from island to island in the Andaman Sea west of Thailand are also stateless; stateless Rohingya refugees from Burma are considered illegal migrants by Thai authorities and are detained in inhumane conditions or expelled; stateless persons are denied access to voting, property, education, employment, healthcare, and driving

note: Thai nationality was granted to more than 23,000 stateless persons between 2012 and 2016; in 2016, the Government of Thailand approved changes to its citizenship laws that could make 80,000 stateless persons eligible for citizenship, as part of its effort to achieve zero statelessness by 2024 (2018)

Trafficking in persons: *current situation:* Thailand is a source, transit, and destination country for men, women, and children subjected to forced labor and sex trafficking; victims from Burma, Cambodia, Laos, China, Vietnam, Uzbekistan, and India, migrate to Thailand in search of jobs but are forced, coerced, or defrauded into labor in commercial fishing, fishing-related industries, factories, domestic work, street begging, or the sex trade; some Thai, Burmese, Cambodian, and Indonesian men forced to work on fishing boats are kept at sea for years; sex trafficking of adults and children from Thailand, Laos, Vietnam, and Burma remains a significant problem; Thailand is a transit country for victims from China, Vietnam,

Bangladesh, and Burma subjected to sex trafficking and forced labor in Malaysia, Indonesia, Singapore, Russia, South Korea, the US, and countries in Western Europe; Thai victims are also trafficked in North America, Europe, Africa, Asia, and the Middle East

tier rating: Tier 2 Watch List - Thailand does not fully comply with the minimum standards for the elimination of trafficking, and is not making significant efforts to do so; in 2014, authorities investigated, prosecuted, and convicted fewer traffickers and identified fewer victims; some cases of official complicity were investigated and prosecuted, but trafficking-related corruption continues to hinder progress in combatting trafficking; authorities' efforts to screen for victims among vulnerable populations remained inadequate due to a poor understanding of trafficking indicators, a failure to recognize non-physical forms of coercion, and a shortage of language interpreters; the government passed new labor laws increasing the minimum age in the fishing industry to 18 years old, guaranteeing the minimum wage, and requiring work contracts, but weak law enforcement and poor coordination among regulatory agencies enabled exploitive labor practices to continue; the government increased efforts to raise public awareness to the dangers of human trafficking and to deny entry to foreign sex tourists (2015)

Illicit drugs: a minor producer of opium, heroin, and marijuana; transit point for illicit heroin en route to the international drug market from Burma and Laos; eradication efforts have reduced the area of cannabis cultivation and shifted some production to neighboring countries; opium poppy cultivation has been reduced by eradication efforts; also a drug money-laundering center; minor role in methamphetamine production for regional consumption; major consumer of methamphetamine since the 1990s despite a series of government crackdowns

TIMOR-LESTE

INTRODUCTION

Background: Timor was actively involved in Southeast Asian trading networks for centuries and by the 14th century exported aromatic sandalwood, slaves, honey, and wax. A number of local chiefdoms ruled the island in the early 16th century when Portuguese traders arrived, chiefly attracted by the relative abundance of sandalwood on Timor; by mid century, the Portuguese had colonized the island. Skirmishing with the Dutch in the region eventually resulted in an 1859 treaty in which Portugal ceded the western portion of the island. Imperial Japan occupied Portuguese Timor from 1942 to 1945, but Portugal resumed colonial authority after the Japanese defeat in World War II. East Timor declared itself independent from Portugal on 28 November 1975 and was invaded and occupied by Indonesian forces nine days later. It was incorporated into Indonesia in July 1976 as the province of Timor Timur (East Timor). An unsuccessful campaign of pacification followed over the next two decades, during which an estimated 100,000 to 250,000 people died. In an August 1999 UNsupervised popular referendum, an overwhelming majority of the people of Timor-Leste voted for independence from Indonesia. However, in the next three weeks, anti-independence Timorese militias - organized and supported by the Indonesian military - commenced a large-scale, scorched-earth campaign of retribution. The militias killed approximately 1,400 Timorese and forced 300,000 people into western Timor as refugees. Most of the country's infrastructure, including homes, irrigation systems, water supply systems, and schools, and nearly all of the country's electrical grid were destroyed. On 20 September 1999, Australian-led peacekeeping troops deployed to the country and brought the violence to an end. On 20 May 2002, Timor-Leste was internationally recognized as an independent state.

In 2006, internal tensions threatened the new nation's security when a military strike led to violence and a breakdown of law and order.

At Dili's request, an Australian-led International Stabilization Force (ISF) deployed to Timor-Leste, and the UN Security Council established the UN Integrated Mission in Timor-Leste (UNMIT), which included an authorized police presence of over 1,600 personnel. The ISF and UNMIT restored stability, allowing for presidential and parliamentary elections in 2007 in a largely peaceful atmosphere. In February 2008, a rebel group staged an unsuccessful attack against the president and prime minister. The ringleader was killed in the attack, and most of the rebels surrendered in April 2008. Since the attack, the government has enjoyed one of its longest periods of post-independence stability, including successful 2012 elections for both the parliament and president and a successful transition of power in February 2015. In late 2012, the UN Security Council ended its peacekeeping mission in Timor-Leste and both the ISF and UNMIT departed the country. Early parliamentary elections in the spring of 2017 finally produced a majority government after months of impasse. Currently, the government is a coalition of three parties and the president is a member of the opposition party. In 2018 and 2019, this configuration stymied nominations for key ministerial positions and slowed progress on certain policy issues.

GEOGRAPHY

Location: Southeastern Asia, northwest of Australia in the Lesser Sunda Islands at the eastern end of the Indonesian archipelago; note Timor-Leste includes the eastern half of the island of Timor, the Oecussi (Ambeno) region on the northwest portion of the island of Timor, and the islands of Pulau Atauro and Pulau Jaco

Geographic coordinates: 8 50 S, 125 55 E

Map references: Southeast Asia

Area: *total:* 14,874 sq km
land: 14,874 sq km
water: 0 sq km
country comparison to the world: 160

Area—comparative: slightly larger than Connecticut; almost half the size of Maryland

Area comparison map: [INSERT IMAGE: TIMOR-LESTE-Area comparison map]

Land boundaries: *total:* 253 km
border countries (1): Indonesia 253 km

Coastline: 706 km

Maritime claims: *territorial sea:* 12 nm
contiguous zone: 24 nm
exclusive fishing zone: 200 nm

Climate: tropical; hot, humid; distinct rainy and dry seasons

Terrain: mountainous

Elevation: *lowest point:* Timor Sea, Savu Sea, and Banda Sea 0 m
highest point: Foho Tatamailau 2,963 m

Natural resources: gold, petroleum, natural gas, manganese, marble

Land use: *agricultural land:* 25.1% (2011 est.)
arable land: 10.1% (2011 est.) / permanent crops: 4.9% (2011 est.) / permanent pasture: 10.1% (2011 est.)
forest: 49.1% (2011 est.)
other: 25.8% (2011 est.)

Irrigated land: 350 sq km (2012)

Population distribution: most of the population concentrated in the western third of the country, particularly around Dili

Natural hazards: floods and landslides are common; earthquakes; tsunamis; tropical cyclones

Environment—current issues: air pollution and deterioration of air quality; greenhouse gas emissions; water quality, scarcity, and access; land and soil degradation; forest depletion; widespread use of slash and burn agriculture has led to deforestation and soil erosion; loss of biodiversity

Environment—international agreements: *party to:* Biodiversity, Climate Change, Climate Change-Kyoto Protocol, Desertification
signed, but not ratified: none of the selected agreements

Geography—note: Timor comes from the Malay word for "east"; the island of Timor is part of the Malay Archipelago and is the largest and easternmost of the Lesser Sunda Islands; the district of Oecussi is an exclave separated from Timor-Leste proper by Indonesia; Timor-Leste has the unique distinction of being the only Asian country located completely in the Southern Hemisphere

PEOPLE AND SOCIETY

Population: 1,383,723 (July 2020 est.)
country comparison to the world: 155

Nationality: *noun:* Timorese
adjective: Timorese

Ethnic groups: Austronesian (Malayo-Polynesian) (includes Tetun, Mambai, Tokodede, Galoli, Kemak, Baikeno), Melanesian-Papuan (includes Bunak, Fataluku, Bakasai), small Chinese minority

Languages: Tetun Prasa 30.6%, Mambai 16.6%, Makasai 10.5%, Tetun Terik 6.1%, Baikenu 5.9%, Kemak 5.8%, Bunak 5.5%, Tokodede 4%, Fataluku 3.5%, Waima'a 1.8%, Galoli 1.4%, Naueti 1.4%, Idate 1.2%, Midiki 1.2%, other 4.5%
note: data represent population by mother tongue; Tetun and Portuguese are official languages; Indonesian and English are working languages; there are about 32 indigenous languages

Religions: Roman Catholic 97.6%, Protestant/Evangelical 2%, Muslim 0.2%, other 0.2% (2015 est.)

Age structure: *0-14 years:* 39.96% (male 284,353/female 268,562)

15-24 years: 20.32% (male 142,693/female 138,508)

25-54 years: 30.44% (male 202,331/female 218,914)

55-64 years: 5.22% (male 34,956/female 37,229)

65 years and over: 4.06% (male 27,153/female 29,024) (2020 est.)

population pyramid: [INSERT IMAGE: TIMOR-LESTE-population pyramid]

Dependency ratios: *total dependency ratio:* 90.3
youth dependency ratio: 83.7
elderly dependency ratio: 6.6
potential support ratio: 15.2 (2020 est.)

Median age: *total:* 19.6 years
male: 18.9 years
female: 20.2 years (2020 est.)
country comparison to the world: 201

Population growth rate: 2.27% (2020 est.)
country comparison to the world: 32

Birth rate: 32 births/1,000 population (2020 est.)
country comparison to the world: 27

Death rate: 5.7 deaths/1,000 population (2020 est.)
country comparison to the world: 178
Net migration rate: -3.9 migrant(s)/1,000 population (2020 est.)
country comparison to the world: 189

Population distribution: most of the population concentrated in the western third of the country, particularly around Dili

Urbanization: *urban population:* 31.3% of total population (2020)
rate of urbanization: 3.35% annual rate of change (2015-20 est.)

total population growth rate v. urban population growth rate, 2000-2030: *Major urban areas - population:* 281,000 DILI (capital) (2018)

Sex ratio: *at birth:* 1.07 male(s)/female
0-14 years: 1.06 male(s)/female
15-24 years: 1.03 male(s)/female
25-54 years: 0.92 male(s)/female
55-64 years: 0.94 male(s)/female
65 years and over: 0.94 male(s)/female
total population: 1 male(s)/female (2020 est.)

Mother's mean age at first birth: 22.1 years (2009/10 est.)
note: median age at first birth among women 25-29

Maternal mortality rate: 142 deaths/100,000 live births (2017 est.)
country comparison to the world: 60

Infant mortality rate: *total:* 31.7 deaths/1,000 live births
male: 34.3 deaths/1,000 live births
female: 28.9 deaths/1,000 live births (2020 est.)
country comparison to the world: 51

Life expectancy at birth: *total population:* 69.3 years
male: 67.6 years
female: 71.1 years (2020 est.)
country comparison to the world: 170

Total fertility rate: 4.44 children born/woman (2020 est.)
country comparison to the world: 25

Contraceptive prevalence rate: 26.1% (2016)

Drinking water source:

improved: *urban:* 100% of population
rural: 72.3% of population
total: 80.7% of population

unimproved: *urban:* 0% of population
rural: 27.7% of population
total: 19.3% of population (2017 est.)

Current Health Expenditure: 3.9% (2017)

Physicians density: 0.75 physicians/1,000 population (2017)

Sanitation facility access:

improved: *urban:* 90.9% of population
rural: 50.3% of population
total: 62.6% of population

unimproved: *urban:* 9.1% of population
rural: 49.7% of population
total: 57.4% of population (2017 est.)

HIV/AIDS—adult prevalence rate: 0.2% (2019)
country comparison to the world: 114

HIV/AIDS—people living with HIV/AIDS: 1,500 (2019)
country comparison to the world: 138

HIV/AIDS—deaths: <100 (2019)

Major infectious diseases: *degree of risk:* very high (2020)
food or waterborne diseases: bacterial diarrhea, hepatitis A, and typhoid fever
vectorborne diseases: dengue fever and malaria

Obesity—adult prevalence rate: 3.8% (2016)
country comparison to the world: 190

Children under the age of 5 years underweight: 37.5% (2013)
country comparison to the world: 3

Education expenditures: 3.8% of GDP (2017)
country comparison to the world: 116

Literacy: *definition:* age 15 and over can read and write
total population: 68.1%
male: 71.9%
female: 64.2% (2018)
School life expectancy (primary to tertiary education): total: 13 years
male: 14 years
female: 13 years (2010)
Unemployment, youth ages 15-24: total: 13.2%
male: 10.9%
female: 15.9% (2016 est.)
country comparison to the world: 105

People—note: one of only two predominantly Christian nations in Southeast Asia, the other being the Philippines

GOVERNMENT

Country name: *conventional long form:* Democratic Republic of Timor-Leste
conventional short form: Timor-Leste
local long form: Republika Demokratika Timor Lorosa'e [Tetum]; Republica Democratica de Timor-Leste [Portuguese]

local short form: Timor Lorosa'e [Tetum]; Timor-Leste [Portuguese]
former: East Timor, Portuguese Timor
etymology: timor' derives from the Indonesian and Malay word "timur" meaning "east"; "leste" is the Portuguese word for "east", so "Timor-Leste" literally means "Eastern-East"; the local [Tetum] name "Timor Lorosa'e" translates as "East Rising Sun"
note: pronounced TEE-mor LESS-tay

Government type: semi-presidential republic

Capital: *name:* Dili

geographic coordinates: 8 35 S, 125 36 E
time difference: UTC+9 (14 hours ahead of Washington, DC, during Standard Time)

Administrative divisions: 12 municipalities (municipios, singular municipio) and 1 special administrative region* (regiao administrativa especial); Aileu, Ainaro, Baucau, Bobonaro (Maliana), Covalima (Suai), Dili, Ermera (Gleno), Lautem (Lospalos), Liquica, Manatuto, Manufahi (Same), Oe-Cusse Ambeno* (Pante Macassar), Viqueque
note: administrative divisions have the same names as their administrative centers (exceptions have the administrative center name following in parentheses)

Independence: 20 May 2002 (from Indonesia); note - 28 November 1975 was the date independence was proclaimed from Portugal; 20 May 2002 was the date of international recognition of Timor-Leste's independence from Indonesia

National holiday: Restoration of Independence Day, 20 May (2002); Proclamation of Independence Day, 28 November (1975)

Constitution: *history:* drafted 2001, approved 22 March 2002, entered into force 20 May 2002

amendments: proposed by Parliament and parliamentary groups; consideration of amendments requires at least four- fifths majority approval by Parliament; passage requires two-thirds majority vote by Parliament and promulgation by the president of the republic; passage of amendments to the republican form of government and the flag requires approval in a referendum

Legal system: civil law system based on the Portuguese model; note - penal and civil law codes to replace the Indonesian codes were passed by Parliament and promulgated in 2009 and 2011, respectively

International law organization participation: accepts compulsory ICJ jurisdiction with reservations; accepts ICCt jurisdiction

Citizenship: *citizenship by birth:* no
citizenship by descent only: at least one parent must be a citizen of Timor-Leste
dual citizenship recognized: no
residency requirement for naturalization: 10 years

Suffrage: 17 years of age; universal

Executive branch: *chief of state:* President Francisco GUTERRES (since 20 May 2017); note - the president is commander in chief of the military and is able to veto legislation, dissolve parliament, and call national elections

head of government: Prime Minister Taur Matan RUAK (since 22 June 2018); note - President GUTERRES dissolved parliament because of an impasse over passing the country's budget on 26 January 2018, with then Prime Minister Mari ALKATIRI assuming the role of caretaker prime minister until a new prime minister was appointed; note - on 25 February 2020, Prime Minister RUAK offered his resignation due to inability to pass 2020 budget in parliament, but the president refused his offer; on 8 April, RUAK withdrew his resignation

cabinet: the governing coalition in the Parliament proposes cabinet member candidates to the Prime Minister, who presents these recommendations to the President of the Republic for swearing in

elections/appointments: president directly elected by absolute majority popular vote in 2 rounds if needed for a 5-year term (eligible for a second term); election last held on 20 March 2017 (next to be held in 2022); following parliamentary elections, the president appoints the leader of the majority party or majority coalition as the prime minister

election results: Francisco GUTERRES elected president; percent of vote - Francisco GUTERRES (FRETILIN) 57.1%, Antonio DA CONCEICAO (PD) 32.5%, Jose Luis GUTERRES (Frenti-Mudanca) 2.6%, Jose NEVES (independent) 2.3%, Luis Alves TILMAN (independent) 2.2%, other 3.4%

Legislative branch: *description:* unicameral National Parliament (65 seats; members directly elected in a single nationwide constituency by proportional representation vote to serve 5-year terms)

elections: last held on 12 May 2018 (next to be held in July 2023)

election results: percent of vote by party - AMP - 49.6%, FRETILIN 34.2%, PD 8.1%, DDF 5.5%, other 2.6%; seats by party - AMP 34, FRETILIN 23, PD 5, DDF 3; composition - men 39, women 26, percent of women 40%

Judicial branch: *highest courts:* Court of Appeals (consists of the court president and NA judges)

judge selection and term of office: court president appointed by the president of the republic from among the other court judges to serve a 4-year term; other court judges appointed - 1 by the Parliament and the others by the Supreme Council for the Judiciary, a body chaired by the court president and that includes mostly presidential and parliamentary appointees; other judges serve for life

subordinate courts: Court of Appeal; High Administrative, Tax, and Audit Court; district courts; magistrates' courts; military courts

note: the UN Justice System Programme, launched in 2003 and being rolled out in 4 phases through 2018, is helping strengthen the country's justice system; the Programme is aligned with the country's long-range Justice Sector Strategic Plan, which includes legal reforms

Political parties and leaders: Alliance for Change and Progress or AMP [Xanana GUSMAO] (alliance includes CNRT, KHUNTO, PLP)
Democratic Development Forum or DDF
Democratic Party or PD
Frenti-Mudanca [Jose Luis GUTERRES]
Kmanek Haburas Unidade Nasional Timor Oan or KHUNTO
National Congress for Timorese Reconstruction or CNRT [Kay Rala Xanana GUSMAO]
People's Liberation Party or PLP [Taur Matan RUAK]
Revolutionary Front of Independent Timor-Leste or FRETILIN [Mari ALKATIRI]

International organization participation: ACP, ADB, AOSIS, ARF, ASEAN (observer), CPLP, EITI (compliant country), FAO, G-77, IBRD, ICAO, ICCt, ICRM, IDA, IFAD, IFC, IFRCS, ILO, IMF, IMO, Interpol, IOC, IOM, IPU, ITU, MIGA, NAM, OPCW, PIF (observer), UN, UNCTAD, UNESCO, UNIDO, Union Latina, UNWTO, UPU, WCO, WHO, WMO

Diplomatic representation in the US: *chief of mission:* Ambassador Isilio Antonio De Fatima COELHO DA SILVA (since 6 January 2020)

chancery: 4201 Connecticut Avenue NW, Suite 504, Washington, DC 20008

telephone: [1] (202) 966-3202

FAX: [1] (202) 966-3205

Diplomatic representation from the US: *chief of mission:* Ambassador Kathleen FITZPATRICK (since 19 January 2018)

telephone: (670) 332-4684, EMER: +(670) 7723-1328

embassy: Avenida de Portugal, Praia dos Coqueiros, Dili

mailing address: US Department of State, 8250 Dili Place, Washington, DC 20521-8250

FAX: (670) 331-3206

Flag description: red with a black isosceles triangle (based on the hoist side) superimposed on a slightly longer yellow arrowhead that extends to the center of the flag; a white star - pointing to the upper hoist-side corner of the flag - is in the center of the black triangle; yellow denotes the colonialism in Timor-Leste's past, black represents the obscurantism that needs to be overcome, red stands for the national liberation struggle; the white star symbolizes peace and serves as a guiding light

National symbol(s): *Mount Ramelau; national colors:* red, yellow, black, white

National anthem: *name:* "Patria" (Fatherland)

lyrics/music: Fransisco Borja DA COSTA/Afonso DE ARAUJO

note: adopted 2002; the song was first used as an anthem when Timor-Leste declared its independence from Portugal in 1975; the lyricist, Francisco Borja DA COSTA, was killed in the Indonesian invasion just days after independence was declared
0:00 / 2:09

ECONOMY

Economy—overview: Since independence in 1999, Timor-Leste has faced great challenges in rebuilding its infrastructure, strengthening the civil administration, and generating jobs for young people entering the work force. The development of offshore oil and gas resources has greatly supplemented government revenues. This technology-intensive industry, however, has done little to create jobs in part because there are no production facilities in Timor-Leste. Gas is currently piped to Australia for processing, but Timor-Leste has expressed interest in developing a domestic processing capability.

In June 2005, the National Parliament unanimously approved the creation of the Timor-Leste Petroleum Fund to serve as a repository for all petroleum revenues and to preserve the value of Timor-Leste's petroleum wealth for future generations. The Fund held assets of $16 billion, as of mid-2016. Oil accounts for over 90% of government revenues, and the drop in the price of oil in 2014-16 has led to concerns about the long-term sustainability of government spending. Timor-Leste compensated for the decline in price by exporting more oil. The Ministry of Finance maintains that the Petroleum Fund is sufficient to sustain government operations for the foreseeable future.

Annual government budget expenditures increased markedly between 2009 and 2012 but dropped significantly through 2016. Historically, the government failed to spend as much as its budget allowed. The government has focused significant resources on basic infrastructure, including electricity and roads, but limited experience in procurement and infrastructure building has hampered these projects. The underlying economic policy challenge the country faces remains how best to use oil-and-gas wealth to lift the non-oil economy onto a higher growth path and to reduce poverty.

GDP (purchasing power parity): $7.426 billion (2017 est.)
$7.784 billion (2016 est.)
$7.391 billion (2015 est.)
note: data are in 2017 dollars
country comparison to the world: 166

GDP (official exchange rate): $2.775 billion (2017 est.)
note: non-oil GDP

GDP—real growth rate: -4.6% (2017 est.)
5.3% (2016 est.)
4% (2015 est.)
country comparison to the world: 216

GDP—per capita (PPP): $6,000 (2017 est.)
$6,400 (2016 est.)
$6,200 (2015 est.)
note: data are in 2017 dollars
country comparison to the world: 164

GDP—composition, by end use: *household consumption:* 33% (2017 est.)
government consumption: 30% (2017 est.)
investment in fixed capital: 10.6% (2017 est.)
investment in inventories: 0% (2017 est.)
exports of goods and services: 78.4% (2017 est.)
imports of goods and services: -52% (2017 est.)

GDP—composition, by sector of origin: *agriculture:* 9.1% (2017 est.)
industry: 56.7% (2017 est.)
services: 34.4% (2017 est.)

Agriculture—products: coffee, rice, corn, cassava (manioc, tapioca), sweet potatoes, soybeans, cabbage, mangoes, bananas, vanilla

Industries: printing, soap manufacturing, handicrafts, woven cloth

Industrial production growth rate: 2% (2017 est.)
country comparison to the world: 133

Labor force: 286,700 (2016 est.)
country comparison to the world: 164

Labor force—by occupation: agriculture: 41%
industry: 13%
services: 45.1% (2013)

Unemployment rate: 4.4% (2014 est.)
3.9% (2010 est.)
country comparison to the world: 65

Population below poverty line: 41.8% (2014 est.)

Household income or consumption by percentage share: *lowest 10%:* 4%
highest 10%: 27% (2007)

Budget: *revenues:* 300 million (2017 est.)
expenditures: 2.4 billion (2017 est.)

Taxes and other revenues: 10.8% (of GDP) (2017 est.)
country comparison to the world: 213

Budget surplus (+) or deficit (-): -75.7% (of GDP) (2017 est.)
country comparison to the world: 222

Public debt: 3.8% of GDP (2017 est.)
3.1% of GDP (2016 est.)
country comparison to the world: 206

Fiscal year: calendar year

Inflation rate (consumer prices): 0.6% (2017 est.)
-1.3% (2016 est.)
country comparison to the world: 34

Current account balance: -$284 million (2017 est.)
-$544 million (2016 est.)
country comparison to the world: 106

Exports: $16.7 million (2017 est.)
$18 million (2015 est.)
country comparison to the world: 214

Exports—commodities: oil, coffee, sandalwood, marble
note: potential for vanilla exports

Imports: $681.2 million (2017 est.)
$558.6 million (2016 est.)
country comparison to the world: 193

Imports—commodities: food, gasoline, kerosene, machinery

Reserves of foreign exchange and gold: $544.4 million (31 December 2017 est.)
$437.8 million (31 December 2015 est.)
note: excludes assets of approximately $9.7 billion in the Petroleum Fund (31 December 2010)
country comparison to the world: 150

Debt—external: $311.5 million (31 December 2014 est.)
$687 million (31 December 2013 est.)
country comparison to the world: 183

Exchange rates: the US dollar is used

ENERGY

Electricity access: *electrification—total population:* 63.4% (2016)

electrification—urban areas: 91.7% (2016)

electrification—rural areas: 49.2% (2016)

Electricity—production: 0 kWh NA (2016 est.)
country comparison to the world: 219

Electricity—consumption: 0 kWh (2016 est.)
country comparison to the world: 218

Electricity—exports: 0 kWh (2017 est.)
country comparison to the world: 207

Electricity—imports: 0 kWh (2016 est.)
country comparison to the world: 209

Electricity—installed generating capacity: 600 kW NA (2016 est.)
country comparison to the world: 215

Electricity—from fossil fuels: 0% of total installed capacity (2016 est.)
country comparison to the world: 215

Electricity—from nuclear fuels: 0% of total installed capacity (2017 est.)
country comparison to the world: 194

Electricity—from hydroelectric plants: 0% of total installed capacity (2017 est.)
country comparison to the world: 204

Electricity—from other renewable sources: 100% of total installed capacity (2017 est.)
country comparison to the world: 1

Crude oil—production: 33,000 bbl/day (2018 est.)
country comparison to the world: 60

Crude oil—exports: 62,060 bbl/day (2015 est.)
country comparison to the world: 39

Crude oil—imports: 0 bbl/day (2015 est.)
country comparison to the world: 203

Crude oil—proved reserves: 0 bbl (1 January 2018 est.)
country comparison to the world: 203

Refined petroleum products—production: 0 bbl/day (2015 est.)
country comparison to the world: 207

Refined petroleum products—consumption: 3,500 bbl/day (2016 est.)
country comparison to the world: 186

Refined petroleum products— exports: 0 bbl/day (2015 est.)
country comparison to the world: 208

Refined petroleum products—imports: 3,481 bbl/day (2015 est.)
country comparison to the world: 182

Natural gas—production: 5.776 billion cu m (2017 est.)
country comparison to the world: 48

Natural gas—consumption: 0 cu m (2017 est.)
country comparison to the world: 205

Natural gas—exports: 5.776 billion cu m (2017 est.)
country comparison to the world: 27

Natural gas—imports: 0 cu m (2017 est.)
country comparison to the world: 199

Natural gas—proved reserves: 200 billion cu m (1 January 2006 est.)
country comparison to the world: 42

Carbon dioxide emissions from consumption of energy: 533,400 Mt (2017 est.)
country comparison to the world: 184

COMMUNICATIONS

Telephones—fixed lines: *total subscriptions:* 2,164

subscriptions per 100 inhabitants: less than 1 (2019 est.)
country comparison to the world: 217

Telephones—mobile cellular: total subscriptions: 1,490,966

subscriptions per 100 inhabitants: 110.22 (2019 est.)
country comparison to the world: 158

Telecommunication systems: *general assessment:* service in urban and some rural areas, which is expanding with the entrance of new competitors; 4G LTE service, with about 97% of population having access, among 3 mobile operators; increase in mobile broadband penetration; govt. aims to boost e-govt. services with new national terrestrial optical fiber network; the launch in 2019 of the Kacific-1 satellite is important to the telecom sector for the entire region (2020)

domestic: system suffered significant damage during the violence associated with independence; limited fixed-line services, less than 1 per 100 and mobile-cellular services have been expanding and are now available in urban and most rural areas with teledensity of 110 per 100 (2019)

international: country code - 670; international service is available; partnership with Australia telecom companies for potential deployment of a submarine fiber-optic link (NWCS); geostationary earth orbit satellite
note: the COVID-19 outbreak is negatively impacting telecommunications production and supply chains globally; consumer spending on telecom devices and services has also slowed due to the pandemic's effect on economies worldwide; overall progress towards improvements in all facets of the telecom industry - mobile, fixed-line, broadband, submarine cable and satellite - has moderated

Broadcast media: 7 TV stations (3 nationwide satellite coverage; 2 terrestrial coverage, mostly in Dili; 2 cable) and 21 radio stations (3 nationwide coverage) (2019)

Internet country code: .tl

Internet users: *total:* 363,398

percent of population: 27.49% (July 2018 est.)
country comparison to the world: 162

Broadband—fixed subscriptions: *total:* 603

subscriptions per 100 inhabitants: less than 1 (2018 est.)
country comparison to the world: 200

TRANSPORTATION

National air transport system: *number of registered air carriers:* 2 (2020)

inventory of registered aircraft operated by air carriers: 2

Civil aircraft registration country code prefix: 4W (2016)

Airports: 6 (2013)
country comparison to the world: 177

Airports—with paved runways: *total:* 2 (2013)
2,438 to 3,047 m: 1 (2013)
1,524 to 2,437 m: 1 (2013)

Airports—with unpaved runways: *total:* 4 (2013)
914 to 1,523 m: 2 (2013)
under 914 m: 2 (2013)

Heliports: 8 (2013)

Roadways: *total:* 6,040 km (2008)

paved: 2,600 km (2008)

unpaved: 3,440 km (2008)
country comparison to the world: 143

Merchant marine: *total:* 1

by type: other 1 (2019)
country comparison to the world: 180

Ports and terminals: *major seaport(s):* Dili

MILITARY AND SECURITY

Military and security forces: *Timor-Leste Defense Force (Falintil-Forcas de Defesa de Timor-L'este, Falintil (F-FDTL)):* Headquarters with Land and Naval components (2019)

Military expenditures: 1% of GDP (2019)
0.7% of GDP (2018)
0.9% of GDP (2017)
1% of GDP (2016)
1.2% of GDP (2015)
country comparison to the world: 120

Military and security service personnel strengths: the Timor-Leste Defense Force (F-FDLT) is comprised of approximately 2,000 troops (2019 est.)

Military equipment inventories and acquisitions: Timor-Leste Defense Force's limited inventory consists of equipment donated by other countries; the only known deliveries of major arms to Timor-Leste since 2010 are naval patrol craft from China and South Korea (2019 est.)

Military service age and obligation: 18 years of age for voluntary military service; 18-month service obligation (2019)

TRANSPORTATION

Disputes—international: three stretches of land borders with Indonesia have yet to be delimited, two of which are in the Oecussi exclave area, and no maritime or Economic Exclusion Zone boundaries have been established between the countries; maritime boundaries with Indonesia remain unresolved; Timor-Leste and Australia reached agreement on a treaty delimiting a permanent maritime boundary in March 2018; the treaty will enter into force once ratified by the two countries' parliaments

Trafficking in persons: *current situation:* Timor-Leste is a source and destination country for men, women, and children subjected to forced labor and sex trafficking; Timorese women and girls from rural areas are lured to the capital with promises of legitimate jobs or education prospects and are then forced into prostitution or domestic servitude, and other women and girls may be sent to Indonesia for domestic servitude; Timorese family members force children into bonded domestic or agricultural labor to repay debts; foreign migrant women are vulnerable to sex trafficking in Timor-Leste, while men and boys from Burma, Cambodia, and Thailand are forced to work on fishing boats in Timorese waters under inhumane conditions

tier rating: Tier 2 Watch List – Timor-Leste does not fully comply with the minimum standards for the elimination of trafficking; however, it is making significant efforts to do so; in 2014, legislation was drafted but not finalized or implemented that outlines procedures for screening potential trafficking victims; law enforcement made modest progress, including one conviction for sex trafficking, but efforts are hindered by prosecutors' and judges' lack of expertise in applying anti-trafficking laws effectively; the government rescued two child victims with support from an NGO but did not provide protective services (2015)

Illicit drugs: NA

TOGO

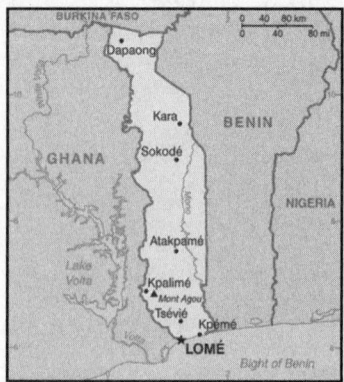

INTRODUCTION

Background: From the 11th to the 16th centuries, various ethnic groups settled the Togo region. From the 16th to the 18th centuries, the coastal region became a major slave trading center and the surrounding region took on the name of "The Slave Coast." In 1884, Germany declared a region including present-day Togo as a protectorate called Togoland. After World War I, rule over Togo was transferred to France. French Togoland became Togo upon independence in 1960. Gen. Gnassingbe EYADEMA, installed as military ruler in 1967, ruled Togo with a heavy hand for almost four decades. Despite the facade of multiparty elections instituted in the early 1990s, the government was largely dominated by President EYADEMA, whose Rally of the Togolese People (RPT) party has been in power almost continually since 1967 and its successor, the Union for the Republic, maintains a majority of seats in today's legislature. Upon EYADEMA's death in February 2005, the military installed the president's son, Faure GNASSINGBE, and then engineered his formal election two months later. Democratic gains since then allowed Togo to hold its first relatively free and fair legislative elections in October 2007. Since 2007, President GNASSINGBE has started the country along a gradual path to democratic reform. Togo has since held multiple presidential and legislative elections deemed generally free and fair by international observers. Despite those positive moves, political reconciliation has moved slowly, and the country experiences periodic outbursts of violent protest by frustrated citizens. Recent constitutional changes to institute a runoff system in presidential elections and establish term limits has done little to reduce the resentment many Togolese feel after over 50 years of one-family rule.

GEOGRAPHY

Location: Western Africa, bordering the Bight of Benin, between Benin and Ghana

Geographic coordinates: 8 00 N, 1 10 E

Map references: Africa

Area: *total:* 56,785 sq km
land: 54,385 sq km
water: 2,400 sq km
country comparison to the world: 127

Area—comparative: slightly smaller than West Virginia

Land boundaries: *total:* 1,880 km
border countries (3): Benin 651 km, Burkina Faso 131 km, Ghana 1098 km

Coastline: 56 km

Maritime claims: *territorial sea:* 30 nm
exclusive economic zone: 200 nm

Climate: tropical; hot, humid in south; semiarid in north

Terrain: gently rolling savanna in north; central hills; southern plateau; low coastal plain with extensive lagoons and marshes

Elevation: *mean elevation:* 236 m
lowest point: Atlantic Ocean 0 m
highest point: Mont Agou 986 m

Natural resources: phosphates, limestone, marble, arable land

Land use: *agricultural land:* 67.4% (2011 est.)
arable land: 45.2% (2011 est.) / permanent crops: 3.8% (2011 est.) / permanent pasture: 18.4% (2011 est.)
forest: 4.9% (2011 est.)
other: 27.7% (2011 est.)

Irrigated land: 70 sq km (2012)

Population distribution: one of the more densely populated African nations with most of the population residing in rural communities, density is highest in the south on or near the Atlantic coast as shown in this population distribution map

Natural hazards: hot, dry harmattan wind can reduce visibility in north during winter; periodic droughts

Environment—current issues: deforestation attributable to slash-and-burn agriculture and the use of wood for fuel; very little rain forest still present and what remains is highly degraded; desertification; water pollution presents health hazards and hinders the fishing industry; air pollution increasing in urban areas

Environment—international agreements:

party to: Biodiversity, Climate Change, Climate Change-Kyoto Protocol, Desertification, Endangered Species, Hazardous Wastes, Law of the Sea, Ozone Layer Protection, Ship Pollution, Tropical Timber 83, Tropical Timber 94, Wetlands, Whaling
signed, but not ratified: none of the selected agreements

Geography—note: the country's length allows it to stretch through six distinct geographic regions; climate varies from tropical to savanna

PEOPLE AND SOCIETY

Population: 8,608,444 (July 2020 est.)
note: estimates for this country explicitly take into account the effects of excess mortality due to AIDS; this can result in lower life expectancy, higher infant mortality, higher death rates, lower population growth rates, and changes in the distribution of population by age and sex than would otherwise be expected
country comparison to the world: 99

Nationality: *noun:* Togolese (singular and plural)
adjective: Togolese

Ethnic groups: Adja-Ewe/Mina 42.4%, Kabye/Tem 25.9%, Para-Gourma/Akan 17.1%, Akposso/Akebu 4.1%, Ana-Ife 3.2%, other Togolese 1.7%, foreigners 5.2%, no response .4% (2013-14 est.)
note: Togo has an estimated 37 ethnic groups

Languages: French (official, the language of commerce), Ewe and Mina (the two major African languages in the south), Kabye (sometimes spelled Kabiye) and Dagomba (the two major African languages in the north)

Religions: Christian 43.7%, folk 35.6%, Muslim 14%, Hindu <.1%, Buddhist <.1%, Jewish <.1%, other .5%, none 6.2% (2010 est.)

Demographic profile: Togo's population is estimated to have grown to four times its size between 1960 and 2010. With nearly 60% of its populace under the age of 25 and a high annual growth rate attributed largely to high fertility, Togo's population is likely to continue to expand for the foreseeable future. Reducing fertility, boosting job creation, and improving education will be essential to reducing the country's high poverty rate. In 2008, Togo eliminated primary school enrollment fees, leading to higher enrollment but increased pressure on limited classroom space, teachers, and materials. Togo has a good chance of achieving universal primary education, but educational quality, the underrepresentation of girls, and the low rate of enrollment in secondary and tertiary schools remain concerns.

Togo is both a country of emigration and asylum. In the early 1990s, southern Togo suffered from the economic decline of the phosphate sector and ethnic and political repression at the hands of dictator Gnassingbe EYADEMA and his northern, Kabye-dominated administration. The turmoil led 300,000 to 350,000 predominantly southern Togolese to flee to Benin and Ghana, with most not returning home until relative stability was restored in 1997. In 2005, another outflow of 40,000 Togolese to Benin and Ghana occurred when violence broke out between the opposition and security forces over the disputed election of EYADEMA's son Faure GNASSINGBE to the presidency. About half of the refugees reluctantly returned home in 2006, many still fearing for their safety. Despite ethnic tensions and periods of political unrest, Togo in September 2017 was home to more than 9,600 refugees from Ghana.

Age structure: *0-14 years:* 39.73% (male 1,716,667/female 1,703,230)
15-24 years: 19.03% (male 817,093/female 820,971)
25-54 years: 33.26% (male 1,423,554/female 1,439,380)
55-64 years: 4.42% (male 179,779/female 200,392)
65 years and over: 3.57% (male 132,304/female 175,074) (2020 est.)

Dependency ratios: *total dependency ratio:* 77.1
youth dependency ratio: 72
elderly dependency ratio: 5.1
potential support ratio: 19.4 (2020 est.)

Median age:
total: 20 years
male: 19.7 years
female: 20.3 years (2020 est.)
country comparison to the world: 197

Population growth rate: 2.56% (2020 est.)
country comparison to the world: 21

Birth rate: 32 births/1,000 population (2020 est.)
country comparison to the world: 28

Death rate: 6.5 deaths/1,000 population (2020 est.)
country comparison to the world: 145

Net migration rate: 0 migrant(s)/1,000 population (2020 est.)
country comparison to the world: 97

Population distribution: one of the more densely populated African nations with most of the population residing in rural communities, density is highest in the south on or near the Atlantic coast as shown in this population distribution map

Urbanization: *urban population:* 42.8% of total population (2020)
rate of urbanization: 3.76% annual rate of change (2015-20 est.)
total population growth rate v. urban population growth rate, 2000-2030:

Major urban areas—population: 1.828 million LOME (capital) (2020)

Sex ratio: *at birth:* 1.03 male(s)/female
0-14 years: 1.01 male(s)/female
15-24 years: 1 male(s)/female
25-54 years: 0.99 male(s)/female
55-64 years: 0.9 male(s)/female
65 years and over: 0.76 male(s)/female
total population: 0.98 male(s)/female (2020 est.)

Mother's mean age at first birth: 21 years (2013/14 est.)

note: median age at first birth among women 25-29

Maternal mortality rate: 396 deaths/100,000 live births (2017 est.)
country comparison to the world: 28

Infant mortality rate:
total: 38.5 deaths/1,000 live births
male: 44.5 deaths/1,000 live births
female: 32.3 deaths/1,000 live births (2020 est.)
country comparison to the world: 41

Life expectancy at birth: *total population:* 66.6 years
male: 63.9 years
female: 69.3 years (2020 est.)
country comparison to the world: 185

Total fertility rate: 4.22 children born/woman (2020 est.)
country comparison to the world: 26

Contraceptive prevalence rate: 23.9% (2017)

Drinking water source:
improved:
urban: 92.3% of population
rural: 56% of population
total: 70.9% of population
unimproved:
urban: 7.7% of population
rural: 44% of population
total: 29.1% of population (2017 est.)

Current Health Expenditure: 6.2% (2017)

Physicians density: 0.03 physicians/1,000 population (2017)

Hospital bed density: 0.7 beds/1,000 population (2011)

Sanitation facility access:
improved:
urban: 80.4% of population (2015 est.)
rural: 16.2% of population
total: 41.6% of population
unimproved:
urban: 19.6% of population
rural: 83.8% of population
total: 57.4% of population (2017 est.)

HIV/AIDS—adult prevalence rate: 2.3% (2019 est.)
country comparison to the world: 22

HIV/AIDS—people living with HIV/AIDS: 120,000 (2019 est.)
country comparison to the world: 41

HIV/AIDS—deaths: 3,000 (2019 est.)
country comparison to the world: 36

Major infectious diseases: *degree of risk:* very high (2020)
food or waterborne diseases: bacterial and proto- zoal diarrhea, hepatitis A, and typhoid fever
vectorborne diseases: malaria, dengue fever, and yellow fever
water contact diseases: schistosomiasis
animal contact diseases: rabies
respiratory diseases: meningococcal meningitis

Obesity—adult prevalence rate: 8.4% (2016)
country comparison to the world: 152

Children under the age of 5 years underweight: 15.2% (2017)
country comparison to the world: 39

Education expenditures: 5% of GDP (2016)
country comparison to the world: 63

Literacy: *definition:* age 15 and over can read and write
total population: 63.7%
male: 77.3%
female: 51.2% (2015)

School life expectancy (primary to tertiary education):
total: 13 years
male: 14 years NA
female: 12 years NA (2017)

Unemployment, youth ages 15-24:
total: 3.9%
male: 3.7%
female: 4.1% (2015 est.)
country comparison to the world: 168

GOVERNMENT

Country name: *conventional long form:* Togolese Republic
conventional short form: Togo
local long form: Republique Togolaise
local short form: none
former: French Togoland
etymology: derived from the Ewe words "to" (river) and "godo" (on the other side) to give the sense of "on the other side of the river"; originally, this designation applied to the town of Togodo (now Togoville) on the northern shore of Lake Togo, but the name was eventually extended to the entire nation

Government type: presidential republic

Capital: *name:* Lome
geographic coordinates: 6 07 N, 1 13 E
time difference: UTC 0 (5 hours ahead of Washington, DC, during Standard Time)
etymology: Lome comes from "alotime" which in the native Ewe language means "among the alo plants"; alo trees dominated the city's original founding site

Administrative divisions: 5 regions (regions, singu- lar—region); Centrale, Kara, Maritime, Plateaux, Savanes

Independence: 27 April 1960 (from French- administered UN trusteeship)

National holiday: Independence Day, 27 April (1960)

Constitution: *history:* several previous; latest adopted 27 September 1992, effective 14 October 1992
amendments: proposed by the president of the republic or supported by at least one fifth of the National Assembly membership; passage requires four-fifths majority vote by the Assembly; a refer- endum is required if approved by only two- thirds majority of the Assembly or if requested by the president; constitutional articles on the repub- lican and secular form of government cannot be amended; amended 2002, 2007, 2019 when the National Assembly unanimously approved a pack- age of amendments, including setting presidential term limits of two 5-year mandates

Legal system: customary law system

International law organization participation: accepts compulsory ICJ jurisdiction with reserva- tions; non-party state to the ICCt

Citizenship: *citizenship by birth:* no
citizenship by descent only: at least one parent must be a citizen of Togo
dual citizenship recognized: yes
residency requirement for naturalization: 5 years

Suffrage: 18 years of age; universal

Executive branch: *chief of state:* President Faure GNASSINGBE (since 4 May 2005)
head of government: Prime Minister Victoire Tomegah DOGBE (since 28 September 2020)
cabinet: Council of Ministers appointed by the president on the advice of the prime minister
elections/appointments: president directly elected by simple majority popular vote for a 5-year term (no term limits); election last held on 22 February 2020 (next to be held February 2025); prime minister appointed by the president

election results: Faure GNASSINGBE reelected president; percent of vote—Faure GNASSINGBE (UNIR) 72.4%, Agbeyome KODJO (MPDD) 18.4%, Jean-Pierre FABRE (ANC) 4.4%, other 5%

Legislative branch: *description:* unicameral National Assembly or Assemblee Nationale (91 seats; members directly elected in multi-seat constituencies by closed, party-list proportional representation vote to serve 5-year terms)

elections: last held on 20 December 2018 (next to be held in 2023)
election results: percent of vote by coalition/ party—NA; seats by party—UNIR 59, UFC 6, NET 3, MPDD 3, other 2, independent 18; com- position—men 75, women 16, percent of women 17.6%

Judicial branch: highest courts: Supreme Court or Cour Supreme (organized into criminal and administrative chambers, each with a chamber president and advisors); Constitutional Court (consists of 9 judges, including the court president)
judge selection and term of office: Supreme Court president appointed by decree of the pres- ident of the republic upon the proposal of the Supreme Council of the Magistracy, a 9-member judicial, advisory, and disciplinary body; other judicial appointments and judge tenure NA; Constitutional Court judges appointed by the National Assembly; judge tenure NA
subordinate courts: Court of Assizes (sessions court); Appeal Court; tribunals of first instance (divided into civil, commercial, and correctional chambers; Court of State Security; military tribunal

Political parties and leaders:
Action Committee for Renewal or CAR [Yaovi AGBOYIBO]
Alliance of Democrats for Integral Development or ADDI [Tchaboure GOGUE]
Democratic Convention of African Peoples or CDPA [Brigitte ADJAMAGBO-JOHNSON]
Democratic Forces for the Republic or FDR [Dodji APEVON]
National Alliance for Change or ANC [Jean- Pierre FABRE]
New Togolese Commitment [Gerry TAAMA]
Pan-African National Party or PNP [Tikpi ATCHADAM]
Pan-African Patriotic Convergence or CPP [Edem KODJO]
Patriotic Movement for Democracy and Development or MPDD [Agbeyome KODJO]
Socialist Pact for Renewal or PSR [Abi TCHESSA]
The Togolese Party [Nathaniel OLYMPIO]
Union of Forces for Change or UFC [Gilchrist OLYMPIO]
Union for the Republic or UNIR [Faure GNASSINGBE]

International organization participation: ACP, AfDB, AU, ECOWAS, EITI (compliant coun- try), Entente, FAO, FZ, G-77, IAEA, IBRD, ICAO, ICRM, IDA, IDB, IFAD, IFC, IFRCS, ILO, IMF, IMO, Interpol, IOC, IOM, IPU, ISO (correspondent), ITSO, ITU, ITUC (NGOs), MIGA, MINURSO, MINUSMA, NAM, OIC, OIF, OPCW, PCA, UN, UNAMID, UNCTAD, UNESCO, UNHCR, UNIDO, UNMIL, UNOCI, UNWTO, UPU, WADB (regional), WAEMU, WCO, WFTU (NGOs), WHO, WIPO, WMO, WTO

Diplomatic representation in the US: *chief of mission:* Ambassador Frederic Edem HEGBE (since 24 April 2017)

chancery: 2208 Massachusetts Avenue NW, Washington, DC 20008
telephone: [1] (202) 234-4212
FAX: [1] (202) 232-3190

Diplomatic representation from the US: *chief of mission:* Ambassador Eric W. STROHMAYER (since 11 April 2019)
telephone: [228] 2261-5470
embassy: 4332 Blvd. Eyadema, Lome
mailing address: B.P. 852, Lome; 2300 Lome Place, Washington, DC 20521-2300
FAX: [228] 2261-5501

Flag description: five equal horizontal bands of green (top and bottom) alternating with yellow; a white five-pointed star on a red square is in the upper hoist-side corner; the five horizontal stripes stand for the five different regions of the country; the red square is meant to express the loyalty and patriotism of the people, green symbolizes hope, fertility, and agriculture, while yellow represents mineral wealth and faith that hard work and strength will bring prosperity; the star symbolizes life, purity, peace, dignity, and Togo's independence
note: uses the popular Pan-African colors of Ethiopia

National symbol(s): *lion; national colors:* green, yellow, red, white

National anthem: *name:* "Salut a toi, pays de nos aieux" (Hail to Thee, Land of Our Forefathers)
lyrics/music: Alex CASIMIR-DOSSEH
note: adopted 1960, restored 1992; this anthem was replaced by another during one-party rule between 1979 and 1992
0:00/ 0:51

ECONOMY

Economy—overview: Togo has enjoyed a period of steady economic growth fueled by political stability and a concerted effort by the government to modernize the country's commercial infrastructure, but discontent with President Faure GNASSINGBE has led to a rapid rise in protests, creating downside risks. The country completed an ambitious large-scale infrastructure improvement program, including new principal roads, a new airport terminal, and a new seaport. The economy depends heavily on both commercial and subsistence agriculture, providing employment for around 60% of the labor force. Some basic foodstuffs must still be imported. Cocoa, coffee, and cotton and other agricultural products generate about 20% of export earnings with cotton being the most important cash crop. Togo is among the world's largest producers of phosphate and seeks to develop its carbonate phosphate reserves, which provide more than 20% of export earnings.

Supported by the World Bank and the IMF, the government's decade-long effort to implement economic reform measures, encourage foreign investment, and bring revenues in line with expenditures has moved slowly. Togo completed its IMF Extended Credit Facility in 2011 and reached a Heavily Indebted Poor Country debt relief completion point in 2010 at which 95% of the country's debt was forgiven. Togo continues to work with the IMF on structural reforms, and in January 2017, the IMF signed an Extended Credit Facility arrangement consisting of a three-year $238 million loan package. Progress depends on follow through on privatization, increased transparency in government financial operations, progress toward legislative elections, and continued support from foreign donors.

Togo's 2017 economic growth probably remained steady at 5.0%, largely driven by infusions of foreign aid, infrastructure investment in its port and mineral industry, and improvements in the business climate. Foreign direct investment inflows have slowed in recent years.

GDP (purchasing power parity):
$12.97 billion (2017 est.)
$12.42 billion (2016 est.)
$11.82 billion (2015 est.)
note: data are in 2017 dollars
country comparison to the world: 155

GDP (official exchange rate):
$4.767 billion (2017 est.)

GDP—real growth rate:
4.4% (2017 est.)
5.1% (2016 est.)
5.7% (2015 est.)
country comparison to the world: 63

GDP—per capita (PPP):
$1,700 (2017 est.)
$1,600 (2016 est.)
$1,600 (2015 est.)
note: data are in 2017 dollars
country comparison to the world: 215

Gross national saving:
16.1% of GDP (2017 est.)
21.8% of GDP (2016 est.)
21.2% of GDP (2015 est.)
country comparison to the world: 129

GDP—composition, by end use:
household consumption: 84.5% (2017 est.)
government consumption: 11.4% (2017 est.)
investment in fixed capital: 23.4% (2017 est.)
investment in inventories: -1.4% (2017 est.)
exports of goods and services: 43.1% (2017 est.)
imports of goods and services: -61% (2017 est.)

GDP—composition, by sector of origin:
agriculture: 28.8% (2017 est.)
industry: 21.8% (2017 est.)
services: 49.8% (2017 est.)

Agriculture—products: coffee, cocoa, cotton, yams, cassava (manioc, tapioca), corn, beans, rice, millet, sorghum; livestock; fish

Industries: phosphate mining, agricultural processing, cement, handicrafts, textiles, beverages

Industrial production growth rate: 5% (2017 est.)
country comparison to the world: 58

Labor force: 2.595 million (2007 est.)
country comparison to the world: 112

Labor force—by occupation:
agriculture: 65%

industry: 5%
services: 30% (1998 est.)

Unemployment rate: 6.9% (2016 est.)
country comparison to the world: 110

Population below poverty line: 55.1% (2015 est.)

Household income or consumption by percentage share: *lowest 10%:* 3.3%
highest 10%: 27.1% (2006)

Budget: *revenues:* 1.023 billion (2017 est.)
expenditures: 1.203 billion (2017 est.)

Taxes and other revenues: 21.5% (of GDP) (2017 est.)
country comparison to the world: 137

Budget surplus (+) or deficit (-): -3.8% (of GDP) (2017 est.)
country comparison to the world: 154

Public debt:
75.7% of GDP (2017 est.)
81.6% of GDP (2016 est.)
country comparison to the world: 40

Fiscal year: calendar year

Inflation rate (consumer prices):
-0.7% (2017 est.)
0.9% (2016 est.)
country comparison to the world: 4

Current account balance:
-$383 million (2017 est.)
-$416 million (2016 est.)
country comparison to the world: 114

Exports:
$1.046 billion (2017 est.)
$967.4 million (2016 est.)
country comparison to the world: 158

Exports—partners: Benin 16.7%, Burkina Faso 15.2%, Niger 8.9%, India 7.3%, Mali 6.7%, Ghana 5.5%, Cote dIvoire 5.4%, Nigeria 4.1% (2017)

Exports—commodities: reexports, cotton, phosphates, coffee, cocoa

Imports:
$1.999 billion (2017 est.)
$2 billion (2016 est.)
country comparison to the world: 167

Imports—commodities: machinery and equipment, foodstuffs, petroleum products

Imports—partners: China 27.5%, France 9.1%, Netherlands 4.4%, Japan 4.3% (2017)

Reserves of foreign exchange and gold:
$77.8 million (31 December 2017 est.)
$42.6 million (31 December 2016 est.)
country comparison to the world: 182

Debt—external:
$1.442 billion (31 December 2017 est.)
$1.22 billion (31 December 2016 est.)
country comparison to the world: 160

Exchange rates: Communaute Financiere Africaine francs (XOF) per US dollar -
617.4 (2017 est.)
593.01 (2016 est.)
593.01 (2015 est.)
591.45 (2014 est.)
494.42 (2013 est.)

ENERGY

Electricity access:
population without electricity: 5 million (2019)
electrification—total population: 43% (2019)
electrification—urban areas: 77% (2019)
electrification—rural areas: 19% (2019)

Electricity—production: 232.6 million kWh (2016 est.)
country comparison to the world: 189

Electricity—consumption: 1.261 billion kWh (2016 est.)
country comparison to the world: 151

Electricity—exports: 0 kWh (2016 est.)
country comparison to the world: 208

Electricity—imports: 1.14 billion kWh (2016 est.)
country comparison to the world: 67

Electricity—installed generating capacity: 230,000 kW (2016 est.)
country comparison to the world: 164

Electricity—from fossil fuels: 70% of total installed capacity (2016 est.)
country comparison to the world: 110

Electricity—from nuclear fuels: 0% of total installed capacity (2017 est.)
country comparison to the world: 195

Electricity—from hydroelectric plants: 29% of total installed capacity (2017 est.)
country comparison to the world: 71

Electricity—from other renewable sources: 1% of total installed capacity (2017 est.)
country comparison to the world: 169

Crude oil—production: 0 bbl/day (2018 est.)
country comparison to the world: 207

Crude oil—exports: 0 bbl/day (2015 est.)
country comparison to the world: 206

Crude oil—imports: 0 bbl/day (2015 est.)
country comparison to the world: 204

Crude oil—proved reserves: 0 bbl (1 January 2018 est.)
country comparison to the world: 204

Refined petroleum products—production: 0 bbl/day (2015 est.)
country comparison to the world: 208

Refined petroleum products—consumption: 15,000 bbl/day (2016 est.)
country comparison to the world: 152

Refined petroleum products—exports: 0 bbl/day (2015 est.)
country comparison to the world: 209

Refined petroleum products—imports: 13,100 bbl/day (2015 est.)
country comparison to the world: 142

Natural gas—production: 0 cu m (2017 est.)
country comparison to the world: 206

Natural gas—consumption: 0 cu m (2017 est.)
country comparison to the world: 206

Natural gas—exports: 0 cu m (2017 est.)
country comparison to the world: 200

Natural gas—imports: 0 cu m (2017 est.)
country comparison to the world: 200

Natural gas—proved reserves: 0 cu m (1 January 2014 est.)
country comparison to the world: 200

Carbon dioxide emissions from consumption of energy: 2.651 million Mt (2017 est.)
country comparison to the world: 151

COMMUNICATIONS

Telephones—fixed lines: *total subscriptions:* 45,311
subscriptions per 100 inhabitants: less than 1 (2019 est.)
country comparison to the world: 159

Telephones—mobile cellular: *total subscriptions:* 6,477,816
subscriptions per 100 inhabitants: 77.2 (2019 est.)
country comparison to the world: 109

Telecommunication systems: *general assessment:* system based on a network of microwave radio relay routes supplemented by open-wire lines and a mobile-cellular system; telecoms supply 8% of GDP; 3 mobile operators; 12% of residents have access to the Internet; mobile subscribers and mobile broadband both increasing (2020)
domestic: fixed-line less than 1 per 100 and mobile-cellular 77 telephones per 100 persons with mobile-cellular use predominating (2019)
international: country code—228; landing point for the WACS submarine cable, linking countries along the west coast of Africa with each other and with Portugal; satellite earth stations—1 Intelsat (Atlantic Ocean), 1 Symphonie (2020)
note: the COVID-19 outbreak is negatively impacting telecommunications production and supply chains globally; consumer spending on telecom devices and services has also slowed due to the pandemic's effect on economies worldwide; overall progress towards improvements in all facets of the telecom industry—mobile, fixed-line, broadband, submarine cable and satellite—has moderated

Broadcast media: 1 state-owned TV station with multiple transmission sites; five private TV stations broadcast locally; cable TV service is available; state-owned radio network with two stations (in Lome and Kara); several dozen private radio stations and a few community radio stations; transmissions of multiple international broadcasters available (2019)

Internet country code: .tg

Internet users: *total:* 1,010,609
percent of population: 12.36% (July 2018 est.)
country comparison to the world: 141

Broadband—fixed subscriptions: *total:* 26,156
subscriptions per 100 inhabitants: less than 1 (2018 est.)
country comparison to the world: 146

TRANSPORTATION

National air transport system: *number of registered air carriers:* 1 (2020)

inventory of registered aircraft operated by air carriers: 8
annual passenger traffic on registered air carriers: 566,295 (2018)
annual freight traffic on registered air carriers: 10.89 million mt-km (2018)

Civil aircraft registration country code prefix: 5V (2016)

Airports: 8 (2013)
country comparison to the world: 163

Airports—with paved runways: *total:* 2 (2019)
2,438 to 3,047 m: 2

Airports—with unpaved runways: *total:* 6 (2013)
914 to 1,523 m: 4 (2013)
under 914 m: 2 (2013)

Pipelines: 62 km gas

Railways: *total:* 568 km (2014)
narrow gauge: 568 km 1.000-m gauge (2014)
country comparison to the world: 111

Roadways:
total: 11,734 km (2081)
paved: 1,794 km (2018)
unpaved: 8,157 km (2018)
urban: 1,783 km (2018)
country comparison to the world: 133

Waterways: 50 km (seasonally navigable by small craft on the Mono River depending on rainfall) (2011)
country comparison to the world: 102

Merchant marine: *total:* 405
by type: bulk carrier 5, container ship 5, general cargo 266, oil tanker 50, other 79 (2019)
country comparison to the world: 44

Ports and terminals: *major seaport(s):* Kpeme, Lome

MILITARY AND SECURITY

Military and security forces: *Togolese Armed Forces (Forces Armees Togolaise, FAT):* Togolese Army (l'Armee de Terre), Togolese Navy (Forces Naval Togolaises), Togolese Air Force (Armee de l'Air), National Gendarmerie (2020)

Military expenditures:
3.1% of GDP (2019)
2% of GDP (2018)
1.9% of GDP (2017)
1.9% of GDP (2016)
1.7% of GDP (2015)
country comparison to the world: 26

Military and security service personnel strengths: the Togolese Armed Forces (FAT) are comprised of approximately 9,100 personnel (8,000 Army; 200 Navy; 200 Navy; 750 Gendarmerie) (2019 est.)

Military equipment inventories and acquisitions: the FAT's small inventory is a mix of older Brazilian, British, French, German, Russian/Soviet, and US equipment; since 2010, France is the leading supplier of military hardware to Togo (2020)

Military deployments: 920 Mali (MINUSMA) (2020)

Military service age and obligation: 18 years of age for military service; 2-year service obligation; currently the military is only an all-volunteer force (2017)

Disputes—international: in 2001, Benin claimed Togo moved boundary monuments—joint commission continues to resurvey the boundary; talks continue between Benin and Togo on funding the Adjrala hydroelectric dam on the Mona River

Refugees and internally displaced persons: *refugees (country of origin):* 9,556 (Ghana) (2020)

Illicit drugs: transit hub for Nigerian heroin and cocaine traffickers; money laundering not a significant problem

TOKELAU

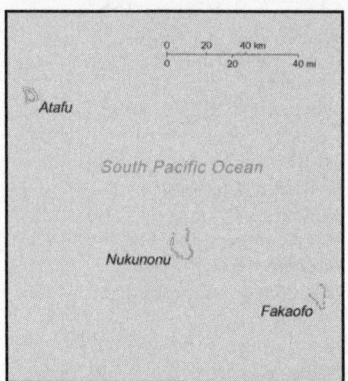

INTRODUCTION

Background: Originally settled by Polynesian emigrants from surrounding island groups, the Tokelau Islands were made a British protectorate in 1889. They were transferred to New Zealand administration in 1925. Referenda held in 2006 and 2007 to change the status of the islands from that of a New Zealand territory to one of free association with New Zealand did not meet the needed threshold for approval.

GEOGRAPHY

Location: Oceania, group of three atolls in the South Pacific Ocean, about one-half of the way from Hawaii to New Zealand

Geographic coordinates: 9 00 S, 172 00 W

Map references: Oceania

Area: *total:* 12 sq km
land: 12 sq km
water: 0 sq km
country comparison to the world: 242

Area—comparative: about 17 times the size of the National Mall in Washington, DC

Land boundaries: 0 km

Coastline: 101 km

Maritime claims: *territorial sea:* 12 nm
exclusive economic zone: 200 nm

Climate: tropical; moderated by trade winds (April to November)

Terrain: low-lying coral atolls enclosing large lagoons

Elevation: *lowest point:* Pacific Ocean 0 m
highest point: unnamed location 5 m

Natural resources: NEGL

Land use: *agricultural land:* 60% (2011 est.)
arable land: 0% (2011 est.) / permanent crops: 60% (2011 est.) / permanent pasture: 0% (2011 est.)
forest: 0% (2011 est.)
other: 40% (2011 est.)

Irrigated land: 0 sq km (2012)

Population distribution: the country's small population is fairly evenly distributed amongst the three atolls

Natural hazards: lies in Pacific cyclone belt

Environment—current issues: overexploitation of certain fish and other marine species, coastal sand, and forest resources; pollution of freshwater lenses and coastal waters from improper disposal of chemicals

Geography—note: consists of three atolls (Atafu, Fakaofo, Nukunonu), each with a lagoon surrounded by a number of reef-bound islets of varying length and rising to over 3 m above sea level

PEOPLE AND SOCIETY

Population: 1,647 (2019 est. est.)
country comparison to the world: 234

Nationality: *noun:* Tokelauan(s)
adjective: Tokelauan

Ethnic groups: Tokelauan 64.5%, part Tokelauan/Samoan 9.7%, part Tokelauan/Tuvaluan 2.8%, Tuvaluan 7.5%, Samoan 5.8%, other Pacific Islander 3.4%, other 5.6%, unspecified 0.8% (2016 est.)

Languages: Tokelauan 88.1% (a Polynesian language), English 48.6%, Samoan 26.7%, Tuvaluan 11.2%, Kiribati 1.5%, other 2.8%, none 2.8%, unspecified 0.8% (2016 ests.)
note: shares sum to more than 100% because some respondents gave more than one answer on the census

Religions: Congregational Christian Church 50.4%, Roman Catholic 38.7%, Presbyterian 5.9%, other Christian 4.2%, unspecified 0.8% (2016 est.)

Population growth rate: -0.01% (2014 est.)
country comparison to the world: 196

Population distribution: the country's small population is fairly evenly distributed amongst the three atolls

Urbanization: *urban population:* 0% of total population (2020)
rate of urbanization: 0% annual rate of change (2015-20 est.)

total population growth rate v. urban population growth rate, 2000-2030: Sex ratio: NA

Infant mortality rate: *total:* NA (2018)
male: NA
female: NA

Life expectancy at birth: *total population:* NA (2017 est.)
male: NA
female: NA

Total fertility rate: NA

Drinking water source:
improved: rural: 100% of population
total: 100% of population
unimproved: rural: 0% of population
total: 0% of population (2017 est.)

Physicians density: 2.72 physicians/1,000 population (2010)

Sanitation facility access:
improved: rural: 100% of population
total: 100% of population
unimproved: rural: 0% of population
total: 0% of population (2017 est.)

HIV/AIDS—adult prevalence rate: NA

HIV/AIDS—people living with HIV/AIDS: NA

HIV/AIDS—deaths: NA

Major infectious diseases: *degree of risk:* high (2020)
food or waterborne diseases: bacterial diarrhea
vectorborne diseases: malaria

Education expenditures: NA

GOVERNMENT

Country name: *conventional long form:* none
conventional short form: Tokelau
former: Union Islands, Tokelau Islands
etymology: "tokelau" is a Polynesian word meaning "north wind"

Dependency status: self-administering territory of New Zealand; note - Tokelau and New Zealand have agreed to a draft constitution as Tokelau moves toward free association with New Zealand; a UN-sponsored referendum on self governance in October 2007 did not meet the two-thirds majority vote necessary for changing the political status

Government type: parliamentary democracy under a constitutional monarchy

Capital: UTC+13 (18 hours ahead of Washington, DC during Standard Time)

Administrative divisions: none (territory of New Zealand)

Independence: none (territory of New Zealand)

National holiday: Waitangi Day (Treaty of Waitangi established British sovereignty over New Zealand), 6 February (1840)

Constitution: *history:* many previous; latest effective 1 January 1949 (Tokelau Islands Act 1948)

amendments: proposed as a resolution by the General Fono; passage requires support by each village and approval by the General Fono; amended many times, last in 2007

Legal system: common law system of New Zealand

Citizenship: see New Zealand

Suffrage: 21 years of age; universal

Executive branch: *chief of state:* Queen ELIZABETH II (since 6 February 1952); represented by Governor General of New Zealand Governor General Dame Patricia Lee REDDY (since 28 September 2016); New Zealand is represented by Administrator Jonathan KINGS (since 30 August 2017)

head of government: Afega GAULOFA (since 10 March 2016); note - position rotates annually among the three Faipule (village leaders)

cabinet: Council for the Ongoing Government of Tokelau (or Tokelau Council) functions as a cabinet; consists of 3 Faipule (village leaders) and 3 Pulenuku (village mayors)

elections/appointments: the monarchy is hereditary; governor general appointed by the monarch; administrator appointed by the Minister of Foreign Affairs and Trade in New Zealand; head of government chosen from the Council of Faipule to serve a 1-year term

note: the meeting place of the Tokelau Council rotates annually among the three atolls; this tradition has given rise to the somewhat misleading description that the capital rotates yearly between the three atolls; in actuality, it is the seat of the government councilors that rotates since Tokelau has no capital

Legislative branch: *description:* unicameral General Fono (20 seats apportioned by island - Atafu 7, Fakaofo 7, Nukunonu 6; members directly elected by simple majority vote to serve 3-year terms); note - the Tokelau Amendment Act of 1996 confers limited legislative power to the General Fono

elections: last held on 23, 27, and 31 January 2017 depending on island (next to be held in 2020)

election results: percent of vote by party - NA; seats by party - independent 20; composition - men 17, women 3, percent of women 15%

Judicial branch: *highest courts:* Court of Appeal (in New Zealand) (consists of the court president and 8 judges sitting in 3- or 5-judge panels, depending on the case)

judge selection and term of office: judges nominated by the Judicial Selection Committee and approved by three- quarters majority of the Parliament; judges serve for life

subordinate courts: High Court (in New Zealand); Council of Elders or Taupulega

Political parties and leaders: none

International organization participation: PIF (associate member), SPC, UNESCO (associate), UPU

Diplomatic representation in the US: none (territory of New Zealand)

Diplomatic representation from the US: none (territory of New Zealand)

Flag description: a yellow stylized Tokelauan canoe on a dark blue field sails toward the manu - the Southern Cross constellation of four, white, five-pointed stars at the hoist side; the Southern Cross represents the role of Christianity in Tokelauan culture and, in conjunction with the canoe, symbolizes the country navigating into the future; the color yellow indicates happiness and peace, and the blue field represents the ocean on which the community relies

National symbol(s): *tuluma (fishing tackle box);* national colors: blue, yellow, white

National anthem: *name:* "Te Atua" (For the Almighty)

lyrics/music: unknown/Falani KALOLO

note: adopted 2008; in preparation for eventual self governance, Tokelau held a national contest to choose an anthem; as a territory of New Zealand, "God Defend New Zealand" and "God Save the Queen" are official (see New Zealand)

ECONOMY

Economy—overview: Tokelau's small size (three villages), isolation, and lack of resources greatly restrain economic development and confine agriculture to the subsistence level. The principal sources of revenue are from sales of copra, postage stamps, souvenir coins, and handicrafts. Money is also remitted to families from relatives in New Zealand.

The people rely heavily on aid from New Zealand - about $15 million annually in FY12/13 and FY13/14 - to maintain public services. New Zealand's support amounts to 80% of Tokelau's recurrent government budget. An international trust fund, currently worth nearly $32 million, was established in 2004 by New Zealand to provide Tokelau an independent source of revenue.

GDP (purchasing power parity): $1.5 million (1993 est.)

country comparison to the world: 229

GDP (official exchange rate): NA

GDP—real growth rate: NA

GDP—per capita (PPP): $1,000 (1993 est.)

country comparison to the world: 225

GDP—composition, by sector of origin: *agriculture:* NA

industry: NA

services: NA

Agriculture—products: coconuts, copra, breadfruit, papayas, bananas; pigs, poultry, goats; fish

Industries: small-scale enterprises for copra production, woodworking, plaited craft goods; stamps, coins; fishing

Labor force: 440 (2001)

country comparison to the world: 231

Unemployment rate: NA

Population below poverty line: NA

Budget: *revenues:* 430,800 (1987 est.)

expenditures: 2.8 million (1987 est.)

Fiscal year: 1 April - 31 March

Inflation rate (consumer prices): NA

Exports: $0 (2002 est.)

country comparison to the world: 224

Exports—commodities: stamps, copra, handicrafts

Imports: $969,200 (2002 est.)

country comparison to the world: 224

Imports—commodities: foodstuffs, building materials, fuel

Exchange rates: New Zealand dollars (NZD) per US dollar—

1.416 (2017 est.)
1.4279 (2016 est.)
1.4279 (2015)
1.4279 (2014 est.)
1.2039 (2013 est.)

ENERGY

Crude oil—proved reserves: 0 bbl (1 January 2010 est.)

country comparison to the world: 205

COMMUNICATIONS

Telephones—fixed lines: *total subscriptions:* 300

subscriptions per 100 inhabitants: 21 (July 2016 est.)

country comparison to the world: 220

Telecommunication systems: *general assessment:* modern satellite-based communications system; demand for mobile broadband increasing due to mobile services being the method of access for Internet across the region; 2G widespread with some 4G LTE service; satellite services has improved with the launch of the Kacific-1 satellite launched in 2019 (2020)

domestic: radiotelephone service between islands; fixed-line 21 per 100 persons (2019)

international: country code - 690; landing point for the Southern Cross NEXT submarine cable linking Australia, Tokelau, Samoa, Kiribati, Fiji, New Zealand and Los Angeles, CA (USA); radiotelephone service to Samoa; government-regulated telephone service (TeleTok); satellite earth stations - 3 (2020)

note: the COVID-19 outbreak is negatively impacting telecommunications production and supply chains globally; consumer spending on telecom devices and services has also slowed due to the pandemic's effect on economies worldwide; overall progress towards improvements in all facets of the

telecom industry - mobile, fixed-line, broadband, submarine cable and satellite - has moderated

Broadcast media: Sky TV access for around 30% of the population; each atoll operates a radio service that provides shipping news and weather reports (2019)

Internet country code: .tk

Internet users: total: 805

percent of population: 60.2% (July 2016 est.)
country comparison to the world: 226

TRANSPORTATION

Roadways: *total:* 10 km (2019)

country comparison to the world: 222

Ports and terminals: none; offshore anchorage only

MILITARY AND SECURITY

Military—note: defense is the responsibility of New Zealand

TRANSPORTATION

Disputes—international: Tokelau included American Samoa's Swains Island (Olosega) in its 2006 draft independence constitution

---

TONGA

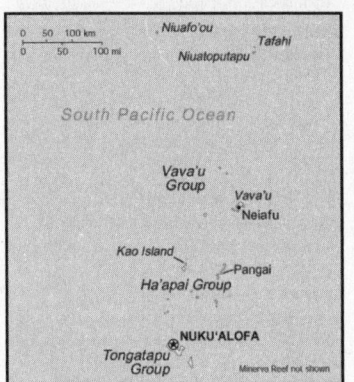

INTRODUCTION

Background: Tonga - unique among Pacific nations - never completely lost its indigenous governance. The archipelagos of "The Friendly Islands" were united into a Polynesian kingdom in 1845. Tonga became a constitutional monarchy in 1875 and a British protectorate in 1900; it withdrew from the protectorate and joined the Commonwealth of Nations in 1970. Tonga remains the only monarchy in the Pacific; in 2008, King George TUPOU V announced he was relinquishing most of his powers leading up to parliamentary elections in 2010. TUPOU died in 2012 and was succeeded by his brother 'Aho'eitu TUPOU VI. Tropical Cyclone Gita, the strongest-ever recorded storm to impact Tonga, hit the islands in February 2018 causing extensive damage.

GEOGRAPHY

Location: Oceania, archipelago in the South Pacific Ocean, about two-thirds of the way from Hawaii to New Zealand

Geographic coordinates: 20 00 S, 175 00 W

Map references: Oceania

Area: *total:* 747 sq km
land: 717 sq km
water: 30 sq km
country comparison to the world: 190

Area—comparative: four times the size of Washington, DC

Area comparison map: [INSERT IMAGE: TONGA-Area comparison map]

Land boundaries: 0 km

Coastline: 419 km

Maritime claims: *territorial sea:* 12 nm
exclusive economic zone: 200 nm
continental shelf: 200-m depth or to the depth of exploitation

Climate: tropical; modified by trade winds; warm season (December to May), cool season (May to December)

Terrain: mostly flat islands with limestone bedrock formed from uplifted coral formation; others have limestone overlying volcanic rock

Elevation: *lowest point:* Pacific Ocean 0 m
highest point: Kao Volcano on Kao Island 1,046 m

Natural resources: arable land, fish

Land use: *agricultural land:* 43.1% (2011 est.)
arable land: 22.2% (2011 est.) / permanent crops: 15.3% (2011 est.) / permanent pasture: 5.6% (2011 est.)
forest: 12.5% (2011 est.)
other: 44.4% (2011 est.)

Irrigated land: 0 sq km (2012)

Population distribution: over two-thirds of the population lives on the island of Tongatapu; only 45 of the nation's 171 islands are occupied

Natural hazards: cyclones (October to April); earthquakes and volcanic activity on Fonuafo'ou

volcanism: moderate volcanic activity; Fonualei (180 m) has shown frequent activity in recent years, while Niuafo'ou (260 m), which last erupted in 1985, has forced evacuations; other historically active volcanoes include Late and Tofua

Environment—current issues: deforestation from land being cleared for agriculture and settlement; soil exhaustion; water pollution due to salinization, sewage, and toxic chemicals from farming activities; coral reefs and marine populations threatened

Environment—international agreements: *party to:* Biodiversity, Climate Change, Climate Change-Kyoto Protocol, Desertification, Hazardous Wastes, Law of the Sea, Marine Dumping, Marine Life Conservation, Ozone Layer Protection, Ship Pollution
signed, but not ratified: none of the selected agreements

Geography—note: the western islands (making up the Tongan Volcanic Arch) are all of volcanic origin; the eastern islands are nonvolcanic and are composed of coral limestone and sand

PEOPLE AND SOCIETY

Population: 106,095 (July 2020 est.)
country comparison to the world: 193

Nationality: *noun:* Tongan(s)
adjective: Tongan

Ethnic groups: Tongan 97%, part-Tongan 0.8%, other 2.2%, unspecified (2016 est.)

Languages: Tongan and English 76.8%, Tongan, English, and other language 10.6%, Tongan only (official) 8.7%, English only (official) 0.7%, other 1.7%, none 2.2% (2016 est.)

note: data represent persons aged 5 and older who can read and write a simple sentence in Tongan, English, or another language

Religions: Protestant 64.1% (includes Free Wesleyan Church 35%, Free Church of Tonga 11.9%, Church of Tonga 6.8%, Assembly of God 2.3%, Seventh Day Adventist 2.2%, Tokaikolo Christian Church 1.6%, other 4.3%), Mormon 18.6%, Roman Catholic 14.2%, other 2.4%, none 0.5%, unspecified 0.1% (2016 est.)

Age structure: 0-14 years: 32% (male 17,250/female 16,698)
15-24 years: 19.66% (male 10,679/female 10,175)
25-54 years: 35.35% (male 18,701/female 18,802)
55-64 years: 6.17% (male 3,345/female 3,202)
65 years and over: 6.83% (male 3,249/female 3,994) (2020 est.)

population pyramid: [INSERT IMAGE: TONGA-population pyramid]

Dependency ratios: *total dependency ratio:* 68.6
youth dependency ratio: 58.6
elderly dependency ratio: 10
potential support ratio: 10 (2020 est.)

Median age: *total:* 24.1 years
male: 23.6 years
female: 24.5 years (2020 est.)
country comparison to the world: 169

Population growth rate: -0.16% (2020 est.)
country comparison to the world: 206

Birth rate: 21 births/1,000 population (2020 est.)
country comparison to the world: 71

Death rate: 4.9 deaths/1,000 population (2020 est.)
country comparison to the world: 203

Net migration rate: -17.9 migrant(s)/1,000 population (2020 est.)
country comparison to the world: 224

Population distribution: over two-thirds of the population lives on the island of Tongatapu; only 45 of the nation's 171 islands are occupied

Urbanization: *urban population:* 23.1% of total population (2020)
rate of urbanization: 0.71% annual rate of change (2015-20 est.)

total population growth rate v. urban population growth rate, 2000-2030: *Major urban areas - population:* 23,000 NUKU'ALOFA (2018)

Sex ratio: *at birth:* 1.03 male(s)/female
0-14 years: 1.03 male(s)/female
15-24 years: 1.05 male(s)/female
25-54 years: 0.99 male(s)/female
55-64 years: 1.04 male(s)/female
65 years and over: 0.81 male(s)/female
total population: 1.01 male(s)/female (2020 est.)

Mother's mean age at first birth: 24.9 years (2012 est.)
note: median age at first birth among women 25-49

Maternal mortality rate: 52 deaths/100,000 live births (2017 est.)
country comparison to the world: 94

Infant mortality rate: *total:* 10.3 deaths/1,000 live births
male: 10.7 deaths/1,000 live births

female: 9.9 deaths/1,000 live births (2020 est.)
country comparison to the world: 129

Life expectancy at birth: *total population:* 77 years
male: 75.4 years
female: 78.8 years (2020 est.)
country comparison to the world: 87

Total fertility rate: 2.87 children born/woman (2020 est.)
country comparison to the world: 57

Contraceptive prevalence rate: 34.1% (2012)

Drinking water source:

improved: *urban:* 100% of population
rural: 100% of population
total: 100% of population

unimproved: *urban:* 0% of population
rural: 0% of population
total: 0% of population (2017 est.)

Current Health Expenditure: 5.3% (2017)

Physicians density: 0.54 physicians/1,000 population (2013)

Hospital bed density: 2.6 beds/1,000 population (2010)

Sanitation facility access:

improved: *urban:* 96.6% of population
rural: 93.6% of population
total: 94.5% of population

unimproved: *urban:* 3.4% of population
rural: 6.4% of population
total: 5.5% of population (2017 est.)

HIV/AIDS—adult prevalence rate: NA

HIV/AIDS—people living with HIV/AIDS: NA

HIV/AIDS—deaths: NA

Major infectious diseases: *degree of risk:* high (2020)
food or waterborne diseases: bacterial diarrhea
vectorborne diseases: malaria

Obesity—adult prevalence rate: 48.2% (2016)
country comparison to the world: 7

Children under the age of 5 years underweight: 1.9% (2012)
country comparison to the world: 114

Education expenditures: NA

Literacy: *definition:* can read and write Tongan and/or English
total population: 99.4%
male: 99.4%
female: 99.5% (2015)

GOVERNMENT

Country name: *conventional long form:* Kingdom of Tonga
conventional short form: Tonga
local long form: Pule'anga Fakatu'i 'o Tonga
local short form: Tonga
former: Friendly Islands
etymology: "tonga" means "south" in the Tongan language and refers to the country's geographic position in relation to central Polynesia

Government type: constitutional monarchy

Capital: *name:* Nuku'alofa

geographic coordinates: 21 08 S, 175 12 W
time difference: UTC+13 (18 hours ahead of Washington, DC, during Standard Time)

daylight saving time: +1hr, begins first Sunday in November; ends second Sunday in January
etymology: composed of the words "nuku," meaning "residence or abode," and "alofa," meaning "love," to signify "abode of love"

Administrative divisions: 5 island divisions; 'Eua, Ha'apai, Ongo Niua, Tongatapu, Vava'u

Independence: 4 June 1970 (from UK protectorate status)

National holiday: Official Birthday of King TUPOU VI, 4 July (1959); note - actual birthday of the monarch is 12 July 1959, 4 July (2015) is the day the king was crowned; Constitution Day (National Day), 4 November (1875)

Constitution: *history:* adopted 4 November 1875

amendments: proposed by the Legislative Assembly; passage requires approval by the Assembly in each of three readings, the unanimous approval of the Privy Council (a high-level advisory body to the monarch), the Cabinet, and assent to by the monarch; revised 1988; amended many times, last in 2016

Legal system: English common law

International law organization participation: has not submitted an ICJ jurisdiction declaration; non-party state to the ICCt

Citizenship: *citizenship by birth:* no
citizenship by descent only: the father must be a citizen of Tonga; if a child is born out of wedlock, the mother must be a citizen of Tonga
dual citizenship recognized: yes
residency requirement for naturalization: 5 years

Suffrage: 21 years of age; universal

Executive branch: *chief of state:* King TUPOU VI (since 18 March 2012); Heir Apparent Crown Prince Siaosi Manumataogo 'Alaivahamama'o 'Ahoeitu Konstantin Tuku'aho, son of the king (born 17 September 1985); note - on 18 March 2012, King George TUPOU V died and his brother, Crown Prince TUPOUTO'A Lavaka, assumed the throne as TUPOU VI

head of government: Prime Minister Pohiva TU'I'ONETOA (since 27 September 2019)
cabinet: Cabinet nominated by the prime minister and appointed by the monarch
elections/appointments: the monarchy is hereditary; prime minister and deputy prime minister indirectly elected by the Legislative Assembly and appointed by the monarch; election last held on 27 September 2019 (next to be held in November 2020)
election results: Pohiva TU'I'ONETOA (Peoples Party) elected prime minister by parliament receiving 15 of 23 votes cast
note: a Privy Council advises the monarch

Legislative branch: *description:* unicameral Legislative Assembly or Fale Alea (up to 30 seats; - 26 for the 2017-19 term); 17 people's representatives directly elected in single-seat constituencies by simple majority vote, and 9 indirectly elected by hereditary leaders; members serve 3-year terms)

elections: last held on 16 November 2017 (next to be held in 2020)

election results: percent of vote - NA; seats by party - Democratic Party 14, nobles' representatives 9, independent 3; composition - men 24, women 2, percent of women 7.7%

Judicial branch: *highest courts:* Court of Appeal (consists of the court president and a number of judges determined by the monarch); note - appeals beyond the Court of Appeal are brought before the King in Privy Council, the monarch's advisory organ that has both judicial and legislative powers

judge selection and term of office: judge appointments and tenures made by the King in Privy Council and subject to consent of the Legislative Assembly

subordinate courts: Supreme Court; Magistrates' Courts; Land Courts

Political parties and leaders: Democratic Party of the Friendly Islands [Samuela 'Akilisi POHIVA] People's Democratic Party or PDP [Tesina FUKO] Sustainable Nation-Building Party [Sione FONUA]

Tonga Democratic Labor Party

Tonga Human Rights and Democracy Movement or THRDM

International organization participation: ACP, ADB, AOSIS, C, FAO, G-77, IBRD, ICAO, ICRM, IDA, IFAD, IFC, IFRCS, IHO, IMF, IMO, IMSO, Interpol, IOC, IPU, ITU, ITUC (NGOs), OPCW, PIF, Sparteca, SPC, UN, UNCTAD, UNESCO, UNIDO, UPU, WCO, WHO, WIPO, WMO, WTO

Diplomatic representation in the US: *chief of mission:* Ambassador Mahe'uli'uli Sandhurst TUPOUNIUA (since 17 September 2013)

chancery: 250 East 51st Street, New York, NY 10022

telephone: [1] (917) 369-1025

FAX: [1] (917) 369-1024

consulate(s) general: San Francisco

Diplomatic representation from the US: the US does not have an embassy in Tonga; the US Ambassador to Fiji is accredited to Tonga

Flag description: red with a bold red cross on a white rectangle in the upper hoist-side corner; the cross reflects the deep-rooted Christianity in Tonga, red represents the blood of Christ and his sacrifice, and white signifies purity

National symbol(s): *red cross on white field, arms equal length; national colors:* red, white

National anthem: *name:* "Ko e fasi 'o e tu''i 'o e 'Otu Tonga" (Song of the King of the Tonga Islands)

lyrics/music: Uelingatoni Ngu TUPOUMALOHI/ Karl Gustavus SCHMITT

note: in use since 1875; more commonly known as "Fasi Fakafonua" (National Song)

ECONOMY

Economy—overview: Tonga has a small, open island economy and is the last constitutional monarchy among the Pacific Island countries. It has a narrow export base in agricultural goods. Squash, vanilla beans, and yams are the main crops. Agricultural exports, including fish, make up two-thirds of total exports. Tourism is the second-largest source of hard currency earnings following remittances. Tonga had 53,800 visitors in 2015. The country must import a high proportion of its food, mainly from New Zealand.

The country remains dependent on external aid and remittances from overseas Tongans to offset its trade deficit. The government is emphasizing the development of the private sector, encouraging investment, and is committing increased funds for health care and education. Tonga's English-speaking and educated workforce offers a viable labor market, and the tropical climate provides fertile soil. Renewable energy and deep-sea mining also offer opportunities for investment.

Tonga has a reasonably sound basic infrastructure and well developed social services. But the government faces high unemployment among the young, moderate inflation, pressures for democratic reform, and rising civil service expenditures.

GDP (purchasing power parity): $591 million (2017 est.)

$576.6 million (2016 est.)

$553.6 million (2015 est.)

note: data are in 2017 dollars

country comparison to the world: 211

GDP (official exchange rate): $455 million (2017 est.)

GDP—real growth rate: 2.5% (2017 est.)

4.2% (2016 est.)

3.5% (2015 est.)

country comparison to the world: 115

GDP—per capita (PPP): $5,900 (2017 est.)

$5,700 (2016 est.)

$5,400 (2015 est.)

note: data are in 2017 dollars

country comparison to the world: 167

GDP—composition, by end use: *household consumption:* 99.4% (2017 est.)

government consumption: 21.9% (2017 est.)

investment in fixed capital: 24.1% (2017 est.)

investment in inventories: 0% (2017 est.)

exports of goods and services: 22.8% (2017 est.)

imports of goods and services: -68.5% (2017 est.)

GDP—composition, by sector of origin: *agriculture:* 19.9% (2017 est.)

industry: 20.3% (2017 est.)

services: 59.8% (2017 est.)

Agriculture—products: squash, coconuts, copra, bananas, vanilla beans, cocoa, coffee, sweet potatoes, cassava, taro, and kava

Industries: tourism, construction, fishing

Industrial production growth rate: 5% (2017 est.)

country comparison to the world: 59

Labor force: 33,800 (2011 est.)

country comparison to the world: 201

Labor force—by occupation: *agriculture:* 2,006% (2006 est.)

industry: 27.5% (2006 est.)

services: 2,006% (2006 est.)

Unemployment rate: 1.1% (2011 est.)

1.1% (2006)

country comparison to the world: 13

Population below poverty line: 22.5% (2010 est.)

Household income or consumption by percentage share: *lowest 10%:* NA

highest 10%: NA

Budget: *revenues:* 181.2 million (2017 est.)

expenditures: 181.2 million (2017 est.)

Taxes and other revenues: 39.8% (of GDP) (2017 est.)

country comparison to the world: 42

Budget surplus (+) or deficit (-): 0% (of GDP) (2017 est.)

country comparison to the world: 47

Public debt: 48% of GDP (FY2017 est.)

51.8% of GDP (FY2016 est.)

country comparison to the world: 109

Fiscal year: 1 July - 30 June

Inflation rate (consumer prices): 7.4% (2017 est.)

2.6% (2016 est.)

country comparison to the world: 195

Current account balance: -$53 million (2017 est.)

-$30 million (2016 est.)

country comparison to the world: 81

Exports: $18.4 million (2017 est.)

$19.4 million (2016 est.)

country comparison to the world: 213

Exports—partners: Hong Kong 25.1%, NZ 22.6%, US 14.3%, Japan 12.8%, Australia 10.5% (2017)

Exports—commodities: squash, fish, vanilla beans, root crops, kava

Imports: $250.2 million (2017 est.)

$269.8 million (2016 est.)

country comparison to the world: 208

Imports—commodities: foodstuffs, machinery and transport equipment, fuels, chemicals

Imports—partners: NZ 33.3%, Fiji 11.7%, US 9.8%, Singapore 9%, Australia 8.9%, China 7.9%, Japan 5.9% (2017)

Reserves of foreign exchange and gold: $198.5 million (31 December 2017 est.)

$176.5 million (31 December 2016 est.)

country comparison to the world: 176

Debt—external: $189.9 million (31 December 2017 est.)

$198.2 million (31 December 2016 est.)

country comparison to the world: 190

Exchange rates: pa'anga (TOP) per US dollar—
2.228 (2017 est.)
2.216 (2016 est.)
2.216 (2015 est.)
2.106 (2014 est.)
1.847 (2013 est.)

ENERGY

Electricity access: *electrification—total population:* 97% (2016)

electrification—urban areas: 98.6% (2016)

electrification—rural areas: 96.6% (2016)

Electricity—production: 52 million kWh (2016 est.)
country comparison to the world: 205

Electricity—consumption: 48.36 million kWh (2016 est.)
country comparison to the world: 205

Electricity—exports: 0 kWh (2016)
country comparison to the world: 209

Electricity—imports: 0 kWh (2016 est.)
country comparison to the world: 210

Electricity—installed generating capacity: 20,300 kW (2016 est.)
country comparison to the world: 204

Electricity—from fossil fuels: 74% of total installed capacity (2016 est.)
country comparison to the world: 98

Electricity—from nuclear fuels: 0% of total installed capacity (2017 est.)
country comparison to the world: 196

Electricity—from hydroelectric plants: 0% of total installed capacity (2017 est.)
country comparison to the world: 205

Electricity—from other renewable sources: 26% of total installed capacity (2017 est.)
country comparison to the world: 28

Crude oil—production: 0 bbl/day (2018 est.)
country comparison to the world: 208

Crude oil—exports: 0 bbl/day (2015 est.)
country comparison to the world: 207

Crude oil—imports: 0 bbl/day (2015 est.)
country comparison to the world: 205

Crude oil—proved reserves: 0 bbl (1 January 2018 est.)
country comparison to the world: 206

Refined petroleum products—production: 0 bbl/day (2017 est.)
country comparison to the world: 209

Refined petroleum products—consumption: 900 bbl/day (2016 est.)
country comparison to the world: 207

Refined petroleum products— exports: 0 bbl/day (2015 est.)
country comparison to the world: 210

Refined petroleum products—imports: 910 bbl/day (2015 est.)
country comparison to the world: 203

Natural gas—production: 0 cu m (2017 est.)
country comparison to the world: 207

Natural gas—consumption: 0 cu m (2017 est.)
country comparison to the world: 207

Natural gas—exports: 0 cu m (2017 est.)
country comparison to the world: 201

Natural gas—imports: 0 cu m (2017 est.)
country comparison to the world: 201

Natural gas—proved reserves: 0 cu m (1 January 2014 est.)
country comparison to the world: 201

Carbon dioxide emissions from consumption of energy: 139,700 Mt (2017 est.)
country comparison to the world: 205

COMMUNICATIONS

Telephones—fixed lines: *total subscriptions:* 6,748

subscriptions per 100 inhabitants: 6.35 (2019 est.)
country comparison to the world: 200

Telephones—mobile cellular: total subscriptions: 63,156

subscriptions per 100 inhabitants: 59.43 (2019 est.)
country comparison to the world: 203

Telecommunication systems: *general assessment:* high speed Internet provided by 3 MNOs, has subsequently allowed for better health care services, faster connections for education and growing e-commerce services; in 2018 new 4G LTE network; fixed-line teledensity has dropped given mobile subscriptions; mobile technology dominates given the island's geography; satellite technology is widespread and is important especially in areas away from the city; the launch in 2019 of the Kacific-1 broadband satellite has made broadband more widely available for around 89 remote communities (2020)

domestic: fixed-line 6 per 100 persons and mobile-cellular teledensity 59 telephones per 100; fully automatic switched network (2019)

international: country code - 676; landing point for the Tonga Cable and the TDCE connecting to Fiji and 3 separate Tonga islands; satellite earth station - 1 Intelsat (Pacific Ocean) (2020)
note: the COVID-19 outbreak is negatively impacting telecommunications production and supply chains globally; consumer spending on telecom devices and services has also slowed due to the pandemic's effect on economies worldwide; overall progress towards improvements in all facets of the telecom industry - mobile, fixed-line, broadband, submarine cable and satellite - has moderated

Broadcast media: 1 state-owned TV station and 3 privately owned TV stations; satellite and cable TV services are available; 1 state-owned and 5 privately owned radio stations; Radio Australia broadcasts available via satellite (2019)

Internet country code: .to

Internet users: *total:* 43,889

percent of population: 41.25% (July 2018 est.)
country comparison to the world: 199

Broadband—fixed subscriptions: *total:* 2,519

subscriptions per 100 inhabitants: 2 (2018 est.)
country comparison to the world: 186

TRANSPORTATION

National air transport system: *number of registered air carriers:* 1 (2020)

inventory of registered aircraft operated by air carriers: 1

Civil aircraft registration country code prefix: A3 (2016)

Airports: 6 (2013)
country comparison to the world: 178

Airports—with paved runways: *total:* 1 (2019)
2,438 to 3,047 m: 1

Airports—with unpaved runways: *total:* 5 (2013)
1,524 to 2,437 m: 1 (2013)
914 to 1,523 m: 3 (2013)
under 914 m: 1 (2013)

Roadways: *total:* 680 km (2011)
paved: 184 km (2011)
unpaved: 496 km (2011)
country comparison to the world: 189

Merchant marine: *total:* 36

by type: bulk carrier 4, general cargo 14, oil tanker 1, other 17 (2019)
country comparison to the world: 127

Ports and terminals: *major seaport(s):* Nuku'alofa, Neiafu, Pangai

MILITARY AND SECURITY

Military and security forces: *Tonga Defense Services:* Joint Force headquarters, Territorial Forces, Land Force, Tonga Navy, Training Wing, Air Wing, and Support Unit (2020)

Military and security service personnel strengths: the Tonga Defense Services have approximately 500 personnel (2020)

Military equipment inventories and acquisitions: the Tonga military's inventory includes mostly light weapons and equipment from European (primarily the UK) countries and the US, as well as naval patrol vessels from Australia; Australia is the only supplier of military systems since 2010 (2019)

Military service age and obligation: Volunteers, 18-25; no conscription (2014)

TRANSPORTATION

Disputes—international: maritime boundary dispute with Fiji

TRINIDAD AND TOBAGO

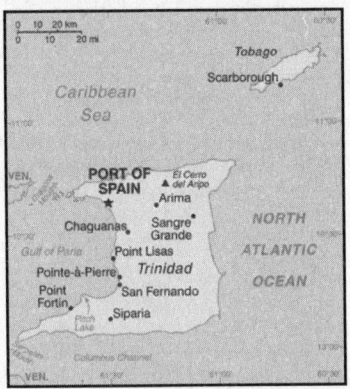

INTRODUCTION

Background: First colonized by the Spanish, the islands came under British control in the early 19th century. The islands' sugar industry was hurt by the emancipation of the slaves in 1834. Manpower was replaced with the importation of contract laborers from India between 1845 and 1917, which boosted sugar production as well as the cocoa industry. The discovery of oil on Trinidad in 1910 added another important export. Independence was attained in 1962. The country is one of the most prosperous in the Caribbean thanks largely to petroleum and natural gas production and processing. Tourism, mostly in Tobago, is targeted for expansion and is growing. The government is struggling to reverse a surge in violent crime.

GEOGRAPHY

Location: Caribbean, islands between the Caribbean Sea and the North Atlantic Ocean, northeast of Venezuela

Geographic coordinates: 11 00 N, 61 00 W

Map references: Central America and the Caribbean

Area: *total:* 5,128 sq km
land: 5,128 sq km
water: 0 sq km
country comparison to the world: 174

Area—comparative: slightly smaller than Delaware

Land boundaries: 0 km

Coastline: 362 km

Maritime claims:
territorial sea: 12 nm
exclusive economic zone: 200 nm
contiguous zone: 24 nm
continental shelf: 200 nm or to the outer edge of the continental margin measured from claimed archipelagic baselines

Climate: tropical; rainy season (June to December)

Terrain: mostly plains with some hills and low mountains

Elevation: *mean elevation:* 83 m
lowest point: Caribbean Sea 0 m
highest point: El Cerro del Aripo 940 m

Natural resources: petroleum, natural gas, asphalt

Land use: *agricultural land:* 10.6% (2011 est.)
arable land: 4.9% (2011 est.) / *permanent crops:* 4.3% (2011 est.) / *permanent pasture:* 1.4% (2011 est.)
forest: 44% (2011 est.)
other: 45.4% (2011 est.)

Irrigated land: 70 sq km (2012)

Population distribution: population on Trinidad is concentrated in the western half of the island, on Tobago in the southern half

Natural hazards: outside usual path of hurricanes and other tropical storms

Environment—current issues: water pollution from agricultural chemicals, industrial wastes, and raw sewage; widespread pollution of waterways and coastal areas; illegal dumping; deforestation; soil erosion; fisheries and wildlife depletion

Environment—international agreements: *party to:* Biodiversity, Climate Change, Climate Change-Kyoto Protocol, Desertification, Endangered Species, Hazardous Wastes, Law of the Sea, Marine Dumping, Marine Life Conservation, Ozone Layer Protection, Ship Pollution, Tropical Timber 83, Tropical Timber 94, Wetlands
signed, but not ratified: none of the selected agreements

Geography—note: Pitch Lake, on Trinidad's southwestern coast, is the world's largest natural reservoir of asphalt

PEOPLE AND SOCIETY

Population: 1,208,789 (July 2020 est.)
country comparison to the world: 159

Nationality: *noun:* Trinidadian(s), Tobagonian(s)
adjective: Trinidadian, Tobagonian
note: Trinbagonian is used on occasion to describe a citizen of the country without specifying the island of origin

Ethnic groups: East Indian 35.4%, African descent 34.2%, mixed—other 15.3%, mixed—African/East Indian 7.7%, other 1.3%, unspecified 6.2% (2011 est.)

Languages: English (official), Trinidadian Creole English, Tobagonian Creole English, Caribbean Hindustani (a dialect of Hindi), Trinidadian Creole French, Spanish, Chinese

Religions: Protestant 32.1% (Pentecostal/Evangelical/Full Gospel 12%, Baptist 6.9%, Anglican 5.7%, Seventh-Day Adventist 4.1%, Presbyterian/Congregational 2.5%, other Protestant 0.9%), Roman Catholic 21.6%, Hindu 18.2%, Muslim 5%, Jehovah's Witness 1.5%, other 8.4%, none 2.2%, unspecified 11.1% (2011 est.)

Age structure: *0-14 years:* 19.01% (male 116,953/female 112,805)
15-24 years: 11.28% (male 70,986/female 65,389)
25-54 years: 43.77% (male 276,970/female 252,108)
55-64 years: 13.83% (male 83,650/female 83,585)
65 years and over: 12.11% (male 64,092/female 82,251) (2020 est.)

Dependency ratios:
total dependency ratio: 46.1
youth dependency ratio: 29.3
elderly dependency ratio: 16.8
potential support ratio: 7.4 (2020 est.)

Median age:
total: 37.8 years
male: 37.3 years
female: 38.3 years (2020 est.)
country comparison to the world: 67

Population growth rate: -0.3% (2020 est.)
country comparison to the world: 219

Birth rate: 11.4 births/1,000 population (2020 est.)
country comparison to the world: 172

Death rate: 9.1 deaths/1,000 population (2020 est.)
country comparison to the world: 60

Net migration rate: -5.4 migrant(s)/1,000 population (2020 est.)
country comparison to the world: 200

Population distribution: population on Trinidad is concentrated in the western half of the island, on Tobago in the southern half

Urbanization: *urban population:* 53.2% of total population (2020)
rate of urbanization: 0.22% annual rate of change (2015-20 est.)
total population growth rate v. urban population growth rate, 2000-2030: Major urban areas—population: 544,000 PORT-OF-SPAIN (capital) (2020)

Sex ratio: *at birth:* 1.03 male(s)/female
0-14 years: 1.04 male(s)/female
15-24 years: 1.09 male(s)/female
25-54 years: 1.1 male(s)/female
55-64 years: 1 male(s)/female
65 years and over: 0.78 male(s)/female
total population: 1.03 male(s)/female (2020 est.)

Maternal mortality rate: 67 deaths/100,000 live births (2017 est.)
country comparison to the world: 85

Infant mortality rate: *total:* 20.1 deaths/1,000 live births
male: 21.3 deaths/1,000 live births
female: 18.8 deaths/1,000 live births (2020 est.)
country comparison to the world: 75

Life expectancy at birth: *total population:* 73.9 years
male: 70.9 years

female: 76.9 years (2020 est.)
country comparison to the world: 138

Total fertility rate: 1.7 children born/woman (2020 est.)
country comparison to the world: 174

Contraceptive prevalence rate: 40.3% (2011)

Drinking water source:
improved: total: 99.3% of population
unimproved: total: 0% of population (2017 est.)

Current Health Expenditure: 7% (2017)

Physicians density: 3.36 physicians/1,000 population (2017)

Hospital bed density: 3 beds/1,000 population (2017)

Sanitation facility access: *improved:* total: 99.3% of population
unimproved: total: 0.7% of population (2017 est.)

HIV/AIDS—adult prevalence rate: 0.9% (2019 est.)
country comparison to the world: 50

HIV/AIDS—people living with HIV/AIDS: 11,000 (2019 est.)
country comparison to the world: 101

HIV/AIDS—deaths: <200 (2019 est.)

Obesity—adult prevalence rate: 18.6% (2016)
country comparison to the world: 116

Children under the age of 5 years underweight: 4.9% (2011)
country comparison to the world: 84

Education expenditures: NA

Literacy: *definition:* age 15 and over can read and write
total population: 99%
male: 99.2%
female: 98.7% (2015)

Unemployment, youth ages 15-24:
total: 8.7%
male: 8.9%
female: 8.4% (2016 est.)
country comparison to the world: 139

GOVERNMENT

Country name: *conventional long form:* Republic of Trinidad and Tobago
conventional short form: Trinidad and Tobago
etymology: explorer Christopher COLUMBUS named the larger island "La Isla de la Trinidad" (The Island of the Trinity) on 31 July 1498 on his third voyage; the tobacco grown and smoked by the natives of the smaller island or its elongated cigar shape may account for the "tobago" name, which is spelled "tobaco" in Spanish

Government type: parliamentary republic

Capital: *name:* Port of Spain
geographic coordinates: 10 39 N, 61 31 W
time difference: UTC-4 (1 hour ahead of Washington, DC, during Standard Time)
etymology: the name dates to the period of Spanish colonial rule (16th to late 18th centuries) when the city was referred to as "Puerto de Espana"; the name was anglicized following the British capture of Trinidad in 1797

Administrative divisions: 9 regions, 3 boroughs, 2 cities, 1 ward
regions: Couva/Tabaquite/Talparo, Diego Martin, Mayaro/Rio Claro, Penal/Debe, Princes Town, Sangre Grande, San Juan/Laventille, Siparia, Tunapuna/Piarco
borough: Arima, Chaguanas, Point Fortin
cities: Port of Spain, San Fernando
ward: Tobago

Independence: 31 August 1962 (from the UK)

National holiday: Independence Day, 31 August (1962)

Constitution: *history:* previous 1962; latest 1976
amendments: proposed by Parliament; passage of amendments affecting constitutional provisions, such as human rights and freedoms or citizenship, requires at least two-thirds majority vote by the membership of both houses and assent of the president; passage of amendments, such as the powers and authorities of the executive, legislative, and judicial branches of government, and the procedure for amending the constitution, requires at least three-quarters majority vote by the House membership, two-thirds majority vote by the Senate membership, and assent of the president; amended many times, last in 2007

Legal system: English common law; judicial review of legislative acts in the Supreme Court

International law organization participation: has not submitted an ICJ jurisdiction declaration; accepts ICCt jurisdiction

Citizenship: *citizenship by birth:* yes
citizenship by descent only: yes
dual citizenship recognized: yes
residency requirement for naturalization: 8 years

Suffrage: 18 years of age; universal

Executive branch: *chief of state:* President Paula-Mae WEEKES (since 19 March 2018)
head of government: Prime Minister Keith ROWLEY (since 9 September 2015)
cabinet: Cabinet appointed from among members of Parliament
elections/appointments: president indirectly elected by an electoral college of selected Senate and House of Representatives members for a 5-year term (eligible for a second term); election last held on 19 January 2018 (next to be held by February 2023); the president usually appoints the leader of the majority party in the House of Representatives as prime minister
election results: Paula-Mae WEEKES (independent) elected president; ran unopposed and was elected without a vote; she is Trinidad and Tabago's first female head of state

Legislative branch: *description:* bicameral Parliament consists of: Senate (31 seats; 16 members appointed by the ruling party, 9 by the president, and 6 by the opposition party; members serve 5-year terms;)
House of Representatives 42 seats; 41 members directly elected in single-seat constituencies by simple majority vote and the house speaker—usually designated from outside Parliament; members serve 5-year terms)

elections: Senate—last appointments on 23 September 2015 (next in 2020)
House of Representatives—last held on 10 August 2020 (next to be held in 2025)
election results: Senate—percent by party—NA; seats by party—NA; composition—men 21, women 10, percent of women 32.3%
House of Representatives—percent by party—NA; seats by party—PNM 22, UNC 19; composition—NA
note: Tobago has a unicameral House of Assembly (16 seats; 12 assemblymen directly elected by simple majority vote and 4 appointed councillors—3 on the advice of the chief secretary and 1 on the advice of the minority leader; members serve 4-year terms)

Judicial branch: *highest courts:* Supreme Court of the Judicature (consists of a chief justice for both the Court of Appeal with 12 judges and the High Court with 24 judges); note—Trinidad and Tobago can file appeals beyond its Supreme Court to the Caribbean Court of Justice, with final appeal to the Judicial Committee of the Privy Council (in London)
judge selection and term of office: Supreme Court chief justice appointed by the president after consultation with the prime minister and the parliamentary leader of the opposition; other judges appointed by the Judicial Legal Services Commission, headed by the chief justice and 5 members with judicial experience; all judges serve for life with mandatory retirement normally at age 65
subordinate courts: Courts of Summary Criminal Jurisdiction; Petty Civil Courts; Family Court

Political parties and leaders: Congress of the People or COP [Carolyn SEEPERSAD-BACHAN]
People's National Movement or PNM [Keith ROWLEY]
Progressive Democratic Patriots (Tobago)
United National Congress or UNC [Kamla PERSAD-BISSESSAR]

International organization participation: ACP, AOSIS, C, Caricom, CDB, CELAC, EITI (compliant country), FAO, G-24, G-77, IADB, IAEA, IBRD, ICAO, ICC (NGOs), ICCt, ICRM, IDA, IFAD, IFC, IFRCS, IHO, ILO, IMF, IMO, Interpol, IOC, IOM, IPU, ISO, ITSO, ITU, ITUC (NGOs), LAES, MIGA, NAM, OAS, OPANAL, OPCW, Pacific Alliance (observer), Paris Club (associate), UN, UNCTAD, UNESCO, UNIDO, UPU, WCO, WFTU (NGOs), WHO, WIPO, WMO, WTO

Diplomatic representation in the US: *chief of mission:* Ambassador Anthony Wayne Jerome PHILLIPS-SPENCER, Brig. Gen. (Ret.) (since 27 June 2016)
chancery: 1708 Massachusetts Avenue NW, Washington, DC 20036
telephone: [1] (202) 467-6490
FAX: [1] (202) 785-3130
consulate(s) general: Miami, New York

Diplomatic representation from the US: *chief of mission:* Ambassador Joseph MONDELLO (since 22 October 2018)

telephone: [1] (868) 622-6371 through 6376
embassy: 15 Queen's Park West, Port of Spain
mailing address: P. O. Box 752, Port of Spain
FAX: [1] (868) 822-5905

Flag description: red with a white-edged black diagonal band from the upper hoist side to the lower fly side; the colors represent the elements of earth, water, and fire; black stands for the wealth of the land and the dedication of the people; white symbolizes the sea surrounding the islands, the purity of the country's aspirations, and equality; red symbolizes the warmth and energy of the sun, the vitality of the land, and the courage and friendliness of its people

National symbol(s): *scarlet ibis (bird of Trinidad), cocrico (bird of Tobago), Chaconia flower; national colors:* red, white, black

National anthem: *name:* Forged From the Love of Liberty
lyrics/music: Patrick Stanislaus CASTAGNE
note: adopted 1962; song originally created to serve as an anthem for the West Indies Federation; adopted by Trinidad and Tobago following the Federation's dissolution in 1962

ECONOMY

Economy—overview: Trinidad and Tobago relies on its energy sector for much of its economic activity, and has one of the highest per capita incomes in Latin America. Economic growth between 2000 and 2007 averaged slightly over 8% per year, significantly above the regional average of about 3.7% for that same period; however, GDP has slowed down since then, contracting during 2009-12, making small gains in 2013 and contracting again in 2014-17. Trinidad and Tobago is buffered by considerable foreign reserves and a sovereign wealth fund that equals about one-and-a-half times the national budget, but the country is still in a recession and the government faces the dual challenge of gas shortages and a low price environment. Large-scale energy projects in the last quarter of 2017 are helping to mitigate the gas shortages.

Energy production and downstream industrial use dominate the economy. Oil and gas typically account for about 40% of GDP and 80% of exports but less than 5% of employment. Trinidad and Tobago is home to one of the largest natural gas liquefaction facilities in the Western Hemisphere. The country produces about nine times more natural gas than crude oil on an energy equivalent basis with gas contributing about two-thirds of energy sector government revenue. The US is the country's largest trading partner, accounting for 28% of its total imports and 48% of its exports.

Economic diversification is a longstanding government talking point, and Trinidad and Tobago has much potential due to its stable, democratic government and its educated, English speaking workforce. The country is also a regional financial center with a well-regulated and stable financial system. Other sectors the Government of Trinidad and Tobago has targeted for increased investment

and projected growth include tourism, agriculture, information and communications technology, and shipping. Unfortunately, a host of other factors, including low labor productivity, inefficient government bureaucracy, and corruption, have hampered economic development.

GDP (purchasing power parity): $42.85 billion (2017 est.)
$43.99 billion (2016 est.)
$46.83 billion (2015 est.)
note: data are in 2017 dollars
country comparison to the world: 115

GDP (official exchange rate): $22.78 billion (2017 est.)

GDP—real growth rate: -2.6% (2017 est.)
-6.1% (2016 est.)
1.7% (2015 est.)
country comparison to the world: 210

GDP—per capita (PPP): $31,300 (2017 est.)
$32,200 (2016 est.)
$34,400 (2015 est.)
note: data are in 2017 dollars
country comparison to the world: 65

Gross national saving: 26.4% of GDP (2017 est.)
16.8% of GDP (2016 est.)
29% of GDP (2015 est.)
country comparison to the world: 50

GDP—composition, by end use:
household consumption: 78.9% (2017 est.)
government consumption: 16.4% (2017 est.)
investment in fixed capital: 8.2% (2017 est.)
investment in inventories: 0.6% (2017 est.)
exports of goods and services: 45.4% (2017 est.)
imports of goods and services: -48.7% (2017 est.)

GDP—composition, by sector of origin:
agriculture: 0.4% (2017 est.)
industry: 47.8% (2017 est.)
services: 51.7% (2017 est.)

Agriculture—products: cocoa, dasheen, pumpkin, cassava, tomatoes, cucumbers, eggplant, hot pepper, pommecythere, coconut water, poultry

Industries: petroleum and petroleum products, liquefied natural gas, methanol, ammonia, urea, steel products, beverages, food processing, cement, cotton textiles

Industrial production growth rate: -4.3% (2017 est.)
country comparison to the world: 195

Labor force: 629,400 (2017 est.)
country comparison to the world: 152

Labor force—by occupation: *agriculture:* 3.1%
industry: 11.5%
services: 85.4% (2016 est.)

Unemployment rate: 4.9% (2017 est.)
4% (2016 est.)
country comparison to the world: 72

Population below poverty line: 20% (2014 est.)

Household income or consumption by percentage share: *lowest 10%:* NA
highest 10%: NA

Budget: *revenues:* 5.581 billion (2017 est.)
expenditures: 7.446 billion (2017 est.)

Taxes and other revenues: 24.5% (of GDP) (2017 est.)
country comparison to the world: 121

Budget surplus (+) or deficit (-): -8.2% (of GDP) (2017 est.)
country comparison to the world: 200

Public debt: 41.8% of GDP (2017 est.)
37% of GDP (2016 est.)
country comparison to the world: 120

Fiscal year: 1 October—30 September

Inflation rate (consumer prices): 1.9% (2017 est.)
3.1% (2016 est.)
country comparison to the world: 101

Current account balance: $2.325 billion (2017 est.)
-$653 million (2016 est.)
country comparison to the world: 37

Exports: $9.927 billion (2017 est.)
$8.714 billion (2016 est.)
country comparison to the world: 92

Exports—partners: US 34.8%, Argentina 9% (2017)

Exports—commodities: petroleum and petroleum products, liquefied natural gas, methanol, ammonia, urea, steel products, beverages, cereal and cereal products, cocoa, fish, preserved fruits, cosmetics, household cleaners, plastic packaging

Imports: $6.105 billion (2017 est.)
$6.858 billion (2016 est.)
country comparison to the world: 120

Imports—commodities: mineral fuels, lubricants, machinery, transportation equipment, manufactured goods, food, chemicals, live animals

Imports—partners: US 23.8%, Russia 15.3%, Colombia 11.1%, Gabon 10.5%, China 7.3% (2017)

Reserves of foreign exchange and gold: $8.892 billion (31 December 2017 est.)
$9.995 billion (31 December 2016 est.)
country comparison to the world: 77

Debt—external: $8.238 billion (31 December 2017 est.)
$8.746 billion (31 December 2016 est.)
country comparison to the world: 120

Exchange rates: Trinidad and Tobago dollars (TTD) per US dollar—6.78 (2017 est.)
6.669 (2016 est.)
6.669 (2015 est.)
6.4041 (2014 est.)
6.4041 (2013 est.)

ENERGY

Electricity access: *electrification—total population:* 100% (2020)

Electricity—production: 10.07 billion kWh (2016 est.)
country comparison to the world: 103

Electricity—consumption: 9.867 billion kWh (2016 est.)
country comparison to the world: 97

Electricity—exports: 0 kWh (2016 est.)

country comparison to the world: 210

Electricity—imports: 0 kWh (2016 est.)
country comparison to the world: 211

Electricity—installed generating capacity: 2.608 million kW (2016 est.)
country comparison to the world: 104

Electricity—from fossil fuels: 100% of total installed capacity (2016 est.)
country comparison to the world: 19

Electricity—from nuclear fuels: 0% of total installed capacity (2017 est.)
country comparison to the world: 197

Electricity—from hydroelectric plants: 0% of total installed capacity (2017 est.)
country comparison to the world: 206

Electricity—from other renewable sources: 0% of total installed capacity (2017 est.)
country comparison to the world: 209

Crude oil—production: 63,000 bbl/day (2018 est.)
country comparison to the world: 48

Crude oil—exports: 31,030 bbl/day (2015 est.)
country comparison to the world: 45

Crude oil—imports: 80,860 bbl/day (2015 est.)
country comparison to the world: 47

Crude oil—proved reserves: 243 million bbl (1 January 2018 est.)
country comparison to the world: 53

Refined petroleum products—production: 134,700 bbl/day (2015 est.)
country comparison to the world: 63

Refined petroleum products—consumption: 51,000 bbl/day (2016 est.)
country comparison to the world: 105

Refined petroleum products—exports: 106,100 bbl/day (2015 est.)
country comparison to the world: 40

Refined petroleum products—imports: 0 bbl/day (2015 est.)
country comparison to the world: 213

Natural gas—production: 36.73 billion cu m (2017 est.)
country comparison to the world: 23

Natural gas—consumption: 21.24 billion cu m (2017 est.)
country comparison to the world: 37

Natural gas—exports: 15.49 billion cu m (2017 est.)
country comparison to the world: 14

Natural gas—imports: 0 cu m (2017 est.)
country comparison to the world: 202

Natural gas—proved reserves: 447.4 billion cu m (1 January 2018 est.)
country comparison to the world: 33

Carbon dioxide emissions from consumption of energy: 48.92 million Mt (2017 est.)
country comparison to the world: 61

COMMUNICATIONS

Telephones—fixed lines:
total subscriptions: 298,493

subscriptions per 100 inhabitants: 24.62 (2019 est.)
country comparison to the world: 109

Telephones—mobile cellular: *total subscriptions:* 1,880,555

subscriptions per 100 inhabitants: 155.11 (2019 est.)
country comparison to the world: 153

Telecommunication systems: *general assessment:* excellent international service; good local service; broadband access; expanded FttP (Fiber to the Home) markets; LTE launch; regulatory development; major growth in mobile telephony and data segments which attacks operation investment in fiber infrastructure; moves to end roaming charges (2020)
domestic: fixed-line 25 per 100 persons and mobile-cellular teledensity 155 per 100 persons (2019)
international: country code—1-868; landing points for the EC Link, ECFS, Southern Caribbean Fiber, SG-SCS and Americas II submarine cable systems provide connectivity to US, parts of the Caribbean and South America; satellite earth station—1 Intelsat (Atlantic Ocean); tropospheric scatter to Barbados and Guyana (2020)
note: the COVID-19 outbreak is negatively impacting telecommunications production and supply chains globally; consumer spending on telecom devices and services has also slowed due to the pandemic's effect on economies worldwide; overall progress towards improvements in all facets of the telecom industry—mobile, fixed-line, broadband, submarine cable and satellite—has moderated

Broadcast media: 6 free-to-air TV networks, 2 of which are state-owned; 24 subscription providers (cable and satellite); over 36 radio frequencies (2019)

Internet country code: .tt

Internet users: *total:* 939,967
percent of population: 77.33% (July 2018 est.)
country comparison to the world: 142

Broadband—fixed subscriptions:
total: 341,045
subscriptions per 100 inhabitants: 28 (2018 est.)
country comparison to the world: 98

TRANSPORTATION

National air transport system: *number of registered air carriers:* 1 (2020)
inventory of registered aircraft operated by air carriers: 19
annual passenger traffic on registered air carriers: 2,525,130 (2018)
annual freight traffic on registered air carriers: 41.14 million mt-km (2018)

Civil aircraft registration country code prefix: 9Y (2016)

Airports: 4 (2013)
country comparison to the world: 190

Airports—with paved runways: *total:* 2 (2019)
over 3,047 m: 1
2,438 to 3,047 m: 1

Airports—with unpaved runways: *total:* 2 (2013)
914 to 1,523 m: 1 (2013)
under 914 m: 1 (2013)

Pipelines: 257 km condensate, 11 km condensate/gas, 1567 km gas, 587 km oil (2013)

Merchant marine:
total: 105
by type: general cargo 1, other 104 (2019)
country comparison to the world: 88

Ports and terminals: *major seaport(s):* Point Fortin, Point Lisas, Port of Spain, Scarborough
oil terminal(s): Galeota Point terminal

LNG terminal(s) (export): Port Fortin

MILITARY AND SECURITY

Military and security forces: *Trinidad and Tobago Defense Force (TTDF):* Trinidad and Tobago Regiment (Land Forces), Coast Guard, Air Guard, Defense Force Reserves (2019)

Military expenditures: 0.7% of GDP (2019)
0.8% of GDP (2018)
1% of GDP (2017)
1% of GDP (2016)
0.9% of GDP (2015)
country comparison to the world: 142

Military and security service personnel strengths: the Trinidad and Tobago Defense Force (TTDF) has approximately 4,000 active troops, including Army, Coast Guard, and Air Guard personnel (2019 est.)

Military equipment inventories and acquisitions: the TTDF's ground force inventory includes only light weapons; the Coast Guard and Air Guard field mostly second-hand equipment from a mix of countries, including Australia, China, the Netherlands, the UK, and the US (2019 est.)

Military service age and obligation: 18-25 years of age for voluntary military service (some age variations between services, reserves); no conscription (2019)

TRANSNATIONAL ISSUES

Disputes—international: Barbados and Trinidad and Tobago abide by the April 2006 Permanent Court of Arbitration decision delimiting a maritime boundary and limiting catches of flying fish in Trinidad and Tobago's EEZ; in 2005, Barbados and Trinidad and Tobago agreed to compulsory international arbitration under UN Convention on the Law of the Sea challenging whether the northern limit of Trinidad and Tobago's and Venezuela's maritime boundary extends into Barbadian waters; Guyana has expressed its intention to include itself in the arbitration, as the Trinidad and Tobago-Venezuela maritime boundary may also extend into its waters

Refugees and internally displaced persons: *refugees (country of origin):* 18,587 (Venezuela) (economic and political crisis; includes Venezuelans who have claimed asylum, are recognized as refugees, or have received alternative legal stay) (2020)

Trafficking in persons: *current situation:* Trinidad and Tobago is a destination, transit, and possible source country for adults and children subjected to sex trafficking and forced labor; women and girls from Venezuela, the Dominican Republic, Guyana, and Colombia have been subjected to sex trafficking in Trinidad and Tobago's brothels and clubs; some economic migrants from the Caribbean region and Asia are vulnerable to forced labor in domestic service and the retail sector; the steady flow of vessels transiting Trinidad and Tobago's territorial waters may also increase opportunities for forced labor for fishing; international crime organizations are increasingly involved in trafficking, and boys are coerced to sell drugs and guns; corruption among police and immigration officials impedes anti-trafficking efforts

tier rating: Tier 2 Watch List – Trinidad and Tobago does not fully comply with the minimum standards for the elimination of trafficking; however, it is making significant efforts to do so; anti-trafficking law enforcement efforts decreased from the initiation of 12 prosecutions in 2013 to 1 in 2014; the government has yet to convict anyone under its 2011 anti-trafficking law, and all prosecutions from previous years remain pending; the government sustained efforts to identify victims and to refer them for care at NGO facilities, which it provided with funding; the government failed to draft a national action plan as mandated under the 2011 anti-trafficking law and did not launch a sufficiently robust awareness campaign to educate the public and officials (2015)

Illicit drugs: transshipment point for South American drugs destined for the US and Europe; producer of cannabis

TUNISIA

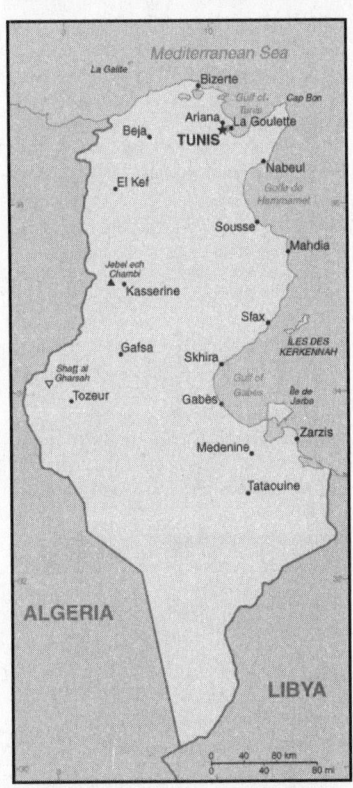

INTRODUCTION

Background: Tunisia has been the nexus of many different colonizations including those of the Phoenicians (as early as the 12 century B.C.), the Carthaginians, Romans, Vandals, Byzantines, various Arab and Berber kingdoms, and the Ottomans (16th to late 19th centuries). Rivalry between French and Italian interests in Tunisia culminated in a French invasion in 1881 and the creation of a protectorate. Agitation for independence in the decades following World War I was finally successful in convincing the French to recognize Tunisia as an independent state in 1956. The country's first president, Habib BOURGUIBA, established a strict one-party state. He dominated the country for 31 years, repressing Islamic fundamentalism and establishing rights for women unmatched by any other Arab nation. In November 1987, BOURGUIBA was removed from office and replaced by Zine el Abidine BEN ALI in a bloodless coup. Street protests that began in Tunis in December 2010 over high unemployment, corruption, widespread poverty, and high food prices escalated in January 2011, culminating in rioting that led to hundreds of deaths. On 14 January 2011, the same day BEN ALI dismissed the government, he fled the country, and by late January 2011, a "national unity government" was formed. Elections for the new Constituent Assembly were held in late October 2011, and in December, it elected human rights activist Moncef MARZOUKI as interim president. The Assembly began drafting a new constitution in February 2012 and, after several iterations and a months-long political crisis that stalled the transition, ratified the document in January 2014.

Parliamentary and presidential elections for a permanent government were held at the end of 2014. Beji CAID ESSEBSI was elected as the first president under the country's new constitution. Following ESSEBSI's death in office in July 2019, Tunisia moved its scheduled presidential election forward two months and after two rounds of voting, Kais SAIED was sworn in as president in October 2019. Tunisia also held legislative elections on schedule in October 2019. SAIED's term, as well as that of Tunisia's 217-member parliament, expires in 2024.

GEOGRAPHY

Location: Northern Africa, bordering the Mediterranean Sea, between Algeria and Libya

Geographic coordinates: 34 00 N, 9 00 E

Map references: Africa

Area: *total:* 163,610 sq km
land: 155,360 sq km
water: 8,250 sq km
country comparison to the world: 94

Area—comparative: slightly larger than Georgia

Land boundaries: *total:* 1,495 km
border countries (2): Algeria 1034 km, Libya 461 km

Coastline: 1,148 km

Maritime claims: *territorial sea:* 12 nm
exclusive economic zone: 12 nm
contiguous zone: 24 nm

Climate: temperate in north with mild, rainy winters and hot, dry summers; desert in south

Terrain: mountains in north; hot, dry central plain; semiarid south merges into the Sahara

Elevation: *mean elevation:* 246 m
lowest point: Shatt al Gharsah -17 m
highest point: Jebel ech Chambi 1,544 m

Natural resources: petroleum, phosphates, iron ore, lead, zinc, salt

Land use: *agricultural land:* 64.8% (2011 est.)
arable land: 18.3% (2011 est.) / permanent crops: 15.4% (2011 est.) / permanent pasture: 31.1% (2011 est.)
forest: 6.6% (2011 est.)
other: 28.6% (2011 est.)

Irrigated land: 4,590 sq km (2012)

Population distribution: the overwhelming majority of the population is located in the northern half of the country; the south remains largely underpopulated as shown in this population distribution map

Natural hazards: flooding; earthquakes; droughts

Environment—current issues: toxic and hazardous waste disposal is ineffective and poses health risks; water pollution from raw sewage; limited natural freshwater resources; deforestation; overgrazing; soil erosion; desertification

Environment—international agreements: *party to:* Biodiversity, Climate Change, Climate Change-Kyoto Protocol, Desertification, Endangered Species, Environmental Modification, Hazardous

Wastes, Law of the Sea, Marine Dumping, Ozone Layer Protection, Ship Pollution, Wetlands
signed, but not ratified: Marine Life Conservation

Geography—note: strategic location in central Mediterranean; Malta and Tunisia are discussing the commercial exploitation of the continental shelf between their countries, particularly for oil exploration

PEOPLE AND SOCIETY

Population: 11, 721,177 (July 2020 est.)
country comparison to the world: 79

Nationality: *noun:* Tunisian(s)
adjective: Tunisian

Ethnic groups: Arab 98%, European 1%, Jewish and other 1%

Languages: Arabic (official, one of the languages of commerce), French (commerce), Berber (Tamazight)
note: despite having no official status, French plays a major role in the country and is spoken by about two thirds of the population

Religions: Muslim (official; Sunni) 99.1%, other (includes Christian, Jewish, Shia Muslim, and Baha'i) 1%

MENA religious affiliation:

Demographic profile: The Tunisian Government took steps in the 1960s to decrease population growth and gender inequality in order to improve socioeconomic development. Through its introduction of a national family planning program (the first in Africa) and by raising the legal age of marriage, Tunisia rapidly reduced its total fertility rate from about 7 children per woman in 1960 to 2 today. Unlike many of its North African and Middle Eastern neighbors, Tunisia will soon be shifting from being a youthbulge country to having a transitional age structure, characterized by lower fertility and mortality rates, a slower population growth rate, a rising median age, and a longer average life expectancy.

Currently, the sizable young working-age population is straining Tunisia's labor market and education and health care systems. Persistent high unemployment among Tunisia's growing workforce, particularly its increasing number of university graduates and women, was a key factor in the uprisings that led to the overthrow of the BEN ALI regime in 2011. In the near term, Tunisia's large number of jobless young, working-age adults; deficiencies in primary and secondary education; and the ongoing lack of job creation and skills mismatches could contribute to future unrest. In the longer term, a sustained low fertility rate will shrink future youth cohorts and alleviate demographic pressure on Tunisia's labor market, but employment and education hurdles will still need to be addressed.

Tunisia has a history of labor emigration. In the 1960s, workers migrated to European countries to escape poor economic conditions and to fill Europe's need for low-skilled labor in construction and manufacturing. The Tunisian Government signed bilateral labor agreements with France, Germany, Belgium, Hungary, and the Netherlands, with the expectation that Tunisian workers would eventually return home. At the same time, growing numbers of Tunisians headed to Libya, often illegally, to work in the expanding oil industry. In the mid-1970s, with European countries beginning to restrict immigration and Tunisian-Libyan tensions brewing, Tunisian economic migrants turned toward the Gulf countries. After mass expulsions from Libya in 1983, Tunisian migrants increasingly sought family reunification in Europe or moved illegally to southern Europe, while Tunisia itself developed into a transit point for Sub-Saharan migrants heading to Europe.

Following the ousting of BEN ALI in 2011, the illegal migration of unemployed Tunisian youths to Italy and onward to France soared into the tens of thousands. Thousands more Tunisian and foreign workers escaping civil war in Libya flooded into Tunisia and joined the exodus. A readmission agreement signed by Italy and Tunisia in April 2011 helped stem the outflow, leaving Tunisia and international organizations to repatriate, resettle, or accommodate some 1 million Libyans and third-country nationals.

Age structure: *0-14 years:* 25.28% (male 1,529,834/female 1,433,357)
15-24 years: 12.9% (male 766,331/female 745,888)
25-54 years: 42.85% (male 2,445,751/female 2,576,335)
55-64 years: 10.12% (male 587,481/female 598,140)
65 years and over: 8.86% (male 491,602/female 546,458) (2020 est.)

Dependency ratios: *total dependency ratio:* 49.6
youth dependency ratio: 36.3
elderly dependency ratio: 13.3
potential support ratio: 7.5 (2020 est.)

Median age: *total:* 32.7 years
male: 32 years
female: 33.3 years (2020 est.)
country comparison to the world: 107

Population growth rate: 0.85% (2020 est.)
country comparison to the world: 124

Birth rate: 15.9 births/1,000 population (2020 est.)
country comparison to the world: 112

Death rate: 6.4 deaths/1,000 population (2020 est.)
country comparison to the world: 146

Net migration rate: -1.4 migrant(s)/1,000 population (2020 est.)
country comparison to the world: 154

Population distribution: the overwhelming majority of the population is located in the northern half of the country; the south remains largely underpopulated as shown in this population distribution map

Urbanization: *urban population:* 69.6% of total population (2020)
rate of urbanization: 1.53% annual rate of change (2015-20 est.)
total population growth rate v. urban population growth rate, 2000-2030:

Major urban areas—population: 2.365 million TUNIS (capital) (2020)

Sex ratio: *at birth:* 1.06 male(s)/female
0-14 years: 1.07 male(s)/female
15-24 years: 1.03 male(s)/female
25-54 years: 0.95 male(s)/female
55-64 years: 0.98 male(s)/female
65 years and over: 0.9 male(s)/female
total population: 0.99 male(s)/female (2020 est.)

Maternal mortality rate: 43 deaths/100,000 live births (2017 est.)
country comparison to the world: 99

Infant mortality rate:
total: 11 deaths/1,000 live births
male: 12 deaths/1,000 live births
female: 9.8 deaths/1,000 live births (2020 est.)
country comparison to the world: 124

Life expectancy at birth: *total population:* 76.3 years
male: 74.6 years
female: 78.1 years (2020 est.)
country comparison to the world: 100

Total fertility rate: 2.06 children born/woman (2020 est.)
country comparison to the world: 105

Contraceptive prevalence rate: 50.7% (2018)

Drinking water source:
improved:
urban: 100% of population
rural: 94.3% of population
total: 98.2% of population
unimproved:
urban: 0% of population
rural: 5.7% of population
total: 1.8% of population (2017 est.)

Current Health Expenditure: 7.2% (2017)

Physicians density: 1.3 physicians/1,000 population (2017)

Hospital bed density: 2.2 beds/1,000 population (2017)

Sanitation facility access:
improved:
urban: 97.6% of population
rural: 92.4% of population
total: 95.9% of population
unimproved:
urban: 2.4% of population
rural: 7.6% of population
total: 4.1% of population (2017 est.)

HIV/AIDS—adult prevalence rate: <.1% (2019 est.)

HIV/AIDS—people living with HIV/AIDS: 6,500 (2019 est.)
country comparison to the world: 116

HIV/AIDS—deaths: <500 (2019 est.)

Obesity—adult prevalence rate: 26.9% (2016)
country comparison to the world: 40

Children under the age of 5 years underweight: 1.6% (2018)
country comparison to the world: 119

Education expenditures: 6.6% of GDP (2015)
country comparison to the world: 19

Literacy: *definition:* age 15 and over can read and write

total population: 81.8%
male: 89.6%
female: 74.2% (2015)

School life expectancy (primary to tertiary education):
total: 15 years
male: 14 years NA
female: 16 years NA (2016)

Unemployment, youth ages 15-24:
total: 35%
male: 34%
female: 37.4% (2015 est.)
country comparison to the world: 22

GOVERNMENT

Country name: *conventional long form:* Republic of Tunisia
conventional short form: Tunisia
local long form: Al Jumhuriyah at Tunisiyah
local short form: Tunis
etymology: the country name derives from the capital city of Tunis

Government type: parliamentary republic

Capital: *name:* Tunis
geographic coordinates: 36 48 N, 10 11 E
time difference: UTC+1 (6 hours ahead of Washington, DC, during Standard Time)
etymology: three possibilities exist for the derivation of the name; originally a Berber settlement (earliest reference 4th century B.C.), the strategic site fell to the Carthaginians (Phoenicians) and the city could be named after the Punic goddess Tanit, since many ancient cities were named after patron deities; alternatively, the Berber root word "ens," which means "to lie down" or "to pass the night," may indicate that the site was originally a camp or rest stop; finally, the name may be the same as the city of Tynes, mentioned in the writings of some ancient authors

Administrative divisions: 24 governorates (wilayat, singular—wilayah); Beja (Bajah), Ben Arous (Bin 'Arus), Bizerte (Banzart), Gabes (Qabis), Gafsa (Qafsah), Jendouba (Jundubah), Kairouan (Al Qayrawan), Kasserine (Al Qasrayn), Kebili (Qibili), Kef (Al Kaf), L'Ariana (Aryanah), Mahdia (Al Mahdiyah), Manouba (Manubah), Medenine (Madanin), Monastir (Al Munastir), Nabeul (Nabul), Sfax (Safaqis), Sidi Bouzid (Sidi Bu Zayd), Siliana (Silyanah), Sousse (Susah), Tataouine (Tatawin), Tozeur (Tawzar), Tunis, Zaghouan (Zaghwan)

Independence: 20 March 1956 (from France)

National holiday: Independence Day, 20 March (1956); Revolution and Youth Day, 14 January (2011)

Constitution: *history:* several previous; latest approved by Constituent Assembly 26 January 2014, signed by the president, prime minister, and Constituent Assembly speaker 27 January 2014
amendments: proposed by the president of the republic or by one third of the Assembly of the Representatives of the People membership; following review by the Constitutional Court, approval to proceed requires an absolute majority vote by the Assembly and final passage requires a two-thirds majority vote by the Assembly; the president can opt to submit an amendment to a referendum, which requires an absolute majority of votes cast for passage

Legal system: mixed legal system of civil law, based on the French civil code and Islamic (sharia) law; some judicial review of legislative acts in the Supreme Court in joint session

International law organization participation: has not submitted an ICJ jurisdiction declaration; accepts ICCt jurisdiction

Citizenship: *citizenship by birth:* no
citizenship by descent only: at least one parent must be a citizen of Tunisia
dual citizenship recognized: yes
residency requirement for naturalization: 5 years

Suffrage: 18 years of age; universal except for active government security forces (including the police and the military), people with mental disabilities, people who have served more than three months in prison (criminal cases only), and people given a suspended sentence of more than six months

Executive branch: *chief of state:* President Kais SAIED (elected 13 October, sworn in 23 October 2019)
head of government: Prime Minister Hichem MECHICHI (since 2 September 2020)
cabinet: selected by the prime minister and approved by the Assembly of the Representatives of the People
elections/appointments: president directly elected by absolute majority popular vote in 2 rounds if needed for a 5-year term (eligible for a second term); last held on 15 September 2019 with a runoff on 13 October 2019 (next to be held in 2024); following legislative elections, the prime minister is selected by the winning party or winning coalition and appointed by the president
election results: first round—Kais SAIED (independent) 18.4%, Nabil KAROUI (Heart of Tunisia) 15.6%, Abdelfattah MOUROU (Nahda Movement) 12.9%, Abdelkrim ZBIDI(independent) 10.7%,Youssef CHAHED (Long Live Tunisia) 7.4%, Safi SAID (independent) 7.1%, Lotfi MRAIHI (Republican People's Union) 6.6%, other 21.3%; runoff—Kais SAIED elected president; Kais SAIED 72.7%, Nabil KAROUI 27.3%

Legislative branch: *description:* unicameral Assembly of the Representatives of the People or Majlis Nuwwab ash-Sha'b (Assemblee des representants du peuple) (217 seats; 199 members directly elected in Tunisian multi-seat constituencies and 18 members in multi-seat constituencies abroad by party-list proportional representation vote; members serve 5-year terms)
elections: initial election held on 6 October 2019 (next to be held in October 2024)
election results: percent of vote by party—Ennahdha 19.6%, Heart of Tunisia 14.6%, Free Destourian Party 6.6%, Democratic Current 6.4%, Dignity Coalition 5.9%, People's Movement 4.5%, TahyaTounes 4.1%, other 35.4%, independent 2.9%;seats by party—Ennahdha 52, Heart of Tunisia 38, Free Destourian Party 17, Democratic Current 22, Dignity Coalition 21, People's Movement 16, Tahya Tounes 14, other 25, independent 12; composition—men 139, women 78, percent of women 35.9%

Judicial branch: *highest courts:* Court of Cassation (consists of the first president, chamber presidents, and magistrates and organized into 27 civil and 11 criminal chambers)
judge selection and term of office: Supreme Court judges nominated by the Supreme Judicial Council, an independent 4-part body consisting mainly of elected judges and the remainder legal specialists; judge tenure based on terms of appointment; Constitutional Court NA
subordinate courts: Courts of Appeal; administrative courts; Court of Audit; Housing Court; courts of first instance; lower district courts; military courts
note: the new Tunisian constitution of January 2014 called for the creation of a constitutional court by the end of 2015, but as of November 2018, the court had not been appointed; the court to consist of 12 members—4 each to be appointed by the president, the Supreme Judicial Council (an independent 4-part body consisting mainly of elected judges and the remainder are legal specialists), and the Chamber of the People's Deputies (parliament); members are to serve 9-year terms with one-third of the membership renewed every 3 years

Political parties and leaders:
Afek Tounes [Yassine BRAHIM]Al Badil Al-Tounisi (The Tunisian Alternative) [Mehdi JOMAA]
Call for Tunisia Party (Nidaa Tounes) [Hafedh CAID ESSEBSI]
Congress for the Republic Party or CPR [Imed DAIMI]
Current of Love [Hachemi HAMDI] (formerly the Popular Petition party)
Democratic Alliance Party [Mohamed HAMDI]
Democratic Current [Mohamed ABBOU]
Democratic Patriots' Unified Party [Zied LAKHDHAR]
Dignity Coalition [Seifeddine MAKHIOUF]
Free Destourian Party [Abir MOUSSI]
Free Patriotic Union (Union patriotique libre) or UPL [Slim RIAHI]
Green Tunisia Party [Abdelkader ZITOUNI]
Heart of Tunisia (Qalb Tounes)
Irada Movement
Long Live Tunisia (Tahya Tounes) [Youssef CHAHED]
Machrou Tounes (Tunisia Project) [Mohsen MARZOUK]
Movement of Socialist Democrats or MDS [Ahmed KHASKHOUSSI]

Ennahda Movement (The Renaissance) [Rachid GHANNOUCHI]

National Destourian Initiative or El Moubadra [Kamel MORJANE]

Party of the Democratic Arab Vanguard [Ahmed JEDDICK, Kheireddine SOUABNI]

People's Movement [Zouheir MAGHZAOUI]

Popular Front (coalition includes Democratic Patriots' Unified Party, Workers' Party, Green Tunisia, Tunisian Ba'ath Movement, Party of the Democratic Arab Vanguard)

Republican Party [Maya JRIBI]

Tunisian Ba'ath Movement [OMAR Othman BELHADJ]

Tunisia First (Tunis Awlan) [Ridha BELHAJ]

Workers' Party [Hamma HAMMAMI]

International organization participation: ABEDA, AfDB, AFESD, AMF, AMU, AU, BSEC (observer), CAEU, CD, EBRD, FAO, G-11, G-77, IAEA, IBRD, ICAO, ICC (national committees), ICCt, ICRM, IDA, IDB, IFAD, IFC, IFRCS, IHO, ILO, IMF, IMO, IMSO, Interpol, IOC, IOM, IPU, ISO, ITSO, ITU, ITUC (NGOs), LAS, MIGA, MONUSCO, NAM, OAS (observer), OIC, OIF, OPCW, OSCE (partner), UN, UNCTAD, UNESCO, UNHCR, UNIDO, UNOCI, UNWTO, UPU, WCO, WFTU (NGOs), WHO, WIPO, WMO, WTO

Diplomatic representation in the US: *chief of mission:* Charge d'Affaires Abdeljelil Ben RABEH (since 24 August 2020)

chancery: 1515 Massachusetts Avenue NW, Washington, DC 20005

telephone: [1] (202) 862-1850

FAX: [1] (202) 862-1858

Diplomatic representation from the US: *chief of mission:* Ambassador Donald A. BLOME (since 21 February 2019)

telephone: [216] 71 107-000

embassy: Les Berges du Lac, 1053 Tunis

mailing address: Zone Nord-Est des Berges du Lac Nord de Tunis 1053

FAX: [216] 71 107-090

Flag description: red with a white disk in the center bearing a red crescent nearly encircling a red five-pointed star; resembles the Ottoman flag (red banner with white crescent and star) and recalls Tunisia's history as part of the Ottoman Empire; red represents the blood shed by martyrs in the struggle against oppression, white stands for peace; the crescent and star are traditional symbols of Islam

note: the flag is based on that of Turkey, itself a successor state to the Ottoman Empire

National symbol(s): encircled red crescent moon and five-pointed star; national colors: red, white

National anthem: *name:* "Humat Al Hima" (Defenders of the Homeland)

lyrics/music: Mustafa Sadik AL-RAFII and Aboul-Qacem ECHEBBI/Mohamad Abdel WAHAB

note: adopted 1957, replaced 1958, restored 1987; Mohamad Abdel WAHAB also composed the music for the anthem of the United Arab Emirates 0:00/ 0:52

ECONOMY

Economy—overview: Tunisia's economy – structurally designed to favor vested interests – faced an array of challenges exposed by the 2008 global financial crisis that helped precipitate the 2011 Arab Spring revolution. After the revolution and a series of terrorist attacks, including on the country's tourism sector, barriers to economic inclusion continued to add to slow economic growth and high unemployment.

Following an ill-fated experiment with socialist economic policies in the 1960s, Tunisia focused on bolstering exports, foreign investment, and tourism, all of which have become central to the country's economy. Key exports now include textiles and apparel, food products, petroleum products, chemicals, and phosphates, with about 80% of exports bound for Tunisia's main economic partner, the EU. Tunisia's strategy, coupled with investments in education and infrastructure, fueled decades of 4-5% annual GDP growth and improved living standards. Former President Zine el Abidine BEN ALI (1987-2011) continued these policies, but as his reign wore on cronyism and corruption stymied economic performance, unemployment rose, and the informal economy grew. Tunisia's economy became less and less inclusive. These grievances contributed to the January 2011 overthrow of BEN ALI, further depressing Tunisia's economy as tourism and investment declined sharply.

Tunisia's government remains under pressure to boost economic growth quickly to mitigate chronic socio-economic challenges, especially high levels of youth unemployment, which has persisted since the 2011 revolution. Ongoing terrorist attacks against the tourism sector and worker strikes in the phosphate sector, which combined account for nearly 15% of GDP, slowed growth from 2015 to 2017. Tunis is seeking increased foreign investment and working with the IMF through an Extended Fund Facility agreement to fix fiscal deficiencies.

GDP (purchasing power parity):
$137.7 billion (2017 est.)
$135 billion (2016 est.)
$133.5 billion (2015 est.)
note: data are in 2017 dollars
country comparison to the world: 78

GDP (official exchange rate):
$39.96 billion (2017 est.)

GDP—real growth rate:
2% (2017 est.)
1.1% (2016 est.)
1.2% (2015 est.)
country comparison to the world: 140

GDP—per capita (PPP):
$11,900 (2017 est.)
$11,800 (2016 est.)
$11,800 (2015 est.)
note: data are in 2017 dollars
country comparison to the world: 131

Gross national saving:
12% of GDP (2017 est.)

13.4% of GDP (2016 est.)
12.5% of GDP (2015 est.)
country comparison to the world: 152

GDP—composition, by end use:
household consumption: 71.7% (2017 est.)
government consumption: 20.8% (2017 est.)
investment in fixed capital: 19.4% (2017 est.)
investment in inventories: 0% (2017 est.)
exports of goods and services: 43.2% (2017 est.)
imports of goods and services: -55.2% (2017 est.)

GDP—composition, by sector of origin:
agriculture: 10.1% (2017 est.)
industry: 26.2% (2017 est.)
services: 63.8% (2017 est.)

Agriculture—products: olives, olive oil, grain, tomatoes, citrus fruit, sugar beets, dates, almonds; beef, dairy products

Industries: petroleum, mining (particularly phosphate, iron ore), tourism, textiles, footwear, agribusiness, beverages

Industrial production growth rate: 0.5% (2017 est.)
country comparison to the world: 166

Labor force: 4.054 million (2017 est.)
country comparison to the world: 89

Labor force—by occupation:
agriculture: 14.8%
industry: 33.2%
services: 51.7% (2014 est.)

Unemployment rate:
15.5% (2017 est.)
15.5% (2016 est.)
country comparison to the world: 176

Population below poverty line: 15.5% (2010 est.)

Household income or consumption by percentage share: *lowest 10%:* 2.6%
highest 10%: 27% (2010 est.)

Budget: *revenues:* 9.876 billion (2017 est.)
expenditures: 12.21 billion (2017 est.)

Taxes and other revenues: 24.7% (of GDP) (2017 est.)
country comparison to the world: 120

Budget surplus (+) or deficit (-): -5.8% (of GDP) (2017 est.)
country comparison to the world: 180

Public debt:
70.3% of GDP (2017 est.)
62.3% of GDP (2016 est.)
country comparison to the world: 51

Fiscal year: calendar year

Inflation rate (consumer prices):
5.3% (2017 est.)
3.7% (2016 est.)
country comparison to the world: 176

Current account balance:
-$4.191 billion (2017 est.)
-$3.694 billion (2016 est.)
country comparison to the world: 180

Exports:
$13.82 billion (2017 est.)
$13.57 billion (2016 est.)
country comparison to the world: 78

Exports—partners: France 32.1%, Italy 17.3%, Germany 12.4% (2017)

Exports—commodities: clothing, semi-finished goods and textiles, agricultural products, mechanical goods, phosphates and chemicals, hydrocarbons, electrical equipment

Imports:
$19.09 billion (2017 est.)
$18.37 billion (2016 est.)
country comparison to the world: 79

Imports—commodities: textiles, machinery and equipment, hydrocarbons, chemicals, foodstuffs

Imports—partners: Italy 15.8%, France 15.1%, China 9.2%, Germany 8.1%, Turkey 4.8%, Algeria 4.7%, Spain 4.5% (2017)

Reserves of foreign exchange and gold:
$5.594 billion (31 December 2017 est.)
$5.941 billion (31 December 2016 est.)
country comparison to the world: 93

Debt—external:
$30.19 billion (31 December 2017 est.)
$28.95 billion (31 December 2016 est.)
country comparison to the world: 79

Exchange rates: Tunisian dinars (TND) per US dollar -
2.48 (2017 est.)
2.148 (2016 est.)
2.148 (2015 est.)
1.9617 (2014 est.)
1.6976 (2013 est.)

ENERGY

Electricity access: electrification—total population: 100% (2020)

Electricity—production: 18.44 billion kWh (2016 est.)
country comparison to the world: 78

Electricity—consumption: 15.27 billion kWh (2016 est.)
country comparison to the world: 79

Electricity—exports: 500 million kWh (2015 est.)
country comparison to the world: 68

Electricity—imports: 134 million kWh (2016 est.)
country comparison to the world: 96

Electricity—installed generating capacity: 5.768 million kW (2016 est.)
country comparison to the world: 77

Electricity—from fossil fuels: 94% of total installed capacity (2016 est.)
country comparison to the world: 49

Electricity—from nuclear fuels: 0% of total installed capacity (2017 est.)
country comparison to the world: 198

Electricity—from hydroelectric plants: 1% of total installed capacity (2017 est.)
country comparison to the world: 151

Electricity—from other renewable sources: 5% of total installed capacity (2017 est.)
country comparison to the world: 109

Crude oil—production: 39,000 bbl/day (2018 est.)
country comparison to the world: 59

Crude oil—exports: 39,980 bbl/day (2015 est.)
country comparison to the world: 42

Crude oil—imports: 17,580 bbl/day (2015 est.)
country comparison to the world: 66

Crude oil—proved reserves: 425 million bbl (1 January 2018 est.)
country comparison to the world: 48

Refined petroleum products—production: 27,770 bbl/day (2015 est.)
country comparison to the world: 85

Refined petroleum products—consumption: 102,000 bbl/day (2016 est.)
country comparison to the world: 79

Refined petroleum products—exports: 13,660 bbl/day (2015 est.)
country comparison to the world: 75

Refined petroleum products—imports: 85,340 bbl/day (2015 est.)
country comparison to the world: 58

Natural gas—production: 1.274 billion cu m (2017 est.)
country comparison to the world: 64

Natural gas—consumption: 5.125 billion cu m (2017 est.)
country comparison to the world: 59

Natural gas—exports: 0 cu m (2017 est.)
country comparison to the world: 202

Natural gas—imports: 3.851 billion cu m (2017 est.)
country comparison to the world: 41

Natural gas—proved reserves: 65.13 billion cu m (1 January 2018 est.)
country comparison to the world: 58

Carbon dioxide emissions from consumption of energy: 23.42 million Mt (2017 est.)
country comparison to the world: 82

COMMUNICATIONS

Telephones—fixed lines: *total subscriptions:* 1,444,631
subscriptions per 100 inhabitants: 12.43 (2019 est.)
country comparison to the world: 66

Telephones—mobile cellular: *total subscriptions:* 14,679,917
subscriptions per 100 inhabitants: 126.31 (2019 est.)
country comparison to the world: 68

Telecommunication systems: *general assessment:* above the African average and continuing to be upgraded; key centers are Sfax, Sousse, Bizerte, and Tunis; telephone network is completely digitized; Internet access available throughout the country; penetration rates for mobile and Internet services are among the highest in the region; 3 MNOs (mobile network operator); government Internet censorship abolished in 2013; telecom invests in LTE network and fiber infrastructure with FttP (fiber to the premises) services; 5G license expected to be launched soon; auction of spectrum in the 800MHz band IoT (location of Things) and mobile services; use of Chinese company Huawei to develop LTE network (2020)

domestic: in an effort to jumpstart expansion of the fixed-line network, the government awarded a concession to build and operate a VSAT network with international connectivity; rural areas are served by wireless local loops; competition between several mobile-cellular service providers has resulted in lower activation and usage charges and a strong surge in subscribership; fixed-line is 12 per 100 and mobile-cellular teledensity has reached about 126 telephones per 100 persons (2019)

international: country code—216; landing points for the SEA-ME-WE-4, Didon, HANNIBAL System and TrapaniKelibia submarine cable systems that provides links to Europe, Africa, the Middle East, Asia and Southeast Asia; satellite earth stations—1 Intelsat (Atlantic Ocean) and 1 Arabsat; coaxial cable and microwave radio relay to Algeria and Libya; participant in Medarabtel; 2 international gateway digital switches (2020)

note: the COVID-19 outbreak is negatively impacting telecommunications production and supply chains globally; consumer spending on telecom devices and services has also slowed due to the pandemic's effect on economies worldwide; overall progress towards improvements in all facets of the telecom industry—mobile, fixed-line, broadband, submarine cable and satellite—has moderated

Broadcast media: 1 state-owned TV station with multiple transmission sites; 5 private TV stations broadcast locally; cable TV service is available; state-owned radio network with 2 stations (in Lome and Kara); several dozen private radio stations and a few community radio stations; transmissions of multiple international broadcasters available (2019)

Internet country code: .tn

Internet users: *total:* 7,392,242
percent of population: 64.19% (July 2018 est.)
country comparison to the world: 70

Broadband—fixed subscriptions: *total:* 1,014,395
subscriptions per 100 inhabitants: 9 (2018 est.)
country comparison to the world: 69

TRANSPORTATION

National air transport system: *number of registered air carriers:* 7 (2020)
inventory of registered aircraft operated by air carriers: 53
annual passenger traffic on registered air carriers: 4,274,199 (2018)
annual freight traffic on registered air carriers: 13.23 million mt-km (2018)

Civil aircraft registration country code prefix: TS (2016)

Airports: 29 (2013)
country comparison to the world: 119

Airports—with paved runways:
total: 15 (2013)
over 3,047 m: 4 (2013)
2,438 to 3,047 m: 6 (2013)
1,524 to 2,437 m: 2 (2013)
914 to 1,523 m: 3 (2013)

Airports—with unpaved runways:
total: 14 (2013)
1,524 to 2,437 m: 1 (2013)
914 to 1,523 m: 5 (2013)
under 914 m: 8 (2013)

Pipelines: 68 km condensate, 3111 km gas, 1381 km oil, 453 km refined products (2013)

Railways: *total:* 2,173 km (1,991 in use) (2014)
standard gauge: 471 km 1.435-m gauge (2014)
narrow gauge: 1,694 km 1.000-m gauge (65 km electrified) (2014)
dual gauge: 8 km 1.435-1.000-m gauge (2014)
country comparison to the world: 70

Roadways: *paved:* 20,000 km (2015)

Merchant marine: *total:* 67
by type: general cargo 9, oil tanker 1, other 57 (2019)
country comparison to the world: 105

Ports and terminals: *major seaport(s):* Bizerte, Gabes, Rades, Sfax, Skhira

MILITARY AND SECURITY

Military and security forces: *Tunisian Armed Forces (Forces Armees Tunisiens, FAT):* Tunisian Army (includes Tunisian Air Defense Force), Tunisian Navy, Republic of Tunisia Air Force; Ministry of Interior: Tunisian National Guard (2020)

Military expenditures:
2.6% of GDP (2019)
2.1% of GDP (2018)
2.1% of GDP (2017)

2.4% of GDP (2016)
2.3% of GDP (2015)
country comparison to the world: 35

Military and security service personnel strengths: the Tunisian Armed Forces (FAT) have approximately 36,000 active personnel (27,000 Army; 5,000 Navy; 4,000 Air Force); est. 12,000 National Guard (2019)

Military equipment inventories and acquisitions: the Tunisian military's inventory includes mostly older or secondhand US and European equipment; since 2010, the Netherlands and US are the leading suppliers of arms to Tunisia (2019 est.)

Military service age and obligation: 20-23 years of age for compulsory service, 1-year service obligation; 18-23 years of age for voluntary service (2019)

TERRORISM

Terrorist group(s): Ansar al-Sharia in Tunisia; Islamic State of Iraq and ash-Sham (ISIS) network in Tunisia; al-Qa'ida in the Islamic Maghreb (2019)
note: details about the history, aims, leadership, organization, areas of operation, tactics, targets, weapons, size, and sources of support of the group(s) appear(s) in Appendix-T

TRANSNATIONAL ISSUES

Disputes—international: none

Trafficking in persons: *current situation:* Tunisia is a source, destination, and possible transit country for men, women, and children subjected to forced labor and sex trafficking; Tunisia's increased number of street children, rural children working to support their families, and migrants who have fled unrest in neighboring countries are vulnerable to human trafficking; organized gangs force street children to serve as thieves, beggars, and drug transporters; Tunisian women have been forced into prostitution domestically and elsewhere in the region under false promises of legitimate work; East and West African women may be subjected to forced labor as domestic workers

tier rating: Tier 2 Watch List – Tunisia does not fully comply with the minimum standards for the elimination of trafficking; however, it is making significant efforts to do so; in 2014, Tunisia was granted a waiver from an otherwise required downgrade to Tier 3 because its government has a written plan that, if implemented would constitute making significant efforts to bring itself into compliance with the minimum standards for the elimination of trafficking; in early 2015, the government drafted a national anti-trafficking action plan outlining proposals to raise awareness and enact draft antitrafficking legislation; authorities did not provide data on the prosecution and conviction of offenders but reportedly identified 24 victims, as opposed to none in 2013, and operated facilities specifically dedicated to trafficking victims, regardless of nationality and gender; the government did not fully implement its national victim referral mechanism; some unidentified victims were not protected from punishment for unlawful acts directly resulting from being trafficked (2015)

TURKEY

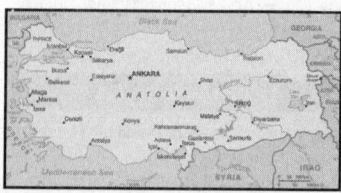

INTRODUCTION

Background: Modern Turkey was founded in 1923 from the remnants of the defeated Ottoman Empire by national hero Mustafa KEMAL, who was later honored with the title Ataturk or "Father of the Turks." Under his leadership, the country adopted radical social, legal, and political reforms. After a period of one-party rule, an experiment with multi-party politics led to the 1950 election victory of the opposition Democrat Party and the peaceful transfer of power. Since then, Turkish political parties have multiplied, but democracy has been fractured by periods of instability and military coups (1960, 1971, 1980), which in each case eventually resulted in a return of formal political power to civilians. In 1997, the military again helped engineer the ouster - popularly dubbed a "post-modern coup" - of the then Islamic-oriented government. An unsuccessful coup attempt was made in July 2016 by a faction of the Turkish Armed Forces.

Turkey intervened militarily on Cyprus in 1974 to prevent a Greek takeover of the island and has since acted as patron state to the "Turkish Republic of Northern Cyprus," which only Turkey recognizes. A separatist insurgency begun in 1984 by the Kurdistan Workers' Party (PKK), a US-designated terrorist organization, has long dominated the attention of Turkish security forces and claimed more than 40,000 lives. In 2013, the Turkish Government and the PKK conducted negotiations aimed at ending the violence, however intense fighting resumed in 2015. Turkey joined the UN in 1945 and in 1952 it became a member of NATO. In 1963, Turkey became an associate member of the European Community; it began accession talks with the EU in 2005. Over the past decade, economic reforms, coupled with some political reforms, have contributed to a growing economy, although economic growth slowed in recent years.

From 2015 and continuing through 2016, Turkey witnessed an uptick in terrorist violence, including major attacks in Ankara, Istanbul, and throughout the predominantly Kurdish southeastern region of Turkey. On 15 July 2016, elements of the Turkish Armed forces attempted a coup that ultimately failed following widespread popular resistance. More than 240 people were killed and over 2,000 injured when Turkish citizens took to the streets en masse to confront the coup forces. The government accused followers of the Fethullah Gulen transnational religious and social movement ("Hizmet") for allegedly instigating the failed coup and designates the movement's followers as terrorists. Since the attempted coup, Turkish Government authorities arrested, suspended, or dismissed more than 130,000 security personnel, journalists, judges, academics, and civil servants due to their alleged connection to Gulen's movement. Following the failed coup, the Turkish Government instituted a State of Emergency from July 2016 to July 2018. The Turkish Government conducted a referendum on 16 April 2017 in which voters approved constitutional amendments changing Turkey from a parliamentary to a presidential system. The amendments went into effect fully following the presidential and parliamentary elections in June 2018.

GEOGRAPHY

Location: Southeastern Europe and Southwestern Asia (that portion of Turkey west of the Bosporus is geographically part of Europe), bordering the Black Sea, between Bulgaria and Georgia, and bordering the Aegean Sea and the Mediterranean Sea, between Greece and Syria

Geographic coordinates: 39 00 N, 35 00 E

Map references: Middle East

Area: *total:* 783,562 sq km
land: 769,632 sq km
water: 13,930 sq km
country comparison to the world: 38

Area—comparative: slightly larger than Texas

Area comparison map: [INSERT IMAGE: TURKEY-Area comparison map]

Land boundaries: *total:* 2,816 km
border countries (8): Armenia 311 km, Azerbaijan 17 km, Bulgaria 223 km, Georgia 273 km, Greece 192 km, Iran 534 km, Iraq 367 km, Syria 899 km

Coastline: 7,200 km

Maritime claims: *territorial sea:* 6 nm in the Aegean Sea
exclusive economic zone: in Black Sea only: to the maritime boundary agreed upon with the former USSR
12 nm in Black Sea and in Mediterranean Sea

Climate: temperate; hot, dry summers with mild, wet winters; harsher in interior

Terrain: high central plateau (Anatolia); narrow coastal plain; several mountain ranges

Elevation: *mean elevation:* 1,132 m
lowest point: Mediterranean Sea 0 m
highest point: Mount Ararat 5,137 m

Natural resources: coal, iron ore, copper, chromium, antimony, mercury, gold, barite, borate, celestite (strontium), emery, feldspar, limestone, magnesite, marble, perlite, pumice, pyrites (sulfur), clay, arable land, hydropower

Land use: *agricultural land:* 49.7% (2011 est.)
arable land: 26.7% (2011 est.) / permanent crops: 4% (2011 est.) / permanent pasture: 19% (2011 est.)
forest: 14.9% (2011 est.)
other: 35.4% (2011 est.)

Irrigated land: 52,150 sq km (2012)

Population distribution: the most densely populated area is found around the Bosporus in the northwest where 20% of the population lives in Istanbul; with the exception of Ankara, urban centers remain small and scattered throughout the interior of Anatolia; an overall pattern of peripheral development exists, particularly along the Aegean Sea coast in the west, and the Tigris and Euphrates River systems in the southeast

Natural hazards: severe earthquakes, especially in northern Turkey, along an arc extending from the Sea of Marmara to Lake Van; landslides; flooding
volcanism: limited volcanic activity; its three historically active volcanoes; Ararat, Nemrut Dagi, and Tendurek Dagi have not erupted since the 19th century or earlier

Environment—current issues: water pollution from dumping of chemicals and detergents; air pollution, particularly in urban areas; deforestation; land degradation; concern for oil spills from increasing Bosporus ship traffic; conservation of biodiversity

Environment—international agreements: *party to:* Air Pollution, Antarctic Treaty, Biodiversity, Climate Change, Desertification, Endangered Species, Hazardous Wastes, Ozone Layer Protection, Ship Pollution, Wetlands
signed, but not ratified: Environmental Modification

Geography—note: strategic location controlling the Turkish Straits (Bosporus, Sea of Marmara, Dardanelles) that link the Black and Aegean Seas; the 3% of Turkish territory north of the Straits lies in Europe and goes by the names of European Turkey, Eastern Thrace, or Turkish Thrace; the 97% of the country in Asia is referred to as Anatolia; Istanbul, which straddles the Bosporus, is the only metropolis in the world located on two continents; Mount Ararat, the legendary landing place of Noah's ark, is in the far eastern portion of the country

PEOPLE AND SOCIETY

Population: 82,017,514 (July 2020 est.)
country comparison to the world: 18

Nationality: *noun:* Turk(s)
adjective: Turkish

Ethnic groups: Turkish 70-75%, Kurdish 19%, other minorities 7-12% (2016 est.)

Languages: Turkish (official), Kurdish, other minority languages

Religions: Muslim 99.8% (mostly Sunni), other 0.2% (mostly Christians and Jews)

Age structure: *0-14 years:* 23.41% (male 9,823,553/female 9,378,767)
15-24 years: 15.67% (male 6,564,263/female 6,286,615)
25-54 years: 43.31% (male 17,987,103/female 17,536,957)
55-64 years: 9.25% (male 3,764,878/female 3,822,946)
65 years and over: 8.35% (male 3,070,258/female 3,782,174) (2020 est.)

population pyramid: [INSERT IMAGE: TURKEY-population pyramid]

Dependency ratios: *total dependency ratio:* 49.1
youth dependency ratio: 35.7
elderly dependency ratio: 13.4
potential support ratio: 7.5 (2020 est.)

Median age: *total:* 32.2 years
male: 31.7 years
female: 32.8 years (2020 est.)

country comparison to the world: 111

Population growth rate: 0.45% (2020 est.)
country comparison to the world: 159

Birth rate: 14.8 births/1,000 population (2020 est.)
country comparison to the world: 123

Death rate: 6.1 deaths/1,000 population (2020 est.)
country comparison to the world: 161

Net migration rate: -4.3 migrant(s)/1,000 population (2020 est.)
country comparison to the world: 191

Population distribution: the most densely populated area is found around the Bosporus in the northwest where 20% of the population lives in Istanbul; with the exception of Ankara, urban centers remain small and scattered throughout the interior of Anatolia; an overall pattern of peripheral development exists, particularly along the Aegean Sea coast in the west, and the Tigris and Euphrates River systems in the southeast

Urbanization: *urban population:* 76.1% of total population (2020)
rate of urbanization: 2.04% annual rate of change (2015-20 est.)

total population growth rate v. urban population growth rate, 2000-2030: *Major urban areas - population:* 15.190 million Istanbul, 5.118 million ANKARA (capital), 2.993 million Izmir, 1.986 million Bursa, 1.771 million Adana, 1.704 million Gaziantep (2020)

Sex ratio: *at birth:* 1.05 male(s)/female
0-14 years: 1.05 male(s)/female
15-24 years: 1.04 male(s)/female
25-54 years: 1.03 male(s)/female
55-64 years: 0.98 male(s)/female
65 years and over: 0.81 male(s)/female
total population: 1.01 male(s)/female (2020 est.)

Mother's mean age at first birth: 22.3 years (2010 est.)

Maternal mortality rate: 17 deaths/100,000 live births (2017 est.)
country comparison to the world: 133

Infant mortality rate: *total:* 15.8 deaths/1,000 live births
male: 16.9 deaths/1,000 live births
female: 14.6 deaths/1,000 live births (2020 est.)
country comparison to the world: 95

Life expectancy at birth: *total population:* 75.7 years
male: 73.3 years
female: 78.2 years (2020 est.)
country comparison to the world: 110

Total fertility rate: 1.96 children born/woman (2020 est.)
country comparison to the world: 118

Contraceptive prevalence rate: 69.8% (2018)

Drinking water source:

improved: *urban:* 98.6% of population
rural: 100% of population
total: 98.9% of population

unimproved: *urban:* 1.4% of population
rural: 0% of population

total: 1.1% of population (2017 est.)

Current Health Expenditure: 4.2% (2017)

Physicians density: 1.85 physicians/1,000 population (2017)

Hospital bed density: 2.8 beds/1,000 population (2017)

Sanitation facility access:

improved: *urban:* 100% of population
rural: 91.6% of population
total: 97.3% of population

unimproved: *urban:* 0% of population
rural: 8.4% of population
total: 2.7% of population (2017 est.)

HIV/AIDS—adult prevalence rate: NA

HIV/AIDS—people living with HIV/AIDS: NA

HIV/AIDS—deaths: NA

Major infectious diseases: *Covid-19 (2020)*
note: widespread ongoing transmission of a respiratory illness caused by the novel coronavirus (COVID-19) is occurring throughout Turkey; as of 10 November 2020, Turkey has reported a total of 391,739 cases of COVID-19 or 4,645 cumulative cases of COVID-19 per 1 million population with 128 cumulative deaths per 1 million population

Obesity—adult prevalence rate: 32.1% (2016)
country comparison to the world: 17

Children under the age of 5 years underweight: 1.5% (2018/19)
country comparison to the world: 121

Education expenditures: 4.3% of GDP (2015)
country comparison to the world: 91

Literacy: *definition:* age 15 and over can read and write
total population: 96.2%
male: 98.8%
female: 93.5% (2017)
School life expectancy (primary to tertiary education): total: 18 years
male: 19 years
female: 18 years (2018)
Unemployment, youth ages 15-24: total: 20.2%
male: 17.5%
female: 25% (2018 est.)
country comparison to the world: 68

GOVERNMENT

Country name: *conventional long form:* Republic of Turkey
conventional short form: Turkey
local long form: Turkiye Cumhuriyeti
local short form: Turkiye
etymology: the name means "Land of the Turks"

Government type: presidential republic

Capital: *name:* Ankara

geographic coordinates: 39 56 N, 32 52 E
time difference: UTC+2 (7 hours ahead of Washington, DC, during Standard Time)
etymology: Ankara has been linked with a second millennium B.C. Hittite cult center of Ankuwash, although this connection is uncertain; in classical and medieval times, the city was known as Ankyra

(meaning "anchor" in Greek and reflecting the city's position as a junction for multiple trade and military routes); by about the 13th century the city began to be referred to as Angora; following the establishment of the Republic of Turkey in 1923, the city's name became Ankara

Administrative divisions: 81 provinces (iller, singular - ili); Adana, Adiyaman, Afyonkarahisar, Agri, Aksaray, Amasya, Ankara, Antalya, Ardahan, Artvin, Aydin, Balikesir, Bartin, Batman, Bayburt, Bilecik, Bingol, Bitlis, Bolu, Burdur, Bursa, Canakkale, Cankiri, Corum, Denizli, Diyarbakir, Duzce, Edirne, Elazig, Erzincan, Erzurum, Eskisehir, Gaziantep, Giresun, Gumushane, Hakkari, Hatay, Igdir, Isparta, Istanbul, Izmir (Smyrna), Kahramanmaras, Karabuk, Karaman, Kars, Kastamonu, Kayseri, Kilis, Kirikkale, Kirklareli, Kirsehir, Kocaeli, Konya, Kutahya, Malatya, Manisa, Mardin, Mersin, Mugla, Mus, Nevsehir, Nigde, Ordu, Osmaniye, Rize, Sakarya, Samsun, Sanliurfa, Siirt, Sinop, Sirnak, Sivas, Tekirdag, Tokat, Trabzon (Trebizond), Tunceli, Usak, Van, Yalova, Yozgat, Zonguldak

Independence: 29 October 1923 (republic proclaimed, succeeding the Ottoman Empire)

National holiday: Republic Day, 29 October (1923)

Constitution: *history:* several previous; latest ratified 9 November 1982

amendments: proposed by written consent of at least one third of Grand National Assembly of Turkey (TBMM) members; adoption of draft amendments requires two debates in plenary TBMM session and three-fifths majority vote of all GNA members; the president of the republic can request TBMM reconsideration of the amendment and, if readopted by two- thirds majority TBMM vote, the president may submit the amendment to a referendum; passage by referendum requires absolute majority vote; amended several times, last in 2017

Legal system: civil law system based on various European legal systems, notably the Swiss civil code

International law organization participation: has not submitted an ICJ jurisdiction declaration; non-party state to the ICCt

Citizenship: *citizenship by birth:* no
citizenship by descent only: at least one parent must be a citizen of Turkey
dual citizenship recognized: yes, but requires prior permission from the government
residency requirement for naturalization: 5 years

Suffrage: 18 years of age; universal

Executive branch: *chief of state:* President Recep Tayyip ERDOGAN (chief of state since 28 August 2014; head of government since 9 July 2019); Vice President Fuat OKTAY (since 9 July 2018); note - the president is both chief of state and head of government

head of government: President Recep Tayyip ERDOGAN (head of government since 9 July

2019; chief of state since 28 August 2014); note - a 2017 constitutional referendum eliminated the post of prime minister after the 2018 general election

cabinet: Council of Ministers appointed by the president

elections/appointments: president directly elected by absolute majority popular vote in 2 rounds if needed for a 5-year term (eligible for a second term); election last held on 24 June 2018 (next scheduled for June 2023)

election results: Recep Tayyip ERDOGAN reelected president in the first round; Recep Tayyip ERDOGAN (AKP) 52.6%, Muharrem INCE (CHP) 30.6%, Selahattin DEMIRTAS (HDP) 8.4%, Meral AKSENER (IYI) 7.3%, other 1.1%

Legislative branch: *description:* unicameral Grand National Assembly of Turkey or Turkiye Buyuk Millet Meclisi (600 seats - increased from 550 seats beginning with June 2018 election; members directly elected in multi-seat constituencies by proportional representation vote to serve 5-year terms - increased from 4 to 5 years beginning with June 2018 election)

elections: last held on 24 June 2018 (next to be held in June 2023)

election results: percent of vote by party - People's Alliance 53.7% (AKP 42.6%, MHP 11.1%), Nation Alliance 33.9% (CHP 22.6%, IYI 10%, SP 1.3%), HDP 11.7%, other 0.7%; seats by party - People's Alliance 344 (AKP 295, MHP 49), National Alliance 189 (CHP 146, IYI 43), HDP 67; composition - men 496, women 104, percent of women 17.3%; note - only parties surpassing a 10% threshold can win parliamentary seats

Judicial branch: *highest courts:* Constitutional Court or Anayasa Mahkemesi (consists of the president, 2 vice presidents, and 12 judges); Court of Cassation (consists of about 390 judges and is organized into civil and penal chambers); Council of State (organized into 15 divisions - 14 judicial and 1 consultative - each with a division head and at least 5 members)

judge selection and term of office: Constitutional Court members - 3 appointed by the Grand National Assembly and 12 by the president of the republic; court president and 2 deputy court presidents appointed from among its members for 4 year terms; judges serve 12-year, nonrenewable terms with mandatory retirement at age 65; Court of Cassation judges appointed by the Board of Judges and Prosecutors, a 13-member body of judicial officials; Court of Cassation judges serve until retirement at age 65; Council of State members appointed by the Board and by the president of the republic; members serve renewable, 4-year terms

subordinate courts: regional appeals courts; basic (first instance) courts; peace courts; aggravated crime courts; specialized courts, including administrative and audit; note - a constitutional amendment in 2017 abolished military courts unless established to investigate military personnel actions during war conditions

Political parties and leaders: Democrat Party or DP [Gultekin UYSAL]

Democratic Regions Party or DBP [Sebahat TUNCEL, Mehmet ARSLAN]
Felicity Party or SP [Temel KARAMOLLAOGLU]
Free Cause Party or HUDAPAR [Ishak SAGLAM]
Good Party or TYIi [Meral AKSENER]
Grand Unity Party or BBP [Mustafa DESTICI]
Justice and Development Party or AKP [Recep Tayyip ERDOGAN]
Nation Alliance (CHP, IYI, SP) (electoral alliance)
Nationalist Movement Party or MHP [Devlet BAHCELI]
People's Alliance (AKP, MHP) (electoral alliance)
Patriotic Party or VP [Dogu PERINCEK]
People's Democratic Party or HDP [Pervin BULDAN, Sezai TEMELLI]
Republican People's Party or CHP [Kemal KILICDAROGLU]
note: as of December 2018, 83 political parties were legally registered

International organization participation: ADB (nonregional member), Australia Group, BIS, BSEC, CBSS (observer), CD, CE, CERN (observer), CICA, CPLP (associate observer), D-8, EAPC, EBRD, ECO, EU (candidate country), FAO, FATF, G-20, IAEA, IBRD, ICAO, ICC (national committees), ICRM, IDA, IDB, IEA, IFAD, IFC, IFRCS, IHO, ILO, IMF, IMO, IMSO, Interpol, IOC, IOM, IPU, ISO, ITSO, ITU, ITUC (NGOs), MIGA, NATO, NEA, NSG, OAS (observer), OECD, OIC, OPCW, OSCE, Pacific Alliance (observer), Paris Club (associate), PCA, PIF (partner), SCO (dialogue member), SELEC, UN, UNCTAD, UNESCO, UNHCR, UNIDO, UNIFIL, UNRWA, UNWTO, UPU, WCO, WFTU (NGOs), WHO, WIPO, WMO, WTO, ZC

Diplomatic representation in the US: *chief of mission:* Ambassador Serdar KILIC (since 21 May 2014)
chancery: 2525 Massachusetts Avenue NW, Washington, DC 20008
telephone: [1] (202) 612-6700
FAX: [1] (202) 612-6744

consulate(s) general: Boston, Chicago, Houston, Los Angeles, Miami, New York

Diplomatic representation from the US: *chief of mission:* Ambassador David M. SATTERFIELD (since 28 August 2019)
telephone: [90] (312) 455-5555
embassy: 110 Ataturk Boulevard, Kavaklidere, 06100 Ankara
mailing address: PSC 93, Box 5000, APO AE 09823
FAX: [90] (312) 467-0019

consulate(s) general: Istanbul

consulate(s): Adana

Flag description: red with a vertical white crescent moon (the closed portion is toward the hoist side) and white five-pointed star centered just outside the crescent opening; the flag colors and designs closely resemble those on the banner of the Ottoman Empire, which preceded modern-day Turkey; the crescent moon and star serve as insignia for Turkic peoples; according to one interpretation, the flag represents the reflection of the moon and a star in a pool of blood of Turkish warriors

National symbol(s): *vertical crescent moon with adjacent five-pointed star; national colors:* red, white

National anthem: *name:* "Istiklal Marsi" (Independence March)

lyrics/music: Mehmet Akif ERSOY/Zeki UNGOR *note:* lyrics adopted 1921, music adopted 1932; the anthem's original music was adopted in 1924; a new composition was agreed upon in 1932
0:00 / 1:23

ECONOMY

Economy—overview: Turkey's largely free-market economy is driven by its industry and, increasingly, service sectors, although its traditional agriculture sector still accounts for about 25% of employment. The automotive, petrochemical, and electronics industries have risen in importance and surpassed the traditional textiles and clothing sectors within Turkey's export mix. However, the recent period of political stability and economic dynamism has given way to domestic uncertainty and security concerns, which are generating financial market volatility and weighing on Turkey's economic outlook.

Current government policies emphasize populist spending measures and credit breaks, while implementation of structural economic reforms has slowed. The government is playing a more active role in some strategic sectors and has used economic institutions and regulators to target political opponents, undermining private sector confidence in the judicial system. Between July 2016 and March 2017, three credit ratings agencies downgraded Turkey's sovereign credit ratings, citing concerns about the rule of law and the pace of economic reforms.

Turkey remains highly dependent on imported oil and gas but is pursuing energy relationships with a broader set of international partners and taking steps to increase use of domestic energy sources including renewables, nuclear, and coal. The joint Turkish-Azerbaijani Trans-Anatolian Natural Gas Pipeline is moving forward to increase transport of Caspian gas to Turkey and Europe, and when completed will help diversify Turkey's sources of imported gas.

After Turkey experienced a severe financial crisis in 2001, Ankara adopted financial and fiscal reforms as part of an IMF program. The reforms strengthened the country's economic fundamentals and ushered in an era of strong growth, averaging more than 6% annually until 2008. An aggressive privatization program also reduced state involvement in basic industry, banking, transport, power generation, and communication. Global economic conditions and tighter fiscal policy caused GDP to contract in 2009, but Turkey's well-regulated financial markets and banking system helped the country weather the global financial crisis, and GDP growth rebounded to around 9% in 2010 and 2011, as exports and investment recovered following the crisis.

The growth of Turkish GDP since 2016 has revealed the persistent underlying imbalances in the Turkish economy. In particular, Turkey's large current account deficit means it must rely on external investment inflows to finance growth, leaving the economy vulnerable to destabilizing shifts in investor confidence. Other troublesome trends include rising unemployment and inflation, which increased in 2017, given the Turkish lira's continuing depreciation against the dollar. Although government debt remains low at about 30% of GDP, bank and corporate borrowing has almost tripled as a percent of GDP during the past decade, outpacing its emerging-market peers and prompting investor concerns about its long-term sustainability.

GDP (purchasing power parity): $2.186 trillion (2017 est.)
$2.034 trillion (2016 est.)
$1.972 trillion (2015 est.)
note: data are in 2017 dollars
country comparison to the world: 13

GDP (official exchange rate): $851.5 billion (2017 est.)

GDP—real growth rate: 0.98% (2019 est.)
3.04% (2018 est.)
7.54% (2017 est.)
country comparison to the world: 174

GDP—per capita (PPP): $27,000 (2017 est.)
$25,500 (2016 est.)
$25,000 (2015 est.)
note: data are in 2017 dollars
country comparison to the world: 77

Gross national saving: 25.5% of GDP (2017 est.)
24.5% of GDP (2016 est.)
24.8% of GDP (2015 est.)
country comparison to the world: 57

GDP—composition, by end use: *household consumption:* 59.1% (2017 est.)
government consumption: 14.5% (2017 est.)
investment in fixed capital: 29.8% (2017 est.)
investment in inventories: 1.1% (2017 est.)
exports of goods and services: 24.9% (2017 est.)
imports of goods and services: -29.4% (2017 est.)

GDP—composition, by sector of origin: *agriculture:* 6.8% (2017 est.)
industry: 32.3% (2017 est.)
services: 60.7% (2017 est.)

Agriculture—products: tobacco, cotton, grain, olives, sugar beets, hazelnuts, pulses, citrus; livestock

Industries: textiles, food processing, automobiles, electronics, mining (coal, chromate, copper, boron), steel, petroleum, construction, lumber, paper

Industrial production growth rate: 9.1% (2017 est.)
country comparison to the world: 18

Labor force: 25.677 million (2020 est.)
note: this number is for the domestic labor force only; number does not include about 1.2 million Turks working abroad, nor refugees

985

country comparison to the world: 20

Labor force—by occupation: agriculture: 18.4%
industry: 26.6%
services: 54.9% (2016)

Unemployment rate: 13.68% (2019 est.)
11% (2018 est.)
country comparison to the world: 169

Population below poverty line: 21.9% (2015 est.)

Household income or consumption by percentage share: lowest 10%: 2.1%
highest 10%: 30.3% (2008)

Budget: revenues: 172.8 billion (2017 est.)
expenditures: 185.8 billion (2017 est.)

Taxes and other revenues: 20.3% (of GDP) (2017 est.)
country comparison to the world: 151

Budget surplus (+) or deficit (-): -1.5% (of GDP) (2017 est.)
country comparison to the world: 91

Public debt: 28.3% of GDP (2017 est.)
28.3% of GDP (2016 est.)
country comparison to the world: 169

Fiscal year: calendar year

Inflation rate (consumer prices): 11.1% (2017 est.)
7.8% (2016 est.)
country comparison to the world: 204

Current account balance: $8.561 billion (2019 est.)
-$20.745 billion (2018 est.)
country comparison to the world: 26

Exports: $166.2 billion (2017 est.)
$150.2 billion (2016 est.)
country comparison to the world: 30

Exports—partners: Germany 9.6%, UK 6.1%, UAE 5.9%, Iraq 5.8%, US 5.5%, Italy 5.4%, France 4.2%, Spain 4% (2017)

Exports—commodities: apparel, foodstuffs, textiles, metal manufactures, transport equipment

Imports: $225.1 billion (2017 est.)
$191.1 billion (2016 est.)
country comparison to the world: 22

Imports—commodities: machinery, chemicals, semi-finished goods, fuels, transport equipment

Imports—partners: China 10%, Germany 9.1%, Russia 8.4%, US 5.1%, Italy 4.8% (2017)

Reserves of foreign exchange and gold: $107.7 billion (31 December 2017 est.)
$106.1 billion (31 December 2016 est.)
country comparison to the world: 24

Debt—external: $452.4 billion (31 December 2017 est.)
$404.9 billion (31 December 2016 est.)
country comparison to the world: 26

Exchange rates: Turkish liras (TRY) per US dollar—
3.628 (2017 est.)
3.0201 (2016 est.)
3.0201 (2015 est.)
2.72 (2014 est.)
2.1885 (2013 est.)

ENERGY

Electricity access: electrification—total population: 100% (2020)

Electricity—production: 261.9 billion kWh (2016 est.)
country comparison to the world: 16

Electricity—consumption: 231.1 billion kWh (2016 est.)
country comparison to the world: 18

Electricity—exports: 1.442 billion kWh (2016 est.)
country comparison to the world: 49

Electricity—imports: 6.33 billion kWh (2016 est.)
country comparison to the world: 31

Electricity—installed generating capacity: 78.5 million kW (2016 est.)
country comparison to the world: 15

Electricity—from fossil fuels: 53% of total installed capacity (2016 est.)
country comparison to the world: 144

Electricity—from nuclear fuels: 0% of total installed capacity (2017 est.)
country comparison to the world: 199

Electricity—from hydroelectric plants: 33% of total installed capacity (2017 est.)
country comparison to the world: 64

Electricity—from other renewable sources: 14% of total installed capacity (2017 est.)
country comparison to the world: 64

Crude oil—production: 55,000 bbl/day (2018 est.)
country comparison to the world: 51

Crude oil—exports: 0 bbl/day (2017 est.)
country comparison to the world: 120

Crude oil—imports: 521,500 bbl/day (2017 est.)
country comparison to the world: 17

Crude oil—proved reserves: 341.6 million bbl (1 January 2018 est.)
country comparison to the world: 51

Refined petroleum products—production: 657,900 bbl/day (2017 est.)
country comparison to the world: 27

Refined petroleum products—consumption: 989,900 bbl/day (2017 est.)
country comparison to the world: 21

Refined petroleum products— exports: 141,600 bbl/day (2017 est.)
country comparison to the world: 37

Refined petroleum products—imports: 560,000 bbl/day (2017 est.)
country comparison to the world: 16

Natural gas—production: 368.1 million cu m (2017 est.)
country comparison to the world: 75

Natural gas—consumption: 53.6 billion cu m (2017 est.)
country comparison to the world: 15

Natural gas—exports: 622.9 million cu m (2017 est.)
country comparison to the world: 42

Natural gas—imports: 55.13 billion cu m (2017 est.)
country comparison to the world: 6

Natural gas—proved reserves: 5.097 billion cu m (1 January 2018 est.)
country comparison to the world: 92

Carbon dioxide emissions from consumption of energy: 379.5 million Mt (2017 est.)
country comparison to the world: 17

COMMUNICATIONS

Telephones—fixed lines: total subscriptions: 11,283,768

subscriptions per 100 inhabitants: 13.82 (2019 est.)
country comparison to the world: 17

Telephones—mobile cellular: total subscriptions: 79,068,023

subscriptions per 100 inhabitants: 96.84 (2019 est.)
country comparison to the world: 20

Telecommunication systems: general assessment: comprehensive telecommunications network undergoing rapid modernization and expansion, especially in mobile-cellular services; rise in subscribers and increase in bundled packages; while mobile broadband becoming increasingly popular DSL has largest share of fixed broadband technologies, but fiber-optic is growing with significant investment; 4G LTE networks well incorporated in Turkey, 93% coverage of the population; strides made with 5G trials with help from Chinese company Huawei (2020)

domestic: additional digital exchanges are permitting a rapid increase in subscribers; the construction of a network of technologically advanced intercity trunk lines, using both fiber-optic cable and digital microwave radio relay, is facilitating communication between urban centers; remote areas are reached by a domestic satellite system; fixed-line 14 per 100 and mobile-cellular teledensity is 97 telephones per 100 persons (2019)

international: country code - 90; landing points for the SeaMeWe-3 & -5, MedNautilus Submarine System, Turcyos-1 & -2 submarine cables providing connectivity to Europe, Africa, the Middle East, Asia, Southeast Asia and Australia ; satellite earth stations - 12 Intelsat; mobile satellite terminals - 328 in the Inmarsat and Eutelsat systems (2020)

note: the COVID-19 outbreak is negatively impacting telecommunications production and supply chains globally; consumer spending on telecom devices and services has also slowed due to the pandemic's effect on economies worldwide; overall progress towards improvements in all facets of the telecom industry - mobile, fixed-line, broadband, submarine cable and satellite - has moderated

Broadcast media: Turkish Radio and Television Corporation (TRT) operates multiple TV and radio networks and stations; multiple privately owned national television stations and 567 private regional and local television stations;

multi-channel cable TV subscriptions available; 1,007 private radio broadcast stations

(2019)

Internet country code: .tr

Internet users: *total:* 57,725,143

percent of population: 71.04% (July 2018 est.) *country comparison to the world:* 15

Broadband—fixed subscriptions: *total:* 13,407,226

subscriptions per 100 inhabitants: 16 (2018 est.) *country comparison to the world:* 15

TRANSPORTATION

National air transport system: *number of registered air carriers:* 11 (2020)

inventory of registered aircraft operated by air carriers: 618

annual passenger traffic on registered air carriers: 115,595,495 (2018)

annual freight traffic on registered air carriers: 5,949,210,000 mt-km (2018)

Civil aircraft registration country code prefix: TC (2016)

Airports: 98 (2013) *country comparison to the world:* 58

Airports—with paved runways: *total:* 91 (2013) *over 3,047 m:* 16 (2013) *2,438 to 3,047 m:* 38 (2013) *1,524 to 2,437 m:* 17 (2013) *914 to 1,523 m:* 16 (2013) *under 914 m:* 4 (2013)

Airports—with unpaved runways: *total:* 7 (2013) *1,524 to 2,437 m:* 1 (2013) *914 to 1,523 m:* 4 (2013) *under 914 m:* 2 (2013)

Heliports: 20 (2013)

Pipelines: 14,666 km gas, 3,293 km oil (2017)

Railways: *total:* 12,710 km (2018)

standard gauge: 11,497 km 1.435-m gauge (1.435 km high speed train) (2018) *country comparison to the world:* 21

Roadways: *total:* 67,333 km (2018) *paved:* 24,082 km (includes 2,159 km of expressways) (2018) *unpaved:* 43,251 km (2018) *country comparison to the world:* 73

Waterways: 1,200 km (2010) *country comparison to the world:* 60

Merchant marine: *total:* 1,234 *by type:* bulk carrier 57, container ship 54, general cargo 363, oil tanker 124, other 636 (2019) *country comparison to the world:* 23

Ports and terminals: *major seaport(s):* Aliaga, Ambarli, Diliskelesi, Eregli, Izmir, Kocaeli (Izmit), Mersin (Icel), Limani, Yarimca

container port(s) (TEUs): Ambarli (3,131,621), Mersin (Icel) (1,592,000) (2017)

LNG terminal(s) (import): Izmir Aliaga, Marmara Ereglisi

MILITARY AND SECURITY

Military and security forces: *Turkish Armed Forces (TSK):* Turkish Land Forces (Turk Kara Kuvvetleri), Turkish Naval Forces (Turk Deniz Kuvvetleri; includes naval air and naval infantry), Turkish Air Forces (Turk Hava Kuvvetleri); Ministry of Interior: Gendarmerie of the Turkish Republic, Turkish Coast Guard Command (2019)

Military expenditures: 1.89% of GDP (2019 est.) 1.85% of GDP (2018) 1.52% of GDP (2017) 1.46% of GDP (2016) 1.39% of GDP (2015) *country comparison to the world:* 57

Military and security service personnel strengths: size assessments for the Turkish Armed Forces (TSK) vary; approximately 375,000 total active duty personnel (280,000 Army; 45,000 Navy; 50,000 Air Force); approximately 150,000 Gendarmerie (2020 est.)

Military equipment inventories and acquisitions: the Turkish Armed Forces inventory is mostly comprised of a mix of domestically-produced and Western weapons systems, although in recent years, Turkey has also acquired some Chinese, Russian, and South Korean equipment; since 2010, the US is the leading provider of armaments to Turkey, followed by Italy, South Korea, and Spain (2019 est.)

Military deployments: 600 Afghanistan (NATO); 250 Bosnia-Herzegovina (EUFOR); est. 25-35,000 Cyprus; 300 Kosovo (NATO); 170 Lebanon (UNIFIL); est. 200 Qatar; est. 200 Somalia; est. 5-10,000 Syria (2020)

note: Turkey has deployed troops into northern Iraq on numerous occasions to combat the Kurdistan Worker's Party (PKK), including operations involving thousands of troops in 2007, 2011, and 2018; its most recent incursion was in June 2020; in 2020, Turkey deployed Turkish troops and an estimated 3,500 Syrian civil war veterans to Libya to support the Libyan Government of National Accord (GNA)

Military service age and obligation: President Erdogan on 25 June 2019 signed a new law cutting the men's mandatory military service period in half, as well as making paid military service permanent; with the new system, the period of conscription was reduced from 12 months to six months for private and non-commissioned soldiers (the service term for reserve officers chosen among university or college graduates will remain 12 months); after completing six months of service, if a conscripted soldier wants to and is suitable for extending his military service, he may do so for an additional six months in return for a monthly salary; under the new law, all male Turkish citizens over the age of 20 will be required to undergo a one month military training period, but they can obtain an exemption from the remaining five months of their mandatory service by paying 31,000 Turkish Liras (2019)

Military—note: the ruling Justice and Development Party (AKP) has actively pursued the goal of asserting civilian control over the military since first taking power in 2002; the Turkish Armed Forces (TSK) role in internal security has been significantly reduced; the TSK leadership continues to be an influential institution within Turkey, but plays a much smaller role in politics; the Turkish military remains focused on the threats emanating from the Syrian civil war, Russia's actions in Ukraine, and the PKK insurgency; primary domestic threats are listed as fundamentalism (with the definition in some dispute with the civilian government), separatism (Kurdish discontent), and the extreme left wing; Ankara strongly opposed establishment of an autonomous Kurdish region in Iraq; an overhaul of the Turkish Land Forces Command (TLFC) taking place under the "Force 2014" program is to produce 20-30% smaller, more highly trained forces characterized by greater mobility and firepower and capable of joint and combined operations; the TLFC has taken on increasing international peacekeeping responsibilities including in Afghanistan; the Turkish Navy is a regional naval power that wants to develop the capability to project power beyond Turkey's coastal waters; the Navy is heavily involved in NATO, multinational, and UN operations; its roles include control of territorial waters and security for sea lines of communications; the Turkish Air Force adopted an "Aerospace and Missile Defense Concept" in 2002 and has initiated project work on an integrated missile defense system; in a controversial move, it purchased the Russian S-400 air defense system for an estimated $2.5 billion in July 2019; Air Force priorities include attaining a modern deployable, survivable, and sustainable force structure, and establishing a sustainable command and control system; Turkey is a NATO ally and hosts NATO's Land Forces Command in Izmir, as well as the AN/TPY-2 radar as part of NATO Missile Defense (2019)

TERRORISM

Terrorist group(s): Islamic State of Iraq and ash-Sham – Turkey; Islamic Movement of Uzbekistan; Islamic Revolutionary Guard Corps/Qods Force; Kurdistan Workers' Party; al-Qa'ida; Revolutionary People's Liberation Party/Front (2019)

note: details about the history, aims, leadership, organization, areas of operation, tactics, targets, weapons, size, and sources of support of the group(s) appear(s) in Appendix-T

TRANSPORTATION

Disputes—international: complex maritime, air, and territorial disputes with Greece in the Aegean Sea; status of north Cyprus question remains; Turkey has expressed concern over the status of Kurds in Iraq; in 2009, Swiss mediators facilitated an accord reestablishing diplomatic ties between Armenia and Turkey, but neither side has ratified the agreement and the rapprochement effort has faltered; Turkish authorities have complained that blasting from quarries in Armenia might be damaging the medieval ruins of Ani, on the other side of the Arpacay valley

Refugees and internally displaced persons: *refugees (country of origin):* 3,630,702 (Syria), 170,000 (Afghanistan), 142,000 (Iraq), 39,000 (Iran), 5,700 (Somalia) (2020)

IDPs: 1.099 million (displaced from 1984-2005 because of fighting between the Kurdish PKK and Turkish military; most IDPs are Kurds from eastern and southeastern provinces; no information available on persons displaced by development projects) (2018)

stateless persons: 117 (2018)

Illicit drugs: key transit route for Southwest Asian heroin to Western Europe and, to a lesser extent, the US - via air, land, and sea routes; major Turkish and other international trafficking organizations operate out of Istanbul; laboratories to convert imported morphine base into heroin exist in remote regions of Turkey and near Istanbul; government maintains strict controls over areas of legal opium poppy cultivation and over output of poppy straw concentrate; lax enforcement of money-laundering controls

TURKMENISTAN

INTRODUCTION

Background: Present-day Turkmenistan covers territory that has been at the crossroads of civilizations for centuries. The area was ruled in antiquity by various Persian empires, and was conquered by Alexander the Great, Muslim armies, the Mongols, Turkic warriors, and eventually the Russians. In medieval times, Merv (located in present-day Mary province) was one of the great cities of the Islamic world and an important stop on the Silk Road. Annexed by Russia in the late 1800s, Turkmenistan later figured prominently in the anti-Bolshevik movement in Central Asia. In 1924, Turkmenistan became a Soviet republic; it achieved independence upon the dissolution of the USSR in 1991. President for Life Saparmyrat NYYAZOW died in December 2006, and Gurbanguly BERDIMUHAMEDOW, a deputy chairman under NYYAZOW, emerged as the country's new president. BERDIMUHAMEDOW won Turkmenistan's first multi-candidate presidential election in February 2007, and again in 2012 and in 2017 with over 97% of the vote in both instances, in elections widely regarded as undemocratic.

Turkmenistan has sought new export markets for its extensive hydrocarbon/natural gas reserves, which have yet to be fully exploited. As of late 2019, Turkmenistan exported the majority of its gas to China and small levels of gas were also being sent to Russia. Turkmenistan's reliance on gas exports has made the economy vulnerable to fluctuations in the global energy market, and economic hardships since the drop in energy prices in 2014 have led many Turkmenistanis to emigrate, mostly to Turkey.

GEOGRAPHY

Location: Central Asia, bordering the Caspian Sea, between Iran and Kazakhstan

Geographic coordinates: 40 00 N, 60 00 E

Map references: Asia

Area: *total:* 488,100 sq km
land: 469,930 sq km
water: 18,170 sq km
country comparison to the world: 54

Area—comparative: slightly more than three times the size of Georgia; slightly larger than California

Land boundaries: *total:* 4,158 km
border countries (4): Afghanistan 804 km, Iran 1148 km, Kazakhstan 413 km, Uzbekistan 1793 km

Coastline: 0 km (landlocked); note - Turkmenistan borders the Caspian Sea (1,768 km)

Maritime claims: none (landlocked)

Climate: subtropical desert

Terrain: flat-to-rolling sandy desert with dunes rising to mountains in the south; low mountains along border with Iran; borders Caspian Sea in west

Elevation: *mean elevation:* 230 m
lowest point: Vpadina Akchanaya (Sarygamysh Koli is a lake in northern Turkmenistan with a water level that fluctuates above and below the elevation of Vpadina Akchanaya, the lake has dropped as low as -110 m) -81 m
highest point: Gora Ayribaba 3,139 m

Natural resources: petroleum, natural gas, sulfur, salt

Land use: *agricultural land:* 72% (2011 est.)
arable land: 4.1% (2011 est.) / permanent crops: 0.1% (2011 est.) / permanent pasture: 67.8% (2011 est.)
forest: 8.8% (2011 est.)
other: 19.2% (2011 est.)

Irrigated land: 19,950 sq km (2012)

Population distribution: the most densely populated areas are the southern, eastern, and northeastern oases; approximately 50% of the population lives in and around the capital of Ashgabat

Natural hazards: earthquakes; mudslides; droughts; dust storms; floods

Environment—current issues: contamination of soil and groundwater with agricultural chemicals, pesticides; salination, water logging of soil due to poor irrigation methods; Caspian Sea pollution; diversion of a large share of the flow of the Amu Darya into irrigation contributes to that river's inability to replenish the Aral Sea; soil erosion; desertification

Environment—international agreements: *party to:* Biodiversity, Climate Change, Climate Change-Kyoto Protocol, Desertification, Hazardous Wastes, Ozone Layer Protection
signed, but not ratified: none of the selected agreements

Geography—note: landlocked; the western and central low-lying desolate portions of the country make up the great Garagum (Kara-Kum) desert, which occupies over 80% of the country; eastern part is plateau

PEOPLE AND SOCIETY

Population: 5,528,627 (July 2020 est.)
some sources suggest Turkmenistan's population could be as much as 1 to 2 million people lower than available estimates because of large-scale emigration during the last 10 years
country comparison to the world: 117

Nationality: *noun:* Turkmenistani(s)
adjective: Turkmenistani

Ethnic groups: Turkmen 85%, Uzbek 5%, Russian 4%, other 6% (2003)

Languages: Turkmen (official) 72%, Russian 12%, Uzbek 9%, other 7%

Religions: Muslim 89%, Eastern Orthodox 9%, unknown 2%

Age structure: *0-14 years:* 25.44% (male 713,441/ female 693,042)
15-24 years: 16.48% (male 458,566/female 452,469)
25-54 years: 44.14% (male 1,214,581/female 1,226,027)

55-64 years: 8.56% (male 221,935/female 251,238)
65 years and over: 5.38% (male 129,332/female 167,996) (2020 est.)

Dependency ratios: *total dependency ratio:* 55.2
youth dependency ratio: 47.8
elderly dependency ratio: 7.4
potential support ratio: 13.5 (2020 est.)

Median age: *total:* 29.2 years
male: 28.7 years
female: 29.7 years (2020 est.)
country comparison to the world: 135

Population growth rate: 1.06% (2020 est.)
country comparison to the world: 100

Birth rate: 18.3 births/1,000 population (2020 est.)
country comparison to the world: 86

Death rate: 6.1 deaths/1,000 population (2020 est.)
country comparison to the world: 162

Net migration rate: -1.7 migrant(s)/1,000 population (2020 est.)
country comparison to the world: 161

Population distribution: the most densely populated areas are the southern, eastern, and northeastern oases; approximately 50% of the population lives in and around the capital of Ashgabat

Urbanization: *urban population:* 52.5% of total population (2020)
rate of urbanization: 2.46% annual rate of change (2015-20 est.)
total population growth rate v. urban population growth rate, 2000-2030: Major urban areas - population: 846,000 ASHGABAT (capital) (2020)

Sex ratio: *at birth:* 1.05 male(s)/female
0-14 years: 1.03 male(s)/female
15-24 years: 1.01 male(s)/female
25-54 years: 0.99 male(s)/female
55-64 years: 0.88 male(s)/female
65 years and over: 0.77 male(s)/female
total population: 0.98 male(s)/female (2020 est.)

Maternal mortality rate: 7 deaths/100,000 live births (2017 est.)
country comparison to the world: 157

Infant mortality rate: *total:* 30.8 deaths/1,000 live births
male: 37.2 deaths/1,000 live births
female: 24.2 deaths/1,000 live births (2020 est.)
country comparison to the world: 53

Life expectancy at birth: *total population:* 71.3 years
male: 68.2 years
female: 74.5 years (2020 est.)
country comparison to the world: 162

Total fertility rate: 2.04 children born/woman (2020 est.)
country comparison to the world: 108

Contraceptive prevalence rate: 50.2% (2015/16)

Drinking water source: *improved:* urban: 100% of population
rural: 100% of population
total: 100% of population
unimproved: urban: 0% of population
rural: 0% of population

total: 0% of population (2017 est.)

Current Health Expenditure: 6.9% (2017)

Physicians density: 2.22 physicians/1,000 population (2014)

Hospital bed density: 4 beds/1,000 population (2014)

Sanitation facility access: *improved:* urban: 100% of population
rural: 100% of population
total: 100% of population
unimproved: urban: 0% of population
rural: 0% of population
total: 0% of population (2017 est.)

HIV/AIDS—adult prevalence rate: NA

HIV/AIDS—people living with HIV/AIDS: NA

HIV/AIDS—deaths: NA

Obesity—adult prevalence rate: 18.6% (2016)
country comparison to the world: 117

Children under the age of 5 years underweight: 3.2% (2015)
country comparison to the world: 98

Education expenditures: 3.1% of GDP (2012)
country comparison to the world: 138

Literacy: *definition:* age 15 and over can read and write
total population: 99.7%
male: 99.8%
female: 99.6% (2015)

School life expectancy (primary to tertiary education): *total:* 13 years
male: 13 years
female: 13 years (2019)

GOVERNMENT

Country name: *conventional long form:* none
conventional short form: Turkmenistan
local long form: none
local short form: Turkmenistan
former: Turkmen Soviet Socialist Republic
etymology: the suffix "-stan" means "place of" or "country," so Turkmenistan literally means the "Land of the Turkmen [people]"

Government type: presidential republic; authoritarian

Capital: *name:* Ashgabat (Ashkhabad)
geographic coordinates: 37 57 N, 58 23 E
time difference: UTC+5 (10 hours ahead of Washington, DC, during Standard Time)
etymology: derived from the Persian words "eshq" meaning "love" and "abad" meaning "inhabited place" or "city," and so loosely translates as "the city of love"

Administrative divisions: *5 provinces (welayatlar, singular - welayat) and 1 independent city*:*
Ahal Welayaty (Anew), Ashgabat*, Balkan Welayaty (Balkanabat), Dasoguz Welayaty, Lebap Welayaty (Turkmenabat), Mary Welayaty
note: administrative divisions have the same names as their administrative centers (exceptions have the administrative center name following in parentheses)

Independence: 27 October 1991 (from the Soviet Union)

National holiday: Independence Day, 27 October (1991)

Constitution: *history:* several previous; latest adopted 14 September 2016
amendments: proposed by the National Assembly; passage requires two-thirds majority vote of the total Assembly membership or absolute majority approval in a referendum; amended 2017

Legal system: civil law system with Islamic (sharia) law influences

International law organization participation: has not submitted an ICJ jurisdiction declaration; non-party state to the ICCt

Citizenship: *citizenship by birth:* no
citizenship by descent only: at least one parent must be a citizen of Turkmenistan
dual citizenship recognized: yes
residency requirement for naturalization: 7 years

Suffrage: 18 years of age; universal

Executive branch: *chief of state:* President Gurbanguly BERDIMUHAMEDOW (since 14 February 2007); note - the president is both chief of state and head of government
head of government: President Gurbanguly BERDIMUHAMEDOW (since 14 February 2007)
cabinet: Cabinet of Ministers appointed by the president
elections/appointments: president directly elected by absolute majority popular vote in 2 rounds if needed for a 7-year term (no term limits); election last held on 12 February 2017 (next to be held in February 2024)
election results: Gurbanguly BERDIMUHAMEDOW reelected president in the first round; percent of vote - Gurbanguly BERDIMUHAMEDOW (DPT) 97.7%, other 2.3%

Legislative branch: *description:* unicameral National Assembly or Mejlis (125 seats; members directly elected from single-seat constituencies by absolute majority vote; members serve 5-year terms)
elections: last held on 25 March 2018, although interim elections are held on an ad hoc basis to fill vacant sets
election results: percent of vote by party - NA; seats by party - DPT 55, APT 11, PIE 11, independent 48 (individuals nominated by citizen groups); composition - men 94, women 31, percent of women 24.8%

Judicial branch: *highest courts:* Supreme Court of Turkmenistan (consists of the court president and 21 associate judges and organized into civil, criminal, and military chambers)
judge selection and term of office: judges appointed by the president for 5-year terms
subordinate courts: High Commercial Court; appellate courts; provincial, district, and city courts; military courts

Political parties and leaders: Agrarian Party of Turkmenistan or APT [Basim ANNAGURBANOW]

989

Democratic Party of Turkmenistan or DPT [Ata SERDAROW]

Party of Industrialists and Entrepreneurs or PIE [Saparmyrat OWGANOW]

note: all of these parties support President BERDIMUHAMEDOW; a law authorizing the registration of political parties went into effect in January 2012; unofficial, small opposition movements exist abroad

International organization participation: ADB, CIS (associate member, has not ratified the 1993 CIS charter although it participates in meetings and held the chairmanship of the CIS in 2012), EAPC, EBRD, ECO, FAO, G-77, IBRD, ICAO, ICRM, IDA, IDB, IFC, IFRCS, ILO, IMF, IMO, Interpol, IOC, IOM (observer), ISO (correspondent), ITU, MIGA, NAM, OIC, OPCW, OSCE, PFP, UN, UNCTAD, UNESCO, UNHCR, UNIDO, UNWTO, UPU, WCO, WFTU (NGOs), WHO, WIPO, WMO

Diplomatic representation in the US: *chief of mission:* Ambassador Meret ORAZOW (since 14 February 2001)

chancery: 2207 Massachusetts Avenue NW, Washington, DC 20008

telephone: [1] (202) 588-1500

FAX: [1] (202) 588-0697

Diplomatic representation from the US: *chief of mission:* Ambassador Matthew S. KLIMOW (since 26 June 2019)

telephone: [993] (12) 94-00-45

embassy: No. 9 1984 Street (formerly Pushkin Street), Ashgabat 744000

mailing address: 7070 Ashgabat Place, Washington DC 20521-7070

FAX: [993] (12) 94-26-14

Flag description: green field with a vertical red stripe near the hoist side, containing five tribal guls (designs used in producing carpets) stacked above two crossed olive branches; five white, five-pointed stars and a white crescent moon appear in the upper corner of the field just to the fly side of the red stripe; the green color and crescent moon represent Islam; the five stars symbolize the regions or welayats of Turkmenistan; the guls reflect the national identity of Turkmenistan where carpet-making has long been a part of traditional nomadic life

note: the flag of Turkmenistan is the most intricate of all national flags

National symbol(s): *Akhal-Teke horse; national colors:* green, white

National anthem: *name:* "Garassyz, Bitarap Turkmenistanyn" (Independent, Neutral, Turkmenistan State Anthem)

lyrics/music: collective/Veli MUKHATOV

note: adopted 1997, lyrics revised in 2008, to eliminate references to deceased President Saparmurat NYYAZOW

ECONOMY

Economy—overview: *Turkmenistan is largely a desert country with intensive agriculture in irrigated oases and significant natural gas and oil resources. The two largest crops are cotton, most of which is produced for export, and wheat, which is domestically consumed. Although agriculture accounts for almost 8% of GDP, it continues to employ nearly half of the country's workforce. Hydrocarbon exports, the bulk of which is natural gas going to China, make up 25% of Turkmenistan's GDP. Ashgabat has explored two initiatives to bring gas to new markets:* a trans-Caspian pipeline that would carry gas to Europe and the Turkmenistan-Afghanistan-Pakistan-India gas pipeline. Both face major financing, political, and security hurdles and are unlikely to be completed soon.

Turkmenistan's autocratic governments under presidents NIYAZOW (1991-2006) and BERDIMUHAMEDOW (since 2007) have made little progress improving the business climate, privatizing state-owned industries, combatting corruption, and limiting economic development outside the energy sector. High energy prices in the mid-2000s allowed the government to undertake extensive development and social spending, including providing heavy utility subsidies.

Low energy prices since mid-2014 are hampering Turkmenistan's economic growth and reducing government revenues. The government has cut subsidies in several areas, and wage arrears have increased. In January 2014, the Central Bank of Turkmenistan devalued the manat by 19%, and downward pressure on the currency continues. There is a widening spread between the official exchange rate (3.5 TMM per US dollar) and the black market exchange rate (approximately 14 TMM per US dollar). Currency depreciation and conversion restrictions, corruption, isolationist policies, and declining spending on public services have resulted in a stagnate economy that is nearing crisis. Turkmenistan claims substantial foreign currency reserves, but non-transparent data limit international institutions' ability to verify this information.

GDP (purchasing power parity): $103.7 billion (2017 est.)

$97.41 billion (2016 est.)

$91.72 billion (2015 est.)

note: data are in 2017 dollars

country comparison to the world: 84

GDP (official exchange rate): $37.93 billion (2017 est.)

GDP—real growth rate: 6.5% (2017 est.)

6.2% (2016 est.)

6.5% (2015 est.)

country comparison to the world: 24

GDP—per capita (PPP): $18,200 (2017 est.)

$17,300 (2016 est.)

$16,500 (2015 est.)

note: data are in 2017 dollars

country comparison to the world: 97

Gross national saving: 23.9% of GDP (2017 est.)

24.3% of GDP (2016 est.)

18.9% of GDP (2015 est.)

country comparison to the world: 70

GDP—composition, by end use: *household consumption:* 50% (2017 est.)

government consumption: 10% (2017 est.)

investment in fixed capital: 28.2% (2017 est.)

investment in inventories: 0% (2017 est.)

exports of goods and services: 26.2% (2017 est.)

imports of goods and services: -14.3% (2017 est.)

GDP—composition, by sector of origin: *agriculture:* 7.5% (2017 est.)

industry: 44.9% (2017 est.)

services: 47.7% (2017 est.)

Agriculture—products: cotton, grain, melons; livestock

Industries: natural gas, oil, petroleum products, textiles, food processing

Industrial production growth rate: 1% (2017 est.)

country comparison to the world: 160

Labor force: 2.305 million (2013 est.)

country comparison to the world: 115

Labor force—by occupation: *agriculture:* 48.2%

industry: 14%

services: 37.8% (2004 est.)

Unemployment rate: 11% (2014 est.)

10.6% (2013)

country comparison to the world: 155

Population below poverty line: 0.2% (2012 est.)

Household income or consumption by percentage share: *lowest 10%:* 2.6%

highest 10%: 31.7% (1998)

Budget: *revenues:* 5.657 billion (2017 est.)

expenditures: 6.714 billion (2017 est.)

Taxes and other revenues: 14.9% (of GDP) (2017 est.)

country comparison to the world: 197

Budget surplus (+) or deficit (-): -2.8% (of GDP) (2017 est.)

country comparison to the world: 126

Public debt: 28.8% of GDP (2017 est.)

24.1% of GDP (2016 est.)

country comparison to the world: 167

Fiscal year: calendar year

Inflation rate (consumer prices): 8% (2017 est.)

3.6% (2016 est.)

country comparison to the world: 198

Current account balance: -$4.359 billion (2017 est.)

-$7.207 billion (2016 est.)

country comparison to the world: 182

Exports: $7.458 billion (2017 est.)

$6.987 billion (2016 est.)

country comparison to the world: 99

Exports—partners: China 83.7%, Turkey 5.1% (2017)

Exports—commodities: gas, crude oil, petrochemicals, textiles, cotton fiber

Imports: $4.571 billion (2017 est.)

$5.215 billion (2016 est.)

country comparison to the world: 135

Imports—commodities: machinery and equipment, chemicals, foodstuffs

Imports—partners: Turkey 24.2%, Algeria 14.4%, Germany 9.8%, China 8.9%, Russia 8%, US 6.6% (2017)

Reserves of foreign exchange and gold: $24.91 billion (31 December 2017 est.)
$25.05 billion (31 December 2016 est.)
country comparison to the world: 56

Debt—external: $539.4 million (31 December 2017 est.)
$425.3 million (31 December 2016 est.)
country comparison to the world: 176

Exchange rates: Turkmenistani manat (TMM) per US dollar -
4.125 (2017 est.)
3.5 (2016 est.)
3.5 (2015 est.)
3.5 (2014 est.)
2.85 (2013 est.)

ENERGY

Electricity access: *electrification - total population:* 100% (2020)

Electricity—production: 21.18 billion kWh (2016 est.)
country comparison to the world: 74

Electricity—consumption: 15.09 billion kWh (2016 est.)
country comparison to the world: 80

Electricity—exports: 3.201 billion kWh (2015 est.)
country comparison to the world: 41

Electricity—imports: 0 kWh (2016 est.)
country comparison to the world: 212

Electricity—installed generating capacity: 4.001 million kW (2016 est.)
country comparison to the world: 88

Electricity—from fossil fuels: 100% of total installed capacity (2016 est.)
country comparison to the world: 20

Electricity—from nuclear fuels: 0% of total installed capacity (2017 est.)
country comparison to the world: 200

Electricity—from hydroelectric plants: 0% of total installed capacity (2017 est.)
country comparison to the world: 207

Electricity—from other renewable sources: 0% of total installed capacity (2017 est.)
country comparison to the world: 210

Crude oil—production: 244,000 bbl/day (2018 est.)
country comparison to the world: 32

Crude oil—exports: 67,790 bbl/day (2015 est.)
country comparison to the world: 38

Crude oil—imports: 0 bbl/day (2015 est.)
country comparison to the world: 206

Crude oil—proved reserves: 600 million bbl (1 January 2018 est.)
country comparison to the world: 43

Refined petroleum products—production: 191,100 bbl/day (2015 est.)
country comparison to the world: 52

Refined petroleum products—consumption: 160,000 bbl/day (2016 est.)
country comparison to the world: 63

Refined petroleum products—exports: 53,780 bbl/day (2015 est.)
country comparison to the world: 53

Refined petroleum products—imports: 0 bbl/day (2015 est.)
country comparison to the world: 214

Natural gas—production: 77.45 billion cu m (2017 est.)
country comparison to the world: 11

Natural gas—consumption: 39.31 billion cu m (2017 est.)
country comparison to the world: 27

Natural gas—exports: 38.14 billion cu m (2017 est.)
country comparison to the world: 10

Natural gas—imports: 0 cu m (2017 est.)
country comparison to the world: 203

Natural gas—proved reserves: 7.504 trillion cu m (1 January 2018 est.)
country comparison to the world: 5

Carbon dioxide emissions from consumption of energy: 100.5 million Mt (2017 est.)
country comparison to the world: 44

COMMUNICATIONS

Telephones—fixed lines: *total subscriptions:* 648,223
subscriptions per 100 inhabitants: 11.85 (2019 est.)
country comparison to the world: 87

Telephones—mobile cellular: *total subscriptions:* 8,908,821
subscriptions per 100 inhabitants: 162.86 (2019 est.)
country comparison to the world: 93

Telecommunication systems: *general assessment:* telecommunications network is gradually improving from the former Soviet republic; state control over most economic activities has not helped growth; in cooperation with foreign partners, the telecom sector has installed high-speed fiber-optic lines and has upgraded most of the country's telephone exchanges and switching centers with new digital technology; the mobile market will see slow growth; some rural areas are still without telephones; mobile broadband is in the early stages of development; in 2019 Russia-based operator said to be leaving the country and leaving only 1 public operator (2020)
domestic: fixed-line 12 per 100 and mobile-cellular teledensity is about 163 per 100 persons; first telecommunication satellite was launched in 2015 (2019)
international: country code - 993; linked by fiber-optic cable and microwave radio relay to other CIS republics and to other countries by leased connections to the Moscow international gateway switch; an exchange in Ashgabat switches international traffic through Turkey via Intelsat; satellite earth stations - 1 Orbita and 1 Intelsat (2018)
note: the COVID-19 outbreak is negatively impacting telecommunications production and supply chains globally; consumer spending on telecom devices and services has also slowed due to the pandemic's effect on economies worldwide; overall progress towards improvements in all facets of the telecom industry - mobile, fixed-line, broadband, submarine cable and satellite - has moderated

Broadcast media: broadcast media is government controlled and censored; 7 state-owned TV and 4 state-owned radio networks; satellite dishes and programming provide an alternative to the state-run media; officials sometimes limit access to satellite TV by removing satellite dishes

Internet country code: .tm

Internet users: *total:* 1,149,840
percent of population: 21.25% (July 2018 est.)
country comparison to the world: 135

Broadband—fixed subscriptions: *total:* 4,000
subscriptions per 100 inhabitants: less than 1 (2017 est.)
country comparison to the world: 183

TRANSPORTATION

National air transport system: *number of registered air carriers:* 1 (2020)
inventory of registered aircraft operated by air carriers: 27
annual passenger traffic on registered air carriers: 2,457,474 (2018)
annual freight traffic on registered air carriers: 16.92 million mt-km (2018)

Civil aircraft registration country code prefix: EZ (2016)

Airports: 26 (2013)
country comparison to the world: 126

Airports—with paved runways: *total:* 21 (2013)
over 3,047 m: 1 (2013)
2,438 to 3,047 m: 9 (2013)
1,524 to 2,437 m: 9 (2013)
914 to 1,523 m: 2 (2013)

Airports—with unpaved runways: *total:* 5 (2013)
1,524 to 2,437 m: 1 (2013)
under 914 m: 4 (2013)

Heliports: 1 (2013)

Pipelines: 7500 km gas, 1501 km oil (2013)

Railways: *total:* 5,113 km (2017)
broad gauge: 5,113 km 1.520-m gauge (2017)
country comparison to the world: 38

Roadways: *total:* 58,592 km (2002)
paved: 47,577 km (2002)
unpaved: 11,015 km (2002)
country comparison to the world: 78

Waterways: 1,300 km (Amu Darya River and Kara Kum Canal are important inland waterways) (2011)
country comparison to the world: 55

Merchant marine: *total:* 71
by type: general cargo 6, oil tanker 8, other 57 (2019)
country comparison to the world: 102

Ports and terminals: *major seaport(s):* Caspian Sea - Turkmenbasy

MILITARY AND SECURITY

Military and security forces: *Armed Forces of Turkmenistan:* National Army, Navy, Air and Air Defense Forces; Federal Border Guard Service (2019)

Military and security service personnel strengths: the Armed Forces of Turkmenistan have approximately 37,000 active troops (est. 33,000 National Army; 500 Navy, 3,500 Air and Air Defense Forces) (2019 est.)

Military equipment inventories and acquisitions: the inventory for Turkmenistan's military is comprised almost entirely of older Russian and Soviet-era weapons systems, although in recent years, Turkmenistan has opened itself up to Chinese and Western equipment; since 2010, China, Italy, Russia, and Turkey are the leading arms suppliers to Turkmenistan (2019 est.)

Military service age and obligation: 18-27 years of age for compulsory male military service; 2-year conscript service obligation; 20 years of age for voluntary service; males may enroll in military schools from age 15 (2013)

TRANSNATIONAL ISSUES

Disputes—international: cotton monoculture in Uzbekistan and Turkmenistan creates water-sharing difficulties for Amu Darya river states; field demarcation of the boundaries with Kazakhstan commenced in 2005; bilateral talks continue with Azerbaijan on dividing the seabed and contested oilfields in the middle of the Caspian

Refugees and internally displaced persons: *stateless persons:* 3,688 (2019)

Trafficking in persons: *current situation:* Turkmenistan is a source country for men, women, and children subjected to forced labor and sex trafficking; Turkmenistanis who migrate abroad are forced to work in the textile, agriculture, construction, and domestic service industries, while women and girls may also be sex trafficked; in 2014, men surpassed women as victims; Turkey and Russia are primary trafficking destinations, followed by the Middle East, South and Central Asia, and other parts of Europe; Turkmenistanis also experience forced labor domestically in the informal construction industry; participation in the cotton harvest is still mandatory for some public sector employees

tier rating: Tier 2 Watch List – Turkmenistan does not fully comply with the minimum standards for the elimination of trafficking; however, it is making significant efforts to do so; in 2014, Turkmenistan was granted a waiver from an otherwise required downgrade to Tier 3 because its government has a written plan that, if implemented, would constitute making significant efforts to bring itself into compliance with the minimum standards for the elimination of trafficking; the government made some progress in its law enforcement efforts in 2014, convicting more offenders than in 2013; authorities did not make adequate efforts to identify and protect victims and did not fund international organizations or NGOs that offered protective services; some victims were punished for crimes as a result of being trafficked (2015)

Illicit drugs: transit country for Afghan narcotics bound for Russian and Western European markets; transit point for heroin precursor chemicals bound for Afghanistan

TURKS AND CAICOS ISLANDS

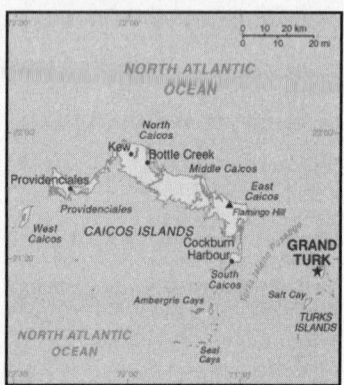

INTRODUCTION

Background: The islands were part of the UK's Jamaican colony until 1962, when they assumed the status of a separate Crown colony upon Jamaica's independence. The governor of The Bahamas oversaw affairs from 1965 to 1973. With Bahamian independence, the islands received a separate governor in 1973. Although independence was agreed upon for 1982, the policy was reversed and the islands remain a British overseas territory. Grand Turk island suffered extensive damage from Hurricane Maria on 22 September 2017 resulting in loss of power and communications as well as damage to housing and businesses.

GEOGRAPHY

Location: two island groups in the North Atlantic Ocean, southeast of The Bahamas, north of Haiti

Geographic coordinates: 21 45 N, 71 35 W

Map references: Central America and the Caribbean

Area: *total:* 948 sq km
land: 948 sq km
water: 0 sq km
country comparison to the world: 186

Area—comparative: 2.5 times the size of Washington, DC

Land boundaries: 0 km

Coastline: 389 km

Maritime claims: *territorial sea:* 12 nm
exclusive fishing zone: 200 nm

Climate: tropical; marine; moderated by trade winds; sunny and relatively dry

Terrain: low, flat limestone; extensive marshes and mangrove swamps

Elevation: *lowest point:* Caribbean Sea 0 m
highest point: Blue Hill on Providenciales and Flamingo Hill on East Caicos 48 m

Natural resources: spiny lobster, conch

Land use: *agricultural land:* 1.1% (2011 est.)
arable land: 1.1% (2011 est.) / permanent crops: 0% (2011 est.) / permanent pasture: 0% (2011 est.)
forest: 36.2% (2011 est.)
other: 62.7% (2011 est.)

Irrigated land: 0 sq km (2012)

Population distribution: eight of the thirty islands are inhabited; the island of Providenciales is the most populated, but the most densely populated is Grand Turk

Natural hazards: frequent hurricanes

Environment—current issues: limited natural freshwater resources, private cisterns collect rainwater

Geography—note: include eight large islands and numerous smaller cays, islets, and reefs; only two of the Caicos Islands and six of the Turks group are inhabited

PEOPLE AND SOCIETY

Population: 55,926 (July 2020 est.)
country comparison to the world: 207

Nationality: *noun:* none
adjective: none

Ethnic groups: black 87.6%, white 7.9%, mixed 2.5%, East Indian 1.3%, other 0.7% (2006 est.)

Languages: English (official)

Religions: Protestant 72.8% (Baptist 35.8%, Church of God 11.7%, Anglican 10%, Methodist 9.3%, Seventh-Day Adventist 6%), Roman Catholic 11.4%, Jehovah's Witness 1.8%, other 14% (2006 est.)

Age structure: *0-14 years:* 21.33% (male 6,077/female 5,852)
15-24 years: 13.19% (male 3,689/female 3,687)
25-54 years: 52.51% (male 14,729/female 14,637)
55-64 years: 7.81% (male 2,297/female 2,069)

65 years and over: 5.17% (male 1,364/female 1,525) (2020 est.)

Median age: total: 34.6 years
male: 34.9 years
female: 34.4 years (2020 est.)
country comparison to the world: 90

Population growth rate: 2% (2020 est.)
country comparison to the world: 48

Birth rate: 14.1 births/1,000 population (2020 est.)
country comparison to the world: 134

Death rate: 3.4 deaths/1,000 population (2020 est.)
country comparison to the world: 222

Net migration rate: 8.9 migrant(s)/1,000 population (2020 est.)
country comparison to the world: 8

Population distribution: eight of the thirty islands are inhabited; the island of Providenciales is the most populated, but the most densely populated is Grand Turk

Urbanization: urban population: 93.6% of total population (2020)
rate of urbanization: 1.77% annual rate of change (2015-20 est.)
total population growth rate v. urban population growth rate, 2000-2030: Major urban areas—population: 5,000 GRAND TURK (capital) (2018)

Sex ratio: at birth: 1.05 male(s)/female
0-14 years: 1.04 male(s)/female
15-24 years: 1 male(s)/female
25-54 years: 1.01 male(s)/female
55-64 years: 1.11 male(s)/female
65 years and over: 0.89 male(s)/female
total population: 1.01 male(s)/female (2020 est.)

Infant mortality rate: total: 9.3 deaths/1,000 live births
male: 11.5 deaths/1,000 live births
female: 7 deaths/1,000 live births (2020 est.)
country comparison to the world: 137

Life expectancy at birth: total population: 80.3 years
male: 77.6 years
female: 83.3 years (2020 est.)
country comparison to the world: 44

Total fertility rate: 1.7 children born/woman (2020 est.)
country comparison to the world: 175

Drinking water source: improved: total: 94.3% of population
unimproved: total: 5.7% of population (2017 est.)

Sanitation facility access: improved: total: 88% of population
unimproved: total: 12% of population (2017)

HIV/AIDS—adult prevalence rate: NA

HIV/AIDS—people living with HIV/AIDS: NA

HIV/AIDS—deaths: NA

Education expenditures: 2.8% of GDP (2018)
country comparison to the world: 149

School life expectancy (primary to tertiary education): total: 9 years
male: NA
female: NA (2015)

People—note: destination and transit point for illegal Haitian immigrants bound for the Bahamas and the US

GOVERNMENT

Country name: conventional long form: none
conventional short form: Turks and Caicos Islands
abbreviation: TCI
etymology: the Turks Islands are named after the Turk's cap cactus (native to the islands and appearing on the flag and coat of arms), while the Caicos Islands derive from the native term "caya hico" meaning "string of islands"

Dependency status: overseas territory of the UK

Government type: parliamentary democracy

Capital: name: Grand Turk (Cockburn Town)
geographic coordinates: 21 28 N, 71 08 W
time difference: UTC-5 (same time as Washington, DC, during Standard Time)
etymology: named after Sir Francis Cockburn, who served as governor of the Bahamas from 1837 to 1844

Administrative divisions: none (overseas territory of the UK)

Independence: none (overseas territory of the UK)

National holiday: Birthday of Queen ELIZABETH II, usually celebrated the Monday after the second Saturday in June

Constitution: history: several previous; latest signed 7 August 2012, effective 15 October 2012 (The Turks and Caicos Constitution Order 2011)
amendments: NA

Legal system: mixed legal system of English common law and civil law

Citizenship: see United Kingdom

Suffrage: 18 years of age; universal

Executive branch: chief of state: Queen ELIZABETH II (since 6 February 1952); represented by Governor Nigel DAKIN (since 15 July 2019)
head of government: Premier Sharlene CARTWRIGHT-ROBINSON (since 20 December 2016); first female Premier of Turks and Caicos
cabinet: Cabinet appointed by the governor from among members of the House of Assembly
elections/appointments: the monarch is hereditary; governor appointed by the monarch; following legislative elections, the leader of the majority party is appointed premier by the governor

Legislative branch: description: unicameral House of Assembly (19 seats; 15 members in multi-seat constituencies and a single all-islands constituency directly elected by simple majority vote, 1 member nominated by the premier and appointed by the governor, 1 nominated by the opposition party leader and appointed by the governor, and 2 from the Turks and Caicos Islands Civic Society directly appointed by the governor; members serve 4-year terms)
elections: last held on 15 December 2016 (next to be held in 2020)

election results: percent of vote—NA; seats by party—PDM 10, PNP 5; composition—men 15, women 6, percent of women 28.6%

Judicial branch: highest courts: Supreme Court (consists of the chief justice and other judges, as determined by the governor); Court of Appeal (consists of the court president and 2 justices); note—appeals beyond the Supreme Court are referred to the Judicial Committee of the Privy Council (in London)
judge selection and term of office: Supreme Court and Appeals Court judges appointed by the governor in accordance with the Judicial Service Commission, a 3-member body of high-level judicial officials; Supreme Court judges serve until mandatory retirement at age 65, but terms can be extended to age 70; Appeals Court judge tenure determined by individual terms of appointment
subordinate courts: magistrates' courts

Political parties and leaders: People's Democratic Movement or PDM [Sharlene CARTWRIGHT-ROBINSON]
Progressive National Party or PNP [Washington MISICK]

International organization participation: Caricom (associate), CDB, Interpol (subbureau), UPU

Diplomatic representation in the US: none (overseas territory of the UK)

Diplomatic representation from the US: none (overseas territory of the UK)

Flag description: blue with the flag of the UK in the upper hoist-side quadrant and the colonial shield centered on the outer half of the flag; the shield is yellow and displays a conch shell, a spiny lobster, and Turk's cap cactus—three common elements of the islands' biota

National symbol(s): conch shell, Turk's cap cactus

National anthem: name: This Land of Ours
lyrics/music: Conrad HOWELL
note: serves as a local anthem; as a territory of the UK, "God Save the Queen" is the official anthem (see United Kingdom)

ECONOMY

Economy—overview: The Turks and Caicos economy is based on tourism, offshore financial services, and fishing. Most capital goods and food for domestic consumption are imported. The US is the leading source of tourists, accounting for more than three-quarters of the more than 1 million visitors that arrive annually. Three-quarters of the visitors come by ship. Major sources of government revenue also include fees from offshore financial activities and customs receipts.

GDP (purchasing power parity): $632 million (2007 est.)
$568.3 million (2006 est.)
country comparison to the world: 210

GDP (official exchange rate): NA

GDP—real growth rate: 11.2% (2007 est.)
country comparison to the world: 4

GDP—per capita (PPP): $29,100 (2007 est.)
country comparison to the world: 72

GDP—composition, by end use: *household consumption:* 49% (2017 est.)
government consumption: 21.5% (2017 est.)
investment in fixed capital: 16.5% (2017 est.)
investment in inventories: -0.1% (2017 est.)
exports of goods and services: 69.5% (2017 est.)
imports of goods and services: -56.4% (2017 est.)

GDP—composition, by sector of origin: *agriculture:* 0.5% (2017 est.)
industry: 8.9% (2017 est.)
services: 90.6% (2017 est.)

Agriculture—products: corn, beans, cassava (manioc, tapioca), citrus fruits; fish

Industries: tourism, offshore financial services

Industrial production growth rate: 3% (2017 est.)
country comparison to the world: 106

Labor force: 4,848 (1990 est.)
country comparison to the world: 220

Labor force—by occupation: *note:* about 33% in government and 20% in agriculture and fishing; significant numbers in tourism, financial, and other services

Unemployment rate: 10% (1997 est.)
country comparison to the world: 146

Population below poverty line: NA

Household income or consumption by percentage share: *lowest 10%:* NA
highest 10%: NA

Budget: *revenues:* 247.3 million (2017 est.)
expenditures: 224.3 million (2017 est.)

Fiscal year: calendar year

Inflation rate (consumer prices): 4% (2017 est.)
0.7% (2016 est.)
country comparison to the world: 158

Exports: $24.77 million (2008 est.)
country comparison to the world: 208

Exports—commodities: lobster, dried and fresh conch, conch shells

Imports: $591.3 million (2008 est.)
country comparison to the world: 197

Imports—commodities: food and beverages, tobacco, clothing, manufactures, construction materials

Debt—external: NA

Exchange rates: the US dollar is used

ENERGY

Electricity access: *electrification - total population:* 100% (2020)

Electricity—production: 235 million kWh (2016 est.)
country comparison to the world: 188

Electricity—consumption: 218.6 million kWh (2016 est.)
country comparison to the world: 191

Electricity—exports: 0 kWh (2016 est.)
country comparison to the world: 211

Electricity—imports: 0 kWh (2016 est.)
country comparison to the world: 213

Electricity—installed generating capacity: 82,000 kW (2016 est.)

country comparison to the world: 183

Electricity—from fossil fuels: 100% of total installed capacity (2016 est.)
country comparison to the world: 21

Electricity—from nuclear fuels: 0% of total installed capacity (2017 est.)
country comparison to the world: 201

Electricity—from hydroelectric plants: 0% of total installed capacity (2017 est.)
country comparison to the world: 208

Electricity—from other renewable sources: 0% of total installed capacity (2017 est.)
country comparison to the world: 211

Crude oil—production: 0 bbl/day (2018 est.)
country comparison to the world: 209

Crude oil—exports: 0 bbl/day (2015 est.)
country comparison to the world: 209

Crude oil—imports: 0 bbl/day (2015 est.)
country comparison to the world: 207

Crude oil—proved reserves: 0 bbl (1 January 2018 est.)
country comparison to the world: 207

Refined petroleum products—production: 0 bbl/day (2015 est.)
country comparison to the world: 210

Refined petroleum products—consumption: 1,420 bbl/day (2016 est.)
country comparison to the world: 201

Refined petroleum products—exports: 0 bbl/day (2015 est.)
country comparison to the world: 211

Refined petroleum products—imports: 1,369 bbl/day (2015 est.)
country comparison to the world: 197

Natural gas—production: 0 cu m (2017 est.)
country comparison to the world: 208

Natural gas—consumption: 0 cu m (2017 est.)
country comparison to the world: 208

Natural gas—exports: 0 cu m (2017 est.)
country comparison to the world: 203

Natural gas—imports: 0 cu m (2017 est.)
country comparison to the world: 204

Natural gas—proved reserves: 0 cu m (1 January 2014 est.)
country comparison to the world: 202

Carbon dioxide emissions from consumption of energy: 221,800 Mt (2017 est.)
country comparison to the world: 198

COMMUNICATIONS

Telephones—fixed lines: *total subscriptions:* 6,096
subscriptions per 100 inhabitants: 11.12 (2019 est.)
country comparison to the world: 203

Telecommunication systems: *general assessment:* fully digital system with international direct dialing; broadband access; expanded FttP (Fiber to the Home) markets; LTE expansion points to investment and focus on data; regulatory development; telecommunication contributes to greatly to GDP (2020)

domestic: full range of services available; GSM wireless service available; fixed-line teledensity 11 per 100 persons (2019)

international: country code—1-649; landing point for the ARCOS fiber-optic telecommunications submarine cable providing connectivity to South and Central America, parts of the Caribbean, and the US; satellite earth station—1 Intelsat (Atlantic Ocean) (2020)

note: the COVID-19 outbreak is negatively impacting telecommunications production and supply chains globally; consumer spending on telecom devices and services has also slowed due to the pandemic's effect on economies worldwide; overall progress towards improvements in all facets of the telecom industry—mobile, fixed-line, broadband, submarine cable and satellite—has moderated

Broadcast media: no local terrestrial TV stations, broadcasts from the Bahamas can be received and multi-channel cable and satellite TV services are available; government-run radio network operates alongside private broadcasters with a total of about 15 stations

Internet country code: .tc

TRANSPORTATION

National air transport system: *number of registered air carriers:* 3 (2020)
inventory of registered aircraft operated by air carriers: 22

Civil aircraft registration country code prefix: VQ-T (2016)

Airports: 8 (2013)
country comparison to the world: 164

Airports—with paved runways: *total:* 6 (2013)
2,438 to 3,047 m: 1 (2013)
1,524 to 2,437 m: 3 (2013)
914 to 1,523 m: 1 (2013)
under 914 m: 1 (2013)

Airports—with unpaved runways: *total:* 2 (2013)
under 914 m: 2 (2013)

Roadways: *total:* 121 km (2003)
paved: 24 km (2003)
unpaved: 97 km (2003)
country comparison to the world: 212

Merchant marine: *total:* 4
by type: general cargo 1, other 3 (2019)
country comparison to the world: 169

Ports and terminals: *major seaport(s):* Cockburn Harbour, Grand Turk, Providenciales

MILITARY AND SECURITY

Military—note: defense is the responsibility of the UK

TRANSNATIONAL ISSUES

Disputes—international: have received Haitians fleeing economic and civil disorder

Illicit drugs: transshipment point for South American narcotics destined for the US and Europe

TUVALU

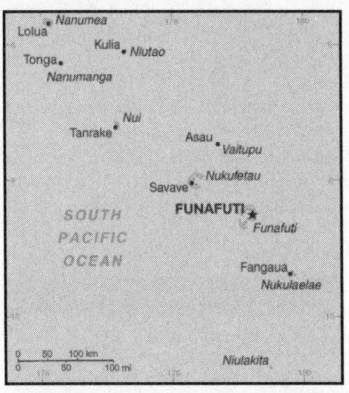

INTRODUCTION

Background: In 1974, ethnic differences within the British colony of the Gilbert and Ellice Islands caused the Polynesians of the Ellice Islands to vote for separation from the Micronesians of the Gilbert Islands. The following year, the Ellice Islands became the separate British colony of Tuvalu. Independence was granted in 1978. In 2000, Tuvalu negotiated a contract leasing its Internet domain name ".tv" for $50 million in royalties over a 12-year period. The agreement was subsequently renegotiated but details were not disclosed. Tuvalu hosted the Pacific Islands Forum Leaders Meeting in August 2019.

GEOGRAPHY

Location: Oceania, island group consisting of nine coral atolls in the South Pacific Ocean, about half way from Hawaii to Australia

Geographic coordinates: 8 00 S, 178 00 E

Map references: Oceania

Area: *total:* 26 sq km
land: 26 sq km
water: 0 sq km
country comparison to the world: 238

Area—comparative: 0.1 times the size of Washington, DC

Area comparison map: [INSERT IMAGE: TUVALU-Area comparison map]

Land boundaries: 0 km

Coastline: 24 km

Maritime claims: *territorial sea:* 12 nm
exclusive economic zone: 200 nm
contiguous zone: 24 nm

Climate: tropical; moderated by easterly trade winds (March to November); westerly gales and heavy rain (November to March)

Terrain: low-lying and narrow coral atolls

Elevation: *mean elevation:* 2 m

lowest point: Pacific Ocean 0 m
highest point: unnamed location 5 m

Natural resources: fish, coconut (copra)

Land use: *agricultural land:* 60% (2011 est.)
arable land: 0% (2011 est.) / permanent crops: 60% (2011 est.) / permanent pasture: 0% (2011 est.)
forest: 33.3% (2011 est.)
other: 6.7% (2011 est.)

Irrigated land: 0 sq km (2012)

Population distribution: over half of the population resides on the atoll of Funafuti

Natural hazards: severe tropical storms are usually rare, but in 1997 there were three cyclones; low levels of islands make them sensitive to changes in sea level

Environment—current issues: water needs met by catchment systems; the use of sand as a building material has led to beachhead erosion; deforestation; damage to coral reefs from increasing ocean temperatures and acidification; rising sea levels threaten water table; in 2000, the government appealed to Australia and New Zealand to take in Tuvaluans if rising sea levels should make evacuation necessary

Environment—international agreements: *party to:* Biodiversity, Climate Change, Climate Change-Kyoto Protocol, Desertification, Law of the Sea, Ozone Layer Protection, Ship Pollution, Whaling
signed, but not ratified: none of the selected agreements

Geography—note: one of the smallest and most remote countries on Earth; six of the nine coral atolls - Nanumea, Nui, Vaitupu, Nukufetau, Funafuti, and Nukulaelae - have lagoons open to the ocean; Nanumaya and Niutao have landlocked lagoons; Niulakita does not have a lagoon

PEOPLE AND SOCIETY

Population: 11,342 (July 2020 est.)
country comparison to the world: 222

Nationality: *noun:* Tuvaluan(s)
adjective: Tuvaluan

Ethnic groups: Tuvaluan 86.8%, Tuvaluan/I-Kiribati 5.6%, Tuvaluan/other 6.7%, other 0.9% (2012 est.)

Languages: Tuvaluan (official), English (official), Samoan, Kiribati (on the island of Nui)

Religions: Protestant 92.4% (Congregational Christian Church of Tuvalu 85.7%, Brethren 3%, Seventh Day Adventist 2.8%, Assemblies of God .9%), Baha'i 2%, Jehovah's Witness 1.3%, Mormon 1%, other 3.1%, none 0.2% (2012 est.)

Age structure: *0-14 years:* 29.42% (male 1,711/female 1,626)
15-24 years: 17.61% (male 1,031/female 966)
25-54 years: 37.17% (male 2,157/female 2,059)

55-64 years: 9.12% (male 427/female 607)
65 years and over: 6.68% (male 289/female 469) (2020 est.)

population pyramid: [INSERT IMAGE: AUSTRALIA -population pyramid]

Median age: *total:* 26.6 years
male: 25.6 years
female: 27.6 years (2020 est.)
country comparison to the world: 152

Population growth rate: 0.87% (2020 est.)
country comparison to the world: 120

Birth rate: 23.4 births/1,000 population (2020 est.)
country comparison to the world: 54

Death rate: 8.2 deaths/1,000 population (2020 est.)
country comparison to the world: 87

Net migration rate: -6.5 migrant(s)/1,000 population (2020 est.)
country comparison to the world: 208

Population distribution: over half of the population resides on the atoll of Funafuti

Urbanization: *urban population:* 64% of total population (2020)
rate of urbanization: 2.27% annual rate of change (2015-20 est.)

total population growth rate v. urban population growth rate, 2000-2030: *Major urban areas - population:* 7,000 FUNAFUTI (capital) (2018)

Sex ratio: *at birth:* 1.05 male(s)/female
0-14 years: 1.05 male(s)/female
15-24 years: 1.07 male(s)/female
25-54 years: 1.05 male(s)/female
55-64 years: 0.7 male(s)/female
65 years and over: 0.62 male(s)/female
total population: 0.98 male(s)/female (2020 est.)

Infant mortality rate: *total:* 26.6 deaths/1,000 live births
male: 28.8 deaths/1,000 live births
female: 24.2 deaths/1,000 live births (2020 est.)
country comparison to the world: 67

Life expectancy at birth: *total population:* 67.9 years
male: 65.6 years
female: 70.2 years (2020 est.)
country comparison to the world: 178

Total fertility rate: 2.88 children born/woman (2020 est.)
country comparison to the world: 56

Drinking water source:

improved: *urban:* 100% of population
rural: 98.8% of population
total: 99% of population

unimproved: *urban:* 0% of population
rural: 1.2% of population
total: 1% of population (2017 est.)

Current Health Expenditure: 17.1% (2017)

Physicians density: 0.91 physicians/1,000 population (2014)

Sanitation facility access:
improved:
urban: 91.8% of population
rural: 91% of population
total: 91.5% of population
unimproved:
urban: 9.2% of population
rural: 9% of population
total: 8.5% of population (2017 est.)

HIV/AIDS—adult prevalence rate: NA

HIV/AIDS—people living with HIV/AIDS: NA

HIV/AIDS—deaths: NA

Obesity—adult prevalence rate: 51.6% (2016)
country comparison to the world: 5

Education expenditures: NA
Unemployment, youth ages 15-24: total: 20.6%
male: 9.8%
female: 45.9% (2016)
country comparison to the world: 65

GOVERNMENT

Country name: *conventional long form:* none
conventional short form: Tuvalu
local long form: none
local short form: Tuvalu
former: Ellice Islands
etymology: "tuvalu" means "group of eight" or "eight standing together" referring to the country's eight traditionally inhabited islands

Government type: parliamentary democracy under a constitutional monarchy; a Commonwealth realm

Capital: *name:* Funafuti; note - the capital is an atoll of some 29 islets; administrative offices are in Vaiaku Village on Fongafale Islet

geographic coordinates: 8 31 S, 179 13 E
time difference: UTC+12 (17 hours ahead of Washington, DC, during Standard Time)
etymology: the atoll is named after a founding ancestor chief, Funa, from the island of Samoa

Administrative divisions: 7 island councils and 1 town council*; Funafuti*, Nanumaga, Nanumea, Niutao, Nui, Nukufetau, Nukulaelae, Vaitupu

Independence: 1 October 1978 (from the UK)

National holiday: Independence Day, 1 October (1978)

Constitution: *history:* previous 1978 (at independence); latest effective 1 October 1986

amendments: proposed by the House of Assembly; passage requires at least two-thirds majority vote by the Assembly membership in the final reading; amended 2007, 2010, 2013

Legal system: mixed legal system of English common law and local customary law

International law organization participation: has not submitted an ICJ jurisdiction declaration; non-party state to the ICCt

Citizenship: *citizenship by birth:* yes
citizenship by descent only: yes; for a child born abroad, at least one parent must be a citizen of Tuvalu

dual citizenship recognized: yes
residency requirement for naturalization: na

Suffrage: 18 years of age; universal

Executive branch: *chief of state:* Queen ELIZABETH II (since 6 February 1952); represented by Governor General Iakoba TAEIA Italeli (since 16 April 2010)

head of government: Prime Minister Kausea NATANO (since 19 September 2019)
cabinet: Cabinet appointed by the governor general on recommendation of the prime minister
elections/appointments: the monarchy is hereditary; governor general appointed by the monarch on recommendation of the prime minister; prime minister and deputy prime minister elected by and from members of House of Assembly following parliamentary elections
election results: Kausea NATANO elected prime minister by House of Assembly; House of Assembly vote count on 19 September 2019 - 10 to 6

Legislative branch: *description:* unicameral House of Assembly or Fale I Fono (16 seats; members directly elected in single- and multi-seat constituencies by simple majority vote to serve 4-year terms)
elections: last held on 9 September 2019 (next to be held on September 2023)
election results: percent of vote - NA; seats - independent 16 (9 members reelected)

Judicial branch: *highest courts:* Court of Appeal (consists of the chief justice and not less than 3 appeals judges); High Court (consists of the chief justice); appeals beyond the Court of Appeal are heard by the Judicial Committee of the Privy Council (in London)
judge selection and term of office: Court of Appeal judges appointed by the governor general on the advice of the Cabinet; judge tenure based on terms of appointment; High Court chief justice appointed by the governor general on the advice of the Cabinet; chief justice serves for life; other judges appointed by the governor general on the advice of the Cabinet after consultation with chief justice; judge tenure set by terms of appointment
subordinate courts: magistrates' courts; island courts; land courts

Political parties and leaders: there are no political parties but members of parliament usually align themselves in informal groupings

International organization participation: ACP, ADB, AOSIS, C, FAO, IBRD, IDA, IFAD, IFRCS (observer), ILO, IMF, IMO, IOC, ITU, OPCW, PIF, Sparteca, SPC, UN, UNCTAD, UNESCO, UNIDO, UPU, WHO, WIPO, WMO

Diplomatic representation in the US: none; the Tuvalu Permanent Mission to the UN serves as the Embassy; it is headed by Samuelu LALONIU (since 21 July 2017); address: 685 Third Avenue, Suite 1104, New York, NY 10017; telephone: [1] (212) 490-0534; FAX: [1] (212) 8084975

Diplomatic representation from the US: the US does not have an embassy in Tuvalu; the US Ambassador to Fiji is accredited to Tuvalu

Flag description: light blue with the flag of the UK in the upper hoist-side quadrant; the outer half of the flag represents a map of the country with nine yellow, five-pointed stars on a blue field symbolizing the nine atolls in the ocean

National symbol(s): *maneapa (native meeting house); national colors:* light blue, yellow

National anthem: *name:* "Tuvalu mo te Atua" (Tuvalu for the Almighty)
lyrics/music: Afaese MANOA
note: adopted 1978; the anthem's name is also the nation's motto

ECONOMY

Economy—overview: Tuvalu consists of a densely populated, scattered group of nine coral atolls with poor soil. Only eight of the atolls are inhabited. It is one of the smallest countries in the world, with its highest point at 4.6 meters above sea level. The country is isolated, almost entirely dependent on imports, particularly of food and fuel, and vulnerable to climate change and rising sea levels, which pose significant challenges to development.

The public sector dominates economic activity. Tuvalu has few natural resources, except for its fisheries. Earnings from fish exports and fishing licenses for Tuvalu's territorial waters are a significant source of government revenue. In 2013, revenue from fishing licenses doubled and totaled more than 45% of GDP.

Official aid from foreign development partners has also increased. Tuvalu has substantial assets abroad. The Tuvalu Trust Fund, an international trust fund established in 1987 by development partners, has grown to $104 million (A$141 million) in 2014 and is an important cushion for meeting shortfalls in the government's budget. While remittances are another substantial source of income, the value of remittances has declined since the 2008-09 global financial crisis, but has stabilized at nearly $4 million per year. The financial impact of climate change and the cost of climate related adaptation projects is one of many concerns for the nation.

GDP (purchasing power parity): $42 million (2017 est.)
$40.68 million (2016 est.)
$39.48 million (2015 est.)
note: data are in 2017 dollars
country comparison to the world: 226

GDP (official exchange rate): $40 million (2017 est.)

GDP—real growth rate: 3.2% (2017 est.)
3% (2016 est.)
9.1% (2015 est.)
country comparison to the world: 94

GDP—per capita (PPP): $3,800 (2017 est.)
$3,700 (2016 est.)
$3,600 (2015 est.)
note: data are in 2017 dollars

country comparison to the world: 180

GDP—composition, by end use: *government consumption:* 87% (2016 est.)
investment in fixed capital: 24.3% (2016 est.)
exports of goods and services: 43.7% (2016 est.)
imports of goods and services: -66.1% (2016 est.)

GDP—composition, by sector of origin: *agriculture:* 24.5% (2012 est.)
industry: 5.6% (2012 est.)
services: 70% (2012 est.)

Agriculture—products: coconuts; fish

Industries: Fishing

Industrial production growth rate: -26.1% (2012 est.)
country comparison to the world: 202

Labor force: 3,615 (2004 est.)
country comparison to the world: 225

Labor force—by occupation: note: most people make a living through exploitation of the sea, reefs, and atolls - and through overseas remittances (mostly from workers in the phosphate industry and sailors)

Unemployment rate: NA

Population below poverty line: 26.3% (2010 est.)

Household income or consumption by percentage share: *lowest 10%:* NA
highest 10%: NA

Budget: *revenues:* 42.68 million (2013 est.)
expenditures: 32.46 million (2012 est.)
note: revenue data include Official Development Assistance from Australia

Taxes and other revenues: 106.7% (of GDP) (2013 est.)
note: revenue data include Official Development Assistance from Australia
country comparison to the world: 1

Budget surplus (+) or deficit (-): 25.6% (of GDP) (2013 est.)
country comparison to the world: 1

Public debt: 37% of GDP (2017 est.)
47.2% of GDP (2016 est.)
country comparison to the world: 142

Fiscal year: calendar year

Inflation rate (consumer prices): 4.1% (2017 est.)
3.5% (2016 est.)
country comparison to the world: 162

Current account balance: $2 million (2017 est.)
$8 million (2016 est.)
country comparison to the world: 64

Exports: $600,000 (2010 est.)
$1 million (2004 est.)
country comparison to the world: 220

Exports—partners: US 18.2%, Bosnia and Herzegovina 17%, Fiji 14.8%, Nigeria 14.2%, Germany 8.2%, South Africa 5.9%, Colombia 5.1% (2017)

Exports—commodities: copra, fish

Imports: $20.69 billion (2018 est.)
$19.09 billion (2017 est.)
country comparison to the world: 74

Imports—commodities: food, animals, mineral fuels, machinery, manufactured goods

Imports—partners: Singapore 33.4%, South Korea 11.5%, Australia 10.8%, NZ 8%, Fiji 7.5%, Chile 6.1%, South Africa 5%, Japan 5% (2017)

Debt—external: NA

Exchange rates: Tuvaluan dollars or Australian dollars (AUD) per US dollar—
1.311 (2017 est.)
1.3442 (2016 est.)

ENERGY

Electricity access: *electrification—total population:* 100% (2020)

Electricity—production: 11.8 million kWh (2011 est.)
country comparison to the world: 214

Electricity—exports: 0 kWh (2014 est.)
country comparison to the world: 212

Electricity—imports: 0 kWh (2014 est.)
country comparison to the world: 214

Electricity—installed generating capacity: 5,100 kW (2011 est.)
country comparison to the world: 212

Electricity—from fossil fuels: 96% of total installed capacity (2015 est.)
country comparison to the world: 43

Electricity—from nuclear fuels: 0% of total installed capacity (2014)
country comparison to the world: 202

Electricity—from hydroelectric plants: 0% of total installed capacity (2014)
country comparison to the world: 209

Crude oil—production: 0 bbl/day (2014 est.)
country comparison to the world: 210

Crude oil—exports: 0 bbl/day (2014 est.)
country comparison to the world: 210

Crude oil—imports: 0 bbl/day (2014 est.)
country comparison to the world: 208

Crude oil—proved reserves: 0 bbl (2014 est.)
country comparison to the world: 208

Refined petroleum products—production: 0 bbl/day (2014 est.)
country comparison to the world: 211

Refined petroleum products— exports: 0 bbl/day
country comparison to the world: 212

Natural gas—production: 0 cu m (2014 est.)
country comparison to the world: 209

Natural gas—consumption: 0 cu m (2014)
country comparison to the world: 209

Natural gas—exports: 0 cu m (2014 est.)
country comparison to the world: 204

Natural gas—imports: 0 cu m (2014 est.)
country comparison to the world: 205

COMMUNICATIONS

Telephones—fixed lines: *total subscriptions:* 1,978

subscriptions per 100 inhabitants: 17.59 (2019 est.)
country comparison to the world: 218

Telephones—mobile cellular: *total subscriptions:* 7,911

subscriptions per 100 inhabitants: 70.36 (2019 est.)
country comparison to the world: 215

Telecommunication systems: *general assessment:* internal communications needs met; small global scale of over 11,000 people on 9 inhabited islands; mobile subscriber penetration about 40% and broadband about 10% penetration; govt. owned and sole provider of telecommunications services; 2G widespread; the launch in 2019 of the Kacific-1 satellite will improve the telecommunication sector for the Asia Pacific region (2020)

domestic: radiotelephone communications between islands; fixed-line 18 per 100 and mobile-cellular 70 per 100 (2019)

international: country code - 688; international calls can be made by satellite
note: the COVID-19 outbreak is negatively impacting telecommunications production and supply chains globally; consumer spending on tele-com devices and services has also slowed due to the pandemic's effect on economies worldwide; overall progress towards improvements in all facets of the telecom industry - mobile, fixed-line, broadband, submarine cable and satellite - has moderated

Broadcast media: no TV stations; many households use satellite dishes to watch foreign TV stations; 1 government-owned radio station, Radio Tuvalu, includes relays of programming from international broadcasters (2019)

Internet country code: .tv

Internet users: *total:* 5,498

percent of population: 49.32% (July 2018 est.)
country comparison to the world: 216

Broadband—fixed subscriptions: *total:* 1,000

subscriptions per 100 inhabitants: 9 (2017 est.)
country comparison to the world: 197

TRANSPORTATION

Civil aircraft registration country code prefix: T2 (2016)

Airports: 1 (2013)
country comparison to the world: 237

Airports—with unpaved runways: *total:* 1 (2013)
1,524 to 2,437 m: 1 (2013)

Roadways: *total:* 8 km (2011)
paved: 8 km (2011)
country comparison to the world: 223

Merchant marine: *total:* 243
by type: bulk carrier 20, container ship 3, general cargo 39, oil tanker 24, other 157 (2019)
country comparison to the world: 60

Ports and terminals: *major seaport(s):* Funafuti

MILITARY AND SECURITY

Military and security forces: *no regular military forces; Tuvalu Police Force (2012)*

TRANSPORTATION

Disputes—international: none

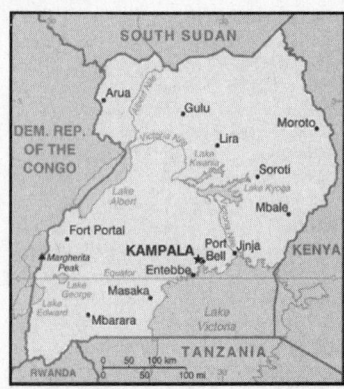

INTRODUCTION

Background: British influence in Uganda began in the 1860s with explorers seeking the source of the Nile and expanded in subsequent decades with various trade agreements and the establishment of the Uganda Protectorate in 1894. The colonial boundaries created by Britain to delimit Uganda grouped together a wide range of ethnic groups with different political systems and cultures. These differences complicated the establishment of a working political community after independence was achieved in 1962. The dictatorial regime of Idi AMIN (1971-79) was responsible for the deaths of some 300,000 opponents; guerrilla war and human rights abuses under Milton OBOTE (1980-85) claimed at least another 100,000 lives. The rule of Yoweri MUSEVENI since 1986 has brought relative stability and economic growth to Uganda. In December 2017, parliament approved the removal of presidential age limits, thereby making it possible for MUSEVENI to continue standing for office. Uganda faces numerous challenges, however, that could affect future stability, including explosive population growth, power and infrastructure constraints, corruption, underdeveloped democratic institutions, and human rights deficits.

GEOGRAPHY

Location: East-Central Africa, west of Kenya, east of the Democratic Republic of the Congo

Geographic coordinates: 1 00 N, 32 00 E

Map references: Africa

Area: *total:* 241,038 sq km
land: 197,100 sq km
water: 43,938 sq km
country comparison to the world: 82

Area—comparative: slightly more than two times the size of Pennsylvania; slightly smaller than Oregon

Land boundaries: *total:* 2,729 km
border countries (5): Democratic Republic of the Congo 877 km, Kenya 814 km, Rwanda 172 km, South Sudan 475 km, Tanzania 391 km

Coastline: 0 km (landlocked)

Maritime claims: none (landlocked)

Climate: tropical; generally rainy with two dry seasons (December to February, June to August); semiarid in northeast

Terrain: mostly plateau with rim of mountains

Elevation: *lowest point:* Albert Nile 614 m
highest point: Margherita Peak on Mount Stanley 5,110 m

Natural resources: copper, cobalt, hydropower, limestone, salt, arable land, gold

Land use: *agricultural land:* 71.2% (2011 est.)
arable land: 34.3% (2011 est.) / *permanent crops:* 11.3% (2011 est.) / *permanent pasture:* 25.6% (2011 est.)
forest: 14.5% (2011 est.)
other: 14.3% (2011 est.)
Irrigated land: 140 sq km (2012)

Population distribution: population density is relatively high in comparison to other African nations; most of the population is concentrated in the central and southern parts of the country, particularly along the shores of Lake Victoria and Lake Albert; the northeast is least populated as shown in this population distribution map

Natural hazards: droughts; floods; earthquakes; landslides; hailstorms

Environment—current issues: draining of wetlands for agricultural use; deforestation; overgrazing; soil erosion; water pollution from industrial discharge and water hyacinth infestation in Lake Victoria; widespread poaching

Environment—international agreements: *party to:* Biodiversity, Climate Change, Climate Change-Kyoto Protocol, Desertification, Endangered Species, Hazardous Wastes, Law of the Sea, Marine Life Conservation, Ozone Layer Protection, Wetlands
signed, but not ratified: Environmental Modification

Geography—note: landlocked; fertile, well-watered country with many lakes and rivers; Lake Victoria, the world's largest tropical lake and the second largest fresh water lake, is shared among three countries: Kenya, Tanzania, and Uganda

PEOPLE AND SOCIETY

Population: 43,252,966 (July 2020 est.)
note: estimates for this country explicitly take into account the effects of excess mortality due to AIDS; this can result in lower life expectancy, higher infant mortality, higher death rates, lower population growth rates, and changes in the distribution of population by age and sex than would otherwise be expected
country comparison to the world: 34

Nationality: *noun:* Ugandan(s)
adjective: Ugandan

Ethnic groups: Baganda 16.5%, Banyankole 9.6%, Basoga 8.8%, Bakiga 7.1%, Iteso 7%, Langi 6.3%, Bagisu 4.9%, Acholi 4.4%, Lugbara 3.3%, other 32.1% (2014 est.)

Languages: English (official language, taught in schools, used in courts of law and by most newspapers and some radio broadcasts), Ganda or Luganda (most widely used of the Niger-Congo languages and the language used most often in the capital), other Niger-Congo languages, Nilo-Saharan languages, Swahili (official), Arabic

Religions: Protestant 45.1% (Anglican 32.0%, Pentecostal/Born Again/Evangelical 11.1%, Seventh Day Adventist 1.7%, Baptist .3%), Roman Catholic 39.3%, Muslim 13.7%, other 1.6%, none 0.2% (2014 est.)

Demographic profile: Uganda has one of the youngest and most rapidly growing populations in the world; its total fertility rate is among the world's highest at 5.8 children per woman. Except in urban areas, actual fertility exceeds women's desired fertility by one or two children, which is indicative of the widespread unmet need for contraception, lack of government support for family planning, and a cultural preference for large families. High numbers of births, short birth intervals, and the early age of childbearing contribute to Uganda's high maternal mortality rate. Gender inequities also make fertility reduction difficult; women on average are less-educated, participate less in paid employment, and often have little say in decisions over childbearing and their own reproductive health. However, even if the birth rate were significantly reduced, Uganda's large pool of women entering reproductive age ensures rapid population growth for decades to come.

Unchecked, population increase will further strain the availability of arable land and natural resources and overwhelm the country's limited means for providing food, employment, education, health care, housing, and basic services. The country's north and northeast lag even further behind developmentally than the rest of the country as a result of long-term conflict (the Ugandan Bush War 1981-1986 and more than 20 years of fighting between the Lord's Resistance Army (LRA) and Ugandan Government forces), ongoing inter-communal violence, and periodic natural disasters.

Uganda has been both a source of refugees and migrants and a host country for refugees. In 1972, then President Idi AMIN, in his drive to return Uganda to Ugandans, expelled the South Asian population that composed a large share of the country's business people and bankers. Since the 1970s, thousands of Ugandans have emigrated, mainly to southern Africa or the West, for security reasons, to escape poverty, to search for jobs, and for access to natural resources. The emigration of Ugandan doctors and nurses due to low wages is a particular concern given the country's shortage of skilled health care workers. Africans escaping conflicts in neighboring states have found refuge in Uganda since the 1950s; the country currently struggles to host tens of thousands from the Democratic Republic of the Congo, South Sudan, and other nearby countries.

Age structure: *0-14 years:* 48.21% (male 10,548,913/female 10,304,876)
15-24 years: 20.25% (male 4,236,231/female 4,521,698)
25-54 years: 26.24% (male 5,202,570/female 6,147,304)
55-64 years: 2.91% (male 579,110/female 681,052)
65 years and over: 2.38% (male 442,159/female 589,053) (2020 est.)

Dependency ratios: *total dependency ratio:* 92.3
youth dependency ratio: 88.5
elderly dependency ratio: 3.8
potential support ratio: 26.2 (2020 est.)

Median age: *total:* 15.7 years
male: 14.9 years
female: 16.5 years (2020 est.)
country comparison to the world: 227

Population growth rate: 3.34% (2020 est.)
country comparison to the world: 5

Birth rate: 42.3 births/1,000 population (2020 est.)
country comparison to the world: 3

Death rate: 5.3 deaths/1,000 population (2020 est.)
country comparison to the world: 193

Net migration rate: -3.5 migrant(s)/1,000 population (2020 est.)
country comparison to the world: 184

Population distribution: population density is relatively high in comparison to other African nations; most of the population is concentrated in the central and southern parts of the country, particularly along the shores of Lake Victoria and Lake Albert; the northeast is least populated as shown in this population distribution map

Urbanization: *urban population:* 25% of total population (2020)
rate of urbanization: 5.7% annual rate of change (2015-20 est.)

total population growth rate v. urban population growth rate, 2000-2030: Major urban areas—population: 3.298 million KAMPALA (capital) (2020)

Sex ratio: *at birth:* 1.03 male(s)/female
0-14 years: 1.02 male(s)/female
15-24 years: 0.94 male(s)/female
25-54 years: 0.85 male(s)/female
55-64 years: 0.85 male(s)/female
65 years and over: 0.75 male(s)/female
total population: 0.94 male(s)/female (2020 est.)

Mother's mean age at first birth: 18.9 years (2011 est.)
note: median age at first birth among women 25-29

Maternal mortality rate: 375 deaths/100,000 live births (2017 est.)
country comparison to the world: 30

Infant mortality rate: *total:* 32.6 deaths/1,000 live births
male: 36.1 deaths/1,000 live births
female: 28.9 deaths/1,000 live births (2020 est.)
country comparison to the world: 47

Life expectancy at birth: *total population:* 68.2 years
male: 66 years
female: 70.5 years (2020 est.)
country comparison to the world: 177

Total fertility rate: 5.54 children born/woman (2020 est.)
country comparison to the world: 7

Contraceptive prevalence rate: 41.8% (2018)

Drinking water source:

improved: *urban:* 92.9% of population
rural: 77.2% of population
total: 80.8% of population

unimproved: *urban:* 7.1% of population
rural: 22.8% of population
total: 19.2% of population (2017 est.)

Current Health Expenditure: 6.3% (2017)

Physicians density: 0.17 physicians/1,000 population (2017)

Hospital bed density: 0.5 beds/1,000 population (2010)

Sanitation facility access:

improved: *urban:* 67.8% of population
rural: 26.6% of population
total: 36.2% of population

unimproved: *urban:* 32.2% of population
rural: 73.4% of population
total: 63.8% of population (2017 est.)

HIV/AIDS—adult prevalence rate: 6.1% (2019 est.)
country comparison to the world: 11

HIV/AIDS—people living with HIV/AIDS: 1.5 million (2019 est.)
country comparison to the world: 7

HIV/AIDS—deaths: 21,000 (2019 est.)
country comparison to the world: 8

Major infectious diseases: *degree of risk:* very high (2020)
food or waterborne diseases: bacterial diarrhea, hepatitis A and E, and typhoid fever
vectorborne diseases: malaria, dengue fever, and Trypanosomiasis-Gambiense (African sleeping sickness)
water contact diseases: schistosomiasis
animal contact diseases: rabies

Obesity—adult prevalence rate: 5.3% (2016)
country comparison to the world: 181

Children under the age of 5 years underweight: 10.4% (2016)
country comparison to the world: 62
Education expenditures: 2.6% of GDP (2017)
country comparison to the world: 157

Literacy: *definition:* age 15 and over can read and write
total population: 76.5%
male: 82.7%
female: 70.8% (2018)

School life expectancy (primary to tertiary education): *total:* 10 years
male: 10 years
female: 10 years (2011)

Unemployment, youth ages 15-24: *total:* 14.8%
male: 12.7%
female: 17.3% (2017 est.)
country comparison to the world: 92

GOVERNMENT

Country name: *conventional long form:* Republic of Uganda
conventional short form: Uganda
etymology: from the name "Buganda," adopted by the British as the designation for their East African colony in 1894; Buganda had been a powerful East African state during the 18th and 19th centuries

Government type: presidential republic

Capital: *name:* Kampala
geographic coordinates: 0 19 N, 32 33 E
time difference: UTC+3 (8 hours ahead of Washington, DC, during Standard Time)
etymology: the site of the original British settlement was referred to by its native name as Akasozi ke'Empala ("hill of the impala" [plural]); over time this designation was shortened to K'empala and finally Kampala

Administrative divisions: 134 districts and 1 capital city*; Abim, Adjumani, Agago, Alebtong, Amolatar, Amudat, Amuria, Amuru, Apac, Arua, Budaka, Bududa, Bugiri, Bugweri, Buhweju, Buikwe, Bukedea, Bukomansimbi, Bukwo, Bulambuli, Buliisa, Bundibugyo, Bunyangabu, Bushenyi, Busia, Butaleja, Butambala, Butebo, Buvuma, Buyende, Dokolo, Gomba, Gulu, Hoima, Ibanda, Iganga, Isingiro, Jinja, Kaabong, Kabale, Kabarole, Kaberamaido, Kagadi, Kakumiro, Kalaki, Kalangala, Kaliro, Kalungu, Kampala*, Kamuli, Kamwenge, Kanungu, Kapchorwa, Kapelebyong, Karenga, Kasese, Kasanda, Katakwi, Kayunga, Kazo, Kibaale, Kiboga, Kibuku, Kikuube, Kiruhura, Kiryandongo, Kisoro, Kitagwenda, Kitgum, Koboko, Kole, Kotido, Kumi, Kwania, Kween, Kyankwanzi, Kyegegwa, Kyenjojo, Kyotera, Lamwo, Lira, Luuka, Luwero, Lwengo, Lyantonde, Madi- Okollo, Manafwa, Maracha, Masaka, Masindi, Mayuge, Mbale, Mbarara, Mitooma, Mityana, Moroto, Moyo, Mpigi, Mubende, Mukono, Nabilatuk, Nakapiripirit, Nakaseke, Nakasongola, Namayingo, Namisindwa, Namutumba, Napak, Nebbi, Ngora, Ntoroko, Ntungamo, Nwoya, Obongi, Omoro,

Otuke, Oyam, Pader, Pakwach, Pallisa, Rakai, Rubanda, Rubirizi, Rukiga, Rukungiri, Rwampara, Sembabule, Serere, Sheema, Sironko, Soroti, Tororo, Wakiso, Yumbe, Zombo

Independence: 9 October 1962 (from the UK)

National holiday: Independence Day, 9 October (1962)

Constitution: *history:* several previous; latest adopted 27 September 1995, promulgated 8 October 1995
amendments: proposed by the National Assembly; passage requires at least two-thirds majority vote of the Assembly membership in the second and third readings; proposals affecting "entrenched clauses," including the sovereignty of the people, supremacy of the constitution, human rights and freedoms, the democratic and multiparty form of government, presidential term of office, independence of the judiciary, and the institutions of traditional or cultural leaders, also requires passage by referendum, ratification by at least two-thirds majority vote of district council members in at least two thirds of Uganda's districts, and assent of the president of the republic; amended several times, last in 2017

Legal system: mixed legal system of English common law and customary law

International law organization participation: accepts compulsory ICJ jurisdiction; accepts ICCt jurisdiction

Citizenship: *citizenship by birth:* no
citizenship by descent only: at least one parent or grandparent must be a native-born citizen of Uganda
dual citizenship recognized: yes
residency requirement for naturalization: an aggregate of 20 years and continuously for the last 2 years prior to applying for citizenship

Suffrage: 18 years of age; universal

Executive branch: head of government: President Yoweri Kaguta MUSEVENI (since seizing power on 26 January 1986); Vice President Edward SSEKANDI (since 24 May 2011); Prime Minister Ruhakana RUGUNDA (since 19 September 2014); First Deputy Prime Minister Moses ALI (since 6 June 2016); Second Deputy Prime Minister Kirunda KIVEJINJA (since 6 June 2016)
cabinet: Cabinet appointed by the president from among elected members of the National Assembly or persons who qualify to be elected as members of the National Assembly
elections/appointments: president directly elected by absolute majority popular vote in 2 rounds if needed for a 5-year term (no term limits); election last held on 18 February 2016 (next scheduled to be held February 2021)
election results: Yoweri Kaguta MUSEVENI reelected president in the first round; percent of vote—Yoweri Kaguta MUSEVENI (NRM) 60.6%, Kizza BESIGYE (FDC) 35.6%, other 3.8%
head of state: President Yoweri Kaguta MUSEVENI (since seizing power on 26 January

1986); Vice President Edward SSEKANDI (since 24 May 2011); note—the president is both head of state and head of government

Legislative branch: *description:* unicameral National Assembly or Parliament (445 seats; 290 members directly elected in single-seat constituencies by simple majority vote, 112 for women directly elected in single-seat districts by simple majority vote, and 25 "representatives" reserved for special interest groups—army 10, disabled 5, youth 5, labor 5; up to 18 ex officio members appointed by the president; members serve 5-year terms)
elections: last held on 18 February 2016 (next to be held in February 2021)
election results: percent of vote by party—NA; seats by party—NRM 292, FDC 37, DP 5, UPDF 10, UPC 6, independent 66 (excludes 19 ex-officio members)

Judicial branch: *highest courts:* Supreme Court of Uganda (consists of the chief justice and at least 6 justices)
judge selection and term of office: justices appointed by the president of the republic in consultation with the Judicial Service Commission, an 8-member independent advisory body, and approved by the National Assembly; justices serve until mandatory retirement at age 70
subordinate courts: Court of Appeal (also acts as the Constitutional Court); High Court (includes 12 High Court Circuits and 8 High Court Divisions); Industrial Court; Chief Magistrate Grade One and Grade Two Courts throughout the country; qadhis courts; local council courts; family and children courts

Political parties and leaders: Alliance for National Transformation or ANT [Ms. Alice ALASO, acting national coordinator]; note—Mugisha MUNTU resigned his position as ANT national coordinator in late June 2020 to run in the 2021 presidential election Democratic Party or DP [Norbert MAO]
Forum for Democratic Change or FDC [Patrick Oboi AMURIAT]
Justice Forum or JEEMA [Asuman BASALIRWA]
National Resistance Movement or NRM [Yoweri MUSEVENI]
Uganda People's Congress or UPC [James AKENA]

International organization participation: ACP, AfDB, AU, C, COMESA, EAC, EADB, FAO, G-77, IAEA, IBRD, ICAO, ICC (national committees), ICCt, IDA, IDB, IFAD, IFC, IFRCS, IGAD, ILO, IMF, IMO, Interpol, IOC, IOM, IPU, ISO (correspondent), ITSO, ITU, ITUC (NGOs), MIGA, NAM, OIC, OPCW, PCA, UN, UNCTAD, UNESCO, UNHCR, UNIDO, UNOCI, UNWTO, UPU, WCO, WFTU (NGOs), WHO, WIPO, WMO, WTO

Diplomatic representation in the US: *chief of mission:* Ambassador Mull Sebujja KATENDE (since 8 September 2017)
chancery: 5911 16th Street NW, Washington, DC 20011

telephone: [1] (202) 726-7100
FAX: [1] (202) 726-1727

Diplomatic representation from the US: *chief of mission:* Ambassador Natalie E. BROWN (since 17 November 2020)
telephone: (256)-414-259791
embassy: 1577 Ggaba Road, Kampala
mailing address: P.O. Box 7007, Kampala
FAX: [256] 414-306-009

Flag description: six equal horizontal bands of black (top), yellow, red, black, yellow, and red; a white disk is superimposed at the center and depicts a grey crowned crane (the national symbol) facing the hoist side; black symbolizes the African people, yellow sunshine and vitality, red African brotherhood; the crane was the military badge of Ugandan soldiers under the UK

National symbol(s): grey crowned crane; national colors: black, yellow, red

National anthem: *name:* Oh Uganda, Land of Beauty!
lyrics/music: George Wilberforce KAKOMOA
note: adopted 1962

ECONOMY

Economy—overview: Uganda has substantial natural resources, including fertile soils, regular rainfall, substantial reserves of recoverable oil, and small deposits of copper, gold, and other minerals. Agriculture is one of the most important sectors of the economy, employing 72% of the work force. The country's export market suffered a major slump following the outbreak of conflict in South Sudan, but has recovered lately, largely due to record coffee harvests, which account for 16% of exports, and increasing gold exports, which account for 10% of exports. Uganda has a small industrial sector that is dependent on imported inputs such as refined oil and heavy equipment. Overall, productivity is hampered by a number of supply-side constraints, including insufficient infrastructure, lack of modern technology in agriculture, and corruption.

Uganda's economic growth has slowed since 2016 as government spending and public debt has grown. Uganda's budget is dominated by energy and road infrastructure spending, while Uganda relies on donor support for long-term drivers of growth, including agriculture, health, and education. The largest infrastructure projects are externally financed through concessional loans, but at inflated costs. As a result, debt servicing for these loans is expected to rise.

Oil revenues and taxes are expected to become a larger source of government funding as oil production starts in the next three to 10 years. Over the next three to five years, foreign investors are planning to invest $9 billion in production facilities projects, $4 billion in an export pipeline, as well as in a $2-3 billion refinery to produce petroleum products for the domestic and East African Community markets. Furthermore, the government is looking to build several hundred million dollars' worth of highway projects to the oil region.

Uganda faces many economic challenges. Instability in South Sudan has led to a sharp increase in Sudanese refugees and is disrupting Uganda's main export market. Additional economic risks include: poor economic management, endemic corruption, and the government's failure to invest adequately in the health, education, and economic opportunities for a burgeoning young population. Uganda has one of the lowest electrification rates in Africa—only 22% of Ugandans have access to electricity, dropping to 10% in rural areas.

GDP (purchasing power parity): $89.19 billion (2017 est.)
$85.07 billion (2016 est.)
$83.14 billion (2015 est.)
note: data are in 2017 dollars
country comparison to the world: 89

GDP (official exchange rate): $26.62 billion (2017 est.)

GDP—real growth rate: 4.8% (2017 est.)
2.3% (2016 est.)
5.7% (2015 est.)
country comparison to the world: 56

GDP—per capita (PPP): $2,400 (2017 est.)
$2,300 (2016 est.)
$2,300 (2015 est.)
note: data are in 2017 dollars
country comparison to the world: 200

Gross national saving: 20.6% of GDP (2017 est.)
21.5% of GDP (2016 est.)
17.7% of GDP (2015 est.)
country comparison to the world: 93

GDP—composition, by end use: *household consumption:* 74.3% (2017 est.)
government consumption: 8% (2017 est.)
investment in fixed capital: 23.9% (2017 est.)
investment in inventories: 0.3% (2017 est.)
exports of goods and services: 18.8% (2017 est.)
imports of goods and services: -25.1% (2017 est.)

GDP—composition, by sector of origin: *agriculture:* 28.2% (2017 est.)
industry: 21.1% (2017 est.)
services: 50.7% (2017 est.)

Agriculture—products: coffee, tea, cotton, tobacco, cassava (manioc, tapioca), potatoes, corn, millet, pulses, cut flowers; beef, goat meat, milk, poultry, and fish

Industries: sugar processing, brewing, tobacco, cotton textiles; cement, steel production

Industrial production growth rate: 4.4% (2017 est.)
country comparison to the world: 69

Labor force: 15.84 million (2015 est.)
country comparison to the world: 34

Labor force—by occupation: *agriculture:* 71%
industry: 7%
services: 22% (2013 est.)

Unemployment rate: 9.4% (2014 est.)
country comparison to the world: 142

Population below poverty line: 21.4% (2017 est.)

Household income or consumption by percentage share: *lowest 10%:* 2.4%
highest 10%: 36.1% (2009 est.)

Budget: *revenues:* 3.848 billion (2017 est.)
expenditures: 4.928 billion (2017 est.)

Taxes and other revenues: 14.5% (of GDP) (2017 est.)
country comparison to the world: 199

Budget surplus (+) or deficit (-): -4.1% (of GDP) (2017 est.)
country comparison to the world: 158

Public debt: 40% of GDP (2017 est.)
37.4% of GDP (2016 est.)
country comparison to the world: 127

Fiscal year: 1 July–30 June

Inflation rate (consumer prices): 5.6% (2017 est.)
5.5% (2016 est.)
country comparison to the world: 181

Current account balance: -$1.212 billion (2017 est.)
-$707 million (2016 est.)
country comparison to the world: 151

Exports: $3.339 billion (2017 est.)
$2.921 billion (2016 est.)
country comparison to the world: 124

Exports—partners: Kenya 17.7%, UAE 16.7%, Democratic Republic of the Congo 6.6%, Rwanda 6.1%, Italy 4.8% (2017)

Exports—commodities: coffee, fish and fish products, tea, cotton, flowers, horticultural products; gold

Imports: $5.036 billion (2017 est.)
$4.424 billion (2016 est.)
country comparison to the world: 127

Imports—commodities: capital equipment, vehicles, petroleum, medical supplies; cereals

Imports—partners: China 17.4%, India 13.4%, UAE 12.2%, Kenya 7.9%, Japan 6.4%, Saudi Arabia 6.3%, Indonesia 4.4%, South Africa 4.1% (2017)

Reserves of foreign exchange and gold: $3.654 billion (31 December 2017 est.)
$3.034 billion (31 December 2016 est.)
note: excludes gold
country comparison to the world: 101

Debt—external: $10.8 billion (22 March 2018 est.)
$11.54 billion (31 December 2017 est.)
$6.241 billion (31 December 2016 est.)
country comparison to the world: 112

Exchange rates: Ugandan shillings (UGX) per US dollar—
3,695 (2017 est.)
3,420.1 (2016 est.)
3,420.1 (2015 est.)
3,234.1 (2014 est.)
2,599.8 (2013 est.)

ENERGY

Electricity access: *population without electricity:* 32 million (2019)

electrification—total population: 29% (2019)
electrification—urban areas: 66% (2019)
electrification—rural areas: 17% (2019)

Electricity—production: 3.463 billion kWh (2016 est.)
country comparison to the world: 130

Electricity—consumption: 3.106 billion kWh (2016 est.)
country comparison to the world: 135

Electricity—exports: 121 million kWh (2015 est.)
country comparison to the world: 81

Electricity—imports: 50 million kWh (2016 est.)
country comparison to the world: 107

Electricity—installed generating capacity: 1.02 million kW (2016 est.)
country comparison to the world: 127

Electricity—from fossil fuels: 19% of total installed capacity (2016 est.)
country comparison to the world: 195

Electricity—from nuclear fuels: 0% of total installed capacity (2017 est.)
country comparison to the world: 203

Electricity—from hydroelectric plants: 68% of total installed capacity (2017 est.)
country comparison to the world: 19

Electricity—from other renewable sources: 12% of total installed capacity (2017 est.)
country comparison to the world: 75

Crude oil—production: 0 bbl/day (2018 est.)
country comparison to the world: 211

Crude oil—exports: 0 bbl/day (2015 est.)
country comparison to the world: 211

Crude oil—imports: 0 bbl/day (2015 est.)
country comparison to the world: 209

Crude oil—proved reserves: 2.5 billion bbl (1 January 2018 est.)
country comparison to the world: 31

Refined petroleum products—production: 0 bbl/day (2015 est.)
country comparison to the world: 212

Refined petroleum products—consumption: 32,000 bbl/day (2016 est.)
country comparison to the world: 119

Refined petroleum products—exports: 0 bbl/day (2015 est.)
country comparison to the world: 213

Refined petroleum products—imports: 31,490 bbl/day (2015 est.)
country comparison to the world: 99

Natural gas—production: 0 cu m (2017 est.)
country comparison to the world: 210

Natural gas—consumption: 0 cu m (2017 est.)
country comparison to the world: 210

Natural gas—exports: 0 cu m (2017 est.)
country comparison to the world: 205

Natural gas—imports: 0 cu m (2017 est.)
country comparison to the world: 206

Natural gas—proved reserves: 14.16 billion cu m (1 January 2018 est.)
country comparison to the world: 76

Carbon dioxide emissions from consumption of energy: 4.703 million Mt (2017 est.)
country comparison to the world: 135

COMMUNICATIONS

Telephones—fixed lines: *total subscriptions:* 184,065
subscriptions per 100 inhabitants: less than 1 (2019 est.)
country comparison to the world: 123

Telephones—mobile cellular: *total subscriptions:* 23,957,740
subscriptions per 100 inhabitants: 57.27 (2019 est.)
country comparison to the world: 51

Telecommunication systems: *general assessment:* in recent years, telecommunications infrastructure has developed through private partnerships; as of 2018, fixed fiber backbone infrastructure is available in over half of Uganda's districts; mobile phone companies now provide 4G networks across all major cities and national parks, while offering 3G coverage in second-tier cities and most rural areas with road access; between 2016 and 2018, commercial Internet services dropped in price from $300/Mbps to $80/Mbps; consumers rely on mobile infrastructure to provide voice and broadband services as fixed-line infrastructure is poor; 5G migration is a few years off; govt. commissions broadband satellite services for rural areas (2020)
domestic: fixed-line 1 per 100 and mobile- cellular systems teledensity about 57 per 100 persons; intercity traffic by wire, microwave radio relay, and radiotelephone communication stations (2019)
international: country code—256; satellite earth stations—1 Intelsat (Atlantic Ocean) and 1 Inmarsat; analog and digital links to Kenya and Tanzania
note: the COVID-19 outbreak is negatively impacting telecommunications production and supply chains globally; consumer spending on telecom devices and services has also slowed due to the pandemic's effect on economies worldwide; overall progress towards improvements in all facets of the telecom industry—mobile, fixed-line, broadband, submarine cable and satellite—has moderated

Broadcast media: public broadcaster, Uganda Broadcasting Corporation (UBC), operates radio and TV networks; 31 Free-To-Air (FTA) TV stations, 2 digital terrestrial TV stations, 3 cable TV stations, and 5 digital satellite TV stations; 258 operational FM stations

Internet country code: .ug

Internet users: *total:* 9,620,681

percent of population: 23.71% (July 2018 est.)
country comparison to the world: 53

Broadband—fixed subscriptions: *total:* 9,485
subscriptions per 100 inhabitants: less than 1 (2018 est.)
country comparison to the world: 170

TRANSPORTATION

National air transport system: *number of registered air carriers:* 6 (2020)
inventory of registered aircraft operated by air carriers: 26
annual passenger traffic on registered air carriers: 21,537 (2018)

Civil aircraft registration country code prefix: 5X (2016)

Airports: 47 (2013)
country comparison to the world: 93

Airports—with paved runways: *total:* 5 (2019)
over 3,047 m: 3
1,524 to 2,437 m: 1
914 to 1,523 m: 1

Airports—with unpaved runways: *total:* 42 (2013)
over 3,047 m: 1 (2013)
1,524 to 2,437 m: 8 (2013)
914 to 1,523 m: 26 (2013)
under 914 m: 7 (2013)

Railways: *total:* 1,244 km (2014)
narrow gauge: 1,244 km 1.000-m gauge (2014)
country comparison to the world: 85

Roadways: *total:* 20,544 km (excludes local roads) (2017)
paved: 4,257 km (2017)
unpaved: 16,287 km (2017)
country comparison to the world: 112

Waterways: (there are no long navigable stretches of river in Uganda; parts of the Albert Nile that flow out of Lake Albert in the northwestern part of the country are navigable; several lakes including Lake Victoria and Lake Kyoga have substantial traffic; Lake Albert is navigable along a 200-km stretch from its northern tip to its southern shores) (2011)

Merchant marine: *total:* 1
by type: bulk carrier 1 (2019)
country comparison to the world: 181

Ports and terminals: *lake port(s):* Entebbe, Jinja, Port Bell (Lake Victoria)

MILITARY AND SECURITY

Military and security forces: *Uganda People's Defense Force (UPDF):* Land Forces, Air Forces, Marine Forces, Special Operations Command, Reserve Force (2019)
Military expenditures: 2.1% of GDP (2019)
1.4% of GDP (2018)
1.3% of GDP (2017)

1.3% of GDP (2016)
1.2% of GDP (2015)
country comparison to the world: 46

Military and security service personnel strengths: size estimates for the Uganda People's Defense Force (UPDF) vary; approximately 50,000 troops, including about 1,000 air and marine personnel (2019 est.)

Military equipment inventories and acquisitions: the UPDF's inventory is mostly older Russian/Soviet-era equipment with a limited mix of more modern Russian- and Western-origin arms; since 2010, the leading suppliers of arms to the UPDF are Russia and Ukraine (2019)

Military deployments: 6,200 Somalia (AMISOM); 620 Somalia (UNSOM); 250 Equatorial Guinea (2020)

Military service age and obligation: 18-25 years of age for voluntary military duty (must be single, no children); 9-year service obligation (2019)

TERRORISM

Terrorist group(s): al-Shabaab; Islamic State of Iraq and ash-Sham—Central Africa (2020)
note: details about the history, aims, leadership, organization, areas of operation, tactics, targets, weapons, size, and sources of support of the group(s) appear(s) in Appendix-T

TRANSNATIONAL ISSUES

Disputes—international: Uganda is subject to armed fighting among hostile ethnic groups, rebels, armed gangs, militias, and various government forces that extend across its borders; Ugandan refugees as well as members of the Lord's Resistance Army (LRA) seek shelter in southern Sudan and the Democratic Republic of the Congo's Garamba National Park; LRA forces have also attacked Kenyan villages across the border

Refugees and internally displaced persons: *refugees (country of origin):* 885,171 (South Sudan) (refugees and asylum seekers), 418,369 (Democratic Republic of the Congo) (refugees and asylum seekers), 49,082 (Burundi), 41,850 (Somalia) (refugees and asylum seekers), 17,239 (Rwanda) (refugees and asylum seekers), 14,865 (Eritrea) (refugees and asylum seekers) (2020)
IDPs: 32,000 (displaced in northern Uganda because of fighting between government forces and the Lord's Resistance Army; as of 2011, most of the 1.8 million people displaced to IDP camps at the height of the conflict had returned home or resettled, but many had not found durable solutions; intercommunal violence, land disputes, and cattle raids) (2019)

UKRAINE

INTRODUCTION

Background: Ukraine was the center of the first eastern Slavic state, Kyivan Rus, which during the 10th and 11th centuries was the largest and most powerful state in Europe. Weakened by internecine quarrels and Mongol invasions, Kyivan Rus was incorporated into the Grand Duchy of Lithuania and eventually into the Polish-Lithuanian Commonwealth. The cultural and religious legacy of Kyivan Rus laid the foundation for Ukrainian nationalism through subsequent centuries. A new Ukrainian state, the Cossack Hetmanate, was established during the mid-17th century after an uprising against the Poles. Despite continuous Muscovite pressure, the Hetmanate managed to remain autonomous for well over 100 years. During the latter part of the 18th century, most Ukrainian ethnographic territory was absorbed by the Russian Empire. Following the collapse of czarist Russia in 1917, Ukraine achieved a short-lived period of independence (1917-20), but was reconquered and endured a brutal Soviet rule that engineered two forced famines (1921-22 and 1932-33) in which over 8 million died. In World War II, German and Soviet armies were responsible for 7 to 8 million more deaths. Although Ukraine achieved independence in 1991 with the dissolution of the USSR, democracy and prosperity remained elusive as the legacy of state control and endemic corruption stalled efforts at economic reform, privatization, and civil liberties.

A peaceful mass protest referred to as the "Orange Revolution" in the closing months of 2004 forced the authorities to overturn a rigged presidential election and to allow a new internationally monitored vote that swept into power a reformist slate under Viktor YUSHCHENKO. Subsequent internal squabbles in the YUSHCHENKO camp allowed his rival Viktor YANUKOVYCH to stage a comeback in parliamentary (Rada) elections, become prime minister in August 2006, and be elected president in February 2010. In October 2012, Ukraine held Rada elections, widely criticized by Western observers as flawed due to use of government resources to favor ruling party candidates, interference with media access, and harassment of opposition candidates. President YANUKOVYCH's backtracking on a trade and cooperation agreement with the EU in November 2013 - in favor of closer economic ties with Russia

- and subsequent use of force against students, civil society activists, and other civilians in favor of the agreement led to a three-month protest occupation of Kyiv's central square. The government's use of violence to break up the protest camp in February 2014 led to all out pitched battles, scores of deaths, international condemnation, a failed political deal, and the president's abrupt departure for Russia. New elections in the spring allowed pro-West president Petro POROSHENKO to assume office in June 2014; he was succeeded by Volodymyr ZELENSKY in May 2019.

Shortly after YANUKOVYCH's departure in late February 2014, Russian President PUTIN ordered the invasion of Ukraine's Crimean Peninsula falsely claiming the action was to protect ethnic Russians living there. Two weeks later, a "referendum" was held regarding the integration of Crimea into the Russian Federation. The "referendum" was condemned as illegitimate by the Ukrainian Government, the EU, the US, and the UN General Assembly (UNGA). In response to Russia's illegal annexation of Crimea, 100 members of the UN passed UNGA resolution 68/262, rejecting the "referendum" as baseless and invalid and confirming the sovereignty, political independence, unity, and territorial integrity of Ukraine. In mid-2014, Russia began supplying proxies in two of Ukraine's eastern provinces with manpower, funding, and materiel driving an armed conflict with the Ukrainian Government that continues to this day. Representatives from Ukraine, Russia, and the unrecognized Russian proxy republics signed the Minsk Protocol and Memorandum in September 2014 to end the conflict. However, this agreement failed to stop the fighting or find a political solution. In a renewed attempt to alleviate ongoing clashes, leaders of Ukraine, Russia, France, and Germany negotiated a follow-on Package of Measures in February 2015 to implement the Minsk agreements. Representatives from Ukraine, Russia, the unrecognized Russian proxy republics, and the Organization for Security and Cooperation in Europe also meet regularly to facilitate implementation of the peace deal. More than 13,000 civilians have been killed or wounded as a result of the Russian intervention in eastern Ukraine.

GEOGRAPHY

Location: Eastern Europe, bordering the Black Sea, between Poland, Romania, and Moldova in the west and Russia in the east

Geographic coordinates: 49 00 N, 32 00 E

Map references: AsiaEurope

Area: *total:* 603,550 sq km
land: 579,330 sq km
water: 24,220 sq km
note: approximately 43,133 sq km, or about 7.1% of Ukraine's area, is Russian occupied; the seized

area includes all of Crimea and about one-third of both Luhans'k and Donets'k oblasts
country comparison to the world: 47

Area—comparative: almost four times the size of Georgia; slightly smaller than Texas

Area comparison map: [INSERT IMAGE: Ukraine- Area comparison map]

Land boundaries: *total:* 5,618 km
border countries (7): Belarus 1111 km, Hungary 128 km, Moldova 1202 km, Poland 535 km, Romania 601 km, Russia 1944 km, Slovakia 97 km

Coastline: 2,782 km

Maritime claims: *territorial sea:* 12 nm
exclusive economic zone: 200 nm
continental shelf: 200 m or to the depth of exploitation

Climate: temperate continental; Mediterranean only on the southern Crimean coast; precipitation disproportionately distributed, highest in west and north, lesser in east and southeast; winters vary from cool along the Black Sea to cold farther inland; warm summers across the greater part of the country, hot in the south

Terrain: mostly fertile plains (steppes) and plateaus, with mountains found only in the west (the Carpathians) or in the extreme south of the Crimean Peninsula

Elevation: *mean elevation:* 175 m
lowest point: Black Sea 0 m
highest point: Hora Hoverla 2,061 m

Natural resources: iron ore, coal, manganese, natural gas, oil, salt, sulfur, graphite, titanium, magnesium, kaolin, nickel, mercury, timber, arable land

Land use: *agricultural land:* 71.2% (2011 est.)
arable land: 56.1% (2011 est.) / permanent crops: 1.5% (2011 est.) / permanent pasture: 13.6% (2011 est.)
forest: 16.8% (2011 est.)
other: 12% (2011 est.)

Irrigated land: 21,670 sq km (2012)

Population distribution: densest settlement in the eastern (Donbas) and western regions; noteable concentrations in and around major urban areas of Kyiv, Kharkiv, Donets'k, Dnipropetrovs'k, and Odesa

Natural hazards: occasional floods; occasional droughts

Environment—current issues: air and water pollution; land degradation; solid waste management; biodiversity loss; deforestation; radiation contamination in the northeast from 1986 accident at Chornobyl' Nuclear Power Plant

Environment—international agreements: *party to:* Air Pollution, Air Pollution-Nitrogen Oxides, Air Pollution-Sulfur 85, Antarctic-Environmental Protocol, Antarctic-Marine Living Resources, Antarctic Treaty, Biodiversity, Climate Change, Climate Change-Kyoto Protocol, Desertification,

Endangered Species, Environmental Modification, Hazardous Wastes, Law of the Sea, Marine Dumping, Ozone Layer Protection, Ship Pollution, Wetlands

signed, but not ratified: Air Pollution-Persistent Organic Pollutants, Air Pollution-Sulfur 94, Air Pollution-Volatile Organic Compounds

Geography—note: strategic position at the crossroads between Europe and Asia; second-largest country in Europe after Russia

PEOPLE AND SOCIETY

Population: 43,922,939 (July 2020 est.)
country comparison to the world: 33

Nationality: *noun:* Ukrainian(s)
adjective: Ukrainian

Ethnic groups: Ukrainian 77.8%, Russian 17.3%, Belarusian 0.6%, Moldovan 0.5%, Crimean Tatar 0.5%, Bulgarian 0.4%, Hungarian 0.3%, Romanian 0.3%, Polish 0.3%, Jewish 0.2%, other 1.8% (2001 est.)

Languages: Ukrainian (official) 67.5%, Russian (regional language) 29.6%, other (includes small Crimean Tatar-, Moldovan/Romanian-, and Hungarian-speaking minorities) 2.9% (2001 est.)
note: in February 2018, the Constitutional Court ruled that 2012 language legislation entitling a language spoken by at least 10% of an oblast's population to be given the status of "regional language" - allowing for its use in courts, schools, and other government institutions - was unconstitutional, thus making the law invalid; Ukrainian remains the country's only official nationwide language

Religions: Orthodox (includes the Orthodox Church of Ukraine (OCU) and the Ukrainian Orthodox - Moscow Patriarchate (UOC-MP)), Ukrainian Greek Catholic, Roman Catholic, Protestant, Muslim, Jewish (2013 est.)
note: Ukraine's population is overwhelmingly Christian; the vast majority - up to two thirds - identify themselves as Orthodox, but many do not specify a particular branch; the OCU and the UOC-MP each represent less than a quarter of the country's population, the Ukrainian Greek Catholic Church accounts for 8-10%, and the UAOC accounts for 1-2%; Muslim and Jewish adherents each compose less than 1% of the total population

Age structure: *0-14 years:* 16.16% (male 3,658,127/female 3,438,887)
15-24 years: 9.28% (male 2,087,185/female 1,987,758)
25-54 years: 43.66% (male 9,456,905/female 9,718,758)
55-64 years: 13.87% (male 2,630,329/female 3,463,851)
65 years and over: 17.03% (male 2,523,600/female 4,957,539) (2020 est.)

population pyramid: [INSERT IMAGE: UKRAINE -population pyramid]

Dependency ratios: *total dependency ratio:* 49.1
youth dependency ratio: 23.8

elderly dependency ratio: 25.3
potential support ratio: 4 (2020 est.)
note: data include Crimea

Median age: *total:* 41.2 years
male: 38.2 years
female: 44.3 years (2020 est.)
country comparison to the world: 45

Population growth rate: -0.1% (2020 est.)
country comparison to the world: 204

Birth rate: 9.6 births/1,000 population (2020 est.)
country comparison to the world: 194

Death rate: 14 deaths/1,000 population (2020 est.)
country comparison to the world: 5

Net migration rate: 2.3 migrant(s)/1,000 population (2020 est.)
country comparison to the world: 46

Population distribution: densest settlement in the eastern (Donbas) and western regions; notable concentrations in and around major urban areas of Kyiv, Kharkiv, Donets'k, Dnipropetrovs'k, and Odesa

Urbanization: *urban population:* 69.6% of total population (2020)
rate of urbanization: -0.33% annual rate of change (2015-20 est.)

total population growth rate v. urban population growth rate, 2000-2030: *Major urban areas - population:* 2.988 million KYIV (capital), 1.429 million Kharkiv, 1.009 million Odesa, 957,000 Dnipropetrovsk, 906,000 Donetsk (2020)

Sex ratio: *at birth:* 1.06 male(s)/female
0-14 years: 1.06 male(s)/female
15-24 years: 1.05 male(s)/female
25-54 years: 0.97 male(s)/female
55-64 years: 0.76 male(s)/female
65 years and over: 0.51 male(s)/female
total population: 0.86 male(s)/female (2020 est.)

Mother's mean age at first birth: 25.6 years (2017 est.)

Maternal mortality rate: 19 deaths/100,000 live births (2017 est.)
country comparison to the world: 128

Infant mortality rate: *total:* 7.4 deaths/1,000 live births
male: 8.3 deaths/1,000 live births
female: 6.4 deaths/1,000 live births (2020 est.)
country comparison to the world: 158

Life expectancy at birth: *total population:* 72.9 years
male: 68.2 years
female: 77.9 years (2020 est.)
country comparison to the world: 150

Total fertility rate: 1.56 children born/woman (2020 est.)
country comparison to the world: 194

Contraceptive prevalence rate: 65.4% (2012)

Drinking water source:
improved: *urban:* 99.5% of population
rural: 100% of population
total: 99.4% of population

unimproved: *urban:* 0.5% of population
rural: 0% of population

total: 0.6% of population (2017 est.)

Current Health Expenditure: 7% (2017)

Physicians density: 2.99 physicians/1,000 population (2014)

Hospital bed density: 7.5 beds/1,000 population (2014)

Sanitation facility access:

improved: *urban:* 99.4% of population
rural: 96.3% of population
total: 98.4% of population

unimproved: *urban:* 0.6% of population
rural: 3.7% of population
total: 1.6% of population (2017 est.)

HIV/AIDS—adult prevalence rate: 0.7% (2019 est.)
country comparison to the world: 57

HIV/AIDS—people living with HIV/AIDS: 250,000 (2019 est.)
country comparison to the world: 23

HIV/AIDS—deaths: 5,900 (2019 est.)
country comparison to the world: 23

Obesity—adult prevalence rate: 24.1% (2016)
country comparison to the world: 61

Education expenditures: 5.4% of GDP (2017)
country comparison to the world: 44

Literacy: *definition:* age 15 and over can read and write
total population: 99.8%
male: 99.8%
female: 99.7% (2015)
School life expectancy (primary to tertiary education): total: 15 years
male: 15 years
female: 15 years (2014)
Unemployment, youth ages 15-24: total: 17.9%
male: 16.9%
female: 19.3% (2018 est.)
country comparison to the world: 74

GOVERNMENT

Country name: *conventional long form:* none
conventional short form: Ukraine
local long form: none
local short form: Ukraina
former: Ukrainian National Republic, Ukrainian State, Ukrainian Soviet Socialist Republic
etymology: name derives from the Old East Slavic word "ukraina" meaning "borderland or march (militarized border region)" and began to be used extensively in the 19th century; originally Ukrainians referred to themselves as Rusyny (Rusyns, Ruthenians, or Ruthenes), an endonym derived from the medieval Rus state (Kyivan Rus)

Government type: semi-presidential republic

Capital: *name:* Kyiv (Kiev)

geographic coordinates: 50 26 N, 30 31 E
time difference: UTC+2 (7 hours ahead of Washington, DC, during Standard Time)

daylight saving time: +1hr, begins last Sunday in March; ends last Sunday in October
note: pronounced KAY-yiv

etymology: the name is associated with that of Kyi, who along with his brothers Shchek and Khoryv, and their sister Lybid, are the legendary founders of the medieval city of Kyiv; Kyi being the eldest brother, the city was named after him

Administrative divisions: 24 provinces (oblasti, singular - oblast'), 1 autonomous republic* (avtonomna respublika), and 2 municipalities** (mista, singular - misto) with oblast status; Cherkasy, Chernihiv, Chernivtsi, Crimea or Avtonomna Respublika Krym* (Simferopol), Dnipropetrovsk (Dnipro), Donetsk, Ivano-Frankivsk, Kharkiv, Kherson, Khmelnytskyi, Kirovohrad (Kropyvnytskyi), Kyiv**, Kyiv, Luhansk, Lviv, Mykolaiv, Odesa, Poltava, Rivne, Sevastopol**, Sumy, Ternopil, Vinnytsia, Volyn (Lutsk), Zakarpattia (Uzhhorod), Zaporizhzhia, Zhytomyr

note: administrative divisions have the same names as their administrative centers (exceptions have the administrative center name following in parentheses); plans include the eventual renaming of Dnipropetrovsk and Kirovohrad oblasts, but because these names are mentioned in the Constitution of Ukraine, the change will require a constitutional amendment

note: the US Government does not recognize Russia's illegal annexation of Ukraine's Autonomous Republic of Crimea and the municipality of Sevastopol, nor their redesignation as the "Republic of Crimea" and the "Federal City of Sevastopol"

Independence: 24 August 1991 (from the Soviet Union); notable earlier dates: ca. 982 (VOLODYMYR I consolidates Kyivan Rus); 1199 (Principality (later Kingdom) of Ruthenia formed; 1648 (establishment of the Cossack Hetmanate); 22 January 1918 (from Soviet Russia)

National holiday: Independence Day, 24 August (1991); note - 22 January 1918, the day Ukraine first declared its independence from Soviet Russia, and the date the short-lived Western and Greater (Eastern) Ukrainian republics united (1919), is now celebrated as Unity Day

Constitution: *history:* several previous; latest adopted and ratified 28 June 1996

amendments: proposed by the president of Ukraine or by at least one third of the Supreme Council members; adoption requires simple majority vote by the Council and at least two-thirds majority vote in its next regular session; adoption of proposals relating to general constitutional principles, elections, and amendment procedures requires two-thirds majority vote by the Council and approval in a referendum; constitutional articles on personal rights and freedoms, national independence, and territorial integrity cannot be amended; amended several times, last in 2019

Legal system: civil law system; judicial review of legislative acts

International law organization participation: has not submitted an ICJ jurisdiction declaration; non-party state to the ICCt

Citizenship: *citizenship by birth:* no

citizenship by descent only: at least one parent must be a citizen of Ukraine
dual citizenship recognized: no
residency requirement for naturalization: 5 years

Suffrage: 18 years of age; universal

Executive branch: *chief of state:* President Volodymyr ZELENSKYY (since 20 May 2019)

head of government: Prime Minister Denys SHMYHAL (since 4 March 2020)
cabinet: Cabinet of Ministers nominated by the prime minister, approved by the Verkhovna Rada
elections/appointments: president directly elected by absolute majority popular vote in 2 rounds if needed for a 5-year term (eligible for a second term); election last held on 31 March and 21 April 2019 (next to be held in March 2024); prime minister selected by the Verkhovna Rada
election results: first round results: percent of vote - Volodymyr ZELENSKYY (Servant of the People) 30.2%, Petro POROSHENKO (BPP-Solidarity) 15.6%, Yuliya TYMOSHENKO (Fatherland) 13.4%, Yuriy BOYKO (Opposition Platform-For Life) 11.7%, 35 other candidates 29.1%; second round results: percent of vote - Volodymyr ZELENSKYY (Servant of the People) 73.2%, Petro POROSHENKO (BPP-Solidarity) 24.5%; Denys SHMYHAL (independent) elected prime minister; Verkhovna Rada vote - 291-59
note: there is also a National Security and Defense Council or NSDC originally created in 1992 as the National Security Council; the NSDC staff is tasked with developing national security policy on domestic and international matters and advising the president; a presidential administration helps draft presidential edicts and provides policy support to the president

Legislative branch: *description:* unicameral Supreme Council or Verkhovna Rada (450 seats; 225 members directly elected in single-seat constituencies by simple majority vote and 225 directly elected in a single nationwide constituency by closed, party-list proportional representation vote; members serve 5-year terms)
elections: last held on 21 July 2019 (next to be held in July 2024)
election results: percent of vote by party - Servant of the People 43.2%, Opposition Platform-For Life 13.1%, Batkivshchyna 8.2%, European Solidarity 8.1%, Voice 5.8%, other 21.6%; seats by party (preliminary) - Servant of the People 254, Oposition Platform for Life 43, Batkivshchyna 26, European Solidarity 25, Voice 20, Opposition Bloc 6, Samopomich 1, Svoboda 1, other parties 2, independent 46; note - voting not held in Crimea and parts of two Russianoccupied eastern oblasts leaving 26 seats vacant; although this brings the total to 424 elected members (of 450 potential), article 83 of the constitution mandates that a parliamentary majority consists of 226 seats

Judicial branch: *highest courts:* Supreme Court of Ukraine or SCU (consists of 100 judges, organized into civil, criminal, commercial and administrative chambers, and a grand chamber); Constitutional Court (consists of 18 justices);

High Anti-Corruption Court (consists of 39 judges, including 12 in the Appeals Chamber)
judge selection and term of office: Supreme Court judges recommended by the High Qualification Commission of Judges (a 16-member state body responsible for judicial candidate testing and assessment and judicial administration), submitted to the High Council of Justice, a 21-member independent body of judicial officials responsible for judicial selfgovernance and administration, and appointed by the president; judges serve until mandatory retirement at age 65; High Anti-Corruption Court judges are selected by the same process as Supreme Court justices, with one addition – a majority of a combined High Qualification Commission of Judges and a 6-member Public Council of International Experts must vote in favor of potential judges in order to recommend their nomination to the High Council of Justice; this majority must include at least 3 members of the Public Council of International Experts; Constitutional Court justices appointed - 6 each by the president, by the Congress of Judges, and by the Verkhovna Rada; judges serve 9-year non-renewable terms
subordinate courts: Courts of Appeal; district courts
note: specialized courts were abolished as part of Ukraine's judicial reform program; in November 2019, President ZELENSKYY signed a bill on legal reforms

Political parties and leaders: Batkivshchyna (Fatherland) [Yuliya TYMOSHENKO]
European Solidarity (BPP-Solidarity) [Petro POROSHENKO]
Holos (Voice) [Sviatoslav VAKARCHUK]
Opposition Bloc or OB [Evgeny MURAYEV]
Opposition Platform-For Life [Yuriy BOYKO, Vadim RABINOVICH]
Radical Party [Oleh LYASHKO]
Samopomich (Self Reliance) [Andriy SADOVYY]
Servant of the People [Oleksandr KORNIENKO]
Svoboda (Freedom) [Oleh TYAHNYBOK]

International organization participation: Australia Group, BSEC, CBSS (observer), CD, CE, CEI, CICA (observer), CIS (participating member, has not signed the 1993 CIS charter), EAEC (observer), EAPC, EBRD, FAO, GCTU, GUAM, IAEA, IBRD, ICAO, ICC (national committees), ICRM, IDA, IFC, IFRCS, IHO, ILO, IMF, IMO, IMSO, Interpol, IOC, IOM, IPU, ISO, ITU, ITUC (NGOs), LAIA (observer), MIGA, MONUSCO, NAM (observer), NSG, OAS (observer), OIF (observer), OPCW, OSCE, PCA, PFP, SELEC (observer), UN, UNCTAD, UNESCO, UNFICYP, UNIDO, UNISFA, UNMIL, UNMISS, UNOCI, UNWTO, UPU, WCO, WFTU (NGOs), WHO, WIPO, WMO, WTO, ZC

Diplomatic representation in the US: *chief of mission:* Ambassador Volodymyr YELCHENKO (since 6 January 2020)
chancery: 3350 M Street NW, Washington, DC 20007
telephone: [1] (202) 349-2920

FAX: [1] (202) 333-0817

consulate(s) general: Chicago, New York, San Francisco, Seattle

Diplomatic representation from the US: *chief of mission:* Ambassador (vacant); Charge d'Affaires Kristina KVIEN (since January 2020)
telephone: [380] (44) 521-5000
embassy: 4 A. I. Igor Sikorsky Street, 04112 Kyiv
mailing address: 5850 Kyiv Place, Washington, DC 20521-5850
FAX: [380] (44) 521-5155

Flag description: two equal horizontal bands of azure (top) and golden yellow represent grain fields under a blue sky

National symbol(s): *tryzub (trident); national colors:* blue, yellow

National anthem: *name:* "Shche ne vmerla Ukraina" (Ukraine Has Not Yet Perished)
lyrics/music: Paul CHUBYNSKYI/Mikhail VERBYTSKYI
note: music adopted 1991, lyrics adopted 2003; song first performed in 1864 at the Ukraine Theatre in Lviv; the lyrics, originally written in 1862, were revised in 2003
0:00 / 1:19

ECONOMY

Economy—overview: After Russia, the Ukrainian Republic was the most important economic component of the former Soviet Union, producing about four times the output of the next-ranking republic. Its fertile black soil accounted for more than one fourth of Soviet agricultural output, and its farms provided substantial quantities of meat, milk, grain, and vegetables to other republics. Likewise, its diversified heavy industry supplied unique equipment such as large diameter pipes and vertical drilling apparatus, and raw materials to industrial and mining sites in other regions of the former USSR.

Shortly after independence in August 1991, the Ukrainian Government liberalized most prices and erected a legal framework for privatization, but widespread resistance to reform within the government and the legislature soon stalled reform efforts and led to some backtracking. Output by 1999 had fallen to less than 40% of the 1991 level. Outside institutions - particularly the IMF encouraged Ukraine to quicken the pace and scope of reforms to foster economic growth. Ukrainian Government officials eliminated most tax and customs privileges in a March 2005 budget law, bringing more economic activity out of Ukraine's large shadow economy. From 2000 until mid-2008, Ukraine's economy was buoyant despite political turmoil between the prime minister and president. The economy contracted nearly 15% in 2009, among the worst economic performances in the world. In April 2010, Ukraine negotiated a price discount on Russian gas imports in exchange for extending Russia's lease on its naval base in Crimea.

Ukraine's oligarch-dominated economy grew slowly from 2010 to 2013 but remained behind peers in the region and among Europe's poorest. After former President YANUKOVYCH fled the country during the Revolution of Dignity, Ukraine's economy fell into crisis because of Russia's annexation of Crimea, military conflict in the eastern part of the country, and a trade war with Russia, resulting in a 17% decline in GDP, inflation at nearly 60%, and dwindling foreign currency reserves. The international community began efforts to stabilize the Ukrainian economy, including a March 2014 IMF assistance package of $17.5 billion, of which Ukraine has received four disbursements, most recently in April 2017, bringing the total disbursed as of that date to approximately $8.4 billion. Ukraine has made progress on reforms designed to make the country prosperous, democratic, and transparent, including creation of a national anti-corruption agency, overhaul of the banking sector, establishment of a transparent VAT refund system, and increased transparency in government procurement. But more improvements are needed, including fighting corruption, developing capital markets, improving the business environment to attract foreign investment, privatizing state-owned enterprises, and land reform. The fifth tranche of the IMF program, valued at $1.9 billion, was delayed in mid-2017 due to lack of progress on outstanding reforms, including adjustment of gas tariffs to import parity levels and adoption of legislation establishing an independent anti-corruption court.

Russia's occupation of Crimea in March 2014 and ongoing Russian aggression in eastern Ukraine have hurt economic growth. With the loss of a major portion of Ukraine's heavy industry in Donbas and ongoing violence, the economy contracted by 6.6% in 2014 and by 9.8% in 2015, but it returned to low growth in in 2016 and 2017, reaching 2.3% and 2.0%, respectively, as key reforms took hold. Ukraine also redirected trade activity towards the EU following the implementation of a bilateral Deep and Comprehensive Free Trade Agreement, displacing Russia as its largest trading partner. A prohibition on commercial trade with separatist-controlled territories in early 2017 has not impacted Ukraine's key industrial sectors as much as expected, largely because of favorable external conditions. Ukraine returned to international debt markets in September 2017, issuing a $3 billion sovereign bond.

GDP (purchasing power parity): $369.6 billion (2017 est.)
$360.5 billion (2016 est.)
$351.9 billion (2015 est.)
note: data are in 2017 dollars
country comparison to the world: 50

GDP (official exchange rate): $112.1 billion (2017 est.)

GDP—real growth rate: 3.24% (2019 est.)
3.41% (2018 est.)
2.48% (2017 est.)
country comparison to the world: 93

GDP—per capita (PPP): $8,800 (2017 est.)
$8,500 (2016 est.)
$8,300 (2015 est.)

note: data are in 2017 dollars
country comparison to the world: 146

Gross national saving: 18.9% of GDP (2017 est.)
20.2% of GDP (2016 est.)
17.7% of GDP (2015 est.)
country comparison to the world: 105

GDP—composition, by end use: *household consumption:* 66.5% (2017 est.)
government consumption: 20.4% (2017 est.)
investment in fixed capital: 16% (2017 est.)
investment in inventories: 4.7% (2017 est.)
exports of goods and services: 47.9% (2017 est.)
imports of goods and services: -55.6% (2017 est.)

GDP—composition, by sector of origin: *agriculture:* 12.2% (2017 est.)
industry: 28.6% (2017 est.)
services: 60% (2017 est.)

Agriculture—products: grain, sugar beets, sunflower seeds, vegetables; beef, milk

Industries: coal, electric power, ferrous and nonferrous metals, machinery and transport equipment, chemicals, food processing

Industrial production growth rate: 3.1% (2017 est.)
country comparison to the world: 100

Labor force: 17.99 million (2017 est.)
country comparison to the world: 31

Labor force—by occupation: agriculture: 5.8%
industry: 26.5%
services: 67.8% (2014)

Unemployment rate: 8.89% (2019 est.)
9.42% (2018 est.)
note: officially registered workers; large number of unregistered or underemployed workers
country comparison to the world: 134

Population below poverty line: 3.8% (2016 est.)

Household income or consumption by percentage share: *lowest 10%:* 4.2%
highest 10%: 21.6% (2015 est.)

Budget: *revenues:* 29.82 billion (2017 est.)
expenditures: 31.55 billion (2017 est.)
note: this is the planned, consolidated budget

Taxes and other revenues: 26.6% (of GDP) (2017 est.)
country comparison to the world: 108

Budget surplus (+) or deficit (-): -1.5% (of GDP) (2017 est.)
country comparison to the world: 92

Public debt: 71% of GDP (2017 est.)
81.2% of GDP (2016 est.)
note: the total public debt of $64.5 billion consists of: domestic public debt ($23.8 billion); external public debt ($26.1 billion); and sovereign guarantees ($14.6 billion)
country comparison to the world: 48

Fiscal year: calendar year

Inflation rate (consumer prices): 14.4% (2017 est.)
13.9% (2016 est.)
note: Excluding the temporarily occupied territories of the Autonomous Republic of Crimea, the city of Sevastopol and part of the anti-terrorist operation zone

country comparison to the world: 210

Current account balance: -$4.124 billion (2019 est.)
-$6.432 billion (2018 est.)
country comparison to the world: 179

Exports: $39.69 billion (2017 est.)
$33.56 billion (2016 est.)
country comparison to the world: 54

Exports—partners: Russia 9.2%, Poland 6.5%, Turkey 5.6%, India 5.5%, Italy 5.2%, China 4.6%, Germany 4.3% (2017)

Exports—commodities: ferrous and nonferrous metals, fuel and petroleum products, chemicals, machinery and transport equipment, foodstuffs

Imports: $49.06 billion (2017 est.)
$40.5 billion (2016 est.)
country comparison to the world: 53

Imports—commodities: energy, machinery and equipment, chemicals

Imports—partners: Russia 14.5%, China 11.3%, Germany 11.2%, Poland 7%, Belarus 6.7%, US 5.1% (2017)

Reserves of foreign exchange and gold: $18.81 billion (31 December 2017 est.)
$15.54 billion (31 December 2016 est.)
country comparison to the world: 61

Debt—external: $130 billion (31 December 2017 est.)
$121.1 billion (31 December 2016 est.)
country comparison to the world: 45

Exchange rates: hryvnia (UAH) per US dollar—
26.71 (2017 est.)
25.5513 (2016 est.)
25.5513 (2015 est.)
21.8447 (2014 est.)
11.8867 (2013 est.)

ENERGY

Electricity access: *electrification—total population:* 100% (2020)

Electricity—production: 153.6 billion kWh (2016 est.)
country comparison to the world: 26

Electricity—consumption: 133.2 billion kWh (2016 est.)
country comparison to the world: 28

Electricity—exports: 3.591 billion kWh (2015 est.)
country comparison to the world: 39

Electricity—imports: 77 million kWh (2016 est.)
country comparison to the world: 103

Electricity—installed generating capacity: 57.28 million kW (2016 est.)
country comparison to the world: 20

Electricity—from fossil fuels: 65% of total installed capacity (2016 est.)
country comparison to the world: 120

Electricity—from nuclear fuels: 23% of total installed capacity (2017 est.)
country comparison to the world: 4

Electricity—from hydroelectric plants: 8% of total installed capacity (2017 est.)
country comparison to the world: 125

Electricity—from other renewable sources: 3% of total installed capacity (2017 est.)
country comparison to the world: 129

Crude oil—production: 32,000 bbl/day (2018 est.)
country comparison to the world: 61

Crude oil—exports: 413 bbl/day (2015 est.)
country comparison to the world: 79

Crude oil—imports: 4,720 bbl/day (2015 est.)
country comparison to the world: 76

Crude oil—proved reserves: 395 million bbl (1 January 2018 est.)
country comparison to the world: 49

Refined petroleum products—production: 63,670 bbl/day (2017 est.)
country comparison to the world: 77

Refined petroleum products—consumption: 233,000 bbl/day (2016 est.)
country comparison to the world: 53

Refined petroleum products— exports: 1,828 bbl/day (2015 est.)
country comparison to the world: 105

Refined petroleum products—imports: 167,000 bbl/day (2015 est.)
country comparison to the world: 37

Natural gas—production: 19.73 billion cu m (2017 est.)
country comparison to the world: 31

Natural gas—consumption: 30.92 billion cu m (2017 est.)
country comparison to the world: 30

Natural gas—exports: 0 cu m (2017 est.)
country comparison to the world: 206

Natural gas—imports: 12.97 billion cu m (2017 est.)
country comparison to the world: 25

Natural gas—proved reserves: 1.104 trillion cu m (1 January 2018 est.)
country comparison to the world: 24

Carbon dioxide emissions from consumption of energy: 238.9 million Mt (2017 est.)
country comparison to the world: 28

COMMUNICATIONS

Telephones—fixed lines: *total subscriptions:* 4,378,911

subscriptions per 100 inhabitants: 9.96 (2019 est.)
country comparison to the world: 32

Telephones—mobile cellular: total subscriptions: 57,431,439

subscriptions per 100 inhabitants: 130.63 (2019 est.)
country comparison to the world: 28

Telecommunication systems: *general assessment:* telecommunication development plan emphasizes improving domestic trunk lines, international connections, and the mobile-cellular system; Turkey and Russia have made investments to Ukraine's telecom market; competition available between 3 alternative operators moving from 3G services, but some areas still use 2G; LTE services available in some areas; FttP networks taking over DSL platforms; political tensions have not added to growth and telecom regulators must not count Crimea numbers (Annexed by Russia in 2014); mobile broadband services present a growth opportunity (2020)

domestic: fixed-line teledensity is 10 per 100; the mobile-cellular telephone system's expansion has slowed, largely due to saturation of the market that is now 131 mobile phones per 100 persons (2019)

international: country code - 380; landing point for the Kerch Strait Cable connecting Ukraine to Russia; 2 new domestic trunk lines are a part of the fiber-optic TAE system and 3 Ukrainian links have been installed in the fiber-optic TEL project that connects 18 countries; additional international service is provided by the Italy-Turkey-Ukraine-Russia (ITUR) fiberoptic submarine cable and by an unknown number of earth stations in the Intelsat, Inmarsat, and Intersputnik satellite systems

note: the COVID-19 outbreak is negatively impacting telecommunications production and supply chains globally; consumer spending on telecom devices and services has also slowed due to the pandemic's effect on economies worldwide; overall progress towards improvements in all facets of the telecom industry - mobile, fixed-line, broadband, submarine cable and satellite - has moderated

Broadcast media: Ukraine's media landscape is dominated by oligarch-owned news outlets, which are often politically motivated and at odds with one another and/or the government; while polls suggest most Ukrainians still receive news from traditional media sources, social media is a crucial component of information dissemination in Ukraine; almost all Ukrainian politicians and opinion leaders communicate with the public via social media and maintain at least one social media page, if not more; this allows them direct communication with audiences, and news often breaks on Facebook or Twitter before being picked up by traditional news outlets

Ukraine television serves as the principal source of news; the largest national networks are controlled by oligarchs: TRK Ukraina is owned by Rinat Akhmetov; Studio 1+1 is owned by Ihor Kolomoyskyy; Inter is owned by Dmytro Firtash and Serhiy Lyovochkin; and StarlightMedia channels (ICTV, STB, and Novyi Kanal) are owned by Victor Pinchuk; a set of 24- hour news channels also have clear political affiliations: 112-Ukraine and NewsOne tacitly support pro-Russian opposition and are believed to be controlled by political and business tycoon Viktor Medvedchuk; pro-Ukrainian government Channel 5 and Pryamyi are linked to President Petro Poroshenko; 24 and ZIK are owned by opposition, but not pro-Russian, politicians; UA: Suspilne is a public television station under the umbrella of the National Public Broadcasting Company of Ukraine; while it is often praised by media experts for balanced coverage, it lags in popularity; Ukrainian Radio, institutionally linked to UA: Suspilne, is one of only two national talk radio networks, with the other being the privately owned Radio NV

(2019)

Internet country code: .ua

Internet users: total: 25,883,509

percent of population: 58.89% (July 2018 est.)
country comparison to the world: 29

Broadband—fixed subscriptions: total: 5,405,125

subscriptions per 100 inhabitants: 12 (2018 est.)
country comparison to the world: 29

Communications—note: a sorting code to expeditiously handle large volumes of mail was first set up in Ukraine (then part of the Soviet Union) in the 1930s; the sophisticated, three-part (number-letter-number) postal code system, referred to as an "index," was the world's first postal zip code; the system functioned well and was in use from 1932 to 1939 when it was abruptly discontinued

TRANSPORTATION

National air transport system: *number of registered air carriers:* 14 (2020)

inventory of registered aircraft operated by air carriers: 126

annual passenger traffic on registered air carriers: 7,854,842 (2018)

annual freight traffic on registered air carriers: 75.26 million mt-km (2018)

Civil aircraft registration country code prefix: UR (2016)

Airports: 187 (2013)
country comparison to the world: 31

Airports—with paved runways: total: 108 (2013)
over 3,047 m: 13 (2013)
2,438 to 3,047 m: 42 (2013)
1,524 to 2,437 m: 22 (2013)
914 to 1,523 m: 3 (2013)
under 914 m: 28 (2013)

Airports—with unpaved runways: total: 79 (2013)
1,524 to 2,437 m: 5 (2013)
914 to 1,523 m: 5 (2013)
under 914 m: 69 (2013)

Heliports: 9 (2013)

Pipelines: 36720 km gas, 4514 km oil, 4363 km refined products (2013)

Railways: total: 21,733 km (2014)
standard gauge: 49 km 1.435-m gauge (49 km electrified) (2014)
broad gauge: 21,684 km 1.524-m gauge (9,250 km electrified) (2014)
country comparison to the world: 12

Roadways: total: 169,694 km (2012)
paved: 166,095 km (includes 17 km of expressways) (2012)
unpaved: 3,599 km (2012)
country comparison to the world: 32

Waterways: 1,672 km (most on Dnieper River) (2012)
country comparison to the world: 46

Merchant marine: total: 408
by type: bulk carrier 1, general cargo 84, oil tanker 15, other 308 (2019)
country comparison to the world: 43

Ports and terminals: *major seaport(s):* Feodosiya (Theodosia), Chornomosk (Illichivsk), Mariupol, Mykolayiv, Odesa, Yuzhnyy

MILITARY AND SECURITY

Military and security forces: *Armed Forces of Ukraine (Zbroyny Syly Ukrayiny, ZSU):* Ground Forces (Sukhoputni Viys'ka), Naval Forces (Viys'kovo-Mors'ki Syly, VMS), Air Forces (Povitryani Syly, PS), Air Assault Forces (Desantno-shturmovi Viyska, DShV); Ministry of Internal Affairs: National Guard of Ukraine, State Border Guard Service of Ukraine (includes Maritime Border Guard) (2020)

Military expenditures: 3.9% of GDP (2019)
3.7% of GDP (2018)
2.9% of GDP (2017)
3.2% of GDP (2016)
3.3% of GDP (2015)
country comparison to the world: 15

Military and security service personnel strengths: size estimates for the Armed Forces of Ukraine (Zbroyni Syly Ukrayiny, ZSU) vary; approximately 215,000 active troops (160,000 Army, including Airborne/Air Assault Forces; 13,000 Navy; 42,000 Air Force); est. 50,000 National Guard (2019 est.)

Military equipment inventories and acquisitions: the Ukrainian military is equipped mostly with older Russian and Soviet-era weapons systems; since 2010, it has imported limited quantities of weapons from several European countries, as well as Canada, the US, and the United Arab Emirates; Ukraine has a broad defense industry capable of building Soviet-era land systems and maintaining and upgrading Sovietera combat aircraft, as well as missile and air defense systems (2019 est.)

Military deployments: 250 Democratic Republic of the Congo (MONUSCO); contributes about 550 troops to the Lithuania, Poland, and Ukraine joint military brigade (LITPOLUKRBRIG), which was established in 2014; the brigade is headquartered in Warsaw and is comprised of an international staff, three battalions, and specialized units (2020)

Military service age and obligation: 20-27 years of age for compulsory military service; conscript service obligation is 12 months (2019)

TRANSPORTATION

Disputes—international: 1997 boundary delimitation treaty with Belarus remains unratified due to unresolved financial claims, stalling demarcation and reducing border security; delimitation of land boundary with Russia is complete and demarcation began in 2012; the dispute over the boundary between Russia and Ukraine through the Kerch Strait and Sea of Azov is suspended due to the occupation of Crimea by Russia; Ukraine and Moldova signed an agreement officially delimiting their border in 1999, but the border has not been demarcated due to Moldova's difficulties with the break-away region of Transnistria; Moldova and Ukraine operate joint customs posts to monitor transit of people and commodities

through Moldova's Transnistria Region, which remains under the auspices of an Organization for Security and Cooperation in Europe-mandated peacekeeping mission comprised of Moldovan, Transnistrian, Russian, and Ukrainian troops; the ICJ ruled largely in favor of Romania in its dispute submitted in 2004 over Ukrainian-administered Zmiyinyy/Serpilor (Snake) Island and Black Sea maritime boundary delimitation; Romania opposes Ukraine's reopening of a navigation canal from the Danube border through Ukraine to the Black Sea

Refugees and internally displaced persons: *IDPs:* 734,000 (Russian-sponsored separatist violence in Crimea and eastern Ukraine) (2020)

stateless persons: 35,642 (2019); note - citizens of the former USSR who were permanently resident in Ukraine were granted citizenship upon Ukraine's independence in 1991, but some missed this window of opportunity; people arriving after 1991, Crimean Tatars, ethnic Koreans, people with expired Soviet passports, and people with no documents have difficulty acquiring Ukrainian citizenship; following the fall of the Soviet Union in 1989, thousands of Crimean Tatars and their descendants deported from Ukraine under the STALIN regime returned to their homeland, some being stateless and others holding the citizenship of Uzbekistan or other former Soviet republics; a 1998 bilateral agreement between Ukraine and Uzbekistan simplified the process of renouncing Uzbek citizenship and obtaining Ukrainian citizenship

Trafficking in persons: *current situation:* Ukraine is a source, transit, and destination country for men, women, and children subjected to forced labor and sex trafficking; Ukrainian victims are sex trafficked within Ukraine as well as in Russia, Poland, Iraq, Spain, Turkey, Cyprus, Greece, Seychelles, Portugal, the Czech Republic, Israel, Italy, South Korea, Moldova, China, the United Arab Emirates, Montenegro, UK, Kazakhstan, Tunisia, and other countries; small numbers of foreigners from Moldova, Russia, Vietnam, Uzbekistan, Pakistan, Cameroon, and Azerbaijan were victims of labor trafficking in Ukraine; Ukrainian recruiters most often target Ukrainians from rural areas with limited job prospects using fraud, coercion, and debt bondage

tier rating: Tier 2 Watch List – Ukraine does not fully comply with the minimum standards for the elimination of trafficking; however, it is making significant efforts to do so; the government's focus on its security situation constrained its anti-trafficking capabilities; law enforcement efforts to pursue trafficking cases weakened in 2014, continuing a multiyear decline, and no investigations, prosecutions, or convictions of government officials were made, despite reports of official complicity in the sex and labor trafficking of children living in state-run institutions; fewer victims were identified and referred to NGOs, which continued to provide and to fund the majority of victims' services (2015)

Illicit drugs: limited cultivation of cannabis and opium poppy, mostly for CIS consumption;

some synthetic drug production for export to the West; limited government eradication program; used as transshipment point for opiates and other illicit drugs from Africa, Latin America, and Turkey to Europe and Russia; Ukraine has improved anti-money-laundering controls, resulting in its removal from the Financial Action Task Force's (FATF's) Noncooperative Countries and Territories List in February 2004; Ukraine's anti-money-laundering regime continues to be monitored by FATF

UNITED ARAB EMIRATES

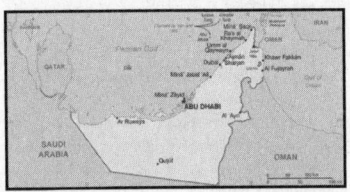

INTRODUCTION

Background: The Trucial States of the Persian Gulf coast granted the UK control of their defense and foreign affairs in 19th century treaties. In 1971, six of these states - Abu Dhabi, 'Ajman, Al Fujayrah, Ash Shariqah, Dubayy, and Umm al Qaywayn - merged to form the United Arab Emirates (UAE). They were joined in 1972 by Ra's al Khaymah. The UAE's per capita GDP is on par with those of leading West European nations. For more than three decades, oil and global finance drove the UAE's economy. In 2008-09, the confluence of falling oil prices, collapsing real estate prices, and the international banking crisis hit the UAE especially hard. The UAE did not experience the "Arab Spring" unrest seen elsewhere in the Middle East in 2010-11, partly because of the government's multi-year, $1.6-billion infrastructure investment plan for the poorer northern emirates, and its aggressive pursuit of advocates of political reform. The UAE in recent years has played a growing role in regional affairs. In addition to donating billions of dollars in economic aid to help stabilize Egypt, the UAE was one of the first countries to join the Defeat-ISIS coalition, and to participate as a key partner in a Saudi-led military campaign in Yemen. On 15 September 2020, the United Arab Emirates and Bahrain signed a peace accord with Israel – brokered by the US – in Washington DC. Referred to as the Abraham Accords, the United Arab Emirates and Bahrain are the two latest Middle Eastern countries, along with Egypt and Jordan, to recognize Israel.

GEOGRAPHY

Location: Middle East, bordering the Gulf of Oman and the Persian Gulf, between Oman and Saudi Arabia

Geographic coordinates: 24 00 N, 54 00 E

Map references: Middle East

Area: *total:* 83,600 sq km
land: 83,600 sq km

water: 0 sq km
country comparison to the world: 116

Area—comparative: slightly larger than South Carolina; slightly smaller than Maine

Area comparison map: [INSERT IMAGE: UNITED ARAB EMIRATES-Area comparison map]

Land boundaries: *total:* 1,066 km
border countries (2): Oman 609 km, Saudi Arabia 457 km

Coastline: 1,318 km

Maritime claims: *territorial sea:* 12 nm
exclusive economic zone: 200 nm
contiguous zone: 24 nm
continental shelf: 200 nm or to the edge of the continental margin

Climate: desert; cooler in eastern mountains

Terrain: flat, barren coastal plain merging into rolling sand dunes of vast desert; mountains in east

Elevation: *mean elevation:* 149 m
lowest point: Persian Gulf 0 m
highest point: Jabal Yibir 1,527 m

Natural resources: petroleum, natural gas

Land use: *agricultural land:* 4.6% (2011 est.)
arable land: 0.5% (2011 est.) / permanent crops: 0.5% (2011 est.) / permanent pasture: 3.6% (2011 est.)
forest: 3.8% (2011 est.)
other: 91.6% (2011 est.)

Irrigated land: 923 sq km (2012)

Population distribution: population is heavily concentrated to the northeast on the Musandam Peninsula; the three largest emirates - Abu Dhabi, Dubai, and Sharjah - are home to nearly 85% of the population

Natural hazards: frequent sand and dust storms

Environment—current issues: air pollution; rapid population growth and high energy demand contribute to water scarcity; lack of natural freshwater resources compensated by desalination plants; land degradation and desertification; waste generation; beach pollution from oil spills

Environment—international agreements: *party to:* Biodiversity, Climate Change, Climate Change-Kyoto Protocol, Desertification, Endangered Species, Hazardous Wastes, Marine Dumping, Ozone Layer Protection
signed, but not ratified: Law of the Sea

Geography—note: strategic location along southern approaches to Strait of Hormuz, a vital transit point for world crude oil

PEOPLE AND SOCIETY

Population: 9,992,083 (July 2020 est.)
note: the UN estimated the country's total population was 9,771,000 as of mid-year 2019; immigrants make up 87.9% of the total population, according to UN data (2019)
country comparison to the world: 92

Nationality: *noun:* Emirati(s)
adjective: Emirati

Ethnic groups: Emirati 11.6%, South Asian 59.4% (includes Indian 38.2%, Bangladeshi 9.5%, Pakistani 9.4%, other 2.3%), Egyptian 10.2%, Filipino 6.1%, other 12.8% (2015 est.)

Languages: Arabic (official), English, Hindi, Malayam, Urdu, Pashto, Tagalog, Persian

Religions: Muslim (official) 76%, Christian 9%, other (primarily Hindu and Buddhist, less than 5% of the population consists of Parsi, Baha'i, Druze, Sikh, Ahmadi, Ismaili, Dawoodi Bohra Muslim, and Jewish) 15% (2005 est.)
note: data represent the total population; as of 2019, immigrants make up about 87.9% of the total population, according to UN data

MENA religious affiliation:

Age structure: *0-14 years:* 14.45% (male 745,492/ female 698,330)
15-24 years: 7.94% (male 431,751/female 361,804)
25-54 years: 68.03% (male 5,204,618/female 1,592,987)
55-64 years: 7.68% (male 658,892/female 108,850)
65 years and over: 1.9% (male 146,221/female 43,138) (2020 est.)

population pyramid: [INSERT IMAGE: UNITED ARAB EMIRATES-population pyramid]

Dependency ratios: *total dependency ratio:* 19.2
youth dependency ratio: 17.7
elderly dependency ratio: 1.5
potential support ratio: 66.4 (2020 est.)

Median age: *total:* 38.4 years
male: 40.4 years
female: 31.5 years (2020 est.)
country comparison to the world: 63

Population growth rate: 1.49% (2020 est.)
country comparison to the world: 71

Birth rate: 9.5 births/1,000 population (2020 est.)
country comparison to the world: 198

Death rate: 2 deaths/1,000 population (2020 est.)
country comparison to the world: 228

Net migration rate: 7.6 migrant(s)/1,000 population (2020 est.)
country comparison to the world: 14

Population distribution: population is heavily concentrated to the northeast on the Musandam Peninsula; the three largest emirates - Abu Dhabi, Dubai, and Sharjah - are home to nearly 85% of the population

Urbanization: *urban population:* 87% of total population (2020)
rate of urbanization: 1.71% annual rate of change (2015-20 est.)

total population growth rate v. urban population growth rate, 2000-2030: *Major urban areas - population:* 2.878 million Dubai, 1.685 million Sharjah, 1.483 million ABU DHABI (capital) (2020)

Sex ratio: *at birth:* 1.06 male(s)/female
0-14 years: 1.07 male(s)/female
15-24 years: 1.19 male(s)/female
25-54 years: 3.27 male(s)/female
55-64 years: 6.05 male(s)/female
65 years and over: 3.39 male(s)/female
total population: 2.56 male(s)/female (2020 est.)

Maternal mortality rate: 3 deaths/100,000 live births (2017 est.)
country comparison to the world: 180

Infant mortality rate: *total:* 5.3 deaths/1,000 live births
male: 5.8 deaths/1,000 live births
female: 4.7 deaths/1,000 live births (2020 est.)
country comparison to the world: 173

Life expectancy at birth: *total population:* 79 years
male: 77.6 years
female: 80.5 years (2020 est.)
country comparison to the world: 63

Total fertility rate: 1.73 children born/woman (2020 est.)
country comparison to the world: 168

Drinking water source:

improved: total: 100% of population
unimproved: total: 0% of population (2017 est.)

Current Health Expenditure: 3.3% (2017)

Physicians density: 2.44 physicians/1,000 population (2017)

Hospital bed density: 1.4 beds/1,000 population (2017)

Sanitation facility access:

improved: total: 98.6% of population
unimproved: total: 1.4% of population (2017 est.)

HIV/AIDS—adult prevalence rate: NA

HIV/AIDS—people living with HIV/AIDS: NA

HIV/AIDS—deaths: NA

Obesity—adult prevalence rate: 31.7% (2016)
country comparison to the world: 20

Education expenditures: NA

Literacy: *definition:* age 15 and over can read and write
total population: 93.8%
male: 93.1%
female: 95.8% (2015)
School life expectancy (primary to tertiary education): total: 14 years
male: 14 years

female: 15 years (2017)
Unemployment, youth ages 15-24: total: 6.9%
male: 5%
female: 12.8% (2018 est.)
country comparison to the world: 153

GOVERNMENT

Country name: *conventional long form:* United Arab Emirates
conventional short form: none
local long form: Al Imarat al Arabiyah al Muttahidah
local short form: none
former: Trucial Oman, Trucial States

abbreviation: UAE
etymology: self-descriptive country name; the name "Arabia" can be traced back many centuries B.C., the ancient Egyptians referred to the region as "Ar Rabi"; "emirates" derives from "amir" the Arabic word for "commander," "lord," or "prince"

Government type: federation of monarchies

Capital: *name:* Abu Dhabi

geographic coordinates: 24 28 N, 54 22 E
time difference: UTC+4 (9 hours ahead of Washington, DC, during Standard Time)
etymology: in Arabic, "abu" means "father" and "dhabi" refers to "gazelle"; the name may derive from an abundance of gazelles that used to live in the area, as well as a folk tale involving the "Father of the Gazelle," Shakhbut bin Dhiyab al Nahyan, whose hunting party tracked a gazelle to a spring on the island where Abu Dhabi was founded

Administrative divisions: 7 emirates (imarat, singular - imarah); Abu Zaby (Abu Dhabi), 'Ajman, Al Fujayrah, Ash Shariqah (Sharjah), Dubayy (Dubai), Ra's al Khaymah, Umm al Qaywayn

Independence: 2 December 1971 (from the UK)

National holiday: Independence Day (National Day), 2 December (1971)

Constitution: *history:* previous 1971 (provisional); latest drafted in 1979, became permanent May 1996

amendments: proposed by the Supreme Council and submitted to the Federal National Council; passage requires at least a two-thirds majority vote of Federal National Council members present and approval of the Supreme Council president; amended 2009

Legal system: mixed legal system of Islamic (sharia) law and civil law

International law organization participation: has not submitted an ICJ jurisdiction declaration; non-party state to the ICCt

Citizenship: *citizenship by birth:* no
citizenship by descent only: the father must be a citizen of the United Arab Emirates; if the father is unknown, the mother must be a citizen
dual citizenship recognized: no
residency requirement for naturalization: 30 years

Suffrage: limited; note - rulers of the seven emirates each select a proportion of voters for the

Federal National Council (FNC) that together account for about 12 percent of Emirati citizens

Executive branch: *chief of state:* President KHALIFA bin Zayid Al-Nuhayyan (since 2 November 2004), ruler of Abu Zaby (Abu Dhabi) (since 4 November 2004); Vice President and Prime Minister MUHAMMAD BIN RASHID Al-Maktum (since 5 January 2006)

head of government: Prime Minister Vice President MUHAMMAD BIN RASHID Al-Maktum (since 5 January 2006); Deputy Prime Ministers SAIF bin Zayid Al-Nuhayyan, MANSUR bin Zayid Al-Nuhayyan (both since 11 May 2009)
cabinet: Council of Ministers announced by the prime minister and approved by the president
elections/appointments: president and vice president indirectly elected by the Federal Supreme Council - composed of the rulers of the 7 emirates - for a 5-year term (no term limits); election last held 3 November 2009 (next election NA); prime minister and deputy prime minister appointed by the president
election results: KHALIFA bin Zayid Al-Nuhayyan reelected president; FSC vote NA
note: there is also a Federal Supreme Council (FSC) composed of the 7 emirate rulers; the FSC is the highest constitutional authority in the UAE; establishes general policies and sanctions federal legislation; meets 4 times a year; Abu Zaby (Abu Dhabi) and Dubayy (Dubai) rulers have effective veto power

Legislative branch: *description:* unicameral Federal National Council (FNC) or Majlis al-Ittihad al-Watani (40 seats; 20 members indirectly elected using single non-transferable vote by an electoral college whose members are selected by each emirate ruler proportional to its FNC membership, and 20 members appointed by the rulers of the 7 constituent states; members serve 4-year terms)
elections: last held for indirectly elected members on 5 October 2019 (next to be held in October 2023)
election results: all candidates ran as independents; seats by emirate - Abu Dhabi 4, Dubai 4, Sharjah 3, Ras al-Khaimah 3, Ajman 2, Fujairah 2, Umm al-Quwain 2; composition (preliminary) - 13 men, 7 women, percent of elected women 35%; note - to attain overall FNC gender parity, 13 women and 7 men will be appointed; overall FNC percent of women 50%

Judicial branch: *highest courts:* Federal Supreme Court (consists of the court president and 4 judges; jurisdiction limited to federal cases)
judge selection and term of office: judges appointed by the federal president following approval by the Federal Supreme Council, the highest executive and legislative authority consisting of the 7 emirate rulers; judges serve until retirement age or the expiry of their appointment terms
subordinate courts: Federal Court of Cassation (determines the constitutionality of laws promulgated at the federal and emirate level; federal level courts of first instance and appeals courts); the emirates of Abu Dhabi, Dubai, and Ra's al

Khaymah have parallel court systems; the other 4 emirates have incorporated their courts into the federal system; note - the Abu Dhabi Global Market Courts and the Dubai International Financial Center Courts, the country's two largest financial free zones, both adjudicate civil and commercial disputes.

Political parties and leaders: none; political parties are banned

International organization participation: ABEDA, AfDB (nonregional member), AFESD, AMF, BIS, CAEU, CICA, FAO, G-77, GCC, IAEA, IBRD, ICAO, ICC (national committees), ICRM, IDA, IDB, IFAD, IFC, IFRCS, IHO, ILO, IMF, IMO, IMSO, Interpol, IOC, IPU, ISO, ITSO, ITU, LAS, MIGA, NAM, OAPEC, OIC, OIF (observer), OPCW, OPEC, PCA, UN, UNCTAD, UNESCO, UNIDO, UNRWA, UNWTO, UPU, WCO, WHO, WIPO, WMO, WTO

Diplomatic representation in the US: *chief of mission:* Ambassador Yusif bin Mani bin Said al-UTAYBA (since 28 July 2008)
chancery: 3522 International Court NW, Suite 400, Washington, DC 20008
telephone: [1] (202) 243-2400
FAX: [1] (202) 243-2432

consulate(s) general: Boston, Los Angeles, New York

Diplomatic representation from the US: *chief of mission:* Ambassador John RAKOLTA Jr. (since 27 October 2019)
telephone: [971] (2) 414-2200
embassy: Embassies District, Plot 38, Sector W59-02, Street No. 4, Abu Dhabi
mailing address: P. O. Box 4009, Abu Dhabi
FAX: [971] (2) 414-2603

consulate(s) general: Dubai

Flag description: three equal horizontal bands of green (top), white, and black with a wider vertical red band on the hoist side; the flag incorporates all four Pan-Arab colors, which in this case represent fertility (green), neutrality (white), petroleum resources (black), and unity (red); red was the traditional color incorporated into all flags of the emirates before their unification

National symbol(s): *golden falcon; national colors:* green, white, black, red

National anthem: *name:* "Nashid al-watani al-imarati" (National Anthem of the UAE)
lyrics/music: AREF Al Sheikh Abdullah Al Hassan/Mohamad Abdel WAHAB
note: music adopted 1971, lyrics adopted 1996; Mohamad Abdel WAHAB also composed the music for the anthem of Tunisia
0:00 / 0:48

ECONOMY

Economy—overview: The UAE has an open economy with a high per capita income and a sizable annual trade surplus. Successful efforts at economic diversification have reduced the portion of GDP from the oil and gas sector to 30%.

Since the discovery of oil in the UAE nearly 60 years ago, the country has undergone a profound transformation from an impoverished region of small desert principalities to a modern state with a high standard of living. The government has increased spending on job creation and infrastructure expansion and is opening up utilities to greater private sector involvement. The country's free trade zones - offering 100% foreign ownership and zero taxes - are helping to attract foreign investors.

The global financial crisis of 2008-09, tight international credit, and deflated asset prices constricted the economy in 2009. UAE authorities tried to blunt the crisis by increasing spending and boosting liquidity in the banking sector. The crisis hit Dubai hardest, as it was heavily exposed to depressed real estate prices. Dubai lacked sufficient cash to meet its debt obligations, prompting global concern about its solvency and ultimately a $20 billion bailout from the UAE Central Bank and Abu Dhabi Government that was refinanced in March 2014.

The UAE's dependence on oil is a significant long-term challenge, although the UAE is one of the most diversified countries in the Gulf Cooperation Council. Low oil prices have prompted the UAE to cut expenditures, including on some social programs, but the UAE has sufficient assets in its sovereign investment funds to cover its deficits. The government reduced fuel subsidies in August 2015, and introduced excise taxes (50% on sweetened carbonated beverages and 100% on energy drinks and tobacco) in October 2017. A five-percent value-added tax was introduced in January 2018. The UAE's strategic plan for the next few years focuses on economic diversification, promoting the UAE as a global trade and tourism hub, developing industry, and creating more job opportunities for nationals through improved education and increased private sector employment.

GDP (purchasing power parity): $696 billion (2017 est.)
$690.5 billion (2016 est.)
$670.5 billion (2015 est.)
note: data are in 2017 dollars
country comparison to the world: 32

GDP (official exchange rate): $382.6 billion (2017 est.)

GDP—real growth rate: 0.8% (2017 est.)
3% (2016 est.)
5.1% (2015 est.)
country comparison to the world: 179

GDP—per capita (PPP): $68,600 (2017 est.)
$70,100 (2016 est.)
$70,000 (2015 est.)
note: data are in 2017 dollars
country comparison to the world: 13

Gross national saving: 28.5% of GDP (2017 est.)
30.9% of GDP (2016 est.)
30.7% of GDP (2015 est.)
country comparison to the world: 39

GDP—composition, by end use: *household consumption:* 34.9% (2017 est.)

government consumption: 12.3% (2017 est.)
investment in fixed capital: 23% (2017 est.)
investment in inventories: 1.8% (2017 est.)
exports of goods and services: 100.4% (2017 est.)
imports of goods and services: -72.4% (2017 est.)

GDP—composition, by sector of origin:
agriculture: 0.9% (2017 est.)
industry: 49.8% (2017 est.)
services: 49.2% (2017 est.)

Agriculture—products: dates, vegetables, watermelons; poultry, eggs, dairy products; fish

Industries: petroleum and petrochemicals; fishing, aluminum, cement, fertilizer, commercial ship repair, construction materials, handicrafts, textiles

Industrial production growth rate: 1.8% (2017 est.)
country comparison to the world: 138

Labor force: 5.344 million (2017 est.)
note: expatriates account for about 85% of the workforce
country comparison to the world: 74

Labor force—by occupation: agriculture: 7%
industry: 15%
services: 78% (2000 est.)

Unemployment rate: 1.6% (2016 est.)
3.6% (2014 est.)
country comparison to the world: 15

Population below poverty line: 19.5% (2003 est.)

Household income or consumption by percentage share: *lowest 10%:* NA
highest 10%: NA

Budget: *revenues:* 110.2 billion (2017 est.)
expenditures: 111.1 billion (2017 est.)
note: the UAE federal budget does not account for emirate-level spending in Abu Dhabi and Dubai

Taxes and other revenues: 28.8% (of GDP) (2017 est.)
country comparison to the world: 90

Budget surplus (+) or deficit (-): -0.2% (of GDP) (2017 est.)
country comparison to the world: 51

Public debt: 19.7% of GDP (2017 est.)
20.2% of GDP (2016 est.)
country comparison to the world: 190

Fiscal year: calendar year

Inflation rate (consumer prices): 2% (2017 est.)
1.6% (2016 est.)
country comparison to the world: 108

Current account balance: $26.47 billion (2017 est.)
$13.23 billion (2016 est.)
country comparison to the world: 14

Exports: $308.5 billion (2017 est.)
$298.6 billion (2016 est.)
country comparison to the world: 18

Exports—partners: India 10.1%, Iran 9.9%, Japan 9.3%, China 5.4%, Oman 5%, Switzerland 4.4%, South Korea 4.1% (2017)

Exports—commodities: crude oil 45%, natural gas, reexports, dried fish, dates (2012 est.)

Imports: $229.2 billion (2017 est.)
$226.5 billion (2016 est.)

country comparison to the world: 21

Imports—commodities: machinery and transport equipment, chemicals, food

Imports—partners: China 8.5%, US 6.8%, India 6.6% (2017)

Reserves of foreign exchange and gold: $95.37 billion (31 December 2017 est.)
$85.39 billion (31 December 2016 est.)
country comparison to the world: 27

Debt—external: $237.6 billion (31 December 2017 est.)
$218.7 billion (31 December 2016 est.)
country comparison to the world: 32

Exchange rates: Emirati dirhams (AED) per US dollar—
3.673 (2017 est.)
3.673 (2016 est.)
3.673 (2015 est.)
3.673 (2014 est.)
3.673 (2013 est.)

ENERGY

Electricity access: *electrification—total population:* 100% (2020)

Electricity—production: 121.8 billion kWh (2016 est.)
country comparison to the world: 31

Electricity—consumption: 113.2 billion kWh (2016 est.)
country comparison to the world: 31

Electricity—exports: 0 kWh (2016 est.)
country comparison to the world: 213

Electricity—imports: 1.111 billion kWh (2016 est.)
country comparison to the world: 66

Electricity—installed generating capacity: 28.91 million kW (2016 est.)
country comparison to the world: 33

Electricity—from fossil fuels: 99% of total installed capacity (2016 est.)
country comparison to the world: 26

Electricity—from nuclear fuels: 0% of total installed capacity (2017 est.)
country comparison to the world: 204

Electricity—from hydroelectric plants: 0% of total installed capacity (2017 est.)
country comparison to the world: 210

Electricity—from other renewable sources: 1% of total installed capacity (2017 est.)
country comparison to the world: 170

Crude oil—production: 3.216 million bbl/day (2018 est.)
country comparison to the world: 8

Crude oil—exports: 2.552 million bbl/day (2015 est.)
country comparison to the world: 5

Crude oil—imports: 0 bbl/day (2015 est.)
country comparison to the world: 210

Crude oil—proved reserves: 97.8 billion bbl (1 January 2018 est.)
country comparison to the world: 7

Refined petroleum products—production: 943,500 bbl/day (2017 est.)
country comparison to the world: 19

Refined petroleum products—consumption: 896,000 bbl/day (2016 est.)
country comparison to the world: 24

Refined petroleum products— exports: 817,700 bbl/day (2015 est.)
country comparison to the world: 10

Refined petroleum products—imports: 392,000 bbl/day (2015 est.)
country comparison to the world: 23

Natural gas—production: 62.01 billion cu m (2017 est.)
country comparison to the world: 14

Natural gas—consumption: 74.48 billion cu m (2017 est.)
country comparison to the world: 12

Natural gas—exports: 7.504 billion cu m (2017 est.)
country comparison to the world: 25

Natural gas—imports: 20.22 billion cu m (2017 est.)
country comparison to the world: 16

Natural gas—proved reserves: 6.091 trillion cu m (1 January 2018 est.)
country comparison to the world: 6

Carbon dioxide emissions from consumption of energy: 289.4 million Mt (2017 est.)
country comparison to the world: 24

COMMUNICATIONS

Telephones—fixed lines: *total subscriptions:* 2,380,238

subscriptions per 100 inhabitants: 24.18 (2019 est.)
country comparison to the world: 51

Telephones—mobile cellular: total subscriptions: 19,749,674

subscriptions per 100 inhabitants: 200.63 (2019 est.)
country comparison to the world: 62

Telecommunication systems: *general assessment:* modern fiber-optic integrated services; digital network with rapidly growing use of mobile-cellular telephones; key centers are Abu Dhabi and Dubai; 5G capabilities launched in 2019; two operators are competitive, but majority owned by the government; HSPA (high speed packet access) + LTE networks cover most of the population; low cost smart phones readily available; mobile penetration levels among the world's highest; well-established fiberbroadband network provides future growth (2020)

domestic: microwave radio relay, fiber-optic and coaxial cable; fixed-line 24 per 100 and mobile-cellular 201 per 100 (2019)

international: country code - 971; landing points for the FLAG, SEA-ME-WE-3 ,-4 & -5, Qater UAE Submarine Cable System, FALCON, FOG, Tat TGN-Gulf, OMRAN/EPEG Cable System, AAE-1, BBG, EIG, FEA, GBICS/MENA,

IMEWE, Orient Express, TEAMS, TW1 and the UAE-Iran submarine cables, linking to Europe, Africa, the Middle East, Asia, Southeast Asia and Australia; satellite earth stations - 3 Intelsat (1 Atlantic Ocean and 2 Indian) (2020)

note: the COVID-19 outbreak is negatively impacting telecommunications production and supply chains globally; consumer spending on telecom devices and services has also slowed due to the pandemic's effect on economies worldwide; overall progress towards improvements in all facets of the telecom industry - mobile, fixed-line, broadband, submarine cable and satellite - has moderated

Broadcast media: except for the many organizations now operating in media free zones in Abu Dhabi and Dubai, most TV and radio stations remain government-owned; widespread use of satellite dishes provides access to pan-Arab and other international broadcasts; restrictions since June 2017 on some satellite channels and websites originating from or otherwise linked to Qatar (2018)

Internet country code: .ae

Internet users: total: 9,550,945

percent of population: 98.45% (July 2018 est.)
country comparison to the world: 54

Broadband—fixed subscriptions: total: 3,024,565

subscriptions per 100 inhabitants: 31 (2018 est.)
country comparison to the world: 44

TRANSPORTATION

National air transport system: *number of registered air carriers:* 10 (2020)

inventory of registered aircraft operated by air carriers: 497

annual passenger traffic on registered air carriers: 95,533,069 (2018)

annual freight traffic on registered air carriers: 15,962,900,000 mt-km (2018)

Civil aircraft registration country code prefix: A6 (2016)

Airports: 43 (2013)
country comparison to the world: 99

Airports—with paved runways: *total:* 25 (2013)
over 3,047 m: 12 (2013)
2,438 to 3,047 m: 3 (2013)
1,524 to 2,437 m: 5 (2013)
914 to 1,523 m: 3 (2013)
under 914 m: 2 (2013)

Airports—with unpaved runways: *total:* 18 (2013)
over 3,047 m: 1 (2013)
2,438 to 3,047 m: 1 (2013)
1,524 to 2,437 m: 4 (2013)
914 to 1,523 m: 6 (2013)
under 914 m: 6 (2013)

Heliports: 5 (2013)

Pipelines: 533 km condensate, 3277 km gas, 300 km liquid petroleum gas, 3287 km oil, 24 km oil/gas/water, 218 km refined products, 99 km water (2013)

Roadways: *total:* 4,080 km (2008)
paved: 4,080 km (includes 253 km of expressways) (2008)

country comparison to the world: 154

Merchant marine: *total:* 637
by type: bulk carrier 2, general cargo 113, oil tanker 17, other 505 (2019)
country comparison to the world: 36

Ports and terminals: *major seaport(s):* Al Fujayrah, Mina' Jabal 'Ali (Dubai), Khor Fakkan (Khawr Fakkan) (Sharjah), Mubarraz Island (Abu Dhabi), Mina' Rashid (Dubai), Mina' Saqr (Ra's al Khaymah)

container port(s) (TEUs): Dubai Port (15,368,000), Khor Fakkan (Khawr Fakkan) (Sharjah) (2,321,000) (2017)

LNG terminal(s) (export): Das Island

MILITARY AND SECURITY

Military and security forces: *United Arab Emirates Armed Forces:* Land Forces, Navy, Air Force, Presidential Guard, Joint Aviation Command; Ministry of Interior: Critical Infrastructure Coastal Patrol Agency (CICPA) (2020)

Military expenditures: 5.7% of GDP (2016)
5.6% of GDP (2014)
6% of GDP (2013)
5.1% of GDP (2012)
5.5% of GDP (2011)

no public data available for 2015 or after 2016
country comparison to the world: 4

Military and security service personnel strengths: the United Arab Emirates Armed Forces have approximately 63,000 total active personnel (44,000 Land Forces; 2,500 Navy; 4,500; 12,000 Presidential Guard) (2019)

Military equipment inventories and acquisitions: the UAE Armed Forces inventory is comprised of mostly modern imported equipment; since 2010, the UAE has acquired military equipment from more than 20 countries with the US as the leading supplier, followed by France and Russia (2019 est.)

Military deployments: est. 1,000 Eritrea; est. 3-4,000 Yemen; maintains a military base in the Eritrean port of Assab (2019)

Military service age and obligation: 18-30 years of age for compulsory military service for men;

17 years of age for male volunteers with parental approval; 24-month general service obligation, 16 months for secondary school graduates; women can volunteer to serve for 9 months regardless of education (2018)

TRANSPORTATION

Disputes—international: boundary agreement was signed and ratified with Oman in 2003 for entire border, including Oman's Musandam Peninsula and Al Madhah enclaves, but contents of the agreement and detailed maps showing the alignment have not been published; Iran and UAE dispute Tunb Islands and Abu Musa Island, which Iran occupies

Illicit drugs: the UAE is a drug transshipment point for traffickers given its proximity to Southwest Asian drug-producing countries; the UAE's position as a major financial center makes it vulnerable to money laundering; anti-money-laundering controls improving, but informal banking remains unregulated

UNITED KINGDOM

INTRODUCTION

Background: The United Kingdom has historically played a leading role in developing parliamentary democracy and in advancing literature and science. At its zenith in the 19th century, the British Empire stretched over one-fourth of the earth's surface. The first half of the 20th century saw the UK's strength seriously depleted in two world wars and the Irish Republic's withdrawal from the union. The second half witnessed the dismantling of the Empire and the UK rebuilding itself into a modern and prosperous European nation. As one of five permanent members of the UN Security Council and a founding member of NATO and the Commonwealth, the UK pursues a global approach to foreign policy. The Scottish Parliament, the National Assembly for Wales, and the Northern Ireland Assembly were established in 1998.

The UK has been an active member of the EU since its accession in 1973, although it chose to remain outside the Economic and Monetary Union. However, motivated in part by frustration at a remote bureaucracy in Brussels and massive migration into the country, UK citizens on 23 June 2016 narrowly voted to leave the EU. The UK is scheduled to depart the EU on 31 January 2020, but negotiations on the future EU-UK economic and security relationship will continue throughout 2020 and potentially beyond.

GEOGRAPHY

Location: Western Europe, islands - including the northern one-sixth of the island of Ireland -

between the North Atlantic Ocean and the North Sea; northwest of France

Geographic coordinates: 54 00 N, 2 00 W

Map references: Europe

Area: *total:* 243,610 sq km
land: 241,930 sq km
water: 1,680 sq km
note 1: the percentage area breakdown of the four UK countries is: England 53%, Scotland 32%, Wales 9%, and Northern Ireland 6%
note 2: includes Rockall and the Shetland Islands, which are part of Scotland
country comparison to the world: 81

Area—comparative: twice the size of Pennsylvania; slightly smaller than Oregon

Area comparison map: [INSERT IMAGE: UNITED KINGDOM -Area comparison map]

Land boundaries: *total:* 499 km
border countries (1): Ireland 499 km

Coastline: 12,429 km

Maritime claims: *territorial sea:* 12 nm
continental shelf: as defined in continental shelf orders or in accordance with agreed upon boundaries
exclusive fishing zone: 200 nm

Climate: temperate; moderated by prevailing southwest winds over the North Atlantic Current; more than one-half of the days are overcast

Terrain: mostly rugged hills and low mountains; level to rolling plains in east and southeast

Elevation: *mean elevation:* 162 m
lowest point: The Fens -4 m

highest point: Ben Nevis 1,345 m

Natural resources: coal, petroleum, natural gas, iron ore, lead, zinc, gold, tin, limestone, salt, clay, chalk, gypsum, potash, silica sand, slate, arable land

Land use: *agricultural land:* 71% (2011 est.)
arable land: 25.1% (2011 est.) / permanent crops: 0.2% (2011 est.) / permanent pasture: 45.7% (2011 est.)
forest: 11.9% (2011 est.)
other: 17.1% (2011 est.)

Irrigated land: 950 sq km (2012)

Population distribution: the core of the population lies in and around London, with significant clusters found in central Britain around Manchester and Liverpool, in the Scottish lowlands between Endinburgh and Glasgow, southern Wales in and around Cardiff, and far eastern Northern Ireland centered on Belfast

Natural hazards: winter windstorms; floods

Environment—current issues: air pollution improved but remains a concern, particularly in the London region; soil pollution from pesticides and heavy metals; decline in marine and coastal habitats brought on by pressures from housing, tourism, and industry

Environment—international agreements: *party to:* Air Pollution, Air Pollution-Nitrogen Oxides, Air Pollution-Persistent Organic Pollutants, Air Pollution-Sulfur 94, Air Pollution-Volatile Organic Compounds, Antarctic-Environmental Protocol, Antarctic-Marine Living Resources, Antarctic Seals, Antarctic Treaty, Biodiversity, Climate Change, Climate Change-Kyoto Protocol, Desertification, Endangered Species, Environmental Modification, Hazardous Wastes, Law of the Sea, Marine Dumping, Marine Life Conservation, Ozone Layer Protection, Ship Pollution, Tropical Timber 83, Tropical Timber 94, Wetlands, Whaling
signed, but not ratified: none of the selected agreements

Geography—note: lies near vital North Atlantic sea lanes; only 35 km from France and linked by tunnel under the English Channel (the Channel Tunnel or Chunnel); because of heavily indented coastline, no location is more than 125 km from tidal waters

PEOPLE AND SOCIETY

Population: 65,761,117 United Kingdom (July 2020 est.)

constituent countries by percentage of total population: England 84%
Scotland 8%
Wales 5%
Northern Ireland 3%
country comparison to the world: 22

Nationality: *noun:* Briton(s), British (collective plural)
adjective: British

Ethnic groups: white 87.2%, black/African/Caribbean/black British 3%, Asian/Asian British:

Indian 2.3%, Asian/Asian British: Pakistani 1.9%, mixed 2%, other 3.7% (2011 est.)

Languages: English
note: the following are recognized regional languages: Scots (about 30% of the population of Scotland), Scottish Gaelic (about 60,000 speakers in Scotland), Welsh (about 20% of the population of Wales), Irish (about 10% of the population of Northern Ireland), Cornish (some 2,000 to 3,000 people in Cornwall) (2012 est.)

Religions: Christian (includes Anglican, Roman Catholic, Presbyterian, Methodist) 59.5%, Muslim 4.4%, Hindu 1.3%, other 2%, unspecified 7.2%, none 25.7% (2011 est.)

Age structure: *0-14 years:* 17.63% (male 5,943,435/female 5,651,780)
15-24 years: 11.49% (male 3,860,435/female 3,692,398)
25-54 years: 39.67% (male 13,339,965/female 12,747,598)
55-64 years: 12.73% (male 4,139,378/female 4,234,701)
65 years and over: 18.48% (male 5,470,116/female 6,681,311) (2020 est.)

population pyramid: [INSERT IMAGE: UNITED KINGDOM -population pyramid]

Dependency ratios: *total dependency ratio:* 57.1
youth dependency ratio: 27.8
elderly dependency ratio: 29.3
potential support ratio: 3.4 (2020 est.)

Median age: *total:* 40.6 years
male: 39.6 years
female: 41.7 years (2020 est.)
country comparison to the world: 50

Population growth rate: 0.49% (2020 est.)
country comparison to the world: 156

Birth rate: 11.9 births/1,000 population (2020 est.)
country comparison to the world: 166

Death rate: 9.5 deaths/1,000 population (2020 est.)
country comparison to the world: 47

Net migration rate: 2.5 migrant(s)/1,000 population (2020 est.)
country comparison to the world: 41

Population distribution: the core of the population lies in and around London, with significant clusters found in central Britain around Manchester and Liverpool, in the Scottish lowlands between Endinburgh and Glasgow, southern Wales in and around Cardiff, and far eastern Northern Ireland centered on Belfast

Urbanization: *urban population:* 83.9% of total population (2020)
rate of urbanization: 0.89% annual rate of change (2015-20 est.)

total population growth rate v. urban population growth rate, 2000-2030: *Major urban areas - population:* 9.304 million LONDON (capital), 2.730 million Manchester, 2.607 million Birmingham, 1.889 million West Yorkshire, 1.663 million Glasgow, 928,000 Southampton/Portsmouth (2020)

Sex ratio: *at birth:* 1.05 male(s)/female

0-14 years: 1.05 male(s)/female
15-24 years: 1.05 male(s)/female
25-54 years: 1.05 male(s)/female
55-64 years: 0.98 male(s)/female
65 years and over: 0.82 male(s)/female
total population: 0.99 male(s)/female (2020 est.)

Mother's mean age at first birth: 28.8 years (2017 est.)
note: data represent England and Wales only

Maternal mortality rate: 7 deaths/100,000 live births (2017 est.)
country comparison to the world: 158

Infant mortality rate: *total:* 4.1 deaths/1,000 live births
male: 4.5 deaths/1,000 live births
female: 3.7 deaths/1,000 live births (2020 est.)
country comparison to the world: 190

Life expectancy at birth: *total population:* 81.1 years
male: 78.8 years
female: 83.5 years (2020 est.)
country comparison to the world: 40

Total fertility rate: 1.86 children born/woman (2020 est.)
country comparison to the world: 140

Drinking water source:

improved: *urban:* 100% of population
rural: 100% of population
total: 100% of population

unimproved: *urban:* 0% of population
rural: 0% of population
total: 0% of population (2017 est.)

Current Health Expenditure: 9.6% (2017)

Physicians density: 2.79 physicians/1,000 population (2017)

Hospital bed density: 2.5 beds/1,000 population (2017)

Sanitation facility access:

improved: *urban:* 100% of population
rural: 100% of population
total: 100% of population

unimproved: *urban:* 0% of population
rural: 0% of population
total: 0% of population (2017 est.)

HIV/AIDS—adult prevalence rate: NA

HIV/AIDS—people living with HIV/AIDS: NA

HIV/AIDS—deaths: NA

Major infectious diseases: *Covid-19 (see note)* (2020)
note: widespread ongoing transmission of a respiratory illness caused by the novel coronavirus (COVID-19) is occurring throughout the UK; as of 10 November 2020, the UK has reported a total of 1,171,445 cases of COVID-19 or 17,256 cumulative cases of COVID-19 per 1 million population with 720 cumulative deaths per 1 million population; individuals arriving in the UK must self-isolate for 14 days and may be contacted to verify compliance; new arrivals will be required to provide UK officials with contact and travel information prior to arrival; the US Department of Homeland Security has issued instructions

requiring US passengers who have been in the UK to travel through select airports where the US Government has implemented enhanced screening procedures

Obesity—adult prevalence rate: 27.8% (2016)
country comparison to the world: 36

Education expenditures: 5.5% of GDP (2016)
country comparison to the world: 40
School life expectancy (primary to tertiary education): total: 17 years
male: 17 years
female: 18 years (2018)
Unemployment, youth ages 15-24: total: 11.3%
male: 12.2%
female: 10.3% (2018 est.)
country comparison to the world: 115

GOVERNMENT

Country name: *conventional long form:* United Kingdom of Great Britain and Northern Ireland; note - the island of Great Britain includes England, Scotland, and Wales
conventional short form: United Kingdom

abbreviation: UK
etymology: self-descriptive country name; the designation "Great Britain," in the sense of "Larger Britain," dates back to medieval times and was used to distinguish the island from "Little Britain," or Brittany in modern France; the name Ireland derives from the Gaelic "Eriu," the matron goddess of Ireland (goddess of the land)

Government type: parliamentary constitutional monarchy; a Commonwealth realm

Capital: *name:* London
geographic coordinates: 51 30 N, 0 05 W
time difference: UTC 0 (5 hours ahead of Washington, DC, during Standard Time)
daylight saving time: +1hr, begins last Sunday in March; ends last Sunday in October
note: the time statements apply to the United Kingdom proper, not to its crown dependencies or overseas territories
etymology: the name derives from the Roman settlement of Londinium, established on the current site of London around A.D. 43; the original meaning of the name is uncertain

Administrative divisions: England: 26 two-tier counties, 32 London boroughs and 1 City of London or Greater London, 36 metropolitan districts, 56 unitary authorities (including 4 single-tier counties*);

two-tier counties: Buckinghamshire, Cambridgeshire, Cumbria, Derbyshire, Devon, Dorset, East Sussex, Essex, Gloucestershire, Hampshire, Hertfordshire, Kent, Lancashire, Leicestershire, Lincolnshire, Norfolk, Northamptonshire, North Yorkshire, Nottinghamshire, Oxfordshire, Somerset, Staffordshire, Suffolk, Surrey, Warwickshire, West Sussex, Worcestershire

London boroughs and City of London or Greater London: Barking and Dagenham, Barnet, Bexley, Brent, Bromley, Camden, Croydon, Ealing, Enfield, Greenwich, Hackney, Hammersmith and Fulham, Haringey, Harrow, Havering, Hillingdon, Hounslow, Islington, Kensington and Chelsea, Kingston upon Thames, Lambeth, Lewisham, City of London, Merton, Newham, Redbridge, Richmond upon Thames, Southwark, Sutton, Tower Hamlets, Waltham Forest, Wandsworth, Westminster

metropolitan districts: Barnsley, Birmingham, Bolton, Bradford, Bury, Calderdale, Coventry, Doncaster, Dudley, Gateshead, Kirklees, Knowsley, Leeds, Liverpool, Manchester, Newcastle upon Tyne, North Tyneside, Oldham, Rochdale, Rotherham, Salford, Sandwell, Sefton, Sheffield, Solihull, South Tyneside, St. Helens, Stockport, Sunderland, Tameside, Trafford, Wakefield, Walsall, Wigan, Wirral, Wolverhampton

unitary authorities: Bath and North East Somerset; Bedford; Blackburn with Darwen; Blackpool; Bournemouth, Christchurch and Poole; Bracknell Forest; Brighton and Hove; City of Bristol; Central Bedfordshire; Cheshire East; Cheshire West and Chester; Cornwall; Darlington; Derby; Dorset; Durham County*; East Riding of Yorkshire; Halton; Hartlepool; Herefordshire*; Isle of Wight*; Isles of Scilly; City of Kingston upon Hull; Leicester; Luton; Medway; Middlesbrough; Milton Keynes; North East Lincolnshire; North Lincolnshire; North Somerset; Northumberland*; Nottingham; Peterborough; Plymouth; Portsmouth; Reading; Redcar and Cleveland; Rutland; Shropshire; Slough; South Gloucestershire; Southampton; Southend-on-Sea; Stockton-on-Tees; Stoke-on-Trent; Swindon; Telford and Wrekin; Thurrock; Torbay; Warrington; West Berkshire; Wiltshire; Windsor and Maidenhead; Wokingham; York

Northern Ireland: 5 borough councils, 4 district councils, 2 city councils;

borough councils: Antrim and Newtownabbey; Ards and North Down; Armagh City, Banbridge, and Craigavon; Causeway Coast and Glens; Mid and East Antrim

district councils: Derry City and Strabane; Fermanagh and Omagh; Mid Ulster; Newry, Murne, and Down

city councils: Belfast; Lisburn and Castlereagh

Scotland: 32 council areas;

council areas: Aberdeen City, Aberdeenshire, Angus, Argyll and Bute, Clackmannanshire, Dumfries and Galloway, Dundee City, East Ayrshire, East Dunbartonshire, East Lothian, East Renfrewshire, City of Edinburgh, Eilean Siar (Western Isles), Falkirk, Fife, Glasgow City, Highland, Inverclyde, Midlothian, Moray, North Ayrshire, North Lanarkshire, Orkney Islands, Perth and Kinross, Renfrewshire, Shetland Islands, South Ayrshire, South Lanarkshire, Stirling, The Scottish Borders, West Dunbartonshire, West Lothian

Wales: 22 unitary authorities;

unitary authorities: Blaenau Gwent, Bridgend, Caerphilly, Cardiff, Carmarthenshire, Ceredigion, Conwy, Denbighshire, Flintshire, Gwynedd, Isle of Anglesey, Merthyr Tydfil, Monmouthshire, Neath Port Talbot, Newport, Pembrokeshire, Powys, Rhondda Cynon Taff, Swansea, The Vale of Glamorgan, Torfaen, Wrexham

Dependent areas: Anguilla; Bermuda; British Indian Ocean Territory; British Virgin Islands; Cayman Islands; Falkland Islands; Gibraltar; Montserrat; Pitcairn Islands; Saint Helena, Ascension, and Tristan da Cunha; South Georgia and the South Sandwich Islands; Turks and Caicos Islands

Independence: no official date of independence: 927 (minor English kingdoms unite); 3 March 1284 (enactment of the Statute of Rhuddlan uniting England and Wales); 1536 (Act of Union formally incorporates England and Wales); 1 May 1707 (Acts of Union formally unite England, Scotland, and Wales as Great Britain); 1 January 1801 (Acts of Union formally unite Great Britain and Ireland as the United Kingdom of Great Britain and Ireland); 6 December 1921 (Anglo-Irish Treaty formalizes partition of Ireland; six counties remain part of the United Kingdom and Northern Ireland); 12 April 1927 (Royal and Parliamentary Titles Act establishes current name of the United Kingdom of Great Britain and Northern Ireland)

National holiday: the UK does not celebrate one particular national holiday

Constitution: *history:* unwritten; partly statutes, partly common law and practice
amendments: proposed as a bill for an Act of Parliament by the government, by the House of Commons, or by the House of Lords; passage requires agreement by both houses and by the monarch (Royal Assent); note - additions include the Human Rights Act of 1998, the Constitutional Reform and Governance Act 2010, the Parliamentary Voting System and Constituencies Act 2011, the Fixed-term Parliaments Act 2011, and the House of Lords (Expulsion and Suspension) Act 2015

Legal system: common law system; has nonbinding judicial review of Acts of Parliament under the Human Rights Act of 1998

International law organization participation: accepts compulsory ICJ jurisdiction with reservations; accepts ICCt jurisdiction

Citizenship: *citizenship by birth:* no
citizenship by descent only: at least one parent must be a citizen of the United Kingdom
dual citizenship recognized: yes
residency requirement for naturalization: 5 years

Suffrage: 18 years of age; universal

Executive branch: *chief of state:* Queen ELIZABETH II (since 6 February 1952); Heir Apparent Prince CHARLES (son of the queen, born 14 November 1948)

head of government: Prime Minister Boris JOHNSON (Conservative) (since 24 July 2019)
cabinet: Cabinet appointed by the prime minister
elections/appointments: the monarchy is hereditary; following legislative elections, the leader of the majority party or majority coalition usually

becomes the prime minister; election last held on 12 December 2019 (next to be held by 2 May 2024) *note:* in addition to serving as the UK head of state, the British sovereign is the constitutional monarch for 15 additional Commonwealth countries (these 16 states are each referred to as a Commonwealth realm)

Legislative branch: *description:* bicameral Parliament consists of:

House of Lords (membership not fixed; as of December 2019, 796 lords were eligible to participate in the work of the House of Lords - 679 life peers, 91 hereditary peers, and 26 clergy; members are appointed by the monarch on the advice of the prime minister and non-party political members recommended by the House of Lords Appointments Commission); note - House of Lords total does not include ineligible members or members on leave of absence

House of Commons (650 seats; members directly elected in single-seat constituencies by simple majority popular vote to serve 5-year terms unless the House is dissolved earlier)

elections: House of Lords - no elections; note - in 1999, as provided by the House of Lords Act, elections were held in the House of Lords to determine the 92 hereditary peers who would remain; elections held only as vacancies in the hereditary peerage arise)

House of Commons—last held on 12 December 2019 (next to be held by 2 May 2024)

election results: House of Lords—composition—men 579, women 217, percent of women 27.3%

House of Commons—percent of vote by party—Conservative 43.6%, Labor 32.1%, Lib Dems 11.6%, SNP 3.9%, Greens 2.7%, Brexit Party 2.0%, other 4.1%; seats by party—Conservative 365, Labor 202, SNP 48, Lib Dems 11, DUP 8, Sinn Fein 7, Plaid Cymru 4, other 9; composition—men 430, women 220, percent of women 34%; total Parliament percent of women 30.2%

Judicial branch: *highest courts:* Supreme Court (consists of 12 justices, including the court president and deputy president); note—the Supreme Court was established by the Constitutional Reform Act 2005 and implemented in 2009, replacing the Appellate Committee of the House of Lords as the highest court in the United Kingdom

judge selection and term of office: judge candidates selected by an independent committee of several judicial commissions, followed by their recommendations to the prime minister, and appointed by the monarch; justices serve for life

subordinate courts: England and Wales: Court of Appeal (civil and criminal divisions); High Court; Crown Court; County Courts; Magistrates' Courts; Scotland: Court of Sessions; Sheriff Courts; High Court of Justiciary; tribunals; Northern Ireland: Court of Appeal in Northern Ireland; High Court; county courts; magistrates' courts; specialized tribunals

Political parties and leaders: Alliance Party (Northern Ireland) [Naomi LONG]

Brexit Party [Nigel FARAGE]
Conservative and Unionist Party [Boris JOHNSON]
Democratic Unionist Party or DUP (Northern Ireland) [Arlene FOSTER]
Green Party of England and Wales or Greens [Sian BERRY and Jonathan BARTLEY]
Labor (Labour) Party [Sir Keir STARMER]
Liberal Democrats (Lib Dems) [Ed Davey]
Party of Wales (Plaid Cymru) [Adam PRICE]
Scottish National Party or SNP [Nicola STURGEON]
Sinn Fein (Northern Ireland) [Mary Lou MCDONALD]
Social Democratic and Labor Party or SDLP (Northern Ireland) [Colum EASTWOOD]
Ulster Unionist Party or UUP (Northern Ireland) [Robin SWANN]
UK Independence Party or UKIP [Pat MOUNTAIN, interim leader]

International organization participation: ADB (nonregional member), AfDB (nonregional member), Arctic Council (observer), Australia Group, BIS, C, CBSS (observer), CD, CDB, CE, CERN, EAPC, EBRD, ECB, EIB, EITI (implementing country), ESA, EU, FAO, FATF, G-5, G-7, G-8, G-10, G-20, IADB, IAEA, IBRD, ICAO, ICC (national committees), ICCt, ICRM, IDA, IEA, IFAD, IFC, IFRCS, IGAD (partners), IHO, ILO, IMF, IMO, IMSO, Interpol, IOC, IOM, IPU, ISO, ITSO, ITU, ITUC (NGOs), MIGA, MINUSMA, MONUSCO, NATO, NEA, NSG, OAS (observer), OECD, OPCW, OSCE, Pacific Alliance (observer), Paris Club, PCA, PIF (partner), OELEC (observer), SICA (observer), UN, UNCTAD, UNESCO, UNFICYP, UNHCR, UNMISS, UNRWA, UN Security Council (permanent), UPU, WCO, WHO, WIPO, WMO, WTO, ZC

Diplomatic representation in the US: *chief of mission:* Ambassador Karen Elizabeth PIERCE (since 8 April 2020)

chancery: 3100 Massachusetts Avenue NW, Washington, DC 20008

telephone: [1] (202) 588-6500

FAX: [1] (202) 588-7870

consulate(s) general: Atlanta, Boston, Chicago, Denver, Houston, Los Angeles, Miami, New York, San Francisco

consulate(s): Orlando (FL), San Juan (Puerto Rico)

Diplomatic representation from the US: *chief of mission:* Ambassador Robert Wood "Woody" JOHNSON IV (since 29 August 2017)

telephone: [44] 20-7499-9000

embassy: 33 Nine Elms Lane, London, SW11 7US or SW8 5DB (driving/GPS postcode)

mailing address: PSC 801, Box 40, FPO AE 09498-4040

FAX: [44] 20-7891-3151

consulate(s) general: Belfast, Edinburgh

Flag description: blue field with the red cross of Saint George (patron saint of England) edged in white superimposed on the diagonal red cross of Saint Patrick (patron saint of Ireland), which is superimposed on the diagonal white cross of Saint Andrew (patron saint of Scotland); properly known as the Union Flag, but commonly called the Union Jack; the design and colors (especially the Blue Ensign) have been the basis for a number of other flags including other Commonwealth countries and their constituent states or provinces, and British overseas territories

National symbol(s): *lion (Britain in general); lion, Tudor rose, oak (England); lion, unicorn, thistle (Scotland); dragon, daffodil, leek (Wales); shamrock, flax (Northern Ireland); national colors:* red, white, blue (Britain in general); red, white (England); blue, white (Scotland); red, white, green (Wales)

National anthem: *name:* God Save the Queen

lyrics/music: unknown

note: in use since 1745; by tradition, the song serves as both the national and royal anthem of the UK; it is known as either "God Save the Queen" or "God Save the King," depending on the gender of the reigning monarch; it also serves as the royal anthem of many Commonwealth nations

0:00 / 1:02

ECONOMY

Economy—overview: The UK, a leading trading power and financial center, is the third largest economy in Europe after Germany and France. Agriculture is intensive, highly mechanized, and efficient by European standards, producing about 60% of food needs with less than 2% of the labor force. The UK has large coal, natural gas, and oil resources, but its oil and natural gas reserves are declining; the UK has been a net importer of energy since 2005. Services, particularly banking, insurance, and business services, are key drivers of British GDP growth. Manufacturing, meanwhile, has declined in importance but still accounts for about 10% of economic output.

In 2008, the global financial crisis hit the economy particularly hard, due to the importance of its financial sector. Falling home prices, high consumer debt, and the global economic slowdown compounded the UK's economic problems, pushing the economy into recession in the latter half of 2008 and prompting the then BROWN (Labour) government to implement a number of measures to stimulate the economy and stabilize the financial markets. Facing burgeoning public deficits and debt levels, in 2010 the then CAMERON-led coalition government (between Conservatives and Liberal Democrats) initiated an austerity program, which has continued under the Conservative government. However, the deficit still remains one of the highest in the G7, standing at 3.6% of GDP as of 2017, and the UK has pledged to lower its corporation tax from 20% to 17% by 2020. The UK had a debt burden of 90.4% GDP at the end of 2017.

The UK economy has begun to slow since the referendum vote to leave the EU in June 2016. A sustained depreciation of the British pound has increased consumer and producer prices, weighing on consumer spending without spurring a meaningful increase in exports. The UK has an extensive trade relationship with other EU members through its single market membership, and economic observers have warned the exit will jeopardize its position as the central location for European financial services. The UK is slated to leave the EU at the end of January 2020.

GDP (purchasing power parity): $2.925 trillion (2017 est.)
$2.877 trillion (2016 est.)
$2.827 trillion (2015 est.)
note: data are in 2017 dollars
country comparison to the world: 9

GDP (official exchange rate): $2.628 trillion (2017 est.)

GDP—real growth rate: 1.26% (2019 est.)
1.25% (2018 est.)
1.74% (2017 est.)
country comparison to the world: 164

GDP—per capita (PPP): $44,300 (2017 est.)
$43,800 (2016 est.)
$43,400 (2015 est.)
note: data are in 2017 dollars
country comparison to the world: 39

Gross national saving: 13.6% of GDP (2017 est.)
12% of GDP (2016 est.)
12.3% of GDP (2015 est.)
country comparison to the world: 142

GDP—composition, by end use: *household consumption:* 65.8% (2017 est.)
government consumption: 18.3% (2017 est.)
investment in fixed capital: 17.2% (2017 est.)
investment in inventories: 0.2% (2017 est.)
exports of goods and services: 30.2% (2017 est.)
imports of goods and services: -31.5% (2017 est.)

GDP—composition, by sector of origin: *agriculture:* 0.7% (2017 est.)
industry: 20.2% (2017 est.)
services: 79.2% (2017 est.)

Agriculture—products: cereals, oilseed, potatoes, vegetables; cattle, sheep, poultry; fish; milk, eggs

Industries: machine tools, electric power equipment, automation equipment, railroad equipment, shipbuilding, aircraft, motor vehicles and parts, electronics and communications equipment, metals, chemicals, coal, petroleum, paper and paper products, food processing, textiles, clothing, other consumer goods

Industrial production growth rate: 3.4% (2017 est.)
country comparison to the world: 93

Labor force: 16.033 million (2020 est.)
country comparison to the world: 33

Labor force—by occupation: agriculture: 1.3%
industry: 15.2%
services: 83.5% (2014 est.)

Unemployment rate: 3.17% (2019 est.)
2.51% (2018 est.)
country comparison to the world: 42

Population below poverty line: 15% (2013 est.)

Household income or consumption by percentage share: *lowest 10%:* 1.7%

highest 10%: 31.1% (2012)

Budget: *revenues:* 1.028 trillion (2017 est.)
expenditures: 1.079 trillion (2017 est.)

Taxes and other revenues: 39.1% (of GDP) (2017 est.)
country comparison to the world: 49

Budget surplus (+) or deficit (-): -1.9% (of GDP) (2017 est.)
country comparison to the world: 102

Public debt: 87.5% of GDP (2017 est.)
87.9% of GDP (2016 est.)
note: data cover general government debt and include debt instruments issued (or owned) by government entities other than the treasury; the data include treasury debt held by foreign entities; the data include debt issued by subnational entities, as well as intragovernmental debt; intragovernmental debt consists of treasury borrowings from surpluses in the social funds, such as for retirement, medical care, and unemployment; debt instruments for the social funds are not sold at public auctions
country comparison to the world: 29

Fiscal year: 6 April - 5 April

Inflation rate (consumer prices): 2.7% (2017 est.)
0.7% (2016 est.)
country comparison to the world: 126

Current account balance: -$121.921 billion (2019 est.)
-$104.927 billion (2018 est.)
country comparison to the world: 205

Exports: $441.2 billion (2017 est.)
$407.3 billion (2016 est.)
country comparison to the world: 10

Exports—partners: US 13.2%, Germany 10.5%, France 7.4%, Netherlands 6.2%, Ireland 5.6%, China 4.8%, Switzerland 4.5% (2017)

Exports—commodities: manufactured goods, fuels, chemicals; food, beverages, tobacco

Imports: $615.9 billion (2017 est.)
$591 billion (2016 est.)
country comparison to the world: 5

Imports—commodities: manufactured goods, machinery, fuels; foodstuffs

Imports—partners: Germany 13.7%, US 9.5%, China 9.3%, Netherlands 8%, France 5.4%, Belgium 5% (2017)

Reserves of foreign exchange and gold: $150.8 billion (31 December 2017 est.)
$129.6 billion (31 December 2015 est.)
country comparison to the world: 17

Debt—external: $8.126 trillion (31 March 2016 est.)
$8.642 trillion (31 March 2015 est.)
country comparison to the world: 2

Exchange rates: British pounds (GBP) per US dollar—
0.7836 (2017 est.)
0.738 (2016 est.)
0.738 (2015 est.)
0.607 (2014 est.)
0.6391 (2013 est.)

Electricity access: *electrification—total population:* 100% (2020)

Electricity—production: 318.2 billion kWh (2016 est.)
country comparison to the world: 12

Electricity—consumption: 309.2 billion kWh (2016 est.)
country comparison to the world: 11

Electricity—exports: 2.153 billion kWh (2016 est.)
country comparison to the world: 45

Electricity—imports: 19.7 billion kWh (2016 est.)
country comparison to the world: 12

Electricity—installed generating capacity: 97.06 million kW (2016 est.)
country comparison to the world: 13

Electricity—from fossil fuels: 50% of total installed capacity (2016 est.)
country comparison to the world: 152

Electricity—from nuclear fuels: 9% of total installed capacity (2017 est.)
country comparison to the world: 17

Electricity—from hydroelectric plants: 2% of total installed capacity (2017 est.)
country comparison to the world: 143

Electricity—from other renewable sources: 39% of total installed capacity (2017 est.)
country comparison to the world: 7

Crude oil—production: 1 million bbl/day (2018 est.)
country comparison to the world: 20

Crude oil—exports: 710,600 bbl/day (2017 est.)
country comparison to the world: 20

Crude oil—imports: 907,100 bbl/day (2017 est.)
country comparison to the world: 11

Crude oil—proved reserves: 2.069 billion bbl (1 January 2018 est.)
country comparison to the world: 33

Refined petroleum products—production: 1.29 million bbl/day (2017 est.)
country comparison to the world: 16

Refined petroleum products—consumption: 1.584 million bbl/day (2017 est.)
country comparison to the world: 15

Refined petroleum products— exports: 613,800 bbl/day (2017 est.)
country comparison to the world: 14

Refined petroleum products—imports: 907,500 bbl/day (2017 est.)
country comparison to the world: 7

Natural gas—production: 42.11 billion cu m (2017 est.)
country comparison to the world: 19

Natural gas—consumption: 79.17 billion cu m (2017 est.)
country comparison to the world: 10

Natural gas—exports: 11.27 billion cu m (2017 est.)
country comparison to the world: 19

Natural gas—imports: 47 billion cu m (2017 est.) *country comparison to the world:* 11

Natural gas—proved reserves: 176 billion cu m (1 January 2018 est.) *country comparison to the world:* 46

Carbon dioxide emissions from consumption of energy: 424 million Mt (2017 est.) *country comparison to the world:* 16

COMMUNICATIONS

Telephones—fixed lines: *total subscriptions:* 31,160,866

subscriptions per 100 inhabitants: 47.62 (2019 est.) *country comparison to the world:* 8

Telephones—mobile cellular: total subscriptions: 76,920,618

subscriptions per 100 inhabitants: 117.55 (2019 est.) *country comparison to the world:* 21

Telecommunication systems: *general assessment:* technologically advanced domestic and international system; one of the largest mobile and telecom markets in Europe for revenue and subscribers; will complete the switch to fiber by 2033; mobile penetration above the EU average; govt. to invest in fiber infrastructure and 5G technologies; operators expanded the reach of 5G services; FttP provided to over a million customers; super-fast broadband available to about 95% of customers (2020)

domestic: equal mix of buried cables, microwave radio relay, and fiber-optic systems, fixed line 48 per 100 and mobilecellular 118 per 100 (2019)

international: country code - 44; Landing points for the GTT Atlantic, Scotland-Northern Ireland -1, & -2, Lanis 1,-2, &-3, Sirius North, BT-MT-1, SHEFA-2, BT Highlands and Islands Submarine Cable System, Northern Lights, FARICE-1, Celtic Norse, Tampnet Offshore FOC Network, England Cable, CC-2, E-LLan, Sirius South, ESAT -1 & -2, Rockabill, Geo-Eirgrid, UK-Netherlands-14, Circle North & South, Ulysses2, Conceto, Farland North, Pan European Crossing, Solas, Swansea- Bream, GTT Express, Tata TGN-Atlantic & -Western Europe, Apollo, EIG, Glo-1, TAT-14, Yellow, Celtic, FLAG Atlantic-1, FEA, Isle of Scilly Cable, UK-Channel Islands-8 and SeaMeWe-3 submarine cables providing links throughout Europe, Asia, Africa, the Middle East, Southeast Asia, Australia, and US; satellite earth stations - 10 Intelsat (7 Atlantic Ocean and 3 Indian Ocean), 1 Inmarsat (Atlantic Ocean region), and 1 Eutelsat; at least 8 large international switching centers (2018)

note: the COVID-19 outbreak is negatively impacting telecommunications production and supply chains globally; consumer spending on telecom devices and services has also slowed due to the pandemic's effect on economies worldwide; overall progress towards improvements in all facets of the telecom industry - mobile, fixed-line, broadband, submarine cable and satellite - has moderated

Broadcast media: public service broadcaster, British Broadcasting Corporation (BBC), is the largest broadcasting corporation in the world; BBC operates multiple TV networks with regional and local TV service; a mixed system of public and commercial TV broadcasters along with satellite and cable systems provide access to hundreds of TV stations throughout the world; BBC operates multiple national, regional, and local radio networks with multiple transmission sites; a large number of commercial radio stations, as well as satellite radio services are available (2018)

Internet country code: .uk

Internet users: *total:* 61,784,878

percent of population: 94.9% (July 2018 est.) *country comparison to the world:* 13

Broadband—fixed subscriptions: *total:* 26,586,110

subscriptions per 100 inhabitants: 41 (2018 est.) *country comparison to the world:* 8

Communications—note: note 1: the British Library claims to be the largest library in the world with well over 150 million items and in most known languages; it receives copies of all books produced in the UK or Ireland, as well as a significant proportion of overseas titles distributed in the UK; in addition to books (print and digital), holdings include: journals, manuscripts, newspapers, magazines, sound and music recordings, videos, maps, prints, patents, and drawings

note 2: on 1 May 1840, the United Kingdom led the world with the introduction of postage stamps; the Austrian Empire had examined the idea of an "adhesive tax postmark" for the prepayment of postage in 1835; while the suggestion was reviewed in detail, it was rejected for the time being; other countries (including Austria) soon followed the UK's example with their own postage stamps; by the 1860s, most countries were issuing stamps; originally, stamps had to be cut from sheets; the UK issued the first postage stamps with perforations in 1854

TRANSPORTATION

National air transport system: *number of registered air carriers:* 20 (2020)

inventory of registered aircraft operated by air carriers: 794

annual passenger traffic on registered air carriers: 165,388,610 (2018)

annual freight traffic on registered air carriers: 6,198,370,000 mt-km (2018)

Civil aircraft registration country code prefix: G (2016)

Airports: 460 (2013) *country comparison to the world:* 16

Airports—with paved runways: *total:* 271 (2013) *over 3,047 m:* 7 (2013) *2,438 to 3,047 m:* 29 (2013) *1,524 to 2,437 m:* 89 (2013) *914 to 1,523 m:* 80 (2013) *under 914 m:* 66 (2013)

Airports—with unpaved runways: *total:* 189 (2013) *1,524 to 2,437 m:* 3 (2013) *914 to 1,523 m:* 26 (2013) *under 914 m:* 160 (2013)

Heliports: 9 (2013)

Pipelines: 502 km condensate, 9 km condensate/gas, 28603 km gas, 59 km liquid petroleum gas, 5256 km oil, 175 km oil/gas/water, 4919 km refined products, 255 km water (2013)

Railways: *total:* 16,837 km (2015) *standard gauge:* 16,534 km 1.435-m gauge (5,357 km electrified) (2015) *broad gauge:* 303 km 1.600-m gauge (in Northern Ireland) (2015) *country comparison to the world:* 17

Roadways: *total:* 394,428 km (2009) *paved:* 394,428 km (includes 3,519 km of expressways) (2009) *country comparison to the world:* 19

Waterways: 3,200 km (620 km used for commerce) (2009) *country comparison to the world:* 31

Merchant marine: *total:* 1,426 *by type:* bulk carrier 143, container ship 108, general cargo 125, oil tanker 137, other 913 (2019) *country comparison to the world:* 19

Ports and terminals: *major seaport(s):* Dover, Felixstowe, Immingham, Liverpool, London, Southampton, Teesport (England); Forth Ports (Scotland); Milford Haven (Wales)

oil terminal(s): Fawley Marine terminal, Liverpool Bay terminal (England); Braefoot Bay terminal, Finnart oil terminal, Hound Point terminal (Scotland)

container port(s) (TEUs): Felixstowe (3,849,700), London (2,431,000), Southampton (2,040,000) (2017)

LNG terminal(s) (import): Isle of Grain, Milford Haven, Teesside

Transportation—note: begun in 1988 and completed in 1994, the Channel Tunnel (nicknamed the Chunnel) is a 50.5-km (31.4-mi) rail tunnel beneath the English Channel at the Strait of Dover that runs from Folkestone, Kent, England to Coquelles, Pas-de-Calais in northern France; it is the only fixed link between the island of Great Britain and mainland Europe

MILITARY AND SECURITY

Military and security forces: British Army, Royal Navy (includes Royal Marines), Royal Air Force (2019)

Military expenditures: 2.14% of GDP (est) (2019 est.)
2.13% of GDP (2018)
2.11% of GDP (2017)
2.11% of GDP (2016)
2.05% of GDP (2015)
country comparison to the world: 45

Military and security service personnel strengths: the British military has approximately 149,000

total active duty troops (83,000 Army; 33,000 Navy, including 7,000 Marines; 33,000 Air Force) (April 2020)

Military equipment inventories and acquisitions: the inventory of the British military is comprised of a mix of domestically-produced and imported Western weapons systems; the US is the leading supplier of armaments to the UK since 2010; the UK defense industry is capable of producing a wide variety of air, land, and sea weapons systems (2019 est.)

Military deployments: 950 Afghanistan (NATO); approx. 1,000 Brunei; more than 400 Canada (BATUS); est. 2,200 Cyprus; 250 Cyprus (UNFICYP); 900 Estonia (NATO); approx. 1,200 Falkland Islands; est. 200 Germany (note - previously about 2,500, but the UK pledged to remove all but 200 troops by 2020); 570 Gibraltar; approx. 1,300 Middle East (coalition against ISIS; NATO); up to 350 Kenya (BATUK); 100 Mali (EUTM, MINUSMA, and Operation Barkhane; note - the UK has pledged to send an additional 250 troops to the MINUSMA mission for three years beginning in 2020); 150 Poland (NATO) (2020)

Military service age and obligation: slight variations by service, but generally 16-36 years of age for enlisted (with parental consent under 18) and 18-29 for officers; minimum length of service 4 years; women serve in military services including ground combat roles (2019)

Terrorist group(s): Continuity Irish Republican Army; Islamic State of Iraq and ash-Sham; New Irish Republican Army (2019)

note: details about the history, aims, leadership, organization, areas of operation, tactics, targets, weapons, size, and sources of support of the group(s) appear(s) in Appendix-T

Disputes—international: in 2002, Gibraltar residents voted overwhelmingly by referendum to reject any "shared sovereignty" arrangement between the UK and Spain; the Government of Gibraltar insisted on equal participation in talks between the two countries; Spain disapproved of UK plans to grant Gibraltar greater autonomy; Mauritius and Seychelles claim the Chagos Archipelago (British Indian Ocean Territory); in 2001, the former inhabitants of the archipelago, evicted 1967 - 1973, were granted UK citizenship and the right of return, followed by Orders in Council in 2004 that banned rehabitation, a High Court ruling reversed the ban, a Court of Appeal refusal to hear the case, and a Law Lords' decision in 2008 denied the right of return; in addition, the UK created the world's largest marine protection area around the Chagos islands prohibiting the extraction of any natural resources therein; UK rejects sovereignty talks requested by Argentina, which still claims the Falkland Islands (Islas Malvinas) and South Georgia and the South Sandwich Islands; territorial claim in Antarctica (British Antarctic Territory) overlaps Argentine claim and partially overlaps Chilean claim; Iceland, the UK, and Ireland dispute Denmark's claim that the Faroe Islands' continental shelf extends beyond 200 nm

Refugees and internally displaced persons: *refugees (country of origin):* 19,744 (Iran), 13,755 (Eritrea), 10,575 (Sudan), 10,389 (Syria), 9,513 (Afghanistan), 8,164 (Pakistan), 5,522 (Sri Lanka) (2019)

stateless persons: 161 (2019)

Illicit drugs: producer of limited amounts of synthetic drugs and synthetic precursor chemicals; major consumer of Southwest Asian heroin, Latin American cocaine, and synthetic drugs; money-laundering center

UNITED STATES

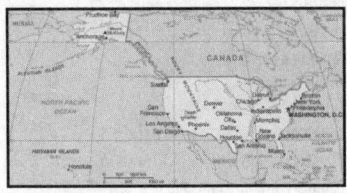

INTRODUCTION

Background: Britain's American colonies broke with the mother country in 1776 and were recognized as the new nation of the United States of America following the Treaty of Paris in 1783. During the 19th and 20th centuries, 37 new states were added to the original 13 as the nation expanded across the North American continent and acquired a number of overseas possessions. The two most traumatic experiences in the nation's history were the Civil War (1861-65), in which a northern Union of states defeated a secessionist Confederacy of 11 southern slave states, and the Great Depression of the 1930s, an economic downturn during which about a quarter of the labor force lost its jobs. Buoyed by victories in World Wars I and II and the end of the Cold War in 1991, the US remains the world's most powerful nation state. Since the end of World War II, the economy has achieved relatively steady growth, low unemployment and inflation, and rapid advances in technology.

GEOGRAPHY

Location: North America, bordering both the North Atlantic Ocean and the North Pacific Ocean, between Canada and Mexico

Geographic coordinates: 38 00 N, 97 00 W

Map references: North America

Area: *total:* 9,833,517 sq km
land: 9,147,593 sq km
water: 685,924 sq km
note: includes only the 50 states and District of Columbia, no overseas territories

country comparison to the world: 4

Area—comparative: about half the size of Russia; about three-tenths the size of Africa; about half the size of South America (or slightly larger than Brazil); slightly larger than China; more than twice the size of the European Union

Land boundaries: *total:* 12,048 km
border countries (2): Canada 8893 km (including 2477 km with Alaska), Mexico 3155 km
note: US Naval Base at Guantanamo Bay, Cuba is leased by the US and is part of Cuba; the base boundary is 28.5 km

Coastline: 19,924 km

Maritime claims: territorial sea: 12 nm

exclusive economic zone: 200 nm

contiguous zone: 24 nm

continental shelf: not specified

Climate: mostly temperate, but tropical in Hawaii and Florida, arctic in Alaska, semiarid in the great plains west of the Mississippi River, and arid in the Great Basin of the southwest; low winter temperatures in the northwest are ameliorated occasionally in January and February by warm chinook winds from the eastern slopes of the Rocky Mountains

Terrain: vast central plain, mountains in west, hills and low mountains in east; rugged mountains and broad river valleys in Alaska; rugged, volcanic topography in Hawaii

Elevation: *mean elevation:* 760 m
lowest point: Death Valley (lowest point in North America) -86 m
highest point: Denali 6,190 m (Mount McKinley) (highest point in North America)
note: the peak of Mauna Kea (4,207 m above sea level) on the island of Hawaii rises about 10,200 m above the Pacific Ocean floor; by this measurement, it is the world's tallest mountain - higher than Mount Everest (8,850 m), which is recognized as the tallest mountain above sea level

Natural resources: coal, copper, lead, molybdenum, phosphates, rare earth elements, uranium, bauxite, gold, iron, mercury, nickel, potash, silver, tungsten, zinc, petroleum, natural gas, timber, arable land, note, the US has the world's largest coal reserves with 491 billion short tons accounting for 27% of the world's total

Land use: *agricultural land:* 44.5% (2011 est.)

arable land: 16.8% (2011 est.) / *permanent crops:* 0.3% (2011 est.) / *permanent pasture:* 27.4% (2011 est.)

forest: 33.3% (2011 est.)

other: 22.2% (2011 est.)

Irrigated land: 264,000 sq km (2012)

Population distribution: large urban clusters are spread throughout the eastern half of the US (particularly the Great Lakes area, northeast, east, and southeast) and the western tier states; mountainous areas, principally the Rocky Mountains and Appalachian chain, deserts in the southwest, the dense boreal forests in the extreme north, and the central prarie states are less densely populated; Alaska's population is concentrated along its southern coast - with particular emphasis on the city of Anchorage - and Hawaii's is centered on the island of Oahu

Natural hazards: tsunamis; volcanoes; earthquake activity around Pacific Basin; hurricanes along the Atlantic and Gulf of Mexico coasts; tornadoes in the Midwest and Southeast; mud slides in California; forest fires in the west; flooding; permafrost in northern Alaska, a major impediment to development

volcanism: volcanic activity in the Hawaiian Islands, Western Alaska, the Pacific Northwest, and in the Northern Mariana Islands; both Mauna Loa (4,170 m) in Hawaii and Mount Rainier (4,392 m) in Washington have been deemed Decade Volcanoes by the International Association of Volcanology and Chemistry of the Earth's Interior, worthy of study due to their explosive history and close proximity to human populations; Pavlof (2,519 m) is the most active volcano in Alaska's Aleutian Arc and poses a significant threat to air travel since the area constitutes a major flight path between North America and East Asia; St. Helens (2,549 m), famous for the devastating 1980 eruption, remains active today; numerous other historically active volcanoes exist, mostly concentrated in the Aleutian arc and Hawaii; they include: in Alaska: Aniakchak, Augustine, Chiginagak, Fourpeaked, Iliamna, Katmai, Kupreanof, Martin, Novarupta, Redoubt, Spurr, Wrangell, Trident, Ugashik-Peulik, Ukinrek Maars, Veniaminof; in Hawaii: Haleakala, Kilauea, Loihi; in the Northern Mariana Islands: Anatahan; and in the Pacific Northwest: Mount Baker, Mount Hood; see note 2 under "Geography - note"

Environment—current issues: air pollution; large emitter of carbon dioxide from the burning of fossil fuels; water pollution from runoff of pesticides and fertilizers; limited natural freshwater resources in much of the western part of the country require careful management; deforestation; mining; desertification; species conservation; invasive species (the Hawaiian Islands are particularly vulnerable)

Environment—international agreements: *party to:* Air Pollution, Air Pollution-Nitrogen Oxides, Antarctic-Environmental Protocol, Antarctic-Marine Living Resources, Antarctic Seals, Antarctic Treaty, Climate Change, Desertification, Endangered Species, Environmental Modification, Marine Dumping, Marine Life Conservation, Ozone Layer Protection, Ship Pollution, Tropical Timber 83, Tropical Timber 94, Wetlands, Whaling

signed, but not ratified: Air Pollution-Persistent Organic Pollutants, Air Pollution-Volatile Organic Compounds, Biodiversity, Climate Change-Kyoto Protocol, Hazardous Wastes

Geography—note: *note 1:* world's third-largest country by size (after Russia and Canada) and by population (after China and India); Denali (Mt. McKinley) is the highest point in North America and Death Valley the lowest point on the continent

note 2: the western coast of the United States and southern coast of Alaska lie along the Ring of Fire, a belt of active volcanoes and earthquake epicenters bordering the Pacific Ocean; up to 90% of the world's earthquakes and some 75% of the world's volcanoes occur within the Ring of Fire

note 3: the Aleutian Islands are a chain of volcanic islands that divide the Bering Sea (north) from the main Pacific Ocean (south); they extend about 1,800 km westward from the Alaskan Peninsula; the archipelago consists of 14 larger islands, 55 smaller islands, and hundreds of islets; there are 41 active volcanoes on the islands, which together form a large northern section of the Ring of Fire

note 4: Mammoth Cave, in west-central Kentucky, is the world's longest known cave system with more than 650 km (405 miles) of surveyed passageways, which is nearly twice as long as the second-longest cave system, the Sac Actun underwater cave in Mexico - the world's longest underwater cave system (see "Geography - note" under Mexico);

note 5: Kazumura Cave on the island of Hawaii is the world's longest and deepest lava tube cave; it has been surveyed at 66 km (41 mi) long and 1,102 m (3,614 ft) deep

note 6: Bracken Cave outside of San Antonio, Texas is the world's largest bat cave; it is the summer home to the largest colony of bats in the world; an estimated 20 million Mexican free-tailed bats roost in the cave from March to October making it the world's largest known concentration of mammals

PEOPLE AND SOCIETY

Population: 332,639,102 (July 2020 est.)

country comparison to the world: 3

Nationality: *noun:* American(s)

adjective: American

Ethnic groups: white 72.4%, black 12.6%, Asian 4.8%, Amerindian and Alaska native 0.9%, native Hawaiian and other Pacific islander 0.2%, other 6.2%, two or more races 2.9% (2010 est.)

note: a separate listing for Hispanic is not included because the US Census Bureau considers Hispanic to mean persons of Spanish/Hispanic/Latino origin including those of Mexican, Cuban, Puerto Rican, Dominican Republic, Spanish, and Central or South American origin living in the US who may be of any race or ethnic group (white, black, Asian, etc.); an estimated 16.3% of the total US population is Hispanic as of 2010

Languages: English only 78.2%, Spanish 13.4%, Chinese 1.1%, other 7.3% (2017 est.)

note: data represent the language spoken at home; the US has no official national language, but English has acquired official status in 32 of the 50 states; Hawaiian is an official language in the state of Hawaii, and 20 indigenous languages are official in Alaska

Religions: Protestant 46.5%, Roman Catholic 20.8%, Jewish 1.9%, Mormon 1.6%, other Christian 0.9%, Muslim 0.9%, Jehovah's Witness 0.8%, Buddhist 0.7%, Hindu 0.7%, other 1.8%, unaffiliated 22.8%, don't know/refused 0.6% (2014 est.)

Age structure: *0-14 years:* 18.46% (male 31,374,555/female 30,034,371)

15-24 years: 12.91% (male 21,931,368/female 21,006,463)

25-54 years: 38.92% (male 64,893,670/female 64,564,565)

55-64 years: 12.86% (male 20,690,736/female 22,091,808)

65 years and over: 16.85% (male 25,014,147/ female 31,037,419) (2020 est.)

population pyramid: [INSERT IMAGE: UNITED STATES-population pyramid]

Dependency ratios: *total dependency ratio:* 53.9

youth dependency ratio: 28.3

elderly dependency ratio: 25.6

potential support ratio: 3.9 (2020 est.)

Median age: *total:* 38.5 years

male: 37.2 years

female: 39.8 years (2020 est.)

country comparison to the world: 61

Population growth rate: 0.72% (2020 est.)

country comparison to the world: 137

Birth rate: 12.4 births/1,000 population (2020 est.)

country comparison to the world: 157

Death rate: 8.3 deaths/1,000 population (2020 est.)

country comparison to the world: 83

Net migration rate: 3 migrant(s)/1,000 population (2020 est.)

country comparison to the world: 37

Population distribution: large urban clusters are spread throughout the eastern half of the US (particularly the Great Lakes area, northeast, east, and southeast) and the western tier states; mountainous areas, principally the Rocky Mountains and Appalachian chain, deserts in the southwest, the dense boreal forests in the extreme north, and the central prarie states are less densely populated; Alaska's population is concentrated along its southern coast - with particular emphasis on the city of Anchorage - and Hawaii's is centered on the island of Oahu

Urbanization: *urban population:* 82.7% of total population (2020)

rate of urbanization: 0.95% annual rate of change (2015-20 est.)

total population growth rate v. urban population growth rate, 2000-2030: *Major urban areas - population:* 18.804 million New York-Newark, 12.447 million Los Angeles-Long Beach-Santa Ana, 8.865 million Chicago, 6.371 million Houston, 6.301 million Dallas-Fort Worth, 5.322 million WASHINGTON, D.C. (capital) (2020)

Sex ratio: *at birth:* 1.05 male(s)/female NA
0-14 years: 1.04 male(s)/female
15-24 years: 1.04 male(s)/female
25-54 years: 1.01 male(s)/female
55-64 years: 0.94 male(s)/female
65 years and over: 0.81 male(s)/female
total population: 0.97 male(s)/female (2020 est.)

Mother's mean age at first birth: 26.4 years (2015 est.)

Maternal mortality rate: 19 deaths/100,000 live births (2017 est.)

country comparison to the world: 129

Infant mortality rate: *total:* 5.3 deaths/1,000 live births
male: 5.7 deaths/1,000 live births
female: 4.9 deaths/1,000 live births (2020 est.)
country comparison to the world: 174

Life expectancy at birth: *total population:* 80.3 years
male: 78 years
female: 82.5 years (2020 est.)
country comparison to the world: 45

Total fertility rate: 1.84 children born/woman (2020 est.)

country comparison to the world: 142

Contraceptive prevalence rate: 75.9% (2015/17)

Drinking water source:

improved: *urban:* 100% of population
rural: 97% of population
total: 99% of population

unimproved: *urban:* 0% of population
rural: 3% of population
total: 1% of population (2017 est.)

Current Health Expenditure: 17.1% (2017)

Physicians density: 2.61 physicians/1,000 population (2017)

Sanitation facility access:

improved: *urban:* 100% of population
rural: 100% of population
total: 100% of population

unimproved: *urban:* 0% of population
rural: 0% of population
total: 0% of population (2017 est.)

HIV/AIDS—adult prevalence rate: NA

HIV/AIDS—people living with HIV/AIDS: NA

HIV/AIDS—deaths: NA

Obesity—adult prevalence rate: 36.2% (2016)

country comparison to the world: 12

Children under the age of 5 years underweight: 0.5% (2012)

country comparison to the world: 131

Education expenditures: 5% of GDP (2014)

country comparison to the world: 64

School life expectancy (primary to tertiary education): *total:* 16 years
male: 16 years
female: 17 years (2018)

Unemployment, youth ages 15-24: *total:* 8.6%
male: 9.5%
female: 7.7% (2018 est.)
country comparison to the world: 140

GOVERNMENT

Country name: *conventional long form:* United States of America
conventional short form: United States
abbreviation: US or USA
etymology: the name America is derived from that of Amerigo VESPUCCI (1454-1512) - Italian explorer, navigator, and cartographer - using the Latin form of his name, Americus, feminized to America

Government type: constitutional federal republic

Capital: *name:* Washington, DC
geographic coordinates: 38 53 N, 77 02 W
time difference: UTC-5 (during Standard Time)
daylight saving time: +1hr, begins second Sunday in March; ends first Sunday in November
note: the 50 United States cover six time zones
etymology: named after George Washington (1732-1799), the first president of the United States

Administrative divisions: 50 states and 1 district*; Alabama, Alaska, Arizona, Arkansas, California, Colorado, Connecticut, Delaware, District of Columbia*, Florida, Georgia, Hawaii, Idaho, Illinois, Indiana, Iowa, Kansas, Kentucky, Louisiana, Maine, Maryland, Massachusetts, Michigan, Minnesota, Mississippi, Missouri, Montana, Nebraska, Nevada, New Hampshire, New Jersey, New Mexico, New York, North Carolina, North Dakota, Ohio, Oklahoma, Oregon, Pennsylvania, Rhode Island, South Carolina, South Dakota, Tennessee, Texas, Utah, Vermont, Virginia, Washington, West Virginia, Wisconsin, Wyoming

Dependent areas: American Samoa, Baker Island, Guam, Howland Island, Jarvis Island, Johnston Atoll, Kingman Reef, Midway Islands, Navassa Island, Northern Mariana Islands, Palmyra Atoll, Puerto Rico, Virgin Islands, Wake Island
note: from 18 July 1947 until 1 October 1994, the US administered the Trust Territory of the Pacific Islands; it entered into a political relationship with all four political entities: the Northern Mariana Islands is a commonwealth in political union with the US (effective 3 November 1986); the Republic of the Marshall Islands signed a Compact of Free Association with the US (effective 21 October 1986); the Federated States of Micronesia signed a Compact of Free Association with the US

(effective 3 November 1986); Palau concluded a Compact of Free Association with the US (effective 1 October 1994)

Independence: 4 July 1776 (declared independence from Great Britain); 3 September 1783 (recognized by Great Britain)

National holiday: Independence Day, 4 July (1776)

Constitution: *history:* previous 1781 (Articles of Confederation and Perpetual Union); latest drafted July - September 1787, submitted to the Congress of the Confederation 20 September 1787, submitted for states' ratification 28 September 1787, ratification completed by nine of the 13 states 21 June 1788, effective 4 March 1789

amendments: proposed as a "joint resolution" by Congress, which requires a two-thirds majority vote in both the House of Representatives and the Senate or by a constitutional convention called for by at least two thirds of the state legislatures; passage requires ratification by three fourths of the state legislatures or passage in state-held constitutional conventions as specified by Congress; the US president has no role in the constitutional amendment process; amended many times, last in 1992

Legal system: common law system based on English common law at the federal level; state legal systems based on common law, except Louisiana, where state law is based on Napoleonic civil code; judicial review of legislative acts

International law organization participation: withdrew acceptance of compulsory ICJ jurisdiction in 2005; withdrew acceptance of ICCt jurisdiction in 2002

Citizenship: *citizenship by birth:* yes

citizenship by descent only: yes

dual citizenship recognized: no, but the US government acknowledges such situations exist; US citizens are not encouraged to seek dual citizenship since it limits protection by the US

residency requirement for naturalization: 5 years

Suffrage: 18 years of age; universal

Executive branch: *chief of state:* President Donald J. TRUMP (since 20 January 2017); Vice President Michael R. PENCE (since 20 January 2017); note - the president is both chief of state and head of government
head of government: President Donald J. TRUMP (since 20 January 2017); Vice President Michael R. PENCE (since 20 January 2017)
cabinet: Cabinet appointed by the president, approved by the Senate
elections/appointments: president and vice president indirectly elected on the same ballot by the Electoral College of 'electors' chosen from each state; president and vice president serve a 4-year term (eligible for a second term); election last held on 8 November 2016 (next to be held on 3 November 2020)
election results: Donald J. TRUMP elected president; electoral vote - Donald J. TRUMP (Republican Party) 304, Hillary D. CLINTON (Democratic Party) 227, other 7; percent of direct

1021

popular vote - Hillary D. CLINTON 48.2%, Donald J. TRUMP 46.1%, other 5.7%

Legislative branch: *description:* bicameral Congress consists of: Senate (100 seats; 2 members directly elected in each of the 50 state constituencies by simple majority vote except in Georgia and Louisiana which require an absolute majority vote with a second round if needed; members serve 6-year terms with one-third of membership renewed every 2 years)

House of Representatives (435 seats; members directly elected in single-seat constituencies by simple majority vote except in Georgia which requires an absolute majority vote with a second round if needed; members serve 2-year terms)

elections: Senate—last held on 6 November 2018 (next to be held on 3 November 2020)

House of Representatives—last held on 6 November 2018 (next to be held on 3 November 2020)

election results: Senate—percent of vote by party—NA; seats by party—Republican Party 53, Democratic Party 45, independent 2; composition—men 75, women 25, percent of women 25%

House of Representatives—percent of vote by party—NA; seats by party—Democratic Party 234, Republican Party 200, 1 seat still undecided; composition—men 328, women 106, percent of women 24.4%; note—total US Congress percent of women 24.5%

note: in addition to the regular members of the House of Representatives there are 6 non-voting delegates elected from the District of Columbia and the US territories of American Samoa, Guam, Puerto Rico, the Northern Mariana Islands, and the Virgin Islands; these are single seat constituencies directly elected by simple majority vote to serve a 2-year term (except for the resident commissioner of Puerto Rico who serves a 4-year term); the delegate can vote when serving on a committee and when the House meets as the Committee of the Whole House, but not when legislation is submitted for a "full floor" House vote; election of delegates last held on 6 November 2018 (next to be held on 3 November 2020)

Judicial branch: *highest courts:* US Supreme Court (consists of 9 justices - the chief justice and 8 associate justices)

judge selection and term of office: president nominates and, with the advice and consent of the Senate, appoints Supreme Court justices; justices serve for life

subordinate courts: Courts of Appeal (includes the US Court of Appeal for the Federal District and 12 regional appeals courts); 94 federal district courts in 50 states and territories

note: the US court system consists of the federal court system and the state court systems; although each court system is responsible for hearing certain types of cases, neither is completely independent of the other, and the systems often interact

Political parties and leaders: Democratic Party [Tom PEREZ]

Green Party [collective leadership]

Libertarian Party [Nicholas SARWARK] Republican Party [Ronna Romney MCDANIEL]

International organization participation: ADB (nonregional member), AfDB (nonregional member), ANZUS, APEC, Arctic Council, ARF, ASEAN (dialogue partner), Australia Group, BIS, BSEC (observer), CBSS (observer), CD, CE (observer), CERN (observer), CICA (observer), CP, EAPC, EAS, EBRD, EITI (implementing country), FAO, FATF, G-5, G-7, G-8, G-10, G-20, IADB, IAEA, IBRD, ICAO, ICC (national committees), ICRM, IDA, IEA, IFAD, IFC, IFRCS, IGAD (partners), IHO, ILO, IMF, IMO, IMSO, Interpol, IOC, IOM, ISO, ITSO, ITU, ITUC (NGOs), MIGA, MINUSMA, MINUSTAH, MONUSCO, NAFTA, NATO, NEA, NSG, OAS, OECD, OPCW, OSCE, Pacific Alliance (observer), Paris Club, PCA, PIF (partner), SAARC (observer), SELEC (observer), SICA (observer), SPC, UN, UNCTAD, UNESCO, UNHCR, UNITAR, UNMIL, UNMISS, UNRWA, UN Security Council (permanent), UNTSO, UPU, WCO, WIPO, WMO, WTO, ZC

Flag description: 13 equal horizontal stripes of red (top and bottom) alternating with white; there is a blue rectangle in the upper hoist-side corner bearing 50 small, white, five-pointed stars arranged in nine offset horizontal rows of six stars (top and bottom) alternating with rows of five stars; the 50 stars represent the 50 states, the 13 stripes represent the 13 original colonies; blue stands for loyalty, devotion, truth, justice, and friendship, red symbolizes courage, zeal, and fervency, while white denotes purity and rectitude of conduct; commonly referred to by its nickname of Old Glory

note: the design and colors have been the basis for a number of other flags, including Chile, Liberia, Malaysia, and Puerto Rico

National symbol(s): bald eagle; national colors: red, white, blue

National anthem: *name:* The Star-Spangled Banner

lyrics/music: Francis Scott KEY/John Stafford SMITH

note: adopted 1931; during the War of 1812, after witnessing the successful American defense of Fort McHenry in Baltimore following British naval bombardment, Francis Scott KEY wrote the lyrics to what would become the national anthem; the lyrics were set to the tune of "The Anacreontic Song"; only the first verse is sung

0:00 / 1:18

ECONOMY

Economy—overview: The US has the most technologically powerful economy in the world, with a per capita GDP of $59,500. US firms are at or near the forefront in technological advances, especially in computers, pharmaceuticals, and medical, aerospace, and military equipment; however, their advantage has narrowed since the end of World War II. Based on a comparison of GDP measured at purchasing power parity conversion rates, the US economy in 2014, having stood as the largest

in the world for more than a century, slipped into second place behind China, which has more than tripled the US growth rate for each year of the past four decades.

In the US, private individuals and business firms make most of the decisions, and the federal and state governments buy needed goods and services predominantly in the private marketplace. US business firms enjoy greater flexibility than their counterparts in Western Europe and Japan in decisions to expand capital plant, to lay off surplus workers, and to develop new products. At the same time, businesses face higher barriers to enter their rivals' home markets than foreign firms face entering US markets.

Long-term problems for the US include stagnation of wages for lower-income families, inadequate investment in deteriorating infrastructure, rapidly rising medical and pension costs of an aging population, energy shortages, and sizable current account and budget deficits.

The onrush of technology has been a driving factor in the gradual development of a "two-tier" labor market in which those at the bottom lack the education and the professional/technical skills of those at the top and, more and more, fail to get comparable pay raises, health insurance coverage, and other benefits. But the globalization of trade, and especially the rise of low-wage producers such as China, has put additional downward pressure on wages and upward pressure on the return to capital. Since 1975, practically all the gains in household income have gone to the top 20% of households. Since 1996, dividends and capital gains have grown faster than wages or any other category of after-tax income.

Imported oil accounts for more than 50% of US consumption and oil has a major impact on the overall health of the economy. Crude oil prices doubled between 2001 and 2006, the year home prices peaked; higher gasoline prices ate into consumers' budgets and many individuals fell behind in their mortgage payments. Oil prices climbed another 50% between 2006 and 2008, and bank foreclosures more than doubled in the same period. Besides dampening the housing market, soaring oil prices caused a drop in the value of the dollar and a deterioration in the US merchandise trade deficit, which peaked at $840 billion in 2008. Because the US economy is energy-intensive, falling oil prices since 2013 have alleviated many of the problems the earlier increases had created.

The sub-prime mortgage crisis, falling home prices, investment bank failures, tight credit, and the global economic downturn pushed the US into a recession by mid-2008. GDP contracted until the third quarter of 2009, the deepest and longest downturn since the Great Depression. To help stabilize financial markets, the US Congress established a $700 billion Troubled Asset Relief Program in October 2008. The government used some of these funds to purchase equity in US banks and industrial corporations, much of which had been returned to the government by early 2011. In January 2009, Congress passed and former President Barack OBAMA signed a bill providing an additional $787 billion fiscal stimulus to be used over 10 years - two-thirds on additional spending and one-third on tax cuts - to create jobs and to help the economy recover. In 2010 and 2011, the federal budget deficit reached nearly 9% of GDP. In 2012, the Federal Government reduced the growth of spending and the deficit shrank to 7.6% of GDP. US revenues from taxes and other sources are lower, as a percentage of GDP, than those of most other countries.

Wars in Iraq and Afghanistan required major shifts in national resources from civilian to military purposes and contributed to the growth of the budget deficit and public debt. Through FY 2018, the direct costs of the wars will have totaled more than $1.9 trillion, according to US Government figures.

In March 2010, former President OBAMA signed into law the Patient Protection and Affordable Care Act (ACA), a health insurance reform that was designed to extend coverage to an additional 32 million Americans by 2016, through private health insurance for the general population and Medicaid for the impoverished. Total spending on healthcare - public plus private - rose from 9.0% of GDP in 1980 to 17.9% in 2010.

In July 2010, the former president signed the DODD-FRANK Wall Street Reform and Consumer Protection Act, a law designed to promote financial stability by protecting consumers from financial abuses, ending taxpayer bailouts of financial firms, dealing with troubled banks that are "too big to fail," and improving accountability and transparency in the financial system - in particular, by requiring certain financial derivatives to be traded in markets that are subject to government regulation and oversight.

The Federal Reserve Board (Fed) announced plans in December 2012 to purchase $85 billion per month of mortgage- backed and Treasury securities in an effort to hold down long-term interest rates, and to keep short-term rates near zero until unemployment dropped below 6.5% or inflation rose above 2.5%. The Fed ended its purchases during the summer of 2014, after the unemployment rate dropped to 6.2%, inflation stood at 1.7%, and public debt fell below 74% of GDP. In December 2015, the Fed raised its target for the benchmark federal funds rate by 0.25%, the first increase since the recession began. With continued low growth, the Fed opted to raise rates several times since then, and in December 2017, the target rate stood at 1.5%.

In December 2017, Congress passed and President Donald TRUMP signed the Tax Cuts and Jobs Act, which, among its various provisions, reduces the corporate tax rate from 35% to 21%; lowers the individual tax rate for those with the highest incomes from 39.6% to 37%, and by lesser percentages for those at lower income levels; changes many deductions and credits used to calculate taxable income; and eliminates in 2019 the penalty imposed on taxpayers who do not obtain the minimum amount of health insurance required under the ACA. The new taxes took effect on 1 January 2018; the tax cut for corporations are permanent, but those for individuals are scheduled to expire after 2025. The Joint Committee on Taxation (JCT) under the Congressional Budget Office estimates that the new law will reduce tax revenues and increase the federal deficit by about $1.45 trillion over the 2018-2027 period. This amount would decline if economic growth were to exceed the JCT's estimate.

GDP (purchasing power parity): $19.49 trillion (2017 est.)
$19.06 trillion (2016 est.)
$18.77 trillion (2015 est.)
note: data are in 2017 dollars

country comparison to the world: 2

GDP (official exchange rate): $19.49 trillion (2017 est.)

GDP—real growth rate: 2.16% (2019 est.)
3% (2018 est.)
2.33% (2017 est.)

country comparison to the world: 130

GDP—per capita (PPP): $59,800 (2017 est.)
$58,900 (2016 est.)
$58,400 (2015 est.)
note: data are in 2017 dollars

country comparison to the world: 19

Gross national saving: 18.9% of GDP (2017 est.)
18.6% of GDP (2016 est.)
20.1% of GDP (2015 est.)

country comparison to the world: 106

GDP—composition, by end use: *household consumption:* 68.4% (2017 est.)
government consumption: 17.3% (2017 est.)
investment in fixed capital: 17.2% (2017 est.)
investment in inventories: 0.1% (2017 est.)
exports of goods and services: 12.1% (2017 est.)
imports of goods and services: -15% (2017 est.)

GDP—composition, by sector of origin: *agriculture:* 0.9% (2017 est.)
industry: 19.1% (2017 est.)
services: 80% (2017 est.)

Agriculture—products: wheat, corn, other grains, fruits, vegetables, cotton; beef, pork, poultry, dairy products; fish; forest products

Industries: highly diversified, world leading, high-technology innovator, second-largest industrial output in the world; petroleum, steel, motor vehicles, aerospace, telecommunications, chemicals, electronics, food processing, consumer goods, lumber, mining

Industrial production growth rate: 2.3% (2017 est.)

country comparison to the world: 122

Labor force: 35.412 million (2020 est.)
note: includes unemployed

country comparison to the world: 16

Labor force—by occupation: *agriculture:* 0.7% (2009)
industry: 20.3% (2009)
services: 37.3% (2009)
industry and services: 24.2% (2009)

manufacturing: 17.6% (2009)

farming, forestry, and fishing: 0.7% (2009)

manufacturing, extraction, transportation, and crafts: 20.3% (2009)

managerial, professional, and technical: 37.3% (2009)

sales and office: 24.2% (2009)

other services: 17.6% (2009)
note: figures exclude the unemployed

Unemployment rate: 3.89% (2018 est.)
4.4% (2017 est.)

country comparison to the world: 57

Population below poverty line: 15.1% (2010 est.)

Household income or consumption by percentage share: *lowest 10%:* 2%

highest 10%: 30% (2007 est.)

Budget: *revenues:* 3.315 trillion (2017 est.)
expenditures: 3.981 trillion (2017 est.)
note: revenues exclude social contributions of approximately $1.0 trillion; expenditures exclude social benefits of approximately $2.3 trillion

Taxes and other revenues: 17% (of GDP) (2017 est.)
note: excludes contributions for social security and other programs; if social contributions were added, taxes and other revenues would amount to approximately 22% of GDP

country comparison to the world: 172

Budget surplus (+) or deficit (-): -3.4% (of GDP) (2017 est.)

country comparison to the world: 145

Public debt: 78.8% of GDP (2017 est.)
81.2% of GDP (2016 est.)
note: data cover only what the United States Treasury denotes as "Debt Held by the Public," which includes all debt instruments issued by the Treasury that are owned by non-US Government entities; the data include Treasury debt held by foreign entities; the data exclude debt issued by individual US states, as well as intragovernmental debt; intragovernmental debt consists of Treasury borrowings from surpluses in the trusts for Federal Social Security, Federal Employees, Hospital and Supplemental Medical Insurance (Medicare), Disability and Unemployment, and several other smaller trusts; if data for intragovernment debt were added, "gross debt" would increase by about one-third of GDP

country comparison to the world: 36

Fiscal year: 1 October–30 September

Inflation rate (consumer prices): 2.1% (2017 est.)
1.3% (2016 est.)

country comparison to the world: 110

Current account balance: -$480.225 billion (2019 est.)
-$449.694 billion (2018 est.)

country comparison to the world: 206

Exports: $1.553 trillion (2017 est.)
$1.456 trillion (2016 est.)

country comparison to the world: 2

Exports—partners: Canada 18.3%, Mexico 15.7%, China 8.4%, Japan 4.4% (2017)

Exports—commodities: agricultural products (soybeans, fruit, corn) 9.2%, industrial supplies (organic chemicals) 26.8%, capital goods (transistors, aircraft, motor vehicle parts, computers, telecommunications equipment) 49.0%, consumer goods (automobiles, medicines) 15.0% (2008 est.)

Imports: $2.361 trillion (2017 est.) $2.208 trillion (2016 est.)

country comparison to the world: 1

Imports—commodities: agricultural products 4.9%, industrial supplies 32.9% (crude oil 8.2%), capital goods 30.4% (computers, telecommunications equipment, motor vehicle parts, office machines, electric power machinery), consumer goods 31.8% (automobiles, clothing, medicines, furniture, toys) (2008 est.)

Imports—partners: China 21.6%, Mexico 13.4%, Canada 12.8%, Japan 5.8%, Germany 5% (2017)

Reserves of foreign exchange and gold: $123.3 billion (31 December 2017 est.) $117.6 billion (31 December 2015 est.)

country comparison to the world: 20

Debt—external: $17.91 trillion (31 March 2016 est.) $17.85 trillion (31 March 2015 est.)

note: approximately 4/5ths of US external debt is denominated in US dollars; foreign lenders have been willing to hold US dollar denominated debt instruments because they view the dollar as the world's reserve currency

country comparison to the world: 1

Exchange rates: *British pounds per US dollar:* 0.7836 (2017 est.), 0.730 (2016 est.), 0.738 (2015 est.), 0.607 (2014 est), 0.6391 (2013 est.)

Canadian dollars per US dollar: 1, 1.308 (2017 est.), 1.3256 (2016 est.), 1.3256 (2015 est.), 1.2788 (2014 est.), 1.0298 (2013 est.)

Chinese yuan per US dollar: 1, 6.7588 (2017 est.), 6.6445 (2016 est.), 6.2275 (2015 est.), 6.1434 (2014 est.), 6.1958 (2013 est.)

euros per US dollar: 0.885 (2017 est.), 0.903 (2016 est.), 0.9214 (2015 est.), 0.885 (2014 est.), 0.7634 (2013 est.)

Japanese yen per US dollar: 111.10 (2017 est.), 108.76 (2016 est.), 108.76 (2015 est.), 121.02 (2014 est.), 97.44 (2013 est.)

ENERGY

Electricity access: electrification—total population: 100% (2020)

Electricity—production: 4.095 trillion kWh (2016 est.)

country comparison to the world: 2

Electricity—consumption: 3.902 trillion kWh (2016 est.)

country comparison to the world: 2

Electricity—exports: 9.695 billion kWh (2016 est.)

country comparison to the world: 22

Electricity—imports: 72.72 billion kWh (2016 est.)

country comparison to the world: 1

Electricity—installed generating capacity: 1.087 billion kW (2016 est.)

country comparison to the world: 2

Electricity—from fossil fuels: 70% of total installed capacity (2016 est.)

country comparison to the world: 111

Electricity—from nuclear fuels: 9% of total installed capacity (2017 est.)

country comparison to the world: 18

Electricity—from hydroelectric plants: 7% of total installed capacity (2017 est.)

country comparison to the world: 128

Electricity—from other renewable sources: 14% of total installed capacity (2017 est.)

country comparison to the world: 65

Crude oil—production: 10.962 million bbl/day (2018 est.)

country comparison to the world: 1

Crude oil—exports: 1.158 million bbl/day (2017 est.)

country comparison to the world: 12

Crude oil—imports: 7.969 million bbl/day (2017 est.)

country comparison to the world: 1

Crude oil—proved reserves: NA bbl (1 January 2018 est.)

Refined petroleum products—production: 20.3 million bbl/day (2017 est.)

country comparison to the world: 1

Refined petroleum products—consumption: 19.96 million bbl/day (2017 est.)

country comparison to the world: 1

Refined petroleum products—exports: 5.218 million bbl/day (2017 est.)

country comparison to the world: 1

Refined petroleum products—imports: 2.175 million bbl/day (2017 est.)

country comparison to the world: 2

Natural gas—production: 772.8 billion cu m (2017 est.)

country comparison to the world: 1

Natural gas—consumption: 767.6 billion cu m (2017 est.)

country comparison to the world: 1

Natural gas—exports: 89.7 billion cu m (2017 est.)

country comparison to the world: 4

Natural gas—imports: 86.15 billion cu m (2017 est.)

country comparison to the world: 4

Natural gas—proved reserves: 0 cu m (1 January 2017 est.)

country comparison to the world: 203

Carbon dioxide emissions from consumption of energy: 5.242 billion Mt (2017 est.)

country comparison to the world: 2

COMMUNICATIONS

Telephones—fixed lines: total subscriptions: 107,667,642

subscriptions per 100 inhabitants: 32.6 (2019 est.)

country comparison to the world: 2

Telephones—mobile cellular: total subscriptions: 408,509,528

subscriptions per 100 inhabitants: 123.69 (2019 est.)

country comparison to the world: 3

Telecommunication systems: general assessment: a large, technologically advanced, multipurpose communications system; mobile subscriber penetration rate of about 129%; national LTE-M services, closes down 2G infrastructure and reassigns spectrum for 5G; FttP rather than FttN efforts (2020)

domestic: a large system of fiber-optic cable, microwave radio relay, coaxial cable, and domestic satellites carries every form of telephone traffic; a rapidly growing cellular system carries mobile telephone traffic throughout the country; fixed-line 33 per 100 and mobile-cellular 124 per 100 (2019)

international: country code—1; landing points for the Quintillion Subsea Cable Network, TERRA SW, AU-Aleutian, KKFL, AKORN, Alaska United -West, & -East & -Southeast, North Star, Lynn Canal Fiber, KetchCan 1, PC-1, SCCN, Tat TGN- Pacific & -Atlantic, Jupiter, Hawaiki, NCP, FASTER, HKA, JUS, AAG, BtoBE, Currie, Southern Cross NEXT, SxS, PLCN, Utility EAO Pacific, SEA-US, Paniolo Cable Network, HICS, HIFN, ASH, Telstra Endeavor, Honotua, AURORA, ARCOS, AMX-1, Americas -I & -II, Columbus IIb & -III, Maya-1, MAC, GTMO-1, BICS, CFX-1, GlobeNet, Monet, SAm-1, Bahamas 2, PCCS, BRUSA, Dunant, MAREA, SAE x1, TAT 14, Apollo, Gemini Bermuda, Havfrue/AEC-2, Seabras-1, WALL-LI, NYNJ-1, FLAG Atalantic-1, Yellow, Atlantic Crossing-1, AE Connect -1, sea2shore, Challenger Bermuda-1, and GTT Atlantic submarine cable systems providing international connectivity to Europe, Africa, the Middle East, Asia, Southeast Asia, Australia, New Zealand, Pacific, & Atlantic, and Indian Ocean Islands, Central and South America, Caribbean, Canada and US; satellite earth stations—61 Intelsat (45 Atlantic Ocean and 16 Pacific Ocean), 5 Intersputnik (Atlantic Ocean region), and 4 Inmarsat (Pacific and Atlantic Ocean regions) (2020)

note: the COVID-19 outbreak is negatively impacting telecommunications production and supply chains globally; consumer spending on telecom devices and services has also slowed due to the pandemic's effect on economies worldwide; overall progress towards improvements in all facets of the telecom industry—mobile, fixed-line, broadband, submarine cable and satellite—has moderated

Broadcast media: 4 major terrestrial TV networks with affiliate stations throughout the country, plus cable and satellite networks, independent stations, and a limited public broadcasting sector that is largely supported by private grants; overall, thousands of TV stations broadcasting; multiple national radio networks with many affiliate stations; while most stations are commercial, National Public Radio (NPR) has a network of some 900 member stations; satellite radio available; in total, over 15,000 radio stations operating (2018)

Internet country code: .us

Internet users: *total:* 285,519,020
percent of population: 87.27% (July 2018 est.)
country comparison to the world: 3

Broadband—fixed subscriptions: *total:* 110.568 million
subscriptions per 100 inhabitants: 34 (2018 est.)
country comparison to the world: 2

Communications—note: note 1: The Library of Congress, Washington DC, USA, claims to be the largest library in the world with more than 167 million items (as of 2018); its collections are universal, not limited by subject, format, or national boundary, and include materials from all parts of the world and in over 450 languages; collections include: books, newspapers, magazines, sheet music, sound and video recordings, photographic images, artwork, architectural drawings, and copyright data

note 2: Cape Canaveral, Florida, USA, hosts one of four dedicated ground antennas that assist in the operation of the Global Positioning System (GPS) navigation system (the others are on Ascension (Saint Helena, Ascension, and Tistan da Cunha), Diego Garcia (British Indian Ocean Territory), and at Kwajalein (Marshall Islands)

TRANSPORTATION

National air transport system: number of registered air carriers: 99 (2020)

inventory of registered aircraft operated by air carriers: 7,249

annual passenger traffic on registered air carriers: 889.022 million (2018)

annual freight traffic on registered air carriers: 42,985,300,000 mt-km (2018)

Civil aircraft registration country code prefix: N (2016)

Airports: 13,513 (2013)

country comparison to the world: 1

Airports—with paved runways: *total:* 5,054 (2013)
over 3,047 m: 189 (2013)
2,438 to 3,047 m: 235 (2013)
1,524 to 2,437 m: 1,478 (2013)
914 to 1,523 m: 2,249 (2013)
under 914 m: 903 (2013)

Airports—with unpaved runways: *total:* 8,459 (2013)
over 3,047 m: 1 (2013)
2,438 to 3,047 m: 6 (2013)
1,524 to 2,437 m: 140 (2013)
914 to 1,523 m: 1,552 (2013)
under 914 m: 6,760 (2013)

Heliports: 5,287 (2013)

Pipelines: 1,984,321 km natural gas, 240,711 km petroleum products (2013)

Railways: *total:* 293,564 km (2014)
standard gauge: 293,564.2 km 1.435-m gauge (2014)
country comparison to the world: 1

Roadways: *total:* 6,586,610 km (2012)
paved: 4,304,715 km (includes 76,334 km of expressways) (2012)
unpaved: 2,281,895 km (2012)
country comparison to the world: 1

Waterways: 41,009 km (19,312 km used for commerce; Saint Lawrence Seaway of 3,769 km, including the Saint Lawrence River of 3,058 km, is shared with Canada) (2012)

country comparison to the world: 5

Merchant marine: *total:* 3,673
by type: bulk carrier 5, container ship 60, general cargo 104, oil tanker 68, other 3,436 (2019)
country comparison to the world: 5

Ports and terminals: *oil terminal(s):* LOOP terminal, Haymark terminal

container port(s) (TEUs): Charleston (2,177,000), Hampton Roads (2,841,000), Houston (2,459,000), Long Beach (7,544,000), Los Angeles (9,343,000), New York/New Jersey (6,710,000), Oakland (2,420,000), Savannah (4,046,000), Seattle/Tacoma (3,665,000) (2017)

LNG terminal(s) (export): Cameron (LA), Corpus Christi (TX), Cove Point (MD), Elba Island (GA), Freeport (TX), Sabine Pass (LA)
note—two additional export facilities are under construction and expected to begin commercial operations in 2023-2024

LNG terminal(s) (import): Cove Point (MD), Elba Island (GA), Everett (MA), Freeport (TX), Golden Pass (TX), Hackberry (LA), Lake Charles (LA), Neptune (offshore), Northeast Gateway (offshore), Pascagoula (MS), Sabine Pass (TX)

cargo ports: Baton Rouge, Corpus Christi, Hampton Roads, Houston, Long Beach, Los Angeles, New Orleans, New York, Plaquemines (LA), Tampa, Texas City

cruise departure ports (passengers): Miami (2,032,000), Port Everglades (1,277,000), Port Canaveral (1,189,000), Seattle (430,000), Long Beach (415,000) (2009)

MILITARY AND SECURITY

Military and security forces: United States Armed Forces: US Army, US Navy (includes Marine Corps), US Air Force, US Space Force; US Coast Guard (administered in peacetime by the Department of Homeland Security, but in wartime reports to the Department of the Navy); National Guard (Army National Guard and Air National Guard); Reserves (all services) (2020)

Military expenditures: 3.42% of GDP (2019 est.)
3.3% of GDP (2018)
3.31% of GDP (2017)
3.52% of GDP (2016)
3.52% of GDP (2015)

country comparison to the world: 20

Military and security service personnel strengths: the US Armed Forces have approximately 1.372 million active duty personnel (475,000 Army; 340,000 Navy; 330,000 Air Force; 185,000 Marine Corps); 42,000 Coast Guard; 335,000 Army National Guard; 105,000 Air National Guard (June 2020)

Military equipment inventories and acquisitions: the US military's inventory is comprised almost entirely of domestically-produced weapons systems (some assembled with foreign components) along with a smaller mix of imported equipment from a variety of Western countries; since 2010, Germany and the UK are the leading suppliers, followed by Australia, Canada, France, the Netherlands, and Norway; the US defense industry is capable of designing, developing, maintaining, and producing the full spectrum of weapons systems (2019 est.)

Military deployments: 4,500 Afghanistan (NATO; note—the US has pledged to further reduce the number of troops in Afghanistan); 5,000 Africa; 300 Australia; 1,150 Belgium; 150 Bulgaria; 250 Diego Garcia; 150 Canada; 800 Cuba; 290 Egypt (MFO); 35,000 Germany (note—in July 2020, the US pledged to reduce the number of troops in Germany by about 12,000); 400 Greece; 150 Greenland; 5,600 Guam; 380 Honduras; 12,300 Italy; 55,200 Japan; 660 Kosovo (KFOR); approximately 10-15,000 assigned with an additional estimated 20-30,000 deployed Middle East (Bahrain/Iraq/Israel/Jordan/Kuwait/Oman/Qatar/Saudi Arabia/Syria/United Arab Emirates); 400 Netherlands; 670 Norway; 200 Philippines; 4,500 Poland; 250 Portugal; 26,500 Republic of Korea; 1,100 Romania; 200 Singapore; 3,200 Spain; 100 Thailand; 1,700 Turkey; 9,400 United Kingdom (2020)

US military rotational policies affect deployed numbers; for example, the US deploys ground and air units to select countries for 6-12 month rotational assignments on a continuous basis; in South Korea, for example, the US continuously rotates combat brigades (3,000-4,000 personnel) for 9 months at a time; contingencies also affect US troop deployments; for example, since May 2019, the US has deployed more than 15,000 additional military personnel to the Middle East for an undetermined period of time; in addition, some overseas US naval bases, such as the headquarters of US Naval Forces Central Command (USNAVCENT) in Manama, Bahrain, are frequented by the crews of US ships on 6-9 month deployments; a US carrier strike group with an air wing and supporting ships typically includes over 6,000 personnel (2020)

Military service age and obligation: 18 years of age (17 years of age with parental consent) for male and female voluntary service; no conscription; maximum enlistment age 34 (Army), 39 (Air Force), 39 (Navy), 28 (Marines), 31 (Coast Guard); 8-year service obligation, including 2-5 years active duty (Army), 2 years active (Navy), 4 years active (Air Force, Marines, Coast Guard); all military occupations and positions open to women (2019)

TERRORISM

Terrorist group(s): Islamic Revolutionary Guard Corps/Qods Force; Islamic State of Iraq and ash-Sham; al-Qa'ida (2020)

note: details about the history, aims, leadership, organization, areas of operation, tactics, targets, weapons, size, and sources of support of the group(s) appear(s) in Appendix-T

TRANSNATIONAL ISSUES

Disputes—international: the US has intensified domestic security measures and is collaborating closely with its neighbors, Canada and Mexico, to monitor and control legal and illegal personnel, transport, and commodities across the international borders; abundant rainfall in recent years along much of the Mexico-US border region has ameliorated periodically strained water-sharing arrangements; 1990 Maritime Boundary Agreement in the Bering Sea still awaits Russian Duma ratification; Canada and the United States dispute how to divide the Beaufort Sea and the status of the Northwest Passage but continue to work cooperatively to survey the Arctic continental shelf; The Bahamas and US have not been able to agree on a maritime boundary; US Naval Base at Guantanamo Bay is leased from Cuba and only mutual agreement or US abandonment of the area can terminate the lease; Haiti claims US-administered Navassa Island; US has made no territorial claim in Antarctica (but has reserved the right to do so) and does not recognize the claims of any other states; Marshall Islands claims Wake Island; Tokelau included American Samoa's Swains Island among the islands listed in its 2006 draft constitution

Refugees and internally displaced persons: *refugees (country of origin):* the US admitted 11,814 refugees during FY2020 including: 2,868 (Democratic Republic of the Congo), 2,115 (Burma), 1,927 (Ukraine), 604 (Afghanistan), 537 (Iraq)

note: 72,722 Venezuelans have claimed asylum since 2014 because of the economic and political crisis (2018)

Illicit drugs: world's largest consumer of cocaine (shipped from Colombia through Mexico and the Caribbean), Colombian heroin, and Mexican heroin and marijuana; major consumer of ecstasy and Mexican methamphetamine; minor consumer of high-quality Southeast Asian heroin; illicit producer of cannabis, marijuana, depressants, stimulants, hallucinogens, and methamphetamine; money-laundering center

UNITED STATES PACIFIC ISLAND WILDLIFE REFUGES

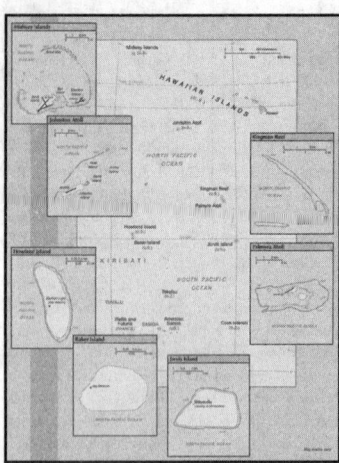

INTRODUCTION

Background: All of the following US Pacific island territories except Midway Atoll constitute the Pacific Remote Islands National Wildlife Refuge (NWR) Complex and as such are managed by the Fish and Wildlife Service of the US Department of the Interior. Midway Atoll NWR has been included in a Refuge Complex with the Hawaiian Islands NWR and also designated as part of Papahanaumokuakea Marine National Monument. These remote refuges are the most widespread collection of marine and terrestrial-life protected areas on the planet under a single country's jurisdiction. They sustain many endemic species including corals, fish, shellfish, marine mammals, seabirds, water birds, land birds, insects, and vegetation not found elsewhere.

Baker Island: The US took possession of the island in 1857. Its guano deposits were mined by US and British companies during the second half of the 19th century. In 1935, a short-lived attempt at colonization began on this island but was disrupted by World War II and thereafter abandoned. The island was established as a NWR in 1974.;

Howland Island: Discovered by the US early in the 19th century, the uninhabited atoll was officially claimed by the US in 1857. Both US and British companies mined for guano deposits until about 1890. In 1935, a short-lived attempt at colonization began on this island, similar to the effort on nearby Baker Island, but was disrupted by World War II and thereafter abandoned. The famed American aviatrix Amelia EARHART disappeared while seeking out Howland Island as a refueling stop during her 1937 round-the-world flight; Earhart Light, a day beacon near the middle of the west coast, was named in her memory. The island was established as a NWR in 1974.;

Jarvis Island: First discovered by the British in 1821, the uninhabited island was annexed by the US in 1858 but abandoned in 1879 after tons of guano had been removed. The UK annexed the island in 1889 but never carried out plans for further exploitation. The US occupied and reclaimed the island in 1935. It was abandoned in 1942 during World War II. The island was established as a NWR in 1974.;

Johnston Atoll: Both the US and the Kingdom of Hawaii annexed Johnston Atoll in 1858, but it was the US that mined the guano deposits until the late 1880s. Johnston and Sand Islands were designated wildlife refuges in 1926. The US Navy took over the atoll in 1934. Subsequently, the US Air Force assumed control in 1948. The site was used for high-altitude nuclear tests in the 1950s and 1960s. Until late in 2000 the atoll was maintained as a storage and disposal site for chemical weapons. Munitions destruction, cleanup, and closure of the facility were completed by May 2005. The Fish and Wildlife Service and the US Air Force are currently discussing future management options; in the interim, Johnston Atoll and the three-mile Naval Defensive Sea around it remain under the jurisdiction and administrative control of the US Air Force.;

Kingman Reef: The US annexed the reef in 1922. Its sheltered lagoon served as a way station for flying boats on Hawaii-to-American Samoa flights during the late 1930s. There are no terrestrial plants on the reef, which is frequently awash, but it does support abundant and diverse marine fauna and flora. In 2001, the waters surrounding the reef out to 12 nm were designated a NWR.;

Midway Islands: The US took formal possession of the islands in 1867. The laying of the transpacific cable, which passed through the islands, brought the first residents in 1903. Between 1935 and 1947, Midway was used as a refueling stop for transpacific flights. The US naval victory over a Japanese fleet off Midway in 1942 was one of the turning points of World War II. The islands continued to serve as a naval station until closed in 1993. Today the islands are a NWR and are the site of the world's largest Laysan albatross colony.;

Palmyra Atoll: The Kingdom of Hawaii claimed the atoll in 1862, and the US included it among the Hawaiian Islands when it annexed the archipelago in 1898. The Hawaii Statehood Act of 1959 did not include Palmyra Atoll, which is now partly privately owned by the Nature Conservancy with the rest owned by the Federal government and managed by the US Fish and Wildlife Service. These organizations are managing the atoll as a wildlife refuge. The lagoons and surrounding waters within the 12-nm US territorial seas were transferred to the US Fish and Wildlife Service and designated a NWR in January 2001.

GEOGRAPHY

Location: Oceania

Baker Island: atoll in the North Pacific Ocean 3,390 km southwest of Honolulu, about halfway between Hawaii and Australia;

Howland Island: island in the North Pacific Ocean 3,360 km southwest of Honolulu, about halfway between Hawaii and Australia;

Jarvis Island: island in the South Pacific Ocean 2,415 km south of Honolulu, about halfway between Hawaii and Cook Islands;

Johnston Atoll: atoll in the North Pacific Ocean 1,330 km southwest of Honolulu, about one-third of the way from Hawaii to the Marshall Islands;

Kingman Reef: reef in the North Pacific Ocean 1,720 km south of Honolulu, about halfway between Hawaii and American Samoa;

Midway Islands: atoll in the North Pacific Ocean 2,335 km northwest of Honolulu near the end of the Hawaiian Archipelago, about one-third of the way from Honolulu to Tokyo;

Palmyra Atoll: atoll in the North Pacific Ocean 1,780 km south of Honolulu, about halfway between Hawaii and American Samoa

Geographic coordinates: *Baker Island:* 0 13 N, 176 28 W;
Howland Island: 0 48 N, 176 38 W;
Jarvis Island: 0 23 S, 160 01 W;
Johnston Atoll: 16 45 N, 169 31 W;
Kingman Reef: 6 23 N, 162 25 W;
Midway Islands: 28 12 N, 177 22 W;
Palmyra Atoll: 5 53 N, 162 05 W
Map references: Oceania

Area: *land:* 6,959.41 sq km (emergent land - 22.41 sq km; submerged - 6,937 sq km)
Baker Island: total - 129.1 sq km; emergent land - 2.1 sq km; submerged - 127 sq km
Howland Island: total - 138.6 sq km; emergent land - 2.6 sq km; submerged - 136 sq km
Jarvis Island: total - 152 sq km; emergent land - 5 sq km; submerged - 147 sq km
Johnston Atoll: total - 276.6 sq km; emergent land - 2.6 sq km; submerged - 274 sq km
Kingman Reef: total - 1,958.01 sq km; emergent land - 0.01 sq km; submerged - 1,958 sq km
Midway Islands: total - 2,355.2 sq km; emergent land - 6.2 sq km; submerged - 2,349 sq km
Palmyra Atoll: total - 1,949.9 sq km; emergent land - 3.9 sq km; submerged - 1,946 sq km
Area—comparative: Baker Island: about 2.5 times the size of the National Mall in Washington, DC;
Howland Island: about three times the size of the National Mall in Washington, DC;
Jarvis Island: about eight times the size of the National Mall in Washington, DC;
Johnston Atoll: about 4.5 times the size of the National Mall in Washington, DC;
Kingman Reef: a little more than 1.5 times the size of the National Mall in Washington, DC;

Midway Islands: about nine times the size of the National Mall in Washington, DC;
Palmyra Atoll: about 20 times the size of the National Mall in Washington, DC
Land boundaries: 0 km

Coastline: *Baker Island:* 4.8 km
Howland Island: 6.4 km
Jarvis Island: 8 km
Johnston Atoll: 34 km
Kingman Reef: 3 km
Midway Islands: 15 km
Palmyra Atoll: 14.5 km

Maritime claims: *territorial sea:* 12 nm
exclusive economic zone: 200 nm

Climate: Baker, Howland, and Jarvis Islands: equatorial; scant rainfall, constant wind, burning sun;

Johnston Atoll and Kingman Reef: tropical, but generally dry; consistent northeast trade winds with little seasonal temperature variation;

Midway Islands: subtropical with cool, moist winters (December to February) and warm, dry summers (May to October); moderated by prevailing easterly winds; most of the 107 cm of annual rainfall occurs during the winter;

Palmyra Atoll: equatorial, hot; located within the low pressure area of the Intertropical Convergence Zone (ITCZ) where the northeast and southeast trade winds meet, it is extremely wet with between 400-500 cm of rainfall each year

Terrain: low and nearly flat sandy coral islands with narrow fringing reefs that have developed at the top of submerged volcanic mountains, which in most cases rise s-teeply from the ocean floor

Elevation: *lowest point:* Pacific Ocean 0 m
highest point: Baker Island, unnamed location 8 m; Howland Island, unnamed location 3 m; Jarvis Island, unnamed location 7 m; Johnston Atoll, Sand Island 10 m; Kingman Reef, unnamed location 2 m; Midway Islands, unnamed location less than 13 m; Palmyra Atoll, unnamed location 3 m

Natural resources: terrestrial and aquatic wildlife

Land use: *agricultural land:* 0% (2011 est.)
arable land: 0% (2011 est.) / *permanent crops:* 0% (2011 est.) / *permanent pasture:* 0% (2011 est.)
forest: 0% (2011 est.)
other: 100% (2011 est.)

Natural hazards: Baker, Howland, and Jarvis Islands: the narrow fringing reef surrounding the island poses a maritime hazard;

Kingman Reef: wet or awash most of the time, maximum elevation of less than 2 m makes Kingman Reef a maritime hazard;

Midway Islands, Johnston, and Palmyra Atolls: NA

Environment—current issues: Baker Island: no natural freshwater resources; feral cats, introduced in 1937 during a short-lived colonization effort, ravaged the avian population and were eradicated in 1965

Howland Island: no natural freshwater resources; the island habitat has suffered from invasive exotic species; black rats, introduced in 1854,

were eradicated by feral cats within a year of their introduction in 1937; the cats preyed on the bird population and were eliminated by 1985

Jarvis Island: no natural freshwater resources; feral cats, introduced in the 1930s during a short-lived colonization venture, were not completely removed until 1990

Johnston Atoll: no natural freshwater resources; the seven decades under US military administration (1934-2004) left the atoll environmentally degraded and required large-scale remediation efforts; a swarm of Anoplolepis (crazy) ants invaded the island in 2010 damaging native wildlife; eradication has been largely, but not completely, successful

Midway Islands: many exotic species introduced, 75% of the roughly 200 plant species on the island are non-native; plastic pollution harms wildlife, via entanglement, ingestion, and toxic contamination

Kingman Reef: none

Palmyra Atoll: black rats, believed to have been introduced to the atoll during the US military occupation of the 1940s, severely degraded the ecosystem outcompeting native species (seabirds, crabs); following a successful rat removal project in 2011, native flora and fauna have begun to recover

Geography—note: Baker, Howland, and Jarvis Islands: scattered vegetation consisting of grasses, prostrate vines, and low growing shrubs; primarily a nesting, roosting, and foraging habitat for seabirds, shorebirds, and marine wildlife; closed to the public;

Johnston Atoll: Johnston Island and Sand Island are natural islands, which have been expanded by coral dredging; North Island (Akau) and East Island (Hikina) are manmade islands formed from coral dredging; the egg-shaped reef is 34 km in circumference; closed to the public;

Kingman Reef: barren coral atoll with deep interior lagoon; closed to the public;

Midway Islands: a coral atoll managed as a National Wildlife Refuge and open to the public for wildlife-related recreation in the form of wildlife observation and photography;

Palmyra Atoll: the high rainfall and resulting lush vegetation make the environment of this atoll unique among the US Pacific Island territories; supports a large undisturbed stand of Pisonia beach forest

PEOPLE AND SOCIETY

Population: no indigenous inhabitants
note: public entry is only by special-use permit from US Fish and Wildlife Service and generally restricted to scientists and educators; visited annually by US Fish and Wildlife Service

Jarvis Island: Millersville settlement on western side of island occasionally used as a weather station from 1935 until World War II, when it was abandoned; reoccupied in 1957 during the International Geophysical Year by scientists who left in 1958; currently unoccupied

Johnston Atoll: in previous years, an average of 1,100 US military and civilian contractor personnel were present; as of May 2005, all US Government personnel had left the island

Midway Islands: approximately 40 people make up the staff of US Fish and Wildlife Service and their services contractor living at the atoll

Palmyra Atoll: four to 20 Nature Conservancy, US Fish and Wildlife staff, and researchers

GOVERNMENT

Country name: *conventional long form:* none
conventional short form: Baker Island, Howland Island, Jarvis Island, Johnston Atoll, Kingman Reef, Midway Islands, Palmyra Atoll
etymology: self-descriptive name specifying the territories' affiliation and location

Dependency status: with the exception of Palmyra Atoll, the constituent islands are unincorporated, unorganized territories of the US; administered from Washington, DC, by the Fish and Wildlife Service of the US Department of the Interior as part of the National Wildlife Refuge System
note: Palmyra Atoll is partly privately owned and partly federally owned; the federally owned portion is administered from Washington, DC, by the Fish and Wildlife Service of the US Department of the Interior as an incorporated, unorganized territory of the US; the Office of Insular Affairs of the US Department of the Interior continues to administer nine excluded areas comprising certain tidal and submerged lands within the 12 nm territorial sea or within the lagoon

Legal system: the laws of the US apply where applicable

Diplomatic representation from the US: none (territories of the US)

Flag description: the flag of the US is used

ECONOMY

Economy—overview: no economic activity

ENERGY

Crude oil—production: 0 bbl/day (2018 est.)
country comparison to the world: 212

TRANSPORTATION

Airports: Baker Island: one abandoned World War II runway of 1,665 m covered with vegetation and unusable (2013)

Howland Island: airstrip constructed in 1937 for scheduled refueling stop on the round-the-world flight of Amelia EARHART and Fred NOONAN; the aviators left Lae, New Guinea, for Howland Island but were never seen again; the airstrip is no longer serviceable (2013)

Johnston Atoll: one closed and not maintained (2013)

Kingman Reef: lagoon was used as a halfway station between Hawaii and American Samoa by Pan American Airways for flying boats in 1937 and 1938 (2013)

Midway Islands: 3 - one operational (2,377 m paved); no fuel for sale except emergencies (2013)

Palmyra Atoll: 1 - 1,846 m unpaved runway; privately owned (2013)

Airports—with paved runways: *2,438 to 3,047 m:* 1 - Johnston Atoll; (2016)

note - abandoned but usable

Airports—with unpaved runways: 1 - Palmyra Atoll (2016)

Ports and terminals: *major seaport(s):* Baker, Howland, and Jarvis Islands, and Kingman Reef

Baker, Howland, and Jarvis Islands, and Kingman Reef: none; offshore anchorage only

Johnston Atoll: Johnston Island

Midway Islands: Sand Island

Palmyra Atoll: West Lagoon

MILITARY AND SECURITY

Military—note: defense is the responsibility of the US

TRANSPORTATION

Disputes—international: none

URUGUAY

INTRODUCTION

Background: Montevideo, founded by the Spanish in 1726 as a military stronghold, soon took advantage of its natural harbor to become an important commercial center. Claimed by Argentina but annexed by Brazil in 1821, Uruguay declared its independence four years later and secured its freedom in 1828 after a three-year struggle. The administrations of President Jose BATLLE in the early 20th century launched widespread political, social, and economic reforms that established a statist tradition. A violent Marxist urban guerrilla movement named the Tupamaros, launched in the late 1960s, led Uruguay's president to cede control of the government to the military in 1973. By yearend, the rebels had been crushed, but the military continued to expand its hold over the government. Civilian rule was restored in 1985. In 2004, the left-of-center Frente Amplio Coalition won national elections that effectively ended 170 years of political control previously held by the Colorado and National (Blanco) parties. Uruguay's political and labor conditions are among the freest on the continent.

GEOGRAPHY

Location: Southern South America, bordering the South Atlantic Ocean, between Argentina and Brazil

Geographic coordinates: 33 00 S, 56 00 W

Map references: South America

Area: *total:* 176,215 sq km
land: 175,015 sq km
water: 1,200 sq km

country comparison to the world: 92

Area—comparative: about the size of Virginia and West Virginia combined; slightly smaller than the state of Washington

Area comparison map: [INSERT IMAGE: URUGUAY-Area comparison map]

Land boundaries: *total:* 1,591 km
border countries (2): Argentina 541 km, Brazil 1050 km

Coastline: 660 km

Maritime claims: *territorial sea:* 12 nm
exclusive economic zone: 200 nm
contiguous zone: 24 nm
continental shelf: 200 nm or the edge of continental margin

Climate: warm temperate; freezing temperatures almost unknown

Terrain: mostly rolling plains and low hills; fertile coastal lowland

Elevation: *mean elevation:* 109 m
lowest point: Atlantic Ocean 0 m
highest point: Cerro Catedral 514 m

Natural resources: arable land, hydropower, minor minerals, fish

Land use: *agricultural land:* 87.2% (2011 est.)

arable land: 10.1% (2011 est.) / permanent crops: 0.2% (2011 est.) / permanent pasture: 76.9% (2011 est.)

forest: 10.2% (2011 est.)

other: 2.6% (2011 est.)

Irrigated land: 2,380 sq km (2012)

Population distribution: most of the country's population resides in the southern half of the country; approximately 80% of the populace is urban, living in towns or cities; nearly half of the population lives in and around the capital of Montevideo

Natural hazards: seasonally high winds (the pampero is a chilly and occasional violent wind that blows north from the Argentine pampas), droughts, floods; because of the absence of mountains, which act as weather barriers, all locations are particularly vulnerable to rapid changes from weather fronts

Environment—current issues: water pollution from meat packing/tannery industry; heavy metal pollution; inadequate solid/hazardous waste disposal; deforestation

Environment—international agreements: *party to:* Antarctic-Environmental Protocol, Antarctic-Marine Living Resources, Antarctic Treaty, Biodiversity, Climate Change, Climate Change-Kyoto Protocol, Desertification, Endangered Species, Environmental Modification, Hazardous Wastes, Law of the Sea, Ozone Layer Protection, Ship Pollution, Wetlands

signed, but not ratified: Marine Dumping, Marine Life Conservation

Geography—note: second-smallest South American country (after Suriname); most of the low-lying landscape (three-quarters of the country) is grassland, ideal for cattle and sheep raising

PEOPLE AND SOCIETY

Population: 3,387,605 (July 2020 est.)
country comparison to the world: 132

Nationality: *noun:* Uruguayan(s)
adjective: Uruguayan

Ethnic groups: white 87.7%, black 4.6%, indigenous 2.4%, other 0.3%, none or unspecified 5% (2011 est.)

note: data represent primary ethnic identity

Languages: Spanish (official)

Religions: Roman Catholic 47.1%, non-Catholic Christians 11.1%, nondenominational 23.2%, Jewish 0.3%, atheist or agnostic 17.2%, other 1.1% (2006 est.)

Demographic profile: Uruguay rates high for most development indicators and is known for its secularism, liberal social laws, and well- developed social security, health, and educational systems. It is one of the few countries in Latin America and the Caribbean where the entire population has access to clean water. Uruguay's provision of free primary through university education has contributed to the country's high levels of literacy and educational attainment. However, the emigration of human capital has diminished the state's return on its investment in education. Remittances from

the roughly 18% of Uruguayans abroad amount to less than 1 percent of national GDP. The emigration of young adults and a low birth rate are causing Uruguay's population to age rapidly.

In the 1960s, Uruguayans for the first time emigrated en masse - primarily to Argentina and Brazil - because of economic decline and the onset of more than a decade of military dictatorship. Economic crises in the early 1980s and 2002 also triggered waves of emigration, but since 2002 more than 70% of Uruguayan emigrants have selected the US and Spain as destinations because of better job prospects. Uruguay had a tiny population upon its independence in 1828 and welcomed thousands of predominantly Italian and Spanish immigrants, but the country has not experienced large influxes of new arrivals since the aftermath of World War II. More recent immigrants include Peruvians and Arabs.

Age structure: *0-14 years:* 19.51% (male 336,336/female 324,563)
15-24 years: 15.14% (male 259,904/female 252,945)
25-54 years: 39.86% (male 670,295/female 679,850)
55-64 years: 10.79% (male 172,313/female 193,045)
65 years and over: 14.71% (male 200,516/female 297,838) (2020 est.)

population pyramid: [INSERT IMAGE: URUGUAY-population pyramid]

Dependency ratios: *total dependency ratio:* 54.9
youth dependency ratio: 31.5
elderly dependency ratio: 23.4
potential support ratio: 4.3 (2020 est.)

Median age: *total:* 35.5 years
male: 33.8 years
female: 37.3 years (2020 est.)
country comparison to the world: 85

Population growth rate: 0.27% (2020 est.)
country comparison to the world: 175

Birth rate: 12.9 births/1,000 population (2020 est.)
country comparison to the world: 146

Death rate: 9.3 deaths/1,000 population (2020 est.)
country comparison to the world: 52

Net migration rate: -0.9 migrant(s)/1,000 population (2020 est.)
country comparison to the world: 141

Population distribution: most of the country's population resides in the southern half of the country; approximately 80% of the populace is urban, living in towns or cities; nearly half of the population lives in and around the capital of Montevideo

Urbanization: *urban population:* 95.5% of total population (2020)
rate of urbanization: 0.46% annual rate of change (2015-20 est.)

total population growth rate v. urban population growth rate, 2000-2030: *Major urban areas - population:* 1.752 million MONTEVIDEO (capital) (2020)

Sex ratio: *at birth:* 1.04 male(s)/female
0-14 years: 1.04 male(s)/female
15-24 years: 1.03 male(s)/female
25-54 years: 0.99 male(s)/female

55-64 years: 0.89 male(s)/female
65 years and over: 0.67 male(s)/female
total population: 0.94 male(s)/female (2020 est.)

Maternal mortality rate: 17 deaths/100,000 live births (2017 est.)
country comparison to the world: 134

Infant mortality rate: *total:* 7.8 deaths/1,000 live births
male: 8.6 deaths/1,000 live births
female: 6.9 deaths/1,000 live births (2020 est.)
country comparison to the world: 152

Life expectancy at birth: *total population:* 77.9 years
male: 74.8 years
female: 81.2 years (2020 est.)
country comparison to the world: 73

Total fertility rate: 1.77 children born/woman (2020 est.)
country comparison to the world: 156

Contraceptive prevalence rate: 79.6% (2015)
note: percent of women aged 15-44

Drinking water source:

improved: *urban:* 100% of population
rural: 93.9% of population
total: 99.7% of population

unimproved: *urban:* 100% of population
rural: 95% of population
total: 100% of population (2017 est.)

Current Health Expenditure: 9.3% (2017)

Physicians density: 5.08 physicians/1,000 population (2017)

Hospital bed density: 2.4 beds/1,000 population (2017)

Sanitation facility access:

improved: *urban:* 99% of population
rural: 98.3% of population
total: 98.9% of population

unimproved: *urban:* 1% of population
rural: 1.7% of population
total: 2.1% of population (2017 est.)

HIV/AIDS—adult prevalence rate: 0.6% (2018 est.)
country comparison to the world: 65

HIV/AIDS—people living with HIV/AIDS: 14,000 (2018 est.)
country comparison to the world: 95

HIV/AIDS—deaths: <200 (2018 est.)

Obesity—adult prevalence rate: 27.9% (2016)
country comparison to the world: 34

Children under the age of 5 years underweight: 4% (2011)
country comparison to the world: 90

Education expenditures: 4.9% of GDP (2017)
country comparison to the world: 69

Literacy: *definition:* age 15 and over can read and write
total population: 98.7%
male: 98.4%
female: 99% (2018)

School life expectancy (primary to tertiary education): total: 17 years
male: NA

female: NA (2017)
Unemployment, youth ages 15-24: total: 25.9%
male: 22.4%
female: 30.7% (2018 est.)
country comparison to the world: 46

GOVERNMENT

Country name: *conventional long form:* Oriental Republic of Uruguay
conventional short form: Uruguay
local long form: Republica Oriental del Uruguay
local short form: Uruguay
former: Banda Oriental, Cisplatine Province
etymology: name derives from the Spanish pronunciation of the Guarani Indian designation of the Uruguay River, which makes up the western border of the country and whose name later came to be applied to the entire country

Government type: presidential republic

Capital: *name:* Montevideo

geographic coordinates: 34 51 S, 56 10 W
time difference: UTC-3 (2 hours ahead of Washington, DC, during Standard Time)
etymology: the name "Montevidi" was originally applied to the hill that overlooked the bay upon which the city of Montevideo was founded; the earliest meaning may have been "[the place where we] saw the hill"

Administrative divisions: 19 departments (departamentos, singular - departamento); Artigas, Canelones, Cerro Largo, Colonia, Durazno, Flores, Florida, Lavalleja, Maldonado, Montevideo, Paysandu, Rio Negro, Rivera, Rocha, Salto, San Jose, Soriano, Tacuarembo, Treinta y Tres

Independence: 25 August 1825 (from Brazil)

National holiday: Independence Day, 25 August (1825)

Constitution: *history:* several previous; latest approved by plebiscite 27 November 1966, effective 15 February 1967
amendments: initiated by public petition of at least 10% of qualified voters, proposed by agreement of at least two fifths of the General Assembly membership, or by existing "constitutional laws" sanctioned by at least two thirds of the membership in both houses of the Assembly; proposals can also be submitted by senators, representatives, or by the executive power and require the formation of and approval in a national constituent convention; final passage by either method requires approval by absolute majority of votes cast in a referendum; amended many times, last in 2004

Legal system: civil law system based on the Spanish civil code

International law organization participation: accepts compulsory ICJ jurisdiction; accepts ICCt jurisdiction

Citizenship: *citizenship by birth:* yes
citizenship by descent only: yes
dual citizenship recognized: yes
residency requirement for naturalization: 3-5 years

Suffrage: 18 years of age; universal and compulsory

Executive branch: *chief of state:* President Luis Alberto LACALLE POU (since 1 March 2020); Vice President Beatriz ARGIMON Cedeira (since 1 March 2020); the president is both chief of state and head of government

head of government: President Luis Alberto LACALLE POU (since 1 March 2020); Vice President Beatriz ARGIMON Cedeira (since 1 March 2020)
cabinet: Council of Ministers appointed by the president with approval of the General Assembly
elections/appointments: president and vice president directly elected on the same ballot by absolute majority vote in 2 rounds if needed for a 5-year term (eligible for nonconsecutive terms); election last held on 27 October 2019 with a runoff election on 24 November 2019 (next to be held in October 2024, and a runoff if needed in November 2024)
election results: Luis Alberto LACALLE POU elected president - results of the first round of presidential elections: percent of vote - Daniel MARTINEZ (FA) 40.7%, Luis Alberto LACALLE POU (Blanco) 29.7%, Ernesto TALVI (Colorado Party) 12.8%, and Guido MANINI RIOS (Open Cabildo) 11.3%, other 5.5%; results of the second round: percent of vote - Luis Alberto LACALLE POU (Blanco) 50.6%, Daniel MARTINEZ (FA) 49.4%

Legislative branch: *description:* bicameral General Assembly or Asamblea General consists of:

Chamber of Senators or Camara de Senadores (31 seats; members directly elected in a single nationwide constituency by proportional representation vote; the vice-president serves as the presiding ex-officio member; elected members serve 5 year terms)

Chamber of Representatives or Camara de Representantes (99 seats; members directly elected in multi-seat constituencies by proportional representation vote to serve 5-year terms)
elections: Chamber of Senators - last held on 27 October 2019 (next to be held in October 2024)

Chamber of Representatives—last held on 27 October 2019 (next to be held in October 2024)
election results: Chamber of Senators - percent of vote by coalition/party - na; seats by coalition/party - Frente Amplio 13, National Party 10, Colorado Party 4, Open Cabildo 3;
Chamber of Representatives—percent of vote by coalition/party—na; seats by coalition/party—Frente Amplio 42, National Party 30, Colorado Party 13, Open Cabildo 11, Independent Party 1, other 2

Judicial branch: *highest courts:* Supreme Court of Justice (consists of 5 judges)
judge selection and term of office: judges nominated by the president and appointed in joint conference of the General Assembly; judges serve 10-year terms, with reelection possible after a lapse of 5 years following the previous term
subordinate courts: Courts of Appeal; District Courts (Juzgados Letrados); Peace Courts (Juzgados de Paz); Rural Courts (Juzgados Rurales)

Political parties and leaders: Broad Front or FA (Frente Amplio) [Javier MIRANDA] - (a broad governing coalition that includes Uruguay Assembly [Danilo ASTORI], Progressive Alliance [Rodolfo NIN NOVOA], New Space [Rafael MICHELINI], Socialist Party [Monica XAVIER], Vertiente Artiguista [Enrique RUBIO], Christian Democratic Party [Jorge RODRIGUEZ], For the People's Victory [Luis PUIG], Popular Participation Movement (MPP) [Jose MUJICA], Broad Front Commitment [Raul SENDIC], Big House [Constanza MOREIRA], Communist Party [Marcos CARAMBULA], The Federal League [Dario PEREZ]
Colorado Party (including Vamos Uruguay (or Let's Go Uruguay), Open Space [Tabare VIERA], and Open Batllism [Ope PASQUET])
Independent Party [Pablo MIERES]
National Party or Blanco (including Everyone [Luis LACALLE POU] and National Alliance [Jorge LARRANAGA])
Popular Unity [Gonzalo ABELLA]
Open Cabildo [Guido MANINI RIOS]

International organization participation: CAN (associate), CD, CELAC, FAO, G-77, IADB, IAEA, IBRD, ICAO, ICC (national committees), ICCt, ICRM, IDA, IFAD, IFC, IFRCS, IHO, ILO, IMF, IMO, Interpol, IOC, IOM, IPU, ISO, ITSO, ITU, LAES, LAIA, Mercosur, MIGA, MINUSTAH, MONUSCO, NAM (observer), OAS, OIF (observer), OPANAL, OPCW, Pacific Alliance (observer), PCA, SICA (observer), UN, UNASUR, UNCTAD, UNESCO, UNIDO, Union Latina, UNMOGIP, UNOCI, UNWTO, UPU, WCO, WFTU (NGOs), WHO, WIPO, WMO, WTO

Diplomatic representation in the US: *chief of mission:* Charge d'Affaires Alejandro Ramon RODRIGUEZ COTRO (since 15 July 2020)
chancery: 1913 I Street NW, Washington, DC 20006
telephone: [1] (202) 331-1313
FAX: [1] (202) 331-8142

consulate(s) general: Chicago, Los Angeles, Miami, New York

Diplomatic representation from the US: *chief of mission:* Ambassador Kenneth S. GEORGE (since 2 September 2019)
telephone: (+598) 1770-2000
embassy: Laura Muller 1776, Montevideo 11200
mailing address: APO AA 34035
FAX: [598] (2) 1770-2128

Flag description: nine equal horizontal stripes of white (top and bottom) alternating with blue; a white square in the upper hoist-side corner with a yellow sun bearing a human face (delineated in black) known as the Sun of May with 16 rays that alternate between triangular and wavy; the stripes represent the nine original departments of Uruguay; the sun symbol evokes the legend of the sun breaking through the clouds on 25 May 1810 as independence was first declared from Spain (Uruguay subsequently won its independence from Brazil); the sun features are said to represent those of Inti, the Inca god of the sun *note:* the banner was inspired by the national colors of Argentina and by the design of the US flag

National symbol(s): Sun of May (a sun-with-face symbol); national colors: blue, white, yellow

National anthem: *name:* "Himno Nacional" (National Anthem of Uruguay)

lyrics/music: Francisco Esteban ACUNA de Figueroa/Francisco Jose DEBALI

note: adopted 1848; the anthem is also known as "Orientales, la Patria o la tumba!" ("Uruguayans, the Fatherland or Death!"); it is the world's longest national anthem in terms of music (105 bars; almost five minutes); generally only the first verse and chorus are sung

0:00 / 1:54

ECONOMY

Economy—overview: Uruguay has a free market economy characterized by an export-oriented agricultural sector, a well-educated workforce, and high levels of social spending. Uruguay has sought to expand trade within the Common Market of the South (Mercosur) and with non-Mercosur members, and President VAZQUEZ has maintained his predecessor's mix of pro-market policies and a strong social safety net.

Following financial difficulties in the late 1990s and early 2000s, Uruguay's economic growth averaged 8% annually during the 2004-08 period. The 2008-09 global financial crisis put a brake on Uruguay's vigorous growth, which decelerated to 2.6% in 2009. Nevertheless, the country avoided a recession and kept growth rates positive, mainly through higher public expenditure and investment; GDP growth reached 8.9% in 2010 but slowed markedly in the 2012-16 period as a result of a renewed slowdown in the global economy and in Uruguay's main trade partners and Mercosur counterparts, Argentina and Brazil. Reforms in those countries should give Uruguay an economic boost. Growth picked up in 2017.

GDP (purchasing power parity): $78.16 billion (2017 est.)
$76.14 billion (2016 est.)
$74.87 billion (2015 est.)
note: data are in 2017 dollars
country comparison to the world: 96

GDP (official exchange rate): $59.18 billion (2017 est.)

GDP—real growth rate: 2.7% (2017 est.)
1.7% (2016 est.)
0.4% (2015 est.)
country comparison to the world: 108

GDP—per capita (PPP): $22,400 (2017 est.)
$21,900 (2016 est.)
$21,600 (2015 est.)
note: data are in 2017 dollars
country comparison to the world: 85

Gross national saving: 17.2% of GDP (2017 est.)
18.6% of GDP (2016 est.)
18.7% of GDP (2015 est.)
country comparison to the world: 119

GDP—composition, by end use: *household consumption:* 66.8% (2017 est.)
government consumption: 14.3% (2017 est.)
investment in fixed capital: 16.7% (2017 est.)
investment in inventories: -1% (2017 est.)
exports of goods and services: 21.6% (2017 est.)
imports of goods and services: -18.4% (2017 est.)

GDP—composition, by sector of origin: *agriculture:* 6.2% (2017 est.)
industry: 24.1% (2017 est.)
services: 69.7% (2017 est.)

Agriculture—products: Cellulose, beef, soybeans, rice, wheat; dairy products; fish; lumber, tobacco, wine

Industries: food processing, electrical machinery, transportation equipment, petroleum products, textiles, chemicals, beverages

Industrial production growth rate: -3.6% (2017 est.)
country comparison to the world: 190

Labor force: 1.748 million (2017 est.)
country comparison to the world: 124

Labor force—by occupation: agriculture: 13%
industry: 14%
services: 73% (2010 est.)

Unemployment rate: 7.6% (2017 est.)
7.9% (2016 est.)
country comparison to the world: 117

Population below poverty line: 9.7% (2015 est.)

Household income or consumption by percentage share: *lowest 10%:* 1.9%
highest 10%: 30.8% (2014 est.)

Budget: *revenues:* 17.66 billion (2017 est.)
expenditures: 19.72 billion (2017 est.)

Taxes and other revenues: 29.8% (of GDP) (2017 est.)
country comparison to the world: 79

Budget surplus (+) or deficit (-): -3.5% (of GDP) (2017 est.)
country comparison to the world: 149

Public debt: 65.7% of GDP (2017 est.)
61.6% of GDP (2016 est.)
note: data cover general government debt and include debt instruments issued (or owned) by government entities other than the treasury; the data include treasury debt held by foreign entities; the data include debt issued by subnational entities, as well as intragovernmental debt; intragovernmental debt consists of treasury borrowings from surpluses in the social funds, such as for retirement, medical care, and unemployment; debt instruments for the social funds are not sold at public auctions.
country comparison to the world: 57

Fiscal year: calendar year

Inflation rate (consumer prices): 6.2% (2017 est.)
9.6% (2016 est.)
country comparison to the world: 189

Current account balance: $879 million (2017 est.)
$410 million (2016 est.)
country comparison to the world: 51

Exports: $11.41 billion (2017 est.)
$8.387 billion (2016 est.)

country comparison to the world: 86

Exports—partners: China 19%, Brazil 16.1%, US 5.7%, Argentina 5.4% (2017)

Exports—commodities: beef, soybeans, cellulose, rice, wheat, wood, dairy products, wool

Imports: $8.607 billion (2017 est.)
$8.463 billion (2016 est.)
country comparison to the world: 106

Imports—commodities: refined oil, crude oil, passenger and other transportation vehicles, vehicle parts, cellular phones

Imports—partners: China 20%, Brazil 19.5%, Argentina 12.6%, US 10.9% (2017)

Reserves of foreign exchange and gold: $15.96 billion (31 December 2017 est.)
$13.47 billion (31 December 2016 est.)
country comparison to the world: 66

Debt—external: $28.37 billion (31 December 2017 est.)
$27.9 billion (31 December 2016 est.)
country comparison to the world: 84

Exchange rates: Uruguayan pesos (UYU) per US dollar—
28.77 (2017 est.)
30.16 (2016 est.)
30.16 (2015 est.)
27.52 (2014 est.)
23.25 (2013 est.)

ENERGY

Electricity access: *electrification—total population:* 100% (2020)

Electricity—production: 13.13 billion kWh (2016 est.)
country comparison to the world: 92

Electricity—consumption: 10.77 billion kWh (2016 est.)
country comparison to the world: 93

Electricity—exports: 1.321 billion kWh (2015 est.)
country comparison to the world: 53

Electricity—imports: 24 million kWh (2016 est.)
country comparison to the world: 111

Electricity—installed generating capacity: 4.808 million kW (2016 est.)
country comparison to the world: 81

Electricity—from fossil fuels: 29% of total installed capacity (2016 est.)
country comparison to the world: 185

Electricity—from nuclear fuels: 0% of total installed capacity (2017 est.)
country comparison to the world: 205

Electricity—from hydroelectric plants: 29% of total installed capacity (2017 est.)
country comparison to the world: 72

Electricity—from other renewable sources: 42% of total installed capacity (2017 est.)
country comparison to the world: 5

Crude oil—production: 0 bbl/day (2018 est.)
country comparison to the world: 213

Crude oil—exports: 0 bbl/day (2015 est.)
country comparison to the world: 212

Crude oil—imports: 40,200 bbl/day (2015 est.)
country comparison to the world: 57

Crude oil—proved reserves: 0 bbl (1 January 2018 est.)
country comparison to the world: 209

Refined petroleum products—production: 42,220 bbl/day (2015 est.)
country comparison to the world: 81

Refined petroleum products—consumption: 53,000 bbl/day (2016 est.)
country comparison to the world: 101

Refined petroleum products— exports: 0 bbl/day (2015 est.)
country comparison to the world: 214

Refined petroleum products—imports: 9,591 bbl/day (2015 est.)
country comparison to the world: 150

Natural gas—production: 0 cu m (2017 est.)
country comparison to the world: 211

Natural gas—consumption: 70.79 million cu m (2017 est.)
country comparison to the world: 110

Natural gas—exports: 0 cu m (2017 est.)
country comparison to the world: 207

Natural gas—imports: 70.79 million cu m (2017 est.)
country comparison to the world: 75

Natural gas—proved reserves: 0 cu m (1 January 2014 est.)
country comparison to the world: 204

Carbon dioxide emissions from consumption of energy: 7,554 million Mt (2017 est.)
country comparison to the world: 122

COMMUNICATIONS

Telephones—fixed lines: *total subscriptions:* 1,137,193

subscriptions per 100 inhabitants: 33.66 (2019 est.)
country comparison to the world: 72

Telephones—mobile cellular: total subscriptions: 4,664,993

subscriptions per 100 inhabitants: 138.08 (2019 est.)
country comparison to the world: 122

Telecommunication systems: *general assessment:* fully digitalized; one of the highest broadband penetrations in Latin America; high fixed-line and mobile penetrations as well; FttP coverage by 2022; nationwide 3G coverage and LTE networks; limited 5G commercial reach; strong focus on fiber infrastructure with 70% residential fixed-broadband connections and all business connections (2020)

domestic: most modern facilities concentrated in Montevideo; nationwide microwave radio relay network; overall fixed- line 34 per 100 and mobile-cellular teledensity 138 per 100 persons (2019)

international: country code - 598; landing points for the Unisor, Tannat, and Bicentenario submarine cable system providing direct connectivity to Brazil and Argentina; Bicentenario 2012 and Tannat 2017 cables helped end-users with Internet bandwidth; satellite earth stations - 2 Intelsat (Atlantic Ocean) (2020)

note: the COVID-19 outbreak is negatively impacting telecommunications production and supply chains globally; consumer spending on telecom devices and services has also slowed due to the pandemic's effect on economies worldwide; overall progress towards improvements in all facets of the telecom industry - mobile, fixed-line, broadband, submarine cable and satellite - has moderated

Broadcast media: mixture of privately owned and state-run broadcast media; more than 100 commercial radio stations and about 20 TV channels; cable TV is available; many community radio and TV stations; adopted the hybrid Japanese/Brazilian HDTV standard (ISDB-T) in December 2010 (2019)

Internet country code: .uy

Internet users: *total:* 2,300,557

percent of population: 68.28% (July 2018 est.)
country comparison to the world: 115

Broadband—fixed subscriptions: *total:* 977,390

subscriptions per 100 inhabitants: 29 (2018 est.)
country comparison to the world: 70

TRANSPORTATION

National air transport system: *number of registered air carriers:* 2 (2020)

inventory of registered aircraft operated by air carriers: 5

Civil aircraft registration country code prefix: CX (2016)

Airports: 133 (2013)
country comparison to the world: 42

Airports—with paved runways: *total:* 11 (2013)
over 3,047 m: 1 (2013)
1,524 to 2,437 m: 4 (2013)
914 to 1,523 m: 4 (2013)
under 914 m: 2 (2013)

Airports—with unpaved runways: *total:* 122 (2013)
1,524 to 2,437 m: 3 (2013)
914 to 1,523 m: 40 (2013)
under 914 m: 79 (2013)

Pipelines: 257 km gas, 160 km oil (2013)

Railways: *total:* 1,673 km (operational; government claims overall length is 2,961 km) (2016)
standard gauge: 1,673 km 1.435-m gauge (2016)
country comparison to the world: 80

Roadways: *total:* 77,732 km (2010)
paved: 7,743 km (2010)
unpaved: 69,989 km (2010)
country comparison to the world: 64

Waterways: 1,600 km (2011)
country comparison to the world: 50

Merchant marine: *total:* 60
by type: bulk carrier 1, general cargo 5, oil tanker 3, other 51 (2019)

country comparison to the world: 109

Ports and terminals: *major seaport(s):* Montevideo

MILITARY AND SECURITY

Military and security forces: *Armed Forces of Uruguay (Fuerzas Armadas del Uruguay):* National Army (Ejercito Nacional), National Navy (Armada Nacional, includes Maritime National Prefecture (Coast Guard)), Uruguayan Air Force (Fuerza Aerea); Guardia Nacional Republicana (paramilitary regiment of the National Police) (2020)

Military expenditures: 2% of GDP (2019)
2.1% of GDP (2018)
2% of GDP (2017)
1.9% of GDP (2016)
1.8% of GDP (2015)
country comparison to the world: 53

Military and security service personnel strengths: the Armed Forces of Uruguay have approximately 22,000 active personnel (14,500 Army; 5,000 Navy; 2,500 Air Force); est. 1,400 Guardia Nacional Republicana (2019 est.)

Military equipment inventories and acquisitions: the Armed Forces of Uruguay inventory includes a wide variety of older or second-hand equipment imported from a range of suppliers, including Brazil, Chile, Ecuador, France, Germany, Italy, Portugal, Russia, Spain, and the US (2019 est.)

Military deployments: 900 Democratic Republic of the Congo (MINUSCO); 210 Golan Heights (UNDOF) (2019)

Military service age and obligation: 18-30 years of age (18-22 years of age for Navy) for male or female voluntary military service; up to 40 years of age for specialists; enlistment is voluntary in peacetime, but the government has the authority to conscript in emergencies (2013)

TRANSPORTATION

Disputes—international: in 2010, the ICJ ruled in favor of Uruguay's operation of two paper mills on the Uruguay River, which forms the border with Argentina; the two countries formed a joint pollution monitoring regime; uncontested boundary dispute between Brazil and Uruguay over Braziliera/Brasiliera Island in the Quarai/Cuareim River leaves the tripoint with Argentina in question; smuggling of firearms and narcotics continues to be an issue along the Uruguay-Brazil border

Refugees and internally displaced persons: *refugees (country of origin):* 19,713 (Venezuela) (economic and political crisis; includes Venezuelans who have claimed asylum or have received alternative legal stay) (2020)

Illicit drugs: small-scale transit country for drugs mainly bound for Europe, often through sea-borne containers; law enforcement corruption; money laundering because of strict banking secrecy laws; weak border control along Brazilian frontier; increasing consumption of cocaine base and synthetic drugs

UZBEKISTAN

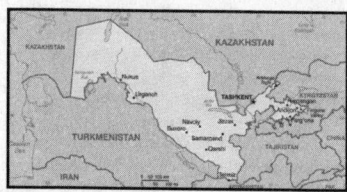

INTRODUCTION

Background: Uzbekistan is the geographic and population center of Central Asia. The country has a diverse economy and a relatively young population. Russia conquered and united the disparate territories of present-day Uzbekistan in the late 19th century. Stiff resistance to the Red Army after the Bolshevik Revolution was eventually suppressed and a socialist republic established in 1924. During the Soviet era, intensive production of "white gold" (cotton) and grain led to the overuse of agrochemicals and the depletion of water supplies, leaving the land degraded and the Aral Sea and certain rivers half-dry. Independent since the dissolution of the USSR in 1991, the country has diversified agricultural production while developing its mineral and petroleum export capacity and increasing its manufacturing base, although cotton remains a major part of its economy. Uzbekistan's first president, Islam KARIMOV, led Uzbekistan for 25 years until his death in September 2016. His successor, former Prime Minister Shavkat MIRZIYOYEV, has improved relations with Uzbekistan's neighbors and introduced wide-ranging economic, judicial, and social reforms.

GEOGRAPHY

Location: Central Asia, north of Turkmenistan, south of Kazakhstan

Geographic coordinates: 41 00 N, 64 00 E

Map references: Asia

Area: *total:* 447,400 sq km
land: 425,400 sq km
water: 22,000 sq km
country comparison to the world: 58

Area—comparative: about four times the size of Virginia; slightly larger than California

Land boundaries: *total:* 6,893 km
border countries (5): Afghanistan 144 km, Kazakhstan 2330 km, Kyrgyzstan 1314 km, Tajikistan 1312 km, Turkmenistan 1793 km

Coastline: 0 km (doubly landlocked); note - Uzbekistan includes the southern portion of the Aral Sea with a 420 km shoreline

Maritime claims: none (doubly landlocked)

Climate: mostly mid-latitude desert, long, hot summers, mild winters; semiarid grassland in east

Terrain: mostly flat-to-rolling sandy desert with dunes; broad, flat intensely irrigated river valleys along course of Amu Darya, Syr Darya (Sirdaryo), and Zarafshon; Fergana Valley in east surrounded by mountainous Tajikistan and Kyrgyzstan; shrinking Aral Sea in west

Elevation: *lowest point:* Sariqamish Kuli -12 m
highest point: Adelunga Toghi 4,301 m

Natural resources: natural gas, petroleum, coal, gold, uranium, silver, copper, lead and zinc, tungsten, molybdenum

Land use: *agricultural land:* 62.6% (2011 est.)
arable land: 10.1% (2011 est.) / permanent crops: 0.8% (2011 est.) / permanent pasture: 51.7% (2011 est.)
forest: 7.7% (2011 est.)
other: 29.7% (2011 est.)

Irrigated land: 42,150 sq km (2012)

Population distribution: most of the population is concentrated in the fertile Fergana Valley in the easternmost arm of the country; the south has significant clusters of people, while the central and western deserts are sparsely populated

Natural hazards: earthquakes; floods; landslides or mudslides; avalanches; droughts

Environment—current issues: shrinkage of the Aral Sea has resulted in growing concentrations of chemical pesticides and natural salts; these substances are then blown from the increasingly exposed lake bed and contribute to desertification and respiratory health problems; water pollution from industrial wastes and the heavy use of fertilizers and pesticides is the cause of many human health disorders; increasing soil salination; soil contamination from buried nuclear processing and agricultural chemicals, including DDT

Environment—international agreements: *party to:* Biodiversity, Climate Change, Climate Change-Kyoto Protocol, Desertification, Endangered Species, Environmental Modification, Hazardous Wastes, Ozone Layer Protection, Wetlands
signed, but not ratified: none of the selected agreements

Geography—note: along with Liechtenstein, one of the only two doubly landlocked countries in the world

PEOPLE AND SOCIETY

Population: 30,565,411 (July 2020 est.)
country comparison to the world: 45

Nationality: *noun:* Uzbekistani
adjective: Uzbekistani

Ethnic groups: Uzbek 83.8%, Tajik 4.8%, Kazakh 2.5%, Russian 2.3%, Karakalpak 2.2%, Tatar 1.5%, other 4.4% (2017 est.)

Languages: Uzbek (official) 74.3%, Russian 14.2%, Tajik 4.4%, other 7.1%

note: in the autonomous Karakalpakstan Republic, both the Karakalpak language and Uzbek have official status

Religions: Muslim 88% (mostly Sunni), Eastern Orthodox 9%, other 3%

Age structure: *0-14 years:* 23.19% (male 3,631,693/female 3,456,750)
15-24 years: 16.63% (male 2,601,803/female 2,481,826)
25-54 years: 45.68% (male 6,955,260/female 7,006,172)
55-64 years: 8.63% (male 1,245,035/female 1,392,263)
65 years and over: 5.87% (male 768,769/female 1,025,840) (2020 est.)

Dependency ratios: *total dependency ratio:* 50.6
youth dependency ratio: 43.4
elderly dependency ratio: 7.2
potential support ratio: 13.9 (2020 est.)

Median age: *total:* 30.1 years
male: 29.4 years
female: 30.7 years (2020 est.)
country comparison to the world: 123

Population growth rate: 0.88% (2020 est.)
country comparison to the world: 118

Birth rate: 16.1 births/1,000 population (2020 est.)
country comparison to the world: 109

Death rate: 5.4 deaths/1,000 population (2020 est.)
country comparison to the world: 188

Net migration rate: -1.9 migrant(s)/1,000 population (2020 est.)
country comparison to the world: 168

Population distribution: most of the population is concentrated in the fertile Fergana Valley in the easternmost arm of the country; the south has significant clusters of people, while the central and western deserts are sparsely populated

Urbanization: *urban population:* 50.4% of total population (2020)
rate of urbanization: 1.28% annual rate of change (2015-20 est.)
total population growth rate v. urban population growth rate, 2000-2030: Major urban areas - population: 2.517 million TASHKENT (capital) (2020)

Sex ratio: *at birth:* 1.06 male(s)/female
0-14 years: 1.05 male(s)/female
15-24 years: 1.05 male(s)/female
25-54 years: 0.99 male(s)/female
55-64 years: 0.89 male(s)/female
65 years and over: 0.75 male(s)/female
total population: 0.99 male(s)/female (2020 est.)

Mother's mean age at first birth: 23.8 years (2017 est.)

Maternal mortality rate: 29 deaths/100,000 live births (2017 est.)
country comparison to the world: 114

Infant mortality rate: *total:* 16.3 deaths/1,000 live births

male: 19.4 deaths/1,000 live births

female: 13.1 deaths/1,000 live births (2020 est.)

country comparison to the world: 93

Life expectancy at birth: *total population:* 74.8 years

male: 71.7 years

female: 78 years (2020 est.)

country comparison to the world: 125

Total fertility rate: 1.74 children born/woman (2020 est.)

country comparison to the world: 164

Drinking water source: *improved:* urban: 100% of population

rural: 96.1% of population

total: 97.8% of population

unimproved: urban: 0% of population

rural: 3.9% of population

total: 2.2% of population (2017 est.)

Current Health Expenditure: 6.4% (2017)

Physicians density: 2.37 physicians/1,000 population (2014)

Hospital bed density: 4 beds/1,000 population (2014)

Sanitation facility access: *improved:* urban: 100% of population

rural: 100% of population

total: 100% of population

unimproved: urban: 0% of population

rural: 0% of population

total: 0% of population (2017 est.)

HIV/AIDS— adult prevalence rate: 0.2% (2019 est.)

country comparison to the world: 115

HIV/AIDS—people living with HIV/AIDS: 50,000 (2019 est.)

country comparison to the world: 61

HIV/AIDS—deaths: <1000 (2019 est.)

Obesity—adult prevalence rate: 16.6% (2016)

country comparison to the world: 123

Children under the age of 5 years underweight: 2.9% (2017)

country comparison to the world: 102

Education expenditures: 6.3% of GDP (2017)

country comparison to the world: 26

Literacy: *definition:* age 15 and over can read and write

total population: 100%

male: 100%

female: 100% (2016)

School life expectancy (primary to tertiary education): *total:* 13 years

male: 13 years

female: 12 years (2019)

GOVERNMENT

Country name: *conventional long form:* Republic of Uzbekistan

conventional short form: Uzbekistan

local long form: O'zbekiston Respublikasi

local short form: O'zbekiston

former: Uzbek Soviet Socialist Republic

etymology: a combination of the Turkic words "uz" (self) and "bek" (master) with the Persian suffix "-stan" (country) to give the meaning "Land of the Free"

Government type: presidential republic; highly authoritarian

Capital: *name:* Tashkent (Toshkent)

geographic coordinates: 41 19 N, 69 15 E

time difference: UTC+5 (10 hours ahead of Washington, DC, during Standard Time)

etymology: "tash" means "stone" and "kent" means "city" in Turkic languages, so the name simply denotes "stone city"

Administrative divisions: 12 provinces (viloyatlar, singular - viloyat), 1 autonomous republic* (avtonom respublikasi), and 1 city** (shahar); Andijon Viloyati, Buxoro Viloyati [Bukhara Province], Farg'ona Viloyati [Fergana Province], Jizzax Viloyati, Namangan Viloyati, Navoiy Viloyati, Qashqadaryo Viloyati (Qarshi), Qoraqalpog'iston Respublikasi [Karakalpakstan Republic]* (Nukus), Samarqand Viloyati [Samarkand Province], Sirdaryo Viloyati (Guliston), Surxondaryo Viloyati (Termiz), Toshkent Shahri [Tashkent City]**, Toshkent Viloyati [Tashkent Province], Xorazm Viloyati (Urganch)

note: administrative divisions have the same names as their administrative centers (exceptions have the administrative center name following in parentheses)

Independence: 1 September 1991 (from the Soviet Union)

National holiday: Independence Day, 1 September (1991)

Constitution: *history:* several previous; latest adopted 8 December 1992

amendments: proposed by the Supreme Assembly or by referendum; passage requires two-thirds majority vote of both houses of the Assembly or passage in a referendum; amended several times, last in 2017 (2018)

Legal system: civil law system; note - in early 2020, the president signed an amendment to the criminal code, criminal procedure code, and code of administrative responsibility

International law organization participation: has not submitted an ICJ jurisdiction declaration; non-party state to the ICCt

Citizenship: *citizenship by birth:* no

citizenship by descent only: at least one parent must be a citizen of Uzbekistan

dual citizenship recognized: no

residency requirement for naturalization: 5 years

Suffrage: 18 years of age; universal

Executive branch: *chief of state:* President Shavkat MIRZIYOYEV (interim president from 8 September 2016; formally elected president on 4 December 2016 to succeed longtime President Islom KARIMOV, who died on 2 September 2016)

head of government: Prime Minister Abdulla ARIPOV (since 14 December 2016); First Deputy Prime Minister/Minister of Transport Achilbay RAMATOV (since 15 December 2016)

cabinet: Cabinet of Ministers appointed by the president with most requiring approval of the Senate chamber of the Supreme Assembly (Oliy Majlis)

elections/appointments: president directly elected by absolute majority popular vote in 2 rounds if needed for a 5-year term (eligible for a second term; previously a 5-year term, extended by a 2002 constitutional amendment to 7 years, and reverted to 5 years in 2011); election last held on 4 December 2016 (next to be held in 2021); prime minister nominated by majority party in legislature since 2011, but appointed along with the ministers and deputy ministers by the president

election results: Shavkat MIRZIYOYEV elected president in first round; percent of vote - Shavkat MIRZIYOYEV (LDPU) 88.6%, Hotamjon KETMONOV (NDP) 3.7%, Narimon UMAROV (Adolat) 3.5%, Sarvar OTAMURODOV (Milliy Tiklanish/National Revival) 2.4%, other 1.8%

Legislative branch: *description:* bicameral Supreme Assembly or Oliy Majlis consists of: Senate (100 seats; 84 members indirectly elected by regional governing councils and 16 appointed by the president; members serve 5-year terms)

Legislative Chamber or Qonunchilik Palatasi (150 seats; members directly elected in single-seat constituencies by absolute majority vote with a second round, if needed; members serve 5-year terms)

elections: Senate - last held 13-14 January 2015 (next to be held in 2020)

Legislative Chamber—last held on 22 December 2019 and 5 January 2020 (next to be held in December 2024)

election results: Senate - percent of vote by party - NA; seats by party - NA; composition - men 83, women 17, percent of women 17% Legislative Chamber - percent of vote by party - NA; seats by party - LDPU 53, National Revival Democratic Party 36, Adolat 24, PDP 22, Ecological Movement 15; composition - NA

note: all parties in the Supreme Assembly support President Shavkat MIRZIYOYEV

Judicial branch: *highest courts:* Supreme Court (consists of 67 judges organized into administrative, civil, criminal, and economic sections); Constitutional Court (consists of 7 judges)

judge selection and term of office: judges of the highest courts nominated by the president and confirmed by the Senate of the Oliy Majlis; judges appointed for initial 5-year term and can be reappointed for subsequent 10-year and lifetime terms

subordinate courts: regional, district, city, and town courts

Political parties and leaders: Ecological Party of Uzbekistan (O'zbekiston Ekologik Partivasi) [Boriy ALIKHANOV]

Justice (Adolat) Social Democratic Party of Uzbekistan [Narimon UMAROV]

Liberal Democratic Party of Uzbekistan (O'zbekiston Liberal-Demokratik Partiyasi) or LDPU [Aktam HAITOV] National Revival Democratic Party of Uzbekistan (O'zbekiston Milliy Tiklanish Demokratik Partiyasi) [Sarvar OTAMURATOV]

People's Democratic Party of Uzbekistan (Xalq Demokratik Partiyas) or PDP [Hotamjon KETMONOV] (formerly Communist Party)

International organization participation: ADB, CICA, CIS, EAPC, EBRD, ECO, FAO, IAEA, IBRD, ICAO, ICC (national committees), ICCt, ICRM, IDA, IDB, IFAD, IFC, IFRCS, ILO, IMF, Interpol, IOC, ISO, ITSO, ITU, MIGA, NAM, OIC, OPCW, OSCE, PFP, SCO, UN, UNCTAD, UNESCO, UNIDO, UNWTO, UPU, WCO, WFTU (NGOs), WHO, WIPO, WMO, WTO (observer)

Diplomatic representation in the US: *chief of mission:* Ambassador Javlon VAHOBOV (since 29 November 2017)

chancery: 1746 Massachusetts Avenue NW, Washington, DC 20036

telephone: [1] (202) 887-5300

FAX: [1] (202) 293-6804

consulate(s) general: New York

Diplomatic representation from the US: *chief of mission:* Ambassador Daniel ROSENBLUM (since 24 May 2019)

telephone: [998] (71) 120-5450

embassy: 3 Moyqo'rq'on, 5th Block, Yunusobod District, Tashkent 100093

mailing address: use embassy street address

FAX: [998] (71) 120-6335

Flag description: three equal horizontal bands of blue (top), white, and green separated by red fimbriations with a vertical, white crescent moon (closed side to the hoist) and 12 white, five-pointed stars shifted to the hoist on the top band; blue is the color of the Turkic peoples and of the sky, white signifies peace and the striving for purity in thoughts and deeds, while green represents nature and is the color of Islam; the red stripes are the vital force of all living organisms that links good and pure ideas with the eternal sky and with deeds on earth; the crescent represents Islam and the 12 stars the months and constellations of the Uzbek calendar

National symbol(s): *khumo (mythical bird);* *national colors:* blue, white, red, green

National anthem: *name:* "O'zbekiston Respublikasining Davlat Madhiyasi" (National Anthem of the Republic of Uzbekistan)

lyrics/music: Abdulla ARIPOV/Mutal BURHANOV

note: adopted 1992; after the fall of the Soviet Union, Uzbekistan kept the music of the anthem from its time as a Soviet Republic but adopted new lyrics

ECONOMY

Economy—overview: Uzbekistan is a doubly land-locked country in which 51% of the population lives in urban settlements; the agriculture-rich Fergana Valley, in which Uzbekistan's eastern borders are situated, has been counted among the most densely populated parts of Central Asia. Since its independence in September 1991, the government has largely maintained its Soviet-style command economy with subsidies and tight controls on production, prices, and access to foreign currency. Despite ongoing efforts to diversify crops, Uzbek agriculture remains largely centered on cotton; Uzbekistan is the world's fifth-largest cotton exporter and seventh-largest producer. Uzbekistan's growth has been driven primarily by state-led investments, and export of natural gas, gold, and cotton provides a significant share of foreign exchange earnings.

Recently, lower global commodity prices and economic slowdowns in neighboring Russia and China have hurt Uzbekistan's trade and investment and worsened its foreign currency shortage. Aware of the need to improve the investment climate, the government is taking incremental steps to reform the business sector and address impediments to foreign investment in the country. Since the death of first President Islam KARIMOV and election of President Shavkat MIRZIYOYEV, emphasis on such initiatives and government efforts to improve the private sector have increased. In the past, Uzbek authorities accused US and other foreign companies operating in Uzbekistan of violating Uzbek laws and have frozen and seized their assets.

As a part of its economic reform efforts, the Uzbek Government is looking to expand opportunities for small and medium enterprises and prioritizes increasing foreign direct investment. In September 2017, the government devalued the official currency rate by almost 50% and announced the loosening of currency restrictions to eliminate the currency black market, increase access to hard currency, and boost investment.

GDP (purchasing power parity): $223 billion (2017 est.)

$211.8 billion (2016 est.)

$196.5 billion (2015 est.)

note: data are in 2017 dollars

country comparison to the world: 63

GDP (official exchange rate): $48.83 billion (2017 est.)

GDP—real growth rate: 5.3% (2017 est.)

7.8% (2016 est.)

7.9% (2015 est.)

country comparison to the world: 41

GDP—per capita (PPP): $6,900 (2017 est.)

$6,700 (2016 est.)

$6,300 (2015 est.)

note: data are in 2017 dollars

country comparison to the world: 158

Gross national saving: 32.7% of GDP (2017 est.)

25.4% of GDP (2016 est.)

27.6% of GDP (2015 est.)

country comparison to the world: 24

GDP—composition, by end use: *household consumption:* 59.5% (2017 est.)

government consumption: 16.3% (2017 est.)

investment in fixed capital: 25.3% (2017 est.)

investment in inventories: 3% (2017 est.)

exports of goods and services: 19% (2017 est.)

imports of goods and services: -20% (2017 est.)

GDP—composition, by sector of origin:

agriculture: 17.9% (2017 est.)

industry: 33.7% (2017 est.)

services: 48.5% (2017 est.)

Agriculture—products: cotton, vegetables, fruits, grain; livestock

Industries: textiles, food processing, machine building, metallurgy, mining, hydrocarbon extraction, chemicals

Industrial production growth rate: 4.5% (2017 est.)

country comparison to the world: 67

Labor force: 13.273 million (2018 est.)

country comparison to the world: 41

Labor force—by occupation: *agriculture:* 25.9%

industry: 13.2%

services: 60.9% (2012 est.)

Unemployment rate: 5% (2017 est.)

5.1% (2016 est.)

note: official data; another 20% are underemployed

country comparison to the world: 79

Population below poverty line: 14% (2016 est.)

Household income or consumption by percentage share: *lowest 10%:* 2.8%

highest 10%: 29.6% (2003)

Budget: *revenues:* 15.22 billion (2017 est.)

expenditures: 15.08 billion (2017 est.)

Taxes and other revenues: 31.2% (of GDP) (2017 est.)

country comparison to the world: 74

Budget surplus (+) or deficit (-): 0.3% (of GDP) (2017 est.)

country comparison to the world: 42

Public debt: 24.3% of GDP (2017 est.)

10.5% of GDP (2016 est.)

country comparison to the world: 178

Fiscal year: calendar year

Inflation rate (consumer prices): 12.5% (2017 est.)

8% (2016 est.)

note: official data; based on independent analysis of consumer prices, inflation reached 22% in 2012

country comparison to the world: 208

Current account balance: $1.713 billion (2017 est.)

$384 million (2016 est.)

country comparison to the world: 43

Exports: $11.48 billion (2017 est.)

$11.2 billion (2016 est.)

country comparison to the world: 84

Exports—partners: Switzerland 38.7%, China 15.5%, Russia 10.7%, Turkey 8.6%, Kazakhstan 7.7%, Afghanistan 4.7% (2017)

Exports—commodities: energy products, cotton, gold, mineral fertilizers, ferrous and non-ferrous metals, textiles, foodstuffs, machinery, automobiles

Imports: $11.42 billion (2017 est.)

$10.92 billion (2016 est.)

country comparison to the world: 93

1035

Imports—commodities: machinery and equipment, foodstuffs, chemicals, ferrous and nonferrous metals

Imports—partners: China 23.7%, Russia 22.5%, Kazakhstan 10.7%, South Korea 9.8%, Turkey 5.8%, Germany 5.6% (2017)

Reserves of foreign exchange and gold: $16 billion (31 December 2017 est.)
$14 billion (31 December 2016 est.)
country comparison to the world: 65

Debt—external: $16.9 billion (31 December 2017 est.)
$16.76 billion (31 December 2016 est.)
country comparison to the world: 101

Exchange rates: Uzbekistani soum (UZS) per US dollar -
3,906.1 (2017 est.)
2,966.6 (2016 est.)
2,966.6 (2015 est.)
2,569.6 (2014 est.)
2,311.4 (2013 est.)

ENERGY

Electricity access: *electrification - total population:* 100% (2020)

Electricity—production: 55.55 billion kWh (2016 est.)
country comparison to the world: 52

Electricity—consumption: 49.07 billion kWh (2016 est.)
country comparison to the world: 50

Electricity—exports: 13 billion kWh (2014 est.)
country comparison to the world: 15

Electricity—imports: 10.84 billion kWh (2016 est.)
country comparison to the world: 23

Electricity—installed generating capacity: 12.96 million kW (2016 est.)
country comparison to the world: 54

Electricity—from fossil fuels: 86% of total installed capacity (2016 est.)
country comparison to the world: 69

Electricity—from nuclear fuels: 0% of total installed capacity (2017 est.)
country comparison to the world: 206

Electricity—from hydroelectric plants: 14% of total installed capacity (2017 est.)
country comparison to the world: 109

Electricity—from other renewable sources: 0% of total installed capacity (2017 est.)
country comparison to the world: 212

Crude oil—production: 41,000 bbl/day (2018 est.)
country comparison to the world: 57

Crude oil—exports: 27,000 bbl/day (2015 est.)
country comparison to the world: 46

Crude oil—imports: 420 bbl/day (2015 est.)
country comparison to the world: 81

Crude oil—proved reserves: 594 million bbl (1 January 2018 est.)
country comparison to the world: 44

Refined petroleum products—production: 61,740 bbl/day (2015 est.)

country comparison to the world: 78

Refined petroleum products—consumption: 60,000 bbl/day (2016 est.)
country comparison to the world: 95

Refined petroleum products—exports: 3,977 bbl/day (2015 est.)
country comparison to the world: 95

Refined petroleum products—imports: 0 bbl/day (2015 est.)
country comparison to the world: 215

Natural gas—production: 52.1 billion cu m (2017 est.)
country comparison to the world: 15

Natural gas—consumption: 43.07 billion cu m (2017 est.)
country comparison to the world: 22

Natural gas—exports: 9.401 billion cu m (2017 est.)
country comparison to the world: 22

Natural gas—imports: 0 cu m (2017 est.)
country comparison to the world: 207

Natural gas—proved reserves: 1.841 trillion cu m (1 January 2018 est.)
country comparison to the world: 18

Carbon dioxide emissions from consumption of energy: 95.58 million Mt (2017 est.)
country comparison to the world: 46

COMMUNICATIONS

Telephones—fixed lines: *total subscriptions:* 3,262,896
subscriptions per 100 inhabitants: 10.77 (2019 est.)
country comparison to the world: 40

Telephones—mobile cellular: *total subscriptions:* 30,662,740
subscriptions per 100 inhabitants: 101.21 (2019 est.)
country comparison to the world: 45

Telecommunication systems: *general assessment:* digital exchanges in large cities and in rural areas; increased investment in infrastructure and growing subscriber base; fixed-line is underdeveloped due to preeminence of mobile market; introduction of prepaid Internet has contributed to home Internet usage; increase in mobile broadband penetration yet still early stages; Wi-Fi hotspot in the city of Tashkent in the future (2020)
domestic: fixed-line 11 per 100 person and mobile-cellular 101 per 100; the state-owned telecommunications company, Uzbek Telecom, owner of the fixed-line telecommunications system, has used loans from the Japanese government and the China Development Bank to upgrade fixed-line services including conversion to digital exchanges; mobile-cellular services are provided by 2 private and 3 state-owned operators with a total subscriber base of 22.8 million as of January 2018 (2019)
international: country code - 998; linked by fiber-optic cable or microwave radio relay with CIS member states and to other countries by leased connection via the Moscow international gateway switch; the country also has a link to the Trans-Asia-Europe (TAE) fiber-optic cable;

Uzbekistan has supported the national fiber-optic backbone project of Afghanistan since 2008
note: the COVID-19 outbreak is negatively impacting telecommunications production and supply chains globally; consumer spending on telecom devices and services has also slowed due to the pandemic's effect on economies worldwide; overall progress towards improvements in all facets of the telecom industry - mobile, fixed-line, broadband, submarine cable and satellite - has moderated

Broadcast media: the government controls media; 17 state-owned broadcasters - 13 TV and 4 radio - provide service to virtually the entire country; about 20 privately owned TV stations, overseen by local officials, broadcast to local markets; privately owned TV stations are required to lease transmitters from the government-owned Republic TV and Radio Industry Corporation; in 2019, the Uzbek Agency for Press and Information was reorganized into the Agency of Information and Mass Communications and became part of the Uzbek Presidential Administration with recent appointment of the Uzbek President's elder daughter as it deputy director (2019)

Internet country code: .uz

Internet users: *total:* 15,705,402
percent of population: 52.31% (July 2018 est.)
country comparison to the world: 42

Broadband—fixed subscriptions: *total:* 4,123,508
subscriptions per 100 inhabitants: 14 (2018 est.)
country comparison to the world: 33

TRANSPORTATION

National air transport system: *number of registered air carriers:* 2 (2020)
inventory of registered aircraft operated by all carriers: 34
annual passenger traffic on registered air carriers: 3,056,558 (2018)
annual freight traffic on registered air carriers: 89.43 million mt-km (2018)

Civil aircraft registration country code prefix: UK (2016)

Airports: 53 (2013)
country comparison to the world: 88

Airports—with paved runways: *total:* 33 (2013)
over 3,047 m: 6 (2013)
2,438 to 3,047 m: 13 (2013)
1,524 to 2,437 m: 6 (2013)
914 to 1,523 m: 4 (2013)
under 914 m: 4 (2013)

Airports—with unpaved runways: *total:* 20 (2013)
2,438 to 3,047 m: 2 (2013)
under 914 m: 18 (2013)

Pipelines: 13,700 km gas, 944 km oil (2016)

Railways: *total:* 4,642 km (2018)
broad gauge: 4,642 km 1.520-m gauge (1,684 km electrified) (2018)
country comparison to the world: 42

Roadways: *total:* 86,496 km (2000)
paved: 75,511 km (2000)
unpaved: 10,985 km (2000)
country comparison to the world: 58

Waterways: 1,100 km (2012)
country comparison to the world: 62

Ports and terminals: *river port(s):* Termiz (Amu Darya)

MILITARY AND SECURITY

Military and security forces: *Armed Forces of Uzbekistan:* Army, Air and Air Defense Forces; National Guard; Ministry of Internal Affairs: Internal Security Troops (2019)

Military expenditures: 4% of GDP (2018)
3.5% of GDP (2010)
country comparison to the world: 12

Military and security service personnel strengths: assessments for the size of the Armed Forces of Uzbekistan vary; approximately 55,000 total active troops (est. 45,000 Army; est. 10,000 Air and Air Defense Forces) (2019 est.)

Military equipment inventories and acquisitions: the Uzbek Armed Forces use mainly Soviet-era equipment, although since 2010 they have received weapons and aircraft from a variety of sources, including China, France, Russia, Spain, Turkey, and the US (2019 est.)

Military service age and obligation: 18-27 years of age for compulsory military service; 1-year conscript service obligation for males (conscripts have the option of paying for a shorter service of one month while remaining in the reserves until the age of 27); Uzbek citizens who have completed their service terms in the armed forces have privileges in employment and admission to higher educational institutions (2016)

TERRORISM

Terrorist group(s): Islamic Jihad Union; Islamic Movement of Uzbekistan (2019)
note: details about the history, aims, leadership, organization, areas of operation, tactics, targets, weapons, size, and sources of support of the group(s) appear(s) in Appendix-T

TRANSNATIONAL ISSUES

Disputes—international: prolonged drought and cotton monoculture in Uzbekistan and Turkmenistan created water-sharing difficulties for Amu Darya river states; field demarcation of the boundaries with Kazakhstan commenced in 2004; border delimitation of 130 km of border with Kyrgyzstan is hampered by serious disputes around enclaves and other areas

Refugees and internally displaced persons: *stateless persons:* 97,346 (2019)

Trafficking in persons: *current situation:* Uzbekistan is a source country for men, women, and children subjected to forced labor and women and children subjected to sex trafficking; government-compelled forced labor of adults remained endemic during the 2014 cotton harvest; despite a decree banning the use of persons under 18, children were mobilized to harvest cotton by local officials in some districts; in some regions, local officials forced teachers, students, private business employees, and others to work in construction, agriculture, and cleaning parks; Uzbekistani women and children are victims of sex trafficking domestically and in the Middle East, Eurasia, and Asia; Uzbekistani men and, to a lesser extent, women are subjected to forced labor in Kazakhstan, Russia, and Ukraine in the construction, oil, agriculture, retail, and food sectors

tier rating: Tier 2 Watch List – Uzbekistan does not fully comply with the minimum standards for the elimination of trafficking; however, it is making significant efforts to do so; law enforcement efforts in 2014 were mixed; the government made efforts to combat sex and transnational labor trafficking, but government-compelled forced labor of adults in the cotton harvest went unaddressed, and the decree prohibiting forced child labor was not applied universally; official complicity in human trafficking in the cotton harvest remained prevalent; authorities made efforts to identify and protect sex and transnational labor victims, although a systematic process is still lacking; minimal efforts were made to assist victims of forced labor in the cotton harvest, as the government does not openly acknowledge the existence of this forced labor; the ILO did not have permission or funding to monitor the 2014 harvest, but the government authorized the UN's International Labour Organization to conduct a survey on recruitment practices and working conditions in agriculture, particularly the cotton sector, and to monitor the 2015-17 cotton harvests for child and forced labor in project areas (2015)

Illicit drugs: transit country for Afghan narcotics bound for Russian and, to a lesser extent, Western European markets; limited illicit cultivation of cannabis and small amounts of opium poppy for domestic consumption; poppy cultivation almost wiped out by government crop eradication program; transit point for heroin precursor chemicals bound for Afghanistan

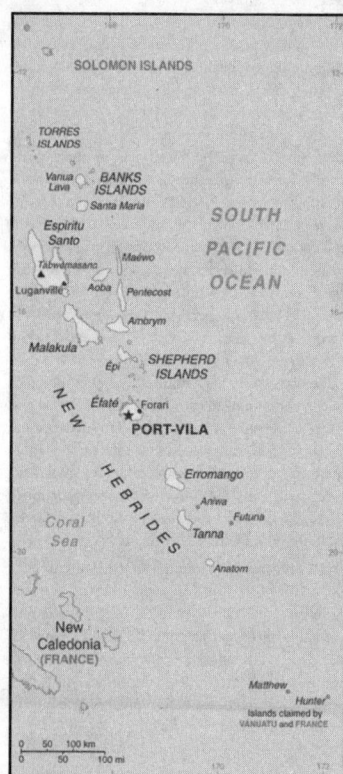

SOLOMON ISLANDS

TORRES ISLANDS
Vanua Lava
Santa Maria
BANKS ISLANDS
Espiritu Santo
Tabwemasana
Maéwo
Luganville
Aoba
Pentecost
Ambrym
Malakula
Epi
SHEPHERD ISLANDS
Éfaté
Forari
PORT-VILA

SOUTH PACIFIC OCEAN

Erromango
Aniwa
Tanna
Futuna
Anatom

Coral Sea

NEW HEBRIDES

New Caledonia (FRANCE)

Matthew
Hunter
Islands claimed by VANUATU and FRANCE

0 50 100 km
0 50 100 mi

INTRODUCTION

Background: Multiple waves of colonizers, each speaking a distinct language, migrated to the New Hebrides in the millennia preceding European exploration in the 18th century. This settlement pattern accounts for the complex linguistic diversity found on the archipelago to this day. The British and French, who settled the New Hebrides in the 19th century, agreed in 1906 to an Anglo-French Condominium, which administered the islands until independence in 1980, when the new name of Vanuatu was adopted. Politics and society continue to be divided along linguistic lines, although those divisions are lessening over time. Coalition governments tend to be weak, and since 2008, prime ministers have been ousted through no-confidence motions or temporary procedural issues 10 times. Prime Minister Charlot SALAWI has survived at least five no-confidence motions since taking office in 2016.

GEOGRAPHY

Location: Oceania, group of islands in the South Pacific Ocean, about three-quarters of the way from Hawaii to Australia

Geographic coordinates: 16 00 S, 167 00 E

Map references: Oceania

Area: *total:* 12,189 sq km
land: 12,189 sq km
water: 0 sq km
note: includes more than 80 islands, about 65 of which are inhabited
country comparison to the world: 163

Area—comparative: slightly larger than Connecticut

Land boundaries: 0 km

Coastline: 2,528 km

Maritime claims: *territorial sea:* 12 nm
exclusive economic zone: 200 nm
contiguous zone: 24 nm
continental shelf: 200 nm or to the edge of the continental margin

measured from claimed archipelagic baselines

Climate: tropical; moderated by southeast trade winds from May to October; moderate rainfall from November to April; may be affected by cyclones from December to April

Terrain: mostly mountainous islands of volcanic origin; narrow coastal plains

Elevation: *lowest point:* Pacific Ocean 0 m
highest point: Tabwemasana 1,877 m

Natural resources: manganese, hardwood forests, fish

Land use: *agricultural land:* 15.3% (2011 est.)
arable land: 1.6% (2011 est.) / *permanent crops:* 10.3% (2011 est.) / *permanent pasture:* 3.4% (2011 est.)
forest: 36.1% (2011 est.)
other: 48.6% (2011 est.)

Irrigated land: 0 sq km (2012)

Population distribution: three-quarters of the population lives in rural areas; the urban populace lives primarily in two cities, Port-Vila and Lugenville; three largest islands - Espiritu Santo, Malakula, and Efate - accomodate over half of the populace

Natural hazards: tropical cyclones (January to April); volcanic eruption on Aoba (Ambae) island began on 27 November 2005, volcanism also causes minor earthquakes; tsunamis
volcanism: significant volcanic activity with multiple eruptions in recent years; Yasur (361 m), one of the world's most active volcanoes, has experienced continuous activity in recent centuries; other historically active volcanoes include Aoba, Ambrym, Epi, Gaua, Kuwae, Lopevi, Suretamatai, and Traitor's Head

Environment—current issues: population growth; water pollution, most of the population does not have access to a reliable supply of potable water; inadequate sanitation; deforestation

Environment—international agreements: *party to:* Antarctic-Marine Living Resources, Biodiversity, Climate Change, Climate Change-Kyoto Protocol, Desertification, Endangered Species, Law of the Sea, Marine Dumping, Ozone Layer Protection, Ship Pollution, Tropical Timber 94
signed, but not ratified: none of the selected agreements

Geography—note: a Y-shaped chain of four main islands and 80 smaller islands; several of the islands have active volcanoes and there are several underwater volcanoes as well

POPULATION: PEOPLE AND SOCIETY

298,333 (July 2020 est.)
country comparison to the world: 180

Nationality: *noun:* Ni-Vanuatu (singular and plural)
adjective: Ni-Vanuatu

Ethnic groups: Melanesian 99.2%, non-Melanesian 0.8% (2016 est.)

Languages: local languages (more than 100) 63.2%, Bislama (official; creole) 33.7%, English (official) 2%, French (official) 0.6%, other 0.5% (2009 est.)

Religions: Protestant 70% (includes Presbyterian 27.9%, Anglican 15.1%, Seventh Day Adventist 12.5%, Assemblies of God 4.7%, Church of Christ 4.5%, Neil Thomas Ministry 3.1%, and Apostolic 2.2%), Roman Catholic 12.4%, customary beliefs 3.7% (including Jon Frum cargo cult), other 12.6%, none 1.1%, unspecified 0.2% (2009 est.)

Age structure: *0-14 years:* 33.65% (male 51,267/ female 49,111)
15-24 years: 19.99% (male 29,594/female 30,050)
25-54 years: 36.09% (male 52,529/female 55,130)
55-64 years: 5.89% (male 8,666/female 8,904)
65 years and over: 4.39% (male 6,518/female 6,564) (2020 est.)

population pyramid: [INSERT IMAGE: VANUATU - population pyramid]

Dependency ratios: *total dependency ratio:* 72.5
youth dependency ratio: 66.2
elderly dependency ratio: 12.3
potential support ratio: 8.1 (2020 est.)

Median age: *total:* 23 years
male: 22.6 years
female: 23.5 years (2020 est.)
country comparison to the world: 179

Population growth rate: 1.73% (2020 est.)
country comparison to the world: 60

Birth rate: 22.4 births/1,000 population (2020 est.)
country comparison to the world: 65

Death rate: 4 deaths/1,000 population (2020 est.)
country comparison to the world: 214

Net migration rate: -1.3 migrant(s)/1,000 population (2020 est.)
country comparison to the world: 151

Population distribution: three-quarters of the population lives in rural areas; the urban populace lives primarily in two cities, Port-Vila and Lugenville; three largest islands - Espiritu Santo,

Malakula, and Efate - accomodate over half of the populace

Urbanization: *urban population:* 25.5% of total population (2020)

rate of urbanization: 2.55% annual rate of change (2015-20 est.)

total population growth rate v. urban population growth rate, 2000-2030: *Major urban areas - population:* 53,000 PORT-VILA (capital) (2018)

Sex ratio: *at birth:* 1.05 male(s)/female
0-14 years: 1.04 male(s)/female
15-24 years: 0.98 male(s)/female
25-54 years: 0.95 male(s)/female
55-64 years: 0.97 male(s)/female
65 years and over: 0.99 male(s)/female
total population: 0.99 male(s)/female (2020 est.)

Maternal mortality rate: 72 deaths/100,000 live births (2017 est.)
country comparison to the world: 82

Infant mortality rate: *total:* 12.7 deaths/1,000 live births
male: 13.6 deaths/1,000 live births
female: 11.9 deaths/1,000 live births (2020 est.)
country comparison to the world: 105

Life expectancy at birth: *total population:* 74.6 years
male: 72.9 years
female: 76.4 years (2020 est.)
country comparison to the world: 130

Total fertility rate: 2.79 children born/woman (2020 est.)
country comparison to the world: 61

Contraceptive prevalence rate: 49% (2013)

Drinking water source:

improved: *urban:* 100% of population
rural: 89.7% of population
total: 92.3% of population

unimproved: *urban:* 0% of population
rural: 10.3% of population
total: 7.7% of population (2017 est.)

Current Health Expenditure: 3.3% (2017)

Physicians density: 0.17 physicians/1,000 population (2016)

Sanitation facility access:

improved: *urban:* 91.6% of population
rural: 60.9% of population
total: 68.6% of population

unimproved: *urban:* 8.4% of population
rural: 39.1% of population
total: 31.4% of population (2017 est.)

HIV/AIDS—adult prevalence rate: NA

HIV/AIDS—people living with HIV/AIDS: NA

HIV/AIDS—deaths: NA

Major infectious diseases: *degree of risk:* high (2020)
food or waterborne diseases: bacterial diarrhea
vectorborne diseases: malaria

Obesity—adult prevalence rate: 25.2% (2016)
country comparison to the world: 52

Children under the age of 5 years underweight: 11.7% (2013)
country comparison to the world: 57

Education expenditures: 4.7% of GDP (2017)
country comparison to the world: 81

Literacy: *definition:* age 15 and over can read and write
total population: 87.5%
male: 88.3%
female: 86.7% (2018)

GOVERNMENT

Country name: *conventional long form:* Republic of Vanuatu
conventional short form: Vanuatu
local long form: Ripablik blong Vanuatu
local short form: Vanuatu
former: New Hebrides
etymology: derived from the words "vanua" (home or land) and "tu" (stand) that occur in several of the Austonesian languages spoken on the islands and which provide a meaning of "the land remains" but which also convey a sense of "independence" or "our land"

Government type: parliamentary republic

Capital: *name:* Port-Vila (on Efate)

geographic coordinates: 17 44 S, 168 19 E
time difference: UTC+11 (16 hours ahead of Washington, DC, during Standard Time)
etymology: there are two possibilities for the origin of the name: early European settlers were Portuguese and "vila" means "village or town" in Portuguese, hence "Port-Vila" would mean "Port Town"; alternatively, the site of the capital is referred to as "Efil" or "Ifira" in native languages, "Vila" is a likely corruption of these names

Administrative divisions: 6 provinces; Malampa, Penama, Sanma, Shefa, Tafea, Torba

Independence: 30 July 1980 (from France and the UK)

National holiday: Independence Day, 30 July (1980)

Constitution: *history:* draft completed August 1979, finalized by constitution conference 19 September 1979, ratified by French and British Governments 23 October 1979, effective 30 July 1980 at independence
amendments: proposed by the prime minister or by the Parliament membership; passage requires at least two-thirds majority vote by Parliament in special session with at least three fourths of the membership; passage of amendments affecting the national and official languages, or the electoral and parliamentary system also requires approval in a referendum; amended several times, last in 2013

Legal system: mixed legal system of English common law, French law, and customary law

International law organization participation: has not submitted an ICJ jurisdiction declaration; accepts ICCt jurisdiction

Citizenship: *citizenship by birth:* no
citizenship by descent only: both parents must be citizens of Vanuatu; in the case of only one parent, it must be the father who is a citizen
dual citizenship recognized: no

residency requirement for naturalization: 10 years

Suffrage: 18 years of age; universal

Executive branch: *chief of state:* President Tallis Obed MOSES (since 6 July 2017)

head of government: Prime Minister Bob LOUGHMAN (since 20 April 2020)
cabinet: Council of Ministers appointed by the prime minister, responsible to Parliament
elections/appointments: president indirectly elected by an electoral college consisting of Parliament and presidents of the 6 provinces; Vanuatu president serves a 5-year term; election last held on 17 June 2017 (next to be held in 2022); following legislative elections, the leader of the majority party or majority coalition usually elected prime minister by Parliament from among its members; election for prime minister last held on 20 April 2020 (next to be held following general elections in 2024)
election results: Bob LOUGHMAN elected prime minister on 20 April 2020; Bob LOUGHMAN 31 votes, Ralph REGENVANU 21 votes

Legislative branch: *description:* unicameral Parliament (52 seats; members directly elected in 8 single-seat and 9 multi-seat constituencies by single non-transferable vote to serve 4-year terms (candidates in multi-seat constituencies can be elected with only 4% of the vote)
elections: last held on 19–20 March 2020 (next to be held in 2024)
election results: percent of vote by party - NA; seats by party - GJP 9, RMC 7, VP 7, LPV 5, UMP 5, NUP 4, other 15; composition - men 52, women 0; percent of women 0%; note - political party associations are fluid
note: the National Council of Chiefs advises on matters of culture and language

Judicial branch: *highest courts:* Court of Appeal (consists of 2 or more judges of the Supreme Court designated by the chief justice); Supreme Court (consists of the chief justice and 6 puisne judges - 3 local and 3 expatriate)
judge selection and term of office: Supreme Court chief justice appointed by the president after consultation with the prime minister and the leader of the opposition; other judges appointed by the president on the advice of the Judicial Service Commission, a 4-member advisory body; judges serve until the age of retirement
subordinate courts: Magistrates Courts; Island Courts

Political parties and leaders: Greens Confederation or GC [Moana CARCASSES Kalosil]
Iauko Group or IG [Tony NARI]
Land and Justice Party (Graon mo Jastis Pati) or GJP [Ralph REGENVANU]
Melanesian Progressive Party or MPP [Barak SOPE]
Nagriamel movement or NAG [Frankie STEVENS]
Natatok Indigenous People's Democratic Party or (NATATOK) or NIPDP [Alfred Roland CARLOT]
National United Party or NUP [Ham LINI]

People's Progressive Party or PPP [Sato KILMAN]

People's Service Party or PSP [Don KEN]

Reunification of Movement for Change or RMC [Charlot SALWAI]

Rural Development Party or RDP [Jay NGWELE, spokesman]

Union of Moderate Parties or UMP [Serge VOHOR]

Vanua'aku Pati (Our Land Party) or VP [Edward NATAPEI]

Vanuatu Democratic Party [Maxime Carlot KORMAN]

Vanuatu First or Vanuatu [Russel NARI]

Vanuatu Liberal Movement or VLM [Gaetan PIKIOUNE]

Vanuatu Liberal Democratic Party or VLDP [Tapangararua WILLIE]

Vanuatu National Party or VNP [Issac HAMARILIU]

Vanuatu National Development Party or VNDP [Robert Bohn SIKOL]

Vanuatu Republican Party or VRP [Marcellino PIPITE]

International organization participation: ACP, ADB, AOSIS, C, FAO, G-77, IBRD, ICAO, ICRM, IDA, IFC, IFRCS, ILO, IMF, IMO, IMSO, IOC, IOM, ITU, ITUC (NGOs), MIGA, NAM, OAS (observer), OIF, OPCW, PIF, Sparteca, SPC, UN, UNCTAD, UNESCO, UNIDO, UNWTO, UPU, WCO, WFTU (NGOs), WHO, WIPO, WMO, WTO

Diplomatic representation in the US: none; the Vanuatu Permanent Mission to the UN serves as the embassy; it is headed by Odo TEVI (since 8 September 2017); address 000 Grand Avenue Suite 400B, New York, NY 10017; telephone: [1] (212) 661-4303; FAX: [1] (212) 422-2437

Diplomatic representation from the US: the US does not have an embassy in Vanuatu; the US Ambassador to Papua New Guinea is accredited to Vanuatu

Flag description: two equal horizontal bands of red (top) and green with a black isosceles triangle (based on the hoist side) all separated by a black-edged yellow stripe in the shape of a horizontal Y (the two points of the Y face the hoist side and enclose the triangle); centered in the triangle is a boar's tusk encircling two crossed namele fern fronds, all in yellow; red represents the blood of boars and men, as well as unity, green the richness of the islands, and black the ni-Vanuatu people; the yellow Y-shape - which reflects the pattern of the islands in the Pacific Ocean - symbolizes the light of the Gospel spreading through the islands; the boar's tusk is a symbol of prosperity frequently worn as a pendant on the islands; the fern fronds represent peace

note: one of several flags where a prominent component of the design reflects the shape of the country; other such flags are those of Bosnia and Herzegovina, Brazil, and Eritrea

National symbol(s): boar's tusk with crossed fern fronds; national colors: red, black, green, yellow

National anthem: *name:* "Yumi, Yumi, Yumi" (We, We, We)

lyrics/music: Francois Vincent AYSSAV

note: adopted 1980; the anthem is written in Bislama, a Creole language that mixes Pidgin English and French

0:00 / 0:59

ECONOMY

Economy—overview: This South Pacific island economy is based primarily on small-scale agriculture, which provides a living for about two thirds of the population. Fishing, offshore financial services, and tourism, with more than 330,000 visitors in 2017, are other mainstays of the economy. Tourism has struggled after Efate, the most populous and most popular island for tourists, was damaged by Tropical Cyclone Pam in 2015. Ongoing infrastructure difficulties at Port Vila's Bauerfield Airport have caused air travel disruptions, further hampering tourism numbers. Australia and New Zealand are the main source of tourists and foreign aid. A small light industry sector caters to the local market. Tax revenues come mainly from import duties. Mineral deposits are negligible; the country has no known petroleum deposits.

Economic development is hindered by dependence on relatively few commodity exports, vulnerability to natural disasters, and long distances from main markets and between constituent islands. In response to foreign concerns, the government has promised to tighten regulation of its offshore financial center.

Since 2002, the government has stepped up efforts to boost tourism through improved air connections, resort development, and cruise ship facilities. Agriculture, especially livestock farming, is a second target for growth.

GDP (purchasing power parity): $772 million (2017 est.)

$740.9 million (2016 est.)

$716.1 million (2015 est.)

note: data are in 2017 dollars

country comparison to the world: 207

GDP (official exchange rate): $870 million (2017 est.)

GDP—real growth rate: 4.2% (2017 est.)

3.5% (2016 est.)

0.2% (2015 est.)

country comparison to the world: 70

GDP—per capita (PPP): $2,700 (2017 est.)

$2,700 (2016 est.)

$2,700 (2015 est.)

note: data are in 2017 dollars

country comparison to the world: 196

GDP—composition, by end use: *household consumption:* 59.9% (2017 est.)

government consumption: 17.4% (2017 est.)

investment in fixed capital: 28.7% (2017 est.)

investment in inventories: 0% (2017 est.)

exports of goods and services: 42.5% (2017 est.)

imports of goods and services: -48.5% (2017 est.)

GDP—composition, by sector of origin: *agriculture:* 27.3% (2017 est.)

industry: 11.8% (2017 est.)

services: 60.8% (2017 est.)

Agriculture—products: copra, coconuts, cocoa, coffee, taro, yams, fruits, vegetables; beef; fish

Industries: food and fish freezing, wood processing, meat canning

Industrial production growth rate: 4.5% (2017 est.)

country comparison to the world: 68

Labor force: 115,900 (2007 est.)

country comparison to the world: 182

Labor force—by occupation: agriculture: 65%

industry: 5%

services: 30% (2000 est.)

Unemployment rate: 1.7% (1999 est.)

country comparison to the world: 17

Population below poverty line: NA

Household income or consumption by percentage share: *lowest 10%:* NA

highest 10%: NA

Budget: *revenues:* 236.7 million (2017 est.)

expenditures: 244.1 million (2017 est.)

Taxes and other revenues: 27.2% (of GDP) (2017 est.)

country comparison to the world: 102

Budget surplus (+) or deficit (-): -0.9% (of GDP) (2017 est.)

country comparison to the world: 73

Public debt: 48.4% of GDP (2017 est.)

46.1% of GDP (2016 est.)

country comparison to the world: 107

Fiscal year: calendar year

Inflation rate (consumer prices): 3.1% (2017 est.)

0.8% (2016 est.)

country comparison to the world: 134

Current account balance: -$13 million (2017 est.)

-$37 million (2016 est.)

country comparison to the world: 70

Exports: $44.7 million (2017 est.)

$53.5 million (2016 est.)

country comparison to the world: 204

Exports—partners: Philippines 23.9%, Australia 16.5%, US 10.4%, Japan 8.8%, Venezuela 8%, France 4.8%, Fiji 4.5%, Hong Kong 4.4% (2017)

Exports—commodities: copra, beef (veal), cocoa, timber, kava, coffee, coconut oil, shell, cowhides, coconut meal, fish

Imports: $273.7 million (2017 est.)

$308.5 million (2016 est.)

country comparison to the world: 207

Imports—commodities: machinery and equipment, foodstuffs, fuels

Imports—partners: Russia 35.2%, Australia 19.8%, NZ 9.8%, China 6.3%, Fiji 5.5% (2017)

Reserves of foreign exchange and gold: $395.1 million (31 December 2017 est.)

$267.4 million (31 December 2016 est.)

country comparison to the world: 160

Debt—external: $200.5 million (31 December 2017 est.)

$182.5 million (31 December 2016 est.)

country comparison to the world: 188

Exchange rates: vatu (VUV) per US dollar—
109.7 (2017 est.)
112.28 (2016 est.)
108.48 (2015 est.)
108.99 (2014 est.)
97.07 (2013 est.)

ENERGY

Electricity access: *electrification—total population:* 57.8% (2016)

electrification—urban areas: 91.4% (2016)

electrification—rural areas: 46.4% (2016)

Electricity—production: 63 million kWh (2016 est.)
country comparison to the world: 204

Electricity—consumption: 58.59 million kWh (2016 est.)
country comparison to the world: 204

Electricity—exports: 0 kWh (2016 est.)
country comparison to the world: 214

Electricity—imports: 0 kWh (2016 est.)
country comparison to the world: 215

Electricity—installed generating capacity: 37,000 kW (2016 est.)
country comparison to the world: 199

Electricity—from fossil fuels: 71% of total installed capacity (2016 est.)
country comparison to the world: 107

Electricity—from nuclear fuels: 0% of total installed capacity (2017 est.)
country comparison to the world: 207

Electricity—from hydroelectric plants: 0% of total installed capacity (2017 est.)
country comparison to the world: 211

Electricity—from other renewable sources: 29% of total installed capacity (2017 est.)
country comparison to the world: 21

Crude oil—production: 0 bbl/day (2018 est.)
country comparison to the world: 214

Crude oil—exports: 0 bbl/day (2015 est.)
country comparison to the world: 213

Crude oil—imports: 0 bbl/day (2015 est.)
country comparison to the world: 211

Crude oil—proved reserves: 0 bbl (1 January 2018 est.)
country comparison to the world: 210

Refined petroleum products—production: 0 bbl/day (2015 est.)
country comparison to the world: 213

Refined petroleum products—consumption: 1,100 bbl/day (2016 est.)
country comparison to the world: 205

Refined petroleum products— exports: 0 bbl/day (2015 est.)
country comparison to the world: 215

Refined petroleum products—imports: 1,073 bbl/day (2015 est.)
country comparison to the world: 201

Natural gas—production: 0 cu m (2017 est.)
country comparison to the world: 212

Natural gas—consumption: 0 cu m (2017 est.)
country comparison to the world: 211

Natural gas—exports: 0 cu m (2017 est.)
country comparison to the world: 208

Natural gas—imports: 0 cu m (2017 est.)
country comparison to the world: 208

Natural gas—proved reserves: 0 cu m (1 January 2014 est.)
country comparison to the world: 205

Carbon dioxide emissions from consumption of energy: 164,800 Mt (2017 est.)
country comparison to the world: 203

COMMUNICATIONS

Telephones—fixed lines: *total subscriptions:* 3,724

subscriptions per 100 inhabitants: 1.27 (2019 est.)
country comparison to the world: 209

Telephones—mobile cellular: total subscriptions: 259,317

subscriptions per 100 inhabitants: 88.44 (2019 est.)
country comparison to the world: 181

Telecommunication systems: *general assessment:* telecom services have progressed significantly in recent years; mobile phones are now the primary means of communication and more than 92% of the population is covered by a mobile network; 2016 saw the launch of LTE services and the introduction of rural satellite broadband services; mobile phone use in some rural areas is constrained by electricity shortages; investment in fixed broadband saw recent growth with fiber-optic cables; mobile broadband infrastructure also expanded with a reduction in prices; general broadband penetration is at 45%; Kacific-1 broadband satellite launch in 2019 will change telecommunications for the region (2020)

domestic: fixed-line 1 per 100 and mobile-cellular 88 per 100 (2019)

international: country code - 678; landing points for the ICN1 & ICN2 submarine cables providing connectivity to the Solomon Islands and Fiji; cables helped end-users with Internet bandwidth; satellite earth station - 1 Intelsat (Pacific Ocean) (2020)

note: the COVID-19 outbreak is negatively impacting telecommunications production and supply chains globally; consumer spending on telecom devices and services has also slowed due to the pandemic's effect on economies worldwide; overall progress towards improvements in all facets of the telecom industry - mobile, fixed-line, broadband, submarine cable and satellite - has moderated

Broadcast media: 1 state-owned TV station; multi-channel pay TV is available; state-owned Radio Vanuatu operates 2 radio stations; 2 privately owned radio broadcasters; programming from multiple international broadcasters is available

Internet country code: .vu

Internet users: total: 74,083

percent of population: 25.72% (July 2018 est.)
country comparison to the world: 184

Broadband—fixed subscriptions: total: 4,718

subscriptions per 100 inhabitants: 2 (2018 est.)
country comparison to the world: 179

TRANSPORTATION

National air transport system: *number of registered air carriers:* 1 (2020)

inventory of registered aircraft operated by air carriers: 8

annual passenger traffic on registered air carriers: 374,603 (2018)

annual freight traffic on registered air carriers: 1.66 million mt-km (2018)

Civil aircraft registration country code prefix: YJ (2016)

Airports: 31 (2013)
country comparison to the world: 113

Airports—with paved runways: *total:* 3 (2019)
2,438 to 3,047 m: 1
1,524 to 2,437 m: 1
914 to 1,523 m: 1

Airports—with unpaved runways: *total:* 28 (2013)

914 to 1,523 m: 7 (2013)
under 914 m: 21 (2013)

Roadways: *total:* 1,070 km (2000)
paved: 256 km (2000)
unpaved: 814 km (2000)
country comparison to the world: 183

Merchant marine: *total:* 369
by type: bulk carrier 26, container ship 1, general cargo 45, other 297 (2019)
country comparison to the world: 49

Ports and terminals: *major seaport(s):* Forari Bay, Luganville (Santo, Espiritu Santo), Port-Vila

MILITARY AND SECURITY

Military and security forces: no regular military forces; Vanuatu Police Force (VPF; includes Vanuatu Mobile Force (VMF) and Police Maritime Wing (VPMW)) (2019)

TRANSPORTATION

Disputes—international: Matthew and Hunter Islands east of New Caledonia claimed by Vanuatu and France

VENEZUELA

INTRODUCTION

Background: Venezuela was one of three countries that emerged from the collapse of Gran Colombia in 1830 (the others being Ecuador and New Granada, which became Colombia). For most of the first half of the 20th century, Venezuela was ruled by generally benevolent military strongmen who promoted the oil industry and allowed for some social reforms. Democratically elected governments have held sway since 1959, although the re-election of current disputed President Nicolás MADURO in an election boycotted by most opposition parties was widely viewed as fraudulent. Under Hugo CHAVEZ, president from 1999 to 2013, and his hand-picked successor, MADURO, the executive branch has exercised increasingly authoritarian control over other branches of government. National Assembly President Juan GUAIDO is currently recognized by more than 50 countries - including the United States - as the interim president while MADURO retains control of all other institutions within the country and has the support of security forces. Venezuela is currently authoritarian with only one democratic institution - the National Assembly - and strong restrictions on freedoms of expression and the press. The ruling party's economic policies expanded the state's role in the economy through expropriations of major enterprises, strict currency exchange and price controls that discourage private sector investment and production, and overdependence on the petroleum industry for revenues, among others. However, Caracas in 2019 relaxed some economic controls to mitigate some impacts of the economic crisis driven by a drop in oil production. Current concerns include human rights abuses, rampant violent crime, high inflation, and widespread shortages of basic consumer goods, medicine, and medical supplies.

GEOGRAPHY

Location: Northern South America, bordering the Caribbean Sea and the North Atlantic Ocean, between Colombia and Guyana

Geographic coordinates: 8 00 N, 66 00 W

Map references: South America

Area: *total:* 912,050 sq km
land: 882,050 sq km
water: 30,000 sq km
country comparison to the world: 34

Area—comparative: almost six times the size of Georgia; slightly more than twice the size of California

Area comparison map: [INSERT IMAGE: VENEZUELA-Area comparison map]

Land boundaries: *total:* 5,267 km
border countries (3): Brazil 2137 km, Colombia 2341 km, Guyana 789 km

Coastline: 2,800 km

Maritime claims: *territorial sea:* 12 nm
exclusive economic zone: 200 nm
contiguous zone: 15 nm
continental shelf: 200-m depth or to the depth of exploitation

Climate: tropical; hot, humid; more moderate in highlands

Terrain: Andes Mountains and Maracaibo Lowlands in northwest; central plains (llanos); Guiana Highlands in southeast

Elevation: *mean elevation:* 450 m
lowest point: Caribbean Sea 0 m
highest point: Pico Bolivar 4,978 m

Natural resources: petroleum, natural gas, iron ore, gold, bauxite, other minerals, hydropower, diamonds

Land use: *agricultural land:* 24.5% (2011 est.)
arable land: 3.1% (2011 est.) / permanent crops: 0.8% (2011 est.) / permanent pasture: 20.6% (2011 est.)
forest: 52.1% (2011 est.)
other: 23.4% (2011 est.)

Irrigated land: 10,550 sq km (2012)

Population distribution: most of the population is concentrated in the northern and western highlands along an eastern spur at the northern end of the Andes, an area that includes the capital of Caracas

Natural hazards: subject to floods, rockslides, mudslides; periodic droughts

Environment—current issues: sewage pollution of Lago de Valencia; oil and urban pollution of Lago de Maracaibo; deforestation; soil degradation; urban and industrial pollution, especially along the Caribbean coast; threat to the rainforest ecosystem from irresponsible mining operations

Environment—international agreements: *party to:* Antarctic Treaty, Biodiversity, Climate Change, Climate Change-Kyoto Protocol, Desertification, Endangered Species, Hazardous Wastes, Marine Life Conservation, Ozone Layer Protection, Ship Pollution, Tropical Timber 83, Tropical Timber 94, Wetlands
signed, but not ratified: none of the selected agreements

Geography—note: note 1: the country lies on major sea and air routes linking North and South America
note 2: Venezuela has some of the most unique geology in the world; tepuis are massive table-top mountains of the western Guiana Highlands that tend to be isolated and thus support unique endemic plant and animal species; their sheer cliffsides account for some of the most spectacular waterfalls in the world including Angel Falls, the world's highest (979 m) that drops off Auyan Tepui

PEOPLE AND SOCIETY

Population: 28,644,603 (July 2020 est.)
country comparison to the world: 50

Nationality: *noun:* Venezuelan(s)
adjective: Venezuelan

Ethnic groups: unspecified Spanish, Italian, Portuguese, Arab, German, African, indigenous people

Languages: Spanish (official), numerous indigenous dialects

Religions: nominally Roman Catholic 96%, Protestant 2%, other 2%

Demographic profile: Social investment in Venezuela during the CHAVEZ administration reduced poverty from nearly 50% in 1999 to about 27% in 2011, increased school enrollment, substantially decreased infant and child mortality, and improved access to potable water and sanitation through social investment. "Missions" dedicated to education, nutrition, healthcare, and sanitation were funded through petroleum revenues. The sustainability of this progress remains questionable, however, as the continuation of these social programs depends on the prosperity of Venezuela's oil industry. In the long-term, education and health care spending may increase economic growth and reduce income inequality, but rising costs and the staffing of new health care jobs with foreigners are slowing development.

While CHAVEZ was in power, more than one million predominantly middle- and upper-class Venezuelans are estimated to have emigrated. The brain drain is attributed to a repressive political system, lack of economic opportunities, steep inflation, a high crime rate, and corruption. Thousands of oil engineers emigrated to Canada, Colombia, and the United States following

CHAVEZ's firing of over 20,000 employees of the state-owned petroleum company during a 2002-03 oil strike. Additionally, thousands of Venezuelans of European descent have taken up residence in their ancestral homelands. Nevertheless, Venezuela has attracted hundreds of thousands of immigrants from South America and southern Europe because of its lenient migration policy and the availability of education and health care. Venezuela also has been a fairly accommodating host to Colombian refugees, numbering about 170,000 as of year-end 2016. However, since 2014, falling oil prices have driven a major economic crisis that has pushed Venezuelans from all walks of life to migrate or to seek asylum abroad to escape severe shortages of food, water, and medicine; soaring inflation; unemployment; and violence. As of November 2019, an estimated 4.6 million Venezuelans were refugees or migrants worldwide, with almost 80% taking refuge in Latin America and the Caribbean (notably Colombia, Peru, Chile, Ecuador, Argentina, and Brazil, as well as the Dominican Republic, Aruba, and Curacao). Asylum applications increased significantly in the US and Brazil in 2016 and 2017. Several receiving countries are making efforts to increase immigration restrictions and to deport illegal Venezuelan migrants - Ecuador and Peru in August 2018 began requiring valid passports for entry, which are difficult to obtain for Venezuelans. Nevertheless, Venezuelans continue to migrate to avoid economic collapse at home.

Age structure: *0-14 years:* 25.66% (male 3,759,280/female 3,591,897)

15-24 years: 16.14% (male 2,348,073/female 2,275,912)

25-54 years: 41.26% (male 5,869,736/female 5,949,082)

55-64 years: 8.76% (male 1,203,430/female 1,305,285)

65 years and over: 8.18% (male 1,069,262/female 1,272,646) (2020 est.)

population pyramid: [INSERT IMAGE: VENEZUELA-population pyramid]

Dependency ratios: *total dependency ratio:* 54.4

youth dependency ratio: 42.1

elderly dependency ratio: 12.3

potential support ratio: 8.1 (2020 est.)

Median age: *total:* 30 years

male: 29.4 years

female: 30.7 years (2020 est.)

country comparison to the world: 124

Population growth rate: -0.18% (2020 est.)

country comparison to the world: 207

Birth rate: 17.9 births/1,000 population (2020 est.)

country comparison to the world: 92

Death rate: 7.5 deaths/1,000 population (2020 est.)

country comparison to the world: 106

Net migration rate: -3.4 migrant(s)/1,000 population (2020 est.)

country comparison to the world: 183

Population distribution: most of the population is concentrated in the northern and western highlands along an eastern spur at the northern end of the Andes, an area that includes the capital of Caracas

Urbanization: *urban population:* 88.3% of total population (2020)

rate of urbanization: 1.28% annual rate of change (2015-20 est.)

total population growth rate v. urban population growth rate, 2000-2030: *Major urban areas - population:* 2.939 million CARACAS (capital), 2.258 million Maracaibo, 1.910 million Valencia, 1.214 million Barquisimeto, 1.203 million Maracay (2020)

Sex ratio: *at birth:* 1.05 male(s)/female

0-14 years: 1.05 male(s)/female

15-24 years: 1.03 male(s)/female

25-54 years: 0.99 male(s)/female

55-64 years: 0.92 male(s)/female

65 years and over: 0.84 male(s)/female

total population: 0.99 male(s)/female (2020 est.)

Maternal mortality rate: 125 deaths/100,000 live births (2017 est.)

country comparison to the world: 63

Infant mortality rate: *total:* 27.9 deaths/1,000 live births

male: 31.1 deaths/1,000 live births

female: 24.5 deaths/1,000 live births (2020 est.)

country comparison to the world: 62

Life expectancy at birth: *total population:* 71 years

male: 67.5 years

female: 74.7 years (2020 est.)

country comparison to the world: 163

Total fertility rate: 2.26 children born/woman (2020 est.)

country comparison to the world: 87

Contraceptive prevalence rate: 75% (2010)

Drinking water source:

improved: total: 95.7% of population

unimproved: total: 4.3% of population (2017 est.)

Current Health Expenditure: 1.2% (2017)

Hospital bed density: 0.9 beds/1,000 population (2017)

Sanitation facility access:

improved: total: 93.9% of population

unimproved: total: 6.4% of population (2017 est.)

HIV/AIDS—adult prevalence rate: 0.5% (2019 est.)

country comparison to the world: 72

HIV/AIDS—people living with HIV/AIDS: 110,000 (2019 est.)

country comparison to the world: 43

HIV/AIDS—deaths: NA

Major infectious diseases: *degree of risk:* high (2020)

food or waterborne diseases: bacterial diarrhea and hepatitis A

vectorborne diseases: dengue fever and malaria

Obesity—adult prevalence rate: 25.6% (2016)

country comparison to the world: 50

Children under the age of 5 years underweight: 2.9% (2009)

country comparison to the world: 103

Education expenditures: 6.9% of GDP (2009)

country comparison to the world: 15

Literacy: *definition:* age 15 and over can read and write

total population: 97.1%

male: 97%

female: 97.2% (2016)

School life expectancy (primary to tertiary education): total: 14 years

male: NA

female: NA (2009)

Unemployment, youth ages 15-24: total: 14.6%

male: NA

female: NA (2015 est.)

country comparison to the world: 94

GOVERNMENT

Country name: *conventional long form:* Bolivarian Republic of Venezuela

conventional short form: Venezuela

local long form: Republica Bolivariana de Venezuela

local short form: Venezuela

former: State of Venezuela, Republic of Venezuela, United States of Venezuela

etymology: native stilt-houses built on Lake Maracaibo reminded early explorers Alonso de OJEDA and Amerigo VESPUCCI in 1499 of buildings in Venice and so they named the region "Venezuola," which in Italian means "Little Venice"

Government type: federal presidential republic

Capital: *name:* Caracas

geographic coordinates: 10 29 N, 66 52 W

time difference: UTC-4 (1 hour ahead of Washington, DC, during Standard Time)

etymology: named for the native Caracas tribe that originally settled in the city's valley site near the Caribbean coast

Administrative divisions: 23 states (estados, singular - estado), 1 capital district* (distrito capital), and 1 federal dependency** (dependencia federal); Amazonas, Anzoategui, Apure, Aragua, Barinas, Bolivar, Carabobo, Cojedes, Delta Amacuro, Dependencias Federales (Federal Dependencies)**, Distrito Capital (Capital District)*, Falcon, Guarico, La Guaira, Lara, Merida, Miranda, Monagas, Nueva Esparta, Portuguesa, Sucre, Tachira, Trujillo, Yaracuy, Zulia

note: the federal dependency consists of 11 federally controlled island groups with a total of 72 individual islands

Independence: 5 July 1811 (from Spain)

National holiday: Independence Day, 5 July (1811)

Constitution: *history:* many previous; latest adopted 15 December 1999, effective 30 December 1999

amendments: proposed through agreement by at least 39% of the National Assembly membership, by the president of the republic in session with the cabinet of ministers, or by petition of at least 15% of registered voters; passage requires simple majority vote by the Assembly and simple majority approval in a referendum; amended 2009; note - in 2016, President MADURO issued a decree to hold an election to form a constituent assembly to change the constitution; the election in July 2017 approved the formation of a 545-member constituent assembly and elected its delegates, empowering them to change the constitution and dismiss government institutions and officials

Legal system: civil law system based on the Spanish civil code

International law organization participation: has not submitted an ICJ jurisdiction declaration; accepts ICCt jurisdiction

Citizenship: *citizenship by birth:* yes
citizenship by descent only: yes
dual citizenship recognized: yes
residency requirement for naturalization: 10 years; reduced to five years in the case of applicants from Spain, Portugal, Italy, or a Latin American or Caribbean country

Suffrage: 18 years of age; universal

Executive branch: *chief of state:* Notification Statement: the United States recognizes Juan GUAIDO as the Interim President of Venezuela
    President Nicolas MADURO Moros (since 19 April 2013); Executive Vice President Delcy RODRIGUEZ Gomez (since 14 June 2018); note - the president is both chief of state and head of government

head of government: President Nicolas MADURO Moros (since 19 April 2013); Executive Vice President Delcy RODRIGUEZ Gomez (since 14 June 2018)
cabinet: Council of Ministers appointed by the president
elections/appointments: president directly elected by simple majority popular vote for a 6-year term (no term limits); election last held on 20 May 2018 (next election scheduled for 2024)
election results: Nicolas MADURO Moros reelected president; percent of vote - Nicolas MADURO Moros (PSUV) 68%, Henri FALCON (AP) 21%, Javier BERTUCCI 11%; note - the election was marked by serious shortcomings and electoral fraud; voter turnout was approximately 46% due largely to an opposition boycott of the election

Legislative branch: *description:* unicameral National Assembly or Asamblea Nacional (167 seats; 113 members directly elected in single- and multi-seat constituencies by simple majority vote, 51 directly elected in multi-seat constituencies by closed, party-list proportional representation vote, and 3 seats reserved for indigenous peoples of Venezuela; members serve 5-year terms)
elections: last held on 6 December 2015 (next to be held on 6 December 2020)

election results: percent of vote by party - MUD (opposition coalition) 56.2%, PSUV (pro-government) 40.9%, other 2.9%; seats by party - MUD 109, PSUV 55, indigenous peoples 3; composition - men 143, women 24, percent of women 14.4%

Judicial branch: *highest courts:* Supreme Tribunal of Justice (consists of 32 judges organized into constitutional, political-administrative, electoral, civil appeals, criminal appeals, and social divisions)
judge selection and term of office: judges proposed by the Committee of Judicial Postulation (an independent body of organizations dealing with legal issues and of the organs of citizen power) and appointed by the National Assembly; judges serve nonrenewable 12-year terms; note - in July 2017, the National Assembly named 33 judges to the court to replace a series of judges, it argued, had been illegally appointed in late 2015 by the outgoing, socialist-party-led Assembly; the Government of President MADURO and the Socialist Party-appointed judges refused to recognize these appointments, however, and many of the new judges have since been imprisoned or forced into exile
subordinate courts: Superior or Appeals Courts (Tribunales Superiores); District Tribunals (Tribunales de Distrito); Courts of First Instance (Tribunales de Primera Instancia); Parish Courts (Tribunales de Parroquia); Justices of the Peace (Justicia de Paz) Network

Political parties and leaders: A New Era or UNT [Manuel ROSALES]
Brave People's Alliance or ABP [Richard BLANCO]
Christian Democrats or COPEI [Roberto ENRIQUEZ]
Clear Accounts or CC [Enzo SCARENO]
Coalition of parties loyal to Hugo CHAVEZ -- Great Patriotic Pole or GPP [Nicolas MADURO]
Coalition of opposition parties -- The Democratic Unity Table or MUD [Jose Luis CARTAYA]
Come On Venezuela or VV [Maria MACHADO]
Communist Party of Venezuela or PCV [Oscar FIGUERA] Democratic Action or AD [Henry RAMOS ALLUP]
Justice First or PJ [Julio BORGES]
Popular Will or VP [Leopoldo LOPEZ]
Progressive Wave or AP [Henri FALCON]
The Radical Cause or La Causa R [Andres VELAZQUEZ]
United Socialist Party of Venezuela or PSUV [Nicolas MADURO]
Venezuelan Progressive Movement or MPV [Simon CALZADILLA] Venezuela Project or PV [Henrique Fernando SALAS FEO]

International organization participation: Caricom (observer), CD, CDB, CELAC, FAO, G-15, G-24, G-77, IADB, IAEA, IBRD, ICAO, ICC (national committees), ICCt (signatory), ICRM, IDA, IFAD, IFC, IFRCS, IHO, ILO, IMF, IMO, IMSO, Interpol, IOC, IOM, IPU, ITSO, ITU, ITUC (NGOs), LAES, LAIA, LAS (observer),

MIGA, NAM, OAS, OPANAL, OPCW, OPEC, PCA, Petrocaribe, UN, UNASUR, UNCTAD, UNESCO, UNHCR, UNIDO, Union Latina, UNWTO, UPU, WCO, WFTU (NGOs), WHO, WIPO, WMO, WTO

Diplomatic representation in the US: *chief of mission:* Ambassador (vacant); Charge d'Affaires Carlos Lissett M. HERNANDEZ Marquez (since May 2018)
chancery: 1099 30th Street NW, Washington, DC 20007
telephone: [1] (202) 342-2214
FAX: [1] (202) 342-6820

consulate(s) general: Boston, Chicago, Houston, New Orleans, New York, San Francisco, San Juan (Puerto Rico)

Diplomatic representation from the US: *chief of mission:* Ambassador (vacant); Charge d'Affaires James "Jimmy" STORY (since July 2018); note - on 11 March 2019, the Department of State announced the temporary suspension of operations of the US Embassy in Caracas and the withdrawal of diplomatic personnel; all consular services, routine and emergency, are suspended
telephone: [58] (212) 975-6411, 907-8400 (after hours)
embassy: now operating from Bogota, Colombia

previously—F St. and Suapure St.; Urb . Colinas de Valle Arriba; Caracas 1080
mailing address: P. O. Box 62291, Caracas 1060-A; APO AA 34037
FAX: [58] (212) 907-8106

Flag description: three equal horizontal bands of yellow (top), blue, and red with the coat of arms on the hoist side of the yellow band and an arc of eight white five-pointed stars centered in the blue band; the flag retains the three equal horizontal bands and three main colors of the banner of Gran Colombia, the South American republic that broke up in 1830; yellow is interpreted as standing for the riches of the land, blue for the courage of its people, and red for the blood shed in attaining independence; the seven stars on the original flag represented the seven provinces in Venezuela that united in the war of independence; in 2006, then President Hugo CHAVEZ ordered an eighth star added to the star arc - a decision that sparked much controversy - to conform with the flag proclaimed by Simon Bolivar in 1827 and to represent the historic province of Guayana

National symbol(s): troupial (bird); national colors: yellow, blue, red

National anthem: *name:* "Gloria al bravo pueblo" (Glory to the Brave People)
lyrics/music: Vicente SALIAS/Juan Jose LANDAETA
note: adopted 1881; lyrics written in 1810, the music some years later; both SALIAS and LANDAETA were executed in 1814 during Venezuela's struggle for independence
0:00 / 1:30

ECONOMY

Economy—overview: Venezuela remains highly dependent on oil revenues, which account for almost all export earnings and nearly half of the government's revenue, despite a continued decline in oil production in 2017. In the absence of official statistics, foreign experts estimate that GDP contracted 12% in 2017, inflation exceeded 2000%, people faced widespread shortages of consumer goods and medicine, and the central bank's international reserves dwindled. In late 2017, Venezuela also entered selective default on some of its sovereign and state oil company, Petroleos de Venezuela, S.A., (PDVSA) bonds. Domestic production and industry continues to severely underperform and the Venezuelan Government continues to rely on imports to meet its basic food and consumer goods needs.

Falling oil prices since 2014 have aggravated Venezuela's economic crisis. Insufficient access to dollars, price controls, and rigid labor regulations have led some US and multinational firms to reduce or shut down their Venezuelan operations. Market uncertainty and PDVSA's poor cash flow have slowed investment in the petroleum sector, resulting in a decline in oil production.

Under President Nicolas MADURO, the Venezuelan Government's response to the economic crisis has been to increase state control over the economy and blame the private sector for shortages. MADURO has given authority for the production and distribution of basic goods to the military and to local socialist party member committees. The Venezuelan Government has maintained strict currency controls since 2003. The government has been unable to sustain its mechanisms for distributing dollars to the private sector, in part because it needed to withhold some foreign exchange reserves to make its foreign bond payments. As a result of price and currency controls, local industries have struggled to purchase production inputs necessary to maintain their operations or sell goods at a profit on the local market. Expansionary monetary policies and currency controls have created opportunities for arbitrage and corruption and fueled a rapid increase in black market activity.

GDP (purchasing power parity): $381.6 billion (2017 est.)
$443.7 billion (2016 est.)
$531.1 billion (2015 est.)
note: data are in 2017 dollars
country comparison to the world: 47

GDP (official exchange rate): $210.1 billion (2017 est.)

GDP—real growth rate: -19.67% (2018 est.)
-14% (2017 est.)
-15.76% (2017 est.)
country comparison to the world: 223

GDP—per capita (PPP): $12,500 (2017 est.)
$14,400 (2016 est.)

$17,300 (2015 est.)
note: data are in 2017 dollars
country comparison to the world: 126

Gross national saving: 12.1% of GDP (2017 est.)
8.6% of GDP (2016 est.)
31.8% of GDP (2015 est.)
country comparison to the world: 150

GDP—composition, by end use: household consumption: 68.5% (2017 est.)
government consumption: 19.6% (2017 est.)
investment in fixed capital: 13.9% (2017 est.)
investment in inventories: 1.7% (2017 est.)
exports of goods and services: 7% (2017 est.)
imports of goods and services: -10.7% (2017 est.)

GDP—composition, by sector of origin: agriculture: 4.7% (2017 est.)
industry: 40.4% (2017 est.)
services: 54.9% (2017 est.)

Agriculture—products: corn, sorghum, sugarcane, rice, bananas, vegetables, coffee; beef, pork, milk, eggs; fish

Industries: agricultural products, livestock, raw materials, machinery and equipment, transport equipment, construction materials, medical equipment, pharmaceuticals, chemicals, iron and steel products, crude oil and petroleum products

Industrial production growth rate: -2% (2017 est.)
country comparison to the world: 183

Labor force: 14.21 million (2017 est.)
country comparison to the world: 37

Labor force—by occupation: agriculture: 7.3%
industry: 21.8%
services: 70.9% (4th quarter, 2011 est.)

Unemployment rate: 6.9% (2018 est.)

27.1% (2017 est.)
country comparison to the world: 111

Population below poverty line: 19.7% (2015 est.)

Household income or consumption by percentage share: lowest 10%: 1.7%
highest 10%: 32.7% (2006)

Budget: revenues: 92.8 billion (2017 est.)
expenditures: 189.7 billion (2017 est.)

Taxes and other revenues: 44.2% (of GDP) (2017 est.)
country comparison to the world: 25

Budget surplus (+) or deficit (-): -46.1% (of GDP) (2017 est.)
country comparison to the world: 220

Public debt: 38.9% of GDP (2017 est.)
31.3% of GDP (2016 est.)
note: data cover central government debt, as well as the debt of state-owned oil company PDVSA; the data include treasury debt held by foreign entities; the data include some debt issued by subnational entities, as well as intragovernmental debt; intragovernmental debt consists of treasury borrowings from surpluses in the social funds, such as for retirement, medical care, and unemployment; some debt instruments for the social funds are sold at public auctions

country comparison to the world: 135

Fiscal year: calendar year

Inflation rate (consumer prices): 1,087.5% (2017 est.)
254.4% (2016 est.)
country comparison to the world: 226

Current account balance: $4.277 billion (2017 est.)
-$3.87 billion (2016 est.)
country comparison to the world: 32

Exports: $32.06 billion (2017 est.)
$27.2 billion (2016 est.)
country comparison to the world: 62

Exports—partners: US 34.8%, India 17.2%, China 16%, Netherlands Antilles 8.2%, Singapore 6.3%, Cuba 4.2% (2017)

Exports—commodities: petroleum and petroleum products, bauxite and aluminum, minerals, chemicals, agricultural products

Imports: $11 billion (2017 est.)
$16.34 billion (2016 est.)
country comparison to the world: 99

Imports—commodities: agricultural products, livestock, raw materials, machinery and equipment, transport equipment, construction materials, medical equipment, petroleum products, pharmaceuticals, chemicals, iron and steel products

Imports—partners: US 24.8%, China 14.2%, Mexico 9.5% (2017)

Reserves of foreign exchange and gold: $9.661 billion (31 December 2017 est.)
$11 billion (31 December 2016 est.)
country comparison to the world: 75

Debt—external: $100.3 billion (31 December 2017 est.)
$109.8 billion (31 December 2016 est.)
country comparison to the world: 47

Exchange rates: bolivars (VEB) per US dollar—
3,345 (2017 est.)
673.76 (2016 est.)
48.07 (2015 est.)
13.72 (2014 est.)
6.284 (2013 est.)

ENERGY

Electricity access: electrification—total population: 99.6% (2016)

electrification—urban areas: 100% (2016)

electrification—rural areas: 96.4% (2016)

Electricity—production: 109.3 billion kWh (2016 est.)
country comparison to the world: 34

Electricity—consumption: 71.96 billion kWh (2016 est.)
country comparison to the world: 39

Electricity—exports: 0 kWh (2015 est.)
country comparison to the world: 215

Electricity—imports: 0 kWh (2016 est.)
country comparison to the world: 216

Electricity—installed generating capacity: 31 million kW (2016 est.)
country comparison to the world: 32

Electricity—from fossil fuels: 51% of total installed capacity (2016 est.)
country comparison to the world: 150

Electricity—from nuclear fuels: 0% of total installed capacity (2017 est.)
country comparison to the world: 208

Electricity—from hydroelectric plants: 49% of total installed capacity (2017 est.)
country comparison to the world: 43

Electricity—from other renewable sources: 0% of total installed capacity (2017 est.)
country comparison to the world: 213

Crude oil—production: 1.484 million bbl/day (2018 est.)
country comparison to the world: 16

Crude oil—exports: 1.656 million bbl/day (2015 est.)
country comparison to the world: 8

Crude oil—imports: 0 bbl/day (2015 est.)
country comparison to the world: 212

Crude oil—proved reserves: 302.3 billion bbl (1 January 2018 est.)
country comparison to the world: 1

Refined petroleum products—production: 926,300 bbl/day (2015 est.)
country comparison to the world: 20

Refined petroleum products—consumption: 659,000 bbl/day (2016 est.)
country comparison to the world: 29

Refined petroleum products— exports: 325,800 bbl/day (2015 est.)
country comparison to the world: 27

Refined petroleum products—imports: 20,640 bbl/day (2015 est.)
country comparison to the world: 117

Natural gas—production: 27.07 billion cu m (2017 est.)
country comparison to the world: 28

Natural gas—consumption: 24.21 billion cu m (2017 est.)
country comparison to the world: 33

Natural gas—exports: 0 cu m (2017 est.)
country comparison to the world: 209

Natural gas—imports: 0 cu m (2017 est.)
country comparison to the world: 209

Natural gas—proved reserves: 5.739 trillion cu m (1 January 2018 est.)
country comparison to the world: 7

Carbon dioxide emissions from consumption of energy: 129.9 million Mt (2017 est.)
country comparison to the world: 36

COMMUNICATIONS

Telephones—fixed lines: *total subscriptions:* 5,501,135

subscriptions per 100 inhabitants: 19.17 (2019 est.)
country comparison to the world: 26

Telephones—mobile cellular: total subscriptions: 16,664,106

subscriptions per 100 inhabitants: 58.07 (2019 est.)
country comparison to the world: 64

Telecommunication systems: *general assessment:* by late 2018 teledensity has fallen due to political upheaval in the country with people holding on to mobile service, but cancelling fixed-line telecom services; poor quality of service in many areas of the country due to financial concerns of customers, decrepit sate of fixed-line network and difficulty to pay for equipment from foreign vendors; popularity of social networks has given growth to mobile data traffic; LTE population coverage about 46%; govt. launches National Fiber Optic backbone project; mobile penetration below average for South America; MNO suffering from stolen or damaged infrastructure (2020)

domestic: two domestic satellite systems with three earth stations; recent substantial improvement in telephone service in rural areas; 3 major providers operate in the mobile market and compete with state-owned company; fixed-line 19 per 100 and mobile-cellular telephone subscribership about 58 per 100 persons (2019)

international: country code - 58; landing points for the Venezuela Festoon, ARCOS, PAN-AM, SAC, GlobeNet, ALBA-1 and Americas II submarine cable system providing connectivity to the Caribbean, Central and South America, and US; satellite earth stations - 1 Intelsat (Atlantic Ocean) and 1 PanAmSat (2020)

Note: the COVID-19 outbreak is negatively impacting telecommunications production and supply chains globally; consumer spending on telecom devices and services has also slowed due to the pandemic's effect on economies worldwide; overall progress towards improvements in all facets of the telecom industry - mobile, fixed-line, broadband, submarine cable and satellite - has moderated

Broadcast media: government supervises a mixture of state-run and private broadcast media; 13 public service networks, 61 privately owned TV networks, a privately owned news channel with limited national coverage, and a government-backed Pan-American channel; state-run radio network includes roughly 65 news stations and another 30 stations targeted at specific audiences; state-sponsored community broadcasters include 235 radio stations and 44 TV stations; the number of private broadcast radio stations has been declining, but many still remain in operation

Internet country code: .ve

Internet users: *total:* 21,354,499

percent of population: 72% (July 2018 est.)
country comparison to the world: 36

Broadband—fixed subscriptions: *total:* 2,604,578
subscriptions per 100 inhabitants: 9 (2018 est.)
country comparison to the world: 47

TRANSPORTATION

National air transport system: *number of registered air carriers:* 12 (2020)

inventory of registered aircraft operated by air carriers: 75

annual passenger traffic on registered air carriers: 2,137,771 (2018)

annual freight traffic on registered air carriers: 1.55 million mt-km (2018)

Civil aircraft registration country code prefix: YV (2016)

Airports: 444 (2013)
country comparison to the world: 17

Airports—with paved runways: *total:* 127 (2013)
over 3,047 m: 6 (2013)
2,438 to 3,047 m: 9 (2013)
1,524 to 2,437 m: 33 (2013)
914 to 1,523 m: 62 (2013)
under 914 m: 17 (2013)

Airports—with unpaved runways: *total:* 317 (2013)
2,438 to 3,047 m: 3 (2013)
1,524 to 2,437 m: 57 (2013)
914 to 1,523 m: 127 (2013)
under 914 m: 130 (2013)

Heliports: 3 (2013)

Pipelines: 981 km extra heavy crude, 5941 km gas, 7588 km oil, 1778 km refined products (2013)

Railways: *total:* 447 km (2014)
standard gauge: 447 km 1.435-m gauge (41.4 km electrified) (2014)
country comparison to the world: 115

Roadways: *total:* 96,189 km (2014)
country comparison to the world: 50

Waterways: 7,100 km (Orinoco River (400 km) and Lake de Maracaibo navigable by oceangoing vessels) (2011)
country comparison to the world: 20

Merchant marine: *total:* 289
by type: bulk carrier 4, container ship 1, general cargo 27, oil tanker 23, other 234 (2019)
country comparison to the world: 54

Ports and terminals: *major seaport(s):* La Guaira, Maracaibo, Puerto Cabello, Punta Cardon

oil terminal(s): Jose terminal

MILITARY AND SECURITY

Military and security forces: *Bolivarian National Armed Forces (Fuerza Armada Nacional Bolivariana, FANB):* Bolivarian Army (Ejercito Bolivariano, EB), Bolivarian Navy (Armada Bolivariana, AB; includes Marines, Coast Guard), Bolivarian Military Aviation (Aviacion Militar Bolivariana, AMB), Integral Aerospace Defense

Command (Comando de Defensa Aeroespacial Integral, CODAI), Bolivarian National Guard (Guardia Nacional Bolivara, GNB); Bolivarian Militia (Milicia Bolivariana, NMB) (2019) note: the CODAI is a joint service command with personnel drawn from other services

Military expenditures: 0.5% of GDP (2017)
0.5% of GDP (2016)
1% of GDP (2015)
1.2% of GDP (2014)
1.7% of GDP (2013)
country comparison to the world: 150

Military and security service personnel strengths: the Bolivarian National Armed Forces (FANB) have approximately 125,000 active personnel (62,000 Army; 25,000 Navy; 11,000 Air Force; 27,000 National Guard) (2019 est.)

Military equipment inventories and acquisitions: the FANB inventory is mainly of Chinese and Russian origin with a smaller mix of equipment from Western countries such as France, Germany, Italy, the Netherlands, Spain, the UK, and the US; since 2010, China and Russia are the top suppliers of military hardware to Venezuela (2019 est.)

Military service age and obligation: all citizens of military service age (18-60 years old) are obligated to register for military service and subject to military training, though mandatory recruitment is forbidden; the minimum service obligation is 24-30 months (2016)

Maritime threats: The International Maritime Bureau continues to report the territorial and offshore waters in the Caribbean Sea as at risk for piracy and armed robbery against ships; numerous vessels, including commercial shipping and pleasure craft, have been attacked and hijacked both at anchor and while underway; crews have been robbed and stores or cargoes stolen; in 2018, 11 attacks were reported which was a slight decrease from the 12 attacks in 2017. Nevertheless, the waters off Venezuela continue to be the fourth most dangerous area for mariners in the world. (2018)

TERRORISM

Terrorist group(s): National Liberation Army; Revolutionary Armed Forces of Colombia (2019) *note:* details about the history, aims, leadership, organization, areas of operation, tactics, targets, weapons, size, and sources of support of the group(s) appear(s) in Appendix-T

TRANSPORTATION

Disputes—international: claims all of the area west of the Essequibo River in Guyana, preventing any discussion of a maritime boundary; Guyana has expressed its intention to join Barbados in asserting claims before the UN Convention on the Law of the Sea that Trinidad and Tobago's maritime boundary with Venezuela extends into their waters; dispute with Colombia over maritime boundary and Venezuelan administered Los Monjes Islands near the Gulf of Venezuela; Colombian organized illegal narcotics and paramilitary activities penetrate Venezuela's shared border region; US, France, and the Netherlands recognize Venezuela's granting full effect to Aves Island, thereby claiming a Venezuelan Economic Exclusion Zone/continental shelf extending over a large portion of the eastern Caribbean Sea; Dominica, Saint Kitts and Nevis, Saint Lucia, and Saint Vincent and the Grenadines protest Venezuela's full effect claim

Refugees and internally displaced persons: *refugees (country of origin):* 67,622 (Colombia) (2019)

Trafficking in persons: *current situation:* Venezuela is a source and destination country for men, women, and children subjected to sex trafficking and forced labor; Venezuelan women and girls, sometimes lured from poor interior regions to urban and tourist areas, are trafficked for sexual exploitation within the country, as well as in the Caribbean; Venezuelan children are exploited, frequently by their families, in domestic servitude; people from South America, the Caribbean, Asia, and Africa are sex and labor trafficking victims in Venezuela; thousands of Cuban citizens, particularly doctors, who work in Venezuela on government social programs in exchange for the provision of resources to the Cuban Government experience conditions of forced labor

tier rating: Tier 3—Venezuela does not fully comply with the minimum standards for the elimination of trafficking and is not making significant efforts to do so; in 2014, the government appeared to increase efforts to hold traffickers criminally accountable, but a lack of government data made anti-trafficking law enforcement efforts difficult to assess; publically available information indicated many cases pursued under anti-trafficking law involved illegal adoption rather than sex and labor trafficking; authorities identified a small number of trafficking victims, and victim referrals to limited government services were made on an ad hoc basis; because no specialized facilities are available for trafficking victims, women and child victims accessed centers for victims of domestic violence or at-risk youth, and services for men were virtually nonexistent; NGOs provided some services to sex and labor trafficking victims; Venezuela has no permanent anti-trafficking interagency body, no national anti-trafficking plan, and still has not passed anti-trafficking legislation drafted in 2010 (2015)

Illicit drugs: small-scale illicit producer of opium and coca for the processing of opiates and coca derivatives; however, large quantities of cocaine, heroin, and marijuana transit the country from Colombia bound for US and Europe; significant narcotics- related money-laundering activity, especially along the border with Colombia and on Margarita Island; active eradication program primarily targeting opium; increasing signs of drug-related activities by Colombian insurgents on border

VIETNAM

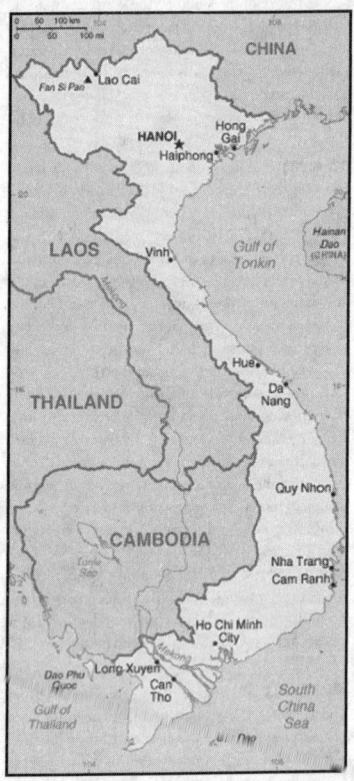

INTRODUCTION

Background: The conquest of Vietnam by France began in 1858 and was completed by 1884. It became part of French Indochina in 1887. Vietnam declared independence after World War II, but France continued to rule until its 1954 defeat by communist forces under Ho Chi MINH. Under the Geneva Accords of 1954, Vietnam was divided into the communist North and anti-communist South. US economic and military aid to South Vietnam grew through the 1960s in an attempt to bolster the government, but US armed forces were withdrawn following a cease-fire agreement in 1973. Two years later, North Vietnamese forces overran the South reuniting the country under communist rule. Despite the return of peace, for over a decade the country experienced little economic growth because of conservative leadership policies, the persecution and mass exodus of individuals - many of them successful South Vietnamese merchants - and growing international isolation. However, since the enactment of Vietnam's "doi moi" (renovation) policy in 1986, Vietnamese authorities have committed to increased economic liberalization and enacted structural reforms needed to modernize the economy and to produce more competitive, export-driven industries. The communist leaders maintain tight control on political expression but have demonstrated some modest steps toward better protection of human rights. The country continues to experience smallscale protests, the vast majority connected to either land-use issues, calls for increased political space, or the lack of equitable mechanisms for resolving disputes. The small-scale protests in the urban areas are often organized by human rights activists, but many occur in rural areas and involve various ethnic minorities such as the Montagnards of the Central Highlands, Hmong in the Northwest Highlands, and the Khmer Krom in the southern delta region.

GEOGRAPHY

Location: Southeastern Asia, bordering the Gulf of Thailand, Gulf of Tonkin, and South China Sea, as well as China, Laos, and Cambodia

Geographic coordinates: 16 10 N, 107 50 E

Map references: Southeast Asia

Area: *total:* 331,210 sq km
land: 310,070 sq km
water: 21,140 sq km
country comparison to the world: 67

Area—comparative: about three times the size of Tennessee; slightly larger than New Mexico

Area comparison map: [INSERT IMAGE: VIETNAM-Area comparison map]

Land boundaries: *total:* 4,616 km
border countries (3): Cambodia 1158 km, China 1297 km, Laos 2161 km

Coastline: 3,444 km (excludes islands)

Maritime claims: *territorial sea:* 12 nm
exclusive economic zone: 200 nm
contiguous zone: 24 nm
continental shelf: 200 nm or to the edge of the continental margin

Climate: tropical in south; monsoonal in north with hot, rainy season (May to September) and warm, dry season (October to March)

Terrain: low, flat delta in south and north; central highlands; hilly, mountainous in far north and northwest

Elevation: *mean elevation:* 398 m
lowest point: South China Sea 0 m
highest point: Fan Si Pan 3,144 m

Natural resources: antimony, phosphates, coal, manganese, rare earth elements, bauxite, chromate, offshore oil and gas deposits, timber, hydropower, arable land

Land use: *agricultural land:* 34.8% (2011 est.)
arable land: 20.6% (2011 est.) / permanent crops: 12.1% (2011 est.) / permanent pasture: 2.1% (2011 est.)
forest: 45% (2011 est.)
other: 20.2% (2011 est.)

Irrigated land: 46,000 sq km (2012)

Population distribution: though it has one of the highest population densities in the world, the population is not evenly dispersed; clustering is heaviest along the South China Sea and Gulf of Tonkin, with the Mekong Delta (in the south) and the Red River Valley (in the north) having the largest concentrations of people

Natural hazards: occasional typhoons (May to January) with extensive flooding, especially in the Mekong River delta

Environment—current issues: logging and slash-and-burn agricultural practices contribute to deforestation and soil degradation; water pollution and overfishing threaten marine life populations; groundwater contamination limits potable water supply; air pollution; growing urban industrialization and population migration are rapidly degrading environment in Hanoi and Ho Chi Minh City

Environment—international agreements: *party to:* Biodiversity, Climate Change, Climate Change-Kyoto Protocol, Desertification, Endangered Species, Environmental Modification, Hazardous Wastes, Law of the Sea, Ozone Layer Protection, Ship Pollution, Wetlands
signed, but not ratified: none of the selected agreements

Geography—note: note 1: extending 1,650 km north to south, the country is only 50 km across at its narrowest point
note 2: Son Doong in Phong Nha-Ke Bang National Park is the world's largest cave (greatest cross sectional area) and is the largest known cave passage in the world by volume; it currently measures a total of 38.5 million cu m (about 1.35 billion cu ft); it connects to Thung cave (but not yet officially); when recognized, it will add an additional 1.6 million cu m in volume; Son Doong is so massive that it contains its own jungle, underground river, and localized weather system; clouds form inside the cave and spew out from its exits and two dolines (openings (sinkhole skylights) created by collapsed ceilings that allow sunlight to stream in)

PEOPLE AND SOCIETY

Population: 98,721,275 (July 2020 est.)
country comparison to the world: 16

Nationality: *noun:* Vietnamese (singular and plural)
adjective: Vietnamese

Ethnic groups: Kinh (Viet) 85.7%, Tay 1.9%, Thai 1.8%, Muong 1.5%, Khmer 1.5%, Mong 1.2%, Nung 1.1%, Hoa 1%, other 4.3% (2009 est.)
note: 54 ethnic groups are recognized by the Vietnamese Government

Languages: Vietnamese (official), English (increasingly favored as a second language), some French, Chinese, and Khmer, mountain area languages (Mon-Khmer and Malayo-Polynesian)

Religions: Buddhist 7.9%, Catholic 6.6%, Hoa Hao 1.7%, Cao Dai 0.9%, Protestant 0.9%, Muslim 0.1%, none 81.8% (2009 est.)

Age structure: *0-14 years:* 22.61% (male 11,733,704/female 10,590,078)
15-24 years: 15.22% (male 7,825,859/female 7,202,716)
25-54 years: 45.7% (male 22,852,429/female 22,262,566)
55-64 years: 9.55% (male 4,412,111/female 5,016,880)
65 years and over: 6.91% (male 2,702,963/female 4,121,969) (2020 est.)

population pyramid: [INSERT IMAGE: VIETNAM-population pyramid]

Dependency ratios: *total dependency ratio:* 45.1
youth dependency ratio: 33.6
elderly dependency ratio: 11.4
potential support ratio: 8.8 (2020 est.)

Median age: *total:* 31.9 years
male: 30.8 years
female: 33 years (2020 est.)
country comparison to the world: 112

Population growth rate: 0.84% (2020 est.)
country comparison to the world: 125

Birth rate: 14.5 births/1,000 population (2020 est.)
country comparison to the world: 130

Death rate: 6 deaths/1,000 population (2020 est.)
country comparison to the world: 167

Net migration rate: -0.3 migrant(s)/1,000 population (2020 est.)
country comparison to the world: 121

Population distribution: though it has one of the highest population densities in the world, the population is not evenly dispersed; clustering is heaviest along the South China Sea and Gulf of Tonkin, with the Mekong Delta (in the south) and the Red River Valley (in the north) having the largest concentrations of people

Urbanization: *urban population:* 37.3% of total population (2020)
rate of urbanization: 2.98% annual rate of change (2015-20 est.)

total population growth rate v. urban population growth rate, 2000-2030: *Major urban areas - population:* 8.602 million Ho Chi Minh City, 4.678 million HANOI (capital), 1.618 million Can Tho, 1.300 million Hai Phong, 1.125 million Da Nang, 1.013 million Bien Hoa (2020)

Sex ratio: *at birth:* 1.09 male(s)/female
0-14 years: 1.11 male(s)/female
15-24 years: 1.09 male(s)/female
25-54 years: 1.03 male(s)/female
55-64 years: 0.88 male(s)/female
65 years and over: 0.66 male(s)/female
total population: 1.01 male(s)/female (2020 est.)

Maternal mortality rate: 43 deaths/100,000 live births (2017 est.)
country comparison to the world: 100

Infant mortality rate: *total:* 15.7 deaths/1,000 live births
male: 16 deaths/1,000 live births

female: 15.3 deaths/1,000 live births (2020 est.)
country comparison to the world: 96

Life expectancy at birth: *total population:* 74.4 years
male: 71.9 years
female: 77.1 years (2020 est.)
country comparison to the world: 132

Total fertility rate: 1.77 children born/woman (2020 est.)
country comparison to the world: 157

Contraceptive prevalence rate: 77.5% (2016)

Drinking water source:

improved: *urban:* 98.6% of population
rural: 92.6% of population
total: 94.7% of population

unimproved: *urban:* 1.4% of population
rural: 7.4% of population
total: 5.3% of population (2017 est.)

Current Health Expenditure: 5.5% (2017)

Physicians density: 0.83 physicians/1,000 population (2016)

Hospital bed density: 2.6 beds/1,000 population (2014)

Sanitation facility access:

improved: *urban:* 96.9% of population
rural: 82.1% of population
total: 87.3% of population

unimproved: *urban:* 3.1% of population
rural: 17.9% of population
total: 12.7% of population (2017 est.)

HIV/AIDS—adult prevalence rate: 0.3% (2019 est.)
country comparison to the world: 94

HIV/AIDS—people living with HIV/AIDS: 230,000 (2019 est.)
country comparison to the world: 27

HIV/AIDS—deaths: 5,000 (2019 est.)
country comparison to the world: 26

Major infectious diseases: *degree of risk:* very high (2020)
food or waterborne diseases: bacterial diarrhea, hepatitis A, and typhoid fever
vectorborne diseases: dengue fever, malaria, and Japanese encephalitis

Obesity—adult prevalence rate: 2.1% (2016)
country comparison to the world: 192

Children under the age of 5 years underweight: 13.4% (2017)
country comparison to the world: 45

Education expenditures: 5.7% of GDP (2013)
country comparison to the world: 35

Literacy: *definition:* age 15 and over can read and write
total population: 95%
male: 96.5%
female: 93.6% (2018)
Unemployment, youth ages 15-24: total: 6.9%
male: 6%
female: 7.9% (2018 est.)
country comparison to the world: 154

Country name: *conventioçnal long form:* Socialist Republic of Vietnam
conventional short form: Vietnam
local long form: Cong Hoa Xa Hoi Chu Nghia Viet Nam
local short form: Viet Nam
abbreviation: SRV
etymology: "Viet nam" translates as "Viet south," where "Viet" is an ethnic self identification dating to a second century B.C. kingdom and "nam" refers to its location in relation to other Viet kingdoms

Government type: communist state

Capital: *name:* Hanoi (Ha Noi)
geographic coordinates: 21 02 N, 105 51 E
time difference: UTC+7 (12 hours ahead of Washington, DC, during Standard Time)
etymology: the city has had many names in its history going back to A.D. 1010 when it first became the capital of imperial Vietnam; in 1831, it received its current name of Ha Noi, meaning "between the rivers," which refers to its geographic location

Administrative divisions: 58 provinces (tinh, singular and plural) and 5 municipalities (thanh pho, singular and plural)

provinces: An Giang, Bac Giang, Bac Kan, Bac Lieu, Bac Ninh, Ba Ria-Vung Tau, Ben Tre, Binh Dinh, Binh Duong, Binh Phuoc, Binh Thuan, Ca Mau, Cao Bang, Dak Lak, Dak Nong, Dien Bien, Dong Nai, Dong Thap, Gia Lai, Ha Giang, Ha Nam, Ha Tinh, Hai Duong, Hau Giang, Hoa Binh, Hung Yen, Khanh Hoa, Kien Giang, Kon Tum, Lai Chau, Lam Dong, Lang Son, Lao Cai, Long An, Nam Dinh, Nghe An, Ninh Binh, Ninh Thuan, Phu Tho, Phu Yen, Quang Binh, Quang Nam, Quang Ngai, Quang Ninh, Quang Tri, Soc Trang, Son La, Tay Ninh, Thai Binh, Thai Nguyen, Thanh Hoa, Thua Thien-Hue, Tien Giang, Tra Vinh, Tuyen Quang, Vinh Long, Vinh Phuc, Yen Bai

municipalities: Can Tho, Da Nang, Ha Noi (Hanoi), Hai Phong, Ho Chi Minh City (Saigon)

Independence: 2 September 1945 (from France)

National holiday: Independence Day (National Day), 2 September (1945)

Constitution: *history:* several previous; latest adopted 28 November 2013, effective 1 January 2014

amendments: proposed by the president, by the National Assembly's Standing Committee, or by at least two thirds of the National Assembly membership; a decision to draft an amendment requires approval by at least a two-thirds majority vote of the Assembly membership, followed by the formation of a constitutional drafting committee to write a draft and collect citizens' opinions; passage requires at least two-thirds majority of the Assembly membership; the Assembly can opt to conduct a referendum

Legal system: civil law system; note - the civil code of 2005 reflects a European-style civil law

International law organization participation: has not submitted an ICJ jurisdiction declaration; non-party state to the ICCt

Citizenship: *citizenship by birth:* no
citizenship by descent only: at least one parent must be a citizen of Vietnam
dual citizenship recognized: no
residency requirement for naturalization: 5 years

Suffrage: 18 years of age; universal

Executive branch: *chief of state:* President Nguyen Phu TRONG (since ·23 October 2018); note - President Tran Dai QUANG (since 2 April 2016) died on 21 September 2018

head of government: Prime Minister Nguyen Xuan PHUC (since 7 April 2016); Deputy Prime Ministers Truong Hoa BINH (since 9 April 2016), Vuong Dinh HUE (since 9 April 2016), Vu Duc DAM (since 13 November 2013), Trinh Dinh DUNG (since 9 April 2016), Pham Binh MINH (since 13 November 2013)
cabinet: Cabinet proposed by prime minister confirmed by the National Assembly and appointed by the president
elections/appointments: president indirectly elected by National Assembly from among its members for a single 5-year term; election last held on 2 April 2016 (next to be held in spring 2021); prime minister recommended by the president and confirmed by National Assembly; deputy prime ministers confirmed by the National Assembly and appointed by the president
election results: Nguyen Phu TRONG (CPV) elected president; percent of National Assembly vote - 99.8%; Nguyen Xuan PHUC elected prime minister; percent of National Assembly vote - 91%

Legislative branch: *description:* unicameral National Assembly or Quoc Hoi (500 seats - number following 2016 election - 494; number of current serving members - 484; members directly elected in multi-seat constituencies by absolute majority vote; members serve 5-year terms)
elections: last held on 22 May 2016 (next to be held in May 2021)
election results: percent of vote by party -CPV 95.8%, non-party members 4.2%; seats by party - CPV 474, non-party CPV-approved 20, self-nominated 2; note - 494 candidates elected, 2 CPV candidates-elect were disqualified; composition - men 364, women 122, percent of women 26.6%

Judicial branch: *highest courts:* Supreme People's Court (consists of the chief justice and 13 judges)
judge selection and term of office: chief justice elected by the National Assembly upon the recommendation of the president for a 5-year, renewable term; deputy chief justice appointed by the president from among the judges for a 5-year term; judges appointed by the president and confirmed by the National Assembly for 5 year terms
subordinate courts: High Courts (administrative, civil, criminal, economic, labor, family, juvenile); provincial courts; district courts; Military Court; note - the National Assembly Standing Committee can establish special tribunals upon the recommendation of the chief justice

Political parties and leaders: Communist Party of Vietnam or CPV [Nguyen Phu TRONG]
note: other parties proscribed

International organization participation: ADB, APEC, ARF, ASEAN, CICA, CP, EAS, FAO, G-77, IAEA, IBRD, ICAO, ICC (NGOs), ICRM, IDA, IFAD, IFC, IFRCS, ILO, IMF, IMO, IMSO, Interpol, IOC, IOM, IPU, ISO, ITSO, ITU, MIGA, NAM, OIF, OPCW, PCA, UN, UNCTAD, UNESCO, UNIDO, UNWTO, UPU, WCO, WFTU (NGOs), WHO, WIPO, WMO, WTO

Diplomatic representation in the US: *chief of mission:* Ambassador Ha Kim NGOC (since 17 September 2018)
chancery: 1233 20th Street NW, Suite 400, Washington, DC 20036
telephone: [1] (202) 861-0737
FAX: [1] (202) 861-0917

consulate(s) general: Houston, San Francisco

consulate(s): New York

Diplomatic representation from the US: *chief of mission:* Ambassador Daniel KRITENBRINK (since 6 November 2017)
telephone: [84] (24) 3850-5000
embassy: 7 Lang Ha Street, Hanoi
mailing address: 7 Lang Ha Street, Ba Dinh District, Hanoi; 4550 Hanoi Place, Washington, DC 20521-4550
FAX: [84] (24) 3850-5010

consulate(s) general: Ho Chi Minh City

Flag description: red field with a large yellow five-pointed star in the center; red symbolizes revolution and blood, the five-pointed star represents the five elements of the populace - peasants, workers, intellectuals, traders, and soldiers - that unite to build socialism

National symbol(s): yellow, five-pointed star on red field; lotus blossom; national colors: red, yellow

National anthem: *name:* "Tien quan ca" (The Song of the Marching Troops)
lyrics/music: Nguyen Van CAO
note: adopted as the national anthem of the Democratic Republic of Vietnam in 1945; it became the national anthem of the unified Socialist Republic of Vietnam in 1976; although it consists of two verses, only the first is used as the official anthem
0:00 / 1:11

ECONOMY

Economy—overview: Vietnam is a densely populated developing country that has been transitioning since 1986 from the rigidities of a centrally planned, highly agrarian economy to a more industrial and market based economy, and it has raised incomes substantially. Vietnam exceeded its 2017 GDP growth target of 6.7% with growth of 6.8%, primarily due to unexpected increases in domestic demand, and strong manufacturing exports.

Vietnam has a young population, stable political system, commitment to sustainable growth, relatively low inflation, stable currency, strong FDI inflows, and strong manufacturing sector. In addition, the country is committed to continuing its global economic integration. Vietnam joined the WTO in January 2007 and concluded several free trade agreements in 2015-16, including the EU-Vietnam Free Trade Agreement (which the EU has not yet ratified), the Korean Free Trade Agreement, and the Eurasian Economic Union Free Trade Agreement. In 2017, Vietnam successfully chaired the Asia- Pacific Economic Cooperation (APEC) Conference with its key priorities including inclusive growth, innovation, strengthening small and medium enterprises, food security, and climate change. Seeking to diversify its opportunities, Vietnam also signed the Comprehensive and Progressive Agreement for the Transpacific Partnership in 2018 and continued to pursue the Regional Comprehensive Economic Partnership.

To continue its trajectory of strong economic growth, the government acknowledges the need to spark a 'second wave' of reforms, including reforming state-owned-enterprises, reducing red tape, increasing business sector transparency, reducing the level of non-performing loans in the banking sector, and increasing financial sector transparency. Vietnam's public debt to GDP ratio is nearing the government mandated ceiling of 65%.

In 2016, Vietnam cancelled its civilian nuclear energy development program, citing public concerns about safety and the high cost of the program; it faces growing pressure on energy infrastructure. Overall, the country's infrastructure fails to meet the needs of an expanding middle class. Vietnam has demonstrated a commitment to sustainable growth over the last several years, but despite the recent speed-up in economic growth the government remains cautious about the risk of external shocks.

GDP (purchasing power parity): $648.7 billion (2017 est.)
$607.4 billion (2016 est.)
$571.9 billion (2015 est.)
note: data are in 2017 dollars
country comparison to the world: 35

GDP (official exchange rate): $220.4 billion (2017 est.)

GDP—real growth rate: 6.8% (2017 est.)
7.16% (2017 est.)
6.2% (2016 est.)
country comparison to the world: 22

GDP—per capita (PPP): $6,900 (2017 est.)
$6,600 (2016 est.)
$6,200 (2015 est.)
note: data are in 2017 dollars
country comparison to the world: 159

Gross national saving: 29% of GDP (2017 est.)
29.5% of GDP (2016 est.)
27.5% of GDP (2015 est.)
country comparison to the world: 33

GDP—composition, by end use: *household consumption:* 66.9% (2017 est.)
government consumption: 6.5% (2017 est.)
investment in fixed capital: 24.2% (2017 est.)
investment in inventories: 2.8% (2017 est.)

exports of goods and services: 100% (2017 est.)
imports of goods and services: -101% (2017 est.)

GDP—composition, by sector of origin:
agriculture: 15.3% (2017 est.)
industry: 33.3% (2017 est.)
services: 51.3% (2017 est.)

Agriculture—products: rice, coffee, rubber, tea, pepper, soybeans, cashews, sugar cane, peanuts, bananas; pork; poultry; seafood

Industries: food processing, garments, shoes, machine-building; mining, coal, steel; cement, chemical fertilizer, glass, tires, oil, mobile phones

Industrial production growth rate: 8% (2017 est.)
country comparison to the world: 24

Labor force: 54.659 million (2019 est.)
country comparison to the world: 10

Labor force—by occupation: agriculture: 40.3%
industry: 25.7%
services: 34% (2017)

Unemployment rate: 3.11% (2018 est.)
2.2% (2017 est.)
country comparison to the world: 41

Population below poverty line: 8% (2017 est.)

Household income or consumption by percentage share: *lowest 10%:* 2.7%
highest 10%: 26.8% (2014)

Budget: *revenues:* 54.59 billion (2017 est.)
expenditures: 69.37 billion (2017 est.)

Taxes and other revenues: 24.8% (of GDP) (2017 est.)
country comparison to the world: 119

Budget surplus (+) or deficit (-): -6.7% (of GDP) (2017 est.)
country comparison to the world: 191

Public debt: 58.5% of GDP (2017 est.)
59.9% of GDP (2016 est.)
note: official data; data cover general government debt and include debt instruments issued (or owned) by government entities other than the treasury; the data include treasury debt held by foreign entities; the data include debt issued by subnational entities, as well as intragovernmental debt; intragovernmental debt consists of treasury borrowings from surpluses in the social funds, such as for retirement, medical care, and unemployment; debt instruments for the social funds are not sold at public auctions
country comparison to the world: 76

Fiscal year: calendar year

Inflation rate (consumer prices): 3.5% (2017 est.)
2.7% (2016 est.)
country comparison to the world: 142

Current account balance: $12.478 billion (2019 est.)
$5.769 billion (2018 est.)
country comparison to the world: 22

Exports: $214.1 billion (2017 est.)
$176.6 billion (2016 est.)
country comparison to the world: 27

Exports—partners: US 20.1%, China 14.5%, Japan 8%, South Korea 6.8% (2017)

Exports—commodities: clothes, shoes, electronics, seafood, crude oil, rice, coffee, wooden products, machinery

Imports: $202.6 billion (2017 est.)
$162.6 billion (2016 est.)
country comparison to the world: 26

Imports—commodities: machinery and equipment, petroleum products, steel products, raw materials for the clothing and shoe industries, electronics, plastics, automobiles

Imports—partners: China 25.8%, South Korea 20.5%, Japan 7.8%, Thailand 4.9% (2017)

Reserves of foreign exchange and gold: $49.5 billion (31 December 2017 est.)
$36.91 billion (31 December 2016 est.)
country comparison to the world: 40

Debt—external: $96.58 billion (31 December 2017 est.)
$84.34 billion (31 December 2016 est.)
country comparison to the world: 48

Exchange rates: dong (VND) per US dollar—
22,425 (2017 est.)
22,159 (2016 est.)
22,355 (2015 est.)
21,909 (2014 est.)
21,189 (2013 est.)

ENERGY

Electricity access: *electrification—total population:* 100% (2019)

Electricity—production: 158.2 billion kWh (2016 est.)
country comparison to the world: 24

Electricity—consumption: 143.2 billion kWh (2016 est.)
country comparison to the world: 25

Electricity—exports: 713 million kWh (2017 est.)
country comparison to the world: 63

Electricity—imports: 2.733 billion kWh (2016 est.)
country comparison to the world: 51

Electricity—installed generating capacity: 40.77 million kW (2016 est.)
country comparison to the world: 25

Electricity—from fossil fuels: 56% of total installed capacity (2016 est.)
country comparison to the world: 140

Electricity—from nuclear fuels: 0% of total installed capacity (2017 est.)
country comparison to the world: 209

Electricity—from hydroelectric plants: 43% of total installed capacity (2017 est.)
country comparison to the world: 47

Electricity—from other renewable sources: 1% of total installed capacity (2017 est.)
country comparison to the world: 171

Crude oil—production: 242,000 bbl/day (2018 est.)
country comparison to the world: 33

Crude oil—exports: 324,600 bbl/day (2015 est.)
country comparison to the world: 25

Crude oil—imports: 0 bbl/day (2015 est.)
country comparison to the world: 213

Crude oil—proved reserves: 4.4 billion bbl (1 January 2018 est.)
country comparison to the world: 25

Refined petroleum products—production: 153,800 bbl/day (2015 est.)
country comparison to the world: 58

Refined petroleum products—consumption: 438,000 bbl/day (2016 est.)
country comparison to the world: 35

Refined petroleum products— exports: 25,620 bbl/day (2015 est.)
country comparison to the world: 67

Refined petroleum products—imports: 282,800 bbl/day (2015 est.)
country comparison to the world: 25

Natural gas—production: 8.098 billion cu m (2017 est.)
country comparison to the world: 44

Natural gas—consumption: 8.098 billion cu m (2017 est.)
country comparison to the world: 52

Natural gas—exports: 0 cu m (2017 est.)
country comparison to the world: 210

Natural gas—imports: 0 cu m (2017 est.)
country comparison to the world: 210

Natural gas—proved reserves: 699.4 billion cu m (1 January 2018 est.)
country comparison to the world: 27

Carbon dioxide emissions from consumption of energy: 235.3 million Mt (2017 est.)
country comparison to the world: 29

COMMUNICATIONS

Telephones—fixed lines: *total subscriptions:* 3,710,210

subscriptions per 100 inhabitants: 3.79 (2019 est.)
·*country comparison to the world:* 37

Telephones—mobile cellular: total subscriptions: 138,256,733

subscriptions per 100 inhabitants: 141.23 (2019 est.)
country comparison to the world: 12

Telecommunication systems: *general assessment:* despite being a communist country there are plans to part privatize the state's holdings in telecom companies as well as a large number of other enterprises; competition is thriving in the market place; mobile dominates over fixed-line; FttH market growing, as is e-commerce; govt. is the driving force for growth and moving towards commercializing 5G services with test licenses issued in 2019; 5 major operators; Ho Chi Minh City to become the first smart city in Vietnam with cloud computing infrastructure, big data, data centers and security-monitoring centers (2020)

domestic: all provincial exchanges are digitalized and connected to Hanoi, Da Nang, and Ho Chi Minh City by fiber-optic cable or microwave radio relay networks; main lines have been increased,

1051

and the use of mobile telephones is growing rapidly; fixed-line 4 per 100 and mobile-cellular 141 per 100 (2019)

international: country code - 84; landing points for the SeaMeWe-3, APG, SJC2, AAE-1, AAG and the TGN-IA submarine cable system providing connectivity to Europe, Africa, the Middle East, Asia, Southeast Asia, Australia, and the US; satellite earth stations - 2 Intersputnik (Indian Ocean region) (2020)

note: the COVID-19 outbreak is negatively impacting telecommunications production and supply chains globally; consumer spending on telecom devices and services has also slowed due to the pandemic's effect on economies worldwide; overall progress towards improvements in all facets of the telecom industry - mobile, fixed-line, broadband, submarine cable and satellite - has moderated

Broadcast media: government controls all broadcast media exercising oversight through the Ministry of Information and Communication (MIC); government-controlled national TV provider, Vietnam Television (VTV), operates a network of several channels with regional broadcasting centers; programming is relayed nationwide via a network of provincial and municipal TV stations; law limits access to satellite TV but many households are able to access foreign programming via home satellite equipment; government-controlled Voice of Vietnam, the national radio broadcaster, broadcasts on several channels and is repeated on AM, FM, and shortwave stations throughout Vietnam (2018)

Internet country code: .vn

Internet users: *total:* 68,267,875

percent of population: 70.35% (July 2018 est.)
country comparison to the world: 11

Broadband—fixed subscriptions: *total:* 12,994,451

subscriptions per 100 inhabitants: 13 (2018 est.)
country comparison to the world: 16

TRANSPORTATION

National air transport system: *number of registered air carriers:* 5 (2020)

inventory of registered aircraft operated by air carriers: 224

annual passenger traffic on registered air carriers: 47,049,671 (2018)

annual freight traffic on registered air carriers: 481.37 million mt-km (2018)

Civil aircraft registration country code prefix: VN (2016)

Airports: 45 (2013)
country comparison to the world: 96

Airports—with paved runways: *total:* 38 (2013)
over 3,047 m: 10 (2013)
2,438 to 3,047 m: 6 (2013)
1,524 to 2,437 m: 13 (2013)
914 to 1,523 m: 9 (2013)

Airports—with unpaved runways: *total:* 7 (2013)
1,524 to 2,437 m: 1 (2013)
914 to 1,523 m: 3 (2013)

under 914 m: 3 (2013)

Heliports: 1 (2013)

Pipelines: 72 km condensate, 398 km condensate/gas, 955 km gas, 128 km oil, 33 km oil/gas/water, 206 km refined products, 13 km water (2013)

Railways: *total:* 2,600 km (2014)
standard gauge: 178 km 1.435-m gauge; 253 km mixed gauge (2014)

narrow gauge: 2,169 km 1.000-m gauge (2014)
country comparison to the world: 66

Roadways: *total:* 195,468 km (2013)
paved: 148,338 km (2013)
unpaved: 47,130 km (2013)
country comparison to the world: 28

Waterways: 47,130 km (30,831 km weight under 50 tons) (2011)
country comparison to the world: 4

Merchant marine: *total:* 1,863
by type: bulk carrier 83, container ship 38, general cargo 1266, oil tanker 114, other 362 (2019)
country comparison to the world: 14

Ports and terminals: *major seaport(s):* Cam Pha Port, Da Nang, Haiphong, Phu My, Quy Nhon

container port(s) (TEUs): Saigon (6,155,535), Cai Mep (3,065,014) (2017)

river port(s): Ho Chi Minh (Mekong)

MILITARY AND SECURITY

Military and security forces: *People's Army of Vietnam (PAVN):* PAVN Ground Forces, PAVN Navy (includes naval infantry), PAVN Air Force and Air Defense, Border Defense Force, and Vietnam Coast Guard; Vietnam People's Public Security; Vietnam Civil Defense Force (2019)

Military expenditures: 2.3% of GDP (2018)
2.3% of GDP (2017)
2.5% of GDP (2016)
2.4% of GDP (2015)
2.3% of GDP (2014)
country comparison to the world: 40

Military and security service personnel strengths: information is limited and estimates of the size of the People's Army of Vietnam (PAVN) vary; approximately 475,000 active duty troops (405,000 ground; 40,000 naval; 30,000 air); est. 40,000 Border Defense Force and Coast Guard (2019)

Military equipment inventories and acquisitions: the PAVN is armed largely with weapons and equipment from Russia and the former Soviet Union; Russia remains the main supplier of newer PAVN military equipment, although in recent years Vietnam has begun diversifying its procurement with purchases from other countries including Belarus, India, Israel, and Ukraine (2019 est.)

Military service age and obligation: 18-27 years of age for compulsory and voluntary military service (females eligible for conscription, but in practice only males are drafted); conscription typically takes place twice annually and service obligation is 2 years (Army, Air Defense) and 3 years (Navy and Air Force) (2019)

Maritime threats: the International Maritime Bureau reports the territorial and offshore waters in the South China Sea as high risk for piracy and armed robbery against ships; numerous commercial vessels have been attacked and hijacked both at anchor and while underway; hijacked vessels are often disguised and cargo diverted to ports in East Asia; crews have been murdered or cast adrift; the number of reported incidents increased from two in 2017 to four in 2018, primarily near the port of Vung Tau

TRANSPORTATION

Disputes—international: southeast Asian states have enhanced border surveillance to check the spread of Asian swine fever; Cambodia and Laos protest Vietnamese squatters and armed encroachments along border; Cambodia accuses Vietnam of a wide variety of illicit cross-border activities; progress on a joint development area with Cambodia is hampered by an unresolved dispute over sovereignty of offshore islands; an estimated 300,000 Vietnamese refugees reside in China; establishment of a maritime boundary with Cambodia is hampered by unresolved dispute over the sovereignty of offshore islands; the decade-long demarcation of the China-Vietnam land boundary was completed in 2009; China occupies the Paracel Islands also claimed by Vietnam and Taiwan; Brunei claims a maritime boundary extending beyond as far as a median with Vietnam, thus asserting an implicit claim to Lousia Reef; the 2002 "Declaration on the Conduct of Parties in the South China Sea" eased tensions but differences between the parties negotiating the Code of Conduct continue; Vietnam continues to expand construction of facilities in the Spratly Islands; in March 2005, the national oil companies of China, the Philippines, and Vietnam signed a joint accord to conduct marine seismic activities in the Spratly Islands; Economic Exclusion Zone negotiations with Indonesia are ongoing, and the two countries in Fall 2011 agreed to work together to reduce illegal fishing along their maritime boundary; in May 2018, Russia's Rosneft Vietnam unit started drilling at a block southeast of Vietnam which is within the area outlined by China's nine-dash line and Beijing issued a warning

Refugees and internally displaced persons: stateless persons: 30,581 (2019); note - Vietnam's stateless ethnic Chinese Cambodian population dates to the 1970s when thousands of Cambodians fled to Vietnam to escape the Khmer Rouge and were no longer recognized as Cambodian citizens; Vietnamese women who gave up their citizenship to marry foreign men have found themselves stateless after divorcing and returning home to Vietnam; the government addressed this problem in 2009, and Vietnamese women are beginning to reclaim their citizenship

Illicit drugs: minor producer of opium poppy; probable minor transit point for Southeast Asian heroin; government continues to face domestic opium/heroin/methamphetamine addiction problems despite longstanding crackdowns; enforces the death penalty for drug trafficking

VIRGIN ISLANDS

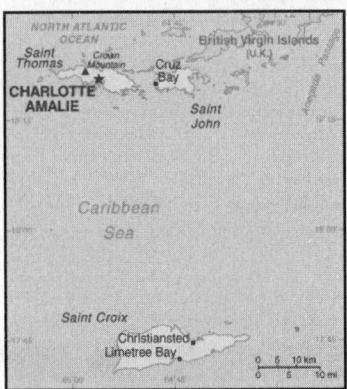

INTRODUCTION

Background: The Danes secured control over the southern Virgin Islands of Saint Thomas, Saint John, and Saint Croix during the 17th and early 18th centuries. Sugarcane, produced by African slave labor, drove the islands' economy during the 18th and early 19th centuries. In 1917, the US purchased the Danish holdings, which had been in economic decline since the abolition of slavery in 1848. On 6 September 2017, Hurricane Irma passed over the northern Virgin Islands of Saint Thomas and Saint John and inflicted severe damage to structures, roads, the airport on Saint Thomas, communications, and electricity. Less than two weeks later, Hurricane Maria passed over the island of Saint Croix in the southern Virgin Islands, inflicting considerable damage with heavy winds and flooding rains.

GEOGRAPHY

Location: Caribbean, islands between the Caribbean Sea and the North Atlantic Ocean, east of Puerto Rico

Geographic coordinates: 18 20 N, 64 50 W

Map references: Central America and the Caribbean

Area: *total:* 1,910 sq km
land: 346 sq km
water: 1,564 sq km
country comparison to the world: 182

Area—comparative: twice the size of Washington, DC

Land boundaries: 0 km

Coastline: 188 km

Maritime claims: *territorial sea:* 12 nm
exclusive economic zone: 200 nm

Climate: subtropical, tempered by easterly trade winds, relatively low humidity, little seasonal temperature variation; rainy season September to November

Terrain: mostly hilly to rugged and mountainous with little flat land

Elevation: *lowest point:* Caribbean Sea 0 m
highest point: Crown Mountain 474 m

Natural resources: pleasant climate, beaches foster tourism

Land use: *agricultural land:* 11.5% (2011 est.)
arable land: 2.9% (2011 est.) / *permanent crops:* 2.9% (2011 est.) / *permanent pasture:* 5.7% (2011 est.)
forest: 57.4% (2011 est.)
other: 31.1% (2011 est.)

Irrigated land: 1 sq km (2012)

Population distribution: while overall population density throughout the islands is relatively low, concentrations appear around Charlotte Amalie on St. Thomas and Christiansted on St. Croix

Natural hazards: several hurricanes in recent years; frequent and severe droughts and floods; occasional earthquakes

Environment—current issues: lack of natural freshwater resources; protection of coral reefs; solid waste management; coastal development; increased boating and overfishing

Geography—note: important location along the Anegada Passage—a key shipping lane for the Panama Canal; Saint Thomas has one of the best natural deepwater harbors in the Caribbean

PEOPLE AND SOCIETY

Population: 106,235 (July 2020 est.)
country comparison to the world: 192

Nationality: *noun:* Virgin Islander(s) (US citizens)
adjective: Virgin Islander

Ethnic groups: black 76%, white 15.6%, Asian 1.4%, other 4.9%, mixed 2.1% (2010 est.)
note: 17.4% self-identify as latino

Languages: English 71.6%, Spanish or Spanish Creole 17.2%, French or French Creole 8.6%, other 2.5% (2010 est.)

Religions: Protestant 59% (Baptist 42%, Episcopalian 17%), Roman Catholic 34%, other 7%

Age structure: *0-14 years:* 19.9% (male 10,820/female 10,322)
15-24 years: 10.32% (male 5,329/female 5,632)
25-54 years: 36.43% (male 18,127/female 20,576)
55-64 years: 14.16% (male 7,177/female 7,864)
65 years and over: 19.19% (male 9,153/female 11,235) (2020 est.)

Dependency ratios: *total dependency ratio:* 66
youth dependency ratio: 32
elderly dependency ratio: 34
potential support ratio: 2.9 (2020 est.)

Median age: *total:* 41.8 years
male: 40.6 years
female: 42.8 years (2020 est.)

country comparison to the world: 42

Population growth rate: -0.37% (2020 est.)
country comparison to the world: 223

Birth rate: 12.1 births/1,000 population (2020 est.)
country comparison to the world: 162

Death rate: 8.5 deaths/1,000 population (2020 est.)
country comparison to the world: 74

Net migration rate: -7.5 migrant(s)/1,000 population (2020 est.)
country comparison to the world: 211

Population distribution: while overall population density throughout the islands is relatively low, concentrations appear around Charlotte Amalie on St. Thomas and Christiansted on St. Croix

Urbanization: *urban population:* 95.9% of total population (2020)
rate of urbanization: 0.1% annual rate of change (2015-20 est.)
total population growth rate v. urban population growth rate, 2000-2030: Major urban areas—population: 52,000 CHARLOTTE AMALIE (capital) (2018)

Sex ratio: *at birth:* 1.06 male(s)/female
0-14 years: 1.05 male(s)/female
15-24 years: 0.95 male(s)/female
25-54 years: 0.88 male(s)/female
55-64 years: 0.91 male(s)/female
65 years and over: 0.81 male(s)/female
total population: 0.91 male(s)/female (2020 est.)

Infant mortality rate: *total:* 7.4 deaths/1,000 live births
male: 8.4 deaths/1,000 live births
female: 6.3 deaths/1,000 live births (2020 est.)
country comparison to the world: 159

Life expectancy at birth: *total population:* 79.8 years
male: 76.6 years
female: 83.2 years (2020 est.)
country comparison to the world: 51

Total fertility rate: 2.03 children born/woman (2020 est.)
country comparison to the world: 109

Drinking water source: *improved:* total: 98.7% of population
unimproved: total: 1.3% of population (2017 est.)

Sanitation facility access: *improved:* total: 100% of population
unimproved: total: 100% of population (2017 est.)

HIV/AIDS—adult prevalence rate: NA

HIV/AIDS—people living with HIV/AIDS: NA

HIV/AIDS—deaths: NA

Education expenditures: NA

GOVERNMENT

Country name: *conventional long form:* none
conventional short form: Virgin Islands

former: Danish West Indies

abbreviation: VI

etymology: the myriad islets, cays, and rocks surrounding the major islands reminded Christopher COLUMBUS in 1493 of Saint Ursula and her 11,000 virgin followers (Santa Ursula y las Once Mil Virgenes), which over time shortened to the Virgins (las Virgenes)

Dependency status: unincorporated organized territory of the US with policy relations between the Virgin Islands and the federal government under the jurisdiction of the Office of Insular Affairs, US

DEPARTMENT OF THE INTERIOR

Government type: republican form of government with separate executive, legislative, and judicial branches; unincorporated organized territory of the US with local self-government

Capital: *name:* Charlotte Amalie

geographic coordinates: 18 21 N, 64 56 W

time difference: UTC-4 (1 hour ahead of Washington, DC, during Standard Time)

etymology: originally called Taphus in Danish—meaning "tap house" or "beer house" because of its many beer halls—the town received a more dignified name in 1691 when it was named Charlotte Amalie in honor of Danish King Christian V's wife, Charlotte Amalie of Hesse-Kassel (1650–1714)

Administrative divisions: none (territory of the US); there are no first-order administrative divisions as defined by the US Government, but there are 3 islands at the second order; Saint Croix, Saint John, Saint Thomas

Independence: none (territory of the US)

National holiday: Transfer Day (from Denmark to the US), 31 March (1917)

Constitution: *history:* 22 July 1954—the Revised Organic Act of the Virgin Islands functions as a constitution for this US territory

amendments: revised 1962, 2000

Legal system: US common law

Citizenship: see United States

Suffrage: 18 years of age; universal; note—island residents are US citizens but do not vote in US presidential elections

Executive branch: *chief of state:* President Donald J. TRUMP (since 20 January 2017); Vice President Michael R. PENCE (since 20 January 2017)

head of government: Governor Albert BRYAN, Jr. (since 7 January 2019), Lieutenant Governor Tregenza ROACH (since 7 January 2019)

cabinet: Territorial Cabinet appointed by the governor and confirmed by the Senate

elections/appointments: president and vice president indirectly elected on the same ballot by an Electoral College of 'electors' chosen from each state; president and vice president serve a 4-year term (eligible for a second term); under the US Constitution, residents of the Virgin Islands do not vote in elections for US president and vice president; however, they may vote in the Democratic and Republican presidential primary elections; governor and lieutenant governor directly elected

on the same ballot by absolute majority vote in 2 rounds if needed for a 4-year term (eligible for a second term); election last held on 6 November 2018 with a runoff on 20 November 2018 (next to be held in November 2022)

election results: Albert BRYAN, Jr. elected governor in the second round; percent of vote in first round—Albert BRYAN, Jr. (Democratic Party) 38.1%, Kenneth MAPP (independent) 33.5%, Adlah "Foncie" DONASTORG, Jr. (independent) 16.5%, other 11.9%; percent of vote in second round- Albert BRYAN, Jr. (Democratic Party) 54.5%, Kenneth MAPP (independent) 45.2%, other .3%

Legislative branch: *description:* unicameral Legislature of the Virgin Islands (15 seats; senators directly elected in single- and multi-seat constituencies by simple majority popular vote to serve 2-year terms) the Virgin Islands directly elects 1 delegate to the US House of Representatives by simple majority vote to serve a 2-year term

elections: Legislature of the Virgin Islands last held on 6 November 2018 (next to be held in November 2020)

US House of Representatives last held on 6 November 2018 (next to be held in November 2020)

election results: Legislature of the Virgin Islands—percent of vote by party—NA; seats by party—Democratic Party 13, independents 2; composition—men 11, women 4, percent of women 26.7% delegate to US House of Representatives—seat by party—Democratic Party 1; composition—1 woman note: the Virgin Islands to the US House of Representatives can vote when serving on a committee and when the House meets as the Committee of the Whole House, but not when legislation is submitted for a "full floor" House vote

Judicial branch: *highest courts:* Supreme Court of the Virgin Islands (consists of the chief justice and 2 associate justices); note—court established by the US Congress in 2004 and assumed appellate jurisdiction in 2007

judge selection and term of office: justices appointed by the governor and confirmed by the Virgin Islands Senate; justices serve initial 10-year terms and upon reconfirmation, during the extent of good behavior; chief justice elected to position by peers for a 3-year term

subordinate courts: Superior Court (Territorial Court renamed in 2004); US Court of Appeals for the Third Circuit (has appellate jurisdiction over the District Court of the Virgin Islands; it is a territorial court and is not associated with a US federal judicial district); District Court of the Virgin Islands

Political parties and leaders: Democratic Party [Stacey PLASKELL]

Independent Citizens' Movement or ICM [Dale BLYDEN]

Republican Party [John CANEGATA]

International organization participation: AOSIS (observer), Interpol (subbureau), IOC, UPU, WFTU (NGOs)

Diplomatic representation in the US: none (territory of the US)

Diplomatic representation from the US: none (territory of the US)

Flag description: white field with a modified US coat of arms in the center between the large blue initials V and I; the coat of arms shows a yellow eagle holding an olive branch in its right talon and three arrows in the left with a superimposed shield of seven red and six white vertical stripes below a blue panel; white is a symbol of purity, the letters stand for the Virgin Islands

National anthem: *name:* Virgin Islands March

lyrics/music: multiple/Alton Augustus ADAMS, Sr.

note: adopted 1963; serves as a local anthem; as a territory of the US, "The Star-Spangled Banner" is official (see United States)

ECONOMY

Economy—overview: Tourism, trade, other services, and rum production are the primary economic activities of the US Virgin Islands (USVI), accounting for most of its GDP and employment. The USVI receives between 2.5 and 3 million tourists a year, mostly from visiting cruise ships. The islands are vulnerable to damage from storms, as evidenced by the destruction from two major hurricanes in 2017. Recovery and rebuilding have continued, but full recovery from these back-to-back hurricanes is years away. The USVI government estimates it will need $7.5 billion, almost twice the territory's GDP, to rebuild the territory.

The agriculture sector is small and most food is imported. In 2016, government spending (both federal and territorial together) accounted for about 27% of GDP while exports of goods and services, including spending by tourists, accounted for nearly 47%. Federal programs and grants, including rum tax cover-over totaling $482.3 million in 2016, contributed 32.2% of the territory's total revenues. The economy picked up 0.9% in 2016 and had appeared to be progressing before the 2017 hurricanes severely damaged the territory's infrastructure and the economy.

GDP (purchasing power parity): $3.872 billion (2016 est.)

$3.759 billion (2015 est.)

$3.622 billion (2014 est.)

note: data are in 2013 dollars

country comparison to the world: 181

GDP (official exchange rate): $5.182 billion (2016 est.)

GDP—real growth rate: 0.9% (2016 est.)

0.3% (2015 est.)

-1% (2014 est.)

country comparison to the world: 175

GDP—per capita (PPP): $37,000 (2016 est.)

$35,800 (2015 est.)

$34,500 (2014 est.)

country comparison to the world: 54

GDP—composition, by end use: *household consumption:* 68.2% (2016 est.)

government consumption: 26.8% (2016 est.)
investment in fixed capital: 7.5% (2016 est.)
investment in inventories: 15% NA (2016 est.)
exports of goods and services: 46.7% (2016 est.)
imports of goods and services: -64.3% (2016 est.)

GDP—composition, by sector of origin:
agriculture: 2% (2012 est.)
industry: 20% (2012 est.)
services: 78% (2012 est.)

Agriculture—products: fruit, vegetables, sorghum; Senepol cattle

Industries: tourism, watch assembly, rum distilling, construction, pharmaceuticals, electronics

Industrial production growth rate: NA

Labor force: 48,550 (2016 est.)
country comparison to the world: 194

Labor force—by occupation: *agriculture:* 1%
industry: 19%
services: 80% (2003 est.)

Unemployment rate: 10.4% (2017 est.)
11% (2016 est.)
country comparison to the world: 151

Population below poverty line: 28.9% (2002 est.)

Household income or consumption by percentage share: *lowest 10%:* NA
highest 10%: NA

Budget: *revenues:* 1.496 billion (2016 est.)
expenditures: 1.518 billion (2016 est.)

Taxes and other revenues: 28.9% (of GDP) (2016 est.)
country comparison to the world: 89

Budget surplus (+) or deficit (-): -0.4% (of GDP) (2016 est.)
country comparison to the world: 59

Public debt: 53.3% of GDP (2016 est.)
45.9% of GDP (2014 est.)
country comparison to the world: 91

Fiscal year: 1 October–30 September

Inflation rate (consumer prices): 1% (2016 est.)
2.6% (2015 est.)
country comparison to the world: 54

Exports: $1.81 billion (2016 est.)
$1.537 billion (2015 est.)
country comparison to the world: 146

Exports—commodities: rum

Imports: $2.489 billion (2016 est.)
$1.549 billion (2015 est.)
country comparison to the world: 158

Imports—commodities: foodstuffs, consumer goods, building materials

Debt—external: NA

Exchange rates: the US dollar is used

ENERGY

Electricity access: *electrification - total population:* 100% (2020)

Electricity—production: 704 million kWh (2016 est.)
country comparison to the world: 157

Electricity—consumption: 654.7 million kWh (2016 est.)

country comparison to the world: 162

Electricity—exports: 0 kWh (2016 est.)
country comparison to the world: 216

Electricity—imports: 0 kWh (2016 est.)
country comparison to the world: 217

Electricity—installed generating capacity:
325,000 kW (2016 est.)
country comparison to the world: 157

Electricity—from fossil fuels: 98% of total installed capacity (2016 est.)
country comparison to the world: 30

Electricity—from nuclear fuels: 0% of total installed capacity (2017 est.)
country comparison to the world: 210

Electricity—from hydroelectric plants: 0% of total installed capacity (2017 est.)
country comparison to the world: 212

Electricity—from other renewable sources: 2% of total installed capacity (2017 est.)
country comparison to the world: 146

Crude oil—production: 0 bbl/day (2018 est.)
country comparison to the world: 215

Crude oil—exports: 0 bbl/day (2015 est.)
country comparison to the world: 214

Crude oil—imports: 0 bbl/day (2015 est.)
country comparison to the world: 214

Crude oil—proved reserves: 0 bbl (1 January 2018 est.)
country comparison to the world: 211

Refined petroleum products—production: 0 bbl/day (2015 est.)
country comparison to the world: 214

Refined petroleum products—consumption: 1,240 bbl/day (2016 est.)
country comparison to the world: 204

Refined petroleum products—exports: 3,285 bbl/day (2015 est.)
country comparison to the world: 97

Refined petroleum products—imports: 23,480 bbl/day (2015 est.)
country comparison to the world: 112

Natural gas—production: 0 cu m (2017 est.)
country comparison to the world: 213

Natural gas—consumption: 0 cu m (2017 est.)
country comparison to the world: 212

Natural gas—exports: 0 cu m (2017 est.)
country comparison to the world: 211

Natural gas—imports: 0 cu m (2017 est.)
country comparison to the world: 211

Natural gas—proved reserves: 0 cu m (1 January 2014 est.)
country comparison to the world: 206

Carbon dioxide emissions from consumption of energy: 2.764 million Mt (2017 est.)
country comparison to the world: 150

COMMUNICATIONS

Telephones—fixed lines: *total subscriptions:* 77,212

subscriptions per 100 inhabitants: 72.41 (2019 est.)
country comparison to the world: 145

Telecommunication systems: *general assessment:* modern system with total digital switching, uses fiber-optic cable and microwave radio relay; good interisland and international connections; broadband access; expansion of FttP (Fiber to the Home) markets; LTE launches; regulatory development and expansion in several markets point to investment and focus on data (2020)
domestic: full range of services available; fixed-line 72 per 100 persons and mobile-cellular 75 per 100 (2019)
international: country code—1-340; landing points for the BSCS, St Thomas-ST Croix System, Southern Caribbean Fiber, Americas II, GCN, MAC, PAN-AM and SAC submarine cable connections to US, the Caribbean, Central and South America; satellite earth stations—NA (2020)
note: the COVID-19 outbreak is negatively impacting telecommunications production and supply chains globally; consumer spending on telecom devices and services has also slowed due to the pandemic's effect on economies worldwide; overall progress towards improvements in all facets of the telecom industry—mobile, fixed-line, broadband, submarine cable and satellite—has moderated

Broadcast media: about a dozen TV broadcast stations including 1 public TV station; multi-channel cable and satellite TV services are available; 24 radio stations

Internet country code: .vi

Internet users: *total:* 68,872
percent of population: 64.38% (July 2018 est.)
country comparison to the world: 189

TRANSPORTATION

Airports: 2 (2013)
country comparison to the world: 207

Airports—with paved runways: *total:* 2 (2019)
over 3,047 m: 1
1,524 to 2,437 m: 1

Roadways: *total:* 1,260 km (2008)
country comparison to the world: 178

Merchant marine: *total:* 1,868
by type: bulk carrier 91, container ship 39, general cargo 1,205, oil tanker 118, other 415 (2019)
country comparison to the world: 13

Ports and terminals: *major seaport(s):* Charlotte Amalie, Christiansted, Cruz Bay, Frederiksted, Limetree Bay

MILITARY AND SECURITY

Military—note: defense is the responsibility of the US

TRANSNATIONAL ISSUES

Disputes—international: none

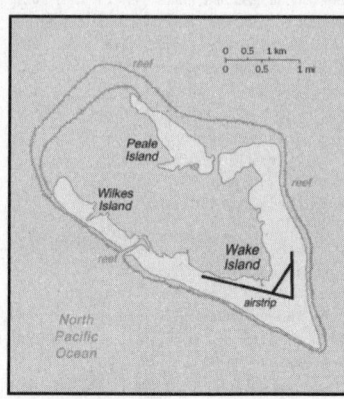

Coastline: 19.3 km

Maritime claims: *territorial sea:* 12 nm
exclusive economic zone: 200 nm

Climate: tropical

Terrain: atoll of three low coral islands, Peale, Wake, and Wilkes, built up on an underwater volcano; central lagoon is former crater, islands are part of the rim

Elevation: *lowest point:* Pacific Ocean 0 m
highest point: unnamed location 8 m

Natural resources: none

Land use: *agricultural land:* 0% (2011 est.)
arable land: 0% (2011 est.) / permanent crops: 0% (2011 est.) / permanent pasture: 0% (2011 est.)
forest: 0% (2011 est.)
other: 100% (2011 est.)

Irrigated land: 0 sq km (2012)

Natural hazards: subject to occasional typhoons

Environment—current issues: potable water obtained through a catchment rainwater system and a desalinization plant for brackish ground water; hazardous wastes moved to an accumulation site for storage and eventual transport off site via barge

Geography—note: strategic location in the North Pacific Ocean; emergency landing location for transpacific flights

PEOPLE AND SOCIETY

Population: no indigenous inhabitants (2018 est.)
note: approximately 100 military personnel and civilian contractors maintain and operate the airfield and communications facilities

GOVERNMENT

Country name: *conventional long form:* none
conventional short form: Wake Island
etymology: although first discovered by British Captain William WAKE in 1792, the island is named after British Captain Samuel WAKE, who rediscovered the island in 1796

Dependency status: unincorporated unorganized territory of the US; administered from Washington, DC, by the Department of the Interior; activities in the atoll are currently conducted by the 11th US Air Force and managed from Pacific Air Force Support Center

Independence: none (territory of the US)

Legal system: US common law

Citizenship: see United States

Flag description: the flag of the US is used

ECONOMY

Economy—overview: Economic activity is limited to providing services to military personnel and contractors located on the island. All food and manufactured goods must be imported.

ENERGY

Electricity access: *electrification—total population:* 100% (2020)

Crude oil—production: 0 bbl/day (2018 est.)
country comparison to the world: 216

COMMUNICATIONS

Telecommunication systems: *general assessment:* satellite communications; 2 Defense Switched Network circuits off the Overseas Telephone System (OTS); located in the Hawaii area code - 808 (2018)

Broadcast media: American Armed Forces Radio and Television Service (AFRTS) provides satellite radio/TV broadcasts (2018)

TRANSPORTATION

Airports: 1 (2018)
country comparison to the world: 238

Airports —with paved runways: *total:* 1 (2019) 2,438 to 3,047 m: 1

Ports and terminals: none; two offshore anchorages for large ships

Transportation - note: there are no commercial or civilian flights to and from Wake Island, except in direct support of island missions; emergency landing is available

MILITARY AND SECURITY

Military—note: defense is the responsibility of the US; the US Air Force is responsible for overall administration and operation of the island facilities; the launch support facility is administered by the US Missile Defense Agency (MDA)

TRANSPORTATION

Disputes—international: claimed by Marshall Islands

INTRODUCTION

Background: The US annexed Wake Island in 1899 for a cable station. An important air and naval base was constructed in 1940-41. In December 1941, the island was captured by the Japanese and held until the end of World War II. In subsequent years, Wake became a stopover and refueling site for military and commercial aircraft transiting the Pacific. Since 1974, the island's airstrip has been used by the US military, as well as for emergency landings. Operations on the island were temporarily suspended and all personnel evacuated in 2006 with the approach of super typhoon IOKE (category 5), but resultant damage was comparatively minor. A US Air Force repair team restored full capability to the airfield and facilities, and the island remains a vital strategic link in the Pacific region.

GEOGRAPHY

Location: Oceania, atoll in the North Pacific Ocean, about two-thirds of the way from Hawaii to the Northern Mariana Islands

Geographic coordinates: 19 17 N, 166 39 E

Map references: Oceania

Area: *total:* 7 sq km
land: 6.5 sq km
water: 0 sq km
country comparison to the world: 246

Area—comparative: about 11 times the size of the National Mall in Washington, DC

Land boundaries: 0 km

WALLIS AND FUTUNA

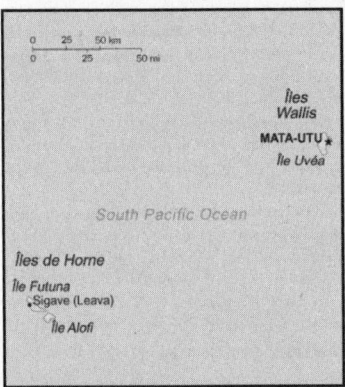

INTRODUCTION

Background: The Futuna island group was discovered by the Dutch in 1616 and Wallis by the British in 1767, but it was the French who declared a protectorate over the islands in 1842, and took official control of them between 1886 and 1888. Notably, Wallis and Futuna was the only French colony to side with the Vichy regime during World War II, a phase that ended in May of 1942 with the arrival of 2,000 American troops. In 1959, the inhabitants of the islands voted to become a French overseas territory and officially assumed that status in 1961. In 2003, Wallis and Futuna's designation changed to that of an overseas collectivity.

GEOGRAPHY

Location: Oceania, islands in the South Pacific Ocean, about two-thirds of the way from Hawaii to New Zealand

Geographic coordinates: 13 18 S, 176 12 W

Map references: Oceania

Area: *total:* 142 sq km
land: 142 sq km
water: 0 sq km
note: includes Ile Uvea (Wallis Island), Ile Futuna (Futuna Island), Ile Alofi, and 20 islets
country comparison to the world: 221

Area—comparative: 1.5 times the size of Washington, DC

Land boundaries: 0 km

Coastline: 129 km

Maritime claims: *territorial sea:* 12 nm
exclusive economic zone: 200 nm

Climate: tropical; hot, rainy season (November to April); cool, dry season (May to October); rains 250-300 cm per year (80% humidity); average temperature 26.6 degrees Celsius

Terrain: volcanic origin; low hills

Elevation: *lowest point:* Pacific Ocean 0 m
highest point: Mont Singavi (on Futuna) 522 m

Natural resources: NEGL

Land use: *agricultural land:* 42.8% (2011 est.)
arable land: 7.1% (2011 est.) / permanent crops: 35.7% (2011 est.) / permanent pasture: 0% (2011 est.)
forest: 41.9% (2011 est.)
other: 15.3% (2011 est.)

Irrigated land: 0 sq km (2012)

Natural hazards: cyclones; tsunamis

Environment—current issues: deforestation (only small portions of the original forests remain) largely as a result of the continued use of wood as the main fuel source; as a consequence of cutting down the forests, the mountainous terrain of Futuna is particularly prone to erosion; there are no permanent settlements on Alofi because of the lack of natural freshwater resources; lack of soil fertility on the islands of Uvea and Futuna negatively impacts agricultural producitivity

Geography—note: both island groups have fringing reefs; Wallis contains several prominent crater lakes

PEOPLE AND SOCIETY

Population: 15,854 (July 2020 est.)
country comparison to the world: 221

Nationality: *noun:* Wallisian(s), Futunan(s), or Wallis and Futuna Islanders
adjective: Wallisian, Futunan, or Wallis and Futuna Islander

Ethnic groups: Polynesian

Languages: Wallisian (indigenous Polynesian language) 58.9%, Futunian 30.1%, French (official) 10.8%, other 0.2% (2003 census)

Religions: Roman Catholic 99%, other 1%

Age structure: *0-14 years:* 20.58% (male 1,702/female 1,561)
15-24 years: 14.72% (male 1,238/female 1,095)
25-54 years: 43.55% (male 3,529/female 3,376)
55-64 years: 9.92% (male 742/female 830)
65 years and over: 11.23% (male 856/female 925) (2020 est.)

population pyramid: [INSERT IMAGE: WALLIS AND FUTUNA - population pyramid]

Median age: *total:* 34 years
male: 33.1 years
female: 35.1 years (2020 est.)
country comparison to the world: 93

Population growth rate: 0.28% (2020 est.)
country comparison to the world: 174

Birth rate: 12.7 births/1,000 population (2020 est.)
country comparison to the world: 151

Death rate: 5.7 deaths/1,000 population (2020 est.)
country comparison to the world: 179

Net migration rate: -4.3 migrant(s)/1,000 population (2020 est.)
note: there has been steady emigration from Wallis and Futuna to New Caledonia
country comparison to the world: 192

Urbanization: *urban population:* 0% of total population (2020)
rate of urbanization: 0% annual rate of change (2015-20 est.)

total population growth rate v. urban population growth rate, 2000-2030: *Major urban areas - population:* 1,000 MATA-UTU (capital) (2018)

Sex ratio: *at birth:* 1.05 male(s)/female
0-14 years: 1.09 male(s)/female
15-24 years: 1.13 male(s)/female
25-54 years: 1.05 male(s)/female
55-64 years: 0.89 male(s)/female
65 years and over: 0.93 male(s)/female
total population: 1.04 male(s)/female (2020 est.)

Infant mortality rate: *total:* 4.2 deaths/1,000 live births
male: 4.4 deaths/1,000 live births
female: 3.9 deaths/1,000 live births (2020 est.)
country comparison to the world: 189

Life expectancy at birth: *total population:* 80.2 years
male: 77.2 years
female: 83.4 years (2020 est.)
country comparison to the world: 48

Total fertility rate: 1.71 children born/woman (2020 est.)
country comparison to the world: 172

Drinking water source:
improved: *rural:* 100% of population
total: 100% of population
unimproved: *rural:* 0% of population
total: 0% of population (2017)

Sanitation facility access:
improved: *rural:* 100% of population
total: 100% of population
unimproved: *rural:* 0% of population
total: 0% of population (2017 est.)

HIV/AIDS—adult prevalence rate: NA

HIV/AIDS—people living with HIV/AIDS: NA

HIV/AIDS—deaths: NA

Major infectious diseases: *degree of risk:* high (2020)
food or waterborne diseases: bacterial diarrhea
vectorborne diseases: malaria

Education expenditures: NA

GOVERNMENT

Country name: *conventional long form:* Territory of the Wallis and Futuna Islands
conventional short form: Wallis and Futuna
local long form: Territoire des Iles Wallis et Futuna

local short form: Wallis et Futuna

former: Hoorn Islands is the former name of the Futuna Islands

etymology: Wallis Island is named after British Captain Samuel WALLIS, who discovered it in 1767; Futuna is derived from the native word "futu," which is the name of the fish-poison tree found on the island

Dependency status: overseas collectivity of France

Government type: parliamentary democracy (Territorial Assembly); overseas collectivity of France

Capital: *name:* Mata-Utu (on Ile Uvea)

geographic coordinates: 13 57 S, 171 56 W

time difference: UTC+12 (17 hours ahead of Washington, DC, during Standard Time)

Administrative divisions: 3 administrative precincts (circonscriptions, singular - circonscription) Alo, Sigave, Uvea

Independence: none (overseas collectivity of France)

National holiday: Bastille Day, 14 July (1789)

Constitution: *history:* 4 October 1958 (French Constitution)

amendments: French constitution amendment procedures apply

Legal system: French civil law

Citizenship: see France

Suffrage: 18 years of age; universal

Executive branch: *chief of state:* President Emmanuel MACRON (since 14 May 2017); represented by High Administrator Thierry QUEFFELEC (since 7 January 2019)

head of government: President of the Territorial Assembly David VERGE (since 4 April 2017)

cabinet: Council of the Territory appointed by the high administrator on the advice of the Territorial Assembly

elections/appointments: French president elected by absolute majority popular vote in 2 rounds if needed for a 5-year term (eligible for a second term); high administrator appointed by the French president on the advice of the French Ministry of the Interior; the presidents of the Territorial Government and the Territorial Assembly elected by assembly members

note: there are 3 traditional kings with limited powers

Legislative branch: *description:* unicameral Territorial Assembly or Assemblee Territoriale (20 seats - Wallis 13, Futuna 7; members directly elected in multi-seat constituencies by party-list proportional representation vote to serve 5-year terms)

Wallis and Futuna indirectly elects 1 senator to the French Senate by an electoral college by absolute majority vote in 2 rounds if needed for a 6 year term, and directly elects 1 deputy to the French National Assembly by absolute majority vote for a 5-year term

elections: Territorial Assembly - last held on 26 March 2017 (next to be held in March 2022)

French Senate—last held on 28 September 2014 (next to be held by September 2020)

French National Assembly—last held on 11 June 2017 (next to be held in June 2022)

election results: Territorial Assembly - percent of vote by party - NA; seats by party - 2 members are elected from the list Fia gaue fakatahi kihe kaha'u e lelei and 1 each from 18 other lists; composition - men 14, women 6, percent of women 30%

French Senate—LR 1

French National Assembly—independent 1

Judicial branch: *highest courts:* Court of Assizes or Cour d'Assizes (consists of 1 judge; court hears primarily serious criminal cases); note - appeals beyond the Court of Assizes are heard before the Court of Appeal or Cour d'Appel (in Noumea, New Caledonia)

judge selection and term of office: NA

subordinate courts: courts of first instance; labor court; note - justice generally administered under French law by the high administrator, but the 3 traditional kings administer customary law, and there is a magistrate in Mata-Utu

Political parties and leaders: Left Radical Party or PRG [Guillaume LACROIX] (formerly Radical Socialist Party or PRS and the Left Radical Movement or MRG)

Lua Kae Tahi (Giscardians) (leader NA)

Rally for Wallis and Futuna-The Republicans (Rassemblement pour Wallis and Futuna) or RPWF-LR [Clovis LOGOLOGOFOLAU]

Socialist Party or PS

Taumu'a Lelei [Soane Muni UHILA]

Union Pour la Democratie Francaise or UDF

International organization participation: PIF (observer), SPC, UPU

Diplomatic representation in the US: none (overseas collectivity of France)

Diplomatic representation from the US: none (overseas collectivity of France)

Flag description: unofficial, local flag has a red field with four white isosceles triangles in the middle, representing the three native kings of the islands and the French administrator; the apexes of the triangles are oriented inward and at right angles to each other; the flag of France, outlined in white on two sides, is in the upper hoist quadrant

note: the design is derived from an original red banner with a white cross pattee that was introduced in the 19th century by French missionaries; the flag of France is used for official occasions

National symbol(s): red saltire (Saint Andrew's Cross) on a white square on a red field; national colors: red, white

National anthem: note: as a territory of France, "La Marseillaise" is official (see France) 0:00 / 1:19

ECONOMY

Economy—overview: The economy is limited to traditional subsistence agriculture, with about 80% of labor force earnings coming from agriculture (coconuts and vegetables), livestock (mostly pigs), and fishing. However, roughly 70% of the labor force is employed in the public sector, although only about a third of the population is in salaried employment.

Revenues come from French Government subsidies, licensing of fishing rights to Japan and South Korea, import taxes, and remittances from expatriate workers in New Caledonia. France directly finances the public sector and health-care and education services. It also provides funding for key development projects in a range of areas, including infrastructure, economic development, environmental management, and health-care facilities.

A key concern for Wallis and Futuna is an aging population with consequent economic development issues. Very few people aged 18-30 live on the islands due to the limited formal employment opportunities. Improving job creation is a current priority for the territorial government.

GDP (purchasing power parity): $60 million (2004 est.)

country comparison to the world: 225

GDP (official exchange rate): $195 million (2005) (2005)

GDP—real growth rate: NA

GDP—per capita (PPP): $3,800 (2004 est.)

country comparison to the world: 181

GDP—composition, by end use: *household consumption:* 26% (2005)

government consumption: 54% (2005)

GDP—composition, by sector of origin: *agriculture:* NA

industry: NA

services: NA

Agriculture—products: coconuts, breadfruit, yams, taro, bananas; pigs, goats; fish

Industries: copra, handicrafts, fishing, lumber

Industrial production growth rate: NA

Labor force: 4,482 (2013)

country comparison to the world: 223

Labor force—by occupation: agriculture: 74%

industry: 3%

services: 23% (2015 est.)

Unemployment rate: 8.8% (2013 est.)

12.2% (2008 est.)

country comparison to the world: 133

Population below poverty line: NA

Household income or consumption by percentage share: *lowest 10%:* NA

highest 10%: NA

Budget: *revenues:* 32.54 million NA (2015 est.)

expenditures: 34.18 million NA (2015 est.)

Taxes and other revenues: 16.7% (of GDP) NA (2015 est.)

country comparison to the world: 175

Budget surplus (+) or deficit (-): -0.8% (of GDP) NA (2015 est.)

country comparison to the world: 70

Public debt: 5.6% of GDP (2004 est.)

note: offical data; data cover general government debt and include debt instruments issued (or owned) by government entities other than the

treasury; the data include treasury debt held by foreign entities; the data include debt issued by subnational entities, as well as intragovernmental debt; intragovernmental debt consists of treasury borrowings from surpluses in the social funds, such as for retirement, medical care, and unemployment; debt instruments for the social funds are not sold at public auctions
country comparison to the world: 204

Fiscal year: calendar year

Inflation rate (consumer prices): 0.9% (2015)
2.8% (2005)
country comparison to the world: 48

Exports: $47,450 (2004 est.)
country comparison to the world: 222

Exports—commodities: copra, chemicals, construction materials

Imports: $61.17 million (2004 est.)
country comparison to the world: 220

Imports—commodities: chemicals, machinery, consumer goods

Debt—external: $3.67 million (2004)
country comparison to the world: 201

Exchange rates: Comptoirs Francais du Pacifique francs (XPF) per US dollar—
110.2 (2015 est.)
89.8 (2014 est.)
89.85 (2013 est.)
90.56 (2012 est.)

COMMUNICATIONS

Telephones—fixed lines: *total subscriptions:* 4,012

subscriptions per 100 inhabitants: 25.38 (2019 est.)
country comparison to the world: 207

Telecommunication systems: *general assessment:* 2G widespread; bandwidth is limited; mobile subscriber numbers are higher than fixed-line and better suited for islands; good mobile coverage in the capital cities and also reasonable coverage across more remote atolls; recent international interest in infrastructure development; increase in demand for mobile broadband as mobile services serve as primary source for Internet access; Kacific-1 broadband satellite launched in 2019 to improve costs and capability (2020)

domestic: fixed-line teledensity 25 per 100 persons (2019)

international: country code - 681; landing point for the Tui-Samoa submarine cable network connecting Wallis & Futuna, Samoa and Fiji (2020)
note: the COVID-19 outbreak is negatively impacting telecommunications production and supply chains globally; consumer spending on telecom devices and services has also slowed due to the pandemic's effect on economies worldwide; overall progress towards improvements in all facets of the telecom industry - mobile, fixed-line, broadband, submarine cable and satellite - has moderated

Broadcast media: the publicly owned French Overseas Network (RFO), which broadcasts to France's overseas departments, collectivities, and territories, is carried on the RFO Wallis and Fortuna TV and radio stations (2019)

Internet country code: .wf

Internet users: *total:* 3,450

percent of population: 22.1% (July 2016 est.)
country comparison to the world: 220

TRANSPORTATION

Airports: 2 (2013)
country comparison to the world: 208

Airports—with paved runways: *total:* 2 (2019)
1,524 to 2,437 m: 1
914 to 1,523 m: 1

Ports and terminals: *major seaport(s):* Leava, Mata-Utu

MILITARY AND SECURITY

Military—note: defense is the responsibility of France

TRANSPORTATION

Disputes—international: none

WEST BANK

INTRODUCTION

Background: Inhabited since at least the 15th century B.C., the West Bank has been dominated by many different peoples throughout its history; it was incorporated into the Ottoman Empire in the early 16th century. The West Bank fell to British forces during World War I, becoming part of the British Mandate of Palestine. Following the 1948 Arab-Israeli War, the West Bank was captured by Transjordan (later renamed Jordan), which annexed the West Bank in 1950; it was captured by Israel in the Six-Day War in 1967. Under a series of agreements known as the Oslo accords signed between 1993 and 1999, Israel transferred to the newly created Palestinian Authority (PA) security and civilian responsibility for many Palestinian-populated areas of the West Bank as well as the Gaza Strip. In 2000, a violent intifada or uprising began, and in 2001 negotiations to determine the permanent status of the West Bank and Gaza Strip stalled. Subsequent attempts to re-start direct negotiations have not resulted in progress toward determining final status of the area.

Roughly 60% of the West Bank, remains under Israeli civil and military control. In early 2006, the Islamic Resistance Movement (HAMAS) won a majority in the Palestinian Legislative Council (PLC) election. Attempts to form a unity government between Fatah, the dominant Palestinian political faction in the West Bank, and HAMAS failed, leading to violent clashed between their respective supporters and HAMAS's violent siezure of all military and governmental institutions in the Gaza Strip in June 2007. Since 2007, the PA has administered parts of the West Bank under its control, mainly the major Palestinian population centers and areas immediately surrounding them. Fatah and HAMAS have made several attempts at reconciliation, but the factions have been unable to implement agreements including the latest agreement signed in October 2017. In December 2018, the Palestinian Constitutional Court dissolved the PLC. In 2019, PA President ABBAS renewed his calls for PLC elections.

GEOGRAPHY

Location: Middle East, west of Jordan, east of Israel

Geographic coordinates: 32 00 N, 35 15 E

Map references: Middle East

Area: *total:* 5,860 sq km
land: 5,640 sq km
water: 220 sq km
note: includes West Bank, Latrun Salient, and the northwest quarter of the Dead Sea, but excludes Mt. Scopus; East Jerusalem and Jerusalem No Man's Land are also included only as a means of depicting the entire area occupied by Israel in 1967
country comparison to the world: 172

Area—comparative: slightly smaller than Delaware

Land boundaries: *total:* 478 km
border countries (2): Israel 330 km, Jordan 148 km

Coastline: 0 km (landlocked)

Maritime claims: none (landlocked)

Climate: temperate; temperature and precipitation vary with altitude, warm to hot summers, cool to mild winters

Terrain: mostly rugged, dissected upland in west, flat plains descending to Jordan River Valley to the east

Elevation: *lowest point:* Dead Sea -431 m
highest point: Khallat al Batrakh 1,020 m

Natural resources: arable land

Land use: *agricultural land:* 43.3% (2011 est.)
arable land: 7.4% (2011 est.) / permanent crops: 11% (2011 est.) / permanent pasture: 24.9% (2011 est.)
forest: 1.5% (2011 est.)
other: 55.2% (2011 est.)
note: includes Gaza Strip

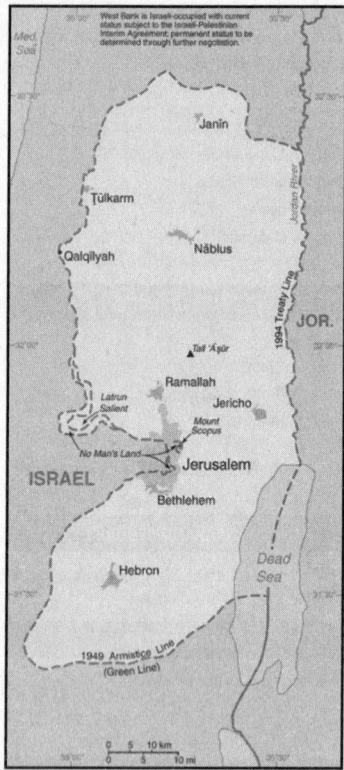

West Bank is Israeli-occupied with current status subject to the Israeli-Palestinian Interim Agreement; permanent status to be determined through further negotiation.

Irrigated land: 240 sq km; note - includes Gaza Strip (2012)

Population distribution: Palestinian settlements are primarily located in the central to western half of the territory; Jewish settlements are found in pockets throughout, particularly in the northeast, north-central, and around Jerusalem

Natural hazards: droughts

Environment—current issues: adequacy of freshwater supply; sewage treatment

Geography—note: landlocked; highlands are main recharge area for Israel's coastal aquifers; there are about 380 Israeli civilian sites, including about 213 settlements and 132 small outpost communities in the West Bank and 35 sites in East Jerusalem (2017)

PEOPLE AND SOCIETY

Population: 2,900,034 (July 2020 est.)
note: approximately 418,600 Israeli settlers live in the West Bank (2018); approximately 215,900 Israeli settlers live in East Jerusalem (2014)
country comparison to the world: 139

Nationality: *noun:* NA
adjective: NA

Ethnic groups: Palestinian Arab, Jewish, other

Languages: Arabic, Hebrew (spoken by Israeli settlers and many Palestinians), English (widely understood)

Religions: Muslim 80-85% (predominantly Sunni), Jewish 12-14%, Christian 1-2.5% (mainly Greek Orthodox), other, unaffiliated, unspecified <1% (2012 est.)

MENA religious affiliation: *Age structure:* 0-14 years: 35.31% (male 525,645/female 498,458)
15-24 years: 20.75% (male 307,420/female 294,469)
25-54 years: 35.19% (male 516,758/female 503,626)
55-64 years: 5.12% (male 76,615/female 72,006)
65 years and over: 3.62% (male 48,387/female 56,650) (2020 est.)

population pyramid: [INSERT IMAGE: WEST BANK-population pyramid]

Dependency ratios: *total dependency ratio:* 71.2
youth dependency ratio: 65.7
elderly dependency ratio: 5.5
potential support ratio: 18.2 (2020 est.)
note: data represent Gaza Strip and the West Bank

Median age: *total:* 21.9 years
male: 21.7 years
female: 22.2 years (2020 est.)
country comparison to the world: 181

Population growth rate: 1.77% (2020 est.)
country comparison to the world: 58

Birth rate: 25.2 births/1,000 population (2020 est.)
country comparison to the world: 49

Death rate: 3.4 deaths/1,000 population (2020 est.)
country comparison to the world: 223

Net migration rate: -4.2 migrant(s)/1,000 population (2020 est.)
country comparison to the world: 190

Population distribution: Palestinian settlements are primarily located in the central to western half of the territory; Jewish settlements are found in pockets throughout, particularly in the northeast, north-central, and around Jerusalem

Urbanization: *urban population:* 76.7% of total population (2020)
rate of urbanization: 3% annual rate of change (2015-20 est.)
note: data represent Gaza Strip and the West Bank

Sex ratio: *at birth:* 1.06 male(s)/female
0-14 years: 1.05 male(s)/female
15-24 years: 1.04 male(s)/female
25-54 years: 1.03 male(s)/female
55-64 years: 1.06 male(s)/female
65 years and over: 0.85 male(s)/female
total population: 1.04 male(s)/female (2020 est.)

Maternal mortality rate: 27 deaths/100,000 live births (2017 est.)
note: data represent Gaza Strip and the West Bank
country comparison to the world: 118

Infant mortality rate: *total:* 12.8 deaths/1,000 live births
male: 14.4 deaths/1,000 live births
female: 11.1 deaths/1,000 live births (2020 est.)
country comparison to the world: 104

Life expectancy at birth: *total population:* 75.9 years
male: 73.8 years

female: 78.1 years (2020 est.)
country comparison to the world: 109

Total fertility rate: 3.07 children born/woman (2020 est.)
country comparison to the world: 50

Contraceptive prevalence rate: 57.2% (2014)
note: includes Gaza Strip and the West Bank

Drinking water source:
improved: *urban:* 97.1% of population
rural: 97.1% of population
total: 96.8% of population
unimproved: *urban:* 2.9% of population
rural: 2.9% of population
total: 3.2% of population (2017 est.)
note: includes Gaza Strip and the West Bank

Physicians density: 1.45 physicians/1,000 population (2017)

Hospital bed density: 1.3 beds/1,000 population (2018)

Sanitation facility access:
improved: *urban:* 100% of population
rural: 99.3% of population
total: 99.8% of population
unimproved: *urban:* 0% of population
rural: 0.7% of population
total: 0.2% of population (2017 est.)
note: note includes Gaza Strip and the West Bank

HIV/AIDS—adult prevalence rate: NA

HIV/AIDS—people living with HIV/AIDS: NA

HIV/AIDS—deaths: NA

Children under the age of 5 years underweight: 1.4% (2014)
note: estimate is for Gaza Strip and the West Bank
country comparison to the world: 123

Education expenditures: 5.3% of GDP (2017)
note: includes Gaza Strip and the West Bank
country comparison to the world: 52

Literacy: *definition:* age 15 and over can read and write
total population: 97.2%
male: 98.7%
female: 95.7% (2018)
note: estimates are for Gaza and the West Bank
School life expectancy (primary to tertiary education): total: 13 years
male: 13 years
female: 14 years (2019)
note: data represent Gaza Strip and the West Bank
Unemployment, youth ages 15-24: total: 42.2%
male: 37%
female: 69.4% (2018 est.)
note: includes Gaza Strip
country comparison to the world: 9

GOVERNMENT

Country name: *conventional long form:* none
conventional short form: West Bank
etymology: name refers to the location of the region - occupied and administered by Jordan after 1948 - that fell on the far side (west bank) of the Jordan River in relation to Jordan proper;

the designation was retained following the 1967 Six- Day War and the subsequent changes in government

ECONOMY

Economy—overview: In 2017, the economic outlook in the West Bank - the larger of the two areas comprising the Palestinian Territories – remained fragile, as security concerns and political friction slowed economic growth. Unemployment in the West Bank remained high at 19.0% in the third quarter of 2017, only slightly better than 19.6% at the same point the previous year, while the labor force participation rate remained flat, year-on-year.

Longstanding Israeli restrictions on imports, exports, and movement of goods and people continue to disrupt labor and trade flows and the territory's industrial capacity, and constrain private sector development. The PA's budget benefited from an effort to improve tax collection, coupled with lower spending in 2017, but the PA for the foreseeable future will continue to rely heavily on donor aid for its budgetary needs and infrastructure development.

GDP (purchasing power parity): $21.22 billion (2014 est.)
$20.15 billion (2013 est.)
$19.95 billion (2012 est.)
note: data are in 2014 US dollars; includes Gaza Strip
country comparison to the world: 147

GDP (official exchange rate): $9.828 billion (2014 est.)
note: excludes Gaza Strip

GDP—real growth rate: 5.3% (2014 est.)
1% (2013 est.)
6% (2012 est.)
note: excludes Gaza Strip
country comparison to the world: 42

GDP—per capita (PPP): $4,300 (2014 est.)
$4,400 (2013 est.)
$4,600 (2012 est.)
note: includes Gaza Strip
country comparison to the world: 175

Gross national saving: 7.8% of GDP (2014 est.)
9.5% of GDP (2013 est.)
5% of GDP (2012 est.)
note: includes Gaza Strip
country comparison to the world: 170

GDP—composition, by end use: *household consumption:* 91.3% (2017 est.)
government consumption: 26.7% (2017 est.)
investment in fixed capital: 23% (2017 est.)
investment in inventories: 0% (2017 est.)
exports of goods and services: 20% (2017 est.)
imports of goods and services: -61% (2017 est.)
note: excludes Gaza Strip

GDP—composition, by sector of origin:
agriculture: 2.9% (2017 est.)
industry: 19.5% (2017 est.)
services: 77.6% (2017 est.)
note: excludes Gaza Strip

Agriculture—products: olives, citrus fruit, vegetables; beef, dairy products

Industries: small-scale manufacturing, quarrying, textiles, soap, olive-wood carvings, and mother-of-pearl souvenirs

Industrial production growth rate: 2.2% (2017 est.)
note: includes Gaza Strip
country comparison to the world: 127

Labor force: 1.24 million (2017 est.)
note: excludes Gaza Strip
country comparison to the world: 134

Labor force—by occupation: agriculture: 11.5%
industry: 34.4%
services: 54.1% (2013 est.)
note: excludes Gaza Strip

Unemployment rate: 27.9% (2017 est.)
27% (2016 est.)
note: excludes Gaza Strip
country comparison to the world: 200

Population below poverty line: 18% (2011 est.)

Household income or consumption by percentage share: *lowest 10%:* 3.2%
highest 10%: 28.2% (2009 est.)
note: includes Gaza Strip

Budget: *revenues:* 1.314 billion (2017 est.)
expenditures: 1.278 billion (2017 est.)
note: includes Palestinian Authority expenditures in the Gaza Strip

Taxes and other revenues: 13.4% (of GDP) (2017 est.)
country comparison to the world: 207

Budget surplus (+) or deficit (-): 0.4% (of GDP) (2017 est.)
country comparison to the world: 39

Public debt: 24.4% of GDP (2014 est.)
23.8% of GDP (2013 est.)
country comparison to the world: 177

Fiscal year: calendar year

Inflation rate (consumer prices): 0.2% (2017 est.)
-0.2% (2016 est.)
note: excludes Gaza Strip
country comparison to the world: 18

Current account balance: -$1.444 billion (2017 est.)
-$1.348 billion (2016 est.)
country comparison to the world: 158

Exports: $2.126 billion (2017 est.)
$1.827 billion (2016 est.)
note: excludes Gaza Strip
country comparison to the world: 138

Exports—commodities: stone, olives, fruit, vegetables, limestone

Imports: $6.565 billion (2017 est.)
$6.207 billion (2016 est.)
note: data include the Gaza Strip
country comparison to the world: 117

Imports—commodities: food, consumer goods, construction materials, petroleum, chemicals

Reserves of foreign exchange and gold: $0 (31 December 2017 est.)
$583 million (31 December 2015 est.)

country comparison to the world: 195

Debt—external: $1.662 billion (31 March 2016 est.)
$1.467 billion (31 March 2015 est.)
note: data include the Gaza Strip
country comparison to the world: 157

Exchange rates: new Israeli shekels (ILS) per US dollar—
3.606 (2017 est.)
3.841 (2016 est.)
3.841 (2015 est.)
3.8869 (2014 est.)
3.5779 (2013 est.)

ENERGY

Electricity access: *electrification—total population:* 100% (2020)
note: data for West Bank and Gaza Strip combined

Electricity—production: 1.093 billion kWh (2016 est.)
country comparison to the world: 148

Electricity—consumption: 6.489 billion kWh (2016 est.)
country comparison to the world: 110

Electricity—exports: 0 kWh (2016)
country comparison to the world: 217

Electricity—imports: 5.473 billion kWh (2016 est.)
country comparison to the world: 36

Electricity—installed generating capacity: 170,000 kW (2016 est.)
note: includes Gaza Strip
country comparison to the world: 170

Electricity—from fossil fuels: 78% of total installed capacity (2016 est.)
country comparison to the world: 91

Electricity—from nuclear fuels: 0% of total installed capacity (2017 est.)
country comparison to the world: 211

Electricity—from hydroelectric plants: 0% of total installed capacity (2017 est.)
country comparison to the world: 213

Electricity—from other renewable sources: 22% of total installed capacity (2017 est.)
country comparison to the world: 34

Crude oil—production: 0 bbl/day (2018 est.)
country comparison to the world: 217

Crude oil—exports: 0 bbl/day (2015 est.)
country comparison to the world: 215

Crude oil—imports: 0 bbl/day (2015 est.)
country comparison to the world: 215

Crude oil—proved reserves: 0 bbl (1 January 2018)
country comparison to the world: 212

Refined petroleum products—production: 0 bbl/day (2015 est.)
country comparison to the world: 215

Refined petroleum products—consumption: 24,000 bbl/day (2016 est.)
country comparison to the world: 131

Refined petroleum products— exports: 19 bbl/day (2015 est.)

country comparison to the world: 123

Refined petroleum products—imports: 22,740 bbl/day (2015 est.)
country comparison to the world: 113

Natural gas—production: 0 cu m (2017 est.)
country comparison to the world: 214

Natural gas—consumption: 0 cu m (2017 est.)
country comparison to the world: 213

Natural gas—exports: 0 cu m (2017 est.)
country comparison to the world: 212

Natural gas—imports: 0 cu m (2017 est.)
country comparison to the world: 212

Natural gas—proved reserves: 0 cu m (1 January 2014 est.)
country comparison to the world: 207

Carbon dioxide emissions from consumption of energy: 3.113 million Mt (2017 est.)
country comparison to the world: 146

COMMUNICATIONS

Telephones—fixed lines: *total subscriptions:* 472,293 (includes Gaza Strip) (2017 est.)

subscriptions per 100 inhabitants: 9 (includes Gaza Strip) (2016 est.)
country comparison to the world: 93

Telephones—mobile cellular: total subscriptions: 4,135,363 (includes Gaza Strip) (2017 est.)

subscriptions per 100 inhabitants: 76 (includes Gaza Strip) (2017 est.)
country comparison to the world: 129

Telecommunication systems: *general assessment:* continuing political and economic instability has impeded liberalization of the telecommunications industry (2018)

domestic: Israeli company BEZEK and the Palestinian company PALTEL are responsible for fixed-line services; two Palestinian cellular providers, JAWWAL and WATANIYA MOBILE, launched 3G mobile networks in the West Bank in January 2018 after Israel lifted its ban; fixed-line 9 per 100 and mobile-cellular 76 per 100 (includes Gaza Strip) (2019)

international: country code 970 or 972; 1 international switch in Ramallah
note: the COVID-19 outbreak is negatively impacting telecommunications production and supply chains globally; consumer spending on telecom devices and services has also slowed due to the pandemic's effect on economies worldwide; overall progress towards improvements in all facets of the telecom industry - mobile, fixed-line, broadband, submarine cable and satellite - has moderated

Broadcast media: the Palestinian Authority operates 1 TV and 1 radio station; about 20 private TV and 40 radio stations; both Jordanian TV and satellite TV are accessible

Internet country code: .ps note - same as Gaza Strip

Internet users: *total:* 2.673 million (includes Gaza Strip)

percent of population: 57.4% (July 2016 est.)
country comparison to the world: 106

Broadband—fixed subscriptions: *total:* 371,299

subscriptions per 100 inhabitants: 14 (2017 est.)
note: includes Gaza Strip
country comparison to the world: 93

TRANSPORTATION

Airports: 2 (2013)
country comparison to the world: 209

Airports—with paved runways: *total:* 2 (2013)
1,524 to 2,437 m: 1 (2013)
under 914 m: 1 (2013)

Heliports: 1 (2013)

Roadways: *total:* 4,686 km (2010)
paved: 4,686 km (2010)
note: includes Gaza Strip
country comparison to the world: 149

MILITARY AND SECURITY

Military and security forces: per the Oslo Accords, the PA is not permitted a conventional military but maintains security and police forces; PA security personnel have operated almost exclusively in the West Bank since HAMAS seized power in the Gaza Strip in 2007; PA forces include National Security Forces, Presidential Guard, Civil Police, Civil Defense, Preventative Security Organization, the General Intelligence Organization, and the Military Intelligence Organization (2020)

Military and security service personnel strengths: the Palestinian Authority Security Forces have approximately 30,000 active personnel (2019 est.)

Military equipment inventories and acquisitions: the Palestinian Authority Security Forces are armed mostly with small arms and light weapons, although since 2007, they have received limited amounts of heavier equipment from Jordan (armored personnel carriers) and Russia (armored personnel carriers and transport helicopters) (2019 est.)

TERRORISM

Terrorist group(s): Al-Aqsa Martyrs Brigade; HAMAS; Islamic Revolutionary Guard Corps/Qods Force; Kahane Chai; Palestine Islamic Jihad; Palestine Liberation Front; Popular Front for the Liberation of Palestine (2019)
note: details about the history, aims, leadership, organization, areas of operation, tactics, targets, weapons, size, and sources of support of the group(s) appear(s) in Appendix-T

TRANSPORTATION

Disputes—international: the current status of the West Bank is subject to the Israeli-Palestinian Interim Agreement - permanent status to be determined through further negotiation; Israel continues construction of a "seam line" separation barrier along parts of the Green Line and within the West Bank; Israel withdrew from Gaza and four settlements in the northern West Bank in August 2005; since 1948, about 350 peacekeepers from the UN Truce Supervision Organization (UNTSO), headquartered in Jerusalem, monitor ceasefires, supervise armistice agreements, prevent isolated incidents from escalating, and assist other UN personnel in the region

Refugees and internally displaced persons: *refugees (country of origin):* 858,758 (Palestinian refugees) (2020)

IDPs: 243,000 (includes persons displaced within the Gaza strip due to the intensification of the Israeli-Palestinian conflict since June 2014 and other Palestinian IDPs in the Gaza Strip and West Bank who fled as long ago as 1967, although confirmed cumulative data do not go back beyond 2006) (2019)

WESTERN SAHARA

INTRODUCTION

Background: Western Sahara is a non-self-governing territory on the northwest coast of Africa bordered by Morocco, Mauritania, and Algeria. After Spain withdrew from its former colony of Spanish Sahara in 1976, Morocco annexed the northern two-thirds of Western Sahara and claimed the rest of the territory in 1979, following Mauritania's withdrawal. A guerrilla war with the Polisario Front contesting Morocco's sovereignty ended in a 1991 cease-fire and the establishment of a UN peacekeeping operation. As part of this effort, the UN sought to offer a choice to the peoples of Western Sahara between independence (favored by the Polisario Front) or integration into Morocco. A proposed referendum on the question of independence never took place due to lack of agreement on voter eligibility. The approximately 1,600 km- (almost 1,000 mi-) long defensive sand berm, built by the Moroccans from 1980 to 1987 and running the length of the territory, continues to separate the opposing forces, with Morocco controlling the roughly three-quarters of the territory west of the berm. There are periodic ethnic tensions between the native Sahrawi population and Moroccan immigrants. Morocco maintains a

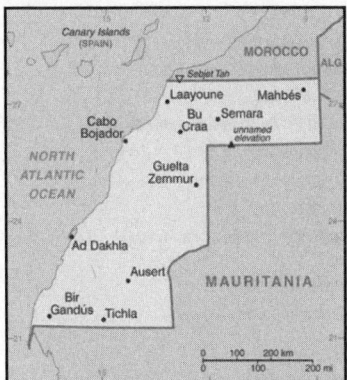

heavy security presence in the territory. The UN revived direct talks about the territory between Morocco, the Polisario Front, Algeria, and Mauritania in December 2018.

GEOGRAPHY

Location: Northern Africa, bordering the North Atlantic Ocean, between Mauritania and Morocco

Geographic coordinates: 24 30 N, 13 00 W

Map references: Africa

Area: *total:* 266,000 sq km
land: 266,000 sq km
water: 0 sq km
country comparison to the world: 79

Area—comparative: about the size of Colorado

Land boundaries: *total:* 2,049 km
border countries (3): Algeria 41 km, Mauritania 1564 km, Morocco 444 km

Coastline: 1,110 km

Maritime claims: contingent upon resolution of sovereignty issue

Climate: hot, dry desert; rain is rare; cold offshore air currents produce fog and heavy dew

Terrain: mostly low, flat desert with large areas of rocky or sandy surfaces rising to small mountains in south and northeast

Elevation: *mean elevation:* 256 m
lowest point: Sebjet Tah—55 m
highest point: unnamed elevation 805 m

Natural resources: phosphates, iron ore

Land use: *agricultural land:* 18.8% (2011 est.)
arable land: 0% (2011 est.) / permanent crops: 0% (2011 est.) / permanent pasture: 18.8% (2011 est.)
forest: 2.7% (2011 est.)
other: 78.5% (2011 est.)
Irrigated land: 0 sq km (2012)

Population distribution: most of the population lives in the two-thirds of the area west of the berm (Moroccan-occupied) that divides the territory; about 40% of that populace resides in Laayoune as shown in this population distribution map

Natural hazards: hot, dry, dust/sand-laden sirocco wind can occur during winter and spring;

widespread harmattan haze exists 60% of time, often severely restricting visibility

Environment—current issues: desertification; overgrazing; sparse water and lack of arable land

Geography—note: the waters off the coast are particularly rich fishing areas

PEOPLE AND SOCIETY

Population: 652,271 (July 2020 est.)
note: estimate is based on projections by age, sex, fertility, mortality, and migration; fertility and mortality are based on data from neighboring countries
country comparison to the world: 168

Nationality: *noun:* Sahrawi(s), Sahraoui(s)
adjective: Sahrawi, Sahrawian, Sahraouian

Ethnic groups: Arab, Berber

Languages: Standard Arabic, Hassaniya Arabic, Moroccan Arabic, Berber, Spanish, French

Religions: Muslim

Demographic profile: Western Sahara is a non-self governing territory; approximately 75% is under Moroccan control. It was inhabited almost entirely by Sahrawi pastoral nomads until the mid-20th century. Their traditional vast migratory ranges, based on following unpredictable rainfall, did not coincide with colonial and later international borders. Since the 1930s, most Sahrawis have been compelled to adopt a sedentary lifestyle and to live in urban settings as a result of fighting, the presence of minefields, job opportunities in the phosphate industry, prolonged drought, the closure of Western Sahara's border with Mauritania from 1979-2002, and the construction of the defensive berm separating Moroccan- and Polisario-controlled (Sahrawi liberalization movement) areas. Morocco supported rapid urbanization to facilitate surveillance and security.

Today more than 80% of Western Sahara's population lives in urban areas; more than 40% live in the administrative center Laayoune. Moroccan immigration has altered the composition and dramatically increased the size of Western Sahara's population. Morocco maintains a large military presence in Western Sahara and has encouraged its citizens to settle there, offering bonuses, pay raises, and food subsidies to civil servants and a tax exemption, in order to integrate Western Sahara into the Moroccan Kingdom and, Sahrawis contend, to marginalize the native population.

Western Saharan Sahrawis have been migrating to Europe, principally to former colonial ruler Spain, since the 1950s. Many who moved to refugee camps in Tindouf, Algeria, also have migrated to Spain and Italy, usually alternating between living in cities abroad with periods back at the camps. The Polisario claims that the population of the Tindouf camps is about 155,000, but this figure may include thousands of Arabs and Tuaregs from neighboring countries. Because international organizations have been unable to conduct an independent census in Tindouf, the UNHCR bases its aid on a figure of 90,000 refugees. Western Saharan coastal towns emerged as key migration transit

points (for reaching Spain's Canary Islands) in the mid-1990s, when Spain's and Italy's tightening of visa restrictions and EU pressure on Morocco and other North African countries to control illegal migration pushed Sub-Saharan African migrants to shift their routes to the south.

Age structure: *0-14 years:* 36.29% (male 119,719/ female 116,997)
15-24 years: 19.44% (male 63,852/female 62,954)
25-54 years: 34.9% (male 112,301/female 115,313)
55-64 years: 5.27% (male 16,095/female 18,292)
65 years and over: 4.1% (male 11,802/female 14,946) (2020 est.)

Dependency ratios: *total dependency ratio:* 44.1
youth dependency ratio: 39.2
elderly dependency ratio: 4.9
potential support ratio: 20.4 (2020 est.)

Median age: *total:* 21.8 years
male: 21.4 years
female: 22.3 years (2020 est.)
country comparison to the world: 184

Population growth rate: 2.54% (2020 est.)
country comparison to the world: 22

Birth rate: 28 births/1,000 population (2020 est.)
country comparison to the world: 38

Death rate: 7.7 deaths/1,000 population (2020 est.)
country comparison to the world: 100

Net migration rate: 4.9 migrant(s)/1,000 population (2020 est.)
country comparison to the world: 25

Population distribution: most of the population lives in the two-thirds of the area west of the berm (Moroccan-occupied) that divides the territory; about 40% of that populace resides in Laayoune as shown in this population distribution map
Urbanization: urban population: 86.8% of total population (2020)
rate of urbanization: 2.61% annual rate of change (2015-20 est.)

total population growth rate v. urban population growth rate, 2000-2030: Major urban areas—population: 232,000 Laayoune (2018)

Sex ratio: *at birth:* 1.04 male(s)/female
0-14 years: 1.02 male(s)/female
15-24 years: 1.01 male(s)/female
25-54 years: 0.97 male(s)/female
55-64 years: 0.88 male(s)/female
65 years and over: 0.79 male(s)/female
total population: 0.99 male(s)/female (2020 est.)

Infant mortality rate: *total:* 47.9 deaths/1,000 live births
male: 52.5 deaths/1,000 live births
female: 43.1 deaths/1,000 live births (2020 est.)
country comparison to the world: 26

Life expectancy at birth: *total population:* 64.5 years
male: 62.1 years
female: 67 years (2020 est.)
country comparison to the world: 202

Total fertility rate: 3.65 children born/woman (2020 est.)
country comparison to the world: 38

HIV/AIDS—adult prevalence rate: NA

HIV/AIDS—people living with HIV/AIDS: NA

HIV/AIDS—deaths: NA

Education expenditures: NA

GOVERNMENT

Country name: *conventional long form: none conventional short form:* Western Sahara *former:* Rio de Oro, Saguia el Hamra, Spanish Sahara

etymology: self-descriptive name specifying the territory's western location on the African continent's vast desert

Government type: legal status of territory and issue of sovereignty unresolved; territory contested by Morocco and Polisario Front (Popular Front for the Liberation of the Saguia el Hamra and Rio de Oro), which in February 1976 formally proclaimed a government-in-exile of the Sahrawi Arab Democratic Republic (SADR), near Tindouf, Algeria, led by President Mohamed ABDELAZIZ until his death in May 2016; current President Brahim GHALI elected in July 2016; territory partitioned between Morocco and Mauritania in April 1976 when Spain withdrew, with Morocco acquiring northern two-thirds; Mauritania, under pressure from Polisario guerrillas, abandoned all claims to its portion in August 1979; Morocco moved to occupy that sector shortly thereafter and has since asserted administrative control; the Polisario's government-in-exile was seated as an Organization of African Unity (OAU) member in 1984; Morocco between 1980 and 1987 built a fortified sand berm delineating the roughly 75% of Western Sahara west of the barrier that currently is controlled by Morocco; guerrilla activities continued sporadically until a UN-monitored cease-fire was implemented on 6 September 1991 (Security Council Resolution 690) by the United Nations Mission for the Referendum in Western Sahara (MINURSO)

Capital: time difference: UTC 0 (5 hours ahead of Washington, DC, during Standard Time)

daylight saving time: +1hr, begins last Sunday in March; ends last Sunday in October

Administrative divisions: none officially; the territory west of the Moroccan berm falls under de facto Moroccan control; Morocco claims the territory of Western Sahara, the political status of which is considered undetermined by the US Government; portions of the regions Guelmim-Es Smara and Laayoune-Boujdour-Sakia El Hamra, as claimed by Morocco, lie within Western Sahara; Morocco also claims Oued Eddahab-Lagouira, another region that falls entirely within Western Sahara

Suffrage: none; (residents of Moroccan-controlled Western Sahara participate in Moroccan elections)

Executive branch: none

International organization participation: AU, CAN (observer), WFTU (NGOs)

Diplomatic representation in the US: none

Diplomatic representation from the US: none

ECONOMY

Economy—overview: Western Sahara has a small market-based economy whose main industries are fishing, phosphate mining, tourism, and pastoral nomadism. The territory's arid desert climate makes sedentary agriculture difficult, and much of its food is imported. The Moroccan Government administers Western Sahara's economy and is a key source of employment, infrastructure development, and social spending in the territory.

Western Sahara's unresolved legal status makes the exploitation of its natural resources a contentious issue between Morocco and the Polisario. Morocco and the EU in December 2013 finalized a four-year agreement allowing European vessels to fish off the coast of Morocco, including disputed waters off the coast of Western Sahara. As of April 2018, Moroccan and EU authorities were negotiating an amendment to renew the agreement.

Oil has never been found in Western Sahara in commercially significant quantities, but Morocco and the Polisario have quarreled over rights to authorize and benefit from oil exploration in the territory. Western Sahara's main long-term economic challenge is the development of a more diverse set of industries capable of providing greater employment and income to the territory. However, following King MOHAMMED VI's November 2015 visit to Western Sahara, the Government of Morocco announced a series of investments aimed at spurring economic activity in the region, while the General Confederation of Moroccan Enterprises announced a $609 million investment initiative in the region in March 2015.

GDP (purchasing power parity): $906.5 million (2007 est.)

country comparison to the world: 205

GDP (official exchange rate): NA

GDP—real growth rate: NA

GDP—per capita (PPP): $2,500 (2007 est.)

country comparison to the world: 198

GDP—composition, by sector of origin:

agriculture: NA (2007 est.)

industry: NA (2007 est.)

services: 40% (2007 est.)

Agriculture—products: fruits and vegetables (grown in the few oases); camels, sheep, goats (kept by nomads); fish

Industries: phosphate mining, handicrafts

Industrial production growth rate: NA

Labor force: 144,000 (2010 est.)

country comparison to the world: 177

Labor force—by occupation: *agriculture:* 50%

industry: 50%

industry and services: 50% (2005 est.)

Unemployment rate: NA

Population below poverty line: NA

Household income or consumption by percentage share: *lowest 10%:* NA

highest 10%: NA

Budget: *revenues:* NA

expenditures: NA

Taxes and other revenues: NA

Budget surplus (+) or deficit (-): NA

Fiscal year: calendar year

Inflation rate (consumer prices): NA

Exports: NA

Exports—commodities: phosphates 62% (2012 est.)

Imports: NA

Imports—commodities: fuel for fishing fleet, foodstuffs

Debt—external: NA

Exchange rates: Moroccan dirhams (MAD) per US dollar—

9.639 (2017 est.)

9.7351 (2016 est.)

9.7351 (2015)

9.7351 (2014 est.)

8.3798 (2013 est.)

ENERGY

Electricity—production: 0 kWh NA (2016 est.)

country comparison to the world: 220

Electricity—consumption: 0 kWh (2016 est.)

country comparison to the world: 219

Electricity—exports: 0 kWh (2016 est.)

country comparison to the world: 218

Electricity—imports: 0 kWh (2016 est.)

country comparison to the world: 218

Electricity—installed generating capacity: 58,000 kW (2016 est.)

country comparison to the world: 188

Electricity—from fossil fuels: 100% of total installed capacity (2016 est.)

country comparison to the world: 22

Electricity—from nuclear fuels: 0% of total installed capacity (2017 est.)

country comparison to the world: 212

Electricity—from hydroelectric plants: 0% of total installed capacity (2017 est.)

country comparison to the world: 214

Electricity—from other renewable sources: 0% of total installed capacity (2017 est.)

country comparison to the world: 214

Crude oil—production: 0 bbl/day (2018 est.)

country comparison to the world: 218

Crude oil—exports: 0 bbl/day (2015 est.)

country comparison to the world: 216

Crude oil—imports: 0 bbl/day (2015 est.)

country comparison to the world: 216

Crude oil—proved reserves: 0 bbl (1 January 2018 est.)

country comparison to the world: 213

Refined petroleum products—production: 0 bbl/day (2015 est.)

country comparison to the world: 216

Refined petroleum products—consumption: 1,700 bbl/day (2016 est.)

country comparison to the world: 197

Refined petroleum products—exports: 0 bbl/day (2015 est.)
country comparison to the world: 216

Refined petroleum products—imports: 1,702 bbl/day (2015 est.)
country comparison to the world: 193

Natural gas—production: 0 cu m (2017 est.)
country comparison to the world: 215

Natural gas—consumption: 0 cu m (2017 est.)
country comparison to the world: 214

Natural gas—exports: 0 cu m (2017 est.)
country comparison to the world: 213

Natural gas—imports: 0 cu m (2017 est.)
country comparison to the world: 213

Natural gas—proved reserves: 0 cu m (1 January 2014 est.)
country comparison to the world: 208

Carbon dioxide emissions from consumption of energy: 268,400 Mt (2017 est.)
country comparison to the world: 194

COMMUNICATIONS

Telecommunication systems: *general assessment:* sparse and limited system
international: country code—212; tied into Morocco's system by microwave radio relay, tropospheric scatter, and satellite; satellite earth stations—2 Intelsat (Atlantic Ocean) linked to Rabat, Morocco
note: the COVID-19 outbreak is negatively impacting telecommunications production and supply chains globally; consumer spending on telecom devices and services has also slowed due to the pandemic's effect on economies worldwide; overall progress towards improvements in all facets of the telecom industry—mobile, fixed-line, broadband, submarine cable and satellite—has moderated

Broadcast media: Morocco's state-owned broadcaster, Radio-Television Marocaine (RTM), operates a radio service from Laayoune and relays TV service; a Polisario-backed radio station also broadcasts

Internet country code: .eh

TRANSPORTATION

Airports: 6 (2013)
country comparison to the world: 179

Airports—with paved runways: *total:* 3 (2019)
2,438 to 3,047 m: 3

Airports—with unpaved runways: *total:* 3 (2013)
1,524 to 2,437 m: 1 (2013)
914 to 1,523 m: 1 (2013)
under 914 m: 1 (2013)

Ports and terminals: *major seaport(s):* Ad Dakhla, Laayoune (El Aaiun)

MILITARY AND SECURITY

Military—note: the United Nations Mission for the Referendum in Western Sahara (MINURSO) has operated in the Western Sahara since 1991 in accordance with settlement proposals accepted in 1988 by Morocco and the Frente Popular para la Liberación de Saguia el-Hamra y de Río de Oro (Frente POLISARIO); the Mission's responsibilities include monitoring the ceasefire, reducing the threat of mines and unexploded ordnance, and providing logistic support to the UNHCR-led Confidence Building Measures pending an agreement to resume those activities, which were suspended in June 2014; as of November 2019, MINURSO had about 460 personnel deployed

TRANSNATIONAL ISSUES

Disputes—international: many neighboring states reject Moroccan administration of Western Sahara; several states have extended diplomatic relations to the "Sahrawi Arab Democratic Republic" represented by the Polisario Front in exile in Algeria, while others support Morocco's proposal to grant the territory autonomy as part of Morocco, although no state recognizes Moroccan sovereignty over Western Sahara; an estimated 100,000 Sahrawi refugees continue to be sheltered in camps in Tindouf, Algeria, which has hosted Sahrawi refugees since the 1980s

WORLD

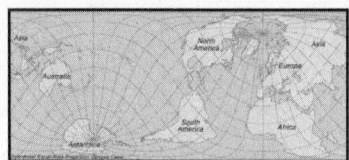

INTRODUCTION

Background: Globally, the 20th century was marked by: (a) two devastating world wars; (b) the Great Depression of the 1930s; (c) the end of vast colonial empires; (d) rapid advances in science and technology, from the first airplane flight at Kitty Hawk, North Carolina (US) to the landing on the moon; (e) the Cold War between the Western alliance and the Warsaw Pact nations; (f) a sharp rise in living standards in North America, Europe, and Japan; (g) increased concerns about environmental degradation including deforestation, energy and water shortages, declining biological diversity, and air pollution; (h) the onset of the AIDS epidemic; and (i) the ultimate emergence of the US as the only world superpower. The planet's population continues to explode: from 1 billion in 1820 to 2 billion in 1930, 3 billion in 1960, 4 billion in 1974, 5 billion in 1987, 6 billion in 1999, and 7 billion in 2012. For the 21st century, the continued exponential growth in science and technology raises both hopes (e. g., advances in medicine and agriculture) and fears (e. g., development of even more lethal weapons of war).

GEOGRAPHY

Geographic overview: The surface of the Earth is approximately 70.9% water and 29.1% land. The former portion is divided into large bodies termed oceans. The World Factbook recognizes and describes five oceans, which are in decreasing order of size: the Pacific Ocean, Atlantic Ocean, Indian Ocean, Southern Ocean, and Arctic Ocean. Because of their immense size, the Pacific and Atlantic Oceans are generally divided at the equator into the North and South Pacific Oceans and the North and South Atlantic Oceans, thus creating seven major water bodies - the so-called "Seven Seas."

Some 97.5% of the Earth's water is saltwater. Of the 2.5% that is fresh, about two-thirds is frozen mostly locked up in the Antarctic ice sheets and mountain glaciers worldwide. If all the surface ice on earth fully melted, the sea level would rise about 70 m (230 ft).

In a 100-year period, a water molecule spends 98 years in the ocean, 20 months as ice, about two weeks in lakes and rivers, and less than a week in the atmosphere. Groundwater can take 50 years to just traverse 1 km (0.6 mi).

Earth's land portion is generally divided into several, large, discrete landmasses termed continents. Depending on the convention used, the number of continents can vary from five to seven. The most common classification recognizes seven, which are (from largest to smallest): Asia, Africa, North America, South America, Antarctica, Europe, and Australia. Asia and Europe are sometimes lumped together into a Eurasian continent resulting in six continents. Alternatively, North and South America are sometimes grouped as simply the Americas, resulting in a continent total of six (or five, if the Eurasia designation is used).

North America is commonly understood to include the island of Greenland, the isles of the Caribbean, and to extend south all the way to the Isthmus of Panama. The easternmost extent of Europe is generally defined as being the Ural Mountains and the Ural River; on the southeast the Caspian Sea; and on the south the Caucasus Mountains, the Black Sea, and the Mediterranean. Portions of five countries - Azerbaijan, Georgia, Kazakhstan, Russia, and Turkey - fall within both

Europe and Asia, but in every instance the larger section is in Asia. These countries are considered part of both continents.

Armenia and Cyprus, which lie completely in Western Asia, are geopolitically European countries.

Asia usually incorporates all the islands of the Philippines, Malaysia, and Indonesia. The islands of the Pacific are often lumped with Australia into a "land mass" termed Oceania or Australasia. Africa's northeast extremity is frequently delimited at the Isthmus of Suez, but for geopolitical purposes, the Egyptian Sinai Peninsula is often included as part of Africa.

Although the above groupings are the most common, different continental dispositions are recognized or taught in certain parts of the world, with some arrangements more heavily based on cultural spheres rather than physical geographic considerations.

Based on the seven-continent model, and grouping islands with adjacent continents, Africa has the most countries with 54. Europe contains 49 countries and Asia 48, but these two continents share five countries: Azerbaijan, Georgia, Kazakhstan, Russia, and Turkey. North America consists of 23 sovereign states, Oceania has 14, and South America 12.

countries by continent: Africa (54): Algeria, Angola, Benin, Botswana, Burkina Faso, Burundi, Cabo Verde, Cameroon, Central African Republic, Chad, Comoros, Democratic Republic of the Congo, Republic of the Congo, Cote d'Ivoire, Djibouti, Egypt, Equatorial Guinea, Eritrea, Eswatini, Ethiopia, Gabon, The Gambia, Ghana, Guinea, Guinea-Bissau, Kenya, Lesotho, Liberia, Libya, Madagascar, Malawi, Mali, Mauritania, Mauritius, Morocco, Mozambique, Namibia, Niger, Nigeria, Rwanda, Sao Tome and Principe, Senegal, Seychelles, Sierra Leone, Somalia, South Africa, South Sudan, Sudan, Tanzania, Togo, Tunisia, Uganda, Zambia, Zimbabwe;

Europe (49): Albania, Andorra, Austria, Azerbaijan*, Belarus, Belgium, Bosnia and Herzegovina, Bulgaria, Croatia, Czech Republic, Denmark, Estonia, Finland, France, Georgia*, Germany, Greece, Holy See (Vatican City), Hungary, Iceland, Ireland, Italy, Kazakhstan*, Kosovo, Latvia, Liechtenstein, Lithuania, Luxembourg, Malta, Moldova, Monaco, Montenegro, Netherlands, North Macedonia, Norway, Poland, Portugal, Romania, Russia*, San Marino, Serbia, Slovakia, Slovenia, Spain, Sweden, Switzerland, Turkey*, Ukraine, United Kingdom (* indicates part of the country is also in Asia);

Asia (48): Afghanistan, Armenia, Azerbaijan*, Bahrain, Bangladesh, Bhutan, Brunei, Burma, Cambodia, China, Cyprus, Georgia*, India, Indonesia, Iran, Iraq, Israel, Japan, Jordan, Kazakhstan*, North Korea, South Korea, Kuwait, Kyrgyzstan, Laos, Lebanon, Malaysia, Maldives, Mongolia, Nepal, Oman, Pakistan, Philippines, Qatar, Russia*, Saudi Arabia,

Singapore, Sri Lanka, Syria, Tajikistan, Thailand, Timor-Leste, Turkey*, Turkmenistan, United Arab Emirates, Uzbekistan, Vietnam, Yemen (* indicates part of the country is also in Europe);

North America (23): Antigua and Barbuda, The Bahamas, Barbados, Belize, Canada, Costa Rica, Cuba, Dominica, Dominican Republic, El Salvador, Grenada, Guatemala, Haiti, Honduras, Jamaica, Mexico, Nicaragua, Panama, Saint Kitts and Nevis, Saint Lucia, Saint Vincent and the Grenadines, Trinidad and Tobago, United States;

Oceania (14): Australia, Fiji, Kiribati, Marshall Islands, Federated States of Micronesia, Nauru, New Zealand, Palau, Papua New Guinea, Samoa, Solomon Islands, Tonga, Tuvalu, Vanuatu;

South America (12): Argentina, Bolivia, Brazil, Chile, Colombia, Ecuador, Guyana, Paraguay, Peru, Suriname, Uruguay, Venezuela

the world from space: Earth is the only planet in the Solar System to have water in its three states of matter: liquid (oceans, lakes, and rivers), solid (ice), and gas (water vapor in clouds); from a distance, Earth would be the brightest of the eight planets in the Solar System; this luminous effect would be because of the sunlight reflected by the planet's water

Earth is also the only planet in the Solar System known to be active with earthquakes and volcanoes; these events form the landscape, replenish carbon dioxide into the atmosphere, and erase impact craters caused by meteors

Map references: Physical Map of the World

Area: total: 510.072 million sq km

land: 148.94 million sq km

water: 361.132 million sq km

note: 70.9% of the world's surface is water, 29.1% is land

Area—comparative: land area about 16 times the size of the US

Area—rankings: top fifteen World Factbook entities ranked by size: Pacific Ocean 155,557,000 km; Atlantic Ocean 76,762,000 sq km; Indian Ocean 68,556,000 sq km; Southern Ocean 20,327,000 sq km; Russia 17,098,242 sq km; Antarctica 14,200,000 sq km; Arctic Ocean 14,056,000 sq km; Canada 9,984,670 sq km; United States 9,826,675 sq km; China 9,596,960 sq km; Brazil 8,515,770 sq km; Australia 7,741,220 sq km; European Union 4,324,782 sq km; India 3,287,263 sq km; Argentina 2,780,400 sq km

top ten largest water bodies: Pacific Ocean 155,557,000 sq km; Atlantic Ocean 76,762,000 sq km; Indian Ocean 68,556,000 sq km; Southern Ocean 20,327,000 sq km; Arctic Ocean 14,056,000 sq km; Coral Sea 4,184,100 sq km; South China Sea 3,595,900 sq km; Caribbean Sea 2,834,000 sq km; Bering Sea 2,520,000 sq km; Mediterranean Sea 2,469,000 sq km

top ten largest landmasses: Asia 44,568,500 sq km; Africa 30,065,000 sq km; North America 24,473,000 sq km; South America 17,819,000 sq km; Antarctica 14,200,000 sq km; Europe 9,948,000 sq km; Australia 7,741,220 sq km;

Greenland 2,166,086 sq km; New Guinea 785,753 sq km; Borneo 751,929 sq km

top ten largest islands: Greenland 2,166,086 sq km; New Guinea (Indonesia, Papua New Guinea) 785,753 sq km; Borneo (Brunei, Indonesia, Malaysia) 751,929 sq km; Madagascar 587,713 sq km; Baffin Island (Canada) 507,451 sq km; Sumatra (Indonesia) 472,784 sq km; Honshu (Japan) 227,963 sq km; Victoria Island (Canada) 217,291 sq km; Great Britain (United Kingdom) 209,331 sq km; Ellesmere Island (Canada) 196,236 sq km

top ten longest mountain ranges (land-based): Andes (Venezuela, Colombia, Ecuador, Peru, Bolivia, Chile, Argentina) 7,000 km; Rocky Mountains (Canada, US) 4,830 km; Great Dividing Range (Australia) 3,700 km; Transantarctic Mountains (Antarctica) 3,500 km; Kunlun Mountains (China) 3,000 km; Ural Mountains (Russia, Kazakhstan) 2,640 km; Atlas Mountains (Morocco, Algeria, Tunisia) 2,500 km; Appalachian Mountains (Canada, US) 2,400 km; Himalayas (Pakistan, Afghanistan, India, China, Nepal, Bhutan) 2,300 km; Altai Mountains (Kazakhstan, Russia, Mongolia) 2,000 km; note - lengths are approximate; if oceans are included, the Mid- Ocean Ridge is by far the longest mountain range at 40,389 km

top ten largest forested countries (sq km and percent of land): Russia 8,149,310 (49.8%); Brazil 4,935,380 (58.9%); Canada 3,470,690 (38.2%); United States 3,103,700 (33.9%); China 2,098,640 (22.3%); Democratic Republic of the Congo 1,522,670 (67.2%); Australia 1,250,590 (16.3%); Indonesia 903,250 (49.9%); Peru 738,054 (57.7%); India 708,600 (23.8%) (2016 est.)

top ten most densely forested countries (percent of land): Suriname (98.3%), Federated States of Micronesia (91.9%), Gabon (90%), Seychelles (88.4%), Palau (87.6%), Guyana (83.9%), Laos (82.1%), Solomon Islands (77.9%), Papua New Guinea (74.1%), Finland (73.1%) (2016 est.)

top ten largest (non-polar) deserts: Sahara (Algeria, Chad, Egypt, Libya, Mali, Mauritania, Niger, Western Sahara, Sudan, Tunisia) 9,200,000 sq km; Arabian (Saudi Arabia, Iraq, Jordan, Kuwait, Oman, Qatar, United Arab Emirates, Yemen) 2,330,000 sq km; Gobi (China, Mongolia) 1,295,000 sq km; Kalahari (Botswana, Namibia, South Africa) 900,000 sq km; Patagonian (Argentina) 673,000 sq km; Syrian (Syria, Iraq, Jordan, Saudi Arabia) 500,000 sq km; Chihuahuan (Mexico) 362,000 sq km; Kara-Kum (Turkmenistan) 350,000 sq km; Great Victoria (Australia) 348,750 sq km; Great Basin (United States) 343,169 sq km; note - if the two polar deserts are included, they would rank first and second: Antarctic Desert 14,200,000 sq km and Arctic Desert 13,900,000 sq km

ton smallest independent countries: Holy See (Vatican City) 0.44 sq km; Monaco 2 sq km; Nauru 21 sq km; Tuvalu 26 sq km; San Marino 61 sq km; Liechtenstein 160 sq km; Marshall Islands

181 sq km; Saint Kitts and Nevis 261 sq km; Maldives 298 sq km; Malta 316 sq km

Land boundaries: the land boundaries in the world total 251,060 km (not counting shared boundaries twice); two nations, China and Russia, each border 14 other countries

note: 46 nations and other areas are landlocked, these include: Afghanistan, Andorra, Armenia, Austria, Azerbaijan, Belarus, Bhutan, Bolivia, Botswana, Burkina Faso, Burundi, Central African Republic, Chad, Czechia, Eswatini, Ethiopia, Holy See (Vatican City), Hungary, Kazakhstan, Kosovo, Kyrgyzstan, Laos, Lesotho, Liechtenstein, Luxembourg, Macedonia, Malawi, Mali, Moldova, Mongolia, Nepal, Niger, Paraguay, Rwanda, San Marino, Serbia, Slovakia, South Sudan, Switzerland, Tajikistan, Turkmenistan, Uganda, Uzbekistan, West Bank, Zambia, Zimbabwe; two of these, Liechtenstein and Uzbekistan, are doubly landlocked

Coastline: 356,000 km

note: 95 nations and other entities are islands that border no other countries, they include: American Samoa, Anguilla, Antigua and Barbuda, Aruba, Ashmore and Cartier Islands, The Bahamas, Bahrain, Baker Island, Barbados, Bermuda, Bouvet Island, British Indian Ocean Territory, British Virgin Islands, Cabo Verde, Cayman Islands, Christmas Island, Clipperton Island, Cocos (Keeling) Islands, Comoros, Cook Islands, Coral Sea Islands, Cuba, Curacao, Cyprus, Dominica, Falkland Islands (Islas Malvinas), Faroe Islands, Fiji, French Polynesia, French Southern and Antarctic Lands, Greenland, Grenada, Guam, Guernsey, Heard Island and McDonald Islands, Howland Island, Iceland, Isle of Man, Jamaica, Jan Mayen, Japan, Jarvis Island, Jersey, Johnston Atoll, Kingman Reef, Kiribati, Madagascar, Maldives, Malta, Marshall Islands, Mauritius, Mayotte, Federated States of Micronesia, Midway Islands, Montserrat, Nauru, Navassa Island, New Caledonia, New Zealand, Niue, Norfolk Island, Northern Mariana Islands, Palau, Palmyra Atoll, Paracel Islands, Philippines, Pitcairn Islands, Puerto Rico, Saint Barthelemy, Saint Helena, Saint Kitts and Nevis, Saint Lucia, Saint Pierre and Miquelon, Saint Vincent and the Grenadines, Samoa, Sao Tome and Principe, Seychelles, Singapore, Sint Maarten, Solomon Islands, South Georgia and the South Sandwich Islands, Spratly Islands, Sri Lanka, Svalbard, Taiwan, Tokelau, Tonga, Trinidad and Tobago, Turks and Caicos Islands, Tuvalu, Vanuatu, Virgin Islands, Wake Island, Wallis and Futuna

Maritime claims: a variety of situations exist, but in general, most countries make the following claims measured from the mean low-tide baseline as described in the 1982 UN Convention on the Law of the Sea: territorial sea - 12 nm, contiguous zone - 24 nm, and exclusive economic zone - 200 nm; additional zones provide for exploitation of continental shelf resources and an exclusive fishing zone; boundary situations with neighboring states prevent many countries from extending their fishing or economic zones to a full 200 nm

Climate: a wide equatorial band of hot and humid tropical climates, bordered north and south by subtropical temperate zones that separate two large areas of cold and dry polar climates

Ten Driest Places on Earth (Average Annual Precipitation): McMurdo Dry Valleys, Antarctica 0 mm (0 in)
Arica, Chile 0.76 mm (0.03 in)
Al Kufrah, Libya 0.86 mm (0.03 in)
Aswan, Egypt 0.86 mm (0.03 in)
Luxor, Egypt 0.86 mm (0.03 in)
Ica, Peru 2.29 mm (0.09 in)
Wadi Halfa, Sudan 2.45 mm (0.1 in)
Iquique, Chile 5.08 mm (0.2 in)
Pelican Point, Namibia 8.13 mm (0.32 in)
El Arab (Aoulef), Algeria 12.19 mm (0.48 in)

Ten Wettest Places on Earth (Average Annual Precipitation): Mawsynram, India 11,871 mm (467.4 in)
Cherrapunji, India 11,777 mm (463.7 in)
Tutunendo, Colombia 11,770 mm (463.4 in)
Cropp River, New Zealand 11,516 mm (453.4 in)
San Antonia de Ureca, Equatorial Guinea 10,450 mm (411.4 in)
Debundsha, Cameroon 10,299 mm (405.5 in)
Big Bog, US (Hawaii) 10,272 mm (404.4 in)
Mt Waialeale, US (Hawaii) 9,763 mm (384.4 in)
Kukui, US (Hawaii) 9,293 mm (365.9 in)
Emeishan, China 8,169 mm (321.6 in)

Ten Coldest Places on Earth (Lowest Average Monthly Temperature): Verkhoyansk, Russia (Siberia) -47°C (-53°F) January
Oymyakon, Russia (Siberia) -46°C (-52°F) January
Eureka, Canada -38.4°C (-37.1°F) February
Isachsen, Canada -36°C (- 32.8°F) February
Alert, Canada -34°C (-28°F) February
Kap Morris Jesup, Greenland -34°C (-29°F) March
Cornwallis Island, Canada -33.5°C (-28.3°F) February
Cambridge Bay, Canada -33.5°C (28.3°F) February
Ilirnej, Russia -33°C (-28°F) January
Resolute, Canada -33°C (-27.4°F) February

Ten Hottest Places on Earth (Highest Average Monthly Temperature): Death Valley, US (California) 39°C (101°F) July
Iranshahr, Iran 38.3°C (100.9°F) June
Ouallene, Algeria 38°C (100.4°F) July
Kuwait City, Kuwait 37.7°C (100°F) July
Medina, Saudi Arabia 36°C (97°F) July
Buckeye, US (Arizona) 34°C (93°F) July
Jazan, Saudi Arabia 33°C (91°F) June
Al Kufrah, Libya 31°C (87°F) July
Alice Springs, Australia 29°C (84°F) January
Tamanrasset, Algeria 29°C (84°F) June

Terrain: tremendous variation of terrain on each of the continents; check the World 'Elevation' entry for a compilation of terrain extremes; the world's ocean floors are marked by mid- ocean ridges while the ocean surfaces form a dynamic, continuously changing environment; check the 'Terrain' field and its 'major surface currents' subfield under each of the five ocean (Arctic, Atlantic, Indian, Pacific, and Southern) entries for further information on oceanic environs

Ten Cave Superlatives: compiled from "Geography - note(s)" under various country entries where more details may be found

largest cave: Son Doong in Phong Nha- Ke Bang National Park, Vietnam is the world's largest cave (greatest cross sectional area) and is the largest known cave passage in the world by volume; it currently measures a total of 38.5 million cu m (about 1.35 billion cu ft); it connects to Thung cave (but not yet officially); when recognized, it will add an additional 1.6 million cu m in volume

largest ice cave: the Eisriesenwelt (Ice Giants World) inside the Hochkogel mountain near Werfen, Austria is the world's largest and longest ice cave system at 42 km (26 mi)

longest cave: Mammoth Cave, in west-central Kentucky, is the world's longest known cave system with more than 650 km (405 mi) of surveyed passageways

longest salt cave: the Malham Cave in Mount Sodom in Israel is the world's longest salt cave at 10 km (6 mi); its survey is not complete and its length will undoubtedly increase

longest underwater cave: the Sac Actun cave system in Mexico at 348 km (216 mi) is the longest underwater cave in the world and the second longest cave worldwide

longest lava tube cave: Kazumura Cave on the island of Hawaii is the world's longest and deepest lava tube cave; it has been surveyed at 66 km (41 mi) long and 1,102 m (3,614 ft) deep

deepest cave: Veryovkina Cave in the Caucasus country of Georgia is the world's deepest cave, plunging down 2,212 m (7,257 ft)

deepest underwater cave: the Hranice Abyss in Czechia is the world's deepest surveyed underwater cave at 404 m (1,325 ft); its survey is not complete and it could end up being some 800-1,200 m deep

largest cave chamber: the Miao Room in the Gebihe cave system at China's Ziyun Getu He Chuandong National Park encloses some 10.78 million cu m (380.7 million cu ft) of volume

largest bat cave: Bracken Cave outside of San Antonio, Texas is the world's largest bat cave; it is the summer home to the largest colony of bats in the world; an estimated 20 million Mexican free-tailed bats roost in the cave from March to October making it the world's largest known concentration of mammals

Elevation: mean elevation: 840 m

lowest point: Denman Glacier (Antarctica) more than -3,500 m (in the oceanic realm, Challenger Deep in the Mariana Trench is the lowest point, lying -10,924 m below the surface of the Pacific Ocean)

highest point: Mount Everest 8,848 m

top ten highest mountains (measured from sea level): Mount Everest (China-Nepal) 8,848 m; K2 (Pakistan) 8,611 m; Kanchenjunga (India-Nepal) 8,598 m; Lhotse (Nepal) 8,516 m; Makalu (China- Nepal) 8,463 m; Cho Oyu (China-Nepal) 8,201 m; Dhaulagiri (Nepal) 8,167 m; Manaslu (Nepal) 8,163 m; Nanga Parbat (Pakistan) 8,125 m; Anapurna (Nepal) 8,091 m; note - Mauna Kea (United States) is the world's tallest mountain as measured from base to summit; the peak of this volcanic colossus lies on the island of Hawaii, but its base begins more than 70 km offshore and at a depth of about 6,000 m; total height estimates range from 9,966 m to 10,203 m

top ten highest island peaks: Puncak Jaya (New Guinea) 4,884 m (Indonesia)*; Mauna Kea (Hawaii) 4,207 m (United States); Gunung Kinabalu (Borneo) 4,095 m (Malaysia)*; Yu Shan (Taiwan) 3,952 (Taiwan)*; Mount Kerinci (Sumatra) 3,805 m (Indonesia); Mount Erebus (Ross Island) 3,794 (Antarctica); Mount Fuji (Honshu) 3,776 m (Japan)*; Mount Rinjani (Lombok) 3,726 m (Indonesia); Aoraki-Mount Cook (South Island) 3,724 m (New Zealand)*; Pico de Teide (Tenerife) 3,718 m (Spain)*; note - * indicates the highest peak for that Factbook entry

highest point on each continent: Asia - Mount Everest (China-Nepal) 8,848 m; South America - Cerro Aconcagua (Argentina) 6,960 m; North America - Denali (Mount McKinley) (United States) 6,190 m; Africa - Kilimanjaro (Tanzania) 5,895 m; Europe - El'brus (Russia) 5,633 m; Antarctica - Vinson Massif 4,897 m; Australia - Mount Kosciuszko 2,229 m

highest capital on each continent: South America - La Paz (Bolivia) 3,640 m; Africa - Addis Ababa (Ethiopia) 2,355 m; Asia - Thimphu (Bhutan) 2,334 m; North America - Mexico City (Mexico) 2,240 m; Europe - Andorra la Vella (Andorra) 1,023 m; Australia - Canberra (Australia) 605 m

lowest point on each continent: Antarctica - Denman Glacier more than - 3, 500 m; Asia - Dead Sea (Israel- Jordan) - 431 m; Africa - Lac Assal (Djibouti) - 155 m; South America - Laguna del Carbon (Argentina) - 105 m; North America - Death Valley (United States) - 86 m; Europe - Caspian Sea (Azerbaijan- Kazakhstan- Russia) - 28 m; Australia - Lake Eyre - 15 m

lowest capital on each continent: Asia - Baku (Azerbaijan) - 28 m; Europe - Amsterdam (Netherlands) - 2 m; Africa - Banjul (Gambia); Bissau (Guinea-Bissau), Conakry (Guinea), Djibouti (Djibouti), Libreville (Gabon), Male (Maldives), Monrovia (Liberia), Tunis (Tunisia), Victoria (Seychelles) 0 m; North America - Basseterre (Saint Kitts and Nevis), Kingstown (Saint Vincent and the Grenadines), Panama City (Panama), Port of Spain (Trinidad and Tobago), Roseau (Dominica), Saint John's (Antigua and Barbuda), Santo Domingo (Dominican Republic) 0 m; South America - Georgetown (Guyana) 0 m; Australia - Canberra (Australia) 605 m

Natural resources: the rapid depletion of nonrenewable mineral resources, the depletion of forest areas and wetlands, the extinction of animal and plant species, and the deterioration in air and water quality pose serious long-term problems

Irrigated land: 3,242,917 sq km (2012 est.)

Population distribution: six of the world's seven continents are widely and permanently inhabited;

Asia is easily the most populous continent with about 60% of the world's population (China and India together account for over 35%); Africa comes in second with over 15% of the earth's populace, Europe has about 10%, North America 8%, South America almost 6%, and Oceania less than 1%; the harsh conditions on Antarctica prevent any permanent habitation

Natural hazards: large areas subject to severe weather (tropical cyclones); natural disasters (earthquakes, landslides, tsunamis, volcanic eruptions)

volcanism: volcanism is a fundamental driver and consequence of plate tectonics, the physical process reshaping the Earth's lithosphere; the world is home to more than 1,500 potentially active volcanoes, with over 500 of these having erupted in historical times; an estimated 500 million people live near these volcanoes; associated dangers include lava flows, lahars (mudflows), pyroclastic flows, ash clouds, ash fall, ballistic projectiles, gas emissions, landslides, earthquakes, and tsunamis; in the 1990s, the International Association of Volcanology and Chemistry of the Earth's Interior, created a list of 16 Decade Volcanoes worthy of special study because of their great potential for destruction: Avachinsky-Koryaksky (Russia), Colima (Mexico), Etna (Italy), Galeras (Colombia), Mauna Loa (United States), Merapi (Indonesia), Nyiragongo (Democratic Republic of the Congo), Rainier (United States), Sakurajima (Japan), Santa Maria (Guatemala), Santorini (Greece), Taal (Philippines), Teide (Spain), Ulawun (Papua New Guinea), Unzen (Japan), Vesuvius (Italy); see second note under "Geography - note"

Environment—current issues: large areas subject to overpopulation, industrial disasters, pollution (air, water, acid rain, toxic substances), loss of vegetation (overgrazing, deforestation, desertification), loss of biodiversity; soil degradation, soil depletion, erosion; ozone layer depletion; waste disposal; global warming becoming a greater concern

Geography—note: note 1: the world is now thought to be about 4.55 billion years old, just about one-third of the 13.8-billion-year age estimated for the universe; the earliest widely accepted date for life appearing on earth is 3.48 billion years ago, but this date is conservative and may get pushed back further

note 2: although earthquakes can strike anywhere at any time, the vast majority occur in three large zones of the earth; the world's greatest earthquake belt, the Circum- Pacific Belt (popularly referred to as the Ring of Fire), is the zone of active volcanoes and earthquake epicenters bordering the Pacific Ocean; about 90% of the world's earthquakes (81% of the largest earthquakes) and some 75% of the world's volcanoes occur within the Ring of Fire; the belt extends northward from Chile, along the South American coast, through Central America, Mexico, the western US, southern Alaska and the Aleutian Islands, to Japan, the Philippines,

Papua New Guinea, island groups in the southwestern Pacific, and New Zealand the second prominent belt, the Alpide, extends from Java to Sumatra, northward along the mountains of Burma, then eastward through the Himalayas, the Mediterranean, and out into the Atlantic Ocean; it accounts for about 17% of the world's largest earthquakes; the third important belt follows the long Mid- Atlantic Ridge

PEOPLE AND SOCIETY

Population: 7,684,292,383 (July 2020 est.)

top ten most populous countries (in millions): China 1394.02; India 1326.1; United States 332.64; Indonesia 267.03;Pakistan 233.5; Nigeria 214.03; Brazil 211.72; Bangladesh 162.65; Russia 141.72; Japan 128.65

ten least populous countries: Holy See (Vatican City) 1,000; Saint Pierre and Miquelon 5,347; Montserrat 5,373; Saint Barthelemy 7,122; Saint Helena, Ascension, and Tristan de Cunha 7,862; Cook Islands 8,574; Nauru 9,785; Tuvalu 11,342; Wallis and Futuna 15,854; Anguilla 18,090

ten most densely populated countries (population per sq km): Macau 21,789; Monaco 15,470; Singapore 8,756; Hong Kong 6,757; Gaza Strip 5,328; Gibraltar 4,551; Bahrain 1,980; Malta 1,447; Bermuda 1,329; Maldives 1,315

ten least densely populated countries (population per sq km): Greenland less than 1; Mongolia 2; Western Sahara 2.5; Namibia 3.2; Australia 3.3; Iceland 3.5; Guyana 3.8; Libya 3.9; Mauritania 3.9; Suriname 3.9

Languages: most-spoken language: English 16.5%, Mandarin Chinese 14.6%, Hindi 8.3%, Spanish 7%, French 3.6%, Arabic 3.6%, Bengali 3.4%, Russian 3.4%, Portuguese 3.3%, Indonesian 2.6% (2020 est.)

most-spoken first language: Mandarin Chinese 12.3%, Spanish 6%, English 5.1%, Arabic 5.1%, Hindi 3.5%, Bengali 3.3%, Portuguese 3%, Russian 2.1%, Japanese 1.7%, Punjabi, Western 1.3%, Javanese 1.1% (2018 est.)

note 1: the six UN languages - Arabic, Chinese (Mandarin), English, French, Russian, and Spanish (Castilian) - are the mother tongue or second language of about 45% of the world's population, and are the official languages in more than half the states in the world; some 400 languages have more than a million first-language speakers (2018)

note 2: all told, there are estimated to be just over 7,115 languages spoken in the world (2020); approximately 80% of these languages are spoken by less than 100,000 people; about 150 languages are spoken by less than 10 people; communities that are isolated from each other in mountainous regions often develop multiple languages; Papua New Guinea, for example, boasts about 840 separate languages (2018)

note 3: approximately 2,300 languages are spoken in Asia, 2,140, in Africa, 1,310 in the Pacific, 1,060 in the Americas, and 290 in Europe (2020)

Religions: Christian 31.2%, Muslim 24.1%, Hindu 15.1%, Buddhist 6.9%, folk religions 5.7%, Jewish 0.2%, other 0.8%, unaffiliated 16% (2015 est.)

Age structure: *0-14 years:* 25.33% (male 1,005,229,963/female 941,107,507)

15-24 years: 15.42% (male 612,094,887/female 572,892,123)

25-54 years: 40.67% (male 1,582,759,769/female 1,542,167,537)

55-64 years: 9.09% (male 341,634,893/female 357,176,983)

65 years and over: 9.49% (male 326,234,036/female 402,994,685) (2020 est.)

World population pyramid: [INSERT IMAGE: The World Factbook pyramid]

Dependency ratios: *total dependency ratio:* 53.3
youth dependency ratio: 39
elderly dependency ratio: 14.3
potential support ratio: 7 (2020 est.)

Median age: *total:* 31 years
male: 30.3 years
female: 31.8 years (2020 est.)
Population growth rate: 1.03% (2020 est.)
note: this rate results in about 149 net additions to the worldwide population every minute or 2.5 every second

Birth rate: 18.1 births/1,000 population (2020 est.)
note: this rate results in about 259 worldwide births per minute or 4.3 births every second

Death rate: 7.7 deaths/1,000 population (2020 est.)
note: this rate results in about 108 worldwide deaths per minute or 1.8 deaths every second

Population distribution: six of the world's seven continents are widely and permanently inhabited; Asia is easily the most populous continent with about 60% of the world's population (China and India together account for over 35%); Africa comes in second with over 15% of the earth's populace, Europe has about 10%, North America 8%, South America almost 6%, and Oceania less than 1%; the harsh conditions on Antarctica prevent any permanent habitation

Urbanization: *urban population:* 56.2% of total population (2020)
rate of urbanization: 1.9% annual rate of change (2015-20 est.)

Major urban areas—population: ten largest urban agglomerations: Tokyo (Japan) - 37,393,000; New Delhi (India) - 30,291,000; Shanghai (China) - 27,058,000; Sao Paulo (Brazil) - 22,043,000; Mexico City (Mexico) - 21,782,000; Dhaka (Bangladesh) - 21,006,000; Cairo (Egypt) - 20,901,000; Beijing (China) - 20,463,000; Mumbai (India) - 20,411,000; Osaka (Japan) - 19,165,000 (2020)

ten largest urban agglomerations, by continent: Africa - Cairo (Egypt) - 20,485,000; Lagos (Nigeria) - 13,904,000; Kinshasha (DRC) - 13,743,000; Luanda (Angola) - 8,045,000; Dar Es Salaam (Tanzania) - 6,368,000; Khartoum (Sudan) - 5,678,000; Johannesburg (South Africa) -5,635,000; Alexandria (Egypt) - 5,182,000;

Abidjan (Cote d'Ivoire) - 5,059,000; Addis Ababa (Ethiopia) - 4,592,000

Asia - Tokyo (Japan) - 37,435,000; New Delhi (India) - 29,399,000; Shanghai (China) - 26,317,000; Dhaka (Bangladesh) - 20,284,000; Mumbai (India) - 20,185,000; Beijing (China) - 20,035,000; Osaka (Japan) - 19,223,000; Karachi (Pakistan) - 15,741,000; Chongqing (China) - 15,354,000; Istanbul (Turkey) - 14,968,000

Europe - Moscow (Russia) - 12,476,000; Paris (France) - 10,958,000; London (United Kingdom) - 9,177,000; Madrid (Spain) - 6,559,000; Barcelona (Spain) - 5,541,000, Saint Petersburg (Russia) - 5,427,000; Rome (Italy) - 4,234,000; Berlin (Germany) - 3,557,000; Athens (Greece) - 3,154,000; Milan (Italy) - 3,136,000

North America - Mexico City (Mexico) - 21,672,000; New York- Newark (United States) - 18,805,000; Los Angeles- Long Beach- Santa Ana (United States) - 12,448,000; Chicago (United States) - 8,862,000; Houston (United States) - 6,245,000; Dallas- Fort Worth (United States) - 6,201,000; Toronto (Canada) - 6,139,000; Miami (United States) - 6,079,000; Philadelphia (United States) - 5,705,000; Atlanta (United States) - 5,689,000

Oceania - Melbourne (Australia) - 4,870,000, Sydney (Australia) - 4,859,000; Brisbane (Australia) - 2,372,000; Perth (Australia) - 2,016,000; Auckland (New Zealand) - 1,582,000; Adelaide (Australia) - 1,328,000; Gold Coast- Tweed Head (Australia) - 687,000; Canberra (Australia) - 452,000; Newcastle- Maitland (Australia) - 447,000; Wellington (New Zealand) - 413,000

South America - Sao Paulo (Brazil) - 21,847,000; Buenos Aires (Argentina) - 15,057,000; Rio de Janeiro (Brazil) - 13,374,000; Bogota (Colombia) - 10,779,000; Lima (Peru) - 10,555,000; Santiago (Chile) - 6,724,000; Belo Horizonte (Brazil) - 6,028,000; Brasilia (Brazil) - 4,559,000; Porto Alegre (Brazil) - 4,115,000; Recife (Brazil) - 4,078,000 (2019)

Sex ratio: *at birth:* 1.07 male(s)/female
0-14 years: 1.07 male(s)/female
15-24 years: 1.07 male(s)/female
25-54 years: 1.03 male(s)/female
55-64 years: 0.96 male(s)/female
65 years and over: 0.81 male(s)/female
total population: 1.01 male(s)/female (2020 est.)

Maternal mortality rate: 211 deaths/100,000 live births (2017 est.)

Infant mortality rate: *total:* 30.8 deaths/1,000 live births
male: 32.8 deaths/1,000 live births
female: 28.6 deaths/1,000 live births (2020 est.)

Life expectancy at birth: *total population:* 70.5 years
male: 68.4 years
female: 72.6 years (2020 est.)

Total fertility rate: 2.42 children born/woman (2020 est.)

Drinking water source:

improved: *urban:* 96.5% of population

rural: 84.7% of population
total: 91.1% of population

unimproved: *urban:* 3.5% of population
rural: 15.3% of population
total: 8.9% of population (2015 est.)

Current Health Expenditure: 10% (2016)

Sanitation facility access:

improved: *urban:* 82.3% of population (2015 est.)
rural: 50.5% of population (2015 est.)
total: 67.7% of population (2015 est.)

unimproved: *urban:* 17.7% of population (2015 est.)
rural: 49.5% of population (2015 est.)
total: 32.3% of population (2015 est.)

HIV/AIDS—adult prevalence rate: 0.6% (2019 est.)

HIV/AIDS—people living with HIV/AIDS: 38 million (2019 est.)

HIV/AIDS—deaths: 690,000 (2019 est.)

Major infectious diseases: *note:* widespread ongoing transmission of a respiratory illness caused by the novel coronavirus (COVID-19) is occurring globally; older adults and people of any age with serious chronic medical conditions are at increased risk for severe disease; some health care systems are becoming overwhelmed and there may be limited access to adequate medical care in affected areas; many countries are implementing travel restrictions and mandatory quarantines, closing borders, and prohibiting non-citizens from entry with little advance notice; US residents may have difficulty returning to the United States; as of 10 November 2020,49,727,316 confirmed cases of COVID- 19 and 1,248,373 deaths have been reported to the World Health Organization

Education expenditures: NA

Literacy: *definition:* age 15 and over can read and write
total population: 86.3%
male: 89.8%
female: 82.8% (2016 est.) (2018)
note: more than three-quarters of the world's 750 million illiterate adults are found in South Asia and sub-Saharan Africa; of all the illiterate adults in the world, almost two-thirds are women (2016)

School life expectancy (primary to tertiary education): *total:* 12 years
male: 12 years
female: 12 years (2019)

GOVERNMENT

Country name: *note:* countries with names connected to animals include: Albania "Land of the Eagles," Anguilla (the name means "eel"), Bhutan "Land of the Thunder Dragon," Cameroon (the name derives from "prawns"), Cayman Islands (named after the caiman, a marine crocodile), Faroe Islands (from Old Norse meaning "sheep"), Georgia "Land of the Wolves," Italy "Land of Young Cattle," Kosovo "Field of Blackbirds," Sierra Leone "Lion Mountains," Singapore "Lion City"

Capital: *time difference:* there are 21 World entities (20 countries and 1 dependency) with multiple time zones: Australia, Brazil, Canada, Chile, Democratic Republic of Congo, Ecuador, France, Greenland (part of the Danish Kingdom), Indonesia, Kazakhstan, Kiribati, Mexico, Micronesia, Mongolia, Netherlands, New Zealand, Papua New Guinea, Portugal, Russia, Spain, United States

note 1: in some instances, the time zones pertain to portions of a country that lie overseas

note 2: in 1851, the British set their prime meridian (0° longitude) through the Royal Observatory at Greenwich, England; this meridian became the international standard in 1884 and thus the basis for the standard time zones of the world; today, GMT is officially known as Coordinated Universal Time (UTC) and is also referred to as "Zulu time"; UTC is the basis for all civil time, with the world divided into time zones expressed as positive or negative differences from UTC

note 3: each time zone is based on 15° starting from the prime meridian; in theory, there are 24 time zones based on the solar day, but there are now upward of 40 because of fractional hour offsets that adjust for various political and physical geographic realities; see the Standard Time Zones of the World map included with the Reference Maps

daylight saving time: some 67 countries - including most of the world's leading industrialized nations - use daylight savings time (DST) in at least a portion of the country; China, Japan, India, and Russia are major industrialized countries that do not use DST; Iran and Africa generally do not observe DST and it is generally not observed near the equator, where sunrise and sunset times do not vary enough to justify it; some countries observe DST only in certain regions; for example, only southeastern Australia observes it; in fact, only a minority of the world's population - about 20% - uses DST

Administrative divisions: 195 countries, 72 dependent areas and other entities

Legal system: the legal systems of nearly all countries are generally modeled upon elements of five main types: civil law (including French law, the Napoleonic Code, Roman law, Roman-Dutch law, and Spanish law); common law (including English and US law); customary law; mixed or pluralistic law; and religious law (including Islamic sharia law); an additional type of legal system - international law - governs the conduct of independent nations in their relationships with one another

International law organization participation: all members of the UN are parties to the statute that established the International Court of Justice (ICJ) or World Court; states parties to the Rome Statute of the International Criminal Court (ICCt) are those countries that have ratified or acceded to the Rome Statute, the treaty that established the Court; as of May 2019, a total of 122 countries have accepted jurisdiction of the ICCt (see Appendix B for a clarification on the differing mandates of the ICJ and ICCt)

Executive branch: there are 27 countries with royal families in the world, most are in Asia (13) and Europe (10), three are in Africa, and one in Oceania; monarchies by continent are as follows: Asia (Bahrain, Bhutan, Brunei, Cambodia, Japan, Jordan, Kuwait, Malaysia, Oman, Qatar, Saudi Arabia, Thailand, United Arab Emirates); Europe (Belgium, Denmark, Liechtenstein, Luxembourg, Monaco, Netherlands, Norway, Spain, Sweden, United Kingdom); Africa (Eswatini, Lesotho, Morocco); Oceania (Tonga); note that Andorra and the Holy See (Vatican) are also monarchies of a sort, but they are not ruled by royal houses; Andorra has two co- princes (the president of France and the bishop of Urgell) and the Holy See is ruled by an elected pope; note too that the sovereign of Great Britain is also the monarch for many of the countries (including Australia, Canada, Jamaica, New Zealand) that make up the Commonwealth

Legislative branch: there are 230 political entities with legislative bodies; of these 144 are unicameral (a single "house") and 86 are bicameral (both upper and lower houses); note - while there are 195 countries in the world, 35 territories, possessions, or other special administrative units also have their own governing bodies

Flag description: while a "World" flag does not exist, the flag of the United Nations (UN) - adopted on 7 December 1946 - has been used on occasion to represent the entire planet; technically, however, it only represents the international organization itself; the flag displays the official emblem of the UN in white on a blue background; the emblem design shows a map of the world in an azimuthal equidistant projection centered on the North Pole, the image is flanked by two olive branches crossed below; blue was selected as the color to represent peace, in contrast to red usually associated with war; the map projection chosen includes all of the continents except Antarctica

note: the flags of 12 nations: Austria, Botswana, Georgia, Jamaica, Japan, Laos, Latvia, Macedonia, Micronesia, Nigeria, Switzerland, and Thailand have no top or bottom and may be flown with either long edge on top without any notice being taken

United Nations flag: [INSERT IMAGE: United Nations-Flag]

ECONOMY

Economy—overview: The international financial crisis of 2008-09 led to the first downturn in global output since 1946 and presented the world with a major new challenge: determining what mix of fiscal and monetary policies to follow to restore growth and jobs, while keeping inflation and debt under control. Financial stabilization and stimulus programs that started in 2009-11, combined with lower tax revenues in 2009-10, required most countries to run large budget deficits. Treasuries issued new public debt - totaling $9.1 trillion since 2008 - to pay for the additional expenditures. To keep interest rates low, most central banks monetized that debt, injecting large sums of money into

their economies - between December 2008 and December 2013 the global money supply increased by more than 35%. Governments are now faced with the difficult task of spurring current growth and employment without saddling their economies with so much debt that they sacrifice long-term growth and financial stability. When economic activity picks up, central banks will confront the difficult task of containing inflation without raising interest rates so high they snuff out further growth.

Fiscal and monetary data for 2013 are currently available for 180 countries, which together account for 98.5% of world GDP. Of the 180 countries, 82 pursued unequivocally expansionary policies, boosting government spending while also expanding their money supply relatively rapidly - faster than the world average of 3.1%; 28 followed restrictive fiscal and monetary policies, reducing government spending and holding money growth to less than the 3.1% average; and the remaining 70 followed a mix of counterbalancing fiscal and monetary policies, either reducing government spending while accelerating money growth, or boosting spending while curtailing money growth.

(For more information, see attached spreadsheet.)

In 2013, for many countries the drive for fiscal austerity that began in 2011 abated. While 5 out of 6 countries slowed spending in 2012, only 1 in 2 countries slowed spending in 2013. About 1 in 3 countries actually lowered the level of their expenditures. The global growth rate for government expenditures increased from 1.6% in 2012 to 5.1% in 2013, after falling from a 10.1% growth rate in 2011. On the other hand, nearly 2 out of 3 central banks tightened monetary policy in 2013, decelerating the rate of growth of their money supply, compared with only 1 out of 3 in 2012. Roughly 1 of 4 central banks actually withdrew money from circulation, an increase from 1 out of 7 in 2012. Growth of the global money supply, as measured by the narrowly defined M1, slowed from 8.7% in 2009 and 10.4% in 2010 to 5.2% in 2011, 4.8% in 2012, and 3.1% in 2013. Several notable shifts occurred in 2013. By cutting government expenditures and expanding money supplies, the US and Canada moved against the trend in the rest of the world. France reversed course completely. Rather than reducing expenditures and money as it had in 2012, it expanded both. Germany reversed its fiscal policy, sharply expanding federal spending, while continuing to grow the money supply. South Korea shifted monetary policy into high gear, while maintaining a strongly expansionary fiscal policy. Japan, however, continued to pursue austere fiscal and monetary policies.

Austere economic policies have significantly affected economic performance. The global budget deficit narrowed to roughly $2.7 trillion in 2012 and $2.1 trillion in 2013, or 3.8% and 2.5% of World GDP, respectively. But growth of the world economy slipped from 5.1% in 2010 and 3.7% in 2011, to just 3.1% in 2012, and 2.9% in 2013.

Countries with expansionary fiscal and monetary policies achieved significantly higher rates of growth, higher growth of tax revenues, and greater success reducing the public debt burden than those countries that chose contractionary policies. In 2013, the 82 countries that followed a pro- growth approach achieved a median GDP growth rate of 4.7%,compared to 1.7% for the 28 countries with restrictive fiscal and monetary policies, a difference of 3 percentage points. Among the 82, China grew 7.7%, Philippines 6.8%, Malaysia 4.7%, Pakistan and Saudi Arabia 3.6%, Argentina 3.5%, South Korea 2.8%, and Russia 1.3%, while among the 28, Brazil grew 2.3%, Japan 2.0%, South Africa 2.0%, Netherlands -0.8%, Croatia -1.0%, Iran -1.5%, Portugal -1.8%, Greece -3.8%, and Cyprus -8.7%.

Faster GDP growth and lower unemployment rates translated into increased tax revenues and a less cumbersome debt burden. Revenues for the 82 expansionary countries grew at a median rate of 10.7%, whereas tax revenues fell at a median rate of 6.8% for the 28 countries that chose austere economic policies. Budget balances improved for about three-quarters of the 28, but, for most, debt grew faster than GDP, and the median level of their public debt as a share of GDP increased 9.1 percentage points, to 59.2%. On the other hand, budget balances deteriorated for most of the 82 pro- growth countries, but GDP growth outpaced increases in debt, and the median level of public debt as a share of GDP increased just 1.9%, to 39.8%.

The world recession has suppressed inflation rates - world inflation declined 1.0 percentage point in 2012 to about 4.1% and 0.2 percentage point to 3.9% in 2013. In 2013 the median inflation rate for the 82 pro-growth countries was 1.3 percentage points higher than that for the countries that followed more austere fiscal and monetary policies. Overall, the latter countries also improved their current account balances by shedding imports; as a result, current account balances deteriorated for most of the countries that pursued pro-growth policies. Slow growth of world income continued to hold import demand in check and crude oil prices fell. Consequently, the dollar value of world trade grew just 1.3% in 2013.

Beyond the current global slowdown, the world faces several long standing economic challenges. The addition of 80 million people each year to an already overcrowded globe is exacerbating the problems of pollution, waste-disposal, epidemics, water-shortages, famine, over-fishing of oceans, deforestation, desertification, and depletion of non-renewable resources. The nation- state, as a bedrock economic-political institution, is steadily losing control over international flows of people, goods, services, funds, and technology. The introduction of the euro as the common currency of much of Western Europe in January 1999, while paving the way for an integrated economic powerhouse, has created economic risks because the participating nations have varying income levels and growth rates, and hence, require a different mix of monetary and fiscal policies. Governments, especially in Western Europe, face the difficult political problem of channeling resources away from welfare programs in order to increase investment

and strengthen incentives to seek employment. Because of their own internal problems and priorities, the industrialized countries are unable to devote sufficient resources to deal effectively with the poorer areas of the world, which, at least from an economic point of view, are becoming further marginalized. The terrorist attacks on the US on 11 September 2001 accentuated a growing risk to global prosperity - the diversion of resources away from capital investments to counter- terrorism programs.

Despite these vexing problems, the world economy also shows great promise. Technology has made possible further advances in a wide range of fields, from agriculture, to medicine, alternative energy, metallurgy, and transportation. Improved global communications have greatly reduced the costs of international trade, helping the world gain from the international division of labor, raise living standards, and reduce income disparities among nations. Much of the resilience of the world economy in the aftermath of the financial crisis resulted from government and central bank leaders around the globe working in concert to stem the financial onslaught, knowing well the lessons of past economic failures.

Fiscal and Monetary Data, 2008-2012: [INSERT IMAGE: The World Factbook pyramid]

GDP (purchasing power parity): $127.8 trillion (2017 est.)
$123.3 trillion (2016 est.)
$119.5 trillion (2015 est.)
note: data are in 2017 dollars

GDP (official exchange rate): $80.27 trillion SGWP (gross world product) (2017 est.)

GDP—real growth rate: 3.7% (2017 est.)
3.2% (2016 est.)
3.3% (2014 est.)

GDP—per capita (PPP): $17,500 (2017 est.)
$17,000 (2016 est.)
$16,800 (2015 est.)
note: data are in 2017 dollars

Gross national saving: 27.9% of GDP (2017 est.)
27.4% of GDP (2016 est.)
27.8% of GDP (2015 est.)

GDP—composition, by end use: *household consumption:* 56.4% (2017 est.)
government consumption: 16.1% (2017 est.)
investment in fixed capital: 25.7% (2017 est.)
investment in inventories: 1.4% (2017 est.)
exports of goods and services: 28.8% (2017 est.)
imports of goods and services: -28.3% (2017 est.)

GDP—composition, by sector of origin: *agriculture:* 6.4% (2017 est.)
industry: 30% (2017 est.)
services: 63% (2017 est.)

Industries: dominated by the onrush of technology, especially in computers, robotics, telecommunications, and medicines and medical equipment; most of these advances take place in OECD nations; only a small portion of non- OECD countries have succeeded in rapidly adjusting to these technological forces; the accelerated development

of new technologies is complicating already grim environmental problems

Industrial production growth rate: 3.2% (2017 est.)

Labor force: 3.432 billion (2017 est.)

Labor force—by occupation: agriculture: 31%
industry: 23.5%
services: 45.5% (2014 est.)

Unemployment rate: 7.7% (2017 est.)
7.5% (2016 est.)
note: combined unemployment and underemployment in many non-industrialized countries; developed countries typically 4%-12% unemployment (2007 est.)

Household income or consumption by percentage share: *lowest 10%:* 2.6%
highest 10%: 30.2% (2008 est.)

Budget: *revenues:* 21.68 trillion (2017 est.)
expenditures: 23.81 trillion (2017 est.)

Taxes and other revenues: 26.7% (of GDP) (2016 est.)

Budget surplus (+) or deficit (-): -3% (of GDP) (2016 est.)

Public debt: 67.2% of GDP (2017 est.)
67.2% of GDP (2016 est.)

Inflation rate (consumer prices): 6.4% (2017 est.)
3.7% (2016 est.)

developed countries: 1.9% (2017 est.) 0.9% (2016 est.)

developing countries: 8.8% (2017 est.) 3.7% (2016 est.)
note: the above estimates are weighted averages; inflation in developed countries is 0% to 4% typically, in developing countries, 4% to 10% typically; national inflation rates vary widely in individual cases; inflation rates have declined for most countries for the last several years, held in check by increasing international competition from several low wage countries and by soft demand due to the world financial crisis

Exports: $17.31 trillion (2017 est.)
$15.82 trillion (2016 est.)

Exports—commodities: the whole range of industrial and agricultural goods and services

top ten—share of world trade: 14.8 electrical machinery, including computers; 14.4 mineral fuels, including oil, coal, gas, and refined products; 14.2 nuclear reactors, boilers, and parts; 8.9 cars, trucks, and buses; 3.5 scientific and precision instruments; 3.4 plastics; 2.7 iron and steel; 2.6 organic chemicals; 2.6 pharmaceutical products; 1.9 diamonds, pearls, and precious stones (2007 est.)

Imports: $20.01 trillion (2018 est.)
$16.02 trillion (2017 est.)

Imports—commodities: the whole range of industrial and agricultural goods and services

top ten—share of world trade: see listing for exports

Debt—external: $76.56 trillion (31 December 2017 est.)
$75.09 trillion (31 December 2016 est.)

note: this figure is the sum total of all countries' external debt, both public and private

ENERGY

Electricity access: population without electricity: 1.201 billion (2013)

electrification—total population: 83% (2013)

electrification—urban areas: 95% (2013)

electrification—rural areas: 70% (2013)

Electricity—production: 23.65 trillion kWh (2015 est.)

Electricity—consumption: 21.78 trillion kWh (2015 est.)

Electricity—exports: 696.1 billion kWh (2016)

Electricity—imports: 721.9 billion kWh (2016 est.)

Electricity—installed generating capacity: 6.386 billion kW (2015 est.)

Electricity—from fossil fuels: 63% of total installed capacity (2015 est.)

Electricity—from nuclear fuels: 6% of total installed capacity (2015 est.)

Electricity—from hydroelectric plants: 18% of total installed capacity (2015 est.)

Electricity—from other renewable sources: 14% of total installed capacity (2015 est.)

Crude oil—production: 80.77 million bbl/day (2016 est.)

Crude oil—exports: 43.57 million bbl/day (2014 est.)

Crude oil—imports: 11.50 million bbl/day (2014 est.)

Crude oil—proved reserves: 1.665 trillion bbl (1 January 2017 est.)

Refined petroleum products—production: 88.4 million bbl/day (2014 est.)

Refined petroleum products—consumption: 96.26 million bbl/day (2015 est.)

Refined petroleum products— exports: 29.66 million bbl/day (2014 est.)

Refined petroleum products—imports: 28.62 million bbl/day (2014 est.)

Natural gas—production: 3.481 trillion cu m (2015 est.)

Natural gas—consumption: 3.477 trillion cu m (2015 est.)

Natural gas—exports: 1.156 trillion cu m (2013 est.)

Natural gas—imports: 1.496 trillion cu m (2013 est.)

Natural gas—proved reserves: 196.1 trillion cu m (1 January 2016 est.)

Carbon dioxide emissions from consumption of energy: 33.62 billion Mt (2013 est.)

COMMUNICATIONS

Telephones—fixed lines: *total subscriptions:* 984,289,950

subscriptions per 100 inhabitants: 1 (2017 est.)

Telephones—mobile cellular: total subscriptions: 7,806,142,681

subscriptions per 100 inhabitants: 105 (2017 est.)

Internet users: *total:* 4.1 billion

percent of population: 53.9% (2019)

top ten countries by Internet usage (in millions): 730.7 China; 374.3 India; 246.8 United States; 122.8 Brazil; 116.6 Japan; 108.8 Russia; 73.3 Mexico; 72.3 Germany; 65.5 Indonesia; 61 United Kingdom (2017)

Broadband—fixed subscriptions:
total: 1,002,793,951

subscriptions per 100 inhabitants: 14 (2017)

Communications—note: note 1: three major data centers - which provide colocation, telecommunications, cloud services, and content ecosystems - compete to be called the world's biggest in terms of physical space occupied:

no. 1. - a data farm in Langfang, Hebei Province, China, identified as the Range International Information Group, claims to be the largest with 585,000 sq m (6.3 million sq ft),

no. 2. - a data farm in Las Vegas, Nevada, USA, known as the Switch SuperNAP data center, comes in second with over 325,000 sq m (3.5 million sq ft); it intends to expand to over 1.615 million sq m (17.4 million sq ft) by 2020,

no. 3. - a data farm in Ashburn, Virginia, USA, referred to as the DFT Data Center, is a transit point for 70% of the world's Internet traffic; it includes 150,000 sq m (1.6 million sq ft) spread out over six separate buildings

note 2. estimates are that by the end of 2019, 53.9% of the global population (4.1 billion people) were using the Internet

TRANSPORTATION

Airports: 41,820 (2016)

top ten by passengers: Atlanta (ATL) - 103,902,992; Beijing (PEK) - 95,786,442; Dubai (DXB) - 88,242,099; Tokyo (HND) - 85,408,975; Los Angeles (LAX) - 84,557,968; Chicago (ORD) - 79,828,183; London (LHR) - 78,014,598; Hong Kong (HKG) 72,664,075; Shanghai (PVG) 70,001,237; Paris (CDG) - 69,471,442 (2017)

top ten by cargo (metric tons): Hong Kong (HKG) - 5,049,898; Memphis, TN (MEM) - 4,336,752; Shanghai (PVG) - 3,824,280; Incheon (ICN) - 2,921,691; Anchorage, AK (ANC) - 2,713,230; Dubai (DXB) - 2,654,494; Louisville, KY (SDF) - 2,602,695; Tokyo (NRT) - 2,336,427; Taipei (TPE) - 2,269,585; Paris (CDG) - 2,195,229 (2017)

Heliports: 6,524 (2013)

Railways: *total:* 1,148,186 km (2013)

Roadways: *total:* 64,785,009 km (2013)

Waterways: 2,293,412 km (2017)

top ten longest rivers: Nile (Africa) 6,693 km; Amazon (South America) 6,436 km; Mississippi-Missouri (North America) 6,238 km; Yenisey-Angara (Asia) 5,981 km; Ob-Irtysh (Asia) 5,569

km; Yangtze (Asia) 5,525 km; Yellow (Asia) 4,671 km; Amur (Asia) 4,52 km; Lena (Asia) 4,345 km; Congo (Africa) 4,344 km

note: rivers are not necessarily navigable along the entire length; if measured by volume, the Amazon is the largest river in the world, responsible for about 20% of the Earth's freshwater entering the ocean

top ten largest natural lakes (by surface area): Caspian Sea (Azerbaijan, Iran, Kazakhstan, Russia, Turkmenistan) 372,960 sq km; Lake Superior (Canada, United States) 82,414 sq km; Lake Victoria (Kenya, Tanzania, Uganda) 69,490 sq km; Lake Huron (Canada, United States) 59,596 sq km; Lake Michigan (United States) 57,441 sq km; Lake Tanganyika (Burundi, Democratic Republic of the Congo, Tanzania, Zambia) 32,890 sq km; Great Bear Lake (Canada) 31,800 sq km; Lake Baikal (Russia) 31,494 sq km; Lake Nyasa (Malawi, Mozambique, Tanzania) 30,044 sq km; Great Slave Lake (Canada) 28,400 sq km

note 1: the areas of the lakes are subject to seasonal variation; only the Caspian Sea is saline, the rest are fresh water

note 2: Lakes Huron and Michigan are technically a single lake because the flow of water between the Straits of Mackinac that connects the two lakes keeps their water levels at near-equilibrium; combined, Lake Huron-Michigan is the largest freshwater lake by surface area in the world

note 3: the deepest lake in the world (1,620 m), and also the largest freshwater lake by volume (23,600 cu km), is Lake Baikal in Russia.

Merchant marine: *total:* 94,980

by type: bulk carrier 11,369, container ship 5,265, general cargo 10,611, oil tanker 10,619, other 49,113 (2019)

Ports and terminals: top twenty container ports as measured by Twenty-Foot Equivalent Units (TEUs) throughput: Shanghai (China) - 40,233,000; Singapore (Singapore) - 33,666,000; Shenzhen (China) - 25,208,000; Ningbo (China) - 24,607,000; Hong Kong (China) - 20,770,000; Busan (South Korea) - 20,493,000; Guangzhou (China) - 18,858,000; Qingdao (China) - 18,262,000; Dubai (UAE) - 15,368,000; - Tianjin (China) - 15,040,000; Rotterdam (Netherlands) - 13,734,000; Port Kelang (Malaysia) - 11,978,000; Antwerp (Belgium) - 10,450,000; Xiamen (China) - 10,380,000; Kaohsiung (Taiwan) - 10,271,000; Dalian (China) - 9,707,000; Los Angeles (US) - 9,343,000; Hamburg (Germany) - 8,860,000; Tanjung Pelepas (Malaysia) - 8,260,000; Laem Chabang (Thailand) - 7,227,000 (2017)

note - it was estimated that in 2017 60% of global sea- borne trade by value moved by container

MILITARY AND SECURITY

Military expenditures: 2.14% of GDP (2018)
2.16% of GDP (2017)
2.2% of GDP (2016)
2.25% of GDP (2015)
2.24% of GDP (2014)

Maritime threats: The International Maritime Bureau (IMB) reports that 2018 saw an increase in global pirate activities; in 2018, pirates attacked a total of 201 ships worldwide including boarding 143 ships, hijacking six ships, and firing on 18; this activity is an increase from 180 incidents in 2017; in 2018, the number of hostages increased to 141, however, the number of seafarers kidnapped for ransom increased to 83 compared with 75 in 2017, with nearly all taken off West Africa

Operation Ocean Shield, the NATO naval task force established in 2009 to combat Somali piracy, concluded its operations in December 2016 as a result of the drop in reported incidents over the last few years; the EU naval mission, Operation ATALANTA, continues its operations in the Gulf of Aden and Indian Ocean through 2020; naval units from Japan, India, and China also operate in conjunction with EU forces; China has established a logistical base in Djibouti to support its deployed naval units in the Horn of Africa

The Horn of Africa continued to see pirate activities with three incidents in 2018, a slight decrease over 2017; the decrease in successful pirate attacks off the Horn of Africa since the peak in 2007 was due, in part, to anti-piracy operations by international naval forces, the hardening of vessels, and the increased use of armed security teams aboard merchant ships; despite these preventative measures, the assessed risk remains high

West African piracy more than doubled in 2018 to become the most dangerous area in the world with 85 attacks in 2018 compared to 33 in 2017; Nigerian pirates are very aggressive, operating as far as 200 nm offshore and boarding 29 ships in 2018; attacks in South Asian waters remain at low levels with 12 attacks off Bangladesh in 2018, up from 11 in 2017; Peru reported four incidents in 2018, up from two in 2017; attacks in Vietnam rose from two in 2017 to four in 2018; the majority of global attacks against shipping have occurred in the offshore waters of five countries - Nigeria, Indonesia, Philippines, Venezuela, and Bangladesh. (2018)

TRANSPORTATION

Disputes—international: stretching over 250,000 km, the world's 325 international land boundaries separate 195 independent states and 71 dependencies, areas of special sovereignty, and other miscellaneous entities; ethnicity, culture, race, religion, and language have divided states into separate political entities as much as history, physical terrain, political fiat, or conquest, resulting in sometimes arbitrary and imposed boundaries; most maritime states have claimed limits that include territorial seas and exclusive economic zones; overlapping limits due to adjacent or opposite coasts create the potential for 430 bilateral maritime boundaries of which 209 have agreements that include contiguous and non-contiguous segments; boundary, borderland/resource, and territorial disputes vary in intensity from managed or dormant to violent or militarized; undemarcated, indefinite, porous, and unmanaged boundaries tend to encourage illegal cross- border activities, uncontrolled migration, and confrontation; territorial disputes may evolve from historical and/ or cultural claims, or they may be brought on by resource competition; ethnic and cultural clashes continue to be responsible for much of the territorial fragmentation and internal displacement of the estimated 45.7 million people and cross- border displacements of approximately 30.2 million refugees and asylum seekers around the world as of yearend 2019; approximately 317,200 refugees were repatriated during 2019; other sources of contention include access to water and mineral (especially hydrocarbon) resources, fisheries, and arable land; armed conflict prevails not so much between the uniformed armed forces of independent states as between stateless armed entities that detract from the sustenance and welfare of local populations, leaving the community of nations to cope with resultant refugees, hunger, disease, impoverishment, and environmental degradation

Refugees and internally displaced persons: the UN High Commissioner for Refugees (UNHCR) estimated that as of year-end 2019 there were 79.5 million people forcibly displaced worldwide; this includes 45.7 million conflict IDPs, 26 million refugees, 4.2 million asylum seekers, and 3.6 million Venezuelans displaced abroad; the UNHCR estimates there are currently at least 10 million stateless persons

Trafficking in persons: *current situation:* the International Labour Organization conservatively estimated that 20.9 million people in 2012 were victims of forced labor, representing the full range of human trafficking (also referred to as ' modern-day slavery') for labor and sexual exploitation; about one-third of reported cases involved crossing international borders, which is often associated with sexual exploitation; trafficking in persons is most prevalent in southeastern Europe, Eurasia, and Africa and least frequent in EU member states, Canada, the US, and other developed countries (2012)

tier rating: (2015)

Tier 2 Watch List: countries that do not fully comply with the minimum standards for the elimination of trafficking but are making significant efforts to do so; (44 countries) Antigua and Barbuda, Bolivia, Botswana, Bulgaria, Burkina Faso, Burma, Cambodia, China, Democratic Republic of the Congo, Republic of the Congo, Costa Rica, Cuba, Djibouti, Egypt, Gabon, Ghana, Guinea, Guyana, Haiti, Jamaica, Laos, Lebanon, Lesotho, Malaysia, Maldives, Mali, Mauritius, Namibia, Pakistan, Papua New Guinea, Qatar, Saudi Arabia, Saint Vincent and the Grenadines, Solomon Islands, Sri Lanka, Sudan, Suriname, Tanzania, Timor-Leste, Trinidad and Tobago, Tunisia, Turkmenistan, Ukraine, Uzbekistan

Tier 3: countries that neither satisfy the minimum standards for the elimination of trafficking nor demonstrate a significant effort to do so; (23 countries) Algeria, Belarus, Belize, Burundi, Central African Republic, Comoros, Equatorial Guinea, Eritrea, The Gambia, Guinea-Bissau, Iran, North Korea, Kuwait, Libya, Marshall Islands, Mauritania, Russia, South Sudan, Syria, Thailand, Venezuela, Yemen, Zimbabwe

Illicit drugs: cocaine: worldwide coca leaf cultivation in 2013 likely amounted to 165,000 hectares, assuming a stable crop in Bolivia; Colombia produced slightly less than half of the worldwide crop, followed by Peru and Bolivia; potential pure cocaine production increased 7% to 640 metric tons in 2013; Colombia conducts an aggressive coca eradication campaign, Peru has increased its eradication efforts, but remains hesitant to eradicate coca in key growing areas;

opiates: worldwide illicit opium poppy cultivation increased in 2013, with potential opium production reaching 6,800 metric tons; Afghanistan is world's primary opium producer, accounting for 82% of the global supply; Southeast Asia was responsible for 12% of global opium; Pakistan produced 3% of global opium; Latin America produced 4% of global opium, and most was refined into heroin destined for the US market (2015)

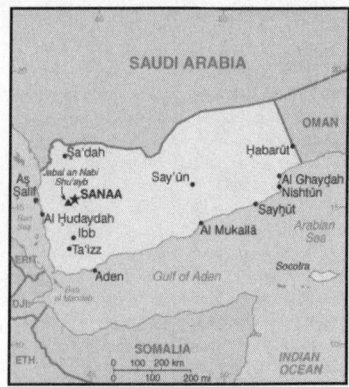

INTRODUCTION

Background: The Kingdom of Yemen (colloquially known as North Yemen) became independent from the Ottoman Empire in 1918 and in 1962 became the Yemen Arab Republic. The British, who had set up a protectorate area around the southern port of Aden in the 19th century, withdrew in 1967 from what became the People's Republic of Southern Yemen (colloquially known as South Yemen). Three years later, the southern government adopted a Marxist orientation and changed the country's name to the People's Democratic Republic of Yemen. The massive exodus of hundreds of thousands of Yemenis from the south to the north contributed to two decades of hostility between the states. The two countries were formally unified as the Republic of Yemen in 1990. A southern secessionist movement and brief civil war in 1994 was quickly subdued. In 2000, Saudi Arabia and Yemen agreed to delineate their border.

Fighting in the northwest between the government and the Huthis, a Zaydi Shia Muslim minority, continued intermittently from 2004 to 2010, and then again from 2014-present. The southern secessionist movement was revitalized in 2007.

Public rallies in Sana'a against then President Ali Abdallah SALIH - inspired by similar demonstrations in Tunisia and Egypt - slowly built momentum starting in late January 2011 fueled by complaints over high unemployment, poor economic conditions, and corruption. By the following month, some protests had resulted in violence, and the demonstrations had spread to other major cities. By March the opposition had hardened its demands and was unifying behind calls for SALIH's immediate ouster. In April 2011, the Gulf Cooperation Council (GCC), in an attempt to mediate the crisis in Yemen, proposed the GCC Initiative, an agreement in which the president would step down in exchange for immunity from prosecution. SALIH's refusal to sign an agreement

led to further violence. The UN Security Council passed Resolution 2014 in October 2011 calling for an end to the violence and completing a power transfer deal. In November 2011, SALIH signed the GCC Initiative to step down and to transfer some of his powers to Vice President Abd Rabuh Mansur HADI. Following HADI's uncontested election victory in February 2012, SALIH formally transferred all presidential powers. In accordance with the GCC Initiative, Yemen launched a National Dialogue Conference (NDC) in March 2013 to discuss key constitutional, political, and social issues. HADI concluded the NDC in January 2014 and planned to begin implementing subsequent steps in the transition process, including constitutional drafting, a constitutional referendum, and national elections.

The Huthis, perceiving their grievances were not addressed in the NDC, joined forces with SALIH and expanded their influence in northwestern Yemen, which culminated in a major offensive against military units and rival tribes and enabled their forces to overrun the capital, Sanaa, in September 2014. In January 2015, the Huthis surrounded the presidential palace, HADI's residence, and key government facilities, prompting HADI and the cabinet to submit their resignations. HADI fled to Aden in February 2015 and rescinded his resignation. He subsequently escaped to Oman and then moved to Saudi Arabia and asked the GCC to intervene militarily in Yemen to protect the legitimate government from the Huthis. In March, Saudi Arabia assembled a coalition of Arab militaries and began airstrikes against the Huthis and Huthi-affiliated forces. Ground fighting between Huthi-aligned forces and anti-Huthi groups backed by the Saudi-led coalition continued through 2016. In 2016, the UN brokered a months-long cessation of hostilities that reduced airstrikes and fighting, and initiated peace talks in Kuwait. However, the talks ended without agreement. The Huthis and SALIH's political party announced a Supreme Political Council in August 2016 and a National Salvation Government, including a prime minister and several dozen cabinet members, in November 2016, to govern in Sanaa and further challenge the legitimacy of HADI's government. However, amid rising tensions between the Huthis and SALIH, sporadic clashes erupted in mid-2017, and escalated into open fighting that ended when Huthi forces killed SALIH in early December 2017. In 2018, anti-Huthi forces made the most battlefield progress in Yemen since early 2016, most notably in Al Hudaydah Governorate. In December 2018, the Huthis and Yemeni Government participated in the first UN-brokered peace talks since 2016, agreeing to a limited ceasefire in Al Hudaydah Governorate and the establishment of a UN Mission to monitor the agreement. In April 2019, Yemen's parliament convened in Say'un for the first time since the conflict broke out in 2014. In

August 2019, violence erupted between HADI's government and the pro-secessionist Southern Transition Council (STC) in southern Yemen. In November 2019, HADI's government and the STC signed a power-sharing agreement to end the fighting between them.

GEOGRAPHY

Location: Middle East, bordering the Arabian Sea, Gulf of Aden, and Red Sea, between Oman and Saudi Arabia

Geographic coordinates: 15 00 N, 48 00 E

Map references: Middle East

Area: *total:* 527,968 sq km
land: 527,968 sq km
water: 0 sq km
note: includes Perim, Socotra, the former Yemen Arab Republic (YAR or North Yemen), and the former People's Democratic Republic of Yemen (PDRY or South Yemen)
country comparison to the world: 51

Area—comparative: almost four times the size of Alabama; slightly larger than twice the size of Wyoming

Area comparison map: [INSERT IMAGE: YEMEN-Area comparison map]

Land boundaries: *total:* 1,601 km
border countries (2): Oman 294 km, Saudi Arabia 1307 km

Coastline: 1,906 km

Maritime claims: *territorial sea:* 12 nm
exclusive economic zone: 200 nm
contiguous zone: 24 nm
continental shelf: 200 nm or to the edge of the continental margin

Climate: mostly desert; hot and humid along west coast; temperate in western mountains affected by seasonal monsoon; extraordinarily hot, dry, harsh desert in east

Terrain: narrow coastal plain backed by flat-topped hills and rugged mountains; dissected upland desert plains in center slope into the desert interior of the Arabian Peninsula

Elevation: *mean elevation:* 999 m
lowest point: Arabian Sea 0 m
highest point: Jabal an Nabi Shu'ayb 3,666 m

Natural resources: petroleum, fish, rock salt, marble; small deposits of coal, gold, lead, nickel, and copper; fertile soil in west

Land use: *agricultural land:* 44.5% (2011 est.)
arable land: 2.2% (2011 est.) / permanent crops: 0.6% (2011 est.) / permanent pasture: 41.7% (2011 est.)
forest: 1% (2011 est.)
other: 54.5% (2011 est.)

Irrigated land: 6,800 sq km (2012)

Population distribution: the vast majority of the population is found in the Asir Mountains (part of the larger Sarawat Mountain system), located in the far western region of the country

Natural hazards: sandstorms and dust storms in summer

volcanism: limited volcanic activity; Jebel at Tair (Jabal al-Tair, Jebel Teir, Jabal al-Tayr, Jazirat at-Tair) (244 m), which forms an island in the Red Sea, erupted in 2007 after awakening from dormancy; other historically active volcanoes include Harra of Arhab, Harras of Dhamar, Harra es-Sawad, and Jebel Zubair, although many of these have not erupted in over a century

Environment—current issues: limited natural freshwater resources; inadequate supplies of potable water; overgrazing; soil erosion; desertification

Environment—international agreements: *party to:* Biodiversity, Climate Change, Climate Change-Kyoto Protocol, Desertification, Endangered Species, Environmental Modification, Hazardous Wastes, Law of the Sea, Ozone Layer Protection *signed, but not ratified:* none of the selected agreements

Geography—note: strategic location on Bab el Mandeb, the strait linking the Red Sea and the Gulf of Aden, one of world's most active shipping lanes

PEOPLE AND SOCIETY

Population: 29,884,405 (July 2020 est.)
country comparison to the world: 48

Nationality: *noun:* Yemeni(s)
adjective: Yemeni

Ethnic groups: predominantly Arab; but also Afro-Arab, South Asian, European

Languages: Arabic (official)
note: a distinct Socotri language is widely used on Socotra Island and Archipelago; Mahri is still fairly widely spoken in eastern Yemen

Religions: Muslim 99.1% (official; virtually all are citizens, an estimated 65% are Sunni and 35% are Shia), other 0.9% (includes Jewish, Baha'i, Hindu, and Christian; many are refugees or temporary foreign residents) (2010 est.)

MENA religious affiliation:

Age structure: *0-14 years:* 39.16% (male 5,711,709 /female 5,513,526)
15-24 years: 21.26% (male 3,089,817 /female 3,005,693)
25-54 years: 32.78% (male 4,805,059 /female 4,591,811)
55-64 years: 4% (male 523,769 /female 623,100)
65 years and over: 2.8% (male 366,891 /female 435,855) (2018 est.)

population pyramid: [INSERT IMAGE: YEMEN-population pyramid]

Dependency ratios: *total dependency ratio:* 71.7
youth dependency ratio: 66.7
elderly dependency ratio: 5
potential support ratio: 19.9 (2020 est.)

Median age: *total:* 19.8 years (2018 est.)
male: 19.6 years
female: 19.9 years

country comparison to the world: 199

Population growth rate: 2.04% (2020 est.)
country comparison to the world: 46

Birth rate: 25.8 births/1,000 population (2020 est.)
country comparison to the world: 46

Death rate: 5.6 deaths/1,000 population (2020 est.)
country comparison to the world: 181

Net migration rate: -0.2 migrant(s)/1,000 population (2020 est.)
country comparison to the world: 111

Population distribution: the vast majority of the population is found in the Asir Mountains (part of the larger Sarawat Mountain system), located in the far western region of the country

Urbanization: *urban population:* 37.9% of total population (2020)
rate of urbanization: 4.06% annual rate of change (2015-20 est.)

total population growth rate v. urban population growth rate, 2000-2030: *Major urban areas - population:* 2.973 million SANAA (capital), 980,000 Aden (2020)

Sex ratio: *at birth:* 1.05 male(s)/female
0-14 years: 1.04 male(s)/female
15-24 years: 1.03 male(s)/female
25-54 years: 1.05 male(s)/female
55-64 years: 0.84 male(s)/female
65 years and over: 0.84 male(s)/female
total population: 1.02 male(s)/female (2018 est.)

Mother's mean age at first birth: 21.4 years (2013 est.)
median age at first birth among women 25-29

Maternal mortality rate: 164 deaths/100,000 live births (2017 est.)
country comparison to the world: 54

Infant mortality rate: *total:* 41.9 deaths/1,000 live births
male: 45.7 deaths/1,000 live births
female: 37.9 deaths/1,000 live births (2020 est.)
country comparison to the world: 34

Life expectancy at birth: *total population:* 66.9 years
male: 64.7 years
female: 69.3 years (2020 est.)
country comparison to the world: 183

Total fertility rate: 3.2 children born/woman (2020 est.)
country comparison to the world: 47

Contraceptive prevalence rate: 33.5% (2013)

Drinking water source:

improved: *urban:* 100% of population
rural: 87.6% of population
total: 92% of population

unimproved: *urban:* 0% of population
rural: 12.4% of population
total: 8% of population (2017 est.)

Current Health Expenditure: 5.6% (2015)

Physicians density: 0.53 physicians/1,000 population (2014)

Hospital bed density: 0.7 beds/1,000 population (2017)

Sanitation facility access:

improved: *urban:* 93.1% of population
rural: 48.5% of population
total: 64.6% of population

unimproved: *urban:* 6.9% of population
rural: 51.5% of population
total: 35.4% of population (2017 est.)

HIV/AIDS—adult prevalence rate: <.1% (2019 est.)

HIV/AIDS—people living with HIV/AIDS: 11,000 (2019 est.)
country comparison to the world: 102

HIV/AIDS—deaths: <500 (2019 est.)

Major infectious diseases: *degree of risk:* high (2020)
food or waterborne diseases: bacterial diarrhea, hepatitis A, and typhoid fever
vectorborne diseases: dengue fever and malaria

water contact diseases: schistosomiasis

Obesity—adult prevalence rate: 17.1% (2016)
country comparison to the world: 120

Children under the age of 5 years underweight: 39.9% (2013)
country comparison to the world: 1

Education expenditures: NA

Literacy: *definition:* age 15 and over can read and write
total population: 70.1%
male: 85.1%
female: 55% (2015)
School life expectancy (primary to tertiary education): total: 9 years
male: 11 years
female: 8 years (2011)
Unemployment, youth ages 15-24: total: 24.5%
male: 23.5%
female: 34.6% (2014 est.)
country comparison to the world: 50

GOVERNMENT

Country name: *conventional long form:* Republic of Yemen
conventional short form: Yemen
local long form: Al Jumhuriyah al Yamaniyah
local short form: Al Yaman
former: Yemen Arab Republic [Yemen (Sanaa) or North Yemen] and People's Democratic Republic of Yemen [Yemen (Aden) or South Yemen]
etymology: name derivation remains unclear but may come from the Arab term "yumn" (happiness) and be related to the region's classical name "Arabia Felix" (Fertile or Happy Arabia); the Romans referred to the rest of the peninsula as "Arabia Deserta" (Deserted Arabia)

Government type: in transition

Capital: *name:* Sanaa

geographic coordinates: 15 21 N, 44 12 E
time difference: UTC+3 (8 hours ahead of Washington, DC, during Standard Time)

etymology: the name is reputed to mean "well-fortified" in Sabaean, the South Arabian language that went extinct in Yemen in the 6th century A.D.

Administrative divisions: 22 governorates (muhafazat, singular - muhafazah); Abyan, 'Adan (Aden), Ad Dali', Al Bayda', Al Hudaydah, Al Jawf, Al Mahrah, Al Mahwit, Amanat al 'Asimah (Sanaa City), 'Amran, Arkhabil Suqutra (Socotra Archipelago), Dhamar, Hadramawt, Hajjah, Ibb, Lahij, Ma'rib, Raymah, Sa'dah, San'a' (Sanaa), Shabwah, Ta'izz

Independence: 22 May 1990 (Republic of Yemen was established with the merger of the Yemen Arab Republic [Yemen (Sanaa) or North Yemen] and the Marxist-dominated People's Democratic Republic of Yemen [Yemen (Aden) or South Yemen]); notable earlier dates: North Yemen became independent on 1 November 1918 (from the Ottoman Empire) and became a republic with the overthrow of the theocratic Imamate on 27 September 1962; South Yemen became independent on 30 November 1967 (from the UK)

National holiday: Unification Day, 22 May (1990)

Constitution: *history:* adopted by referendum 16 May 1991 (following unification); note - after the National Dialogue ended in January 2015, a Constitutional Drafting Committee appointed by the president worked to prepare a new draft constitution that was expected to be put to a national referendum before being adopted; however, the start of the current conflict in early 2015 interrupted the process

amendments: amended several times, last in 2009

Legal system: mixed legal system of Islamic (sharia) law, Napoleonic law, English common law, and customary law

International law organization participation: has not submitted an ICJ jurisdiction declaration; non-party state to the ICCt

Citizenship: *citizenship by birth:* no
citizenship by descent only: the father must be a citizen of Yemen; if the father is unknown, the mother must be a citizen
dual citizenship recognized: no
residency requirement for naturalization: 10 years

Suffrage: 18 years of age; universal

Executive branch: *chief of state:* President Abd Rabuh Mansur HADI (since 21 February 2012); Vice President ALI MUHSIN al-Ahmar, Lt. Gen. (since 3 April 2016)

head of government: Prime Minister Maeen Abd al-Malik SAEED (since 15 October 2018)
cabinet: appointed by the president
elections/appointments: president directly elected by absolute majority popular vote in 2 rounds if needed for a 7-year term (eligible for a second term); election last held on 21 February 2012 (next election NA); note - a special election was held on 21 February 2012 to remove Ali Abdallah SALIH under the terms of a Gulf Cooperation Council-mediated deal during the

political crisis of 2011; vice president appointed by the president; prime minister appointed by the president

election results: Abd Rabuh Mansur HADI (GPC) elected as a consensus president with about 50% popular participation; no other candidates

Legislative branch: *description:* bicameral Parliament or Majlis consists of:
Shura Council or Majlis Alshoora (111 seats; members appointed by the president; member tenure NA) House of Representatives or Majlis al Nuwaab (301 seats; members directly elected in single-seat constituencies by simple majority vote to serve 6-year terms)
elections: House of Representatives - last held on 27 April 2003 (next scheduled for April 2009 but postponed indefinitely)
election results: percent of vote by party - GPC 58.0%, Islah 22.6%, YSP 3.8%, Unionist Party 1.9%, other 13.7%; seats by party - GPC 238, Islah 46, YSP 8, Nasserist Unionist Party 3, National Arab Socialist Ba'ath Party 2, independent 4

Judicial branch: *highest courts:* Supreme Court (consists of the president, 2 deputies, and nearly 50 judges; court organized into constitutional, civil, commercial, family, administrative, criminal, military, and appeals scrutiny divisions)
judge selection and term of office: judges appointed by the Supreme Judicial Council, which is chaired by the president of the republic and includes 10 high-ranking judicial officers; judges serve for life with mandatory retirement at age 65
subordinate courts: appeal courts; district or first instance courts; commercial courts

Political parties and leaders: General People's Congress or GPC (3 factions: pro-Hadi [Abdrabbi Mansur HADI], pro-Houthi [Sadeq Ameen Abu RAS], pro-Saleh [Ahmed SALEH]
National Arab Socialist Ba'ath Party [Qassem Salam SAID]
Nasserist Unionist People's Organization [Abdulmalik al-MEKHLAFI]
Southern Transitional Council or STC [Aidarus al-ZOUBAIDA]
Yemeni Reform Grouping or Islah [Muhammed Abdallah al-YADUMI]
Yemeni Socialist Party or YSP [Dr. Abd al-Rahman Umar al-SAQQAF]

International organization participation: AFESD, AMF, CAEU, CD, EITI (temporarily suspended), FAO, G-77, IAEA, IBRD, ICAO, ICRM, IDA, IDB, IFAD, IFC, IFRCS, ILO, IMF, IMO, IMSO, Interpol, IOC, IOM, IPU, ISO, ITSO, ITU, ITUC (NGOs), LAS, MIGA, MINURSO, MINUSMA, MONUSCO, NAM, OAS (observer), OIC, OPCW, UN, UNAMID, UNCTAD, UNESCO, UNHCR, UNIDO, UNISFA, UNMIL, UNMIS, UNOCI, UNWTO, UPU, WCO, WFTU (NGOs), WHO, WIPO, WMO, WTO

Diplomatic representation in the US: *chief of mission:* Ambassador Ahmad Awadh BIN MUBARAK (since 3 August 2015)
chancery: 2319 Wyoming Avenue NW, Washington, DC 20008

telephone: [1] (202) 965-4760
FAX: [1] (202) 337-2017

Diplomatic representation from the US: *chief of mission:* Ambassador Christopher HENZEL (since 20 May 2019); note - the embassy closed in March 2015;

Yemen Affairs Unit currently operates out of US Embassy Riyadh
telephone: US Embassy Riyadh [966] 11-488-3800

previously—[967] 1 755-2000
embassy: previously - Sa'awan Street, Sanaa
mailing address: US Embassy Riyadh

previously—US Embassy in Sana'a
Address: Sa'awan Street
P.O. Box 22347
FAX: US Embassy Riyadh [966] 11-488-7360

Flag description: three equal horizontal bands of red (top), white, and black; the band colors derive from the Arab Liberation flag and represent oppression (black), overcome through bloody struggle (red), to be replaced by a bright future (white)
note: similar to the flag of Syria, which has two green stars in the white band, and of Iraq, which has an Arabic inscription centered in the white band; also similar to the flag of Egypt, which has a heraldic eagle centered in the white band

National symbol(s): golden eagle; national colors: red, white, black

National anthem: *name:* "al-qumhuriyatu l-muttahida" (United Republic)
lyrics/music: Abdullah Abdulwahab NOA'MAN/ Ayyoab Tarish ABSI
note: adopted 1990; the music first served as the anthem for South Yemen before unification with North Yemen in 1990
0:00 / 0:48

ECONOMY

Economy—overview: Yemen is a low-income country that faces difficult long-term challenges to stabilizing and growing its economy, and the current conflict has only exacerbated those issues. The ongoing war has halted Yemen's exports, pressured the currency's exchange rate, accelerated inflation, severely limited food and fuel imports, and caused widespread damage to infrastructure. The conflict has also created a severe humanitarian crisis - the world's largest cholera outbreak currently at nearly 1 million cases, more than 7 million people at risk of famine, and more than 80% of the population in need of humanitarian assistance.

Prior to the start of the conflict in 2014, Yemen was highly dependent on declining oil and gas resources for revenue. Oil and gas earnings accounted for roughly 25% of GDP and 65% of government revenue. The Yemeni Government regularly faced annual budget shortfalls and tried to diversify the Yemeni economy through a reform program designed to bolster non-oil sectors of the economy and foreign investment. In July 2014, the government continued reform efforts by eliminating some fuel subsidies and in August 2014, the

IMF approved a three-year, $570 million Extended Credit Facility for Yemen.

However, the conflict that began in 2014 stalled these reform efforts and ongoing fighting continues to accelerate the country's economic decline. In September 2016, President HADI announced the move of the main branch of Central Bank of Yemen from Sanaa to Aden where his government could exert greater control over the central bank's dwindling resources. Regardless of which group controls the main branch, the central bank system is struggling to function. Yemen's Central Bank's foreign reserves, which stood at roughly $5.2 billion prior to the conflict, have declined to negligible amounts. The Central Bank can no longer fully support imports of critical goods or the country's exchange rate. The country also is facing a growing liquidity crisis and rising inflation. The private sector is hemorrhaging, with almost all businesses making substantial layoffs. Access to food and other critical commodities such as medical equipment is limited across the country due to security issues on the ground. The Social Welfare Fund, a cash transfer program for Yemen's neediest, is no longer operational and has not made any disbursements since late 2014.

Yemen will require significant international assistance during and after the protracted conflict to stabilize its economy. Long-term challenges include a high population growth rate, high unemployment, declining water resources, and severe food scarcity.

GDP (purchasing power parity): $73.63 billion (2017 est.)
$78.28 billion (2016 est.)
$90.63 billion (2015 est.)
note: data are in 2017 dollars
country comparison to the world: 98

GDP (official exchange rate): $31.27 billion (2017 est.)

GDP—real growth rate: -5.9% (2017 est.)
-13.6% (2016 est.)
-16.7% (2015 est.)
country comparison to the world: 219

GDP—per capita (PPP): $2,500 (2017 est.)
$2,700 (2016 est.)
$3,200 (2015 est.)
note: data are in 2017 dollars
country comparison to the world: 199

Gross national saving: -1.9% of GDP (2017 est.)
-3.7% of GDP (2016 est.)
-4.5% of GDP (2015 est.)
country comparison to the world: 182

GDP—composition, by end use: household consumption: 116.6% (2017 est.)
government consumption: 17.6% (2017 est.)
investment in fixed capital: 2.2% (2017 est.)
investment in inventories: 0% (2017 est.)
exports of goods and services: 7.5% (2017 est.)
imports of goods and services: -43.9% (2017 est.)

GDP—composition, by sector of origin:
agriculture: 20.3% (2017 est.)
industry: 11.8% (2017 est.)
services: 67.9% (2017 est.)

Agriculture—products: grain, fruits, vegetables, pulses, qat, coffee, cotton; dairy products, livestock (sheep, goats, cattle, camels), poultry; fish

Industries: crude oil production and petroleum refining; small-scale production of cotton textiles, leather goods; food processing; handicrafts; aluminum products; cement; commercial ship repair; natural gas production

Industrial production growth rate: 8.9% (2017 est.)
country comparison to the world: 21

Labor force: 7.425 million (2017 est.)
country comparison to the world: 62

Labor force—by occupation: note: most people are employed in agriculture and herding; services, construction, industry, and commerce account for less than one-fourth of the labor force

Unemployment rate: 27% (2014 est.)
35% (2003 est.)
country comparison to the world: 198

Population below poverty line: 54% (2014 est.)
Household income or consumption by percentage share: lowest 10%: 2.6%
highest 10%: 30.3% (2008 est.)

Budget: revenues: 2.821 billion (2017 est.)
expenditures: 4.458 billion (2017 est.)

Taxes and other revenues: 9% (of GDP) (2017 est.)
country comparison to the world: 216
Budget surplus (+) or deficit (-): -5.2% (of GDP) (2017 est.)
country comparison to the world: 171

Public debt: 74.5% of GDP (2017 est.)
68.1% of GDP (2016 est.)
country comparison to the world: 41

Fiscal year: calendar year

Inflation rate (consumer prices): 24.7% (2017 est.)
-12.6% (2016 est.)
country comparison to the world: 218

Current account balance: -$1.236 billion (2017 est.)
-$1.868 billion (2016 est.)
country comparison to the world: 153

Exports: $384.5 million (2017 est.)
$940 million (2016 est.)
country comparison to the world: 183

Exports—partners: Egypt 29.4%, Thailand 16.7%, Belarus 13.5%, Oman 10.5%, UAE 6.5%, Saudi Arabia 5% (2017)

Exports—commodities: crude oil, coffee, dried and salted fish, liquefied natural gas

Imports: $4.079 billion (2017 est.)
$3.117 billion (2016 est.)
country comparison to the world: 139

Imports—commodities: food and live animals, machinery and equipment, chemicals

Imports—partners: UAE 12.2%, China 12.1%, Turkey 8.7%, Brazil 7.3%, Saudi Arabia 6.5%, Argentina 5.5%, India 4.7% (2017)

Reserves of foreign exchange and gold: $245.4 million (31 December 2017 est.)
$592.6 million (31 December 2016 est.)
country comparison to the world: 170

Debt—external: $7.068 billion (31 December 2017 est.)
$7.181 billion (31 December 2016 est.)
country comparison to the world: 124

Exchange rates: Yemeni rials (YER) per US dollar—
275 (2017 est.)
214.9 (2016 est.)
214.9 (2015 est.)
228 (2014 est.)
214.89 (2013 est.)

ENERGY

Electricity access: population without electricity: 16 million (2019)

electrification—total population: 47% (2019)

electrification—urban areas: 72% (2019)

electrification—rural areas: 31% (2019)

Electricity—production: 4.784 billion kWh (2016 est.)
country comparison to the world: 122

Electricity—consumption: 3.681 billion kWh (2016 est.)
country comparison to the world: 130

Electricity—exports: 0 kWh (2016 est.)
country comparison to the world: 219

Electricity—imports: 0 kWh (2016 est.)
country comparison to the world: 219

Electricity—installed generating capacity: 1.819 million kW (2016 est.)
country comparison to the world: 116

Electricity—from fossil fuels: 79% of total installed capacity (2016 est.)
country comparison to the world: 89

Electricity—from nuclear fuels: 0% of total installed capacity (2017 est.)
country comparison to the world: 213

Electricity—from hydroelectric plants: 0% of total installed capacity (2017 est.)
country comparison to the world: 215

Electricity—from other renewable sources: 21% of total installed capacity (2017 est.)
country comparison to the world: 37

Crude oil—production: 61,000 bbl/day (2018 est.)
country comparison to the world: 49

Crude oil—exports: 8,990 bbl/day (2015 est.)
country comparison to the world: 60

Crude oil—imports: 0 bbl/day (2015 est.)
country comparison to the world: 217

Crude oil—proved reserves: 3 billion bbl (1 January 2018 est.)
country comparison to the world: 29

Refined petroleum products—production: 20,180 bbl/day (2015 est.)
country comparison to the world: 89

Refined petroleum products—consumption: 104,000 bbl/day (2016 est.)
country comparison to the world: 78

Refined petroleum products— exports: 12,670 bbl/day (2015 est.)
country comparison to the world: 78

Refined petroleum products—imports: 75,940 bbl/day (2015 est.)
country comparison to the world: 65

Natural gas—production: 481.4 million cu m (2017 est.)
country comparison to the world: 72

Natural gas—consumption: 481.4 million cu m (2017 est.)
country comparison to the world: 99

Natural gas—exports: 0 cu m (2017 est.)
country comparison to the world: 214

Natural gas—imports: 0 cu m (2017 est.)
country comparison to the world: 214

Natural gas—proved reserves: 478.5 billion cu m (1 January 2018 est.)
country comparison to the world: 31

Carbon dioxide emissions from consumption of energy: 13.68 million Mt (2017 est.)
country comparison to the world: 95

COMMUNICATIONS

Telephones—fixed lines: *total subscriptions:* 1,253,287

subscriptions per 100 inhabitants: 4.28 (2019 est.)
country comparison to the world: 69

Telephones—mobile cellular: total subscriptions: 16,158,028

subscriptions per 100 inhabitants: 55.18 (2019 est.)
country comparison to the world: 66

Telecommunication systems: *general assessment:* large percent of the population is in need of humanitarian assistance, and given the civil conflict, telecommunications services are vital but disrupted; mobile towers are often deliberately targeted; maintenance is dangerous to staff; aid organization rely on satellite and radio communications; there is a scarcity of telecommunications equipment in rural areas; ownership of telecommunications services and the related revenues and taxes have become a political issue; Chinese company Huawei helping with rebuilding and moving some equipment; little progress in the near future until civil unrest stabilizes; earlier damage to the FALCON submarine cable, left Internet service interrupted for a month until repaired (2020)

domestic: the national network consists of micro-wave radio relay, cable, tropospheric scatter, GSM and CDMA mobile- cellular telephone systems; fixed-line teledensity remains low by regional standards at 4 per 100 but mobile cellular use expanding at 55 per 100 (2019)

international: country code - 967; landing points for the FALCON, SeaMeWe-5, Aden-Djibouti, and the AAE-1 international submarine cable connecting Europe, Africa, the Middle East, Asia and Southeast Asia; satellite earth stations - 3 Intelsat (2 Indian Ocean and 1 Atlantic Ocean), 1 Intersputnik (Atlantic Ocean region), and 2 Arabsat; microwave radio relay to Saudi Arabia and Djibouti (2020)

note: the COVID-19 outbreak is negatively impacting telecommunications production and supply chains globally; consumer spending on tele-com devices and services has also slowed due to the pandemic's effect on economies worldwide; overall progress towards improvements in all facets of the telecom industry - mobile, fixed-line, broadband, submarine cable and satellite - has moderated

Broadcast media: state-run TV with 2 stations; state-run radio with 2 national radio stations and 5 local stations; stations from Oman and Saudi Arabia can be accessed

Internet country code: ye

Internet users: total: 7,659,884

percent of population: 26.72% (July 2018 est.)
country comparison to the world: 66

Broadband—fixed subscriptions: total: 386,330

subscriptions per 100 inhabitants: 1 (2018 est.)
country comparison to the world: 91

TRANSPORTATION

National air transport system: *number of registered air carriers:* 2 (2020)

inventory of registered aircraft operated by air carriers: 8

annual passenger traffic on registered air carriers: 556,518 (2018)

annual freight traffic on registered air carriers: 3.27 million mt-km (2018)

Civil aircraft registration country code prefix: 7O (2016)

Airports: 57 (2013)
country comparison to the world: 82

Airports—with paved runways: *total:* 17 (2013)
over 3,047 m: 4 (2013)
2,438 to 3,047 m: 9 (2013)
1,524 to 2,437 m: 3 (2013)
914 to 1,523 m: 1 (2013)

Airports—with unpaved runways: *total:* 40 (2013)
over 3,047 m: 3 (2013)
2,438 to 3,047 m: 5 (2013)
1,524 to 2,437 m: 7 (2013)
914 to 1,523 m: 16 (2013)
under 914 m: 9 (2013)

Pipelines: 641 km gas, 22 km liquid petroleum gas, 1370 km oil (2013)

Roadways: *total:* 71,300 km (2005)
paved: 6,200 km (2005)
unpaved: 65,100 km (2005)
country comparison to the world: 69

Merchant marine: *total:* 34
by type: general cargo 3, oil tanker 4, other 27 (2019)
country comparison to the world: 128

Ports and terminals: *major seaport(s):* Aden, Al Hudaydah, Al Mukalla

MILITARY AND SECURITY

Military and security forces: Land Forces (includes seven Military Regional Commands, supported by Strategic Reserve Forces), Naval and Coastal Defense Forces (includes naval infantry/marines and Coast Guard), Air and Air Defense Force (although it still exists in name, in practice many of the officers and soldiers in this branch have been distributed to other military branches and jobs), Border Guards, Strategic Reserve Forces (supports the Land Forces at the discretion of the Armed Forces Commander-in-Chief; includes a Missile Group, Presidential Protection Brigades, and Special Operations Forces) (2018)

Military expenditures: 3.98% of GDP (2014)
4.08% of GDP (2013)
4.57% of GDP (2012)
4.19% of GDP (2011)
note—no reliable information exists following the start of renewed conflict in 2015
country comparison to the world: 13

Military and security service personnel strengths: N/A; note: prior to the civil war, Yemeni Government armed forces had approximately 70,000 active personnel, including about 60,000 Army (2019 est.)

Military equipment inventories and acquisitions: the inventory of the Yemeni Government forces consists primarily of Russian and Soviet-era equip-ment, although much of it has been lost in the cur-rent conflict; since 2010, it has received limited amounts of equipment from a variety of countries, including Belarus, Czechia, Jordan, Russia, Saudi Arabia, South Africa, Spain, UAE, Ukraine, and the US (2019 est.)

Military service age and obligation: 18 is the legal minimum age for voluntary military service; no conscription; 2-year service obligation (2018)

Maritime threats: the International Maritime Bureau reports offshore waters in the Gulf of Aden are high risk for piracy; numerous vessels, includ-ing commercial shipping and pleasure craft, have been attacked and hijacked both at anchor and while underway; crew, passengers, and cargo have been held for ransom; the presence of several naval task forces in the Gulf of Aden and additional anti-piracy measures on the part of ship operators reduced the incidence of piracy in that body of water; one attack was reported in 2016 while three ships reported being fired upon in 2017

TERRORISM

Terrorist group(s): Islamic Revolutionary Guard Corps/Qods Force; Islamic State of Iraq and ash-Sham - Yemen; al-Qa'ida in the Arabian Peninsula (2020)

note: details about the history, aims, leader-ship, organization, areas of operation, tactics, targets, weapons, size, and sources of support of the group(s) appear(s) in Appendix-T

TRANSPORTATION

Disputes—international: Saudi Arabia has reinforced its concrete-filled security barrier along sections of the fully demarcated border with Yemen to stem illegal cross-border activities

Refugees and internally displaced persons: *refugees (country of origin):* 14,638 (Ethiopia) (2019); 220,753 (Somalia) (2020)

IDPs: 3,635,000 (conflict in Sa'ada Governorate; clashes between al-Qa'ida in the Arabian Peninsula and government forces) (2019)

Trafficking in persons: *current situation:* Yemen is a source and, to a lesser extent, transit and destination country for men, women, and children subjected to forced labor and women and children subjected to sex trafficking; trafficking activities grew in Yemen in 2014, as the country's security situation deteriorated and poverty worsened; armed groups increased their recruitment of Yemeni children as combatants or checkpoint guards, and the Yemeni military and security forces continue to use child soldiers; some other Yemeni children, mostly boys, migrate to Yemeni cities or Saudi Arabia and, less frequently Oman, where they end up as beggars, drug smugglers, prostitutes, or forced laborers in domestic service or small shops; Yemeni children increasingly are also subjected to sex trafficking in country and in Saudi Arabia; tens of thousands of Yemeni migrant workers deported from Saudi Arabia and thousands of Syrian refugees are vulnerable to trafficking; additionally, Yemen is a destination and transit country for women and children from the Horn of Africa who are looking for work or receive fraudulent job offers in the Gulf states but are subjected to sexual exploitation or forced labor upon arrival; reports indicate that adults and children are still sold or inherited as slaves in Yemen

tier rating: Tier 3 – Yemen does not fully comply with the minimum standards for the elimination of trafficking and is not making significant efforts to do so; weak government institutions, corruption, economic problems, security threats, and poor law enforcement capabilities impeded the government's ability to combat human trafficking; not all forms of trafficking are criminalized, and officials continue to conflate trafficking and smuggling; the status of an anti-trafficking law drafted with assistance from an international organization remains unknown following the dissolution of the government in January 2015; the government did not report efforts to investigate, prosecute, or convict anyone of trafficking or slavery offenses, including complicit officials, despite reports of officials willfully ignoring trafficking crimes and using child soldiers in the government's armed forces; the government acknowledged the use of child soldiers and signed a UN action plan to end the practice in 2014 but made no efforts to release child soldiers from the military and provide them with rehabilitative services; authorities failed to identify victims and refer them to protective services; the status of a draft national anti-trafficking strategy remains unknown (2015)

Background: Multiple waves of Bantu-speaking groups moved into and through what is now Zambia over the past thousand years. In the 1880s, the British began securing mineral and other economic concessions from various local leaders and the territory that is now Zambia eventually came under the control of the former British South Africa Company and was incorporated as the protectorate of Northern Rhodesia in 1911. Administrative control was taken over by the UK in 1924. During the 1920s and 1930s, advances in mining spurred development and immigration.

The name was changed to Zambia upon independence in 1964. In the 1980s and 1990s, declining copper prices, economic mismanagement, and a prolonged drought hurt the economy. Elections in 1991 brought an end to one-party rule and propelled the Movement for Multiparty Democracy (MMD) to government. The subsequent vote in 1996, however, saw increasing harassment of opposition parties and abuse of state media and other resources. The election in 2001 was marked by administrative problems, with three parties filing a legal petition challenging the election of ruling party candidate Levy MWANAWASA. MWANAWASA was reelected in 2006 in an election that was deemed free and fair. Upon his death in August 2008, he was succeeded by his vice president, Rupiah BANDA, who won a special presidential byelection later that year. The MMD and BANDA lost to the Patriotic Front (PF) and Michael SATA in the 2011 general elections. SATA, however, presided over a period of haphazard economic management and attempted to silence opposition to PF policies. SATA died in October 2014 and was succeeded by his vice president, Guy SCOTT, who served as interim president until January 2015, when Edgar LUNGU won the presidential byelection and completed SATA's term. LUNGU then won a full term in August 2016 presidential elections.

Location: Southern Africa, east of Angola, south of the Democratic Republic of the Congo

Geographic coordinates: 15 00 S, 30 00 E

Map references: Africa

Area: *total:* 752,618 sq km
land: 743,398 sq km
water: 9,220 sq km
country comparison to the world: 40

Area—comparative: almost five times the size of Georgia; slightly larger than Texas

Land boundaries: *total:* 6,043.15 km
border countries (8): Angola 1065 km, Botswana 0.15 km, Democratic Republic of the Congo 2332 km, Malawi 847 km, Mozambique 439 km, Namibia 244 km, Tanzania 353 km, Zimbabwe 763 km

Coastline: 0 km (landlocked)

Maritime claims: none (landlocked)

Climate: tropical; modified by altitude; rainy season (October to April)

Terrain: mostly high plateau with some hills and mountains

Elevation: *mean elevation:* 1,138 m
lowest point: Zambezi river 329 m
highest point: unnamed elevation in Mafinga Hills 2,301 m

Natural resources: copper, cobalt, zinc, lead, coal, emeralds, gold, silver, uranium, hydropower

Land use: *agricultural land:* 31.7% (2011 est.)
arable land: 4.8% (2011 est.) / permanent crops: 0% (2011 est.) / permanent pasture: 26.9% (2011 est.)
forest: 66.3% (2011 est.)
other: 2% (2011 est.)
Irrigated land: 1,560 sq km (2012)

Population distribution: one of the highest levels of urbanization in Africa; high density in the central area, particularly around the cities of Lusaka, Ndola, Kitwe, and Mufulira as shown in this population distribution map

Natural hazards: periodic drought; tropical storms (November to April)

Environment—current issues: air pollution and resulting acid rain in the mineral extraction and refining region; chemical runoff into watersheds; loss of biodiversity; poaching seriously threatens rhinoceros, elephant, antelope, and large cat populations; deforestation; soil erosion; desertification; lack of adequate water treatment presents human health risks

Environment—international agreements: *party to:* Biodiversity, Climate Change, Climate Change-Kyoto Protocol, Desertification, Endangered Species, Hazardous Wastes, Law of the Sea, Ozone Layer Protection, Wetlands

signed, but not ratified: none of the selected agreements

Geography—note: landlocked; the Zambezi forms a natural riverine boundary with Zimbabwe; Lake Kariba on the Zambia-Zimbabwe border forms the world's largest reservoir by volume (180 cu km; 43 cu mi)

Population: 17,426,623 (July 2020 est.)
note: estimates for this country explicitly take into account the effects of excess mortality due to AIDS; this can result in lower life expectancy, higher infant mortality, higher death rates, lower population growth rates, and changes in the distribution of population by age and sex than would otherwise be expected
country comparison to the world: 66

Nationality: *noun:* Zambian(s)
adjective: Zambian

Ethnic groups: Bemba 21%, Tonga 13.6%, Chewa 7.4%, Lozi 5.7%, Nsenga 5.3%, Tumbuka 4.4%, Ngoni 4%, Lala 3.1%, Kaonde 2.9%, Namwanga 2.8%, Lunda (north Western) 2.6%, Mambwe 2.5%, Luvale 2.2%, Lamba 2.1%, Ushi 1.9%, Lenje 1.6%, Bisa 1.6%, Mbunda 1.2%, other 13.8%, unspecified 0.4% (2010 est.)

Languages: Bemba 33.4%, Nyanja 14.7%, Tonga 11.4%, Lozi 5.5%, Chewa 4.5%, Nsenga 2.9%, Tumbuka 2.5%, Lunda (North Western) 1.9%, Kaonde 1.8%, Lala 1.8%, Lamba 1.8%, English (official) 1.7%, Luvale 1.5%, Mambwe 1.3%, Namwanga 1.2%, Lenje 1.1%, Bisa 1%, other 9.7%, unspecified 0.2% (2010 est.)
note: Zambia is said to have over 70 languages, although many of these may be considered dialects; all of Zambia's major languages are members of the Bantu family; Chewa and Nyanja are mutually intelligible dialects

Religions: Protestant 75.3%, Roman Catholic 20.2%, other 2.7% (includes Muslim Buddhist, Hindu, and Baha'i), none 1.8% (2010 est.)

Demographic profile: Zambia's poor, youthful population consists primarily of Bantu-speaking people representing nearly 70 different ethnicities. Zambia's high fertility rate continues to drive rapid population growth, averaging almost 3 percent annually between 2000 and 2010. The country's total fertility rate has fallen by less than 1.5 children per woman during the last 30 years and still averages among the world's highest, almost 6 children per woman, largely because of the country's lack of access to family planning services, education for girls, and employment for women. Zambia also exhibits wide fertility disparities based on rural or urban location, education, and income. Poor, uneducated women from rural areas are more likely to marry young, to give birth early, and to have more children, viewing children as a sign of prestige and recognizing that not all of their children will live to adulthood. HIV/AIDS is prevalent in Zambia and contributes to its low life expectancy.

Zambian emigration is low compared to many other African countries and is comprised predominantly of the well- educated. The small amount of brain drain, however, has a major impact in Zambia because of its limited human capital and lack of educational infrastructure for developing skilled professionals in key fields. For example, Zambia has few schools for training doctors, nurses, and other health care workers. Its spending on education is low compared to other Sub-Saharan countries.

Age structure: *0-14 years:* 45.74% (male 4,005,134/female 3,964,969)

15-24 years: 20.03% (male 1,744,843/female 1,746,561)

25-54 years: 28.96% (male 2,539,697/female 2,506,724)

55-64 years: 3.01% (male 242,993/female 280,804)

65 years and over: 2.27% (male 173,582/female 221,316) (2020 est.)

Dependency ratios: *total dependency ratio:* 85.7
youth dependency ratio: 81.7
elderly dependency ratio: 4
potential support ratio: 25.3 (2020 est.)

Median age: *total:* 16.9 years
male: 16.7 years
female: 17 years (2020 est.)
country comparison to the world: 221

Population growth rate: 2.89% (2020 est.)
country comparison to the world: 10

Birth rate: 40.4 births/1,000 population (2020 est.)
country comparison to the world: 8

Death rate: 11. 6 deaths/1,000 population (2020 est.)
country comparison to the world: 19

Net migration rate: 0 migrant(s)/1,000 population (2020 est.)
country comparison to the world: 98

Population distribution: one of the highest levels of urbanization in Africa; high density in the central area, particularly around the cities of Lusaka, Ndola, Kitwe, and Mufulira as shown in this population distribution map

Urbanization: urban population: 44.6% of total population (2020)
rate of urbanization: 4.23% annual rate of change (2015-20 est.)

total population growth rate v. urban population growth rate, 2000-2030: Major urban areas—population: 2.774 million LUSAKA (capital) (2020)

Sex ratio: *at birth:* 1.03 male(s)/female
0-14 years: 1.01 male(s)/female
15-24 years: 1 male(s)/female
25-54 years: 1.01 male(s)/female
55-64 years: 0.87 male(s)/female
65 years and over: 0.78 male(s)/female
total population: 1 male(s)/female (2020 est.)

Mother's mean age at first birth: 19.2 years (2013/14 est.)
note: median age at first birth among women 25-29

Maternal mortality rate: 213 deaths/100,000 live births (2017 est.)
country comparison to the world: 46

Infant mortality rate: *total:* 56 deaths/1,000 live births
male: 61.1 deaths/1,000 live births
female: 50.7 deaths/1,000 live births (2020 est.)
country comparison to the world: 16

Life expectancy at birth: *total population:* 53.6 years
male: 51.9 years
female: 55.3 years (2020 est.)
country comparison to the world: 226

Total fertility rate: 5.49 children born/woman (2020 est.)
country comparison to the world: 10

Contraceptive prevalence rate: 49.5% (2018)

Drinking water source:

improved: *urban:* 89.5% of population
rural: 50.9% of population
total: 67.5% of population

unimproved: *urban:* 10.5% of population
rural: 49.1% of population
total: 32.5% of population (2017 est.)

Current Health Expenditure: 4.5% (2017)

Physicians density: 0.16 physicians/1,000 population (2016)

Hospital bed density: 2 beds/1,000 population (2010)

Sanitation facility access:

improved: *urban:* 69.6% of population
rural: 24.8% of population
total: 44.1% of population

unimproved: *urban:* 31.4% of population
rural: 75.2% of population
total: 55.9% of population (2017 est.)

HIV/AIDS—adult prevalence rate: 12.1% (2019 est.)
country comparison to the world: 8

HIV/AIDS—people living with HIV/AIDS: I. 2 million (2019 est.)
country comparison to the world: 9

HIV/AIDS—deaths: 17,000 (2019 est.)
country comparison to the world: 10

Major infectious diseases: *degree of risk:* very high (2020)
food or waterborne diseases: bacterial and protozoal diarrhea, hepatitis A, and typhoid fever
vectorborne diseases: malaria and dengue fever
water contact diseases: schistosomiasis
animal contact diseases: rabies

Obesity—adult prevalence rate: 8.1% (2016)
country comparison to the world: 155

Children under the age of 5 years underweight: II. 8% (2018/19)
country comparison to the world: 55
Education expenditures: NA

Literacy: *definition:* age 15 and over can read and write English
total population: 86.7%
male: 90.6%

female: 83.1% (2018)

Unemployment, youth ages 15-24: *total:* 24%
male: 23.6%
female: 24.4% (2017 est.)
country comparison to the world: 52

GOVERNMENT

Country name: *conventional long form:* Republic of Zambia
conventional short form: Zambia
former: Northern Rhodesia
etymology: name derived from the Zambezi River, which flows through the western part of the country and forms its southern border with neighboring Zimbabwe

Government type: presidential republic

Capital: *name:* Lusaka; note—a proposal to build a new capital city in Ngabwe was announced in May 2017

geographic coordinates: 15 25 S, 28 17 E
time difference: UTC+2 (7 hours ahead of Washington, DC, during Standard Time)
etymology: named after a village called Lusaka, located at Manda Hill, near where Zambia's National Assembly building currently stands; the village was named after a headman (chief) Lusakasa

Administrative divisions: 10 provinces; Central, Copperbelt, Eastern, Luapula, Lusaka, Muchinga, Northern, North-Western, Southern, Western

Independence: 24 October 1964 (from the UK)

National holiday: Independence Day, 24 October (1964)

Constitution: *history:* several previous; latest adopted 24 August 1991, promulgated 30 August 1991
amendments: proposed by the National Assembly; passage requires two-thirds majority vote by the Assembly in two separate readings at least 30 days apart; passage of amendments affecting fundamental rights and freedoms requires approval by at least one half of votes cast in a referendum prior to consideration and voting by the Assembly; amended 1996, 2015, 2016

Legal system: mixed legal system of English common law and customary law

International law organization participation: has not submitted an ICJ jurisdiction declaration; accepts ICCt jurisdiction

Citizenship: *citizenship by birth:* only if at least one parent is a citizen of Zambia
citizenship by descent only: yes, if at least one parent was a citizen of Zambia
dual citizenship recognized: yes
residency requirement for naturalization: 5 years for those with an ancestor who was a citizen of Zambia, otherwise 10 years residency is required

Suffrage: 18 years of age; universal

Executive branch: *chief of state:* President Edgar LUNGU (since 25 January 2015); Vice President Inonge WINA (since 26 January 2015);

note—the president is both chief of state and head of government

head of government: President Edgar LUNGU (since 25 January 2015); Vice President Inonge WINA (since 26 January 2015)

cabinet: Cabinet appointed by president from among members of the National Assembly

elections/appointments: president directly elected by absolute majority popular vote in 2 rounds if needed for a 5-year term (eligible for a second term); last held on 11 August 2016 (next to be held in 2021)

election results: Edgar LUNGU reelected president in the first round; percent of vote—Edgar LUNGU (PF) 50.4%, Hakainde HICHILEMA (UPND) 47.6%, other 2.0%

Legislative branch: *description:* unicameral National Assembly (165 seats; 156 members directly elected in single-seat constituencies by simple majority vote in 2 rounds if needed, and up to 8 appointed by the president; members serve 5-year terms); note—6 additional electoral seats were added for the 11 August 2016 election, up from 150 electoral seats in the 2011 election

elections: last held on 11 August 2016 (next to be held in 2021)

election results: percent of vote by party—PF 42%, UPND 41.7%, MMD 2.7%, FDD 2.2%, other 1.9%,independent 9.5%; seats by party—PF 89, UPND 54, MMD 5, FDD 1, NDC 1, independent 14; composition—men 135, women 30, percent of women 18.2%

Judicial branch: *highest courts:* Supreme Court (consists of the chief justice, deputy chief justice, and at least 11 judges); Constitutional Court (consists of the court president, vice president, and 11 judges); note—the Constitutional Court began operation in June 2016

judge selection and term of office: Supreme Court and Constitutional Court judges appointed by the president of the republic upon the advice of the 9-member Judicial Service Commission, which is headed by the chief justice, and ratified by the National Assembly; judges normally serve until age 65

subordinate courts: Court of Appeal; High Court; Industrial Relations Court; subordinate courts (3 levels, based on upper limit of money involved); Small Claims Court; local courts (2 grades, based on upper limit of money involved)

Political parties and leaders: Alliance for Democracy and Development or ADD [Charles MILUPI]

Forum for Democracy and Development or FDD [Edith NAWAKWI]

Movement for Multiparty Democracy or MMD [Felix MUTATI]

National Democratic Congress or NDC [Chishimba KAMBWILI]

Patriotic Front or PF [Edgar LUNGU]

United Party for National Development or UPND [Hakainde HICHILEMA]

International organization participation: ACP, AfDB, AU, C, COMESA, EITI (compliant country), FAO, G-77, IAEA, IBRD, ICAO, ICCt, ICRM, IDA, IFAD, IFC, IFRCS, ILO, IMF, Interpol, IOC, IOM, IPU, ISO (correspondent), ITSO, ITU, ITUC (NGOs), MIGA, MONUSCO, NAM, OPCW, PCA, SADC, UN, UNAMID, UNCTAD, UNESCO, UNHCR, UNIDO, UNISFA, UNMIL, UNMISS, UNOCI, UNWTO, UPU, WCO, WHO, WIPO, WMO, WTO

Diplomatic representation in the US: *chief of mission:* Ambassador Lazarous KAPAMBWE (since 8 April 2020)

chancery: 2200 R Street NW, Washington, DC 20008

telephone: [1] (202) 265-9717 through 9719

FAX: [1] (202) 332-0826

Diplomatic representation from the US: *chief of mission:* Charge d'Affaires David J. YOUNG (since 2 March 2020)

telephone: [260] (0) 211-357-000

embassy: Eastern end of Kabulonga Road, Ibex Hill, Lusaka

mailing address: P. O. Box 320065, Lusaka

FAX: [260] 211-357-224

Flag description: green field with a panel of three vertical bands of red (hoist side), black, and orange below a soaring orange eagle, on the outer edge of the flag; green stands for the country's natural resources and vegetation, red symbolizes the struggle for freedom, black the people of Zambia, and orange the country's mineral wealth; the eagle represents the people's ability to rise above the nation's problems

National symbol(s): African fish eagle; national colors: green, red, black, orange

National anthem: *name:* "Lumbanyeni Zambia" (Stand and Sing of Zambia, Proud and Free)

lyrics/music: multiple/Enoch Mankayi SONTONGA

note: adopted 1964; the melody, from the popular song "God Bless Africa," is the same as that of Tanzania but with different lyrics; the melody is also incorporated into South Africa's anthem

ECONOMY

Economy—overview: Zambia had one of the world's fastest growing economies for the ten years up to 2014, with real GDP growth averaging roughly 6.7% per annum, though growth slowed during the period 2015 to 2017, due to falling copper prices, reduced power generation, and depreciation of the kwacha. Zambia's lack of economic diversification and dependency on copper as its sole major export makes it vulnerable to fluctuations in the world commodities market and prices turned downward in 2015 due to declining demand from China; Zambia was overtaken by the Democratic Republic of Congo as Africa's largest copper producer. GDP growth picked up in 2017 as mineral prices rose.

Despite recent strong economic growth and its status as a lower middle-income country, widespread and extreme rural poverty and high unemployment levels remain significant problems, made worse by a high birth rate, a relatively high HIV/AIDS burden, by market-distorting agricultural and energy policies, and growing government debt. Zambia raised $7 billion from international investors by issuing separate sovereign bonds in 2012, 2014, and 2015. Concurrently, it issued over $4 billion in domestic debt and agreed to Chinese-financed infrastructure projects, significantly increasing the country's public debt burden to more than 60% of GDP. The government has considered refinancing $3 billion worth of Eurobonds and significant Chinese loans to cut debt servicing costs.

GDP (purchasing power parity): $68.93 billion (2017 est.)

$66.66 billion (2016 est.)

$64.25 billion (2015 est.)

note: data are in 2017 dollars

country comparison to the world: 102

GDP (official exchange rate): $25.71 billion (2017 est.)

GDP—real growth rate: 3.4% (2017 est.)

3.8% (2016 est.)

2.9% (2015 est.)

country comparison to the world: 88

GDP—per capita (PPP): $4,000 (2017 est.)

$4,000 (2016 est.)

$4,000 (2015 est.)

note: data are in 2017 dollars

country comparison to the world: 178

Gross national saving: 38.3% of GDP (2017 est.)

37.3% of GDP (2016 est.)

38.9% of GDP (2015 est.)

country comparison to the world: 11

GDP—composition, by end use: *household consumption:* 52.6% (2017 est.)

government consumption: 21% (2017 est.)

investment in fixed capital: 27.1% (2017 est.)

investment in inventories: 1.2% (2017 est.)

exports of goods and services: 43% (2017 est.)

imports of goods and services: -44.9% (2017 est.)

GDP—composition, by sector of origin: *agriculture:* 7.5% (2017 est.)

industry: 35.3% (2017 est.)

services: 57% (2017 est.)

Agriculture—products: corn, sorghum, rice, peanuts, sunflower seeds, vegetables, flowers, tobacco, cotton, sugarcane, cassava (manioc, tapioca), coffee; cattle, goats, pigs, poultry, milk, eggs, hides

Industries: copper mining and processing, emerald mining, construction, foodstuffs, beverages, chemicals, textiles, fertilizer, horticulture

Industrial production growth rate: 4.7% (2017 est.)

country comparison to the world: 62

Labor force: 6.898 million (2017 est.)

country comparison to the world: 66

Labor force—by occupation: *agriculture:* 54.8%

industry: 9.9%

services: 35.3% (2017 est.)

Unemployment rate: 15% (2008 est.)

50% (2000 est.)

country comparison to the world: 175

Population below poverty line: 54.4% (2015 est.)

Household income or consumption by percentage share: *lowest 10%:* 1.5%
highest 10%: 47.4% (2010)

Budget: *revenues:* 4.473 billion (2017 est.)
expenditures: 6.357 billion (2017 est.)

Taxes and other revenues: 17.4% (of GDP) (2017 est.)
country comparison to the world: 169

Budget surplus (+) or deficit (-): -7.3% (of GDP) (2017 est.)
country comparison to the world: 195

Public debt: 63.1% of GDP (2017 est.)
60.7% of GDP (2016 est.)
country comparison to the world: 66

Fiscal year: calendar year

Inflation rate (consumer prices): 6.6% (2017 est.)
17.9% (2016 est.)
country comparison to the world: 192

Current account balance: -$1.006 billion (2017 est.)
-$934 million (2016 est.)
country comparison to the world: 145

Exports: $8.216 billion (2017 est.)
$6.514 billion (2016 est.)
country comparison to the world: 96

Exports—partners: Switzerland 44.8%, China 16.1%, Democratic Republic of the Congo 6.2%, Singapore 6%, South Africa 5.9% (2017)

Exports—commodities: copper/cobalt, cobalt, electricity; tobacco, flowers, cotton

Imports: $7.852 billion (2017 est.)
$6.539 billion (2016 est.)
country comparison to the world: 112

Imports—commodities: machinery, transportation equipment, petroleum products, electricity, fertilizer, foodstuffs, clothing

Imports—partners: South Africa 28.2%, Democratic Republic of the Congo 20.8%, China 12.9%, Kuwait 5.4%, UAE 4.6% (2017)

Reserves of foreign exchange and gold: $2.082 billion (31 December 2017 est.)
$2.353 billion (31 December 2016 est.)
country comparison to the world: 121

Debt—external: $11.66 billion (31 December 2017 est.)
$9.562 billion (31 December 2016 est.)
country comparison to the world: 108

Exchange rates: Zambian kwacha (ZMK) per US dollar—
9.2 (2017 est.)
10.3 (2016 est.)
10.3 (2015 est.)
8.6 (2014 est.)
6.2 (2013 est.)

ENERGY

Electricity access: *population without electricity:* 11 million (2019)
electrification—total population: 37% (2019)

electrification—urban areas: 76% (2019)
electrification—rural areas: 6% (2019)

Electricity—production: 11. 55 billion kWh (2016 est.)
country comparison to the world: 98

Electricity—consumption: 11.04 billion kWh (2016 est.)
country comparison to the world: 91

Electricity—exports: 1.176 billion kWh (2015 est.)
country comparison to the world: 56

Electricity—imports: 2.185 billion kWh (2016 est.)
country comparison to the world: 56

Electricity—installed generating capacity: 2.573 million kW (2016 est.)
country comparison to the world: 107

Electricity—from fossil fuels: 5% of total installed capacity (2016 est.)
country comparison to the world: 205

Electricity—from nuclear fuels: 0% of total installed capacity (2017 est.)
country comparison to the world: 214

Electricity—from hydroelectric plants: 93% of total installed capacity (2017 est.)
country comparison to the world: 9

Electricity—from other renewable sources: 2% of total installed capacity (2017 est.)
country comparison to the world: 147

Crude oil—production: 0 bbl/day (2018 est.)
country comparison to the world: 219

Crude oil—exports: 0 bbl/day (2015 est.)
country comparison to the world: 217

Crude oil—imports: 12,860 bbl/day (2015 est.)
country comparison to the world: 70

Crude oil—proved reserves: 0 bbl (1 January 2018 est.)
country comparison to the world: 214

Refined petroleum products—production: 13,120 bbl/day (2015 est.)
country comparison to the world: 98

Refined petroleum products—consumption: 23,000 bbl/day (2016 est.)
country comparison to the world: 133

Refined petroleum products—exports: 371 bbl/day (2015 est.)
country comparison to the world: 113

Refined petroleum products—imports: 10,150 bbl/day (2015 est.)
country comparison to the world: 149

Natural gas—production: 0 cu m (2017 est.)
country comparison to the world: 216

Natural gas—consumption: 0 cu m (2017 est.)
country comparison to the world: 215

Natural gas—exports: 0 cu m (2017 est.)
country comparison to the world: 215

Natural gas—imports: 0 cu m (2017 est.)
country comparison to the world: 215

Natural gas—proved reserves: 0 cu m (1 January 2014 est.)
country comparison to the world: 209

Carbon dioxide emissions from consumption of energy: 3.777 million Mt (2017 est.)
country comparison to the world: 140

COMMUNICATIONS

Telephones—fixed lines: *total subscriptions:* 91,422
subscriptions per 100 inhabitants: less than 1 (2019 est.)
country comparison to the world: 140

Telephones—mobile cellular: *total subscriptions:* 16,322,168
subscriptions per 100 inhabitants: 96.41 (2019 est.)
country comparison to the world: 65

Telecommunication systems: *general assessment:* service is among the best in Sub-Saharan Africa; regulatory promotes competition and is a partner to private sector service providers, offering mobile voice and Internet at some of the lowest prices in the region; investment made in data centers, education centers and computer assembly training plants; operators invest in 3G and LTE-based services; Chinese company Huawei is helping to upgrade state-owned mobile infrastructure for 5G services; 3 cellular telephone providers currently in operation, plus several data only ISPs; 1,010 towers project to soon be completed (2020)
domestic: fiber optic connections are available between most larger towns and cities with microwave radio relays serving more rural areas; 3G and LTE with FttX in limited urban areas and private Ku or Ka band VSAT terminals in remote locations; fixed-line 1 per 100 and mobile-cellular 96 per 100 (2019)
international: country code—260; multiple providers operate overland fiber optic routes via Zimbabwe/South Africa, Botswana/Namibia and Tanzania provide access to the major undersea cables
note: the COVID-19 outbreak is negatively impacting telecommunications production and supply chains globally; consumer spending on telecom devices and services has also slowed due to the pandemic's effect on economies worldwide; overall progress towards improvements in all facets of the telecom industry—mobile, fixed-line, broadband, submarine cable and satellite—has moderated

Broadcast media: according to the Independent Broadcast Authority, there are 137 radio stations and 47 television stations in Zambia; out of the 137 radio stations, 133 are private (categorized as either commercial or community radio stations), while 4 are public- owned; state-owned Zambia National Broadcasting Corporation (ZNBC) operates 2 television channels and 3 radio stations; ZNBC owns 75% shares in GoTV, 40% in MultiChoice, and 40% in TopStar Communications Company, all of which operate in-country (2019)

Internet country code: .zm

Internet users: *total:* 2,351,646

percent of population: 14.3% (July 2018 est.)
country comparison to the world: 113

Broadband—fixed subscriptions: *total:* 43,365
subscriptions per 100 inhabitants: less than
1 (2018 est.)
country comparison to the world: 137

TRANSPORTATION

National air transport system: *number of
registered air carriers:* 3 (2020)
*inventory of registered aircraft operated by air
carriers:* 6
annual passenger traffic on registered air carriers: 8,904 (2018)

annual freight traffic on registered air carriers:
75.08 million mt-km (2018)

Civil aircraft registration country code prefix: 9J
(2016)

Airports: 88 (2013)
country comparison to the world: 64

Airports—with paved runways: *total:* 8 (2013)
over 3,047 m: 1 (2013)
2,438 to 3,047 m: 3 (2013)
1,524 to 2,437 m: 3 (2013)
914 to 1,523 m: 1 (2013)

Airports—with unpaved runways: *total:* 80 (2013)
2,438 to 3,047 m: 1 (2013)
1,524 to 2,437 m: 5 (2013)
914 to 1,523 m: 53 (2013)
under 914 m: 21 (2013)
Pipelines: 771 km oil (2013)

Railways: *total:* 3,126 km (2014)

narrow gauge: 3,126 km 1.067-m gauge (2014)

note: includes 1,860 km of the Tanzania-Zambia
Railway Authority (TAZARA)
country comparison to the world: 59

Roadways: *total:* 67,671 km (2018)
paved: 14,888 km (2018)
unpaved: 52,783 km (2018)
country comparison to the world: 72

Waterways: 2,250 km (includes Lake Tanganyika
and the Zambezi and Luapula Rivers) (2010)
country comparison to the world: 38

Merchant marine: *total:* 1
by type: other 1 (2019)
country comparison to the world: 182

Ports and terminals: *river port(s):* Mpulungu
(Zambezi)

MILITARY AND SECURITY

Military and security forces: Zambia Defense
Force (ZDF): Zambia Army, Zambia Air Force,
Zambia National Service (support organization);
the Zambia Police includes a paramilitary battalion (2019)

Military expenditures: 1.2% of GDP (2019)
1.3% of GDP (2018)
1.3% of GDP (2017)
1.4% of GDP (2016)
1.8% of GDP (2015)
country comparison to the world: 108

Military and security service personnel strengths:
the Zambia Defense Force (ZDF) has an estimated
17,000 active troops (15,500 Army; 1,500 Air);
1,400 paramilitary Police (2019 est.)

Military equipment inventories and acquisitions:
the ZDF's inventory is largely comprised of Soviet-
era and older Chinese- and Russian-origin equipment; since 2010, China is the leading supplier of
arms to Zambia (2019 est.)

Military deployments: 920 Central African
Republic (MINUSCA) (2020)

Military service age and obligation: 18-25 years
of age for male and female voluntary military service; no conscription; 12-year enlistment period (7
years active, 5 in the Reserves) (2019)

TRANSNATIONAL ISSUES

Disputes—international: in 2004, Zimbabwe
dropped objections to plans between Botswana
and Zambia to build a bridge over the Zambezi
River, thereby de facto recognizing a short, but not
clearly delimited, Botswana-Zambia boundary in
the river

Refugees and internally displaced persons:
refugees (country of origin): 55,523 (Democratic
Republic of the Congo) (refugees and asylum
seekers), 18,815 (Angola), 7,997 (Burundi), 5,982
(Rwanda) (2020)

Illicit drugs: transshipment point for moderate
amounts of methaqualone, small amounts of heroin, and cocaine bound for southern Africa and
possibly Europe; a poorly developed financial
infrastructure coupled with a government commitment to combating money laundering make it
an unattractive venue for money launderers; major
consumer of cannabis

ZIMBABWE

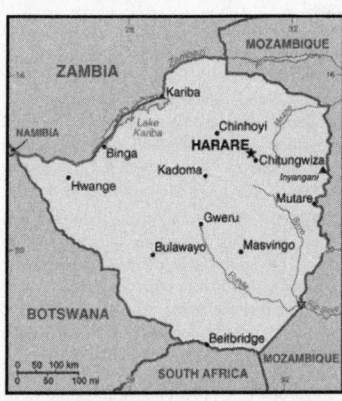

INTRODUCTION

Background: A series of trading states developed
in the area of Zimbabwe prior to the arrival of
the first European explorers; the largest of these
was the Kingdom of Zimbabwe (ca. 1220-1450).

In the 1880s, European colonists arrived with the
British South Africa Company (BSAC), which
obtained mining rights and established company
rule over the area. The southern portion of BSAC
holdings were annexed by the UK in 1923 and
became the British colony of Southern Rhodesia.
A 1961 constitution was formulated that favored
whites in power. In 1965 the government unilaterally declared its independence, but the UK
did not recognize the act and demanded more
complete voting rights for the black African
majority in the country (then called Rhodesia).
UN sanctions and a guerrilla uprising finally
led to free elections in 1979 and independence
(as Zimbabwe) in 1980. Robert MUGABE, the
nation's first prime minister, was the country's
only ruler (as president since 1987) from independence until his resignation in November 2017.
His chaotic land redistribution campaign, which
began in 1997 and intensified after 2000, caused
an exodus of white farmers, crippled the economy,
and ushered in widespread shortages of basic commodities. Ignoring international condemnation,
MUGABE rigged the 2002 presidential election
to ensure his reelection.

In 2005, the capital city of Harare embarked
on Operation Restore Order, ostensibly an
urban rationalization program, which resulted
in the destruction of the homes or businesses of
700,000 mostly poor supporters of the opposition.
MUGABE in 2007 instituted price controls on
all basic commodities causing panic buying and
leaving store shelves empty for months. General
elections in both 2008 and 2013 were severely
flawed and widely condemned, but allowed
MUGABE to remain president. As a prerequisite
to holding the 2013 election, Zimbabwe enacted
a new constitution by referendum, although many
provisions in the new constitution have yet to be
codified in law. In November 2017, Vice President
Emmerson MNANGAGWA took over following
a military intervention that forced MUGABE to
resign. MNANGAGWA was inaugurated president days later, promising to hold presidential
elections in 2018. In July 2018, MNANGAGWA
won the presidential election after a close contest
with Movement for Democratic Change Alliance
candidate Nelson CHAMISA. MNANGAGWA
has since resorted to the government's longstanding practice of violently disrupting protests or

opposition rallies. Official inflation rates soared in 2019, approaching 500% by the end of the year. MUGABE died in September 2019.

GEOGRAPHY

Location: Southern Africa, between South Africa and Zambia

Geographic coordinates: 20 00 S, 30 00 E

Map references: Africa

Area: *total:* 390,757 sq km
land: 386,847 sq km
water: 3,910 sq km
country comparison to the world: 62

Area—comparative: about four times the size of Indiana; slightly larger than Montana

Land boundaries: *total:* 3,229 km
border countries (4): Botswana 834 km, Mozambique 1402 km, South Africa 230 km, Zambia 763 km

Coastline: 0 km (landlocked)

Maritime claims: none (landlocked)

Climate: tropical; moderated by altitude; rainy season (November to March)

Terrain: mostly high plateau with higher central plateau (high veld); mountains in east

Elevation: *mean elevation:* 961 m
lowest point: junction of the Runde and Save Rivers 162 m
highest point: Inyangani 2,592 m

Natural resources: coal, chromium ore, asbestos, gold, nickel, copper, iron ore, vanadium, lithium, tin, platinum group metals

Land use: *agricultural land:* 42.5% (2011 est.)
arable land: 10.9% (2011 est.) / permanent crops: 0.3% (2011 est.) / permanent pasture: 31.3% (2011 est.)
forest: 39.5% (2011 est.)
other: 18% (2011 est.)
Irrigated land: 1,740 sq km (2012)

Population distribution: Aside from major urban agglomerations in Harare and Bulawayo, population distribution is fairly even, with slightly greater overall numbers in the eastern half as shown in this population distribution map

Natural hazards: recurring droughts; floods and severe storms are rare

Environment—current issues: deforestation; soil erosion; land degradation; air and water pollution; the black rhinoceros herd—once the largest concentration of the species in the world—has been significantly reduced by poaching; poor mining practices have led to toxic waste and heavy metal pollution

Environment—international agreements: *party to:* Biodiversity, Climate Change, Desertification, Endangered Species, Hazardous Wastes, Law of the Sea, Ozone Layer Protection
signed, but not ratified: none of the selected agreements

Geography—note: landlocked; the Zambezi forms a natural riverine boundary with Zambia; in full flood (February-April) the massive Victoria Falls on the river forms the world's largest curtain of falling water; Lake Kariba on the Zambia-Zimbabwe border forms the world's largest reservoir by volume (180 cu km; 43 cu mi)

PEOPLE AND SOCIETY

Population: 14,546,314 (July 2020 est.)
note: estimates for this country explicitly take into account the effects of excess mortality due to AIDS; this can result in lower life expectancy, higher infant mortality, higher death rates, lower population growth rates, and changes in the distribution of population by age and sex than would otherwise be expected
country comparison to the world: 73

Nationality: *noun:* Zimbabwean(s)
adjective: Zimbabwean

Ethnic groups: African 99.4% (predominantly Shona; Ndebele is the second largest ethnic group), other 0.4%, unspecified 0.2% (2012 est.)

Languages: Shona (official; most widely spoken), Ndebele (official, second most widely spoken), English (official; traditionally used for official business), 13 minority languages (official; includes Chewa, Chibarwe, Kalanga, Koisan, Nambya, Ndau, Shangani, sign language, Sotho, Tonga, Tswana, Venda, and Xhosa)

Religions: Protestant 74.8% (includes Apostolic 37.5%, Pentecostal 21.8%, other 15.5%), Roman Catholic 7.3%, other Christian 5.3%, traditional 1.5%, Muslim 0.5%, other 0.1%, none 10.5% (2015 est.)

Demographic profile: Zimbabwe's progress in reproductive, maternal, and child health has stagnated in recent years. According to a 2010 Demographic and Health Survey, contraceptive use, the number of births attended by skilled practitioners, and child mortality have either stalled or somewhat deteriorated since the mid-2000s. Zimbabwe's total fertility rate has remained fairly stable at about 4 children per woman for the last two decades, although an uptick in the urban birth rate in recent years has caused a slight rise in the country's overall fertility rate. Zimbabwe's HIV prevalence rate dropped from approximately 29% to 15% since 1997 but remains among the world's highest and continues to suppress the country's life expectancy rate. The proliferation of HIV/AIDS information and prevention programs and personal experience with those suffering or dying from the disease have helped to change sexual behavior and reduce the epidemic.

Historically, the vast majority of Zimbabwe's migration has been internal – a rural-urban flow. In terms of international migration, over the last 40 years Zimbabwe has gradually shifted from being a destination country to one of emigration and, to a lesser degree, one of transit (for East African illegal migrants traveling to South Africa). As a British colony, Zimbabwe attracted significant numbers of permanent immigrants from the UK and other European countries, as well as temporary economic migrants from Malawi, Mozambique, and Zambia.

Although Zimbabweans have migrated to South Africa since the beginning of the 20th century to work as miners, the first major exodus from the country occurred in the years before and after independence in 1980. The outward migration was politically and racially influenced; a large share of the white population of European origin chose to leave rather than live under a new black-majority government.

In the 1990s and 2000s, economic mismanagement and hyperinflation sparked a second, more diverse wave of emigration. This massive out migration – primarily to other southern African countries, the UK, and the US – has created a variety of challenges, including brain drain, illegal migration, and human smuggling and trafficking. Several factors have pushed highly skilled workers to go abroad, including unemployment, lower wages, a lack of resources, and few opportunities for career growth.

Age structure: *0-14 years:* 38.32% (male 2,759,155/female 2,814,462)
15-24 years: 20.16% (male 1,436,710/female 1,495,440)
25-54 years: 32.94% (male 2,456,392/female 2,334,973)
55-64 years: 4.07% (male 227,506/female 363,824)
65 years and over: 4.52% (male 261,456/female 396,396) (2020 est.)

Dependency ratios: *total dependency ratio:* 81.6
youth dependency ratio: 76.1
elderly dependency ratio: 5.5
potential support ratio: 18.3 (2020 est.)

Median age: *total:* 20.5 years
male: 20.3 years
female: 20.6 years (2020 est.)
country comparison to the world: 190

Population growth rate: 1.87% (2020 est.)
country comparison to the world: 53

Birth rate: 33.6 births/1,000 population (2020 est.)
country comparison to the world: 24

Death rate: 9.3 deaths/1,000 population (2020 est.)
country comparison to the world: 53

Net migration rate: -5 migrant(s)/1,000 population (2020 est.)
country comparison to the world: 198

Population distribution: Aside from major urban agglomerations in Harare and Bulawayo, population distribution is fairly even, with slightly greater overall numbers in the eastern half as shown in this population distribution map
Urbanization: urban population: 32.2% of total population (2020)
rate of urbanization: 2.19% annual rate of change (2015-20 est.)

total population growth rate v. urban population growth rate, 2000-2030: Major urban areas—population: 1.530 million HARARE (capital) (2020)

Sex ratio: *at birth:* 1.03 male(s)/female
0-14 years: 0.98 male(s)/female
15-24 years: 0.96 male(s)/female
25-54 years: 1.05 male(s)/female
55-64 years: 0.63 male(s)/female

65 years and over: 0.66 male(s)/female

total population: 0.96 male(s)/female (2020 est.)

Mother's mean age at first birth: 20 years (2015 est.)

note: median age at first birth among women 25-29

Maternal mortality rate: 458 deaths/100,000 live births (2017 est.)

country comparison to the world: 24

Infant mortality rate: total: 30.3 deaths/1,000 live births

male: 34.2 deaths/1,000 live births

female: 26.4 deaths/1,000 live births (2020 est.)

country comparison to the world: 55

Life expectancy at birth: total population: 62.3 years

male: 60.2 years

female: 64.5 years (2020 est.)

country comparison to the world: 210

Total fertility rate: 3.93 children born/woman (2020 est.)

country comparison to the world: 31

Contraceptive prevalence rate: 66.8% (2015)

Drinking water source:

improved: urban: 98% of population

rural: 67.4% of population

total: 77.3% of population

unimproved: urban: 2% of population

rural: 32.6% of population

total: 22.7% of population (2017 est.)

Current Health Expenditure: 6.6% (2017)

Physicians density: 0.19 physicians/1,000 population (2017)

Hospital bed density: 1.7 beds/1,000 population (2011)

Sanitation facility access:

improved: urban: 96.1% of population

rural: 49% of population

total: 64.2% of population

unimproved: urban: 3.9% of population

rural: 51% of population

total: 35.8% of population (2017 est.)

HIV/AIDS—adult prevalence rate: 13.4% (2019 est.)

country comparison to the world: 5

HIV/AIDS—people living with HIV/AIDS: 1.4 million (2019 est.)

country comparison to the world: 8

HIV/AIDS—deaths: 20,000 (2019 est.)

country comparison to the world: 9

Major infectious diseases: degree of risk: high (2020)

food or waterborne diseases: bacterial and protozoal diarrhea, hepatitis A, and typhoid fever

vectorborne diseases: malaria and dengue fever

water contact diseases: schistosomiasis

animal contact diseases: rabies

Obesity—adult prevalence rate: 15.5% (2016)

country comparison to the world: 126

Children under the age of 5 years underweight: 9.7% (2019)

country comparison to the world: 64

Education expenditures: 6.1% of GDP (2014)

country comparison to the world: 31

Literacy: definition: age 15 and over can read and write English

total population: 86.5%

male: 88.5%

female: 84.6% (2015)

School life expectancy (primary to tertiary education): total: 11 years

male: 12 years

female: 11 years (2013)

Unemployment, youth ages 15-24: total: 16.5%

male: 11.6%

female: 21.2% (2014 est.)

country comparison to the world: 83

GOVERNMENT

Country name: conventional long form: Republic of Zimbabwe

conventional short form: Zimbabwe

former: Southern Rhodesia, Rhodesia, Zimbabwe-Rhodesia

etymology: takes its name from the Kingdom of Zimbabwe (13th-15th century) and its capital of Great Zimbabwe, the largest stone structure in pre-colonial southern Africa

Government type: presidential republic

Capital: name: Harare

geographic coordinates: 17 49 S, 31 02 E

time difference: UTC+2 (7 hours ahead of Washington, DC, during Standard Time)

etymology: named after a village of Harare at the site of the present capital; the village name derived from a Shona chieftain, Ne-haraya, whose name meant "he who does not sleep"

Administrative divisions: 8 provinces and 2 cities* with provincial status; Bulawayo*, Harare*, Manicaland, Mashonaland Central, Mashonaland East, Mashonaland West, Masvingo, Matabeleland North, Matabeleland South, Midlands

Independence: 18 April 1980 (from the UK)

National holiday: Independence Day, 18 April (1980)

Constitution: history: previous 1965 (at Rhodesian independence), 1979 (Lancaster House Agreement), 1980 (at Zimbabwean independence); latest final draft completed January 2013, approved by referendum 16 March 2013, approved by Parliament 9 May 2013, effective 22 May 2013

amendments: proposed by the Senate or by the National Assembly; passage requires two-thirds majority vote by the membership of both houses of Parliament and assent of the president of the republic; amendments to constitutional chapters on fundamental human rights and freedoms and on agricultural lands also require approval by a majority of votes cast in a referendum; amended many times, last in 2017

Legal system: mixed legal system of English common law, Roman-Dutch civil law, and customary law

International law organization participation: has not submitted an ICJ jurisdiction declaration; non-party state to the ICCt

Citizenship: citizenship by birth: no

citizenship by descent only: the father must be a citizen of Zimbabwe; in the case of a child born out of wedlock, the mother must be a citizen

dual citizenship recognized: no

residency requirement for naturalization: 5 years

Suffrage: 18 years of age; universal

Executive branch: chief of state: President Emmerson Dambudzo MNANGAGWA (since 24 November 2017); First Vice President Constantino CHIWENGA (since 28 December 2017); note—Robert Gabriel MUGABE resigned on 21 November 2017, after ruling for 37 years

head of government: President Emmerson Dambudzo MNANGAGWA (since 24 November 2017); Vice President Constantino CHIWENGA (since 28 December 2017); Vice President Kembo MOHADI (since 28 December 2017)

cabinet: Cabinet appointed by president, responsible to National Assembly

elections/appointments: each presidential candidate nominated with a nomination paper signed by at least 10 registered voters (at least 1 candidate from each province) and directly elected by absolute majority popular vote in 2 rounds if needed for a 5-year term (no term limits); election last held on 3 July 2018 (next to be held in 2023); co-vice presidents drawn from party leadership

election results: Emmerson MNANGAGWA reelected president in 1st round of voting; percent of vote—Emmerson MNANGAGWA (ZANU-PF) 50.8%, Nelson CHAMISA (MDC-T) 44.3%, Thokozani KHUPE (MDC-N) .9%, other 3%

Legislative branch: description: bicameral Parliament consists of:

Senate (80 seats; 60 members directly elected in multi-seat constituencies—6 seats in each of the 10 provinces—by proportional representation vote, 16 indirectly elected by the regional governing councils, 2 reserved for the National Council Chiefs, and 2 reserved for members with disabilities; members serve 5-year terms)

National Assembly (270 seats; 210 members directly elected in single-seat constituencies by simple majority vote and 60 seats reserved for women directly elected by proportional representation vote; members serve 5-year terms)

elections: Senate—last held for elected member on 30 July 2018 (next to be held in 2023)

National Assembly—last held on 30 July 2018 (next to be held in 2023)

election results: Senate—percent of vote by party—NA; seats by party—ZANU-PF 34, MDC Alliance 25, Chiefs 18, people with disabilities 2, MDC-T 1; composition—men 45, women 35, percent of women 43.8%

National Assembly—percent of vote by party—NA; seats by party—ZANU PF 179, MDC Alliance 88, MDC-T 1, NPF 1, independent 1; composition—men 185, women 25, percent of women 31.5%; note—total Parliament percent of women 34.3%

Judicial branch: *highest courts:* Supreme Court (consists of the chief justice and 4 judges); Constitutional Court (consists of the chief and deputy chief justices and 9 judges)

judge selection and term of office: Supreme Court judges appointed by the president upon recommendation of the Judicial Service Commission, an independent body consisting of the chief justice, Public Service Commission chairman, attorney general, and 2-3 members appointed by the president; judges normally serve until age 65 but can elect to serve until age 70; Constitutional Court judge appointment NA; judges serve nonrenewable 15-year terms

subordinate courts: High Court; Labor Court; Administrative Court; regional magistrate courts; customary law courts; special courts

Political parties and leaders: MDC Alliance [Thokozane KHUPEIS] (acting)

Movement for Democratic Change—MDC-T [Thokozani KHUPE]

National People's Party or NPP [Joyce MUJURU] (formerly Zimbabwe People First or ZimPF)

National Patriotic Front or NPF [Ambrose MUTINHIRI]

Zimbabwe African National Union-Patriotic Front or ZANU-PF [Emmerson Dambudzo MNANGAGWA]

Zimbabwe African Peoples Union or ZAPU [Isaac MABUKA]

International organization participation: ACP, AfDB, AU, COMESA, FAO, G-15, G-77, IAEA, IBRD, ICAO, ICRM, IDA, IFAD, IFC, IFRCS, ILO, IMF, IMO, Interpol, IOC, IOM, IPU, ISO, ITSO, ITU, ITUC (NGOs), MIGA, NAM, OPCW, PCA, SADC, UN, UNAMID, UNCTAD, UNESCO, UNIDO, UNMIL, UNMISS, UNOCI, UNWTO, UPU, WCO, WFTU (NGOs), WHO, WIPO, WMO, WTO

Diplomatic representation in the US: *chief of mission:* Ambassador Ammon MUTEMBWA (since 18 November 2014)

chancery: 1608 New Hampshire Avenue NW, Washington, DC 20009

telephone: [1] (202) 332-7100

FAX: [1] (202) 483-9326

Diplomatic representation from the US: *chief of mission:* Ambassador Brian A. NICHOLS (since 19 July 2018)

telephone: [263] (0) 867-701-1000

embassy: 2 Lorraine Drive, Bluffhill, Harare

mailing address: P.O. Box 3340, Harare

FAX: [263] (4) 796-488

Flag description: seven equal horizontal bands of green (top), yellow, red, black, red, yellow, and green with a white isosceles triangle edged in black with its base on the hoist side; a yellow Zimbabwe bird representing the long history of the country is superimposed on a red five-pointed star in the center of the triangle, which symbolizes peace; green represents agriculture, yellow mineral wealth, red the blood shed to achieve

independence, and black stands for the native people

National symbol(s): Zimbabwe bird symbol, African fish eagle, flame lily; national colors: green, yellow, red, black, white

National anthem: *name:* "Kalibusiswe Ilizwe leZimbabwe" [Northern Ndebele language] "Simudzai Mureza WeZimbabwe" [Shona] (Blessed Be the Land of Zimbabwe)

lyrics/music: Solomon MUTSWAIRO/Fred Lecture CHANGUNDEGA

note: adopted 1994

ECONOMY

Economy—overview: Zimbabwe's economy depends heavily on its mining and agriculture sectors. Following a contraction from 1998 to 2008, the economy recorded real growth of more than 10% per year in the period 2010-13, before falling below 3% in the period 2014-17, due to poor harvests, low diamond revenues, and decreased investment. Lower mineral prices, infrastructure and regulatory deficiencies, a poor investment climate, a large public and external debt burden, and extremely high government wage expenses impede the country's economic performance.

Until early 2009, the Reserve Bank of Zimbabwe (RBZ) routinely printed money to fund the budget deficit, causing hyperinflation. Adoption of a multi-currency basket in early 2009—which allowed currencies such as the Botswana pula, the South Africa rand, and the US dollar to be used locally—reduced inflation below 10% per year. In January 2015, as part of the government's effort to boost trade and attract foreign investment, the RBZ announced that the Chinese renminbi, Indian rupee, Australian dollar, and Japanese yen would be accepted as legal tender in Zimbabwe, though transactions were predominantly carried out in US dollars and South African rand until 2016, when the rand's devaluation and instability led to near-exclusive use of the US dollar. The government in November 2016 began releasing bond notes, a parallel currency legal only in Zimbabwe which the government claims will have a one-to-one exchange ratio with the US dollar, to ease cash shortages. Bond notes began trading at a discount of up to 10% in the black market by the end of 2016.

Zimbabwe's government entered a second Staff Monitored Program with the IMF in 2014 and undertook other measures to reengage with international financial institutions. Zimbabwe repaid roughly $108 million in arrears to the IMF in October 2016, but financial observers note that Zimbabwe is unlikely to gain new financing because the government has not disclosed how it plans to repay more than $1.7 billion in arrears to the World Bank and African Development Bank. International financial institutions want Zimbabwe to implement significant fiscal and structural reforms before granting new loans. Foreign and domestic investment continues to be hindered by the lack of land tenure and titling, the inability to repatriate dividends to investors

overseas, and the lack of clarity regarding the government's Indigenization and Economic Empowerment Act.

GDP (purchasing power parity): $34.27 billion (2017 est.)

$33.04 billion (2016 est.)

$32.82 billion (2015 est.)

note: data are in 2017 dollars

country comparison to the world: 127

GDP (official exchange rate): $17.64 billion (2017 est.)

GDP—real growth rate: 3.7% (2017 est.)

0.7% (2016 est.)

1.4% (2015 est.)

country comparison to the world: 81

GDP—per capita (PPP): $2,300 (2017 est.)

$2,300 (2016 est.)

$2,300 (2015 est.)

note: data are in 2017 dollars

country comparison to the world: 203

Gross national saving: 23.3% of GDP (2017 est.)

19.1% of GDP (2016 est.)

8% of GDP (2015 est.)

country comparison to the world: 74

GDP—composition, by end use: *household consumption:* 77.6% (2017 est.)

government consumption: 24% (2017 est.)

investment in fixed capital: 12.6% (2017 est.)

investment in inventories: 0% (2017 est.)

exports of goods and services: 25.6% (2017 est.)

imports of goods and services: -39.9% (2017 est.)

GDP—composition, by sector of origin:

agriculture: 12% (2017 est.)

industry: 22.2% (2017 est.)

services: 65.8% (2017 est.)

Agriculture—products: tobacco, corn, cotton, wheat, coffee, sugarcane, peanuts; sheep, goats, pigs

Industries: mining (coal, gold, platinum, copper, nickel, tin, diamonds, clay, numerous metallic and nonmetallic ores), steel; wood products, cement, chemicals, fertilizer, clothing and footwear, foodstuffs, beverages

Industrial production growth rate: 0.3% (2017 est.)

country comparison to the world: 167

Labor force: 7.907 million (2017 est.)

country comparison to the world: 61

Labor force—by occupation: *agriculture:* 67.5%

industry: 7.3%

services: 25.2% (2017 est.)

Unemployment rate: 11.3% (2014 est.)

80% (2005 est.)

note: data include both unemployment and underemployment; true unemployment is unknown and, under current economic conditions, unknowable

country comparison to the world: 158

Population below poverty line: 72.3% (2012 est.)

Household income or consumption by percentage share: *lowest 10%:* 2%

highest 10%: 40.4% (1995)

Budget: *revenues:* 3.8 billion (2017 est.)

expenditures: 5.5 billion (2017 est.)

Taxes and other revenues: 21.5% (of GDP) (2017 est.)
country comparison to the world: 138

Budget surplus (+) or deficit (-): -9.6% (of GDP) (2017 est.)
country comparison to the world: 208

Public debt: 82.3% of GDP (2017 est.)
69.9% of GDP (2016 est.)
country comparison to the world: 34

Fiscal year: calendar year

Inflation rate (consumer prices): 0.9% (2017 est.)
-1.6% (2016 est.)
country comparison to the world: 49

Current account balance: -$716 million (2017 est.)
-$553 million (2016 est.)
country comparison to the world: 134

Exports: $4.353 billion (2017 est.)
$3.366 billion (2016 est.)
country comparison to the world: 113

Exports—partners: South Africa 50.3%, Mozambique 22.5%, UAE 9.8%, Zambia 4.9% (2017)

Exports—commodities: platinum, cotton, tobacco, gold, ferroalloys, textiles/clothing

Imports: $5.472 billion (2017 est.)
$5.236 billion (2016 est.)
country comparison to the world: 122

Imports—commodities: machinery and transport equipment, other manufactures, chemicals, fuels, food products

Imports—partners: South Africa 47.8%, Zambia 20.5% (2017)

Reserves of foreign exchange and gold: $431.8 million (31 December 2017 est.)
$407.2 million (31 December 2016 est.)
country comparison to the world: 157

Debt—external: $9.357 billion (31 December 2017 est.)
$10.14 billion (31 December 2016 est.)
country comparison to the world: 116

Exchange rates: Zimbabwean dollars (ZWD) per US dollar—
1 (2017 est.)
1 (2016 est.)
(2013)
234.25 (2010)
note: the dollar was adopted as a legal currency in 2009; since then the Zimbabwean dollar has experienced hyperinflation and is essentially worthless

ENERGY

Electricity access: *population without electricity:* 7 million (2019)
electrification—total population: 53% (2019)
electrification—urban areas: 89% (2019)
electrification—rural areas: 36% (2019)

Electricity—production: 6.8 billion kWh (2016 est.)
country comparison to the world: 113

Electricity—consumption: 7.118 billion kWh (2016 est.)

country comparison to the world: 108

Electricity—exports: 1.239 billion kWh (2015 est.)
country comparison to the world: 54

Electricity—imports: 2.22 billion kWh (2016 est.)
country comparison to the world: 54

Electricity—installed generating capacity: 2.122 million kW (2016 est.)
country comparison to the world: 111

Electricity—from fossil fuels: 58% of total installed capacity (2016 est.)
country comparison to the world: 136

Electricity—from nuclear fuels: 0% of total installed capacity (2017 est.)
country comparison to the world: 215

Electricity—from hydroelectric plants: 37% of total installed capacity (2017 est.)
country comparison to the world: 58

Electricity—from other renewable sources: 5% of total installed capacity (2017 est.)
country comparison to the world: 110

Crude oil—production: 0 bbl/day (2018 est.)
country comparison to the world: 220

Crude oil—exports: 0 bbl/day (2015 est.)
country comparison to the world: 218

Crude oil—imports: 0 bbl/day (2015 est.)
country comparison to the world: 218

Crude oil—proved reserves: 0 bbl (1 January 2018 est.)
country comparison to the world: 215

Refined petroleum products—production: 0 bbl/day (2015 est.)
country comparison to the world: 217

Refined petroleum products—consumption: 27,000 bbl/day (2016 est.)
country comparison to the world: 127

Refined petroleum products—exports: 0 bbl/day (2015 est.)
country comparison to the world: 217

Refined petroleum products—imports: 26,400 bbl/day (2015 est.)
country comparison to the world: 104

Natural gas—production: 0 cu m (2017 est.)
country comparison to the world: 217

Natural gas—consumption: 0 cu m (2017 est.)
country comparison to the world: 216

Natural gas—exports: 0 cu m (2017 est.)
country comparison to the world: 216

Natural gas—imports: 0 cu m (2017 est.)
country comparison to the world: 216

Natural gas—proved reserves: 0 cu m (1 January 2014 est.)
country comparison to the world: 210

Carbon dioxide emissions from consumption of energy: 12.06 million Mt (2017 est.)
country comparison to the world: 100

COMMUNICATIONS

Telephones—fixed lines: *total subscriptions:* 258,419

subscriptions per 100 inhabitants: 1.81 (2019 est.)
country comparison to the world: 117

Telephones—mobile cellular: *total subscriptions:* 12,863,830
subscriptions per 100 inhabitants: 90.1 (2019 est.)
country comparison to the world: 72

Telecommunication systems: *general assessment:* competition has driven the expansion of the telecommunications sector, particularly cellular voice and mobile broadband, in recent years; 3 mobile network operators continue to invest in M-commerce and M-banking facilities; continued advancement with national and international fiber backbone network as well as 3G and LTE mobile broadband services; mobile Internet connections make up 98% of all Internet connections (2020)
domestic: consists of microwave radio relay links, open-wire lines, radiotelephone communication stations, fixed wireless local loop installations, fiber-optic cable, VSAT terminals, and a substantial mobile-cellular network; Internet connection is most readily available in Harare and major towns; two government owned and two private cellular providers; fixed-line 2 per 100 and mobile-cellular 90 per 100 (2019)
international: country code—263; fiber-optic connections to neighboring states provide access to international networks via undersea cable; satellite earth stations—2 Intelsat; 5 international digital gateway exchanges
note: the COVID-19 outbreak is negatively impacting telecommunications production and supply chains globally; consumer spending on telecom devices and services has also slowed due to the pandemic's effect on economies worldwide; overall progress towards improvements in all facets of the telecom industry—mobile, fixed-line, broadband, submarine cable and satellite—has moderated

Broadcast media: government owns all local radio and TV stations; foreign shortwave broadcasts and satellite TV are available to those who can afford antennas and receivers; in rural areas, access to TV broadcasts is extremely limited; analog TV only, no digital service (2017)

Internet country code: .zw

Internet users: *total:* 3,796,618

percent of population: 27.06% (July 2018 est.)
country comparison to the world: 93

Broadband—fixed subscriptions: *total:* 203,056
subscriptions per 100 inhabitants: 1 (2018 est.)
country comparison to the world: 107

TRANSPORTATION

National air transport system: *number of registered air carriers:* 2 (2020)
inventory of registered aircraft operated by air carriers: 12
annual passenger traffic on registered air carriers: 285,539 (2018)

annual freight traffic on registered air carriers: 670,000 mt-km (2018)

Civil aircraft registration country code prefix: Z (2016)

Airports: 196 (2013)
country comparison to the world: 29

Airports—with paved runways: *total:* 17 (2013)
over 3,047 m: 3 (2013)
2,438 to 3,047 m: 2 (2013)
1,524 to 2,437 m: 5 (2013)
914 to 1,523 m: 7 (2013)

Airports—with unpaved runways: *total:* 179 (2013)
1,524 to 2,437 m: 3 (2013)
914 to 1,523 m: 104 (2013)
under 914 m: 72 (2013)
Pipelines: 270 km refined products (2013)

Railways: *total:* 3,427 km (2014)

narrow gauge: 3,427 km 1.067-m gauge (313 km electrified) (2014)
country comparison to the world: 58

Roadways: *total:* 97,267 km (2019)
paved: 18,481 km (2019)
unpaved: 78,786 km (2019)
country comparison to the world: 49

Waterways: (some navigation possible on Lake Kariba) (2011)

Ports and terminals: *river port(s):* Binga, Kariba (Zambezi)

MILITARY AND SECURITY

Military and security forces: Zimbabwe Defense Forces (ZDF): Zimbabwe National Army (ZNA), Air Force of Zimbabwe (AFZ) (2020)
Military expenditures: 1% of GDP (2019)
1.2% of GDP (2018)
1.5% of GDP (2017)
1.7% of GDP (2016)
1.9% of GDP (2015)
country comparison to the world: 121

Military and security service personnel strengths: size estimates for the Zimbabwe Defense Forces (ZDF) vary; approximately 30,000 active duty troops, including about 4,000 serving in the Air Force (2019 est.)

Military equipment inventories and acquisitions: the ZDF inventory is comprised mostly of older Chinese- and Russian-origin equipment; since 2000, China is the leading arms supplier to the ZDF, although there are no recorded deliveries of weapons since 2006 (2019 est.)

Military service age and obligation: 18-22 years of age for voluntary military service (18-24 for officer cadets; 18-30 for technical/specialist personnel); no conscription; women are eligible to serve (2019)

Military—note: the ZDF was formed after independence from the former Rhodesian Army and the two guerrilla forces that opposed it during the Rhodesian Civil War (aka "Bush War") of the 1970s, the Zimbabwe African National Liberation Army (ZANLA) and the Zimbabwe People's Revolutionary Army (ZIPRA); internal security is a key current responsibility, and the military continues to play an active role in the country's politics since the coup of 2017 (2020)

TRANSNATIONAL ISSUES

Disputes—international: Namibia has supported, and in 2004 Zimbabwe dropped objections to, plans between Botswana and Zambia to build a bridge over the Zambezi River, thereby de facto recognizing a short, but not clearly delimited, Botswana-Zambia boundary in the river; South Africa has placed military units to assist police operations along the border of Lesotho, Zimbabwe, and Mozambique to control smuggling, poaching, and illegal migration

Refugees and internally displaced persons: *refugees (country of origin):* 116,237 (Nigeria), 10,901 (Democratic Republic of Congo) (refugees and asylum seekers), 8,133 (Mozambique) (2020)
IDPs: 25,517 (tropical cyclone, 2019) (2020)

stateless persons: 300,000 (2016)

Trafficking in persons: *current situation:* Zimbabwe is a source, transit, and destination country for men, women, and children subjected to forced labor and sex trafficking; Zimbabwean women and girls from towns bordering South Africa, Mozambique, and Zambia are subjected to forced labor, including domestic servitude, and prostitution catering to long-distance truck drivers; Zimbabwean men, women, and children experience forced labor in agriculture and domestic servitude in rural areas; family members may recruit children and other relatives from rural areas with promises of work or education in cities and towns where they end up in domestic servitude and sex trafficking; Zimbabwean women and men are lured into exploitative labor situations in South Africa and other neighboring countries

tier rating: Tier 3—Zimbabwe does not fully comply with the minimum standards for the elimination of trafficking and is not making significant efforts to do so; the government passed an anti-trafficking law in 2014 defining trafficking in persons as a crime of transportation and failing to capture the key element of the international definition of human trafficking – the purpose of exploitation – which prevents the law from being comprehensive or consistent with the 2000 UN TIP Protocol that Zimbabwe acceded to in 2013; the government did not report on anti-trafficking law enforcement efforts during 2014, and corruption in law enforcement and the judiciary remain a concern; authorities made minimal efforts to identify and protect trafficking victims, relying on NGOs to identify and assist victims; Zimbabwe's 2014 antitrafficking law required the opening of 10 centers for trafficking victims, but none were established during the year; five existing shelters for vulnerable children and orphans may have accommodated child victims; in January 2015, an inter- ministerial anti-trafficking committee was established, but it is unclear if the committee ever met or initiated any activities (2015)

Illicit drugs: transit point for cannabis and South Asian heroin, mandrax, and methamphetamines en route to South Africa

APPENDIX A

ABBREVIATIONS

°C	degree(s) Celsius, degree(s) centigrade
°F	degree(s) Fahrenheit
ABEDA	Arab Bank for Economic Development in Africa
ACP Group	African, Caribbean, and Pacific Group of States
ADB	Asian Development Bank
AfDB	African Development Bank
AFESD	Arab Fund for Economic and Social Development
AG	Australia Group
Air Pollution	Convention on Long-Range Transboundary Air Pollution
Air Pollution-Nitrogen Oxides	Protocol to the 1979 Convention on Long-Range Transboundary Air Pollution Concerning the Control of Emissions of Nitrogen Oxides or Their Transboundary Fluxes
Air Pollution-Persistent Organic Pollutants	Protocol to the 1979 Convention on Long-Range Transboundary Air Pollution on Persistent Organic Pollutants
Air Pollution-Sulphur 85	Protocol to the 1979 Convention on Long-Range Transboundary Air Pollution on the Reduction of Sulphur Emissions or Their Transboundary Fluxes by at Least 30%
Air Pollution-Sulphur 94	Protocol to the 1979 Convention on Long-Range Transboundary Air Pollution on Further Reduction of Sulphur Emissions
Air Pollution-Volatile Organic Compounds	Protocol to the 1979 Convention on Long-Range Transboundary Air Pollution Concerning the Control of Emissions of Volatile Organic Compounds or Their Transboundary Fluxes
AMF	Arab Monetary Fund
AMISOM	African Union Mission in Somalia
AMU	Arab Maghreb Union
Antarctic Marine Living Resources	Convention on the Conservation of Antarctic Marine Living Resources
Antarctic Seals	Convention for the Conservation of Antarctic Seals
Antarctic-Environmental Protocol	Protocol on Environmental Protection to the Antarctic Treaty
ANZUS	Australia-New Zealand-United States Security Treaty
AOSIS	Alliance of Small Island States
APEC	Asia-Pacific Economic Cooperation
Arabsat	Arab Satellite Communications Organization
ARF	ASEAN Regional Forum
ASEAN	Association of Southeast Asian Nations
AU	African Union
Autodin	Automatic Digital Network
BA	Baltic Assembly
bbl/day	barrels per day
BCIE	Central American Bank for Economic Integration
BDEAC	Central African States Development Bank
Benelux	Benelux Union
BGN	United States Board on Geographic Names
BIMSTEC	Bay of Bengal Initiative for Multi-sectoral Technical and Economic Cooperation
Biodiversity	Convention on Biological Diversity
BIS	Bank for International Settlements
BRICS	(Brazil, Russia, India, China, and South Africa)
BSEC	Black Sea Economic Cooperation Zone
C	Commonwealth
c.i.f.	cost, insurance, and freight
CACM	Central American Common Market
CAEU	Council of Arab Economic Unity
CAN	Andean Community
Caricom	Caribbean Community and Common Market
CB	citizen's band mobile radio communications

CBSS	Council of the Baltic Sea States
CCC	Customs Cooperation Council
CD	Community of Democracies
CDB	Caribbean Development Bank
CE	Council of Europe
CEI	Central European Initiative
CELAC	Community of Latin America and Caribbean States
CEMA	Council for Mutual Economic Assistance
CEMAC	Economic and Monetary Community of Central Africa
CEPGL	Economic Community of the Great Lakes Countries
CERN	European Organization for Nuclear Research
CIA	Central Intelligence Agency
CICA	Conference of Interaction and Confidence-Building Measures in Asia
CIS	Commonwealth of Independent States
CITES	see Endangered Species
Climate Change	United Nations Framework Convention on Climate Change
Climate Change-Kyoto Protocol	Kyoto Protocol to the United Nations Framework Convention on Climate Change
COCOM	Coordinating Committee on Export Controls
COMESA	Common Market for Eastern and Southern Africa
Comsat	Communications Satellite Corporation
CP	Colombo Plan
CPLP	Comunidade dos Paises de Lingua Portuguesa
CSN	South American Community of Nations became UNASUL—Union of South American Nations
CSTO	Collective Security Treaty Organization
CTBTO	Preparatory Commission for the Nuclear-Test-Ban Treaty Organization
CY	calendar year
D-8	Developing Eight
DC	developed country
DDT	dichloro-diphenyl-trichloro-ethane
Desertification	United Nations Convention to Combat Desertification in Those Countries Experiencing Serious Drought and/or Desertification, Particularly in Africa
DIA	United States Defense Intelligence Agency
DSN	Defense Switched Network
DST	daylight savings time
DWT	deadweight ton
EAC	East African Community
EADB	East African Development Bank
EAEC	Eurasian Economic Community
EAPC	Euro-Atlantic Partnership Council
EAS	East Asia Summit
EBRD	European Bank for Reconstruction and Development
EC	European Community or European Commission
ECA	Economic Commission for Africa
ECB	European Central Bank
ECE	Economic Commission for Europe
ECLAC	Economic Commission for Latin America and the Caribbean
ECO	Economic Cooperation Organization
ECOMIG	ECOWAS Mission in The Gambia
ECOSOC	Economic and Social Council
ECOWAS	Economic Community of West African States
ECSC	European Coal and Steel Community
EE	Eastern Europe
EEC	European Economic Community
EEZ	exclusive economic zone
EFTA	European Free Trade Association

EIB	European Investment Bank
EITI	Extractive Industry Transparency Initiative
EMU	European Monetary Union
Endangered Species	Convention on the International Trade in Endangered Species of Wild Flora and Fauna (CITES)
Entente	Council of the Entente
Environmental Modification	Convention on the Prohibition of Military or Any Other Hostile Use of Environmental Modification Techniques
ESA	European Space Agency
ESCAP	Economic and Social Commission for Asia and the Pacific
ESCWA	Economic and Social Commission for Western Asia
est.	estimate
EU	European Union
EUFOR	European Union Force in Bosnia and Herzegovina
Euratom	European Atomic Energy Community
Eutelsat	European Telecommunications Satellite Organization
EUTM	European Union Training Mission (military force deployed to provide advice, operational training, and education to security forces in Bosnia-Herzegovina, Central African Republic, Mali, and Somalia)
Ex-Im	Export-Import Bank of the United States
f.o.b.	free on board
FAO	Food and Agriculture Organization
FATF	Financial Action Task Force
FAX	facsimile
FLS	Front Line States
FOC	flags of convenience
FSU	former Soviet Union
ft	foot
FttP	FttP: Fiber to the Home (FttP) is a pure fiber-optic cable connection running from an Internet Service Provider (ISP) directly to the user's home or business
FY	fiscal year
FZ	Franc Zone
G-10	Group of 10
G-11	Group of 11
G-15	Group of 15
G-20	Group of 20
G-24	Group of 24
G-3	Group of 3
G-5	Group of 5
G-6	Group of 6
G-7	Group of 7
G-77	Group of 77
G-8	Group of 8
G-9	Group of 9
GATT	General Agreement on Tariffs and Trade; now WTO
GCC	Gulf Cooperation Council
GCN	Global Caribbean Network
GCTU	General Confederation of Trade Unions
GDP	gross domestic product
GMT	Greenwich Mean Time
GNP	gross national product
GRT	gross register ton
GSM	global system for mobile cellular communications
GUAM	Organization for Democracy and Economic Development; acronym for member states— Georgia, Ukraine, Azerbaijan, Moldova
GWP	gross world product

Hazardous Wastes	Basel Convention on the Control of Transboundary Movements of Hazardous Wastes and Their Disposal
HF	high-frequency
HIV/AIDS	human immunodeficiency virus/acquired immune deficiency syndrome
IADB	Inter-American Development Bank
IAEA	International Atomic Energy Agency
IANA	Internet Assigned Numbers Authority
IBRD	International Bank for Reconstruction and Development (World Bank)
ICAO	International Civil Aviation Organization
ICC	International Chamber of Commerce
ICCt	International Criminal Court
ICJ	International Court of Justice (World Court)
ICRC	International Committee of the Red Cross
ICRM	International Red Cross and Red Crescent Movement
ICSID	International Center for Settlement of Investment Disputes
ICTR	International Criminal Tribunal for Rwanda
ICTY	International Criminal Tribunal for the former Yugoslavia
IDA	International Development Association
IDB	Islamic Development Bank
IDP	Internally Displaced Person
IEA	International Energy Agency
IFAD	International Fund for Agricultural Development
IFC	International Finance Corporation
IFRCS	International Federation of Red Cross and Red Crescent Societies
IGAD	Inter-Governmental Authority on Development
IHO	International Hydrographic Organization
ILO	International Labor Organization
IMF	International Monetary Fund
IMO	International Maritime Organization
IMSO	International Mobile Satellite Organization
in	inch
Inmarsat	International Maritime Satellite Organization
InOC	Indian Ocean Commission
Intelsat	International Telecommunications Satellite Organization
Interpol	International Criminal Police Organization
Intersputnik	International Organization of Space Communications
IOC	International Olympic Committee
IOM	International Organization for Migration
IPU	Inter-Parliamentary Union
ISO	International Organization for Standardization
ISP	Internet Service Provider
ITC	International Trade Center
ITSO	International Telecommunications Satellite Organization
ITU	International Telecommunication Union
ITUC	International Trade Union Confederation, the successor to ICFTU (International Confederation of Free Trade Unions) and the WCL (World Confederation of Labor)
kg	kilogram
kHz	kilohertz
km	kilometer
kW	kilowatt
kWh	kilowatt-hour
LAES	Latin American and Caribbean Economic System
LAIA	Latin American Integration Association
LAS	League of Arab States
Law of the Sea	United Nations Convention on the Law of the Sea (LOS)
LDC	less developed country

LLDC	least developed country
LNG	liquefied natural gas
London Convention	see Marine Dumping
LOS	see Law of the Sea
m	meter
Marecs	Maritime European Communications Satellite
Marine Dumping	Convention on the Prevention of Marine Pollution by Dumping Wastes and Other Matter
Marine Life Conservation	Convention on Fishing and Conservation of Living Resources of the High Seas
MARPOL	see Ship Pollution
Medarabtel	Middle East Telecommunications Project of the International Telecommunications Union
Mercosur	Southern Cone Common Market
MFO	Multinational Force & Observers--Sinai
MHz	megahertz
mi	mile
MICAH	International Civilian Support Mission in Haiti
MIGA	Multilateral Investment Guarantee Agency
MINURCAT	United Nations Mission in the Central African Republic and Chad
MINURSO	United Nations Mission for the Referendum in Western Sahara
MINUSCA	United Nations Multidimensional Integrated Stabilization Mission in the Central African Republic
MINUSMA	United Nations Multidimensional Integrated Stabilization Mission in Mali
MINUSTAH	United Nations Stabilization Mission in Haiti
mm	millimeter
MONUSCO	United Nations Organization Stabilization Mission in the Democratic Republic of the Congo
mt	metric ton
Mt	Mount
NA	not available
NAFTA	North American Free Trade Agreement
NAM	Nonaligned Movement
NATO	North Atlantic Treaty Organization
NC	Nordic Council
NEA	Nuclear Energy Agency
NEGL	negligible
NGA	National Geospatial-Intelligence Agency
NGO	nongovernmental organization
NIB	Nordic Investment Bank
NIC	newly industrializing country
NIE	newly industrializing economy
NIS	new independent states
nm	nautical mile
NMT	Nordic Mobile Telephone
NSG	Nuclear Suppliers Group
Nuclear Test Ban	Treaty Banning Nuclear Weapons Tests in the Atmosphere, in Outer Space, and Under Water
NZ	New Zealand
OAPEC	Organization of Arab Petroleum Exporting Countries
OAS	Organization of American States
OAU	Organization of African Unity; see African Union
ODA	official development assistance
OECD	Organization for Economic Cooperation and Development
OECS	Organization of Eastern Caribbean States
OHCHR	Office of the United Nations High Commissioner for Human Rights
OIC	Organization of the Islamic Conference
OIF	International Organization of the French-speaking World

OOF	other official flows
OPANAL	Agency for the Prohibition of Nuclear Weapons in Latin America and the Caribbean
OPCW	Organization for the Prohibition of Chemical Weapons
OPEC	Organization of Petroleum Exporting Countries
OSCE	Organization for Security and Cooperation in Europe
Ozone Layer Protection	Montreal Protocol on Substances That Deplete the Ozone Layer
PCA	Permanent Court of Arbitration
PFP	Partnership for Peace
PIF	Pacific Islands Forum
PPP	purchasing power parity
Ramsar	see Wetlands
RG	Rio Group
SAARC	South Asian Association for Regional Cooperation
SACEP	South Asia Co-operative Environment Program
SACU	Southern African Customs Union
SADC	Southern African Development Community
SAFE	South African Far East Cable
SCO	Shanghai Cooperation Organization
SECI	Southeast European Cooperative Initiative
SELEC	Convention of the Southeast European Law Enforcement Centers (successor to SECI)
SHF	super-high-frequency
Ship Pollution	Protocol of 1978 Relating to the International Convention for the Prevention of Pollution From Ships, 1973 (MARPOL)
SICA	Central American Integration System
Sparteca	South Pacific Regional Trade and Economic Cooperation Agreement
SPC	Secretariat of the Pacific Communities
SPF	South Pacific Forum
sq km	square kilometer
sq mi	square mile
TAT	Trans-Atlantic Telephone
TEU	Twenty-Foot Equivalent Unit, a unit of measure for containerized cargo capacity
Tropical Timber 83	International Tropical Timber Agreement, 1983
Tropical Timber 94	International Tropical Timber Agreement, 1994
UAE	United Arab Emirates
UDEAC	Central African Customs and Economic Union
UHF	ultra-high-frequency
UK	United Kingdom
UN	United Nations
UN-AIDS	Joint United Nations Program on HIV/AIDS
UN-Habitat	United Nations Center for Human Settlements
UNAMA	United Nations Assistance Mission in Afghanistan
UNAMID	African Union/United Nations Hybrid Operation in Darfur
UNASUR	Union of South American Nations
UNCLOS	United Nations Convention on the Law of the Sea, also known as LOS
UNCTAD	United Nations Conference on Trade and Development
UNDCP	United Nations Drug Control Program
UNDEF	United Nations Democracy Fund
UNDOF	United Nations Disengagement Observer Force
UNDP	United Nations Development Program
UNEP	United Nations Environment Program
UNESCO	United Nations Educational, Scientific, and Cultural Organization
UNFICYP	United Nations Peacekeeping Force in Cyprus
UNFPA	United Nations Population Fund
UNHCR	United Nations High Commissioner for Refugees
UNICEF	United Nations Children's Fund
UNICRI	United Nations Interregional Crime and Justice Research Institute

UNIDIR	United Nations Institute for Disarmament Research
UNIDO	United Nations Industrial Development Organization
UNIFIL	United Nations Interim Force in Lebanon
UNISFA	United Nations Interim Force for Abyei
UNITAR	United Nations Institute for Training and Research
UNMIK	United Nations Interim Administration Mission in Kosovo
UNMIL	United Nations Mission in Liberia
UNMIS	United Nations Mission in the Sudan
UNMISS	United Nations Mission in South Sudan
UNMIT	United Nations Integrated Mission in Timor-Leste
UNMOGIP	United Nations Military Observer Group in India and Pakistan
UNOCI	United Nations Operation in Cote d'Ivoire
UNODC	United Nations Office of Drugs and Crime
UNOPS	United Nations Office of Project Services
UNRISD	United Nations Research Institute for Social Development
UNRWA	United Nations Relief and Works Agency for Palestine Refugees in the Near East
UNSC	United Nations Security Council
UNSOM	United Nations Assistance Mission in Somalia
UNSSC	United Nations System Staff College
UNTSO	United Nations Truce Supervision Organization
UNU	United Nations University
UNWTO	World Tourism Organization
UPU	Universal Postal Union
US	United States
USSR	Union of Soviet Socialist Republics (Soviet Union); used for information dated before 25 December 1991
UTC	Coordinated Universal Time
UV	ultraviolet
VHF	very-high-frequency
VSAT	very small aperture terminal
WADB	West African Development Bank
WAEMU	West African Economic and Monetary Union
WCL	World Confederation of Labor
WCO	World Customs Organization
Wetlands	Convention on Wetlands of International Importance Especially As Waterfowl Habitat
WEU	Western European Union
WFP	World Food Program
WFTU	World Federation of Trade Unions
Whaling	International Convention for the Regulation of Whaling
WHO	World Health Organization
WIPO	World Intellectual Property Organization
WMO	World Meteorological Organization
WP	Warsaw Pact
WTO	World Trade Organization
ZC	Zangger Committee

APPENDIX B

SELECTED INTERNATIONAL ENVIRONMENTAL AGREEMENTS

Air Pollution
see Convention on Long-Range Transboundary Air Pollution

Air Pollution-Heavy Metals
see Protocol to the 1979 Convention on Long-Range Transboundary Air Pollution on Heavy Metals

Air Pollution-Nitrogen Oxides
see Protocol to the 1979 Convention on Long-Range Transboundary Air Pollution Concerning the Control of Emissions of Nitrogen Oxides or Their Transboundary Fluxes

Air Pollution-Persistent Organic Pollutants
see Protocol to the 1979 Convention on Long-Range Transboundary Air Pollution on Persistent Organic Pollutants

Air Pollution-Sulphur 85
see Protocol to the 1979 Convention on Long-Range Transboundary Air Pollution on the Reduction of Sulphur Emissions or Their Transboundary Fluxes by at least 30%

Air Pollution-Sulphur 94
see Protocol to the 1979 Convention on Long-Range Transboundary Air Pollution on Further Reduction of Sulphur Emissions

Air Pollution-Volatile Organic Compounds
see Protocol to the 1979 Convention on Long-Range Transboundary Air Pollution Concerning the Control of Emissions of Volatile Organic Compounds or Their Transboundary Fluxes

Antarctic—Environmental Protocol
see Protocol on Environmental Protection to the Antarctic Treaty

Antarctic Treaty
opened for signature—1 December 1959
entered into force—23 June 1961
objective—to ensure that Antarctica is used for peaceful purposes only (such as international cooperation in scientific research); to defer the question of territorial claims asserted by some nations and not recognized by others; to provide an international forum for management of the region; applies to land and ice shelves south of 60 degrees south latitude
parties—(53) Argentina, Australia, Austria, Belarus, Belgium, Brazil, Bulgaria, Canada, Chile, China, Colombia, Cuba, Czechia, Denmark, Ecuador, Estonia, Finland, France, Germany, Greece, Guatemala, Hungary, Iceland, India, Italy, Japan, Kazakhstan, North Korea, South Korea, Malaysia, Monaco, Mongolia, Netherlands, NZ, Norway, Pakistan, Papua New Guinea, Peru, Poland, Portugal, Romania, Russia, Slovakia, South Africa, Spain, Sweden, Switzerland, Turkey, Ukraine, UK, US, Uruguay, Venezuela

Basel Convention on the Control of Transboundary Movements of Hazardous Wastes and Their Disposal
note—abbreviated as Hazardous Wastes
opened for signature—22 March 1989
entered into force—5 May 1992
objective—to reduce transboundary movements of wastes subject to the Convention to a minimum consistent with the environmentally sound and efficient management of such wastes; to minimize the amount and toxicity of wastes generated and ensure their environmentally sound management as closely as possible to the source of generation; and to assist LDCs in environmentally sound management of the hazardous and other wastes they generate
parties—(186 and the Palestine Liberation Organization) Afghanistan, Albania, Algeria, Andorra, Angola, Antigua and Barbuda, Argentina, Armenia, Australia, Austria, Azerbaijan, The Bahamas, Bahrain, Bangladesh, Barbados, Belarus, Belgium, Belize, Benin, Bhutan, Bolivia, Bosnia and Herzegovina, Botswana, Brazil, Brunei, Bulgaria, Burkina Faso,
Burma, Burundi, Cambodia, Cameroon, Canada, Cape Verde, Central African Republic, Chad, Chile, China, Colombia, Comoros, Democratic Republic of the Congo, Republic of the Congo, Cook Islands, Costa Rica, Cote d'Ivoire, Croatia, Cuba, Cyprus, Czechia, Denmark, Djibouti, Dominica, Dominican Republic, Ecuador, Egypt, El Salvador, Equatorial Guinea, Eritrea, Estonia, Eswatini, Ethiopia, EU, Finland, France, Gabon, The Gambia, Georgia, Germany, Ghana, Greece, Guatemala, Guinea, Guinea-Bissau, Guyana, Haiti, Honduras, Hungary, Iceland, India, Indonesia, Iran, Iraq, Ireland, Israel, Italy, Jamaica, Japan, Jordan, Kazakhstan, Kenya, Kiribati, North Korea, South Korea, Kuwait, Kyrgyzstan, Latvia, Laos, Lebanon, Lesotho, Liberia, Libya, Liechtenstein, Lithuania, Luxembourg, Madagascar, Malawi, Malaysia, Maldives, Mali, Malta, Marshall Islands, Mauritania, Mauritius, Mexico, Federated States of Micronesia, Moldova, Monaco, Mongolia, Montenegro, Morocco, Mozambique, Namibia, Nauru, Nepal, Netherlands, NZ, Nicaragua, Niger, Nigeria, North Macedonia, Norway, Oman, Pakistan, Palau, Panama, Papua New Guinea, Paraguay, Peru, Philippines, Poland, Portugal, Qatar, Romania, Russia, Rwanda, Saint Kitts and Nevis, Saint Lucia, Saint Vincent and the Grenadines, Samoa, Sao Tome and Principe, Saudi Arabia, Senegal, Serbia, Seychelles, Sierra Leone, Singapore, Slovakia, Slovenia, Somalia, South Africa, Spain, Sri Lanka, Sudan, Suriname, Sweden, Switzerland,

Syria, Tajikistan, Tanzania, Thailand, Togo, Tonga, Trinidad and Tobago, Tunisia, Turkey, Turkmenistan, Uganda, Ukraine, UAE, UK, Uruguay, Uzbekistan, Venezuela, Vietnam, Yemen, Zambia, Zimbabwe, Palestine Liberation Organization
countries that have signed, but not yet ratified—(1) US

Biodiversity
see Convention on Biological Diversity

Climate Change
see United Nations Framework Convention on Climate Change

Climate Change-Kyoto Protocol
see Kyoto Protocol to the United Nations Framework Convention on Climate Change

Convention for the Conservation of Antarctic Seals
note—abbreviated as Antarctic Seals
opened for signature—1 June 1972
entered into force—11 March 1978
objective—to promote and achieve the protection, scientific study, and rational use of Antarctic seals, and to maintain a satisfactory balance within the ecological system of Antarctica
parties—(17) Argentina, Australia, Belgium, Brazil, Canada, Chile, France, Germany, Italy, Japan, Norway, Pakistan, Poland, Russia, South Africa, UK, US
countries that have signed, but not yet ratified—(1) NZ

Convention on Biological Diversity
note—abbreviated as Biodiversity
opened for signature—5 June 1992
entered into force—29 December 1993
objective—to develop national strategies for the conservation and sustainable use of biological diversity and to address the fair and equitable sharing of benefits arising out of the utilization of genetic resources
parties—(195 and the Palestine Liberation Organization) Afghanistan, Albania, Algeria, Andorra, Angola, Antigua and Barbuda, Argentina, Armenia, Australia, Austria, Azerbaijan, The Bahamas, Bahrain, Bangladesh, Barbados, Belarus, Belgium, Belize, Benin, Bhutan, Bolivia, Bosnia and Herzegovina, Botswana, Brazil, Brunei, Bulgaria, Burkina Faso, Burma, Burundi, Cambodia, Cameroon, Canada, Cape Verde, Central African Republic, Chad, Chile, China, Colombia, Comoros, Democratic Republic of the Congo, Republic of the Congo, Cook Islands, Costa Rica, Cote d'Ivoire, Croatia, Cuba, Cyprus, Czechia, Denmark, Djibouti, Dominica, Dominican Republic, Ecuador, Egypt, El Salvador, Equatorial Guinea, Eritrea, Estonia, Eswatini, Ethiopia, EU, Fiji, Finland, France, Gabon, The Gambia, Georgia, Germany, Ghana, Greece, Grenada, Guatemala, Guinea, Guinea-Bissau, Guyana, Haiti, Honduras, Hungary, Iceland, India, Indonesia, Iran, Iraq, Ireland, Israel, Italy, Jamaica, Japan, Jordan, Kazakhstan, Kenya, Kiribati, North Korea, South Korea, Kuwait, Kyrgyzstan, Laos, Latvia, Lebanon, Lesotho, Liberia, Libya, Liechtenstein, Lithuania, Luxembourg, Madagascar, Malawi, Malaysia, Maldives, Mali, Malta, Marshall Islands, Mauritania, Mauritius, Mexico, Federated States of Micronesia, Moldova, Monaco, Mongolia, Montenegro, Morocco, Mozambique, Namibia, Nauru, Nepal, Netherlands, NZ, Nicaragua, Niger, Nigeria, Niue, North Macedonia, Norway, Oman, Pakistan, Palau, Panama, Papua New Guinea, Paraguay, Peru, Philippines, Poland, Portugal, Qatar, Romania, Russia, Rwanda, Saint Kitts and Nevis, Saint Lucia, Saint Vincent and the Grenadines, Samoa, San Marino, Sao Tome and Principe, Saudi Arabia, Senegal, Serbia, Seychelles, Sierra Leone, Singapore, Slovakia, Slovenia, Solomon Islands, Somalia, South Africa, South Sudan, Spain, Sri Lanka, Sudan, Suriname, Sweden, Switzerland, Syria, Tajikistan, Tanzania, Thailand, Timor-Leste, Togo, Tonga, Trinidad and Tobago, Tunisia, Turkey, Turkmenistan, Tuvalu, Uganda, Ukraine, UAE, UK, Uruguay, Uzbekistan, Vanuatu, Venezuela, Vietnam, Yemen, Zambia, Zimbabwe, Palestine Liberation Organization
countries that have signed, but not yet ratified—(1) US

Convention on Fishing and Conservation of Living Resources of the High Seas
note—abbreviated as Marine Life Conservation
opened for signature—29 April 1958
entered into force—20 March 1966
objective—to solve through international cooperation the problems involved in the conservation of living resources of the high seas, considering that because of the development of modern technology some of these resources are in danger of being overexploited
parties—(39) Australia, Belgium, Bosnia and Herzegovina, Burkina Faso, Cambodia, Colombia, Republic of the Congo, Denmark, Dominican Republic, Fiji, Finland, France, Haiti, Jamaica, Kenya, Lesotho, Madagascar, Malawi, Malaysia, Mauritius, Mexico, Montenegro, Netherlands, Nigeria, Portugal, Senegal, Serbia, Sierra Leone, Solomon Islands, South Africa, Spain, Switzerland, Thailand, Tonga, Trinidad and Tobago, Uganda, UK, US, Venezuela
countries that have signed, but not yet ratified—(21) Afghanistan, Argentina, Bolivia, Canada, Costa Rica, Cuba, Ghana, Iceland, Indonesia, Iran, Ireland, Israel, Lebanon, Liberia, Nepal, NZ, Pakistan, Panama, Sri Lanka, Tunisia, Uruguay

Convention on Long-Range Transboundary Air Pollution
note—abbreviated as Air Pollution
opened for signature—13 November 1979
entered into force—16 March 1983

objective—to protect the human environment against air pollution and, as far as possible, to gradually reduce and prevent air pollution, including long-range transboundary air pollution

parties—(51) Albania, Armenia, Austria, Azerbaijan, Belarus, Belgium, Bosnia and Herzegovina, Bulgaria, Canada, Croatia, Cyprus, Czechia, Denmark, Estonia, EU, Finland, France, Georgia, Germany, Greece, Hungary, Iceland, Ireland, Italy, Kazakhstan, Kyrgyzstan, Latvia, Liechtenstein, Lithuania, Luxembourg, Malta, Moldova, Monaco, Montenegro, Netherlands, North Macedonia, Norway, Poland, Portugal, Romania, Russia, Serbia, Slovakia, Slovenia, Spain, Sweden, Switzerland, Turkey, Ukraine, UK, US

countries that have signed, but not yet ratified—(1) Holy See, San Marino

Convention on the Conservation of Antarctic Marine Living Resources

note—abbreviated as Antarctic-Marine Living Resources

opened for signature—5 May 1980

entered into force—7 April 1982

objective—to safeguard the environment and protect the integrity of the ecosystem of the seas surrounding Antarctica, and to conserve Antarctic marine living resources

members—(25) Argentina, Australia, Belgium, Brazil, Chile, China, EU, France, Germany, India, Italy, Japan, South Korea, Namibia, NZ, Norway, Poland, Russia, South Africa, Spain, Sweden, Ukraine, UK, US, Uruguay

acceding states—(11) Bulgaria, Canada, Cook Islands, Finland, Greece, Mauritius, Netherlands, Pakistan, Panama, Peru, Vanuatu

Convention on the International Trade in Endangered Species of Wild Flora and Fauna (CITES)

note—abbreviated as Endangered Species

opened for signature—3 March 1973

entered into force—1 July 1975

objective—to protect certain endangered species from overexploitation by means of a system of import/export permits

parties—(183) Afghanistan, Albania, Algeria, Angola, Antigua and Barbuda, Argentina, Armenia, Australia, Austria, Azerbaijan, The Bahamas, Bahrain, Bangladesh, Barbados, Belarus, Belgium, Belize, Benin, Bhutan, Bolivia, Bosnia and Herzegovina, Botswana, Brazil, Brunei, Bulgaria, Burkina Faso, Burma, Burundi, Cambodia, Cameroon, Canada, Cape Verde, Central African Republic, Chad, Chile, China, Colombia, Comoros, Democratic Republic of the Congo, Republic of the Congo, Costa Rica, Cote d'Ivoire, Croatia, Cuba, Cyprus, Czechia, Denmark, Djibouti, Dominica, Dominican Republic, Ecuador, Egypt, El Salvador, Equatorial Guinea, Eritrea, Estonia, Eswatini, Ethiopia, EU, Fiji, Finland, France, Gabon, The Gambia, Georgia, Germany, Ghana, Greece, Grenada, Guatemala, Guinea, Guinea-Bissau, Guyana, Honduras, Hungary, Iceland, India, Indonesia, Iran, Iraq, Ireland, Israel, Italy, Jamaica, Japan, Jordan, Kazakhstan, Kenya, South Korea, Kuwait, Kyrgyzstan, Laos, Latvia, Lebanon, Lesotho, Liberia, Libya, Liechtenstein, Lithuania, Luxembourg, Madagascar, Malawi, Malaysia, Maldives, Mali, Malta, Mauritania, Mauritius, Mexico, Moldova, Monaco, Mongolia, Montenegro, Morocco, Mozambique, Namibia, Nepal, Netherlands, NZ, Nicaragua, Niger, Nigeria, North Macedonia, Norway, Oman, Palau, Pakistan, Panama, Papua New Guinea, Paraguay, Peru, Philippines, Poland, Portugal, Qatar, Romania, Russia, Rwanda, Saint Kitts and Nevis, Saint Lucia, Saint Vincent and the Grenadines, Samoa, San Marino, Sao Tome and Principe, Saudi Arabia, Senegal, Serbia, Seychelles, Sierra Leone, Singapore, Slovakia, Slovenia, Solomon Islands, Somalia, South Africa, Spain, Sri Lanka, Sudan, Suriname, Sweden, Switzerland, Syria, Tajikistan, Tanzania, Thailand, Togo, Tonga, Trinidad and Tobago, Tunisia, Turkey, Uganda, Ukraine, UAE, UK, US, Uruguay, Uzbekistan, Vanuatu, Venezuela, Vietnam, Yemen, Zambia, Zimbabwe

Convention on the Prevention of Marine Pollution by Dumping Wastes and Other Matter (London Convention)

note—abbreviated as Marine Dumping

opened for signature—29 December 1972

entered into force—30 August 1975

objective—to promote effective control of all sources of marine pollution and to take all practicable steps to prevent pollution of the sea by dumping and to encourage regional agreements supplementary to the Convention

parties—(87) Afghanistan, Antigua and Barbuda, Argentina, Australia, Azerbaijan, Barbados, Belarus, Belgium, Benin, Bolivia, Brazil, Bulgaria, Canada, Cape Verde, Chile, China, Democratic Republic of the Congo, Costa Rica, Cote d'Ivoire, Croatia, Cuba, Denmark, Dominican Republic, Egypt, Equatorial Guinea, Finland, France, Gabon, Germany, Greece, Guatemala, Haiti, Honduras, Hungary, Iceland, Iran, Ireland, Italy, Jamaica, Japan, Jordan, Kenya, Kiribati, South Korea, Libya, Luxembourg, Malta, Mexico, Monaco, Montenegro, Morocco, Nauru, Netherlands, NZ, Nigeria, Norway, Oman, Pakistan, Panama, Papua New Guinea, Peru, Philippines, Poland, Portugal, Russia, Saint Lucia, Saint Vincent and the Grenadines, Serbia, Seychelles, Sierra Leon, Slovenia, Solomon Islands, South Africa, Spain, Suriname, Sweden, Switzerland, Syria, Tanzania, Tonga, Tunisia, Ukraine, UAE, UK, US, Vanuatu

associate members to the London Convention—(3) Faroe Islands, Hong Kong, Macau i>countries that have signed, but not yet ratified*—(3) Chad, Kuwait, Uruguay

Convention on the Prohibition of Military or Any Other Hostile Use of Environmental Modification Techniques

note—abbreviated as Environmental Modification

opened for signature—18 May 1977

entered into force—5 October 1978

objective—to prohibit the military or other hostile use of environmental modification techniques in order to further world peace and trust among nations

parties—(77 and the Palestine Liberation Organization) Afghanistan, Algeria, Antigua and Barbuda, Argentina, Armenia, Australia, Austria, Bangladesh, Belarus, Belgium, Benin, Brazil, Bulgaria, Canada, Cameroon, Cape Verde, Chile, China, Costa Rica, Cuba, Cyprus, Czechia, Denmark, Dominica, Egypt, Estonia, Finland, Germany, Ghana, Greece, Guatemala, Honduras, Hungary, India, Ireland, Italy, Japan, Kazakhstan, North Korea, South Korea, Kuwait, Kyrgyzstan, Laos, Lithuania, Malawi, Mauritius, Mongolia, Netherlands, NZ,

Nicaragua, Niger, Norway, Pakistan, Panama, Papua New Guinea, Poland, Romania, Russia, Saint Lucia, Saint Vincent and the Grenadines, Sao Tome and Principe, Slovakia, Slovenia, Solomon Islands, Spain, Sri Lanka, Sweden, Switzerland, Tajikistan, Tunisia, Ukraine, UK, US, Uruguay, Uzbekistan, Vietnam, Yemen, Palestine Liberation Organization
countries that have signed, but not yet ratified—(16) Bolivia, Democratic Republic of the Congo, Ethiopia, Holy See, Iceland, Iran, Iraq, Lebanon, Liberia, Luxembourg, Morocco, Portugal, Sierra Leone, Syria, Turkey, Uganda

Convention on Wetlands of International Importance Especially as Waterfowl Habitat (Ramsar)
note—abbreviated as Wetlands
opened for signature—2 February 1971
entered into force—21 December 1975
objective—to stem the progressive encroachment on and loss of wetlands now and in the future
parties—(170) Albania, Algeria, Andorra, Antigua and Barbuda, Argentina, Armenia, Australia, Austria, Azerbaijan, The Bahamas, Bahrain, Bangladesh, Barbados, Belarus, Belgium, Belize, Benin, Bhutan, Bolivia, Bosnia and Herzegovina, Botswana, Brazil, Bulgaria, Burkina Faso, Burma, Burundi, Cabo Verde, Cambodia, Cameroon, Canada, Central African Republic, Chad, Chile, China, Colombia, Comoros, Democratic Republic of the Congo, Republic of the Congo, Costa Rica, Cote d'Ivoire, Croatia, Cuba, Cyprus, Czechia, Denmark, Djibouti, Dominican Republic, Ecuador, Egypt, El Salvador, Equatorial Guinea, Estonia, Eswatini, Fiji, Finland, France, Gabon, The Gambia, Georgia, Germany, Ghana, Greece, Grenada, Guatemala, Guinea, Guinea-Bissau, Honduras, Hungary, Iceland, India, Indonesia, Iran, Iraq, Ireland, Israel, Italy, Jamaica, Japan, Jordan, Kazakhstan, Kenya, Kiribati, South Korea, Kyrgyzstan, Kuwait, Laos, Latvia, Lebanon, Lesotho, Liberia, Libya, Liechtenstein, Lithuania, Luxembourg, Madagascar, Malawi, Malaysia, Mali, Malta, Marshall Islands, Mauritania, Mauritius, Mexico, Moldova, Monaco, Mongolia, Montenegro, Morocco, Mozambique, Namibia, Nepal, Netherlands, NZ, Nicaragua, Niger, Nigeria, North Macedonia, Norway, Oman, Pakistan, Palau, Panama, Papua New Guinea, Paraguay, Peru, Philippines, Poland, Portugal, Romania, Russia, Rwanda, Saint Lucia, Samoa, Sao Tome and Principe, Senegal, Serbia, Seychelles, Sierra Leone, Slovakia, Slovenia, South Africa, South Sudan, Spain, Sri Lanka, Sudan, Suriname, Sweden, Switzerland, Syria, Tanzania, Tajikistan, Thailand, Togo, Trinidad and Tobago, Tunisia, Turkey, Turkmenistan, Uganda, Ukraine, UAE, UK, US, Uruguay, Uzbekistan, Venezuela, Vietnam, Yemen, Zambia, Zimbabwe

Desertification
see United Nations Convention to Combat Desertification in those Countries Experiencing Serious Drought and/or Desertification, Particularly in Africa

Endangered Species
see Convention on the International Trade in Endangered Species of Wild Flora and Fauna (CITES)

Environmental Modification
see Convention on the Prohibition of Military or Any Other Hostile Use of Environmental Modification Techniques

Hazardous Wastes
see Basel Convention on the Control of Transboundary Movements of Hazardous Wastes and Their Disposal

International Convention for the Regulation of Whaling
note—abbreviated as Whaling
opened for signature—2 December 1946
entered into force—10 November 1948
objective—to protect all species of whales from overhunting; to establish a system of international regulation for the whale fisheries to ensure proper conservation and development of whale stocks; and to safeguard for future generations the great natural resources represented by whale stocks
parties—(89) Antigua and Barbuda, Argentina, Australia, Austria, Belgium, Belize, Benin, Brazil, Bulgaria, Cambodia, Cameroon, Chile, China, Colombia, Republic of the Congo, Costa Rica, Cote D'Ivoire, Croatia, Cyprus, Czechia, Denmark, Dominica, Dominican Republic, Ecuador, Eritrea, Estonia, Finland, France, Gabon, The Gambia, Germany, Ghana, Grenada, Guinea, Guinea-Bissau, Hungary, Iceland, India, Ireland, Israel, Italy, Japan, Kenya, Kiribati, South Korea, Laos, Liberia, Lithuania, Luxembourg, Mali, Marshall Islands, Mauritania, Mexico, Monaco, Mongolia, Morocco, Nauru, Netherlands, NZ, Nicaragua, Norway, Oman, Palau, Panama, Peru, Poland, Portugal, Romania, Russia, Saint Kitts and Nevis, Saint Lucia, Saint Vincent and the Grenadines, San Marino, Sao Tome and Principe, Senegal, Slovakia, Slovenia, Solomon Islands, South Africa, Spain, Suriname, Sweden, Switzerland, Tanzania, Togo, Tuvalu, UK, US, Uruguay

International Tropical Timber Agreement, 2006
note—abbreviated as Tropical Timber, 2006; ITTA, 2006; or ITTA3
opened for signature—3 April 2006
entered into force—7 December 2011; note—superseded the International Tropical Timber Agreement, 1994, which itself superseded the International Tropical Timber Agreement, 1983
objective—to promote the expansion and diversification of international trade in tropical timber from sustainably managed and legally harvested forests and to promote the sustainable management of tropical timber producing forests
parties—(74) Abania, Australia, Austria, Belgium, Benin, Brazil, Bulgaria, Burma, Cambodia, Cameroon, Central African Republic, China, Colombia, Democratic Republic of the Congo, Republic of the Congo, Costa Rica, Cote d'Ivoire, Croatia, Cyprus, Czechia, Denmark, Ecuador, Estonia, EU, Fiji, Finland, France, Gabon, Germany, Ghana, Greece, Guatemala, Guyana, Honduras, Hungary, India, Indonesia, Ireland, Italy, Japan, South Korea, Latvia, Liberia, Lithuania, Luxembourg, Madagascar, Malaysia, Mali, Malta, Mexico, Mozambique,

Netherlands, NZ, Norway, Panama, Papua New Guinea, Peru, Philippines, Poland, Portugal, Romania, Slovakia, Slovenia, Spain, Suriname, Sweden, Switzerland, Thailand, Togo, Trinidad and Tobago, UK, US, Venezuela, Vietnam
countries that have signed, but not yet ratified—(2) Nigeria, Paraguay

Kyoto Protocol to the United Nations Framework Convention on Climate Change

note—abbreviated as Climate Change-Kyoto Protocol
opened for signature—16 March 1998
entered into force—16 February 2005
objective—to further reduce greenhouse gas emissions by enhancing the national programs of developed countries aimed at this goal and by establishing percentage reduction targets for the developed countries
parties—(192) Afghanistan, Albania, Algeria, Angola, Antigua and Barbuda, Argentina, Armenia, Australia, Austria, Azerbaijan, The Bahamas, Bahrain, Bangladesh, Barbados, Belarus, Belgium, Belize, Benin, Bhutan, Bolivia, Bosnia and Herzegovina, Botswana, Brazil, Brunei, Bulgaria, Burkina Faso, Burma, Burundi, Cabo Verde, Cambodia, Cameroon, Central African Republic, Chad, Chile, China, Colombia, Comoros, Democratic Republic of the Congo, Republic of the Congo, Cook Islands, Costa Rica, Cote d'Ivoire, Croatia, Cuba, Cyprus, Czechia, Denmark, Djibouti, Dominica, Dominican Republic, Ecuador, Egypt, El Salvador, Equatorial Guinea, Eritrea, Estonia, Eswatini, Ethiopia, EU, Fiji, Finland, France, Gabon, The Gambia, Georgia, Germany, Ghana, Greece, Grenada, Guatemala, Guinea, Guinea-Bissau, Guyana, Haiti, Honduras, Hungary, Iceland, India, Indonesia, Iran, Iraq, Ireland, Israel, Italy, Jamaica, Japan, Jordan, Kazakhstan, Kenya, Kiribati, North Korea, South Korea, Kuwait, Kyrgyzstan, Laos, Latvia, Lebanon, Lesotho, Liberia, Libya, Liechtenstein, Lithuania, Luxembourg, Madagascar, Malawi, Malaysia, Maldives, Mali, Malta, Marshall Islands, Mauritania, Mauritius, Mexico, Federated States of Micronesia, Moldova, Monaco, Mongolia, Montenegro, Morocco, Mozambique, Namibia, Nauru, Nepal, Netherlands, NZ, Nicaragua, Niger, Nigeria, Niue, North Macedonia, Norway, Oman, Pakistan, Palau, Panama, Papua New Guinea, Paraguay, Peru, Philippines, Poland, Portugal, Qatar, Romania, Russia, Rwanda, Saint Kitts and Nevis, Saint Lucia, Saint Vincent and the Grenadines, Samoa, San Marino, Sao Tome and Principe, Saudi Arabia, Senegal, Serbia, Seychelles, Sierra Leone, Singapore, Slovakia, Slovenia, Solomon Islands, Somalia, South Africa, Spain, Sri Lanka, Sudan, Suriname, Sweden, Switzerland, Syria, Tajikistan, Tanzania, Thailand, Timor-Leste, Togo, Tonga, Trinidad and Tobago, Tunisia, Turkey, Turkmenistan, Tuvalu, Uganda, Ukraine, UAE, UK, Uruguay, Uzbekistan, Vanuatu, Venezuela, Vietnam, Yemen, Zambia, Zimbabwe countries that have signed, but not yet ratified—(1) US

Law of the Sea

see United Nations Convention on the Law of the Sea (LOS)

Marine Dumping

see Convention on the Prevention of Marine Pollution by Dumping Wastes and Other Matter (London Convention)

Marine Life Conservation

see Convention on Fishing and Conservation of Living Resources of the High Seas

Montreal Protocol on Substances That Deplete the Ozone Layer

note—abbreviated as Ozone Layer Protection
opened for signature—16 September 1987
entered into force—1 January 1989
objective—to protect the ozone layer by controlling emissions of substances that deplete it
parties—(197) Afghanistan, Albania, Algeria, Andorra, Angola, Antigua and Barbuda, Argentina, Armenia, Australia, Austria, Azerbaijan, The Bahamas, Bahrain, Bangladesh, Barbados, Belarus, Belgium, Belize, Benin, Bhutan, Bolivia, Bosnia and Herzegovina, Botswana, Brazil, Brunei, Bulgaria, Burkina Faso, Burma, Burundi, Cambodia, Cameroon, Canada, Cape Verde, Central African Republic, Chad, Chile, China, Colombia, Comoros, Democratic Republic of the Congo, Republic of the Congo, Cook Islands, Costa Rica, Cote d'Ivoire, Croatia, Cuba, Cyprus, Czechia, Denmark, Djibouti, Dominica, Dominican Republic, Ecuador, Egypt, El Salvador, Equatorial Guinea, Eritrea, Estonia, Eswatini, Ethiopia, EU, Fiji, Finland, France, Gabon, The Gambia, Georgia, Germany, Ghana, Greece, Grenada, Guatemala, Guinea, Guinea-Bissau, Guyana, Haiti, Holy See, Honduras, Hungary, Iceland, India, Indonesia, Iran, Iraq, Ireland, Israel, Italy, Jamaica, Japan, Jordan, Kazakhstan, Kenya, Kiribati, North Korea, South Korea, Kuwait, Kyrgyzstan, Laos, Latvia, Lebanon, Lesotho, Liberia, Libya, Liechtenstein, Lithuania, Luxembourg, Madagascar, Malawi, Malaysia, Maldives, Mali, Malta, Marshall Islands, Mauritania, Mauritius, Mexico, Federated States of Micronesia, Moldova, Monaco, Mongolia, Montenegro, Morocco, Mozambique, Namibia, Nauru, Nepal, Netherlands, NZ, Nicaragua, Niger, Nigeria, Niue, North Macedonia, Norway, Oman, Pakistan, Palau, Panama, Papua New Guinea, Paraguay, Peru, Philippines, Poland, Portugal, Qatar, Romania, Russia, Rwanda, Saint Kitts and Nevis, Saint Lucia, Saint Vincent and the Grenadines, Samoa, San Marino, Sao Tome and Principe, Saudi Arabia, Senegal, Serbia, Seychelles, Sierra Leone, Singapore, Slovakia, Slovenia, Solomon Islands, Somalia, South Africa, South Sudan, Spain, Sri Lanka, Sudan, Suriname, Sweden, Switzerland, Syria, Tajikistan, Tanzania, Thailand, Timor-Leste, Togo, Tonga, Trinidad and Tobago, Tunisia, Turkey, Turkmenistan, Tuvalu, Uganda, Ukraine, UAE, UK, US, Uruguay, Uzbekistan, Vanuatu, Venezuela, Vietnam, Yemen, Zambia, Zimbabwe

Multi-effect Protocol

see Protocol to the 1979 Convention on Long-Range Transboundary Air Pollution to Abate Acidification, Eutrophication, and Ground-Level Ozone

Nuclear Test Ban

see Treaty Banning Nuclear Weapons Tests in the Atmosphere, in Outer Space, and Under Water

Ozone Layer Protection
see Montreal Protocol on Substances That Deplete the Ozone Layer

Protocol of 1978 Relating to the International Convention for the Prevention of Pollution From Ships, 1973 (MARPOL)
note—abbreviated as Ship Pollution
opened for signature—1 June 1978
entered into force—2 October 1983
objective—to preserve the marine environment in an attempt to completely eliminate pollution by oil and other harmful substances and to minimize accidental spillage of such substances
parties—(153) Albania, Algeria, Angola, Antigua and Barbuda, Argentina, Australia, Austria, Azerbaijan, The Bahamas, Bahrain, Bangladesh, Barbados, Belarus, Belgium, Belize, Benin, Bolivia, Brazil, Brunei, Bulgaria, Burma, Cabo Verde, Cambodia, Cameroon, Canada, Chile, China, Colombia, Comoros, Republic of Congo, Cook Islands, Cote d'Ivoire, Croatia, Cuba, Cyprus, Czechia, Denmark, Djibouti, Dominica, Dominican Republic, Ecuador, Egypt, El Salvador, Equatorial Guinea, Estonia, Fiji, Finland, France, Gabon, The Gambia, Georgia, Germany, Ghana, Greece, Guatemala, Guinea, Guinea-Bissau, Guyana, Honduras, Hungary, Iceland, India, Indonesia, Iran, Iraq, Ireland, Israel, Italy, Jamaica, Japan, Jordan, Kazakhstan, Kenya, Kiribati, North Korea, South Korea, Kuwait, Latvia, Lebanon, Liberia, Libya, Lithuania, Luxembourg, Madagascar, Malawi, Malaysia, Maldives, Malta, Marshall Islands, Mauritania, Mauritius, Mexico, Moldova, Monaco, Mongolia, Montenegro, Morocco, Mozambique, Namibia, Netherlands, NZ, Nicaragua, Nigeria, Niue, Norway, Oman, Pakistan, Palau, Panama, Papua New Guinea, Peru, Philippines, Poland, Portugal, Qatar, Romania, Russia, Saint Kitts and Nevis, Saint Lucia, Saint Vincent and the Grenadines, Samoa, Sao Tome and Principe, Saudi Arabia, Senegal, Serbia, Sierra Leone, Singapore, Slovakia, Slovenia, Solomon Islands, South Africa, Spain, Sri Lanka, Sudan, Suriname, Sweden, Switzerland, Syria, Tanzania, Togo, Tonga, Trinidad and Tobago, Tunisia, Turkey, Tuvalu, Ukraine, UAE, UK, US, Uruguay, Vanuatu, Venezuela, Vietnam

Protocol on Environmental Protection to the Antarctic Treaty
note—abbreviated as Antarctic-Environmental Protocol
opened for signature—4 October 1991
entered into force—14 January 1998
objective—to provide for comprehensive protection of the Antarctic environment and dependent and associated ecosystems; applies to the area covered by the Antarctic Treaty
consultative parties—(29) Argentina, Australia, Belgium, Brazil, Bulgaria, Chile, China, Czechia, Ecuador, Finland, France, Germany, India, Italy, Japan, South Korea, Netherlands, NZ, Norway, Peru, Poland, Russia, South Africa, Spain, Sweden, Ukraine, UK, US, Uruguay
non consultative parties—(24) Austria, Belarus, Canada, Colombia, Cuba, Denmark, Estonia, Greece, Guatemala, Hungary, Iceland, Kazakhstan, North Korea, Malaysia, Monaco, Mongolia, Pakistan, Papua New Guinea, Portugal, Romania, Slovakia, Switzerland, Turkey, Venezuela

Protocol to the 1979 Convention on Long-Range Transboundary Air Pollution Concerning the Control of Emissions of Nitrogen Oxides or Their Transboundary Fluxes
note—abbreviated as Air Pollution-Nitrogen Oxides
opened for signature—31 October 1988
entered into force—14 February 1991
objective—to provide for the control or reduction national of nitrogen oxide emissions and their transboundary fluxes
parties—(35) Albania, Austria, Belarus, Belgium, Bulgaria, Canada, Croatia, Cyprus, Czechia, Denmark, Estonia, EU, Finland, France, Germany, Greece, Hungary, Ireland, Italy, Liechtenstein, Lithuania, Luxembourg, Netherlands, North Macedonia, Norway, Poland, Russia, Slovakia, Slovenia, Spain, Sweden, Switzerland, Ukraine, UK, US

Protocol to the 1979 Convention on Long-Range Transboundary Air Pollution Concerning the Control of Emissions of Volatile Organic Compounds or Their Transboundary Fluxes
note—abbreviated as Air Pollution-Volatile Organic Compounds
opened for signature—18 November 1991
entered into force—29 September 1997
objective—to provide for the control and reduction of national emissions of volatile organic compounds in order to reduce their transboundary fluxes
parties—(24) Austria, Belgium, Bulgaria, Croatia, Czechia, Denmark, Estonia, Finland, France, Germany, Hungary, Italy, Liechtenstein, Lithuania, Luxembourg, Monaco, Netherlands, North Macedonia, Norway, Slovakia, Spain, Sweden, Switzerland, UK
countries that have signed, but not yet ratified—(6) Canada, EU, Greece, Portugal, Ukraine, US

Protocol to the 1979 Convention on Long-Range Transboundary Air Pollution on Further Reduction of Sulphur Emissions
note—abbreviated as Air Pollution-Sulphur 94
opened for signature—14 June 1994
entered into force—5 August 1998
objective—to provide for a further reduction in national sulfur emissions or transboundary fluxes on a regional basis within Europe
parties—(30) Austria, Belgium, Bulgaria, Canada, Croatia, Cyprus, Czechia, Denmark, EU, Finland, France, Germany, Greece, Hungary, Ireland, Italy, Liechtenstein, Lithuania, Luxembourg, Monaco, Netherlands, North Macedonia, Norway, Poland, Slovakia, Slovenia, Spain, Sweden, Switzerland, UK
countries that have signed, but not yet ratified—(2) Russia, Ukraine

Protocol to the 1979 Convention on Long-Range Transboundary Air Pollution on Heavy Metals

note—abbreviated as Air Pollution-Heavy Metals
opened for signature—24 June 1998
entered into force—29 December 2003
objective—to reduce emissions of the heavy metals cadmium, lead, and mercury from industrial sources, combustion processes, and waste incineration
parties—(34) Austria, Belgium, Bulgaria, Canada, Croatia, Cyprus, Czechia, Denmark, Estonia, EU, Finland, France, Germany, Hungary, Iceland, Latvia, Liechtenstein, Lithuania, Luxembourg, Moldova, Monaco, Montenegro, Netherlands, North Macedonia, Norway, Portugal, Romania, Serbia, Slovakia, Slovenia, Spain, Sweden, Switzerland, UK, US
countries that have signed, but not yet ratified—(7) Armenia, Greece, Iceland, Ireland, Italy, Poland, Ukraine

Protocol to the 1979 Convention on Long-Range Transboundary Air Pollution on Persistent Organic Pollutants

note—abbreviated as Air Pollution-Persistent Organic Pollutants
opened for signature—24 June 1998
entered into force—23 October 2003
objective—to provide for the control, reduction, or elimination of discharges, emissions of persistent organic pollutants
parties—(33) Austria, Belgium, Bulgaria, Canada, Croatia, Cyprus, Czechia, Denmark, EU, Finland, France, Germany, Hungary, Iceland, Italy, Latvia, Liechtenstein, Lithuania, Luxembourg, Moldova, Montenegro, Netherlands, North Macedonia, Norway, Romania, Serbia, Slovakia, Slovenia, Spain, Sweden, Switzerland, UK
countries that have signed, but not yet ratified—(7) Armenia, Greece, Ireland, Poland, Portugal, Ukraine, US

Protocol to the 1979 Convention on Long-Range Transboundary Air Pollution on the Reduction of Sulphur Emissions or Their Transboundary Fluxes by at Least 30%

note—abbreviated as Air Pollution-Sulphur 85
opened for signature—8 July 1985
entered into force—2 September 1987
objective—to provide for national reductions in sulfur emissions or transboundary fluxes by 30% of 1980 emission or transboundary flux levels by no later than 1993
parties—(25) Albania, Austria, Belarus, Belgium, Bulgaria, Canada, Czechia, Denmark, Estonia, Finland, France, Germany, Hungary, Italy, Liechtenstein, Lithuania, Luxembourg, Netherlands, North Macedonia, Norway, Russia, Slovakia, Sweden, Switzerland, Ukraine
note—Albania and Macedonia listed as non-compliant

Protocol to the 1979 Convention on Long-Range Transboundary Air Pollution to Abate Acidification, Eutrophication, and Ground-Level Ozone

note—abbreviated as Multi-effect Protocol
opened for signature—30 November 1999
entered into force—17 May 2005
objective—to reduce acidification, eutrophication, and ground-level ozone by setting emissions ceilings for sulphur dioxide, nitrogen oxides, volatile organic compounds, and ammonia
parties—(27) Belgium, Bulgaria, Canada, Croatia, Cyprus, Czechia, Denmark, EU, Finland, France, Germany, Hungary, Latvia, Lithuania, Luxembourg, Netherlands, North Macedonia, Norway, Portugal, Romania, Slovakia, Slovenia, Spain, Sweden, Switzerland, UK, US
countries that have signed, but not yet ratified—(8) Armenia, Austria, Greece, Ireland, Italy, Liechtenstein, Moldova, Poland

Ship Pollution

see Protocol of 1978 Relating to the International Convention for the Prevention of Pollution From Ships, 1973 (MARPOL)

Treaty Banning Nuclear Weapon Tests in the Atmosphere, in Outer Space, and Under Water

note—abbreviated as Nuclear Test Ban
opened for signature—5 August 1963
entered into force—10 October 1963
objective—to ban nuclear weapons testing in the atmosphere, outer space, or under water
parties—(125) Afghanistan, Antigua and Barbuda, Argentina, Armenia, Australia, Austria, The Bahamas, Bangladesh, Belarus, Belgium, Benin, Bhutan, Bolivia, Bosnia and Herzegovina, Botswana, Brazil, Bulgaria, Burma, Cabo Verde, Canada, Central African Republic, Chad, Chile, Colombia, Democratic Republic of the Congo, Costa Rica, Cote d'Ivoire, Croatia, Cyprus, Czechia, Denmark, Dominican Republic, Ecuador, Egypt, El Salvador, Eswatini, Equatorial Guinea, Fiji, Finland, Gabon, The Gambia, Germany, Ghana, Greece, Guatemala, Guinea-Bissau, Honduras, Hungary, Iceland, India, Indonesia, Iran, Iraq, Ireland, Israel, Italy, Jamaica, Japan, Jordan, Kenya, South Korea, Kuwait, Laos, Lebanon, Liberia, Libya, Luxembourg, Madagascar, Malawi, Malaysia, Malta, Mauritania, Mauritius, Mexico, Mongolia, Montenegro, Morocco, Nepal, Netherlands, NZ, Nicaragua, Niger, Nigeria, Norway, Pakistan, Panama, Papua New Guinea, Peru, Philippines, Poland, Romania, Russia, Rwanda, Samoa, San Marino, Senegal, Serbia, Seychelles, Sierra Leone, Singapore, Slovakia, Slovenia, South Africa, Spain, Sri Lanka, Sudan, Suriname, Sweden, Switzerland, Syria, Tanzania, Thailand, Togo, Tonga, Trinidad and Tobago, Tunisia, Turkey, Uganda, Ukraine, UK, US, Uruguay, Venezuela, Yemen, Zambia
countries that have signed, but not yet ratified—(10) Algeria, Burkina-Faso, Burundi, Cameroon, Ethiopia, Haiti, Mali, Paraguay, Portugal, Somalia

Tropical Timber, 2006
see International Tropical Timber Agreement, 2006

United Nations Convention on the Law of the Sea (LOS)
note—abbreviated as Law of the Sea
opened for signature—10 December 1982
entered into force—16 November 1994
objective—to provide a comprehensive legal regime for the sea and oceans
parties—(167 and the Palestine Liberation Organization) Albania, Algeria, Angola, Antigua and Barbuda, Argentina, Armenia, Australia, Austria, Azerbaijan, The Bahamas, Bahrain, Bangladesh, Barbados, Belarus, Belgium, Belize, Benin, Bolivia, Bosnia and Herzegovina, Botswana, Brazil, Brunei, Bulgaria, Burkina Faso, Burma, Burundi, Cabo Verde, Cameroon, Canada, Chad, Chile, China, Comoros, Democratic Republic of the Congo, Republic of the Congo, Cook Islands, Costa Rica, Cote d'Ivoire, Croatia, Cuba, Cyprus, Czechia, Denmark, Djibouti, Dominica, Dominican Republic, Ecuador, Egypt, Equatorial Guinea, Estonia, Eswatini, EU, Fiji, Finland, France, Gabon, The Gambia, Georgia, Germany, Ghana, Greece, Grenada, Guatemala, Guinea, Guinea-Bissau, Guyana, Haiti, Honduras, Hungary, Iceland, India, Indonesia, Iraq, Ireland, Italy, Jamaica, Japan, Jordan, Kenya, Kiribati, South Korea, Kuwait, Laos, Latvia, Lebanon, Lesotho, Liberia, Lithuania, Luxembourg, Madagascar, Malawi, Malaysia, Maldives, Mali, Malta, Marshall Islands, Mauritania, Mauritius, Mexico, Federated States of Micronesia, Moldova, Monaco, Mongolia, Montenegro, Morocco, Mozambique, Namibia, Nauru, Nepal, Netherlands, NZ, Nicaragua, Niger, Nigeria, Niue, North Macedonia, Norway, Oman, Pakistan, Palau, Panama, Papua New Guinea, Paraguay, Philippines, Poland, Portugal, Qatar, Romania, Russia, Rwanda, Saint Kitts and Nevis, Saint Lucia, Saint Vincent and the Grenadines, Samoa, Sao Tome and Principe, Saudi Arabia, Senegal, Serbia, Seychelles, Sierra Leone, Singapore, Slovakia, Slovenia, Solomon Islands, Somalia, South Africa, Spain, Sri Lanka, Sudan, Suriname, Sweden, Switzerland, Tanzania, Thailand, Timor-Leste, Togo, Tonga, Trinidad and Tobago, Tunisia, Tuvalu, Uganda, Ukraine, UK, Uruguay, Vanuatu, Vietnam, Yemen, Zambia, Zimbabwe, Palestine Liberation Organization
countries that have signed, but not yet ratified—(14) Afghanistan, Bhutan, Burundi, Cambodia, Central African Republic, Colombia, El Salvador, Ethiopia, Iran, North Korea, Libya, Liechtenstein, Rwanda, UAE

United Nations Convention to Combat Desertification in Those Countries Experiencing Serious Drought and/or Desertification, Particularly in Africa
note—abbreviated as Desertification
opened for signature—14 October 1994
entered into force—26 December 1996
objective—to combat desertification and mitigate the effects of drought through an integrated framework that is consistent with Agenda 21, employing international cooperation and partnership arrangements, and effective action at all levels
parties—(196 and the Palestine Liberation Organization) Afghanistan, Albania, Algeria, Andorra, Angola, Antigua and Barbuda, Argentina, Armenia, Australia, Austria, Azerbaijan, The Bahamas, Bahrain, Bangladesh, Barbados, Belarus, Belgium, Belize, Benin, Bhutan, Bolivia, Bosnia and Herzegovina, Botswana, Brazil, Brunei, Bulgaria, Burkina Faso, Burma, Burundi, Cambodia, Cameroon, Canada, Cape Verde, Central African Republic, Chad, Chile, China, Colombia, Comoros, Democratic Republic of the Congo, Republic of the Congo, Cook Islands, Costa Rica, Cote d'Ivoire, Croatia, Cuba, Cyprus, Czechia, Denmark, Djibouti, Dominica, Dominican Republic, Ecuador, Egypt, El Salvador, Equatorial Guinea, Eritrea, Estonia, Eswatini, Ethiopia, EU, Fiji, Finland, France, Gabon, The Gambia, Germany, Ghana, Greece, Grenada, Guatemala, Guinea, Guinea-Bissau, Guyana, Haiti, Honduras, Hungary, Iceland, India, Indonesia, Iran, Iraq, Ireland, Israel, Italy, Jamaica, Japan, Jordan, Kazakhstan, Kenya, Kiribati, North Korea, South Korea, Kuwait, Kyrgyzstan, Laos, Latvia, Lebanon, Lesotho, Liberia, Libya, Liechtenstein, Lithuania, Luxembourg, Madagascar, Malawi, Malaysia, Maldives, Mali, Malta, Marshall Islands, Mauritania, Mauritius, Mexico, Federated States of Micronesia, Moldova, Monaco, Mongolia, Montenegro, Morocco, Mozambique, Namibia, Nauru, Nepal, Netherlands, NZ, Nicaragua, Niger, Nigeria, Niue, North Macedonia, Norway, Oman, Pakistan, Palau, Panama, Papua New Guinea, Paraguay, Peru, Philippines, Poland, Portugal, Qatar, Romania, Russia, Rwanda, Saint Kitts and Nevis, Saint Lucia, Saint Vincent and the Grenadines, Samoa, San Marino, Sao Tome and Principe, Saudi Arabia, Senegal, Serbia, Seychelles, Sierra Leone, Singapore, Slovakia, Slovenia, Solomon Islands, Somalia, South Africa, South Sudan, Spain, Sri Lanka, Sudan, Suriname, Sweden, Switzerland, Syria, Tajikistan, Thailand, Tanzania, Timor-Leste, Togo, Tonga, Trinidad and Tobago, Tunisia, Turkey, Turkmenistan, Tuvalu, Uganda, Ukraine, UAE, UK, US, Uruguay, Uzbekistan, Vanutu, Venezuela, Vietnam, Yemen, Zambia, Zimbabwe, Palestine Liberation Organization

United Nations Framework Convention on Climate Change
note—abbreviated as Climate Change
opened for signature—9 May 1992
entered into force—21 March 1994
objective—to achieve stabilization of greenhouse gas concentrations in the atmosphere at a low enough level to prevent dangerous anthropogenic interference with the climate system
parties—(196 and the Palestine Liberation Organization) Afghanistan, Albania, Algeria, Andorra, Angola, Antigua and Barbuda, Argentina, Armenia, Australia, Austria, Azerbaijan, The Bahamas, Bahrain, Bangladesh, Barbados, Belarus, Belgium, Belize, Benin, Bhutan, Bolivia, Bosnia and Herzegovina, Botswana, Brazil, Brunei, Bulgaria, Burkina Faso, Burma, Burundi, Cambodia, Cameroon, Canada, Cape Verde, Central African Republic, Chad, Chile, China, Colombia, Comoros, Democratic Republic of the Congo, Republic of the Congo, Cook Islands, Costa Rica, Cote d'Ivoire, Croatia, Cuba, Cyprus, Czechia, Denmark, Djibouti, Dominica, Dominican Republic, Ecuador, Egypt, El Salvador, Equatorial Guinea, Eritrea, Estonia, Eswatini,Ethiopia, EU, Fiji, Finland, France, Gabon, The Gambia, Georgia, Germany, Ghana, Greece, Grenada, Guatemala, Guinea, Guinea-Bissau, Guyana, Haiti, Honduras, Hungary, Iceland, India, Indonesia, Iran, Iraq, Ireland, Israel, Italy, Jamaica, Japan, Jordan, Kazakhstan, Kenya, Kiribati, North Korea, South Korea, Kuwait, Kyrgyzstan, Laos, Latvia, Lebanon, Lesotho, Liberia, Libya, Liechtenstein, Lithuania, Luxembourg, Madagascar, Malawi, Malaysia, Maldives, Mali, Malta, Marshall

Islands, Mauritania, Mauritius, Mexico, Federated States of Micronesia, Moldova, Monaco, Mongolia, Montenegro, Morocco, Mozambique, Namibia, Nauru, Nepal, Netherlands, NZ, Nicaragua, Niger, Nigeria, Niue, North Macedonia, Norway, Oman, Pakistan, Palau, Panama, Papua New Guinea, Paraguay, Peru, Philippines, Poland, Portugal, Qatar, Romania, Russia, Rwanda, Saint Kitts and Nevis, Saint Lucia, Saint Vincent and the Grenadines, Samoa, San Marino, Sao Tome and Principe, Saudi Arabia, Senegal, Serbia, Seychelles, Sierra Leone, Singapore, Slovakia, Slovenia, Solomon Islands, Somalia, South Africa, South Sudan, Spain, Sri Lanka, Sudan, Suriname, Sweden, Switzerland, Syria, Tajikistan, Tanzania, Thailand, Timor-Leste, Togo, Tonga, Trinidad and Tobago, Tunisia, Turkey, Turkmenistan, Tuvalu, Uganda, Ukraine, UAE, UK, US, Uruguay, Uzbekistan, Vanuatu, Venezuela, Vietnam, Yemen, Zambia, Zimbabwe, Palestine Liberation Organization

Wetlands
see Convention on Wetlands of International Importance Especially As Waterfowl Habitat (Ramsar)

Whaling
see International Convention for the Regulation of Whaling

POLITICAL MAP OF AFRICA

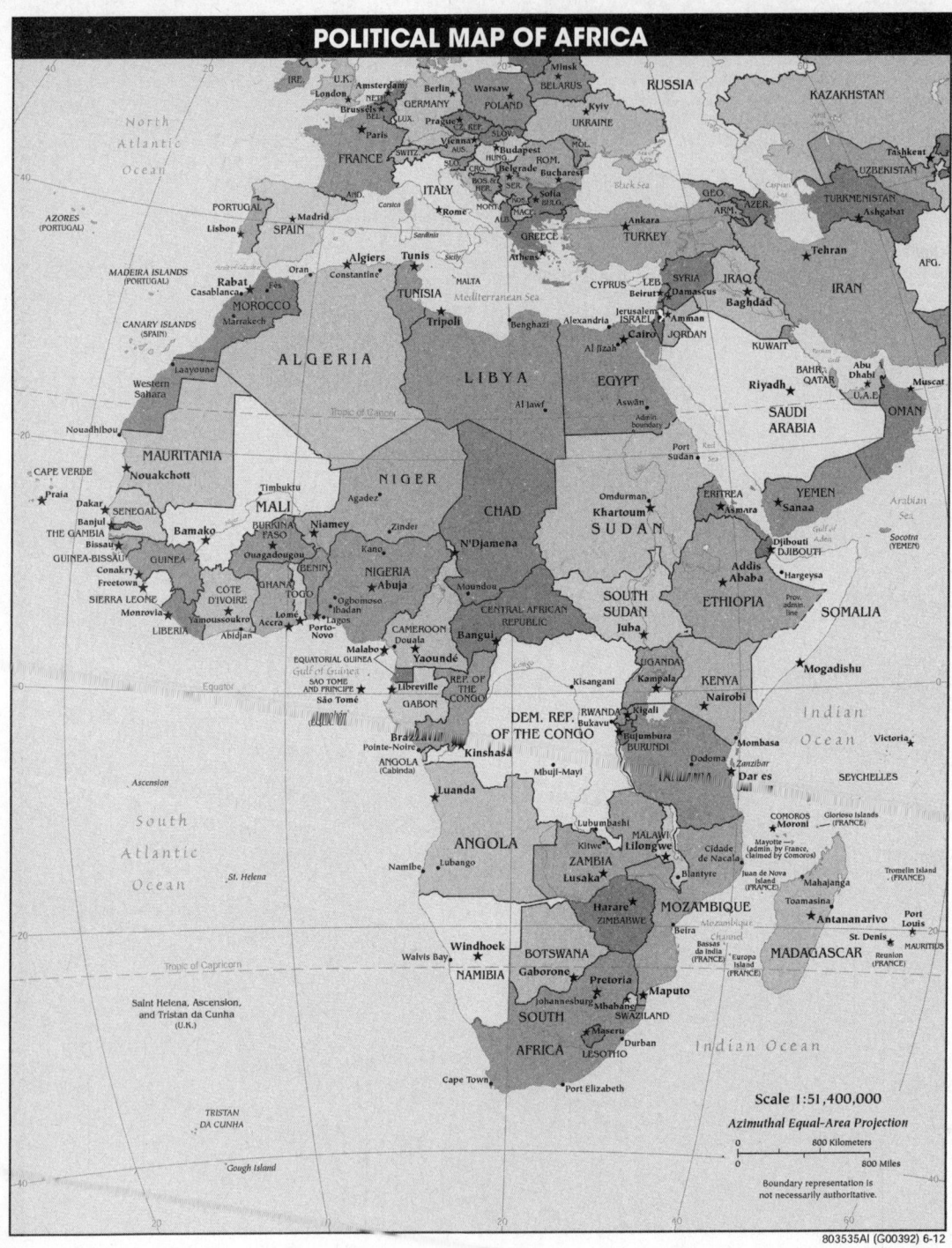

Scale 1:51,400,000

Azimuthal Equal-Area Projection

0   800 Kilometers

0   800 Miles

Boundary representation is
not necessarily authoritative.

803535AI (G00392) 6-12

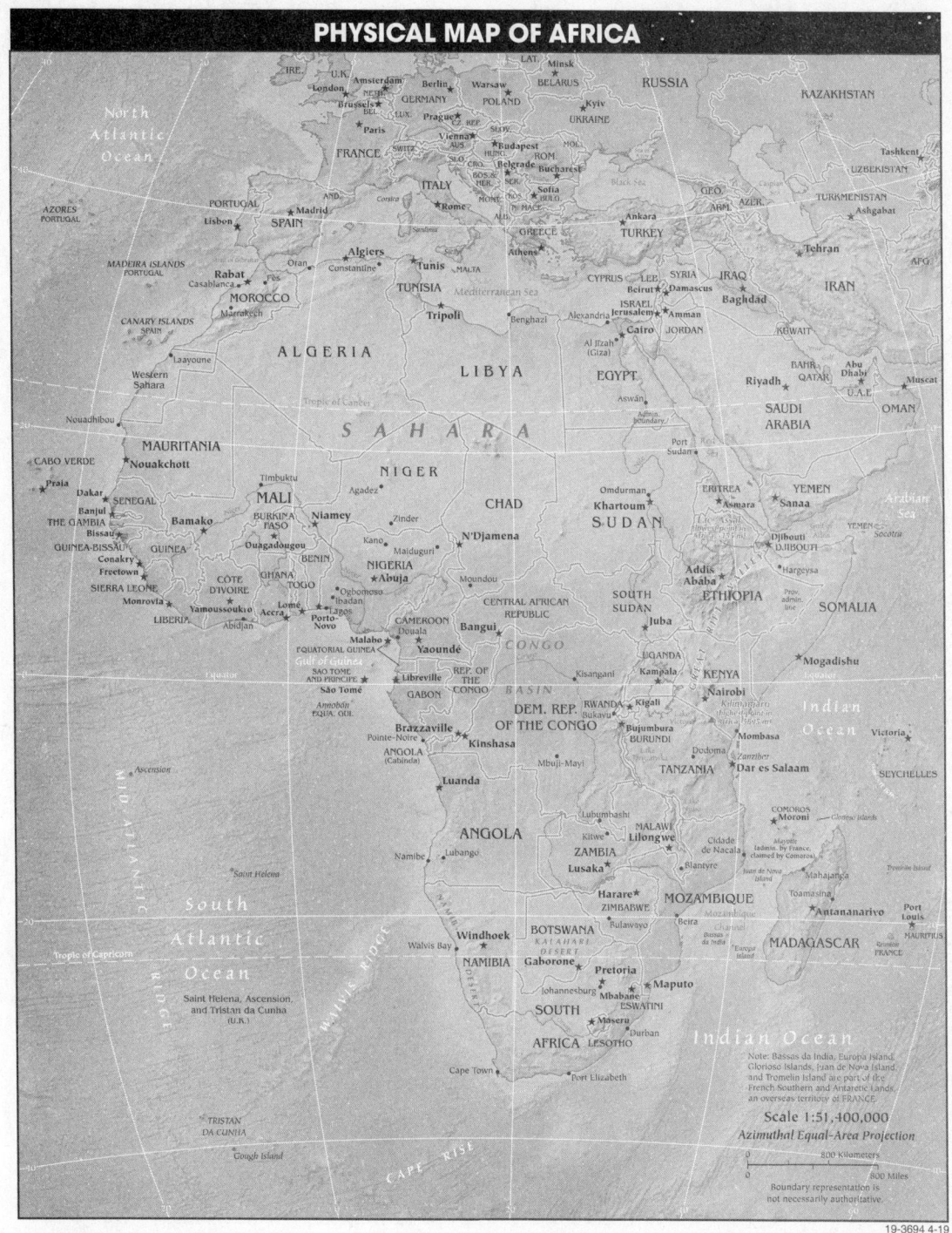

PHYSICAL MAP OF AFRICA

Note: Bassas da India, Europa Island, Glorioso Islands, Juan de Nova Island, and Tromelin Island are part of the French Southern and Antarctic Lands, an overseas territory of FRANCE.

Scale 1:51,400,000
Azimuthal Equal-Area Projection

Boundary representation is not necessarily authoritative.

19-3694 4-19

PHYSICAL MAP OF ANTARCTIC REGION

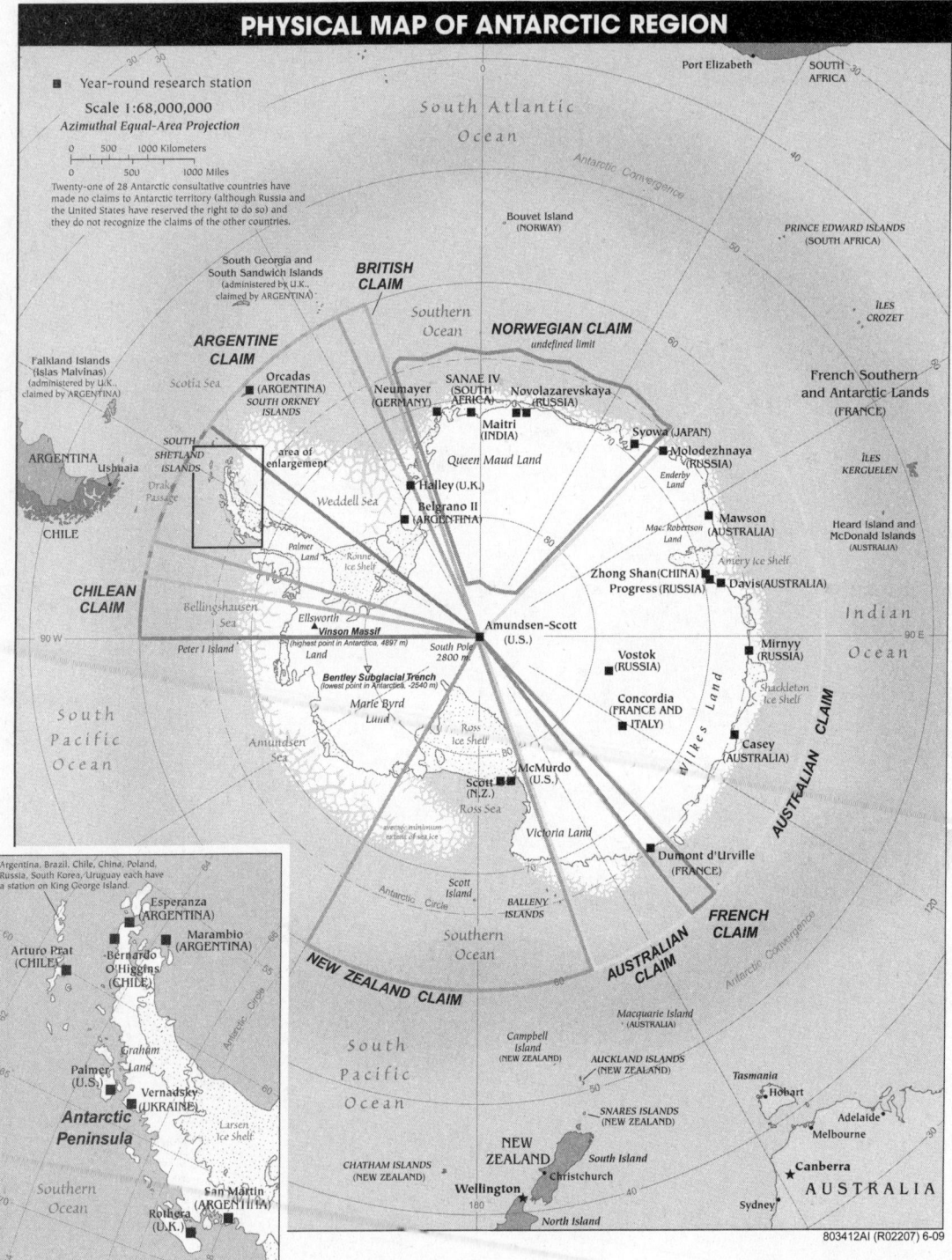

■ Year-round research station

Scale 1:68,000,000
Azimuthal Equal-Area Projection

0   500   1000 Kilometers
0   500   1000 Miles

Twenty-one of 28 Antarctic consultative countries have
made no claims to Antarctic territory (although Russia and
the United States have reserved the right to do so) and
they do not recognize the claims of the other countries.

Port Elizabeth
SOUTH AFRICA

South Atlantic Ocean

Antarctic Convergence

Bouvet Island (NORWAY)

PRINCE EDWARD ISLANDS (SOUTH AFRICA)

ÎLES CROZET

South Georgia and South Sandwich Islands (administered by U.K., claimed by ARGENTINA)

BRITISH CLAIM

Southern Ocean

NORWEGIAN CLAIM
undefined limit

French Southern and Antarctic Lands (FRANCE)

ARGENTINE CLAIM

Falkland Islands (Islas Malvinas) (administered by U.K., claimed by ARGENTINA)

Scotia Sea

Orcadas (ARGENTINA) SOUTH ORKNEY ISLANDS

Neumayer (GERMANY)   SANAE IV (SOUTH AFRICA)   Novolazarevskaya (RUSSIA)

Maitri (INDIA)

Syowa (JAPAN)
Molodezhnaya (RUSSIA)

ÎLES KERGUELEN

ARGENTINA   Ushuaia

SOUTH SHETLAND ISLANDS

Drake Passage

area of enlargement

Queen Maud Land

Enderby Land

Heard Island and McDonald Islands (AUSTRALIA)

CHILE

Weddell Sea

Halley (U.K.)

Belgrano II (ARGENTINA)

Mawson (AUSTRALIA)

Mac. Robertson Land

CHILEAN CLAIM

Palmer Land

Ronne Ice Shelf

Amery Ice Shelf

Zhong Shan (CHINA)
Progress (RUSSIA)

Davis (AUSTRALIA)

Indian Ocean

Bellingshausen Sea

Ellsworth
Vinson Massif
(highest point in Antarctica, 4897 m)
Land

90 W

Peter I Island

Amundsen-Scott (U.S.)

South Pole 2800 m

Bentley Subglacial Trench
(lowest point in Antarctica, -2540 m)

Marie Byrd Land

Vostok (RUSSIA)

Mirnyy (RUSSIA)

90 E

Shackleton Ice Shelf

Concordia (FRANCE AND ITALY)

Casey (AUSTRALIA)

AUSTRALIAN CLAIM

South Pacific Ocean

Amundsen Sea

Ross Ice Shelf

Wilkes Land

McMurdo (U.S.)

Scott (N.Z.)

Ross Sea

Victoria Land

Dumont d'Urville (FRANCE)

FRENCH CLAIM

Scott Island

BALLENY ISLANDS

average minimum extent of sea ice

Antarctic Circle

Southern Ocean

NEW ZEALAND CLAIM

AUSTRALIAN CLAIM

Macquarie Island (AUSTRALIA)

Campbell Island (NEW ZEALAND)

AUCKLAND ISLANDS (NEW ZEALAND)

Tasmania
Hobart

South Pacific Ocean

SNARES ISLANDS (NEW ZEALAND)

Adelaide
Melbourne

CHATHAM ISLANDS (NEW ZEALAND)

NEW ZEALAND   South Island
Christchurch

Canberra

AUSTRALIA

Wellington

North Island

Sydney

803412AI (R02207) 6-09

Argentina, Brazil, Chile, China, Poland, Russia, South Korea, Uruguay each have a station on King George Island.

Esperanza (ARGENTINA)

Marambio (ARGENTINA)

Arturo Prat (CHILE)

Bernardo O'Higgins (CHILE)

Palmer (U.S.)

Vernadsky (UKRAINE)

Graham Land

Antarctic Peninsula

Larsen Ice Shelf

San Martín (ARGENTINA)

Rothera (U.K.)

Southern Ocean

POLITICAL MAP OF ARCTIC REGION

North Pacific Ocean

ALEUTIAN ISLANDS

Bering Sea

Petropavlovsk-Kamchatskiy

KURIL ISLANDS

occupied by the Soviet Union in 1945; administered by Russia; claimed by Japan

JAPAN

Sea of Japan

Kodiak

Bristol Bay

Bethel

Gulf of Alaska

Anchorage

Valdez

Juneau

Prince George

Whitehorse

Fairbanks

Dawson

Fort Nelson

Inuvik

Fort McMurray

Yellowknife

Nome

Providenlya

Anadyr'

Magadan

Sea of Okhotsk

Khabarovsk

Arctic Circle

Okhotsk

Oymyakon

CHINA

UNITED STATES

Chukchi Sea

Peyek

Cherskiy

Yakutsk

Verkhoyansk

120

Prudhoe Bay

Barrow

Wrangel Island

East Siberian Sea

Beaufort Sea

Banks Island

sea ice extent summer average 2000-2006

80

NEW SIBERIAN ISLANDS

Tiksi

Laptev Sea

Cambridge Bay

Victoria Island

CANADA

Churchill

Arviat

Gjoa Haven

Resolute

QUEEN ELIZABETH ISLANDS

Ellesmere Island

Arctic Ocean

SEVERNAYA ZEMLYA

RUSSIA

Noril'sk

90W

Rankin Inlet

Repulse Bay

Pond Inlet

Baffin Island

Alert

Kara Sea

Dikson

90E

Yenisey

Hudson Bay

Qaanaaq (Thule)

FRANZ IOSEF LAND

NOVAYA ZEMLYA

Baffin Bay

Nord

Iqaluit

Svalbard (NORWAY)

Longyearbyen

Greenland (DENMARK)

70

Kuujjuaq

Davis Strait

Julissat (Jakobshavn)

Sisimiut (Holsteinsborg)

Bjørnøya (NORWAY)

Barents Sea

Yekaterinburg

Perm'

60

Nuuk (Godthåb)

Labrador Sea

Ittoqqortoormiit (Scoresbysund)

Greenland Sea

Murmansk

Arkhangel'sk

Lake Onega

Severnaya Dvina

Qaqortoq (Julianehåb)

Tasiilaq

Jan Mayen (NORWAY)

White Sea

Kazan'

Samara

Denmark Strait

Norwegian Sea

Lake Ladoga

Nizhniy Novgorod

KAZ.

Arctic Circle

Reykjavik

ICELAND

NORWAY

FINLAND

Saint Petersburg

Moscow

Saratov

North Atlantic Ocean

Tórshavn

Faroe Islands (DENMARK)

SHETLAND ISLANDS

SWEDEN

Helsinki

Oslo

Stockholm

Tallinn EST.

Volgograd

Voronezh

LATVIA

Riga

Baltic Sea

LITH. Vilnius

Minsk

BELARUS

Kharkiv

Rostov

Krasnodar

Copenhagen

Kaliningrad (RUS.)

North Sea

Belfast

Dublin IRE.

London

U.K.

Amsterdam

NETH.

DENMARK

GERMANY

Berlin

Warsaw

POLAND

Prague CZECH REP.

SLOV.

HUNG.

Kyiv

UKRAINE

Chisinau

MOL.

ROMANIA

Black Sea

Sochi

TURKEY

Scale 1:39,000,000

Lambert Azimuthal Equal-Area Projection

0   500 Kilometers

0   30   500 Miles

PHYSICAL MAP OF ARCTIC REGION

Scale 1:39,000,000
Lambert Azimuthal Equal-Area Projection

803400AI (G01486) 6-0

POLITICAL MAP OF ASIA

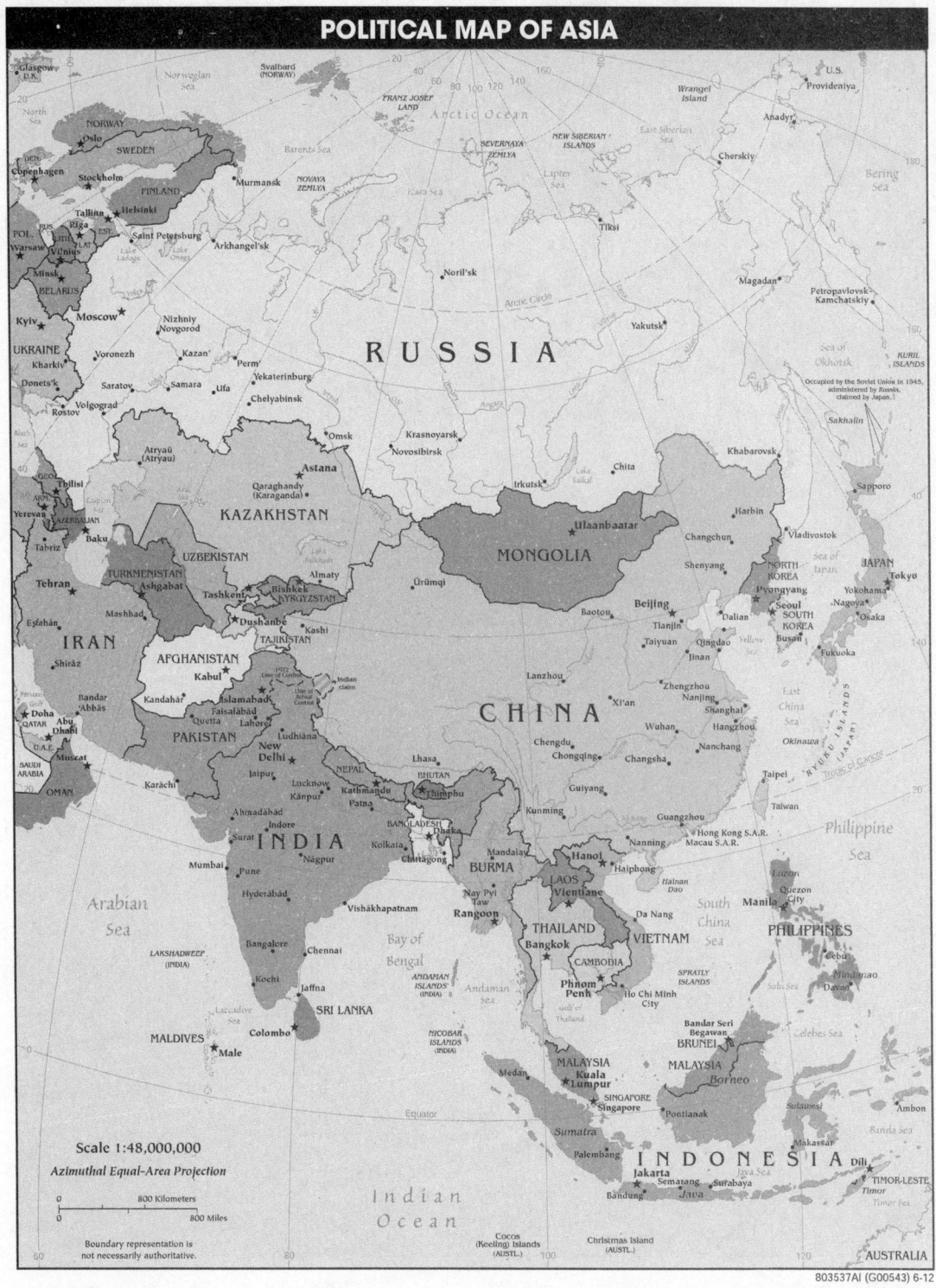

Scale 1:48,000,000

Azimuthal Equal-Area Projection

	800 Kilometers
0	
0	800 Miles

Boundary representation is
not necessarily authoritative.

803537AI (G00543) 6-12

PHYSICAL MAP OF ASIA

Scale 1:48,000,000

Azimuthal Equal-Area Projection

0   800 Kilometers

0   800 Miles

Boundary representation is
not necessarily authoritative

803623AI (G00543) 8-13

POLITICAL MAP OF CENTRAL AMERICA AND THE CARIBBEAN

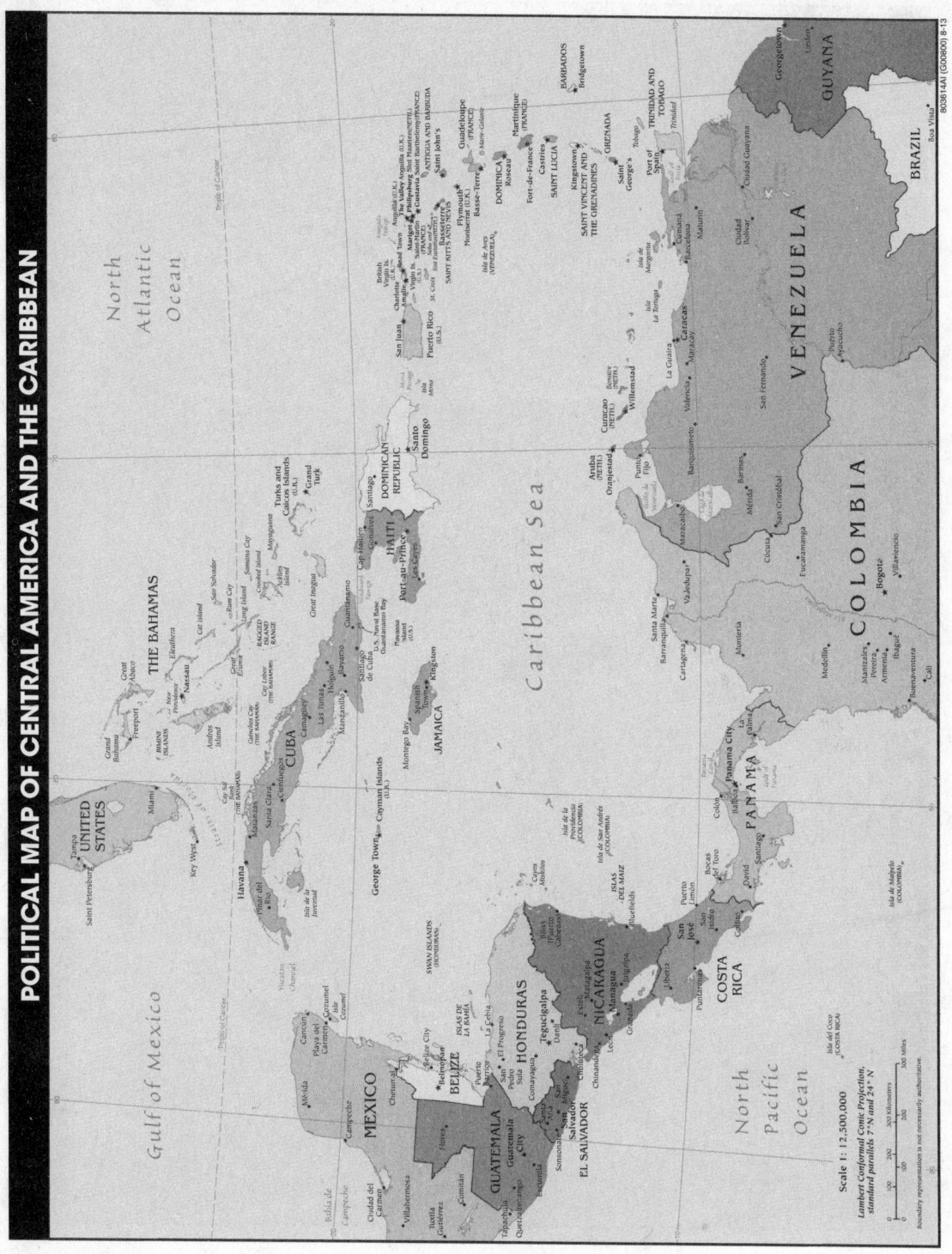

Scale 1: 12,500,000

Lambert Conformal Conic Projection,
standard parallels 7°N and 24°N

Boundary representation is not necessarily authoritative.

PHYSICAL MAP OF CENTRAL AMERICA AND THE CARIBBEAN

POLITICAL MAP OF EUROPE

Scale 1:19,300,000
Lambert Conformal Conic Projection,
standard parallels 40°N and 68°N

803539AI (G00772) 6-12

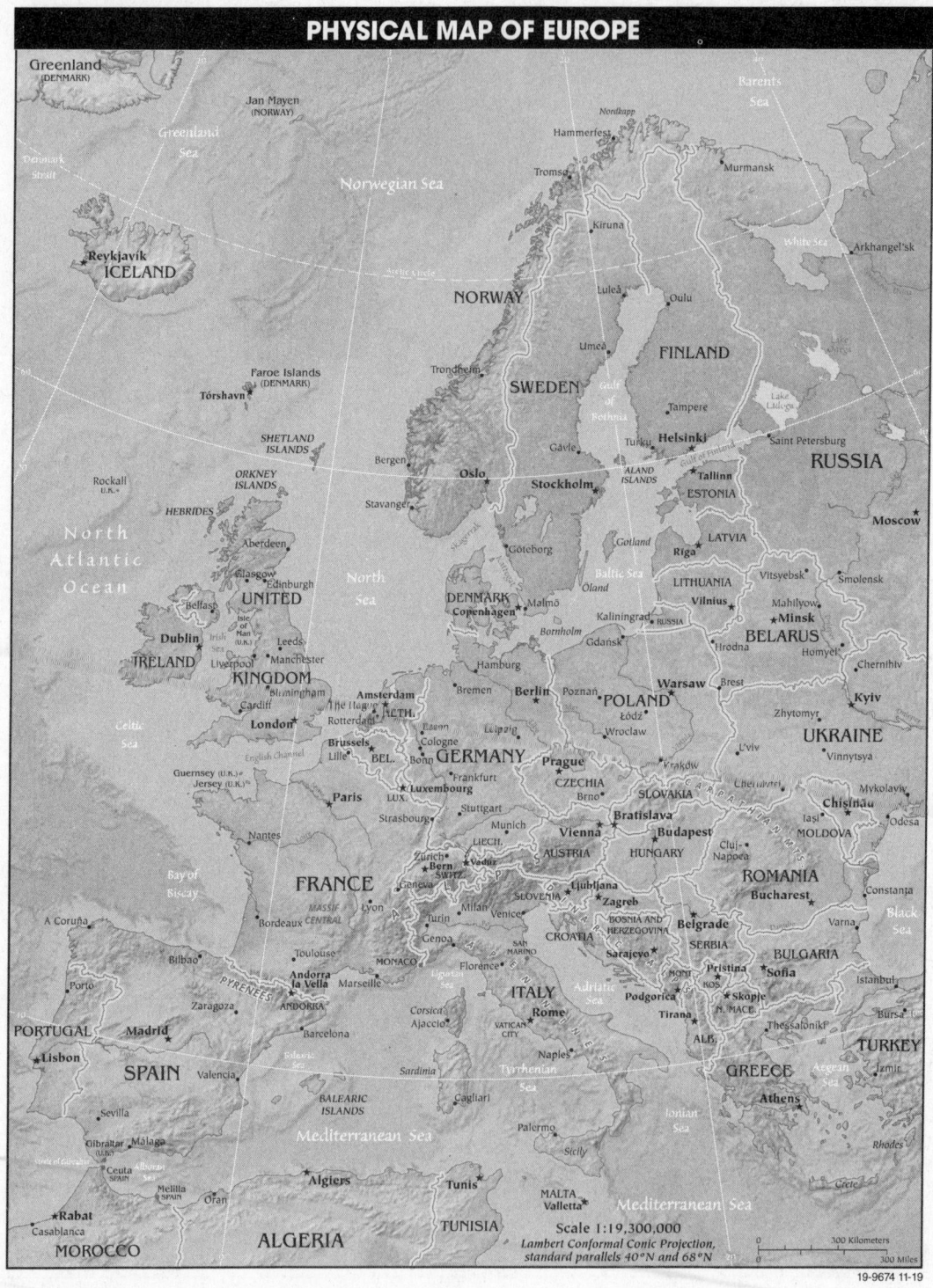

PHYSICAL MAP OF EUROPE

Scale 1:19,300,000
Lambert Conformal Conic Projection,
standard parallels 40°N and 68°N

19-9674 11-19

POLITICAL MAP OF MIDDLE EAST

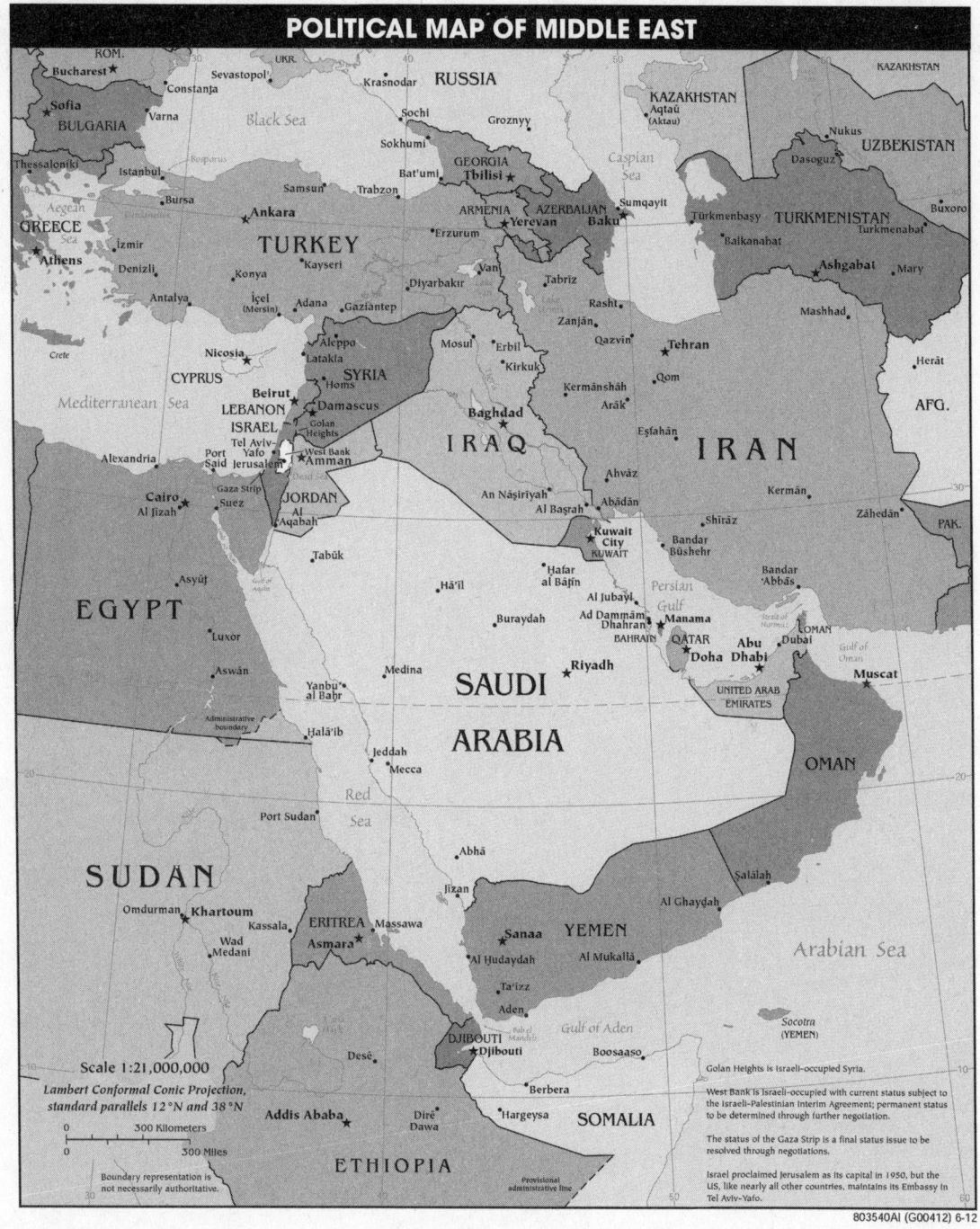

Scale 1:21,000,000

*Lambert Conformal Conic Projection,
standard parallels 12°N and 38°N*

| 0 | 300 Kilometers |
| 0 | 300 Miles |

Boundary representation is
not necessarily authoritative.

Golan Heights is Israeli-occupied Syria.

West Bank is Israeli-occupied with current status subject to
the Israeli-Palestinian Interim Agreement; permanent status
to be determined through further negotiation.

The status of the Gaza Strip is a final status issue to be
resolved through negotiations.

Israel proclaimed Jerusalem as its capital in 1950, but the
US, like nearly all other countries, maintains its Embassy in
Tel Aviv-Yafo.

803540AI (G00412) 6-12

PHYSICAL MAP OF MIDDLE EAST

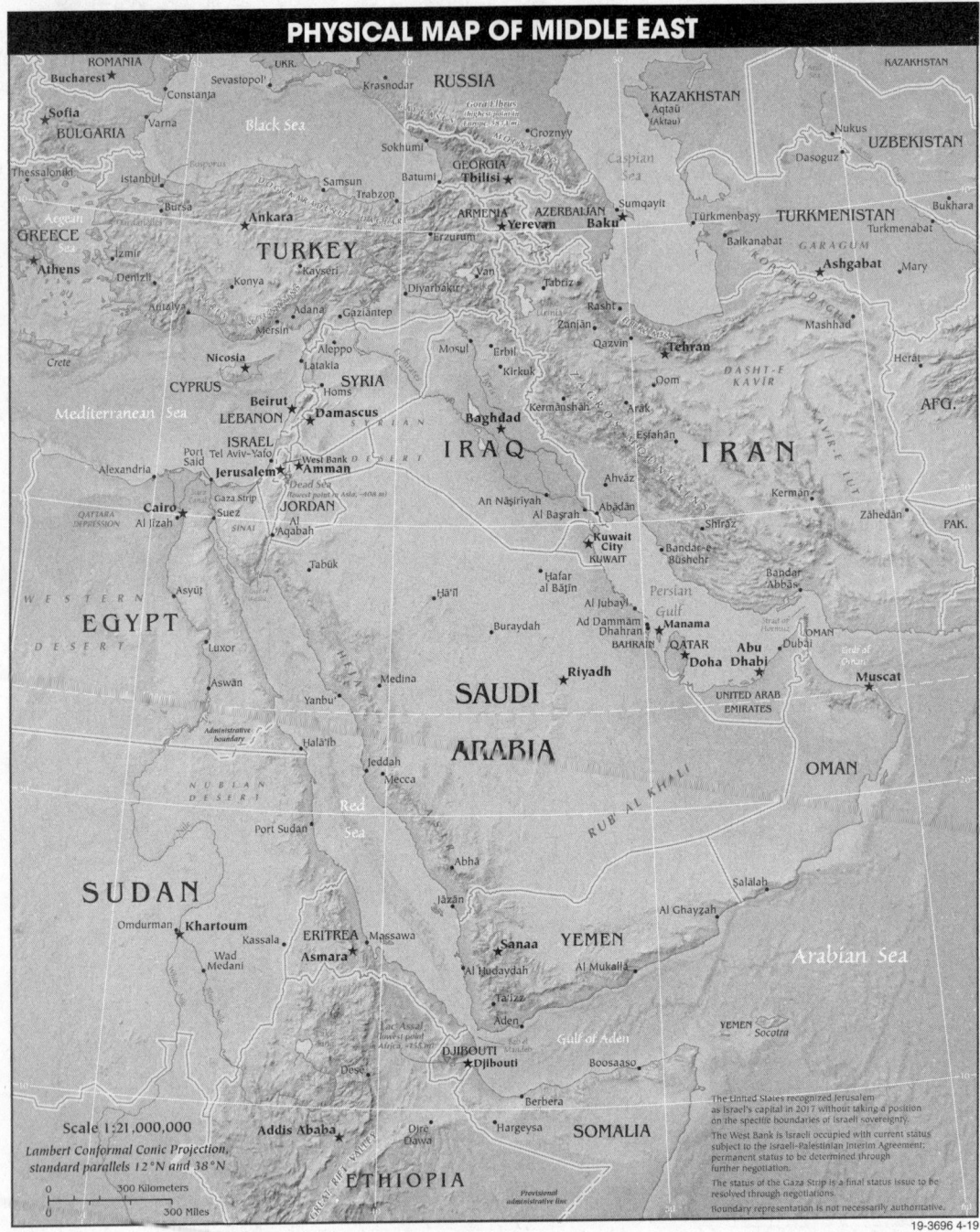

Scale 1:21,000,000
Lambert Conformal Conic Projection,
standard parallels 12°N and 38°N

0   300 Kilometers
0   300 Miles

The United States recognized Jerusalem
as Israel's capital in 2017 without taking a position
on the specific boundaries of Israeli sovereignty.

The West Bank is Israeli occupied with current status
subject to the Israeli-Palestinian Interim Agreement;
permanent status to be determined through
further negotiation.

The status of the Gaza Strip is a final status issue to be
resolved through negotiations.

Boundary representation is not necessarily authoritative.

19-3696 4-19

POLITICAL MAP OF NORTH AMERICA

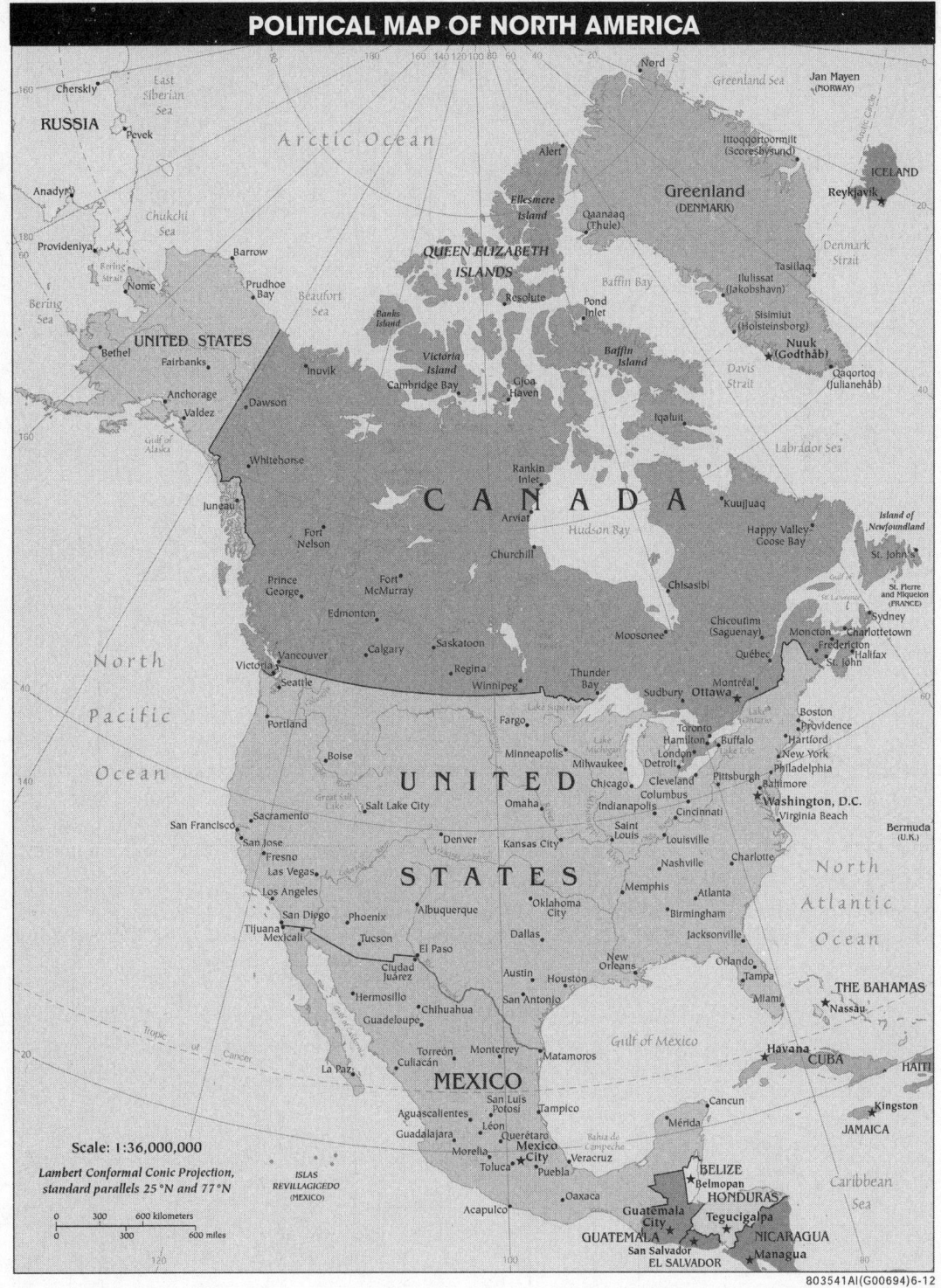

Scale: 1:36,000,000

Lambert Conformal Conic Projection,
standard parallels 25°N and 77°N

0   300   600 kilometers
0   300   600 miles

803541AI(G00694)6-12

PHYSICAL MAP OF NORTH AMERICA

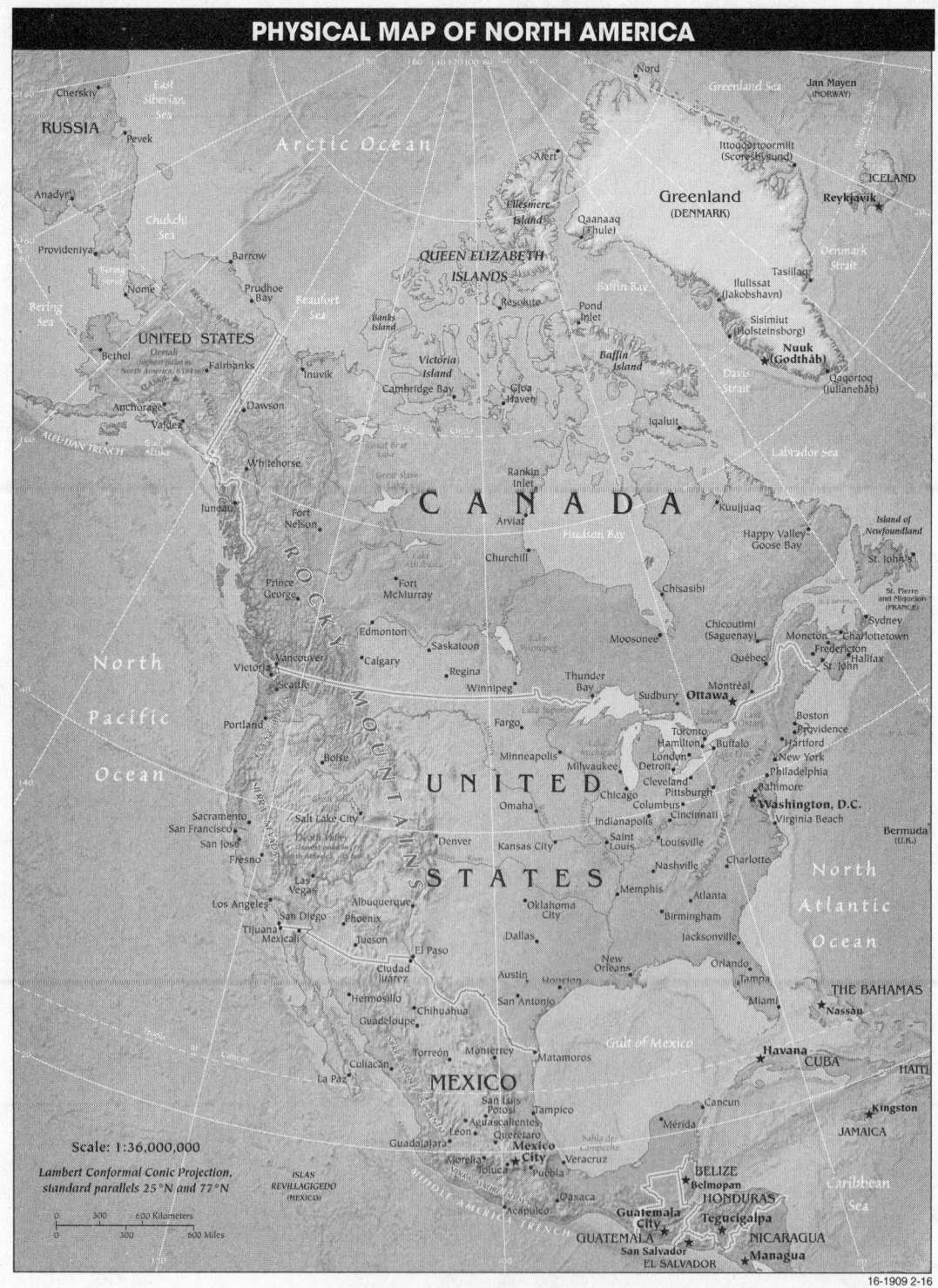

Scale: 1:36,000,000

Lambert Conformal Conic Projection,
standard parallels 25°N and 77°N

0   300   600 Kilometers
0   300   600 Miles

16-1909 2-16

POLITICAL MAP OF OCEANIA

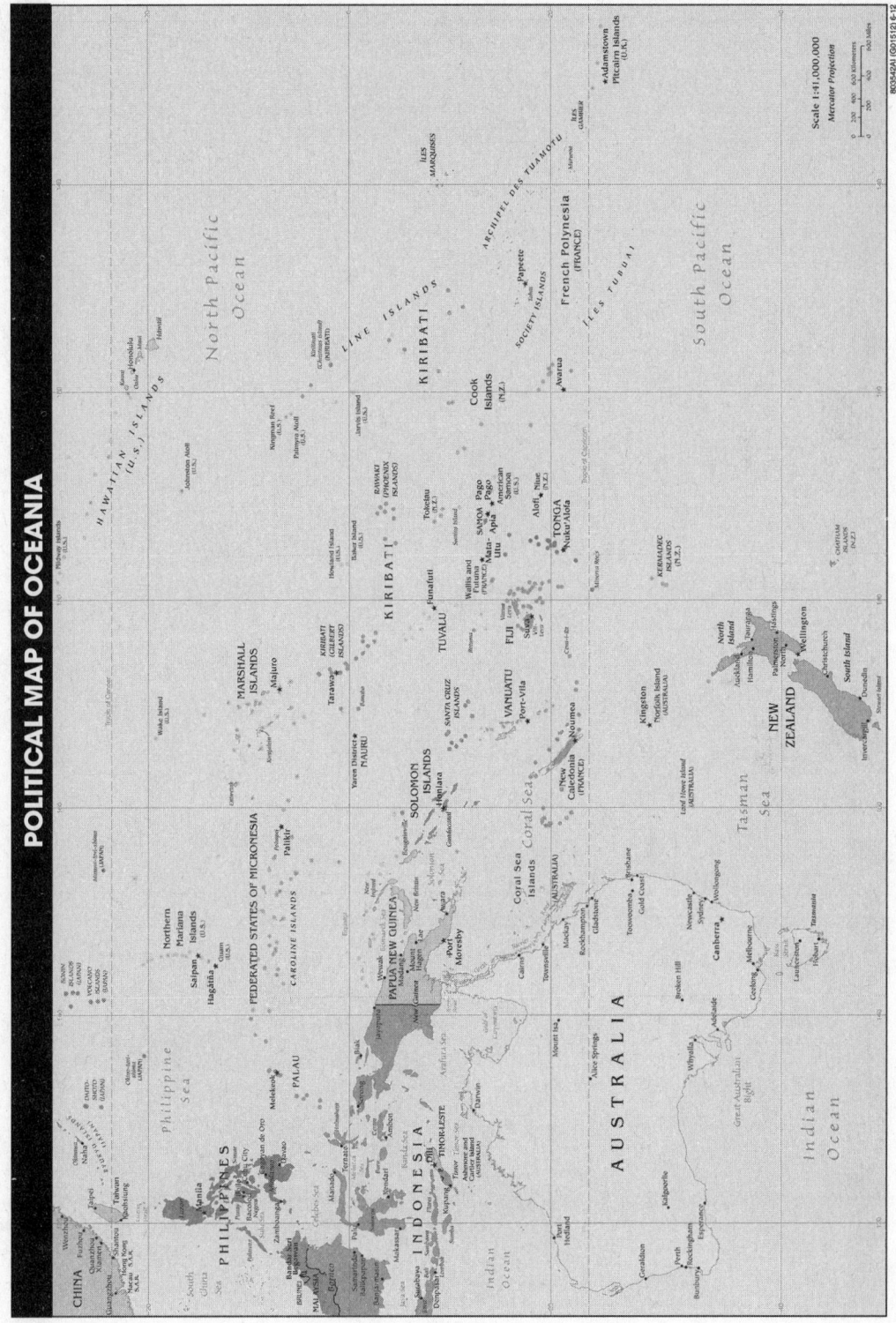

Scale 1:41,000,000
Mercator Projection

PHYSICAL MAP OF OCEANIA

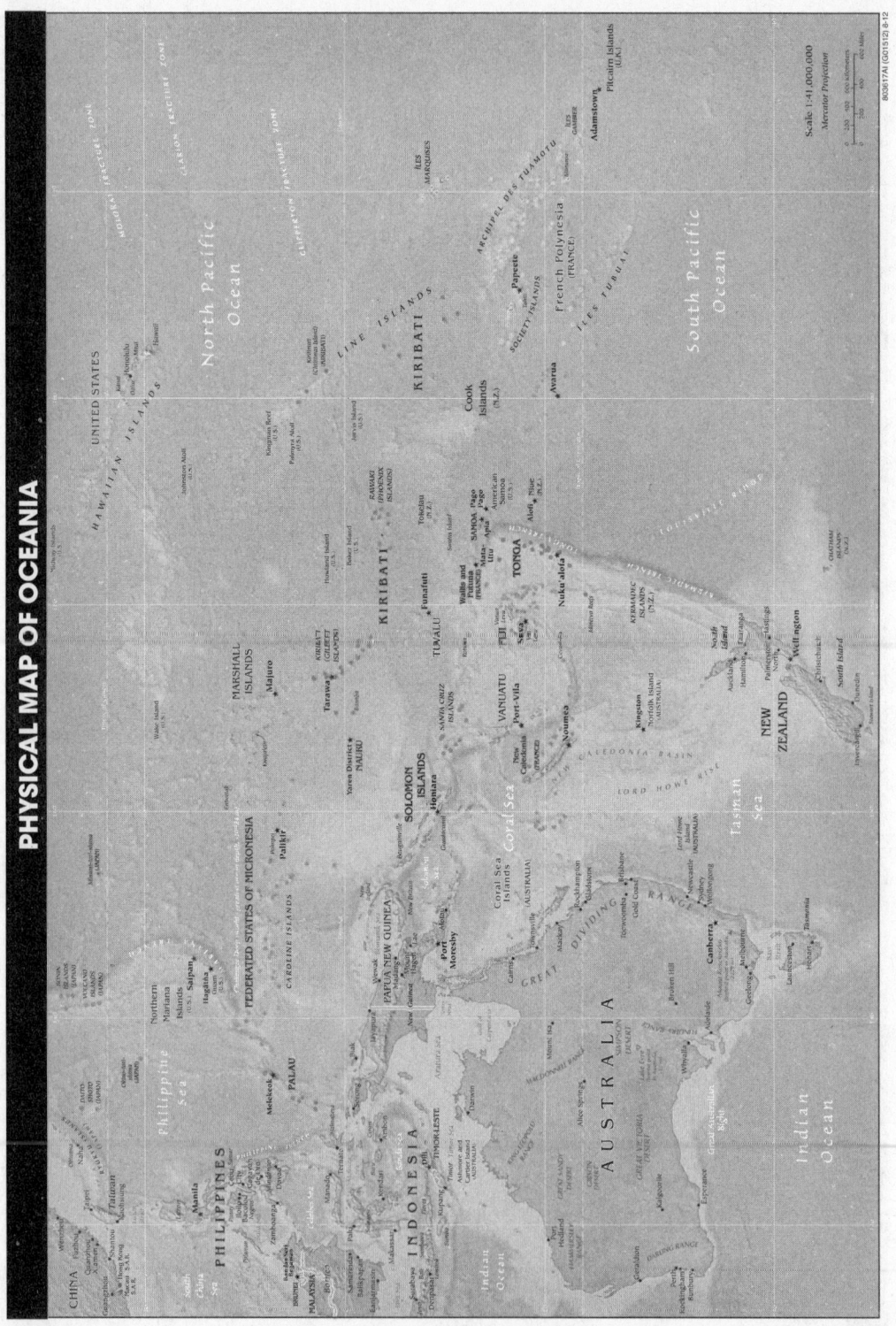

Scale 1:41,000,000
Mercator Projection

PHYSICAL MAP OF THE WORLD

Physical Map of the World, January 2015

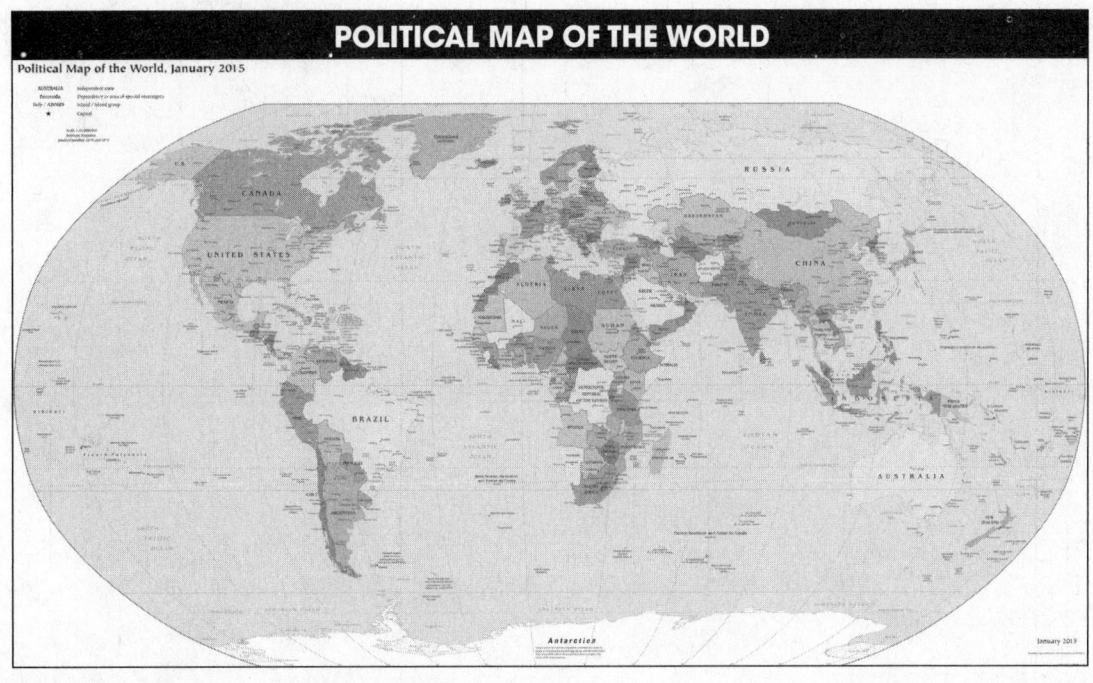

POLITICAL MAP OF THE WORLD

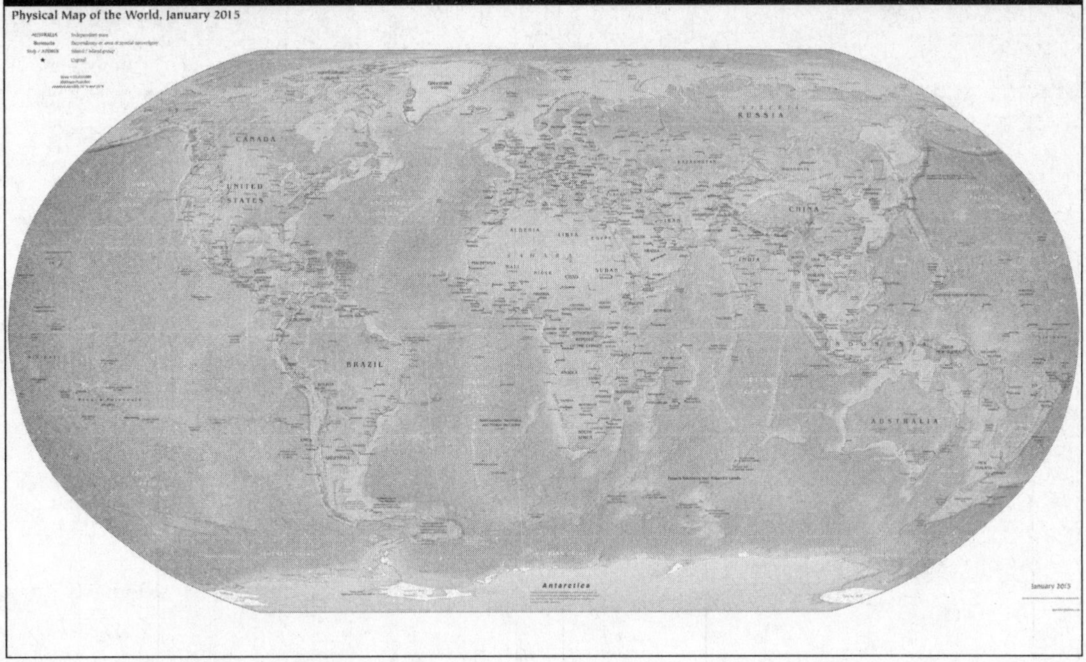

Political Map of the World, January 2015

POLITICAL MAP OF SOUTH AMERICA

Scale 1:35,000,000
Azimuthal Equal-Area Projection

500 Kilometers

500 Miles

Boundary representation is
not necessarily authoritative.

Stanley
Falkland Islands
(Islas Malvinas)
(administered by U.K.,
claimed by ARGENTINA)

South Georgia and
South Sandwich Islands
(administered by U.K.,
claimed by ARGENTINA)

803543AI (G00186) 6-12

PHYSICAL MAP OF SOUTH AMERICA

Scale 1:35,000,000
Azimuthal Equal-Area Projection

0   500 Kilometers
0   500 Miles

Boundary representation is
not necessarily authoritative.

803521AI (G00186) 11-11

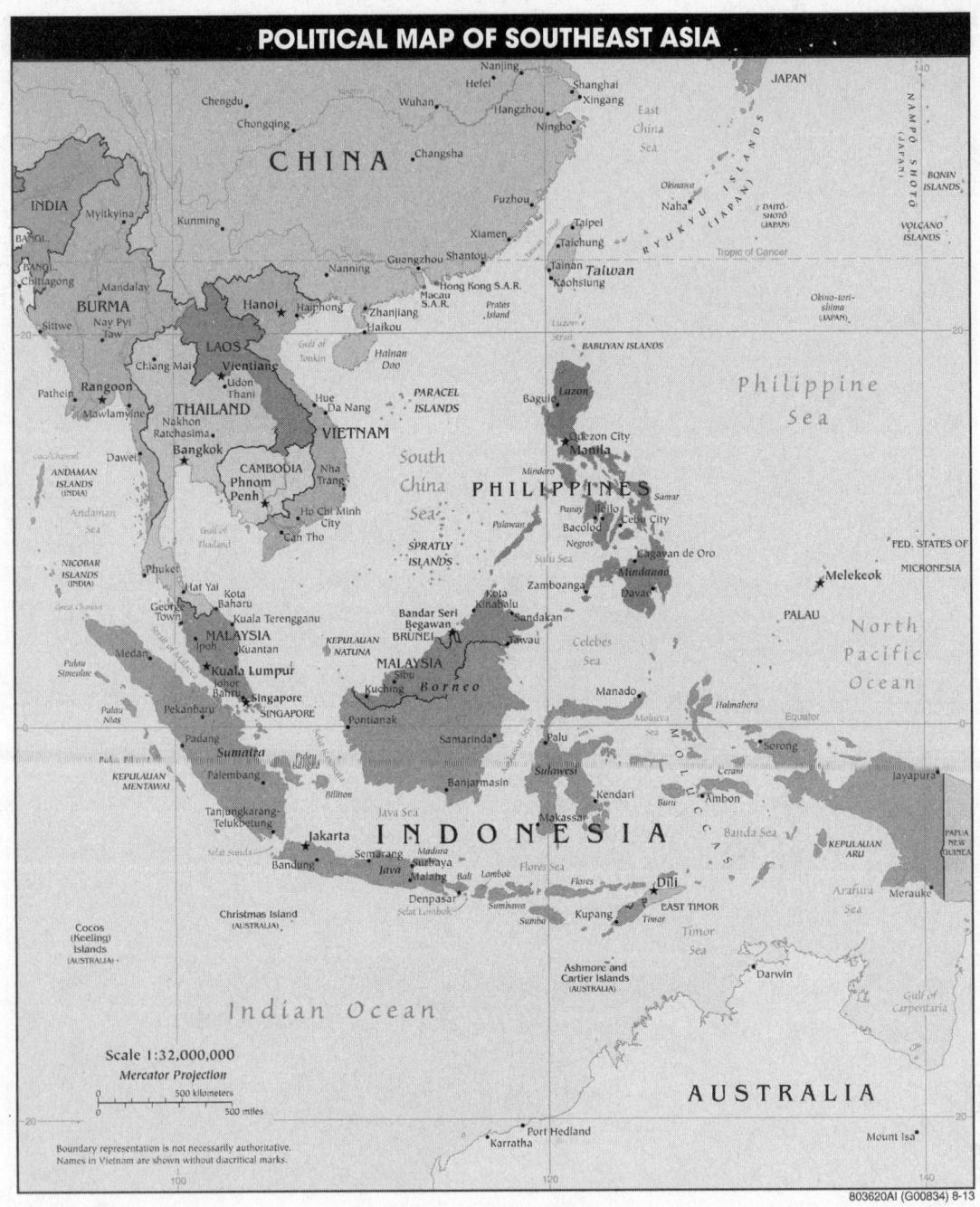

POLITICAL MAP OF SOUTHEAST ASIA

PHYSICAL MAP OF SOUTHEAST ASIA

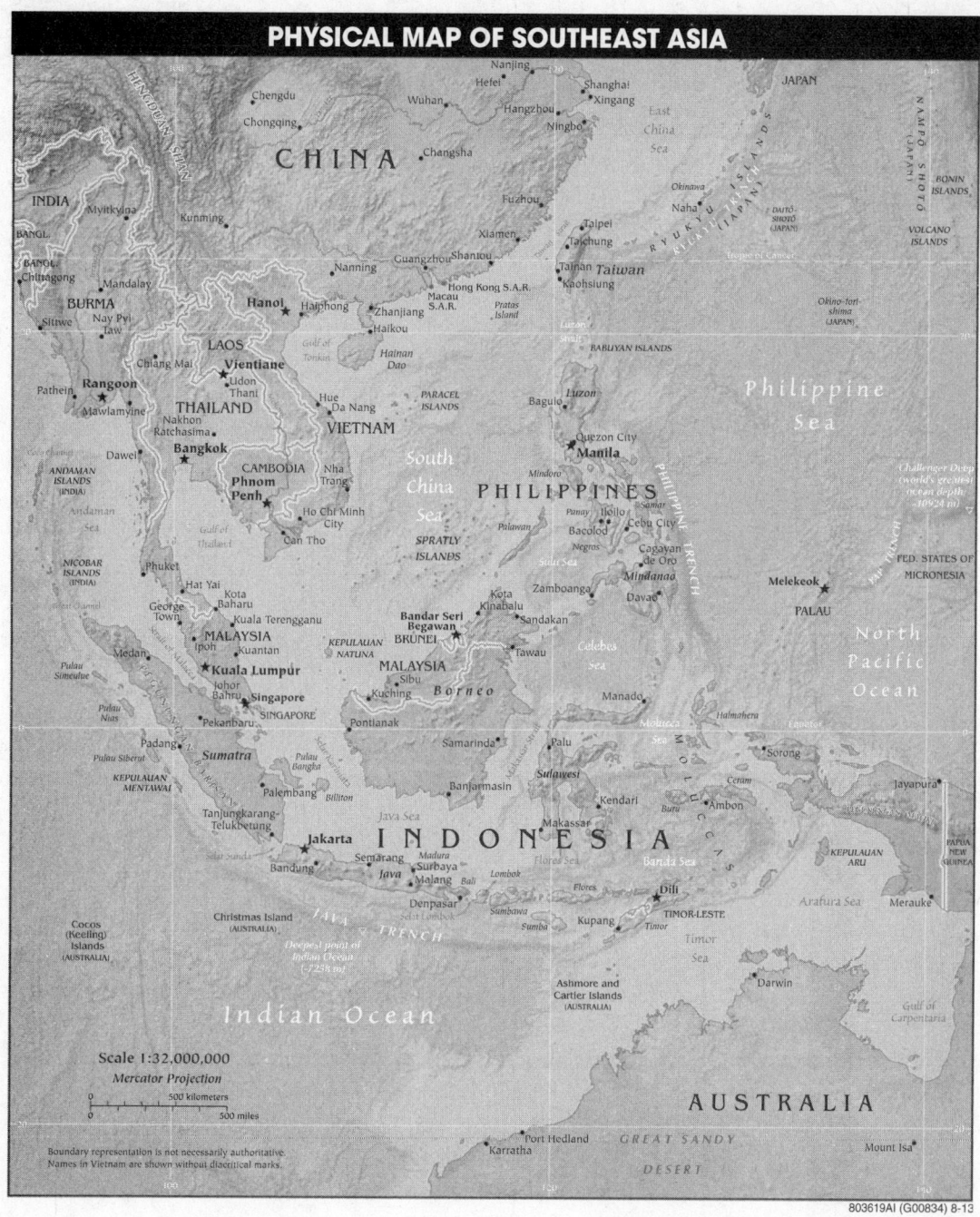

Scale 1:32,000,000
Mercator Projection

0   500 kilometers
0   500 miles

Boundary representation is not necessarily authoritative.
Names in Vietnam are shown without diacritical marks.

803619AI (G00834) 8-13

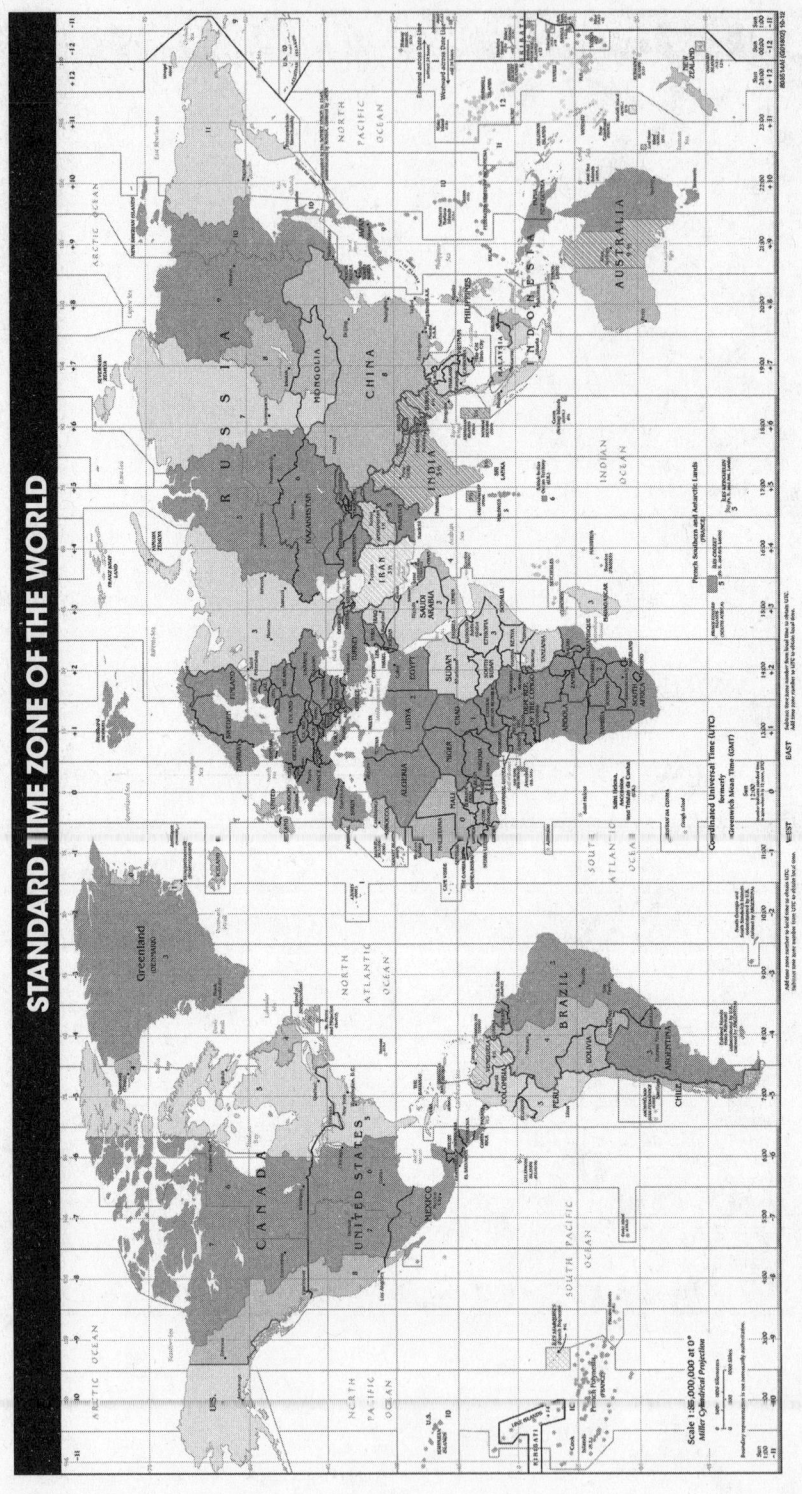

STANDARD TIME ZONE OF THE WORLD

UNITED STATES

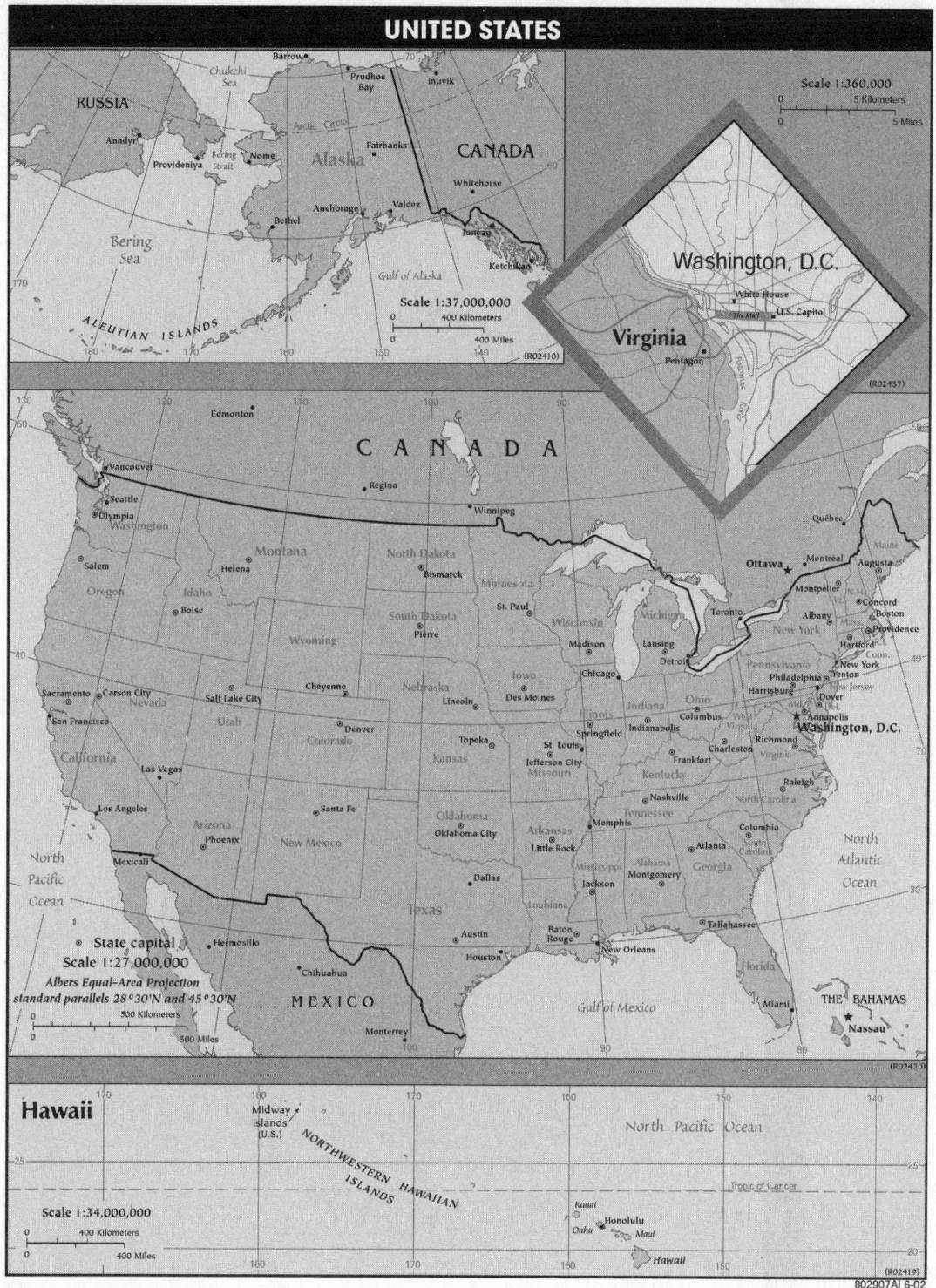

Scale 1:360,000
0   5 Kilometers
0   5 Miles

RUSSIA

CANADA

Alaska

Chukchi Sea

Barrow
Prudhoe Bay
Inuvik
Anadyr'
Providéniya
Nome
Bering Strait
Fairbanks
Whitehorse
Anchorage
Valdez
Bethel
Juneau
Ketchikan
Bering Sea
Gulf of Alaska
ALEUTIAN ISLANDS
Arctic Circle

Washington, D.C.
White House
The Mall
U.S. Capitol
Virginia
Pentagon
Potomac River

Scale 1:37,000,000
0   400 Kilometers
0   400 Miles
(R02418)

(R02437)

CANADA

Edmonton
Vancouver
Seattle
Olympia
Washington
Salem
Oregon
Helena
Montana
Boise
Idaho
Regina
North Dakota
Bismarck
Winnipeg
Minnesota
St. Paul
Wisconsin
Madison
Michigan
Lansing
Québec
Maine
Ottawa
Montréal
Augusta
Montpelier
N.H.
Concord
Vt.
Boston
Mass.
Albany
Providence
Conn.
Hartford
New York
R.I.
Toronto
Detroit
New York
Pennsylvania
New Jersey
Philadelphia
Trenton
South Dakota
Pierre
Wyoming
Cheyenne
Nebraska
Iowa
Des Moines
Chicago
Illinois
Indiana
Ohio
Columbus
Indianapolis
West Virginia
Harrisburg
Dover
Md.
Del.
Annapolis
Washington, D.C.
Sacramento
Carson City
Nevada
Salt Lake City
Utah
Denver
Colorado
Lincoln
Topeka
Kansas
St. Louis
Springfield
Missouri
Jefferson City
Frankfort
Kentucky
Richmond
Virginia
Raleigh
North Carolina
San Francisco
California
Las Vegas
Los Angeles
Arizona
Phoenix
Santa Fe
New Mexico
Oklahoma
Oklahoma City
Arkansas
Little Rock
Nashville
Tennessee
Memphis
Columbia
South Carolina
Atlanta
Georgia
Alabama
Montgomery
Mississippi
Jackson
Mexicali
North Pacific Ocean
Hermosillo
Chihuahua
MEXICO
Texas
Dallas
Austin
Houston
Louisiana
Baton Rouge
New Orleans
Tallahassee
Florida
Miami
North Atlantic Ocean
THE BAHAMAS
Nassau
Monterrey
Gulf of Mexico

State capital
Scale 1:27,000,000
Albers Equal-Area Projection
standard parallels 28°30'N and 45°30'N
0   500 Kilometers
0   500 Miles

(R02420)

Hawaii

Midway Islands (U.S.)
NORTHWESTERN HAWAIIAN ISLANDS
North Pacific Ocean
Tropic of Cancer
Kauai
Oahu
Honolulu
Maui
Hawaii

Scale 1:34,000,000
0   400 Kilometers
0   400 Miles

(R02419)

802907AI 6-02

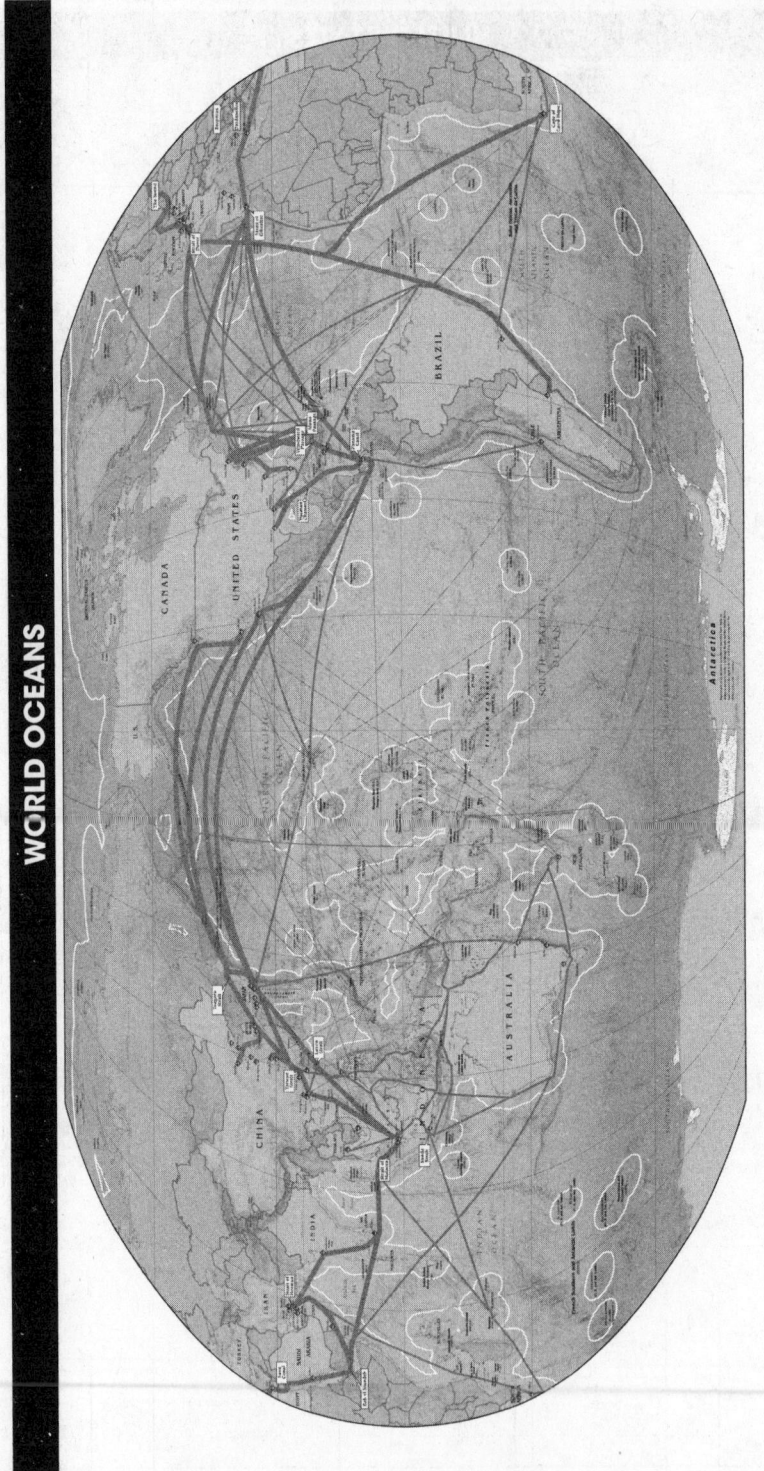

WORLD OCEANS